A Note from the Authors

You are about to embark on an amazing journey of discovery. The study of life spans from the inner workings of cells to the complex interactions of entire ecosystems, through the information stored in DNA to the ways genetic information evolves over time. At the same time that our understanding of biology is growing in leaps and bounds, so too are great insights into how learners acquire new knowledge and skills. We are thrilled to join Scott Freeman on *Biological Science*, a book dedicated to active, research-based learning and to exploring the experimental evidence that informs what we know about biology. The next few pages highlight the features in this book and in MasteringBiology® that will help you succeed.

From left to right: Michael Black, Emily Taylor, Jon Monroe,
Lizabeth Allison, Greg Podgorski, Kim Quillin

To the Student: How to Use This Book

New chapter-opening Roadmaps visually group and organize information to help you anticipate key ideas as well as recognize meaningful relationships and connections between them.

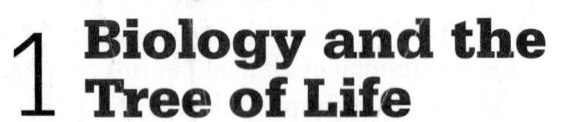

1 Biology and the Tree of Life

In this chapter you will learn about

Key themes to structure your thinking about Biology

starting with

What does it mean to say that something is alive? 1.1

including

Two of the greatest unifying ideas in Biology

including

The process of doing Biology 1.5

first
The cell theory 1.2

then
The theory of evolution by natural selection 1.3

predicts
The tree of life 1.4

Each Roadmap begins with a statement of why the chapter topic is important.

Key topics from each chapter are previewed, and related ideas are connected through blue linking words.

Chapter section numbers help you find key ideas easily in the chapter.

These Chinese Water Dragon hatchlings are exploring their new world and learning how to find food and stay alive. They represent one of the key characteristics of life introduced in this chapter: replication.

In essence, biological science is a search for ideas and observations that unify our understanding of the diversity of life, from bacteria living in rocks a mile underground to humans and majestic sequoia trees. This chapter is an introduction to this search.

The goals of this chapter are to introduce the nature of life and explore how biologists go about studying it. The chapter also introduces themes that will resonate throughout this book:

* Analyzing how organisms work at the molecular level.
* Understanding organisms in terms of their evolutionary history.
* Helping you learn to think like a biologist.

Let's begin with what may be the most fundamental question of all: What is life?

BIG PICTURE
This chapter is part of the Big Picture. See how on pages 16–17.

✔ When you see this checkmark, stop and test yourself. Answers are available in Appendix A.

1

Big Picture Concept Maps are referenced on the opening page of related chapters, pointing you to summary pages that help you synthesize challenging topics.

Big Picture Concept Maps integrate visuals and words to help you synthesize information about challenging topics in biology that span multiple chapters and units.

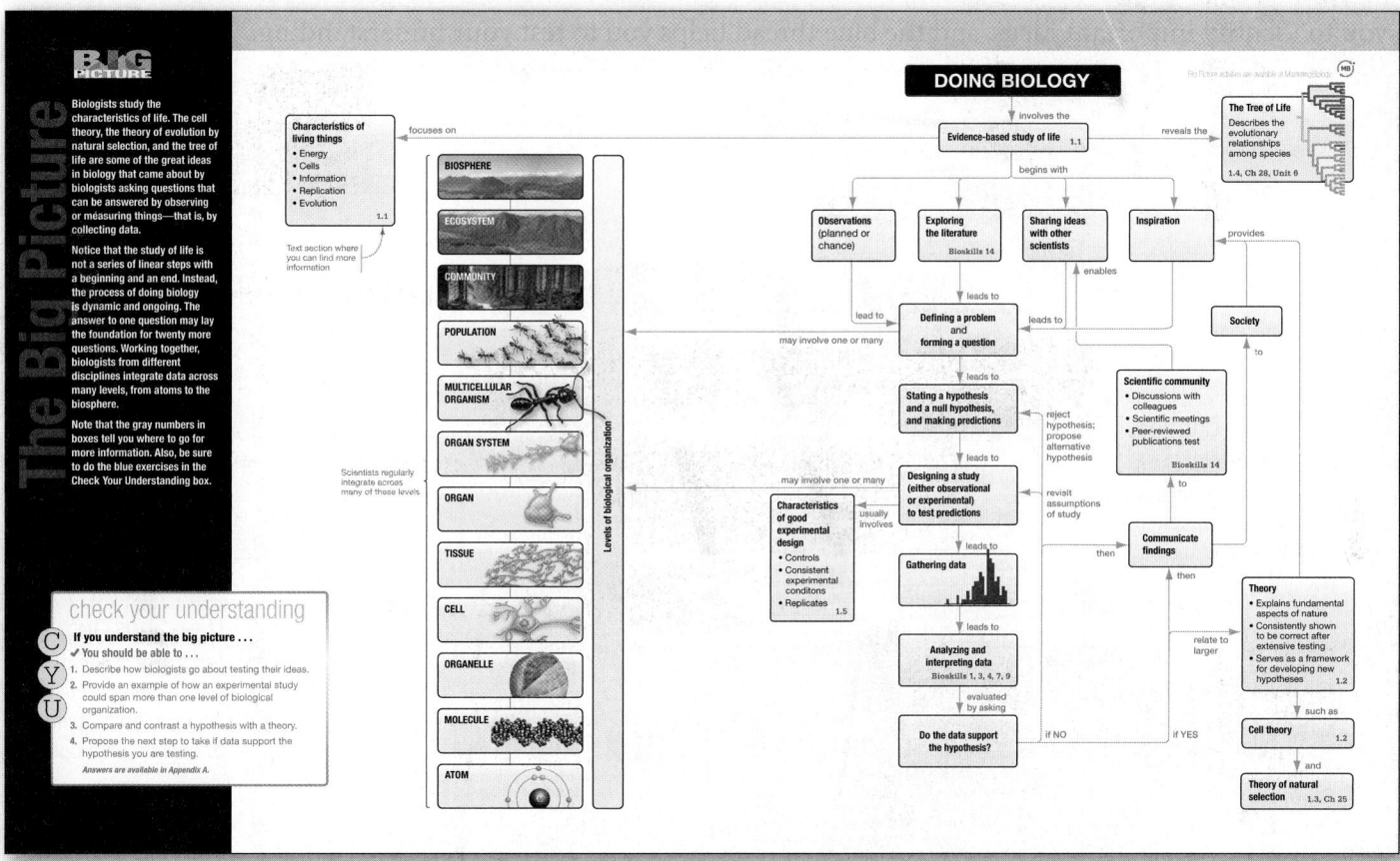

Three New Big Picture topics have been added to the Fifth Edition:

- NEW! Doing Biology
- NEW! The Chemistry of Life
- Energy for Life
- Genetic Information
- Evolution
- NEW! Plant and Animal Form and Function
- Ecology

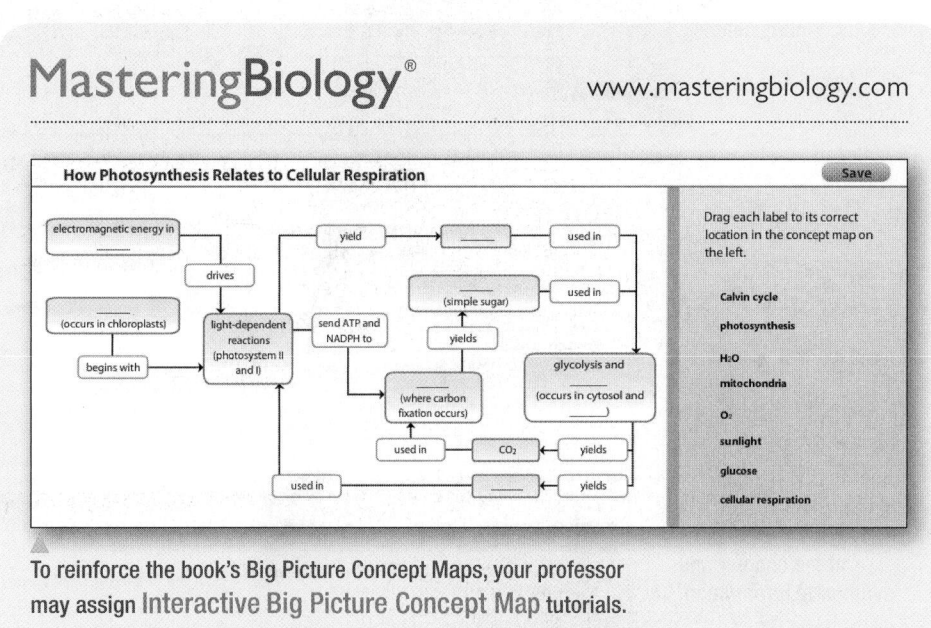

To reinforce the book's Big Picture Concept Maps, your professor may assign Interactive Big Picture Concept Map tutorials.

Practice for success on tests and exams

Intertwined color-coded "active learning threads" are embedded in the text. The gold thread helps you to identify important ideas, and the blue thread helps you to test your understanding.

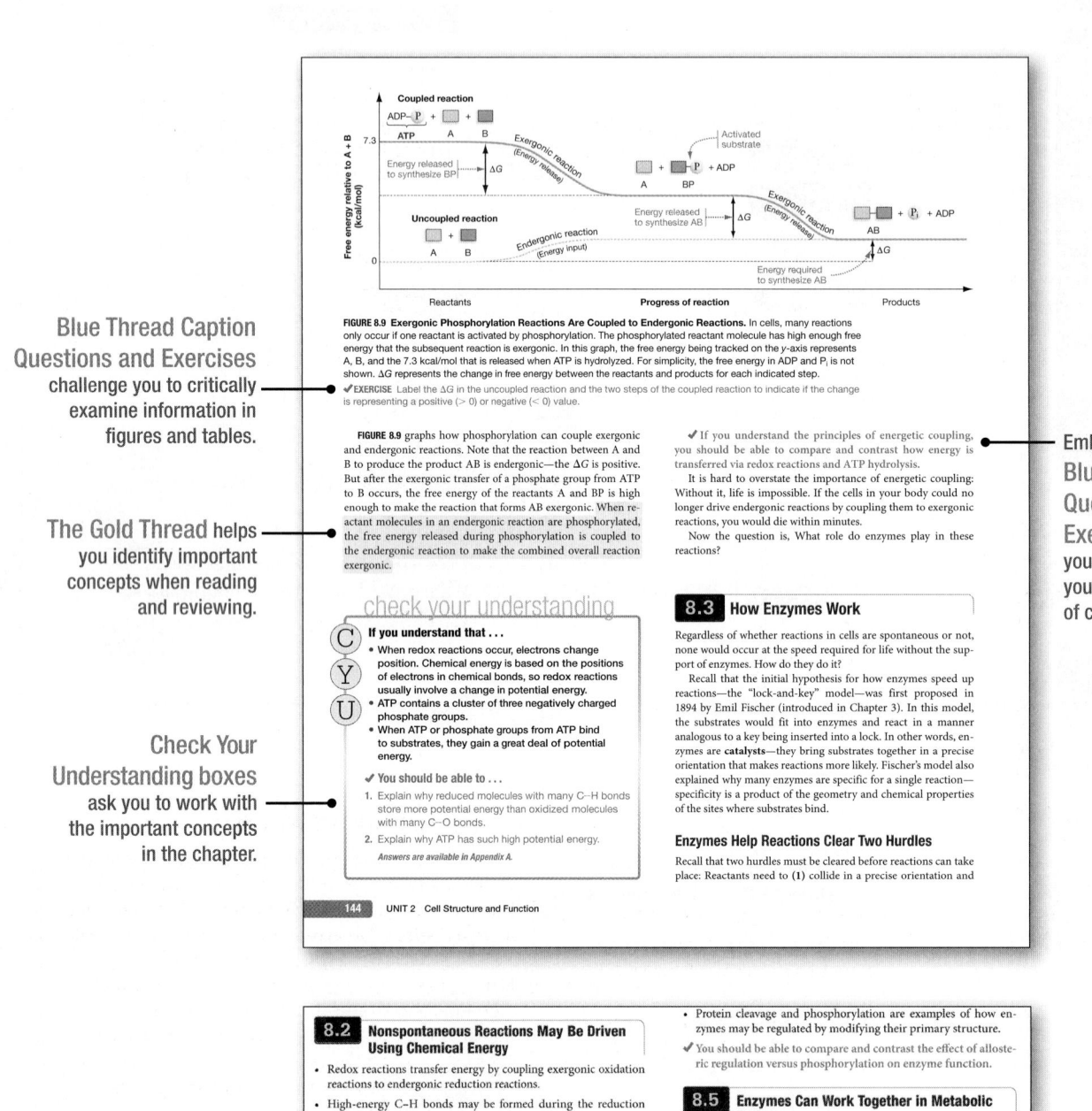

FIGURE 8.9 Exergonic Phosphorylation Reactions Are Coupled to Endergonic Reactions. In cells, many reactions only occur if one reactant is activated by phosphorylation. The phosphorylated reactant molecule has high enough free energy that the subsequent reaction is exergonic. In this graph, the free energy being tracked on the y-axis represents A, B, and the 7.3 kcal/mol that is released when ATP is hydrolyzed. For simplicity, the free energy in ADP and P_i is not shown. ΔG represents the change in free energy between the reactants and products for each indicated step.

✓ **EXERCISE** Label the ΔG in the uncoupled reaction and the two steps of the coupled reaction to indicate if the change is representing a positive (> 0) or negative (< 0) value.

Blue Thread Caption Questions and Exercises challenge you to critically examine information in figures and tables.

FIGURE 8.9 graphs how phosphorylation can couple exergonic and endergonic reactions. Note that the reaction between A and B to produce the product AB is endergonic—the ΔG is positive. But after the exergonic transfer of a phosphate group from ATP to B occurs, the free energy of the reactants A and BP is high enough to make the reaction that forms AB exergonic. When reactant molecules in an endergonic reaction are phosphorylated, the free energy released during phosphorylation is coupled to the endergonic reaction to make the combined overall reaction exergonic.

The Gold Thread helps you identify important concepts when reading and reviewing.

check your understanding

Check Your Understanding boxes ask you to work with the important concepts in the chapter.

If you understand that . . .

- When redox reactions occur, electrons change position. Chemical energy is based on the positions of electrons in chemical bonds, so redox reactions usually involve a change in potential energy.
- ATP contains a cluster of three negatively charged phosphate groups.
- When ATP or phosphate groups from ATP bind to substrates, they gain a great deal of potential energy.

✓ You should be able to . . .

1. Explain why reduced molecules with many C—H bonds store more potential energy than oxidized molecules with many C—O bonds.
2. Explain why ATP has such high potential energy.

Answers are available in Appendix A.

✓ If you understand the principles of energetic coupling, you should be able to compare and contrast how energy is transferred via redox reactions and ATP hydrolysis.

It is hard to overstate the importance of energetic coupling: Without it, life is impossible. If the cells in your body could no longer drive endergonic reactions by coupling them to exergonic reactions, you would die within minutes.

Now the question is, What role do enzymes play in these reactions?

Embedded Blue Thread Questions and Exercises encourage you to stop and test your understanding of challenging topics.

8.3 How Enzymes Work

Regardless of whether reactions in cells are spontaneous or not, none would occur at the speed required for life without the support of enzymes. How do they do it?

Recall that the initial hypothesis for how enzymes speed up reactions—the "lock-and-key" model—was first proposed in 1894 by Emil Fischer (introduced in Chapter 3). In this model, the substrates would fit into enzymes and react in a manner analogous to a key being inserted into a lock. In other words, enzymes are **catalysts**—they bring substrates together in a precise orientation that makes reactions more likely. Fischer's model also explained why many enzymes are specific for a single reaction—specificity is a product of the geometry and chemical properties of the sites where substrates bind.

Enzymes Help Reactions Clear Two Hurdles

Recall that two hurdles must be cleared before reactions can take place: Reactants need to (**1**) collide in a precise orientation and

8.2 Nonspontaneous Reactions May Be Driven Using Chemical Energy

- Redox reactions transfer energy by coupling exergonic oxidation reactions to endergonic reduction reactions.
- High-energy C—H bonds may be formed during the reduction step of a redox reaction when an H^+ is combined with a transferred electron.
- The hydrolysis of ATP is an exergonic reaction and may be used to drive a variety of cellular processes.
- When a phosphate group from ATP is added to a substrate that participates in an endergonic reaction, the potential energy of the substrate is raised enough to make the reaction exergonic and thus spontaneous.

✓ You should be able to explain what energetic coupling means, and why life would not exist without it.

8.3 How Enzymes Work

- Enzymes are protein catalysts. They speed reaction rates but do not affect the change in free energy of the reaction.
- The structure of an enzyme has an active site that brings sub-

End-of-Chapter Blue Thread Exercises, integrated in the chapter summary, help you review the major themes of the chapter and synthesize information.

- Protein cleavage and phosphorylation are examples of how enzymes may be regulated by modifying their primary structure.

✓ You should be able to compare and contrast the effect of allosteric regulation versus phosphorylation on enzyme function.

8.5 Enzymes Can Work Together in Metabolic Pathways

- In cells, enzymes often work together in metabolic pathways that sequentially modify a substrate to make a product.
- A pathway may be regulated by controlling the activity of one enzyme, often the first in the series of reactions. Feedback inhibition results from the accumulation of a product that binds to an enzyme in the pathway and inactivates it.
- Metabolic pathways were vital to the evolution of life, and new pathways continue to evolve in cells.

✓ You should be able to predict how the removal of the intermediate in a two-step metabolic pathway would affect the enzymatic rates of the first and last.

MB www.masteringbiology.com

1. MasteringBiology Assignments

Identify gaps in your understanding, then fill them

The Fifth Edition provides many opportunities for you to test yourself and offers **helpful learning strategies.**

Analyze: Can I recognize underlying patterns and structure?

Evaluate: Can I make judgments on the relative value of ideas and information?

Create: Can I put ideas and information together to generate something new?

Apply: Can I use these ideas in the same way or in a new situation?

Understand: Can I explain this concept in my own words?

Remember: Can I recall the key terms and ideas?

◀ **Bloom's Taxonomy** describes six learning levels: Remember, Understand, Apply, Analyze, Evaluate, and Create. Questions in the book span all levels, including self-testing at the higher levels to help you develop higher-order thinking skills that will prepare you for exams.

Steps to Building Understanding
Each chapter ends with three groups of questions that build in difficulty:

✔ **TEST YOUR KNOWLEDGE**

Begin by testing your basic knowledge of new information.

✔ **TEST YOUR UNDERSTANDING**

Once you're confident with the basics, demonstrate your deeper understanding of the material.

✔ **TEST YOUR PROBLEM-SOLVING SKILLS**

Work towards mastery of the content by answering questions that challenge you at the highest level of competency.

BIOSKILL 16 using Bloom's taxonomy

Most students have at one time or another wondered why a particular question on an exam seemed so hard, while others seemed easy. The explanation lies in the type of cognitive skills required to answer the question. Let's take a closer look.

NEW! BioSkill Covering Bloom's Taxonomy helps you to recognize question types using the Bloom's cognitive hierarchy, and it provides specific strategies to help you study for questions at all six levels.

Answer Appendix Includes Bloom's Taxonomy Information
Answers to all questions in the text now include the Bloom's level being tested. You can simultaneously practice assessing your understanding of content and recognizing Bloom's levels. Combining this information with the guidance in the BioSkill on Bloom's Taxonomy will help you form a plan to improve your study skills.

▶ ✔ **Test Your Problem-Solving Skills**

13. analyze A scientific theory is not a guess—it is an idea whose validity can be tested with data. Both the cell theory and the theory of evolution have been validated by large bodies of observational and experimental data.
14. apply If all eukaryotes living today have a nucleus, then it is logical to conclude that the nucleus arose in a common ancestor of all eukaryotes, indicated by the arrow you should have added to the figure. See **FIGURE A1.2.** If it had arisen in a common ancestor of Bacteria or Archaea, then species in those groups would have had to lose the trait—an unlikely event.
15. evaluate The data set was so large and diverse that it was no longer reasonable to argue that noncellular life-forms would be discovered. **16.** apply b

MasteringBiology®
www.masteringbiology.com

NEW! End-of-chapter questions from the book are now available for your professor to assign as homework in MasteringBiology.

Practice scientific thinking and scientific skills

A unique emphasis on the process of scientific discovery and experimental design teaches you how to think like a scientist as you learn fundamental biology concepts.

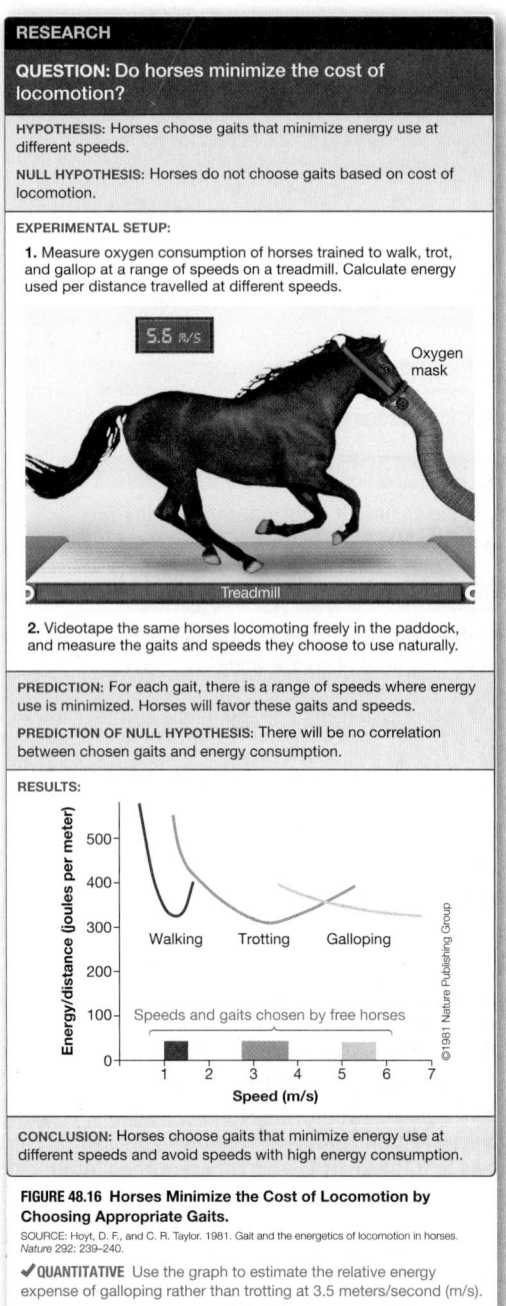

RESEARCH

QUESTION: Do horses minimize the cost of locomotion?

HYPOTHESIS: Horses choose gaits that minimize energy use at different speeds.

NULL HYPOTHESIS: Horses do not choose gaits based on cost of locomotion.

EXPERIMENTAL SETUP:

1. Measure oxygen consumption of horses trained to walk, trot, and gallop at a range of speeds on a treadmill. Calculate energy used per distance travelled at different speeds.

5.8 m/s — Oxygen mask — Treadmill

2. Videotape the same horses locomoting freely in the paddock, and measure the gaits and speeds they choose to use naturally.

PREDICTION: For each gait, there is a range of speeds where energy use is minimized. Horses will favor these gaits and speeds.

PREDICTION OF NULL HYPOTHESIS: There will be no correlation between chosen gaits and energy consumption.

RESULTS:

Energy/distance (joules per meter) vs Speed (m/s): Walking, Trotting, Galloping. Speeds and gaits chosen by free horses. ©1981 Nature Publishing Group

CONCLUSION: Horses choose gaits that minimize energy use at different speeds and avoid speeds with high energy consumption.

FIGURE 48.16 Horses Minimize the Cost of Locomotion by Choosing Appropriate Gaits.
SOURCE: Hoyt, D. F., and C. R. Taylor. 1981. Gait and the energetics of locomotion in horses. *Nature* 292: 239–240.

✓**QUANTITATIVE** Use the graph to estimate the relative energy expense of galloping rather than trotting at 3.5 meters/second (m/s).

Research Boxes explain how research studies are designed and give you additional practice interpreting data. Each Research Box consistently models the scientific method, presenting the research question, hypotheses, experimental setup, predictions, results, and conclusion. 15 Research Boxes are new to the Fifth Edition.

All of the Research Boxes cite the original research paper and include a question that asks you to analyze the design of the experiment or study.

MasteringBiology®
www.masteringbiology.com

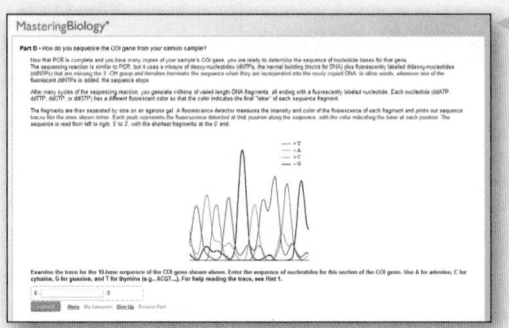

NEW! Solve It Tutorials are available for homework assignments in MasteringBiology and give you an opportunity to work like a scientist through a simulated investigation that requires you to analyze and interpret data.

Experimental Inquiry Tutorials based on some of biology's most seminal experiments give you a chance to analyze data and the reasoning that led scientists from the data to their conclusions.

Experimental Inquiry tutorial topics include:

- What Can You Learn About the Process of Science from Investigating a Cricket's Chirp?
- Which Wavelengths of Light Drive Photosynthesis?
- What Is the Inheritance Pattern of Sex-Linked Traits?
- Does DNA Replication Follow the Conservative, Semiconservative, or Dispersive Model?
- How Do Calcium Ions Help to Prevent Polyspermy During Egg Fertilization?

- Did Natural Selection of Ground Finches Occur When the Environment Changed?
- What Effect Does Auxin Have on Coleoptile Growth?
- What Role Do Genes Play in Appetite Regulation?
- Can a Species' Niche Be Influenced by Interspecific Competition?
- What Factors Influence the Loss of Nutrients from a Forest Ecosystem?

Build important skills scientists use to perform, evaluate, and communicate scientific research.

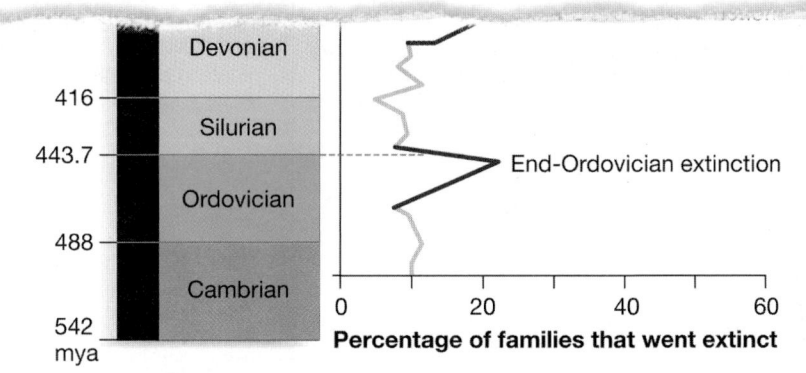

FIGURE 28.14 The Big Five Mass Extinction Events. This graph shows the percentage of lineages called families that went extinct over each interval in the fossil record since the Cambrian explosion. Over 50 percent of families and 90 percent of species went extinct during the end-Permian extinction.

DATA: Benton, M. J., 1995. *Science* 268: 52–58.

✔**QUANTITATIVE** Which extinction event ended the era of the dinosaurs 65 million years ago? About what percentage of families went extinct?

NEW! Graphs and tables now include their data sources, emphasizing the research process that leads to our understanding of biological ideas.

NEW! Quantitative questions are identified throughout the text, helping you practice computational problem solving and data analysis.

Expanded BioSkills Appendix helps you build skills that will be important to your success in biology. At relevant points in the text, you'll find references to the BioSkills appendix that will help you learn and practice foundational skills.

BioSkills Topics include:

- The Metric System and Significant Figures
- Some Common Latin and Greek Roots Used in Biology
- Reading Graphs
- Using Statistical Tests and Interpreting Standard Error Bars
- Combining Probabilities

- Using Logarithms
- Reading a Phylogenetic Tree
- Reading Chemical Structures
- Separating and Visualizing Molecules
- Separating Cell Components by Centrifugation
- Biological Imaging: Microscopy and X-ray Crystallography

- Cell and Tissue Culture Methods
- Model Organisms
- NEW! Primary Literature and Peer Review
- Making Concept Maps
- NEW! Using Bloom's Taxonomy

MasteringBiology®
www.masteringbiology.com

You can access self-paced BioSkills activities in the Study Area, and your instructor can assign additional activities in MasteringBiology.

Visualize biology
processes and structures

A carefully crafted visual program helps you gain a better understanding of biology through accurate, appropriately detailed figures.

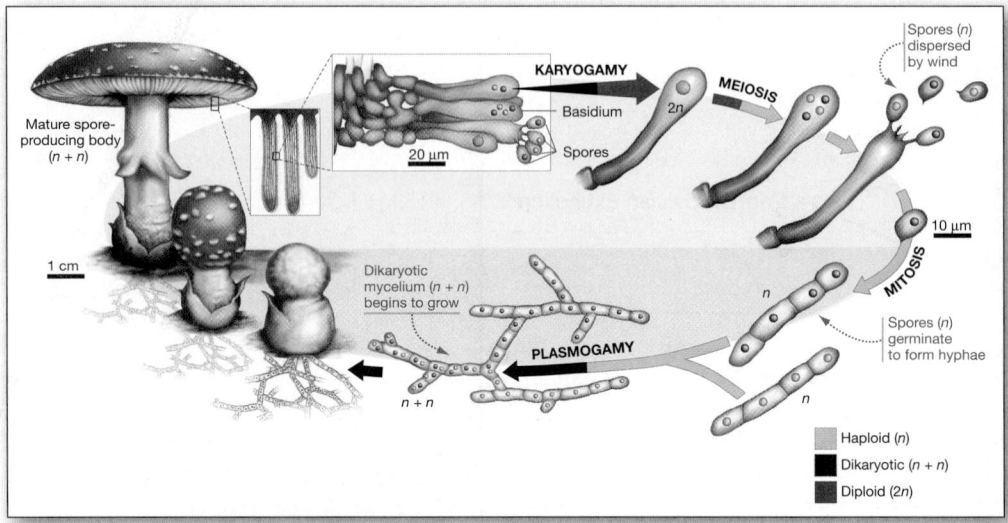

◄ **NEW! Redesigned Life Cycle diagrams** in Unit 6 and 7 help you compare and contrast processes among different organisms.

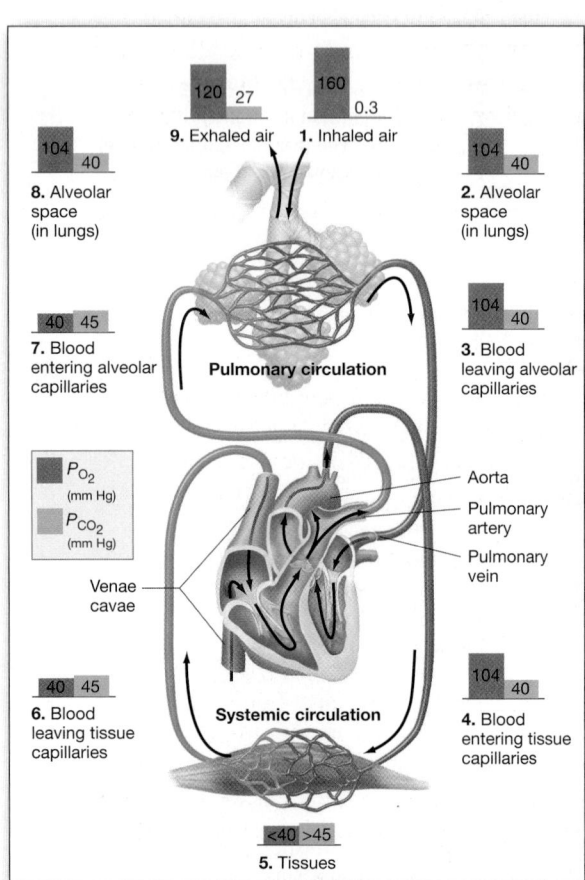

▲ **Informative figures** help you think through complex biological processes in manageable steps.

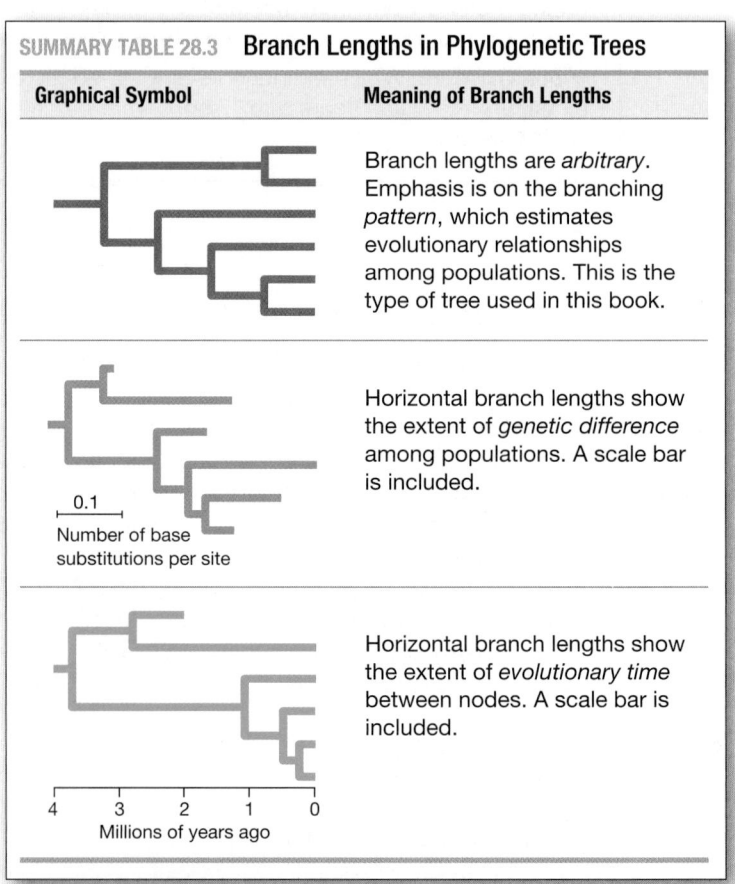

SUMMARY TABLE 28.3 **Branch Lengths in Phylogenetic Trees**

Graphical Symbol	Meaning of Branch Lengths
	Branch lengths are *arbitrary*. Emphasis is on the branching *pattern*, which estimates evolutionary relationships among populations. This is the type of tree used in this book.
0.1 — Number of base substitutions per site	Horizontal branch lengths show the extent of *genetic difference* among populations. A scale bar is included.
4 3 2 1 0 — Millions of years ago	Horizontal branch lengths show the extent of *evolutionary time* between nodes. A scale bar is included.

▲ **Visual Summary Tables** pull together important information in a format that allows for easy comparison and review.

Instructor and Student Resources

For Instructors

Instructor Resource DVD-ROM
978-0-321-86112-2 • 0-321-86112-4
Everything you need for lectures in one place, including video segments that demonstrate how to incorporate active-learning techniques into your own classroom. Enhanced menus make locating and assessing the digital resources for each chapter easy. The Instructor Resource CD/DVD-ROM includes PowerPoint® Lecture Outlines that integrate figures and animations for classroom presentations. All textbook figures, art, and photos are in JPEG format, and all PowerPoint slides and JPEGs have editable labels. Over 300 Instructor Animations accurately depict complex topics and dynamic processes described in the book.

Instructor Guide (Download only)
Available in the instructor resource area of MasteringBiology®

TestGen® (Download only)
All of the exam questions in the Test Bank have been peer reviewed and student tested, providing questions that set the standard for quality and accuracy. To improve the Test Bank, Metadata from MasteringBiology users has been incorporated directly into the software. Test questions that are ranked according to Bloom's taxonomy and improved TestGen® software makes assembling tests that much easier.

For Students

Study Guide
978-0-321-85832-0 • 0-321-85832-8
The Study Guide presents a breakdown of key biological concepts, difficult topics, and quizzes to help students prepare for exams. Unique to this study guide are four introductory, stand-alone chapters that introduce students to foundational ideas and skills necessary for classroom success: Introduction to Experimentation and Research in the Biological Sciences, Presenting Biological Data, Understanding Patterns in Biology and Improving Study Techniques, and Reading and Writing to Understand Biology. "Looking Forward" and "Looking Back" sections help students make connections across the chapters instead of viewing them as discrete entities.

Practicing Biology: A Student Workbook
978-0-321-88647-7 • 0-321-88647-X
This workbook focuses on key ideas, principles, and concepts that are fundamental to understanding biology. A variety of hands-on activities such as mapping and modeling suit different learning styles and help students discover which topics they need more help on. Students learn biology by doing biology. An instructors guide can be downloaded from the Instructor Area of MasteringBiology.

MasteringBiology®   www.masteringbiology.com

MasteringBiology is an online homework, tutorial, and assessment system that delivers self-paced tutorials that provide individualized coaching, focus on your course objectives, and respond to each student's progress. The Mastering system helps instructors maximize class time with customizable, easy-to-assign, and automatically graded assessments that motivate students to learn outside of class and arrive prepared for lecture. MasteringBiology is also available with a complete Pearson eText edition of *Biological Science*.

Highlights of the Fifth Edition Item Library include:

* NEW! **assignment options** include Solve It activities, end-of-chapter problems, and questions that accompany new BioSkills and new Big Picture Interactive Concept Maps.
* NEW! **"best of" homework pre-built assignments** help professors assign popular, key content quickly, including a blend of tutorials, end-of-chapter problems, and test bank questions.

* *Get Ready for Biology* and **Chemistry Review assignment options** help students get up to speed with activities that review chemistry, mathematics, and basic biology.

MasteringBiology® Virtual Labs

978-0-321-88644-6 • 0-321-88644-5
MasteringBiology: Virtual Labs is an online environment that promotes critical-thinking skills using virtual experiments and explorations that might be difficult to perform in a wet-lab environment due to time, cost, or safety concerns. MasteringBiology: Virtual Labs offers unique learning experiences in the areas of microscopy, molecular biology, genetics, ecology, and systematics.

For more information, please visit
www.pearsonhighered.com/virtualbiologylabs

BIOLOGICAL SCIENCE

After you explore this book . . . you should be able to . . .

• Pose an evolutionary hypothesis to explain why meter-long male water dragons are larger and have more colorful throats than the females.

• Propose how DNA sequences could be used to determine the relationships among populations of these lizards in China, India, and Southeast Asia.

• Design an experiment to study the relative importance of swimming, tree climbing, and running to the ability of these semi-aquatic lizards to find food and escape from predators.

• Create questions of your own and suggest methods for finding the answers!

Chinese Water Dragon,
Physignathus cocincinus

BIOLOGICAL SCIENCE

FIFTH EDITION

SCOTT FREEMAN
University of Washington

LIZABETH ALLISON
College of William & Mary

MICHAEL BLACK
California Polytechnic State University in San Luis Obispo

GREG PODGORSKI
Utah State University

KIM QUILLIN
Salisbury University

JON MONROE
James Madison University

EMILY TAYLOR
California Polytechnic State University in San Luis Obispo

PEARSON

Boston   Columbus   Indianapolis   New York   San Francisco   Upper Saddle River
Amsterdam   Cape Town   Dubai   London   Madrid   Milan   Munich   Paris   Montréal   Toronto
Delhi   Mexico City   São Paulo   Sydney   Hong Kong   Seoul   Singapore   Taipei   Tokyo

Editor-in-Chief: Beth Wilbur
Senior Acquisitions Editor: Michael Gillespie
Executive Director of Development: Deborah Gale
Senior Development Editor: Sonia DiVittorio
Project Editor: Anna Amato
Development Editors: Mary Catherine Hager,
  Moira Lerner Nelson, Bill O'Neal
Art Development Editors: Fernanda Oyarzun,
  Adam Steinberg
Associate Editor: Brady Golden
Assistant Editor: Leslie Allen
Editorial Assistant: Eddie Lee
Executive Media Producer: Laura Tommasi
Media Producer: Joseph Mochnick
Associate Media Producer: Daniel Ross
Associate Media Project Manager: David Chavez
Text Permissions Project Manager: Joseph Croscup
Text Permissions Specialist: Sheri Gilbert
Director of Production: Erin Gregg
Managing Editor: Michael Early

Production Project Manager: Lori Newman
Production Management: S4Carlisle Publishing Services
Copyeditor: Christianne Thillen
Compositor: S4Carlisle Publishing Services
Design Manager: Mark Ong
Interior Designer: Integra Software Services
Cover Designer: tt eye
Illustrators: Imagineering Media Services
Photo Permissions Management: Phutu
Photo Researchers: Kristin Piljay, Eric Schrader,
  Maureen Spuhler
Senior Photo Editor: Travis Amos
Manufacturing Buyer: Michael Penne
Director of Marketing: Christy Lesko
Executive Marketing Manager: Lauren Harp
Sales Director for Key Markets: David Theisen

Cover Photo Credit: *Physignathus cocincinus*
  Eric Isselée/Fotolia

Library of Congress Cataloging-in-Publication Data
Freeman, Scott, 1955- Biological science / Scott Freeman.—Fifth edition.
     pages cm
   ISBN-13: 978-0-321-74367-1 (student edition)
   ISBN-10: 0-321-74367-9 (student edition)
   ISBN-13: 978-0-321-84159-9 (instructors review copy)
   ISBN-10: 0-321-84159-X (instructors review copy) [etc.]
  1. Biology—Textbooks.   I. Title.
   QH308.2.F73 2014
   570—dc23

2 3 4 5 6 7 8 9 10—V011—16 15 14 13

www.pearsonhighered.com

ISBN 10: 0-321-74367-9; ISBN 13: 978-0-321-74367-1 (Student Edition)
ISBN 10: 0-321-84159-X; ISBN 13: 978-0-321-84159-9 (Instructor's Review Copy)
ISBN 10: 0-321-86216-3; ISBN 13: 978-0-321-86216-7 (Books a la Carte Edition)
ISBN 10: 0-321-84180-8; ISBN 13: 978-0-321-84180-3 (Volume 1)
ISBN 10: 0-321-84181-6; ISBN 13: 978-0-321-84181-0 (Volume 2)
ISBN 10: 0-321-84182-4; ISBN 13: 978-0-321-84182-7 (Volume 3)

Detailed Contents

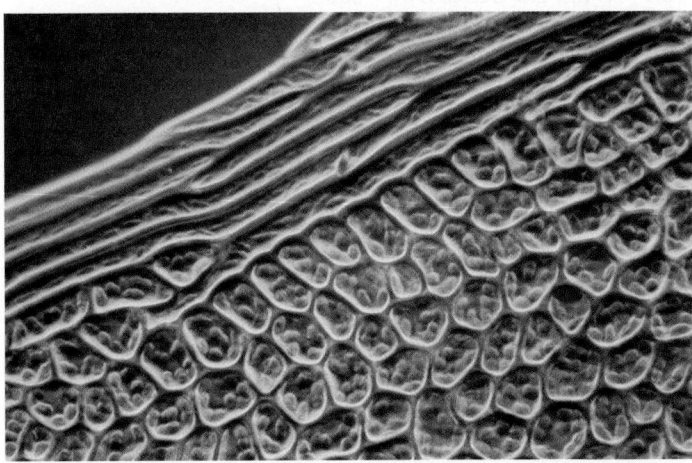

30 Protists   552

31 Green Algae and Land Plants   577

36 Viruses 711

UNIT **7** HOW PLANTS WORK 731

37 Plant Form and Function 731

38 Water and Sugar Transport in Plants 754

39 Plant Nutrition 775

About the Authors

A Letter from Scott:

I started working on *Biological Science* in 1997 with a simple goal: To help change the way biology is taught. After just shy of 20,000 hours of work on four editions of this text, that goal still gets me out of bed in the morning. But instead of focusing my energies on textbook writing, I've decided to devote myself full-time to research on student learning and developing new courses for undergraduate and graduate students at the University of Washington.

So with this edition I am passing the torch—to an all-star cast of leading scientists and educators who have enthusiastically taught from, and contributed to, previous editions of *Biological Science*. Working with them, I have seen the new team bring their passion, talent, and creativity to the book, with expertise that spans the breadth of the life sciences. Just as important, they work beautifully together because they think alike. They are driven by a shared concern for student learning, a commitment to the craft of writing, and a background in evidence-based teaching.

These pages provide a brief introduction to Liz Allison, Michael Black, Greg Podgorski, Kim Quillin, Jon Monroe, and Emily Taylor. As a group, they've built on the book's existing strengths and infused this edition with fresh energy, perspective, and ideas. I'm full of admiration for what they have accomplished, and excited about the impact this edition will have on biology students from all over the world.—*Scott Freeman*

Lizabeth A. Allison is professor and chair of the Biology Department at the College of William & Mary. She received her Ph.D. in Zoology from the University of Washington, specializing in molecular and cellular biology. Before coming to William & Mary, she spent eight years as a faculty member at the University of Canterbury in New Zealand. Liz teaches introductory biology for majors and upper-division molecular biology courses. She has mentored graduate students and more than 80 undergraduate research students, many of them coauthoring papers with her on intracellular trafficking of the thyroid hormone receptor in normal and cancer cells. The recipient of numerous awards, including a State Council for Higher Education in Virginia (SCHEV) Outstanding Faculty Award in 2009, Liz received one of the three inaugural Arts & Sciences Faculty Awards for Teaching Excellence in 2011, and a Plumeri Award for Faculty Excellence in 2012. In addition to her work on this text, she is author of *Fundamental Molecular Biology*, now in its second edition.
Lead Author; Chapter 1 and BioSkills
laalli@wm.edu

Scott Freeman received a Ph.D. in Zoology from the University of Washington and was subsequently awarded an Alfred P. Sloan Postdoctoral Fellowship in Molecular Evolution at Princeton University. He has done research in evolutionary biology on topics ranging from nest parasitism to the molecular systematics of the blackbird family and is coauthor, with Jon Herron, of the standard-setting undergraduate text *Evolutionary Analysis*. Scott is the recipient of a Distinguished Teaching Award from the University of Washington and is currently a Senior Lecturer in the UW Department of Biology, where he teaches introductory biology for majors, a writing-intensive course for majors called The Tree of Life, and a graduate seminar in college science teaching. Scott's current research focuses on how active learning affects student learning and academic performance.

Michael Black received his Ph.D. in Microbiology & Immunology from Stanford University School of Medicine as a Howard Hughes Predoctoral Fellow. After graduation, he studied cell biology as a Burroughs Wellcome Postdoctoral Fellow at the MRC Laboratory of Molecular Biology in Cambridge, England. His current research focuses on the use of molecules to identify and track the transmission of microbes in the environment. Michael is a professor of Cell & Molecular Biology at California Polytechnic State University in San Luis Obispo, where he teaches introductory and advanced classes for majors in cell biology and microbiology. In addition to his teaching and research activities, Michael serves as the director of the Undergraduate Biotechnology Lab, where he works alongside undergraduate technicians to integrate research projects and inquiry-based activities into undergraduate classes.
Chapters 2–12, 36, and 51
mblack@calpoly.edu

Greg Podgorski received his Ph.D. in Molecular and Cellular Biology from Penn State University and has been a postdoctoral fellow at the Max Plank Institute for Biochemistry and Columbia University. His research interests are in biology education, developmental genetics, and computational biology. Greg's most recent work has been in mathematical modeling of how patterns of different cell types emerge during development and how tumors recruit new blood vessels in cancer. Greg has been teaching at Utah State University for more than 20 years in courses that include introductory biology for majors and for nonmajors, genetics, cell biology, developmental biology, and microbiology, and he has offered courses in nonmajors biology in Beijing and Hong Kong. He's won teaching awards at Utah State University and has been recognized by the National Academies as a Teaching Fellow and a Teaching Mentor.

Chapters 13–24
greg.podgorski@usu.edu

Jon Monroe is professor of Biology at James Madison University in Harrisonburg, Virginia. Jon completed his undergraduate work in Botany at the University of Michigan and his graduate work in Plant Physiology at Cornell University. He began his current position after a postdoc in biochemistry at Michigan State University. He currently teaches Plant Biology, and Cell and Molecular Biology. Jon's interest in plants is broad, ranging from systematics and taxonomy to physiology and biochemistry. His research, mostly with undergraduates, uses Arabidopsis thaliana to study the functions of a family of β-amylase genes in starch metabolism. Jon has been active in promoting undergraduate research through his work with the American Society of Plant Biologists (ASPB) and the Council on Undergraduate Research. He has received ASPB's Excellence in Teaching award and James Madison University Alumni Association's Distinguished Faculty Award.

Chapters 29–32; 37–41
monroejd@jmu.edu

Kim Quillin received her B.A. in Biology at Oberlin College *summa cum laude* and her Ph.D. in Integrative Biology from the University of California, Berkeley (as a National Science Foundation Graduate Fellow). Kim has worked in the trenches with Scott Freeman on every edition of *Biological Science*, starting with the ground-up development of the illustrations in the first edition in 1999 and expanding her role in each edition, always with the focus of helping students to think like biologists. Kim currently teaches introductory biology at Salisbury University, a member of the University System of Maryland, where she is actively involved in the ongoing student-centered reform of the concepts-and-methods course for biology majors. Her current research focuses on the scholarship of teaching and learning with an emphasis on measuring science process skills and the advantages and pitfalls of active multimedia learning.

Chapters 25–28; 33–35; 48; 52–57
kxquillin@salisbury.edu

Emily Taylor earned a B.A. in English at the University of California, Berkeley followed by a Ph.D. in Biological Sciences from Arizona State University, where she conducted research in the field of environmental physiology as a National Science Foundation Graduate Research Fellow. She is currently an associate professor of Biological Sciences at the California Polytechnic State University in San Luis Obispo, California. Her student-centered research program focuses on the endocrine and reproductive physiology of free-ranging reptiles, especially rattlesnakes. She teaches numerous undergraduate and graduate courses, including introductory biology, anatomy and physiology, and herpetology, and received the California Faculty Association's Distinguished Educator Award in 2010 and Cal Poly's Distinguished Teaching Award in 2012. Her revision of Unit 8 is her first foray into textbook writing.

Chapters 42–50
etaylor@calpoly.edu

Preface to Instructors

The first edition of *Biological Science* was visionary in its unique emphasis on the process of scientific discovery and experimental design—teaching how we know what we know. The goal was for students not only to learn the language of biology and understand fundamental concepts but also to begin to apply those concepts in new situations, analyze experimental design, synthesize results, and evaluate hypotheses and data—to learn how to think like biologists. Each edition since has proudly expanded on this vision. The Fifth Edition is no exception.

A team of six dedicated teacher-scholars has joined Scott to build on and refine the original vision, and by so doing, make the book an even better teaching and learning tool. The pace of biological discovery is rapid, and with each novel breakthrough it becomes even more challenging to decide what is essential to include in an introductory biology text. Pulling together an author team with firsthand expertise from molecules to ecosystems has ensured that the content of the Fifth Edition reflects cutting-edge biology that is pitched at the right level for introductory students and is as accurate and as exciting as ever for instructors and students alike.

New findings from education research continue to inform and inspire the team's thinking about *Biological Science*—we know more today than ever before about how students learn. These findings demand that we constantly look for new ways to increase student engagement in the learning process, and to help instructors align course activities and learning goals with testing strategies.

The New Coauthors

The new coauthor team brings a broad set of talents and interests to the project, motivated by a deep commitment to undergraduate teaching, whether at a small liberal arts college or a large university. Kim Quillin has been a partner in this textbook in every edition. For the Fifth Edition, she revised chapters across three units in addition to spearheading the continued effort to enhance the visual-teaching program. Michael Black, Greg Podgorski, Jon Monroe, and Emily Taylor, who served as unit advisors on the Fourth Edition, were already familiar with the book. And most of the authorial team have been avid users of previous editions for many years.

Core Values

Together, the coauthor team has worked to extend the vision and maintain the core values of *Biological Science*—to provide a book for instructors who embrace the challenge of boosting students to higher levels of learning, and to provide a book for students that helps them each step of the way in learning to think like scientists. Dedicated instructors have high expectations of their students—the Fifth Edition provides scaffolding to help students learn at the level called for by the National Academy of Sciences, the Howard Hughes Medical Institute, the American Association of Medical Academies, and the National Science Foundation.

What's New in This Edition

The Fifth Edition contains many new or expanded features, all of them targeted at ways to help students learn to construct their own knowledge and think like biologists.

- **Road Maps** The new Road Maps at the beginning of each chapter pair with the Big Picture concept maps introduced in the Fourth Edition. Together they help students navigate chapter content and see the forest for the trees. Each Road Map starts with a purpose statement that tells students what they can expect to learn from each chapter. It then goes on to visually group and organize information to help students anticipate key ideas as well as recognize meaningful relationships and connections between the ideas.

- **The Big Picture** Introduced in the Fourth Edition, Big Picture concept maps integrate words and visuals to help students synthesize information about challenging topics that span multiple chapters and units. In response to requests from instructors and students, three new Big Pictures focused on additional tough topics have been added: Doing Biology, The Chemistry of Life, and Plant and Animal Form and Function. In addition, the Ecology Big Picture is completely revised to reflect changes to that unit.

- **New Chapters** Two new chapters are added to better serve instructors and students. Unit 2 now contains a new Chapter 8, Energy and Enzymes: An Introduction to Metabolic Pathways. This chapter consolidates these critical topics in a place where students and instructors need it most—right before the chapters on cellular respiration and photosynthesis. In the Fourth Edition, animal movement was discussed in a chapter largely focused on animal sensory systems. In the Fifth Edition, this important topic is treated in depth in a new Chapter 48, Animal Movement, that explores how muscle and skeletal systems work together to produce locomotion.

- **New BioSkills** Instructors recognize that biology students need to develop foundational science skills in addition to content knowledge. While these skills are emphasized throughout the book, *Biological Science*, beginning with the Third

Edition, has provided a robust set of materials and activities to guide students who need extra help. To promote even fuller use of this resource, the BioSkills are now updated, expanded, and reorganized. New in this edition are a discussion of significant figures within the BioSkills on the Metric System, and two new BioSkills on Primary Literature and Peer Review and Using Bloom's Taxonomy. BioSkills are located in Appendix B, and practice activities can be assigned online in MasteringBiology®.

- **Promotion of Quantitative Skills** Reports like *Biology 2010, Scientific Foundations for Future Physicians,* and *Vision and Change* all place a premium on quantitative skills. To infuse a quantitative component throughout the text, new and existing quantitative questions are flagged in each chapter to encourage students to work on developing their ability to read or create a graph, perform or interpret a calculation, or use other forms of quantitative reasoning.

- **Bloom's Taxonomy** In the Fifth Edition, all questions in the text are assigned a Bloom's Taxonomy level to help both students and instructors understand whether a question requires higher-order or lower-order cognitive skills. Questions span all six Bloom's levels. (Bloom's levels are identified in Appendix A: Answers.) The coauthors were trained by experts Mary Pat Wenderoth and Clarissa Dirks[1] to ensure we followed a process that would result in high inter-rater reliability—or agreement among raters—in assigning Bloom's levels to questions. The new BioSkill, Using Bloom's Taxonomy, explains the six Bloom's levels to students and offers a practical guide to the kinds of study activities best suited for answering questions at each level.

- **Expanded Emphasis on "Doing Biology"** A constant hallmark of this text is its emphasis on experimental evidence—on teaching how we know what we know. To reflect the progress of science, in the Fifth Edition, the coauthor team replaced many experiments with fresh examples and added new Research Boxes. And as noted earlier, they added a new Big Picture on Doing Biology, focusing on the process of science and the organizational levels of biology. Data sources are now cited for all graphs and data tables to model the importance of citing data sources to students. Updated Research Box questions continue to encourage students to analyze some aspect of experimental design. Also new to this edition is a BioSkill on Primary Literature and Peer Review.

- **Art Program** The art program is further enhanced in this edition by the addition of more illustrated summary tables. These tables make subject areas more accessible to visual learners and reinforce key concepts of the chapter. Many of the life-cycle figures in Unit 6 are significantly overhauled.

[1] Crowe, A., C. Dirks, and M. P. Wenderoth. 2008. Biology in Bloom: Implementing Bloom's Taxonomy to enhance student learning in biology. *CBE–Life Sciences Education* 7: 368–381.

Updated Blue Thread Scaffolding

In the Third and Fourth editions of *Biological Science*, a metacognitive tool was formulated as the now popular feature known as "Blue Thread"—sets of questions designed to help students identify what they do and don't understand. The fundamental idea is that if students really understand a piece of information or a concept, they should be able to do something with it.

In the Fifth Edition, the Blue Thread is revised to reflect changes in chapter content, and to incorporate user feedback. Blue-Thread questions appear in the following locations:

- **In-text "You should be able to's"** offer exercises on topics that professors and students have identified as the most difficult concepts in each chapter.

- **Caption questions and exercises** challenge students to examine the information in a figure or table critically—not just absorb it.

- **Check Your Understanding boxes** present two to three tasks that students should be able to complete in order to demonstrate a mastery of summarized key ideas.

- **Chapter summaries** include "You should be able to" problems or exercises related to each key concept.

- **End-of-chapter** questions are organized in three levels of increasing difficulty so students can build from lower to higher-order cognitive questions.

Integration of Media

The textbook continues to be supported by MasteringBiology®, the most powerful online homework, tutorial, and assessment system available. Tutorials follow the Socratic method, coaching students to the correct answer by offering feedback specific to a student's misconceptions as well as providing hints students can access if they get stuck. Instructors can associate content with publisher-provided learning outcomes or create their own. Content highlights include the following:

- **NEW! Solve It Tutorials** These activities allow students to act like scientists in simulated investigations. Each tutorial presents an interesting, real-world question that students will answer by analyzing and interpreting data.

- **Experimental Inquiry Tutorials** The call to teach students about the process of science has never been louder. To support such teaching, there are 10 interactive tutorials on classic scientific experiments—ranging from Meselson–Stahl on DNA replication to the Grants' work on Galápagos finches and Connell's work on competition. Students who use these tutorials should be better prepared to think critically about experimental design and evaluate the wider implications of the data—preparing them to do the work of real scientists in the future.

- **BioFlix® Animations and Tutorials** BioFlix are movie-quality, 3-D animations that focus on the most difficult core topics and are accompanied by in-depth, online tutorials that provide hints and feedback to guide student learning. Eighteen BioFlix animations and tutorials tackle topics such as meiosis, mitosis, DNA replication, photosynthesis, homeostasis, and the carbon cycle.

- **NEW! End-of-Chapter Questions** Multiple choice end-of-chapter questions are now available to assign in MasteringBiology.

- **Blue-Thread Questions** Over 500 questions based on the Blue-Thread Questions in the textbook are assignable in MasteringBiology.

- **Big Picture Tutorials** Interactive concept map activities based on the Big Picture figures in the textbook are assignable in MasteringBiology, including tutorials to support the three new Big Pictures: Doing Biology, The Chemistry of Life, and Plant and Animal Form and Function.

- **BioSkills Activities** Activities based on the BioSkills content in the textbook are assignable in MasteringBiology, including activities to support the new BioSkills on Primary Literature and Peer Review and Using Bloom's Taxonomy.

- **Reading Quiz Questions** Every chapter includes reading quiz questions you can assign to ensure students read the textbook and understand the basics. These quizzes are perfect as a pre-lecture assignment to get students into the content before class, allowing you to use class time more effectively.

Serving a Community of Teachers

All of us on the coauthor team are deeply committed to students and to supporting the efforts of dedicated teachers. Doing biology is what we love. At various points along our diverse paths, we have been inspired by our own teachers when we were students, and now are inspired by our colleagues as we strive to become even better teacher-scholars. In the tradition of all previous editions of *Biological Science*, we have tried to infuse this textbook with the spirit and practice of evidence-based teaching. We welcome your comments, suggestions, and questions.

Thank you for your work on behalf of your students.

Content Highlights of the Fifth Edition

As discussed in the preface, a major focus of this revision is to enhance the pedagogical utility of *Biological Science*. Another major goal is to ensure that the content reflects the current state of science and is accurate. The expanded author team has scrutinized every chapter to add new, relevant content, update descriptions when appropriate, and adjust the approach to certain topics to enhance student comprehension. In this section, some of the key content improvements to the textbook are highlighted.

Chapter 1 Biology and the Tree of Life A concept map summarizing the defining characteristics of life is added. The process of doing biology coverage is expanded to include discussion of both experimental and descriptive studies, and more rigorous definitions of the terms hypothesis and theory.

Chapter 2 Water and Carbon: The Chemical Basis of Life A stronger emphasis on chemical evolution is threaded throughout the chapter to bring chemistry to life for the student reader. Two prominent models for chemical evolution are introduced; the historic Miller prebiotic soup experiment was moved here. Advanced discussion of energy and chemical reactions was moved to a new chapter (see Chapter 8).

Chapter 3 Protein Structure and Function The chapter is reorganized to emphasize the link between structure and function, from amino acids to folded proteins. Updated content illustrates that protein shapes are flexible and dynamic, and may remain incompletely folded until the protein interacts with other molecules or ions. Details of how enzymes work were moved to Chapter 8.

Chapter 4 Nucleic Acids and the RNA World New experimental results concerning the synthesis of nucleotides and nucleic acids in a prebiotic environment are discussed. The section on the RNA world is expanded to include the artificial evolution of a novel ribozyme involved in nucleotide synthesis.

Chapter 5 An Introduction to Carbohydrates The molecular basis for resistance of structural polymers, such as cellulose, to degradation is clarified. A new research box illustrates the role of carbohydrates in cellular recognition and attachment using the egg and sperm of mice as a model system.

Chapter 6 Lipids, Membranes, and the First Cells New content on lipid and membrane evolution and the proposed characteristics of the first protocell is introduced. The aquaporin and potassium channel figures are updated; how key amino acids serve as selectivity filters is now highlighted.

Chapter 7 Inside the Cell Several new electron micrographs were selected to more clearly illustrate cell component structure and function. A new figure is added to better depict the pulse–chase assay used to identify the secretory pathway. Coverage of nuclear transport is expanded to differentiate between passive diffusion and active nuclear import. Updated content emphasizes the role of the cytoskeleton in localizing organelles, and how polarity of microtubules and microfilaments influences their growth rate.

Chapter 8 Energy and Enzymes: An Introduction to Pathways This new chapter pulls together concepts in energy, chemical reactions, and enzymes that previously were covered in three different chapters. Oxidation and reduction reactions are emphasized to prepare students for Chapters 9 and 10. The energetics behind ATP hydrolysis and its role in driving endergonic reactions is discussed, and figures are revised to better illustrate the process. Updated content on enzyme regulation and a new process figure show a model for how metabolic pathways may have evolved.

Chapter 9 Cellular Respiration and Fermentation Two new summary tables for glycolysis and the citric acid cycle are added that provide the names of the enzymes and the reaction each catalyzes. New content is introduced to propose a connection between the universal nature of the proton motive force and the story of the chemical evolution of life.

Chapter 10 Photosynthesis More extensive comparison between the chemical reactions in mitochondria and chloroplasts is added. A new figure is introduced to illustrate noncyclic electron flow in the context of the thylakoid membrane. Greater emphasis is placed on the number of ATPs and NADPHs required for each cycle of carbon fixation and reduction.

Chapter 11 Cell–Cell Interactions Coverage of extracellular matrix structure and function is expanded, including its role in intercellular adhesions and cell signaling. The plant apoplast and symplast are now introduced as key terms in the text and illustrated in a new figure. New content and a new figure on unicellular models for intercellular communication via pheromone sensing (yeast) and quorum sensing (slime mold) are added.

Chapter 12 The Cell Cycle A new figure helps explain the pulse–chase assay for identifying phases of the cell cycle. Content is added to the text and to a figure that illustrates the similarities between chromosome segregation in eukaryotes and prokaryotes. A revised description of anaphase emphasizes how microtubule fraying at the kinetochore can drive chromosome movement. The explanation of how phosphorylation and dephosphorylation turns on MPF activity is updated to reflect current research.

Chapter 13 Meiosis To improve the flow of the chapter, the section on advantages of sexual reproduction was moved to before mistakes in meiosis. The discussion of the role and timing of crossing over during meiosis I is updated. A new study that supports the hypothesis that sex evolved in response to the selective pressure of pathogens is introduced.

Chapter 14 Mendel and the Gene   Material on gene linkage is revised to emphasize the importance of genetic mapping. A new matched set of figures on pedigree analysis brings together the various modes of transmission that were previously shown in four individual figures. A new summary table on characteristics of different patterns of inheritance is added.

Chapter 15 DNA and the Gene: Synthesis and Repair   A new research figure is added that focuses on the relationship between telomere length and senescence in cultured somatic cells.

Chapter 16 How Genes Work   Coverage of the evolving concept of the gene and of different types of RNA is expanded. A figure showing the karyotype of a cancer cell is revised to improve clarity.

Chapter 17 Transcription, RNA Processing, and Translation   The sections on transcription in bacteria and eukaryotes are now separated, and content on charging tRNAs was moved to a new section. The discussion of translation is reorganized, first to emphasize the process in bacteria and then to highlight differences in eukaryotes.

Chapter 18 Control of Gene Expression in Bacteria   Coverage of *lac* operon positive regulation is updated to reflect current research. A new section and new process figure on global gene regulation are added, using the *lexA* regulon as an example.

Chapter 19 Control of Gene Expression in Eukaryotes   Extensive updates to the discussion of epigenetics include a new research box and a section on DNA methylation. Coverage of transcription initiation is updated to reflect current science. A new figure illustrates the role of p53 in the cell cycle in normal and cancerous cells.

Chapter 20 Analyzing and Engineering Genes   The material on sequencing the Neanderthal genome is updated, including evidence of limited Neanderthal genetic material in some modern human populations. New information on current generation sequencing technologies and massive parallelism is added. Recent advances in gene therapy are highlighted.

Chapter 21 Genomics and Beyond   Extensive updates throughout reflect recent advances in genomics. Changes include sequence database statistics, genomes that have been sequenced to study evolutionary relationships, and new figures illustrating gene count versus genome size in prokaryotes and eukaryotes and functional classes of human DNA sequences. A new section on systems biology is added. Also included are notes on the discovery of widespread transcription of eukaryotic genomes, deep sequencing, and the spectrum of mutations in human tumors.

Chapter 22 Principles of Development   New information is added on dedifferentiation in induced pluripotent stem cells, maternal genes in *Drosophila* development, how morphogens work, and tool-kit genes. The order of topics in the discussion of developmental principles is reorganized. The figure on *Hox* genes in *Drosophila* and the mouse is updated.

Chapter 23 An Introduction to Animal Development   The chapter is streamlined by focusing on principles of animal development. The discussion of gametogenesis was moved to Chapter 50 (Animal Reproduction). The presentation of fertilization is simplified, and a new figure summarizing steps of fertilization is added. The figure on gastrulation is modified to better depict the arrangement of the germ layers and their movement.

Chapter 24 An Introduction to Plant Development   The chapter is modified to impart an evolutionary perspective on the similarities and differences in plant and animal development. The chapter also was streamlined by removing material such as details of gametogenesis, which now appears in Chapter 41 (Plant Reproduction).

Chapter 25 Evolution by Natural Selection   Several new key passages are included, among them the use of the Grand Canyon as a context for understanding relative dating of fossils, Darwin's artificial selection experiments with fancy pigeons, and Malthus's concept of struggle for existence. A new example of people living at high altitude in Tibet clarifies the difference between acclimatization and adaptation. An illustrated summary table of common misconceptions is added.

Chapter 26 Evolutionary Processes   Discussion of sexual selection now falls within the section on natural selection, and the terms intersexual and intrasexual selection are added. Several new examples replace those in the Fourth Edition, including inbreeding depression in Florida panthers, gene flow in Oregon steelhead trout, and lateral gene transfer in aphids. An illustrated summary table is added on modes of selection.

Chapter 27 Speciation   Several points are clarified, such as the gradient-like (rather than all-or-nothing) nature of reproductive isolation. The section on polyploidy is reorganized and the figures revised, including a side-by-side comparison of autopolyploidy and allopolyploidy.

Chapter 28 Phylogenies and the History of Life   The phylogenetics section is reorganized and expanded to include three illustrated summary tables and updated life-history timelines. New content is added, including the concept of the Anthropocene, the calendar analogy to the history of the Earth, the Chengjiang fossils, and a Life-in-the-Cambrian illustration. Evidence for the impact hypothesis is combined into an illustrated summary table.

Chapter 29 Bacteria and Archaea   The chapter is updated to include a description of metagenomic experiments with an emphasis on the role of gut bacteria in digestion. A newly recognized phylum of Archaea, the Thaumarchaeota, is included, and the table comparing key characteristics of the Bacteria, Archaea, and Eukarya is streamlined.

Chapter 30 Protists   For simplicity, protist lineages are now referred to throughout the chapter by their more familiar common names. Also, some key lineage boxes were consolidated to trim the number to one box per major lineage. Discussion of the origin of the nuclear envelope and mitochondria is expanded to reflect new thinking on the evolution of eukaryotic cells. Protist life cycle figures are significantly overhauled.

Chapter 31 Green Algae and Land Plants Coverage of the evolution of land plants is expanded to include the importance of UV light and UV-absorbing molecules on the colonization of land. The "Redwood group" is now referred to as the Cupressophyta. Updates emphasize the role(s) of each stage of a life cycle in dispersal and in increasing genetic variation and individual numbers. A new research box is added, showing the importance of flower color to pollinator preference. Plant life-cycle figures are significantly overhauled.

Chapter 32 Fungi Coenocytic fungal hyphae are illustrated with a new image showing GFP-labeled nuclei in *Neurospora crassa*. The chapter now points out the similarity between fungal and animal modes of nutrition, in terms of extracellular digestion and absorption of small molecules. The discussion of lignin degradation is updated, and new descriptions of mutualisms of fungi with animals are included. Fungal life cycle figures are significantly overhauled.

Chapter 33 An Introduction to Animals The chapter is extensively revised to streamline and modernize the presentation, including emphasis on genetic tool kits and symmetry in the phylogeny of animals. The "Themes of Diversification" section is reorganized around five illustrated summary tables. The discussion of life cycles is revised to be more general. Insect metamorphosis has moved to Chapter 34, and a sea urchin life cycle replaces *Obelia*.

Chapter 34 Protostome Animals Two themes are threaded throughout the chapter: the water-to-land transition and modular body plans. The section on lophotrochozoans emphasizes spiral cleavage, indirect versus direct development, hemocoels, and radulas. The section on ecdysozoans highlights segmentation and *Hox* genes, including discussion of the origin of the wing and metamorphosis. Key lineage boxes include new phylogenies for annelids, crustaceans, and chelicerates.

Chapter 35 Deuterostome Animals Updates to reflect current research include revised phylogenies, evolution of flight and feathers, *Australopithecus sediba*, human migration out of Africa, and genetic evidence for interbreeding of *Homo neanderthalensis* and *Homo sapiens*.

Chapter 36 Viruses New content focuses on how viruses contribute to evolution via lateral gene transfer and direct addition of genes to cellular genomes. Content is updated and expanded on viral structure and function, and on lytic and latent infections. Three new figures are added, including a comparison of replication of viruses and cells, how pandemic strains of influenza arise via reassortment, and the devastating impact of the 1918 influenza pandemic.

Chapter 37 Plant Form and Function The use of terminology is streamlined for consistency and clarity. For example,

"lateral meristem" is replaced with "vascular cambium" to avoid confusion with lateral buds, and the description of bark is clarified to avoid using the term phelloderm. Several complex figures were converted to illustrated summary tables.

Chapter 38 Water and Sugar Transport in Plants The chapter is revised to improve accuracy, and points out that water loss is also a means for transporting minerals from roots to shoots. Updated content clarifies the role of energy expenditure in moving water across roots, and the Casparian strip as a barrier to the back diffusion of ions and water out of the root. A new research figure shows the importance of the sucrose proton symporter in long-distance transport in *Arabidopsis*.

Chapter 39 Plant Nutrition In this chapter the coverage of mycorrhizae is modified to emphasize their overall role in nutrient acquisition. In the section on nitrogen fixation, a description of the worldwide practice of crop rotation involving legumes and grains is added.

Chapter 40 Plant Sensory Systems, Signals, and Responses A new research box is added that reveals the essential role of PHOT1 phosphorylation in phototropism. A section on the effect of day length on flowering was moved here from Chapter 41 and is integrated with the discussion of phytochromes. New content on the role of plasmodesmata in plant action potentials is added. The section on how plants respond to pathogens is simplified and updated with an example of control of stomata during a bacterial infection.

Chapter 41 Plant Reproduction To provide a clearer example of a gametophyte-dominant life cycle, the liverwort life cycle has been replaced with a moss life cycle. The term "pollination syndrome" is clarified. Also, a research box on how capsaicin prevents seed predation and facilitates dispersal was reinstated from an earlier edition of *Biological Science*.

Chapter 42 Animal Form and Function The chapter includes a new experiment illustrating physiological trade-offs, along with improved examples of thermoregulatory strategies in animals. Several complex figures were converted to illustrated summary tables.

Chapter 43 Water and Electrolyte Balance in Animals The chapter is reorganized to better integrate the relationship between excretion and water and electrolyte balance. Osmoregulatory strategies are now organized according to the challenges presented by marine, freshwater, and terrestrial habitats. Coverage of osmoregulation in bony fishes versus cartilaginous fishes, mammalian kidney function, and how nonmammalian vertebrates concentrate their urine is expanded and clarified.

Chapter 44 Animal Nutrition This chapter contains new information on nutritional imbalances, including diabetes and obesity.

Chapter 45 Gas Exchange and Circulation Discussion of the insect tracheal system is expanded, including new content on how respiration restricts upper limits of body size of insects.

Details regarding the lymphatic system and heart anatomy in vertebrates are updated, and a new section on cardiovascular disease is added.

Chapter 46 Electrical Signals in Animals This chapter is greatly expanded to reflect recent research and growing interest in neuroscience. New information includes comparative anatomy of vertebrate brains, more case studies of brain injuries or dysfunctions that have led to major discoveries in neuroscience, and the concept of neuroplasticity—especially neurogenesis.

Chapter 47 Animal Sensory Systems Content from the Fourth Edition has been split into two chapters (47, Animal Sensory Systems; 48, Animal Movement). The chapter on sensory systems is now organized by type of sensory reception: mechanoreception (with new coverage on the lateral line system of fishes), photoreception, chemoreception (with new coverage of pheromones), and a new section introducing thermoreception, electroreception, and magnetoreception.

Chapter 48 Animal Movement This new chapter introduces the importance of movement in animals, building from small to large scale. The mechanism of muscle contraction (with revised figures) is covered, followed by discussions of types of muscle tissue (with new content on skeletal-muscle fiber types and parallel- versus pennate-muscle fiber orientation), and skeletal systems (hydrostatic skeletons, exoskeletons, endoskeletons). A completely new final section discusses how biologists study locomotion on land, in the air, and in the water.

Chapter 49 Chemical Signals in Animals Figures and content are updated for clarity. The chapter includes a new section on endocrine disruptors.

Chapter 50 Animal Reproduction New content includes details of sperm and egg structure and function, reproduction in the spotted hyena, and human contraceptive methods.

Chapter 51 The Immune System in Animals Coverage of the innate immune response is expanded to include more detail on Toll-like receptors and how they transmit signals. The section on adaptive immunity is reorganized to improve flow. Updated content on inappropriate immune responses (autoimmunity and allergies) and inadequate responses (immunodeficiency) is grouped together in one section. The hygiene hypothesis is introduced to explain the growing trend of inappropriate immune responses in populations that have reduced exposure to common pathogens and parasites.

Chapter 52 An Introduction to Ecology The first section on levels of ecological study is expanded to include global ecology. The rest of the chapter is reorganized, beginning with the factors that determine the distribution and abundance of organisms (including a new Argentine ant case study) and ending with biomes. The biome section is streamlined with a new emphasis on human impacts, including an introduction to anthropogenic biomes and the Anthropocene.

Chapter 53 Behavioral Ecology The introduction is revised to provide a clearer framework for types of behavior. Sections are now organized around proximate versus ultimate causation. These new examples replace those in the Fourth Edition: Argentine ant territorial behavior (replacing spiny lobsters), optimal foraging in gerbils (replacing white-fronted bee-eaters), sexual selection in *Anolis* lizards (replacing barn swallows), and map orientation in green sea turtles.

Chapter 54 Population Ecology A new introductory section focuses on the distribution of organisms in populations, including dispersion patterns and a consolidated discussion of measurement methods. The human population content is updated and separated into a new section. The quantitative methods boxes and life table are now more student friendly.

Chapter 55 Community Ecology Several changes to content are made, such as a clarification of competitive exclusion versus niche differentiation. New content includes a summary table on constitutive defenses, a discovery story on mimicry (including Bates and Müller), an introduction to food webs, and the process of soil formation in primary succession.

Chapter 56 Ecosystems The title and scope of the chapter are updated to include global ecology. Extensive revisions include many content updates and new figures on the relationship between GPP and NPP, the one-way flow of energy and cycling of nutrients, the food web, open versus closed aquifers, the High Plains Aquifer, the biomagnification of DDT, and the greenhouse effect. The climate change section is expanded and updated, including an illustrated summary table.

Chapter 57 Biodiversity and Conservation Biology Throughout the chapter, there is more emphasis on conserving ecosystem function. Many new examples are added, including Smits's restoration project in Borneo, the Census of Marine Life, the IUCN Red List, the Sinervo lizard extinction experiment, and Florida panther genetic restoration. New summary tables highlight ecosystem services and threats to biodiversity.

Acknowledgments

Reviewers

The peer review system is the key to quality and clarity in science publishing. In addition to providing a filter, the investment that respected individuals make in vetting the material—catching errors or inconsistencies and making suggestions to improve the presentation—gives authors, editors, and readers confidence that the text meets rigorous professional standards.

Peer review plays the same role in textbook publishing. The time and care that this book's reviewers have invested is a tribute to their professional integrity, their scholarship, and their concern for the quality of teaching. Virtually every paragraph in this edition has been revised and improved based on insights from the following individuals.

Tamarah Adair, *Baylor University*
Sandra D. Adams, *Montclair State University*
Marc Albrecht, *University of Nebraska at Kearney*
Larry Alice, *Western Kentucky University*
Leo M. Alves, *Manhattan College*
David R. Angelini, *American University*
Dan Ardia, *Franklin & Marshall College*
Paul Arriola, *Elmhurst College*
Davinderjit K. Bagga, *University of Montevallo*
Susan Barrett, *Wheaton College*
Donald Baud, *University of Memphis*
Vernon W. Bauer, *Francis Marion University*
Robert Bauman, *Amarillo College*
Christopher Beck, *Emory University*
Vagner Benedito, *West Virginia University*
Scott Bingham, *Arizona State University*
Stephanie Bingham, *Barry University*
Wendy Birky, *California State University, Northridge*
Jason Blank, *California Polytechnic State University*
Kristopher A. Blee, *California State University, Chico*
Margaret Bloch-Qazi, *Gustavus Adolphus College*
Lanh Bloodworth, *Florida State College at Jacksonville*
Catherine H. Borer, *Berry College*
James Bottesch, *Brevard Community College*
Jacqueline K. Bowman, *Arkansas Tech University*
John Bowman, *University of California, Davis*
Chris Brochu, *University of Iowa*
Matthew Brown, *Dalhousie University*
Mark Browning, *Purdue University*
Carolyn J. W. Bunde, *Idaho State University*
David Byres, *Florida State College at Jacksonville*
Michael Campbell, *Penn State Erie*
Manel Camps, *University of California, Santa Cruz*
Geralyn M. Caplan, *Owensboro Community and Technical College*
Richard Cardullo, *University of California, Riverside*

David Carlini, *American University*
Dale Casamatta, *University of North Florida*
Deborah Chapman, *University of Pittsburgh*
Joe Coelho, *Quincy University*
Allen Collins, *Smithsonian Museum of Natural History*
Robert A. Colvin, *Ohio University*
Kimberly L. Conner, *Florida State College at Jacksonville*
Karen Curto, *University of Pittsburgh*
Clarissa Dirks, *Evergreen State College*
Peter Ducey, *SUNY Cortland*
Erastus Dudley, *Huntingdon College*
Jeffrey P. Duguay, *Delta State University*
Tod Duncan, *University of Colorado, Denver*
Joseph Esdin, *University of California, Los Angeles*
Brent Ewers, *University of Wyoming*
Amy Farris, *Ivy Tech Community College*
Bruce Fisher, *Roane State Community College*
Ryan Fisher, *Salem State University*
David Fitch, *New York University*
Elizabeth Fitch, *Motlow State Community College*
Michael P. Franklin, *California State University, Northridge*
Susannah French, *Utah State University*
Caitlin Gabor, *Texas State University*
Matthew Gilg, *University of North Florida*
Kendra Greenlee, *North Dakota State University*
Patricia A. Grove, *College of Mount Saint Vincent*
Nancy Guild, *University of Colorado, Boulder*
Cynthia Hemenway, *North Carolina State University*
Christopher R. Herlihy, *Middle Tennessee State University*
Kendra Hill, *South Dakota State University*
Sara Hoot, *University of Wisconsin, Milwaukee*
Kelly Howe, *University of New Mexico*
Robin Hulbert, *California Polytechnic State University*
Rick Jellen, *Brigham Young University*
Russell Johnson, *Colby College*
William Jira Katembe, *Delta State University*
Elena K. Keeling, *California Polytechnic State University*
Jill B. Keeney, *Juniata College*
Greg Kelly, *University of Western Ontario*
Scott L. Kight, *Montclair State University*
Charles Knight, *California Polytechnic State University*
Jenny Knight, *University of Colorado, Boulder*
William Kroll, *Loyola University Chicago*
Dominic Lannutti, *El Paso Community College*
Brenda Leady, *University of Toledo*
David Lindberg, *University of California, Berkeley*
Barbara Lom, *Davidson College*
Robert Maxwell, *Georgia State University*
Marshall D. McCue, *St. Mary's University*
Kurt A. McKean, *SUNY Albany*
Michael Meighan, *University of California, Berkeley*
John Merrill, *Michigan State University*

Richard Merritt, *Houston Community College*
Alan Molumby, *University of Illinois at Chicago*
Jeremy Montague, *Barry University*
Chad E. Montgomery, *Truman State University*
Kimberly D. Moore, *Lone Star College System, North Harris*
Michael Morgan, *Berry College*
James Mulrooney, *Central Connecticut State University*
John D. Nagy, *Scottsdale Community College*
Margaret Olney, *St. Martin's University*
Nathan Okia, *Auburn University at Montgomery*
Robert Osuna, *SUNY Albany*
Daniel Panaccione, *West Virginia University*
Stephanie Pandolfi, *Michigan State University*
Michael Rockwell Parker, *Monell Chemical Senses Center*
Lisa Parks, *North Carolina State University*
Nancy Pelaez, *Purdue University*
Shelley W. Penrod, *Lone Star College System, North Harris*
Andrea Pesce, *James Madison University*
Raymond Pierotti, *University of Kansas*
Melissa Ann Pilgrim, *University of South Carolina Upstate*
Paul Pillitteri, *Southern Utah University*
Debra Pires, *University of California, Los Angeles*
P. David Polly, *Indiana University, Bloomington*
Vanessa Quinn, *Purdue University North Central*
Stacey L. Raimondi, *Elmhurst College*
Stephanie Randell, *McLennan Community College*
Marceau Ratard, *Delgado Community College*
Flona Redway, *Barry University*
Srebrenka Robic, *Agnes Scott College*
Dave Robinson, *Bellarmine University*
George Robinson, *SUNY Albany*
Adam W. Rollins, *Lincoln Memorial University*
Amanda Rosenzweig, *Delgado Community College*
Leonard C. Salvatori, *Indian River State College*
Dee Ann Sato, *Cypress College*
Leena Sawant, *Houston Community College*
Jon Scales, *Midwestern State University*
Oswald Schmitz, *Yale University*
Joan Sharp, *Simon Fraser University*
Julie Schroer, *North Dakota State University*
Timothy E. Shannon, *Francis Marion University*
Lynnette Sievert, *Emporia State University*
Susan Skambis, *Valencia College*
Ann E. Stapleton, *University of North Carolina, Wilmington*
Mary-Pat Stein, *California State University, Northridge*
Christine Strand, *California Polytechnic State University*
Denise Strickland, *Midlands Technical College*
Jackie Swanik, *Wake Technical Community College*
Billie J. Swalla, *University of Washington*
Zuzana Swigonova, *University of Pittsburgh*
Briana Timmerman, *University of South Carolina*
Catherine Ueckert, *Northern Arizona University*
Sara Via, *University of Maryland, College Park*
Thomas J. Volk, *University of Wisconsin–La Crosse*
Jeffrey Walck, *Middle Tennessee State University*
Andrea Weeks, *George Mason University*
Margaret S. White, *Scottsdale Community College*
Steven D. Wilt, *Bellarmine University*
Candace Winstead, *California Polytechnic State University*
James A. Wise, *Hampton University*

Correspondents

One of the most enjoyable interactions we have as textbook authors is correspondence or conversations with researchers and teachers who take the time and trouble to contact us to discuss an issue with the book, or who respond to our queries about a particular data set or study. We are always amazed and heartened by the generosity of these individuals. They care, deeply.

Lawrence Alice, *Western Kentucky University*
David Baum, *University of Wisconsin–Madison*
Meredith Blackwell, *Louisiana State University*
Nancy Burley, *University of California, Irvine*
Thomas Breithaupt, *University of Hull*
Philip Cantino, *Ohio University*
Allen Collins, *Smithsonian Museum of Natural History*
Robert Full, *University of California, Berkeley*
Arundhati Ghosh, *University of Pittsburgh*
Jennifer Gottwald, *University of Wisconsin–Madison*
Jon Harrison, *Arizona State University*
David Hawksworth, *Natural History Museum, London*
Jim Herrick, *James Madison University*
John Hunt, *University of Exeter*
Doug Jensen, *Converse College*
Scott Kight, *Montclair State University*
Scott Kirkton, *Union College*
Mimi Koehl, *University of California, Berkeley*
Rodger Kram, *University of Colorado*
Matthew McHenry, *University of California, Irvine*
Alison Miyamoto, *California State University, Fullerton*
Sean Menke, *Lake Forest College*
Rich Mooi, *California Academy of Sciences*
Michael Oliver, *MalawiCichlids.com*
M. Rockwell Parker, *Monell Chemical Senses Center*
Andrea Pesce, *James Madison University*
Chris Preston, *Monterey Bay Aquarium Research Institute*
Scott Sakaluk, *Illinois State University*
Kyle Seifert, *James Madison University*
Jos Snoeks, *Royal Museum for Central Africa*
Jeffrey Spring, *University of Louisiana*
Christy Strand, *California Polytechnic State University, San Luis Obispo*
Torsten Struck, *University of Osnabrueck, Germany*
Oswald Schmitz, *Yale University*
Ian Tattersal, *American Museum of Natural History*
Robert Turgeon, *Cornell University*
Tom Volk, *University of Wisconsin–La Crosse*
Naomi Wernick, *University of Massachusetts, Lowell*

Supplements Contributors

Instructors depend on an impressive array of support materials—in print and online—to design and deliver their courses. The student experience would be much weaker without the study guide, test bank, activities, animations, quizzes, and tutorials written by the following individuals.

Brian Bagatto, *University of Akron*
Scott Bingham, *Arizona State University*
Jay L. Brewster, *Pepperdine University*

Mirjana Brockett, *Georgia Institute of Technology*
Warren Burggren, *University of North Texas*
Jeff Carmichael, *University of North Dakota*
Tim Christensen, *East Carolina University*
Erica Cline, *University of Washington—Tacoma*
Patricia Colberg, *University of Wyoming*
Elia Crisucci, *University of Pittsburgh*
Elizabeth Cowles, *Eastern Connecticut State University*
Clarissa Dirks, *Evergreen State College*
Lisa Elfring, *University of Arizona, Tucson*
Brent Ewers, *University of Wyoming*
Rebecca Ferrell, *Metropolitan State University of Denver*
Miriam Ferzli, *North Carolina State University*
Cheryl Frederick, *University of Washington*
Cindee Giffen, *University of Wisconsin–Madison*
Kathy M. Gillen, *Kenyon College*
Linda Green, *Georgia Institute of Technology*
Christopher Harendza, *Montgomery County Community College*
Cynthia Hemenway, *North Carolina State University*
Laurel Hester, *University of South Carolina*
Jean Heitz, *University of Wisconsin–Madison*
Tracey Hickox, *University of Illinois, Urbana–Champaign*
Jacob Kerby, *University of South Dakota*
David Kooyman, *Brigham Young University*
Barbara Lom, *Davidson College*
Cindy Malone, *California State University, Northridge*
Jim Manser, retired, *Harvey Mudd College*
Jeanette McGuire, *Michigan State University*
Mark Music, *Indian River State College*
Jennifer Nauen, *University of Delaware*
Chris Pagliarulo, *University of California, Davis*
Stephanie Scher Pandolfi, *Michigan State University*
Lisa Parks, *North Carolina State University*
Debra Pires, *University of California, Los Angeles*
Carol Pollock, *University of British Columbia*
Jessica Poulin, *University at Buffalo, the State University of New York*
Vanessa Quinn, *Purdue University North Central*
Eric Ribbens, *Western Illinois University*
Christina T. Russin, *Northwestern University*
Leonard Salvatori, *Indian River State College*
Joan Sharp, *Simon Fraser University*
Chrissy Spencer, *Georgia Institute of Technology*
Mary-Pat Stein, *California State University, Northridge*
Suzanne Simon-Westendorf, *Ohio University*
Fred Wasserman, *Boston University*
Cindy White, *University of Northern Colorado*
Edward Zalisko, *Blackburn College*

Book Team

Anyone who has been involved in producing a textbook knows that many people work behind the scenes to make it all happen. The coauthor team is indebted to the many talented individuals who have made this book possible.

Development editors Mary Catherine Hager, Moira Lerner-Nelson, and Bill O'Neal provided incisive comments on the revised manuscript. Fernanda Oyarzun and Adam Steinberg used their artistic sense, science skills, and love of teaching to hone the figures for many chapters.

The final version of the text was copyedited by Chris Thillen and expertly proofread by Pete Shanks. The final figure designs were rendered by Imagineering Media Services and carefully proofread by Frank Purcell. Maureen Spuhler, Eric Schrader, and Kristen Piljay researched images for the Fifth Edition.

The book's clean, innovative design was developed by Mark Ong and Emily Friel. Text and art were skillfully set in the design by S4Carlisle Publishing Services. The book's production was supervised by Lori Newman and Mike Early.

The extensive supplements program was managed by Brady Golden and Katie Cook. All of the individuals mentioned—and more—were supported with cheerful, dedicated efficiency by Editorial Assistant Leslie Allen for the first half of the project; Eddie Lee has since stepped in to skillfully fill this role.

Creating MasteringBiology® tutorials and activities also requires a team. Media content development was overseen by Tania Mlawer and Sarah Jensen, who benefited from the program expertise of Caroline Power and Caroline Ross. Joseph Mochnick and Daniel Ross worked together as media producers. Lauren Fogel (VP, Director, Media Development), Stacy Treco (VP, Director, Media Product Strategy), and Laura ensured that the complete media program that accompanies the Fifth Edition, including MasteringBiology, will meet the needs of the students and professors who use our offerings.

Pearson's talented sales reps, who listen to professors, advise the editorial staff, and get the book in students' hands, are supported by tireless Executive Marketing Manager Lauren Harp and Director of Marketing Christy Lesko. The marketing materials that support the outreach effort were produced by Lillian Carr and her colleagues in Pearson's Marketing Comunications group. David Theisen, national director for Key Markets, tirelessly visits countless professors each year, enthusiastically discussing their courses and providing us with meaningful editorial guidance.

The vision and resources required to run this entire enterprise are the responsibility of Vice President and Editor-in-Chief Beth Wilbur, who provided inspirational and focused leadership, and President of Pearson Science Paul Corey, who displays unwavering commitment to high-quality science publishing.

Becky Ruden recruited the coauthor team, drawing us to the project with her energy and belief in this book. The editorial team was skillfully directed by Executive Director of Development Deborah Gale. Finally, we are deeply grateful for three key drivers of the Fifth Edition. Project Editor Anna Amato's superb organizational skills and calm demeanor assured that all the wheels and cogs of the process ran smoothly to keep the mammoth project steadily rolling forward. Supervising Development Editor Sonia DiVittorio's deep expertise, creative vision, keen attention to detail, level, and clarity, and inspiring insistence on excellence kept the bar high for everyone on every aspect of the project. Lastly, Senior Acquisitions Editor Michael Gillespie's boundless energy and enthusiasm, positive attitude, and sharp intellect have fueled and united the team and also guided the book through the hurdles to existence. The coauthor team thanks these exceptional people for making the art and science of book writing a productive and exhilarating process.

1 Biology and the Tree of Life

In this chapter you will learn about

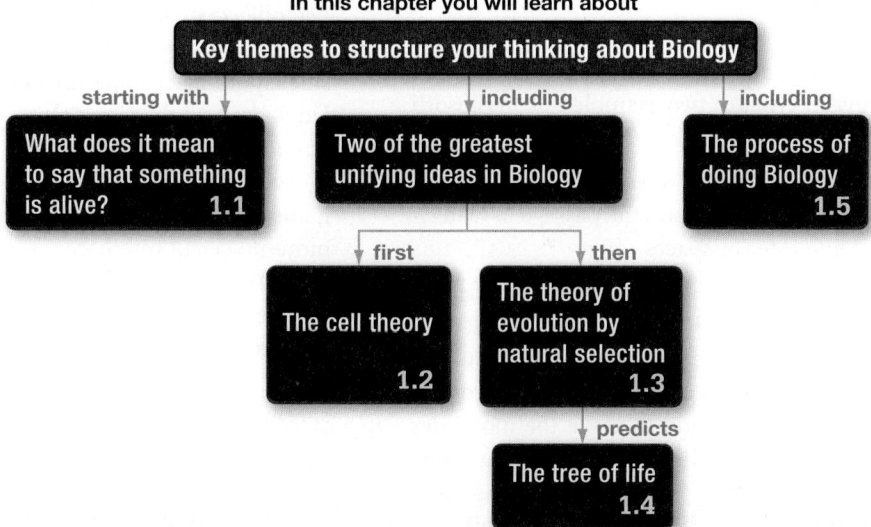

Key themes to structure your thinking about Biology

starting with → **What does it mean to say that something is alive? 1.1**

including → **Two of the greatest unifying ideas in Biology**

including → **The process of doing Biology 1.5**

first → **The cell theory 1.2**

then → **The theory of evolution by natural selection 1.3**

predicts → **The tree of life 1.4**

These Chinese water dragon hatchlings are exploring their new world and learning how to find food and stay alive. They represent one of the key characteristics of life introduced in this chapter—replication.

n essence, biological science is a search for ideas and observations that unify our understanding of the diversity of life, from bacteria living in rocks a mile underground to humans and majestic sequoia trees. This chapter is an introduction to this search.

The goals of this chapter are to introduce the nature of life and explore how biologists go about studying it. The chapter also introduces themes that will resonate throughout this book:

- Analyzing how organisms work at the molecular level.
- Understanding organisms in terms of their evolutionary history.
- Helping you learn to think like a biologist.

Let's begin with what may be the most fundamental question of all: What is life?

This chapter is part of the Big Picture. See how on pages 16–17.

✔ When you see this checkmark, stop and test yourself. Answers are available in Appendix A.

1.1 What Does It Mean to Say That Something Is Alive?

An **organism** is a life-form—a living entity made up of one or more cells. Although there is no simple definition of life that is endorsed by all biologists, most agree that organisms share a suite of five fundamental characteristics.

- *Energy*   To stay alive and reproduce, organisms have to acquire and use energy. To give just two examples: plants absorb sunlight; animals ingest food.

- *Cells*   Organisms are made up of membrane-bound units called cells. A cell's membrane regulates the passage of materials between exterior and interior spaces.

- *Information*   Organisms process hereditary, or genetic, information encoded in units called genes. Organisms also respond to information from the environment and adjust to maintain stable internal conditions. Right now, cells throughout your body are using information to make the molecules that keep you alive; your eyes and brain are decoding information on this page that will help you learn some biology, and if your room is too hot you might be sweating to cool off.

- *Replication*   One of the great biologists of the twentieth century, François Jacob, said that the "dream of a bacterium is to become two bacteria." Almost everything an organism does contributes to one goal: replicating itself.

- *Evolution*   Organisms are the product of evolution, and their populations continue to evolve.

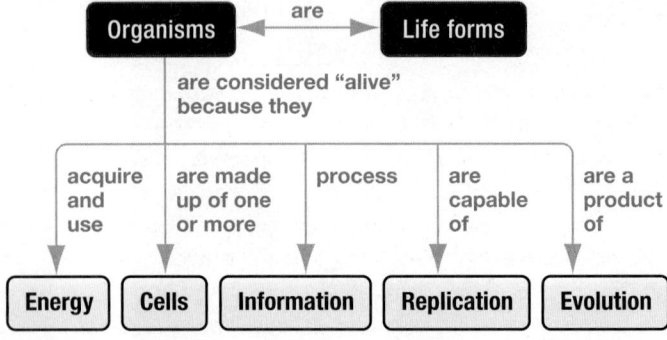

You can think of this text as one long exploration of these five traits. Here's to life!

1.2 The Cell Theory

Two of the greatest unifying ideas in all of science laid the groundwork for modern biology: the cell theory and the theory of evolution by natural selection. Formally, scientists define a **theory** as an explanation for a very general class of phenomena or observations that are supported by a wide body of evidence. The cell theory and theory of evolution address fundamental questions: What are organisms made of? Where do they come from?

When these concepts emerged in the mid-1800s, they revolutionized the way biologists think about the world. They established

two of the five attributes of life: Organisms are cellular, and their populations change over time.

Neither insight came easily, however. The cell theory, for example, emerged after some 200 years of work. In 1665 the Englishman Robert Hooke devised a crude microscope to examine the structure of cork (a bark tissue) from an oak tree. The instrument magnified objects to just 30× (30 times) their normal size, but it allowed Hooke to see something extraordinary. In the cork he observed small, pore-like compartments that were invisible to the naked eye. Hooke coined the term "cells" for these structures because of their resemblance to the cells inhabited by monks in a monastery.

Soon after Hooke published his results, a Dutch scientist named Anton van Leeuwenhoek succeeded in developing much more powerful microscopes, some capable of magnifications up to 300×. With these instruments, van Leeuwenhoek inspected samples of pond water and made the first observations of a dazzling collection of single-celled organisms that he called "animalcules." He also observed and described human blood cells and sperm cells, shown in **FIGURE 1.1**.

In the 1670s an Italian researcher who was studying the leaves and stems of plants with a microscope concluded that plant tissues were composed of many individual cells. By the early 1800s, enough data had accumulated for a German biologist to claim that *all* organisms consist of cells. Did this claim hold up?

All Organisms Are Made of Cells

Advances in microscopy have made it possible to examine the amazing diversity and complexity of cells at higher and higher magnifications. Biologists have developed microscopes that are tens of thousands of times more powerful than van Leeuwenhoek's and have described over a million new species. The basic conclusion made in the 1800s remains intact, however: All organisms are made of cells.

The smallest organisms known today are bacteria that are barely 200 nanometers wide, or 200 *billionths* of a meter. (See **BioSkills 1** in Appendix B to review the metric system and its prefixes.[1]) It would take 5000 of these organisms lined up side by side to span a millimeter. This is the distance between the smallest hash marks on a metric ruler. In contrast, sequoia trees can be over 100 meters tall. This is the equivalent of a 20-story building. Bacteria and sequoias are composed of the same fundamental building block, however—the cell. Bacteria consist of a single cell; sequoias are made up of many cells.

Today a **cell** is defined as a highly organized compartment that is bounded by a thin, flexible structure called a plasma membrane and that contains concentrated chemicals in an aqueous (watery) solution. The chemical reactions that sustain life take place inside cells. Most cells are also capable of reproducing by dividing—in effect, by making a copy of themselves.

The realization that all organisms are made of cells was fundamentally important, but it formed only the first part of the cell

---

[1]BioSkills are located in the second appendix at the back of the book. They focus on general skills that you'll use throughout this course. More than a few students have found them to be a life-saver. Please use them!

(a) van Leeuwenhoek built his own microscopes—which, while small, were powerful. They allowed him to see, for example . . .

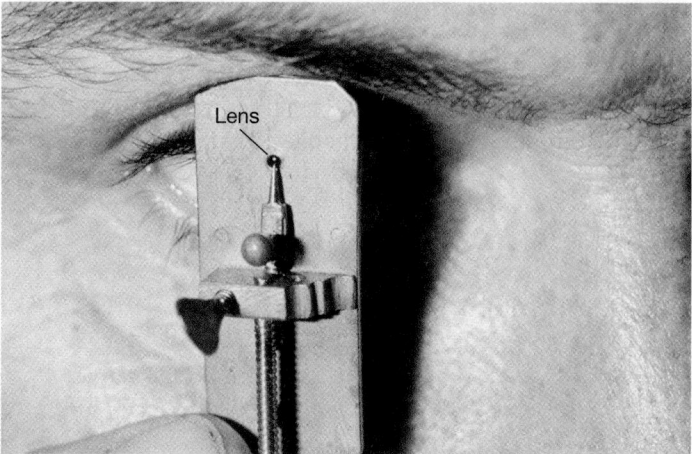

Lens

(b) . . . human blood cells (this modern photo was shot through one of van Leeuwenhoek's original microscopes) . . .

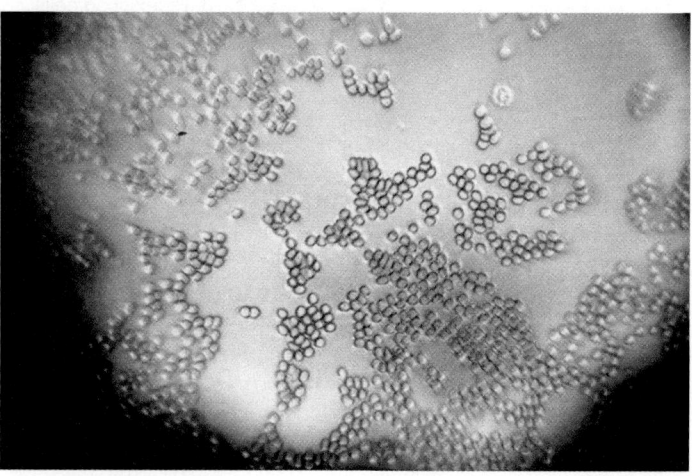

(c) . . . and animal sperm (drawing by van Leeuwenhoek of canine sperm cells on left, human on right).

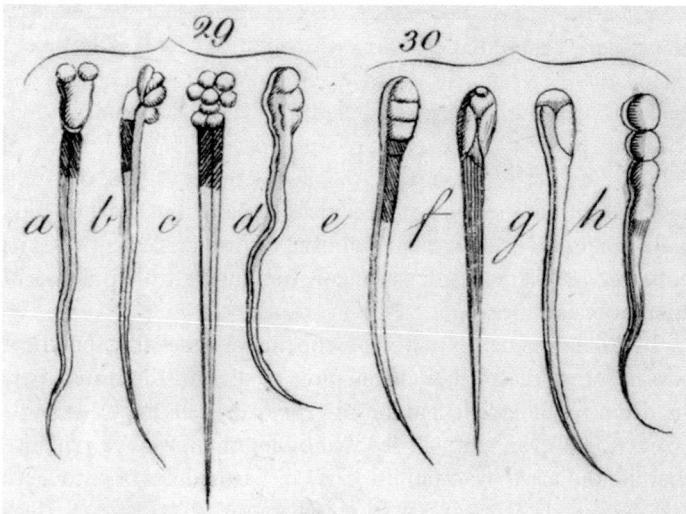

FIGURE 1.1 Van Leeuwenhoek's Microscope Made Cells Visible.

theory. In addition to understanding what organisms are made of, scientists wanted to understand how cells come to be.

Where Do Cells Come From?

Most scientific theories have two components: The first describes a pattern in the natural world; the second identifies a mechanism or process that is responsible for creating that pattern. Hooke and his fellow scientists articulated the pattern component of the cell theory. In 1858, a German scientist named Rudolph Virchow added the process component by stating that all cells arise from preexisting cells.

The complete **cell theory** can be stated as follows: All organisms are made of cells, and all cells come from preexisting cells.

Two Hypotheses The cell theory was a direct challenge to the prevailing explanation of where cells come from, called spontaneous generation. In the mid-1800s, most biologists believed that organisms could arise spontaneously under certain conditions. For example, the bacteria and fungi that spoil foods such as milk and wine were thought to appear in these nutrient-rich media of their own accord—springing to life from nonliving materials. In contrast, the cell theory maintained that cells do not spring to life spontaneously but are produced only when preexisting cells grow and divide. The all-cells-from-cells explanation was a **hypothesis:** a testable statement to explain a phenomenon or a set of observations.

Biologists usually use the word theory to refer to proposed explanations for broad patterns in nature and prefer hypothesis to refer to explanations for more tightly focused questions. A theory serves as a framework for the development of new hypotheses.

An Experiment to Settle the Question Soon after Virchow's all-cells-from-cells hypothesis appeared in print, a French scientist named Louis Pasteur set out to test its predictions experimentally. An experimental **prediction** describes a measurable or observable result that must be correct if a hypothesis is valid.

Pasteur wanted to determine whether microorganisms could arise spontaneously in a nutrient broth or whether they appear only when a broth is exposed to a source of preexisting cells. To address the question, he created two treatment groups: a broth that was not exposed to a source of preexisting cells and a broth that was.

The spontaneous generation hypothesis predicted that cells would appear in both treatment groups. The all-cells-from-cells hypothesis predicted that cells would appear only in the treatment exposed to a source of preexisting cells.

FIGURE 1.2 (on page 4) shows Pasteur's experimental setup. Note that the two treatments are identical in every respect but one. Both used glass flasks filled with the same amount of the same nutrient broth. Both were boiled for the same amount of time to kill any existing organisms such as bacteria or fungi. But because the flask pictured in Figure 1.2a had a straight neck, it was exposed to preexisting cells after sterilization by the heat treatment. These preexisting cells are the bacteria and fungi that cling to dust particles in the air. They could drop into the nutrient broth because the neck of the flask was straight.

In contrast, the flask drawn in Figure 1.2b had a long swan neck. Pasteur knew that water would condense in the crook of the swan neck after the boiling treatment and that this pool of water

QUESTION: Do cells arise spontaneously or from other cells?

SPONTANEOUS GENERATION HYPOTHESIS: Cells arise spontaneously from nonliving materials.

ALL-CELLS-FROM-CELLS HYPOTHESIS: Cells are produced only when preexisting cells grow and divide.

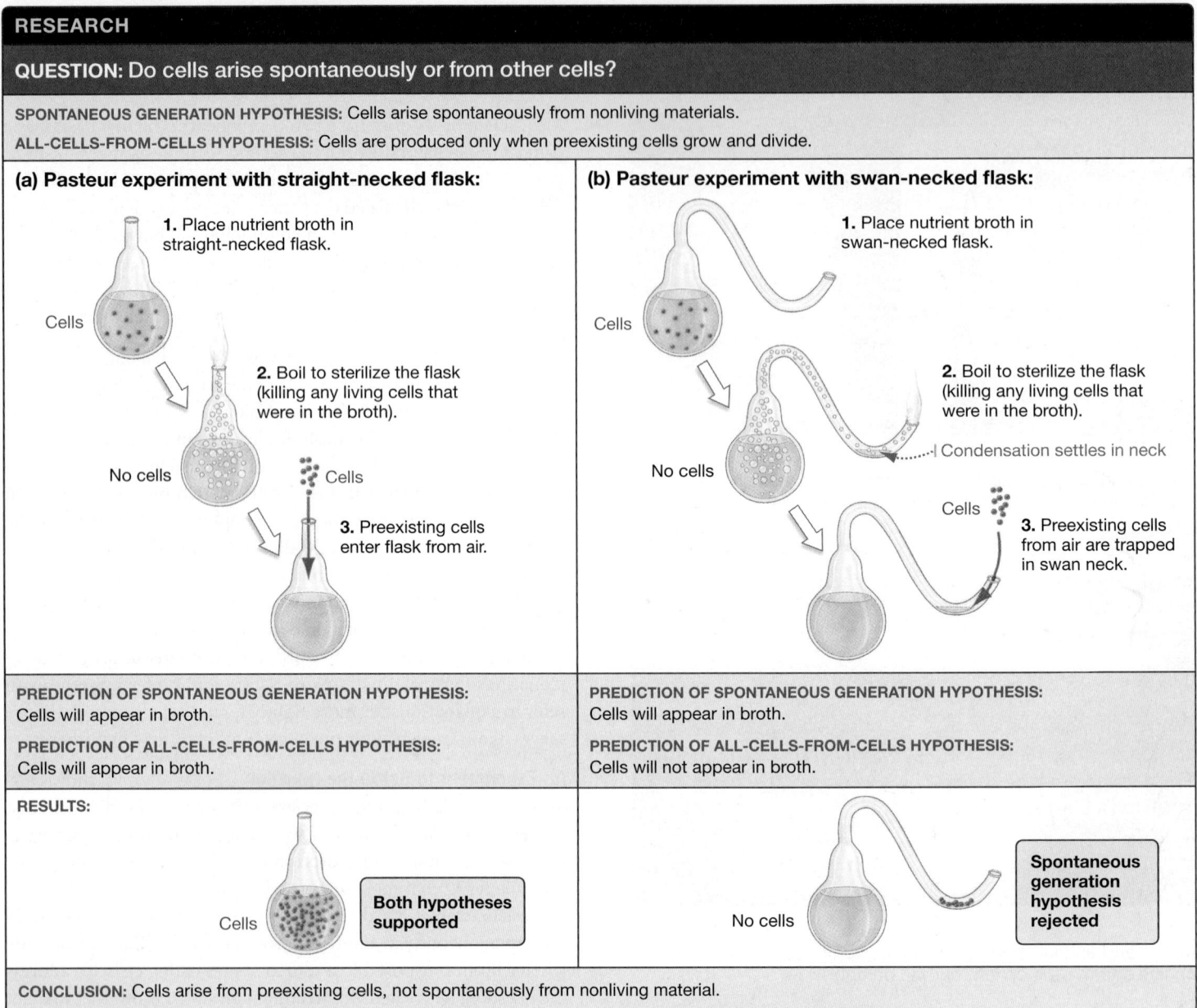

(a) Pasteur experiment with straight-necked flask:

1. Place nutrient broth in straight-necked flask.

Cells

2. Boil to sterilize the flask (killing any living cells that were in the broth).

No cells

Cells

3. Preexisting cells enter flask from air.

(b) Pasteur experiment with swan-necked flask:

1. Place nutrient broth in swan-necked flask.

Cells

2. Boil to sterilize the flask (killing any living cells that were in the broth).

No cells

Condensation settles in neck

Cells

3. Preexisting cells from air are trapped in swan neck.

PREDICTION OF SPONTANEOUS GENERATION HYPOTHESIS: Cells will appear in broth.

PREDICTION OF ALL-CELLS-FROM-CELLS HYPOTHESIS: Cells will appear in broth.

PREDICTION OF SPONTANEOUS GENERATION HYPOTHESIS: Cells will appear in broth.

PREDICTION OF ALL-CELLS-FROM-CELLS HYPOTHESIS: Cells will not appear in broth.

RESULTS:

Cells

Both hypotheses supported

No cells

Spontaneous generation hypothesis rejected

CONCLUSION: Cells arise from preexisting cells, not spontaneously from nonliving material.

FIGURE 1.2 The Spontaneous Generation and All-Cells-from-Cells Hypotheses Were Tested Experimentally.

✔**QUESTION** What problem would arise in interpreting the results of this experiment if Pasteur had (1) put different types of broth in the two treatments, (2) heated them for different lengths of time, or (3) used a ceramic flask for one treatment and a glass flask for the other?

would trap any bacteria or fungi that entered on dust particles. Thus, the contents of the swan-necked flask were isolated from any source of preexisting cells even though still open to the air.

Pasteur's experimental setup was effective because there was only one difference between the two treatments and because that difference was the factor being tested—in this case, a broth's exposure to preexisting cells.

One Hypothesis Supported And Pasteur's results? As Figure 1.2 shows, the treatment exposed to preexisting cells quickly filled with bacteria and fungi. This observation was important because it showed that the heat sterilization step had not altered the nutrient broth's capacity to support growth.

The broth in the swan-necked flask remained sterile, however. Even when the flask was left standing for months, no organisms appeared in it. This result was inconsistent with the hypothesis of spontaneous generation.

Because Pasteur's data were so conclusive—meaning that there was no other reasonable explanation for them—the results persuaded most biologists that the all-cells-from-cells hypothesis was correct. However, you will see that biologists now have evidence that life did arise from nonlife early in Earth's history, through a process called chemical evolution (Chapters 2–6).

The success of the cell theory's process component had an important implication: If all cells come from preexisting cells, it follows that all individuals in an isolated population of single-celled organisms are related by common ancestry. Similarly, in you and most other multicellular individuals, all the cells present are descended from preexisting cells, tracing back to a fertilized egg. A fertilized egg is a cell created by the fusion of sperm and egg—cells that formed in individuals of the previous generation. In this way, all the cells in a multicellular organism are connected by common ancestry.

The second great founding idea in biology is similar, in spirit, to the cell theory. It also happened to be published the same year as the all-cells-from-cells hypothesis. This was the realization, made independently by the English scientists Charles Darwin and Alfred Russel Wallace, that all species—all distinct, identifiable types of organisms—are connected by common ancestry.

1.3 The Theory of Evolution by Natural Selection

In 1858 short papers written separately by Darwin and Wallace were read to a small group of scientists attending a meeting of the Linnean Society of London. A year later, Darwin published a book that expanded on the idea summarized in those brief papers. The book was called *The Origin of Species*. The first edition sold out in a day.

What Is Evolution?

Like the cell theory, the theory of evolution by natural selection has a pattern and a process component. Darwin and Wallace's theory made two important claims concerning patterns that exist in the natural world.

1. Species are related by common ancestry. This contrasted with the prevailing view in science at the time, which was that species represent independent entities created separately by a divine being.

2. In contrast to the accepted view that species remain unchanged through time, Darwin and Wallace proposed that the characteristics of species can be modified from generation to generation. Darwin called this process descent with modification.

Evolution is a change in the characteristics of a population over time. It means that species are not independent and unchanging entities, but are related to one another and can change through time.

What Is Natural Selection?

This pattern component of the theory of evolution was actually not original to Darwin and Wallace. Several scientists had already come to the same conclusions about the relationships between species. The great insight by Darwin and Wallace was in proposing a process, called **natural selection,** that explains *how* evolution occurs.

Two Conditions of Natural Selection Natural selection occurs whenever two conditions are met.

1. Individuals within a population vary in characteristics that are **heritable**—meaning, traits that can be passed on to offspring.

A **population** is defined as a group of individuals of the same species living in the same area at the same time.

2. In a particular environment, certain versions of these heritable traits help individuals survive better or reproduce more than do other versions.

If certain heritable traits lead to increased success in producing offspring, then those traits become more common in the population over time. In this way, the population's characteristics change as a result of natural selection acting on individuals. This is a key insight: Natural selection acts on individuals, but evolutionary change occurs in populations.

Selection on Maize as an Example To clarify how selection works, consider an example of **artificial selection**—changes in populations that occur when *humans* select certain individuals to produce the most offspring. Beginning in 1896, researchers began a long-term selection experiment on maize (corn).

1. In the original population, the percentage of protein in maize kernels was variable among individuals. Kernel protein content is a heritable trait—parents tend to pass the trait on to their offspring.

2. Each year for many years, researchers chose individuals with the highest kernel protein content to be the parents of the next generation. In this environment, individuals with high kernel protein content produced more offspring than individuals with low kernel protein content.

FIGURE 1.3 shows the results. Note that this graph plots generation number on the *x*-axis, starting from the first generation (0 on the graph) and continuing for 100 generations. The average percentage of protein in a kernel among individuals in this population is plotted on the *y*-axis.

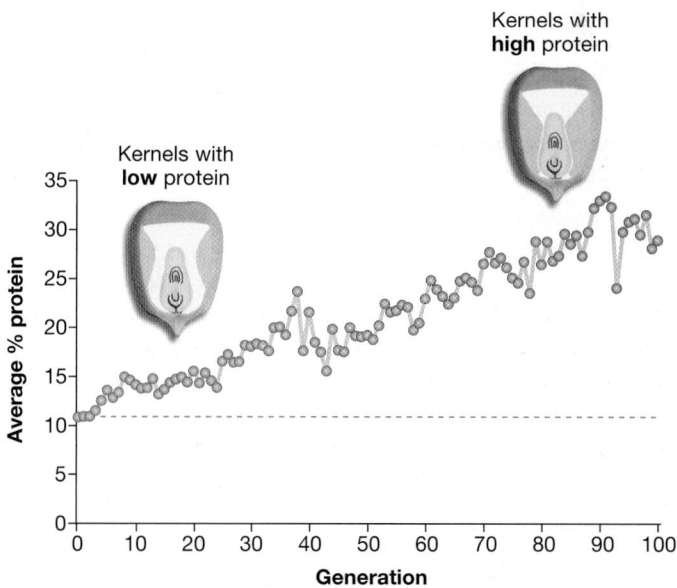

FIGURE 1.3 Response to Selection for High Kernel Protein Content in Maize.

DATA: Moose, S. P., J. W. Dudley, and T. R. Rocheford. 2004. *Trends in Plant Sciences* 9: 358–364; and the Illinois long-term selection experiment for oil and protein in corn (University of Illinois at Urbana–Champaign).

To read this graph, put your finger on the x-axis at generation 0. Then read up the y-axis, and note that kernels averaged about 11 percent protein at the start of the experiment. Now read the graph to the right. Each dot is a data point, representing the average kernel protein concentration in a particular generation. (A generation in maize is one year.) The lines on this graph simply connect the dots, to make the pattern in the data easier to see. During a few years the average protein content goes down, because of poor growing conditions or chance changes in how the many genes responsible for this trait interact. However, at the end of the graph, after 100 generations of selection, average kernel protein content is about 29 percent. (For more help with reading graphs, see **BioSkills 3** in Appendix B.)

This sort of change in the characteristics of a population, over time, is evolution. Humans have been practicing artificial selection for thousands of years, and biologists have now documented evolution by *natural* selection—where humans don't do the selecting—occurring in thousands of different populations, including humans. Evolution occurs when heritable variation leads to differential success in reproduction.

✔ **QUANTITATIVE** If you understand the concepts of selection and evolution, you should be able to describe how protein content in maize kernels changed over time, using the same x-axis and y-axis as in Figure 1.3, when researchers selected individuals with the *lowest* kernel protein content to be the parents of the next generation. (This experiment was actually done, starting with the same population at the same time as selection for high protein content.)

Fitness and Adaptation Darwin also introduced some new terminology to identify what is happening during natural selection.

- In everyday English, fitness means health and well-being. But in biology, **fitness** means the ability of an individual to produce viable offspring. Individuals with high fitness produce many surviving offspring.

- In everyday English, adaptation means that an individual is adjusting and changing to function in new circumstances. But in biology, an **adaptation** is a trait that increases the fitness of an individual in a particular environment.

Once again, consider kernel protein content in maize: In the environment of the experiment graphed in Figure 1.3, individuals with high kernel protein content produced more offspring and had higher fitness than individuals with lower kernel protein content. In this population and this environment, high kernel protein content was an adaptation that allowed certain individuals to thrive.

Note that during this process, the amount of protein in the kernels of any individual maize plant did not change within its lifetime—the change occurred in the characteristics of the population over time.

Together, the cell theory and the theory of evolution provided the young science of biology with two central, unifying ideas:

1. The cell is the fundamental structural unit in all organisms.

2. All species are related by common ancestry and have changed over time in response to natural selection.

check your understanding

If you understand that . . .

- Natural selection occurs when heritable variation in certain traits leads to improved success in reproduction. Because individuals with these traits produce many offspring with the same traits, the traits increase in frequency and evolution occurs.
- Evolution is a change in the characteristics of a population over time.

✔ **You should be able to . . .**

Using the graph you just analyzed in Figure 1.3, describe the average kernel protein content over time in a maize population where *no* selection occurred.

Answers are available in Appendix A.

1.4 The Tree of Life

Section 1.3 focuses on how individual populations change through time in response to natural selection. But over the past several decades, biologists have also documented dozens of cases in which natural selection has caused populations of one species to diverge and form new species. This divergence process is called **speciation.**

Research on speciation has two important implications: All species come from preexisting species, and all species, past and present, trace their ancestry back to a single common ancestor.

The theory of evolution by natural selection predicts that biologists should be able to construct a **tree of life**—a family tree of organisms. If life on Earth arose just once, then such a diagram would describe the genealogical relationships between species with a single, ancestral species at its base.

Has this task been accomplished? If the tree of life exists, what does it look like?

Using Molecules to Understand the Tree of Life

One of the great breakthroughs in research on the tree of life occurred when American biologist Carl Woese (pronounced *woze*) and colleagues began analyzing the chemical components of organisms as a way to understand their evolutionary relationships. Their goal was to understand the **phylogeny** of all organisms—their actual genealogical relationships. Translated literally, phylogeny means "tribe-source."

To understand which organisms are closely versus distantly related, Woese and co-workers needed to study a molecule that is found in all organisms. The molecule they selected is called small subunit ribosomal RNA (rRNA). It is an essential part of the machinery that all cells use to grow and reproduce.

Although rRNA is a large and complex molecule, its underlying structure is simple. The rRNA molecule is made up of sequences of four smaller chemical components called ribonucleotides. These ribonucleotides are symbolized by the letters A, U, C, and G. In rRNA, ribonucleotides are connected to one another linearly, like the boxcars of a freight train.

Analyzing rRNA Why might rRNA be useful for understanding the relationships between organisms? The answer is that the ribonucleotide sequence in rRNA is a trait that can change during the course of evolution. Although rRNA performs the same function in all organisms, the sequence of ribonucleotide building blocks in this molecule is not identical among species.

In land plants, for example, the molecule might start with the sequence A-U-A-U-C-G-A-G (**FIGURE 1.4**). In green algae, which are closely related to land plants, the same section of the molecule might contain A-U-A-U-G-G-A-G. But in brown algae, which are not closely related to green algae or to land plants, the same part of the molecule might consist of A-A-A-U-G-G-A-C.

The research that Woese and co-workers pursued was based on a simple premise: If the theory of evolution is correct, then rRNA sequences should be very similar in closely related organisms but less similar in organisms that are less closely related. Species that are part of the same evolutionary lineage, like the plants, should share certain changes in rRNA that no other species have.

To test this premise, the researchers determined the sequence of ribonucleotides in the rRNA of a wide array of species. Then they considered what the similarities and differences in the sequences implied about relationships between the species. The goal was to produce a diagram that described the phylogeny of the organisms in the study.

A diagram that depicts evolutionary history in this way is called a phylogenetic tree. Just as a family tree shows relationships between individuals, a phylogenetic tree shows relationships between species. On a phylogenetic tree, branches that share a recent common ancestor represent species that are closely related; branches that don't share recent common ancestors represent species that are more distantly related.

The Tree of Life Estimated from Genetic Data To construct a phylogenetic tree, researchers use a computer to find the arrangement of branches that is most consistent with the similarities and differences observed in the data.

Although the initial work was based only on the sequences of ribonucleotides observed in rRNA, biologists now use data sets that include sequences from a wide array of genetic material. **FIGURE 1.5** shows a recent tree produced by comparing these sequences. Because this tree includes such a diverse array of

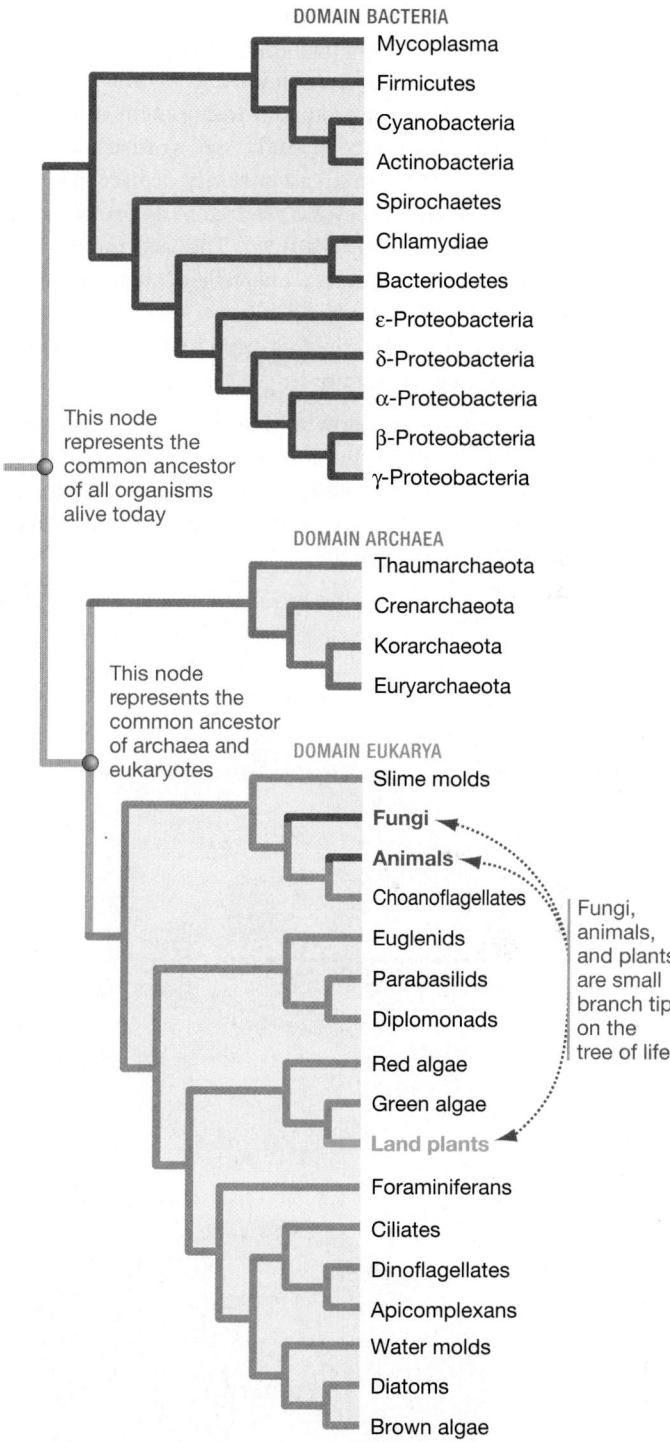

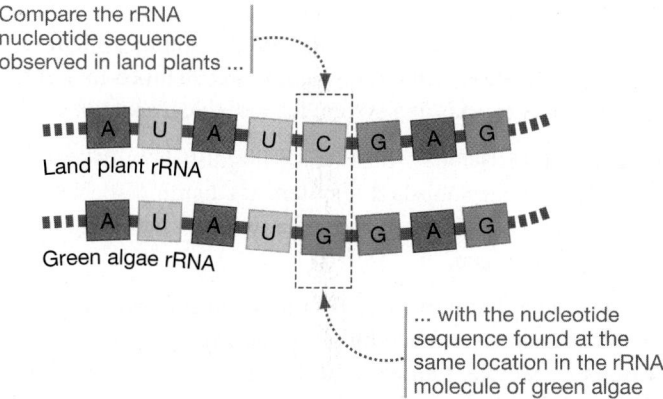

Compare the rRNA nucleotide sequence observed in land plants ...

Land plant rRNA

Green algae rRNA

... with the nucleotide sequence found at the same location in the rRNA molecule of green algae

FIGURE 1.4 RNA Molecules Are Made Up of Smaller Molecules. A complete small subunit rRNA molecule contains about 2000 ribonucleotides; just 8 are shown in this comparison.

✔**QUESTION** Suppose that in the same section of rRNA, molds and other fungi have the sequence A-U-A-U-G-G-A-C. Are fungi more closely related to green algae or to land plants? Explain your logic.

FIGURE 1.5 The Tree of Life. A phylogenetic tree estimated from a large amount of genetic sequence data. The three domains of life revealed by the analysis are labeled. Common names are given for lineages in the domains Bacteria and Eukarya. Phyla names are given for members of the domain Archaea, because most of these organisms have no common names.

species, it is often called the universal tree, or the tree of life. (For help in learning how to read a phylogenetic tree, see **BioSkills 7** in Appendix B.) Notice that the tree's main node is the common ancestor of all living organisms. Researchers who study the origin of life propose that the tree's root extends even further back to the "*last universal common ancestor*" of cells, or **LUCA.**

The tree of life implied by rRNA and other genetic data established that there are three fundamental groups or lineages of organisms: (1) the Bacteria, (2) the Archaea, and (3) the Eukarya. In all **eukaryotes,** cells have a prominent component called the nucleus (**FIGURE 1.6a**). Translated literally, the word eukaryotes means "true kernel." Because the vast majority of bacterial and archaeal cells lack a nucleus, they are referred to as **prokaryotes** (literally, "before kernel"; see **FIGURE 1.6b**). The vast majority of bacteria and archaea are unicellular ("one-celled"); many eukaryotes are multicellular ("many-celled").

When results based on genetic data were first published, biologists were astonished. For example:

- Prior to Woese's work and follow-up studies, biologists thought that the most fundamental division among organisms was between prokaryotes and eukaryotes. The Archaea were virtually unknown—much less recognized as a major and highly distinctive branch on the tree of life.

- Fungi were thought to be closely related to plants. Instead, they are actually much more closely related to animals.

(a) Eukaryotic cells have a membrane-bound nucleus.

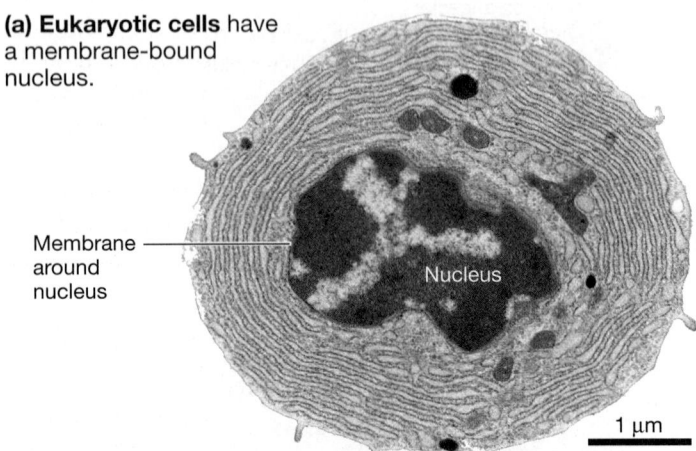

Membrane around nucleus

Nucleus

1 μm

(b) Prokaryotic cells do *not* have a membrane-bound nucleus.

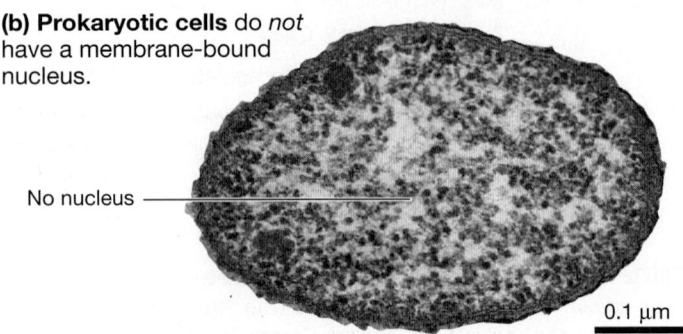

No nucleus

0.1 μm

FIGURE 1.6 Eukaryotes and Prokaryotes.

✔**QUANTITATIVE** How many times larger is the eukaryotic cell in this figure than the prokaryotic cell? (Hint: Study the scale bars.)

- Traditional approaches for classifying organisms—including the system of five kingdoms divided into various classes, orders, and families that you may have learned in high school— are inaccurate in many cases, because they do not reflect the actual evolutionary history of the organisms involved.

The Tree of Life Is a Work in Progress   Just as researching your family tree can help you understand who you are and where you came from, so the tree of life helps biologists understand the relationships between organisms and the history of species. The discovery of the Archaea and the accurate placement of lineages such as the fungi qualify as exciting breakthroughs in our understanding of evolutionary history and life's diversity.

Work on the tree of life continues at a furious pace, however, and the location of certain branches on the tree is hotly debated. As databases expand and as techniques for analyzing data improve, the shape of the tree of life presented in Figure 1.5 will undoubtedly change. Our understanding of the tree of life, like our understanding of every other topic in biological science, is dynamic.

How Should We Name Branches on the Tree of Life?

In science, the effort to name and classify organisms is called **taxonomy.** Any named group is called a **taxon** (plural: **taxa**). Currently, biologists are working to create a taxonomy, or naming system, that accurately reflects the phylogeny of organisms.

Based on the tree of life implied by genetic data, Woese proposed a new taxonomic category called the **domain.** The three domains of life are the Bacteria, Archaea, and Eukarya.

Biologists often use the term **phylum** (plural: **phyla**) to refer to major lineages within each domain. Although the designation is somewhat arbitrary, each phylum is considered a major branch on the tree of life. Within the lineage called animals, biologists currently name 30–35 phyla—each of which is distinguished by distinctive aspects of its body structure as well as by distinctive gene sequences. For example, the mollusks (clams, squid, octopuses) constitute a phylum, as do chordates (the vertebrates and their close relatives).

Because the tree of life is so new, though, naming systems are still being worked out. One thing that hasn't changed for centuries, however, is the naming system for individual species.

Scientific (Latin) Names   In 1735, a Swedish botanist named Carolus Linnaeus established a system for naming species that is still in use today. Linnaeus created a two-part name unique to each type of organism.

- *Genus*   The first part indicates the organism's **genus** (plural: **genera**). A genus is made up of a closely related group of species. For example, Linnaeus put humans in the genus *Homo*. Although humans are the only living species in this genus, at least six extinct organisms, all of which walked upright and made extensive use of tools, were later also assigned to *Homo*.

- *Species*   The second term in the two-part name identifies the organism's species. Linnaeus gave humans the species name *sapiens*.

An organism's genus and species designation is called its **scientific name** or Latin name. Scientific names are always italicized. Genus names are always capitalized, but species names are not—as in *Homo sapiens*.

Scientific names are based on Latin or Greek word roots or on words "Latinized" from other languages. Linnaeus gave a scientific name to every species then known, and also Latinized his own name—from Karl von Linné to Carolus Linnaeus.

Linnaeus maintained that different types of organisms should not be given the same genus and species names. Other species may be assigned to the genus *Homo*, and members of other genera may be named *sapiens*, but only humans are named *Homo sapiens*. Each scientific name is unique.

Scientific Names Are Often Descriptive Scientific names and terms are often based on Latin or Greek word roots that are descriptive. For example, *Homo sapiens* is derived from the Latin *homo* for "man" and *sapiens* for "wise" or "knowing." The yeast that bakers use to produce bread and that brewers use to brew beer is called *Saccharomyces cerevisiae*. The Greek root *saccharo* means "sugar," and *myces* refers to a fungus. *Saccharomyces* is aptly named "sugar fungus" because yeast is a fungus and because the domesticated strains of yeast used in commercial baking and brewing are often fed sugar. The species name of this organism, *cerevisiae*, is Latin for "beer." Loosely translated, then, the scientific name of brewer's yeast means "sugar-fungus for beer."

Scientific names and terms often seem daunting at first glance. So, most biologists find it extremely helpful to memorize some of the common Latin and Greek roots. To aid you in this process, new terms in this text are often accompanied by a translation of their Latin or Greek word roots in parentheses. (A glossary of common root words with translations and examples is also provided in **BioSkills 2** in Appendix B.)

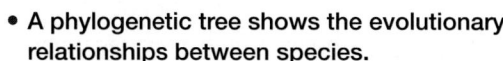

If you understand that . . .

- A phylogenetic tree shows the evolutionary relationships between species.
- To infer where species belong on a phylogenetic tree, biologists examine genetic and other characteristics of the species involved. Closely related species should have similar characteristics, while less closely related species should be less similar.

✔ **You should be able to . . .**

Examine the following rRNA ribonucleotide sequences and draw a phylogenetic tree showing the relationships between species A, B, and C that these data imply:

Species A: A A C T A G C G C G A T

Species B: A A C T A G C G C C A T

Species C: T T C T A G C G G T A T

Answers are available in Appendix A.

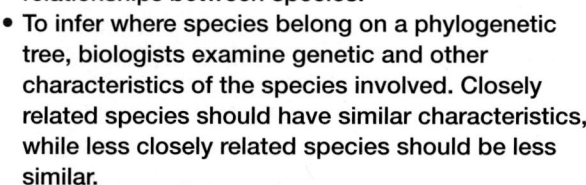

1.5 Doing Biology

This chapter has introduced some of the great ideas in biology. The development of the cell theory and the theory of evolution by natural selection provided cornerstones when the science was young; the tree of life is a relatively recent insight that has revolutionized our understanding of life's diversity.

These theories are considered great because they explain fundamental aspects of nature, and because they have consistently been shown to be correct. They are considered correct because they have withstood extensive testing.

How do biologists go about testing their ideas? Before answering this question, let's step back a bit and consider the types of questions that researchers can and cannot ask.

The Nature of Science

Biologists ask questions about organisms, just as physicists and chemists ask questions about the physical world or geologists ask questions about Earth's history and the ongoing processes that shape landforms.

No matter what their field, all scientists ask questions that can be answered by observing or measuring things—by collecting data. Conversely, scientists cannot address questions that can't be answered by observing or measuring things.

This distinction is important. It is at the root of continuing controversies about teaching evolution in publicly funded schools. In the United States and in Turkey, in particular, some Christian and Islamic leaders have been particularly successful in pushing their claim that evolution and religious faith are in conflict. Even though the theory of evolution is considered one of the most successful and best-substantiated ideas in the history of science, they object to teaching it.

The vast majority of biologists and many religious leaders reject this claim; they see no conflict between evolution and religious faith. Their view is that science and religion are compatible because they address different types of questions.

- Science is about formulating hypotheses and finding evidence that supports or conflicts with those hypotheses.

- Religious faith addresses questions that cannot be answered by data. The questions addressed by the world's great religions focus on why we exist and how we should live.

Both types of questions are seen as legitimate and important.

So how do biologists go about answering questions? After formulating hypotheses, biologists perform experimental studies, or studies that yield descriptive data, such as observing a behavior, characterizing a structure within a cell by microscopy, or sequencing rRNA. Let's consider two recent examples of this process.

Why Do Giraffes Have Long Necks? An Introduction to Hypothesis Testing

If you were asked why giraffes have long necks, you might say based on your observations that long necks enable giraffes to reach food that is unavailable to other mammals. This hypothesis is expressed in African folktales and has traditionally been accepted by many biologists. The food competition hypothesis is so plausible, in fact, that for decades no one thought to test it.

In the mid-1990s, however, Robert Simmons and Lue Scheepers assembled data suggesting that the food competition hypothesis is only part of the story. Their analysis supports an alternative hypothesis—that long necks allow giraffes to use their heads as effective weapons for battering their opponents, and that longer-necked giraffes would have a competitive advantage in fights.

Before exploring these alternative explanations, it's important to recognize that hypothesis testing is a two-step process:

Step 1 State the hypothesis as precisely as possible and list the predictions it makes.

Step 2 Design an observational or experimental study that is capable of testing those predictions.

If the predictions are accurate, the hypothesis is supported. If the predictions are not met, then researchers do further tests, modify the original hypothesis, or search for alternative explanations. But the process does not end here. Biologists also talk to other researchers. Over coffee, at scientific meetings, or through publications, biologists communicate their results to the scientific community and beyond. (You can see the Big Picture of the process of doing biology on pages 16–17.)

Now that you understand more about hypothesis testing, let's return to the giraffes. How did biologists test the food competition hypothesis? What data support their alternative explanation?

The Food Competition Hypothesis: Predictions and Tests The food competition hypothesis claims that giraffes compete for food with other species of mammals. When food is scarce, as it is during the dry season, giraffes with longer necks can reach food that is unavailable to other species and to giraffes with shorter necks. As a result, the longest-necked individuals in a giraffe population survive better and produce more young than do shorter-necked individuals, and average neck length of the population increases with each generation.

To use the terms introduced earlier, long necks are adaptations that increase the fitness of individual giraffes during competition for food. This type of natural selection has gone on so long that the population has become extremely long necked.

The food competition hypothesis makes several explicit predictions. For example, the food competition hypothesis predicts that:

- neck length is variable among giraffes;
- neck length in giraffes is heritable; and
- giraffes feed high in trees, especially during the dry season, when food is scarce and the threat of starvation is high.

The first prediction is correct. Studies in zoos and natural populations confirm that neck length is variable among individuals.

The researchers were unable to test the second prediction, however, because they studied giraffes in a natural population and were unable to do breeding experiments. As a result, they simply had to accept this prediction as an assumption. In

(a) Most feeding is done at about shoulder height.

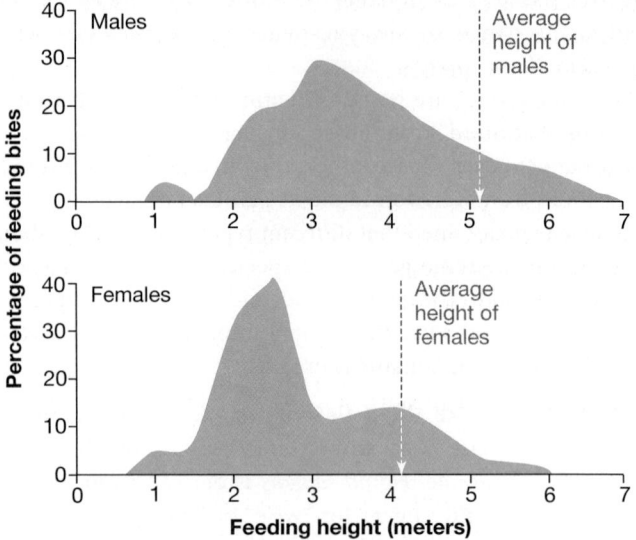

(b) Typical feeding posture in giraffes

FIGURE 1.7 Giraffes Do Not Usually Extend Their Necks Upward to Feed.
DATA: Young, T. P., L. A. Isbell. 1991. *Ethology* 87: 79–89.

general, though, biologists prefer to test every assumption behind a hypothesis.

What about the prediction regarding feeding high in trees? According to Simmons and Scheepers, this is where the food competition hypothesis breaks down.

Consider, for example, data collected by a different research team on the amount of time that giraffes spend feeding in vegetation of different heights. **FIGURE 1.7a** plots the height of vegetation versus the percentage of bites taken by a giraffe, for males and for females from the same population in Kenya. The dashed line on each graph indicates the average height of a male or female in this population.

Note that the average height of a giraffe in this population is much greater than the height where most feeding takes place. In this population, both male and female giraffes spend most of their feeding time eating vegetation that averages just 60 percent of their full height. Studies on other populations of giraffes, during both the wet and dry seasons, are consistent with these data. Giraffes usually feed with their necks bent (**FIGURE 1.7b**).

These data cast doubt on the food competition hypothesis, because one of its predictions does not appear to hold. Biologists have not abandoned this hypothesis completely, though, because feeding high in trees may be particularly valuable during extreme droughts, when a giraffe's ability to reach leaves far above the ground could mean the difference between life and death. Still, Simmons and Scheepers have offered an alternative explanation for why giraffes have long necks. The new hypothesis is based on the mating system of giraffes.

The Sexual Competition Hypothesis: Predictions and Tests

Giraffes have an unusual mating system. Breeding occurs year round rather than seasonally. To determine when females are coming into estrus or "heat" and are thus receptive to mating, the males nuzzle the rumps of females. In response, the females urinate into the males' mouths. The males then tip their heads back and pull their lips to and fro, as if tasting the liquid. Biologists who have witnessed this behavior have proposed that the males taste the females' urine to detect whether estrus has begun.

Once a female giraffe enters estrus, males fight among themselves for the opportunity to mate. Combat is spectacular. The bulls stand next to one another, swing their necks, and strike thunderous blows with their heads. Researchers have seen males knocked unconscious for 20 minutes after being hit and have cataloged numerous instances in which the loser died. Giraffes are not the only animals known to fight in this way—male giraffe weevils also use enormously long necks to fight for mating rights.

These observations inspired a new explanation for why giraffes have long necks. The sexual competition hypothesis is based on the idea that longer-necked giraffes are able to strike harder blows during combat than can shorter-necked giraffes. In engineering terms, longer necks provide a longer "moment arm." A long moment arm increases the force of an impact. (Think about the type of sledgehammer you'd use to bash down a concrete wall—one with a short handle or one with a long handle?)

The idea here is that longer-necked males should win more fights and, as a result, father more offspring than shorter-necked males do. If neck length in giraffes is inherited, then the average neck length in the population should increase over time. Under the sexual competition hypothesis, long necks are adaptations that increase the fitness of males during competition for females.

Although several studies have shown that long-necked males are more successful in fighting and that the winners of fights gain access to estrous females, the question of why giraffes have long necks is not closed. With the data collected to date, most biologists would probably concede that the food competition hypothesis needs further testing and refinement and that the sexual competition hypothesis appears promising. It could also be true that both hypotheses are correct. For our purposes, the important take-home message is that all hypotheses must be tested rigorously.

In many cases in biological science, testing hypotheses rigorously involves experimentation. Experimenting on giraffes is difficult. But in the case study considered next, biologists were able to test an interesting hypothesis experimentally.

How Do Ants Navigate? An Introduction to Experimental Design

Experiments are a powerful scientific tool because they allow researchers to test the effect of a single, well-defined factor on a particular phenomenon. Because experiments testing the effect of neck length on food and sexual competition in giraffes haven't been done yet, let's consider a different question: When ants leave their nest to search for food, how do they find their way back?

The Saharan desert ant lives in colonies and makes a living by scavenging the dead carcasses of insects. Individuals leave the burrow and wander about searching for food at midday, when temperatures at the surface can reach 60°C (140°F) and predators are hiding from the heat.

Foraging trips can take the ants hundreds of meters—an impressive distance when you consider that these animals are only about a centimeter long. But when an ant returns, it doesn't follow the same long, wandering route it took on its way away from the nest. Instead, individuals return in a straight line (**FIGURE 1.8**).

Once individuals are close to the nest, they engage in a characteristic set of back-and-forth U-turns until they find their nest hole. How do they do know how far they are from the nest?

The Pedometer Hypothesis

Early work on navigation in desert ants showed that they use the Sun's position as a compass—meaning that they always know the approximate direction of the nest relative to the Sun. But how do they know how far to go?

After experiments had shown that the ants do not use landmarks to navigate, Matthias Wittlinger and co-workers set out to test a novel idea. The biologists proposed that Saharan desert ants know how far they are from the nest by integrating information from leg movements.

According to this pedometer hypothesis, the ants always know how far they are from the nest because they track the number

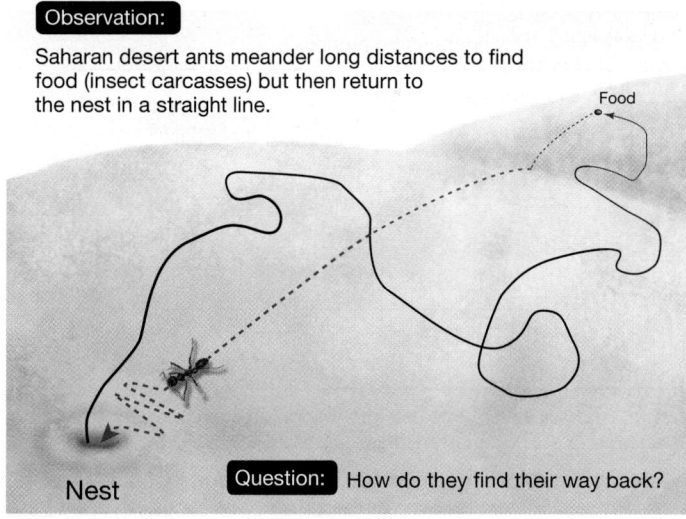

Observation:

Saharan desert ants meander long distances to find food (insect carcasses) but then return to the nest in a straight line.

Food

Question: How do they find their way back?

Nest

FIGURE 1.8 Navigation in Foraging Desert Ants.

of steps they have taken and their stride length. The idea is that they can make a beeline back toward the burrow because they integrate information on the angles they have traveled *and* the distance they have gone—based on step number and stride length.

If the pedometer hypothesis is wrong, however, then stride length and step number should have no effect on the ability of an ant to get back to its nest. This latter possibility is called a **null hypothesis.** A null hypothesis specifies what should be observed when the hypothesis being tested isn't correct.

Testing the Hypothesis To test their idea, Wittlinger's group allowed ants to walk from a nest to a feeder through a channel—a distance of 10 m. Then they caught ants at the feeder and created three test groups, each with 25 individuals (**FIGURES 1.9** and **1.10**):

- *Stumps* By cutting the lower legs of some individuals off, they created ants with shorter-than-normal legs.

- *Normal* Some individuals were left alone, meaning that they had normal leg length.

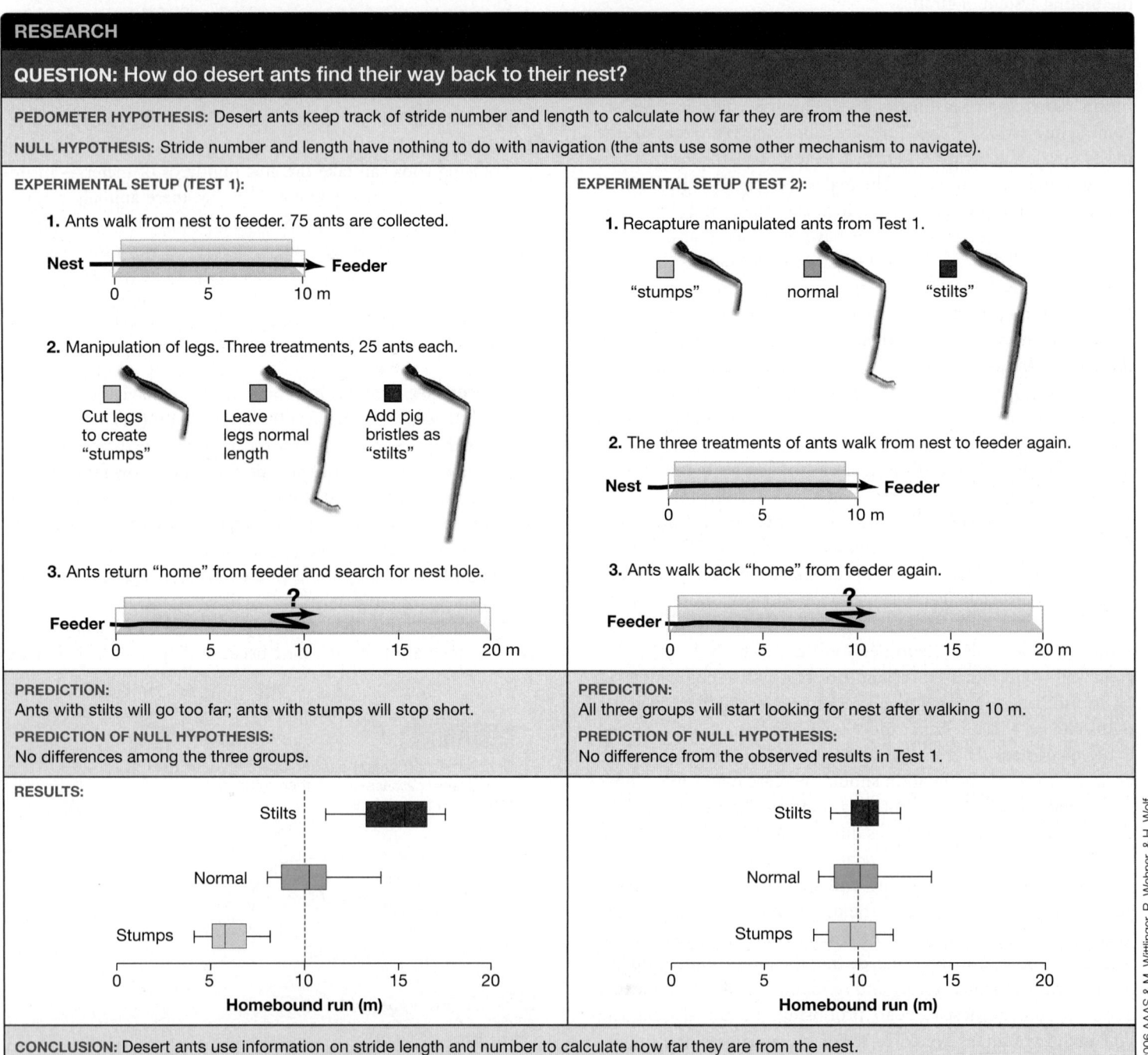

RESEARCH

QUESTION: How do desert ants find their way back to their nest?

PEDOMETER HYPOTHESIS: Desert ants keep track of stride number and length to calculate how far they are from the nest.

NULL HYPOTHESIS: Stride number and length have nothing to do with navigation (the ants use some other mechanism to navigate).

EXPERIMENTAL SETUP (TEST 1):

1. Ants walk from nest to feeder. 75 ants are collected.

Nest — Feeder
0   5   10 m

2. Manipulation of legs. Three treatments, 25 ants each.

Cut legs to create "stumps"   Leave legs normal length   Add pig bristles as "stilts"

3. Ants return "home" from feeder and search for nest hole.

Feeder
0   5   10   15   20 m

EXPERIMENTAL SETUP (TEST 2):

1. Recapture manipulated ants from Test 1.

"stumps"   normal   "stilts"

2. The three treatments of ants walk from nest to feeder again.

Nest — Feeder
0   5   10 m

3. Ants walk back "home" from feeder again.

Feeder
0   5   10   15   20 m

PREDICTION:
Ants with stilts will go too far; ants with stumps will stop short.

PREDICTION OF NULL HYPOTHESIS:
No differences among the three groups.

PREDICTION:
All three groups will start looking for nest after walking 10 m.

PREDICTION OF NULL HYPOTHESIS:
No difference from the observed results in Test 1.

RESULTS:

Stilts
Normal
Stumps
0   5   10   15   20
Homebound run (m)

Stilts
Normal
Stumps
0   5   10   15   20
Homebound run (m)

CONCLUSION: Desert ants use information on stride length and number to calculate how far they are from the nest.

FIGURE 1.9 An Experimental Test: Do Desert Ants Use a "Pedometer"?

SOURCE: Wittlinger, M., R. Wehner, and H. Wolf. 2006. The ant odometer: Stepping on stilts and stumps. *Science* 312: 1965–1967.

✓**QUESTION** What is the advantage of using 25 ants in each group instead of just one?

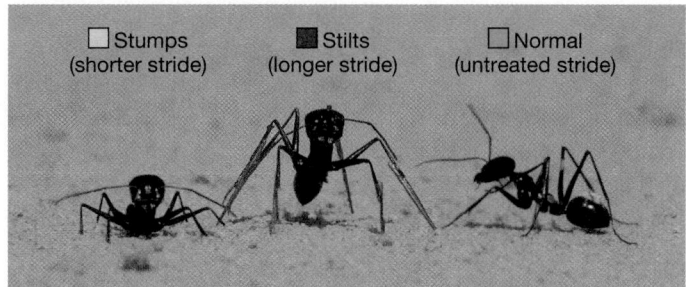

□ Stumps   ■ Stilts   □ Normal
(shorter stride) (longer stride) (untreated stride)

FIGURE 1.10 Manipulation of Desert Ant Stride Length.

- *Stilts*   By gluing pig bristles onto each leg, the biologists created ants with longer-than-normal legs.

Next they put the ants in a different channel and recorded how far they traveled in a direct line before starting their nest-searching behavior. To see the data they collected, look at the graph on the left side of the "Results" section in Figure 1.9.

- *Stumps*   The ants with stumps stopped short, by about 5 m, before starting to search for the nest opening.

- *Normal*   The normal ants walked the correct distance—about 10 m.

- *Stilts*   The ants with stilts walked about 5 m too far before starting to search for the nest opening.

To check the validity of this result, the researchers put the test ants back in the nest and recaptured them one to several days later, when they had walked to the feeder on their stumps, normal legs, or stilts. Now when the ants were put into the other channel to "walk back," they all traveled the correct distance—10 m—before starting to search for the nest (see the graph on the right side of the "Results" section in Figure 1.9).

The graphs in the "Results" display "box-and-whisker" plots that allow you to easily see where most of the data fall. Each box indicates the range of distances where 50 percent of the ants stopped to search for the nest. The whiskers indicate the lower extreme (stopping short of the nest location) and the upper extreme (going too far) of where the ants stopped to search. The vertical line inside each box indicates the median—meaning that half the ants stopped above this distance and half below. (For more details on how biologists report medians and indicate the variability and uncertainty in data, see **BioSkills 4** in Appendix B.)

Interpreting the Results   The pedometer hypothesis predicts that an ant's ability to walk home depends on the number and length of steps taken on its outbound trip. Recall that a prediction specifies what we should observe if a hypothesis is correct. Good scientific hypotheses make testable predictions—predictions that can be supported or rejected by collecting and analyzing data. In this case, the researchers tested the prediction by altering stride length and recording the distance traveled on the return trip. Under the null hypothesis in this experiment, all the ants—altered and unaltered—should have walked 10 m in the first test before they started looking for their nest.

Important Characteristics of Good Experimental Design   In relation to designing effective experiments, this study illustrates several important points:

- It is critical to include **control** groups. A control checks for factors, other than the one being tested, that might influence the experiment's outcome. In this case, there were two controls. Including a normal, unmanipulated individual controlled for the possibility that switching the individuals to a new channel altered their behavior. In addition, the researchers had to control for the possibility that the manipulation itself—and not the change in leg length—affected the behavior of the stilts and stumps ants. This is why they did the second test, where the outbound and return runs were done with the same legs.

- The experimental conditions must be as constant or equivalent as possible. The investigators used ants of the same species, from the same nest, at the same time of day, under the same humidity and temperature conditions, at the same feeders, in the same channels. Controlling all the variables except one—leg length in this case—is crucial because it eliminates alternative explanations for the results.

- Repeating the test is essential. It is almost universally true that larger sample sizes in experiments are better. By testing many individuals, the amount of distortion or "noise" in the data caused by unusual individuals or circumstances is reduced.

✔ If you understand these points, you should be able to explain: (1) What you would conclude if in the first test, the normal individual had not walked 10 m on the return trip before

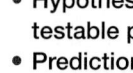

looking for the nest; and (2) What you would conclude if the stilts and stumps ants had not navigated normally during the second test.

From the outcomes of these experiments, the researchers concluded that desert ants use stride length and number to measure how far they are from the nest. They interpreted their results as strong support for the pedometer hypothesis.

The giraffe and ant studies demonstrate a vital point: Biologists practice evidence-based decision making. They ask questions about how organisms work, pose hypotheses to answer those questions, and use experimental or observational evidence to decide which hypotheses are correct.

The data on giraffes and ants are a taste of things to come. In this text you will encounter hypotheses and research on questions ranging from how water gets to the top of 100-meter-tall sequoia trees to how the bacterium that causes tuberculosis has become resistant to antibiotics. As you work through this book, you'll get lots of practice thinking about hypotheses and predictions, analyzing the nature of control treatments, and interpreting graphs.

A commitment to tough-minded hypothesis testing and sound experimental design is a hallmark of biological science. Understanding their value is an important first step in becoming a biologist.

CHAPTER 1 REVIEW

For media, go to MasteringBiology

If you understand . . .

1.1 What Does It Mean to Say That Something Is Alive?

- There is no single, well-accepted definition of life. Instead, biologists point to five characteristics that organisms share.

✔ You should be able to explain why the cells in a dead organism are different from the cells in a live organism.

1.2 The Cell Theory

- The cell theory identified the fundamental structural unit common to all life.

✔ You should be able to describe the evidence that supported the pattern and the process components of the cell theory.

1.3 The Theory of Evolution by Natural Selection

- The theory of evolution states that all organisms are related by common ancestry.

- Natural selection is a well-tested explanation for why species change through time and why they are so well adapted to their habitats.

✔ You should be able to explain why the average protein content of seeds in a natural population of a grass species would increase over time, if seeds with higher protein content survive better and grow into individuals that produce many seeds with high protein content when they mature.

1.4 The Tree of Life

- The theory of evolution predicts that all organisms are part of a genealogy of species, and that all species trace their ancestry back to a single common ancestor.

- To construct this phylogeny, biologists have analyzed the sequences in rRNA and in an array of genetic material found in all cells.

- A tree of life, based on similarities and differences in these molecules, has three fundamental lineages, or domains: the Bacteria, the Archaea, and the Eukarya.

✔ You should be able to explain how biologists can determine which of the three domains a newly discovered species belongs to by analyzing its rRNA.

1.5 Doing Biology

- Biology is a hypothesis-driven, experimental science.

✔ You should be able to explain (1) the relationship between a hypothesis and a prediction and (2) why experiments are convincing ways to test predictions.

MasteringBiology

1. **MasteringBiology Assignments**

   Tutorials and Activities An Introduction to Graphing; Experimental Inquiry: What Can You Learn about the Process of Science from Investigating a Cricket's Chirp?; Introduction to Experimental Design; Levels of Life Card Game; Metric System Review; The Scientific Method

   Questions Reading Quizzes, Blue-Thread Questions, Test Bank

2. **eText** Read your book online, search, take notes, highlight text, and more.

3. **The Study Area** Practice Test, Cumulative Test, BioFlix® 3-D Animations, Videos, Activities, Audio Glossary, Word Study Tools, Art

You should be able to . . .

1. Anton van Leeuwenhoek made an important contribution to the development of the cell theory. How?
   a. He articulated the pattern component of the theory—that all organisms are made of cells.
   b. He articulated the process component of the theory—that all cells come from preexisting cells.
   c. He invented the first microscope and saw the first cell.
   d. He invented more powerful microscopes and was the first to describe the diversity of cells.

2. What does it mean to say that experimental conditions are controlled?
   a. The test groups consist of the same individuals.
   b. The null hypothesis is correct.
   c. There is no difference in outcome between the control and experimental treatment.
   d. All physical conditions except for one are identical for all groups tested.

3. The term *evolution* means that _____ change through time.

4. What does it mean to say that a characteristic of an organism is heritable?
   a. The characteristic evolves.
   b. The characteristic can be passed on to offspring.
   c. The characteristic is advantageous to the organism.
   d. The characteristic does not vary in the population.

5. In biology, to what does the term *fitness* refer?

6. Could *both* the food competition hypothesis and the sexual competition hypothesis explain why giraffes have long necks? Why or why not?
   a. No. In science, only one hypothesis can be correct.
   b. No. Observations have shown that the food competition hypothesis cannot be correct.
   c. Yes. Long necks could be advantageous for more than one reason.
   d. Yes. All giraffes have been shown to feed at the highest possible height and fight for mates.

7. What would researchers have to demonstrate to convince you that they had discovered life on another planet?

8. What did Linnaeus's system of naming organisms ensure?
   a. Two different organisms never end up with the same genus and species name.
   b. Two different organisms have the same genus and species name if they are closely related.
   c. The genus name is different for closely related species.
   d. The species name is the same for each organism in a genus.

9. What does it mean to say that a species is adapted to a particular habitat?

10. Explain how selection occurs during natural selection. What is selected, and why?

11. The following two statements explain the logic behind the use of molecular sequence data to estimate evolutionary relationships:

    "If the theory of evolution is true, then rRNA sequences should be very similar in closely related organisms but less similar in organisms that are less closely related."

    "On a phylogenetic tree, branches that share a recent common ancestor represent species that are closely related; branches that don't share recent common ancestors represent species that are more distantly related."

    Is the logic of these statements sound? Why or why not?

12. Explain why researchers formulate a null hypothesis in addition to a hypothesis when designing an experimental study.

13. A scientific theory is a set of propositions that defines and explains some aspect of the world. This definition contrasts sharply with the everyday usage of the word theory, which often carries meanings such as "speculation" or "guess." Explain the difference between the two definitions, using the cell theory and the theory of evolution by natural selection as examples.

14. Turn back to the tree of life shown in Figure 1.5. Note that Bacteria and Archaea are prokaryotes, while Eukarya are eukaryotes. On the simplified tree below, draw an arrow that points to the branch where the structure called the nucleus originated. Explain your reasoning.

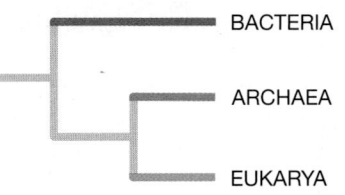

BACTERIA

ARCHAEA

EUKARYA

15. The proponents of the cell theory could not "prove" that it was correct in the sense of providing incontrovertible evidence that all organisms are made up of cells. They could state only that all organisms examined to date were made of cells. Why was it reasonable for them to conclude that the theory was valid?

16. Some humans have heritable traits that make them resistant to infection by HIV. In areas of the world where HIV infection rates are high, are human populations evolving? Explain your logic.
    a. No. HIV infection rates would not affect human evolution.
    b. Yes. The heritable traits that confer resistance to HIV should increase over time.
    c. No. The heritable traits that confer resistance to HIV should decrease over time.
    d. Yes. The heritable traits that confer resistance to HIV should decrease over time.

Biologists study the characteristics of life. The cell theory, the theory of evolution by natural selection, and the tree of life are some of the great ideas in biology that came about by biologists asking questions that can be answered by observing or measuring things—that is, by collecting data.

Notice that the study of life is not a series of linear steps with a beginning and an end. Instead, the process of doing biology is dynamic and ongoing. The answer to one question may lay the foundation for twenty more questions. Working together, biologists from different disciplines integrate data across many levels, from atoms to the biosphere.

Note that the gray numbers in boxes tell you where to go for more information. Also, be sure to do the blue exercises in the Check Your Understanding box.

Characteristics of living things
- Energy
- Cells
- Information
- Replication
- Evolution

1.1

Text section where you can find more information

focuses on

Scientists regularly integrate across many of these levels

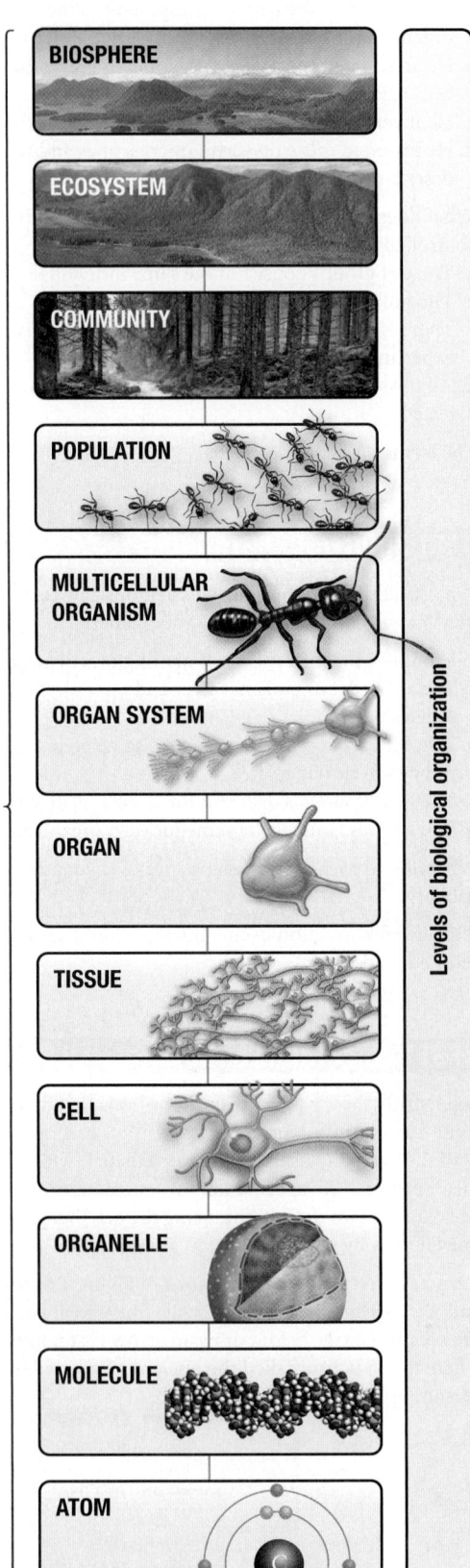

BIOSPHERE

ECOSYSTEM

COMMUNITY

POPULATION

MULTICELLULAR ORGANISM

ORGAN SYSTEM

ORGAN

TISSUE

CELL

ORGANELLE

MOLECULE

ATOM

Levels of biological organization

check your understanding

Ⓒ Ⓨ Ⓤ

If you understand the big picture . . .

✔ You should be able to . . .

1. Describe how biologists go about testing their ideas.
2. Provide an example of how an experimental study could span more than one level of biological organization.
3. Compare and contrast a hypothesis with a theory.
4. Propose the next step to take if data support the hypothesis you are testing.

Answers are available in Appendix A.

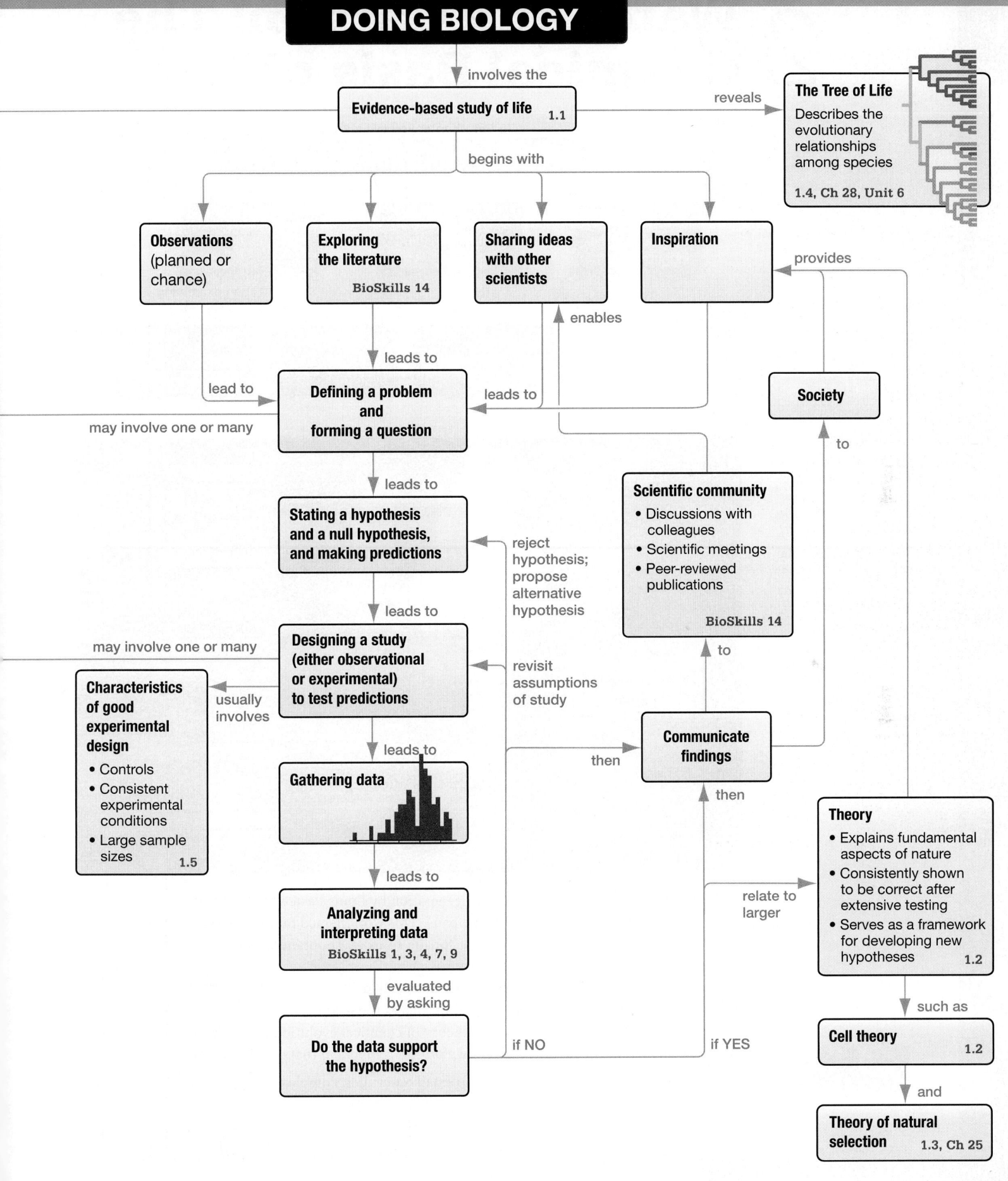

DOING BIOLOGY

involves the

Evidence-based study of life 1.1

reveals

The Tree of Life
Describes the evolutionary relationships among species

1.4, Ch 28, Unit 6

begins with

Observations (planned or chance)

Exploring the literature BioSkills 14

Sharing ideas with other scientists

Inspiration

provides

enables

leads to

lead to

leads to

may involve one or many

Defining a problem and forming a question

Society

to

leads to

Stating a hypothesis and a null hypothesis, and making predictions

reject hypothesis; propose alternative hypothesis

Scientific community
• Discussions with colleagues
• Scientific meetings
• Peer-reviewed publications

BioSkills 14

leads to

may involve one or many

revisit assumptions of study

to

Designing a study (either observational or experimental) to test predictions

Characteristics of good experimental design
• Controls
• Consistent experimental conditions
• Large sample sizes 1.5

usually involves

Communicate findings

then

leads to

Gathering data

then

Theory
• Explains fundamental aspects of nature
• Consistently shown to be correct after extensive testing
• Serves as a framework for developing new hypotheses 1.2

relate to larger

leads to

Analyzing and interpreting data BioSkills 1, 3, 4, 7, 9

evaluated by asking

such as

Cell theory 1.2

if NO

if YES

Do the data support the hypothesis?

and

Theory of natural selection 1.3, Ch 25

2 Water and Carbon: The Chemical Basis of Life

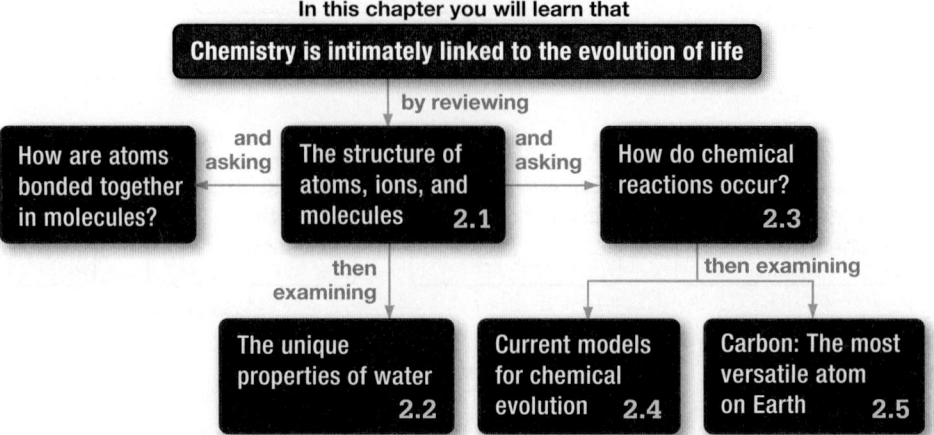

In this chapter you will learn that

Chemistry is intimately linked to the evolution of life

by reviewing

How are atoms bonded together in molecules? ← and asking — The structure of atoms, ions, and molecules **2.1** — and asking → How do chemical reactions occur? **2.3**

then examining

The unique properties of water **2.2**

then examining

Current models for chemical evolution **2.4**

Carbon: The most versatile atom on Earth **2.5**

These deep-sea hydrothermal vents produce hydrogen-rich, highly basic fluids at temperatures that range from 40° to 90°C. It has been proposed that life emerged from similar seafloor chimneys early in Earth's history via chemical evolution.

A classic experiment on spontaneous generation by Louis Pasteur tested the idea that life arises from nonliving materials (see Chapter 1). This work helped build a consensus that spontaneous generation does not occur. But for life to exist, spontaneous generation must have occurred at least once, early in Earth's history.

How did life begin? This simple query has been called "the mother of all questions." This chapter examines a theory, called **chemical evolution,** that is the leading scientific explanation for the origin of life. Like all scientific theories, the theory of chemical evolution has a *pattern component* that makes a claim about the natural world and a *process component* that explains that pattern.

- *The pattern component* In addition to small molecules, complex carbon-containing substances exist and are required for life.

- *The process component* Early in Earth's history, simple chemical compounds combined to form more complex carbon-containing substances before the evolution of life.

This chapter is part of the Big Picture. See how on pages 104–105.

✔ When you see this checkmark, stop and test yourself. Answers are available in Appendix A.

The theory maintains that inputs of energy led to the formation of increasingly complex carbon-containing substances, culminating in a compound that could replicate itself. At this point, there was a switch from chemical evolution to biological evolution.

As the original molecule multiplied, the process of evolution by natural selection took over. Eventually a descendant of the original molecule became metabolically active and acquired a membrane. When this occurred, the five attributes of life (discussed in Chapter 1) were fulfilled. Life had begun.

At first glance, the theory of chemical evolution may seem implausible. But is it? What evidence do biologists have that chemical evolution occurred? What approaches do they take to gathering this evidence?

Let's start with the fundamentals—the atoms and molecules that would have combined to get chemical evolution started.

2.1 Atoms, Ions, and Molecules: The Building Blocks of Chemical Evolution

Just four types of atoms—hydrogen, carbon, nitrogen, and oxygen—make up 96 percent of all matter found in organisms today. Many of the molecules found in your cells contain thousands, or even millions, of these atoms bonded together. But early in Earth's history, these elements existed only in simple substances such as water and carbon dioxide, which contain just three atoms apiece.

Two questions are fundamental to understanding how elements could have evolved into the more complex substances found in living cells:

1. What is the physical structure of the hydrogen, carbon, nitrogen, and oxygen atoms found in living cells?

2. What is the structure of the simple molecules—water, carbon dioxide, and others—that served as the building blocks of chemical evolution?

The focus on structure follows from one of the most central themes in biology: *Structure affects function.* To understand how a molecule affects your body or the role it played in chemical evolution, you have to understand how it is put together.

Basic Atomic Structure

FIGURE 2.1a shows a simple way of depicting the structure of an atom, using hydrogen and carbon as examples. Extremely small particles called electrons orbit an atomic nucleus made up of larger particles called protons and neutrons. **FIGURE 2.1b** provides a sense of scale at the atomic level.

Protons have a positive electric charge (+1), neutrons are electrically neutral, and electrons have a negative electric charge (−1). When the number of protons and the number of electrons in an atom are the same, the charges balance and the atom is electrically neutral.

(a) Diagrams of atoms

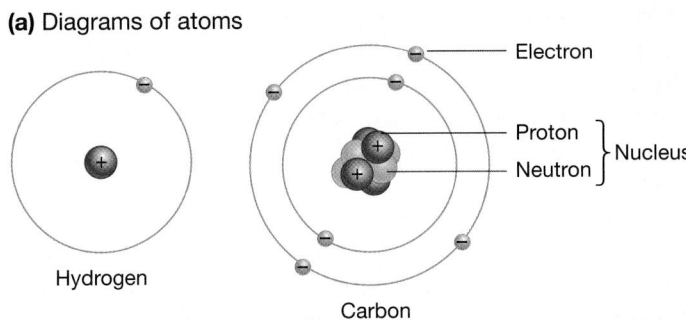

Hydrogen

Carbon

(b) Most of an atom's volume is empty space.

If an atom occupied the same volume as this stadium, the nucleus would be about the size of a pea

FIGURE 2.1 Parts of an Atom. The atomic nucleus, made up of protons and neutrons, is surrounded by orbiting electrons. In reality, electrons do not orbit the nucleus in circles; their actual orbits are complex.

FIGURE 2.2 shows a segment of the periodic table of the elements. Notice that each atom of a given **element** contains a characteristic number of protons, called its **atomic number.** The atomic number is written as a subscript to the left of an element's symbol in Figure 2.2. The sum of the protons and neutrons in an atom is called its **mass number** and is written as a superscript to the left of its symbol.

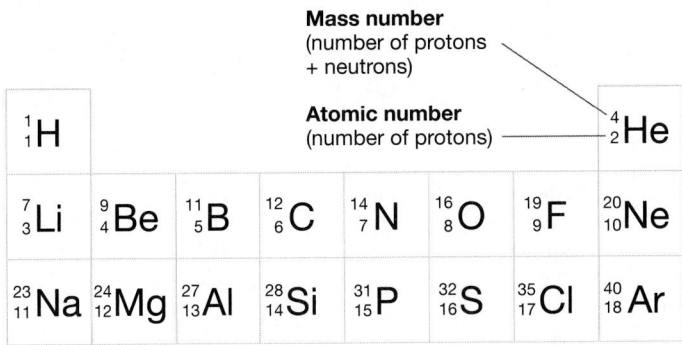

FIGURE 2.2 A Portion of the Periodic Table. Each element has a unique atomic number and is represented by a unique one- or two-letter symbol. The mass numbers given here are the most common for each element. (Appendix C provides a complete periodic table of elements.)

The number of protons in an element does not vary—if the atomic number of an atom changes, then it is no longer the same element. The number of neutrons present in an element can vary, however. Forms of an element with different numbers of neutrons are known as **isotopes** (literally, "equal-places" in regard to position in the periodic table).

Different isotopes have different masses, yet are the same element. For example, all atoms of the element carbon have 6 protons. But naturally occurring isotopes of carbon can have 6, 7, or even 8 neutrons, giving them a mass number of 12, 13, or 14, respectively. The **atomic weight** of an element is an average of all the mass numbers of the naturally occurring isotopes based on their abundance. This is why the atomic weights for elements are often slightly different from the mass numbers—the atomic weight of carbon, for example, is 12.01.

Most isotopes are stable, but not all. For example, ^{14}C, with 6 protons and 8 neutrons, represents an unstable **radioactive isotope.** Its nucleus will eventually decay and release energy (radiation). When ^{14}C decays, one of its neutrons changes into a proton, converting ^{14}C to the stable ^{14}N isotope of nitrogen, with 7 protons and 7 neutrons. Timing of decay is specific to each radioisotope, a fact that has been very useful in estimating the dates of key events in the fossil record (see Chapter 25).

Although the masses of protons, neutrons, and electrons can be measured in grams, the numbers involved are so small that biologists prefer to use a special unit called the **dalton.** The masses of protons and neutrons are virtually identical and are routinely rounded to 1 dalton. A carbon atom that contains 6 protons and 6 neutrons has a mass of 12 daltons, while a carbon atom with 6 protons and 7 neutrons would have a mass of 13 daltons. These isotopes would be written as ^{12}C and ^{13}C,

respectively. The mass of an electron is so small that it is normally ignored.

To understand how the atoms involved in chemical evolution behave, focus on how electrons are arranged around the nucleus:

- Electrons move around atomic nuclei in specific regions called **orbitals.**

- Each orbital can hold up to two electrons.

- Orbitals are grouped into levels called **electron shells.**

- Electron shells are numbered 1, 2, 3, and so on, to indicate their relative distance from the nucleus. Smaller numbers are closer to the nucleus.

- Each electron shell contains a specific number of orbitals. An electron shell comprising a single orbital can hold up to two electrons; a shell with four orbitals can contain up to eight electrons.

- The electrons of an atom fill the innermost shells first, before filling outer shells.

To understand how the structures of atoms differ, take a moment to study **FIGURE 2.3**. This chart highlights the elements that are most abundant in living cells. The gray ball in the center of each box represents an atomic nucleus, and the orange circle or circles represent the electron shells around that nucleus. The small orange balls on the circles indicate the number of electrons that are distributed in the shells of each element. Electrons shown as pairs share the same orbital within a shell.

Now focus on the outermost shell of each atom. This is the element's **valence shell.** The electrons found in this shell are referred to as **valence electrons.** Two observations are important:

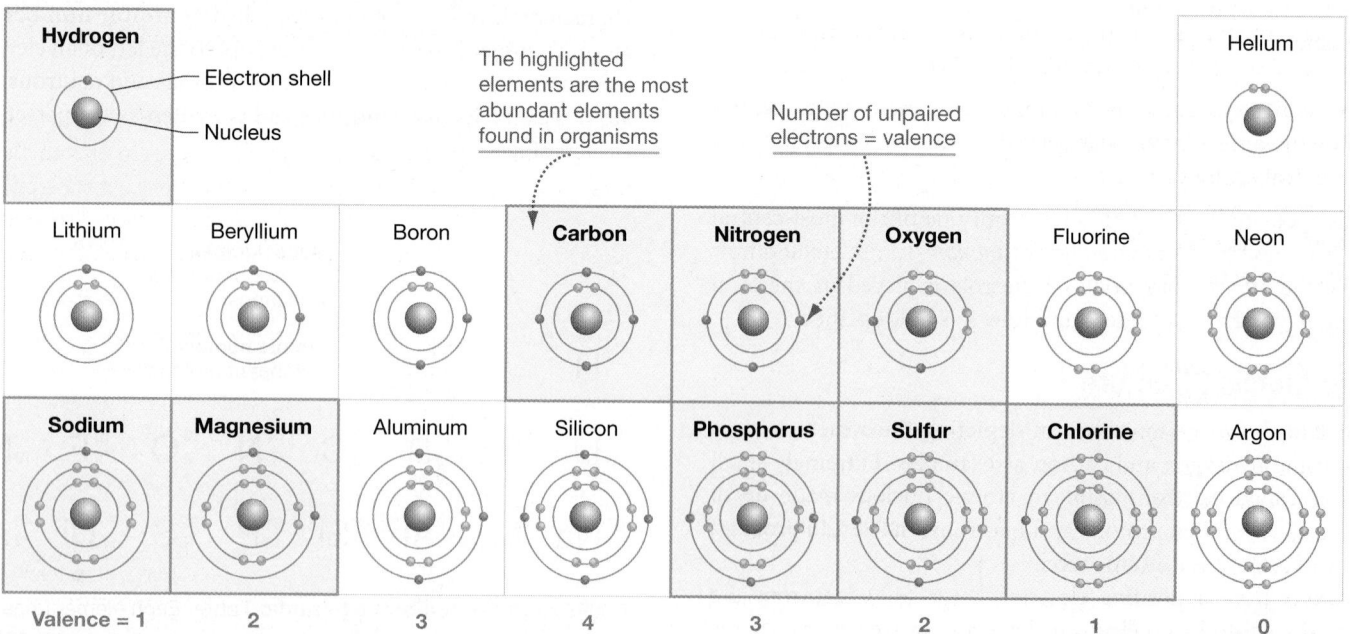

FIGURE 2.3 The Structure of Atoms Found in Organisms.

✔**QUESTION** If the mass number of phosphorus is 31, how many neutrons exist in the most common isotope of phosphorus?

1. In each of the highlighted elements, the outermost electron shell is not full—not all orbitals in the valence shell have two electrons. The highlighted elements have at least one unpaired valence electron—meaning at least one unfilled valence shell orbital.

2. The number of unpaired valence electrons varies among elements. Carbon, for example, has four valence electrons, all unpaired. Oxygen has six valence electrons; four are paired, two are not. The number of unpaired electrons found in an atom is called its **valence.** Carbon's valence is four, oxygen's is two.

These observations are significant because an atom is most stable when its valence shell is filled. One way that shells can be filled is through the formation of strong **chemical bonds**—attractions that bind atoms together. A strong attraction where two atoms share one or more pairs of electrons is called a **covalent bond.**

How Does Covalent Bonding Hold Molecules Together?

To understand how atoms can become more stable by making covalent bonds, consider hydrogen. The hydrogen atom has just one electron, which resides in a shell that can hold two electrons.

Because it has an unpaired valence electron, the hydrogen atom is not very stable. But when two atoms of hydrogen come into contact, the two electrons become shared by the two nuclei (**FIGURE 2.4**). Both atoms now have a completely filled outer shell. Together, the hydrogen atoms are more stable than the two individual hydrogen atoms.

Shared electrons "glue" two hydrogen atoms together. Substances held together by covalent bonds are called **molecules.** In the case of two hydrogen atoms, the bonded atoms form a single molecule of hydrogen, written as H—H or H_2.

It can also be helpful to think about covalent bonding as electrical attraction and repulsion. Opposite charges attract; like charges repel. As two hydrogen atoms move closer together, their positively charged nuclei repel each other and their negatively charged electrons repel each other. But each proton attracts both electrons, and each electron attracts both protons. Covalent bonds form when the attractive forces overcome the repulsive forces. This is the case when hydrogen atoms interact to form the hydrogen molecule (H_2).

Nonpolar and Polar Bonds In **FIGURE 2.5a**, the covalent bond between hydrogen atoms is represented by a dash and the electrons are drawn as dots halfway between the two nuclei. This depiction shows that the electrons are shared equally between the two hydrogen atoms, resulting in a covalent bond that is symmetrical.

It's important to note, though, that the electrons participating in a covalent bond are not always shared equally between the atoms involved. This happens because some atoms hold the electrons in covalent bonds much more tightly than do other atoms. Chemists call this property **electronegativity.**

What is responsible for an atom's electronegativity? It's a combination of two things—the number of protons in the nucleus and the distance between the nucleus and the valence shell. If

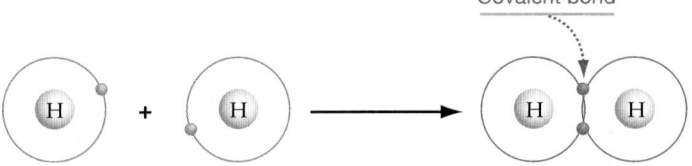

Hydrogen atoms each have one unpaired electron

H_2 molecule has two shared electrons

FIGURE 2.4 Covalent Bonds Result from Electron Sharing. When two hydrogen atoms come into contact, their electrons are attracted to the positive charge in each nucleus. As a result, their orbitals overlap, the electrons are shared by each nucleus, and a covalent bond forms.

you return to Figure 2.3 and move your finger along a row from left to right, you will be moving toward elements that increase in protons and in electronegativity (ignoring the elements in the far right column, which have full outer shells). Each row, however, represents shells of electrons, so if your finger moved down the table, the elements would decrease in electronegativity.

Oxygen, which has eight protons and only two electron shells, is among the most electronegative of all elements. It attracts covalently bonded electrons more strongly than does any other atom commonly found in organisms. Nitrogen's electronegativity is somewhat lower than oxygen's. Carbon and hydrogen, in turn, have relatively low and approximately equal electronegativities. Thus, the electronegativities of the four most abundant elements in organisms are related as follows: O > N > C ≅ H.

Because carbon and hydrogen have approximately equal electronegativity, the electrons in a C—H bond are shared equally or symmetrically. The result is a **nonpolar covalent bond.** In contrast, asymmetric sharing of electrons results in a **polar covalent bond.** The electrons in a polar covalent bond spend most of their time close to the nucleus of the more electronegative atom. Why is this important?

Polar Bonds Produce Partial Charges on Atoms To understand the consequences of differences in electronegativity and the formation of polar covalent bonds, consider the water molecule.

Water consists of an oxygen bonded to two hydrogen atoms, and is written H_2O. As **FIGURE 2.5b** illustrates, the electrons

(a) Nonpolar covalent bond in hydrogen molecule

Electrons are halfway between the two atoms, shared equally

(b) Polar covalent bonds in water molecule

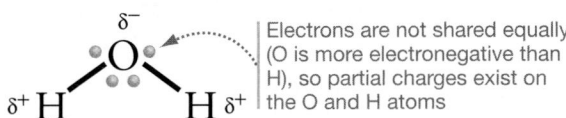

Electrons are not shared equally (O is more electronegative than H), so partial charges exist on the O and H atoms

FIGURE 2.5 Electron Sharing and Bond Polarity. Delta (δ) symbols in polar covalent bonds refer to partial positive and negative charges that arise owing to unequal electron sharing.

involved in the covalent bonds in water are not shared equally but are held much more tightly by the oxygen nucleus than by the hydrogen nuclei. Hence, water has two polar covalent bonds—one between the oxygen atom and each of the hydrogen atoms.

Here's the key observation: Because electrons are shared unequally in each O—H bond, they spend more time near the oxygen atom, giving it a partial negative charge, and less time near the hydrogen atoms, giving them a partial positive charge. These partial charges are symbolized by the lowercase Greek letter delta, δ.

As Section 2.2 shows, the partial charges on water molecules—due simply to the difference in electronegativity between oxygen and hydrogen—are one of the primary reasons that life exists.

Ionic Bonding, Ions, and the Electron-Sharing Continuum

Ionic bonds are similar in principle to covalent bonds, but instead of being shared between two atoms, the electrons in ionic bonds are completely transferred from one atom to the other. The electron transfer occurs because it gives the resulting atoms a full outermost shell.

Sodium atoms (Na), for example, tend to lose an electron, leaving them with a full second shell. This is a much more stable arrangement, energetically, than having a lone electron in their third shell (**FIGURE 2.6a**). The atom that results has a net electric charge of +1, because it has one more proton than it has electrons.

An atom or molecule that carries a full charge, rather than the partial charges that arise from polar covalent bonds, is called an **ion.** The sodium ion is written Na^+ and, like other positively charged ions, is called a **cation** (pronounced *KAT-eye-un*).

Chlorine atoms (Cl), in contrast, tend to gain an electron, filling their outermost shell (**FIGURE 2.6b**). The ion has a net charge of −1, because it has one more electron than protons. This

negatively charged ion, or **anion** (pronounced *AN-eye-un*), is written Cl^- and is called chloride.

When sodium and chlorine combine to form sodium chloride (NaCl, common table salt), they pack into a crystal structure consisting of sodium cations and chloride anions (**FIGURE 2.6c**). The electrical attraction between the ions is so strong that salt crystals are difficult to break apart.

This discussion of covalent and ionic bonding supports an important general observation: The degree to which electrons are shared in chemical bonds forms a continuum from equal sharing in nonpolar covalent bonds to unequal sharing in polar covalent bonds to the transfer of electrons in ionic bonds.

As the left-hand side of **FIGURE 2.7** shows, covalent bonds between atoms with exactly the same electronegativity—for example, between the atoms of hydrogen in H_2—represent one end of the continuum. The electrons in these nonpolar bonds are shared equally.

In the middle of the continuum are bonds where one atom is much more electronegative than the other. In these asymmetric bonds, substantial partial charges exist on each of the atoms. These types of polar covalent bonds occur when a highly electronegative atom such as oxygen or nitrogen is bonded to an atom with a lower affinity for electrons, such as carbon or hydrogen. Ammonia (NH_3) and water (H_2O) are examples of molecules with polar covalent bonds.

At the right-hand side of the continuum are molecules made up of atoms with extreme differences in their electronegativities. In this case, electrons are transferred rather than shared, the atoms have full charges, and the bonding is ionic. Sodium chloride (NaCl) is a familiar example of a molecule formed by ionic bonds.

Most chemical bonds that occur in biological molecules are on the left-hand side and the middle of the continuum; in the molecules found in organisms, ionic bonding is less common.

(a) A sodium ion being formed

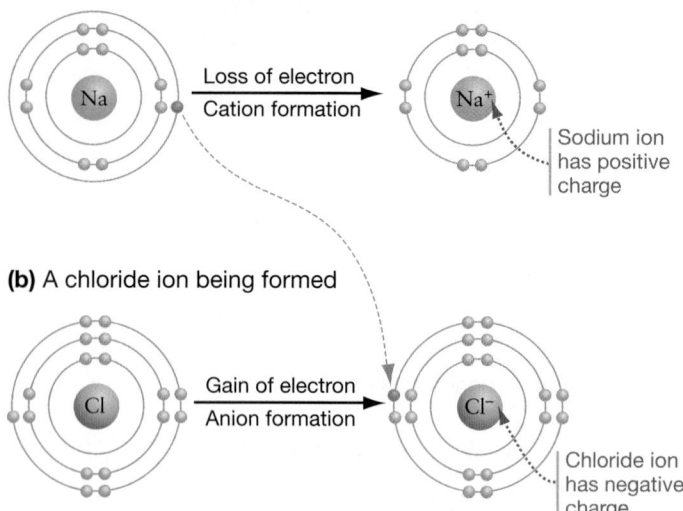

Loss of electron
Cation formation

Sodium ion has positive charge

(b) A chloride ion being formed

Gain of electron
Anion formation

Chloride ion has negative charge

(c) Table salt (NaCl) is a crystal composed of two ions.

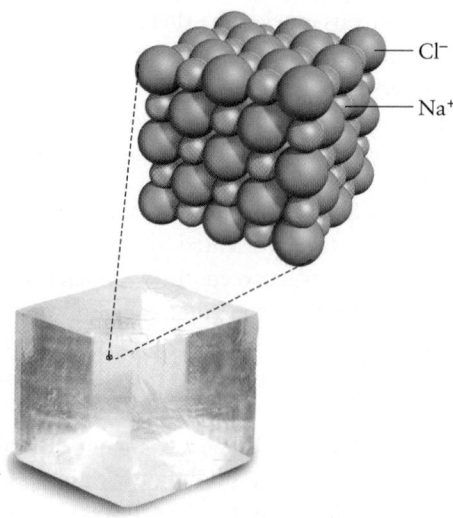

Cl⁻
Na⁺

FIGURE 2.6 Ion Formation and Ionic Bonding. The sodium ion (Na^+) and the chloride ion (Cl^-) are stable because they have full valence shells. In table salt (NaCl), sodium and chloride ions pack into a crystal structure held together by electrical attraction between their positive and negative charges.

Equal sharing of electrons ◄──────────────────────────────────► Transfer of electrons

| Nonpolar covalent bonds (atoms have no charge) | Polar covalent bonds (atoms have partial charge) | Ionic bonds (atoms have full charge) |

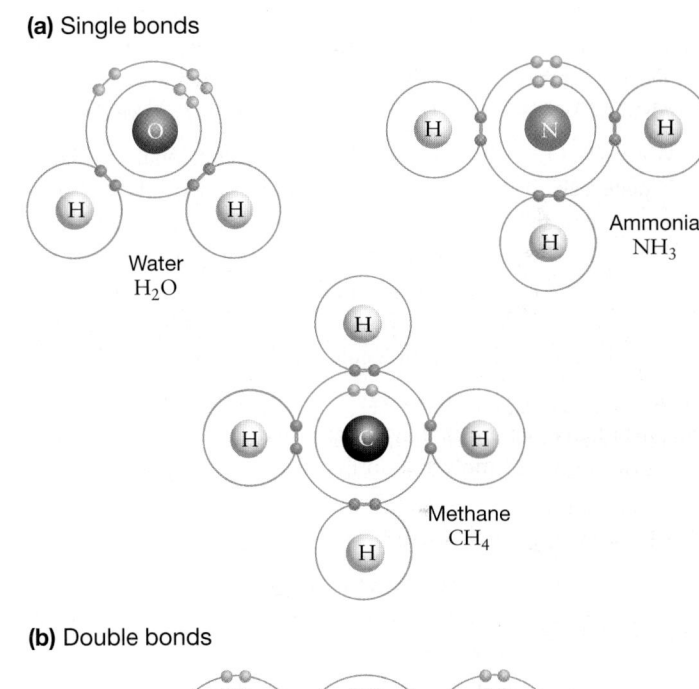

| Hydrogen | Methane | Ammonia | Water | Sodium chloride |

FIGURE 2.7 The Electron-Sharing Continuum. The degree of electron sharing in chemical bonds can be thought of as a continuum, from equal sharing in nonpolar covalent bonds to no sharing in ionic bonds.

✔QUESTION Why do most polar covalent bonds involve nitrogen or oxygen?

Some Simple Molecules Formed from C, H, N, and O

Look back at Figure 2.3 and count the number of unpaired electrons in the valence shells of carbon, nitrogen, oxygen, and hydrogen atoms. Each unpaired electron in a valence shell can make up half of a covalent bond. It should make sense to you that a carbon atom can form a total of four covalent bonds; nitrogen can form three; oxygen can form two; and hydrogen, one.

When each of the four unpaired electrons of a carbon atom covalently bonds with a hydrogen atom, the molecule that results is written CH_4 and is called methane (**FIGURE 2.8a**). Methane is the most common molecule found in natural gas. When a nitrogen atom's three unpaired electrons bond with three hydrogen atoms, the result is NH_3, or ammonia. Similarly, an atom of oxygen can form covalent bonds with two atoms of hydrogen, resulting in a water molecule (H_2O). As Figure 2.4 showed, a hydrogen atom can bond with another hydrogen atom to form hydrogen gas (H_2).

In addition to forming more than one single bond, atoms with more than one unpaired electron in the valence shell can form double bonds or triple bonds. **FIGURE 2.8b** shows how carbon forms double bonds with oxygen atoms to produce carbon dioxide (CO_2). Triple bonds result when three pairs of electrons are shared. **FIGURE 2.8c** shows the structure of molecular nitrogen (N_2), which forms when two nitrogen atoms establish a triple bond.

✔ If you understand how electronegativity affects covalent bonds, you should be able to draw arrows between the atoms in each molecule shown in Figure 2.8 to indicate the relative position of the shared electrons. If they are equally shared, then draw a double-headed arrow.

The Geometry of Simple Molecules

In many cases, the overall shape of a molecule dictates how it behaves. In chemistry and in biology, function is based on structure.

The shapes of the simple molecules you've just learned about are governed by the geometry of their bonds. Nitrogen (N_2) and

(a) Single bonds

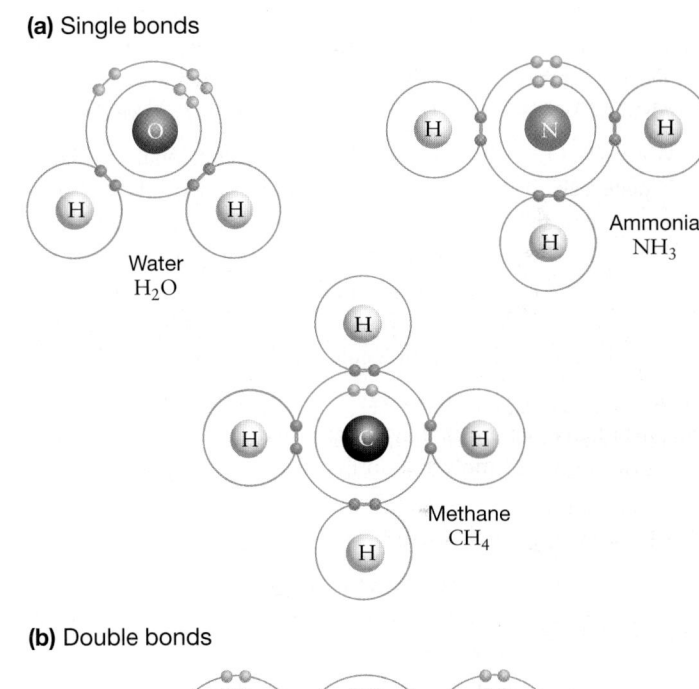

Water H_2O

Ammonia NH_3

Methane CH_4

(b) Double bonds

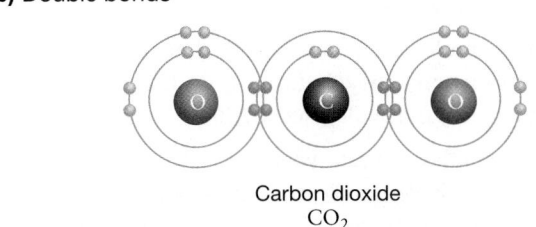

Carbon dioxide CO_2

(c) Triple bonds

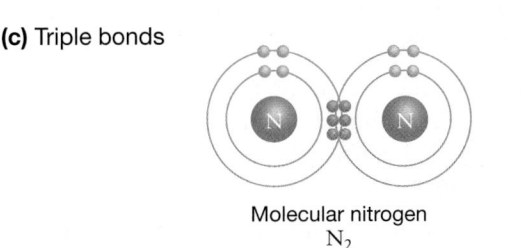

Molecular nitrogen N_2

FIGURE 2.8 Unpaired Electrons in the Valence Shell Participate in Covalent Bonds. Covalent bonding is based on sharing of electrons in the outermost shell. Covalent bonds can be **(a)** single, **(b)** double, or **(c)** triple.

(a) Methane (CH_4) **(b)** Water (H_2O)

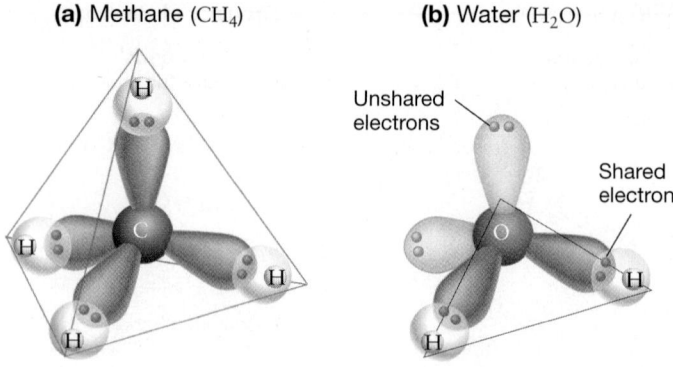

Unshared electrons

Shared electrons

FIGURE 2.9 The Geometry of Methane and Water.

	Methane	Ammonia	Water	Oxygen
(a) Molecular formulas:	CH_4	NH_3	H_2O	O_2
(b) Structural formulas:	$H-\underset{\displaystyle H}{\overset{\displaystyle H}{C}}-H$	$H-\underset{\displaystyle H}{N}-H$	$\underset{H \quad H}{\overset{O}{\diagup \diagdown}}$	$O{=}O$
(c) Ball-and-stick models:				
(d) Space-filling models:				

FIGURE 2.10 Molecules Can Be Represented Several Ways. Each method of representing a molecule has particular advantages.

carbon dioxide (CO_2), for example, have linear structures (see Figure 2.8). Molecules with more complex geometries include

- Methane (CH_4)—which is tetrahedral, a structure with four tri-angular faces like a pyramid (**FIGURE 2.9a**). The tetrahedron forms because the electrons in the four C—H bonds repulse each other equally. The electron pairs are as far apart as they can get.

- Water (H_2O)—which is bent and two-dimensional, or planar (**FIGURE 2.9b**). Why? The electrons in the four orbitals of oxy-gen's valence shell repulse each other, just as they do in meth-ane. But in water, two of the orbitals are filled with electron pairs from the oxygen atom, and two are filled with electron pairs from covalent bonds between oxygen and hydrogen. The shared electrons form a molecule that is V-shaped and flat.

Section 2.2 explores how water's shape, in combination with the partial charges on the oxygen and hydrogen atoms, makes it the most important molecule on Earth.

Representing Molecules

Molecules can be represented in a variety of increasingly com-plex ways—only some of which reflect their actual shape. Each method has advantages and disadvantages.

- **Molecular formulas** are compact, but don't contain a great deal of information—they indicate only the numbers and types of atoms in a molecule (**FIGURE 2.10a**).

- **Structural formulas** indicate which atoms in a molecule are bonded together. Single, double, and triple bonds are represented by single, double, and triple dashes, respectively. Structural for-mulas also indicate geometry in two dimensions (**FIGURE 2.10b**). This method is useful for planar molecules such as water and O_2.

- **Ball-and-stick models** take up more space than structural formulas, but provide information on the three-dimensional shape of molecules and indicate the relative sizes of the atoms involved (**FIGURE 2.10c**).

- **Space-filling models** are more difficult to read than ball-and-stick models but more accurately depict the spatial relation-ships between atoms. (**FIGURE 2.10d**).

In both ball-and-stick and space-filling models, biologists use certain colors to represent certain atoms. A black ball, for

example, always symbolizes carbon. (For more information on interpreting chemical structures, see **BioSkills 8** in Appendix B.)

Some of the small molecules you've just learned about are found in volcanic gases, the atmospheres of nearby planets, and in deep-sea hydrothermal vents, like those shown in the photo-graph at the start of this chapter. Based on these observations, re-searchers claim that they were important components of Earth's ancient atmosphere and oceans. If so, then they provided the building blocks for chemical evolution. The question is: How did these simple building blocks combine to form more complex products, early in Earth's history?

Researchers postulate that most of the critical reactions in chemical evolution occurred in an aqueous, or water-based, en-vironment. To understand what happened and why, let's delve into the properties of water and then turn to analyzing the reac-tions that triggered chemical evolution.

check your understanding

C Y U

If you understand that . . .

- Covalent bonds are based on electron sharing. Electron sharing allows atoms to fill all the orbitals in their valence shell, making them more stable.
- Covalent bonds can be polar or nonpolar, depending on whether the electronegativities of the two atoms involved are the same or different.

✔ You should be able to . . .

Draw the structural formula of formaldehyde (CH_2O) and add dots to indicate the relative locations of the electrons being shared in each covalent bond, based on the relative electronegativities of C, H, and O.

Answers are available in Appendix A.

2.2 Properties of Water and the Early Oceans

Life is based on water. It arose in an aqueous environment and remains dependent on water today. In fact, 75 percent of the volume in a typical cell is water; water is the most abundant molecule in organisms (**FIGURE 2.11**). You can survive for weeks without eating, but you aren't likely to live more than 3 or 4 days without drinking.

Water is vital for a simple reason: It is an excellent **solvent**—that is, an agent for dissolving substances and getting them into **solution.** The reactions that were responsible for chemical evolution some 3.5 billion years ago, like those occurring inside your body right now, depend on direct, physical interaction between molecules. Substances are most likely to come into contact with one another and react as **solutes**—meaning, when they are dissolved in a solvent like water. The formation of Earth's first ocean, about 3.8 billion years ago, was a turning point in chemical evolution because it gave the process a place to happen.

Why Is Water Such an Efficient Solvent?

To understand why water is such an effective solvent, recall that

1. Both of the O—H bonds in a water molecule are polar, owing to the difference in the electronegativities of hydrogen and oxygen. As a result, the oxygen atom has a partial negative charge and each hydrogen atom has a partial positive charge.

2. The molecule is bent. Consequently, the partial negative charge on the oxygen atom sticks out, away from the partial positive charges on the hydrogen atoms, giving a water molecule an overall polarity (**FIGURE 2.12a**).

FIGURE 2.12b illustrates how water's structure affects its interactions with other water molecules. When two water molecules approach each other, the partial positive charge on hydrogen attracts the partial negative charge on oxygen. This weak electrical attraction forms a **hydrogen bond** between the molecules.

✔ If you understand how water's structure makes hydrogen bonding possible, you should be able to (1) draw a fictional version of Figure 2.12b that shows water as a linear (not bent)

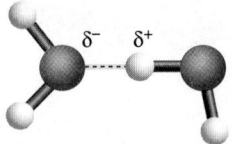

(a) Water is polar.

Electrons are pulled toward oxygen

(b) Hydrogen bonds form between water molecules.

FIGURE 2.12 Water Is Polar and Participates in Hydrogen Bonds. **(a)** Because of oxygen's high electronegativity, the electrons that are shared between hydrogen and oxygen spend more time close to the oxygen nucleus, giving the oxygen atom a partial negative charge and the hydrogen atom a partial positive charge. **(b)** The electrical attraction that occurs between the partial positive and negative charges on water molecules forms a hydrogen bond.

molecule with partial charges on the oxygen and hydrogen atoms; and (2) explain why electrostatic attractions between such water molecules would be much weaker as a result.

In an aqueous solution, hydrogen bonds also form between water molecules and other polar molecules. Similar interactions occur between water and ions. Ions and polar molecules stay in solution because of their interactions with water's partial charges (**FIGURE 2.13**). Substances that interact with water in this way are said to be **hydrophilic** ("water-loving"). Hydrogen bonding makes it possible for almost any charged or polar molecule to dissolve in water.

In contrast, compounds that are uncharged and nonpolar do not interact with water through hydrogen bonding and do not dissolve in water. Substances that do not interact with water are said to be **hydrophobic** ("water-fearing"). Because their interactions with water are minimal or nonexistent, they are forced to interact with each other (**FIGURE 2.14**, see page 26). The water molecules surrounding nonpolar molecules form hydrogen bonds with one another and increase the stability of these **hydrophobic interactions.**

Although individual hydrogen bonds are not as strong as covalent or ionic bonds, many of them occur in a solution. Hydrogen bonding is extremely important in biology owing to the

FIGURE 2.11 Fruits Shrink When They Are Dried Because They Consist Primarily of Water.

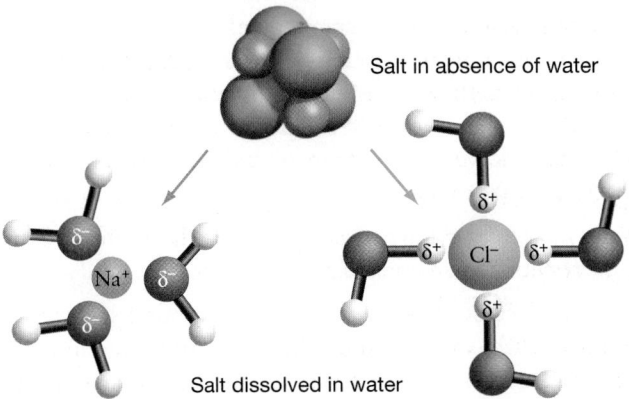

Salt in absence of water

Salt dissolved in water

FIGURE 2.13 Polar Molecules and Ions Dissolve Readily in Water. Water's polarity makes it a superb solvent for polar molecules and ions.

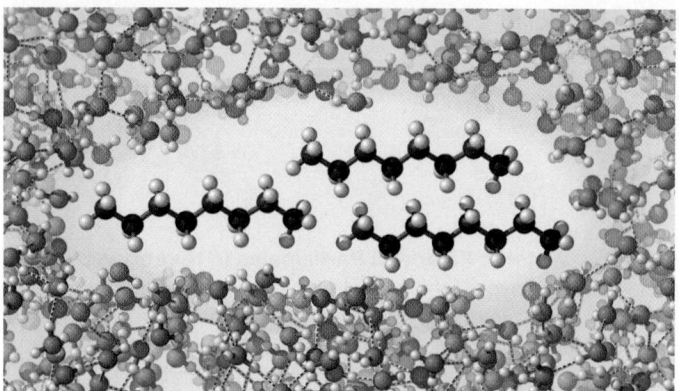

FIGURE 2.14 Nonpolar Molecules Do Not Dissolve in Water.
In aqueous solution, nonpolar molecules and compounds are forced to interact with each other. This occurs because water is much more stable when it interacts with itself rather than with the nonpolar molecules.

✔**QUESTION** What is the physical basis of the expression, "Oil and water don't mix"?

(a) A meniscus forms where water meets a solid surface, as a result of two forces.

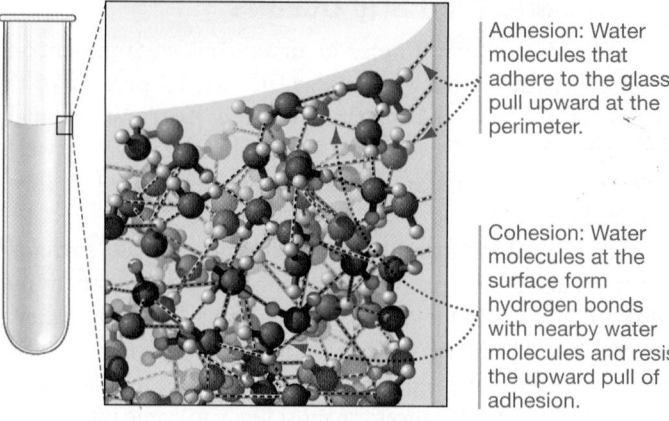

Adhesion: Water molecules that adhere to the glass pull upward at the perimeter.

Cohesion: Water molecules at the surface form hydrogen bonds with nearby water molecules and resist the upward pull of adhesion.

(b) Water has high surface tension.

Because of surface tension, light objects do not fall through the water's surface

FIGURE 2.15 Cohesion, Adhesion, and Surface Tension.
(a) Meniscus formation is based on hydrogen bonding. **(b)** Water resists forces—like the weight of a spider—that increase its surface area. The resistance is great enough that light objects do not break the surface.

sheer number of hydrogen bonds that form between water and hydrophilic molecules.

What Properties Are Correlated with Water's Structure?

Water's small size, highly polar covalent bonds, and bent shape resulting in overall polarity are unique among molecules. Because the structure of molecules routinely correlates with their function, it's not surprising that water has some remarkable properties, in addition to its extraordinary capacity to act as a solvent.

Cohesion, Adhesion, and Surface Tension   Attraction between like molecules is called **cohesion.** Water is cohesive—meaning that it stays together—because of the hydrogen bonds that form between individual molecules.

Attraction between unlike molecules, in contrast, is called **adhesion.** Adhesion is usually analyzed in regard to interactions between a liquid and a solid surface. Water adheres to surfaces that have any polar or charged components.

Cohesion and adhesion are important in explaining how water can move from the roots of plants to their leaves against the force of gravity (see Chapter 38). But you can also see them in action in the concave surface, or meniscus, that forms in a glass tube (**FIGURE 2.15a**). A meniscus forms as a result of two forces:

1. Water molecules at the perimeter of the surface adhere to the glass, resulting in an upward pull.

2. Water molecules at the surface hydrogen-bond with water molecules next to them and below them, resulting in a net lateral and downward pull that resists the upward pull of adhesion.

Cohesion is also instrumental in the phenomenon known as **surface tension.** When water molecules are at the surface, there

are no water molecules above them for hydrogen bonding. As a result, they exhibit stronger attractive forces between their nearest neighboring molecules. This enhanced attraction between the surface water molecules results in tension that minimizes the total surface area.

This fact has an important consequence: Water resists any force that increases its surface area. More specifically, any force that depresses a water surface meets with resistance. This resistance makes a water surface act like an elastic membrane (**FIGURE 2.15b**).

In water, the "elastic membrane" is stronger than it is in other liquids. Water's surface tension is extraordinarily high because of the stronger hydrogen bonding that occurs between molecules at the surface. This explains why it is better to cut the water's surface with your fingertips when you dive into a pool, instead of doing a belly flop.

Water Is Denser as a Liquid than as a Solid   When factory workers pour molten metal or plastic into a mold and allow it to cool to the solid state, the material shrinks. When molten lava pours out of a volcano and cools to solid rock, it shrinks. But when you fill an ice tray with water and put it in the freezer to make ice, the water expands.

Unlike most substances, water is denser as a liquid than it is as a solid. In other words, there are more molecules of water in a given volume of liquid water than there are in the same volume of solid water, or ice. **FIGURE 2.16** illustrates why this is so.

Note that in ice, each water molecule participates in four hydrogen bonds. These hydrogen bonds cause the water molecules to form a regular and repeating structure, or crystal (Figure 2.16a). The crystal structure of ice is fairly open, meaning that there is a relatively large amount of space between molecules.

Now compare the extent of hydrogen bonding and the density of ice with that of liquid water, illustrated in Figure 2.16b. Note that the extent of hydrogen bonding in liquid water is much less than that found in ice, and that the hydrogen bonds in liquid water are constantly being formed and broken. As a result, molecules in the liquid phase are packed much more closely together than in the solid phase.

Normally, heating a substance causes it to expand because molecules begin moving faster and colliding more often and with greater force. But heating ice causes hydrogen bonds to break and the open crystal structure to collapse. In this way, hydrogen bonding explains why water is denser as a liquid than as a solid.

This property of water has an important result: Ice floats (Figure 2.16c). If it didn't, ice would sink to the bottom of lakes, ponds, and oceans soon after it formed. The ice would stay frozen in the cold depths. Instead, ice serves as a blanket, insulating the liquid below from the cold air above. If water weren't so unusual, it is almost certain that Earth's oceans would have frozen solid before life had a chance to start.

Water Has a High Capacity for Absorbing Energy Hydrogen bonding is also responsible for another of water's remarkable physical properties: Water has a high capacity for absorbing energy.

Specific heat, for example, is the amount of energy required to raise the temperature of 1 gram of a substance by 1°C. Water has a high specific heat because when a source of energy hits it, hydrogen bonds must be broken before heat can be transferred and the water molecules begin moving faster. As **TABLE 2.1** indicates, as molecules increase in overall polarity, and thus in their ability to form hydrogen bonds, it takes an extraordinarily large amount of energy to change their temperature.

TABLE 2.1 Specific Heats of Some Liquids

The specific heats reported in this table were measured at 25°C and are given in units of joules per gram of substance per degree Celsius. (The joule is a unit of energy.)

With extensive hydrogen bonding	Specific Heat
Water (H_2O)	4.18
With some hydrogen bonding	
Ethanol (C_2H_6O)	2.44
Glycerol ($C_3H_8O_3$)	2.38
With little or no hydrogen bonding	
Benzene (C_6H_6)	1.74
Xylene (C_8H_{10})	1.72

DATA: D. R. Lide (editor). 2008. Standard Thermodynamic Properties of Chemical Substances, in *CRC Handbook of Physics and Chemistry.* 89th ed. Boca Raton, FL: CRC Press.

(a) In ice, water molecules form a crystal lattice.

(b) In liquid water, no crystal lattice forms.

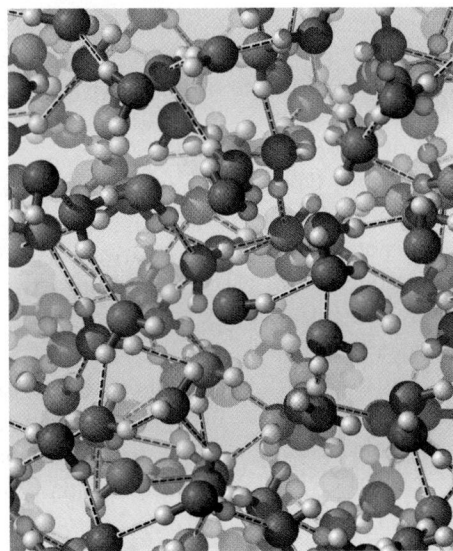

(c) Liquid water is denser than ice. As a result, ice floats.

FIGURE 2.16 Hydrogen Bonding Forms the Crystal Structure of Ice. In ice, each molecule can form four hydrogen bonds at one time. Each oxygen atom can form two; each hydrogen atom can form one.

Property	Cause	Biological Consequences
Solvent for charged or polar compounds	_____ _____ _____ _____	Most chemical reactions important for life take place in aqueous solution.
Denser as a liquid than a solid	As water freezes, each molecule forms a total of four hydrogen bonds, leading to the formation of the low-density crystal structure called ice.	_____ _____ _____
High specific heat	Water molecules must absorb lots of heat energy to break hydrogen bonds and experience increased movement (and thus temperature).	Oceans absorb and release heat slowly, moderating coastal climates.
High heat of vaporization	_____ _____	Evaporation of water from an organism cools the body.

✔ **EXERCISE** You should be able to fill in the missing cells in this table.

Similarly, it takes a large amount of energy to break the hydrogen bonds in liquid water and change the molecules from the liquid phase to the gas phase. Water's **heat of vaporization**—the energy required to change 1 gram of it from a liquid to gas—is higher than that of most molecules that are liquid at room temperature. As a result, water has to absorb a great deal of energy to evaporate. Water's high heat of vaporization is the reason that sweating or dousing yourself with water is an effective way to cool off on a hot day. Water molecules absorb a great deal of energy from your body before they evaporate, so you lose heat.

Water's ability to absorb energy is critical to the theory of chemical evolution. Molecules that were formed in the ocean were well protected from sources of energy that could break them apart, such as intense sunlight. As a result, they would have persisted and slowly increased in concentration over time, making them more likely to react and continue the process.

TABLE 2.2 summarizes some of the key properties of water.

The Role of Water in Acid–Base Reactions

You've seen that water's high specific heat and heat of vaporization tend to keep its temperature and liquid form stable. One other aspect of water's chemistry is important for understanding chemical evolution and how organisms work: Water is not a completely stable molecule. In reality, water molecules continually undergo a chemical reaction with themselves. When a **chemical reaction** occurs, one substance is combined with others or broken down into another substance. Atoms may also be rearranged; in most cases, chemical bonds are broken and new bonds form. The chemical reaction that takes place between water molecules is called a "dissociation" reaction. It can be written as follows:

$$H_2O \rightleftharpoons H^+ + OH^-$$

The double arrow indicates that the reaction proceeds in both directions.

The substances on the right-hand side of the expression are the **hydrogen ion** (H^+) and the **hydroxide ion** (OH^-). A hydrogen ion is simply a proton. In reality, however, protons do not exist by themselves. In water, for example, protons associate with water molecules to form hydronium ions (H_3O^+). Thus, the dissociation of water is more accurately written as:

$$H_2O + H_2O \rightleftharpoons H_3O^+ + OH^-$$

One of the water molecules on the left-hand side of the expression has given up a proton, while the other water molecule has accepted a proton.

Substances that give up protons during chemical reactions and raise the hydronium ion concentration of water are called **acids;** molecules or ions that acquire protons during chemical reactions and lower the hydronium ion concentration of water are called **bases.** Most acids act only as acids, and most bases act only as bases; but water can act as both an acid and a base.

A chemical reaction that involves a transfer of protons is called an acid–base reaction. Every acid–base reaction requires a proton donor and a proton acceptor—an acid and a base, respectively.

Water is an extremely weak acid—very few water molecules dissociate to form hydronium ions and hydroxide ions. In contrast, strong acids like the hydrochloric acid (HCl) in your stomach readily give up a proton when they react with water.

$$HCl + H_2O \rightleftharpoons H_3O^+ + Cl^-$$

Strong bases readily acquire protons when they react with water. For example, sodium hydroxide (NaOH, commonly called lye) dissociates completely in water to form Na^+ and OH^-. The hydroxide ion produced by that reaction then accepts a proton from a hydronium ion in the water, forming two water molecules.

$$NaOH(aq) \longrightarrow Na^+ + OH^-$$
$$OH^- + H_3O^+ \rightleftharpoons 2 H_2O$$

(The "*aq*" in the first expression indicates that NaOH is in aqueous solution.)

To summarize, adding an acid to a solution increases the concentration of protons; adding a base to a solution lowers the concentration of protons. Water is both a weak acid and a weak base.

Determining the Concentration of Protons In a solution, the tendency for acid–base reactions to occur is largely a function of the number of protons present. The problem is, there's no simple way to count the actual number of protons present in a sample. Researchers solve this problem using the mole concept.

A **mole** refers to the number 6.022×10^{23}—just as the unit called the dozen refers to the number 12 or the unit million refers to the number 1×10^6. The mole is a useful unit because the mass of one mole of any substance is the same as its molecular weight expressed in grams. **Molecular weight** is the sum of the atomic weights of all the atoms in a molecule.

For example, to get the molecular weight of H_2O, you sum the atomic weights of two atoms of hydrogen and one atom of oxygen. Since the atomic weights of hydrogen and oxygen are very close to their mass numbers (see Figure 2.2), the molecular weight of water would be $1 + 1 + 16$, or a total of 18. Thus, if you weighed a sample of 18 grams of water, it would contain around 6×10^{23} water molecules, or about 1 mole of water molecules.

When substances are dissolved in water, their concentration is expressed in terms of molarity (symbolized by "M"). **Molarity** is the number of moles of the substance present per liter of solution. A 1-molar solution of protons in water, for example, means that 1 mole of protons is contained in 1 liter of solution.

Chemists can measure the concentration of protons in a solution directly using molarity and an instrument called a pH meter. In a sample of pure water at 25°C, the concentration of H^+ is 1.0×10^{-7}M, or 1 ten-millionth molar.

The pH of a Solution Reveals Whether It Is Acidic or Basic Because the concentration of protons in water is such a small number, exponential notation is cumbersome. So chemists and biologists prefer to express the concentration of protons in a solution, and thus whether it is acidic or basic, with a logarithmic notation called **pH**.[1]

By definition, the pH of a solution is the negative of the base-10 logarithm, or log, of the hydrogen ion concentration:

$$pH = -\log[H^+]$$

(To review logarithms, see **BioSkills 6** in Appendix B. The square brackets are a standard notation for indicating "concentration" of a substance in solution.)

Taking antilogs gives

$$[H^+] = \text{antilog}(-pH) = 10^{-pH}$$

Solutions that contain acids have a proton concentration larger than 1×10^{-7}M and thus a pH < 7. This is because acidic

molecules tend to release protons into solution. In contrast, solutions that contain bases have a proton concentration less than 1×10^{-7}M and thus a pH > 7. This is because basic molecules tend to accept protons from solution.

pH is a convenient way to indicate the concentration of protons in a solution, but take note of what the number represents. For example, if the concentration of H^+ in a sample of water is 1.0×10^{-7}M, then its pH is 7. If the pH changes to 5, then the sample contains 100 times more protons and has become 100 times more acidic. ✔ **QUANTITATIVE** If you understand how pH is related to $[H^+]$, you should be able to calculate the concentration of protons in a solution that has a pH of 8.5.

FIGURE 2.17 shows the pH scale and reports the pH of some selected solutions. Pure water is used as a standard, or point of reference, for pH 7 on the pH scale. The solution inside living cells is about pH 7, which is considered neutral—neither acidic nor basic. The normal function of a cell is dependent on maintaining this neutral internal environment. What is responsible for regulating pH?

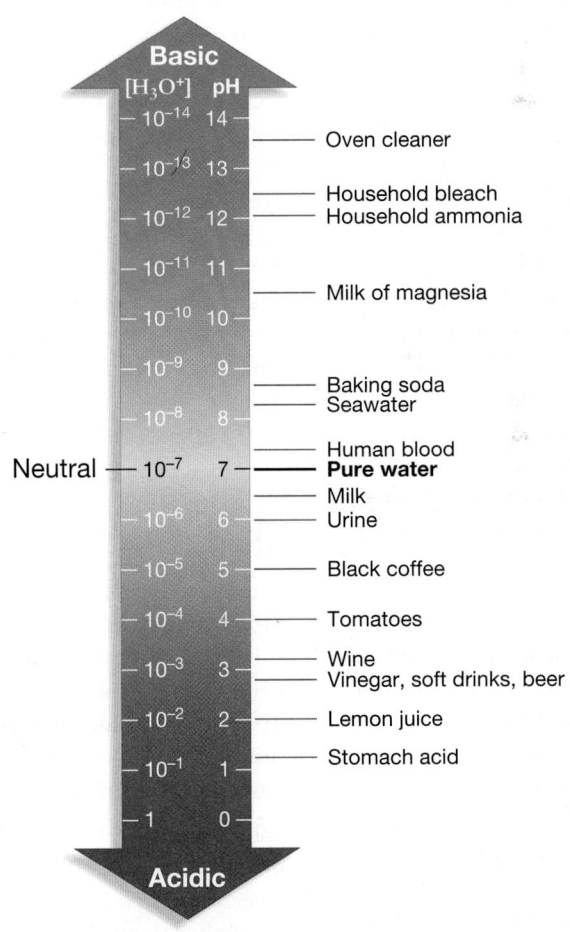

FIGURE 2.17 The pH Scale. Because the pH scale is logarithmic, a change in one unit of pH represents a change in the concentration of hydrogen ions equal to a factor of 10. Coffee has a hundred times more H^+ than pure water has.

✔ **QUESTION** What happens to the concentration of protons in black coffee after you add milk?

---

[1]The term pH is derived from the French *puissance d'hydrogéne*, or "power of hydrogen."

Buffers Protect Against Damaging Changes in pH Life is sensitive to changes in pH. Changes in proton concentration affect the structure and function of polar or charged substances as well as the tendency of acid–base reactions to occur.

Compounds that minimize changes in pH are called **buffers** because they reduce the impact of adding acids or bases on the overall pH of a solution. Buffers are important in maintaining relatively constant conditions, or **homeostasis,** in cells and tissues. In cells, a wide array of naturally occurring molecules act as buffers.

Most buffers are weak acids, meaning that they are somewhat likely to give up a proton in solution, but once the proton concentration rises, the acid is regenerated. To see how buffers work, consider the disassociation of carbonic acid in water to form bicarbonate ions and protons:

$$CH_2O_3 \rightleftharpoons CHO_3^- + H^+$$
$$\text{carbonic acid} \qquad \text{bicarbonate}$$

When carbonic acid and bicarbonate are present in about equal concentrations in a solution, they function as a buffering system. If the concentration of protons increases slightly, the protons react with bicarbonate ions to form carbonic acid and pH does not change. If the concentration of protons decreases slightly, carbonic acid gives up protons and pH does not change. ✔ If you understand this concept, you should be able to predict what would happen to the concentration of bicarbonate ions if a base like sodium hydroxide (NaOH) were added to the solution of carbonic acid.

As chemical evolution began, then, water provided the physical environment for key reactions to take place. In some cases water also acted as an important reactant. Although acid–base reactions were not critical to the initial stages of chemical evolution, they became extremely important once the process was under way. Now let's consider what happened in solution, some 3.5 billion years ago.

2.3 Chemical Reactions, Energy, and Chemical Evolution

Proponents of the theory of chemical evolution contend that simple molecules present in the atmosphere and oceans of early Earth participated in chemical reactions that eventually produced larger, more complex organic (carbon-containing) molecules—such as the proteins, nucleic acids, sugars, and lipids introduced in the next four chapters. Currently, researchers are investigating two environments where these reactions may have occurred:

1. *The atmosphere*, which was probably dominated by gases ejected from volcanoes. Water vapor, carbon dioxide (CO_2), and nitrogen (N_2) are the dominant gases ejected by volcanoes today; a small amount of molecular hydrogen (H_2) and carbon monoxide (CO) may also be present.

2. *Deep-sea hydrothermal vents*, where extremely hot rocks contact deep cracks in the seafloor. In addition to gases such as CO_2 and H_2, certain deep-sea vents are rich in minerals containing reactive metals such as nickel and iron.

When gases like CO_2, N_2, H_2, and CO are put together and allowed to interact on their own, however, very little happens. They do not suddenly link together to create large, complex substances like those found in living cells. Instead, their bonds remain intact. To understand why the bonds of these molecules remain unchanged, you must first learn about how chemical reactions proceed.

How Do Chemical Reactions Happen?

Chemical reactions are written in a format similar to mathematical equations: The initial, or **reactant,** molecules are shown on the left and the resulting reaction **product(s)** shown on the right. For example, the most common reaction in the mix of gases and water vapor that emerges from volcanoes results in the production of carbonic acid, which can be precipitated with water as acid rain:

$$CO_2(g) + H_2O(l) \rightleftharpoons CH_2O_3(aq)$$
$$\text{carbonic acid}$$

The physical state of each reactant and product is indicated as gas (*g*), liquid (*l*), solid (*s*), or in aqueous solution (*aq*).

Note that the expression is balanced; that is, 1 carbon, 3 oxygen, and 2 hydrogen atoms are present on each side of the expression. This illustrates the conservation of mass in closed systems—mass cannot be created or destroyed, but it may be rearranged through chemical reactions.

Note also that the expression contains a double arrow, meaning that the reaction is reversible. When the forward and reverse reactions proceed at the same rate, the quantities of reactants and products remain constant, although not necessarily equal. A dynamic but stable state such as this is termed a **chemical equilibrium.**

Changing the concentration of reactants or products can disturb a chemical equilibrium. For example, adding more CO_2 to the mixture would drive the reaction to the right, creating more CH_2O_3 until the equilibrium proportions of reactants and products are reestablished. Removing CO_2 or adding more CH_2O_3 would drive the reaction to the left.

A chemical equilibrium can also be altered by changes in temperature. For example, the water molecules in the following set of interacting elements, or **system,** would be present as a combination of liquid water and water vapor:

$$H_2O(l) \rightleftharpoons H_2O(g)$$

If liquid water molecules absorb enough energy, like the heat released from a volcano, they transform to the gaseous state. (You may recall that water has a high heat of vaporization and requires a large amount of energy to change its state from liquid to gas.) As a result, this change is termed **endothermic** ("within heating") because heat is absorbed during the process. In contrast, the transformation of water vapor to liquid water releases heat and is **exothermic** ("outside heating"). Raising the temperature of this system drives the equilibrium to the right; cooling the system drives it to the left.

In relation to chemical evolution, though, these reactions and changes of physical state are not particularly interesting. Carbonic acid is not an important intermediate in the formation of more complex molecules. However, interesting things do begin to happen when energy is added to mixtures of volcanic gases.

What Is Energy?

Energy can be defined as the capacity to do work or to supply heat. This capacity exists in one of two ways—as a stored potential or as an active motion.

Stored energy is called **potential energy.** An object gains or loses its ability to store energy because of its position. An electron that resides in an outer electron shell will, if the opportunity arises, fall into a lower electron shell closer to the positive charges on the protons in the nucleus. Because of its position farther from the positive charges in the nucleus, an electron in an outer electron shell has more potential energy than does an electron in an inner shell (**FIGURE 2.18**). When stored in chemical bonds, this form of potential energy is called **chemical energy.**

Kinetic energy is energy of motion. Molecules have kinetic energy because they are constantly in motion.

- The kinetic energy of molecular motion is called **thermal energy.**

- The **temperature** of an object is a measure of how much thermal energy its molecules possess. If an object has a low temperature, its molecules are moving slowly. (We perceive this as "cold.") If an object has a high temperature, its molecules are moving rapidly. (We perceive this as "hot.")

- When two objects with different temperatures come into contact, thermal energy is transferred between them. This transferred energy is called **heat.**

There are many forms of potential energy and kinetic energy, and energy can change from one form into another. However, according to the **first law of thermodynamics,** energy is conserved—it cannot be created or destroyed, but only transferred and transformed. (A more thorough explanation of energy transformation is provided in Chapter 8 in the context of cellular metabolism.)

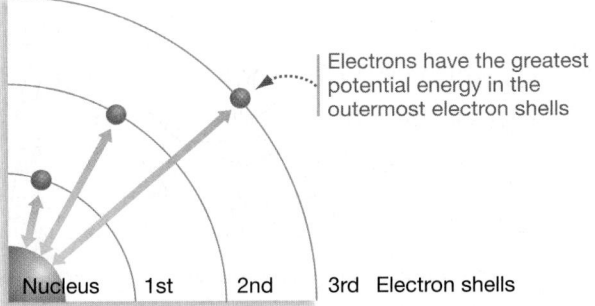

Electrons have the greatest potential energy in the outermost electron shells

Nucleus | 1st | 2nd | 3rd Electron shells

FIGURE 2.18 Potential Energy as a Function of Electron Shells. Electrons in outer shells have more potential energy than do electrons in inner shells, because the negative charges of the electrons in outer shells are farther from the positive charges of the protons in the nucleus. Each shell represents a distinct level of potential energy.

Energy transformation is the heart of chemical evolution. According to the best data available, molecules that were part of the early Earth were exposed to massive inputs of energy. Kinetic energy, in the form of heat, was present in the gradually cooling molten mass that initially formed the planet. The atmosphere and surface of the early Earth were also bombarded with electricity from lightening and radiation from the Sun. Energy stored as potential energy in the chemical bonds of small molecules was also abundant.

Now that you understand that energy transformations are involved in chemical reactions, a big question remains: What determines if a reaction will take place?

What Makes a Chemical Reaction Spontaneous?

When chemists say that a reaction is spontaneous, they have a precise meaning in mind: Chemical reactions are spontaneous if they are able to proceed on their own, without any continuous external influence, such as added energy. Two factors determine if a reaction will proceed spontaneously:

1. Reactions tend to be spontaneous when the product molecules are less ordered than the reactant molecules. For example, nitroglycerin is a single, highly ordered molecule. But when nitroglycerin explodes, it breaks up into gases like carbon dioxide, nitrogen, oxygen, and water vapor. These molecules are much less ordered than the reactant nitroglycerin molecules. The heat that is given off from this explosion also contributes to increasing disorder in the environment. The amount of disorder in a system is called **entropy.** When the products of a chemical reaction are less ordered than the reactant molecules are, entropy increases and the reaction tends to be spontaneous. The **second law of thermodynamics,** in fact, states that entropy always increases in an isolated system.

2. Reactions tend to be spontaneous if the products have lower potential energy than the reactants. If the electrons in the reaction products are held more tightly than those in the reactants, then they have lower potential energy. Recall that highly electronegative atoms such as oxygen or nitrogen hold electrons in covalent bonds much more tightly than do atoms with a lower electronegativity, such as hydrogen or carbon. For example, when hydrogen and oxygen gases react, water is produced spontaneously:

$$2\,H_2(g) + O_2(g) \longrightarrow 2\,H_2O(g)$$

The electrons involved in the O—H bonds of water are held much more tightly by the more electronegative oxygen atom than when they were shared equally in the H—H and O=O bonds of hydrogen and oxygen (see **FIGURE 2.19a** on page 32). As a result, the products have much lower potential energy than the reactants. The difference in chemical energy between reactants and products is given off as heat, so the reaction is exothermic. And although the reaction between hydrogen and oxygen results in less entropy—three molecules of gas produce two molecules of water vapor—the reaction is still spontaneous due to the large drop in potential energy

(a) When hydrogen and oxygen gas react, the product has much lower potential energy than the reactants.

Electrons are held "loosely" in bonds between atoms with equal electronegativities

Electrons are held tightly by highly electronegative atoms

H—H   +   O=O   →   H—O—H   +

2 Hydrogens (H_2)   1 Oxygen (O_2)   *Potential energy drops*   2 Waters (H_2O)

(b) The difference in potential energy is released as heat and light, which vaporizes the water produced.

Heat and light

Released energy

FIGURE 2.19 Potential Energy May Change during Chemical Reactions. In the Hindenburg disaster of 1937, the hydrogen gas from this lighter-than-air craft reacted with oxygen in the atmosphere, with devastating results.

✔**EXERCISE** Label which electrons have relatively low potential energy and which electrons have relatively high potential energy.

released as heat. Since heat increases disorder in the environment, the second law of thermodynamics remains intact. The Hindenburg disaster of 1937 illustrates the large and terrifying amount of heat energy that is given off from this relatively simple reaction (**FIGURE 2.19b**).

To summarize: In general, physical and chemical processes proceed in the direction that results in increased entropy and lower potential energy (**FIGURE 2.20**). These two factors—potential energy and entropy—are used to figure out whether a reaction is spontaneous (see Chapter 8 for more detail). Were the reactions that led to chemical evolution spontaneous? Section 2.4 explores how researchers address this question.

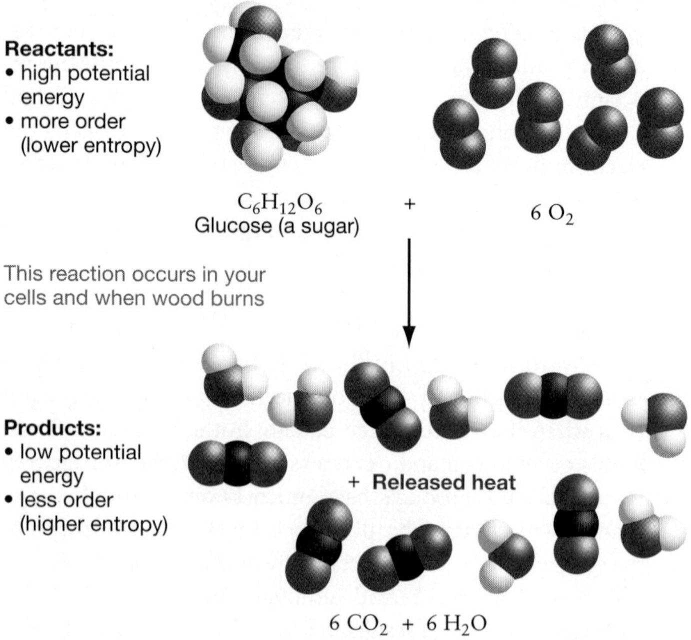

Reactants:
• high potential energy
• more order (lower entropy)

    +

$C_6H_{12}O_6$
Glucose (a sugar)   +   6 O_2

This reaction occurs in your cells and when wood burns

Products:
• low potential energy
• less order (higher entropy)

+ **Released heat**

6 CO_2 + 6 H_2O

FIGURE 2.20 Spontaneous Processes Result in Lower Potential Energy, Increased Disorder, or Both.

---

check your understanding

C Y U

If you understand that . . .

• Chemical reactions result in the transformation of energy, either through the release of energy stored in chemical bonds or the uptake of energy from external sources.

• Chemical reactions tend to be spontaneous if they lead to lower potential energy and higher entropy (more disorder).

✔ **You should be able to . . .**

1. Determine if the reaction between methane (CH_4) and oxygen (O_2) shown here is spontaneous or not, addressing both potential energy and entropy:

$$CH_4 + 2 O_2 \longrightarrow CO_2 + 2 H_2O$$

2. Explain how the positions of the valence electrons in carbon and hydrogen change as methane is converted into carbon dioxide and water.

Answers are available in Appendix A.

2.4 Investigating Chemical Evolution: Approaches and Model Systems

To probe the kinds of reactions that may have set chemical evolution in motion, researchers have used two different approaches—one looking from the "top down" and the other from the "bottom up."

1. In the top-down approach, researchers examine modern cells to identify chemistry that is shared throughout the tree of life. Such ancient reactions are prime candidates for being involved in the chemical evolution that led up to **LUCA,** or last universal common ancestor (introduced in Chapter 1).

2. In the bottom-up approach, the primary focus is on the small molecules and environmental conditions that were present in

early Earth. Here, researchers attempt to identify reactions that could build the molecules found in life using only what was available at the time, without regard to reactions used by modern cells.

These approaches have been used to investigate two different model systems that attempt to explain the process component of the theory of chemical evolution:

1. The **prebiotic soup model** proposes that certain molecules were synthesized from gases in the atmosphere or arrived via meteorites. Afterward they would have condensed with rain and accumulated in oceans. This process would result in an "organic soup" that allowed for continued construction of larger, even more complex molecules.

2. The **surface metabolism model** suggests that dissolved gases came in contact with minerals lining the walls of deep-sea vents and formed more complex, organic molecules.

Since it is impossible to directly examine how and where chemical evolution occurred, the next best thing is to re-create the conditions in the lab and test predictions made by these models. In the following sections, you will learn about how biologists used the top-down and bottom-up approaches to identify reactions that support each of these models for chemical evolution.

Early Origin-of-Life Experiments

Chemical evolution was first taken seriously in 1953 when a graduate student named Stanley Miller performed a breakthrough experiment in the study of the prebiotic soup model.

Miller wanted to answer a simple question: Can complex organic compounds be synthesized from the simple molecules present in Earth's early atmosphere? In other words, is it possible to re-create the first steps in chemical evolution by simulating early-Earth conditions in the laboratory?

Miller's experimental setup (**FIGURE 2.21**) was designed to produce a microcosm of early Earth. The large glass flask represented the atmosphere and contained the gases methane (CH_4), ammonia (NH_3), and hydrogen (H_2), all of which have high potential energy. This large flask was connected to a smaller flask by glass tubing. The small flask held a tiny ocean—200 milliliters (mL) of liquid water.

To connect the mini-atmosphere with the mini-ocean, Miller boiled the water constantly. This added water vapor to the mix of gases in the large flask. As the vapor cooled and condensed, it flowed back into the smaller flask, where it boiled again. In this way, water vapor circulated continuously through the system. This was important: If the molecules in the simulated atmosphere reacted with one another, the "rain" would carry them into the mini-ocean, forming a simulated version of the prebiotic soup.

Had Miller stopped at merely boiling the molecules, little or nothing would have happened. Even at the boiling point of water (100°C), the starting molecules used in the experiment are stable and do not undergo spontaneous chemical reactions.

Something did start to happen in the apparatus, however, when Miller sent electrical discharges across the electrodes he'd inserted into the atmosphere. These miniature lightning bolts

RESEARCH

QUESTION: Can simple molecules and kinetic energy lead to chemical evolution?

HYPOTHESIS: If kinetic energy is added to a mix of simple molecules, reactions will occur that produce more complex molecules, perhaps including some with C–C bonds.

NULL HYPOTHESIS: Chemical evolution will not occur, even with an input of energy.

EXPERIMENTAL SETUP:

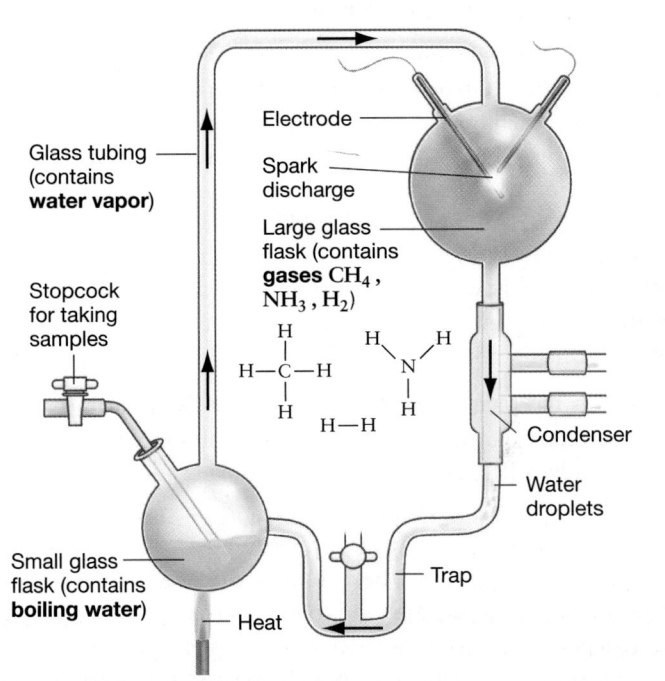

PREDICTION: Complex organic compounds will be found in the liquid water.

PREDICTION OF NULL HYPOTHESIS: Only the starting molecules will be found in the liquid water.

RESULTS

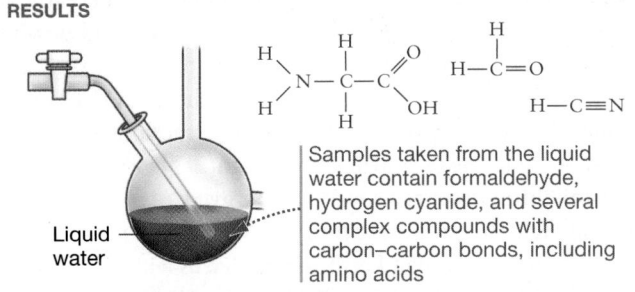

Samples taken from the liquid water contain formaldehyde, hydrogen cyanide, and several complex compounds with carbon–carbon bonds, including amino acids

CONCLUSION: Chemical evolution occurs readily if simple molecules with high free energy are exposed to a source of kinetic energy.

FIGURE 2.21 Miller's Spark-Discharge Experiment. The arrows in the "Experimental setup" diagram indicate the flow of water vapor or liquid. The condenser is a jacket with cold water flowing through it.

SOURCE: Miller, S. L. 1953. A production of amino acids under possible primitive Earth conditions. *Science* 117: 528–529.

✔ **QUESTION** Which parts of the apparatus mimic the ocean, atmosphere, rain, and lightning?

added a crucial element to the reaction mix—pulses of intense electrical energy. After a day of continuous boiling and sparking, the solution in the boiling flask began to turn pink. After a week, it was deep red and cloudy.

When Miller analyzed samples from the mini-ocean, he found large quantities of hydrogen cyanide and formaldehyde. Even more exciting, the sparks and heating had led to the synthesis of additional, more complex organic compounds, including amino acids, which are the building blocks of proteins (see Chapter 3).

Recent Origin-of-Life Experiments

The production of more complex molecules from simple molecules in Miller's experiment supported his claim that the formation of a prebiotic soup was possible. The results came under fire, however, when other researchers pointed out that the early atmosphere was dominated by volcanic gases like CO, CO_2, and H_2, not the CH_4 and NH_3 used in Miller's experiment.

This controversy stimulated a series of follow-up experiments, which showed that the assembly of small molecules into more complex molecules can also occur under more realistic early Earth conditions.

Synthesis of Precursors Using Light Energy One such reaction that may have played a role in chemical evolution is the synthesis of formaldehyde (CH_2O) from carbon dioxide and hydrogen:

$$CO_2(g) + 2\,H_2(g) \longrightarrow \underset{\text{formaldehyde}}{CH_2O(g)} + H_2O(g)$$

This reaction has not been observed in cells—like Miller's experiment, it represents the bottom-up approach. But researchers have shown that when molecules of formaldehyde are heated, they react with one another to produce larger organic compounds, including energy-rich molecules like sugars (see Chapter 5). Note, however, that this reaction does not occur spontaneously—a large input of energy is required.

To explore the possibility of early formaldehyde synthesis, a research group constructed a computer model of the early atmosphere of Earth. The model consisted of a list of all possible chemical reactions that can occur among the molecules now thought to have dominated the early atmosphere: CO_2, H_2O, N_2, CO, and H_2. In this model, they included reactions that occur when these molecules are struck by sunlight. This was crucial because sunlight represents a source of energy.

The sunlight that strikes Earth is made up of packets of light energy called **photons.** Today, Earth is protected by a blanket of ozone (O_3) in the upper atmosphere that absorbs most of the higher-energy photons in sunlight. But since Earth's early atmosphere was filled with volcanic gases released as the molten planet cooled, and ozone is not among these gases, it is extremely unlikely that appreciable quantities of ozone existed. Based on this logic, researchers infer that when chemical evolution was occurring, large quantities of high-energy photons bombarded the planet.

To understand why this energy source was so important, recall that the atoms in hydrogen and carbon dioxide molecules have full valence shells through covalent bonding. This arrangement makes these molecules largely unreactive. However, energy

from photons can break up molecules by knocking apart shared electrons. The fragments that result, called **free radicals,** have unpaired electrons in their outermost shells and are extremely reactive (**FIGURE 2.22**). To mimic the conditions on early Earth more accurately, the computer model included several reactions that produce highly reactive free radicals.

The result? The researchers calculated that, under conditions accepted as reasonable approximations of early Earth by most scientists, appreciable quantities of formaldehyde would have been produced. The energy in sunlight was converted to chemical energy in the form of new bonds in formaldehyde.

The complete reaction that results in the formation of formaldehyde is written as

$$CO_2(g) + 2\,H_2(g) + \text{sunlight} \longrightarrow CH_2O(g) + H_2O(g)$$

Notice that the reaction is balanced in terms of the atoms *and* the energy involved. The sunlight on the reactant side balances the higher energy required for the formation of formaldehyde and water. This result makes sense if you take a moment to think about it. Energy is the capacity to do work, and building larger, more complex molecules requires work to be done.

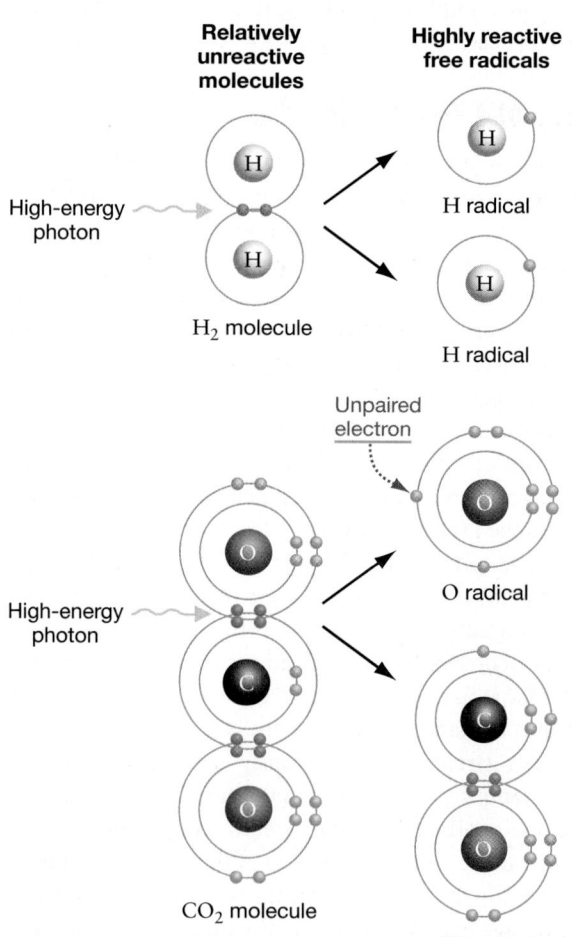

FIGURE 2.22 Free Radicals Are Extremely Reactive. When high-energy photons or pulses of intense electrical energy, such as lightning, strike molecules of hydrogen or carbon dioxide, free radicals can be created. Formation of free radicals is thought to be responsible for some key reactions in chemical evolution.

Using a similar model, other researchers have shown that hydrogen cyanide (HCN)—another important precursor of molecules required for life—could also have been produced in the early atmosphere. According to this research, large quantities of potential precursors for chemical evolution would have formed in the atmosphere and rained out into the early oceans. As a result, organic compounds with relatively high potential energy could have accumulated, and the groundwork would have been in place for the prebiotic soup model of chemical evolution to take off (**FIGURE 2.23a**).

Concentration and Catalysis in Hydrothermal Vents A major stumbling block in the prebiotic soup model is that precursor molecules would have become diluted when they entered the early oceans. Without some means of localized concentration, the formaldehyde and hydrogen cyanide mentioned in the previous section would have been unlikely to meet and react to form larger, more complex molecules. The surface metabolism model offers one possible solution to this dilution effect.

In the surface metabolism model, reactants are recruited to a defined space—a layer of reactive minerals deposited on the

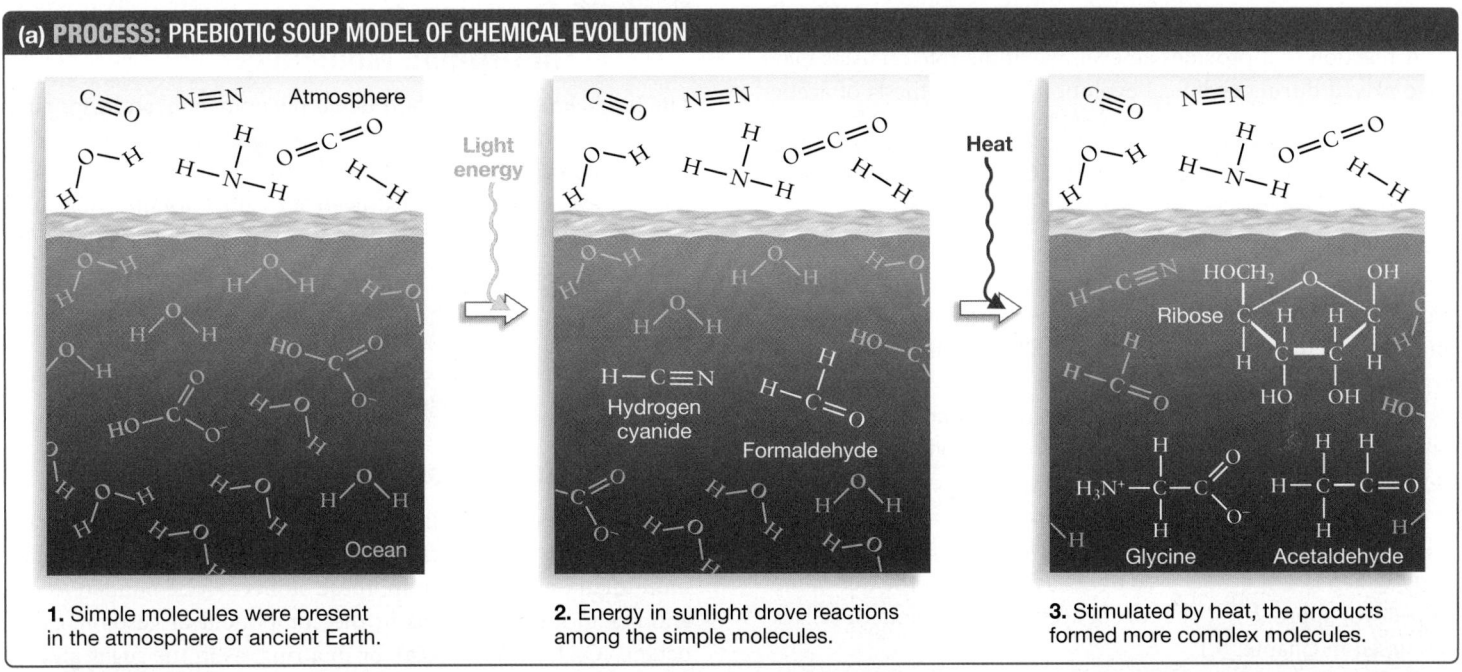

(a) PROCESS: PREBIOTIC SOUP MODEL OF CHEMICAL EVOLUTION

1. Simple molecules were present in the atmosphere of ancient Earth.

2. Energy in sunlight drove reactions among the simple molecules.

3. Stimulated by heat, the products formed more complex molecules.

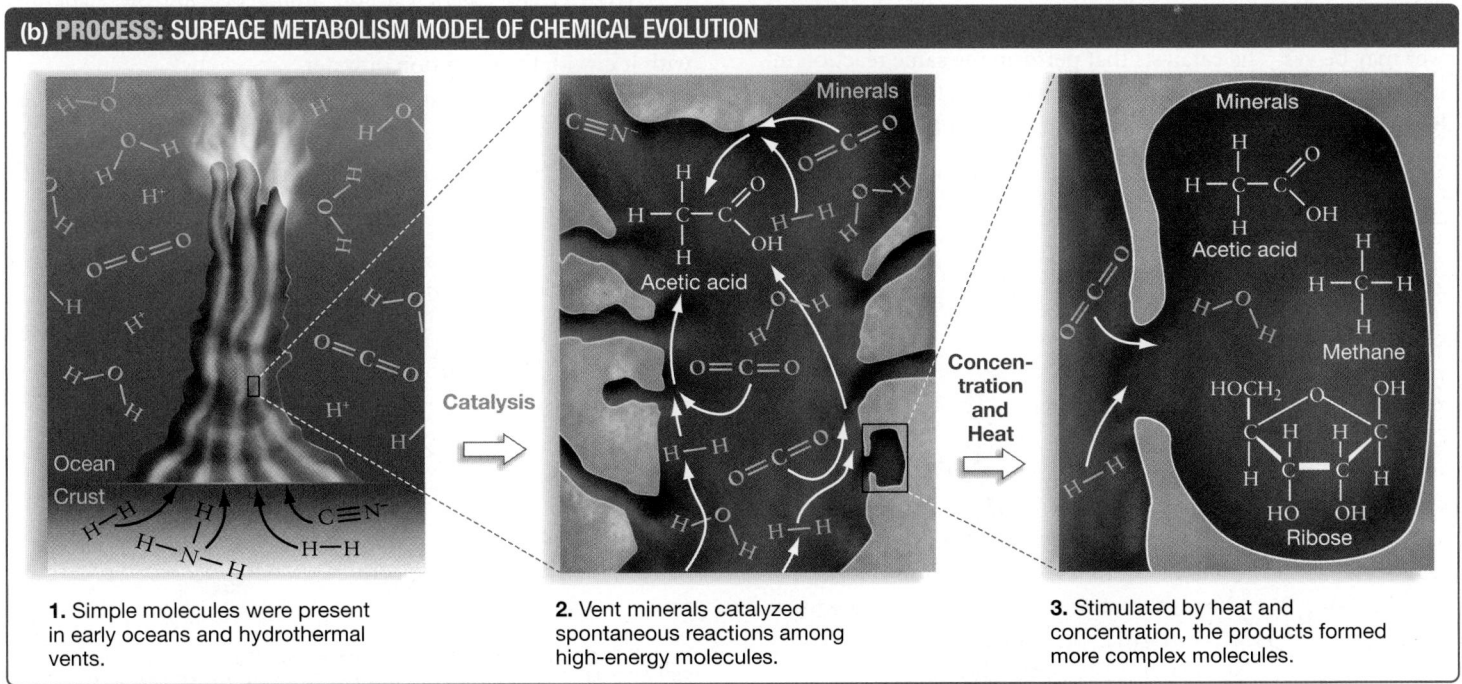

(b) PROCESS: SURFACE METABOLISM MODEL OF CHEMICAL EVOLUTION

1. Simple molecules were present in early oceans and hydrothermal vents.

2. Vent minerals catalyzed spontaneous reactions among high-energy molecules.

3. Stimulated by heat and concentration, the products formed more complex molecules.

FIGURE 2.23 The Start of Chemical Evolution—Two Models. The prebiotic soup and surface metabolism models illustrate how simple molecules containing C, H, O, and N reacted to form organic compounds that served as building blocks for more complex molecules.

walls of deep-sea vent chimneys. Dissolved gases would be attracted by the minerals and concentrated on vent-wall surfaces (**FIGURE 2.23b**).

Here's a key point of this model: Not only would vent-wall minerals bring reactants together, they would also be critical to the rate at which reaction products formed. Even if a potential reaction were spontaneous, it would probably not occur at a level useful for chemical evolution without the support of a **catalyst.** A catalyst provides the appropriate chemical environment for reactants to interact with one another effectively. (You will learn in Chapter 8 that a catalyst only influences the rate of a reaction—it does not provide energy or alter spontaneity.)

A reaction that provides an example of the role catalysts may have played during chemical evolution is the synthesis of acetic acid (CH_3COOH) from carbon dioxide and hydrogen:

$$2\,CO_2(aq) + 4\,H_2(aq) \longrightarrow CH_3COOH(aq) + 2\,H_2O(l)$$
$$\text{acetic acid}$$

The reaction is driven by chemical energy stored in one of the reactants—H_2—and is spontaneous despite the apparent decrease in entropy. It is employed by certain groups of Bacteria and Archaea today as a step toward building even more complex organic molecules.

This reaction has grabbed wide attention among the chemical evolution research community, for two reasons in particular: (**1**) Acetic acid can be formed under conditions that simulate a hydrothermal vent environment (bottom-up approach). (**2**) It is a key intermediate in an ancient pathway that produces acetyl CoA, which is a molecule used by cells throughout the tree of life (top-down approach). (The role of acetyl CoA in modern cells is discussed in Chapter 9.)

Did vent minerals serve as catalysts in the synthesis of acetic acid in early Earth? Evidence from modern cells suggests the answer may be yes. The catalysts that perform the same reaction in modern cells contain minerals similar to those found in hydrothermal vents. These minerals may represent a form of molecular luggage taken from the deep-sea hydrothermal vents as LUCA evolved its independence.

Research is currently under way to establish laboratory systems to more closely mimic surface metabolism conditions in hydrothermal vents. Preliminary results show that in addition to the production of acetic acid, a variety of larger carbon-based molecules can be formed under early Earth conditions. Among these are precursors for the synthesis of nucleotides, the building blocks for the molecules of inheritance used by every living organism on Earth (see Chapter 4).

2.5 The Importance of Organic Molecules

Life has been called a carbon-based phenomenon, and with good reason. Except for water, almost all of the molecules found in organisms contain this atom. Molecules that contain carbon bonded to other elements, such as hydrogen, are called **organic** molecules. (Other types of molecules are referred to as *inorganic* compounds.)

Carbon has great importance in biology because it is the most versatile atom on Earth. Because of its four valence electrons, it will form four covalent bonds. This results in an almost limitless array of molecular shapes, made possible by different combinations of single and double bonds.

Linking Carbon Atoms Together

You have already examined the tetrahedral structure of methane and the linear shape of carbon dioxide. When molecules contain more than one carbon atom, these atoms can be bonded to one another in long chains, as in the component of gasoline called octane (C_8H_{18}; **FIGURE 2.24a**), or in a ring, as in the sugar glucose ($C_6H_{12}O_6$; **FIGURE 2.24b**). Carbon atoms provide the structural framework for virtually all the important compounds associated with life, with the exception of water.

The formation of carbon–carbon bonds was an important event in chemical evolution: It represented a crucial step toward the production of the types of molecules found in living organisms.

(a) Carbons linked in a chain

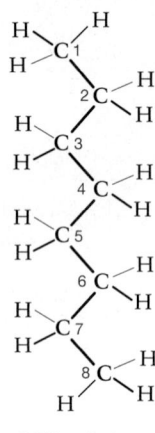

C_8H_{18} Octane

(b) Carbons linked in a ring

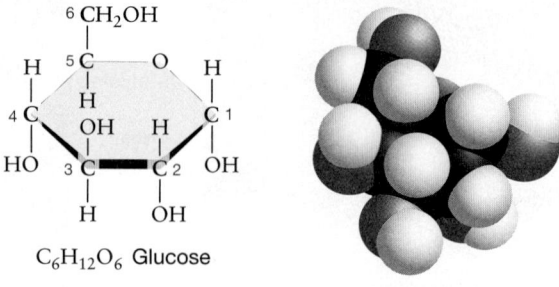

$C_6H_{12}O_6$ Glucose

FIGURE 2.24 The Shapes of Carbon-Containing Molecules. **(a)** Octane is a hydrocarbon chain, and one of the primary ingredients in gasoline. **(b)** Glucose is a sugar that can form a ring-like structure.

Functional Groups

In general, the carbon atoms in an organic molecule furnish a skeleton that gives the molecule its overall shape. But the chemical behavior of the compound—meaning the types of reactions that it participates in—is dictated by groups of H, N, O, P, or S atoms that are bonded to one of the carbon atoms in a specific way.

The critically important H-, N-, O-, P-, and S-containing groups found in organic compounds are called **functional groups**. The composition and properties of six prominent functional groups that are commonly found in organic molecules and recognized by organic chemists are summarized in **TABLE 2.3**. To understand the role that organic compounds play in organisms, it is important to analyze how these functional groups behave.

SUMMARY TABLE 2.3   **Six Functional Groups Commonly Attached to Carbon Atoms**

Functional Group	Formula*	Family of Molecules	Properties of Functional Group	Example
Amino		Amines	Acts as a base—tends to attract a proton to form:	Glycine (an amino acid)
Carboxyl		Carboxylic acids	Acts as an acid—tends to lose a proton in solution to form:	Acetic acid
Carbonyl		Aldehydes	Aldehydes, especially, react with certain compounds to produce larger molecules to form:	Acetaldehyde
		Ketones		Acetone
Hydroxyl	R—OH	Alcohols	Highly polar, so makes compounds more soluble through hydrogen bonding with water; may also act as a weak acid and drop a proton	Ethanol
Phosphate		Organic phosphates	Molecules with more than one phosphate linked together store large amounts of chemical energy	3–Phosphoglyceric acid
Sulfhydryl	R—SH	Thiols	When present in proteins, can form disulfide (S–S) bonds that contribute to protein structure	Cysteine

*In these structural formulas, "R" stands for the rest of the molecule.

✔**EXERCISE** Based on the electronegativities of the atoms involved, predict whether each functional group is polar or nonpolar.

- *Amino and carboxyl functional groups* tend to attract or drop a proton, respectively, when in solution. Amino groups function as bases; carboxyl groups act as acids. During chemical evolution and in organisms today, the most important types of amino- and carboxyl-containing molecules are the amino acids (which Chapter 3 analyzes in detail). Amino acids contain both an amino group and a carboxyl group. (It's common for organic compounds to contain more than one functional group.) Amino acids can be linked together by covalent bonds that form between amino and carboxyl groups. In addition, both of these functional groups participate in hydrogen bonding.

- *Carbonyl groups* are found on aldehyde and ketone molecules such as formaldehyde, acetaldehyde, and acetone. This functional group is the site of reactions that link these molecules into larger, more complex organic compounds.

- *Hydroxyl groups* are important because they act as weak acids. In many cases, the protons involved in acid–base reactions that occur in cells come from hydroxyl groups on organic compounds. Because hydroxyl groups are polar, molecules containing hydroxyl groups will form hydrogen bonds and tend to be soluble in water.

- *Phosphate groups* carry two negative charges. When phosphate groups are transferred from one organic compound to another, the change in charge often dramatically affects the structure of the recipient molecule. In addition, phosphates that are bonded together store chemical energy that can be used in chemical reactions (some of these are discussed in Chapter 3).

- *Sulfhydryl groups* consist of a sulfur atom bonded to a hydrogen atom. They are important because sulfhydryl groups can link to one another via disulfide (S–S) bonds.

To summarize, functional groups make things happen. The number and types of functional groups attached to a framework of carbon atoms imply a great deal about how that molecule is going to behave.

When you encounter an organic compound that is new to you, it's important to do the following three things:

1. Examine the overall size and shape provided by the carbon framework.

2. Identify the types of covalent bonds present based on the electronegativities of the atoms. Use this information to estimate the polarity of the molecule and the amount of potential energy stored in its chemical bonds.

3. Locate any functional groups and note the properties these groups give to the molecule.

Understanding these three features will help you to predict the molecule's role in the chemistry of life.

Once carbon-containing molecules with functional groups had appeared early in Earth's history, what happened next? For chemical evolution to continue, small carbon-based molecules had to form still larger, more complex molecules like those found in living cells. How were the molecules of life—proteins, nucleic acids, carbohydrates, and lipids—formed, and how do they function in organisms today? The rest of this unit explores the next steps in chemical evolution, culminating in the formation of the first living cell.

CHAPTER 2   REVIEW

For media, go to MasteringBiology

If you understand . . .

2.1 | Atoms, Ions, and Molecules: The Building Blocks of Chemical Evolution

- When atoms participate in chemical bonds to form molecules, the shared electrons give the atoms full valence shells and thus contribute to the atoms' stability.

- The electrons in a chemical bond may be shared equally or unequally, depending on the relative electronegativities of the two atoms involved.

- Nonpolar covalent bonds result from equal sharing; polar covalent bonds are due to unequal sharing. Ionic bonds form when an electron is completely transferred from one atom to another.

✔ You should be able to compare and contrast the types of bonds found in methane (CH_4), ammonia (NH_3), and sodium chloride (NaCl).

2.2 | Properties of Water and the Early Oceans

- The chemical reactions required for life take place in water.

- Water is polar—meaning that it has partial positive and negative charges—because it is bent and has two polar covalent bonds.

- Polar molecules and charged substances, including ions, interact with water and stay in solution via hydrogen bonding and electrostatic attraction.

- Water's ability to participate in hydrogen bonding also gives it an extraordinarily high capacity to absorb heat and cohere to other water molecules.

- Water spontaneously dissociates into hydrogen ions (or protons, H^+) and hydroxide ions (OH^-). The concentration of protons in a solution determines the pH, which can be altered by acids and bases or stabilized by buffers.

✔ You should be able to predict what part of water molecules would interact with amino, carboxyl, and hydroxyl functional groups in solution and the types of bonds that would be involved.

2.3 Chemical Reactions, Energy, and Chemical Evolution

- The first step in chemical evolution was the formation of small organic compounds from molecules such as molecular hydrogen (H_2) and carbon dioxide (CO_2).

- Chemical reactions typically involve bonds being broken, atoms being rearranged, and new bonds being formed. This process involves energy, either from the reactants or external sources (e.g., heat).

- Energy comes in different forms. Although energy cannot be created or destroyed, one form of energy can be transformed into another.

✓ You should be able to explain how the energy in electricity can drive a reaction that is nonspontaneous.

2.4 Investigating Chemical Evolution: Approaches and Model Systems

- Experiments suggest that early in Earth's history, external sources of energy, such as sunlight or lightning, could have driven chemical reactions between simple molecules to form molecules with higher potential energy. In this way, energy in the form of radiation or electricity was transformed into chemical energy.

- The prebiotic soup and surface metabolism models for chemical evolution have been supported by the synthesis of organic molecules in laboratory simulations of the early Earth environment.

✓ You should be able to explain how the surface metabolism model is supported by both the top-down and bottom-up approaches used to investigate reactions involved in chemical evolution.

2.5 The Importance of Organic Molecules

- Carbon is the foundation of organic molecules based on its valence, which allows for the construction of molecules with complex shapes.

- Organic molecules are critical to life because they possess versatility of chemical behavior due to the presence of functional groups.

✓ You should be able to predict how adding hydroxyl groups to the octane molecule in Figure 2.24 would affect the properties of the molecule.

(MB) MasteringBiology

1. MasteringBiology Assignments

Tutorials and Activities Acids, Bases, and pH; Anatomy of Atoms; Atomic Number and Mass Number; BioSkill: Using Logarithms; Carbon Bonding and Functional Groups; Cohesion of Water; Covalent Bonds; Dissociation of Water Molecules; Diversity of Carbon-Based Molecules; Electron Arrangement; Energy Transformations; Functional Groups; Hydrogen Bonding and Water; Hydrogen Bonds; Ionic Bonds; Nonpolar and Polar Molecules; pH Scale; Polarity of Water; Properties of Water; Structure of the Atomic Nucleus

Questions Reading Quizzes, Blue-Thread Questions, Test Bank

2. eText Read your book online, search, take notes, highlight text, and more.

3. The Study Area Practice Test, Cumulative Test, BioFlix® 3-D Animations, Videos, Activities, Audio Glossary, Word Study Tools, Art

You should be able to . . .

✓ TEST YOUR KNOWLEDGE

Answers are available in Appendix A

1. Which of the following occurs when a covalent bond forms?
   a. The potential energy of electrons drops.
   b. Electrons in valence shells are shared between nuclei.
   c. Ions of opposite charge interact.
   d. Polar molecules interact.

2. If a reaction is exothermic, then which of the following statements is true?
   a. The products have lower potential energy than the reactants.
   b. Energy must be added for the reaction to proceed.
   c. The products have lower entropy (are more ordered) than the reactants.
   d. It occurs extremely quickly.

3. Which of the following is most likely to have been the energy source responsible for the formation of acetic acid in deep-sea hydrothermal vents?
   a. heat released from the vents
   b. solar radiation that passed through the ocean water
   c. chemical energy present in the reactants
   d. the increase in entropy in the products

4. What is thermal energy?
   a. a form of potential energy
   b. the temperature increase that occurs when any form of energy is added to a system
   c. mechanical energy
   d. the kinetic energy of molecular motion, measured as heat

5. What factors determine whether a chemical reaction is spontaneous or not?

6. What are the two models that have been proposed to explain the process component of chemical evolution?

7. Which of the following molecules would you predict to have the largest number of polar covalent bonds based on their molecular formulas?
   a. C_2H_6O (ethanol)
   b. C_2H_6 (ethane)
   c. $C_2H_4O_2$ (acetic acid)
   d. C_3H_8O (propanol)

8. Locate fluorine (F) on the partial periodic table provided in Figure 2.2. Predict its relative electronegativity compared to hydrogen, sodium, and oxygen. State the number and type of bond(s) you expect it would form if it reacted with sodium.

9. Oxygen is extremely electronegative, meaning that its nucleus pulls electrons shared in covalent bonds very strongly. Explain the changes in electron position that are illustrated in Figure 2.19 based on oxygen's electronegativity.

10. Draw the electron-sharing continuum and place molecular oxygen (O_2), magnesium chloride ($MgCl_2$), and carbon dioxide (CO_2) on it.

11. Consider the reaction between carbon dioxide and water, which forms carbonic acid:

$$CO_2(g) + H_2O(l) \rightleftharpoons CH_2O_3(aq)$$

In aqueous solution, carbonic acid immediately dissociates to form a proton and bicarbonate ion, as follows:

$$CH_2O_3(aq) \rightleftharpoons H^+(aq) + CHO_3^-(aq)$$

If an underwater volcano bubbled additional CO_2 into the ocean, would this sequence of reactions be driven to the left or the right? How would this affect the pH of the ocean?

12. What is the relationship between the carbon framework in an organic molecule (the "R" in Table 2.3) and its functional groups?

13. When H_2 and CO_2 react, acetic acid can be formed spontaneously while the production of formaldehyde requires an input of energy. Which of the following conclusions may be drawn from this observation?
    a. More heat is released when formaldehyde is produced compared to the production of acetic acid.
    b. Compared to the reactants from which it is formed, formaldehyde has more potential energy than does acetic acid.
    c. Entropy decreases when acetic acid is produced and increases when formaldehyde is produced.
    d. The mineral catalyst involved in acetic acid production provides energy to make the reaction spontaneous.

14. When chemistry texts introduce the concept of electron shells, they emphasize that shells represent distinct potential energy levels.

In introducing electron shells, this chapter also emphasizes that they represent distinct distances from the positive charges in the nucleus. Are these two points of view in conflict? Why or why not?

15. Draw a concept map relating water's structure to its properties. (For an introduction to concept mapping, see **BioSkills 15** in Appendix B.) Your concept map should include the following terms or phrases: polar covalent bonds, polarity (on the water molecule), hydrogen bonding, high heat of vaporization, high specific heat, less dense as a solid, effective solvent, unequal sharing of electrons, high energy input required to break bonds, high electronegativity of oxygen.

16. From what you have learned about water, why do coastal regions tend to have climates with lower annual variation in temperature than do inland areas at the same latitude?

3 Protein Structure and Function

In this chapter you will learn that

Proteins are the most abundant and versatile macromolecules in life

↓ *composed of*

20 amino acids with unique side chains
3.1

↓ *polymerize to form*

Protein structure:
Primary – Secondary – Tertiary – Quaternary
3.2

determines ↓   ↓ *determines*

Protein function
3.3

informs →

Importance of proteins for life
3.4

A space-filling model of hemoglobin—a protein that is carrying oxygen in your blood right now.

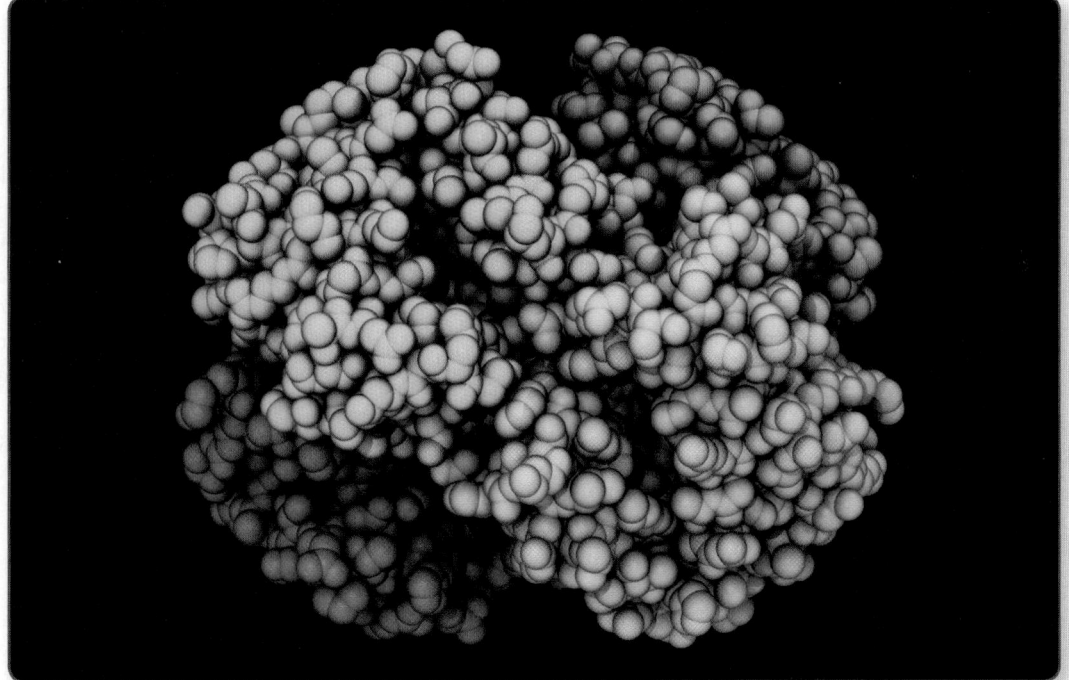

This chapter is part of the Big Picture. See how on pages 104–105.

Chemical reactions in the atmosphere and ocean of ancient Earth are thought to have led to the formation of the first complex carbon-containing compounds. This idea, called chemical evolution, was first proposed by Alexander I. Oparin in 1924. The hypothesis was published again—independently and five years later—by J. B. S. Haldane.

Today, the Oparin–Haldane proposal is considered a formal scientific theory (see Chapter 1). Scientific theories are continuously refined as new information comes to light, and many of Oparin

✔ When you see this checkmark, stop and test yourself. Answers are available in Appendix A.

and Haldane's original ideas have been revised. In its current form, the theory can be broken into four steps.

Step 1 Chemical evolution began with the production of small organic compounds from reactants such as H_2, N_2, NH_3, and CO_2. (Chapter 2 focuses on this step.)

Step 2 These small, simple organic compounds reacted to form mid-sized molecules, such as amino acids, nucleotides, and sugars. (Amino acids are introduced in this chapter. Nucleotides and sugars are discussed in Chapters 4 and 5, respectively.)

Step 3 Mid-sized, building-block molecules linked to form the types of large molecules found in cells today, including proteins, nucleic acids, and complex carbohydrates. Each of these large molecules is composed of distinctive chemical subunits that join together: Proteins are composed of amino acids, nucleic acids are composed of nucleotides, and complex carbohydrates are composed of sugars.

Step 4 Life became possible when one of these large, complex molecules acquired the ability to replicate itself. By increasing in copy number, this molecule would then emerge from the pool of chemicals. At that point, life had begun—chemical evolution gave way to biological evolution.

What type of molecule was responsible for the origin of life? Answering this question is a recurring theme in this and the next three chapters.

To address this question, researchers first designed experiments to identify the types of molecules that could be produced in the waters of prebiotic Earth (Chapter 2). One series of results sparked particular excitement for origin-of-life researchers—the repeated discovery of amino acids among the products of early Earth simulations.

Amino acids have also been found in meteorites and produced in experiments that approximate the environment of interstellar space. Taken together, these observations have led researchers to conclude that amino acids were present and probably abundant during chemical evolution. Since amino acids are the building blocks of proteins, many researchers have therefore asked, Could a protein have been the initial spark of life?

For this question to be valid, proteins would need to possess three of the fundamental attributes of life, namely: information, replication, and evolution. To determine if they do, let's look at the molecules themselves. What are amino acids, and how are they linked to form proteins?

3.1 Amino Acids and Their Polymerization

Modern cells, such as those that make up your body, produce tens of thousands of distinct proteins. Most of these molecules are composed of just 20 different building blocks, called **amino acids.** All 20 of these building blocks have a common structure.

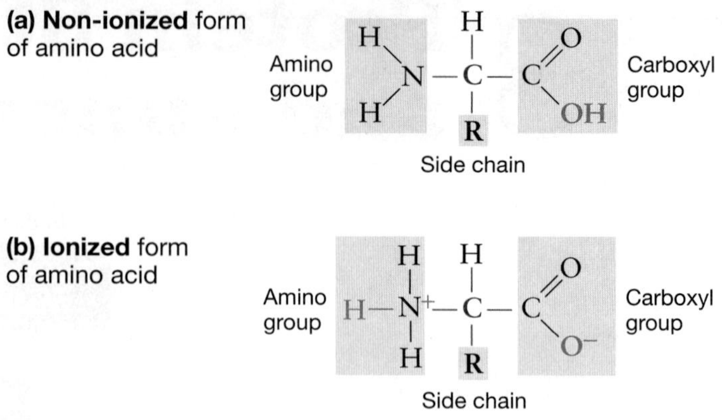

(a) Non-ionized form of amino acid

(b) Ionized form of amino acid

FIGURE 3.1 All Amino Acids Have the Same General Structure. The central α-carbon is shown in red.

The Structure of Amino Acids

To understand how amino acids are put together, recall that carbon atoms have a valence of four—they form four covalent bonds (Chapter 2). All 20 amino acids thus have a common core structure—with a central carbon atom (referred to as the α-carbon) bonded to the four different atoms or groups of atoms diagrammed in **FIGURE 3.1a**:

1. H—a hydrogen atom
2. NH_2—an amino functional group
3. COOH—a carboxyl functional group
4. a distinctive "R-group" (often referred to as a "side chain")

The combination of amino and carboxyl groups not only inspired the name amino acid, but is key to how these molecules behave. In water at pH 7, the concentration of protons causes the amino group to act as a base. It attracts a proton to form NH_3^+ (**FIGURE 3.1b**). The carboxyl group, in contrast, is acidic because its two oxygen atoms are highly electronegative. They pull electrons away from the hydrogen atom, which means that it is relatively easy for this group to lose a proton to form COO^-.

The charges on these functional groups are important for two reasons: **(1)** They help amino acids stay in solution, where they can interact with one another and with other solutes, and **(2)** they affect the amino acid's chemical reactivity.

The Nature of Side Chains

What about the R-group? The R-groups, or side chains, on amino acids vary from a single hydrogen atom to large structures containing carbon atoms linked into rings. While all amino acids share the same core structure, each of the 20 R-groups is unique. The properties of amino acids vary because their R-groups vary.

FIGURE 3.2 highlights the R-groups on the 20 most common amino acids found in cells.[1] As you examine these side chains,

---

[1]There are actually 22 amino acids found in proteins that occur in organisms, but two are very rare.

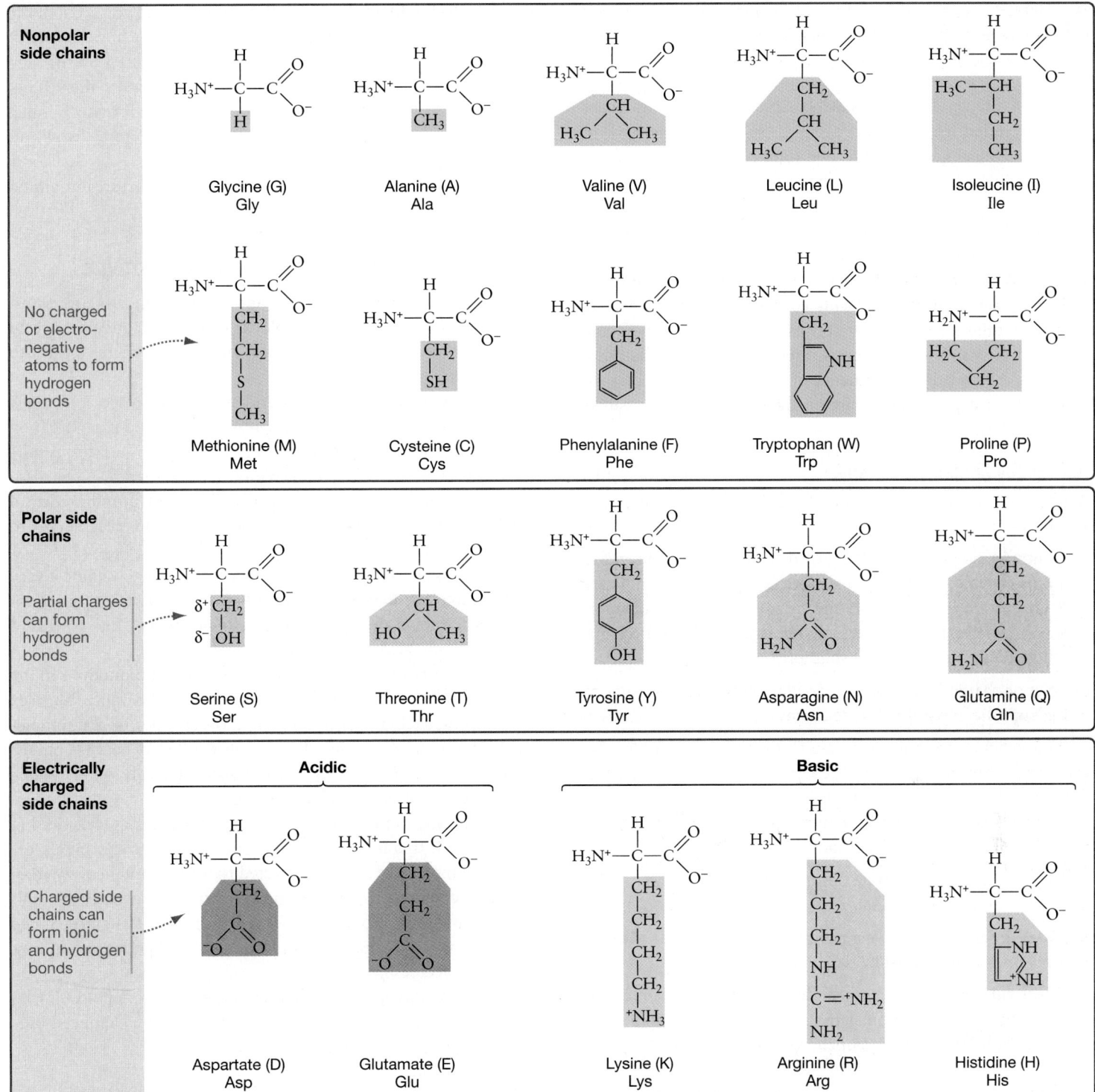

FIGURE 3.2 The 20 Major Amino Acids Found in Organisms. At the pH (about 7.0) found in cells, the 20 major amino acids found in organisms have the structural formulas shown here. The side chains are highlighted, and standard single-letter and three-letter abbreviations for each amino acid are given. For clarity, the carbon atoms in the ring structures of phenylalanine, tyrosine, tryptophan, and histidine are not shown; each bend in a ring is the site of a carbon atom. The hydrogen atoms in these structures are also not shown. A double line inside a ring indicates a double bond.

✔**EXERCISE** Explain why the green R-groups are nonpolar and why the pink R-groups are polar, based on the relative electronegativities of O, N, C, and H (see Chapter 2). Note that sulfur (S) has an electronegativity almost equal to that of carbon and slightly higher than that of hydrogen, making cysteine's side chain mildly hydrophobic.

ask yourself two questions (while referring to Table 2.3): Is this R-group likely to participate in chemical reactions? Will it help this amino acid stay in solution?

Functional Groups Affect Reactivity   Several of the side chains found in amino acids contain carboxyl, sulfhydryl, hydroxyl, or amino functional groups. Under the right conditions, these functional groups can participate in chemical reactions. For example, amino acids with a sulfhydryl group (SH) in their side chains can form disulfide (S—S) bonds that help link different parts of large proteins. Such bonds naturally form between the proteins in your hair; curly hair contains many cross-links and straight hair far fewer.

In contrast, some amino acids contain side chains that are devoid of functional groups—consisting solely of carbon and hydrogen atoms. These R-groups rarely participate in chemical reactions. As a result, the influence of these amino acids on protein function depends primarily on their size and shape rather than reactivity.

The Polarity of Side Chains Affects Solubility   The nature of its R-group affects the polarity, and thus the solubility, of an amino acid in water.

- Nonpolar side chains lack charged or highly electronegative atoms capable of forming hydrogen bonds with water. These R-groups are **hydrophobic,** meaning that they do not interact with water. Instead of dissolving, hydrophobic side chains tend to coalesce in aqueous solution.

- Polar or charged side chains interact readily with water and are **hydrophilic.** Hydrophilic side chains dissolve in water easily.

Amino acid side chains distinguish the different amino acids and can be grouped into four general types: acidic, basic, uncharged polar, and nonpolar. If given a structural formula for an amino acid, as in Figure 3.2, you can determine which type of amino acid it is by asking three questions:

1. Does the side chain have a negative charge? If so, it has lost a proton, so it must be acidic.

2. Does the side chain have a positive charge? If so, it has taken on a proton, so it must be basic.

3. If the side chain is uncharged, does it have an oxygen atom? If so, the highly electronegative oxygen will result in a polar covalent bond and thus is uncharged polar.

If the answers to all three questions are no, then you are looking at a nonpolar amino acid. ✔ If you understand how the interaction between amino acids and water is affected by the side chains, you should be able to use Figure 3.2 to order the following amino acids from most hydrophilic to most hydrophobic: valine, aspartate, asparagine, and tyrosine. Explain why you have chosen this order.

Now that you have seen the diversity of structures in amino acids, let's put them together to make a protein.

How Do Amino Acids Link to Form Proteins?

Amino acids are linked to one another to form proteins. Similarly, the molecular building blocks called nucleotides attach to one another to form nucleic acids, and simple sugars connect to form complex carbohydrates.

In general, a molecular subunit such as an amino acid, a nucleotide, or a sugar is called a **monomer** ("one-part"). When a large number of monomers are bonded together, the resulting structure is called a **polymer** ("many-parts"). The process of linking monomers together is called **polymerization** (**FIGURE 3.3**). Thus, amino acid monomers can polymerize to form proteins.

Biologists also use the word **macromolecule** to denote a very large molecule that is made up of smaller molecules joined together. Proteins are macromolecules—polymers—that consist of linked amino acid monomers.

The theory of chemical evolution states that monomers in the prebiotic soup polymerized to form larger and more complex molecules, such as the proteins and other types of macromolecules found in organisms. This is a difficult step, because monomers such as amino acids do not spontaneously self-assemble into macromolecules such as proteins.

According to the second law of thermodynamics (reviewed in Chapter 2), this fact is not surprising. Complex and highly organized molecules are not expected to form spontaneously from simpler constituents, because polymerization organizes the molecules involved into a more complex, ordered structure. Stated another way, polymerization decreases the disorder, or entropy, of the molecules involved.

For monomers to link together and form macromolecules, an input of energy is required. How could this have happened during chemical evolution?

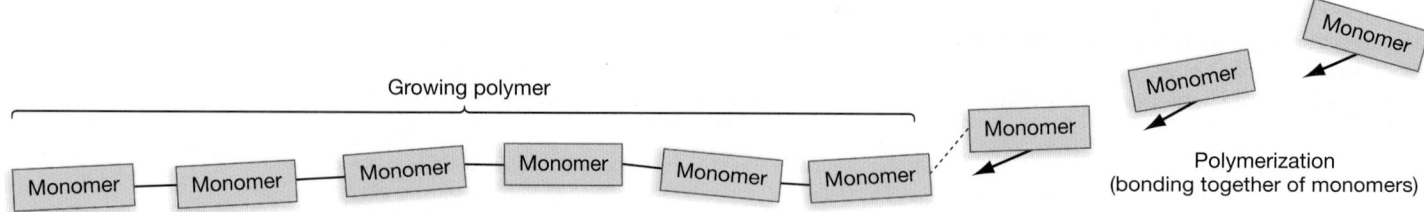

FIGURE 3.3 Monomers Are the Building Blocks of Polymers.

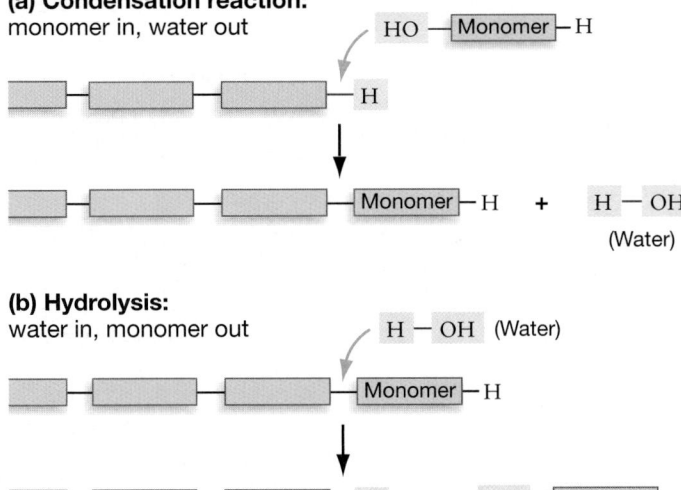

(a) Condensation reaction:
monomer in, water out

HO — Monomer — H

— — — H

Monomer — H + H — OH
(Water)

(b) Hydrolysis:
water in, monomer out

H — OH (Water)

Monomer — H

— — — H + HO — Monomer — H

FIGURE 3.4 Polymers Can Be Extended or Broken Apart.

Could Polymerization Occur in the Energy-Rich Environment of Early Earth?

Monomers polymerize through **condensation reactions,** also known as **dehydration reactions.** These reactions are aptly named because the newly formed bond results in the loss of a water molecule (**FIGURE 3.4a**). The reverse reaction, called **hydrolysis,** breaks polymers apart by adding a water molecule (**FIGURE 3.4b**). The water molecule reacts with the bond linking the monomers, separating one monomer from the polymer chain.

In a solution such as the prebiotic soup, condensation and hydrolysis represent the forward and reverse reactions of a chemical equilibrium:

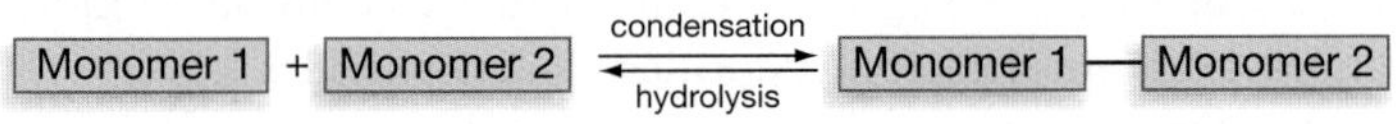

Monomer 1 + Monomer 2 $\underset{\text{hydrolysis}}{\overset{\text{condensation}}{\rightleftarrows}}$ Monomer 1 — Monomer 2

Hydrolysis dominates because it increases entropy and is favorable energetically.

This means that, in the prebiotic soup, polymerization would occur only if there were a very high concentration of amino acids to push the reaction toward condensation. Since the equilibrium favors free monomers over polymers even under concentrated conditions, a polymer is unlikely to have grown much beyond a short chain.

According to recent experiments, though, there are several ways that amino acids could have polymerized early in chemical evolution.

- Researchers evaluating the surface metabolism model of chemical evolution have been able to generate stable polymers by mixing free amino acids with a source of chemical energy and tiny mineral particles. Apparently, growing macromolecules are protected from hydrolysis if they cling, or adsorb, to a mineral surface. One such experiment produced polymers that were 55 amino acids long.

- In conditions that simulate the hot, metal-rich environments of undersea volcanoes, researchers have observed not only amino acid formation but also their polymerization.

- Amino acids have also joined into polymers in experiments in cooler water if a carbon- and sulfur-containing gas—one that is commonly ejected from undersea volcanoes—is present.

The current consensus is that several mechanisms could have led to polymerization reactions between amino acids, early in chemical evolution. What kind of bond is responsible for linking these monomers?

The Peptide Bond As **FIGURE 3.5** shows, amino acids polymerize when a bond forms between the carboxyl group of one amino acid and the amino group of another. The C—N covalent bond that results from this condensation reaction is called a **peptide bond.** When a water molecule is removed in the condensation reaction, the carboxyl group is converted to a carbonyl functional group (C=O) in the resulting polymer, and the amino group is reduced to an N—H.

Peptide bonds are unusually stable compared to linkages in other types of macromolecules. This is because a pair of valence electrons on the nitrogen is partially shared in the C—N bond (see Figure 3.5). The degree of electron sharing is great enough that peptide bonds actually have some of the characteristics of a double bond. For example, the peptide bond is planar, limiting the movement of the atoms participating in the peptide bond.

When amino acids are linked by peptide bonds into a chain, the amino acids are referred to as residues to distinguish them from free monomers.

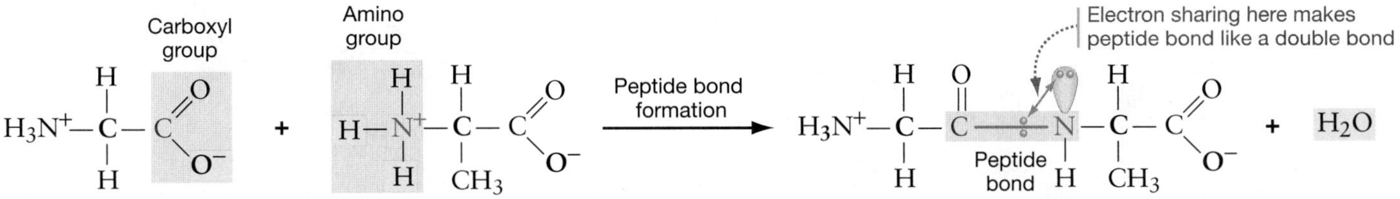

Carboxyl group / Amino group

$$H_3N^+ - \overset{\overset{\displaystyle H}{|}}{\underset{\underset{\displaystyle H}{|}}{C}} - \overset{\overset{\displaystyle O}{\parallel}}{\underset{\underset{\displaystyle O^-}{}}{C}} \quad + \quad H - \overset{\overset{\displaystyle H}{|}}{\underset{\underset{\displaystyle H}{|}}{N^+}} - \overset{\overset{\displaystyle H}{|}}{\underset{\underset{\displaystyle CH_3}{|}}{C}} - \overset{\overset{\displaystyle O}{\parallel}}{\underset{\underset{\displaystyle O^-}{}}{C}}$$

Peptide bond formation →

Electron sharing here makes peptide bond like a double bond

$$H_3N^+ - \overset{\overset{\displaystyle H}{|}}{\underset{\underset{\displaystyle H}{|}}{C}} - \overset{\overset{\displaystyle O}{\parallel}}{C} - N - \overset{\overset{\displaystyle H}{|}}{\underset{\underset{\displaystyle CH_3}{|}}{C}} - \overset{\overset{\displaystyle O}{\parallel}}{\underset{\underset{\displaystyle O^-}{}}{C}} \quad + \quad H_2O$$

Peptide bond

FIGURE 3.5 Peptide Bonds Form When the Carboxyl Group of One Amino Acid Reacts with the Amino Group of a Second Amino Acid.

(a) Peptide chain

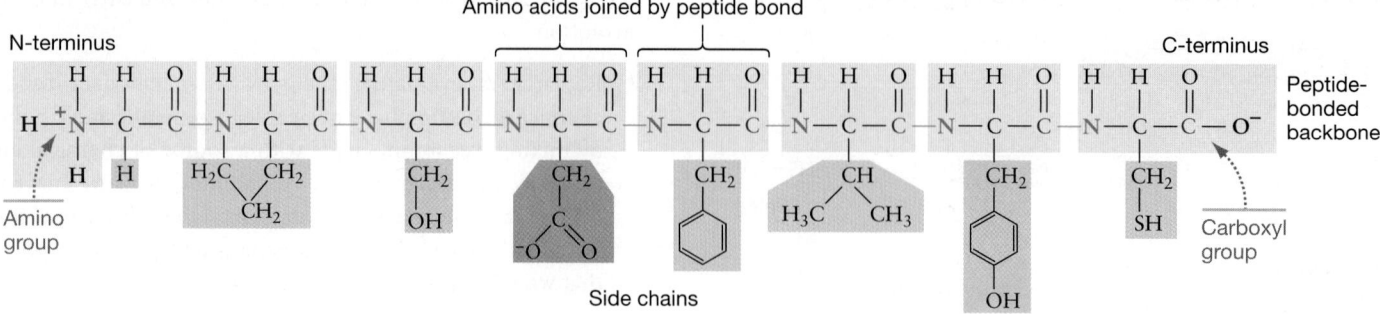

(b) Numbering system

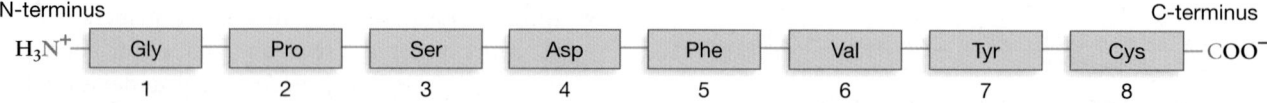

FIGURE 3.6 Amino Acids Polymerize to Form Chains.

FIGURE 3.6a shows how the chain of peptide bonds in a short polymer gives the molecule a structural framework, or a "backbone." There are three key points to note about the peptide-bonded backbone:

1. **R-group orientation** The side chains present in each residue extend out from the backbone, making it possible for them to interact with each other and with water.

2. **Directionality** There is an amino group (NH_3^+) on one end of the backbone and a carboxyl group (COO^-) on the other. The end of the sequence that has the free amino group is called the N-terminus, or amino-terminus, and the end with the free carboxyl group is called the C-terminus, or carboxy-terminus. By convention, biologists always write amino acid sequences

from the N-terminus to the C-terminus (**FIGURE 3.6b**), because the N-terminus is the start of the chain when proteins are synthesized in cells.

3. **Flexibility** Although the peptide bond itself cannot rotate because of its double-bond nature, the single bonds on either side of the peptide bond can rotate. As a result, the structure as a whole is flexible (**FIGURE 3.7**).

When fewer than 50 amino acids are linked together in this way, the resulting polymer is called an **oligopeptide** ("few peptides") or simply a **peptide**. Polymers that contain 50 or more amino acids are called **polypeptides** ("many peptides").

The term **protein** is often used to describe any chain of amino acid residues, but formally protein refers to the complete, often

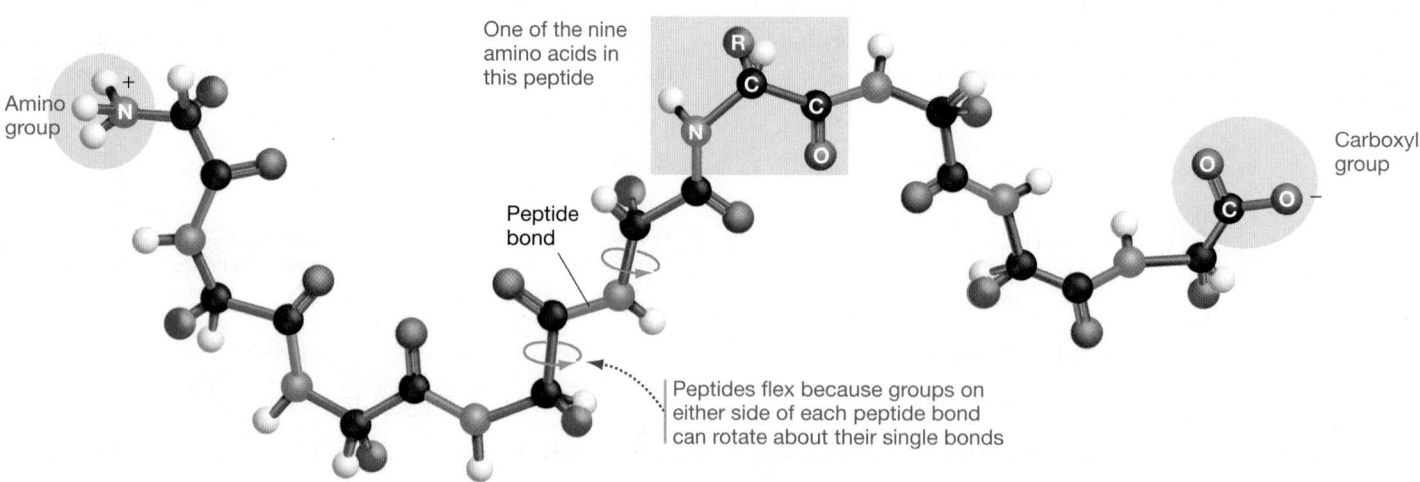

FIGURE 3.7 Peptide Chains Are Flexible.

functional form of the molecule. In Section 3.2, you'll see that some proteins consist of a single polypeptide while others are functional only when multiple polypeptides are bonded to one another.

Proteins are the stuff of life. Let's take a look at how they are put together and then see what they do.

check your understanding

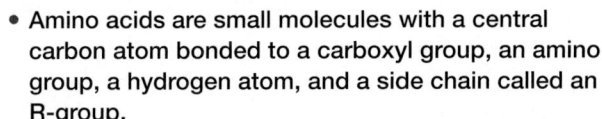

If you understand that . . .

- Amino acids are small molecules with a central carbon atom bonded to a carboxyl group, an amino group, a hydrogen atom, and a side chain called an R-group.
- Each amino acid has distinctive chemical properties because each has a unique R-group.
- Proteins are polymers made up of amino acids.
- When the carboxyl group of one amino acid reacts with the amino group of another amino acid, a strong covalent bond called a peptide bond forms. Small chains are called oligopeptides; large chains are called polypeptides, or proteins.

✔ **You should be able to . . .**

Draw the structural formulas of two glycine residues (glycine's R-group is an H) linked by a peptide bond, and label the amino- and carboxy-terminus.

Answers are available in Appendix A.

3.2 What Do Proteins Look Like?

The unparalleled diversity of proteins—in size, shape, and other aspects of structure—is important because function follows from structure. Proteins can serve diverse functions in cells because they are diverse in size and shape as well as in the chemical properties of their amino acid residues.

FIGURE 3.8 illustrates some of the variety in the sizes and shapes observed in proteins. In the case of the TATA box–binding protein in **FIGURE 3.8a** and the porin protein in **FIGURE 3.8b**, the shape of the molecule has a clear correlation with its function. The TATA box–binding protein has a groove where DNA molecules fit; porin has a hole that forms a pore. The groove in the TATA box–binding protein interacts with specific regions of a DNA molecule, while porin fits in cell membranes and allows certain hydrophilic molecules to pass through. Proteins that provide structural support for cells or tissues, such as the collagen triple helix in **FIGURE 3.8d**, often form long, cable-like fibers.

But many of the proteins found in cells do not have shapes that are noticeably correlated with their functions. For example, the trypsin protein in **FIGURE 3.8c** has an overall globular shape that tells little about its function, which is to bind and cleave peptide bonds of other proteins.

How can biologists make sense of this diversity of protein size and shape? Initially, the amount of variation seems overwhelming. Fortunately, it is not. No matter how large or complex a protein may be, its underlying structure can be broken down into just four basic levels of organization.

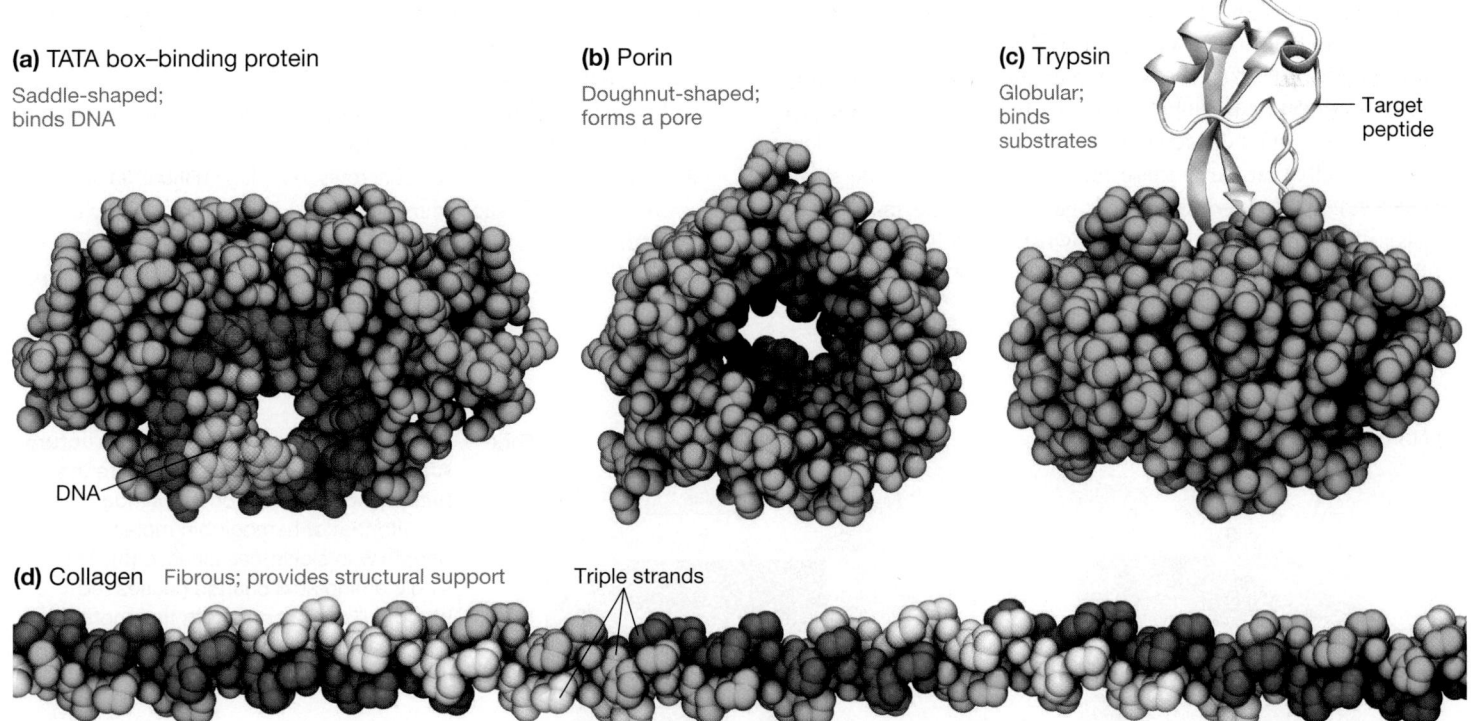

(a) TATA box–binding protein

Saddle-shaped; binds DNA

DNA

(b) Porin

Doughnut-shaped; forms a pore

(c) Trypsin

Globular; binds substrates

Target peptide

(d) Collagen   Fibrous; provides structural support

Triple strands

FIGURE 3.8 In Overall Shape, Proteins Are the Most Diverse Class of Molecules Known.

Primary Structure

Each protein has a unique sequence of amino acids. That simple conclusion was the culmination of 12 years of study by Frederick Sanger and co-workers during the 1940s and 1950s. Sanger's group worked out the first techniques for determining the amino acid sequence of insulin, a hormone that helps regulate sugar concentrations in the blood of humans and other mammals. When other proteins were analyzed, it rapidly became clear that each protein has a definite and distinct amino acid sequence.

Biochemists call the unique sequence of amino acids in a protein the **primary structure** of that protein. The sequence of amino acid residues in Figure 3.6, for example, defines the peptide's primary structure.

With 20 types of amino acids available and length ranging from two amino acid residues to tens of thousands, the number of primary structures that are possible is practically limitless. There may, in fact, be 20^n different combinations of amino acid residues for a polymer with a given length of n. For example, a peptide that is just 10 amino acids long has 20^{10} possible sequences. This is over 10,000 billion.

Why is the order and type of residues in the primary structure of a protein important? Recall that the R-groups present on each amino acid affect its chemical reactivity and solubility. It's therefore reasonable to predict that the R-groups present in a polypeptide will affect that molecule's properties and function.

This prediction is correct. In some cases, even a single change in the sequence of amino acids can cause radical changes in the way the protein as a whole behaves.

As an example, consider hemoglobin, an oxygen-binding protein in human red blood cells. In some individuals, hemoglobin has a valine instead of a glutamate at the 6th position of a strand containing 146 amino acid residues (**FIGURE 3.9a**). Valine's side chain is radically different from the R-group in glutamate. The change in R-group produces hemoglobin molecules that stick to one another and form fibers when oxygen concentrations in the blood are low. Red blood cells that carry these fibers adopt a sickle-like shape (**FIGURE 3.9b**). Sickled red blood cells get stuck in the small blood vessels called capillaries and starve downstream cells of oxygen. A debilitating illness called sickle-cell disease results.

A protein's primary structure is fundamental to its function. Primary structure is also fundamental to the higher levels of protein structure: secondary, tertiary, and quaternary.

Secondary Structure

Even though variation in the amino acid sequence of a protein is virtually limitless, it is only the tip of the iceberg in terms of generating structural diversity.

The next level of organization in proteins—**secondary structure**—is created in part by hydrogen bonding between components of the peptide-bonded backbone. Secondary structures are distinctively shaped sections of proteins that are stabilized largely by hydrogen bonding that occurs between the oxygen on the C=O group of one amino acid residue and the hydrogen on the N—H groups of another (**FIGURE 3.10a**). The oxygen atom in the C=O group has a partial negative charge due to its high electronegativity, while the hydrogen atom in the N—H group has a partial positive charge because it is bonded to nitrogen, which has high electronegativity.

Note a key point: Hydrogen bonding between sections of the same backbone is possible only when a polypeptide bends in a way that puts C=O and N—H groups close together. In most proteins, these polar groups are aligned and form hydrogen bonds with one another when the backbone bends to form one of two possible structures (**FIGURE 3.10b**):

1. an α-**helix** (alpha-helix), in which the polypeptide's backbone is coiled; or

2. a β-**pleated sheet** (beta-pleated sheet), in which segments of a peptide chain bend 180° and then fold in the same plane.

In both structures, the distance between residues that hydrogen-bond to one another is small. In an α-helix, for example, H-bonds form between residues that are just four linear positions apart in the polypeptide's primary sequence (Figure 3.10a).

When biologists use illustrations called ribbon diagrams to represent the shape of a protein, α-helices are shown as coils; β-pleated sheets are shown by groups of arrows in a plane (**FIGURE 3.10c**).

In most cases, secondary structure consists of α-helices and β-pleated sheets. Which one forms, if either, depends on the molecule's primary structure—specifically, the geometry and

(a) Normal amino acid sequence

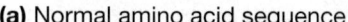

| Pro | **Glu** | Glu |
| 5 | 6 | 7 |

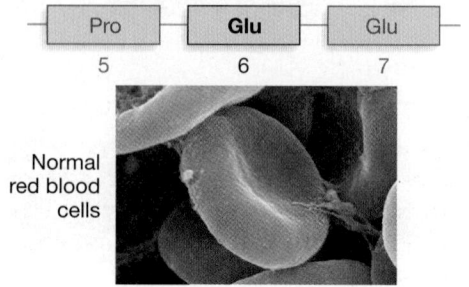

Normal red blood cells

(b) Single change in amino acid sequence

| Pro | **Val** | Glu |
| 5 | 6 | 7 |

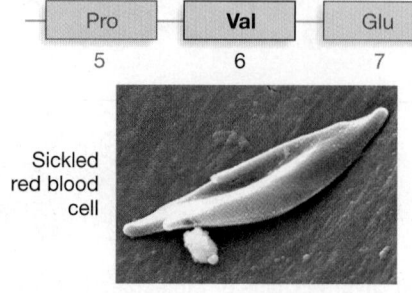

Sickled red blood cell

FIGURE 3.9 Changes in Primary Structure Affect Protein Function. Compare the primary structure of normal hemoglobin **(a)** with that of hemoglobin molecules in people with sickle-cell disease **(b)**. The single amino acid change causes red blood cells to change from their normal disc shape in (a) to a sickled shape in (b) when oxygen concentrations are low.

properties of the amino acids in the sequence. Certain amino acids are more likely to be involved in α-helices than in β-pleated sheets, and vice versa, due to the specific geometry of their side chains. Proline, for example, may be present in β-pleated sheets,

(a) Hydrogen bonds can form between nearby amino and carbonyl groups on the same polypeptide chain.

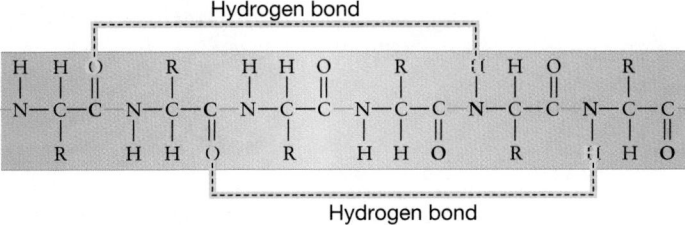

Hydrogen bond

Hydrogen bond

(b) Secondary structures of proteins result.

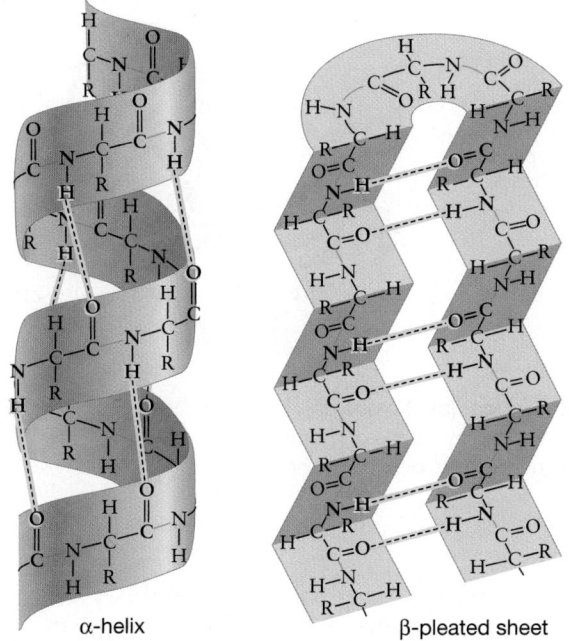

α-helix   β-pleated sheet

(c) Ribbon diagrams of secondary structure

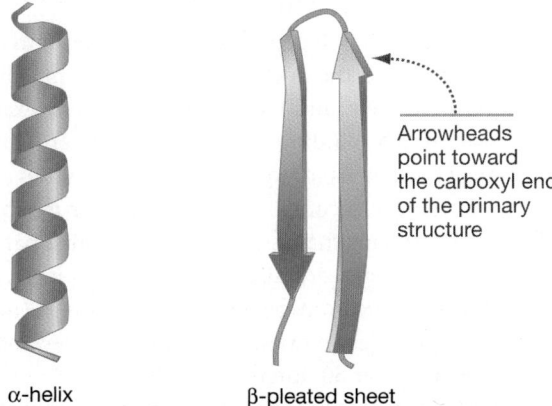

α-helix   β-pleated sheet

Arrowheads point toward the carboxyl end of the primary structure

FIGURE 3.10 Secondary Structures of Proteins. A polypeptide chain can coil or fold in on itself when hydrogen bonds form between N—H and C=O groups on its peptide-bonded backbone.

but it will terminate α-helices due to its unusual side chain. The bond formed between proline's R-group and the nitrogen of the core amino group introduces kinks in the backbone that do not conform to the shape of the helix.

Although each of the hydrogen bonds in an α-helix or a β-pleated sheet is weak relative to a covalent bond, the large number of hydrogen bonds in these structures makes them highly stable. As a result, they increase the stability of the molecule as a whole and help define its shape. In terms of overall shape and stability, though, the tertiary structure of a protein is even more important.

Tertiary Structure

Most of the overall shape, or **tertiary structure,** of a polypeptide results from interactions between R-groups or between R-groups and the backbone. In contrast to the secondary structures, where hydrogen bonds link backbone components together, these side chains can be involved in a wide variety of bonds and interactions. In addition, the amino acid residues that interact with one another are often far apart in the linear sequence. Because each contact between R-groups causes the peptide-bonded backbone to bend and fold, each contributes to the distinctive three-dimensional shape of a polypeptide.

Five types of interactions involving side chains are particularly important:

1. *Hydrogen bonding* Hydrogen bonds form between polar R-groups and opposite partial charges either in the peptide backbone or other R-groups.

2. *Hydrophobic interactions* In an aqueous solution, water molecules interact with the hydrophilic polar side chains of a polypeptide and force the hydrophobic nonpolar side chains to coalesce into globular masses. When these nonpolar R-groups come together, the surrounding water molecules form more hydrogen bonds with each other, increasing the stability of their own interactions.

3. *van der Waals interactions* Once hydrophobic side chains are close to one another, their association is further stabilized by electrical attractions known as **van der Waals interactions.** These weak attractions occur because the constant motion of electrons gives molecules a tiny asymmetry in charge that changes with time. If nonpolar molecules get extremely close to each other, the minute partial charge on one molecule induces an opposite partial charge in the nearby molecule and causes an attraction. Although the interaction is very weak relative to covalent bonds or even hydrogen bonds, a large number of van der Waals attractions can significantly increase the stability of the structure.

4. *Covalent bonding* Covalent bonds can form between the side chains of two cysteines through a reaction between the sulfhydryl groups. These **disulfide ("two-sulfur") bonds** are frequently referred to as bridges, because they create strong links between distinct regions of the same polypeptide or two separate polypeptides.

(a) Interactions that determine the tertiary structure of proteins

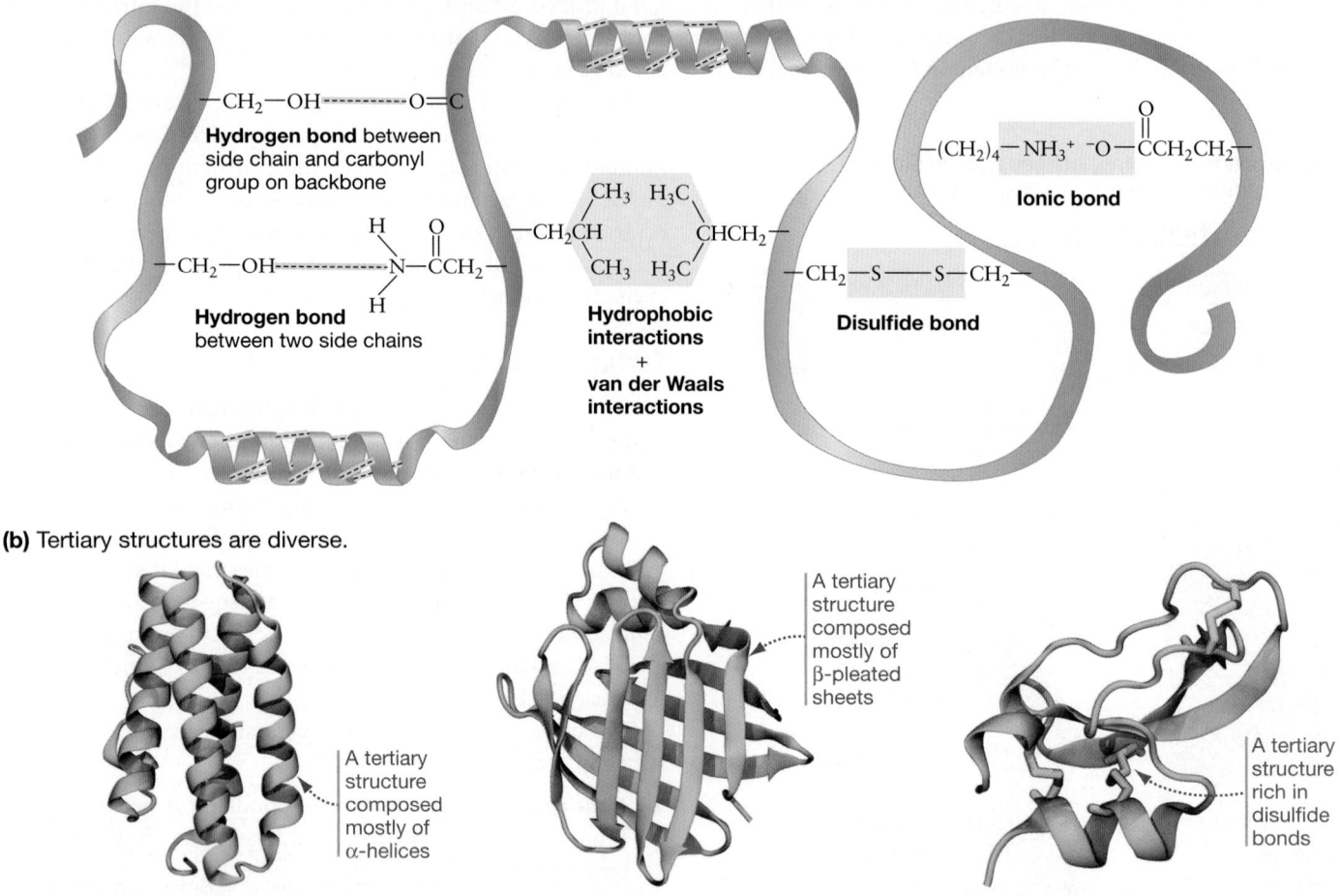

(b) Tertiary structures are diverse.

A tertiary structure composed mostly of β-pleated sheets

A tertiary structure composed mostly of α-helices

A tertiary structure rich in disulfide bonds

FIGURE 3.11 Tertiary Structure of Proteins Results from Interactions Involving R-Groups. (a) The overall shape of a single polypeptide is called its tertiary structure. This level of structure is created by bonds and other interactions that cause it to fold. **(b)** The tertiary structure of these proteins includes interactions between α-helices and β-pleated sheets.

5. *Ionic bonding* Ionic bonds form between groups that have full and opposing charges, such as the ionized acidic and basic side chains highlighted on the right in **FIGURE 3.11a**.

In addition, the overall shape of many proteins depends in part on the presence of secondary structures like α-helices and β-pleated sheets. Thus, tertiary structure depends on both primary and secondary structures.

With so many interactions possible between side chains and peptide-bonded backbones, it's not surprising that polypeptides vary in shape from rod-like filaments to ball-like masses. (See **FIGURE 3.11b**, and look again at Figure 3.8.)

Quaternary Structure

The first three levels of protein structure involve individual polypeptides. But some proteins contain multiple polypeptides that interact to form a single structure. The combination of polypeptides, referred to as subunits, gives a protein **quaternary structure.** The individual polypeptides are held together by the same types of bonds and interactions found in the tertiary level of structure.

In the simplest case, a protein with quaternary structure can consist of just two subunits that are identical. The Cro protein found in a virus called bacteriophage λ (pronounced *LAMB-da*) is an example (**FIGURE 3.12a**). Proteins with two polypeptide subunits are called dimers ("two-parts").

More than two polypeptides can be linked into a single protein, however, and the polypeptides involved may be distinct in primary, secondary, and tertiary structure. For example hemoglobin, an oxygen-binding protein, is a tetramer ("four-parts"). It consists of two copies of two different polypeptides (**FIGURE 3.12b**).

In addition, cells contain **macromolecular machines:** groups of multiple proteins that assemble to carry out a particular function. Some proteins are also found in complexes that include other types of macromolecules. The ribosome (introduced in Chapter 7) provides an example; it consists of several nucleic acid molecules and over 50 different proteins.

TABLE 3.1 summarizes the four levels of protein structure, using hemoglobin as an example. The key thing to note is that protein structure is hierarchical. Quaternary structure is based on tertiary structure, which is based in part on secondary

(a) Cro protein, a dimer

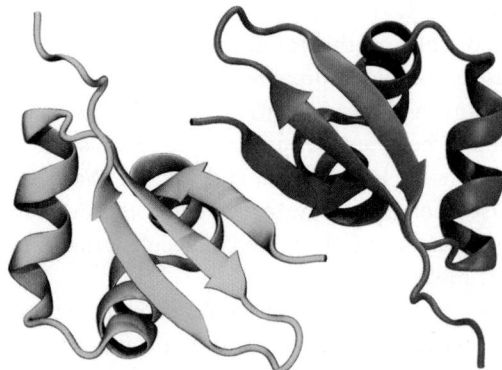

(b) Hemoglobin, a tetramer

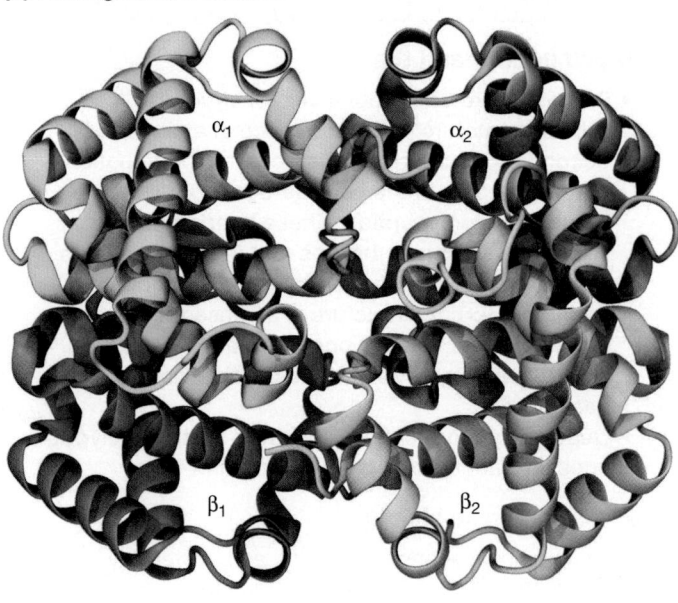

FIGURE 3.12 Quaternary Structures of Proteins Are Created by Multiple Polypeptides. These diagrams represent primary sequences as ribbons. **(a)** The Cro protein is a dimer—it consists of two polypeptide subunits, colored light and dark green. The subunits are identical in this case. **(b)** Hemoglobin is a tetramer—it consists of four polypeptide subunits. The α subunits (light and dark green) are identical; so are the β subunits (light and dark blue).

structure. All three of the higher-level structures are based on primary structure.

The summary table and preceding discussion convey three important messages:

1. The combination of primary, secondary, tertiary, and quaternary levels of structure is responsible for the fantastic diversity of sizes and shapes observed in proteins.

2. Protein folding is directed by the sequence of amino acids present in the primary structure.

3. Most elements of protein structure are based on folding of polypeptide chains.

Does protein folding occur spontaneously? What happens to the function of a protein if normal folding is disrupted? Let's use these questions as a guide to dig deeper into how proteins fold.

SUMMARY TABLE 3.1 **Protein Structure**

Level	Description	Stabilized by	Example: Hemoglobin
Primary	The sequence of amino acids in a polypeptide	Peptide bonds	Gly — Ser — Asp — Cys
Secondary	Formation of α-helices and β-pleated sheets in a polypeptide	Hydrogen bonding between groups along the peptide-bonded backbone; thus, depends on primary structure	One α-helix
Tertiary	Overall three-dimensional shape of a polypeptide (includes contribution from secondary structures)	Bonds and other interactions between R-groups, or between R-groups and the peptide-bonded backbone; thus, depends on primary structure	One of hemoglobin's subunits
Quaternary	Shape produced by combinations of polypeptides (thus, combinations of tertiary structures)	Bonds and other interactions between R-groups, and between peptide backbones of different polypeptides; thus, depends on primary structure	Hemoglobin consists of four polypeptide subunits

check your understanding

C Y U

If you understand that . . .

- Proteins have up to four levels of structure.
- Primary structure is the sequence of amino acids.
- Secondary structure results from hydrogen bonds between atoms in the peptide-bonded backbone of the same polypeptide. These bonds produce structures such as α-helices and β-pleated sheets.
- Tertiary structure is the overall shape of a polypeptide. Most tertiary structure is a consequence of bonds or other interactions between R-groups or between R-groups and the peptide-bonded backbone.
- Quaternary structure occurs when multiple polypeptides interact to form a single protein.

✔ You should be able to . . .

1. Explain how secondary, tertiary, and quaternary levels of structure depend on primary structure.

2. **QUANTITATIVE** Calculate the number of different primary sequences that could be generated by randomly assembling amino acids into peptides that are five residues long.

Answers are available in Appendix A.

3.3 Folding and Function

If you were able to synthesize one of the polypeptides in hemoglobin from individual amino acids, and you then placed the resulting chain in an aqueous solution, it would spontaneously fold into the shape of the tertiary structure shown in Table 3.1.

In terms of entropy, this result probably seems counterintuitive. Because an unfolded protein has many more ways to move about, it has much higher entropy than the folded version. Folding *does* tend to be spontaneous, however, because the chemical bonds, hydrophobic interactions, and van der Waals forces that occur release enough energy to overcome the decrease in entropy. In terms of energy, the folded molecule is more stable than the unfolded molecule.

Folding is crucial to the function of a completed protein. This relationship between protein structure and function was hammered home in a set of classic experiments by Christian Anfinsen and colleagues during the 1950s.

Normal Folding Is Crucial to Function

Anfinsen studied a protein called ribonuclease that is found in many organisms. Ribonuclease is an enzyme that cleaves ribonucleic acid polymers. Anfinsen found that ribonuclease could be unfolded, or **denatured,** by treating it with compounds that break hydrogen bonds and disulfide bonds. The denatured ribonuclease was unable to function normally—it could no longer break apart nucleic acids (**FIGURE 3.13**).

When the denaturing agents were removed, however, the molecule refolded and began to function normally again. These experiments confirmed that ribonuclease folds spontaneously and that folding is essential for normal function.

More recent work has shown that in cells, folding is often facilitated by specific proteins called **molecular chaperones.** Many molecular chaperones belong to a family of molecules called the heat-shock proteins. Heat-shock proteins are produced in large quantities after cells experience high temperatures or other

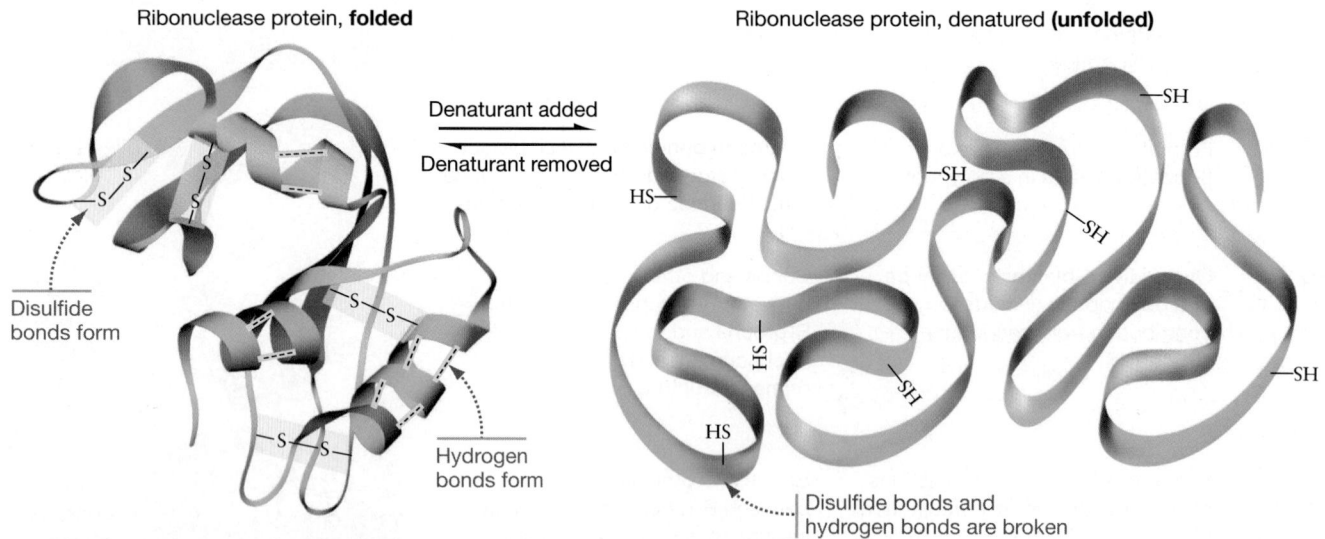

FIGURE 3.13 Protein Structure Determines Function. (left) Ribonuclease is functional when properly folded via hydrogen and disulfide bonds. **(right)** When the disulfide and various noncovalent bonds are broken, ribonuclease is no longer able to function. The double arrow indicates that this process is reversible.

treatments that make other proteins lose their tertiary structure. Heat-shock proteins recognize denatured proteins by binding to hydrophobic patches that would not normally be exposed in properly folded proteins. This interaction blocks inappropriate interactions with other molecules and allows the proteins to refold.

So what is the "normal shape" of a protein? Is only one shape possible for each protein, or could there be several different folded shapes with only one serving as the functional form?

Protein Shape Is Flexible

Although each protein has a characteristic folded shape that is necessary for its function, most proteins are flexible and dynamic, not rigid and static. As it turns out, many polypeptides are unable to fold into their active shape on their own. Over half of the proteins that have been analyzed to date have been found to contain disordered regions lacking any apparent structure. These proteins exist in an assortment of shapes. Only when they interact with particular ions or molecules, or are chemically modified, will they adopt the shape, or conformation, that allows them to perform their function in the cell.

Protein Folding Is Often Regulated   Since the function of a protein is dependent on its shape, controlling when or where it is folded will regulate the protein's activity.

For example, proteins involved in cell signaling are often regulated by controlling their shape. The inactive form of calmodulin—a protein that helps maintain normal blood pressure—has a disordered shape. When the concentration of calcium ions increases in the cell, calmodulin binds these ions, folds into an ordered, active conformation, and sends a signal to increase the diameter of blood vessels. **FIGURE 3.14** illustrates the major shape change that is induced in calmodulin when it binds to calcium.

Misfolding Can Be "Infectious"   In 1982, Stanley Prusiner published what may be the most surprising result to emerge from research on protein folding: Certain proteins can be folded into infectious, disease-causing agents. These proteins are called **prions** (pronounced *PREE-ons*), or proteinaceous infectious particles.

Infectious prions are alternate forms of normal proteins that are present in healthy individuals. The infectious and normal forms do not necessarily differ in amino acid sequence, but their *shapes* are radically different. The infectious form propagates by inducing conformational changes in normal proteins that cause them to adopt the alternate, infectious shape.

FIGURE 3.15 illustrates the differences in shape observed between the normal and infectious forms of the prion responsible for "mad cow disease" in cattle. The molecule in Figure 3.15a is called the prion protein (PrP) and is a normal component of mammalian cells. The improperly folded version of this protein, like the one in Figure 3.15b, represents the infectious form of the prion.

Prions cause a family of diseases known as the spongiform encephalopathies—literally, "sponge-brain-illnesses." Sheep, cows, goats, and humans afflicted with these diseases undergo massive degeneration of the brain. Although some spongiform encephalopathies

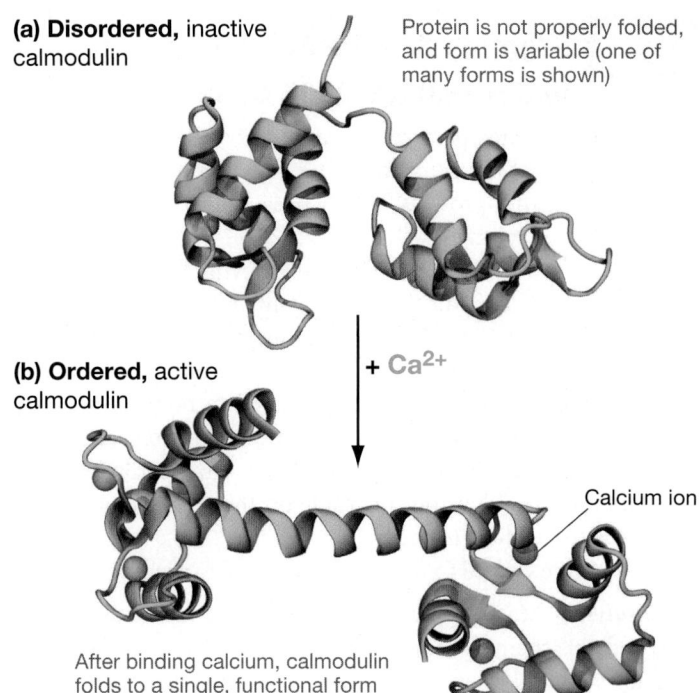

(a) Disordered, inactive calmodulin

Protein is not properly folded, and form is variable (one of many forms is shown)

(b) Ordered, active calmodulin

+ Ca²⁺

Calcium ion

After binding calcium, calmodulin folds to a single, functional form

FIGURE 3.14 Calmodulin Requires Calcium to Fold Properly. Many proteins, like calmodulin, do not complete their folding until after interacting with ions or other molecules. Once calmodulin binds to calcium, it assumes its functional shape.

can be inherited, in many cases the disease is transmitted when individuals eat tissues containing the infectious form of PrP. All the prion illnesses are fatal.

Prions are a particularly dramatic example of how a protein's function depends on its shape as well as how the final shape of a protein depends on folding.

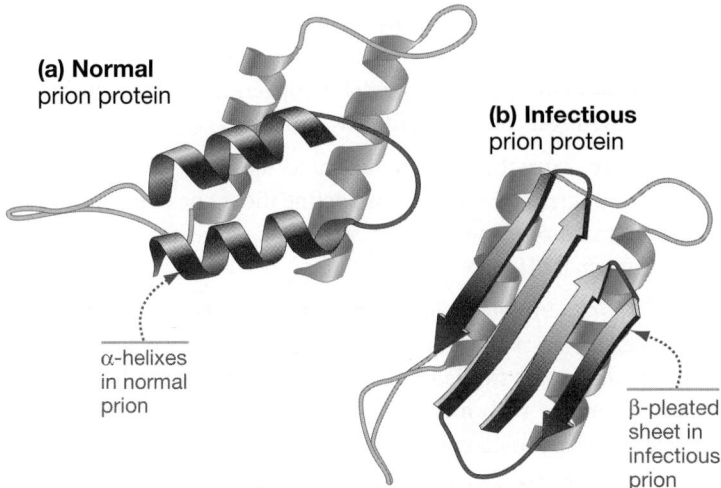

(a) Normal prion protein

(b) Infectious prion protein

α-helixes in normal prion

β-pleated sheet in infectious prion

FIGURE 3.15 Prion Infectivity Is Linked to Structure. Ribbon model of **(a)** a normal, noninfectious prion protein; and **(b)** the infectious form that causes mad cow disease in cattle. Secondary structure is represented by coils (α-helices) and arrows (β-pleated sheets).

3.4 Proteins Are the Most Versatile Macromolecules in Cells

As a group, proteins perform more types of cell functions than any other type of molecule does. It makes sense to hypothesize that life began with proteins, simply because proteins are so vital to the life of today's cells.

Consider the red blood cells that are moving through your arteries right now. Each of these cells contains about 300 million copies of hemoglobin. Hemoglobin carries oxygen from your lungs to cells throughout the body. But every red blood cell also has thousands of copies of a protein called carbonic anhydrase, which is important for moving carbon dioxide from cells back to the lungs, where it can be breathed out. Other proteins form the cell's internal "skeleton" or reside on the cell's membrane to interact with neighboring cells.

Proteins are crucial to most tasks required for cells to exist:

- *Catalysis* Many proteins are specialized to **catalyze,** or speed up, chemical reactions. A protein that functions as a catalyst is called an **enzyme.** The carbonic anhydrase molecules in red blood cells are catalysts. So is the protein called salivary amylase, found in your mouth. Salivary amylase helps begin the digestion of starch and other complex carbohydrates into simple sugars. Most chemical reactions that make life possible depend on enzymes.

- *Defense* Proteins called antibodies and complement proteins attack and destroy viruses and bacteria that cause disease.

- *Movement* Motor proteins and contractile proteins are responsible for moving the cell itself, or for moving large molecules and other types of cargo inside the cell. As you turn this page, for example, specialized proteins called actin and myosin will slide past one another to flex or extend muscle cells in your fingers and arm.

- *Signaling* Proteins are involved in carrying and receiving signals from cell to cell inside the body. If sugar levels in your blood are low, a small protein called glucagon will bind to receptor proteins on your liver cells, triggering enzymes inside to release sugar into your bloodstream.

- *Structure* Structural proteins make up body components such as fingernails and hair, and define the shape of individual cells. Structural proteins keep red blood cells flexible and in their normal disc-like shape.

- *Transport* Proteins allow particular molecules to enter and exit cells or carry them throughout the body. Hemoglobin is a particularly well-studied transport protein, but virtually every cell is studded with membrane proteins that control the passage of specific molecules and ions.

Of all the functions that proteins perform in cells, catalysis may be the most important. The reason is speed. Life, at its most basic level, consists of chemical reactions. But most don't occur fast enough to support life unless a catalyst is present. Enzymes are the most effective catalysts on Earth. Why is this so?

Why Are Enzymes Good Catalysts?

Part of the reason enzymes are such effective catalysts is that they bring reactant molecules—called **substrates**—together in a precise orientation so the atoms involved in the reaction can interact.

The initial hypothesis for how enzymes work was proposed by Emil Fischer in 1894. According to Fischer's "lock-and-key" model, enzymes are analogous to a lock and the keys are substrates that fit into the lock and then react.

Several important ideas in this model have stood the test of time. For example, Fischer was correct in proposing that enzymes bring substrates together in a precise orientation that makes reactions more likely. His model also accurately explained why most enzymes catalyze one specific reaction effectively. Enzyme specificity is a product of the geometry and types of functional groups in the sites where substrates bind.

As researchers began to test Fischer's model, the location where substrates bind and react became known as the enzyme's **active site.** The active site is where catalysis actually occurs.

When techniques for solving the three-dimensional structure of enzymes became available, the active sites were identified as clefts or cavities within the globular shapes. The digestive enzyme trypsin, which is at work in your body now, is a good example. As **FIGURE 3.16** shows, the active site in trypsin is a small notch that contains three key amino acid residues with functional groups that catalyze the cleavage of peptide bonds in other proteins. No other class of macromolecule can match proteins for their catalytic potential. The variety of reactive functional groups present in amino acids is much better suited for this activity than those found in nucleotides or sugars.

The role of enzymes in catalyzing reactions is discussed in more detail in the next unit (see Chapter 8). There you will see that Fischer's model had to be modified as research on enzyme action progressed.

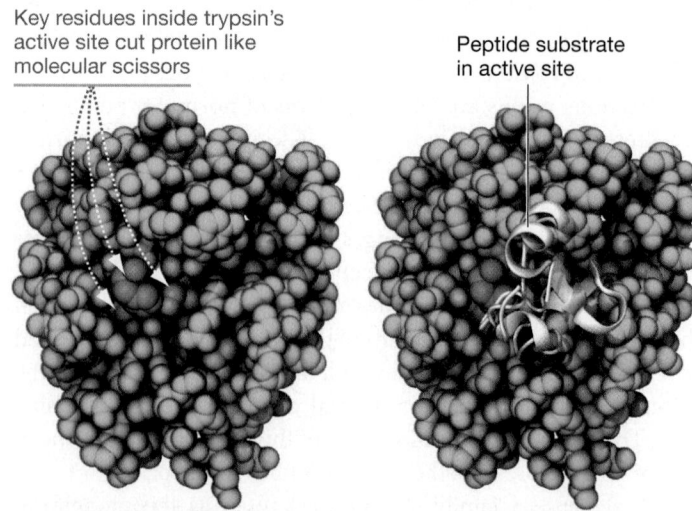

Key residues inside trypsin's active site cut protein like molecular scissors

Peptide substrate in active site

FIGURE 3.16 Substrates Bind to a Specific Location in an Enzyme Called the Active Site. The active site in trypsin, as in many enzymes, is a cleft that contains key amino acid residues that bind substrates and catalyze a reaction.

Was the First Living Entity a Protein Catalyst?

Several observations in the preceding sections could argue that a protein was the first molecule capable of replication. Experimental studies have shown that amino acids were likely abundant during chemical evolution, and that they could have polymerized to form small proteins. In addition, proteins are the most efficient catalysts known.

To date, however, attempts to simulate the origin of life with proteins have not been successful. The only experimental glimpse of a protein's potential to replicate involved an enzyme that could link two oligopeptides together to form a functional duplicate of itself. However, this result required a high concentration of preformed, specific oligopeptides that would not have been present during chemical evolution.

Although it is too early to arrive at definitive conclusions, most origin-of-life researchers are increasingly skeptical that life began with a protein. Their reasoning is that to make a copy of something, a mold or template is required. Proteins cannot furnish this information. Nucleic acids, in contrast, *can*. How they do so is the subject of the next chapter.

check your understanding

If you understand that . . .

- Proteins are the most versatile large molecules in cells, and each function is directly connected to structure.
- Enzymes speed up chemical reactions by binding substrates at their active site, where catalysis takes place.

✔ You should be able to . . .

Predict where amino acid changes would most likely occur in an enzyme that would result in it catalyzing a reaction with a new substrate, and explain how these changes could result in the new activity.

Answers are available in Appendix A.

CHAPTER 3 REVIEW

For media, go to MasteringBiology

If you understand . . .

3.1 Amino Acids and Their Polymerization

- Amino acids have a central carbon bonded to an amino group, a hydrogen atom, a carboxyl group, and an R-group.
- The structure of the R-group affects the chemical reactivity and solubility of the amino acid.
- In proteins, amino acids are joined by a peptide bond between the carboxyl group of one amino acid and the amino group of another amino acid.

✔ You should be able to explain how you could use the structural formula of an amino acid to determine if it is acidic, basic, uncharged polar, or nonpolar.

3.2 What Do Proteins Look Like?

- A protein's primary structure, or sequence of amino acids, is responsible for most of its chemical properties.
- Interactions that take place between C=O and N−H groups in the same peptide-bonded backbone create secondary structures, which are stabilized primarily by hydrogen bonding.
- Tertiary structure results from interactions between R-groups—or R-groups and the peptide-bonded backbone—that stabilize a folded protein into a characteristic overall shape.
- In many cases, a complete protein consists of several different polypeptides, bonded together. The combination of polypeptides represents the protein's quaternary structure.

✔ You should be able to predict where nonpolar amino acid residues would be found in a globular protein, such as the trypsin molecule shown in Figure 3.8c.

3.3 Folding and Function

- Protein folding is a spontaneous process.
- A protein's normal folded shape is essential to its function.
- Many proteins must first bind to other molecules or ions before they can adopt their active conformation.
- Improperly folded proteins can be detrimental to life, and certain proteins even cause deadly infectious diseases.

✔ You should be able to identify one way in which the process of folding in calmodulin and infectious prions is similar.

3.4 Proteins Are the Most Versatile Macromolecules in Cells

- In organisms, proteins function in catalysis, defense, movement, signaling, structural support, and transport of materials.
- Proteins can have diverse functions in cells because they have such diverse structures and chemical properties.
- Catalysis takes place at the enzyme's active site, which has unique chemical properties and a distinctive size and shape.

✔ You should be able to provide the characteristics of proteins that make them especially useful for the following cellular activities: catalysis, defense, and signaling.

You should be able to . . .

✓ TEST YOUR KNOWLEDGE
Answers are available in Appendix A

1. What two functional groups are present on every amino acid?
   a. a carbonyl (C=O) group and a carboxyl group
   b. an N–H group and a carbonyl group
   c. an amino group and a hydroxyl group
   d. an amino group and a carboxyl group

2. Twenty different amino acids are found in the proteins of cells. What distinguishes these molecules?

3. By convention, biologists write the sequence of amino acids in a polypeptide in which direction?
   a. carboxy- to amino-terminus
   b. amino- to carboxy-terminus
   c. polar residues to nonpolar residues
   d. charged residues to uncharged residues

4. In a polypeptide, what bonds are responsible for the secondary structure called an α-helix?

   a. peptide bonds
   b. hydrogen bonds that form between the core C=O and N–H groups on different residues
   c. hydrogen bonds and other interactions between side chains
   d. disulfide bonds that form between cysteine residues

5. Where is the information stored that directs different polypeptides to fold into different shapes?

6. What is an active site?
   a. the position in an enzyme where substrates bind
   b. the place where a molecule or ion binds to a protein to induce a shape change
   c. the portion of a motor protein that is involved in moving cargo in a cell
   d. the site on an antibody where it binds to bacterial cells or viruses

✓ TEST YOUR UNDERSTANDING
Answers are available in Appendix A

7. Explain how water participates in the development of the interactions that glue nonpolar amino acids together in the interior of globular proteins.

8. If amino acids were mixed together in a solution, resembling the prebiotic soup, would they spontaneously polymerize into polypeptides? Why or why not?

9. Provide an example of how a specific shape of a protein is correlated with its function.

10. A major theme in this chapter is that the structure of molecules correlates with their function. Use this theme to explain why proteins can perform so many different functions in organisms and why enzymes are such effective catalysts.

11. Why are proteins not considered to be a good candidate for the first living molecule?
    a. Their catalytic capability is insufficient.
    b. Their amino acid monomers were likely not present during chemical evolution.
    c. They cannot serve as a template for replication.
    d. They could not have polymerized on their own from amino acids during chemical evolution.

12. If proteins folded only into rigid, inflexible structures, how might this affect the cell's ability to regulate protein function?

✓ TEST YOUR PROBLEM-SOLVING SKILLS
Answers are available in Appendix A

13. Based on what you know of the peptide bonds that link together amino acid residues, why would proline's side chain reduce the flexibility of the backbone?

14. Make a concept map (see **BioSkills 15** in Appendix B) that relates the four levels of protein structure and shows how they can contribute to the formation of an active site. Your map should include the following boxed terms: Primary structure, Secondary structure, Tertiary structure, Quaternary structure, Active site, Amino acid sequence, R-groups, Helices and sheets, 3-D shape.

15. Proteins that interact with DNA often interact with the phosphates that are part of this molecule. Which of the following types of

amino acids would you predict to be present in the DNA binding sites of these proteins?
    a. acidic amino acids
    b. basic amino acids
    c. uncharged polar amino acids
    d. nonpolar amino acids

16. Some prion-associated diseases are inherited, such as fatal familial insomnia. What is likely to be different between the infectious forms of these inherited prions compared to those that arise via transmission from one animal to another?

4 Nucleic Acids and the RNA World

In this chapter you will learn that

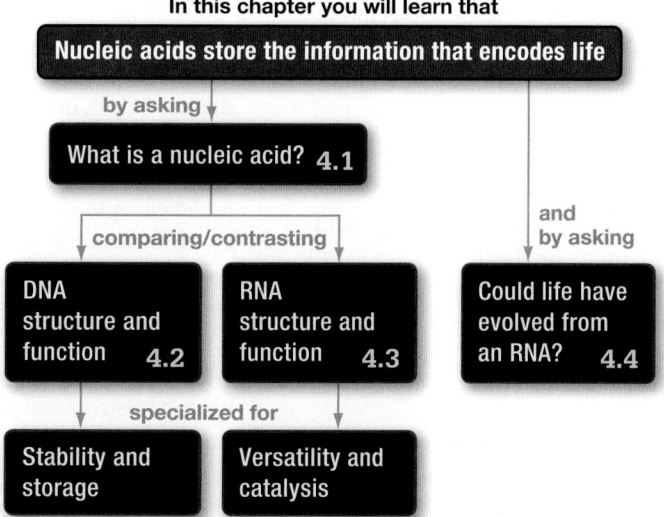

Nucleic acids store the information that encodes life

by asking

What is a nucleic acid? **4.1**

comparing/contrasting

DNA structure and function **4.2**

RNA structure and function **4.3**

and by asking

Could life have evolved from an RNA? **4.4**

specialized for

Stability and storage

Versatility and catalysis

This is part of the sheet-metal-and-wire model that James Watson and Francis Crick used to figure out the secondary structure of DNA. The large "T" stands for the nitrogen-containing base thymine.

This chapter is part of the Big Picture. See how on pages 104–105.

life began when chemical evolution led to the production of a molecule that could promote its own replication. The nature of this first "living molecule," however, has been the subject of many investigations and heated debates. Even though proteins are the workhorse molecules of today's cells, relatively few researchers favor the hypothesis that life began as a protein molecule. Instead, the vast majority of biologists contend that life began as a polymer called a nucleic acid—specifically, a molecule of ribonucleic acid (RNA). This proposal is called the **RNA world hypothesis.**

The RNA world hypothesis contends that chemical evolution led to the existence of an RNA molecule that could replicate itself. Once this molecule existed, chance errors in the copying process

✔ When you see this checkmark, stop and test yourself. Answers are available in Appendix A.

created variations that would undergo natural selection—the evolutionary process by which individuals with certain attributes are selectively reproduced (see Chapter 1). At this point, chemical evolution was over and biological evolution was off and running.

To test this hypothesis, several groups around the world have been working to synthesize a self-replicating RNA molecule in the laboratory. If they ever succeed, they will have created a life-form in a test tube.

This chapter focuses on the structure and function of nucleic acids. Let's begin with an analysis of nucleic acid monomers and how they are linked together into polymers. Afterwards, you will learn about the experiments used to determine if a nucleic acid could have triggered the evolution of life on Earth.

4.1 What Is a Nucleic Acid?

Nucleic acids are polymers, just as proteins are polymers. But instead of being made up of monomers called amino acids, **nucleic acids** are made up of monomers called **nucleotides.**

FIGURE 4.1a diagrams the three components of a nucleotide: (1) a phosphate group, (2) a five-carbon sugar, and (3) a nitrogenous (nitrogen-containing) base. The phosphate is bonded to the sugar molecule, which in turn is bonded to the nitrogenous base.

The sugar component of a nucleotide is an organic compound bearing reactive hydroxyl (−OH) functional groups. Notice that the prime symbols (′) in Figure 4.1 indicate that the carbon being referred to is part of the sugar—not of the attached nitrogenous base. The phosphate group in a nucleotide is attached to the 5′ carbon.

Although a wide variety of nucleotides are found in living cells, origin-of-life researchers concentrate on two types: **ribonucleotides,** the monomers of **ribonucleic acid (RNA)**, and **deoxyribonucleotides,** the monomers of **deoxyribonucleic acid (DNA)**. In ribonucleotides, the sugar is ribose; in deoxyribonucleotides, it is deoxyribose (*deoxy* means "lacking oxygen"). As **FIGURE 4.1b** shows, these two sugars differ by a single oxygen atom. Ribose has an −OH group bonded to the 2′ carbon. Deoxyribose has an H instead at the same location. In both of these sugars, an −OH group is bonded to the 3′ carbon.

In addition to the type of sugar, nucleotides also differ in the type of nitrogenous base. These bases, diagrammed in **FIGURE 4.1c**, belong to structural groups called **purines** and **pyrimidines.** The purines are adenine (A) and guanine (G); the pyrimidines are cytosine (C), uracil (U), and thymine (T). Note that the two rings in adenine and guanine are linked together by nine atoms, compared to the six atoms that make a single ring in each pyrimidine. This makes remembering which bases are purines easy, since both adenine and guanine include "nine" in their names.

As Figure 4.1c shows, ribonucleotides and deoxyribonucleotides also differ in one of their pyrimidine bases. Ribonucleotides

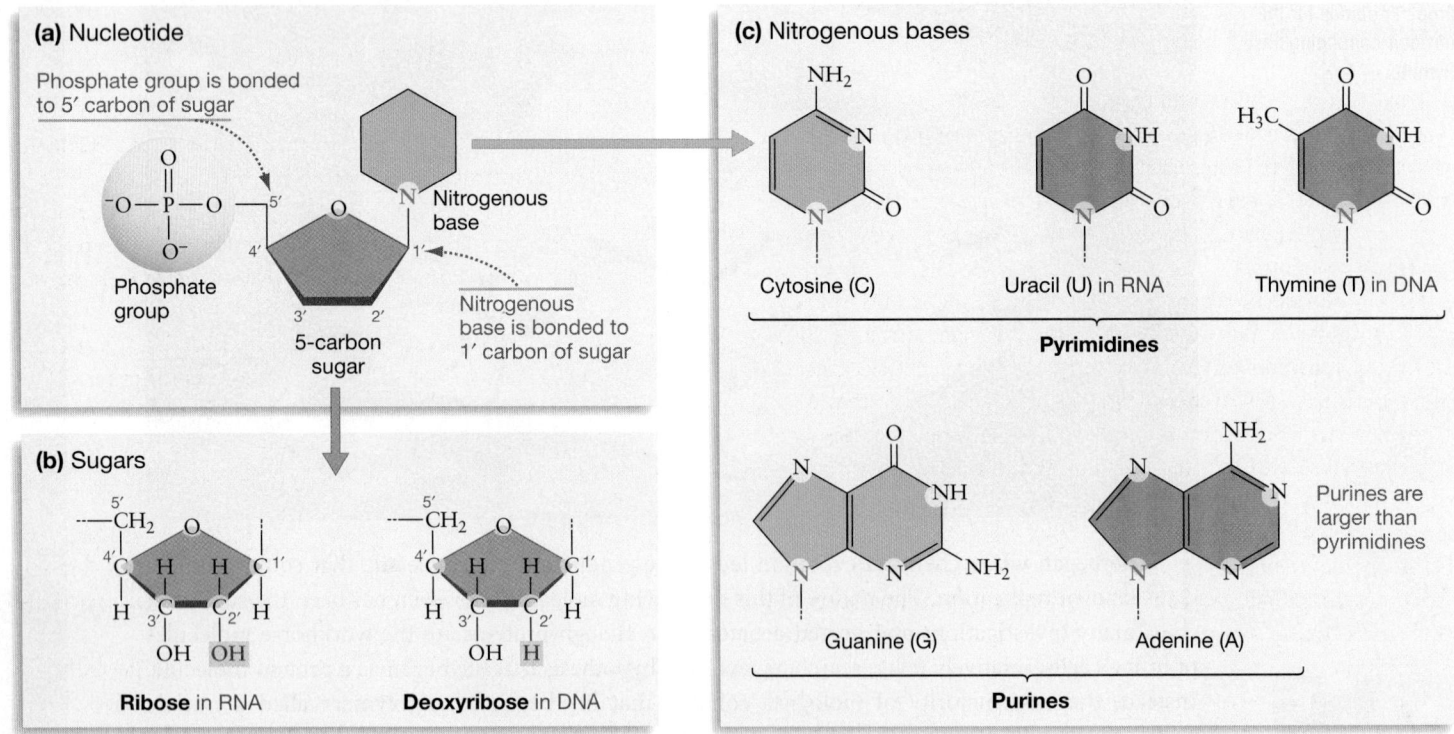

FIGURE 4.1 The General Structure of a Nucleotide. Note that in the bases, the nitrogen that bonds to the sugar is colored blue.

use uracil (U) while deoxyribonucleotides use the closely related base thymine (T).

✔ You should be able to diagram a ribonucleotide and a deoxyribonucleotide. Use a ball for the phosphate group, a pentagon to represent the sugar subunit, and a hexagon to represent the nitrogenous base. Label the 2′, 3′, and 5′ carbons on the sugar molecule, and add the atoms or groups that are bonded to each.

To summarize: After the different sugars and bases are taken into account, eight different nucleotides are used to build nucleic acids—four ribonucleotides (A, G, C, and U) and four deoxyribonucleotides (A, G, C, and T). If nucleic acids played any role in the chemical evolution of life, then at least some of these nucleotides must have been present in the prebiotic oceans. Is there any evidence to suggest that this was possible?

Could Chemical Evolution Result in the Production of Nucleotides?

Based on data from Stanley Miller and researchers who followed (Chapter 2), most biologists contend that amino acids could have been synthesized early in Earth's history. The reactions behind the prebiotic synthesis of nucleotides, however, have been more difficult to identify.

Miller-like laboratory simulations have shown that nitrogenous bases and many different types of sugars can be synthesized readily under conditions that mimic the prebiotic soup. In these experiments, almost all the sugars that have five or six carbons—called pentoses and hexoses, respectively—are produced in approximately equal amounts. If nucleic acids were to form in the prebiotic soup, however, ribose would have had to predominate.

How ribose came to be the dominant sugar during chemical evolution (i.e., what selective process was at work) is still a mystery. Origin-of-life researchers refer to this issue as the "ribose problem." Recent work focusing on the conditions that exist in deep-sea hydrothermal vent systems (see Chapter 2) may point to a possible solution.

Here's the line of reasoning researchers are currently pursuing: Ribose molecules may have been selectively enriched from the mix of sugars in certain early Earth deep-sea vent systems. In one experiment, researchers simulated the conditions that exist in these vents. Then they tested whether minerals that are predicted to have existed in the vent chimneys are able to bind sugars. What they found was striking—the minerals preferentially bound to ribose over other pentoses and hexoses. Did this occur in the ancient vents? If so, the implications are exciting: A high concentration of ribose would be present in the same deep-sea vent environment where chemical evolution is thought to have taken place.

Despite the observed synthesis of nitrogenous bases and the recent discovery of ribose enrichment, the production of nucleotides remains a serious challenge for the theory of chemical evolution. At this time, experiments that attempt to simulate early

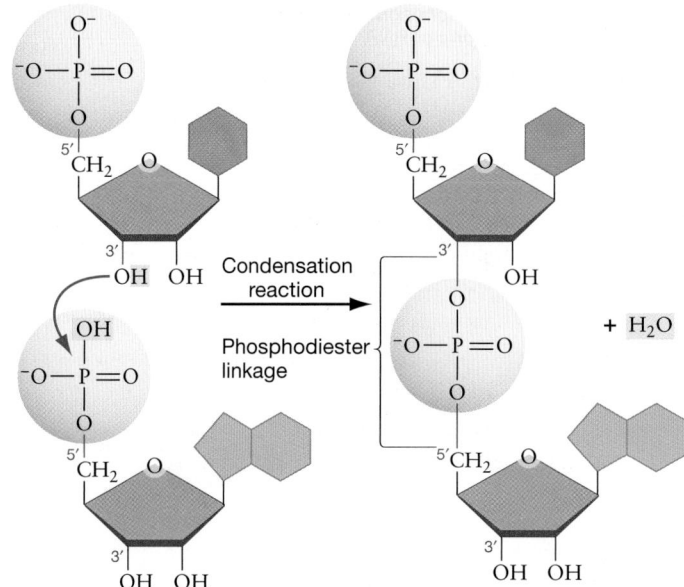

FIGURE 4.2 Nucleotides Polymerize via Phosphodiester Linkages. Ribonucleotides can polymerize via condensation reactions. The resulting phosphodiester linkage connects the 3′ carbon of one ribonucleotide and the 5′ carbon of another ribonucleotide.

Earth environments have yet to synthesize complete nucleotides. But research on this issue continues.

In the meantime, let's consider the next question: Once nucleotides formed, how would they polymerize to form RNA and DNA? This question has an answer.

How Do Nucleotides Polymerize to Form Nucleic Acids?

Nucleic acids form when nucleotides polymerize. As **FIGURE 4.2** shows, the polymerization reaction involves the formation of a bond between a hydroxyl on the sugar component of one nucleotide and the phosphate group of another nucleotide. The result of this condensation reaction is called a **phosphodiester linkage,** or a phosphodiester bond.

A phosphodiester linkage joins the 5′ carbon on the sugar of one nucleotide to the 3′ carbon on the sugar of another. When the nucleotides involved contain the sugar ribose, the polymer that is produced is RNA. If the nucleotides contain the sugar deoxyribose instead, then the resulting polymer is DNA.

DNA and RNA Strands Are Directional **FIGURE 4.3** (see page 60) shows how the chain of phosphodiester linkages in a nucleic acid acts as a backbone, analogous to the peptide-bonded backbone found in proteins.

Like the peptide-bonded backbone of a polypeptide, the sugar-phosphate backbone of a nucleic acid is directional. In a strand of RNA or DNA, one end has an unlinked 5′ phosphate while the other end has an unlinked 3′ hydroxyl—meaning the groups are not linked to another nucleotide. By convention, the

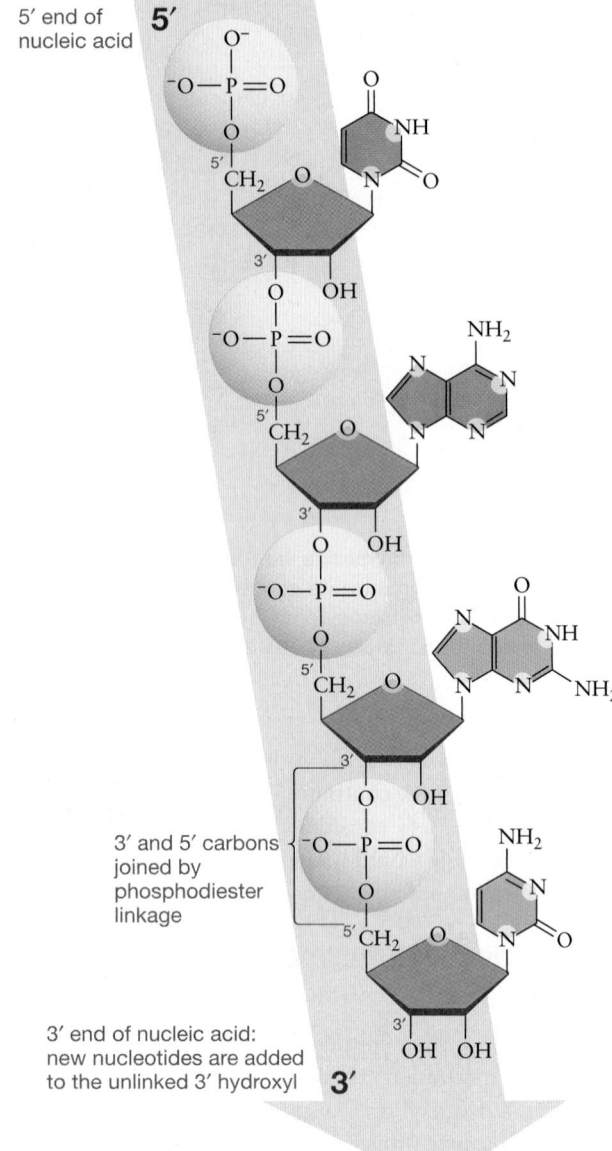

The sugar-phosphate
backbone of RNA

5′ end of
nucleic acid

5′

3′ and 5′ carbons
joined by
phosphodiester
linkage

3′ end of nucleic acid:
new nucleotides are added
to the unlinked 3′ hydroxyl

3′

FIGURE 4.3 RNA Has a Sugar-Phosphate Backbone.

✔**EXERCISE** Identify the four bases in this RNA strand, using Figure 4.1c as a key. Then write down the base sequence, starting at the 5′ end.

sequence of bases found in an RNA or DNA strand is always written in the 5′→3′ direction. (This system is logical because in cells, RNA and DNA are always synthesized in this direction. Bases are added only at the 3′ end of the growing molecule.)

The order of the different nitrogenous bases in a nucleic acid forms the primary structure of the molecule. When biologists write the primary structure of a stretch of DNA or RNA, they simply list the sequence of nucleotides in the 5′→3′ direction, using their single-letter abbreviations. For example, a six-base-long DNA sequence might be ATTAGC. It would take roughly

6 billion of these letters to write the primary structure of the DNA in most of your cells.

Polymerization Requires an Energy Source In cells, the polymerization reactions that join nucleotides into nucleic acids are catalyzed by enzymes. Like other polymerization reactions, the process is not spontaneous. An input of energy is needed to tip the energy balance in favor of the process.

Polymerization can take place in cells because the potential energy of the nucleotide monomers is first raised by reactions that add two phosphate groups to the ribonucleotides or deoxyribonucleotides, creating nucleoside triphosphates.[1] In the case of nucleic acid polymerization, researchers refer to these nucleotides as "activated." **FIGURE 4.4a** shows an example of an activated nucleotide; this molecule is called **adenosine triphosphate,** or **ATP.**

Why do added phosphate groups raise the energy content of a molecule? Recall that phosphates are negatively charged and that like charges repel (Chapter 2). Linking two or more phosphates with covalent bonds generates strong repulsive forces. These bonds therefore carry a large amount of potential energy, which can be harvested to power other chemical reactions (**FIGURE 4.4b**). You will see in later chapters that the potential energy stored in ATP is used to drive other cellular activities, independent of nucleotide polymerization.

This is a key point, and one that you will encounter again and again in this text: The addition of one or more phosphate groups raises the potential energy of substrate molecules enough to make an otherwise nonspontaneous reaction possible. (Chapter 8 explains how this happens in more detail.)

Could Nucleic Acids Have Formed in the Absence of Cellular Enzymes? Accumulating data suggest that the answer is yes.

Activation of nucleotides has been observed when prebiotic conditions are simulated experimentally. In a suite of follow-up experiments, researchers have produced RNA molecules by incubating activated nucleotides with tiny mineral particles—in one case, molecules up to 50 nucleotides long were observed. These results support the hypothesis that polymerization of activated nucleotides in the prebiotic world may have been catalyzed by minerals. This model would be in line with the surface metabolism model for chemical evolution (introduced in Chapter 2).

More recent work has shown that under certain conditions, up to 100 nucleotides can be linked together, even without first being activated. To accomplish this, heat was introduced as a source of energy and small nonpolar molecules, called lipids, were added to help the monomers interact. This experiment is particularly interesting with respect to the setting for chemical evolution, because both of these factors—heat and lipids—are thought to have been present in prebiotic hydrothermal vents. (The chemical origins and properties of lipids are covered in Chapter 6.)

---

[1] A molecule consisting of a sugar and one of the bases in Figure 4.1c is called a nucleoside (a nucleotide is a sugar, a base, and one or more phosphate groups). Thus, a sugar attached to a base and three phosphate groups is called a nucleoside triphosphate.

(a) ATP is an example of an activated nucleotide.

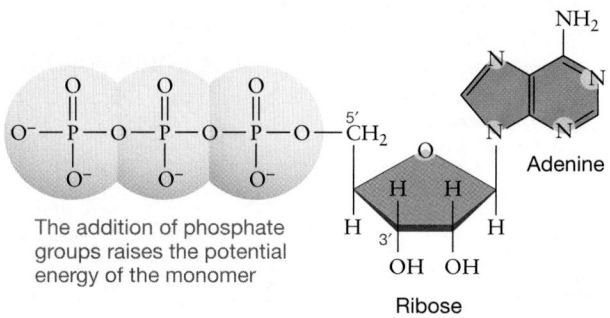

The addition of phosphate groups raises the potential energy of the monomer

Adenine

Ribose

FIGURE 4.4 Activated Monomers Drive Polymerization Reactions. Polymerization reactions are generally nonspontaneous, but those reactions involving nucleoside triphosphates, such as ATP, are spontaneous. The potential energy stored in activated nucleotides is released when the pyrophosphate (PP$_i$) is removed before the polymerizing condensation reaction shown in Figure 4.2.

(b) Energy is released when phosphates are removed by hydrolysis.

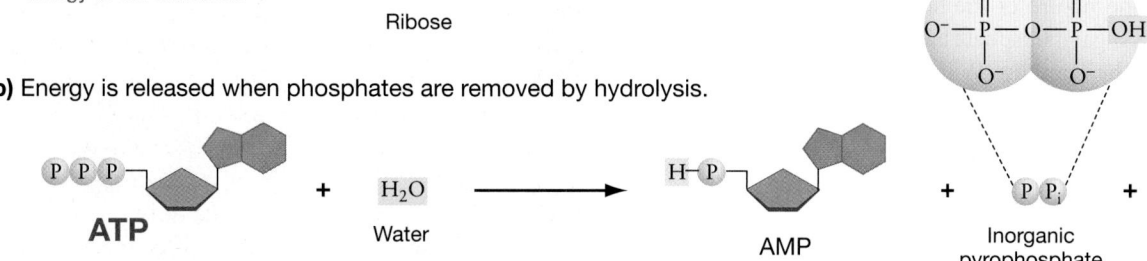

ATP

Water

AMP

Inorganic pyrophosphate

Energy used to link nucleotide to RNA

Based on these results, there is a strong consensus that if ribonucleotides and deoxyribonucleotides were able to form during chemical evolution, they would be able to polymerize into DNA and RNA. Now, what do these nucleic acids look like, and what can they do?

check your understanding

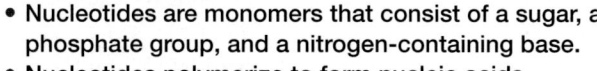

If you understand that . . .

- Nucleotides are monomers that consist of a sugar, a phosphate group, and a nitrogen-containing base.
- Nucleotides polymerize to form nucleic acids through formation of phosphodiester linkages between the 3′ hydroxyl on one nucleotide and the 5′ phosphate on another.
- During polymerization, nucleotides are added only to the 3′ end of a nucleic acid strand.

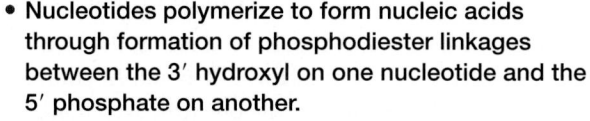

✔ **You should be able to . . .**

Draw a simplified diagram of the phosphodiester linkage between two nucleotides, indicate the 5′→3′ polarity, and mark where the next nucleotide would be added to the growing chain.

Answers are available in Appendix A.

4.2 DNA Structure and Function

The primary structure of nucleic acids is somewhat similar to the primary structure of proteins. Proteins have a peptide-bonded backbone with a series of R-groups that extend from it. DNA and RNA molecules have a sugar-phosphate backbone, created by phosphodiester linkages, and a sequence of any of four nitrogenous bases that extend from it.

Like proteins, DNA and RNA also have secondary structure. While the α-helices and β-pleated sheets of proteins are formed by hydrogen bonding between groups in the backbone, the secondary structure of nucleic acids is formed by hydrogen bonding between the nitrogenous bases.

Let's analyze the secondary structure and function of DNA first, and then dig into the secondary structure and function of RNA.

What Is the Nature of DNA's Secondary Structure?

The solution to DNA's secondary structure, announced in 1953, ranks among the great scientific breakthroughs of the twentieth century. James Watson and Francis Crick presented a model for the secondary structure of DNA in a one-page paper published in the scientific journal *Nature*.

Early Data Provided Clues Watson and Crick's finding was a hypothesis based on a series of results from other laboratories. They were trying to propose a secondary structure that could explain several important observations about the DNA found in cells:

- Chemists had worked out the structure of nucleotides and knew that DNA polymerized through the formation of phosphodiester linkages. Thus, Watson and Crick knew that the molecule had a sugar-phosphate backbone.

- By analyzing the nitrogenous bases in DNA samples from different organisms, Erwin Chargaff had established two empirical rules: **(1)** The number of purines in a given DNA molecule is equal to the number of pyrimidines, and **(2)** the number of T's and A's in DNA are equal, and the number of C's and G's in DNA are equal.

- By bombarding DNA with X-rays and analyzing how it scattered the radiation, Rosalind Franklin and Maurice Wilkins had calculated the distances between groups of atoms in the molecule (see **BioSkills 11** in Appendix B for an introduction to this technique, called **X-ray crystallography**). The scattering patterns showed that three distances were repeated many times: 0.34 nanometer (nm), 2.0 nm, and 3.4 nm. Because the measurements repeated, the researchers inferred that DNA molecules had a regular and repeating structure. The pattern of X-ray scattering suggested that the molecule was helical, or spiral, in nature.

Based on this work, understanding DNA's structure boiled down to understanding the nature of the helix involved. What type of helix would have a sugar-phosphate backbone and explain both Chargaff's rules and the Franklin–Wilkins measurements?

DNA Strands Are Antiparallel   Watson and Crick began by analyzing the size and geometry of deoxyribose, phosphate groups, and nitrogenous bases. The bond angles and measurements suggested that the distance of 2.0 nm probably represented the width of the helix and that 0.34 nm was likely to be the distance between bases stacked in a spiral.

How could they make sense of Chargaff's rules and the 3.4-nm distance, which appeared to be exactly 10 times the distance between a single pair of bases?

To solve this problem, Watson and Crick constructed a series of physical models like the one pictured in **FIGURE 4.5**. The models allowed them to tinker with different types of helical configurations. After many false starts, something clicked:

- They arranged two strands of DNA side by side and running in opposite directions—meaning that one strand ran in the $5' \rightarrow 3'$ direction while the other strand was oriented $3' \rightarrow 5'$. Strands with this orientation are said to be **antiparallel.**

- If the antiparallel strands are twisted together to form a **double helix,** the coiled sugar-phosphate backbones end up on the outside of the spiral and the nitrogenous bases on the inside.

FIGURE 4.5 Building a Physical Model of DNA Structure. Watson (left) and Crick (right) represented the arrangement of the four deoxyribonucleotides in a double helix, using metal plates and wires with precise lengths and geometries.

- For the bases from each backbone to fit in the interior of the 2.0-nm-wide structure, they have to form purine-pyrimidine pairs (see **FIGURE 4.6a**). This is a key point: The pairing allows hydrogen bonds to form between certain purines and pyrimidines. Adenine forms hydrogen bonds with thymine, and guanine forms hydrogen bonds with cytosine (**FIGURE 4.6b**).

- The A-T and G-C bases were said to be complementary. Two hydrogen bonds form when A and T pair, and three hydrogen bonds form when G and C pair. As a result, the G-C interaction is slightly stronger than the A-T bond. In contrast, A-C and G-T pairs allowed no or only one hydrogen bond.

(a) Only purine-pyrimidine pairs fit inside the double helix.

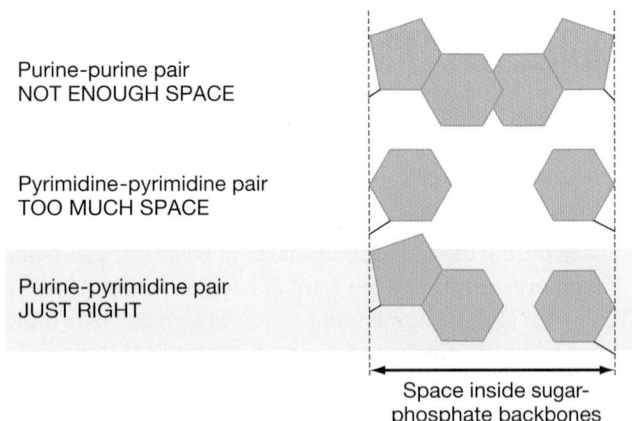

Purine-purine pair
NOT ENOUGH SPACE

Pyrimidine-pyrimidine pair
TOO MUCH SPACE

Purine-pyrimidine pair
JUST RIGHT

Space inside sugar-
phosphate backbones

(b) Hydrogen bonds form between G-C pairs and A-T pairs.

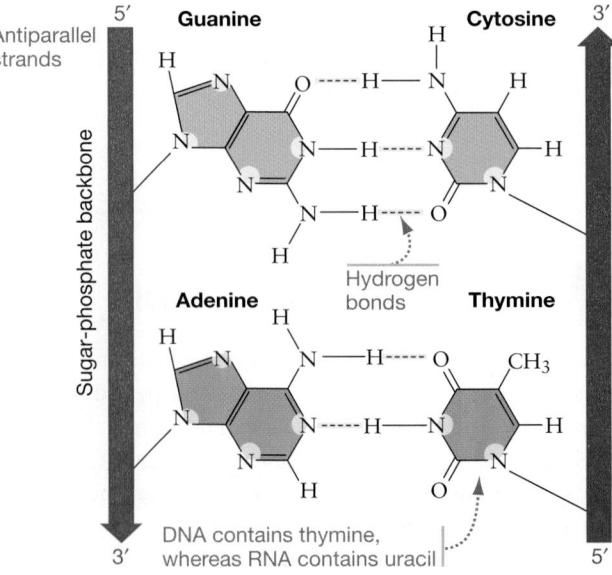

FIGURE 4.6 Complementary Base Pairing Is Based on Hydrogen Bonding.

Watson and Crick had discovered **complementary base pairing.** In fact, the term **Watson–Crick pairing** is now used interchangeably with the phrase complementary base pairing. The physical restraints posed by these interactions resulted in a full helical twist every 10 bases, or 3.4 nm.

The Double Helix FIGURE 4.7a shows how antiparallel strands of DNA form when complementary bases line up and form hydrogen bonds. As you study the figure, notice that DNA is put together like a ladder whose ends have been twisted in opposite directions. The sugar-phosphate backbones form the supports of the ladder; the base pairs represent the rungs of the ladder. The twisting allows the nitrogenous bases to line up in a way that makes hydrogen bonding between them possible.

The nitrogenous bases in the middle of the molecule are hydrophobic. This is a key point, because twisting into a double helix minimizes contact between the bases and surrounding water molecules. In addition to hydrogen bonding, van der Waals interactions between the tightly stacked bases in the interior further contribute to the stability of the helix. You see the same forces—hydrogen bonding, hydrophobicity, and van der Waals interactions—play similar roles in protein folding (Chapter 3). But DNA as a whole is hydrophilic and water soluble because the backbones, which face the exterior of the molecule, contain negatively charged phosphate groups that interact with water.

FIGURE 4.7b highlights additional features of DNA's secondary structure. It's important to note that the outside of the helical DNA molecule forms two types of grooves. The larger of the two is known as the major groove, and the smaller one is known as the minor groove. From this figure, you can identify how DNA's secondary structure explains the measurements observed by Franklin and Wilkins.

Since the model of the double helix was published, experimental tests have shown that the hypothesis is correct in almost every detail. To summarize:

- DNA's secondary structure consists of two antiparallel strands twisted into a double helix.

- The molecule is stabilized by hydrophobic interactions in its interior and by hydrogen bonding between the complementary base pairs A-T and G-C.

✔ You should be able to explain why complementary base pairing would not be possible if two DNA strands were aligned in a parallel fashion—instead of the antiparallel alignment shown in Figure 4.6b.

Now the question is, how does this secondary structure affect the molecule's function?

(a) Cartoons of DNA structure

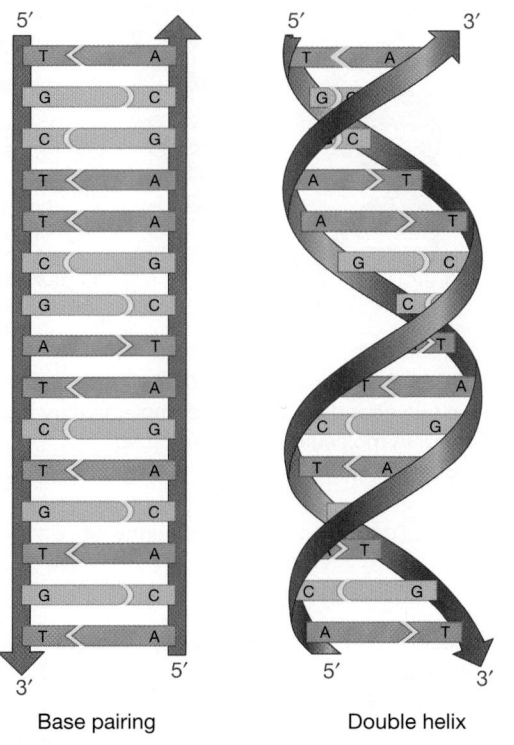

Base pairing   Double helix

(b) Space-filling model of DNA double helix

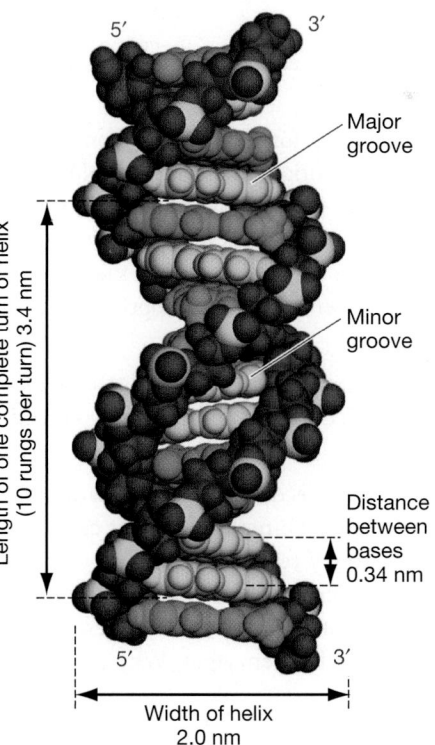

Length of one complete turn of helix (10 rungs per turn) 3.4 nm

Major groove

Minor groove

Distance between bases 0.34 nm

Width of helix 2.0 nm

FIGURE 4.7 The Secondary Structure of DNA Is a Double Helix. (a) The cartoons illustrate complementary base pairing and how strands are twisted into a double helix. **(b)** The space-filling model shows tight packing of the bases inside the double helix. The double-helix structure explains the measurements inferred from X-ray analysis of DNA molecules.

DNA Functions as an Information-Containing Molecule

Watson and Crick's model created a sensation for a simple reason: It revealed how DNA could store and transmit biological information. In literature, information consists of letters on a page. In music, information is composed of the notes on a staff. But inside cells, information consists of a sequence of nucleotides in a nucleic acid. The four nitrogenous bases function like letters of the alphabet. A sequence of bases is like the sequence of letters in a word—it has meaning.

In all organisms that have been examined to date, from tiny bacteria to gigantic redwood trees, DNA carries the information required for the organism's growth and reproduction. Exploring how hereditary information is encoded and translated into action is the heart of several later chapters (Chapters 16 through 19).

Here, however, our focus is on how life began. The theory of chemical evolution holds that life began once a molecule emerged that could make a copy of itself. Does the information contained within DNA allow it to be replicated?

Watson and Crick ended their paper on the double helix with one of the classic understatements in the scientific literature: "It has not escaped our notice that the specific pairing we have postulated immediately suggests a possible copying mechanism." Here's the key insight: DNA's primary structure serves as a mold or template for the synthesis of a complementary strand. DNA contains the information required for a copy of itself to be made. **FIGURE 4.8** shows how a copy of DNA can be made by complementary base pairing.

Step 1 Heating or enzyme-catalyzed reactions can cause the double helix to separate.

Step 2 Free deoxyribonucleotides form hydrogen bonds with complementary bases on the original strand of DNA—also called a **template strand.** As they do, their sugar-phosphate groups form phosphodiester linkages to create a new strand—also called a **complementary strand.** Note that the $5' \rightarrow 3'$ directionality of the complementary strand is opposite that of the template strand.

Step 3 Complementary base pairing allows each strand of a DNA double helix to be copied exactly, producing two identical daughter molecules.

DNA copying is the basis for a second of the five characteristics of life (introduced in Chapter 1): replication. But can DNA catalyze the reactions needed to *self*-replicate? In today's cells and in laboratory experiments, the answer is no. Instead, the molecule is copied through a complicated series of energy-demanding reactions, catalyzed by a large suite of enzymes. Why can't DNA catalyze these reactions itself?

Is DNA a Catalytic Molecule?

The DNA double helix is highly structured. It is regular, symmetric, and held together by hydrogen bonding, hydrophobic

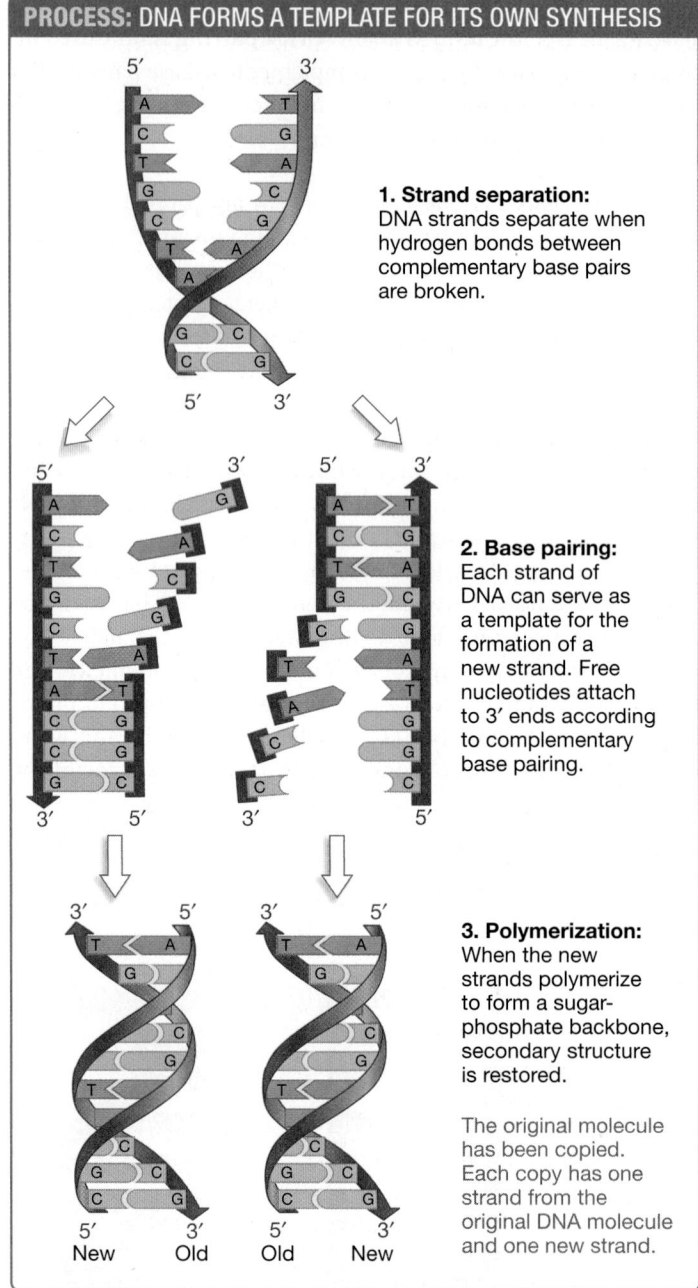

PROCESS: DNA FORMS A TEMPLATE FOR ITS OWN SYNTHESIS

1. Strand separation: DNA strands separate when hydrogen bonds between complementary base pairs are broken.

2. Base pairing: Each strand of DNA can serve as a template for the formation of a new strand. Free nucleotides attach to 3' ends according to complementary base pairing.

3. Polymerization: When the new strands polymerize to form a sugar-phosphate backbone, secondary structure is restored.

The original molecule has been copied. Each copy has one strand from the original DNA molecule and one new strand.

FIGURE 4.8 Making a Copy of DNA. If new bases are added to each of the two strands of DNA via complementary base pairing, a copy of the DNA molecule can be produced.

✔**QUESTION** When double-stranded DNA is heated to 95°C, the bonds between complementary base pairs break and single-stranded DNA results. Considering this observation, is the reaction shown in step 1 spontaneous?

interactions, and phosphodiester linkages. In addition, the molecule has few functional groups exposed that can participate in chemical reactions. For example, the lack of a 2' hydroxyl group on each deoxyribonucleotide makes the polymer much less reactive than RNA, and thus much more resistant to degradation.

Intact stretches of DNA have been recovered from fossils that are tens of thousands of years old. The molecules have the same sequence of bases as the organisms had when they were alive, despite death and exposure to a wide array of pH, temperature, and chemical conditions. DNA's stability is the key to its effectiveness as a reliable information-bearing molecule. DNA's structure is consistent with its function in cells.

The orderliness and stability that make DNA such a dependable information repository also make it extraordinarily inept at catalysis, however. Recall that enzyme function is based on a specific binding event between a substrate and a protein catalyst (Chapter 3). Thanks to variation in reactivity among R-groups in amino acids, and the enormous diversity of shapes found in proteins, a wide array of catalytic activities can be generated. In comparison, DNA's primary and secondary structures are simple. It is not surprising, then, that DNA has never been observed to catalyze any reaction in any organism. Although researchers have been able to construct single-stranded DNA molecules that can catalyze some reactions in the laboratory, the number and diversity of reactions involved is a minute fraction of the activity catalyzed by enzymes.

In short, DNA furnishes an extraordinarily stable template for copying itself and for storing information encoded in a sequence of bases. But owing to its inability to act as an effective catalyst, there is virtually no support for the hypothesis that the first lifeform consisted of DNA. Instead, most biologists who are working on the origin of life support the hypothesis that life began with RNA. How does the structure of RNA differ from DNA?

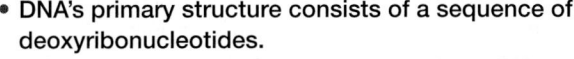
check your understanding

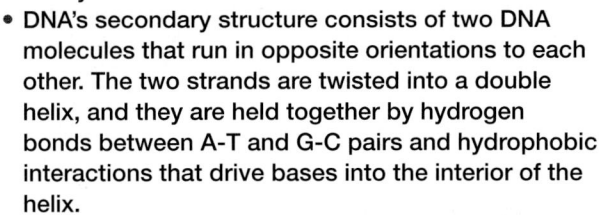

If you understand that . . .

- DNA's primary structure consists of a sequence of deoxyribonucleotides.
- DNA's secondary structure consists of two DNA molecules that run in opposite orientations to each other. The two strands are twisted into a double helix, and they are held together by hydrogen bonds between A-T and G-C pairs and hydrophobic interactions that drive bases into the interior of the helix.
- The sequence of deoxyribonucleotides in DNA contains information. Owing to complementary base pairing, each DNA strand also contains the information required to form its complementary strand.

✔ You should be able to . . .

Make a sketch of a double-stranded DNA molecule in the form of a ladder with the sequence of A-G-C-T. Label the 5′ and 3′ ends, the sugar-phosphate backbones, the hydrogen bonds between complementary bases, and the location of hydrophobic interactions.

Answers are available in Appendix A.

The first living molecule would have needed to perform two key functions: carry information and catalyze reactions that promoted its own replication. At first glance, these two functions appear to conflict. Information storage requires regularity and stability; catalysis requires variation in chemical composition and flexibility in shape. How is it possible for a molecule to do both? The answer lies in structure.

Structurally, RNA Differs from DNA

Recall that proteins can have up to four levels of structure. Single-chain proteins possess a primary sequence of amino acids, secondary folds that are stabilized by hydrogen bonding between atoms in the peptide-bonded backbone, and tertiary folds that are stabilized by interactions involving R-groups. Quaternary structure is found in proteins consisting of multiple polypeptides.

DNA has only primary and secondary structure. But RNA, like single-chained proteins, can have up to three levels of structure.

Primary Structure   Like DNA, RNA has a primary structure consisting of a sugar-phosphate backbone formed by phosphodiester linkages and, extending from that backbone, a sequence of four types of nitrogenous bases. But it's important to recall two significant differences between these nucleic acids:

1. The sugar in the sugar-phosphate backbone of RNA is ribose, not deoxyribose as in DNA.

2. The pyrimidine base thymine does not exist in RNA. Instead, RNA contains the closely related pyrimidine base uracil.

The first point is critical. Look back at Figure 4.1b and compare the functional groups attached to ribose and deoxyribose. Notice the hydroxyl (−OH) group on the 2′ carbon of ribose. This additional hydroxyl is much more reactive than the hydrogen atom on the 2′ carbon of deoxyribose. When RNA molecules fold in certain ways, the hydroxyl group can attack the phosphate linkage between nucleotides, which would generate a break in the sugar-phosphate backbone. While this −OH group makes RNA much less stable than DNA, it can also support catalytic activity by the molecule.

Secondary Structure   Like DNA molecules, most RNA molecules have secondary structure that results from complementary base pairing between purine and pyrimidine bases. In RNA, adenine forms hydrogen bonds only with uracil, and guanine again forms hydrogen bonds with cytosine. (Other, non-Watson–Crick base pairs occur, although less frequently.) Three hydrogen bonds form between guanine and cytosine, but only two form between adenine and uracil.

This hydrogen bonding should seem familiar, since DNA bonds together in a similar manner—so how do the secondary structures of RNA and DNA differ? In the vast majority of cases, the purine and pyrimidine bases in RNA undergo hydrogen

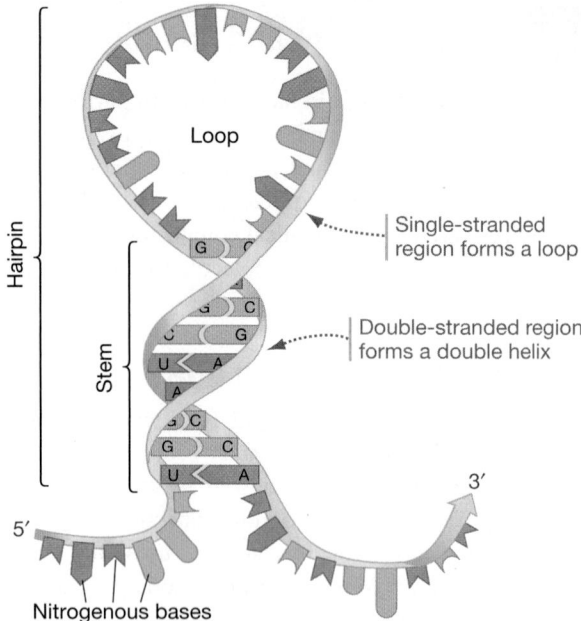

FIGURE 4.9 Complementary Base Pairing and Secondary Structure in RNA: Stem-and-Loop Structures. This RNA molecule has secondary structure. The double-stranded "stem" and single-stranded "loop" form a hairpin. The bonded bases in the stem are antiparallel, meaning that they are oriented in opposite directions.

bonding with complementary bases on the *same strand*, rather than forming hydrogen bonds with complementary bases on a different strand, as in DNA.

FIGURE 4.9 shows how within-strand base pairing works. The key is that when bases on one part of an RNA strand fold over and align with ribonucleotides on another part of the same strand, the two sugar-phosphate strands are antiparallel. In this orientation, hydrogen bonding between complementary bases results in a stable double helix.

If the section where the fold occurs includes unpaired bases, then the stem-and-loop configuration shown in Figure 4.9 results. This type of secondary structure is called a **hairpin.** Several other types of RNA secondary structures are possible, each involving a different length and arrangement of base-paired segments.

Like the α-helices and β-pleated sheets observed in many proteins, RNA secondary structures can form spontaneously. They are directed by hydrophobic interactions and stabilized by hydrogen bonding between the bases. Even though hairpins and other types of secondary structure reduce the entropy of RNA molecules, the energy released in these interactions makes the overall process favorable.

Tertiary Structure   RNA molecules can also have tertiary structure, which arises when secondary structures fold into more complex shapes. As a result, RNA molecules with different base sequences can have very different overall shapes and chemical properties. RNA molecules are much more diverse in size, shape, and reactivity than DNA molecules are. Structurally and chemically, RNA is intermediate between the complexity of proteins and the simplicity of DNA.

TABLE 4.1 summarizes the similarities and differences in the structures of RNA and DNA.

RNA's Structure Makes It an Extraordinarily Versatile Molecule

In terms of structure, you've seen that RNA is intermediate between DNA and proteins. RNA is intermediate in terms of function as well. RNA molecules cannot archive information nearly as efficiently as DNA molecules do, but they do perform key functions in information processing. Likewise, they cannot catalyze as many reactions as proteins do. But as it turns out, the reactions they do catalyze are particularly important.

In cells, RNA molecules function like a jackknife or a pocket tool with an array of attachments: They perform a wide variety of

SUMMARY TABLE 4.1   **DNA and RNA Structure**

Level of Structure	DNA		RNA	
Primary	Sequence of deoxyribonucleotides; bases are A, T, G, C	5′ AATGTGCCG 3′	Sequence of ribonucleotides; bases are A, U, G, C	5′ UUACACGGC 3′
Secondary	Two antiparallel strands twist into a double helix, stabilized by hydrogen bonding between complementary bases (A-T, G-C) and hydrophobic interactions		Most common are hairpins, formed when a single strand folds back on itself to form a double-helix "stem" and a single-stranded "loop"	
Tertiary	None*		Folds that form distinctive three-dimensional shapes	Example: tRNA

*In cells, DNA coils around proteins that bind to the double helix. In many cases the DNA-protein complex folds into highly organized, compact structures. But DNA does not form tertiary structure on its own.

tasks reasonably well. Some of the most surprising results in the last decade of biological science, in fact, involve new insights into the diversity of roles that RNAs play in cells. These molecules process information stored in DNA, synthesize proteins, and defend against attack by viruses, among other things.

Next let's focus on the roles that RNA could have played in the origin of life—as an information-containing entity and as a catalyst.

RNA Is an Information-Containing Molecule

Because RNA contains a sequence of bases analogous to the letters in a word, it can function as an information-containing molecule. And because hydrogen bonding occurs specifically between A-U pairs and G-C pairs in RNA, it is possible for RNA to furnish the information required to make a copy of itself.

FIGURE 4.10 illustrates how the information stored in an RNA molecule can be used to direct its own replication.

First, a complementary copy of the RNA is made when free ribonucleotides form hydrogen bonds with complementary bases on the original strand of RNA—the template strand. As they do, their sugar-phosphate groups form phosphodiester linkages to produce a double-stranded RNA molecule (steps 1 and 2).

To make a copy of the original single-stranded RNA, the hydrogen bonds between the double-stranded product must first be broken by heating or by a catalyzed reaction (step 3). The newly made complementary RNA molecule now exists independently of the original template strand. If steps 1–3 were repeated with the new strand serving as a template (steps 4–6), the resulting molecule would be a copy of the original. In this way, the primary sequence of an RNA serves as a mold.

RNA Can Function as a Catalytic Molecule

In terms of diversity in chemical reactivity and overall shape, RNA molecules are no match for proteins. The primary structure of RNA molecules is much more restricted because RNA has only four types of nucleotides versus the 20 types of amino acids found in proteins. Secondary through tertiary structure is more limited as a result, meaning that RNA cannot form the wide array of catalysts observed among proteins.

But because RNA has a degree of structural and chemical complexity, it is capable of catalyzing a number of chemical reactions. Sidney Altman and Thomas Cech shared the 1989 Nobel Prize in chemistry for showing that catalytic RNAs, or **ribozymes,** exist in organisms.

FIGURE 4.11 (on page 68) shows the structure of a ribozyme Cech isolated from a single-celled organism called *Tetrahymena*. This ribozyme catalyzes both the hydrolysis and the condensation of phosphodiester linkages in RNA. Researchers have since discovered a variety of ribozymes that catalyze an array of reactions in cells. For example, ribozymes catalyze the formation of peptide bonds when amino acids polymerize to form polypeptides. Ribozymes are at work in your cells right now.

The three-dimensional nature of ribozymes is vital to their catalytic activity. To catalyze a chemical reaction, substrates must

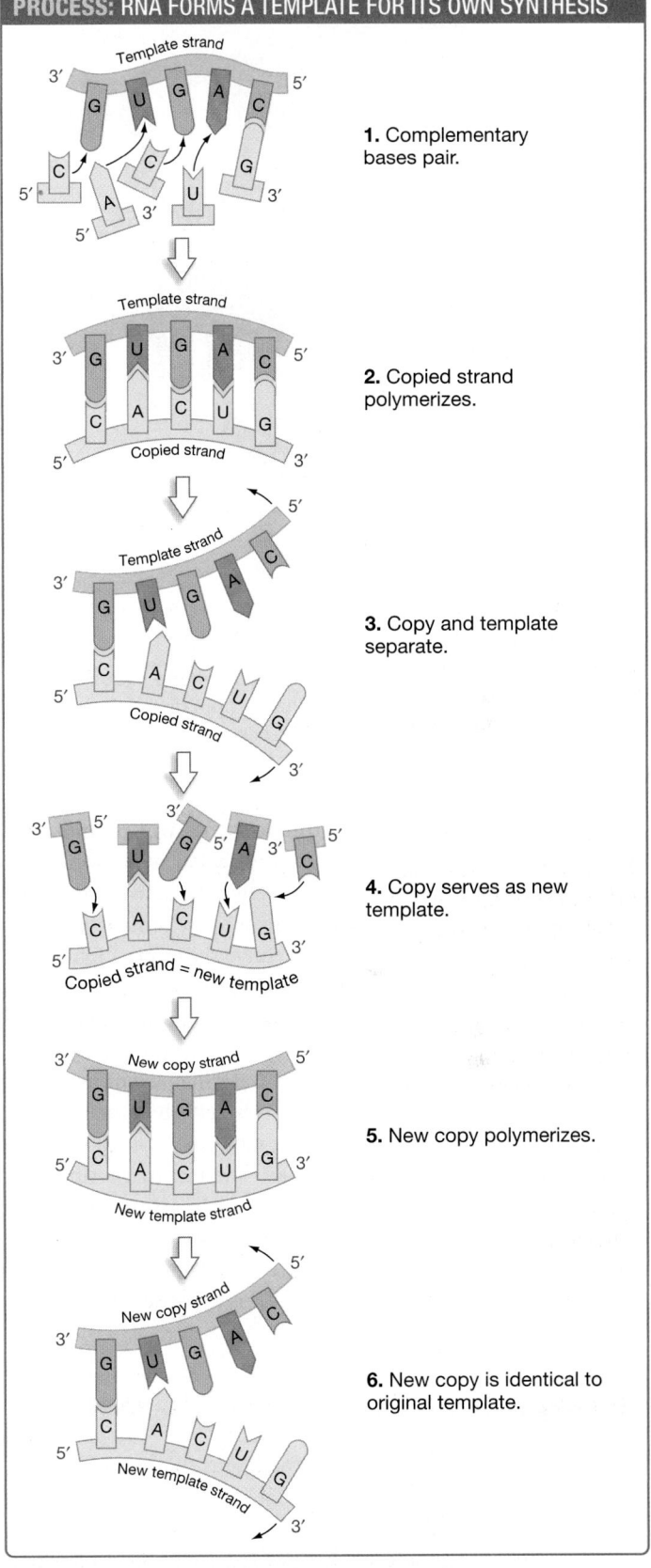

PROCESS: RNA FORMS A TEMPLATE FOR ITS OWN SYNTHESIS

1. Complementary bases pair.

2. Copied strand polymerizes.

3. Copy and template separate.

4. Copy serves as new template.

5. New copy polymerizes.

6. New copy is identical to original template.

FIGURE 4.10 RNA Molecules Contain Information That Allows Them to Be Replicated. For a single-stranded RNA to be copied, it must pass through double-stranded RNA intermediates.

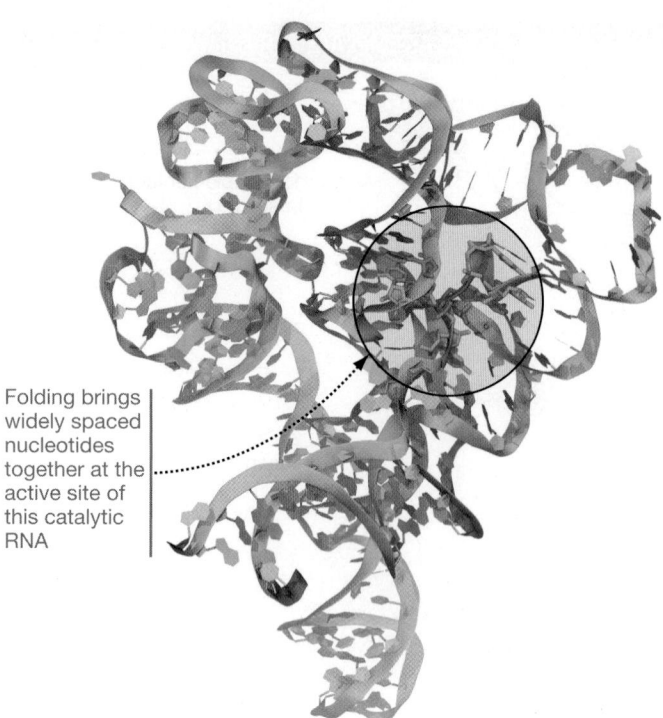

Folding brings widely spaced nucleotides together at the active site of this catalytic RNA

FIGURE 4.11 Tertiary Structure of the *Tetrahymena* Ribozyme. The folded structure brings together bases from distant locations in the primary structure to form the active site.

be brought together in an environment that will promote the reaction. As with protein enzymes, the region of the ribozyme that is responsible for this activity is called the active site. When the *Tetrahymena* ribozyme was compared to protein enzymes that catalyze similar reactions, their active sites were found to be similar in structure. This observation about two very different molecules demonstrates the critical relationship between structure and function

The discovery of ribozymes was a watershed event in origin-of-life research. Before Altman and Cech published their results, most biologists thought that the only molecules capable of catalyzing reactions in cells were proteins. The fact that a ribozyme in *Tetrahymena* catalyzed a condensation reaction raised the possibility that an RNA molecule could make a copy of itself. Such a molecule could qualify as the first living entity. Is there any experimental evidence to support this hypothesis?

4.4 In Search of the First Life-Form

The theory of chemical evolution maintains that life began as a naked self-replicator—a molecule that existed by itself in solution, without being enclosed in a membrane. To make a copy of itself, that first living molecule had to **(1)** provide a template that could be copied, and **(2)** catalyze polymerization reactions that would link monomers into a copy of that template. Because RNA is capable of both processes, most origin-of-life researchers propose that the first life-form was made of RNA.

No self-replicating RNA molecules have been discovered in nature, however, so researchers test the hypothesis by trying to simulate the RNA world in the laboratory. The eventual goal is to create an RNA molecule that can catalyze its own replication.

How Biologists Study the RNA World

To understand how researchers go about testing the RNA world hypothesis, consider two recent experiments by researchers in David Bartel's laboratory. In one study, the team attempted to generate an RNA molecule that could catalyze the kind of template-directed polymerization needed for RNA replication—an RNA "replicase." Starting with a ribozyme capable of joining two ribonucleotides together, they generated billions of copies into which random mutations were introduced.

Next they incubated the mutants with free ribonucleotides and began selecting for replicase activity. Molecules that exhibited such activity were isolated and copied. After two weeks and 18 rounds of selection, the team succeeded in isolating a ribozyme that could add 14 nucleotides to an existing RNA strand.

Note that the team's experimental protocol was designed to mimic the process of natural selection introduced in Chapter 1. The population of RNAs from each round had variable characteristics that could be replicated and passed on to the next generation of ribozymes. In addition, the researchers were able to select the most efficient RNAs to be the "parents" of the next generation—and in the process introduce new mutations that potentially could make some of the "offspring" even better ribozymes.

This research created considerable excitement among biologists interested in the origin of life, because adding ribonucleotides to a growing strand is a key attribute of an RNA replicase. However, since the maximum product length generated was less than 10 percent of the ribozyme's own length, an RNA replicase capable of making a full-length copy of itself was far from being discovered. In fact, the difficulty in creating an effective RNA replicase has led many researchers to question the idea of a replicase being the first ribozyme to emerge in the RNA world.

In another study, Bartel's group asked a different question: Would it be possible to select for a ribozyme that could make ribonucleotides? This type of ribozyme is not known to exist in nature but would be a key component in the RNA world.

Recall that the direction of a chemical reaction and how much product it makes is influenced by the amount of reactants present (Chapter 2). Since the chemical evolution of nucleotides is thought to have been inefficient, nucleotides would have been a scarce resource on early Earth. Ribozymes that could catalyze the production of nucleotides would be more likely to be copied due to local accumulation of monomers.

Starting with a large pool of randomly generated RNA sequences, the researchers selected for RNAs that could catalyze the addition of a uracil base to a ribose sugar. By round 11, the group had recovered ribozymes that were 50,000 times better at catalyzing the reaction than those found in the fourth round and over 1 million times more efficient than the uncatalyzed reaction. In effect, molecular evolution had occurred in the reaction tubes.

Thanks to similar efforts at other laboratories around the world, biologists have produced an increasingly impressive set of catalytic activities from RNA molecules. The results from each of these studies help clarify our view of what occurred in the RNA world. If a living ribozyme ever existed, then each round of simulated molecular evolution brings us closer to resurrecting it.

The RNA World May Have Sparked the Evolution of Life

Although ribozymes like these lab-generated molecules may have been present in the RNA world, they have not been observed in nature. Of those that have been discovered in modern cells, most play key roles in the synthesis of proteins. This relationship suggests the order of events in chemical evolution—the RNA world preceded proteins.

The evolution of protein enzymes would have marked the end of the RNA world—providing the means for catalyzing reactions necessary for life to emerge in a cellular form. After this milestone, three of the five fundamental characteristics of life (see Chapter 1) were solidly in place:

1. *Information*   Proteins and ribozymes were processing information stored in nucleic acids for the synthesis of more proteins.

2. *Replication*   Enzymes, and possibly ribozymes, were replicating the nucleic acids that stored the hereditary information.

3. *Evolution*   Random changes in the synthesis of proteins, and selective advantages resulting from some of these changes, allowed for the evolution of new proteins and protein families.

If these events occurred in a hydrothermal vent, the molecular assemblages of nucleic acids and proteins would have been constantly fed with thermal and chemical energy. To gain independence from their undersea hatchery, enzymes would have evolved to store this energy as something more portable—carbohydrates. The structure and function of carbohydrates will be the focus of the next chapter.

CHAPTER 4   REVIEW

For media, go to MasteringBiology

If you understand . . .

4.1   What Is a Nucleic Acid?

- Nucleic acids are polymers of nucleotide monomers, which consist of a sugar, a phosphate group, and a nitrogenous base. Ribonucleotide monomers polymerize to form RNA. Deoxyribonucleotide monomers polymerize to form DNA.

- Ribonucleotides have a hydroxyl (—OH) group on their 2′ carbon; deoxyribonucleotides do not.

- Nucleic acids polymerize when condensation reactions join nucleotides together via phosphodiester linkages.

- Nucleic acids are directional: they have a 5′ end and a 3′ end. During polymerization, new nucleotides are added only to the 3′ end.

✔ You should be able to state what cells do to activate nucleotides for incorporation into a polymer and explain why activation is required.

4.2   DNA Structure and Function

- DNA's primary structure consists of a sequence of linked nitrogenous bases. Its secondary structure consists of two DNA strands running in opposite directions that are twisted into a double helix.

- DNA is an extremely stable molecule that serves as a superb archive for information in the form of base sequences. It lacks a reactive 2′ hydroxyl group, and its secondary structure is stabilized by hydrophobic interactions and hydrogen bonds that form between complementary bases stacked on the inside of the helix.

- DNA is readily copied via complementary base pairing. Complementary base pairing occurs between A-T and G-C pairs in DNA.

- DNA's structural stability and regularity are advantageous for information storage, but they make DNA an ineffective catalyst.

✔ You should be able to explain why DNA molecules with a high percentage of guanine and cytosine are particularly stable.

4.3   RNA Structure and Function

- Like DNA, RNA's primary structure consists of a sequence of linked nitrogenous bases. RNA's secondary structure includes short regions of double helices and looped structures called hairpins.

- RNA molecules are usually single stranded. They have secondary structure because of complementary base pairing between A-U and G-C pairs on the same strand.

- Unlike DNA, the secondary structures of RNA can fold into more complex shapes, stabilized by hydrogen bonding, which give the molecule tertiary structure.

- RNA is versatile. The primary function of proteins is to catalyze chemical reactions, and the primary function of DNA is to carry information. But RNA is an "all-purpose" macromolecule that can do both.

✔ You should be able to explain why many RNA molecules exhibit tertiary structure, while most DNA molecules do not.

4.4 In Search of the First Life-Form

- To test the RNA world hypothesis, researchers are attempting to synthesize new ribozymes in the laboratory. Using artificial selection strategies, they have succeeded in identifying RNAs that catalyze several different reactions.

- Ribozymes that catalyze reactions necessary for the production of nucleotides may have preceded the evolution of RNA replicases.

✔ You should be able to provide two examples of activities in the RNA world you expect would benefit from catalysis and justify your choices.

You should be able to . . .

✔ TEST YOUR KNOWLEDGE
Answers are available in Appendix A

1. What are the four nitrogenous bases found in RNA?
   a. uracil, guanine, cytosine, thymine (U, G, C, T)
   b. adenine, guanine, cytosine, thymine (A, G, C, T)
   c. adenine, uracil, guanine, cytosine (A, U, G, C)
   d. alanine, threonine, glycine, cysteine (A, T, G, C)

2. What determines the primary structure of a DNA molecule?
   a. the sugar-phosphate backbone
   b. complementary base pairing and the formation of hairpins
   c. the sequence of deoxyribonucleotides
   d. the sequence of ribonucleotides

3. DNA attains a secondary structure when hydrogen bonds form between the nitrogenous bases called purines and pyrimidines. What are the complementary base pairs that form in DNA?
   a. A-T and G-C
   b. A-U and G-C
   c. A-G and T-C
   d. A-T and G-U

4. Which of the following rules apply to the synthesis of nucleic acids?
   a. Nucleotides are added to the 5′ end of nucleic acids.
   b. The synthesis of nucleic acids cannot occur without the presence of an enzyme to catalyze the reaction.
   c. Strands are synthesized in a parallel direction such that one end of the double-stranded product has the 3′ ends and other has the 5′ ends.
   d. Complementary pairing between bases is required for copying nucleic acids.

5. Nucleic acids are directional, meaning that there are two different ends. What functional groups define the two different ends of a DNA strand?

6. What is responsible for the increased stability of DNA compared to RNA?

✔ TEST YOUR UNDERSTANDING
Answers are available in Appendix A

7. Explain how Chargaff's rules relate to the complementary base pairing seen in the secondary structure of DNA. Would you expect these rules to apply to RNA as well? Explain why or why not.

8. **QUANTITATIVE** If nucleotides from the DNA of a human were quantified and 30 percent of them consisted of adenine, what percentage of guanine nucleotides would be present?
   a. 20 percent
   b. 30 percent
   c. 40 percent
   d. 70 percent

9. What would be the sequence of the strand of DNA that is made from the following template: 5′-GATATCGAT-3′ (Your answer must be written 5′→3′.) How would this sequence be different if RNA were made from this DNA template?

10. A major theme in this chapter is that the structure of molecules correlates with their function. Explain how DNA's secondary structure limits its catalytic abilities compared with that of RNA. Why is it expected that RNA molecules can catalyze a modest but significant array of reactions?

11. To replicate a ribozyme, a complete complementary copy must be made. Would you expect the double-stranded intermediate to maintain its catalytic activity? Justify your answer with an explanation.

12. Suppose that Bartel's research group succeeded in producing a molecule that could make a copy of itself. Which of the five fundamental characteristics of life (provided in Chapter 1) would support the claim that this molecule is alive?

13. Make a concept map (see **BioSkills 15** in Appendix B) that relates DNA's primary structure to its secondary structure. Your diagram should include deoxyribonucleotides, hydrophobic interactions, purines, pyrimidines, phosphodiester linkages, DNA primary structure, DNA secondary structure, complementary base pairing, and antiparallel strands.

14. Viruses are particles that infect cells. In some viruses, the genetic material consists of two strands of RNA, bonded together via complementary base pairing. Would these antiparallel strands form a double helix? Explain why or why not.

15. Before Watson and Crick published their model of the DNA double helix, Linus Pauling offered a model based on a triple helix. If the three sugar-phosphate backbones were on the outside of such a molecule, would hydrogen bonding or hydrophobic interactions be more important in keeping such a secondary structure together?

16. How would you expect the structure of ribozymes in organisms that grow in very hot environments, such as hot springs or deep-sea vents, to differ from those in organisms that grow in cooler environments?
  a. These ribozymes would have more hairpin secondary structures.
  b. The hairpins would have more G's and C's in the primary structure.
  c. The hairpins would have more A's and U's in the primary structure.
  d. These ribozymes would exhibit no tertiary structure.

5 An Introduction to Carbohydrates

In this chapter you will learn that

The role carbohydrates play in life is based on how they are linked together

by examining

The structure of monosaccharides
5.1

and how they link to form

Polymers called polysaccharides
5.2

then asking

What major roles do carbohydrates play? **5.3**

and looking at

Cell structure

Cell identity

Energy storage

A cross section through a buttercup root. Cellulose-rich cell walls are stained green; starch-filled structures are stained purple. Cellulose is a structural carbohydrate; starch is an energy-storage carbohydrate.

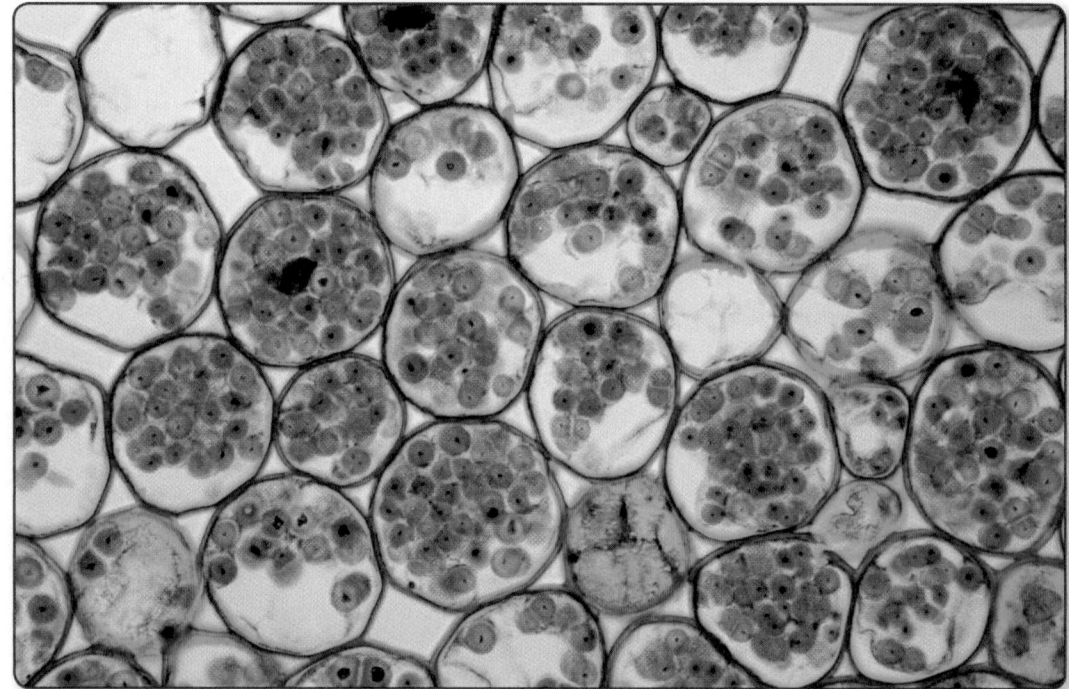

This unit highlights the four types of macromolecules that were key to the evolution of the cell: proteins, nucleic acids, carbohydrates, and lipids. Understanding the structure and function of macromolecules is a basic requirement for exploring how life began and how organisms work. Recall that proteins and nucleic acids could satisfy only three of the five fundamental characteristics of life: information, replication, and evolution (Chapter 4). Carbohydrates, the subject of this chapter, play an important role in a fourth characteristic—energy.

The term **carbohydrate,** or **sugar,** encompasses the monomers called **monosaccharides** (literally, "one-sugar"), small polymers called **oligosaccharides** ("few-sugars"), and the large polymers called **polysaccharides** ("many-sugars"). The name carbohydrate is logical because the molecular formula of many of these molecules is $(CH_2O)_n$, where the n refers to the number of "carbohydrate" groups. The value of n can vary from 3, for the smallest sugar, to well over a thousand for some of the large polymers.

This chapter is part of the Big Picture. See how on pages 104–105.

✔ When you see this checkmark, stop and test yourself. Answers are available in Appendix A.

An aldose
Carbonyl group at end of carbon chain

A ketose
Carbonyl group in middle of carbon chain

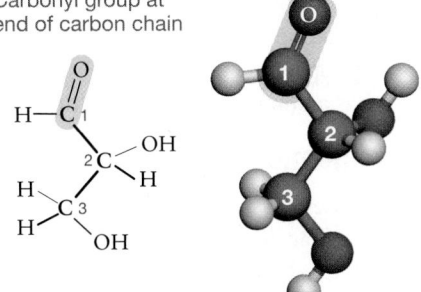

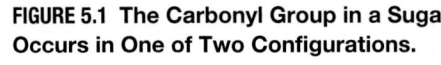

The name can also be misleading, though, because carbohydrates do not consist of carbon atoms bonded to water molecules. Instead, they are molecules with a carbonyl (C=O) and several hydroxyl (−OH) functional groups, along with several to many carbon–hydrogen (C–H) bonds. Consider formaldehyde, which was introduced as one of the molecules present in early Earth (Chapter 2). Even though formaldehyde has the same molecular formula as the one given above (CH_2O), it is not a carbohydrate since it does not contain a hydroxyl group.

Let's begin with monosaccharides, put them together into polysaccharides, and then explore how carbohydrates figured in the origin of life and what they do in cells today. As you study this material, be sure to ask yourself the central question of biological chemistry: How does this molecule's structure relate to its properties and function?

5.1 Sugars as Monomers

Sugars are fundamental to life. They provide chemical energy in cells and furnish some of the molecular building blocks required for the synthesis of larger, more complex compounds. Monosaccharides were important during chemical evolution, early in Earth's history. For example, as you've seen, the sugar called ribose is required for the formation of the nucleotides that make up nucleic acids (Chapter 4).

What Distinguishes One Monosaccharide from Another?

Monosaccharides, or simple sugars, are the monomers of carbohydrates. **FIGURE 5.1** illustrates two of the smallest monosaccharides. Although their molecular formulas are identical ($C_3H_6O_3$), their molecular structures are different. The carbonyl group that serves as one of monosaccharides' distinguishing features can be found either at the end of the molecule, forming an aldehyde sugar (an aldose), or within the carbon chain, forming a ketone sugar (a ketose). The presence of a carbonyl group along with multiple hydroxyl groups provides sugars with an array of reactive and hydrophilic functional groups. Based on this observation, it's not surprising that sugars are able to participate in a large number of chemical reactions.

The number of carbon atoms present also varies in monosaccharides. By convention, the carbons in a monosaccharide are numbered consecutively, starting with the end nearest the carbonyl group. Figure 5.1 features three-carbon sugars, or **trioses.** Ribose, which acts as a building block for nucleotides, has five carbons and is called a **pentose;** the glucose that is coursing through your bloodstream right now is a six-carbon sugar, or a **hexose.**

Besides varying in the location of the carbonyl group and the total number of carbon atoms present, monosaccharides can vary in the spatial arrangement of their atoms. There is, for example, a wide array of pentoses and hexoses. Each is distinguished by the configuration of its hydroxyl functional groups. **FIGURE 5.2**

Glucose

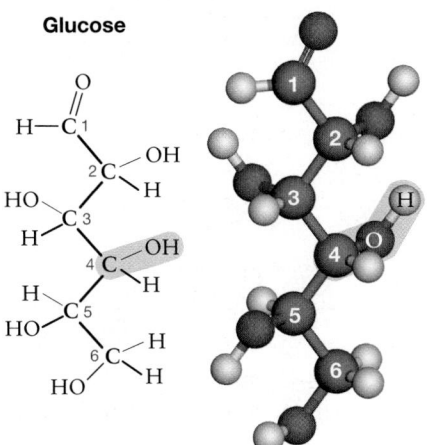

Galactose

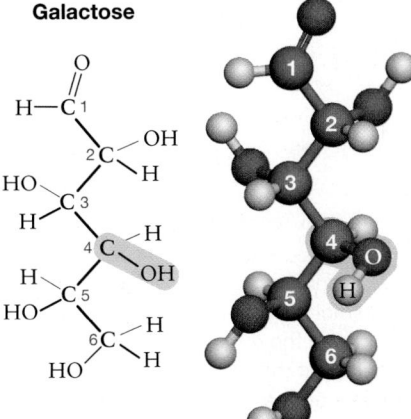

FIGURE 5.2 **Sugars May Vary in the Configuration of Their Hydroxyl Groups.** The two six-carbon sugars shown here vary only in the spatial orientation of their hydroxyl groups on carbon number 4.

✔**EXERCISE** Mannose is a six-carbon sugar that is identical to glucose, except that the hydroxyl (–OH) group on carbon number 2 is switched in orientation. Circle carbon number 2 in glucose and galactose; then draw the structural formula of mannose.

illustrates glucose and galactose, which are six-carbon sugars. Notice that the two molecules have the same molecular formula ($C_6H_{12}O_6$) but not the same structure. Both are aldose sugars with six carbons, but they differ in the spatial arrangement of the hydroxyl group at the fourth carbon (highlighted in Figure 5.2).

This is a key point: Because the structures of glucose and galactose differ, their functions differ. In cells, glucose is used as a source of carbons for the construction of other molecules and chemical energy that sustains life. But for galactose to be used in these roles, it first has to be converted to glucose via an enzyme-catalyzed reaction. This example underscores a general theme: Even seemingly simple changes in structure—like the location of a single hydroxyl group—can have enormous consequences for function. This is because molecules interact in precise ways, based on their shape.

It's rare for sugars consisting of five or more carbons to exist in the form of the linear chains illustrated in Figure 5.2, however. In aqueous solution they tend to form ring structures. The bond responsible for ring formation occurs only between the carbon containing the carbonyl group and one of the carbons with a hydroxyl group. Glucose serves as the example in **FIGURE 5.3**. When the cyclic structure forms in glucose, the C-1 carbon (the carbon numbered 1 in the linear chain) forms a bond with the oxygen atom of the C-5 hydroxyl and transfers its hydrogen to the C-1 carbonyl, turning it into a hydroxyl group.

Transfer of hydrogen between the C-5 and C-1 functional groups preserves the number of atoms and hydroxyls found in the ring and linear forms. The newly formed C-1 hydroxyl group can be oriented in two distinct ways: above or below the plane of the ring. The different configurations produce the molecules α-glucose and β-glucose.

To summarize, many distinct monosaccharides exist because so many aspects of their structure are variable: aldose or ketose placement of the carbonyl group, variation in carbon number, different arrangements of hydroxyl groups in space, and alternative ring forms. Each monosaccharide has a unique structure and function.

Monosaccharides and Chemical Evolution

Laboratory simulations, like those you read about in Chapter 2, have shown that most monosaccharides are readily synthesized under conditions that mimic the conditions of early Earth. For example, when formaldehyde (CH_2O) molecules are heated in solution, they react with one another to form almost all the pentoses and hexoses.

In addition, researchers have discovered the three-carbon ketose illustrated in Figure 5.1, along with a wide array of compounds closely related to sugars, on a meteorite that struck Murchison, Australia, in 1969. Based on these observations, investigators suspect that sugars are synthesized on dust particles and other debris in interstellar space and could have rained down onto Earth as the planet was forming, as well as being synthesized in the hot water near ancient undersea volcanoes.

More recent evidence suggests that synthesis of sugars could have been catalyzed by minerals found in the walls of deep-sea hydrothermal vents. Most researchers interested in chemical evolution maintain that one or more of the above mechanisms led to the accumulation of monosaccharides in the early oceans.

Modern cells display a wide range of carbohydrates beyond monosaccharides. How do these monomers join together to form polymers? Is the process similar to how amino acids link together to form proteins and nucleotides join to form nucleic acids? Let's explore how the array of functional groups in monosaccharides influences the polymerization of carbohydrates.

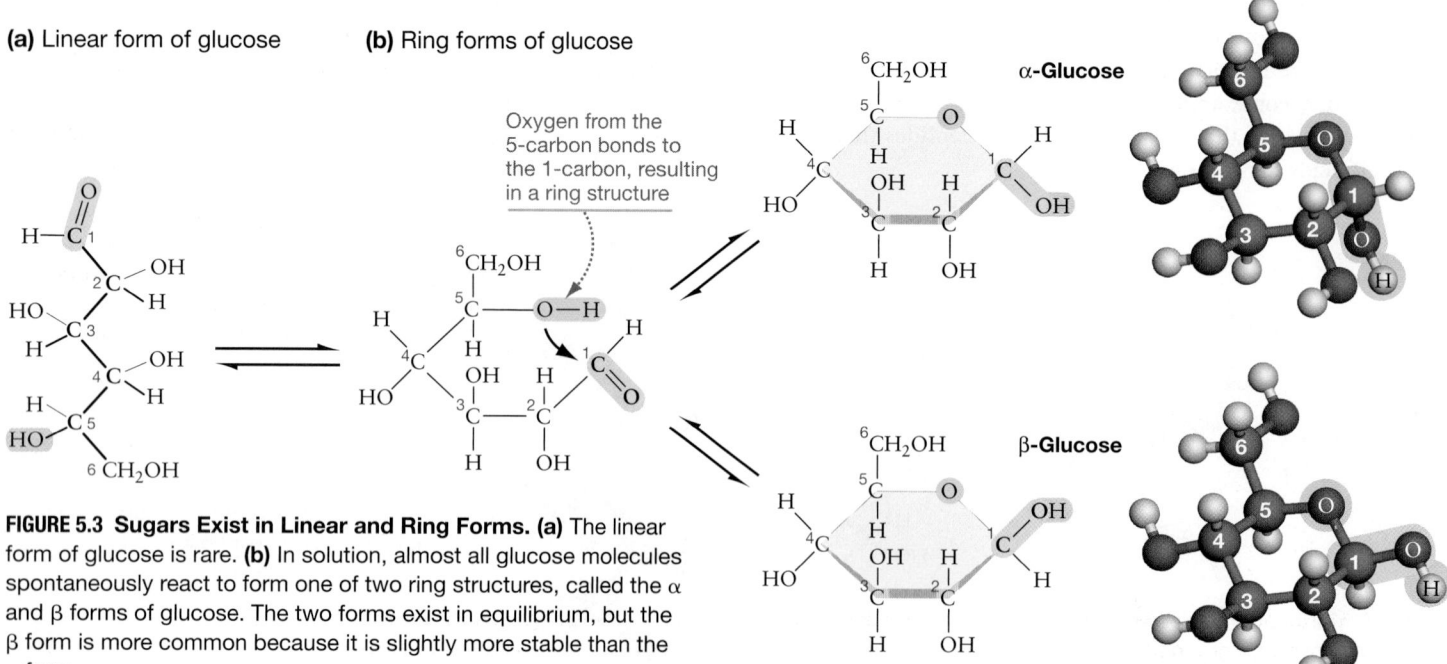

(a) Linear form of glucose

(b) Ring forms of glucose

Oxygen from the 5-carbon bonds to the 1-carbon, resulting in a ring structure

α-Glucose

β-Glucose

FIGURE 5.3 Sugars Exist in Linear and Ring Forms. (a) The linear form of glucose is rare. **(b)** In solution, almost all glucose molecules spontaneously react to form one of two ring structures, called the α and β forms of glucose. The two forms exist in equilibrium, but the β form is more common because it is slightly more stable than the α form.

If you understand that . . .

- Simple sugars differ from each other in three respects:
  1. the location of their carbonyl group,
  2. the number of carbon atoms present, and
  3. the spatial arrangement of their atoms— particularly the relative positions of hydroxyl (−OH) groups.

✔ **You should be able to . . .**

Draw the structural formula of a three-carbon monosaccharide ($C_3H_6O_3$) in linear form and then draw three other sugars that illustrate the three differences listed above.

Answers are available in Appendix A.

5.2 The Structure of Polysaccharides

Simple sugars can be covalently linked into chains of varying lengths, also known as complex carbohydrates. These chains range in size from small oligomers, or oligosaccharides, to the large polymers called polysaccharides. When only two sugars are linked together, they are known as **disaccharides.**

Similar to proteins and nucleic acids, the structure and function of larger carbohydrates depends on the types of monomers involved and how they are linked together. For example, maltose, also known as malt sugar, and lactose, an important sugar in milk, are two disaccharides that differ by just one monosaccharide. Maltose consists of two identical glucose molecules (**FIGURE 5.4a**), while lactose is made up of glucose and galactose (**FIGURE 5.4b**).

Monosaccharides polymerize when a condensation reaction occurs between two hydroxyl groups, resulting in a covalent interaction called a **glycosidic linkage.** The inverse reaction, hydrolysis, cleaves these linkages. (To review condensation and hydrolysis reactions, see Chapter 3.)

(a) Formation of α-glycosidic linkage

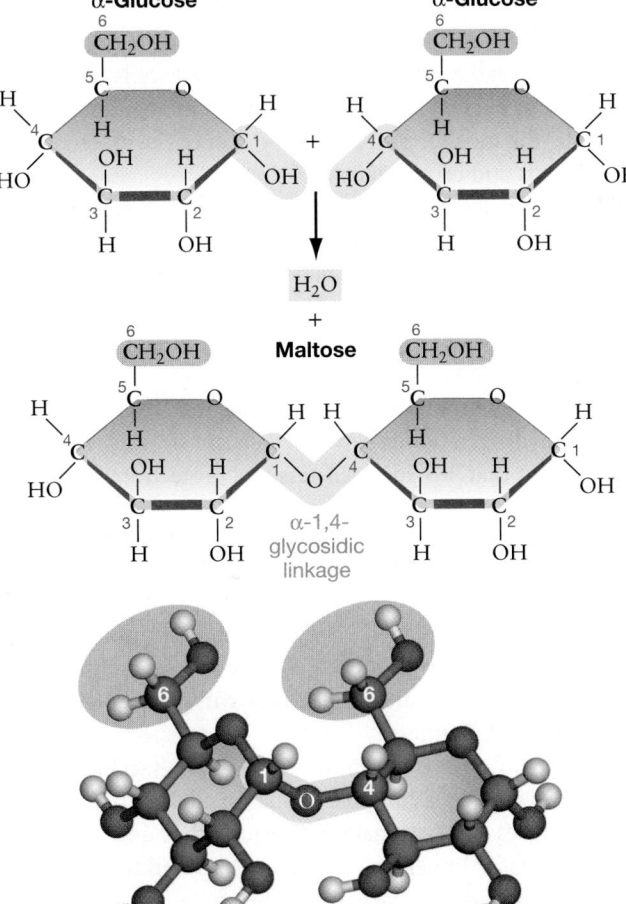

(b) Formation of β-glycosidic linkage

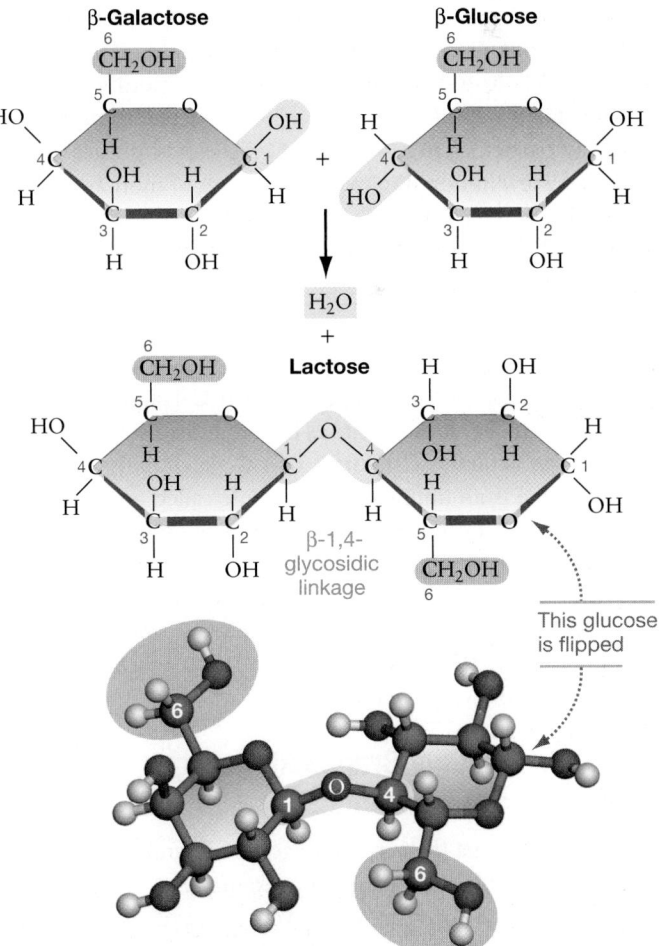

FIGURE 5.4 Monosaccharides Polymerize through Formation of Glycosidic Linkages. A glycosidic linkage occurs when hydroxyl groups on two monosaccharides undergo a condensation reaction. Maltose and lactose are disaccharides.

In that they hold monomers together, glycosidic linkages are analogous to the peptide bonds and phosphodiester linkages in proteins and nucleic acids. There is an important difference, however. Peptide bonds and phosphodiester linkages form between the same locations in their monomers, giving proteins and nucleic acids a standard backbone structure, but this is not the case for polysaccharides. Because glycosidic linkages form between hydroxyl groups, and because every monosaccharide contains at least two hydroxyls, the location and geometry of glycosidic linkages can vary widely among polysaccharides.

Maltose and lactose illustrate two of the most common glycosidic linkages, called the α-1,4-glycosidic linkage and the β-1,4-glycosidic linkage. The numbers refer to the carbons on either side of the linkage, indicating that both linkages are between the C-1 and C-4 carbons. Their geometry, however, is different: α and β refer to the contrasting orientations of the C-1 hydroxyls—on opposite sides of the plane of the glucose rings (i.e., "above" versus "below" the plane).

As Section 5.3 explains, the orientation of this hydroxyl in glycosidic linkages is particularly important in the structure, function, and durability of the molecules. In essence, the difference between polysaccharides used for storage and structural polysaccharides is a simple twist of a link.

To drive this point home, let's consider the structures of the most common polysaccharides found in organisms today: starch, glycogen, cellulose, and chitin, along with a modified polysaccharide called peptidoglycan. Each of these macromolecules is joined by particular α-1,4- or β-1,4-glycosidic linkages and can consist of a few hundred to many thousands of monomers.

Starch: A Storage Polysaccharide in Plants

In plant cells, some monosaccharides are stored for later use in the form of starch. **Starch** consists entirely of α-glucose monomers joined by glycosidic linkages. As the top panel in **TABLE 5.1** shows, the angle of the linkages between C-1 and C-4 carbons causes a chain of glucose subunits to coil into a helix.

Starch is actually a mixture of two such polysaccharides, however. One is an unbranched molecule called amylose, which contains only α-1,4-glycosidic linkages. The other is a branched molecule called amylopectin. The branching in amylopectin occurs when glycosidic linkages form between the C-1 carbon of a glucose monomer on one strand and the C-6 carbon of a glucose monomer on another strand. In amylopectin, branches occur at about one out of every 30 monomers.

Glycogen: A Highly Branched Storage Polysaccharide in Animals

Glycogen performs the same storage role in animals that starch performs in plants. In humans, for example, glycogen is stored in the liver and in muscles. When you start exercising, enzymes begin breaking glycogen into glucose monomers, which are then processed in muscle cells to supply energy. Glycogen is a polymer of α-glucose and is nearly identical to the branched form of starch. However, instead of an α-1,6-glycosidic linkage occurring

in about 1 out of every 30 monomers, a branch occurs in about 1 out of every 10 glucose subunits (see Table 5.1).

Cellulose: A Structural Polysaccharide in Plants

All cells are enclosed by a membrane (Chapter 1). In most organisms living today, the cell is also surrounded by a layer of material called a wall. A **cell wall** is a protective sheet that occurs outside the membrane. In plants, bacteria, fungi, and many other groups, the cell wall is composed primarily of one or more polysaccharides.

In plants, cellulose is the major component of the cell wall. **Cellulose** is a polymer of β-glucose monomers, joined by β-1,4-glycosidic linkages. As Table 5.1 shows, the geometry of the linkage is such that each glucose monomer in the chain is flipped in relation to the adjacent monomer. The flipped orientation is important because (**1**) it generates a linear molecule, rather than the helix seen in starch; and (**2**) it permits multiple hydrogen bonds to form between adjacent, parallel strands of cellulose. As a result, cellulose forms long, parallel strands that are joined by hydrogen bonds. The linked cellulose fibers are strong and give the cell structural support.

Chitin: A Structural Polysaccharide in Fungi and Animals

Chitin is a polysaccharide that stiffens the cell walls of fungi. It is also found in a few types of protists and in many animals. It is, for example, the most important component of the external skeletons of insects and crustaceans.

Chitin is similar to cellulose, but instead of consisting of glucose monomers, the monosaccharide involved is one called *N*-acetylglucosamine (abbreviated as NAG). These NAG monomers are joined by β-1,4-glycosidic linkages (see Table 5.1). As in cellulose, the geometry of these bonds results in every other residue being flipped in orientation.

Like the glucose monomers in cellulose, the NAG subunits in chitin form hydrogen bonds between adjacent strands. The result is a tough sheet that provides stiffness and protection.

Peptidoglycan: A Structural Polysaccharide in Bacteria

Most bacteria, like all plants, have cell walls. But unlike plants, in bacteria the ability to produce cellulose is extremely rare. Instead, a polysaccharide called **peptidoglycan** gives bacterial cell walls strength and firmness.

Peptidoglycan is the most complex of the polysaccharides discussed thus far. It has a long backbone formed by two types of monosaccharides that alternate with each other and are linked by β-1,4-glycosidic linkages (see Table 5.1). In addition, a short chain of amino acids is attached to one of the two sugar types. When molecules of peptidoglycan align, peptide bonds link the amino acid chains on adjacent strands. These links serve the same purpose as the hydrogen bonds between the parallel strands of cellulose and chitin in the cell walls of other organisms.

Polysaccharides Differ in Structure

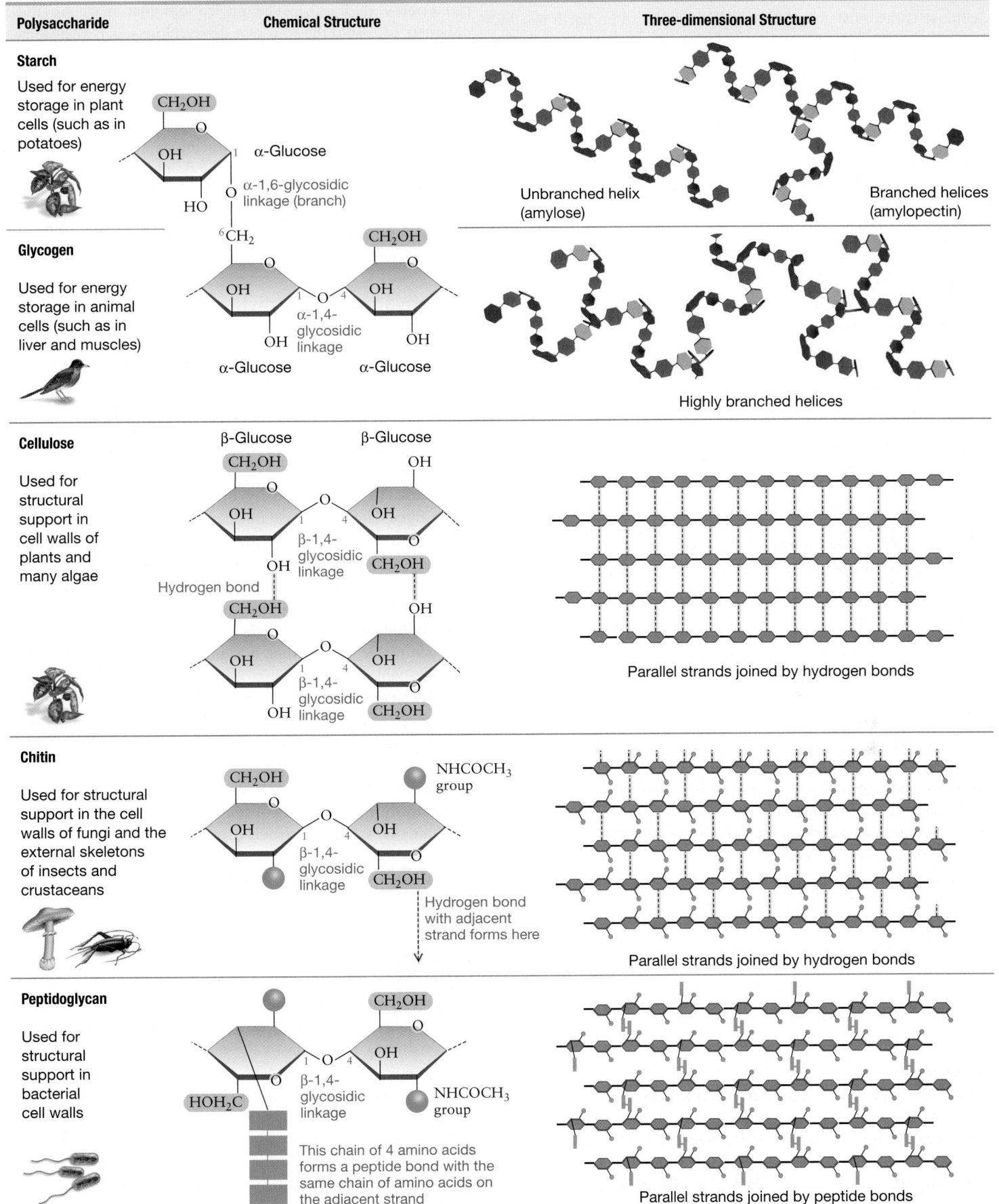

Polysaccharide	Chemical Structure	Three-dimensional Structure

Starch

Used for energy storage in plant cells (such as in potatoes)

CH₂OH / α-Glucose / α-1,6-glycosidic linkage (branch)

Unbranched helix (amylose) Branched helices (amylopectin)

Glycogen

Used for energy storage in animal cells (such as in liver and muscles)

CH₂OH / α-1,4-glycosidic linkage / α-Glucose / α-Glucose

Highly branched helices

Cellulose

Used for structural support in cell walls of plants and many algae

β-Glucose β-Glucose / CH₂OH / OH / β-1,4-glycosidic linkage / Hydrogen bond / β-1,4-glycosidic linkage / CH₂OH

Parallel strands joined by hydrogen bonds

Chitin

Used for structural support in the cell walls of fungi and the external skeletons of insects and crustaceans

CH₂OH / NHCOCH₃ group / β-1,4-glycosidic linkage / CH₂OH / Hydrogen bond with adjacent strand forms here

Parallel strands joined by hydrogen bonds

Peptidoglycan

Used for structural support in bacterial cell walls

CH₂OH / OH / β-1,4-glycosidic linkage / NHCOCH₃ group / HOH₂C / This chain of 4 amino acids forms a peptide bond with the same chain of amino acids on the adjacent strand

Parallel strands joined by peptide bonds

Polysaccharides and Chemical Evolution

Cellulose is the most abundant organic compound on Earth today, and chitin is probably the second most abundant by weight. Virtually all organisms depend on glycogen or starch as an energy source. But despite their current importance to life, polysaccharides probably played little to no role in the origin of life. This conclusion is supported by several observations:

- *No plausible mechanism exists for the polymerization of monosaccharides under conditions that prevailed early in Earth's history.* In cells and in laboratory experiments, the glycosidic linkages illustrated in Figure 5.4 and Table 5.1 form only with the aid of protein enzymes. No enzyme-like RNAs are known to catalyze these reactions.

- *To date, no polysaccharide has been discovered that can catalyze polymerization reactions.* Even though polysaccharides contain reactive hydroxyl and carbonyl groups, they lack the structural and chemical complexity that makes proteins, and to a lesser extent RNA, effective catalysts.

- *The monomers in polysaccharides are not capable of complementary base pairing.* Like proteins, but unlike nucleic acids, polysaccharides cannot act as templates for their own replication.

Even though polysaccharides probably did not play a significant role in the earliest forms of life, they became enormously important once cellular life evolved. In the next section, let's take a detailed look at how they function in today's cells.

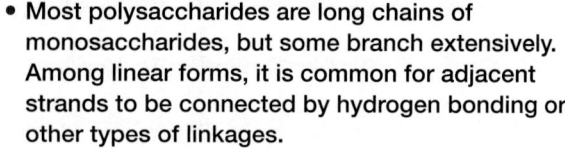
check your understanding

C Y U

If you understand that . . .

- Polysaccharides form when enzymes catalyze the formation of glycosidic linkages between monosaccharides that are in the α or β form.
- Most polysaccharides are long chains of monosaccharides, but some branch extensively. Among linear forms, it is common for adjacent strands to be connected by hydrogen bonding or other types of linkages.

✔ **You should be able to . . .**

Provide four structural differences that could result in different oligosaccharides consisting of two glucose monomers and two galactose monomers.

Answers are available in Appendix A.

5.3 What Do Carbohydrates Do?

One of the basic functions that carbohydrates perform in organisms is to serve as a substrate for synthesizing more-complex molecules. For example, recall that RNA contains the five-carbon sugar ribose ($C_5H_{10}O_5$) and DNA contains the modified sugar deoxyribose ($C_5H_{10}O_4$). The nucleotides that make up these polymers consist of the ribose or deoxyribose sugar, a phosphate group, and a nitrogenous base (Chapter 4). The sugar itself acts as a subunit of each of these monomers.

In addition, sugars frequently furnish the raw "carbon skeletons" that are used as building blocks in the synthesis of important molecules. Your cells are producing amino acids right now, for example, using sugars as a starting point.

Carbohydrates have diverse functions in cells: In addition to serving as precursors to larger molecules, they **(1)** provide fibrous structural materials, **(2)** indicate cell identity, and **(3)** store chemical energy. Let's look at each function in turn.

Carbohydrates Can Provide Structural Support

Cellulose and chitin, along with the modified polysaccharide peptidoglycan, are key structural compounds. They form fibers that give cells and organisms strength and elasticity.

To appreciate why cellulose, chitin, and peptidoglycan are effective structural molecules, recall that they form long strands and that bonds can form between adjacent strands. In the cell walls of plants, for example, a collection of about 80 cellulose molecules are cross-linked by hydrogen bonding to create a tough fiber. These cellulose fibers, in turn, crisscross to form a tough sheet that is able to withstand pulling and pushing forces—what an engineer would call tension and compression.

Besides being stiff and strong, the structural carbohydrates are durable. Almost all organisms have the enzymes required to break the various α-glycosidic linkages that hold starch and glycogen molecules together, but only a few organisms have enzymes capable of hydrolyzing cellulose, chitin, and peptidoglycan. Due to the strong interactions between strands consisting of β-1,4-glycosidic linkages, water is excluded and the fibers tend to be insoluble. The absence of water within these fibers makes their hydrolysis more difficult. As a result, the structural polysaccharides are resistant to degradation and decay.

Ironically, the durability of cellulose supports digestion. The cellulose that you ingest when you eat plant cells—what biologists call dietary fiber—forms a porous mass that absorbs and retains water. This sponge-like mass adds moisture and bulk that helps fecal material move through the intestinal tract more quickly, preventing constipation and other problems.

The Role of Carbohydrates in Cell Identity

Structural polymers tend to be repetitive, with only one or two types of monosaccharides. The same is not true for all complex carbohydrates. Some types exhibit enormous structural diversity, because their component monomers—and the linkages between them—vary a lot. As a result, they are capable of displaying information to other cells through their structure. More specifically, polysaccharides act as an identification badge on the outer surface of the plasma membrane that surrounds a cell. (Chapter 6 describes plasma membranes in detail.)

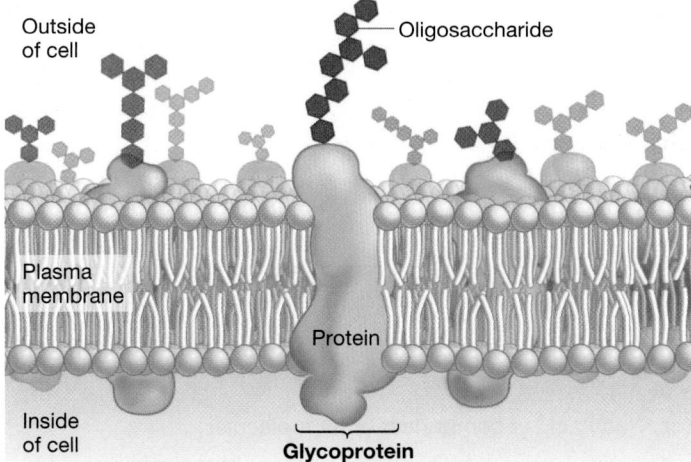

FIGURE 5.5 Carbohydrates Are an Identification Badge for Cells. Glycoproteins contain sugar groups that project outside the cell from the surface of the plasma membrane enclosing the cell. These sugar groups have distinctive structures that identify the type or species of the cell.

FIGURE 5.5 shows how this information about cell identity is displayed. Molecules called glycoproteins project outward from the cell surface into the surrounding environment. A **glycoprotein** is a protein that has one or more carbohydrates covalently bonded to it—usually relatively short oligosaccharides.

Glycoproteins are key molecules in what biologists call cell–cell recognition and cell–cell signaling. Each cell in your body has glycoproteins on its surface that identify it as part of your body. Immune system cells use these glycoproteins to distinguish your body's cells from foreign cells, such as bacteria. In addition, each distinct type of cell in a multicellular organism—for example, the nerve cells and muscle cells in your body—displays a different set of glycoproteins on its surface.

The identification information displayed by glycoproteins helps cells recognize and communicate with each other.

The key point here is to recognize that the variety in the types of monosaccharides and how they can be linked together makes it possible for an enormous number of unique oligosaccharides to exist. As a result, each cell type and each species can display a unique identity.

During the 1980s, Paul Wassarman and colleagues investigated the role of glycoproteins in one of the most important cell–cell recognition events in the life of a plant or animal—the attachment of sperm to eggs during fertilization. This step guarantees specificity—sperm recognize and bind only to eggs of their own species.

In one experiment, the researchers mixed sperm with purified egg-surface glycoproteins and discovered that most of the sperm lost their ability to attach to eggs (**FIGURE 5.6**). Such loss of function is an example of what researchers call competitive inhibition. The glycoproteins had bound to—and thus blocked—the same structure on the sperm that it uses to bind to eggs. This result showed that sperm attach to eggs via egg glycoproteins.

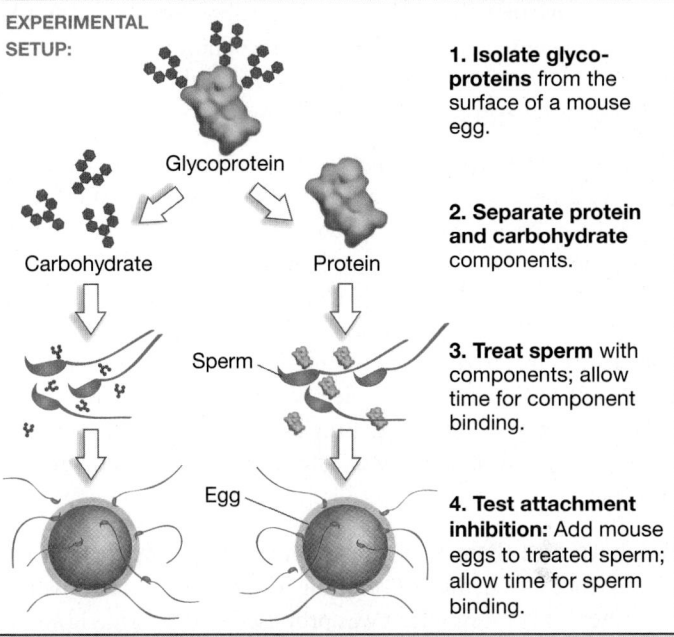

RESEARCH

QUESTION: What part of surface glycoproteins do sperm recognize when they attach to eggs?

HYPOTHESIS: Sperm attach to the carbohydrate component.

NULL HYPOTHESIS: Sperm attach to the protein component.

EXPERIMENTAL SETUP:

Glycoprotein

Carbohydrate — Protein

Sperm

Egg

1. **Isolate glyco-proteins** from the surface of a mouse egg.

2. **Separate protein and carbohydrate** components.

3. **Treat sperm** with components; allow time for component binding.

4. **Test attachment inhibition:** Add mouse eggs to treated sperm; allow time for sperm binding.

PREDICTION: The carbohydrate component of the glycoprotein will bind to sperm and block their attachment to eggs.

PREDICTION OF NULL HYPOTHESIS: The protein component of the glycoprotein will block sperm attachment to eggs.

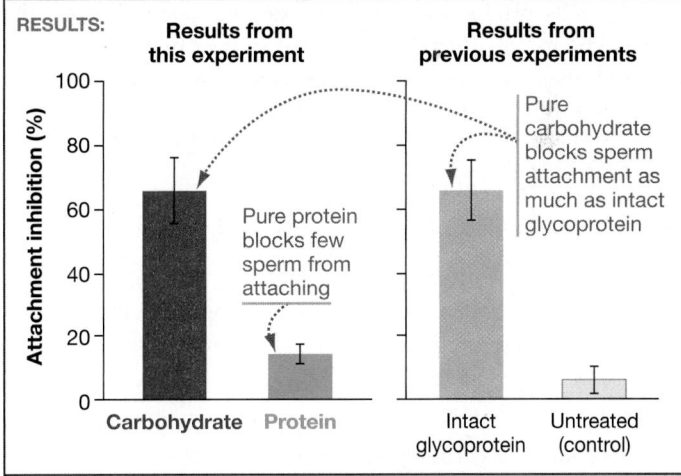

RESULTS:

Results from this experiment | Results from previous experiments

Pure protein blocks few sperm from attaching

Pure carbohydrate blocks sperm attachment as much as intact glycoprotein

(Attachment inhibition (%): axis 0, 20, 40, 60, 80, 100)

Carbohydrate — Protein | Intact glycoprotein — Untreated (control)

CONCLUSION: Sperm recognize and bind to the carbohydrates of egg-surface glycoproteins when they attach to egg cells.

FIGURE 5.6 Carbohydrates Are Required for Cellular Recognition and Attachment.

SOURCES: Florman, H. M., K. B. Bechtol, and P. M. Wassarman. 1984. Enzymatic dissection of the functions of the mouse egg's receptor for sperm. *Developmental Biology* 106: 243–255. Also Florman, H. M., and P. M. Wassarman. 1985. O-linked oligosaccharides of mouse egg ZP3 account for its sperm receptor activity. *Cell* 41: 313–324.

✔**QUANTITATIVE** How would the bars change in the graph if sperm attachment required only the protein portion of egg glycoproteins?

But which part of the egg glycoproteins is essential for recognition and attachment—the protein or the carbohydrate? In follow-up experiments, Wassarman's group used the same type of competitive-binding assay to answer this question. When sperm were mixed with purified carbohydrates alone, most were unable to attach to eggs. In contrast, most sperm treated with purified protein alone were not inhibited and still attached to eggs. Both results show that the carbohydrate component plays a fundamental role in the process of egg-cell recognition.

Carbohydrates and Energy Storage

Candy-bar wrappers promise a quick energy boost, and ads for sports drinks claim that their products provide the "carbs" needed for peak activity. If you were to ask friends or family members what carbohydrates do in your body, they would probably say something like "They give you energy." And after pointing out that carbohydrates are also used in cell identity, as a structural material, and as a source of carbon skeletons for the synthesis of other complex molecules, you'd have to agree.

Carbohydrates store and provide chemical energy in cells. What aspect of carbohydrate structure makes this function possible?

Carbohydrates Store Sunlight as Chemical Energy Recall that the essence of chemical evolution was energy transformations (Chapter 2). For example, it was proposed that the kinetic energy in sunlight may have been converted into chemical energy and stored in bonds of molecules such as formaldehyde (CH_2O).

This same type of transformation from light energy to chemical energy occurs in cells today, but instead of making formaldehyde, cells produce sugars. For example, plants harvest the kinetic energy in sunlight and store it in the bonds of carbohydrates by the process known as **photosynthesis.** (Photosynthesis is the focus of Chapter 10.)

Photosynthesis entails a complex set of reactions that can be summarized most simply as follows:

$$CO_2 + H_2O + sunlight \longrightarrow (CH_2O)_n + O_2$$

where $(CH_2O)_n$ represents a carbohydrate. The key to understanding the energy conversion that is taking place in this reaction is to compare the positions of the electrons in the reactants to those in the products.

1. The electrons in the C=O bonds of carbon dioxide and the C—O bonds of carbohydrates are held tightly because of oxygen's high electronegativity. Thus, they have relatively low potential energy.

2. The electrons involved in the C—H bonds of carbohydrates are shared equally because the electronegativity of carbon and hydrogen is about the same. Thus, these electrons have relatively high potential energy.

3. Electrons are also shared equally in the carbon–carbon C—C bonds of carbohydrates—meaning that they, too, have relatively high potential energy.

(a) Carbon dioxide

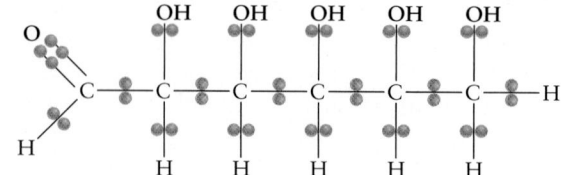

(b) A carbohydrate

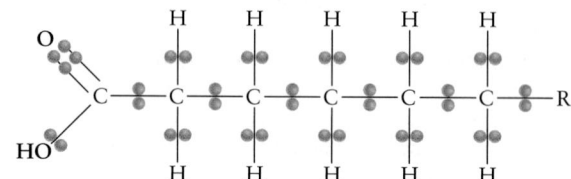

(c) A fatty acid (a component of fat molecules)

FIGURE 5.7 **In Organisms, Potential Energy Is Stored in C—H and C—C Bonds. (a)** In carbon dioxide, the electrons involved in covalent bonds are held tightly by oxygen atoms. **(b)** In carbohydrates such as the sugar shown here, many of the covalently bonded electrons are held equally between C and H atoms. **(c)** The fatty acids found in fat molecules have more C—H bonds and fewer C—O bonds than carbohydrates do. ("R" stands for the rest of the molecule.)

✔**EXERCISE** Circle the bonds in this diagram that have high potential energy.

C—C and C—H bonds have much higher potential energy than C—O bonds have. As a result, carbohydrates have much more chemical energy than carbon dioxide has.

FIGURE 5.7 summarizes and extends these points. Start by comparing the structure of carbon dioxide in Figure 5.7a with the carbohydrate in Figure 5.7b. The main difference is the presence of C—C and C—H bonds in the carbohydrate. Now compare the carbohydrate in Figure 5.7b with the fatty acid—a subunit of a fat molecule—in Figure 5.7c. Compared with carbohydrates, fats contain many more C—C and C—H bonds and many fewer C—O bonds.

This point is important. C—C and C—H bonds have high potential energy because the electrons are shared equally by atoms with low electronegativities. C—O bonds, in contrast, have low potential energy because the highly electronegative oxygen atom holds the electrons so tightly. Both carbohydrates and fats are used as fuel in cells, but fats store twice as much energy per gram compared with carbohydrates. (Fats are discussed in more detail in Chapter 6.)

Enzymes Hydrolyze Polysaccharides to Release Glucose Starch and glycogen are efficient energy-storage molecules because they polymerize via α-glycosidic linkages instead of the β-glycosidic linkages observed in the structural polysaccharides. The α-linkages in storage polysaccharides are readily hydrolyzed to release glucose, while the structural polysaccharides resist enzymatic degradation.

The most important enzyme involved in catalyzing the hydrolysis of α-glycosidic linkages in glycogen molecules is a protein called **phosphorylase.** Many of your cells contain phosphorylase, so they can break down glycogen to provide glucose on demand.

The enzymes involved in breaking the α-glycosidic linkages in starch are called **amylases.** Your salivary glands and pancreas produce amylases that are secreted into your mouth and small intestine, respectively. These amylases are responsible for digesting the starch that you eat.

The glucose subunits that are hydrolyzed from glycogen and starch are processed in reactions that result in the production of chemical energy that can be used in the cell. Glycogen and starch are like a candy bar that has segments, so you can break off chunks whenever you need a boost.

Energy Stored in Glucose is Used to Make ATP When a cell needs energy, reactions lead to the breakdown of the glucose and capture of the released energy through synthesis of the nucleotide adenosine triphosphate (ATP) (introduced in Chapter 4).

More specifically, the energy that is released when sugars are processed is used to synthesize ATP from a precursor called adenosine diphosphate (ADP) plus a free inorganic phosphate (P_i) molecule. The overall reaction can be written as follows:

$$(CH_2O)_n + O_2 + ADP + P_i \longrightarrow CO_2 + H_2O + ATP$$

To put this in words, the chemical energy stored in the C−H and C−C bonds of carbohydrate is transferred to a new bond linking a third phosphate group to ADP to form ATP.

How much energy does it take to form ATP? Consider this example: A cell can use the 10 calories of energy stored in a LifeSavers candy to produce approximately 2×10^{23} molecules of ATP. Although this sounds like a lot of ATP, an average human's energy needs would burn through all of this ATP energy in a little over a minute! The energy in ATP drives reactions like polymerization and cellular processes like moving your muscles.

Carbohydrates are like the water that piles up behind a dam; ATP is like the electricity generated at a dam, which lights up your home. Carbohydrates store chemical energy; ATP makes chemical energy useful to the cell.

Later chapters analyze in detail how cells capture and store energy in sugars and how these sugars are then broken down to provide cells with usable chemical energy in the form of ATP (Chapters 8, 9, and 10). For both of these processes to occur, however, a selectively permeable membrane barrier is required. The following chapter introduces the lipids needed to build these membranes and the role they played in the evolution of the first cell.

check your understanding

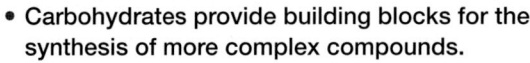
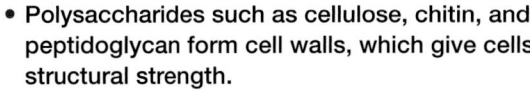
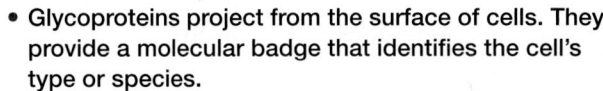

If you understand that . . .

- Carbohydrates provide building blocks for the synthesis of more complex compounds.
- Polysaccharides such as cellulose, chitin, and peptidoglycan form cell walls, which give cells structural strength.
- Glycoproteins project from the surface of cells. They provide a molecular badge that identifies the cell's type or species.
- Starch and glycogen store sugars for later use in reactions that produce ATP. Sugars contain large amounts of chemical energy because they contain carbon atoms that are bonded to hydrogen atoms or other carbon atoms. The C−H and C−C bonds have high potential energy because the electrons are shared equally by atoms with low electronegativity.

✔ **You should be able to . . .**

1. Identify two aspects of the structures of cellulose, chitin, and peptidoglycan that correlate with their function as structural molecules.

2. Describe how the carbohydrates you ate during breakfast today are functioning in your body right now.

Answers are available in Appendix A.

CHAPTER 5 REVIEW

For media, go to MasteringBiology

If you understand . . .

5.1 Sugars as Monomers

- Monosaccharides are organic compounds that have a carbonyl group and several hydroxyl groups. The molecular formula for a sugar is typically $(CH_2O)_n$, but the number of "carbon-hydrate" groups may vary between sugars, as indicated by the n.

- Although some monosaccharides may have the same molecular formula, the arrangement of functional groups can lead to differences in the molecular structure of the sugars.

- Individual monosaccharides may form ring structures that differ from one another in the orientation of a hydroxyl group.

✔ You should be able to explain how a relatively small difference in the location of a carbonyl or hydroxyl group can lead to dramatic changes in the properties and function of a monosaccharide.

5.2 The Structure of Polysaccharides

- Monosaccharides can be covalently bonded to one another via glycosidic linkages, which join hydroxyl groups on adjacent molecules.

- In contrast to proteins and nucleic acids, polysaccharides do not always form a single uniform backbone structure. The numerous hydroxyls found in each monosaccharide allow glycosidic linkages to form at different sites and new strands to branch from existing chains.

- The types of monomers involved and the geometries of the glycosidic linkages between monomers distinguish different polysaccharides from one another.

- The most common polysaccharides in organisms today are starch, glycogen, cellulose, and chitin; peptidoglycan is an abundant polysaccharide that has short chains of amino acids attached.

✔ You should be able to compare and contrast glycosidic linkages in polysaccharides with the linkages between monomers in proteins and nucleic acids.

5.3 What Do Carbohydrates Do?

- In carbohydrates, as in proteins and nucleic acids, structure correlates with function.

- Cellulose, chitin, and peptidoglycan are polysaccharides that function in support. They are made up of monosaccharide monomers joined by β-1,4-glycosidic linkages. When individual molecules of these polysaccharides align side by side, bonds form between them—resulting in strong, flexible fibers or sheets that resist hydrolysis.

- The oligosaccharides on cell-surface glycoproteins can function as specific signposts or identity tags because their constituent monosaccharides are so diverse in geometry and composition.

- Both starch and glycogen function as energy-storage molecules. They are made up of glucose molecules that are joined by α-glycosidic linkages. These linkages are readily hydrolyzed to release glucose for the production of ATP.

✔ You should be able to describe four key differences in the structure of polysaccharides that function in energy storage versus structural support.

You should be able to . . .

✔ **TEST YOUR KNOWLEDGE**   *Answers are available in Appendix A*

1. What is the difference between a monosaccharide, an oligosaccharide, and a polysaccharide?
   a. the number of carbon atoms in the molecule
   b. the type of glycosidic linkage between monomers
   c. the spatial arrangement of the various hydroxyl residues in the molecule
   d. the number of monomers in the molecule

2. What are three ways monosaccharides differ from one another?

3. What type of bond is formed between two sugars in a disaccharide?
   a. glycosidic linkage
   b. phosphodiester bond
   c. peptide bond
   d. hydrogen bond

4. What holds cellulose molecules together in bundles large enough to form fibers?
   a. the cell wall
   b. peptide bonds
   c. hydrogen bonds
   d. hydrophobic interactions between different residues in the cellulose helix

5. What are the primary functions of carbohydrates in cells?
   a. energy storage, cell identity, structure, and building blocks for synthesis
   b. catalysis, structure, and energy storage
   c. information storage and catalysis
   d. source of carbon, information storage, and energy storage

6. What is responsible for the difference in potential energy between carbohydrates and carbon dioxide?

7. Which of the differences listed here could be found in the same monosaccharide?

   a. different orientation of a hydroxyl in the linear form
   b. different number of carbons
   c. different orientation of a hydroxyl in the ring form
   d. different position of the carbonyl group in the linear form

8. What would most likely occur if the galactose in lactose were replaced with glucose?
   a. It would not be digested by human infants or adults.
   b. It would be digested by most adult humans.
   c. It would be digested by human infants, but not adults.
   d. It would be digested by human adults, but not infants.

9. Explain how the structure of carbohydrates supports their function in displaying the identity of a cell.

10. What is the difference between linking glucose molecules with α-1,4-glycosidic linkages versus β-1,4-glycosidic linkages? What are the consequences?

11. Give three reasons why researchers have concluded that polysaccharides were unlikely to play a large role in the origin of life.

12. Compare and contrast the structures and functions of starch and glycogen. How are these molecules similar? How are they different?

13. A weight-loss program for humans that emphasized minimal consumption of carbohydrates was popular in some countries

in the early 2000s. What was the logic behind this diet? (Note: This diet plan caused controversy and is not endorsed by some physicians and researchers).

14. Galactosemia is a potentially fatal disease that occurs in humans who lack the enzyme that converts galactose to glucose. To treat this disease, physicians exclude the monosaccharide galactose from the diet. Which of the following would you also predict to be excluded from the diet?
   a. maltose   **b.** starch   **c.** mannose   **d.** lactose

15. If you hold a salty cracker in your mouth long enough, it will begin to taste sweet. What is responsible for this change in taste?

16. Lysozyme, an enzyme found in human saliva, tears, and other secretions, catalyzes the hydrolysis of the β-1,4-glycosidic linkages in peptidoglycan. Predict the effect of this enzyme on bacteria, and explain the role its activity plays in human health.

6 Lipids, Membranes, and the First Cells

In this chapter you will learn how

Life's defining barrier—the plasma membrane— is built of lipids and proteins

by looking at ↓

Lipid structure and function **6.1**

How do substances move across bilayers?

via

Diffusion and osmosis **6.3**

Membrane proteins **6.4**

and how ↓

Lipids spontaneously form bilayers **6.2**

then asking

A space-filling model of a phospholipid bilayer. In single-celled organisms, this cluster of molecules forms part of the boundary between life (inside the cell) and nonlife (outside the cell)—the cell membrane.

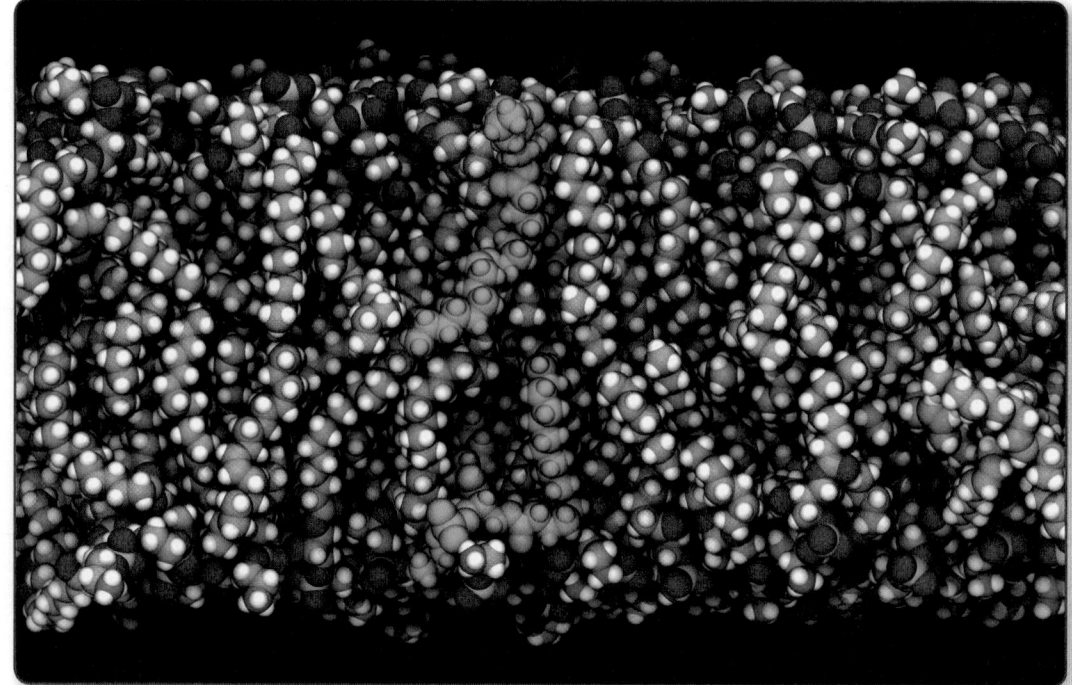

Currently, most biologists support the hypothesis that biological evolution began with a catalytic RNA molecule that could replicate itself. As the offspring of this molecule multiplied, natural selection would have favored the most efficient versions. A second great milestone in the history of life occurred when descendants of these replicators became enclosed within a membrane.

Why is the presence of a membrane so important? The **plasma membrane,** or **cell membrane,** separates life from nonlife. It is a layer of molecules that surrounds the cell interior and separates it from the environment.

BIG PICTURE

This chapter is part of the Big Picture. See how on pages 104–105.

84

- The plasma membrane serves as a selective barrier: It keeps damaging compounds out of the cell and allows entry of compounds needed by the cell.

- Because the plasma membrane sequesters the appropriate chemicals in an enclosed area, reactants collide more frequently—the chemical reactions necessary for life occur much more efficiently.

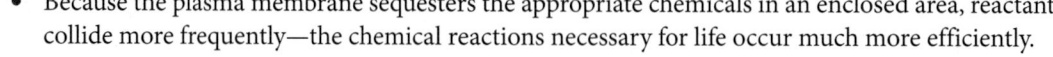

✔ When you see this checkmark, stop and test yourself. Answers are available in Appendix A.

While researchers of chemical evolution are currently debating when membranes arose—whether early or late during the emergence of life—there is little argument about the importance of this event. After life secured a membrane, it continued to evolve into an efficient and dynamic reaction vessel—the cell.

How do membranes form? Which ions and molecules can pass through a membrane and which cannot, and why? These are some of the most fundamental questions in all of biological science. Let's delve into them, beginning with the membrane's foundation—lipids.

6.1 Lipid Structure and Function

Lipid is a catchall term for carbon-containing compounds that are found in organisms and are largely nonpolar and hydrophobic—meaning that they do not dissolve readily in water. (Recall from Chapter 2 that water is a polar solvent.) Lipids do dissolve, however, in liquids consisting of nonpolar organic compounds.

To understand why lipids are insoluble in water, examine the five-carbon compound called isoprene, illustrated in **FIGURE 6.1a**. Note that isoprene consists of carbon atoms bonded to hydrogen atoms. The figure also shows the structural formula of a chain of linked isoprenes, called an isoprenoid.

Molecules that contain only carbon and hydrogen are known as **hydrocarbons.** Hydrocarbons are nonpolar because electrons are shared equally in C–H bonds—owing to the approximately equal electronegativity of carbon and hydrogen. Since these bonds form no partial charges, hydrocarbons are hydrophobic. Thus lipids do not dissolve in water, because they have a significant hydrocarbon component.

Bond Saturation Is an Important Aspect of Hydrocarbon Structure

FIGURE 6.1b gives the structural formula of a **fatty acid,** a simple lipid consisting of a hydrocarbon chain bonded to a carboxyl (–COOH) functional group. Fatty acids and isoprenes are key building blocks of important lipids found in organisms. Just as subtle differences in the orientation of hydroxyls in sugars can lead to dramatic effects in their structure and function, the type of C–C bond used in hydrocarbon chains is a key factor in lipid structure.

When two carbon atoms form a double bond, the attached atoms are found in a plane instead of a three-dimensional tetrahedron. The carbon atoms involved are also locked into place. They cannot rotate freely, as they do in carbon–carbon single bonds. As a result, certain double bonds between carbon atoms produce a "kink" in an otherwise straight hydrocarbon chain (Figure 6.1b, left).

Hydrocarbon chains that consist of only single bonds between the carbons are called **saturated.** If one or more double bonds exist in the hydrocarbon chains, then they are **unsaturated.** The choice of terms is logical. If a hydrocarbon chain does not contain a double bond, it is saturated with the maximum number of hydrogen atoms that can attach to the carbon skeleton. If it is unsaturated, then a C–H bond is removed to form a C=C double bond, resulting in fewer than the maximum number of attached hydrogen atoms.

Foods that contain lipids with many double bonds are said to be polyunsaturated and are advertised as healthier than foods with saturated fats. Recent research suggests that polyunsaturated fats help protect the heart from disease. Exactly how this occurs is under investigation.

(a) Isoprenes can be linked into chains called isoprenoids.

(b) Fatty acids can be saturated or unsaturated.

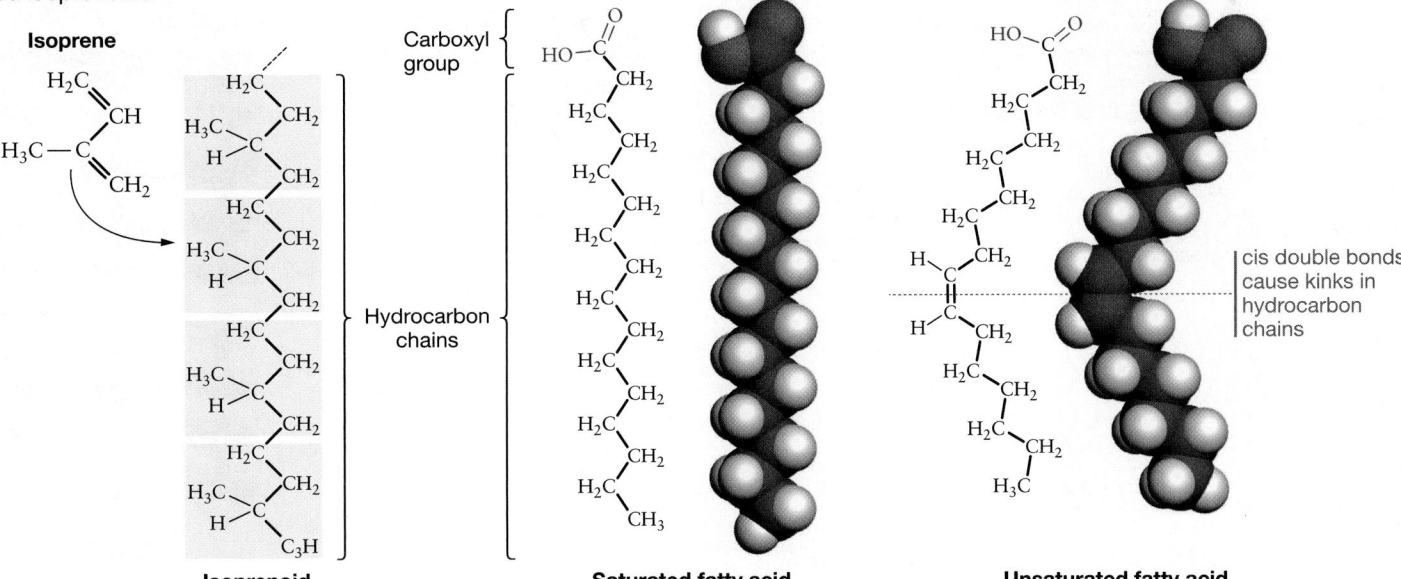

FIGURE 6.1 Hydrocarbon Structure. (a) Isoprene subunits, like the one shown to the left, can be linked to each other, end to end, to form long hydrocarbon chains called isoprenoids. **(b)** Fatty acids typically contain a total of 14–20 carbon atoms, most found in their long hydrocarbon "tails." Unsaturated hydrocarbons contain carbon–carbon double bonds; saturated hydrocarbons do not.

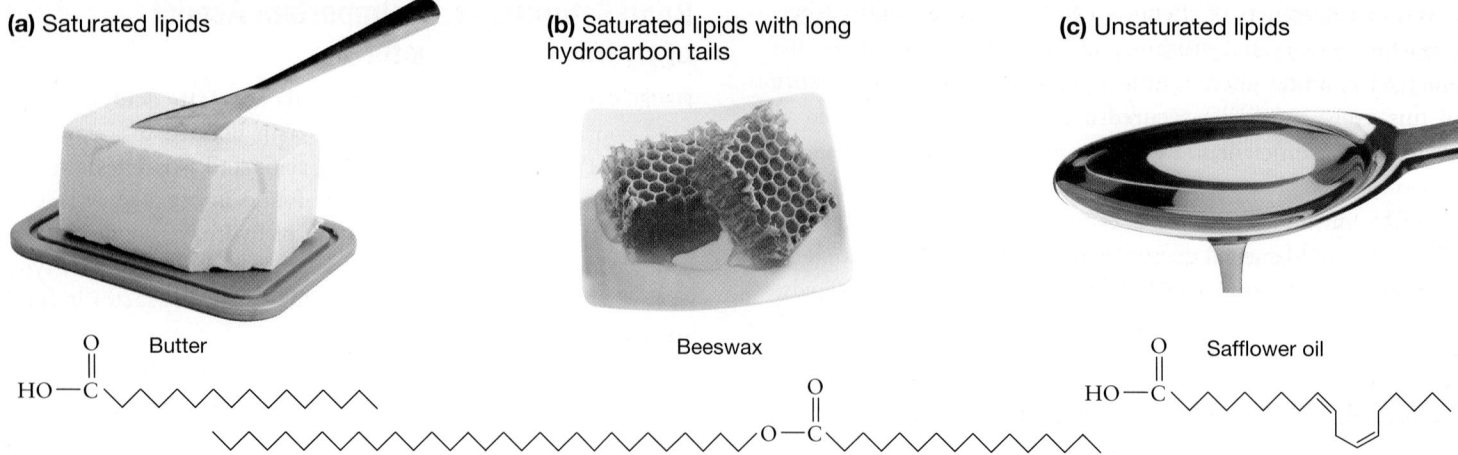

(a) Saturated lipids

(b) Saturated lipids with long hydrocarbon tails

(c) Unsaturated lipids

Butter

Beeswax

Safflower oil

FIGURE 6.2 The Fluidity of Lipids Depends on the Length and Saturation of Their Hydrocarbon Chains. (a) Butter consists primarily of saturated lipids. **(b)** Waxes are lipids with extremely long saturated hydrocarbon chains. **(c)** Oils are dominated by "polyunsaturates"—lipids with hydrocarbon chains that contain multiple C=C double bonds.

Bond saturation also profoundly affects the physical state of lipids. Highly saturated fats, such as butter, are solid at room temperature (**FIGURE 6.2a**). Saturated lipids that have extremely long hydrocarbon tails, like **waxes** do, form particularly stiff solids at room temperature (**FIGURE 6.2b**). Highly unsaturated fats are liquid at room temperature (**FIGURE 6.2c**).

A Look at Three Types of Lipids Found in Cells

Unlike amino acids, nucleotides, and monosaccharides, lipids are characterized by a physical property—their insolubility in water—instead of a shared chemical structure. This insolubility is based on the high proportion of nonpolar C–C and C–H bonds relative to polar functional groups. As a result, the structure of lipids varies widely. For example, consider the most important types of lipids found in cells: fats, steroids, and phospholipids.

Fats Fats are nonpolar molecules composed of three fatty acids that are linked to a three-carbon molecule called **glycerol.** Because of this structure, fats are also called triacylglycerols or triglycerides. When the fatty acids are polyunsaturated, they form liquid triacylglycerols called **oils.** In organisms, the primary role of fats is energy storage.

As **FIGURE 6.3a** shows, fats form when a dehydration reaction occurs between a hydroxyl group of glycerol and the carboxyl

(a) Fats form via dehydration reactions.

(b) Fats consist of glycerol linked by ester linkages to three fatty acids.

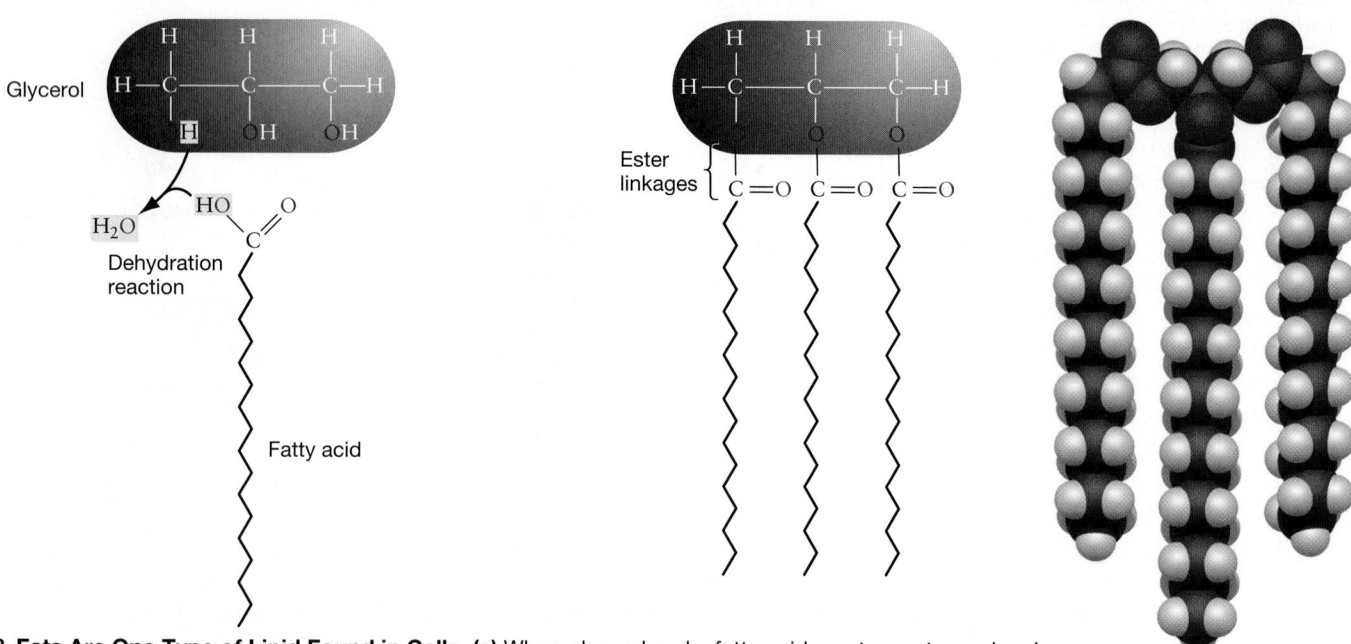

FIGURE 6.3 Fats Are One Type of Lipid Found in Cells. (a) When glycerol and a fatty acid react, a water molecule leaves. The covalent bond that results from this reaction is termed an ester linkage. **(b)** The structural formula and a space-filling model of tristearin, the most common type of fat in beef.

group of a fatty acid. The glycerol and fatty acid molecules become joined by an **ester linkage.** Fats are not polymers, however, and fatty acids are not monomers. As **FIGURE 6.3b** shows, fatty acids are not linked together to form a macromolecule in the way that amino acids, nucleotides, and monosaccharides are.

Steroids **Steroids** are a family of lipids distinguished by the bulky, four-ring structure shown in orange in **FIGURE 6.4a**. The various steroids differ from one another by the functional groups or side groups attached to different carbons in those hydrophobic rings. The steroid shown in the figure is cholesterol, which has a hydrophilic hydroxyl group attached to the top ring and an isoprenoid "tail" attached at the bottom. Cholesterol is an important component of plasma membranes in many organisms.

Phospholipids **Phospholipids** consist of a glycerol that is linked to a phosphate group and two hydrocarbon chains of either isoprenoids or fatty acids. The phosphate group is also bonded to a small organic molecule that is charged or polar (**FIGURE 6.4b**).

Phospholipids composed of fatty acids are found in the domains Bacteria and Eukarya; phospholipids with isoprenoid chains are found in the domain Archaea. (The domains of life were introduced in Chapter 1.) In all three domains, phospholipids are crucial components of the plasma membrane.

The lipids found in organisms have a wide array of structures and functions. In addition to storing chemical energy, lipids act as pigments that capture or respond to sunlight, serve as signals between cells, form waterproof coatings on leaves and skin, and act as vitamins used in many cellular processes. The most prominent function of lipids, however, is their role in cell membranes.

The Structures of Membrane Lipids

Not all lipids can form membranes. Membrane-forming lipids have a polar, hydrophilic region—in addition to the nonpolar, hydrophobic region found in all lipids.

To better understand this structure, take another look at the phospholipid illustrated in Figure 6.4b. Notice that the molecule has a "head" region containing highly polar covalent bonds as well as a negatively charged phosphate attached to a polar or charged group. The charges and polar bonds in the head region interact with water molecules when a phospholipid is placed in solution. In contrast, the long hydrocarbon tails of a phospholipid are nonpolar and hydrophobic. Water molecules cannot form hydrogen bonds with the hydrocarbon tail, so they do not interact extensively with this part of the molecule.

Compounds that contain both hydrophilic and hydrophobic elements are **amphipathic** (literally, "dual-sympathy"). Phospholipids are amphipathic. As Figure 6.4a shows, cholesterol is also amphipathic. Because it has a hydroxyl functional group attached to its rings, it has both hydrophilic and hydrophobic regions. ✔ If you understand these concepts, you should be able to look back at Figure 6.1b and explain why fatty acids are also amphipathic.

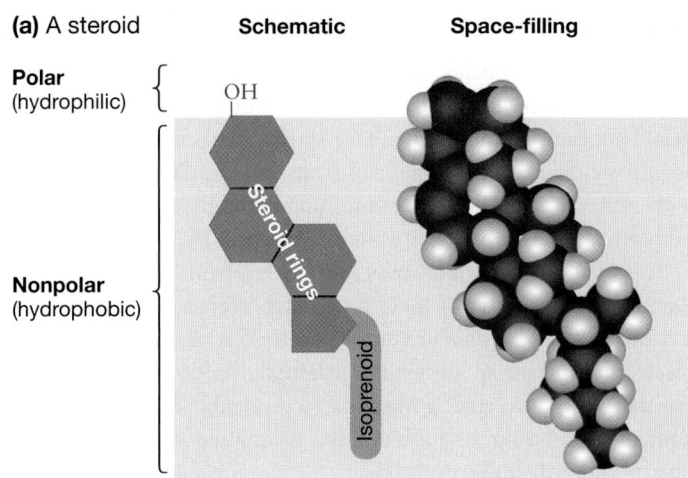

(a) A steroid

Schematic   Space-filling

Polar (hydrophilic)

OH

Steroid rings

Nonpolar (hydrophobic)

Isoprenoid

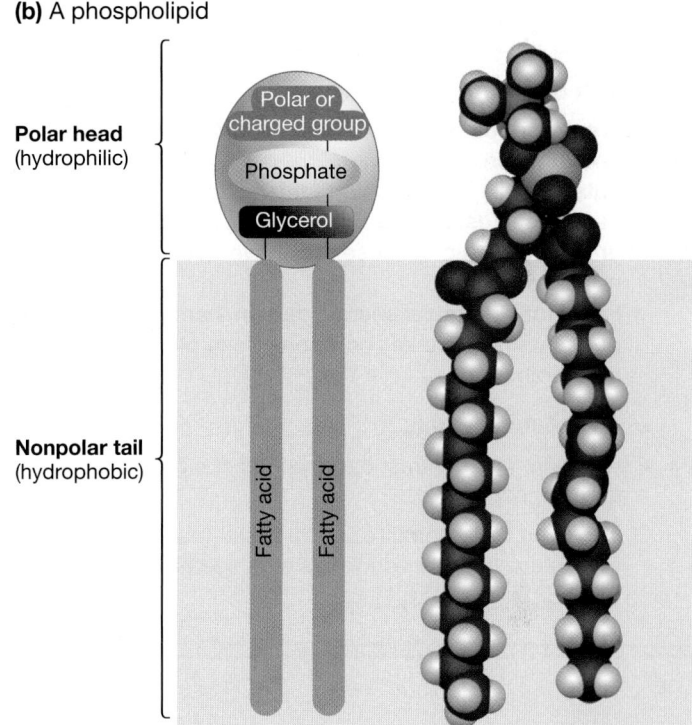

(b) A phospholipid

Polar head (hydrophilic)

Polar or charged group

Phosphate

Glycerol

Nonpolar tail (hydrophobic)

Fatty acid   Fatty acid

FIGURE 6.4 Some Lipids Contain Hydrophilic and Hydrophobic Regions. (a) All steroids have the distinctive four-ring structure shown in orange. Cholesterol has a polar hydroxyl group and an isoprenoid chain attached to these rings. **(b)** Most phospholipids consist of two fatty acid or isoprenoid chains that are linked to glycerol, which is linked to a phosphate group, which is linked to a small organic molecule that is polar or charged.

✔**QUESTION** If cholesterol and phospholipids were in solution, which part of the molecules would interact with water molecules?

The amphipathic nature of phospholipids is far and away their most important feature biologically. It is responsible for life's defining barrier—the plasma membrane. If the membrane defines life, then amphipathic lipids must have existed when life first originated during chemical evolution. Was that possible?

Were Lipids Present during Chemical Evolution?

Like amino acids, nucleic acids, and carbohydrates (Chapters 3–5), there is evidence that lipids were present during chemical evolution. Laboratory experiments have shown that simple lipids, such as fatty acids, can be synthesized from H_2 and CO_2 via reactions with mineral catalysts under conditions thought to be present in prebiotic hydrothermal vent systems (Chapter 2).

It is also possible that lipids literally fell from the sky early in Earth's history. Modern meteorites have been found to contain not only amino acids and carbohydrates but also lipids that exhibit amphipathic qualities. For example, lipids extracted from the meteorite that struck Murchison, Australia, in 1969 spontaneously formed lipid "bubbles" that resembled small cells. Why do amphipathic lipids do this?

check your understanding

If you understand that . . .

- Fats, steroids, and phospholipids differ in structure and function.
- Fats and oils are nonpolar; fatty acids, phospholipids, and certain steroids, like cholesterol, are amphipathic because they have both polar and nonpolar regions.
- Fats store chemical energy; certain steroids and phospholipids are key components of plasma membranes.

✔ You should be able to . . .

1. Compare and contrast the structure of a fat, a steroid, and a phospholipid.
2. Based on their structure, explain what makes cholesterol and phospholipids amphipathic.

Answers are available in Appendix A.

6.2 Phospholipid Bilayers

Amphipathic lipids do not dissolve when they are placed in water. Their hydrophilic heads interact with water, but their hydrophobic tails do not. Instead of dissolving in water, then, amphipathic lipids assume one of two types of structures: micelles or lipid bilayers.

- Micelles (**FIGURE 6.5a**) are tiny droplets created when the hydrophilic heads of a set of lipids face the water and form hydrogen bonds, while the hydrophobic tails interact with each other in the interior, away from the water.

- A **lipid bilayer** is created when two sheets of lipid molecules align. As **FIGURE 6.5b** shows, the hydrophilic heads in each layer face the surrounding solution while the hydrophobic tails face one another inside the bilayer. In this way, the hydrophilic heads interact with water while the hydrophobic tails interact with one another.

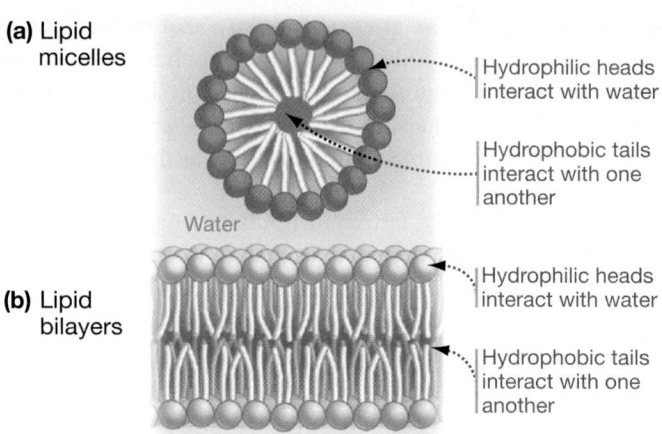

(a) Lipid micelles

Hydrophilic heads interact with water

Hydrophobic tails interact with one another

Water

(b) Lipid bilayers

Hydrophilic heads interact with water

Hydrophobic tails interact with one another

FIGURE 6.5 Lipids Form Micelles and Bilayers in Solution. In **(a)** a micelle or **(b)** a lipid bilayer, the hydrophilic heads of lipids face out, toward water; the hydrophobic tails face in, away from water. Lipid bilayers are the foundation of plasma membranes.

Micelles tend to form from fatty acids or other simple amphipathic hydrocarbon chains. Bilayers tend to form from phospholipids that contain two hydrocarbon tails. For this reason, bilayers are often called phospholipid bilayers.

It's critical to recognize that micelles and phospholipid bilayers form spontaneously—no input of energy is required. This concept can be difficult to grasp because entropy clearly decreases when these structures form. The key is to recognize that micelles and lipid bilayers are much more stable energetically than are independent phospholipids in solution.

Independent lipids are unstable in water because their hydrophobic tails disrupt hydrogen bonds that could otherwise form between water molecules. As a result, the tails of amphipathic molecules are forced together and participate in hydrophobic interactions (introduced in Chapter 2). This point should also remind you of the aqueous behavior of hydrophobic side chains in proteins and bases in nucleic acids.

Artificial Membranes as an Experimental System

When phospholipids are added to an aqueous solution and agitated, lipid bilayers spontaneously form small spherical structures. The hydrophilic heads on both sides of the bilayer remain in contact with the aqueous solution—water is present both inside and outside the vesicle. Artificial membrane-bound vesicles like these are called liposomes (**FIGURE 6.6**).

To explore how membranes work, researchers began creating and experimenting with liposomes and planar bilayers—lipid bilayers constructed across a hole in a glass or plastic wall separating two aqueous solutions (**FIGURE 6.7a**). Some of the first questions they posed concerned the permeability of lipid bilayers. The **permeability** of a structure is its tendency to allow a given substance to pass through it.

Using liposomes and planar bilayers, researchers can study what happens when a known ion or molecule is added to one side of a lipid bilayer (**FIGURE 6.7b**). Does the substance cross the

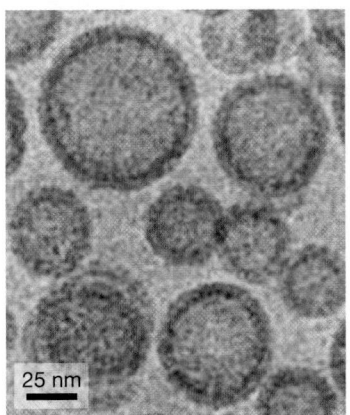
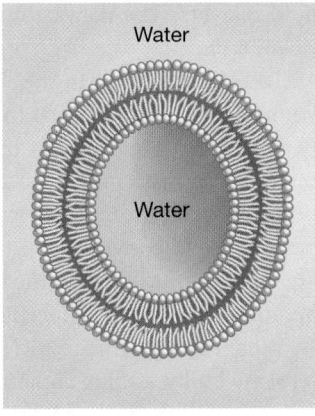

FIGURE 6.6 Liposomes Are Artificial Membrane-Bound Vesicles. Electron micrograph of liposomes in cross section (left) and a cross-sectional diagram of the lipid bilayer in a liposome (right).

(a) Planar bilayers: Artificial membranes

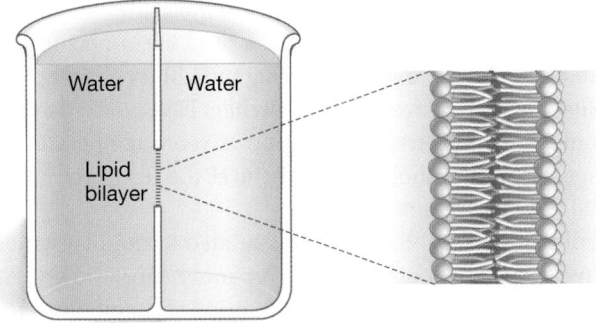

(b) Artificial-membrane experiments

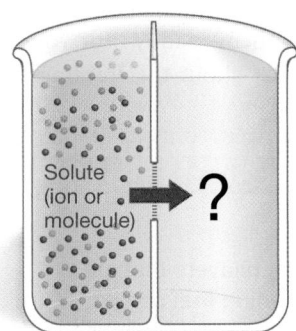

How rapidly can different solutes cross the membrane (if at all) when …

1. Different types of phospholipids are used to make the membrane?

2. Proteins or other molecules are added to the membrane?

FIGURE 6.7 Use of Planar Bilayers in Experiments. (a) The construction of a planar bilayer across a hole in a wall separating two water-filled compartments. **(b)** A wide variety of experiments are possible with planar bilayers; a few are suggested here.

membrane and show up on the other side? If so, how rapidly does the movement take place? What happens when a different type of phospholipid is used to make the artificial membrane? Does the membrane's permeability change when proteins or other types of molecules become part of it?

Biologists describe such an experimental system as elegant and powerful because it gives them precise control over which factor changes from one experimental treatment to the next.

Control, in turn, is why experiments are such an effective way to explore scientific questions. Recall that good experimental design allows researchers to alter one factor at a time and determine what effect, if any, each has on the process being studied (Chapter 1).

Selective Permeability of Lipid Bilayers

When researchers put molecules or ions on one side of a liposome or planar bilayer and measure the rate at which the molecules arrive on the other side, a clear pattern emerges: Lipid bilayers are highly selective.

Selective permeability means that some substances cross a membrane more easily than other substances do. Small nonpolar molecules move across bilayers quickly. In contrast, large molecules and charged substances cross the membrane slowly, if at all. This difference in membrane permeability is a critical issue because controlling what passes between the exterior and interior environments is a key characteristic of cells.

According to the data in **FIGURE 6.8**, small nonpolar molecules such as oxygen (O_2) move across selectively permeable membranes more than a billion times faster than do chloride ions (Cl^-). In essence, ions cannot cross membranes at all—unless they have "help" in the form of membrane proteins introduced later in the chapter. Very small and uncharged molecules such as water (H_2O) can cross membranes relatively rapidly, even if they are polar. Small polar molecules such as glycerol have intermediate permeability.

The leading hypothesis to explain this pattern is that charged compounds and large polar molecules are more stable dissolved in water than they are in the nonpolar interior of membranes. ✔ If you understand this hypothesis, you should be able to predict where amino acids and nucleotides would be placed in Figure 6.8 and explain your reasoning.

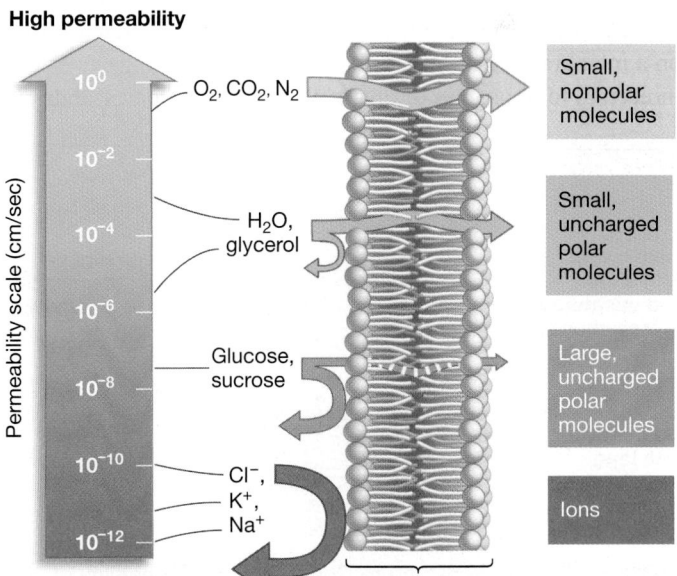

FIGURE 6.8 Lipid Bilayers Show Selective Permeability. Only certain substances cross lipid bilayers readily. Size and polarity or charge affect the rate of diffusion across a membrane.

How Does Lipid Structure Affect Membrane Permeability?

The amphipathic nature of phospholipids allows them to spontaneously form membranes. But not all phospholipid bilayers are the same. The nature of the hydrocarbon tails, in addition to the presence of cholesterol molecules, profoundly influences how a membrane behaves.

Bond Saturation and Hydrocarbon Chain Length Affect Membrane Fluidity and Permeability The degree of saturation in a phospholipid—along with the length of its hydrocarbon tails—affects key aspects of a lipid's behavior in a membrane.

- When unsaturated hydrocarbon tails are packed into a lipid bilayer, kinks created by double bonds produce spaces among the tails. These spaces reduce the strength of the van der Waals interactions (see Chapter 3) that hold the hydrophobic tails together, weakening the barrier to solutes.

- Packed saturated hydrocarbon tails have fewer spaces and stronger van der Waals interactions. As the length of saturated hydrocarbon tails increases, the forces that hold them together also grow stronger, making the membrane even denser.

These observations have profound impacts on membrane fluidity and permeability—two closely related properties. As **FIGURE 6.9** shows, lipid bilayers are more permeable as well as more fluid when they contain short, kinked, unsaturated hydrocarbon tails. An unsaturated membrane allows more materials to pass because its interior is held together less tightly. Bilayers containing long, straight, saturated hydrocarbon tails are much less permeable and fluid. Experiments on liposomes have shown exactly these patterns.

Cholesterol Reduces Membrane Permeability Cholesterol molecules are present, to varying extents, in the membranes of every cell in your body. What effect does adding cholesterol have on a membrane? Researchers have found that adding cholesterol molecules to liposomes dramatically reduces the permeability of

lipid bilayers. The data behind this conclusion are presented in **FIGURE 6.10**.

To read the graph in the "Results" section of Figure 6.10, put your finger on the x-axis at the point marked 20°C, and note that permeability to glycerol is much higher at this temperature in membranes that contain no cholesterol versus 20 percent or 50 percent cholesterol. Using this procedure at other temperature points should convince you that membranes lacking cholesterol are more permeable than the other two membranes at every temperature tested in the experiment.

What explains this result? Because the steroid rings in cholesterol are bulky, adding cholesterol fills gaps that would otherwise be present in the hydrophobic section of the membrane.

How Does Temperature Affect the Fluidity and Permeability of Membranes?

At about 25°C—or "room temperature"—the phospholipids in a plasma membrane have a consistency resembling olive oil. This fluid physical state allows individual lipid molecules to move laterally within each layer (**FIGURE 6.11**), a little like a person moving about in a dense crowd. By tagging individual phospholipids and following their movement, researchers have clocked average speeds of 2 micrometers (μm)/second at room temperature. At these speeds, a phospholipid could travel the length of a small bacterial cell in a second.

Recall that permeability is closely related to fluidity. As temperature drops, molecules in a bilayer move more slowly. As a result, the hydrophobic tails in the interior of membranes pack together more tightly. At very low temperatures, lipid bilayers even begin to solidify. As the graph in Figure 6.10 indicates, low temperatures can make membranes impervious to molecules that would normally cross them readily. Put your finger on the x-axis of that graph, just about the freezing point

Lipid bilayer with **short** and **unsaturated** hydrocarbon tails

Higher permeability and fluidity

Lipid bilayer with **long** and **saturated** hydrocarbon tails

Lower permeability and fluidity

FIGURE 6.9 Fatty Acid Structure Changes the Permeability of Membranes. Lipid bilayers consisting of phospholipids containing unsaturated fatty acids should have more gaps and be more permeable than those with saturated fatty acids.

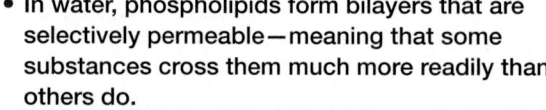

QUESTION: Does adding cholesterol to a membrane affect its permeability?

HYPOTHESIS: Cholesterol reduces permeability because it fills spaces in phospholipid bilayers.

NULL HYPOTHESIS: Cholesterol has no effect on permeability.

EXPERIMENTAL SETUP:

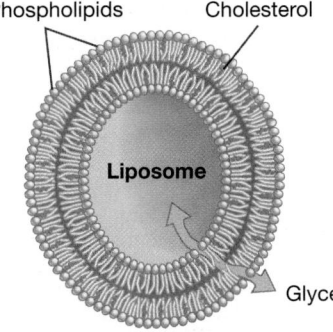

Phospholipids   Cholesterol

Liposome

Glycerol

1. Construct liposomes: Create with no cholesterol, 20% cholesterol, and 50% cholesterol.

2. Measure glycerol movement: Record how quickly glycerol moves across each type of membrane at different temperatures.

PREDICTION: Liposomes with higher cholesterol levels will have reduced permeability.

PREDICTION OF NULL HYPOTHESIS: All liposomes will have the same permeability.

RESULTS:

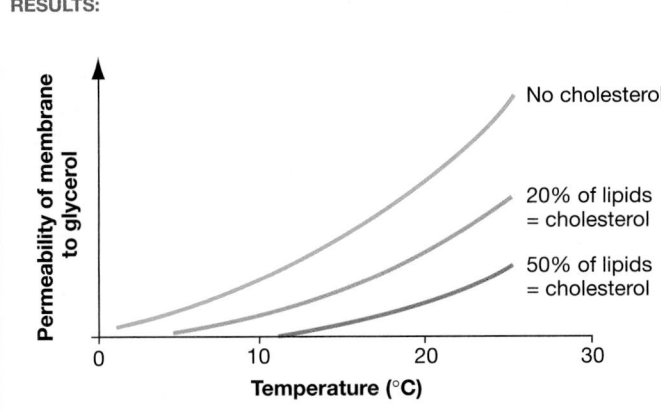

No cholesterol

20% of lipids = cholesterol

50% of lipids = cholesterol

Permeability of membrane to glycerol

Temperature (°C)

CONCLUSION: Adding cholesterol to membranes decreases their permeability to glycerol. The permeability of all membranes analyzed in this experiment increases with increasing temperature.

FIGURE 6.10 The Permeability of a Membrane Depends on Its Composition.

SOURCE: de Gier, J., et al. (1968). Lipid composition and permeability of liposomes. *Biochimica et Biophysica Acta* 150: 666–675.

✔**QUANTITATIVE** Suppose the investigators had instead created liposomes using phospholipids with fully saturated tails and compared them to two other sets of liposomes where either 20 percent or 50 percent of the phospholipids contained polyunsaturated tails. Label the three lines on the graph above with your prediction for the three different liposomes in this new experiment.

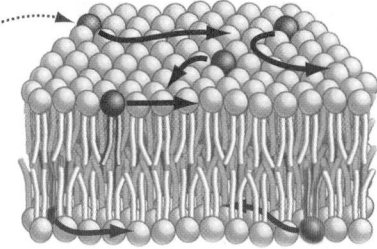

Phospholipids are in constant lateral motion, but rarely flip to the other side of the bilayer

FIGURE 6.11 Phospholipids Move within Membranes. Membranes are dynamic—in part because phospholipid molecules randomly move laterally within each layer in the structure.

of water (0°C), and note that even membranes that lack cholesterol are almost completely impermeable to glycerol. Indeed, trace any of the three lines in Figure 6.10, and as you move to the right (increasing temperature), you also move up (increasing permeability).

These observations on glycerol and lipid movement demonstrate that membranes are dynamic. Phospholipid molecules whiz around each layer, while water and small nonpolar molecules shoot in and out of the membrane. How quickly molecules move within and across membranes is a function of temperature, the structure of hydrocarbon tails, and the number of cholesterol molecules in the bilayer.

6.3 How Molecules Move across Lipid Bilayers: Diffusion and Osmosis

Small uncharged molecules and hydrophobic compounds can cross membranes readily and spontaneously—without an input of energy. The question now is: How is this possible? What process is responsible for movement of molecules across lipid bilayers?

Diffusion

A thought experiment can help explain how substances can cross membranes spontaneously. Suppose you rack up a set of billiard balls in the middle of a pool table and then begin to vibrate the table.

1. Because of the vibration, the billiard balls will move about randomly. They will also bump into one another.

2. After these collisions, some balls will move outward—away from their original position.

3. As movement and collisions continue, the overall or net movement of balls will be outward. This occurs because the random motion of the balls disrupts their original, nonrandom position. As the balls move at random, they are more likely to move away from one another than to stay together.

4. Eventually, the balls will be distributed randomly across the table. The entropy of the billiard balls has increased. Recall that entropy is a measure of the randomness or disorder in

a system (Chapter 2). The second law of thermodynamics states that in an isolated system, entropy always increases.

This hypothetical example illustrates how vibrating billiard balls move at random. More to the point, it also explains how substances located on one side of a lipid bilayer can move to the other side spontaneously. All dissolved molecules and ions, or **solutes,** have thermal energy and are in constant, random motion. Movement of molecules and ions that results from their kinetic energy is known as **diffusion.**

A difference in solute concentrations creates what is called a **concentration gradient.** Solutes move randomly in all directions, but when a concentration gradient exists, there is a net movement from regions of high concentration to regions of low concentration. Diffusion down a concentration gradient, or away from the higher concentration, is a spontaneous process because it results in an increase in entropy.

Once the molecules or ions are randomly distributed throughout a solution, a chemical equilibrium is established. For example, consider two aqueous solutions separated by a lipid bilayer. **FIGURE 6.12** shows how molecules that can pass through the bilayer diffuse to the other side. At equilibrium, these molecules continue to move back and forth across the membrane, but at equal rates—simply because they are equally likely to move in any direction. This means that there is no longer a net movement of molecules across the membrane. ✔ If you understand diffusion, you should be able to predict how a difference in temperature across a membrane would affect the concentration of a solute at equilibrium.

Osmosis

What about water? As the data in Figure 6.8 show, water moves across lipid bilayers relatively quickly. The movement of water is a special case of diffusion that is given its own name: **osmosis.** Osmosis occurs only when solutions are separated by a membrane that permits water to cross, but holds back some or all of the solutes—that is, a selectively permeable membrane.

It's important to note that some of the water molecules in a solution are unavailable to diffuse across the membrane. Recall that solutes form ionic or hydrogen bonds with water molecules (Chapter 2). Water molecules that are bound to a solute that can't cross the membrane are themselves prevented from crossing.

Only unbound water molecules are able to diffuse across the membrane during osmosis. When these unbound water molecules move across a membrane, they flow from the solution with the lower solute concentration into the solution with the higher solute concentration.

To drive this point home, let's suppose the concentration of a particular solute is higher on one side of a selectively permeable membrane than it is on the other side (**FIGURE 6.13**, step 1). Further, suppose that this solute cannot diffuse through the membrane to establish equilibrium. What happens? Water will move from the side with a lower concentration of solute to the side with a higher concentration of solute (Figure 6.13, step 2). Osmosis dilutes the higher concentration and equalizes the concentrations

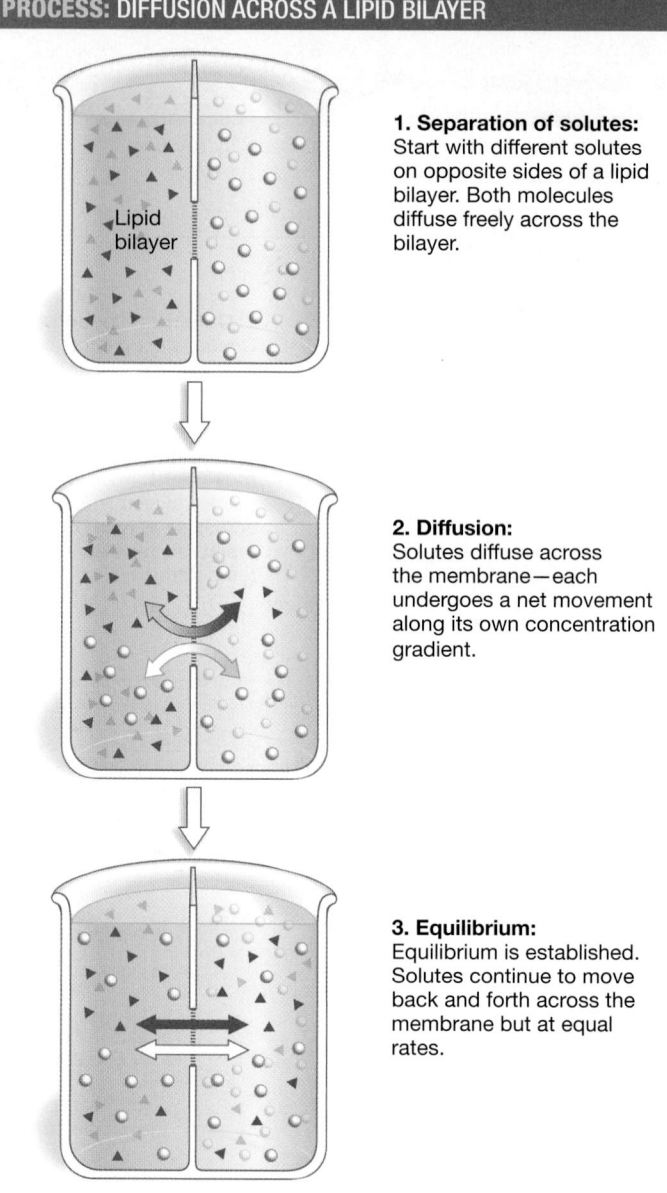

PROCESS: DIFFUSION ACROSS A LIPID BILAYER

Lipid bilayer

1. Separation of solutes: Start with different solutes on opposite sides of a lipid bilayer. Both molecules diffuse freely across the bilayer.

2. Diffusion: Solutes diffuse across the membrane—each undergoes a net movement along its own concentration gradient.

3. Equilibrium: Equilibrium is established. Solutes continue to move back and forth across the membrane but at equal rates.

FIGURE 6.12 Diffusion across a Selectively Permeable Membrane Establishes an Equilibrium.

on both sides. The movement of water is spontaneous. It is driven by the increase in entropy achieved when solute concentrations are equal on both sides of the membrane.

Movement of water by osmosis is important because it can swell or shrink a membrane-bound vesicle. Consider the liposomes illustrated in **FIGURE 6.14.** (Remember that osmosis occurs only when a solute cannot pass through a separating membrane.)

- *Left* If the solution inside the membrane has a lower concentration of solutes than the exterior has, water moves out of the vesicle into the solution outside. The solution inside is said to be **hypotonic** ("lower-tone") relative to the outside of the vesicle. As water leaves, the vesicle shrinks and the membrane shrivels, resulting in lower vesicle firmness.

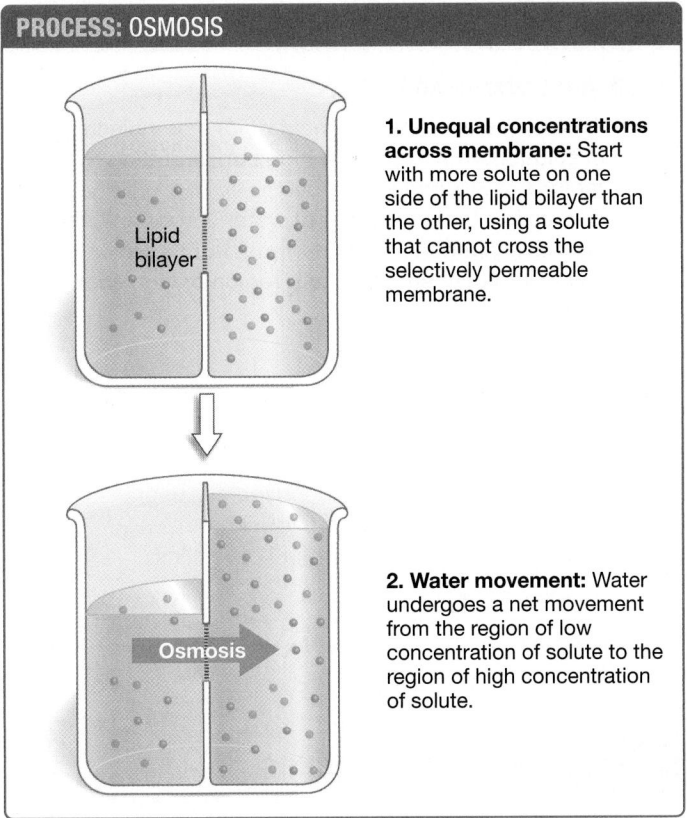

1. Unequal concentrations across membrane: Start with more solute on one side of the lipid bilayer than the other, using a solute that cannot cross the selectively permeable membrane.

Lipid bilayer

Osmosis

2. Water movement: Water undergoes a net movement from the region of low concentration of solute to the region of high concentration of solute.

FIGURE 6.13 Osmosis Is the Diffusion of Water.

✔**QUESTION** Suppose you doubled the number of solute molecules on the left side of the membrane (at the start). At equilibrium, would the water level on the left side be higher or lower than what is shown in the second drawing?

• *Middle* If the solution inside the membrane has a higher concentration of solutes than the exterior has, water moves into the vesicle via osmosis. The inside solution is said to be **hypertonic** ("excess-tone") relative to the outside of the vesicle. The incoming water causes the vesicle to swell and increase in firmness, or even burst.

• *Right* If solute concentrations are equal on both sides of the membrane, the liposome maintains its size. When the inside solution does not affect the membrane's shape, that solution is called **isotonic** ("equal-tone").

Note that the terms hypertonic, hypotonic, and isotonic are relative—they can be used only to express the relationship between a given solution and another solution separated by a membrane. Biologists also commonly use these terms to describe the solution that is exterior to the cells or vesicles.

Membranes and Chemical Evolution

What do diffusion and osmosis have to do with the first membranes floating in the prebiotic soup? Both processes tend to *reduce* differences in chemical composition between the inside and outside of membrane-bound compartments.

If liposome-like structures first arose in the oceans of early Earth, their interiors probably didn't offer a radically different environment from the surrounding solution. In all likelihood, the primary importance of the first lipid bilayers was simply to provide a container for replicating RNA, the macromolecule most likely to have been the first "living" molecule (see Chapter 4). But ribonucleotide monomers would need to be available for these

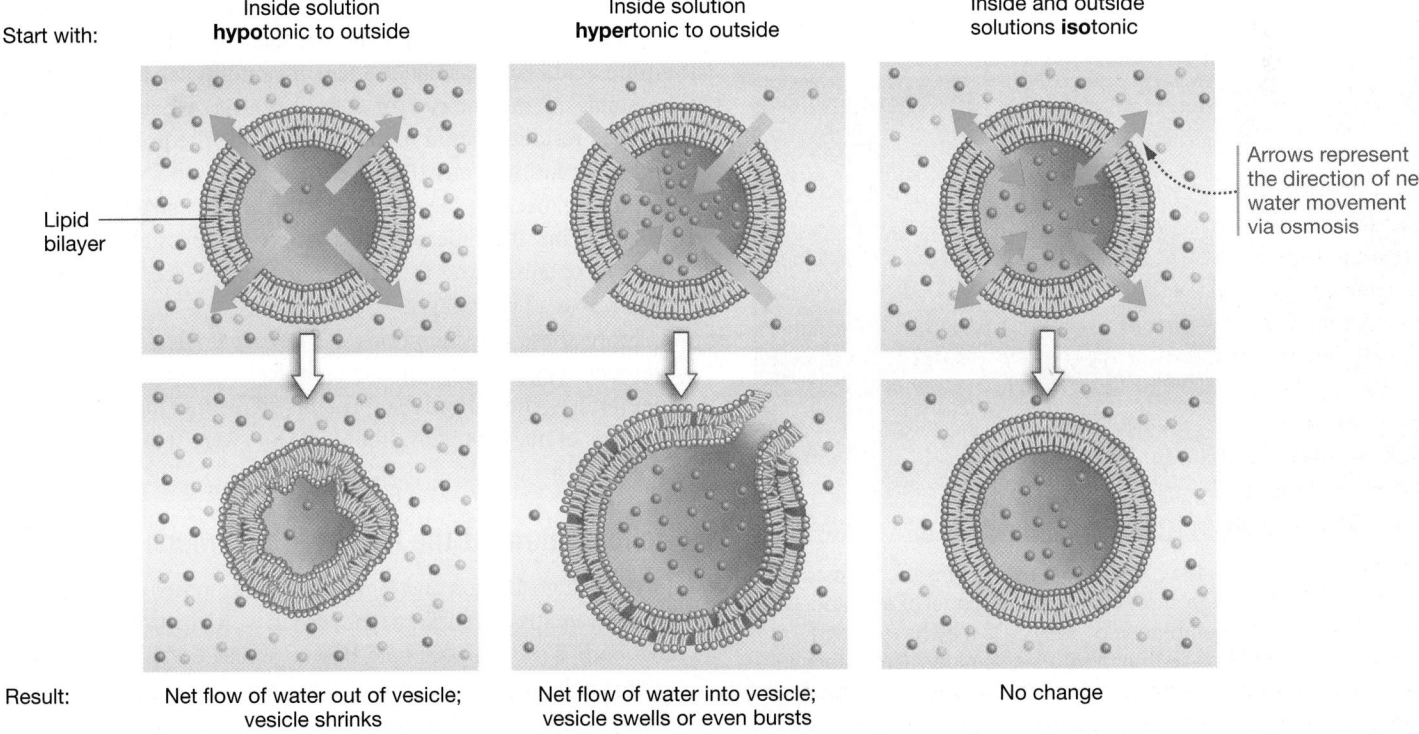

Start with:

Inside solution **hypo**tonic to outside

Inside solution **hyper**tonic to outside

Inside and outside solutions **iso**tonic

Lipid bilayer

Arrows represent the direction of net water movement via osmosis

Result: Net flow of water out of vesicle; vesicle shrinks

Net flow of water into vesicle; vesicle swells or even bursts

No change

FIGURE 6.14 Osmosis Can Shrink or Burst Membrane-Bound Vesicles.

RNAs to replicate. Can negatively charged ribonucleotides get across lipid bilayers and inside lipid-bounded vesicles?

The answer is yes. Jack Szostak and colleagues first set out to study the permeability of membranes consisting of fatty acids and other simple amphipathic lipids thought to be present in the early oceans. Like phospholipids, fatty acids will spontaneously assemble into lipid bilayers and water-filled vesicles. Their experiments showed that ions, and even ribonucleotides, can diffuse across the fatty acid vesicle membranes—meaning that monomers could have been available for RNA synthesis.

Lending support to this hypothesis, the same minerals found to catalyze the polymerization of RNA from activated nucleotides (see Chapter 4) will also promote the formation of fatty acid vesicles—and in the process, often incorporate themselves and RNA inside. Simple vesicle-like structures that harbor nucleic acids are referred to as **protocells** (**FIGURE 6.15**). Most origin-of-life researchers view protocells as possible intermediates in the evolution of the cell.

Laboratory simulations also showed that free lipids and micelles can become incorporated into fatty acid bilayers, causing protocells to grow. Shearing forces, as from bubbling, shaking, or wave action, cause protocells to divide. Based on these observations, it is reasonable to hypothesize that once replicating RNAs became surrounded by a lipid bilayer, this simple life-form and its descendants would occupy cell-like structures that grew and divided.

Now let's investigate the next great innovation in the evolution of the cell: the ability to create and maintain a specialized internal environment that is conducive to life. What is necessary to construct an effective plasma membrane—one that imports ions and molecules needed for life while excluding ions and molecules that might damage it?

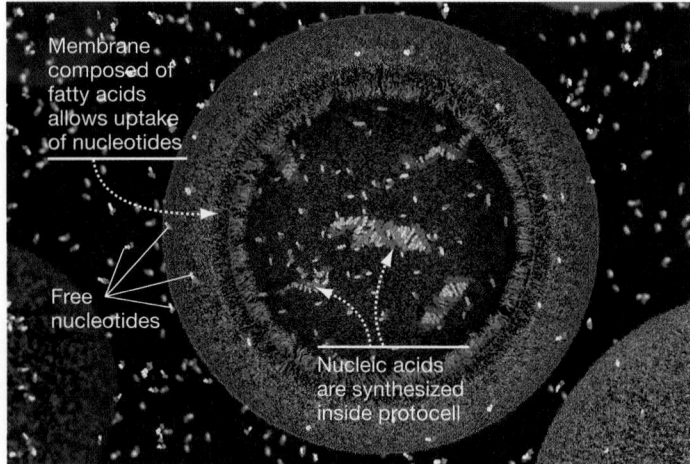

FIGURE 6.15 Protocells May Have Possessed Simple, Permeable Membranes. This image shows a computer model of a protocell. Like this model, the membranes of early cells may have been built of fatty acids. Passive transport of nucleotides across these membranes, as well as replication of nucleic acids inside, has been observed in the laboratory.

If you understand that . . .

- Diffusion is the net movement of ions or molecules in solution from regions of high concentration to regions of low concentration.
- Osmosis is the movement of water across a selectively permeable membrane, from a region of low solute concentration to a region of high solute concentration.

✔ You should be able to . . .

Make a concept map (see **BioSkills 15** in Appendix B) that includes the boxed terms water molecules, solute molecules, osmosis, diffusion, areas of high-to-low concentration, selectively permeable membranes, concentration gradients, hypertonic solutions, hypotonic solutions, and isotonic solutions.

Answers are available in Appendix A.

6.4 Membrane Proteins

What sort of molecule could become incorporated into a lipid bilayer and affect the bilayer's permeability? The title of this section gives the answer away. Proteins that are amphipathic can be inserted into lipid bilayers.

Proteins can be amphipathic because their monomers, amino acids, have side chains that range from highly nonpolar to highly polar or charged (see Figure 3.2). It's conceivable, then, that a protein could have a series of nonpolar amino acid residues in the middle of its primary structure flanked by polar or charged amino acid residues (**FIGURE 6.16a**). The nonpolar residues would be stable in the interior of a lipid bilayer, while the polar or charged residues would be stable alongside the polar lipid heads and surrounding water (**FIGURE 6.16b**).

Further, because the secondary and tertiary structures of proteins are almost limitless in their variety, it is possible for proteins to form openings and thus function as some sort of channel or pore across a lipid bilayer.

From these considerations, it's not surprising that when researchers began analyzing the chemical composition of plasma membranes, they found that proteins were often just as common, in terms of mass, as phospholipids. How were these two types of molecules arranged?

Development of the Fluid-Mosaic Model

In 1935 Hugh Davson and James Danielli proposed that cell membranes were structured like a sandwich in which hydrophilic proteins coat both sides of a pure lipid bilayer (**FIGURE 6.17a**). Early electron micrographs of plasma membranes seemed to be consistent with the sandwich model, and for decades it was widely accepted.

(a) Proteins can be amphipathic.

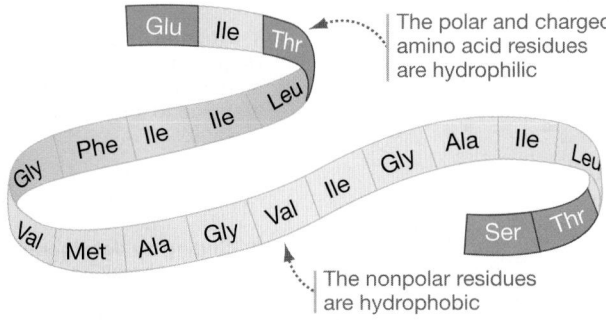

The polar and charged amino acid residues are hydrophilic

The nonpolar residues are hydrophobic

(b) Amphipathic proteins can integrate into lipid bilayers.

Outside cell

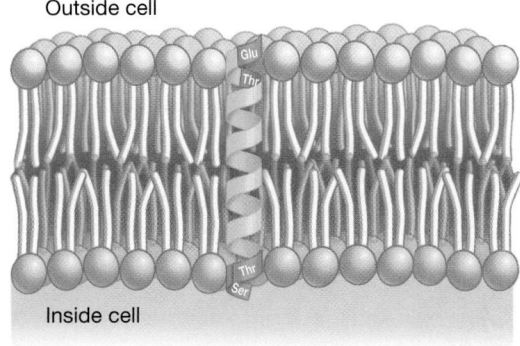

Inside cell

FIGURE 6.16 Amphipathic Proteins Are Anchored in Lipid Bilayers.

(a) Sandwich model

Cell exterior

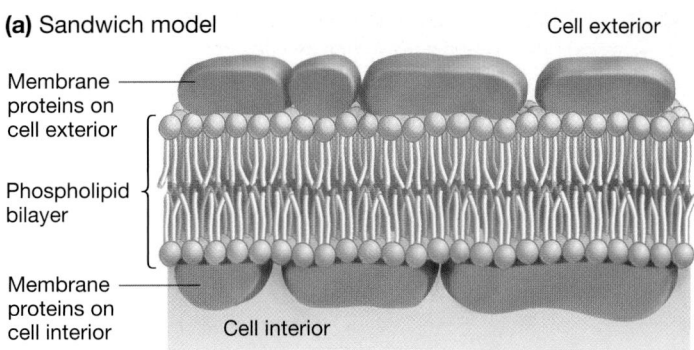

Membrane proteins on cell exterior

Phospholipid bilayer

Membrane proteins on cell interior

Cell interior

(b) Fluid-mosaic model

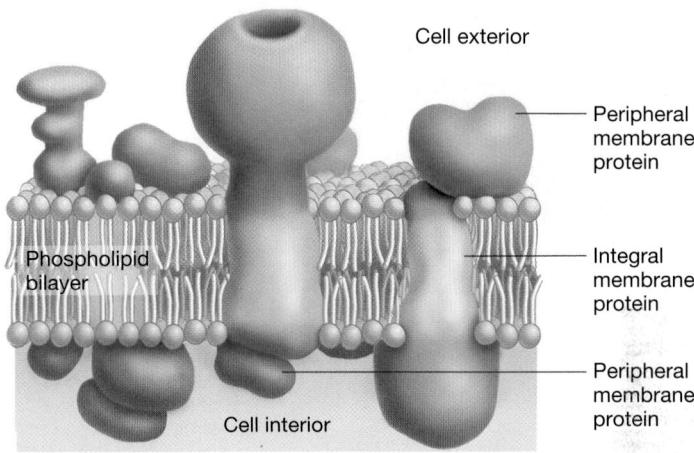

Cell exterior

Peripheral membrane protein

Phospholipid bilayer

Integral membrane protein

Peripheral membrane protein

Cell interior

FIGURE 6.17 Past and Current Models of Membrane Structure Differ in Where Membrane Proteins Reside. (a) The protein-lipid-lipid-protein sandwich model was the first hypothesis for the arrangement of lipids and proteins in cell membranes. **(b)** The fluid-mosaic model was a radical departure from the sandwich hypothesis.

The realization that membrane proteins could be amphipathic, however, led S. Jon Singer and Garth Nicolson to suggest an alternative hypothesis. In 1972, they proposed that at least some proteins span the membrane instead of being found only outside the lipid bilayer. Their hypothesis was called the **fluid-mosaic model** (**FIGURE 6.17b**). Singer and Nicolson suggested that membranes are a mosaic of phospholipids and different types of proteins. The overall structure was proposed to be dynamic and fluid.

The controversy over the nature of the cell membrane was resolved in the early 1970s with the development of an innovative technique for visualizing the surface of plasma membranes. The method is called freeze-fracture electron microscopy because the steps involve freezing and fracturing the membrane before examining it with a **scanning electron microscope (SEM)**, which produces images of an object's surface (see **BioSkills 11** in Appendix B).

As **FIGURE 6.18** (see page 96) shows, the freeze-fracture technique allows researchers to split cell membranes and view the middle of the structure. The scanning electron micrographs that result show pits and mounds studding the inner surfaces of the lipid bilayer. Researchers interpreted these structures as the locations of membrane proteins. As step 4 in the figure shows, the mounds represent proteins that remained attached to one side of the split lipid bilayer and the pits are the holes they left behind.

These observations conflicted with the sandwich model but were consistent with the fluid-mosaic model. Based on these and subsequent observations, the fluid-mosaic model is now widely accepted.

Notice in Figure 6.17b that some proteins span the membrane and have segments facing both the interior and the exterior surfaces. Proteins like these are called **integral membrane proteins, or transmembrane proteins.** Proteins that bind to the membrane without passing through it are called **peripheral membrane proteins.**

Certain peripheral proteins are found only on the interior surface of a cellular membrane, while others are found only on the exterior surface. As a result, the interior and exterior surfaces of the plasma membrane are distinct—the peripheral proteins and the ends of transmembrane proteins differ. Peripheral membrane proteins are often attached to transmembrane proteins.

How do these proteins affect the permeability of membranes? The answer to this question starts with an investigation of the structure of proteins involved in the transport of molecules and ions across the plasma membrane.

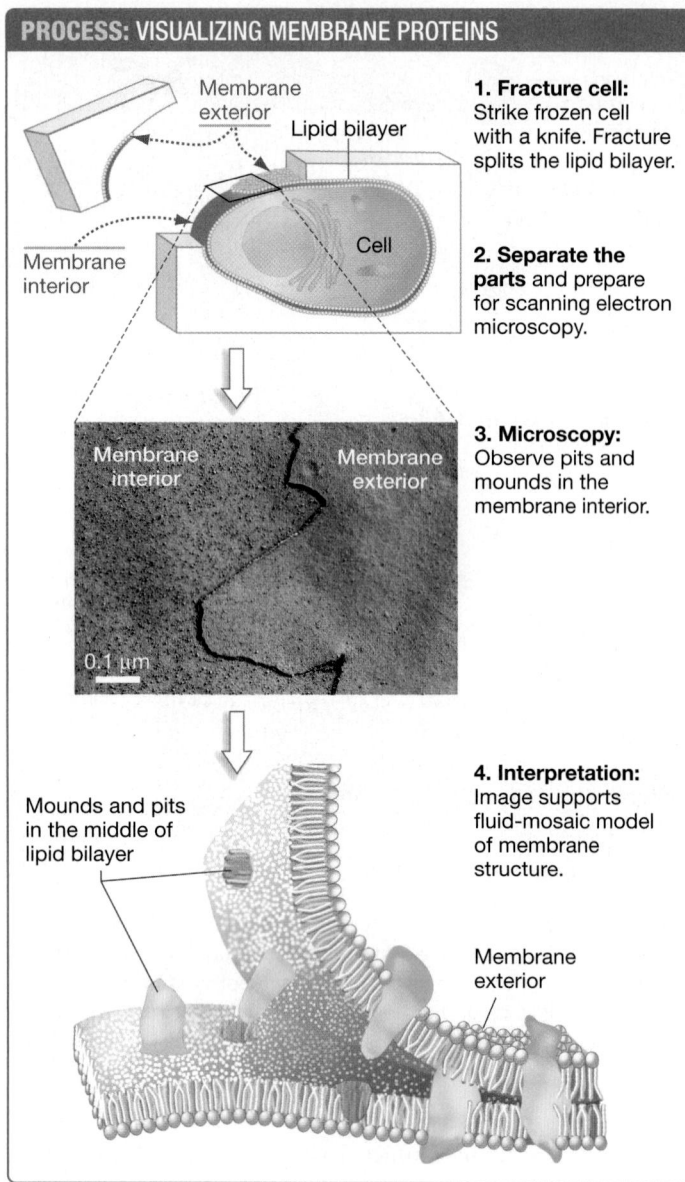

Membrane exterior

Lipid bilayer

Cell

Membrane interior

1. Fracture cell: Strike frozen cell with a knife. Fracture splits the lipid bilayer.

2. Separate the parts and prepare for scanning electron microscopy.

Membrane interior

Membrane exterior

0.1 μm

3. Microscopy: Observe pits and mounds in the membrane interior.

Mounds and pits in the middle of lipid bilayer

Membrane exterior

4. Interpretation: Image supports fluid-mosaic model of membrane structure.

FIGURE 6.18 Freeze-Fracture Preparations Allow Biologists to View Membrane Proteins.

✔ QUESTION What would be an appropriate control to show that the pits and mounds were not simply irregularities in the lipid bilayer caused by the freeze-fracture process?

Systems for Studying Membrane Proteins

The discovery of transmembrane proteins was consistent with the hypothesis that proteins affect membrane permeability. To test this hypothesis, researchers needed some way to isolate and purify membrane proteins.

FIGURE 6.19 outlines one method that researchers developed to separate proteins from membranes. The key to the technique is the use of detergents. A **detergent** is a small amphipathic molecule. When detergents are added to the solution surrounding a lipid bilayer, the hydrophobic tails of the detergent molecule interact with the hydrophobic tails of the lipids and with the

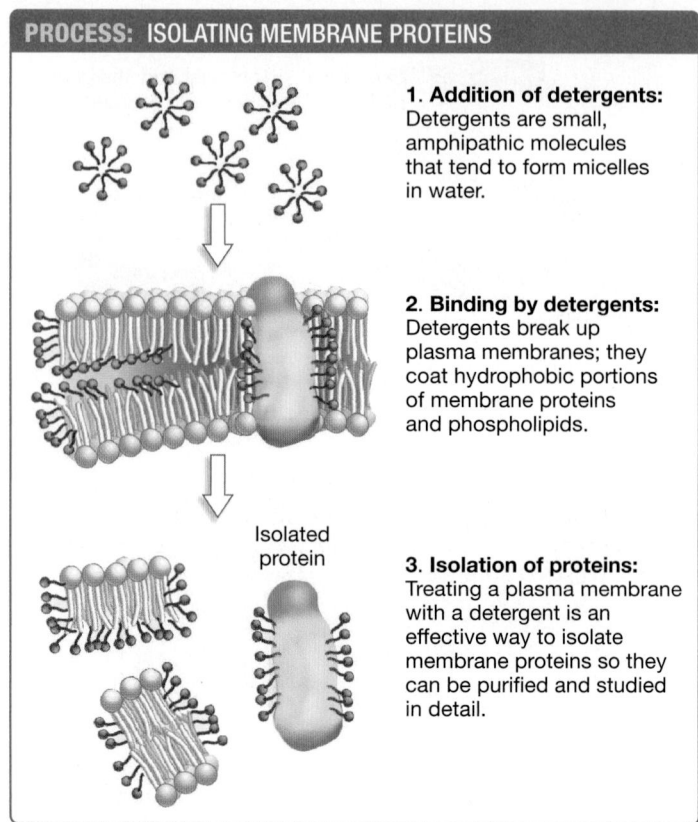

1. Addition of detergents: Detergents are small, amphipathic molecules that tend to form micelles in water.

2. Binding by detergents: Detergents break up plasma membranes; they coat hydrophobic portions of membrane proteins and phospholipids.

Isolated protein

3. Isolation of proteins: Treating a plasma membrane with a detergent is an effective way to isolate membrane proteins so they can be purified and studied in detail.

FIGURE 6.19 Detergents Can Be Used to Isolate Proteins from Membranes.

hydrophobic portions of transmembrane proteins. These interactions displace the membrane phospholipids and end up forming water-soluble detergent–protein complexes that can be isolated.

Since intensive experimentation on membrane proteins began, researchers have identified three broad classes of proteins that affect membrane permeability: channels, carriers, and pumps. Let's consider each class in turn.

Facilitated Diffusion via Channel Proteins

As the data in Figure 6.8 show, ions almost never cross pure phospholipid bilayers on their own. But in cells, ions routinely cross membranes through specialized membrane proteins called **ion channels.**

Ion channels form pores, or openings, in a membrane. Ions move through these pores in a predictable direction: from regions of high concentration to regions of low concentration and from areas of like charge to areas of unlike charge.

In **FIGURE 6.20**, for example, a large concentration gradient favors the movement of sodium ions from the outside of a membrane to the inside. But in addition, the inside of this cell has a net negative charge while the outside has a net positive charge. As a result, the combination of these two factors influences the final concentration of sodium ions inside the cell once equilibrium has been established.

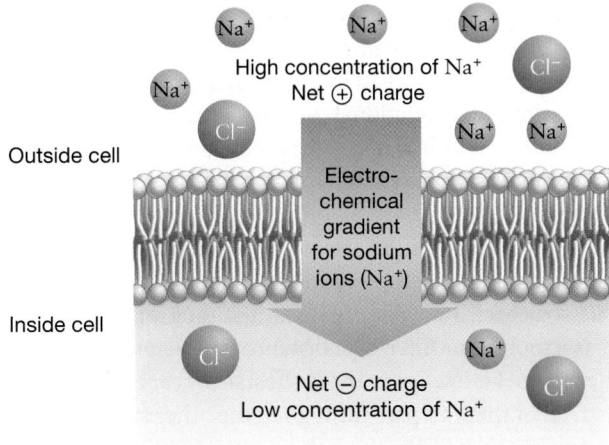

FIGURE 6.20 An Electrochemical Gradient Is a Combined Concentration and Electrical Gradient. Electrochemical gradients are established when ions build up on one side of a membrane.

Ions move in response to a combined concentration and electrical gradient, or what biologists call an **electrochemical gradient.** ✔ If you understand this concept, you should be able to add an arrow to Figure 6.20 indicating the electrochemical gradient for chloride ions.

Is an Ion Channel Involved in Cystic Fibrosis? To understand the types of experiments that biologists do to confirm that a membrane protein is an ion channel, consider work on the cause of cystic fibrosis.

Cystic fibrosis (CF) is the most common genetic disease in humans of Northern European descent. It affects cells that produce mucus, sweat, and digestive juices. Normally these secretions are thin and slippery and act as lubricants. In individuals with CF, however, the secretions become abnormally concentrated and sticky and clog passageways in organs like the lungs.

Experiments published in 1983 suggested that cystic fibrosis is caused by defects in a membrane protein that allow chloride ions (Cl⁻) to move across plasma membranes. It was proposed that reduced chloride ion transport would account for the thick mucus.

How is the transport of chloride ions involved in mucus consistency? Water movement across cell membranes is largely determined by the presence of extracellular ions like chloride. If a defective channel prevents chloride ions from leaving cells, water isn't pulled from cells by osmosis to maintain the proper mucus consistency. In effect, the disease results from the mismanagement of osmosis.

Using molecular techniques introduced in Unit 3 (see Chapter 20), biologists were able to **(1)** find the gene that is defective in people suffering from CF and **(2)** use the gene to produce copies of the normal protein, which was called CFTR (short for cystic fibrosis transmembrane conductance regulator).

Is CFTR a chloride channel? To answer this question, researchers inserted purified CFTR into planar bilayers and

RESEARCH

QUESTION: Is CFTR a chloride channel?

HYPOTHESIS: CFTR increases the flow of chloride ions across a membrane.

NULL HYPOTHESIS: CFTR has no effect on membrane permeability.

EXPERIMENTAL SETUP:

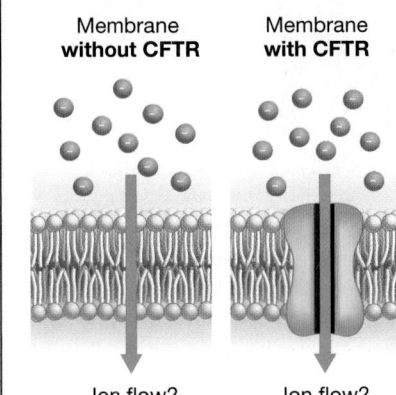

Membrane **without CFTR**   Membrane **with CFTR**

Ion flow?   Ion flow?

1. **Create planar bilayers** with and without CFTR.

2. **Add chloride ions** to one side of the planar bilayer to create an electrochemical gradient.

3. **Record electrical currents** to measure ion flow across the planar bilayers.

PREDICTION: Ion flow will be higher in membrane with CFTR.

PREDICTION OF NULL HYPOTHESIS: Ion flow will be the same in both membranes.

RESULTS:

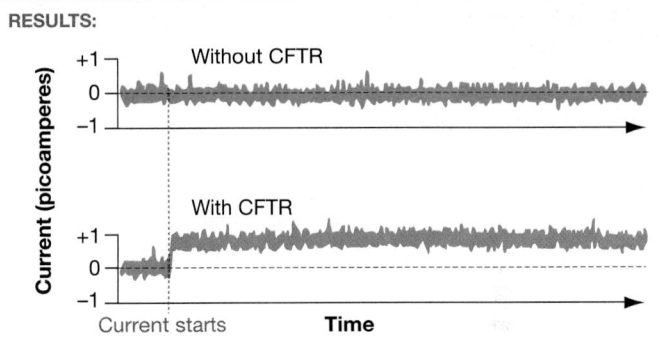

CONCLUSION: CFTR facilitates diffusion of chloride ions along an electrochemical gradient. CFTR is a chloride channel.

FIGURE 6.21 Electric Current Measurements Indicate that Chloride Flows through CFTR.

SOURCE: Bear, C. A., et al. (1992). Purification and functional reconstitution of the cystic fibrosis transmembrane conductance regulator (CFTR). *Cell* 68: 809–818.

✔**QUESTION** The researchers repeated the "with CFTR" treatment 45 times, but recorded a current in only 35 of the replicates. Does this observation negate the conclusion? Explain why or why not.

measured the flow of electric current across the membrane. Because ions carry a charge, ion movement across a membrane produces an electric current.

The graphs in **FIGURE 6.21**, which plot the amount of current flowing across the membrane over time, show the results from this experiment. Notice that when CFTR was absent, no electric current passed through the membrane. But when CFTR was

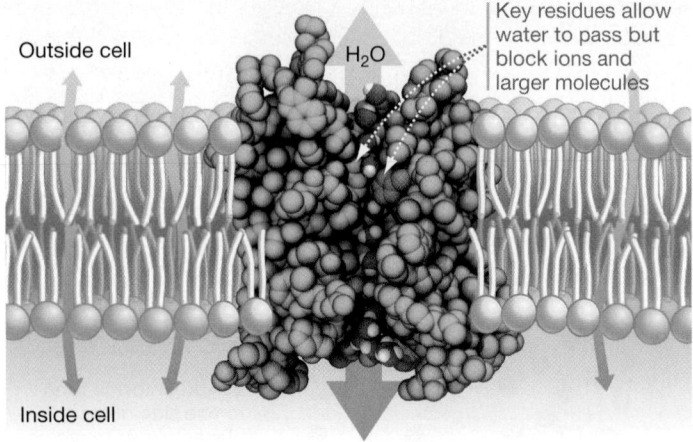

Outside cell

H₂O

Key residues allow water to pass but block ions and larger molecules

Inside cell

FIGURE 6.22 Membrane Channels Are Highly Selective. A cutaway view looking at the inside of a membrane channel, aquaporin. The key residues identified in the space-filling model selectively filter ions and other small molecules, allowing only water (red and white structures) to pass through.

inserted into the membrane, current began to flow. This was strong evidence that CFTR was indeed a chloride ion channel.

Protein Structure Determines Channel Selectivity Subsequent research has shown that cells have many different types of pore-like **channel proteins** in their membranes, including ion channels like CFTR. Channel proteins are selective. Each channel protein has a structure that permits only a particular type of ion or small molecule to pass through it.

For example, Peter Agre and co-workers discovered channels called **aquaporins** ("water-pores") that allow water to cross the plasma membrane over 10 times faster than it does in the absence of aquaporins. Aquaporins admit water but not other small molecules or ions.

FIGURE 6.22 shows a cutaway view from the side of an aquaporin, indicating how it fits in a plasma membrane. Like other channels that have been studied in detail, aquaporins have a pore that is lined with polar functional groups—in this case, carbonyl groups that interact with water. A channel's pore is hydrophilic relative to the hydrophobic residues facing the phospholipid tails of the membrane.

But how can aquaporin be selective for water and not other polar molecules? The answer was found when researchers examined its structure. Key side chains in the interior of the pore function as a molecular filter. The distance between these groups across the channel allows only those substances capable of interacting with all of them to pass through to the other side.

Movement Through Many Membrane Channels Is Regulated Recent research has shown that many aquaporins and ion channels are **gated channels**—meaning that they open or close in response to a signal, such as the binding of a particular molecule or a change in the electrical voltage across the membrane.

As an example of how voltage-gated channels work, **FIGURE 6.23** shows a potassium channel in closed and open configurations. The electrical charge on the membrane is normally negative on the inside relative to the outside, which causes the channel to adopt a closed shape that prevents potassium ions from passing through. When this charge asymmetry is reversed, the shape changes in a way that opens the channel and allows potassium ions to cross. The key point here is that in almost all cases, the flow of ions and small molecules through membrane channels is carefully controlled.

In all cases, however, the movement of substances through channels is passive—meaning it does not require an input of energy. **Passive transport** is powered by diffusion along an electrochemical gradient. Channel proteins simply enable ions or polar molecules to move across lipid bilayers efficiently, in response to

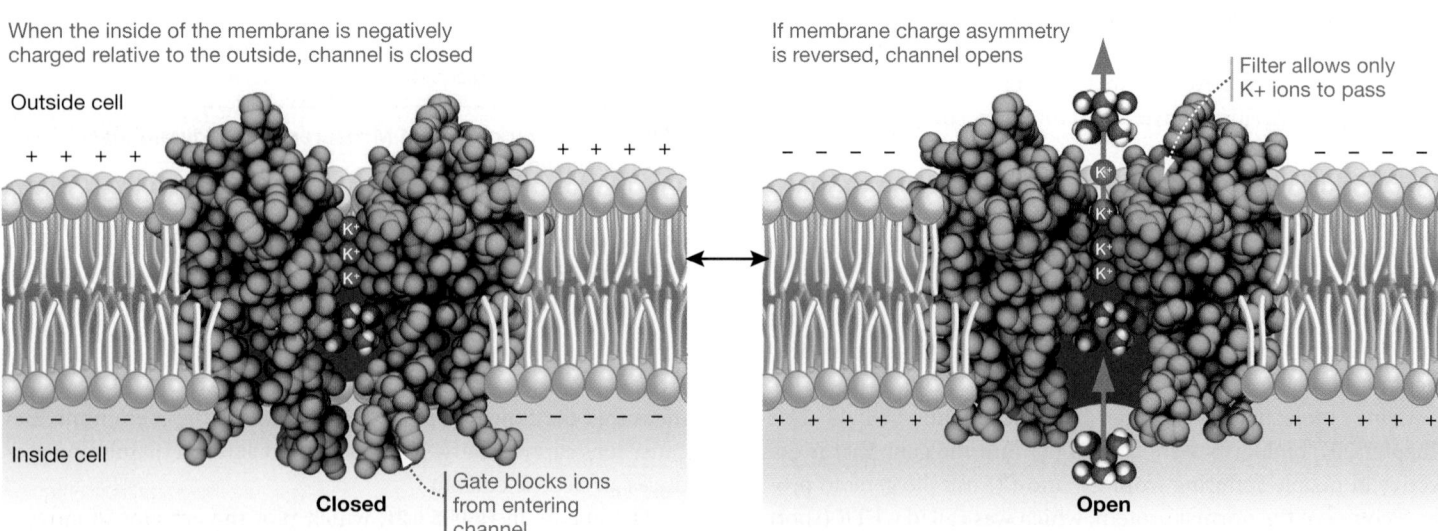

When the inside of the membrane is negatively charged relative to the outside, channel is closed

Outside cell

Inside cell

Closed

Gate blocks ions from entering channel

If membrane charge asymmetry is reversed, channel opens

Filter allows only K+ ions to pass

Open

FIGURE 6.23 Some Membrane Channels Are Highly Regulated. A model of a voltage-gated K⁺ channel in the closed and open configurations. The channel filter displaces water molecules that normally surround the K⁺ ions in an aqueous solution.

an existing gradient. They are responsible for **facilitated diffusion:** the passive transport of substances that otherwise would not cross a membrane readily.

Facilitated Diffusion via Carrier Proteins

Facilitated diffusion can also occur through **carrier proteins**—specialized membrane proteins that change shape during the transport process. Perhaps the best-studied carrier protein is one that is involved in transporting glucose into cells.

The Search for a Glucose Carrier
Next to ribose, the six-carbon sugar glucose is the most prevalent sugar found in organisms. Virtually all cells alive today use glucose as a building block for important macromolecules and as a source of stored chemical energy (Chapter 5). But as Figure 6.8 shows, lipid bilayers are only moderately permeable to glucose. It is reasonable to expect, then, that plasma membranes have some mechanism for increasing their permeability to this sugar.

This prediction was supported in experiments on pure preparations of plasma membranes from human red blood cells. These plasma membranes turned out to be much more permeable to glucose than are pure lipid bilayers. Why?

After isolating and analyzing many proteins from red blood cell membranes, researchers found one protein that specifically increases membrane permeability to glucose. When they added this purified protein to liposomes, the artificial membrane transported glucose at the same rate as a membrane from a living cell. This experiment convinced biologists that the membrane protein—now called GLUT-1 (short for glucose transporter 1)—was indeed responsible for transporting glucose across plasma membranes.

How Does GLUT-1 Work?
Recall that proteins frequently change shape when they bind to other molecules and that such conformational changes are often a critical step in their function (Chapter 3).

FIGURE 6.24 illustrates the current hypothesis for how GLUT-1 works to facilitate the movement of glucose. The idea is that when glucose binds to GLUT-1, it changes the shape of the protein in a way that moves the sugar through the hydrophobic region of the membrane and releases it on the other side.

What powers the movement of molecules through carriers? The answer is diffusion. GLUT-1 facilitates diffusion by allowing glucose to enter the carrier from either side of the membrane. Glucose will pass through the carrier in the direction dictated by its concentration gradient. A large variety of molecules move across plasma membranes via specific carrier proteins.

Pumps Perform Active Transport

Diffusion—whether it is facilitated by proteins or not—is a passive process that will move substances in either direction across a membrane to make the cell interior and exterior more similar. But it is also possible for cells to move molecules or ions in a directed manner, often *against* their electrochemical gradient. Accomplishing this task requires an input of energy, because the cell must counteract the decrease in entropy that occurs when molecules or ions are concentrated. It makes sense, then, that transport against an electrochemical gradient is called **active transport.**

In cells, ATP (adenosine triphosphate) often provides the energy for active transport by transferring a phosphate group (HPO_4^{2-}) to an active transport protein called a **pump.** Recall that ATP contains three phosphate groups (Chapter 4), and that phosphate groups carry two negative charges (Chapter 2). When a phosphate group leaves ATP and binds to a pump, its negative charges interact with charged amino acid residues in the protein. As a result, the protein's potential energy increases and its shape changes.

The Sodium–Potassium Pump
A classic example of how structural change leads to active transport is provided in the **sodium–potassium pump,** or more formally, Na^+/K^+-ATPase. The Na^+/K^+ part of the name refers to the ions that are transported, ATP indicates that adenosine triphosphate is used, and *–ase* identifies the molecule as an enzyme.

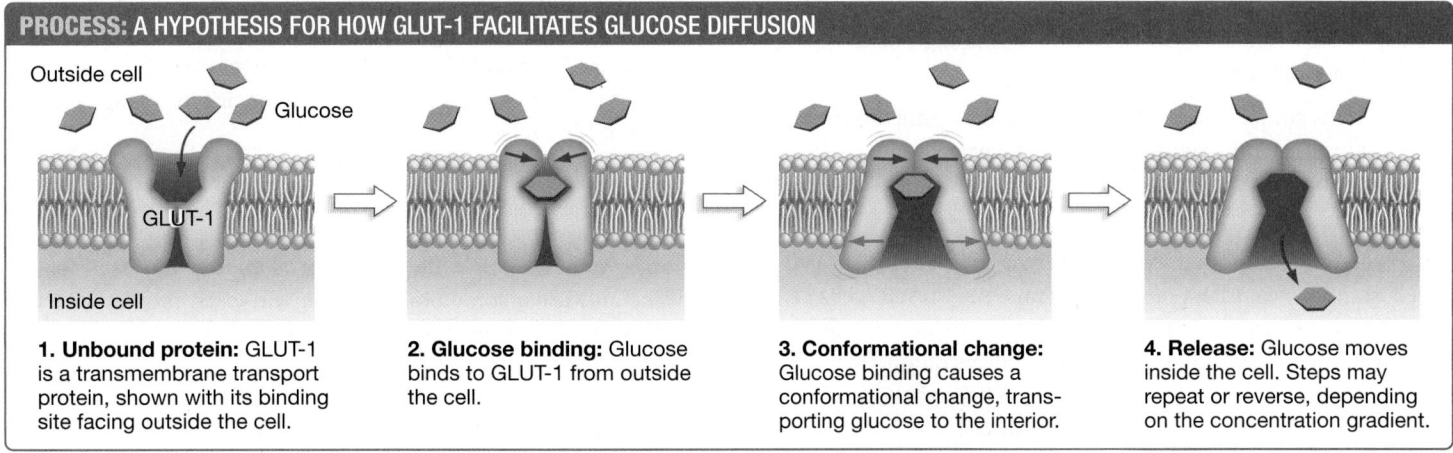

PROCESS: A HYPOTHESIS FOR HOW GLUT-1 FACILITATES GLUCOSE DIFFUSION

1. Unbound protein: GLUT-1 is a transmembrane transport protein, shown with its binding site facing outside the cell.

2. Glucose binding: Glucose binds to GLUT-1 from outside the cell.

3. Conformational change: Glucose binding causes a conformational change, transporting glucose to the interior.

4. Release: Glucose moves inside the cell. Steps may repeat or reverse, depending on the concentration gradient.

FIGURE 6.24 Carrier Proteins Undergo Structural Changes to Move Substances. This model suggests that GLUT-1 binds a glucose molecule, undergoes a conformational change, and releases glucose on the other side of the membrane.

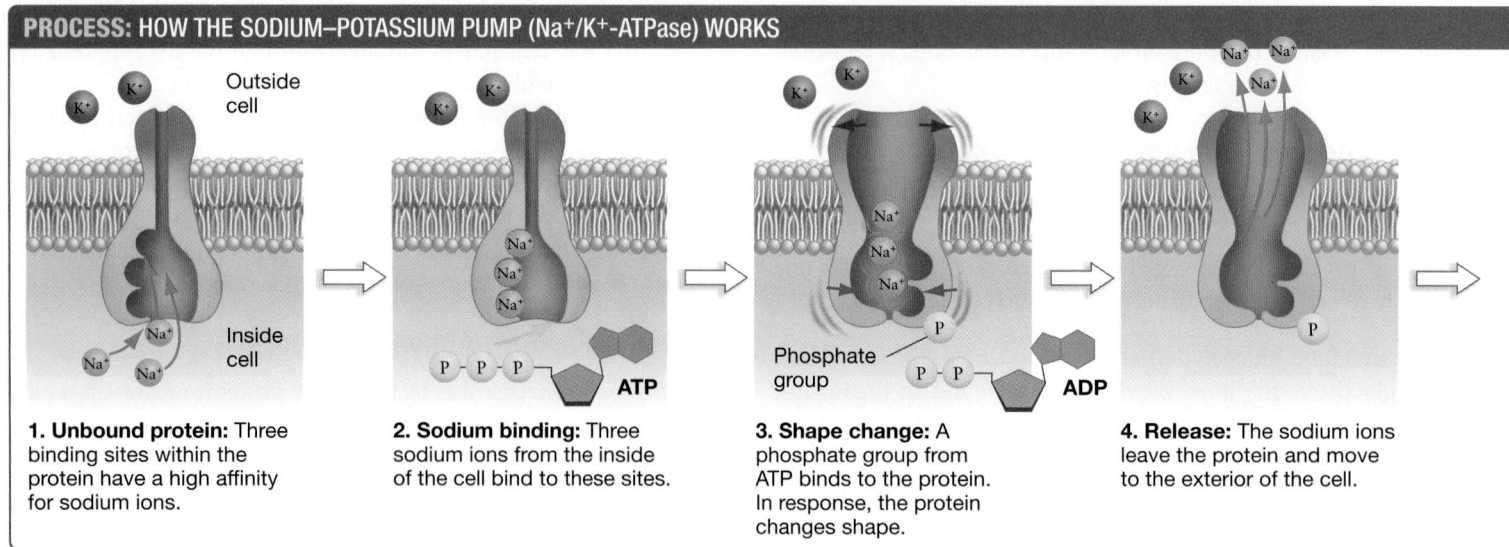

1. **Unbound protein:** Three binding sites within the protein have a high affinity for sodium ions.

2. **Sodium binding:** Three sodium ions from the inside of the cell bind to these sites.

3. **Shape change:** A phosphate group from ATP binds to the protein. In response, the protein changes shape.

4. **Release:** The sodium ions leave the protein and move to the exterior of the cell.

FIGURE 6.25 The Sodium–Potassium Pump Depends on an Input of Chemical Energy Stored in ATP.

As shown in **FIGURE 6.25**, sodium and potassium ions move in a multistep process:

Step 1 When Na+/K+-ATPase is in the conformation shown here, binding sites with a high affinity for sodium ions are available.

Step 2 Three sodium ions from the inside of the cell bind to these sites and activate the ATPase activity in the pump.

Step 3 A phosphate group from ATP is transferred to the pump. When the phosphate group attaches, the pump changes its shape in a way that opens the ion-binding pocket to the external environment and reduces its affinity for sodium ions.

Step 4 The sodium ions leave the protein and move to the exterior of the cell.

Step 5 In this conformation, the pump has binding sites with a high affinity for potassium ions facing the external environment.

Step 6 Two potassium ions from outside the cell bind to the pump.

Step 7 When the potassium is bound, the phosphate group is cleaved from the protein and its structure changes in response—back to the original shape with the ion-binding pocket facing the interior of the cell.

Step 8 In this conformation, the pump has low affinity for potassium ions. The potassium ions leave the protein and move to the interior of the cell. The cycle then repeats.

Other types of pumps move protons (H^+), calcium ions (Ca^{2+}), or other ions or molecules across membranes in a directed manner, regardless of the gradients. As a result, cells can import and concentrate valuable nutrients and ions inside the cell despite their relatively low external concentration. They can also expel molecules or ions, even when a concentration gradient favors diffusion of these substances into the cell.

Secondary Active Transport Approximately 30 percent of all the ATP generated in your body is used to drive the Na+/K+-ATPase cycle. Each cycle exports three Na^+ ions for every two K^+ ions it

imports. In this way, the sodium–potassium pump converts energy from ATP to an electrochemical gradient across the membrane. The outside of the membrane becomes positively charged relative to the inside. This gradient favors a flow of anions out of the cell and a flow of cations into the cell.

The electrochemical gradients established by the Na+/K+-ATPase represent a form of stored energy, much like the electrical energy stored in a battery. How do cells use this energy?

Gradients are crucial to the function of the cell, in part because they make it possible for cells to engage in **secondary active transport**—also known as cotransport. When cotransport occurs, a gradient set up by a pump provides the energy required to power the movement of a different molecule against its particular gradient.

Recall that GLUT-1 facilitates the movement of glucose into or out of cells in the direction of its gradient. Can glucose be moved against its gradient? The answer is yes—a cotransport protein in your gut cells uses the Na^+ gradient created by Na+/K+-ATPases to import glucose against its chemical gradient. When Na^+ ions bind to this cotransporter, its shape changes in a way that allows glucose to bind. Once glucose binds, another shape transports both the sodium and glucose to the inside of the cell. After dropping off sodium and glucose, the protein's original shape returns to repeat the cycle.

In this way, glucose present in the food you are digesting is actively transported into your body. The glucose molecules eventually diffuse into your bloodstream and are transported to your brain, where they provide the chemical energy you need to stay awake and learn some biology. (You will learn more about secondary active transport in Units 7 and 8.)

Plasma Membranes and the Intracellular Environment

Taken together, the selective permeability of the lipid bilayer and the specificity of the proteins involved in passive transport and

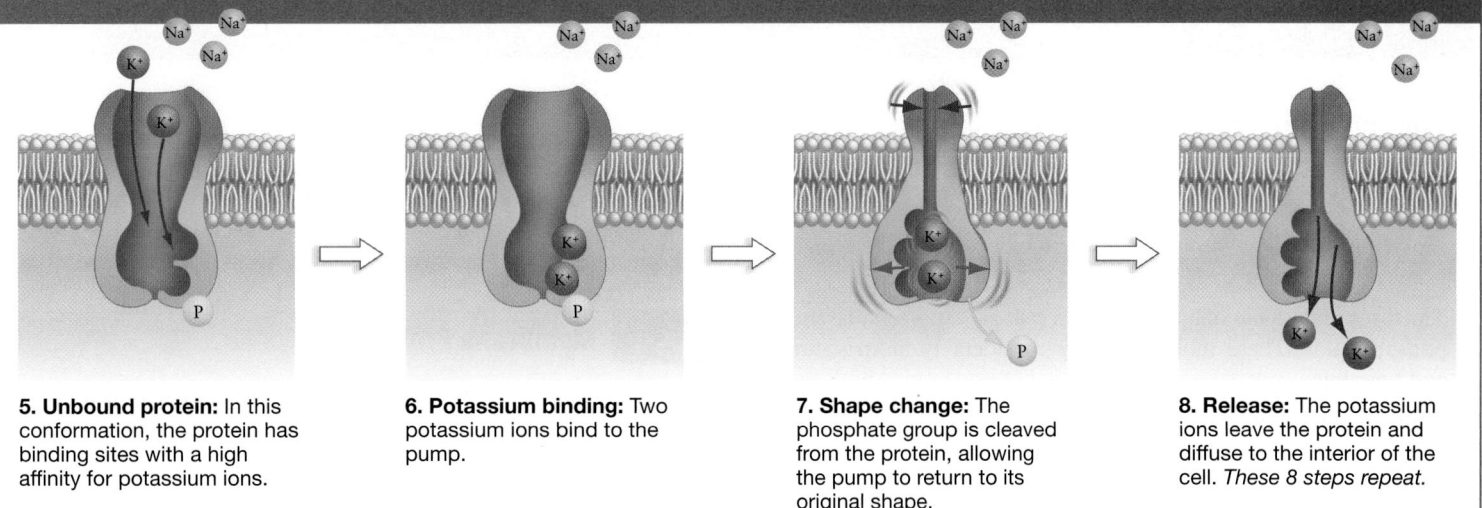

5. Unbound protein: In this conformation, the protein has binding sites with a high affinity for potassium ions.

6. Potassium binding: Two potassium ions bind to the pump.

7. Shape change: The phosphate group is cleaved from the protein, allowing the pump to return to its original shape.

8. Release: The potassium ions leave the protein and diffuse to the interior of the cell. *These 8 steps repeat.*

active transport enable cells to create an internal environment that is much different from the external one (**FIGURE 6.26**).

With the evolution of membrane proteins, the early cells acquired the ability to create an internal environment that was conducive to life—one that contained the substances required for manufacturing ATP and copying ribozymes. Cells with particularly efficient and selective membrane proteins would be favored by natural selection and would come to dominate the population. Cellular life had begun.

Some 3.5 billion years later, cells continue to evolve. What do today's cells look like, and how do they produce and store the chemical energy that makes life possible? Answering these and related questions is the focus of the following unit.

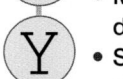

check your understanding

If you understand that . . .

- Membrane proteins allow substances that ordinarily do not readily cross lipid bilayers to enter or exit cells.
- Substances may move across a membrane along an electrochemical gradient, via facilitated diffusion through channel or carrier proteins. Or, they may move against a gradient in response to work done by pumps.

✔ **You should be able to . . .**

Explain what is passive about passive transport, active about active transport, and "co" about cotransport.

Answers are available in Appendix A.

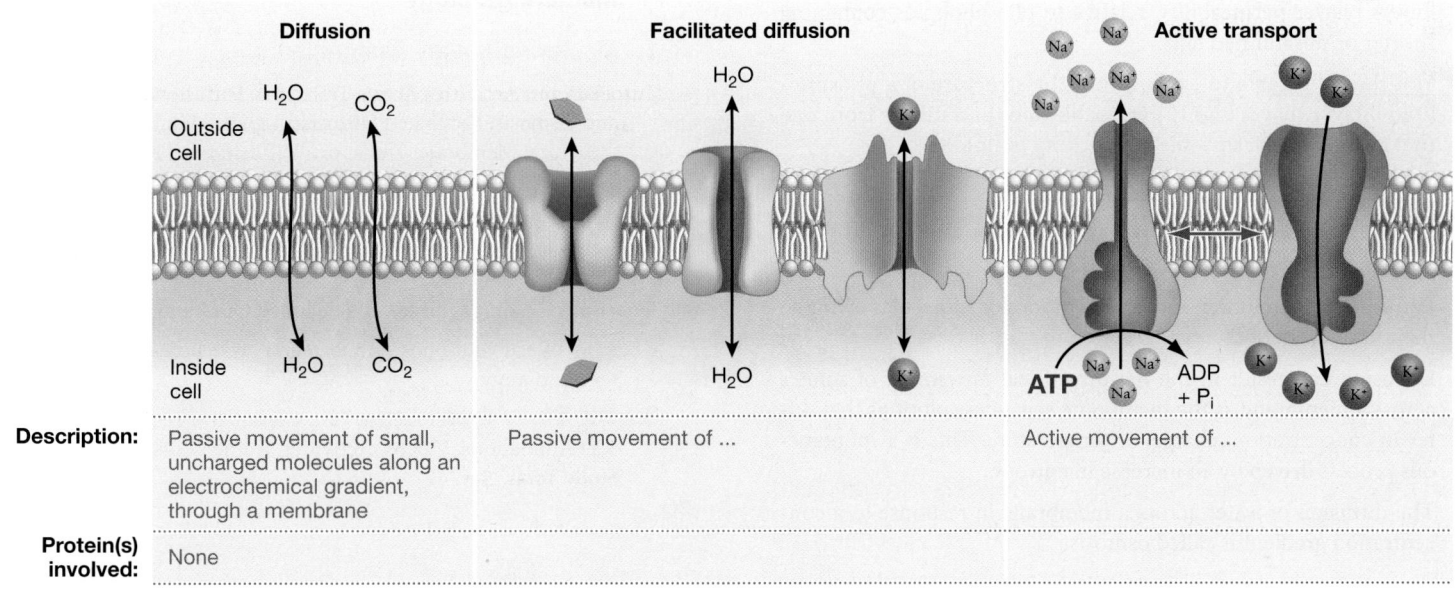

Description:	Passive movement of small, uncharged molecules along an electrochemical gradient, through a membrane	Passive movement of ...	Active movement of ...
Protein(s) involved:	None		

FIGURE 6.26 Summary of the Passive and Active Mechanisms of Membrane Transport.

✔**EXERCISE** Complete the chart.

If you understand . . .

6.1   Lipid Structure and Function

- Lipids are largely hydrophobic compounds due to their high number of nonpolar C–H bonds.

- The three main types of lipids found in cells are fats, steroids, and phospholipids. These molecules vary considerably in structure and function.

- In hydrocarbon chains, the length and degree of saturation have a profound effect on their physical properties.

- Amphipathic lipids possess a distinct hydrophilic region containing polar or charged groups. Phospholipids have a polar head and a nonpolar tail. The nonpolar tail usually consists of fatty acids or isoprenoids.

✔ You should be able to explain how adding hydrogen (H_2) to vegetable oil, a process called hydrogenation, results in a butter-like solid called margarine.

6.2   Phospholipid Bilayers

- In solution, phospholipids spontaneously assemble into bilayers that can serve as a physical barrier between an internal and external environment.

- Small nonpolar molecules tend to move across lipid bilayers readily; ions cross rarely, if at all.

- The permeability and fluidity of lipid bilayers depend on temperature, on the concentration of steroids, and on the chemical structure of the lipids present, such as the saturation status and length of the hydrocarbon chains. Phospholipids with longer or saturated tails form a dense and highly hydrophobic interior that lowers bilayer permeability, relative to phospholipids containing shorter or unsaturated tails.

✔ You should be able to explain how the structure of a phospholipid bilayer that is highly permeable and fluid differs from one that is highly impermeable and lacking in fluidity.

6.3   How Molecules Move across Lipid Bilayers: Diffusion and Osmosis

- Diffusion is the random movement of ions or molecules owing to their kinetic energy.

- Diffusion can result in the net directional movement of solutes across a membrane, if the membrane separates solutions that differ in concentration, charge, or temperature. This is a spontaneous process driven by an increase in entropy.

- The diffusion of water across a membrane in response to a concentration gradient is called osmosis.

✔ You should be able to imagine a beaker with solutions separated by a planar membrane and then predict what will happen after addition of a solute to one side if the solute (1) crosses the membrane readily or (2) is incapable of crossing the membrane.

6.4   Membrane Proteins

- The permeability of lipid bilayers can be altered significantly by membrane proteins.

- Channel proteins provide pores in the membrane and facilitate the diffusion of specific solutes into and out of the cell.

- Carriers undergo conformational changes that facilitate the diffusion of specific molecules into and out of the cell.

- Pumps use energy to actively move ions or molecules in a single direction, often against the electrical or chemical gradient.

- In combination, the selective permeability of phospholipid bilayers and the specificity of transport proteins make it possible to create an environment inside a cell that is radically different from the exterior.

✔ You should be able to draw and label the membrane of a cell that is placed in a solution containing calcium ions and lactose and show the activity of the following membrane proteins: (1) an H^+ pump that exports protons; (2) a calcium channel; and (3) a lactose carrier. Your drawing should include arrows and labels indicating the direction of solute movement and the direction of the appropriate electrochemical gradients.

(MB) ## MasteringBiology

1. MasteringBiology Assignments

Tutorials and Activities Active Transport; Diffusion, Diffusion and Osmosis; Facilitated Diffusion Lipids; Membrane Structure; Membrane Transport: Diffusion and Passive Transport; Membrane Transport: The Sodium–Potassium Pump; Membrane Transport Proteins; Osmosis; Membrane Transport: Cotransport; Osmosis and Water Balance in Cells; Selective Permeability of Membranes

Questions Reading Quizzes, Blue-Thread Questions, Test Bank

2. eText Read your book online, search, take notes, highlight text, and more.

3. The Study Area Practice Test, Cumulative Test, BioFlix® 3-D Animations, Videos, Activities, Audio Glossary, Word Study Tools, Art

You should be able to . . .

1. How is the structure of saturated fats different from that of unsaturated fats?
   a. All of the carbons in the hydrocarbon tails of saturated fats are bonded to one another with double bonds.
   b. Saturated fats have three hydrocarbon tails bonded to the glycerol molecule instead of just two.
   c. The hydrocarbon tails in a saturated fat have the maximum number of hydrogens possible.
   d. Saturated fats have no oxygens present.

2. What distinguishes amphipathic lipids from other lipids?
   a. Amphipathic lipids have polar and nonpolar regions.
   b. Amphipathic lipids have saturated and unsaturated regions.
   c. Amphipathic lipids are steroids.
   d. Amphipathic lipids dissolve in water.

3. If a solution surrounding a cell is hypertonic relative to the inside of the cell, how will water move?

   a. It will move into the cell via osmosis.
   b. It will move out of the cell via osmosis.
   c. It will not move, because equilibrium exists.
   d. It will evaporate from the cell surface more rapidly.

4. When does a concentration gradient exist?
   a. when membranes rupture
   b. when solute concentrations are high
   c. when solute concentrations are low
   d. when solute concentrations differ on the two sides of a membrane

5. What two conditions must be present for the effects of osmosis to occur?

6. In terms of structure, how do channel proteins differ from carrier proteins?

7. If a cell were placed in a solution with a high potassium concentration and no sodium, what would happen to the sodium–potassium pump's activity?
   a. It would stop moving ions across the membrane.
   b. It would continue using ATP to pump sodium out of the cell and potassium into the cell.
   c. It would move sodium and potassium ions across the membrane, but no ATP would be used.
   d. It would reverse the direction of sodium and potassium ions to move them against their gradients.

8. Cooking oil lipids consist of long, unsaturated hydrocarbon chains. Would you expect these molecules to form membranes spontaneously? Why or why not? Describe, on a molecular level, how you would expect these lipids to interact with water.

9. Explain why phospholipids form a bilayer in solution, and why the process is spontaneous.

10. Ethanol (C_2H_5OH) is the active ingredient in alcoholic beverages. Would you predict that this molecule crosses lipid bilayers quickly, slowly, or not at all? Explain your reasoning.

11. Integral membrane proteins are anchored in lipid bilayers. Of the following four groups of amino acids—nonpolar, polar, charged/acidic, charged/basic (see Figure 3.2)—which would likely be found in the portion that crosses the lipid bilayer? Explain your reasoning.

12. Examine the experimental chamber in Figure 6.7a. If the lipid bilayer were to contain the CFTR molecule, what would pass through the membrane if you added a 1-molar solution of sodium chloride on the left side and a 1-molar solution of potassium ions on the right? Assume that there is an equal amount of water on each side at the start of the experiment.

13. In an experiment, you create two groups of liposomes—one made from red blood cell membranes and the other from frog egg cell membranes. When placed in water, those made with red blood cell membranes burst more rapidly than those made from frog membranes. What is the best explanation for these results?
    a. The red blood cell liposomes are more hypertonic relative to water than the frog egg liposomes.
    b. The red blood cell liposomes are more hypotonic relative to water than the frog egg liposomes.
    c. The red blood cell liposomes contain aquaporins, which are not abundant in the frog egg liposomes.
    d. The frog egg liposomes contain ion channels, which are not present in the red blood cell liposomes.

14. When phospholipids are arranged in a bilayer, it is theoretically possible for individual molecules in the bilayer to flip-flop. That is, a phospholipid could turn 180° and become part of the membrane's other surface. From what you know about the

behavior of polar heads and nonpolar tails, predict whether flip-flops are frequent or rare. Then design an experiment, using a planar bilayer with one side made up of phospholipids that contain a dye molecule on their hydrophilic head, to test your prediction.

15. Unicellular organisms live in a wide range of habitats, from the hot springs in Yellowstone National Park to the freezing temperatures of the Antarctic. Make a prediction about the saturation status of membrane phospholipids in organisms that live in extremely cold environments versus those that live in extremely hot environments. Explain your reasoning.

16. When biomedical researchers design drugs, they sometimes add methyl (CH_3) groups or charged groups to the molecules. If these groups are not directly involved in the activity of the drug, predict the purpose of these modifications and explain why these strategies are necessary.

The Big Picture

The first spark of life ignited when simple chemical reactions began to convert small molecules into larger, more complex molecules with novel 3-D structures and activities. According to the theory of chemical evolution, these reactions eventually led to the formation of the four types of macromolecules characteristic of life—proteins, nucleic acids, carbohydrates, and lipids.

As you look through this concept map, consider how the functions of the four types of macromolecules are determined by their structures, and how these structures stem from the chemical properties of the atoms and bonds used to build them.

Note that each box in the concept map indicates the chapters and sections where you can go for review. Also, be sure to do the blue exercises in the Check Your Understanding box below.

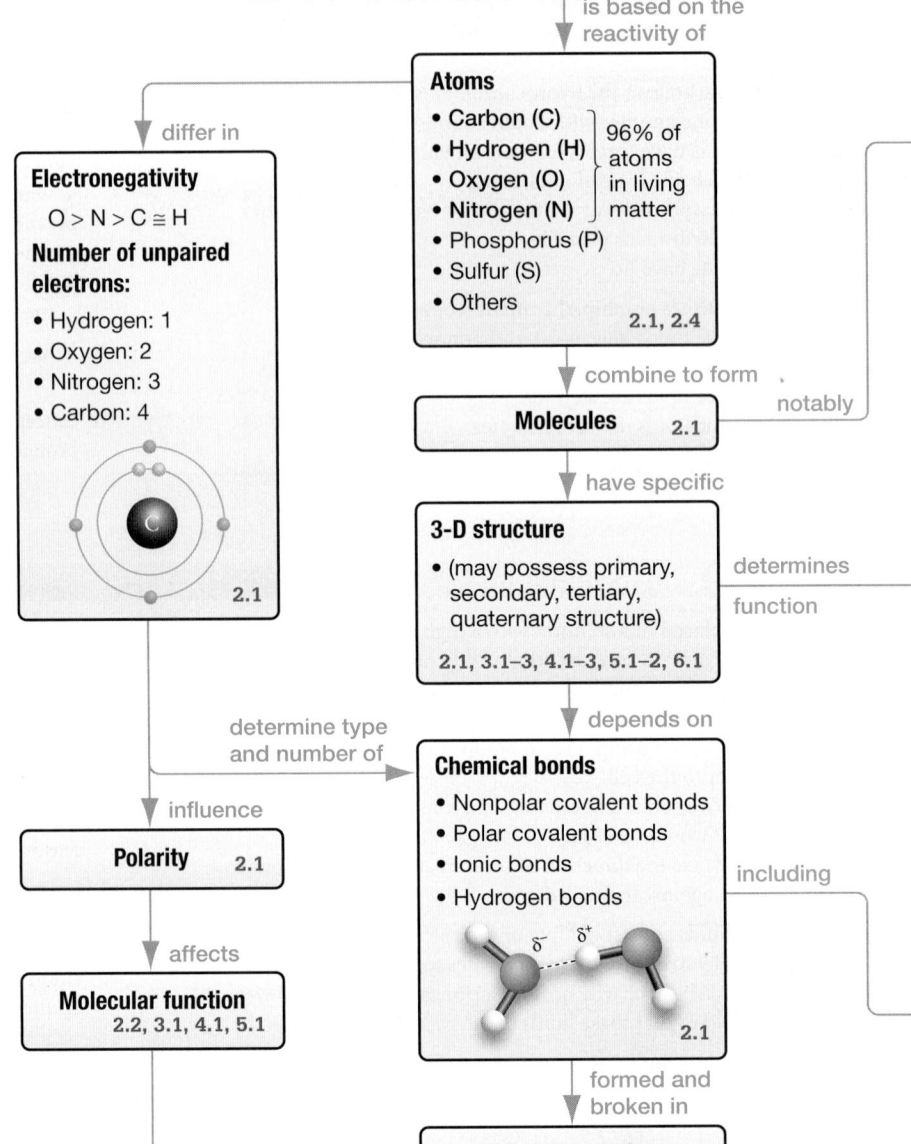

is based on the reactivity of

Atoms
- Carbon (C)
- Hydrogen (H) } 96% of atoms in living matter
- Oxygen (O)
- Nitrogen (N)
- Phosphorus (P)
- Sulfur (S)
- Others

2.1, 2.4

differ in

Electronegativity

$O > N > C \cong H$

Number of unpaired electrons:
- Hydrogen: 1
- Oxygen: 2
- Nitrogen: 3
- Carbon: 4

2.1

combine to form

Molecules 2.1 notably

have specific

3-D structure
- (may possess primary, secondary, tertiary, quaternary structure)

2.1, 3.1–3, 4.1–3, 5.1–2, 6.1

determines function

determine type and number of

depends on

Chemical bonds
- Nonpolar covalent bonds
- Polar covalent bonds
- Ionic bonds
- Hydrogen bonds

2.1 including

influence

Polarity 2.1

affects

Molecular function
2.2, 3.1, 4.1, 5.1

formed and broken in

Chemical reactions 2.3

as demonstrated by

WATER

is essential for life due to its

- Efficiency as a solvent
- Cohesion, adhesion and surface tension properties
- Higher density as a liquid than as a solid
- High capacity for energy absorption

2.2

Basic

14

H_2O 7 **pH** 2.2 has neutral

0

Acidic

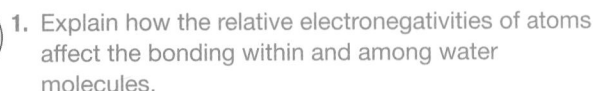

check your understanding

C Y U

If you understand the big picture . . .

✔ **You should be able to . . .**

1. Explain how the relative electronegativities of atoms affect the bonding within and among water molecules.

2. Describe the attributes of RNA that make it a candidate for the origin of life. Why isn't DNA considered a candidate?

3. Circle the atoms in amino acids and nucleotides that engage in creating bonds with other monomers.

4. Draw a protein in the lipid bilayer. What role might it play?

Answers are available in Appendix A.

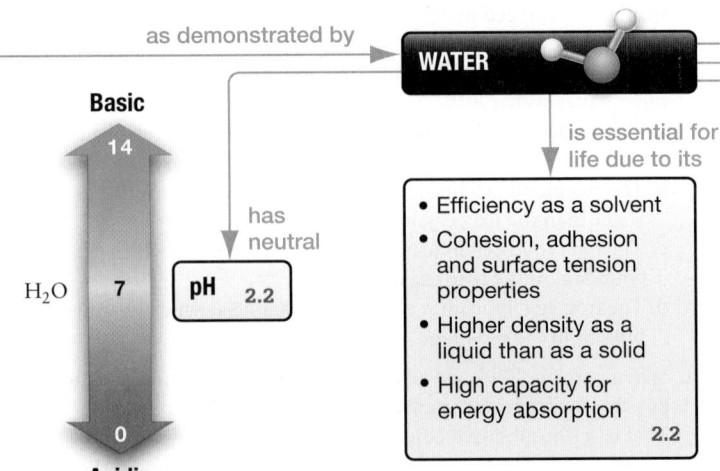

Biological macromolecules

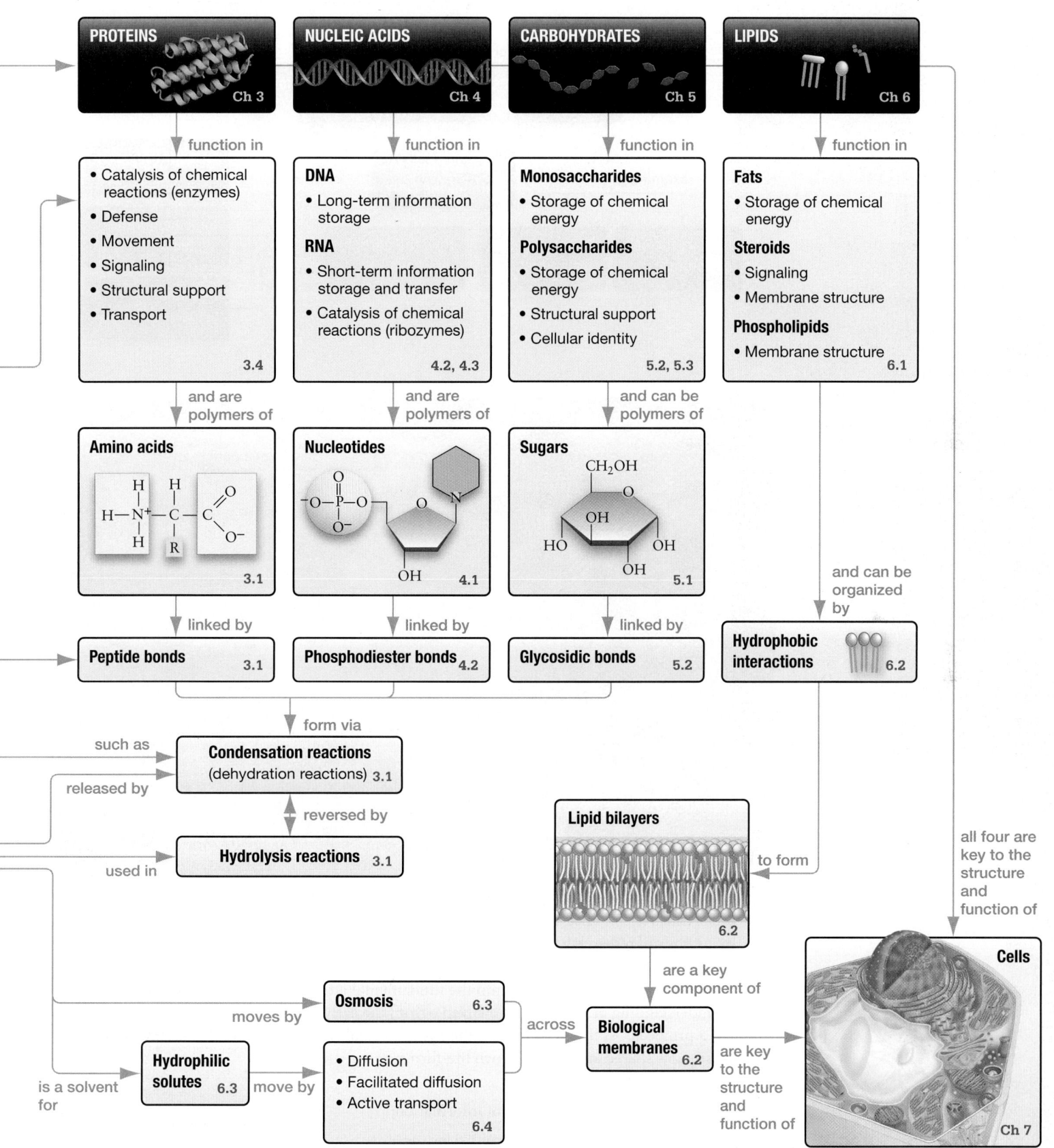

PROTEINS — Ch 3

function in
- Catalysis of chemical reactions (enzymes)
- Defense
- Movement
- Signaling
- Structural support
- Transport

3.4

and are polymers of

Amino acids

3.1

linked by

Peptide bonds 3.1

NUCLEIC ACIDS — Ch 4

function in

DNA
- Long-term information storage

RNA
- Short-term information storage and transfer
- Catalysis of chemical reactions (ribozymes)

4.2, 4.3

and are polymers of

Nucleotides

4.1

linked by

Phosphodiester bonds 4.2

CARBOHYDRATES — Ch 5

function in

Monosaccharides
- Storage of chemical energy

Polysaccharides
- Storage of chemical energy
- Structural support
- Cellular identity

5.2, 5.3

and can be polymers of

Sugars

5.1

linked by

Glycosidic bonds 5.2

LIPIDS — Ch 6

function in

Fats
- Storage of chemical energy

Steroids
- Signaling
- Membrane structure

Phospholipids
- Membrane structure

6.1

and can be organized by

Hydrophobic interactions 6.2

form via

such as

released by

Condensation reactions
(dehydration reactions) 3.1

reversed by

used in

Hydrolysis reactions 3.1

Lipid bilayers

6.2

to form

all four are key to the structure and function of

Osmosis 6.3

moves by

is a solvent for

Hydrophilic solutes 6.3

move by

- Diffusion
- Facilitated diffusion
- Active transport

6.4

across

are a key component of

Biological membranes 6.2

are key to the structure and function of

Cells — Ch 7

7 Inside the Cell

In this chapter you will learn that

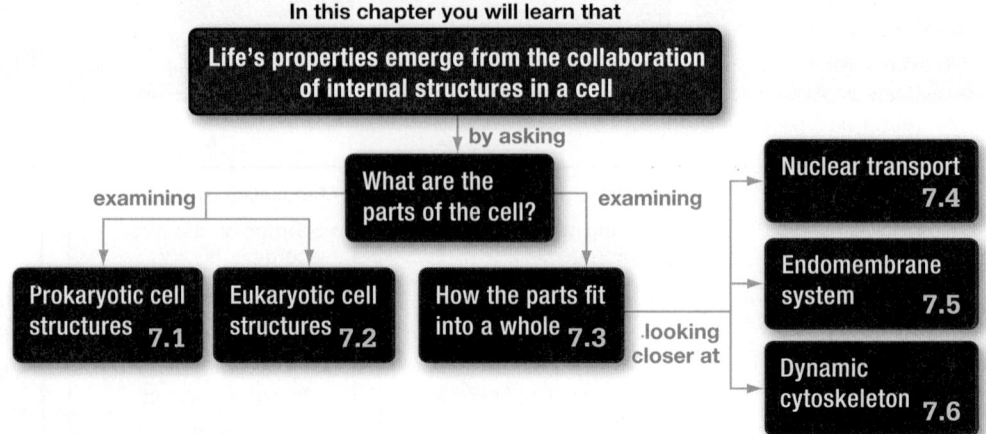

Life's properties emerge from the collaboration of internal structures in a cell

by asking

What are the parts of the cell?

examining — Prokaryotic cell structures **7.1** — Eukaryotic cell structures **7.2**

examining — How the parts fit into a whole **7.3**

looking closer at — Nuclear transport **7.4** — Endomembrane system **7.5** — Dynamic cytoskeleton **7.6**

This cell has been treated with fluorescing molecules that bind to its fibrous cytoskeleton. Microtubules (large protein fibers) are yellow; actin filaments (smaller fibers) are blue. The cell's nucleus has been stained green.

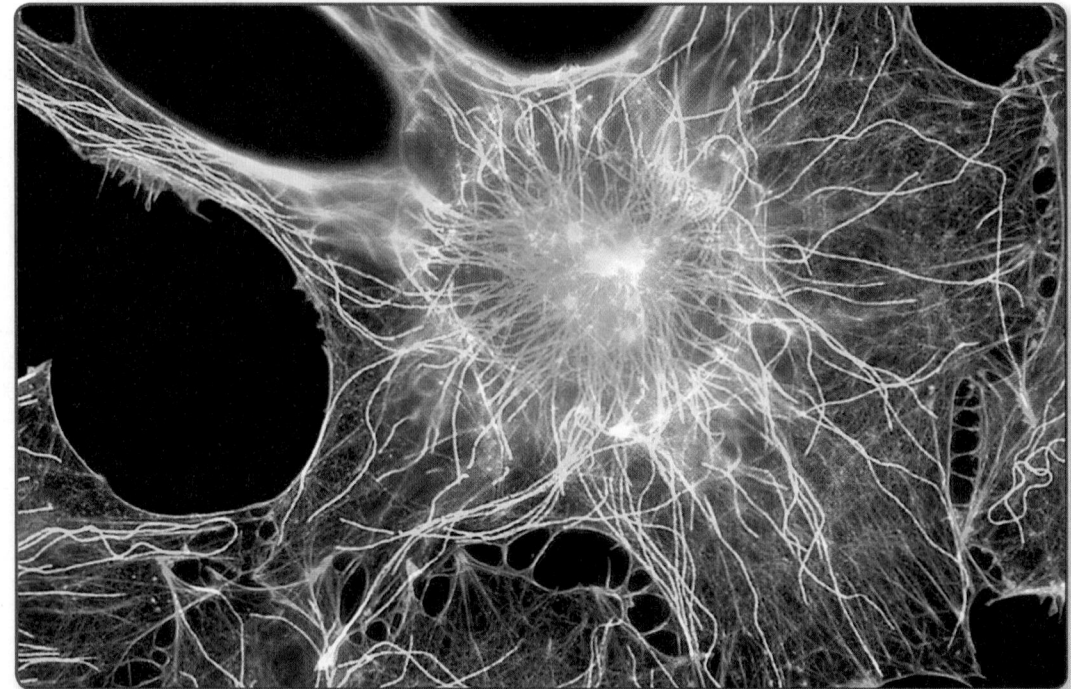

The cell theory states that all organisms consist of cells and all cells are derived from preexisting cells (Chapter 1). Since this theory was initially developed and tested in the 1850s, an enormous body of research has confirmed that the cell is the fundamental structural and functional unit of life. Life on Earth is cellular.

Previous chapters (Unit 1) delved into the fundamental attributes of life by looking at biologists' current understanding of how the cell evolved—from the early chemistry to the assembly and replication of a protocell. As the first cells left the hydrothermal vents, they took with them characteristics that are now shared among all known life-forms.

All cells have

1. nucleic acids that store and transmit information;

2. proteins that perform most of the cell's functions;

✔ When you see this checkmark, stop and test yourself. Answers are available in Appendix A.

3. carbohydrates that provide chemical energy, carbon, support, and identity; and

4. a plasma membrane, which serves as a selectively permeable membrane barrier.

Thanks to the selective permeability of phospholipid bilayers and the activity of membrane transport proteins, the plasma membrane creates an internal environment that differs from conditions outside the cell. Our task now is to explore the structures inside the cell to understand how the properties of life emerged from the combination of these characteristics.

Let's begin by analyzing how the parts inside a cell function individually and then exploring how they work as a unit. This approach is analogous to studying individual organs in the body and then analyzing how they work together to form the nervous system or digestive system. As you study this material, keep asking yourself some key questions: How does the structure of this part or group of parts correlate with its function? What problem does it solve?

7.1 Bacterial and Archaeal Cell Structures and Their Functions

Cells are divided into two fundamental types called eukaryotes and prokaryotes (see Chapter 1). This division is mostly based on cell **morphology** ("form-science")—eukaryotic cells have a membrane-bound compartment called a nucleus, and prokaryotic cells do not.

But according to **phylogeny** ("tribe-source"), or evolutionary history, organisms are divided into three broad domains called **(1)** Bacteria, **(2)** Archaea, and **(3)** Eukarya. Members of the Bacteria and Archaea are prokaryotic; members of the Eukarya—including algae, fungi, plants, and animals—are eukaryotic.

A Revolutionary New View

For almost 200 years, biologists thought that prokaryotic cells were simple in terms of their morphology and that there was little structural diversity among species. This conclusion was valid at the time, given the resolution of the microscopes that were available and the number of species that had been studied.

Things have changed. Recent improvements in microscopy and other research tools have convinced biologists that prokaryotic cells, among which bacteria are the best understood, possess an array of distinctive structures and functions found among millions of species. This conclusion represents one of the most exciting discoveries in cell biology over the past 10 years.

To keep things simple at the start, though, **FIGURE 7.1** offers a low-magnification, stripped-down diagram of a bacterial cell.

Prokaryotic Cell Structures: A Parts List

The labels in Figure 7.1 highlight the components common to all or most bacteria studied to date. Let's explore these elements one by one, and also look at more specialized structures found in particular species, starting from the inside and working out.

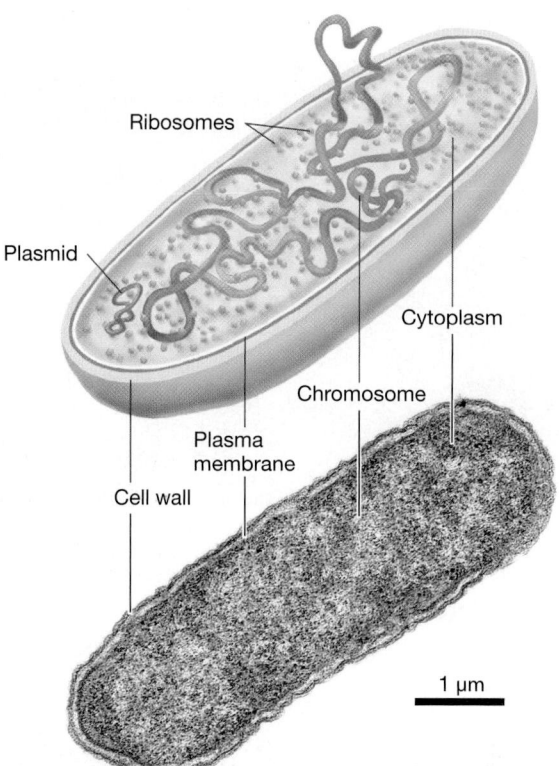

FIGURE 7.1 Overview of a Prokaryotic Cell. Prokaryotic cells are identified by a negative trait—the absence of a membrane-bound nucleus. Although there is wide variation in the size and shape of bacterial and archaeal cells, they all contain a plasma membrane, a chromosome, and protein-synthesizing ribosomes.

The Chromosome Is Organized in a Nucleoid The most prominent structure inside a bacterial cell is the **chromosome.** Most bacterial species have a single, circular chromosome that consists of a large DNA molecule associated with a small number of proteins. The DNA molecule contains information, and the proteins provide structural support for the DNA.

Recall that the information in DNA is encoded in its sequence of nitrogenous bases. Segments of DNA that contain information for building functional RNAs, some of which may be used to make polypeptides, are called **genes** (Chapter 4). Thus, chromosomes contain DNA, which contains genes.

In the well-studied bacterium *Escherichia coli*, the circular chromosome would be over 1 mm long if it were linear—500 times longer than the cell itself (**FIGURE 7.2a**; see page 108). This situation is typical in prokaryotes. To fit into the cell, the DNA double helix coils on itself with the aid of enzymes to form a compact, "supercoiled" structure. Supercoiled regions of DNA resemble a rubber band that has been held at either end and then twisted until it coils back upon itself.

The location and structural organization of the circular chromosome is called the **nucleoid** (pronounced *NEW-klee-oyd*). The genetic material in the nucleoid is often organized by clustering loops of DNA into distinct domains, but it is not separated from the rest of the cell interior by a membrane. The functional role of this organization of the bacterial chromosome and how it changes over time is currently the subject of intense research.

(a) Compared to the cell, chromosomal DNA is very long.

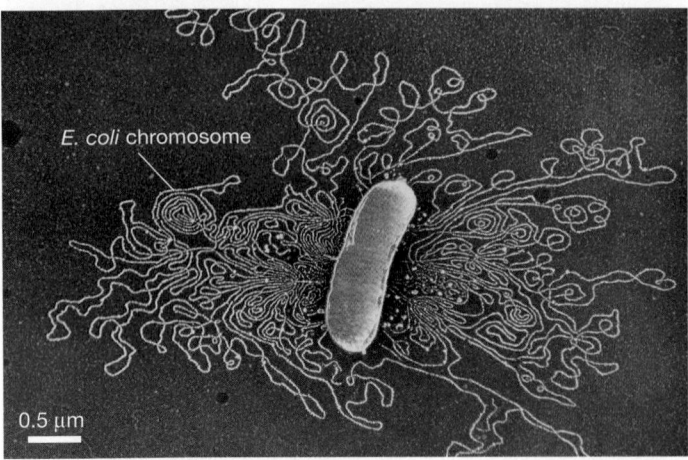

E. coli chromosome

0.5 μm

(b) DNA is packaged by supercoiling.

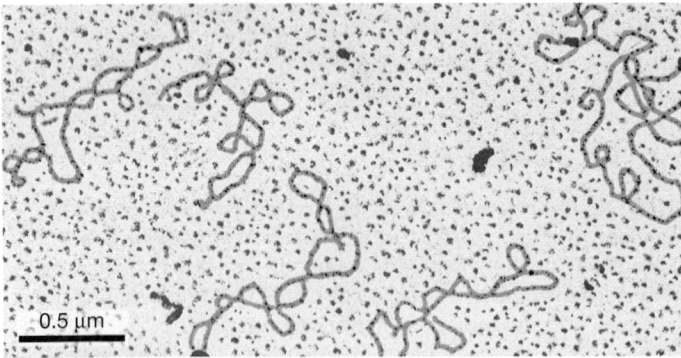

0.5 μm

FIGURE 7.2 Bacterial DNA Is Supercoiled. (a) The chromosomes of bacteria and archaea are often over 1000 times the length of the cell, as shown in this micrograph of *E. coli* that has been treated to release its DNA. To fit inside cells, this DNA must be highly compacted by supercoiling. **(b)** A colorized electron micrograph showing the effect of supercoiling on the DNA of isolated plasmids.

In addition to one or more chromosomes, bacterial cells may contain from one to about a hundred small, usually circular, supercoiled DNA molecules called **plasmids** (**FIGURE 7.2b**). Plasmids contain genes but are physically independent of the cellular chromosome. In many cases the genes carried by plasmids are not required under normal conditions; instead, they help cells adapt to unusual circumstances, such as the sudden presence of a poison in the environment. As a result, plasmids can be considered auxiliary genetic elements.

Ribosomes Manufacture Proteins   **Ribosomes** are observed in all prokaryotic cells and are found throughout the cell interior. It is not unusual for a single cell to contain 10,000 ribosomes, each functioning as a protein-manufacturing center.

Ribosomes are complex structures composed of large and small subunits, each of which contains RNA and protein molecules. Biologists often refer to ribosomes, along with other multicomponent complexes that perform specialized tasks, as "macromolecular machines." (Chapter 17 analyzes the structure and function of ribosomes in detail.)

Photosynthetic Species Have Internal Membrane Complexes   In addition to the nucleoid and ribosomes found in all bacteria and archaea studied to date, it is common to observe extensive internal membranes in prokaryotes that perform photosynthesis. Photosynthesis is the suite of chemical reactions responsible for converting the energy in sunlight into chemical energy stored in sugars.

The photosynthetic membranes observed in prokaryotes contain the enzymes and pigment molecules required for these reactions to occur and develop as infoldings of the plasma membrane. In some cases, vesicles pinch off as the plasma membrane folds in. In other cases, flattened stacks of photosynthetic membrane remain connected to the plasma membrane, like those shown in **FIGURE 7.3**. The extensive surface area provided by these internal membranes makes it possible for more photosynthetic reactions to occur and thus increases the cell's ability to make food.

Organelles Perform Specialized Functions   Recent research indicates that several bacterial species have internal compartments that qualify as **organelles** ("little organs"). An organelle is a membrane-bound compartment inside the cell that contains enzymes or structures specialized for a particular function.

Bacterial organelles perform an array of tasks, including

- storing calcium ions or other key molecules;

- holding crystals of the mineral magnetite, which function like a compass needle to help cells sense a magnetic field and swim in a directed way;

- organizing enzymes responsible for synthesizing complex carbon compounds from carbon dioxide; and

- sequestering enzymes that generate chemical energy from ammonium ions.

The Cytoskeleton Structures the Cell Interior   Recent research has also shown that bacteria and archaea contain long, thin fibers that serve a variety of roles inside the cell. All bacterial species,

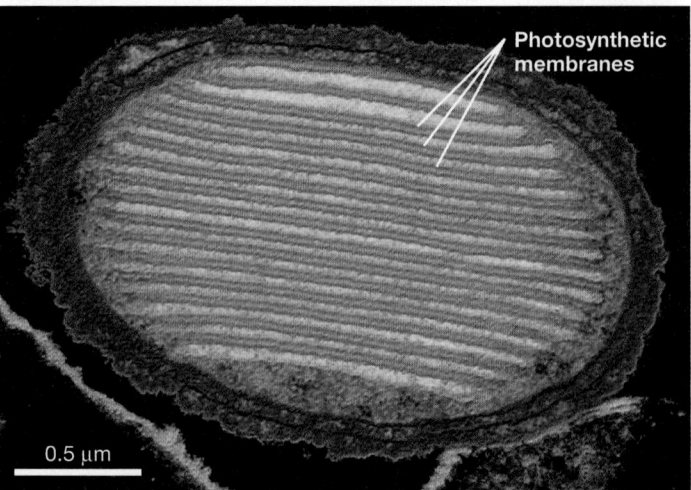

Photosynthetic membranes

0.5 μm

FIGURE 7.3 Photosynthetic Membranes in Bacteria. The green stripes in this photosynthetic bacterium are membranes that contain the pigments and enzymes required for photosynthesis. This photo has been colorized to enhance the membranes.

for example, contain protein fibers that are essential for cell division to take place. Some species also have protein filaments that help maintain cell shape. Protein filaments such as these form the basis of the **cytoskeleton** ("cell skeleton").

The discovery of bacterial cytoskeletal elements is so new that much remains to be learned. Currently, researchers are working to understand how the different cytoskeletal elements enable cells to divide and if they play a role in organizing the cell interior into distinctive regions.

The Plasma Membrane Separates Life from Nonlife

The plasma membrane consists of a phospholipid bilayer and proteins that either span the bilayer or attach to one side. Inside the membrane, all the contents of a cell, excluding the nucleus, are collectively termed the **cytoplasm** ("cell-formed").

Because all archaea and virtually all bacteria are unicellular, the plasma membrane creates an internal environment that is distinct from the outside, nonliving environment. The combined effect of a lipid bilayer and membrane proteins prohibits the entry of many substances that would be dangerous to life while allowing the passage of molecules and ions required for life (see Chapter 6).

The Cell Wall Forms a Protective "Exoskeleton"

Because the cytoplasm contains a high concentration of solutes, in most habitats it is hypertonic relative to the surrounding environment. When this is the case, water enters the cell via osmosis and makes the cell's volume expand. In virtually all bacteria and archaea, this pressure is resisted by a stiff **cell wall.**

Bacterial and archaeal cell walls are a tough, fibrous layer that surrounds the plasma membrane. In prokaryotes, the pressure of the plasma membrane against the cell wall is about the same as the pressure in an automobile tire.

The cell wall protects the organism and gives it shape and rigidity, much like the exoskeleton (external skeleton) of a crab or insect. In addition, many bacteria have another protective layer outside the cell wall that consists of lipids with polysaccharides attached. Lipids that contain carbohydrate groups are termed **glycolipids.**

External Structures Enable Movement and Attachment

Besides having a cell wall to provide protection, as just described, many bacteria also interact with their environment via structures that grow from the plasma membrane. The flagella and fimbriae shown in **FIGURE 7.4** are examples that are commonly found on bacterial surfaces.

Bacterial **flagella** (singular: **flagellum**) are assembled from over 40 different proteins at the cell surface of certain species. The base of this structure is embedded in the plasma membrane, and its rotation spins a long, helical filament that propels cells through water. At top speed, flagellar movement can drive a bacterial cell through water at 60 cell lengths per second. In contrast, the fastest animal in the ocean—the sailfish—can swim at a mere 10 body lengths per second.

Fimbriae (singular: **fimbria**) are needlelike projections that extend from the plasma membrane of some bacteria and promote attachment to other cells or surfaces. These structures are

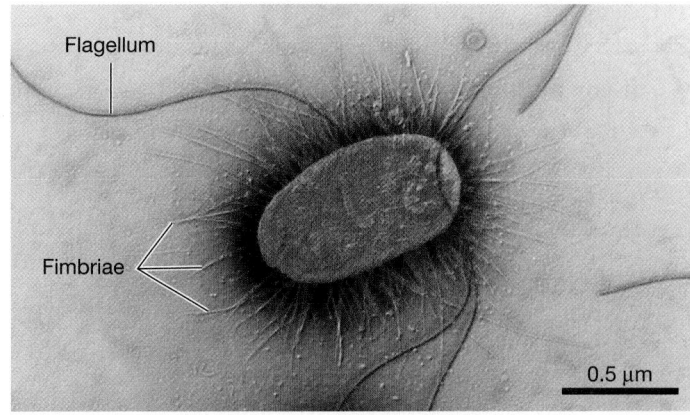

FIGURE 7.4 Extracellular Appendages Found on Bacteria. Some species of bacteria, such as the *E. coli* shown here, assemble large protein structures used for swimming through liquid (flagella) or adhering to surfaces (fimbriae).

more numerous than flagella and are often distributed over the entire surface of the cell. Fimbriae are crucial to the establishment of many infections based on their ability to glue bacteria to the surface of tissues.

The painting in **FIGURE 7.5** shows a cross section of a bacterial cell and provides a close-up view of the internal and external structures introduced in this section. One feature that prokaryotic and eukaryotic cells have in common: They are both packed with dynamic, highly integrated structures.

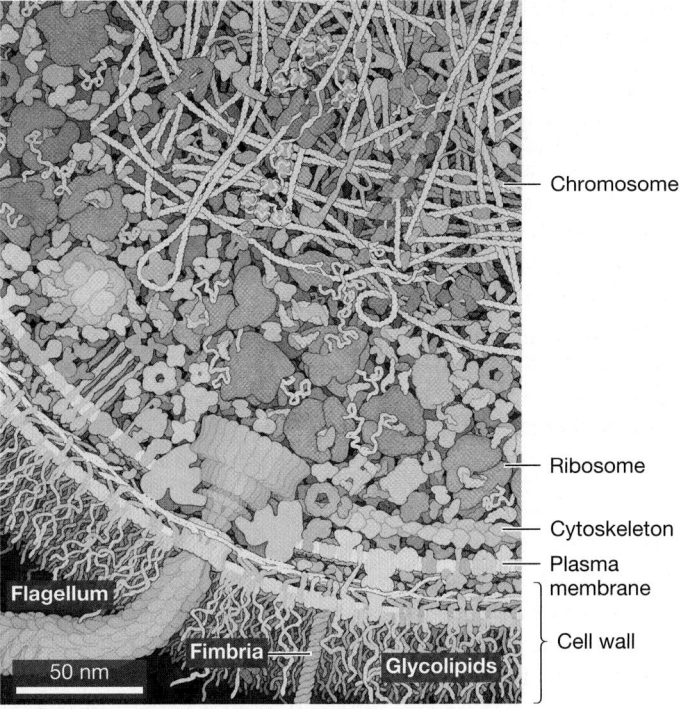

FIGURE 7.5 Close-up View of a Prokaryotic Cell. This painting is David Goodsell's representation of a cross section through part of a bacterial cell. It is based on electron micrographs of bacterial cells and is drawn to scale. Note that the cell is packed with proteins, DNA, ribosomes, and other molecular machinery.

If you understand that . . .

- Each structure in a prokaryotic cell performs a function vital to the cell.

✔ **You should be able to . . .**

Describe the structure and function of (1) the nucleoid, (2) photosynthetic membranes, (3) flagella, and (4) the cell wall.

Answers are available in Appendix A.

7.2 Eukaryotic Cell Structures and Their Functions

The Eukarya includes species that range from microscopic algae to 100-meter-tall redwood trees. Brown algae, red algae, fungi, amoebae, slime molds, green plants, and animals are all eukaryotic. Although multicellularity has evolved several times among eukaryotes (see Chapter 30), many species are unicellular.

The first thing that strikes biologists about eukaryotic cells is how much larger they are on average than bacteria and archaea. Most prokaryotic cells measure 1 to 10 μm in diameter, while most eukaryotic cells range from about 5 to 100 μm in diameter. A micrograph of an average eukaryotic cell, at the same scale as the bacterial cell in Figure 7.3, would fill this page. For many species of unicellular eukaryotes, this size difference allows them to make a living by ingesting bacteria and archaea whole.

Large size has a downside, however. As a cell increases in diameter, its volume increases more than its surface area. In other words, the relationship between them—the surface-area-to-volume ratio—changes. Since the surface is where the cell exchanges substances with its environment, the reduction in this ratio decreases the rate of exchange: Diffusion only allows for rapid movement across very small distances.

Prokaryotic cells tend to be small enough so that ions and small molecules arrive where they are needed via diffusion. The random movement of diffusion alone, however, is insufficient for this type of transport as the cell's diameter increases.

The Benefits of Organelles

How do eukaryotic cells solve the problems that size can engender? The answer lies in their numerous organelles. In effect, the huge volume inside a eukaryotic cell is compartmentalized into many small bins. Because eukaryotic cells are subdivided, the **cytosol**—the fluid portion between the plasma membrane and these organelles—is only a fraction of the total cell volume. This relatively small volume of cytosol reduces the effect of the total cell surface-area-to-volume ratio with respect to the exchange of nutrients and waste products.

Compartmentalization also offers two key advantages:

1. Incompatible chemical reactions can be separated. For example, new fatty acids can be synthesized in one organelle while

	Bacteria and Archaea	Eukaryotes
Location of DNA	In nucleoid (not membrane bound); plasmids also common	Inside nucleus (membrane bound); plasmids extremely rare
Internal Membranes and Organelles	Extensive internal membranes only in photosynthetic species; limited types and numbers of organelles	Large numbers of organelles; many types of organelles
Cytoskeleton	Limited in extent, relative to eukaryotes	Extensive—usually found throughout volume of cell
Overall Size	Usually small relative to eukaryotes	Most are larger than prokaryotes

excess or damaged fatty acids are degraded and recycled in a different organelle.

2. Chemical reactions become more efficient. First, the substrates required for particular reactions can be localized and maintained at high concentrations within organelles. Second, if substrates are used up in a particular part of the organelle, they can be replaced by substrates that have only a short distance to diffuse. Third, groups of enzymes that work together can be clustered within or on the membranes of organelles instead of floating free in the cytosol. When the product of one reaction is the substrate for a second reaction catalyzed by another enzyme, clustering the enzymes increases the speed and efficiency of both reaction sequences.

If bacteria and archaea can be compared to small, specialized machine shops, then eukaryotic cells resemble sprawling industrial complexes. The organelles and other structures found in eukaryotes are analogous to highly specialized buildings that act as administrative centers, factories, transportation corridors, waste and recycling facilities, warehouses, and power stations.

When typical prokaryotic and eukaryotic cells are compared, four key differences, identified in **TABLE 7.1**, stand out:

1. Eukaryotic chromosomes are found inside a membrane-bound compartment called the **nucleus.**

2. Eukaryotic cells are often much larger than prokaryotes.

3. Eukaryotic cells contain extensive amounts of internal membrane.

4. Eukaryotic cells feature a particularly diverse and dynamic cytoskeleton.

Eukaryotic Cell Structures: A Parts List

FIGURE 7.6 provides a simplified view of a typical animal cell and a plant cell. The artist has removed most of the cytoskeletal elements

(a) Generalized animal cell

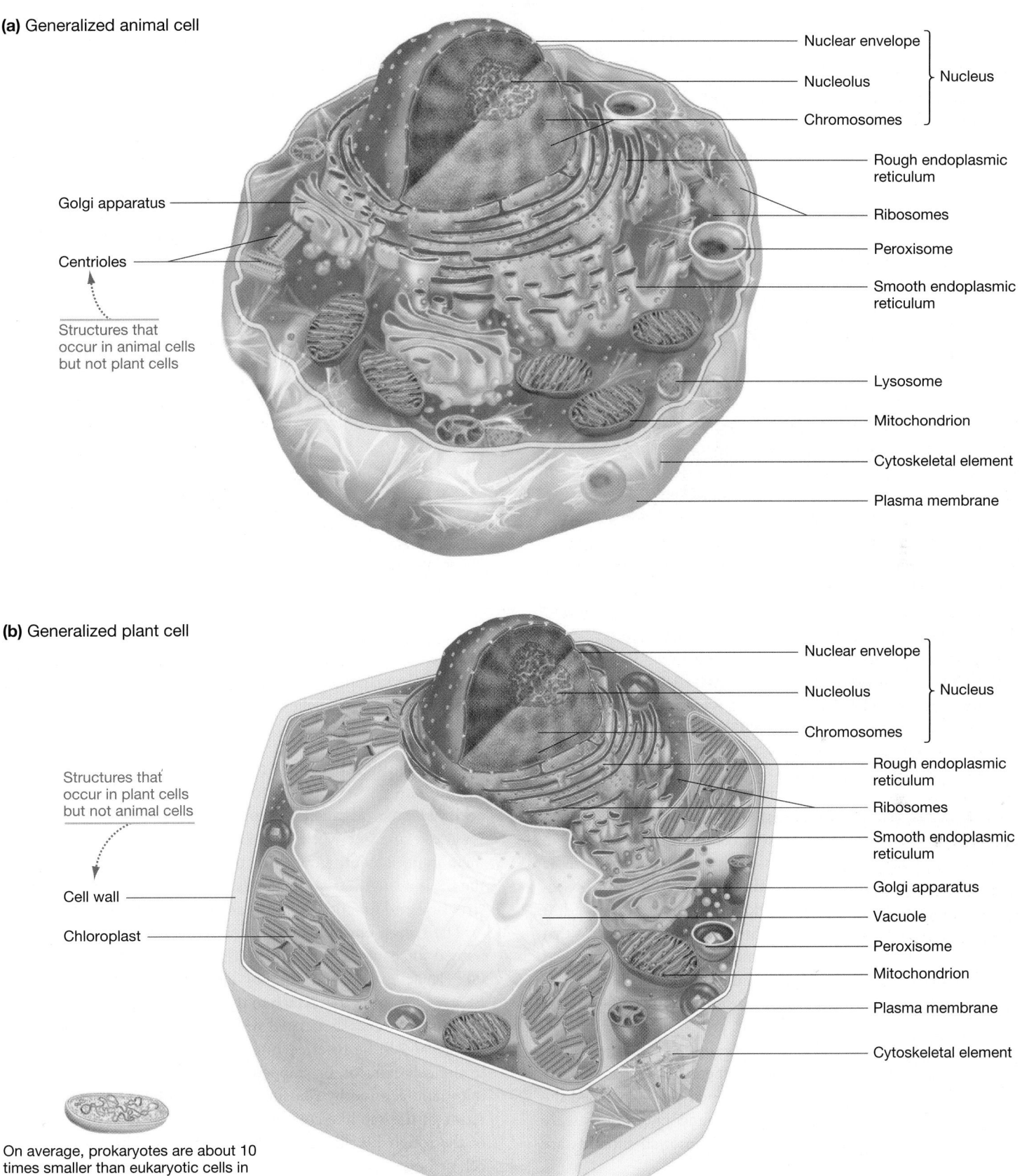

Nuclear envelope ⎫
Nucleolus   ⎬ Nucleus
Chromosomes   ⎭

Rough endoplasmic reticulum

Ribosomes

Peroxisome

Smooth endoplasmic reticulum

Golgi apparatus

Centrioles

Structures that
occur in animal cells
but not plant cells

Lysosome

Mitochondrion

Cytoskeletal element

Plasma membrane

(b) Generalized plant cell

Nuclear envelope ⎫
Nucleolus   ⎬ Nucleus
Chromosomes   ⎭

Rough endoplasmic reticulum

Ribosomes

Smooth endoplasmic reticulum

Structures that
occur in plant cells
but not animal cells

Golgi apparatus

Vacuole

Cell wall

Peroxisome

Chloroplast

Mitochondrion

Plasma membrane

Cytoskeletal element

On average, prokaryotes are about 10
times smaller than eukaryotic cells in
diameter and about 1000 times smaller
than eukaryotic cells in volume.

FIGURE 7.6 Overview of Eukaryotic Cells. Generalized images of **(a)** animal and **(b)** plant cells that illustrate the
cellular structures in the "typical" eukaryote. The structures have been color-coded for clarity. Compare with the
prokaryotic cell, shown at true relative size at bottom left.

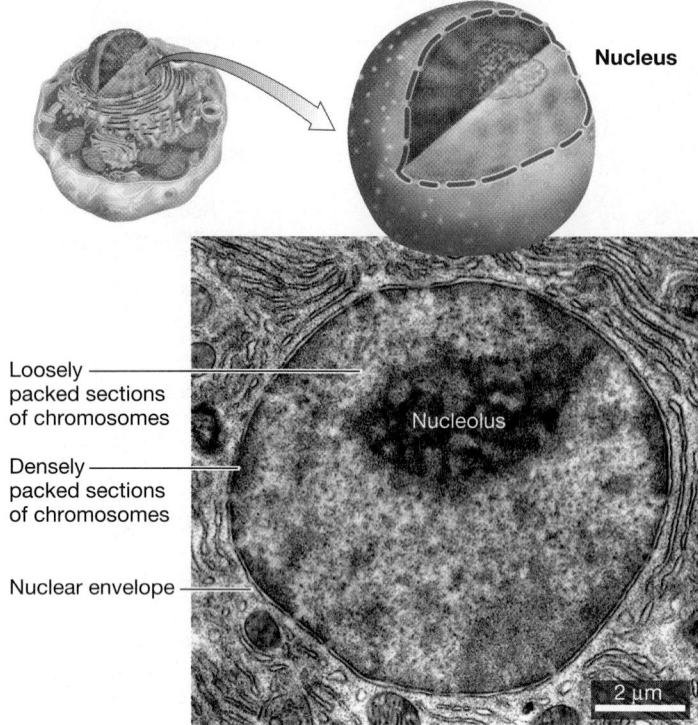

FIGURE 7.7 The Nucleus Stores and Transmits Information. The genetic, or hereditary, information is encoded in DNA, which is a component of the chromosomes inside the nucleus.

Loosely packed sections of chromosomes

Densely packed sections of chromosomes

Nuclear envelope

Nucleolus

2 μm

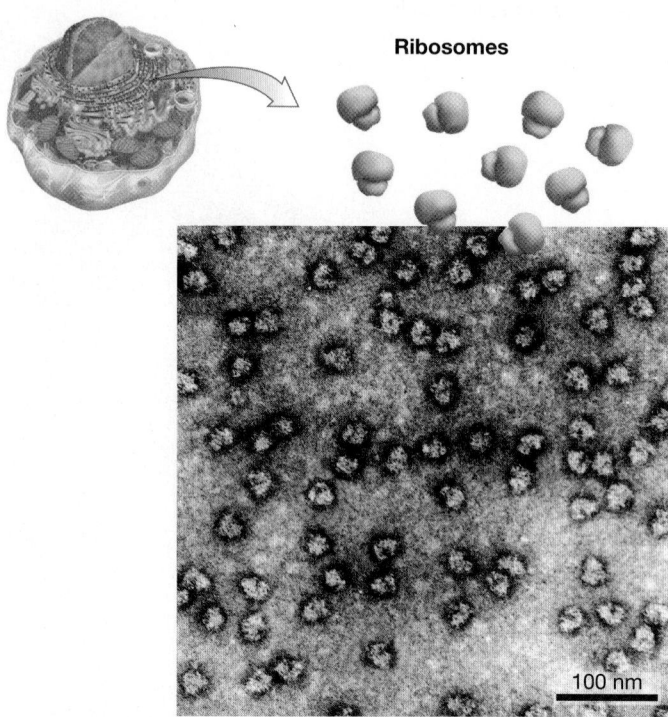

FIGURE 7.8 Ribosomes Are the Site of Protein Synthesis. Eukaryotic ribosomes are larger than bacterial and archaeal ribosomes, but similar in overall structure.

100 nm

to make the organelles and other cellular parts easier to see. As you read about each cell component in the pages that follow, focus on identifying how its structure correlates with its function. Then use **TABLE 7.2** (see page 117) as a study guide. As with bacterial cells, let's start from the inside and move to the outside.

The Nucleus The nucleus contains the chromosomes and functions as an administrative center for information storage and processing. Among the largest and most highly organized of all organelles (**FIGURE 7.7**), it is enclosed by a unique structure—a complex double membrane called the **nuclear envelope.** As Section 7.4 will detail, the nuclear envelope is studded with pore-like openings, and the inside surface is linked to fibrous proteins that form a lattice-like sheet called the **nuclear lamina.** The nuclear lamina stiffens the structure and maintains its shape.

Chromosomes do not float freely inside the nucleus—instead, each chromosome occupies a distinct area, which may vary in different cell types and over the course of cell replication. The nucleus also contains specific sites where gene products are processed and includes at least one distinctive region called the **nucleolus,** where the RNA molecules found in ribosomes are manufactured and the large and small ribosomal subunits are assembled.

Ribosomes In eukaryotes, the cytoplasm consists of everything inside the plasma membrane excluding the nucleus. Scattered throughout this cytoplasm are millions of ribosomes (**FIGURE 7.8**).

Like bacterial ribosomes, eukaryotic ribosomes are complex macromolecular machines that manufacture proteins. They are not classified as organelles because they are not surrounded by membranes.

Endoplasmic Reticulum The portions of the nuclear envelope extend into the cytoplasm to form an extensive membrane-enclosed factory called the **endoplasmic reticulum** (literally, "inside-formed-network"), or ER. As Figure 7.6 shows, the ER membrane is continuous with the nuclear envelope. Although the ER is a single structure, it has two regions that are distinct in structure and function. Let's consider each region in turn.

The **rough endoplasmic reticulum (RER),** or **rough ER,** is named for its appearance in transmission electron micrographs (see **FIGURE 7.9**, left). The knobby-looking structures in the rough ER are ribosomes that attach to the membrane.

The ribosomes associated with the rough ER synthesize proteins that will be inserted into the plasma membrane, secreted to the cell exterior, or shipped to an organelle. As they are being manufactured by ribosomes, these proteins move to the interior of the sac-like component of the rough ER. The interior of the rough ER, like the interior of any sac-like structure in a cell or body, is called the **lumen.** In the lumen of the rough ER, newly manufactured proteins undergo folding and other types of processing.

The proteins produced in the rough ER have a variety of functions. Some carry messages to other cells; some act as membrane

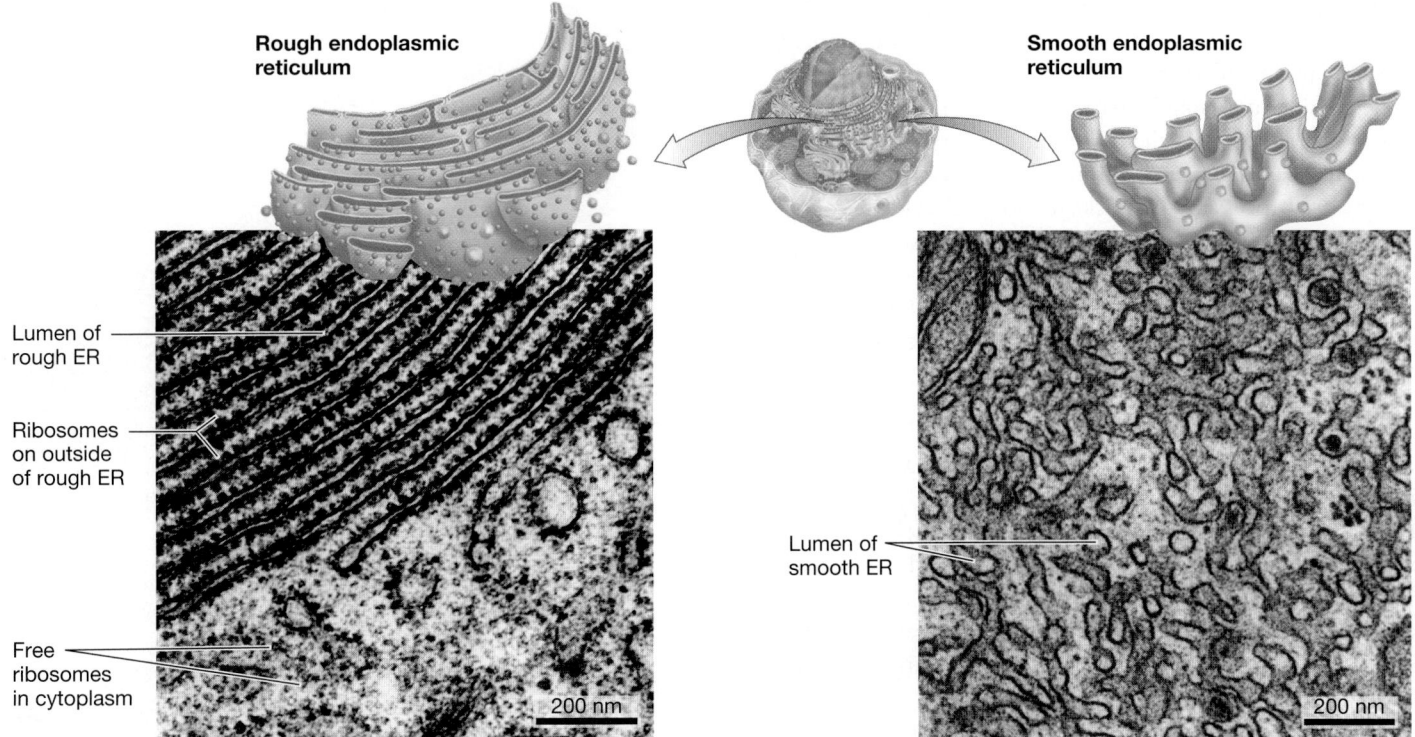

Rough endoplasmic reticulum

Lumen of rough ER

Ribosomes on outside of rough ER

Free ribosomes in cytoplasm

200 nm

Smooth endoplasmic reticulum

Lumen of smooth ER

200 nm

FIGURE 7.9 The Endoplasmic Reticulum Is a Site of Synthesis, Processing, and Storage. The ER is continuous with the nuclear envelope and possesses two distinct regions: on the left, the rough ER is a system of membrane-bound sacs and tubules with ribosomes attached; on the right, the smooth ER is a system of membrane-bound sacs and tubules that lacks ribosomes.

transport proteins or pumps; others are enzymes. The common theme is that many of the rough ER products are packaged into vesicles and transported to various distant destinations—often to the surface of the cell or beyond.

In electron micrographs, parts of the ER that are free of ribosomes appear smooth and even. Appropriately, these parts of the ER are called the **smooth endoplasmic reticulum (SER),** or **smooth ER** (see **FIGURE 7.9**, right).

The smooth ER contains enzymes that catalyze reactions involving lipids. Depending on the type of cell, these enzymes may synthesize lipids needed by the organism or break down lipids and other molecules that are poisonous. For example, the smooth ER is the manufacturing site for phospholipids used in plasma membranes. In addition, the smooth ER functions as a reservoir for calcium ions (Ca^{2+}) that act as a signal triggering a wide array of activities inside the cell.

The structure of the endoplasmic reticulum correlates closely with its function. The rough ER has ribosomes and functions primarily as a protein-manufacturing center; the smooth ER lacks ribosomes and functions primarily as a lipid-processing center.

Golgi Apparatus In many cases, the products of the rough ER pass through the Golgi apparatus before they reach their final destination. The **Golgi apparatus** consists of discrete flattened, membranous sacs called **cisternae** (singular: **cisterna**), which are stacked on top of one another (**FIGURE 7.10**). The organelle also

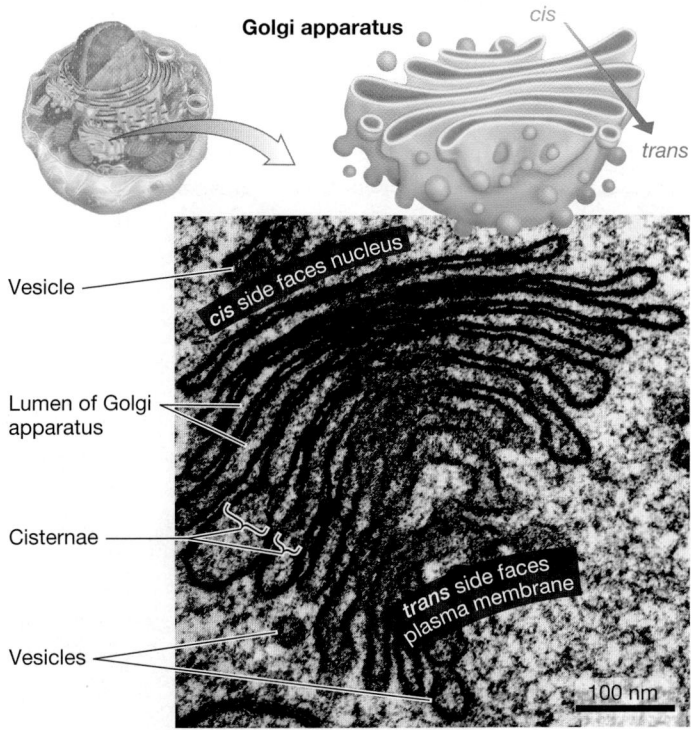

Golgi apparatus

cis

trans

Vesicle

cis side faces nucleus

Lumen of Golgi apparatus

Cisternae

trans side faces plasma membrane

Vesicles

100 nm

FIGURE 7.10 The Golgi Apparatus Is a Site of Protein Processing, Sorting, and Shipping. The Golgi apparatus is a collection of flattened vesicles called cisternae.

has a distinct polarity, or sidedness. The *cis* ("this side") surface is closest to the nucleus, and the *trans* ("across") surface is oriented toward the plasma membrane.

The *cis* side of a Golgi apparatus receives products from the rough ER, and the *trans* side ships them out to other organelles or the cell surface. In between, within the cisternae, the rough ER's products are processed and packaged for delivery. Micrographs often show "bubbles" on either side of the Golgi stack. These are membrane-bound vesicles that carry proteins or other products to and from the organelle. Section 7.5 analyzes the intracellular movement of molecules from the rough ER to the Golgi apparatus and beyond in more detail.

Lysosomes   Animal cells contain organelles called **lysosomes** that function as recycling centers (**FIGURE 7.11**). Lysosomes contain about 40 different enzymes, each specialized for hydrolyzing different types of macromolecules—proteins, nucleic acids, lipids, or carbohydrates. The amino acids, nucleotides, sugars, and other molecules that result from acid hydrolysis leave the lysosome via transport proteins in the organelle's membrane. Once in the cytosol, they can be used as sources of energy or building blocks for new molecules.

These digestive enzymes are collectively called acid hydrolases because under acidic conditions (pH of 5.0), they use water to break monomers from macromolecules. In the cytosol, where the pH is about 7.2, acid hydrolases are less active. Proton pumps in the lysosomal membrane maintain an acidic pH in the lumen of the lysosome by importing hydrogen ions.

Even though lysosomes are physically separated from the Golgi apparatus and the endoplasmic reticulum, these various

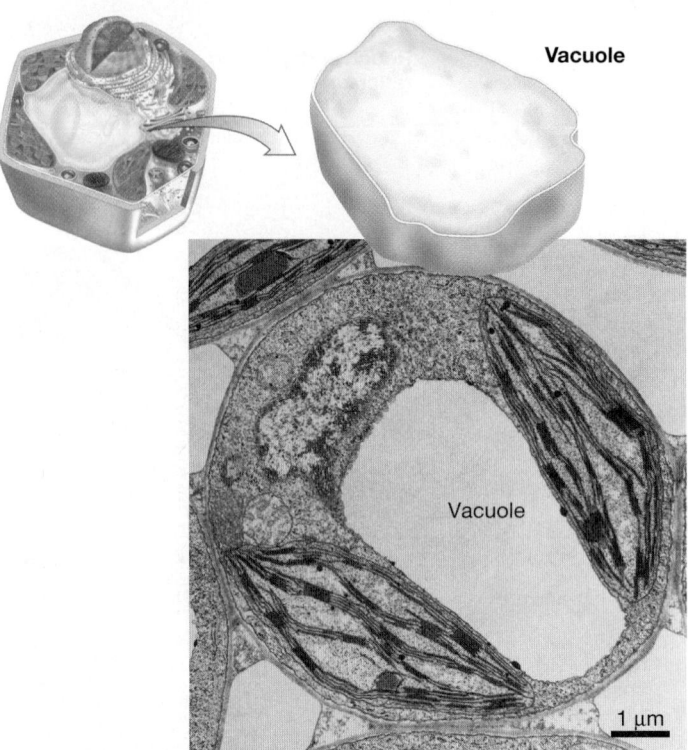

Vacuole

Vacuole

FIGURE 7.12 Vacuoles Are Generally Storage Centers in Plant and Fungal Cells. Vacuoles vary in size and function. Some contain digestive enzymes and serve as recycling centers; most are large storage containers.

✔**QUESTION** Why are toxins like nicotine, cocaine, and caffeine stored in vacuoles instead of the cytosol?

organelles jointly form a key functional grouping referred to as the **endomembrane system.** The endomembrane ("inner-membrane") system is a center for producing, processing, and transporting proteins and lipids in eukaryotic cells. For example, acid hydrolases are synthesized in the ER, processed in the Golgi, and then shipped to the lysosome.

Vacuoles   The cells of plants, fungi, and certain other groups lack lysosomes. Instead, they contain a prominent organelle called a vacuole. Compared with the lysosomes of animal cells, the **vacuoles** of plant and fungal cells are large—sometimes taking up as much as 80 percent of a plant cell's volume (**FIGURE 7.12**).

Although some vacuoles contain enzymes that are specialized for digestion, most of the vacuoles observed in plant and fungal cells act as storage depots. In many cases, ions such as potassium (K^+) and chloride (Cl^-), among other solutes, are stored at such high concentrations they draw water in from the environment. As the vacuole expands in volume, the cytoplasm pushes the plasma membrane against the cell wall, which maintains the plant cell's shape. In other cells, vacuoles have more specialized storage functions:

- Inside seeds, cells may contain a large vacuole filled with proteins. When the embryonic plant inside the seed begins to grow, enzymes begin digesting these proteins to provide amino acids for the growing individual.

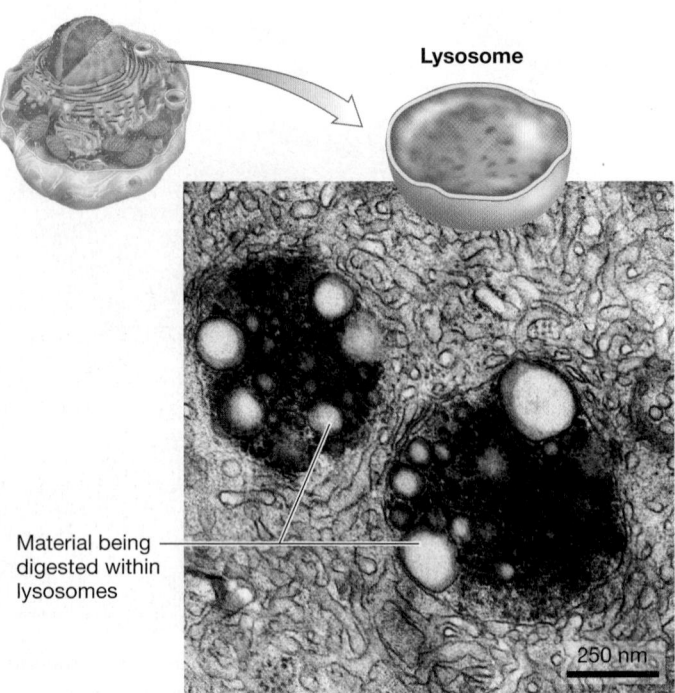

Lysosome

Material being digested within lysosomes

250 nm

FIGURE 7.11 Lysosomes Are Recycling Centers. Lysosomes are usually oval or globular and have a single membrane.

- In cells that make up flower petals or fruits, vacuoles are filled with colorful pigments.

- Elsewhere, vacuoles may be packed with noxious compounds that protect leaves and stems from being eaten by predators. The type of chemical involved varies by species, ranging from bitter-tasting tannins to toxins such as nicotine, morphine, caffeine, or cocaine.

Peroxisomes Virtually all eukaryotic cells contain globular organelles called **peroxisomes** (**FIGURE 7.13**). These organelles have a single membrane and originate as buds from the ER.

Although different types of cells from the same individual may have distinct types of peroxisomes, these organelles all share a common function: Peroxisomes are centers for reduction–oxidation (redox) reactions. (Chapter 8 explains in detail how redox reactions transfer electrons between atoms and molecules.) For example, the peroxisomes in your liver cells contain enzymes that remove electrons from, or oxidize, the ethanol in alcoholic beverages.

Different types of peroxisomes contain different suites of redox enzymes. In the leaves of plants, specialized peroxisomes called **glyoxysomes** are packed with enzymes that oxidize fats to form a compound that can be used to store energy for the cell. But plant seeds have a different type of peroxisome—one that is packed with enzymes responsible for releasing energy from stored fatty acids. The young plant uses this energy as it begins to grow.

In animals and plants, the products of these reactions often include hydrogen peroxide (H_2O_2), which is highly reactive.

If hydrogen peroxide escaped from the peroxisome, it would quickly react with and damage DNA, proteins, and cellular membranes. This event is rare, however, because inside the peroxisome, the enzyme catalase quickly "detoxifies" hydrogen peroxide by catalyzing its oxidation to form water and oxygen. The enzymes found inside the peroxisome make a specialized set of oxidation reactions possible and safe for the cell.

Mitochondria The energy required to build these organelles and do other types of work comes from adenosine triphosphate (ATP), most of which is produced in the cell's **mitochondria** (singular: **mitochondrion**).

As **FIGURE 7.14** shows, each mitochondrion has two membranes. The outer membrane defines the organelle's surface, while the inner membrane is connected to a series of sac-like **cristae.** The solution enclosed within the inner membrane is called the **mitochondrial matrix.** In eukaryotes, most of the enzymes and molecular machines responsible for synthesizing ATP are embedded in the membranes of the cristae or suspended in the matrix (see Chapter 9). Depending on the type of cell, from 50 to more than a million mitochondria may be present.

Each mitochondrion has many copies of a small, circular chromosome that is independent of the nuclear chromosomes. This mitochondrial DNA contains only around 37 genes in most eukaryotes—most of the genes responsible for the function of the organelle reside in the nuclear DNA.

Among the genes present in mitochondrial DNA are those that encode RNAs for mitochondrial ribosomes. These ribosomes are

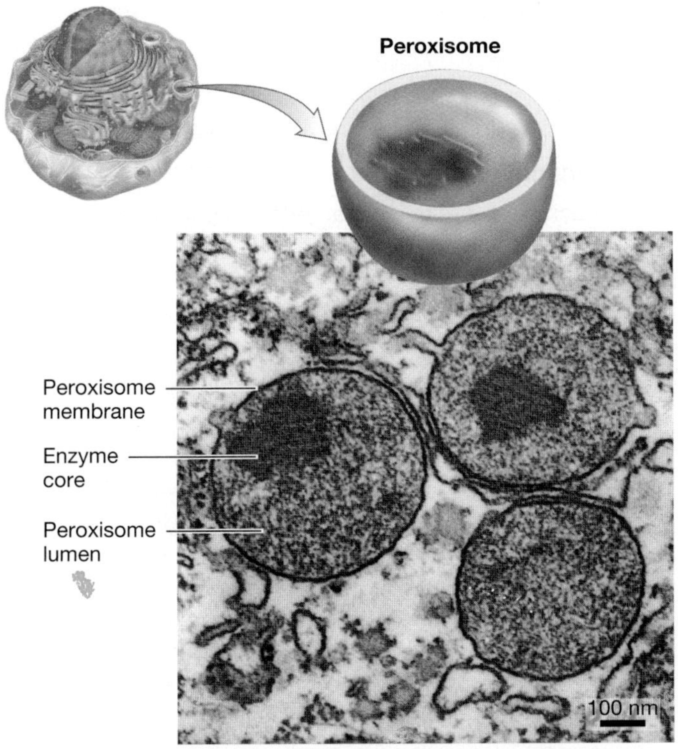

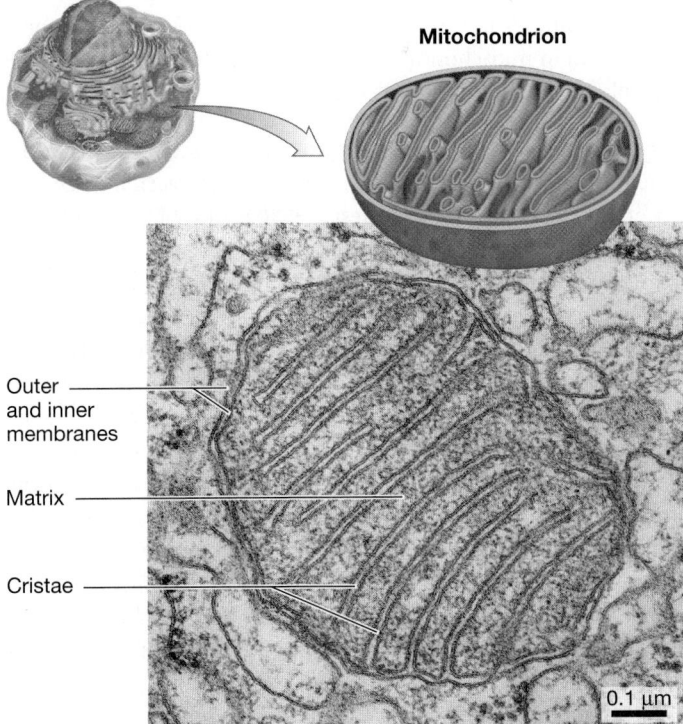

FIGURE 7.13 Peroxisomes Are the Site of Oxidation Reactions.
Peroxisomes are globular organelles that are defined by a single membrane.

Labels (left figure): Peroxisome / Peroxisome membrane / Enzyme core / Peroxisome lumen / 100 nm

FIGURE 7.14 Mitochondria Are Power-Generating Stations.
Mitochondria vary in size and shape, but all have a double membrane with sac-like cristae inside.

Labels (right figure): Mitochondrion / Outer and inner membranes / Matrix / Cristae / 0.1 μm

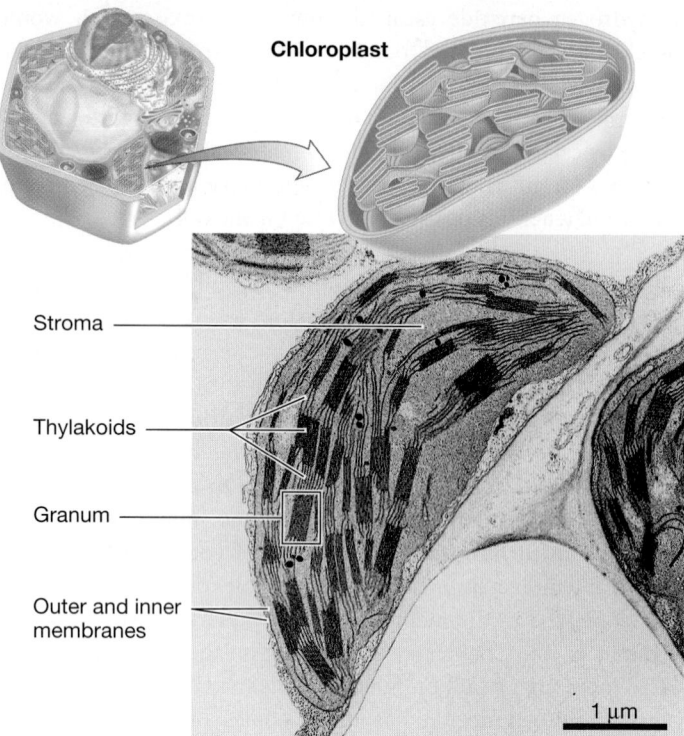

Chloroplast

Stroma

Thylakoids

Granum

Outer and inner membranes

1 μm

FIGURE 7.15 Chloroplasts Are Sugar-Manufacturing Centers in Plants and Algae. Many of the enzymes and other molecules required for photosynthesis are located in membranes inside the chloroplast. These membranes form thylakoids that consist of discs stacked into grana.

smaller than those found in the cytosol, yet they still function to produce some of the mitochondrial proteins. (Most of the proteins found in mitochondria are produced from ribosomes in the cytosol and imported into the organelle.)

Chloroplasts   Most algal and plant cells possess an organelle called the **chloroplast,** in which sunlight is converted to chemical energy during photosynthesis (**FIGURE 7.15**). The number of chloroplasts per cell varies from none to several dozen.

The chloroplast has a double membrane around its exterior, analogous to the structure of a mitochondrion. Instead of featuring sac-like cristae that connect to the inner membrane, though, the interior of the chloroplast is dominated by a network of hundreds of membrane-bound, flattened, sac-like structures called **thylakoids,** which are independent of the inner membrane.

Thylakoids have stacks, like pancakes, that are called **grana** (singular: **granum**). Many of the pigments, enzymes, and macromolecular machines responsible for converting light energy into chemical energy are embedded in the thylakoid membranes (see Chapter 10). The region outside the thylakoids, called the **stroma,** contains enzymes that use this chemical energy to produce sugars.

Like mitochondria, each chloroplast contains copies of a circular chromosome and small ribosomes that manufacture some, but not all, of the organelle's proteins. Both mitochondria

and chloroplasts also grow and divide independently of cell division through a process that resembles bacterial fission (see Chapter 12).

These attributes are odd compared with those of the other organelles and have led biologists to propose that mitochondria and chloroplasts were once free-living bacteria. According to the **endosymbiosis theory,** the ancestors of modern eukaryotes ingested these bacteria, but instead of destroying them, established a mutually beneficial relationship with them. (In Chapter 30, you will learn more about the origins of these eukaryotic organelles.)

Cytoskeleton   The final major structural feature that is common to all eukaryotic cells is the cytoskeleton, an extensive system of protein fibers. In addition to giving the cell its shape and structural stability, cytoskeletal proteins are involved in moving the cell itself and moving materials within the cell. In essence, the cytoskeleton organizes all the organelles and other cellular structures into a cohesive whole. Section 7.6 will analyze the structure and functions of the cytoskeleton in detail.

The Cell Wall   In fungi, algae, and plants, cells possess an outer cell wall in addition to their plasma membrane. The cell wall is located outside the plasma membrane and furnishes a durable, outer layer that provides structural support for the cell. The cells of animals, amoebae, and other groups lack a cell wall—their exterior surface consists of the plasma membrane only.

Although the composition of the cell wall varies among species and even among types of cells in the same individual, the general plan is similar: Rods or fibers composed of a carbohydrate run through a stiff matrix made of other polysaccharides and proteins (see Chapter 11 for details).

check your understanding

 If you understand that . . .

- Each structure in a eukaryotic cell performs a function vital to the cell.
- In eukaryotes, many of the cellular functions are compartmentalized into organelles.

✔ **You should be able to . . .**

1. Explain how the structure of lysosomes and peroxisomes correlates with their function.

2. In Table 7.2, label each component with one of the following roles: administrative/information hub, power station, warehouse, large molecule manufacturing and shipping facility (with subtitles for lipid factory, protein finishing and shipping line, protein synthesis and folding center, waste processing and recycling center), support beams, perimeter fencing with secured gates, protein factory, food-manufacturing facility, and fatty-acid processing and detox center.

Answers are available in Appendix A.

SUMMARY TABLE 7.2 Eukaryotic Cell Components

Icons Not to Scale		Structure		
		Membrane	**Components**	**Function**
	Nucleus	Double ("envelope"); openings called nuclear pores	Chromosomes	Information storage and transmission
			Nucleolus	Ribosome subunit assembly
			Nuclear lamina	Structural support
	Ribosomes	None	Complex of RNA and proteins	Protein synthesis
	Endomembrane system			
	Endoplasmic reticulum: rough	Single; contains receptors for entry of selected proteins	Network of branching sacs Ribosomes associated	Protein synthesis and processing
	Endoplasmic reticulum: smooth	Single; contains enzymes for synthesizing phospholipids	Network of branching sacs Enzymes for synthesizing or breaking down lipids	Lipid synthesis and processing
	Golgi apparatus	Single; contains receptors for products of rough ER	Stack of flattened, distinct cisternae	Protein, lipid, and carbohydrate processing
	Lysosomes	Single; contains proton pumps	Acid hydrolases (catalyze hydrolysis reactions)	Digestion and recycling
	Vacuoles	Single; contains transporters for selected molecules	Varies—pigments, oils, carbohydrates, water, or toxins	Varies—coloration, storage of oils, carbohydrates, water, or toxins
	Peroxisomes	Single; contains transporters for selected macromolecules	Enzymes that catalyze oxidation reactions Catalase (processes peroxide)	Oxidation of fatty acids, ethanol, or other compounds
	Mitochondria	Double; inner contains enzymes for ATP production	Enzymes that harvest energy from molecules to make ATP	ATP production
	Chloroplasts	Double; plus membrane-bound sacs in interior	Pigments Enzymes that use light energy to make sugars	Production of sugars via photosynthesis
	Cytoskeleton	None	Actin filaments Intermediate filaments Microtubules	Structural support; movement of materials; in some species, movement of whole cell
	Plasma membrane	Single; contains transport and receptor proteins	Phospholipid bilayer with transport and receptor proteins	Selective permeability—maintains intracellular environment
	Cell wall	None	Carbohydrate fibers running through carbohydrate or protein matrix	Protection, structural support

7.3 Putting the Parts into a Whole

Within a cell, the structure of each component correlates with its function. In the same way, the overall size, shape, and composition of a cell correlate with its function.

Cells might be analogous to machine shops or industrial complexes, but clothing manufacturing centers are very different in layout and composition from airplane production facilities. How does the physical and chemical makeup of a cell correlate with its function?

Structure and Function at the Whole-Cell Level

Inside an individual plant or animal, cells are specialized for certain tasks and have a structure that correlates with those tasks. For example, the muscle cells in your upper leg are extremely long, tube-shaped structures. They are filled with protein fibers that slide past one another as the entire muscle flexes or relaxes. It is this sliding motion that allows your muscles to contract or extend as you run. Muscle cells are also packed with mitochondria, which produce the ATP required for the sliding motion to occur.

In contrast, nearby fat cells are rounded, globular structures that store fatty acids. They consist of little more than a plasma membrane, a nucleus, and a fat droplet. Neither cell bears a close resemblance to the generalized animal cell pictured in Figure 7.6a.

To drive home the correlation between the overall structure and function of a cell, examine the transmission electron micrographs in **FIGURE 7.16**.

- The animal cell in Figure 7.16a, located in the pancreas, manufactures and exports digestive enzymes. It is packed with rough ER and Golgi, which make these functions possible.

- The animal cell in Figure 7.16b, from the testis, synthesizes and exports the steroid hormone testosterone—a lipid-soluble signal. This cell is dominated by smooth ER, where processing of steroids and other lipids takes place.

- The plant cell in Figure 7.16c, from the leaf of a potato, has hundreds of chloroplasts and is specialized for absorbing light and manufacturing sugar.

- The animal cells in Figure 7.16d come from brown fat. The cells have numerous mitochondria that have been altered so they convert energy stored in fat into heat instead of ATP.

In each case, the types of organelles in each cell and their size and number correlate with the cell's specialized function.

The Dynamic Cell

Biologists study the structure and function of organelles and cells with a combination of tools and approaches. For several decades, a technique called **differential centrifugation** was particularly important because it allowed researchers to isolate particular cell components and analyze their chemical composition. Differential centrifugation is based on breaking cells apart to create a complex mixture and separating components in a centrifuge (see **BioSkills 10** in Appendix B). The individual parts of the cell

(a) Animal pancreatic cell: Exports digestive enzymes.

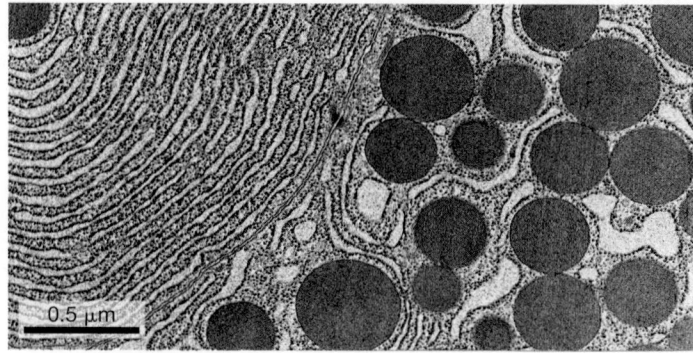

(b) Animal testis cell: Exports lipid-soluble signals.

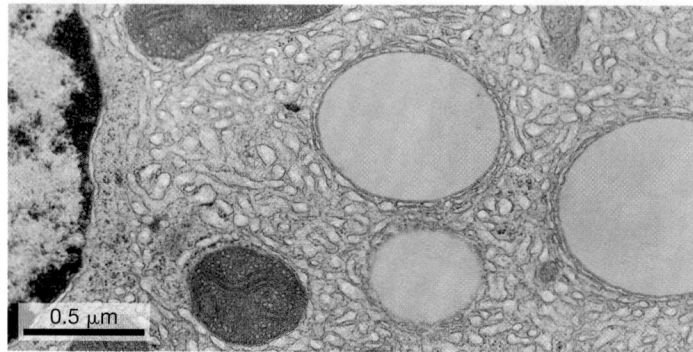

(c) Plant leaf cell: Manufactures ATP and sugar.

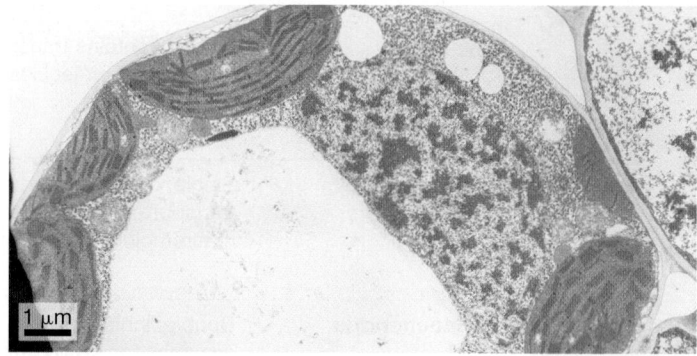

(d) Brown fat cells: Burn fat to generate heat in lieu of ATP.

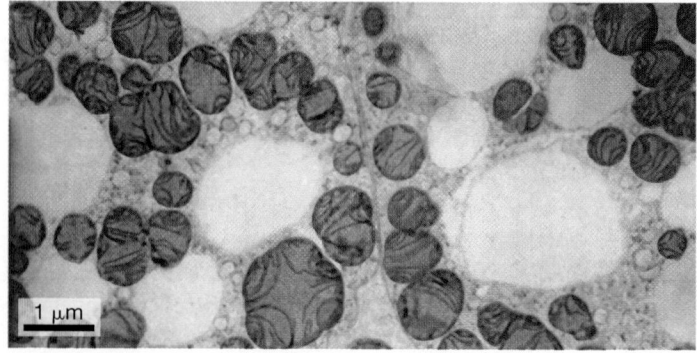

FIGURE 7.16 Cell Structure Correlates with Function.

✔**EXERCISE** In part (a), label the rough ER and the dark, round secretory vesicles. In (b), label the smooth ER. In (c), label the chloroplasts, vacuole, and nucleus. In (d), label the mitochondria.

can then be purified and studied in detail, in isolation from other parts of the cell.

Historically and currently, however, the most important research in cell biology is based on imaging—simply looking at cells. Recent innovations allow biologists to put fluorescing tags or other types of markers on particular cell components and then look at them with increasingly sophisticated light microscopes and electron microscopes. Advances in microscopy provide increasingly high magnification and better resolution.

It's important to recognize, though, that some of these techniques have limitations. Differential centrifugation splits cells into parts that are analyzed independently, and electron microscopy gives a fixed "snapshot" of the cell or organisms being observed. Neither technique allows investigators to explore directly how things move from place to place in the cell or how parts interact. The information gleaned from these techniques can make cells seem static. In reality, cells are dynamic.

The amount of chemical activity and the speed of molecular movement inside cells are nothing short of fantastic. Bacterial ribosomes add up to 20 amino acids per second to a growing polypeptide, and eukaryotic ribosomes typically add 2 per second. Given that there are about 15,000 ribosomes in each bacterium and possibly a million in an average eukaryotic cell, hundreds or even thousands of new protein molecules can be produced each second in every cell. Here are some other remarkable cellular feats:

- In an average second, a typical cell in your body uses an average of 10 million ATP molecules and synthesizes just as many.

- It's not unusual for a cellular enzyme to catalyze 25,000 or more reactions per second; most cells contain hundreds or thousands of different enzymes.

- A minute is more than enough time for each membrane phospholipid in your body to travel the breadth of the organelle or cell where it resides.

- The hundreds of trillions of mitochondria inside you are completely replaced about every 10 days, for as long as you live.

Because humans are such large organisms, it's impossible for us to imagine what life is really like inside a cell. At the scale of a ribosome or an organelle or a cell, gravity is inconsequential. Instead, the dominant forces are the charge- or polarity-based electrostatic attractions between molecules and their energy of motion. At this level, events take nanoseconds, and speeds are measured in micrometers per second. This is the speed of life.

Contemporary methods for studying cells (including some of the imaging techniques featured in **BioSkills 11** in Appendix B) capture this dynamism by tracking how organelles and molecules move and interact over time. The ability to digitize video images of live cells, or take time-lapse photographs of living cells, is allowing researchers to see and study dynamic processes.

The rest of this chapter focuses on this theme of cellular dynamism and movement. Its goal is to put some of the individual pieces of a cell together and ask how they work as systems to accomplish key tasks.

To begin, let's first look at how molecules move into and out of the cell's control center—the nucleus—and then consider how proteins move from ribosomes into the lumen of the rough ER and then to the Golgi apparatus and beyond. The chapter closes by introducing the cytoskeletal elements and their associated motor proteins and how they are used to transport cargo inside the cell or move the cell itself.

7.4 Cell Systems I: Nuclear Transport

The nucleus is the information center of eukaryotic cells—a corporate headquarters, design center, and library all rolled into one. Appropriately enough, its interior is highly organized.

The organelle's overall shape and structure are defined by the mesh-like nuclear lamina. The nuclear lamina provides an attachment point for the chromosomes, each of which occupies a well-defined region in the nucleus.

In addition, specific centers exist where the genetic information in DNA is decoded and processed. At these locations, large suites of enzymes interact to produce RNA messages from specific genes at specific times. Meanwhile, the nucleolus functions as the site of ribosome assembly.

Structure and Function of the Nuclear Envelope

The nuclear envelope separates the nucleus from the rest of the cell. Starting in the 1950s, transmission electron micrographs of cross sections through the nuclear envelope showed that the structure is supported by the fibrous nuclear lamina and bounded by two membranes, each consisting of a lipid bilayer. How does this administrative center communicate with the rest of the cell across the double membrane barrier?

Micrographs like the one in **FIGURE 7.17** (see page 120) show that the nuclear envelope is broken with openings, approximately 60 nanometers (nm) in diameter, called **nuclear pores.** Because these pores extend through both the inner and outer nuclear membranes, they connect the inside of the nucleus with the cytosol. Follow-up research showed that each pore consists of over 50 different proteins. As the diagram on the right side of Figure 7.17 shows, these protein molecules form an elaborate structure called the **nuclear pore complex.**

What substances traverse nuclear pores? Chromosomal DNA clearly does not—it remains in the nucleus as long as the nuclear envelope remains intact. But DNA is used to synthesize RNA inside the nucleus, most of which is exported through nuclear pores to the cytoplasm.

Several types of RNA molecules are produced, each distinguished by size and function. For example, **ribosomal RNAs** are manufactured in the nucleolus, where they bind to proteins to form ribosomes. Molecules called **messenger RNAs (mRNA)** carry the information required to manufacture proteins. Both the newly assembled ribosomes and the mRNAs must be transported from the nucleus to the cytoplasm, where protein synthesis takes place.

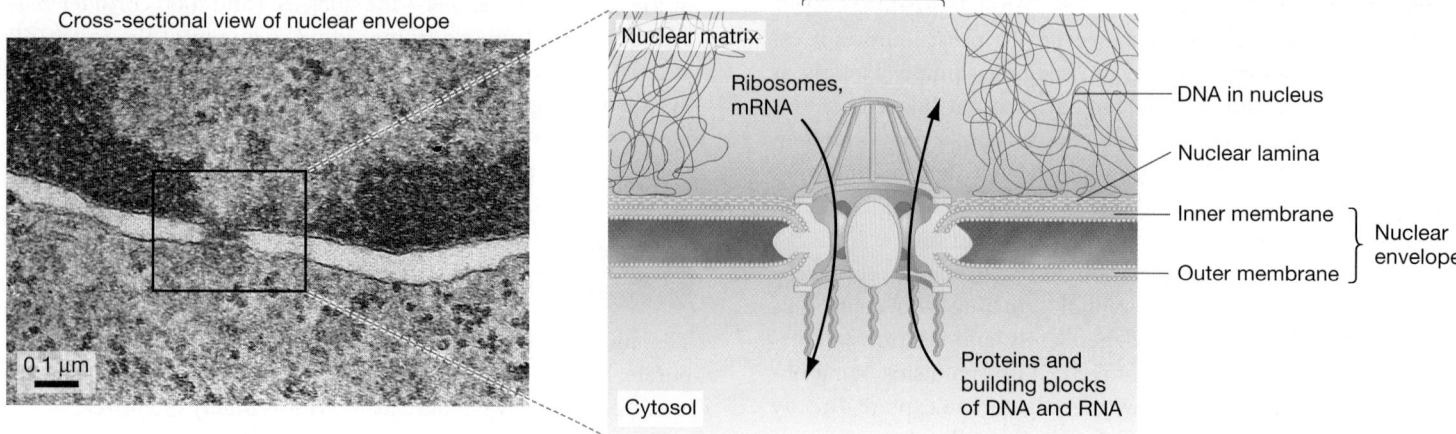

FIGURE 7.17 Structure of the Nuclear Envelope and Nuclear Pore Complex.

Inbound traffic is also impressive. Nucleoside triphosphates that act as building blocks for DNA and RNA enter the nucleus, as do a variety of proteins responsible for copying DNA, synthesizing RNAs, extending the nuclear lamina, or assembling ribosomes.

To summarize, ribosomal subunits and various types of RNAs exit the nucleus; nucleotides and certain proteins enter it. In a typical cell, over 500 molecules pass through each of the 3000–4000 nuclear pores every second. The scale of traffic through the nuclear pores is mind-boggling. How is it regulated and directed?

Experiments in the early 1960s showed that size matters in the passage of molecules through nuclear pores. This conclusion was based on the results from injecting tiny gold particles that varied in diameter and tracking their movement across the pores. In electron micrographs, gold particles show up as defined black dots that can be easily distinguished from cellular structures. Immediately after injection, most of the gold particles were observed in the cytoplasm, and only a few were closely associated with nuclear pores. Ten minutes after injection, only the small particles ($<$ 12.5 nm in diameter) appeared to be distributed throughout both the nucleus and the cytoplasm, and the larger particles were excluded from entering the nucleus.

The fact that the pore opening is almost 5 times larger than this 12.5-nm size limit supports the hypothesis that the nuclear pore complex serves as a gate to control passage through the envelope. If this is the case, then what is required to open these gates so that proteins larger than the size limit, like those responsible for replicating DNA, may pass?

How Do Large Molecules Enter the Nucleus?

It was clear to researchers that size was not the sole factor in selective transport across the nuclear envelope. Certain proteins were concentrated in the nucleus, while others were completely excluded—even if they were similar in size.

A series of experiments on a protein called nucleoplasmin helped researchers understand the nature of nuclear import. Nucleoplasmin is strictly found in the nucleus and plays an important role in the assembly of chromatin. When researchers labeled nucleoplasmin with a radioactive atom and injected it into the cytoplasm of living cells, they found that the radioactive protein was quickly concentrated into the nucleus. Is there a "send-to-nucleus" signal within the nucleoplasmin protein that is responsible for this directed transport?

As shown in **FIGURE 7.18**, the distinctive structure of nucleoplasmin was used to further investigate this process. First, researchers used enzymes called proteases to cleave the core sections of nucleoplasmin from the tails. After separating the tails from the core fragments, they labeled each component with radioactive atoms and injected them into the cytoplasm of different cells.

At various times after the injections, researchers examined the nuclei and cytoplasm of the cells to track down the radioactive label. The results were striking. They found that tail fragments were rapidly transported from the cytoplasm into the nucleus. Core fragments, in contrast, were not allowed to pass through the nuclear envelope and remained in the cytoplasm.

These data led to a key hypothesis: Nuclear proteins are synthesized by ribosomes in the cytosol and contain a "zip code"—a molecular address tag—that marks them for transport through the nuclear pore complex. This zip code allows the nuclear pore complex to open in some way that permits larger proteins and RNA molecules to pass through.

By analyzing different stretches of the tail, the biologists eventually found a 17-amino-acid-long section that had to be present to direct nucleoplasmin to the nucleus. Follow-up work confirmed that other proteins bound for the nucleus, even those expressed by some viruses, have similar amino acid sequences directing their transport. This common sequence came to be called the **nuclear localization signal (NLS).** Proteins that leave the nucleus have a different signal, required for nuclear export.

QUESTION: Does the nucleoplasmin protein contain a "Send to nucleus" signal?

HYPOTHESIS: Nucleoplasmin contains a discrete "Send to nucleus" signal that resides in either the tail or core region.

NULL HYPOTHESIS: Nucleoplasmin does not require a signal to enter the nucleus, or the entire protein serves as the signal.

EXPERIMENTAL SETUP:

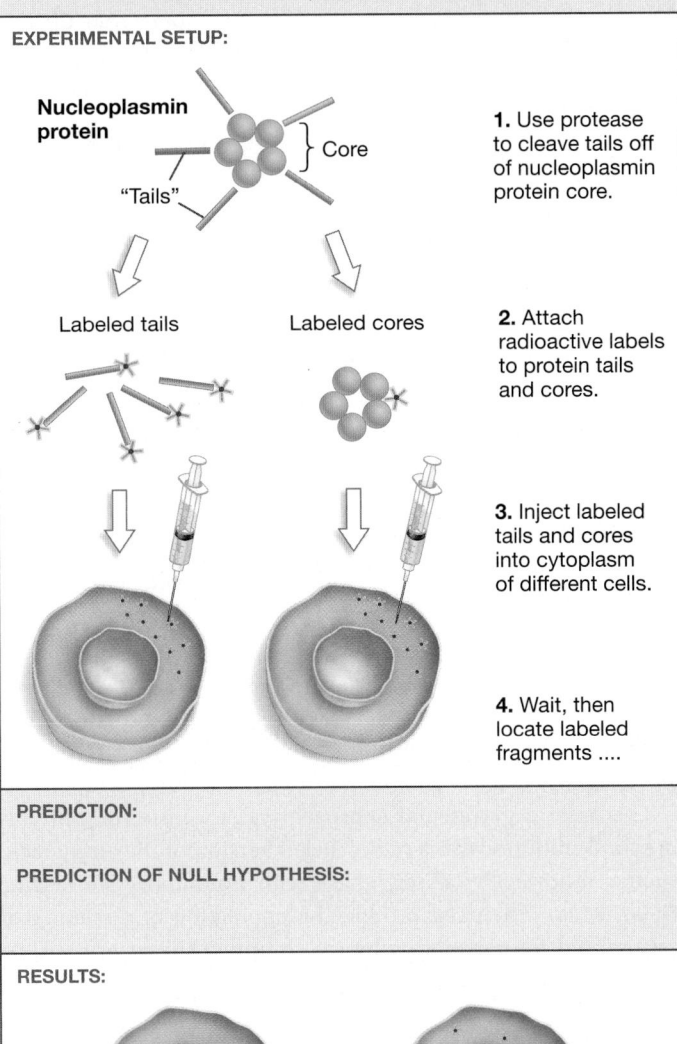

1. Use protease to cleave tails off of nucleoplasmin protein core.

2. Attach radioactive labels to protein tails and cores.

3. Inject labeled tails and cores into cytoplasm of different cells.

4. Wait, then locate labeled fragments ....

PREDICTION:

PREDICTION OF NULL HYPOTHESIS:

RESULTS:

Labeled tail fragments **located in nucleus**

Labeled core fragments still **located in cytoplasm**

CONCLUSION:

FIGURE 7.18 Does the Nucleoplasmin Protein Contain a "Send to Nucleus" Signal?

SOURCES: Mills, A. D., R. A. Laskey, P. Black, et al. 1980. *Journal of Molecular Biology* 139: 561–568; Dingwall, C., S. V. Sharnick, and R. A. Laskey. 1982. *Cell* 30: 449–458.

✔**EXERCISE** Without looking at the text, fill in the prediction(s) and conclusion(s) in this experiment.

More recent research has shown that the movement of proteins and other large molecules into and out of the nucleus is an energy-demanding process that involves special transport proteins. These nuclear transport proteins function like trucks that haul cargo into or out of the nucleus through the nuclear pore complex, depending on whether they have an import or export zip code. Currently, biologists are trying to unravel how all this traffic in and out of the nucleus is regulated to avoid backups and head-on collisions.

✔ If you understand the process of nuclear transport, you should be able to compare and contrast the movement of (1) nucleotides and (2) large proteins through the nuclear pore complex. Which would you expect to require the input of energy?

7.5 Cell Systems II: The Endomembrane System Manufactures, Ships, and Recycles Cargo

The nuclear membrane is not the only place in cells where cargo moves in a regulated and energy-demanding fashion. Most of the proteins found in peroxisomes, mitochondria, and chloroplasts are also actively imported from the cytosol. These proteins contain special signal sequences, like the nuclear localization signal, that target them to the appropriate organelles.

If you think about it for a moment, the need to sort proteins and ship them to specific destinations should be clear. Proteins are produced by ribosomes that are either free in the cytosol or on the surface of the ER. Many of these proteins must be transported to a compartment inside the eukaryotic cell. Acid hydrolases must be shipped to lysosomes and catalase to peroxisomes. To get to the right location, each protein has to have an address tag and a transport and delivery system.

To get a better understanding of protein sorting and transport in eukaryotic cells, let's consider perhaps the most intricate of all manufacturing and shipping complexes: the endomembrane system. In this system, proteins that are synthesized in the rough ER move to the Golgi apparatus for processing, and from there they travel to the cell surface or other destinations.

Studying the Pathway through the Endomembrane System

The idea that materials move through the endomembrane system in an orderly way was inspired by a simple observation. According to electron micrographs, cells that secrete digestive enzymes, hormones, or other products have particularly large amounts of rough ER and Golgi. This correlation led to the idea that these organelles may participate in a "secretory pathway" that starts in the rough ER and ends with products leaving the cell (**FIGURE 7.19**, see page 122). How does this hypothesized pathway work?

Tracking Protein Movement via Pulse–Chase Assay George Palade and colleagues did pioneering research on the secretory

FIGURE 7.19 The Secretory Pathway Hypothesis. The secretory pathway hypothesis proposes that proteins intended for secretion from the cell are synthesized and processed in a highly prescribed series of steps. Note that proteins are packaged into vesicles when they move from the RER to the Golgi and from the Golgi to the cell surface.

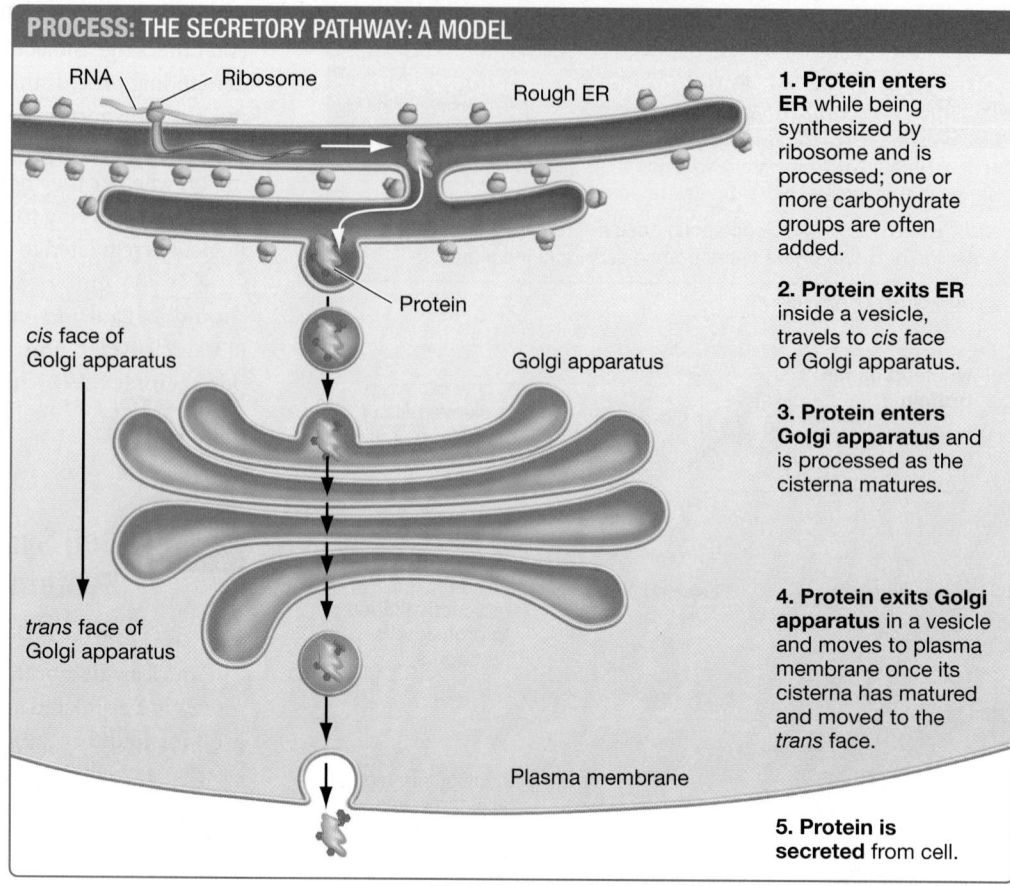

PROCESS: THE SECRETORY PATHWAY: A MODEL

RNA — Ribosome — Rough ER

Protein

cis face of Golgi apparatus

Golgi apparatus

trans face of Golgi apparatus

Plasma membrane

1. Protein enters ER while being synthesized by ribosome and is processed; one or more carbohydrate groups are often added.

2. Protein exits ER inside a vesicle, travels to *cis* face of Golgi apparatus.

3. Protein enters Golgi apparatus and is processed as the cisterna matures.

4. Protein exits Golgi apparatus in a vesicle and moves to plasma membrane once its cisterna has matured and moved to the *trans* face.

5. Protein is secreted from cell.

pathway using a **pulse–chase experiment** to track protein movement. This strategy is based on two steps:

1. *The "Pulse"* Expose experimental cells to a high concentration of a modified amino acid for a short time. For example, if a cell is briefly exposed to a large amount of radioactively labeled amino acid, virtually all the proteins synthesized during that interval will be radiolabeled.

2. *The "Chase"* The pulse ends by washing away the modified amino acid and replacing it with the normal version of the same molecule. The time following the end of the pulse is referred to as the chase. If the chase consists of unlabeled amino acid, then the proteins synthesized during the chase period will *not* be radiolabeled.

The idea is to mark a population of molecules at a particular interval and then follow their fate over time. This approach is analogous to adding a small amount of dye to a stream and then following the movement of the dye molecules.

To understand why the chase is necessary in these experiments, imagine what would happen if you added dye to a stream continuously. Soon the entire stream would be dyed—you could no longer tell where a specific population of dye molecules were moving.

In testing the secretory pathway hypothesis, Palade's team focused on pancreatic cells that were growing in **culture,** or

in vitro.[1] These cells are specialized for secreting digestive enzymes into the small intestine and are packed with rough ER and Golgi.

The basic experimental approach was to pulse the cell culture for 3 minutes with a radiolabeled version of the amino acid leucine, followed by a long chase with nonradioactive leucine (**FIGURE 7.20a**). The pulse produced a population of proteins that were related to one another by the timing of their synthesis. At different points during the chase, the researchers tracked the movement of these proteins by preparing samples of the cells for autoradiography and electron microscopy (see **BioSkills 10** and **11** in Appendix B). The drawings in **FIGURE 7.20b** illustrate what the researchers would have seen in micrographs taken at different times before and after the start of the chase.

Results of the Pulse–Chase Experiment The graph in Figure 7.20b was based on the electron microscopy results, which showed that proteins are trafficked through the secretory pathway in a highly organized and directed manner. Track the movement of proteins through the cell during the chase by covering the graph with a piece of paper and then slowly sliding it off from

---

[1]The term in vitro is Latin for "in glass." Experiments that are performed outside living organisms are done in vitro. The term in vivo, in contrast, is Latin for "in life." Experiments performed with living organisms are done in vivo.

(a) Setup for a pulse-chase experiment

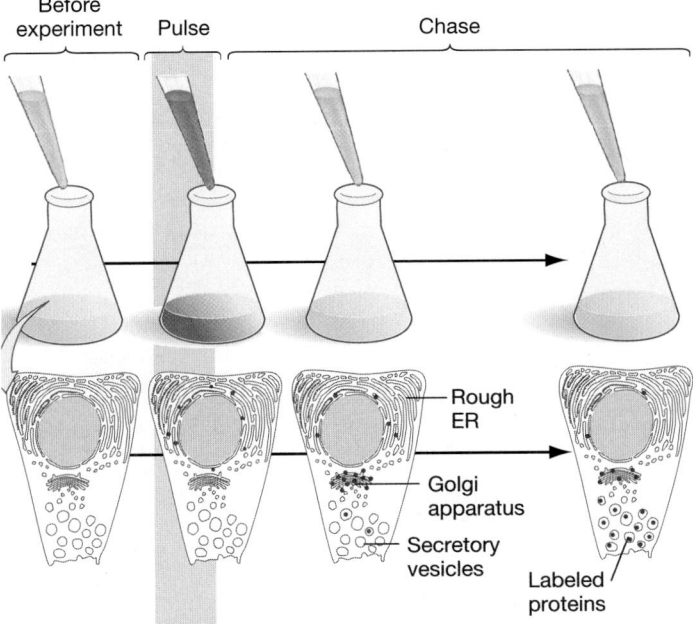

(b) Tracking pulse-labeled proteins during the chase

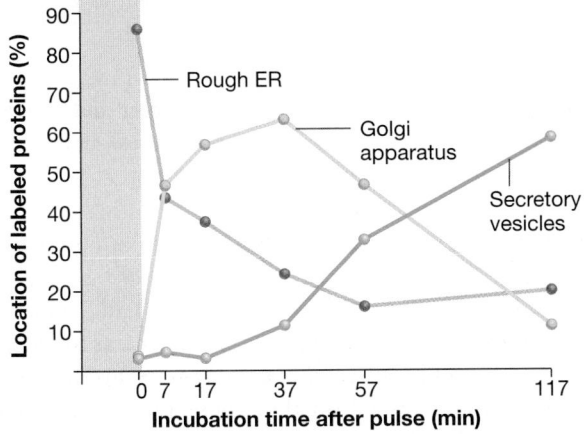

FIGURE 7.20 Tracking Protein Movement in a Pulse–Chase Experiment. Part **(a)** shows how investigators label newly synthesized proteins during the pulse with radioactive amino acids (red). At the start of the chase, this medium is replaced with non-radioactive amino acids (yellow) so only those proteins labeled in the pulse will be tracked. Part **(b)** provides the results of a pulse–chase experiment. The drawings represent micrographs taken that show the radiolabeled proteins (red dots) in the cells. The graph shows the relative abundance of radiolabeled proteins in three different organelles during the chase.

left to right. Notice what is happening to each line at the following three time points:

1. Immediately after the pulse, most of the newly synthesized proteins are inside this cell's rough ER.

2. At 37 minutes into the chase, the situation has changed. Most of the labeled proteins have left the rough ER and entered the

Golgi apparatus, and some of them have accumulated inside structures called secretory vesicles.

3. By the end of the chase, at 117 minutes, most of the labeled proteins have left the Golgi and are either in secretory vesicles or were secreted from the cells.

Over a period of two hours, the labeled population of proteins moved along a defined trail through the rough ER, Golgi apparatus, and secretory vesicles to reach the exterior of the cell. ✔ **QUANTITATIVE** If you understand how the pulse–chase experiment is used to track proteins, use the graph in Figure 7.20b to estimate the time it takes for proteins to pass through the Golgi apparatus.

The results support the hypotheses that a secretory pathway exists and that the rough ER and Golgi apparatus function together as an integrated endomembrane system. Next, let's break this secretory pathway down to examine four of the steps in more detail:

1. How do proteins enter the lumen of the ER?

2. How do the proteins move from the ER to the Golgi apparatus?

3. Once they're inside the Golgi, what happens to them?

4. And finally, how does the Golgi sort out the proteins so each will end up going to the appropriate place?

Entering the Endomembrane System: The Signal Hypothesis

The synthesis of proteins destined to be secreted or embedded in membranes begins in ribosomes free in the cytosol. Günter Blobel and colleagues proposed that at some point these ribosomes become attached to the outside of the ER. But what directs these ribosomes to the ER? The signal hypothesis predicts that proteins bound for the endomembrane system have a molecular zip code analogous to the nuclear localization signal. Blobel proposed that the first few amino acids in the growing polypeptide act as a signal that marks the ribosome for transport to the ER membrane.

This hypothesis received important support when researchers made a puzzling observation: When proteins that are normally synthesized in the rough ER are instead manufactured by isolated ribosomes in vitro—with *no* ER present—they are 20 amino acids longer, on average, than usual.

Blobel seized on these data. He claimed that the extra amino acids are the "send-to-ER" signal, and that the signal is removed inside the organelle. When the same protein is synthesized in vitro, the signal is not removed.

Blobel's group went on to produce convincing data that supported the hypothesis: They identified a sequence of amino acids that will move proteins into the ER lumen, called the **ER signal sequence.**

More recent work has documented the mechanisms responsible for receiving this send-to-ER signal and inserting the

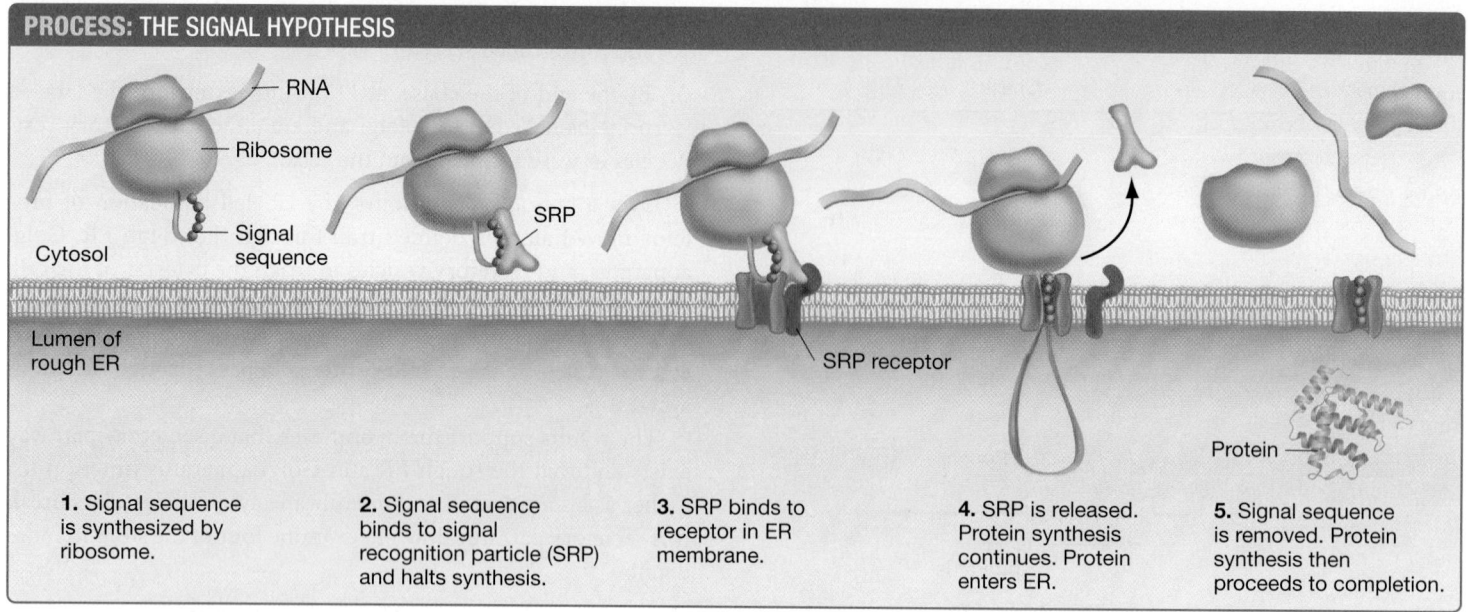

RNA

Ribosome

Signal
sequence

Cytosol

SRP

Lumen of
rough ER

SRP receptor

Protein

1. Signal sequence is synthesized by ribosome.

2. Signal sequence binds to signal recognition particle (SRP) and halts synthesis.

3. SRP binds to receptor in ER membrane.

4. SRP is released. Protein synthesis continues. Protein enters ER.

5. Signal sequence is removed. Protein synthesis then proceeds to completion.

FIGURE 7.21 The Signal Hypothesis Explains How Proteins Destined for Secretion Enter the Endomembrane System. According to the signal hypothesis, proteins destined for secretion contain a short stretch of amino acids that interact with a signal recognition particle (SRP) in the cytoplasm. This interaction directs the synthesis of the remaining protein into the ER.

protein into the rough ER. **FIGURE 7.21** illustrates the key steps involved.

Step 1 Protein synthesis begins on a free ribosome in the cytosol. The ribosome synthesizes the ER signal sequence.

Step 2 The signal sequence binds to a **signal recognition particle (SRP)**—a complex of RNA and protein. The attached SRP causes protein synthesis to stop.

Step 3 The ribosome + signal sequence + SRP complex moves to the ER membrane, where it attaches to the SRP receptor. Think of the SRP as a key that is activated by an ER signal sequence. The SRP receptor in the ER membrane is the lock.

Step 4 Once the lock (the receptor) and key (the SRP) connect, the SRP is released and protein synthesis continues.

Step 5 The growing protein is fed into the ER lumen through a channel, and the signal sequence is removed.

If the protein will eventually be shipped to the inside of an organelle or secreted from the cell, it is completely transferred into the lumen of the rough ER. If it is an integral membrane protein, part of it remains in the cytosol and rough ER membrane while it is being processed.

Once proteins are inside the rough ER or inserted into its membrane, they fold into their three-dimensional shape with the help of chaperone proteins (see Chapter 3). In addition, proteins that enter the ER lumen interact with enzymes that catalyze the addition of carbohydrate side chains (Figure 7.19). Because carbohydrates are polymers of sugar monomers, the addition of one or more carbohydrate groups is called **glycosylation** ("sugar-together"). The resulting molecule is a **glycoprotein**

("sugar-protein"; see Chapter 5). The number and arrangement of these sugars changes as the protein matures, serving as an indicator for shipment to the next destination.

Moving from the ER to the Golgi

How do proteins travel from the ER to the Golgi apparatus? In Palade's pulse–chase experiment, labeled proteins found between the rough ER and the Golgi apparatus were inside membrane-bound structures. Based on these observations, Palade's group suggested that proteins are transported in vesicles that bud off from the ER, move away, fuse with the membrane on the *cis* face of the Golgi apparatus, and dump their cargo inside.

This hypothesis was supported when other researchers used differential centrifugation to isolate and characterize the vesicles that contained labeled proteins. They found that a distinctive type of vesicle carries proteins from the rough ER to the Golgi apparatus. Ensuring that only appropriate cargo is loaded into these vesicles and that the vesicles dock and fuse only with the *cis* face of the Golgi involves a complex series of events and is an area of active research.

What Happens Inside the Golgi Apparatus?

Section 7.2 indicated that the Golgi apparatus consists of a stack of flattened vesicles called cisternae, and that cargo enters one side of the organelle and exits the other. Recent research has shown that the composition of the Golgi apparatus is dynamic. New cisternae constantly form at the *cis* face of the Golgi, while old cisternae break apart at the *trans* face, to be replaced by the cisternae behind it. In this way a new cisterna follows those formed earlier, advancing toward the *trans* face of the Golgi. As it does, it changes in composition and activity through a process called **cisternal maturation.**

By separating individual cisternae and analyzing their contents, researchers have found that cisternae at various stages of maturation contain different suites of enzymes. Many of these enzymes catalyze glycosylation reactions that further modify the oligosaccharides that were attached to the protein in the ER. As the cisternae slowly move from *cis* to *trans*, these enzymes are replaced with those representing more mature cisternae. The result is that proteins are modified in a stepwise manner as they slowly move through the Golgi.

If the rough ER is like a foundry and stamping plant where rough parts are manufactured, then the Golgi can be considered a finishing area where products are polished, painted, and readied for shipping.

How Do Proteins Reach Their Destinations?

The rough ER and Golgi apparatus constitute an impressive assembly line. Certain proteins manufactured by this process remain in these organelles, replacing worn-out resident molecules. But those proteins that are simply passing through as cargo must be sorted and sent to their intended destination as the *trans* cisterna they are in breaks up into vesicles.

How are these finished products put into the right shipping containers, and how are the different containers addressed?

Studies on enzymes that are shipped to lysosomes have provided some answers to both questions. A key finding was that lysosome-bound proteins have a phosphate group attached to a specific sugar subunit on their surface, forming the compound mannose-6-phosphate. If mannose-6-phosphate is removed from these proteins, they are not transported to a lysosome.

This is strong evidence that the phosphorylated sugar serves as a zip code, analogous to the nuclear localization signal and ER signal sequence discussed earlier. Data indicate that mannose-6-phosphate binds to a receptor protein in the membrane of the *trans*-Golgi cisterna. Regions that are enriched with these receptor–cargo complexes will form vesicles that, in turn, have proteins on their cytosolic surfaces that direct their transport and fusion with pre-lysosomal compartments. In this way, the presence of mannose-6-phosphate targets proteins for vesicles that deliver their contents to organelles that eventually become lysosomes.

FIGURE 7.22 presents a simplified model of how cargo is sorted and loaded into specific vesicles that are shipped to different destinations. Each cargo protein has a molecular tag that directs it to particular vesicle budding sites by interacting with receptors in the *trans* cisterna. These receptors, along with other cytosolic proteins that are not shown, direct the transport vesicles to the correct destinations.

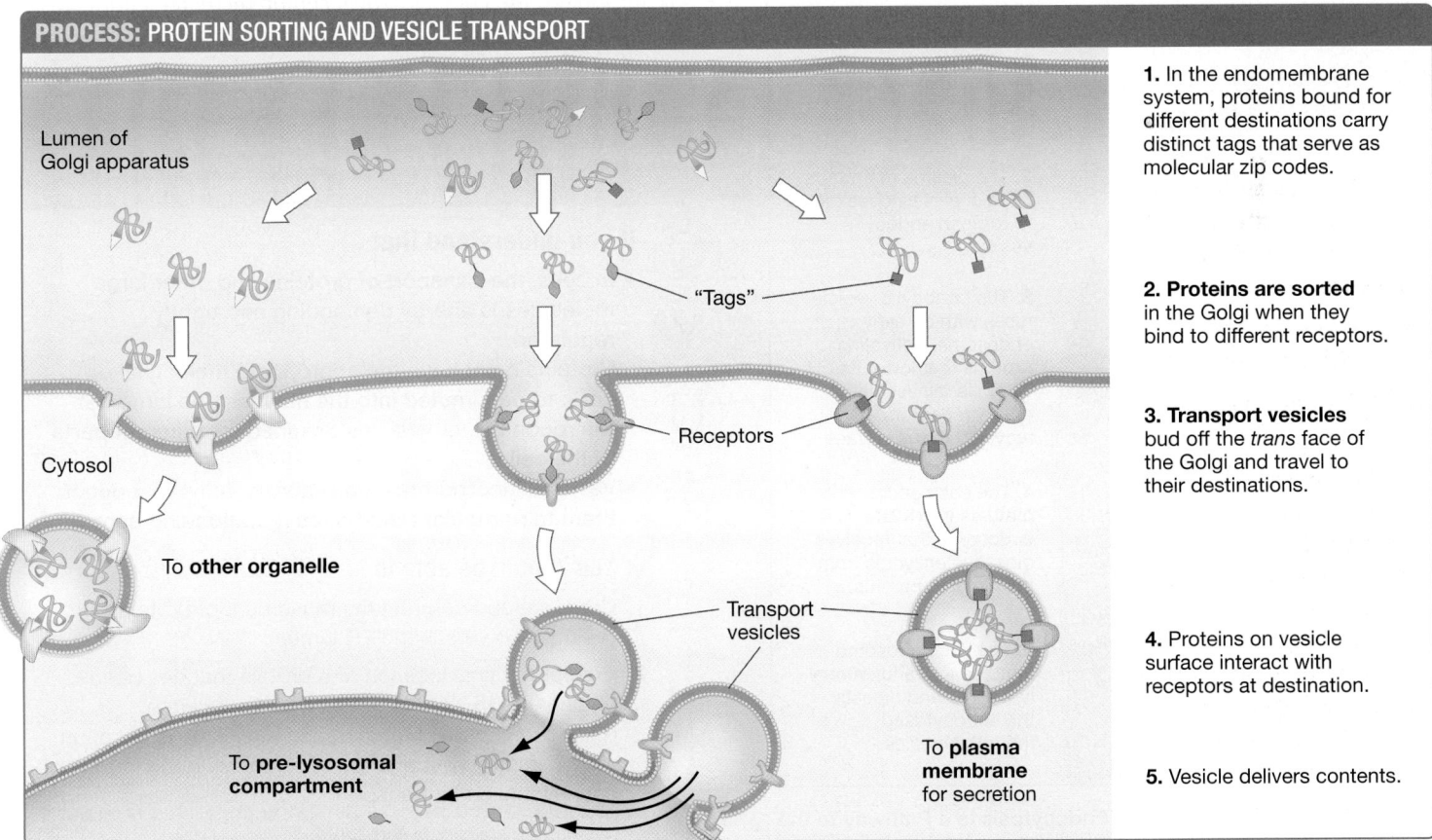

PROCESS: PROTEIN SORTING AND VESICLE TRANSPORT

Lumen of Golgi apparatus

"Tags"

Receptors

Cytosol

To **other organelle**

To **pre-lysosomal compartment**

Transport vesicles

To **plasma membrane** for secretion

1. In the endomembrane system, proteins bound for different destinations carry distinct tags that serve as molecular zip codes.

2. **Proteins are sorted** in the Golgi when they bind to different receptors.

3. **Transport vesicles** bud off the *trans* face of the Golgi and travel to their destinations.

4. Proteins on vesicle surface interact with receptors at destination.

5. Vesicle delivers contents.

FIGURE 7.22 In the Golgi Apparatus, Proteins Are Sorted into Vesicles That Are Targeted to a Destination.

In particular, notice that the transport vesicle shown on the right of Figure 7.22 is bound for the plasma membrane, where it will secrete its contents to the outside. This process is called **exocytosis** ("outside-cell-act"). When exocytosis occurs, the vesicle membrane and plasma membrane make contact. As the two membranes fuse, their lipid bilayers rearrange in a way that exposes the interior of the vesicle to the outside of the cell. The vesicle's contents then diffuse into the space outside the cell. This is how cells in your pancreas deliver digestive enzymes to the duct that leads to your small intestine—where food is digested.

Recycling Material in the Lysosome

Now that you have seen how cargo moves out of the cell, let's look at how cargo is brought into the cell. Previously, you learned about how cells import small molecules across lipid bilayers (see Chapter 6), but this is not possible for large molecules like proteins and complex carbohydrates. For these molecules to be recycled and used by the cell, they must first be digested in the lysosome—but how do they get there?

Endocytosis ("inside-cell-act") refers to any pinching off of the plasma membrane that results in the uptake of material from outside the cell. **Receptor-mediated endocytosis** is illustrated in **FIGURE 7.23**. As its name implies, the sequence of events begins when macromolecules outside the cell bind to receptors on the plasma membrane. More than 25 distinct receptors have now been characterized, each specialized for binding to different cargo.

Once receptor binding occurs, the plasma membrane folds in and pinches off to form an endocytic vesicle. These vesicles then drop off their cargo in a transient organelle called the **early endosome** ("inside-body"). The activity of proton pumps in the membrane of this organelle acidifies its lumen, which causes the cargo to be released from their receptors. Many of these emptied cargo receptors are then repackaged into vesicles and returned to the plasma membrane.

As proton pumps continue to lower the early endosome's pH, it undergoes a series of processing steps that cause it to mature into a **late endosome.** The late endosome is the pre-lysosomal compartment introduced earlier (Figure 7.22), where the acid hydrolases from the Golgi apparatus are dropped off. As before, the emptied cargo receptors transported from the Golgi are removed from the late endosome as it matures into a fully active lysosome.

In addition to receptor-mediated endocytosis, the lysosome is involved in recycling material via autophagy and phagocytosis (see **FIGURE 7.24**). During **autophagy** (literally, "same-eating"), damaged organelles are enclosed within an internal membrane and delivered to a lysosome. There the components are digested and recycled. In **phagocytosis** ("eat-cell-act"), the plasma membrane of a cell surrounds a smaller cell or food particle and engulfs it, forming a structure called a phagosome. This structure is delivered to a lysosome, where it is taken in and digested.

Regardless of whether the materials in lysosomes originate via autophagy, phagocytosis, or receptor-mediated endocytosis, the result is similar: Molecules are hydrolyzed. ✔ If you understand the interaction between the endomembrane system

PROCESS: RECEPTOR-MEDIATED ENDOCYTOSIS

Recycling of membrane proteins

Endocytic vesicle

H^+

Early endosome

H^+ H^+

Vesicle from Golgi apparatus

Late endosome

Lysosome

1. Macromolecules outside the cell bind to membrane proteins that act as receptors.

2. The plasma membrane folds in and pinches off to form an endocytic vesicle.

3. The endocytic vesicle fuses with an early endosome, activating protons that lower its pH. Cargo is released and empty receptors are recycled to the surface.

4. The early endosome matures into a late endosome that receives digestive enzymes from the Golgi apparatus.

5. The late endosome matures into a functional lysosome and digests the endocytosed macromolecules.

FIGURE 7.23 Receptor-Mediated Endocytosis Is a Pathway to the Lysosome. Endosomes created by receptor-mediated endocytosis will mature into lysosomes.

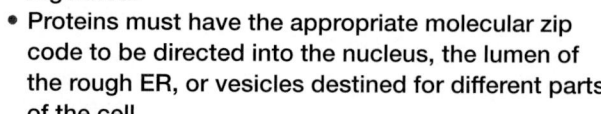
check your understanding

C Y U If you understand that . . .

- In cells, the transport of proteins and other large molecules is energy demanding and tightly regulated.
- Proteins must have the appropriate molecular zip code to be directed into the nucleus, the lumen of the rough ER, or vesicles destined for different parts of the cell.
- Vesicles incorporate membrane proteins that direct them to particular target sites for unloading cargo.

✔ You should be able to . . .

1. Compare and contrast the movement of proteins into the nucleus versus the ER lumen.

2. Predict the final location of a protein that has been engineered to include an ER signal sequence, mannose-6-phosphate tag, and a nuclear localization signal. Justify your answer by addressing the impact of each signal on its transport.

Answers are available in Appendix A.

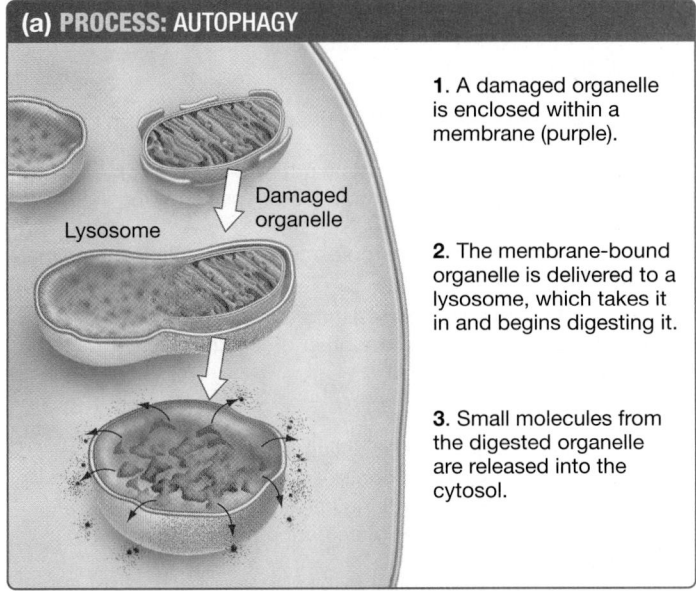

(a) PROCESS: AUTOPHAGY

Damaged organelle

Lysosome

1. A damaged organelle is enclosed within a membrane (purple).

2. The membrane-bound organelle is delivered to a lysosome, which takes it in and begins digesting it.

3. Small molecules from the digested organelle are released into the cytosol.

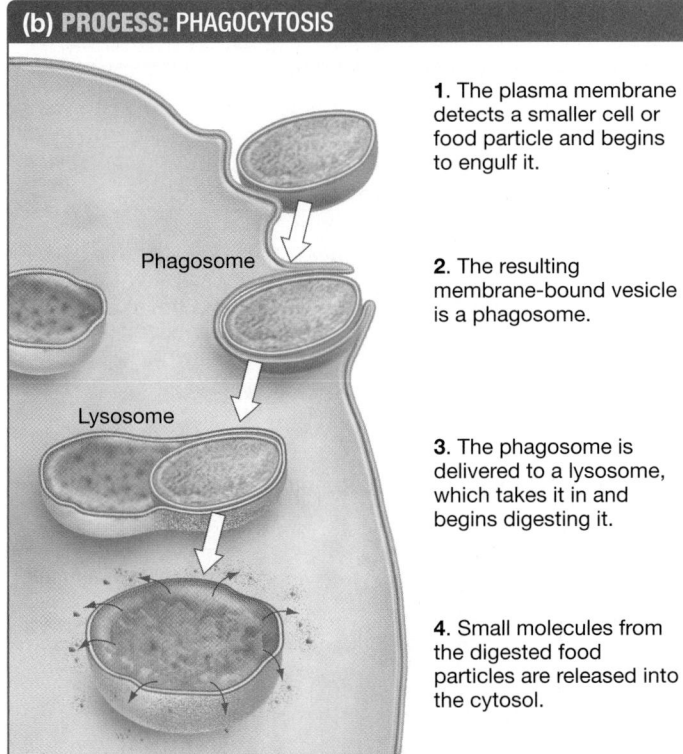

(b) PROCESS: PHAGOCYTOSIS

Phagosome

Lysosome

1. The plasma membrane detects a smaller cell or food particle and begins to engulf it.

2. The resulting membrane-bound vesicle is a phagosome.

3. The phagosome is delivered to a lysosome, which takes it in and begins digesting it.

4. Small molecules from the digested food particles are released into the cytosol.

FIGURE 7.24 Two More Ways to Deliver Materials to Lysosomes. Materials can be transported to lysosomes **(a)** via autophagy or **(b)** after phagocytosis.

and endocytosis, you should be able to predict how the loss of the mannose-6-phosphate receptor would affect receptor-mediated endocytosis.

It is important to note, however, that not all the materials that are surrounded by membrane and taken into a cell end up in lysosomes. In addition to receptor-mediated endocytosis and phagocytosis, small fluid-filled vesicles can be brought into a cell via **bulk-phase endocytosis.** There does not appear to be any cargo selection in bulk-phase endocytosis, and the vesicles are

not transported to lysosomes. These tiny vesicles are used elsewhere in the cell and are likely involved in recycling lipids deposited on the plasma membrane during exocytosis.

Throughout this section, vesicles have been key to the transport of cargo. If these transport steps depended on the random movement of diffusion alone, however, then the vesicles and their cargo might never reach their intended destinations. Are there instead defined tracks that direct the movement of these shipping containers? If so, what are these tracks, and what molecule or molecules function to transport the vesicles along them? Let's delve into these questions in the next section.

7.6 Cell Systems III: The Dynamic Cytoskeleton

The endomembrane system may be the best-studied example of how individual organelles work together in a dynamic, highly integrated way. This integration depends in part on the physical relationship of organelles, which is organized by the cytoskeletal system.

The cytoskeleton is a dense and complex network of fibers that helps maintain cell shape by providing structural support. However, the cytoskeleton is not a static structure like the scaffolding used at construction sites. Its fibrous proteins move and change to alter the cell's shape, shift its contents, and even move the cell itself. Like the rest of the cell, the cytoskeleton is dynamic.

As **TABLE 7.3** (see page 128) shows, there are three distinct cytoskeletal elements in eukaryotic cells: actin filaments, intermediate filaments, and microtubules. Recent research has shown structural and functional relationships between these three eukaryotic filaments and cytoskeletal elements in bacteria.

Each of the three cytoskeletal elements found in eukaryotes has a distinct size, structure, and function. Let's look at each one in turn.

Actin Filaments

Sometimes called **microfilaments** because they are the cytoskeletal element with the smallest diameter, **actin filaments** are fibrous structures made of the globular protein actin (Table 7.3). In animal cells, actin is often the most abundant of all proteins—typically it represents 5–10 percent of the total protein in the cell. Each of your liver cells contains about half a billion of these molecules.

Actin Filament Structure A completed actin filament resembles two long strands that coil around each other. Actin filaments form when individual actin protein subunits assemble, or polymerize, from head to tail through the formation of noncovalent bonds.

Because the actin proteins are not symmetrical, this head-to-tail arrangement of actin subunits results in filaments that have two different ends, or polarity. The two distinct ends of an actin filament are referred to as plus and minus ends. The structural difference between these two ends results in different rates of

Filament	Structure	Subunits	Functions
The three types of filaments that make up the cytoskeleton are distinguished by their size, structure, and type of protein subunit.			
Actin filaments (microfilaments)	Strands in double helix − end + end 7 nm	Actin	• maintain cell shape by resisting tension (pull) • move cells via muscle contraction or cell crawling • divide animal cells in two • move organelles and cytoplasm in plants, fungi, and animals
Intermediate filaments	Fibers wound into thicker cables 10 nm	Keratins, lamins, or others	• maintain cell shape by resisting tension (pull) • anchor nucleus and some other organelles
Microtubules	Hollow tube 25 nm − end + end	α- and β-tubulin dimers	• maintain cell shape by resisting compression (push) • move cells via flagella or cilia • move chromosomes during cell division • assist formation of cell plate during plant cell division • move organelles • provide tracks for intracellular transport

assembling new actin subunits: The plus end grows faster than the minus end.

Each filament is generally unstable and will grow or shrink depending on the concentration of free actin subunits. In addition to controlling the availability of free actin, cells regulate the length and longevity of microfilaments via actin-binding proteins that either stabilize or destabilize their structure.

In animal cells, actin filaments are particularly abundant just under the plasma membrane. They are organized into long, parallel bundles or dense, crisscrossing networks in which individual actin filaments are linked to one another by other proteins. The reinforced bundles and networks of actin filaments help stiffen the cell and define its shape.

Actin Filament Function   In addition to providing structural support, actin filaments are involved in movement. In several cases, actin's role in movement depends on the protein myosin. Myosin is a **motor protein:** a protein that converts the potential energy in ATP into the kinetic energy of mechanical work, just as a car's motor converts the chemical energy in gasoline into spinning wheels.

The interaction between actin and myosin is frequently presented in the context of how it produces muscle contraction and movement (Chapter 48). For now, it's enough to recognize that when myosin binds and hydrolyzes ATP to ADP, it undergoes a series of shape changes that extends the "head" region, attaches it to actin, and then contracts to pull itself along the actin filament. The shape change of this protein causes the actin and myosin to slide past each other. After repeated rounds of this contraction cycle, the myosin progressively moves toward the plus end of the actin filament (**FIGURE 7.25a**). This type of movement is analogous to an inchworm contracting its body as it moves along a stick.

(a) Actin and myosin interact to cause movement.

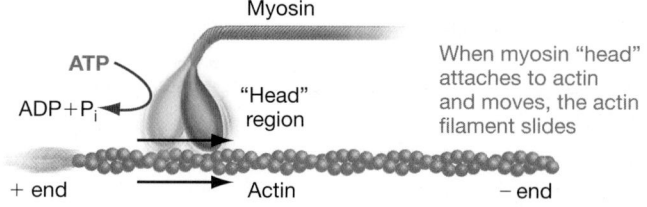

When myosin "head" attaches to actin and moves, the actin filament slides

(b) Examples of movement caused by actin–myosin interactions

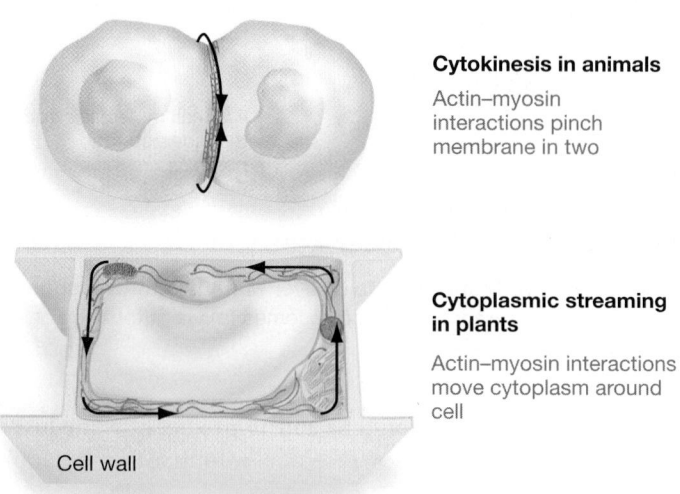

Cytokinesis in animals

Actin–myosin interactions pinch membrane in two

Cytoplasmic streaming in plants

Actin–myosin interactions move cytoplasm around cell

FIGURE 7.25 Many Cellular Movements Are Based on Actin–Myosin Interactions. (a) ATP hydrolysis in the "head" region of myosin causes the protein to attach to actin and change shape. The movement slides the myosin toward the plus end of actin. **(b)** Actin–myosin interactions can divide cells and move organelles and cytoplasm.

As **FIGURE 7.25b** shows, the ATP-powered interaction between actin and myosin is the basis for an array of cell movements:

- **Cytokinesis** ("cell-moving") is the process of cell division. In animals, this occurs by the use of actin filaments that are connected to the plasma membrane and arranged in a ring around the circumference of the cell. Myosin causes the filaments to slide past one another, drawing in the membrane and pinching the cell in two.

- **Cytoplasmic streaming** is the directed flow of cytosol and organelles within plant cells. The movement occurs along actin filaments and is powered by myosin. It is especially common in large cells, where the circulation of cytoplasm facilitates material transport.

In addition, the movement called **cell crawling** occurs when groups of actin filaments grow, creating bulges in the plasma membrane that extend and move the cell. Cell crawling occurs in a wide range of organisms and cell types, including amoebae, slime molds, and certain animal cells.

Intermediate Filaments

Many types of **intermediate filament** exist, each consisting of a different—though similar in size and structure—type of protein (Table 7.3). Humans, for example, have 70 genes that code for intermediate filament proteins. This is in stark contrast to actin filaments and microtubules, which are made from the same protein subunits in all eukaryotic cells.

Moreover, intermediate filaments are not polar; instead, each end of these filaments is identical. They are not involved in directed movement driven by myosin or other motor proteins, but instead serve a purely structural role in eukaryotic cells.

The intermediate filaments that you are most familiar with belong to a family of molecules called the keratins. The cells that make up your skin and line surfaces inside your body contain about 20 types of keratin. These intermediate filaments provide the mechanical strength required for these cells to resist pressure and abrasion. Certain cells in the skin can also produce secreted forms of keratin. Depending on the location of the cell and keratins involved, the secreted filaments form fingernails, toenails, or hair.

Nuclear lamins, which make up the nuclear lamina layer introduced in Section 7.4, also qualify as intermediate filaments. Nuclear lamins form a dense mesh under the nuclear envelope. Recall that in addition to giving the nucleus its shape, they anchor the chromosomes. They are also involved in the breakup and reassembly of the nuclear envelope when cells divide.

Some intermediate filaments project from the nucleus through the cytoplasm to the plasma membrane, where they are linked to intermediate filaments that run parallel to the cell surface. In this way, intermediate filaments form a flexible skeleton that helps shape the cell surface and hold the nucleus in place.

Microtubules

Microtubules are the largest cytoskeletal components in terms of diameter. As Table 7.3 shows, they are assembled from subunits consisting of two polypeptides, called α-tubulin and β-tubulin, that exist as stable protein **dimers** ("two-parts").

Tubulin dimers polymerize from head to tail to form filaments that interact with one another to create relatively large, hollow tubes. Because of this polarity, these microtubules have α-tubulin polypeptides at one end (the minus end) and β-tubulins at the other end (the plus end). Like actin filaments, microtubules are dynamic and grow faster at their plus ends compared with their minus ends.

Microtubules originate from a structure called the **microtubule organizing center (MTOC).** Their plus ends grow outward, radiating throughout the cell. Although plant cells typically have hundreds of sites where microtubules start growing, most animal and fungal cells have just one site that is near the nucleus.

In animals, the microtubule organizing center has a distinctive structure and is called a **centrosome.** As **FIGURE 7.26** shows, animal centrosomes contain two bundles of microtubules called **centrioles.** Although additional microtubules emanate from these structures in animals, they do not grow directly from the centrioles.

(a) In animals, microtubules originate from centrosomes.

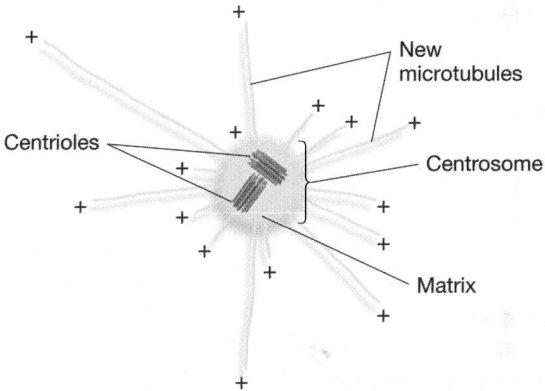

(b) Centrioles consist of microtubules.

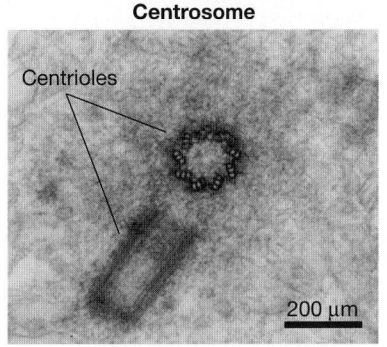

Centrosome

Centrioles

200 μm

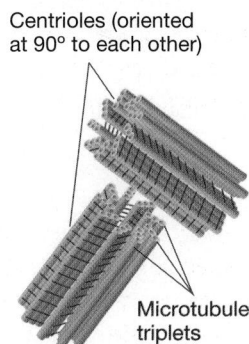

Centrioles (oriented at 90° to each other)

Microtubule triplets

FIGURE 7.26 Centrosomes Are a Type of Microtubule-Organizing Center. (a) Microtubule-organizing centers, such as the centrosomes of animal cells, are the sites where new microtubules are made. Microtubules grow from the matrix surrounding the centrioles, and their positive ends point away from the centrosomes. **(b)** The two centrioles inside a centrosome consist of microtubules as triplets arranged in a circle.

In function, microtubules are similar to actin filaments: They provide stability and are involved in movement. Like steel girders in a skyscraper, the microtubules that radiate from an organizing center stiffen the cell by resisting compression forces. Microtubules also provide a structural framework for organelles. If microtubules are prevented from forming, the network-like configuration of the ER collapses and the Golgi apparatus disappears into vesicles.

Microtubules are best known for their role in separating chromosomes during mitosis and meiosis (see Chapters 12 and 13). But microtubules are involved in many other types of cellular movement as well. Let's first consider their role in moving materials inside cells and then explore how microtubules can help cells to swim.

Microtubules Serve as Tracks for Vesicle Transport Recall from Section 7.5 that vesicles are used to transport materials to a wide array of destinations inside cells. To study how this movement happens, Ronald Vale and colleagues focused on the giant axon, an extremely large nerve cell in squid that runs the length of the animal's body. If the squid is disturbed, the cell signals muscles to contract so it can jet away to safety. The researchers decided to study this particular cell for three reasons.

1. The giant axon is so large that it is relatively easy to see and manipulate.

2. Large numbers of vesicles are transported down the length of the cell. As a result, a large amount of cargo moves a long distance.

(a) Electron micrograph

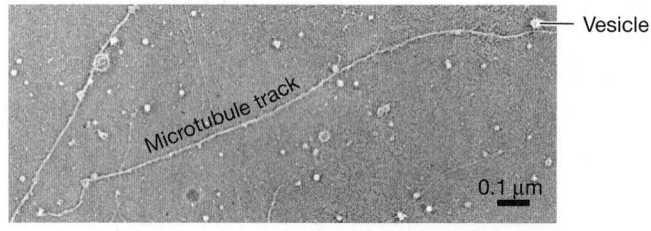

(b) Video image

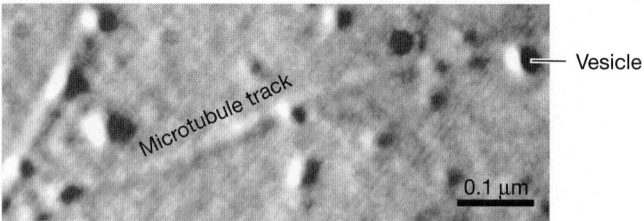

FIGURE 7.27 Transport Vesicles Move along Microtubule Track. The images show extruded cytoplasm from a squid giant axon. **(a)** An electron micrograph that allowed researchers to measure the diameter of the filaments and confirm that they are microtubules. In the upper part of this image, you can see a vesicle on a "track." **(b)** A video microscope image using enhanced contrast that allowed researchers to watch vesicles move in real time.

3. The researchers found that if they gently squeezed the cytoplasm out of the cell, vesicle transport still occurred in the extracellular cytoplasmic material. This allowed them to do experiments on vesicle transport without the plasma membrane being in the way.

In short, the squid giant axon provided a system that could be observed and manipulated efficiently in the lab. To watch vesicle transport in action, the researchers mounted a video camera to a microscope. As **FIGURE 7.27** shows, this technique allowed them to document that vesicle transport occurred along filamentous tracks.

To identify the filament involved, the biologists measured the diameter of the tracks and analyzed their chemical composition. Both types of data indicated that the tracks consist of microtubules. Microtubules also appear to be required for movement of materials elsewhere in the cell. For instance, if experimental cells are treated with a drug that disrupts microtubules, the movement of vesicles from the rough ER to the Golgi apparatus is impaired.

The general message of these experiments is that transport vesicles move through the cell along microtubules. How? Do the tracks themselves move, like a conveyer belt, or are vesicles carried along on some sort of molecular vehicle?

Motor Proteins Pull Vesicles Along the Tracks To study the way vesicles move along microtubules, Vale's group took the squid axon's transport system apart and then determined what components were required to put it back together. A simple experiment convinced the group that this movement is an energy-dependent process: If they depleted the amount of ATP in the cytoplasm, vesicle transport stopped.

To examine this process further, they mixed purified microtubules and vesicles with ATP, but no transport occurred. Something had been left out—but what? To find the missing element or elements, the researchers purified one subcellular part after another and added it to the microtubule + vesicle + ATP system.

Through trial and error, and further purification steps, the researchers finally succeeded in isolating a protein that generated vesicle movement. They named the molecule **kinesin,** from the Greek word *kinein* ("to move").

Like myosin, kinesin is a motor protein. Kinesin converts the chemical energy in ATP into mechanical energy in the form of movement. More specifically, when ATP is hydrolyzed by kinesin, the protein moves along microtubules in a directional manner: toward the plus end.

Biologists began to understand how kinesin works when X-ray diffraction studies showed that it has three major regions: a head section with two globular pieces, a tail associated with small polypeptides, and a stalk that connects the head and tail (**FIGURE 7.28a**). Follow-up studies confirmed that the head region binds to the microtubule while the tail region binds to the transport vesicle. Recent work has shown that kinesin uses these domains to "walk" along the microtubule through a series of conformational changes as it hydrolyzes ATP (**FIGURE 7.28b**). Amazingly, these motors have been found to "walk" up to 375 steps per second.

Cells contain several different versions of the kinesin motor, each specialized for a different role in the cell. If kinesins move

(a) Structure of kinesin

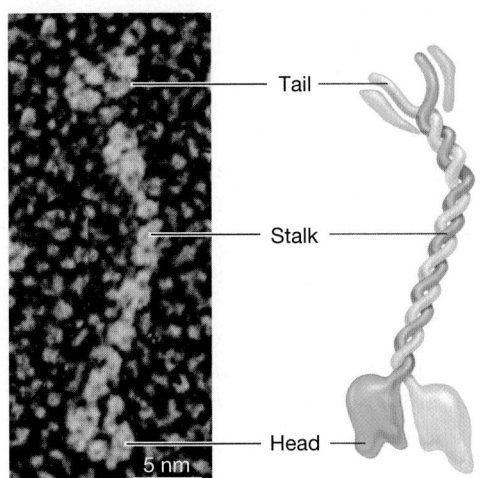

Tail

Stalk

Head

5 nm

(b) Kinesin "walks" along a microtubule track.

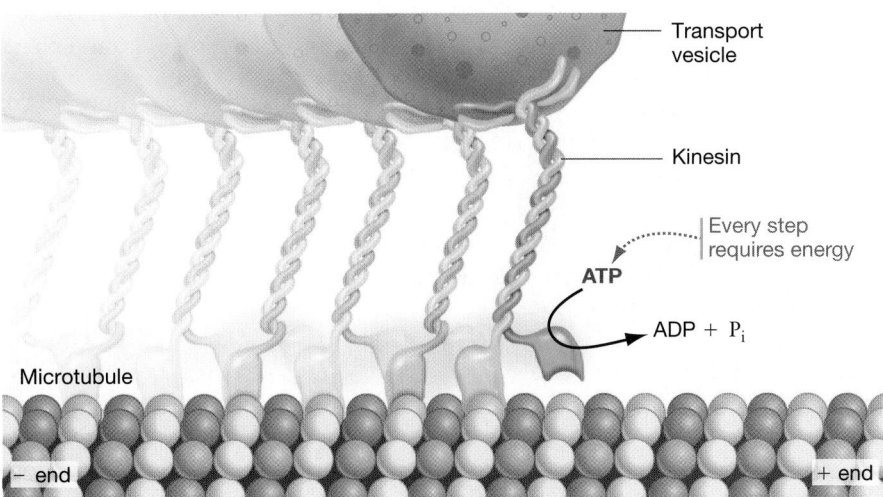

Transport vesicle

Kinesin

Every step requires energy

ATP

ADP + P$_i$

Microtubule

− end

+ end

FIGURE 7.28 Motor Proteins Move Vesicles along Microtubules. (a) Kinesin has three distinct regions. **(b)** The current model depicting how kinesin "walks" along a microtubule track to transport vesicles. The two head segments act like feet that alternately attach, pivot, and release in response to the gain or loss of a phosphate group from ATP.

only toward the plus ends of microtubules, then what is responsible for moving the cargo in the opposite direction? By studying whole-cell locomotion, researchers discovered a motor that could move toward the minus end of microtubules.

Flagella and Cilia: Moving the Entire Cell

Flagella are long, whiplike projections from the cell surface that function in movement. While many bacteria and eukaryotes have flagella, the structure is completely different in the two groups.

- Bacterial flagella are helical rods made of a protein called flagellin; eukaryotic flagella consist of several microtubules constructed from tubulin dimers.

- Bacterial flagella move the cell by rotating the rod like a ship's propeller; eukaryotic flagella move the cell by undulating—they whip back and forth.

- Eukaryotic flagella are surrounded by the plasma membrane and are considered organelles; bacterial flagella are not.

Based on these observations, biologists conclude that the two structures evolved independently, even though their function is similar.

To understand how some cells move, let's focus on eukaryotic flagella. Eukaryotic flagella are closely related to structures called **cilia** (singular: **cilium**), which are short, hairlike projections that are also found in some eukaryotic cells (**FIGURE 7.29**). Flagella are generally much longer than cilia, and the two structures differ in

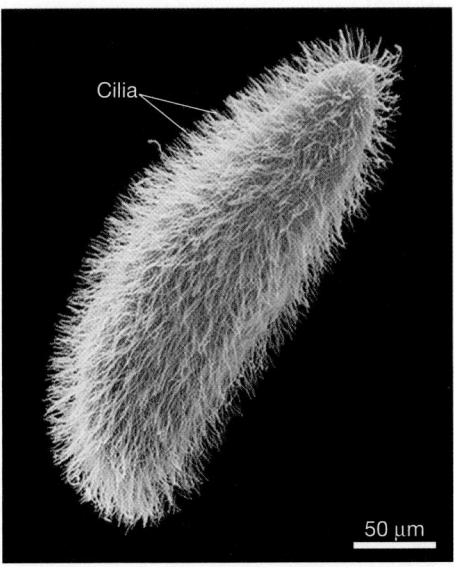

Cilia

50 μm

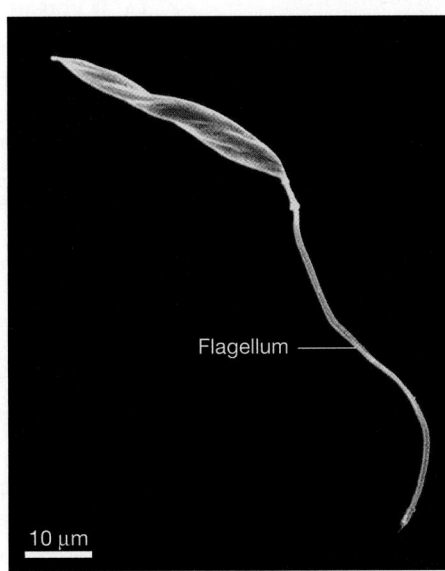

Flagellum

10 μm

FIGURE 7.29 Cilia and Flagella Differ in Length and Number. Cells typically only have 1–4 flagella but may have up to 14,000 cilia. The cells in these scanning electron micrographs have been colorized.

(a) Transmission electron micrograph of axoneme

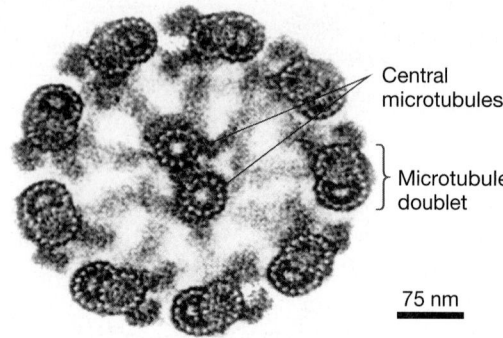

Central microtubules

Microtubule doublet

75 nm

(b) Structure of axoneme

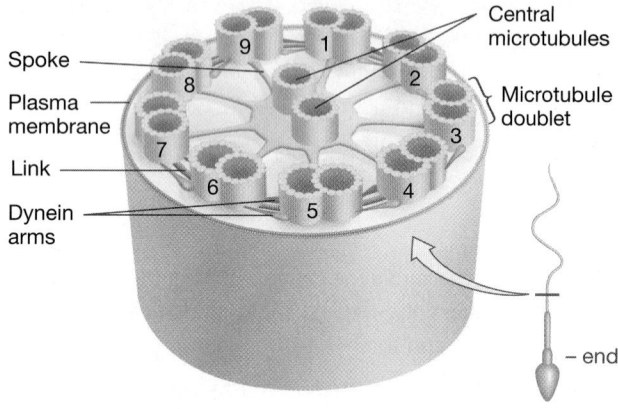

Spoke

Plasma membrane

Link

Dynein arms

Central microtubules

Microtubule doublet

9 1 2 3 4 5 6 7 8

– end

(c) Mechanism of axoneme bending

Microtubule doublet

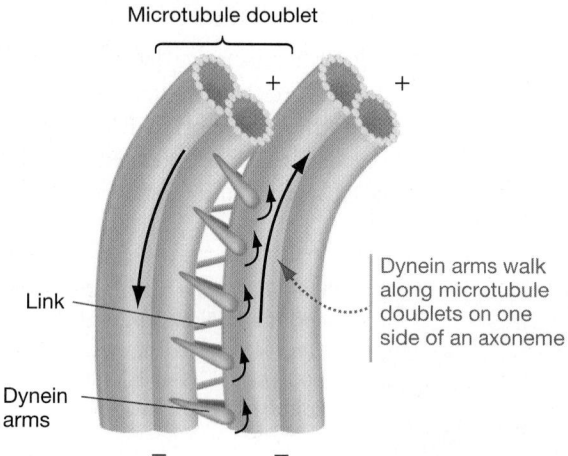

+ +

Link

Dynein arms

Dynein arms walk along microtubule doublets on one side of an axoneme

– –

+ **ATP**: Causes dynein to walk toward minus end and pull toward plus end

FIGURE 7.30 The Structure and Function of Cilia and Flagella.
(a) Transmission electron micrograph of a cross section through an axoneme. **(b)** The microtubules in cilia and flagella are connected by links and spokes, and the entire structure is surrounded by the plasma membrane. **(c)** When dynein arms walk along the microtubule doublets on one side of a flagellum, force is transmitted to these links and spokes, causing the entire axoneme to bend.

✔**QUESTION** If the links and spokes were removed from the microtubule doublets, what would happen to the axoneme after adding ATP?

their abundance and pattern of movement. But when researchers examined the two structures with an electron microscope, they found that their underlying organization is identical.

How are Cilia and Flagella Constructed? In the 1950s, anatomical studies established that most cilia and flagella have a characteristic "9 + 2" arrangement of microtubules. As **FIGURE 7.30a** shows, nine microtubule pairs, or doublets, surround two central microtubules. The doublets consist of one complete and one incomplete microtubule and are arranged around the periphery of the structure.

The entire 9 + 2 structure is called the **axoneme** ("axle-thread"). The nine doublets of the axoneme originate from a structure called the **basal body.** The basal body is identical in structure with a centriole and plays a central role in the growth of the axoneme.

Through further study, biologists gained a more detailed view of the axoneme's structure. Spoke-like proteins connect each doublet to the central pair of microtubules, and molecular links connect the nine doublets to one another (**FIGURE 7.30b**). Each doublet also has a set of arms that project toward an adjacent doublet.

Axonemes are complex. How do their components interact to generate motion?

What Provides the Force Required for Movement? In the 1960s Ian Gibbons began studying the cilia of a common unicellular eukaryote called *Tetrahymena*. Gibbons found that he could isolate axonemes by using a detergent to remove the plasma membrane that surrounds cilia and then subjecting the resulting solution to differential centrifugation. These steps gave Gibbons a cell-free system for studying how the axonemes in cilia and flagella work. He found that the isolated structures would beat only if he supplied them with ATP, confirming that the beating of cilia is an energy-demanding process.

check your understanding

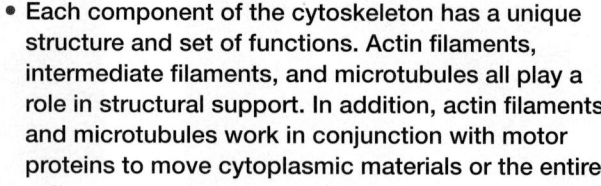

If you understand that . . .

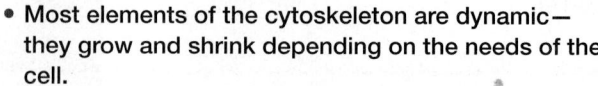

- Each component of the cytoskeleton has a unique structure and set of functions. Actin filaments, intermediate filaments, and microtubules all play a role in structural support. In addition, actin filaments and microtubules work in conjunction with motor proteins to move cytoplasmic materials or the entire cell.
- Most elements of the cytoskeleton are dynamic— they grow and shrink depending on the needs of the cell.

✔ **You should be able to . . .**

Compare and contrast the structure and function of actin filaments, intermediate filaments, and microtubules.

Answers are available in Appendix A.

In another experiment, Gibbons treated the isolated axonemes with a molecule that disrupts interactions between proteins. The resulting axonemes could not beat even after being supplied with ATP. When Gibbons examined them in the electron microscope, he found that the treatment had removed the arms from the doublets. This result suggested that the arms are required for movement. Follow-up work showed that the arms are made of a large protein that Gibbons named **dynein** (from the Greek word *dyne*, meaning "force").

Like myosin and kinesin, dynein is a motor protein that uses ATP to undergo conformational changes. These shape changes move dynein along microtubules toward the minus end. Note that dynein moves in the opposite direction from the kinesin motor, which moves toward the plus end. In the cytoplasm, dynein motors are known to play various roles similar to the other motors, including the transport of vesicles. In the context of the axoneme, however, the outcome of dynein walking in the axoneme is very different.

So what is special about the axoneme? Remember that each of the nine doublets in the axoneme is connected to the central pair

of microtubules by a spoke, and all the doublets are connected to each other by molecular links (Figure 7.30b). As a result, the sliding motion produced by dynein walking is constrained—if one doublet slides, it transmits force to the rest of the axoneme via the links and spokes (**FIGURE 7.30c**). If the dynein arms on just one side of the axoneme are activated, then the localized movement results in bending. The bending of cilia or flagella results in a swimming motion.

Scaled for size, flagella-powered swimming can be rapid. In terms of the number of body or cell lengths traveled per second, a sperm cell from a bull moves faster than a human world-record-holder does when swimming freestyle. At the cellular level, life is fast paced.

Taken together, the data reviewed in this chapter can be summed up in six words: Cells are dynamic, highly integrated structures. To maintain the level of organization that is required for life, chemical reactions must take place at mind-boggling speeds. How cells accomplish this feat is taken up elsewhere (see Chapter 8).

If you understand . . .

7.1 Bacterial and Archaeal Cell Structures and Their Functions

- There are two basic cellular designs: prokaryotic and eukaryotic. The single defining characteristic that differentiates prokaryotes from eukaryotes is the absence of a nucleus.

- Structures common to most, if not all, prokaryotes are ribosomes, a cell wall, a plasma membrane, an interior cytoskeleton, and a nucleoid.

- Many prokaryotes also possess flagella, fimbriae, and internal membrane structures, some of which are considered organelles.

✓ You should be able to predict what would happen to cells that are exposed to (1) a drug that prevents ribosomes from functioning, (2) an enzyme that degrades the cell wall, or (3) a drug that prevents the assembly of the cytoskeleton.

7.2 Eukaryotic Cell Structures and Their Functions

- Eukaryotic cells are usually much larger and more structurally complex than prokaryotic cells.

- Eukaryotic cells contain numerous specialized organelles, which allow eukaryotic cells to compartmentalize functions and grow to a large size. Organelles common to most, if not all, eukaryotes are as follows:

  1. The nucleus, which contains the cell's chromosomes and serves as its control center.

  2. The endomembrane system, which consists of a diverse group of interrelated organelles, including the endoplasmic reticulum, Golgi apparatus, lysosomes or vacuoles, and endosomes. These organelles work together to synthesize, process, sort, transport, and recycle material.

  3. Peroxisomes, which are organelles where key reactions take place that often result in the generation of toxic by-products. Specialized enzymes are included that safely disarm these by-products soon after they are generated.

  4. Mitochondria and chloroplasts, which have extensive internal membrane systems where the enzymes responsible for ATP generation and photosynthesis reside.

✓ You should be able to predict what would happen to a plant cell that is exposed to (1) a drug that poisons mitochondria, (2) a drug that inhibits catalase in the peroxisome, or (3) a drug that inhibits the formation of centrioles.

7.3 Putting the Parts into a Whole

- Cells have a tightly organized interior, where the presence and quantity of organelles often reflect the function of the cell.

- The activity in a cell illustrates the dynamic nature of life. Organelles and cytosolic proteins continually bustle about with a seemingly nonstop rush hour.

- Much of what is known about cellular activity has come from advances in cell imaging and techniques for isolating cellular components.

✔ You should be able to predict how a liver cell would differ compared with a salivary gland cell in terms of organelles.

7.4 Cell Systems I: Nuclear Transport

- Cells have sophisticated systems for making sure that proteins and other products end up in the right place.

- Traffic across the nuclear envelope occurs through nuclear pores, which contain a multiprotein nuclear pore complex that serves as gatekeeper.

- Small molecules can passively diffuse through the nuclear pore. Larger molecules enter the nucleus only if they contain a specific molecular signal that directs them through the pore via nuclear transport proteins.

✔ You should be able to propose a hypothesis that would address how certain cytoplasmic proteins can be induced to enter the nucleus by either the addition or the removal of phosphates.

7.5 Cell Systems II: The Endomembrane System Manufactures, Ships, and Recycles Cargo

- Molecules synthesized in the ER may be transported as cargo to the Golgi apparatus and then to a number of different sites, depending on the cargo.

- Before products leave the Golgi, they are sorted by their molecular "zip codes" that direct them to specific vesicles. The vesicles interact with receptor proteins at the target location so that the contents are delivered correctly.

- The lysosome is built from enzymes and membranes that are made and processed through the endomembrane system. These organelles are involved in recycling products via autophagy, phagocytosis, and receptor-mediated endocytosis.

✔ You should be able to justify why proteins (see Chapter 3)—and not RNA, DNA, carbohydrates, or lipids—are the molecules responsible for "reading" the array of molecular zip codes in cells.

7.6 Cell Systems III: The Dynamic Cytoskeleton

- The cytoskeleton is an extensive system of fibers that provides (1) structural support and a framework for arranging and organizing organelles and other cell components; (2) paths for moving vesicles inside cells; and (3) machinery for moving the cell as a whole through the beating of flagella or cilia, or through cell crawling.

- Subunits are constantly being added to or removed from cytoskeletal filaments. Actin filaments and microtubules are polarized, meaning different ends of the filaments are designated as plus or minus ends. The plus ends have a higher growth rate than the minus ends.

- Movement often depends on motor proteins, which use chemical energy stored in ATP to change shape and position. Myosin motors move toward the plus ends of actin filaments. Kinesin and dynein motors move along microtubules toward the plus and minus ends, respectively.

- A specific type of dynein is found in the axonemes of eukaryotic cilia and flagella. These motors move microtubules to generate forces that bend the structures and enable cells to swim or generate water currents.

✔ You should be able to predict which of the three motors presented in this section would be responsible for transporting vesicles from the Golgi to the plasma membrane.

(MB) **MasteringBiology**

1. MasteringBiology Assignments

Tutorials and Activities Cilia and Flagella; Endomembrane System; Exocytosis and Endocytosis; Form Fits Function: Cells; Membrane Transport: Bulk Transport; Prokaryotic Cell Structure and Function; Pulse–Chase Experiment; Review: Animal Cell Structure and Function; Tour of a Plant Cell: Structures and Functions; Tour of an Animal Cell: Structures and Functions; Tour of an Animal Cell: The Endomembrane System; Transport into the Nucleus

Questions Reading Quizzes, Blue-Thread Questions, Test Bank

2. eText Read your book online, search, take notes, highlight text, and more.

3. The Study Area Practice Test, Cumulative Test, BioFlix® 3-D Animations, Videos, Activities, Audio Glossary, Word Study Tools, Art

You should be able to . . .

✔ TEST YOUR KNOWLEDGE   *Answers are available in Appendix A*

1. Which of the following accurately describes a difference between prokaryotic and eukaryotic cells?
   a. Prokaryotic cells have fimbriae that allow the cell to swim whereas eukaryotic cells have flagella.
   b. Eukaryotic cells are generally larger than prokaryotic cells. -
   c. Eukaryotic cells have organelles.
   d. Prokaryotic cells have nuclei and eukaryotic cells have nucleoids.

2. What are three attributes of mitochondria and chloroplasts that suggest they were once free-living bacteria?

3. Which of the following is *not* true of secreted proteins?
   a. They are synthesized using ribosomes.
   b. They enter the ER lumen during translation.
   c. They contain a signal that directs them into the lysosome. ⌐
   d. They are transported between organelles in membrane-bound vesicles.

4. Which of the following results provided evidence of a nuclear localization signal in the nucleoplasmin protein?
   a. The protein was small and easily slipped through the nuclear pore complex.
   —b. After cleavage of the protein, only the tail segments appeared in the nucleus.
   c. Removing the tail allowed the core segment to enter the nucleus.
   d. The SRP bound only to the tail, not the core segment.

5. Molecular zip codes direct molecules to particular destinations in the cell. How are these signals read?
   a. They bind to receptor proteins. ⌐
   b. They enter transport vesicles.
   c. They bind to motor proteins.
   d. They are glycosylated by enzymes in the Golgi apparatus.

6. How does the hydrolysis of ATP result in the movement of a motor protein along a cytoskeletal filament?

✔ TEST YOUR UNDERSTANDING   *Answers are available in Appendix A*

7. Compare and contrast the structure of a generalized plant cell, animal cell, and prokaryotic cell. Which features are common to all cells? Which are specific to just prokaryotes, or just plants, or just animals?

8. Cells that line your intestines are known to possess a large number of membrane proteins that transport small molecules and ions across the plasma membrane. Which of the following cell structures would you expect to be required for this function of the cells?
   a. the endoplasmic reticulum ⌐
   b. peroxisomes
   c. lysosomes
   d. the cell wall

9. Most of the proteins that reside in the nucleus possess a nuclear localization signal (NLS), even if they are small enough to pass

through the pore complex unhindered. Why would a small protein have an NLS, when it naturally diffuses across the pore without one?

10. Make a flowchart that traces the movement of a secreted protein from its site of synthesis to the outside of a eukaryotic cell. Identify all the organelles that the protein passes through. Add notes indicating what happens to the protein at each step.

11. Although all three cytoskeletal fibers constantly replace their subunits, only actin filaments and microtubules demonstrate differences in the rate of growth between the two ends. What is responsible for this difference, and why is this not observed in intermediate filaments?

12. Describe how vesicles move in a directed manner between organelles of the endomembrane system. Explain why this movement requires ATP.

✔ TEST YOUR PROBLEM-SOLVING SKILLS   *Answers are available in Appendix A*

13. Which of the following cell structures would you expect to be most important in the growth of bacteria on the surface of your teeth?
    a. cell wall
    b. fimbriae
    c. flagella
    d. cilia

14. The enzymes found in peroxisomes are synthesized by cytosolic ribosomes. Suggest a hypothesis for how these proteins find their way to the peroxisomes.

15. Propose an experiment that would determine if the NLS in nucleoplasmin is limited to this protein only or if it could direct other structures into the nucleus.

16. George Palade's research group used the pulse–chase assay to dissect the secretory pathway in pancreatic cells. If they had instead performed this assay on muscle cells, which have high energy demands and primarily consist of actin and myosin filaments, where would you expect the labeled proteins to go during the chase?

8 Energy and Enzymes: An Introduction to Metabolic Pathways

In this chapter you will learn how

Enzymes use energy to drive the chemistry of life

looking at energy, asking

What happens to energy in chemical reactions? **8.1**

Can chemical energy drive nonspontaneous reactions? **8.2**

looking at enzymes, asking

How do enzymes help speed chemical reaction rates? **8.3**

What factors affect enzyme function? **8.4**

How do enzymes work together in metabolic pathways? **8.5**

When table sugar is heated in the presence of oxygen, it undergoes the uncontrolled oxidation reaction known as burning. The heat energy in the flame is released as electrons are transferred from sugar to oxygen. Cells use the energy released from this type of reaction to drive the energy-demanding processes required for life.

This chapter is part of the Big Picture. See how on pages 198–199.

Cells are dynamic. Vesicles move cargo from the Golgi apparatus to the plasma membrane and other destinations, enzymes catalyze the synthesis of a complex array of macromolecules, and millions of proteins transport ions and molecules across cellular membranes. These activities change constantly in response to signals from other cells or the environment.

What drives all this action? The answer is twofold—energy and enzymes. Because staying alive takes work, there is no life without energy. Life, at its most basic level, consists of chemical reactions catalyzed by enzymes. By using enzymes to direct which reactions occur and which do not, life possesses the distinguishing feature of creating order from a naturally disordered environment.

✔ When you see this checkmark, stop and test yourself. Answers are available in Appendix A.

This chapter is about how enzymes work to help cells acquire and use energy. It is also your introduction to metabolic pathways—the ordered series of chemical reactions that build up or break down a particular molecule.

Let's begin by reviewing some fundamental concepts about energy and how it is used in cells.

<table>
<tr><td>8.1</td><td>What Happens to Energy in Chemical Reactions?</td></tr>
</table>

When biologists consider energy in chemical reactions, they often use the term **free energy** to describe the amount of energy that is available to do work. Recall that two types of energy exist: kinetic energy or potential energy (Chapter 2). **Kinetic energy** is energy of motion. There are several different forms of kinetic energy—at the molecular level, the energy of motion is called thermal energy. **Potential energy** is energy that is associated with position or configuration. In molecules, this is referred to as chemical energy and is stored in the position of electrons.

Chemical Reactions Involve Energy Transformations

The existence of two types of energy does not mean that energy is locked into either the kinetic or the potential type. Energy is often transformed from one type to the other. To drive this point home, consider a water molecule sitting at the top of a waterfall, as in **FIGURE 8.1**.

Step 1 The molecule has potential energy (E_p) because of its position.

Step 2 As the molecule passes over the waterfall, its potential energy is converted to the kinetic energy (E_k) of motion.

Step 3 When the molecule reaches the rocks below, it undergoes a change in potential energy because it has changed position. The difference in potential energy is transformed into an equal amount of kinetic energy that is manifested in a variety of forms: mechanical energy, which tends to break up the rocks; heat (thermal energy), which raises the temperature of the rocks and the water itself; and sound.

The amount of potential energy in an electron is based on its position relative to other electrons and the protons in the nuclei of nearby atoms (see **FIGURE 8.2a** on page 138). If an electron is close to negative charges on other electrons and far from the positive charges in nuclei, it has high potential energy. In general, the potential energy of a molecule is a function of the way its electrons are configured or positioned.

An electron in an outer electron shell is analogous to the water molecule at the top of a waterfall (**FIGURE 8.2b**). If the electron falls to a lower shell, its potential energy is converted to the kinetic energy of motion. After the electron occupies the lower electron shell, it undergoes a change in potential energy. As panel 3 in Figure 8.1b shows, the change in potential energy is transformed

into an equal amount of kinetic energy—usually thermal energy, but sometimes light.

These examples illustrate the **first law of thermodynamics,** which states that energy is conserved. Energy cannot be created or destroyed, but only transferred and transformed.

The total energy in a molecule is referred to as its **enthalpy** (represented by H). Enthalpy includes the potential energy of the molecule, often referred to as heat content, plus the effect of the molecule on its surroundings in terms of pressure and volume.

1. Potential energy
A water molecule sitting at the top of a waterfall has a defined amount of potential energy, E_p.

2. Kinetic energy
As the molecule falls, some of this stored energy is converted to kinetic energy (the energy of motion), E_k.

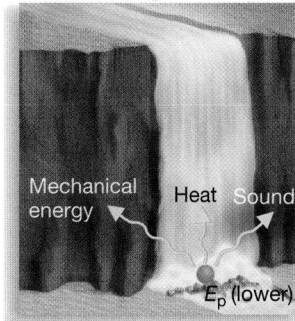

3. Other forms of kinetic energy
When the molecule strikes the rocks below, its energy of motion is converted to thermal, mechanical, and sound energy. The molecule's potential energy is now much lower. The change in potential energy has been transformed into an equal amount of other forms of kinetic energy.

Conclusion: Energy is neither created nor destroyed; it simply changes form.

FIGURE 8.1 Energy Transformations. During an energy transformation, the total amount of energy in the system remains constant.

(a) The potential energy of an electron is related to its position.

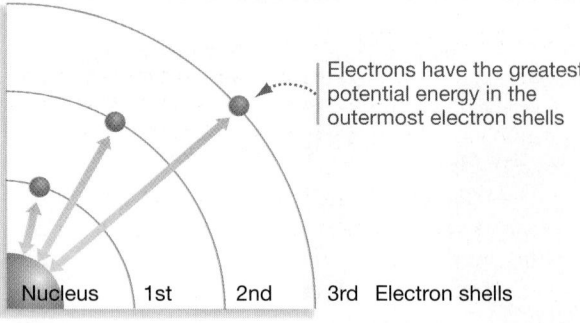

Electrons have the greatest potential energy in the outermost electron shells

Nucleus   1st   2nd   3rd   Electron shells

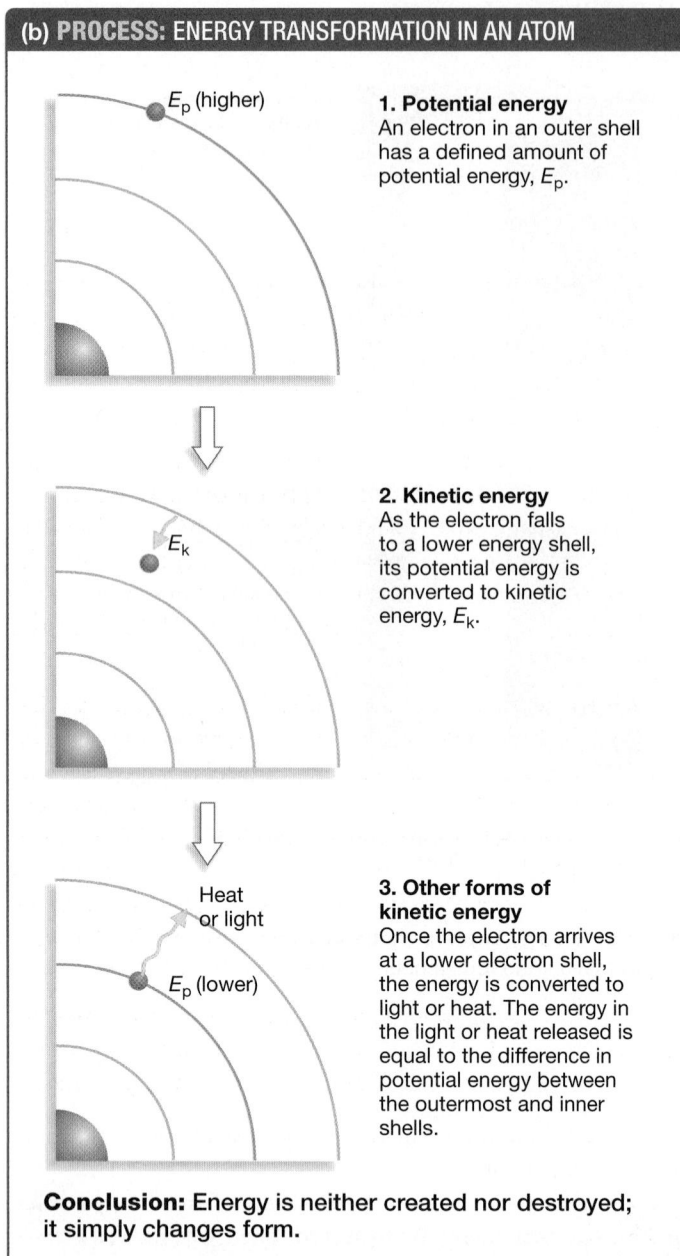

(b) PROCESS: ENERGY TRANSFORMATION IN AN ATOM

E_p (higher)

1. Potential energy
An electron in an outer shell has a defined amount of potential energy, E_p.

E_k

2. Kinetic energy
As the electron falls to a lower energy shell, its potential energy is converted to kinetic energy, E_k.

Heat or light

E_p (lower)

3. Other forms of kinetic energy
Once the electron arrives at a lower electron shell, the energy is converted to light or heat. The energy in the light or heat released is equal to the difference in potential energy between the outermost and inner shells.

Conclusion: Energy is neither created nor destroyed; it simply changes form.

FIGURE 8.2 Chemical energy transformations Potential energy energy stored in atoms or molecules may be transformed into kinetic energy by changes in electron position.

The contributions of heat, pressure, and volume to the enthalpy of a molecule are best understood by observing the changes in enthalpy in a chemical reaction. For example, let's examine the reaction responsible for the explosive bursts of scalding hot liquid a bombardier beetle can produce when provoked, as seen in **FIGURE 8.3**:

$$2 H_2O_2(aq) \longrightarrow 2 H_2O(l) + O_2(g)$$

In this reaction, hydrogen peroxide (H_2O_2) is broken down into water and O_2 gas, which expands to over 500 times the original volume of the H_2O_2. Heat given off from the reaction also increases the temperature of the liquid dramatically. These massive increases in temperature and volume generate the pressure that propels the boiling liquid out of an opening at the tip of the beetle's abdomen.

Changes in enthalpy in chemical reactions can be measured and are represented by ΔH. (The uppercase Greek letter delta, Δ, is often used in chemical and mathematical notation to represent change.) The value of ΔH is primarily based on the difference in heat content, since—apart from the reaction in the bombardier beetle—most biological reactions do not result in substantial changes in pressure and volume. When a reaction releases heat energy (products have less potential energy than the reactants), it is **exothermic** and the ΔH is negative. If heat energy is taken up during the reaction, generating products that have higher potential energy than the reactants, the reaction is **endothermic** and ΔH is positive.

Another factor that changes during a chemical reaction is the amount of disorder or **entropy** (symbolized by ΔS). When the products of a chemical reaction become less ordered than the reactant molecules were, entropy increases and ΔS is positive. The **second law of thermodynamics,** in fact, states that total entropy always increases in an isolated system. Keep in mind that the isolated system in this case is the universe, which includes the surroundings as well as the products of the reaction.

To determine whether a chemical reaction is spontaneous, it's necessary to assess the combined contributions of changes in heat and disorder. Chemists do this with a quantity called the **Gibbs free-energy change,** symbolized by ΔG.

$$\Delta G = \Delta H - T\Delta S$$

Here, T stands for temperature measured on the Kelvin scale (see **BioSkills 1**, in Appendix B). Water freezes at 273.15 K and boils at 373.15 K.

In words, the free-energy change in a reaction is equal to the change in enthalpy minus the change in entropy multiplied by the temperature. The $T\Delta S$ term simply means that entropy becomes more important in determining free-energy change as the temperature of the molecules increases. Thermal energy increases the amount of disorder in the system, so the faster molecules are moving, the more important entropy becomes in determining the overall free-energy change.

Chemical reactions are spontaneous when ΔG is less than zero. Such reactions are said to be **exergonic.** Reactions are

FIGURE 8.3 Reactions May Be Explosive due to Changes in Enthalpy. When provoked, the bombardier beetle mixes reactants with enzymes in a special chamber near the tip of its abdomen. The enzyme-catalyzed reaction releases heat energy and oxygen gas. The result is the projection of boiling hot liquid at a predator.

If you understand that . . .

- The ability of chemical reactions to proceed without an input of energy depends on the difference in enthalpy and entropy between the products and reactants.
- The combined effects of enthalpy and entropy changes are summarized in the equation for the Gibbs free-energy change.

✔ **You should be able to . . .**

1. Write out the Gibbs equation and define each of the components.
2. State when the overall free energy change in a reaction is most likely to be negative (meaning that the reaction is exergonic).

Answers are available in Appendix A.

nonspontaneous when ΔG is greater than zero. Such reactions are termed **endergonic.** When ΔG is equal to zero, reactions are at equilibrium. ✔ If you understand these concepts, you should be able to explain (1) why the same reaction can be nonspontaneous at low temperature but spontaneous at high temperature, and (2) why some exothermic reactions are nonspontaneous.

Free energy changes when the potential energy and/or entropy of substances change. Spontaneous chemical reactions run in the direction that lowers the free energy of the system. Exergonic reactions are spontaneous and release energy; endergonic reactions are nonspontaneous and require an input of energy to proceed.

Temperature and Concentration Affect Reaction Rates

Even if a chemical reaction occurs spontaneously, it may not happen quickly. The reactions that convert iron to rust or sugar molecules to carbon dioxide and water are spontaneous, but at room temperature they occur very slowly, if at all.

For most reactions to proceed, one or more chemical bonds have to break and others have to form. For this to happen, the substances involved must collide in a specific orientation that brings the electrons involved near each other. (See Chapter 2 to review the forces involved in bond formation.)

The number of collisions occurring between the substances in a mixture depends on their temperature and concentration:

- When the concentration of reactants is high, more collisions should occur and reactions should proceed more quickly.
- When their temperature is high, reactants should move faster and collide more frequently.

Higher concentrations and higher temperatures should speed up chemical reactions. To test this hypothesis, students at Parkland College in Champaign, Illinois, performed the experiments shown in **FIGURE 8.4** (see page 140). Pay special attention to the two graphs in the "Results" section:

- *Temperature versus reaction rate* The graph on the left is based on experiments where the concentration of the reactants was the same, but the temperature varied. Each data point represents one experiment. Notice that the points represent a trend that rises from left to right—meaning, in this case, that the reaction rate speeded up when the temperature of the reaction mixture was higher.

- *Concentration versus reaction rate* The graph on the right is based on experiments where the temperature was constant, but the concentration of reactants varied. Each bar represents the average reaction rate over many replicates of each treatment, or set of concentrations. The thin lines at the top of each bar indicate the standard error of the mean—a measure of variability (see **BioSkills 4** in Appendix B). The take-home message of this graph is that reaction rates are higher when reactant concentrations are higher.

The reactions shown in Figure 8.4 were exergonic, meaning that the products had lower free energy than the reactants, so no input of energy was required. But, what drives nonspontaneous, endergonic reactions? Let's take a closer look.

8.2 Nonspontaneous Reactions May Be Driven Using Chemical Energy

By definition, endergonic reactions require an input of energy to proceed. Recall that radiation from the Sun and electricity from lightning could have driven nonspontaneous reactions during

RESEARCH

QUESTION: Do chemical reaction rates increase with increased temperature and concentration?

RATE INCREASE HYPOTHESIS: Chemical reaction rates increase with increased temperature. They also increase with increased concentration of reactants.

NULL HYPOTHESIS: Chemical reaction rates are not affected by increases in temperature or concentration of reactants.

EXPERIMENTAL SETUP:

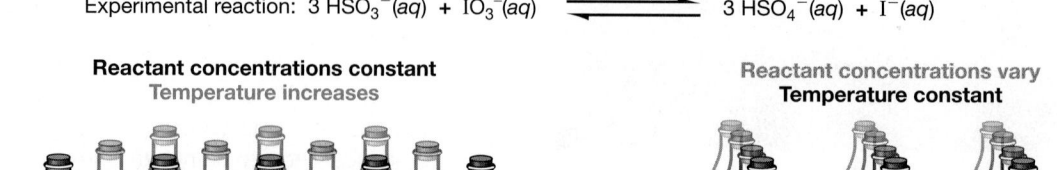

Experimental reaction: $3\ HSO_3^-(aq) + IO_3^-(aq) \rightleftharpoons 3\ HSO_4^-(aq) + I^-(aq)$

Reactant concentrations constant
Temperature increases

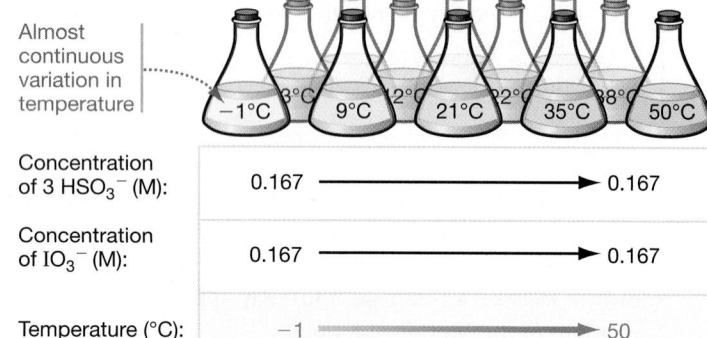

Almost continuous variation in temperature

−1°C 3°C 9°C 12°C 21°C 32°C 35°C 38°C 50°C

Concentration of 3 HSO_3^- (M):	0.167 ⟶ 0.167	
Concentration of IO_3^- (M):	0.167 ⟶ 0.167	
Temperature (°C):	−1 ⟶ 50	

Reactant concentrations vary
Temperature constant

Treatment 1   Treatment 2   Treatment 3

Many replicates at each concentration

	Treatment 1	Treatment 2	Treatment 3
Concentration of 3 HSO_3^- (M):	0.167	0.167	0.333
Concentration of IO_3^- (M):	0.167	0.333	0.333
Temperature (°C):	23	23	23

PREDICTION: Reaction rate, measured as 1/(time for reaction to go to completion), will increase with increased concentrations of reactants and increased temperature of reaction mix.

PREDICTION OF NULL HYPOTHESIS: There will be no difference in reaction rates among treatments in each setup.

RESULTS:

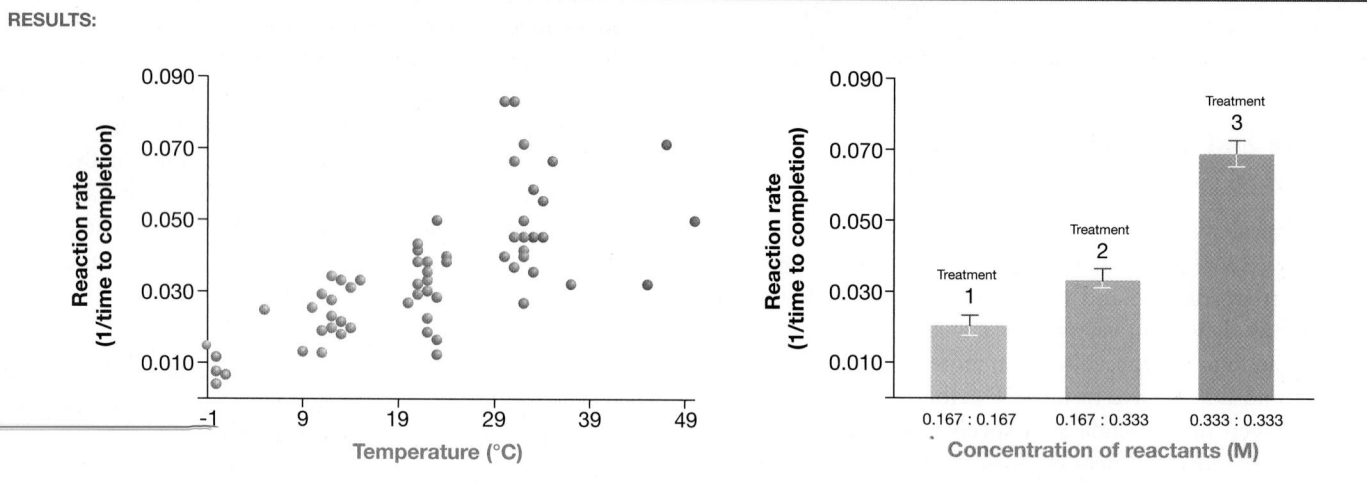

CONCLUSION: Chemical reaction rates increase with increased temperature or concentration.

FIGURE 8.4 Testing the Hypothesis that Reaction Rates Are Sensitive to Changes in Temperature and Concentration.

✓**QUESTION** Use **BioSkills 4** in Appendix B to explain why no error bars are used for the points shown on the graph on the left side of the "Results" section.

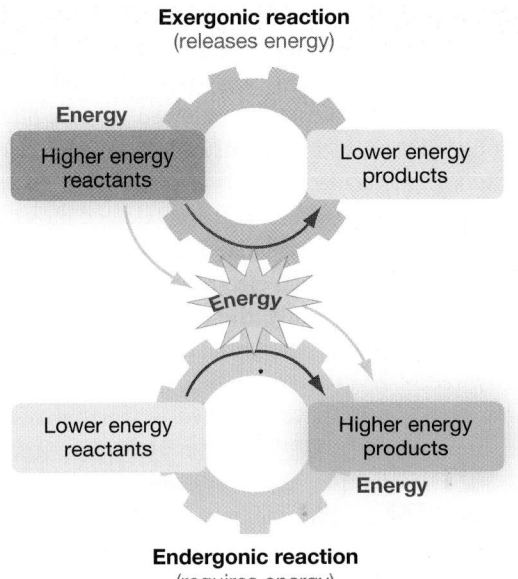

Exergonic reaction
(releases energy)

Energy

Higher energy reactants → Lower energy products

Energy

Lower energy reactants → Higher energy products

Energy

Endergonic reaction
(requires energy)

FIGURE 8.5 Energetic Coupling Allows Endergonic Reactions to Proceed Using the Energy Released from Exergonic Reactions.

chemical evolution (Chapter 2). What source of energy drives these reactions inside cells?

Exergonic reactions release free energy. **FIGURE 8.5** shows how **energetic coupling** between exergonic and endergonic reactions allows chemical energy released from one reaction to drive another. In cells, this process generally occurs in one of two ways, either through the transfer of high-energy electrons or the transfer of a phosphate group.

Redox Reactions Transfer Energy via Electrons

Chemical reactions that involve the loss or gain of one or more electrons are called **reduction–oxidation reactions,** or **redox reactions.** When an atom or molecule loses one or more electrons, it is oxidized. This makes sense if you notice that the term

oxidized sounds as if oxygen has done something to an atom or molecule. Recall that oxygen is highly electronegative and often pulls electrons from other atoms (Chapter 2). On the other hand, when an atom or molecule gains one or more electrons, it is reduced. To keep these terms straight, students often use the mnemonic "OIL RIG"—**Oxidation** *Is* *L*oss of electrons; **Reduction** *Is* *Gain* of electrons.

Oxidation events are always paired with a reduction; if one atom loses an electron, another has to gain it, and vice versa. Since electron position is related to energy levels, redox reactions represent the energetic coupling of two half-reactions, one exergonic and one endergonic. Oxidation is the exergonic half-reaction, and reduction is the endergonic half-reaction. Some of the energy that is lost by the oxidized molecule is used to increase potential energy of the reduced molecule. In cases where more free energy is released by the oxidation step than is necessary for the reduction step, the overall reaction is exergonic.

The gain or loss of an electron can be relative, however. During a redox reaction, an electron can be transferred completely from one atom to another, or an electron can simply shift its position in a covalent bond.

An Example of Redox in Action To see how redox reactions work, consider the spontaneous reaction that occurs when reduced carbons in glucose ($C_6H_{12}O_6$) are oxidized as the sugar is burned in the presence of oxygen (O_2) (**FIGURE 8.6**). The orange dots in the illustration represent the positions of the electrons involved in covalent bonds.

Now compare the position of the electrons in the first reactant, glucose, with their position in the first product, carbon dioxide. Notice that many of the electrons have moved farther from the carbon nucleus in carbon dioxide. This means that carbon has been oxidized: it has "lost" electrons. The change occurred because the carbon and hydrogen atoms in glucose share electrons equally, while the carbon and oxygen atoms in CO_2 don't. In CO_2, the high electronegativity of the oxygen atoms pulled electrons away from the carbon atom.

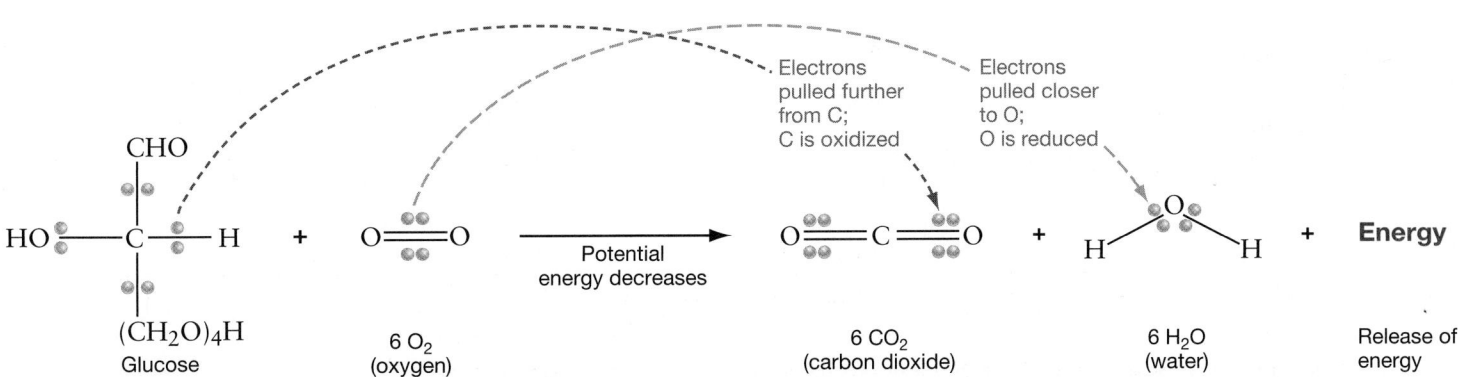

FIGURE 8.6 Redox Reactions Involve the Gain or Loss of One or More Electrons. This diagram shows how the position of electrons changes when glucose reacts with oxygen. The carbons of glucose are oxidized while the oxygen atoms of O_2 are reduced.

Now compare the position of the electrons in the reactant O_2 molecules with their position in the product water molecules. In water, the electrons have moved closer to the oxygen nuclei than they were in the O_2 molecules, meaning that the oxygen atoms have been reduced. Oxygen has "gained" electrons. Thus, when glucose burns, carbon atoms are oxidized while oxygen atoms are reduced.

These shifts in electron position change the amount of chemical energy in the reactants and products. When glucose reacts with oxygen, electrons are held much tighter in the product molecules than in the reactant molecules. This means their potential energy has decreased. The entropy of the products is also much higher than that of the reactants, as indicated by the increase in the number of molecules. As a result, this reaction is exergonic. It releases energy in the form of heat and light.

Another Approach to Understanding Redox During the redox reactions that occur in cells, electrons (e^-) may also be transferred from an atom in one molecule, called the **electron donor,** to an atom in a different molecule, the **electron acceptor.** When this occurs, the electron may be accompanied by a proton (H^+), which would result in the addition of a neutral hydrogen (H) atom to the electron acceptor.

Molecules that obtain hydrogens via redox reactions tend to gain potential energy because the electrons in C—H bonds are equally shared and hence relatively far from the positive charges on the C and H nuclei. This observation should sound familiar,

from what you have learned about carbohydrates (see Chapter 5). Molecules that have a large number of C—H bonds, such as carbohydrates and fats, store a great deal of potential energy.

Conversely, molecules that are oxidized in cells often lose a proton along with an electron. Instead of having many C—H bonds, oxidized molecules in cells tend to have an increased number of C—O bonds (see Figure 8.6). Oxidized molecules tend to lose potential energy. To understand why, remember that oxygen atoms have extremely high electronegativity. Because oxygen atoms hold electrons so tightly, the electrons involved in bonds with oxygen atoms have low potential energy.

In many redox reactions in biology, understanding where oxidation and reduction have occurred is a matter of following hydrogen atoms—reduction often "adds Hs" and oxidation often "removes Hs." For example, **flavin adenine dinucleotide (FAD)** is a cellular electron acceptor that is reduced by two electrons accompanied by two protons to form $FADH_2$ (**FIGURE 8.7a**). $FADH_2$ readily donates these high-energy electrons to other molecules. As a result, it is called an **electron carrier** and is said to have "reducing power."

Another common electron carrier is **nicotinamide adenine dinucleotide (NAD$^+$),** which is reduced to form **NADH.** Like FAD, two electrons reduce NAD$^+$. These two carriers differ, however, in the number of hydrogen atoms transferred. NAD$^+$ acquires only one of the two hydrogens and releases the second into the environment as H^+ (**FIGURE 8.7b**).

(a) Flavin adenine dinucleotide

$$AH_2 + FAD \longrightarrow A + FADH_2$$

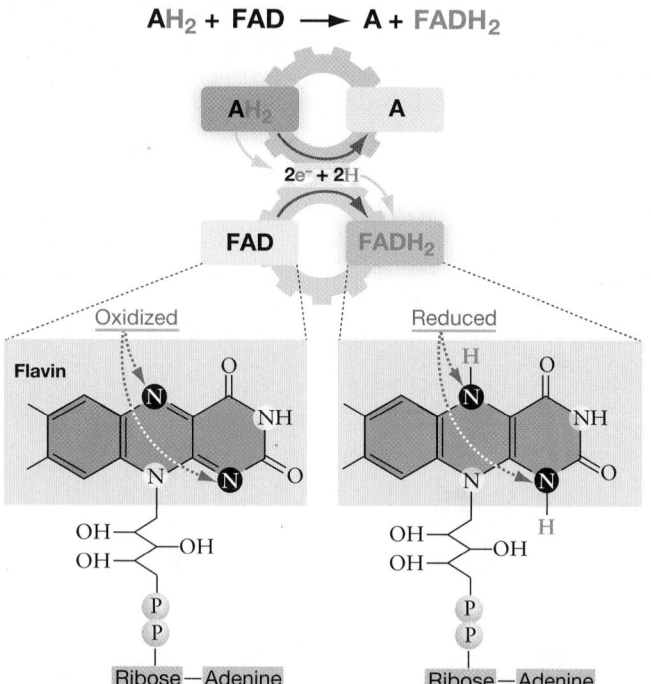

(b) Nicotinamide adenine dinucleotide

$$BH_2 + NAD^+ \longrightarrow B + NADH + H^+$$

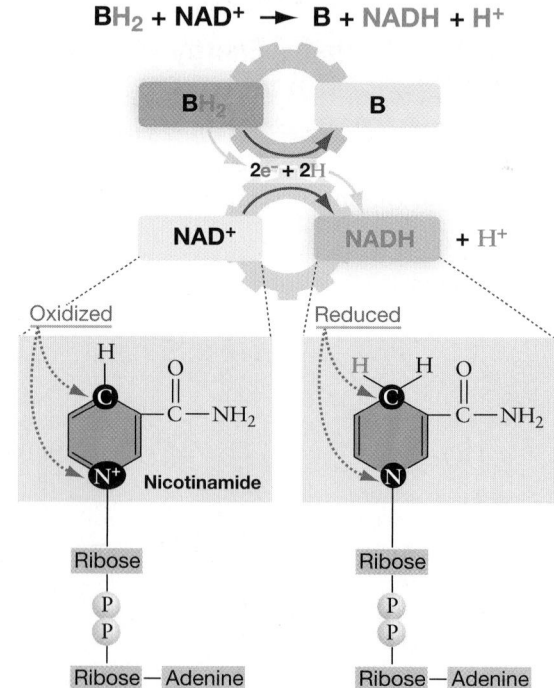

FIGURE 8.7 Redox Reactions May Transfer Protons Along with Electrons. The potential energy of NAD$^+$ and FAD is increased by redox reactions that transfer high-energy electrons, which may or may not be accompanied by protons. The products FADH$_2$ and NADH are important electron carriers.

The two examples in Figure 8.6 illustrate an important point—all redox reactions involve the transfer of electrons, but they do not always involve the transfer of hydrogens. Redox reactions are central in biology—they transfer energy via electrons. The energy released from certain key redox reactions (see Chapters 9 and 10) is used to drive the endergonic formation of the nucleotide ATP from ADP and P_i. How is the energy stored in ATP used by the cell?

ATP Transfers Energy via Phosphate Groups

Adenosine triphosphate (ATP) (introduced in Chapter 4) makes things happen in cells because it has a great deal of potential energy. As **FIGURE 8.8a** shows, four negative charges are confined to a small area in the three phosphate groups in ATP. In part because these negative charges repel each other, the potential energy of the electrons in the phosphate groups is extraordinarily high.

ATP Hydrolysis Releases Free Energy When ATP reacts with water during a hydrolysis reaction, the bond between ATP's outermost phosphate group and its neighbor is broken, resulting in the formation of ADP and inorganic phosphate, P_i, which has the formula $H_2PO_4^-$ (**FIGURE 8.8b**). This reaction is highly exergonic. Under standard conditions of temperature and pressure in the laboratory, a total of 7.3 kilocalories of energy per mole of ATP (or 7.3 kcal/mol), is released during the reaction. A **kilocalorie (kcal)** of energy raises 1 kilogram (kg) of water 1°C.

ATP hydrolysis is exergonic because the entropy of the product molecules is higher than that of the reactants, and because there is a large drop in potential energy when ATP breaks down into ADP and P_i. The change in potential energy occurs in part because the electrons from ATP's phosphate groups are now spread across two molecules instead of being clustered on one molecule—meaning that there is now less electrical repulsion.

In addition, the destabilizing effect of the negative charges is reduced in ADP and P_i since these products interact with the partial positive charges on surrounding water molecules more efficiently than the clustered negative charges on ATP did.

How Does ATP Drive Endergonic Reactions? In the time it takes to read this sentence, millions of endergonic reactions have occurred in your cells. This chemical activity is possible, in part, because cells are able to use the energy released from the exergonic hydrolysis of ATP.

If the reaction diagrammed in Figure 8.8b occurred in a test tube, the energy released would be lost as heat. But cells don't lose that 7.3 kcal/mole as heat. Instead, they use it to make things happen. Specifically, the energy that is released when ATP is hydrolyzed may be used to transfer the cleaved phosphate to a target molecule, called a **substrate.**

The addition of a phosphate group to a substrate is called **phosphorylation.** When ATP is used as the phosphate donor, phosphorylation is exergonic because the electrons in ADP and the phosphate added to the substrate have much less potential energy than they did in ATP.

To see how this process works, consider an endergonic reaction between two reactant molecules—compound A and compound B—that results in a product AB needed by your cells. For this reaction to proceed, an input of energy is required.

When a phosphate group from ATP is added to one or both of the reactant molecules, the potential energy of the reactant is increased. This phosphorylated intermediate is referred to as an activated substrate. This is the critical point: Activated substrates have high enough potential energy that the reaction between compound A and, for example, the activated form of compound B is now exergonic. The two compounds then go on to react and form the product molecule AB.

(a) ATP stores a large amount of potential energy.

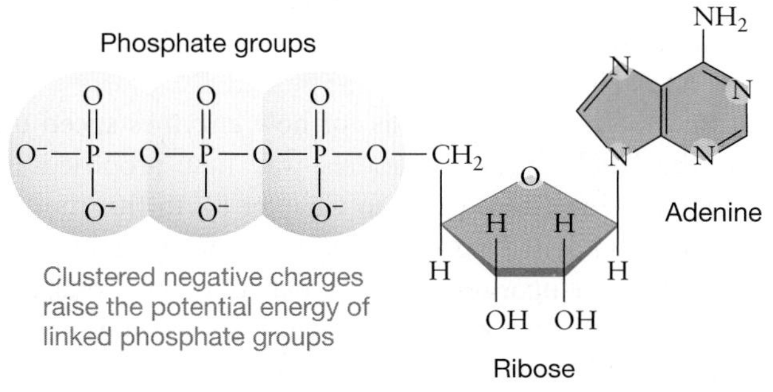

Phosphate groups

Clustered negative charges raise the potential energy of linked phosphate groups

Adenine

Ribose

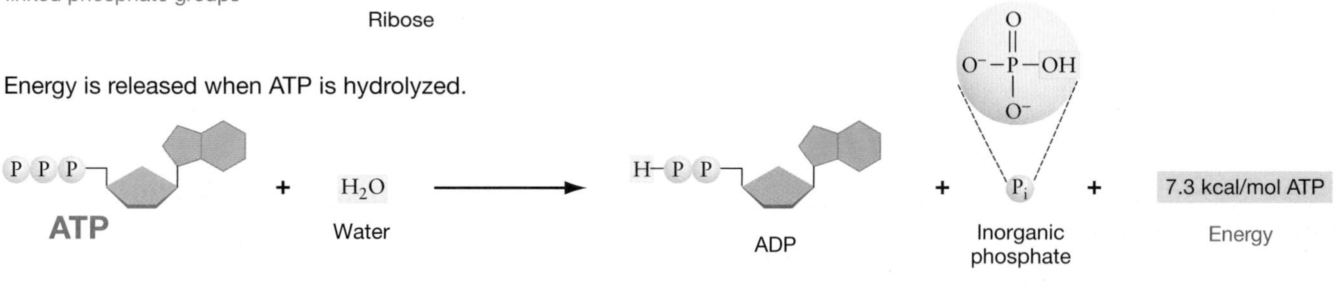

FIGURE 8.8 Adenosine Triphosphate (ATP) Has High Potential Energy. (a) ATP's high potential energy results, in part, from the four negative charges clustered in its three phosphate groups. The negative charges repel each other, raising the potential energy of the electrons. **(b)** When ATP is hydrolyzed to ADP and inorganic phosphate, a large free-energy change occurs.

(b) Energy is released when ATP is hydrolyzed.

ATP   + H_2O ⟶   ADP   + P_i +   7.3 kcal/mol ATP

         Water   Inorganic   Energy
                                        phosphate

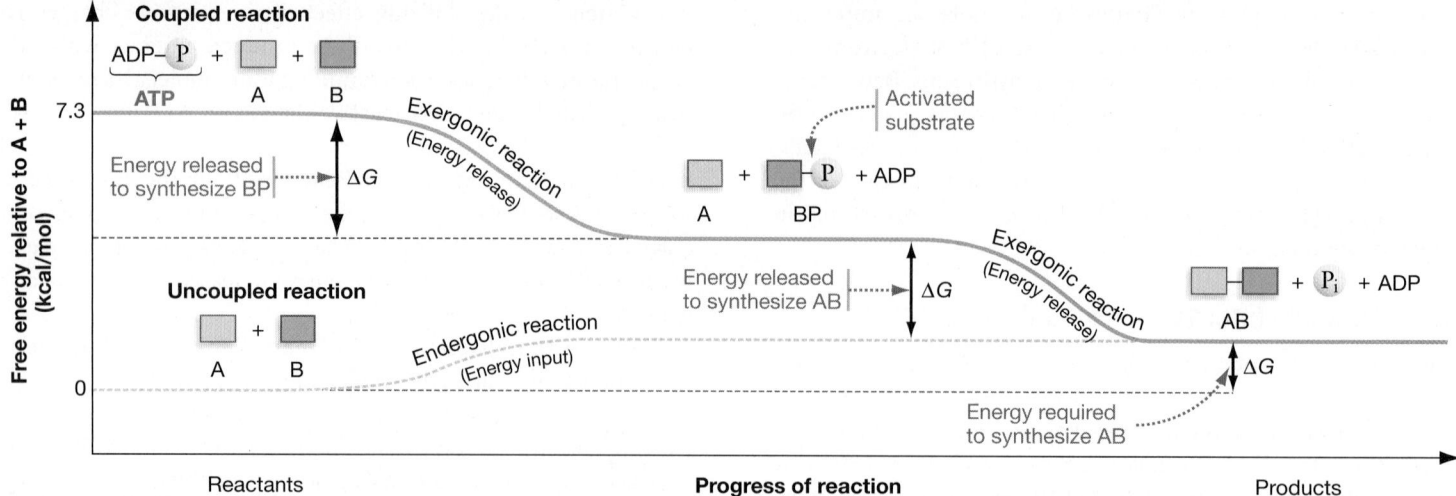

FIGURE 8.9 Exergonic Phosphorylation Reactions Are Coupled to Endergonic Reactions. In cells, many reactions only occur if one reactant is activated by phosphorylation. The phosphorylated reactant molecule has high enough free energy that the subsequent reaction is exergonic. In this graph, the free energy being tracked on the y-axis represents A, B, and the 7.3 kcal/mol that is released when ATP is hydrolyzed. For simplicity, the free energy in ADP and P_i is not shown. ΔG represents the change in free energy between the reactants and products for each indicated step.

✔**EXERCISE** Label the ΔG in the uncoupled reaction and the two steps of the coupled reaction to indicate if the change is representing a positive (> 0) or negative (< 0) value.

FIGURE 8.9 graphs how phosphorylation can couple exergonic and endergonic reactions. Note that the reaction between A and B to produce the product AB is endergonic—the ΔG is positive. But after the exergonic transfer of a phosphate group from ATP to B occurs, the free energy of the reactants A and BP is high enough to make the reaction that forms AB exergonic. When reactant molecules in an endergonic reaction are phosphorylated, the free energy released during phosphorylation is coupled to the endergonic reaction to make the combined overall reaction exergonic.

✔ If you understand the principles of energetic coupling, you should be able to compare and contrast how energy is transferred via redox reactions and ATP hydrolysis.

It is hard to overstate the importance of energetic coupling: Without it, life is impossible. If the cells in your body could no longer drive endergonic reactions by coupling them to exergonic reactions, you would die within minutes.

Now the question is, What role do enzymes play in these reactions?

check your understanding

If you understand that . . .

- When redox reactions occur, electrons change position. Chemical energy is based on the positions of electrons in chemical bonds, so redox reactions usually involve a change in potential energy.
- ATP contains a cluster of three negatively charged phosphate groups.
- When ATP or phosphate groups from ATP bind to substrates, they gain a great deal of potential energy.

✔ You should be able to . . .

1. Explain why reduced molecules with many C–H bonds store more potential energy than oxidized molecules with many C–O bonds.
2. Explain why ATP has such high potential energy.

Answers are available in Appendix A.

8.3 How Enzymes Work

Regardless of whether reactions in cells are spontaneous or not, none would occur at the speed required for life without the support of enzymes. How do they do it?

Recall that the initial hypothesis for how enzymes speed up reactions—the "lock-and-key" model—was first proposed in 1894 by Emil Fischer (introduced in Chapter 3). In this model, the substrates would fit into enzymes and react in a manner analogous to a key being inserted into a lock. In other words, enzymes are **catalysts**—they bring substrates together in a precise orientation that makes reactions more likely. Fischer's model also explained why many enzymes are specific for a single reaction—specificity is a product of the geometry and chemical properties of the sites where substrates bind.

Enzymes Help Reactions Clear Two Hurdles

Recall that two hurdles must be cleared before reactions can take place: Reactants need to **(1)** collide in a precise orientation and

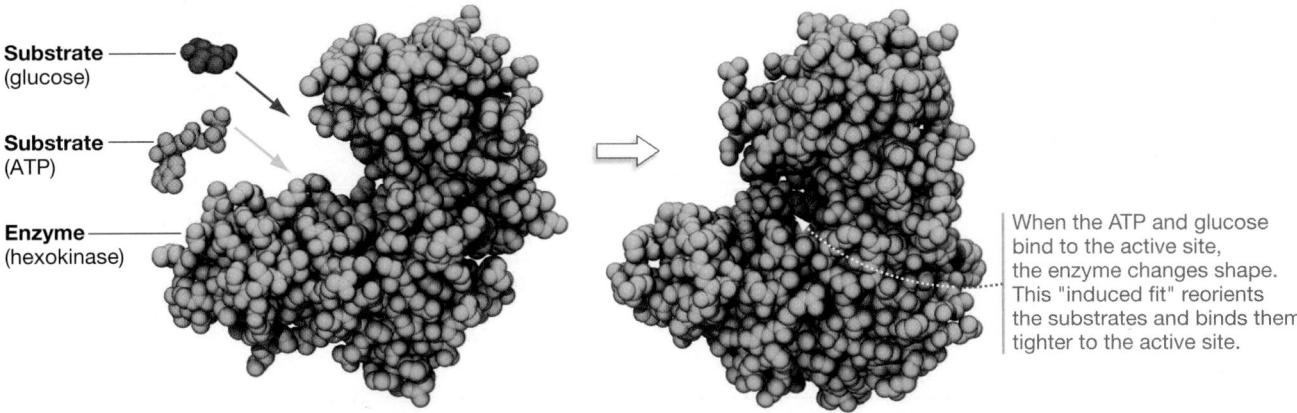

Substrate
(glucose)

Substrate
(ATP)

Enzyme
(hexokinase)

When the ATP and glucose
bind to the active site,
the enzyme changes shape.
This "induced fit" reorients
the substrates and binds them
tighter to the active site.

FIGURE 8.10 Reactant Molecules Bind to Specific Locations in an Enzyme. The reactant molecules, shown in red and yellow, fit into a precise location, called the active site, in the green enzyme. In this enzyme and in many others, the binding event causes the protein to change shape.

(2) have enough kinetic energy to overcome repulsion between electrons that come into contact as a bond forms (Chapter 2). To appreciate how enzymes work, let's consider each hurdle in turn.

Enzymes Bring Substrates Together

Part of the reason enzymes are such effective catalysts is that they bring substrate molecules together in a substrate binding site known as the enzyme's **active site** (Chapter 3). In this way, enzymes help substrates collide in a precise orientation so that the electrons involved in the reaction can interact.

Enzymes generally are very large relative to substrates and roughly globular. The active site is in a cleft or cavity within the globular shape. A good example can be seen in the enzyme glucokinase, which catalyzes the phosphorylation of the sugar glucose. (Many enzymes have names that hint at the identity of the substrate and end with -*ase*.) As the left side of **FIGURE 8.10** shows, the active site in glucokinase is a small notch in an otherwise large, crescent-shaped enzyme.

In Fischer's original lock-and-key model, enzymes were conceived of as being rigid—almost literally as rigid as a lock. As research on enzyme action progressed, however, Fischer's model had to be modified. Perhaps the most important realization was that enzymes are not rigid and static, but flexible and dynamic. In fact, many enzymes undergo a significant change in shape, or conformation, when reactant molecules bind to the active site. You can see this conformational change, called an **induced fit,** in the glucokinase molecule on the right side of Figure 8.10. Once glucokinase binds its substrates—ATP and glucose—the enzyme rocks forward over the active site to bring the two substrates together.

In addition, recent research has clarified the nature of Fischer's key. When one or more substrate molecules enter the active site, they are held in place through hydrogen bonding or other weak interactions with amino acids in the active site. Once the substrate is bound, one or more R-groups in the active site come into play. The degree of interaction between the substrate and enzyme increases and reaches a maximum when a temporary,

unstable, intermediate condition called the **transition state** is formed. When Fischer's key is in its lock, it represents the transition state of the substrate.

There is more to achieving this transition state than simply an enzyme binding to its substrates, however. Even if the reaction is spontaneous, a certain amount of kinetic energy is required to strain the chemical bonds in substrates so they can achieve this transition state—called the **activation energy.** How do enzymes help clear the activation energy hurdle?

Enzymes Lower the Activation Energy

Reactions happen when reactants have enough kinetic energy to reach the transition state. The kinetic energy of molecules, in turn, is a function of their temperature. (This is why reactions tend to proceed faster at higher temperatures.)

FIGURE 8.11 (see page 146) graphs the changes in free energy that take place during the course of a chemical reaction. As you read along the x-axis from left to right, note that a dramatic rise in free energy occurs when the reactants combine to form the transition state—followed by a dramatic drop in free energy when products form. The free energy of the transition state is high because the bonds that existed in the substrates are destabilized—it is the transition point between breaking old bonds and forming new ones.

The ΔG label on the graph indicates the overall change in free energy in the reaction—that is, the energy of the products minus the energy of the reactants. In this particular case, the products have lower free energy than the reactants, meaning that the reaction is exergonic. But because the activation energy for this reaction, symbolized by E_a, is high, the reaction would proceed slowly—even at high temperature.

This is an important point: The more unstable the transition state, the higher the activation energy and the less likely a reaction is to proceed quickly.

Reaction rates, then, depend on both the kinetic energy of the reactants and the activation energy of the particular reaction—meaning the free energy of the transition state. If the kinetic

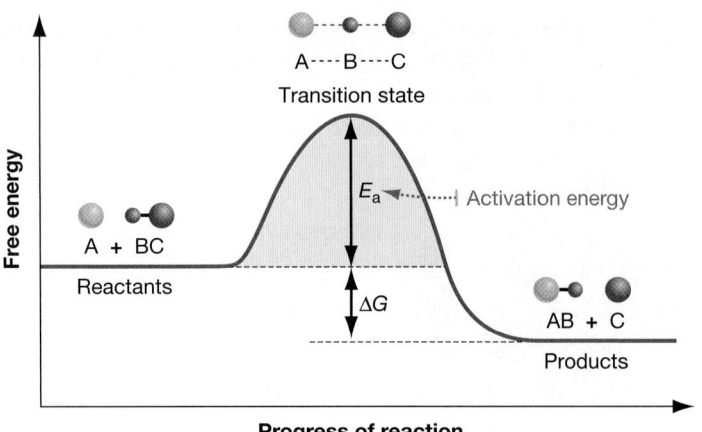

FIGURE 8.11 Changes in Free Energy during a Chemical Reaction. The energy profile shows changes in free energy that occur over the course of a hypothetical reaction between a molecule A and a molecule containing parts B and C. The overall reaction would be written as $A + BC \rightarrow AB + C$. E_a is the activation energy of the reaction.

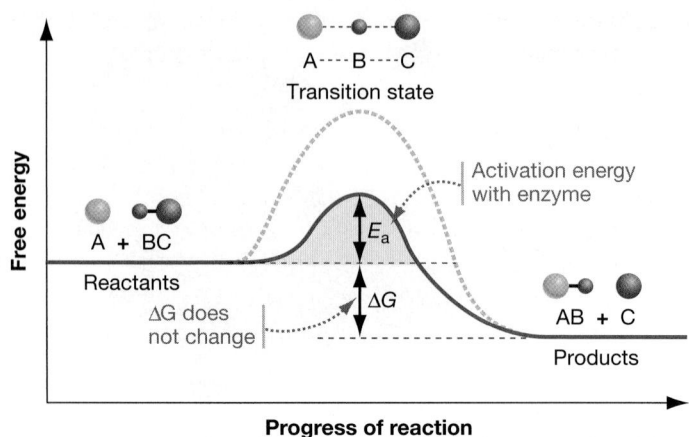

FIGURE 8.12 A Catalyst Changes the Activation Energy of a Reaction. The energy profile for the same reaction diagrammed in Figure 8.11, but now with a catalyst present. Even though the energy barrier to the reaction, E_a, is much lower, ΔG does not change.

✔**QUESTION** Can a catalyst make a nonspontaneous reaction occur spontaneously? Explain why or why not.

energy of the participating molecules is high, such as at high temperatures, then molecular collisions are more likely to overcome the activation energy barrier. At this point, the transition state is formed and the reaction takes place.

Enzymes don't change the temperature of a solution, though. How do they fit in?

Interactions with amino acid R-groups at the enzyme active site stabilize the transition state and thus lower the activation energy required for the reaction to proceed. At the atomic level, R-groups that line the active site may form short-lived covalent bonds that assist with the transfer of atoms or groups of atoms from one reactant to another. More commonly, the presence of acidic or basic R-groups allows the reactants to lose or gain a proton more readily.

FIGURE 8.12 diagrams how enzymes lower the activation energy for a reaction by lowering the free energy of the transition state. Note that the presence of an enzyme does not affect the overall energy change, ΔG, or change the energy of the reactants or the products. An enzyme changes only the free energy of the transition state.

Most enzymes are specific in their activity—they catalyze just a single reaction by lowering the activation energy that is required—and many are astonishingly efficient. Most of the important reactions in biology would not occur at all, or else proceed at imperceptible rates, without a catalyst. It's not unusual for enzymes to speed up reactions by a factor of a million; some enzymes make reactions go many *trillions* of times faster than they would without a catalyst.

It's also important to note that an enzyme is not consumed in a chemical reaction, even though it participates in the reaction. The composition of an enzyme is exactly the same after the reaction as it was before.

Enzyme catalysis can be analyzed as a three-step process. **FIGURE 8.13** summarizes this model:

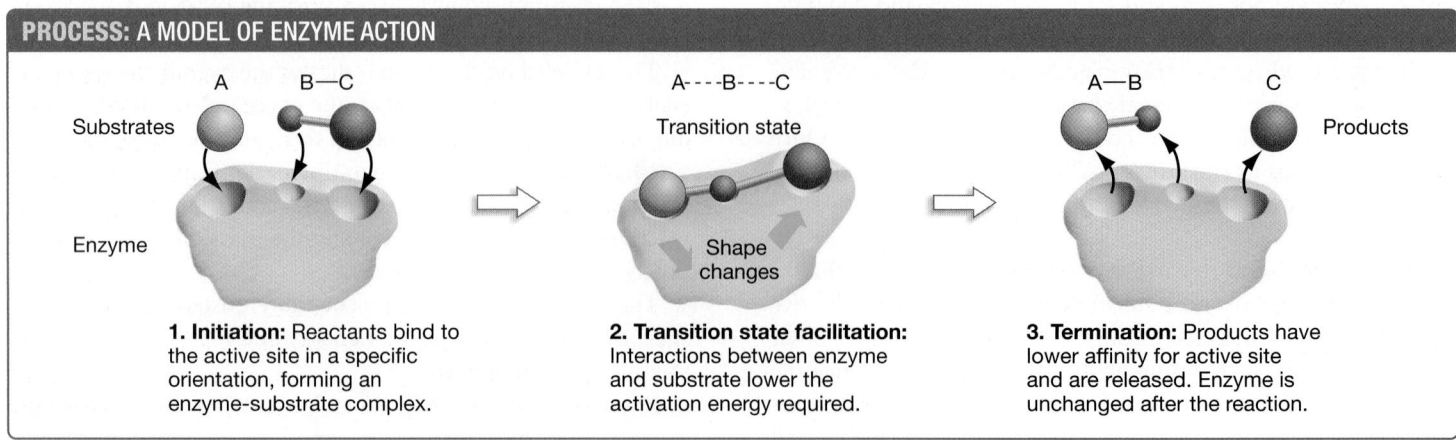

PROCESS: A MODEL OF ENZYME ACTION

Substrates / Enzyme

1. Initiation: Reactants bind to the active site in a specific orientation, forming an enzyme-substrate complex.

Transition state / Shape changes

2. Transition state facilitation: Interactions between enzyme and substrate lower the activation energy required.

Products

3. Termination: Products have lower affinity for active site and are released. Enzyme is unchanged after the reaction.

FIGURE 8.13 Enzyme Action Can Be Analyzed as a Three-Step Process.

1. *Initiation*   Instead of reactants occasionally colliding in a random fashion, enzymes orient reactants precisely as they bind at specific locations within the active site.

2. *Transition state facilitation*   Inside a catalyst's active site, reactant molecules are more likely to reach their transition state. In some cases the transition state is stabilized by a change in the enzyme's shape. Interactions between the substrate and R-groups in the enzyme's active site lower the activation energy required for the reaction. Thus, the catalyzed reaction proceeds much more rapidly than the uncatalyzed reaction.

3. *Termination*   The reaction products have less affinity for the active site than the transition state does. Binding ends, the enzyme returns to its original conformation, and the products are released.

✔ If you understand the basic principles of enzyme catalysis, you should be able to complete the following sentences: (1) Enzymes speed reaction rates by _____ and lowering activation energy. (2) Activation energies drop because enzymes destabilize bonds in the substrates, forming the _____. (3) Enzyme specificity is a function of the active site's shape and the chemical properties of the _____ at the active site. (4) In enzymes, as in many molecules, function follows from _____.

What Limits the Rate of Catalysis?

For several decades after Fischer's model was published, most research on enzymes focused on rates of enzyme action, or what biologists call enzyme kinetics. Researchers observed that, when the amount of product produced per second—indicating the speed of the reaction—is plotted as a function of substrate concentration, a graph like that shown in **FIGURE 8.14** results.

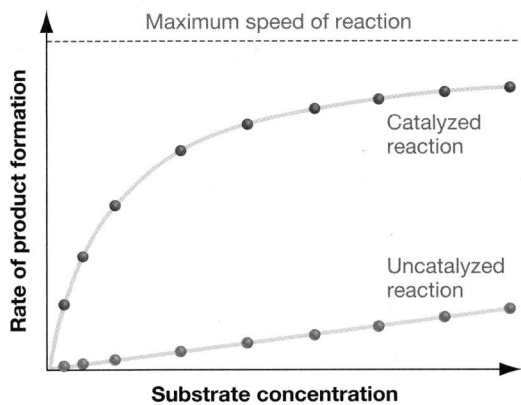

FIGURE 8.14 Enzyme-Catalyzed Reactions Can Be Saturated. At high substrate concentration, enzyme-catalyzed reactions reach a maximum rate. Uncatalyzed reactions slowly increase as substrate concentration increases.

✔**EXERCISE** Label the parts of the *catalyzed reaction curve* that represent where (1) the reaction rate is most sensitive to changes in substrate concentration and (2) most or all of the active sites present are occupied.

In this graph, each data point represents an experiment where reaction rate was measured when substrates were at various concentrations. The two lines represent two series of experiments: one with the reactions catalyzed by an enzyme and the other uncatalyzed. As you read the curve for the catalyzed reaction from left to right, note that it has three basic sections:

1. When substrate concentrations are low, the speed of an enzyme-catalyzed reaction increases in a steep, linear fashion.

2. At intermediate substrate concentrations, the increase in speed begins to slow.

3. At high substrate concentration, the reaction rate plateaus at a maximum speed.

This pattern is in striking contrast to the situation for the uncatalyzed reactions, where the reaction speed is far slower, but tends to show a continuing linear increase with substrate concentration. The "saturation kinetics" of enzyme-catalyzed reactions were taken as strong evidence that the enzyme–substrate complex proposed by Fischer actually exists. The idea was that, at some point, active sites cannot accept substrates any faster, no matter how large the concentration of substrates gets. Stated another way, reaction rates level off because all available enzyme molecules are being used.

Do Enzymes Work Alone?

The answer to this question, in many cases, is no. Atoms or molecules that are not part of an enzyme's primary structure are often required for an enzyme to function normally. These enzyme "helpers" can be divided into three different types:

1. **Cofactors:** Inorganic ions, such as the metal ions Zn^{2+} (zinc), Mg^{2+} (magnesium), and Fe^{2+} (iron), which reversibly interact with enzymes. Cofactors that now participate in key reactions in virtually all living cells are thought to have been involved in catalysis early on in chemical evolution (see Chapter 2).

2. **Coenzymes:** Organic molecules that reversibly interact with enzymes, such as the electron carriers NADH or $FADH_2$.

3. **Prosthetic groups:** Non-amino acid atoms or molecules that are permanently attached to proteins, such as the molecule retinal. Retinal is involved in converting light energy into chemical energy (see Chapter 47).

In many cases, these enzyme helpers are part of the active site and play a key role in stabilizing the transition state. Their presence is therefore essential for the catalytic activity of many enzymes.

To appreciate why this is important, consider that many of the vitamins in your diet are required for the production of coenzymes. Vitamin deficiencies result in coenzyme deficiencies. Lack of coenzymes, in turn, disrupts normal enzyme function and causes disease. For example, thiamine (vitamin B_1) is required for the production of a coenzyme called thiamine pyrophosphate, which is required by three different enzymes. Lack of thiamine in the diet dramatically reduces the activity of these enzymes and causes an array of nervous system and heart disorders collectively known as beriberi.

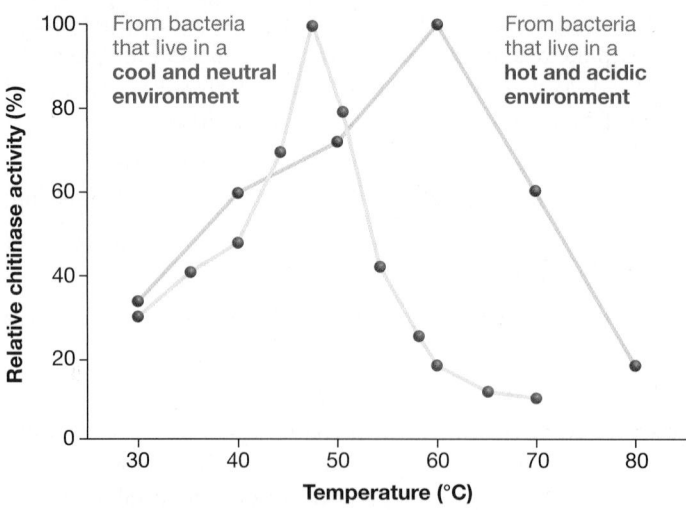

(a) Enzymes from different organisms may function best at different temperatures.

From bacteria that live in a **cool and neutral environment**

From bacteria that live in a **hot and acidic environment**

Relative chitinase activity (%)

Temperature (°C)

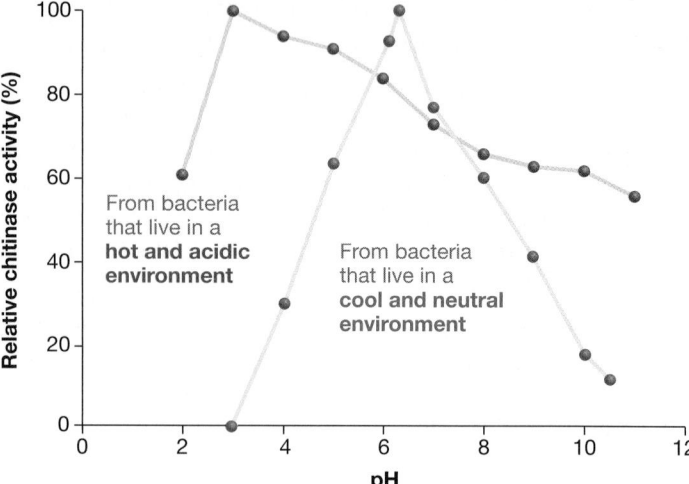

(b) Enzymes from different organisms may function best at different pHs.

From bacteria that live in a **hot and acidic environment**

From bacteria that live in a **cool and neutral environment**

Relative chitinase activity (%)

pH

FIGURE 8.15 Enzymes Have an Optimal Temperature and pH. The activity of enzymes is sensitive to changes in temperature **(a)** and pH **(b)**.

DATA: Nawani, N., B. P. Kapadnis, A. D. Das, et al. 2002. *Journal of Applied Microbiology* 93: 865–975. Also Nawani, N., and B. P. Kapadnis. 2001. *Journal of Applied Microbiology* 90: 803–808.

8.4 What Factors Affect Enzyme Function?

Given that an enzyme's structure is critical to its function, it's not surprising that an enzyme's activity is sensitive to conditions that alter protein shape. Recall that protein structure is dependent on the sequence of amino acids and a variety of chemical bonds and interactions that fold the polypeptide into its functional form (Chapter 3).

In particular, the activity of an enzyme often changes drastically as a function of temperature, pH, interactions with other molecules, and modifications of its primary structure. Let's take a look at how enzyme function is affected by, and sometimes even regulated by, each of these factors.

Enzymes Are Optimized for Particular Environments

Temperature affects the folding and movement of an enzyme as well as the kinetic energy of its substrates. The concentration of protons in a solution, as measured by pH, also affects enzyme structure and function. pH affects the charge on carboxyl and amino groups in residue side chains, and also the active site's ability to participate in reactions that involve the transfer of protons or electrons.

Do data support these assertions? **FIGURE 8.15a** shows how the activity of an enzyme, plotted on the *y*-axis, changes as a function of temperature, plotted on the *x*-axis. These data were collected for an enzyme called chitinase, which is used by bacteria to digest cell walls of fungi. In this graph, each data point represents the enzyme's relative activity—meaning the rate of the enzyme-catalyzed reaction, scaled relative to the highest rate observed—in

experiments conducted under conditions that differed only in temperature. Results are shown for two types of bacteria.

Note that, in both bacterial species, the enzyme has a distinct optimum or peak—a temperature at which it functions best. One of the bacterial species lives in the cool soil under palm trees, where the temperature is about 25°C, while the other lives in hot springs, where temperatures can be close to 100°C. The temperature optimum for the enzyme reflects these environments.

The two types of bacteria have different versions of the enzyme that differ in primary structure. Natural selection (introduced in Chapter 1) has favored a structure in each species that is best suited for its distinct environment. The two versions are adaptations that allow each species to thrive at different temperatures.

FIGURE 8.15b makes the same point for pH. The effect of pH on enzyme activity was tested on the same chitinases used in Figure 8.15a, but this time using conditions that varied only in pH. The soil-dwelling bacteria described earlier grow in a neutral pH environment, but the species that lives in hot springs is also exposed to acidic conditions.

Note that the organism that thrives in a hot, acidic environment has a version of the enzyme that performs best at high temperatures and low pH; the organism that lives in the cool soil has a version of the enzyme that functions best at cooler temperatures and nearly neutral pH. Each enzyme is sensitive to changes in temperature and pH, but each species' version of the enzyme has a structure that allows it to function best in its particular environment.

To summarize, the rate of an enzyme-catalyzed reaction depends not only on substrate concentration and the enzyme's intrinsic affinity for the substrate but also on temperature and pH (among other factors). Temperature affects the movement of the substrates and enzyme; pH affects the enzyme's shape and reactivity.

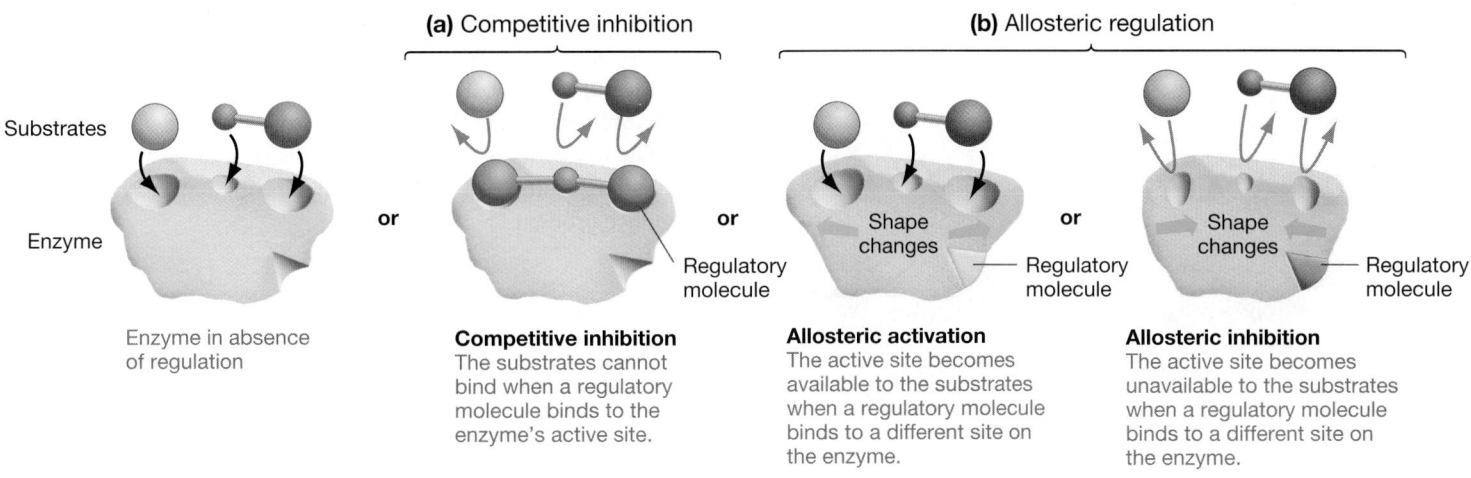

(a) Competitive inhibition

(b) Allosteric regulation

Substrates

Enzyme

or

or

or

Regulatory molecule

Shape changes

Regulatory molecule

Shape changes

Regulatory molecule

Enzyme in absence of regulation

Competitive inhibition
The substrates cannot bind when a regulatory molecule binds to the enzyme's active site.

Allosteric activation
The active site becomes available to the substrates when a regulatory molecule binds to a different site on the enzyme.

Allosteric inhibition
The active site becomes unavailable to the substrates when a regulatory molecule binds to a different site on the enzyme.

FIGURE 8.16 An Enzyme's Activity Is Precisely Regulated. Enzymes are turned on or off when specific regulatory molecules bind to them.

Most Enzymes Are Regulated

Controlling when and where enzymes will function is vital to the work of a cell. While temperature and pH affect the activity of enzymes, they are not often used as a means of regulating enzyme function. Instead, other molecules, in some cases other enzymes, regulate most of the cell's enzymatic activity. These regulatory molecules often change the enzyme's structure in some way, and their activity either activates or inactivates the enzyme.

Regulating Enzymes via Noncovalent Modifications Many molecules that regulate enzyme activity bind non-covalently to the enzyme to either activate or inactivate it. Since the interaction does not alter the enzyme's primary structure, it is often referred to as a "reversible" modification.

Reversible modifications affect enzyme function in one of two ways:

1. The regulatory molecule is similar in size and shape to the enzyme's natural substrate and inhibits catalysis by binding to the enzyme's active site. This event is called **competitive inhibition** because the molecule involved competes with the substrate for access to the enzyme's active site (**FIGURE 8.16a**).

2. The regulatory molecule binds at a location other than the active site and changes the shape of the enzyme. This type of regulation is called **allosteric** ("different-structure") **regulation** because the binding event changes the shape of the enzyme in a way that makes the active site available or unavailable (**FIGURE 8.16b**).

Both strategies depend on the concentration of the regulatory molecule—the more regulatory molecule present, the more likely it will be to bind to the enzyme and affect its activity. The amount of regulatory molecule is often tightly controlled and, as you'll see in Section 8.5, the regulatory molecules themselves often manage the enzymes that produce them.

Regulating Enzymes via Covalent Modifications In some cases, the function of an enzyme is altered by a chemical change in its primary structure. This change may be reversible or irreversible, depending on the type of modification.

Irreversible changes often result from the cleavage of peptide bonds that make up the primary structure of the enzyme. The enzyme trypsin, for example, is not functional until a small section of the protein is removed by a specific protease.

The most common modification of enzymes is the addition of one or more phosphate groups, similar to what was described for activated substrates in Section 8.2. In this case, however, the enzyme is phosphorylated instead of the substrate molecule. The transfer of a phosphate from ATP to the enzyme may be catalyzed by the enzyme itself or by a different enzyme.

When phosphorylation adds a negative charge to one or more amino acid residues in a protein, the electrons in that part of the protein change configuration. The enzyme's conformation

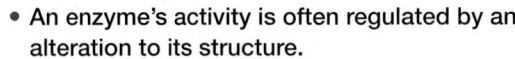

check your understanding

If you understand that . . .

- An enzyme's function is dependent on its folded structure.
- Enzymes have been optimized to fold into functional structures at particular environmental conditions, such as temperature and pH.
- An enzyme's activity is often regulated by an alteration to its structure.

✔ **You should be able to . . .**

1. Explain why the relative activity appears to drop off in Figure 8.15b, when it has been shown that reaction rates tend to increase at higher temperatures (Figure 8.4).

2. Predict how the shape change that occurs when an enzyme is phosphorylated would affect its catalytic activity.

Answers are available in Appendix A.

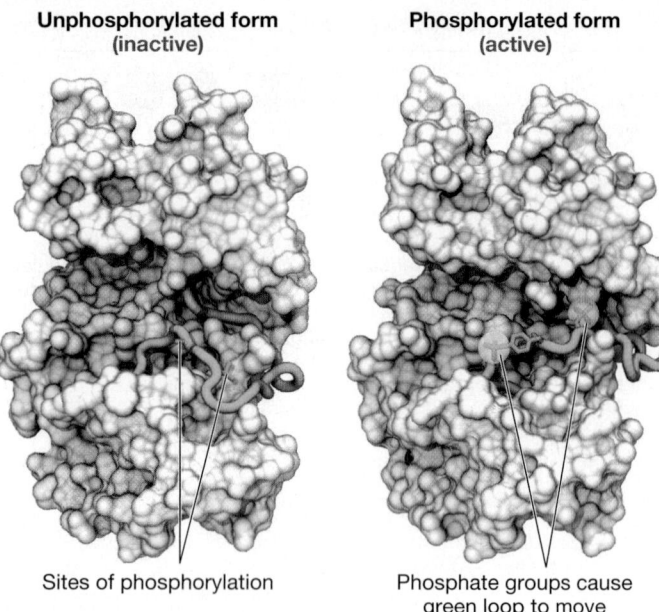

Unphosphorylated form
(inactive)

Phosphorylated form
(active)

Sites of phosphorylation

Phosphate groups cause
green loop to move

FIGURE 8.17 Phosphorylation Changes the Shape and Activity of Proteins. When proteins are phosphorylated, they often change shape in a way that alters their activity. The figure shows the subtle structural change that occurs when mitogen-activated protein kinase (MAPK) is activated by adding two phosphate groups (yellow) to the enzyme.

usually changes as well, which may activate or inactivate its function. Note that although a substrate or an enzyme may be "activated" via phosphorylation, this activation does not represent the same effect. When a substrate is activated, its potential energy has increased, and this energy is used to convert an endergonic reaction to one that is exergonic. When an enzyme is activated, its catalytic function has been turned on—any change in the potential energy of the enzyme is not directly used in driving the reaction.

To see how phosphorylation affects the shape and activity of an enzyme, let's look at an enzyme called mitogen-activated protein kinase (MAPK), which is involved in cell signaling (see Chapter 11). As shown in **FIGURE 8.17**, phosphorylation of amino acid residues in a particular loop of the primary sequence causes a shape change, which functions like a switch to activate the enzyme.

Phosphorylation of an enzyme is a reversible modification to the protein's structure. Dephosphorylation—removal of phosphates—can quickly return the protein to its previous shape. The relative abundance of enzymes that catalyze phosphorylation and dephosphorylation, then, regulates the function of the protein.

8.5 Enzymes Can Work Together in Metabolic Pathways

The eukaryotic cell has been compared to an industrial complex, where distinct organelles are functionally integrated into a cooperative network with a common goal—life (see Chapter 7). Similarly, enzymes often work together in a manner resembling an

assembly line in a factory. Each of the molecules of life presented in this book is built by a series of reactions, each catalyzed by a different enzyme. These multistep processes are referred to as **metabolic pathways.**

The following is an example of this type of teamwork, where an initial substrate A is sequentially modified by enzymes 1–3 to produce product D:

$$A \xrightarrow{\text{enzyme 1}} B \xrightarrow{\text{enzyme 2}} C \xrightarrow{\text{enzyme 3}} D$$

The B and C molecules are referred to as intermediates in the pathway—they serve as both a product and a reactant. For example, molecule B is the product of reaction 1 and the reactant for reaction 2.

Although these reactions have been written in a single direction, from left to right, the directionality is dependent on the relative concentrations of the reactants and products. At equilibrium, however, the concentration of the product for each reaction will be higher than the concentration of its respective reactant. Since D is the overall product for this pathway, it will have the highest concentration at equilibrium.

Metabolic Pathways Are Regulated

Since enzymes catalyze the reactions in metabolic pathways, the mechanisms that regulate enzyme function introduced in Section 8.4 also apply to the individual steps in a pathway. For example, to understand how blocking an individual reaction can affect an entire pathway, go back to the three-step model presented earlier and inactivate enzyme 2 by crossing it out. ✔ If you understand the assembly-line behavior of enzymes in a metabolic pathway, you should be able to predict how inactivating enzyme 2 would affect the concentration of molecules A, B, C, and D relative to what they would be if the pathway were fully functional.

When an enzyme in a pathway is inhibited by the product of the reaction sequence, **feedback inhibition** occurs. This is a convenient way for pathways to shut themselves down when their activity is no longer needed. As the concentration of the product molecule becomes abundant, it "feeds back" to stop the reaction sequence (**FIGURE 8.18**). By inhibiting a step early in the pathway, the amount of the initial substrate is not depleted unnecessarily, allowing it to be stored or used for other reactions.

Metabolic Pathways Evolve

While many enzymes are extraordinarily specific, some can catalyze a range of reactions and are able to interact with a family of related substrates. Research suggests that this flexibility allowed new enzymes to evolve and that enzymes specialized for catalyzing key reactions provided cells with a selective advantage. Could the same flexibility also help explain the evolution of the stepwise series of reactions seen in metabolic pathways?

In 1945, Norman Horowitz proposed a simple, stepwise process that could have directed pathway evolution. In Horowitz's model, enzymes first would have evolved to make the building blocks of life from readily available substrates, such as small organic compounds (see Chapter 2).

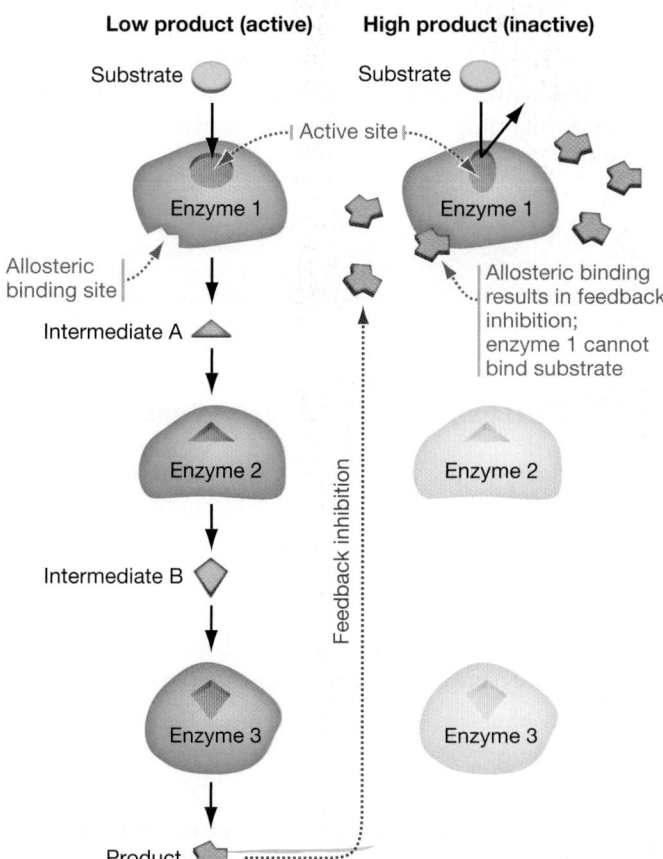

Low product (active)   **High product (inactive)**

Substrate

Active site

Enzyme 1

Allosteric binding site

Allosteric binding results in feedback inhibition; enzyme 1 cannot bind substrate

Intermediate A

Enzyme 2

Feedback inhibition

Intermediate B

Enzyme 3

Product

FIGURE 8.18 Feedback Inhibition May Regulate Metabolic Pathways. Feedback inhibition occurs when the product of a metabolic pathway inhibits an enzyme that functions early in the pathway.

If an original substrate were depleted, natural selection would next favor the evolution of a new enzyme to make more of it from other existing molecules. By evolving a new reaction step to produce the original substrate—now serving as an intermediate in a two-step pathway—the original enzyme would have been able to continue its work. **FIGURE 8.19** illustrates this model—referred to as retro-evolution—in which repetition of this backward process produces a multistep metabolic pathway.

Researchers also speculate that as early pathways emerged, early enzymes may have been recruited to new pathways, where they evolved new catalytic activities that performed new tasks. This hypothesis is called patchwork evolution, since the new reaction series would consist of enzymes brought together from different pathways.

Evidence of patchwork evolution has been observed in modern organisms, where new metabolic activities have emerged in response to human-made chemicals. For example, a novel pathway has recently evolved in one species of bacterium to break down the pesticide pentachlorophenol, for use as a source of energy and carbon building blocks. Pentachlorophenol was first introduced into the environment in the 1930s as a timber preservative. The new pathway uses enzymes from two preexisting pathways, which had evolved the ability to work together. The metabolic activity of microbes is now being scrutinized and engineered to clean up a variety of human-made pollutants—giving rise to a new technology called **bioremediation** (see Chapter 29).

Regardless of how they evolved, metabolic pathways are now vital to the function of all cells. Those that break down molecules for sources of energy and carbon building blocks are called **catabolic pathways;** those that use energy and carbon building blocks to synthesize molecules are called **anabolic pathways.**

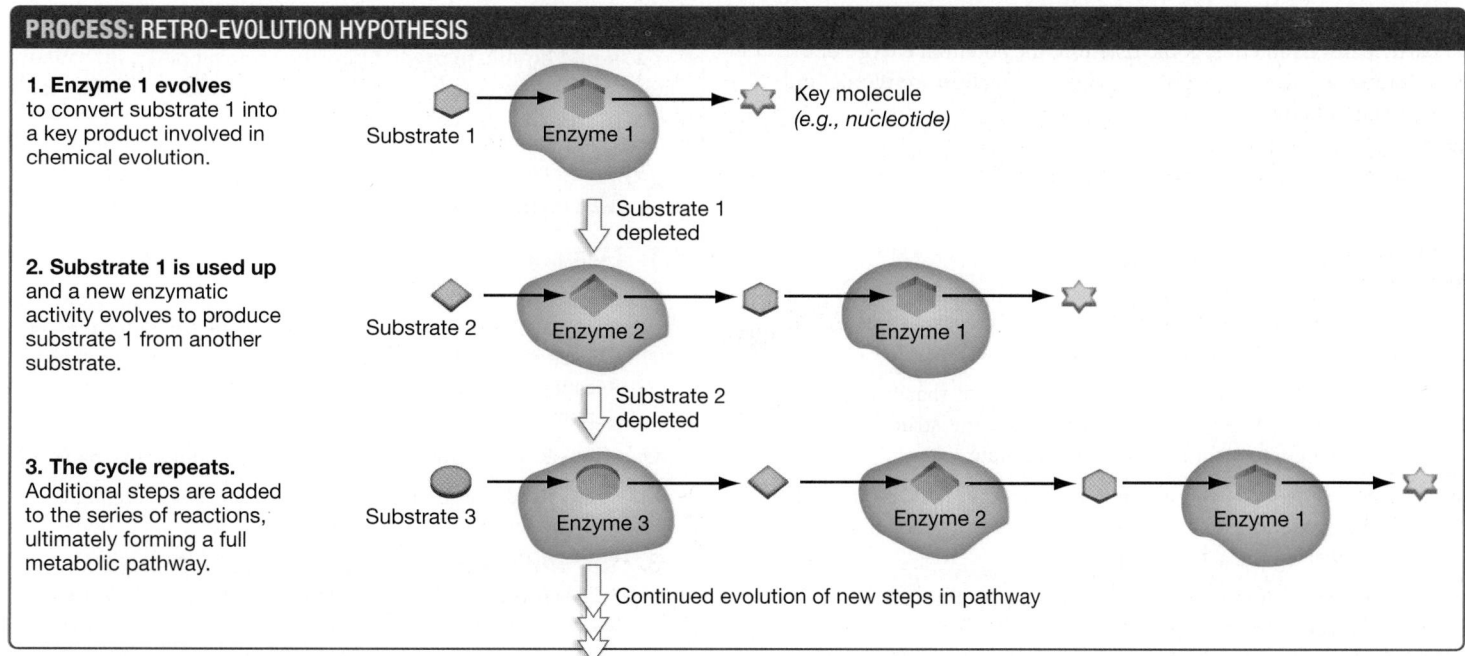

PROCESS: RETRO-EVOLUTION HYPOTHESIS

1. Enzyme 1 evolves to convert substrate 1 into a key product involved in chemical evolution.

Substrate 1   Enzyme 1   Key molecule *(e.g., nucleotide)*

Substrate 1 depleted

2. Substrate 1 is used up and a new enzymatic activity evolves to produce substrate 1 from another substrate.

Substrate 2   Enzyme 2   Enzyme 1

Substrate 2 depleted

3. The cycle repeats. Additional steps are added to the series of reactions, ultimately forming a full metabolic pathway.

Substrate 3   Enzyme 3   Enzyme 2   Enzyme 1

Continued evolution of new steps in pathway

FIGURE 8.19 A Hypothetical Model for Metabolic Pathway Evolution.

You are being kept alive by key catabolic and anabolic pathways. The catabolic pathways of cellular respiration (introduced in Chapter 9) harvest high-energy electrons from reduced carbons (from foods such as starch and sugar) and pass them through redox reactions to generate ATP. These reduced carbons are produced by the anabolic pathways of photosynthesis that are driven by light energy (introduced in Chapter 10). The reactions involved in cellular respiration and photosynthesis perform the most important energy transformations to life on Earth.

If you understand . . .

8.1 What Happens to Energy in Chemical Reactions?

- Spontaneous reactions do not require an input of energy to occur.
- The Gibbs free energy change, ΔG, summarizes the combined effects of changes in enthalpy and entropy during a chemical reaction.
- Spontaneous reactions have a negative ΔG and are said to be exergonic; nonspontaneous reactions have a positive ΔG and are said to be endergonic.

✔ You should be able to explain why changes in enthalpy and entropy are used to determine whether a reaction is spontaneous.

8.2 Nonspontaneous Reactions May Be Driven Using Chemical Energy

- Redox reactions transfer energy by coupling exergonic oxidation reactions to endergonic reduction reactions.
- High-energy C−H bonds may be formed during the reduction step of a redox reaction when an H^+ is combined with a transferred electron.
- The hydrolysis of ATP is an exergonic reaction and may be used to drive a variety of cellular processes.
- When a phosphate group from ATP is added to a substrate that participates in an endergonic reaction, the potential energy of the substrate is raised enough to make the reaction exergonic and thus spontaneous.

✔ You should be able to explain what energetic coupling means, and why life would not exist without it.

8.3 How Enzymes Work

- Enzymes are protein catalysts. They speed reaction rates but do not affect the change in free energy of the reaction.
- The structure of an enzyme has an active site that brings substrates together. After binding to substrates, the structure of the enzyme changes to stabilize the transition state.
- Activation energy is the amount of kinetic energy required to reach the transition state of a reaction. Enzymes speed up a reaction by lowering the activation energy.
- Many enzymes function only with the help of cofactors, coenzymes, or prosthetic groups.

✔ You should be able to explain how an enzyme's active site can reduce the activation energy of a reaction.

8.4 What Factors Affect Enzyme Function?

- Enzymes are proteins, and thus their activity can be directly influenced by modifications or environmental factors, such as temperature and pH, that alter their three-dimensional structure.
- Most enzymes are regulated by molecules that either compete with substrates to occupy the active site, or alter enzyme shape.
- Protein cleavage and phosphorylation are examples of how enzymes may be regulated by modifying their primary structure.

✔ You should be able to compare and contrast the effect of allosteric regulation versus phosphorylation on enzyme function.

8.5 Enzymes Can Work Together in Metabolic Pathways

- In cells, enzymes often work together in metabolic pathways that sequentially modify a substrate to make a product.
- A pathway may be regulated by controlling the activity of one enzyme, often the first in the series of reactions. Feedback inhibition results from the accumulation of a product that binds to an enzyme in the pathway and inactivates it.
- Metabolic pathways were vital to the evolution of life, and new pathways continue to evolve in cells.

✔ You should be able to predict how the removal of the intermediate in a two-step metabolic pathway would affect the enzymatic rates of the first and last.

MB MasteringBiology

1. **MasteringBiology Assignments**

   Tutorials and Activities ATP and Energy; Chemical Reactions and ATP; Energy Transformations; Enzyme and Substrate Concentrations; Enzyme Inhibition; Factors That Affect Reaction Rate; How Enzymes Function; Regulating Enzyme Action; Redox Reactions

   Questions Reading Quizzes, Blue-Thread Questions, Test Bank

2. **eText** Read your book online, search, take notes, highlight text, and more.

3. **The Study Area** Practice Test, Cumulative Test, BioFlix® 3-D Animations, Videos, Activities, Audio Glossary, Word Study Tools, Art

You should be able to . . .

1. The first law of thermodynamics states which of the following?
   a. Energy exists in two forms: kinetic and potential.
   b. Reactions will take place only if energy is released.
   c. Energy is conserved: it cannot be created or destroyed.
   d. Disorder always increases in the universe.

2. If a reaction is exergonic, then which of these statements is true?
   a. The products have lower free energy than the reactants.
   b. Energy must be added for the reaction to proceed.
   c. The products have lower entropy (are more ordered) than the reactants.
   d. The reaction occurs extremely quickly.

3. What is a transition state?
   a. the complex formed as covalent bonds are being broken and re-formed during a reaction
   b. the place where an allosteric regulatory molecule binds to an enzyme

   c. an interaction between reactants with high kinetic energy, due to high temperature
   d. the shape adopted by an enzyme that has an inhibitory molecule bound at its active site

4. What often happens to an enzyme after it binds to its substrate? Is this a permanent change?

5. How does pH affect enzyme-catalyzed reactions?
   a. Protons serve as substrates for most reactions.
   b. Energy stored in protons is used to drive endergonic reactions.
   c. Proton concentration increases the kinetic energy of the reactants, allowing them to reach their transition state.
   d. The concentration of protons affects the folded structure of the enzyme.

6. When does feedback inhibition occur?

7. Explain the lock-and-key model of enzyme activity. What was incorrect about this model?

8. If you were to expose glucose to oxygen on your lab bench, why would you not expect to see it burn as shown in Figure 8.6?
   a. The reaction is endergonic and requires an input of energy.
   b. The reaction is not spontaneous unless an enzyme is added to the substrates.
   c. The sugar must first be phosphorylated to increase its potential energy.
   d. Energy is required for the sugar and oxygen to reach their transition state.

9. Explain why substrate phosphorylation using ATP is an exergonic reaction. How does the phosphorylation of reactants result in driving reactions that would normally be endergonic?

10. QUANTITATIVE In Figure 8.9, the energetic coupling of ATP hydrolysis and an endergonic reaction are shown. If the hydrolysis of ATP releases 7.3 kcal of free energy, use the graph in this figure to estimate what you would expect the ΔG values to be for the uncoupled reaction and the two steps in the coupled reaction.

11. Compare and contrast competitive inhibition and allosteric regulation.

12. Using what you have learned about changes in free energy, would you predict the ΔG value of catabolic reactions to be positive or negative? What about anabolic reactions? Justify your answers using the terms enthalpy and entropy.

13. Draw a redox reaction that occurs between compounds AH_2 and B^+ to form A, BH, and H^+. On the drawing, connect the reactant and product forms of each compound and state if it is the reduction or oxidation step and how many electrons are transferred. If this represents an exergonic reaction, identify which of the five substances would have the highest-energy electrons.

14. Researchers can analyze the atomic structure of enzymes during catalysis. In one recent study, investigators found that the transition state included the formation of a free radical (see Chapter 2) and that a coenzyme bound to the active site donated an electron to help stabilize the free radical. How would the reaction rate and the stability of the transition state change if the coenzyme were not available?

15. Recently, researchers were able to measure movement that occurred in a single amino acid in an enzyme as reactions were taking place in its active site. The amino acid that moved was

located in the active site, and the rate of movement correlated closely with the rate at which the reaction was taking place. Discuss the significance of these findings, using the information in Figures 8.10 and 8.13.

16. You have discovered an enzyme that appears to function only when a particular sugar accumulates. Which of the following scenarios would you predict to be responsible for activating this enzyme?
   a. The sugar cleaves the enzyme so it is now in an active conformation.
   b. The sugar binds to the enzyme and changes the conformation of the active site.
   c. The sugar binds to the active site and competes with the normal substrate.
   d. The sugar phosphorylates the enzyme, triggering a conformational change.

9 Cellular Respiration and Fermentation

In this chapter you will learn how

Cells make ATP starting from sugars and other high potential energy compounds

by examining ↓

How cells produce ATP when oxygen is present   **9.1**

by examining ↓

How cells produce ATP when oxygen is absent

looking closer at

Glycolysis **9.2**

Pyruvate oxidation **9.3**

Citric acid cycle **9.4**

Electron transport and chemiosmosis **9.5**

focusing on

Fermentation **9.6**

This hydroelectric dam on the Duero river between Spain and Portugal uses pumps to move water from the lower reservoir to the upper reservoir. During periods of high energy demand, the potential energy stored by this activity is used to generate electricity. A similar process is used by cells to produce ATP during cellular respiration.

This chapter is part of the Big Picture. See how on pages 198–199.

ife requires energy. From the very start, chemical evolution was driven by energy from chemicals, radiation, heat, or other sources (see Chapter 2). Harnessing energy and controlling its flow has been the single most important step in the evolution of life.

What fuels life in cells? The answer is the nucleotide adenosine triphosphate (ATP). ATP has high potential energy and allows cells to overcome life's energy barriers (see Chapter 8).

This chapter investigates how cells make ATP, starting with an introduction to the metabolic pathways that harvest energy from high-energy molecules like **glucose**—the most common source of chemical energy used by organisms. The four central pathways of cellular respiration will be

✔ When you see this checkmark, stop and test yourself. Answers are available in Appendix A.

presented with emphasis on how the oxidation of glucose leads to ATP production. Fermentation will also be introduced as an alternative pathway used to make ATP when key reactions in cellular respiration are either shut down or not available.

As cells process sugar, the energy that is released is used to transfer a phosphate group to adenosine diphosphate (ADP), generating ATP. (You can see the Big Picture of how the production of glucose in photosynthesis is related to its catabolism in cellular respiration on pages 198–199.)

9.1 An Overview of Cellular Respiration

In general, a cell contains only enough ATP to last from 30 seconds to a few minutes. Because it has such high potential energy, ATP is unstable and is not stored. Like many other cellular processes, the production and use of ATP is fast. Most cells are making ATP all the time.

Most of the glucose that is used to make ATP is produced by plants and other photosynthetic species. These organisms use the energy in sunlight to reduce carbon dioxide (CO_2) to glucose and other carbohydrates. While they are alive, photosynthetic species use the glucose that they produce to make ATP for themselves. When photosynthetic species decompose or are eaten, they provide glucose to animals, fungi, and many bacteria and archaea.

All organisms use glucose in the synthesis of complex carbohydrates, fats, and other energy-rich compounds. Storage carbohydrates, such as starch and glycogen, act like savings accounts for chemical energy. ATP, in contrast, is like cash. To withdraw chemical energy from the accounts to get cash, storage carbohydrates are first hydrolyzed into their glucose monomers. The glucose is then used to produce ATP through one of two general processes: cellular respiration or fermentation (**FIGURE 9.1**). The primary difference between these two processes lies in the degree to which glucose is oxidized.

What Happens When Glucose Is Oxidized?

When glucose undergoes the uncontrolled oxidation reaction called burning, some of the potential energy stored in its chemical bonds is converted to kinetic energy in the form of heat:

$$C_6H_{12}O_2 + 6\,O_2 \longrightarrow 6\,CO_2 + 6\,H_2O + \text{Heat}$$

glucose   oxygen   carbon dioxide water

More specifically, a total of about 685 kilocalories (kcal) of heat is released when one mole of glucose is oxidized. To put this in perspective, if you burned this amount of glucose, it would give off enough heat to bring almost 2.5 gallons of room-temperature water to a boil.

Glucose does not burn in cells, however. Instead, the glucose in cells is oxidized through a long series of carefully controlled redox reactions. These reactions are occurring, millions of

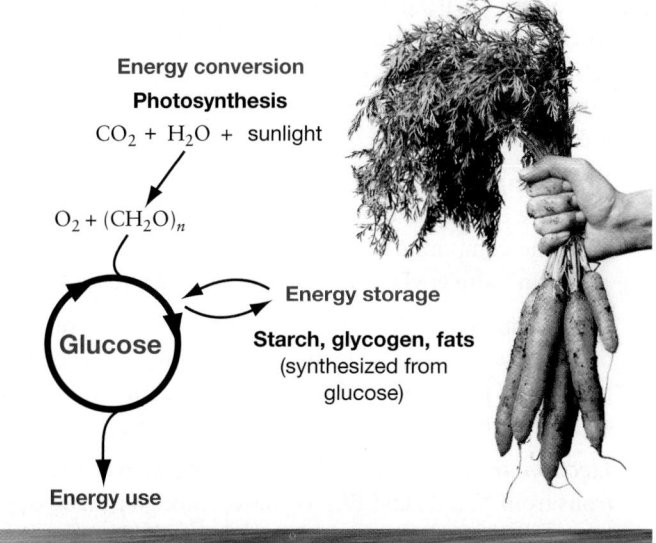

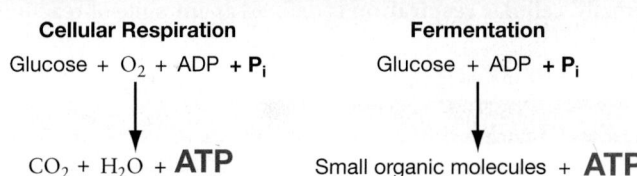

FIGURE 9.1 Glucose Is the Hub of Energy Processing in Cells. Glucose is a product of photosynthesis. Both plants and animals store glucose and oxidize it to provide chemical energy in the form of ATP.

times per minute, in your cells right now. Instead of releasing all of this energy as heat, much of it is being used to make the ATP you need to read, think, move, and stay alive. In cells, the change in free energy (Chapter 8) that occurs during the oxidation of glucose is used to synthesize ATP from ADP and P_i.

So how does fermentation differ from cellular respiration? Respiration, like burning, results in the complete oxidation of glucose into CO_2 and water. Fermentation, on the other hand, does not fully oxidize glucose. Instead, small, reduced organic molecules are produced as waste. As a result, cellular respiration releases more energy from glucose than fermentation.

The complete oxidation of glucose via cellular respiration can be thought of as a four-step process used to convert the chemical energy in glucose to chemical energy in ATP. Each of the four steps consists of a series of chemical reactions, and each step has a distinctive starting molecule and a characteristic set of products.

1. *Glycolysis* During **glycolysis,** one 6-carbon molecule of glucose is broken into two molecules of the three-carbon compound pyruvate. During this process, ATP is produced from ADP, and nicotinamide adenine dinucleotide (NAD^+) is reduced to form NADH.

2. *Pyruvate processing* Pyruvate is processed to release one molecule of CO_2, and the remaining two carbons are used to form the compound acetyl CoA. The oxidation of pyruvate results in more NAD^+ being reduced to NADH.

3. *Citric acid cycle* Acetyl CoA is oxidized to two molecules of CO_2. During this sequence of reactions, more ATP and NADH are produced, and flavin adenine dinucleotide (FAD) is reduced to form $FADH_2$.

4. *Electron transport and oxidative phosphorylation* Electrons from NADH and $FADH_2$ move through a series of proteins called an electron transport chain (ETC). The energy released in this chain of redox reactions is used to create a proton gradient across a membrane; the ensuing flow of protons back across the membrane is used to make ATP. Because this mode of ATP production links the phosphorylation of ADP with the oxidation of NADH and $FADH_2$, it is called **oxidative phosphorylation.**

FIGURE 9.2 summarizes the four steps in cellular respiration. Formally, **cellular respiration** is defined as any suite of reactions that uses electrons harvested from high-energy molecules to produce ATP via an electron transport chain.

The enzymes, products, and intermediates involved in cellular respiration and fermentation do not exist in isolation. Instead, they are part of a huge and dynamic inventory of chemicals inside the cell.

This complexity can be boiled down to a simple essence, however. Two of the most fundamental requirements of a cell are energy and carbon. They need a source of high-energy electrons for generating chemical energy in the form of ATP, and a source of carbon-containing molecules that can be used to synthesize DNA, RNA, proteins, fatty acids, and other molecules. Let's take a closer look at the central role cellular respiration plays in metabolic pathways as a whole.

Cellular Respiration Plays a Central Role in Metabolism

Recall that sets of reactions that break down molecules are called catabolic pathways (Chapter 8). These reactions often harvest stored chemical energy to produce ATP. On the other hand, sets of reactions that synthesize larger molecules from smaller components are called anabolic pathways. Anabolic reactions often use energy in the form of ATP.

How does the process of cellular respiration interact with other catabolic and anabolic pathways? Let's first consider how

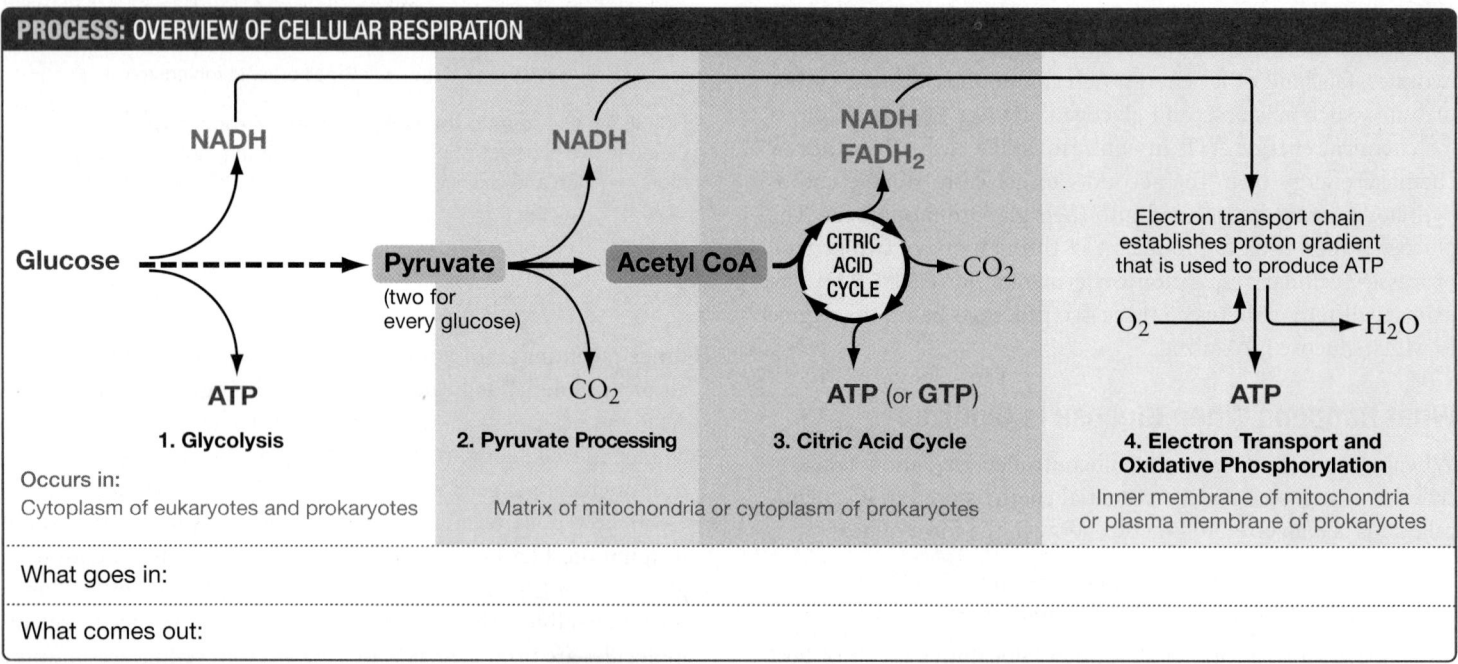

FIGURE 9.2 **Cellular Respiration Oxidizes Glucose to Make ATP.** Cells produce ATP from glucose via a series of processes: (1) glycolysis, (2) pyruvate processing, (3) the citric acid cycle, and (4) electron transport and oxidative phosphorylation. Each component produces high-energy molecules in the form of nucleotides (ATP or GTP) or electron carriers (NADH or $FADH_2$). Because the four components are connected, glucose oxidation is an integrated metabolic pathway. The first three steps oxidize glucose to produce NADH and $FADH_2$, which then feed the electron transport chain.

✔EXERCISE Fill in the chart along the bottom.

eukaryotes extract energy from molecules other than glucose and then examine how intermediates produced in glycolysis and the citric acid cycle are used as building blocks to synthesize cell components.

Catabolic Pathways Break Down a Variety of Molecules Most organisms ingest, absorb, or synthesize many different carbohydrates. These molecules range from sucrose, maltose, and other simple sugars to large polymers such as glycogen and starch (see Chapter 5).

Recall that both glycogen and starch are polymers of glucose, but differ in the way their long chains of glucose branch. Using enzyme-catalyzed reactions, cells can produce glucose from glycogen, starch, and most simple sugars. Glucose and fructose can then be processed in glycolysis.

Carbohydrates are not the only important source of carbon compounds used in catabolic pathways, however. Fats are highly reduced macromolecules consisting of glycerol bonded to chains of fatty acids (see Chapter 6). In cells, enzymes routinely break down fats to release the glycerol and convert the fatty acids into acetyl CoA molecules. Glycerol can be further processed and enter glycolysis. Acetyl CoA enters the citric acid cycle.

Proteins can also be catabolized, meaning that they can be broken down and used to produce ATP. Once they are hydrolyzed to their constituent amino acids, enzyme-catalyzed reactions remove the amino ($-NH_2$) groups. The amino groups are excreted in urine as waste. The carbon compounds that remain are converted to pyruvate, acetyl CoA, and other intermediates in glycolysis and the citric acid cycle.

The top half of **FIGURE 9.3** summarizes the catabolic pathways of carbohydrates, fats, and proteins and shows how their breakdown products feed an array of steps in glucose oxidation and cellular respiration. When all three types of molecules are available in the cell to generate ATP, carbohydrates are used up first, then fats, and finally proteins.

Catabolic Intermediates Are Used in Anabolic Pathways Where do cells get the precursor molecules required to synthesize amino acids, RNA, DNA, phospholipids, and other cell components? Not surprisingly, the answer often involves intermediates in carbohydrate metabolism. For example,

- In humans, about half the required amino acids can be synthesized from molecules siphoned from the citric acid cycle.
- Acetyl CoA is the starting point for anabolic pathways that result in the synthesis of fatty acids. Fatty acids can then be used to build phospholipid membranes or fats.
- Intermediates in glycolysis can be oxidized to start the synthesis of the sugars in ribonucleotides and deoxyribonucleotides. Nucleotides, in turn, are building blocks used in RNA and DNA synthesis.
- If ATP is abundant, pyruvate and lactate (from fermentation) can be used in the synthesis of glucose. Excess glucose may be converted to glycogen or starch and stored.

The bottom half of Figure 9.3 summarizes how intermediates in carbohydrate metabolism are drawn off to synthesize macromolecules. The take-home message is that the same molecule can serve many different functions in the cell. As a result, catabolic and anabolic pathways are closely intertwined.

Metabolism comprises thousands of different chemical reactions, yet the amounts and identities of molecules inside cells are relatively constant. By regulating key reactions involved in catabolic and anabolic pathways, the cell is able to maintain its internal environment even under different environmental conditions—a process referred to as **homeostasis.** Cellular respiration and

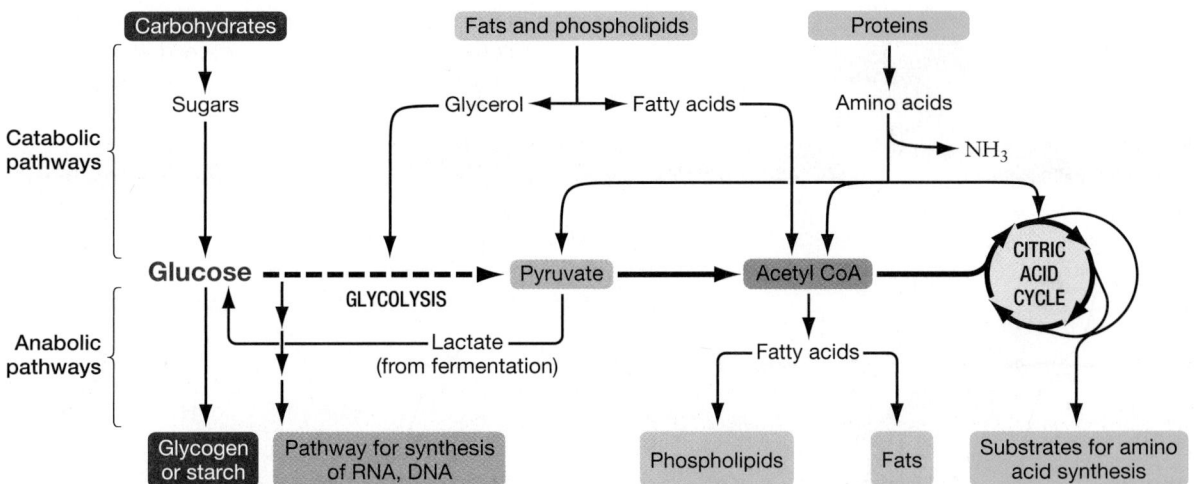

FIGURE 9.3 Cellular Respiration Interacts with Other Catabolic and Anabolic Pathways. A variety of high-energy compounds from carbohydrates, fats, or proteins can be broken down in catabolic reactions and used by cellular respiration for ATP production. Several of the intermediates in carbohydrate metabolism act as precursor molecules in anabolic reactions leading to the synthesis of glycogen or starch, RNA, DNA, fatty acids, and amino acids.

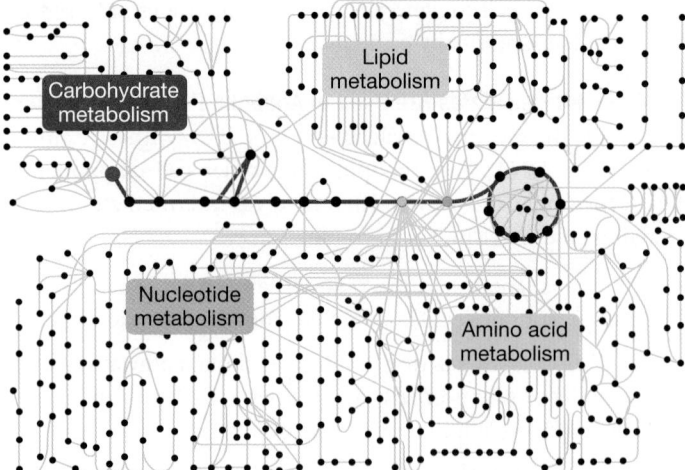

FIGURE 9.4 Pathways of Cellular Respiration Play a Central Role in the Metabolic Activity of Cells. A representation of a few of the thousands of chemical reactions that occur in cells. The dots represent molecules, and the lines represent enzyme-catalyzed reactions. At the center of all this, the first three steps of cellular respiration are emphasized by bold dots and thick lines. For reference, glucose, pyruvate, and acetyl CoA are represented by the distinctive colors used in Figure 9.3.

fermentation pathways may be crucial to the life of a cell, but they also have to be seen as central parts of a whole (**FIGURE 9.4**).

Once you've filled in the chart at the bottom of Figure 9.2, you'll be ready to analyze each of the four steps of cellular respiration in detail. As you delve in, keep asking yourself the same key questions: What goes in and what comes out? What happens to the potential energy that is released? Where does each step occur, and how is it regulated? Then take a look in the mirror. All these processes are occurring right now, in virtually all your cells.

Because the enzymes responsible for glycolysis have been observed in nearly every bacterium, archaean, and eukaryote, it is logical to infer that the ancestor of all organisms living today made ATP by glycolysis. It's ironic, then, that the process was discovered by accident.

In the 1890s Hans and Edward Buchner were working out techniques for breaking open baker's yeast cells and extracting the contents for commercial and medicinal use. (Yeast extracts are still added to some foods as a flavor enhancer or nutritional supplement.) In one set of experiments, the Buchners added sucrose to their extracts. At the time, sucrose was commonly used as a preservative—a substance used to preserve food from decay.

Instead of preserving the yeast extracts, though, the sucrose was quickly broken down and fermented, and alcohol appeared as a by-product. This was a key finding: It showed that metabolic pathways like fermentation could be studied in vitro—outside the organism. Until then, researchers thought that metabolism could take place only in intact organisms.

When researchers studied how the sugar was being processed, they found that the reactions could go on much longer than normal if inorganic phosphate were added to the mixture. This result implied that some of the compounds involved were being phosphorylated. Soon after, a molecule called fructose bisphosphate was isolated. (The prefix *bis*– means that the phosphate groups are attached to the fructose molecule at two different locations.) Subsequent work showed that all but the starting

FIGURE 9.5 Glycolysis Pathway. This sequence of 10 reactions oxidizes glucose to pyruvate. Each reaction is catalyzed by a different enzyme to produce two net ATP (4 ATP are produced, but 2 are invested), two molecules of NADH, and two molecules of pyruvate. In step 4, fructose-1,6-bisphosphate is divided into two products that both proceed through steps 6–10. The amounts for "What goes in" and "What goes out" are the combined totals for both molecules.

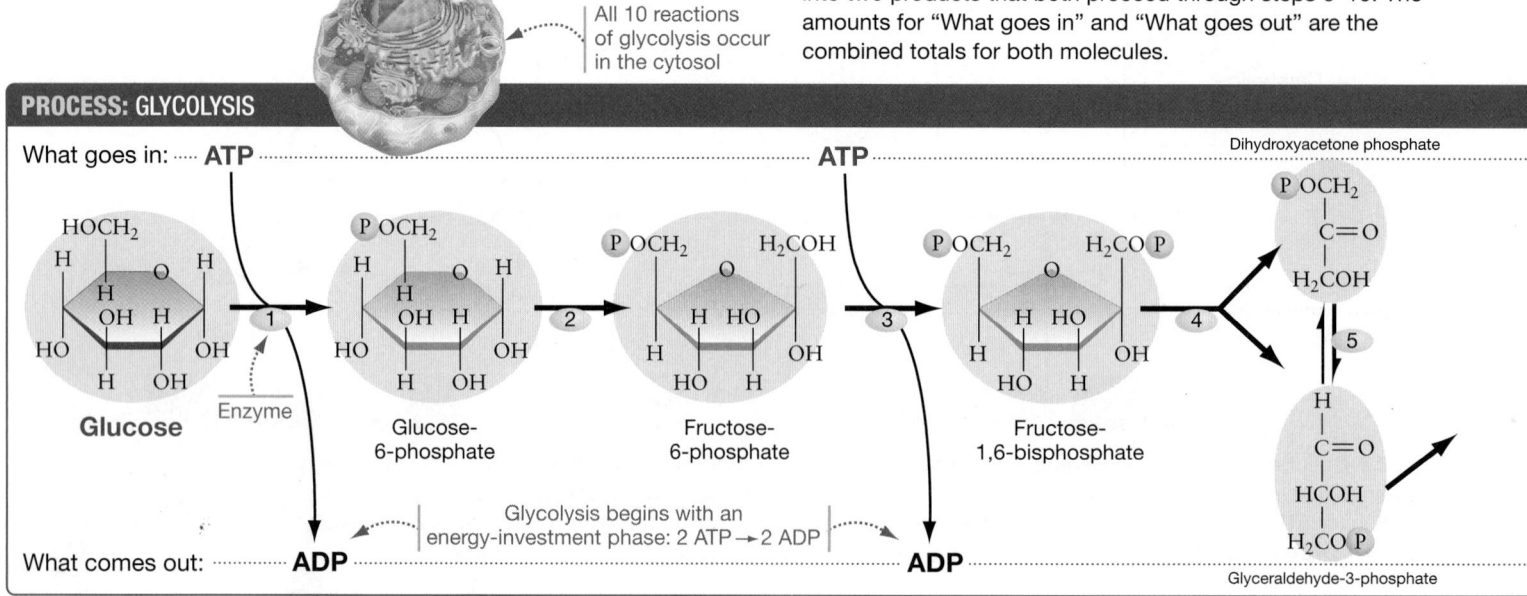

PROCESS: GLYCOLYSIS

All 10 reactions of glycolysis occur in the cytosol

What goes in: ATP ⋯⋯⋯⋯ ATP

Glucose → (1) → Glucose-6-phosphate → (2) → Fructose-6-phosphate → (3) → Fructose-1,6-bisphosphate → (4) → Dihydroxyacetone phosphate / Glyceraldehyde-3-phosphate (5)

Enzyme

Glycolysis begins with an energy-investment phase: 2 ATP → 2 ADP

What comes out: ADP ⋯⋯⋯⋯ ADP

and ending molecules in glycolysis—glucose and pyruvate—are phosphorylated.

In 1905 researchers found that the processing of sugar by yeast extracts stopped if they boiled the reaction mix. Because it was known that enzymes could be inactivated by heat, their discovery suggested that enzymes were involved in at least some of the processing steps. Years later, investigators realized that each step in glycolysis is catalyzed by a different enzyme. Eventually, each of the reactions and enzymes involved was worked out.

Glycolysis Is a Sequence of 10 Reactions

In both eukaryotes and prokaryotes, all 10 reactions of glycolysis occur in the cytosol (**FIGURE 9.5**). Note three key points about this reaction sequence:

1. Glycolysis starts by *using* ATP, not producing it. In the initial step, glucose is phosphorylated to form glucose-6-phosphate. After the second reaction rearranges the sugar to form fructose-6-phosphate, the third reaction adds a second phosphate group, forming the fructose-1,6-bisphosphate observed by early researchers. Thus, two ATP molecules are used up before any ATP is produced.

2. Once the energy-investment phase of glycolysis is complete, the subsequent reactions represent an energy-payoff phase. The sixth reaction in the sequence results in the reduction of two molecules of NAD^+; the seventh produces two molecules of ATP. This is where the energy "debt"—of two molecules of ATP invested early in glycolysis—is paid off. The final reaction in the sequence produces another two ATPs. For each molecule of glucose processed, the net yield is two molecules of NADH, two of ATP, and two of pyruvate.

3. In reactions 7 and 10 of Figure 9.5, an enzyme catalyzes the transfer of a phosphate group from a phosphorylated substrate to ADP, forming ATP. Enzyme-catalyzed reactions that result in

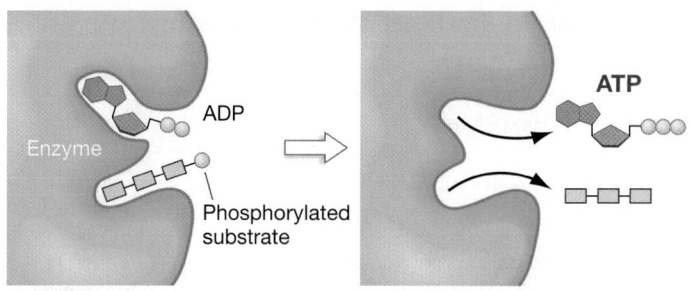

FIGURE 9.6 Substrate-Level Phosphorylation Involves an Enzyme and a Phosphorylated Substrate. Substrate-level phosphorylation occurs when an enzyme catalyzes the transfer of a phosphate group from a phosphorylated substrate to ADP, forming ATP.

ATP production are termed **substrate-level phosphorylation** (**FIGURE 9.6**). The key idea to note here is that the energy to produce the ATP comes from the phosphorylated substrate—not from a proton gradient, as it does when ATP is produced by oxidative phosphorylation.

The discovery and elucidation of the glycolytic pathway ranks as one of the great achievements in the history of biochemistry. For more detail concerning the enzymes that catalyze each step, see **TABLE 9.1** (on page 160). While the catabolism of glucose can occur via other pathways, this set of reactions is among the most ancient and fundamental of all life processes.

How Is Glycolysis Regulated?

An important advance in understanding how glycolysis is regulated occurred when biologists observed that high levels of ATP inhibit a key glycolytic enzyme called phosphofructokinase. **Phosphofructokinase** catalyzes reaction 3 in Figure 9.5—the synthesis of fructose-1,6-bisphosphate from fructose-6-phosphate. This is a crucial step in the sequence.

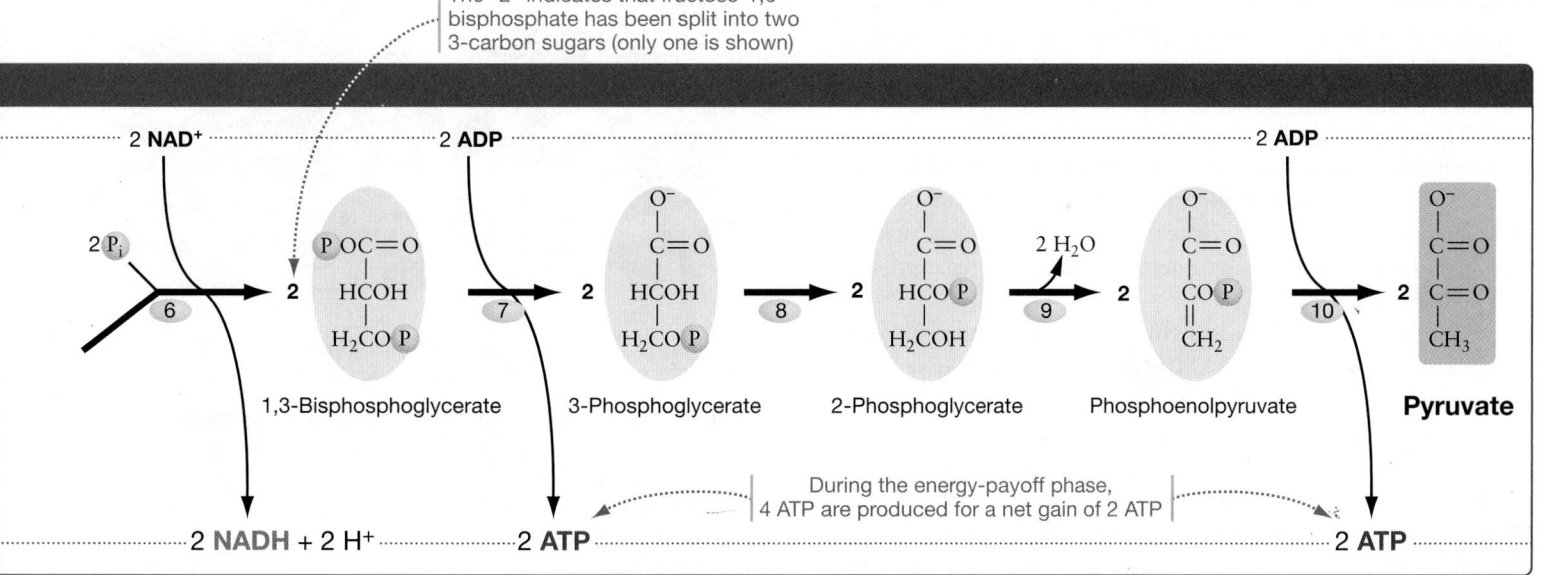

The "2" indicates that fructose-1,6-bisphosphate has been split into two 3-carbon sugars (only one is shown)

During the energy-payoff phase, 4 ATP are produced for a net gain of 2 ATP

Step	Enzyme	Reaction
1	Hexokinase	Transfers a phosphate from **ATP** to glucose, increasing its potential energy.
2	Phosphoglucose isomerase	Converts glucose-6-phosphate to fructose-6-phosphate; referred to as an isomer of glucose-6-phosphate.
3	Phosphofructokinase	Transfers a phosphate from **ATP** to the opposite end of fructose-6-phosphate, increasing its potential energy.
4	Fructose-bis-phosphate aldolase	Cleaves fructose-1,6-bisphosphate into two different 3-carbon sugars.
5	Triose phosphate isomerase	Converts dihydroxyacetone phosphate (DAP) to glyceraldehyde-3-phosphate (G3P). Although the reaction is fully reversible, the DAP-to-G3P reaction is favored because G3P is immediately used as a substrate for step 6.
6	Glyceraldehyde-3-phosphate dehydrogenase	A two-step reaction that first oxidizes G3P using the **NAD$^+$** coenzyme to produce **NADH**. Energy from this reaction is used to attach a P$_i$ to the oxidized product to form 1,3-bisphosphoglycerate.
7	Phosphoglycerate kinase	Transfers a phosphate from 1,3-bisphosphoglycerate to **ADP** to make 3-phosphoglycerate and **ATP**.
8	Phosphoglycerate mutase	Rearranges the phosphate in 3-phosphoglycerate to make 2-phosphoglycerate.
9	Enolase	Removes a water molecule from 2-phosphoglycerate to form a C=C double bond and produce phosphoenolpyruvate.
10	Pyruvate kinase	Transfers a phosphate from phosphoenolpyruvate to **ADP** to make pyruvate and **ATP**.

After reactions 1 and 2 occur, an array of enzymes can reverse the process and regenerate glucose for use in other pathways. Before step 3, then, the sequence is not committed to glycolysis. But once fructose-1,6-bisphosphate is synthesized, there is no point in stopping the process. Based on these observations, it makes sense that the pathway is regulated at step 3. How do cells do it?

As shown in Figure 9.5, ATP serves as a substrate for the addition of a phosphate to fructose-6-phosphate. In the vast majority of cases, increasing the concentration of a substrate would *speed* the rate of a chemical reaction, but in this case, it inhibits it. Why would ATP—a substrate that is required for the reaction—also serve as an inhibitor of the reaction? The answer lies in the fact that ATP is also the end product of the overall catabolic pathway.

Recall that when an enzyme in a pathway is inhibited by the product of the reaction sequence, feedback inhibition occurs (see Chapter 8). When the product molecule is abundant, it can inhibit its own production by interfering with the reaction sequence used to create it.

Feedback inhibition increases efficiency. Cells that are able to stop glycolytic reactions when ATP is abundant can conserve their stores of glucose for times when ATP is scarce. As a result, natural selection should favor individuals who have phosphofructokinase molecules that are inhibited by high concentrations of ATP.

How do high levels of the substrate inhibit the enzyme? As **FIGURE 9.7** shows, phosphofructokinase has two distinct binding sites for ATP. ATP can bind at the enzyme's active site, where it

is used to phosphorylate fructose-6-phosphate, or at a regulatory site, where it turns off the enzyme's activity.

The key to feedback inhibition lies in the ability of the two sites to bind to ATP. When concentrations are low, ATP binds

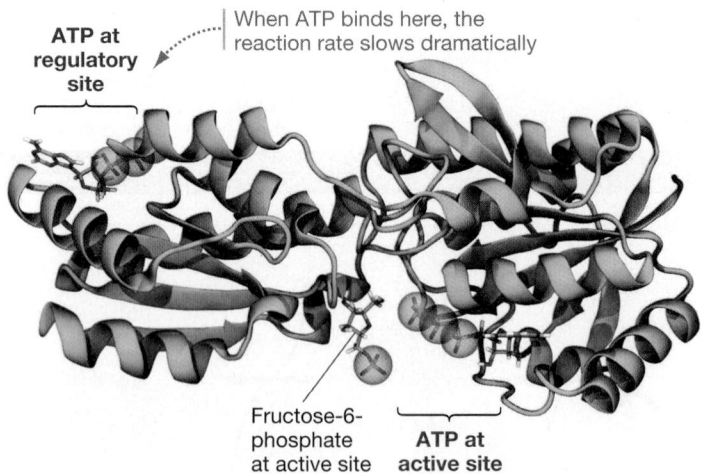

ATP at regulatory site

When ATP binds here, the reaction rate slows dramatically

Fructose-6-phosphate at active site

ATP at active site

FIGURE 9.7 Phosphofructokinase Has Two Binding Sites for ATP. A model of one of the four identical subunits of phosphofructokinase. In the active site, ATP is used as a substrate to transfer one of its phosphate groups to fructose-6-phosphate. In the regulatory site, ATP binding inhibits the reaction by changing the shape of the enzyme.

only to the active site, which has a greater affinity for ATP than does the regulatory site. As ATP concentrations increase, however, it also binds at the regulatory site on phosphofructokinase. When ATP binds at this second location, the enzyme's conformation changes in a way that dramatically lowers the reaction rate at the active site. In phosphofructokinase, ATP acts as an allosteric regulator (see Chapter 8). ✔ **If you understand the principle behind the difference in affinity between the two ATP binding sites, you should be able to predict the consequences if the regulatory site had higher affinity for ATP than the active site did.**

To summarize, glycolysis starts with one 6-carbon glucose molecule and ends with two 3-carbon pyruvate molecules. The reactions occur in the cytoplasm, and the energy that is released is used to produce a net total of two ATP and two NADH. Now the question is, what happens to the pyruvate?

9.3 Processing Pyruvate to Acetyl CoA

In eukaryotes, the pyruvate produced by glycolysis is transported from the cytosol to mitochondria. Mitochondria are organelles found in virtually all eukaryotes (see Chapter 7).

As shown in **FIGURE 9.8**, mitochondria have two membranes, called the inner membrane and outer membrane. The interior of the organelle is filled with layers of sac-like structures called **cristae.** Short tubes connect the cristae to the main part of the inner membrane. The region inside the inner membrane but outside the cristae is the **mitochondrial matrix.**

Pyruvate moves across the mitochondrion's outer membrane through small pores, but how it is transported across the inner membrane is still unclear. Current research suggests that either pyruvate is transported directly into the matrix using an unknown

transporter, or it is converted first into lactate, transported across the membrane, and then converted back into pyruvate.

Inside the mitochondrion, pyruvate reacts with a compound called **coenzyme A (CoA).** Coenzyme A is sometimes abbreviated as CoA-SH to call attention to its key sulfhydryl functional group. In this and many other reactions, CoA acts as a coenzyme by accepting and then transferring an acetyl group ($-COCH_3$) to a substrate (the A stands for acetylation). Pyruvate reacts with CoA, through a series of steps, to produce **acetyl CoA.**

The reaction sequence occurs inside an enormous and intricate enzyme complex called **pyruvate dehydrogenase.** In eukaryotes, pyruvate dehydrogenase is located in the mitochondrial matrix. In bacteria and archaea, pyruvate dehydrogenase is located in the cytosol.

As pyruvate is being processed, one of the carbons in the pyruvate is oxidized to CO_2 and NAD^+ is reduced to NADH. The remaining two-carbon acetyl unit is transferred to CoA (**FIGURE 9.9**).

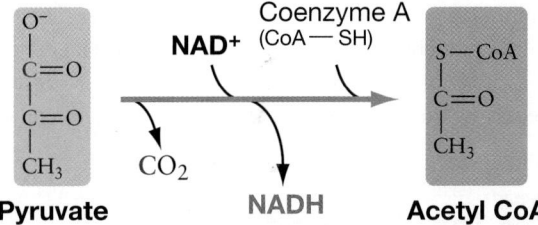

FIGURE 9.9 Pyruvate Is Oxidized to Acetyl CoA. The reaction shown here is catalyzed by pyruvate dehydrogenase.

✔**EXERCISE** Above the reaction arrow, list three molecules whose presence speeds up the reaction. Label them "Positive control." Below the reaction arrow, list three molecules whose presence slows down the reaction. Label them "Negative control by feedback inhibition."

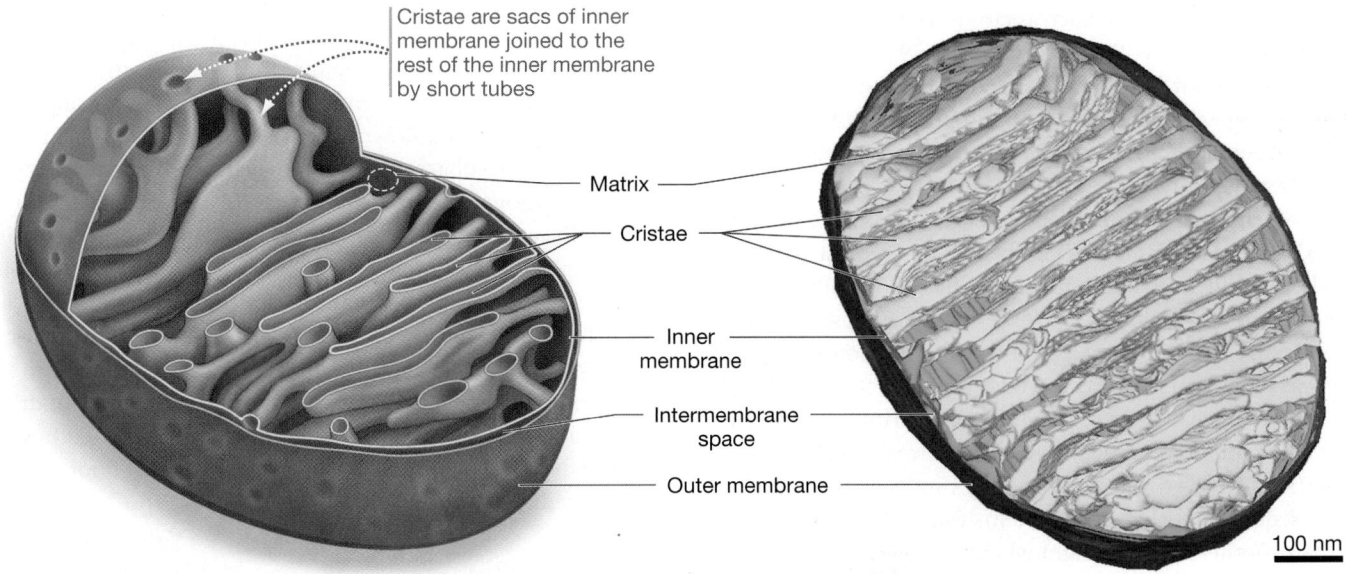

FIGURE 9.8 The Structure of the Mitochondrion. Mitochondria have outer and inner membranes that define the intermembrane space and matrix. Pyruvate processing occurs within the mitochondrial matrix. Recent research using cryo-electron tomography (the colorized micrograph on the right) shows the inner membrane is connected by short tubes to sac-like cristae.

Acetyl CoA is the final product of the pyruvate-processing step in glucose oxidation. Pyruvate, NAD^+, and CoA go in; CO_2, NADH, and acetyl CoA come out.

When supplies of ATP are abundant, however, the process shuts down. Pyruvate processing stops when the pyruvate dehydrogenase complex becomes phosphorylated and changes shape. The rate of phosphorylation increases when other products—specifically acetyl CoA and NADH—are at high concentration.

These regulatory changes are more examples of feedback inhibition. Reaction products feed back to stop or slow down the pathway.

On the contrary, high concentrations of NAD^+, CoA, or adenosine monophosphate (AMP)—which indicates low ATP supplies—*speed up* the reactions catalyzed by the pyruvate dehydrogenase complex.

Pyruvate processing is under both positive and negative control. Large supplies of products inhibit the enzyme complex; large supplies of reactants and low supplies of products stimulate it.

To summarize, pyruvate processing starts with the three-carbon pyruvate molecule and ends with one carbon released as CO_2 and the remaining two carbons in the form of acetyl CoA. The reactions occur in the mitochondrial matrix, and the potential energy that is released is used to produce one NADH for each pyruvate that is processed. Now the question is, what happens to the acetyl CoA?

9.4 The Citric Acid Cycle: Oxidizing Acetyl CoA to CO_2

While researchers were working out the sequence of reactions in glycolysis, biologists in other laboratories were focusing on redox reactions that oxidize small organic acids called **carboxylic acids.** Note that carboxylic acids all have carboxyl functional groups (R-COOH), hence the name.

A key finding emerged from their studies: Redox reactions that involve carboxylic acids such as citrate, malate, and succinate produce carbon dioxide. Recall from Section 9.1 that carbon dioxide is the endpoint of glucose oxidation via cellular respiration. Thus, it was logical for researchers to propose that the oxidation of small carboxylic acids could be an important component of glucose catabolism.

Early researchers identified eight small carboxylic acids that are rapidly oxidized in sequence, from least to most oxidized. What they found next was puzzling. When they added one of the eight carboxylic acids to cells, the rate of glucose oxidation increased, suggesting that the reactions are somehow connected to pathways involved in glucose catabolism. But, the added molecules did not appear to be used up. Instead, virtually all the carboxylic acids added were recovered later. How is this possible?

Hans Krebs solved the mystery when he proposed that the reaction sequence occurs in a cycle instead of a linear pathway. Krebs had another crucial insight when he suggested that the reaction sequence was directly tied to the processing of pyruvate—the endpoint of the glycolytic pathway.

To test these hypotheses, Krebs and a colleague set out to determine if adding pyruvate could link the two ends of the sequence

of eight carboxylic acids. If pyruvate is the key link in forming a cycle, it would need to be involved in the conversion of oxaloacetate, the most oxidized of the eight carboxylic acids, to citrate, the most reduced carboxylic acid. When Krebs added pyruvate, the series of redox reactions occurred. The conclusion? The sequence of eight carboxylic acids is indeed arranged in a cycle (**FIGURE 9.10**).

Many biologists now refer to the cycle as the **citric acid cycle** because it starts with citrate, which is the salt of citric acid after the protons are released. The citric acid cycle is also known as the tricarboxylic acid (TCA) cycle, because citrate has three carboxyl groups, and also as the Krebs cycle, after its discoverer.

When radioactive isotopes of carbon became available in the early 1940s, researchers showed that carbon atoms cycle through the reactions just as Krebs had proposed. For more detail concerning the enzymes that catalyze each step, see **TABLE 9.2** (on page 164). In each cycle, the energy released by the oxidation of one molecule of acetyl CoA is used to produce three molecules of NADH, one of $FADH_2$, and one of **guanosine triphosphate (GTP),** or ATP, through substrate-level phosphorylation. Whether GTP or ATP is produced depends on the type of cell being considered.[1] For example, GTP appears to be produced in the liver cells of mammals, while ATP is produced in muscle cells.

In bacteria and archaea, the enzymes responsible for the citric acid cycle are located in the cytosol. In eukaryotes, most of the enzymes responsible for the citric acid cycle are located in the mitochondrial matrix. Because glycolysis produces two molecules of pyruvate, the cycle turns twice for each molecule of glucose processed in cellular respiration.

How Is the Citric Acid Cycle Regulated?

By now, it shouldn't surprise you to learn that the citric acid cycle is carefully regulated. The citric acid cycle can be turned off at multiple points, via several different mechanisms of feedback inhibition. Reaction rates are high when ATP is scarce; reaction rates are low when ATP is abundant.

FIGURE 9.11 highlights the major control points. Notice that in step 1, the enzyme that combines acetyl CoA and oxaloacetate to form citrate is shut down when ATP binds to it. This is another example of feedback inhibition, which also regulates enzymes at two additional points in the cycle. In step 3, NADH interferes with the reaction by binding to the enzyme's active site. This is an example of competitive inhibition (see Chapter 8). In step 4, ATP binds to the enzyme at an allosteric regulatory site.

To summarize, the citric acid cycle starts with the two-carbon acetyl molecule in the form of acetyl CoA and ends with the release of two CO_2. The reactions occur in the mitochondrial matrix, and the potential energy that is released is used to produce three NADH, one $FADH_2$, and one ATP or GTP for each acetyl oxidized. But a major question remains.

---

[1]Traditionally it was thought that the citric acid cycle produced GTP, which was later converted to ATP in the same cell. Recent work suggests that ATP is produced directly in some cell types, while GTP is produced in other cells. See C. O. Lambeth, Reconsideration of the significance of substrate-level phosphorylation in the citric acid cycle. *Biochemistry and Molecular Biology Education* 34 (2006): 21–29.

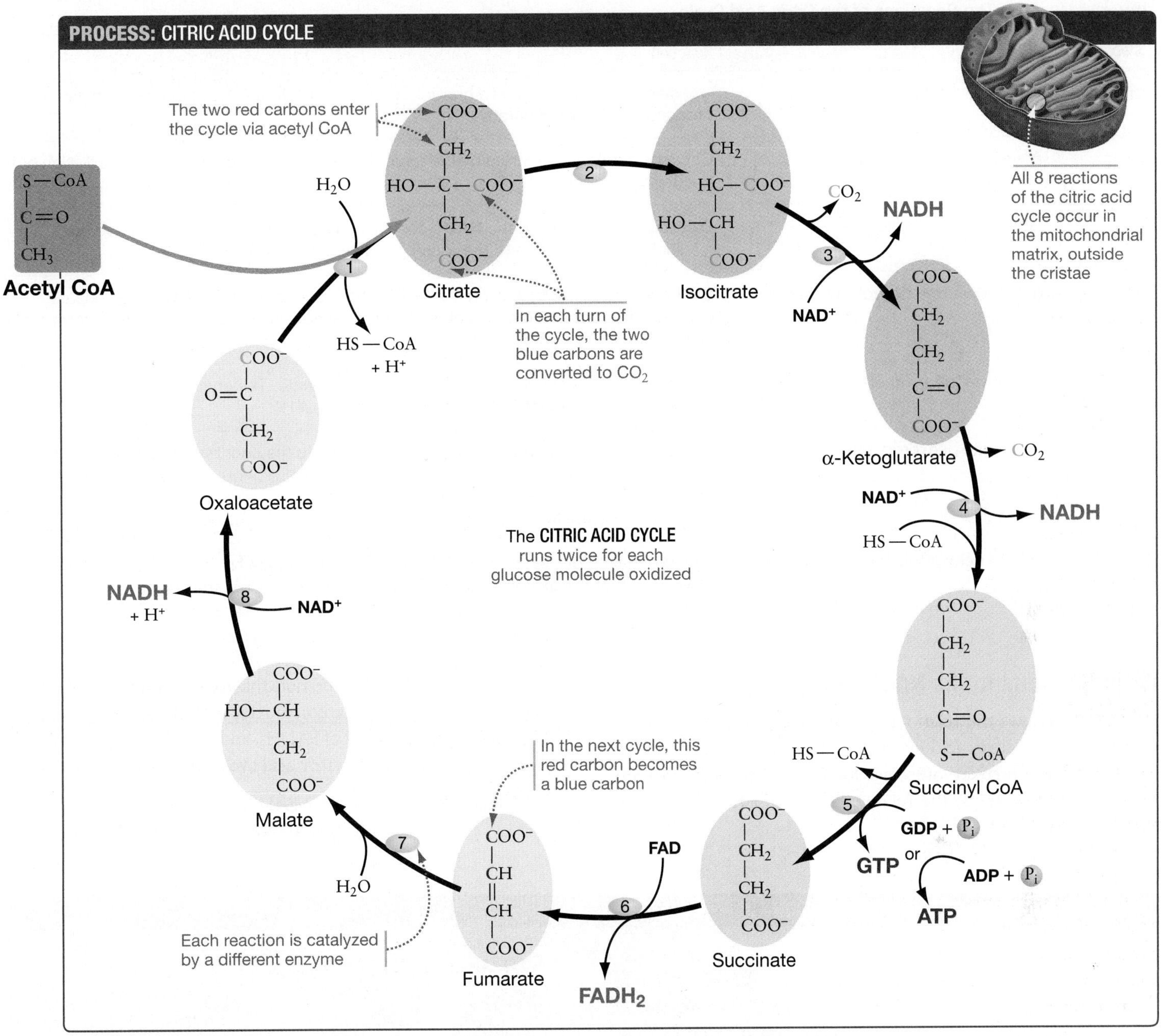

The two red carbons enter the cycle via acetyl CoA

H_2O

S—CoA
|
C=O
|
CH_3

Acetyl CoA

1

HS—CoA
+ H^+

COO⁻
|
CH_2
|
HO—C—COO⁻
|
CH_2
|
COO⁻

Citrate

In each turn of the cycle, the two blue carbons are converted to CO_2

2

COO⁻
|
CH_2
|
HC—COO⁻
|
HO—CH
|
COO⁻

Isocitrate

CO_2

NADH

3

NAD^+

All 8 reactions of the citric acid cycle occur in the mitochondrial matrix, outside the cristae

COO⁻
|
CH_2
|
CH_2
|
C=O
|
COO⁻

α-Ketoglutarate

CO_2

NAD^+

4

NADH

HS—CoA

COO⁻
|
O=C
|
CH_2
|
COO⁻

Oxaloacetate

NADH
+ H^+

8

NAD^+

The **CITRIC ACID CYCLE** runs twice for each glucose molecule oxidized

COO⁻
|
HO—CH
|
CH_2
|
COO⁻

Malate

7

H_2O

In the next cycle, this red carbon becomes a blue carbon

Each reaction is catalyzed by a different enzyme

COO⁻
|
CH
||
CH
|
COO⁻

Fumarate

FAD

6

FADH₂

COO⁻
|
CH_2
|
CH_2
|
COO⁻

Succinate

HS—CoA

5

COO⁻
|
CH_2
|
CH_2
|
C=O
|
S—CoA

Succinyl CoA

GDP + P_i

GTP or

ATP

ADP + P_i

FIGURE 9.10 The Citric Acid Cycle Completes the Oxidation of Glucose. Acetyl CoA goes into the citric acid cycle, and carbon dioxide, NADH, FADH₂, and GTP or ATP come out. GTP or ATP is produced by substrate-level phosphorylation. If you follow individual carbon atoms around the cycle several times, you'll come to an important conclusion: each of the carbons in the cycle is eventually a "blue carbon" that is released as CO_2.

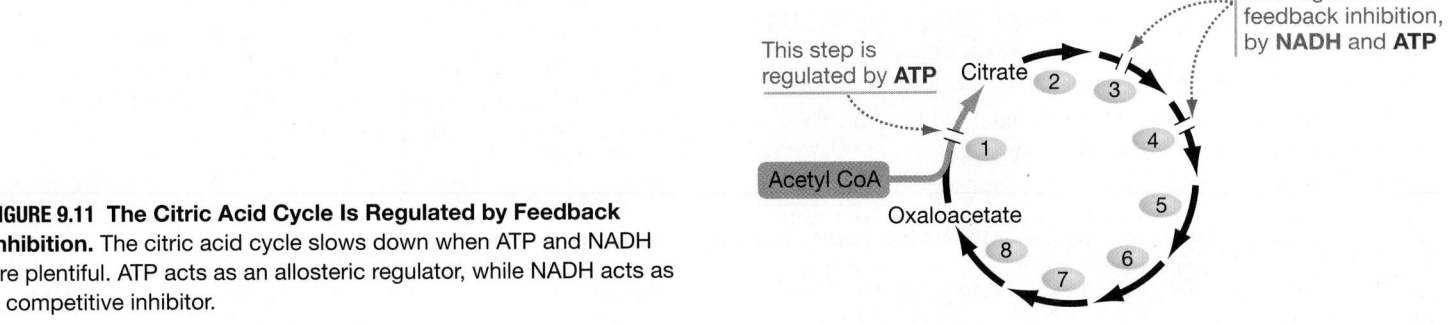

FIGURE 9.11 The Citric Acid Cycle Is Regulated by Feedback Inhibition. The citric acid cycle slows down when ATP and NADH are plentiful. ATP acts as an allosteric regulator, while NADH acts as a competitive inhibitor.

This step is regulated by **ATP**

These steps are also regulated via feedback inhibition, by **NADH** and **ATP**

Citrate

Acetyl CoA

Oxaloacetate

Step	Enzyme	Reaction
1	Citrate synthase	Transfers the 2-carbon acetyl group from acetyl CoA to the 4-carbon oxaloacetate to produce the 6-carbon citrate.
2	Aconitase	Converts citrate to isocitrate by the removal of one water molecule and the addition of another water molecule.
3	Isocitrate dehydrogenase	Oxidizes isocitrate using the **NAD$^+$** coenzyme to produce **NADH** and release one CO_2, resulting in the formation of the 5-carbon molecule α-ketoglutarate.
4	α-Ketoglutarate dehydrogenase	Oxidizes α-ketoglutarate using the **NAD$^+$** coenzyme to produce **NADH** and release one CO_2. The remaining 4-carbon molecule is added to coenzyme A (CoA) to form succinyl CoA.
5	Succinyl-CoA synthetase	CoA is removed, converting succinyl CoA to succinate. The energy released is used to transfer P_i to GDP to form **GTP**, or to ADP to form **ATP**.
6	Succinate dehydrogenase	Oxidizes succinate by transferring two hydrogens to the coenzyme **FAD** to produce **FADH$_2$**, resulting in the formation of fumarate.
7	Fumarase	Converts fumarate to malate by the addition of one water molecule.
8	Malate dehydrogenase	Oxidizes malate by using the **NAD$^+$** coenzyme to produce **NADH**, resulting in the regeneration of the oxaloacetate that will be used in step 1 of the cycle.

What Happens to the NADH and FADH$_2$?

FIGURE 9.12 reviews the relationships of glycolysis, pyruvate processing, and the citric acid cycle and identifies where each process takes place in eukaryotic cells. As the carbons in glucose are oxidized in these steps, the relative changes in free energy are shown in **FIGURE 9.13**.

As you study these figures, note that for each molecule of glucose that is fully oxidized to 6 carbon dioxide molecules, the cell produces 10 molecules of NADH, 2 of FADH$_2$, and 4 of ATP. The overall reaction for glycolysis and the citric acid cycle can be written as

$$C_6H_{12}O_2 + 10\ NAD^+ + 2\ FAD + 4\ ADP + 4\ P_i \longrightarrow$$
$$6\ CO_2 + 10\ NADH + 2\ FADH_2 + 4\ ATP$$

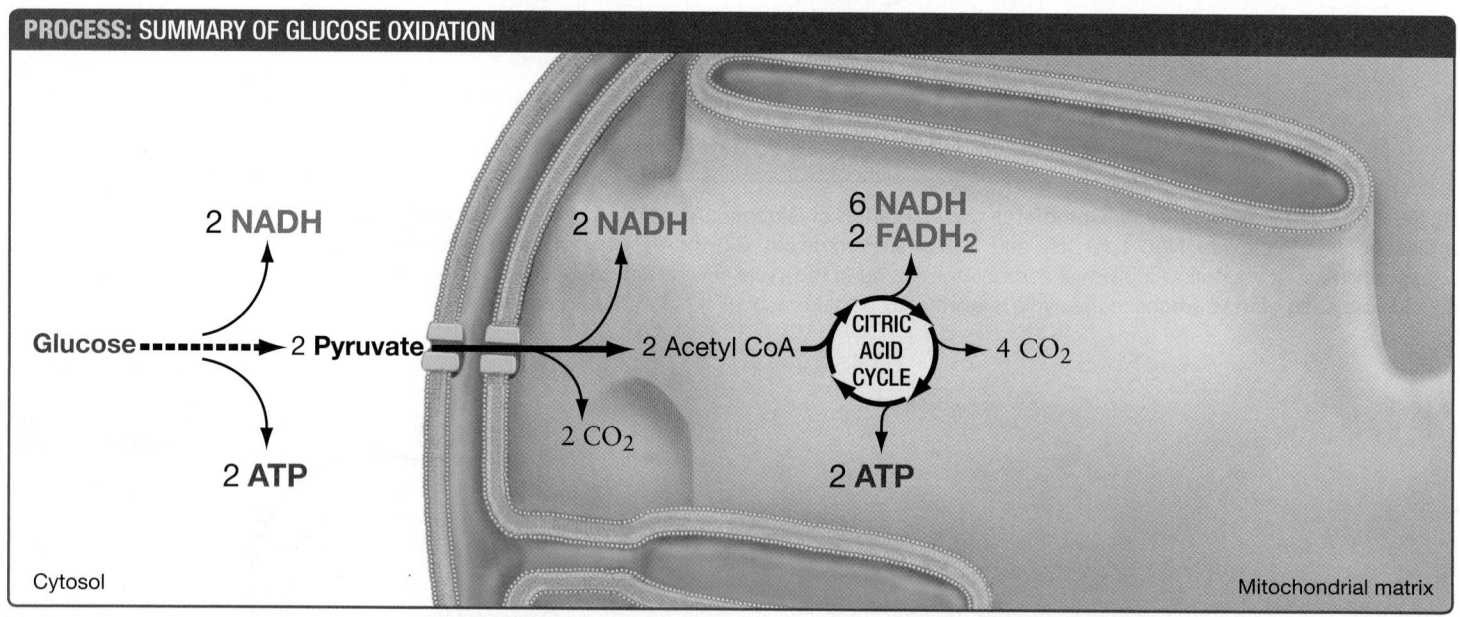

FIGURE 9.12 Glucose Oxidation Produces ATP, NADH, FADH$_2$, and CO$_2$. Glucose is completely oxidized to carbon dioxide via glycolysis, pyruvate processing, and the citric acid cycle. In eukaryotes, glycolysis occurs in the cytosol; pyruvate oxidation and the citric acid cycle take place in the mitochondrial matrix.

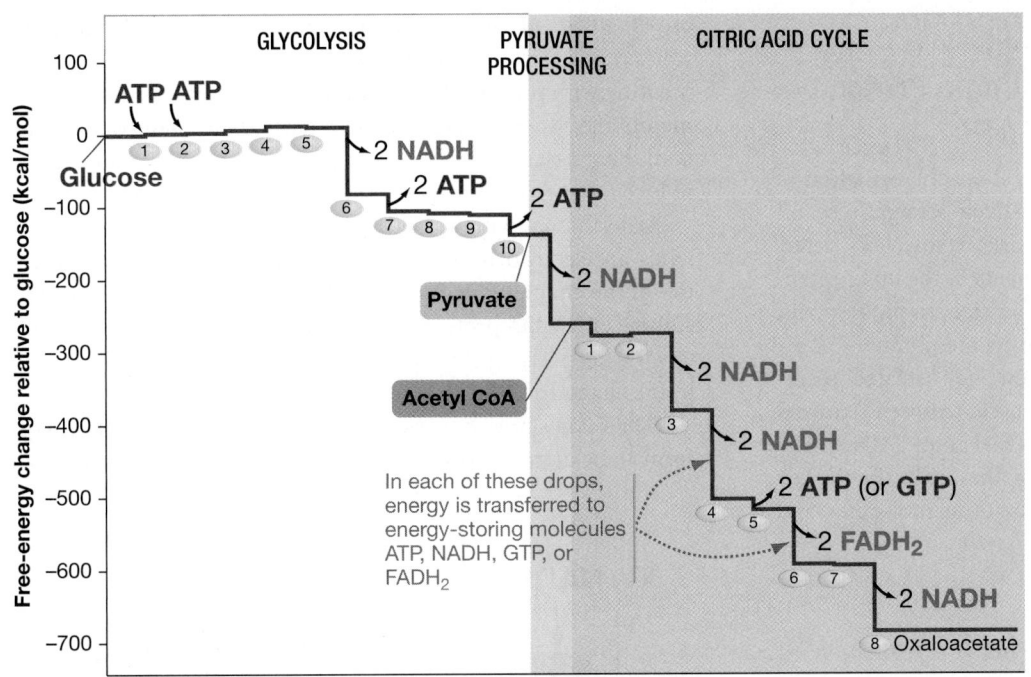

GLYCOLYSIS   PYRUVATE   CITRIC ACID CYCLE
              PROCESSING

ATP ATP

Glucose

2 NADH

2 ATP

2 ATP

Pyruvate

2 NADH

Acetyl CoA

2 NADH

In each of these drops,
energy is transferred to
energy-storing molecules
ATP, NADH, GTP, or
$FADH_2$

2 NADH

2 ATP (or GTP)

2 $FADH_2$

2 NADH

8 Oxaloacetate

Oxidation of glucose →

FIGURE 9.13 Free Energy Changes as Glucose Is Oxidized. If you read the vertical axis of this graph carefully, it should convince you that about 685 kcal/mol of free energy is released from the oxidation of glucose. Much of the energy is harnessed in the form of ATP, NADH, and $FADH_2$. The numbered green ovals identify the reaction steps in glycolysis and the citric acid cycle (see Tables 9.1 and 9.2).

DATA: Li, X., R. K. Dash, R. K. Pradhan, et al. 2010. *Journal of Physical Chemistry B.* 114: 16068–16082.

✔ **QUANTITATIVE** Based on the data in this graph, which of the three high-energy molecules produced during glucose oxidation would you expect to carry the highest amount of chemical energy? Justify your answer.

The ATP molecules are produced by substrate-level phosphorylation and can be used to drive endergonic reactions. The CO_2 molecules are a gas that is disposed of as waste—you exhale it; plants release it or use it as a reactant in photosynthesis.

What happens to the NADH and $FADH_2$ produced by glycolysis, pyruvate processing, and the citric acid cycle? Recall that the overall reaction for glucose oxidation is

$$C_6H_{12}O_6 + 6\,O_2 \longrightarrow 6\,CO_2 + 6\,H_2O + Energy$$

These three steps account for the glucose, the CO_2, and—because ATP is produced—some of the chemical energy that results from the overall reaction. But the O_2 and the H_2O are still unaccounted for. As it turns out, so is much of the chemical energy. The reaction that has yet to occur is

$$NADH + FADH_2 + O_2 + ADP + P_i \longrightarrow$$
$$NAD^+ + FAD + 2\,H_2O + ATP$$

In the above reaction, the electrons from NADH and $FADH_2$ are transferred to oxygen. NADH and $FADH_2$ are oxidized to NAD^+ and FAD, and oxygen is reduced to form water.

In effect, glycolysis, pyruvate processing, and the citric acid cycle transfer electrons from glucose to NAD^+ and FAD to form NADH and $FADH_2$. When oxygen accepts electrons from these reduced molecules, water is produced.

At this point, all the components of the overall reaction for glucose oxidation are accounted for, except for the energy. What happens to the energy that is released as electrons are transferred from NADH and $FADH_2$ to the highly electronegative oxygen atoms?

Specifically, how is the transfer of electrons linked to the production of ATP? In the 1960s—decades after the details of glycolysis and the citric acid cycle had been worked out—a startling answer to these questions emerged.

check your understanding

C Y U

If you understand that . . .

- During glycolysis, glucose is oxidized to pyruvate, in the cytosol.
- During pyruvate processing, pyruvate is oxidized to acetyl CoA, in the mitochondrial matrix.
- In the citric acid cycle, the acetyl from acetyl CoA is oxidized to carbon dioxide (CO_2), in the mitochondrial matrix.
- Glycolysis, pyruvate processing, and the citric acid cycle are all regulated processes. The cell produces ATP only when ATP is needed.

✔ **You should be able to . . .**

Model the following components of cellular respiration by pretending that a large piece of paper is a cell. Draw a large mitochondrion inside it. Cut out small circles of paper and label them glucose, pyruvate, acetyl CoA, CO_2, ADP → ATP, NAD^+ → NADH, and FAD → $FADH_2$. Cut out small squares of paper and label them as glycolytic reactions, citric acid cycle reactions, and pyruvate dehydrogenase complex.

1. Put each of the squares in the appropriate location in the cell.

2. Add the circles and draw arrows to connect the appropriate molecules and reactions.

3. Using 12 paper triangles for pairs of electrons, show how electrons from glucose are transferred to NADH or $FADH_2$ (one pair should go to each NADH or $FADH_2$ formed) as glucose is oxidized to CO_2.

4. Label points where regulation occurs.

Answers are available in Appendix A.

9.5 Electron Transport and Chemiosmosis: Building a Proton Gradient to Produce ATP

The answer to one fundamental question about the oxidation of NADH and FADH$_2$ turned out to be relatively straightforward. By isolating different parts of mitochondria, researchers determined that NADH is oxidized by components in the inner membrane of the mitochondria, including the cristae. In prokaryotes, the oxidation of NADH occurs in the plasma membrane.

Biologists made a key discovery when they isolated membrane components—they were found to cycle between oxidized and reduced states after the addition of NADH and FADH$_2$. The membrane-associated molecules were hypothesized to be the key to processing NADH and FADH$_2$. What are these molecules, and how do they work?

The Electron Transport Chain

Collectively, the molecules responsible for the oxidation of NADH and FADH$_2$ are designated the **electron transport chain (ETC).** As electrons are passed from one molecule to another in the chain, the energy released by the redox reactions is used to move protons across the inner membrane of mitochondria.

Several points are fundamental to understanding how the ETC works:

- Most of the molecules are proteins that contain distinctive cofactors and prosthetic groups where the redox events take place (see Chapter 8). They include iron–sulfur complexes, ring-containing structures called flavins, or iron-containing heme groups called cytochromes. Each of these groups is readily reduced or oxidized.

- The inner membrane of the mitochondrion also contains a molecule called **ubiquinone,** which is not a protein. Ubiquinone got its name because it is nearly ubiquitous in organisms and belongs to a family of compounds called quinones. Also called **coenzyme Q** or simply **Q,** ubiquinone is lipid soluble and moves efficiently throughout the hydrophobic interior of the inner mitochondrial membrane.

- The molecules involved in processing NADH and FADH$_2$ differ in electronegativity, or their tendency to hold electrons. Some of the molecules pick up a proton with each electron, forming hydrogen atoms, while others obtain only electrons.

Because Q and the ETC proteins can cycle between a reduced state and an oxidized state, and because they differ in electronegativity, investigators realized that it should be possible to arrange them into a logical sequence. The idea was that electrons would pass from a molecule with lower electronegativity to one with higher electronegativity, via a redox reaction.

As electrons moved through the chain, they would be held more and more tightly. A small amount of energy would be released in each reaction, and the potential energy in each successive bond would lessen.

Organization of the Electron Transport Chain Researchers worked out the sequence of compounds in the ETC by experimenting with poisons that inhibit particular proteins in the inner membrane. It was expected that if part of the chain were inhibited, then the components upstream of the block would become reduced, but those downstream would remain oxidized.

Experiments with various poisons showed that NADH donates an electron to a flavin-containing protein (FMN) at the top of the chain, while FADH$_2$ donates electrons to an iron- and sulfur-containing protein (Fe·S) that then passes them directly to Q. After passing through each of the remaining components in the chain, the electrons are finally accepted by oxygen.

FIGURE 9.14 shows how electrons step down in potential energy from the electron carriers NADH and FADH$_2$ to O$_2$. The x-axis

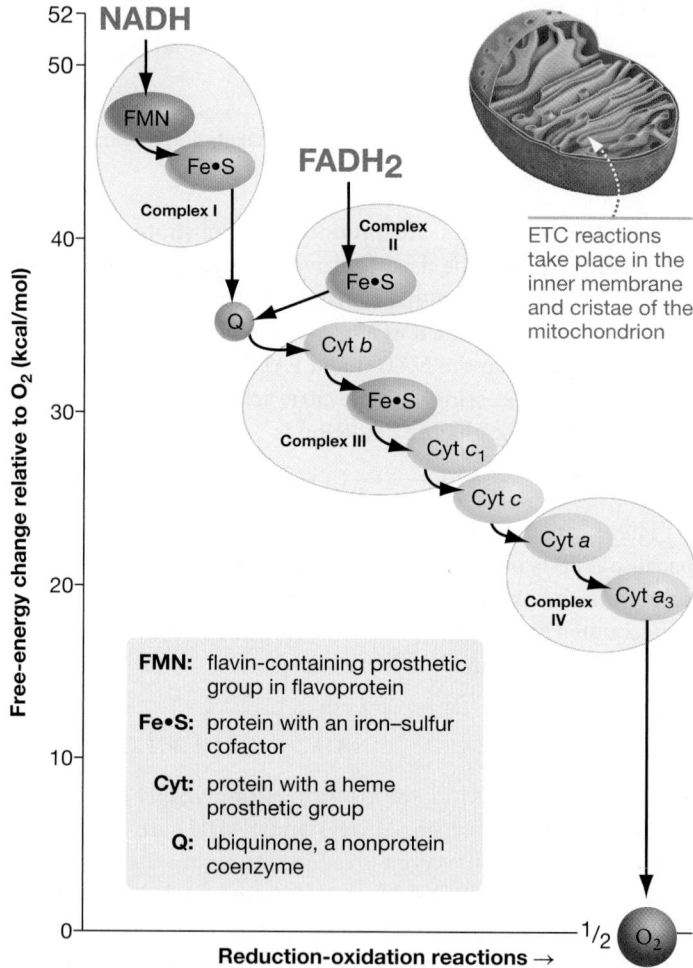

ETC reactions take place in the inner membrane and cristae of the mitochondrion

FMN: flavin-containing prosthetic group in flavoprotein

Fe•S: protein with an iron–sulfur cofactor

Cyt: protein with a heme prosthetic group

Q: ubiquinone, a nonprotein coenzyme

FIGURE 9.14 A Series of Reduction–Oxidation Reactions Occur in an Electron Transport Chain. Electrons step down in potential energy from the electron carriers NADH and FADH$_2$ through an electron transport chain to a final electron acceptor. When oxygen is the final electron acceptor, water is formed. The overall free-energy change of 52 kcal/mol (from NADH to oxygen) is broken into small steps.

DATA: Wilson D. F., M. Erecinska, and P. L. Dutton. 1974. *Annual Review of Biophysics and Bioengineering* 3: 203–230. Also Sled, V. D., N. I. Rudnitzky, Y. Hatefi, et al. 1994. *Biochemistry* 33: 10069–10075.

plots the sequence of redox reactions in the ETC; the *y*-axis plots the free-energy changes that occur.

The components of the electron transport chain are organized into four large complexes of proteins, often referred to as simply complexes I–IV. Q and the protein **cytochrome *c*** act as shuttles that transfer electrons between these complexes. Once the electrons at the bottom of the ETC are accepted by oxygen to form water, the oxidation of glucose is complete. Details on the names of the complexes and their role in the electron transport chain are provided in **TABLE 9.3** (on page 168).

Under controlled conditions in the laboratory, the total potential energy difference from NADH to oxygen is a whopping 53 kilocalories/mole (kcal/mol). Oxidation of the 10 molecules of NADH produced from each glucose accounts for almost 80 percent of the total energy released from the sugar. What does the ETC do with all this energy?

Role of the Electron Transport Chain Throughout the 1950s most biologists working on cellular respiration assumed that electron transport chains include enzymes that catalyze substrate-level phosphorylation. Recall that when substrate-level

phosphorylation occurs, a phosphate group is transferred from a phosphorylated substrate to ADP, forming ATP. Despite intense efforts, however, no one was able to find an enzyme among the components of the ETC that would catalyze the phosphorylation of ADP to produce ATP.

What researchers did find, however, is that the movement of electrons through the ETC actively transports protons from the matrix, across the inner membrane, and into the intermembrane space (see **FIGURE 9.15**). The exact route and mechanism used to pump protons is still being worked out. In some cases, it is not clear how the redox reactions taking place inside each complex result in the movement of protons.

The best-understood interaction between electron transport and proton transport takes place in complex III. Research has shown that when Q accepts electrons from complex I or complex II, it picks up protons from the matrix side of the inner membrane. The reduced form of Q then diffuses through the inner

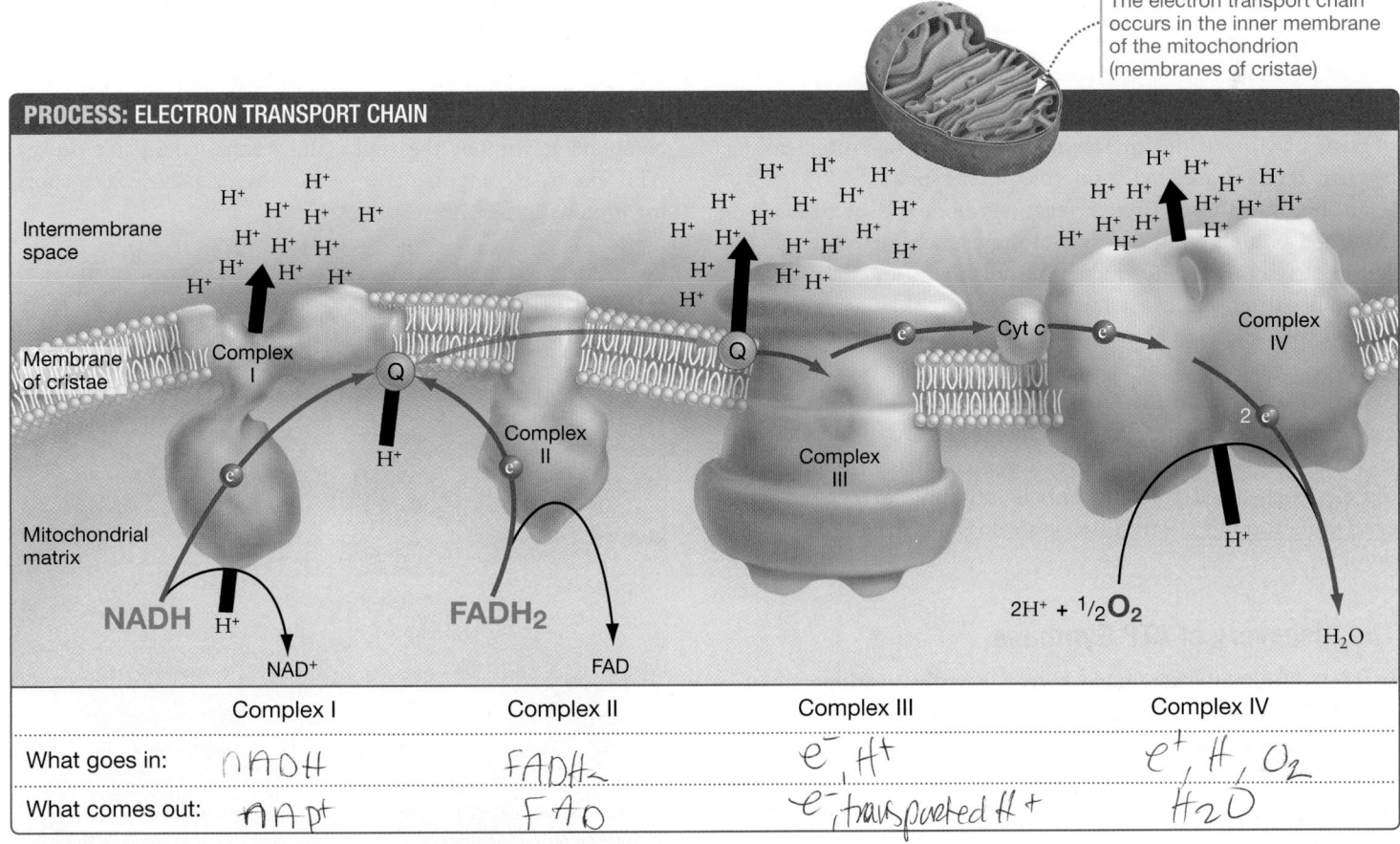

The electron transport chain occurs in the inner membrane of the mitochondrion (membranes of cristae)

PROCESS: ELECTRON TRANSPORT CHAIN

	Complex I	Complex II	Complex III	Complex IV
What goes in:	NADH	FADH₂	e⁻, H⁺	e⁺, H, O₂
What comes out:	NADp⁺	FAO	e⁻ transported H⁺	H₂O

FIGURE 9.15 How Does the Electron Transport Chain Work? The individual components of the electron transport chain diagrammed in Figure 9.14 are found in the inner membrane of mitochondria. Electrons are carried from one complex to another by Q and by cytochrome *c*; Q also shuttles protons across the membrane. The orange arrow indicates Q moving back and forth. Complexes I and IV use the potential energy released by the redox reactions to pump protons from the mitochondrial matrix to the intermembrane space.

✔**EXERCISE** Add an arrow across the membrane and label it "Proton gradient." In the boxes at the bottom, list "What goes in" and "What comes out" for each complex.

ETC Component	Descriptive Name	Reaction
Complex I	NADH dehydrogenase	Oxidizes **NADH** and transfers the two electrons through proteins containing FMN prosthetic groups and Fe·S cofactors to reduce an oxidized form of ubiquinone (Q). Four **H$^+$** are pumped out of the matrix to the intermembrane space.
Complex II	Succinate dehydrogenase	Oxidizes **FADH$_2$** and transfers the two electrons through proteins containing Fe·S cofactors to reduce an oxidized form of Q. This complex is also used in step 6 of the citric acid cycle.
Q	Ubiquinone	Reduced by complexes I and II and moves throughout the hydrophobic interior of the ETC membrane, where it is oxidized by complex III.
Complex III	Cytochrome c reductase	Oxidizes Q and transfers one electron at a time through proteins containing heme prosthetic groups and Fe·S cofactors to reduce an oxidized form of cytochrome c (cyt c). A total of four **H$^+$** for each pair of electrons is transported from the matrix to the intermembrane space.
Cyt c	Cytochrome c	Reduced by accepting a single electron from complex III and moves along the surface of ETC membrane, where it is oxidized by complex IV.
Complex IV	Cytochrome c oxidase	Oxidizes cyt c and transfers each electron through proteins containing heme prosthetic groups to reduce oxygen gas (O$_2$), which picks up two **H$^+$** from the matrix to produce water. Two additional **H$^+$** are pumped out of the matrix to the intermembrane space.

membrane, where its electrons are used to reduce a component of complex III near the intermembrane space. The protons held by Q are then released to the intermembrane space.

In this way, through redox reactions alone, Q shuttles electrons and protons from one side of the membrane to the other. The electrons proceed down the transport chain, and the protons contribute to an electrochemical gradient as they are released into the intermembrane space.

Once the nature of the electron transport chain became clear, biologists understood the fate of the electrons and the energy carried by NADH and FADH$_2$. Much of the chemical energy that was originally present in glucose is now accounted for in the proton electrochemical gradient. This is satisfying, except for one crucial question: If electron transport does not make ATP, what does?

The Discovery of ATP Synthase

In 1960 Efraim Racker made several key observations about how ATP is synthesized in mitochondria. When he used mitochondrial membranes to make vesicles, Racker noticed that some vesicles formed with their membrane inside out. Electron microscopy revealed that the inside-out membranes had many large proteins studded along their surfaces. Each protein appeared to have a base in the membrane, from which a lollipop-shaped stalk and a knob project (**FIGURE 9.16**). If the solution was vibrated or treated with a compound called urea, the stalks and knobs fell off.

Racker seized on this technique to isolate the stalks and knobs and do experiments with them. For example, he found that isolated stalks and knobs could hydrolyze ATP, forming ADP and

inorganic phosphate. The vesicles that contained just the base component, without the stalks and knobs, could not process ATP. The base components were, however, capable of transporting protons across the membrane.

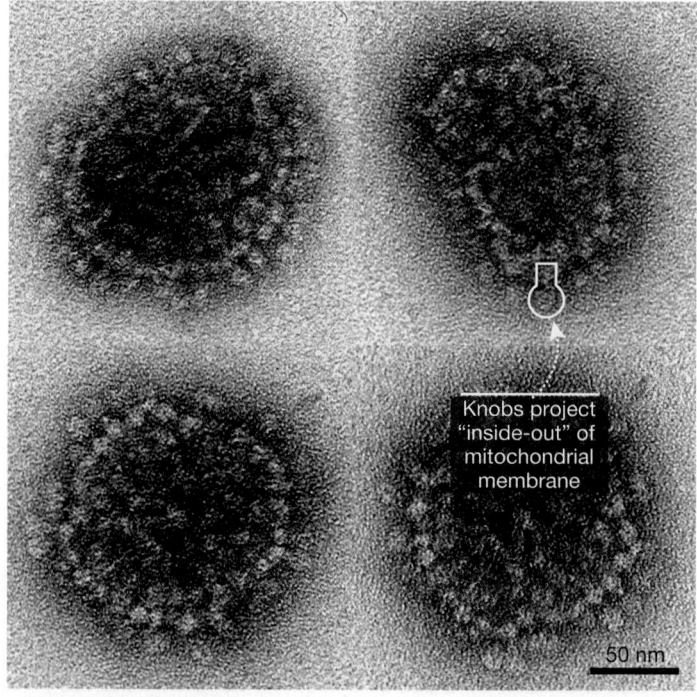

Knobs project "inside-out" of mitochondrial membrane

50 nm

FIGURE 9.16 The Discovery of ATP Synthase. When patches of mitochondrial membrane turn inside out and form vesicles, the lollipop-shaped stalk-and-knob structures of ATP synthase proteins face outward. Normally, the stalk and knob face inward, toward the mitochondrial matrix.

Based on these observations, Racker proposed that the stalk-and-knob component of the protein was an enzyme that both hydrolyzes and synthesizes ATP. To test his idea, Racker added the stalk-and-knob components back to vesicles that had been stripped of them and confirmed that the vesicles regained the ability to synthesize ATP. The entire complex is known as **ATP synthase.** Follow-up work also confirmed his hypothesis that the membrane-bound base component of ATP synthase is a proton channel. Is there a connection between proton transport and ATP synthesis?

The Chemiosmosis Hypothesis

In 1961 Peter Mitchell broke with the prevailing ideas that electron transport produces ATP via substrate phosphorylation. Instead, he proposed something completely new—an indirect connection between electron transport and ATP production. Mitchell's novel hypothesis? The real job of the electron transport chain is to pump protons across the inner membrane of mitochondria from the matrix to the intermembrane space. After a proton gradient is established, an enzyme in the inner membrane, like Racker's ATP synthase, would synthesize ATP from ADP and P_i.

Mitchell introduced the term **chemiosmosis** to describe the use of a proton gradient to drive energy-requiring processes, like the production of ATP. Here, osmosis refers to the force generated from the proton gradient rather than the transport of water. Although proponents of a direct link between electron transport and substrate-level phosphorylation objected vigorously to Mitchell's idea, several key experiments supported it.

FIGURE 9.17 illustrates how the existence of a key element in Mitchell's hypothesis was confirmed: A proton gradient alone can be used to synthesize ATP via ATP synthase. The researchers made vesicles from artificial membranes that contained Racker's ATP synthase isolated from mitochondria. Along with this enzyme, they inserted bacteriorhodopsin, a well-studied membrane protein that acts as a light-activated proton pump.

When light strikes bacteriorhodopsin, it absorbs some of the light energy and changes conformation in a way that pumps protons from the interior of a membrane to the exterior. As a result, the experimental vesicles established a strong electrochemical gradient favoring proton movement to the interior. When the vesicles were illuminated to initiate proton pumping, ATP began to be produced from ADP and P_i inside the vesicles.

Mitchell's prediction was correct: In this situation, ATP production depended solely on the existence of a **proton-motive force,** which is based on a proton electrochemical gradient. It could occur in the *absence* of an electron transport chain. This result, along with many others, has provided strong support for the hypothesis of chemiosmosis. Most of the ATP produced by cellular respiration is made by a flow of protons.

✔ If you understand chemiosmosis, you should be able to explain why ATP production during cellular respiration is characterized as indirect. More specifically, you should be able to explain the relationship between glucose oxidation, the proton gradient, and ATP synthase.

RESEARCH

QUESTION: How are the electron transport chain and ATP production linked?

CHEMIOSMOTIC HYPOTHESIS: The linkage is indirect. The ETC creates a proton-motive force that drives ATP synthesis by the mitochondrial ATP synthase.

ALTERNATIVE HYPOTHESIS: The linkage is direct. The ETC is associated with enzymes that perform substrate-level phosphorylation.

EXPERIMENTAL SETUP:

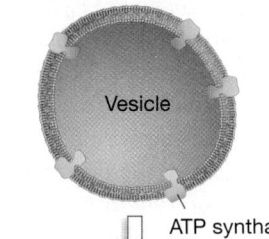

1. **Produce vesicles from artificial membranes;** add ATP synthase, an enzyme found in mitochondria.

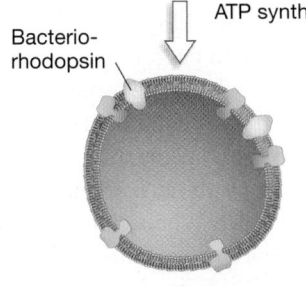

2. **Add bacteriorhodopsin,** a protein that acts as a light-activated proton pump.

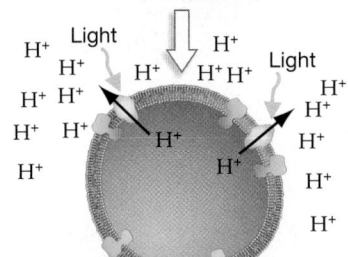

3. **Illuminate vesicle** so that bacteriorhodopsin pumps protons out of vesicle, creating a proton gradient.

PREDICTION OF CHEMIOSMOTIC HYPOTHESIS: ATP will be produced within the vesicle.

PREDICTION OF ALTERNATIVE HYPOTHESIS: No ATP will be produced.

RESULTS:

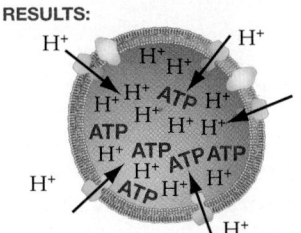

ATP is produced within the vesicle, in the absence of the electron transport chain.

CONCLUSION: The linkage between electron transport and ATP synthesis is indirect; the movement of protons drives the synthesis of ATP.

FIGURE 9.17 Evidence for the Chemiosmotic Hypothesis.

Racker, E., and W. Stoeckenius. 1974. Reconstitution of purple membrane vesicles catalyzing light-driven proton uptake and adenosine triphosphate formation. *Journal of Biological Chemistry.* 249: 662–663.

✔**QUESTION** If bacteriorhodopsin were not available, what could the researchers have done with the ATP synthase vesicles to test their hypothesis?

Electron transport chains and ATP synthases are used by organisms throughout the tree of life. They are humming away in your cells now. Let's look in more detail at how they function.

The Proton-Motive Force Couples Electron Transport to ATP Synthesis

As **FIGURE 9.18** shows, the structure of ATP synthase is now well understood. The ATP synthase "knob" component is called the F_1 unit; the membrane-bound, proton-transporting base component is the F_o unit. The F_1 and F_o units are connected by a shaft, as well as by a stator, which holds the two units in place.

The F_o unit serves as a rotor, whose turning is conveyed to the F_1 unit via the shaft. A flow of protons through the F_o unit causes the rotor and shaft to spin. By attaching long actin filaments to the shaft and examining them with a videomicroscope, researchers have been able to see the rotation, which can reach speeds of 350 revolutions per second. As the shaft spins within the F_1 unit, it is thought to change the conformation of the F_1 subunits in a way that catalyzes the phosphorylation of ADP to ATP.

Chemiosmosis is like the process of generating electricity in a hydroelectric dam (like the one pictured on page 154). The ETC is analogous to a series of gigantic pumps that force water up and behind the dam. The inner mitochondrial membrane functions as the dam, with ATP synthase spinning and generating electricity inside as water passes through—like a turbine. In a mitochondrion, protons are pumped instead of water. When protons move through ATP synthase, the protein spins and generates ATP.

It has been determined that the ETC transports enough protons to produce approximately three ATP for each NADH and

two for each $FADH_2$, depending on the type of ATP synthase used. These yields, however, are not observed in cells, since the proton-motive force is also used to drive other processes, such as the import of phosphates into the mitochondrial matrix.

Unlike the turbines in a hydroelectric dam, however, ATP synthase can reverse its direction and hydrolyze ATP to build a proton gradient. If the proton gradient dissipates, the direction of the spin is reversed and ATP is hydrolyzed to pump protons from the matrix to the intermembrane space. Understanding how these reactions occur is currently the focus of intense research. ATP synthase makes most of the ATP that keeps you alive.

The Proton-Motive Force and Chemical Evolution

How was energy first transformed into a usable form during the evolution of life? Since chemiosmosis is responsible for most of the ATP produced by cells throughout the tree of life, it is likely to have arisen early in evolution. But how could a complex electron transport chain evolve to produce the proton-motive force without a proton-motive force to supply the energy?

This apparent conundrum left many of the chemical evolution theorists perplexed until a key discovery was made deep in the ocean along the Mid-Atlantic Ridge—the Lost City hydrothermal vents (see Chapter 2). Researchers propose that the alkaline fluid (low proton concentration) released from these vents in the acidic oceans (high proton concentration) of early Earth may have provided such a gradient.

While there is still considerable debate concerning the role hydrothermal vents may have played in chemical evolution, their discovery has generated much excitement. By harnessing the natural electrochemical gradient deep in the early oceans, the proton-motive force of life may have evolved to mimic the environment of its origin.

Organisms Use a Diversity of Electron Acceptors

FIGURE 9.19 summarizes glucose oxidation and cellular respiration by tracing the fate of the carbon atoms and electrons in glucose. Notice that electrons from glucose are transferred to NADH and $FADH_2$, passed through the electron transport chain, and accepted by oxygen. Proton pumping during electron transport creates the proton-motive force that drives ATP synthesis.

The diagram also indicates the approximate yield of ATP from each component of the process. Recent research shows that about 29 ATP molecules are produced from each molecule of glucose.[2] Of these, 25 ATP molecules are produced by ATP synthase. The fundamental message here? The vast majority of the "payoff" from the oxidation of glucose occurs via oxidative phosphorylation.

Aerobic Versus Anaerobic Respiration

During cellular respiration, oxygen is the electron acceptor used by all eukaryotes and

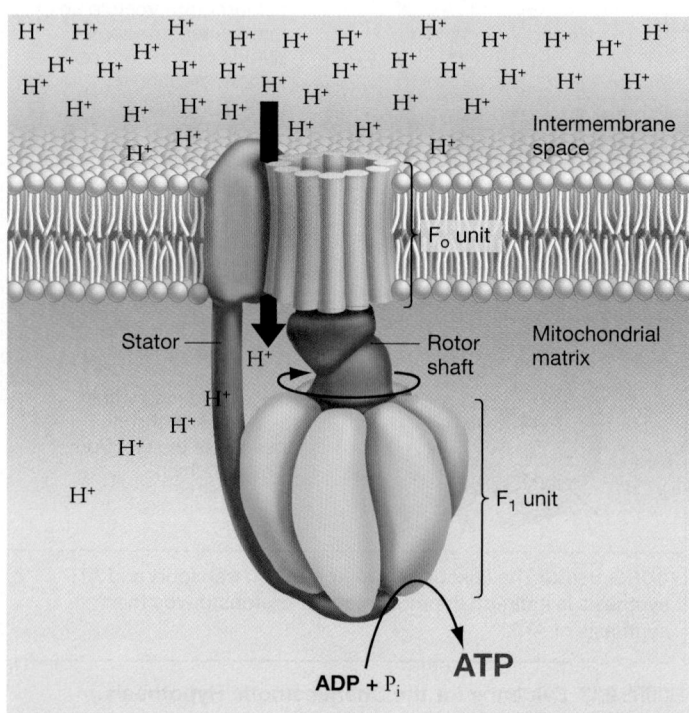

FIGURE 9.18 ATP Synthase Is a Motor. ATP synthase has two major components, designated F_o and F_1, connected by a shaft. The F_o unit spins as protons pass through. The shaft transmits the rotation to the F_1 unit, causing it to make ATP from ADP and P_i.

---

[2]Traditionally, biologists thought that 36 ATP would be synthesized for every molecule of glucose oxidized in eukaryotic cells. More recent work has shown that actual yield is only about 29 ATP [see M. Brand, Approximate yield of ATP from glucose, designed by Donald Nicholson. *Biochemistry and Molecular Biology Education* 31 (2003): 2–4]. Also, it's important to note that yield varies with conditions in the cell.

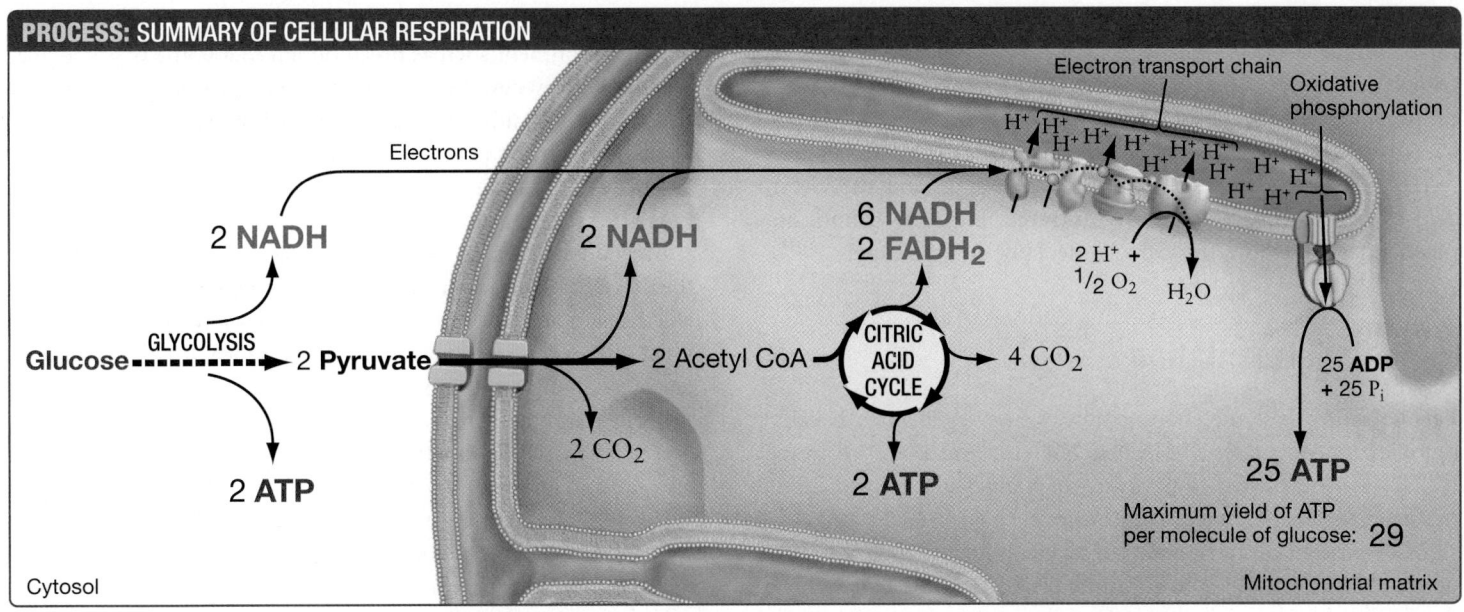

PROCESS: SUMMARY OF CELLULAR RESPIRATION

FIGURE 9.19 ATP Yield during Cellular Respiration. The actual yield of ATP per glucose (29 ATP) is lower than the theoretical calculation (38 ATP) because of energy required for the import of NADH from the cytoplasm and the use of the proton-motive force to actively transport P_i into the mitochondrial matrix.

a wide diversity of bacteria and archaea. Species that depend on oxygen as an electron acceptor for the ETC use **aerobic** respiration and are called aerobic organisms. (The Latin root *aero* means "air.")

It is important to recognize, though, that cellular respiration can occur without oxygen. Many thousands of bacterial and archaeal species rely on electron acceptors other than oxygen, and electron donors other than glucose. For example, nitrate (NO_3^-) and sulfate (SO_4^{2-}) are particularly common electron acceptors in species that live in oxygen-poor environments (see Chapter 29). In addition, many bacteria and archaea use H_2, H_2S, CH_4, or other inorganic compounds as electron donors—not glucose.

Cells that depend on electron acceptors other than oxygen are said to use **anaerobic** ("no air") respiration. Even though the starting and ending points of cellular respiration differ, aerobic and anaerobic cells still use electron transport chains to create a proton-motive force that drives the synthesis of ATP. In bacteria and archaea, the ETC and ATP synthase are located in the plasma membrane.

Aerobic Respiration Is Most Efficient Even though an array of compounds can serve as the final electron acceptor in cellular respiration, oxygen is the most efficient. Because oxygen holds electrons so tightly, the potential energy of electrons in a bond between an oxygen atom and a non-oxygenic atom, such as hydrogen, is low. As a result, there is a large difference between the potential energy of electrons in NADH and the potential energy of electrons bonded to an oxygen atom, such as found in water (see Figure 9.14). The large differential in potential energy means that the electron transport chain can generate a large proton-motive force.

Cells that do not use oxygen as an electron acceptor cannot generate such a large potential energy difference. As a result, they make less ATP from each glucose molecule than cells that use aerobic respiration. This finding is important: It means that anaerobic organisms tend to grow much more slowly than aerobic organisms. If cells that use anaerobic respiration compete with cells using aerobic respiration, those that use oxygen as an electron acceptor almost always grow faster and reproduce more.

What happens when oxygen or other electron acceptors get used up? When there is no terminal electron acceptor, the electrons in

check your understanding

 If you understand that . . .

- As electrons from NADH and $FADH_2$ move through the electron transport chain, protons are pumped into the intermembrane space of mitochondria.
- The electrochemical gradient across the inner mitochondrial membrane drives protons through ATP synthase, resulting in the production of ATP from ADP.

 ✔ **You should be able to . . .**

Add paper squares labeled ETC and ATP synthase and a paper circle labeled $\frac{1}{2} O_2 \rightarrow H_2O$ to the model you made in Section 9.4. Explain the steps in electron transport and chemiosmosis using paper triangles to represent electron pairs and dimes to represent protons.

Answers are available in Appendix A.

each of the complexes of the electron transport chain have no place to go and the electron transport chain stops. Without an oxidized complex I, NADH remains reduced. The concentration of NAD^+ drops rapidly as cells continue to convert NAD^+ to NADH.

This situation is life threatening. When there is no longer any NAD^+ to drive glycolysis, pyruvate processing, and the citric acid cycle, then no ATP can be produced. If NAD^+ cannot be regenerated somehow, the cell will die. How do cells cope?

9.6 Fermentation

Fermentation is a metabolic pathway that regenerates NAD^+ by oxidizing stockpiles of NADH. The electrons removed from NADH are transferred to pyruvate, or a molecule derived from pyruvate, instead of an electron transport chain (**FIGURE 9.20**).

In respiring cells, fermentation serves as an emergency backup that allows glycolysis to continue producing ATP even when the ETC is shut down. It allows the cell to survive and even grow in the absence of electron transport chains.

In many cases, the cell cannot use the molecule that is formed when pyruvate (or another electron acceptor) accepts electrons from NADH. This by-product may even be toxic and excreted from the cell as waste even though it has not been fully oxidized.

Many Different Fermentation Pathways Exist When you run up a long flight of stairs, your muscles begin metabolizing glucose so fast that the supply of oxygen is rapidly used up by their mitochondria. When oxygen is absent, the electron transport chains shut down and NADH cannot donate its electrons there. The pyruvate produced by glycolysis then begins to accept electrons from NADH, and fermentation takes place. This process, called **lactic acid fermentation,** regenerates NAD^+ by forming a product molecule called lactate: a deprotonated form of lactic acid (**FIGURE 9.21a**). Your body reacts by making you breathe faster and increasing your heart rate. By getting more oxygen to your muscle cells, the electron transport chain is revived.

FIGURE 9.21b illustrates a different fermentation pathway, **alcohol fermentation,** which occurs in the fungus *Saccharomyces*

cerevisiae—baker's and brewer's yeast. When yeast cells are placed in an environment such as bread dough or a bottle of grape juice and begin growing, they quickly use up all the available oxygen. Instead of depositing the electrons from NADH into pyruvate, yeast first convert pyruvate to the two-carbon compound acetaldehyde. This reaction gives off carbon dioxide, which causes bread to rise and produces the bubbles in champagne and beer.

Acetaldehyde then accepts electrons from NADH, forming the NAD^+ required to keep glycolysis going. The addition of electrons to acetaldehyde forms ethanol as a waste product. The yeast cells excrete ethanol as waste. In essence, the active ingredient in alcoholic beverages is like yeast urine.

Cells that employ other types of fermentation are used commercially in the production of soy sauce, tofu, yogurt, cheese, vinegar, and other products.

Bacteria and archaea that exist exclusively through fermentation are present in phenomenal numbers in the oxygen-free environment of your small intestine and in the first compartment of a cow's stomach, called the rumen. The rumen is a specialized digestive organ that contains over 10^{10} (10 billion) bacterial and archaeal cells per *milliliter* of fluid. The fermentations that occur in these cells produce an array of fatty acids. Cattle don't actually live off grass directly—they eat it to feed their bacteria and archaea and then use the fermentation by-products from these organisms as a source of energy.

Fermentation as an Alternative to Cellular Respiration Even though fermentation is a widespread type of metabolism, it is extremely inefficient compared with aerobic cellular respiration. Fermentation produces just 2 molecules of ATP for each molecule of glucose metabolized, while cellular respiration produces about 29—almost 15 times more energy per glucose molecule than fermentation. The reason for the disparity is that oxygen has much higher electronegativity than electron acceptors such as pyruvate and acetaldehyde. As a result, the potential energy drop between the start and end of fermentation is a tiny fraction of the potential energy change that occurs during cellular respiration.

Based on these observations, it is not surprising that organisms capable of both processes almost never use fermentation

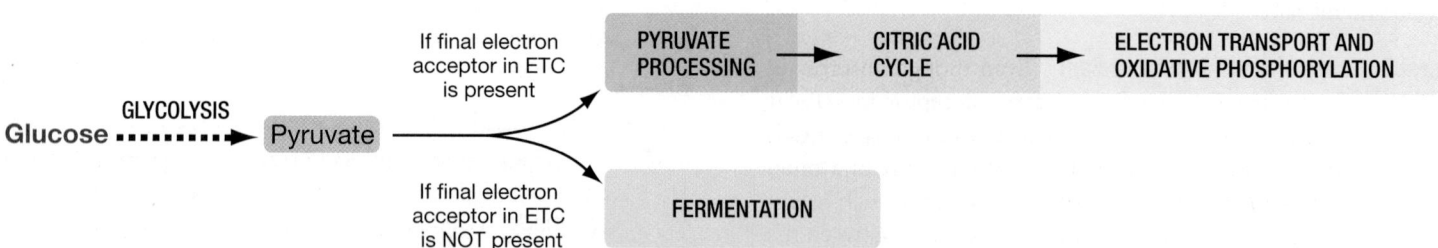

FIGURE 9.20 Cellular Respiration and Fermentation Are Alternative Pathways for Producing Energy. When oxygen or another electron acceptor used by the ETC is present in a cell, the pyruvate produced by glycolysis enters the citric acid cycle and the electron transport system is active. But if no electron acceptor is available to keep the ETC running, the pyruvate undergoes reactions known as fermentation.

(a) Lactic acid fermentation occurs in humans.

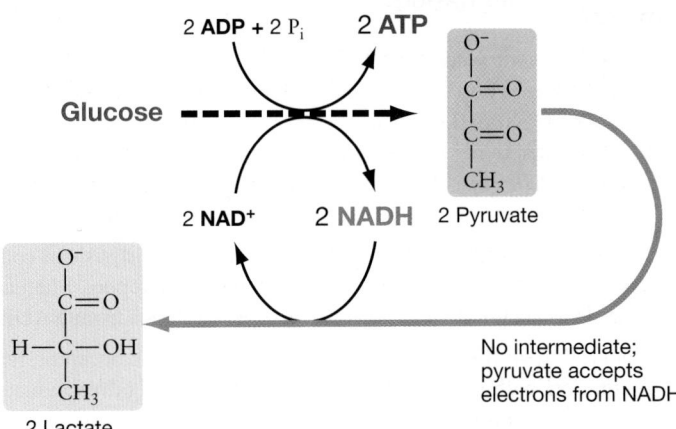

2 Lactate

No intermediate; pyruvate accepts electrons from NADH

(b) Alcohol fermentation occurs in yeast.

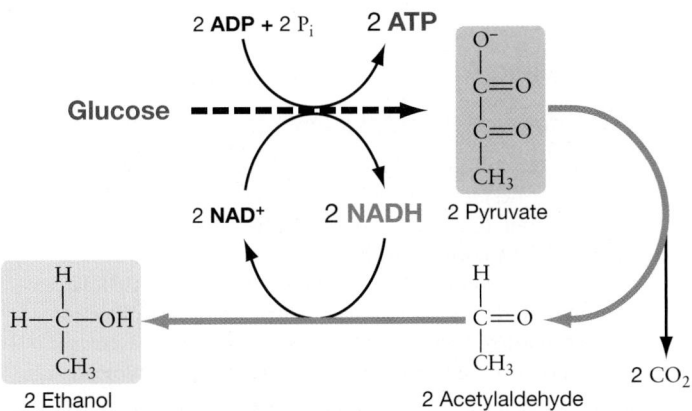

2 Ethanol

2 Acetylaldehyde

2 CO_2

FIGURE 9.21 Fermentation Regenerates NAD^+ So That Glycolysis Can Continue. These are just two examples of the many types of fermentation that occur among bacteria, archaea, and eukaryotes.

when an appropriate electron acceptor is available for cellular respiration. In organisms that usually use oxygen as an electron acceptor, fermentation is an alternative mode of ATP production when oxygen supplies temporarily run out.

Organisms that can switch between fermentation and cellular respiration that uses oxygen as an electron acceptor are called **facultative anaerobes.** The adjective facultative reflects the ability to use cellular respiration when oxygen is present and fermentation when it is absent (anaerobic). Many of your cells can function as facultative anaerobes to a certain extent; however, you cannot survive for long without oxygen. To make this point clear, try holding your breath—it should take only a minute for you to realize how important electron transport is to your cells.

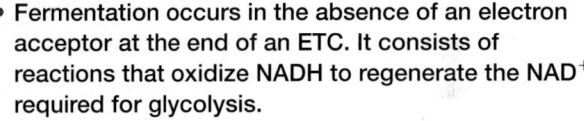

If you understand that . . .

• Fermentation occurs in the absence of an electron acceptor at the end of an ETC. It consists of reactions that oxidize NADH to regenerate the NAD^+ required for glycolysis.

✓ **You should be able to . . .**

Explain why organisms that have an ETC as well as fermentation pathways seldom ferment pyruvate if an electron acceptor at the end of the ETC is readily available.

Answers are available in Appendix A.

CHAPTER 9   REVIEW

For media, go to MasteringBiology (MB)

If you understand . . .

9.1   An Overview of Cellular Respiration

• Cellular respiration is based on redox reactions that transfer electrons from a compound with high free energy, such as glucose, to a molecule with lower free energy, such as oxygen, through an electron transport chain.

• In eukaryotes, cellular respiration consists of four steps: glycolysis, pyruvate processing, the citric acid cycle, and electron transport coupled to oxidative phosphorylation.

• Glycolysis, pyruvate processing, and the citric acid cycle are central to the metabolism of most cells. Other catabolic pathways feed into them, and the intermediates of the central pathways are used in the synthesis of many key molecules.

✓ You should be able to explain why many different molecules—including lipids, amino acids, and CO_2—are radiolabeled when cells are fed glucose with ^{14}C radioactive carbons.

9.2   Glycolysis: Processing Glucose to Pyruvate

• The glycolytic pathway is a 10-step reaction sequence in which glucose is broken down into two molecules of pyruvate. It takes place in the cytosol, where ATP and NADH are produced.

• Glycolysis slows when ATP binds to phosphofructokinase.

✓ **QUANTITATIVE** You should be able to draw a graph predicting how the rate of ATP production in glycolysis changes as a function of ATP concentration. (Write "ATP concentration" on the *x*-axis and "ATP production" on the *y*-axis.)

9.3 Processing Pyruvate to Acetyl CoA

- During pyruvate processing, a series of reactions convert pyruvate to acetyl CoA. NADH is produced and CO_2 is released.

- The pyruvate dehydrogenase complex is inhibited when it is phosphorylated by ATP. It speeds up in the presence of substrates like NAD and ADP.

- ✔ You should be able to explain why it is not surprising that pyruvate dehydrogenase consists of a large, multi-enzyme complex.

9.4 The Citric Acid Cycle: Oxidizing Acetyl CoA to CO_2

- The citric acid cycle is an eight-step reaction cycle that begins with acetyl CoA. $FADH_2$, NADH, and GTP or ATP are produced; CO_2 is released. By the end of the citric acid cycle, glucose is completely oxidized to CO_2.

- Certain enzymes in the citric acid cycle are inhibited when NADH or ATP binds to them.

- ✔ You should be able to describe what would happen to NADH levels in a cell in the first few seconds after a drug has poisoned the enzyme that combines acetyl CoA and oxaloacetate to form citrate.

9.5 Electron Transport and Chemiosmosis: Building a Proton Gradient to Produce ATP

- NADH and $FADH_2$ donate electrons to an electron transport chain that resides in the inner membrane of mitochondria and the plasma membrane of many bacteria. The series of redox reactions in these chains gradually steps the electrons down in potential energy until they are transferred to a final electron acceptor (often O_2).

- The energy released from redox reactions in the electron transport chain is used to move protons across the inner mitochondrial membrane, creating an electrochemical gradient. ATP synthase uses the energy stored in this gradient to produce ATP via chemiosmosis—a process called oxidative phosphorylation.

- ✔ You should be able to predict the effect of a drug that inhibits ATP synthase on the pH in the mitochondrial matrix.

9.6 Fermentation

- In many eukaryotes and bacteria, fermentation occurs when cellular respiration slows down or stops due to an insufficient amount of the final electron acceptor. If the final electron acceptor is absent, then the electron transport chain would no longer oxidize NADH to NAD^+ and ATP could no longer be produced by glycolysis, the citric acid cycle, or oxidative phosphorylation.

- Fermentation pathways regenerate NAD^+, so glycolysis can continue to make ATP and keep the cell alive. This happens when an organic molecule such as pyruvate accepts electrons from NADH.

- Depending on the molecule that acts as an electron acceptor, fermentation pathways produce lactate, ethanol, or other reduced organic compounds as a by-product.

- ✔ You should be able to explain why you would expect organisms that produce ATP only via fermentation to grow much more slowly than organisms that produce ATP via cellular respiration.

(MB) **MasteringBiology**

1. MasteringBiology Assignments

Tutorials and Activities Build a Chemical Cycling System; Cellular Respiration (1 of 5): Inputs and Outputs; Cellular Respiration (2 of 5): Glycolysis; Cellular Respiration (3 of 5): Acetyl CoA Formation and the Citric Acid Cycle; Cellular Respiration (4 of 5): Oxidative Phosphorylation; Cellular Respiration (5 of 5): Summary; Citric Acid Cycle; Electron Transport; Fermentation; Glucose Metabolism; Glycolysis; Overview of Cellular Respiration; Pathways for Pyruvate

Questions Reading Quizzes, Blue-Thread Questions, Test Bank

2. eText Read your book online, search, take notes, highlight text, and more.

3. The Study Area Practice Test, Cumulative Test, BioFlix® 3-D animations, Videos, Activities, Audio Glossary, Word Study Tools, Art

You should be able to . . .

✔ **TEST YOUR KNOWLEDGE**   *Answers are available in Appendix A*

1. Make a flowchart indicating the relationships among the four steps of cellular respiration. Which steps are responsible for glucose oxidation? Which produce the most ATP?

2. Where does the citric acid cycle occur in eukaryotes?
   a. in the cytosol
   b. in the matrix of mitochondria
   c. in the inner membrane of mitochondria
   d. in the intermembrane space of mitochondria

3. What does the chemiosmotic hypothesis claim?
   a. Substrate-level phosphorylation occurs in the electron transport chain.
   b. Substrate-level phosphorylation occurs in glycolysis and the citric acid cycle.
   c. The electron transport chain is located in the inner membrane of mitochondria.
   d. Electron transport chains generate ATP indirectly, by the creation of a proton-motive force.

4. After glucose is fully oxidized by glycolysis, pyruvate processing, and the citric acid cycle, where is most of the energy stored?

5. What is the function of the reactions in a fermentation pathway?
   a. to generate NADH from NAD^+, so electrons can be donated to the electron transport chain
   b. to synthesize pyruvate from lactate
   c. to generate NAD^+ from NADH, so glycolysis can continue
   d. to synthesize electron acceptors, so that cellular respiration can continue

6. Which of the following would cause cells to switch from cellular respiration to fermentation?
   a. The final electron acceptor in the ETC is not available.
   b. The proton-motive force runs down.
   c. NADH and $FADH_2$ supplies are low.
   d. Pyruvate is not available.

✔ **TEST YOUR UNDERSTANDING**   *Answers are available in Appendix A*

7. Describe the relationship between carbohydrate metabolism, the catabolism of proteins and fats, and anabolic pathways.

8. Compare and contrast substrate-level phosphorylation and oxidative phosphorylation.

9. Why does aerobic respiration produce much more ATP than anaerobic respiration?

10. If you were to expose cells that are undergoing cellular respiration to a radioactive oxygen isotope in the form of O_2, which of the following molecules would you expect to be radiolabeled?
    a. pyruvate
    b. water
    c. NADH
    d. CO_2

11. In step 3 of the citric acid cycle, the enzyme isocitrate dehydrogenase is regulated by NADH. Compare and contrast the regulation of this enzyme with what you have learned about phosphofructokinase in glycolysis.

12. Explain the relationship between electron transport and oxidative phosphorylation. What does ATP synthase look like, and how does it work?

✔ **TEST YOUR PROBLEM-SOLVING SKILLS**   *Answers are available in Appendix A*

13. Cyanide ($C\equiv N^-$) blocks complex IV of the electron transport chain. Suggest a hypothesis for what happens to the ETC when complex IV stops working. Your hypothesis should explain why cyanide poisoning in humans is fatal.

14. The presence of many sac-like cristae results in a large amount of membrane inside mitochondria. Suppose that some mitochondria had few cristae. How would their output of ATP compare with that of mitochondria with many cristae? Justify your answer.

15. **QUANTITATIVE** Early estimates suggested that the oxidation of glucose via aerobic respiration would produce 38 ATP. Based on what you know of the theoretical yields of ATP from each step, show how this total was determined. Why do biologists now think this amount of ATP/glucose is not achieved in cells?

16. Suppose a drug were added to mitochondria that allowed protons to freely pass through the inner membrane. Which of the following mitochondrial activities would most likely be inhibited?
    a. the citric acid cycle
    b. oxidative phosphorylation
    c. substrate-level phosphorylation
    d. the electron transport chain

10 Photosynthesis

In this chapter you will learn how

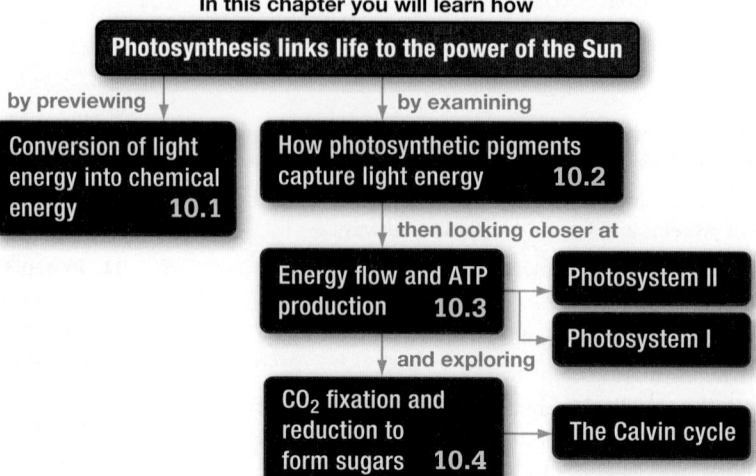

Photosynthesis links life to the power of the Sun

by previewing

Conversion of light energy into chemical energy   10.1

by examining

How photosynthetic pigments capture light energy   10.2

then looking closer at

Energy flow and ATP production   10.3

Photosystem II

Photosystem I

and exploring

CO_2 fixation and reduction to form sugars   10.4

The Calvin cycle

A close-up of moss cells filled with chloroplasts, where photosynthesis converts the energy in sunlight to chemical energy in the bonds of sugar. The sugar produced by photosynthetic organisms fuels cellular respiration and growth. Photosynthetic organisms, in turn, are consumed by other organisms, including you. Directly or indirectly, most organisms on Earth get their energy from photosynthesis.

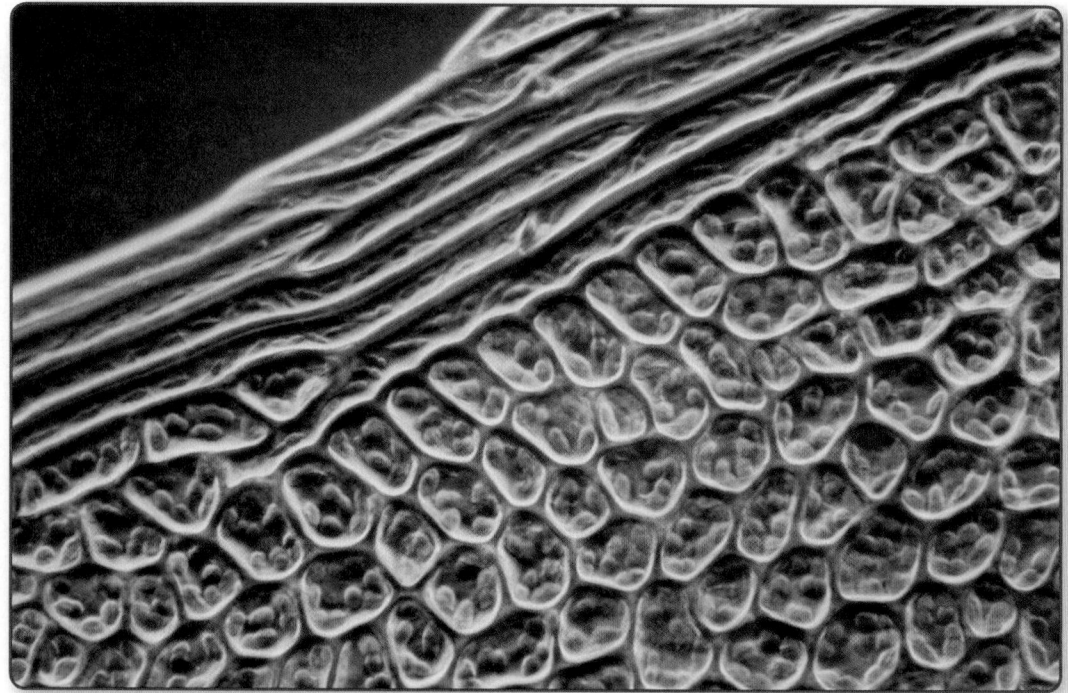

This chapter is part of the Big Picture. See how on pages 198–199.

Some 3 billion years ago, a novel combination of light-absorbing molecules and enzymes gave a bacterial cell the capacity to convert light energy into chemical energy in the C−C and C−H bonds of sugar. The origin of **photosynthesis**—the use of sunlight to manufacture carbohydrate—ranks as one of the great events in the history of life.

The vast majority of organisms alive today rely on photosynthesis, either directly or indirectly, to stay alive. Maples, mosses, and other photosynthetic organisms are termed **autotrophs** (literally, "self-feeders") because they make all their own food from ions and simple molecules. Humans, houseflies, and other non-photosynthetic organisms are called **heterotrophs** ("different-feeders") because they have to obtain the sugars and many of the other macromolecules they need from other organisms.

✔ When you see this checkmark, stop and test yourself. Answers are available in Appendix A.

Because there could be no heterotrophs without autotrophs, photosynthesis is fundamental to almost all life. Glycolysis may qualify as the most ancient set of energy-related chemical reactions from an evolutionary viewpoint, but ecologically—meaning, in terms of how organisms interact with one another—photosynthesis is easily the most important.

How does photosynthesis work? Let's begin with an overview and then delve into a step-by-step analysis of some of the most remarkable chemistry on Earth.

10.1 Photosynthesis Harnesses Sunlight to Make Carbohydrate

Research on photosynthesis began early in the history of biological science. In the 1770s, Joseph Priestley performed a series of experiments showing that the green parts of plants would "restore air" that had been consumed by animals or fire. This work led to the discovery of oxygen (O_2) and the finding that plants produce it in the presence of sunlight, carbon dioxide (CO_2), and water (H_2O).

By the 1840s, enough was known about the process for biologists to propose that photosynthesis allows plants to convert the electromagnetic energy of sunlight into chemical energy in the $C-C$ and $C-H$ bonds of carbohydrates. When glucose is the carbohydrate that is eventually produced, the overall reaction—the sum of many independent reactions—can be simplified and written as

$$6\,CO_2 + 6\,H_2O + light\,energy \longrightarrow C_6H_{12}O_6 + 6\,O_2$$
$$glucose$$

Now read the reaction again, and note the contrast with cellular respiration. Photosynthesis is an endergonic suite of redox reactions that produce sugars from carbon dioxide and light energy. Cellular respiration is an exergonic suite of redox reactions that produces carbon dioxide and ATP from sugars.

FIGURE 10.1 provides an incomplete electron-sharing diagram for the reaction shown above. ✔ If you understand the fundamental principles of reduction–oxidation (see Chapter 8), you should be able to complete Figure 10.1 (following the instructions in the caption exercise) and then use the data from the figure to explain why the reaction is endergonic.

So how does photosynthesis produce O_2 and glucose? Early investigators assumed that CO_2 and H_2O react directly to produce the CH_2O found in carbohydrates and release O_2 as a by-product. This idea, however, turned out to be incorrect. Instead, CO_2 and H_2O participate in entirely different reactions, and the oxygen atoms in O_2 come from water. How was this discovered?

Photosynthesis: Two Linked Sets of Reactions

Starting in the 1930s, two independent lines of research on photosynthesis converged, leading to a major advance in biologists' understanding of how oxygen gas and carbohydrates are produced.

The first research program, led by Cornelius van Niel, focused on photosynthesis in organisms called purple sulfur bacteria. Van Niel and his group found that these cells are autotrophs that manufacture their own carbohydrates from CO_2, sunlight, and hydrogen sulfide (H_2S).

Van Niel also showed that these cells did not produce oxygen as a by-product of photosynthesis. Instead, elemental sulfur (S) accumulated in their medium. In these organisms, the overall reaction for photosynthesis was

$$CO_2 + 2\,H_2S + light\,energy \longrightarrow (CH_2O)_n + H_2O + 2\,S$$

Van Niel's work was crucial for two reasons:

1. It showed that H_2S, the equivalent of H_2O in the plant reactions, and CO_2 do *not* combine directly during photosynthesis.

2. It showed that the oxygen atoms in CO_2 are *not* released as oxygen gas (O_2). The purple sulfur bacteria produced no oxygen, even though carbon dioxide participated in the reaction—just as it did in plants.

Based on these findings, biologists hypothesized that the oxygen atoms released during plant photosynthesis must come from H_2O. The hypothesis was confirmed when heavy isotopes

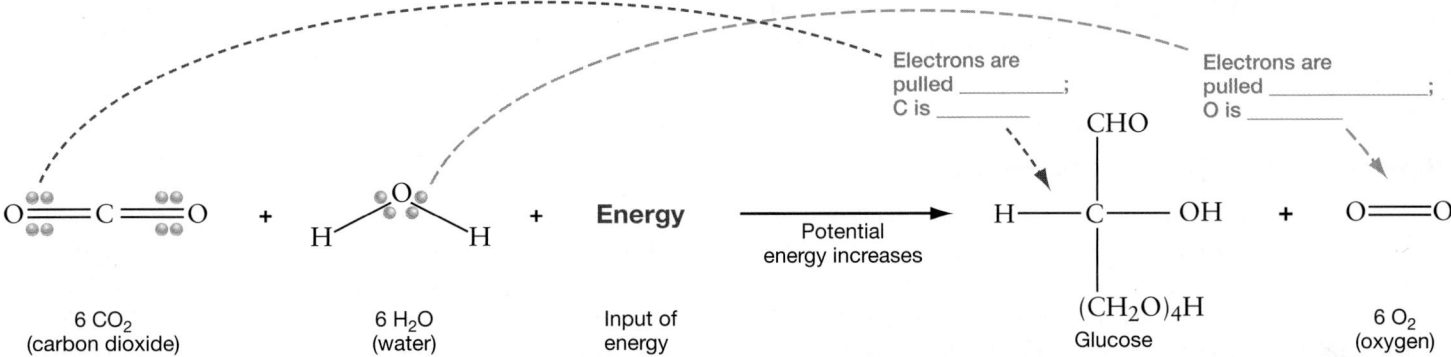

FIGURE 10.1 Electron Transfer during the Reduction of Glucose.

✔ **EXERCISE** Fill in the electron positions for each bond in the reaction products, and complete the labels explaining which product is reduced and which is oxidized.

of oxygen—^{18}O in contrast to the normal isotope, ^{16}O—became available to researchers. They observed the ^{18}O in oxygen gas only when algae or plants were exposed to ^{18}O-labeled H_2O, not the ^{18}O-labeled CO_2.

In addition, the reactions responsible for producing oxygen gas occurred only in the presence of sunlight, but did not require the presence of CO_2. These data suggested that there were two distinct sets of reactions: one that uses light to produce O_2 from H_2O and one that converts CO_2 into sugars.

A second major line of research supported the idea of two sets of reactions. Between 1945 and 1955, a team led by Melvin Calvin began introducing radioactively labeled carbon dioxide ($^{14}CO_2$) to algae and identifying the molecules that subsequently became labeled with the radioisotope. These experiments allowed researchers to identify the sequence of reactions involved in reducing CO_2 to sugars.

Because Calvin played an important role in this research, the reactions that reduce carbon dioxide and produce sugar came to be known as the **Calvin cycle.** Later research showed that the Calvin cycle can function only if the light-capturing reactions are occurring.

To summarize: Early research showed that photosynthesis consists of two linked sets of reactions. One set is triggered by light; the other set—the Calvin cycle—requires the products of the light-capturing reactions. The light-capturing reactions produce oxygen from water; the Calvin cycle produces sugar from carbon dioxide.

The two reactions are linked by electrons that are released when water is split to form oxygen gas. During the light-capturing reactions, these electrons are promoted to a high-energy state by light and then transferred through a series of redox reactions to a phosphorylated version of NAD$^+$, called **NADP$^+$ (nicotinamide adenine dinucleotide phosphate).** This reaction forms **NADPH,** which functions as a reducing agent similar to the NADH produced in cellular respiration. Some of the energy released from these redox reactions is also used to produce ATP (**FIGURE 10.2**).

During the Calvin cycle, the electrons in NADPH and the potential energy in ATP are used to reduce CO_2 to carbohydrate. The resulting sugars are used in cellular respiration to produce ATP for the cell. Plants oxidize sugars in their mitochondria and consume O_2 in the process, just as animals and other eukaryotes do.

Where does all this activity take place?

Photosynthesis Occurs in Chloroplasts

Once experiments had established that photosynthesis takes place only in the green portions of plants, biologists focused on the bright green organelles called **chloroplasts** ("green-formed elements"). One leaf cell typically contains 40 to 50 chloroplasts, and a square millimeter of leaf averages about 500,000 (**FIGURE 10.3**).

When membranes derived from chloroplasts were found to release oxygen after exposure to sunlight, the hypothesis that chloroplasts are the site of photosynthesis became widely accepted.

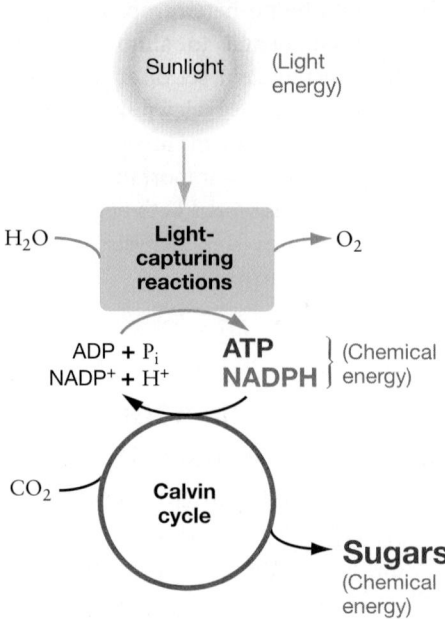

FIGURE 10.2 Photosynthesis Has Two Linked Components. In the light-capturing reactions of photosynthesis, light energy is transformed to chemical energy in the form of ATP and NADPH. During the Calvin cycle, the ATP and NADPH produced in the light-capturing reactions are used to reduce carbon dioxide to carbohydrate.

As Figure 10.3 shows, a chloroplast is enclosed by an outer membrane and an inner membrane (see Chapter 7). The interior is dominated by flattened, sac-like structures called **thylakoids,** which often occur in interconnected stacks called **grana** (singular: **granum**). The space inside a thylakoid is its **lumen.** (Recall that lumen is a general term for the interior of any sac-like structure. Your stomach and intestines have a lumen.) The fluid-filled space between the thylakoids and the inner membrane is the **stroma.**

When researchers analyzed the chemical composition of thylakoid membranes, they found huge quantities of pigments. **Pigments** are molecules that absorb only certain wavelengths of light—other wavelengths are either reflected or transmitted (pass through). Pigments have colors because we see the wavelengths that they do *not* absorb.

The most abundant pigment in the thylakoid membranes turned out to be chlorophyll ("green-leaf"), which reflects or transmits green light. As a result, chlorophyll is responsible for the green color of plants, some algae, and many photosynthetic bacteria.

Before plunging into the details of how photosynthesis occurs, take a moment to consider just how astonishing the process is. Chemists have synthesized an amazing diversity of compounds from relatively simple starting materials, but their achievements pale in comparison to a cell that can synthesize sugar from just carbon dioxide, water, and sunlight. If photosynthesis is not *the* most sophisticated chemistry on Earth, it is certainly a contender.

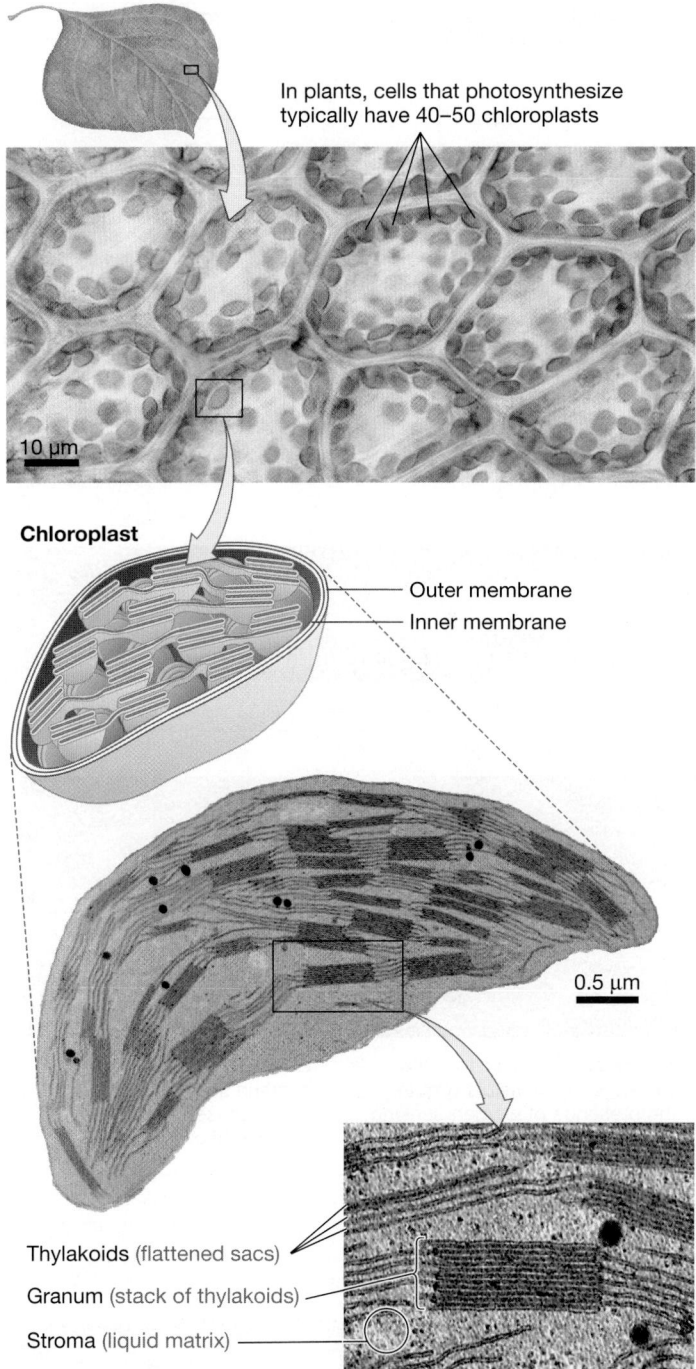

In plants, cells that photosynthesize typically have 40–50 chloroplasts

10 µm

Chloroplast

Outer membrane
Inner membrane

0.5 µm

Thylakoids (flattened sacs)
Granum (stack of thylakoids)
Stroma (liquid matrix)

FIGURE 10.3 Photosynthesis Takes Place in Chloroplasts.

10.2 How Do Pigments Capture Light Energy?

The light-capturing reactions of photosynthesis begin with the simple act of sunlight striking chlorophyll. To understand the consequences of this event, it's helpful to review the nature of light.

Light is a type of electromagnetic radiation, a form of energy. Photosynthesis converts electromagnetic energy in the form of

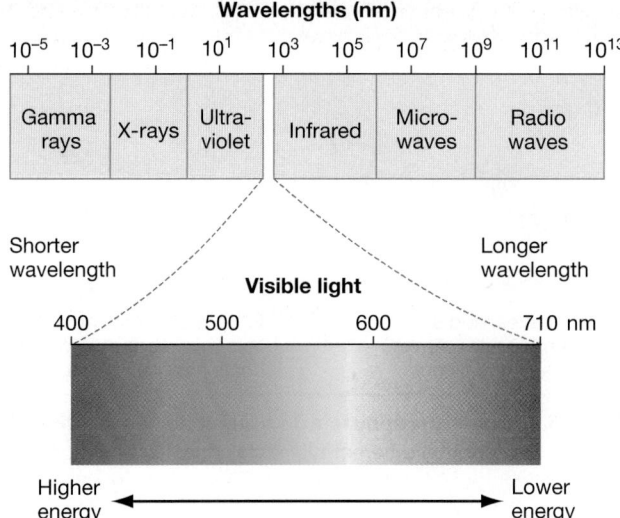

Wavelengths (nm)

10^{-5} 10^{-3} 10^{-1} 10^{1} 10^{3} 10^{5} 10^{7} 10^{9} 10^{11} 10^{13}

| Gamma rays | X-rays | Ultra-violet | Infrared | Micro-waves | Radio waves |

Shorter wavelength

Longer wavelength

Visible light

400   500   600   710 nm

Higher energy

Lower energy

FIGURE 10.4 The Electromagnetic Spectrum. Electromagnetic energy radiates through space in the form of waves. Humans can see radiation at wavelengths between about 400 nanometers (nm) to 710 nm. The shorter the wavelength of electromagnetic radiation, the higher its energy.

sunlight into chemical energy in the C—C and C—H bonds of sugar.

Physicists describe light's behavior as both wavelike and particle-like. Like water waves or airwaves, electromagnetic radiation is characterized by its **wavelength**—the distance between two successive wave crests (or wave troughs). The wavelength determines the type of electromagnetic radiation.

FIGURE 10.4 illustrates the range of wavelengths of electromagnetic radiation—the **electromagnetic spectrum.** The electromagnetic radiation that humans can see, the **visible light,** ranges in wavelength from about 400 to about 710 nanometers (nm, or 10^{-9} m). Shorter wavelengths of electromagnetic radiation contain more energy than longer wavelengths do. Thus, there is more energy in blue light than in red light.

To emphasize the particle-like nature of light, physicists point out that it exists in discrete packets called **photons.** Each photon of light has a characteristic wavelength and energy level. Pigment molecules absorb the energy of some of these photons. How?

Photosynthetic Pigments Absorb Light

When a photon strikes an object, the photon may be absorbed, transmitted, or reflected. A pigment molecule absorbs photons of particular wavelengths. Sunlight includes white light, which consists of all wavelengths in the visible portion of the electromagnetic spectrum at once.

If a pigment absorbs all the visible wavelengths, the pigment appears black because no visible wavelength of light is reflected back to your eye. If a pigment absorbs many or most of the wavelengths in the blue and green parts of the spectrum but transmits or reflects longer wavelengths, it appears red.

What wavelengths do various plant pigments absorb? In one approach to answering this question, researchers grind up leaves

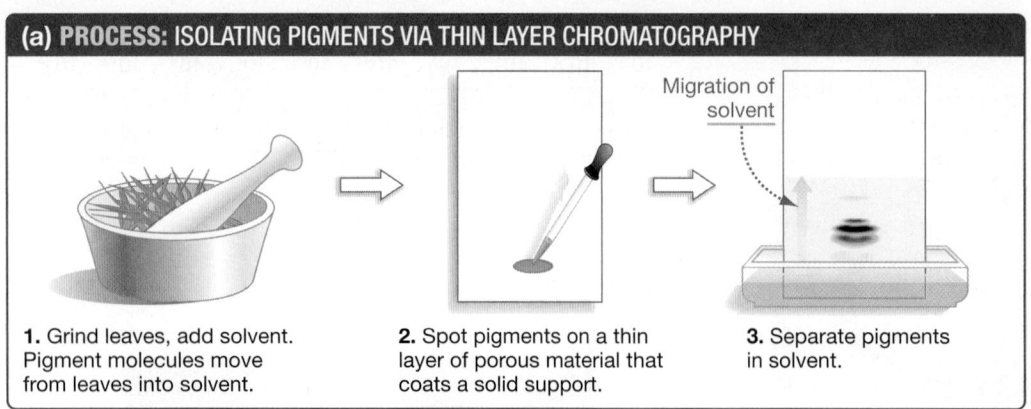

Migration of solvent

1. Grind leaves, add solvent. Pigment molecules move from leaves into solvent.

2. Spot pigments on a thin layer of porous material that coats a solid support.

3. Separate pigments in solvent.

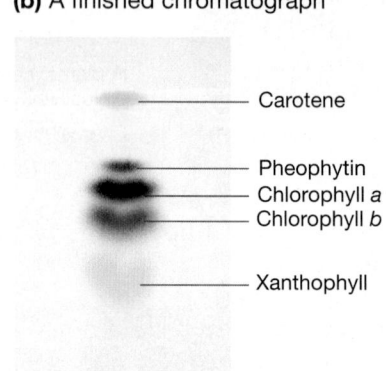

(b) A finished chromatograph

Carotene

Pheophytin
Chlorophyll *a*
Chlorophyll *b*

Xanthophyll

FIGURE 10.5 Chromatography Is a Technique for Separating Molecules. Different species of photosynthetic organisms may contain different types and quantities of pigments. This example shows grass leaves.

in a liquid that acts as a solvent to extract pigment molecules from the leaf mixture. A technique called thin layer chromatography separates the pigments in the extract (**FIGURE 10.5a**).

To begin, spots of raw leaf extract are placed near the bottom of a stiff support that is coated with a thin layer of silica gel, cellulose, or similar porous material. The coated support is then placed in a solvent solution. As the solvent wicks upward through the coating, it carries the pigment molecules in the mixture with it. Because the pigment molecules vary in size, solubility, or both, they are carried at different rates.

FIGURE 10.5b shows a chromatograph from a grass-leaf extract. Notice that this leaf contains an array of pigments. To find out which wavelengths are absorbed by each of these molecules, researchers cut out a single region (color band) of the porous material, extract the pigment, and use an instrument to record the wavelengths absorbed.

Different Pigments Absorb Different Wavelengths of Light   Research based on the techniques shown in Figure 10.5 has confirmed that there are two major pigment classes in plant leaves: chlorophylls and carotenoids.

1. **Chlorophylls,** designated chlorophyll *a* and chlorophyll *b*, absorb strongly in the blue and red regions of the visible spectrum. The presence of chlorophylls makes plants look green because they reflect green light, which they do not absorb.

2. **Carotenoids** absorb in the blue and green parts of the visible spectrum. Thus, carotenoids appear yellow, orange, or red. The carotenoids found in plants belong to two classes, called carotenes and xanthophylls.

FIGURE 10.6 Certain Wavelengths of Light Are Used to Drive Photosynthesis.

SOURCE: Engelmann, T. W. 1882. Oxygen excretion from plant cells in a microspectrum. *Botanische Zeitung* 40: 419–426.

✔**EXERCISE** Draw what you expect the results of this experiment would look like if the pigments that drive photosynthesis in the algae were to absorb most strongly at 500 nm and 560 nm.

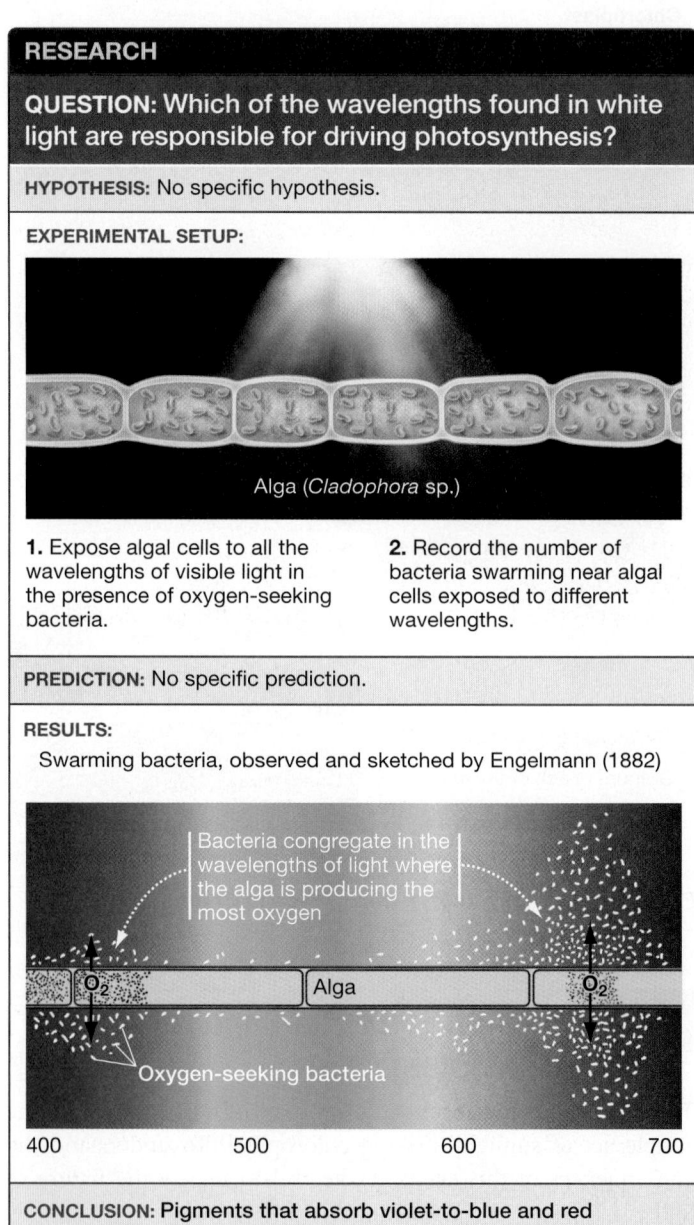

RESEARCH

QUESTION: Which of the wavelengths found in white light are responsible for driving photosynthesis?

HYPOTHESIS: No specific hypothesis.

EXPERIMENTAL SETUP:

Alga (*Cladophora* sp.)

1. Expose algal cells to all the wavelengths of visible light in the presence of oxygen-seeking bacteria.

2. Record the number of bacteria swarming near algal cells exposed to different wavelengths.

PREDICTION: No specific prediction.

RESULTS:

Swarming bacteria, observed and sketched by Engelmann (1882)

Bacteria congregate in the wavelengths of light where the alga is producing the most oxygen

O_2 Alga O_2

Oxygen-seeking bacteria

400   500   600   700

CONCLUSION: Pigments that absorb violet-to-blue and red wavelengths are most effective at triggering photosynthesis.

Which of these wavelengths drive photosynthesis?

In 1882, T. W. Engelmann answered this question by laying a filamentous alga across a microscope slide that was illuminated with a spectrum of colors generated by passing light through a prism to separate the wavelengths (**FIGURE 10.6**). The idea was that the algal cells would begin performing photosynthesis in response to the various wavelengths of light and produce oxygen as a by-product. To determine exactly where oxygen was being produced, Engelmann added bacterial cells from a species that is attracted to oxygen.

As the drawing in the "Results" section of Figure 10.6 shows, most of the bacteria congregated in the violet-to-blue and red regions of the slide. Because wavelengths in these parts of the spectrum were associated with high oxygen concentrations, Engelmann concluded that they defined the **action spectrum** for photosynthesis—the wavelengths that drive the light-capturing reactions. Engelmann's data indicate that violet-to-blue and red photons are the most effective at driving photosynthesis. Because the chlorophylls absorb these wavelengths, this early experiment showed that chlorophylls are the main photosynthetic pigments.

Using thin layer chromatography, and more advanced techniques to evaluate photosynthetic activity, biologists have produced data like those shown in **FIGURE 10.7**. This graph shows the action spectrum and the absorption spectra for three different pigments found in chloroplasts. An **absorption spectrum** measures how the wavelength of photons influences the amount of light absorbed by a pigment. In the combined graph, peaks indicate wavelengths where absorbance or photosynthetic activity is high; troughs indicate wavelengths where absorbance or photosynthetic activity is low.

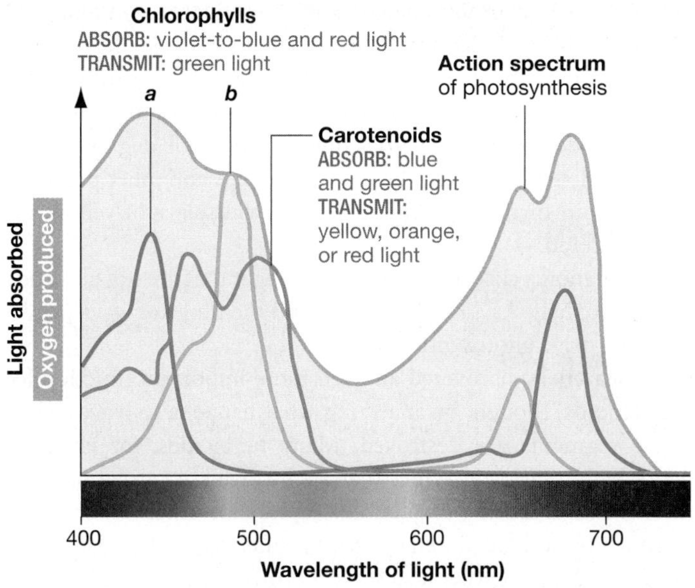

Which Part of a Pigment Absorbs Light?

As **FIGURE 10.8a** shows, chlorophyll *a* and chlorophyll *b* are similar in structure. Both have two fundamental parts: a long isoprenoid "tail" (introduced in Chapter 6) and a "head" consisting of a large ring structure

Chlorophylls
ABSORB: violet-to-blue and red light
TRANSMIT: green light

Action spectrum of photosynthesis

Carotenoids
ABSORB: blue and green light
TRANSMIT: yellow, orange, or red light

Light absorbed
Oxygen produced

400   500   600   700
Wavelength of light (nm)

FIGURE 10.7 There Is a Strong Correlation between the Absorption Spectra of Pigments and the Action Spectrum for Photosynthesis.

DATA: Singhal, G. S., et al. 1999. *Concepts in Photobiology: Photosynthesis and Photomorphogenesis*. Dordrecht: Kluwer Academic; co-published with Narosa Publishing House (New Delhi), 11–51.

with a magnesium atom in the middle. The tail interacts with proteins embedded in the thylakoid membrane; the head is where light is absorbed.

The structure of β-carotene, shown in **FIGURE 10.8b**, has an isoprenoid chain connecting two rings that are responsible for absorbing light. This pigment is what gives carrots their orange color. A xanthophyll called zeaxanthin, which gives corn kernels their bright yellow color, is nearly identical to β-carotene, except that the ring structures on either end of the molecule contain a hydroxyl (−OH) group.

Researchers had shown that chlorophylls are the main photosynthetic pigments, but carotenoids also absorb light. What do they do? Before analyzing what happens when chlorophyll pigments absorb light, let's first look at the function of the carotenoids.

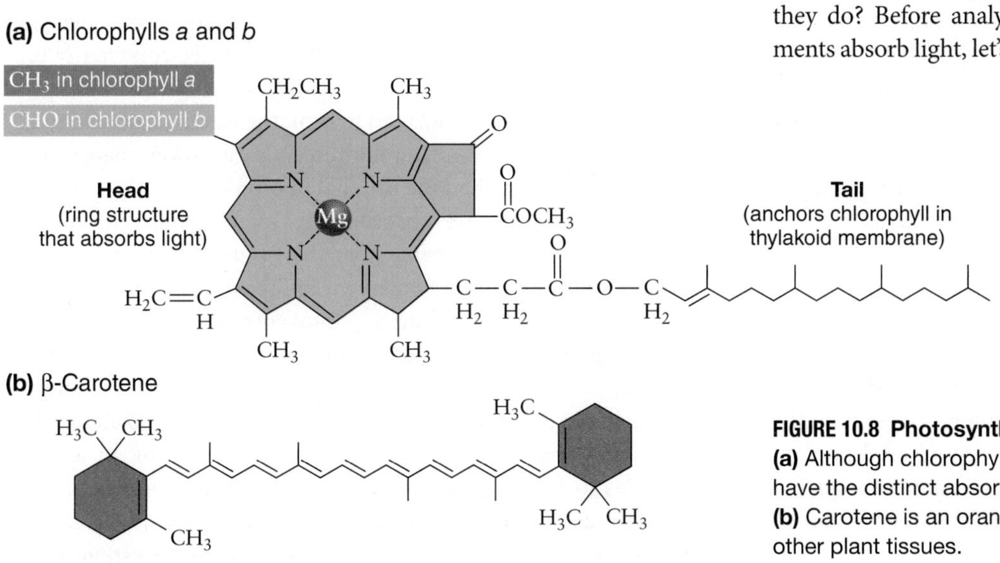

(a) Chlorophylls *a* and *b*

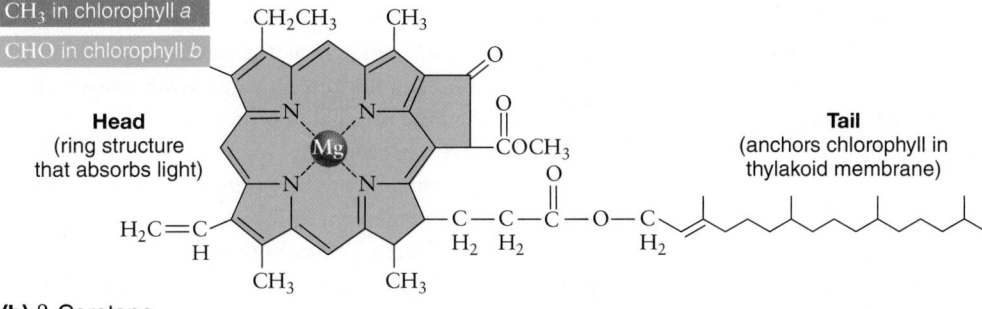

CH₃ in chlorophyll *a*
CHO in chlorophyll *b*

Head (ring structure that absorbs light)

Tail (anchors chlorophyll in thylakoid membrane)

(b) β-Carotene

FIGURE 10.8 Photosynthetic Pigments Contain Ring Structures.
(a) Although chlorophylls *a* and *b* are very similar structurally, they have the distinct absorption spectra shown in Figure 10.7.
(b) Carotene is an orange pigment found in carrot roots and other plant tissues.

What Is the Role of Carotenoids and Other Accessory Pigments?

Carotenoids are called accessory pigments because they absorb light and pass the energy on to chlorophyll. Both xanthophylls and carotenes are found in chloroplasts. In autumn, when the leaves of deciduous trees die, their chlorophyll degrades first. The wavelengths reflected by the carotenoids and other pigments that remain turn forests into spectacular displays of yellow, orange, and red.

Carotenoids absorb wavelengths of light that are not absorbed by chlorophyll. As a result, they extend the range of wavelengths that can drive photosynthesis.

Researchers discovered an even more important function for carotenoids, though, by analyzing what happens to leaves when these pigments are destroyed. Many herbicides, for example, work by inhibiting enzymes that are involved in carotenoid synthesis. Plants lacking carotenoids rapidly lose their chlorophyll, turn white, and die. Based on these results, researchers have concluded that carotenoids also serve a protective function.

To understand why carotenoids are protective, recall that photons—especially the high-energy, short-wavelength photons in the ultraviolet part of the electromagnetic spectrum—contain enough energy to knock electrons out of atoms and create free radicals (see Chapter 2). Free radicals, in turn, trigger reactions that can disrupt and degrade molecules.

Carotenoids "quench" free radicals by accepting or stabilizing unpaired electrons. As a result, they protect chlorophyll molecules from harm. When carotenoids are absent, chlorophyll molecules are destroyed and photosynthesis stops. Starvation and death follow.

When Light Is Absorbed, Electrons Enter an Excited State

Just what is absorption? What happens when a photon of a particular wavelength—say, red light with a wavelength of 680 nm—strikes a chlorophyll molecule?

When a photon strikes a chlorophyll molecule, the photon's energy can be transferred to an electron in the chlorophyll molecule's head region. In response, the electron is "excited," or raised to a higher energy state.

The excited electron states that are possible in a particular pigment are discrete—meaning, incremental rather than continuous—and can be represented as lines on an energy scale. These discrete energy levels are a property of the electron configurations in a particular pigment.

FIGURE 10.9 shows the ground state, or unexcited state, as 0 and the higher energy states as 1 and 2. If the difference between the possible energy states is the same as the energy in the photon, the photon can be absorbed and an electron excited to a higher energy state.

In chlorophyll, for example, the energy difference between the ground state and state 1 is equal to the energy in a red photon, while the energy difference between state 0 and state 2 is equal to the energy in a blue photon. Thus, chlorophyll can readily absorb red photons and blue photons.

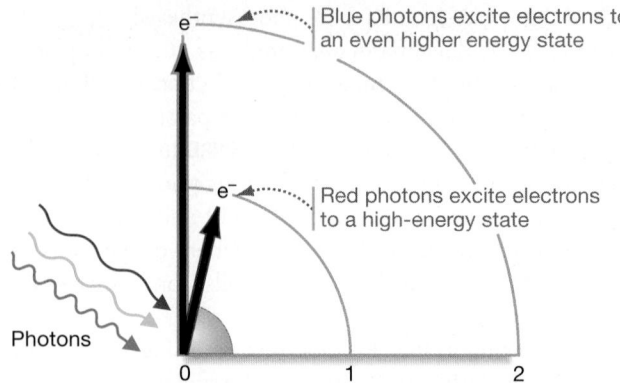

FIGURE 10.9 Electrons Are Promoted to High-Energy States When Photons Strike Chlorophyll. The unexcited, or ground state, is labeled 0, and the discrete energy states are labeled 1 and 2. The wavelength of light that will excite electrons to these energy states is a property of chlorophyll's structure.

✔**QUESTION** Suppose a pigment had a discrete energy state that corresponded to the energy in green light. Where would you draw this energy state on this diagram?

Chlorophyll does not absorb green light well, because there is no discrete step—no difference in possible energy states for its electrons—that corresponds to the amount of energy in a green photon.

Wavelengths in the ultraviolet part of the spectrum have so much energy that they may actually eject electrons from a pigment molecule and create a free radical. In contrast, wavelengths in the infrared regions have so little energy that in most cases they merely increase the movement of atoms in the pigment, generating heat—meaning molecular movement—rather than exciting electrons.

But if a pigment absorbs a photon with the right amount of energy, energy in the form of electromagnetic radiation is transferred to that electron. The electron now has high potential energy. What happens next?

If the excited electron simply falls back to its ground state, the absorbed energy is released as heat or a combination of heat and electromagnetic radiation (light). When the electron energy produces light, it is called **fluorescence.** Because some of the original photon's energy is transformed to heat, the electromagnetic radiation that is given off during fluorescence has lower energy and a longer wavelength than the original photon did.

When photons are absorbed by pigments in chloroplasts though, only about 2 percent of the excited electrons produce fluorescence. The other 98 percent of the energized pigments use their excited electrons to drive photosynthesis.

To understand what happens to these excited electrons, it's important to recognize that chlorophyll molecules work in groups—not individually. In the thylakoid membrane, 200–300 chlorophyll molecules and accessory pigments are organized by an array of proteins to form structures called the **antenna complex** and the reaction center. These complexes, along with the molecules that capture and process excited electrons, form a **photosystem.**

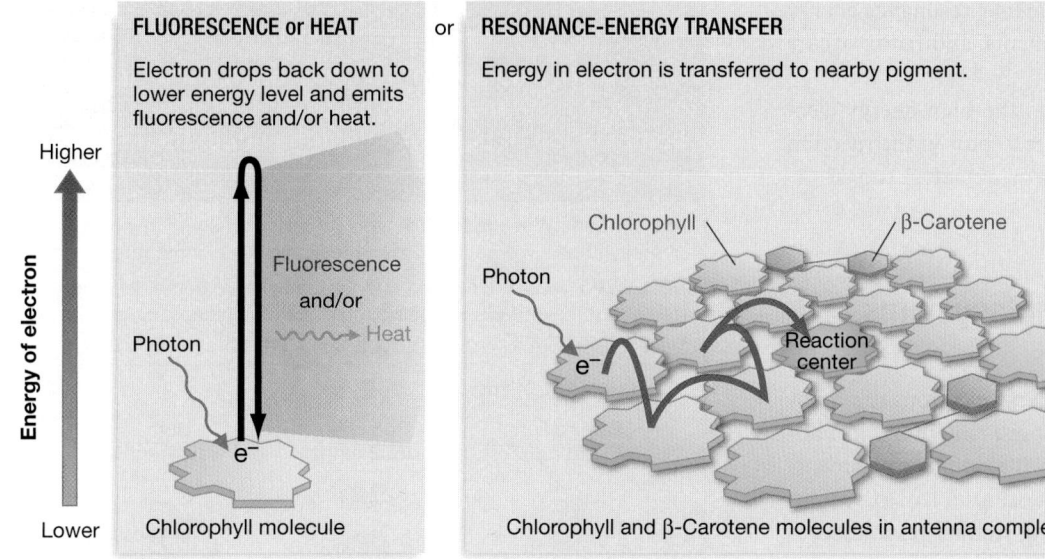

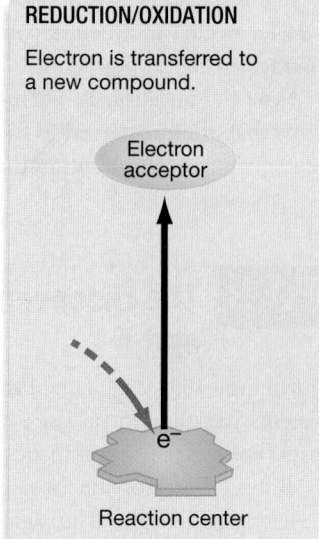

FLUORESCENCE or HEAT

Electron drops back down to lower energy level and emits fluorescence and/or heat.

Higher

Energy of electron

Lower

Photon

e⁻

Fluorescence and/or

Heat

Chlorophyll molecule

or

RESONANCE-ENERGY TRANSFER

Energy in electron is transferred to nearby pigment.

Chlorophyll

β-Carotene

Photon

e⁻

Reaction center

Chlorophyll and β-Carotene molecules in antenna complex

or

REDUCTION/OXIDATION

Electron is transferred to a new compound.

Electron acceptor

e⁻

Reaction center

FIGURE 10.10 Four Fates for Excited Electrons in Photosynthetic Pigments. When sunlight promotes electrons in pigments to a high-energy state, four things can happen: They can fluoresce, release heat, pass energy to a nearby pigment via resonance, or transfer the electron to an electron acceptor.

The Antenna Complex When a red or blue photon strikes a pigment molecule in the antenna complex, the energy is absorbed and an electron is excited in response. This energy—but not the electron itself—is passed to a nearby chlorophyll molecule, where another electron is excited in response. This phenomenon is known as resonance energy transfer.

Resonance energy transfer is possible only between pigments that are able to absorb different wavelengths of photons—from those absorbing higher-energy photons to those absorbing lower-energy photons. The organization of the antenna complex makes it possible for this resonance energy to be efficiently moved between pigments, as the potential energy drops at each step.

Once the energy is transferred, the original excited electron falls back to its ground state. In this way, energy is transferred inside the antenna complex in a manner that may be likened to the transfer of excitement between fans at a sports event during the "wave." But unlike the stadium wave, most of this resonance energy is directed to a particular location in a photosystem, called the reaction center.

The Reaction Center When a chlorophyll molecule is excited in the **reaction center,** its excited electron is transferred to an electron acceptor. When the acceptor becomes reduced, the energy transformation event that started with the absorption of light becomes permanent: Electromagnetic energy is transformed to chemical energy. The redox reaction that occurs in the reaction center results in the production of chemical energy from sunlight.

Note that in the absence of light, the electron acceptor does not accept electrons. It remains in an oxidized state because the redox reaction that transfers an electron to the electron acceptor is endergonic. But when light excites electrons in chlorophyll to a high-energy state, the reaction becomes exergonic. In this way, the energy in light transforms an endergonic reaction to an exergonic one.

FIGURE 10.10 summarizes the four possible fates of electrons in chlorophyll that are excited by photons. The energy released from these electrons can

1. be emitted in the form of light via fluorescence, or

2. be given off as heat alone, or

3. excite an electron in a nearby pigment and induce resonance, or

4. be transferred to an electron acceptor in a redox reaction.

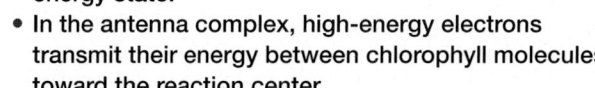

check your understanding

C Y U

If you understand that . . .

- Pigments absorb specific wavelengths of light.
- When a chlorophyll molecule in the antenna complex of a chloroplast membrane absorbs red or blue light, one of its electrons is promoted to a high-energy state.
- In the antenna complex, high-energy electrons transmit their energy between chlorophyll molecules toward the reaction center.
- When energy is transferred to a chlorophyll molecule in the reaction center, the excited electron reduces an electron acceptor. In this way, light energy is transformed to chemical energy.

✔ **You should be able to . . .**

Predict how the pigments of the antenna complex would be organized, with regard to the wavelength of photons absorbed, to allow the directional transport of energy from the outer pigments to the reaction center.

Answers are available in Appendix A.

Fluorescence is typical of isolated pigments, resonance energy transfer occurs in antenna complex pigments, and redox occurs in reaction center pigments.

Now the question is, what happens to the high-energy electrons that are transferred to the electron acceptor in the reaction center? Specifically, how are they used to manufacture sugar?

10.3 The Discovery of Photosystems I and II

During the 1950s, the fate of the high-energy electrons in photosystems was the central issue facing biologists interested in photosynthesis. A key breakthrough began with simple experiments by Robert Emerson on how green algae responded to various wavelengths of light. The algal cells being studied responded to wavelengths in the red and far-red regions of the visible spectrum.

Emerson found that if the algal cells were illuminated with either red or far-red wavelengths of light, the photosynthetic response was moderate. But if cells were exposed to a combination of both wavelengths, the rate of photosynthesis increased more than the sum of the rates produced by each wavelength independently. This phenomenon was called the enhancement effect, and is not limited to algal cells. In follow-up work by other researchers, it was also observed in isolated chloroplasts from plants (**FIGURE 10.11**). Why the enhancement effect occurred was a complete mystery at the time.

A solution to this puzzle was proposed by Robin Hill and Faye Bendall, who hypothesized that this enhancement effect resulted from two distinct types of reaction centers, each absorbing different wavelengths of light. According to the two-photosystem hypothesis, the enhancement effect occurs because photosynthesis is much more efficient when both photosystems operate together.

Subsequent work has shown that the two-photosystem hypothesis is correct for cyanobacteria ("blue-green bacteria") and the chloroplasts of eukaryotes, such as algae and plants. These two photosystems differ in structure and function, but work together in the light-capturing reactions.

To figure out how the two photosystems work, investigators focused on species of photosynthetic bacteria that possess one or the other of the two photosystems, but not both. Once each type of photosystem was understood in isolation, researchers explored how they work in combination. Let's do the same—first let's analyze **photosystem II,** then **photosystem I** (so named because it was discovered first), and then how the two interact.

How Does Photosystem II Work?

To study photosystem II, researchers focused on purple photosynthetic bacteria, including the purple sulfur bacteria that were studied by van Niel (see Section 10.1). These cells have a single photosystem that has many of the same components observed in photosystem II of cyanobacteria and the chloroplasts of algae and plants. (For simplicity, the eukaryotic chloroplast will serve as the model system for the remainder of the chapter.)

RESEARCH

QUESTION: Red and far-red light each stimulate a moderate rate of photosynthesis. How does a combination of both wavelengths affect the rate of photosynthesis?

HYPOTHESIS: When red and far-red light are combined, the rate of photosynthesis will be the sum of the single wavelength rates.

NULL HYPOTHESIS: When red and far-red light are combined, the rate of photosynthesis will be no more than the highest single wavelength rate.

EXPERIMENTAL SETUP:

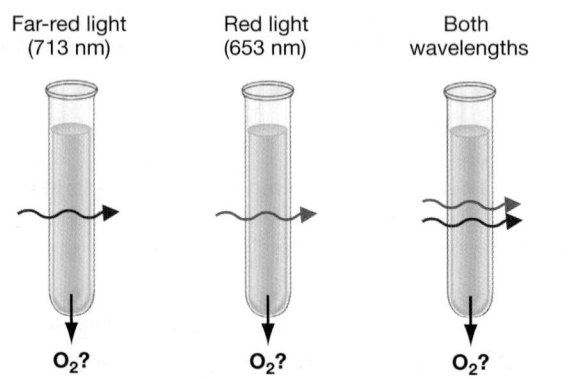

1. Expose algal cells to far-red light and then red light. Record oxygen produced as a measure of rate of photosynthesis.

2. Expose same cells to a combination of both lights.

PREDICTION: When the two wavelengths are combined, the amount of oxygen produced will be the sum of the single wavelength tests.

PREDICTION OF NULL HYPOTHESIS: When the two wavelengths are combined, the amount of oxygen produced will be no more than the single wavelength test that yielded the highest amount of oxygen.

RESULTS:

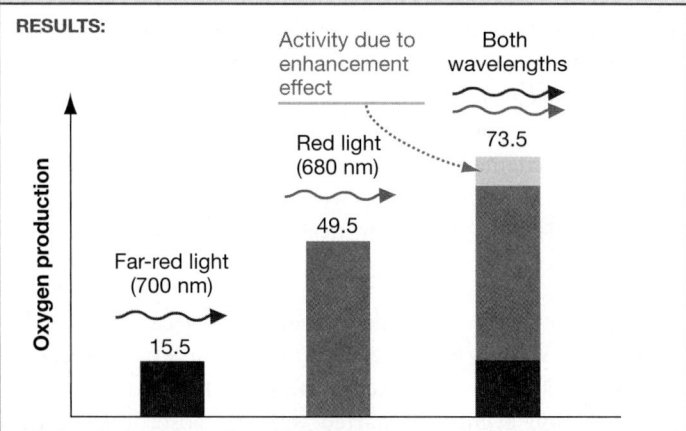

CONCLUSION: Neither hypothesis is correct. The combination of both wavelengths yielded more oxygen than the sum of the single tests. A new hypothesis is required to explain this enhancement effect.

FIGURE 10.11 The "Enhancement Effect" of Two Different Wavelengths in Isolated Chloroplasts.

SOURCE: Govindjee, R., Govindjee, and G. Hoch. 1964. Emerson enhancement effect in chloroplast reactions. *Plant Physiology* 39: 10–14.

✔**QUESTION** Was it important for the researchers to keep the density of chloroplasts fairly constant in each treatment? Explain why or why not.

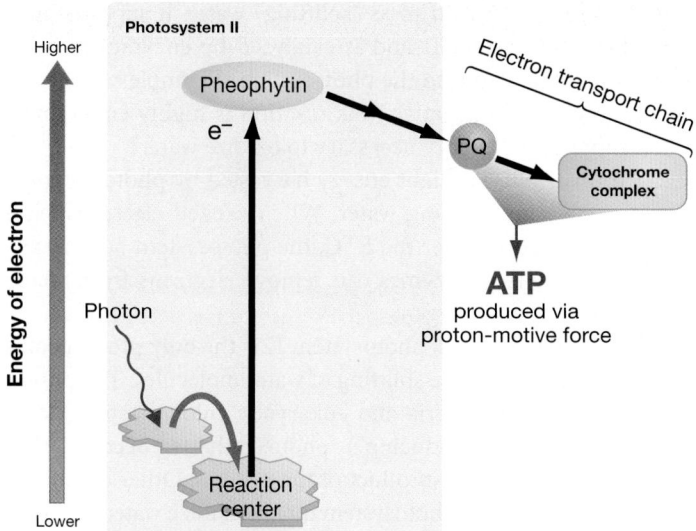

FIGURE 10.12 Photosystem II Feeds High-Energy Electrons to an Electron Transport Chain. When an excited electron leaves the chlorophyll in the reaction center of photosystem II, the electron is accepted by pheophytin, transferred to plastoquinone (PQ), and then stepped down in energy along an electron transport chain.

Converting Light Energy into Chemical Energy In photosystem II, the action begins when the antenna complex transmits resonance energy to the reaction center, where the electron acceptor pheophytin comes into play (**FIGURE 10.12**). Structurally,

pheophytin is identical to chlorophyll except that pheophytin lacks a magnesium atom in its head region. Functionally, the two molecules are extremely different.

Instead of acting as a pigment that energizes an electron when it absorbs a photon, pheophytin accepts high-energy electrons from the excited reaction center chlorophylls. The reduction of pheophytin (and the accompanying oxidation of the reaction center chlorophyll pigment) is a key step in the transformation of light energy into chemical energy.

Electrons that reduce pheophytin are passed through additional carriers to an electron transport chain (ETC) in the thylakoid membrane. In both structure and function, this ETC is similar to components in the mitochondrial ETC (see Chapter 9).

- Structurally, the ETC associated with photosystem II and the ETC in the mitochondrion both contain quinones and cytochromes.

- Functionally, the redox reactions that occur in both ETCs result in protons being actively transported from one side of an internal membrane to the other. The resulting proton electrochemical gradient forms a proton-motive force that drives ATP production via ATP synthase. Photosystem II triggers chemiosmosis and ATP synthesis in the chloroplast.

FIGURE 10.13 explains how the electron transport chain associated with photosystem II works in more detail. Start by focusing on the molecule called **plastoquinone (PQ)**—a quinone similar to ubiquinone in the ETC of cellular respiration. Recall that

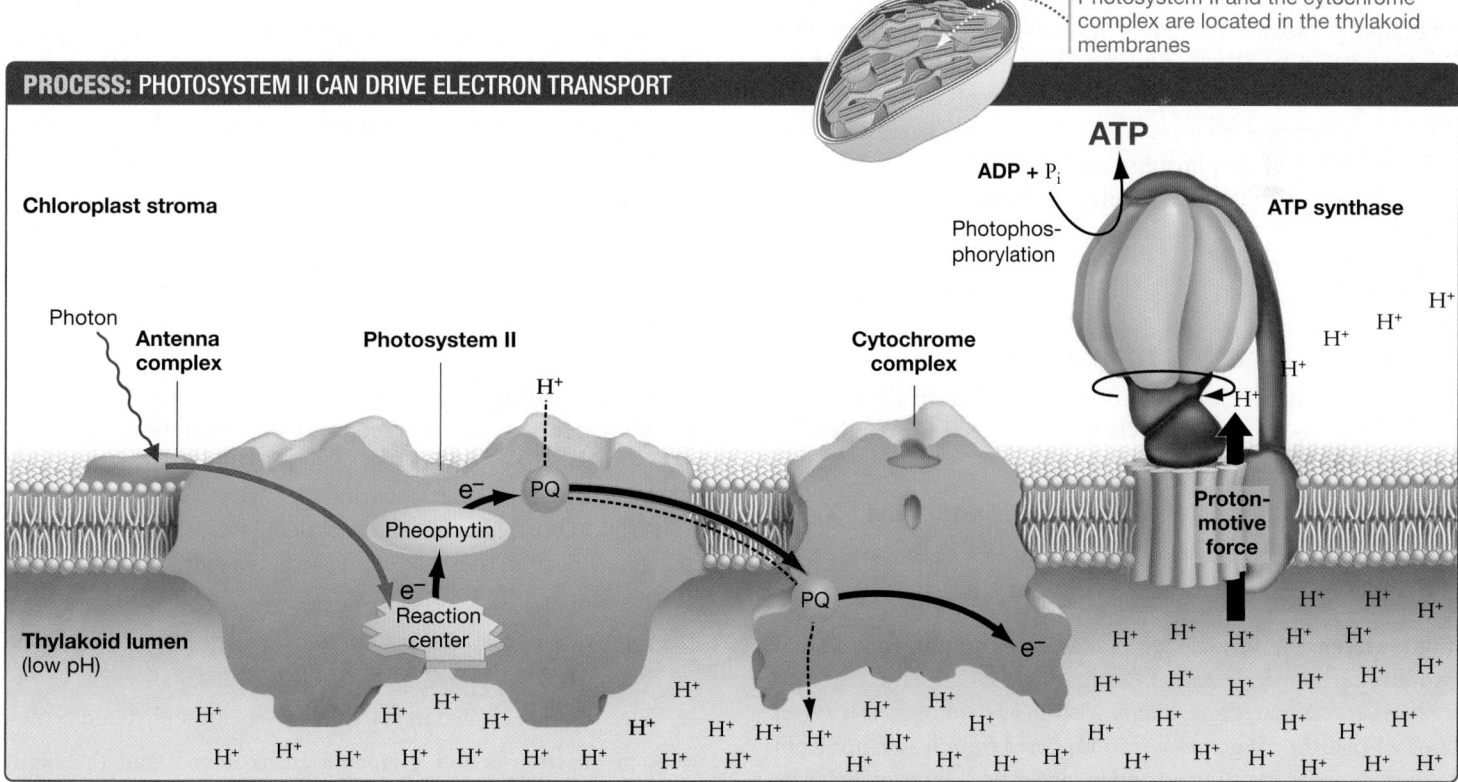

FIGURE 10.13 Electron Transport between Photosystem II and the Cytochrome Complex. Plastoquinone (PQ) carries electrons from photosystem II along with protons from the stroma. The cytochrome complex oxidizes plastoquinone, releasing the protons in the thylakoid lumen that drive ATP synthesis.

quinones are small hydrophobic molecules that can transport electrons between molecules (see Chapter 9). Because plastoquinone is lipid soluble and not anchored to the thylakoid membrane, it is free to move from one side of the thylakoid membrane to the other.

When plastoquinone receives electrons from photosystem II, it carries them across the membrane to the lumen side of the thylakoid and delivers them to more electronegative molecules in the cytochrome complex. In this way, plastoquinone shuttles electrons from photosystem II to the cytochrome complex much like ubiquinone shuttled electrons between complexes I or II and complex III in mitochondria. The potential energy released by these reactions allows protons to be picked up from the stroma and dropped off in the lumen side of the thylakoid membrane.

The protons transported by plastoquinone result in a large concentration of protons in the thylakoid lumen. When photosystem II is active, the pH of the thylakoid interior reaches 5 while the pH of the stroma hovers around 8. Because the pH scale is logarithmic, the difference of 3 units means that the concentration of H^+ is $10 \times 10 \times 10 = 1000$ times higher in the lumen than in the stroma. In addition, the stroma becomes negatively charged relative to the thylakoid lumen.

The net effect of electron transport, then, is to set up a large proton electrochemical gradient, resulting in a proton-motive force that drives H^+ out of the thylakoid lumen and into the stroma. Based on what you know of cellular respiration, it should come as no surprise that this proton-motive force drives the production of ATP.

Specifically, proton flow down the electrochemical gradient is an exergonic process that drives the endergonic synthesis of ATP from ADP and P_i. The stream of protons through ATP synthase causes conformational changes that drive the phosphorylation of ADP. Since this process is initiated by the energy harvested from light, it is called **photophosphorylation.**

Photophosphorylation is similar to the oxidative phosphorylation that occurs in plant and animal mitochondria. Both depend on chemiosmosis.

The photosystem II story is not yet complete, however. The electrons from PQ are passed through the cytochrome complex, but what about the oxidized photosystem II reaction center? To continue this ETC, the electron removed from the reaction center needs to be replaced. Where do the electrons required by photosystem II come from?

Photosystem II Obtains Electrons by Oxidizing Water Think back to the overall reaction for photosynthesis:

$$6 CO_2 + 6 H_2O + \text{light energy} \longrightarrow C_6H_{12}O_6 + 6 O_2$$

In the presence of sunlight, carbon dioxide and water are used to produce carbohydrate and oxygen gas.

Now recall that experiments with heavy isotopes of oxygen showed that the oxygen atoms in O_2 come from water, not from carbon dioxide. For this to happen, water must be oxidized. The oxygen-generating reaction can be written as

$$2 H_2O \longrightarrow 4 H^+ + 4 e^- + O_2$$

This reaction is referred to as "splitting" water. It supplies electrons for photosystem II and is catalyzed by enzymes that are physically integrated into the photosystem II complex. Since oxygen is very electronegative, this reaction is highly endergonic. What supplies the energy necessary to oxidize water?

As it turns out, the light energy harvested by photosystem II is responsible for splitting water. When excited electrons leave photosystem II and enter the ETC, the photosystem becomes so electronegative that enzymes can remove electrons from water, leaving protons and oxygen.

Among all life-forms, photosystem II is the only protein complex that can catalyze the splitting of water molecules. The photosystem II of cyanobacteria and eukaryotic chloroplasts perform **oxygenic** ("oxygen-producing") photosynthesis, because they generate oxygen as a by-product of the process. Other organisms that have only a single photosystem do not oxidize water, and thus do not produce O_2 gas. Instead, these organisms use different electron donors, such as H_2S in the purple sulfur bacteria, to perform **anoxygenic** ("no oxygen-producing") photosynthesis.

✔ If you understand photosystem II, you should be able to make an energy flowchart that includes the antenna complex, ATP synthase, pheophytin, light, the proton gradient, an ETC, and a reaction center and then add notes explaining where the enzyme complex that splits water fits in.

What happens next in green algae and land plants? The answer lies in photosystem I. Let's take a closer look.

How Does Photosystem I Work?

Recall that researchers dissected photosystem II by studying similar, but simpler, photosystems in the purple photosynthetic bacteria. To understand the structure and function of photosystem I, they turned to heliobacteria ("sun-bacteria").

Like the purple bacteria, heliobacteria have only one photosystem that uses the energy in sunlight to promote electrons to a high-energy state. But instead of being passed to an electron transport chain that pumps protons across a membrane, the high-energy electrons in heliobacteria are used to reduce NAD^+. When NAD^+ gains two electrons and a proton, NADH is produced.

In the cyanobacteria and eukaryotic chloroplasts, a similar set of light-capturing reactions reduces a phosphorylated version of NAD^+, symbolized $NADP^+$, to yield NADPH. Both NADH and NADPH function as electron carriers.

FIGURE 10.14 explains how photosystem I works in chloroplasts—put your finger on the "2 photons" arrows and trace the steps that follow.

1. Pigments in the antenna complex absorb photons and pass the energy to the photosystem I reaction center.

2. Electrons are excited in reaction center chlorophyll molecules.

3. The reaction center pigments are oxidized, and the high-energy electrons are passed through a series of carriers inside the photosystem, then to a molecule called **ferredoxin,** and then to the enzyme called $NADP^+$ reductase.

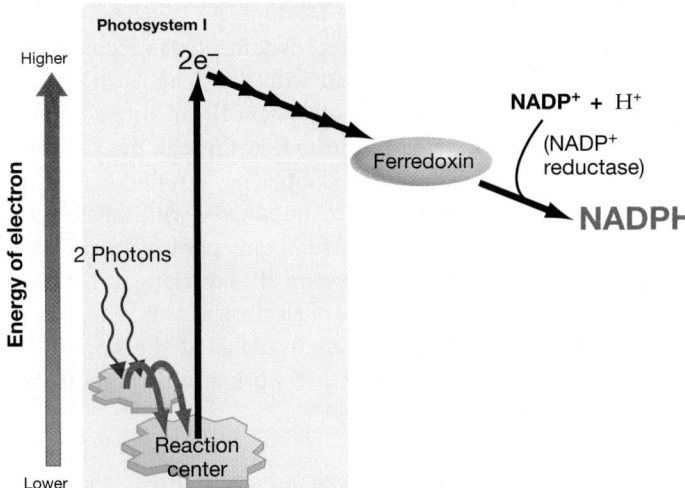

FIGURE 10.14 Photosystem I Produces NADPH. When excited electrons leave the chlorophyll molecule in the reaction center of photosystem I, they pass through a series of iron- and sulfur-containing proteins until they are accepted by ferredoxin. In an enzyme-catalyzed reaction, the reduced form of ferredoxin reacts with NADP$^+$ to produce NADPH.

4. NADP$^+$ reductase transfers two electrons and a proton to NADP$^+$. This reaction forms NADPH.

Photosystem I and NADP$^+$ reductase are anchored in the thylakoid membrane; ferredoxin is in the stroma, but it is closely associated with the thylakoid membrane.

To summarize: Electrons from photosystem I are used to produce NADPH, which is a reducing agent similar in function to the NADH and FADH$_2$ produced by the citric acid cycle. Electrons from photosystem II, in contrast, are used to produce a proton-motive force that drives the synthesis of ATP.

In combination, then, photosystems II and I produce chemical energy stored in ATP and NADPH. But there are still gaps in the flow of electrons through these two photosystems. Where do the electrons from photosystem II end up? How does the oxidized reaction center of photosystem I obtain electrons so NADPH will continue to be made?

The Z Scheme: Photosystems II and I Work Together

FIGURE 10.15 illustrates the **Z-scheme** model for how photosystems II and I interact. The name was inspired by the changes occurring in electron potential energy as plotted on a vertical axis, which takes on the shape of a Z that has fallen over.

To drive home how energy flows through the light-capturing reactions, trace the route of electrons through Figure 10.15 with your finger. Start on the lower left. The process starts when photons excite electrons in the chlorophyll molecules of photosystem II's antenna complex. When the energy in the excited electrons is transferred to the reaction center, a special pair of chlorophyll molecules, called P680, passes excited electrons to pheophytin. These are the same reaction center pigments described previously, and the name represents the wavelength of photons absorbed (680 nm).

When pheophytin is reduced, it transfers the high-energy electron to an electron transport chain. There the electron is gradually stepped down in potential energy through redox reactions among a series of quinones and cytochromes. Using the energy released by the redox reactions, plastoquinone (PQ) carries protons across the thylakoid membrane, from the stroma to the lumen. ATP synthase uses the resulting proton-motive force to phosphorylate ADP, creating ATP.

When electrons reach the end of the cytochrome complex, they are passed to a small diffusible protein called **plastocyanin**

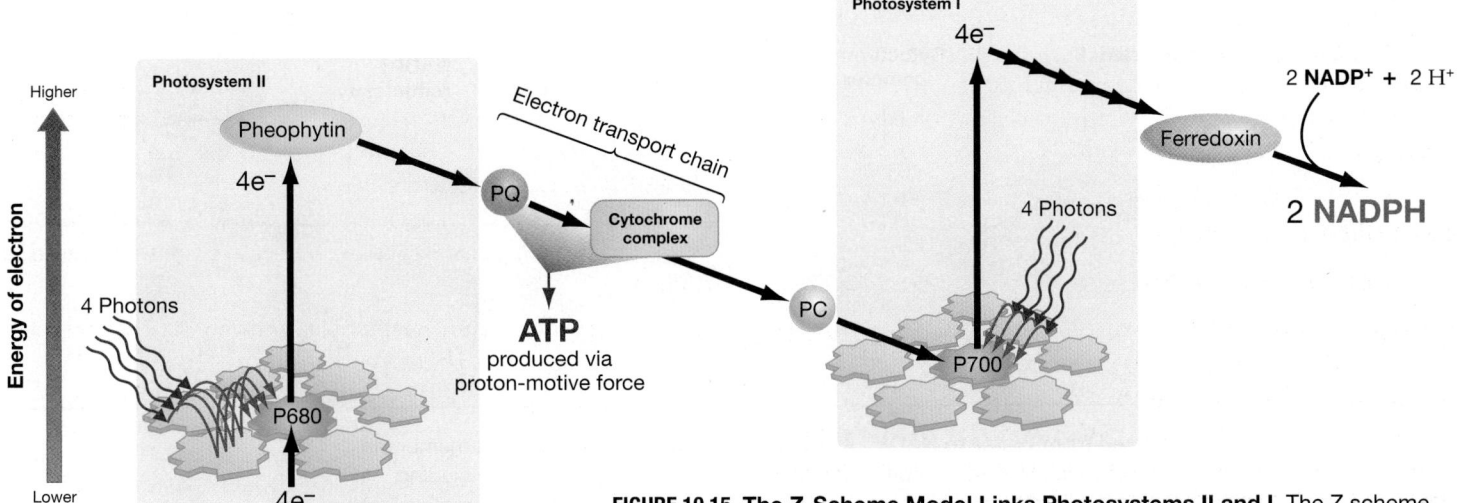

FIGURE 10.15 The Z-Scheme Model Links Photosystems II and I. The Z scheme proposes that electrons from water are first energized by photosystem II to generate ATP and then energized again by photosystem I to reduce NADP$^+$ to NADPH.

(symbolized as PC in Figure 10.15). The reduced plastocyanin diffuses through the lumen of the thylakoid, and donates the electron to an oxidized reaction center pigment in photosystem I.

Stop tracing for a moment, and consider the following:

- Plastocyanin is critical—it forms a physical link between photosystem II and photosystem I.

- A single plastocyanin molecule can shuttle over 1000 electrons per second between the cytochrome complex and photosystem I.

- The flow of electrons between photosystems, by means of plastocyanin, is important because it replaces electrons that are carried away from the pair of pigments in the photosystem I reaction center. This pair of specialized chlorophyll molecules is called P700 (absorbs 700-nm photons).

Now keep going. The electrons that flow from photosystem II to P700, via plastocyanin, are eventually transferred to the protein ferredoxin, which passes electrons to an enzyme that catalyzes the reduction of NADP$^+$ to NADPH.

Finally, direct your attention back to the lower-left portion of the figure. Note that the electrons that initially left photosystem II are replaced by electrons that are stripped away from water, producing oxygen gas as a by-product.

✔ If you understand the Z-scheme model, you should be able to describe (1) the role of plastocyanin in linking the two photosystems, and (2) the point where the electrons that flow through the system have their highest potential energy.

Understanding the Enhancement Effect The Z-scheme model helps explain the enhancement effect documented in Figure 10.11. When chloroplasts are illuminated with wavelengths in the red portion of the spectrum, only photosystem II can run at a maximum rate. The overall rate of electron flow through the Z scheme is moderate because photosystem I's efficiency is reduced.

Similarly, when chloroplasts are illuminated with wavelengths in the far-red portion of the spectrum, only photosystem I is capable of peak efficiency; photosystem II is working at a below-maximum rate, so the overall rate of electron flow is reduced.

But when both wavelengths are available at the same time, both photosystems are activated and work at a maximum rate, leading to enhanced efficiency.

Noncyclic Electron Flow between Water and NADP$^+$ The complete path that electrons follow from photosystem II to photosystem I and how it is oriented in the thylakoid membrane is shown in **FIGURE 10.16**. Note that electrons pass from water to NADP$^+$ through a chain of redox reactions in a linear fashion, referred to as **noncyclic electron flow.**

Compare the movement of electrons and protons in Figure 10.16 with what you have learned about electron transport chains in mitochondria (see Figure 9.15). In both these organelles, the energy released from redox reactions is used to build a proton gradient for ATP production. At the end of the chains, electrons are donated to terminal electron acceptors.

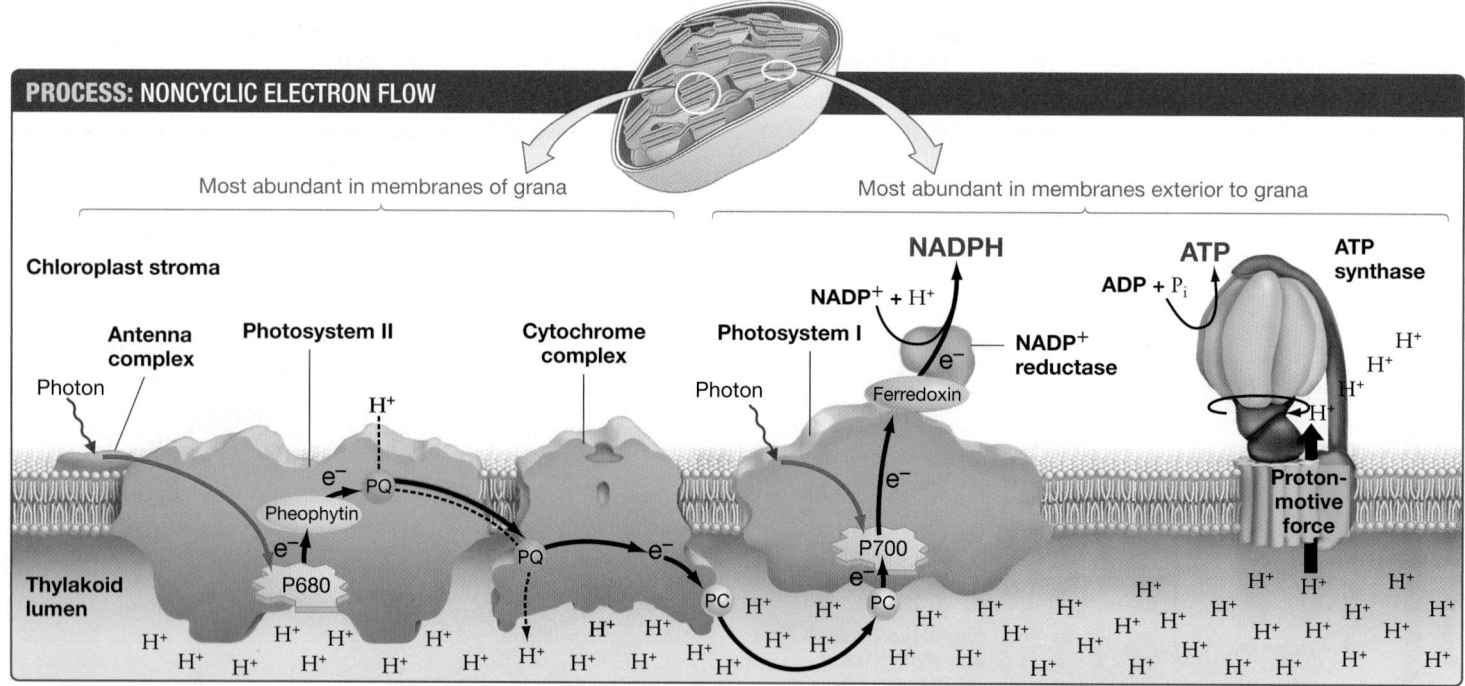

FIGURE 10.16 Electrons Are Passed from Water to NADP$^+$ in a Linear Pathway. In the thylakoid membrane, photosystem II uses light to energize electrons taken from water and pass them through an ETC including plastoquinone (PQ), the cytochrome complex, and plastocyanin (PC). The ETC produces a proton motive force that is used to make ATP. Photosystem I energizes electrons from PC and passes them on to ferredoxin to reduce NADP$^+$ to NADPH.

Chloroplasts and mitochondria differ, however, in how electron potential energy changes between the primary electron donor and the terminal electron acceptor. In the mitochondrial ETC, electron potential energy starts high and then steadily drops as the electrons are transferred to the terminal electron acceptor, which has the lowest potential energy. In chloroplasts, the reduced terminal electron acceptor (NADPH) has higher potential energy than the electron donor (H_2O) (see Figure 10.15).

Cyclic Electron Flow Recycles Electrons and Drives Photophosphorylation

Recent evidence indicates that an alternative electron path, called **cyclic electron flow,** also occurs in green algae and plants (**FIGURE 10.17**). In these organisms, ATP is produced via cyclic as well as noncyclic photophosphorylation.

During cyclic electron flow, photosystem I transfers electrons back to the electron transport chain associated with photosystem II, generating ATP through photophosphorylation instead of reducing $NADP^+$. This "extra" ATP is used for the chemical reactions that reduce carbon dioxide (CO_2) and produce sugars. Cyclic electron flow coexists with the noncyclic electron flow and produces additional ATP.

Where Are Photosystems II and I Located?

Although both photosystems reside in the thylakoid membrane, their distribution is far from random. Photosystem II and the cytochrome complex are much more abundant in the interior, stacked membranes of grana, while photosystem I and ATP synthase are much more common in the exterior, unstacked membranes.

This organization seems appropriate for ATP synthase. As shown in Figure 10.16, this enzyme complex is oriented with its bulky head toward the stroma, so avoiding the tightly stacked grana makes sense. In addition, ATP synthase and the $NADP^+$ reductase that is associated with photosystem I require substrates that are found in the stroma, such as ADP, P_i, and $NADP^+$. These substrates would not be as readily available if the enzymes were buried in the membrane folds of the grana.

The benefit of physically separating the two photosystems is the focus of intense research and debate. Unlike ATP synthase, the functions of photosystems I and II are tightly integrated, requiring electrons to be transported between them in noncyclic electron flow. Compared to other electron transport chains, the distance between where plastocyanin is reduced—the cytochrome complex—and where it is oxidized—the photosystem I reaction center—is huge. This separation between the photosystems is currently thought to be involved in regulating the switch between noncyclic and cyclic electron flow.

Oxygenic Photosynthesis and the Evolution of Earth

Although oxygen is a by-product of photosynthesis, the impact of producing this molecule on the environment of early Earth cannot be overstated. Photosynthesis produces the oxygen that is keeping you alive right now. Biologists rank the evolution of Earth's oxygen-rich atmosphere as one of the most important events in the history of life. Why?

According to the geologic record, oxygen levels in the atmosphere and oceans began to rise only about 2 billion years ago, as organisms that performed oxygenic photosynthesis increased in abundance. O_2 was, in fact, almost nonexistent on Earth before enzymes evolved that could catalyze the oxidation of water. Since ozone is formed from O_2 gas, this protective layer would have arisen in our atmosphere only after the evolution of oxygenic photosynthesis. Without the ozone layer, Earth's surface would have been bombarded continually by the searing intensity of ultraviolet radiation—making the evolution of life on land nearly impossible.

As oxygen became more abundant, bacterial cells that evolved the ability to use it as an electron acceptor via cellular respiration began to dominate. O_2 is so electronegative that it creates a huge potential energy drop for the electron transport chains involved in cellular respiration. As a result, organisms that use O_2 as an electron acceptor in cellular respiration can produce much more ATP than can organisms that use other electron acceptors (see Chapter 9). In addition, this accumulation of oxygen was a disaster for anaerobic organisms because O_2 is such a powerful oxidant it is toxic to them.

Determining exactly how photosystem II splits water and generates oxygen may be the greatest challenge currently facing researchers interested in photosynthesis. This issue has important practical applications: If human chemists could replicate the reaction, it might be possible to produce huge volumes of O_2 and hydrogen gas (H_2) from water. The resulting H_2 could provide a clean, inexpensive fuel for vehicles.

Despite the importance of oxygen in the evolution and maintenance of life, in terms of photosynthesis, it is simply waste. The useful products of the light-capturing reactions are ATP and NADPH, which are required to reduce carbon dioxide to sugar. Your life, and the life of most organisms, also depends on this process. How does it happen?

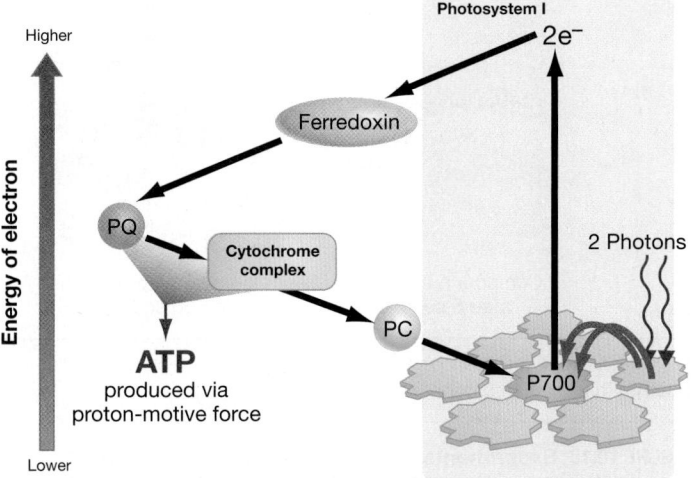

FIGURE 10.17 Cyclic Electron Flow Leads to ATP Production. Cyclic electron flow is an alternative to the Z scheme. Instead of being donated to $NADP^+$, electrons are returned to plastoquinone (PQ) and cycle between photosystem I and the ETC, resulting in the production of additional ATP via photophosphorylation.

check your understanding

If you understand that . . .

- Photosystem II contributes high-energy electrons to an electron transport chain that pumps protons, creating a proton-motive force that drives ATP synthase.
- Photosystem I uses high-energy electrons to make NADPH and can produce additional ATP by building a proton-motive force via cyclic electron flow.

✔ You should be able to . . .

Compare and contrast the flow of electrons in mitochondria and chloroplasts. What are the primary electron donors and terminal electron acceptors, and how do they differ in terms of energy?

Answers are available in Appendix A.

10.4 How Is Carbon Dioxide Reduced to Produce Sugars?

The reactions analyzed in Section 10.3 are triggered by light. This is logical, because their entire function is focused on transforming electromagnetic energy in the form of sunlight into chemical energy in the phosphate bonds of ATP and the electrons of NADPH. The reactions that produce sugar from carbon dioxide, in contrast, are not triggered directly by light. Instead, they depend on the ATP and NADPH produced by the light-capturing reactions of photosynthesis.

The Calvin Cycle Fixes Carbon

Carbon fixation is the addition of carbon dioxide to an organic compound. The word fix is appropriate because the process converts or fixes CO_2 gas to a biologically useful form. Once carbon atoms are fixed, they can be used as sources of energy and as building blocks to construct the molecules found in cells.

Carbon fixation is a redox reaction—the carbon atom in CO_2 is reduced. Research on how this happens in chloroplasts gained momentum just after World War II, when radioactive isotopes of carbon became available for research purposes.

Melvin Calvin's group made great strides early in this effort by tracking the incorporation of $^{14}CO_2$ into molecules during photosynthesis (**FIGURE 10.18**). After injecting $^{14}CO_2$ into a culture of algae that were undergoing photosynthesis, they stopped the reaction after different periods of time by killing the cells in hot alcohol. This treatment immediately denatured the enzymes involved in the reactions, effectively halting any further change in the radiolabeled intermediates.

The molecules labeled with the ^{14}C in this extract were separated by chromatography and detected using X-ray film. If radioactively labeled molecules were present in the chromatograph,

---

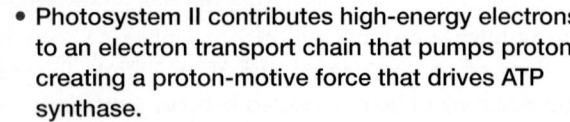

RESEARCH

QUESTION: What intermediates are produced as carbon dioxide is reduced to sugar?

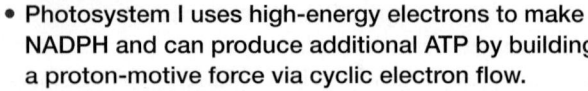

HYPOTHESIS: No specific hypothesis.

EXPERIMENTAL SETUP:

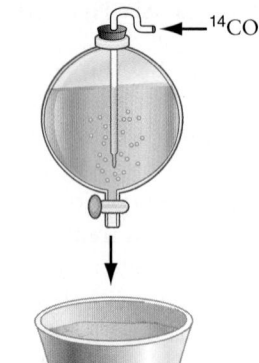

$^{14}CO_2$

1. Add $^{14}CO_2$ to actively photosynthesizing algae.

2. Wait 5–60 seconds; then homogenize cells by immersing in hot alcohol.

3. Separate molecules via chromatography.

4. Lay X-ray film on chromatograph to locate radioactive label.

PREDICTION: No specific prediction.

RESULTS:

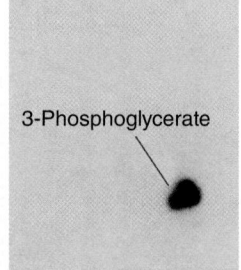

3-Phosphoglycerate

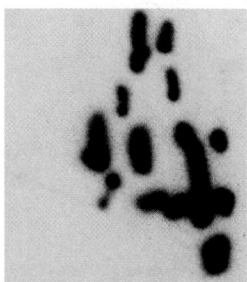

Compounds produced after 5 seconds

Compounds produced after 60 seconds

CONCLUSION: 3-Phosphoglycerate is the first intermediate product. Other intermediates appear later.

FIGURE 10.18 Experiments Revealed the Reaction Pathway Leading to Reduction of CO_2.

SOURCE: Benson, A. A., J. A. Bassham, M. Calvin, et al. 1950. The path of carbon in photosynthesis. V. Paper chromatography and radioautography of the products. *Journal of the American Chemistry Society* 72: 1710–1718.

✔QUESTION Why wasn't this experiment based on a specific hypothesis and set of predictions?

the energy they emitted would expose the film and create a dark spot. The labeled compounds could then be isolated and identified.

By varying the amount of time the algae were exposed to labeled $^{14}CO_2$, Calvin and co-workers pieced together the sequence in which various intermediates formed. For example, when the team analyzed cells almost immediately after adding the $^{14}CO_2$, they found that the ^{14}C was predominantly in a three-carbon compound called 3-phosphoglycerate (3PGA). This result suggested that 3PGA was the initial product of carbon reduction. Stated another way, it appeared that carbon dioxide reacted with some unknown molecule to produce 3PGA.

This was an intriguing result, because 3-phosphoglycerate is also one of the 10 intermediates in glycolysis. The Calvin cycle manufactures carbohydrate; glycolysis breaks it down. Because the two processes are related in this way, it was logical that at least some intermediates in glycolysis and the Calvin cycle are the same.

RuBP Is the Initial Reactant with CO_2   Which compound reacts with CO_2 to produce 3-phosphoglycerate? This was the key, initial step. Calvin's group searched in vain for a two-carbon compound that might serve as the initial carbon dioxide acceptor and yield 3PGA.

Then, while Calvin was running errands one day, it occurred to him that the molecule reacting with carbon dioxide might contain five carbons, not two. Adding CO_2 to a five-carbon molecule would produce a six-carbon compound, which could then split in half to form 2 three-carbon molecules.

Experiments to test this hypothesis confirmed that the five-carbon compound **ribulose bisphosphate (RuBP)** is the initial reactant.

The Calvin Cycle Is a Three-Step Process   The complete Calvin cycle, as it came to be called, has three phases (**FIGURE 10.19**):

1. **Fixation phase**   The Calvin cycle begins when CO_2 reacts with RuBP. This phase fixes carbon and produces two molecules of 3PGA.

2. **Reduction phase**   The 3PGA is phosphorylated by ATP and then reduced by electrons from NADPH. The product is the phosphorylated three-carbon sugar **glyceraldehyde-3-phosphate (G3P)**. Some of the G3P that is synthesized is drawn off to manufacture glucose and fructose.

3. **Regeneration phase**   The rest of the G3P keeps the cycle going by serving as the substrate for the third phase in the cycle: reactions that use additional ATP in the regeneration of RuBP.

All three phases take place in the stroma of chloroplasts. One turn of the Calvin cycle fixes one molecule of CO_2. Three turns of the cycle fix three molecules of CO_2, yielding one molecule of G3P and fully regenerated RuBP (Figure 10.19).

The discovery of the Calvin cycle clarified how the ATP and NADPH produced by light-capturing reactions allow cells to reduce CO_2 gas to carbohydrate $(CH_2O)_n$. Because sugars store a great deal of potential energy, producing them takes a great deal of chemical energy. In the Calvin cycle, each mole of CO_2 requires the energy from 3 moles of ATP and 2 moles of

(a) The Calvin cycle has three phases.

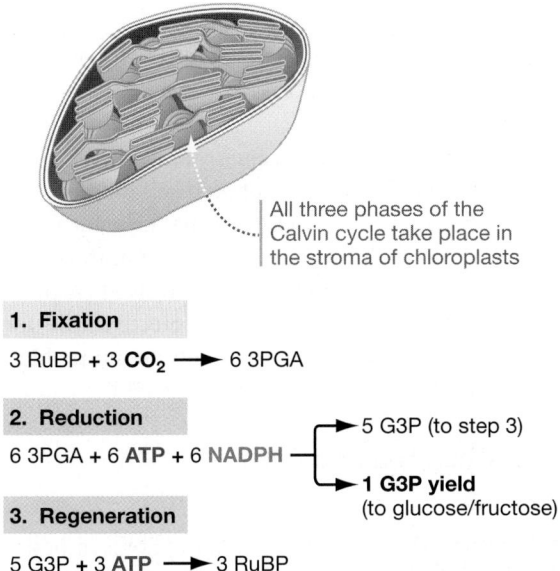

All three phases of the Calvin cycle take place in the stroma of chloroplasts

1. Fixation

3 RuBP + 3 CO_2 ⟶ 6 3PGA

2. Reduction

6 3PGA + 6 **ATP** + 6 **NADPH** ⟶ 5 G3P (to step 3)

⟶ **1 G3P yield** (to glucose/fructose)

3. Regeneration

5 G3P + 3 **ATP** ⟶ 3 RuBP

(b) The reaction occurs in a cycle.

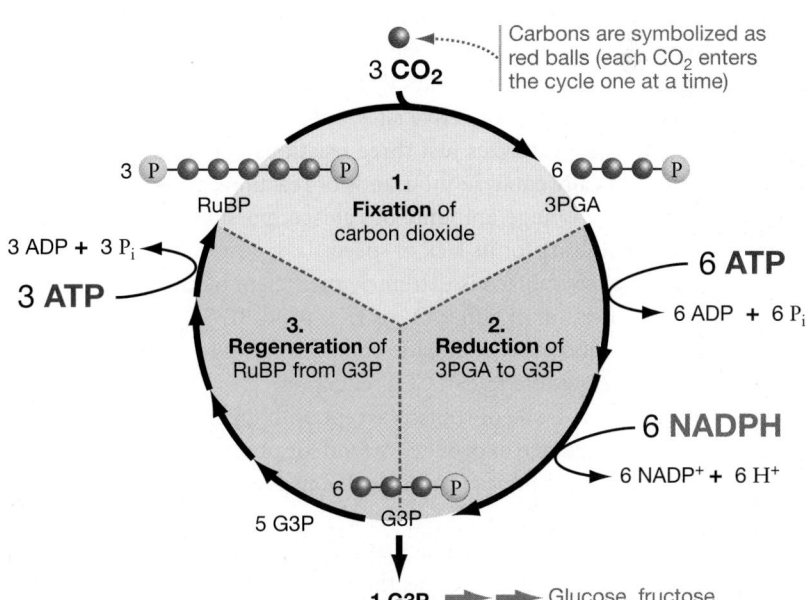

Carbons are symbolized as red balls (each CO_2 enters the cycle one at a time)

3 CO_2

3 P RuBP

1. **Fixation** of carbon dioxide

3 ADP + 3 P_i

3 ATP

6 3PGA

6 ATP

6 ADP + 6 P_i

3. **Regeneration** of RuBP from G3P

2. **Reduction** of 3PGA to G3P

6 NADPH

6 NADP⁺ + 6 H⁺

5 G3P   6 G3P

1 G3P ⟹ Glucose, fructose

FIGURE 10.19 Carbon Dioxide Is Reduced in the Calvin Cycle. The number of reactants and products resulting from three turns of the cycle are shown. Of the six G3Ps that are generated during the reduction phase, one is used in the synthesis of glucose or fructose and the other five are used to regenerate RuBP. The 3 RuBPs that are regenerated participate in fixation reactions for additional turns of the cycle.

NADPH to fix it and reduce it to sugar. ✔ **QUANTITATIVE** If you understand the Calvin cycle, you should be able to provide the *minimum number* of RuBP, ATP, and NADPH molecules that would be required to run through six complete cycles. Explain why you would not need six RuBP molecules to fix and reduce six CO_2.

The conversion of CO_2 gas into carbohydrate is, without doubt, worthy of this energy investment. Plants use sugars to fuel cellular respiration and build leaves and other structures. Millions of non-photosynthetic organisms—from fungi to mammals—also depend on this reaction to provide the sugars they need for cellular respiration.

Ecologically, the addition of CO_2 to RuBP may be the most important chemical reaction on Earth. The enzyme that catalyzes it is fundamental to life. How does this protein work?

The Discovery of Rubisco

Most reactions involved in reducing CO_2 also occur during glycolysis or other metabolic pathways. The initial CO_2 fixation phase of the Calvin cycle, however, is one of only two reactions that are entirely unique to the Calvin cycle.

To find the enzyme that fixes CO_2 to RuBP, Arthur Weissbach and colleagues ground up spinach leaves, purified a large series of proteins from the resulting cell extracts, and tested each protein to see if it could catalyze this step. Eventually they isolated the catalyst, which happens to be the most abundant enzyme in leaf tissue. The researchers' data suggested that it constituted almost 50 percent of the total protein in spinach leaves.

The CO_2-fixing enzyme, ribulose-1,5-bisphosphate carboxylase/oxygenase (commonly referred to as **rubisco**), is found in all photosynthetic organisms that use the Calvin cycle to fix carbon and is thought to be the most abundant enzyme on Earth. As shown in **FIGURE 10.20a**, the rubisco enzyme is cube-shaped and consists of 16 polypeptides that form eight active sites where CO_2 is fixed.

Despite its large number of active sites, rubisco is a slow enzyme. Each active site catalyzes just three reactions per second; other enzymes typically catalyze thousands of reactions per second. Plants synthesize huge amounts of rubisco, possibly as an adaptation compensating for its lack of speed.

Besides being slow, rubisco is extremely inefficient because it will catalyze the addition of either O_2 or CO_2 to RuBP. This is a key point: Oxygen and carbon dioxide compete at the enzyme's active sites, which slows the rate of CO_2 reduction.

Why would an active site of rubisco accept both O_2 and CO_2? Given rubisco's importance in producing food for photosynthetic species, this trait would appear to be **maladaptive**—it reduces the fitness of individuals.

The reaction of O_2 with RuBP actually does more than just compete with the reaction of CO_2 at the same active site. One of the molecules produced from the addition of oxygen to RuBP is processed in reactions that consume ATP and release CO_2 in order to regenerate 3PGA. Part of this pathway occurs in chloroplasts, and part occurs in peroxisomes and mitochondria. The

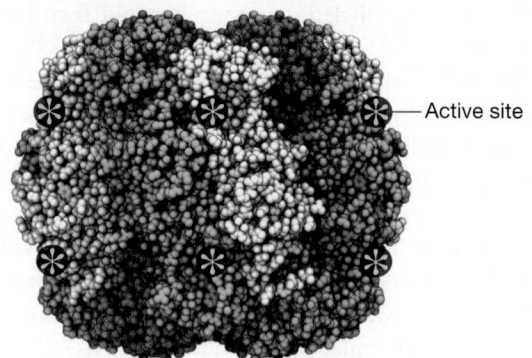

(a) Rubisco has 16 subunits and a total of 8 active sites.

—Active site

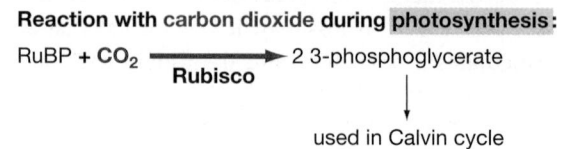

(b) Rubisco's active sites can interact with CO_2 or O_2.

Reaction with carbon dioxide during photosynthesis:

$$RuBP + CO_2 \xrightarrow{\text{Rubisco}} 2 \text{ 3-phosphoglycerate}$$

used in Calvin cycle

Reaction with oxygen during photorespiration:

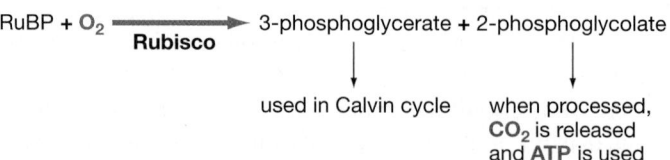

$$RuBP + O_2 \xrightarrow{\text{Rubisco}} \text{3-phosphoglycerate} + \text{2-phosphoglycolate}$$

used in Calvin cycle   when processed, CO_2 is released and **ATP** is used

FIGURE 10.20 Rubisco Is a Large Enzyme Complex That Can React with CO_2 or O_2. (a) The cube shape of rubisco consists of multiple polypeptides that form eight catalytic active sites. **(b)** In addition to fixing CO_2 in photosynthesis, rubisco catalyzes a competing reaction with O_2 with a very different outcome.

reaction sequence resembles respiration, because it consumes oxygen and produces carbon dioxide. As a result, it is called **photorespiration** (**FIGURE 10.20b**).

Because photorespiration consumes energy and releases fixed CO_2, it "undoes" photosynthesis. When photorespiration occurs, the overall rate of photosynthesis declines. This does not mean that there is no benefit to the plant, however. Some of the products from photorespiration are known to be involved in plant signaling and development. In addition, a protective role for photorespiration has been proposed when plants are under high light and low CO_2 conditions.

Oxygen and Carbon Dioxide Pass through Stomata

Atmospheric carbon dioxide is a key reactant in photosynthesizing cells. It would seem straightforward, then, for CO_2 to diffuse directly into plants along a concentration gradient. But the situation is not this simple, because plants are covered with a waxy coating called a cuticle. This lipid layer prevents water from evaporating out of tissues, but it also prevents the transport of gases like CO_2 and O_2.

How does CO_2 get into photosynthesizing tissues? The surface of a leaf is dotted with openings bordered by two distinctively shaped cells called **guard cells** (**FIGURE 10.21a**). The opening between these paired cells is called a pore, and the entire structure is a **stoma** (plural: **stomata**).

An open stoma allows CO_2 from the atmosphere to diffuse into air-filled spaces inside the leaf and excess O_2 to diffuse out (**FIGURE 10.21b**). Eventually the CO_2 diffuses along a concentration gradient into the chloroplasts of photosynthesizing cells. A strong concentration gradient favoring entry of CO_2 is maintained by the Calvin cycle, which constantly uses up the CO_2 in chloroplasts.

Stomata are normally open during the day, when photosynthesis is occurring, and closed at night. But if the daytime is extremely hot and dry, leaf cells may lose a great deal of water to evaporation through their stomata. When this occurs, they must either close the openings and halt photosynthesis or risk death from dehydration.

When conditions are hot and dry, then, stomata must close and CO_2 and O_2 transport stops—meaning that photosynthesis slows and photorespiration increases. How do plants that live in hot, dry environments prevent dehydration while keeping CO_2 supplies high enough to avoid photorespiration?

Mechanisms for Increasing CO_2 Concentration

The oxygenation reaction that triggers photorespiration is favored when oxygen concentrations are high and CO_2 concentrations are low. But even with the stomata open, the atmosphere is 21 percent oxygen and only 0.03 percent carbon dioxide. How can photosynthesizing cells raise CO_2 concentrations to make photosynthesis more efficient? An answer emerged in a surprising experimental result.

The C_4 Pathway After the Calvin cycle had been worked out in algae, researchers in a variety of labs used the same radioactive carbon dioxide tracking approach to investigate how carbon fixation occurs in other species. Hugo Kortschak and colleagues and Y. S. Karpilov and associates exposed leaves of sugarcane and maize (corn) to $^{14}CO_2$ and sunlight; then they isolated and identified the intermediates.

Both research teams expected to find the first of the radioactive carbon atoms in 3-phosphoglycerate—the normal product of carbon fixation by rubisco. Instead, they found that in their species, the radioactive carbon atom ended up in four-carbon compounds such as malate and aspartate.

Instead of creating a three-carbon molecule as in the Calvin cycle, it appeared that these species were able to fix CO_2 to produce four-carbon molecules. This newly identified set of reactions became known as the **C_4 pathway** to distinguish it from Calvin's CO_2 fixation via what is now termed the **C_3 pathway** (**FIGURE 10.22**).

Researchers who followed up on the initial reports found that the C_4 pathway does not replace the Calvin cycle, but serves as an additional fixation step. C_4 plants can actually fix carbon dioxide using both pathways—to a three-carbon compound by an enzyme called **PEP carboxylase** (C_4) and to RuBP by rubisco (C_3). They also showed that the two pathways are found in distinct cell types within the same leaf. PEP carboxylase is common in **mesophyll cells** near the surface of leaves, while rubisco is found in **bundle-sheath cells** that surround the vascular tissue in the

(a) Leaf surfaces contain stomata.

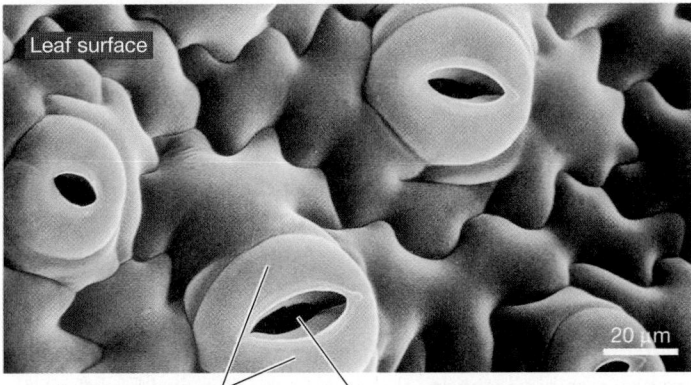

Guard cells + Pore = **Stoma**

(b) Carbon dioxide diffuses into leaves through stomata.

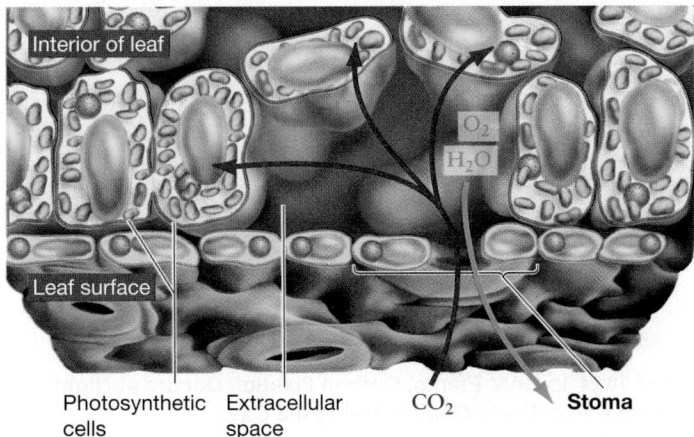

Photosynthetic cells   Extracellular space CO_2 **Stoma**

FIGURE 10.21 Leaf Cells Obtain Carbon Dioxide through Stomata.

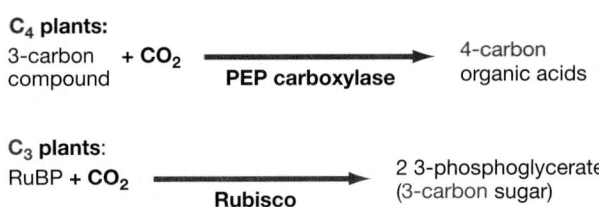

FIGURE 10.22 Initial Carbon Fixation in C_4 Plants Is Different from That in C_3 Plants.

(a) C_4 plant

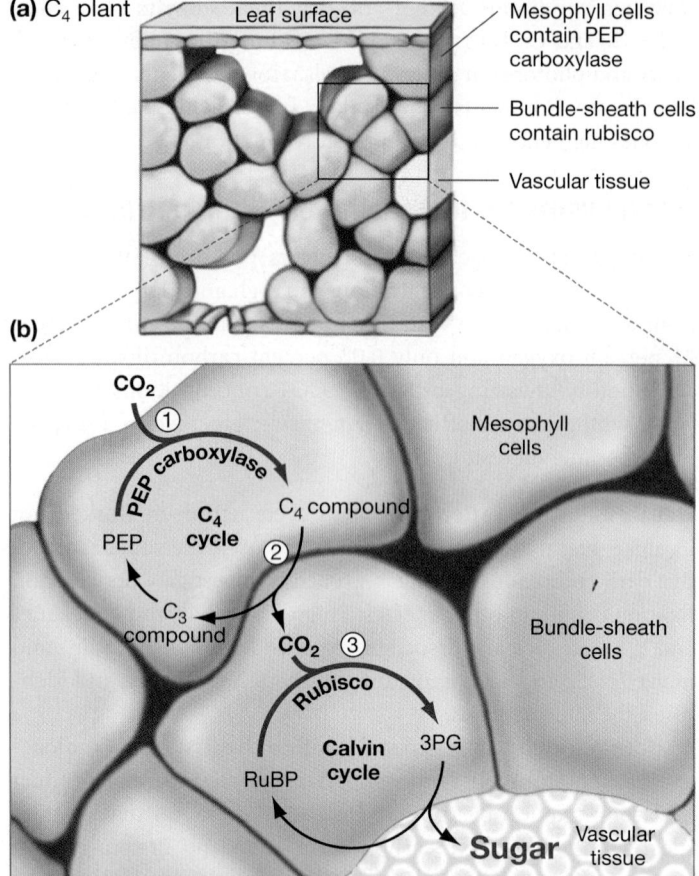

Leaf surface

Mesophyll cells contain PEP carboxylase

Bundle-sheath cells contain rubisco

Vascular tissue

(b)

CO_2

① PEP carboxylase

C_4 cycle

PEP

C_4 compound

②

C_3 compound

CO_2 ③

Rubisco

Calvin cycle

RuBP

3PG

Sugar

Mesophyll cells

Bundle-sheath cells

Vascular tissue

FIGURE 10.23 In C_4 Plants, Carbon Fixation and the Calvin Cycle Occur in Different Cell Types. (a) The carbon-fixing enzyme PEP carboxylase is located in mesophyll cells, while rubisco is in bundle-sheath cells. **(b)** CO_2 is fixed to the three-carbon compound PEP by PEP carboxylase, forming a four-carbon organic acid. A CO_2 molecule from the four-carbon sugar then feeds the Calvin cycle.

interior of the leaf (**FIGURE 10.23a**). Vascular tissue conducts water and nutrients in plants (see Chapter 38).

Based on the observations about C_4 plants, Hal Hatch and Roger Slack proposed a four-step model to explain how CO_2 that is fixed to a four-carbon sugar feeds the Calvin cycle (**FIGURE 10.23b**):

Step 1 PEP carboxylase fixes CO_2 to a three-carbon molecule (phosphoenolpyruvate, or PEP) in mesophyll cells.

Step 2 The four-carbon organic acids that result are transported to bundle-sheath cells via channels called plasmodesmata (see Chapter 11).

Step 3 The four-carbon organic acids release a CO_2 molecule that rubisco uses as a substrate to form 3PGA. This step initiates the Calvin cycle.

Step 4 The three-carbon compound remaining after CO_2 is released is returned to the mesophyll cell to regenerate PEP.

In effect, then, the C_4 pathway acts as a CO_2 concentrator. The reactions that take place in mesophyll cells require energy in the

form of ATP, but they increase CO_2 concentrations in cells where rubisco is active. Because it increases the ratio of carbon dioxide to oxygen in photosynthesizing cells, less O_2 binds to rubisco's active sites. As a result, the C_4 pathway improves the efficiency of the Calvin cycle.

The C_4 pathway is an adaptation that keeps CO_2 concentrations in leaves high, but it comes at a cost. For each glucose molecule generated via photosynthesis, C_4 plants expend 30 ATP molecules compared to the 18 ATP molecules required by C_3 plants. This energy expenditure, however, is justified by the increased efficiency of photosynthesis in conditions where stomata are mostly closed to prevent dehydration. The affinity for CO_2 by PEP carboxylase is also much higher than that of rubisco, which means that stomata can be open for shorter periods in C_4 plants.

This strategy is not the only mechanism that plants use to continue growth in hot, dry climates, however. Some environments are so arid that even C_4 plants are unable to avoid dehydration. Nevertheless, certain plants use the C_4 pathway in a unique way that allows them to thrive in these deserts. How do they do it?

CAM Plants Researchers studying a group of flowering plants called the Crassulaceae discovered a second mechanism for limiting the effects of dehydration and photorespiration. This photosynthetic pathway, **crassulacean acid metabolism,** or **CAM,** resembles the C_4 pathway in a number of ways. It is a CO_2 concentrator that acts as an additional, preparatory step to the Calvin cycle. It also generates an organic acid with four carbons in its first CO_2 fixation step. But unlike the C_4 pathway, CAM occurs at a different time than the Calvin cycle does—not in a different place.

CAM occurs in cacti and other species that routinely keep their stomata closed on hot, dry days. At night, when conditions are cooler and more humid, CAM plants open their stomata and take in huge quantities of CO_2. The CO_2 is temporarily fixed to organic acids and stored in the central vacuoles of photosynthetic cells. During the day, when stomata are closed, these acids are processed in reactions that release the CO_2 and feed the Calvin cycle (**FIGURE 10.24**).

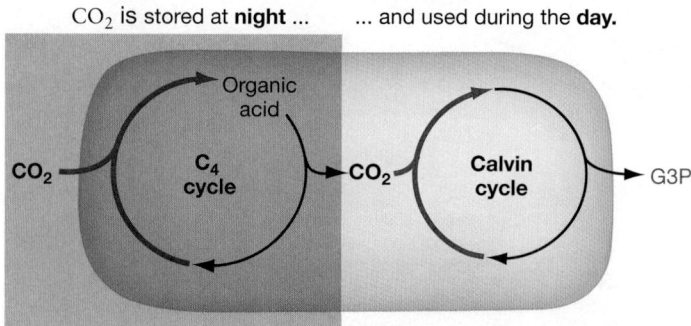

CO_2 is stored at **night** ... ... and used during the **day.**

Organic acid

CO_2

C_4 cycle

CO_2

Calvin cycle

G3P

FIGURE 10.24 In CAM Plants, Carbon Fixation Occurs at Night and the Calvin Cycle Occurs during the Day.

✔**QUESTION** At what part of the day would there be the highest concentration of four-carbon acids in the vacuoles of CAM plants?

The C$_4$ and CAM pathways function as CO$_2$ pumps. They minimize photorespiration when stomata are closed and CO$_2$ cannot diffuse in directly from the atmosphere. Both are found in species that live in hot, dry environments.

But while C$_4$ plants stockpile CO$_2$ by fixing and storing organic acids in cells *where* rubisco is not active, CAM plants store CO$_2$ *when* rubisco is inactive. In C$_4$ plants, the reactions catalyzed by PEP carboxylase and rubisco are separated in space; in CAM plants, the reactions are separated in time.

How Is Photosynthesis Regulated?

Like cellular respiration, photosynthesis is regulated. Although the mechanisms responsible for turning photosynthesis on or off are still under investigation, several patterns have emerged:

- The presence of light triggers the production of proteins required for photosynthesis.

- When sugar supplies are high, the production of proteins required for photosynthesis is inhibited, but the production of proteins required to process and store sugars is stimulated.

- Rubisco is activated by regulatory molecules that are produced when light is available, but inhibited in conditions of low CO$_2$ availability—when photorespiration is favored.

The central message here is that the rate of photosynthesis is finely tuned to use resources efficiently in response to changes in environmental conditions.

What Happens to the Sugar That Is Produced by Photosynthesis?

The products of the Calvin cycle enter one of several reaction pathways. The most important of these reaction sequences produces the monosaccharides glucose and fructose from G3P, a process called **gluconeogenesis.** This glucose is often combined with fructose to form the disaccharide ("two-sugar") **sucrose.**

When photosynthesis is taking place slowly, almost all the glucose that is produced is used to make sucrose. Sucrose is water soluble and readily transported to other parts of the plant. If sucrose is delivered to rapidly growing parts of the plant, it is broken down to fuel cellular respiration and growth.

An alternative pathway occurs when photosynthesis is proceeding rapidly and sucrose is abundant. Under these conditions, the glucose molecules are polymerized to form **starch** in the leaves and in storage cells in the roots. Starch production occurs inside the chloroplast; sucrose synthesis takes place in the cytosol.

In photosynthesizing cells, starch acts as a temporary sugar-storage product. At night, the starch that is stored in leaf cells is broken down and used to manufacture sucrose molecules. The sucrose is then broken down via cellular respiration or transported to other parts of the plant. In this way, chloroplasts provide sugars for cells throughout the plant by day and by night.

If a mouse eats the starch that is stored in the leaves or roots of a plant, however, the chemical energy in the reduced carbons of starch fuels the mouse's growth and reproduction. If an owl eats the mouse, the chemical energy in the mouse's tissues fuels the predator's growth and reproduction. (You can see the Big Picture of how energy is processed via photosynthesis and cellular respiration on pages 198–199.)

In this way, virtually all cell activity can be traced back to the sun's energy that was originally captured by photosynthesis. Photosynthesis is the staff of life.

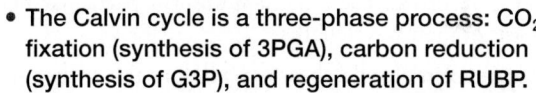

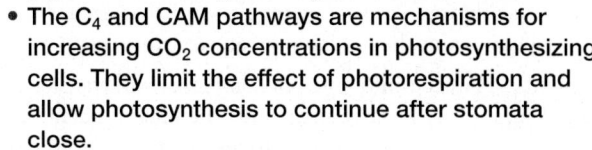

If you understand that . . .

- The Calvin cycle is a three-phase process: CO$_2$ fixation (synthesis of 3PGA), carbon reduction (synthesis of G3P), and regeneration of RUBP.
- The C$_4$ and CAM pathways are mechanisms for increasing CO$_2$ concentrations in photosynthesizing cells. They limit the effect of photorespiration and allow photosynthesis to continue after stomata close.
- In photosynthesizing cells, G3P is used to make sucrose or stored as starch. Sucrose is transported to all plant cells and used to drive cellular respiration.

✔ **You should be able to . . .**

1. Describe how CO$_2$ is delivered to rubisco (a) via organic acids in mesophyll cells, (b) via organic acids stored in vacuoles, and (c) directly.

2. Predict the relative concentration of starch in leaves at the start of the day versus the end of the day.

Answers are available in Appendix A.

If you understand . . .

10.1 Photosynthesis Harnesses Sunlight to Make Carbohydrate

- The light-capturing reactions occur in internal membranes of the chloroplast that are organized into structures called thylakoids in stacks known as grana.

- The Calvin cycle takes place in a fluid portion of the chloroplast called the stroma.

- The CO_2-reduction reactions of photosynthesis depend on the products of the light-capturing reactions.

✔ You should be able to explain why it is not entirely accurate to adopt the common phrase "light-independent reactions" when referring to the Calvin cycle.

10.2 How Do Pigments Capture Light Energy?

- Pigment molecules capture light energy by exciting electrons after a photon is absorbed. Each pigment absorbs particular photons on the basis of their wavelengths.

- After a pigment molecule absorbs a photon, the energy may be released as fluorescence, resonance energy that excites a neighboring pigment, or the reduction of an electron acceptor.

- Pigments organized into antenna complexes will transfer the absorbed light energy via resonance to the reaction center, where an excited electron is transferred to an electron acceptor. The reduction of this electron acceptor completes the transformation of light energy into chemical energy.

✔ You should be able to explain why extracted chlorophyll molecules produce more fluorescence compared to the same number of chlorophyll molecules that remain in chloroplasts.

10.3 The Discovery of Photosystems I and II

- In photosystem II, high-energy electrons are accepted by pheophytin and passed along an electron transport chain, releasing energy that moves protons across the thylakoid membrane. The resulting proton-motive force drives the synthesis of ATP by ATP synthase. Photosystem II takes electrons from water, releasing oxygen and protons.

- In photosystem I, high-energy electrons are passed to ferredoxin. In an enzyme-catalyzed reaction, the reduced form of ferredoxin passes electrons to $NADP^+$, forming NADPH.

- The Z scheme connects photosystems II and I. Plastocyanin carries electrons from the end of photosystem II's ETC to photosystem I. They are promoted to a high-energy state in photosystem I's reaction center, and subsequently used to reduce $NADP^+$.

- Electrons from photosystem I may occasionally be passed back to photosystem II's ETC instead of being used to reduce $NADP^+$. A cyclic flow of electrons between the two photosystems boosts ATP supplies.

✔ You should be able to explain why measuring the rate of oxygen production in chloroplasts is appropriate for estimating the rate of photosynthesis.

10.4 How Is Carbon Dioxide Reduced to Produce Sugars?

- The Calvin cycle starts when CO_2 is attached to a five-carbon compound called ribulose bisphosphate (RuBP) in a reaction catalyzed by the enzyme rubisco.

- The six-carbon compound that results immediately splits in half to form two molecules of 3-phosphoglycerate (3PGA), which is then phosphorylated by ATP and reduced by NADPH to produce a sugar called glyceraldehyde-3-phosphate (G3P).

- Some G3P is used to synthesize glucose and fructose, which combine to form sucrose; the rest are phosphorylated by more ATP in a series of reactions that regenerate RuBP so the cycle can continue.

- Rubisco catalyzes the addition of oxygen as well as carbon dioxide to RuBP. The reaction with oxygen leads to a loss of fixed CO_2 and ATP and is called photorespiration.

- C_4 plants and CAM plants fix CO_2 to organic acids, before it is transferred to rubisco. As a result, they can increase CO_2 levels in their tissues, reducing the effect of photorespiration and allowing photosynthesis to continue when stomata close.

✔ QUANTITATIVE You should be able to connect the light-capturing reactions and Calvin cycle by estimating the number of photons required to produce one glucose molecule from CO_2. How would photorespiration affect the number of photons required per glucose?

MasteringBiology

1. MasteringBiology Assignments

Tutorials and Activities Calvin Cycle; Chemiosmosis; Energy Flow in Plants; Experimental Inquiry: Which Wavelengths of Light Drive Photosynthesis?; Light Energy and Pigments; Light Reactions; Overview of Photosynthesis; Photosynthesis: Inputs, Outputs, and Chloroplast Structure; Photosynthesis: The Light Reactions; Photosynthesis in Dry Climates; Sites of Photosynthesis

Questions Reading Quizzes, Blue-Thread Questions, Test Bank

2. eText Read your book online, search, take notes, highlight text, and more.

3. The Study Area Practice Test, Cumulative Test, BioFlix® 3-D Animations, Videos, Activities, Audio Glossary, Word Study Tools, Art

You should be able to . . .

1. In antenna complexes, how is energy transferred among the pigment molecules?
   a. photophosphorylation
   b. redox reactions
   c. fluorescence
   d. resonance

2. Why is chlorophyll green?
   a. It absorbs all wavelengths in the visible spectrum.
   b. It absorbs wavelengths only in the red portions of the spectrum (680 nm, 700 nm).
   c. It absorbs wavelengths in only the blue and red parts of the visible spectrum.
   d. It absorbs wavelengths only in the blue part of the visible spectrum.

3. What do the light-capturing reactions of photosynthesis produce?
   a. G3P
   b. RuBP
   c. ATP and NADPH
   d. sucrose or starch

4. Why do the absorption spectrum for chlorophyll and the action spectrum for photosynthesis coincide?
   a. Photosystems I and II are activated by different wavelengths of light.
   b. Wavelengths of light that are absorbed by chlorophyll trigger the light-capturing reactions.
   c. Energy from wavelengths absorbed by carotenoids is passed on to chlorophyll.
   d. The rate of photosynthesis depends on the amount of light received.

5. At what point in the light-capturing reactions is the electromagnetic energy of light converted into chemical energy? Where does this occur?

6. In noncyclic electron flow, photosystems I and II function as an integrated unit. What connects the two photosystems?

7. Explain how electrons from water can be used to produce both ATP and NADPH.

8. In addition to their protective function, carotenoids absorb certain wavelengths of light and pass the energy to other pigments via resonance. Based on this function, where would you expect carotenoids to be located in the chloroplast?
   a. the reaction centers of photosystems I and II
   b. the inner membrane of chloroplasts
   c. the antenna complex
   d. the stroma

9. Describe the three phases of the Calvin cycle and how the products of the light-capturing reactions participate in this process.

10. What conditions favor photorespiration? What are its consequences for the plant?

11. Compare and contrast how C_4 plants and CAM plants separate the acquisition of CO_2 from the production of sugar in the Calvin cycle.

12. Why do plants need both chloroplasts and mitochondria? How do their roles differ in the cell?

13. Predict how the following conditions would affect the production of O_2, ATP, and NADPH and state whether noncyclic or cyclic electron flow would occur in each: (1) Only blue photons hit a chloroplast; (2) blue and red photons hit a chloroplast, but no $NADP^+$ is available; (3) blue and red photons hit a chloroplast, but a proton channel has been introduced into the thylakoid membrane, so it is fully permeable to protons.

14. Some biologists claim that photorespiration is an evolutionary "holdover," because rubisco evolved over a billion years ago when O_2 levels were extremely low and CO_2 concentrations relatively high. Do you agree with this hypothesis? Why or why not?

15. An investigator exposes chloroplasts to 700-nm photons and observes low O_2 production, but high ATP production. Which of the following best explains this observation?

   a. The electrons from water are directly transferred to $NADP^+$, which is used to generate ATP.
   b. Photosystem II is not splitting water, and the ATP is being produced by cycling electrons via photosystem I.
   c. The O_2 is being converted to water as a terminal electron acceptor in the production of ATP.
   d. Electron transport has stopped and ATP is being produced by the Calvin cycle.

16. Consider plants that occupy the top, middle, or ground layer of a forest, and algae that live near the surface of the ocean or in deeper water. Would you expect the same photosynthetic pigments to be found in species that live in these different habitats? Why or why not? How would you test your hypothesis?

The Big Picture

It takes energy to stay alive. Use this concept map to study how the information on energy and energetics presented in this book fits together.

As you read the map, remember that chemical energy is potential energy. Potential energy is based on the position of matter in space, and chemical energy is all about the position of electrons in covalent bonds. When hydrogen gas reacts explosively with oxygen, all that's happening is that electrons are moving from high-energy positions to lower-energy positions.

In essence, organisms transform energy from the Sun into chemical energy in the C–C and C–H bonds of glucose, and then into chemical energy in the P–P bonds of ATP.

The potential energy in ATP allows cells to do work: pump ions, synthesize molecules, move cargo, and send and receive signals.

ENERGY FOR LIFE

begins as

Electromagnetic energy in SUNLIGHT 10.2

Text section where you can find more information

drives

PHOTOSYNTHESIS (in chloroplasts) 10.1

begins with

Antenna complex
- Light excites electrons in pigment molecules 10.2

H_2O

donates energy from excited electrons to

donates energy from excited electrons to

enters

Photosystem II
- "Splits" water to yield electrons
- Electron transport chain pumps H^+ 10.3

donates electrons to

Photosystem I
- Electron transport ends with ferredoxin 10.3

Chemiosmosis
- H^+ gradient drives ATP synthase 9.5

releases

yields

O_2

ATP 9.1

NADPH 10.1

used in

Calvin cycle
- Series of enzyme-catalyzed reactions 10.4

CO_2

fixed by rubisco to start

yields substrate for synthesis of

stored as

Glycogen, starch 5.2

broken down to yield

GLUCOSE 5.1

check your understanding

If you understand the big picture . . .
✔ You should be able to . . .

1. Explain how H_2O and O_2 are cycled between photosynthesis and cellular respiration.
2. Explain how CO_2 is cycled between photosynthesis and cellular respiration.
3. Describe what might happen to life on Earth if rubisco were suddenly unable to fix CO_2.
4. Fill in the blue ovals with appropriate linking verbs or phrases.

Answers are available in Appendix A.

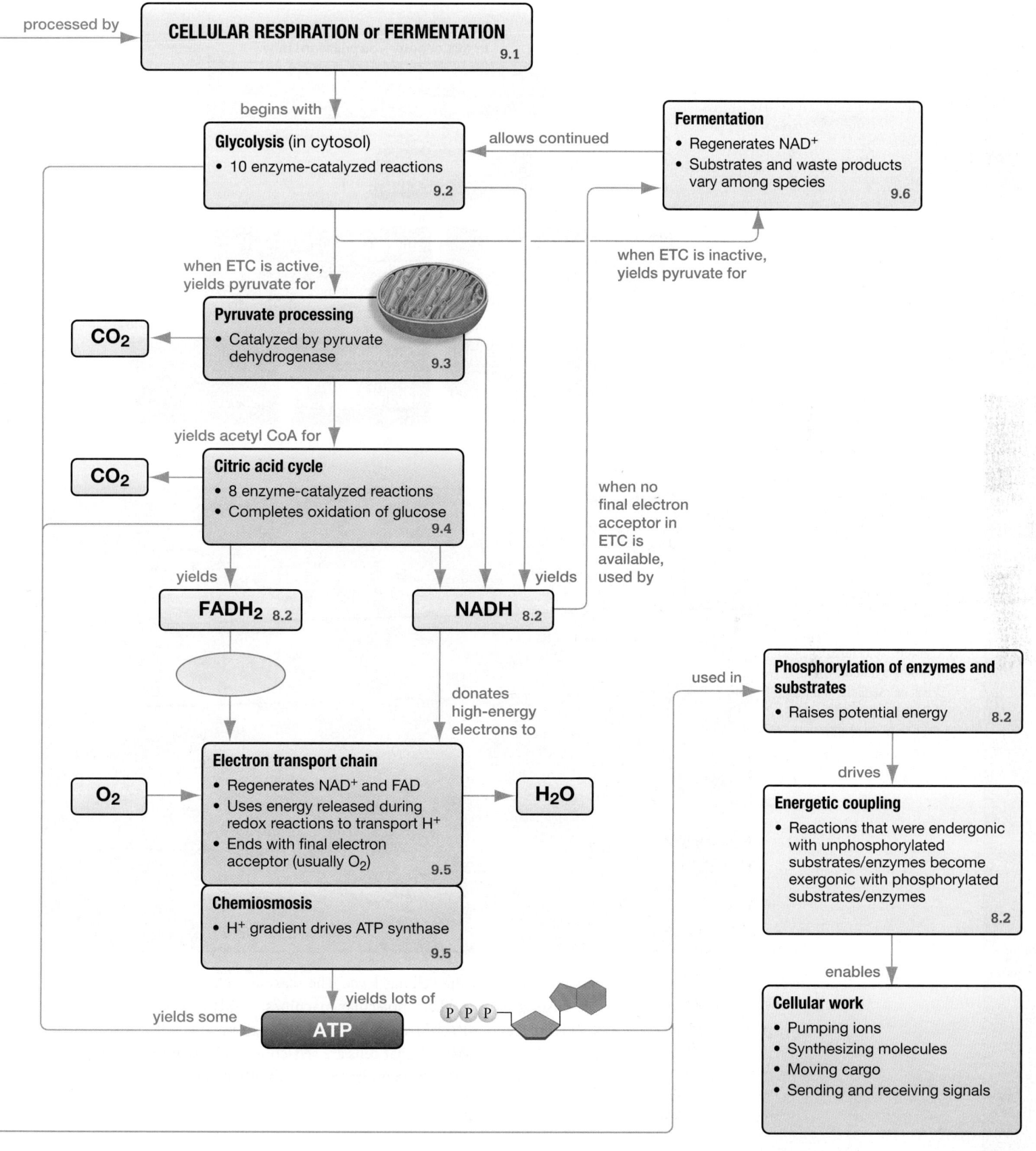

CELLULAR RESPIRATION or FERMENTATION 9.1

processed by →

begins with

Glycolysis (in cytosol)
• 10 enzyme-catalyzed reactions 9.2

Fermentation
• Regenerates NAD^+
• Substrates and waste products vary among species 9.6

allows continued

when ETC is active, yields pyruvate for

when ETC is inactive, yields pyruvate for

Pyruvate processing
• Catalyzed by pyruvate dehydrogenase 9.3

CO_2

yields acetyl CoA for

Citric acid cycle
• 8 enzyme-catalyzed reactions
• Completes oxidation of glucose 9.4

CO_2

when no final electron acceptor in ETC is available, used by

yields

yields

FADH$_2$ 8.2

NADH 8.2

donates high-energy electrons to

Phosphorylation of enzymes and substrates
• Raises potential energy 8.2

used in

Electron transport chain
• Regenerates NAD^+ and FAD
• Uses energy released during redox reactions to transport H^+
• Ends with final electron acceptor (usually O_2) 9.5

O_2

H_2O

Energetic coupling
• Reactions that were endergonic with unphosphorylated substrates/enzymes become exergonic with phosphorylated substrates/enzymes 8.2

drives

Chemiosmosis
• H^+ gradient drives ATP synthase 9.5

enables

yields lots of

yields some

ATP

P P P

Cellular work
• Pumping ions
• Synthesizing molecules
• Moving cargo
• Sending and receiving signals

11 Cell–Cell Interactions

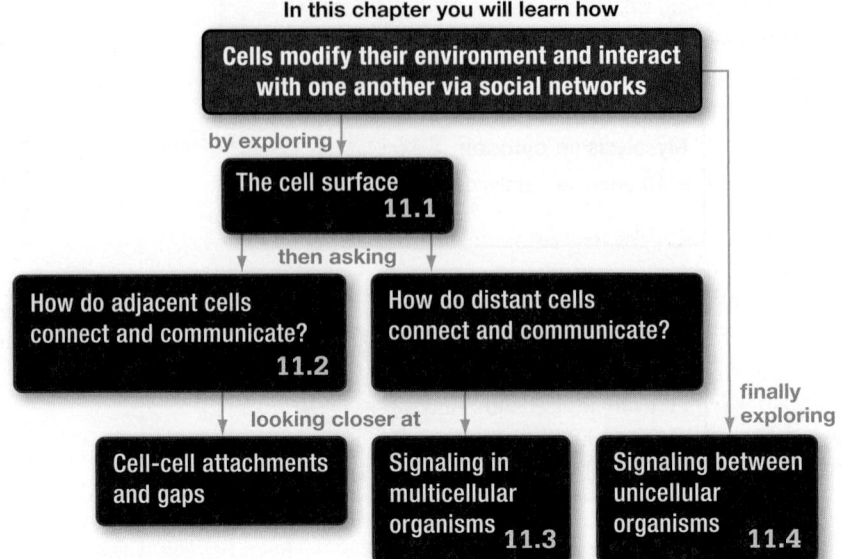

In this chapter you will learn how

Cells modify their environment and interact with one another via social networks

by exploring ↓

The cell surface
11.1

then asking

How do adjacent cells connect and communicate?
11.2

How do distant cells connect and communicate?

looking closer at

Cell-cell attachments and gaps

Signaling in multicellular organisms **11.3**

finally exploring

Signaling between unicellular organisms **11.4**

In this micrograph of cardiac tissue, muscle cells are stained red and their nuclei are stained blue. The green dye highlights a protein called dystrophin, which links the cytoskeleton of muscle cells to proteins that attach to the extracellular matrix. Deficiency in dystrophin leads to muscular dystrophy.

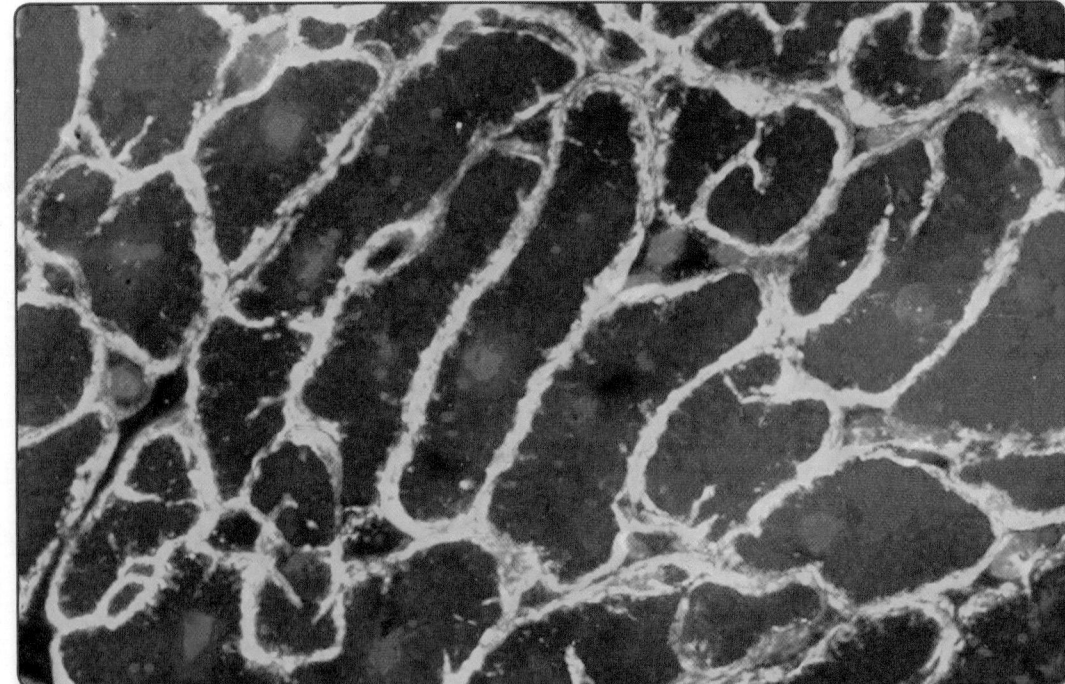

A diversity of events takes place at the cellular level. The plasma membrane surrounds a bustling enterprise consisting of organelles, molecular machines, and cytoskeletal elements (see Chapters 6 and 7). Molecular motors transport cargo throughout the cell at breathtaking speed. It would be a mistake, however, to think that cells are self-contained—that they are worlds in and of themselves. Instead, cells are dependent on interactions with other cells and the surrounding environment.

✔ When you see this checkmark, stop and test yourself. Answers are available in Appendix A.

For most unicellular species, the outside environment is teeming with other organisms. Inside your gut, for example, hundreds of billions of bacterial cells are jostling for space and resources. In addition to interacting with these individuals, every unicellular organism must contend with constant shifts in the physical environment, such as heat, light, ion concentrations, and food supplies. If unicellular organisms are unable to sense these conditions and respond appropriately, they die.

In multicellular species, the environment outside the cell is made up of other cells, both neighboring and distant. The cells that make up a redwood tree, a mushroom, or your body are intensely social. Although biologists often study cells in isolation, an individual tree, fungus, or person is actually an interdependent community of cells. If those cells do not communicate and cooperate, the whole will break into dysfunctional parts and die.

To understand the life of a cell, then, it is critical to analyze how the cell interacts with the world outside its membrane. How do cells obtain information about the world and respond to that information? In particular, how do cells interact with other cells? To answer these questions, let's begin with the cell surface—with the molecules that separate the cell from its environment.

11.1 The Cell Surface

The line between life and nonlife is drawn by the plasma membrane that surrounds every cell. Recall that the structure of this membrane consists of a phospholipid bilayer studded with membrane proteins that are integral, meaning that they are embedded in the bilayer, or peripheral, meaning that they are attached to one surface (see Chapter 6). These proteins participate in the primary function of the plasma membrane: to create an environment inside the cell that is different from conditions outside by regulating the transport of substances.

The plasma membrane does not exist in isolation, however. Cytoskeletal elements attach to the interior face of the bilayer (see Chapter 7), and a complex array of extracellular structures interacts with the membrane's exterior surface. Let's consider the nature of the material outside the cell and then analyze how the cell interacts with it and other cells.

The Structure and Function of an Extracellular Layer

It is actually extremely rare for cells to be bounded simply by a plasma membrane. Most cells secrete products that are assembled into a layer or wall just beyond the membrane. This extracellular material helps define the cell's shape and either attaches it to another cell or acts as a first line of defense against the outside world.

Virtually all types of extracellular structures—from the cell walls of bacteria, algae, fungi, and plants to the extracellular material that surrounds most animal cells—follow the same fundamental design principle. Like reinforced concrete, they are "fiber composites": They consist of a cross-linked network of long

FIGURE 11.1 Fiber Composites Resist Tension and Compression. Fiber composites, such as reinforced concrete, consist of ground substance that fills spaces between cross-linked rods.

filaments embedded in a stiff surrounding material, or ground substance (**FIGURE 11.1**). The molecules that make up the filaments and the encasing material vary from group to group, but the engineering principle is the same. Why?

- The rods or filaments in a fiber composite are extremely effective at withstanding stretching and straining forces, or tension. The fibers present in extracellular material of most cells are functionally similar to the steel rods in reinforced concrete—they resist being pulled or pushed lengthwise.

- The stiff surrounding substance is effective at withstanding the pressing forces called compression. Concrete performs this function in highways, and a gel-forming mixture of polysaccharides achieves the same end in extracellular material.

Thanks to the combination of tension- and compression-resisting elements, fiber composites are particularly rugged. And in many living cells, the fiber and composite elements are flexible as well as strong.

What molecules make up the rods and ground substance found on the surface of plant and animal cells? How are these extracellular layers synthesized, and what do they do?

The Cell Wall in Plants

Virtually all plant cells are surrounded by a cell wall—a fiber composite that is the basis of major industries. The paper in this book, the threads in your cotton clothing, and the wood in your neighborhood's houses are made up primarily of plant cell walls.

Before analyzing the structure of plant cell walls in detail, it's important to note that these structures are dynamic. If they are damaged by attacking insects, they may release signaling molecules that trigger the reinforcement of walls in nearby cells. Cell walls are also degraded in a controlled way as fruits ripen, making the fruits softer and more digestible for the animals that disperse the seeds inside.

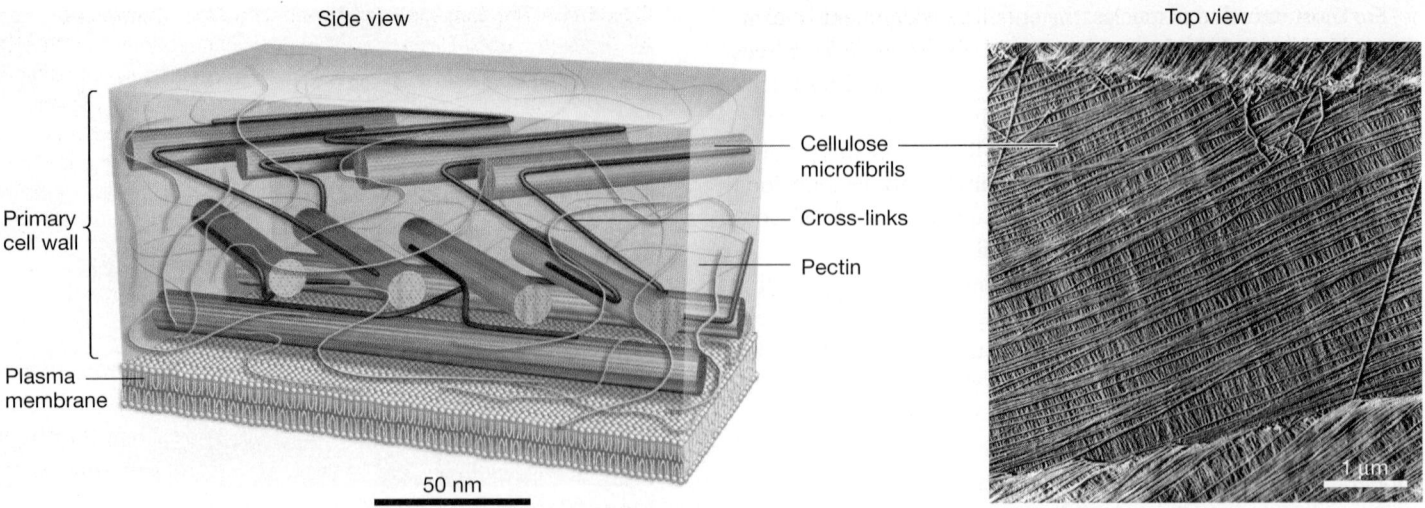

Side view　　　　　　　　　　　　　　　　　　　Top view

Primary cell wall

Plasma membrane

Cellulose microfibrils

Cross-links

Pectin

50 nm

1 μm

FIGURE 11.2 Primary Cell Walls of Plants Are Fiber Composites. In a plant's primary cell wall, cellulose microfibrils are cross-linked by polysaccharide chains. The spaces between the microfibrils are filled with pectin molecules, which form a gelatinous solid.

Primary Cell Walls　When plant cells first form, they secrete an initial fiber composite called a **primary cell wall.**

- The fibrous component of the primary cell wall consists of long strands of cellulose, which are bundled into stout, cable-like structures termed **microfibrils** and then cross-linked by other polysaccharide filaments. The microfibrils are synthesized by a complex of enzymes in the plasma membrane, forming a crisscrossed network (**FIGURE 11.2**).

- The space between microfibrils is filled with gelatinous polysaccharides such as **pectins**—the molecules that are used to thicken jams and jellies. Because the polysaccharides in pectin are hydrophilic, they attract and hold large amounts of water to keep the cell wall moist. The gelatinous components of the cell wall are synthesized in the rough endoplasmic reticulum and Golgi apparatus and secreted to the extracellular space.

The primary cell wall defines the shape of a plant cell. Under normal conditions, the nucleus and cytoplasm fill the entire volume of the cell and push the plasma membrane up against the wall. Because the concentration of solutes is higher inside the cell than outside, water tends to enter the cell via osmosis. The incoming water increases the cell's volume, exerting a force against the wall that is known as **turgor pressure.**

Although plant cells experience turgor pressure throughout their lives, it is particularly important in young cells that are actively growing. Young plant cells secrete proteins named expansins into their cell wall. **Expansins** disrupt hydrogen bonds that cross-link the microfibrils in the wall, allowing them to slide past one another. Turgor pressure then forces the wall to elongate and expand. The result is cell growth (see Chapter 40).

Secondary Cell Walls　As plant cells mature and stop growing, they may secrete an additional layer of material—a **secondary cell wall**—between the plasma membrane and the primary cell

wall. The structure of the secondary cell wall varies from cell to cell in the plant and correlates with that cell's function. Cells on the surface of a leaf have secondary cell walls that are impregnated with waxes that form a waterproof coating; the cells that support the plant's stem have secondary cell walls that contain a great deal of cellulose.

In cells that form wood, the secondary cell wall also includes **lignin,** a complex polymer that forms an exceptionally rigid network. Cells that have thick secondary cell walls of cellulose and lignin help plants withstand the forces of gravity and wind.

Although animal cells do not make a cell wall, they do form a fiber composite outside their plasma membrane. What is this substance, and what does it do?

The Extracellular Matrix in Animals

Most animal cells secrete a fiber composite called the **extracellular matrix (ECM).** Like the extracellular materials found in other organisms, structural support is one of the ECM's most important functions.

ECM design follows the same principles observed in the cell walls of bacteria, archaea, algae, fungi, and plants. There is a key difference, however: The animal ECM contains much more protein relative to carbohydrate than does a cell wall.

- The fibrous component of animal ECM is dominated by a cable-like protein termed **collagen** (**FIGURE 11.3a**).

- The matrix that surrounds collagen and other fibrous components contains gel-forming **proteoglycans** that consist of protein cores with many large polysaccharides attached to them. In some tissues, complexes of proteoglycans may also be produced (**FIGURE 11.3b**).

Most ECM components are synthesized in the rough endoplasmic reticulum (ER), processed in the Golgi apparatus, and

(a) Collagen proteins consist of three polypeptide chains that wind around one another to form the fibrous component of the animal ECM.

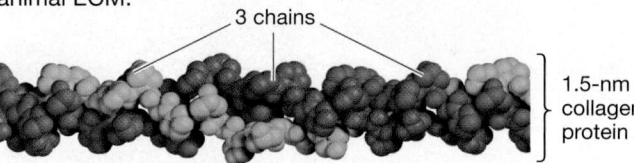

3 chains

1.5-nm collagen protein

(b) Complexes of gelatinous proteoglycans form the ground substance of the animal ECM.

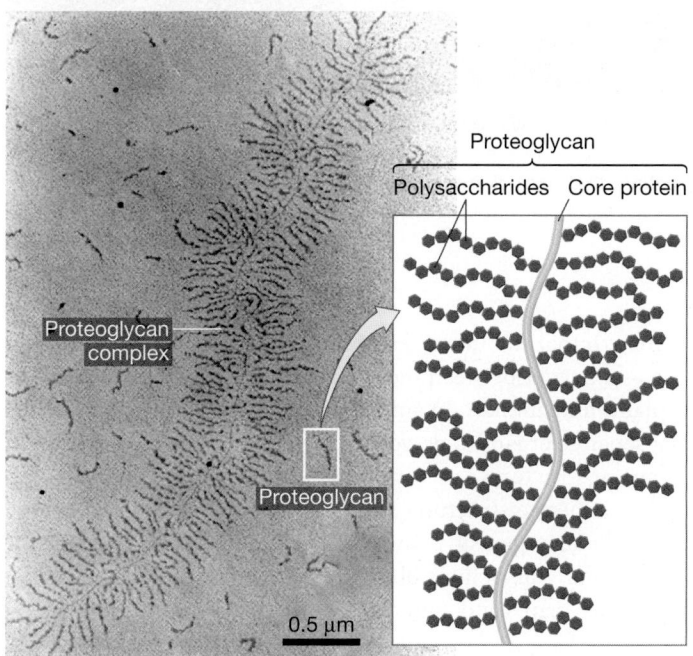

Proteoglycan complex

Proteoglycan

Proteoglycan

Polysaccharides   Core protein

0.5 μm

FIGURE 11.3 The Extracellular Matrix Is a Fiber Composite.
(a) Although several types of fibrous proteins are found in the ECM of animal cells, the most abundant is collagen. Groups of collagen proteins coalesce to form collagen fibrils, and bundles of fibrils link to form collagen fibers. **(b)** The spaces between the collagen fibers are filled with complexes of gelatinous proteoglycans. The proteoglycan subunits consist of a protein core attached to many polysaccharides.

secreted from the cell via exocytosis. After secretion, however, proteins like collagen may then assemble into larger structures, such as the fibrils shown in **FIGURE 11.4**. In addition, the secreted proteoglycans may be attached to long polysaccharides synthesized by cellular enzymes in the extracellular space. These huge complexes, such as the one shown in Figure 11.3b, are responsible for the rubber-like consistency of cartilage.

Even in the same organism, the amount of ECM varies among different types of **tissues,** which consist of similar cells that function as a unit. Bone and cartilage, for example, have relatively few cells surrounded by a large amount of ECM. Skin cells, in contrast, are packed together with a minimal amount of ECM.

The composition of the ECM also varies among tissue types. For example, the ECM surrounding cells in lung tissue contains large amounts of a rubber-like protein called elastin, which allows the ECM to expand and contract during breathing. The structure of a cell's ECM correlates with the function of the tissue.

Although collagen and the other common ECM proteins are much more elastic and bendable than the stiff cell walls of plants, they support cell structure via attachments to the cell surface. As

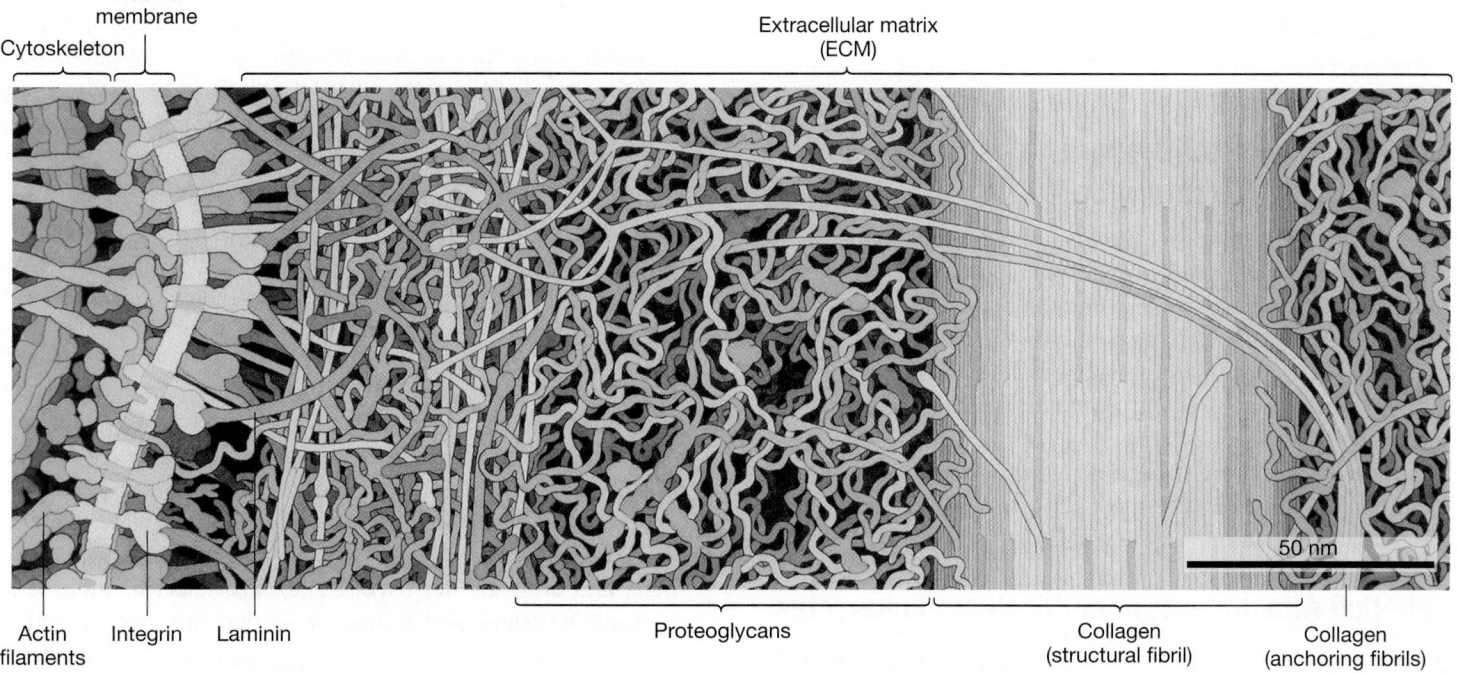

Cytoskeleton

Plasma membrane

Extracellular matrix (ECM)

Actin filaments   Integrin   Laminin

Proteoglycans

Collagen (structural fibril)

Collagen (anchoring fibrils)

50 nm

FIGURE 11.4 Integrins Connect the Extracellular Matrix to the Cytoskeleton.

Figure 11.4 shows, membrane proteins called **integrins** bind to extracellular proteins, including laminins, which in turn bind to other components of the ECM. **Laminins** are ECM crosslinking proteins—not to be confused with lamins, which are intermediate filaments found in the nucleus (see Chapter 7).

The intracellular portions of the integrins also bind to proteins that are connected to the cytoskeleton, effectively forming a bridge between the two support systems. This linkage between the cytoskeleton and ECM is critical. Besides keeping individual cells in place, it helps adjacent cells adhere to each other via their common connection to the ECM.

Cells monitor this cytoskeleton–ECM linkage via signaling pathways that will be introduced in Section 11.3. When integrins bind to the ECM, they transmit signals that inform the cell it is in the right place and properly anchored. If these linkages break down, the signals are not transmitted and cells normally die as a result. For most of the cells in your body, anchorage to the ECM is a matter of life and death.

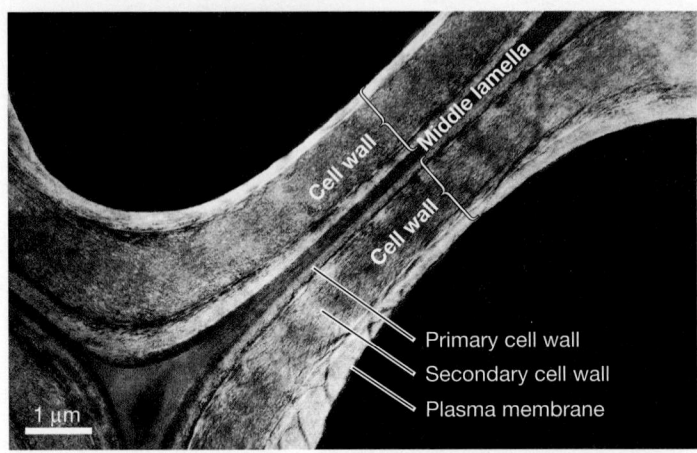

FIGURE 11.5 The Middle Lamella Connects Adjacent Plant Cells. The middle lamella contains gelatinous polysaccharides, called pectins, that help glue together the walls of adjacent cells.

11.2 How Do Adjacent Cells Connect and Communicate?

Intercellular connections are the basis of **multicellularity.** These physical connections between cells—either direct, or indirect via the ECM—are particularly important in the structure and function of tissues. The muscle tissue in your heart, for example, depends on these attachments to support the structure of the cells as they contract and relax with each beat (see the micrograph on the opening page of this chapter).

Let's look first at the structures that attach cells to each other and then examine how they allow adjacent cells to exchange materials and information.

Cell–Cell Attachments in Multicellular Eukaryotes

Materials and structures that bind cells together are particularly important in the **epithelium** (plural: **epithelia**)—a tissue that forms external and internal surfaces. These epithelial layers function as a barrier between the external and internal environments of plants and animals. In animals, epithelial cells also form layers that separate organs to prevent mixing of solutions from adjacent organs or structures.

The adhesive structures that hold cells together vary among multicellular organisms. To illustrate this diversity, consider the intercellular connections observed in the best-studied groups of organisms: plants and animals.

Indirect Intercellular Attachments The extracellular space between adjacent plant cells comprises three layers (**FIGURE 11.5**). The primary cell walls of adjacent plant cells sandwich a central layer designated the middle lamella, which consists primarily of gelatinous pectins. Because this gel layer is continuous with the primary cell walls of the adjacent cells, it serves to glue them together. The two cell walls are like slices of bread; the middle lamella is like a layer of peanut butter. If enzymes degrade the middle lamella, as they do when flower petals and leaves detach and fall, the surrounding cells separate.

In many animal tissues, integrins connect the cytoskeleton of each cell to the extracellular matrix (see Section 11.1). A middle-lamella-like layer of gelatinous polysaccharides and proteoglycans runs between adjacent animal cells. Along with the cytoskeleton–ECM connections, the polysaccharide glue helps hold cells together in tissues. In addition, in certain animal tissues the polysaccharide glue is reinforced by collagen fibrils that span the ECM to connect adjacent cells.

In animals, where cell walls do not exist, a variety of membrane proteins allow for direct cell–cell attachments in epithelia and other tissues (**FIGURE 11.6**). Let's start by looking at the tight junctions and desmosomes that hold cells together and then examine the role of gap junctions in intercellular communication.

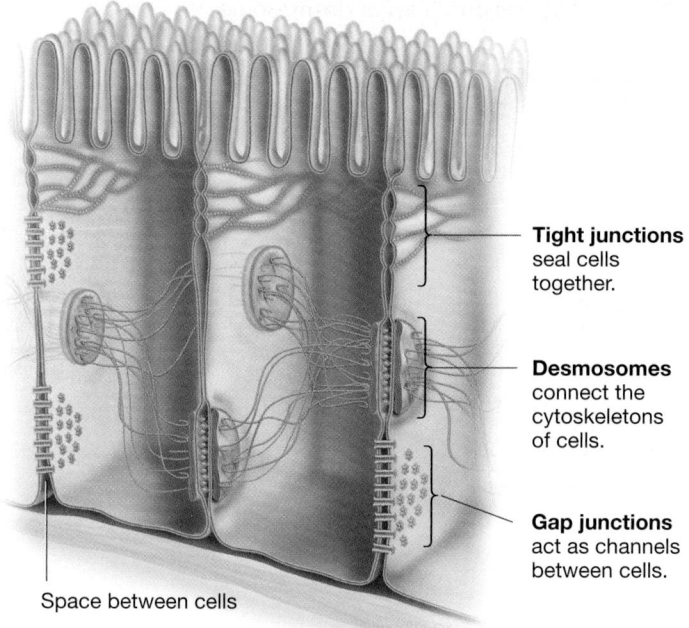

Tight junctions
seal cells
together.

Desmosomes
connect the
cytoskeletons
of cells.

Gap junctions
act as channels
between cells.

Space between cells

FIGURE 11.6 An Array of Structures Are Involved in Cell–Cell Adhesion and Communication between Animal Cells.

Tight Junctions Form a Seal between Cells

A **tight junction** is a cell–cell attachment composed of specialized proteins in the plasma membranes of adjacent animal cells (**FIGURE 11.7a**). As the drawing in **FIGURE 11.7b** indicates, these proteins line up and bind to one another. The resulting structure resembles quilting, where the proteins "stitch" the membranes of two cells together to form a watertight seal. In this way, tight junctions prevent solutions from flowing through the space between the two cells.

Because tight junctions form a watertight seal, this type of junction is commonly found in cells that form a barrier, such as the epithelial cells lining your stomach and intestines. There, they restrict the passive movement of substances between the contents of your gut and the rest of your body. Instead, only selected nutrients enter and leave the epithelia via specialized transport proteins and channels in the plasma membrane (Chapter 6).

Although tight junctions are indeed tight, they are variable. The tight junctions between the cells lining your bladder draw the cells closer together than those between the cells lining your small intestine, because they consist of different proteins. As a result, small ions can pass between the cells lining the surface of the small intestine more easily than between those lining the bladder—helping you absorb ions in your food and eliminate them in your waste.

Tight junctions are also dynamic. For example, they loosen to permit more transport between epithelial cells lining the small intestine after a meal and then "retighten" later. In this way, tight junctions can open and close in response to changes in environmental conditions.

Although tight junctions are very good at holding cells close together, they are weak adhesions that can be easily broken. Since epithelial cells often experience pulling and shearing forces, other intercellular adhesions are required to help hold cells together in a tissue. What are these other adhesions, and how do they resist being pulled apart?

(a) Electron micrograph of a tight junction in longitudinal section

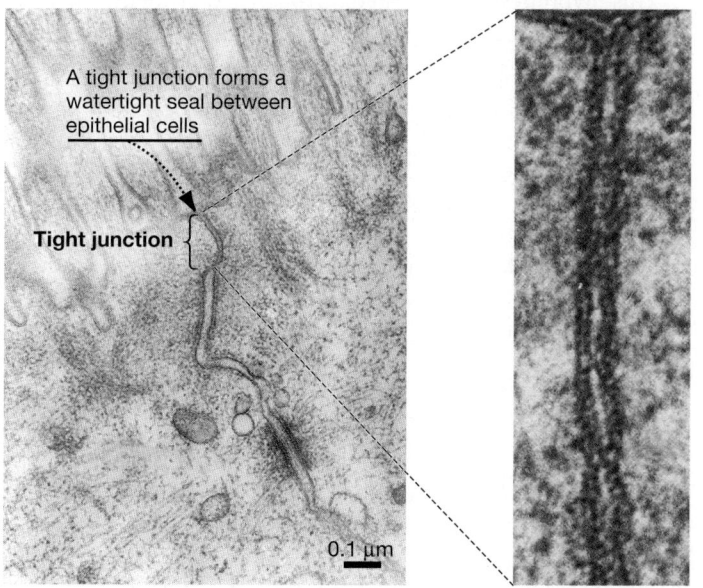

A tight junction forms a
watertight seal between
epithelial cells

Tight junction

0.1 μm

(b) Three-dimensional view of a tight junction

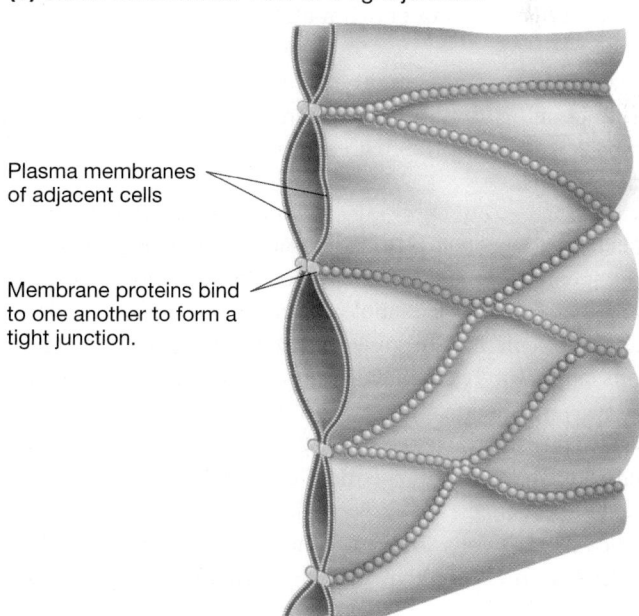

Plasma membranes
of adjacent cells

Membrane proteins bind
to one another to form a
tight junction.

FIGURE 11.7 In Animals, Tight Junctions Form a Seal between Adjacent Cells.

(a) Micrograph of desmosome in longitudinal section

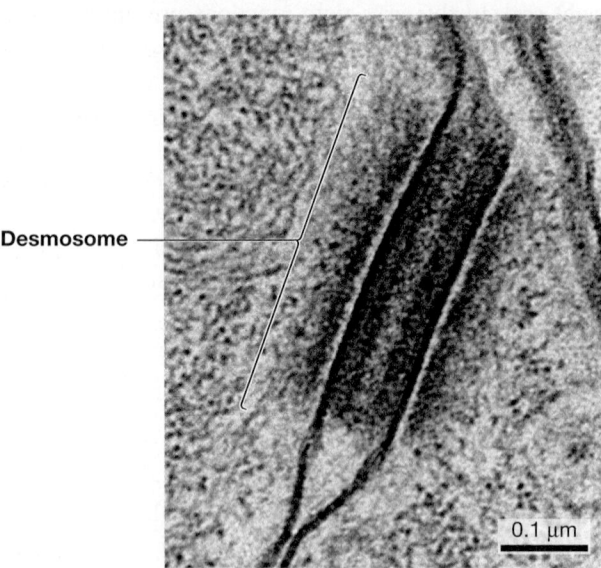

Desmosome

0.1 μm

(b) Three-dimensional view of desmosome

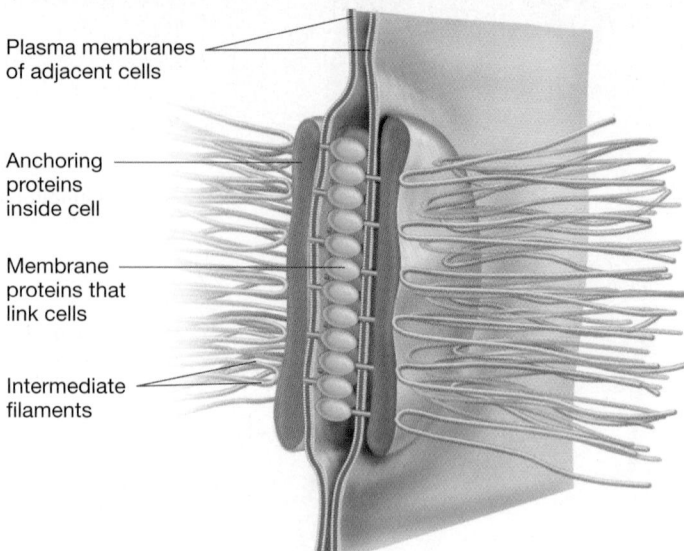

Plasma membranes of adjacent cells

Anchoring proteins inside cell

Membrane proteins that link cells

Intermediate filaments

FIGURE 11.8 Adjacent Animal Cells Are Linked by Desmosomes, Which Bind Cytoskeletons Together.

Desmosomes Form Secure Adhesions **FIGURE 11.8a** illustrates **desmosomes,** cell–cell attachments particularly common in animal epithelial cells and certain muscle cells. The structure and function of a desmosome are analogous to the rivets that hold pieces of sheet metal together.

As **FIGURE 11.8b** indicates, desmosomes are extremely sophisticated cell–cell connections. At their heart are integral membrane attachment proteins that form bridges between anchoring proteins inside adjacent cells. Intermediate filaments help reinforce these connections by attaching to the anchoring proteins in the cytoplasm. In this way, desmosomes help form a continuous structural support system between all the cells in the tissue (see Figure 11.6).

What are the membrane proteins that serve this cell attachment function in desmosomes? The answer to this question traces back to some of the first experiments conducted on cell–cell interactions.

Intercellular Adhesions Are Selective Long before electron micrographs revealed the presence of desmosomes, biologists realized that some sort of molecule must bind animal cells to one another. This insight grew out of experiments from H. V. Wilson's lab that were conducted on sponges in the early 1900s.

Sponges are aquatic animals, and the sponge species used in this study consists of just two basic types of tissues. When Wilson treated adult sponges with chemicals that made the cells separate from one another, the result was a jumbled mass of individual and unconnected cells. But when normal chemical conditions were restored, he noted that the cells gradually began to move and stick together.

As the experiment continued, cells began to aggregate based on their origin—adhering to other cells of the same tissue type.

This phenomenon is now called **selective adhesion.** Eventually the experimental sponge cells re-formed functional adult sponges with two distinct tissues. How could this happen?

The Discovery of Cell–Cell Adhesion Proteins What is the molecular basis of selective adhesion? The initial hypothesis, proposed in the 1970s, was that specialized membrane proteins were involved. The idea was that different types of cells have different types of adhesion proteins in their membranes, and only those with the same or complementary adhesion proteins are able to attach to one another.

This hypothesis was tested through experiments that relied on molecules called antibodies. An **antibody** is a protein produced by an immune response that binds specifically to a unique molecule type, often another protein (see Chapter 51). When an antibody binds to a protein, it can change the target protein's structure or interfere with its ability to interact with other molecules. This property of antibodies was crucial to these experiments.

FIGURE 11.9 shows how researchers tested the hypothesis that cell–cell adhesion takes place via interactions between membrane proteins:

Step 1 Isolate the membrane proteins from a certain cell type. Produce pure preparations of each protein.

Step 2 Inject one of the membrane proteins into a rabbit. The rabbit's immune system cells respond by creating antibodies to the membrane protein, which is recognized as being foreign. Purify those antibodies. Repeat this procedure for the other membrane proteins that were isolated. In this way, obtain a large collection of antibodies—each of which binds specifically to one (and only one) type of membrane protein.

QUESTION: Do animal cells have adhesion proteins on their surfaces?

HYPOTHESIS: Selective adhesion is due to specific membrane proteins.

NULL HYPOTHESIS: Selective adhesion is not due to specific membrane proteins.

EXPERIMENTAL SETUP:

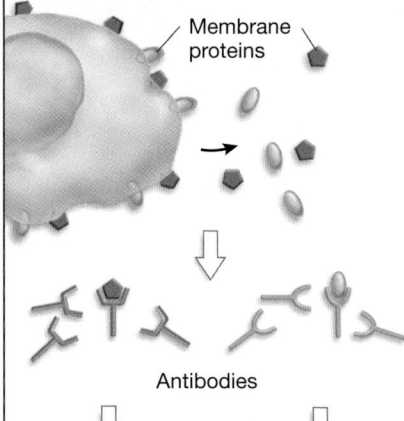

Membrane proteins

1. Isolate the membrane proteins from a certain cell type that adheres to other cells of the same type. (There are many membrane proteins; only two are shown here.)

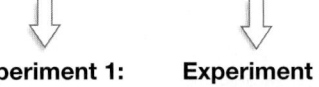

Antibodies

2. Produce antibodies that bind to specific membrane proteins. Purify the antibodies.

Experiment 1:   Experiment 2:

3. Treat cells with an antibody, one type at a time. Wait; then observe whether cells adhere normally.

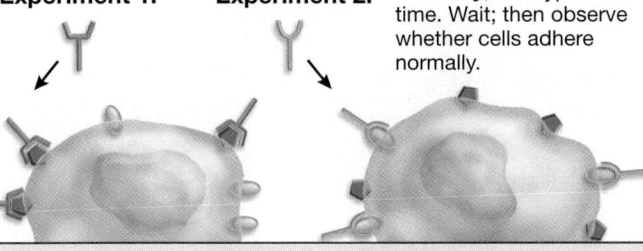

PREDICTION:

PREDICTION OF NULL HYPOTHESIS:

RESULTS:

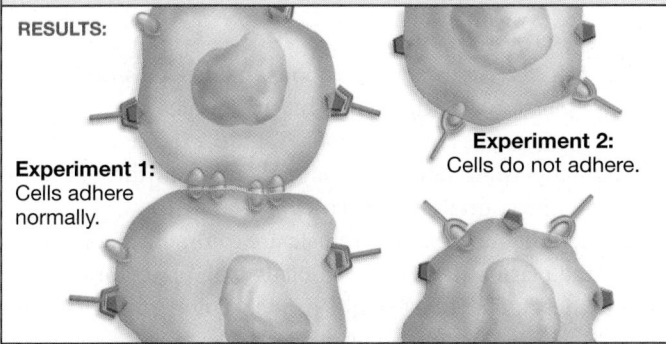

Experiment 1: Cells adhere normally.

Experiment 2: Cells do not adhere.

CONCLUSION: The protein that was blocked in experiment 2 (called a cadherin) is involved in cell–cell adhesion.

FIGURE 11.9 Evidence for Adhesion Proteins on Animal Cells.

SOURCES: Hatta, K., and M. Takeichi. 1986. Expression of N-cadherin adhesion molecules associated with early morphogenetic events in chick development. *Nature* 320: 447–449. Also Takeichi, M. 1988. The cadherins: Cell–cell adhesion molecules controlling animal morphogenesis. *Development* 102: 639–655.

✔**EXERCISE** Fill in the prediction made by each hypothesis.

Step 3 Add one antibody type to a mixture of dissociated cells from a tissue and observe whether the cells reaggregate normally. Repeat this experiment with each of the other antibody types, one type at a time.

If treatment with a particular antibody prevents the cells from attaching to one another, the antibody is probably bound to an adhesion protein. The logic is that if the antibody "shakes hands" with the adhesion protein, the adhesion protein can't shake hands with other adhesion proteins and attach the cells to one another.

This approach allowed biologists to identify several major classes of cell adhesion proteins, including **cadherins**—the attachment molecules in desmosomes. There are various types of cadherins, and cells from different tissues have different forms of cadherin in their plasma membranes. Each cadherin can bind only to cadherins of the same type. In this way, cells of the same tissue type attach specifically to one another.

To summarize: Animal cells attach to one another in a selective manner because different types of cell adhesion proteins can bind and rivet certain cells together. Cadherins provide the physical basis for selective adhesion in many cells and are a critical component of the desmosomes that join mature cells.

✔ If you understand cell–cell attachments, you should be able to predict what would happen if you treated cells in a developing frog embryo with a molecule that blocked a cadherin present in muscle tissue.

In addition to providing structural support to tissues, intercellular connections can direct cell–cell communication. But how can cellular connections pass information between cells?

Cells Communicate via Cell–Cell Gaps

In both plants and animals, direct connections between cells in the same tissue help them to work in a coordinated fashion. One way of accomplishing this is to generate channels in the membranes of adjacent cells, allowing them to communicate via the diffusion of cytosolic ions and small molecules from cell to cell.

How cells respond to this exchange of information varies from signal to signal and from cell to cell, but the result falls into two general categories:

1. Signals may alter which proteins are produced and which are not, by regulating gene expression; or

2. Signals may activate or deactivate particular proteins that already exist in the cell—often those involved in metabolism, membrane transport, secretion, and the cytoskeleton.

Whatever the mechanism, the activity of the cell often changes dramatically after the signal arrives. Let's take a closer look at how these signals are able to travel between adjacent cells connected by gap junctions and plasmodesmata.

Gap Junctions Connect Cells via Protein Channels   In most animal tissues, structures called **gap junctions** connect adjacent cells. The key feature of gap junctions is the specialized proteins that assemble in the membranes of adjacent cells, creating

(a) Gap junctions create gaps that connect animal cells.

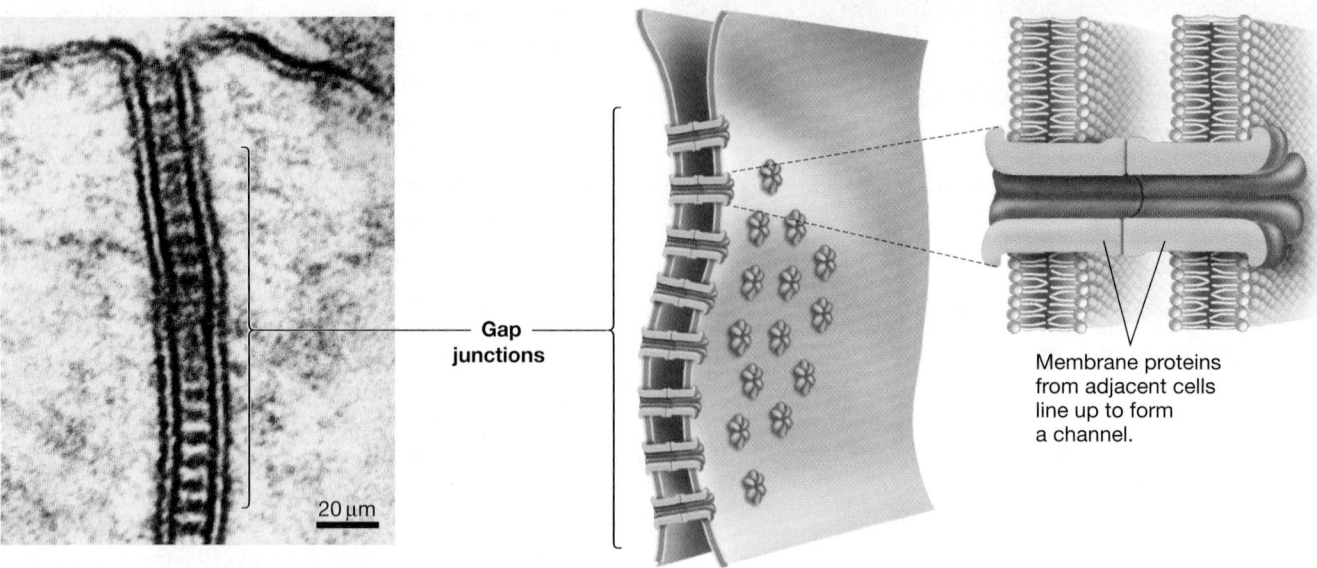

Gap junctions

Membrane proteins from adjacent cells line up to form a channel.

20 μm

(b) Plasmodesmata create gaps that connect plant cells.

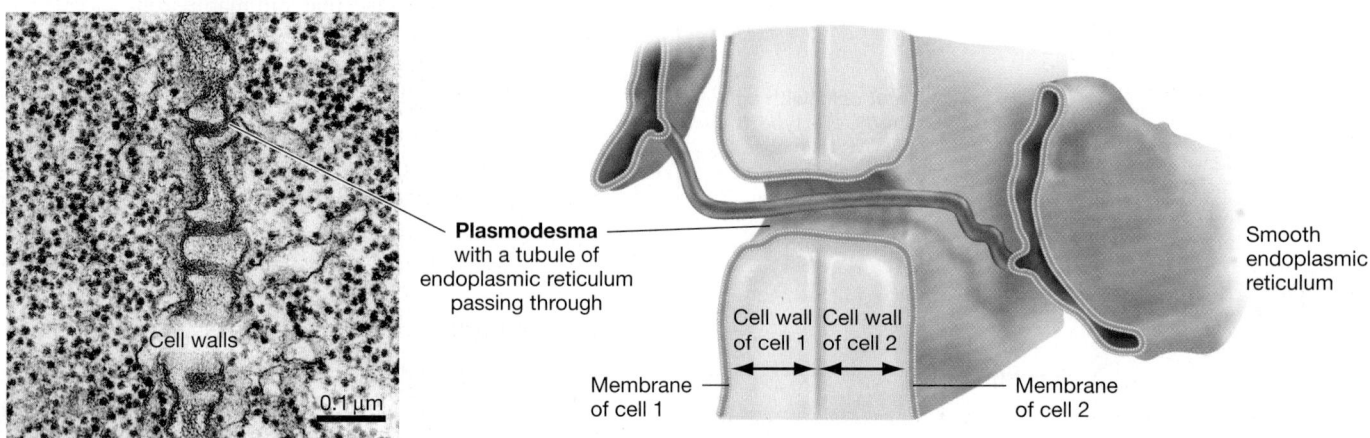

Plasmodesma with a tubule of endoplasmic reticulum passing through

Cell walls

0.1 μm

Cell wall of cell 1   Cell wall of cell 2

Membrane of cell 1

Membrane of cell 2

Smooth endoplasmic reticulum

FIGURE 11.10 **Adjacent Animal Cells and Adjacent Plant Cells Communicate Directly.**

interconnected pores between the cells (**FIGURE 11.10a**). These channels allow water, ions, and small molecules such as amino acids, sugars, and nucleotides to move between adjacent cells.

Gap junctions are communication portals. The flow of small molecules through gap junctions can help adjacent cells coordinate their activities by allowing the rapid passage of regulatory ions or molecules. In the muscle cells of your heart, for example, a flow of ions through gap junctions acts as a signal that coordinates contractions. Without this cell–cell communication, a normal heartbeat would be impossible.

In plants, direct interactions between membrane proteins are impossible due to the presence of cell walls. How do adjacent plant cells communicate?

Plasmodesmata Connect Cells via Membrane Channels   In plants, gaps in cell walls create direct connections between the cytoplasm

of adjacent cells. At these connections, named **plasmodesmata** (singular: **plasmodesma**), the plasma membrane and the cytoplasm of the two cells are continuous. Tubular extensions from the smooth endoplasmic reticulum (smooth ER) run through these membrane-lined portals (**FIGURE 11.10b**).

Like gap junctions, plasmodesmata are communication portals through the plasma membrane. In plants, the plasma membrane separates most tissues into two independent compartments: (**1**) the **symplast,** which is a continuous network of cytoplasm connected by plasmodesmata, and (**2**) the **apoplast,** which is the region outside the plasma membrane (**FIGURE 11.11**). The apoplast consists of cell walls, the middle lamella, and air spaces. Small molecules can move through plant tissues in either of these compartments without ever crossing a membrane (see Chapter 38).

Gap junctions and plasmodesmata allow for adjacent cells to transmit information, like a conversation between neighbors.

But how do multicellular organisms send messages between different tissues, where in most cases there is no direct contact? For example, suppose that the muscle cells in your arm are exercising so hard they run low on sugar or that leaf cells in a maple tree are attacked by caterpillars. How do these cells signal tissues or organs elsewhere in the organism to release materials that are needed to fend off exhaustion or caterpillars? Distant cell communication is the subject of Section 11.3.

11.3 How Do Distant Cells Communicate?

Cells that are not in physical contact communicate with one another. This is true for unicellular organisms, where hundreds or

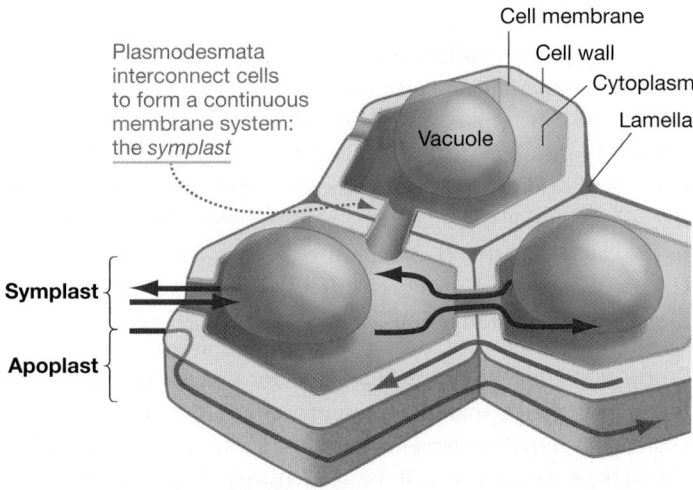

FIGURE 11.11 Most Plant Tissues Are Divided into Two Compartments—Symplast and Apoplast. Small molecules may be transported through plant tissues either within the shared cytoplasm (symplast) or through the extracellular space (apoplast).

thousands of cells may live in close proximity, as well as for multicellular organisms like humans and maple trees, which typically contain trillions of cells and dozens of tissue types.

Cell–cell communication qualifies as one of the most dynamic and important research areas in biology. Let's begin by analyzing how distant cells in humans and other multicellular eukaryotes exchange information, and then in Section 11.4 explore how unicellular organisms communicate.

Cell–Cell Signaling in Multicellular Organisms

Suppose that cells in your brain sense that you are becoming dehydrated. Brain cells can't do much about the water you lose during urination, but kidney cells can. In response to dehydration, certain brain cells release a signaling molecule that travels to the kidneys. The arrival of this molecule activates specialized membrane channels that prevent water from being lost in urine—an important aspect of fighting dehydration.

Thanks to cell–cell signals, the activities of cells in different parts of a multicellular body are coordinated.

Biologists have classified many different types of signaling molecules that keep distant tissues in touch. For example, neurotransmitters activate membrane channel receptors that open to allow a flow of ions into the cytosol of the cell, changing the electrical properties of the membrane. This type of signal is responsible for the transmission of information along neurons, allowing your brain to control the movements of the rest of your body (see Chapters 46 and 48). The best-studied means of distant signaling, however, may be via **hormones**—information-carrying molecules that can act on distant target cells because they are secreted by plant and animal cells into bodily fluids.

Hormones are usually small molecules and are typically present in minute concentrations. Even so, they have a large impact on the activity of target cells and the condition of the body as a whole. Hormones are like a scent or whispered phrase from someone you are attracted to—a tiny signal, but one that makes your cheeks flush and your heart pound.

As **TABLE 11.1** (see page 210) indicates, hormones have a wide array of chemical structures and effects. The important point about a signaling molecule, though, is not whether it is a gas or peptide or steroid, but whether it is lipid soluble or not. The ability of a signaling molecule to pass through lipid bilayers is crucial in determining how a target cell recognizes it. Where does this recognition occur—inside the cell or outside?

- Most lipid-soluble signaling molecules are able to diffuse across the hydrophobic region of the plasma membrane and enter the cytoplasm of their target cells.

- Large or hydrophilic signaling molecules are lipid insoluble and do not cross the plasma membrane. To affect a target cell, they have to be recognized at the cell surface.

How do cells receive and process signals from distant cells? The basic steps are common to all cell–cell signaling systems. Let's consider each step in turn.

TABLE 11.1 Hormones Have Diverse Structures and Functions

Hormone Name	Chemical Structure	Where Is Signal Received?	Function of Signal
Auxin	Small organic compound	At plasma membrane	Signals changes in long axis of plant body
Brassinosteroids	Steroid	At plasma membrane	Stimulate plant cell elongation
Estrogens	Steroid	Inside cell	Stimulate development of female characteristics in animals
Ethylene	C_2H_4 (a gas)	At plasma membrane	Stimulates fruit ripening; regulates aging
FSH	Glycoprotein	At plasma membrane	Stimulates egg maturation, sperm production in animals
Insulin	Protein, 51 amino acids	At plasma membrane	Stimulates glucose uptake in animal bloodstream
Prostaglandins	Modified fatty acid	At plasma membrane	Perform a variety of functions in animal cells
Systemin	Peptide, 111 amino acids	At plasma membrane	Stimulates plant defenses against herbivores
Thyroxine (T4)	Modified amino acid	Inside cell	Regulates metabolism in animals

Signal Reception

Hormones and other types of cell–cell signaling molecules deliver their message by binding to receptor molecules. Even though the molecule that carries the message "We're getting dehydrated—conserve water" is broadcast throughout the body, only certain kidney cells respond because only they have the appropriate receptor. The presence of an appropriate receptor dictates which cells will respond to a particular hormone. Bone and muscle cells don't respond to the "conserve water" message, because they don't have a receptor for it.

Cells in a wide array of tissues may respond to the same signaling molecule, though, if they have the appropriate receptor. If you are startled by a loud noise, cells in your adrenal glands secrete a hormone called adrenaline (also called epinephrine) that carries the message "Get ready to fight or run." In response, your heart rate increases, your breathing rate increases, and cells in your liver release sugars that your muscles can use to power rapid movement. This is the basis of an "adrenaline rush." Heart, lung, and liver cells respond to adrenaline because they each have the receptor that binds to it. Identical receptors in diverse cells and tissues allow long-distance signals to coordinate the activities of cells throughout a multicellular organism.

No matter where signal receptors are located, it's critical to note two important points about these proteins:

1. **Receptors are dynamic.** The number of receptors in a particular cell may decline if hormonal stimulation occurs at high levels over a long period of time. The ability of a receptor molecule to bind tightly to a signaling molecule may also decline in response to intensive stimulation. As a result, the sensitivity of a cell to a particular hormone may change over time.

2. **Receptors can be blocked.** The drugs called beta-blockers, for example, bind to certain receptors for the hormone adrenaline. When adrenaline binds to receptors in heart cells, it stimulates more rapid and forceful contractions. So if a physician wants to reduce a patient's heart cell contraction as a way to lower pressure, she is likely to prescribe a beta-blocker.

Most signal receptors are located in the plasma membrane, where they can bind to signaling molecules that cannot or do not cross the membrane. Other signal receptors exist inside the cell, where they respond to lipid-soluble signaling molecules that readily diffuse through the plasma membrane.

The most important general characteristic of signal receptors, though, is that their conformation—meaning, overall shape—changes when a hormone binds to them. A **signal receptor** is a protein that changes its shape and activity after binding to a signaling molecule.

This is a critical event in cell–cell signaling. The change in receptor structure means that the signal has been received. It's like throwing an "on" switch. What happens next?

Signal Processing

Once a cell receives a signal, something has to happen to initiate the cell's response. This signal processing step happens in one of two ways, depending on whether the signal is received inside the cell or at the membrane surface.

Processing Lipid-Soluble Signals When lipid-soluble signals enter a cell, the information they carry is processed directly—without any intermediate steps. For example, steroid hormones such as testosterone and estradiol (one of a group commonly referred to as estrogens) diffuse through the plasma membrane and enter the cytoplasm, where they bind to a cytosolic receptor protein. The hormone–receptor complex is then transported to the nucleus, where it triggers changes in the genes being expressed in the cell (**FIGURE 11.12**). By altering the expression of genes (see Chapter 17), the cell produces different proteins that will have a direct effect on the function or shape of the cell.

Processing Lipid-Insoluble Signals Hormones that *cannot* diffuse across the plasma membrane and enter the cytoplasm can't change the activity of genes or pumps directly. Instead, the signal that arrives at the surface of the cell has to initiate an intracellular signal—the signal processing step is indirect.

When a signaling molecule binds at the cell surface, it triggers **signal transduction**—the conversion of a signal from one form to another. A long and often complex series of events ensues, collectively called a signal transduction pathway.

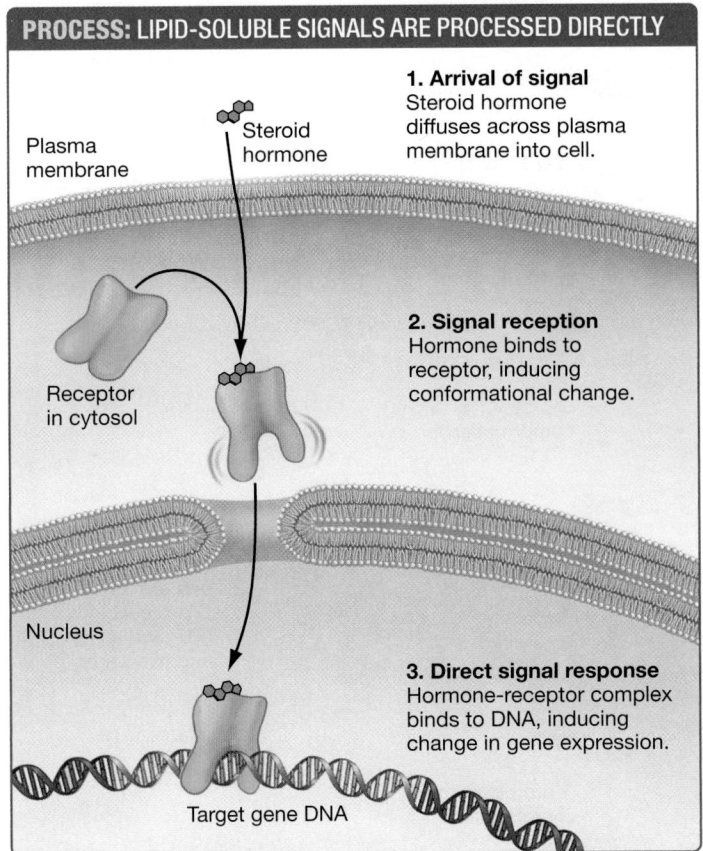

PROCESS: LIPID-SOLUBLE SIGNALS ARE PROCESSED DIRECTLY

Plasma membrane

Steroid hormone

1. Arrival of signal
Steroid hormone diffuses across plasma membrane into cell.

Receptor in cytosol

2. Signal reception
Hormone binds to receptor, inducing conformational change.

Nucleus

3. Direct signal response
Hormone-receptor complex binds to DNA, inducing change in gene expression.

Target gene DNA

FIGURE 11.12 Some Cell–Cell Signaling Molecules Enter the Cell and Bind to Receptors in the Cytoplasm. Because they are lipids, steroid hormones can diffuse across the plasma membrane and bind to signal receptors inside the cell. The hormone–receptor complex is transported to the nucleus and binds to genes, changing their activity.

✔ **QUESTION** Based on what you have learned about nuclear transport (see Chapter 7), what type of signal would you expect to be exposed on the cytosolic receptor after the steroid hormone changes the receptor's conformation?

Signal transduction is a common occurrence in everyday life. For example, the e-mail messages you receive are transmitted from one computer to another over cables or wireless transmissions. These electronic signals can be transmitted efficiently over long distances but would be meaningless to you. Software in your computer has to transduce, or convert, the signals into a form that you can understand and respond to, such as words on the screen.

Signal transduction pathways work the same way (**FIGURE 11.13**). In a cell, signal transduction converts an extracellular signal to an intracellular signal. As in an e-mail transmission, a signal that is easy to transmit is converted to a signal that is easily understood and that triggers a response.

Intracellular Signals May Be Amplified Recall that hormones are present in minuscule concentrations but trigger a large response from cells. Signal amplification is one reason this is possible. When a hormone arrives at the cell surface, the message it transmits may be amplified as it changes form. The amplifier

in your portable music player performs an analogous function: Once it is amplified, a tiny sound signal can get a whole roomful of people dancing.

In cells, signal transduction begins at the plasma membrane; amplification occurs inside. Amplification may occur in a variety of ways. In general, the mechanism of amplification correlates with the mechanism of signal transduction. But the general observation is that the arrival of a single signaling molecule may result in a secondary signal that involves many ions or molecules.

For example, one major type of signal transduction system consists of membrane channel receptors that open to allow a flow of ions into the cytosol of the cell. In muscle cells, this type of amplification occurs when calcium ions flood into the cytosol, activating all the myosin filaments so the entire cell contracts as a whole (see Chapter 48).

Here let's focus on two other major types of signal transduction and amplification systems that are distinguished based on how they are initiated:

1. G-protein-coupled receptors initiate the production of intracellular or "second" messengers that then amplify the signal.

2. Enzyme-linked receptors amplify the signal by triggering the activation of a series of proteins inside the cell, through the addition of phosphate groups.

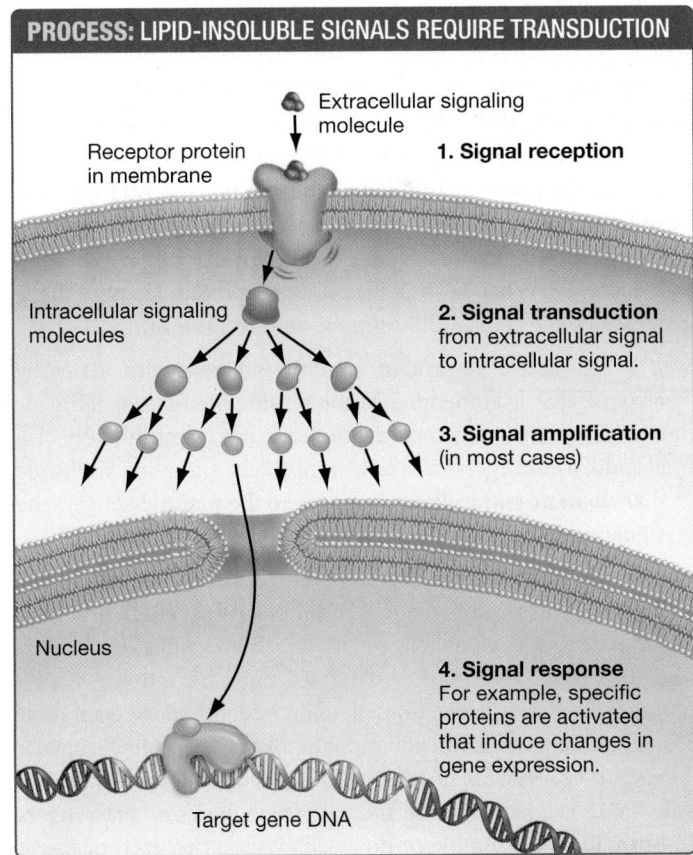

PROCESS: LIPID-INSOLUBLE SIGNALS REQUIRE TRANSDUCTION

Extracellular signaling molecule

Receptor protein in membrane

1. Signal reception

Intracellular signaling molecules

2. Signal transduction
from extracellular signal to intracellular signal.

3. Signal amplification
(in most cases)

Nucleus

4. Signal response
For example, specific proteins are activated that induce changes in gene expression.

Target gene DNA

FIGURE 11.13 Signal Transduction Converts an Extracellular Signal to an Intracellular Signal. Signal transduction is a multistep process.

Although there are many variations in the signaling pathways that fall within these two categories, the common features are emphasized here. Let's look at these two signal transduction systems in turn.

Signal Transduction via G-Protein-Coupled Receptors Many signal receptors span the plasma membrane and are closely associated with membrane-anchored proteins inside the cell called **G proteins.** When G proteins are activated by a signal receptor, they trigger a key step in signal transduction: the production of a messenger inside the cell. They link the receipt of an extracellular signal to the production of an intracellular signal.

G proteins got their name because the type of guanine nucleotide they are bound to regulates their activity: either guanosine triphosphate (GTP) or guanosine diphosphate (GDP). GTP is a nucleoside triphosphate that is similar in structure to adenosine triphosphate (ATP; introduced in Chapter 4). Recall that nucleoside triphosphates have high potential energy because their three phosphate groups have four negative charges close together.

When GTP binds to a G protein, the addition of the negative charges alters the protein's shape. Changes in shape produce changes in activity. G proteins are turned on or activated when they bind GTP; they are turned off or inactivated when a phosphate group, and thus a negative charge, is removed to form GDP. The G protein will remain in this off position until the GDP is removed and the protein binds to a new GTP.

To understand how G proteins fit into an overall signal transduction pathway, follow the events in **FIGURE 11.14**.

Step 1 A hormone arrives and binds to a receptor in the plasma membrane. Notice that the receptor is a transmembrane protein with the intracellular portion coupled to a G protein that is composed of multiple subunits.

Step 2 In response to hormone binding, the receptor changes shape and activates its G protein. Specifically, the receptor kicks out the GDP from the inactive G protein, allowing it to bind to a new GTP. When GTP is bound, the G protein changes shape radically: the active GTP-binding subunit splits off.

Step 3 The active G protein subunit interacts with a nearby enzyme that is embedded in the plasma membrane. This interaction stimulates the enzyme to catalyze production of a **second messenger**—a small, nonprotein signaling molecule that elicits an intracellular response to the first messenger (the signaling molecule that arrived at the cell surface).

Second messengers are effective because they are small and diffuse rapidly to spread the signal throughout the cell. In addition, they can be produced quickly in large quantities. This characteristic is important. Because the arrival of a single hormone molecule can stimulate the production of many second messenger molecules, the signal transduction event amplifies the original signal.

Several types of small molecules act as second messengers in cells. **TABLE 11.2** lists some of the best-studied second messengers and provides an example of how cells respond to each of them. Note that several second messengers activate **protein kinases**—enzymes that activate or inactivate other proteins by adding a phosphate group to them.

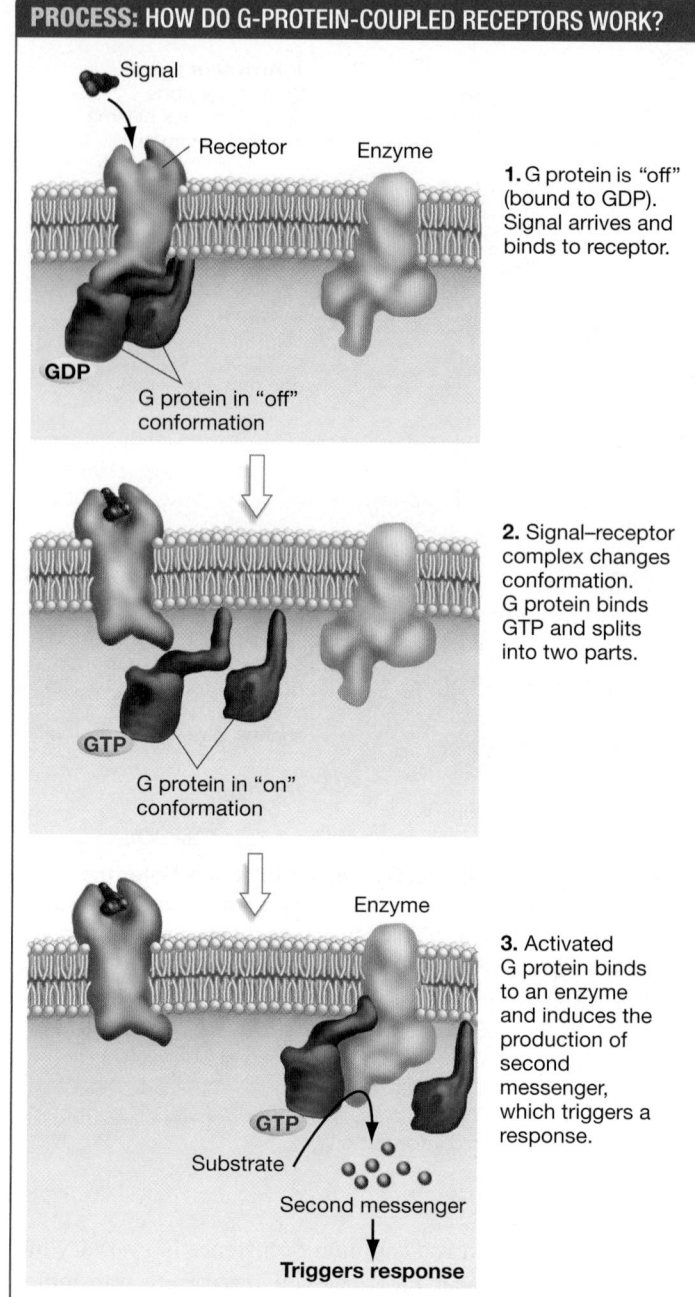

PROCESS: HOW DO G-PROTEIN-COUPLED RECEPTORS WORK?

1. G protein is "off" (bound to GDP). Signal arrives and binds to receptor.

2. Signal–receptor complex changes conformation. G protein binds GTP and splits into two parts.

3. Activated G protein binds to an enzyme and induces the production of second messenger, which triggers a response.

FIGURE 11.14 G-Protein-Coupled Receptors Trigger the Production of a Second Messenger.

It's also important to note two things:

1. Second messengers aren't restricted to a single role or single cell type—the same second messenger can initiate dramatically different events in different cell types; and

2. It is common for more than one second messenger to be involved in triggering a cell's response to the same extracellular signaling molecule.

To make sure that you understand how G proteins and second messengers work, imagine the following movie scene: A spy arrives at a castle gate. The guard receives a note from the spy, but

TABLE 11.2 Examples of Second Messengers

Name	Type of Response
Cyclic guanosine monophosphate (cGMP)	Opens ion channels; activates certain protein kinases
Diacylglycerol (DAG)	Activates certain protein kinases
Inositol trisphosphate (IP$_3$)	Opens calcium channels to transport stored calcium ions
Cyclic adenosine monophosphate (cAMP)	Activates certain protein kinases
Calcium ions (Ca^{2+})	Binds to a receptor called calmodulin; Ca^{2+}/calmodulin complex then activates proteins

he cannot read the coded message. Instead, the guard turns to the queen. She reads the note and summons the commander of the guard, who sends soldiers throughout the castle to warn everyone of approaching danger. ✔ You should be able to identify which characters in the scene correspond to the second messenger, G protein, hormone, receptor, and enzyme activated by the G protein.

It's difficult to overstate the importance of signal transduction by G-protein-coupled receptors. Biomedical researchers estimate that half of human drugs target signal receptors that are associated with G proteins.

Signal Transduction via Enzyme-Linked Receptors Enzyme-linked receptors transduce hormonal signals by directly catalyzing a reaction inside the cell. **FIGURE 11.15** focuses on the best-studied group of enzyme-linked receptors: the **receptor tyrosine kinases (RTKs)**.

Step 1 A hormone binds to an RTK.

Step 2 The protein forms a dimer. In this conformation, the catalytic activity of the receptor is turned on, allowing it to phosphorylate itself using ATP inside the cell.

Step 3 Proteins inside the cell bind to the phosphorylated RTK to form a bridge between the receptor and a peripheral membrane protein called **Ras,** which is a G protein. The formation of the RTK bridge activates Ras by causing it to exchange its GDP for a GTP.

Step 4 When Ras is activated, it triggers the phosphorylation and activation of another protein.

Step 5 The phosphorylated protein is a protein kinase, which then catalyzes the phosphorylation and activation of other kinases, which phosphorylate yet another population of proteins.

This sequence of events is termed a **phosphorylation cascade,** and it culminates in a response by the cell. The enzymes involved are called **mitogen-activated protein kinases (MAPK).** They are so named because many of the signaling molecules that start

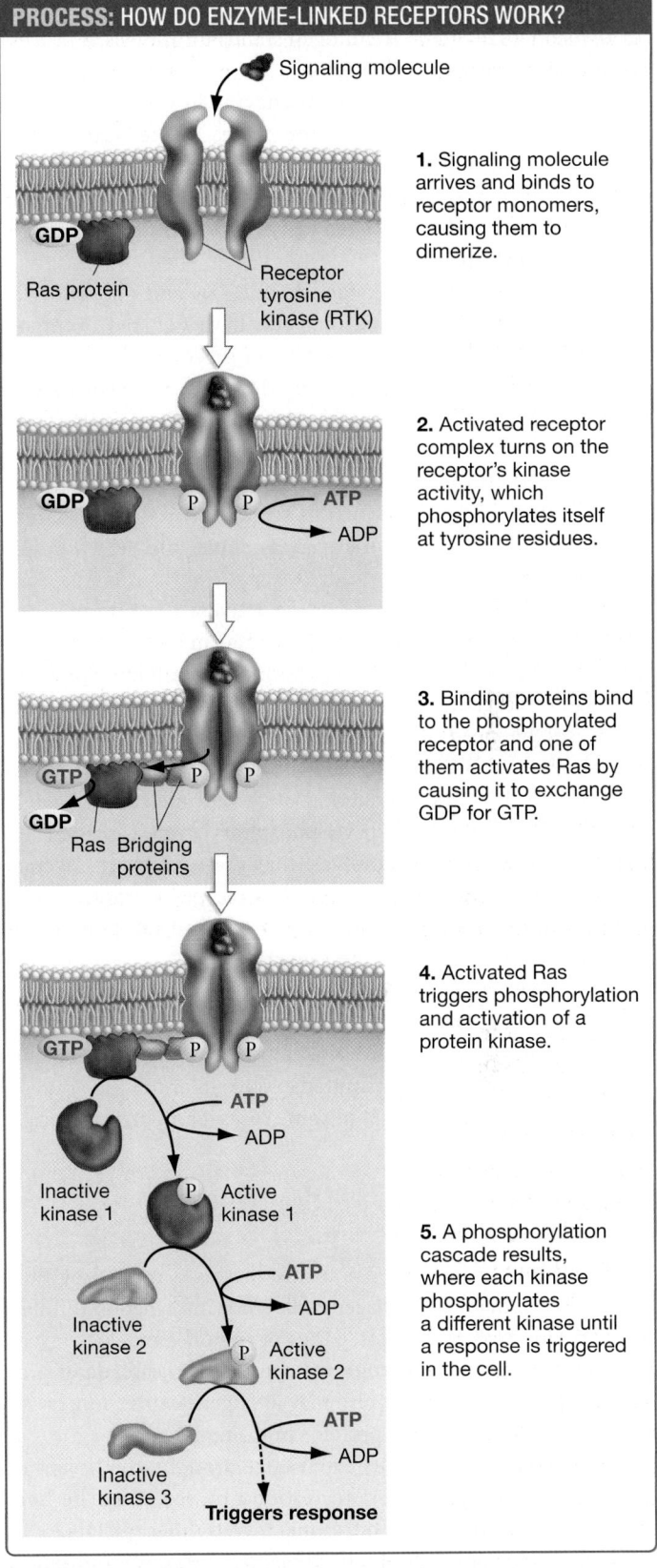

PROCESS: HOW DO ENZYME-LINKED RECEPTORS WORK?

Signaling molecule

GDP

Ras protein

Receptor tyrosine kinase (RTK)

1. Signaling molecule arrives and binds to receptor monomers, causing them to dimerize.

GDP

P   P

ATP

ADP

2. Activated receptor complex turns on the receptor's kinase activity, which phosphorylates itself at tyrosine residues.

GTP

GDP

Ras   Bridging proteins

P   P

3. Binding proteins bind to the phosphorylated receptor and one of them activates Ras by causing it to exchange GDP for GTP.

GTP

P   P

4. Activated Ras triggers phosphorylation and activation of a protein kinase.

ATP

ADP

Inactive kinase 1

P   Active kinase 1

Inactive kinase 2

ATP

ADP

P   Active kinase 2

Inactive kinase 3

ATP

ADP

Triggers response

5. A phosphorylation cascade results, where each kinase phosphorylates a different kinase until a response is triggered in the cell.

FIGURE 11.15 Enzyme-Linked Receptors Trigger a Phosphorylation Cascade.

these pathways are substances called mitogens, which activate cell division. (The *mito–* in mitogen stands for mitosis, a process involved in eukaryotic cell division.)

In some cases, each copy of an enzyme in the cascade catalyzes the phosphorylation of many copies of the next "downstream" protein, and so on. When this occurs, activated enzymes at a given stage in the cascade exist in greater numbers than the activated enzymes that preceded them, and the original signal is amplified many times over.

To make sure you understand how RTKs and phosphorylation cascades work, imagine that you have two red dominos, one black domino, and a large supply of green, blue, and yellow dominos. The red dominos represent the two subunits of an RTK dimer, and the black domino represents Ras. Each of the other colors represents a different protein kinase in a phosphorylation cascade. ✔ QUANTITATIVE You should be able to (1) explain how you would set up the dominos to simulate a phosphorylation cascade, and (2) state how many green, blue, and yellow dominos would be required to model the pathway if Ras and each protein kinase in the cascade were to activate 10 proteins.

In many cases, the proteins that take part in a phosphorylation cascade are held in close physical proximity by scaffolding proteins. Although this organization limits the amplification of the response, it increases the speed and efficiency of the reaction sequence.

In general, intracellular signals initiated by G-protein-coupled receptors result in the production of second messengers, while enzyme-linked receptors drive phosphorylation cascades. It's important to recognize, however, that these pathways overlap significantly. Some G-protein-coupled receptors trigger phosphorylation cascades, and some enzyme-linked receptors result in the production of second messengers.

To summarize: Many of the key signal transduction events observed in cells occur via G-protein-coupled receptors or enzyme-linked receptors. The signal transduction event has two results: (**1**) It converts an easily transmitted extracellular message into an intracellular message, and (**2**) in some cases it amplifies the original message many times over.

Signal Response

What is the ultimate response to the messages carried by hormones? Recall that when adjacent cells share information through cell–cell gaps, two general categories of response may occur (see Section 11.2). Likewise, second messengers or a cascade of protein phosphorylation events also may alter gene expression, or activate or deactivate existing proteins in the target cell.

For example, when plants experience drought, the tissues in the root system are the first to respond by secreting the hormone abscisic acid. This hormone travels huge distances to reach the leaves and eventually bind to its receptors in guard cells that control the stomatal pores that allow for gas exchange (see Chapter 10). When abscisic acid binds to these receptors, a signal transduction pathway ensues that increases the concentration of calcium inside the guard cells. In response, potassium ions (K^+) move out of the guard cells, creating an osmotic gradient that leads to the movement of water out of the guard cells. The guard cells deflate and close the pore, which prevents water loss from the plant.

At this point, you've analyzed the first three steps of cell–cell communication: signal reception, signal processing, and the response. Now the question is, how is the signal turned off? Consider the flush of testosterone and estrogens that you experienced during puberty, and the morphological changes these hormones induced. Abnormalities would result if these changes continued indefinitely. What limits the response to a cell–cell signal?

Signal Deactivation

Cells have built-in systems for turning off intracellular signals. For example, once activated G proteins turn on downstream enzymes, the GTP is hydrolyzed to GDP and P_i. When this reaction occurs, the G protein's conformation changes and it returns to an inactive state. Activation of its downstream target stops, and production of the second messenger ceases.

The presence of second messengers in the cytosol is also short lived. For example, pumps in the membrane of the smooth ER return calcium ions to storage, and enzymes called phosphodiesterases convert active cAMP and cGMP (see Table 11.2) to inactive AMP and GMP. When second messengers are cleared from the cytosol, the response stops.

To continue the response from G-protein-coupled receptors, the G proteins must be reactivated by the activated signal receptor to start the process again. Otherwise, the signal transduction system quickly shuts down.

Phosphorylation cascades are also sensitive to the presence of the external signal. Enzymes, called **phosphatases,** that remove phosphate groups from cascade proteins are always present in the cell. If hormone stimulation of a receptor tyrosine kinase ends, phosphatases will dephosphorylate enough components of the phosphorylation cascade that the response ceases.

Although an array of specific mechanisms are involved, here is the general observation: Signal transduction systems trigger a rapid response and can be shut down quickly. As a result, they are exquisitely sensitive to small changes in the concentration of hormones or in the number and activity of signal receptors.

It is critical, though, to appreciate what happens when a signal transduction system does not shut down properly. For example, recall that Ras is active when it binds GTP, but is deactivated once it has hydrolyzed GTP to GDP and P_i. If this hydrolysis activity were defective, however, Ras would stay in the GTP-bound "on" position and continue stimulating a phosphorylation cascade even when the external signal is absent.

Why is continuously active Ras a problem? Recall that the phosphorylation cascade that is activated by Ras involves mitogen-activated protein kinases, many of which induce cell replication. If cells express this type of defective Ras, they would receive a never-ending "divide now" signal that may lead to the development of cancer. An estimated 25–30 percent of all human cancers express this type of defective Ras. (To learn more about the family of diseases called cancer, see Chapters 12 and 19.)

Crosstalk: Synthesizing Input from Many Signals

Although the preceding discussion focused on how cells respond to individual signals, it's crucial to realize that every cell has an array of signal receptors on its plasma membrane and in its cytoplasm, and many cells receive an almost constant stream of different signals. Just as you receive information about your environment via text messages, e-mails, phone calls, and snail mail, cells get an array of chemical signals about changes in their environment.

The signal transduction pathways that are triggered by these signals and receptors intersect and connect. In reality, they are not strictly linear like the pathways illustrated in Figures 11.12 through 11.15. Signal transduction pathways form a network. This complexity is important: It allows cells to respond to many different signals in an integrated way.

The diverse signals that a cell receives are integrated by what biologists call **crosstalk**—meaning, the signals from different pathways interact to modify the cell response (**FIGURE 11.16**). This would be like getting advice from multiple people before making a decision. Here are three key things to note:

1. Elements or products from one pathway may inhibit steps in a different pathway—reducing the cell's response, even though the appropriate signal is present.

2. A response from one pathway may stimulate a greater response by a protein in a different pathway, increasing the cell's response to the other signal.

3. The presence of multiple steps in each signaling pathway provides a series of points where crosstalk can regulate the flow of information. These interactions are important, because they allow the cell to respond appropriately to many signals at the same time.

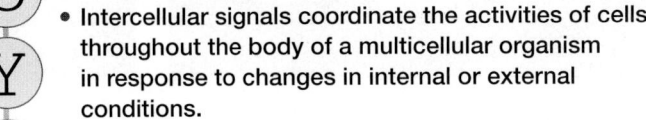
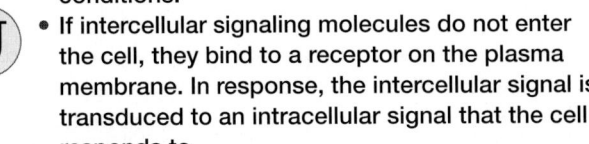
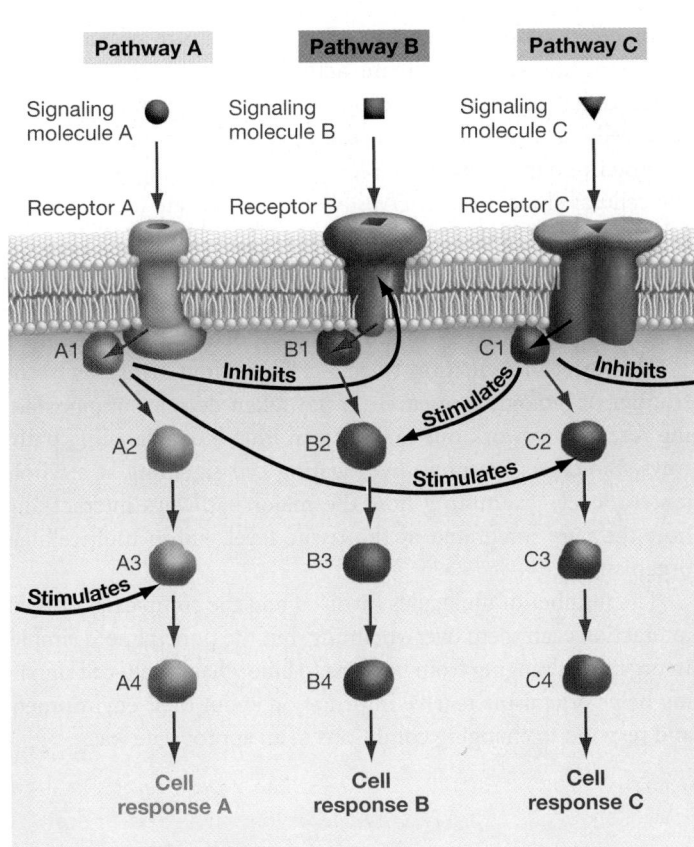

FIGURE 11.16 Signaling Pathways Interact via "Crosstalk."

✔ **EXERCISE** Predict which response would occur in cells exposed to the following signaling molecules: (1) A + B; (2) A + C; (3) C alone.

11.4 Signaling between Unicellular Organisms

Cell–cell signaling has been one of the hottest research areas in biological science over the past two decades. Surprisingly, much of what we know about signal transduction in multicellular organisms has come from the study of unicellular organisms. While single-cell microbes communicate with one another in a manner similar to what is observed in multicellular organisms, the topic of conversation often differs. Rather than asking for help, as when a dehydrated brain asks the kidney to conserve water, the conversations between unicellular microbes are often about sex and environmental change.

Responding to Sex Pheromones

While unicellular eukaryotes generally reproduce by cell division, some also are known to undergo sexual reproduction (see Chapters 12 and 13 to learn more about cellular reproduction). At its most basic level, sex involves the fusion of two cells such that genetic material of the two individuals is combined into one nucleus. What attracts individuals of the opposite sex to each other?

In *Saccharomyces cerevisiae*, or baker's yeast, there are two sexes, or mating types, referred to as "**a**" cells and "**α**" cells. By

(a) Yeast cells alter their growth in response to pheromones of the opposite mating type.

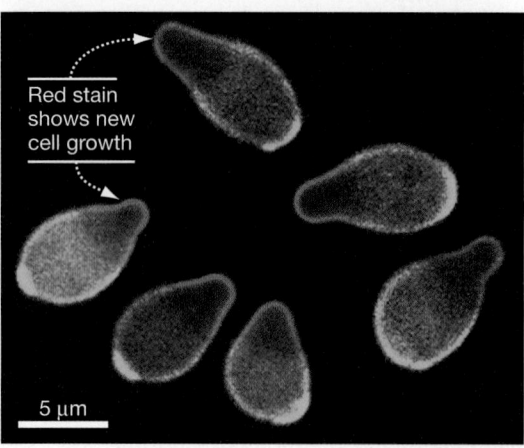

Red stain shows new cell growth

5 μm

(b) Slime mold amoebae aggregate in response to sensing a quorum.

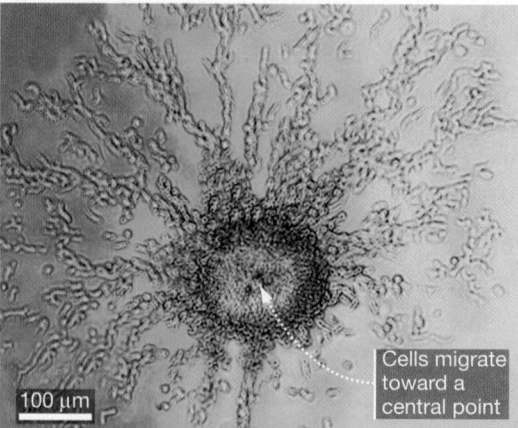

100 μm

Cells migrate toward a central point

FIGURE 11.17 Unicellular Organisms Interact and Respond in a Variety of Ways. (a) Yeast cells respond to pheromone signals from the opposite mating type by growing toward the source of the signal during sexual reproduction. **(b)** Signals secreted by free-living slime mold protists stimulate them to aggregate under high population densities.

studying this eukaryotic model organism, biologists have figured out the cell signaling events that bring yeast of the opposite sex together (to learn more about model organisms, see **BioSkills 13** in Appendix B).

Although the yeast cells are visually indistinguishable, opposite mating types recognize one another via chemical signaling molecules called **pheromones.** The **a** mating type secretes the pheromone **a** factor, and the **α** type secretes the pheromone **α** factor. Receptors on the surface of these cells will bind only to the opposite mating type factor, so when an **a** cell detects the **α** factor, it recognizes that a suitable mate is nearby.

Since yeast cells are not motile, part of the cellular response to the pheromone is to direct new growth toward the signal (**FIGURE 11.17a**). The signaling pathway that is responsible for this morphological change uses both G-protein-coupled receptors and phosphorylation cascades that were presented in Section 11.3. One of the proteins affected by this signaling pathway is actin, which is used to construct new microfilaments at the site where the G proteins have been activated. These new filaments push out new growth toward the highest concentration of the pheromone to allow the cell to find, and eventually fuse with, its mate.

Responding to Population Density

Within a population of unicellular organisms, widespread communication can occur that closely monitors the environment—in particular, the density of the population. Signaling pathways that respond to population density in microbes are collectively referred to as **quorum sensing.** The name was inspired by the observation that cells of the same species may undergo dramatic changes in activity when their numbers reach a threshold, or quorum.

Quorum sensing is based on signaling molecules that are secreted by cells and diffuse through the environment. The response to these molecules varies dramatically between species and ranges from bacterial bioluminescence—or light emission—in the light organs of squid to the secretion of molecules that help glue a community of microbes to a surface in biofilms (see Chapter 29).

In effect, quorum sensing allows unicellular organisms to communicate and coordinate activity. When it occurs, these single-celled organisms take on some of the characteristics of multicellular organisms. For example, quorum sensing via a G-protein-coupled receptor recruits the free-living amoebae of the cellular slime mold *Dictyostelium* to aggregate into multicellular mounds (**FIGURE 11.17b**). Amazingly, the slug-like bodies that are formed from these aggregates can crawl across a surface and eventually organize themselves into a fruiting body that releases spores into the air (see **BioSkills 13** in Appendix B).

This brief introduction to cell–cell signaling brings us to the frontier of biological research. It has taken decades of painstaking research to work out each step in individual signaling pathways. Biologists are now investigating cell signaling at a whole system level—examining how the major pathways interact and how they are integrated at the tissue level within multicellular organisms.

The number of molecules involved and the complexity of their interactions can seem overwhelming, but the punch line is simple: In organisms ranging from bacteria to blue whales, cell–cell signaling helps organisms receive information about their environment and respond to changing conditions in an appropriate way.

If you understand . . .

11.1 The Cell Surface

- The vast majority of cells secrete an extracellular layer.

- In bacteria, archaea, algae, and plants, the extracellular material is stiff and is called a cell wall. In animals, the secreted layer is flexible and is called the extracellular matrix (ECM).

- Extracellular layers are fiber composites. They consist of cross-linked filaments that provide tensile strength and a ground substance that fills space and resists compression.

- In plants, the extracellular filaments are cellulose microfibrils; in animals, the most abundant filaments are made of the protein collagen. In both plants and animals, the ground substance is composed of gel-forming polysaccharides.

✔ You should be able to predict what happens to animal cells when they are treated with an enzyme that (1) cuts integrin molecules, or (2) digests collagen fibrils.

11.2 How Do Adjacent Cells Connect and Communicate?

- In multicellular organisms, molecules in the extracellular layer and plasma membrane mediate interactions between adjacent cells.

- Adjacent cells may be physically bound to one another by glue-like middle lamellae in plants or by tight junctions and desmosomes in animals.

- The cytoplasm of adjacent cells is in direct communication through openings called plasmodesmata in plants and gap junctions in animals.

- Cells may respond to signals by activating certain enzymes, releasing or taking up specific ions or molecules, or changing the activity of target genes. As a result, cells and tissues throughout the body can alter their activity in response to changing conditions, and do so in a coordinated way.

✔ You should be able to predict the consequences of removing the gap junctions between the cells in the cardiac muscle in your heart.

11.3 How Do Distant Cells Communicate?

- Distant cells in multicellular organisms communicate through signaling molecules that bind to receptors found in or on specific target cells.

- Cell–cell signaling molecules that are not lipid soluble bind to receptors in the plasma membrane. The receptor then changes conformation and triggers production of a new type of intracellular signal—a second messenger or phosphorylation cascade.

- Because enzymes inside the cell quickly deactivate the signal and signaling pathways often interact, the cell's response is tightly regulated.

✔ You should be able to explain how the hormone adrenalin can stimulate cells in both the heart and the liver, yet trigger different responses (increasing heart rate versus releasing glucose).

11.4 Signaling between Unicellular Organisms

- Unicellular organisms use chemical signals to sense the pheromones of opposite mating types and population density. Quorum sensing allows closely related cells to coordinate changes in their activity when population density is high.

✔ You should be able to compare and contrast the role of intercellular signaling between unicellular organisms and the cells in a multicellular organism.

(MB) **MasteringBiology**

1. **MasteringBiology Assignments**

   Tutorials and Activities Build a Signaling Pathway; Cell Junctions; Cell Signaling: Reception; Cell Signaling: Transduction and Response; Cellular Responses; Overview of Cell Signaling; Reception; Signal Transduction Pathways

   Questions Reading Quizzes, Blue-Thread Questions, Test Bank

2. **eText** Read your book online, search, take notes, highlight text, and more.

3. **The Study Area** Practice Test, Cumulative Test, BioFlix® 3-D Animations, Videos, Activities, Audio Glossary, Word Study Tools, Art

You should be able to . . .

1. What is a fiber composite? How do cellular fiber composites resemble reinforced concrete?

2. In animals, where are most components of the extracellular material synthesized?
   a. smooth ER
   b. the rough ER
   c. in the extracellular layer itself
   d. in the plasma membrane

3. Treating dissociated cells with certain antibodies makes the cells unable to reaggregate. Why?
   a. The antibodies bind to cell adhesion proteins.
   b. The antibodies bind to the fiber component of the extracellular matrix.
   c. The antibodies bind to the cell surface and inhibit motility.
   d. The antibodies act as enzymes that break down desmosomes.

4. What does it mean to say that a signal is transduced?
   a. The signaling molecule enters the cell directly and binds to a receptor inside.
   b. The physical form of the signal changes between the outside of the cell and the inside.
   c. The signal is amplified, such that even a single molecule evokes a large response.
   d. The signal triggers a sequence of phosphorylation events inside the cell.

5. What characteristics do tight junctions bestow on tissues that use these adhesions to connect adjacent cells?
   a. They allow communication between adjacent cells.
   b. They provide strong connections to resist pulling forces.
   c. They use the extracellular matrix to indirectly connect adjacent cells.
   d. They form a watertight barrier between the cells.

6. What are the two general categories of cellular responses to an intercellular signal?

7. Which of the following statements represents a fundamental difference between the fibers found in the extracellular layers of plants and those of animals?
   a. Plant fibers resist compression forces; animal fibers resist pulling forces.
   b. Animal fibers consist of proteins; plant fibers consist of polysaccharides instead.
   c. Plant extracellular fibers never move; animal fibers can slide past one another.
   d. Cellulose microfibrils run parallel to one another; collagen filaments crisscross.

8. Explain how it is possible for a phosphorylation cascade to amplify an intercellular signal.

9. Compare and contrast the structure and function of tight junctions, desmosomes, and gap junctions.

10. Animal cells adhere to each other selectively. Summarize experimental evidence that supports this statement. Explain the molecular basis of selective adhesion.

11. Make a flowchart summarizing the reception, processing, response, and deactivation steps for a signaling molecule that binds to an intracellular receptor.

12. What is the significance of the observation that many signal transduction pathways create a network, where they intersect or overlap?

13. What would be the impact on the structure of a plant tissue if the cells lacked the ability to modify the extracellular environment?
    a. Cells would swell and burst if placed in a hypotonic environment.
    b. Cells would not be able to adhere to one another.
    c. No defined tissues, consisting of similar cells with coordinated activities, would be possible.
    d. All of the above.

14. Suppose that a particular cell–cell signaling molecule induces a cellular response without requiring signal transduction (i.e., no second messengers or phosphorylation cascades). Compared to the signal transduction pathways you learned about in this chapter, how would an event like this affect (a) the types of responses that are possible, (b) amplification, and (c) regulation?

15. In most species of fungi, chitin is a major polysaccharide found in cell walls. Review the structure of chitin (see Chapter 5), and then describe what would have to take place for the directional growth that occurs when yeast, a type of fungi, respond to sex pheromones.

16. Suppose you created an antibody that bound to the receptor tyrosine kinase illustrated in Figure 11.15. You expected this antibody to inhibit the cell response, but instead it resulted in activating the response, even when no signal was present. Explain this result.

12 The Cell Cycle

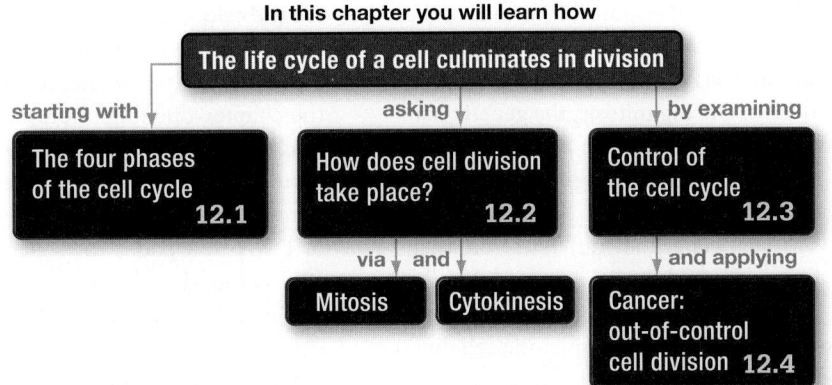

In this chapter you will learn how

The life cycle of a cell culminates in division

starting with

The four phases of the cell cycle 12.1

asking

How does cell division take place? 12.2

by examining

Control of the cell cycle 12.3

via and

Mitosis **Cytokinesis**

and applying

Cancer: out-of-control cell division 12.4

This cell, from a hyacinth plant, is undergoing a type of nuclear division called mitosis. Understanding how mitosis occurs is a major focus of this chapter.

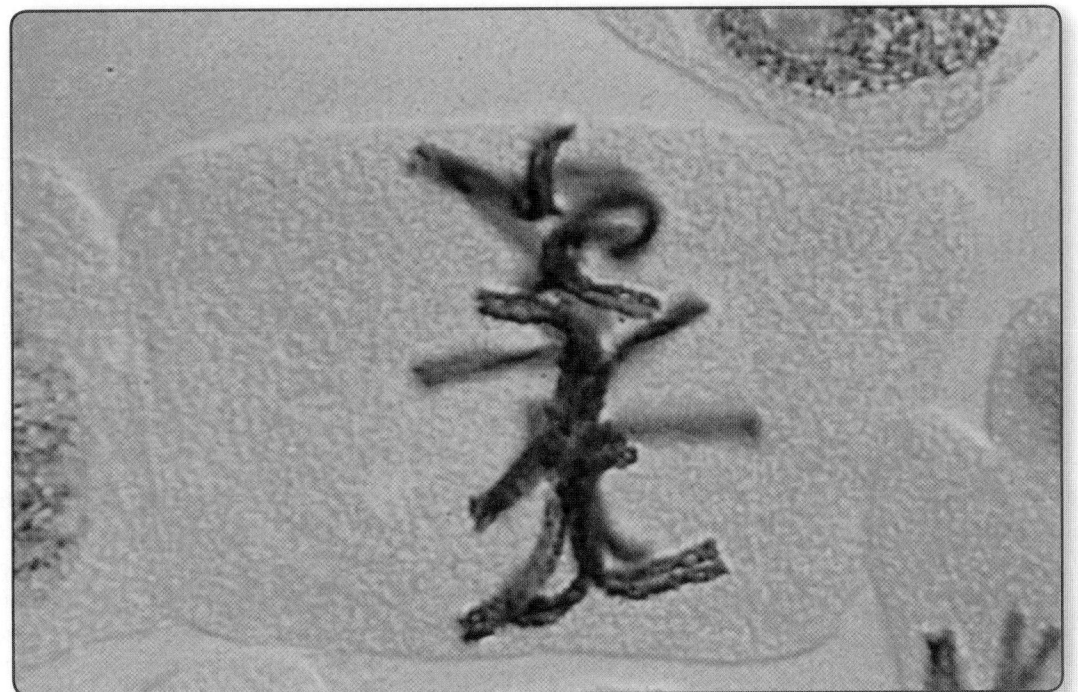

This chapter is part of the Big Picture. See how on pages 366–367.

The cell theory maintains that all organisms are made of cells and all cells arise from preexisting cells (Chapter 1). Although the cell theory was widely accepted among biologists by the 1860s, most believed that new cells arose within preexisting cells by a process that resembled the growth of mineral crystals. But Rudolf Virchow proposed that new cells arise by splitting preexisting cells—that is, by **cell division.**

In the late 1800s, microscopic observations of newly developing organisms, or **embryos,** confirmed Virchow's hypothesis. Multicellular eukaryotes start life as single-celled embryos and grow through a series of cell divisions.

Early studies revealed two fundamentally different ways that nuclei divide before cell division: meiosis and mitosis. In animals, **meiosis** leads to the production of sperm and eggs, which are the male and female reproductive cells termed **gametes. Mitosis** leads to the production of all other cell types, referred to as **somatic** (literally, "body-belonging") **cells.** (You can see the Big Picture of

✔ When you see this checkmark, stop and test yourself. Answers are available in Appendix A.

219

how these two nuclear divisions are related to each other and the transmission of genetic information on pages 366–367.)

Mitosis and meiosis are usually accompanied by **cytokinesis** ("cell movement")—the division of the cytoplasm into two distinct cells. When cytokinesis is complete, a so-called parent cell has given rise to two daughter cells.

Mitotic and meiotic cell divisions are responsible for one of the five fundamental attributes of life: reproduction (see Chapter 1). But even though mitosis and meiosis share many characteristics, they are fundamentally different. During mitosis, the genetic material is copied and then divided equally between two cells. This is referred to as cellular *replication*, since these daughter cells are genetically identical with the original parent cell. In contrast, meiosis results in daughter cells that are genetically different from each other and that have half the amount of hereditary material as the parent cell.

This chapter focuses on mitotic cell division; meiotic cell division is the subject of the next chapter (Chapter 13). Let's begin with a look at the key events in a cell's life cycle, continue with an in-depth analysis of mitosis and the regulation of the cell cycle, and end by examining how uncontrolled cell division can lead to cancer.

12.1 How Do Cells Replicate?

For life on Earth to exist, cells must replicate. The general requirements for cellular replication are to **(1)** copy the DNA (deoxyribonucleic acid), **(2)** separate the copies, and **(3)** divide the cytoplasm to create two complete cells.

This chapter focuses on eukaryotic cell replication, which is responsible for three key events:

1. *Growth*   The trillions of genetically identical cells that make up your body are the product of mitotic divisions that started in a single fertilized egg.

2. *Wound repair*   When you suffer a scrape, cellular replication generates the cells that repair your skin.

3. *Reproduction*   When yeast cells grow in bread dough or in a vat of beer, they are reproducing by cellular replication. In yeasts and other single-cell eukaryotes, mitotic division is the basis of asexual reproduction. **Asexual reproduction** produces offspring that are genetically identical with the parent.

These events are so basic to life that cell replication has been studied for well over a century. Like much work in biology, the research on how cells divide began by simply observing the process.

What Is a Chromosome?

As studies of cell division in eukaryotes began, biologists found that certain chemical dyes made threadlike structures visible within nuclei. In 1879, Walther Flemming used a dye made from a coal tar to observe these threadlike structures and how they changed in the dividing cells of salamander embryos. The threads

first appeared in pairs just before division and then split to produce single, unpaired threads in the daughter cells. Flemming introduced the term mitosis, from the Greek *mitos* ("thread"), to describe this process.

Others studied the roundworm *Ascaris* and noted that the total number of threads in a cell was the same before and after mitosis. All the cells in a roundworm had the same number of threads.

In 1888 Wilhelm Waldeyer coined the term **chromosome** ("colored-body") to refer to these threadlike structures (visible in the chapter-opening photo). A chromosome consists of a single, long DNA double helix that is wrapped around proteins, called **histones,** in a highly organized manner. DNA encodes the cell's hereditary information, or genetic material. A gene is a length of DNA that codes for a particular protein or ribonucleic acid (RNA) found in the cell.

Before mitosis, each chromosome is replicated. As mitosis starts, the chromosomes condense into compact structures that can be moved around the cell efficiently. Then one of the chromosome copies is distributed to each of two daughter cells.

FIGURE 12.1 illustrates unreplicated chromosomes, replicated chromosomes before they have condensed prior to mitosis, and replicated chromosomes that have condensed at the start of mitosis. Each of the DNA copies in a replicated chromosome is called a **chromatid.** Before mitosis, the two chromatids are joined along their entire length by proteins called cohesins. Once mitosis begins, however, many of these connections are removed except for those at a specialized region of the chromosome called the **centromere.** Chromatid copies that remain attached at their centromere are referred to as **sister chromatids.** Even though a replicated chromosome consists of two chromatids, it is still considered a single chromosome.

Cells Alternate between M Phase and Interphase

The division of eukaryotic cells is like a well-choreographed stage performance. The most visually stimulating part of the show occurs when cells are in their dividing phase, called the **M** (*m*itotic or *m*eiotic) **phase.** With a light microscope, chromosomes can be stained and observed as discrete units only during M phase, when they condense into compact structures.

The rest of the time, the cell is in **interphase** ("between-phase"). No dramatic changes in the nucleus are visible by light microscopy during interphase. The chromosomes uncoil into the extremely long, thin structures shown in Figure 12.1 and are no longer stained as individual threads. However, this does not mean that the cell is idle. Interphase is an active time: The cell is either growing and preparing to divide or fulfilling its specialized function in a multicellular individual. Cells actually spend most of their time in interphase.

The Discovery of S Phase

Once M phase and interphase were identified by microscopy, researchers could start assigning roles to these distinct phases. They could see that the separation of chromosomes and cytokinesis

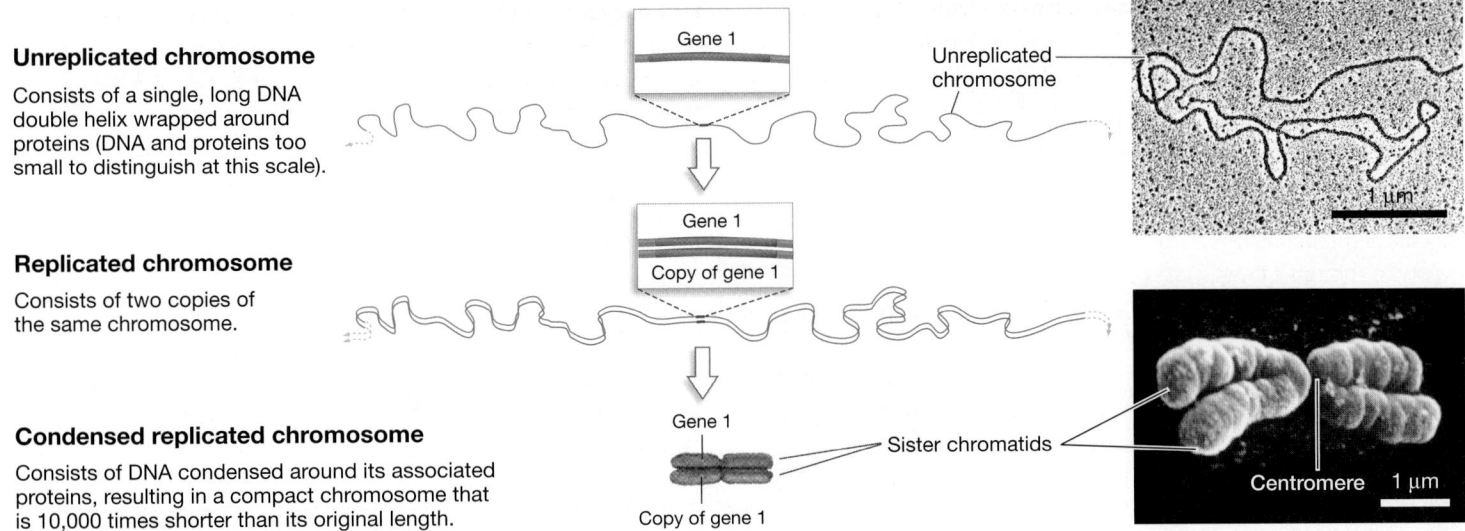

Unreplicated chromosome

Consists of a single, long DNA double helix wrapped around proteins (DNA and proteins too small to distinguish at this scale).

Gene 1

Unreplicated chromosome

1 μm

Replicated chromosome

Consists of two copies of the same chromosome.

Gene 1

Copy of gene 1

Condensed replicated chromosome

Consists of DNA condensed around its associated proteins, resulting in a compact chromosome that is 10,000 times shorter than its original length.

Gene 1

Sister chromatids

Copy of gene 1

Centromere   1 μm

FIGURE 12.1 Changes in Chromosome Morphology. After chromosomes replicate, the two identical copies are attached to each other along their entire length. Early in mitosis, replicated chromosomes condense and sister chromatids remain attached at a region called the centromere.

took place during the M phase, but when are the chromosomes copied?

To answer this question, researchers needed to distinguish cells that were replicating their DNA from those that were not. They were able to do this by adding radioactive phosphorus, in the form of phosphates, to cells. Those cells that were synthesizing DNA would incorporate the radioactive isotope into nucleotides (see Chapter 4 to review where phosphates are in DNA).

The idea was to:

1. label DNA as chromosomes were being copied;

2. wash away any radioactive isotope that hadn't been incorporated and remove RNA, which would also incorporate phosphorus; and then

3. visualize the labeled, newly synthesized DNA by exposing the treated cells to X-ray film. Emissions from radioactive phosphorus create a black dot in the film. This is the technique called autoradiography (see **BioSkills 9** in Appendix B).

In 1951, Alma Howard and Stephen Pelc performed this experiment and looked for cells with black dots—indicating active DNA synthesis—immediately after the exposure to a radioactive isotope ended. They found black dots in some of the interphase cells, but none in M-phase cells. Several years later, these results were verified using radioactive thymidine, which is incorporated into DNA but not RNA. These results were strong evidence that DNA replication occurs during interphase.

Thus, biologists had identified a new stage in the life of a cell. They called it **synthesis (or S) phase.** S phase is part of interphase. Replication of the genetic material is separated, in time, from the partitioning of chromosome copies during M phase.

Howard and Pelc coined the term **cell cycle** to describe the orderly sequence of events that leads a eukaryotic cell through the duplication of its chromosomes to the time it divides.

The Discovery of the Gap Phases

In addition to discovering the S phase, Howard and Pelc made another key observation—not all the interphase cells were labeled. This meant that there was at least one "gap" in interphase when DNA was not being copied.

Howard and Pelc, along with researchers in other labs, followed up on these early results by asking where S phase was positioned in interphase. There were three possible scenarios:

1. The S phase is immediately before M phase, with a single gap between the end of M and start of S phase;

2. the S phase is immediately after M phase, with a gap between the end of S and the start of M phase; or

3. two gaps exist, one before and one after the S phase.

To address which of these models, if any, is correct, many experiments were done using cells in culture. Cultured cells are powerful experimental tools because they can be manipulated much more easily than cells in an intact organism (see **BioSkills 12** in Appendix B). In most of these studies, researchers used cultures that were asynchronous, meaning that the cells were randomly distributed in the cycle.

To understand the value of these asynchronous cultures, imagine the cell cycle were a clock. Every complete rotation around the clock would represent one cell division, and each tick would represent a different point in the cycle. At any given time, an asynchronous culture would have at least one cell present at each of the ticks on the clock. As time passes, these cells would move around this cell-cycle clock at the same rate and in the same direction.

FIGURE 12.2 The Pulse–Chase Assay Reveals a Gap Phase. Cells labeled with radioactive thymidine during the pulse can be tracked during the chase to identify when they enter M phase. In this assay, a gap between the end of S phase and start of M phase was identified based on the delay observed between the pulse and the presence of labeled mitotic cells.

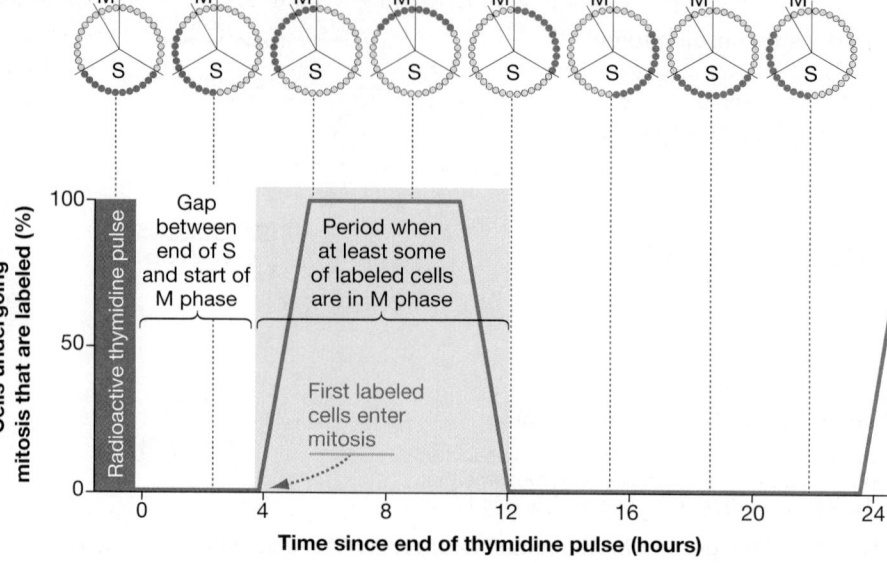

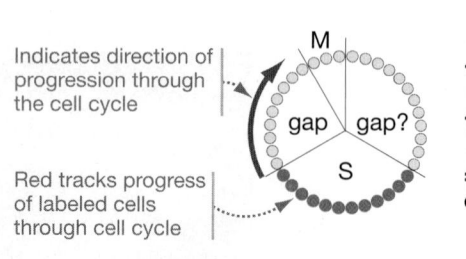

In one experiment, researchers marked the S-phase cells in a human cell culture by exposing it to radioactively labeled thymidine. A short time later, they stopped the labeling by flooding the solution surrounding the cultured cells with nonradioactive thymidine. This pulse–chase approach (introduced in Chapter 7) labeled only those cells that were in S phase during the radioactive pulse. Imagine these marked cells moving like a hand on the clock that could be tracked as they progressed through the cell cycle.

Once the pulse ended, the researchers analyzed samples of the culture at different times during the chase. For each batch of cells, they recorded how many labeled cells were undergoing mitosis, meaning that the cells that were in S phase during the pulse had entered M phase.

One striking result emerged early on: None of the labeled cells started mitosis immediately. Because the cultures were asynchronous, at least some of the cells must have been at the very end of their S phase when exposed to the pulse. If the S phase had been immediately followed by the M phase, some of these labeled cells would have entered M just as the chase began. Instead, it took several hours before any of the labeled cells began mitosis.

The time lag between the end of the pulse and the appearance of the first labeled mitotic nuclei corresponds to a period between the end of S phase and the beginning of M phase. This gap represents the time when chromosome replication is complete, but mitosis has not yet begun. **FIGURE 12.2** shows how cells labeled with radioactive thymidine can be tracked as they progress through the M phase.

After this result, the possibilities for the organization of the cell cycle were narrowed to either one gap between the end of S and start of M phase, or two gaps that flank the S phase. Which of the models best represents the eukaryotic cell cycle? Once researchers determined the lengths of the S and M phases, they found that the combined time, including the gap between these phases, was still short compared with the length of the cell cycle. This discrepancy represents an additional gap phase that is between the end of M and the start of S phase.

The cell cycle was thus finally mapped out. The gap between the end of M and start of S phase is called the **G₁ phase**. The second gap, between the end of S and start of M phase, is called the **G₂ phase**.

The Cell Cycle

FIGURE 12.3 pulls these results together into a comprehensive view of the cell cycle. The cell cycle involves four phases: M phase and an interphase consisting of the G₁, S, and G₂ phases. In the cycle diagrammed here, the G₁ phase is about twice as long as G₂, but the timing of these phases varies depending on the cell type and growth conditions.

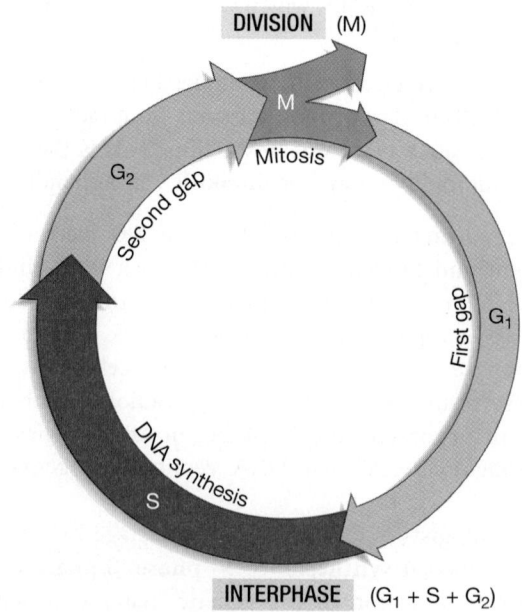

FIGURE 12.3 The Cell Cycle Has Four Phases. A representative cell cycle. The time required for the G₁ and G₂ phases varies dramatically among cells and organisms.

Why do the gap phases exist? Besides needing to copy their chromosomes during interphase, cells also must prepare for division by replicating organelles and increasing in size. Before mitosis can take place, the parent cell must grow large enough to divide into two cells that will be normal in size and function. The two gap phases provide the time required to accomplish these tasks. They allow the cell to complete all the requirements for cell division other than chromosome replication.

Now let's turn to the M phase. Once the genetic material has been copied, how do cells divide it between daughter cells?

12.2 What Happens during M Phase?

The M phase typically consists of two distinct events: the division of the nucleus and the division of the cytoplasm. During cell replication, mitosis divides the replicated chromosomes to form two daughter nuclei with identical chromosomes and genes. Mitosis is usually accompanied by cytokinesis—cytoplasmic division that results in two daughter cells.

FIGURE 12.4 provides an overview of how chromosomes change before, during, and after mitosis and cytokinesis, beginning with a hypothetical plant cell or animal cell in G_1 phase. The first drawing shows a total of four chromosomes in the cell, but chromosome number varies widely among species—chimpanzees and potato plants have a total of 48 chromosomes in each cell; a maize (corn) plant has 20, dogs have 78, and fruit flies have 8.

Eukaryotic chromosomes consist of DNA wrapped around the globular histone proteins. In eukaryotes this DNA–protein material is called **chromatin.** During interphase, the chromatin of each chromosome is in a "relaxed" or uncondensed state, forming long, thin strands (see Figure 12.1, top).

The second drawing in Figure 12.4 shows chromosomes that have been copied before mitosis. Each chromosome now consists of two sister chromatids. Each chromatid contains one long DNA double helix, and sister chromatids represent exact copies of the same genetic information.

At the start of mitosis, then, each chromosome consists of two sister chromatids that are attached to each other at the centromere.

✔ You should be able to explain the relationship between chromosomes and (1) genes, (2) chromatin, and (3) sister chromatids.

Events in Mitosis

As the third drawing in Figure 12.4 indicates, mitosis begins when chromatin condenses to form a much more compact structure. Replicated, condensed chromosomes correspond to the paired threads observed by early biologists.

During mitosis, the two sister chromatids separate to form independent daughter chromosomes. One copy of each chromosome goes to each of the two daughter cells. (See the final drawing in Figure 12.4.) As a result, each cell receives an identical copy of the genetic information that was contained in the parent cell.

Biologists have identified five subphases within M phase based on distinctive events that occur. Interphase is followed by the mitotic subphases of prophase, prometaphase, metaphase, anaphase, and telophase.

Recall that before mitosis begins, chromosomes have already replicated during the S phase of interphase. Now let's look at how cells separate the chromatids in these replicated chromosomes by investigating each subphase of mitosis in turn (**FIGURE 12.5**, on page 224).

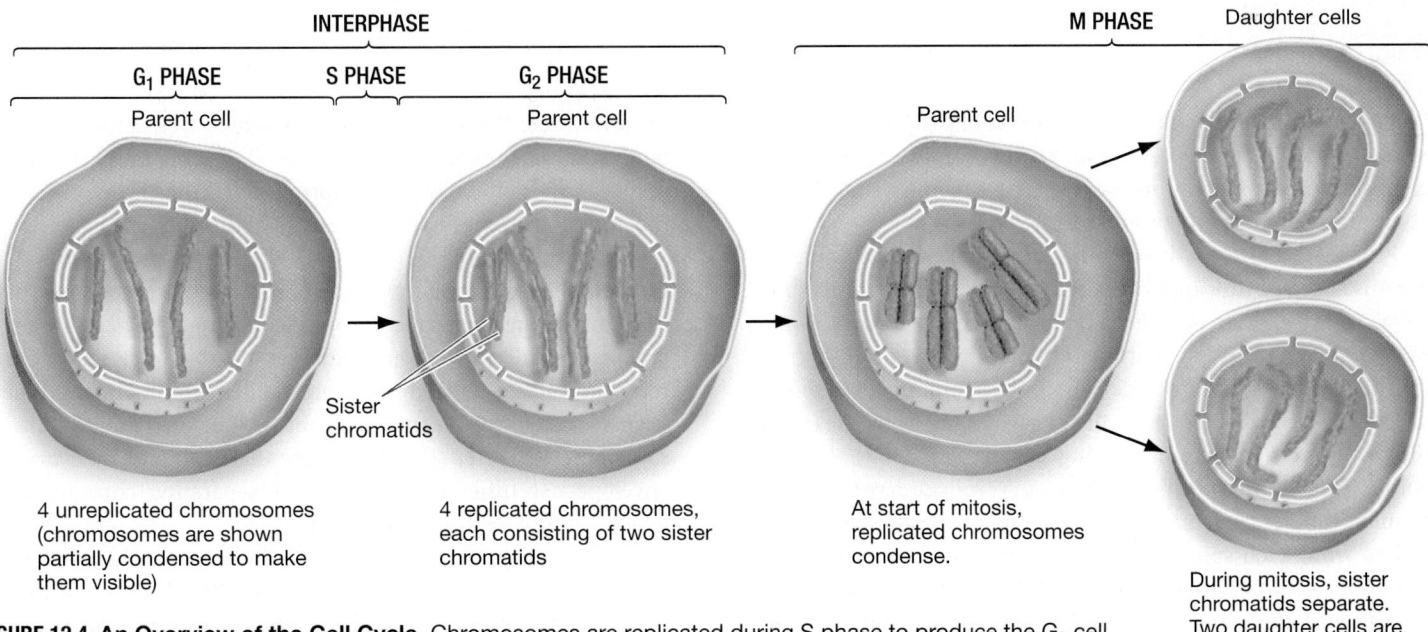

INTERPHASE

G₁ PHASE   S PHASE   G₂ PHASE

Parent cell   Parent cell

M PHASE   Daughter cells

Parent cell

4 unreplicated chromosomes (chromosomes are shown partially condensed to make them visible)

Sister chromatids

4 replicated chromosomes, each consisting of two sister chromatids

At start of mitosis, replicated chromosomes condense.

During mitosis, sister chromatids separate. Two daughter cells are formed by cytokinesis.

FIGURE 12.4 An Overview of the Cell Cycle. Chromosomes are replicated during S phase to produce the G_2 cell. During M phase, the replicated chromosomes are partitioned to the two daughter cells. Each daughter cell contains the same complement of chromosomes as the parent cell.

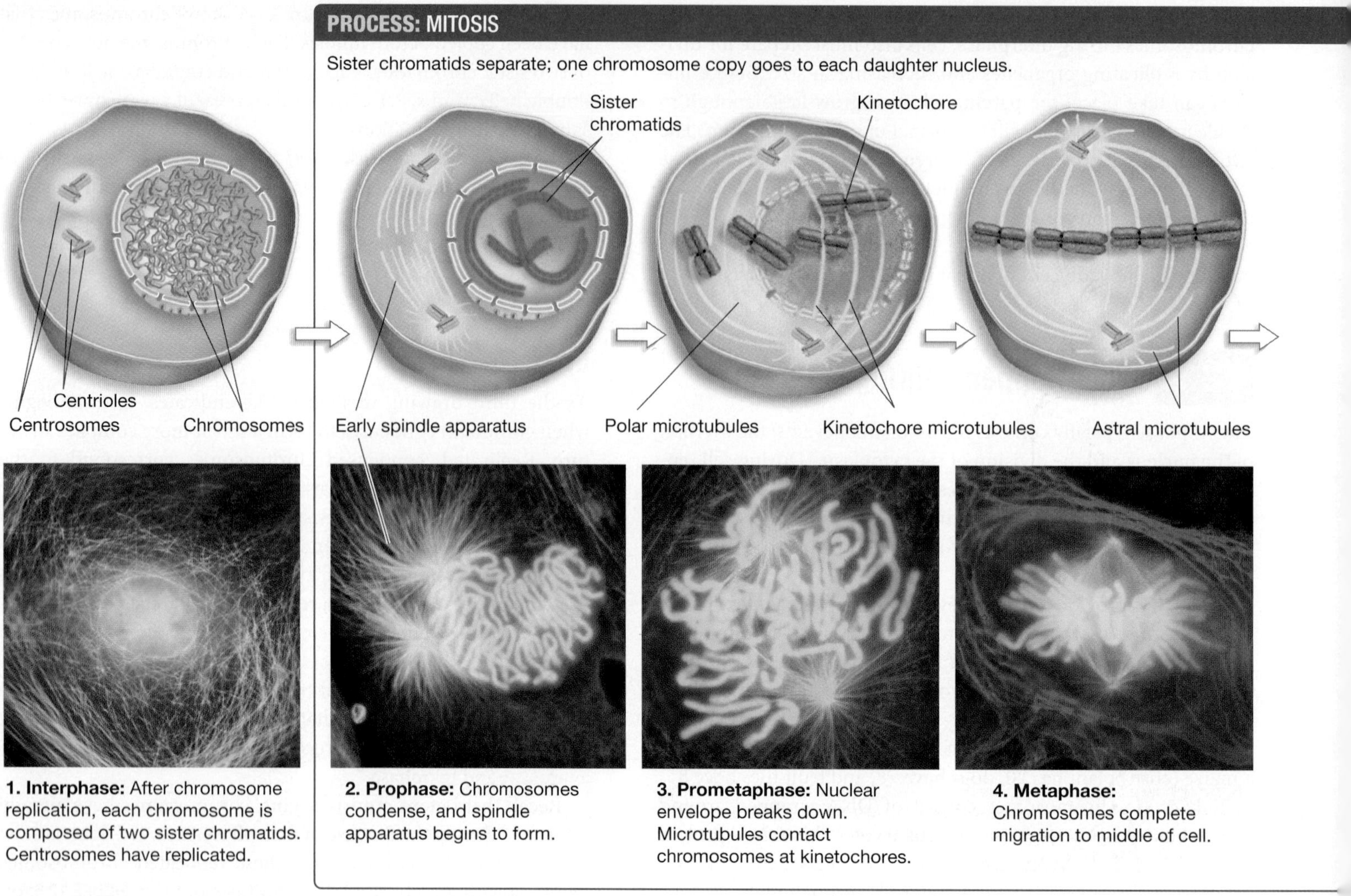

Sister chromatids separate; one chromosome copy goes to each daughter nucleus.

Sister chromatids

Kinetochore

Centrioles
Centrosomes Chromosomes Early spindle apparatus Polar microtubules Kinetochore microtubules Astral microtubules

1. Interphase: After chromosome replication, each chromosome is composed of two sister chromatids. Centrosomes have replicated.

2. Prophase: Chromosomes condense, and spindle apparatus begins to form.

3. Prometaphase: Nuclear envelope breaks down. Microtubules contact chromosomes at kinetochores.

4. Metaphase: Chromosomes complete migration to middle of cell.

FIGURE 12.5 Mitosis and Cytokinesis. In the micrographs, under the drawings, chromosomes are stained blue, microtubules are yellow/green, and intermediate filaments are red.

✔**QUANTITATIVE:** If the model cell in this figure has x amount of DNA and four chromosomes in its G_1 phase, then what is the amount of DNA and number of chromosomes in **(1)** prophase; **(2)** anaphase; **(3)** each daughter cell after division is complete?

Prophase Mitosis begins with the events of **prophase** ("before-phase," Figure 12.5, step 2), when chromosomes condense into compact structures. Chromosomes first become visible in the light microscope during prophase.

Prophase is also marked by the formation of the spindle apparatus. The **spindle apparatus** is a structure that produces mechanical forces that **(1)** move replicated chromosomes during early mitosis and **(2)** pull chromatids apart in late mitosis.

The spindle consists of microtubules—components of the cytoskeleton (see Chapter 7). In all eukaryotes, microtubules originate from microtubule-organizing centers (MTOCs). MTOCs define the two poles of the spindle and produce large numbers of microtubules. During prophase, some of these microtubules extend from each spindle pole and overlap with one another—these are called **polar microtubules.**

Although the nature of this MTOC varies among plants, animals, fungi, and other eukaryotic groups, the spindle apparatus has the same function. Figure 12.5 illustrates an animal cell undergoing

mitosis, where the MTOC is a **centrosome**—a structure that contains a pair of **centrioles** (see Chapter 7). During prophase in animal cells, the spindle begins to form around the chromosomes by moving centrosomes to opposite sides of the nucleus.

Prometaphase In many eukaryotes, once chromosomes have condensed, the nuclear envelope disintegrates. Once the envelope has been removed, microtubules are able to attach to chromosomes at specialized structures called **kinetochores.** These events occur during **prometaphase** ("before middle-phase"; see Figure 12.5, step 3). (Organisms that maintain their nuclear envelope use different strategies for separating chromosomes, which will not be discussed here.)

Each sister chromatid has its own kinetochore, which is assembled at the centromere. Since the centromere is also the attachment site for chromatids, the result is two kinetochores on opposite sides of each replicated chromosome. The microtubules that are attached to these structures are called **kinetochore microtubules.**

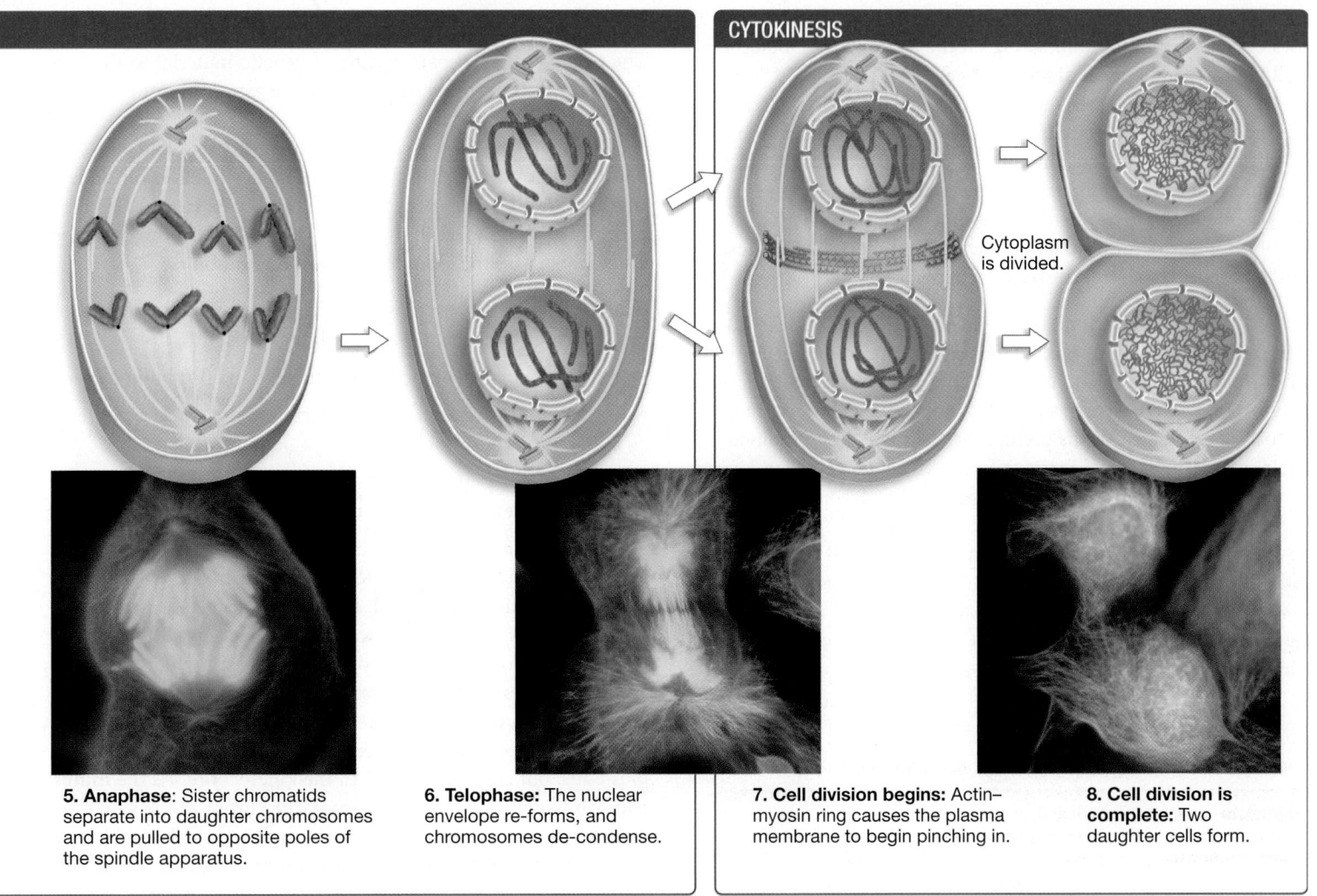

Cytoplasm is divided.

5. Anaphase: Sister chromatids separate into daughter chromosomes and are pulled to opposite poles of the spindle apparatus.

6. Telophase: The nuclear envelope re-forms, and chromosomes de-condense.

7. Cell division begins: Actin–myosin ring causes the plasma membrane to begin pinching in.

8. Cell division is complete: Two daughter cells form.

Early in mitosis, kinesin and dynein motors are recruited to the kinetochore, where they can "walk" the chromosome up and down microtubules. These motors are thought to be very important in the initial attachment of the kinetochore to the plus end of the microtubule. If these ideas are correct, then the process is similar to the way these motors walk along microtubules during vesicle transport (see Chapter 7).

In all eukaryotes, after the kinetochores have attached to microtubules, chromosomes begin to move to the middle of the cell during prometaphase.

Metaphase Once the kinetochore microtubules have moved all the chromosomes to the middle of the spindle (Figure 12.5, step 4), the mitotic cells enter **metaphase** ("middle-phase"). At this point, the chromosomes are lined up along an imaginary plane between the two spindle poles called the **metaphase plate.**

The formation of the spindle apparatus is now complete. The polar microtubules that extend from each spindle pole overlap in the middle of the cell, thereby forming a pole-to-pole connection. Each chromosome is held by kinetochore microtubules reaching out from opposite poles and exerting the same amount of tension, or pull. The spindle poles are held in place partly because of

astral microtubules that extend from the MTOCs and interact with proteins on the cell membrane.

The alignment of these chromosomes results from the growth and shrinkage of the attached kinetochore microtubules. When chromosomes reach the metaphase plate, the shrinkage of these microtubules at the MTOCs is balanced by slow growth of microtubules at the kinetochores. Since the sister chromatids of each chromosome are connected to opposite poles, a tug of war occurs during metaphase that pulls them in opposite directions.

Anaphase At the start of **anaphase** ("against-phase"), the cohesins that are holding sister chromatids together at the centromeres split (Figure 12.5, step 5). Because the chromatids are under tension, each replicated chromosome is pulled apart to create two independent daughter chromosomes. By definition, this separation of chromatids instantly doubles the number of chromosomes in the cell.

Two types of movement occur during anaphase. First, the daughter chromosomes move to opposite poles via the attachment of kinetochore proteins to the shrinking kinetochore microtubules. Second, the two poles of the spindle are pushed and pulled farther apart. The motor proteins in overlapping polar microtubules push the poles away from each other. Different motors on the membrane walk along on the astral microtubules to pull the poles to opposite sides of the cell.

During anaphase, then, replicated chromosomes split into two identical sets of daughter chromosomes. Their separation to opposite poles is a critical step in mitosis because it ensures that each daughter cell receives the same complement of chromosomes.

When anaphase is complete, two complete collections of chromosomes are fully separated, each being identical with those of the parent cell before chromosome replication.

Telophase   During **telophase** ("end-phase"), the nuclear envelope that dissolved in prometaphase reforms around each set of chromosomes, and the chromosomes begin to de-condense (Figure 12.5, step 6). Once two independent nuclei have formed, mitosis is complete. At this point, most cells will go on to divide their cytoplasm via cytokinesis to form two daughter cells.

TABLE 12.1 summarizes the key structures involved in mitosis.

✔ After you've studied the table and reviewed Figure 12.5, you should be able to make a table with rows titled (1) spindle apparatus, (2) nuclear envelope, and (3) chromosomes, and columns titled with the five phases of mitosis. Fill in the table by summarizing what happens to each structure during each phase of mitosis.

SUMMARY TABLE 12.1   **Structures Involved in Mitosis**

Structure	Definition
Chromosome	A structure composed of a DNA molecule and associated proteins
Chromatin	The material that makes up eukaryotic chromosomes; consists of a DNA molecule complexed with histone proteins (see Chapter 19)
Chromatid	One strand of a replicated chromosome, with its associated proteins
Sister chromatids	The two strands of a replicated chromosome. When chromosomes are replicated, they consist of two sister chromatids. The genetic material in sister chromatids is identical. When sister chromatids separate during mitosis, they become independent chromosomes.
Centromere	The structure that joins sister chromatids
Kinetochores	The structures on sister chromatids where microtubules attach
Microtubule-organizing center	Any structure that organizes microtubules (see Chapter 7)
Centrosome	The microtubule-organizing center in animals and some plants
Centrioles	Cylindrical structures that comprise microtubules, located inside animal centrosomes

How Do Chromosomes Move during Anaphase?

The exact and equal partitioning of genetic material to the two daughter nuclei is the most fundamental aspect of mitosis. How does this process occur?

To understand how sister chromatids separate and move to opposite sides of the spindle, biologists have focused on understanding the function of spindle microtubules. How do kinetochore microtubules pull chromatids apart? And how does the kinetochore join the chromosome and microtubules?

Mitotic Spindle Forces   The spindle apparatus is composed of microtubules (see Chapter 7). Recall that:

- microtubules are composed of α-tubulin and β-tubulin dimers,

- microtubules are asymmetric—meaning they have a plus end and a minus end, and

- the plus end is the site where microtubule growth normally occurs while disassembly is more frequent at the minus end.

During mitosis, the microtubules originating from the poles are highly dynamic. Rapid growth and shrinkage ensures that some of the microtubules will be able to attach to kinetochores with their plus ends. Others will be stabilized by different proteins in the cytoplasm and become polar or astral microtubules.

These observations suggest two possible mechanisms for the movement of chromosomes during anaphase. The simplest mechanism would be for microtubules to stop growing at the plus ends, but remain attached to the kinetochore. As the minus ends disassemble at the spindle pole, the chromosome would be reeled in like a hooked fish. An alternative model would have the chromosomes moving along microtubules that are being disassembled at the plus ends at the kinetochore. In this case, the chromosome is like a yo-yo running up a string into your hand.

To test these hypotheses, biologists introduced fluorescently labeled tubulin subunits into prophase or metaphase cells. This treatment made the kinetochore microtubules visible (**FIGURE 12.6**, step 1). Once anaphase began, the researchers marked a region of these microtubules with a bar-shaped beam of laser light. The laser permanently bleached a section of the fluorescently labeled structures, darkening them—although they were still functional (Figure 12.6, step 2).

As anaphase progressed, two things happened: (**1**) The darkened region appeared to remain stationary, and (**2**) the chromosomes moved closer to the darkened regions of the microtubules, eventually overtaking them.

This result suggested that the kinetochore microtubules remain stationary during anaphase, but shorten because tubulin subunits are lost from their plus ends. As microtubule ends shrink back to the spindle poles, the chromosomes are pulled along. But if the microtubule is disassembling at the kinetochore, how does the chromosome remain attached?

Kinetochores Are Linked to Retreating Microtubule Ends   The kinetochore is a complex of many proteins that build a base on the centromere region of the chromosome and a "crown" of fibrous proteins projecting outward. **FIGURE 12.7** shows a current

QUESTION: How do kinetochore microtubules shorten to pull daughter chromosomes apart during anaphase?

HYPOTHESIS: Microtubules shorten at the spindle pole.

ALTERNATIVE HYPOTHESIS: Microtubules shorten at the kinetochore.

EXPERIMENTAL SETUP:

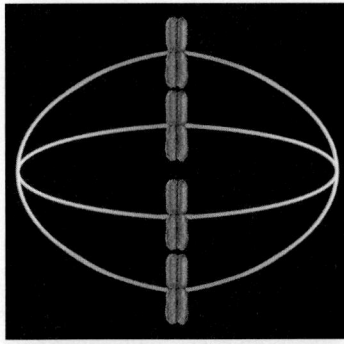

1. Label targets:
Use fluorescent labels to make the metaphase chromosomes fluoresce blue and the microtubules fluoresce yellow.

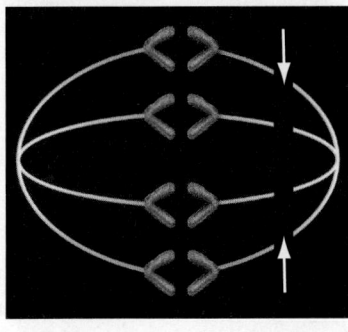

2. Mark microtubules:
At the start of anaphase, darken a section of microtubules to mark them without changing their function.

PREDICTION:

PREDICTION OF ALTERNATIVE HYPOTHESIS: Daughter chromosomes will move toward the pole faster than the darkened section.

RESULTS:

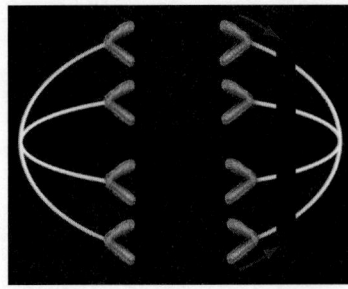

The darkened areas of the microtubules remained stationary as the chromosomes moved through them toward the pole.

CONCLUSION: Kinetochore microtubules shorten at the kinetochore to pull daughter chromosomes apart during anaphase.

FIGURE 12.6 During Anaphase, Microtubules Shorten at the Kinetochore.
SOURCE: Gorbsky, G. J., et al. 1987. Chromosomes move poleward during anaphase along stationary microtubules that coordinately disassemble from their kinetochore ends. *Journal of Cellular Biology* 104: 9–18.

✔**EXERCISE** Complete the prediction for what would occur if chromosome movement were based on microtubules shortening at the spindle pole.

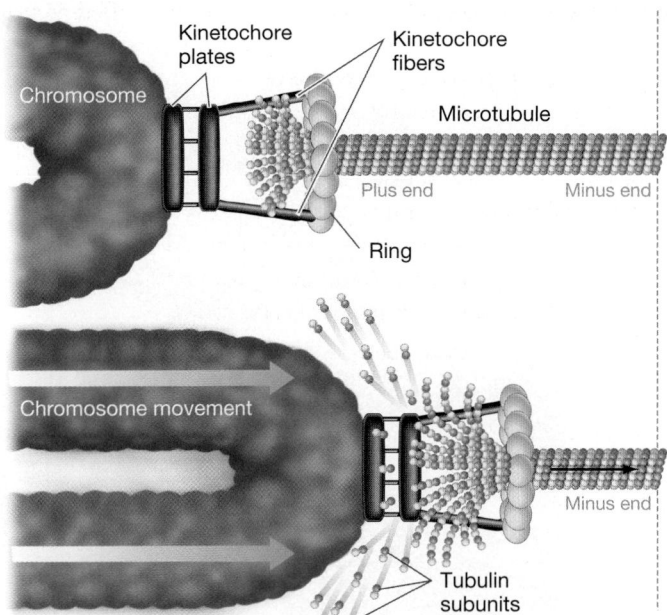

FIGURE 12.7 How Do Microtubules Move Chromosomes during Anaphase? Microtubules are disassembled at the kinetochore during anaphase. In yeast, kinetochore proteins tether the chromosome to a ring that is pushed toward the spindle pole by the fraying plus end of the microtubule.

model for kinetochore structure and function during chromosome movement in anaphase. For simplicity, a yeast kinetochore is shown, which attaches to only one microtubule. (Other eukaryotes can have as many as 30 microtubules attached to each kinetochore.)

Biologists have found that as anaphase gets under way, the plus ends of the kinetochore microtubules begin to fray and disassemble. Fibers that extend from the yeast kinetochore are tethered to this retreating end by attaching to a ring that surrounds the kinetochore microtubule (Figure 12.7, top). As the fraying end widens, its expansion forces the ring, and the attached chromosome, toward the minus end of the microtubule (see Figure 12.7, bottom). The result is that the chromosome is pulled to the spindle pole by the depolymerization of the kinetochore microtubule.

Cytokinesis Results in Two Daughter Cells

At this point, the chromosomes have been replicated in S phase and partitioned to opposite sides of the spindle via mitosis. Now it's time to divide the cell into two daughters that contain identical copies of each chromosome. If these cells are to be viable, however, the parent cell must also ensure that more than just chromosomes make it into each daughter cell.

While the cell is in interphase, the cytoplasmic contents, including the organelles, have increased in number or volume. During cytokinesis (Figure 12.5, steps 7 and 8), the cytoplasm divides to form two daughter cells, each with its own nucleus and complete set of organelles. In most types of cells, cytokinesis directly follows mitosis.

(a) Cytokinesis in plants

Microtubules direct vesicles to center of spindle where they fuse to divide the cell in two

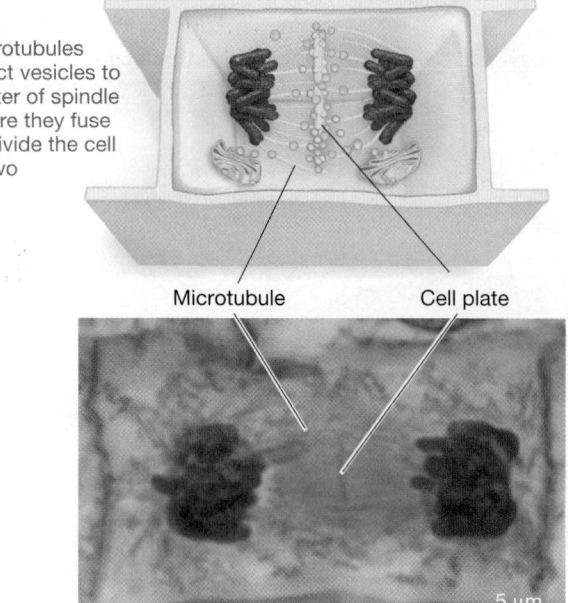

Microtubule   Cell plate

5 μm

(b) Cytokinesis in animals

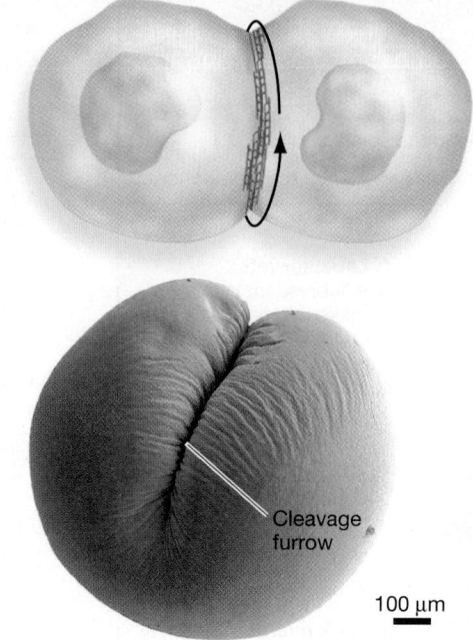

Actin–myosin interactions pinch the membrane in two

Cleavage furrow

100 μm

FIGURE 12.8 The Mechanism of Cytokinesis Varies among Eukaryotes. (a) In plants, the cytoplasm is divided by a cell plate that forms in the middle of the parent cell. **(b)** In animals, the cytoplasm is divided by a cleavage furrow. (The cells in both micrographs have been stained or colorized.)

In plants, polar microtubules left over from the spindle help define and organize the region where the new plasma membranes and cell walls will form. Vesicles from the Golgi apparatus carry components to build a new cell wall to the middle of the dividing cell. These vesicles are moved along the polar microtubules via motor proteins. In the middle of what was the spindle, the vesicles start to fuse together to form a flattened sac-like structure called the **cell plate** (**FIGURE 12.8a**). The cell plate continues to grow as new vesicles fuse with it, eventually contacting the existing plasma membrane. When the cell plate fuses with the existing plasma membrane, it divides the cell into two new daughter cells.

In animals and many other eukaryotes, cytokinesis begins with the formation of a **cleavage furrow** (**FIGURE 12.8b**). The furrow appears because a ring of actin filaments forms just inside the plasma membrane, in the middle of what used to be the spindle. Myosin motor proteins bind to these actin filaments and use adenosine triphosphate (ATP) to contract in a way that causes actin filaments to slide (see Chapter 7).

As myosin moves the ring of actin filaments on the inside of the plasma membrane, the ring shrinks in size and tightens. Because the ring is attached to the plasma membrane, the shrinking ring pulls the membrane with it. As a result, the plasma membrane pinches inward. The actin and myosin filaments continue to slide past each other, tightening the ring further, until the original membrane pinches in two and cell division is complete.

The overall process involved in chromosome separation and cytoplasmic division is a common requirement for all living organisms. The mechanisms involved in accomplishing these events, however, vary depending on the type of cell. What about bacterial cells? How does chromosomal segregation and cytokinesis compare between prokaryotes and eukaryotes?

Bacterial Cell Replication Many bacteria divide using a process called **binary fission.** Recent research has shown that chromosome segregation and cytokinesis in bacterial division is strikingly similar to what occurs in the eukaryotic M phase (**FIGURE 12.9**). As the bacterial chromosome is being replicated, protein filaments attach to the copies and separate them to opposite sides of the cell.

PROCESS: BACTERIAL CELL DIVISION

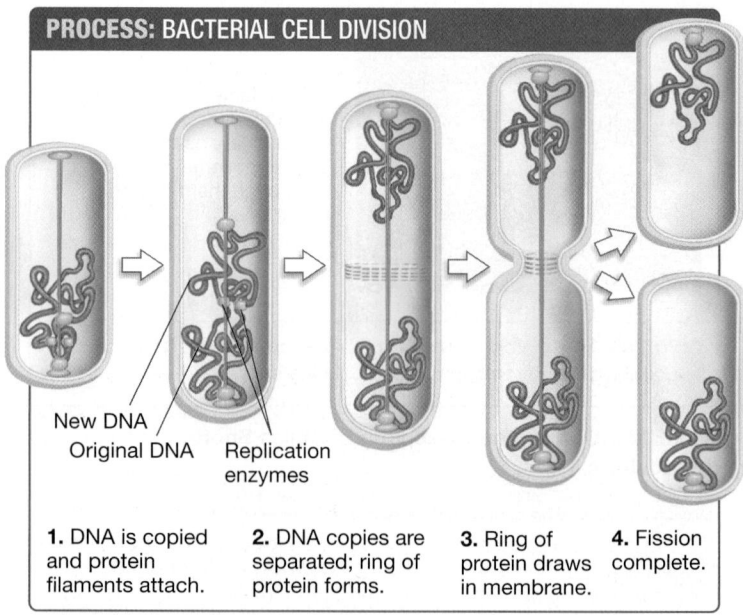

New DNA
Original DNA   Replication enzymes

1. DNA is copied and protein filaments attach.

2. DNA copies are separated; ring of protein forms.

3. Ring of protein draws in membrane.

4. Fission complete.

FIGURE 12.9 Bacterial Cells Divide but Do Not Undergo Mitosis.

Once the copies of the chromosome have been partitioned to opposite sides of the cell, other filaments, made up of proteins that are similar to eukaryotic tubulin, are responsible for dividing the cytoplasm. These filaments attach to the cell membrane to form a ring between the chromosome copies. A signal from the cell causes the filaments to constrict, drawing in the membrane and eventually cleaving the parent into two genetically identical cells.

Having explored what occurs during cell division, let's focus on how it is controlled in eukaryotes. When does a eukaryotic cell divide, and when does it stop dividing?

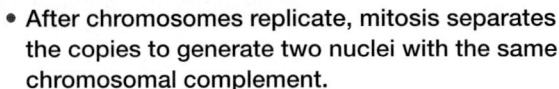
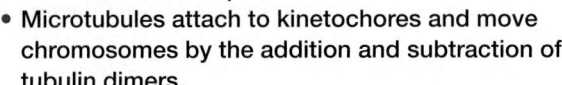

check your understanding

C Y U If you understand that . . .

- After chromosomes replicate, mitosis separates the copies to generate two nuclei with the same chromosomal complement.
- Microtubules attach to kinetochores and move chromosomes by the addition and subtraction of tubulin dimers.
- Cytokinesis divides the nuclei and cytoplasmic components into two daughter cells that are genetically identical with each other and the parent cell.

✔ You should be able to . . .

1. Draw the mitotic spindle for an animal cell that has two chromosomes in metaphase and label the sister chromatids, kinetochores, centrosomes, and the three types of microtubules.
2. Predict how the inhibition of microtubule motors in a plant cell would affect the activities in M phase.

Answers are available in Appendix A.

12.3 Control of the Cell Cycle

Although the events of mitosis are virtually identical in all eukaryotes, other aspects of the cell cycle vary. In humans, for example, intestinal cells routinely divide more than twice a day to replace tissue that is lost during digestion; mature human nerve and muscle cells do not divide at all.

Most of these differences are due to variation in the length of the G_1 phase. In rapidly dividing cells, G_1 is essentially eliminated. Most nondividing cells, in contrast, are permanently stuck in G_1. Researchers refer to this arrested stage as the G_0 state, or simply "G zero." Cells that are in G_0 have effectively exited the cell cycle and are sometimes referred to as post-mitotic. Nerve cells, muscle cells, and many other cell types enter G_0 once they have matured.

A cell's division rate can also vary in response to changing conditions. For example, human liver cells normally replicate about once per year. But if part of the liver is damaged or lost, the remaining cells divide every one or two days until repair is accomplished. Cells of unicellular organisms such as yeasts, bacteria, or archaeans divide rapidly only if the environment is rich in nutrients; otherwise, they enter a quiescent (inactive) state.

To explain these differences, biologists hypothesized that the cell cycle must be regulated in some way. Cell-cycle control is now the most prominent issue in research on cell division—partly because defects in control can lead to uncontrolled, cancerous growth.

The Discovery of Cell-Cycle Regulatory Molecules

The first solid evidence for cell-cycle control molecules came to light in 1970. Researchers found that certain chemicals, viruses, or an electric shock could fuse the membranes of two mammalian cells that were growing in culture, forming a single cell with two nuclei.

How did cell-fusion experiments relate to cell-cycle regulation? When investigators fused cells that were in different stages of the cell cycle, certain nuclei changed phases. For example, when a cell in M phase was fused with one in interphase, the nucleus of the interphase cell immediately initiated mitosis, even if the chromosomes had not been replicated. The biologists hypothesized that the cytoplasm of M-phase cells contains a regulatory molecule that induces interphase cells to enter M phase.

But cell-fusion experiments were difficult to control and left researchers wondering if it was the nucleus or the cytoplasm that was responsible for the induction. To address this issue, they turned to the South African clawed frog, *Xenopus laevis*.

As an egg of these frogs matures, it changes from a cell called an oocyte, which is arrested in a phase similar to G_2, to a mature egg that is arrested in M phase. The large size of these cells—more than 1 mm in diameter—makes them relatively easy to manipulate. Using instruments with extremely fine needles, researchers could specifically examine the effects of the cytoplasm by pulling a sample from a mature egg or an oocyte and injecting it into another.

When biologists purified cytoplasm from M-phase frog eggs and injected it into the cytoplasm of frog oocytes arrested in G_2, the immature oocytes immediately entered M phase (**FIGURE 12.10**, see page 230). But when cytoplasm from interphase cells was injected into G_2 oocytes, the cells remained in the G_2 phase. The researchers concluded that the cytoplasm of M-phase cells—but not the cytoplasm of interphase cells—contains a factor that drives immature oocytes into M phase to complete their maturation.

The factor that initiates M-phase in oocytes was purified and is now called **M phase–promoting factor,** or **MPF.** Subsequent experiments showed that MPF induces M phase in all eukaryotes. For example, injecting M-phase cytoplasm from mammalian cells into immature frog oocytes results in egg maturation, and human MPF can also trigger M phase in yeast cells.

MPF appears to be a general signal that says "Start M phase." How does it work?

QUESTION: Is M phase controlled by regulatory molecules in the cytoplasm?

HYPOTHESIS: Cytoplasmic regulatory molecules control entry into M phase.

NULL HYPOTHESIS: M-phase regulatory molecules are not in the cytoplasm or do not exist.

EXPERIMENTAL SETUP:

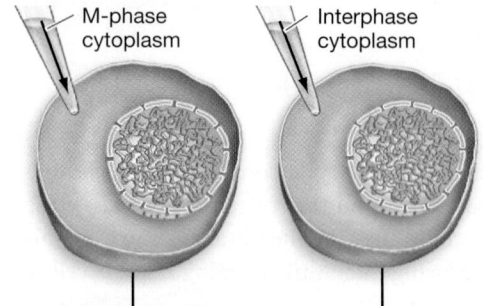

M-phase cytoplasm

Interphase cytoplasm

Microinject cytoplasm from M-phase cell into one frog oocyte and cytoplasm from interphase cell into another frog oocyte.

PREDICTION: Only the oocyte injected with M-phase cytoplasm will begin M phase.

PREDICTION OF NULL HYPOTHESIS: Neither of the frog oocytes will begin M phase.

RESULTS:

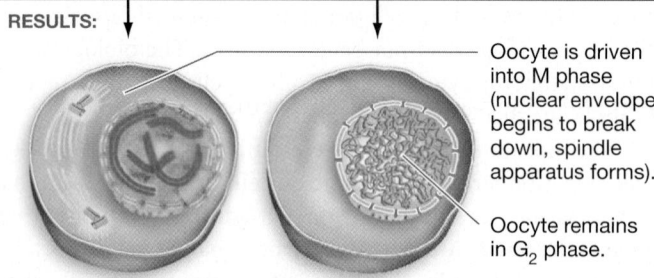

Oocyte is driven into M phase (nuclear envelope begins to break down, spindle apparatus forms).

Oocyte remains in G₂ phase.

CONCLUSION: M-phase cytoplasm contains a regulatory molecule that induces M phase in interphase cells.

FIGURE 12.10 Experimental Evidence for Cell-Cycle Control Molecules. When the cytoplasm from M-phase cells is microinjected into cells in interphase, the interphase chromosomes condense and begin M phase.

SOURCE: Masui, Y., and C. L. Markert. 1971. Cytoplasmic control of nuclear behavior during meiotic maturation of frog oocytes. *Journal of Experimental Zoology* 177: 129–145.

✔ **QUESTION** This experiment was done using cells that were undergoing meiosis. What could the investigators do to show that the factor used in meiotic division is the same as used for mitotic division?

MPF Contains a Protein Kinase and a Cyclin

MPF is made up of two distinct polypeptide subunits. One subunit is a protein kinase—an enzyme that catalyzes the transfer of a phosphate group from ATP to a target protein. Recall that phosphorylation may activate or inactivate the function of proteins by changing their shape (Chapter 8). As a result, kinases frequently act as regulatory proteins in the cell.

These observations suggested that MPF phosphorylates proteins that trigger the onset of M phase. But research showed that

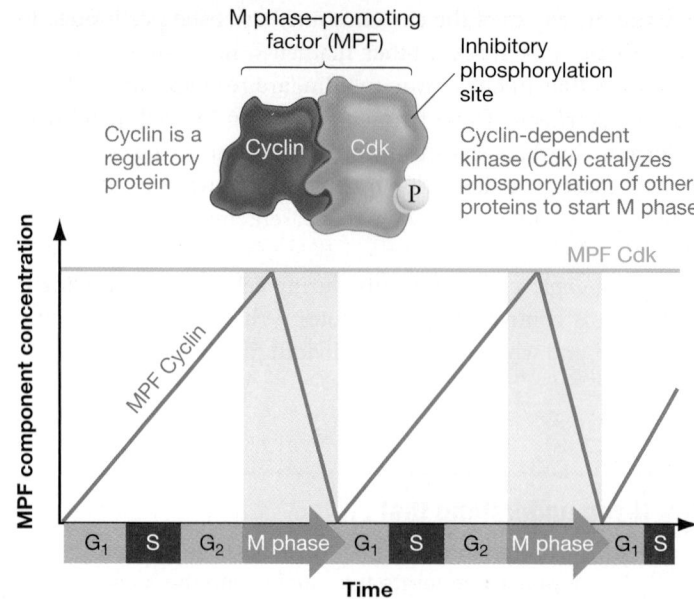

M phase–promoting factor (MPF)

Cyclin is a regulatory protein

Inhibitory phosphorylation site

Cyclin-dependent kinase (Cdk) catalyzes phosphorylation of other proteins to start M phase

FIGURE 12.11 Cyclin Concentration Regulates the Concentration of the MPF Dimer. Cyclin concentrations cycle in dividing cells, reaching a peak in M phase. The activity of MPF, shown in the blue shaded area, requires both cyclin and Cdk components.

✔ **QUESTION** Proteins that degrade cyclin are activated by events that MPF initiates. Why is this important?

the concentration of the protein kinase is more or less constant throughout the cell cycle. How can MPF trigger M phase if the protein kinase subunit is always present?

The answer lies in the second MPF subunit, which belongs to a family of proteins called **cyclins.** Cyclins got their name because their concentrations fluctuate throughout the cell cycle.

As **FIGURE 12.11** shows, the concentration of the cyclin associated with MPF builds during interphase and peaks in M phase. The timing of this increase is important because the protein kinase subunit in MPF is functional only when it is bound to the cyclin subunit. As a result, the protein kinase subunit of MPF is called a **cyclin-dependent kinase,** or **Cdk.**

To summarize, MPF is a dimer consisting of a cyclin and a cyclin-dependent kinase. The cyclin subunit regulates the formation of the MPF dimer; the kinase subunit catalyzes the phosphorylation of other proteins to start M phase.

How Is MPF Turned On? According to Figure 12.11, the number of cyclins builds up steadily during interphase. Why doesn't this increasing concentration of MPF trigger the onset of M phase?

The answer is that the activity of MPF's Cdk subunit is further regulated by two phosphorylation events. The phosphorylation of one site in Cdk activates the kinase, but when the second site is phosphorylated, it is inactivated. Both these sites are phosphorylated after cyclin binds to the Cdk. This allows the concentration of the dimer to increase without prematurely starting M phase. Late in G₂ phase, however, an enzyme removes the inhibitory phosphate. This dephosphorylation reaction, coupled with the addition of the activating phosphate, changes the Cdk's shape in a way that turns on its kinase activity.

Once MPF is active, it triggers a chain of events. Although the exact mechanisms involved are still under investigation, the result is that chromosomes begin to condense and the spindle apparatus starts to form. In this way, MPF triggers the onset of M phase.

How Is MPF Turned Off?   During anaphase, an enzyme complex begins degrading MPF's cyclin subunit. In this way, MPF triggers a chain of events that leads to its own destruction.

MPF deactivation illustrates two key concepts about regulatory systems in cells:

- **Negative feedback** occurs when a process is slowed or shut down by one of its products. Thermostats shut down furnaces when temperatures are high; phosphofructokinase is inhibited by ATP (see Chapter 9); MPF is turned off by an enzyme complex that is activated by events in mitosis.

- Destroying specific proteins is a common way to control cell processes. In this case, the enzyme complex that is activated in anaphase attaches small proteins called ubiquitins to MPF's cyclin subunit. This marks the subunit for destruction by a protein complex called the proteasome.

In response to MPF activity, then, the concentration of cyclin declines rapidly. Slowly, it builds up again during interphase. This sets up an oscillation in cyclin concentration.

✔ If you understand this aspect of cell-cycle regulation, you should be able to explain the relationship between MPF and (1) cyclin, (2) Cdk, and (3) the enzymes that phosphorylate MPF, dephosphorylate MPF, and degrade cyclin.

Cell-Cycle Checkpoints Can Arrest the Cell Cycle

The dramatic changes in cyclin concentrations and Cdk activity drive the ordered events of the cell cycle. These events are occurring in your body right now. Over a 24-hour period, you swallow millions of cheek cells and lose millions of cells from your intestinal lining as waste. To replace them, cells in your cheek and intestinal tissue are making and degrading cyclin and pushing themselves through the cell cycle.

MPF is only one of many protein complexes involved in regulating the cell cycle, however. A different cyclin complex triggers the passage from G_1 phase into S phase, and several regulatory proteins maintain the G_0 state of quiescent cells. An array of regulatory molecules holds cells in particular stages or stimulates passage to the next phase.

To make sense of these observations, Leland Hartwell and Ted Weinert introduced the concept of a **cell-cycle checkpoint.** A cell-cycle checkpoint is a critical point in the cell cycle that is regulated.

Hartwell and Weinert identified checkpoints by analyzing yeast cells with defects in the cell cycle. The defective cells kept dividing under culture conditions when normal cells stopped growing, because the defective cells lacked a specific checkpoint. In multicellular organisms, cells that keep dividing in this way may form a mass of cells called a **tumor.**

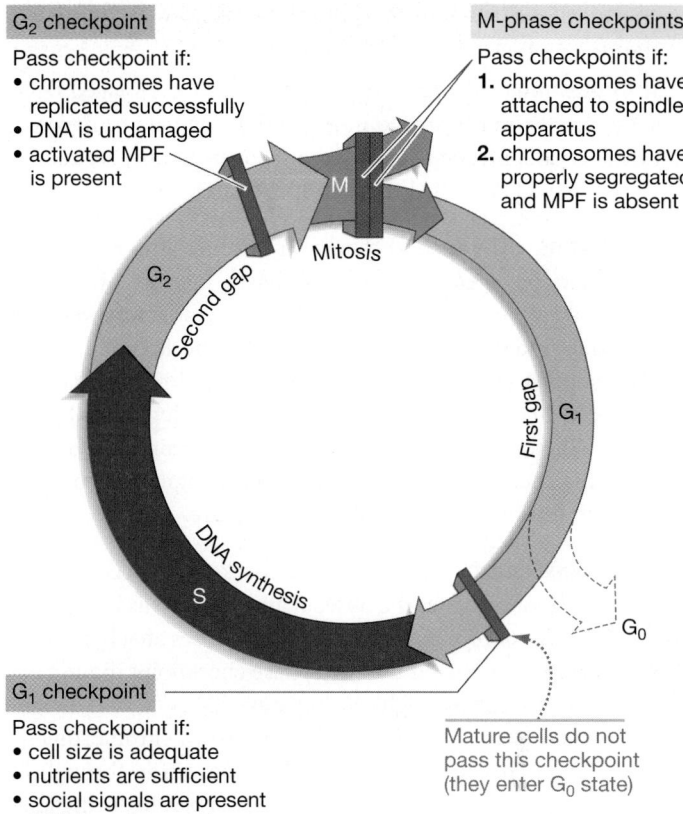

G₂ checkpoint

Pass checkpoint if:
- chromosomes have replicated successfully
- DNA is undamaged
- activated MPF is present

M-phase checkpoints

Pass checkpoints if:
1. chromosomes have attached to spindle apparatus
2. chromosomes have properly segregated and MPF is absent

G₁ checkpoint

Pass checkpoint if:
- cell size is adequate
- nutrients are sufficient
- social signals are present
- DNA is undamaged

Mature cells do not pass this checkpoint (they enter G_0 state)

FIGURE 12.12 The Four Cell-Cycle Checkpoints.

There are distinct checkpoints in three of the four phases of the cell cycle (**FIGURE 12.12**). In effect, interactions among regulatory molecules at each checkpoint allow a cell to "decide" whether to proceed with division or not. If these regulatory molecules are defective, the checkpoint may fail and cells may start dividing in an uncontrolled fashion.

G₁ Checkpoint   The first cell-cycle checkpoint occurs late in G_1. For most cells, this checkpoint is the most important in establishing whether the cell will continue through the cycle and divide, or exit the cycle and enter G_0. What determines whether a cell passes the G_1 checkpoint?

- *Size*   Because a cell must reach a certain size before its daughter cells will be large enough to function normally, biologists hypothesize that some mechanism exists to arrest the cell cycle if the cell is too small.

- *Availability of nutrients*   Unicellular organisms arrest at the G_1 checkpoint if nutrient conditions are poor.

- *Social signals*   Cells in multicellular organisms pass (or do not pass) through the G_1 checkpoint in response to signaling molecules from other cells, which are termed social signals.

- *Damage to DNA*   If DNA is physically damaged, the protein **p53** activates genes that either stop the cell cycle until the damage can be repaired or cause the cell's programmed, controlled destruction—a phenomenon known as **apoptosis.** In this way, p53 acts as a brake on the cell cycle.

If "brake" molecules such as p53 are defective, damaged DNA remains unrepaired. Damage in genes that regulate cell growth can lead to uncontrolled cell division. Consequently, regulatory proteins like p53 are called **tumor suppressors.**

G_2 Checkpoint The second checkpoint occurs after S phase, at the boundary between the G_2 and M phases. Because MPF is the key signal triggering the onset of M phase, investigators were not surprised to find that it is involved in the G_2 checkpoint.

Data suggest that if DNA is damaged or if chromosomes are not replicated correctly, removal of the inactivating phosphate is blocked. When MPF is not turned on, cells remain in G_2 phase. Cells at this checkpoint may also respond to signals from other cells and to internal signals relating to their size.

M-Phase Checkpoints The final two checkpoints occur during mitosis. The first regulates the onset of anaphase. Cells in M phase will not split the chromatids until all kinetochores attach properly to the spindle apparatus. If the metaphase checkpoint did not exist, some chromosomes might not separate correctly, and daughter cells would receive either too many or too few chromosomes.

The second checkpoint regulates the progression through M phase into G_1. If chromosomes do not fully separate during anaphase, MPF will not decline and the cell will be arrested in M phase. The enzymes that are responsible for cyclin destruction are activated only when all the chromosomes have been properly separated. The presence of MPF activity prevents the cell from undergoing cytokinesis and exiting the M phase.

To summarize, the four cell-cycle checkpoints have the same purpose: They prevent the division of cells that are damaged or that have other problems. The G_1 checkpoint also prevents mature cells that are in the G_0 state from dividing.

Understanding cell-cycle regulation is fundamental. If one of the checkpoints fails, the affected cells may begin dividing in an uncontrolled fashion. For the organism as a whole, the consequences of uncontrolled cell division may be dire: cancer.

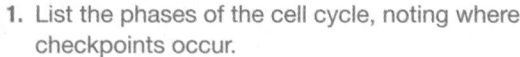

check your understanding

C
If you understand that . . .
- The cell cycle consists of four carefully controlled phases.

Y
✔ **You should be able to . . .**

U
1. List the phases of the cell cycle, noting where checkpoints occur.
2. Summarize how levels of Cdk and cyclin change over time and how this is related to MPF activity, noting the particular phases that are involved.

Answers are available in Appendix A.

12.4 Cancer: Out-of-Control Cell Division

Fifty percent of American men and 33 percent of American women will develop cancer during their lifetime. In the United States, one in four of all deaths is from cancer. It is the second leading cause of death, exceeded only by heart disease.

Cancer is a general term for disease caused by cells that divide in an uncontrolled fashion, invade nearby tissues, and spread to other sites in the body. Cancerous cells cause disease because they use nutrients and space needed by normal cells and disrupt the function of normal tissues.

Humans suffer from at least 200 types of cancer. Stated another way, cancer is not a single illness but a complex family of diseases that affect an array of organs, including the breast, colon, brain, lung, and skin (**FIGURE 12.13**). In addition, several types of cancer can affect the same organ. Skin cancers, for example, come in multiple forms.

Although cancers vary in time of onset, growth rate, seriousness, and cause, they have a unifying feature: Cancers arise from cells in which cell-cycle checkpoints have failed.

Cancerous cells have two types of defects related to cell division: **(1)** defects that make the proteins required for cell growth active when they shouldn't be, and **(2)** defects that prevent tumor suppressor genes from shutting down the cell cycle.

For example, the protein Ras is a key component in signal transduction systems—including phosphorylation cascades that trigger cell growth (see Chapter 11). Many cancers have defective forms of Ras that do not become inactivated. Instead, the defective Ras constantly sends signals that trigger mitosis and cell division.

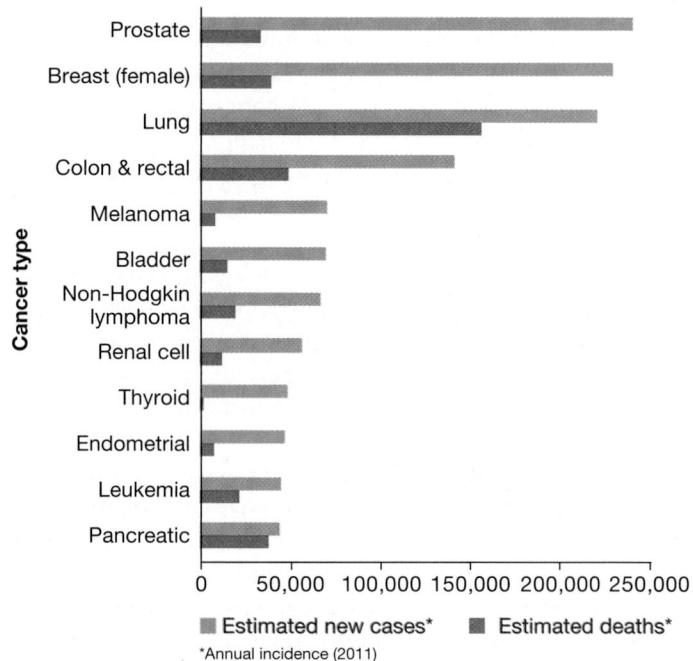

FIGURE 12.13 Cancers Vary in Type and Severity.

DATA: The website of the National Cancer Institute (http://www.cancer.gov), Common Cancer Types, November 2010.

Likewise, a large percentage of cancers have defective forms of the tumor suppressor p53. Instead of being arrested or destroyed, cells with damaged DNA are allowed to continue growing.

Let's review the general characteristics of cancer and then explore how regulatory mechanisms become defective.

Properties of Cancer Cells

When even a single cell in a multicellular organism begins to divide in an uncontrolled fashion, a mass of cells called a tumor may result. If a tumor can be surgically removed without damage to the affected organ, a cure might be achieved. Often, though, surgery doesn't cure cancer. Why?

In addition to uncontrolled replication, cancer cells are invasive—meaning that they are able to spread to adjacent tissues and throughout the body via the bloodstream or the lymphatic vessels (introduced in Chapter 51), which collect excess fluid from tissues and return it to the bloodstream.

Invasiveness is a defining feature of a **malignant tumor**—one that is cancerous. Masses of noninvasive cells are noncancerous and form **benign tumors.** Some benign tumors are largely harmless. Others grow quickly and can cause problems if they are located in the brain or other sensitive parts of the body.

Cells become malignant and cancerous if they gain the ability to detach from the original tumor and invade other tissues. By spreading from the primary tumor site, cancer cells can establish secondary tumors elsewhere in the body (**FIGURE 12.14**). This process is called **metastasis.**

If metastasis has occurred by the time the original tumor is detected, secondary tumors have already formed and surgical removal of the primary tumor will not lead to a cure. This is why early detection is the key to treating cancer most effectively.

Cancer Involves Loss of Cell-Cycle Control

What causes cancer at the molecular level? Recall that when many cells mature, they enter the G_0 phase—meaning their cell cycle is arrested at the G_1 checkpoint. In contrast, cells that do pass through the G_1 checkpoint are irreversibly committed to replicating their DNA and entering G_2.

Based on this observation, biologists hypothesize that many types of cancer involve defects in the G_1 checkpoint. To understand the molecular nature of the disease, then, researchers have focused on understanding the normal mechanisms that operate at that checkpoint. Cancer research and research on the normal cell cycle have become two sides of the same coin.

Social Control In unicellular organisms, passage through the G_1 checkpoint is thought to depend primarily on cell size and the availability of nutrients. If nutrients are plentiful, cells pass through the checkpoint and divide rapidly.

In multicellular organisms, however, cells divide in response to signals from other cells. Biologists refer to this as *social control* over cell division. The general idea is that individual cells should be allowed to divide only when their growth is in the best interests of the organism as a whole.

(a) Benign tumor

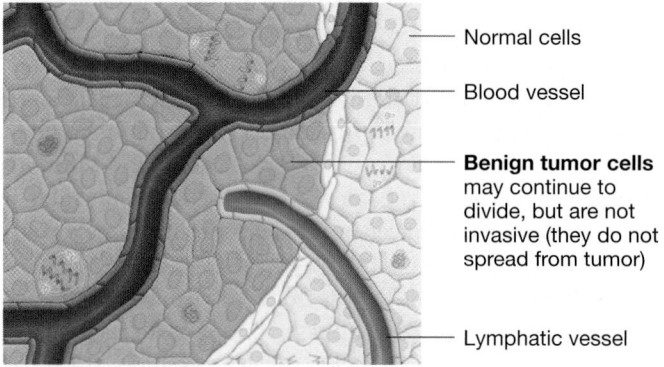

- Normal cells
- Blood vessel
- **Benign tumor cells** may continue to divide, but are not invasive (they do not spread from tumor)
- Lymphatic vessel

(b) Malignant tumor

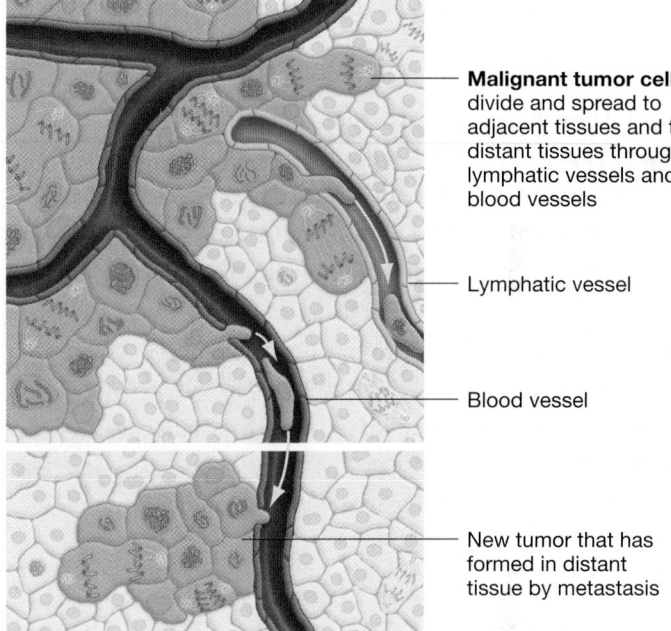

- **Malignant tumor cells** divide and spread to adjacent tissues and to distant tissues through lymphatic vessels and blood vessels
- Lymphatic vessel
- Blood vessel
- New tumor that has formed in distant tissue by metastasis

FIGURE 12.14 Cancers Spread to New Locations in the Body. (a) Benign tumors grow in a single location. **(b)** Malignant tumors are metastatic—meaning that their cells can spread to distant parts of the body and initiate new tumors. Malignant tumors cause cancer.

Social control of the cell cycle is based on **growth factors**—polypeptides or small proteins that stimulate cell division. Many growth factors were discovered by researchers who were trying to grow cells in culture. When isolated mammalian cells were placed in a culture flask and provided with adequate nutrients, they arrested in G_1 phase. The cells began to grow again only when biologists added **serum**—the liquid portion of blood that remains after blood cells and cell fragments have been removed. Researchers identified growth factors as the components in the serum that were responsible for allowing cells to pass through the G_1 checkpoint.

Cancer cells are an exception. They can often be cultured successfully without externally supplied growth factors. This observation suggests that the normal social controls on the G_1 checkpoint have broken down in cancer cells.

How Does the G₁ Checkpoint Work? In G₀ cells, the arrival of growth factors stimulates the production of a key regulatory protein called E2F. When E2F is activated, it triggers the expression of genes required for S phase.

When E2F is first produced, however, its activity is blocked by a tumor suppressor protein called Rb. **Rb protein** is one of the key molecules that enforces the G₁ checkpoint. It is called Rb because a nonfunctional version was first discovered in children with retinoblastoma, a cancer in the light-sensing tissue, or retina, of the eye.

When E2F is bound to Rb, it is in the "off" position—it can't activate the genes required for S phase. As long as Rb stays bound to E2F, the cell remains in G₀. But as **FIGURE 12.15** shows, the situation changes dramatically if growth factors continue to arrive. To understand how growth factors affect E2F activity, think back to how cells progress from G₂ to M phase. As in passage from G₂ to M phase, phosphorylation of other proteins catalyzed by an activated cyclin–Cdk dimer permits passage from G₁ to S.

Step 1 Growth factors arrive from other cells.

Step 2 The growth factors stimulate the production of E2F and of G₁ cyclins, which are different from those used in MPF.

Step 3 Rb binds to E2F, inactivating it. The G₁ cyclins begin forming cyclin–Cdk dimers. Initially, the Cdk component is phosphorylated and inactive.

Step 4 When dephosphorylation turns on the G₁ cyclin–Cdk complexes, they catalyze the phosphorylation of Rb.

Step 5 The phosphorylated Rb changes shape and releases E2F.

Step 6 The unbound E2F is free to activate its target genes. Production of S-phase proteins gets S phase under way.

In this way, growth factors function as a social signal that says, "It's OK to override Rb. Go ahead and pass the G₁ checkpoint and divide."

How Do Social Controls and Cell-Cycle Checkpoints Fail? Cells can become cancerous when social controls fail—meaning, when cells begin dividing in the absence of the go-ahead signal from growth factors. One of two things can go wrong: The G₁ cyclin is overproduced, or Rb is defective.

When cyclins are overproduced and stay at high concentrations, the Cdk that binds to cyclin phosphorylates Rb continuously. This activates E2F and sends the cell into S phase.

Cyclin overproduction results from **(1)** excessive amounts of growth factors or **(2)** cyclin production in the absence of growth signals. Cyclins are produced continuously when a signaling pathway is defective. Because this pathway includes the Ras protein (highlighted in Chapter 11), it is common to find overactive Ras proteins in cancerous cells.

What happens if Rb is defective? When Rb is missing or does not bind normally to E2F, any E2F that is present pushes the cell through the G₁ checkpoint and into S phase, leading to uncontrolled cell division.

Because cancer is actually a family of diseases with a complex and highly variable molecular basis, there will be no "magic bullet," or single therapy, that cures all forms of the illness. Still, recent progress in understanding the cell cycle and the molecular basis of cancer has been dramatic, and cancer prevention and early detection programs are increasingly effective. The prognosis for many cancer patients is remarkably better now than it was even a few years ago. Thanks to research, almost all of us know someone who is a cancer survivor.

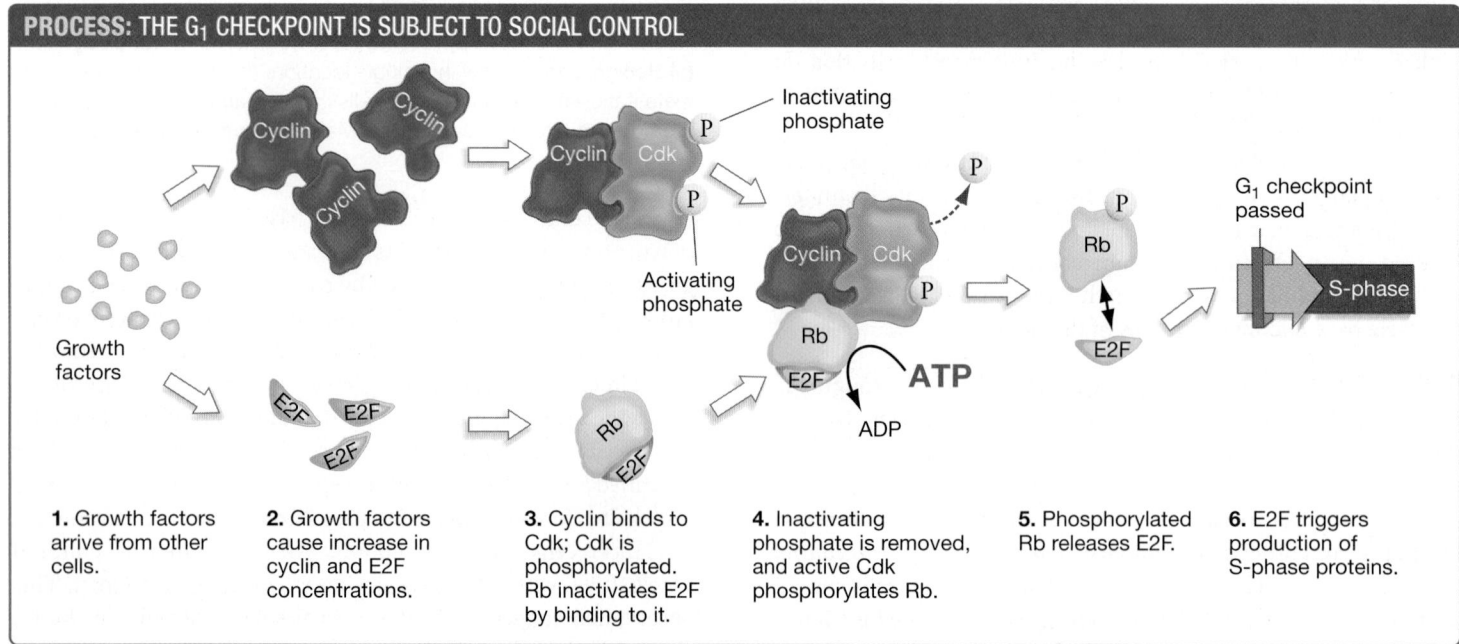

PROCESS: THE G₁ CHECKPOINT IS SUBJECT TO SOCIAL CONTROL

1. Growth factors arrive from other cells.

2. Growth factors cause increase in cyclin and E2F concentrations.

3. Cyclin binds to Cdk; Cdk is phosphorylated. Rb inactivates E2F by binding to it.

4. Inactivating phosphate is removed, and active Cdk phosphorylates Rb.

5. Phosphorylated Rb releases E2F.

6. E2F triggers production of S-phase proteins.

FIGURE 12.15 Growth Factors Move Cells through the G₁ Checkpoint.

If you understand . . .

12.1 How Do Cells Replicate?

- When a cell divides, it must copy its chromosomes, separate the copies, and divide the cytoplasm to generate daughter cells such that each carries the same chromosomal complement as the parent.
- Eukaryotic cells divide by alternating between interphase and M phase.
- Interphase consists of S phase, when chromosomes are replicated, and gap phases called G_1 and G_2, when cells grow and prepare for division.
- Eukaryotic cells divide by cycling through four phases: G_1, S, G_2, and M. Mature cells arrest at G_1 and enter a nonreplicating phase called G_0.

✔ **You should be able to** explain the roles of each of the four stages of the cell cycle.

12.2 What Happens during M Phase?

- Mitosis and cytokinesis are responsible for the partitioning of chromosomes and division of the parent cell into two daughter cells.
- Mitosis can be described as a sequence of five phases:
  1. *Prophase* Chromosomes condense. The spindle apparatus begins to form, and polar microtubules overlap each other.
  2. *Prometaphase* In cells of many organisms, the nuclear envelope disintegrates. Microtubules attach to the kinetochores of chromosomes and begin moving them to the middle of the spindle.
  3. *Metaphase* All the chromosomes are positioned in the middle of the spindle. The spindle is anchored to the cell membrane by astral microtubules.
  4. *Anaphase* Sister chromatids are pulled apart by the disassembly of kinetochore microtubules at the kinetochore. The separated chromatids are now daughter chromosomes. The spindle poles are moved farther apart to fully separate the replicated chromosomes.
  5. *Telophase* Daughter chromosomes are fully separated and are clustered at opposite poles of the spindle. A nuclear envelope forms around each set and the chromosomes de-condense.
- In most cells, mitosis is followed by cytokinesis—division of all cell contents.

✔ **You should be able to** predict how mitosis would be different in cells where the nuclear envelope remains intact (e.g., yeast).

12.3 Control of the Cell Cycle

- The onset of S and M phases is primarily determined by the activity of protein dimers consisting of cyclin and cyclin-dependent kinases (Cdks).

- Cyclin concentrations oscillate during the cell cycle, regulating the formation of the dimer. The activity of the Cdk is further regulated by addition of a phosphate in its activating site and removal of one from its inhibitory site.
- Progression through the cell cycle is controlled by checkpoints in three phases.
  1. The G_1 checkpoint regulates progress based on nutrient availability, cell size, DNA damage, and social signals.
  2. The G_2 checkpoint delays progress until chromosome replication is complete and any damaged DNA that is present is repaired.
  3. The two M-phase checkpoints will **(1)** delay anaphase until all chromosomes are correctly attached to the spindle apparatus and **(2)** delay the onset of cytokinesis and G_1 until all the chromosomes have been properly partitioned.

✔ **You should be able to** predict what would happen if the kinase that adds the inhibitory phosphates to Cdk were defective.

12.4 Cancer: Out-of-Control Cell Division

- Cancer is characterized by **(1)** loss of control at the G_1 checkpoint, resulting in cells that divide in an uncontrolled fashion; and **(2)** metastasis, or the ability of tumor cells to spread throughout the body.
- The G_1 checkpoint depends in part on Rb, which prevents progression to S phase, and G_1 cyclin–Cdk complexes that trigger progression to S phase. Defects in Rb and G_1 cyclin are common in human cancer cells.

✔ **You should be able to** compare and contrast the effect of removing growth factors from asynchronous cultures of human cells that are normal versus those that are cancerous.

(MB) **MasteringBiology**

1. **MasteringBiology Assignments**

   Tutorials and Activities Causes of Cancer; Cell Culture Methods; Four Phases of the Cell Cycle; Mitosis (1 of 3): Mitosis and the Cell Cycle; Mitosis (2 of 3): Mechanism of Mitosis; Mitosis (3 of 3): Comparing Cell Division in Animals, Plants, and Bacteria; Mitosis and Cytokinesis Animation; Roles of Cell Division; The Cell Cycle; The Phases of Mitosis

   Questions Reading Quizzes, Blue-Thread Questions, Test Bank

2. **eText** Read your book online, search, take notes, highlight text, and more.

3. **The Study Area** Practice Test, Cumulative Test, BioFlix® 3-D Animations, Videos, Activities, Audio Glossary, Word Study Tools, Art

You should be able to . . .

1. Which statement about the daughter cells following mitosis and cytokinesis is correct?
   a. They are genetically different from each other and from the parent cell.
   b. They are genetically identical with each other and with the parent cell.
   c. They are genetically identical with each other but different from the parent cell.
   d. Only one of the two daughter cells is genetically identical with the parent cell.

2. Progression through the cell cycle is regulated by oscillations in the concentration of which type of molecule?
   a. p53, Rb, and other tumor suppressors
   b. receptor tyrosine kinases
   c. cyclin-dependent kinases
   d. cyclins

3. After the S phase, what comprises a single chromosome?
   a. two daughter chromosomes
   b. a double-stranded DNA molecule
   c. two single-stranded molecules of DNA
   d. two sister chromatids

4. What major events occur during anaphase of mitosis?
   a. Chromosomes replicate, so each chromosome consists of two identical sister chromatids.
   b. Chromosomes condense and the nuclear envelope disappears.
   c. Sister chromatids separate, and the spindle poles are pushed farther apart.
   d. The chromosomes end up at opposite ends of the cell, and two nuclear envelopes form around them.

5. What evidence suggests that during anaphase, kinetochore microtubules shorten at the kinetochore?

6. Under normal conditions, what happens to the cell cycle if the chromosomes fail to separate properly at anaphase?

7. Identify at least two events in the cell cycle that must be completed successfully for daughter cells to share an identical complement of chromosomes.

8. Make a concept map illustrating normal events at the G_1 checkpoint. Your diagram should include p53, DNA damage, Rb, E2F, social signals, G_1 Cdk, G_1 cyclin, S-phase proteins, phosphorylated (inactivated) cyclin–Cdk, dephosphorylated (activated) cyclin–Cdk, phosphorylated (inactivated) Rb.

9. Explain how microinjection experiments supported the hypothesis that specific molecules in the cytoplasm are involved in the transition from interphase to M phase. What was the control for this experiment?

10. Why are most protein kinases considered regulatory proteins?

11. Why are cyclins called cyclins? Explain their relationship to MPF activity.

12. In multicellular organisms, nondividing cells stay in G_0 phase. For the cell, why is it better to be held in G_1 rather than S, G_2, or M phase?
    a. G_1 cells are larger and more likely to perform the normal functions of the cell.
    b. G_1 cells have not replicated their DNA in preparation for division.
    c. G_1 cells are the only ones that do not have their chromatin in a highly condensed state.
    d. MPF is required to enter S phase, so the cell is committed to entering M phase if the cycle moves beyond G_1.

13. **QUANTITATIVE** A particular cell spends 4 hours in G_1 phase, 2 hours in S phase, 2 hours in G_2 phase, and 30 minutes in M phase. If a pulse–chase assay were performed with radioactive thymidine on an asynchronous culture, what percentage of mitotic cells would be radiolabeled after 9 hours?
    a. 0%
    b. 50%
    c. 75%
    d. 100%

14. When fruit fly embryos first begin to develop, a large cell is generated that contains over 8000 nuclei that are genetically identical with one another. What is most likely responsible for this result?

15. What is most likely responsible for the reduction in death rates over the past several years in cancers of the breast and prostate? How is this related to the development of cancer?

16. Cancer is primarily a disease of older people. Further, a group of individuals may share a genetic predisposition to developing certain types of cancer, yet vary a great deal in time of onset—or not get the disease at all. What conclusion could be drawn based on these observations? How does this relate to the requirements for a cell to become cancerous?

13 Meiosis

In this chapter you will learn how

Meiosis promotes genetic diversity and allows the benefits of sex

starting with

How does meiosis occur? **13.1**

and comparing to

Mitosis **Ch. 12**

by examining

How meiosis produces genetic variation

looking at

Independent assortment, crossing over, and fertilization **13.2**

then asking

What happens when things go wrong? **13.3**

by asking

Why does meiosis exist? **13.4**

Purifying-selection hypothesis

Changing-environment hypothesis

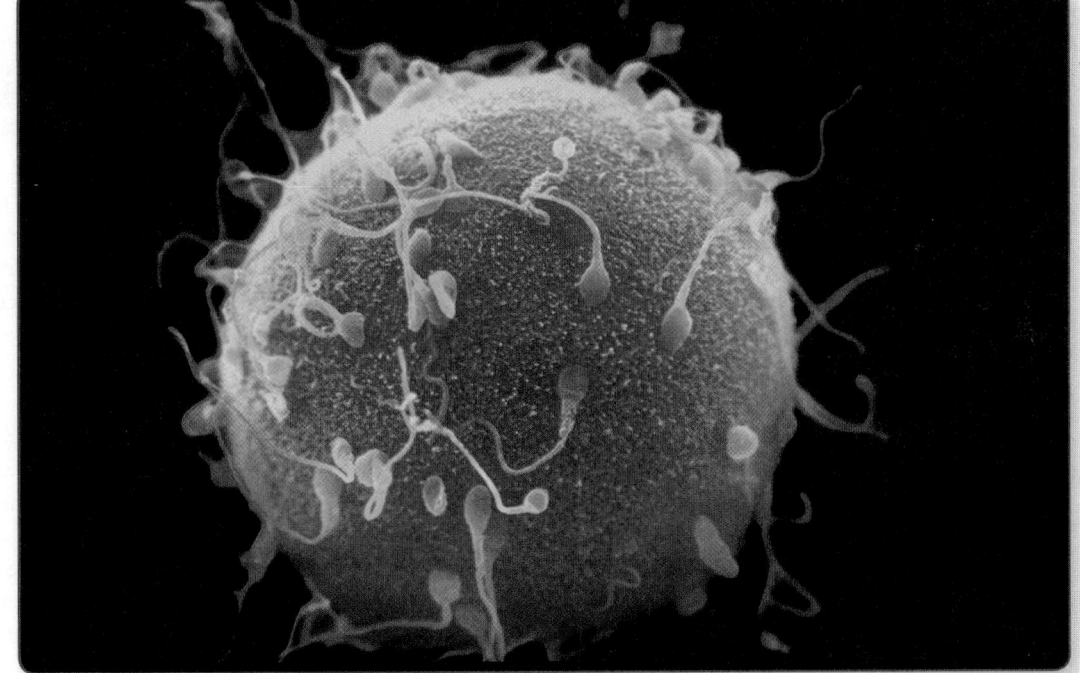

Scanning electron micrograph (with color added) showing human sperm attempting to enter a human egg. This chapter introduces the type of nuclear division called meiosis, which in animals occurs before the formation of sperm and eggs.

This chapter is part of the Big Picture. See how on pages 366–367.

W hy sex?

Simple questions—such as why sexual reproduction exists—are sometimes the best. This chapter asks what sexual reproduction is and why some organisms employ it. The focus here is on how organisms reproduce, or replicate—one of the five fundamental attributes of life introduced in Chapter 1.

For centuries people have known that during sexual reproduction, a male reproductive cell—a **sperm**—and a female reproductive cell—an **egg**—unite to form a new individual. The process of uniting sperm and egg is called **fertilization**. The first biologists to observe fertilization studied the large, translucent eggs of sea urchins. Owing to the semitransparency of the sea urchin egg cell, researchers were able to see the nuclei of a sperm and an egg fuse.

✔ When you see this checkmark, stop and test yourself. Answers are available in Appendix A.

When these observations were published in 1876, they raised an important question, because biologists had already established that the number of chromosomes is constant from cell to cell within a multicellular organism. The question is, How can the chromosomes from a sperm cell and an egg cell combine, but form an offspring that has the same chromosome number as its mother and its father?

A hint at the answer came in 1883, with the observation that cells in the body of roundworms of the genus *Ascaris* have four chromosomes, while their sperm and egg nuclei have only two chromosomes apiece.

Four years later, August Weismann formally proposed a hypothesis to explain the riddle: During the formation of **gametes**—reproductive cells such as sperm and eggs—there must be a distinctive type of cell division that leads to a reduction in chromosome number. Specifically, if the sperm and egg contribute an equal number of chromosomes to the fertilized egg, Weismann reasoned, they must each contain half of the usual number of chromosomes. Then, when sperm and egg combine, the resulting cell has the same chromosome number as its mother's cells and its father's cells have.

In the decades that followed, biologists confirmed this hypothesis by observing gamete formation in a wide variety of plant and animal species. Eventually this form of cell division came to be called meiosis (literally, "lessening-act").

Meiosis is nuclear division that leads to a halving of chromosome number and ultimately to the production of sperm and egg. (Meiosis is an important part of The Big Picture of Genetic Information on pages 366-367). To a biologist, asking "Why sex?" is equivalent to asking "Why meiosis?" Let's delve in by first looking at how meiosis happens.

13.1 How Does Meiosis Occur?

To understand meiosis, it is critical to grasp some key ideas about chromosomes. For example, when cell biologists began to study the cell divisions that lead to gamete formation, they made an important observation: Each organism has a characteristic number of chromosomes.

Consider the drawings in **FIGURE 13.1**, based on a paper published by Nettie Maria Stevens in 1908. They show the chromosomes of the fruit fly *Drosophila melanogaster*, or *Drosophila* for short. This model organism has been a focus of biological research for more than 100 years (see **BioSkills 13** in Appendix B). Stevens was studying the cell divisions leading up to the formation of egg and sperm. In total, she found eight chromosomes in *Drosophila* cells. Your cells have 46 chromosomes, and some ferns have over 1000.

Chromosomes Come in Distinct Sizes and Shapes

Stevens found that each *Drosophila* cell has eight chromosomes, but just five distinct types, distinguished by their size and shape. Three of these chromosomes always occurred in pairs and are labeled chromosomes 2–4 in Figure 13.1. In males, Stevens

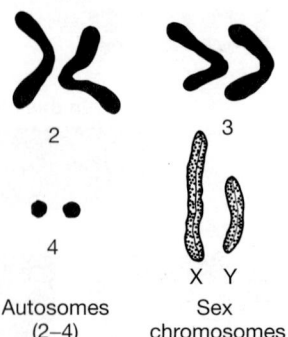 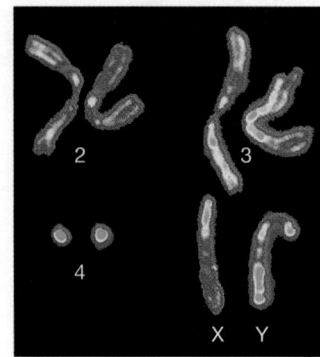

Drosophila chromosomes drawn by Nettie Stevens in 1908 ...

... and photographed through a modern microscope.

Autosomes (2–4) Sex chromosomes

FIGURE 13.1 Cells Contain Different Types of Chromosomes, and in Diploid Organisms, Chromosomes Come in Pairs. Numbers and letters designate the types of *Drosophila* chromosomes. These chromosomes are from the cell of a male, so there is an X and a Y chromosome plus three homologous pairs of autosomes.

observed an unpaired set of chromosomes, which came to be known as the X and the Y chromosomes. Stevens found that females lack a Y chromosome but contain a pair of X chromosomes. This is the same situation in some other insects and in mammals. The X and Y chromosomes are called **sex chromosomes** and are associated with an individual's sex. Non-sex chromosomes, such as chromosomes 2–4 in *Drosophila*, are **autosomes.**

Chromosomes that are the same size and shape are called **homologous** ("same-proportion") **chromosomes,** or **homologs,** and the pair is called a homologous pair. Later work showed that homologous chromosomes are similar in content as well as in size and shape. Homologous chromosomes carry the same genes. A **gene** is a section of DNA that influences some hereditary trait in an individual. For example, each copy of chromosome 2 found in *Drosophila* carries genes that influence eye color, wing size and shape, and bristle size.

The versions of a gene found on homologous chromosomes may differ, however. Biologists use the term **allele** to denote different versions of the same gene. For example, the allele for an eye-color gene on one homolog of chromosome 2 may be associated with red eyes, the normal color in *Drosophila*, whereas the allele of the same eye-color gene on the other homolog may be associated with purple eyes (**FIGURE 13.2**); the particular alleles of the bristle-size gene will influence whether the fly's bristles are long or short, and so on.

Homologous chromosomes carry the same genes, but each homolog may contain different alleles.

The Concept of Ploidy

At this point in her study, Stevens had determined the *Drosophila* **karyotype**—meaning the number and types of chromosomes present. As karyotyping studies became more common, cell biologists realized that, like *Drosophila*, the vast majority of plants and animals have more than one of each type of chromosome.

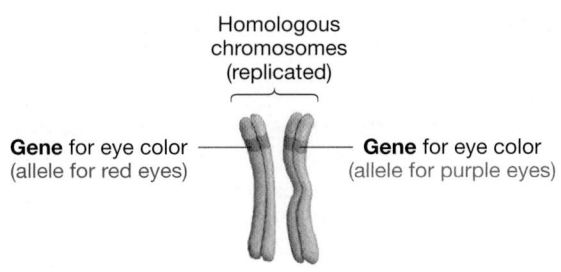

Homologous
chromosomes
(replicated)

Gene for eye color
(allele for red eyes)

Gene for eye color
(allele for purple eyes)

Drosophila autosome 2

FIGURE 13.2 Homologous Chromosomes May Contain Different Alleles of the Same Gene. The homologs of *Drosophila* chromosome 2 are shown; the location of only one of many genes is indicated.

Insects, humans, oak trees, and other organisms that have two versions of each type of chromosome are called **diploid** ("double-form"). Diploid organisms have two alleles of each gene. One allele is carried on each of the homologous pairs of chromosomes. Although a diploid individual can carry only two different alleles of a gene, there can be many different alleles in a population.

Organisms whose cells contain just one of each type of chromosome—for example, bacteria, archaea, and many algae and fungi—are called **haploid** ("single-form"). Haploid organisms have only one copy of each chromosome and just one allele of each gene.

Biologists use a compact notation to indicate the number of chromosomes and chromosome sets in a particular organism or type of cell:

- By convention, the letter n stands for the number of distinct types of chromosomes in a given cell and is called the **haploid number.** If sex chromosomes are present, they are counted as a single type in the haploid number. In humans, n is 23.

- To indicate the number of complete chromosome sets observed, a number is placed before the n. Thus, a cell can be n, or $2n$, or $3n$, and so on.

The combination of the number of sets and n is termed the cell's **ploidy.** Diploid cells or species are designated $2n$, because two chromosomes of each type are present—one from each parent. A **maternal chromosome** comes from the mother, and a **paternal chromosome** comes from the father.

Humans are diploid; $2n$ is 46. Haploid cells or species are labeled simply n, because they have just one set of chromosomes—no homologs are present. In haploid cells, the number 1 in front of n is implied and is not written out.

To summarize, the haploid number n indicates the number of distinct types of chromosomes present. In contrast, a cell's ploidy (n, $2n$, $3n$, etc.) indicates the number of each type of chromosome present. Stating a cell's ploidy is the same as stating the number of haploid chromosome sets present. ✔ If you understand how these terms relate, you should be able to state the haploid number, ploidy, and total number of chromosomes present in a male *Drosophila*.

Later work revealed that it is common for species in some lineages—particularly certain land plants, such as ferns—to contain more than two of each type of chromosome. Instead of having two homologous chromosomes per cell, **polyploid** ("many-form") species have three or more of each type of chromosome in each cell.

Depending on the number of homologs present, polyploid species are called triploid ($3n$), tetraploid ($4n$), hexaploid ($6n$), octoploid ($8n$), and so on.

Stevens and other early cell biologists did more than just describe the karyotypes observed in their study organisms. Through careful examination, they were able to track how chromosome numbers change during meiosis. These studies confirmed Weismann's hypothesis that a special type of cell division occurs during gamete formation.

An Overview of Meiosis

Cells replicate each of their chromosomes before undergoing meiosis. At the start of meiosis, chromosomes are in the same state they are in before mitosis.

Recall that an unreplicated eukaryotic chromosome consists of a single, long DNA double helix organized around proteins called histones (see Chapter 12). When chromosome replication is complete, each chromosome will consist of two identical **sister chromatids.** Sister chromatids contain identical copies of the DNA double helix present in the unreplicated chromosome and therefore the same genetic information. They remain physically joined along their entire length during much of meiosis (**FIGURE 13.3**).

To understand meiosis, it is critical to understand the relationship between chromosomes and sister chromatids. The trick is to recognize that unreplicated and replicated chromosomes are both considered *single* chromosomes—even though the replicated chromosome contains *two* sister chromatids. This makes sense if you consider that a chromosome carries a particular set of genetic information in its DNA and that the amount of *unique* information is the same whether there is one copy of it present or two.

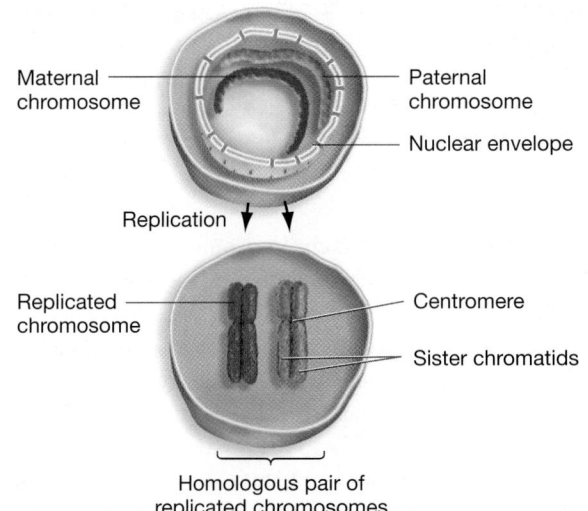

Maternal chromosome

Paternal chromosome

Nuclear envelope

Replication

Replicated chromosome

Centromere

Sister chromatids

Homologous pair of replicated chromosomes

FIGURE 13.3 Each Chromosome Replicates before Undergoing Meiosis.

Term	Definition	Example or Comment
Chromosome	Structure made up of DNA and proteins; carries the cell's hereditary information (genes)	Eukaryotes have linear chromosomes; most bacteria and archaea have just one, circular, chromosome
• **Sex chromosome**	Chromosome associated with an individual's sex	X and Y chromosomes of humans (males are XY, females XX); Z and W chromosomes of birds and butterflies (males are ZZ, females ZW)
• **Autosome**	A non-sex chromosome	Chromosomes 1–22 in humans
Unreplicated chromosome	A chromosome that consists of one double-helical molecule of DNA packaged with proteins	
Replicated chromosome	A chromosome that has been copied; consists of two identical chromatids, each containing one double-helical DNA molecule	Centromere
Sister chromatids	The two identical chromatid copies in a replicated chromosome	Sister chromatids
Homologous chromosomes (homologs)	In a diploid organism, chromosomes that are similar in size, shape, and gene content	You have a chromosome 22 from each parent. Homologous chromosomes
Non-sister chromatids	Chromatids belonging to homologous chromosomes	Non-sister chromatids
Bivalent (or tetrad)	Homologous replicated chromosomes that are joined together during prophase I and metaphase I of meiosis	Bivalent
Haploid number	The number of different types of chromosomes in a cell; symbolized *n*	Humans have 23 different types of chromosomes (*n* = 23)
Diploid number	The number of chromosomes present in a diploid cell (see below); symbolized 2*n*	In humans all cells except gametes are diploid and contain 46 chromosomes (2*n* = 46)
Ploidy	The number of each type of chromosome present	The number of haploid chromosome sets present
• **Haploid**	Having one of each type of chromosome (*n*)	Bacteria and archaea are haploid, as are many algae; plant and animal gametes are haploid
• **Diploid**	Having two of each type of chromosome (2*n*)	Most familiar plants and animals are diploid
• **Polyploid**	Having more than two of each type of chromosome; cells may be triploid (3*n*), tetraploid (4*n*), hexaploid (6*n*), and so on	Seedless bananas are triploid; many ferns are tetraploid; bread wheat is hexaploid

Note that an unreplicated chromosome is never called a chromatid; you can refer to chromatids only as the structures in a replicated chromosome.

TABLE 13.1 summarizes the vocabulary that biologists use to describe chromosomes and illustrates the relationship between chromosomes and chromatids. ✔ If you understand this relationship, you should be able to draw the same chromosome in the unreplicated and replicated state, label the sister chromatids, indicate the number of DNA molecules present in each structure, and explain why both structures represent a single chromosome.

Meiosis Comprises Two Cell Divisions Meiosis consists of two cell divisions, called **meiosis I** and **meiosis II**. As **FIGURE 13.4** shows, the two divisions occur consecutively but differ sharply.

During meiosis I, the homologs in each chromosome pair separate from each other. One homolog goes to one daughter cell; the other homolog goes to the other daughter cell. It is a matter of chance which daughter cell receives which homolog.

At the end of meiosis I, each of the two daughter cells has one of each type of chromosome instead of two, and thus half as many chromosomes as the parent cell had. Put another way: During meiosis I, the diploid (2*n*) parent cell produces two haploid (*n*)

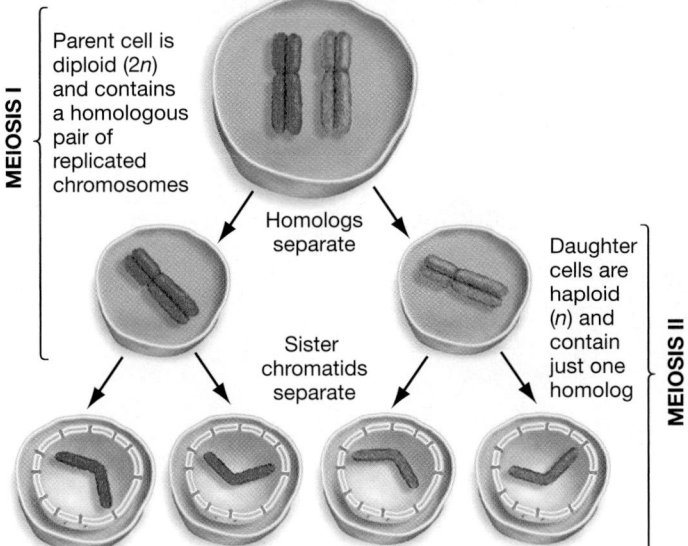

Four daughter cells contain one chromosome each (*n*). In animals, these cells become gametes.

FIGURE 13.4 The Major Events in Meiosis. Before undergoing meiosis, chromosomes are replicated so there are two chromatids per chromosome. Meiosis reduces chromosome number by half. In diploid organisms, the products of meiosis are haploid. Maternal chromosomes are shown in red; paternal chromosomes, blue. Note that in this cell, $2n = 2$.

daughter cells. Notice, however, that each chromosome still consists of two sister chromatids—meaning that chromosomes are still replicated at the end of meiosis I.

During meiosis II, sister chromatids from each chromosome separate. One sister chromatid (now called a daughter chromosome) goes to one daughter cell; the other sister chromatid goes to the other daughter cell as a daughter chromosome. Remember that the cell that started meiosis II had one of each type of chromosome, but each chromosome was still replicated (meaning still consisting of two sister chromatids). The cells produced by meiosis II also have one of each type of chromosome, but now the daughter chromosomes are no longer replicated.

To reiterate, sister chromatids separate into daughter chromosomes during meiosis II. This is just what happens during mitosis. Meiosis II is actually equivalent to mitosis occurring in a haploid cell. In meiosis I, on the other hand, sister chromatids stay together. This sets meiosis I apart from both mitosis and meiosis II.

As in mitosis, chromosome movements during meiosis I and II are coordinated by microtubules of the **spindle apparatus** that attach to **kinetochores** located at the **centromere** of each chromosome. Recall that the centromere is a region on the chromosome; kinetochores are structures in that region (see Chapter 12). Chromosome movement is driven by fraying of the ends of microtubules at each kinetochore, just as it is in mitosis (see Figure 12.7).

Meiosis I is a Reduction Division

A host of early cell biologists worked out this sequence of events through careful observation of cells with the light microscope. Based on these studies, they came to a key realization: The outcome of meiosis I is a reduction

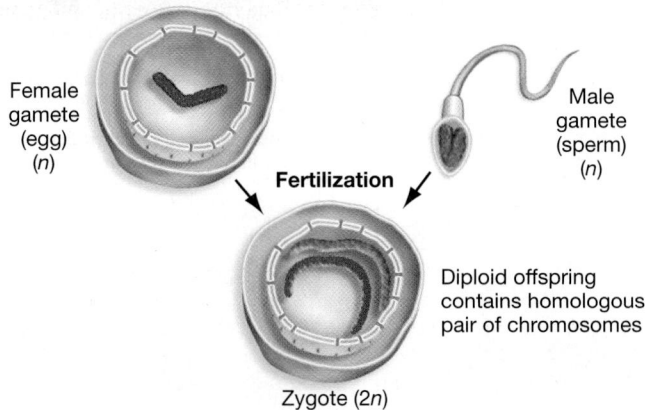

FIGURE 13.5 Fertilization Restores a Full Complement of Chromosomes.

in chromosome number. For this reason, meiosis I is known as a reduction division. Reduction is another important way in which meiosis I differs from meiosis II and mitosis.

In most plants and animals, the original cell entering meiosis is diploid and the four final daughter cells are haploid. In animals, the haploid daughter cells, each containing one of each homologous chromosome, eventually go on to form egg cells or sperm cells via a process called **gametogenesis** ("gamete-origin").

When two haploid gametes fuse during fertilization, a full complement of chromosomes is restored (**FIGURE 13.5**). The cell that results from fertilization is diploid and is called a **zygote**. In this way, each diploid individual receives a haploid chromosome set from its mother and a haploid set from its father.

FIGURE 13.6 puts these events into the context of an animal's **life cycle**—the sequence of events that occurs over the life span of an individual, from fertilization to the production of offspring. As you study the figure, note how ploidy changes as the result of

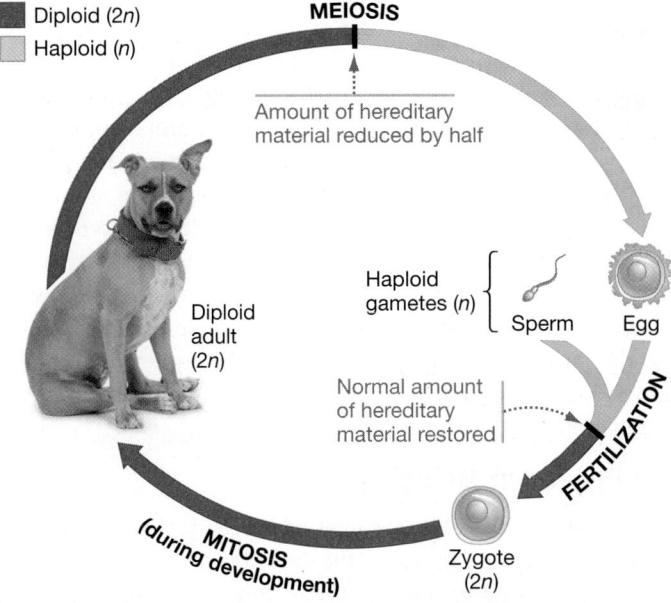

FIGURE 13.6 Ploidy Changes during the Life Cycle of a Dog. Most of the dog life cycle involves diploid cells.

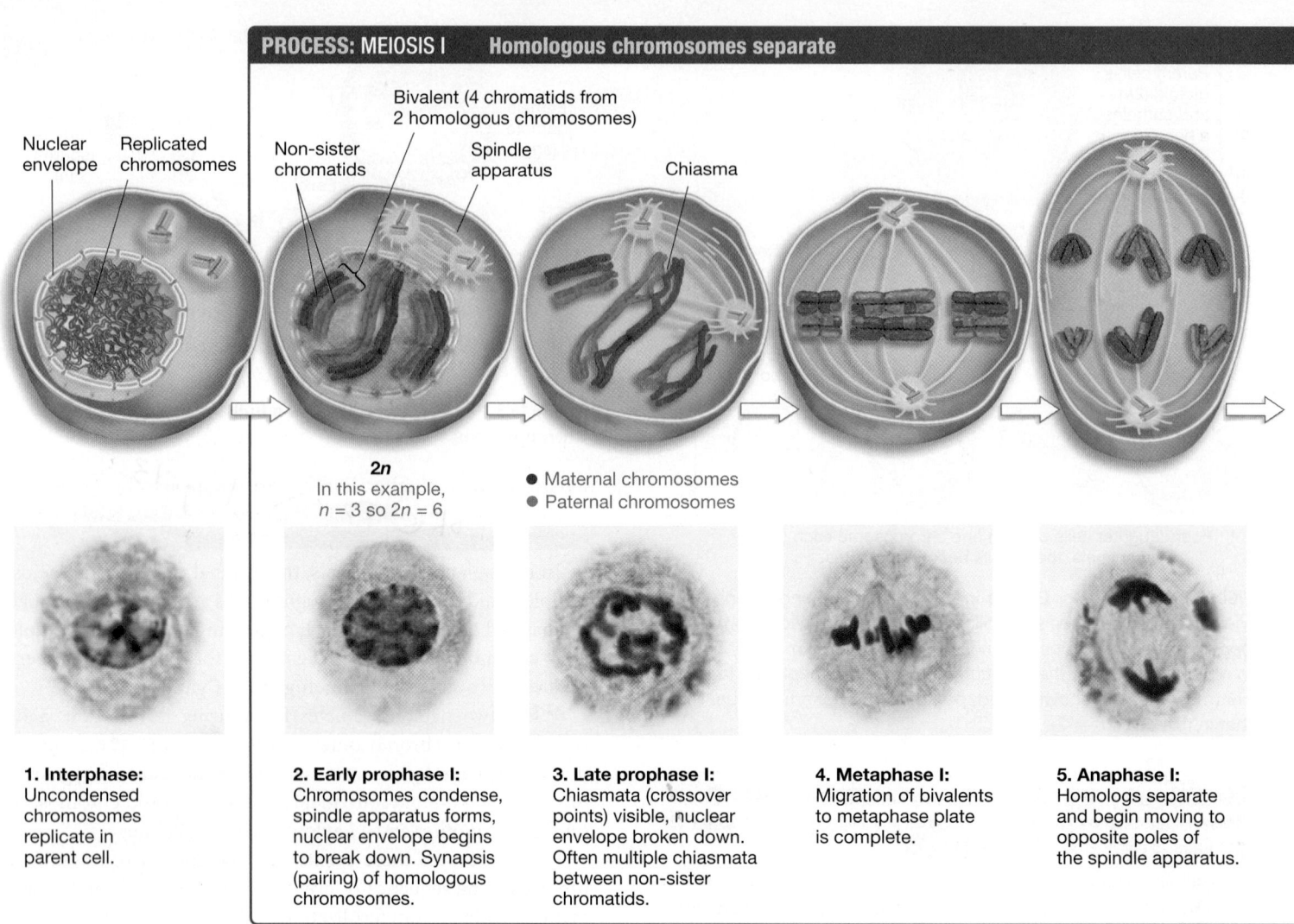

FIGURE 13.7 **The Phases of Meiosis.** The micrographs of each phase are from a species of salamander.

meiosis and fertilization. In the case of the dog illustrated here, meiosis in a diploid adult results in the formation of haploid gametes, which combine to form a diploid zygote. The zygote marks the start of a new generation, and through mitotic divisions during development, the zygote goes on to form an adult dog.

✔ **If you understand the events of meiosis, you should be able to predict how many DNA molecules will be present in the gametes of the fruit fly *Drosophila*, a diploid organism that has eight replicated chromosomes in each cell that enters meiosis.**

Once Stevens and others had published their work on meiosis and the accompanying changes in ploidy, the mystery of fertilization was finally solved. To appreciate the consequences of meiosis fully, let's analyze the events in more detail.

The Phases of Meiosis I

Meiosis begins after chromosomes have been replicated during S phase (see Chapter 12). Before the start of meiosis, chromosomes are extremely long structures, just as they are during interphase of the normal cell cycle. The major steps that occur once meiosis begins are shown in **FIGURE 13.7.**

Early Prophase I   During early prophase I, the nuclear envelope begins to break down, chromosomes condense and the spindle apparatus begins to form. Then a crucial event occurs: Homologous chromosome pairs come together. The end result of this process is called **synapsis** and is illustrated in step 2 of Figure 13.7. Synapsis is possible because regions of homologous chromosomes that are similar at the molecular level come together. In most organisms, synapsis requires breaking and then connecting together DNA of the two homologs at one or more spots along their length.

The structure that results from synapsis is called a **bivalent** (*bi* means "two" in Latin) or **tetrad** (*tetra* means "four" in Greek). A bivalent consists of paired homologous chromosomes, with each homolog consisting of two sister chromatids. Chromatids from different homologs are referred to as **non-sister chromatids.** In the figure, the red-colored chromatids are non-sister chromatids with respect to the blue-colored chromatids.

Late Prophase I   During late prophase I, the nuclear envelope breaks down and microtubules of the spindle apparatus attach to

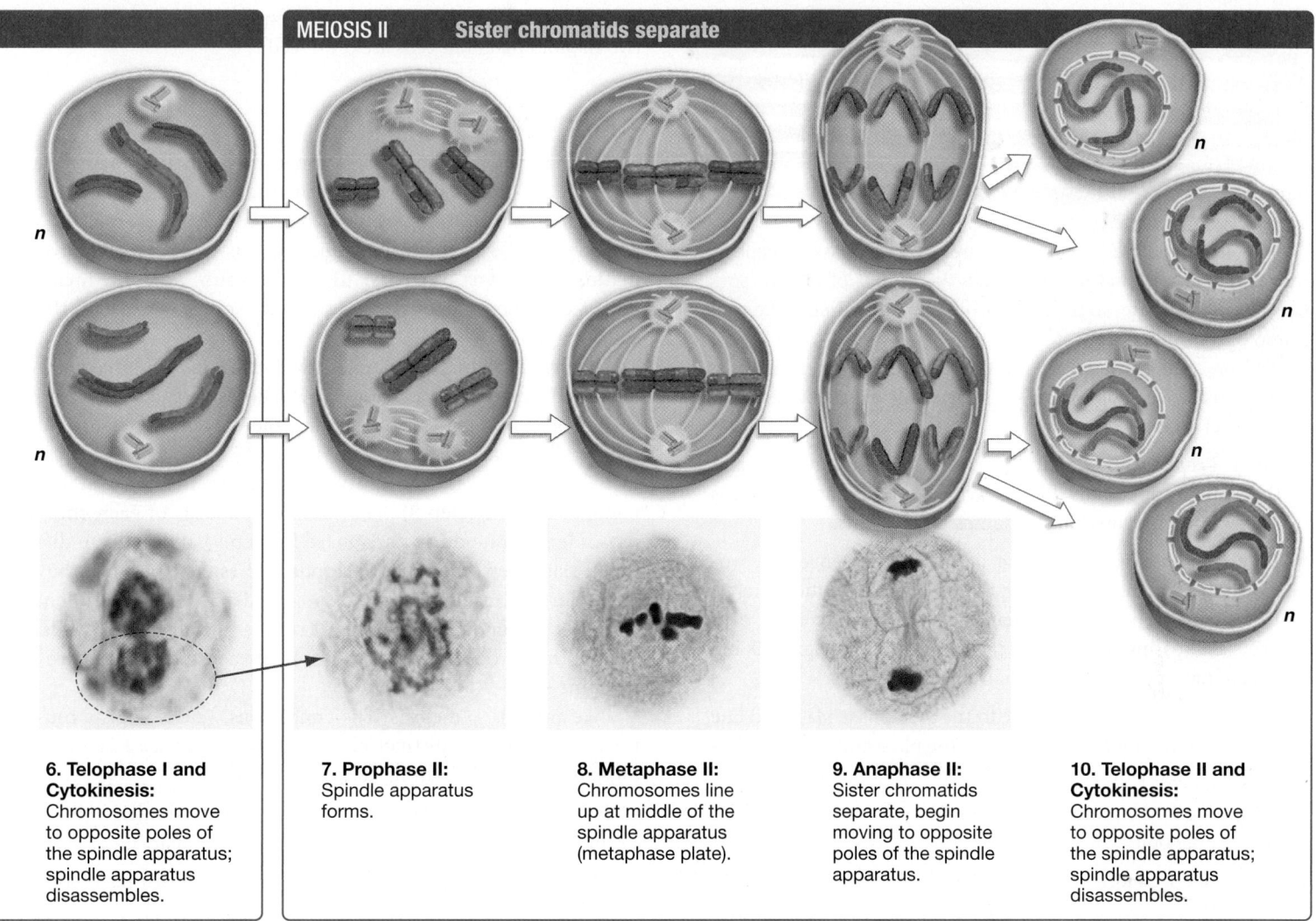

6. Telophase I and Cytokinesis: Chromosomes move to opposite poles of the spindle apparatus; spindle apparatus disassembles.

7. Prophase II: Spindle apparatus forms.

8. Metaphase II: Chromosomes line up at middle of the spindle apparatus (metaphase plate).

9. Anaphase II: Sister chromatids separate, begin moving to opposite poles of the spindle apparatus.

10. Telophase II and Cytokinesis: Chromosomes move to opposite poles of the spindle apparatus; spindle apparatus disassembles.

kinetochores. Non-sister chromatids begin to separate at many points along their length. They stay joined at certain locations, however, each of which forms an X-shaped structure called a **chiasma** (plural: **chiasmata**). (In the Greek alphabet, the letter X is called "chi.") Normally, at least one chiasma forms in every pair of homologous chromosomes; often there are several chiasmata. The chiasmata mark sites where DNA was broken and rejoined between homologs early in prophase I.

As step 3 of Figure 13.7 shows, the chromatids that meet to form a chiasma are non-sister chromatids. At each chiasma there is an exchange of parts of chromosomes between paternal and maternal homologs. These reciprocal exchanges between different homologs create non-sister chromatids that have both paternal and maternal segments. This process of chromosome exchange is called **crossing over.**

In step 4 of Figure 13.7, the result of crossing over is illustrated by chromosomes with a combination of red and blue segments. When crossing over occurs, the chromosomes that result have a mixture of maternal and paternal alleles. Crossing over is a major way that meiosis creates genetic diversity.

Metaphase I The next major stage in meiosis I is metaphase I. This is when kinetochore microtubules move the pairs of homologous chromosomes (bivalents) to a region called the **metaphase plate** in the middle of the spindle apparatus (step 4). The metaphase plate is not a physical structure but an imaginary plane dividing the spindle apparatus.

Here are two key points about chromosome movement: Each bivalent moves to the metaphase plate independently of the other bivalents, and the alignment on one side or the other of the metaphase plate is random for maternal and paternal homologs from each chromosome. This movement explains the most basic principles of genetics (see Chapter 14).

Anaphase and Telophase I Sister chromatids of each chromosome remain together. During anaphase I, the homologous chromosomes in each bivalent separate and begin moving to opposite poles of the spindle apparatus (step 5). Meiosis I concludes with telophase I, when the homologs finish moving to opposite sides of the spindle (step 6). When meiosis I is complete, **cytokinesis** (division of cytoplasm) occurs and two haploid daughter cells form.

Meiosis I: A Recap The end result of meiosis I is that one chromosome of each homologous pair is distributed to a different daughter cell. A reduction division has occurred: The daughter cells of meiosis I are haploid, having only one copy of each type of chromosome. The sister chromatids remain attached in each chromosome, however, meaning that the haploid daughter cells produced by meiosis I still contain replicated chromosomes.

The chromosomes in each cell are a random assortment of maternal and paternal chromosomes as a result of (1) crossing over and (2) the random distribution of maternal and paternal homologs during metaphase.

The preceding discussion shows that although meiosis I is a continuous process, biologists summarize the events by identifying distinct phases:

- *Early Prophase I* Replicated chromosomes condense and the spindle apparatus forms. Synapsis of homologs forms pairs of homologous chromosomes, or bivalents.

- *Late Prophase I* Breakdown of the nuclear envelope. Microtubules of the spindle apparatus attach to kinetochores. Chiasmata become visible, marking sites where crossing over occurs. Crossing over results in an exchange of segments between maternal and paternal chromosomes.

- *Metaphase I* Bivalents move to the metaphase plate and line up. One homolog is on one side of the plate and the other homolog is on the other.

- *Anaphase I* Homologs separate and begin moving to opposite spindle poles.

- *Telophase I* Homologs finish moving to opposite spindle poles. Spindle apparatus disassembles. In some species, a nuclear envelope re-forms around each set of chromosomes.

Throughout, chromosome movement takes place as microtubules that are attached to the kinetochore dynamically assemble and disassemble. When meiosis I is complete, the cell divides and two haploid daughter cells are produced.

The Phases of Meiosis II

Recall that chromosome replication occurred before meiosis I. Throughout meiosis I, sister chromatids remained attached. Because no chromosome replication occurs between meiosis I and meiosis II, each chromosome consists of two sister chromatids at the start of meiosis II. And because only one member of each homologous pair of chromosomes is present, the cell is haploid.

During prophase II, a spindle apparatus forms in both daughter cells. Microtubules attach to kinetochores on each side of every chromosome and begin moving the chromosomes toward the middle of each cell (step 7 of Figure 13.7).

In metaphase II, the chromosomes are lined up at the metaphase plate (step 8). The sister chromatids of each chromosome separate during anaphase II (step 9) and move to different daughter cells during telophase II (step 10). Once they are separated, each chromatid is considered an independent daughter chromosome. Meiosis II results in four haploid cells, each with one daughter chromosome of each type in the chromosome set.

Like meiosis I, meiosis II is continuous, but biologists routinely divide it into distinct phases. These stages are essentially those of mitosis. To summarize,

- *Prophase II* The spindle apparatus forms. If a nuclear envelope formed at the end of meiosis I, it breaks apart.

- *Metaphase II* Replicated chromosomes, consisting of two sister chromatids, are lined up at the metaphase plate.

- *Anaphase II* Sister chromatids separate. The daughter chromosomes that result begin moving to opposite poles of the spindle apparatus.

SUMMARY TABLE 13.2 **Key Differences between Mitosis and Meiosis**

Feature	Mitosis	Meiosis
Number of cell divisions	One	Two
Number of chromosomes in daughter cells compared with parent cell	Same	Half
Synapsis of homologs	No	Yes
Number of crossing-over events	None	One or more per pair of homologous chromosomes
Makeup of chromosomes in daughter cells	Identical	Different—various combinations of maternal and paternal chromosomes, paternal and maternal segments mixed within chromosomes
Role in organism life cycle	Asexual reproduction in some eukaryotes; cell division for growth	Halving of chromosome number in cells that will produce gametes

- *Telophase II*   Chromosomes finish moving to opposite poles of the spindle apparatus. A nuclear envelope forms around each haploid set of chromosomes.

When meiosis II is complete, each cell divides to form two daughter cells. Because meiosis II occurs in both daughter cells of meiosis I, the process results in a total of four daughter cells from each original, parent cell. To describe meiosis in a nutshell, one diploid cell with replicated chromosomes gives rise to four haploid cells with unreplicated chromosomes.

TABLE 13.2 and **FIGURE 13.8** provide a comparison of mitosis and meiosis. A key difference between the two processes is that homologous chromosomes pair early in meiosis but do not pair at all during mitosis. Because homologs pair through synapsis in prophase of meiosis I, they can migrate to the metaphase plate together and then separate during anaphase of meiosis I, resulting in a reduction division. ✔ If you understand this key distinction between meiosis and mitosis, you should be able to describe the consequences for meiosis if homologs do not pair.

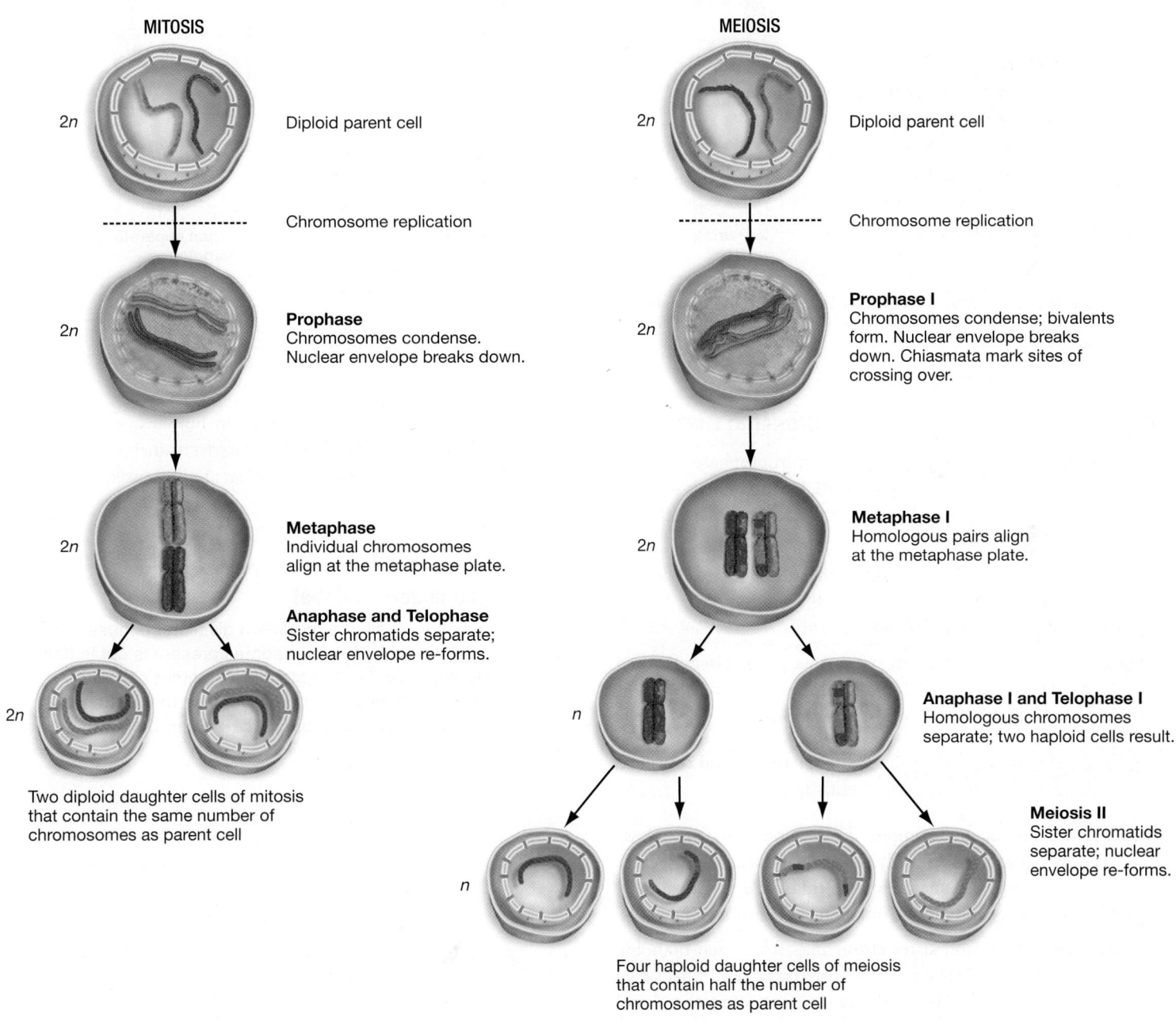

FIGURE 13.8 A Comparison of Mitosis and Meiosis. Mitosis produces two daughter cells with chromosomal complements identical to the parent cell. Meiosis produces four haploid cells with chromosomal complements unlike one another and unlike the diploid parent cell.

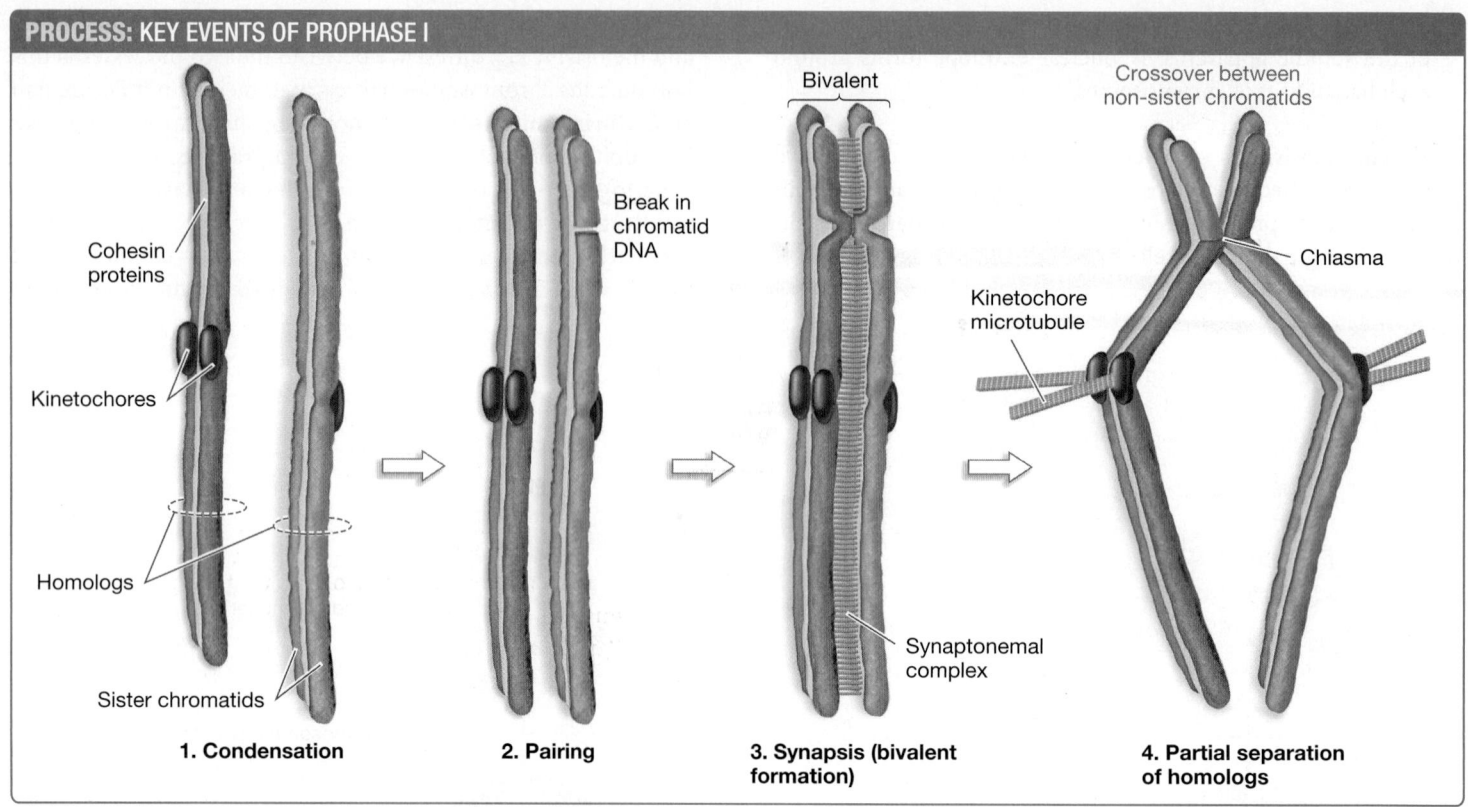

1. Condensation   **2. Pairing**   **3. Synapsis (bivalent formation)**   **4. Partial separation of homologs**

FIGURE 13.9 A Closer Look at Key Events in Prophase of Meiosis I.

A Closer Look at Synapsis and Crossing Over

The pairing of homologs and crossing over in prophase I are important events unique to meiosis. **FIGURE 13.9** takes a closer look at how chromosomes come together and exchange parts during meiosis I.

Step 1 Sister chromatids are held together along their full length by proteins known as cohesins. At the entry to prophase I, chromosomes begin to condense.

Step 2 Homologs pair. In many organisms, pairing begins when a break is made in the DNA of one chromatid. This break initiates a crossover between non-sister chromatids.

Step 3 A network of proteins forms the **synaptonemal complex,** which holds the two homologs tightly together.

Step 4 The synaptonemal complex disassembles in late prophase I. The two homologs partially separate and are held together only at chiasmata. Attachments at chiasmata are eventually broken to restore individual, unconnected chromosomes.

At a chiasma the non-sister chromatids from each homolog have been physically broken at the same point and *attached to each other.* As a result, corresponding segments of maternal and paternal chromosomes are exchanged.

Crossing over can occur at many locations along the length of paired homologs, and it routinely occurs at least once between each pair of non-sister chromatids. In humans, each chromosome undergoes an average of 1½ crossovers during meiosis.

Why does meiosis exist at all? What are its consequences?

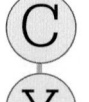

check your understanding

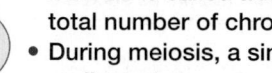

If you understand that . . .
- Meiosis is called a reduction division because the total number of chromosomes present is cut in half.
- During meiosis, a single diploid parent cell with replicated chromosomes gives rise to four haploid daughter cells, with unreplicated chromosomes.

✓ **You should be able to . . .**

1. Demonstrate the phases of meiosis I illustrated in Figure 13.7 by using pipe cleaners or pieces of cooked spaghetti.
2. Identify the event that makes meiosis a reduction division, unlike mitosis, and explain why it is responsible for reduction division.
3. Explain how meiosis generates cells with one of every kind of chromosome rather than random mixtures of different chromosomes.

Answers are available in Appendix A.

13.2 Meiosis Promotes Genetic Variation

The cell biologists who worked out the details of meiosis in the late 1800s and early 1900s realized that the process solved the riddle of fertilization. Weismann's hypothesis—that a reduction division precedes gamete formation in animals—was confirmed.

By now, having come to appreciate that meiosis is an intricate, tightly regulated process, you shouldn't be surprised to learn that it involves dozens, if not hundreds, of different proteins. Given this complexity, it is logical to hypothesize that meiosis accomplishes something extremely important: Thanks to the independent shuffling of maternal and paternal chromosomes and crossing over during meiosis I, the chromosomes in one gamete are different from the chromosomes in another gamete and different from the chromosomes in parental cells. Subsequently, fertilization brings haploid sets of chromosomes from a mother and father together to form a diploid offspring. The chromosome complement of this offspring is unlike that of either parent. It is a random combination of genetic material from each parent.

This change in chromosomal complement is crucial. The critical factor here is that changes in chromosome sets occur only during sexual reproduction—*not* during asexual reproduction.

- **Asexual reproduction** is any mechanism of producing offspring that does not involve the production and fusion of gametes. Asexual reproduction in eukaryotes is based on mitosis. The chromosomes in cells produced by mitosis are identical to the chromosomes in the parental cell.

- **Sexual reproduction** is the production of offspring through the production and fusion of gametes. Sexual reproduction results in offspring that have chromosome complements unlike those of their siblings or their parents.

Why is this difference important?

Chromosomes and Heredity

The changes in chromosomes produced by meiosis and fertilization are significant because chromosomes contain the cell's hereditary material. Stated another way, chromosomes contain the instructions for specifying particular traits. These inherited traits range from eye color and height in humans to the number or shape of the bristles on a fruit fly's leg to the color or shape of the seeds found in pea plants.

In the early 1900s, biologists began using the term gene to refer to the inherited instructions for a particular trait. Recall that the term allele refers to a particular version of a gene and that homologous chromosomes may carry different alleles.

Chromosomes are the repositories of genes, and identical copies of chromosomes are distributed to daughter cells during mitosis. Thus, cells that are produced by mitosis are genetically identical to the parent cell, and offspring produced during asexual reproduction are genetically identical to one another as well as to their parent. Offspring produced by asexual reproduction

are **clones**—or exact copies—of their parent. A familiar example of asexual reproduction is growing a new plant from a cutting.

In contrast, the offspring produced by sexual reproduction are genetically different from one another and unlike either their mother or their father.

Let's begin by analyzing two aspects of meiosis that create variation among chromosomes: (1) separation and distribution of homologous chromosomes and (2) crossing over. We'll then look at how these processes interact with fertilization to produce genetically variable offspring.

The Role of Independent Assortment

Each somatic cell in your body contains 23 homologous pairs of chromosomes and 46 chromosomes in total. Half of these chromosomes came from your mother, and half came from your father. Each chromosome contains genes, and genes influence particular traits. For example, one gene that affects your eye color might be located on one chromosome, while one of the genes that affects your hair color might be located on a different chromosome (**FIGURE 13.10a**).

(a) Example: Individual with different alleles of two genes

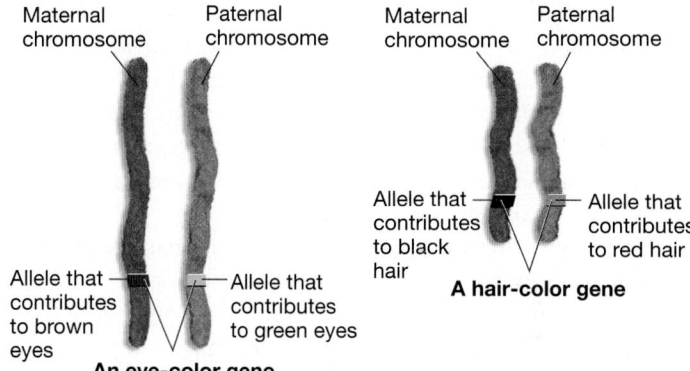

Maternal chromosome Paternal chromosome

Allele that contributes to brown eyes Allele that contributes to green eyes

An eye-color gene

Maternal chromosome Paternal chromosome

Allele that contributes to black hair Allele that contributes to red hair

A hair-color gene

(b) During meiosis I, bivalents can line up two different ways before the homologs separate.

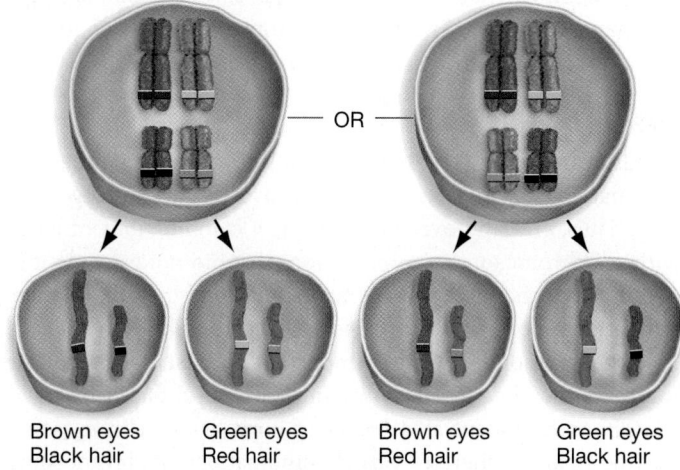

— OR —

Brown eyes Black hair   Green eyes Red hair   Brown eyes Red hair   Green eyes Black hair

FIGURE 13.10 Independent Assortment of Homologous Chromosomes Results in Varied Combinations of Genes.

Suppose that the chromosomes you inherited from your mother contain alleles associated with brown eyes and black hair, but the chromosomes you inherited from your father include alleles associated with green eyes and red hair. (This is a simplification for the purpose of explanation. In reality, several genes with various alleles interact in complex ways to produce human eye color and hair color.)

Will any particular gamete you produce contain the genetic instructions inherited from your mother or the instructions inherited from your father? To answer this question, study the diagram of meiosis in **FIGURE 13.10b**. It shows that when pairs of homologous chromosomes line up during meiosis I and the homologs separate, a variety of combinations of maternal and paternal chromosomes can result. Each daughter cell gets a random assortment of maternal and paternal chromosomes.

This phenomenon is known as the principle of independent assortment. In the example given here, meiosis results in gametes with alleles for brown eyes and black hair, the traits from your mother, and green eyes and red hair, the traits from your father. But two additional combinations also occur: brown eyes and red hair, or green eyes and black hair. The appearance of new combinations of alleles is called **genetic recombination.** Four different combinations of paternal and maternal chromosomes are possible when two chromosomes are distributed to daughter cells during meiosis I.

✔ If you understand how independent assortment produces genetic variation in the daughter cells of meiosis, you should be able to explain how genetic variation would be affected if maternal chromosomes always lined up together on one side of the metaphase plate during meiosis I and paternal chromosomes always lined up on the other side.

How many different combinations of maternal and paternal homologs are possible when more chromosomes are involved? With each additional pair of chromosomes, the number of combinations doubles. In general, a diploid organism can produce 2^n combinations of maternal and paternal chromosomes, where n is the haploid chromosome number. This means that you ($n = 23$) can produce 2^{23}, or about 8.4 million, gametes that differ in their combination of maternal and paternal chromosome sets. The random assortment of whole chromosomes generates an impressive amount of genetic variation among gametes.

The Role of Crossing Over

Recall from Section 13.1 that segments of paternal and maternal chromatids exchange when crossing over occurs during meiosis I. Thus, crossing over produces new combinations of alleles within a chromosome—combinations that did not exist in either parent. This phenomenon is known as recombination. Crossing over is an important source of genetic recombination.

Genetic recombination is important because it dramatically increases the genetic variability of gametes produced by meiosis. The independent assortment of homologous chromosomes during meiosis generates varied combinations of chromosomes in gametes; genetic recombination due to crossing over varies the combinations of alleles along each chromosome that is involved

in a crossover. With crossing over, the number of genetically different gametes that you can produce is much more than the 8.4 million—it is virtually limitless.

How Does Fertilization Affect Genetic Variation?

Crossing over and the independent assortment of maternal and paternal chromosomes ensure that each gamete is genetically unique. Even if two gametes produced by the same individual fuse to form a diploid offspring—in which case **self-fertilization,** or "selfing," is taking place—the offspring are very likely to be genetically different from the parent (**FIGURE 13.11**). Selfing is common in many plant species, and it also occurs in animal species in which single individuals—called hermaphrodites—contain both male and female sex organs.

Self-fertilization, however, is rare or nonexistent in many sexually reproducing species. Instead, gametes from different individuals combine to form offspring. This process is called **outcrossing.** Outcrossing increases the genetic diversity of offspring even further because it combines chromosomes from different individuals. These chromosomes are likely to contain different alleles.

How many genetically distinct offspring can be produced when outcrossing occurs? Let's answer this question using humans as an example. Recall that a single human can produce about 8.4 million different gametes by independent assortment alone. When a sperm and egg come together at fertilization, the number of possible genetic combinations that can result is equal to the product of the numbers of different gametes produced by each parent. (To understand this logic, see **BioSkills 5** in Appendix B.) In humans this means that two parents can potentially produce 8.4 million $\times$ 8.4 million $= 70.6 \times 10^{12}$ genetically distinct offspring, even without crossing over. This number is far greater than the total number of people who have ever lived.

check your understanding

If you understand that . . .

• The daughter cells produced by meiosis are genetically different from the parent cell because (1) maternal and paternal homologs align randomly at metaphase of meiosis I and (2) crossing over leads to recombination within chromosomes.

✔ You should be able to . . .

1. Draw a diploid parent cell with $2n = 6$ (three types of chromosomes) and then sketch the genetically distinct types of daughter cells that may result by independent assortment of these chromosomes at meiosis.

2. Discuss how crossing over would influence the genetic diversity of these gametes.

3. Compare and contrast the degree of genetic variation that results from asexual reproduction, selfing, and outcrossing.

Answers are available in Appendix A.

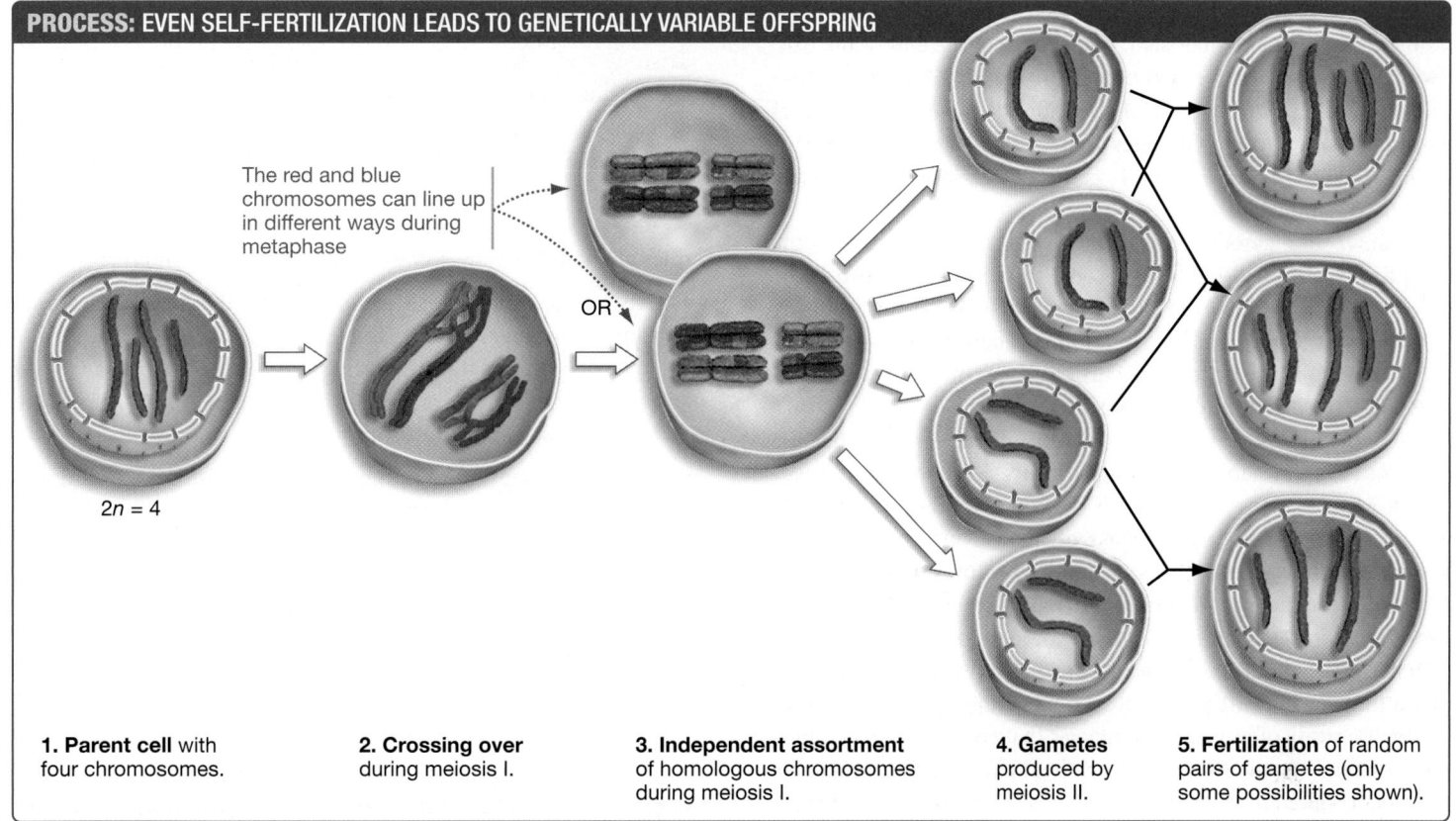

The red and blue chromosomes can line up in different ways during metaphase

OR

$2n = 4$

1. Parent cell with four chromosomes.

2. Crossing over during meiosis I.

3. Independent assortment of homologous chromosomes during meiosis I.

4. Gametes produced by meiosis II.

5. Fertilization of random pairs of gametes (only some possibilities shown).

FIGURE 13.11 Crossing Over, Independent Assortment, and the Random Pairing of Gametes during Fertilization Increase Genetic Variation, Even in Offspring Produced by Self-Fertilization.

✔**EXERCISE** In step 5, only a few of the many types of offspring that could be produced are shown. Sketch two additional types that are different from those shown.

In any complicated process such as meiosis, things can and do go wrong. What happens if there is a mistake, and the chromosomes are not properly distributed?

13.3 What Happens When Things Go Wrong in Meiosis?

Errors in meiosis are surprisingly common. If this were like a spelling mistake, it might be only an annoyance. But in humans, a conservative estimate is that 25 percent of conceptions are spontaneously terminated because of problems in meiosis. What are the consequences for offspring if gametes contain an abnormal set of chromosomes?

In 1866 Langdon Down described a distinctive set of co-occurring conditions observed in some people. The syndrome was characterized by mental retardation, a high risk for heart problems and leukemia, and a degenerative brain disorder similar to Alzheimer's disease. **Down syndrome,** as the disorder came to be called, is observed in about 0.15 percent of live births (1 infant in every 666).

For over 80 years the cause of the syndrome was unknown. Then, in the late 1950s, a study of the chromosome sets of nine Down syndrome children suggested that the condition is associated with the presence of an extra copy of chromosome 21. This situation is called **trisomy** ("three-bodies")—in this case, trisomy-21—because each cell has three copies of the chromosome. The explanation proposed for the trisomy was that a mistake had occurred during meiosis in either the mother or the father.

How Do Mistakes Occur?

For a gamete to get one complete set of chromosomes, two steps in meiosis must be perfectly executed.

1. The chromosomes in each homologous pair must separate from each other during the first meiotic division, so that only one homolog ends up in each daughter cell.

2. Sister chromatids must separate from each other and move to opposite poles of the dividing cell during meiosis II.

If both homologs in meiosis I or both sister chromatids in meiosis II move to the same pole of the parent cell, the products of meiosis will be abnormal. This sort of meiotic error is referred to as **nondisjunction,** because the homologs or sister chromatids fail to separate, or disjoin.

FIGURE 13.12 shows what happens when homologs do not separate correctly during meiosis I. Notice that at the end of meiosis, two daughter cells have two copies of the same chromosome—the smaller one in Figure 13.12—while the other two lack that chromosome entirely. Gametes that contain an extra chromosome are symbolized as $n + 1$; gametes that lack one chromosome are symbolized as $n - 1$.

If an $n + 1$ gamete is fertilized by a normal n gamete, the resulting zygote will be $2n + 1$. This situation is trisomy. If the $n - 1$ gamete is fertilized by a normal n gamete, the resulting zygote will be $2n - 1$. This situation is called **monosomy.** Cells that have too many or too few chromosomes of a particular type are said to be **aneuploid** ("without-form").

Meiotic mistakes occur often. Researchers estimate that 25 percent of all human conceptions produce a zygote that is aneuploid. Most of these errors result from the failure of a homologous pair to separate in anaphase of meiosis I; less often, sister chromatids stay together during anaphase of meiosis II.

The consequences of meiotic mistakes are almost always severe. Trisomy-21 is unusual in allowing development to proceed when there are three copies of chromosome 21. Even for this chromosome, live births are not seen when there is only one copy.

In one study of human pregnancies that ended in early embryonic or fetal death, 38 percent involved abnormal chromosome sets that resulted from mistakes in meiosis. Trisomy accounted for about one-third of the abnormal karyotypes. Three copies of every chromosome ($3n$), a condition called triploidy, was also common. Monosomy and abnormally sized or shaped chromosomes were also seen. Mistakes in meiosis are common and are the leading cause of spontaneous abortion (miscarriage) in humans.

Why Do Mistakes Occur?

Trisomy and other meiotic mistakes are random errors that occur during meiosis. Recent research indicates that problems are especially common in attaching microtubules to kinetochores early in meiosis I and in separating chromosomes that have a single chiasma near their ends or near centromeres.

Researchers see strong patterns in occurrence of human trisomies at birth. Here are some of the patterns that emerge:

- Trisomy is much more common with the smaller chromosomes (numbers 13–22) than it is with the larger chromosomes (numbers 1–12), and trisomy-21 is far and away the most common type of trisomy observed. Chromosome 21 is the smallest human autosome.

- With the exception of trisomy-21, most trisomies and monosomies observed in humans involve the sex chromosomes.

- Errors in meiosis leading to eggs are more common than errors in meiosis leading to sperm.

- Maternal age is an important factor in the occurrence of trisomy. For example, in the case of Down syndrome, as **FIGURE 13.13** shows, the incidence increases dramatically in babies born to mothers over 35 years old.

Why do these patterns occur? One important point to remember is that these observations are made at birth, not fertilization. The frequency of nondisjunction is about equal among chromosomes, but aneuploidy tends to be lethal to embryos if it involves chromosomes that contain a large number of genes. Trisomy-21 is common most likely because it involves a small chromosome with a correspondingly small number of genes.

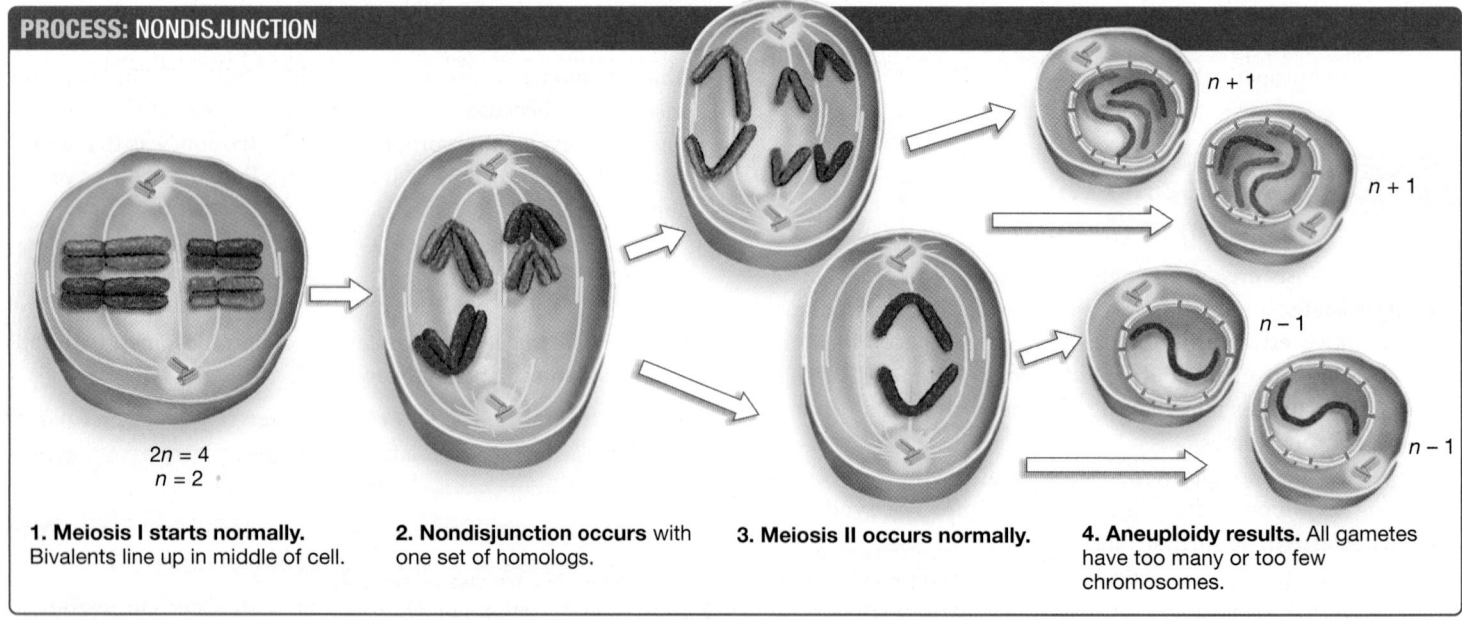

PROCESS: NONDISJUNCTION

$2n = 4$
$n = 2$

$n + 1$

$n + 1$

$n - 1$

$n - 1$

1. Meiosis I starts normally. Bivalents line up in middle of cell.

2. Nondisjunction occurs with one set of homologs.

3. Meiosis II occurs normally.

4. Aneuploidy results. All gametes have too many or too few chromosomes.

FIGURE 13.12 Nondisjunction Leads to Gametes with Abnormal Chromosome Numbers. If homologous chromosomes fail to separate during meiosis I, the gametes that result will have an extra chromosome or will lack a chromosome. Nondisjunction can also occur during meiosis II if sister chromatids fail to separate.

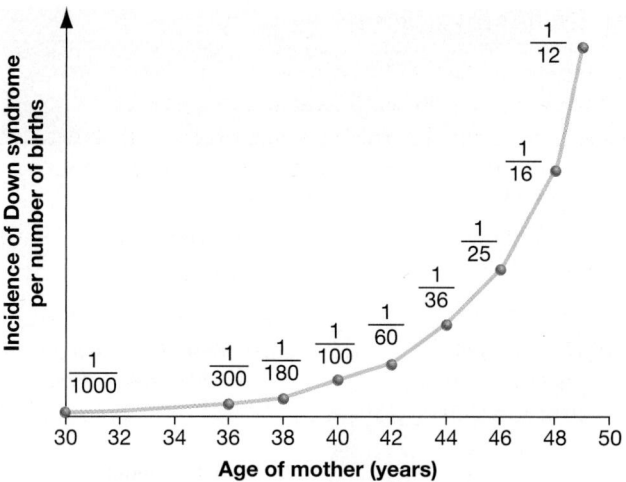

FIGURE 13.13 The Frequency of Down Syndrome Increases as a Function of a Mother's Age.

DATA: National Down Syndrome Society. 2012. www.ndss.org/index.php?option=com_content&view=article&id=61&Itemid=78

There are more questions than answers. Why are there more errors in women than men? Why is there such a strong correlation between maternal age and frequency of trisomy-21?

In part, the answers lie with an unusual feature of human egg development (oogenesis). Diploid precursors to eggs enter meiosis I before birth and arrest in prophase I until ovulation. For some eggs, this is 40 years or more later. Spindle apparatus function and the ability to separate chromosomes properly appear to decline after this long wait. Much remains to be discovered, but one thing is clear: Successful meiosis is critical to the health of offspring.

Why sex and meiosis? Although it seems obvious that sex and meiosis are needed universally for reproduction, that's not the case. Meiosis and sexual reproduction occur in only a small fraction of the lineages on the tree of life. Bacteria and archaea normally undergo only asexual reproduction; most algae, all fungi, and some animals and land plants reproduce asexually as well as sexually. Asexual reproduction is even observed in some vertebrates. Several species of guppy in the genus *Poeciliopsis*, for example, reproduce exclusively by mitosis.

Although sexual reproduction plays a central role in the life of most familiar organisms—including us—until recently, scientists had no clear idea of why it occurs. In fact, on the basis of theory, biologists had good reason to think that sexual reproduction should *not* exist.

The Paradox of Sex

In 1978 John Maynard Smith pointed out that the existence of sexual reproduction presents a paradox. Maynard Smith developed a mathematical model showing that because asexually reproducing individuals do not have to produce male offspring, their progeny on average can produce twice as many offspring as individuals that reproduce sexually. **FIGURE 13.14** diagrams this result by showing the number of females (♀), males (♂), and asexually reproducing organisms (**O**) produced over several generations by asexual versus sexual reproduction.

In this example, each asexually reproducing individual and each sexually reproducing couple produces four offspring over the course of their lifetimes. Note that in the sexual population, it takes two individuals—one male and one female—to produce

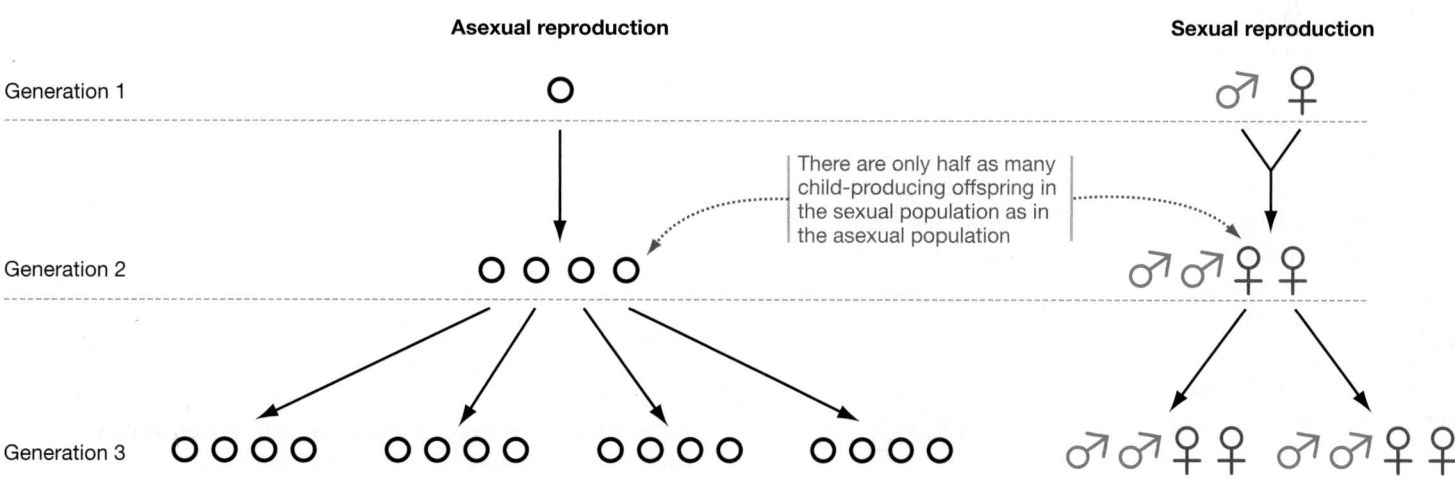

FIGURE 13.14 Asexual Reproduction Produces More Offspring. Each female (♀), male (♂), and asexual (**O**) symbol represents an individual. This hypothetical example assumes that (1) every asexual individual or sexually reproducing couple produces four offspring over the course of a lifetime, (2) sexually reproducing individuals produce half males and half females, and (3) all offspring survive to breed.

✓**QUANTITATIVE** How many asexually produced offspring would be present in generation 4? How many sexually produced offspring?

four offspring. Two out of every four children that each female produces sexually are males who cannot have children of their own. As a result, after one generation (generation 2 in Figure 13.14) the sexual population has just half as many child-producing individuals as the asexual population. Maynard Smith referred to this as the "two-fold cost of males." Asexual reproduction is much more efficient than sexual reproduction because no males are produced.

Based on this analysis, what will happen when asexual and sexual individuals exist in the same population and compete with one another? If all other things are equal, individuals that reproduce asexually should increase in frequency in the population while individuals that reproduce sexually should decline in frequency. In fact, Maynard Smith's model predicts that sexual reproduction is so inefficient that it should be completely eliminated.

To resolve this paradox, biologists began examining the assumption "If all other things are equal." Stated another way, biologists began looking for ways that meiosis and outcrossing could lead to the production of offspring that reproduce more than asexually produced individuals do. After decades of debate and analysis, two solid hypotheses to explain the paradox of sex are beginning to emerge.

The Purifying Selection Hypothesis

The first clue to unraveling the paradox of sex is a simple observation: If a gene is damaged or altered in a way that causes it to function poorly, it will be inherited by *all* of that individual's offspring when asexual reproduction occurs. Suppose the altered gene arose in generation 1 of Figure 13.14. If this gene is important, its alteration might cause the four asexual females present in generation 2 to produce fewer than four offspring apiece—perhaps because the members of generation 2 die young. If so, then generation 3 will not have twice as many individuals in the asexual lineage compared with the sexual lineage.

An allele that functions poorly and lowers the fitness of an individual is said to be deleterious. Asexual individuals are doomed to transmitting all their deleterious alleles to all of their offspring.

Suppose, however, that the same deleterious allele arose in the sexually reproducing female in generation 1 of Figure 13.14. If the female also has a normal copy of the gene, and if she mates with a male that has normal copies of the gene, then on average half her offspring will lack the deleterious allele. Sexual individuals are likely to have some offspring that lack the deleterious alleles that are present in a parent.

Natural selection against deleterious alleles is called purifying selection. Purifying selection should reduce the numerical advantage of asexual reproduction.

To test this hypothesis, researchers recently compared the same genes in two closely related species of *Daphnia,* a tiny crustacean that is a common inhabitant of ponds and lakes (see Chapter 50). One of these species reproduces asexually and the other reproduces sexually. As predicted, the researchers found that individuals in the asexual species contained many more deleterious alleles than individuals in the sexual species. Results like

RESEARCH

QUESTION: Does exposure to evolving pathogens favor outcrossing?

HYPOTHESIS: In environments where evolving pathogens are present, sexual reproduction by outcrossing will be favored.

NULL HYPOTHESIS: The presence of evolving pathogens will not favor outcrossing.

EXPERIMENTAL SETUP:

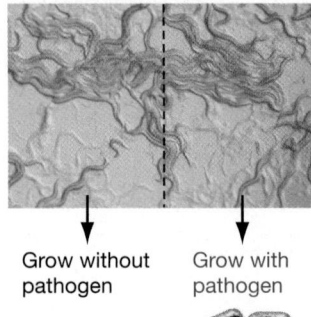

Grow without pathogen   Grow with pathogen

1. **Start with a pathogen free population of roundworms** with a 20% rate of outcrossing.

2. **Divide the population;** grow one subgroup in the absence of a pathogen and another subgroup in the presence of an evolving pathogen.

3. **Assess the rate of outcrossing** over many generations.

PREDICTION: The rate of outcrossing will increase in response to exposure by a pathogen.

PREDICTION OF NULL HYPOTHESIS: The rate of outcrossing will not be influenced by a pathogen.

RESULTS:

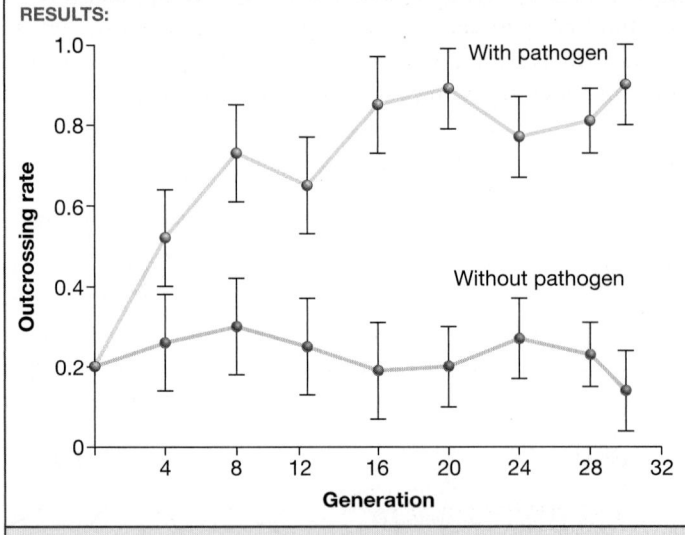

CONCLUSION: Exposure to evolving pathogens favors outcrossing.

FIGURE 13.15 Does Exposure to Pathogens Favor Sexual Reproduction through Outcrossing? Each point in the graph shows the average percentage of reproduction by outcrossing for five populations. The bars indicate the degree of variation in the data (see **BioSkills 4** in Appendix B for a description of error bars).

Morran, L. T., et al. 2011. Running with the red queen: Host-parasite coevolution selects for biparental sex. *Science* 333: 216–218.

✔ QUESTION What would you predict if a non-evolving pathogen were used?

these have convinced biologists that purifying selection is an important factor promoting the success of sexual reproduction.

The Changing-Environment Hypothesis

The second hypothesis to explain sexual reproduction also focuses on the benefits of producing genetically diverse offspring. Here's the key idea: Offspring that are genetic clones of their parents are unlikely to thrive if the environment changes.

What type of environmental change might favor genetically diverse offspring? The leading hypothesis points to pressure put on hosts by rapidly changing pathogens and parasites—bacteria, viruses, fungi, and other entities that cause disease. In your own lifetime, for example, several new disease-causing agents have emerged that afflict humans. These include the SARS virus, HIV, and new strains of the tuberculosis bacterium. Hundreds of genes help defend you against these types of invaders. Certain alleles help you fight off particular strains of disease-causing bacteria, eukaryotes, or viruses. In this evolutionary arms race, pathogens and parasites constantly evolve new ways to infect the most common types of host.

What happens if all the offspring produced by an individual are genetically identical? If a new strain of disease-causing agent evolves, then all the asexually produced offspring are likely to be susceptible to that new strain. But if the offspring are genetically varied, then it is likely that at least some offspring will have combinations of alleles that enable them to fight off the new strain of pathogen or parasite and produce offspring of their own.

Recall from Section 13.2 that over multiple generations, outcrossing—mating between two different individuals—increases the amount of genetic diversity relative to self-fertilization. A logical question to ask is: Does exposure to evolving pathogens favor outcrossing in sexually reproducing organisms? To address this question, Levi Morran, Curtis Lively, and colleagues carried out a pivotal study on a tiny (only about 1 mm long) roundworm named *Caenorhabditis elegans*.

C. elegans is an important model organism (see **BioSkills 13** in Appendix B) that was chosen for this study because it leads an unusual sex life. There are no females, only males and hermaphrodites. Because hermaphrodites have both male and female sex organs, this means that *C. elegans* can reproduce either by self-fertilization or by outcrossing with males. The proportion of roundworms that reproduce by self-fertilization versus outcrossing can vary in different strains or in different environments over time. So, the research team was able to test whether the rate of outcrossing increased in response to intense selection by a pathogen.

The setup of Morran, Lively, and colleagues' experiment is shown in **FIGURE 13.15**. The team began with a population of roundworms that had not been exposed to the pathogen and that reproduced predominantly by self-fertilization. The researchers then split the starting population into different groups. Half the groups were grown in the presence of a pathogen—a deadly bacterium—and the other half were grown without it. Once ingested by a roundworm, the bacterial pathogen could kill a susceptible individual within 24 hours.

At each generation, bacteria were collected from the carcasses of roundworms killed in the previous generation. Companion experiments showed that the pathogen evolved to become even more infectious over the course of the study.

The results are shown at the bottom of Figure 13.15. The rate of outcrossing stayed low over 32 generations in populations that did not encounter the pathogen. In contrast, populations that were exposed to the evolving pathogen showed a rapid increase in the rate of outcrossing.

At the end of the experiment, the roundworms in the pathogen-exposed population were significantly more resistant to the evolved pathogen than their ancestors. In other words, the roundworms in the predominantly outcrossing population had evolved along with the pathogen. In striking contrast, when a parallel experiment was done with a strain of roundworms that could reproduce only by self-fertilization, those populations were unable to evolve resistance to the pathogen. In fact, they became extinct.

These results and many others support the changing-environment hypothesis. Although the paradox of sex remains an active area of research, more biologists are becoming convinced that sexual reproduction is helpful for two reasons: **(1)** Offspring are not doomed to inherit harmful alleles, and **(2)** the production of genetically varied offspring means that at least some may be able to resist rapidly evolving pathogens and parasites.

CHAPTER 13　REVIEW

For media, go to MasteringBiology

If you understand . . .

13.1　How Does Meiosis Occur?

- Meiosis is a nuclear division resulting in cells that have only one of each type of chromosome and half as many chromosomes as the parent cell. In animals it leads to the formation of eggs and sperm.

- In diploid ($2n$) organisms, individuals have two versions of each type of chromosome. The two versions are called homologs. One homolog is inherited from the mother and one from the father. Haploid organisms (n) have just one of each type of chromosome.

- Each chromosome is replicated before meiosis begins. At the start of meiosis I, each chromosome consists of a pair of sister chromatids.

- Homologous pairs of chromosomes synapse early in meiosis I, forming a bivalent—two closely paired homologous chromosomes. Non-sister chromatids undergo crossing over.

- When crossing over is complete, the pair of homologous chromosomes is moved to the metaphase plate.

- At the end of meiosis I, the homologous chromosomes are separated and distributed to two daughter cells. The daughter cells are haploid, because each receives one of each type of chromosome.

- During meiosis II, sister chromatids separate and are distributed to two daughter cells.

- From one diploid cell with replicated chromosomes, meiosis produces four haploid daughter cells with unreplicated chromosomes.

✔ You should be able to diagram a diploid cell with a homologous pair of chromosomes and show when in meiosis this cell produces haploid daughter cells. Be sure to show all chromatids.

13.2 Meiosis Promotes Genetic Variation

- Each cell produced by meiosis receives a different combination of chromosomes. Because genes are located on chromosomes, each cell produced by meiosis receives a different complement of genes. The resulting offspring are genetically distinct from one another and from their parents.

- When meiosis and outcrossing occur, the chromosome complements of offspring differ from one another and from their parents for three reasons:

1. Gametes receive a random assortment of maternal and paternal chromosomes when homologs separate in meiosis I. This is independent assortment.
2. Because of crossing over, each chromosome contains a random assortment of paternal and maternal alleles.
3. Outcrossing results in a combination of chromosome sets from different individuals.

✔ You should be able to draw a diploid cell with four chromosomes entering meiosis and illustrate (a) how different combinations of chromosomes can result from independent assortment; and (b) focusing on a single homologous pair, show how many recombinant chromosomes (chromosomes with mixtures of maternal and paternal segments) are produced when crossing over occurs once along this homologous pair.

13.3 What Happens When Things Go Wrong in Meiosis?

- If mistakes occur during meiosis, the resulting egg and sperm cells may contain the wrong number of chromosomes. It is rare for offspring with an incorrect number of chromosomes to develop normally.

- Mistakes during meiosis lead to gametes and offspring with an abnormal number of chromosomes. Children with Down syndrome, for example, have an extra copy of chromosome 21.

- The leading hypothesis to explain meiotic mistakes is that they are accidental failures of homologous chromosomes or sister chromatids to separate properly during meiosis.

✔ Using pipe cleaners or spaghetti to model a homologous pair of chromosomes, you should be able to demonstrate (a) what happens if one pair of homologous chromosomes fails to separate at anaphase of meiosis I and (b) what happens if sister chromatids of one chromosome fail to separate at anaphase of meiosis II but later separate in a daughter cell.

13.4 Why Does Meiosis Exist?

- Asexual reproduction is much more efficient than sexual reproduction because all individuals produced asexually are capable of bearing progeny. With sexual reproduction, half the offspring (the males) are unable to bear progeny.

- The leading hypotheses to explain the existence of meiosis and sexual reproduction are that:

1. parents can produce offspring that lack harmful alleles; and
2. genetically diverse offspring are likely to include some that are better able to resist evolving pathogens and parasites.

✔ You should be able to predict whether, in species that alternate between asexual and sexual reproduction, sexual reproduction occurs during times when environmental conditions are stable or times when conditions change rapidly.

MB MasteringBiology

1. MasteringBiology Assignments

Tutorials and Activities Asexual and Sexual Life Cycles; Genetic Variation from Sexual Recombination; Meiosis; Meiosis (1 of 3): Genes, Chromosomes, and Sexual Reproduction; Meiosis (2 of 3): The Mechanism; Meiosis (3 of 3): Determinants of Heredity and Genetic Variation; Meiosis Animation; Mistakes in Meiosis; Origins of Genetic Variation

Questions Reading Quizzes, Blue-Thread Questions, Test Bank

2. eText Read your book online, search, take notes, highlight text, and more.

3. The Study Area Practice Test, Cumulative Test, BioFlix® 3-D Animations, Videos, Activities, Audio Glossary, Word Study Tools, Art

You should be able to . . .

1. In the roundworm *Ascaris*, eggs and sperm have two chromosomes, but all other cells have four. Observations such as this inspired which important hypothesis?
   a. Before gamete formation, a special type of cell division leads to a quartering of chromosome number.
   b. Before gamete formation, a special type of cell division leads to a halving of chromosome number.
   c. After gamete formation, half the chromosomes are destroyed.
   d. After gamete formation, either the maternal or the paternal set of chromosomes disintegrates.

2. What are homologous chromosomes?
   a. chromosomes that are similar in their size, shape, and gene content
   b. similar chromosomes that are found in different individuals of the same species
   c. the two "threads" in a replicated chromosome (they are identical copies)
   d. the products of crossing over, which contain a combination of segments from maternal chromosomes and segments from paternal chromosomes

3. What is a bivalent?
   a. the X that forms when chromatids from homologous chromosomes cross over
   b. a group of four chromatids produced when homologs synapse
   c. the four points where homologous chromosomes touch as they synapse
   d. the group of four genetically identical daughter cells produced by mitosis

4. What is an outcome of genetic recombination?
   a. the synapsing of homologs during prophase of meiosis I
   b. the new combination of maternal and paternal chromosome segments that results when homologs cross over
   c. the new combinations of chromosome segments that result when self-fertilization occurs
   d. the combination of a haploid phase *and* a diploid phase in a life cycle

5. What proportion of chromosomes in a human skin cell are paternal chromosomes?

6. Meiosis II is similar to _____.

7. Explain the relationship between homologous chromosomes and the relationship between sister chromatids.

8. Lay four pens and four pencils on a tabletop, and imagine that they represent replicated chromosomes in a diploid cell where $n = 2$. Explain the phases of meiosis II by moving the pens and pencils around. (If you don't have enough pens and pencils, use strips of paper or fabric.)

9. Meiosis is called a reduction division, but all the reduction occurs during meiosis I—no reduction occurs during meiosis II. Explain why meiosis I is a reduction division but meiosis II is not.

10. Dogs have 78 chromosomes in their diploid cells. If a diploid dog cell enters meiosis, how many chromosomes and chromatids will be present in each daughter cell at the end of meiosis I?

   a. 39 chromosomes and 39 chromatids
   b. 39 chromosomes and 78 chromatids
   c. 78 chromosomes and 78 chromatids
   d. 78 chromosomes and 156 chromatids

11. Triploid ($3n$) watermelons are produced by crossing a tetraploid ($4n$) strain with a diploid ($2n$) plant. Briefly explain why this mating produces a triploid individual. Why can mitosis proceed normally in triploid cells, but meiosis cannot?

12. Some plant breeders are concerned about the susceptibility of asexually cultivated plants, such as seedless bananas, to new strains of disease-causing bacteria, viruses, or fungi. Briefly explain their concern by discussing the differences in the genetic "outcomes" of asexual and sexual reproduction.

13. The gibbon has 44 chromosomes per diploid set, and the siamang has 50 chromosomes per diploid set. In the 1970s a chance mating between a male gibbon and a female siamang produced an offspring. Predict how many chromosomes were observed in the somatic cells of the offspring. Do you predict that this individual would be able to form viable gametes? Why or why not?

14. Meiosis results in a reassortment of maternal and paternal chromosomes. If $n = 3$ for a given organism, there are eight different combinations of paternal and maternal chromosomes. If no crossing over occurs, what is the probability that a gamete will receive *only* paternal chromosomes?
   a. 0; b. 1/16; c. 1/8; d. 1/3

15. Some researchers hypothesize that older women are less responsive to triggers of spontaneous abortion than younger women. How could the data shown in Figure 13.13, which graphs a mother's age versus the incidence of Down syndrome, be used to support this hypothesis?

16. A species of rotifer, a small freshwater invertebrate, abandoned sexual reproduction millions of years ago. A remarkable feature of the rotifer's life cycle is its ability to withstand extreme drying. When the rotifer's watery environment dries out, so does the rotifer, and it can be blown in the wind to a new environment. Once blown to water, the rotifer rehydrates and resumes an active life. A major pathogen of these rotifers is a species of fungus. Some scientists hypothesize that fungus-infected rotifers rid themselves of the pathogen when they dry.
   a. Design an experimental study to test this hypothesis.
   b. Provide an explanation for how these asexually reproducing rotifers are able to evade pathogens even though they are genetically identical.

14 Mendel and the Gene

In this chapter you will learn how

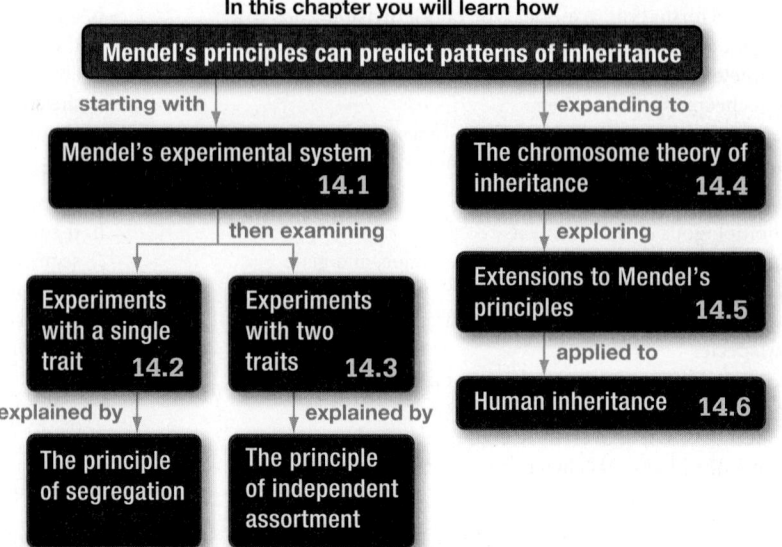

Mendel's principles can predict patterns of inheritance

starting with

Mendel's experimental system **14.1**

then examining

Experiments with a single trait **14.2**

Experiments with two traits **14.3**

explained by

The principle of segregation

explained by

The principle of independent assortment

expanding to

The chromosome theory of inheritance **14.4**

exploring

Extensions to Mendel's principles **14.5**

applied to

Human inheritance **14.6**

Experiments on garden peas and sweet peas (shown here) helped launch the science of genetics.

This chapter is part of the Big Picture. See how on pages 366–367.

The science of biology is built on a series of great ideas. Two of these—the cell theory and the theory of evolution—were introduced in Chapter 1. The cell theory describes the basic structure of organisms; the theory of evolution by natural selection clarifies why species change through time. Life is cellular; populations evolve. These are two of the five fundamental attributes of life.

This chapter introduces a third great idea in biology: the chromosome theory of inheritance. The chromosome theory explained how genetic information is transmitted from one generation to the next. It shed light on a third fundamental attribute of life: Organisms process information.

✔ When you see this checkmark, stop and test yourself. Answers are available in Appendix A.

An Austrian monk named Gregor Mendel laid the groundwork for the theory in 1865, when he announced that he had worked out the rules of inheritance through a series of experiments on garden peas. Another key insight emerged during the final decades of the nineteenth century, when biologists described the details of meiosis (see Chapter 13).

The chromosome theory of inheritance, formulated in 1902 by Walter Sutton and Theodor Boveri, linked these two insights. This theory contends that meiosis causes the patterns of inheritance that Mendel observed. It also asserts that the hereditary factors called genes are located on chromosomes. (Genes determine inherited traits and are center stage in the Big Picture on pages 366–367.)

The chromosome theory launched the study of **genetics,** the branch of biology that focuses on the inheritance of traits. Let's start at the beginning: What are the rules of inheritance that Mendel discovered?

14.1 Mendel's Experimental System

When biological science began to emerge as an important discipline, questions about **heredity**—meaning inheritance, or the transmission of traits from parents to offspring—were primarily the concern of animal breeders and horticulturists. A **trait** is any characteristic of an individual, ranging from height to the primary structure of a particular membrane protein.

In the city where Gregor Mendel lived, there was particular interest in how selective breeding could result in hardier and more productive varieties of sheep, fruit trees, and vines; and an agricultural society had been formed there to promote research into making selective breeding more efficient. Mendel was an active member of this society, and the monastery he belonged to was also devoted to scientific teaching and research.

What Questions Was Mendel Trying to Answer?

Mendel set out to address the most fundamental of all issues concerning heredity: What are the basic patterns in the transmission of traits from parents to offspring?

At the time, two hypotheses had been formulated to answer this question:

1. *Blending inheritance* claimed that the traits observed in a mother and father blend together to form the traits observed in their offspring. As a result, an offspring's traits are intermediate between the mother's and father's traits.

2. *Inheritance of acquired characters* claimed that traits present in parents are modified, through use, and passed on to their offspring in the modified form.

Each of these hypotheses made predictions. Blending inheritance contended that when black sheep and white sheep mate, their hereditary determinants will blend to form a new hereditary determinant for gray wool. Therefore, their offspring should be gray. Inheritance of acquired characters predicts that if giraffes extend their necks by straining to reach leaves high in the tops of trees, they subsequently produce longer-necked offspring.

These hypotheses were being promoted by the greatest scientists of Mendel's time. Are they correct?

The Garden Pea Served as the First Model Organism in Genetics

After investigating and discarding several candidate species to study, Mendel chose the garden pea, *Pisum sativum*. His reasons were practical: Peas are inexpensive and easy to grow from seed, have a relatively short generation time, and produce reasonably large numbers of seeds. These features made it possible for him to continue experiments over several generations and collect data from large numbers of individuals.

Peas served as a **model organism:** a species that is used for research because it is practical and because conclusions drawn from studying it turn out to apply to many other species as well. **BioSkills 13** in Appendix B introduces some of the important model organisms used in biological science today.

Two additional features of the pea made it possible for Mendel to design his experiments: Individuals were available that differed in easily recognizable traits, such as flower color or seed shape, and he could control which parents were involved in a mating.

How Did Mendel Control Matings? **FIGURE 14.1a** shows the male and female reproductive organs of a garden pea flower. Sperm cells are produced in pollen grains, which are small sacs that

(a) Self-pollination

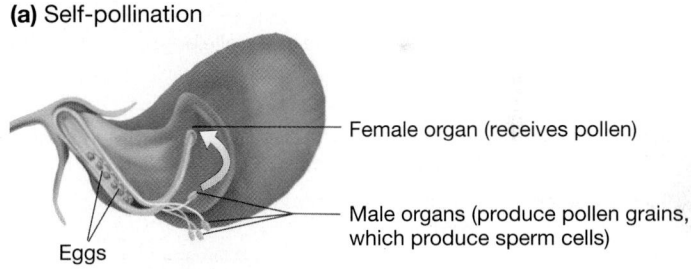

Female organ (receives pollen)

Male organs (produce pollen grains, which produce sperm cells)

Eggs

(b) Cross-pollination

Collect pollen from one individual and transfer it ...

... to the female organs of an individual whose male organs have been removed.

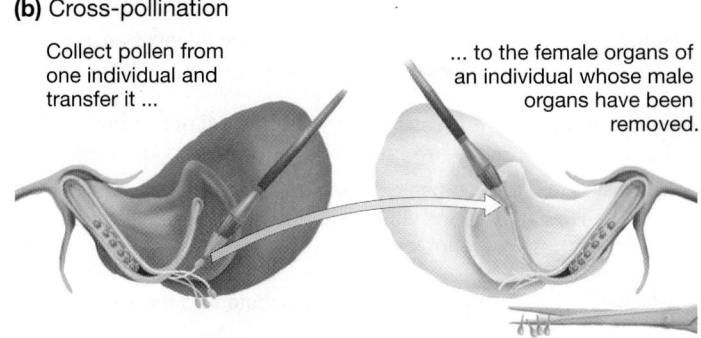

FIGURE 14.1 Peas Can Be Self-Pollinated or Cross-Pollinated.
(a) Under normal conditions, garden peas pollinate themselves.
(b) Mendel developed a method of controlling the matings of his model organism.

mature in the male reproductive structure of the plant. Eggs are produced in the female reproductive structure.

Under normal conditions, garden peas **self-fertilize**: a flower's pollen falls on the female reproductive organ of that same flower. As **FIGURE 14.1b** shows, however, Mendel could prevent self-fertilization by removing the male reproductive organs from a flower before any pollen formed. Later he could transfer pollen from another pea plant to the target flower's female reproductive organ with a brush. This type of mating is referred to as a cross-fertilization, or simply a **cross**. Using this technique, Mendel could control the matings of his model organism.

What Traits Did Mendel Study? Mendel conducted his experiments on varieties of peas that differed in seven traits: seed shape, seed color, pod shape, pod color, flower color, flower and pod position, and stem length. Biologists refer to the observable traits of an individual, such as the shape of a pea seed or the eye color of a person, as its **phenotype** (literally, "show-type"). Phenotype is just one term in the rich vocabulary of genetics. You can review many of these terms in **TABLE 14.1**. In the first pea populations that Mendel studied, two distinct phenotypes existed for each of the seven traits.

Mendel began his work by obtaining individuals from what breeders called pure lines or true-breeding lines. A **pure line** consists of individuals that produce offspring identical to themselves when they are self-pollinated or crossed to another member of the pure-line population. For example, earlier breeders had developed pure lines for wrinkled seeds and round seeds. During two years of trial experiments, Mendel confirmed that individuals that germinated from his wrinkled seeds produced only wrinkled-seeded offspring when they were mated to themselves or to another pure-line individual that germinated from a wrinkled seed, and he confirmed that the same was true for round seeds.

Why is this important? Remember that Mendel wanted to find out how traits are transmitted from parents to offspring. Once he had confirmed that he was working with pure lines, he could compare the results of crosses within a pure line with crosses between individuals from different pure lines.

SUMMARY TABLE 14.1 **Vocabulary Used in Mendelian Genetics**

Term	Definition	Example or Comment
Gene	A hereditary factor that influences a particular trait.	This definition will become more precise in later chapters.
Allele	A particular form of a gene.	The two alleles in a diploid may be the same or different.
Genotype	A listing of the alleles in an individual.	In diploids, the genotype lists two alleles of each gene; in haploids, the genotype lists one allele of each gene.
Phenotype	An individual's observable traits.	Can be observed at levels from molecules to the whole organism; influenced, not dictated, by the genotype.
Homozygous	Having two of the same allele.	Refers to a particular gene.
Heterozygous	Having two different alleles.	Refers to a particular gene.
Dominant allele	An allele that produces its phenotype in heterozygous and homozygous form.	Dominance does not imply high frequency or high fitness.
Recessive allele	An allele that produces its phenotype only in homozygous form.	Phenotype "recedes" or disappears in heterozygous individuals.
Pure line	Individuals of the same phenotype that, when crossed, always produce offspring with the same phenotype.	Pure-line individuals are homozygous for the gene in question.
Hybrid	Offspring from crosses between homozygous parents with different genotypes.	Hybrids are heterozygous.
Reciprocal cross	A cross in which the phenotypes of the male and female are reversed compared with a prior cross.	If reciprocal crosses give identical results, the sex of the parent does not influence transmission of the trait.
Testcross	A cross between a homozygous recessive individual and an individual with the dominant phenotype but an unknown genotype.	Usually used to determine whether a parent with a dominant phenotype is homozygous or heterozygous.
X-linked	Referring to a gene located on the X chromosome.	X-linked genes and traits show different patterns of inheritance in males and females.
Y-linked	Referring to a gene located on the Y chromosome.	In humans, Y-linked genes determine male-specific development.
Autosomal	Referring to a gene located on any non-sex chromosome (an autosome) or a trait determined by an autosomal gene.	Mendel studied only autosomal genes and traits.

Suppose that Mendel arranged matings between a pure-line individual with round seeds and a pure-line individual with wrinkled seeds. He knew that one parent carried a hereditary determinant for round seeds, while the other carried a hereditary determinant for wrinkled seeds. But each offspring from this mating would contain both types of hereditary determinants. They would be **hybrids**—offspring from matings between true-breeding parents that differ in one or more traits.

Would these hybrid offspring have wrinkled seeds, round seeds, or a blended combination of wrinkled and round? What would the seed shape in subsequent generations be when hybrid individuals self-pollinated or were crossed with members of the pure lines?

14.2 Mendel's Experiments with a Single Trait

Mendel's first set of experiments consisted of crossing pure lines that differed in just one trait. This is an important research strategy in biological science: Start with a simple situation. Once you understand what's going on, you can consider more complex questions, such as, What happens in crosses between individuals that differ in two traits?

Mendel began his single-trait crosses by crossing individuals from round-seeded and wrinkled-seeded pure lines. The adults used in an initial experimental cross are the **parental generation.** Their progeny (that is, offspring) are the **F₁ generation.** F_1 stands for "first filial"; the Latin roots *filius* and *filia* mean "son" and "daughter," respectively. Subsequent generations are called the F_2 generation, F_3 generation, and so on.

The Monohybrid Cross

In his first set of crosses, Mendel took pollen from round-seeded plants and placed it on the female reproductive organs of plants from the wrinkled-seeded line. As **FIGURE 14.2a** shows, all the seeds produced by progeny from this cross were round.

This was a remarkable result, for two reasons:

1. The traits did not blend together to form an intermediate phenotype. Instead, the round-seeded form appeared intact. This result was in stark contrast to the predictions of the blending-inheritance hypothesis shown in **FIGURE 14.2b**.

2. The genetic determinant for wrinkled seeds seemed to have disappeared. Where did it go?

Dominant and Recessive Traits To figure out what was going on, Mendel did something that turned out to be brilliant: He planted the F_1 seeds and allowed the individuals to self-pollinate when they matured.

Remember that he knew that each of these individuals had inherited a genetic determinant for round seeds and a genetic determinant for wrinkled seeds. A mating like this—between parents that each carry two different genetic determinants for the same trait—is called a **monohybrid cross.**

When he collected the seeds that were produced by many plants in the resulting F_2 generation, he observed that 5474 were round and 1850 were wrinkled (see Figure 14.2a). This observation was astonishing. The wrinkled seed shape had reappeared in the F_2 generation after disappearing completely in the F_1 generation. No one had observed the phenomenon before, simply because it had been customary for biologists to stop their breeding experiments with F_1 offspring.

Mendel invented some important terms to describe this result.

- He designated wrinkled shape as a **recessive** trait relative to the round-seed trait. This was an appropriate term because none of the F_1 individuals had wrinkled seeds—meaning the wrinkled-seed phenotype appeared to recede or temporarily become latent or hidden.

- He referred to round seeds as **dominant** to the wrinkled-seed trait. This term was apt because the round-seed phenotype appeared to dominate over the wrinkled-seed determinant when both were present.

It's important to note, though, that in genetics the term dominant has nothing to do with its everyday English usage as powerful

(a) Results of Mendel's single-trait (monohybrid) cross

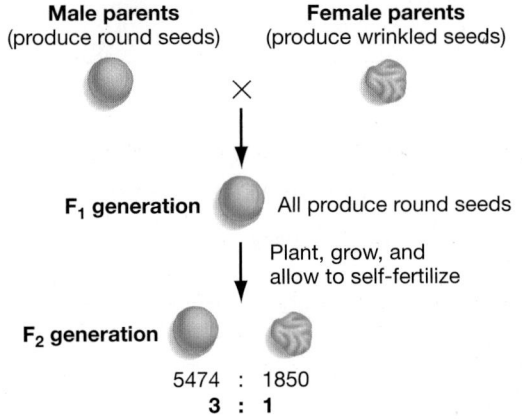

Male parents
(produce round seeds)   ×   Female parents
(produce wrinkled seeds)

F₁ generation   All produce round seeds

Plant, grow, and allow to self-fertilize

F₂ generation

5474 : 1850
3 : 1

(b) Prediction of blending-inheritance hypothesis

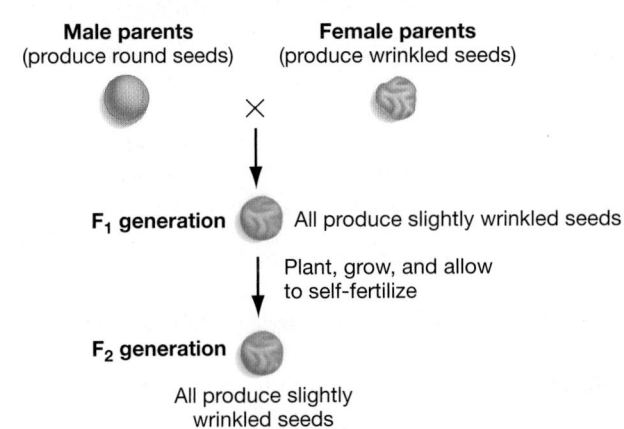

Male parents
(produce round seeds)   ×   Female parents
(produce wrinkled seeds)

F₁ generation   All produce slightly wrinkled seeds

Plant, grow, and allow to self-fertilize

F₂ generation

All produce slightly wrinkled seeds

FIGURE 14.2 A Monohybrid Cross. The results of Mendel's crosses involving a single trait **(a)** contrasted strongly with the predictions of the blending-inheritance hypothesis **(b)**.

or superior. Subsequent research has shown that individuals with the dominant phenotype do not necessarily have higher fitness than individuals with the recessive phenotype. Nor are genetic determinants associated with a dominant phenotype necessarily more common than recessive ones. For example, a fatal illness—a type of brain degeneration called Huntington's disease—is caused by a rare, dominant genetic determinant. In genetics, the terms dominant and recessive identify *only* which phenotype is observed in individuals carrying two different genetic determinants for a given trait.

Mendel also noticed that the round and wrinkled seeds of the F_2 generation were present in a ratio of 2.96 : 1, or essentially 3 : 1. The 3 : 1 ratio means that for every four individuals, on average three had the dominant phenotype and one had the recessive phenotype. The results can also be stated in terms of frequencies or proportions: In this case, about ¾ of the F_2 seeds were round and ¼ were wrinkled.

A Reciprocal Cross   Mendel wanted to test the hypothesis that it mattered which parent and gamete type had a particular genetic determinant—that gender influenced the inheritance of

seed shape. To do this, he performed a second set of crosses between two pure-breeding lines—this time with pollen taken from an individual from a pure line of wrinkled-seeded peas (**FIGURE 14.3**).

These experiments completed a **reciprocal cross**—a set of matings where the mother's phenotype in the initial cross is the father's phenotype in a subsequent cross, and the father's phenotype in the initial cross is the mother's phenotype in a subsequent cross.

In this case the results of the reciprocal crosses were identical: All the F_1 progeny had round seeds, just as in the initial cross. The reciprocal cross established that it does not matter whether the genetic determinants for seed shape are located in the male or female parent.

Do Mendel's Results Hold for Other Traits?   Before he tried to interpret this pattern, it was important for Mendel to establish that the results were not restricted to inheritance of seed shape. So he repeated the experiments with each of the six other traits listed earlier. As **TABLE 14.2** shows, in each case, he obtained similar results:

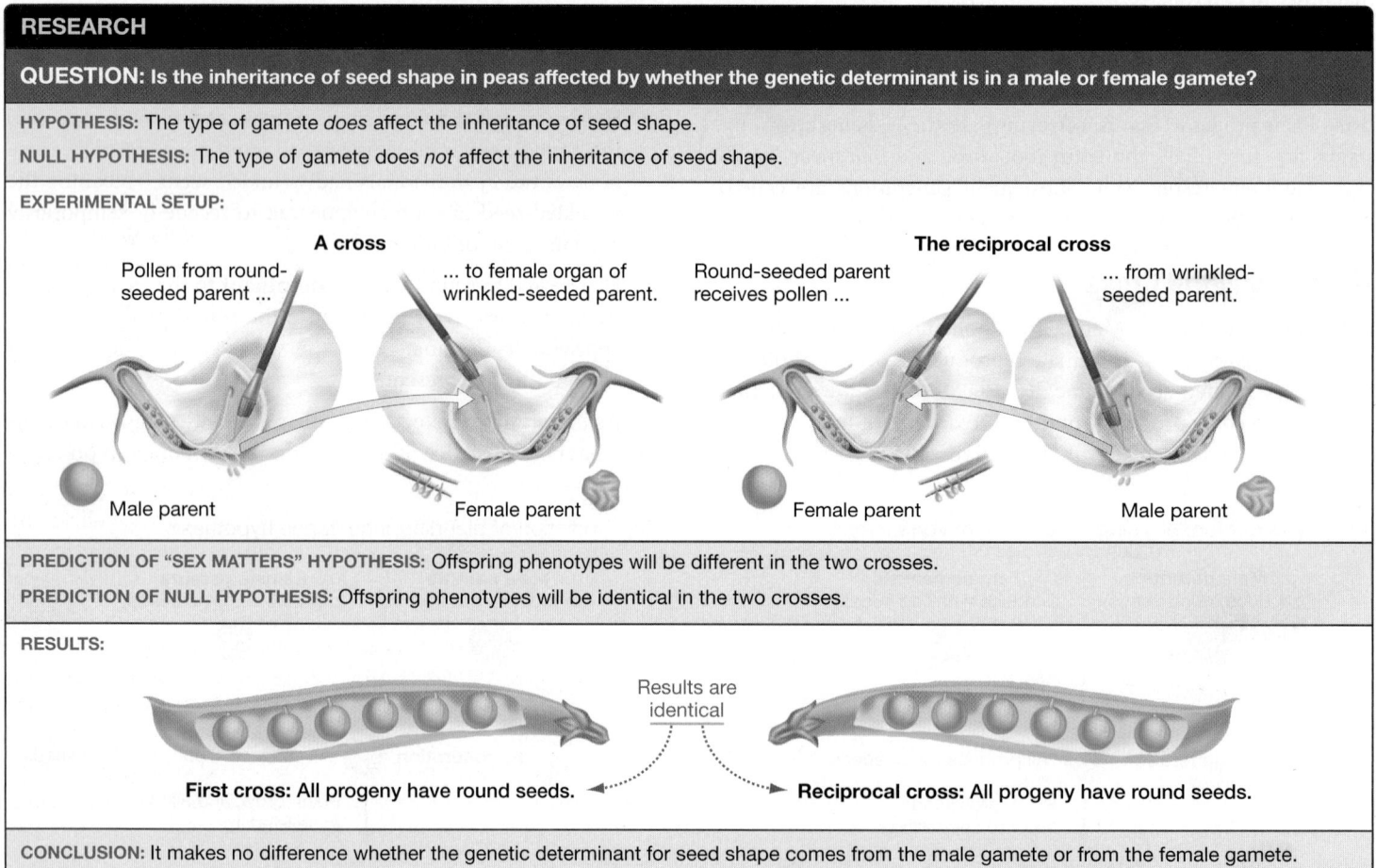

RESEARCH

QUESTION: Is the inheritance of seed shape in peas affected by whether the genetic determinant is in a male or female gamete?

HYPOTHESIS: The type of gamete *does* affect the inheritance of seed shape.

NULL HYPOTHESIS: The type of gamete does *not* affect the inheritance of seed shape.

EXPERIMENTAL SETUP:

A cross

Pollen from round-seeded parent ...

... to female organ of wrinkled-seeded parent.

Male parent   Female parent

The reciprocal cross

Round-seeded parent receives pollen ...

... from wrinkled-seeded parent.

Female parent   Male parent

PREDICTION OF "SEX MATTERS" HYPOTHESIS: Offspring phenotypes will be different in the two crosses.

PREDICTION OF NULL HYPOTHESIS: Offspring phenotypes will be identical in the two crosses.

RESULTS:

Results are identical

First cross: All progeny have round seeds.   **Reciprocal cross:** All progeny have round seeds.

CONCLUSION: It makes no difference whether the genetic determinant for seed shape comes from the male gamete or from the female gamete.

FIGURE 14.3 A Reciprocal Cross.

SOURCE: Mendel, G. 1866. Versuche über Pflanzen-hybriden. *Verhandlungen des naturforschenden Vereines in Brünn.* 4: 3–47. English translation available from ESP: Electronic Scholarly Publishing (www.esp.org).

✔**QUESTION** Some people think that experiments are failures if the hypothesis being tested is not supported. What does it mean to say that an experiment failed? Was this experiment a failure?

TABLE 14.2 F₂ Results from Mendel's Monohybrid Reciprocal Cross Experiments*

Trait	Dominant Phenotype	Recessive Phenotype	Ratio
Seed shape	5474 round	1850 wrinkled	2.96 : 1
Seed color	6022 yellow	2001 green	3.01 : 1
Pod shape	882 inflated	299 constricted	2.95 : 1
Pod color	428 green	152 yellow	2.82 : 1
Flower color	705 purple	224 white	3.15 : 1
Flower and pod position	651 axial	207 terminal	3.14 : 1
Stem length	787 tall	265 dwarf	2.96 : 1

*Mendel pooled the results from the reciprocal crosses for each trait because the results were the same whether the dominant trait originated from the male parent or the female parent.
DATA: Mendel, G. 1866. *Verhandlungen des naturforschenden Vereines in Brünn.* 4: 3–47.

✔ **EXERCISE** Two entries in the last rows are left blank. Fill in these entries with the correct values calculated from the available data.

- The F₁ progeny showed only the dominant trait and did not exhibit an intermediate phenotype.
- Reciprocal crosses produced the same results.
- The ratio of F₂ generation individuals with dominant and recessive phenotypes was about 3 to 1.

How could these patterns be explained? Mendel answered this question with a series of propositions about the nature and behavior of the hereditary determinants. These hypotheses are considered some of the most brilliant insights in the history of biological science.

Particulate Inheritance

Mendel's results were clearly inconsistent with either the hypothesis of blending inheritance or the hypothesis of acquired characters. To explain the patterns that he observed, Mendel proposed a competing hypothesis called **particulate inheritance.** He maintained that the hereditary determinants for traits do not blend together or become modified through use. In fact, hereditary determinants maintain their integrity from generation to generation. Instead of blending together, they act as discrete entities or particles.

Mendel's hypothesis was the only way to explain the observation that phenotypes disappeared in one generation and reappeared intact in the next. It also represented a fundamental break with ideas that had prevailed for hundreds of years.

Genes, Alleles, and Genotypes Today, geneticists use the word **gene** to indicate the hereditary determinant for a trait. For example, the hereditary factor that determines whether the seeds of garden peas are round or wrinkled is referred to as the gene for seed shape.

Mendel also proposed that each individual can have two versions of any gene. Today different versions of the same gene are called **alleles.** Different alleles are responsible for the variation in the traits that Mendel studied. In the case of the gene for seed shape, one allele of this gene is responsible for the round form of the seed while another allele is responsible for the wrinkled form.

The alleles found in a particular individual are called the **genotype.** An individual's genotype has a profound effect on the phenotype—the observable physical traits.

The hypothesis that pea plants have two copies of each gene—either two of the same allele or two different alleles—was important because it gave Mendel a framework for explaining dominance and recessiveness. He proposed that some alleles are dominant and others are recessive. Recall that dominance and recessiveness determine which phenotype appears in an individual when two different alleles are present. In garden peas, the allele for round seeds is dominant; the allele for wrinkled seeds is recessive. Therefore, so long as one allele for round seeds is present, seeds are round. When both alleles present are for wrinkled seeds (thus no allele for round seeds is present), seeds are wrinkled.

These hypotheses explain why the wrinkled-seed phenotype disappeared in the F₁ generation. But why did wrinkled seeds reappear in the F₂, and why was there a 3:1 ratio of round and wrinkled seeds in the F₂ generation?

The Principle of Segregation To explain the reappearance of the recessive phenotype and the characteristic 3:1 ratio of phenotypes in F₂ individuals, Mendel reasoned that the two members of each gene pair must segregate—that is, separate—into different gamete cells during the formation of eggs and sperm. As a result, each gamete contains one allele of each gene. This idea is called the **principle of segregation.**

To show how this principle works, Mendel used a letter to indicate the gene for a particular trait. For example, he used uppercase *R* to symbolize a dominant allele for seed shape and lowercase *r* to symbolize a recessive allele for seed shape. (Notice that the symbols for genes are always italicized.)

Using this notation, Mendel described the genotype of the individuals in the round-seed pure line as *RR* (having two of the dominant allele). The genotype of the wrinkled-seed pure line is *rr* (two of the recessive allele). Because *RR* and *rr* individuals have two copies of the same allele, they are said to be **homozygous** for the seed-shape gene (*homo* is the Greek root for "same," while *zygo* means "yoked"). Crosses of individuals from the same pure line always produce offspring with the same phenotype because they are homozygous—no other allele is present.

FIGURE 14.4a uses a diagram called a Punnett square to show what happened to these alleles when Mendel crossed the *RR* and *rr* pure lines. R. C. Punnett invented this straightforward technique for predicting the genotypes and phenotypes of different crosses years after Mendel published his work. According to Mendel's hypothesis, *RR* parents produce eggs and sperm that all carry the *R* allele, while *rr* parents produce gametes with the *r* allele only. When two gametes—one from each parent—come together at fertilization, they create offspring with the *Rr* genotype. Such individuals, with two different alleles for the same gene, are said to be **heterozygous** (*hetero* is the Greek root for "different"). Heterozygous individuals, or heterozygotes, show that the *R* allele is dominant because only the round phenotype is expressed even though the wrinkled allele is present.

Why do the two phenotypes appear in a 3:1 ratio in the F₂ generation? Mendel proposed that during gamete formation in the F₁ heterozygotes, the paired *Rr* alleles separate into different gamete cells. As a result, and as shown in the Punnett square of **FIGURE 14.4b**, half the gametes should carry the *R* allele and half should carry the *r* allele. A given sperm has an equal chance of fertilizing either an *R*-bearing egg or an *r*-bearing egg.

Predicting Offspring Genotypes and Phenotypes with a Punnett Square

The box you've just studied in Figure 14.4b is an example of a simple Punnett square. To produce a Punnett square, follow these steps:

1. Write each of the *unique* gamete genotypes produced by one parent in a horizontal row along the top of the diagram.

2. Write each of the *unique* gamete genotypes produced by the other parent in a vertical column down the left side of the diagram.

3. Create a table under the horizontal row of gametes and to the right of the vertical column of gametes.

4. Fill in the table with the entries for the parental gamete genotypes that are written at the top and at the left side. This step represents fertilization and produces the offspring genotypes.

(a) A cross between two **homozygotes**

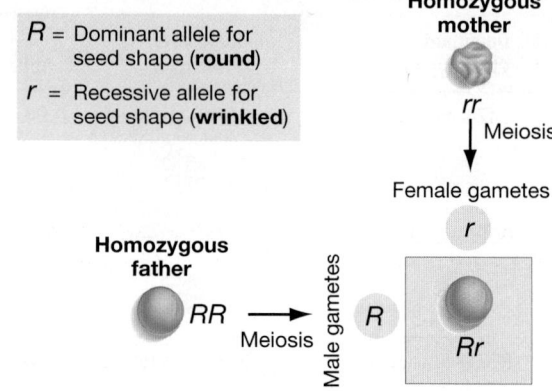

Offspring genotypes: All *Rr* (heterozygous)
Offspring phenotypes: All round seeds

(b) A cross between two **heterozygotes**

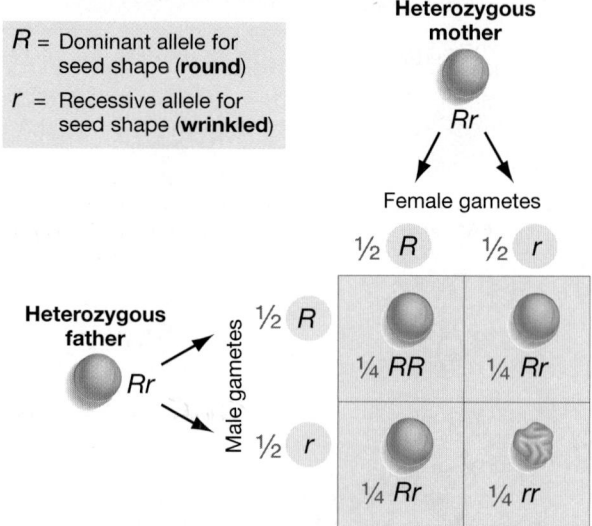

Offspring genotypes: ¼ *RR* : ½ *Rr* : ¼ *rr*
Offspring phenotypes: ¾ round : ¼ wrinkled

FIGURE 14.4 Mendel Analyzed the F₁ and F₂ Offspring of a Cross between Pure Lines. Notice that when you construct a Punnett square, you only need to list each unique type of gamete once at the head of a row or column. For example, even though the *RR* alleles segregate in the male parent of part (a), you have to list just one *R* gamete to represent the male's contribution, not two.

✓**QUESTION** In constructing a Punnett square, does it matter whether the male or female gametes go on the left or across the top? Why or why not?

5. Tally the proportions or ratios of each offspring genotype and phenotype.

✓ If you understand these concepts, you should be able to state how filling in the top and side of the Punnett square is related to the principle of segregation and predict the phenotype and genotype ratios for a cross between *Rr* and *rr* peas.

Mendel's Claims	Comments
1. Peas have two copies of each gene and thus may have two different alleles of the gene.	This also turns out to be true for many other organisms.
2. Genes are particles of inheritance that do not blend together.	Genes maintain their integrity from generation to generation.
3. Each gamete contains one copy of each gene (one allele).	This is because of the principle of segregation—the members of each gene pair segregate during the formation of gametes.
4. Males and females contribute equally to the genotype of their offspring.	When gametes fuse, offspring acquire a total of two of each gene—one from each parent.
5. Some alleles are dominant to other alleles.	When a dominant and a recessive allele for the same gene are found in the same individual (a heterozygote), that individual has the dominant phenotype.

*Mendel did not use these modern terms. He expressed these ideas in different words.

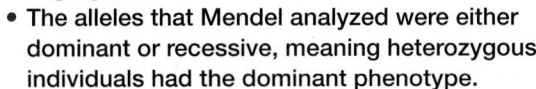

check your understanding

If you understand that . . .

- Mendel discovered that individuals have two alleles of each gene and that these alleles separate (segregate) into gametes. This is the principle of segregation.
- The alleles that Mendel analyzed were either dominant or recessive, meaning heterozygous individuals had the dominant phenotype.

✔ You should be able to . . .

Use the genetic problems at the end of this chapter to practice the following skills:

1. Starting with parents of known genotypes, create and analyze Punnett squares of crosses involving a single trait to predict the genotypes and phenotypes that will occur in their F_1 and F_2 offspring; then use the Punnett square to determine the expected frequency of each genotype and phenotype. (Do Problem 13 in Test Your Problem-Solving Skills.)

2. Given the outcome of a cross, infer the genotypes and phenotypes of the parents. (Do Problem 15 in Test Your Problem-Solving Skills.)

Answers are available in Appendix A.

As an example of the concluding step in analyzing a cross, the Punnett square in Figure 14.4b predicts that ¼ of the F_2 offspring will be *RR*, ½ will be *Rr*, and ¼ will be *rr*. Because the *R* allele is dominant to the *r* allele, ¾ of the offspring should be round-seeded (the sum of the *RR* and the *Rr* offspring) and ¼ should be wrinkled-seeded (the *rr* offspring). These results are what Mendel found in his experiments with peas. In the simplest and most elegant fashion possible, Mendel's interpretation explains the 3 : 1 ratio of round to wrinkled seeds observed in the F_2 offspring and the mysterious reappearance of the wrinkled seeds.

The term genetic model refers to a set of hypotheses that explains how a particular trait is inherited. **TABLE 14.3** summarizes Mendel's model for explaining the basic patterns in the transmission of traits from parents to offspring; these hypotheses are sometimes referred to as Mendel's rules. They represent a radical break from the ideas of blending inheritance and the inheritance of acquired characters that had dominated scientific thinking about heredity.

14.3 Mendel's Experiments with Two Traits

Working with one trait at a time allowed Mendel to establish that blending inheritance does not occur. It also allowed him to infer that each pea plant had two copies of each gene and to recognize the principle of segregation.

Mendel's next step extended these results. The important question now was whether the principle of segregation holds true if individuals differ with respect to two traits, instead of just one. Do different genes segregate together, or independently?

The Dihybrid Cross

Mendel crossed a pure-line parent that produced round, yellow seeds with a pure-line parent that produced wrinkled, green seeds. According to his model, the F_1 offspring of this cross should be heterozygous for both genes. A mating between two such individuals—both heterozygous for two traits—is called a **dihybrid cross.**

Mendel's earlier experiments had established that the allele for yellow seeds was dominant to the allele for green seeds; these alleles were designated *Y* for yellow and *y* for green. As **FIGURE 14.5** (see page 264) indicates, two distinct possibilities existed for how the alleles of these two different genes—the gene for seed shape and the gene for seed color—would be transmitted to offspring.

- The first possibility was that the allele for seed shape and the allele for seed color originally present in each parent would separate from each other and be transmitted independently. This hypothesis is called independent assortment because the two alleles would be sorted into gametes independently of each other (Figure 14.5a).

- The second possibility was that the allele for seed shape and the allele for seed color originally present in each parent

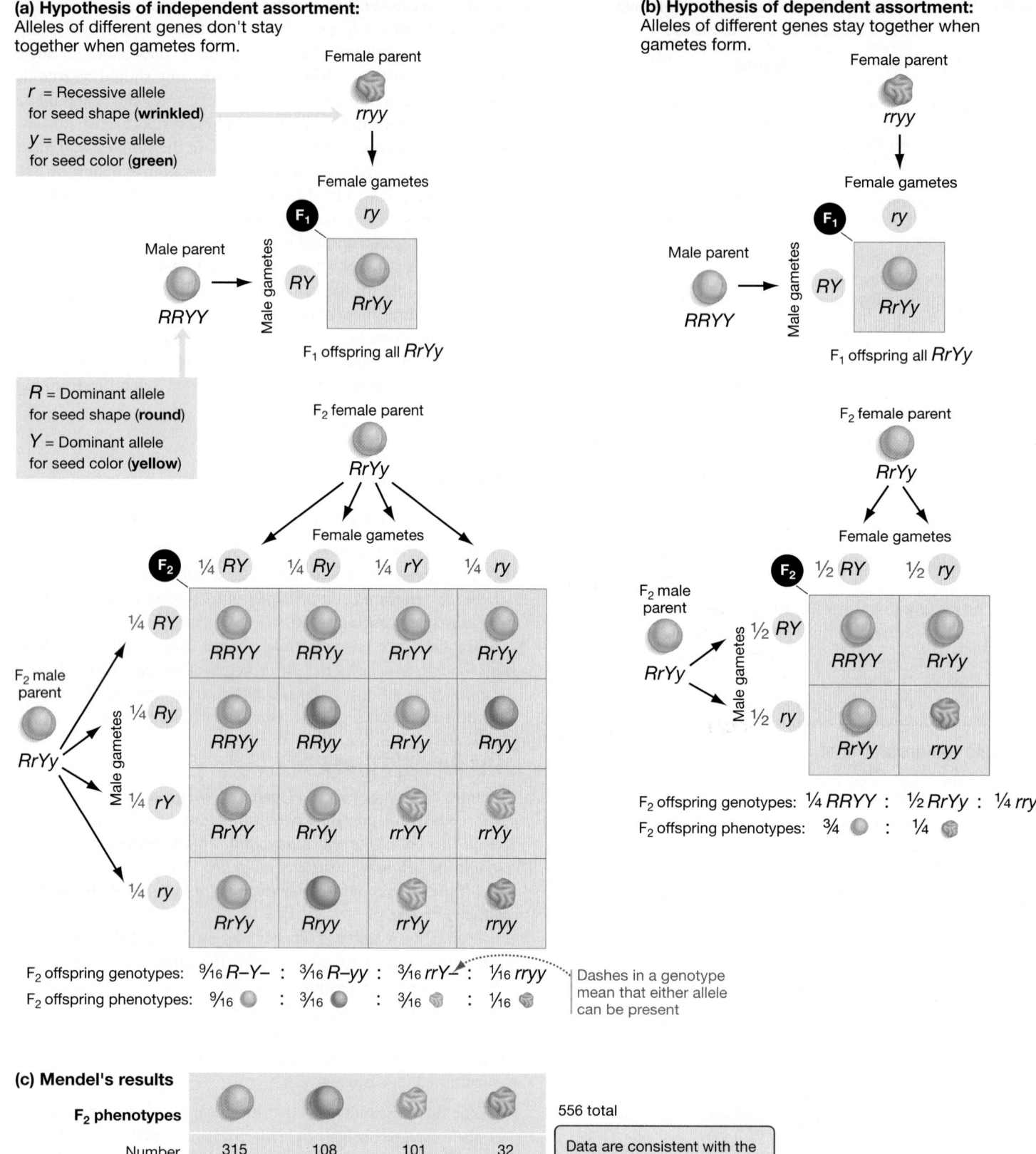

(a) Hypothesis of independent assortment:
Alleles of different genes don't stay together when gametes form.

r = Recessive allele for seed shape (**wrinkled**)

y = Recessive allele for seed color (**green**)

R = Dominant allele for seed shape (**round**)

Y = Dominant allele for seed color (**yellow**)

Female parent

rryy

Male parent

RRYY

F₁ offspring all *RrYy*

Female gametes: *ry*
Male gametes: *RY*
F₁: *RrYy*

F₂ female parent

RrYy

Female gametes: ¼ *RY* ¼ *Ry* ¼ *rY* ¼ *ry*

F₂ male parent

RrYy

Male gametes: ¼ *RY* ¼ *Ry* ¼ *rY* ¼ *ry*

	¼ RY	¼ Ry	¼ rY	¼ ry
¼ RY	RRYY	RRYy	RrYY	RrYy
¼ Ry	RRYy	RRyy	RrYy	Rryy
¼ rY	RrYY	RrYy	rrYY	rrYy
¼ ry	RrYy	Rryy	rrYy	rryy

F₂ offspring genotypes: ⁹⁄₁₆ *R–Y–* : ³⁄₁₆ *R–yy* : ³⁄₁₆ *rrY–* : ¹⁄₁₆ *rryy*
F₂ offspring phenotypes: ⁹⁄₁₆ : ³⁄₁₆ : ³⁄₁₆ : ¹⁄₁₆

Dashes in a genotype mean that either allele can be present

(b) Hypothesis of dependent assortment:
Alleles of different genes stay together when gametes form.

Female parent

rryy

Male parent

RRYY

F₁ offspring all *RrYy*

Female gametes: *ry*
Male gametes: *RY*
F₁: *RrYy*

F₂ female parent

RrYy

Female gametes: ½ *RY* ½ *ry*

F₂ male parent

RrYy

Male gametes: ½ *RY* ½ *ry*

	½ RY	½ ry
½ RY	RRYY	RrYy
½ ry	RrYy	rryy

F₂ offspring genotypes: ¼ *RRYY* : ½ *RrYy* : ¼ *rryy*
F₂ offspring phenotypes: ¾ : ¼

(c) Mendel's results

F₂ phenotypes					556 total
Number	315	108	101	32	
Fraction of offspring	⁹⁄₁₆	³⁄₁₆	³⁄₁₆	¹⁄₁₆	

Data are consistent with the predictions of independent assortment.

FIGURE 14.5 Mendel Analyzed the F₁ and F₂ Offspring of a Cross between Pure Lines for Two Traits. Each of two hypotheses predicted a different pattern for the outcome when alleles of different genes are transmitted to offspring: **(a)** The alleles could be sorted into gametes independently of each other, or **(b)** particular alleles could always be transmitted together. **(c)** Mendel's results supported independent assortment.

would be transmitted to gametes together. This hypothesis can be called dependent assortment because the transmission of one allele would depend on the transmission of another (Figure 14.5b).

As Figure 14.5 shows, the F₁ offspring of Mendel's mating are expected to have the dominant round and yellow phenotypes whether the different genes are transmitted together or independently. When Mendel did the cross and observed the F₁ individuals, this is exactly what he found. All the F₁ offspring had round, yellow seeds.

The two hypotheses make radically different predictions, however, about what will be observed when the F₁ individuals are allowed to self-fertilize and produce an F₂ generation. If the alleles of different genes assort independently to form gametes, then each heterozygous parent should produce four different gamete genotypes, as shown in Figure 14.5a. A 4-row-by-4-column Punnett square results, and it predicts that there should be 9 different offspring genotypes and 4 phenotypes. Further, the yellow-round, green-round, yellow-wrinkled, and green-wrinkled phenotypes should be present in frequencies of ⁹⁄₁₆, ³⁄₁₆, ³⁄₁₆, and ¹⁄₁₆, respectively. This is a ratio of 9 : 3 : 3 : 1.

On the other hand, if the alleles from each parent stay together, then the prediction is for only three possible offspring genotypes and a 3 : 1 ratio of two phenotypes—yellow-round or green-wrinkled—in the F₂, as Figure 14.5b shows.

When Mendel examined the phenotypes of the F₂ offspring, he found that they conformed to the predictions of the hypothesis of independent assortment. Four phenotypes were present in proportions that closely approximated the predicted ratio of 9 : 3 : 3 : 1 (Figure 14.5c). On the basis of these data, Mendel accepted the hypothesis that alleles of different genes are transmitted independently of one another. This result became known as the **principle of independent assortment**.

✔ If you understand the principle of independent assortment, it should make sense to you that an individual with the genotype *AaBb* produces gametes with the genotypes *AB*, *Ab*, *aB*, and *ab*. You should be able to predict the genotypes of the gametes produced by individuals with the genotypes *AABb*, *PpRr*, and *AaPpRr*.

Using a Testcross to Confirm Predictions

Mendel did experiments with combinations of traits other than seed shape and color and obtained results similar to those in Figure 14.5c. Each paired set of traits produced a 9 : 3 : 3 : 1 ratio of progeny phenotypes in the F₂ generation. He even did a limited set of crosses examining three traits at a time. Although all these data were consistent with the principle of independent assortment, his most powerful support for the hypothesis came from a different type of experiment.

In designing this study, Mendel's goal was to test the prediction that an *RrYy* plant produces four different types of gametes in equal proportions. To accomplish this, Mendel invented a technique called a testcross. A **testcross** uses a parent that

contributes only recessive alleles to its offspring and helps to determine the unknown genotype of the second parent.

Testcrosses are useful because the genetic contribution of the homozygous recessive parent is known. As a result, a testcross allows experimenters to test the genetic contribution of the other parent. If the other parent has the dominant phenotype but an unknown genotype, the results of the testcross allow researchers to infer whether that parent is homozygous or heterozygous for the dominant allele.

In this case, Mendel performed a cross between a parent that was homozygous for the recessive green and wrinkled phenotypes (*rryy*) and a parent that had an unknown genotype but was known from its yellow- and round-seeded phenotype to possess the dominant *R* and *Y* alleles. Two (of four) possible genotypes for this yellow- and round-seeded parent are *RrYy* and *RRYY*. The types and proportions of offspring that could result from a testcross involving *RrYy* or *RRYY* pea plants can be predicted with the Punnett square shown in **FIGURE 14.6** (see page 266). If the principle of independent assortment is valid, the testcross should produce four types of offspring in equal proportions if the tested parent is *RrYy*, and only one type of offspring if the tested parent is *RRYY*.

What were the actual proportions observed? Mendel did a testcross using the F₁ offspring of a pure line yellow-, round-seeded parent and a green-, wrinkled-seeded parent. These F₁ offspring were expected to have an *RrYy* genotype as shown in Figure 14.5a. When he examined the seeds produced by the testcross, Mendel found that among the 110 progeny, 31 were round

check your understanding

If you understand that . . .

- Mendel found that alleles of different genes are transmitted to gametes independently of one another. This is the principle of independent assortment.
- The genotype of a strain with a dominant phenotype can be revealed in testcrosses between the dominant strain and a homozygous recessive strain.

✔ **You should be able to . . .**

Use the genetics problems at the end of this chapter to practice the following skills:

1. Starting with parents of known genotypes for two different traits, create and analyze Punnett squares to predict the genotypes and phenotypes that will occur in their F₁ and F₂ offspring and then calculate the expected frequency of each genotype and phenotype. (Do Problem 14 in Test Your Problem-Solving Skills.)

2. Given the outcome of a cross, infer the genotypes and phenotypes of the parents. (Do Problem 16 in Test Your Problem-Solving Skills.)

Answers are available in Appendix A.

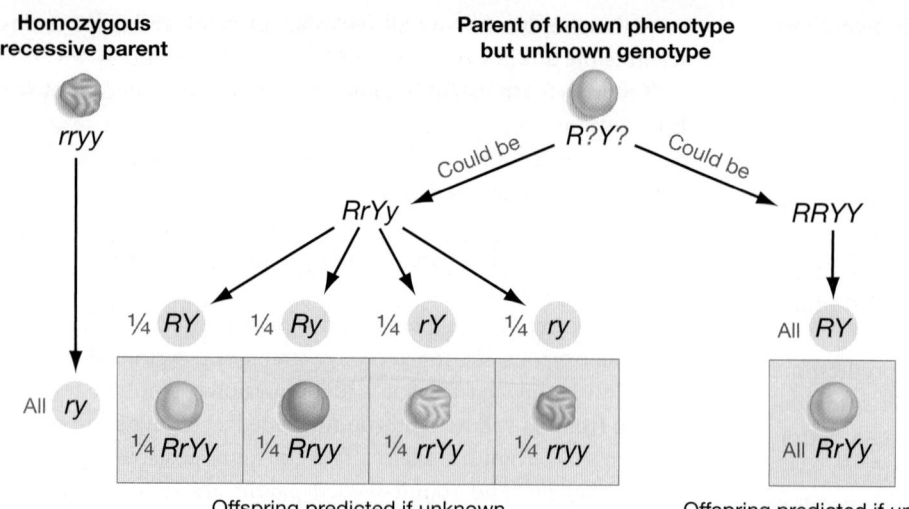

Homozygous recessive parent

rryy

Parent of known phenotype but unknown genotype

R?Y?

Could be → *RrYy*

Could be → *RRYY*

¼ *RY*   ¼ *Ry*   ¼ *rY*   ¼ *ry*

All *RY*

All *ry*

| ¼ *RrYy* | ¼ *Rryy* | ¼ *rrYy* | ¼ *rryy* |

All *RrYy*

Offspring predicted if unknown parent is **heterozygous** at both genes

Offspring predicted if unknown parent is **homozygous dominant** at both genes

FIGURE 14.6 The Predictions Made by the Principle of Independent Assortment Can Be Evaluated in a Testcross. If the principle of independent assortment is correct, and *RrYy* parents produce four types of gametes in equal proportions, then a mating between *RrYy* and *rryy* parents should produce four types of offspring in equal proportions, as shown on the left. Test crosses can also reveal the genotype of a parent with dominant phenotypes, as seen in the example of different results obtained from *RrYy* and *RRYY* parental genotypes.

and yellow, 26 were round and green, 27 were wrinkled and yellow, and 26 were wrinkled and green. This is almost exactly ¼ of each type, which matched the predicted proportions for offspring of an *RrYy* parent. The testcross had confirmed the principle of independent assortment.

Mendel's work provided a powerful conceptual framework for thinking about transmission genetics—the patterns that occur as alleles pass from one generation to the next. This framework was based on **(1)** the segregation of discrete, paired genes into separate gametes, and **(2)** the independent assortment of genes that affect different traits.

The experiments you've just reviewed were brilliant in design, execution, and interpretation. Unfortunately, they were ignored for 34 years.

14.4 The Chromosome Theory of Inheritance

Historians of science debate why Mendel's work was overlooked for so long. It is probably true that his use of ratios and proportions were difficult for biologists of that time to understand and absorb. It may also be true that the theory of blending inheritance was so well entrenched that his results were dismissed as peculiar or unbelievable.

Whatever the reason, Mendel's work was not appreciated until 1900, when three biologists, working with a variety of plants and animals, independently "discovered" Mendel's work and reached the same main conclusions.

The rediscovery of Mendel's work more than three decades after its publication ignited the young field of genetics. Mendel's experiments had established the basic patterns of inheritance, but what process is responsible for these patterns? Two biologists, working independently, came up with the answer. Walter Sutton and Theodor Boveri each realized that meiosis could account for

Mendel's rules. When this hypothesis was published in 1902, research in genetics exploded.

Meiosis Explains Mendel's Principles

What Sutton and Boveri grasped is that meiosis explains the principle of segregation and the principle of independent assortment. The cell at the top of **FIGURE 14.7** illustrates Sutton and Boveri's central insight: Mendel's hereditary determinants, or genes, are located on chromosomes. In this example, the gene for seed shape is shown at a particular position along a certain chromosome. This location is known as a **locus** ("place"; plural, **loci**).

The paternal and maternal chromosomes shown in Figure 14.7 happen to possess different alleles at the seed shape gene locus: One allele specifies round seeds (*R*) and the other specifies wrinkled seeds (*r*).

The subsequent steps in Figure 14.7 show how these alleles segregate into different daughter cells during meiosis I, when homologous chromosomes separate. The physical separation of alleles during anaphase of meiosis I is responsible for Mendel's principle of segregation.

FIGURE 14.8 follows the segregation of two different gene pairs—in this case, for seed shape and seed color—as meiosis proceeds. If the alleles for different genes are located on different chromosomes, they assort independently of one another at meiosis I. This is the physical basis of Mendel's principle of independent assortment. Over many meiotic divisions, four types of gametes will be produced in equal proportions.

Sutton and Boveri formalized these observations in the **chromosome theory of inheritance.** Like other theories in biology, the chromosome theory describes a predictable pattern—a set of observations about the natural world—and a process that explains the pattern. The chromosome theory states that Mendel's rules can be explained by the independent alignment and separation of homologous chromosomes at meiosis I.

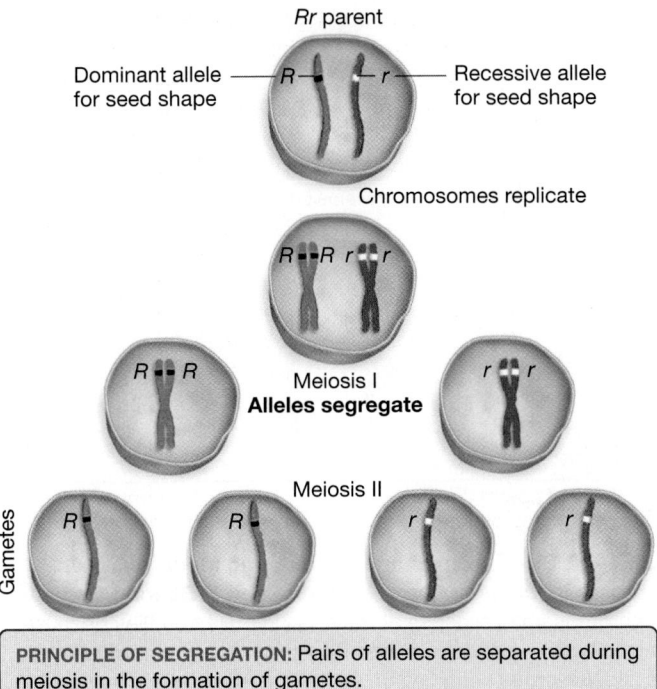

Dominant allele for seed shape — *R* ● *r* — Recessive allele for seed shape

Rr parent

Chromosomes replicate

Meiosis I
Alleles segregate

Meiosis II

Gametes

PRINCIPLE OF SEGREGATION: Pairs of alleles are separated during meiosis in the formation of gametes.

FIGURE 14.7 Meiosis Explains the Principle of Segregation. The two members of a parent's gene pair segregate into different gametes because homologous chromosomes separate during meiosis I.

When Sutton and Boveri published their ideas, however, the hypothesis that genes are located on chromosomes was untested. What experiments confirmed that chromosomes contain genes?

Testing the Chromosome Theory

During the first decade of the twentieth century, an unassuming insect rose to prominence as a model organism for testing the chromosome theory of inheritance. This organism—the fruit fly *Drosophila melanogaster*—has been at the center of genetic studies ever since (see **BioSkills 13** in Appendix B).

Drosophila melanogaster has all the attributes of a useful model organism for studies in genetics: small size, ease of rearing in the lab, a short generation time (about 10 days), and abundant offspring (up to a few hundred per mating). The elaborate external anatomy of this insect also makes it possible to identify interesting phenotypic variation among individuals (**FIGURE 14.9a**; see page 268).

Drosophila was adopted as a model organism by Thomas Hunt Morgan and his students. But because *Drosophila* is not a domesticated species like the garden pea, common phenotypic variants such as Mendel's round and wrinkled seeds were not available to Morgan. Instead, he had access only to flies with the most common phenotype for each trait, phenotypes referred to as **wild type.** Consequently, an early goal of Morgan's research

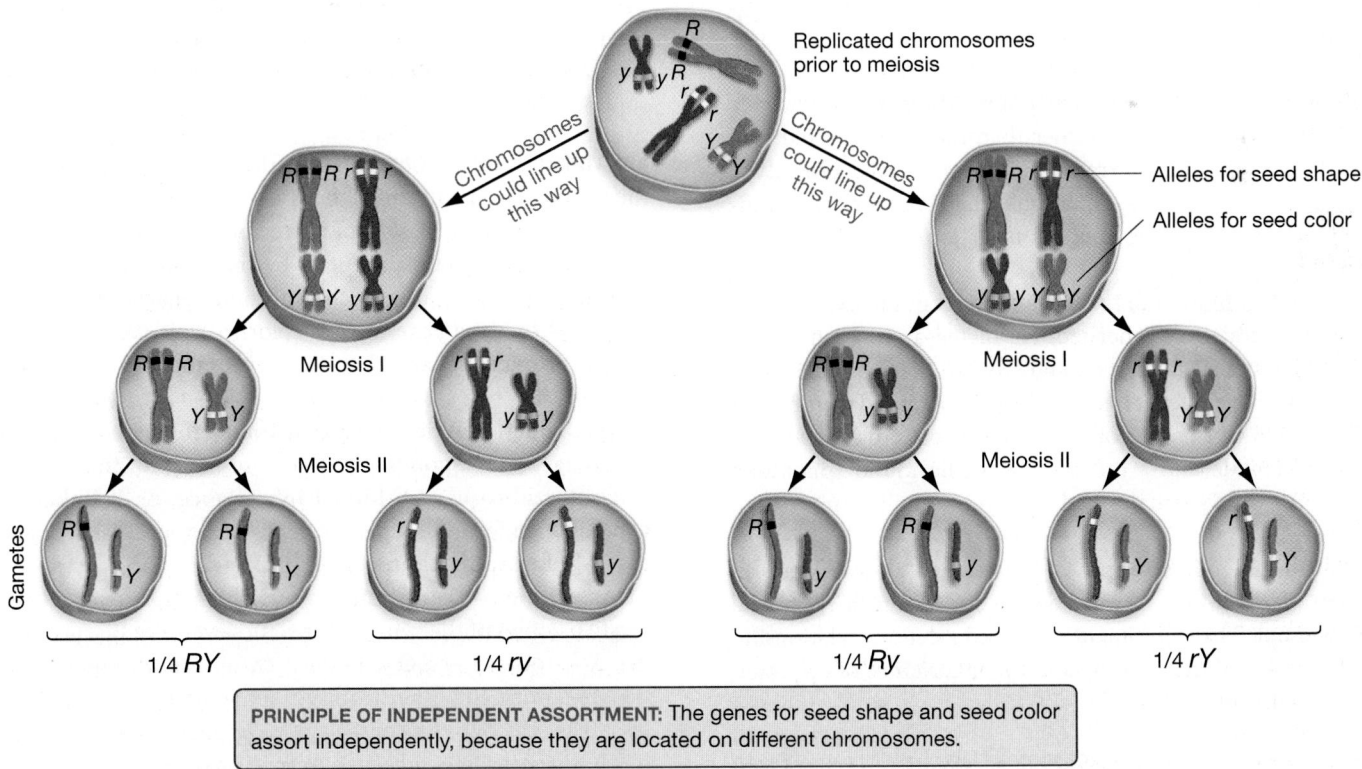

Replicated chromosomes prior to meiosis

Chromosomes could line up this way

Chromosomes could line up this way

Alleles for seed shape

Alleles for seed color

Meiosis I

Meiosis II

Gametes

1/4 *RY* | 1/4 *ry* | 1/4 *Ry* | 1/4 *rY*

PRINCIPLE OF INDEPENDENT ASSORTMENT: The genes for seed shape and seed color assort independently, because they are located on different chromosomes.

FIGURE 14.8 Meiosis Is Responsible for the Principle of Independent Assortment. The genes for different traits assort independently because nonhomologous chromosomes assort independently during meiosis.

(a) The fruit fly *Drosophila melanogaster*

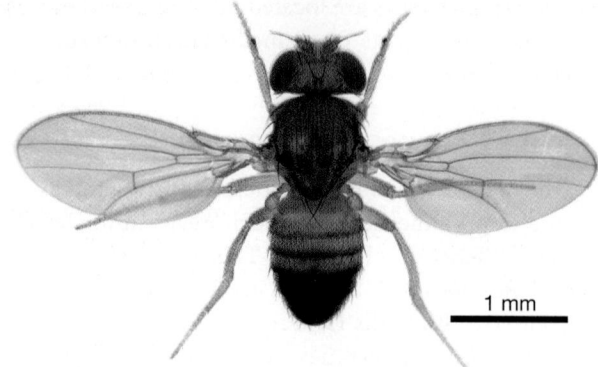

1 mm

(b) Eye color is a variable trait.

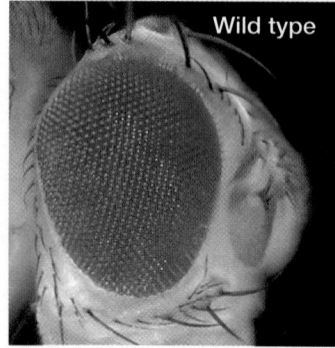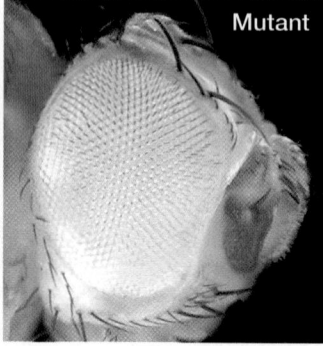

Wild type

Mutant

FIGURE 14.9 The Fruit Fly *Drosophila melanogaster* Is an Important Model Organism in Genetics.

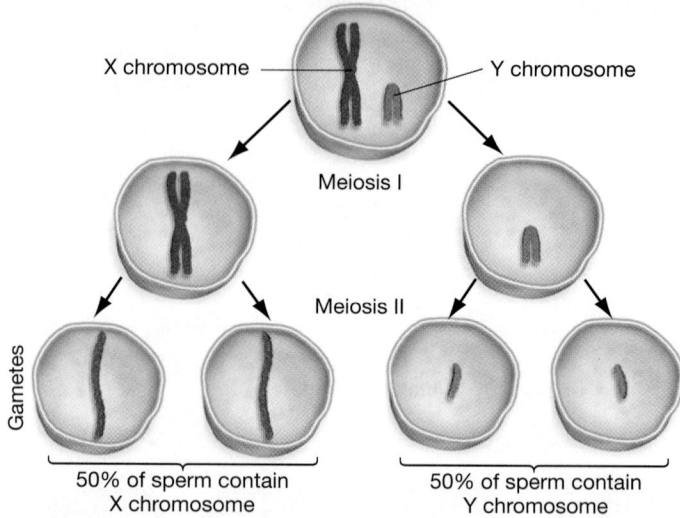

X chromosome

Y chromosome

Meiosis I

Meiosis II

Gametes

50% of sperm contain X chromosome

50% of sperm contain Y chromosome

FIGURE 14.10 Sex Chromosomes Pair during Meiosis I, Then Segregate in Males to Form X-Bearing and Y-Bearing Gametes. Sex chromosomes synapse at meiosis I in male fruit flies because of a small, gene-free region shared by the X and Y chromosomes. This allows normal segregation, so half the sperm cells bear an X chromosome and half have a Y chromosome.

was simply to find and characterize individuals with different phenotypes to use in genetic studies.

The White-Eyed Mutant At one point, Morgan discovered a male fly that had white eyes rather than the wild-type red eyes (**FIGURE 14.9b**). Morgan inferred that the white-eyed phenotype resulted from a **mutation**—a heritable change in a gene. An individual with a phenotype due to a mutation is referred to as a **mutant.**

With his first mutant in hand, Morgan set out to explore how the eye-color trait was inherited. He mated a red-eyed female fly with the mutant white-eyed male fly. All the F₁ progeny had red eyes. By continued crosses, Morgan obtained white-eyed female flies. When he performed a reciprocal cross between a white-eyed female and a red-eyed male, he found something puzzling: All the F₁ females had red eyes, but all F₁ males had white eyes.

Recall that Mendel's reciprocal crosses had always given results that were similar to each other. But Morgan's reciprocal crosses did not. The experiment suggested a definite relationship between the sex of the progeny and the inheritance of eye color. What was going on?

The Discovery of Sex Chromosomes Nettie Stevens began studying the karyotypes of insects about the time that Morgan began his work with *Drosophila*. First, in the beetle *Tenebrio*

molitor, and later in other insects, including *Drosophila*, she noticed a striking difference in the chromosome complements of males and females.

Recall that Stevens and others discovered that there were sex chromosomes (the X and the Y) and autosomes (see Chapter 13). Female flies have a pair of X chromosomes and male flies have an X and a Y chromosome. Morgan's knowledge of Stevens' findings was the key to explaining his puzzling results and in providing support for the chromosome theory.

Sex Linkage and the Chromosome Theory Morgan realized that the transmission pattern of the X chromosome in males and females explained the results of his reciprocal crosses. He reasoned that half the gametes produced by males would contain an X chromosome and half a Y chromosome (**FIGURE 14.10**). Morgan proposed that the gene for eye color in fruit flies is located on the X chromosome and that the Y chromosome does not carry this gene.

This situation is described as **X-linked inheritance,** or simply **X-linkage.** Correspondingly, a gene residing on the Y chromosome is said to have **Y-linked inheritance,** or **Y-linkage.** The general term for inheritance of genes on either sex chromosome is **sex-linked inheritance,** or **sex-linkage.**

According to the hypothesis of X-linkage, a female fruit fly has two copies of the gene that specifies eye color because she has two X chromosomes. One of these chromosomes came from her female parent, the other from her male parent. A male, in contrast, has only one copy of the eye-color gene because he has only one X chromosome, inherited from his mother.

The Punnett squares in **FIGURE 14.11** show that Morgan's experimental results can be explained if the gene for eye color is located

(a) One half of reciprocal cross

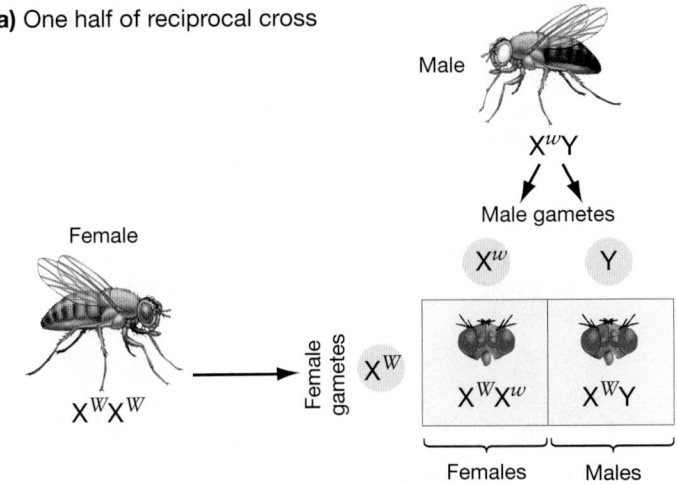

(b) Other half of reciprocal cross

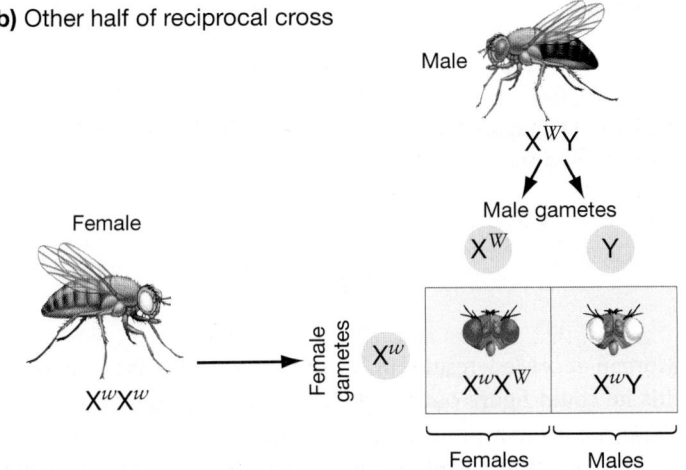

FIGURE 14.11 Reciprocal Crosses Confirm that Eye Color in *Drosophila* Is an X-Linked Trait. When Morgan crossed red-eyed females with white-eyed males **(a)** and then crossed white-eyed females with red-eyed males **(b)**, he observed strikingly different results.

on the X chromosome, and if the allele for red color is dominant to the allele for white color. In this figure, the allele for red eyes is denoted X^W while the allele for white eyes is denoted X^w. The Y chromosome present in males is simply designated by Y. Using this notation,[1] the genotypes used in the experiment are written as $X^W X^W$ for red-eyed females; $X^w Y$ for white-eyed males; $X^w X^w$ for white-eyed females; and $X^W Y$ for red-eyed males.

Notice how the symbol for *Drosophila* alleles represents the mutant phenotype. Instead of showing alleles for the eye color gene with an *R* for wild-type red eyes and an *r* for mutant white eyes, the alleles are shown with a *W* (red eyes) or a *w* (white eyes).

---

[1]Scientific papers on fruit fly genetics use a different notation. The wild-type allele is designated with a superscript +, and no X is used for X-linked traits. The red-eye allele, for example, is denoted w^+; the white-eye allele w. The notation used here is simplified and conforms to conventions used in human genetics.

By applying the principles of segregation and random fertilization, you should see that the results predicted by the hypothesis of X-linkage match the observed results.

When reciprocal crosses give different results, such as those illustrated in Figure 14.11, it is likely that the gene in question is located on a sex chromosome—it is sex-linked. Recall that non-sex chromosomes are called autosomes (see Chapter 13). Genes on non-sex chromosomes are said to show **autosomal inheritance.**

Morgan's discovery of X-linked inheritance carried an even more fundamental message. In *Drosophila*, the gene for white eye color is clearly correlated with inheritance of the X chromosome. This correlation was important evidence in support of the hypothesis that chromosomes contain genes. The discovery of X-linked inheritance convinced most biologists that the chromosome theory of inheritance was correct.

check your understanding

C Y U

If you understand that . . .

- Meiosis is the process responsible for Mendel's principle of segregation. It occurs because alleles on homologous chromosomes separate at anaphase of meiosis I.
- Meiosis is the process responsible for Mendel's principle of independent assortment. Alleles of different genes go to gametes independently because pairs of homologous chromosomes line up randomly at metaphase of meiosis I.

✔ **You should be able to . . .**

1. Draw the chromosomes involved in a cross between *Pp* and *Pp* peas, and use your diagram to explain the segregation of alleles.
2. Draw the chromosomes involved in a cross between *YyRr* and *YyRr* peas, and use your diagram to explain the independent assortment of alleles.

Answers are available in Appendix A.

14.5 Extending Mendel's Rules

Biologists point out that Mendel analyzed the simplest possible genetic system. The traits that he was studying were not sex-linked. Moreover, they were influenced by just two alleles of each gene, and each allele was completely dominant or recessive.

With this well-chosen model system, Mendel was able to discover the most fundamental rules of inheritance. Mendel probably would have failed, as so many others had done before him, had he been trying to analyze more complex patterns of inheritance.

Once Mendel's work was rediscovered, researchers began to analyze traits and alleles whose inheritance was more complicated.

If experimental crosses produced F_2 progeny that did not conform to the expected 3:1 or 9:3:3:1 ratios, researchers had a strong hint that something interesting was going on. The discovery of sex-linkage is a prominent example. How can other traits that don't appear to follow Mendel's rules contribute to a more complete understanding of heredity?

Linkage: What Happens When Genes Are Located on the Same Chromosome?

Once the chromosome theory had been tested and supported, biologists began to reevaluate Mendel's principle of independent assortment. It seemed unlikely that genes on the same chromosome would assort independently.

Linkage is the tendency of particular alleles of different genes to be inherited together. Linkage is seen when genes are on the same chromosome. Notice that the terms linkage and sex-linkage have different meanings. If genes are linked, it means that they are located on the same chromosome. If a gene is sex-linked, it means that it is located on a sex chromosome but says nothing about its location relative to other genes.

The first examples of linked genes were those on the X chromosome of fruit flies. After Morgan established that the *white-eye* gene was located on *Drosophila*'s X chromosome, he and colleagues established that one of the several genes that affects body color is also located on the X. Red eyes and gray body are the wild-type phenotypes; white eyes and a yellow body occur as rare mutant phenotypes. The alleles for red eyes (X^W) and gray body (X^Y) are dominant to the alleles for white eyes (X^w) and yellow body (X^y). (Be sure not to confuse the notation for the Y chromosome in males, Y, with the gray body allele, X^Y.)

Linked Genes Do Not Assort Independently Because linked genes are located on the same chromosome, it is logical to predict that they should always be transmitted together during gamete formation. Stated another way, linked genes should violate the principle of independent assortment.

Recall from Section 14.4 that genes on different chromosomes show independent assortment because different chromosomes assort independently during meiosis I. How will the X-linked body-color and eye-color genes be inherited? **FIGURE 14.12** shows a cell of a female fruit fly with one X chromosome carrying the white eye and gray body alleles, written X^{wY}, and the homologous X chromosome carrying the red eye and yellow body alleles, written X^{Wy}. This female would be expected to generate just two classes of gametes in equal numbers during meiosis, instead of the four classes that are predicted under the principle of independent assortment. Is this what actually occurs?

The Role of Crossing Over To determine whether linked traits always stay linked, Morgan performed crosses like the one described in the "Experimental Setup" section of **FIGURE 14.13**. In this case, $X^{wY}X^{Wy}$ females mated with $X^{wY}Y$ males.

The "Results" table in Figure 14.13 summarizes the phenotypes and genotypes observed in this experimental cross.

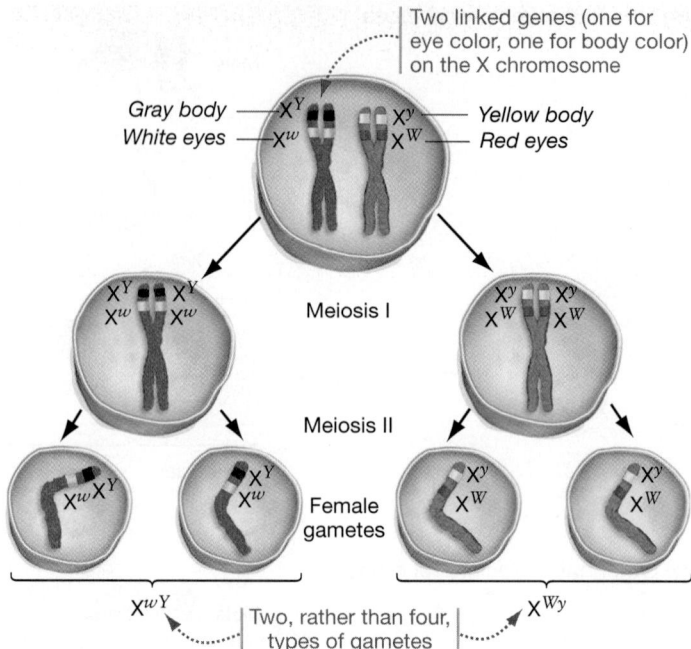

FIGURE 14.12 Linked Genes Are Often Inherited Together. If the eye-color and body-color genes were found on different chromosomes, then this female would generate four different types of gametes instead of just two types as shown here.

✔**EXERCISE** List the four genotypes that would be generated if the white-eye and yellow-body genes were not linked.

Morgan recorded results only from male offspring. By doing this he could figure out which X-linked alleles were present on the chromosomes produced during meiosis in the mother. Since there is a single X chromosome in the male offspring, the phenotype associated with any X-linked allele, dominant or recessive, is expressed:

- Most of these males carried an X chromosome with one of the two combinations of alleles found in the chromosomes of their mothers: X^{wY} or X^{Wy}.

- A small percentage of males had novel combinations of phenotypes and genotypes: X^{wy} and X^{WY}. Morgan referred to these individuals as **recombinant** because the combination of alleles on their X chromosome was different from the combinations of alleles present in their mother.

Morgan concluded that alleles on the same chromosome don't always stay together. To explain the recombinant phenotypes, Morgan proposed that gametes with new combinations of alleles were generated when crossing over occurred during prophase of meiosis I in the females.

Recall that crossing over involves a physical exchange of segments of non-sister chromatids between homologous chromosomes (see Chapter 13). Crossing over typically occurs at least once in every synapsed pair of homologous chromosomes, and usually multiple times. (Male fruit flies are an exception to this rule. For unknown reasons, no crossing over occurs in male fruit flies.)

QUESTION: Will genes undergo independent assortment if they are on the same chromosome?

LINKAGE HYPOTHESIS: Linked genes will violate the principle of independent assortment.

NULL HYPOTHESIS: Linked genes will adhere to the principle of independent assortment.

EXPERIMENTAL SETUP:

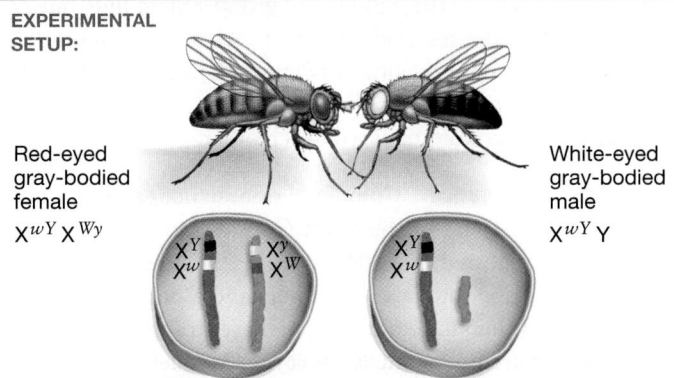

Red-eyed gray-bodied female $X^{wY}X^{Wy}$

White-eyed gray-bodied male $X^{wY}Y$

PREDICTION: Because these two genes are X-linked, male offspring will have only one copy of each gene, from their mother; the two possible male offspring genotypes are $X^{wY}Y$ and $X^{Wy}Y$

PREDICTION OF NULL HYPOTHESIS: Four male genotypes are possible ($X^{wY}Y : X^{Wy}Y : X^{wy}Y : X^{WY}Y$) and will occur with equal frequency.

RESULTS:

Male offspring

Phenotype	Genotype	Number	
	$X^{wY}Y$	4292	Four male genotypes were observed (rather than two), but not the equal frequencies predicted by independent assortment
	$X^{Wy}Y$	4605	
Recombinant genotypes	$X^{wy}Y$	86	
	$X^{WY}Y$	44	

CONCLUSION: Neither hypothesis is fully supported. Independent assortment does not apply to linked genes—linked genes segregate together except when crossing over and genetic recombination have occurred.

FIGURE 14.13 Linked Genes Are Inherited Together Unless Recombination Occurs.

SOURCE: Morgan, T. H. 1911. An attempt to analyze the constitution of the chromosomes on the basis of sex-limited inheritance in *Drosophila. Journal of Experimental Zoology* 11: 365–414.

✔**QUESTION** Why didn't Morgan observe equal numbers of white-eyed, yellow-bodied males and red-eyed, gray-bodied males?

As **FIGURE 14.14** shows, a crossover between the eye-color and body-color genes in the $X^{wY}X^{Wy}$ females can explain the recombinant gametes. Male progeny produced from fertilization with these gametes are predicted to have either yellow bodies and white eyes or gray bodies and red eyes. This is what Morgan observed.

Notice, however, that the results of Figure 14.14 don't fit either the model of independent assortment or complete linkage. Independent assortment predicts a 1:1:1:1 ratio of all four

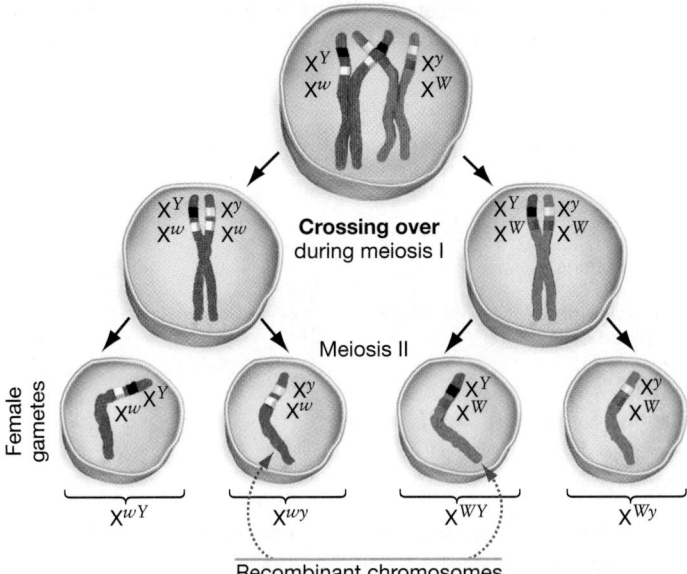

Crossing over during meiosis I

Meiosis II

Female gametes

X^{wY} X^{wy} X^{WY} X^{Wy}

Recombinant chromosomes

FIGURE 14.14 Genetic Recombination Results from Crossing Over. To explain the results in Figure 14.13, Morgan proposed that crossing over occurred between the body color (*y*) and eye color (*w*) genes in a small percentage of meiotic divisions in the female parent. The recombinant chromosomes that resulted would produce the recombinant phenotypes observed in the male offspring.

combinations of phenotypes, while complete linkage would give only the two phenotypes associated with the nonrecombinant or parental chromosomes. Instead, most flies have parental phenotypes and a smaller number have recombinant phenotypes.

As **Quantitative Methods 14.1** (see page 274) explains, the percentage of recombinant offspring that occur in crosses like the one diagrammed in Figure 14.14 can be used to estimate the relative distance between genes. The reasoning is that the farther genes are apart on the same chromosome, the more likely it is that a crossover will occur someplace between these genes. Data on the frequency of crossing over between many genes on the same chromosome can be used to create a **genetic map**—a diagram showing the relative positions of genes along a particular chromosome.

Knowing a gene's locus relative to others can be very useful. An important example is genes involved in human genetic diseases. Most of these genes have been identified on the basis of mapping their location relative to other known genes.

The take-home message of Morgan's experiments is simple: Linked genes are inherited together unless crossing over occurs between them. When crossing over takes place, genetic recombination occurs. Linkage is an important exception to Mendel's rules.

How Many Alleles Can a Gene Have?

Mendel worked with genes that each had two alleles. In most populations, however, it's not unusual to find dozens of alleles of a single gene. The existence of more than two alleles of the same gene is known as **multiple allelism.** When you consider that

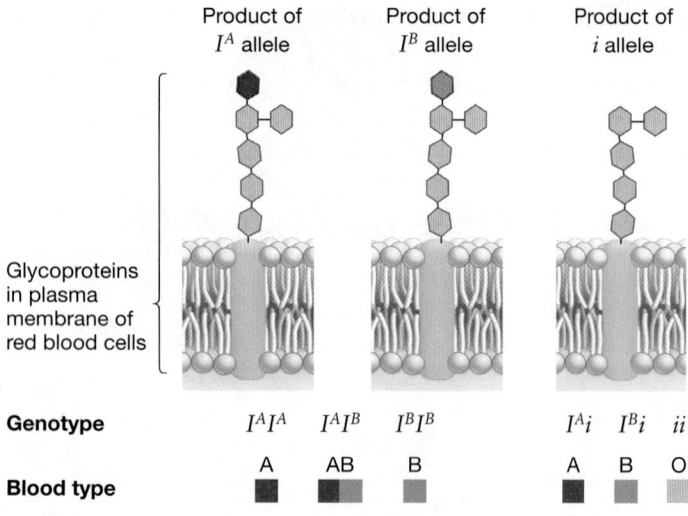

Product of I^A allele Product of I^B allele Product of i allele

Glycoproteins in plasma membrane of red blood cells

Genotype	I^AI^A	I^AI^B	I^BI^B	I^Ai	I^Bi	ii
Blood type	A	AB	B	A	B	O

FIGURE 14.15 Phenotypes Produced by Alleles Responsible for ABO Blood Types. Alleles I^A and I^B produce a phenotype called AB when paired with each other in heterozygotes. Both I^A and I^B produce a dominant phenotype when paired with allele i. The different colored hexagons on the products of the I^A and I^B alleles represent related but distinct sugars.

genes are made of DNA sequences that can change over time, the idea of multiple alleles isn't surprising.

The ABO blood group types in humans are coded for by a gene with three common alleles. The gene is known as I, and it has I^A, I^B, and i alleles. As the number of alleles for a gene increases, the number of possible genotypes rises sharply. When more than two distinct phenotypes are present in a population owing to multiple allelism, the trait is **polymorphic** ("many-formed"). ABO blood type is a polymorphic trait.

Each I gene allele controls the production of a different polysaccharide attached to a glycoprotein (see Chapter 5) found in the plasma membrane of red blood cells. The I^A and I^B alleles code for different forms of an enzyme that adds a different sugar to the end of a core polysaccharide. The i allele codes for a nonfunctional form of this enzyme, so no sugar is added to the core polysaccharide. ABO blood group types are important in blood transfusions. Some mismatches of blood type between a donor and recipient are tolerated, but others can cause fatal reactions.

The type of polysaccharide associated with each allele is shown in **FIGURE 14.15** along with all possible genotypes and phenotypes. As you can see, there are six possible genotypes for the three alleles of the I gene. The I^A allele codes for the type A polysaccharide, and the I^B allele codes for the type B polysaccharide. What happens in a person with both alleles?

Are Alleles Always Dominant or Recessive?

The terms dominant and recessive describe which phenotype is observed when two different alleles of a gene occur in a heterozygous individual. In all traits that Mendel studied, only the phenotype associated with one allele—the "dominant" one—appeared in heterozygotes. However, not all combinations of alleles work this way.

Codominance Many alleles show a relationship that is called **codominance.** The type AB blood group shown in Figure 14.15 is an example of codominance. In this case, an AB heterozygote expresses both the A and the B polysaccharides together on the surface of red blood cells. This is the essence of codominance—the simultaneous expression of the phenotype associated with each allele in a heterozygote. An AB individual expresses both the A and the B phenotypes.

The three alleles of the ABO blood group system illustrate another interesting point—alleles of one gene can show more than one form of dominance. Notice in Figure 14.15 that the I^A and I^B alleles are both completely dominant to the i allele while the I^A and I^B alleles are codominant with each other.

Incomplete Dominance Complete dominance and codominance are not the end of the story. Consider the flowers called four-o'clocks, pictured in **FIGURE 14.16a**. Plant breeders have developed a pure line that has red flowers and a pure line that has white flowers. When individuals from these strains are mated, all their offspring are pink (**FIGURE 14.16b**). In Mendel's peas, crosses between dominant and recessive parents produced only offspring with the dominant phenotype. Why the difference?

Biologists answered this question by allowing the pink flowered F_1 plants to self-fertilize and examining the phenotypes of F_2 four-o'clocks. Of the F_2 plants, ¼ have red flowers, ½ have pink flowers, and ¼ have white flowers. This 1:2:1 ratio of phenotypes exactly matches the 1:2:1 ratio of genotypes that is produced when flower color is controlled by one gene with two alleles.

To convince yourself that this explanation is sound, study the genetic model shown in Figure 14.16b. According to the diagram, the pattern of inheritance of flower color genotypes in four-o'clocks and peas is identical, but the four-o'clock alleles show a different form of dominance known as **incomplete dominance.**

When incomplete dominance occurs, heterozygotes have a phenotype that is between the two different homozygous parents. In the case of four-o'clocks, neither red nor white alleles dominate. Instead, the heterozygous F_1 progeny show a phenotype in between the two parental strains.

By this point, it should be clear that the answer to the question of whether alleles are always dominant or recessive is a resounding no. Instead, there are three possible dominance relationships between different alleles: complete, incomplete, and codominance.

Does Each Gene Affect Just One Trait?

Mendel's results led him to hypothesize that one gene influences one trait. The gene for seed color in garden peas, for example, did not appear to affect other aspects of the individual's phenotype. In reality, however, many genes influence more than one trait.

A gene that influences many traits is said to be **pleiotropic** ("more-turning"). For example, mutations in one gene, *FBN1*, cause Marfan syndrome in humans. In this case the mutant

(a) Flower color is variable in four-o'clocks.

(b) Incomplete dominance in flower color

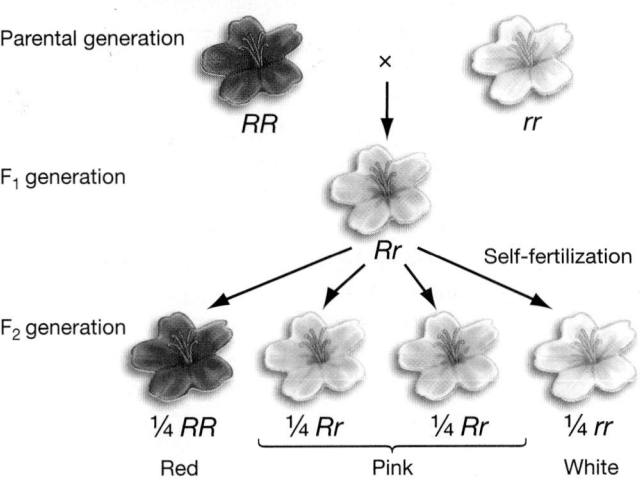

Parental generation

RR   ×   rr

F_1 generation

Rr   Self-fertilization

F_2 generation

¼ RR   ¼ Rr   ¼ Rr   ¼ rr

Red   Pink   White

FIGURE 14.16 When Incomplete Dominance Occurs, Heterozygotes Have Intermediate Phenotypes. The cross in part (b) is explained by hypothesizing that a single gene influences flower color and that alleles *R* and *r* exhibit incomplete dominance.

FBN1 allele is dominant to the wild-type allele. What is the phenotype of an individual with a mutant *FBN1* allele?

People with a mutant *FBN1* allele are tall, have disproportionately long limbs and fingers and an abnormally shaped chest, and often have severe heart problems. Therefore, the gene associated with Marfan syndrome influences many traits and is pleiotropic. On the basis of this set of phenotypes, many medical scientists believe Abraham Lincoln had Marfan syndrome.

Is There More to Phenotype than Genotype?

After analyzing the results of Mendel's experiments, it would be tempting to conclude that *R* alleles dictate that seeds are round and *T* alleles dictate that individual plants are tall—that there is a strict correspondence between alleles and phenotypes.

It's important to recognize, though, that when Mendel analyzed height in his experiments, he ensured that each plant received a similar amount of sunlight and grew in similar soil. This was critical because even individuals with alleles for tallness will be stunted if they are deprived of nutrients, sunlight, or water—so much so that they look similar to individuals with alleles for dwarfing. Mendel also worked with pure lines that had been inbred for many generations. This breeding method reduces the genetic variation in each line.

For Mendel to analyze the hereditary determinants of height, he had to control the environmental determinants of height. Let's consider how the environment and alleles at other gene loci affect phenotype.

The Environment Affects Phenotypes The phenotypes produced by most genes are strongly affected by the individual's environment. Consequently, an individual's phenotype is often as much a product of the environment as it is a product of the genotype. Environmental influences include temperature, sunlight, nutrient availability, competition, and even a mother's hormone levels during development of an embryo. To capture this point, biologists refer to the combined effect of genes and environment as gene-by-environment interaction.

Gene-by-environment interactions have a profound effect on how physicians treat people with the genetic disease phenylketonuria (PKU). These individuals are homozygous for a recessive allele of an enzyme-coding gene. The enzyme helps convert the amino acid phenylalanine to the amino acid tyrosine. In PKU, this enzyme is absent and, as a result, phenylalanine and a related molecule, phenylpyruvic acid, accumulate. These compounds interfere with the development of the nervous system and produce severe mental retardation.

But are people without the ability to metabolize phenylalanine genetically fated to mental retardation?

In many countries, newborns are routinely tested for the defect. If identified at birth, individuals with PKU can be placed on a low-phenylalanine diet. The change in environment—reduced phenylalanine in the diet—has a dramatic influence on phenotype. Treated individuals develop normally. PKU is a genetic disease, but by controlling the environment, it is neither inevitable nor invariant.

Interactions Between Genes Affect Phenotypes In Mendel's pea plants, there was a one-to-one correspondence between genes and traits. The pea seeds he analyzed were round or wrinkled regardless of the types of alleles present at other loci. Only one gene influenced each trait.

In many cases, however, different genes work together to control a single trait. Consider a classic experiment published in 1905 on comb shape in chickens. William Bateson and R. C. Punnett crossed parents from pure lines with comb shapes called rose and pea and found that the F_1 offspring had a new phenotype, called walnut combs, a new phenotype not seen in either parent. When these individuals bred, their offspring had walnut, rose,

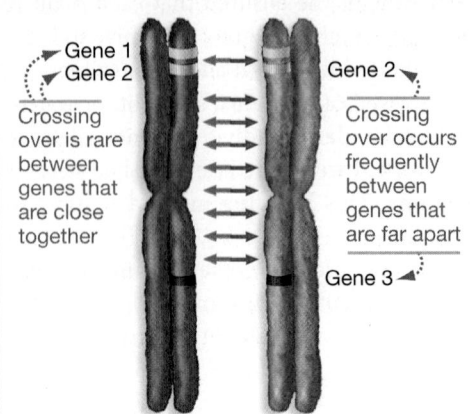

Gene 1
Gene 2

Crossing over is rare between genes that are close together

Gene 2

Crossing over occurs frequently between genes that are far apart

Gene 3

FIGURE 14.17 The Physical Distance between Genes Determines the Frequency of Crossing Over. The arrows show that crossing over is possible at any point between the genes, but the chance of a crossover between a pair of genes increases when the distance between the genes is large.

In experiments like the one diagrammed in Figure 14.13, researchers calculate the recombination frequency as the number of offspring with recombinant phenotypes divided by the total number of offspring. With crosses involving the X-linked traits of white eyes and yellow bodies, 1.4 percent of offspring have recombinant phenotypes and genotypes.

But in crosses with different pairs of X-linked traits, the fraction of recombinant offspring varies, and this variation provides information on the relative distance separating linked genes. For example, in crosses of fruit flies with X-linked genes for white eyes and another mutant phenotype called singed bristles, recombinant phenotypes are seen 19.6 percent of the time.

To explain these observations, Alfred Sturtevant, an undergraduate student working with Morgan, proposed that the physical distance between genes determines how frequently crossing over occurs between them. His idea was that crossing over occurs at random and can take place at any location along a chromosome. The shorter the distance between a pair of genes, the lower the probability that crossing over will take place between them (**FIGURE 14.17**). Using this reasoning, he set out to create a genetic map.

To define the unit of distance on his genetic map, Sturtevant used the percentage of offspring that have recombinant phenotypes with respect to two genes. One map unit (later called 1 centiMorgan [cM]) is the physical distance that produces 1 percent recombinant offspring.

The eye-color and bristle-shape genes of fruit flies are 19.6 cM apart on the X chromosome, because recombination between these genes results in 19.6 percent recombinant offspring, on average. The genes for yellow body and white eye color, in contrast, are just 1.4 cM apart.

Where is the *yellow-body* gene relative to the *singed-bristles* gene? Twenty-one percent of the offspring were recombinant for these traits, meaning that the *yellow-body* and *singed-bristles* genes are 21.0 cM apart. Sturtevant inferred that the gene for white eyes must be located *between* the genes for yellow body and singed bristles, as shown in **FIGURE 14.18a**.

Mapping genes relative to one another is like fitting pieces into a puzzle: Placing *white* between *yellow* and *singed bristles* is the only way to make the distances between each pair sum correctly. The key observation is that 21.0 cM—the distance between *yellow* and *singed bristles*—is equal to 1.4 cM + 19.6 cM, or the sum of the distances between *yellow* and *white* and *white* and *singed bristles*.

FIGURE 14.18b provides a partial genetic map of the X chromosome in *Drosophila melanogaster,* along with the data used to establish the map positions. Using this logic and similar data, Sturtevant assembled the first genetic map.

(a) Mapping genetic distance

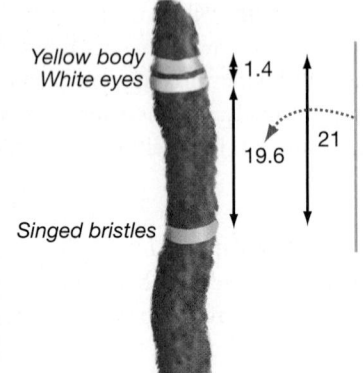

Yellow body
White eyes

1.4

19.6

21

Singed bristles

Frequency of recombinant offspring correlates directly with the distance between two genes; 19.6% recombinant offspring, for example, translates to 19.6 map units (centiMorgans, cM).

(b) Constructing a genetic map

% Frequency of crossing over between some genes on the X chromosome of fruit flies		
	Miniature Wings	Ruby Eyes
Yellow body	36.1	7.5
White eyes	34.7	6.1
Singed bristles	15.1	13.5
Miniature wings	——	28.6

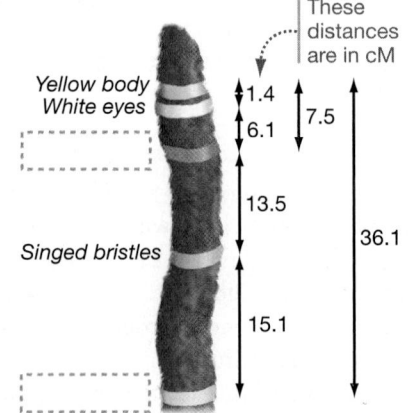

These distances are in cM

Yellow body
White eyes

1.4
6.1
7.5

13.5

Singed bristles

36.1

15.1

FIGURE 14.18 The Locations of Genes Can Be Mapped by Analyzing the Frequency of Recombination. (a) The *yellow body* gene is on the end of the fruit fly X chromosome. To explain the recombination frequencies observed in experimental crosses, the *yellow body, white eyes,* and *singed bristles* genes must be in the locations shown here. **(b)** A partial genetic map of the X chromosome in fruit flies.

✔**EXERCISE** In part (b), label the orange and blue genes. (Which is *ruby* and which is *miniature wings?*)

pea, and a fourth phenotype called single combs in a 9:3:3:1 ratio (**FIGURE 14.19a**).

The genetic model in **FIGURE 14.19b** shows how the interaction between two different genes that control one trait (comb shape) can account for the results. If comb morphology results from interactions between two genes (symbolized *R* and *P*), if a dominant and a recessive allele exist for each gene, and if the four comb phenotypes are associated with the genotypes indicated at the bottom of the figure, then a cross between *RRpp* and *rrPP* parents would give the results that Bateson and Punnett observed.

When gene-by-gene interactions occur, one trait is influenced by the alleles of two or more different genes. If a chicken has an *R* allele, its phenotype depends on the allele present at the *P* gene.

Gene-by-gene interaction is very common and has important implications in human genetics. Imagine that two people have the same genotype at one locus that increases risk for a heart disease. If there is gene-by-gene interaction, then the risk of developing heart disease also depends on the genotype at other loci. Even if they experience identical environments, these two people may have very different overall risks from genetic factors alone.

Can Mendel's Principles Explain Traits That Don't Fall into Distinct Categories?

Mendel worked with **discrete traits**—traits that are clearly different from each other. In garden peas, seed color is either yellow or green—no intermediate phenotypes exist. But many traits in peas and other organisms don't fall into discrete categories. In humans, for example, height, weight, and skin color vary continuously. People are not limited to being either 160 cm tall or 180 cm tall—countless other heights are possible.

For height and many other characteristics, individuals differ by degree. These types of continuously varying traits that don't fall into discrete categories are called **quantitative traits**. Like discrete traits, quantitative traits are greatly influenced by the environment. The effects of nutrition on human height, intelligence, and disease resistance, for example, have been well documented.

Many quantitative traits share a common characteristic: When the frequencies of different trait values observed in a population are plotted on a histogram, or frequency distribution (see **BioSkills 3** in Appendix B), they often form a bell-shaped curve, or normal distribution (**FIGURE 14.20**).

In 1909 Herman Nilsson-Ehle had an important insight: If many genes each contribute a small amount to the value of a quantitative trait, then a normal distribution results for the population as a whole.

(a) Crosses between chickens with different comb phenotypes give odd results.

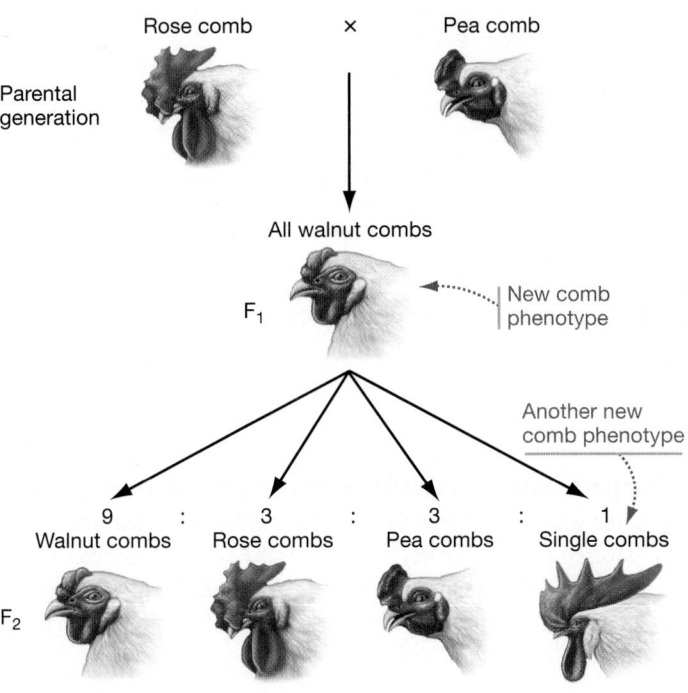

(b) A genetic model based on gene-by-gene interactions can explain the results.

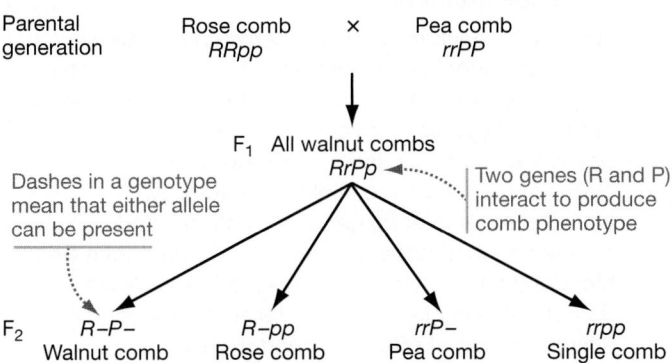

FIGURE 14.19 Genes at Different Loci Can Interact to Influence a Trait. (a) This cross is notable because new phenotypes show up in the F$_1$ and the F$_2$ generation. **(b)** To explain the results, researchers hypothesized that comb shape depends on two genes that interact. The phenotype associated with any one allele of one gene depends on the alleles at a second gene.

✔**EXERCISE** What is different about the 9:3:3:1 ratio in the F$_2$ of this cross compared with the ratio observed in a standard dihybrid cross?

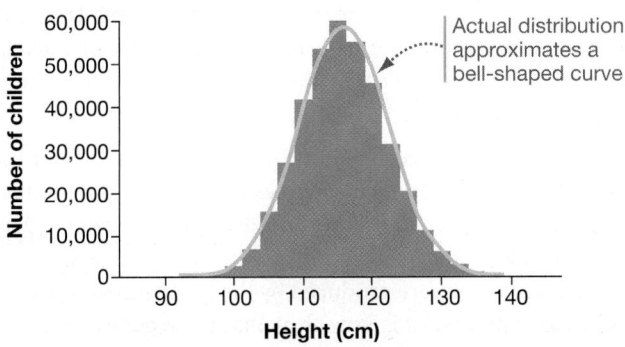

FIGURE 14.20 Quantitative Traits Have a Normal Distribution.
A histogram showing the heights of first-grade schoolchildren in Guatemala in 2001.

DATA: Pan American Health Organization/WHO. 2004. *Epidemiological Bulletin* 25: 9-13, Graph 1.

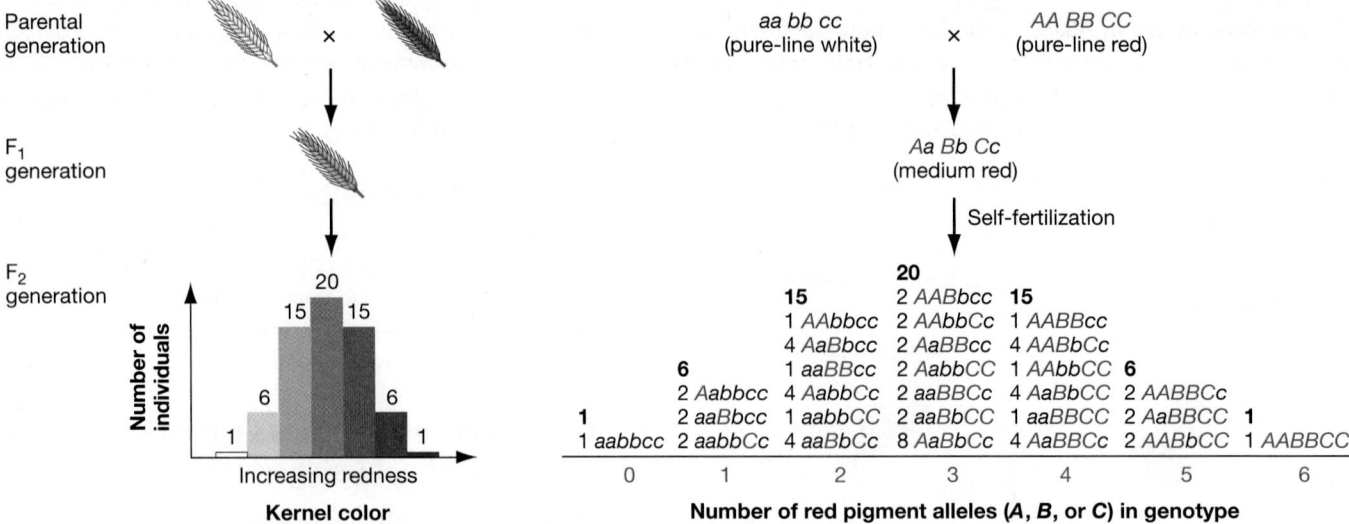

(a) Wheat kernel color is a quantitative trait.

Parental generation ✕

F₁ generation

F₂ generation

Number of individuals

20
15   15
6   6
1   1

Increasing redness

Kernel color

(b) Hypothesis to explain inheritance of kernel color

aa bb cc
(pure-line white)   ✕   *AA BB CC*
(pure-line red)

Aa Bb Cc
(medium red)

Self-fertilization

20
15   2 *AABbcc* **15**
1 *AAbbcc* 2 *AAbbCc* 1 *AABBcc*
4 *AaBbcc* 2 *AaBBcc* 4 *AABbCc*
6   1 *aaBBcc*   2 *AabbCC* 1 *AAbbCC* **6**
2 *Aabbcc* 4 *AabbCc* 2 *aaBBCc* 4 *AaBbCC* 2 *AABBCc*
1   2 *aaBbcc* 1 *aabbCC* 2 *aaBBCC* 1 *aaBBCC* 2 *AaBBCC* **1**
1 *aabbcc* 2 *aabbCc* 4 *aaBbCc* 8 *AaBbCc* 4 *AaBBCc* 2 *AABbCC* 1 *AABBCC*

0   1   2   3   4   5   6

Number of red pigment alleles (*A*, *B*, or *C*) in genotype

FIGURE 14.21 Quantitative Traits Result from the Action of Many Genes. (a) When wheat plants with white kernels were crossed with wheat plants with red kernels, the F₂ offspring showed a range of kernel colors. The frequency of these phenotypes approximates a normal distribution. **(b)** This model can explain the results of part (a). The bold numbers above each genotype match the numbers shown in part (a) and indicate the relative number of plants with each phenotype.

✔**QUESTION** Why are there fewer very dark or very light wheat kernels compared with kernels of intermediate coloration?

Nilsson-Ehle established this finding using strains of wheat that differed in kernel color. **FIGURE 14.21a** includes a histogram showing the distribution of F₂ phenotypes from a cross he performed between pure lines of white wheat and dark-red wheat. Notice that the frequency of colors in F₂ progeny approximates a bell-shaped curve. To explain these results, Nilsson-Ehle proposed the hypotheses illustrated in **FIGURE 14.21b**:

- The parental strains differ with respect to three genes that control kernel color: *AABBCC* produces dark-red kernels, and *aabbcc* produces white kernels.

- The three genes assort independently.

- The *a*, *b*, and *c* alleles do not contribute to pigment production, but the *A*, *B*, and *C* alleles contribute to pigment production in an equal and additive way. This is a form of incomplete dominance. As a result, the degree of red pigmentation is determined by the number of *A*, *B*, or *C* alleles present. Each uppercase (dominant) allele that is present makes a wheat kernel slightly darker red.

Later work supported Nilsson-Ehle's model. He did not have to propose any new genetic principles to explain the inheritance of a quantitative trait. All that was needed was extension of Mendel's hypotheses about segregation and independent assortment.

Quantitative traits are produced by the independent actions of many genes, although it is now clear that some genes have much greater effects on the trait in question than other genes do. As a result, the transmission of quantitative traits is said to result from polygenic ("many-genes") inheritance. In **polygenic inheritance,** each gene adds a small amount to the value of the phenotype.

In the decades immediately after the rediscovery of Mendel's work, the question of why offspring resemble their parents could be answered in more satisfying ways. **TABLE 14.4** summarizes some of the key exceptions and extensions to Mendel's rules and gives you a chance to compare and contrast their effects on patterns of inheritance.

check your understanding

If you understand that . . .

- Genes near each other on the same chromosome violate the principle of independent assortment. They are not transmitted to gametes independently of each other unless crossing over occurs between them.
- Sex linkage, linkage, incomplete dominance, codominance, multiple allelism, pleiotropy, environmental effects, gene interactions, and polygenic inheritance are aspects of inheritance that Mendel did not study. When they occur, crosses do not result in classical Mendelian monohybrid or dihybrid ratios of offspring phenotypes.

✔ **You should be able to . . .**

Explain why the following crosses don't produce a 3:1 phenotype ratio in F₂ offspring:

1. Rose-comb ✕ pea-comb chickens
2. Red-kernel ✕ white-kernel wheat plants

Answers are available in Appendix A.

Type of Inheritance	Definition	Consequences or Comments
Sex linkage	Genes located on sex chromosomes.	Patterns of inheritance in males and females differ.
Linkage	Two genes found on same chromosome.	Linked genes violate principle of independent assortment.
Incomplete dominance	Heterozygotes have intermediate phenotype.	Polymorphism—heterozygotes have unique phenotype.
Codominance	Heterozygotes have phenotype of both alleles.	Polymorphism is possible—heterozygotes have unique phenotype.
Multiple allelism	In a population, more than two alleles present at a locus.	Polymorphism is possible.
Polymorphism	In a population, more than two phenotypes associated with a single gene are present.	Can result from actions of multiple alleles, incomplete dominance, and codominance.
Pleiotropy	A single gene affects many traits.	This is common.
Gene-by-gene interaction	In discrete traits, the phenotype associated with an allele depends on which alleles are present at another gene.	One allele can be associated with different phenotypes.
Gene-by-environment interaction	Phenotype influenced by environment experienced by individual.	Same genotypes can be associated with different phenotypes.
Polygenic inheritance of quantitative traits	Many genes are involved in specifying traits that exhibit continuous variation.	Unlike alleles that determine discrete traits, each allele adds a small amount to phenotype.

14.6 Applying Mendel's Rules to Human Inheritance

When researchers set out to study how a particular gene is transmitted in wheat or fruit flies or garden peas, they begin by making a series of controlled experimental crosses. For obvious reasons, this strategy is not possible with humans. But suppose you are concerned about an illness that runs in your family and go to a genetic counselor to find out how likely your children are to have the disease. To advise you, the counselor needs to know how the trait is transmitted, including whether the gene involved is autosomal or sex-linked and what type of dominance is associated with the disease allele.

To understand the transmission of human traits, investigators have to analyze human genotypes and phenotypes that already exist. A **mode of transmission** describes a trait as autosomal or sex-linked and gives the type of dominance of the allele. To learn the mode of transmission, scientists construct a **pedigree,** or family tree, of affected and unaffected individuals. By analyzing pedigrees, biomedical researchers have been able to discover how more than 2000 human genetic diseases are inherited.

A pedigree records the genetic relationships between the individuals in a family along with each person's sex and phenotype with respect to the trait in question. If the trait is governed by a single gene, then analyzing the pedigree may reveal whether a given phenotype is due to a dominant or recessive allele and whether the gene responsible is located on a sex chromosome or

on an autosome. Let's look at a series of specific case histories to see how this work is done.

Identifying Human Alleles as Recessive or Dominant

To analyze the inheritance of a discrete trait, biologists begin by assuming that a single autosomal gene is responsible and that the alleles present in the population have a simple dominant–recessive relationship.

This is the simplest possible situation. If the pattern of inheritance fits this model, then the assumptions—of inheritance by a single autosomal gene and simple dominance—are supported. Let's first analyze the pattern of inheritance that is typical of autosomal recessive traits and then examine patterns that emerge in pedigrees for autosomal dominant traits.

Patterns of Inheritance: Autosomal Recessive Traits If a phenotype is due to an autosomal recessive allele, then

- Individuals with the trait must be homozygous.

- If the parents of an affected individual do not have the trait, then the parents are heterozygous for the trait.

Heterozygous individuals who carry a recessive allele for an inherited disease are referred to as **carriers** of the disease. These individuals carry the allele and may transmit it even though they do not exhibit signs of the disease. When two carriers mate,

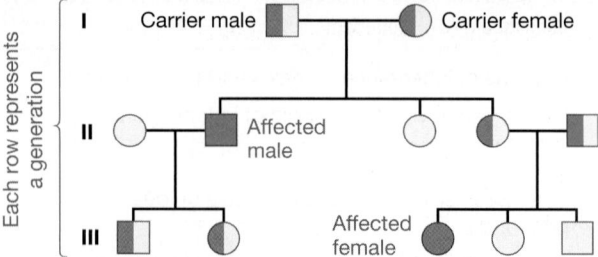

(a) Autosomal recessive trait (e.g., sickle-cell disease)

Each row represents a generation

I — Carrier male — Carrier female

II — (unaffected female) — Affected male — (unaffected female) — (carrier female) — (carrier male)

III — (carrier male) — (carrier female) — Affected female — (carrier female) — (unaffected male)

CHARACTERISTICS:
- Males and females are equally likely to be affected
- Affected offspring often have unaffected parents
- Unaffected parents of affected offspring are heterozygous (carriers)
- Affected offspring are homozygous
- If both parents are heterozygous, about ¼ of the offspring will be affected
- Trait often skips generations

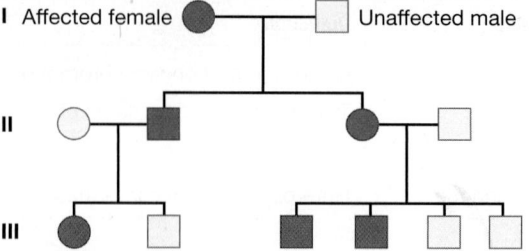

(b) Autosomal dominant trait (e.g., Huntington's disease)

I — Affected female — Unaffected male

II — (unaffected female) — Affected male — Affected female — (unaffected male)

III — Affected female — (unaffected male) — Affected male — Affected male — (unaffected male) — (unaffected male)

CHARACTERISTICS:
- Males and females are equally likely to be affected
- Affected offspring have at least one affected parent
- Affected offspring are heterozygous if only one parent is affected
- Unaffected offspring are homozygous recessive
- If one parent is heterozygous, about ½ of the offspring will be affected
- Trait does not skip generations

FIGURE 14.22 Pedigrees of Families with Autosomal Recessive and Autosomal Dominant Traits. Pedigrees use standard symbols: squares = males, circles = females; unfilled symbols = unaffected individuals (those without the trait), filled symbols = affected individuals, half-filled symbols = heterozygotes for a recessive trait (carriers); horizontal lines connect parents, vertical lines connect parents to children.

about ¼ of their offspring are expected to express the recessive phenotype.

FIGURE 14.22a is a pedigree from a family in which an autosomal recessive trait, such as sickle-cell disease, occurs. The key feature to notice in this pedigree is that both boys and girls can exhibit the trait even though their parents do not. This is the pattern you would expect when the parents of an individual with the trait are heterozygous. It is also logical to observe that when an affected (homozygous) individual has children, those children do not necessarily have the trait. This pattern is predicted if affected people marry individuals who are homozygous for the dominant allele. This is likely to occur if the recessive allele is rare in the population.

In general, a recessive phenotype should show up in offspring only when both parents have that recessive allele and pass it on to their offspring. By definition, a recessive allele produces a given phenotype only when the individual is homozygous for that allele.

Patterns of Inheritance: Autosomal Dominant Traits When a trait is autosomal dominant, individuals who are homozygous or heterozygous for it must have the dominant phenotype. Even if one parent is heterozygous and the other is homozygous recessive, on average half their children should show the dominant phenotype. And unless a new mutation has occurred in a gamete, any child with the trait must have a parent with this trait. The latter observation is in strong contrast to the pattern seen in autosomal recessive traits.

FIGURE 14.22b shows the inheritance of the degenerative brain disorder called Huntington's disease (see Section 14.2). This pedigree has two features that indicate Huntington's disease is passed to the next generation through an autosomal dominant allele.

First, if a child shows the trait, then one of its parents shows the trait as well. Second, if families have a large number of children, the trait usually shows up in every generation—owing to the high probability of heterozygous parents having affected children.

Identifying Human Traits as Autosomal or Sex-Linked

When it is not possible to arrange reciprocal crosses, can data in a pedigree indicate whether a trait is autosomal or sex-linked? The answer is based on a simple premise. If a trait appears about equally often in males and females, then it is likely to be autosomal. But if males express the trait in question more often than females, then the allele responsible is likely to be recessive and found on the X chromosome.

X-linked Recessive Traits X-linked recessive traits are relatively common. They include the form of red–green color blindness that affects about 10 percent of males and the devastating blood clotting disorder hemophilia. A pedigree for red–green color blindness is shown in **FIGURE 14.23a**.

A key characteristic of X-linked recessive traits is that males express the trait more often than females. This is because males have only one copy of the X chromosome. Therefore, any X-linked allele will be expressed in a male.

Because human females have two X chromosomes, they have two copies of each X-linked gene and a recessive allele will not be expressed in heterozygotes. In the pedigree of Figure 14.23a the fact that only males are affected gives an immediate clue that this is an X-linked recessive trait.

Further, the appearance of an X-linked recessive trait usually skips a generation in the pedigree. Notice how on the right-hand

(a) X-linked recessive trait (e.g., red–green color blindness)

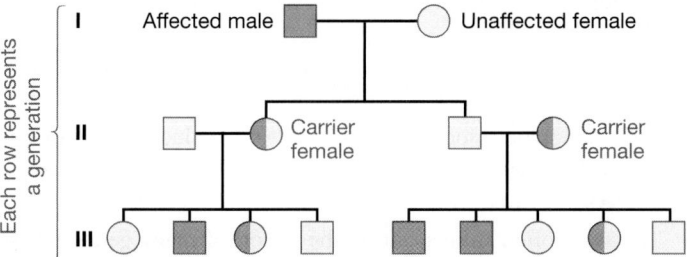

(b) X-linked dominant trait (e.g., hypophosphatemia)

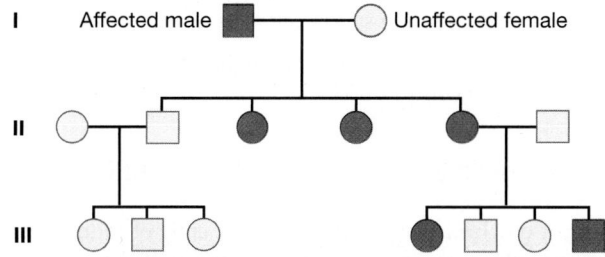

Each row represents a generation

CHARACTERISTICS:
- Males are affected more frequently than females
- Trait is never passed from father to son
- Affected sons are usually born to carrier mothers
- About ½ of the sons of a carrier mother will be affected
- All daughters of affected males and unaffected non-carrier females are carriers
- Trait often skips generations

CHARACTERISTICS:
- Males and females are equally likely to be affected
- All daughters of an affected father are affected, but no sons
- Affected sons always have affected mothers
- About ½ of the offspring of an affected mother will be affected
- Affected daughters can have an affected mother or father
- Trait does not skip generations

FIGURE 14.23 Pedigrees of Families with X-linked Recessive (a) and X-linked Dominant (b) Traits.

✔**QUESTION** What genotype in a mother and a father would be predicted to produce a 1:1 ratio of normal : color-blind offspring? What would the ratio of color-blind male : female be from this mating?

side of the pedigree the trait appears in generations I and III but not in generation II. This pattern occurs because an affected male passes his only X chromosome on to his daughters. But because his daughters almost always receive a wild-type allele from their mother, the daughters don't show the trait. They will pass the defective allele on to about half their sons, however.

Most X-linked traits are recessive. There are only a few rare examples of X-linked dominant traits.

X-linked Dominant Traits One example of an X-linked dominant trait is a bone disorder known as hypophosphatemia, or vitamin D–resistant rickets. What can be predicted about the inheritance of an X-linked dominant trait? As the pedigree in **FIGURE 14.23b** shows, the most telling feature is that an affected male will pass the trait to all his daughters and none of his sons.

This is because daughters receive their father's only X chromosome. In contrast, a heterozygous female will pass the trait to half her daughters and half her sons. This occurs because there is an equal chance that a heterozygous mother will pass on an X chromosome with either the dominant or the recessive trait.

What about Y-linked traits? Although the patterns for Y-linked inheritance can easily be predicted, the reality is that very few genes occur on the Y chromosome. These genes are involved with male-specific sexual development. Except for maleness, there are no known human Y-linked traits.

Within a few decades of the rediscovery of Mendel's work, the burning question in genetics was no longer the nature of inheritance but the nature of the gene. What are genes made of, and how are they copied so that parents pass their alleles on to their offspring? These are the questions turned to in Chapter 15.

CHAPTER 14 REVIEW

For media, go to MasteringBiology

If you understand . . .

14.1 Mendel's Experimental System

- When Mendel began his work, there were two leading hypotheses of inheritance: blending inheritance and the inheritance of acquired characteristics.
- Mendel chose pea plants as a model organism and began his work with pure lines that he crossed to produce hybrids.
- Mendel sought to discover rules of heredity that explained how phenotypes were transmitted from parent to offspring.

✔ You should be able explain why pure lines are homozygous.

14.2 Mendel's Experiments with a Single Trait

- From analysis of the trait expressed in hybrids, Mendel concluded that one trait was dominant and the other recessive.
- From the results of monohybrid crosses, Mendel concluded that inheritance is particulate—genes do not blend together.
- The traits that Mendel studied are specified by paired hereditary determinants (genes) that separate from each other during gamete formation.

- Analysis of monohybrid crosses led to the principle of segregation: Before the formation of gametes, the alleles of each gene separate so that each egg or sperm cell receives only one of them.

✔ You should be able to predict the gamete genotypes generated by a parent with the genotype *Bb*.

14.3 Mendel's Experiments with Two Traits

- Analysis of dihybrid crosses led to the principle of independent assortment: alleles of different genes are transmitted to egg cells and sperm cells independently of each other.

- Testcrosses allow an investigator to determine whether an individual of dominant phenotype is homozygous or heterozygous.

✔ You should be able to predict the gamete genotypes generated by a parent with the genotype *BbRr*.

14.4 The Chromosome Theory of Inheritance

- The chromosome theory states that chromosomes contain genes and that Mendel's rules can be explained by the segregation and independent assortment of homologous chromosomes at meiosis I.

- The chromosome theory was supported by the discovery of sex linkage. Crosses with X-linked traits supported the theory's contention that genes are found on chromosomes.

- X-linked traits give different results in reciprocal crosses.

✔ You should be able to use a Punnett square to show why reciprocal crosses involving an X-linked recessive gene give different results.

14.5 Extending Mendel's Rules

- There are important exceptions and extensions to the basic patterns of inheritance that Mendel discovered.

- All genes follow the principle of segregation, but genes close together on the same chromosome do not follow the principle of independent assortment.

- Crossing over between homologous chromosomes creates new combinations of alleles along each homolog.

- The frequency of crossing over can be used to create genetic maps that show the position and spacing of genes along chromosomes.

- Many genes have more than two alleles; this is multiple allelism. When more than two phenotypes are due to multiple allelism, the trait is polymorphic.

- Not all heterozygotes show a dominant phenotype. In addition to complete dominance, incomplete dominance and codominance are also common.

- Many genes are pleiotropic, meaning that they influence more than one trait.

- The phenotype associated with an allele is influenced by the environment that the individual experiences, and by the actions of alleles of other genes.

- Traits are often influenced by the action of many genes. These polygenic traits show quantitative instead of discrete variation. The frequency of different phenotypes is distributed normally (a plot produces a "bell curve" or normal distribution).

✔ You should be able to explain why crossing over is said to break up linkage between alleles.

14.6 Applying Mendel's Rules to Human Inheritance

- Pedigrees map out the transmission of human traits.

- Applying Mendel's rules to pedigrees can reveal the mode of transmission—whether a trait is dominant or recessive, autosomal or X-linked.

✔ You should be able to draw a pedigree that shows two sons and two daughters produced by a red–green color-blind father and a homozygous mother with normal color vision and explain why all the daughters are expected to be carriers of color blindness and none of the sons are expected to be color-blind.

MB) MasteringBiology

1. **MasteringBiology Assignments**

   Tutorials and Activities Determining Genotype: Pea Pod Color; What Is the Inheritance Pattern of Sex-Linked Traits? Gregor's Garden; Incomplete Dominance; Inheritance of Fur Color in Mice; Linked Genes and Crossing Over; Linked Genes and Linkage Mapping; Mendel's Experiments; Mendel's Law of Independent Assortment; Mendel's Law of Segregation; Pedigree Analysis: Dominant and Recessive Autosomal Conditions; Pedigree Analysis: Galactosemia, Sex Linkage, Sex-Linked Genes

   Questions Reading Quizzes, Blue-Thread Questions, Test Bank

2. **eText** Read your book online, search, take notes, highlight text, and more.

3. **The Study Area** Practice Test, Cumulative Test, BioFlix® 3-D Animations, Videos, Activities, Audio Glossary, Word Study Tools, Art

You should be able to . . .

✔ **TEST YOUR KNOWLEDGE**   *Answers are available in Appendix A*

1. In studies of how traits are inherited, what makes certain species candidates for model organisms?
   a. They are the first organisms to be used in a particular type of experiment, so they are a historical "model" of what researchers expect to find.
   b. They are easy to study because a great deal is already known about them.
   c. They are the best or most fit of their type.
   d. They are easy to maintain, have a short life cycle, produce many offspring, and yield data that are relevant to many other organisms.

2. Why is the allele for wrinkled seed shape in garden peas considered recessive?

    a. It "recedes" in the F_2 generation when homozygous parents are crossed.

    b. The trait associated with the allele is not expressed in heterozygotes.

    c. Individuals with the allele have lower fitness than that of individuals with the dominant allele.

    d. The allele is less common than the dominant allele. (The wrinkled allele is a rare mutant.)

3. The alleles found in haploid organisms cannot be dominant or recessive. Why?

    a. Dominance and recessiveness describe which allele is expressed in phenotype when different alleles occur in the same individual.

    b. Because only one allele is present, alleles in haploid organisms are always dominant.

    c. Alleles in haploid individuals are transmitted like mitochondrial DNA or chloroplast DNA.

    d. Most haploid individuals are bacteria, and bacterial genetics is completely different from eukaryotic genetics.

4. Why can you infer that individuals that are "pure line" are homozygous for the gene in question?

    a. Because they are highly inbred.

    b. Because only two alleles are present at each gene in the populations to which these individuals belong.

    c. Because in a pure line, phenotypes are not affected by environmental conditions or gene interactions.

    d. Because no other phenotype is ever observed in a pure-line population, this implies that only one allele is present.

5. The genes for the traits that Mendel worked with are either located on different chromosomes or so far apart on the same chromosome that crossing over almost always occurs between them. How did this circumstance help Mendel recognize the principle of independent assortment?

    a. Otherwise, his dihybrid crosses would not have produced a $9:3:3:1$ ratio of F_2 phenotypes.

    b. The occurrence of individuals with unexpected phenotypes led him to the discovery of recombination.

    c. It led him to the realization that the behavior of chromosomes during meiosis explained his results.

    d. It meant that the alleles involved were either dominant or recessive, which gave $3:1$ ratios in the F_1 generation.

6. What is meant by the claim that Mendel worked with the simplest possible genetic system?

    a. Discrete traits, two alleles, simple dominance and recessiveness, no sex chromosomes, and unlinked genes are the simplest situation known.

    b. The ability to self-fertilize or cross-pollinate made it simple for Mendel to set up controlled crosses.

    c. Mendel was aware of meiosis and the chromosome theory of inheritance, so it was easy to reach the conclusions he did.

    d. Mendel's experimental designs and his rules of inheritance are actually neither complex nor sophisticated.

7. Mendel's rules do not correctly predict patterns of inheritance for tightly linked genes or the inheritance of alleles that show incomplete dominance. Does this mean that his hypotheses are incorrect?

    a. Yes, because they are relevant to only a small number of organisms and traits.

    b. Yes, because not all data support his hypotheses.

    c. No, because he was not aware of meiosis or the chromosome theory of inheritance.

    d. No, it just means that his hypotheses are limited to certain conditions.

8. The artificial sweetener NutraSweet consists of a phenylalanine molecule linked to aspartic acid. The labels of diet sodas that contain NutraSweet include a warning to people with PKU. Why?

    a. NutraSweet stimulates the same taste receptors that natural sugars do.

    b. People with PKU have to avoid phenylalanine in their diet.

    c. In people with PKU, phenylalanine reacts with aspartic acid to form a toxic compound.

    d. People with PKU cannot lead normal lives, even if their environment is carefully controlled.

9. When Sutton and Boveri published the chromosome theory of inheritance, research on meiosis had not yet established that paternal and maternal homologs of different chromosomes assort independently. Then, in 1913, Elinor Carothers published a paper about a grasshopper species with an unusual karyotype: One chromosome had no homolog (meaning no pairing partner at meiosis I); another chromosome had homologs that could be distinguished under the light microscope. If chromosomes assort independently, how often should Carothers have observed each of the four products of meiosis shown in the following figure?

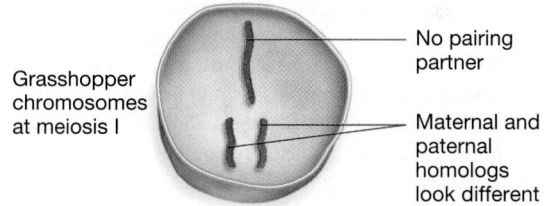

Grasshopper chromosomes at meiosis I

No pairing partner

Maternal and paternal homologs look different

Four types of gametes possible
(each meiotic division can produce only two of the four)

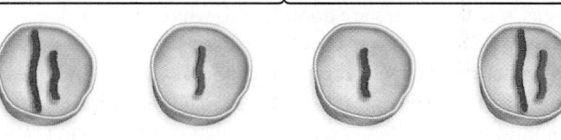

    a. Only the gametes with one of each type of chromosome would occur.

    b. The four types of gametes should be observed to occur at equal frequencies.

    c. The chromosome with no pairing partner would disintegrate, so only gametes with one copy of the other chromosome would be observed.

    d. Gametes with one of each type of chromosome would occur twice as often as gametes with just one chromosome.

10. Which of the following is the strongest evidence that a trait might be influenced by polygenic inheritance?

    a. F_1 offspring of parents with different phenotypes have an intermediate phenotype.

    b. F_1 offspring of parents with different phenotypes have the dominant phenotype.

    c. The trait shows qualitative (discrete) variation.

    d. The trait shows quantitative variation.

The best way to test and extend your knowledge of transmission genetics is to work problems. Most genetics problems are set up as follows: You are given some information about the genotypes or phenotypes of one or both parents, along with data on the phenotypes of F_1 or F_2 offspring. Your task is to generate a set of hypotheses—a genetic model—to explain the results. Your hypotheses should address each of the following questions:

- Is the trait under study discrete or quantitative?
- Is the phenotype a product of one gene or many genes?
- For each gene involved, how many alleles are present—one, two, or many?
- Do the alleles involved show complete dominance, incomplete dominance, or codominance?
- Are the genes involved sex-linked or autosomal?
- If more than one gene is involved, are they linked or unlinked? If they are linked, does crossing over occur frequently?

It's also helpful to ask yourself whether gene interactions or pleiotropy might be occurring and whether it is safe to assume that the experimental design carefully controlled for effects of variation in other genes and in the environment.

In working the problem, be sure to start with the simplest possible explanation. For example, if you are dealing with a discrete trait, you might hypothesize that the cross involves a single autosomal gene with two alleles that show complete dominance. Your next step is to infer what the parental genotypes would be (according to your hypothesis), if they are not already given, and then do a Punnett square to predict what the offspring phenotypes and their frequencies should be based on your hypothesis. Next, check whether these predictions match the observed results given in the problem. If the answer is yes, you have a valid solution. But if the answer is no, you need to go back and change one of your hypotheses, redo the Punnett square, and check to see if the predictions and observations match. Keep repeating these steps until you have a model that fits the data.

11. Example Problem *Plectritis congesta* plants produce fruits that either have or do not have prominent structures called wings. The alleles involved are W^+ = winged fruit; W^- = wingless fruit. Researchers collected an array of individuals from the field and performed a series of crosses. The results are given in the following table. Complete the table by writing down the genotype of the parent or parents involved in each cross.

Parental Phenotype(s)	Number of Offspring with Winged Fruits	Number of Offspring with Wingless Fruits	Parental Genotype(s)
Wingless (self-fertilized)	0	80	
Winged (self-fertilized)	90	30	
Winged × wingless	46	0	
Winged × winged	44	0	

A worked solution is available in Appendix A.

12. Example Problem Two black female mice are crossed with a brown male. In several litters, female I produced 9 blacks and 7 browns; female II produced 57 blacks. What deductions can you make concerning the inheritance of black and brown coat color in mice? What are the genotypes of the parents in this case?

A worked solution is available in Appendix A.

13. In peas, purple flowers are dominant to white. If a purple-flowered, heterozygous plant were crossed with a white-flowered plant, what is the expected ratio of genotypes and phenotypes among the F_1 offspring? If two of the purple-flowered F_1 offspring were randomly selected and crossed, what is the expected ratio of genotypes and phenotypes among the F_2 offspring?

14. In garden peas, yellow seeds (Y) are dominant to green seeds (y), and inflated pods (I) are dominant to constricted pods (i). Suppose you have crossed YYII parents with yyii parents.
- Draw the F_1 Punnett square and predict the expected F_1 phenotype(s).
- List the genotype(s) of gametes produced by F_1 individuals.
- Draw the F_2 Punnett square. Based on this Punnett square, predict the expected phenotype(s) in the F_2 generation and the expected frequency of each phenotype.

15. The smooth feathers on the back of the neck in pigeons can be reversed by a mutation to produce a "crested" appearance in which feathers form a distinctive spike at the back of the head. A pigeon breeder examined offspring produced by a single pair of non-crested birds and recorded the following: 22 non-crested and 7 crested. She then made a series of crosses using offspring from the first cross. When she crossed two of the crested birds, all 20 of the offspring were crested. When she crossed a non-crested bird with a crested bird, 7 offspring were non-crested and 6 were crested.
- For these three crosses, provide genotypes for parents and offspring that are consistent with these results.
- Which allele is dominant?

16. A plant with orange, spotted flowers was grown in the greenhouse from a seed collected in the wild. The plant was self-pollinated and gave rise to the following progeny: 88 orange with spots, 34 yellow with spots, 32 orange with no spots, and 8 yellow with no spots. What can you conclude about the dominance relationships of the alleles responsible for the spotted and unspotted phenotypes? For the orange and yellow phenotypes? What can you conclude about the genotype of the original plant that had orange, spotted flowers?

17. As a genetic counselor, you routinely advise couples about the possibility of genetic disease in their offspring based on their family histories. This morning you met with an engaged couple, both of whom are phenotypically normal. The man, however, has a brother who died of Duchenne-type muscular dystrophy, an X-linked condition that results in death before the age of 20. The allele responsible for this disease is recessive. His prospective bride, whose family has no history of the disease, is worried that the couple's sons or daughters might be afflicted.
- How would you advise this couple?
- The sister of this man is planning to marry his fiancée's brother. How would you advise this second couple?

18. Suppose you are heterozygous for two genes that are located on different chromosomes. You carry alleles *A* and *a* for one gene and alleles *B* and *b* for the other. Draw a diagram illustrating what happens to these genes and alleles when meiosis occurs in your reproductive tissues. Label the stages of meiosis, the homologous chromosomes, sister chromatids, nonhomologous chromosomes, genes, and alleles. Be sure to list all the genetically different gametes that could form and indicate how frequently each type should be observed. On the diagram, identify the events responsible for the principle of segregation and the principle of independent assortment.

19. Review the text's description of ABO blood types. Suppose a woman with blood type O married a man with blood type AB. What phenotypes and genotypes would you expect to observe in their offspring, and in what proportions? Answer the same question for a heterozygous mother with blood type A and a heterozygous father with blood type B.

20. An alien friend named Tukan has two sets of eyes, one set forward-looking and one set backward-looking, and smooth skin. His mate, Valco, lacks eyes but has skin covered with tiny hooks that attract all sorts of debris. Tukan and Valco have thrived on earth and have had four children, all with no eyes and smooth skin. Typical of their ways, the children interbred and produced 32 children of their own.
- Under the models of inheritance proposed by Mendel, identify which alleles are dominant and which are recessive.
- Provide gene symbols that would reflect the dominant–recessive allelic relationships.
- Of the 32 children, how many would you expect to have two sets of eyes and smooth skin?

21. Phenylketonuria (PKU) is a genetic disease caused by homozygosity for a recessive mutation in the enzyme that converts the amino acid phenylalanine to tyrosine. In the absence of this enzyme, phenylalanine and some of its derivatives accumulate in the body and cause mental retardation. If individuals are identified soon enough after birth, they can be treated by a low-phenylalanine diet for the early years of their lives. As adults, though, homozygous recessive individuals are allowed to adopt a diet with normal amounts of phenylalanine. Not long after such treatments were initiated, a troubling phenomenon was observed. A high number of children born to treated mothers were mentally retarded even though the children were heterozygous for the PKU gene. Children born of treated PKU males suffered no ill effects.
- Can you offer an explanation as to why genetically heterozygous children of treated PKU mothers might be prone to mental retardation?
- Propose a solution to reduce the likelihood of mental retardation in children of treated PKU mothers.

22. The blending-inheritance hypothesis proposed that the genetic material from parents is unavoidably and irreversibly mixed in the offspring. As a result, offspring and later descendants should always appear intermediate in phenotype to their forebears. Mendel, in contrast, proposed that genes are discrete and that their integrity is maintained in the offspring and in subsequent generations. Suppose the year is 1890. You are a horse breeder and have just read Mendel's paper. You don't believe his results, however, because you often work with cremello (very light-colored) and chestnut (reddish-brown) horses. You know that if you cross a cremello individual from a pure-breeding line with a chestnut individual from a pure-breeding line, the offspring will be palomino—meaning they have an intermediate (golden-yellow) body color. What additional crosses would you do to test whether Mendel's model is valid in the case of genes for horse color? List the crosses and the offspring genotypes and phenotypes you'd expect to obtain. Explain why these experimental crosses would provide a test of Mendel's model.

23. Two mothers give birth to sons at the same time in a busy hospital. The son of couple 1 is afflicted with hemophilia A, which is a recessive X-linked disease. Neither parent has the disease. Couple 2 has a normal son even though the father has hemophilia A. The two couples sue the hospital in court, claiming that a careless staff member swapped their babies at birth. You appear in court as an expert witness. What do you tell the jury? Make a diagram that you can submit to the jury.

24. You have crossed two *Drosophila melanogaster* individuals that have long wings and red eyes—the wild-type phenotype. In the progeny, the mutant phenotypes called curved wings and lozenge eyes appear as follows:

Females	Males
600 long wings, red eyes	300 long wings, red eyes
200 curved wings, red eyes	100 curved wings, red eyes
	300 long wings, lozenge eyes
	100 curved wings, lozenge eyes

- According to these data, is the curved-wing allele autosomal recessive, autosomal dominant, sex-linked recessive, or sex-linked dominant?
- Is the lozenge-eyed allele autosomal recessive, autosomal dominant, sex-linked recessive, or sex-linked dominant?
- What is the genotype of the female parent?
- What is the genotype of the male parent?

25. In parakeets, two autosomal genes that are located on different chromosomes control the production of feather pigment. Gene *B* codes for an enzyme that is required for the synthesis of a blue pigment, and gene *Y* codes for an enzyme required for the synthesis of a yellow pigment. Recessive mutations that result in no production of the affected pigment are known for both genes. Suppose that a bird breeder has two green parakeets and mates them. The offspring are green, blue, yellow, and albino (unpigmented).
- Based on this observation, what are the genotypes of the green parents? What is the genotype of each type of offspring? What fraction of the total progeny should exhibit each type of color?
- Suppose that the parents were the progeny of a cross between two true-breeding strains. What two types of crosses between true-breeding strains could have produced the green parents? Indicate the genotypes and phenotypes for each cross.

26. Recall that hemophilia is an X-linked recessive disease. If a woman with hemophilia had children with a man without hemophilia, what is the chance that their first child will have the disease? What is the chance that their first child will be a carrier?

15 DNA and the Gene: Synthesis and Repair

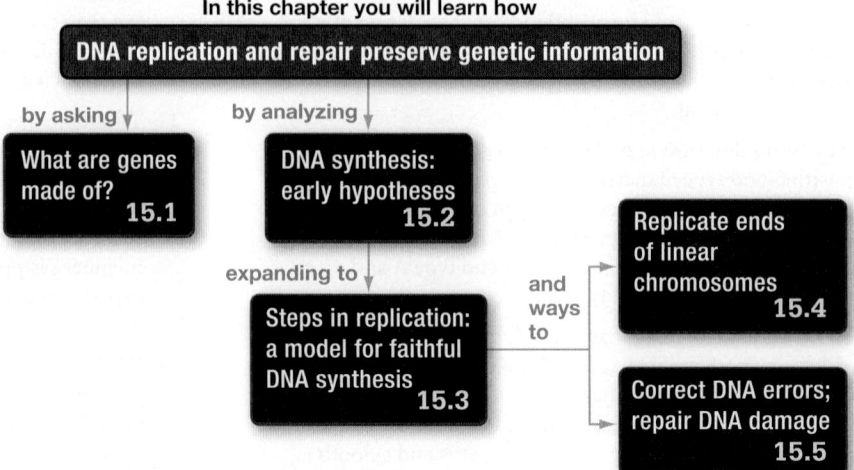

In this chapter you will learn how

DNA replication and repair preserve genetic information

by asking ↓

by analyzing ↓

What are genes made of? 15.1

DNA synthesis: early hypotheses 15.2

expanding to ↓

Steps in replication: a model for faithful DNA synthesis 15.3

and ways to

Replicate ends of linear chromosomes 15.4

Correct DNA errors; repair DNA damage 15.5

Electron micrograph (with color added) showing DNA in the process of replication. The original DNA double helix (far right) is being replicated into two DNA double helices (on the left). The two helices diverge at the replication fork, which is where DNA synthesis is taking place.

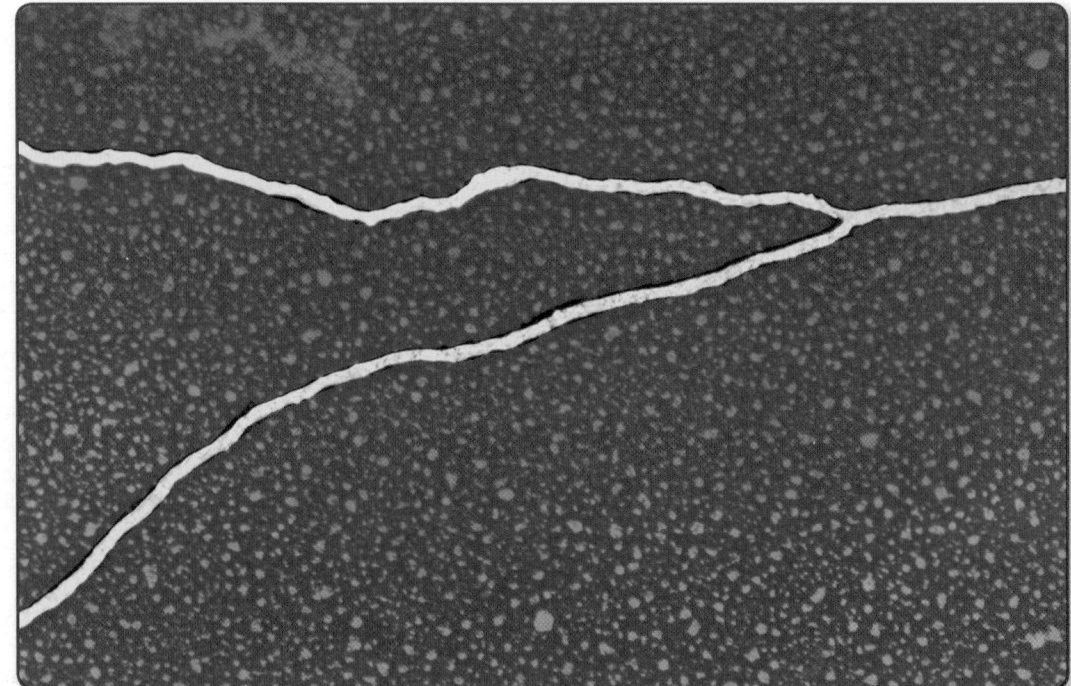

This chapter is part of the Big Picture. See how on pages 366–367.

What are genes made of, and how are they copied so that they are faithfully passed on to offspring? These questions dominated biology during the middle of the twentieth century.

Since Mendel's time, the predominant research strategy in genetics had been to conduct a series of experimental crosses, create a genetic model to explain the types and proportions of phenotypes that resulted, and then test the model's predictions through reciprocal crosses, testcrosses, or other techniques. This strategy led to virtually all the discoveries of classical genetics, including Mendel's rules, sex linkage, linkage, and quantitative inheritance (see Chapter 14).

The chemical composition and molecular structure of Mendel's hereditary factors—which came to be called genes—remained a mystery for the first half of the twentieth century. Although biologists knew that genes and chromosomes were replicated during the cell cycle, with copies distributed

✔ When you see this checkmark, stop and test yourself. Answers are available in Appendix A.

to daughter cells during mitosis and meiosis (see Chapters 12 and 13), no one had the slightest clue about how the copying occurred.

The goal of this chapter is to explore how researchers solved these mysteries. The results provided a link between two of the five attributes of life (introduced in Chapter 1): processing genetic information and replication. (You can see how DNA synthesis and repair fits into the Big Picture of Genetic Information on pages 366–367.)

How are genes copied, so they can be passed on to succeeding generations? Let's begin with studies that identified the nature of the genetic material, then explore how genes are copied during the synthesis phase of the cell cycle, and conclude by analyzing how incorrectly copied or damaged genes are repaired. Once the molecular nature of the gene was known, the nature of biological science changed forever.

15.1 What Are Genes Made Of?

The chromosome theory of inheritance (Chapter 14) proposed that chromosomes contain genes. It had been known since the late 1800s that chromosomes are a complex of DNA and proteins. The question, then, of what genes are made of came down to a simple choice: DNA or protein?

Initially, most biologists backed the hypothesis that genes are made of proteins. The arguments in favor of this hypothesis were compelling. Hundreds, if not thousands, of complex and highly regulated chemical reactions occur in even the simplest living cells. The amount of information required to specify and coordinate these reactions is mind-boggling. With their almost limitless variation in structure and function, proteins are complex enough to contain this much information.

In contrast, DNA was known to be composed of just four types of deoxyribonucleotides (Chapter 4). Early but incorrect evidence suggested that DNA was a simple molecule with some sort of repetitive and uninteresting structure. It seemed impossible that such a simple compound could hold complex information.

DNA or protein? The experiment that settled the question is considered a classic in biological science.

The Hershey–Chase Experiment

In 1952 Alfred Hershey and Martha Chase took up the question of whether genes were made of protein or DNA by studying how a virus called T2 infects and replicates within the bacterium *Escherichia coli*. Nearly 10 years before Hershey and Chase began their study, Oswald Avery and colleagues showed in 1944 that DNA could serve as genetic material, but many scientists remained unconvinced of the finding or its generality. Hershey and Chase knew that T2 infections begin when the virus attaches to the cell wall of *E. coli* and injects its genes into the cell's interior (**FIGURE 15.1a**). These genes then direct the production of a new generation of

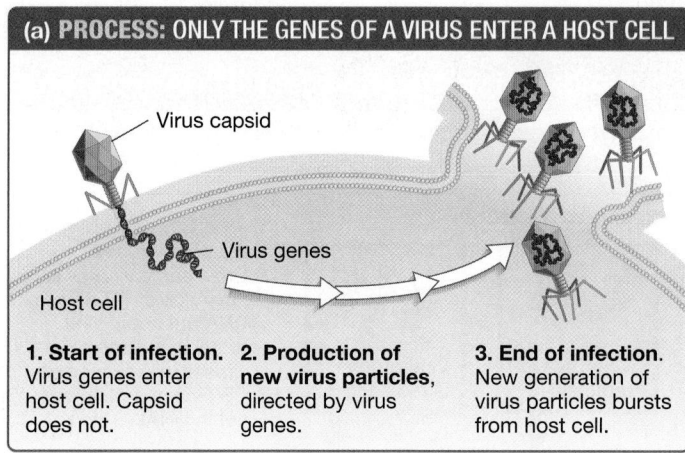

(a) PROCESS: ONLY THE GENES OF A VIRUS ENTER A HOST CELL

Virus capsid
Virus genes
Host cell

1. Start of infection. Virus genes enter host cell. Capsid does not.

2. Production of new virus particles, directed by virus genes.

3. End of infection. New generation of virus particles bursts from host cell.

(b) The virus's capsid stays outside the cell.

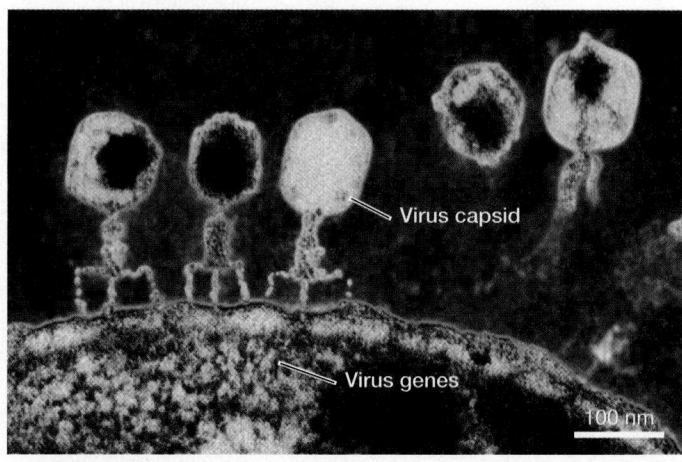

Virus capsid
Virus genes
100 nm

FIGURE 15.1 Viruses Inject Genes into Bacterial Cells and Leave a Capsid Behind. Color has been added to the transmission electron micrograph in (b) to make key structures more visible.

virus particles inside the infected cell, which acts as a host for the virus. (For more information on viruses, see Chapter 36.)

During the infection, the exterior protein coat, or **capsid,** of the original, parent virus is left behind. The capsid remains attached to the exterior of the host cell (**FIGURE 15.1b**). Hershey and Chase also knew that T2 is made up almost exclusively of protein and DNA. Was it protein or DNA that entered the host cell and directed the production of new viruses?

Hershey and Chase's strategy for determining the composition of the viral substance that enters the cell and acts as the hereditary material was based on two biochemical facts: (1) Proteins contain sulfur but not phosphorus, and (2) DNA contains phosphorus but not sulfur.

As **FIGURE 15.2** (see page 286) shows, the researchers began their work by growing viruses in the presence of either a radioactive isotope of sulfur (^{35}S) or a radioactive isotope of phosphorus (^{32}P). Because these isotopes were incorporated into newly synthesized proteins and DNA, this step produced a population of viruses with radioactive proteins and a population with radioactive DNA.

QUESTION: Do viral genes consist of DNA or protein?

DNA HYPOTHESIS: Viral genes consist of DNA.

PROTEIN HYPOTHESIS: Viral genes consist of protein.

EXPERIMENTAL SETUP:

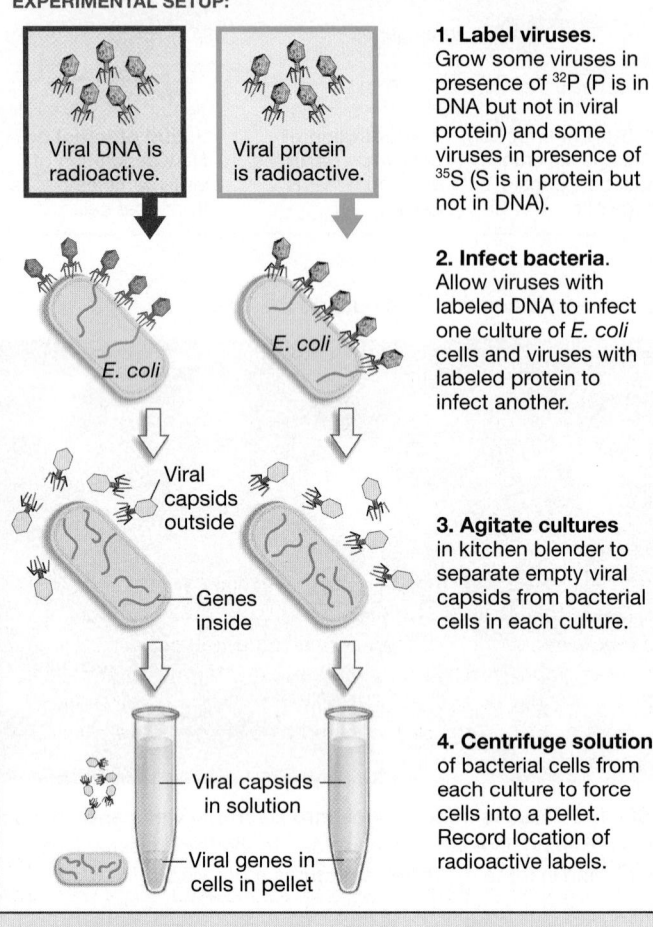

Viral DNA is radioactive.

Viral protein is radioactive.

E. coli

E. coli

Viral capsids outside

Genes inside

Viral capsids in solution

Viral genes in cells in pellet

1. Label viruses. Grow some viruses in presence of ^{32}P (P is in DNA but not in viral protein) and some viruses in presence of ^{35}S (S is in protein but not in DNA).

2. Infect bacteria. Allow viruses with labeled DNA to infect one culture of *E. coli* cells and viruses with labeled protein to infect another.

3. Agitate cultures in kitchen blender to separate empty viral capsids from bacterial cells in each culture.

4. Centrifuge solutions of bacterial cells from each culture to force cells into a pellet. Record location of radioactive labels.

PREDICTION OF DNA HYPOTHESIS: Radioactive DNA will be located within pellet.

PREDICTION OF PROTEIN HYPOTHESIS: Radioactive protein will be located within pellet.

RESULTS:

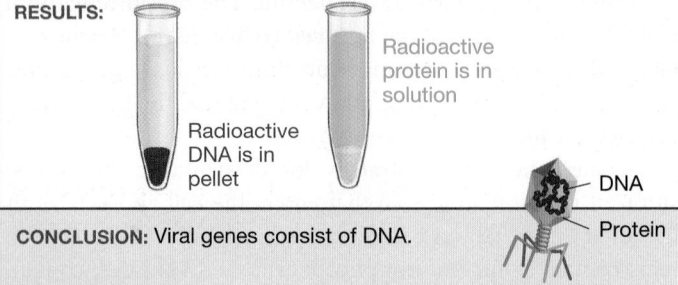

Radioactive DNA is in pellet

Radioactive protein is in solution

DNA

Protein

CONCLUSION: Viral genes consist of DNA.

FIGURE 15.2 Experimental Evidence that DNA Is the Hereditary Material.

SOURCE: Hershey, A. D., and M. Chase. 1952. Independent functions of viral protein and nucleic acid in growth of bacteriophage. *Journal of General Physiology* 36: 39–56.

✔ **QUESTION** What evidence would these investigators have to produce to convince you that the viral capsids were shaken off the bacterial cells by the agitation step?

Hershey and Chase allowed each set of radioactive viruses to infect *E. coli* cells. If genes consist of DNA, then radioactive protein should be found only in the capsids outside the infected host cells, while radioactive DNA should be located inside the cells. But if genes consist of proteins, then radioactive protein—and no radioactive DNA—should be inside the cells.

To test these predictions, Hershey and Chase sheared the capsids off the cells by vigorously agitating each of the cultures in kitchen blenders. When the researchers spun the samples in a centrifuge, the small phage capsids remained in the solution while the cells formed a pellet at the bottom of the centrifuge tube (see **BioSkills 10** in Appendix B to review how centrifugation works).

As predicted by the DNA hypothesis, the biologists found that virtually all the radioactive protein was outside cells in the emptied capsids, while virtually all the radioactive DNA was inside the host cells. Because the injected component of the virus directs the production of a new generation of virus particles, this component must represent the virus's genes.

After these results were published, proponents of the protein hypothesis accepted that DNA, not protein, must be the hereditary material. An astonishing claim—that DNA contained all the information for life's complexity—was correct.

The Secondary Structure of DNA

In 1953, one year after Hershey and Chase's landmark results were published, Watson and Crick proposed a model for the secondary structure of DNA. Recall that DNA is typically double-stranded with each strand consisting of a long, linear polymer made up of monomers called deoxyribonucleotides (Chapter 4).

Each deoxyribonucleotide consists of a deoxyribose sugar, a phosphate group, and a nitrogenous base (**FIGURE 15.3a**). Deoxyribonucleotides link together into a polymer when a phosphodiester bond forms between a hydroxyl group on the 3′ carbon of one deoxyribose and the phosphate group attached to the 5′ carbon of another deoxyribose. The two strands together make up one DNA molecule that functions as the genetic information storage molecule of cells.

As **FIGURE 15.3b** shows, the primary structure of each strand of DNA has two major features: **(1)** a "backbone" made up of the sugar and phosphate groups of deoxyribonucleotides and **(2)** a series of bases that project from the backbone. Each strand of DNA has a directionality, or polarity: One end has an exposed hydroxyl group on the 3′ carbon of a deoxyribose, while the other has an exposed phosphate group on a 5′ carbon. Thus, the molecule has distinctly different 3′ and 5′ ends.

As they explored different models for the secondary structure of DNA, Watson and Crick hit on the idea of lining up two of these long strands in opposite directions, or in what is called antiparallel fashion (**FIGURE 15.4a**). They realized that antiparallel strands will twist around each other into a spiral or helix because certain bases fit together snugly in pairs inside the spiral and form hydrogen bonds (**FIGURE 15.4b**). The double-stranded molecule that results is called a **double helix**.

(a) Structure of a deoxyribonucleotide

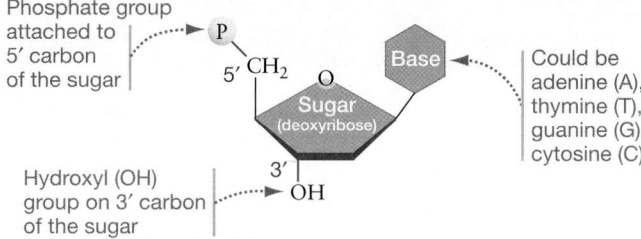

(b) Primary structure of DNA

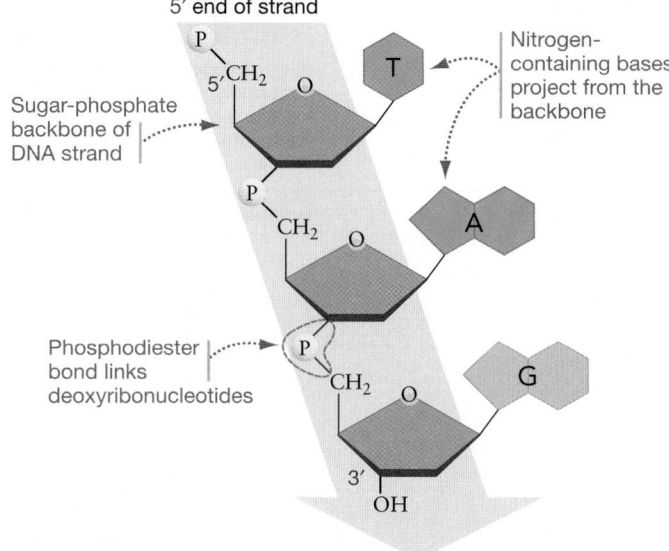

FIGURE 15.3 DNA's Primary Structure. (a) Deoxyribonucleotides are monomers that polymerize to form DNA. **(b)** DNA's primary structure is made up of a sequence of deoxyribonucleotides. Notice that the structure has a sugar–phosphate "backbone" with nitrogen-containing bases attached.

✔ **EXERCISE** Write the base sequence of the DNA in part (b), in the 5′ → 3′ direction.

The double-helical DNA is stabilized by hydrogen bonds that form between the bases adenine (A) and thymine (T) and between the bases guanine (G) and cytosine (C), along with hydrophobic interactions that the bases experience inside the helix. Hydrogen bonding of particular base pairs is **complementary base pairing.**

15.2 Testing Early Hypotheses about DNA Synthesis

Watson and Crick realized that the A-T and G-C pairing rules suggested a way for DNA to be copied when chromosomes are replicated during S phase of the cell cycle, before mitosis and meiosis. They suggested that the existing strands of DNA served as a template (pattern) for the production of new strands and that deoxyribonucleotides were added to the new strands according to complementary base pairing. For example, if the template

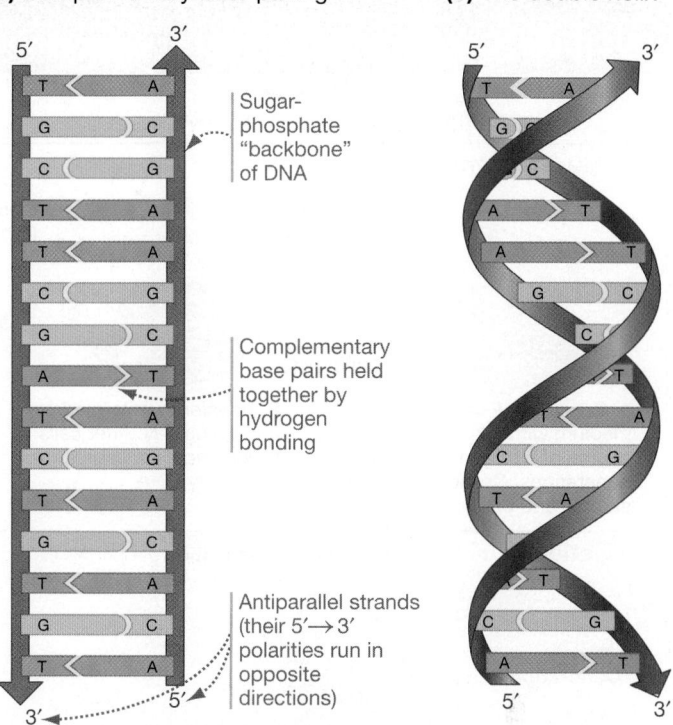

FIGURE 15.4 DNA's Secondary Structure: The Double Helix. (a) DNA normally consists of two strands, each with a sugar–phosphate backbone. Nitrogen-containing bases project from each strand and form hydrogen bonds. Only A-T and G-C pairs fit together in a way that allows hydrogen bonding to occur between the strands. **(b)** Bonding between complementary bases twists the molecule into a double helix.

strand contained a T, then an A would be added to the new strand to pair with that T. Similarly, a G on the template strand would dictate the addition of a C on the new strand.

Complementary base pairing provided a mechanism for DNA to be copied. But many questions remained about how the copying was done.

Three Alternative Hypotheses

Biologists at the time proposed three alternative hypotheses about how the old and new strands might interact during replication:

1. *Semiconservative replication* If the old, **parental strands** of DNA separated, each could then be used as a template for the synthesis of a new, **daughter strand.** This hypothesis is called **semiconservative replication** because each new daughter DNA molecule would consist of one old strand and one new strand.

2. *Conservative replication* If the bases temporarily turned outward so that complementary strands no longer faced each other, they could serve as a template for the synthesis of an entirely new double helix all at once. This hypothesis, called conservative replication, would result in an intact parental

QUESTION: Is replication semiconservative, conservative, or dispersive?

HYPOTHESIS 1:	HYPOTHESIS 2:	HYPOTHESIS 3:
Replication is semiconservative.	Replication is conservative.	Replication is dispersive.

EXPERIMENTAL SETUP:

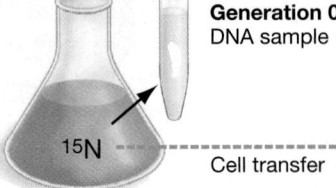 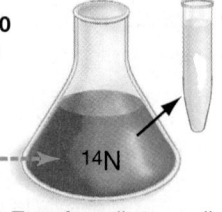 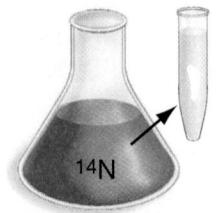 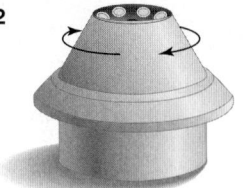

Generation 0
DNA sample

Generation 1
DNA sample

Generation 2
DNA sample

1. Grow *E. coli* cells in medium with ^{15}N as sole source of nitrogen for many generations. Collect sample and purify DNA.

2. Transfer cells to medium containing ^{14}N. After cells divide once, collect sample and purify DNA.

3. After cells have divided a second time in ^{14}N medium, collect sample and purify DNA.

4. Centrifuge the three samples separately. Compare the locations of the DNA bands in each sample.

PREDICTIONS:

Semiconservative replication	Conservative replication	Dispersive replication

Generation 0

Generation 1

Generation 2

Semiconservative replication

Generation 0: ^{15}N
Generation 1: Hybrid — Hybrid
Generation 2: Hybrid ^{14}N Hybrid ^{14}N

After 2 generations:
1/2 low-density DNA (^{14}N)
1/2 intermediate-density DNA (hybrid)

Conservative replication

Generation 0: ^{15}N
Generation 1: ^{15}N — ^{14}N
Generation 2: ^{15}N ^{14}N

After 2 generations:
1/4 high-density DNA (^{15}N)
3/4 low-density DNA (^{14}N)

Dispersive replication

Generation 0: ^{15}N
Generation 1: Hybrid — Hybrid
Generation 2: Hybrid

After 2 generations:
All intermediate-density DNA (hybrid)

RESULTS:

Top of centrifuge tube (lower density)

^{14}N
Hybrid
^{15}N

Bottom of centrifuge tube (higher density)

0   1   2
Generation

After 2 generations:
1/2 low-density DNA
1/2 intermediate-density DNA

CONCLUSION: Data from generation 1 conflict with conservative replication hypothesis. Data from generation 2 conflict with dispersive replication hypothesis. Replication is semiconservative.

FIGURE 15.5 The Meselson–Stahl Experiment.

SOURCE: Meselson, M., and F. W. Stahl. 1958. The replication of DNA in *Escherichia coli*. *Proceedings of the National Academy of Sciences USA* 44: 671–682.

✔**EXERCISE** Meselson and Stahl actually let their experiment run for a fourth generation with cultures growing in the presence of ^{14}N. Explain what data from third- and fourth-generation DNA should look like—that is, where the DNA band(s) should be.

molecule and a daughter DNA molecule consisting entirely of newly synthesized strands.

3. *Dispersive replication*   If the parental double helix were cut wherever one strand crossed over another and DNA was synthesized in short sections by extending each of the cut parental strands to the next strand crossover, then there would be a mix of new and old segments along each replicated molecule. This possibility is called dispersive replication—stretches of old DNA would be interspersed with new DNA down the length of each daughter strand.

Matthew Meselson and Franklin Stahl realized that if they could tag or mark parental and daughter strands of DNA in a way that would make them distinguishable from each other, they could determine whether replication was conservative, semiconservative, or dispersive.

The Meselson–Stahl Experiment

Before Meselson and Stahl could do any tagging to distinguish old DNA from new DNA, they needed to choose an organism to study. They decided to work with the bacterium *Escherichia coli*—the same inhabitant of the human gastrointestinal tract that Hershey and Chase used. Because *E. coli* is small and grows quickly and readily in the laboratory, it had become a favored model organism in studies of biochemistry and molecular genetics. (See **BioSkills 13** in Appendix B for more on *E. coli*.)

Like all organisms, bacterial cells copy their entire complement of DNA, or their **genome,** before every cell division. To distinguish parental strands of DNA from daughter strands when *E. coli* replicated, Meselson and Stahl grew the cells for successive generations in the presence of different isotopes of nitrogen: first ^{15}N and later ^{14}N. Because ^{15}N contains an extra neutron, it is heavier than the normal isotope, ^{14}N.

This difference in mass, which creates a difference in density between ^{14}N-containing and ^{15}N-containing DNA, was the key to the experiment summarized in **FIGURE 15.5**. The logic ran as follows:

- If different nitrogen isotopes were available in the growth medium when different generations of DNA were produced, then the parental and daughter strands would have different densities.

- The technique called density-gradient centrifugation separates molecules based on their density (**BioSkills 10** in Appendix B). Low-density molecules cluster in bands high in the centrifuge tube; higher-density molecules cluster in bands lower in the centrifuge tube.

- When intact, double-stranded DNA molecules are subjected to density-gradient centrifugation, DNA that contains ^{14}N should form a band higher in the centrifuge tube; DNA that contains ^{15}N should form a band lower in the centrifuge tube.

In short, DNA containing ^{14}N and DNA containing ^{15}N should form separate bands. How could this tagging system be used to test whether replication is semiconservative, conservative, or dispersive?

Meselson and Stahl began by growing *E. coli* cells with nutrients that contained only ^{15}N. They purified DNA from a sample of these cells and transferred the rest of the culture to a growth medium containing only the ^{14}N isotope. After enough time had elapsed for these cells to divide once—meaning that the DNA had been copied once—they removed a sample and isolated the DNA. After the remainder of the culture had divided again, they removed another sample and isolated its DNA.

As Figure 15.5 shows, the conservative, semiconservative, and dispersive models make distinct predictions about the makeup of the DNA molecules after replication occurs in the first and second generation. Examine the figure carefully to understand these distinct predictions.

The photograph at the bottom of Figure 15.5 shows the experiment's results. After one generation, the density of the DNA molecules was intermediate. These data suggested that the hypothesis of conservative replication was wrong, since it predicted two different densities in the first generation.

After two generations, a lower-density band appeared in addition to the intermediate-density band. This result offered strong support for the hypothesis that DNA replication is semiconservative. Had dispersive replication occurred, the second generation would have produced only a single, intermediate density band. Each newly made DNA molecule comprises one old strand and one new strand—replication is semiconservative.

15.3   A Model for DNA Synthesis

The DNA inside a cell is like an ancient text that has been painstakingly copied and handed down, generation after generation. But while the most ancient of all human texts contain messages that are thousands of years old, the DNA in living cells has been copied and passed down for billions of years. And instead of being copied by monks or clerks, DNA is replicated by molecular scribes. What molecules are responsible for copying DNA, and how do they work?

Meselson and Stahl showed that each strand of DNA is copied in its entirety each time replication occurs, but how does DNA synthesis proceed? Does it require an input of energy in the form of ATP, or it is spontaneous? Is it catalyzed by an enzyme, or does it occur quickly on its own?

The initial breakthrough on DNA replication came with the discovery of an enzyme called **DNA polymerase,** so named because it polymerizes deoxyribonucleotides into DNA. This protein catalyzes DNA synthesis. Follow-up work showed that there are several types of DNA polymerase. DNA polymerase III, for example, is the enzyme that is primarily responsible for copying *E. coli*'s chromosome before cell division.

FIGURE 15.6 (see page 290) illustrates a critical characteristic of DNA polymerases: They can work in only one direction. DNA polymerases can add deoxyribonucleotides only to the 3' end of a growing DNA chain. As a result, DNA synthesis always proceeds in the $5' \rightarrow 3'$ direction. ✔ If you understand this

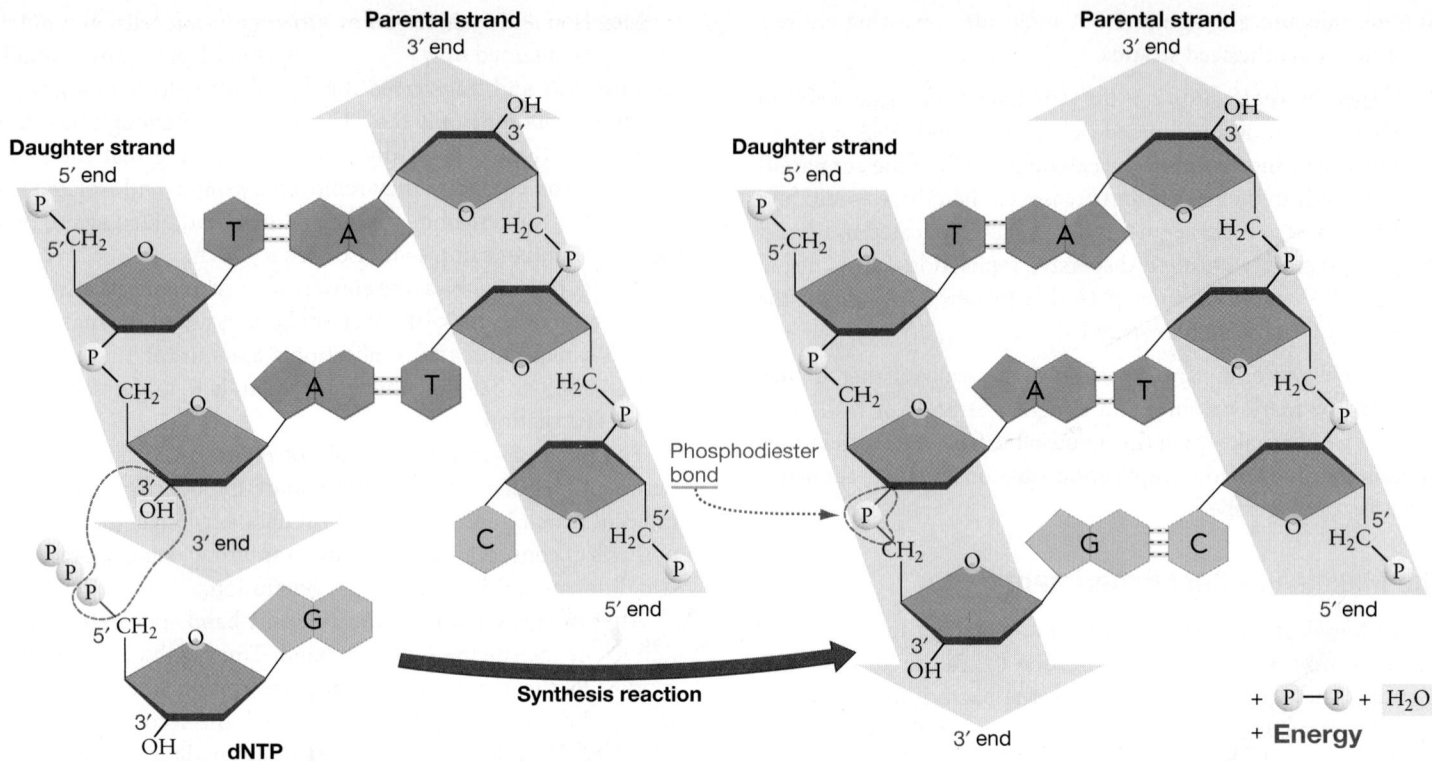

FIGURE 15.6 The DNA Synthesis Reaction. A condensation reaction results in formation of a phosphodiester bond between the 3′ carbon on the end of a DNA strand and the 5′ carbon on an incoming deoxyribonucleoside triphosphate (dNTP) monomer.

concept, you should be able to draw two lines representing a DNA molecule, assign the 3′-to-5′ polarity of each strand, and then label the direction in which DNA synthesis will proceed for each strand.

Figure 15.6 makes another important point about DNA synthesis. You might recall from earlier chapters that polymerization reactions generally are endergonic, meaning they require an input of energy. But for DNA synthesis, the reaction is exergonic (it releases energy) because the monomers that are used in the DNA synthesis reaction are **deoxyribonucleoside triphosphates (dNTPs).** (The *N* in dNTP stands for any of the four bases found in DNA: adenine, thymine, guanine, or cytosine). Because they have three closely spaced phosphate groups, dNTPs have high potential energy—high enough to make the formation of phosphodiester bonds in a growing DNA strand exergonic as two of the phosphates are cleaved off (see Chapter 8).

How Does Replication Get Started?

Another major insight into the mechanism of DNA synthesis emerged when electron microscopy caught DNA replication in action. As **FIGURE 15.7a** shows, a "bubble" forms when DNA is being synthesized. Initially, the replication bubble forms at a specific sequence of bases called the **origin of replication** (**FIGURE 15.7b**). Bacterial chromosomes have only one origin of replication, and thus a single replication bubble forms. Eukaryotes have multiple origins of replication along each chromosome, and thus multiple replication bubbles (**FIGURE 15.7c**).

DNA synthesis is bidirectional—that is, it occurs in both directions at the same time. Therefore, replication bubbles grow in two directions as DNA replication proceeds.

A specific set of proteins are responsible for recognizing sites where replication begins and opening the double helix at those points. These proteins are activated by the proteins that initiate S phase in the cell cycle (see Chapter 12).

Once a replication bubble opens at the origin of replication, a different set of enzymes takes over to start DNA synthesis. Active DNA synthesis takes place at the replication forks of each replication bubble (shown in Figure 15.7c). The **replication fork** is the Y-shaped region where the parent–DNA double helix is split into two single strands and copied.

How Is the Helix Opened and Stabilized?

A large group of enzymes and specialized proteins converge on the point where the double helix opens. The enzyme called **DNA helicase** breaks the hydrogen bonds between the base pairs. This reaction causes the two strands of DNA to separate. **Single-strand DNA-binding proteins (SSBPs)** attach to the separated strands and prevent them from snapping back into a double helix. Working together, DNA helicase and single-strand DNA-binding proteins open up the double helix and maintain the separation of both strands during copying (**FIGURE 15.8**, step 1).

The "unzipping" process that occurs at the replication fork creates tension farther down the helix. To understand why, imagine what would happen if you started to pull apart the twisted

(a) DNA being replicated

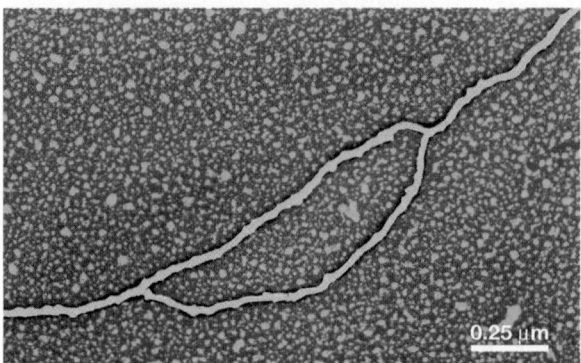

(b) Bacterial chromosomes have a single origin of replication.

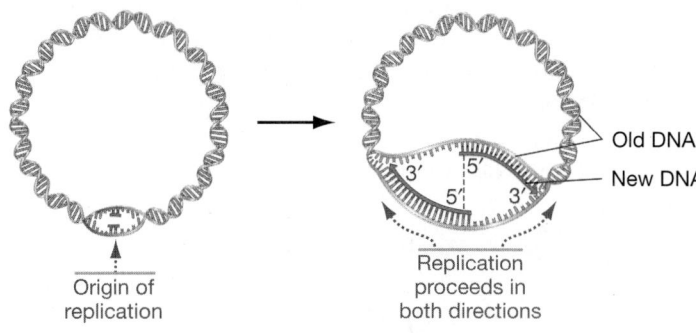

Old DNA

New DNA

Origin of replication

Replication proceeds in both directions

(c) Eukaryotic chromosomes have multiple origins of replication.

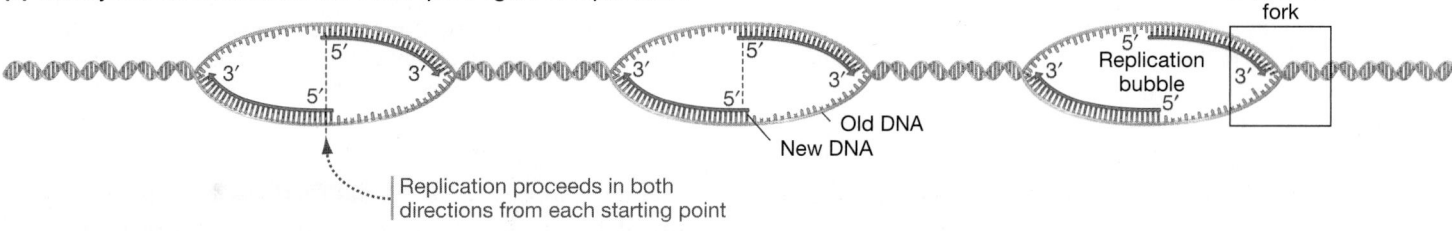

Replication fork

Replication bubble

Old DNA

New DNA

Replication proceeds in both directions from each starting point

FIGURE 15.7 DNA Synthesis Proceeds in Two Directions from an Origin of Replication. Color has been added to the micrograph in part (a).

strands of a rope. The untwisting movements at one end would force the intact section to rotate in response. If the intact end of the rope were fixed in place, it would coil on itself in response to the twisting forces. DNA does not become tightly coiled ahead of the replication fork, because the twisting induced by helicase is relaxed by proteins called topoisomerases. A **topoisomerase** is an enzyme that cuts DNA, allows it to unwind, and rejoins it ahead of the advancing replication fork.

Now, what happens once the DNA helix is open?

How Is the Leading Strand Synthesized?

The keys to understanding what happens at the start of DNA synthesis are to recall that DNA polymerase (1) works only in the $5' \rightarrow 3'$ direction and (2) requires both a $3'$ end to extend from and a single-stranded template. Both of these properties control how synthesis occurs on both template strands of DNA, and as you'll soon see, they significantly complicate copying one of these. The single-stranded template dictates which

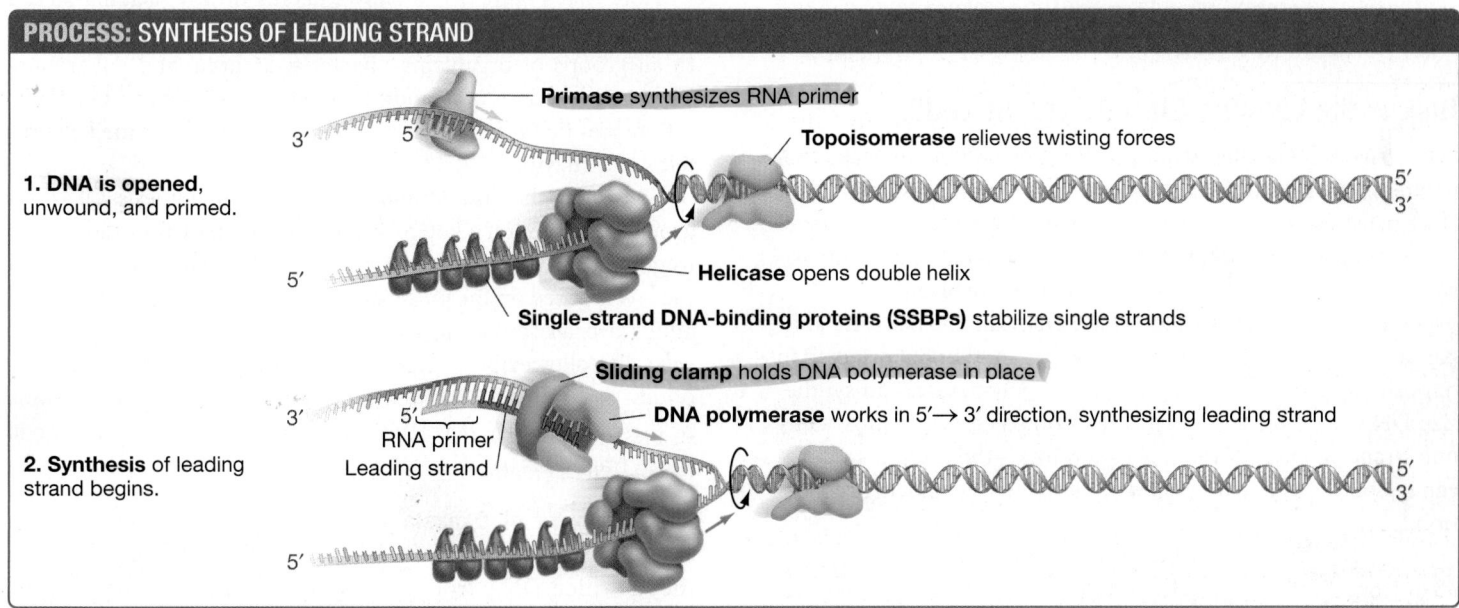

PROCESS: SYNTHESIS OF LEADING STRAND

Primase synthesizes RNA primer

Topoisomerase relieves twisting forces

1. DNA is opened, unwound, and primed.

Helicase opens double helix

Single-strand DNA-binding proteins (SSBPs) stabilize single strands

Sliding clamp holds DNA polymerase in place

DNA polymerase works in $5' \rightarrow 3'$ direction, synthesizing leading strand

2. Synthesis of leading strand begins.

RNA primer
Leading strand

FIGURE 15.8 Leading-Strand Synthesis.

deoxyribonucleotide should be added next. A **primer**—a strand a few nucleotides long that is bonded to the template strand—provides DNA polymerase with a free 3′ hydroxyl (−OH) group that can combine with an incoming deoxyribonucleotide to form a phosphodiester bond. As shown in the figure below and in Figure 15.8, step 2, primers used during cellular DNA synthesis are short RNA strands, not DNA strands.

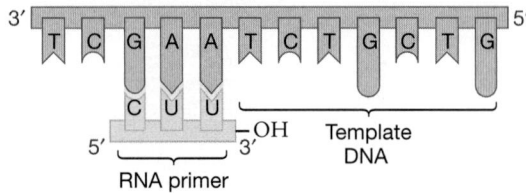

But what adds the primer? Before DNA synthesis can get under way, an enzyme called **primase** synthesizes a short stretch of RNA that acts as a primer for DNA polymerase. The primer is about 10 nucleotides long in *E. coli*. Primase is a type of **RNA polymerase**—an enzyme that catalyzes the polymerization of ribonucleotides into RNA (see Chapter 4 to review RNA's structure). Unlike DNA polymerases, primase and other RNA polymerases do not require a primer to begin synthesis.

Once a primer is present on a single-stranded template, DNA polymerase begins working in the 5′ → 3′ direction and adds deoxyribonucleotides to complete the complementary strand. As Figure 15.8, step 2, shows, DNA polymerase has a shape that grips the DNA strand during synthesis, similar to your hand clasping a rope. Deoxyribonucleotide addition is catalyzed at an active site in a groove between the enzyme's "thumb" and "fingers." As DNA polymerase moves along the DNA molecule, a doughnut-shaped structure behind it, called the sliding clamp, holds the enzyme in place on the template strand.

The enzyme's product is called the **leading strand,** or **continuous strand,** because it leads into the replication fork and is synthesized continuously. ✔ If you understand leading-strand synthesis, you should be able to list the enzymes involved and predict the consequences if any of them are defective.

How Is the Lagging Strand Synthesized?

Synthesis of the leading strand is straightforward. After an RNA primer is in place, DNA polymerase moves along, adding deoxyribonucleotides to the 3′ end of that strand. The enzyme moves into the replication fork, which "unzips" ahead of it. By comparison, events on the opposite strand are more involved.

Recall that the two strands of the DNA double helix are antiparallel—meaning they lie parallel to one another but oriented in opposite directions. The fact that DNA polymerases can synthesize DNA only in the 5′ → 3′ direction creates a paradox. Only one strand of DNA at the replication fork—the leading strand—can be synthesized in a direction that follows the moving replication fork.

The other strand must be synthesized in a direction that runs *away* from the moving replication fork, as illustrated in **FIGURE 15.9**. The strand of DNA that extends in the direction away from the

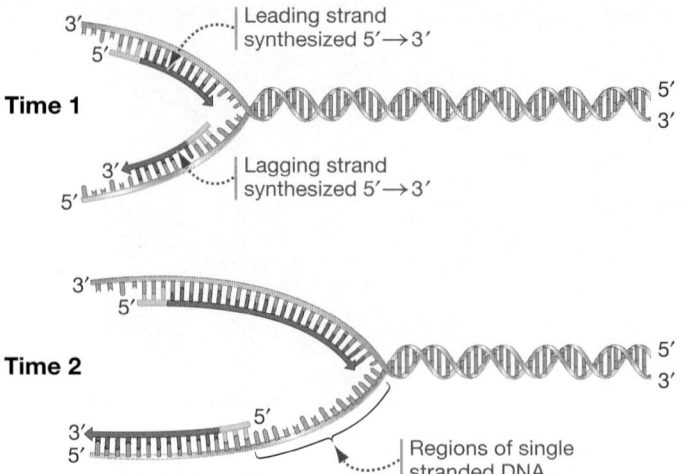

FIGURE 15.9 The Lagging Strand Is Synthesized in a Direction Moving Away from the Replication Fork. This occurs because the DNA strands are antiparallel and DNA polymerase can work only in the 5′ → 3′ direction.

replication fork is called the **lagging strand,** or **discontinuous strand,** because it lags behind the synthesis occurring at the fork. As the replication fork moves, it exposes gaps of single-stranded template DNA (Time 2 in Figure 15.9). How are the growing gaps filled in?

The Discontinuous Replication Hypothesis The puzzle posed by lagging-strand synthesis was resolved when Reiji Okazaki and colleagues tested a hypothesis called discontinuous replication. This hypothesis held that primase synthesizes new RNA primers for lagging strands as the moving replication fork opens single-stranded regions of DNA, and that DNA polymerase uses these primers to synthesize short lagging-strand DNA fragments that are linked together into a continuous strand. These ideas are illustrated in **FIGURE 15.10**.

Note that Figure 15.10 shows details of how lagging-strand synthesis occurs in *E. coli*. The overall process, however, applies to all groups of organisms—bacteria, archaea, and eukaryotes. The basic reactions of lagging-strand synthesis are universal. The differences lie in the names or specific properties of the key proteins and enzymes.

To explore the discontinuous replication hypothesis, Okazaki's group set out to test a key prediction: Could they find short DNA fragments produced during replication? Their critical experiment was based on the pulse–chase strategy (see Chapter 7). They added a brief "pulse" of radioactive deoxyribonucleotides to *E. coli* cells, followed by a "chase" of nonradioactive deoxyribonucleotides. According to the discontinuous replication model, some of these radioactive deoxyribonucleotides should first appear in short, fragments of DNA.

The Discovery of Okazaki Fragments As predicted, the researchers succeeded in finding short DNA fragments when they purified DNA from the experimental cells, separated the two strands of DNA, and analyzed the size of the molecules

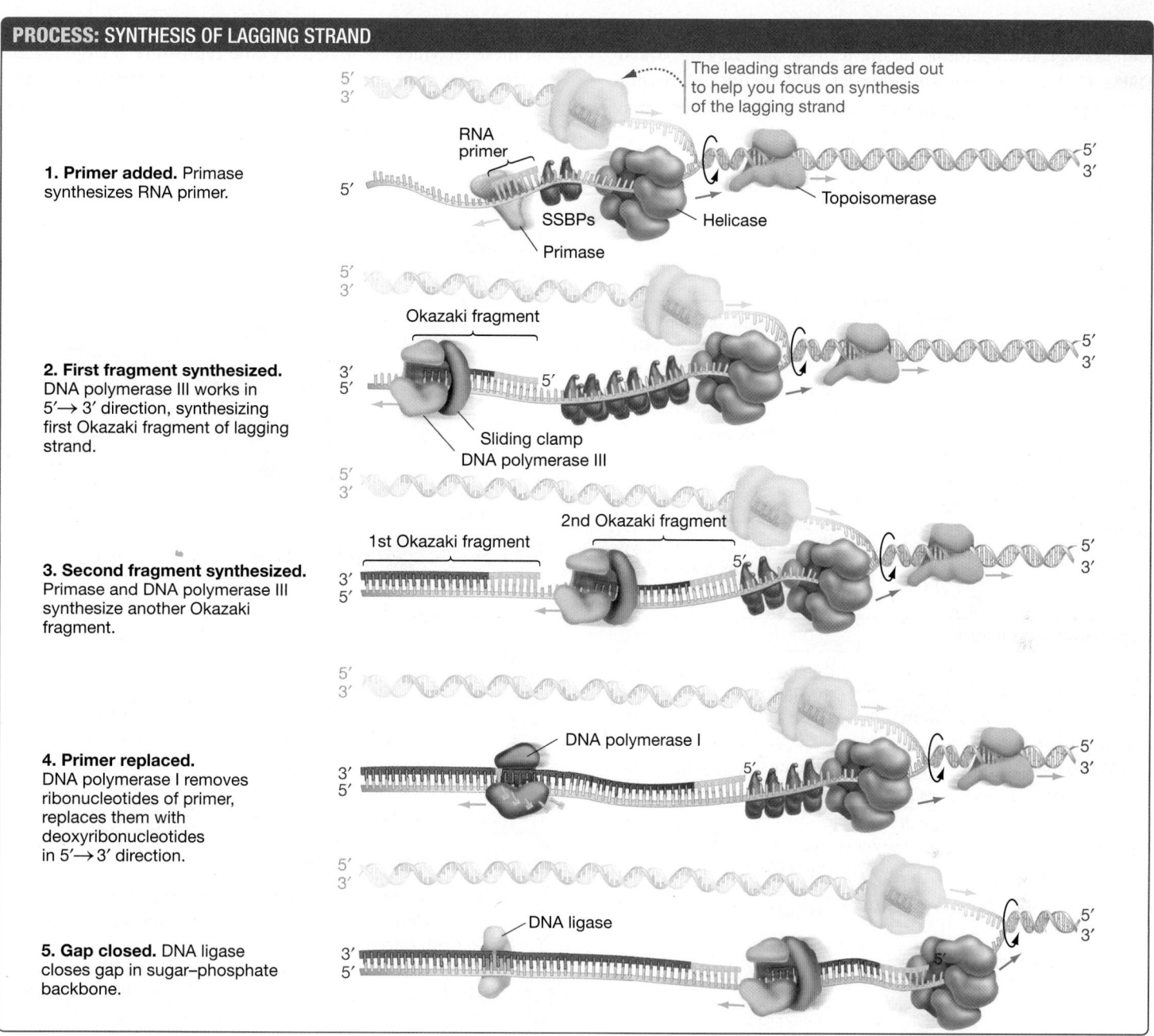

The leading strands are faded out to help you focus on synthesis of the lagging strand

1. Primer added. Primase synthesizes RNA primer.

RNA primer
SSBPs
Primase
Helicase
Topoisomerase

2. First fragment synthesized. DNA polymerase III works in 5′ → 3′ direction, synthesizing first Okazaki fragment of lagging strand.

Okazaki fragment
Sliding clamp
DNA polymerase III

3. Second fragment synthesized. Primase and DNA polymerase III synthesize another Okazaki fragment.

2nd Okazaki fragment
1st Okazaki fragment

4. Primer replaced. DNA polymerase I removes ribonucleotides of primer, replaces them with deoxyribonucleotides in 5′ → 3′ direction.

DNA polymerase I

5. Gap closed. DNA ligase closes gap in sugar–phosphate backbone.

DNA ligase

FIGURE 15.10 Lagging-Strand Synthesis.

by centrifugation. A small number of labeled DNA fragments about 1000 base pairs long were present immediately after the pulse. These short DNAs came to be known as **Okazaki fragments** and are shown in steps 2 and 3 of Figure 15.10. These small DNAs became larger during the chase as they were linked together into longer pieces. Subsequent work showed that Okazaki fragments in eukaryotes are even smaller—just 100 to 200 base pairs long.

How are Okazaki fragments connected? First, as step 4 of Figure 15.10 shows, in *E. coli* a specialized DNA polymerase called DNA polymerase I attaches to the 3′ end of an Okazaki fragment. As DNA polymerase I moves along in the 5′ → 3′ direction, it

removes that RNA primer ahead of it and replaces the ribonucleotides with the appropriate deoxyribonucleotides.

Once the RNA primer is removed and replaced by DNA, an enzyme called **DNA ligase** catalyzes the formation of a phosphodiester bond between the adjacent fragments (Figure 15.10, step 5). ✔ If you understand lagging-strand synthesis, you should be able to draw what the two newly synthesized molecules of DNA at a single replication fork would look like if DNA ligase were defective.

In eukaryotes, the mechanism for primer removal is different, but the mechanism of synthesizing short Okazaki fragments that are later joined into an unbroken chain of DNA is the same.

Working together, the enzymes that open the replication fork and manage the synthesis of the leading and lagging strands (**TABLE 15.1**) produce faithful copies of DNA before cell division. Although separate enzymes are drawn at different locations around the replication fork in Figures 15.8 and 15.10, in reality, all these enzymes are joined into the **replisome,** a large macromolecular machine. In *E. coli*, the replisome contains two copies of DNA polymerase III that are actively engaged in DNA synthesis. As shown in **FIGURE 15.11**, the lagging strand loops out and

SUMMARY TABLE 15.1 **Proteins Required for DNA Synthesis in Bacteria**

Name	Structure	Function
Opening the helix		
Helicase		Catalyzes the breaking of hydrogen bonds between base pairs to open the double helix
Single-strand DNA-binding proteins (SSBPs)		Stabilizes single-stranded DNA
Topoisomerase		Breaks and rejoins the DNA double helix to relieve twisting forces caused by the opening of the helix
Leading strand synthesis		
Primase		Catalyzes the synthesis of the RNA primer
DNA polymerase III		Extends the leading strand
Sliding clamp		Holds DNA polymerase in place during strand extension
Lagging strand synthesis		
Primase		Catalyzes the synthesis of the RNA primer on an Okazaki fragment
DNA polymerase III		Extends an Okazaki fragment
Sliding clamp		Holds DNA polymerase in place during strand extension
DNA polymerase I		Removes the RNA primer and replaces it with DNA
DNA ligase		Catalyzes the joining of Okazaki fragments into a continuous strand

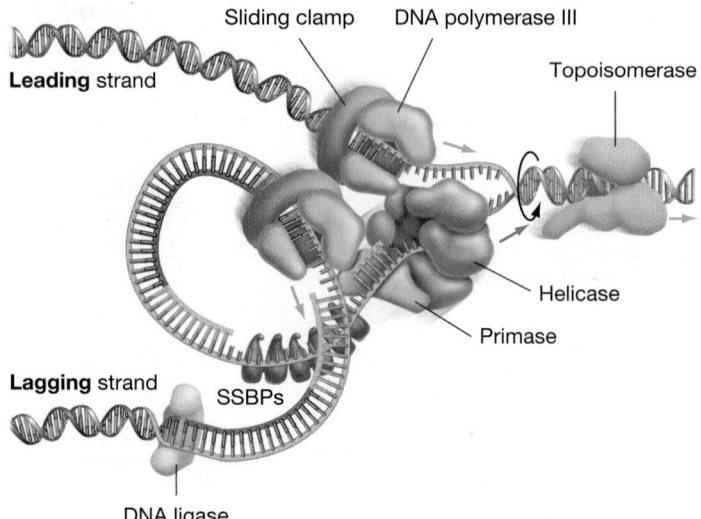

FIGURE 15.11 **The Replisome.** The enzymes required for DNA synthesis are organized into a macromolecular machine. Note how the lagging strand loops out as the leading strand is being synthesized.

check your understanding

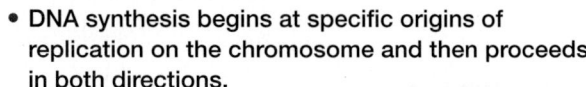

If you understand that . . .

- DNA synthesis begins at specific origins of replication on the chromosome and then proceeds in both directions.
- Synthesis at the replication fork occurs in three steps: (1) Helicase opens the double helix, SSBPs stabilize the exposed single strands, and topoisomerase prevents twists downstream of the fork; (2) DNA polymerase synthesizes the leading strand after primase has added an RNA primer; and (3) A series of enzymes synthesize the lagging strand.
- Lagging-strand synthesis cannot be continuous, because it moves away from the replication fork. In bacteria, enzymes called primase, DNA polymerase III, DNA polymerase I, and ligase work in sequence to synthesize Okazaki fragments and link them into a continuous whole.

✔ **You should be able to . . .**

1. Explain the function of primase.
2. Explain why DNA polymerase I is used predominantly on the lagging strand.

Answers are available in Appendix A.

around the complex, allowing the replisome to move as a single unit as it follows the replication fork. After the DNA polymerase on the lagging strand completes synthesis of an Okazaki fragment, it is released from the DNA and reassembles on the most recently synthesized primer.

15.4 Replicating the Ends of Linear Chromosomes

The circular DNA molecules in bacteria and archaea can be synthesized by the enzymes introduced in Section 15.3, and so can most of the linear DNA molecules found in eukaryotes. But replication at the very ends of linear eukaryotic chromosomes is another story altogether. Replication of chromosome ends requires a specialized DNA replication enzyme that has been the subject of intense research.

The End Replication Problem

The region at the end of a eukaryotic chromosome is called a **telomere** (literally, "end-part"). **FIGURE 15.12** illustrates the problem that arises during the replication of telomeres.

- When the replication fork reaches the end of a linear chromosome, a eukaryotic DNA polymerase synthesizes the leading strand all the way to the end of the parent DNA template (step 1 and step 2, top strand). As a result, leading-strand synthesis results in a double-stranded copy of the DNA molecule.

- On the lagging strand, primase adds an RNA primer close to the tip of the chromosome (see step 2, bottom strand).

- DNA polymerase synthesizes the final Okazaki fragment on the lagging strand (step 3). An enzyme that degrades ribonucleotides removes the primer.

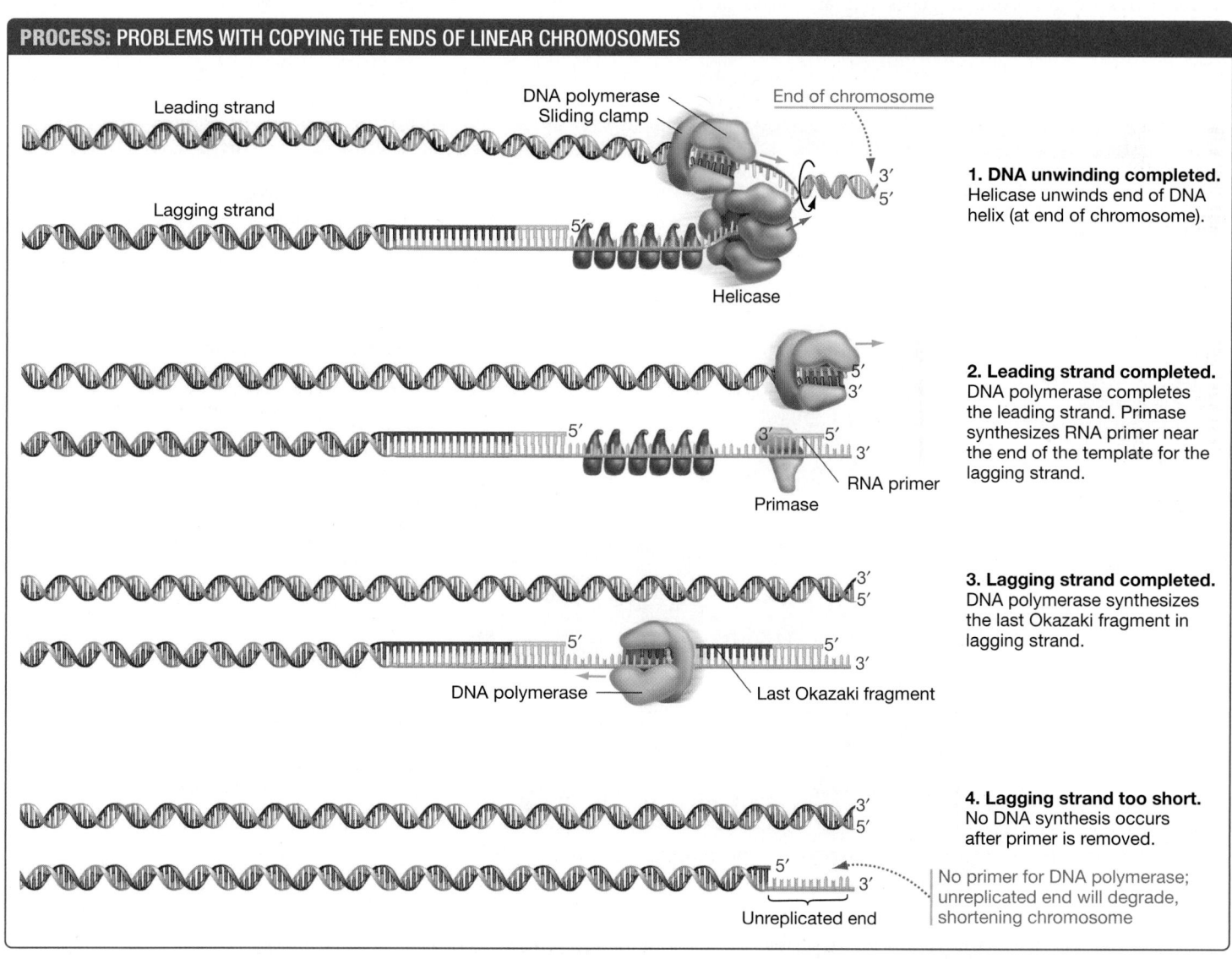

PROCESS: PROBLEMS WITH COPYING THE ENDS OF LINEAR CHROMOSOMES

Leading strand

DNA polymerase
Sliding clamp

End of chromosome

3′
5′

Lagging strand

5′

Helicase

1. DNA unwinding completed.
Helicase unwinds end of DNA helix (at end of chromosome).

5′
3′

5′

3′ 5′
3′

RNA primer

Primase

2. Leading strand completed.
DNA polymerase completes the leading strand. Primase synthesizes RNA primer near the end of the template for the lagging strand.

3′
5′

5′

5′
3′

DNA polymerase

Last Okazaki fragment

3. Lagging strand completed.
DNA polymerase synthesizes the last Okazaki fragment in lagging strand.

3′
5′

5′
3′

Unreplicated end

4. Lagging strand too short.
No DNA synthesis occurs after primer is removed.

No primer for DNA polymerase; unreplicated end will degrade, shortening chromosome

FIGURE 15.12 Chromosomes Shorten during Normal DNA Replication. An RNA primer is added to the lagging strand near the end of the chromosome. Once the primer is removed, it cannot be replaced with DNA. As a result, the chromosome shortens.

- DNA polymerase is unable to add DNA near the tip of the chromosome, because it cannot synthesize DNA without a primer (step 4). As a result, the single-stranded DNA that is left stays single stranded.

The single-stranded DNA at the end of the lagging strand is eventually degraded, which results in the shortening of the chromosome. If this process were to continue unabated, every chromosome would shorten by about 50 to 100 deoxyribonucleotides each time DNA replication occurred. Over time, linear chromosomes would vanish.

Telomerase Solves the End Replication Problem

How do eukaryotes maintain their chromosomes? One answer emerged after Elizabeth Blackburn, Carol Greider, and Jack Szostak reported two striking discoveries:

1. Telomeres do not contain genes but are made of short stretches of bases that are repeated over and over. In human telomeres, for example, the base sequence TTAGGG is repeated thousands of times.

2. A remarkable enzyme called telomerase that carries its own template is involved in replicating telomeres.

Telomerase is extraordinary because it catalyzes the synthesis of DNA from an RNA template that it contains. Telomerase adds DNA onto the end of a chromosome to prevent it from getting shorter.

FIGURE 15.13 shows one model for how telomerase works to maintain the ends of eukaryotic chromosomes.

Step 1 The unreplicated segment of the telomere at the 3′ end of the template for the lagging strand forms a single-stranded "overhang".

Step 2 Telomerase binds to the overhanging single-stranded DNA and begins DNA synthesis. The template for this reaction is a portion of the RNA held within telomerase.

Step 3 Telomerase synthesizes DNA in the 5′ → 3′ direction and catalyzes repeated additions of the same short DNA sequence to the end of the growing single strand.

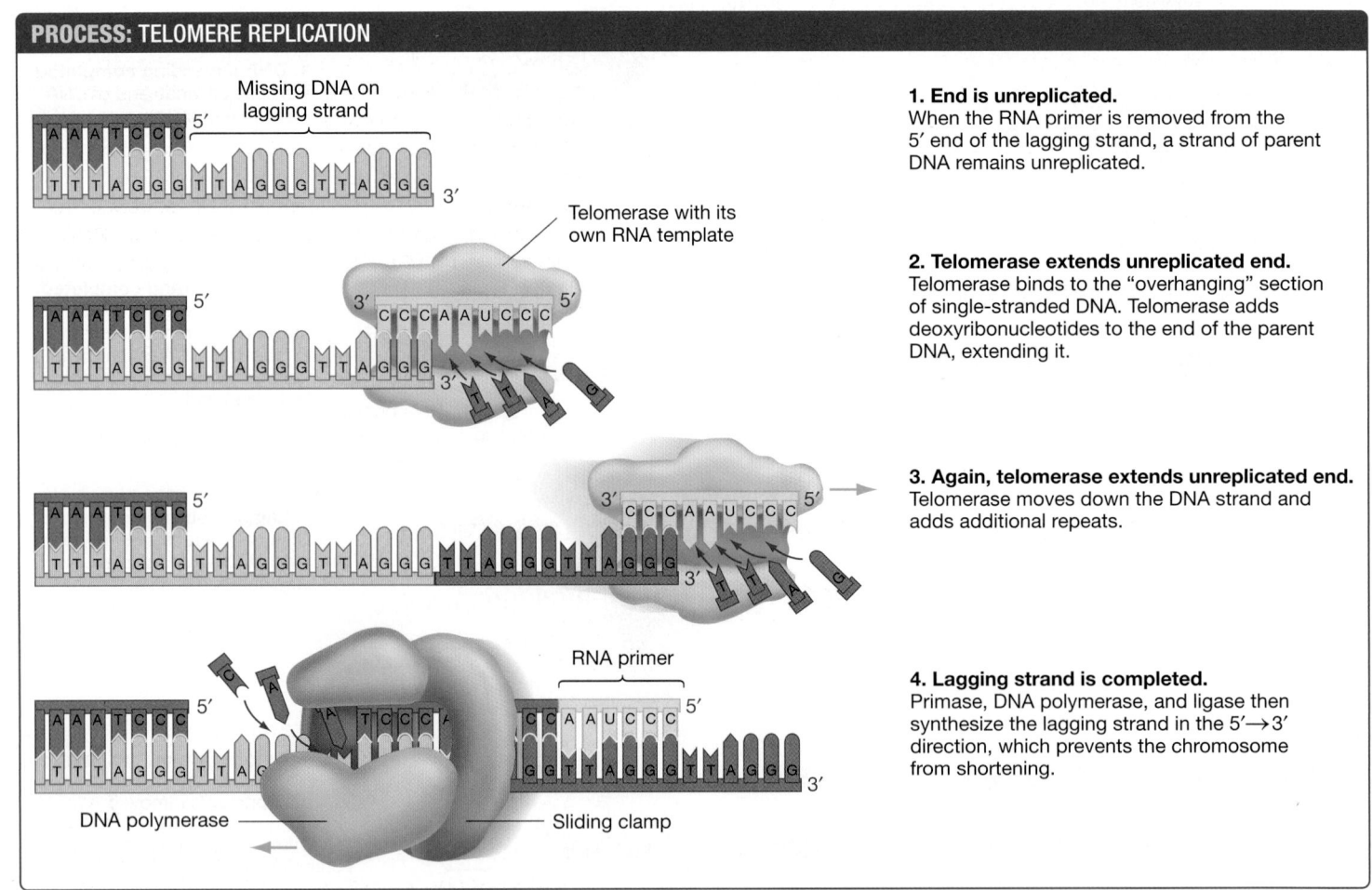

PROCESS: TELOMERE REPLICATION

Missing DNA on lagging strand

Telomerase with its own RNA template

1. End is unreplicated.
When the RNA primer is removed from the 5′ end of the lagging strand, a strand of parent DNA remains unreplicated.

2. Telomerase extends unreplicated end.
Telomerase binds to the "overhanging" section of single-stranded DNA. Telomerase adds deoxyribonucleotides to the end of the parent DNA, extending it.

3. Again, telomerase extends unreplicated end.
Telomerase moves down the DNA strand and adds additional repeats.

RNA primer

4. Lagging strand is completed.
Primase, DNA polymerase, and ligase then synthesize the lagging strand in the 5′→ 3′ direction, which prevents the chromosome from shortening.

DNA polymerase — Sliding clamp

FIGURE 15.13 Telomerase Prevents Shortening of Telomeres during Replication. By extending the number of repeated sequences in the 5′ → 3′ direction, telomerase provides room for enzymes to add an RNA primer to the lagging-strand template. Normal DNA replication enzymes can then fill in the missing section of the lagging strand.

✔ QUESTION Would this telomerase work as well if its RNA template had a different sequence?

Step 4 Once the single-stranded overhang on the parent strand is lengthened, the normal enzymes of DNA synthesis use this strand as a template to synthesize a complementary strand. The result is that the lagging strand becomes slightly longer than it was originally.

Telomerase Regulation

The way telomerase is regulated is just as remarkable as the enzyme itself. Telomerase is active in only a limited number of cell types. In humans, for example, active telomerase is found primarily in the cells that produce gametes. Most **somatic cells,** meaning cells that are not involved in gamete formation, lack telomerase activity. As predicted, the chromosomes of somatic cells gradually shorten with each mitotic division, becoming progressively smaller as an individual ages.

These observations led to the hypothesis that the number of cell divisions possible for a somatic cell would be limited by the initial length of its telomeres. Carol Greider and colleagues tested this hypothesis by obtaining cells with a variety of telomere lengths from donors aged newborn to 90 years old and growing these cells in culture. (For an introduction to cell culture, see **BioSkills 12** in Appendix B.) Results of their study are shown in **FIGURE 15.14**. As predicted, there was a positive relationship between initial telomere length and the number of cell divisions before cells stop dividing—longer initial telomere length allowed a greater number of cell divisions, regardless of the donor's age.

You've probably noticed that the data points in Figure 15.14 do not fall perfectly on a line. This scatter or noise is typical in many studies. In interpreting results like these, researchers must consider what might account for the scatter and use statistical tests (see **BioSkills 4** in Appendix B) to determine how reliable the results are likely to be.

If telomere shortening controls the number of divisions possible for a cell, then a related prediction is that by restoring telomerase in somatic cells, these cells should be freed from growth limitations. As predicted, when researchers added telomerase to human cells growing in culture, the cells continued dividing long past the age when otherwise identical cells stop growing. Most biologists are convinced that telomere shortening has a role in limiting the number of cell divisions for somatic cells.

There is a dark side of telomerase activity, however. Unlike the somatic cells they derive from, most cancer cells have active telomerase. Many cancer biologists have proposed that telomerase activity allows the unlimited divisions of cancer cells. A simple prediction is that by inhibiting telomerase, the progression of cancer can be slowed or stopped. When combined with other approaches, could drugs that knock out telomerase be an effective way to fight cancer? Unfortunately, the complexity of cancer often thwarts such simple predictions. So far, answers to this question are unclear. Research continues.

check your understanding

If you understand that . . .
- Linear chromosomes shorten during replication because the end of the lagging strand lacks a primer and cannot be synthesized.
- Shortening is prevented in certain cells—particularly those that produce sperm and egg—because telomerase adds short, repeated DNA sequences to the template strand. Primase can then add an RNA primer to the lagging strand, and DNA polymerase can fill in the missing sections.

✔ You should be able to . . .
1. Explain why telomerase is not needed by bacterial cells.
2. Explain why telomerase has to have a built-in template.

Answers are available in Appendix A.

15.5 Repairing Mistakes and DNA Damage

DNA polymerases work fast. In *E. coli*, for example, each replication fork advances about 500 nucleotides per second. But the replication process is also astonishingly accurate. In organisms ranging from *E. coli* to animals, the error rate during DNA replication averages about one mistake per *billion* deoxyribonucleotides.

This level of accuracy is critical. Humans, for example, develop from a fertilized egg that has roughly 12 billion deoxyribonucleotides in its DNA. This DNA is replicated over and over to create the trillions of cells that eventually make up the adult body. If more than one or two mutations occurred during each

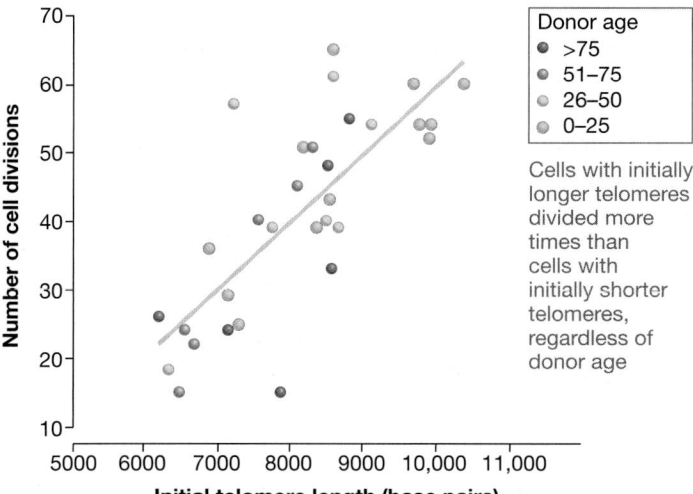

Cells with initially longer telomeres divided more times than cells with initially shorter telomeres, regardless of donor age

FIGURE 15.14 Telomere Length Predicts the Number of Divisions before Cells Stop Dividing.

DATA: Allsopp, R. C., et al. 1992. *Proceedings of the National Academy of Sciences*, 82: 10114–10118.

cell division cycle as a person developed, genes would be riddled with errors by the time the individual reached maturity. Genes that contain errors are often defective.

Based on these observations, it is no exaggeration to claim that the accurate replication of DNA is a matter of life and death. Natural selection favors individuals with enzymes that copy DNA quickly and accurately.

These observations raise a key question. How can the enzymes of DNA replication be as precise as they are?

Correcting Mistakes in DNA Synthesis

As DNA polymerase marches along a DNA template, hydrogen bonding occurs between incoming deoxyribonucleotides and the deoxyribonucleotides on the template strand. DNA polymerases are selective about the bases they add to a growing strand because (1) the correct base pairings (A-T and G-C) are energetically the most favorable, and (2) these correct pairings have a distinct shape. As a result, DNA polymerase inserts an incorrect deoxyribonucleotide (**FIGURE 15.15a**) only about once in every 100,000 bases added.

An error rate of one in 100,000 seems low, but it is much higher than the rate of one in a billion listed at the start of this section. What happens when DNA polymerase makes a mistake?

DNA Polymerase Proofreads Biologists learned more about how DNA synthesis could be so accurate when they found mutant cells in which DNA synthesis was *in*accurate.

Specifically, researchers found *E. coli* mutants with error rates that were 100 times greater than normal. Recall that a mutant is an individual with a novel trait caused by a mutation (see Chapter 14). In the case of *E. coli* mutants with high error rates

(a) DNA polymerase adds a mismatched base...

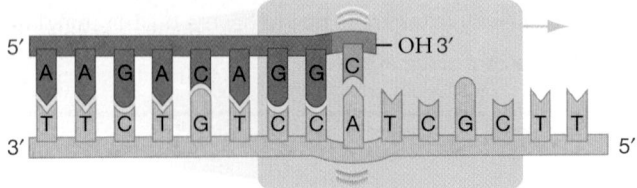

(b) ...but detects the mistake and corrects it.

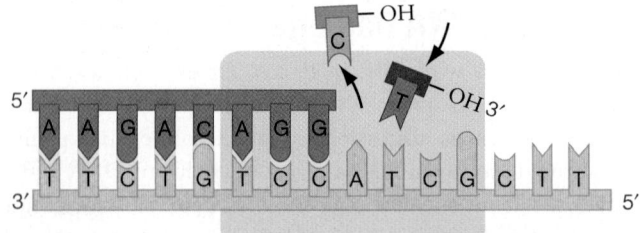

FIGURE 15.15 DNA Polymerase Can Proofread. If a mismatch such as the pairing of A with C occurs **(a)**, DNA polymerase can act as a 3′ → 5′ exonuclease, meaning that it can remove bases in that direction **(b)**. The DNA polymerase then adds the correct base.

in DNA replication, biologists found a defect in a portion of the DNA polymerase III enzyme called the ε (epsilon) subunit. Further analyses showed that the ε subunit acts as an exonuclease—meaning an enzyme that removes deoxyribonucleotides from the ends of DNA strands (**FIGURE 15.15b**).

If a newly added deoxyribonucleotide is not correctly paired with a base on the complementary strand, the positioning of the incorrect deoxyribonucleotide provides a poor substrate for DNA polymerase to extend. This is because the geometry of incorrect base pairs differs from that of the correct A-T and G-C pairs. DNA polymerase's active site can detect these shapes and will add a new deoxyribonucleotide only when the previous base pair is correct. In wild-type *E. coli*, the polymerase pauses when it detects the wrong shape, and the exonuclease activity of the ε subunit removes the mismatched deoxyribonucleotide.

These findings led to the conclusion that DNA polymerase III can **proofread.** If the wrong base is added during DNA synthesis, the enzyme pauses, removes the mismatched deoxyribonucleotide that was just added, and then proceeds again with synthesis.

Eukaryotic DNA polymerases have the same type of proofreading ability. Typically, proofreading reduces the overall error rate of DNA synthesis to about one mistake in 10 million bases added. Is this accurate enough? The answer remains no.

Mismatch Repair If—despite its proofreading ability—DNA polymerase leaves a mismatched base behind in the newly synthesized strand, a battery of enzymes springs into action to correct the problem. **Mismatch repair** occurs when mismatched bases are corrected after DNA synthesis is complete.

The proteins responsible for mismatch repair were discovered in the same way proofreading was—by analyzing *E. coli* mutants. In this case, the mutants had normal DNA polymerase III but abnormally high mutation rates.

The first mutation that caused a deficiency in mismatch repair was identified in the late 1960s and was called *mutS*. (The *mut* is short for "mutator.") Twenty years later, researchers had identified 10 proteins involved in the identification and repair of base-pair mismatches in *E. coli*.

These proteins recognize the mismatched base, remove a section containing the incorrect base from the newly synthesized strand, and fill in the correct bases using the older strand as a template. In *E. coli*, chemical marks on the older strand allow the enzymes to distinguish the original strand from the newly synthesized strand. Eukaryotes use a different scheme to recognize the old and new strands of DNA.

This final layer of error detection and correction brings the overall error rate of DNA synthesis down to roughly one mistake per billion deoxyribonucleotides. The mismatch-repair enzymes are like a copy editor who corrects the errors that a writer—DNA polymerase—did not catch.

The importance of mismatch repair is revealed by grim discoveries: Mutations in components of the mismatch repair system are observed in many common human cancers, where they play an important role in cancer development and progression.

Repairing Damaged DNA

Even after DNA is synthesized and proofread and mismatches repaired, the job of ensuring accuracy doesn't end. Genes are under constant assault. DNA is damaged by sunlight, X-rays, and many chemicals like the hydroxyl (OH) radicals produced during aerobic metabolism, aflatoxin B1 found in moldy peanuts and corn, and benzo[α]pyrene in cigarette smoke. If this damage were ignored, mutations would quickly accumulate to lethal levels. To fix problems caused by chemical attack, radiation, or other events, organisms have evolved a wide array of DNA damage-repair systems. As an example, consider the **nucleotide excision repair** system that works on DNA damage caused by ultraviolet light and many different chemicals.

Ultraviolet (UV) light in sunlight—and tanning booths—can cause a covalent bond to form between adjacent pyrimidine bases within the same DNA strand. The thymine-thymine pair illustrated in **FIGURE 15.16** is a common example. This defect, called a thymine dimer, creates a kink in the structure of DNA. The kink stalls standard DNA polymerases, blocking DNA replication. If the damage is not repaired, the cell may die.

Nucleotide excision repair fixes thymine dimers and many other types of damage that distort the DNA helix. In the first step of excision repair, an enzyme recognizes the kink in the DNA helix (step 1 in **FIGURE 15.17**). Once a damaged region is recognized, another enzyme removes a segment of single-stranded DNA containing the defective sequence (step 2). The intact DNA strand provides a template for synthesis of a corrected strand, and the 3′ hydroxyl of the DNA strand next to the gap serves as a primer (step 3). DNA ligase links the newly synthesized DNA to the original undamaged DNA (step 4). As with mismatch repair, multiple enzymes work together and DNA synthesis plays a central role in repair.

What happens when a human DNA repair system is defective?

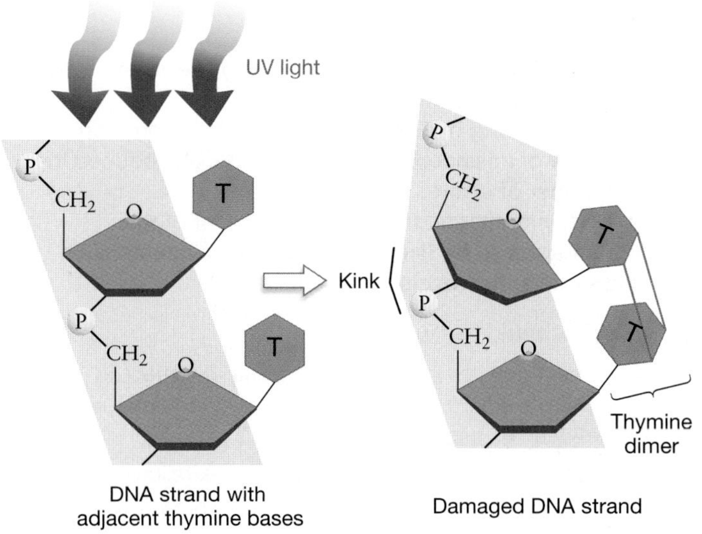

Xeroderma Pigmentosum: A Case Study

Xeroderma pigmentosum (XP) is a rare autosomal recessive disease in humans. Individuals with this condition are extremely sensitive to ultraviolet (UV) light. Their skin develops lesions including rough, scaly patches and irregular dark spots after even slight exposure to sunlight.

In 1968 James Cleaver proposed a connection between XP and DNA nucleotide excision repair. He knew that mutants of *E. coli* had defects in nucleotide excision repair that caused an increased sensitivity to radiation. Cleaver's hypothesis was that people with XP have similar mutations. He proposed that they are extremely sensitive to sunlight because they are unable to repair damage induced by UV light.

Cleaver and other researchers made extensive use of cell cultures (**BioSkills 12**, see Appendix B) to study the hypothesized connection between DNA damage, faulty nucleotide excision repair, and XP. They collected skin cells from people with XP and from people with normal UV light sensitivity. When these cells were grown in culture, the biologists exposed them to increasing amounts of UV radiation and recorded how many survived.

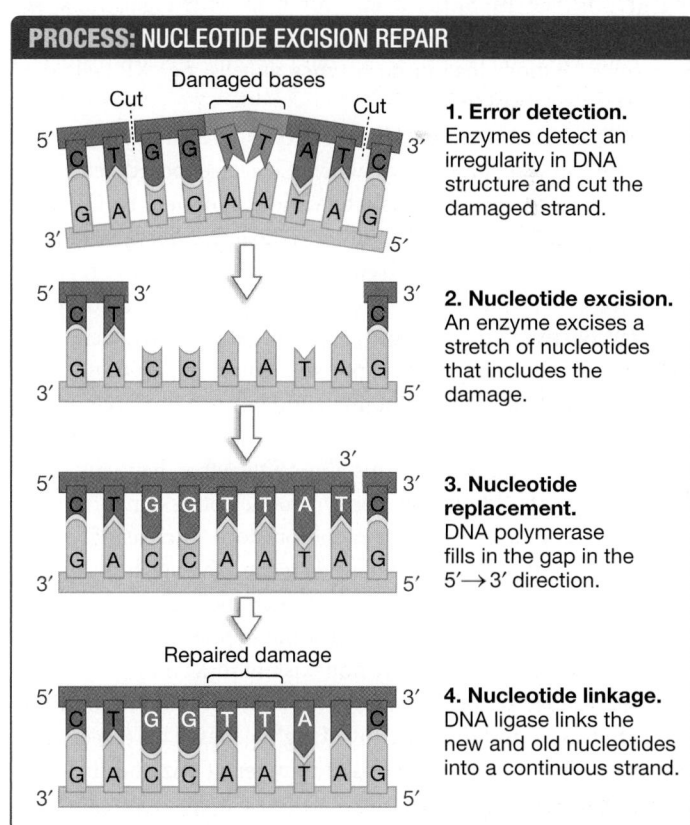

PROCESS: NUCLEOTIDE EXCISION REPAIR

1. Error detection. Enzymes detect an irregularity in DNA structure and cut the damaged strand.

2. Nucleotide excision. An enzyme excises a stretch of nucleotides that includes the damage.

3. Nucleotide replacement. DNA polymerase fills in the gap in the 5′→3′ direction.

4. Nucleotide linkage. DNA ligase links the new and old nucleotides into a continuous strand.

FIGURE 15.17 In Nucleotide Excision Repair, Defective Bases Are Removed and Replaced.

FIGURE 15.16 UV Light Damages DNA. When UV light strikes a section of DNA that has adjacent thymines, the energy can break bonds within each base and lead to the formation of bonds *between* them. The thymine dimer that is produced causes a kink in the DNA.

✔ **QUESTION** Why are infrared wavelengths much less likely than UV to damage DNA? (Hint: See Figure 10.4.)

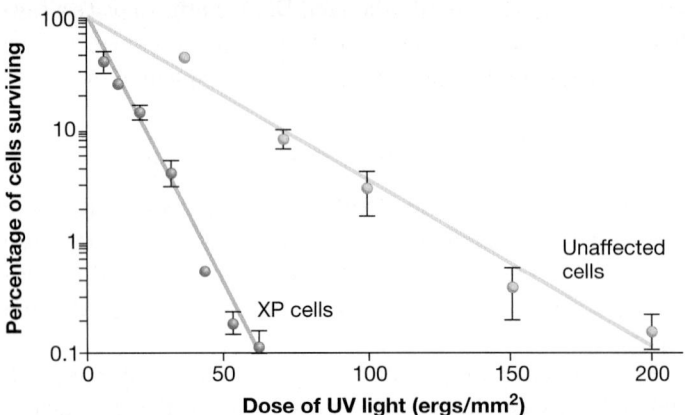

(a) Vulnerability of cells to UV light damage

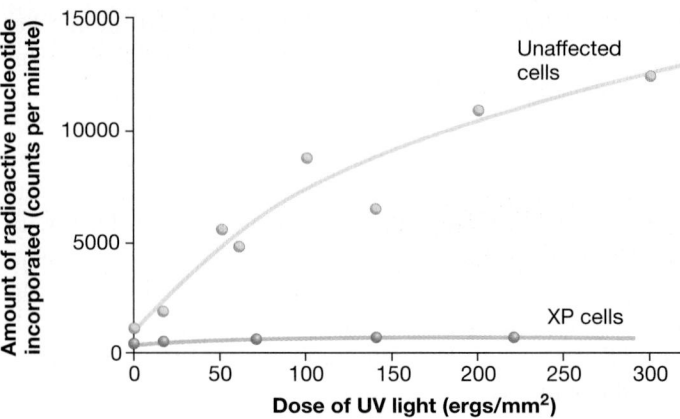

(b) Ability of cells to repair UV light damage

FIGURE 15.18 DNA Damage from UV Light Is Not Repaired Properly in Individuals with XP. (a) When cell cultures from unaffected individuals and XP patients are irradiated with various doses of UV light (expressed here as ergs/mm^2), the percentage of cells that survive is strikingly different. **(b)** When cell cultures from unaffected individuals and XP patients are irradiated with various doses of UV light and then provided with a radioactive deoxyribonucleotide, only cells from unaffected individuals incorporate the labeled deoxyribonucleotide into their DNA.

DATA: (a) Cleaver, J. E. 1970. *Int. J. Rad. Biol.* 18: 577–565, Fig 3. (b) Cleaver, J. E. 1972. *J. Invest. Dermatol.* 58: 124–128, Fig 1.

✔**QUESTION** Why are people who cultivate a sun tan increasing their risk of developing cancer? (Hint: Tanning is a response to UV light.)

FIGURE 15.18a shows the results of one such study by Cleaver. Note that the intensity of the radiation is graphed on the *x*-axis, and the percentage of cells surviving is graphed on the *y*-axis. Note, too, that the *y*-axis is logarithmic. (For help with reading graphs, see **BioSkills 3** and for help with logarithms, see **BioSkills 6**, both in Appendix B.) Cell survival declined with increasing radiation dose in both types of cells, but XP cells died off much more rapidly.

The connection to nucleotide excision repair systems was confirmed in a separate study when Cleaver exposed cells from unaffected and XP individuals to various amounts of UV light and then incubated the cells with a radioactive deoxyribonucleotide to label DNA synthesized during DNA repair. If repair is defective in XP individuals, then their cells should incorporate little radioactive deoxyribonucleotide into their DNA. Cells from unaffected individuals, in contrast, should incorporate large amounts of labeled deoxyribonucleotide into their DNA as it is repaired.

As **FIGURE 15.18b** shows, this is exactly what happens. Here the amount of radioactive deoxyribonucleotides incorporated into DNA is graphed against radiation dose. Increasingly large amounts of radioactivity are found in the DNA of healthy cells as UV dose increases, but almost no such increase occurs in XP cells. These data are consistent with the hypothesis that nucleotide excision repair is virtually nonexistent in XP individuals.

Genetic analyses of XP patients have shown that the condition can result from mutations in any of eight genes. This discovery is not surprising in light of the large number of enzymes involved in repairing damaged DNA.

As you saw for mismatch repair, defects in DNA repair genes are frequently associated with cancer. Individuals with xeroderma pigmentosum, for example, are 1000 to 2000 times more likely to get skin cancer than are individuals with intact excision repair systems. To explain this pattern, biologists suggest that if DNA damage in the genes involved in the cell cycle goes unrepaired, mutations will result that may allow the cell to grow in an uncontrolled manner. Tumor formation could result. Recall

check your understanding

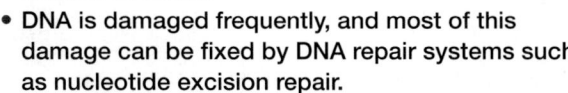

If you understand that . . .

- DNA polymerases occasionally add the wrong base during DNA synthesis.
- Proofreading by DNA polymerase and mismatch repair of misincorporated bases sharply reduces the number of errors.
- DNA is damaged frequently, and most of this damage can be fixed by DNA repair systems such as nucleotide excision repair.

✔ **You should be able to . . .**

1. Predict how the mutation rate would be affected if there were no differences in stability and shape between all possible base pairs.

2. Predict the effect on mutation rate of a failure in the system for distinguishing old and newly synthesized DNA.

3. State which nucleotide excision repair enzymes are specific for DNA repair and which work in both normal DNA replication and in DNA repair.

Answers are available in Appendix A.

that most cancers develop only after several genes have been damaged (see Chapter 12). If the overall mutation rate in a cell is elevated because of defects in DNA repair, then mutations that trigger cancer become more likely.

At this point, it's clear that genes are made of DNA and that DNA is accurately copied and passed on to offspring. How can information be stored in DNA, and how can this information be used? (These are the topics of the next two chapters.)

If you understand . . .

15.1 What Are Genes Made Of?

- Experiments on viruses that had labeled proteins or DNA showed that DNA is the hereditary material.

- DNA's primary structure consists of a sugar–phosphate backbone and a sequence of nitrogen-containing bases.

- DNA's secondary structure consists of two strands in an antiparallel orientation. The strands twist into a helix and are held together by complementary pairing between bases.

✔ You should be able to interpret an imaginary experiment like the one done by Hershey and Chase that shows that ^{32}P is found only in the pellet and that ^{35}S is found in both the pellet and the solution.

15.2 Testing Early Hypotheses about DNA Synthesis

- By labeling DNA with ^{15}N or ^{14}N, researchers were able to validate the hypothesis that DNA replication is semiconservative.

- In semiconservative replication, each strand of a parent DNA molecule provides a template for the synthesis of a daughter strand, resulting in two complete DNA double helices.

✔ You should be able to write a sequence of double-stranded DNA that is 10 base pairs long, separate the strands, and, without comparing them, write in the bases that are added during DNA replication.

15.3 A Model for DNA Synthesis

- DNA synthesis requires many different enzymes, and it occurs in one direction only.

- DNA synthesis requires both a template and a primer sequence. It takes place at the replication fork where the double helix is opened.

- Synthesis of the leading strand in the $5' \rightarrow 3'$ direction is continuous, but synthesis of the lagging strand is discontinuous because on that strand, the DNA polymerase moves away from the replication fork.

- On the lagging strand, short DNA fragments called Okazaki fragments form and are joined together. Okazaki fragments are primed by a short strand of RNA.

✔ You should be able to draw and label a diagram of a replication bubble that shows (1) the $5' \rightarrow 3'$ polarity of the two parental DNA strands and (2) the leading and lagging daughter strands at each replication fork.

15.4 Replicating the Ends of Linear Chromosomes

- At the ends of linear chromosomes in eukaryotes, the enzyme telomerase adds short, repeated sections of DNA so that the lagging strand can be synthesized without shortening the chromosome.

- Telomerase is active in reproductive cells that eventually undergo meiosis. As a result, gametes contain chromosomes of normal length.

- Chromosomes in cells without telomerase shorten with continued cell division until their telomeres reach a critical length at which cell division no longer occurs.

✔ You should be able to explain the significance of telomerase reactivation in cancer cells.

15.5 Repairing Mistakes and DNA Damage

- DNA replication is remarkably accurate because (1) DNA polymerase selectively adds a deoxyribonucleotide that correctly pairs with the template strand; (2) DNA proofreads each added deoxyribonucleotide; and (3) mismatch repair enzymes remove incorrect bases once synthesis is complete and replace them with the correct base.

- DNA repair occurs after DNA has been damaged by chemicals or radiation.

- Nucleotide excision repair cuts out damaged portions of DNA and replaces them with correct sequences.

- If DNA repair enzymes are defective, mutation rate increases. Because of this, several types of human cancers are associated with defects in the genes responsible for DNA repair.

✔ You should be able to explain the logical connections between failure of repair systems, increases in mutation rate, and high likelihood of cancer developing.

MasteringBiology

1. MasteringBiology Assignments

Tutorials and Activities DNA and RNA Structure; DNA Double Helix; DNA Replication; DNA Replication: A Closer Look; DNA Replication: A Review; DNA Replication: An Overview; DNA Synthesis; Experimental Inquiry: Does DNA Replication Follow the Conservative, Semiconservative, or Dispersive Model; Hershey–Chase Experiment

Questions Reading Quizzes, Blue-Thread Questions, Test Bank

2. eText Read your book online, search, take notes, highlight text, and more.

3. The Study Area Practice Test, Cumulative Test, BioFlix® 3-D Animations, Videos, Activities, Audio Glossary, Word Study Tools, Art

You should be able to . . .

✓ TEST YOUR KNOWLEDGE
Answers are available in Appendix A

1. What does it mean to say that strands in a double helix are antiparallel?
   a. Their primary sequences consist of a sequence of *complementary* bases.
   b. They each have a sugar–phosphate backbone.
   c. They each have a $5' \rightarrow 3'$ directionality.
   d. They have opposite directionality, or polarity.

2. Which of the following is *not* a property of DNA polymerase?
   a. It adds dNTPs only in the $5' \rightarrow 3'$ direction.
   b. It requires a primer to work.
   c. It is associated with a sliding clamp only on the leading strand.
   d. Its exonuclease activity is involved in proofreading.

3. The enzyme that removes twists in DNA ahead of the replication fork is _____.

4. What is the function of primase?
   a. synthesis of the short section of double-stranded DNA required by DNA polymerase
   b. synthesis of a short RNA, complementary to single-stranded DNA
   c. closing the gap at the $3'$ end of DNA after excision repair
   d. removing primers and synthesizing a short section of DNA to replace them

5. How are Okazaki fragments synthesized?
   a. using the leading strand template, and synthesizing $5' \rightarrow 3'$
   b. using the leading strand template, and synthesizing $3' \rightarrow 5'$
   c. using the lagging strand template, and synthesizing $5' \rightarrow 3'$
   d. using the lagging strand template, and synthesizing $3' \rightarrow 5'$

6. An enzyme that uses an internal RNA template to synthesize DNA is _____.

✓ TEST YOUR UNDERSTANDING
Answers are available in Appendix A

7. Researchers design experiments so that only one thing is different between the treatments that are being compared. In the Hershey–Chase experiment, what was this single difference?

8. What is the relationship between defective DNA repair and cancer?

9. Why is the synthesis of the lagging strand of DNA discontinuous? How is it possible for the synthesis of the leading strand to be continuous?

10. Explain how telomerase prevents linear chromosomes from shortening during replication.

11. Predict what would occur in a bacterial mutant that lost the ability to chemically mark the template strand of DNA.
    a. The mutation rate would increase.

    b. The ability of DNA polymerase to discriminate between correct and incorrect base pairs would decrease.
    c. The energy differences between correct and incorrect base pairs would decrease.
    d. The energy differences between correct and incorrect base pairs would increase.

12. What aspect of DNA structure makes it possible for the enzymes of nucleotide excision repair to recognize many different types of DNA damage?
    a. the polarity of each DNA strand
    b. the antiparallel orientation of strands in the double helix
    c. the energy differences between correct and incorrect base pairs
    d. the regularity of DNA's overall structure

13. If you could engineer an activity into DNA polymerase to allow both strands to follow the replication fork, what would this additional activity be?
   a. the ability to begin DNA synthesis without a primer
   b. the ability to proofread in the $5' \rightarrow 3'$ direction
   c. the ability to synthesize DNA in the $3' \rightarrow 5'$ direction
   d. the ability to synthesize DNA without using a template

14. In the late 1950s, Herbert Taylor grew bean root-tip cells in a solution of radioactive thymidine and allowed them to undergo one round of DNA replication. He then transferred the cells to a solution without the radioactive deoxyribonucleotide, allowed them to replicate again, and examined their chromosomes for the presence of radioactivity. His results are shown in the following figure, where red indicates a radioactive chromatid.

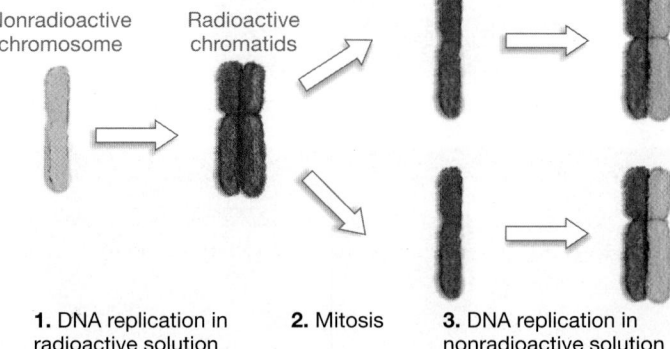

1. DNA replication in radioactive solution   **2.** Mitosis   **3.** DNA replication in nonradioactive solution

   a. Draw diagrams explaining the pattern of radioactivity observed in the sister chromatids after the first and second rounds of replication.
   b. What would the results of Taylor's experiment be if eukaryotes used a conservative mode of DNA replication?

15. The graph that follows shows the survival of four different *E. coli* strains after exposure to increasing doses of ultraviolet light. The wild-type strain is normal, but the other strains have a mutation in either a gene called *uvrA*, a gene called *recA*, or both.

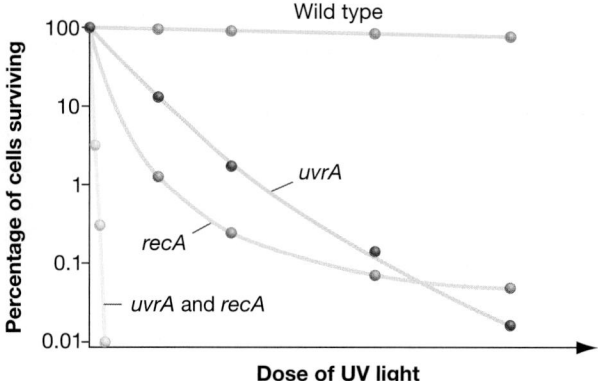

DATA: Howard-Flanders, P., and R. P. Boyce. 1966. *Radiation Research Supplement* 6: 156–184, Fig. 8.

   a. Which strains are most sensitive to UV light? Which strains are least sensitive?
   b. What are the relative contributions of these genes to the repair of UV damage?

16. QUANTITATIVE Assuming that each replication fork moves at a rate of 500 base pairs per second, how long would it take to replicate the *E. coli* chromosome (with 4.6 million base pairs) from a single origin of replication?

16 How Genes Work

In this chapter you will learn how

Genetic information flows from DNA → RNA → proteins

by asking ↓   *then examining* ↓   *and analyzing* ↓

What do genes do?
16.1

The central dogma of molecular biology
16.2

The genetic code, with its 3-letter "words"
16.3

which together explain ↓

How mutations can modify genes and genomes
16.4

This image shows a normal human male spectral karyotype—a micrograph of metaphase chromosomes stained to show different homologous chromosome pairs. This chapter explores how DNA sequences in chromosomes are related to phenotypes.

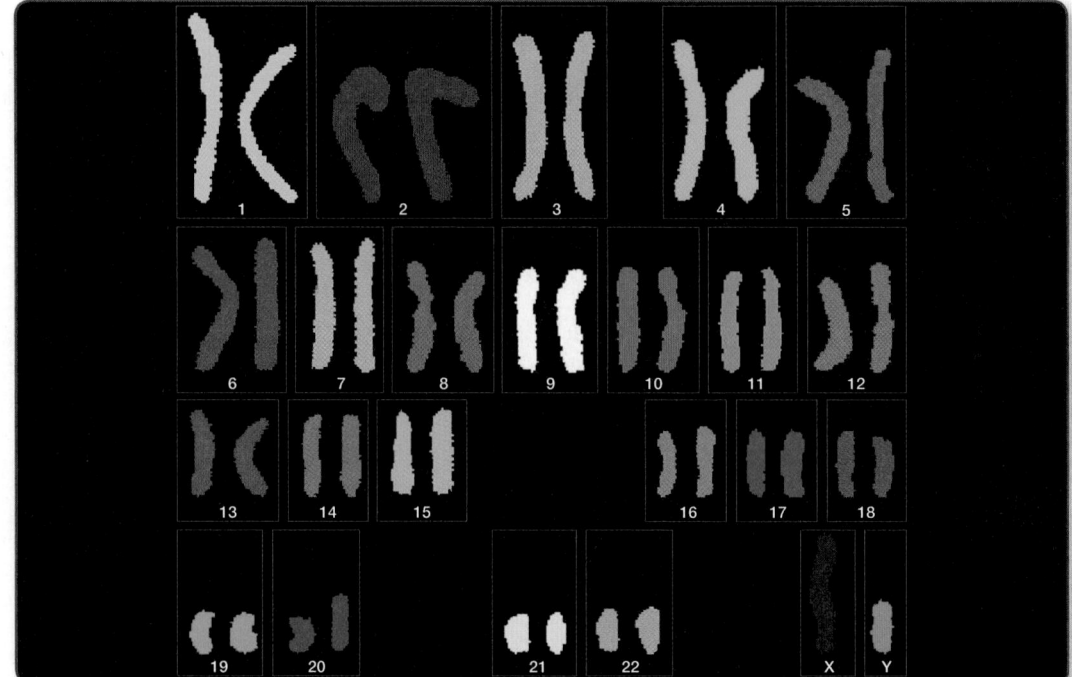

B**G PICTURE**

This chapter is part of the Big Picture. See how on pages 366–367.

DNA has been called the blueprint of life. If an organism's DNA is like a set of blueprints, then its cells are like construction sites, and the enzymes inside a cell are like construction workers. But how does the DNA inside each cell assemble this team of skilled laborers and specify the materials needed to build and maintain the cell, and remodel it when conditions change?

Mendel provided insights that made the study of these questions possible. He discovered that particular alleles are associated with certain phenotypes and that alleles do not change when transmitted from parent to offspring. Later, the chromosome theory of inheritance established that genes are found in chromosomes, whose movement during meiosis explains Mendel's results.

The science of molecular biology began with the discovery that DNA is the hereditary material and that DNA is a double-helical structure containing sequences of four bases. From these early advances, it was clear that genes are made of DNA and that genes carry the instructions for making and maintaining an individual.

✔ When you see this checkmark, stop and test yourself. Answers are available in Appendix A.

But biologists still didn't know how the information in DNA is translated into action. How does **gene expression**—the process of converting archived information into molecules that actually do things in the cell—occur?

This chapter introduces some of the most pivotal ideas in all of biology—ideas that connect genotypes to phenotypes by revealing how genes work at the molecular level. They also speak to the heart of a key attribute of life: processing genetic information to produce a living organism. (You can see how these concepts fit into the Big Picture of Genetic Information on pages 366–367.)

Understanding how genes work triggered a major transition in biological science. Instead of thinking about genes solely in relation to their effects on eye color in fruit flies or on seed shape in garden peas, biologists could begin analyzing the molecular composition of genes and their products. The molecular revolution in biology took flight.

16.1 What Do Genes Do?

Although biologists of the early twentieth century made tremendous progress in understanding how genes are inherited, an explicit hypothesis explaining what genes do did not appear until 1941. That year George Beadle and Edward Tatum published a series of breakthrough experiments on a bread mold called *Neurospora crassa*.

Beadle and Tatum's research was inspired by an idea that was brilliant in its simplicity. As Beadle said: "One ought to be able to discover what genes do by making them defective." The idea was to knock out a gene by damaging it and then infer what the gene does by observing the phenotype of the mutant individual.

Today, alleles that do not function at all are called **knock-out, null,** or **loss-of-function alleles.** Creating knock-out mutant alleles and analyzing their effects is still one of the most common research strategies in studies of gene function. But Beadle and Tatum were the pioneers.

The One-Gene, One-Enzyme Hypothesis

To start their work, Beadle and Tatum exposed a large number of *N. crassa* cells to radiation. As described earlier (Chapter 15), radiation can damage the double-helical structure of DNA—often in a way that makes the affected gene nonfunctional.

Their next step was to examine the mutant cells. Eventually they succeeded in finding *N. crassa* mutants that could not make specific compounds. For example, one of the mutants could not make pyridoxine, also called vitamin B_6, even though normal individuals can. Further, Beadle and Tatum showed that the inability to synthesize pyridoxine was due to a defect in a single gene, and that the inability to synthesize other molecules was due to defects in other genes.

These results inspired their **one-gene, one-enzyme hypothesis.** Beadle and Tatum proposed that the mutant *N. crassa* could not make pyridoxine because it lacked an enzyme required to synthesize the compound. They further proposed that the lack of the enzyme was due to a genetic defect. Based on analyses of knock-out mutants, the one-gene, one-enzyme hypothesis claimed that each gene contains the information needed to make an enzyme.

An Experimental Test of the Hypothesis

Three years later, Adrian Srb and Norman Horowitz published a rigorous test of the one-gene, one-enzyme hypothesis. These biologists focused on the ability of *N. crassa* to synthesize the amino acid arginine. In the lab, normal cells of this bread mold grow well on a laboratory culture medium that lacks arginine. This is possible because *N. crassa* cells are able to synthesize their own arginine.

Previous work had shown that organisms synthesize arginine in a series of steps called a **metabolic pathway.** As **FIGURE 16.1** shows, compounds called ornithine and citrulline are intermediate products in the metabolic pathway leading to arginine. Specific enzymes are required to synthesize ornithine, convert ornithine to citrulline, and change citrulline to arginine. Srb and Horowitz hypothesized that specific *N. crassa* genes are responsible for producing each of the three enzymes involved.

To test this idea, Srb and Horowitz used radiation to create a large number of mutant cells. However, radiation is equally likely to damage DNA and mutate genes in any part of the organism's genome, and most organisms have thousands or tens of thousands of genes. Of the many mutants the biologists created, how could they find the handful that specifically knocked out a step in the pathway for arginine synthesis?

To find the mutants they were looking for, the researchers performed what is now known as a genetic screen. A **genetic screen** is any technique for picking certain types of mutants out of many randomly generated mutants.

Srb and Horowitz began their screen by raising colonies of irradiated cells on a medium that included arginine. Then they transferred a sample of each colony to a medium that *lacked* arginine. If an individual could grow in the presence of arginine but failed to grow without arginine, they concluded that it couldn't make its own arginine.

Metabolic pathway for arginine synthesis: Precursor — Enzyme 1 → Ornithine — Enzyme 2 → Citrulline — Enzyme 3 → Arginine

FIGURE 16.1 Different Enzymes Catalyze Each Step in the Metabolic Pathway for Arginine.

✔**QUESTION** If a cell lacked enzyme 2 but was placed in growth medium with ornithine, could it grow? Could it grow if it received citrulline instead?

The biologists followed up by confirming that the offspring of these cells also had this defect. Based on these data, they were confident that they had isolated individuals with mutations in one or more of the genes for the enzymes shown in Figure 16.1.

To test the one-gene, one-enzyme hypothesis, the biologists grew each mutant under four different conditions: on normal media without added arginine, and on normal medium supplemented with ornithine, citrulline, or arginine.

As **FIGURE 16.2** shows, the results from these growth experiments were dramatic. Some of the mutant cells were able to grow on some of these media but not on others. More specifically, the mutants fell into three distinct classes, which the researchers called *arg1*, *arg2*, and *arg3*.

As the "Interpretation" section of the figure shows, the data make sense if each type of mutant lacked a different, specific step in a metabolic pathway because of a defect in a particular gene. In short, Srb and Horowitz had documented a correlation between a specific genetic defect and a defect at a specific point in a metabolic pathway. This experiment convinced most investigators that the one-gene, one-enzyme hypothesis was correct.

RESEARCH

QUESTION: What do genes do?

HYPOTHESIS: Each gene contains the information required to make one enzyme.

NULL HYPOTHESIS: Genes do not have a one-to-one correspondence with enzymes.

EXPERIMENTAL STRATEGY: Mutate specific genes. Test to see if each mutant also lacks one of the enzymes required for different steps in the pathway for synthesizing arginine.

EXPERIMENTAL SETUP: Isolate mutant *N. crassa* that cannot synthesize arginine. Grow each type of mutant on normal medium that is:

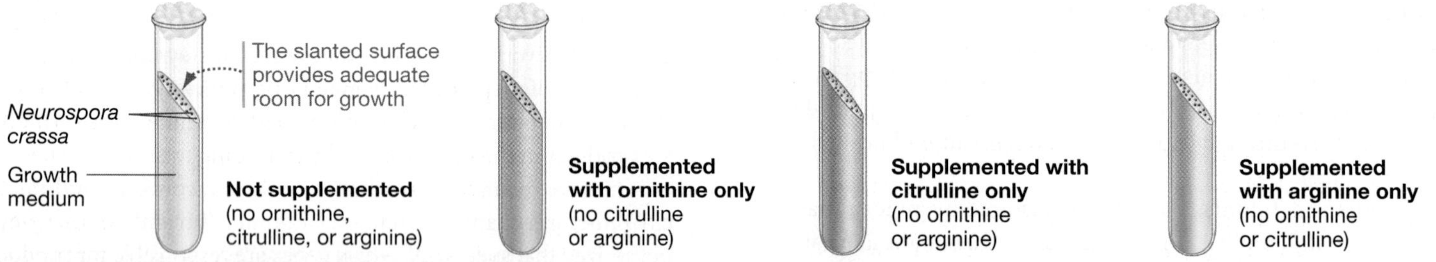

The slanted surface provides adequate room for growth

Neurospora crassa

Growth medium

Not supplemented (no ornithine, citrulline, or arginine)

Supplemented with ornithine only (no citrulline or arginine)

Supplemented with citrulline only (no ornithine or arginine)

Supplemented with arginine only (no ornithine or citrulline)

PREDICTION: There will be three distinct types of mutants, corresponding to defects in enzyme 1, enzyme 2, and enzyme 3 in the pathway for synthesizing arginine. Each type of mutant will be able to grow on different combinations of the four types of media.

PREDICTION OF NULL HYPOTHESIS: There will not be a simple correspondence between a particular mutation and a particular enzyme.

RESULTS: There are three distinct types of mutants, called *arg1*, *arg2*, and *arg3*, each defective in one enzyme.

Mutant type	None	Ornithine only	Citrulline only	Arginine only
arg1	no growth	GROWTH	GROWTH	GROWTH
arg2	no growth	no growth	GROWTH	GROWTH
arg3	no growth	no growth	no growth	GROWTH

(Supplement type)

INTERPRETATION:

Precursor → Ornithine → Citrulline → Arginine

arg1 cells lack enzyme 1

arg2 cells lack enzyme 2

arg3 cells lack enzyme 3

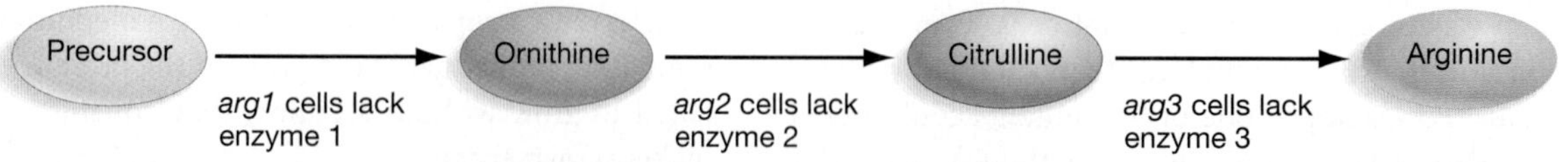

CONCLUSION: The one-gene, one-enzyme hypothesis is supported.

FIGURE 16.2 Experimental Support for the One-Gene, One-Enzyme Hypothesis. The association between specific genetic defects in *N. crassa* and specific defects in the metabolic pathway for arginine synthesis provided evidence that supported the one-gene, one-enzyme hypothesis.

SOURCE: Srb, A. M., and N. H. Horowitz. 1944. The ornithine cycle in *Neurospora* and its genetic control. *Journal of Biological Chemistry* 154: 129–139.

✔**QUESTION** Experimental designs must be repeatable so that other investigators can try the experiment themselves to check the results. Name three things that these researchers would need to describe so that others could repeat this experiment.

Follow-up work showed that genes contain the information for all the proteins produced by an organism—not just enzymes. Biologists finally understood what most genes do: They contain the instructions for making proteins.

In many cases, though, a protein is made up of several different polypeptides, each of which is a product of a different gene. Consequently, for greater accuracy, the one-gene, one-enzyme hypothesis is best called the one-gene, one-polypeptide hypothesis.

16.2 The Central Dogma of Molecular Biology

How does a gene specify the production of a protein? As soon as Beadle and Tatum's hypothesis had been supported in *N. crassa* and a variety of other organisms, this question became a central one.

Part of the answer lay in the molecular structure of the gene. Biochemists knew that the primary components of DNA were four nitrogen-containing bases: the pyrimidines thymine (abbreviated T) and cytosine (C), and the purines adenine (A) and guanine (G). They also knew that these bases were connected in a linear sequence by a sugar–phosphate backbone. Watson and Crick's model for the secondary structure of the DNA molecule (see Chapters 4 and 15) revealed that two strands of DNA are wound into a double helix, held together by hydrogen bonds between the complementary base pairs A-T and G-C.

Given DNA's structure, it appeared extremely unlikely that DNA directly catalyzed the reactions that produce proteins. Its shape was too regular to suggest that it could bind a wide variety of substrate molecules and lower the activation energy for chemical reactions. So how, then, did information translate into action?

The Genetic Code Hypothesis

Crick proposed that the sequence of bases in DNA might act as a code. His idea was that DNA was *only* an information-storage molecule. The instructions it contained would have to be read and then translated into proteins.

Crick offered Morse code as an analogy. Morse code is a message-transmission system using dots and dashes to represent the letters of the alphabet, and in that way it can convey all the complex information of human language. Crick proposed that different combinations of bases could specify the 20 amino acids, just as different combinations of dots and dashes specify the 26 letters of the alphabet. A particular stretch of DNA, then, could contain the information needed to produce the amino acid sequence of a particular polypeptide.

In code form, the tremendous quantity of information required to build and operate a cell could be stored compactly. This information could also be copied through complementary base pairing and transmitted efficiently from one generation to the next.

It soon became apparent, however, that the information encoded in the base sequence of DNA is not translated into the amino acid sequence of proteins directly. Instead, the link between DNA as information repository and proteins as cellular machines is indirect.

RNA as the Intermediary between Genes and Proteins

The first clue that the biological information in DNA must go through an intermediary in order to produce proteins came from knowledge of cell structure. In eukaryotic cells, DNA is enclosed within a membrane-bound organelle called the nucleus (see Chapter 7). But the cells' ribosomes, where protein synthesis takes place, are outside the nucleus, in the cytoplasm.

To make sense of this observation, François Jacob and Jacques Monod suggested that RNA molecules act as a link between genes and the protein-manufacturing centers. Jacob and Monod's hypothesis is illustrated in **FIGURE 16.3**. They predicted that short-lived molecules of RNA, which they called **messenger RNA,** or **mRNA** for short, carry information out of the nucleus from DNA to the site of protein synthesis. Messenger RNA is one of several distinct types of RNA in cells.

Follow-up research confirmed that the messenger RNA hypothesis is correct. One particularly important piece of evidence was the discovery of an enzyme that catalyzes the synthesis of RNA. This protein is called **RNA polymerase** because it polymerizes ribonucleotides into strands of RNA.

RNA polymerase synthesizes RNA molecules according to the information provided by the sequence of bases in a particular stretch of DNA. Unlike DNA polymerase, RNA polymerase

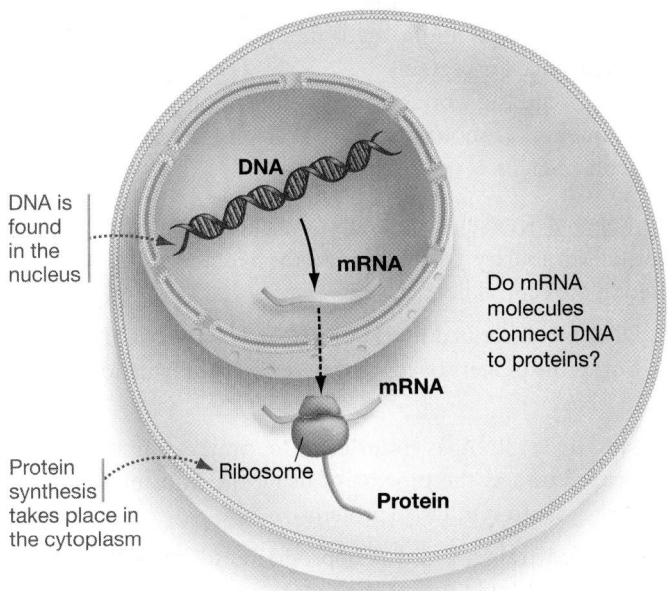

FIGURE 16.3 The Messenger RNA Hypothesis. In cells of eukaryotes such as plants, animals, and fungi, most DNA is found in the nucleus, but proteins are manufactured using ribosomes in the cytoplasm outside the nucleus. Biologists proposed that the information coded in DNA is carried from inside the nucleus out to the ribosomes by messenger RNA (mRNA).

does not require a primer to begin connecting ribonucleotides together to produce a strand of RNA.

To test the mRNA hypothesis, researchers created a reaction mix containing three critical elements: **(1)** the enzyme RNA polymerase; **(2)** ribonucleotides containing the bases adenine (A), uracil (U), guanine (G), and cytosine (C); and **(3)** strands of synthetic DNA that contained deoxyribonucleotides in which the only base was thymine (T).

After allowing the polymerization reaction to proceed, the biologists isolated RNA molecules that contained only the base adenine.

This result supported the hypothesis that RNA polymerase synthesizes RNA according to the rules of complementary base pairing (introduced in Chapter 4), because thymine pairs with adenine. Similar experiments showed that synthetic DNAs containing only cytosine result in the production of RNA molecules containing only guanine.

Dissecting the Central Dogma

Once the mRNA hypothesis was accepted, Francis Crick articulated what became known as the central dogma of molecular biology. The **central dogma** summarizes the flow of information in cells. It simply states that DNA codes for RNA, which codes for proteins:

$$DNA \longrightarrow RNA \longrightarrow proteins$$

Crick's simple statement encapsulates much of the research reviewed in this chapter and the preceding one. DNA is the hereditary material. Genes consist of specific stretches of DNA that code for products used in the cell. The sequence of bases in DNA specifies the sequence of bases in an RNA molecule, which specifies the sequence of amino acids in a protein. In this way, genes ultimately code for proteins.

Proteins are the workers of cells, functioning not only as enzymes but also as motors, structural elements, transporters, and molecular signals.

The Roles of Transcription and Translation Biologists use specialized vocabulary to summarize the sequence of events captured in the central dogma.

1. DNA is transcribed to RNA by RNA polymerase. **Transcription** is the process of copying hereditary information in DNA to RNA.

2. Messenger RNA is translated to proteins in ribosomes. **Translation** is the process of using the information in nucleic acids to synthesize proteins.

The term transcription is appropriate. In everyday English, transcription simply means making a copy of information. The scientific use is similar because it conveys the idea that DNA acts as a permanent record—an information archive or blueprint. This permanent record is copied, during transcription, to produce the short-lived form called mRNA.

Translation is also an appropriate term. In everyday English, translation refers to converting information from one language to another. In biology, translation is the transfer of information from one type of molecule to another—from the "language" of nucleic acids to the "language" of proteins. Translation is also referred to simply as protein synthesis.

The following equation summarizes the relationship between transcription and translation as well as the relationships between DNA, RNA, and proteins:

Gene expression occurs via transcription and translation.

Linking Genotypes and Phenotypes An organism's genotype is determined by the sequence of bases in its DNA, while its phenotype is a product of the proteins it produces.

To appreciate this point, consider that the proteins encoded by genes are what make the "stuff" of the cell and dictate which chemical reactions occur inside. For example, in populations of the oldfield mouse native to southeastern North America, individuals have a gene for a protein called the melanocortin receptor. Melanocortin is a hormone—an important type of molecular signal (discussed in Chapter 11)—that works through the melanocortin receptor to influence how much dark pigment is deposited in fur. An important aspect of a mouse's phenotype—its coat color—is determined in part by the DNA sequence at the gene for this receptor (**FIGURE 16.4a**).

Later work revealed that alleles of a gene differ in their DNA sequence. As a result, the proteins produced by different alleles of the gene may differ in their amino acid sequence. If the primary structures of proteins vary, their functions are likely to vary as well.

To drive this point home, look at the DNA sequence in the portion of the melanocortin receptor gene shown in **FIGURE 16.4b**, and compare it with the sequence in Figure 16.4a. The sequences differ—meaning that they are different alleles. Now look at the protein products of each allele, and note that one of the amino acids in the protein's primary structure differs—one allele specifies an arginine residue; the other specifies a cysteine residue.

At the protein level, the phenotypes associated with these alleles differ. The consequences for the mouse are striking: Melanocortin receptors that have arginine in this location deposit a large amount of pigment, but receptors that have cysteine in this location deposit small amounts of pigment. Whether a mouse is dark or light depends, largely, on a single base change in its DNA sequence. In this case, a tiny difference in genotype produces a large change in phenotype. The central dogma links genotypes to phenotypes.

Exceptions to the Central Dogma The central dogma provided an important conceptual framework for the burgeoning field of molecular genetics and inspired a series of fundamental questions about how genes and cells work. But important modifications to the central dogma have occurred in the decades since Frances Crick first proposed it:

(a) Genetic information flows from DNA to RNA to proteins.

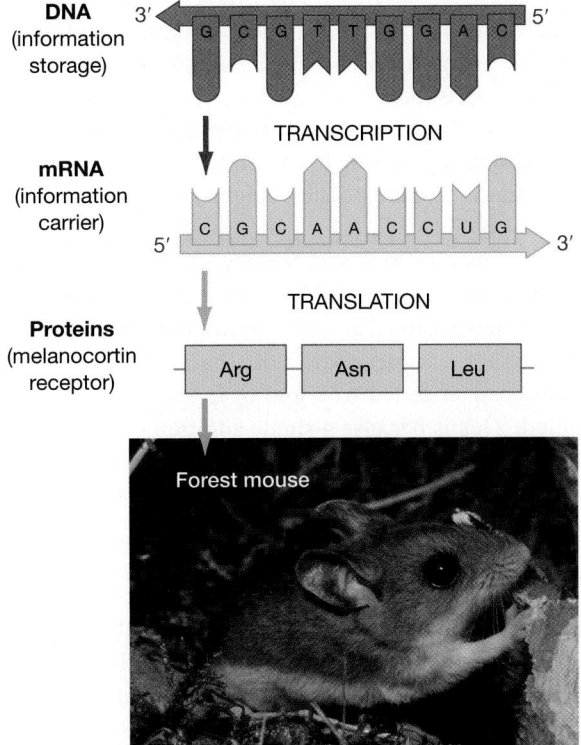

Mice with this DNA sequence have **dark** coats.

(b) Differences in genotype may cause differences in phenotype.

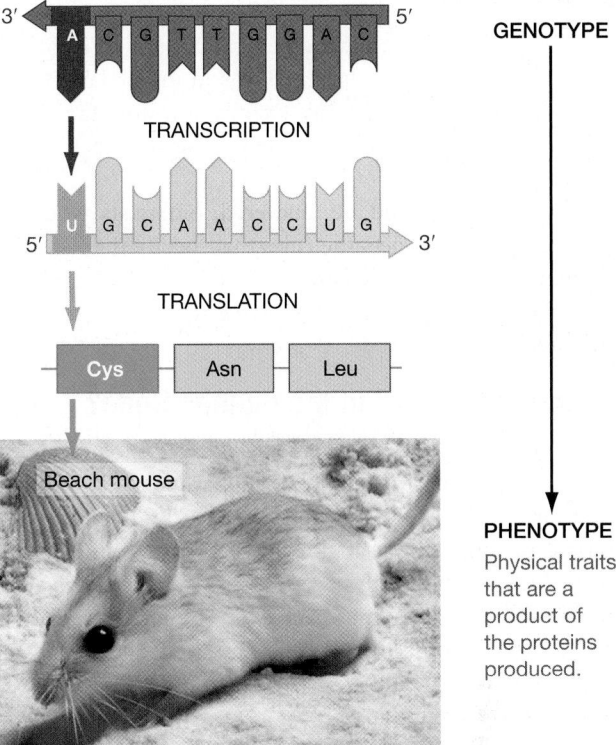

Mice with this DNA sequence have **light** coats.

FIGURE 16.4 The Relationship between Genotype and Phenotype. The central dogma revealed the flow of information within the cell. The DNA sequences given in parts **(a)** and **(b)** are from different alleles (genotypes) that influence coat color (phenotypes) in oldfield mice. Forest-dwelling mice are dark, which camouflages them in their forested habitats. Beach-dwelling mice are light, which camouflages them in their sandy habitat.

- Many genes code for RNA molecules that do not function as mRNAs—they are not translated into proteins.
- In some cases, information flows from RNA back to DNA.

The discovery of a wide array of different RNA types ranks among the most profound advances in the past decade of biological science. Some RNAs form major parts of the ribosome, others help to form mRNA from a much longer precursor RNA (Chapter 17), and yet others regulate which genes are expressed (see Chapter 19). New types of RNA are still being discovered. For the genes coding for these types of RNA, information flow would be diagrammed as simply DNA → RNA.

In the early 1970s, the discovery of "reverse" information flow created the kind of excitement now being generated by the discovery of so many kinds of RNA. Some viruses, for example, have genes consisting of RNA. When some RNA viruses infect a cell, a specialized viral polymerase called **reverse transcriptase** synthesizes a DNA version of the RNA genes. In these viruses, information flows from RNA to DNA.

The human immunodeficiency virus (HIV), which causes AIDS, is an RNA virus that uses reverse transcriptase. Several of the most commonly prescribed drugs for AIDS patients fight the infection by poisoning the HIV reverse transcriptase. The drugs prevent viruses from replicating efficiently by disrupting reverse information flow.

The punch line? Crick's hypothesis is a central concept in biology, but cells, viruses, and researchers aren't dogmatic about it.

check your understanding

If you understand that . . .

- Genes code for proteins, but they do so indirectly.
- The sequence of bases in DNA is used to produce RNA, including messenger RNA (mRNA), via transcription. The sequence of bases in an RNA molecule is complementary to one of the DNA strands of a gene.
- Messenger RNAs are translated into proteins.
- Differences in DNA sequence can lead to differences in the amino acid sequence of proteins.

✔ You should be able to . . .

List the steps that link a change in the base sequence of a gene to a change in the phenotype of an organism.

Answers are available in Appendix A.

16.3 The Genetic Code

Once biologists understood the general pattern of information flow in the cell, the next challenge was to understand the final link between DNA and proteins. Exactly how does the sequence of bases in a strand of mRNA code for the sequence of amino acids in a protein?

If this question could be answered, biologists would have cracked the **genetic code**—the rules that specify the relationship between a sequence of nucleotides in DNA or RNA and the sequence of amino acids in a protein. Researchers from all over the world took up the challenge. A race was on.

How Long Is a Word in the Genetic Code?

The first step in cracking the genetic code was to determine how many bases make up a "word." In a sequence of mRNA, how long is a message that specifies one amino acid?

Based on some simple logic, George Gamow suggested that each code word contains three bases. His reasoning derived from the observation that 20 amino acids are commonly used in cells and from the hypothesis that each amino acid must be specified by a particular sequence of mRNA. **FIGURE 16.5** illustrates Gamow's reasoning:

- There are only four different bases in ribonucleotides (A, U, G, and C), so a one-base code could specify only four different amino acids.

- A two-base code could represent just 4 × 4, or 16, different amino acids.

- A three-base code could specify 4 × 4 × 4, or 64, different amino acids.

A three-base code provides more than enough words to code for all 20 amino acids. A three-base code is known as a **triplet code**.

Gamow's hypothesis suggested that the genetic code could be redundant. That is, more than one triplet of bases might specify the same amino acid. As a result, different three-base sequences in an mRNA—say, AAA and AAG—might code for the same amino acid—say, lysine.

The group of three bases that specifies a particular amino acid is called a **codon.** According to the triplet code hypothesis, many of the 64 codons that are possible might specify the same amino acids.

Work by Francis Crick and Sydney Brenner confirmed that codons are three bases long. Their experiments used chemicals that caused an occasional addition or deletion of a base in DNA. As predicted for a triplet code, a one-base addition or deletion in the base sequence led to a loss of function in the gene being studied. This is because a single addition or deletion mutation throws the sequence of codons, or the **reading frame,** out of register. To understand how a reading frame works, consider the sentence

"The fat cat ate the rat."

The reading frame of this sentence is a three-letter word and a space. If the fourth letter in this sentence—the *f* in *fat*—were deleted, the reading frame would transform the sentence into

"The atc ata tet her at."

This is gibberish.

When the reading frame in a DNA sequence is thrown out of register by the addition or deletion of a base, the composition of each codon changes just like the letters in each word of the example sentence above. The protein produced from the altered DNA sequence has a completely different sequence of amino acids. In terms of its normal function, this protein is gibberish.

Crick and Brenner were also able to produce DNA sequences that had deletions or additions of two base pairs or three base

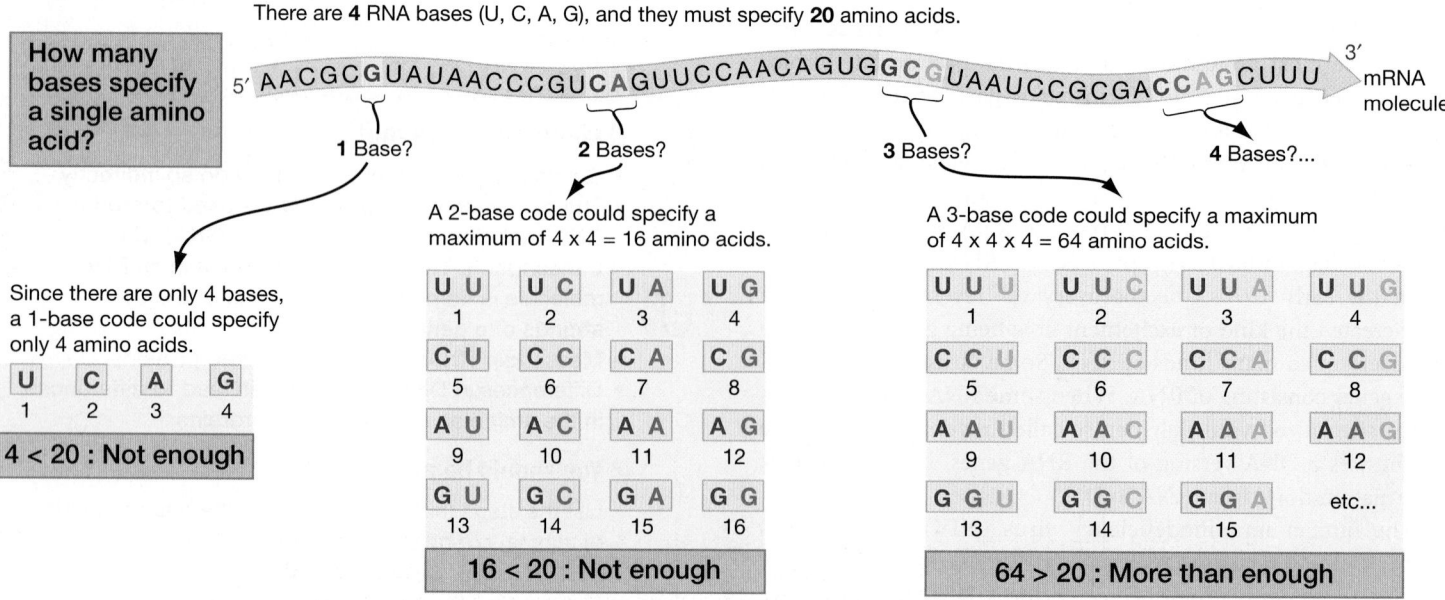

FIGURE 16.5 How Many Bases Form a "Word" in the Genetic Code?

pairs. The only time functional proteins were produced was when three bases were added or removed. In the sentence

"The fat cat ate the rat."

the combination of removing one letter from each of the first three words might result in

"Tha tca ate the rat."

Just as the altered sentence still conveys some meaning, genes with three deletion mutations were able to produce a functional protein.

The researchers interpreted these results as strong evidence in favor of the triplet code hypothesis. Most other biologists agreed.

The confirmation of the triplet code launched an effort to determine which amino acid is specified by each of the 64 codons. Ultimately, it was successful.

How Did Researchers Crack the Code?

The initial advance in deciphering the genetic code came in 1961, when Marshall Nirenberg and Heinrich Matthaei developed a method for synthesizing RNAs of known sequence. They began by creating a long polymer of uracil-containing ribonucleotides. These synthetic RNAs were added to an in vitro system for synthesizing proteins. The researchers analyzed the resulting amino acid chain and determined that it was polyphenylalanine—a polymer consisting of the amino acid phenylalanine.

This result provided evidence that the RNA triplet UUU codes for the amino acid phenylalanine. By complementary base pairing, it was clear that the corresponding DNA sequence would be AAA. This initial work was followed by experiments using RNAs consisting of only A or C. RNAs with only AAAAA . . . produced

polypeptides composed of only lysine; poly-C RNAs (RNAs consisting of only CCCCC . . .) produced polypeptides composed entirely of proline.

Nirenberg and Philip Leder later devised a system for synthesizing specific codons. With these they performed a series of experiments in which they added each codon to a cell extract containing the 20 different amino acids, ribosomes, and other molecules required for protein synthesis. Recall that ribosomes are macromolecular machines that synthesize proteins (Chapter 7). Then the researchers determined which amino acid became bound to the ribosomes when a particular codon was present. For example, when the codon CAC was in the reaction mix, the amino acid histidine would bind to the ribosomes. This result indicated that CAC codes for histidine.

These ribosome-binding experiments allowed Nirenberg and Leder to determine which of the 64 codons coded for each of the 20 amino acids.

Researchers also discovered that certain codons are punctuation marks signaling "start of message" or "end of message." These codons indicate that protein synthesis should start at a given codon or that the protein chain is complete.

- There is one **start codon** (AUG), which signals that protein synthesis should begin at that point on the mRNA molecule. The start codon specifies the amino acid methionine.

- There are three **stop codons,** also called termination codons (UAA, UAG, and UGA). The stop codons signal that the protein is complete, they do not code for any amino acid, and they end translation.

The complete genetic code is given in **FIGURE 16.6**. Deciphering it was a tremendous achievement, requiring more than five years of work by several teams of researchers.

SECOND BASE

		U	C	A	G	
FIRST BASE	**U**	UUU UUC — Phenylalanine (Phe) UUA UUG — Leucine (Leu)	UCU UCC UCA UCG — Serine (Ser)	UAU UAC — Tyrosine (Tyr) UAA — Stop codon UAG — Stop codon	UGU UGC — Cysteine (Cys) UGA — Stop codon UGG — Tryptophan (Trp)	U C A G
	C	CUU CUC CUA CUG — Leucine (Leu)	CCU CCC CCA CCG — Proline (Pro)	CAU CAC — Histidine (His) CAA CAG — Glutamine (Gln)	CGU CGC CGA CGG — Arginine (Arg)	U C A G
	A	AUU AUC AUA — Isoleucine (Ile) AUG — Methionine (Met) Start codon	ACU ACC ACA ACG — Threonine (Thr)	AAU AAC — Asparagine (Asn) AAA AAG — Lysine (Lys)	AGU AGC — Serine (Ser) AGA AGG — Arginine (Arg)	U C A G
	G	GUU GUC GUA GUG — Valine (Val)	GCU GCC GCA GCG — Alanine (Ala)	GAU GAC — Aspartic acid (Asp) GAA GAG — Glutamic acid (Glu)	GGU GGC GGA GGG — Glycine (Gly)	U C A G

(Right side label: **THIRD BASE**)

FIGURE 16.6 The Genetic Code. To read a codon in mRNA, locate its first base in the red band on the left; then move rightward to the box under the codon's second base in the blue band along the top. Finally, locate the codon's third base in the green band on the right side to learn the amino acid. By convention, codons are always written in the $5' \rightarrow 3'$ direction.

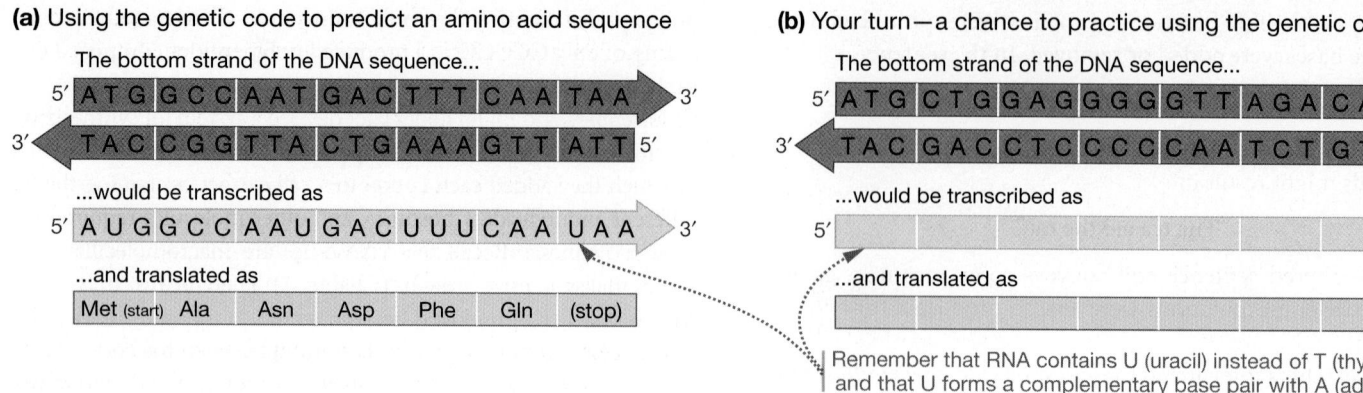

(a) Using the genetic code to predict an amino acid sequence

The bottom strand of the DNA sequence...

5′ | A | T | G | G | C | C | A | A | T | G | A | C | T | T | T | C | A | A | T | A | A | 3′

3′ | T | A | C | C | G | G | T | T | A | C | T | G | A | A | A | G | T | T | A | T | T | 5′

...would be transcribed as

5′ | A | U | G | G | C | C | A | A | U | G | A | C | U | U | U | C | A | A | U | A | A | 3′

...and translated as

| Met (start) | Ala | Asn | Asp | Phe | Gln | (stop) |

(b) Your turn—a chance to practice using the genetic code

The bottom strand of the DNA sequence...

5′ | A | T | G | C | T | G | G | A | G | G | G | G | G | T | T | A | G | A | C | A | T | 3′

3′ | T | A | C | G | A | C | C | T | C | C | C | C | C | A | A | T | C | T | G | T | A | 5′

...would be transcribed as

5′ | | | | | | | | 3′

...and translated as

| | | | | | | |

Remember that RNA contains U (uracil) instead of T (thymine), and that U forms a complementary base pair with A (adenine)

FIGURE 16.7 Using the Genetic Code.

✔**EXERCISE** Fill in the mRNA and amino acid sequences in part (b).

Analyzing the Code Once biologists had cracked the genetic code, they realized that it has a series of important properties.

- *The code is redundant.* All amino acids except methionine and tryptophan are coded by more than one codon.

- *The code is unambiguous.* A single codon never codes for more than one amino acid.

- *The code is non-overlapping.* Once the ribosome locks onto the first codon, it then reads each separate codon one after another.

- *The code is nearly universal.* With a few minor exceptions, all codons specify the same amino acids in all organisms.

- *The code is conservative.* When several codons specify the same amino acid, the first two bases in those codons are almost always identical.

The last point is subtle, but important. Here's the key: If a mutation in DNA or an error in transcription or translation affects the third position in a codon, it is less likely to change the amino acid in the final protein. This feature makes individuals less vulnerable to small, random changes or errors in their DNA sequences. Compared with randomly generated codes, the existing genetic code minimizes the phenotypic effects of small changes in DNA sequence and errors during translation. Stated another way, the genetic code does not represent a random assemblage of bases, like letters drawn from a hat. It has been honed by natural selection and is remarkably efficient.

Using the Code Using the genetic code and the central dogma, biologists can:

1. Predict the codons and amino acid sequence encoded by a particular DNA sequence (see **FIGURE 16.7a**).

2. Determine the set of mRNA and DNA sequences that would code for a particular sequence of amino acids.

Why is a *set* of mRNA or DNA sequences predicted from a given amino acid sequence? The answer lies in the code's redundancy. If a polypeptide contains phenylalanine, you don't know if the codon responsible is UUU or UUC.

✔If you understand how to read the genetic code, you should be able to do the following tasks: (1) Identify the codons in Figure 16.4 and decide whether they are translated correctly. (2) Complete the exercise for **FIGURE 16.7b**. (3) Write an mRNA that codes for the amino acid sequence Ala-Asn-Asp-Phe-Gln yet is different from the one given in Figure 16.7a. Indicate the mRNA's 5′ → 3′ polarity. Then write the double-stranded DNA that corresponds to this mRNA. Indicate the 5′ → 3′ polarity of both DNA strands.

Once they understood the central dogma and genetic code, biologists were able to explore and eventually understand the molecular basis of mutation. How do novel traits—such as dwarfing in garden peas and white eye color in fruit flies—come to be?

check your understanding

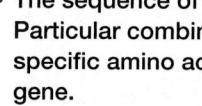

If you understand that . . .

- The sequence of bases in mRNA constitutes a code. Particular combinations of three bases specify specific amino acids in the protein encoded by the gene.

- The genetic code is redundant. It consists of 64 combinations of bases, but only 20 amino acids plus start and stop "punctuation marks" need to be specified.

✔ **You should be able to . . .**

Consider the consequences of a mutation in the DNA template sequence ATA to one of the following sequences: GTA, TTA, or GCA.

1. In each case, specify the resulting change in the mRNA codon.

2. In each case, describe the effect on the resulting protein.

Answers are available in Appendix A.

16.4 How Can Mutation Modify Genes and Chromosomes?

This chapter has explored how the information archived in DNA is put into action in the form of working RNAs and proteins. Now the questions are, what happens if the information in DNA changes? In what ways can this information be changed? What are the consequences for the cell and organism?

A **mutation** is any permanent change in an organism's DNA. It is a modification in a cell's information archive—a change in its genotype. Mutations create new alleles.

Mutations can alter DNA sequences that range in size from a single base pair in DNA to whole sets of chromosomes. Let's look at these different types of mutation and their consequences.

Point Mutation

FIGURE 16.8 shows how a common type of mutation occurs. If a mistake is made during DNA synthesis or DNA repair, a change in the sequence of bases in DNA results. A single-base change such as this is called a **point mutation.**

What happens when point mutations occur in regions of DNA that code for proteins? To answer this question, look back at Figure 16.4 and recall that a change in a single base in DNA is associated with a difference in coat color in populations of oldfield mice. The DNA sequence in Figure 16.4a is found in dark-colored mice that live in forest habitats; the sequence in Figure 16.4b is found in light-colored mice that live in beach habitats.

Because beach-dwelling populations are evolutionarily younger than the nearby forest-dwelling populations, researchers hypothesize the following sequence of events:

1. Forest mice colonized beach habitats.

2. Either before or after the colonization event, a random point mutation occurred in a mouse that altered the melanocortin receptor gene and resulted in some offspring with light coats.

3. Light-colored mice are camouflaged in beach habitats; in sandy environments, they suffer lower predation than dark-colored mice.

4. Over time, the allele created by the point mutation increased in frequency in beach-dwelling populations.

Point mutations that cause these types of changes in the amino acid sequence of proteins are called **missense mutations.** But note that if the same G-to-A change had occurred in the third position of the same DNA codon, instead of the first position, there would have been no change in the protein produced. The mRNA codons CGC and CGU both code for arginine. A point mutation that does not change the amino acid sequence of the gene product is called a **silent mutation.**

Some point mutations disrupt major portions of a protein. Recall that a single addition or deletion mutation throws the sequence of codons out of register and alters the meaning of all subsequent codons. Such mutations are called **frameshift mutations.** Another type of point mutation with a large effect is a **nonsense mutation.** Nonsense mutations occur when a codon that specifies an amino acid is changed by mutation to one that specifies a stop codon. This causes early termination of the polypeptide chain and often results in a non-functional protein.

In terms of the impact on organisms, biologists divide mutations into three categories:

1. **Beneficial** Some mutations increase the fitness of the organism—meaning, its ability to survive and reproduce—in certain environments. The G-to-A mutation is beneficial in beach habitats because it camouflages mice.

2. **Neutral** If a mutation has no effect on fitness, it is termed neutral. Silent mutations are usually neutral.

3. **Deleterious** Because organisms tend to be well adapted to their current habitat, and because mutations are random changes in the genotype, many mutations lower fitness. These mutations are termed harmful or deleterious. The G-to-A mutation would be deleterious in the forest habitat.

Recent studies indicate that the majority of point mutations are slightly deleterious or neutral. **TABLE 16.1** (see page 314) summarizes the types of point mutations that occur in protein-coding sequences of a gene and reviews their consequences for the amino acid sequences of proteins and for fitness.

Point mutations can and do occur in DNA sequences that do not code for proteins. These mutations, however, are not referred to as missense, silent, frameshift, or nonsense mutations

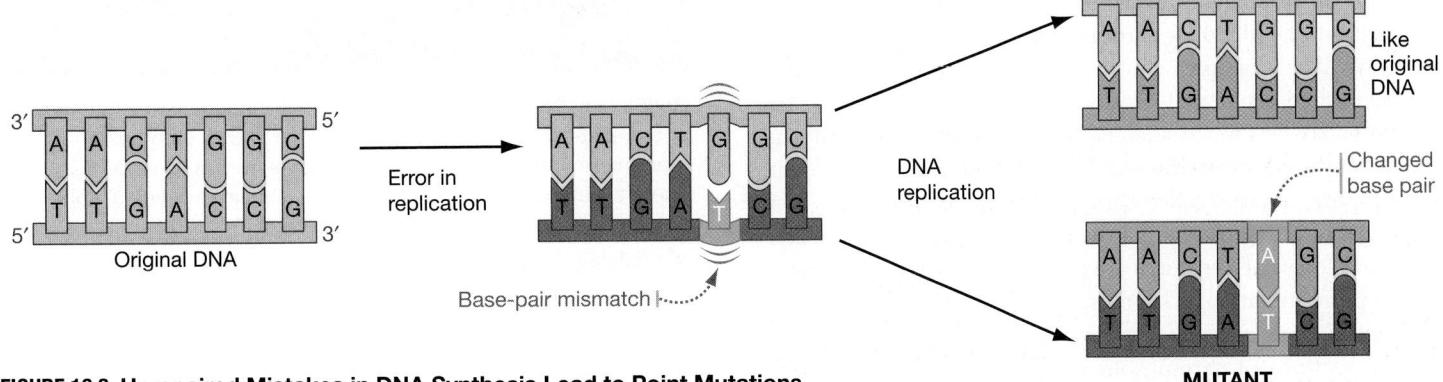

FIGURE 16.8 Unrepaired Mistakes in DNA Synthesis Lead to Point Mutations.

Name	Definition	Example	Consequence
		Original DNA sequence —— TAT TGG CTA GTA CAT	
		Original mRNA transcript —— UAU UGG CUA GUA CAU	
		Tyr – Trp – Leu – Val – His —— Original polypeptide	
Silent	Change in nucleotide sequence that does not change the amino acid specified by a codon	TAC TGG CTA GTA CAT UAC UGG CUA GUA CAU Tyr – Trp – Leu – Val – His	No change in phenotype; neutral with respect to fitness
Missense	Change in nucleotide sequence that changes the amino acid specified by codon	TAT TGT CTA GTA CAT UAU UGU CUA GUA CAU Tyr – Cys – Leu – Val – His	Change in primary structure of protein; may be beneficial, neutral, or deleterious
Nonsense	Change in nucleotide sequence that results in an early stop codon	TAT TGA CTA GTA CAT UAU UGA CUA GUA CAU Tyr – STOP	Leads to mRNA breakdown or a shortened polypeptide; usually deleterious
Frameshift	Addition or deletion of a nucleotide	TAT TCG GCT AGT ACA T UAU UCG GCU AGU ACA U Tyr – Ser – Ala – Ser – Thr	Reading frame is shifted, altering the meaning of all subsequent codons; almost always deleterious

because these terms apply only to mutations that can change the protein-coding potential of a gene. If point mutations alter DNA sequences that are important for *gene expression*, they can have important effects on phenotype even though they do not change the amino acid sequence of a protein.

Chromosome Mutations

Besides documenting various types of point mutations, biologists study larger-scale mutations that change chromosomes. You might recall, for example, that polyploidy is an increase in the number of each type of chromosome, while aneuploidy is the addition or deletion of individual chromosomes (Chapter 13).

Polyploidy, aneuploidy, and other changes in chromosome number result from chance mistakes in moving chromosomes into daughter cells during meiosis or mitosis. Polyploidy and aneuploidy are forms of mutation that don't change DNA sequences, but alter the number of chromosome copies.

In addition to changes in overall chromosome number, the composition of individual chromosomes can change in important ways. For example, chromosome segments can become detached when accidental breaks in chromosomes occur. The segments may be flipped and rejoined—a phenomenon known as a chromosome **inversion**—or become attached to a different chromosome, an event called chromosome **translocation.** When a segment of chromosome is lost, this is a **deletion,** and when additional copies of a segment are present, this is a **duplication.**

Like point mutations, chromosome mutations can be beneficial, neutral, or deleterious. For example, more than 200 different inverted sections of chromosomes were found in comparisons

of the DNA from eight phenotypically normal people. These mutations appear to be neutral. Not all chromosome mutations are so harmless, however. Chromosomes of cancer cells exhibit deleterious chromosome mutations that include aneuploidy, inversions, translocations, deletions, and duplications. **FIGURE 16.9**

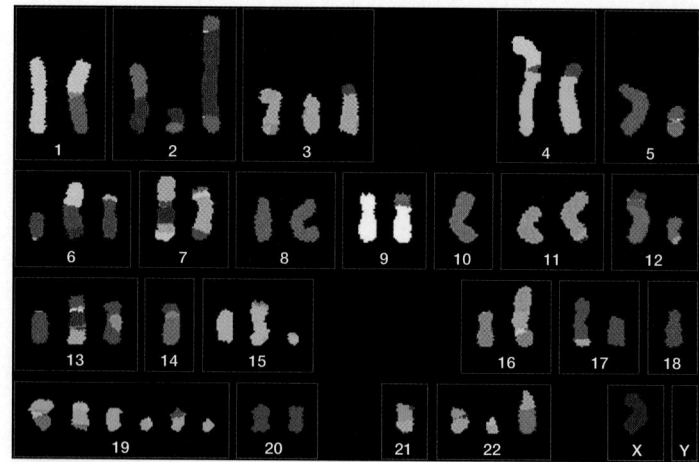

FIGURE 16.9 Chromosome-Level Mutations. A spectral karyotype of a breast cancer cell from a female that shows chromosome rearrangements and aneuploidy typical of cancer. In a normal spectral karyotype, each chromosome is stained a single, solid color, which varies for each chromosome pair.

✔ **EXERCISE** Compare this karyotype to the one shown in the chapter-opening image. Remember that females normally have two X chromosomes and males normally have one X chromosome. Which chromosomes show evidence of aneuploidy? Which chromosomes show evidence of rearrangements?

drives this point home by showing the **karyotype**—the complete set of chromosomes in a cell—of a cancerous human cell.

To summarize, point mutations and chromosome mutations are random changes in DNA that can produce new genes, new alleles, and new traits. At the level of individuals, mutations can cause disease or death or lead to increases in fitness. At the level of populations, mutations furnish the heritable variation that Mendel and Morgan analyzed and that makes evolution possible. The central role of mutation in evolution is explored in depth in Unit 5.

If you understand . . .

16.1 What Do Genes Do?

- Experiments with mutants of the bread mold *N. crassa* led to the one-gene, one-enzyme hypothesis.

- The original one-gene, one-enzyme hypothesis has been broadened to account for genes that code for proteins other than enzymes and for genes that have RNA as a final product.

- ✔ You should be able to use Figure 16.1 to explain what compounds could be added to the medium to allow the growth of a mutant unable to synthesize citrulline because of a mutation in the gene for enzyme 1.

16.2 The Central Dogma of Molecular Biology

- DNA is transcribed to messenger RNA (mRNA) by RNA polymerase, and then mRNA is translated to proteins by ribosomes. In this way, genetic information is converted from DNA to RNA to protein.

- The flow of information from DNA to RNA to protein is called the central dogma of molecular biology.

- Many RNAs do not code for proteins. Instead, these RNAs perform other important functions in the cell.

- Reverse transcriptase reverses information flow by copying RNA into DNA. Some viruses with an RNA genome use this enzyme during their replication.

- ✔ You should be able to explain how a compound that blocks RNA synthesis will affect protein synthesis.

16.3 The Genetic Code

- Each amino acid in a protein is specified by a codon—a group of three bases in mRNA.

- By synthesizing RNAs of known base composition and then observing the results of translation, researchers were able to decipher the genetic code.

- The genetic code is redundant—meaning that most of the 20 amino acids are specified by more than one codon.

- Certain codons signal where translation starts and stops.

- ✔ Using the genetic code shown in Figure 16.6, you should be able to write all possible mRNA sequences that would produce the following sequence of amino acids: Met-Trp-Lys-Gln.

16.4 How Can Mutation Modify Genes and Chromosomes?

- Mutations are random, heritable changes in DNA that range from changes in a single base to changes in the structure and number of chromosomes.

- Point mutations in protein-coding regions may have no effect on the protein (silent mutation), may change a single amino acid (missense mutation), may shorten the protein (nonsense mutation), or may shift the reading frame and cause many amino acids to be wrong (frameshift mutation).

- Mutations can have beneficial, neutral, or harmful effects on organisms.

- ✔ You should be able to explain how redundancy in the genetic code allows for silent mutations and whether a silent mutation is likely to be beneficial, neutral, or harmful.

(MB) MasteringBiology

1. **MasteringBiology Assignments**

   Tutorials and Activities Genetic Code; One-Gene One-Enzyme Hypothesis; Overview of Protein Synthesis; Role of the Nucleus and Ribosomes in Protein Synthesis; Triplet Nature of the Genetic Code

   Questions Reading Quizzes, Blue-Thread Questions, Test Bank

2. **eText** Read your book online, search, take notes, highlight text, and more.

3. **The Study Area** Practice Test, Cumulative Test, BioFlix® 3-D Animations, Videos, Activities, Audio Glossary, Word Study Tools, Art

You should be able to . . .

1. What does the one-gene, one-enzyme hypothesis state?
   a. Genes are composed of stretches of DNA.
   b. Genes are made of protein.
   c. Genes code for ribozymes.
   d. A single gene codes for a single protein.

2. Which of the following is an important exception to the central dogma of molecular biology?
   a. Many genes code for RNAs that function directly in the cell.
   b. DNA is the repository of genetic information in all cells.
   c. Messenger RNA is a short-lived "information carrier."
   d. Proteins are responsible for most aspects of the phenotype.

3. DNA's primary structure is made up of just four different bases, and its secondary structure is regular and highly stable. How can a molecule with these characteristics hold all the information required to build and maintain a cell?
   a. The information is first transcribed, then translated.
   b. The messenger RNA produced from DNA has much more complex secondary structures, allowing mRNA to hold much more information.
   c. A protein coded for in DNA has much more complex primary and secondary structures, allowing it to hold much more information.
   d. The information in DNA is in a code form that is based on the sequence of bases.

4. Why did researchers suspect that DNA does not code for proteins directly?

5. Which of the following describes an important experimental strategy in deciphering the genetic code?
   a. comparing the amino acid sequences of proteins with the base sequence of their genes
   b. analyzing the sequence of RNAs produced from known DNA sequences
   c. analyzing mutants that changed the code
   d. examining the polypeptides produced when RNAs of known sequence were translated

6. What is a stop codon?

7. Explain why Morse code is an appropriate analogy for the genetic code.

8. Draw a hypothetical metabolic pathway in *Neurospora crassa* composed of five substrates, five enzymes, and a product called Biological Sciazine. Number the substrates 1–5, and label the enzymes A–E, in order. (For instance, enzyme A catalyzes the reaction between substrates 1 and 2.)
   - Suppose a mutation made the gene for enzyme C nonfunctional. What molecule would accumulate in the affected cells?
   - Suppose a mutant strain can survive if substrate 5 is added to the growth medium but it cannot grow if substrates 1, 2, 3, or 4 are added. Which enzyme in the pathway is affected in this mutant?

9. How did experiments with *Neurospora crassa* mutants support the one-gene, one-enzyme hypothesis?

10. Why does a single-base deletion mutation within a protein-coding sequence usually have a more severe effect than a deletion of three adjacent bases?
    a. because single-base deletions prevent the ribosome from binding to mRNA
    b. because single-base deletions stabilize mRNA
    c. because single-base deletions change the reading frame
    d. because single-base deletions alter the meaning of individual codons

11. When researchers discovered that a combination of three deletion mutations or three addition mutations would restore the function of a gene, most biologists were convinced that the genetic code was read in triplets. Explain the logic behind this conclusion.

12. Explain why all point mutations change the genotype, but why only some point mutations change the phenotype.

13. Recall that DNA and RNA are synthesized only in the $5' \rightarrow 3'$ direction and that DNA and RNA sequences are written in the $5' \rightarrow 3'$ direction, unless otherwise noted. Consider the following DNA sequence:

    5′ TTGAAATGCCCGTTTGGAGATCGGGTTACAGCTAGTCAAAG 3′

    3′ AACTTTACGGGCAAACCTCTAGCCCAATGTCGATCAGTTTC 5′

    - Identify bases in the bottom strand that can be transcribed into start and stop codons.
    - Write the mRNA sequence that would be transcribed between start and stop codons if the bottom strand served as the template for RNA polymerase.
    - Write the amino acid sequence that would be translated from the mRNA sequence you just wrote.

14. What problems would arise if the genetic code contained only 22 codons—one for each amino acid, a start signal, and a stop signal?

15. Scientists say that a phenomenon is a "black box" if they can describe it and study its effects but don't know the underlying mechanism that causes it. In what sense was genetics—meaning the transmission of heritable traits—a black box before the central dogma of molecular biology was understood?

16. **QUANTITATIVE** One of the possibilities that researchers interested in the genetic code considered was that the code was overlapping, meaning that a single base could be part of up to three codons. How many amino acids would be encoded in the sequence 5′ AUGUUACGGAAU 3′ by a non-overlapping and maximally overlapping code?
    a. 4 (non-overlapping) and 16 (overlapping)
    b. 4 and 12
    c. 4 and 10
    d. 12 and 4

17 Transcription, RNA Processing, and Translation

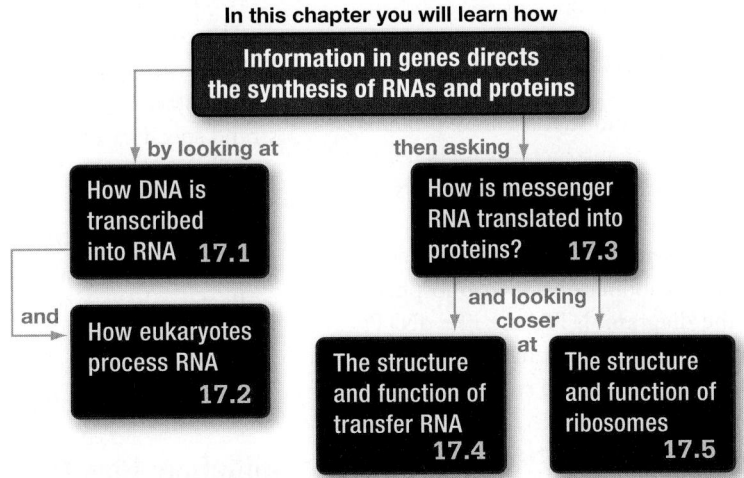

In this chapter you will learn how

Information in genes directs the synthesis of RNAs and proteins

by looking at

How DNA is transcribed into RNA   17.1

and

How eukaryotes process RNA   17.2

then asking

How is messenger RNA translated into proteins?   17.3

and looking closer at

The structure and function of transfer RNA   17.4

The structure and function of ribosomes   17.5

Extensive transcription is occurring along this gene within a frog cell. The horizontal strand in the middle of this micrograph is DNA; the strands that have been colored yellow and red, and that are coming off on either side, are RNA molecules.

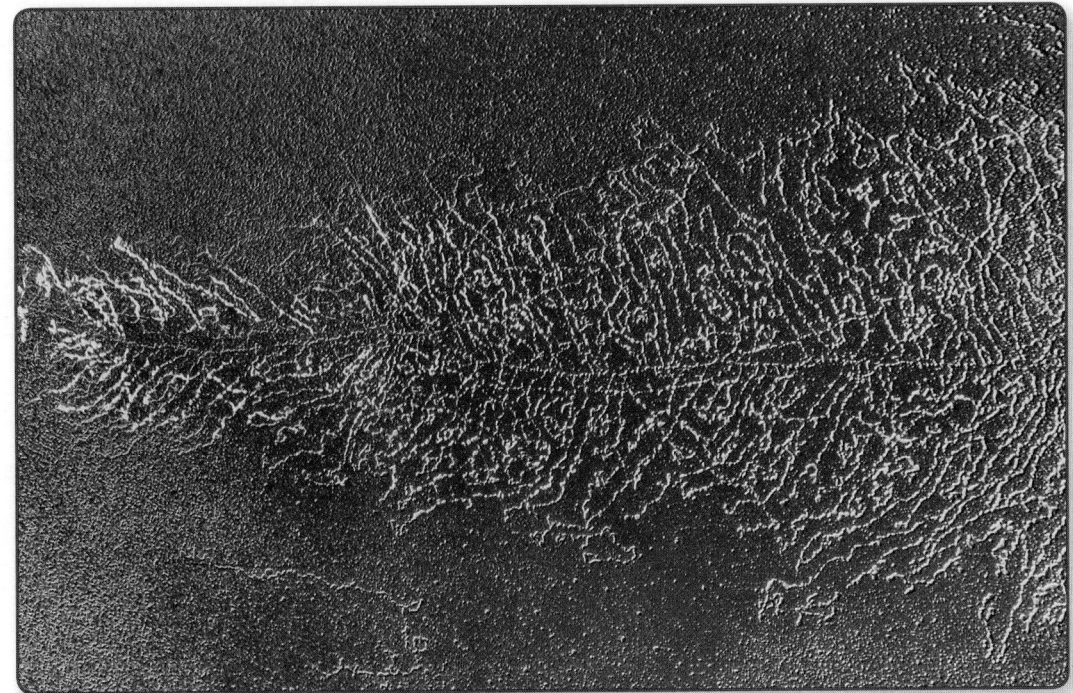

Proteins are the stuff of life. They give shape to our cells, control the chemical reactions that go on inside them, and regulate how materials move into, out of, and through them. Some of these proteins may not be produced at all in some types of cells; others may be present in quantities ranging from millions of copies to fewer than a dozen.

A cell builds the proteins it needs from instructions encoded in its DNA. The central dogma of molecular biology states that the flow of information in cells is from DNA to mRNA to protein (Chapter 16). Once this pattern of information flow had been established, biologists puzzled over how cells actually accomplish the two major steps of the central dogma: transcription and

This chapter is part of the Big Picture. See how on pages 366–367.

✔ When you see this checkmark, stop and test yourself. Answers are available in Appendix A.

translation. Specifically, how does RNA polymerase know where to start transcribing a gene, and where to end? And once an RNA message is produced, how is the linear sequence of ribonucleotides translated into the linear sequence of amino acids in a protein?

This chapter delves into the molecular mechanisms of gene expression—the blood and guts of the central dogma. It starts with the monomers that build RNA and ends with a protein.

17.1 An Overview of Transcription

The first step in converting genetic information into proteins is to synthesize an RNA version of the instructions archived in DNA. Enzymes called **RNA polymerases** are responsible for synthesizing mRNA (see Chapter 15).

FIGURE 17.1 shows how the polymerization reaction occurs. Note the incoming monomer—a ribonucleoside triphosphate, or NTP—at the far right of the diagram. NTPs are like dNTPs (introduced in Chapter 15), except that they have a hydroxyl (−OH) group on the 2′ carbon. This makes the sugar in an NTP a ribose instead of the deoxyribose sugar of DNA.

Once an NTP that matches a base on the DNA template is in place, RNA polymerase cleaves off two phosphates and catalyzes the formation of a phosphodiester linkage between the 3′ end of the growing mRNA chain and the new ribonucleoside monophosphate. As this 5′ → 3′ matching-and-catalysis process continues, an RNA that is complementary to the gene is synthesized. This is transcription.

Notice that only one of the two DNA strands is used as a template and transcribed, or "read," by RNA polymerase.

- The strand that is read by the enzyme is the **template strand.**
- The other strand is called the **non-template strand** or **coding strand.** Coding strand is an appropriate name, because, with one exception, its sequence matches the sequence of the RNA that is transcribed from the template strand and codes for a polypeptide.

The coding strand and the RNA don't match exactly, because RNA has uracil (U) rather than the thymine (T) found in the coding strand. Likewise, an adenine (A) in the DNA template strand specifies a U in the complementary RNA strand.

Like DNA polymerases (see Chapter 15), an RNA polymerase performs a template-directed synthesis in the 5′ → 3′ direction. But unlike DNA polymerases, RNA polymerases do not require a primer to begin transcription.

Bacteria have a single RNA polymerase. In contrast, eukaryotes, have at least three distinct types. Let's first take a look at general principles of transcription using bacteria as an example and then examine things that differ in eukaryotes.

Initiation: How Does Transcription Begin in Bacteria?

How does RNA polymerase know where and in which direction to start transcription on the DNA template? The answer to this question defined what biologists call the **initiation** phase of transcription.

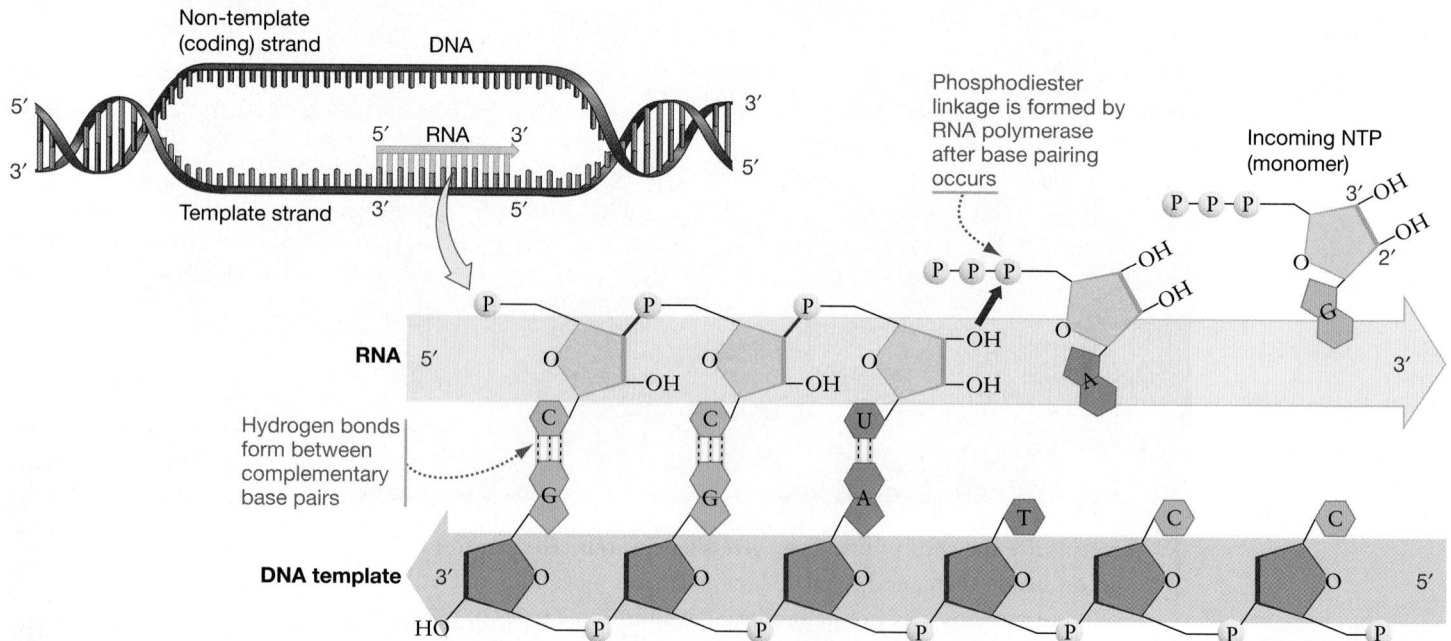

FIGURE 17.1 Transcription Is the Synthesis of RNA from a DNA Template. The reaction catalyzed by RNA polymerase (not shown) results in the formation of a phosphodiester linkage between ribonucleotides. RNA polymerase produces an RNA strand whose sequence is complementary to the bases in the DNA template.

✔**QUESTION** In which direction is RNA synthesized, 5′ → 3′ or 3′ → 5′? In which direction is the DNA template "read"?

Soon after the discovery of bacterial RNA polymerase, researchers found that the enzyme cannot initiate transcription on its own. Instead, a detachable protein subunit called **sigma** must bind to the polymerase before transcription can begin.

Bacterial RNA polymerase and sigma form a **holoenzyme** (literally, "whole enzyme"; **FIGURE 17.2a**). A holoenzyme consists of a **core enzyme** (RNA polymerase, in this case), which contains the active site for catalysis, and other required proteins (such as sigma).

What does sigma do? When researchers mixed the polymerase and DNA together, they found that the core enzyme could bind to any sequence of DNA. When sigma was added to this mixture, the holoenzyme formed and bound only to specific sections of DNA. These binding sites were named **promoters,** because they are sections of DNA that promote the start of transcription.

Most bacteria have alternative sigma proteins that bind to promoters with slightly different DNA base sequences, and may activate a group of genes in response to environmental change. For example, one type of sigma initiates the transcription of genes that help the cell cope with high temperatures. Controlling which sigma proteins are used is one of the ways that bacterial cells regulate which groups of genes are expressed.

The discovery of promoters suggested that sigma was responsible for guiding RNA polymerase to specific locations where transcription should begin. What is the nature of these specific locations? What do promoters look like, and what do they do?

Bacterial Promoters David Pribnow offered an initial answer to these questions in the mid-1970s. When Pribnow analyzed the base sequence of promoters from various bacteria and from viruses that infect bacteria, he found that the promoters were 40–50 base pairs long and had a particular section in common: a series of bases identical or similar to TATAAT. This six-base-pair sequence is now known as the −10 box, because it is centered about 10 bases from the point where bacterial RNA polymerase starts transcription (**FIGURE 17.2b**).

DNA that is located in the direction RNA polymerase moves during transcription is said to be **downstream** from the point of reference; DNA located in the opposite direction is said to be **upstream.** Thus, the −10 box is centered about 10 bases upstream from the transcription start site. The place where transcription begins is called the +1 site.

Soon after the discovery of the −10 box, researchers recognized that the sequence TTGACA also occurred in promoters and was about 35 bases upstream from the +1 site. This second key sequence is called the −35 box. Although all bacterial promoters have a −10 box and a −35 box, the sequences within the promoter but outside these boxes vary.

Events inside the Holoenzyme In bacteria, transcription begins when sigma, as part of the holoenzyme complex, binds to the −35 and −10 boxes. Sigma, and not RNA polymerase, makes the initial contact with DNA of the promoter. Sigma's binding to a promoter determines where and in which direction RNA polymerase will start synthesizing RNA.

Once the holoenzyme is bound to a promoter for a bacterial gene, the DNA helix is opened by RNA polymerase, creating two

(a) RNA polymerase and sigma form a holoenzyme.

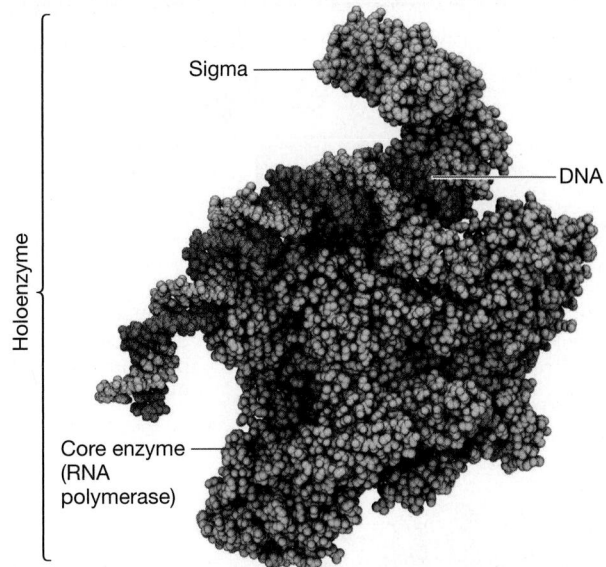

(b) Sigma recognizes and binds to the promoter.

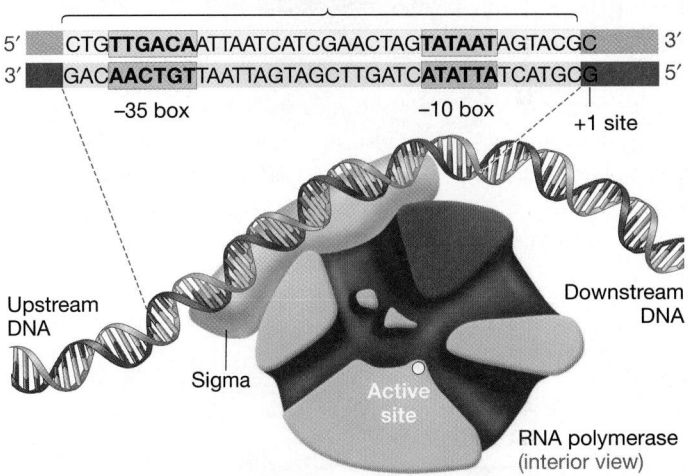

FIGURE 17.2 Sigma Is the Promoter-Recognizing Subunit of Bacterial RNA Polymerase Holoenzyme. (a) A space-filling model of bacterial RNA polymerase holoenzyme. **(b)** A cartoon of bacterial RNA polymerase, showing that sigma binds to the −35 box and −10 box of the promoter.

separated strands of DNA as shown in **FIGURE 17.3** (see page 320), steps 1 and 2. As step 2 shows, the template strand is threaded through a channel that leads to the active site inside RNA polymerase. Ribonucleoside triphosphates (NTPs)—the RNA building blocks—enter a channel in the enzyme and diffuse to the active site.

When an incoming NTP pairs with a complementary base on the template strand of DNA, RNA polymerization begins. The reaction catalyzed by RNA polymerase is exergonic and spontaneous because NTPs have significant potential energy, owing to their three phosphate groups. As step 3 of Figure 17.3 shows, the initiation phase of transcription is complete as RNA polymerase extends the mRNA from the +1 site.

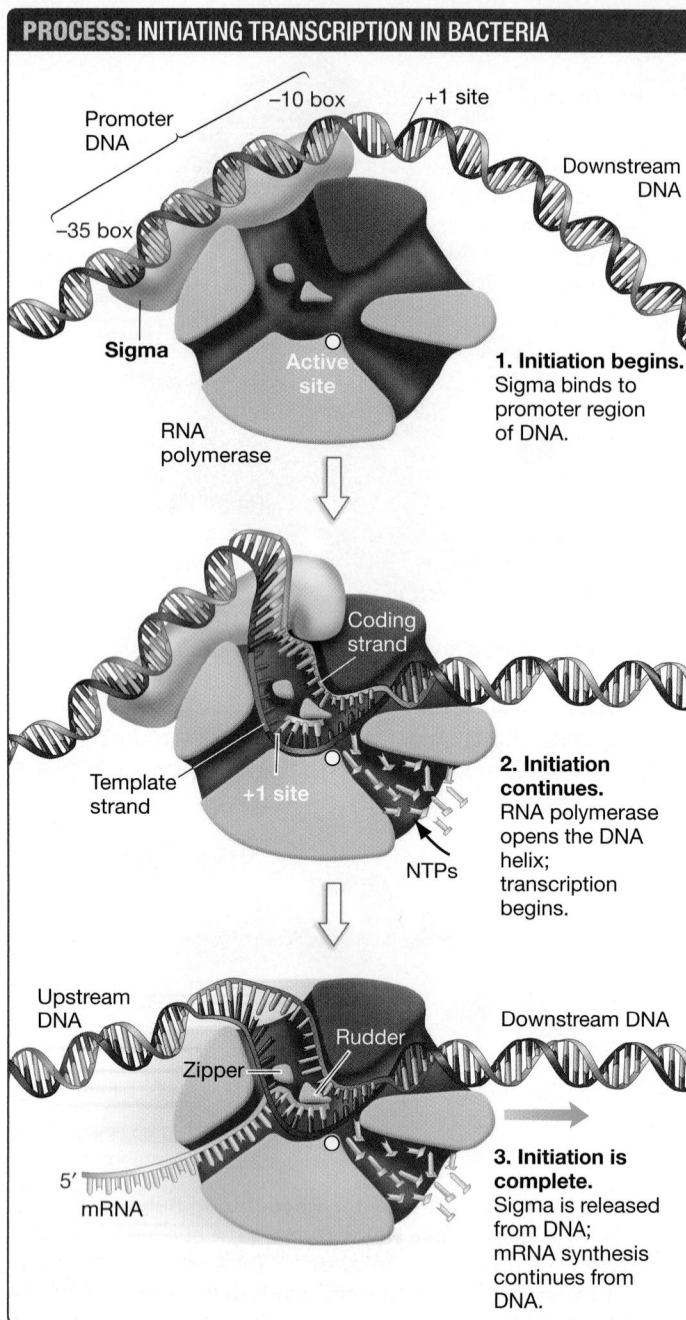

PROCESS: INITIATING TRANSCRIPTION IN BACTERIA

Promoter DNA

−10 box

+1 site

Downstream DNA

−35 box

Sigma

Active site

RNA polymerase

1. Initiation begins. Sigma binds to promoter region of DNA.

Coding strand

Template strand

+1 site

NTPs

2. Initiation continues. RNA polymerase opens the DNA helix; transcription begins.

Upstream DNA

Rudder

Downstream DNA

Zipper

5′

mRNA

3. Initiation is complete. Sigma is released from DNA; mRNA synthesis continues from DNA.

FIGURE 17.3 Sigma Orients the DNA Template inside RNA Polymerase. Sigma binds to the promoter, and RNA polymerase opens the DNA helix and threads the template strand through the active site.

Elongation and Termination

Once RNA polymerase begins moving along the DNA template synthesizing RNA, the **elongation** phase of transcription is under way. RNA polymerase is a macromolecular machine with different parts. In the interior of the enzyme, a group of amino acids forms a rudder to help steer the template and non-template strands through channels inside the enzyme (see Figure 17.3, step 3). Meanwhile, the enzyme's active site catalyzes the addition

of nucleotides to the 3′ end of the growing RNA molecule at the rate of about 50 nucleotides per second. A group of projecting amino acids forms a region called the zipper to help separate the newly synthesized RNA from the DNA template.

During the elongation phase of transcription, all the prominent channels and grooves in the enzyme are filled (Figure 17.3, step 3). Double-stranded DNA goes into and out of one groove, ribonucleoside triphosphates enter another, and the growing RNA strand exits to the rear. The enzyme's structure is critical for its function.

Termination ends transcription. In bacteria, transcription stops when RNA polymerase transcribes a DNA sequence that functions as a transcription-termination signal.

The bases that make up the termination signal in bacteria are transcribed into a stretch of RNA with an important property: As soon as it is synthesized, this portion of the RNA folds back on itself and forms a short double helix that is held together by complementary base pairing (**FIGURE 17.4**). Recall that this type of RNA secondary structure is called a hairpin (Chapter 4). The hairpin structure disrupts the interaction between RNA polymerase and the RNA transcript, resulting in the physical separation of the enzyme and its product.

Transcription in Eukaryotes

Fundamental features of transcription are the same in bacteria and eukaryotes. In fact, these similarities provide compelling evidence for a common ancestor of all cells. There are, however, some differences that are worth noting:

- Eukaryotes have three polymerases—RNA polymerase I, II, and III—that are often referred to as pol I, pol II, and pol III. Each polymerase transcribes only certain types of RNA in eukaryotes. RNA pol II is the only polymerase that transcribes protein-coding genes.

- Promoters in eukaryotic DNA are more diverse than bacterial promoters. Most eukaryotic promoters include a sequence called the **TATA box,** centered about 30 base pairs upstream of the transcription start site, and other important sequences that vary more widely.

- Instead of using a sigma protein, eukaryotic RNA polymerases recognize promoters using a group of proteins called **basal transcription factors.** Basal transcription factors assemble at the promoter, and RNA polymerase follows. (This idea, as well as the extensive use of other types of transcription factors to regulate transcription, will be covered in Chapter 19.)

- Termination of eukaryotic protein-coding genes involves a short sequence called the polyadenylation signal or **poly(A) signal.** Soon after the signal is transcribed, the RNA is cut by an enzyme downstream of the poly(A) signal as the polymerase continues to transcribe the DNA template. Eventually RNA polymerase falls off the DNA template and terminates transcription. Bacteria end transcription at a distinct site for each gene, but in eukaryotes, transcription ends variable distances from the poly(A) signal.

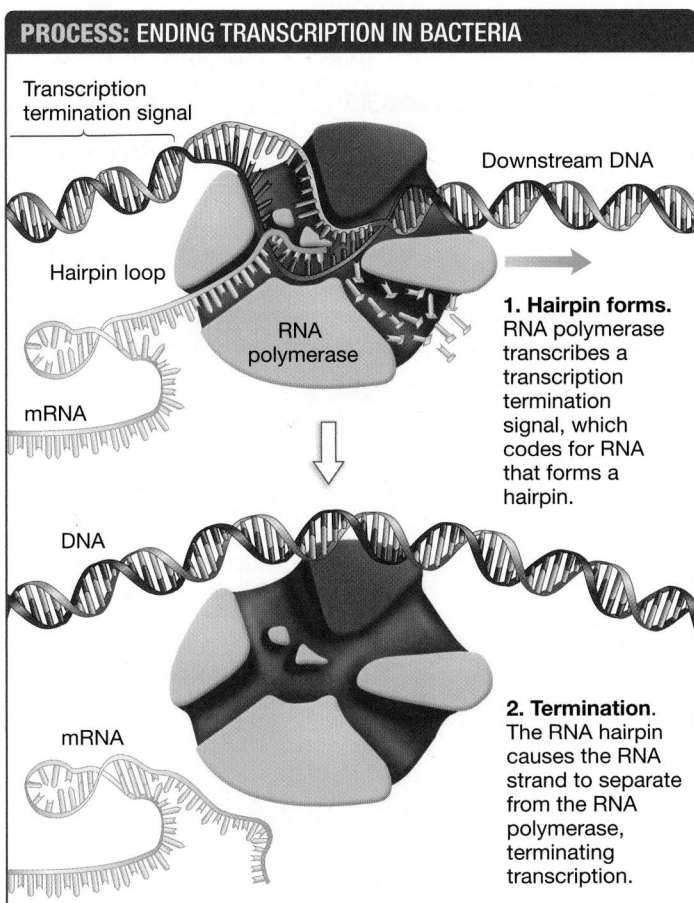

PROCESS: ENDING TRANSCRIPTION IN BACTERIA

Transcription termination signal

Downstream DNA

Hairpin loop

RNA polymerase

mRNA

1. Hairpin forms. RNA polymerase transcribes a transcription termination signal, which codes for RNA that forms a hairpin.

DNA

mRNA

2. Termination. The RNA hairpin causes the RNA strand to separate from the RNA polymerase, terminating transcription.

FIGURE 17.4 Transcription Terminates When an RNA Hairpin Forms.

 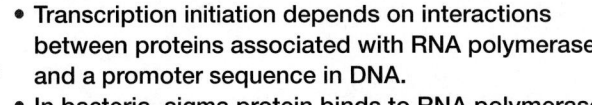
17.2 RNA Processing in Eukaryotes

The molecular machinery required for transcription is much more complex in eukaryotes than in bacteria. But these differences are minor when compared with what happens to the eukaryotic RNA after transcription. In bacteria, when transcription terminates, the result is a mature mRNA that's ready to be translated into a protein. In fact, translation often begins while the mRNA is still being transcribed.

The fate of the transcript in eukaryotes is more complicated. When eukaryotic genes of any type are transcribed, the initial product is termed a **primary transcript.** This RNA must undergo multistep processing before it is functional. For protein-coding genes, the primary transcript is called a **pre-mRNA.**

The processing of primary transcripts has important consequences for gene expression in eukaryotes. Let's delve in to see how and why.

The Startling Discovery of Split Eukaryotic Genes

Eukaryotic genes do not consist of one continuous DNA sequence that codes for a product, as do bacterial genes. Instead, the regions in a eukaryotic gene that code for proteins are intermittently interrupted by stretches of hundreds or even thousands of intervening bases.

Although these intervening bases are part of the gene, they do not code for a product. To make a functional RNA, eukaryotic cells must dispose of certain sequences inside the primary transcript and then combine the separated sections into an integrated whole.

What sort of data would provoke such a startling claim? The first evidence came from work that Phillip Sharp and colleagues carried out in the late 1970s to determine the location of genes within the DNA of a virus that infects mammalian cells. Viruses are often used as tools to provide insights into fundamental processes of the cells they infect.

They began one of their experiments by heating the virus' DNA sufficiently to break the hydrogen bonds between complementary bases. This treatment separated the two strands. The single-stranded DNA was then incubated with the mRNA encoded by the virus. The team's intention was to promote base pairing between the mRNA and the single-stranded DNA.

The researchers expected that the mRNA would form base pairs with the DNA sequence that acted as the template for its synthesis—that the mRNA and DNA would match up exactly.

(a) Micrograph of DNA-RNA hybrid

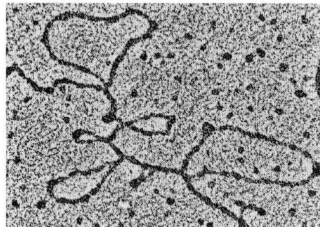

(b) Interpretation of micrograph

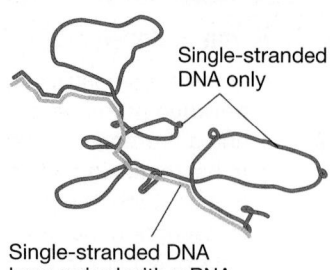

Single-stranded DNA only

Single-stranded DNA base paired with mRNA

FIGURE 17.5 The Discovery of Introns. The loops in the micrograph and drawing represent regions of DNA that are transcribed but are not found in the final mRNA. These regions are introns.

✔**QUESTION** If the noncoding regions of the gene did not exist, what would the micrograph in part (a) look like?

But when the team examined the DNA–RNA hybrid molecules using an electron microscope, they observed the structure shown in **FIGURE 17.5a**. Instead of matching up exactly, parts of the DNA formed loops.

What was going on? As **FIGURE 17.5b** shows, Sharp's group interpreted these loops as stretches of DNA that are present in the template strand but are *not* in the corresponding mRNA.

Sharp's group and a team headed by Richard Roberts then carried out similar studies on eukaryotic genes. The results were the same as for the viral genes. They went on to propose that there is not a one-to-one correspondence between the nucleotide sequence of a eukaryotic gene and its mRNA. As an analogy, it could be said that eukaryotic genes do not carry messages such as "Biology is my favorite course of all time." Instead, they carry messages that read something like:

BIOLτηεπροτεινχοδινγρεγιονσοφγενεσOGY IS MY
FAVORαρειντερρθπτεθβυνονψοθινγθITE COURSE
OF ανθηαωετοβεσπλιχεθτογετηερ ALL TIME

Here the sections of noncoding sequence are represented with Greek letters. They must be removed from the mRNA before it can carry an intelligible message to the translation machinery.

When it became clear that the genes-in-pieces hypothesis was correct, Walter Gilbert suggested that regions of eukaryotic genes that are part of the final mRNA be referred to as **exons** (because they are *ex*pressed) and the sections of primary transcript not in mRNA be referred to as **introns** (because they are *inter*vening). Introns are sections of genes that are not represented in the final RNA product. Because of introns, eukaryotic genes are much larger than their corresponding mature RNAs. Introns were first discovered in genes that produce mRNA, but researchers later found that genes for other types of RNA also could be split.

RNA Splicing

The transcription of eukaryotic genes by RNA polymerase generates a primary transcript (**FIGURE 17.6a**) that contains both exons and introns. As transcription proceeds, the introns are removed

(a) Introns must be removed from eukaryotic RNA transcripts.

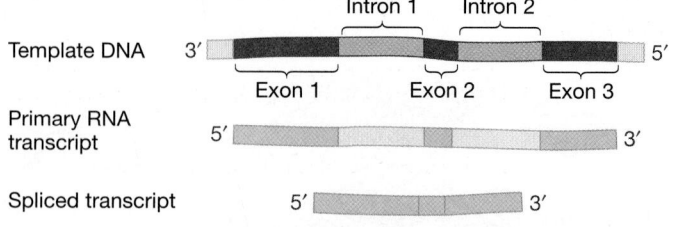

(b) PROCESS: snRNPs SPLICE RNA WITHIN THE NUCLEUS

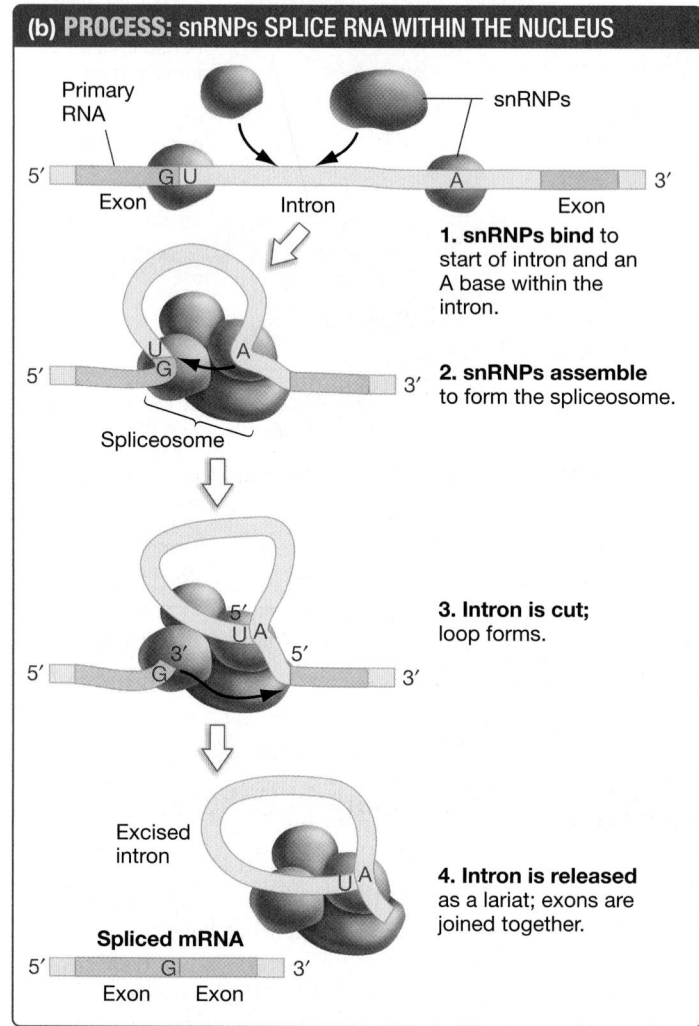

1. snRNPs bind to start of intron and an A base within the intron.

2. snRNPs assemble to form the spliceosome.

3. Intron is cut; loop forms.

4. Intron is released as a lariat; exons are joined together.

Spliced mRNA

FIGURE 17.6 Introns Are Spliced Out of the Primary Transcript.

from the growing RNA strand by a process known as **splicing**. In this phase of information processing, pieces of the primary transcript are removed and the remaining segments are joined together. Splicing occurs within the nucleus while transcription is still under way and results in an RNA that contains an uninterrupted genetic message.

FIGURE 17.6b provides more detail about how introns are removed from primary transcripts to form mRNA. Splicing of primary transcripts is catalyzed by RNAs called small nuclear RNAs (snRNAs) working with a complex of proteins. These protein-plus-RNA macromolecular machines are known as **small nuclear ribonucleoproteins,** or **snRNPs** (pronounced "snurps").

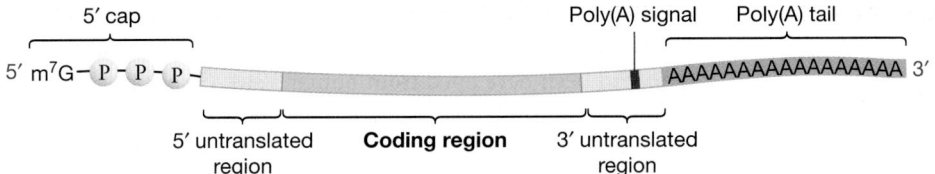

5′ cap
Poly(A) signal   Poly(A) tail

5′ m⁷G– P – P – P –[   ]AAAAAAAAAAAAAAAAAAAAA 3′

5′ untranslated   **Coding region**   3′ untranslated
region   region

FIGURE 17.7 In Eukaryotes, a Cap and a Tail Are Added to mRNAs. As part of eukaryotic pre-mRNA processing, a cap consisting of a modified guanine (G) nucleotide (symbolized as m⁷G) bonded to three phosphate groups is added to the 5′ end, and a tail made up of a long series of adenine (A) residues is added to the 3′ end after cleavage of the primary transcript.

The snRNAs of the snRNPs recognize RNA sequences critical for splicing. Splicing can be broken into four steps:

1. The process begins when snRNPs bind to the 5′ exon–intron boundary, which is marked by the bases GU, and to a key adenine ribonucleotide (A) near the end the intron.

2. Once the initial snRNPs are in place, other snRNPs arrive to form a multipart complex called a **spliceosome.** The spliceosomes found in human cells contain about 145 different proteins and RNAs, making them the most complex macromolecular machines known.

3. The intron forms a loop plus a single-stranded stem (a lariat) with the adenine at its connecting point.

4. The lariat is cut out, and a phosphodiester linkage links the exons on either side, producing a continuous coding sequence—the mRNA.

Splicing is now complete. In most cases, the excised intron is degraded to ribonucleoside monophosphates.

As you'll see later (Chapter 19), for many genes, the RNA can be spliced in more than one way. This allows the production of different, related mRNAs and proteins from one gene.

Current data suggest that both the cutting and rejoining reactions that occur during splicing are catalyzed by the snRNA molecules in the spliceosome—meaning that the reactions are catalyzed by a ribozyme. Section 17.5 will demonstrate that ribozymes also play a key role in translation. As the RNA world hypothesis (Chapter 4) predicts, proteins are not the only important catalysts in cells.

What is the origin of introns? One hypothesis is that introns in eukaryotes arose from an ancient type of DNA sequence that is present in many bacteria and archaea. These ancient sequences are related to viruses (Chapter 36) and, like viruses, can infect cells and insert into their genomes. Remarkably, when this DNA sequence inserts into a gene and is transcribed, the RNA catalyzes its own splicing out of the primary transcript. This is possible in part because these virus-like elements have sequences similar to snRNAs.

The bacterium that was the source of mitochondria (see Chapter 30) likely carried some of these virus-like DNA sequences. When this bacterium was taken up by an ancestral eukaryote, the virus-like sequences are hypothesized to have spread rapidly. Later in evolution, the portion of the sequence that was a precursor to today's snRNA may have separated from the portion of the sequence that was spliced. This spliced sequence is hypothesized to be the ancestor of the modern eukaryotic intron.

Adding Caps and Tails to Transcripts

For pre-mRNAs, intron splicing is accompanied by other important processing steps.

- As soon as the 5′ end of a eukaryotic pre-mRNA emerges from RNA polymerase, enzymes add a structure called the **5′ cap** (**FIGURE 17.7**). The cap consists of a modified guanine (7-methylguanylate) nucleotide with three phosphate groups.

- An enzyme cleaves the 3′ end of the pre-mRNA downstream of the poly(A) signal (introduced in Section 17.1). Another enzyme adds a long row of 100–250 adenine nucleotides that are not encoded on the DNA template strand. This string of adenines is known as the **poly(A) tail.**

With the addition of the cap and tail and completion of splicing, processing of the pre-mRNA is complete. The product is a mature mRNA.

Figure 17.7 also shows that in the mature RNA molecule, the coding sequence for the polypeptide is flanked by sequences that are not destined to be translated. These 5′ and 3′ untranslated regions (or UTRs) help stabilize the mature RNA and regulate its translation. The mRNAs in bacteria also possess 5′ and 3′ UTRs.

Not long after the caps and tails on eukaryotic mRNAs were discovered, evidence began to accumulate that they protect mRNAs from degradation by ribonucleases—enzymes that

check your understanding

If you understand that . . .

- Eukaryotic genes consist of exons, which are parts of the primary transcript that remain in mature RNA, and introns, which are regions of the primary transcript that are removed in forming mature RNA.
- Macromolecular machines, called spliceosomes, splice introns out of pre-mRNAs.
- Enzymes add a 5′ cap and a poly(A) tail to spliced transcripts, producing a mature mRNA that is ready to be translated.

✔ **You should be able to . . .**

1. Explain why ribonucleoprotein is an appropriate name for the subunits of the spliceosome.
2. Explain the function of the 5′ cap and the poly(A) tail.

Answers are available in Appendix A.

degrade RNA—and enhance the efficiency of translation. For example:

- Experimental mRNAs that have a cap and a tail last longer when they are introduced into cells than do experimental mRNAs that lack a cap or a tail.

- Experimental mRNAs with caps and tails produce more proteins than do experimental mRNAs without caps and tails.

Follow-up work has shown that the 5′ cap and the poly(A) tail are bound by proteins that prevent ribonucleases in the cytoplasm from recognizing and destroying the mRNA. The 5′ cap and the poly(A) tail also are important for initiating translation.

RNA processing is the general term for any of the modifications, such as splicing or poly(A) tail addition, needed to convert a primary transcript into a mature RNA. It is summarized in **TABLE 17.1** along with other important differences in how RNAs are produced in eukaryotes as compared with bacteria.

17.3 An Introduction to Translation

To synthesize a protein, the sequence of bases in a messenger RNA molecule is translated into a sequence of amino acids in a polypeptide. The genetic code specifies the correspondence between each triplet codon in mRNA and the amino acid it codes for (see Chapter 16). But how are the amino acids assembled into a polypeptide according to the information in messenger RNA?

Studies of translation in cell-free systems proved extremely effective in answering this question. Once in vitro translation systems had been developed from human cells, *E. coli*, and a variety of other organisms, biologists could see that the sequence of events is similar in bacteria, archaea, and eukaryotes. As with similarities in transcription across the domains of life, the shared mechanisms of translation argue for a common ancestor of all cells living today.

Ribosomes Are the Site of Protein Synthesis

The first question that biologists answered about translation concerned where it occurs. The answer grew from a simple observation: There is a strong correlation between the number of **ribosomes** in a given type of cell and the rate at which that cell synthesizes proteins. Based on this observation, investigators proposed that ribosomes are the site of protein synthesis.

To test this hypothesis, Roy Britten and collaborators did a pulse–chase experiment similar in design to experiments introduced earlier (Chapter 7). Recall that a pulse–chase experiment labels a population of molecules as they are being produced. The location of the tagged molecules is then followed over time.

In this case, the tagging was done by supplying a pulse of radioactive sulfur atoms that would be incorporated into the amino acids methionine and cysteine, followed by a chase of unlabeled sulfur atoms. If the ribosome hypothesis were correct, the radioactive signal should be associated with ribosomes for a short period of time—when the amino acids are being polymerized into proteins. Later, when translation was complete, all the radioactivity should be found in proteins that are not associated with ribosomes.

This is exactly what the researchers found. Based on these data, biologists concluded that proteins are synthesized at ribosomes and then released.

Translation in Bacteria and Eukaryotes

About a decade after the ribosome hypothesis was confirmed, electron micrographs showed bacterial ribosomes in action (**FIGURE 17.8a**). The images showed that in bacteria, ribosomes

(a) Bacterial ribosomes during translation

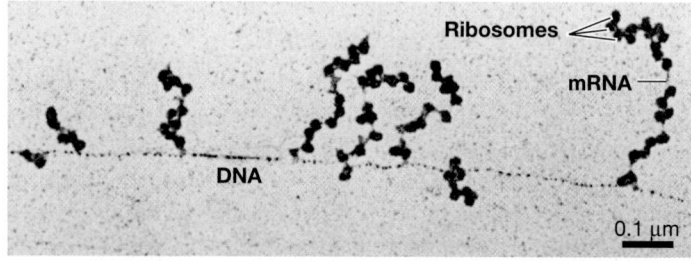

(b) In bacteria, transcription and translation are tightly coupled.

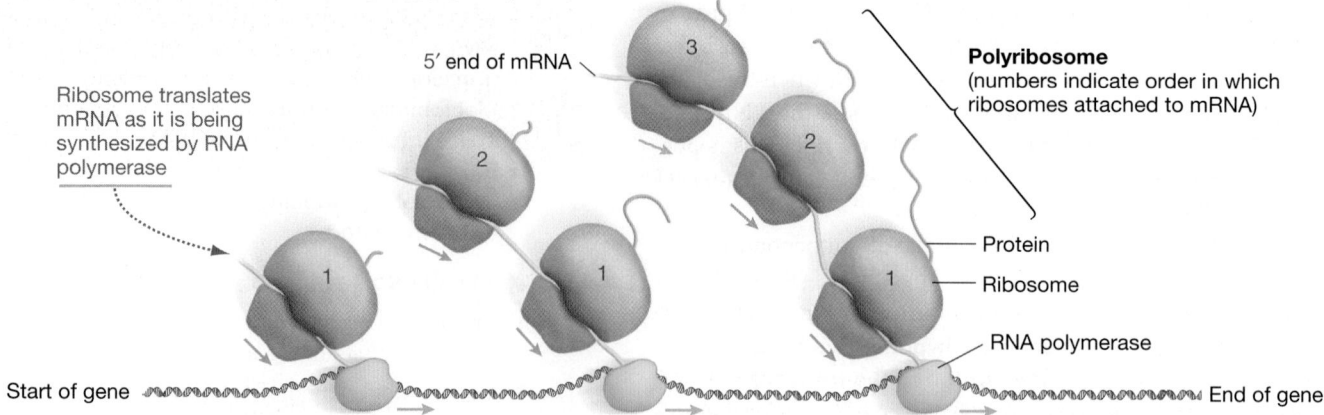

FIGURE 17.8 Transcription and Translation Occur Simultaneously in Bacteria. In bacteria, ribosomes attach to mRNA transcripts and begin translation while RNA polymerase is still transcribing the DNA template strand.

TABLE 17.1 Transcription and RNA Processing in Bacteria and Eukaryotes

Point of Comparison	Bacteria	Eukaryotes
RNA polymerase(s)	One	Three; each produces a different class of RNA
Promoter structure	Typically contains a −35 box and a −10 box	More variable; often includes a TATA box about −30 from the transcription start site
Proteins involved in recognizing promoter	Sigma; different versions of sigma bind to different promoters	Many basal transcription factors
RNA processing	None	Extensive; several processing steps occur in the nucleus before RNA is exported to the cytoplasm: **(1)** Enzyme-catalyzed addition of 5′ cap on mRNAs, **(2)** Splicing (intron removal); by spliceosome to produce mRNA, **(3)** Enzyme-catalyzed addition of 3′ poly(A) tail on mRNAs

attach to mRNAs and begin synthesizing proteins even before transcription is complete. In fact, multiple ribosomes attach to each mRNA, forming a **polyribosome** (**FIGURE 17.8b**). In this way, many copies of a protein can be produced from a single mRNA.

Transcription and translation can occur concurrently in bacteria because there is no nuclear envelope to separate the two processes.

The situation is different in eukaryotes. In these organisms, primary transcripts are processed in the nucleus to produce a mature mRNA, which is then exported to the cytoplasm (**FIGURE 17.9**). This means that in eukaryotes, transcription and translation are separated in time and space. Once mRNAs are outside the nucleus, ribosomes can attach to them and begin translation. As in bacteria, polyribosomes form.

How Does an mRNA Triplet Specify an Amino Acid?

When an mRNA interacts with a ribosome, instructions encoded in nucleic acids are translated into a different chemical language—the amino acid sequences found in proteins. The discovery of the genetic code revealed that triplet codons in mRNA specify particular amino acids in a protein. How does this conversion occur?

One early hypothesis was that mRNA codons and amino acids interact directly. This hypothesis proposed that the bases in a particular codon were complementary in shape or charge to the side group of a particular amino acid (**FIGURE 17.10a**, see page 326). But Francis Crick pointed out that the idea didn't make chemical sense. For example, how could the nucleic acid bases interact with a hydrophobic amino acid side group, which does not form hydrogen bonds?

Crick proposed an alternative hypothesis. As **FIGURE 17.10b** shows, he suggested that some sort of adapter molecule holds amino acids in place while interacting directly and specifically with a codon in mRNA by hydrogen bonding. In essence, Crick predicted the existence of a chemical go-between that produced a physical connection between the two types of molecules. As it turns out, Crick was right.

(a) mRNAs are exported to the cytoplasm.

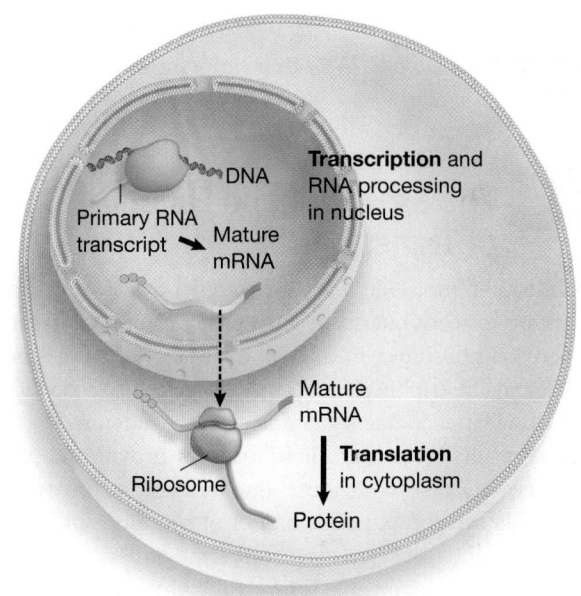

(b) Polypeptides grow from ribosomes translating mRNA.

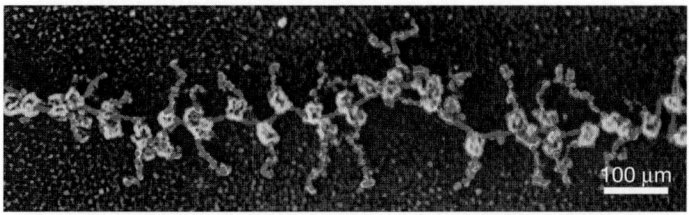

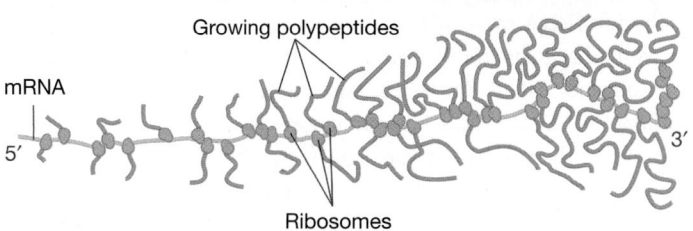

FIGURE 17.9 Transcription and Translation Are Separated in Space and Time in Eukaryotes.

(a) Hypothesis 1: Amino acids interact directly with mRNA codons.

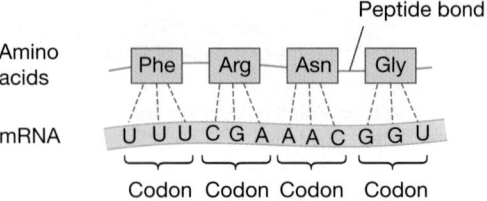

(b) Hypothesis 2: Adapter molecules hold amino acids and interact with mRNA codons.

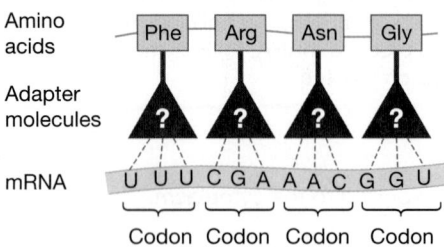

FIGURE 17.10 How Do mRNA Codons Interact with Amino Acids?

17.4 The Structure and Function of Transfer RNA

Crick's adapter molecule was discovered by accident. Biologists were trying to work out an in vitro protein-synthesis system and reasoned that ribosomes, mRNA, amino acids, ATP, and a molecule called guanosine triphosphate, or GTP, would be needed. (GTP is similar to ATP but contains guanine instead of adenine.)

These results were logical: Ribosomes provide the catalytic machinery, mRNAs contribute the message to be translated, amino acids are the building blocks of proteins, and ATP and GTP supply potential energy to drive the endergonic polymerization reactions responsible for forming proteins.

But, in addition, a cellular fraction that contained a previously unknown type of RNA turned out to be indispensable. If this type of RNA is missing, protein synthesis does not occur. What is this mysterious RNA, and why is it essential to translation?

The novel class of RNAs eventually became known as **transfer RNA (tRNA).** The role of tRNA in translation was a mystery until some researchers happened to add a radioactive amino acid—leucine—to an in vitro protein-synthesis system. The treatment was actually done as a control for an unrelated experiment. To the researchers' amazement, some of the radioactive leucine attached to tRNA molecules.

What happens to the amino acids bound to tRNAs? To answer this question, Paul Zamecnik and colleagues tracked the fate of radioactive leucine molecules that were attached to tRNAs. They found that the amino acids are transferred from tRNAs to proteins.

The data supporting this conclusion are shown in the "Results" section of **FIGURE 17.11**. The graph shows that radioactive amino acids are lost from tRNAs and incorporated into polypeptides synthesized by ribosomes. To understand this conclusion:

RESEARCH

QUESTION: What happens to the amino acids attached to tRNAs?

HYPOTHESIS: Aminoacyl tRNAs transfer amino acids to growing polypeptides.

NULL HYPOTHESIS: Aminoacyl tRNAs do not transfer amino acids to growing polypeptides.

EXPERIMENTAL SETUP:

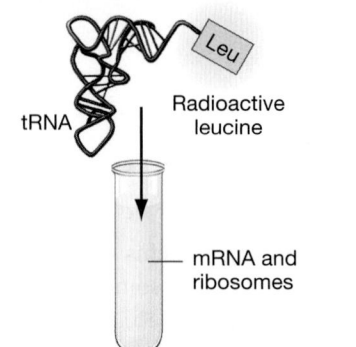

1. Attach radioactive leucine molecules to tRNAs.

2. Add these aminoacyl tRNAs to in vitro translation system. Follow fate of the radioactive amino acids.

PREDICTION: Radioactive amino acids will be found in proteins.

PREDICTION OF NULL HYPOTHESIS: Radioactive amino acids will not be found in proteins.

RESULTS:

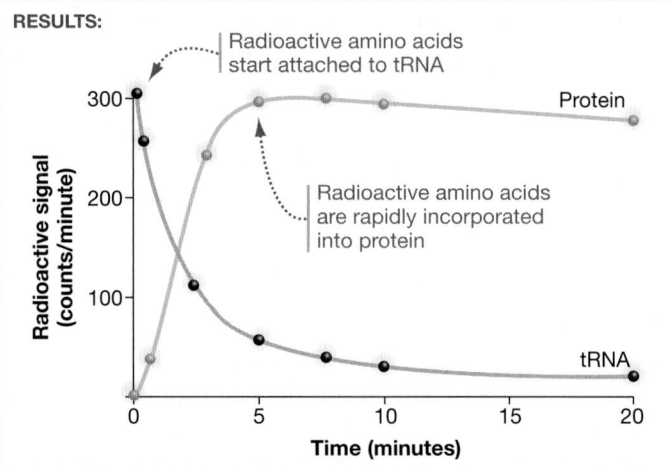

CONCLUSION: Aminoacyl tRNAs transfer amino acids to growing polypeptides.

FIGURE 17.11 Evidence that Amino Acids Are Transferred from tRNAs to Proteins.

SOURCE: Hoagland, M. B., M. L. Stephenson, J. F. Scott, et al. 1958. A soluble ribonucleic acid intermediate in protein synthesis. *Journal of Biological Chemistry* 231: 241–257.

✔ **QUESTION** What would the graphed results look like if the null hypothesis were correct?

1. Put your finger on the point on the *x*-axis that indicates that one minute has passed since the start of the experiment.

2. Read up until you hit the green line and the gray line. The green line represents data from proteins; the gray line represents data from tRNAs.

3. Check the *y*-axis—which indicates the amount of radioactive leucine present—at each point.

4. It should be clear that early in the experiment, almost all the radioactive leucine is attached to tRNA, not protein.

Next, do the same four steps at the point on the *x*-axis labeled 10 minutes (since the start of the experiment). Your conclusion now should be that late in the experiment, almost all the radioactive leucine is attached to proteins, not tRNA.

These results inspired the use of the word transfer in tRNA's name, because amino acids are transferred from the RNA to a growing polypeptide. The experiment also confirmed that tRNAs act as the interpreter during translation: tRNAs are Crick's adapter molecules.

What Do tRNAs Look Like?

Transfer RNAs serve as chemical go-betweens that allow amino acids to interact with an mRNA template. But precisely how does the connection occur?

This question was answered by research on tRNA's molecular structure. The initial studies established the sequence of nucleotides in various tRNAs, or what is termed their primary structure. Transfer RNA sequences are relatively short, ranging from 75 to 85 nucleotides in length.

When biologists studied the primary sequence closely, they noticed that certain parts of the molecules can form secondary structures. Specifically, some sequences of bases in the tRNA molecule can form hydrogen bonds with complementary base sequences elsewhere in the same molecule. As a result, portions of the molecule form stem-and-loop structures (introduced in Chapter 4). The stems are short stretches of double-stranded RNA; the loops are single stranded.

Two aspects of tRNA's secondary structure proved especially important. A CCA sequence at the 3′ end of each tRNA molecule offered a site for amino acid attachment, while a triplet on the loop at the other end of the structure could serve as an anticodon. An **anticodon** is a set of three ribonucleotides that forms base pairs with the mRNA codon.

Later, X-ray crystallography studies revealed the tertiary structure of tRNAs. Recall that the tertiary structure of a molecule is the three-dimensional arrangement of its atoms (Chapter 3). As **FIGURE 17.12** shows, tRNAs fold into an L-shaped molecule. The anticodon is at one end of the structure; the CCA sequence and attached amino acid are at the other end.

All the tRNAs in a cell have the same general structure, shaped like an upside-down L. They vary at the anticodon and attached amino acid. The tertiary structure of tRNAs is important because it maintains a precise physical distance between the anticodon and amino acid. As it turns out, this separation is important in positioning the amino acid and the anticodon within the ribosome.

✔ If you understand the structure of tRNAs, you should be able to (1) describe where on the L-shaped structure the amino acid attaches; and (2) explain the relationship between the anticodon of a tRNA and a codon in an mRNA.

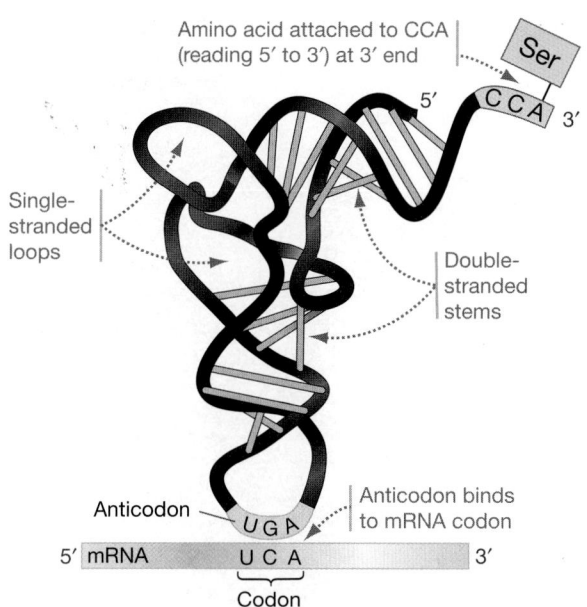

FIGURE 17.12 The Structure of an Aminoacyl Transfer RNA. The anticodon forms complementary base pairs with an mRNA codon.

How Are Amino Acids Attached to tRNAs?

How are amino acids linked to tRNAs? Just as important, what allows the right amino acid for a particular tRNA to be attached?

- An input of energy, in the form of ATP, is required to attach an amino acid to a tRNA.

- Enzymes called **aminoacyl-tRNA synthetases** catalyze the addition of amino acids to tRNAs—what biologists call "charging" a tRNA.

- For each of the 20 major amino acids, there is a different aminoacyl-tRNA synthetase and one or more tRNAs.

Each aminoacyl-tRNA synthetase has a binding site for a particular amino acid and a particular tRNA. Subtle differences in tRNA shape and base sequence allow the enzymes to recognize the correct tRNA for the correct amino acid. The combination of a tRNA molecule covalently linked to an amino acid is called an **aminoacyl tRNA. FIGURE 17.13** (see page 328) shows an aminoacyl-tRNA synthetase bound to a tRNA that has just been charged with an amino acid. Note how tightly the two structures fit together—making it possible for the enzyme and its tRNA and amino acid substrates to interact in a precise way.

How Many tRNAs Are There?

After characterizing all the different types of tRNAs, biologists encountered a paradox. According to the genetic code (Chapter 16), the 20 most common amino acids found in proteins are specified by 61 different mRNA codons. Instead of containing 61 different tRNAs with 61 different anticodons, though, most cells contain only about 40. How can all 61 codons be translated with only two-thirds that number of tRNAs?

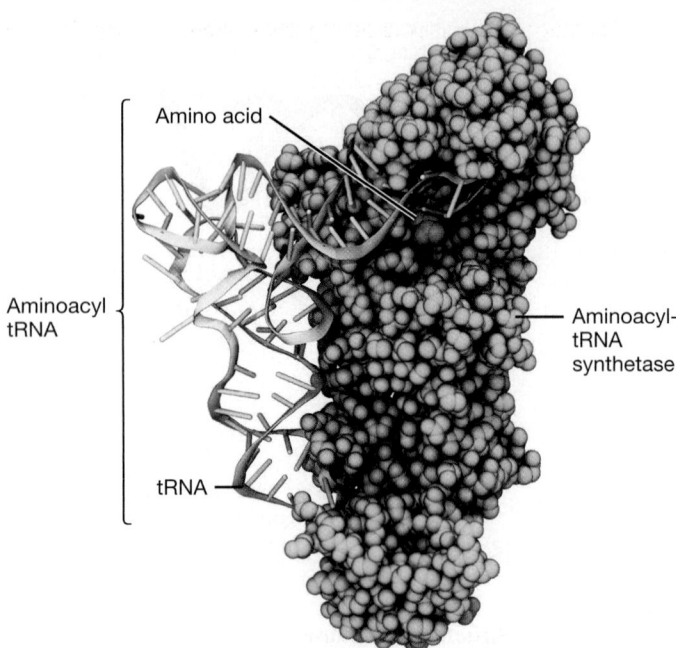

Amino acid

Aminoacyl
tRNA

Aminoacyl-
tRNA
synthetase

tRNA

FIGURE 17.13 Aminoacyl-tRNA Synthetases Couple the Appropriate Amino Acid to the Appropriate tRNA.

To resolve this paradox, Francis Crick proposed what is known as the **wobble hypothesis.** Recall that:

1. Many amino acids are specified by more than one codon.

2. Codons for the same amino acid tend to have the same nucleotides at the first and second positions but a different nucleotide at the third position.

For example, both of the codons CAA and CAG code for the amino acid glutamine. Surprisingly, experimental data have shown that a tRNA with an anticodon of GUU can base-pair with both CAA and CAG in mRNA. The GUU anticodon matches the first two bases (C and A) in both codons, but the U in the anticodon's third position forms a nonstandard base pair with a G in the CAG codon.

Crick proposed that inside the ribosome, certain bases in the third position of tRNA anticodons can bind to bases in the third position of a codon in a manner that does not match Watson–Crick base pairing. If so, this would allow a limited flexibility, or "wobble," in the base pairing.

According to the wobble hypothesis, particular nonstandard base pairs—such as G-U—are acceptable in the third position of a codon and do not change the amino acid that the codon specifies. In this way, wobble in the third position of a codon allows just 40 or so tRNAs to bind to all 61 mRNA codons.

17.5 The Structure and Function of Ribosomes

Recall that protein synthesis occurs when the sequence of bases in an mRNA is translated into a sequence of amino acids in a polypeptide. The translation of each mRNA codon begins when the anticodon of an aminoacyl tRNA binds to the codon. Translation of a codon is complete when a peptide bond forms between the tRNA's amino acid and the growing polypeptide chain.

Both of these events take place inside a ribosome. Biologists have known since the 1930s that ribosomes contain many proteins and **ribosomal RNAs (rRNAs).** Later work showed that ribosomes can be separated into two major substructures, called the large subunit and small subunit. Each ribosome subunit consists of a complex of RNA molecules and proteins. The small subunit holds the mRNA in place during translation; the large subunit is where peptide-bond formation takes place.

FIGURE 17.14 shows two views of how the molecules required for translation fit together. Note that during protein synthesis, three distinct tRNAs are lined up inside the ribosome. All three are bound to their corresponding mRNA codons.

- The tRNA that is on the right in the figure, and colored red, carries an amino acid. This tRNA's position in the ribosome is called the A site—"A" for acceptor or aminoacyl.

- The tRNA that is in the middle (green) holds the growing polypeptide chain and occupies the P site, for peptidyl, inside the ribosome. (Think of "P" for peptide-bond formation.)

- The left-hand (blue) tRNA no longer has an amino acid attached and is about to leave the ribosome. It occupies the ribosome's E site—"E" for exit.

Because all tRNAs have similar secondary and tertiary structure, they all fit equally well in the A, P, and E sites.

The ribosome is a macromolecular machine that synthesizes proteins in a three-step sequence:

1. An aminoacyl tRNA diffuses into the A site; if its anticodon matches a codon in mRNA, it stays in the ribosome.

2. A peptide bond forms between the amino acid held by the aminoacyl tRNA in the A site and the growing polypeptide, which was held by a tRNA in the P site.

3. The ribosome moves down the mRNA by one codon, and all three tRNAs move one position within the ribosome. The tRNA in the E site exits; the tRNA in the P site moves to the E site; and the tRNA in the A site switches to the P site.

The protein that is being synthesized grows by one amino acid each time this three-step sequence repeats. The process occurs up to 20 times per second in bacterial ribosomes and about 2 times per second in eukaryotic ribosomes. Protein synthesis starts at the amino end (N-terminus) of a polypeptide and proceeds to the carboxy end (C-terminus; see Chapter 3).

This introduction to how tRNAs, mRNAs, and ribosomes interact during protein synthesis leaves several key questions unanswered. How do mRNAs and ribosomes get together to start the process? Once protein synthesis is under way, how is peptide-bond formation catalyzed inside the ribosome? And how does protein synthesis conclude when the ribosome reaches the end of the message? Let's consider each question in turn.

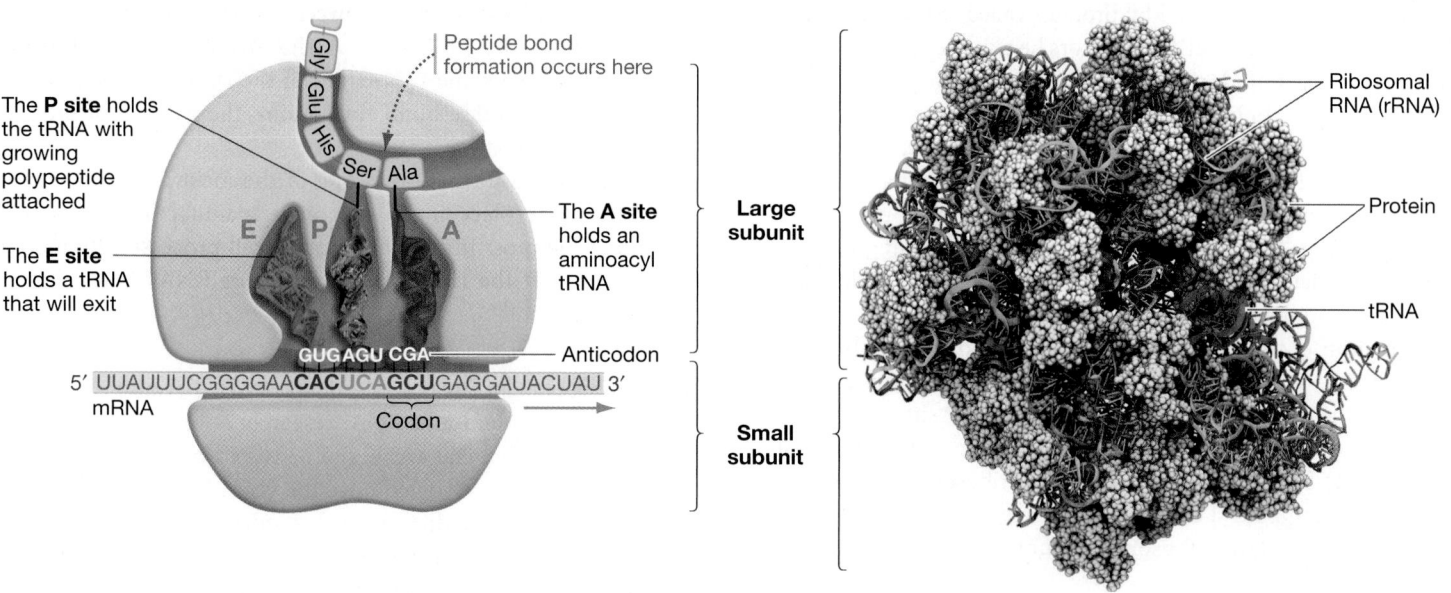

(a) Diagram of ribosome during translation (interior view)

Peptide bond formation occurs here

The **P site** holds the tRNA with growing polypeptide attached

The **E site** holds a tRNA that will exit

The **A site** holds an aminoacyl tRNA

Large subunit

Anticodon

5′ UUAUUUCGGGGAA**CAC**UCA**GCU**GAGGAUACUAU 3′
mRNA

Codon

Small subunit

(b) Model of ribosome during translation (exterior view)

Ribosomal RNA (rRNA)

Protein

tRNA

FIGURE 17.14 The Structure of the Ribosome. Ribosomes have three distinct tRNA binding sites in their interior.

Initiating Translation

To translate an mRNA properly, a ribosome must begin at a specific point in the message, translate the mRNA up to the message's termination codon, and then stop. Using the same terminology that they apply to transcription, biologists call these three phases of protein synthesis initiation, elongation, and termination, respectively.

One key to understanding translation initiation is to recall that a start codon (usually AUG) is found near the 5′ end of

all mRNAs and that it codes for the amino acid methionine (Chapter 16).

FIGURE 17.15 shows how translation gets under way in bacteria. The process begins when a section of rRNA in a small ribosomal subunit binds to a complementary sequence on an mRNA. The mRNA region is called the **ribosome binding site,** or **Shine–Dalgarno sequence,** after the biologists who discovered it. The site is about six nucleotides upstream from the start codon.

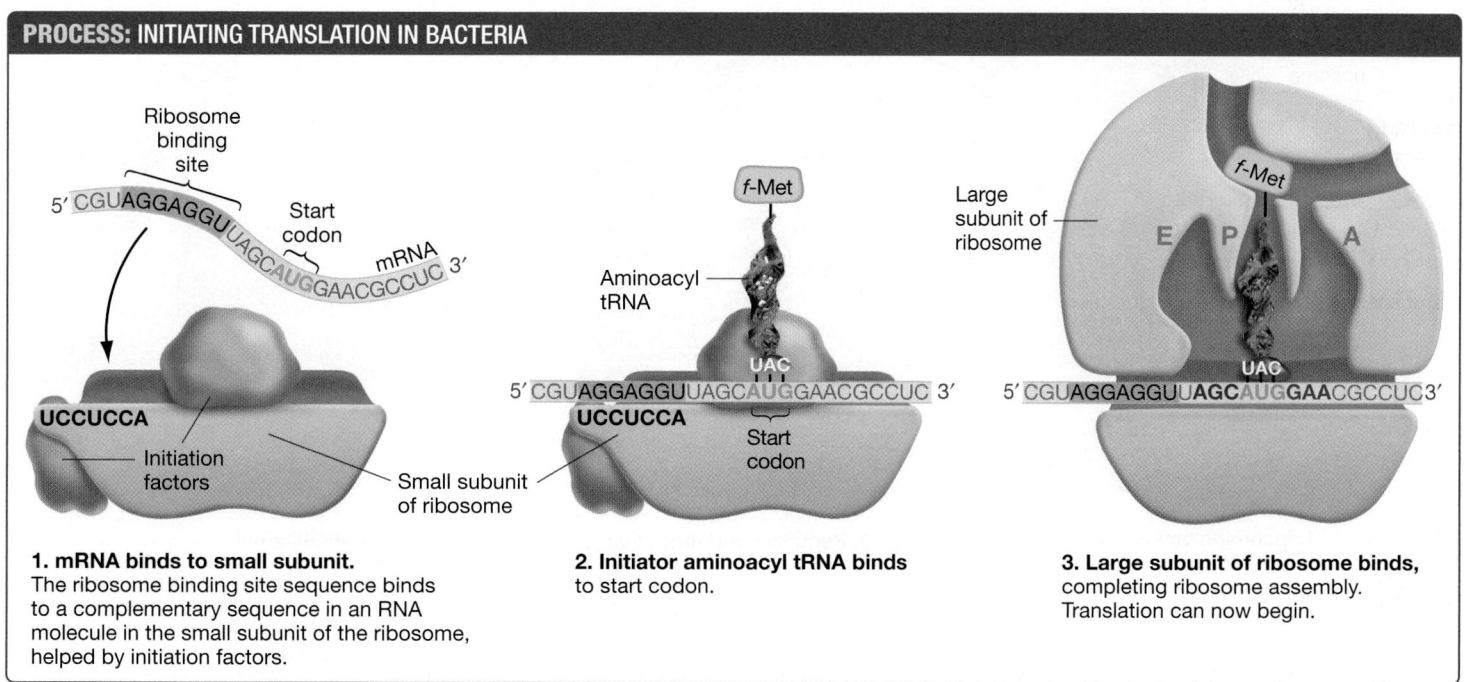

PROCESS: INITIATING TRANSLATION IN BACTERIA

Ribosome binding site

Start codon

5′ CGU**AGGAGGU**UAGCAUGGAACGCCUC 3′

mRNA

UCCUCCA

Initiation factors

Small subunit of ribosome

1. mRNA binds to small subunit.
The ribosome binding site sequence binds to a complementary sequence in an RNA molecule in the small subunit of the ribosome, helped by initiation factors.

f-Met

Aminoacyl tRNA

UAC

5′ CGU**AGGAGGU**UAGC**AUG**GAACGCCUC 3′

UCCUCCA

Start codon

2. Initiator aminoacyl tRNA binds to start codon.

f-Met

Large subunit of ribosome

E P A

UAC

5′ CGU**AGGAGGU**UA**GC**AUG**GAA**CGCCUC3′

3. Large subunit of ribosome binds, completing ribosome assembly. Translation can now begin.

FIGURE 17.15 Initiation of Translation.

The interactions between the small subunit, the message, and the tRNA are mediated by proteins called **initiation factors** (Figure 17.15, step 1). Initiation factors help in preparing the ribosome for translation, including binding the first aminoacyl tRNA to the ribosome. In bacteria this initiator tRNA bears a modified form of methionine called *N*-formylmethionine (abbreviated *f*-met) (Figure 17.15, step 2). In eukaryotes, this initiating tRNA carries a normal methionine. Initiation factors also prevent the small and large subunits of the ribosome from coming together until the initiator tRNA is in place at the AUG start codon, and they help bind the mRNA to the small ribosomal subunit.

Initiation is complete when the large subunit joins the complex (Figure 17.15, step 3). When the ribosome is completely assembled, the tRNA bearing *f*-met occupies the P site.

To summarize, translation initiation is a three-step process in bacteria: **(1)** The mRNA binds to a small ribosomal subunit, **(2)** the initiator aminoacyl tRNA bearing *f*-met binds to the start codon, and **(3)** the large ribosomal subunit binds, completing the complex.

In eukaryotes, the details of initiation are different, but they still involve recognition of a start codon, assembly of the ribosome, assistance from initiation factors, and the positioning of a methionine-carrying tRNA in the P site. The cap and poly(A) tail are also important in assembling the ribosome on the mRNA.

Elongation: Extending the Polypeptide

At the start of elongation, the E and A sites in the ribosome are empty of tRNAs. As a result, an mRNA codon is exposed in the A site. As step 1 in **FIGURE 17.16** illustrates, elongation proceeds when an aminoacyl tRNA binds to the codon in the A site by complementary base pairing between the anticodon and codon.

When both the P site and A site are occupied by tRNAs, the amino acids on the tRNAs are in the ribosome's active site. This is where peptide-bond formation—the essence of protein synthesis—occurs.

Peptide-bond formation is one of the most important reactions that take place in cells because manufacturing proteins is among the most fundamental of all cell processes. Biologists wondered, is it the ribosome's proteins or RNAs that catalyze this reaction?

Is the Ribosome an Enzyme or a Ribozyme? Because ribosomes contain both protein and RNA, researchers had argued for decades over whether the active site consisted of protein or RNA. The debate was not resolved until the year 2000, when researchers completed three-dimensional models that were detailed enough to reveal the structure of the active site. These models confirmed that the active site consists entirely of ribosomal RNA. Based on these results, biologists are now convinced that protein synthesis is catalyzed by RNA. The ribosome is a ribozyme—not a protein-based enzyme.

The observation that protein synthesis is catalyzed by RNA is important because it supports the RNA-world hypothesis (Chapter 4). Recall that proponents of this hypothesis claim that life began with RNA molecules and that the presence of DNA and proteins in cells evolved later. If the RNA-world hypothesis is correct, then it would make sense that the production of proteins is catalyzed by RNA.

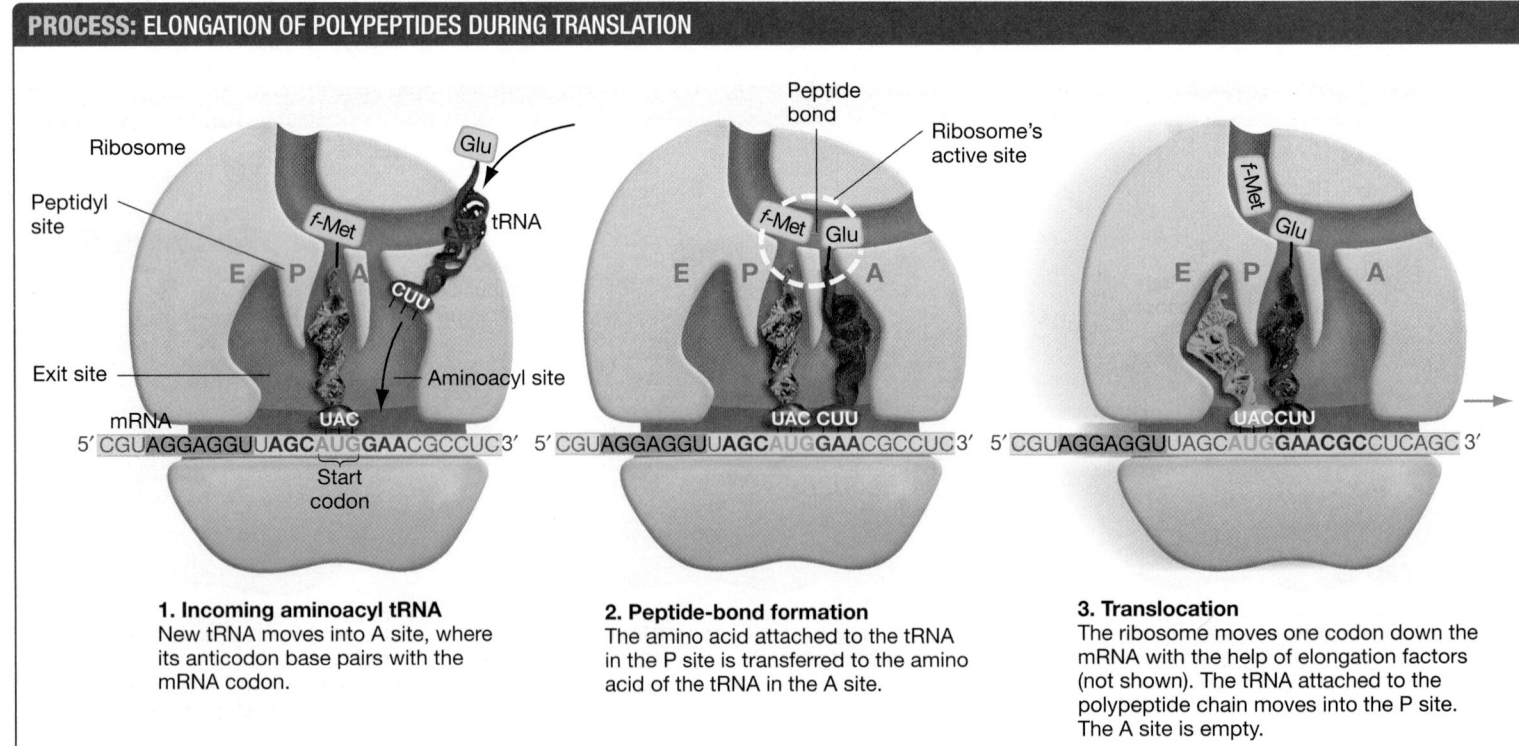

PROCESS: ELONGATION OF POLYPEPTIDES DURING TRANSLATION

1. Incoming aminoacyl tRNA
New tRNA moves into A site, where its anticodon base pairs with the mRNA codon.

2. Peptide-bond formation
The amino acid attached to the tRNA in the P site is transferred to the amino acid of the tRNA in the A site.

3. Translocation
The ribosome moves one codon down the mRNA with the help of elongation factors (not shown). The tRNA attached to the polypeptide chain moves into the P site. The A site is empty.

FIGURE 17.16 The Elongation Phase of Translation.

Moving Down the mRNA What happens after a peptide bond forms? Step 2 in Figure 17.16 shows that when peptide-bond formation is complete, the polypeptide chain is transferred from the tRNA in the P site to the amino acid held by the tRNA in the A site. Step 3 shows the process called **translocation**, which occurs when proteins called **elongation factors** help move the ribosome relative to the mRNA so that translation occurs in the $5' \rightarrow 3'$ direction. Translocation is an energy-demanding event that requires GTP.

Translocation does several things: It moves the uncharged RNA into the E site; it moves the tRNA containing the growing polypeptide into the P site; and it opens the A site and exposes a new mRNA codon. The empty tRNA that finds itself in the E site is ejected into the cytosol.

The three steps in elongation—(**1**) arrival of aminoacyl tRNA, (**2**) peptide-bond formation, and (**3**) translocation—repeat down the length of the mRNA. Recent three-dimensional models of ribosomes in various stages of translation show that the machine is highly dynamic. The ribosome constantly changes shape as tRNAs come and go and catalysis and translocation occur. The ribosome is a complex and dynamic macromolecular machine.

Terminating Translation

How does protein synthesis end? Recall that the genetic code includes three stop codons: UAA, UAG, and UGA (see Chapter 16). In most cells, no aminoacyl tRNA has an anticodon that binds to these sequences. When the translocating ribosome reaches one of the stop codons, a protein called a **release factor** recognizes the stop codon and fills the A site (**FIGURE 17.17**, see page 333).

Release factors fit tightly into the A site because they have the size and shape of a tRNA coming into the ribosome. However, release factors do not carry an amino acid. When a release factor occupies the A site, the protein's active site catalyzes the hydrolysis of the bond that links the tRNA in the P site to the polypeptide chain. This reaction frees the polypeptide.

The newly synthesized polypeptide and uncharged tRNAs are released from the ribosome, the ribosome separates from the mRNA, and the two ribosomal subunits dissociate. The subunits are ready to attach to the start codon of another message and start translation anew. Termination occurs in very similar ways in bacteria and eukaryotes.

Post-Translational Modifications

Proteins are not fully formed and functional when termination occurs. From earlier chapters, it should be clear that most proteins go through an extensive series of processing steps, collectively called post-translational modification, before they are completely functional. These steps require a wide array of molecules and events and take place in many different locations throughout the cell.

Folding Recall that a protein's function depends on its shape, and that a protein's shape depends on how it folds (see Chapter 3). Although folding can occur spontaneously, it is frequently speeded up by proteins called **molecular chaperones.**

Recent data have shown that in some bacteria, chaperone proteins bind to the ribosome near the "tunnel" where the growing polypeptide emerges from the ribosome. This finding suggests that folding occurs as the polypeptide emerges from the ribosome.

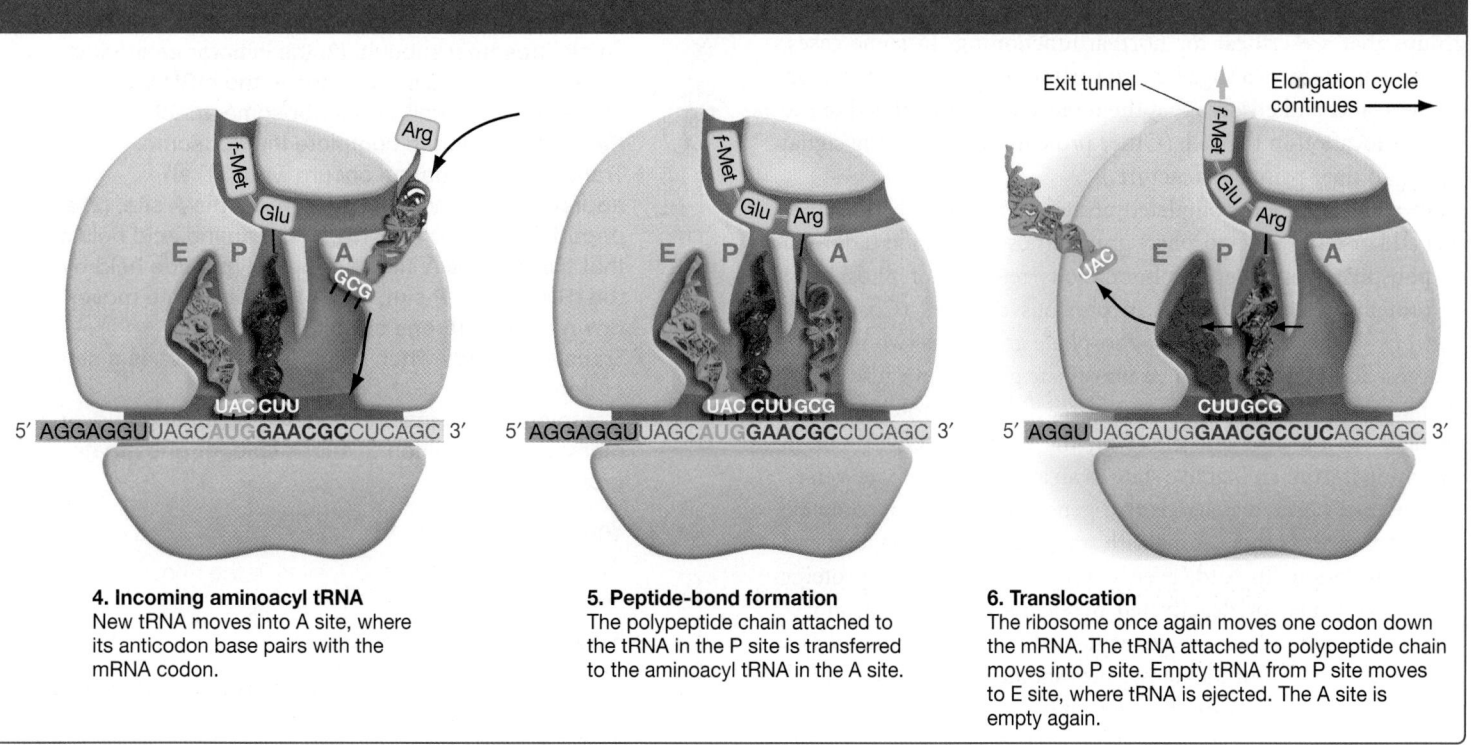

4. Incoming aminoacyl tRNA
New tRNA moves into A site, where its anticodon base pairs with the mRNA codon.

5. Peptide-bond formation
The polypeptide chain attached to the tRNA in the P site is transferred to the aminoacyl tRNA in the A site.

6. Translocation
The ribosome once again moves one codon down the mRNA. The tRNA attached to polypeptide chain moves into P site. Empty tRNA from P site moves to E site, where tRNA is ejected. The A site is empty again.

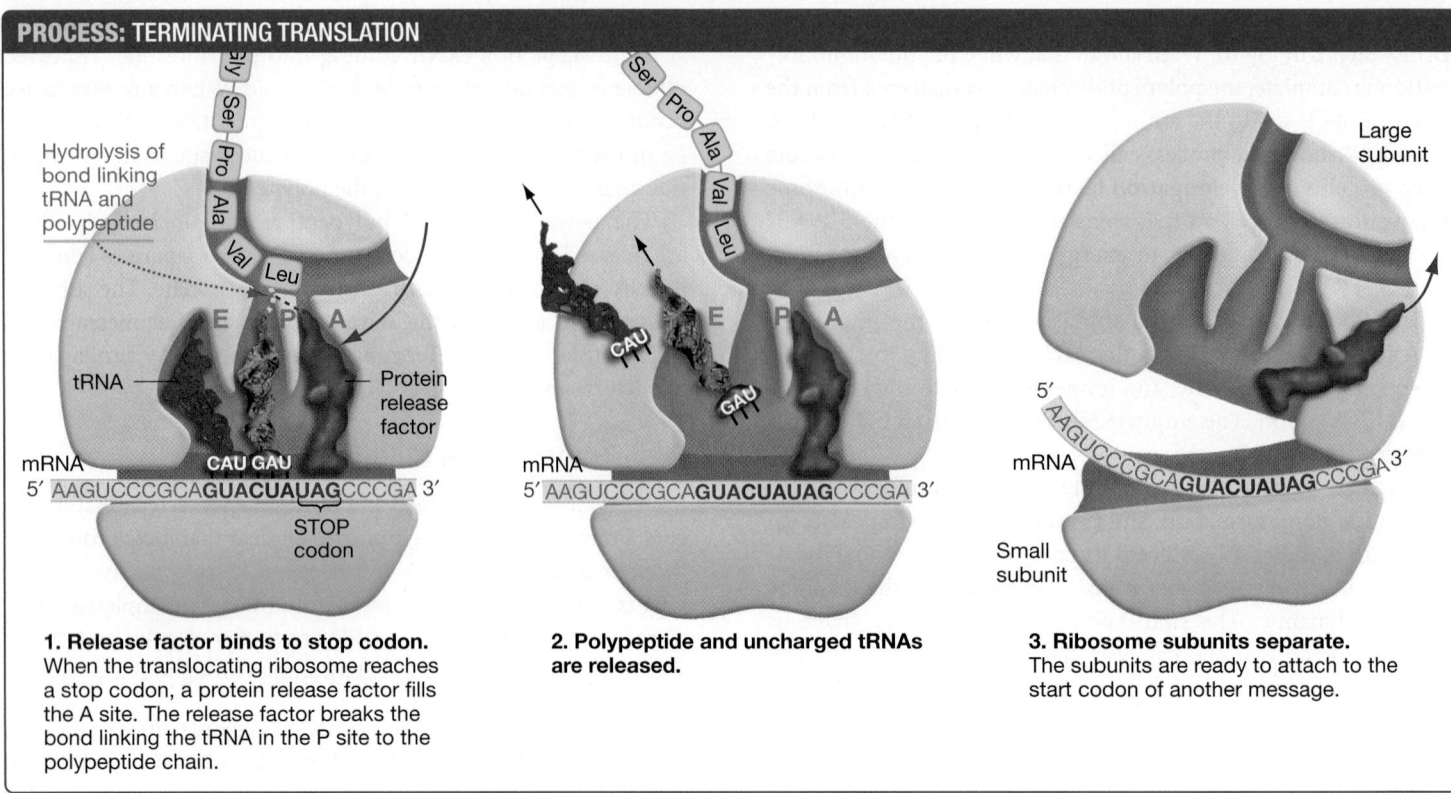

1. Release factor binds to stop codon.
When the translocating ribosome reaches a stop codon, a protein release factor fills the A site. The release factor breaks the bond linking the tRNA in the P site to the polypeptide chain.

2. Polypeptide and uncharged tRNAs are released.

3. Ribosome subunits separate.
The subunits are ready to attach to the start codon of another message.

FIGURE 17.17 Termination of Translation.

Chemical Modifications An earlier chapter pointed out that many eukaryotic proteins are extensively modified after they are synthesized (see Chapter 7). For example, in the organelles called the rough endoplasmic reticulum and the Golgi apparatus, small chemical groups may be added to proteins—often sugar or lipid groups that are critical for normal functioning. In some cases, the proteins receive a sugar-based sorting signal that serves as an address label and ensures that the molecule will be carried to the correct location in the cell. (Other proteins have a sorting signal built into their primary structure.)

In addition, many completed proteins are altered by enzymes that add or remove a phosphate group. Phosphorylation (addition of phosphate) and dephosphorylation (removal of phosphate) of proteins were introduced in previous chapters (Chapters 9 and 11). Recall that because a phosphate group has two negative charges, adding or removing a phosphate group can cause major changes in the shape and chemical reactivity of proteins. These changes have a dramatic effect on the protein's activity—often switching it from an inactive state to an active state, or vice versa.

The take-home message is that gene expression is a complex, multistep process that begins with transcription but may not end with translation. Instead, even completed and folded proteins may be activated or deactivated by events such as phosphorylation. In general, how are genes turned on or off? How does a cell "decide" which of its many genes should be expressed at any time? These questions are the focus of the next two chapters.

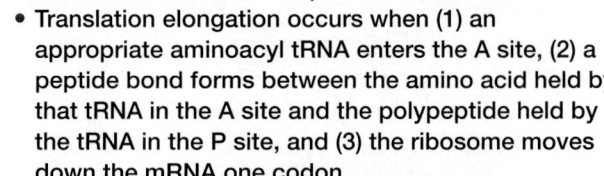

check your understanding

(C)(Y)(U) If you understand that . . .

- Translation begins when (1) the ribosome binding site on an mRNA binds to an rRNA sequence in the small ribosomal subunit, (2) the initiator aminoacyl tRNA binds to the start codon in the mRNA, and (3) the large subunit of the ribosome attaches to the small subunit to complete the ribosome.
- Translation elongation occurs when (1) an appropriate aminoacyl tRNA enters the A site, (2) a peptide bond forms between the amino acid held by that tRNA in the A site and the polypeptide held by the tRNA in the P site, and (3) the ribosome moves down the mRNA one codon.
- Translation ends when the ribosome reaches a stop codon.
- Completed proteins are modified by folding and, in many cases, addition of sugar, lipid, or phosphate groups.

✔ You should be able to . . .

Explain why the E, P, and A sites in the ribosome are appropriately named.

Answers are available in Appendix A.

If you understand . . .

17.1 An Overview of Transcription

- RNA polymerase catalyzes the production of an RNA molecule whose base sequence is complementary to the base sequence of the DNA template strand.

- RNA polymerase binds DNA with the help of other proteins.

- RNA polymerase begins transcription by binding to promoter sequences in DNA.

- In bacteria, this binding occurs in conjunction with a protein called sigma. Sigma associates with RNA polymerase and then recognizes particular sequences within promoters that are centered 10 bases and 35 bases upstream from the start of the actual genetic message.

- Eukaryotic promoters vary more than bacterial promoters.

- In eukaryotes, transcription begins when a large array of proteins called basal transcription factors bind to a promoter. In response, RNA polymerase binds to the site.

- In both bacteria and eukaryotes, the RNA elongates in a $5' \rightarrow 3'$ direction.

- Transcription in bacteria ends when RNA polymerase encounters a stem-loop structure in the just transcribed RNA; in eukaryotes, transcription terminates after the RNA is cleaved downstream of the poly(A) sequence.

✔ You should be able to predict the consequences of a mutation in bacteria that inserts random nucleotides into the hairpin-coding region near the 3' end of a transcribed region.

17.2 RNA Processing in Eukaryotes

- In eukaryotes, the primary (initial) transcript must be processed to produce a mature RNA.

- In primary transcripts, stretches of RNA called introns are spliced out and regions called exons are joined together.

- Complex macromolecular machines called spliceosomes splice introns out of pre-mRNA.

- A "cap" is added to the 5' end of pre-mRNAs, and a poly(A) tail is added to their 3' end.

- The cap and tail serve as recognition signals for translation and protect the message from degradation by ribonucleases.

- RNA processing occurs in the nucleus.

✔ You should be able to predict whether the protein-coding portion of a gene for an identical protein will be of the same or different lengths in a bacterium and a eukaryote.

17.3 An Introduction to Translation

- Ribosomes translate mRNAs into proteins with the help of adaptor molecules called transfer RNAs.

- In bacteria, an RNA is often transcribed and translated at the same time because there is no nucleus.

- In eukaryotes, transcription and translation of an RNA cannot occur together, because transcription and RNA processing occur in the nucleus and translation occurs in the cytoplasm.

- Experiments with radioactively labeled amino acids showed that transfer RNAs (tRNAs) serve as the chemical bridge between the RNA message and the polypeptide product.

✔ You should be able to explain why it is correct to say that transfer RNAs work as molecular adaptors.

17.4 The Structure and Function of Transfer RNA

- Each transfer RNA carries an amino acid corresponding to the tRNA's three-base-long anticodon.

- tRNAs have an L-shaped, tertiary structure. One leg of the L contains the anticodon, which forms complementary base pairs with the mRNA codon. The other leg holds the amino acid appropriate for that codon.

- Enzymes called aminoacyl-tRNA synthetases link the correct amino acid to the correct tRNA.

- Imprecise pairing—or "wobble pairing"—is allowed in the third position of the codon and anticodon, so only about 40 different tRNAs are required to translate the 61 codons that code for amino acids.

✔ You should be able to predict what would occur if a mutation caused an aminoacyl-tRNA synthetase to recognize two different amino acids.

17.5 The Structure and Function of Ribosomes

- Ribosomes are large macromolecular machines made of many proteins and RNAs.
- In the ribosome, the tRNA anticodon binds to a three-base-long mRNA codon to bring the correct amino acid into the ribosome.
- Peptide-bond formation by the ribosome is catalyzed by a ribozyme (RNA), not an enzyme (protein).
- Protein synthesis occurs in three steps: **(1)** an incoming aminoacyl tRNA occupies the A site; **(2)** the growing polypeptide chain is transferred from a peptidyl tRNA in the ribosome's P site to the amino acid bound to the tRNA in the A site, and a peptide bond is formed; and **(3)** the ribosome is translocated to the next codon on the mRNA, accompanied by ejection of the uncharged RNA from the E site.
- Chaperone proteins help fold newly synthesized proteins into their three-dimensional conformation (tertiary structure).
- Most proteins need to be modified after translation (post-translational modification) to activate them or target them to specific locations.

✓ You should be able to create a concept map (see BioSkills 15 in Appendix B) that describes the relationships among the following concepts and structures: translation, initiation, elongation, termination, growing polypeptide in P site, start codon, ribosome subunits.

(MB) MasteringBiology

1. **MasteringBiology Assignments**

   Tutorials and Activities Chromosomal Mutations; Following the Instructions in DNA; Point Mutations Protein Synthesis (1 of 3): Overview; Protein Synthesis (2 of 3): Transcription and RNA Processing; Protein Synthesis (3 of 3): Translation and Protein Targeting Pathways; RNA Processing; RNA Synthesis; Synthesizing Proteins; Transcription; Translation; Types of RNA

   Questions Reading Quizzes, Blue-Thread Questions, Test Bank

2. **eText** Read your book online, search, take notes, highlight text, and more.

3. **The Study Area** Practice Test, Cumulative Test, BioFlix® 3-D Animations, Videos, Activities, Audio Glossary, Word Study Tools, Art

You should be able to . . .

✓ TEST YOUR KNOWLEDGE
Answers are available in Appendix A

1. How did the A site of the ribosome get its name?
   a. It is where amino acids are joined to tRNAs, producing aminoacyl tRNAs.
   b. It is where the amino group on the growing polypeptide chain is available for peptide-bond formation.
   c. It is the site occupied by incoming aminoacyl tRNAs.
   d. It is surrounded by α-helices of ribosomal proteins.

2. Where is the start codon located?
   a. at the very start (5′ end) of the mRNA
   b. at the downstream end of the 3′ untranslated region (UTR)
   c. at the downstream end of the 5′ untranslated region (UTR)
   d. at the upstream end of the 3′ untranslated region (UTR)

3. What is the function of a molecular chaperone?

4. What does a bacterial RNA polymerase produce when it transcribes a protein-coding gene?

   a. rRNA
   b. tRNA
   c. mRNA
   d. pre-mRNA

5. Where is an amino acid attached to a tRNA?

6. Compared with mRNAs that have a cap and tail, what do researchers observe when eukaryotic mRNAs that lack a cap and poly(A) tail are translated within a cell?
   a. The primary transcript cannot be processed properly.
   b. Translation occurs inefficiently.
   c. Enzymes on the ribosome add back a cap and poly(A) tail.
   d. tRNAs become resistant to degradation (being broken down).

✓ TEST YOUR UNDERSTANDING
Answers are available in Appendix A

7. Explain the relationship between eukaryotic promoter sequences, basal transcription factors, and RNA polymerase. Explain the relationship between bacterial promoter sequences, sigma, and RNA polymerase.

8. According to the wobble rules, the correct amino acid can be added to a growing polypeptide chain even if the third base in the mRNA codon is not complementary to the corresponding base in the tRNA anticodon. How do the wobble rules relate to the redundancy of the genetic code?

9. RNases and proteases are enzymes that destroy RNAs and proteins, respectively. Which of the following enzymes when added to a spliceosome is predicted to prevent recognition of pre-mRNA regions critical for splicing?
   a. an RNase specific for tRNAs
   b. an RNase specific for snRNAs
   c. a protease specific for initiation factors
   d. a protease specific for a release factor

10. Describe the sequence of events that occurs during translation as a protein elongates by one amino acid and the ribosome moves down the mRNA. Your answer should specify what is happening in the ribosome's A site, P site, and E site.

11. **QUANTITATIVE** Controlling the rates of transcription and translation is important in bacteria to avoid collisions between ribosomes and RNA polymerases. Calculate the maximum rate of translation by a ribosome in a bacterial cell, provided in units of amino acids per second, so that the ribosome doesn't overtake an RNA polymerase that is transcribing mRNA at a rate of 60 nucleotides per second. How long would it take for this bacterial cell to translate an mRNA containing 1800 codons?

12. In an aminoacyl tRNA, why is the observed distance between the amino acid and the anticodon important?

TEST YOUR PROBLEM-SOLVING SKILLS   *Answers are available in Appendix A*

13. The 5′ cap and poly(A) tail in eukaryotic mRNAs protect the message from degradation by ribonucleases. But why do ribonucleases exist? What function would an enzyme that destroys messages serve? Answer this question using the example of an mRNA for a hormone that causes human heart rate to increase.

14. The nucleotide shown below is called cordycepin triphosphate.

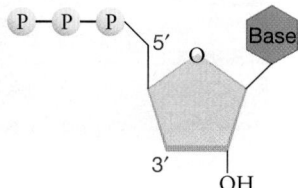

If cordycepin triphosphate is added to a cell-free transcription reaction, the nucleotide is added onto the growing RNA chain but no more nucleotides can be added. The added cordycepin is always found at the 3′ end of an RNA, confirming that synthesis occurs in the 5′ → 3′ direction. Why does cordycepin end transcription?

a. It prevents the association of RNA polymerase and sigma.
b. It irreversibly binds to the active site of RNA polymerase.
c. It cannot be recognized by RNA polymerase.
d. It lacks a 3′ OH.

15. Certain portions of the rRNAs in the large subunit of the ribosome are very similar in all organisms. To make sense of this finding, Carl Woese suggests that the conserved sequences have an important functional role. His logic is that these conserved sequences evolved in a common ancestor of all modern cells and are so important to cell function that any changes in the sequences cause death. In addition to rRNAs, which specific portions of the ribosome would you expect to be identical or nearly identical in all organisms? Explain your logic.

16. Recent structural models show that a poison called α-amanitin inhibits transcription by binding to a site inside eukaryotic RNA polymerase II but not to the active site itself. Based on the model of RNA polymerase in Figure 17.2, predict a place or places where α-amanitin might bind to inhibit transcription.

18 Control of Gene Expression in Bacteria

In this chapter you will learn how

Bacteria turn their genes on and off to adapt to changing environments

surveying ↓

Different ways genes can be regulated **18.1**

and ↓

How mutants help identify regulated genes **18.2**

looking closer at

Negative control of gene expression **18.3**

Positive control of gene expression **18.4**

Ways bacteria regulate many genes together **18.5**

The structures that have been colored blue in this scanning electron micrograph are projections from human intestinal cells; the structures colored yellow are the bacterium *Escherichia coli*. In the intestine, the nutrients available to bacteria constantly change. This chapter explores how changes in gene expression help bacteria respond to environmental changes.

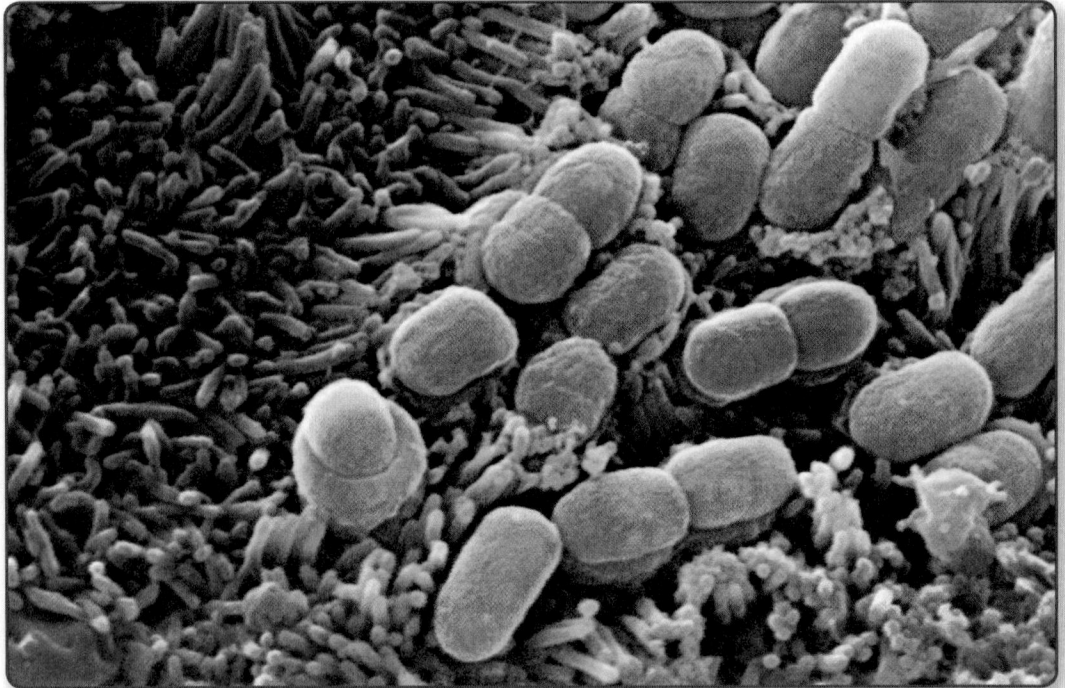

This chapter is part of the Big Picture. See how on pages 366–367.

Imagine waiting eagerly to hear the opening lines of a wonderfully melodic symphony played by a renowned orchestra. The crowd applauds as the celebrated conductor comes onstage and then hushes as he takes the podium. He cocks the baton; the musicians raise their instruments. As the baton comes down, every instrument begins blaring a different tune at full volume. A tuba plays "Dixie," a violinist renders "In-A-Gadda-Da-Vida," a snare drum lays down beats for Hot Chelle Rae's "Tonight, Tonight," while the bass drum simulates cannons in the "1812 Overture." Instead of music, there is pandemonium. The conductor staggers offstage, clutching his heart.

Cacophony like this would result if a bacterial cell "played" all its genes at full volume all the time. The *Escherichia coli* cells living in your gut right now have over 4300 genes. If all those genes were expressed at the fastest possible rate at all times, the *E. coli* cells would stagger off the stage,

✔ When you see this checkmark, stop and test yourself. Answers are available in Appendix A.

too. But this does not happen. Cells are extremely selective about which genes are expressed, in what amounts, and when.

This chapter explores how bacterial cells control the activity, or expression, of their genes. **Gene expression** is the process of converting information that is archived in DNA into molecules that actually do things in the cell. It occurs when a protein or other gene product is synthesized and active. (You can see on pages 366–367 how gene expression fits into the Big Picture of Genetic Information.)

Previous chapters detailed how genetic information is processed in cells; this chapter focuses on ways to control *when* genetic information is used. Let's begin by reviewing some of the environmental challenges that bacterial cells face and then explore how these organisms meet them.

18.1 An Overview of Gene Regulation and Information Flow

The bacteria that live in and on your body vastly outnumber your own cells. Consider just one of the species present: the gut-dwelling *Escherichia coli*. These cells can use a wide array of carbohydrates to supply the carbon and energy they need. But as your diet changes from day to day, the availability of different sugars in your intestines varies. Each type of nutrient requires a different membrane transport protein to bring the molecule into the cell and a different suite of enzymes to process it. Precise control of gene expression gives *E. coli* the ability to use the available sugars efficiently.

To understand why precise control over gene expression is so important, you have to realize that bacterial cells from an array of species can be densely packed along your intestinal walls. All of these organisms are competing for space and nutrients. In an environment like this, a cell has to use resources efficiently if it's going to be able to survive and reproduce. An individual that synthesizes proteins it doesn't need has fewer resources to devote to making the proteins it does need. Such cells are losers—they compete less successfully for the resources that are required to produce offspring.

Realizing this, biologists predicted that most gene expression is triggered by specific signals from the environment, such as the presence of specific sugars. Did you drink milk at your last meal, or eat French fries and a candy bar? Each type of food contains different sugars. Each sugar should induce a different response from the *E. coli* cells in your intestine. Just as a conductor needs to regulate the orchestra's musicians, cells need to regulate which proteins they produce.

Mechanisms of Regulation

The flow of information from DNA to activation of the final gene product occurs in three steps, represented by arrows in the following diagram:

$$\text{DNA} \longrightarrow \text{mRNA} \longrightarrow \text{protein} \longrightarrow \text{activated protein}$$

Gene expression can be controlled at any of these steps. The arrow from DNA to RNA represents transcription—the making of messenger RNA (mRNA). The arrow from RNA to protein represents translation, in which ribosomes read the information in mRNA and use that information to synthesize a protein. The arrow from protein to activated protein represents post-translational modifications that can lead to changes in shape and activity.

How can a bacterial cell avoid producing proteins that are not needed at a particular time, and thus use resources efficiently? A look at the flow of information from DNA to protein suggests three possible mechanisms:

1. The cell could avoid making the mRNAs for particular enzymes. If there is no mRNA, then ribosomes cannot make the gene product. **Transcriptional control** occurs when regulatory proteins affect RNA polymerase's ability to bind to a promoter and initiate transcription:

$$\text{DNA} \xrightarrow{\times} \text{mRNA} \longrightarrow \text{protein} \longrightarrow \text{activated protein}$$

2. If the mRNA for an enzyme has been made, the cell could prevent the mRNA from being translated into protein. **Translational control** occurs when regulatory molecules alter the length of time an mRNA survives, or affect translation initiation or elongation:

$$\text{DNA} \longrightarrow \text{mRNA} \xrightarrow{\times} \text{protein} \longrightarrow \text{activated protein}$$

3. Many proteins have to be activated by chemical modification, such as the addition of a phosphate group. Regulating this final step is **post-translational control:**

$$\text{DNA} \longrightarrow \text{mRNA} \longrightarrow \text{protein} \xrightarrow{\times} \text{activated protein}$$

Which of these three forms of control occur in bacteria? The short answer is all the above. As **FIGURE 18.1** (see page 338) shows, many factors affect how much active protein is produced from a particular gene.

- Transcriptional control is particularly important due to its efficiency—it saves the most energy for the cell, because it stops the process of gene expression at the earliest possible point.

- Translational control allows a cell to make rapid changes in the amounts of different proteins because the mRNA is already present and available for translation.

- Post-translational control provides the most rapid response of all three mechanisms because only one step is needed to activate an existing protein.

Among these mechanisms of gene regulation, there is a clear trade-off between the speed of response and the conservation of ATP, amino acids, and other resources. Transcriptional control is slow but efficient in resource use; post-translational control is fast but energetically expensive.

Although this chapter focuses almost exclusively on mechanisms of transcriptional control, it is important to keep in mind that bacteria also possess translational and post-translational

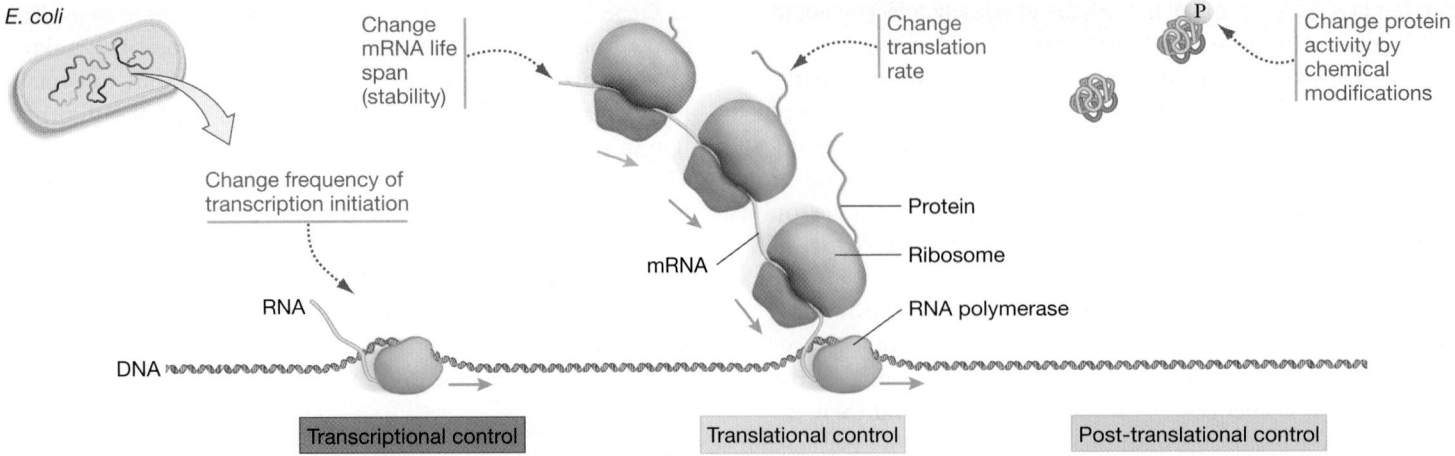

FIGURE 18.1 Gene Expression in Bacteria Can Be Regulated at Three Levels.

✔**EXERCISE** Label the mode of regulation that is slowest in response time and the mode that is fastest. Label the most efficient and least efficient mode in resource use.

controls. Just as important, some genes—such as those that code for the enzymes required for glycolysis—are transcribed all the time, or **constitutively.** Finally, it is critical to realize that gene expression is not an all-or-none proposition. Genes are not just "on" or "off"—instead, the level of expression can vary between these extremes.

The ability to regulate gene expression allows cells to respond to changes in their environment.

Metabolizing Lactose—A Model System

Many of the great advances in genetics have been achieved through the analysis of model systems (see Chapters 14–17). Mendel studied garden peas and discovered the fundamental patterns of gene transmission; Morgan studied fruit flies and confirmed the chromosome theory of inheritance; an array of researchers used *E. coli* and its viruses to work out the mechanisms of DNA synthesis, transcription, and translation. In early studies of gene regulation, a key model system was the metabolism of the sugar lactose in *E. coli*.

Jacques Monod, François Jacob, and many colleagues introduced lactose metabolism in *E. coli* as a model system during the 1950s and 1960s. Although they worked with a single species of bacteria, their results had a profound effect on thinking about gene regulation in all organisms.

E. coli can use a wide variety of sugars for ATP production, via cellular respiration or fermentation. These sugars also serve as raw material in the synthesis of amino acids, vitamins, and other complex compounds. Glucose, however, is *E. coli*'s preferred carbon source—meaning that it is the source of energy and carbon atoms that the organism uses most efficiently.

A preference for glucose makes sense, because glycolysis begins with glucose and is the main pathway for the production of ATP. Lactose, the sugar found in milk, can also be used by *E. coli*, but it is not used until glucose supplies are depleted. Recall that lactose is a disaccharide made up of one molecule of glucose and one molecule of galactose (see Chapter 5).

To use lactose, *E. coli* must first transport the sugar into the cell. Once lactose is inside the cell, the enzyme β-galactosidase catalyzes a reaction that breaks down the disaccharide into glucose and galactose. The glucose released by this reaction goes directly into the glycolytic pathway; other enzymes convert the galactose to a substance that can also be processed in the glycolytic pathway.

In the early 1950s, biologists discovered that *E. coli* produces high levels of β-galactosidase only when lactose is present in the environment. Based on this observation, researchers proposed that lactose itself regulates the gene for β-galactosidase—meaning that lactose acts as an inducer. An **inducer** is a small molecule that triggers transcription of a specific gene.

In the late 1950s, Jacques Monod wondered how the presence of glucose affects the regulation of the β-galactosidase gene. Would *E. coli* produce high levels of β-galactosidase when both glucose and lactose were present in the surrounding environment? As the experiment summarized in **FIGURE 18.2** shows, the answer was no. Significant amounts of β-galactosidase are produced only when lactose is present and glucose is not present.

Monod teamed up with François Jacob to investigate exactly how lactose and glucose regulate the genes responsible for lactose metabolism—the gene for the membrane protein that imports lactose and the gene for β-galactosidase. Discoveries about how these genes are regulated shed light on how genes in all organisms are controlled. Research on this system is still going strong, over 50 years later.

✔ You should be able to make a chart summarizing the molecules involved in regulating lactose use in *E. coli*. There should be 7 rows and 2 columns. Title the first column "Name" and the second column "Function." The rows are *lacZ*, *lacY*, operator, promoter, repressor, lactose, and glucose. As you read this chapter, fill in the "Function" column.

QUESTION: *E. coli* produces β-galactosidase when lactose is present. Does *E. coli* produce β-galactosidase when both glucose and lactose are present?

HYPOTHESIS: *E. coli* does not produce β-galactosidase when glucose is present, even if lactose is present. (Glucose is the preferred food source.)

NULL HYPOTHESIS: *E. coli* produces β-galactosidase whenever lactose is present, regardless of the presence or absence of glucose.

EXPERIMENTAL SETUP:

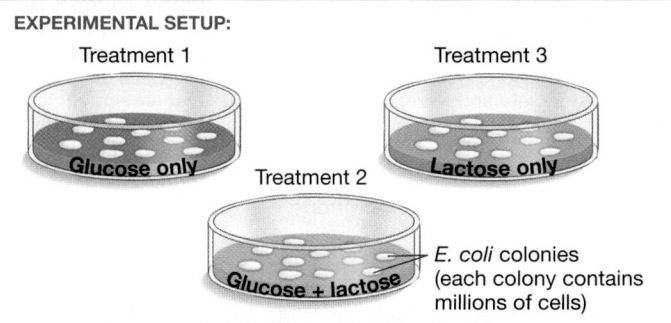

PREDICTION: β-Galactosidase will be produced only in treatment 3.

PREDICTION OF NULL HYPOTHESIS: β-Galactosidase will be produced in treatments 2 and 3.

RESULTS:

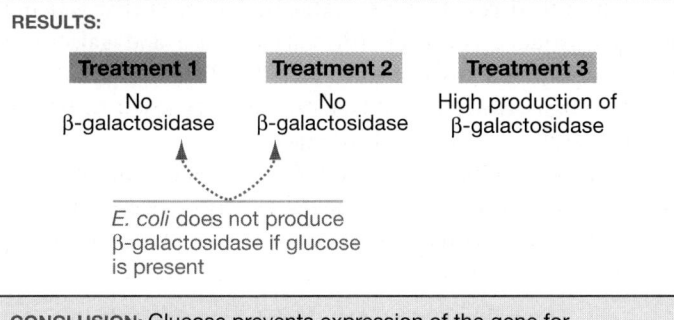

CONCLUSION: Glucose prevents expression of the gene for β-galactosidase. The presence of lactose without glucose stimulates expression of that gene.

FIGURE 18.2 Glucose Affects the Regulation of the β-Galactosidase Gene.

✔**QUESTION** How would you control growth conditions in the three treatments so that the results of this experiment are valid?

SOURCE: Pardee, A. B., F. Jacob, and J. Monod. 1959. The genetic control and cytoplasmic expression of "inducibility" in the synthesis of β-galactosidase by *E. coli. Journal of Molecular Biology* 1: 165–178.

18.2 Identifying Regulated Genes

To understand how *E. coli* controls production of β-galactosidase and the transport protein that brings lactose into the cell, Jacob and Monod first had to find the genes that code for these proteins. To do this, they employed the same tactic used in the pioneering studies of DNA replication, transcription, and translation reviewed in earlier chapters: They isolated and analyzed mutants. In this case, their goal was to find *E. coli* cells that could not metabolize lactose. Cells that can't use lactose must lack either β-galactosidase or the lactose transporter protein.

To find mutants that are associated with a particular trait, a researcher has to complete two steps:

1. Generate a large number of individuals with mutations at random locations in their genomes. Monod and colleagues accomplished this step by exposing *E. coli* populations to **mutagens**—X-rays, UV light, or chemicals that damage DNA and increase mutation rates.

2. Screen the treated individuals for mutants with defects in the process or biochemical pathway in question—in this case, defects in lactose metabolism. Recall that a genetic screen is any technique for selecting individuals with certain types of mutations out of a large population, and that a mutant is an individual with a mutation (see Chapter 16).

The researchers were looking for cells that cannot grow in an environment that contains only lactose as an energy source. Normal cells grow well in this environment. How could the researchers select cells on the basis of *lack* of growth?

Replica Plating to Find Lactose Metabolism Mutants

Replica plating and growth on indicator plates were key techniques in the search for mutants with defects in lactose metabolism. **FIGURE 18.3** (see page 6) shows how **replica plating** works.

Step 1 When mutants with defects in lactose metabolism are desired, mutagenized bacteria are spread on a "master plate" filled with a gelatinous growth **medium** containing glucose but no lactose. Growth medium is any liquid or solid that supports the growth of cells. It is important that the mutant cells are capable of growing on the master plate. The bacteria are then allowed to grow, so that each cell produces a colony—a large number of identical cells descended from a single cell.

Step 2 A block covered with a piece of sterilized velvet is pressed onto the master plate. Some cells from each colony on the master plate are transferred to the velvet.

Step 3 The velvet is pressed onto a plate called a replica plate that contains medium that differs from the master plate by a single component. In this case, the second medium has only lactose and no glucose as the source of carbon and energy. Cells from the velvet stick to the replica plate's surface, producing an exact copy of the locations of the colonies on the master plate.

Step 4 After these transferred cells grow, compare the colonies on the replica plate with those on the master plate. In this example, colonies that grow on the master plate but are missing on the replica plate are mutants that cannot metabolize lactose. By picking cells from these colonies on the master plate, researchers build a collection of lactose metabolism mutants.

Several Genes Are Involved in Lactose Metabolism

The initial mutant screen yielded three types of mutants. In one class, the mutant cells were unable to cleave lactose—even when

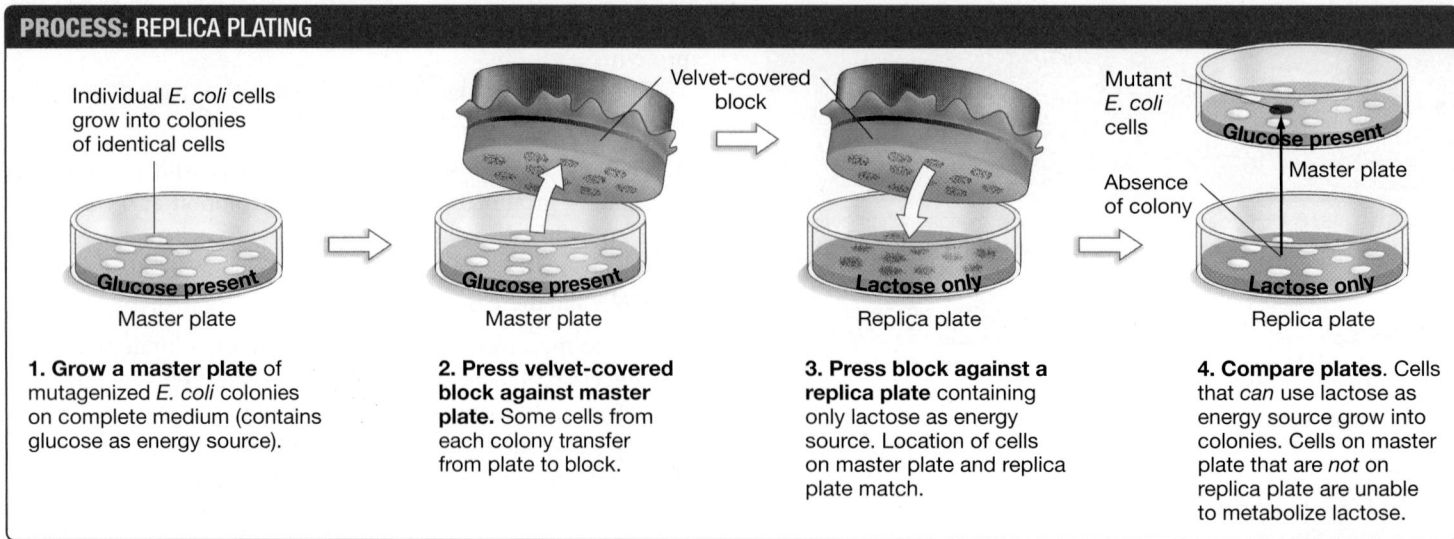

PROCESS: REPLICA PLATING

Individual *E. coli* cells grow into colonies of identical cells

Velvet-covered block

Mutant *E. coli* cells

Glucose present

Master plate

Absence of colony

Glucose present
Master plate

Glucose present
Master plate

Lactose only
Replica plate

Lactose only
Replica plate

1. Grow a master plate of mutagenized *E. coli* colonies on complete medium (contains glucose as energy source).

2. Press velvet-covered block against master plate. Some cells from each colony transfer from plate to block.

3. Press block against a replica plate containing only lactose as energy source. Location of cells on master plate and replica plate match.

4. Compare plates. Cells that *can* use lactose as energy source grow into colonies. Cells on master plate that are *not* on replica plate are unable to metabolize lactose.

FIGURE 18.3 Replica Plating Is a Technique for Identifying Mutants That Cannot Grow in Particular Conditions.
Here, replica plating is used to isolate mutant *E. coli* cells with a deficiency in lactose metabolism.
✔**QUESTION** How would you alter this protocol to isolate mutant cells with a deficiency in the enzymes required to synthesize tryptophan?

lactose was in the medium and transported into cells to induce production of the β-galactosidase protein. Jacob and Monod concluded that these mutants must lack a functioning version of the β-galactosidase protein and, therefore, the gene that encodes β-galactosidase is defective. This gene was designated *lacZ*, and the mutant allele *lacZ⁻*.

In the second class of mutants, the cells failed to accumulate lactose inside the cell. To explain this result, Jacob and Monod hypothesized that the mutant cells had defective copies of the membrane protein responsible for transporting lactose into the cell. This protein was identified and named galactoside permease; the gene that encodes it was designated *lacY*. **FIGURE 18.4** summarizes the functions of β-galactosidase and galactoside permease.

The third and most surprising class of mutants did not show normal regulation of β-galactosidase and galactoside permease expression. Instead, these mutants made the proteins all the

time—even if no lactose was present. **TABLE 18.1** summarizes these three types of mutants.

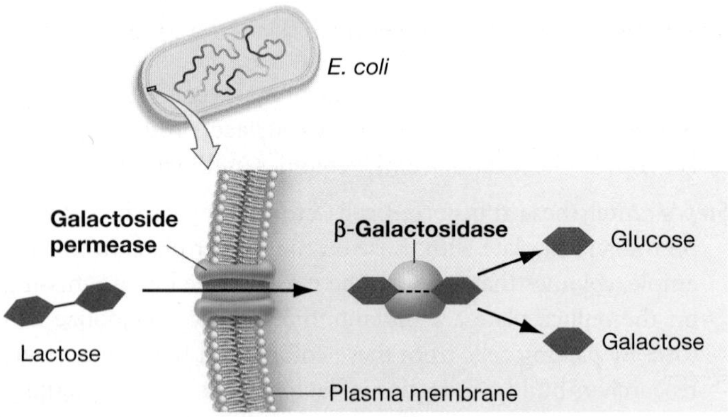

Cells that are abnormal because they produce a product at all times are called **constitutive mutants**. The gene that was mutated to produce constitutive β-galactosidase and galactoside permease expression was named *lacI*. The letter I signified that these mutants did not need an inducer—lactose—to express β-galactosidase or galactoside permease.

To understand the significance of the *lacI* mutation, recall that in normal cells, the expression of the *lacZ* (β-galactosidase) and *lacY* (galactoside permease) genes is induced by lactose. But in

TABLE 18.1 Three Types of Lactose Metabolism Mutants in *E. coli*

Observed Phenotype	Interpretation	Genotype
1. Cells cannot cleave lactose, even in the presence of inducer.	No β-galactosidase; gene for β-galactosidase is defective. Call this gene *lacZ*.	*lacZ⁻*
2. Cells cannot accumulate lactose.	No membrane protein (galactoside permease) to import lactose; gene for galactoside permease is defective. Call this gene *lacY*.	*lacY⁻*
3. Cells cleave lactose even if lactose is absent as an inducer.	Constitutive (constant) expression of *lacZ* and *lacY*; gene for regulatory protein that shuts down *lacZ* and *lacY* is defective—it does not need to be induced by lactose. Call this gene *lacI*.	*lacI⁻*

E. coli

Galactoside permease

β-Galactosidase

Glucose

Lactose

Galactose

Plasma membrane

FIGURE 18.4 Two Proteins *E. coli* Needs for Using Lactose.
The membrane protein galactoside permease brings lactose into the cell, and the enzyme β-galactosidase breaks lactose into its glucose and galactose subunits.

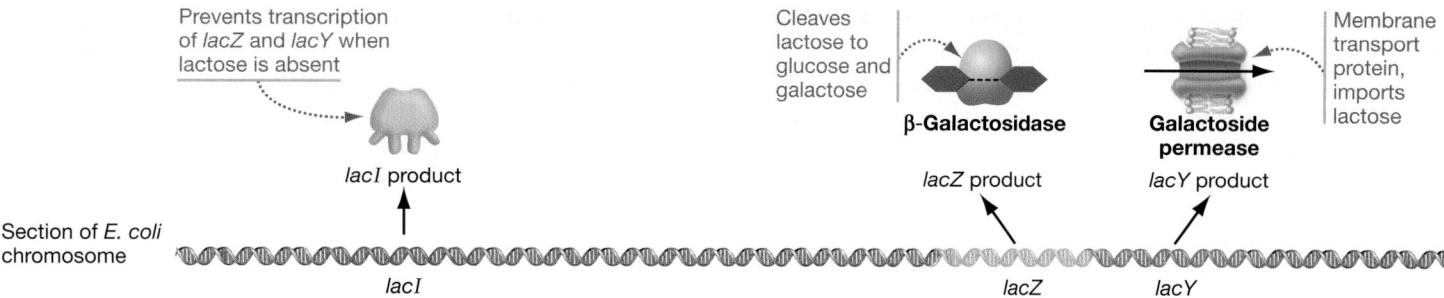

FIGURE 18.5 The *lac* Genes Are in Close Physical Proximity. The associated proteins and their functions are shown above each gene.

cells with a mutant form of *lacI* (*lacI⁻* mutants), gene expression occurs with or without lactose. This means that *lacI⁻* mutants have a defect in gene regulation. In these mutants, the gene remains on when it should be turned off.

To pull these observations together, the researchers hypothesized that the normal product of the *lacI* gene prevents the transcription of *lacZ* and *lacY* when lactose is absent. Because lactose triggers production of β-galactosidase, it was reasonable to expect that the *lacI* gene or gene product interacts with lactose in some way. (Later work showed that the inducer is actually a derivative of lactose called *allolactose*. For historical accuracy and simplicity, however, this discussion refers to lactose itself as the inducer.)

Jacob and Monod had succeeded in identifying three genes involved in lactose metabolism: *lacZ*, *lacY*, and *lacI*. They concluded that *lacZ* and *lacY* code for proteins required for the metabolism and import of lactose, while *lacI* is responsible for some sort of regulatory function. When lactose is absent, the *lacI* gene or gene product shuts down the expression of *lacZ* and *lacY*. But when lactose is present, the opposite occurs—transcription of *lacZ* and *lacY* is induced.

✔ If you understand the genes involved in lactose metabolism, you should be able to describe the specific function of *lacZ* and *lacY*. You should also be able to describe the effect of the *lacI* gene product when lactose is present versus absent and explain why these effects are logical.

Jacob and Monod followed up on these experiments by mapping the location of the three genes on the *E. coli* chromosome (**FIGURE 18.5**). They discovered that the genes are close together. This was a crucial finding because it suggested that *lacZ* and *lacY* might be transcribed together. Could the *lacI* regulatory gene govern both the *lacZ* and *lacY* genes? How does *lacI* actually work? And why do lactose and glucose have opposite effects on gene expression?

18.3 Negative Control of Transcription

In principle, there are two general ways that transcription can be regulated: by negative control or positive control.

1. **Negative control** occurs when a regulatory protein called a **repressor** binds to DNA and shuts down transcription (**FIGURE 18.6a**).

2. **Positive control** occurs when a regulatory protein called an **activator** binds to DNA and triggers transcription (**FIGURE 18.6b**).

When you are driving a car, negative control is exerted by setting the parking brake; positive control occurs when you step on the gas pedal. It turned out that the *lacZ* and *lacY* genes in *E. coli* are controlled by engaging or releasing a parking brake—they are under negative control.

(a) Negative control: Regulatory protein *shuts down* transcription.

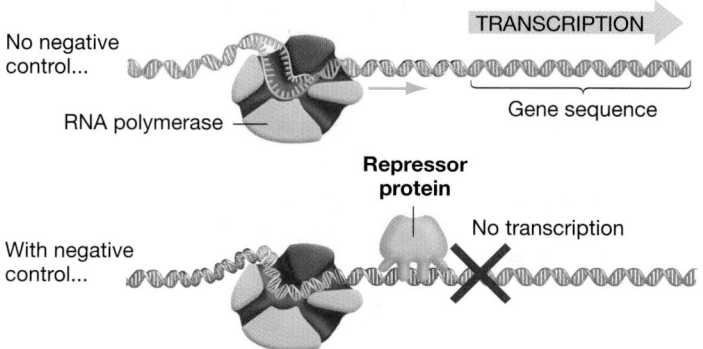

(b) Positive control: Regulatory protein *triggers* transcription.

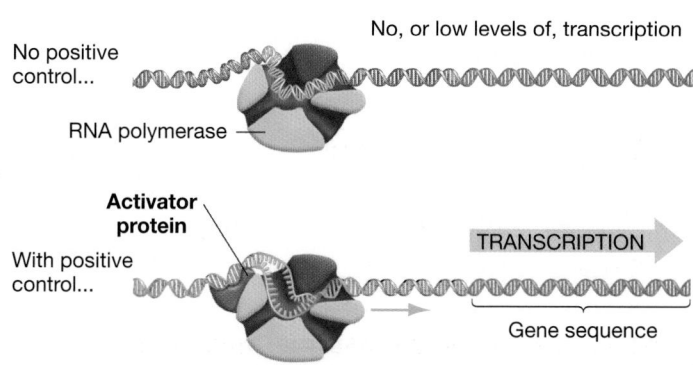

FIGURE 18.6 Genes Are Regulated by Negative Control, Positive Control, or Both. (To review transcription initiation, see Figure 17.3.)

The hypothesis that the *lacZ* and *lacY* genes might be under negative control originated with Leo Szilard in the late 1950s. Szilard suggested to Monod that the *lacI* gene could code for a product that represses transcription of the *lacZ* and *lacY* genes. As it turned out, Szilard was right.

The *lacI* gene produces a repressor protein that exerts negative control over *lacZ* and *lacY* gene transcription. The repressor was proposed to bind directly to DNA at or near the promoter for the *lacZ* and *lacY* genes (**FIGURE 18.7a**).

To explain how lactose triggers transcription, Szilard and Monod proposed that lactose interacts with the repressor in a way that makes the repressor release from its binding site (**FIGURE 18.7b**). In negative control, the repressor is the parking brake; lactose releases the brake.

What about the constitutive mutants? **FIGURE 18.7c** shows that constitutive transcription is observed in *lacI⁻* mutants because a functional repressor is absent—the parking brake is broken.

To test the hypothesis of negative control by a repressor, Jacob, Monod, and co-workers added back a functioning copy of the *lacI* repressor gene to the *lacI⁻* mutants that made β-galactosidase all the time. If these cells were grown using glucose and no lactose, β-galactosidase production declined and then stopped.

This result supported the hypothesis that the repressor codes for a protein that shuts down transcription. Significantly, if the experimental cells were grown using lactose instead of glucose, β-galactosidase activity resumed. This result supported the hypothesis that lactose removes the repressor.

What's the take-home message? The *lacI* gene codes for a repressor protein that exerts negative control on *lacZ* and *lacY*. Lactose acts as an inducer by causing the repressor to release from DNA and ending negative control.

The Operon Model

Jacob and Monod summarized the results of their experiments with a comprehensive model of negative control that was published in 1961. One of their key conclusions was that the genes for β-galactosidase and galactoside permease are controlled together and transcribed into a single mRNA. To encapsulate this idea, they coined the term **operon** for a set of coordinately regulated bacterial genes that are transcribed together into one mRNA. Logically enough, the group of genes involved in lactose metabolism was termed the ***lac* operon.**

Later, a gene called *lacA* was found to be adjacent to *lacY* and *lacZ* and transcribed as part of the same operon. The *lacA* gene

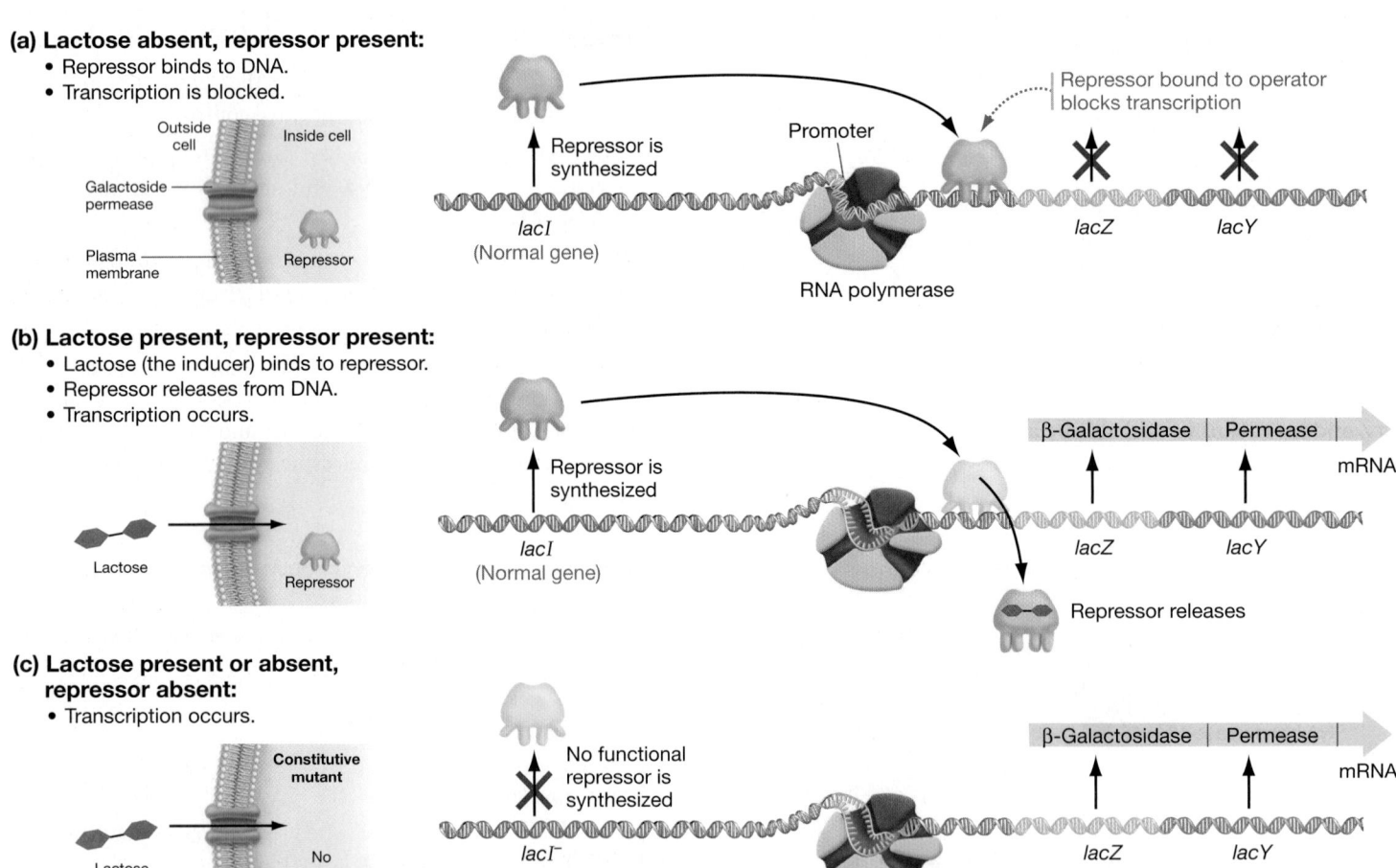

(a) Lactose absent, repressor present:
• Repressor binds to DNA.
• Transcription is blocked.

(b) Lactose present, repressor present:
• Lactose (the inducer) binds to repressor.
• Repressor releases from DNA.
• Transcription occurs.

(c) Lactose present or absent, repressor absent:
• Transcription occurs.

FIGURE 18.7 The Hypothesis of Negative Control of the *lac* Operon. The plasma membrane and galactoside permease are shown as a reminder that lactose comes from outside the cell and controls genes within the *E. coli* chromosome. Repression of the *lac* operon is never complete, so there is always some galactoside permease to transport lactose into the cell and begin induction of the *lac* operon.

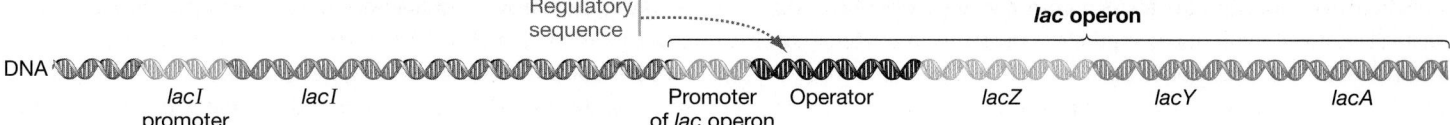

FIGURE 18.8 The *lac* Operon and *lacI* Gene. This view emphasizes the arrangement of genes and regulatory sequences and is not drawn to scale.

✔**EXERCISE** Using small, colored bits of candy or paper, add the repressor protein to the figure. Next, add RNA polymerase; then add lactose. At each step, explain what happens after the molecule is added.

codes for the enzyme transacetylase. This enzyme catalyzes reactions that allow certain types of sugars to be exported from the cell when they are too abundant and could harm the cell. The components of the *lac* operon are summarized in **FIGURE 18.8.**

Three hypotheses are central to the Jacob–Monod model of *lac* operon regulation:

1. The *lacZ*, *lacY*, and *lacA* genes are adjacent and are transcribed into one mRNA initiated from the single promoter of the *lac* operon. This is known as cotranscription, and it results in the coordinated expression of the three genes.

2. The repressor is a protein encoded by *lacI* that binds to DNA and prevents transcription of the *lac* operon genes (*lacZ*, *lacY*, and *lacA)*. Jacob and Monod proposed that *lacI* is expressed constitutively, and that the repressor binds to a section of DNA in the *lac* operon called the **operator.**

3. The inducer (lactose) binds to the repressor. When it does, the repressor changes shape. The shape change causes the repressor to come off the DNA. Recall that this form of control over protein function is **allosteric regulation** (see Chapter 8). In allosteric regulation, a small molecule binds to a protein and causes it to change its shape and activity. When the inducer binds to the repressor, the repressor can no longer bind to DNA and transcription can proceed.

✔ If you understand negative control of the *lac* operon, you should be able to predict the effect of a mutation in the *lacI* gene that alters the repressor so it cannot bind to lactose, and the effect of a mutation in the operator that prevents repressor binding.

How Does Glucose Regulate the *lac* Operon?

The model of *lac* operon control, summarized in Figure 18.7, is elegant and successful in explaining experimental results. But it is not complete. After studying the model, you may think of an important question that it fails to answer: Where does glucose fit in?

Transcription of the *lac* operon is drastically reduced when glucose is present in the environment—even when lactose is available to induce β-galactosidase expression (see Figure 18.2). This makes sense, given that glucose is *E. coli*'s preferred carbon source. When glucose is already present, the cell doesn't need to cleave lactose as a way of acquiring glucose.

How can glucose prevent expression of the *lac* operon? Researchers recently discovered that glucose inhibits the lactose transport activity of galactoside permease through a chain of molecular events. When both glucose and lactose are present in the environment, the transport of lactose into the cell is inhibited. Because lactose does not accumulate in the cytoplasm, the repressor remains bound to the operator. Negative control (as in Figure 18.7a) is in place. In contrast, when glucose levels outside the cell are low, galactoside permease is active. If lactose is present, it is transported into the cell and induces *lac* operon expression (as in Figure 18.7b).

The mechanism of glucose preventing the transport of inducer is known as inducer exclusion. Inducer exclusion affects the activity of many different sugar transporters in addition to galactoside permease. It allows *E. coli* to preferentially use glucose, even when other sugars are also present outside the cell.

This understanding of how glucose regulates the *lac* operon is relatively new. For decades, researchers thought that when glucose levels outside the cell declined, an activator protein called CAP bound to a regulatory sequence in DNA just upstream of the promoter to increase the frequency of transcription initiation. There is strong evidence that binding of CAP to the regulatory sequence is important for efficient transcription of the *lac* operon. However, recent results indicate that CAP is always bound to the regulatory sequence, even in the presence of glucose.

Why Has the *lac* Operon Model Been So Important?

The *lac* operon has been an immensely important model system for two reasons. First, follow-up work showed that many bacterial genes and operons are under negative control by repressor proteins. This means that the findings on the *lac* operon are general. Second, the *lac* operon model introduced a fundamentally important idea: Gene expression is regulated by physical contact between regulatory proteins and specific regulatory sites in DNA. Publication of the *lac* operon model was a watershed event in the history of biological science.

Work on the *lac* operon also offered an important example of post-translational control over gene expression. To understand why, you have to realize that the repressor protein is always present; it is transcribed and translated constitutively. When a rapid change in *lac* operon activity is needed, it does not require changes in the transcription or translation of new repressor proteins. Instead, the activity of *existing* repressor proteins is altered.

This is exactly the prediction made at the beginning of this chapter—post-translational control is best when a rapid response is needed. As it turns out, this is a common type of control. In most cases, the activity of key regulatory proteins is controlled by post-translational modifications.

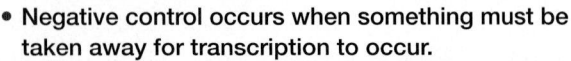

check your understanding

C Y U

If you understand that . . .

- Negative control occurs when something must be taken away for transcription to occur.
- The *lac* operon repressor exerts negative control over three protein-coding genes by binding to the operator site in DNA, near the promoter.
- For transcription to occur in the *lac* operon, an inducer molecule (a derivative of lactose) must bind to the repressor, causing it to change shape and release from the operator.
- The *lac* operon is not transcribed when glucose is available because glucose prevents lactose transport into the cell.

✔ **You should be able to . . .**

1. Explain why lactose should induce transcription of the *lac* operon.
2. Diagram the *lac* operon, showing the relative positions of the operator, the promoter, and the three protein-coding genes; indicate what is happening at the operon in the absence of lactose and in the presence of lactose.

Answers are available in Appendix A.

working on a project for a laboratory course were the first to uncover it.

This operon contains three genes that allow *E. coli* to use the sugar arabinose. Arabinose is found in many plant cell walls. When you eat vegetables, arabinose is available to the bacteria that inhabit your gut. Without arabinose in the environment, the *ara* operon is not transcribed. But when arabinose is present, transcription of the *ara* operon is turned on by an activator protein called **AraC**. The *ara* operon and an adjacent gene, ***araC***, that codes for the araC activator are shown in **FIGURE 18.9**.

FIGURE 18.10 outlines how AraC controls the *ara* operon. The AraC protein is allosterically regulated by arabinose. When bound to arabinose, two copies of the AraC protein attach to a regulatory sequence of DNA called the *ara* initiator that lies just upstream of the promoter (see Figure 18.10a). Once AraC is bound to DNA, it can also bind to RNA polymerase. This interaction between AraC and the RNA polymerase helps to dock the polymerase to the promoter and accelerate the initiation of transcription.

Continued work on the *ara* operon revealed a surprise—AraC is both an activator and a repressor. In the absence of arabinose, the two copies of the AraC protein remain together; but while one araC copy remains bound to the initiator, the other copy now binds to a different regulatory site in DNA, the *ara* operator, as shown in Figure 18.10b. In this configuration, AraC works as a repressor to prevent the transcription of both the *ara* operon and the *araC* gene.

✔ If you understand positive control by the AraC protein, you should be able to predict the effect of a mutation that removes the part of AraC that binds to RNA polymerase.

18.4 Positive Control of Transcription

Positive control is an important way of controlling transcription. In positive control, an activator protein binds to a regulatory sequence in DNA when genes are turned on. When bound to DNA, the activator interacts with RNA polymerase to increase the rate of initiating transcription (see Figure 18.6b).

The ***ara* operon** provides an important example of positive control and of the process of science. The *ara* operon wasn't discovered in the laboratory of a famous scientist. Instead, students

18.5 Global Gene Regulation

A theme of this chapter is that cells respond to changing environments. To compete for resources, bacteria must be able to coordinate the expression of large sets of genes. As you've seen for the *lac* and *ara* operons, an effective way to express sets of genes together is to group them into an operon and transcribe them into a single mRNA. But there are limits to the size of operons. How can bacterial cells manage responses that require the expression of dozens or even hundreds of genes?

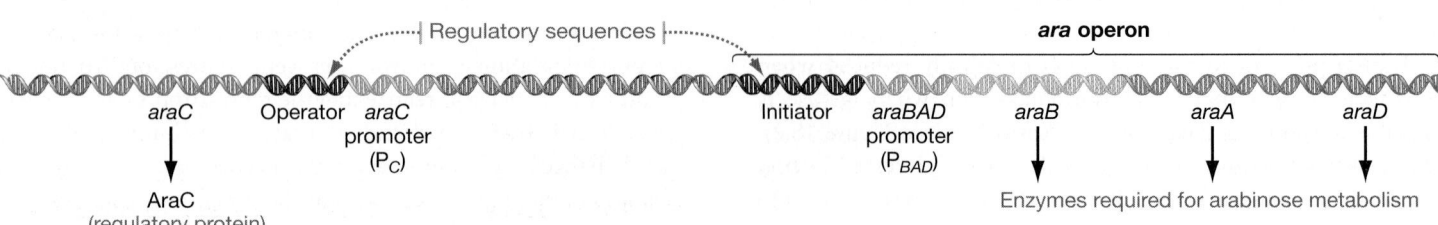

FIGURE 18.9 The *ara* Operon, Regulatory Sequences, and *araC* Gene.

(a) AraC protein is an **activator** when bound to arabinose.

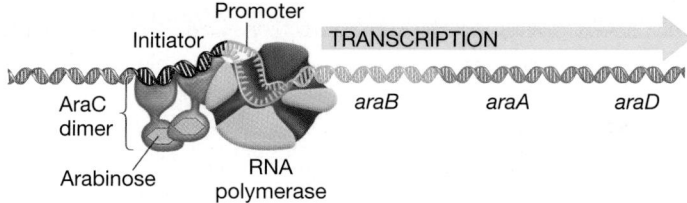

(b) AraC protein is a **repressor** when arabinose is absent.

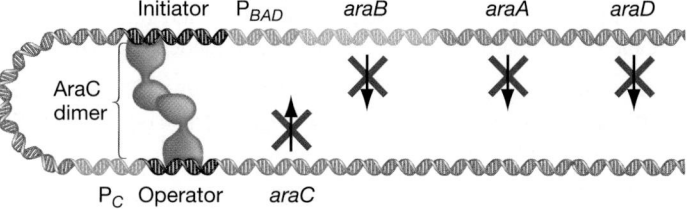

FIGURE 18.10 Positive and Negative Regulation of the *ara* Operon.

Global gene regulation is the coordinated regulation of many genes. You already learned that alternative sigma proteins provide one way for bacteria to turn on large numbers of genes in response to environmental change (Chapter 17). But there are other means of global gene regulation, such as grouping genes into a **regulon**—a set of separate genes or operons that contain the same regulatory sequences and that are controlled by a single type of regulatory protein.

Regulons allow bacteria to respond to challenges that include shortages of nutrients, sudden changes in temperature, exposure to radiation, or shifts in habitat. Let's explore how regulons work in general, and then look at two specific examples.

A general strategy for controlling regulon genes is shown in **FIGURE 18.11**. In this example, the regulon consists of many genes that are scattered across the genome. All of these genes are controlled by the same type of repressor protein that binds to the same operator sequences near the promoter of each gene. When an environmental change triggers the removal of the repressor protein from all the operators, every gene in the regulon is transcribed.

Regulons can be under negative control by a repressor protein or positive control by an activator protein. The regulon in Figure 18.11 is under negative control. The SOS response regulon works exactly this way to allow bacterial cells to repair extensive damage to DNA that can occur when cells are exposed to ultraviolet light, other types of radiation, or some chemicals. Damaged DNA sets off an SOS signal that induces the transcription of more than 40 genes that code for DNA repair enzymes and for DNA polymerases that can use damaged DNA as a template. Without the SOS response, bacteria with massive DNA damage would face almost certain death.

The ToxR regulon of *Vibrio cholera*—the bacterium that causes cholera—is under positive control. This regulon allows *V. cholera* to colonize the human gut and to produce toxins that cause diarrhea. Cholera kills 120,000 people each year and sickens as many as 18 million. ToxR regulon genes are inactive when *V. cholera* lives outside a human host. When bacteria from contaminated drinking water encounter the environment of the human gut, this sets off a signal that activates an activator protein. The activator induces a response by binding to a regulatory DNA sequence near the promoters of all ToxR regulon genes to stimulate their transcription. The diarrhea induced by this regulon is adaptive for *V. cholera* because it spreads more bacteria into the environment to infect new hosts.

What are the general messages of this chapter? Interactions among protein regulators and the DNA sequences they bind produce finely tuned control over gene expression, regulating individual genes, operons, or large sets of genes. With these exquisite controls over gene expression, bacteria have been able to compete, grow, and reproduce for more than 3 billion years of life's history.

Do eukaryotes control their genes the same way as bacteria? If not, what are the differences? These questions are the focus of the next chapter.

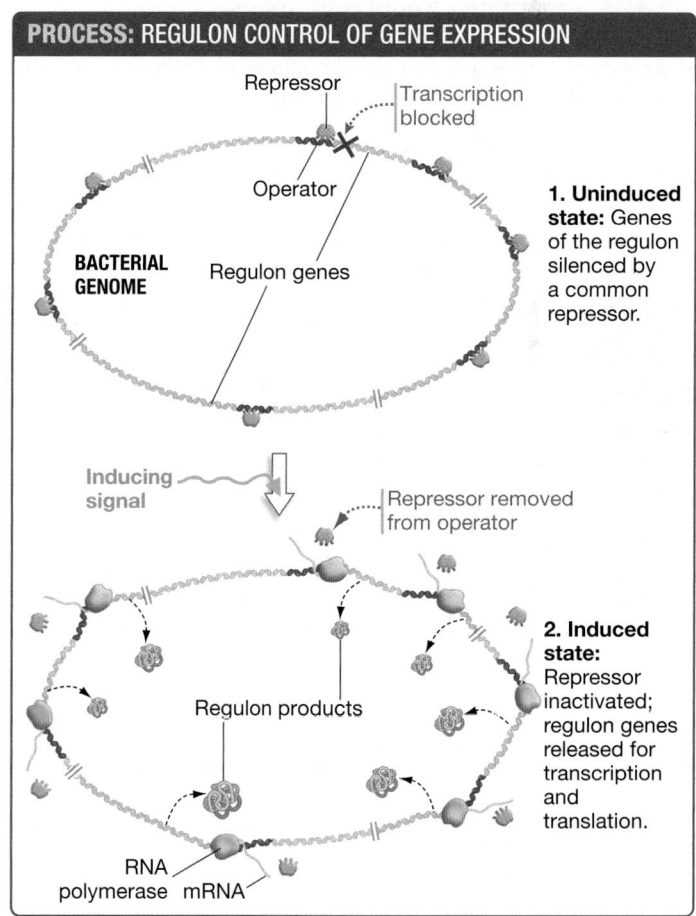

PROCESS: REGULON CONTROL OF GENE EXPRESSION

1. Uninduced state: Genes of the regulon silenced by a common repressor.

2. Induced state: Repressor inactivated; regulon genes released for transcription and translation.

FIGURE 18.11 Genes of a Regulon Are Expressed Together. The "| |" symbols indicate regions of the bacterial genome not shown between regulon genes.

If you understand . . .

18.1 An Overview of Gene Regulation and Information Flow

- Changes in gene expression allow bacterial cells to respond to environmental changes.

- Most gene products are produced or activated only when they are needed.

- Gene expression can be controlled at three levels: transcription, translation, or post-translation (protein activation).

- Transcriptional control can be negative or positive. Negative control occurs when a regulatory protein prevents transcription. Positive control occurs when a regulatory protein increases the frequency of initiating transcription.

✔ You should be able to describe one component of the *lac* operon that is under transcriptional control and one component that is under post-translational control.

18.2 Identifying Regulated Genes

- Replica plating is a technique that allows researchers to identify mutants that cannot grow in a particular condition.

- Replica plating led to the isolation of three types of lactose metabolism mutants.

- Transcription may be constitutive or regulated. Constitutive expression occurs in genes whose products are required at all times, such as genes that encode glycolytic enzymes.

✔ You should be able to propose a strategy to isolate *E. coli* mutants that can grow at 33°C, but not at 42°C.

18.3 Negative Control of Transcription

- The *lac* operon is transcribed efficiently when lactose is present and glucose is absent.

- The *lac* operon is under negative control.

- Negative control occurs because a repressor protein binds to an operator sequence in DNA near the promoter of the protein-encoding genes to prevent their transcription.

- When lactose is present, it binds to the repressor and causes it to fall off the operator, allowing transcription to occur.

- Glucose inhibits transcription of the *lac* operon by inhibiting lactose transport into the cell.

✔ You should be able explain how the operator, repressor, and inducer relate to a car's parking brake.

18.4 Positive Control of Transcription

- Positive control of transcription occurs when a regulatory protein called an activator binds to a regulatory sequence in DNA.

- Activator proteins bind to RNA polymerase in addition to DNA. Binding between the activator and RNA polymerase increases the rate of transcription initiation.

- The *ara* operon codes for genes required for metabolism of the sugar arabinose. The operon is controlled by the AraC regulatory protein. AraC is an activator when bound to arabinose and a repressor when the protein is not bound to arabinose.

✔ You should be able to predict if mutations in the *ara* initiator sequence of the *ara* operon are most likely to affect positive regulation, negative regulation, or both.

18.5 Global Gene Regulation

- Bacterial cells often need to coordinate the expression of large sets of genes in response to changing environments.

- Regulons coordinate the expression of different genes by using a shared regulator that acts on a regulatory sequence found in all genes of the regulon. Regulons can work through negative control using repressors, or through positive control using activators.

✔ You should be able to propose a method that would allow more genes to become part of the SOS regulon.

MasteringBiology

1. **MasteringBiology Assignments**
   Tutorials and Activities The *lac* Operon
   Questions Reading Quizzes, Blue-Thread Questions, Test Bank

2. **eText** Read your book online, search, take notes, highlight text, and more.

3. **The Study Area** Practice Test, Cumulative Test, BioFlix® 3-D animations, Videos, Activities, Audio Glossary, Word Study Tools, Art

You should be able to . . .

✔ TEST YOUR KNOWLEDGE

Answers are available in Appendix A

1. Replica plating is used to isolate mutants that
   a. can produce an enzyme.
   b. cannot grow in a particular condition.
   c. can utilize lactose.
   d. turn yellow when lactose is broken down.

2. Why are the genes involved in lactose metabolism considered to be an operon?
   a. They occupy adjacent locations on the *E. coli* chromosome.
   b. They have a similar function.
   c. They are all required for normal cell function.
   d. They are all controlled by the same promoter.

3. In the *lac* operon, the repressor inhibits transcription when
   a. the repressor is bound to the inducer.
   b. the repressor is not bound to the inducer.
   c. the repressor is bound to glucose.
   d. the repressor is not bound to the operator.

4. Activators bind to regulatory sequences in _____ and to _____ polymerase.

5. How does inducer exclusion control gene expression in the *lac* operon?

6. A regulon is a set of genes controlled by
   a. one type of regulator of transcription.
   b. two or more different alternative sigma proteins.
   c. many different types of promoters.
   d. glucose.

7. *E. coli* expresses genes for glycolytic enzymes constitutively. Why?

8. Explain the difference between positive and negative control over transcription.

9. Predict what would happen if the *lac* repressor protein were altered so it could not bind inducer.
   a. The repressor could not bind to DNA.
   b. The repressor would always be bound to DNA.
   c. The repressor could bind to DNA only when cells were grown with glucose.
   d. The repressor could bind to DNA only when cells were grown without glucose.

10. Predict what would happen to regulation of the *lac* operon if the *lacI* gene were moved 50,000 nucleotides upstream of its normal location.

11. If any of the following hypothetical drugs could be developed, which would be most effective in preventing cholera?
    a. a drug that increased the amount of the ToxR activator
    b. a drug that blocked the DNA-binding activity of the activator
    c. a drug that increased rates of transcription in *V. cholerae*
    d. a drug that increased rates of translation in *V. cholerae*

12. IPTG is a molecule with a structure very similar to lactose. IPTG can be transported into cells by galactoside permease and can bind to the *lac* repressor protein. However, unlike lactose, IPTG is not broken down by β-galactosidase. Predict what would occur regarding *lac* operon regulation if IPTG were added to *E. coli* growth medium containing arabinose and no glucose or lactose.

13. You are interested in using bacteria to metabolize wastes at an old chemical plant and convert them into harmless compounds. You find bacteria that are able to tolerate high levels of the toxic compounds toluene and benzene, and you suspect that this is because the bacteria can break down these compounds into less-toxic products. If that is true, these toluene- and benzene-resistant strains will be valuable for cleaning up toxic sites. How could you find out whether these bacteria are metabolizing toluene as a source of carbon compounds?

14. **QUANTITATIVE** Imagine that you are repeating the replica-plating procedure of Jacob and Monod to find mutants that can't grow using lactose. After treating cells with a mutagen, you anticipate a mutation rate of 1×10^{-4} lactose-nonutilizing mutants per mutagen-treated cell. Based on this estimate, how many cells should you replica-plate to have a good chance of finding one mutant?

15. A type of mutation in the *lac* operator known as *lacOᶜ* prevents repressor binding to DNA and causes constitutive transcription of the *lac* operon. Which of the following secondary mutations might restore normal regulation to the *lac* operon in a *lacOᶜ* mutant?
    a. a *lacI* mutation that decreases the ability of the repressor to bind the inducer
    b. a *lacI* mutation that produces a repressor than can recognize the mutated *lacOᶜ* DNA sequence
    c. a promoter mutation that prevents it from being recognized by sigma
    d. an RNA polymerase mutation that allows it to bind to the promoter without using sigma

16. X-gal is a colorless, lactose-like molecule that can be split into two fragments by β-galactosidase. One of these product molecules is blue. The following photograph shows *E. coli* colonies growing in a medium that contains X-gal.

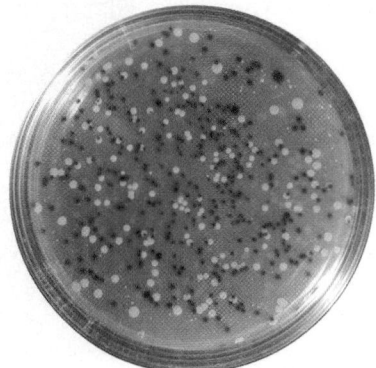

Find three colonies whose cells have functioning copies of β-galactosidase. Find three colonies whose cells might have mutations in the *lacZ* or in the *lacY* genes. Suppose you analyze the protein-coding sequence of the *lacZ* and *lacY* genes of cells from the three mutant colonies and find that these sequences are wild type (normal). What other region of the *lac* operon might be altered to account for the mutant phenotype of these colonies?

19 Control of Gene Expression in Eukaryotes

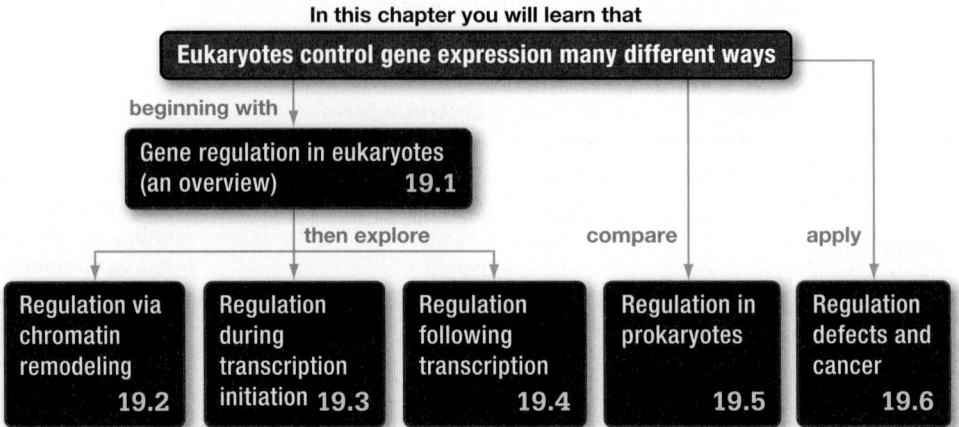

In this chapter you will learn that

Eukaryotes control gene expression many different ways

beginning with

Gene regulation in eukaryotes (an overview)   19.1

then explore — *compare* — *apply*

| Regulation via chromatin remodeling 19.2 | Regulation during transcription initiation 19.3 | Regulation following transcription 19.4 | Regulation in prokaryotes 19.5 | Regulation defects and cancer 19.6 |

A model of eukaryotic DNA in the condensed state. The DNA (shown in red and pink) is wrapped around proteins (in green). The DNA has to be uncoiled before transcription can take place.

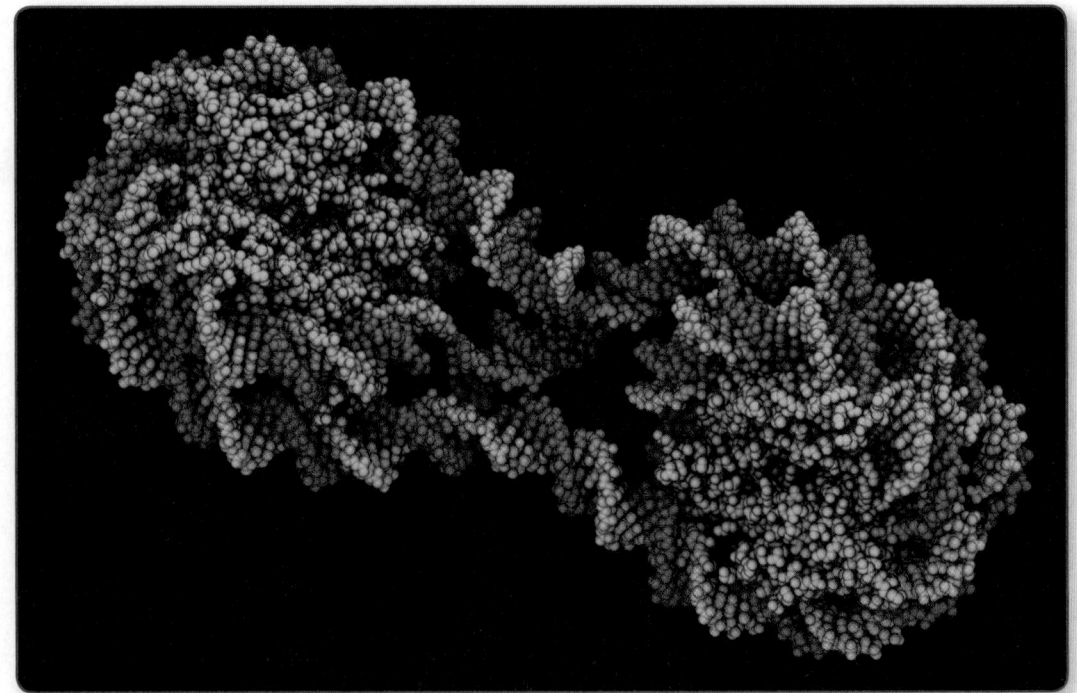

This chapter is part of the Big Picture. See how on pages 366–367.

Bacteria regulate gene expression to respond to changes in their environment. *Escherichia coli* thrive best if the genes that are required to import and cleave lactose are expressed only when the cells are relying on lactose as a source of energy (see Chapter 18).

Unicellular eukaryotes face similar challenges. Consider the yeast *Saccharomyces cerevisiae*, which is used extensively in the production of beer, wine, and bread. In nature this species lives on the skins of grapes and other fruits, where the sugars that the cells use as food vary in type and concentration as the fruit ripens, falls, and rots. For yeast cells to grow and reproduce efficiently, gene expression has to be modified in response to these changes.

The cells that make up multicellular eukaryotes face additional challenges. Consider your body, which contains trillions of cells, each with a specialized structure and function. You have heart

✔ When you see this checkmark, stop and test yourself. Answers are available in Appendix A.

muscle cells, lung cells, nerve cells, skin cells, and so on. Even though these cells are different, they contain the same genes. Your bone cells and blood cells aren't different because of a difference in their genes but because they *express* different genes. Your bone cells have blood-cell genes—they just don't transcribe them.

Why not? The answer is that your cells respond to their environment, just as bacteria and unicellular eukaryotes do. But there's a key difference. The cells in a multicellular eukaryote express different genes in response to changes in the *internal* environment—specifically, to signals from other cells. As a human being or an oak tree develops, cells that are located in different parts of the organism are exposed to different cell–cell signals. As a result, they express different genes. **Differential gene expression** is responsible for creating different cell types, arranging them into tissues, and coordinating their activity to form the multicellular society we call an individual.

How does all of this regulation and differentiation happen? Later chapters introduce the signals that trigger the formation of muscle, bone, leaf, and flower cells (see Unit 4). In contrast, this chapter focuses on what happens after a eukaryotic cell receives such a signal. Let's start with an overview of how gene expression can be controlled, and close with a look at how defects in the process can trigger cancer.

19.1 Gene Regulation in Eukaryotes—An Overview

Like bacteria, eukaryotes can control gene expression at the levels of transcription, translation, and post-translation. But as **FIGURE 19.1** shows, three additional levels of control occur in eukaryotes as genetic information flows from DNA to proteins.

The first additional level of control involves the DNA–protein complex at the top of the figure. In eukaryotes, DNA is wrapped around proteins to create a structure called **chromatin.** Eukaryotic genes have promoters, just as bacterial genes do; but before transcription can begin in eukaryotes, the stretch of DNA containing the promoter must be released from tight interactions with proteins, so that RNA polymerase can make contact with the promoter. To capture this idea, biologists say that **chromatin remodeling** must occur before transcription.

The second level of regulation that is unique to eukaryotes is **RNA processing**—the steps required to produce a mature, processed mRNA from a primary RNA transcript. Recall that introns have to be spliced out of primary transcripts (see Chapter 17). In many cases, carefully orchestrated alternative splicing occurs—meaning that different combinations of exons are included in the mRNA. If different cells use different splicing patterns, different gene products result.

Third, mRNA life span is regulated in eukaryotes: mRNAs that remain in the cell for a long time tend to be translated more than mRNAs that have a short life span.

Each of the six potential control points shown in Figure 19.1 is employed in eukaryotic cells. This chapter explores all six—chromatin

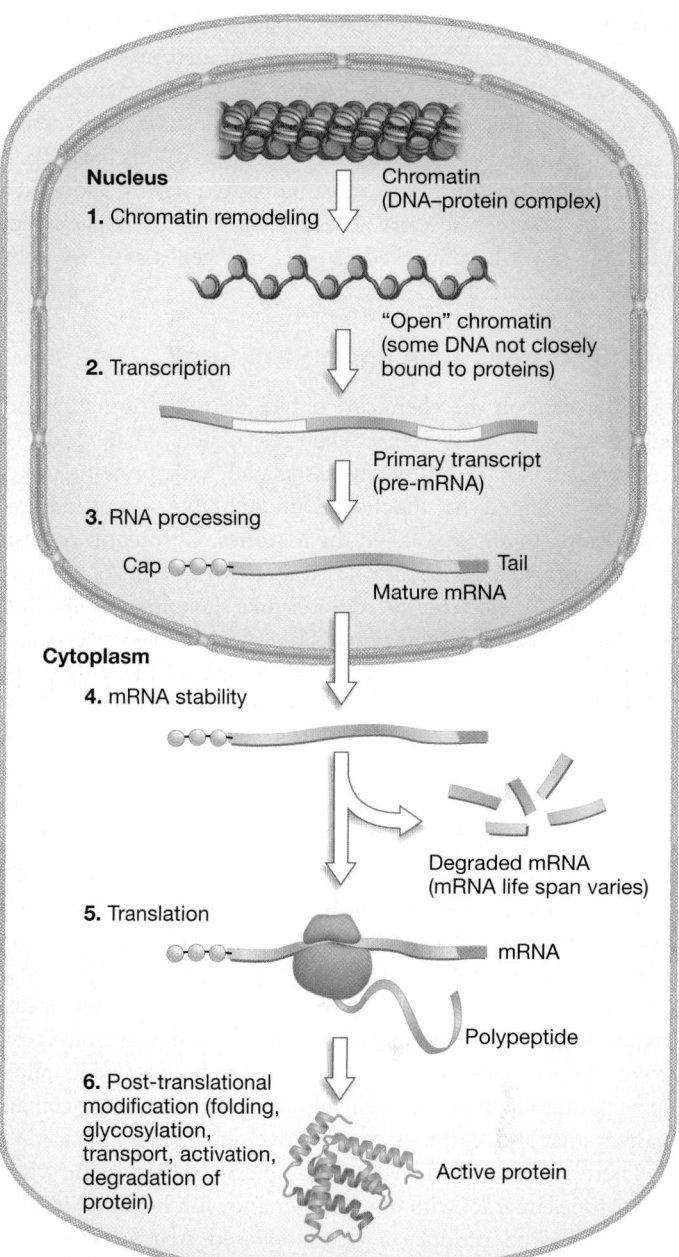

FIGURE 19.1 In Eukaryotes, Gene Expression Can Be Controlled at Many Different Levels.

remodeling, transcription, RNA processing, mRNA stability, translation, and post-translational modification of proteins.

To appreciate the breadth and complexity of gene regulation in eukaryotes, let's follow the series of events that occur as an embryonic cell responds to a developmental signal. Suppose a molecule arrives that specifies the production of a muscle-specific protein. What happens next?

19.2 Chromatin Remodeling

For a molecular signal to trigger the transcription of a particular gene, the chromatin around the target gene must be remodeled.

To appreciate why, consider that a typical cell in your body contains about 6 billion base pairs of DNA. Lined up end to end, these nucleotide pairs would form a double helix about 2 m (6.5 feet) long. But the nucleus that holds this DNA is only about 5 μm in diameter—far less than the thickness of this page. To fit inside the nucleus, the DNA must be packed tightly—so tightly that RNA polymerase can't access it. How is DNA packaged? And how can it be unpacked at particular genes so RNA polymerase can transcribe it?

What Is Chromatin's Basic Structure?

The first data on the chemical composition of chromatin were published in the early 1900s, when researchers established that eukaryotic DNA is intimately associated with proteins. Later work documented that the most abundant DNA-associated proteins belong to a group called the **histones.** Chromatin consists of DNA complexed with histones and other proteins.

In the 1970s electron micrographs like the one in **FIGURE 19.2a** revealed that chromatin has a regular structure. In some preparations for electron microscopy, chromatin looked like beads on a string. The "beads" came to be called **nucleosomes.**

More information emerged in 1984 when researchers determined the three-dimensional structure of eukaryotic chromatin by using X-ray crystallography (see **BioSkills 11** in Appendix B). The X-ray crystallographic data indicated that each nucleosome consists of DNA wrapped almost twice around a core of eight histone proteins. As **FIGURE 19.2b** indicates, a histone called H1 "seals" DNA to each nucleosome. Between each pair of nucleosomes there is a "linker" stretch of DNA.

The intimate association between DNA and histones occurs in part because DNA is negatively charged and histones are positively charged. DNA has a negative charge because of its phosphate groups; histones are positively charged because they contain many lysines and arginines, two positively charged amino acids.

There are additional layers of complexity in packaging DNA. H1 histones interact with one another and with histones in other nucleosomes to produce a tightly packed structure like that shown in Figure 19.2b. Based on its width, this structure is called the 30-nanometer fiber. (Recall that a nanometer is one-billionth of a meter and is abbreviated nm.)

Finally, the 30-nm fibers are attached at intervals along their length to proteins that form a scaffold or framework inside the nucleus. In this way, the entire chromosome is organized and held in place. When chromosomes condense before mitosis or meiosis, the scaffold proteins and 30-nm fibers are folded into still larger and more tightly packed structures.

A eukaryotic chromosome, then, is made up of chromatin that has several layers of organization: The DNA is wrapped around histones to form nucleosomes, nucleosomes are packed into 30-nm fibers, 30-nm fibers are attached to scaffold proteins, and the entire assembly can be folded into the highly condensed structure observed during cell division.

Although research has shown that bacterial DNA interacts with proteins that are organized similarly to nucleosomes,

(a) Nucleosomes in chromatin

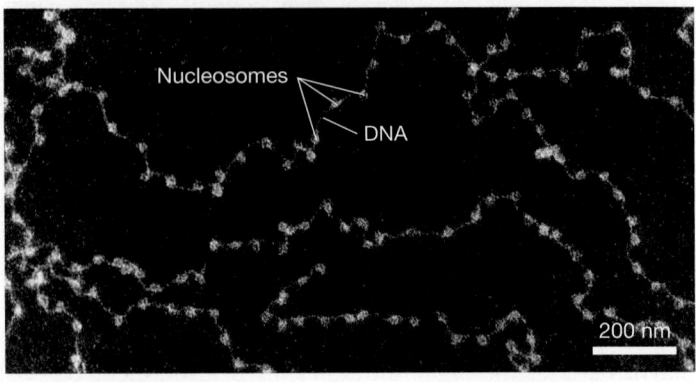

(b) Nucleosome structure

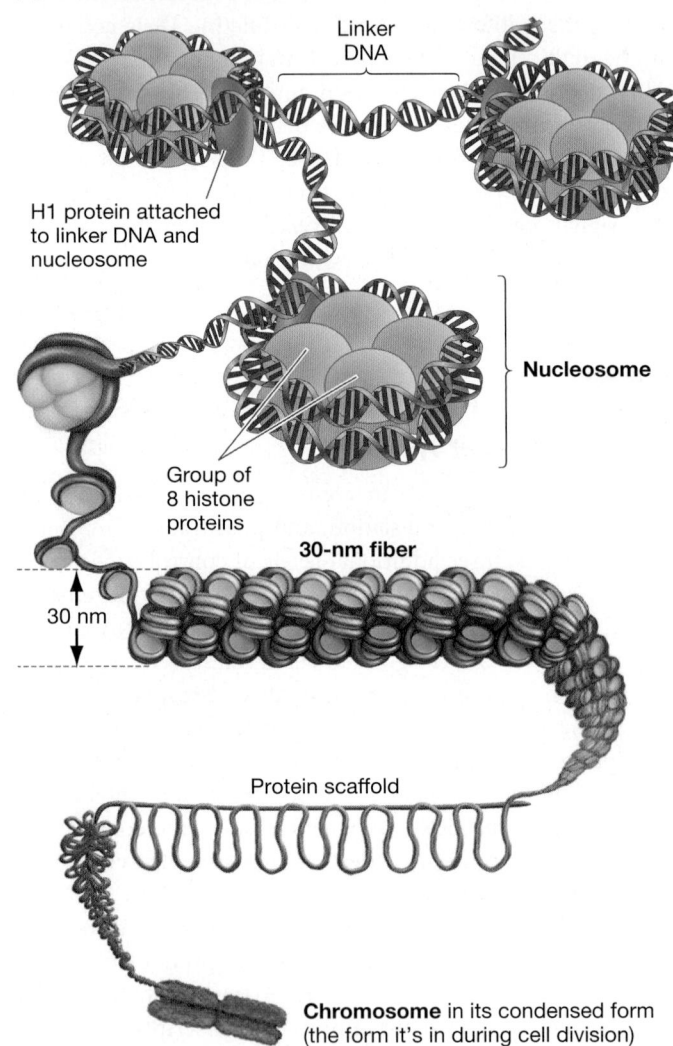

FIGURE 19.2 Chromatin Has Several Levels of Organization.

nothing like the 30-nm fibers or higher-order arrangements has been observed in bacterial chromosomes.

The elaborate structure of eukaryotic chromatin does more than just package DNA so that it fits into the nucleus. Chromatin structure also has profound implications for the control of gene expression. To appreciate this point, consider the 30-nm fiber illustrated in Figure 19.2b. If this tightly packed stretch of DNA contains a promoter, how can RNA polymerase bind to it and initiate transcription?

Evidence that Chromatin Structure Is Altered in Active Genes

Once the nucleosome-based structure of chromatin was established, biologists hypothesized that the close physical interaction between DNA and histones must be altered for RNA polymerase to make contact with DNA. More specifically, biologists hypothesized that a gene could not be transcribed until the condensed chromatin near its promoter was remodeled.

The central idea is that chromatin must be decondensed, to expose the promoter so RNA polymerase can bind to it. If so, then chromatin remodeling would represent the first step in the control of eukaryotic gene expression. Two types of studies have provided strong support for this hypothesis.

DNA in Condensed Chromatin Is Protected from DNase DNases are enzymes that cut DNA. Some DNases cleave DNA at random locations, and these cannot cut efficiently if DNA is tightly wrapped with proteins. As **FIGURE 19.3** shows, this type of DNase works effectively only if DNA is in a decondensed configuration.

Harold Weintraub and Mark Groudine used this observation to test the hypothesis that the DNA of actively transcribed genes is in an open configuration. In chicken blood cells, they compared chromatin structure in two genes: β-globin and ovalbumin. β-globin is a protein that is part of the hemoglobin found in red blood cells; ovalbumin is a protein found in egg white. In blood cells, the β-globin gene is transcribed at high levels, but the ovalbumin gene is not transcribed at all.

After treating blood cells with DNase and then analyzing the state of chromatin at the β-globin and ovalbumin genes, the researchers found that DNase cut the β-globin gene DNA much more readily than DNA of the ovalbumin gene. They interpreted this finding as evidence that in blood cells, chromatin of the

actively transcribed β-globin gene was decondensed; and conversely, chromatin of the non-transcribed ovalbumin gene was condensed. Studies using DNase on different genes in different cell types yielded similar results.

Histone Mutants The second type of evidence in support of the chromatin-remodeling hypothesis comes from studies of mutant brewer's yeast cells that do not produce the usual complement of histones. In these mutant cells, many genes that are normally never transcribed are instead always transcribed at high levels.

To interpret this finding, biologists hypothesized that the lack of histone proteins prevented the assembly of normal chromatin. If the absence of normal histone–DNA interactions promotes transcription, then the presence of normal histone–DNA interactions must prevent it.

Taken together, the data suggest that in their normal, or default, state, eukaryotic genes are turned off. This is a new mechanism of negative control—different from repressor proteins (introduced in Chapter 18). When DNA is wrapped into a 30-nm fiber, the parking brake is on. If so, then gene expression depends on chromatin being opened up in the promoter region.

How Is Chromatin Altered?

Research on chromatin remodeling has been proceeding at a furious pace, and biologists have succeeded in identifying some of the key players that work to change the state of chromatin condensation. These include enzymes that add methyl groups to DNA, enzymes that chemically modify histones, and macromolecular machines that actively reshape chromatin. Let's examine each of these in turn.

DNA Methylation A group of enzymes known as **DNA methyltransferases** add methyl groups ($-CH_3$) to cytosine residues in DNA. In mammals, the sequence recognized by these enzymes is a C next to a G in one strand of the DNA. This sequence is abbreviated CpG and is shown below in methylated form within a stretch of DNA.

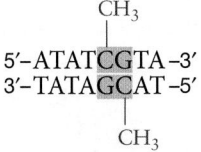

Why is **DNA methylation** important? Methylated CpG sequences are recognized by proteins that trigger chromatin condensation. Actively transcribed genes usually have low levels of methylated CpG near their promoters, and non-transcribed genes usually have high levels of methylated CpG.

Histone Modification DNA methylation is only one part of the chromatin alteration story. A large set of enzymes adds a variety of chemical groups to specific amino acids of histone proteins. These include phosphate groups, methyl groups, short polypeptide chains, and acetyl groups ($-COCH_3$). Addition of these groups to histones promotes condensed or decondensed chromatin depending on the specific set of modifications.

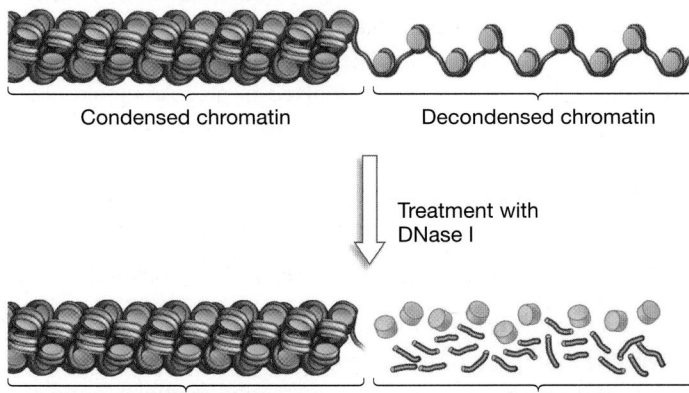

FIGURE 19.3 DNase Assay for Chromatin Structure. DNase is an enzyme that cuts DNA at random locations. It cannot cut condensed chromatin.

To account for these effects, researchers have proposed the existence of a **histone code.** The histone code hypothesis postulates that particular combinations of histone modifications set the state of chromatin condensation for a particular gene. In turn, this has an important role in regulating transcription. Let's take a closer look at one way histone modifications can control chromatin structure.

As shown in **FIGURE 19.4**, two different types of enzymes can add or remove acetyl groups from histones. **Histone acetyltransferases (HATs)** add acetyl groups to the positively charged lysine residues in histones, and **histone deacetylases (HDACs)** remove them. **Acetylation** of histones usually results in decondensed chromatin, a state associated with active transcription. How can acetylation of histones promote chromatin decondensation? When HATs add acetyl groups, the acetyl groups neutralize the positive charge on lysine residues and loosens the close association of nucleosomes with the negatively charged DNA. The addition of acetyl groups also creates a binding site for other proteins that help open the chromatin.

In contrast, when HDACs remove acetyl groups from histones, this process usually leads to condensed chromatin, a state associated with no transcription. HATs are an on switch for transcription, and HDACs are an off switch.

Chromatin-Remodeling Complexes   Other major players in chromatin alteration and gene regulation are enzymes that form macromolecular machines called **chromatin-remodeling complexes.** These machines harness the energy in ATP to reshape chromatin. Chromatin-remodeling complexes cause nucleosomes to slide along the DNA or, in some cases, knock the histones completely off the DNA to open up stretches of chromatin and allow gene transcription.

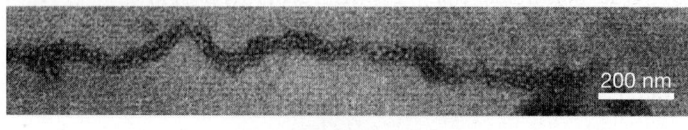

Condensed chromatin

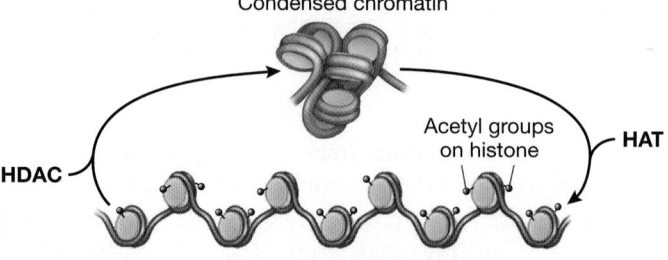

Decondensed chromatin

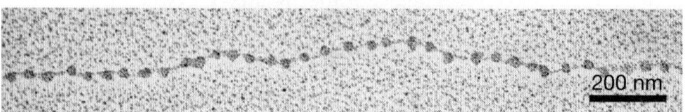

FIGURE 19.4 Acetyl Groups Decondense Chromatin. Histone acetyltransferases (HATs) cause chromatin to decondense; histone deacetylases (HDACs) cause it to condense.

✔QUESTION Are HAT and HDAC elements in positive control or negative control? Explain your reasoning.

DNA methylation, histone modifications, and chromatin-remodeling complexes work together to fine-tune chromatin condensation at specific genes. The take-home message from work on chromatin remodeling is simple: The condensation state of chromatin is critical in determining whether transcription occurs.

Chromatin Modifications Can Be Inherited

The pattern of chromatin modifications varies from one cell type to another. For example, suppose within an individual you analyzed the same gene in a muscle cell and a brain cell. This and other genes would likely have a different pattern of DNA methylation and histone acetylation in the two cell types.

DNA methylation and histone modifications are an example of **epigenetic inheritance,** the collective term for patterns of inheritance that are due to something other than differences in DNA sequences. The *epi–* of epigenetics comes from the Greek word meaning "upon." It implies another level of inheritance that adds to standard DNA-based mechanisms to explain how different phenotypes are transmitted.

With epigenetic inheritance, if a cell received a "become muscle" signal early in development, it would modify its chromatin in distinctive ways and pass those modifications on to its descendants. Muscle cells are different from brain cells not because they contain different genes, but largely because they have inherited different patterns of DNA methylation and histone modifications during the course of their development.

But the story of epigenetic inheritance involves more than just differentiation of cell types during development. Evidence is emerging that epigenetic mechanisms can record early-life events and that this archive can be difficult to erase. This is the case when prenatal conditions alter the patterns of chromatin modification and the later-life phenotypes of a mother's offspring.

For example, biologists have long known that rats born to mothers fed low-protein diets during pregnancy and while nursing have a greatly increased risk of developing disorders similar to type 2 diabetes. Type 2 diabetes is a serious and increasingly common disease that alters the cellular uptake of glucose (Chapter 44). Both genetic factors and environmental factors, such as diet, play important roles in diabetes development.

One significant gene associated with diabetes is *Hnf4a. Hnf4a* codes for a regulator of genes involved in glucose uptake. Rats born to protein-deprived mothers develop symptoms of diabetes later in life, even when these rats are fed a normal, healthy diet from the time they are weaned. These diabetic rats also express the *Hnf4a* regulatory gene at lower levels than normal rats. Could epigenetic mechanisms be at work in silencing *Hnf4a* expression?

One team's approach to probing this question is shown in **FIGURE 19.5**. Using a treatment group and a control group, the researchers measured the types of histone modifications found at a key regulatory region of the *Hnf4a* gene. A regulatory region is a section of DNA that, like prokaryotic operators (Chapter 18), is involved in controlling the activity of a gene. The chromatin at this region has to be opened up for transcription to occur. Levels

QUESTION: Does poor nutrition in a mother produce epigenetic effects in offspring?

HYPOTHESIS: Protein-deprived mothers will produce offspring with abnormal histone modifications.

NULL HYPOTHESIS: Protein-deprived mothers will produce offspring with normal histone modifications.

EXPERIMENTAL SETUP:

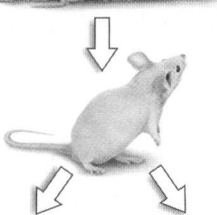

1. Provide rat mothers with a normal or a low-protein diet during pregnancy and while nursing.

2. After weaning, feed rat pups a normal diet and raise to old age.

3. Determine types of histone modifications for a regulatory gene involved in diabetes. (Also measure gene transcription.)

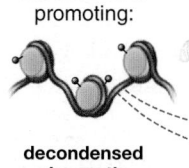

Modifications promoting: **condensed chromatin**

Modifications promoting: **decondensed chromatin**

TRANSCRIPTION

Hnf4a

PREDICTION: Offspring of mothers fed a low-protein diet will have abnormal histone modifications.

PREDICTION OF NULL HYPOTHESIS: Offspring of mothers fed a low-protein diet will have normal histone modifications.

RESULTS:

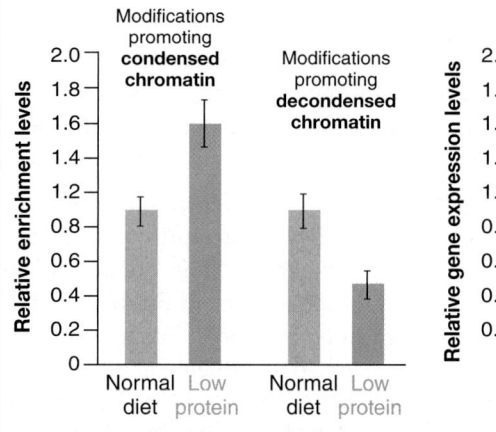

CONCLUSION: A mother's diet influences chromatin modifications and gene expression patterns throughout her offspring's life.

FIGURE 19.5 Events in Early Life Can Be Recorded through Epigenetic Mechanisms.

SOURCE: Sandovici, I., N. H. Smith, and M. D. Nitert. 2011. Maternal diet and aging alter the epigenetic control of a promoter-enhancer interaction at the *Hnf4a* gene in rat pancreatic islets. *Proceedings of the National Academy of Sciences USA* 108: 5449–5454.

✔ **QUESTION** What could researchers do to prove that the histone modifications are causing reduced regulatory gene transcription?

---

of *Hnf4a* transcription in the control and treatment groups were also measured. What did the team learn?

As the graph on the left in the "Results" section of Figure 19.5 shows, they found that histone modifications that lead to condensed chromatin were *elevated* in rats born to malnourished mothers compared to control offspring. Conversely, histone modifications associated with decondensed chromatin were significantly *reduced* in the treatment group. The graph on the right confirms that transcription of *Hnf4a* was much lower in the treatment group than the control group. Together these results demonstrate a correlation between altered histone modifications and decreased levels of *Hnf4a* gene expression.

Remember that in this study, all rats were provided a healthy diet after weaning. This finding implies that a mother's nutritional status during pregnancy and nursing is responsible for the types of chromatin modifications seen in rats that develop diabetes.

Chromatin remodeling must occur before transcription. Now the question is, What happens once a section of chromatin is opened and DNA becomes accessible to RNA polymerase?

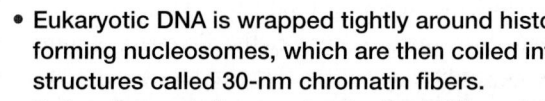
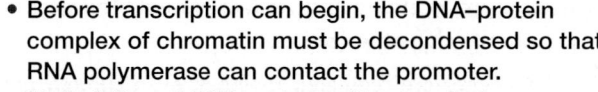

check your understanding

C Y U

If you understand that . . .

- Eukaryotic DNA is wrapped tightly around histones, forming nucleosomes, which are then coiled into structures called 30-nm chromatin fibers.
- Before transcription can begin, the DNA–protein complex of chromatin must be decondensed so that RNA polymerase can contact the promoter.
- Methylation of DNA and specific chemical modifications of histones play a key role in determining whether chromatin is opened and a gene is expressed.
- In many cases, the patterns of DNA methylation and histone modification in a cell are passed on to its daughter cells.

✔ **You should be able to . . .**

1. Predict how gene expression will be affected if a cell is grown with compounds that prevent DNA methylation.

2. Explain how certain patterns of histone acetylation or DNA methylation could influence whether a cell became a muscle cell or a brain cell.

Answers are available in Appendix A.

19.3 Initiating Transcription: Regulatory Sequences and Regulatory Proteins

As in bacteria, the **promoter** is a site in DNA where RNA polymerase binds to initiate transcription. Recent findings from genome sequencing projects (see Chapter 20) have shown that there is still much to be learned about eukaryotic promoters. What is known is that promoters in eukaryotes are more complex than bacterial promoters, often containing two or three conserved sequences that serve as binding sites for proteins needed to start transcription. The most intensively studied of these is a sequence known as the **TATA box.**

Once a promoter that contains a TATA box has been exposed by chromatin remodeling, the first step in initiating transcription is binding of the **TATA-binding protein (TBP).** Proteins related to TBP also work on promoters with other conserved sequences. But the binding of TBP or any of its relatives does not guarantee that a gene will be transcribed. In eukaryotes, a wide array of other DNA sequences and proteins work together to allow transcription.

Promoter-Proximal Elements Are Regulatory Sequences Near the Promoter

The first **regulatory sequences** in eukaryotic DNA were discovered in the late 1970s, when Yasuji Oshima and co-workers set out to understand how yeast cells control the metabolism of the sugar galactose.

When galactose is absent, *S. cerevisiae* cells produce only tiny quantities of the enzymes required to metabolize it. But when galactose is present, transcription of the genes encoding these enzymes increases by a factor of 1000.

The team's first major result was the discovery of mutant cells that failed to produce any of the five enzymes required for galactose metabolism, even if galactose was present. To interpret this observation, they hypothesized that

1. the five genes are regulated together, even though they are not on the same chromosome;

2. normal cells have an activator protein that exerts positive control over the five genes;

3. the mutant cells have a mutation that completely disables the activator protein.

Other researchers were able to isolate the regulatory protein and confirm that it binds to a short stretch of DNA located just upstream from the promoter for all five genes required for galactose use.

In bacteria, genes that need to be regulated together are often clustered into a single operon and transcribed into a single mRNA. In contrast, eukaryotes use the strategy uncovered by Oshima's group for the galactose-metabolizing genes in yeast—co-regulated genes are not clustered together, but instead share a regulatory DNA sequence that binds the same regulatory protein.

Regulatory DNA sequences similar to those first discovered in yeast have now been found in a wide array of eukaryotic genes and species. Regulatory sequences like these that are located close to the promoter and bind regulatory proteins are termed **promoter-proximal elements.**

Unlike the promoter itself, promoter-proximal elements have sequences that are unique to specific sets of genes. In this way, they furnish a mechanism for eukaryotic cells to express certain genes but not others.

The discovery of promoter-proximal elements and a mechanism of positive control suggested a satisfying parallel between gene regulation in bacteria and in eukaryotes. This picture changed, however, when researchers discovered a new class of eukaryotic DNA regulatory sequences—sequences unlike anything in bacteria.

Enhancers Are Regulatory Sequences Far from the Promoter

Susumu Tonegawa and colleagues made a startling discovery while exploring how human cells regulate gene expression. The gene studied by Tonegawa's group was broken into many introns and exons. Recall that introns are transcribed sequences that are spliced out of the primary transcript; exons are transcribed regions that are included in the mature RNA once splicing is complete (Chapter 17). The researchers discovered a regulatory sequence within one of the introns that was required for transcription of the gene.

This finding was remarkable for two reasons: **(1)** The regulatory sequence was thousands of bases away from the promoter, and **(2)** it was downstream of the promoter instead of upstream. Regulatory sequences that are far from the promoter and activate transcription are termed **enhancers.** Follow-up work has shown that enhancers occur in all eukaryotes and that they have several key characteristics:

- Enhancers can be more than 100,000 bases away from the promoter. They can be located in introns or in untranscribed sequences on either the 5′ or 3′ side of the gene (See **FIGURE 19.6**).

- Like promoter-proximal elements, many types of enhancers exist.

- Most genes have more than one enhancer.

- Enhancers usually have binding sites for more than one protein.

- Enhancers can work even if their normal 5′→3′ orientation is flipped, or if they are moved to a new location in the vicinity of the gene.

Enhancers are regulatory DNA sequences unique to eukaryotes. When regulatory proteins called **transcriptional activators** bind to enhancers, transcription begins. Thus, enhancers and activators are like a gas pedal—an element in positive control. Eukaryotes also possess regulatory sequences that are similar in structure and share key characteristics with enhancers but work to inhibit transcription. These DNA sequences are called **silencers.** When regulatory proteins called **repressors** bind to silencers, transcription is shut down. Silencers and repressors are like a brake—an element in negative control.

The Role of Transcription Factors in Differential Gene Expression

Follow-up work supported the hypothesis that enhancers and silencers are binding sites for activators and repressors that regulate transcription. Collectively, these proteins are termed

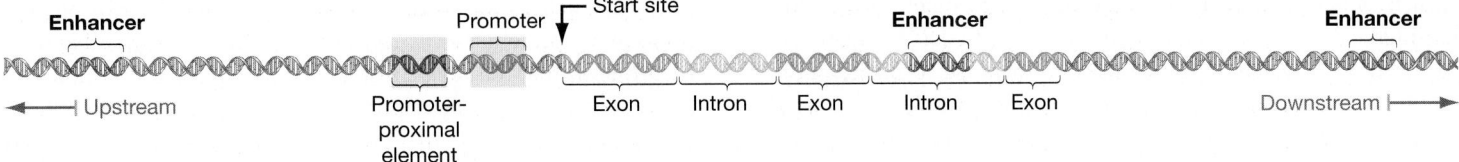

FIGURE 19.6 Enhancers and Promoter-Proximal Elements Regulate the Expression of Eukaryotic Genes. Promoter-proximal elements are near the promoter. Enhancers are located farther away, may be upstream or downstream of the promoter, and may even be within introns. Exons and introns are not drawn to scale. They are typically very large compared with regulatory sequences.

✔ **EXERCISE** Compare and contrast the structure of this typical eukaryotic gene and the structure of a bacterial operon.

regulatory transcription factors, or often **transcription factors** for short. By analyzing mutant yeast, fruit flies, and roundworms that have defects in the expression of particular genes, biologists have identified a large number of transcription factors that bind to enhancers, silencers, and promoter-proximal elements.

These results support one of the most general statements researchers are able to make about gene regulation in eukaryotes: Different types of cells express different genes because they have different transcription factors. In multicellular species, the transcription factors, in turn, are produced in response to signals that arrive from other cells early in embryonic development.

For example, if a signal that says "become a muscle cell" reaches a cell in the early embryo, it triggers a signal transduction cascade (see Chapter 11) that leads to the production of transcription factors specific to muscle cells. Because different transcription factors bind to specific regulatory sequences, they turn on the production of muscle-specific proteins. But if no "become-a-muscle-cell" signal arrives, then no muscle-specific transcription factors are produced and no muscle-specific gene expression takes place.

Differential gene expression is a result of the production or activation of specific transcription factors. Eukaryotic genes are turned on when transcription factors bind to enhancers and promoter-proximal elements; the genes are turned off when transcription factors bind to silencers or when chromatin is condensed. Distinctive transcription factors are what make a muscle cell a muscle cell and a bone cell a bone cell.

How Do Transcription Factors Recognize Specific DNA Sequences?

Each transcription factor must be able to recognize and bind to a specific DNA sequence. How can it do this?

Recall that DNA bases are partially exposed in the major and minor grooves of the DNA double helix (see Figure 4.7 for a review). The edges of an AT base pair and a GC base pair that project into the grooves of the DNA helix contain different sets of atoms and have different surface shapes (**FIGURE 19.7a**). These differences in composition and shape can be recognized by transcription factors.

(a) AT and CG base pairs present different shapes and chemical groups in the grooves of DNA.

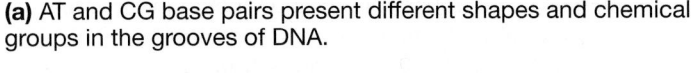

(b) Transcription factors recognize a specific sequence of bases in target DNA.

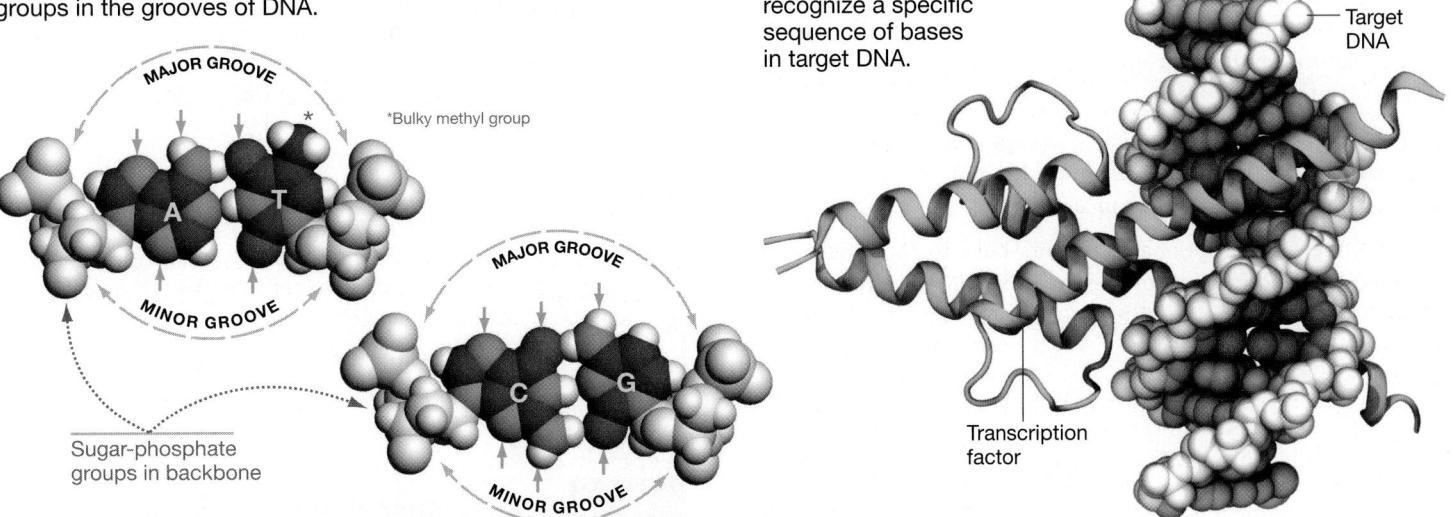

FIGURE 19.7 How Transcription Factors Bind to Regulatory Sequences. (a) The edges of base pairs that project into the major and minor grooves of the double helix present a different structure and set of atoms. Atoms that can participate in hydrogen bonding with amino acids of transcription factors are indicated by arrows. The methyl group on thymine (T) is also important in recognition. **(b)** A transcription factor (green) involved in muscle-cell differentiation binding to a regulatory sequence in DNA. The bases recognized by the protein are highlighted in red.

Just as base pairs come together by complementary molecular interactions, so too can proteins and specific DNA sequences. An example is shown in **FIGURE 19.7b**. In this case, a transcription factor that is essential for the development of muscle cells inserts amino acid side chains into two major grooves of DNA. This particular transcription factor binds to a specific enhancer sequence because of complementary interactions between base pairs and its amino acids. Without such specific interactions between transcription factors and DNA, the development of muscle cells—or any other specialized cell type—would not be possible.

A Model for Transcription Initiation

Although gene expression can be controlled at many levels, regulating the start of transcription is at center stage. For a process so important, many questions remain. What is clear is that transcription factors must interact with regulatory sequences to initiate transcription.

Besides the regulatory transcription factors you've learned about that bind to enhancers, silencers, and promoter-proximal elements, there is another type: **basal transcription factors.** These are proteins that interact with the promoter and are not restricted to particular genes or cell types. The term basal implies that these proteins are necessary for transcription to occur, but they do not provide much in the way of regulation. The promoter-recognized TATA-binding protein (TBP) that you learned about earlier is an example of a basal transcription factor that is common to many genes. ✔ If you understand this concept, you should be able to compare and contrast the regulatory and basal transcription factors found in muscle cells versus nerve cells.

In addition to transcription factors, a large complex of proteins called **Mediator** acts as a bridge between regulatory transcription factors, basal transcription factors, and RNA polymerase II.

FIGURE 19.8 summarizes one model for how transcription is initiated in eukaryotes.

Step 1 Transcriptional activators bind to DNA and recruit chromatin-remodeling complexes and histone acetyltransferases (HATs).

Step 2 The chromatin-remodeling complexes and HATs open a swath of chromatin that includes the promoter, promoter-proximal elements, and enhancers.

Step 3 Other transcriptional activators bind to the newly exposed enhancers and promoter-proximal elements; basal transcription factors bind to the promoter and recruit RNA polymerase II.

Step 4 Mediator connects the transcriptional activators and basal transcription factors that are bound to DNA. This step is made possible through DNA looping. RNA polymerase II can now begin transcription.

✔ If you understand this model, you should be able to explain why DNA forms loops near the promoter in order for transcription to begin.

An important point in this model of transcription initiation is the dual role of transcriptional activators. Activators work not only to stimulate transcription but also to bring chromatin-remodeling proteins to the right place at the right time. None of the proteins that remodel chromatin can recognize specific DNA sequences. It is the transcriptional activators that bind to regulatory sequences at particular genes to recruit the proteins needed to remodel chromatin.

The role of transcriptional activators in bringing in proteins that decondense chromatin leads to a chicken-and-egg paradox: How can an activator bind to DNA in the first place if chromatin is condensed? It turns out that except in its most highly condensed forms, chromatin is dynamic. DNA occasionally dissociates from nucleosomes, exposing regulatory sequences to activators that are present in a particular cell type.

Getting RNA polymerase to initiate transcription requires interactions between many proteins, including transcriptional activators that are bound to enhancers and promoter-proximal elements, Mediator, basal transcription factors, and RNA polymerase itself. The result is a large, macromolecular machine that is positioned at a gene's start site and capable of starting transcription.

Compared with what happens in bacteria, where just three to five proteins may interact at the promoter to initiate transcription, the process in eukaryotes is remarkably complicated.

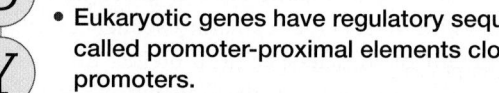

check your understanding

If you understand that . . .

- Eukaryotic genes have regulatory sequences called promoter-proximal elements close to their promoters.
- Eukaryotic genes also have regulatory sequences called enhancers or silencers far from their promoters.
- Transcription initiation is a multistep process that begins when transcriptional activators bind to DNA and recruit proteins that open chromatin.
- Interactions between regulatory transcription factors, Mediator, and basal transcription factors position RNA polymerase II at the gene's start site.

✔ **You should be able to . . .**

1. Compare and contrast the nature of regulatory sequences and regulatory proteins in bacteria versus eukaryotes.

2. Explain why the presence of certain transcription factors could influence whether a cell becomes a muscle cell or a brain cell.

Answers are available in Appendix A.

19.4 Post-Transcriptional Control

Chromatin remodeling and transcription are just the opening to the story of gene regulation. Once a gene is transcribed, a series of events has to occur before a final product appears (see Figure 19.1). Each of these events offers an opportunity

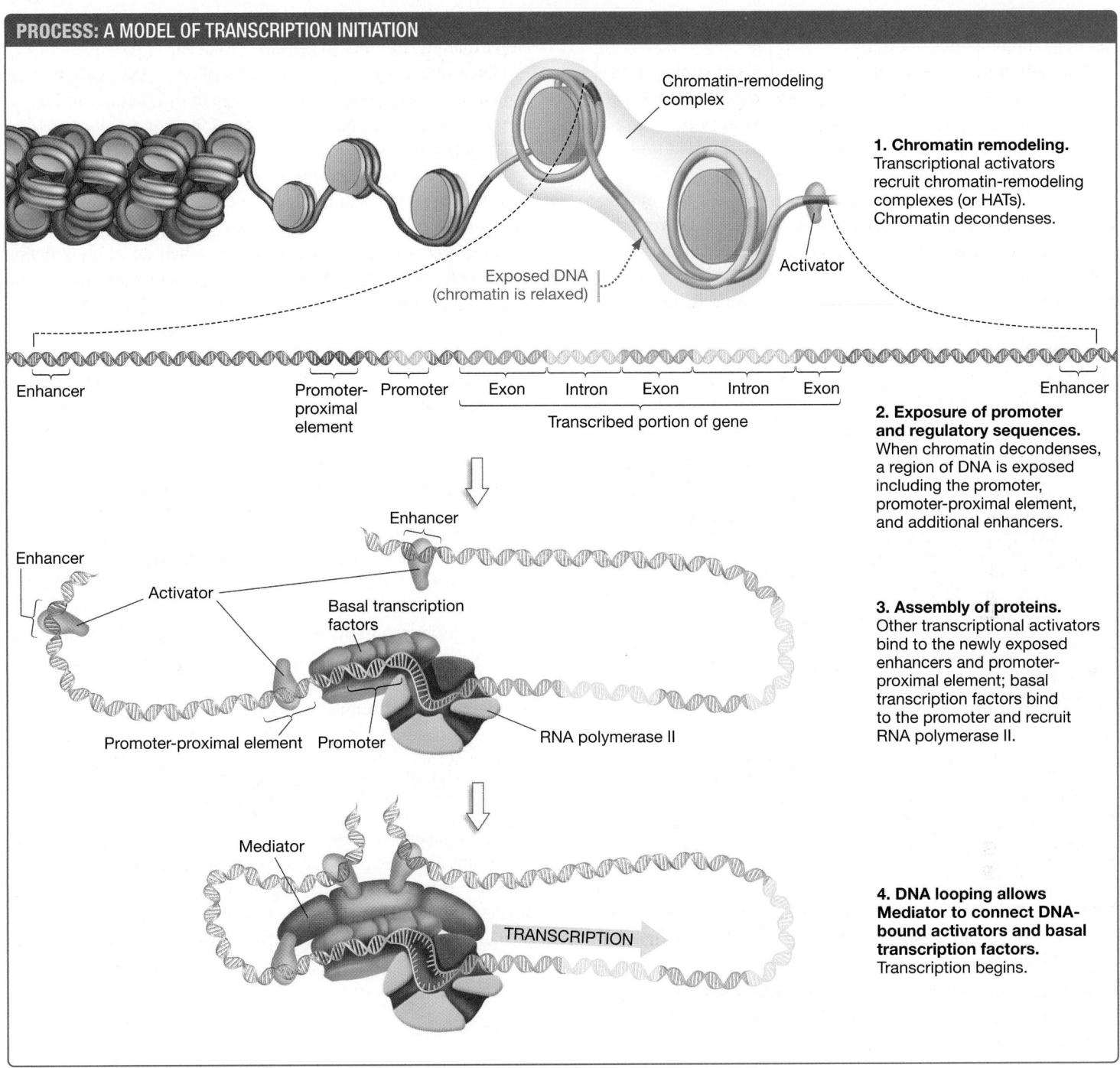

1. Chromatin remodeling. Transcriptional activators recruit chromatin-remodeling complexes (or HATs). Chromatin decondenses.

Chromatin-remodeling complex

Exposed DNA (chromatin is relaxed)

Activator

Enhancer | Promoter-proximal element | Promoter | Exon | Intron | Exon | Intron | Exon | Enhancer

Transcribed portion of gene

2. Exposure of promoter and regulatory sequences. When chromatin decondenses, a region of DNA is exposed including the promoter, promoter-proximal element, and additional enhancers.

Enhancer

Enhancer

Activator

Basal transcription factors

Promoter-proximal element   Promoter

RNA polymerase II

3. Assembly of proteins. Other transcriptional activators bind to the newly exposed enhancers and promoter-proximal element; basal transcription factors bind to the promoter and recruit RNA polymerase II.

Mediator

TRANSCRIPTION

4. DNA looping allows Mediator to connect DNA-bound activators and basal transcription factors. Transcription begins.

FIGURE 19.8 Transcription Initiation in Eukaryotes.

to regulate gene expression, and each is used in some cells at least some of the time. These control mechanisms include (1) splicing RNAs in various ways, (2) modifying the life span of mRNAs, (3) altering the rate at which translation is initiated, and (4) activating or inactivating proteins after translation has occurred. Let's consider each in turn.

Alternative Splicing of mRNAs

Introns are spliced out in the nucleus as the primary RNA is transcribed. Recall that the mRNA that results from splicing consists of sequences encoded by exons, and that it is protected by a cap on the 5′ end and a long poly(A) tail on the 3′ end (see Chapter 17). You may also recall that splicing is accomplished by macromolecular machines called **spliceosomes,** and that many primary transcripts can be spliced in more than one way. This turns out to be a major way of regulating eukaryotic gene expression.

During splicing, gene expression is regulated when selected exons are removed from the primary transcript along with the introns. As a result, the same primary RNA transcript can yield more than one kind of mature, processed mRNA, consisting of different combinations of exons.

This is important. Since these mature mRNAs contain differences in their sequences, the polypeptides translated from them will likewise differ. Splicing the same primary RNA transcript in different ways to produce different mature mRNAs and thus different proteins is referred to as **alternative splicing.**

To see how alternative splicing works, consider the muscle-cell protein tropomyosin. The tropomyosin gene is expressed in both skeletal muscle cells and smooth muscle cells. These cells make up two distinct kinds of muscle tissue. Skeletal muscle is responsible for moving your bones; smooth muscle lines many parts of your gut and certain blood vessels.

As **FIGURE 19.9a** shows, the primary transcript from the tropomyosin gene contains 14 exons. In each type of muscle cell, a different subset of the 14 exons are spliced together to produce two different mRNAs (**FIGURE 19.9b**). As a result of alternative splicing, the tropomyosin proteins found in these two cell types are distinct. One reason skeletal muscle and smooth muscle are different is that they contain different types of tropomyosin.

Alternative splicing is controlled by proteins that bind to RNAs in the nucleus and interact with spliceosomes to influence which sequences are used for splicing. When cells that are destined to become skeletal muscle or smooth muscle are developing, they receive signals leading to the production of specific proteins that are active in the regulation of splicing. Instead of transcribing different tropomyosin genes, the cells transcribe a single gene and splice the same primary RNA transcript in different ways.

Before the importance of alternative splicing was widely appreciated, a gene was considered to be a nucleotide sequence that encodes one specific protein or RNA, along with its regulatory sequences. Based on this view, estimates for the number of genes in the human genome were typically in the range of 60,000 to 100,000. But once the complete human genome sequence became available, researchers realized that we may have as few as 20,000 sequences for primary mRNA transcripts.

Even though our genomes contain a relatively low number of genes, recent data indicate that over 90 percent of them undergo alternative splicing to produce multiple products. Thus, the number of different proteins that your cells can produce is far larger than the number of genes.

Given the extent of alternative splicing, the definition of protein-coding genes has been changed to the coding and regulatory sequences that direct the production of one *or more* related mRNAs and polypeptides.

Many alternatively spliced genes produce just a few different products, but some can produce a bewildering array of mRNAs. The current record is the *Dscam* gene of the fruit fly *Drosophila melanogaster*. The products of this gene help to guide growing nerve cells within the embryo. The primary transcript can be spliced into more than 38,000 distinct forms of mRNA, and the *Dscam* gene can produce thousands of different products.

Alternative splicing is a major mechanism in the control of gene expression in multicellular eukaryotes. ✔ If you understand alternative splicing, you should be able to explain why it does not occur in bacteria and describe where it occurs in Figure 19.1.

mRNA Stability and RNA Interference

Once splicing is complete and processed mRNAs are exported to the cytoplasm, new regulatory mechanisms come into play. For example, it has long been known that the life span of an mRNA in a cell can vary. The mRNA for casein—the major protein in milk—is produced in the mammary gland tissue of female mammals. Normally, casein mRNA persists in cells for only about an hour, and little casein protein is produced. But when a female mouse is lactating, regulatory molecules help the mRNAs persist almost 30 times longer—leading to a huge increase in the production of casein. In this instance, shortening of the poly(A) tail decreases the life-span of the mRNA.

More recent discoveries in a variety of organisms, from worms to plants to people, have uncovered a widespread and important form of post-transcriptional gene regulation known as **RNA interference.** RNA interference occurs when a tiny, single-stranded RNA held by a protein complex binds to a complementary sequence in an mRNA. This event unleashes either the destruction of the mRNA or a block to the mRNA's translation. How does it work?

(a) Tropomyosin gene

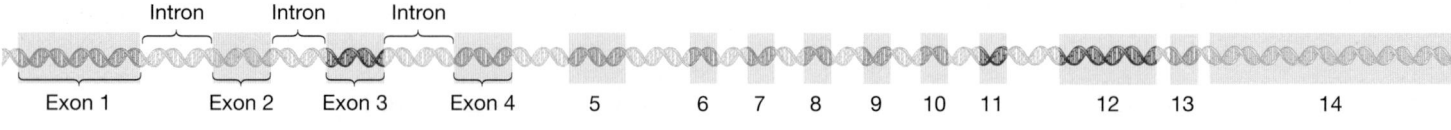

(b) Alternative splicing produces more than one mature mRNA.

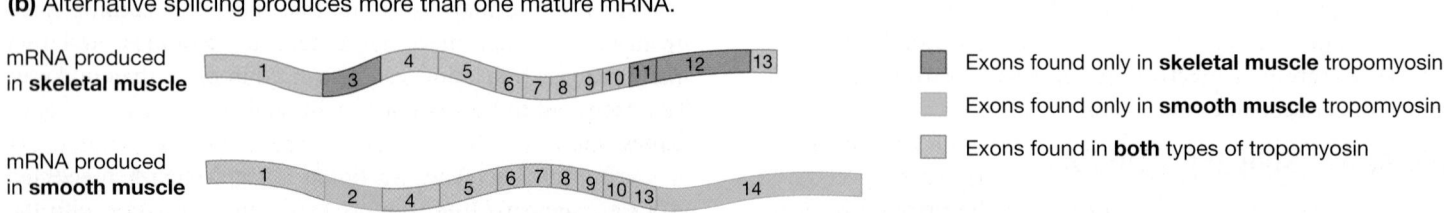

mRNA produced in **skeletal muscle**

mRNA produced in **smooth muscle**

Exons found only in **skeletal muscle** tropomyosin

Exons found only in **smooth muscle** tropomyosin

Exons found in **both** types of tropomyosin

FIGURE 19.9 Alternative Splicing Produces More than One Mature mRNA from the Same Gene.

FIGURE 19.10 walks through the sequence of events.

Step 1 RNA interference begins when RNA polymerase transcribes genes that code for RNAs that double back on themselves to form a hairpin. Hairpin formation occurs because pairs of bases within the RNA transcript are complementary.

Step 2 Some of the RNA is trimmed by enzymes in the nucleus; then the double-stranded hairpin that remains is exported to the cytoplasm.

Step 3 In the cytoplasm, another enzyme cuts out the hairpin loop to form double-stranded RNA molecules that are only about 22 nucleotides long.

Step 4 One of the strands from this short RNA is taken up by a group of proteins called the *R*NA-*i*nduced *s*ilencing *c*omplex, or RISC. The RNA strand held by the RISC is a **microRNA (miRNA).**

Step 5 Once it is part of a RISC, the miRNA binds to its complementary sequences in a target mRNA.

Step 6 If the match between a miRNA and an mRNA is perfect, an enzyme in the RISC destroys the mRNA by cutting it in two. In effect, tight binding by a miRNA is a "kiss of death" for the mRNA. If the match isn't perfect, however, the mRNA is not destroyed. Instead, its translation is inhibited. Either way, miRNAs "interfere" with gene expression.

The first papers on RNA interference were published in the mid-1990s, and the first miRNAs were characterized in 2001. Since then, research on miRNAs and RNA interference has exploded. Current data suggest that a typical animal or plant species has at least 500 genes that code for miRNAs and that each miRNA regulates more than one mRNA. Because of this evidence, it is estimated that a large percentage of all animal and plant genes are controlled by these tiny molecules. miRNAs are critical for normal development, and mutations in miRNA genes are associated with many diseases. RNA interference is increasingly recognized as a key aspect of post-transcriptional control.

Researchers are currently testing whether certain miRNAs could be used to knock out specific genes associated with illness, or to destroy mRNAs produced by viruses during an infection. Research on RNA interference has quickly moved from an exciting new frontier in basic biology to possible applications in medicine.

✔ If you understand RNA interference, you should be able to describe where it occurs in Figure 19.1.

How Is Translation Controlled?

RNA interference is not the only mechanism of gene control that acts on mRNAs and affects whether translation occurs. For example, cells may slow or stop translation in response to a sudden increase in temperature or infection by a virus. The slowdown occurs because regulatory proteins that are activated by the temperature spike or viral invasion add a phosphate group to a protein that is part of the ribosome.

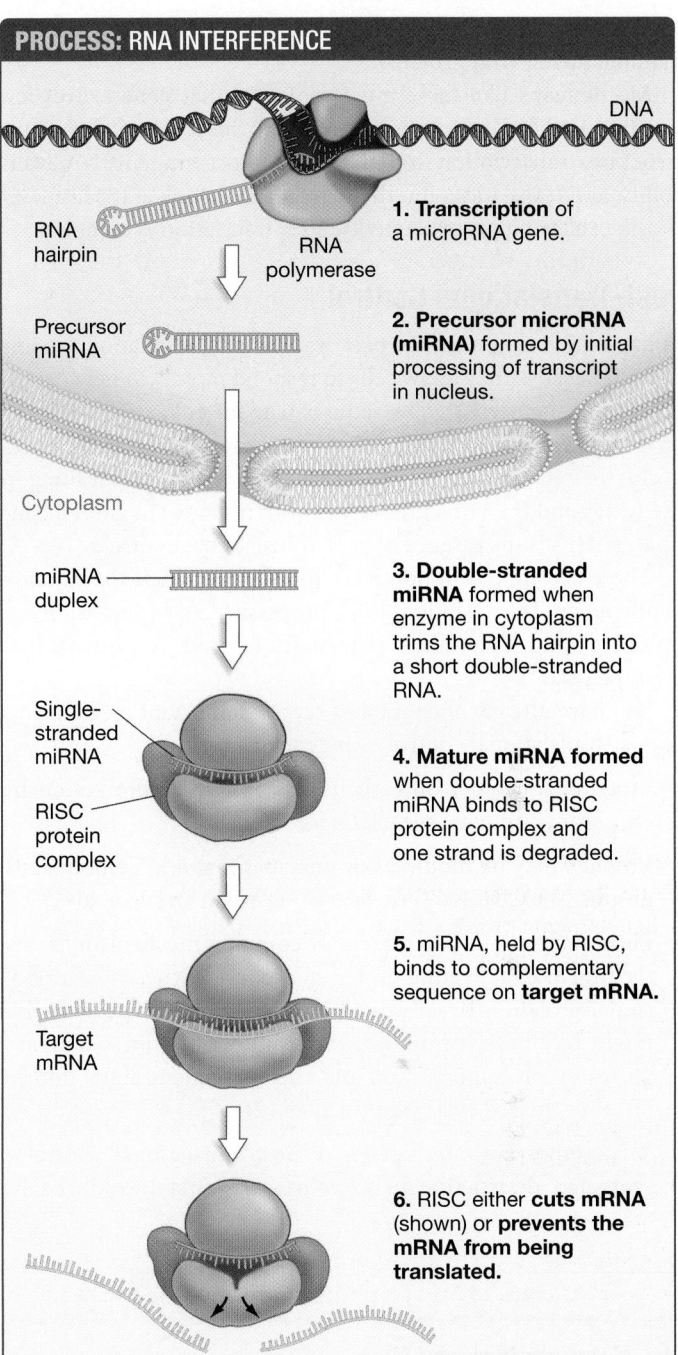

PROCESS: RNA INTERFERENCE

DNA

RNA hairpin

RNA polymerase

1. Transcription of a microRNA gene.

Precursor miRNA

2. Precursor microRNA (miRNA) formed by initial processing of transcript in nucleus.

Cytoplasm

miRNA duplex

3. Double-stranded miRNA formed when enzyme in cytoplasm trims the RNA hairpin into a short double-stranded RNA.

Single-stranded miRNA

RISC protein complex

4. Mature miRNA formed when double-stranded miRNA binds to RISC protein complex and one strand is degraded.

5. miRNA, held by RISC, binds to complementary sequence on **target mRNA.**

Target mRNA

6. RISC either **cuts mRNA** (shown) or **prevents the mRNA from being translated.**

FIGURE 19.10 MicroRNAs Either Target mRNAs for Destruction or Prevent Their Translation. MicroRNAs are held by the RISC protein complex and bind to target mRNAs by complementary base pairing.

You might recall that phosphorylation frequently leads to changes in the shape and chemical reactivity of proteins. In the case of the phosphorylated ribosomal protein, the shape change slows or prevents translation.

For the cell, this dramatic change in gene expression can mean the difference between life and death. High temperatures disrupt protein folding, so shutting down translation prevents the production of improperly folded polypeptides. If the problem is an invading virus, the cell stops the infection because it avoids manufacturing viral proteins.

Mechanisms like these are a reminder that gene expression can be regulated at multiple points: at the level of chromatin structure, transcription initiation, RNA processing, mRNA availability, and translation. But that's not all. Let's look at the last level possible: altering protein activity, after translation is complete.

Post-Translational Control

In bacteria, mechanisms of post-translational regulation are important because they allow cells to respond rapidly to new conditions (see Chapter 18). The same is true for eukaryotes. Instead of waiting for transcription, RNA processing, and translation to occur, the cell can keep an existing but inactive protein waiting in the wings and then quickly activate it in response to altered conditions. This is the essence of post-translational control.

There is a trade-off, however: Speed is gained at the expense of efficiency. Transcription, RNA processing, and translation use up energy and materials; it is wasteful to produce proteins that won't be used.

You have already encountered several important mechanisms of post-translational control over gene expression.

- Proteins are folded into their final, active conformation by chaperone proteins (see Chapters 3 and 17).

- Proteins may be modified by enzymes that add carbohydrate groups (see Chapter 7) or cleave off certain amino acids.

- Phosphorylation is an extremely common mechanism for activating or deactivating proteins. You just learned how a ribosomal protein is deactivated by adding a phosphate. You also might recall discussion of the activation of cyclin–Cdk complexes by phosphorylation and the subsequent entry into M phase of the cell cycle (see Chapter 12).

Yet another key mechanism of post-translational control—the targeted destruction of proteins—was first introduced by

describing the short life span of cyclin proteins. When a protein such as a cyclin needs to be destroyed, enzymes mark it by adding many copies of a small polypeptide called ubiquitin. Ubiquitin got its name because it is ubiquitous in cells. A macromolecular machine called the **proteasome** recognizes proteins that have a ubiquitin tag and cuts them into short segments.

As you can see, the regulation of gene expression in eukaryotes includes everything from opening chromatin in the nucleus to controlling the life span of proteins.

Do bacteria use the same range of regulatory mechanisms? Let's explore this question next.

19.5 How Does Gene Expression Compare in Bacteria and Eukaryotes?

Almost as soon as biologists knew that information in DNA is transcribed into RNA and then translated into proteins, they began asking questions about how that flow of information is regulated. **TABLE 19.1** summarizes what biologists have learned over the past half century about how bacterial and eukaryotic gene expression is controlled—organized by the six steps in gene expression introduced in Figure 19.1.

How does the regulation of gene expression differ in bacteria and eukaryotes? Biologists point to five fundamental differences in the control of gene expression in bacteria and eukaryotes:

1. *DNA Packaging* The chromatin of eukaryotic DNA must be decondensed for basal and regulatory transcription factors to gain access to genes and for RNA polymerase to initiate transcription. The tight packaging of eukaryotic DNA means that the default state of transcription in eukaryotes is "off." In contrast, the default state of transcription in bacteria, which lack histone proteins and have freely accessible promoters, is "on." Chromatin structure provides a mechanism of negative control that does not exist in bacteria.

2. *Complexity of transcription* Transcriptional control is much more complex in eukaryotes than in bacteria. The sheer number of eukaryotic proteins involved in regulating transcription dwarfs that in bacteria, as does the complexity of their interactions.

3. *Coordinated transcription* In bacteria, genes that take part in the same cellular response are often organized into operons and transcribed together from a single promoter. In contrast, operons are rare in eukaryotes. Instead, for coordinated gene expression, eukaryotes rely on the strategy used in bacterial regulons—physically scattered genes are expressed together when the same regulatory transcription factors trigger the transcription of genes with the same DNA regulatory sequences.

4. *Greater reliance on post-transcriptional control* Eukaryotes make greater use of post-transcriptional control, such as alternative splicing. Alternative splicing allows eukaryotes to regulate the production of many proteins from each gene.

Level of Regulation	Bacteria	Eukaryotes
Chromatin remodeling	• Limited packaging of DNA • Remodeling not a major issue in regulating gene expression.	• Extensive packaging of DNA • Chromatin must be decondensed for transcription to begin.
Transcription	• Positive and negative control by regulatory proteins that act at sites close to the promoter • Sigma interacts with promoter.	• Positive and negative control by regulatory proteins that act at sites close to *and* far from promoter • Large set of basal transcription factors interact with promoter. • Mediator required.
RNA processing	• Rare	• Extensive processing: alternative splicing of introns
mRNA stability	• Rarely used for control	• Commonly used: RNA interference limits life span or translation rate of many mRNAs.
Translation	• Regulatory proteins bind to mRNAs and ribosomes and affect translation rate.	• Regulatory proteins bind to mRNAs and ribosomes and affect translation rate.
Post-translational modification	• Folding by chaperone proteins • Chemical modification (e.g., phosphorylation) changes protein activity.	• Folding by chaperone proteins • Chemical modification (e.g., phosphorylation) changes protein activity. • Ubiquitination targets proteins for destruction by proteasome.

Alternative splicing, microRNAs, and regulation of mRNA stability are seldom seen in bacteria, but these constitute major elements of control in eukaryotes.

To date, biologists do not have a good explanation for why gene expression is so much more complex in eukaryotes than it is in bacteria. After decades of research, the debate continues.

For multicellular eukaryotes, one hypothesis is that the requirement for complex regulation of gene expression during development has driven the evolution of such complicated and multilayered means of controlling genes.

Normal regulation of gene expression results in the orderly development of an embryo and appropriate responses to environmental change in adults. What happens when gene expression goes awry? Unfortunately, one answer is uncontrolled cell growth and the set of diseases called cancer. Understanding how changes in gene expression can lead to cancer is one of today's great research frontiers.

19.6 Linking Cancer with Defects in Gene Regulation

All cancers involve uncontrolled cell division. What allows this unbridled increase in cell number? Each type of cancer is caused by a different set of mutations that lead to cancer when they affect one of two classes of genes: (1) genes that stop or slow the cell cycle, and (2) genes that trigger cell growth and division by initiating specific phases in the cell cycle. Many of the genes that are mutated in cancer influence gene regulation. Let's take a closer look at how altered gene regulation can cause uncontrolled cell growth.

The Genetic Basis of Uncontrolled Cell Growth

As you learned in the chapter on the cell cycle (Chapter 12), proteins that stop or slow the cell cycle when conditions are unfavorable for cell division are called tumor suppressors. Logically enough, the genes that code for these proteins are called **tumor suppressor** genes. If the function of a tumor suppressor gene is lost because of mutation, then a key brake on the cell cycle is eliminated.

Genes that stimulate cell division are called **proto-oncogenes** (literally, "first cancer genes"). In normal cells, the proteins produced from proto-oncogenes are active only when conditions are appropriate for growth. In cancerous cells, defects in the regulation of proto-oncogenes can cause these genes to stimulate growth at all times. (In the context of cancer, cell growth refers to an increase in cell numbers, not an increase in the size of individual cells.) In such cases, a mutation has converted the proto-oncogene into an **oncogene**—an allele that promotes cancer development.

For cancers to develop, many mutations are required within a single cell, and these alter both tumor suppressor genes and proto-oncogenes.

The *p53* Tumor Suppressor: A Case Study

To gain a deeper understanding of how defects in gene expression can lead to cancer, consider research on the gene that is most often defective in human cancers. The gene is called *p53* because

(a) Normal cell

UV light

Damaged DNA

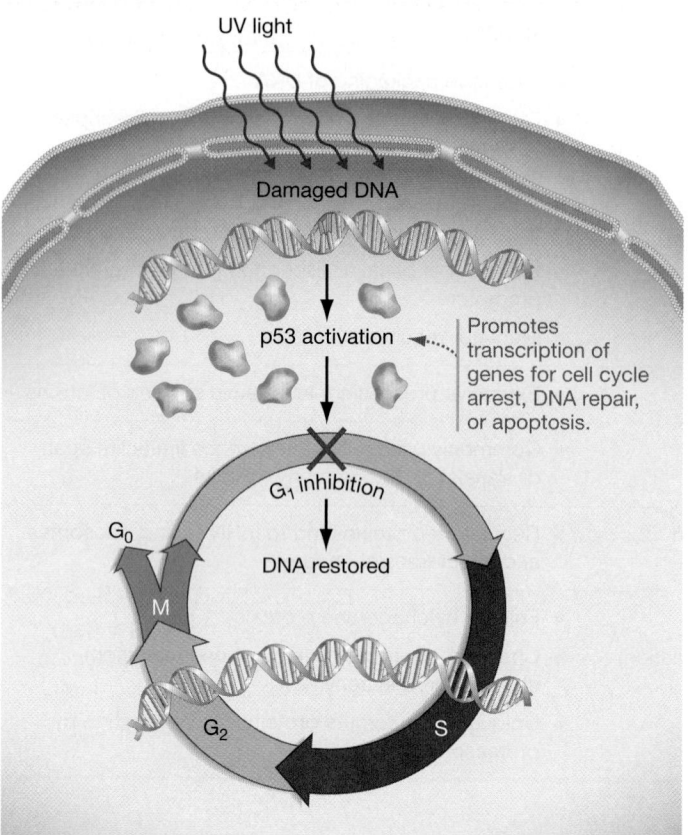

p53 activation ←---- Promotes transcription of genes for cell cycle arrest, DNA repair, or apoptosis.

G₁ inhibition

G₀

DNA restored

M

G₂

S

(b) *p53* mutant cell

UV light

Damaged *p53* gene

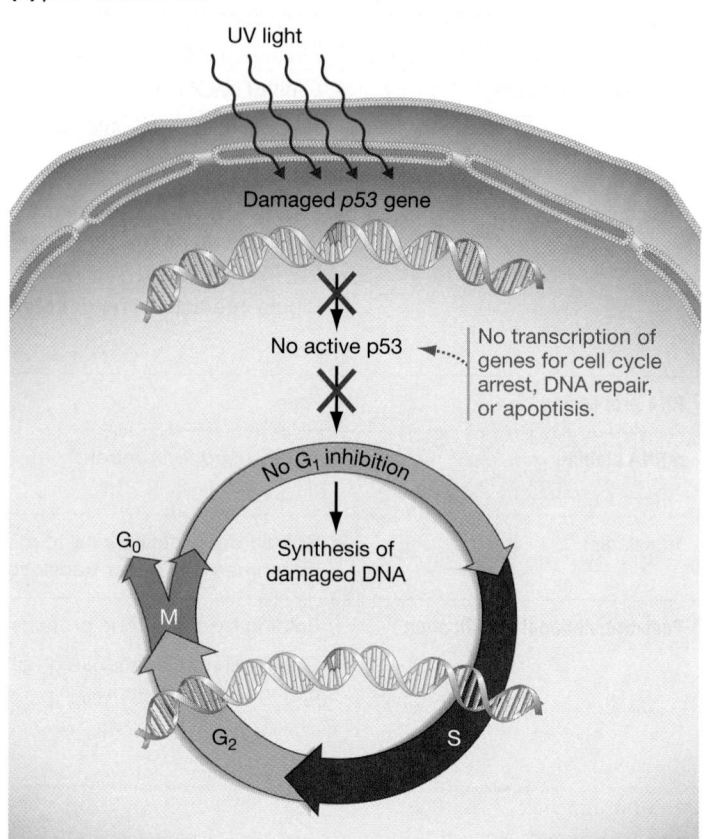

No active p53 ←---- No transcription of genes for cell cycle arrest, DNA repair, or apoptisis.

No G₁ inhibition

G₀

Synthesis of damaged DNA

M

G₂

S

FIGURE 19.11 Consequences of *p53* Mutation.

the protein it codes for has a molecular weight of approximately 53 kilodaltons. Sequencing studies have revealed that mutant, nonfunctional forms of *p53* are found in over half of all human cancers. The *p53* gene codes for a regulatory transcription factor.

Researchers began to understand what *p53* does when they exposed normal, noncancerous human cells to UV radiation and noticed that levels of active *p53* protein increased markedly. Recall that UV radiation damages DNA (see Chapter 15). Follow-up studies confirmed that there is a close correlation between DNA damage and the amount of *p53* in a cell. In addition, analyses of the protein's primary structure suggested that it might contain a DNA-binding region similar to the one shown for the muscle-specific transcription factor in Figure 19.7.

These observations inspired the hypothesis that *p53* is a regulatory transcription factor that serves as a master brake on the cell cycle. In this model, shown in **FIGURE 19.11**, *p53* is activated by DNA damage. Activated *p53* binds to the enhancers of genes that arrest the cell cycle, repair DNA damage, and when all else fails, trigger apoptosis (cell death). Expression of these genes allows the cell to halt the cell cycle in order to repair its DNA, if this is possible, or commit suicide if the DNA damage is too severe. In mutant cells that lack a form of *p53* that can bind to enhancers, DNA damage cannot arrest the cell cycle, the cell cannot kill

check your understanding

C Y U

If you understand that . . .

- Many mutations associated with cancer alter gene regulation.
- Cancer is associated with mutations that lead to loss of control over the cell cycle.
- Uncontrolled cell growth may result when a mutation in a regulatory gene creates a protein that activates the cell cycle constitutively.
- Uncontrolled cell growth may result when a mutation prevents a tumor suppressor protein from shutting down the cell cycle in damaged cells.

✔ **You should be able to . . .**

1. Explain why cancer has a common pattern of uncontrolled cell growth, but not a common cause. Your answer should refer to the six levels of gene regulation outlined in Figure 19.1.

2. Explain why loss-of-function mutations in *p53* are observed in so many cancers.

Answers are available in Appendix A.

itself, and damaged DNA is replicated. This situation leads to mutations that can move the cell farther down the road to cancer. The *p53* protein is like a quality control officer. If it is missing, errors are made and things go downhill.

Here are some of the results that support this model of *p53* function:

- *p53* activates many different genes, including genes for cell cycle regulation, DNA repair, and apoptosis.

- X-ray crystallography studies show that *p53* binds directly to DNA regulatory sequences of the genes it controls.

- Virtually all the *p53* mutations associated with cancer are located in the protein's DNA-binding region and alter amino acids that interfere with *p53*'s ability to bind to regulatory DNA sequences.

The role of *p53* in preventing cancer is so important that biologists call this gene "the guardian of the genome." Today, biologists are searching for molecules that can restore *p53* activity to protect the genome and act against cancer.

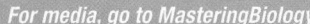

CHAPTER 19 REVIEW

For media, go to MasteringBiology

If you understand . . .

19.1 Gene Regulation in Eukaryotes— An Overview

- Changes in gene expression allow eukaryotic cells to respond to changes in the environment and cause distinct cell types to develop.

- In a multicellular eukaryote, cells are different because they express different genes, not because they have different genes.

- Gene expression is regulated at six levels: Chromatin has to be remodeled, the transcription of specific genes may be initiated or repressed, mRNAs may be spliced in different ways to produce a different product, the life span of specific mRNAs may be extended or shortened, translation rate may be increased or decreased, and the life span or activity of particular proteins may be altered.

✔ You should be able to describe how the presence of the nuclear envelope, and the physical separation of transcription and translation, influences the levels of gene regulation observed in eukaryotes versus bacteria and archaea.

19.2 Chromatin Remodeling

- Eukaryotic DNA is packaged with proteins into chromatin that must be opened before transcription can occur.

- Eukaryotic DNA is wrapped around histone proteins to form bead-like nucleosomes that are then coiled into 30-nm fibers and higher-order chromatin structures.

- Transcription cannot be initiated until chromatin around regulatory regions is decondensed.

- The state of chromatin condensation depends on the methylation of cytosines in DNA, acetylation and other modifications of histones, and the action of molecular machines called chromatin-remodeling complexes.

- Patterns of DNA methylation and histone modifications can be passed from mother cells to daughter cells.

- Epigenetic inheritance is the inheritance of different phenotypes due to anything other than differences in alleles; transmitting patterns of chromatin condensation from mother to daughter cells or from parent to offspring is a mechanism of epigenetic inheritance.

✔ You should be able to explain why chromatin remodeling has to be the first step in gene activation.

19.3 Initiating Transcription: Regulatory Sequences and Regulatory Proteins

- In eukaryotes, transcription is triggered by regulatory proteins called transcription factors that bind to sequences both close to and far from the promoter.

- Regulatory transcription factors can be activators or repressors; these bind to regulatory sequences called (1) promoter-proximal sequences that are near promoters or (2) enhancers and silencers that are often located at a distance from gene promoters.

- Amino acids on regulatory transcription factors interact with the projections of base pairs in the grooves of the DNA helix to allow binding to specific regulatory sequences.

- The first regulatory transcription factors that bind to DNA recruit proteins that loosen the interaction between nucleosomes and DNA, making the promoter, promoter-proximal elements, and enhancers accessible to other transcription factors.

- Interactions between regulatory and basal transcription factors occur through a complex of proteins called Mediator and lead to the positioning of RNA polymerase at the promoter and the start of transcription.

✔ You should be able to draw a model of a eukaryotic gene undergoing transcription. Label enhancers, promoter-proximal elements, the promoter, activators, basal transcription factors, Mediator, and RNA polymerase.

19.4 Post-Transcriptional Control

- Once transcription is complete, gene expression is controlled by (1) alternative splicing, (2) RNA interference, and (3) activation or inactivation of protein products.

- Alternative splicing allows a single gene to produce more than one version of an mRNA and more than one kind of protein. It is regulated by proteins that interact with potential splice sites in the primary transcript to control which ones the spliceosome uses.

- RNA interference occurs when tiny strands of complementary RNA bind to mRNAs in association with the protein complex called RISC. This marks the mRNAs for destruction or prevents their translation. If the short RNA strands come from a transcribed cellular gene, they are known as microRNAs (miRNAs).

- Once translation occurs, proteins may be activated or inactivated by the addition or removal of chemical groups such as phosphates, or marked for destruction in the proteasome by adding polypeptides known as ubiquitin.

✓ You should be able to explain why humans can have so few genes.

19.5 How Does Gene Expression Compare in Bacteria and Eukaryotes?

- Examine Table 19.1 Regulating Gene Expression in Bacteria and Eukaryotes.

✓ You should be able to compare and contrast how bacteria and eukaryotes regulate the transcription of genes that need to be turned on together.

19.6 Linking Cancer with Defects in Gene Regulation

- If mutations alter regulatory proteins that promote or inhibit progression through the cell cycle, then uncontrolled cell growth and tumor formation may result.

✓ You should be able to explain why mutations of *p53* that prevent the protein from binding to DNA set the stage for cancer.

(MB) **MasteringBiology**

1. MasteringBiology Assignments

Tutorials and Activities Control of Gene Expression; Control of Transcription; DNA Packing; Overview: Control of Gene Expression; Post-Transcriptional Control Mechanisms; Regulation of Gene Expression in Eukaryotes; Transcription Initiation in Eukaryotes

Questions Reading Quizzes, Blue-Thread Questions, Test Bank

2. eText Read your book online, search, take notes, highlight text, and more.

3. The Study Area Practice Test, Cumulative Test, BioFlix® 3-D Animations, Videos, Activities, Audio Glossary, Word Study Tools, Art

You should be able to . . .

✓ **TEST YOUR KNOWLEDGE**   *Answers are available in Appendix A*

1. What is chromatin?
   a. the histone-containing protein core of the nucleosome
   b. the 30-nm fiber
   c. the complex of DNA and proteins found in eukaryotes
   d. the histone *and* non-histone proteins in eukaryotic nuclei

2. What is a tumor suppressor?

3. Which of the following statements about enhancers is correct?
   a. They contain a unique base sequence called a TATA box.
   b. They are located only in 5′-flanking regions.
   c. They are located only in introns.
   d. They are found in a variety of locations and are functional in any orientation.

4. In eukaryotes, what allows only certain genes to be expressed in certain types of cells?

5. What is alternative splicing?
   a. phosphorylation events that lead to different types of post-translational regulation
   b. mRNA processing events that lead to different combinations of exons being spliced together
   c. folding events that lead to proteins with alternative conformations
   d. action by regulatory proteins that leads to changes in the life span of an mRNA

6. What types of proteins bind to promoter-proximal elements?
   a. the basal transcription complex
   b. the basal transcription complex plus RNA polymerase
   c. basal transcription factors such as TBP
   d. regulatory transcription factors such as activators

7. Compare and contrast (a) enhancers and the *E. coli araC* binding site (see Chapter 18), (b) promoter-proximal elements and the operator of the *lac* operon, and (c) basal transcription factors and sigma.

8. Explain how alternative splicing could play a role in changing eukaryotic gene expression in response to changes in the environment.

9. Compare and contrast (a) enhancers and silencers; (b) promoter-proximal elements and enhancers; and (c) regulatory transcription factors and Mediator.

10. Predict how a drug that inhibits histone deacetylase will alter gene expression.

11. Relative to the genetic code, the histone code
   a. has more triplets.
   b. is much simpler to read because of complementary base pairing.
   c. does not depend on particular base sequences.
   d. requires methylated Cs rather than standard Cs in DNA.

12. Predict how a mutation that caused continuous production of active *p53* would affect the cell.

13. In the follow-up work to the experiment shown in Figure 19.5, the researchers used a technique that allowed them to see if two DNA sequences are in close physical proximity (association). They applied this method to examine how often an enhancer and the promoter of the *Hnf4a* regulatory gene were near each other. A logical prediction is that compared with rats born to mothers fed a healthy diet, the *Hnf4a* gene in rats born to mothers fed a protein poor diet would
   a. show no difference in how often the promoter and enhancer associated.
   b. never show any promoter–enhancer association.
   c. show a lower frequency of promoter–enhancer association.
   d. show a higher frequency of promoter–enhancer association.

14. **QUANTITATIVE** Imagine repeating the experiment on epigenetic inheritance that is shown in Figure 19.5. You measure the amount of radioactive uridine (U) incorporated into *Hnf4a* mRNA in counts per minute (cpm) to determine the level of *Hnf4a* gene transcription in rats born to mothers fed either a normal diet or a low protein diet. The results are 11,478 cpm for the normal diet

and 7368 cpm for the low-protein diet conditions. You should prepare a graph similar to the one at the bottom of Figure 19.5 that shows the normalized results relative to the normal diet condition. Normalizing values means that value obtained from one condition is expressed as 1.0 (the norm; the normal diet in this case) and the values obtained from any other conditions (low-protein diet in this case) are expressed as decimal values relative to the norm.

15. After DNA damage, levels of activated p53 protein in the cytoplasm increase. Design an experiment to determine whether this increase is due to increased transcription of the *p53* gene or to activation of preexisting p53 proteins by a post-translational mechanism such as phosphorylation.

16. Researchers have discovered that if a single-stranded mRNA produced by a virus is used as a template to create a double-stranded RNA, that this double-stranded RNA often blocks infection by the virus. Propose an explanation for how infection can be blocked by this type of double-stranded RNA.

The Big Picture

Copying, using, and transmitting genetic information is fundamental to life. Cells use the genetic information archived in their DNA to respond to changes in the environment and, in multicellular organisms, to develop into specific cell types.

Hereditary information is transmitted to offspring with random changes called mutations. Thus, genetic information is dynamic—both within generations and between generations.

Note that each box in the concept map indicates the chapter and section where you can go for review. Also, be sure to do the blue exercises in the Check Your Understanding box below.

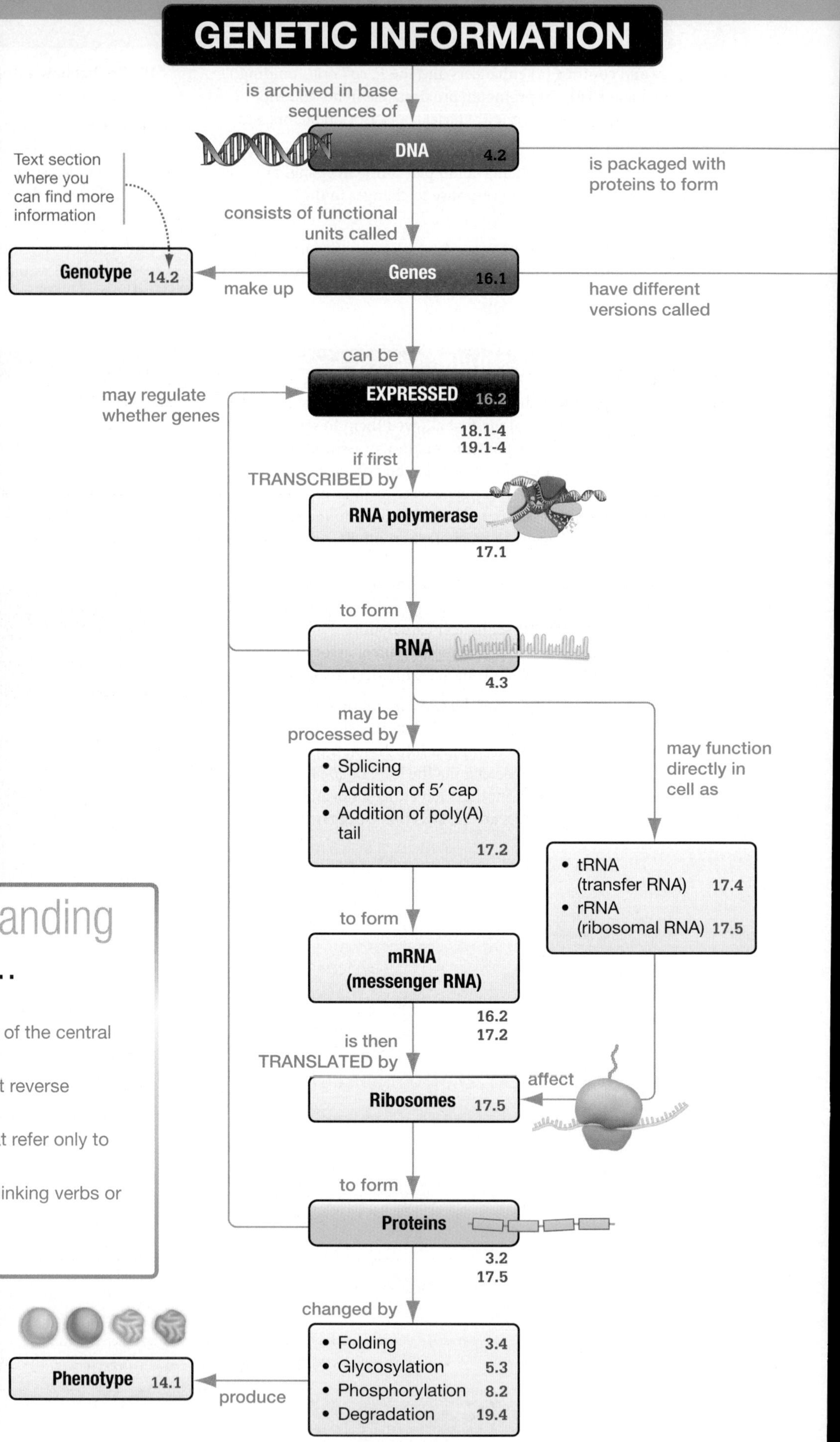

GENETIC INFORMATION

is archived in base sequences of

DNA 4.2 → is packaged with proteins to form

consists of functional units called

Text section where you can find more information ⟶ **Genotype** 14.2 ← make up **Genes** 16.1 → have different versions called

can be

may regulate whether genes → **EXPRESSED** 16.2
18.1-4
19.1-4

if first TRANSCRIBED by

RNA polymerase 17.1

to form

RNA 4.3

may be processed by

- Splicing
- Addition of 5′ cap
- Addition of poly(A) tail 17.2

may function directly in cell as

- tRNA (transfer RNA) 17.4
- rRNA (ribosomal RNA) 17.5

to form

mRNA (messenger RNA) 16.2 17.2

is then TRANSLATED by

Ribosomes 17.5 ← affect

to form

Proteins 3.2 17.5

changed by

- Folding 3.4
- Glycosylation 5.3
- Phosphorylation 8.2
- Degradation 19.4

produce → **Phenotype** 14.1

check your understanding

If you understand the big picture . . .

✔ You should be able to . . .

1. Draw stars next to the three elements of the central dogma of molecular biology.

2. Add arrows and labels indicating what reverse transcriptase does.

3. Draw an E in the corners of boxes that refer only to eukaryotes, not prokaryotes.

4. Fill in the blue ovals with appropriate linking verbs or phrases.

Answers are available in Appendix A.

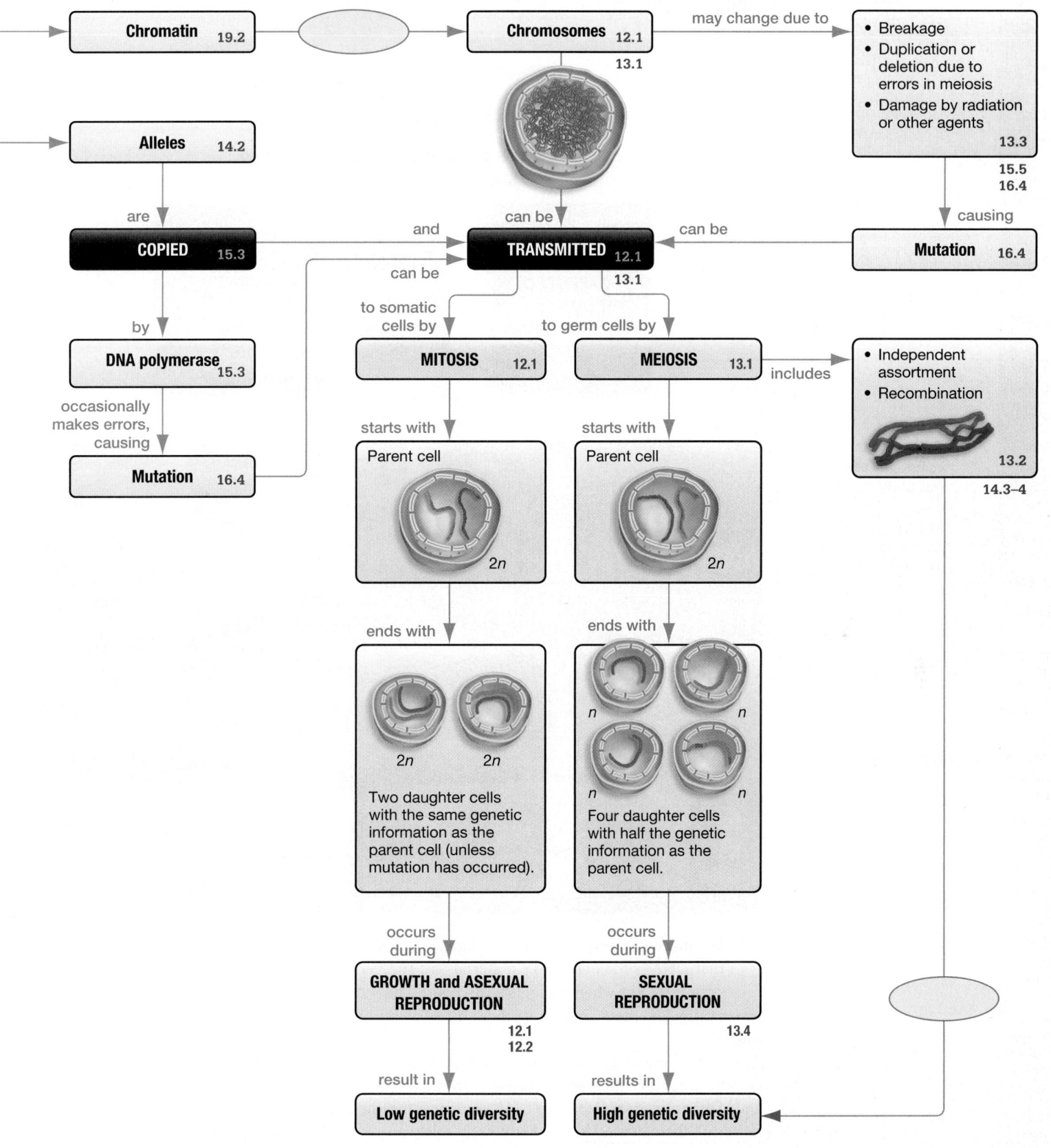

Chromatin 19.2

Chromosomes 12.1 13.1

may change due to

- Breakage
- Duplication or deletion due to errors in meiosis
- Damage by radiation or other agents
  13.3
  15.5
  16.4

Alleles 14.2

are

COPIED 15.3

and

TRANSMITTED 12.1 13.1

can be

Mutation 16.4

causing

by

DNA polymerase 15.3

occasionally makes errors, causing

Mutation 16.4

can be

to somatic cells by

to germ cells by

includes

- Independent assortment
- Recombination
  13.2
  14.3–4

MITOSIS 12.1

MEIOSIS 13.1

starts with

Parent cell

2n

starts with

Parent cell

2n

ends with

2n   2n

Two daughter cells with the same genetic information as the parent cell (unless mutation has occurred).

ends with

n   n

n   n

Four daughter cells with half the genetic information as the parent cell.

occurs during

GROWTH and ASEXUAL REPRODUCTION
12.1
12.2

occurs during

SEXUAL REPRODUCTION
13.4

result in

Low genetic diversity

results in

High genetic diversity

20 Analyzing and Engineering Genes

In this chapter you will learn that

Biotechnology depends on methods to analyze and alter genomes

by exploring

Case studies of key genetic technologies

applied to

Recombinant DNA technologies **20.1**

Finding genes by mapping **20.4**

Gene therapy **20.5**

Genome analysis

Medicine

Agriculture

The polymerase chain reaction **20.2**

Dideoxy DNA sequencing **20.3**

Transgenic crops **20.6**

The rice plants in these bottles have been genetically engineered—using techniques introduced in this chapter—to produce a molecule needed for a key vitamin.

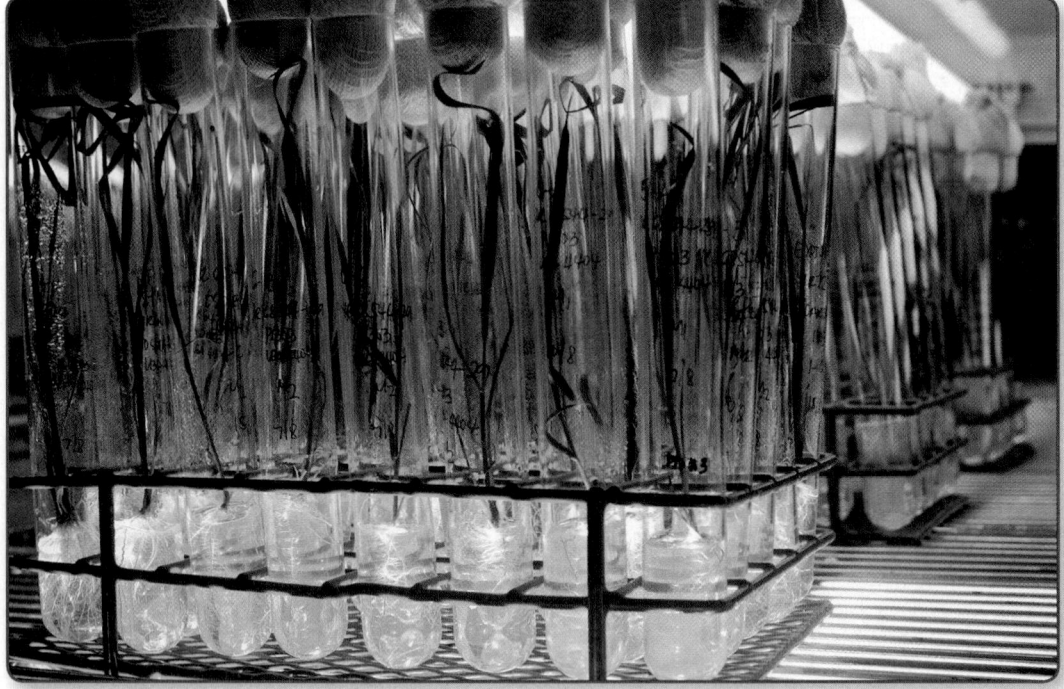

The molecular revolution in biological science got its start when researchers confirmed that DNA is the hereditary material and succeeded in describing the molecule's secondary structure. But when biologists discovered how to remove DNA sequences from an organism, manipulate them, and insert them into different individuals, the molecular revolution really took off.

Efforts to manipulate DNA sequences in organisms are often referred to as genetic engineering. Genetic engineering became possible with the discovery of enzymes that cut DNA at specific sites and of other enzymes that paste DNA sequences together. These new molecular tools were extremely powerful. Biologists no longer had to rely solely on controlled breeding experiments to change the genetic characteristics of individuals. Instead, they could mix and match specific DNA sequences in

✔ When you see this checkmark, stop and test yourself. Answers are available in Appendix A.

the lab. Because successful efforts to manipulate genes usually result in novel combinations of DNA, techniques used to engineer genes are often referred to as **recombinant DNA technology.**

This chapter uses a series of case histories to introduce basic molecular biology techniques in the context of solving problems. It also considers the ethical and economic issues raised by efforts to manipulate genes. What are the potential perils and benefits of introducing recombinant genes into human beings, food plants, and other organisms? This question, one of the great ethical challenges of the twenty-first century, is a recurrent theme in the following pages.

20.1 Case 1—The Effort to Cure Pituitary Dwarfism: Basic Recombinant DNA Technologies

To understand the basic techniques and tools of genetic engineering, let's consider the role they played in developing a treatment for pituitary dwarfism in humans.

The pituitary gland is a structure at the base of the mammalian brain that produces several important biomolecules, including a protein that stimulates growth. This protein, which was found to be just 191 amino acids long, was named human growth hormone (HGH). In humans, the gene that codes for it is called *GH1*.

The discovery of growth hormone led researchers immediately to suspect that at least some forms of inherited dwarfism might be caused by a defect in the *GH1* gene. This hypothesis was confirmed by studies showing that people with certain types of dwarfism produce little growth hormone or none at all. These people have defective copies of *GH1* and exhibit pituitary dwarfism, type I (**FIGURE 20.1a**).

By studying the pedigrees of families in which dwarfism was common, several teams of researchers established that pituitary dwarfism, type I, is an autosomal recessive trait (see Chapter 14). In other words, affected individuals have two copies of the defective allele. Individuals who are affected by pituitary dwarfism have normal body proportions but grow more slowly than average people, reach puberty from two to 10 years later than average, and are short in stature as adults—typically no more than 120 cm (4 feet) tall (**FIGURE 20.1b**).

Why Did Early Efforts to Treat the Disease Fail?

Once the molecular basis of pituitary dwarfism was understood, physicians began treating the disease with injections of naturally produced growth hormone. This approach was inspired by the spectacular success that had been achieved in treating type I diabetes mellitus. This form of diabetes is caused by a deficiency of the peptide hormone insulin, and clinicians had been able to alleviate the disease's symptoms by injecting patients with insulin from pigs.

Early trials showed that people with pituitary dwarfism could be treated successfully with growth hormone therapy, but only if the protein came from humans. Growth hormones isolated from pigs, cows, or other animals were ineffective. Until the 1980s, however, the only source of human growth hormone was pituitary glands from human cadavers. As a result, the drug was extremely scarce and expensive.

It turned out that meeting demand was the least of the problems with growth hormone therapy. To understand why, recall that infectious proteins called prions can cause degenerative brain disorders in mammals (see Chapter 3). When some of the children treated with human growth hormone developed a prion disease in their teens and twenties, physicians realized that the supply of growth hormone was contaminated with a prion protein from

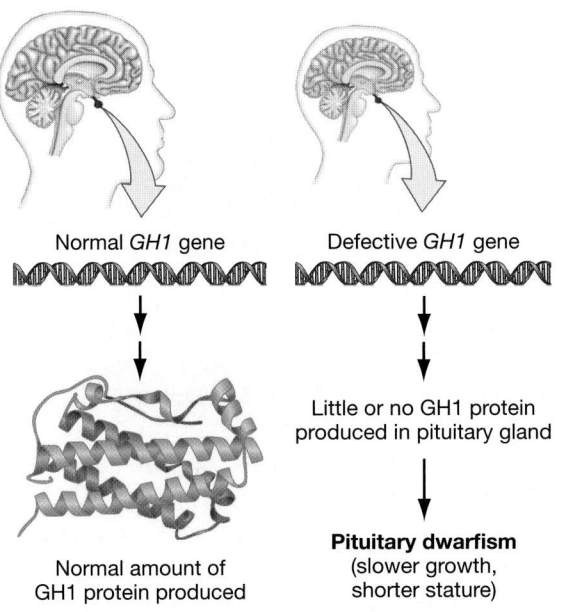

(a) *GH1* codes for a pituitary growth hormone.

Normal *GH1* gene

Defective *GH1* gene

Little or no GH1 protein produced in pituitary gland

Normal amount of GH1 protein produced

Pituitary dwarfism (slower growth, shorter stature)

(b) Normal versus GH1-deficient

FIGURE 20.1 Pituitary Dwarfism Is a Genetic Disease. (a) If mutations in the human *GH1* sequence are severe enough to inactivate the gene, pituitary dwarfism may result. **(b)** William Harrison and Charles Stratton, in a photo taken about 1860. Harrison and Stratton were both celebrated comedians and performers. Stratton, whose stage name was Tom Thumb, enjoyed audiences in the White House with Abraham Lincoln and Buckingham Palace with Queen Victoria. Stratton had pituitary dwarfism; Harrison had normal height.

some of the cadavers supplying the hormone. In 1984, the use of growth hormone isolated from cadavers was banned.

Steps in Engineering a Safe Supply of Growth Hormone

To replace natural sources of growth hormone, researchers turned to genetic engineering. Their plan was to insert fully functional copies of human *GH1* into the bacterium *Escherichia coli*, which they hoped would then produce huge quantities of recombinant progeny. If the plan worked, the recombinant cells would produce uncontaminated growth hormone in sufficient quantities to meet demand at an affordable price.

The plan required investigators to find *GH1*, obtain many copies of the gene, and insert them into *E. coli* cells. Their ability to do these things hinged on using basic tools of molecular biology.

Using Reverse Transcriptase to Produce cDNAs An enzyme called reverse transcriptase (Chapter 16) is responsible for an exception to the central dogma of molecular biology: It allows information to flow from RNA to DNA. More specifically, reverse transcriptase catalyzes the synthesis of DNA from an RNA template.

DNA that is produced from RNA is called **complementary DNA, or cDNA.** Although reverse transcriptase initially produces a single-stranded cDNA, it is also capable of synthesizing the complementary strand to yield a double-stranded DNA. In many cases, however, researchers add a chemically synthesized primer to single-stranded cDNAs and use DNA polymerase to synthesize the second strand (**FIGURE 20.2**).

Reverse transcriptase played a key role in the search for the growth hormone gene. Knowing that *GH1* is actively transcribed in the pituitary gland, researchers isolated mRNAs from pituitary-gland cells and used the enzyme to reverse-transcribe those mRNAs to cDNAs. These reaction products contained double-stranded cDNAs corresponding to each gene that is actively expressed in pituitary cells.

The next move? Isolating each of the cDNAs and making many identical copies of them.

Using Plasmids in Cloning Producing many copies of a gene is referred to as **DNA cloning.** If a researcher says that she has cloned a gene, it means that she has isolated it and then produced many identical copies.

In many cases, researchers can clone a gene by inserting it into a small, circular DNA molecule called a **plasmid.** You might recall that plasmids are common in bacterial cells (see Chapter 7). They are physically separate from the bacterial chromosome and are not required for normal growth and reproduction. Most replicate independently of the chromosome. Some plasmids carry genes for antibiotic resistance or other traits that increase the cell's ability to grow in a particular environment.

Researchers realized that if they could splice a loose piece of DNA into a plasmid and then insert the modified plasmid into a bacterial cell, the engineered plasmid would be replicated and passed on to daughter cells as the bacterium grew and divided. If this recombinant bacterium were then placed in a nutrient broth and allowed to grow and reproduce overnight, billions of copies of the original cell, each containing identical modified plasmid DNA, would result. When a plasmid is used in this way—to make copies of a foreign DNA sequence—it is called a **cloning vector,** or simply a **vector.**

Biologists harvest the recombinant genes by breaking the bacteria open, isolating all the DNA, and then separating the plasmids from the main chromosomes. But how do they insert a gene into a plasmid in the first place?

Using Restriction Endonucleases and DNA Ligase to Cut and Paste DNA To cut a gene out for later insertion into a cloning vector, researchers use enzymes called restriction endonucleases. A **restriction endonuclease** is a bacterial enzyme that cuts DNA molecules at specific base sequences. In bacterial cells, these enzymes cut up DNA from invading viruses and prevent a fatal infection.

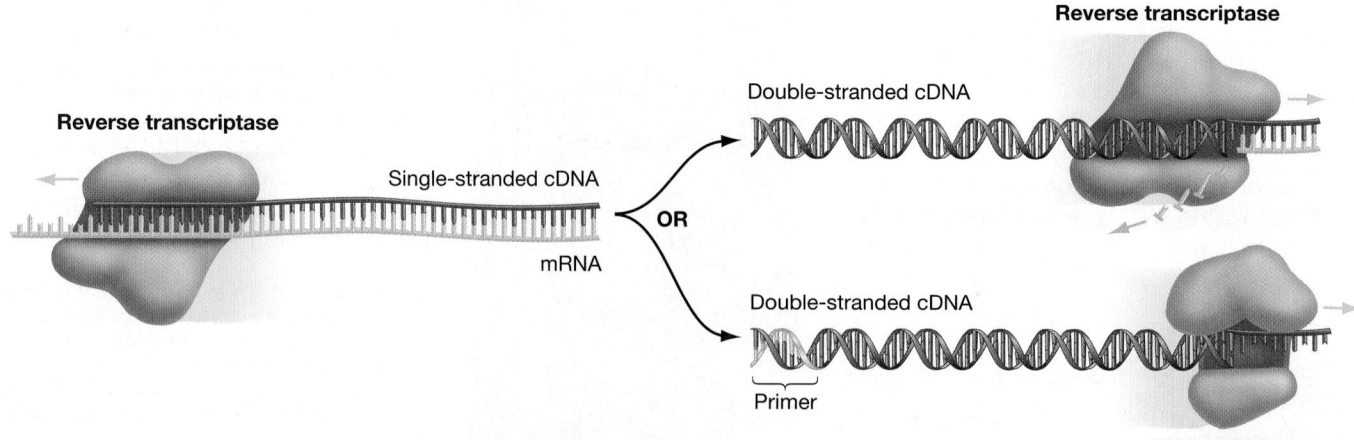

FIGURE 20.2 Reverse Transcriptase Catalyzes the Synthesis of DNA from RNA. The single-stranded DNA produced by reverse transcriptase is complementary to the RNA template. The cDNA can be made double stranded by reverse transcriptase or DNA polymerase. DNA polymerase requires a primer.

More than 800 restriction endonucleases are known, and many of them cut DNA only at sequences that form palindromes. In English, a word or sentence is a palindrome if it reads the same way backward as it does forward. "Madam, I'm Adam" is an example. In biology, a stretch of double-stranded DNA forms a palindrome if the 5′→3′ sequence of one strand is identical to the 5′→3′ sequence on the antiparallel, complementary strand.

To insert the pituitary-gland cDNAs into plasmids, researchers performed the sequence of steps outlined in **FIGURE 20.3**.

Step 1 The left side of the figure shows a plasmid containing a palindromic sequence, or recognition site, that is cut by a specific restriction endonuclease. As the right side of the figure shows, the researchers attached the same palindromic sequence to the ends of each cDNA in their sample.

Step 2 They cut the recognition sites in each plasmid (left) and at the ends of each cDNA (right) with a restriction endonuclease called EcoRI. (The name comes from the fact that this enzyme was the first (Roman numeral *I*) restriction endonuclease discovered in *E. coli* strain <u>R</u>Y13.)

Step 3 Like most restriction endonucleases, EcoRI makes a staggered cut in the palindrome. The resulting DNA fragments are described as having **sticky ends,** because the single-stranded bases on one fragment are complementary to the single-stranded bases on the other fragment. As a result, the two ends can pair up and hydrogen-bond to each other: The complementary sequences in the sticky ends of the plasmid (in grey) will bind to the sticky ends in the cDNA (in red) by complementary base pairing.

Step 4 Finally, researchers used **DNA ligase**—the enzyme that connects Okazaki fragments during DNA replication (see Chapter 15)—to seal the pieces of DNA together at the arrows marked in green.

The creation of sticky ends in DNA is important. If restriction sites in different DNA sequences are cut with the same restriction

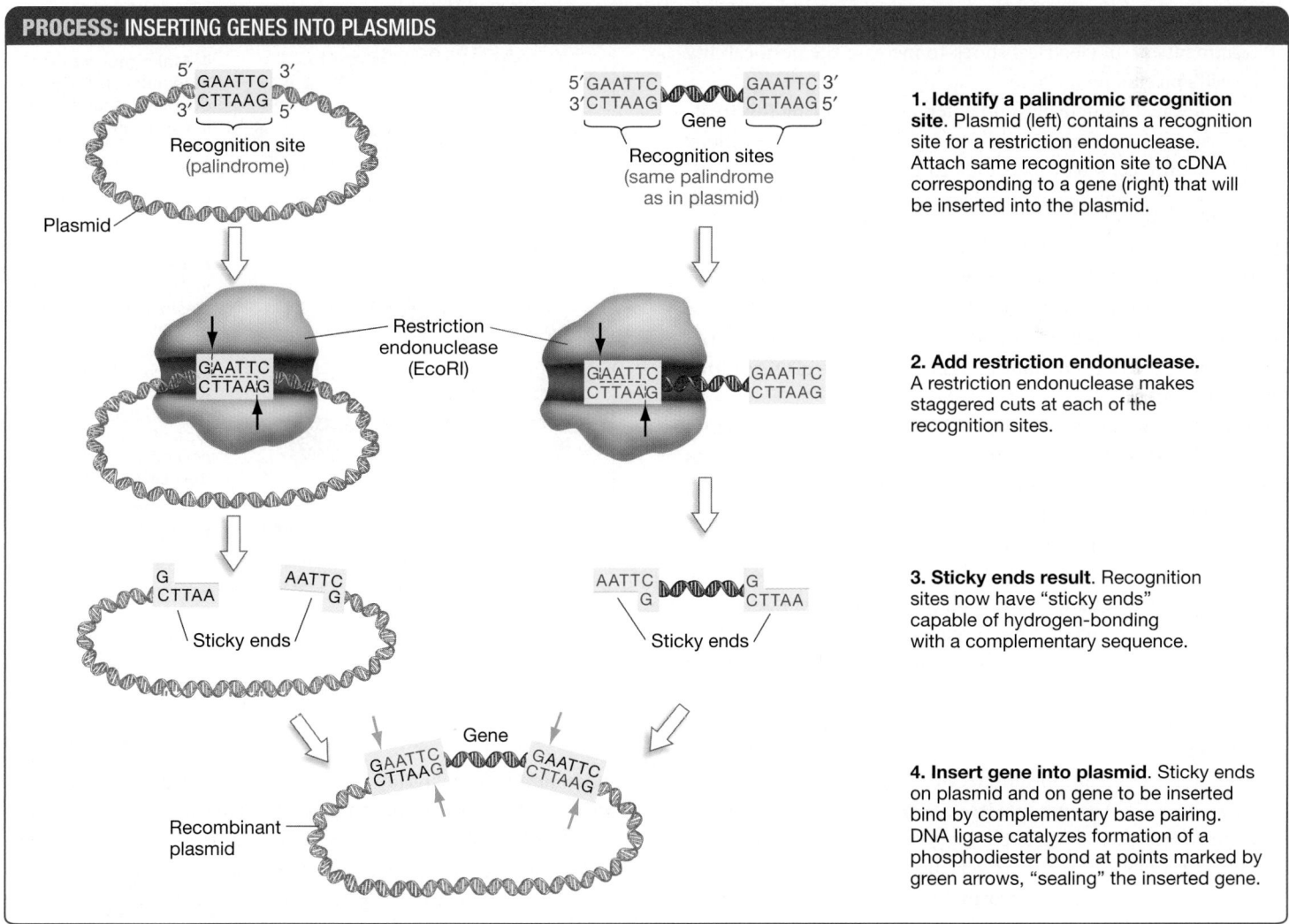

PROCESS: INSERTING GENES INTO PLASMIDS

1. **Identify a palindromic recognition site**. Plasmid (left) contains a recognition site for a restriction endonuclease. Attach same recognition site to cDNA corresponding to a gene (right) that will be inserted into the plasmid.

2. **Add restriction endonuclease.** A restriction endonuclease makes staggered cuts at each of the recognition sites.

3. **Sticky ends result**. Recognition sites now have "sticky ends" capable of hydrogen-bonding with a complementary sequence.

4. **Insert gene into plasmid**. Sticky ends on plasmid and on gene to be inserted bind by complementary base pairing. DNA ligase catalyzes formation of a phosphodiester bond at points marked by green arrows, "sealing" the inserted gene.

FIGURE 20.3 Genes Can Be Inserted into Plasmids in Preparation for Cloning. Once a gene has been inserted into a plasmid, the recombinant plasmid can be introduced into bacterial cells that grow and divide to produce many identical copies of the gene.

endonuclease, the presence of the same sticky ends in both samples of DNA promotes joining of the resulting fragments. This is the essence of recombinant DNA technology—the ability to create novel combinations of DNA sequences by cutting specific sequences and pasting them into new locations.

After performing this procedure, the researchers who were hunting for the growth hormone gene had a set of recombinant plasmids. Each contained a cDNA made from one of the many human pituitary-gland mRNAs. But the *GH1* gene was still not cloned.

Transformation: Introducing Recombinant Plasmids into Bacterial Cells

If a recombinant plasmid can be inserted into a bacterial or yeast cell, the foreign DNA will be copied and transmitted to new cells as the host cell grows and divides. In short, it can be cloned. In this way, researchers can obtain millions or billions of copies of specific genes. How is the insertion brought about?

Cells that take up DNA from the environment and incorporate it into their genomes are said to undergo **transformation.** Most bacterial cells do not take up DNA on their own under laboratory conditions. So, to get plasmid DNA into *E. coli* and other common laboratory species, researchers use simple chemical treatments or an electrical shock to increase the permeability of the cell's plasma membrane.

Typically, just a single plasmid enters the cell during this treatment. The cells are then spread out on plates under conditions that allow only cells with plasmids to grow into colonies. Each colony contains millions of identical cells, each with many identical copies of the recombinant plasmid.

Producing a cDNA Library

FIGURE 20.4 summarizes the steps covered thus far in the hunt for the growth hormone gene. The result, shown in step 5, is a collection of transformed bacterial cells. Each of the cells contains a plasmid with one cDNA from a pituitary gland mRNA.

A collection of DNA sequences, each of which is inserted into a vector, is called a **DNA library.** If the sequences are cDNAs made from a particular cell type or tissue, the library is called a **cDNA library.** If the sequences are fragments of DNA from the genome of an individual, the library is called a **genomic library.**

DNA libraries are made up of cloned genes. Each gene can be produced in large quantity and isolated in pure form. ✔ If you understand this concept, you should be able to describe how you could make a genomic library starting with DNA from your own cells and using the restriction endonuclease EcoRI to cut the genome into fragments that can be inserted into a plasmid vector.

DNA libraries are important because they provide researchers a way to store DNA fragments from a particular cell type or genome in a form that is accessible for gene cloning. But like a college library, a DNA library isn't very useful unless there is a way to retrieve specific pieces of information. At your school's library, you use call numbers or computer searches to retrieve a particular book or article. How do you go about retrieving a particular gene from a DNA library? For example, how did researchers find

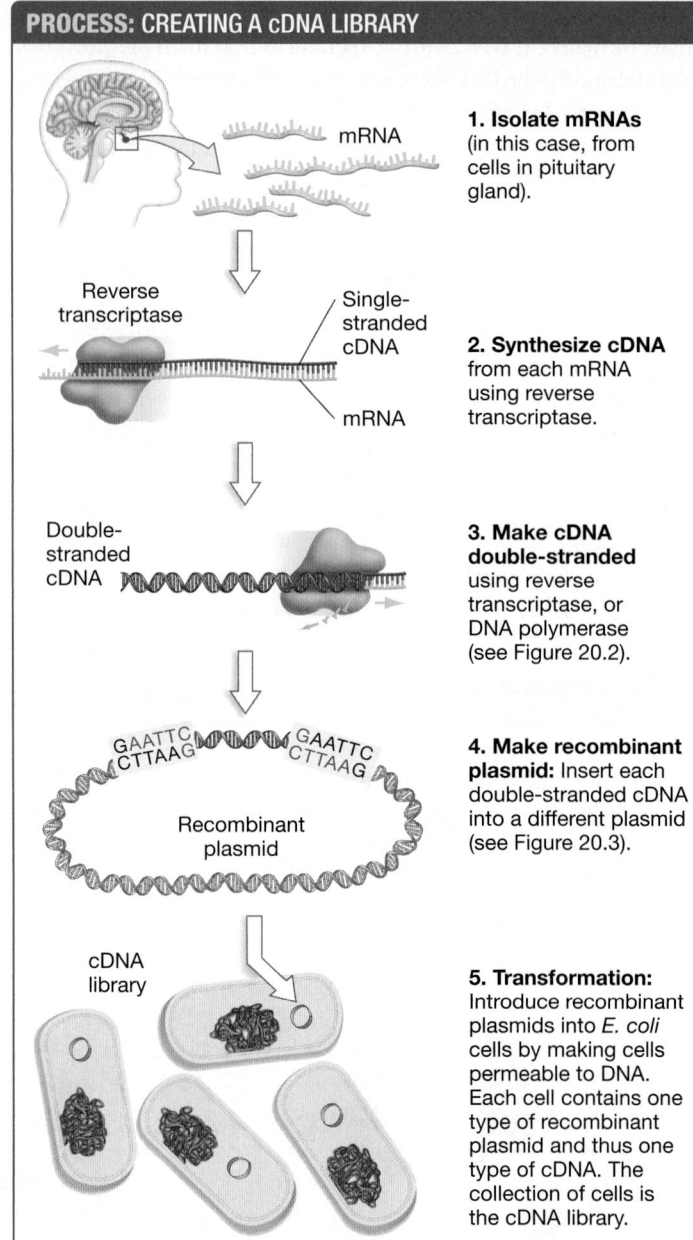

PROCESS: CREATING A cDNA LIBRARY

1. **Isolate mRNAs** (in this case, from cells in pituitary gland).

2. **Synthesize cDNA** from each mRNA using reverse transcriptase.

3. **Make cDNA double-stranded** using reverse transcriptase, or DNA polymerase (see Figure 20.2).

4. **Make recombinant plasmid:** Insert each double-stranded cDNA into a different plasmid (see Figure 20.3).

5. **Transformation:** Introduce recombinant plasmids into *E. coli* cells by making cells permeable to DNA. Each cell contains one type of recombinant plasmid and thus one type of cDNA. The collection of cells is the cDNA library.

FIGURE 20.4 Complementary DNA (cDNA) Libraries Represent a Collection of the mRNAs in a Cell.

✔**QUESTION** Would each type of cDNA in the library be represented just once? Why or why not?

the growth hormone gene in the cDNA library of the human pituitary gland?

Screening a DNA Library

Molecular biologists are often faced with the task of finding one specific gene in a large collection of DNA fragments. To do this requires a **probe**—a marked molecule that binds to the molecule the biologist is looking for.

A DNA probe is a single-stranded fragment that will bind to a particular single-stranded complementary sequence in a mixture of DNAs. By binding to the target sequence, the probe marks the fragment containing that sequence, distinguishing it from all the other DNA fragments in the sample. As **FIGURE 20.5** shows, a DNA probe must be labeled in some way so that it can be found after it has bound to the complementary sequence in the large sample of fragments.

✔ If you understand the concept of a DNA probe, you should be able to explain why the probe must be single stranded and labeled in order to work, and why it binds to just one specific fragment. You should also be able to indicate where a probe with the sequence AATCG (recall that sequences are always written $5' \rightarrow 3'$) will bind to a target DNA with the sequence TTTTACCCATTTACGATTGGCCT (again written $5' \rightarrow 3'$).

To find an appropriate probe for the human growth hormone gene, researchers began by using the genetic code to predict possible DNA sequences of *GH1*. They could do this because they knew the sequence of amino acids in the polypeptide, which they could use to infer the codons that coded for each amino acid. You made similar inferences in some of the exercises in earlier chapters. But recall that there is more than one codon for most amino acids (see Chapter 16). As a result, the researchers could not infer a unique sequence for the growth hormone gene. Instead, they deduced a set of possible sequences that could encode the *GH1* gene.

The next step was to chemically synthesize the set of short, single-stranded DNAs that were complementary to the possible *GH1* sequences. Because one of these sequences would bind to single-stranded fragments from the actual gene by complementary base pairing, it could act as a probe. In this case, the label the researchers attached to the probe was a radioactive atom.

FIGURE 20.6 shows how researchers used this probe to find the plasmid in the cDNA library that contained *GH1*. (For more information on how to use probes, see **BioSkills 9** in Appendix B.) As predicted, the labeled probe bound to its complementary sequence in the cDNA library—identifying the recombinant cell that contained the human growth hormone cDNA.

Mass-Producing Growth Hormone To accomplish their goal of producing large quantities of the human growth hormone, the investigators used recombinant DNA techniques to transfer the growth hormone cDNA to a new plasmid. The plasmid in question contained a promoter sequence recognized by *E. coli*'s RNA polymerase holoenzyme (see Chapter 17). The recombinant plasmids were then introduced into *E. coli* cells.

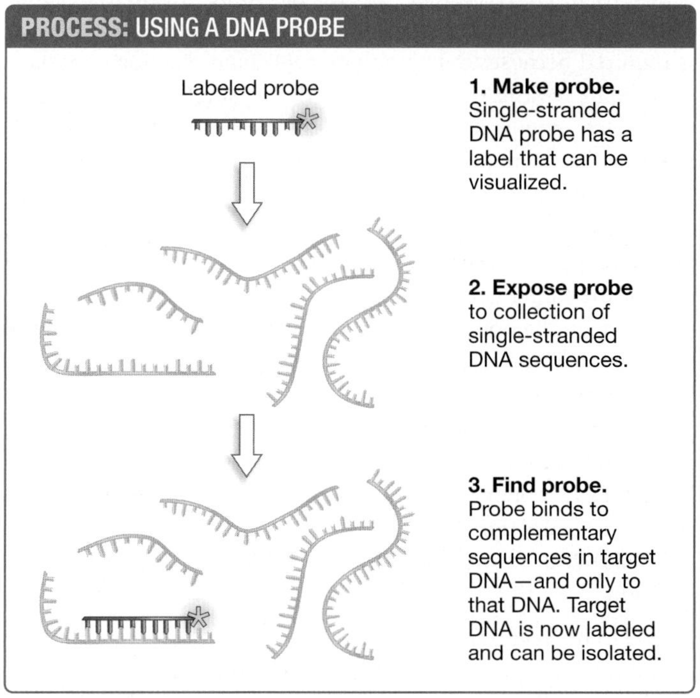

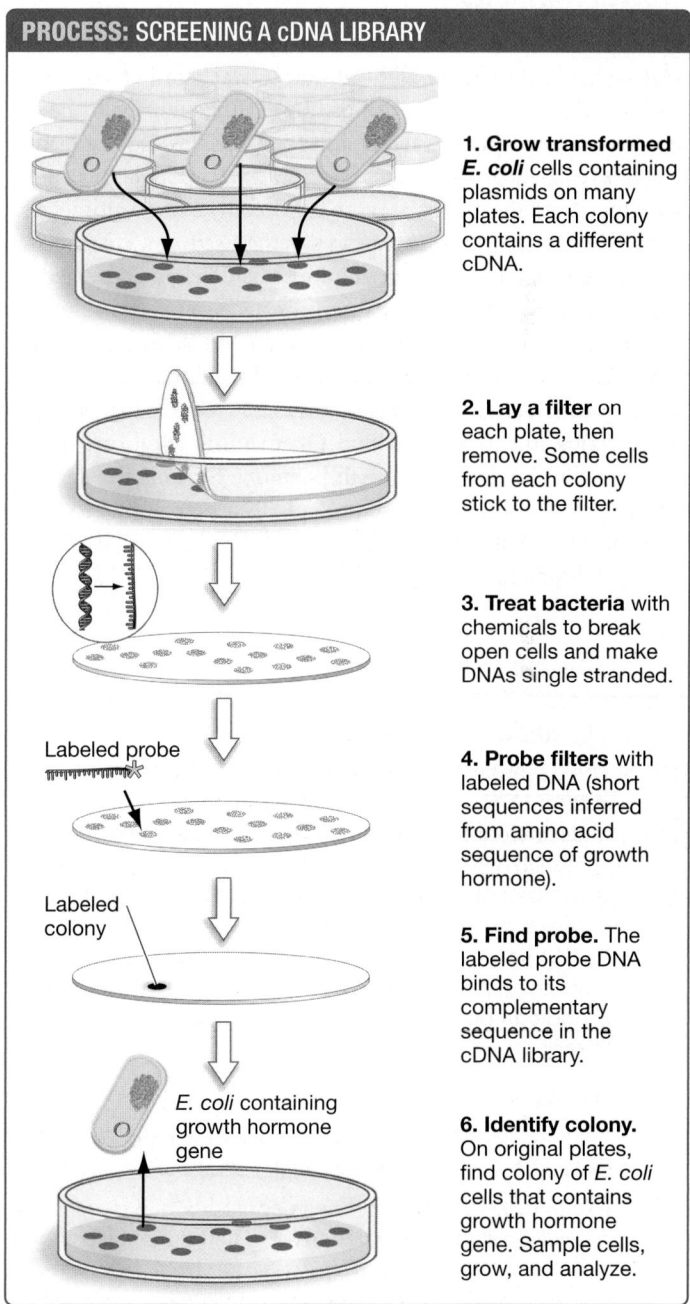

PROCESS: USING A DNA PROBE

Labeled probe

1. Make probe. Single-stranded DNA probe has a label that can be visualized.

2. Expose probe to collection of single-stranded DNA sequences.

3. Find probe. Probe binds to complementary sequences in target DNA—and only to that DNA. Target DNA is now labeled and can be isolated.

FIGURE 20.5 DNA Probes Bind to and Identify Specific Target Sequences.

PROCESS: SCREENING A cDNA LIBRARY

1. Grow transformed *E. coli* cells containing plasmids on many plates. Each colony contains a different cDNA.

2. Lay a filter on each plate, then remove. Some cells from each colony stick to the filter.

3. Treat bacteria with chemicals to break open cells and make DNAs single stranded.

Labeled probe

4. Probe filters with labeled DNA (short sequences inferred from amino acid sequence of growth hormone).

Labeled colony

5. Find probe. The labeled probe DNA binds to its complementary sequence in the cDNA library.

E. coli containing growth hormone gene

6. Identify colony. On original plates, find colony of *E. coli* cells that contains growth hormone gene. Sample cells, grow, and analyze.

FIGURE 20.6 Finding Specific Genes by Probing a cDNA Library.

The resulting transformed *E. coli* cells now contained a gene for human growth hormone attached to a promoter. These cells began to transcribe and translate the human growth hormone gene. Human growth hormone accumulated in the cells and was subsequently isolated and purified.

Today, bacterial cells containing the human growth hormone gene are grown in huge quantities. These cells have proved to be a safe and reliable source of the human growth hormone protein. The effort to cure pituitary dwarfism using recombinant DNA technology was a spectacular success—a triumph of applied biology, or **biotechnology.**

Ethical Concerns over Recombinant Growth Hormone

As supplies of growth hormone increased, physicians used it in treating not only people with pituitary dwarfism but also short children who had no growth hormone deficiency. Even though the treatment requires several injections per week until adult height is reached, growth hormone therapy was popular because it often increased the height of these children by a few centimeters.

In essence, growth hormone was being used as a cosmetic—a way to improve appearance in cultures where height is deemed attractive. But if short people are discriminated against in a culture, is a medical treatment a better solution than education and changes in attitudes? And what if parents wanted a tall child to be even taller, to enhance her potential success as, say, a basketball player?

Currently, the U.S. Food and Drug Administration has approved the use of human growth hormone for only the shortest 1.2 percent of children. These individuals are projected to reach adult heights of less than 160 cm (5′3″) in males and 150 cm (4′11″) in women.

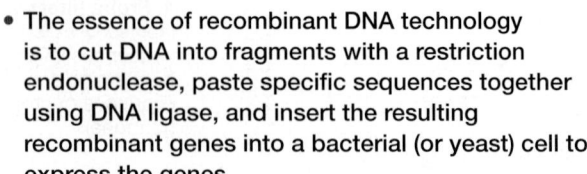

check your understanding

C Y U

If you understand that . . .

- The essence of recombinant DNA technology is to cut DNA into fragments with a restriction endonuclease, paste specific sequences together using DNA ligase, and insert the resulting recombinant genes into a bacterial (or yeast) cell to express the genes.
- A DNA library consists of different sequences that have been inserted into plasmids or other vectors. A probe can be used to find specific sequences in the library.

✔ You should be able to . . .

1. Explain why restriction endonucleases like EcoRI create DNA fragments with sticky ends.
2. Explain why the word probe is appropriate to describe a labeled sequence that is used to find a particular gene in a DNA library.

Answers are available in Appendix A.

Growth hormone has also become a popular performance-enhancing drug for athletes, because it improves the maintenance of bone density and muscle mass. Part of its popularity stems from the fact that it is difficult to detect in current drug tests.

Should athletes be able to enhance their physical skills by taking hormones or other types of drugs? Is the drug safe at the dosages athletes are using? These questions are being debated by physicians, researchers, agencies that govern sports, and legislative bodies.

In the meantime, it is clear that while solving one important problem, recombinant DNA technology created others. One of this chapter's recurring themes is that genetic engineering has costs that must be carefully weighed against its benefits.

20.2 Case 2–Amplification of Fossil DNA: The Polymerase Chain Reaction

Inserting a gene into a bacterial plasmid is one method for cloning DNA. The polymerase chain reaction is another.

The **polymerase chain reaction (PCR)** is an in vitro DNA synthesis reaction that uses DNA polymerase to replicate a specific section of DNA over and over. It generates many identical copies of a particular region of DNA.

Requirements of PCR

Although PCR is much faster and technologically easier than cloning genes into a DNA library, there is a catch: PCR is possible only when a researcher already has some information about DNA sequences near the gene in question. Sequence information is required because to do a polymerase chain reaction, you have to start by synthesizing short lengths of single-stranded DNA that match sequences on either side of the gene. These short segments act as primers for the DNA synthesis reaction.

As **FIGURE 20.7a** shows, the primer sequences must be complementary to bases on either side of the target gene—the DNA you want to copy. One primer is complementary to a sequence on one side of the target gene; the other primer is complementary to a sequence on the opposite strand of DNA, on the other side of the target gene. If the target DNA molecule is made single stranded, then the primers will bind, or anneal, to their complementary sequences, as shown in **FIGURE 20.7b**. You might recall that DNA polymerase cannot work without a primer. Once the primers are bound, DNA polymerase can extend each new strand of DNA in the $5' \rightarrow 3'$ direction.

FIGURE 20.8 shows the sequence of the polymerase chain reaction.

Step 1 The researcher creates a reaction mix containing an abundant supply of the four deoxyribonucleoside triphosphates (dNTPs; see Chapter 15), a DNA sample that includes the gene of interest, many copies of the two primers, and a heat-resistant DNA polymerase called *Taq* polymerase.

(a) PCR primers must bind to sequences on either side of the target sequence, on opposite strands.

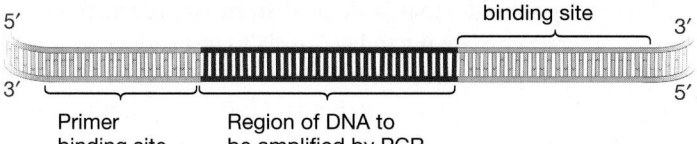

(b) When target DNA is made single stranded, primers bind and allow DNA polymerase to work.

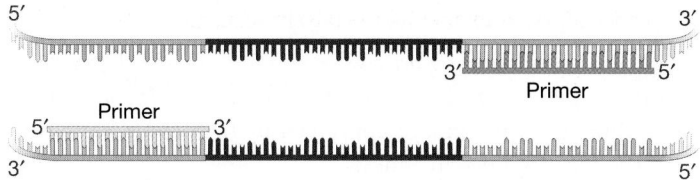

FIGURE 20.7 The Polymerase Chain Reaction Requires Appropriate Primers. (a) To design an appropriate primer, the base sequence at the primer binding sites must be known. **(b)** The primers bind by complementary base pairing to single-stranded target DNA.

✔**EXERCISE** Indicate where DNA polymerase would begin to work on each strand; add an arrow indicating the direction of DNA synthesis.

Step 2 The reaction mix is heated to 95°C. At this temperature, the double-stranded template DNA denatures. This means that the two DNA strands separate, forming single-stranded templates.

Step 3 The mixture is allowed to cool to 50–60°C. In this temperature range, the primers bind, or anneal, to complementary portions of the single-stranded template DNA. This step is called primer annealing.

Step 4 The reaction mix is heated to 72°C. At this temperature, *Taq* polymerase efficiently synthesizes the complementary DNA strand from the dNTPs, starting at the primer. This step is called *extension*.

Step 5 Repeat steps 2 through 4.

Step 6 Continue repeating steps 2 through 4 until the necessary number of copies is obtained.

The temperature changes required in each step are controlled by automated PCR machines, and there is no need to add more components once the reaction starts.

Taq polymerase is a DNA polymerase found in the thermophilic ("heat-loving") bacterium *Thermus aquaticus*, which was discovered in a hot spring in Yellowstone National Park. Researchers use *Taq* polymerase because the PCR mixture has to be heated, and *Taq* polymerase is heat stable. Enzymes from most organisms are destroyed at high temperature, but *Taq* polymerase functions normally even when heated to 95°C.

The denaturation, primer annealing, and extension steps constitute a single PCR cycle. If one copy of the template sequence existed in the original sample, then two copies are present at the

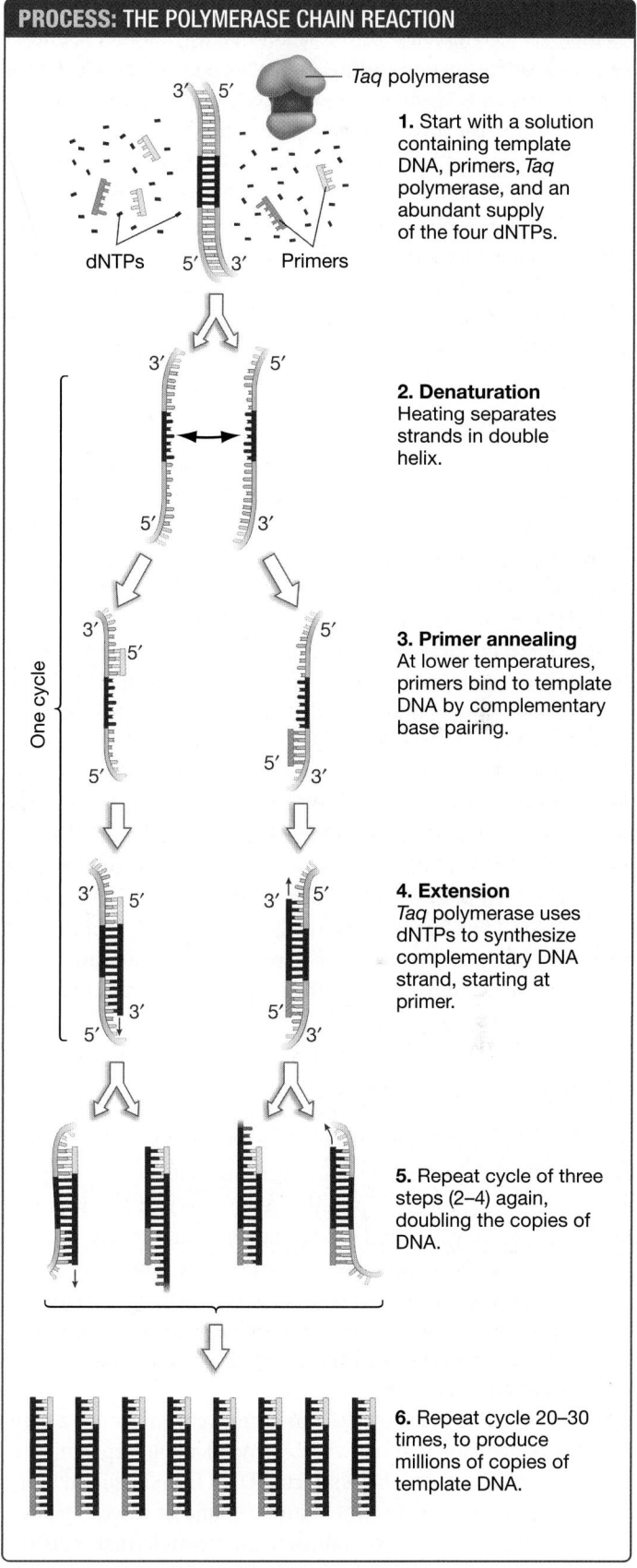

PROCESS: THE POLYMERASE CHAIN REACTION

1. Start with a solution containing template DNA, primers, *Taq* polymerase, and an abundant supply of the four dNTPs.

2. Denaturation Heating separates strands in double helix.

3. Primer annealing At lower temperatures, primers bind to template DNA by complementary base pairing.

4. Extension *Taq* polymerase uses dNTPs to synthesize complementary DNA strand, starting at primer.

5. Repeat cycle of three steps (2–4) again, doubling the copies of DNA.

6. Repeat cycle 20–30 times, to produce millions of copies of template DNA.

FIGURE 20.8 The Polymerase Chain Reaction Produces Many Copies of a Specific Sequence. Each PCR cycle (denaturation, primer annealing, and extension) results in a doubling of the number of target sequences.

end of the first cycle (see step 4 in Figure 20.8). These two copies then act as templates for the second cycle—another round of denaturation, primer annealing, and extension—after which four copies of the target gene are present (see step 5).

Each time the cycle repeats, the amount of template sequence in the reaction mixture doubles (step 6). Doubling occurs because each newly synthesized segment of DNA serves as a template in the subsequent cycle, along with the previously synthesized segments. Starting with a single copy, successive cycles result in the production of 2, 4, 8, 16, 32, 64, 128, 256 copies, and so on. A total of n cycles can generate 2^n copies. In just 20 cycles, one sequence can be amplified to over a million copies.

PCR in Action

To understand why PCR is so valuable, consider a study by biologist Svante Pääbo and colleagues, who wanted to analyze DNA recovered from the 30,000-year-old bones of a fossilized human of the species *Homo neanderthalensis*. Their goal was to determine the sequence of bases in the ancient DNA and compare it with DNA from modern humans (*Homo sapiens*).

If modern humans have sequences that are identical or almost identical to the sequences found in Neanderthals, it would suggest that some of us inherited DNA directly from a Neanderthal ancestor. That could happen only if *H. sapiens* and *H. neanderthalensis* interbred while they coexisted in Europe.

The Neanderthal bone was so old, however, that most of the DNA in it had degraded into tiny fragments. The biologists could recover only a minute amount of DNA that was still in even moderate-sized pieces. Fortunately, the Neanderthal DNA sample included a few fragments of the gene region that Pääbo's team wanted to study. The researchers were able to design primers that bracketed this region, based on the sequence of highly conserved sections of the same gene from *H. sapiens*.

Using PCR, the researchers produced millions of copies of the Neanderthal DNA fragment. After analyzing these sequences, the team found that they differ from the same gene segment found in modern humans. Subsequent work with DNA from 14 other Neanderthal fossils, from locations throughout Europe, gave the same result. These data support the hypothesis that Neanderthals never interbred with modern humans—even though the two species lived in the same areas of Europe at the same time.

But a nagging doubt remained—the conclusion was based on the analysis of a small region of the genome, the best that could be done at the time. Would the conclusion hold if more of the genome was analyzed?

To answer this question, Pääbo's team went on to use a form of DNA sequencing that has a DNA amplification step similar to PCR (see Section 20.3). They extracted DNA from fossilized bones and were able to sequence the entire genome of three Neanderthals. The researchers next compared the Neanderthal sequence with the genome sequences of people from different populations living today.

Their conclusion? For African populations there was no evidence of inbreeding with Neanderthals. The story is different,

however, for non-African populations. People from these groups have a small amount of DNA, roughly 1 percent to 4 percent of each person's genome, that is derived from Neanderthal ancestors. So, many people living today can claim just a touch of Neanderthal in their family tree.

PCR has opened new research possibilities in countless areas beyond ancient DNA. For example:

- Forensic scientists, who use biological analyses to help solve crimes, clone DNA from tiny drops of blood or hair. The copied DNA can then be analyzed to identify victims, implicate perpetrators, or exonerate the falsely accused.

- Genetic counselors, who advise couples on how likely their offspring are to suffer from inherited diseases, can use PCR to find out if an embryo conceived by the couple has alleles associated with deadly illness.

Because the complete genomes of a wide array of organisms have now been sequenced, researchers can easily find appropriate primer sequences to use in cloning almost any target gene by PCR. The polymerase chain reaction is now one of the most basic and widely used techniques in biology.

check your understanding

If you understand that . . .

- PCR is a technique for amplifying a specific region of DNA into millions of copies, which can then be sequenced or used for other types of analyses.

✔ **You should be able to . . .**

1. Explain the purpose of the denaturation, annealing, and extension steps in a PCR cycle, and why "chain reaction" is an appropriate part of the term PCR.

2. Write down the sequence of a double-stranded DNA that is 50 base pairs long. Then design 21-base-pair-long primers that would allow you to amplify the segment by PCR.

Answers are available in Appendix A.

20.3 Case 3–Sanger's Breakthrough: Dideoxy DNA Sequencing

Once researchers have cloned a gene from a DNA library or by PCR, determining the gene's base sequence is usually one of the first things they want to do. Learning a gene's sequence is valuable for a variety of reasons. For example:

- Once a gene's sequence is known, the amino acid sequence of its product can be inferred from the genetic code. Knowing a protein's primary structure often provides clues to its function.

- Comparing sequences is fundamental to understanding why alleles vary in function—for example, why one allele causes disease and another doesn't.

- Researchers can infer evolutionary relationships by comparing the sequences of the same gene in different species (see Chapter 1). This information can be used to study an array of questions, ranging from how new traits evolve to where new diseases come from.

How do researchers sequence DNA? In 1977 Frederick Sanger published a technique called dideoxy sequencing that is still in use today.

The Logic of Dideoxy Sequencing

As **FIGURE 20.9** shows, **dideoxy sequencing** is a clever variation on the basic in vitro DNA synthesis reaction. But saying "clever" is an understatement. Sanger had to link three important insights to make his sequencing strategy work.

Dideoxynucleotides Terminate DNA Synthesis   Sanger's first insight was to use monomers for DNA synthesis called dideoxyribonucleoside triphosphates (ddNTPs) along with the normal deoxyribonucleoside triphosphates (dNTPs) (Chapter 15) in the reaction mix. The ddNTPs are identical to dNTPs except that they lack a hydroxyl group at their 3′ carbon. Four types of ddNTPs are used in dideoxy sequencing, each named according to whether it contains adenine (ddATP), thymine (ddTTP), cytosine (ddCTP), or guanine (ddGTP). The use of ddNTPs inspired the name dideoxy sequencing.

Sanger realized that if a ddNTP were added to a growing DNA strand, it would terminate synthesis. Why? After a ddNTP is added, no hydroxyl group is available on a 3′ carbon to link to the 5′ carbon on an incoming dNTP monomer. As a result, DNA polymerization stops once a ddNTP is added.

Fragment Length Correlates with the Location of Each Base
Sanger linked the ability of ddNTPs to stop DNA synthesis to a second fundamental insight: Every time a ddNTP is added to a growing strand, the result is a fragment with a length corresponding to the position in the template of a base complementary to the ddNTP. To produce these fragments, biologists create a reaction mix containing (**1**) many copies of the template DNA with (**2**) a primer, (**3**) DNA polymerase, (**4**) a large supply of the four dNTPs, and (**5**) a small amount of the four ddNTPs (Figure 20.9, step 1). Each of the four ddNTPs carries a different fluorescent tag. (In the figure, ddGTP is purple, ddCTP is blue, ddATP is green, and ddTTP is orange.) Fluorescent molecules absorb light at one wavelength and reemit the light at a longer wavelength. As described in **BioSkills 10** (see Appendix B), fluorescent tags provide a very sensitive way of detecting molecules.

Under these conditions, many daughter strands of different lengths are synthesized. All fragments that are the same length end in the same kind of ddNTP.

Step 2 in Figure 20.9 shows why:

- DNA polymerase synthesizes a complementary strand from each template in the reaction mix.

- The synthesis of each one of these complementary strands starts at the same point—the primer.

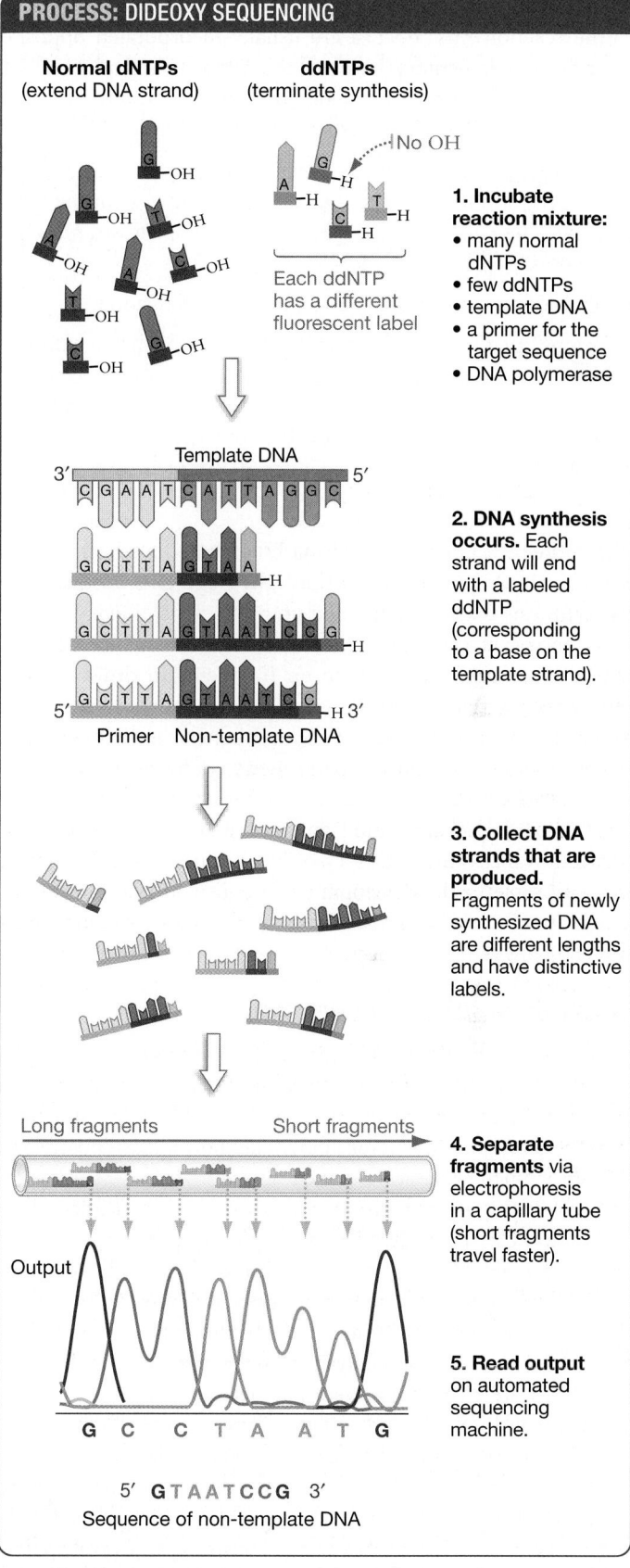

FIGURE 20.9 Dideoxy Sequencing Can Determine the Base Sequence of DNA.

- Because there are many dNTPs and relatively few ddNTPs in the reaction mix, dNTPs are usually incorporated opposite each complementary base on the template strand as DNA polymerase works its way along the template strand. Incorporating a dNTP allows DNA synthesis to continue.

- Occasionally, one of the few ddNTPs is incorporated into the growing strand, opposite the corresponding base in the template. The complementary base in the template strand pairs randomly with either a ddNTP or a dNTP.

- The addition of the ddNTP stops further elongation.

- "Stops" of this kind happen for each base in the template strand. As a result, the overall reaction produces a collection of newly synthesized strands (fragments) whose various lengths correspond to the location of each base in the template strand (see step 3 in Figure 20.9). Each fragment will fluoresce in the color of its ddNTP.

DNA Sequence Can Be Read from Fragments Lined Up by Size
Sanger's third insight? When the DNA fragments produced by the synthesis reactions are lined up by size, the dideoxy monomers on the successive fragments reveal the sequence of bases in the template DNA. To line up fragments in order of size, biologists separate them using gel electrophoresis (step 4 in Figure 20.9). As step 5 shows, a machine can read the pattern of fluorescence, indicating the sequence of bases in the newly synthesized strand.

Dideoxy sequencing ranks among the greatest of all technological advances in the history of biological science. Its impact is comparable to the development of light and electron microscopes, microelectrodes for recording membrane potentials, and recombinant gene technology.

"Next Generation" Sequencing

Sequencing technology is advancing at a blindingly fast pace. Dideoxy sequencing is still performed, particularly when relatively long sequences need to be read with great accuracy. But new, "next generation" sequencing approaches now make it possible to sequence much faster and more cheaply.

Most of the newer methods are based on amplification steps related to PCR. These create many copies of each template DNA molecule and allow sequencing of minute quantities of DNA. Even more important, millions of different DNAs can be amplified and sequenced in a single run. Methods that allow simultaneous sequencing or analysis of huge numbers of different molecules are called massively parallel approaches. These sequencing technologies are opening research possibilities that were barely imaginable only a few years ago. Today, instead of sequencing individual genes, researchers often obtain the sequence of entire genomes. A case in point is the sequencing of the Neanderthal genome—this was made possible using a massively parallel approach.

The project to sequence the human genome for the first time took 10 years and cost $3 billion U.S. dollars. Now researchers could sequence your genome in a day for less than $5,000.

20.4 Case 4–The Huntington's Disease Story: Finding Genes by Mapping

Mendel had no idea what a "hereditary determinant" actually was. But now we know. Biology's molecular revolution has allowed researchers to find and characterize individual genes—to explore the connection between genotype and phenotype as explicitly and directly as possible. The question is, How do researchers find the genes associated with certain traits in the first place? How do you find the gene responsible for seed shape in peas, or white eyes in flies, or DNA polymerase III in *E. coli*?

One widely used approach is conceptually simple: You begin with a map of known sites in the genome and then look for an association between one of those known sites and the phenotype you're interested in. The gene that affects the phenotype is probably close to the known site.

In practice, the process is not so simple. As an example of how this type of gene hunt is done, let's consider one of the first successful searches ever conducted for a human gene—the gene associated with Huntington's disease.

How Was the Huntington's Disease Gene Found?

Huntington's disease is a rare but devastating illness. Typically, affected individuals first show symptoms between the ages of 35 and 45. At onset, an individual appears to be clumsier than normal and tends to develop small tics and abnormal movements. As the disease progresses, uncontrollable movements become more pronounced. Eventually the affected individual twists and writhes involuntarily. Personality and intelligence are also compromised—to the extent that the early stage of this disease is sometimes misdiagnosed as the brain disorder schizophrenia. The illness may continue to progress for 10 to 20 years and is eventually fatal.

Because Huntington's disease runs in families, physicians suspected that it was a genetic disease. An analysis of pedigrees from families affected by Huntington's disease suggested that the trait was due to a single, autosomal dominant allele (see Chapter 14). To understand the molecular basis for the disease, researchers used many of the tools and techniques of genetic engineering that are shown in **TABLES 20.1** and **20.2** (see pages 380–381) as they set out to locate and identify the gene or genes involved. It took over 10 years of intensive effort to reach this goal.

The search for the Huntington's disease gene was led by Nancy Wexler, whose mother had died of the disease. If the trait was indeed due to an autosomal dominant allele, it meant that Wexler had a 50 percent chance of receiving the allele from her mother and would begin to show symptoms when she reached middle age.

Using Genetic Markers To locate the gene or genes associated with a particular phenotype, such as a disease, researchers traditionally start with a **genetic map,** also known as a **linkage map** (see Chapter 14). Recall that a genetic map shows the relative positions of genes on the same chromosome, determined by analyzing the frequency of recombination between pairs of genes. Biologists also use **physical maps** of the genome. A physical map shows the absolute position of a gene—in numbers of base pairs—along a chromosome. Genome sequencing (see Chapter 21) has produced physical maps for a wide array of species.

A genetic map is valuable in gene hunts because it contains **genetic markers**—easily identified genes or DNA sequences that have known locations. Each genetic marker provides a landmark—a position along a chromosome that is known relative to other markers. The key is to find where the unknown gene lies relative to the established genetic markers.

To understand how genetic markers can be used, let's use a hypothetical example. Suppose that you knew the position of a hair-color gene in humans relative to other genetic markers. Suppose too that various alleles of this gene contributed to the development of black hair, red hair, blond hair, and brown hair in the group of people you were studying. This variation in phenotype associated with the marker is crucial. To be useful in a gene hunt, a genetic marker has to be **polymorphic,** meaning that the phenotype associated with the marker varies. In our hypothetical example, hair color is a polymorphic genetic marker.

Now suppose that the genetic disease called cystic fibrosis is common among the individuals you were studying and that your goal is to find the gene associated with cystic fibrosis. Further, suppose that people who have cystic fibrosis almost always have black hair—even though they are just as likely as unaffected individuals to have any other inherited trait observed in the study population, such as the presence or absence of a widow's peak or detached earlobes.

If you observe that a certain form of a marker and a certain disease are almost always inherited together, this means that the marker gene and disease gene are physically close to each other on the same chromosome—they are closely linked. If they were not closely linked, then crossing over in between them would be common and they would *not* be inherited together. In this hypothetical study, you could infer that the gene for cystic fibrosis is very close to the hair-color gene. ✔ **If you understand this concept, you should be able to explain why it's helpful to hunt for genes using a genetic map with many genetic markers rather than only a few.**

Gene hunts in humans boil down to this: Researchers have to find a large number of people who are affected and unaffected. Then they must attempt to locate a genetic marker that almost always occurs in one form in the affected individuals but only rarely in unaffected people. If such a marker is found, the disease gene is almost guaranteed to be nearby.

The types of genetic markers used in gene mapping have changed over time. Today, researchers often have a large catalog of polymorphic genetic markers available, including the particularly abundant markers known as **single nucleotide polymorphisms** (**SNPs,** pronounced *snips*). A SNP is a site in DNA that varies between alleles at a single base pair. Below is an example of a SNP:

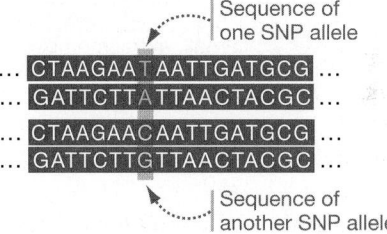

To date, roughly 10 million human SNPs have been identified.

In the late 1970s and early 1980s, when biologists were searching for the Huntington's disease gene, SNPs were unknown. The best genetic markers available were restriction sites—short stretches of DNA where restriction endonucleases cut the double helix. These sequences are also known as restriction endonuclease recognition sites.

The restriction sites that Wexler's team used were polymorphic: Some alleles had a sequence that allowed cuts to occur; but in other alleles, the DNA sequence at the same site varied slightly, and no cuts occurred. Thus, just as an individual might have an A instead of a C at a certain SNP, an individual might have a restriction site allele that allowed cutting or not. Wexler's team was looking for restriction site alleles that were almost always present in diseased individuals but not found any more often than predicted by chance in healthy individuals (**FIGURE 20.10**).

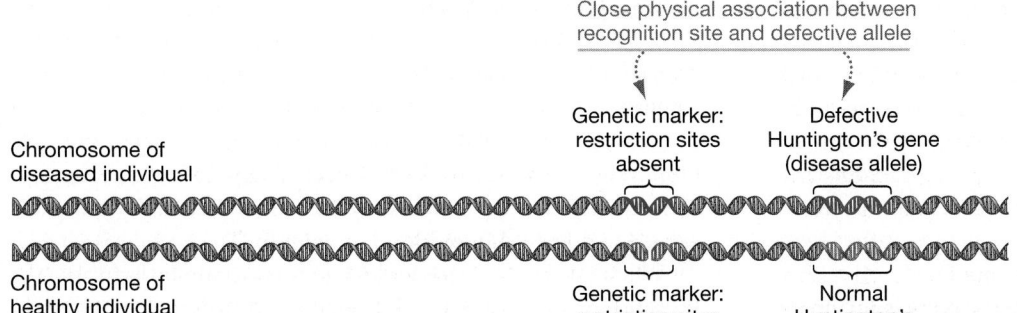

FIGURE 20.10 Genetic Markers Can be Used to Locate Disease Alleles. Because of genetic recombination, genetic markers that are far from the gene of interest are equally likely to be found in both affected and unaffected individuals—there will be no association between particular forms of the marker and either the normal or the disease-causing allele.

Some Common Tools Used in Genetic Engineering

Tool	Description	How Used	Illustration
Reverse transcriptase	Enzyme that catalyzes synthesis of a complementary DNA (cDNA) from an RNA template.	Many applications, including making cDNAs used in constructing a genetic library.	
Restriction endonucleases	Enzymes that cut DNA at a specific sequence—often a palindromic sequence that is six base pairs long.	Allows researchers to cut DNA at specific locations. Cuts in palindromic sites create "sticky ends."	
DNA ligase	An enzyme that catalyzes the formation of a phosphodiester bond between nucleotides on the same DNA strand.	Ligates (joins) sequences that were cut with a restriction endonuclease. Gives researchers the ability to splice fragments of DNA together.	
Plasmids	Small, extrachromosomal circles of DNA found in many bacteria and in some yeast.	After a target gene is inserted into a plasmid, the recombinant plasmid serves as a vector for transferring the gene into a bacterial or yeast cell, so the gene can be cloned.	
Taq polymerase	DNA polymerase from the bacterium *Thermus aquaticus*. Catalyzes synthesis of DNA from a primed DNA template; remains stable at 95°C.	Responsible for the "primer extension" step in the polymerase chain reaction. Heat stability allows enzyme to be active even after the 95°C denaturation step of PCR.	
Single nucleotide polymorphisms (SNPs)	Sites in DNA where the identity of a single base pair varies between alleles.	An important type of polymorphic DNA sequence that is useful in creating the genetic maps required for gene hunts.	

A Linkage Study Once a genetic map containing many genetic markers has been assembled, to find the gene in question, biologists need help from groups of people that include individuals affected by an inherited disease. Recall that the fundamental goal is to find a genetic marker that is almost always inherited along with the disease-causing allele. Biologists call this a linkage study. Gene hunts based on linkage studies are more likely to be successful if large groups are involved. Large sample sizes reduce the possibility that researchers will observe an association between one or more markers and the disease just by chance, rather than because they are closely linked.

Huntington's disease is rare, but Wexler's team was fortunate to find a large, extended family affected with the disease living along the shores of Lake Maracaibo, Venezuela. The researchers followed the inheritance of Huntington's disease and various polymorphic genetic markers within this extended family.

From historical records, the researchers deduced that the Huntington's disease allele was introduced to this family by an English sailor who visited the area in the early 1800s. At the time of the study, over 3000 of his descendants were living in the area. Hundreds of these people had been diagnosed with Huntington's disease. To help in the search for the gene, family members agreed to donate skin or blood samples for DNA analysis.

When Wexler's team looked for associations between the presence or absence of the disease phenotype and genetic markers observed in each family member, they found one marker that

Technique	Description	How Used	Illustration
Recombinant DNA technology	Taking a copy of a gene from one individual and placing it in the genome of a different individual (usually of a different species).	Many applications, including DNA cloning, gene therapy (see Section 20.5), and biotechnology (see Sections 20.1 and 20.6).	Inserted gene
DNA libraries	A collection of all DNA sequences present in a particular source. The library consists of individual DNA fragments that are isolated and inserted into a plasmid or other vector, so they can be cloned.	cDNA libraries allow researchers to catalog the genes being expressed in a particular cell type and to work with coding sequences uninterrupted by introns. Genomic libraries allow researchers to archive all the DNA sequences present in a genome. Libraries can be screened to find a particular target gene.	Stored cDNA
Probing/screening a DNA library	Use of a labeled, known DNA fragment to hybridize (by complementary base pairing) with a collection of unlabeled, unknown fragments.	Allows a researcher to find a particular DNA sequence in a large collection of sequences.	Labeled probe
Polymerase chain reaction (PCR)	A DNA synthesis reaction that uses known primer sequences on either side of a target gene. Reaction is based on many cycles of DNA denaturation, primer annealing, and primer extension.	Produces many identical copies of a target sequence. A shortcut method for DNA cloning.	
Dideoxy sequencing	In vitro DNA synthesis reaction that includes dideoxyribo-nucleoside triphosphates (ddNTPs) as monomers.	Determining the base sequence of a gene or other section of DNA.	G C C T A A T G
Genetic mapping	Creation of a map showing the relative positions of genes or specific DNA sequences on chromosomes. Done by analyzing the frequency of recombination between sequences.	Many applications, including use of mapped genetic markers in genetic association studies to find unknown genes associated with diseases or other distinctive phenotypes.	Yellow body White eyes Ruby eyes 1.4 6.1

turned out to be especially important. Four different restriction site alleles (A, B, C, and D) were present at this location in the genome. The key finding was that the C form of the marker was almost always found in diseased individuals. Almost certainly, the English sailor who introduced the Huntington's disease allele had the C form of the marker in his DNA. The marker and the Huntington's disease gene are so close together that recombination in between them has been extremely rare. No other genetic marker showed this tight association with Huntington's disease.

From the human genetic map that was available at the time, Wexler's team knew that the marker they had identified was on chromosome 4. Eventually the team succeeded in narrowing down the location of the marker, and thus the Huntington's

disease gene, to a region about 500,000 base pairs long. Because the haploid human genome contains over 3 billion base pairs, this was a huge step in focusing the search for the gene.

Pinpointing the Defect Once the general location of the Huntington's disease gene was known, biologists looked in that region for exons that encode an mRNA. Then they used dideoxy sequencing to determine the sequence of exons from diseased and normal individuals, compared the data, and pinpointed specific bases that differed between the two groups.

When this analysis was complete, the research team found that individuals with Huntington's disease have an unusually large number of CAG codons near the 5′ end of one gene. CAG codes for glutamine. Healthy individuals have 11–25 copies of the CAG codon at that location, while affected individuals have 42 or more copies.

When the Huntington's disease research team confirmed that the increased number of CAG codons was always observed in affected individuals, they concluded that the long search for the Huntington's disease gene was over. They named the newly discovered gene *IT15* and its protein product huntingtin. The huntingtin protein is involved in the development of nerve cells. Only later in life do the mutant forms of the protein cause disease.

What Are the Benefits of Finding a Disease Gene?

How have efforts to find disease genes improved human health and welfare? Has the effort to locate the Huntington's disease gene helped researchers and physicians understand and treat the illness? Biomedical researchers point to three major benefits of disease-gene discovery.

Improved Understanding of the Phenotype

Once a disease gene is found and its sequence is known, researchers can usually figure out why its product causes disease. In the case of *IT15*, autopsies of Huntington's patients had shown that their brains decrease in size, and that the brain tissue contains clumps, or aggregates of the protein now called huntingtin.

Huntingtin aggregates are a direct consequence of the increased number of CAG repeats in the *IT15* gene. Long stretches of glutamine are known to promote protein aggregations. The leading hypothesis to explain Huntington's disease proposes that a gradual buildup of huntingtin aggregates triggers neurons to undergo apoptosis, or programmed cell death.

These results explained why Huntington's disease is pleiotropic (see Chapter 14). Patients suffer from abnormal movements *and* personality changes because neurons throughout the brain are killed. The results also help explain why the disease takes so long to appear, and why it is progressive: The defective huntingtin proteins take time to build up to harmful levels and then continue to increase over time. Finally, understanding the molecular mechanism responsible for the illness explained why the disease allele is dominant. One copy of the defective gene is enough to produce fatal concentrations of huntingtin aggregates.

Therapy

Once *IT15* was found, biologists began a search for new therapies for Huntington's disease by introducing the defective allele into mice, using the types of genetic engineering techniques discussed in Section 20.5. These mice with alleles that have been modified by genetic engineering are called **transgenic** (literally, "across-genes").

Transgenic mice that produce defective versions of the huntingtin protein develop a version of Huntington's disease, exhibiting tremors and abnormal movements, higher-than-normal levels of aggression toward litter and cage mates, and a loss of neurons in the brain. Laboratory animals with disease symptoms that parallel those of a human disease provide an **animal model** of the disease (see **BioSkills 13** in Appendix B).

Animal models are valuable in disease research because they can be used to test potential treatments. For example, research groups are using transgenic mice to test drugs that may prevent or reduce the aggregation of the huntingtin protein.

Genetic Testing

When the Huntington's gene was found and sequenced, biologists used the knowledge to develop a test for the presence of the defective allele. The test consists of obtaining a DNA sample from an individual and using the polymerase chain reaction to amplify the chromosome region that contains the CAG repeats responsible for the disease. If the number of CAG repeats is 35 or less, the individual is not considered at risk. Forty or more repeats results in a positive diagnosis for Huntington's.

Thanks to genetic maps based on SNPs, gene hunts are increasingly successful. Biologists have recently documented alleles associated with a predisposition to developing type I and type II diabetes, breast and ovarian cancer, obesity, coronary heart disease, bipolar disorder, Crohn's disease, and rheumatoid arthritis. Genetic testing for these alleles is now available for both prenatal and adult screens.

Ethical Concerns over Genetic Testing

Knowing the genetic basis of human diseases offers hope, but it also raises difficult ethical issues.

Genetic testing, for example, can create serious moral and legal dilemmas as well as harrowing personal choices. Consider that some people maintain that it is morally wrong to terminate any pregnancy, even if the fetus is certain to be born with a debilitating or fatal genetic disease. Think too about Nancy Wexler's position soon after the discovery of *IT15*: Would you choose to be tested for the defective allele and risk finding out that you were almost certain to develop an incurable disease such as Huntington's?

There are other, equally serious, questions. Should people be tested for any disease that has no cure? Should it be legal for insurance companies to test clients? If so, can companies refuse to insure people at risk for diseases that require expensive treatments? What about employers?

These questions are being debated by political and religious leaders, health-care workers, philosophers, and the public at large. In many cases, we've yet to find answers.

check your understanding

If you understand that . . .

- Genes for particular traits can be located if they are inherited together with a known genetic marker.

✓ **You should be able to . . .**

Describe how you would design a study aimed at identifying alleles associated with alcoholism.

Answers are available in Appendix A.

20.5 Case 5–Severe Immune Disorders: The Potential of Gene Therapy

For physicians who treat inherited disorders such as Hunting-ton's disease, sickle-cell anemia, and cystic fibrosis, the ultimate goal is to cure the disease. This may be done by replacing or augmenting defective copies of the gene with normal alleles. This approach to treatment is called **gene therapy.**

For gene therapy to succeed, two crucial requirements must be met. First, the sequence of the allele associated with the healthy phenotype must be known. Second, a method must be available for introducing this allele into affected individuals and having it be expressed in the correct tissues, in the correct amount, and at the correct time. If the defective allele is dominant, then the introduction step may be even more complicated: In at least some cases, the introduced allele must physically replace or block the expression of the undesirable dominant allele.

How Can Genes Be Introduced into Human Cells?

Section 20.1 reviewed how recombinant DNA sequences are packaged into plasmids and taken up by *E. coli* cells. However, humans and other mammals lack plasmids. How can foreign genes be introduced into human cells?

Researchers have focused on packaging foreign genes into viruses for transport into human cells. These viruses have been engineered so they can deliver genes to cells but cannot replicate to produce new viruses. Viral infection begins when a virus particle attaches to a cell and delivers its genome into the cell (Chapter 36). For some viruses, the viral DNA becomes

integrated into a host-cell chromosome. This trait makes it possible to use these viruses as vectors to carry engineered genes into the chromosomes of target cells. Potentially, the genes delivered by the virus could be expressed and produce a product capable of curing a genetic disease.

Vectors used today in gene therapy are often modified retroviruses. Retroviruses have genomes made of single-stranded RNA. When a **retrovirus** infects a human cell, a reverse transcriptase encoded by the virus catalyzes the production of a DNA copy of the virus's RNA genome. Other viral enzymes catalyze the insertion of the viral DNA into a host-cell chromosome. If an RNA version of a human gene can be packaged into a recombinant retrovirus, then the virus will insert a DNA copy of the human gene into a chromosome in a target cell (**FIGURE 20.11**).

Unfortunately, there are problems associated with using retroviruses as agents in gene therapy. For example, if the virus happens to insert the recombinant human gene in a position that disrupts the function of an important gene in the target cell, the consequences may be serious. Despite these risks, modified retroviruses are still among the best vectors currently available for human gene therapy.

Using Gene Therapy to Treat X-Linked Immune Deficiency

In 2000, a research team reported the successful use of gene therapy to treat an illness called severe combined immunodeficiency (SCID). Children who are born with SCID lack a normal immune system and are unable to fight off infections.

The type of SCID the team treated is designated SCID-X1, because it is caused by mutations in a gene on the X chromosome.

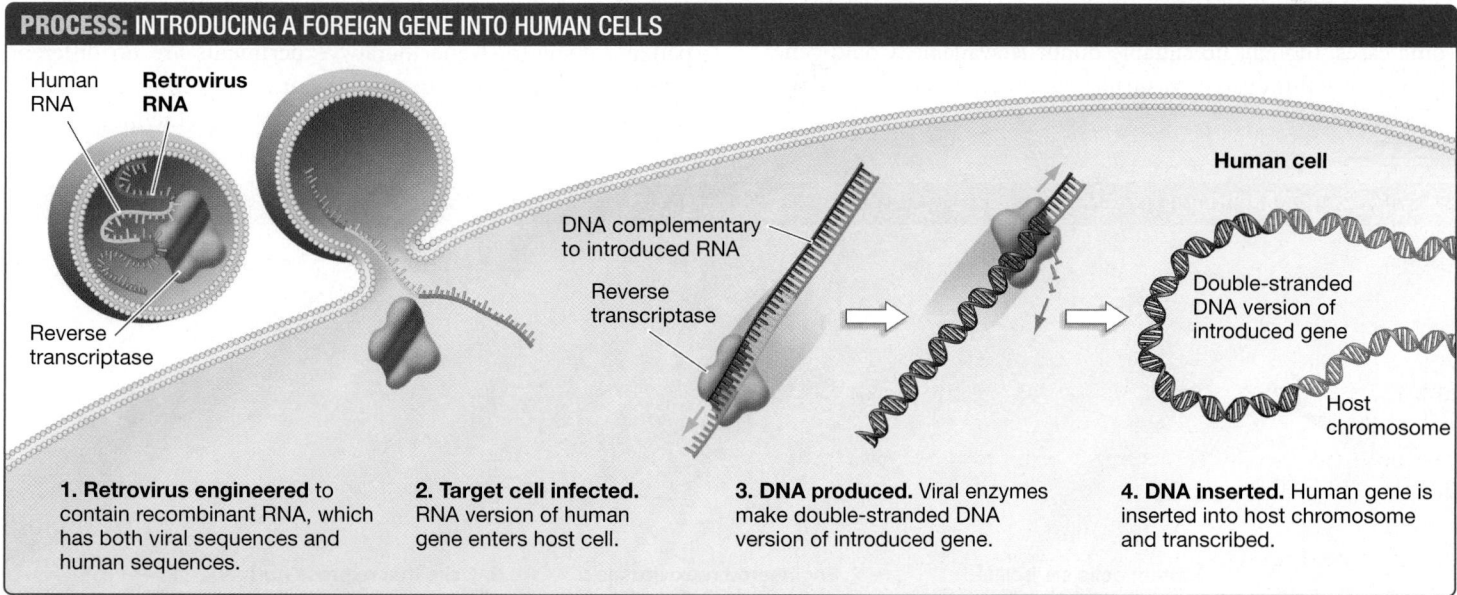

PROCESS: INTRODUCING A FOREIGN GENE INTO HUMAN CELLS

Human RNA
Retrovirus RNA
Reverse transcriptase
DNA complementary to introduced RNA
Reverse transcriptase
Human cell
Double-stranded DNA version of introduced gene
Host chromosome

1. Retrovirus engineered to contain recombinant RNA, which has both viral sequences and human sequences.

2. Target cell infected. RNA version of human gene enters host cell.

3. DNA produced. Viral enzymes make double-stranded DNA version of introduced gene.

4. DNA inserted. Human gene is inserted into host chromosome and transcribed.

FIGURE 20.11 Modified Retroviruses Can Insert a Foreign Gene into a Host-Cell Chromosome. (Many details have been omitted for conceptual clarity.)

✔**QUESTION** What happens if the recombinant DNA is inserted in the middle of a gene that is critical to normal cell function?

FIGURE 20.12 A "Bubble Child." Children with SCID cannot fight off bacterial or viral infections. As a result, such children must live in a sterile environment.

The gene codes for a receptor protein necessary for the development of immune system cells, called T cells, that develop in bone marrow (Chapter 51).

Traditionally, physicians have treated SCID-X1 by keeping the patient in a sterile environment, isolated from any direct human contact, until the person could receive a transplant of bone-marrow tissue from a close relative (**FIGURE 20.12**). In most cases, the T cells that the patient needs are produced by the transplanted bone-marrow cells and allow the individual to live normally. In some cases, though, no suitable donor is available. Could gene therapy cure this disease by furnishing functioning copies of the defective gene?

After extensive testing suggested that their treatment plan was safe and effective, the research team gained approval to treat 10 boys, each less than 1 year old, who had SCID-X1 but no suitable bone-marrow donor. The researchers removed bone marrow from each child, collected the stem cells that produce mature T cells, and infected those cells with an engineered retrovirus that delivered the normal receptor gene. Cells that began to produce normal receptor protein were then isolated and transferred back into the patients (**FIGURE 20.13**).

Within four months after reinsertion of the transformed marrow cells, nine of the boys had normal levels of functioning T cells. These patients were removed from germ-free isolation rooms and began residing at home, where they grew and developed normally.

Subsequently, however, four of the boys developed a cancer characterized by unchecked growth of T cells. Follow-up analyses of their bone-marrow cells showed that the normal receptor gene had been inserted either near a gene for a transcription factor that triggers T-cell growth, or near a gene for a cyclin that drives the cell cycle (see Chapter 12). The inserted receptor gene provided an enhancer that led to constitutive (constant) expression of the transcription factor or cyclin.

Three of the four boys responded to cancer chemotherapy and are healthy. The fourth did not respond to treatment and died of cancer.

The tenth boy to receive gene therapy never produced T cells. For unknown reasons, his recombinant stem cells failed to function normally when they were transplanted back into his bone marrow. Fortunately, physicians were later able to find a bone-marrow donor whose cells matched the boy's closely enough to make a successful transplant possible.

Ethical Concerns over Gene Therapy

Throughout the history of medicine, efforts to test new drugs, vaccines, and surgical protocols have always carried a risk for the patients involved. Gene therapy experiments are no different. The researchers who run gene therapy trials must explain the risks clearly and make every effort to minimize them.

PROCESS: ONE APPROACH TO GENE THERAPY

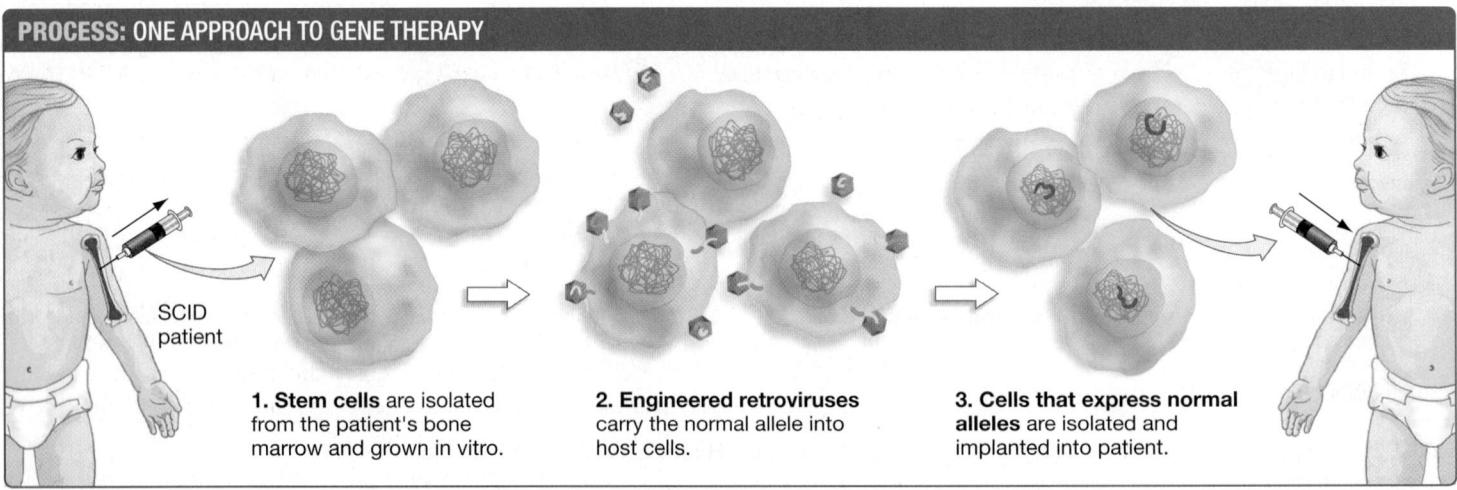

SCID patient

1. **Stem cells** are isolated from the patient's bone marrow and grown in vitro.

2. **Engineered retroviruses** carry the normal allele into host cells.

3. **Cells that express normal alleles** are isolated and implanted into patient.

FIGURE 20.13 Gene Therapy Can Cure a Genetic Disorder. For gene therapy to work in the case of a loss-of-function allele, copies of a normal allele have to be introduced into a patient's cells and be expressed.

The initial report on the development of cancer in the boys who received gene therapy for SCID-X1 concluded with the following statement: "We have proposed . . . a halt to our trial until further evaluation of the causes of this adverse event and a careful reassessment of the risks and benefits of continuing our study of gene therapy."

When recombinant DNA technology first became possible, many researchers thought they would live to see most or all of the serious inherited diseases caused by single-gene mutations cured by gene therapy. After several decades of rare successes punctuated by tragic failures, that optimism was tempered. In the past few years, however, renewed hope has emerged for gene therapy. Improved vectors have been used successfully to treat two different forms of blindness, a brain disorder, and another type of SCID. Perhaps gene therapy is finally poised to deliver on some of its promises.

20.6 Case 6—The Development of Golden Rice: Biotechnology in Agriculture

Progress in transforming crop plants with recombinant genes has been breathtaking. In 2010, a total 10 percent of all farmland worldwide was planted with transgenic crops—and this number is predicted to show double-digit growth over the foreseeable future. In the same year, roughly 90 percent of soybeans, cotton, and corn grown in the United States were genetically engineered. You almost certainly have eaten food from a genetically modified plant sometime, if not today.

Transgenic crops have been engineered largely to meet three objectives:

1. **Reducing losses from herbivore damage** For instance, researchers have transferred a gene from the bacterium *Bacillus thuringiensis* into corn; the presence of the "Bt toxin" encoded by this gene protects the plant from corn borers and other caterpillar pests.

2. **Reducing competition with weeds** An example is the genetic engineering of soybeans for resistance to an herbicide—a molecule that kills plants—called glyphosate. Soybean fields with the engineered strain can be sprayed with glyphosate to kill weeds without harming the soybeans.

3. **Improving food quality** An important example is engineering soybeans and canola to produce a higher percentage of unsaturated fatty acids relative to saturated fatty acids (Chapter 6). Reducing the amounts of saturated fatty acids helps prevent heart disease, so these crops produce healthier vegetable oils.

How is this work done?

Rice as a Target Crop

Almost half the world's population depends on rice as its staple food. Unfortunately, rice is a poor source of some vitamins and essential nutrients—including vitamin A. Vitamin A deficiency causes blindness in 250,000 Southeast Asian children each year. It also increases susceptibility to diarrhea, respiratory infections, and childhood diseases such as measles.

Humans and other mammals synthesize vitamin A from a precursor molecule known as β-carotene (beta-carotene). β-carotene belongs to a family of orange, yellow, and red plant pigments called the carotenoids (Chapter 10).

Rice plants synthesize β-carotene in their chloroplasts but not in the carbohydrate-rich seed tissue called endosperm—the part of the rice seed that you eat. Could genetic engineering produce a strain of rice plants that synthesizes β-carotene in the endosperm?

Synthesizing β-Carotene in Rice

To explore the possibility of genetically engineering rice, a research team searched for compounds in rice endosperm that could serve as precursors for the synthesis of β-carotene. They found that maturing rice endosperm contains a molecule that could be converted to β-carotene in three enzyme-catalyzed reactions. The researchers reasoned that if genes that encode these enzymes could be introduced into rice plants along with regulatory sequences that would trigger their synthesis in endosperm, it should be possible to create a transgenic strain of rice that would contain β-carotene.

Fortunately, genes that encode two of the required enzymes had already been isolated from daffodils, and the gene for the third enzyme had been purified from a bacterium. These genes had been cloned in bacteria. To each of the coding sequences, biologists added regulatory sequences from an endosperm-specific protein. This segment would promote transcription of the recombinant sequences in endosperm cells.

Next, the three sets of sequences had to be inserted into rice plants. How are foreign genes introduced into plants?

The *Agrobacterium* Transformation System

Agrobacterium tumefaciens is a bacterium that infects plant tissues and triggers formation of tumorlike growths called galls. When researchers looked into how these infections occur, they found that a plasmid carried by the *Agrobacterium* cells, called a **Ti (tumor-inducing) plasmid,** plays a key role (**FIGURE 20.14**, see page 386).

Ti plasmids contain several sets of genes. One set encodes products that allow the bacterium to bind to the cell walls of a host. Another set, referred to as the virulence genes, encodes the proteins required to transfer part of the Ti DNA, called T-DNA (transferred DNA), into the plant cell. The T-DNA then travels to the nucleus and integrates into host-cell chromosomes (Figure 20.14, step 1). T-DNA genes are expressed and their products induce the infected cell to grow and divide. This results in the formation of a gall that houses a growing population of *Agrobacterium* cells (Figure 20.14, step 2).

Researchers soon realized that the Ti plasmid offers an efficient way to introduce recombinant genes into plant cells. Follow-up experiments confirmed that recombinant genes could be

PROCESS: Ti PLASMIDS TRANSFER GENES INTO HOST DNA

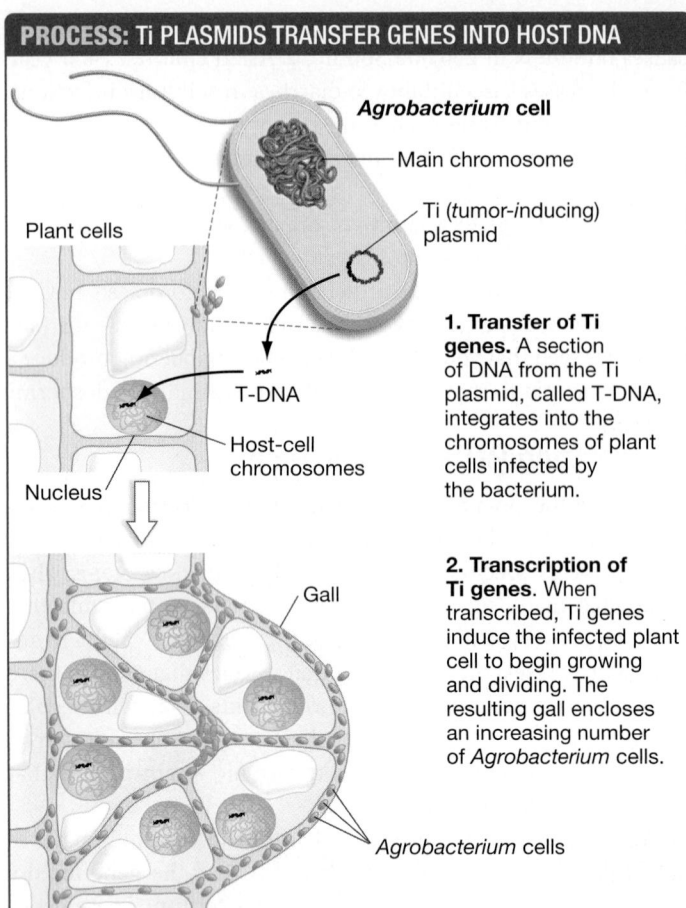

1. Transfer of Ti genes. A section of DNA from the Ti plasmid, called T-DNA, integrates into the chromosomes of plant cells infected by the bacterium.

2. Transcription of Ti genes. When transcribed, Ti genes induce the infected plant cell to begin growing and dividing. The resulting gall encloses an increasing number of *Agrobacterium* cells.

FIGURE 20.14 *Agrobacterium* **Infections Introduce Genes into a Plant Host-Cell Chromosome.** Ti plasmids of *Agrobacterium* cells induce gall formation—a tumorlike growth.

added to the T-DNA that integrates into the host chromosome, that the gall-inducing genes could be removed from the T-DNA, and that the resulting sequence is efficiently transferred and expressed in its new host plant.

Using the Ti Plasmid to Produce Golden Rice

To generate a strain of rice that produces all three enzymes needed to synthesize β-carotene in endosperm, the researchers exposed embryos to *Agrobacterium* cells containing genetically modified Ti plasmids (**FIGURE 20.15**). When the transgenic rice plants grew and produced seeds, the researchers found that some rice grains contained so much β-carotene that they were yellow. The biologists called the engineered plants "golden rice."

Follow-up experiments used gene sequences from corn, rather than daffodil, to produce 23 times more β-carotene in rice than the original transformants contained.

Will golden rice help solve a serious public health problem? The answer is not clear. Many environmental groups are strongly

(a) PROCESS: ENGINEERING OF Ti PLASMIDS

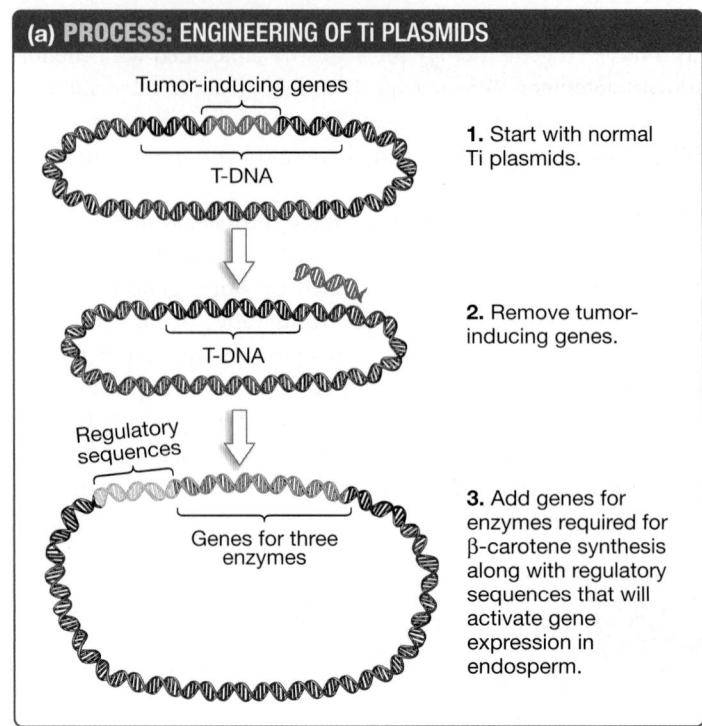

1. Start with normal Ti plasmids.

2. Remove tumor-inducing genes.

3. Add genes for enzymes required for β-carotene synthesis along with regulatory sequences that will activate gene expression in endosperm.

(b) Golden rice (right) is engineered to synthesize β-carotene.

FIGURE 20.15 Constructing a Ti Plasmid to Produce "Golden Rice." Golden rice is a transgenic strain capable of synthesizing β-carotene in the endosperm of its seeds.

opposed to golden rice and any other engineered crops. Regulatory agencies in an array of countries would need to approve the use of golden rice—but there is strong resistance in some nations—and seed would have to be made available to farmers at an affordable price. The barriers to answering the question of whether golden rice can solve health problems are more societal than scientific.

It's important to recognize that each solution offered by genetic engineering introduces new issues to resolve. Biology students and others who are well informed about the techniques and issues involved will be important participants in this debate.

If you understand . . .

20.1 Basic Recombinant DNA Technologies

- In genetic engineering, DNA is added to a cell either to modify the cell's properties or to clone (obtain many identical copies of) the DNA.

- Restriction endonucleases cut DNA at specific locations. The resulting DNA fragments can be inserted into plasmids or other vectors with the help of DNA ligase.

- In many cases, the DNA fragments are inserted into vectors containing regulatory sequences that control expression of inserted genes.

✔ You should be able to explain why a plasmid is needed for gene cloning.

20.2 The Polymerase Chain Reaction

- The polymerase chain reaction (PCR) produces many identical copies of a gene without using cells for cloning.

- PCR depends on having primers that bracket a target stretch of DNA. These allow *Taq* polymerase, a heat-stable DNA polymerase, to amplify a single target DNA sequence to millions of identical copies.

✔ You should be able to list the advantages and disadvantages of cloning in cells versus using PCR to obtain many copies of genes.

20.3 Dideoxy DNA Sequencing

- Dideoxy sequencing determines the sequence of bases in DNA.

- Dideoxy sequencing is based on an in vitro DNA synthesis reaction in which dideoxyribonucleotides stop different DNA replication reactions at different bases in the sequence.

- The DNA fragments of different lengths that are generated by a dideoxy sequencing reaction are separated via gel electrophoresis to determine the sequence of bases.

✔ You should be able to explain how the newly synthesized DNA fragments—when they are lined up by size—can be used to determine the sequence of bases in the template DNA.

20.4 Finding Genes by Mapping

- Genetic maps are often used to find genes associated with phenotypes such as diseases.

- If individuals with a certain phenotype share a particular form of a polymorphic genetic marker (a mapped site with two or more forms in DNA that is unrelated to the phenotype), the gene responsible for the phenotype is likely to be near that marker.

- Once the general area of a gene is known, DNA in the region can be sequenced to determine exactly where the gene is located.

✔ You should be able to explain why genetic markers that are not polymorphic (come in only one form) are not useful in gene hunts.

20.5 The Potential of Gene Therapy

- Researchers are working to cure genetic diseases by gene therapy. This involves inserting normal copies of the defective gene into patients.

- In humans, genes used for gene therapy are often introduced using modified viruses.

- Gene therapy has faced many difficulties but recently has met with some notable successes.

✔ You should be able to describe what makes retroviruses well suited for gene therapy and what the concerns are about their use.

20.6 Biotechnology in Agriculture

- Many important crop plants are genetically engineered for traits that include pest and herbicide resistance, and improved food quality.

- Genes are often introduced into crops by infecting plant cells with bacteria that integrate their plasmid genes into the host-plant genome. By adding recombinant genes to these plasmids, researchers have been able to introduce genes that improve crops.

✔ You should be able to explain how genes are inserted into plants.

(MB) MasteringBiology

1. **MasteringBiology Assignments**

Tutorials and Activities Analyzing DNA Fragments Using Gel Electrophoresis; Cloning a Gene in Bacteria; Gel Electrophoresis of DNA; Making Decisions about DNA Technology: Golden Rice; Producing Human Growth Hormone; Restriction Enzymes; Restriction Enzymes, Recombinant DNA, and Gene Cloning; The Polymerase Chain Reaction

Questions Reading Quizzes, Blue-Thread Questions, Test Bank

2. **eText** Read your book online, search, take notes, highlight text, and more.

3. **The Study Area** Practice Test, Cumulative Test, BioFlix® 3-D Animations, Videos, Activities, Audio Glossary, Word Study Tools, Art

You should be able to . . .

1. What do restriction endonucleases do?

2. What is a plasmid?
   a. an organelle found in many bacteria and certain eukaryotes
   b. a circular DNA molecule that often replicates independently of the main chromosome(s)
   c. a type of virus that has a DNA genome and that infects certain types of human cells, including lung and respiratory tract tissue
   d. a type of virus that has an RNA genome, codes for reverse transcriptase, and inserts a cDNA copy of its genome into host cells

3. When present in a DNA synthesis reaction mixture, a ddNTP molecule is added to the growing chain of DNA. No further nucleotides can be added afterward. Why?

4. Once the gene that causes Huntington's disease was found, researchers introduced the defective allele into mice to create an animal model of the disease. Why was this model valuable?
   a. It allowed the testing of potential drug therapies without endangering human patients.
   b. It allowed the study of how the gene is regulated.
   c. It allowed the production of large quantities of the huntingtin protein.
   d. It allowed the study of how the gene was transmitted from parents to offspring.

5. To begin the hunt for the human growth hormone gene, researchers created a cDNA library from cells in the pituitary gland. What did this library contain?
   a. only the sequence encoding growth hormone
   b. DNA versions of all the mRNAs in the pituitary-gland cells
   c. all the coding sequences in the human genome, but no introns
   d. all the coding sequences in the human genome, including introns

6. What does it mean to say that a genetic marker and a disease gene are closely linked?
   a. The marker lies within the coding region for the disease gene.
   b. The sequence of the marker and the sequence of the disease gene are extremely similar.
   c. The marker and the disease gene are on different chromosomes.
   d. The marker and the disease gene are in close physical proximity and tend to be inherited together.

7. Explain how restriction endonucleases and DNA ligase are used to insert foreign genes into plasmids and create recombinant DNA. Make a drawing that shows why sticky ends are sticky and that identifies the location where DNA ligase catalyzes a key reaction.

8. **QUANTITATIVE** If a particular sequence of DNA were amplified using 25 PCR cycles, then the amount of this DNA would be predicted to increase by _____ -fold.

9. What is a cDNA library? Would you expect the cDNA library from a human muscle cell to be different from the cDNA library from a human nerve cell in the same individual? Explain why or why not.

10. What are genetic markers, and how are they used to create a genetic map?

11. Researchers added regulatory sequences from an endosperm-specific gene to the Ti plasmids used in creating golden rice. This was important to
    a. allow inserted genes to integrate into the plant genome.
    b. increase the endosperm growth rate.
    c. prevent the introduced plasmid from harming the endosperm.
    d. promote expression of introduced genes in the rice grain.

12. Compare and contrast PCR with the DNA synthesis that occurs in cells (see Chapter 15).

13. Suppose you had a large amount of sequence data, similar to the data that Nancy Wexler's team had in the region of the Huntington's disease gene, and that you knew that mRNAs of the species being studied typically contain protein-coding regions about 1500 bases long. How would you use the genetic code (see Chapter 16) and information on the structure of promoters (see Chapters 17 and 18) to find the precise location of one or more genes in your sequence?

14. Modifying germ-line or somatic cells for gene therapy involves the same ethical concerns. True or false?

15. Describe similarities between how researchers screen a DNA library and how they perform a genetic screen—for example, for mutant *E. coli* cells that cannot metabolize lactose (see Chapter 18).

16. A friend of yours is doing a series of PCRs and comes to you for advice. She purchased two sets of primers, hoping that one set would amplify the template sequence shown here. (The dashed lines in the template sequence stand for a long sequence of unspecified bases.) Neither of the primer pairs produced any product DNA, however.

	Primer a		Primer b
Primer Pair 1:	5' CAAGTCC 3'	&	5' GCTGGAC 3'
Primer Pair 2:	5' GGACTTG 3'	&	5' GTCCAGC 3'
Template:	5' ATTCGGACTTG---GTCCAGCTAGAGG 3'		
	3' TAAGCCTGAAC---CAGGTCGATCTCC 5'		

    a. Explain why each primer pair didn't work. Indicate whether both primers are at fault, or just one of them.
    b. Your friend doesn't want to buy new primers. She asks you whether she can salvage this experiment. What do you tell your friend to do?

21 Genomics and Beyond

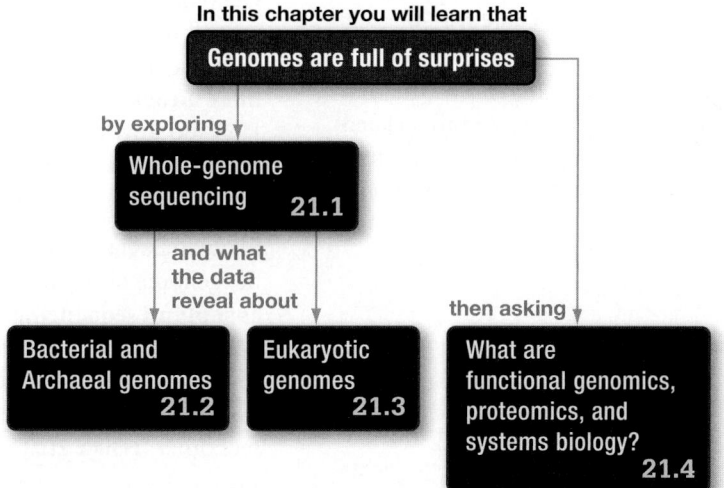

In this chapter you will learn that

Genomes are full of surprises

by exploring

Whole-genome sequencing 21.1

and what the data reveal about

Bacterial and Archaeal genomes 21.2

Eukaryotic genomes 21.3

then asking

What are functional genomics, proteomics, and systems biology? 21.4

A high-throughput robotic genome sequencer. Advances in DNA sequencing technologies are opening new questions to biologists and deepening the understanding of life and its evolution.

The first data sets describing the complete DNA sequence, or **genome,** of humans were published in February 2001. These papers were immediately hailed as a landmark in the history of science. In just 50 years, biologists had gone from not understanding the molecular nature of the gene to knowing the molecular makeup of every gene present in our species.

Years later, knowledge continues to stream from the multinational effort called the **Human Genome Project** and its many spinoffs. It's important to recognize, though, that research on *Homo sapiens* is part of a much larger, ongoing effort to gain insights from the genome sequences of an array of other eukaryotes, bacteria, and archaea. The pace of progress in this field is nothing short of explosive.

The effort to sequence, interpret, and compare whole genomes is **genomics.** While whole-genome sequencing supplies a list of the genes present in an organism, **functional genomics**

✔ When you see this checkmark, stop and test yourself. Answers are available in Appendix A.

UNIT 3

GENE STRUCTURE AND EXPRESSION

389

answers questions about the functioning of that genome, such as what particular genes do and how they're expressed.

Genomics has spawned a host of related fields. These are often referred to as the *–omics*—proteomics, metabolomics, and transcriptomics—but also include emerging areas like systems biology. Like genomics, these fields take a holistic approach to learning about the entire set of proteins, metabolites, or RNA transcripts present in a given cell or tissue type at a given time.

As an introductory biology student, you are part of the first generation trained in the genome era. Genomics and the related fields it has generated are revolutionizing biological science. They will almost certainly be an important part of your personal and professional life. Let's delve in.

21.1 Whole-Genome Sequencing

Genomics has moved to the cutting edge of research in biology, largely because of technological advances. These began with the development of dideoxy sequencing and progressed to next-generation sequencing techniques (introduced in Chapter 20). These technical breakthroughs have enabled obtaining immense quantities of high-quality sequence data rapidly and at low cost.

As technology continues to become faster and less expensive, the pace of genome sequencing accelerates. The result is that an almost mind-boggling number of sequences of genes and whole genomes are now being generated. As this book goes to press, the primary international repository for DNA sequence data contained over 425 *billion* nucleotides. By way of comparison, a haploid human genome contains about 3 billion nucleotides on each strand of the DNA double helix.

FIGURE 21.1 gives a visual sense of the growth in sequence data by plotting the number of nucleotides, in billions, versus time. There are three large international online repositories for

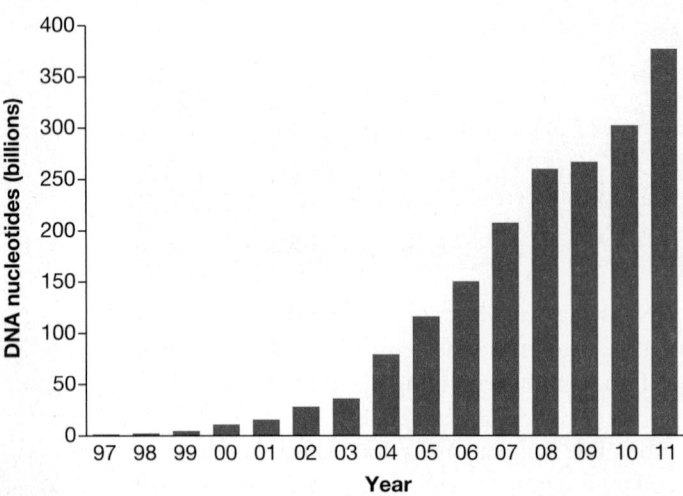

FIGURE 21.1 The Total Number of Bases Sequenced Is Growing Rapidly. Data from the EMBL Nucleotide Sequence Database (also known as EMBL-Bank).

DATA: European Nucleotide Archive/EMBL-Bank Release Notes. Release 110, December 2011. www.ebi.ac.uk/embl/.

sequence data; the numbers plotted here were compiled from one of them. Over 15 years, this database has grown at a staggering average rate of about 46 percent per year.

How Are Complete Genomes Sequenced?

Genomes range in size from about a half million base pairs to several billion. But even under the best conditions, a single dideoxy sequencing reaction can analyze only about 1000 nucleotides. Reads from next-generation sequencing are even shorter. How do investigators break a genome into sequencing-sized pieces and then figure out how the thousands or millions of pieces go back together?

Shotgun Sequencing When researchers first set out to sequence the genome of a species, they usually rely on an approach known as **shotgun sequencing.** In shotgun sequencing, a genome is broken up into a set of overlapping fragments that are small enough to be sequenced completely. The regions of overlap are then used as guides for putting the sequenced fragments back into the correct order (**FIGURE 21.2**).

Step 1 Application of high-frequency sound waves, or sonication, is used to break a genome randomly into pieces about 160 kilobases (kb) long (1 kb = 1000 bases).

Step 2 Each 160-kb piece is inserted into a type of cloning vector called a **bacterial artificial chromosome (BAC).** BACs are able to replicate large segments of DNA. Each BAC is then inserted into a different *Escherichia coli* cell (using techniques introduced in Chapter 20), creating a **BAC library.** A BAC library is a genomic library: a set of all the DNA sequences in a particular genome, split into small segments and inserted into cloning vectors. By allowing each cell to grow into a colony, researchers can isolate large numbers of each 160-kb fragment.

Step 3 After many copies of each 160-kb fragment have been produced, each cloned DNA is again broken into fragments—but this time, into segments about 1 kb long.

Step 4 These small fragments are then inserted into plasmids and placed inside bacterial cells. (Note that by this point the genome has been broken down twice, into increasingly manageable pieces: 160-kb fragments in BACs and 1-kb segments in plasmids.) The plasmids are copied many times as each bacterial cell grows into a large population. Cloned 1-kb fragments are then available for sequencing reactions.

Step 5 Next, the cloned 1-kb fragments from each 160-kb BAC clone are sequenced, and computer programs analyze regions where the ends of different 1-kb segments overlap. Overlaps occur because many copies were made of each 160-kb segment, and these copies were fragmented randomly by sonication.

Step 6 A computer program searches for overlaps between 1-kb fragments from a single BAC clone and stitches the sequences together until a continuous sequence across the BAC has been reconstructed.

Step 7 The ends of the reconstructed BACs are analyzed in a similar way. The goal is to link sequences from each 160-kb

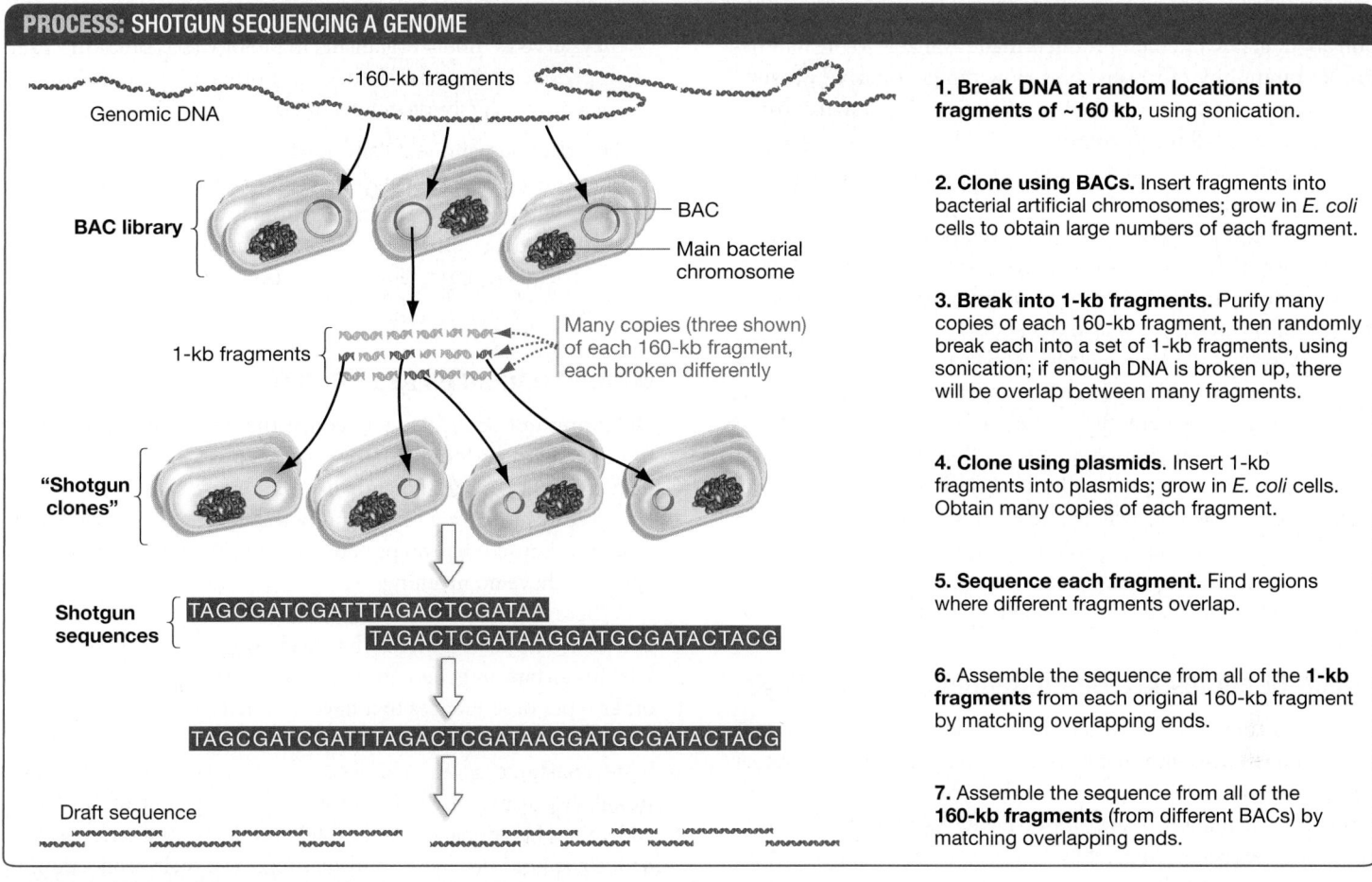

~160-kb fragments

Genomic DNA

BAC library

BAC

Main bacterial chromosome

1-kb fragments

Many copies (three shown) of each 160-kb fragment, each broken differently

"Shotgun clones"

Shotgun sequences

TAGCGATCGATTTAGACTCGATAA

TAGACTCGATAAGGATGCGATACTACG

TAGCGATCGATTTAGACTCGATAAGGATGCGATACTACG

Draft sequence

1. **Break DNA at random locations into fragments of ~160 kb**, using sonication.

2. **Clone using BACs.** Insert fragments into bacterial artificial chromosomes; grow in *E. coli* cells to obtain large numbers of each fragment.

3. **Break into 1-kb fragments.** Purify many copies of each 160-kb fragment, then randomly break each into a set of 1-kb fragments, using sonication; if enough DNA is broken up, there will be overlap between many fragments.

4. **Clone using plasmids.** Insert 1-kb fragments into plasmids; grow in *E. coli* cells. Obtain many copies of each fragment.

5. **Sequence each fragment.** Find regions where different fragments overlap.

6. Assemble the sequence from all of the **1-kb fragments** from each original 160-kb fragment by matching overlapping ends.

7. Assemble the sequence from all of the **160-kb fragments** (from different BACs) by matching overlapping ends.

FIGURE 21.2 Shotgun Sequencing Breaks Large Genomes into Many Short Segments.

✔ **QUESTION** A shotgun blast produces many small, scattered pieces of shot. Why is "shotgun" an appropriate way to describe this sequencing strategy?

segment based on regions of overlap until the sequence of an entire chromosome is assembled.

In essence, the shotgun strategy consists of breaking a genome into many small fragments, sequencing each fragment, and then putting the sequence data back in the correct order.

✔ If you understand shotgun sequencing, you should be able to explain why it is essential for fragments to have regions of overlap.

The Impact of Next-Generation Sequencing Today, there are approaches that are much faster and cheaper than dideoxy sequencing (see Chapter 20). These next-generation methods are massively parallel, meaning that millions of DNA fragments can be sequenced simultaneously in one run of a sequencing machine. The downside of these methods is that they produce sequence reads of only about 50–200 nucleotides, depending on the particular technology. This is in contrast to the roughly 1000 nucleotides obtained by dideoxy sequencing. For piecing together a whole genome, especially one with many repetitive sequences such as those present in most eukaryotes, these read lengths are too short.

But if a complete genome is already available for the organism, next-generation sequencing offers a remarkably quick and inexpensive way to sequence the entire genome from a particular individual—with all the tiny fragments arranged in the correct order by being compared with the "master genome." This sequencing power has opened up possibilities that were unimaginable even a few years ago.

Consider what was involved in sequencing the human genome. The Human Genome Project required more than 15 years and about $3 billion to assemble the first human genome sequence. In 2011, ten years after the first human genome sequence was available, more than 2700 individual human genomes had been sequenced at a cost as low as $5000 per genome. This is a 2700-fold increase in output and a 600,000-fold drop in price.

What's important is not only the numbers, but what can be done with the information. One illustration is an offshoot of the Human Genome Project, the 1000 Genomes Project. This effort has already sequenced the genomes of over 1000 people selected from diverse populations spread across the planet. An important goal is to assess the genetic similarities and differences among people in order to understand our own evolution. Many similarly ambitious and exciting genomics projects are under way.

Bioinformatics How do researchers piece together the millions of fragments produced by shotgun sequencing? Once a complete genome is assembled, how are the raw sequence data and information about genes and their products made available to the international community of researchers? How can genomes of different species be compared to learn about evolutionary relationships?

The answer is **bioinformatics**—a field that fuses mathematics, computer science, and biology in an effort to manage and analyze sequence data. Researchers in bioinformatics have created searchable databases. These vast repositories hold sequence information that allows investigators to evaluate the similarities between newly discovered genes and genes that have been studied previously in the same or other species.

The World Wide Web has put sequence databases at the fingertips of anyone with an Internet connection. For example, the U.S. National Center for Biotechnology Information (NCBI) is only a click away on your computer. At this free and publicly accessible site, you can search billions of nucleotides by using programs such as BLAST, which can quickly find DNA sequences related to any new gene uncovered in a genomics project.

The vast quantity of data generated by genome sequencing centers makes bioinformatics an indispensable element of genomics.

Which Genomes Are Being Sequenced, and Why?

The first genome to be sequenced from an organism—not a virus—came from a bacterium that lives in the human upper respiratory tract. *Haemophilus influenzae* has one circular chromosome and a total of 1.8 million base pairs of DNA. Its genome was small enough to sequence completely in a reasonable amount of time and within a reasonable budget, given the technology available in the early 1990s. *H. influenzae* was an important research subject because it causes earaches and respiratory tract infections in children. One strain is also capable of infecting the membranes surrounding the brain and spinal cord, causing meningitis.

Publication of the *H. influenzae* genome in 1995 was quickly followed by publication of complete genomes sequenced from an assortment of bacteria and archaea. Sequencing of the first eukaryotic genome, from the yeast *Saccharomyces cerevisiae*, was finished in 1996. Today, the genomes of more than 2000 species from all domains of life have been sequenced. That number is certain to continue climbing in the coming years.

Most of the organisms that have been selected for whole-genome sequencing have interesting biological properties, represent a particular branch of life informative for evolutionary investigations, or cause disease. For example:

- Genomes of bacteria and archaea from hot environments have been sequenced in the hopes of discovering enzymes useful for high-temperature industrial applications.

- A set of more than 50 genomes of diverse bacteria and archaea was sequenced to explore patterns of evolution that are impossible to study by other means.

- Genomes such as those of rice and maize (corn) have been sequenced for crop improvement applications.

- The fruit fly *Drosophila melanogaster*, the roundworm *Caenorhabditis elegans*, the house mouse *Mus musculus*, and the mustard plant *Arabidopsis thaliana* were analyzed because they serve as model organisms in biology (see **BioSkills 13** in Appendix B). Data from these and other well-studied organisms have helped researchers interpret the human genome.

- The platypus and African elephant genomes have been sequenced to reveal evolutionary relationships among mammals. Although the elephant genomes are not complete, the available data confirmed that there are two distinct species of African elephant. This information is vital to conservation plans.

Which Sequences Are Genes?

Obtaining raw sequence data is just the beginning of the effort to understand a genome. As researchers point out, raw sequence data are analogous to the parts list for a house. The list, however, would read something like "windowwabeborogovestaircasedoorjubjub" because it has no punctuation and contains portions that appear to have no meaning.

Where do the genes for "window," "staircase," and "door" start and end? Are the segments that read "wabeborogove" and "jubjub" important in gene regulation, or are they simply spacers or other types of sequences that have no function at all?

The most basic task in interpreting a genome is to identify which bases constitute genes. This task is called **genome annotation.** Recall the current definition of a gene: A segment of DNA that codes for a functional RNA or protein product—or a series of alternatively spliced products—and that regulates their production. In bacteria and archaea, identifying genes is relatively straightforward. The task is much more difficult in eukaryotes.

Identifying Genes in Bacterial and Archaeal Genomes To interpret bacterial and archaeal genomes, biologists begin with computer programs that scan the sequence of a genome in both directions. These programs identify each reading frame that is possible on the two strands of the DNA. Recall that a reading frame is a continuous sequence of non-overlapping codons (see Chapter 16).

Codons consist of three bases, so three reading frames are possible on each DNA strand, for a total of six possible reading frames (**FIGURE 21.3**). Because randomly generated sequences contain a stop codon at about 1 in every 20 codons on average, a long stretch of codons that lacks a stop codon is a good indication of a protein-coding sequence. The computer programs draw attention to any "gene-sized" stretches of sequence that lack an internal stop codon and are flanked by a stop codon and a start codon. Because polypeptides range in size from a few dozen amino acids to many hundreds of amino acids, gene-sized stretches of sequence range from several hundred bases to thousands of bases. In addition, the computer programs look for sequences typical of promoters, ribosome binding sites, or other regulatory sites. DNA segments that are identified in this way are called **open reading frames,** or **ORFs.**

Once an ORF is found, a computer program compares its sequence with the sequences of known genes from another species.

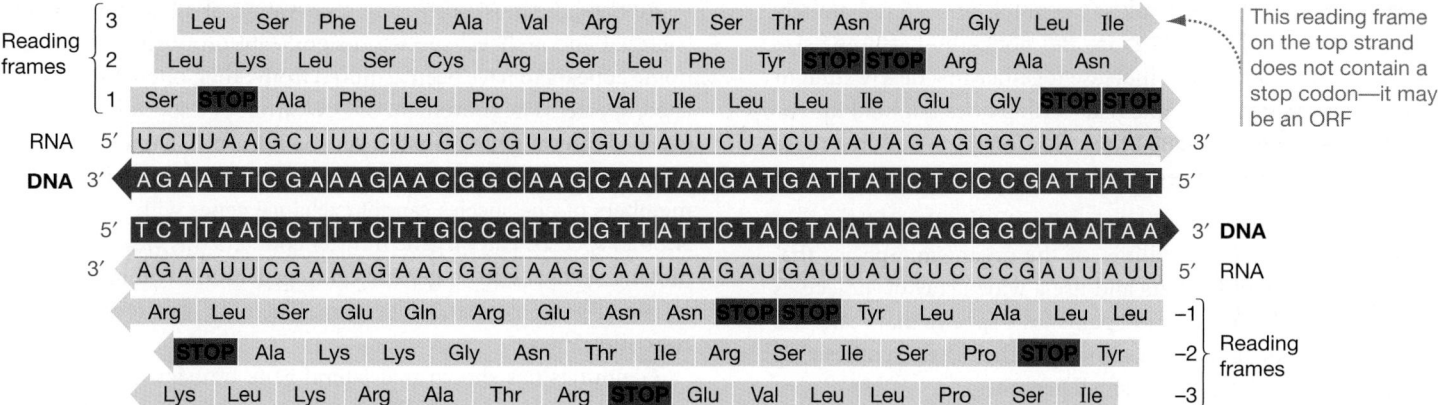

FIGURE 21.3 Open Reading Frames Can Identify Genes. Computer programs use the genetic code to translate the three possible reading frames on each strand of DNA. A long stretch of codons that lacks a stop codon may be an open reading frame (ORF) and identify a gene.

✓**QUESTION** To predict the mRNA codons that would be produced by a particular reading frame, a computer analyzes the DNA in the 3′-to-5′ direction. Why?

If the ORF is unlike any gene that has so far been described in any species, further research is required before it can actually be considered a gene. In contrast, if an ORF shares a significant amount of sequence with a known gene from another species, then it is very likely to be a gene.

Similarities between genes in different species are usually due to **homology.** If genes are homologous, it means they are similar because they are related by descent from a common ancestor. Homologous genes have similar base sequences and frequently the same or a similar function. For example, consider the genes that code for enzymes involved in repairing mismatches in DNA (introduced in Chapter 15). The mismatch-repair genes in *E. coli*, yeast, and humans are similar in DNA sequence and function. To explain this similarity, biologists hypothesize that the common ancestor of all cells living today had mismatch-repair genes—thus, the descendants of this ancestral species also have versions of these genes.

Identifying Genes in Eukaryotic Genomes Mining eukaryotic sequence data for protein-coding genes is complicated. For example, because coding regions are broken up by introns, it is not possible to scan for long ORFs. Instead, researchers combine an array of approaches.

Perhaps the most productive gene-finding strategy has been to isolate mRNAs from cells, use reverse transcriptase to produce a cDNA version of each mRNA, and sequence a portion of the resulting molecule to produce an **expressed sequence tag,** or **EST.** ESTs represent protein-coding genes. To locate the gene, researchers use the EST to find the matching sequence in genomic DNA.

Although ESTs and many other gene-finding strategies have been fruitful, it will likely be many years before biologists are convinced they have identified all the genes in even a single eukaryotic genome. As gene identification efforts continue, researchers are analyzing the data and making some remarkable discoveries. Let's first consider what genome sequencing has revealed about the nature of bacterial and archaeal genomes and then move on to eukaryotes. Is the effort to sequence whole genomes paying off?

21.2 Bacterial and Archaeal Genomes

Biologists have obtained the genome sequences of thousands of distinct bacterial and archaeal species or strains. For example, researchers have sequenced the genome of a laboratory population of *Escherichia coli*—derived from the harmless strain that lives in your gut—as well as the genome of a form that causes severe disease in humans. As a result, researchers can identify genes that differ between these strains and begin experiments to learn what accounts for the infectious properties of some strains.

What general observations have biologists been able to make about the nature of all these bacterial and archaeal genomes?

The Natural History of Prokaryotic Genomes

In a sense, biologists who are working in genomics can be compared to the naturalists of the eighteenth and nineteenth centuries. These early biologists explored the globe, collecting the plants and animals they encountered. Their goals were to describe what existed and identify any patterns. Similarly, the first task of a genome sequencer is to catalog what is in a genome—specifically, the number, type, and organization of genes—and then look for patterns within and between different genomes. Here are some principles that have emerged from analysis of prokaryotic genomes:

- Bacterial and archaeal genomes are compact. They have uninterrupted coding sequences, little space between genes, extensive use of operons, and relatively few regulatory sequences. This structure leads to a linear relationship between genome size and gene number. Look at the graph of bacterial and archaeal genomes in **FIGURE 21.4a**, and notice how genome size and the number of genes increase together in a nearly straight line. In contrast, this simple relationship does not hold for eukaryotic genomes (**FIGURE 21.4b**). Section 21.3 explores these features of eukaryotic genome organization in detail.

- In bacteria, there is a correlation between the size of a genome and the metabolic capabilities of the organism. Species that live in a variety of habitats and use a wide array of molecules for food have large genomes; parasites—species that make use of a host's biochemical machinery rather than synthesizing their own molecules—have small genomes.

- The function of many bacterial and archaeal genes is still unknown. Across a wide range of species, a function cannot yet be assigned to 15 to 30 percent of genes.

- Most of the genes found in one species are not shared widely with others. In a study that sampled the genomes of diverse prokaryotes, every time a new genome was sequenced, more than 1000 new genes were discovered. Only a small set of genes involved in processes such as DNA replication, transcription, and translation are similar across a wide range of bacteria or archaea.

- The content of genomes varies widely even within species. For example, sequencing of 17 different *E. coli* strains revealed a total of about 13,000 different genes. The genome of any one strain has only about 4400 genes, and roughly 2200 of these are shared by all strains. The remaining genes in each genome are found only in some strains but not in others.

- Prokaryotic genomes are frequently rearranged during evolution. Even closely related species show little similarity in gene order.

Perhaps the most surprising observation of all is that in many bacterial and archaeal species, a significant proportion of the genome appears to have been acquired from other, often distantly related, species. The movement of DNA from one species to another is called **lateral gene transfer.** As you'll learn (Chapter 27), there are many ways to define a species. But in all traditional definitions, members of one species cannot exchange genes with members of another species. Lateral gene transfer counters this view: Instead of moving vertically from generation to generation within a species, in lateral gene transfer, genes move "laterally" between different coexisting species.

Genomics has shown that lateral gene transfer is more common than ever imagined. This finding is causing many biologists to wonder if conventional evolutionary trees capture the way bacteria and archaea have evolved.

Lateral Gene Transfer in Bacteria and Archaea

The extent of lateral gene transfer is still debated, but genomic data indicate that all bacterial and archaeal species have experienced lateral gene transfer. Lateral gene transfer is a major force in the evolution of bacteria and archaea.

One illustration is the bacterium *Thermotoga maritima*. This species thrives in high-temperature environments near deep-sea vents. Almost 25 percent of the genes in this species are closely related to genes found in archaea that live in the same habitats. The archaea-like genes occur in well-defined clusters within the *T. maritima* genome. These observations support the hypothesis that these sequences were transferred in large pieces from an archaean to a bacterium—organisms in two different domains of life.

How are laterally transferred genes identified? Biologists primarily use two criteria: (**1**) A gene is much more similar to genes in distantly related species than to those in closely related species,

(a) In bacteria (○) and archaea (◉), genome size and number of genes increase together in a linear relationship.

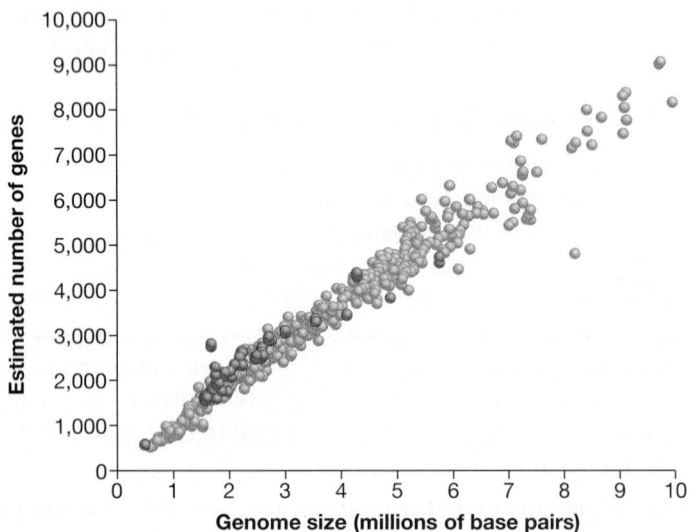

(b) The same relationship does not hold true for eukaryotes (●).

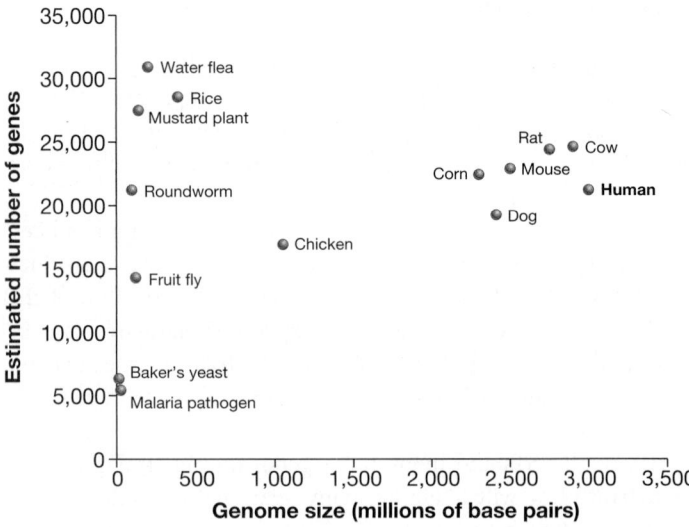

FIGURE 21.4 Relationship Between Genome Size and Gene Number.

DATA: (a) Hou, Y., and S. Lin. 2009. *PLoS ONE* 4(9): e6978, Supplemental Table S1. (b) KEGG: Kyoto Encyclopedia of Genes and Genomes, KEGG Organisms: Complete Genomes. www.genome.jp/kegg/.

and (2) the proportion of G-C base pairs to A-T base pairs in a particular gene or series of genes is markedly different from the base composition of the rest of the genome. This second criterion works because the proportion of G-C base pairs in a genome is characteristic of the particular genus or species.

How can genes move from one species to another? In some cases, plasmids are responsible. For example, most of the genes that are responsible for conferring resistance to antibiotics are found on plasmids. Researchers have documented the transfer of plasmid-borne antibiotic-resistance genes between distantly related bacteria. In many cases of lateral gene transfer, genes from plasmids become integrated into the main chromosome of a bacterium through genetic recombination.

Lateral gene transfer may also occur by transformation—when bacteria and archaea take up raw pieces of DNA from the environment—and by viruses that pick up DNA from one cell and transfer it to another cell.

There is no doubt that lateral gene transfer occurs even between distantly related organisms. What is still debated, however, is how much this shakes the tree of life. If lateral gene transfer is rare, then evolutionary paths in the bacteria and archaea form a set of branches that begin at common ancestors and spread out to descendants. If lateral gene transfer is as widespread as some biologists believe, then evolutionary paths must form a complex interconnected network that links species in a web of vertical and lateral gene transfers. These alternative views are shown in **FIGURE 21.5**.

In light of new genomics findings, will a tree of life stand for bacteria and archaea? Or are the evolutionary relationships in these domains of life best viewed as a web? Stay tuned.

Metagenomics

Biologists continue to gain important insights from sequencing the genomes of individual species and strains. But more recently, some research groups have taken a different approach: cataloging all the genes present in a community of bacteria and archaea. This type of research is called **metagenomics,** or **environmental sequencing.** The subject of these studies is genes—not organisms.

The first environmental sequencing study was conducted in the Sargasso Sea, near the Caribbean island of Bermuda. Researchers chose the spot because it is extremely nutrient poor and species poor—a desert in the ocean. This is a common strategy

in biological science: Start by studying a simple system, then go on to more complex situations. To inventory the complete array of bacterial genes present, the research group collected cells from different water depths and locations, isolated DNA from the samples, and sequenced the DNA.

After analyzing over 1 billion nucleotides, the team concluded that at least 1800 bacterial species were present, of which 148 were previously undiscovered. They also identified more than 1.2 million genes that had never before been characterized. These genes included over 780 sequences that code for proteins similar to the rhodopsin found in the cells of your retina—a molecule that is absorbing the light entering your eye right now. Follow-up work suggests that most of the Sargasso Sea rhodopsin-like molecules are also absorbing light, and that bacterial cells use the energy of the light to pump protons across their plasma membranes—creating a chemiosmotic gradient that can synthesize ATP (see Chapter 9).

Many metagenomics studies are going on today, including one that examined microbes of the human gut (Chapter 29). All these investigations are providing new insights about the living world, from how rhodopsin-like proteins help bacteria thrive in a desert to how your lunch is being digested.

check your understanding

C Y U

If you understand that . . .

- The size of bacterial and archaeal genomes correlates with the cell's metabolic capabilities.
- Lateral gene transfer—movement of DNA from one species to another—is extensive in prokaryotes.
- Environmental sequencing catalogs all the genes found in a particular habitat.

✓ **You should be able to . . .**

1. Explain why it is logical to observe that parasitic bacteria have small genomes.

2. Explain the logic behind claiming that a gene's similarity to a gene in a distantly related species, and dissimilarity to the same gene in closely related species, is evidence for lateral gene transfer.

Answers are available in Appendix A.

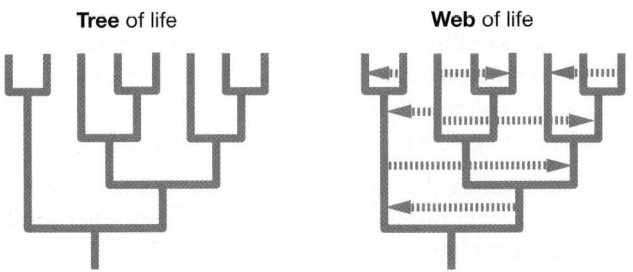

Tree of life   **Web** of life

FIGURE 21.5 Tree of Life and Web of Life Views of Evolution. The broken horizontal arrows indicate lateral gene transfer events.

21.3 Eukaryotic Genomes

DNA sequencing has revealed some extraordinary features of eukaryotic genome organization. Genome size varies widely, but the number of genes is much more similar. For example, the genomes shown in Figure 21.4b vary roughly 250-fold in size (from 12 to 3000 million base pairs) but less than sixfold in the number of genes (from 5400 to 30,900). How can gene number be so similar among organisms with vastly different genome sizes and that range in complexity from single-celled parasites to large multicellular plants and animals?

Before exploring possible answers, let's consider two daunting challenges in sequencing eukaryotic genomes. The first is size. The largest bacterial genome, that of *Sorangium cellulosum,* is slightly more than 13 million base pairs. However, as Figure 21.4b shows, even modest-sized eukaryotic genomes are much larger. The 3-billion-base-pair human genome dwarfs those of bacteria but is miniscule compared with the genome of the Japanese flower *Paris japonica.* This plant's genome contains 149 billion base pairs of DNA—that's 50 times the size of the human genome!

The second challenge in sequencing eukaryotic genes is coping with sequences that are repeated many times. Many eukaryotic genomes are dominated by repeated DNA sequences that occur between genes or inside introns and do not code for products used by the organism. These repeated sequences greatly complicate the work of aligning and interpreting sequence data. They also explain some of the paradox of the immense variation in eukaryotic genome sizes. If repeated sequences don't code for products needed by cells, what do they do?

Transposable Elements and Other Repeated Sequences

In many eukaryotes, the exons and regulatory sequences associated with genes make up a relatively small percentage of the genome. Over 90 percent of a bacterial or archaeal genome consists of genes, but about 50 percent of an average eukaryotic genome consists of repeated sequences that do not code for a product used by the cell.

When repeated sequences were discovered, they were initially considered "junk DNA" that was nonfunctional and probably unimportant and uninteresting. But subsequent work has shown that many of the repeated sequences observed in eukaryotes are actually derived from sequences known as transposable elements.

Transposable elements are segments of DNA that are capable of being inserted into new locations, or transposing, in a genome. They were first discovered in corn by Barbara McClintock and later shown to be present in organisms from every domain of life. Transposable elements behave similarly to some viruses that insert into the genome. In contrast to viruses, however, transposable elements seldom leave their host cell—instead, they make copies of themselves that become inserted in new locations. Transposable elements are passed from mother to daughter cell and from parents to offspring, generation after generation, because they are part of the genome.

A transposable element is an example of what biologists call selfish DNA: a DNA sequence that has invaded a host and persists and reproduces using the resources of the host. Selfish or not, transposable elements play a big role in a species as they move from place to place and cause mutation. Like any mutation, a transposable element insertion into a new site in the genome can have negative, neutral, or positive effects on fitness. Transposable elements are a significant part of almost all cellular genomes and play a major role in evolution. Transposable elements have many effects, disrupting the coding sequence of genes, changing patterns of gene regulation, and promoting gene duplication and loss. They are genome invaders that also shape the structure and function of genomes in profound ways.

How Do Transposable Elements Work? Transposable elements come in a wide variety of types and spread through genomes in a variety of ways. Different organisms—*E. coli,* fruit flies, yeast, and humans, for example—contain distinct types of transposable elements. Bacterial and archaeal genomes, however, have far fewer transposable elements compared with most eukaryotes. This observation has inspired the hypothesis that bacteria and archaea either have efficient means of removing parasitic sequences or can somehow thwart insertion events.

As an example of how these selfish DNA sequences work, consider a well-studied type of transposable element called a **long interspersed nuclear element (LINE).** LINEs are found in humans and other animals. Because LINEs have a reverse transcriptase like retroviruses (introduced in Chapter 20), biologists hypothesize that they are derived from them evolutionarily. Your genome contains nearly 1 million LINEs, each between 1000 and 5000 bases long. Transposable elements of different types make up over 45 percent of the human genome and 85 percent of the corn (maize) genome. **FIGURE 21.6** illustrates the steps that allow an active LINE to transpose.

Most of the LINEs observed in the human genome do not actually function, however, because they don't contain a promoter or the genes for either reverse transcriptase or integrase. To make sense of this observation, researchers hypothesize that the insertion process illustrated in steps 6 and 7 of Figure 21.6 is often disrupted in some way, leaving the inserted replica of the original LINE incomplete.

Research on transposable elements and lateral gene transfer has revolutionized how biologists view the genome. Many genomes are riddled with transposable elements, and others have undergone radical change in response to lateral gene transfer events. In other words, genomes are much more dynamic and complex than previously thought. Their size and composition can change dramatically over time.

Repeated Sequences and DNA Fingerprinting In addition to containing repeated sequences from transposable elements, many eukaryotic genomes have several thousand loci of relatively short DNA sequences. They are repeated in tandem, one after another, contiguously along part of a chromosome.

These tandem repetitive DNA sequences fall into two major classes:

1. Repeating units that are just 2 to about 6 bases long. These are **microsatellites,** also known as **short tandem repeats (STRs)** or **simple sequence repeats.** The most common type of microsatellite in humans is a repeated stretch of the dinucleotide AC, giving the sequence ACACACAC . . . .

2. Repeating units that are longer, from about 6 to 100 bases long. These are **minisatellites,** or **variable number tandem repeats (VNTRs).**

Microsatellite sequences are thought to originate when DNA polymerase skips or mistakenly adds extra bases during replication;

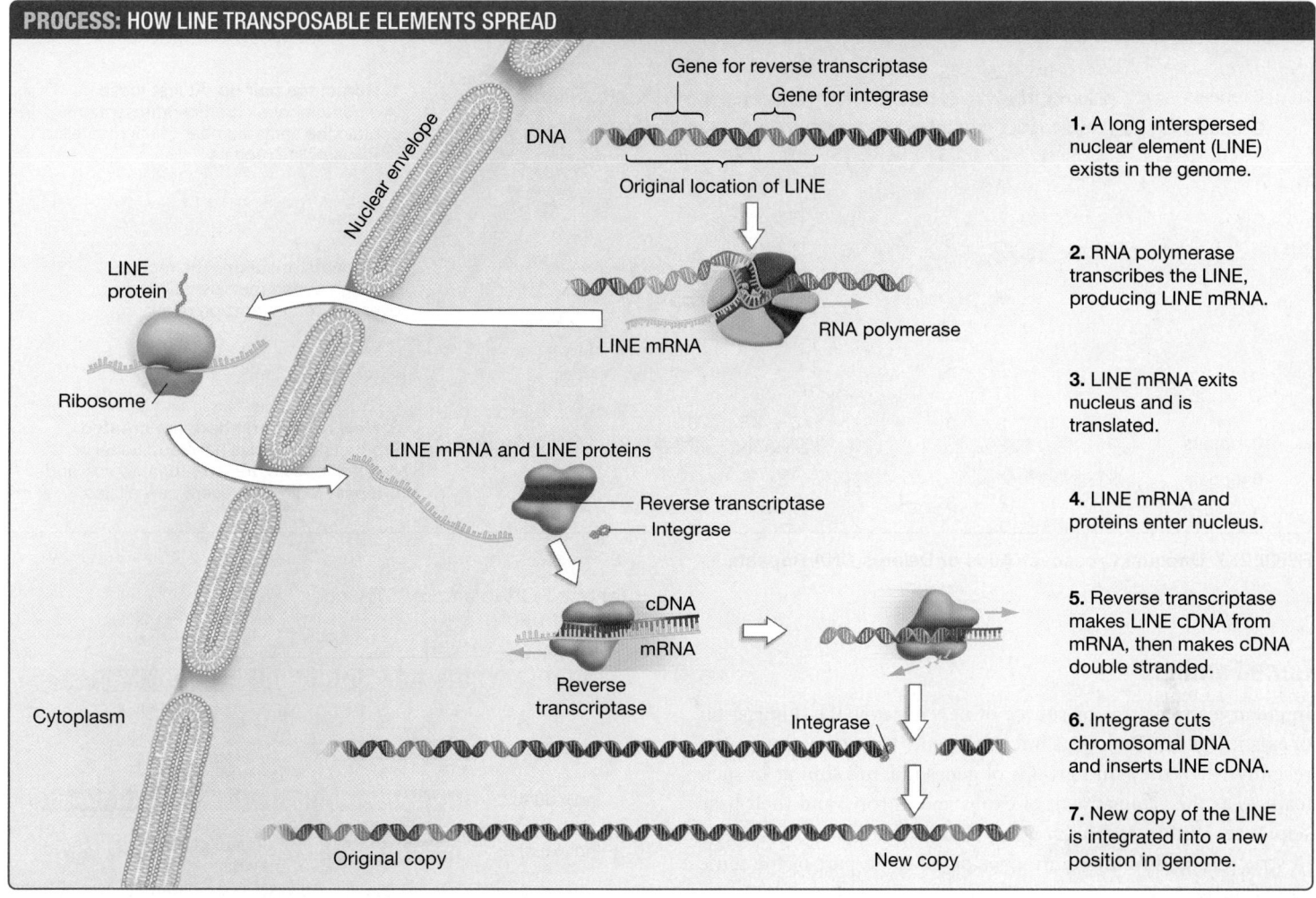

FIGURE 21.6 Transposable Elements Spread within a Genome.

the origin of minisatellites is still unclear. Together, the two types of repeated sequences make up 3 percent of the human genome.

Soon after these sequences were first characterized, Alec Jeffreys and co-workers established that microsatellite and minisatellite loci are "hypervariable," meaning that they vary among individuals much more than any other type of sequence does.

FIGURE 21.7 (see page 398) illustrates one mechanism to explain why microsatellites and minisatellites have so many different alleles: a process called **unequal crossover.** Here's how it works: Homologous chromosomes sometimes align incorrectly during prophase of meiosis I. Instead of lining up in exactly the same location, the two chromosomes pair in a way that matches up bases in different DNA repeats. When crossover occurs, the resulting chromosomes have different numbers of repeats.

Repeated sequences are particularly prone to unequal crossover, because their homologs are so similar that they are likely to misalign. If the region in question has a unique number of repeats, it represents a unique allele. Like any other alleles, microsatellite and minisatellite alleles are transmitted from parents to offspring.

The variation in repeat number among individuals is more than a curious feature of genome organization—it is the basis

of most DNA fingerprinting. **DNA fingerprinting** refers to any technique for identifying individuals based on the unique features of their genomes. Because microsatellite and minisatellite sequences vary so much, they are now the sequences of choice for DNA fingerprinting. How can these sequences be used for DNA fingerprinting?

Investigators obtain a DNA sample and perform the polymerase chain reaction (PCR), using primers that flank a region containing an STR (**FIGURE 21.8**; see page 398). Once the region has been amplified, it can be analyzed to determine the number of repeats present. Primers are now available that allow the analysis of many different STR loci.

These advances have profound impacts on society. Police use DNA fingerprinting to put people behind bars, and DNA fingerprinting has been used to show that people who were accused of crimes were actually innocent. Beyond criminal investigations, DNA fingerprinting has also been used to assign paternity and to identify remains.

Now that some characteristics of eukaryotic genomes have been reviewed, let's consider the genes they contain. Let's start with the most basic question of all: Where do eukaryotic genes come from?

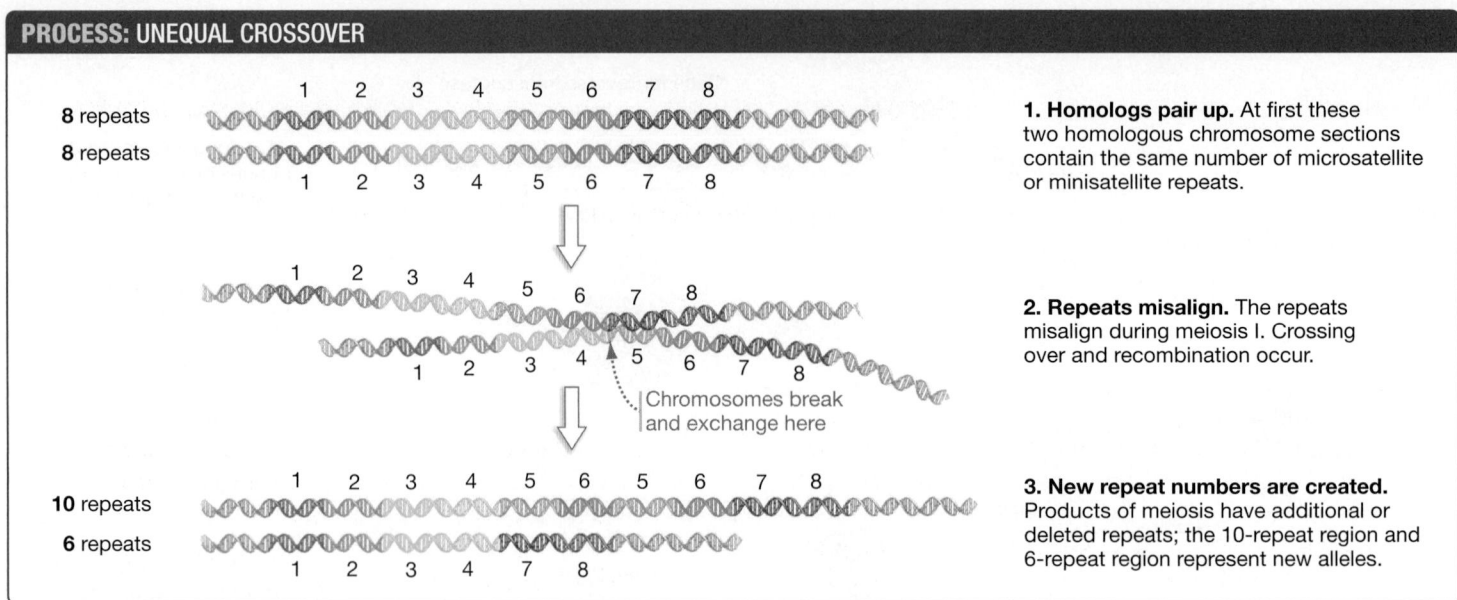

PROCESS: UNEQUAL CROSSOVER

8 repeats
8 repeats

1. Homologs pair up. At first these two homologous chromosome sections contain the same number of microsatellite or minisatellite repeats.

2. Repeats misalign. The repeats misalign during meiosis I. Crossing over and recombination occur.

Chromosomes break and exchange here

10 repeats
6 repeats

3. New repeat numbers are created. Products of meiosis have additional or deleted repeats; the 10-repeat region and 6-repeat region represent new alleles.

FIGURE 21.7 Unequal Crossover Adds or Deletes DNA Repeats.

Gene Families

In eukaryotes, the major source of new genes is the duplication of existing genes. Biologists infer that genes have been duplicated recently when they find groups of genes that are similar in such features as the arrangement of exons and introns, and their base sequence. Within a species, genes that are similar to each other in structure and function are considered to be part of the same **gene family.**

The degree of sequence similarity among members of a gene family varies. In the genes that code for ribosomal RNAs (rRNAs) in vertebrates, the sequences are virtually identical—meaning that each individual has many exact copies of the same gene. In other cases, though, 50 percent or less of the bases are identical.

How Do Gene Families Arise? Genes that make up gene families are hypothesized to have arisen from a common ancestral sequence through gene duplication. When **gene duplication** occurs, an extra copy of a gene is added to the genome.

The most common type of gene duplication results from unequal crossover during meiosis—the same process that resulted in extra microsatellite and minisatellite repeats in Figure 21.7. Gene-sized segments of chromosomes can be deleted or duplicated if homologous chromosomes misalign during prophase of meiosis I and an unequal crossover occurs. Like microsatellites or minisatellites, the duplicated segments are arranged in tandem—one after the other.

New Genes—New Functions? Gene duplication is important because the original gene is still functional and produces a normal product. As a result, the new, duplicated stretches of sequence are redundant. If mutations in the duplicated sequence alter the protein product so that it performs a valuable new function, then an important new gene has been created.

(a) Use PCR to amplify STR loci.

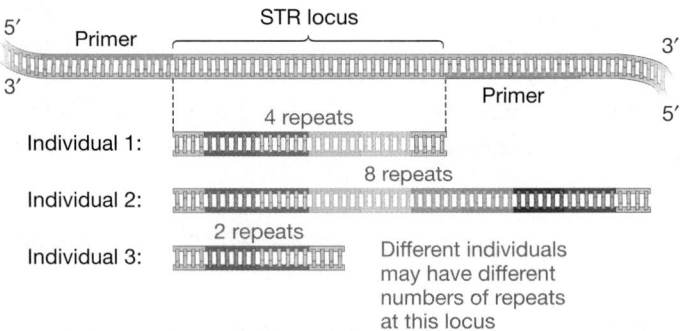

Individual 1:
Individual 2:
Individual 3:

Different individuals may have different numbers of repeats at this locus

(b) Compare number of STR repeats in alleles to test paternity.

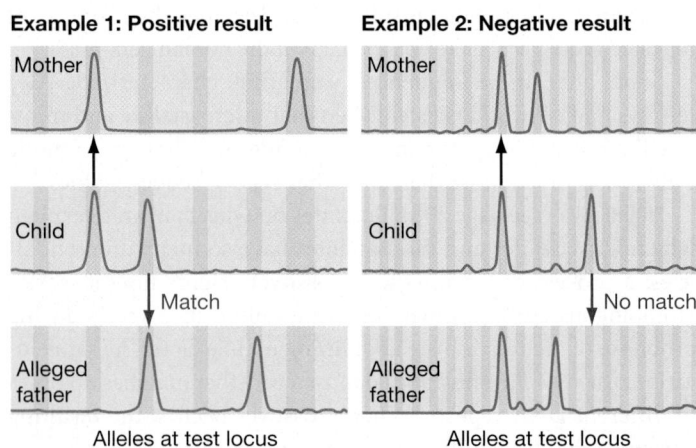

Example 1: Positive result

Mother

Child

Match

Alleged father

Alleles at test locus

Example 2: Negative result

Mother

Child

No match

Alleged father

Alleles at test locus

FIGURE 21.8 DNA Fingerprinting Can Be Used to Identify Parents. **(a)** The lengths of STR loci vary. Only one allele is shown for each individual. Individuals are often heterozygous, so the repeat number varies within and between individuals. **(b)** Here, the position of each peak indicates the number of repeats at a particular locus. Each individual is heterozygous and thus shows two peaks. One of the peaks from the mother and one of the peaks from the father should line up with a peak in the child. Typically 6 to 16 loci are tested to determine paternity.

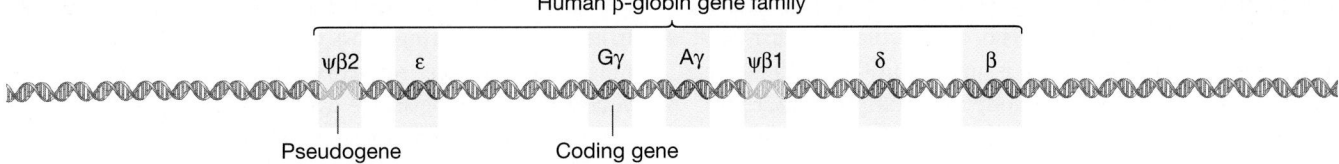

Human β-globin gene family

ψβ2   ε   Gγ   Aγ   ψβ1   δ   β

Pseudogene   Coding gene

FIGURE 21.9 Gene Families Are Closely Related Genes. The β-globin gene family is shown with coding genes in red and pseudogenes in yellow.

✔**EXERCISE** Suppose that during prophase of meiosis I, the β locus on one chromosome aligned with the ψβ2 locus on another chromosome. Then crossing over occurred in the noncoding sequences just to the left (as oriented in the figure) of this β-ψβ2 pairing. List the order of the β-globin-family genes that would result on each chromosome.

As an example of a gene family, consider the human globin genes diagrammed in **FIGURE 21.9**. Collectively, this set of genes is known as the β-globin gene family, and they code for proteins that form part of hemoglobin—the oxygen-carrying molecule in your red blood cells. Each coding gene in the family serves a slightly different function. For example, some genes are transcribed only in the fetus or the adult. The product of the fetal gene binds oxygen more tightly than the proteins expressed in adults. Consequently, oxygen moves from the mother's blood, where it is not as tightly bound to hemoglobin, to the fetus's blood (see Chapter 45).

In addition to creating genes with new functions, mutations in duplicated regions often make gene expression impossible. For example, a mutation could produce a stop codon in the middle of an exon. A member of a gene family that resembles a working gene but does not code for a functional product because of a mutation is a **pseudogene.** Pseudogenes do not function. Note that the β-globin gene family contains pseudogenes along with several genes that code for oxygen-transporting proteins. The number of pseudogenes is remarkable. In the human genome, for example, there are roughly as many pseudogenes as functional genes.

Lateral Gene Transfer in Eukaryotes   Genome sequencing projects are revealing more and more instances of lateral gene transfer in eukaryotes. What is the role of lateral gene transfer in the evolution of eukaryotes?

There are some clear examples of lateral gene transfer playing pivotal roles in eukaryote evolution. One key example is the capture of bacterial cells that were predecessors of today's mitochondria and chloroplasts and the subsequent transfer of many bacterial genes to the host genome (see Chapter 30). But in general, lateral gene transfer seems to be relatively rare in eukaryotes compared with prokaryotes. While biologists debate whether a tree or a web of life best describes evolution in bacteria and archaea, the eukaryotic tree remains rooted in what appear to be more modest rates of lateral gene transfer.

Insights from the Human Genome Project

More than 10 years ago, President Clinton announced the completion of the human genome sequence as "the most wondrous map ever produced by humankind." What has analysis of human genome sequence revealed? Let's first explore what types of DNA sequences make up the human genome and then examine some questions raised by this wondrous map.

Given biologists' focus on protein-coding genes, the composition of the human genome was unexpected. As **FIGURE 21.10** reveals, less than 2 percent of the genome consists of protein-coding exons, and nearly half is made of transposable elements. Introns make up over one-quarter of the genome and are 17 times more abundant than protein-coding exons.

Of all the observations stemming from the human genome, perhaps the most striking is that organisms with complex morphology, biochemistry, and behavior do not have particularly large numbers of genes. Notice in Figure 21.4b that the total number of genes in humans is only 50 percent more than the number in fruit flies, is about the same as in roundworms, and is substantially lower than the number of genes in water fleas, rice, and the mustard plant *Arabidopsis thaliana*.

Before the human genome was sequenced, many biologists expected that humans would have at least 100,000 genes. But we may only have a fifth of that number. How can this be?

In prokaryotes there is a correlation between genome size, gene number, a cell's metabolic capabilities, and the cell's ability to live in a variety of habitats. But why isn't there a stronger correlation between gene number and morphological, biochemical, and behavioral complexity in eukaryotes?

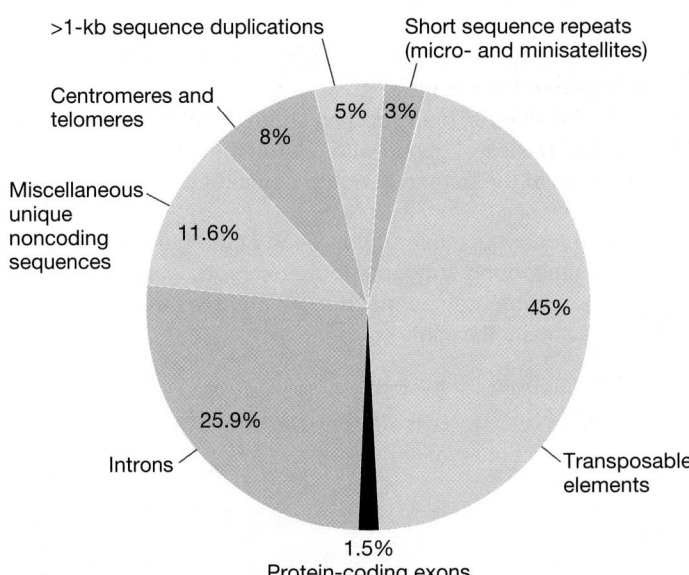

FIGURE 21.10 Composition of the Human Genome.
DATA: Gregory, T. R. 2005. *Nature Reviews Genetics* 6: 699–708.

One hypothesis to explain this observation is based on **alternative splicing.** Recall that the exons of a particular gene can be spliced in ways that produce distinct, mature mRNAs. As a result, a single eukaryotic gene can code for multiple transcripts and thus multiple proteins (see Chapter 19). The alternative-splicing hypothesis claims that multicellular eukaryotes do not need enormous numbers of distinct genes. Instead, more extensive use of alternative splicing in these organisms creates different proteins from the same gene.

In support of the alternative-splicing hypothesis, researchers have analyzed the mRNAs produced by human genes. They estimate that at least 90 percent of genes produce transcripts that are alternatively spliced, with an average of more than three distinct transcripts per gene. This means that the number of different proteins that can be produced is more than triple the gene number. Because of extensive alternative splicing, the number of genes and number of proteins do not have to be tightly linked.

There is likely more to the story, however, than extensive alternative splicing. Accompanying the rapid advances in sequencing technology have been equally dramatic advances in methods for studying how the genome functions. Some of these methods are described later, in Section 21.4. But what's important here are the findings.

Roughly 90 percent of the human genome is transcribed. This is far more than believed even a few years ago. Some of these transcripts code for regulatory RNAs with known roles, such as microRNAs (Chapter 19). Many of these transcripts, however, have no currently known roles. If many of these newly discovered RNAs have a function, then the human genome and the genomes of other complex organisms may not be so gene poor after all. If so,

this would require an adjustment in how we view a typical gene. Instead of considering most genes to code for proteins, perhaps most genes produce regulatory RNAs that are never translated. As you read this, biologists are working hard on this central question that has been opened by functional studies of the human genome.

What are these functional studies, and how are they carried out? The following section considers these questions.

21.4 Functional Genomics, Proteomics, and Systems Biology

Genomics researcher Eric Lander has compared the sequencing of the human genome to the establishment of the periodic table of the elements in chemistry. Once the periodic table was validated, chemists focused on understanding how the elements combine to form molecules. Similarly, biologists now want to understand how the elements of the human genome combine to produce an individual.

Remember that a genome sequence is essentially a parts list. Once that list is assembled, researchers delve deeper to understand how genes interact to produce an organism.

What Is Functional Genomics?

For decades, biologists have worked at understanding how and when individual genes are expressed. Research on the *lac* operon is typical of this type of study (see Chapter 18). But now researchers can ask how and when *all* the genes in an organism are expressed.

Large-scale analysis of gene expression is part of functional genomics—research on how genes work together to produce a phenotype. The effort is motivated by the realization that gene products do not exist in a vacuum. Instead, groups of RNAs and proteins act together to respond to environmental challenges such as extreme heat or drought. Similarly, distinct groups of genes are transcribed at different stages as a multicellular eukaryote grows and develops.

A basic tool of functional genomics is a microarray. A **DNA microarray** consists of as many as 1 million different single-stranded DNA segments that are permanently attached at one end to a glass slide or silicon chip. Each DNA sequence is linked to the slide or chip in a known location and serves as a probe for a specific transcript. For example, the slide pictured in **FIGURE 21.11** contains thousands of spots, each one containing single-stranded DNA from a unique exon found in the human genome.

Microarrays can be used for many applications. The most common use is to learn which genes are expressed as RNAs in a particular cell type under particular conditions. A typical experiment done with a DNA microarray would follow the protocol outlined in **FIGURE 21.12**. For example, if the researchers' goal was to learn how a cell alters its gene expression to meet the challenges of heat stress, the first step would be to isolate mRNAs produced in control cells functioning at normal temperature and in cells of the same kind exposed to high temperatures.

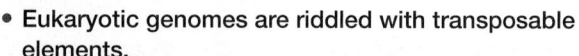

check your understanding

If you understand that . . .

- Eukaryotic genomes are riddled with transposable elements.
- Relatively short repeated sequences are common in eukaryotic genomes.
- In eukaryotes, many coding sequences are organized into families of genes with related functions.
- Much of the human genome does not code for proteins.
- The recent discovery that much of the genome is transcribed suggests that eukaryotic gene expression may be much more complex than previously thought.

✔ **You should be able to . . .**

1. Explain why there is no simple relationship between the size of a eukaryotic genome and the complexity of the organism.

2. Estimate the ratio of protein-coding portions of the human genome to all transcribed regions of the genome.

Answers are available in Appendix A.

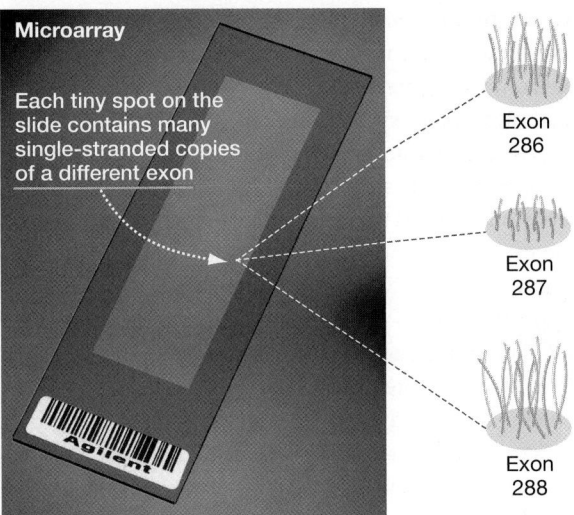

Microarray

Each tiny spot on the slide contains many single-stranded copies of a different exon

Exon 286

Exon 287

Exon 288

FIGURE 21.11 A DNA Microarray. To create this microarray, thousands of short, single-stranded DNA sequences were synthesized in defined positions on a glass plate. In this microarray, the synthesized DNAs represent portions of all exons from a particular species.

Once they had purified mRNAs from the two populations of cells (step 1), investigators would use reverse transcriptase to make a single-stranded cDNA version of each RNA in each of the two samples (see Chapter 20). In addition to the four standard dNTPs, one of the DNA building blocks used in synthesizing the cDNA would carry a fluorescent label (step 2). The label used for the cDNA of the control cells would glow one color (let's say green), while the label chosen for the cDNA of the heat-stressed cells would glow another color (let's say red). The labeled cDNAs of both colors would then be used to bind to the complementary DNA probes on the microarray (step 3). This step is called hybridization because hybrids between probe DNAs and cDNAs will form.

Out of all the exons in the genome, then, only the exons that are being expressed by the two populations of cells will be labeled on the microarray. In this example, genes that are expressed by the control cells under normal conditions will be labeled green, while those expressed by the cells during heat stress will be labeled red. If an exon in the microarray is expressed under both sets of conditions, then both green- and red-labeled cDNAs will bind to the DNA in that spot and make it appear yellow (step 4).

A microarray lets researchers study the expression of thousands of genes at a time. As a result, they can identify which sets of genes are expressed in concert under specific sets of conditions.

Researchers can use microarrays to establish which genes are transcribed in different organs and tissues, in cancers, or—as you saw in Figure 21.12—in response to changes in environmental conditions such as heat stress. ✔ If you understand how microarrays are used, you should be able to explain how you would use a DNA microarray to compare the genes expressed in brain cells versus liver cells of an adult human.

Besides using microarrays, investigators are now able to assess gene expression by directly sequencing cDNAs using next-generation sequencing technologies. For example, if biologists

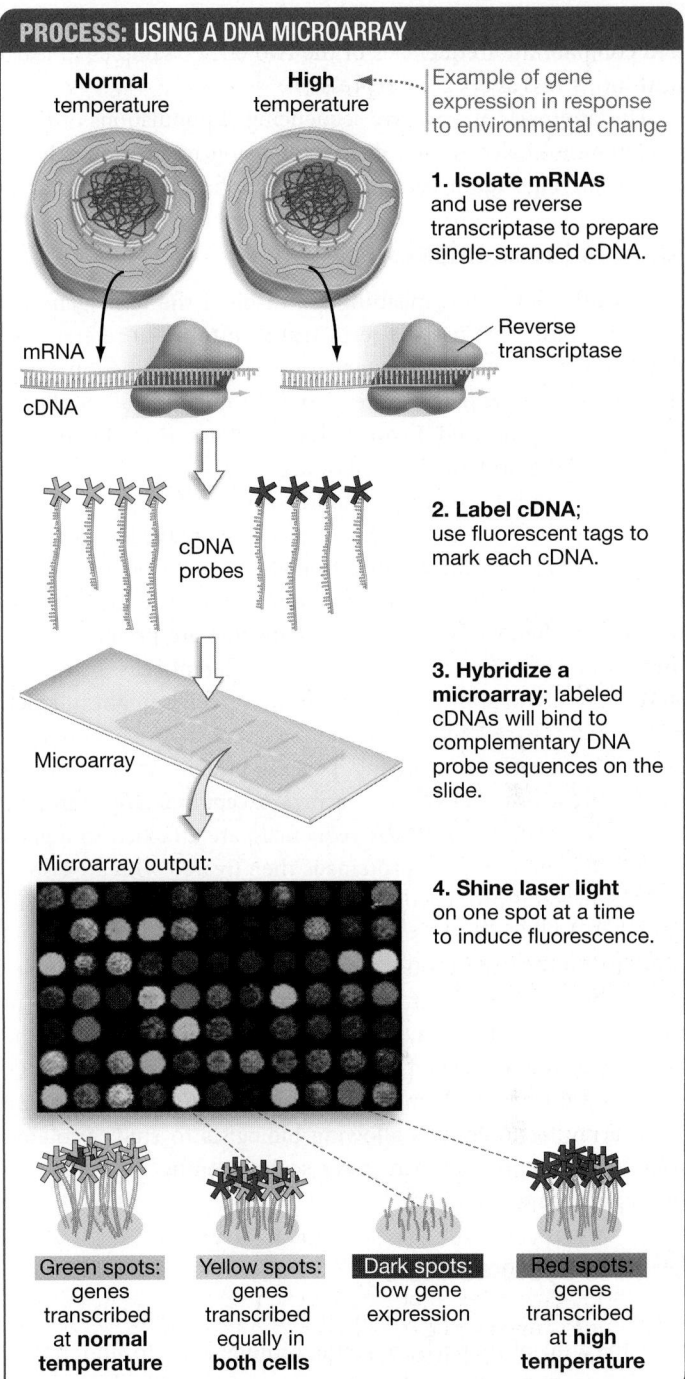

PROCESS: USING A DNA MICROARRAY

Normal temperature   High temperature   Example of gene expression in response to environmental change

1. Isolate mRNAs and use reverse transcriptase to prepare single-stranded cDNA.

mRNA   Reverse transcriptase

cDNA

cDNA probes

2. Label cDNA; use fluorescent tags to mark each cDNA.

Microarray

3. Hybridize a microarray; labeled cDNAs will bind to complementary DNA probe sequences on the slide.

Microarray output:

4. Shine laser light on one spot at a time to induce fluorescence.

Green spots: genes transcribed at **normal temperature**

Yellow spots: genes transcribed equally in **both cells**

Dark spots: low gene expression

Red spots: genes transcribed at **high temperature**

FIGURE 21.12 DNA Microarrays Can Be Used to Study Changes in Gene Expression. By probing a microarray with labeled cDNAs synthesized from mRNAs, researchers can identify which sequences are being transcribed. Here mRNAs from cells growing at normal temperature are detected by green color, while mRNAs from cells growing at high temperature are detected by red color.

wanted to learn about gene expression changes induced by heat stress, they would start as they did for the microarray by preparing two different sets of cDNAs, one from control and one from heat-stressed cells. But instead of using these cDNAs to hybridize with microarray probes, biologists would sequence millions of cDNAs from each treatment type. Using bioinformatics tools, they would then determine the frequency of each type of cDNA and compare the frequencies of the two cDNA samples to learn how heat stress alters gene expression.

This approach of extensive sequencing of populations of DNA or cDNA molecules is called **deep sequencing,** and it is quickly becoming an important research approach for functional genomics.

What Is Proteomics?

The Greek root –*ome,* meaning all, inspired the term genome. Similarly, biologists use the term **transcriptome** in referring to the complete set of DNA sequences that are transcribed in a particular cell, and **proteome** in referring to the complete set of proteins that are produced. **Proteomics,** it follows, is the large-scale study of all the proteins in a cell or organism.

Like genomics, proteomic studies begin with a parts list by identifying the proteins present in a cell or organelle. The techniques used for protein identification are distinct from those used in working with DNA. Once individual proteins are identified, researchers then study how the proteins that are present change through time, interact, or vary between different cells. Instead of studying individual proteins or how two proteins might interact, proteomics is based on studying all the proteins present at once.

One approach to studying protein–protein interactions is similar to the use of DNA microarrays, except that large numbers of proteins, rather than DNA sequences, are attached to a glass plate. This microarray of proteins is then treated with an assortment of proteins produced by the same organism. These proteins are labeled with a fluorescent or radioactive tag. If any labeled proteins bind to the proteins in the microarray, the two molecules may also interact in the cell. In this way, researchers can identify proteins that physically bind to one another—like the G proteins and associated enzymes (introduced in Chapter 11), or the cyclin and Cdk molecules (introduced in Chapter 12). Microarray technology is allowing biologists to study protein–protein interactions on a massive scale, opening the door to a new approach to biology.

What Is Systems Biology?

Systems biology is based on the premise that a whole is greater than the sum of its parts. **Systems biology** aims to understand how interactions between the individual parts of a biological system create new properties. For example, how does metabolism come about from the interaction of proteins within a cell? How does cancer arise from the interplay of individual genes? Complex properties that arise from the interaction of simpler elements are **emergent properties.**

DNA replication, metabolism, cancer, and the development of an organism all are emergent properties because they come about from the interactions of simpler elements—genes and proteins. ✔ If you understand this concept, you should be able to explain why cell replication is an emergent property.

Many systems biology investigations focus on mapping the interactions between genes or proteins. Genomics and proteomics provide the parts list needed to start these systems biology studies. It's the job of a systems biologist to learn how these parts are linked together into networks and how new properties emerge from these interactions.

Let's look at how a systems biology approach was taken to predict all possible interactions between proteins in *Schizosaccharomyces pombe,* a species of yeast. An interaction can mean either that two proteins bind to one another to form a stable complex or that one protein acts on another protein. An example of the first type of interaction is cyclin binding to a cyclin-dependent kinase (Cdk); an example of the second interaction type is when the cyclin–Cdk complex acts as a kinase to add a phosphate to a target protein (see Chapter 12).

To begin their study, the researchers used existing databases of known protein–protein interactions to make predictions about the fission yeast protein interaction network. Then they verified some of the predicted interactions experimentally to show that their prediction method worked. Once they had shown that their prediction method was accurate, they mapped 37,325 possible interactions between 3438 different proteins. If you look at the interaction network in **FIGURE 21.13**, you will see how clusters of interacting proteins emerge from the tangle. These clusters reveal highly connected portions of the network.

The significance? Some of the predicted interactions in this yeast may have implications for human disease. For example, some of the key players in the cluster associated with signal transduction point to previously unexplored relationships that could be important in understanding cancer and neurodegenerative disorders, such as Huntington's disease (see Chapter 20), and for targeted drug design.

Work in genomics, proteomics, and systems biology is opening the door to knowing how cells work in health and disease. These new areas of biology also lay the foundation for understanding one of the most wondrous aspects of life—how a single cell develops into a complete organism. This topic is explored in the unit ahead.

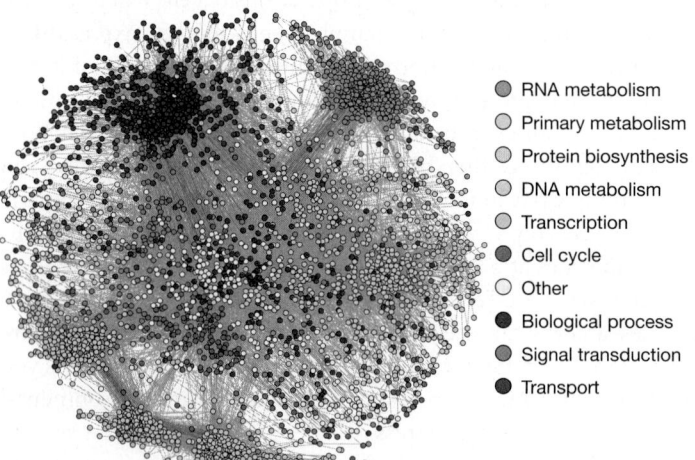

FIGURE 21.13 Yeast Protein Interaction Network. Each circle represents a different protein, with each protein color-coded according to its cellular function. Lines that connect a pair of proteins indicate an interaction.

DATA: V. Pancaldi et al. 2012. *G3: Genes, Genomes, Genetics* 2: 453–467.

If you understand . . .

21.1 Whole-Genome Sequencing

- Advances in DNA sequencing technologies have allowed investigators to sequence DNA more rapidly and cheaply, resulting in a flood of genome data.

- Thousands of genomes have been sequenced to date for many different purposes.

- Bioinformatics is the application of computer science to genome analysis and is essential for genome research.

- Researchers annotate genome sequences by finding genes and determining their function.

- To identify genes in bacteria and archaea, researchers use computers to scan the genome for start and stop codons that are in the same reading frame and that are separated by gene-sized stretches of sequence.

- Genes are also identified based on their homology (similarity due to evolutionary relatedness) to previously identified genes in other species.

- To find genes in eukaryotes, researchers couple the use of computers and study of RNAs to identify transcribed sequences.

✓ You should be able to describe how a research group that discovered a gene for coat color in mice would determine whether a homologous gene exists in the human genome.

21.2 Bacterial and Archaeal Genomes

- Bacterial and archaeal genomes are small relative to many eukaryotic genomes and have tightly spaced genes.

- There is a linear relationship between bacterial and archaeal gene number and genome size, and a similar correspondence in the relationship between metabolic capacity and genome size.

- The function of many of the genes identified in bacteria and archaea is still unknown.

- There is a huge amount of genetic diversity in bacterial and archaeal genomes, even among different strains of the same species.

- Lateral gene transfer is common in bacteria and archaea. It is an important source of new genes in many species.

- Metagenomics allows the analysis of all the genes of all the bacteria or archaea in a community.

✓ You should be able to propose genome characteristics you would look for to distinguish a bacterial parasite from a nonparasitic bacterial species that is found in a range of environments.

21.3 Eukaryotic Genomes

- Eukaryotic genomes tend to be large and complex. They include many sequences that have little to no effect on the fitness of the organism, and many transcribed sequences whose function is not known.

- There is no correlation between morphological complexity and gene number in eukaryotes.

- Because of alternative splicing, the number of distinct transcripts produced in many eukaryotes is much larger than the gene number.

- Gene duplication has been an important source of new genes in eukaryotes.

- Lateral gene transfer occurs in eukaryotes, but appears to play a smaller role than in bacteria and archaea.

- Recent findings show that much more of the eukaryotic genome is transcribed than previously believed, and the function of many noncoding transcripts is unknown.

✓ You should be able to explain what features of eukaryotic genomes result in a lack of correspondence between genome size, gene number, and morphological complexity.

21.4 Functional Genomics, Proteomics, and Systems Biology

- Functional genomics uses tools such as DNA microarrays to learn patterns of gene expression and gene function.

- Proteomics is similar to genomics but works to identify the complete set of proteins expressed in a cell, how this set changes under different conditions, and how it relates to phenotype.

- Systems biology starts with genomics and proteomics data and studies the set of interactions between different genes or proteins to understand how biological systems work.

✓ You should be able to explain how Figure 21.13 indicates that there are networks within networks for interacting proteins associated with particular functions.

(MB) **MasteringBiology**

1. **MasteringBiology Assignments**

   Tutorials and Activities DNA Fingerprinting; Human Genome Project: Genes on Human Chromosome 17; Human Genome Sequencing Strategies; Shotgun Approach to Whole-Genome Sequencing

   Questions Reading Quizzes, Blue-Thread Questions, Test Bank

2. **eText** Read your book online, search, take notes, highlight text, and more.

3. **The Study Area** Practice Test, Cumulative Test, BioFlix® 3-D Animations, Videos, Activities, Audio Glossary, Word Study Tools, Art

You should be able to . . .

1. What is an open reading frame in bacteria?
   a. a gene whose function is already known
   b. a DNA section that is thought to code for a protein because it is similar to a complementary DNA (cDNA)
   c. a DNA section that is thought to code for a protein because it has a start codon and a stop codon flanking hundreds of nucleotides
   d. any member of a gene family

2. What best describes the logic behind shotgun sequencing?
   a. Break the genome into tiny pieces. Sequence each piece. Use overlapping ends to assemble the pieces in the correct order.
   b. Start with one end of each chromosome. Sequence straight through to the other end of the chromosome.
   c. Use a variety of techniques to identify genes and ORFs. Sequence these segments—not the noncoding and repeated sequences.
   d. Break the genome into pieces. Map the location of each piece. Then sequence each piece.

3. A _____ is a 2- to 6-base-pair repeated sequence in DNA.
   a. LINE
   b. restriction site
   c. gene duplication
   d. microsatellite

4. What is a leading hypothesis to explain the paradox that large, morphologically complex eukaryotes such as humans have relatively small numbers of genes?
   a. lateral transfer of genes from other species
   b. alternative splicing of mRNAs
   c. polyploidy, or the doubling of the genome's entire chromosome complement
   d. expansion of gene families through gene duplication

5. What evidence do biologists use to infer that a gene is part of a gene family?

6. What are some characteristics of a pseudogene?

7. Explain how open reading frames are identified in the genomes of bacteria and archaea. Why is it more difficult to find open reading frames in eukaryotes?

8. In a genomics-based search for mutations that caused a patient's cancer, which of the following would provide the most informative comparison with the cancer cell?
   a. the average human DNA sequence available from a database
   b. the DNA sequence of a noncancerous cell from another person
   c. the DNA sequence of a noncancerous cell from the patient
   d. the DNA sequence of a different tumor type from another cancer patient

9. **QUANTITATIVE** Gene density is the number of genes per unit length of DNA. Most often, gene density is expressed as the number of genes per million base pairs (Mbp). Go to Figure 21.4b and find the approximate number of genes estimated in water fleas and in humans and the size of each genome. What is the gene density per Mbp in water fleas? What is the gene density per Mbp in humans? How much greater is the gene density in water fleas relative to humans?

10. In DNA fingerprinting, why is it an advantage to analyze an STR locus with many different repeat length alleles versus an STR locus with only a few different repeat length alleles?

11. Explain how microarrays of short, single-stranded DNAs that represent many or all of the exons in a genome are used to document changes in the transcription of genes over time or in response to environmental challenges.

12. Explain the concept of homology and how identifying homologous genes helps researchers identify the function of unknown genes.

13. Parasites lack genes for many of the enzymes found in their hosts. Most parasites, however, have evolved from free-living ancestors that had larger genomes. Based on these observations, W. Ford Doolittle claims that the loss of genes in parasites represents an evolutionary trend. He summarizes his hypothesis with the quip "use it or lose it." What does he mean?

14. According to eyewitness accounts, communist revolutionaries executed Nicholas II, the last czar of Russia, along with his wife and five children, the family physician, and about a dozen servants. Many decades after this event, a grave purported to hold the remains of the royal family was discovered. Biologists were asked to analyze DNA from each adult and juvenile skeleton and determine whether the bodies were indeed those of several young siblings, two parents, and several unrelated adults. If the remains of the family were in this grave, predict how similar the DNA fingerprints would be between the parents, the children, and the unrelated individuals in the grave.

15. Pleiotropy (a concept presented in Chapter 14) occurs when a mutation in one gene results in many different phenotypes. In a study of a gene interaction network, similar to the protein interaction network shown in Figure 21.13, which type of gene would you predict to exhibit the greatest degree of pleiotropy when mutated?
   a. a gene that interacts with one neighbor
   b. a gene that has a duplicate copy
   c. a gene that is the center of interactions with functionally related genes
   d. a gene that is the center of interactions with genes of many different functions

16. One hypothesis for differences between humans and chimps involves differences in gene regulation. A recent study used microarrays to compare the patterns of expression of genes that are active in the brain, liver, and blood of chimpanzees and humans. The overall patterns of gene expression were similar in the liver and blood of the two species, but the expression patterns were strikingly different in the brain. How do these results relate to the hypothesis?

22 Principles of Development

In this chapter you will learn that

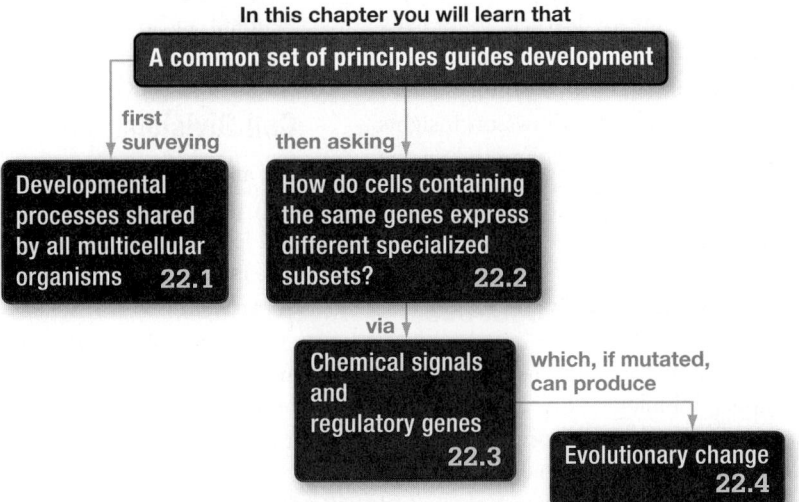

A common set of principles guides development

first surveying

Developmental processes shared by all multicellular organisms **22.1**

then asking

How do cells containing the same genes express different specialized subsets? **22.2**

via

Chemical signals and regulatory genes **22.3**

which, if mutated, can produce

Evolutionary change **22.4**

An embryonic fish, still attached to the nutrient-filled yolk in the egg. This chapter introduces the processes responsible for transforming a fertilized egg into an individual that has specialized cells, tissues, and organs.

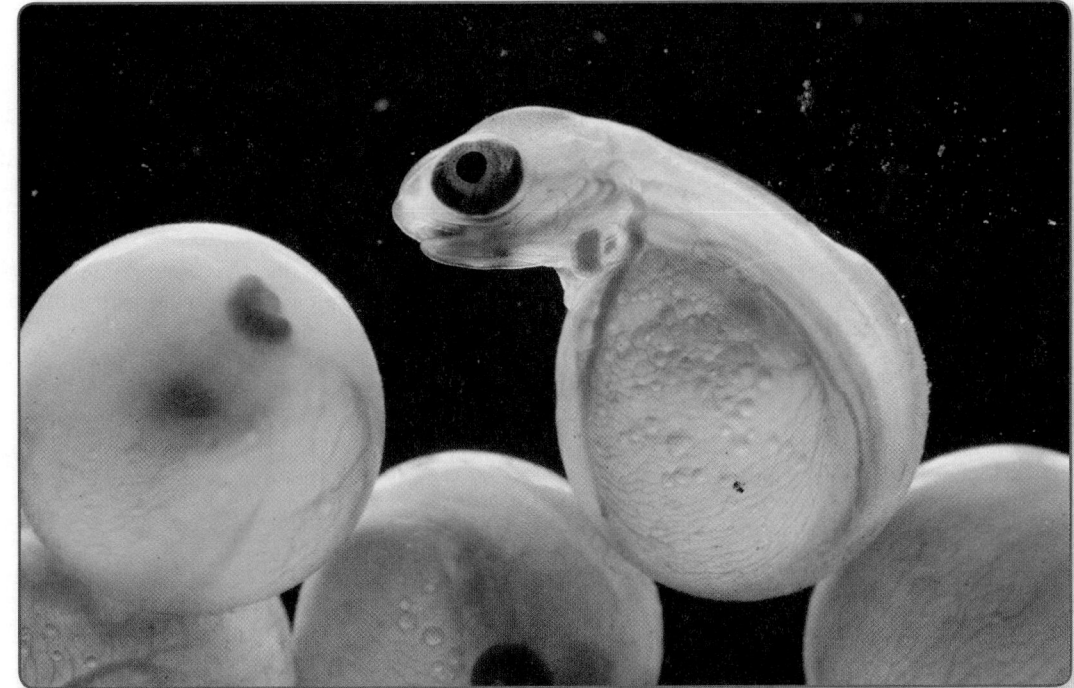

What is today's greatest challenge in biological science? Although there are many candidates, one of the most compelling is the question addressed in Unit 4: How does a multicellular individual develop from a single cell—a fertilized egg?

It's important to pause for a moment and think about the magnitude of this problem. Consider that once you were a single cell. If you could have watched your own development, you would have seen that single cell divide rapidly and form a ball of tiny, identical-looking cells. At that point, the fertilized egg had become an **embryo**—a young, developing organism. After continued cell division, large groups of cells suddenly began moving into the interior of the embryo. Cell division continued at a dizzying pace. After a week or two the embryo elongated, and a recognizable head and tail portion appeared. Tiny precursors of vertebrae became visible, along with rudimentary eyes. Eventually

✔ When you see this checkmark, stop and test yourself. Answers are available in Appendix A.

405

buds emerged and went on to form your limbs. As development continued, the embryo became you.

Biologists who have watched this process in humans or other organisms never cease to marvel at it. How does all the growth and formation of body parts happen?

To understand how researchers are answering this question, you'll need to draw on what you've already learned about cell–cell interactions, the regulation of gene expression, and a host of other topics described in previous units. Part of the excitement surrounding developmental biology is that it draws on insights from biochemistry, cell biology, genetics, and evolution. It is one of the most interdisciplinary fields in all of biology. Let's delve in.

22.1 Shared Developmental Processes

Over one hundred years of research has culminated in one of the great insights of contemporary biology: A few fundamental principles are common to all developmental processes in every multicellular organism. This discovery has brought a unified understanding to how the embryos of fruit flies, oak trees, and humans each grow and develop from a single cell into complex, multicellular individuals.

An individual develops as cells divide; signal to one another about where they are, what they are doing, and what type of cell they are becoming; begin to express certain genes rather than others; move or expand in specific directions; and, in the case of some cells, die (TABLE 22.1). Let's first consider these processes in turn, while keeping in mind that in the embryo they are interdependent and occur together. Then, let's go on to consider more specific questions about how cells interact and specialize.

Cell Division

For an embryo to grow and develop, its cells have to proliferate—they must divide and make more cells. This statement is obvious, but less obvious and equally important is the following point: The location, timing, and extent of cell division have to be tightly controlled.

How is cell division controlled? You might recall that mitosis and cytokinesis are responsible for cell proliferation in eukaryotes (Chapter 12) and that cells initiate mitosis in response to a regulatory protein complex called M phase–promoting factor (MPF). You also may remember that each stage of the cell cycle has carefully regulated checkpoints, and that cells control their division in response to what biologists call "social controls"—meaning, signals from other cells.

SUMMARY TABLE 22.1 **Essential Developmental Processes**

Cell proliferation		Cells divide by mitosis and cytokinesis. The timing, location, and amount of cell division are regulated.
Cell–cell interactions		Signals that are produced by cells influence their neighbors to divide, die, move, or differentiate.
Cell differentiation		Undifferentiated cells specialize at specific times and places in a stepwise fashion.
Cell movement and expansion		Cells can move past one another within a block of animal cells, causing drastic shape changes in the embryo.
		Cells can break away from a block of animal cells and migrate to new locations.
		Plant cells can regulate the plane of cell division and expand in specific directions, causing dramatic changes in shape.
Programmed cell death		The timing, location, and amount of cell death are regulated.

As development proceeds in both plants and animals, most cells stop dividing when they mature. But in later stages of development and in the adult, populations of undifferentiated cells keep proliferating throughout the individual's life. These are **stem cells.** Stem cells divide to produce a daughter cell that remains a stem cell and another daughter cell that ultimately differentiates into a specialized cell type. By producing one daughter that stays a stem cell, populations of stem cells are maintained.

- In plants, stem cells are located in **meristems.** Meristems are present in embryos and adults and produce the stems, roots, leaves, flowers, and other structures that develop throughout a plant's life (see Chapter 24).

- In animals, stem cells exist in specific locations in the body. In adults, stem cells proliferate to replace skin, blood, and gut cells that die; repair wounds; and create a constant supply of disease-fighting cells in the immune system.

Cell–Cell Interactions

Cells interact constantly during development. They engage in "conversations" with their neighbors through a diverse set of chemical signals. This ongoing communication drives most of the cellular behaviors that are the foundation of development.

You might recall that when a signal arrives at the surface of a cell that possesses the appropriate receptor, the message is received and processed (Chapter 11). Signal molecules used in development may diffuse in the watery environment that surrounds cells, but they also can be present on the surface of other cells or bound to the extracellular matrix. These nondiffusible signals are detected by developing cells as they expand or move within the embryo.

In most cases in development, when a signal is bound by a receptor, it activates a signal transduction cascade that triggers the production of transcription factors (Chapter 19). As a result, patterns of gene expression change, and this alters the cell's activity. In response to a chemical signal, embryonic cells may divide, differentiate, move, expand, or die. In this way, the fate of a cell inside an embryo hinges on the signals it receives from other cells.

Cell Differentiation

As development progresses, most cells undergo **differentiation**— the process of becoming a specialized type of cell. Differentiation is a progressive, step-by-step process. As cells differentiate, decisions are made at branch points to follow one developmental path or another. As a result, a fertilized egg may give rise to hundreds of distinctive cell types.

In the case of your own development, some of your embryonic cells differentiated to form muscle cells that contract and relax, while others became nerve cells that conduct electrical signals throughout the body. As an oak tree develops, some cells secrete thickened walls and transport water as part of a woody stem; other cells become flattened and secrete the waxy coating found on the surface of leaves.

Remarkably, sometimes the process of differentiation can be reversed. For example, some plant cells can "de-differentiate" after they have become specialized. When a branch of a western red cedar tree droops low enough to make contact with the soil, cells in the branch de-differentiate. These cells then re-differentiate to form root cells, resulting in the growth of a fully formed root where the branch came to rest on the ground.

For decades, biologists thought that animal cells could not de-differentiate. As you'll see in Section 22.2, however, the cloning of frogs and mammals reversed this view. In a spectacular recent study, scientists found that it is even possible to de-differentiate many adult human cells. Adding as few as three transcription factors to adult cells can cause them to de-differentiate into cells that resemble those of the early embryo. The hope is that these de-differentiated cells can be coaxed to re-differentiate in tissue culture to create any desired cell type that might be needed to replace cells damaged by disease. If this dream becomes reality, it will largely be through understanding cell differentiation during normal development.

Cell Movement and Cell Expansion

Many animal cells have to move to new locations for normal development to occur. In plants, cells cannot move because of the cell wall. However, plants are masters at controlling both the direction of cell expansion—the growth of an individual cell—and the orientation of cell division to determine where cells are added.

Some of the most dramatic cell movements in animals occur early in development once rapid cell divisions have produced a mass of cells. During a sequence of events called **gastrulation,** cells in different parts of the mass rearrange themselves into three distinctive layers, which then give rise to the skin, gut, and other basic parts of the body (gastrulation is described in Chapter 23).

Later in development, some animal cells break away from their original sites and migrate to new locations in the embryo. There they give rise to germ cells (cells that produce sperm or eggs), pigment-containing cells, precursors of blood cells, or certain nerve cells. If these cell movements go awry, the embryo will be deformed or die.

Plant cells, in contrast, are encased in stiff cell walls and do not move. Instead, the direction of cell division and cell expansion is carefully regulated to form stems, leaves, roots, and all other parts of the plant. Differential cell expansion is a key part of plant development, just as regulated cell movement is a key part of animal development.

Cell Death

Cell division is controlled by signals from other cells. But sometimes, signals trigger cell death. Cell death is a highly regulated aspect of plant and animal development. **Programmed cell death** occurs as certain tissues and organs take shape.

There are a few different forms of programmed cell death. In animals, the predominant one used in development is **apoptosis** (literally, "falling away"). Although biologists debate whether to call the form of programmed cell death that occurs in plants apoptosis, an important role for cell death in plant development is undisputed.

Apoptosis plays fascinating roles. For example, the feet of a chicken embryo initially develop with webs. Cells that are present between the toes must die to form separate toes (**FIGURE 22.1a**). The same is true for you—without apoptosis, your hands and feet would be webbed like a duck's feet. A duck has webbed feet because cells between the toes do not undergo apoptosis. Apoptosis doesn't end with digits. Vertebrate nerve cells are overproduced in the embryo. As a normal part of "wiring" an effective nervous system, at least half of embryonic nerve cells are destroyed by apoptosis.

The initial studies on apoptosis were done on the roundworm *Caenorhabditis* (pronounced *see-no-rab-DIE-tiss*) *elegans*. This species is a popular model organism in developmental biology because the adult contains only about a thousand cells and because the embryonic worm is transparent, so biologists can follow individual cells throughout development. (For more on *C. elegans*, see **BioSkills 13** in Appendix B.)

As a *C. elegans* worm matures, exactly 131 of its cells undergo apoptosis. To explore how this happens, Hillary Ellis and Robert Horvitz studied mutant embryos that do *not* exhibit the normal pattern of cell death and then used techniques (introduced in Chapter 20) to locate the mutated genes responsible for the defect. Their early work uncovered two genes required for apoptosis. The researchers proposed that the genes are part of a genetic program for apoptosis—in short, they are cell-death genes.

In follow-up work, researchers searched databases of DNA sequences and discovered that mice have similar genes. When a research team used genetic engineering techniques to produce mice in which both copies of one of their cell-death genes were knocked out, the embryos had severe malformation of the brain (**FIGURE 22.1b**). The defect occurred because nerve cells that normally die during development survived.

The same genes exist in humans and other mammals. Abnormal apoptosis—either too much or too little—occurs in certain diseases. For example, inappropriate activation of programmed cell death is proposed to be involved in some devastating neurodegenerative diseases such as ALS (Lou Gehrig's disease).

(a) Chicken embryo with normal (left) and defective (right) cell-death genes.

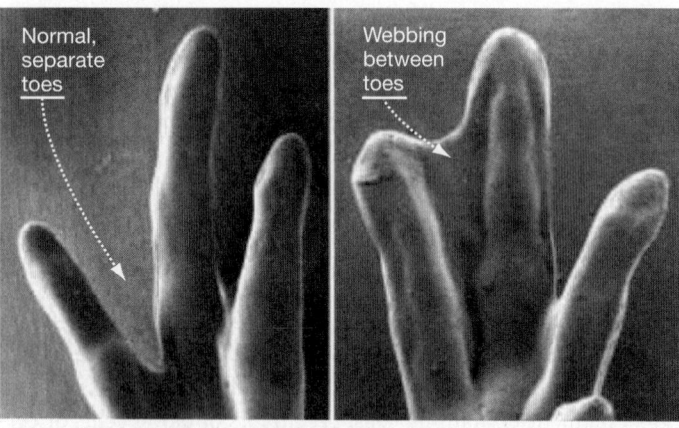

Normal, separate toes

Webbing between toes

(b) Mouse embryo with normal (left) and defective (right) cell-death genes.

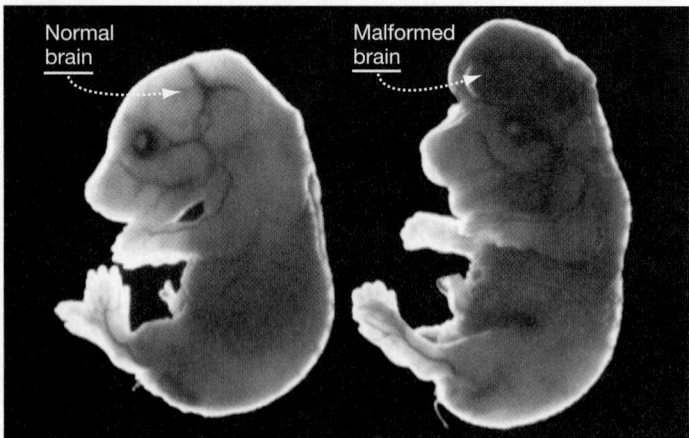

Normal brain

Malformed brain

FIGURE 22.1 Programmed Cell Death Is a Normal Part of Development.

22.2 Genetic Equivalence and Differential Gene Expression in Development

The differentiation of a cell occurs through differential gene expression. The muscle cells in your body are different from your nerve cells because they express different genes and therefore produce different proteins. The water-transport cells in an oak tree are different from its leaf-surface cells for the same reason.

If you think about these statements, you'll realize that they have to be true. The only way that cells can have different structures and functions is if they contain different molecules. What is less obvious is whether cells express different genes because they contain different genes, or whether all the cells in a body contain the same genes but express only a specialized subset.

Evidence that Differentiated Plant Cells Are Genetically Equivalent

If cells from the branch of a cedar tree can de-differentiate to form roots, the branch cells must contain the genes required by root cells. Gardeners and farmers have known for centuries that in many plant species, complete new plants can be produced from a small section of a root or shoot.

From observations like this, researchers strongly suspected that all plant cells contain the same genes—meaning that they show **genetic equivalence.** This notion of genetically equivalent cells was confirmed in the 1950s when biologists were able to grow entire tobacco plants or carrots from a single, differentiated cell taken from an adult. (For more information on how plant cells are grown in culture, see **BioSkills 12** in Appendix B.) These experiments confirmed that differentiated plant cells are genetically equivalent.

Evidence that Differentiated Animal Cells Are Genetically Equivalent

The issue of genetic equivalence proved more difficult to resolve in animals than plants. Serious work began in the 1950s, but the question wasn't settled until the late 1990s.

Early experiments were based on transferring nuclei from differentiated frog cells into unfertilized egg cells whose nuclei had been removed. Some of these transplanted nuclei were able to direct the development of tadpoles. These results provided strong evidence that all cells in the same individual are genetically equivalent.

This conclusion was confirmed in a stunning way in 1997 when Ian Wilmut and colleagues reported their results of nuclear transfer experiments in sheep. As **FIGURE 22.2** shows, the researchers removed mammary-gland cells from a female, grew them in culture, and fused them with eggs whose nuclei had been removed. As the upper drawings show, the eggs came from a black-faced breed of sheep, while the donor nuclei came from a white-faced breed. After developing in culture, the embryos were implanted into surrogate mothers. In one of several hundred such transfer attempts, an apparently normal white-faced lamb named Dolly was born.

Genetic tests showed that Dolly was a **clone**—a genetically identical copy—of the white-faced donor of the mammary-gland cell. Dolly grew into a fertile adult. Soon after, other research groups reported similar results in mice and cows. More recently, a menagerie of animals, including horses, cows, monkeys, and dogs, have been cloned. The success rate of the procedure remains low, however, with the vast majority of nuclear transplants failing to produce healthy newborns.

Taken together, research on cloning plants and animals has shown that cellular differentiation typically does not involve changes in the genetic makeup of cells. Instead, it results from differential gene expression.

How Does Differential Gene Expression Occur?

Eukaryotic cells control gene expression at several different levels: chromatin remodeling and modification, transcriptional regulation, alternative splicing of mRNAs, selective destruction of mRNAs, translation rate, and activation and deactivation of proteins after they are translated (see Chapter 19). All of these processes occur during development. Which is most important in differentiation?

The answer is transcriptional control. To understand why, ask yourself whether a muscle cell should produce mRNAs and proteins that are specifically required by nerve cells. The answer is no. If it did, it would also have to either produce microRNAs that disable nerve-cell mRNAs or expend the energy to translate the mRNAs into nerve-cell proteins that would then have to be maintained in an inactive state. It is more logical to expect that muscle cells transcribe genes required by muscle cells and not genes needed for specialized functions in other types of cells. This is exactly what researchers have found.

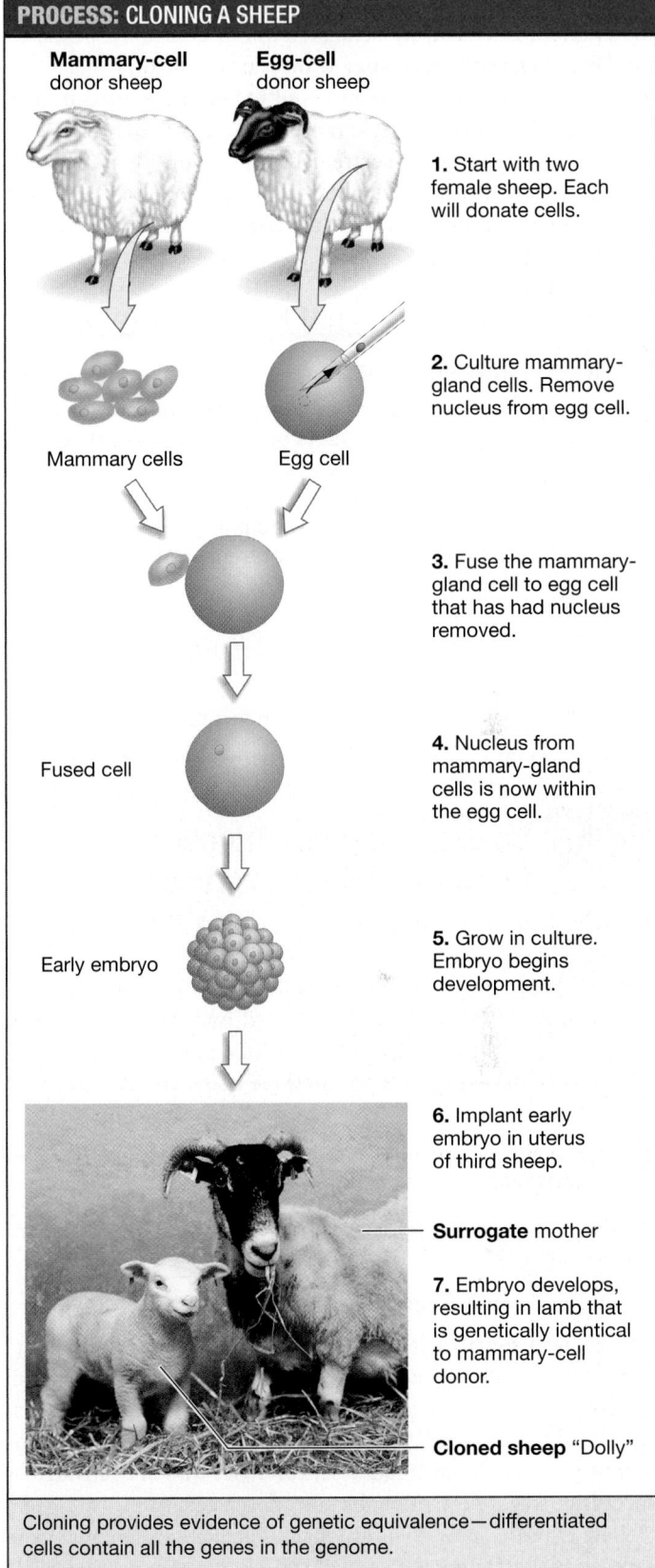

PROCESS: CLONING A SHEEP

Mammary-cell donor sheep   Egg-cell donor sheep

1. Start with two female sheep. Each will donate cells.

Mammary cells   Egg cell

2. Culture mammary-gland cells. Remove nucleus from egg cell.

3. Fuse the mammary-gland cell to egg cell that has had nucleus removed.

Fused cell

4. Nucleus from mammary-gland cells is now within the egg cell.

Early embryo

5. Grow in culture. Embryo begins development.

6. Implant early embryo in uterus of third sheep.

Surrogate mother

7. Embryo develops, resulting in lamb that is genetically identical to mammary-cell donor.

Cloned sheep "Dolly"

Cloning provides evidence of genetic equivalence—differentiated cells contain all the genes in the genome.

FIGURE 22.2 Mammals Can Be Cloned. The lamb that resulted from this experiment was genetically identical to the white-faced individual that donated the nucleus, not the black-faced egg donor or surrogate mother.

Transcription is the fundamental level of control during differentiation. In eukaryotes, transcription is controlled primarily by the presence of **regulatory transcription factors** that influence chromatin remodeling and bind to promoter-proximal elements, enhancers, silencers, and other regulatory sites in DNA.

This simple insight is extremely important. To understand differentiation, researchers have to understand how and why regulatory transcription factors vary among cells.

check your understanding

C Y U

If you understand that . . .

- Differentiation occurs because embryonic cells express distinctive subsets of genes, not because they contain different genes.
- Differential gene expression is predominantly based on transcriptional control. Different types of cells have different combinations of transcription factors.

✔ **You should be able to . . .**

Explain the evidence for genetic equivalence in both plant and animal cells.

Answers are available in Appendix A.

22.3 Chemical Signals Trigger Differential Gene Expression

To understand development, you have to think like a cell. Suppose that you were one of the thousands of cells in a developing animal embryo. Your fate—whether you ended up as part of an arm or a kidney, or whether you differentiated into a nerve cell or a blood cell—would depend on your position in four dimensions—time during development and the three body axes illustrated in **FIGURE 22.3**.

1. One axis runs **anterior** (toward the head) to **posterior** (toward the tail).

2. One axis runs **dorsal** (toward the back) to **ventral** (toward the belly).

3. One axis runs left to right.

Chemical signals tell cells where they are in time and space. This information works through signal receptors and signal transduction cascades to activate transcription factors that turn specific genes on or off, resulting in differentiation. As development proceeds, the distinctive sets of genes that are activated at successive stages determine the fate of each cell.

Let's consider how chemical signals activate differential gene expression, beginning with one of the first developmental signals ever discovered. Although you'll be analyzing what happens as a fruit fly embryo develops, keep an important point in mind: Principles that were discovered in fruit flies are relevant to all multicellular organisms—from mustard plants to humans.

(a) The three body axes observed in humans and other animals...

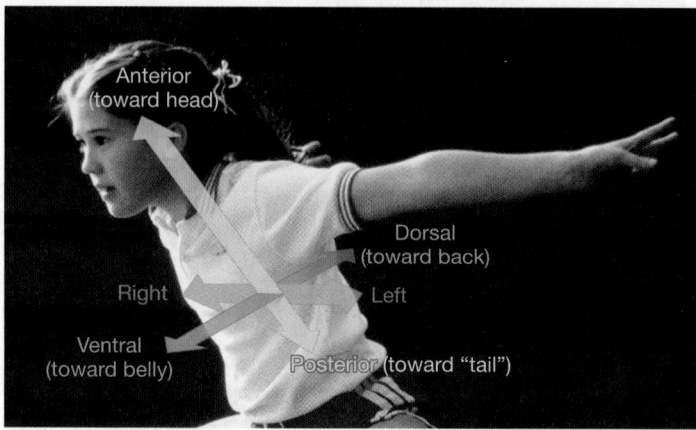

(b) ...are initially established in embryos.

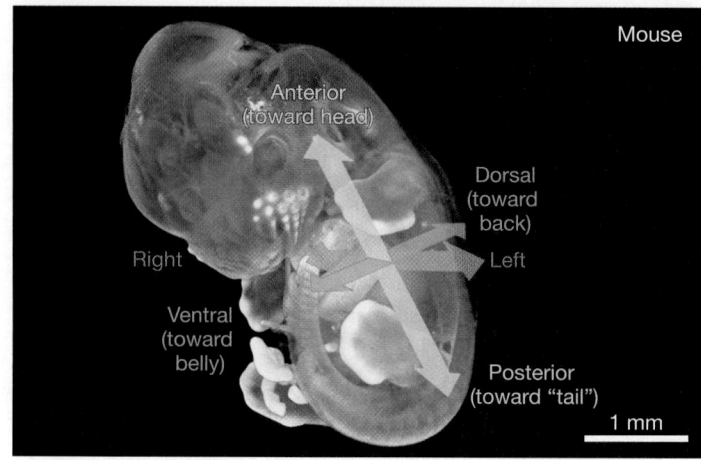

FIGURE 22.3 Most Animals Have Three Major Body Axes.

Morphogens Set Up the Major Body Axes

Biologists use the term **pattern formation** to describe the events that determine the spatial organization of an embryo. If a molecule signals that a target cell is in the head or tail, or dorsal or ventral side of an embryo, then that molecule is involved in pattern formation.

Many molecules that work in pattern formation are present in a concentration gradient, with high concentrations near the source of the molecule and lower concentrations farther away. If cells detect the different concentrations to learn their position, then this molecule is a **morphogen.** Morphogens play central roles in setting up the body axes and at many later stages of development.

Complex patterns don't develop all at once. Instead, pattern formation is progressive. Morphogens set up the major elements of the body—the anterior–posterior, dorsal–ventral, and left–right axes. Genes activated by these morphogens generate signals of their own with more specific information about the location of cells. As development continues, the process repeats: New signals arrive and activate genes that specify finer and finer control over

(a) A normal fruit fly embryo

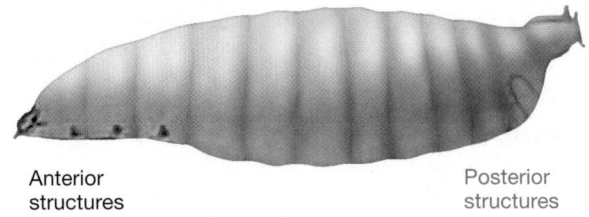

Anterior structures   Posterior structures

(b) A mutant that does not express *bicoid*

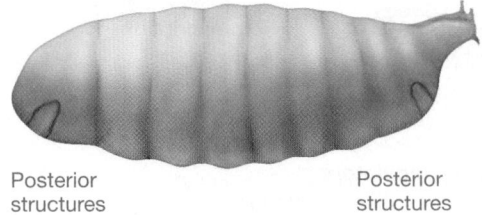

Posterior structures   Posterior structures

FIGURE 22.4 Pattern-Formation Mutants Have Misshapen Bodies.

✔**QUESTION** Predict the phenotype of a mutant that expresses twice the normal levels of Bicoid instead of no Bicoid at all.

what a cell becomes. Development unfolds in a cascade of events unleashed by a few key signals.

The Discovery of Bicoid Morphogens had long been proposed on theoretical grounds, but discovery of the first morphogen emerged from work on the fruit fly *Drosophila melanogaster*—a key model organism in genetics and development (see **BioSkills 13** in Appendix B).

Christiane Nüsslein-Volhard and Eric Wieschaus started this work in the 1970s, taking an approach based on genetics. They began by exposing adult flies to chemicals that cause mutations and examining their offspring for defects in development. One of the most dramatic mutations affected the anterior–posterior axis of the embryos. As **FIGURE 22.4** shows, the mutant embryos were missing all the structures normally found in the anterior end. Instead, the anterior end contained some structures normally found in the posterior.

The mutated gene responsible for this phenotype was dubbed *bi-coid*, meaning two tailed. Based on its phenotype, Nüsslein-Volhard and Wieschaus suspected that the product of the normal *bicoid* gene must provide positional information. In other words, they hypothesized that the *bicoid* gene coded for a morphogen that tells cells where they are located along the anterior–posterior body axis.

The Importance of Morphogen Concentration Gradients To test their hypothesis that *bicoid* encodes a morphogen, Nüsslein-Volhard and colleagues cloned and sequenced the *bicoid* gene (using techniques introduced in Chapter 20). They then used **in situ** (literally, "in place") **hybridization** to find where *bicoid* mRNAs are located. As **FIGURE 22.5** shows, in situ hybridization works by adding a label to single-stranded DNA or RNA molecules to create a probe that is complementary in sequence to the

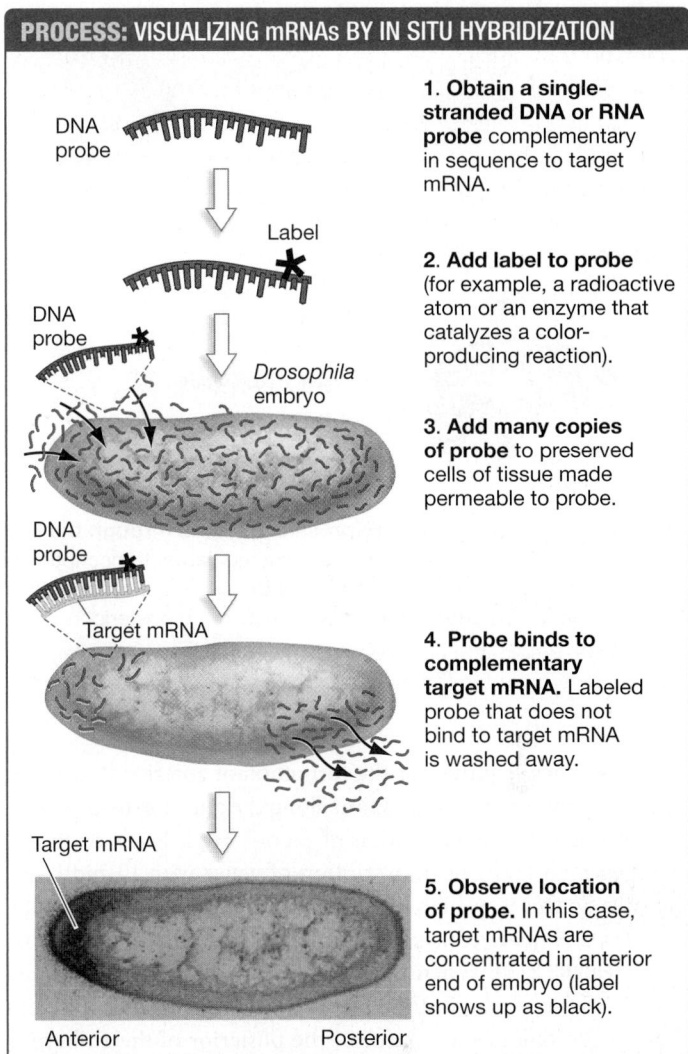

PROCESS: VISUALIZING mRNAs BY IN SITU HYBRIDIZATION

DNA probe

1. Obtain a single-stranded DNA or RNA probe complementary in sequence to target mRNA.

Label

2. Add label to probe (for example, a radioactive atom or an enzyme that catalyzes a color-producing reaction).

DNA probe

Drosophila embryo

3. Add many copies of probe to preserved cells of tissue made permeable to probe.

DNA probe

Target mRNA

4. Probe binds to complementary target mRNA. Labeled probe that does not bind to target mRNA is washed away.

Target mRNA

5. Observe location of probe. In this case, target mRNAs are concentrated in anterior end of embryo (label shows up as black).

Anterior   Posterior

FIGURE 22.5 In Situ Hybridization Allows Researchers to Pinpoint the Location of Specific mRNAs. The micrograph in the last step shows the location of the *bicoid* mRNA in a fruit fly embryo.

mRNA of interest. In this case, the probe sequence was designed to bind to *bicoid* mRNA inside the embryo. As a result, the labeled probes marked the location of the *bicoid* mRNAs.

When Nüsslein-Volhard's group treated eggs and early embryos with labeled probes, they found the *bicoid* mRNA was tightly localized to the anterior end (see step 5 in Figure 22.5). In related studies they also made another surprising discovery: the *bicoid* gene is not expressed in the egg or embryo—it is only transcribed in special cells of the mother fruit fly that are near the egg. The *bicoid* mRNA is then deposited by these maternal cells into the developing egg at the anterior end. Follow-up work showed that when these mRNAs are translated, the protein product forms a steep concentration gradient: Bicoid protein is abundant in the anterior end but declines to progressively lower concentrations in the posterior end.

Later work showed that the Bicoid morphogen is a regulatory transcription factor (see Chapter 19). It binds to DNA and

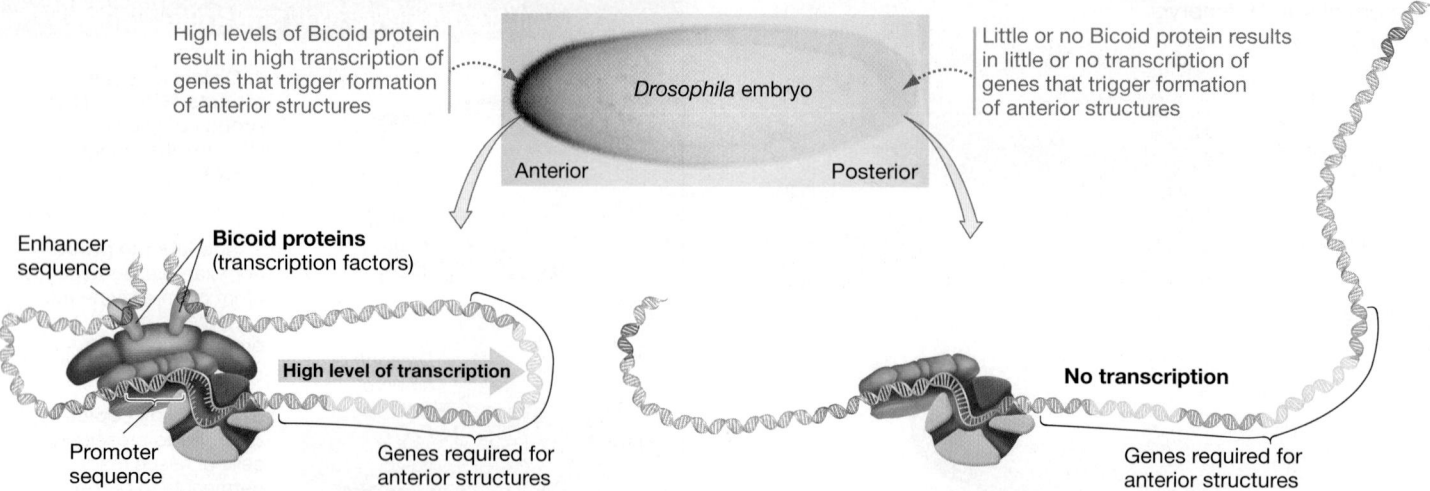

High levels of Bicoid protein result in high transcription of genes that trigger formation of anterior structures

Drosophila embryo

Anterior   Posterior

Little or no Bicoid protein results in little or no transcription of genes that trigger formation of anterior structures

Enhancer sequence

Bicoid proteins (transcription factors)

High level of transcription

Promoter sequence

Genes required for anterior structures

No transcription

Genes required for anterior structures

FIGURE 22.6 Differential Gene Expression Occurs through the Presence or Absence of Regulatory Transcription Factors. Bicoid protein is a regulatory transcription factor that triggers the formation of anterior structures. Because the concentration of Bicoid decreases toward the embryo's posterior, different genes are expressed in the anterior of the embryo than in the posterior.

✔ **QUESTION** Predict the effect on development if some cytoplasm from the anterior of a normal early *Drosophila* embryo were added to the central region of a mutant embryo that lacked *bicoid* mRNA.

activates genes required for the formation of anterior structures. In effect, cells learn their position along the anterior-to-posterior axis through the concentration of Bicoid. High levels of Bicoid result in high levels of transcription of genes with Bicoid binding sites. In this way, high concentrations tell a cell that it's in the anterior, and lower concentrations tell the cell that it's more posterior (**FIGURE 22.6**). When Bicoid is lacking, cells throughout the embryo get a "you're in the posterior" message from a different morphogen that has its source in the posterior of the embryo—leading to the mutant phenotype you saw in Figure 22.4.

Morphogens in Plant Development   Plants also use morphogens to provide positional information. One important example is auxin, a small molecule that is used first in early development and then repeatedly and in different ways throughout the plant's life. In the embryo, auxin enters cells and triggers the production of transcription factors that affect differentiation.

Auxin is produced in the embryo's meristem cells that lie at the tip of what will become the stem. Auxin is actively transported toward the base of the embryo—what will become the root. In the process, an auxin concentration gradient forms. A high concentration of auxin signals, "You're near the top of the stem"; lower auxin concentrations that accumulate at the base of the embryo signal, "You're the root."

Regulatory Genes Provide Increasingly Specific Positional Information

The initial work on *bicoid* illustrated the importance of morphogens as a general theme in development. A second fundamental developmental principle common to both plants and animals soon emerged: Differentiation is a progressive, step-by-step process.

Dividing Up the Embryo   Along with the "two-tailed monster" *bicoid* mutants, Nüsslein-Volhard and Wieschaus found an array of embryos that had defects in smaller regions of their body. Flies and many other animals are constructed from a set of **segments**—portions of the body that are repeated along its length. The mutants with altered segments had defective **segmentation genes.**

Researchers have now identified four classes of segmentation genes that act in a cascade:

1. First, maternally expressed genes deposit morphogen mRNAs in the egg. Morphogens, such as *bicoid*, control the formation of large groups of segments that span the anterior or the posterior halves of the embryo.

2. Gap genes are expressed second, in broad regions along the head-to-tail axis (**FIGURE 22.7a**). Gap genes control the formation of groups of segments that define large body regions.

3. Pair-rule genes are expressed next, in alternating bands along the embryo (**FIGURE 22.7b**). This pattern suggests that these genes control the formation of individual segments.

4. Segment polarity genes are expressed later, in a restricted band within every segment (**FIGURE 22.7c**). This pattern and order of expression imply that segment polarity genes create specific regions within each segment.

Specifying Segment Identity   Once segmentation gene products create segments along the anterior–posterior axis of a fly embryo, development continues with the activation of *Hox* genes. These genes are turned on by products of the segmentation genes. *Hox* genes have spectacular and far-reaching effects. While segmentation gene products establish each segment, *Hox* gene products specify each segment's identity by activating effector genes.

(a) Gap genes

Early in development, gap genes define broad regions that often span several segments.

Anterior   Posterior

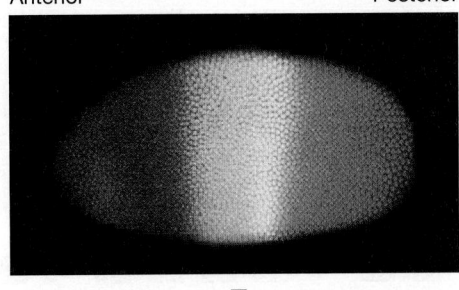

(b) Pair-rule genes

Later in development, pair-rule genes demarcate individual segments.

(c) Segment polarity genes

Still later, segment polarity genes delineate regions within individual segments.

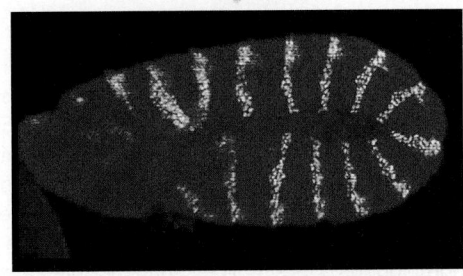

FIGURE 22.7 Domains of Gene Expression Demarcate Body Regions in Fruit Flies. The embryos in **(a)** and **(b)** were stained for two different segmentation gene products. The embryo in **(c)** was stained for one segmentation gene product.

Effector gene expression leads to the production of the specialized proteins and structures of differentiated cells.

The *Hox* genes were discovered when researchers found adult fruit flies with body parts in the wrong place. This bizarre outcome is called homeosis, and the types of mutations that cause it are termed **homeotic mutations.** For example, homeotic mutations in *Hox* genes can turn a segment in the middle part of the body into a segment just like the one that lies in front of it. Instead of bearing the pair of small balancing structures called halteres, the transformed segment now bears a pair of wings—as does the segment in front of it. The mutant has four wings instead of the normal two (**FIGURE 22.8**).

Hox genes code for transcription factors that trigger the development of the structures appropriate to each type of segment, such as antennae, wings, or legs. The *Hox* genes are distinguished by a sequence called the **homeobox** that encodes a DNA-binding domain.

To put all this information into perspective, it helps to realize that the genes involved in early development form a regulatory cascade (**FIGURE 22.9**; see page 414). Morphogens—produced by maternally expressed genes and deposited in the egg—trigger the production of other regulatory signals and transcription factors, which initiate the production of another set of signals and regulatory proteins, and so on down the chain. Ultimately, the genes needed to form antennae, wings, and legs are turned on in the right places and times within the developing embryo. Similar types of regulatory cascades exist in all animal and plant species.

The unfolding regulatory cascades provide progressively detailed information about where cells are located. This positional information, in turn, causes changes in cell proliferation, death, movement, interactions, and differentiation.

- Because chemical signals vary in identity and concentration along the three major body axes, cells in different locations receive unique positional information.

Normal fruit fly

Homeotic mutant

Homeotic mutant

FIGURE 22.8 Homeotic Mutants in *Drosophila* Have Structures in the Wrong Locations. These colorized scanning electron micrographs show homeotic mutants in fruit flies. Wings are growing where small, stabilizing structures called halteres should be, or legs are growing where antennae should be.

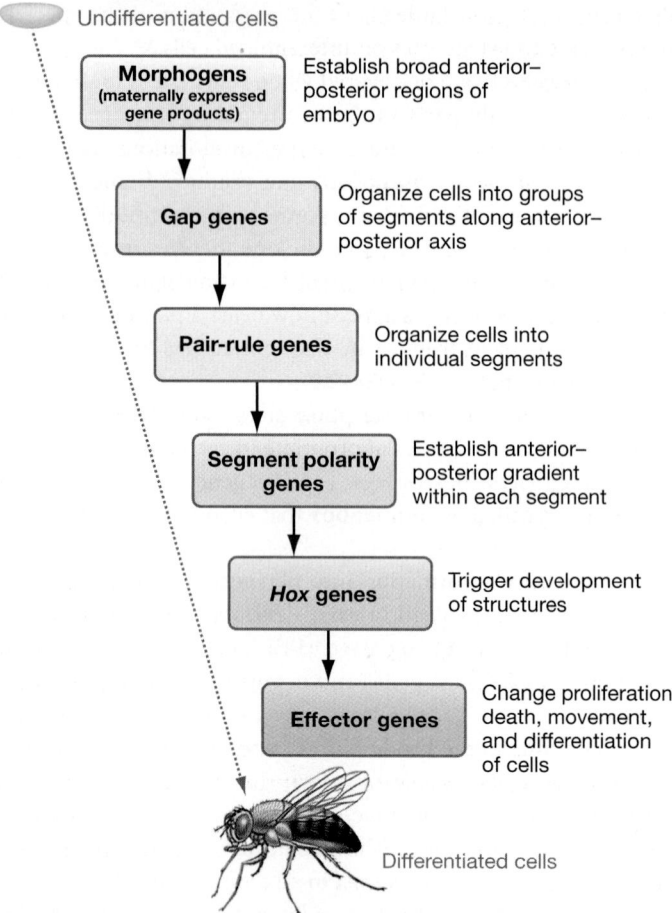

FIGURE 22.9 A Regulatory Gene Cascade in Fruit Flies. This cascade establishes the positions and identities of structures along the anterior-to-posterior axis. Morphogens—products of maternally expressed genes that are deposited in the egg—unleash the sequential expression of genes that work at each level of the cascade. Genes at one level of the cascade control the expression of genes at lower levels. The *Hox* genes are controlled by gap, pair-rule, and segment polarity genes. The effector genes are controlled by both segment polarity and *Hox* genes.

In the flowchart:

Morphogens (maternally expressed gene products) — Establish broad anterior–posterior regions of embryo

Gap genes — Organize cells into groups of segments along anterior–posterior axis

Pair-rule genes — Organize cells into individual segments

Segment polarity genes — Establish anterior–posterior gradient within each segment

Hox genes — Trigger development of structures

Effector genes — Change proliferation, death, movement, and differentiation of cells

Undifferentiated cells → Differentiated cells

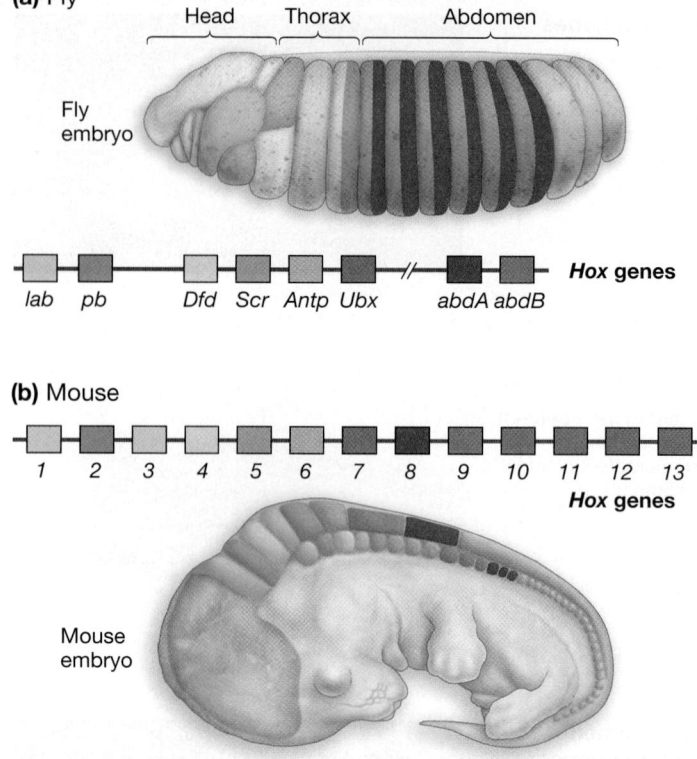

(a) Fly

Head Thorax Abdomen

Fly embryo

Hox genes: lab pb Dfd Scr Antp Ubx abdA abdB

(b) Mouse

Hox genes: 1 2 3 4 5 6 7 8 9 10 11 12 13

Mouse embryo

FIGURE 22.10 Organization and Expression of *Hox* Genes. The location of *Hox* genes on the chromosome correlates with their pattern of expression in the embryos of flies and mice. Matching colors of genes and regions in the embryo indicate where each gene is expressed at its highest levels. The genes represented by same-color boxes in the fly **(a)** and mouse **(b)** are considered homologous. For simplicity, three features are not shown in this figure: the fly genes are present in two separate groups that were created by a relatively recent chromosome breakage in the fly lineage, the mouse has three similar *Hox* gene clusters in addition to the one shown, and there is overlap in the regions in which *Hox* genes are expressed.

✔QUESTION What evidence would support the claim that these genes are homologous?

- Each level in a regulatory cascade provides more specific information about where a cell is located.

- As regulatory cascades are deployed over time, a cell's identity becomes more and more finely specified.

Chemical Signals and Regulatory Genes Are Evolutionarily Conserved

Early genetic mapping studies revealed that the *Drosophila Hox* genes were present in clusters as shown in **FIGURE 22.10a**. Once methods to isolate and sequence DNA were available (see Chapter 20), researchers soon applied them to the *Hox* genes. The results of molecular studies were striking—the order of the *Hox* genes along the chromosome corresponds to the order of where the genes are expressed along the anterior-to-posterior axis.

Conservation of *Hox* Gene Organization With information about fruit fly *Hox* genes in hand, a few biologists thought it was worth looking for similar genes in other animals. This was a long shot, because it seemed unlikely that creatures as different as flies, sea urchins, mice, and humans would use the same genes to build such vastly different bodies. The gamble paid off. Virtually all animals were found to possess related sets of *Hox* genes. Incredibly, these *Hox* genes were organized in clusters in an order that aligned with the *Drosophila* genes. Even more amazing, the curious relationship first seen in *Drosophila* between *Hox* gene order along the chromosome and where a particular *Hox* gene is expressed was shared by all these animals.

A comparison of *Hox* genes in flies and mice (**FIGURE 22.10b**) reveals that there are more *Hox* genes in the mouse, but their chromosomal organization is similar. Notice how genes at the "left-hand" side of the cluster are expressed in the anterior end of both embryos, while genes at the opposite end of the cluster are expressed in posterior regions. The discovery that similar regulatory genes are used to build such a diverse array of animals ranks as one of the greatest surprises in modern biology.

Conservation of *Hox* Gene Function   Is there evidence that *Hox* genes in organisms other than *Drosophila* actually specify parts of the animal? Could it be that these genes are organized and expressed in similar ways but aren't used to specify pattern except in flies? Mutations discovered in many animals, including humans, have shown that when *Hox* genes are altered, defects in pattern formation result. Based on these data, biologists conclude that in flies, mice, humans, and most other animals, *Hox* genes play a key role in specifying which body structures to build.

This conclusion was supported in dramatic fashion when researchers in William McGinnis's lab introduced the *Hoxb6* gene from mice into fruit fly eggs. The *Hoxb6* gene in mice is similar in structure and sequence to the *Antp* gene of flies (Figure 22.10). Because it was introduced without its normal regulatory sequences, the *Hoxb6* gene was expressed throughout the treated fly embryos. The resulting larvae had defects identical to those observed in naturally occurring fly mutants in which the *Antp* gene is improperly expressed throughout the embryo. This is a stunning result: A mouse gene not only affected the development of a fly, but even mimicked the effect of the related fly gene.

To interpret these observations, biologists hypothesize that the genes in *Hox* complexes of animals are **homologous**—meaning that they are similar because they are descended from genes in a common ancestor. This hypothesis implies that the first *Hox* genes arose very early in animal evolution. For the past roughly 600 million years of animal evolution, *Hox* gene products have been helping to direct the development of animals.

The take-home message from these studies is that key molecular mechanisms of pattern formation have been highly conserved during animal evolution. The discovery of these shared mechanisms is one of the most significant results to have emerged from studies of animal development. Although animal bodies are spectacularly diverse in size and shape, the underlying mechanisms responsible for their development are similar.

One Regulator Can Be Used Many Different Ways

Regulatory gene cascades and the evolutionary conservation of key developmental genes are general features of animal and plant development. But biologists have articulated another organizing principle as well: During development, the same regulatory genes, proteins, and chemical signals are used over and over in a variety of contexts.

The Wnt signal protein offers an impressive example of repeated uses of the same molecule to achieve different ends. *Drosophila* has a single *Wnt* gene known as *wingless*. From the name, you can guess that *wingless* is involved in wing development—flies that have a mutated *wingless* gene lack wings. The first role of *wingless*, however, is not in controlling wing formation, but in specifying boundaries between segments. The list doesn't stop here—it turns out that *wingless* is also a regulator of many other

developmental decisions. The *wingless* gene produces a signal protein that binds to specific receptors on different target cells and controls development of the leg, head, and digestive system as well as specifying particular nerve cells.

An analogy for this developmental principle of one signal, many messages is the signal called a pinch. When you were little, a pinch on the cheek from your grandmother indicated something very different from a pinch on the arm from an older sibling. It also means something different from a pinch you might receive now from someone you are romantically attracted to. In development, as in human communication, the context in which a signal is sent and received—its location, timing, and intensity—has a major effect on the signal's meaning and consequences.

In contrast to single genes such as *wingless* that are used in many different ways during the development of one organism, in many different organisms there are genes that control similar events in development. Animals and plants each have a conserved set of **tool-kit genes** that code for signals, signal-transduction pathways, and regulatory proteins that are used to direct related aspects of development in many different species.

Hox genes are an important example of tool-kit genes because they are present in a wide array of animals and are used to specify body structures. Similar tool-kit genes can, however, produce dramatically different structures such as the wing of a bat or the flipper of a whale if they are used at different times, in different places, and in different developmental contexts.

What determines the ways tool-kit genes are used? The answer lies largely in the regulatory sequences of DNA. Mutation of these sequences allows tool-kit genes to be used in new ways in different species. As you'll see in the next section, changes in regulatory sequence have profound effects on evolution.

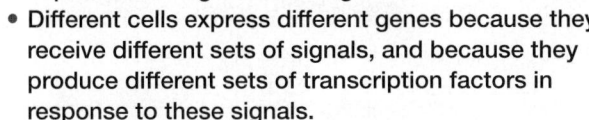

check your understanding

If you understand that . . .

- Chemical signals trigger the production of specific sets of transcription factors that change gene expression in signal-receiving cells.
- Different cells express different genes because they receive different sets of signals, and because they produce different sets of transcription factors in response to these signals.

✔ **You should be able to . . .**

1. Explain how a gradient of Bicoid protein delivers information about where cells are along the anterior–posterior axis of a fly embryo.

2. Describe how the cascade of transcription factors triggered by Bicoid leads to the gradual differentiation of segments along the anterior–posterior axis of a fly embryo.

Answers are available in Appendix A.

22.4 Changes in Developmental Gene Expression Underlie Evolutionary Change

This chapter has shown that for an embryo to develop, cells have to proliferate, move or expand, differentiate, interact in specific ways, and sometimes die. Differentiation is caused by differential gene expression. It results from signals that tell cells where they are in time and space, triggering deployment of a complex cascade of transcription factors.

If any of these processes are disrupted, the embryo is likely to die. But if one of these processes is modified, the effect may be a structure with a different size, shape, or activity. As a result, the embryo will develop new features, and the adult will have a novel phenotype.

Once biologists began working out the regulatory signals and cascades introduced earlier in the chapter, they realized that the genetic changes altering these developmental processes must be the foundation of evolutionary change. The increase in body size that has occurred during human evolution, for example, must have resulted from mutations that altered the signals, regulatory sequences in DNA, or transcription factors that are involved in the amount and timing of cell proliferation throughout the body.

An emerging research field called evolutionary developmental biology, or **evo-devo,** focuses on understanding how changes in developmentally important genes have led to the evolution of new forms such as the flower, the leaf, and animal limbs. As an example of this exciting work, let's consider how snakes *lost* their limbs.

Although some snakes do not develop any sort of forelimb or hindlimb, boas and pythons have tiny pelvic (hip) bones and a rudimentary femur (thigh bone). The fossil record shows that the ancestor of all snakes had four functional legs. How did snakes lose their legs?

Researchers were able to answer this question because the signals and genes responsible for limb development are well understood. In chicken embryos, scientists had learned that *Hox* genes called *Hoxc6* and *Hoxc8* are expressed together in cells where ribs form, but *Hoxc6* is expressed alone in the region that gives rise to the forelimbs (**FIGURE 22.11a**).

Could modifications in where homologous *Hox* genes are expressed in snakes explain their limb loss? As **FIGURE 22.11b** shows, *Hoxc6* and *Hoxc8* are expressed together throughout the snake embryo—including in the region where forelimbs should form. These data suggest that a change in the regulation of where *Hoxc8* is expressed led to the evolutionary loss of the forelimb. Snakes make ribs instead of forelimbs.

It turns out that there's more than one way to lose a limb, even in the same animal. Hindlimb loss in snakes is due to defects in production of a signaling molecule encoded by the gene *sonic hedgehog*. This gene is homologous to a *Drosophila* segment polarity gene, *hedgehog,* which also produces a signal protein. Recent work has shown that defects in *sonic hedgehog* signaling affect animals other than snakes, as well—loss of *sonic hedgehog* signals in the pelvic region of whales led to the disappearance of their hindlimbs. Changes in the regulation of one widespread tool-kit gene—*sonic hedgehog*—have achieved extraordinary outcomes in animals as different as snakes and whales.

In limb loss, biologists can point to alterations in gene expression or in the production of specific signals that explain why an important evolutionary change occurred. Changes in the regulation of tool-kit genes led to developmental changes that build a snake or whale. These creatures have been legless ever since.

(a) Pattern of gene expression in tetrapods.

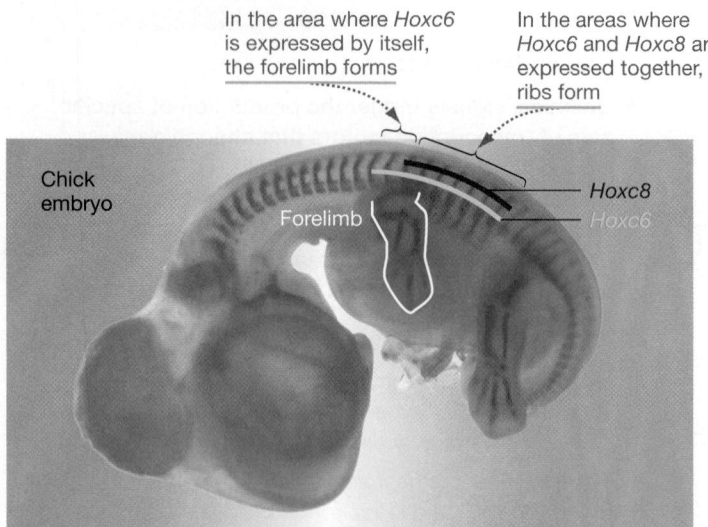

In the area where *Hoxc6* is expressed by itself, the forelimb forms

In the areas where *Hoxc6* and *Hoxc8* are expressed together, ribs form

Chick embryo

Forelimb

Hoxc8
Hoxc6

(b) Pattern of gene expression in snakes.

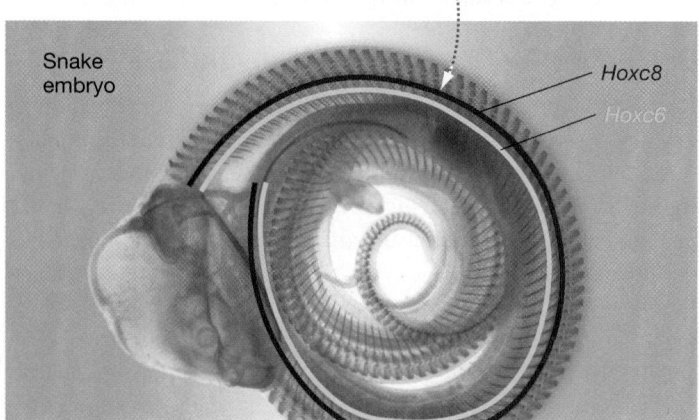

Hoxc6 and *Hoxc8* are always expressed together, so no forelimbs form

Snake embryo

Hoxc8
Hoxc6

FIGURE 22.11 Changes in Homeotic Gene Expression Led to Limb Loss in Snakes. *Hoxc6* and *Hoxc8* code for transcription factors. If both transcription factors are present in a population of cells along the anterior–posterior axis of a tetrapod (a four-limbed vertebrate), genes that lead to the formation of vertebrae and ribs are activated. But if only *Hoxc6* is expressed, then genes that lead to the formation of forelimbs are activated.

If you understand . . .

22.1 Shared Developmental Processes

- During development in all organisms, cells divide, interact, differentiate, move or expand in a directed manner, and sometimes die.

- Cells have to proliferate in a regulated manner to promote growth of the body.

- Chemical signals sent between cells provide a constant flow of information about where cells are in space and time.

- Cells undergo a step-by-step process of differentiation that leads to specialized cell types.

- Development requires the controlled movement and expansion of cells. In animals, many cells move in a coordinated way throughout development. Plant cells cannot move because of the cell wall, but plant cells precisely control the plane of cell cleavage and the direction of cell expansion to shape the embryo and the adult.

- Programmed cell death of some cells is an essential part of normal development.

✓ You should be able to predict how a mutation that disrupts control of the plane of cell cleavage would affect the development of an oak seedling.

22.2 Genetic Equivalence and Differential Gene Expression in Development

- Cells become specialized because they express different genes, not because they contain different genes.

- Cloning of plants and animals from differentiated adult cells indicates the genetic equivalence—that is, all cells of an individual contain the same genetic information.

- An implication of genetic equivalence is that cells are different because they express a common set of genes differently. Creating different cell types by expression of different genes from the same genome is differential gene expression.

- Stem cells in animals and plants do not fully differentiate, but proliferate throughout life to provide a source of cells that can specialize to replace damaged cells or promote growth.

- Transcriptional regulation through the production of different sets of transcription factors in different types of cells is the most important form of gene control during development.

You should be able to explain why it's possible to propagate banana trees from cuttings. Your answer should cover how cells in a stem cutting can give rise to roots.

22.3 Chemical Signals Trigger Differential Gene Expression

- Cells "know" where they are in the body and how far along development has progressed because they receive a steady stream of signals. These signals are produced by other cells and diffuse or are transported throughout the body.

- Morphogens are signals present in a gradient. Cells can learn where they are along a body axis by "reading" the concentration of morphogen.

- Cells respond to chemical signals by activating signal transduction pathways that lead to expression of a distinctive set of proteins. These proteins are expressed because new groups of genes are transcribed in response to the signal.

- Cells gradually acquire positional information, first learning where they are in broad regions and later in smaller parts of each region.

- In some organisms, products of maternally expressed genes (morphogens) are deposited in the egg, and these activate the sequential expression of many other segmentation genes that are linked in regulatory cascades.

- Chemical signals are conserved across species.

- Each specific signaling pathway is used over and over in development in different contexts to achieved different outcomes.

- A set of conserved tool-kit genes is used to control core developmental processes in a wide array of species.

✓ You should be able to explain why a mutation in a *Drosophila* pair-rule gene affects the expression of segment polarity genes but has no effect on the expression of gap genes.

22.4 Changes in Developmental Gene Expression Underlie Evolutionary Change

- Mutations in genes responsible for development can lead to the evolution of new body sizes, shapes, and structures. These changes are important components of evolution.

- Mutations in regulatory sequences of DNA are especially important because they can change when, where, and to what levels key regulators are expressed.

✓ You should be able to use the information in Figure 22.11 to predict the effect on chick development of replacing the *Hoxc8* gene regulatory sequence with the regulatory sequences of *Hoxc6* gene.

(MB) MasteringBiology

1. MasteringBiology Assignments

Tutorials and Activities Early Pattern Formation in *Drosophila*; Pattern Formation; Role of *bicoid* Gene in *Drosophila* Development

Questions Reading Quizzes, Blue-Thread Questions, Test Bank

2. eText Read your book online, search, take notes, highlight text, and more.

3. The Study Area Practice Test, Cumulative Test, BioFlix® 3-D Animations, Videos, Activities, Audio Glossary, Word Study Tools, Art

You should be able to . . .

1. What is apoptosis?
   a. an experimental technique used to kill specific cells
   b. programmed cell death that is required for normal development
   c. a pathological condition observed only in damaged or diseased organisms
   d. a developmental mechanism unique to the roundworm *C. elegans*

2. In adult animals, _____ are a source of undifferentiated cells that can divide to produce cells that can specialize.

3. Why is in situ hybridization such a valuable tool for studying development?
   a. It can identify the location of specific mRNAs to provide information on where particular genes are being expressed.
   b. It allows researchers to understand how cell–cell signals and regulatory transcription factors interact.
   c. It provides information on gene evolution.
   d. It can locate promoters and enhancers of important developmental genes.

4. Embryonic cells can learn their position by "reading" the concentration of a morphogen.
   a. True
   b. False

5. What is a homeotic mutant?
   a. an individual with a structure located in the wrong place
   b. an individual with an abnormal head-to-tail axis
   c. an individual that is missing segments
   d. An individual with double the normal number of structures

6. A tool-kit gene is _____.

7. What does it mean to say that cell proliferation, death, movement, or expansion; differentiation; and interaction are shared developmental processes?

8. Explain the logic of cloning animals to test the hypothesis that all cells in the animal body are genetically equivalent.

9. How did researchers go about looking for genes that have important roles in establishing the anterior–posterior body axis and body segments in *Drosophila*?

10. Why is it significant that many of the genes involved in development encode regulatory transcription factors?

11. Explain the connection between the existence of regulatory cascades and the observation that differentiation is a step-by-step process.

12. Which of the following provides the strongest evidence for the conservation of tool-kit genes?
    a. Bicoid moved from one fly embryo into the posterior of another fly embryo causes the formation of two head regions.
    b. Mutation of an unrelated gene in another species of fly has a similar effect to mutation of *bicoid* in *Drosophila*.
    c. A mouse *Hox* gene can be used to take over the function of a mutated *Drosophila Hox* gene.
    d. Sheep can be cloned by fusing a differentiated adult cell with an enucleated egg.

13. Biologists found that the products of two different *Drosophila* genes are required to keep *bicoid* mRNA concentrated at the anterior end of the egg. In individuals with mutant forms of these proteins, *bicoid* mRNA spreads farther toward the posterior pole than it normally does. Predict what effect these mutations will have on segmentation of the larva.

14. In 1992 David Vaux and colleagues used genetic engineering technology to introduce a human gene for a protein that inhibits apoptosis into embryos of the roundworm *C. elegans*. When the team examined the embryos, they found that cells that normally undergo programmed cell death survived. What is the significance of this observation?

15. QUANTITATIVE Imagine a situation in which a morphogen has its source at the posterior end of a *Drosophila* embryo. Every 100 μm from the posterior pole, the morphogen concentration decreases by half. If a cell required 1/16th the amount of morphogen found at the posterior pole to form part of a leg, how far from the posterior pole would the leg form?
    a. 100 μm
    b. 160 μm
    c. 400 μm
    d. 1600 μm

16. Fruit flies have 6 legs; spiders have 8 legs; centipedes have many; earthworms have none. All these species are segmented and have *Hox* genes. Generate a hypothesis to explain how variation in leg number evolved in these species.

23 An Introduction to Animal Development

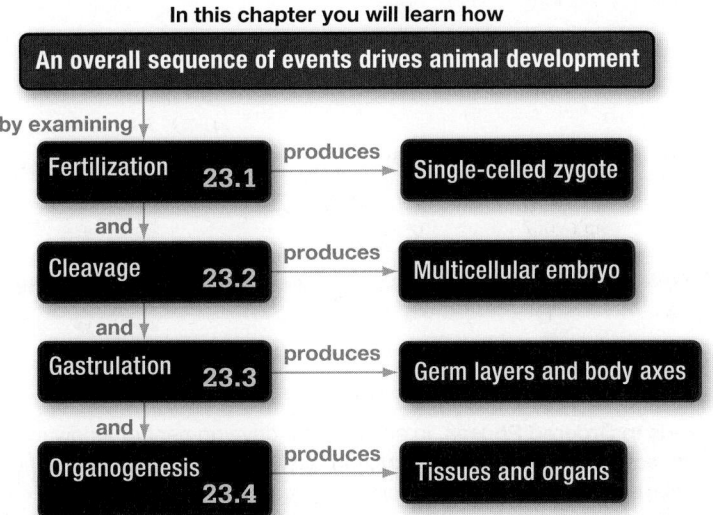

In this chapter you will learn how

An overall sequence of events drives animal development

by examining

Fertilization 23.1 → produces → Single-celled zygote

and

Cleavage 23.2 → produces → Multicellular embryo

and

Gastrulation 23.3 → produces → Germ layers and body axes

and

Organogenesis 23.4 → produces → Tissues and organs

A human embryo, about 3 days old, on the tip of a pin.

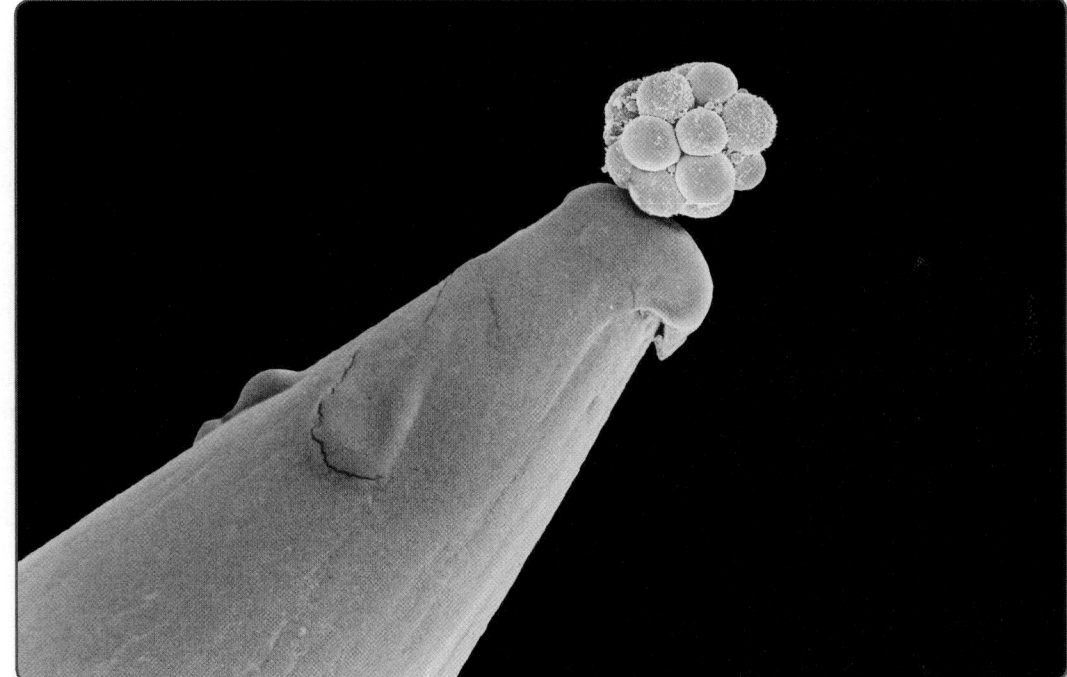

"Making Babies: 25,000 Genes, Some Assembly Required." This wry chapter title written by evolutionary-developmental biologist Sean Carroll highlights what is surely one of the most remarkable events of life. How *is* a baby assembled from a single cell using instructions in the genome?

In a previous chapter (Chapter 22), you learned about the basic cellular processes of development. The central message was that cell–cell signals cause different sets of transcription factors to be produced in various cells throughout the embryo, resulting in differential gene expression and differentiation. For example, muscle cells "know" that they are muscle cells because they have received specific types and amounts of cell–cell signals. In response, they produce muscle-specific proteins.

✔ When you see this checkmark, stop and test yourself. Answers are available in Appendix A.

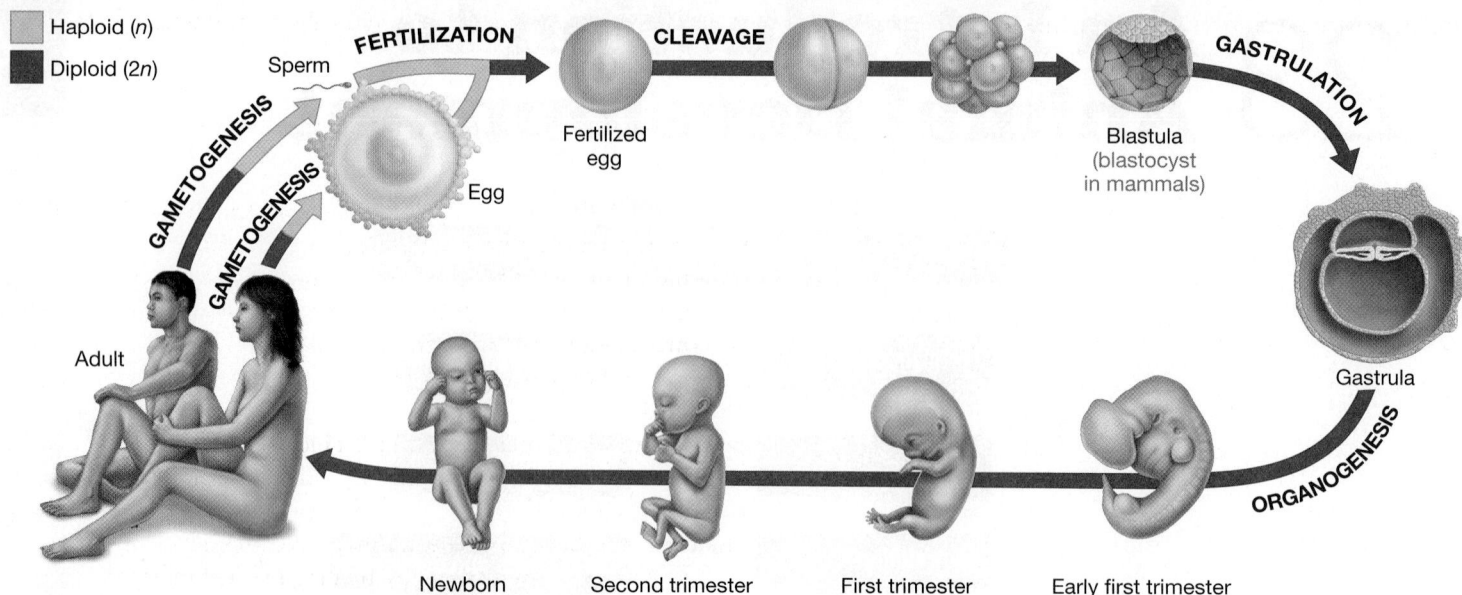

FIGURE 23.1 Development Proceeds in Ordered Phases. In animals, the development of a new individual starts with fertilization and continues with early cell divisions (cleavage) that result in a blastula. Gastrulation then rearranges the blastula into a multilayered structure called a gastrula. Organogenesis follows and leads to the formation of adult tissues and structures. The cycle is continued into a new generation with gametogenesis, the formation of sperm and egg, and their fusion in fertilization.

This chapter's goal is to apply these general developmental principles to the events that are needed to develop an animal; in essence, what's involved in "some assembly required." The major stages in this assembly process for a human are summarized in **FIGURE 23.1**.

Let's begin with the union of sperm and egg, and end with formation of differentiated cells within organs. At each step, remember that it is the readout of the genes that produces the cell–cell signals and regulatory proteins needed to direct developing cells to divide, move, interact, and differentiate—to do the things required to form a new individual.

23.1 Fertilization

Sperm and egg come together to create the **zygote,** the first cell of the new individual. What could be simpler? When this process called **fertilization** is examined closely, however, the question is quickly turned on its head. What could be more complex than two cells finding each other and doing something as odd as fusing together?

Sperm and eggs are among the most specialized of all cells. Recall that sperm and egg are gametes, haploid reproductive cells that are produced by meiosis and differentiation (Chapter 13). Sperm, the male gametes, are stripped-down speedsters designed to deliver the paternal (the father's) haploid genome to the egg. Eggs are huge and specialized to support early—and in some organisms, virtually all—development. A frog egg is more than 400,000 times larger in volume, and an ostrich egg is over

a billion times larger, than other cells in the body. Human eggs, in comparison, are relatively small, but they still contain more than 100 times the volume of a "typical" cell. A later chapter (Chapter 50) explores the process of gametogenesis, how eggs and sperm are formed.

What are the challenges of fertilization? First, sperm need to find the egg (**FIGURE 23.2a**). In many species, this is accomplished by following chemical signals secreted by the egg. Next, sperm and egg must recognize one another. In organisms that use external fertilization—that is, fertilization outside the body—there's an additional challenge. For these organisms, there needs to be a mechanism to ensure that only sperm and egg from the same species come together. Once in contact, sperm and egg have to fuse, something that few other cells in the body ever do. What's more, in most species, fusion must be limited to a single sperm so the egg does not receive extra chromosomes. Finally, the fusion of the two gametes has to trigger the onset of development. There's little simple about fertilization.

Research on fertilization began in earnest early in the twentieth century, when biologists started to study the sperm–egg interaction in sea urchins. Like most aquatic animals, sea urchins shed their gametes into the surrounding water, and fertilization occurs externally. **FIGURE 23.2b** outlines major events of this process.

Sea urchin sperm are drawn to the egg by following a gradient of a chemical secreted by the egg's jelly layer (step 1). The head of the sperm initially binds to the jelly layer that surrounds the egg cell (step 2). This triggers the acrosome reaction, a release of the contents of the sperm's acrosome (step 3). The acrosome is a large

(a) Many sperm compete to fertilize an egg.

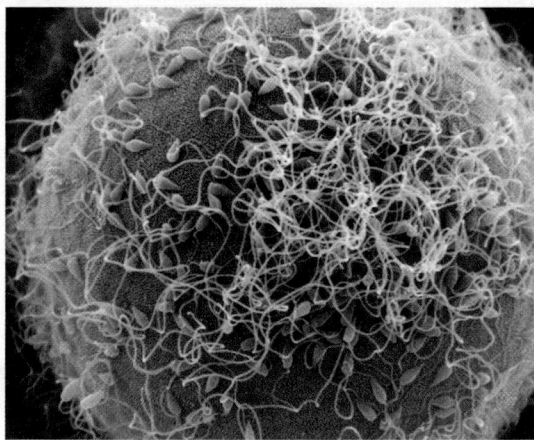

(b) Steps in fertilization

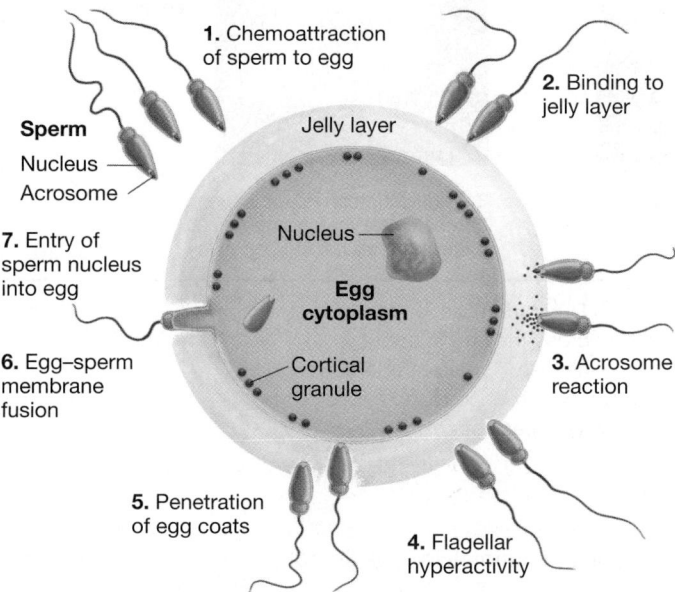

FIGURE 23.2 **Fertilization Is a Complex Process. (a)** Many sperm competing to fertilize an egg. **(b)** Steps of fertilization.

vesicle containing digestive enzymes that bore a hole through the jelly layer and through another protective coat called the vitelline envelope. The initial contact with the egg's jelly layer ramps up the beating of the flagellum to power the sperm through the egg layers (steps 4 and 5).

Once the sperm head contacts the egg-cell surface, specific proteins in the plasma membranes of the egg and sperm induce membrane fusion (step 6). The sperm nucleus then enters the egg (step 7). Contact between the sperm and egg plasma membranes initiates many responses in the egg. Sperm and egg nuclei fuse, and the egg is now an activated machine ready for development. Fertilization is complete.

But for species like the sea urchin that cast their sperm indiscriminately, what prevents cross-species fertilization and the consequent production of dysfunctional hybrid offspring? After

all, in many habitats sperm and eggs from different species are mixed in the same sea water.

How Do Gametes from the Same Species Recognize Each Other?

In the 1970s, Victor Vacquier and co-workers succeeded in identifying a protein on the head of sea urchin sperm that binds to the plasma membrane of sea urchin eggs in a species-specific manner. They called this protein bindin. Follow-up work showed that bindin proteins from even very closely related species are distinct. This led to the proposal that bindin ensures that a sperm binds only to eggs from the same species.

But what binds to bindin? If bindin is a key, what's the lock? Charles Glabe and Vacquier hypothesized that egg-cell membranes contain a receptor for bindin. To test this hypothesis, they set out to isolate the receptor molecule from sea urchin eggs.

The researchers began by treating sea urchin eggs with a set of chemicals that released proteins from the egg cell surface. When the investigators separated the proteins that were released from the egg surface, they found one that bound to isolated bindin molecules. Significantly, further testing by the researchers showed that this binding occurred in a species-specific manner. The egg-cell receptor molecule only bound to bindin isolated from sperm of the same species, not to bindin isolated from sperm of different species. Based on these observations, Glabe and Vacquier concluded that they had found the egg-cell receptor protein for sperm.

During sea urchin fertilization, species-specific bindin molecules on sperm interact with species-specific receptors on the surface of the egg. Similar types of specific protein–protein interactions occur between the sperm and egg cells of mammals.

✔ If you understand the importance of protein–protein interactions for binding sperm and egg, you should be able to explain why adding a molecule that bound to a bindin-like protein on the human sperm head would be an effective contraceptive.

What Prevents More Than One Sperm from Entering the Egg?

In early studies on sea urchin fertilization, researchers noticed that only one sperm fertilized an egg, even when hundreds of sperm were clustered around it. This observation makes sense: Multiple fertilization, or polyspermy, would result in a zygote that had more than two copies of each chromosome. Embryos with more than two copies of each chromosome die. How do animals avoid polyspermy?

In sea urchins, there are two mechanisms that block polyspermy. One works immediately after sperm–egg contact and is only briefly effective. A second mechanism, seen in almost all species, including mammals, results in the erection of a physical barrier to sperm entry. This second process begins when sperm entry causes calcium ions (Ca^{2+}) to be released from storage

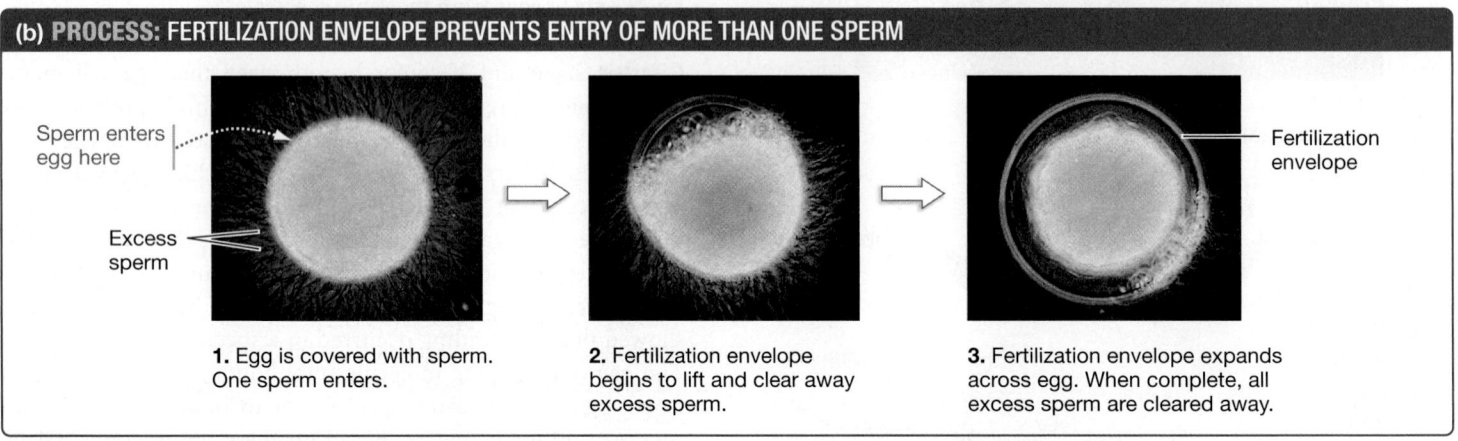

(a) PROCESS: A WAVE OF Ca²⁺ RELEASE SPREADS FROM THE SITE OF SPERM ENTRY

Sperm enters egg here

Ca²⁺

Ca²⁺

A wave of calcium ion release starts at the point of sperm entry and spreads throughout the egg

(b) PROCESS: FERTILIZATION ENVELOPE PREVENTS ENTRY OF MORE THAN ONE SPERM

Sperm enters egg here

Excess sperm

Fertilization envelope

1. Egg is covered with sperm. One sperm enters.

2. Fertilization envelope begins to lift and clear away excess sperm.

3. Fertilization envelope expands across egg. When complete, all excess sperm are cleared away.

FIGURE 23.3 A Physical Barrier Erected after Fertilization Prevents Polyspermy. (a) During fertilization, a wave of Ca²⁺ begins at the point of sperm entry and spreads under the plasma membrane across the egg in about 30 seconds. Colors represent Ca²⁺ concentration, where white is highest and blue lowest. Ca²⁺ concentrations were measured using an injected reagent that fluoresces in the presence of calcium ions. **(b)** In response to increased Ca²⁺ concentrations, a fertilization envelope arises in about 40 seconds and clears away excess sperm.

areas inside the egg. As **FIGURE 23.3a** shows, a wave of calcium ion release starts at the point of sperm entry and spreads throughout the egg.

The egg contains thousands of vesicles called cortical granules that lie just beneath the plasma membrane (see **FIGURE 23.2b**). Ca²⁺ causes the cortical granules to fuse with the plasma membrane and release their contents by exocytosis to the exterior. As you'll see in a later chapter, this same mechanism of Ca²⁺-mediated exocytosis is used to allow nerve cells to communicate with other cells (Chapter 46).

The egg cortical granules release proteases that digest the exterior-facing portion of the receptor for sperm. This prevents any new sperm from binding to the egg surface. In addition, ions and other compounds released by the cortical granules accumulate between the egg cell's plasma membrane and the vitelline envelope. These concentrated solutes cause water to flow into the space between the plasma membrane and vitelline envelope by osmosis (see Chapter 6). The influx of water lifts the vitelline envelope away from the cell. Compounds from the cortical granules cross-link molecules in the vitelline envelope together to form a tough **fertilization envelope** (**FIGURE 23.3b**). This impenetrable barrier keeps additional sperm from reaching the sea urchin egg.

Although mammalian eggs do not produce a fertilization envelope, a wave of Ca²⁺ ions also trigger the release of enzymes from cortical granules that perform a function similar to the sea

check your understanding

If you understand that . . .

- Fertilization depends on specific interactions between proteins on the plasma membranes of animal sperm and egg cells.
- Enzymes that digest the egg-cell receptor for sperm help to prevent polyspermy.

✔ **You should be able to . . .**

1. Predict the consequences of mutations that change the structure of bindin of sea urchin species A to the bindin structure of sea urchin species B.

2. Describe the role of calcium signaling in processes that block polyspermy.

Answers are available in Appendix A.

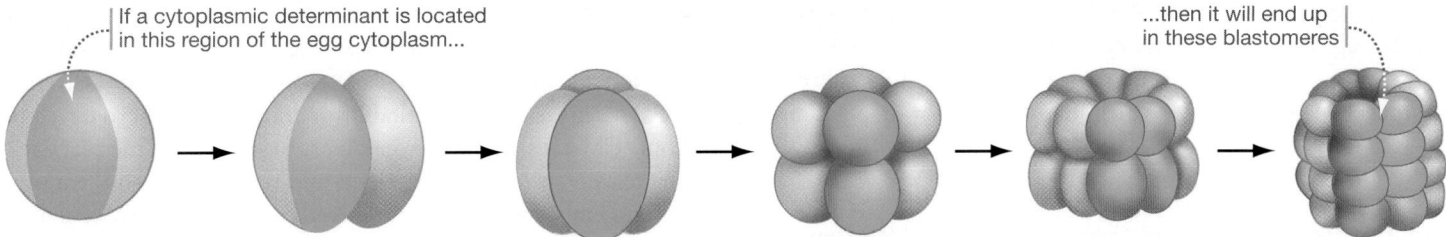

If a cytoplasmic determinant is located in this region of the egg cytoplasm...

...then it will end up in these blastomeres

FIGURE 23.4 Cytoplasmic Determinants Are Partitioned into Certain Blastomeres during Cleavage.

urchin enzymes that destroy the egg's receptor for sperm. These enzymes modify proteins on the egg-cell surface and on a structure similar to the vitelline envelope. This process blocks polyspermy by preventing the binding of any additional sperm once the egg is fertilized.

Once a sperm nucleus enters a sea urchin or human egg, the chromosomes of the two haploid nuclei come together to form a diploid genome. Development is off and running.

23.2 Cleavage

Fertilization activates development. One cell becomes two, two become four, and the beginnings of a multicellular organism are in place. The stage of rapid cell divisions that follows fertilization is called **cleavage.**

In most animals, cleavage divides the egg cytoplasm without any overall growth of the embryo. The zygote simply divides without increasing in size. The key feature of cleavage is that the cytoplasm present in the egg is divided into a larger and larger number of smaller and smaller daughter cells.

Cleavage is fast—the fastest cell divisions recorded over an individual's lifetime. For example, in fruit flies, mitosis can occur every 10 minutes, producing about 50,000 cells in half a day.

The cells that are created during cleavage are called **blastomeres** (literally, "bud-part"). When cleavage is complete, the embryo in many animals consists of a mass of blastomeres called a **blastula** ("little-sprout," "bud").

Partitioning Cytoplasmic Determinants

A key to understanding cleavage is to analyze what is happening to the egg cytoplasm as it is divided into many blastomeres. Many embryos contain **cytoplasmic determinants.** Cytoplasmic determinants are regulatory molecules that are located in specific regions of the egg cytoplasm. Through cleavage, they end up in specific populations of blastomeres (**FIGURE 23.4**). As a result, certain regulatory gene cascades are triggered only in certain blastomeres.

By precisely dividing the egg cytoplasm to distribute (partition) cytoplasmic determinants to certain cells, cleavage initiates the step-by-step process that, in combination with signals received from other cells, results in cell differentiation.

Cleavage in Mammals

In humans and other mammals, cleavage occurs as the embryo moves down the fallopian tube, or oviduct, a structure that connects the ovary and the uterus (**FIGURE 23.5**). The ovary is the organ in which the egg matures, and the uterus is the organ in

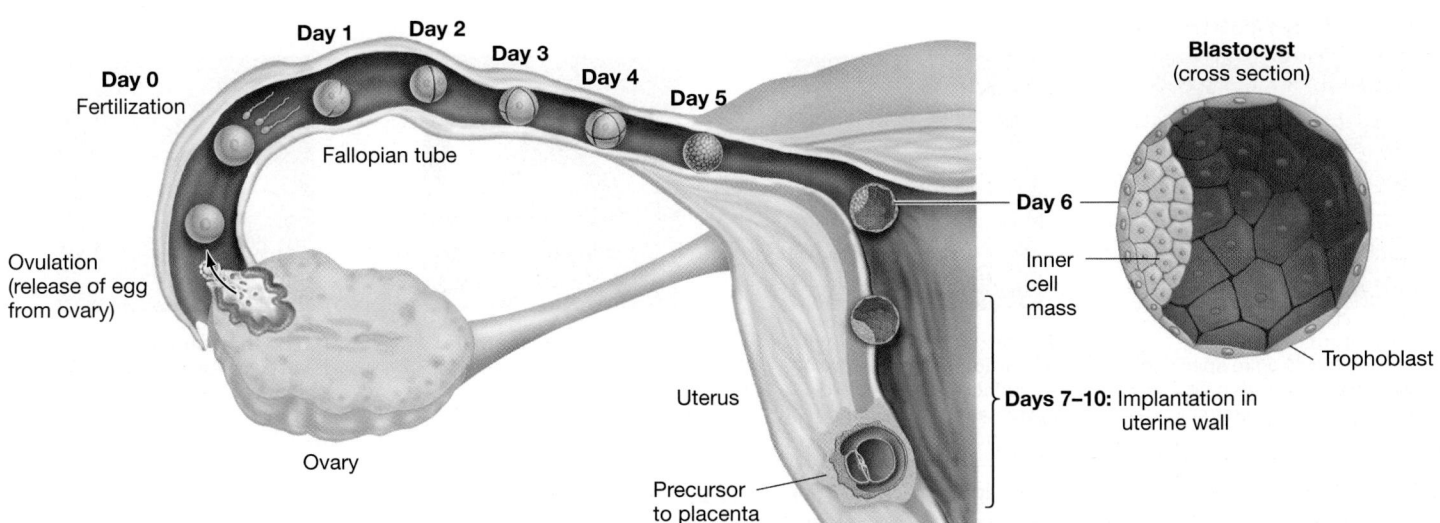

FIGURE 23.5 In Mammals, Cleavage Occurs before Implantation into the Uterus. Human cleavage and the human blastocyst are shown here.

which the embryo develops. Fertilization occurs near the ovary, and cleavage occurs as the embryo travels toward the uterus.

Cleavage in mammals results in a type of blastula called a **blastocyst** ("sprout-bag"), which has two major populations of cells. The exterior of the blastocyst is a thin-walled, hollow structure called the **trophoblast** ("feeding-sprout"). Inside the trophoblast, there is a cluster of cells called the **inner cell mass (ICM).** This is a fundamental distinction—the embryo develops from ICM cells, and the trophoblast forms part of the **placenta.** The placenta exchanges nutrients and wastes between the mother's and embryo's blood. It is derived from a mixture of maternal uterine cells and trophoblast cells.

The placenta is a key evolutionary innovation of mammals that allows internal development—but it isn't part of a newborn or adult.

23.3 Gastrulation

As cleavage nears completion, cell division slows. During the next phase of development, cell proliferation is no longer the most important developmental process. Instead, cell movement takes the leading role. In this stage of development, called **gastrulation,** extensive and highly organized cell movements radically rearrange the embryonic cells into a structure called the gastrula. **FIGURE 23.6** shows the remarkable reshaping of a frog embryo during gastrulation.

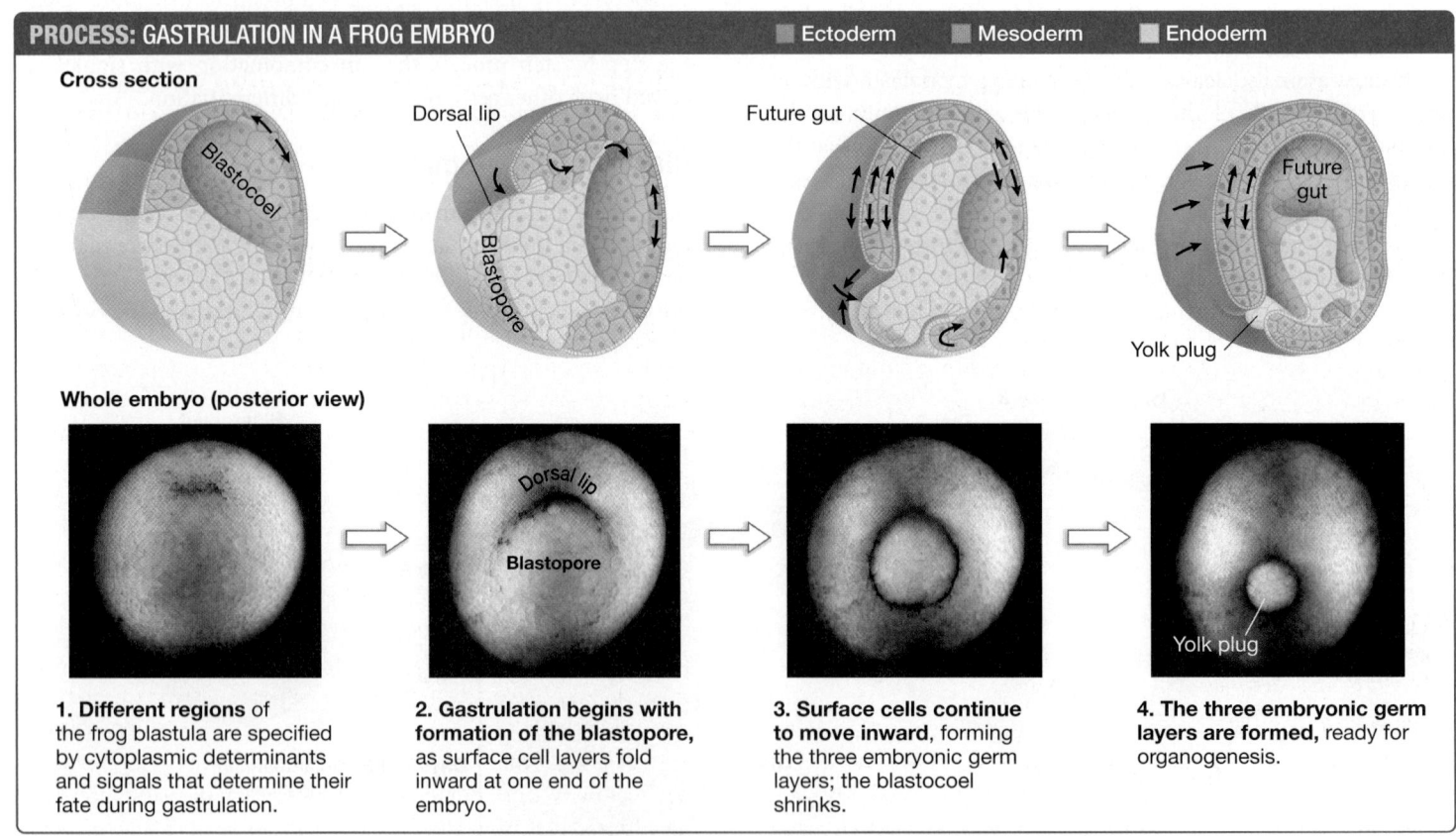

PROCESS: GASTRULATION IN A FROG EMBRYO

■ Ectoderm ■ Mesoderm ■ Endoderm

Cross section

Blastocoel · Dorsal lip · Blastopore · Future gut · Future gut · Yolk plug

Whole embryo (posterior view)

Dorsal lip · Blastopore · Yolk plug

1. Different regions of the frog blastula are specified by cytoplasmic determinants and signals that determine their fate during gastrulation.

2. Gastrulation begins with formation of the blastopore, as surface cell layers fold inward at one end of the embryo.

3. Surface cells continue to move inward, forming the three embryonic germ layers; the blastocoel shrinks.

4. The three embryonic germ layers are formed, ready for organogenesis.

FIGURE 23.6 Gastrulation Creates Head-to-Tail and Back-to-Belly Axes and Three Germ Layers in the Embryo. Gastrulation in the frog is shown here. In all animal embryos, gastrulation requires coordination of a remarkable set of cell movements.

Research on this phase of development started in the 1920s with efforts to follow the movement of individual cells during newt and frog gastrulation. In these early experiments, tiny blocks of agar (a gelatinous compound) were soaked with a non-toxic dye. The dyed blocks were then pressed against the surface of blastula-stage embryos so that a small number of blastomeres became marked with dye. By allowing marked embryos to develop and then examining them at intervals during gastrulation, researchers were able to follow the movement of cells. Today, research on gastrulation continues, using new ways to mark individual cells and track them throughout the developing embryo.

Formation of Germ Layers

The pattern of gastrulation varies widely among animal species, but the general outcome is the same: Gastrulation forms three embryonic tissue layers. A **tissue** is an integrated set of cells that function as a unit.

Most animal embryos have three primary tissue layers: **(1)** ectoderm ("outside skin"), **(2)** mesoderm ("middle skin"), and **(3)** endoderm ("inner skin"). These embryonic tissues are called **germ layers** because they give rise to all the organs and tissues of the adult. It is important to note that each germ layer forms certain tissues and organs, and the correspondence between germ layer and organ type is the same in most animals.

Figure 23.6 shows two views of how the cell movements of gastrulation form the three germ layers in a frog embryo. The cell movements of gastrulation are remarkably complex. Cells that will become ectoderm are shown in blue; cells destined to form mesoderm are shown in pink, and cells that will form endoderm are colored yellow. These cell fates are set by signals between blastomeres and cytoplasmic determinants within blastomeres.

Step 1 The frog blastula contains a fluid-filled interior space called the **blastocoel.** The blastula of most animals contains a blastocoel.

Step 2 As gastrulation begins, an indentation (invagination) forms on the outer surface. In frogs this invagination starts out as a slit that eventually forms a circular opening known as the **blastopore.**

Step 3 Cells from the surface move into the interior of the embryo through the blastopore, forming a tube that will become the gut or digestive tract. The blastocoel is displaced and eventually disappears.

Step 4 The movement of cells into the embryo ultimately results in the formation of three layers—an inside, a middle, and an outside. These are the endoderm, mesoderm, and ectoderm.

FIGURE 23.7 shows what the germ layers produce in a human. Each germ layer gives rise to the same types of organs in virtually all animals. **Ectoderm** forms the outer covering of the body and the nervous system; **mesoderm** produces muscle, most internal organs, and connective tissues such as bone and cartilage; **endoderm** produces the lining of the digestive tract and the many organs that develop from the gut, such as the liver and lungs.

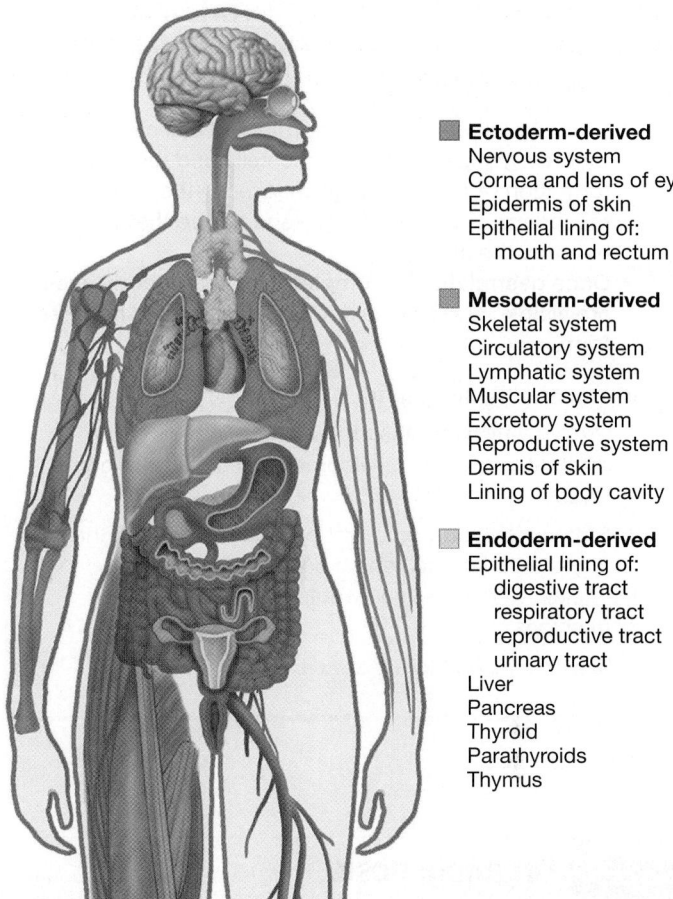

Ectoderm-derived
Nervous system
Cornea and lens of eye
Epidermis of skin
Epithelial lining of:
  mouth and rectum

Mesoderm-derived
Skeletal system
Circulatory system
Lymphatic system
Muscular system
Excretory system
Reproductive system
Dermis of skin
Lining of body cavity

Endoderm-derived
Epithelial lining of:
  digestive tract
  respiratory tract
  reproductive tract
  urinary tract
Liver
Pancreas
Thyroid
Parathyroids
Thymus

FIGURE 23.7 The Three Embryonic Germ Layers Give Rise to Different Adult Tissues and Organs. Each germ layer forms the same types of organs in all animals.

Creating Body Axes

In addition to establishing the germ layers, gastrulation has another major role: Creating the body axes. In frogs, for example, the blastopore becomes the anus (posterior), the region at the opposite end of the gut tube becomes the mouth (anterior), the region where cells first move into the blastopore (as drawn in Figure 23.6) defines the dorsal, or back, side of the embryo, and the region opposite this becomes the ventral region. In this way, the anterior–posterior and dorsal–ventral axes of the body become apparent.

Note, however, that in frogs, the major body axes are partially determined early in development by cytoplasmic determinants that were stored in the egg and partitioned into blastomeres during cleavage. These master regulators start regulatory gene cascades that begin the step-by-step process of differentiation (see Chapter 22).

Current research on gastrulation is focused on understanding how cells can coordinate their behaviors—especially, the mechanisms responsible for cell movement and navigation. Meanwhile, like the embryo itself, let's move along to the next phase in development—the formation of tissues and organs as the body takes shape.

If you understand that . . .

- Gastrulation consists of coordinated cell movements that reorganize the embryonic cells and result in the formation of embryonic germ layers—the ectoderm, mesoderm, and endoderm—and a tube that will become the gut.
- Once gastrulation is complete, the major body axes are visible, and organs and other structures begin to form.

✔ **You should be able to . . .**

1. Predict which germ layers would be present in an animal that during development uses only two germ layers.
2. State when you would first be able to point out the future anterior and ventral portion of the embryo, and explain what clues you would use to identify their positions.

Answers are available in Appendix A.

23.4 Organogenesis

Gastrulation creates the rudiments of an animal. There are outside, inside, and middle layers; a head and a tail; a back and a belly—but much remains to be done. For one thing, there are no organs within the embryo. The heart, brain, liver, and lungs all need to be formed from the germ layers, properly positioned, and connected with other organs. **Organogenesis** ("organ-origin") is the process of tissue and organ formation. During organogenesis, cells proliferate, move, differentiate, and assemble into tissues and organs.

To explore how organogenesis takes place, let's consider one example—how muscle tissue forms in the embryo. Muscle comes from mesoderm, the middle germ layer. To form muscle, mesodermal cells must specialize and organize into muscle tissue. What causes undifferentiated mesodermal cells to form muscle tissue and to begin producing the muscle-specific proteins that make breathing, walking, and turning the pages of textbooks possible?

Organizing Mesoderm into Somites

To understand the formation of any tissue or organ, it is necessary to trace back through many earlier structures, some of which exist only in the embryo. Let's begin with the assembly of key embryonic structures called somites.

Somites are paired blocks of mesodermal tissue that extend along either side of the dorsal midline (top middle) of the embryo. In other words, they run along the "back" of the

(a) Surface view of somites

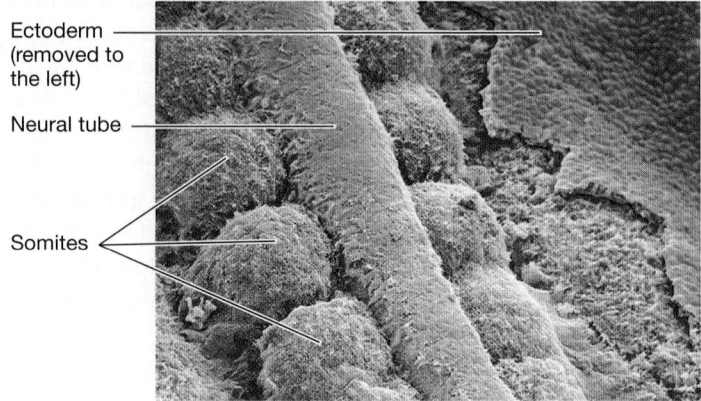

Ectoderm (removed to the left)

Neural tube

Somites

(b) Cross section of somites

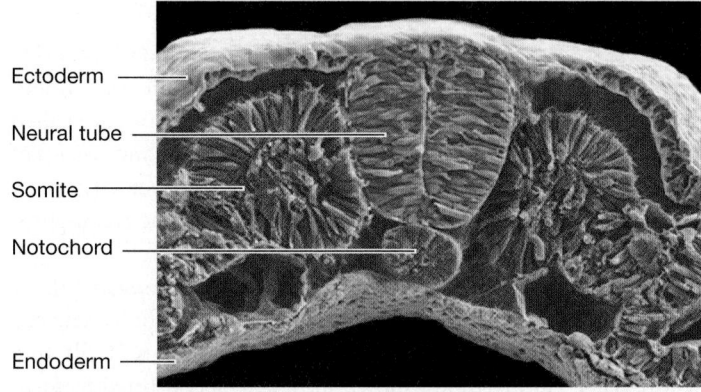

Ectoderm

Neural tube

Somite

Notochord

Endoderm

FIGURE 23.8 Somites Are Made of Mesodermal Cells. Somites (color-coded pink in these chick-embryo micrographs) form on either side of the neural tube, a structure that will develop into the brain and spinal cord and that runs along the middle of the embryo's back.

embryo. **FIGURE 23.8** shows two views of somites and surrounding embryonic structures early in organogenesis of a chick embryo, where somites are easily studied. Note that at this stage of development, a new structure has appeared in the dorsal mesoderm—a rod-like element called the **notochord.** This structure is shown in cross section in the figure and runs the length of the anterior–posterior axis of the embryo, just under the dorsal surface.

The notochord is unique to the group of animals called the **chordates,** which includes humans and all other vertebrates. In some species of chordates, the notochord is a long-lasting structure that functions as a simple internal skeleton—it stiffens the body and makes efficient swimming movements possible. But in vertebrates, such as chicks, frogs, and humans, the notochord is transient. It appears only in embryos. As organogenesis proceeds, many of the cells in the notochord undergo programmed cell death by apoptosis.

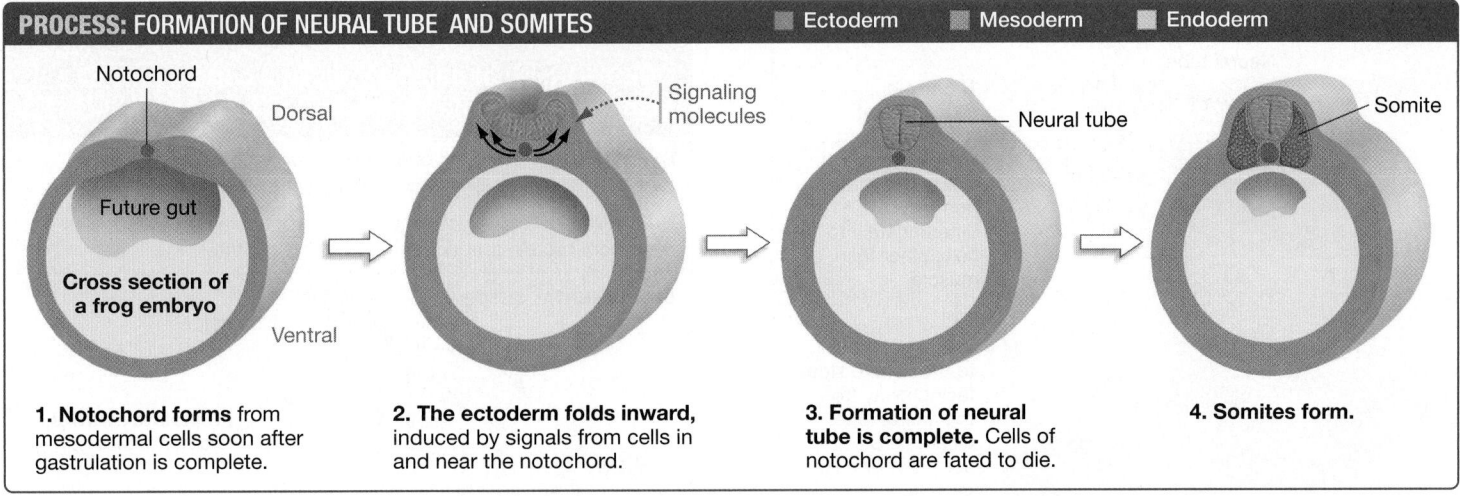

PROCESS: FORMATION OF NEURAL TUBE AND SOMITES ■ Ectoderm ■ Mesoderm ■ Endoderm

Notochord

Dorsal

Signaling molecules

Neural tube

Somite

Future gut

Cross section of a frog embryo

Ventral

1. Notochord forms from mesodermal cells soon after gastrulation is complete.

2. The ectoderm folds inward, induced by signals from cells in and near the notochord.

3. Formation of neural tube is complete. Cells of notochord are fated to die.

4. Somites form.

FIGURE 23.9 The Notochord, Neural Tube, and Somites Form Early in Organogenesis. In vertebrates, the notochord forms from mesodermal cells soon after gastrulation is complete. Molecules produced in the notochord induce the formation of the neural tube.

✔**QUESTION** Predict how the formation of the neural tube would be affected by a drug that locks the cytoskeleton in place.

But where do somites come from? **FIGURE 23.9** illustrates the developmental path to somite formation in a frog embryo. What's shown here differs in detail but not in substance from what occurs in a chick, human, or any other vertebrate embryo. From your study of gastrulation, you should already be familiar with the ectoderm, mesoderm, endoderm, and future gut cavity shown in step 1 of Figure 23.9. The somites form from the mesoderm near the notochord.

Formation of Somites and the Neural Tube Somite formation, the stepping stone to formation of muscle, is coupled to the creation of another major structure in the embryo—the **neural tube.** The neural tube is the precursor of the brain and spinal cord. As step 2 in Figure 23.9 shows, the notochord produces signaling molecules that induce the ectoderm on the dorsal (back) side of the embryo to fold. The notochord is a key organizing element during organogenesis.

Folding of the dorsal ectoderm results from changes in cell shape—analogous to the differential cell expansion that occurs in plant cell development (see Chapter 22). The folding begins when signals from the notochord trigger massive changes in cytoskeletal elements inside each of the dorsal ectodermal cells. As the cytoskeleton is reorganized, the ectodermal cells grow taller and then constrict at their dorsal end and expand at their ventral end. The constriction above and expansion below makes the sheet of cells fold upward. When the folds of dorsal ectoderm come together, they form the neural tube (step 3). This is a monumental event in the development of chordates, because the newly formed neural tube is the beginning of the brain and spinal cord.

As organogenesis continues, cells near the neural tube and notochord becomes organized into blocks of mesodermal tissue. These are the somites (step 4). Somite formation is a response

to changes in the cell adhesion molecules that keep mesodermal cells attached to each other (see Chapter 11). Look back at Figure 23.8 to see how the somites, notochord, and neural tube appear in an actual embryo (in this case, chick).

Somites Are Precursors to Skin, Bone, and Muscle Somites are transient structures, just like the notochord is. But unlike the vertebrate notochord, somites produce many important structures of the adult. By marking somite cells and following them over time, researchers discovered that somites give rise to the lower (dermal) layer of the skin, to much of the skeleton, and to skeletal muscles.

As organogenesis proceeds, somite cells break away in distinct groups that migrate to their final location in the developing embryo. There they continue to proliferate and eventually differentiate. Cell movements like these are critical to organogenesis. As **FIGURE 23.10** shows, the fate of a somite cell depends on its position within the somite. When is the fate determined?

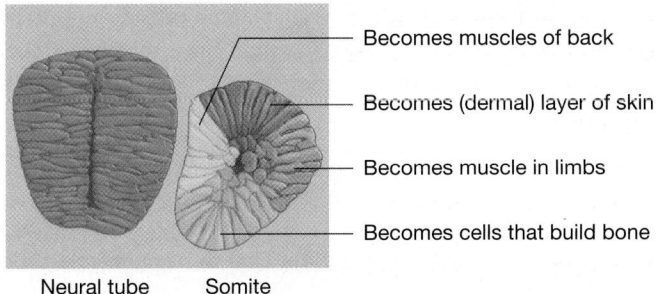

Becomes muscles of back

Becomes (dermal) layer of skin

Becomes muscle in limbs

Becomes cells that build bone

Neural tube   Somite

FIGURE 23.10 A Cell's Position within a Somite Determines Its Fate. Each somite eventually breaks up into distinct populations of cells, each of which gives rise to distinct tissues.

PROCESS: TESTING SOMITE CELL DETERMINATION

Neural tube

Somite

Somite

1. Early in development, remove cells from various locations in somites that normally differentiate into cells other than muscle.

Future limb muscle

Future limb muscle

2. Transplant cells to the somite sides facing away from the neural tube, where cells normally form limb muscle.

3. Cells from the outer side of somites will migrate away and become limb muscle, regardless of their normal fate and original location.

FIGURE 23.11 Evidence for Progressive Determination of Mesodermal Cells in Somites. By transplanting somite cells early in somite formation to new locations in another developing somite, researchers showed that in the early somites, the fate of cells is not yet determined.

Cell Determination By transplanting cells from one location to another within the developing somite, researchers found that initially any of those cells can become any of the somite-derived elements. For example, biologists transplanted cells from various parts of the early somite to the somite region farthest from the neural tube—the region that produces limb muscle. All the transplanted cells eventually formed limb muscles (**FIGURE 23.11**), even though these same cells normally would have formed other structures. Cells transplanted later in development, however, failed to become the cell type associated with their new position. Instead, the transplanted cells differentiated into the cell types they normally would have formed in their original position, even though they were now in a new and inappropriate location.

FIGURE 23.12 The Search for a Protein That Causes Muscle-Cell Differentiation.

Weintraub, H., S. J. Tapscott, R. L. Davis, et al. 1989. Activation of muscle-specific genes in pigment, nerve, fat, liver, and fibroblast cell lines by forced expression of *MyoD*. *Proceedings of the National Academy of Sciences, USA* 86: 5434–5438.

✔**QUESTION** Why did the researchers have to attach a "general purpose" promoter to the cDNAs?

RESEARCH

QUESTION: What causes cells in a certain part of somites to become committed to produce muscle?

HYPOTHESIS: Production of an mRNA for a regulatory protein (or proteins) by myoblasts commits them to their fate.

NULL HYPOTHESIS: Myoblasts do not produce an mRNA for a regulatory protein that commits them to their fate.

EXPERIMENTAL SETUP:

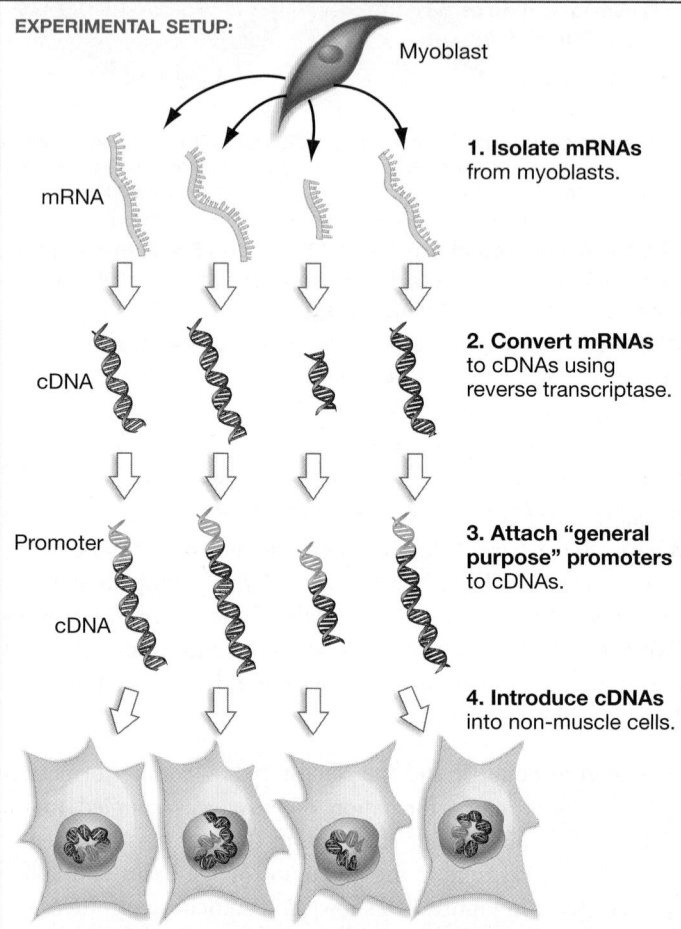

Myoblast

mRNA

cDNA

Promoter

cDNA

1. Isolate mRNAs from myoblasts.

2. Convert mRNAs to cDNAs using reverse transcriptase.

3. Attach "general purpose" promoters to cDNAs.

4. Introduce cDNAs into non-muscle cells.

PREDICTION: One of the myoblast-derived cDNAs will convert non-muscle cells into cells that produce muscle-specific proteins.

PREDICTION OF NULL HYPOTHESIS: None of the myoblast-derived cDNAs will convert non-muscle cells into cells that produce muscle-specific proteins.

RESULTS:

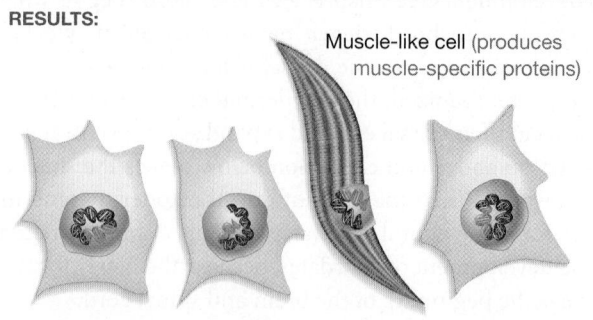

Muscle-like cell (produces muscle-specific proteins)

CONCLUSION: Certain somite cells produce an mRNA for a regulatory protein (later called MyoD) that commits them to differentiate into muscle.

These transplantation studies reveal an important aspect of development: Cells gradually become determined. **Determination** is the irreversible commitment of a cell to a particular developmental fate. Before determination, cells transplanted to a new location can take on the fate appropriate for that location. But once determined, cells do not alter their course, even if they are placed in a new environment and receive signals to become a different type of tissue.

Determination comes about through epigenetic mechanisms that were discussed in the chapter on eukaryotic gene expression (Chapter 19). Signals in the embryo induce the production of new sets of transcription factors. In turn, these transcription factors trigger new and long-lasting patterns of chromatin condensation. Gene expression patterns appropriate for a particular cell type are set in place, and switching to different patterns becomes nearly impossible.

In short, the determination of a cell's fate as the somite matures **(1)** is a step-by-step process guided by regulatory gene cascades and **(2)** is a function of the cell's location in the embryo. Location is important because different positions within the embryo have different molecular signals.

Signals coming from cells in the notochord, the neural tube, and nearby ectoderm and mesoderm set up divisions within the somite. Different combinations of these signals inform cells where they are within the somite and what fate they should adopt. These signals induce new sets of transcription factors and new patterns of gene expression that determine whether a cell becomes muscle, bone, or skin.

Differentiation of Muscle Cells

What makes a cell become muscle? Harold Weintraub and colleagues answered this question in the late 1980s by experimenting with cells called myoblasts. A myoblast is derived from cells in the somite. Myoblasts are destined to become muscle cells but have not yet begun producing muscle-specific proteins.

Weintraub and co-workers hypothesized that myoblasts must express at least one regulatory protein that commits them to their fate. The idea was that myoblasts begin producing this muscle-determining protein after they receive an appropriate set of signals from nearby tissues. In effect, the researchers were looking for a transcription factor in myoblasts that dictates, "Become a muscle cell."

FIGURE 23.12 outlines how the biologists went about searching for this hypothetical protein.

1. They began by isolating mRNAs from myoblasts.

2. They used reverse transcriptase to convert the mRNAs to cDNAs. Because myoblasts transcribe genes required for a muscle-cell fate, they reasoned that the cDNAs must include a muscle-determining gene transcript.

3. They attached a type of promoter to the cDNAs that would ensure expression of the cDNA in any type of cell.

4. They introduced the modified cDNAs into non-muscle cells called fibroblasts and monitored the development of the cells.

As predicted, one of the myoblast-derived cDNAs converted fibroblasts into muscle-like cells. Follow-up experiments showed that the same gene product could convert pigment cells, nerve cells, fat cells, and liver cells into cells that produced muscle-specific proteins.

Weintraub's group called the protein product of this gene **MyoD**, for *myo*blast *d*etermination. Subsequent work showed that the *MyoD* gene encodes a regulatory transcription factor and that the MyoD protein binds to enhancer elements located upstream of muscle-specific genes (see Chapter 19). MyoD is a master regulator of muscle differentiation.

In addition, researchers found that the MyoD protein activates further expression of the *MyoD* gene. This was a key observation because it meant that once *MyoD* is turned on, it triggers its own expression—meaning that the gene continues to be transcribed.

To put this specific example of muscle differentiation into context, think back to the sequence of events occurring in early development and to the principles of development (see Chapter 22). Differentiation is a step-by-step process that is complete when cells begin producing proteins that are specific to a particular cell type. By what path did the cells in your bicep become muscle cells?

Step 1 Fertilization of the egg inside your mother triggered the onset of cleavage, resulting in a blastocyst.

Step 2 Certain cells in the blastocyst began producing signals that triggered regulatory gene cascades that led specific cells during gastrulation to become mesoderm in your back.

Step 3 During organogenesis, signals from cells of the notochord and nearby structures induced the production of MyoD and other muscle-determining proteins in particular somite cells. In response, these target cells became committed to developing into muscle and moved into the regions that formed your arms.

Step 4 Later, these MyoD-expressing cells began producing muscle-specific proteins.

All of these steps made it possible for you to move your arms after birth. The rest, as they say, is history.

If you understand . . .

23.1 Fertilization

- Fertilization creates the zygote, the first cell of the animal.

- Fertilization involves mechanisms to attract the sperm to the egg; bind sperm to egg in a species-specific manner; prevent the entry of more than one sperm; fuse the sperm- and egg-cell plasma membranes; bring the chromosomes of egg and sperm together; and activate metabolism in the egg to support development.

✔ You should be able to explain why bindin and the egg-cell receptor for sperm are said to interact like a lock and key.

23.2 Cleavage

- Cleavage is the set of rapid, early cell divisions that follow fertilization.

- Cleavage divides the fertilized egg into a mass of cells.

- In many species, an array of signals and regulatory transcription factors (cytoplasmic determinants) present in the egg are distributed to different blastomeres. As a result, different blastomeres have different fates.

- Once cleavage is complete, the embryo consists of a mass of cells.

✔ You should be able to describe how cleavage in a frog embryo can result in certain cells containing a cytoplasmic determinant.

23.3 Gastrulation

- Gastrulation is the dramatic, coordinated movement of cells that transform the cleavage-stage embryo into the gastrula.

- Cell movements of gastrulation require changes in cell shape and size and changes in the affinity of one cell for another.

- The gastrula contains the three germ layers of ectoderm, mesoderm, and endoderm and has recognizable embryonic axes (anterior–posterior, dorsal–ventral, right–left).

- The germ layers give rise to distinct organs of the adult, and each germ layer produces organs of similar types in all animals.

✔ You should be able to compare and contrast, in broad terms, a mesodermal cell in the anterior end of an embryo and a mesodermal cell in the posterior end, regarding the cell–cell signals they receive and the regulatory transcription factors they contain.

23.4 Organogenesis

- Organogenesis is the formation of tissues and organs.

- Organogenesis occurs through the coordination of cell–cell signals, cell proliferation, cell movements, and differentiation. Differentiation is complete when cells express tissue-specific proteins and the specialized properties of adult cells.

- The creation of muscle is one example of organogenesis.

- Early in vertebrate organogenesis, cells in the notochord release cell–cell signals that induce the formation of a neural tube—precursor to the brain and spinal cord—from ectoderm that overlies the notochord.

- Blocks of mesodermal cells, called somites, form next to the neural tube.

- In response to cell–cell signals from the notochord, neural tube, and other nearby tissues, cells in the somites are determined, move to new positions, proliferate, and begin expressing tissue-specific proteins.

- Determination of cells occurs during organogenesis and is the generally irreversible commitment of a cell to follow a particular pathway of differentiation.

✔ You should be able to give examples of cell–cell signals, expansion, movement, proliferation, and differentiation during neural tube, somite, and muscle development in mammals.

(MB) MasteringBiology

1. MasteringBiology Assignments

Tutorials and Activities Early Stages of Animal Development; Embryonic Development; Frog Development; How Do Calcium Ions Help to Prevent Polyspermy During Egg Fertilization?; Sea Urchin Development

Questions Reading Quizzes, Blue-Thread Questions, Test Bank

2. eText Read your book online, search, take notes, highlight text, and more.

3. The Study Area Practice Test, Cumulative Test, BioFlix® 3-D Animations, Videos, Activities, Audio Glossary, Word Study Tools, Art

You should be able to . . .

1. What is bindin, and what does it do?

2. How is calcium signaling involved in blocking polyspermy?
   a. It triggers a regulatory gene cascade.
   b. Accumulation of calcium ions causes an outward flow of water, raising the fertilization envelope.
   c. It triggers fusion of cortical granules with the plasma membrane that leads to formation of the fertilization shell.
   d. Calcium ions bind to egg-cell receptors for sperm, blocking binding sites.

3. What happens during cleavage?
   a. The neural tube—precursor of the spinal cord and brain—forms.
   b. The inner cell mass of the blastocyst begins dividing rapidly to produce the placenta.
   c. The fertilized egg divides without growth occurring, forming a mass of cells.
   d. Massive movements of cells make the primary body axes visible and organize the three embryonic tissues.

4. What happens during gastrulation?
   a. The neural tube—precursor of the spinal cord and brain—forms.
   b. The inner cell mass of the blastocyst begins dividing rapidly to produce the placenta.
   c. The fertilized egg divides without growth occurring, forming a ball of cells.
   d. Massive movements of cells make the primary body axes visible and organize the three embryonic tissues.

5. Which of the embryonic germ layers gives rise to the brain?

6. Which of the following does *not* occur during organogenesis?
   a. establishment of the anterior–posterior axis
   b. differentiation of cells
   c. movement of cells into new positions
   d. extensive cell–cell signaling

7. Many frogs and mice are similar in size, yet a frog egg is vastly larger than a mouse egg. Propose a plausible explanation for the difference in the size of these eggs. Explain your reasoning.

8. A variety of agents are known that bind tightly to Ca^{2+} inside of cells. This binding prevents Ca^{2+} from exerting any of its normal effects. Imagine a situation in which a Ca^{2+}-binding agent was injected into a sea urchin egg before thousands of sperm were added to the seawater surrounding the egg. Predict how this treatment would affect fertilization, and provide reasons why.

9. Blastomeres look identical. Describe how they are not.

10. Plants depend on directional cell expansion to create structures. Which of the following events in animal development relies on a mechanism related to directional cell expansion in plants?

   a. formation of the neural tube from a sheet of ectoderm
   b. migration of myoblasts to developing limbs
   c. determination of cells in different regions of the somite
   d. formation of the inner cell mass within the blastocyst

11. Explain how cell-transplantation experiments provided evidence that cells in a somite are determined in a step-by-step fashion.

12. **QUANTITATIVE** In the study that discovered MyoD, Weintraub and his colleagues reasoned that MyoD was likely to be a rare mRNA. If they assumed that MyoD mRNA comprised 2×10^{-5} (0.00002) of all mRNAs, what is the smallest number of fibroblast cells into which they would have needed to introduce cDNAs so they would have a reasonable chance of finding one that expressed MyoD?

13. At the molecular level, explain why bindin on the sperm of the sea urchin *Strongylocentrotus purpuratus* attaches to the egg-cell receptor for sperm on eggs of *S. purpuratus*, but why bindin from other species of *Strongylocentrotus* cannot.

14. A molecule called G418 inhibits translation. When researchers treat fruit fly eggs with G418, the early stages of cleavage proceed normally after fertilization. What can you conclude from this experiment?

15. In the marine organisms called sea squirts, eggs contain a yellow pigment. Blastomeres that contain this yellow pigment become muscle cells. Is the yellow pigment a cytoplasmic determinant? Why or why not?

16. One reason chicks are such a valuable model organism for studying development is that their embryos can be surgically manipulated. Researchers have done experiments that involve transplanting an additional piece of notochord near the upper portion of the neural tube in an embryo that has just formed somites. A reasonable prediction for how this manipulation would affect somite development is that
   a. more bone than normal will be produced from the somite.
   b. more limb muscle than normal will be produced from the somite.
   c. more dermis than normal will be produced from the somite.
   d. somite development will be unchanged, because somite cells are determined as soon as the somite forms.

24 An Introduction to Plant Development

In this chapter you will learn how

Plant development is both similar to and distinct from animal development

by asking what happens | during three key stages

Embryogenesis 24.1	Vegetative development 24.2	Reproductive development 24.3
including	including	including
What genes and proteins set up body axes?	What genes and proteins determine leaf shape?	What genes and proteins control organ patterns in flowers?

The 4700-year old bristlecone pine is Earth's oldest known living organism. This chapter explores how plant embryos can develop into adults that grow and reproduce throughout their lives.

About the year 2750 B.C., perhaps more than a century before the first pyramids were constructed in ancient Egypt, a sperm cell and an egg cell from a bristlecone pine tree fused on a mountainside in today's California. A zygote was created that produced a tree that has been growing and developing ever since—for over 4700 years.

Unlike many animals, plants continue to grow and develop throughout their lives, whether that life lasts two weeks or thousands of years. Continuous development is a hallmark of plant biology. Why the difference?

To begin to answer this question, consider the distinct lifestyles that have evolved in plants and animals. Most animals move to find food and mates, while plants stand their ground, make their own food, and use water, wind, or insects to transport their gametes. Animal cells lack a rigid cell

✔ When you see this checkmark, stop and test yourself. Answers are available in Appendix A.

wall, while plant cells are encased in a protective but constraining wall. These very different ways of life account for many of the contrasts in plant and animal development.

Animals, land plants, two groups of fungi, and the red and brown algae all have evolved independent solutions to the common problem of how to build a multicellular organism. However, despite their separate evolutionary paths, there are abundant similarities in plant and animal development. How can this be explained? Recall that a few fundamental principles are common to the developmental events observed in all multicellular organisms (Chapter 22)—there are a limited number of ways to build a plant or animal.

This chapter is about how plants develop, and how plant development is both similar to and distinct from the development of animals. Let's begin with embryogenesis in plants, the formation of a multicellular embryo from a single-celled zygote.

24.1 Embryogenesis

FIGURE 24.1 provides an overview of development in a flowering plant—the most familiar, species-rich, and abundant group of plants (see Chapter 31). The development of flowering plants is the primary focus of this chapter. In this figure and in many other places in this chapter, a small species called *Arabidopsis thaliana*—the best-studied plant—will serve as a model organism (see **BioSkills 13** in Appendix B).

Note that the sequence of developmental events, or life cycle, diagrammed in Figure 24.1 begins with gametogenesis, or gamete formation. Next come pollination and fertilization. In *Arabidopsis* and other flowering plants, an egg is fertilized inside a protective, womb-like structure called the ovule. You'll learn more about these processes in a later chapter (Chapter 41).

Development continues inside the ovule with **embryogenesis**—literally, "embryo-origin." In flowering plants, embryogenesis ends with the maturation of the ovule into a **seed**—a structure that contains the embryo and a supply of nutrients surrounded by a protective maternally-provided coat.

An embryo may remain in a nongrowing state inside the seed for months, years, or in some cases, centuries. This state of arrested development is rare in animals. When conditions are favorable the seed undergoes **germination**—meaning that it resumes growth—to form a seedling, beginning the process of organogenesis. Here is another contrast with animal development: organogenesis—the development of organs—continues throughout the plant's life.

Vegetative development produces the nonreproductive portions of the plant body—the roots, leaves, and stems. As the plant matures, some cells in the stem will become converted to reproductive structures, a process known as **reproductive development.**

Reproductive development is a distinguishing feature of plant development. In animals, determination of cells that have the potential to form sperm or egg—the **germ line**—is one of the earliest events in development. In plants, there is no determined germ line. A small group of cells in each stem have the potential to switch from vegetative to reproductive development in response to environmental conditions.

What Happens during Plant Embryogenesis?

The sequence of events in plant embryogenesis was worked out through careful observation of developing embryos. Embryogenesis is shown in **FIGURE 24.2** (see page 434) within the overall scheme for development of the model plant *Arabidopsis*.

After fertilization, the zygote undergoes an asymmetric cell division (see Figure 24.2 steps 1 and 2). The cells resulting from this initial division are different in size, content, and fate. Similar asymmetric cell divisions are seen often in animal development.

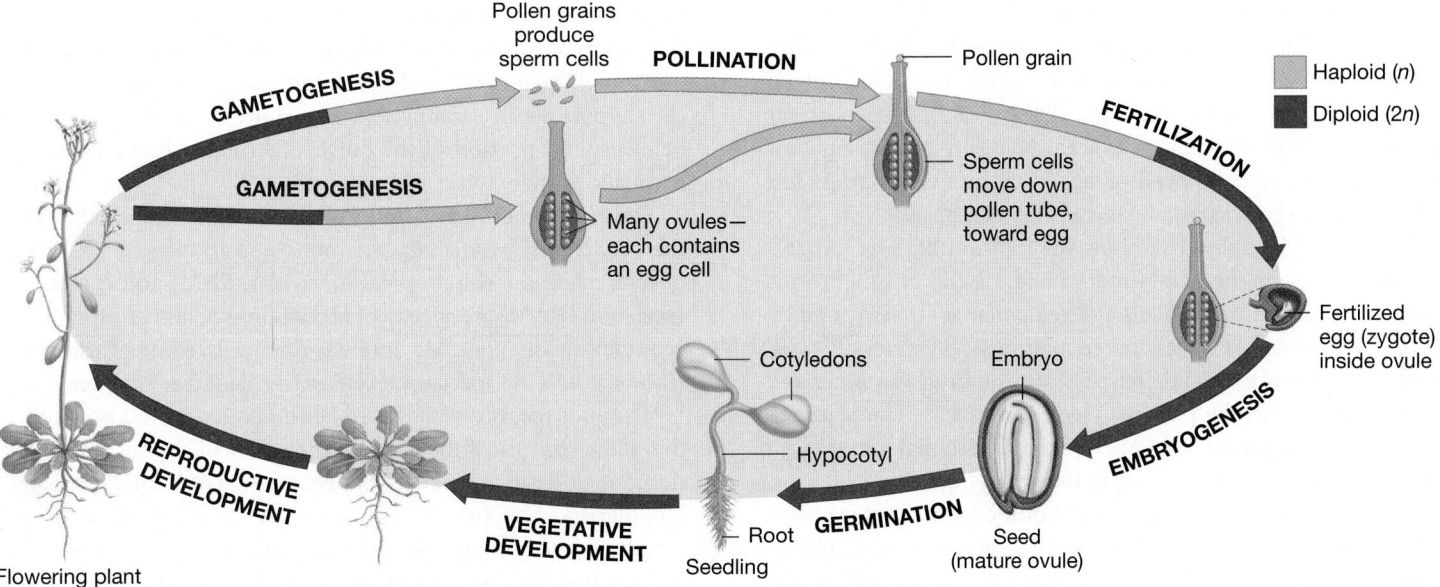

FIGURE 24.1 The Stages of Development in *Arabidopsis*.

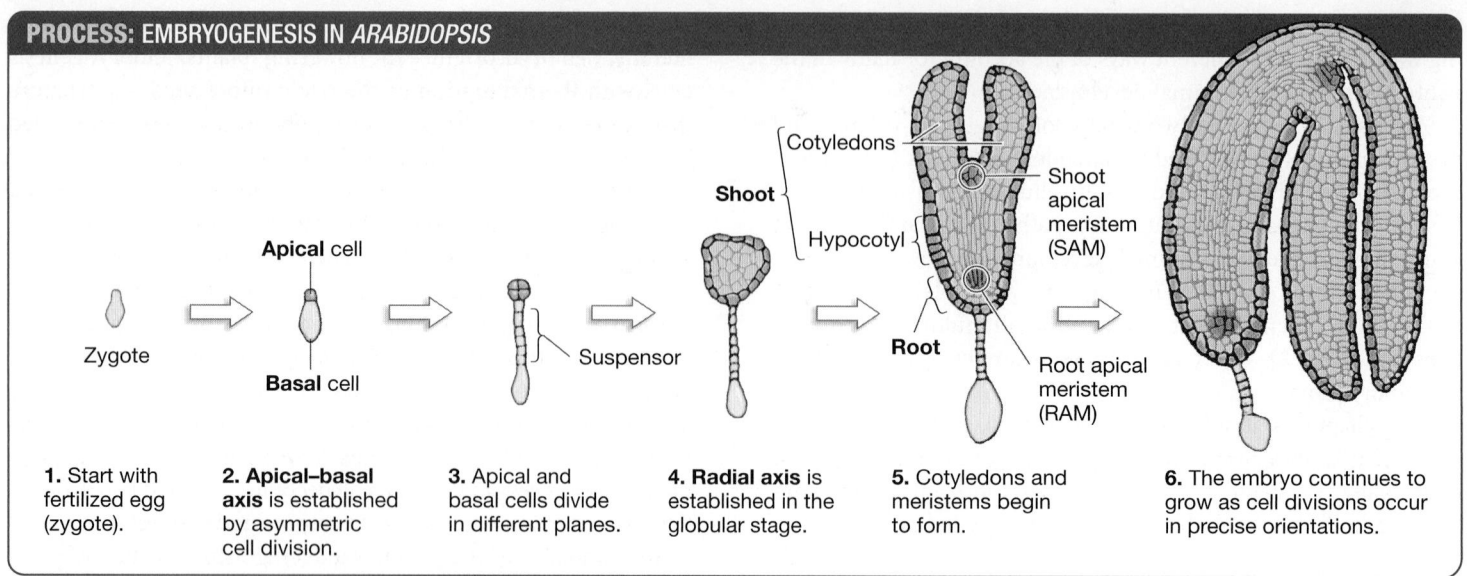

Cotyledons

Shoot apical meristem (SAM)

Shoot

Hypocotyl

Root

Root apical meristem (RAM)

Apical cell

Zygote

Basal cell

Suspensor

1. Start with fertilized egg (zygote).

2. Apical–basal axis is established by asymmetric cell division.

3. Apical and basal cells divide in different planes.

4. Radial axis is established in the globular stage.

5. Cotyledons and meristems begin to form.

6. The embryo continues to grow as cell divisions occur in precise orientations.

FIGURE 24.2 Embryogenesis Takes Place Inside the Developing Seed. Only the embryo is shown in these panels.

✔**EXERCISE** Label the apical–basal and radial axes on the globular-stage embryo.

(For example, think of the distribution of cytoplasmic determinants in Figure 23.4.) The bottom, or basal, cell is large, and it gives rise to a column of cells called the suspensor. The suspensor connects the embryo to surrounding tissues within the seed. The small cell above the basal cell, the apical cell, gives rise to most of the embryo.

Apical refers to the tip; **basal** refers to the base. The asymmetry between the basal and apical cells helps establish one of the primary axes of the plant body: the **apical–basal axis.**

As steps 3 and 4 in Figure 24.2 show, the basal cell divides perpendicularly to the apical–basal axis to produce the suspensor. Only one cell in the suspensor—the one closest to the apical cell—contributes cells to the embryo and thus to the mature adult. The apical cell, in contrast, divides both perpendicularly to the apical–basal axis and parallel to it.

As you learned in an earlier chapter, the regulation of cleavage plane orientation and the direction of cell expansion are critical for creating structures in the plant (Chapter 22). Perpendicular and parallel divisions in the cells at the apical tip of the plant embryo give rise to a ball of cells attached to the suspensor. At this point, the embryo is said to be in the globular stage.

You should be able to see how the division between an apical cell that forms the plant embryo and a basal cell that forms a supportive structure parallels a developmental strategy used by mammalian embryos. Recall that in mammals, the inner cell mass produces the embryo, and the cells surrounding the inner cell mass create part of the placenta (see Figure 23.5). The placenta, like the plant suspensor, is a structure that is essential for development but that doesn't contribute to the embryo.

Another asymmetry arises as cells of the globular-stage plant embryo continue to divide. Cells in the interior are completely surrounded by other cells of the embryo. In contrast, cells in the outermost layer contact surrounding tissues in the seed in addition to underlying embryo cells. Interior and exterior cells

become visibly different. This creates the second major body axis, the radial axis. The radial axis extends from the interior of the plant body out to the exterior. Note that the animals examined in the animal development chapter (Chapter 23) do not have a radial axis.

The initial events in embryogenesis illustrate a general point: Just as for most animal cells, the fate of a plant cell can be summed up in the old quip about the three keys to success in real estate—"location, location, location." Starting with the initial division that creates the apical and basal cells, plant cells differentiate based on where they are in the body.

As the ball of cells continues to grow and develop, embryonic leaves, or **cotyledons,** begin to take shape (step 5). The cotyledons are connected to the developing root by a stem-like structure called the **hypocotyl.** Together, the cotyledons and hypocotyl make up the **shoot,** which will become the aboveground portion of the body. The shoot system functions in photosynthesis, support, and reproduction. The **root,** in contrast, forms the belowground portion of the body. The root system anchors the plant and gathers water and nutrients.

Once the apical–basal and radial axes are established, and as the cotyledons, hypocotyl, and root begin to take shape, groups of cells called the **shoot apical meristem (SAM)** and **root apical meristem (RAM)** are specified. **Meristems** contain groups of stem cells that divide repeatedly into daughter cells, some of which differentiate into the specialized cells of the plant (see Chapter 22).

The root meristem can form all the underground portions of the plant, and the shoot meristem can form all the aerial portions, including reproductive structures. Throughout a plant's life, meristematic tissues continue to produce cells that can differentiate into adult tissues and structures.

Note that all of this growth and development takes place without any of the cell migration seen in gastrulation in animals. Because plant cells have stiff cell walls, they do not move. This is one of the

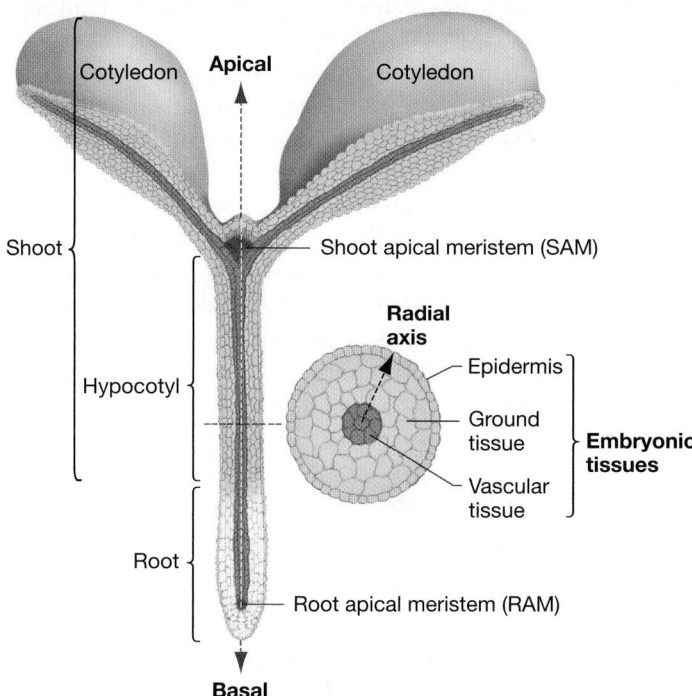

FIGURE 24.3 Embryogenesis Establishes the Two Body Axes and Three Embryonic Tissues.

fundamental differences between plant and animal development. For the cotyledons and other embryonic structures to take shape, it is important that cell divisions occur in precise orientations (step 6 of Figure 24.2). What's more, the resulting cells exhibit differential growth. Some cells grow larger than others, and even more important, the direction of cell expansion along the apical–basal or radial axes is tightly controlled and often radically different.

During embryogenesis, three embryonic tissues are produced and arranged along the radial axis (**FIGURE 24.3**).

1. The **epidermis** (literally, "over-skin") is an outer covering of specialized cells that protects the individual.

2. Inside the epidermal layer of cells is **ground tissue,** a mass of cells that may later differentiate into cells that are specialized for photosynthesis, food storage, or other functions.

3. The **vascular tissue** in the center of the plant will eventually differentiate into specialized cells that transport food and water between root and shoot.

These embryonic tissues are analogous to the ectoderm, mesoderm, and endoderm of developing animals (see Chapter 23). Although the apical–basal and radial axes of plants do not directly correspond to animal axes, note the parallel of creating axes during early embryogenesis in both groups of organisms.

Which Genes and Proteins Set Up Body Axes?

The genetic approach to development that was pioneered with research on *Drosophila melanogaster* has also been a powerful way of studying plant embryogenesis. Although the specific genes involved are different in plants than in animals, the basic mechanism by which genes direct the earliest events in development is similar.

Research on the genetics of early development in plants was pioneered by Gerd Jürgens and colleagues in the 1990s. This research group set out to identify genes that are transcribed in the zygote or embryo of *Arabidopsis* and that are responsible for establishing the apical–basal axis of the plant body. It's no coincidence that this effort was similar to the work on anterior–posterior pattern formation in *Drosophila* (see Chapter 22)—Jürgens had participated in the work with flies.

The biologists' initial goal was to identify individuals with developmental defects at the seedling stage. More specifically, they were looking for patterning mutants that lacked particular regions along the apical–basal axis of the body. The team succeeded in finding several bizarre-looking mutants (**FIGURE 24.4**). Apical mutants lacked the first leaves, or cotyledons. Some mutants, called central mutants, lacked the embryonic stem, or hypocotyl. Other plants, dubbed basal mutants, lacked both hypocotyls and roots.

To interpret these results, the researchers suggested that each type of *Arabidopsis* mutant had a defect in a different gene and that each gene played a role in specifying the position of cells along the apical–basal axis of the body. They hypothesized that these genes are analogous to the segmentation genes of fruit flies, which specify the fate of cells within well-defined regions along the anterior–posterior axis of insects.

What are these *Arabidopsis* genes, and how do they exert their effects? To answer these questions, consider the gene responsible for the basal mutants lacking hypocotyls and roots. This gene has been cloned and sequenced and named *MONOPTEROS*. Because its DNA sequence indicated that its protein product has a DNA-binding domain, *MONOPTEROS* was hypothesized to encode a transcription factor that regulates the activity of target genes. This hypothesis was later shown to be correct.

The *MONOPTEROS* gene is activated in response to signals from auxin—a cell-to-cell signal molecule. Auxin is produced in the shoot apical meristem and transported toward the basal parts of the individual. This results in an auxin concentration gradient along the apical–basal axis of a plant. Much like the Bicoid concentration

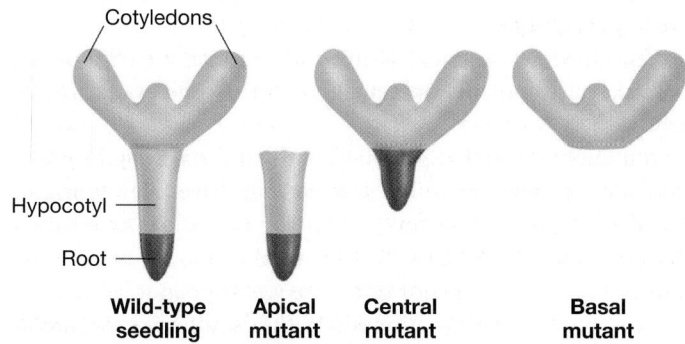

FIGURE 24.4 *Arabidopsis* Developmental Mutations Can Result in Misshapen Embryos. Researchers have identified *Arabidopsis* patterning mutants that are missing specific sections of the body along the apical–basal axis.

gradient in fruit fly embryos (see Chapter 22), the auxin concentration gradient provides positional information.

Together these observations indicate that auxin acts as a morphogen to trigger the production of the regulatory transcription factor MONOPTEROS. In turn, MONOPTEROS unleashes a regulatory cascade that determines which cells in the basal portion of the embryo will form hypocotyl and roots.

The take-home message? Although the specific genes and proteins differ in animals and plants, independent evolution in the two groups has either converged or been constrained to adopt solutions for development that involve common principles. Cell-to-cell signals and regulatory cascades result in the step-by-step specification of a cell's position and fate.

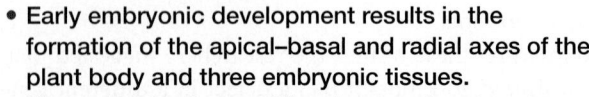
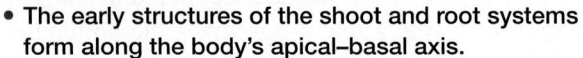

check your understanding

If you understand that . . .

- Early embryonic development results in the formation of the apical–basal and radial axes of the plant body and three embryonic tissues.
- The early structures of the shoot and root systems form along the body's apical–basal axis.
- The genes responsible for setting up the body axes are currently the focus of intense research.

✔ You should be able to. . . .

1. Relate the "location, location, location" quip to the differentiation of epidermal, ground, and vascular tissue.
2. Predict the effect on the *MONOPTEROS* gene and on the embryo of adding auxin to embryonic root cells.

Answers are available in Appendix A.

24.2 Vegetative Development

Whether a plant lives for one season or the 4700 seasons of the oldest bristlecone pine, it continues to develop throughout its life. New roots, stems, and leaves appear and grow. This is vegetative development.

Vegetative development is finely tuned to the environment. Consider just one environmental condition—the availability of light.

You might recall that plants primarily use wavelengths in the blue and red portions of the spectrum to drive photosynthesis (see Chapter 10). Now suppose that you are an oak tree with a life expectancy of 300 years. The quality and quantity of light that your leaves receive depends on where you happen to germinate. Are you growing on flat ground? With a southern exposure in full sun? With a northern exposure rarely exposed to full sun?

In addition, from the time you emerge from an acorn to the time of your death, the light you receive will depend on changes in climate and weather as well as on your size relative to the size

and proximity of plants that compete with you for light and shade. Finally, the leaves in your bottommost branches experience a different light regime from the leaves at your apex.

How do plants cope with all this variation in their living conditions? Unlike most animals, they don't move around to find a place that suits their requirements. Instead, they adjust to their immediate surroundings, in large part through the continuous growth and development of roots, stems, and leaves. If an oak tree is heavily shaded on one side, it stops growing in that direction and extends branches on the other side. If it is heavily shaded on all sides, its growth is directed upward. This constant adjustment to changing environmental conditions is possible because of the meristems that are located at the tips of shoots and roots.

Meristems Provide Lifelong Growth and Development

Just as in animals, when early embryonic development in a plant is complete, the basic body axes are established and the initial structures have formed. For the rest of the individual's life, further development is driven by the groups of stem cells held within meristems (**FIGURE 24.5a**). Meristematic tissue is located at each tip in the shoot and at the tips of the root system. As a result, the individual is capable of growing in any direction aboveground or belowground, depending on conditions.

FIGURE 24.5b provides a close-up view of a shoot apical meristem, or SAM. The cells within the meristem are small and undifferentiated. Within each meristem, the rate of cell division is dictated by cell–cell signals produced in response to environmental cues, such as the arrival of spring, the presence of abundant water, or the amount of light striking the plant. Just below and at the periphery of the meristem, daughter cells produced by mitosis and cytokinesis in the meristem grow in specific directions and initially differentiate into epidermal, ground, or vascular tissue. Eventually these cells will differentiate into more specialized cell types.

Careful microscopy allowed biologists to tease out the sequence of events that occur as meristems grow, and intense research continues to explore how interactions between auxin and other cell–cell signals influence the fate of cells produced by meristems.

Recently researchers have also taken up the question of which genes respond to these cell–cell signals to direct the development of specific structures. Let's consider one example—the genetic control of leaf shape.

Which Genes and Proteins Determine Leaf Shape?

Once a leaf begins to grow, three different axes must be formed: the proximal–distal, mediolateral (middle-to-side), and adaxial–abaxial (upper–lower; **FIGURE 24.6**). Proximal is toward the main body; distal is away from the main body. The mediolateral axis runs from the middle of a leaf toward its margin. Unlike the shoot, but similar to the body of most animals, most leaves are bilaterally symmetric—they have mirror-image left and right

(a) Meristems are located at the tips of shoots and roots.

(b) Cross section of a shoot apical meristem.

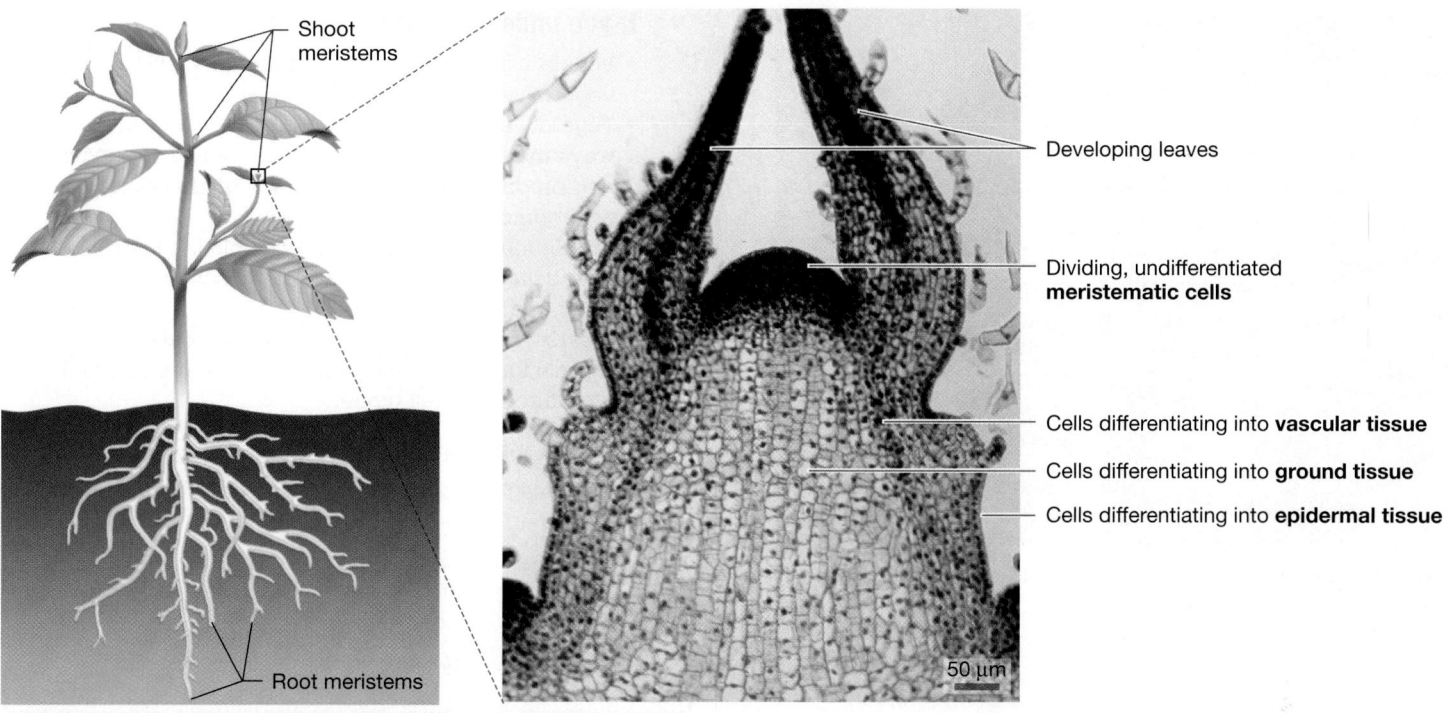

Shoot meristems

Root meristems

Developing leaves

Dividing, undifferentiated **meristematic cells**

Cells differentiating into **vascular tissue**

Cells differentiating into **ground tissue**

Cells differentiating into **epidermal tissue**

50 µm

FIGURE 24.5 Development Takes Place in Meristems. (a) Each tip in the root and shoot systems contains a meristem. The individual can grow in any direction to which a meristem is oriented. **(b)** When stem cells in the meristem divide, the daughter cells either remain undifferentiated and continue to function as stem cells or differentiate into new epidermal, ground, or vascular cells.

sides. The amount of growth along these axes determines the shape of the leaf.

Recent research has begun to identify the genes responsible for specifying the three leaf axes. For example, analyses of mutant snapdragons—a flowering plant you may have seen growing in a garden—and other species has shown that a gene called

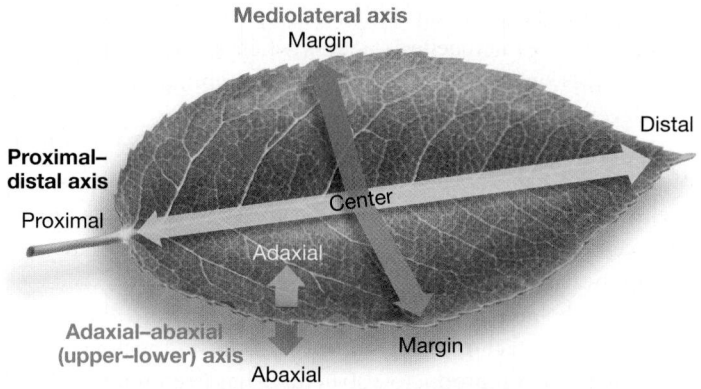

Mediolateral axis

Margin

Proximal–distal axis

Distal

Proximal

Center

Adaxial

Adaxial–abaxial (upper–lower) axis

Margin

Abaxial

FIGURE 24.6 Leaves Have Three Axes. Overall, the plant body has just two axes: apical–basal and radial. Individually, however, every leaf has the three axes shown here. (To remember the difference between adaxial and abaxial, pretend that the *b* in a*b*axial stands for "below.")

✔**QUESTION** How do the body axes in plants compare and contrast with the body axes in animals?

PHANTASTICA (abbreviated *PHAN*) may play a critical role in setting up the adaxial–abaxial axis of leaves.

The protein product of *PHAN* is a regulatory transcription factor. PHAN controls the expression of genes that cause cells to form the upper surface of leaves. It is part of a regulatory cascade that begins with auxin and other cell–cell signals and ends with the growth of a normal-shaped leaf.

Research on tomatoes suggests that changes in *PHAN* expression may also underlie some of the evolutionary changes observed in leaf shape. Leaf shape varies widely among species (see Chapter 37). Simple leaves consist of a single blade, but as **FIGURE 24.7a** (see page 438) shows, tomatoes have compound leaves—each leaf blade is divided into smaller units called leaflets. Other species have palmately compound leaves, meaning that leaflets radiate from a single point.

To explore whether changes in *PHAN* might have a role in the evolution of various leaf shapes, a team of biologists produced transgenic tomato plants using techniques like those that created "golden rice" (Chapter 20). In these tomato plants, the PHAN transcription factor was expressed at reduced levels in the developing leaf.

As **FIGURE 24.7b** shows, leaf shape in the transgenic tomato plants was dramatically altered. Some had simple leaves that were cup shaped, while others had several leaflets emerging from the same point (palmate leaves). Although it is still uncertain exactly how PHAN affects leaf shape, it is clear that *PHAN* expression plays a role.

(a) Normal tomato leaf

(b) Transgenic leaves (reduced *PHAN* expression)

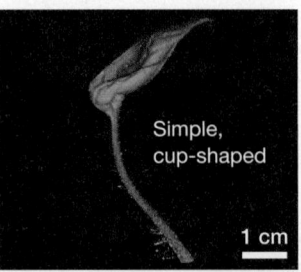

Simple, cup-shaped

1 cm

Palmately compound

1 cm

Compound

1 cm

FIGURE 24.7 Changes in the Expression of Genes That Establish Leaf Axes Can Change Leaf Shape.

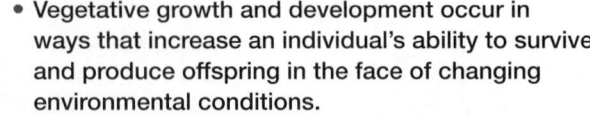
These results have inspired the hypothesis that at least some evolutionary changes in leaf size and shape are due to mutations that alter *PHAN* expression. Alleles that result in lowered *PHAN* expression might lead to simple leaves with a single blade; alleles that increase the extent of *PHAN* expression might result in compound leaves like those of normal tomatoes. By experimentally altering the genes that regulate development, researchers are beginning to understand the genetic changes leading to different types of leaves. Biologists now have another example of how changes in regulatory pathways that direct development could underlie evolutionary change (see Chapter 22). Clearly, evo-devo is not just for animals.

Do Plant Cells Become Irreversibly Determined?

Normal animal cells are irreversibly determined and do not de-differentiate without considerable coaxing. You may recall that producing Dolly, the first cloned mammal, wasn't easy (see Chapter 22). More than 300 embryos were generated from differentiated adult cells, but only one healthy lamb—Dolly—developed. In plants, however, the situation is different.

For millennia, people have grown plants from cuttings. Removing just part of a stem and planting it can produce a new plant with a complete set of roots. What this requires is either **(1)** for plant cells to de-differentiate into an unspecialized cell type and then for the descendants of these cells to repeat development, or **(2)** for one type of specialized cell to differentiate directly into another cell type.

Unlike animal cells, determination in plants either doesn't occur or is readily reversible in at least some cells. This extraordinary flexibility is another hallmark of plant development.

24.3 Reproductive Development

An important difference between animal and plant development is how reproductive tissues and organs form. In animals, the cells that give rise to sperm and egg cells are set aside early in development. These **germ cells** migrate to the ovaries or testes as the reproductive organs begin to develop. As a result, the cells that give rise to animal gametes undergo relatively few rounds of mitosis—perhaps 20 to 50—before meiosis and gametogenesis.

In contrast, plants do not set aside germ cells early in development. Instead, the formation of reproductive structures—flowers—and gametogenesis occurs when a shoot apical meristem (SAM) switches from vegetative development to reproductive development.

As a result, meristematic cells that have divided hundreds of times can give rise to the reproductive organs of plants, and eventually to sperm and eggs. Because mutations occur during each cell cycle, plants generate much more genetic variation by mutation than animals do.

Although biologists have just begun to explore the consequences of this fact, research on the mechanisms responsible for the formation of reproductive structures has been intense. As an introduction to this work, let's consider some highlights of research on *Arabidopsis*.

The Floral Meristem and the Flower

When specialized proteins in *Arabidopsis* sense that nights are getting shorter and the temperature is favorable, they trigger the

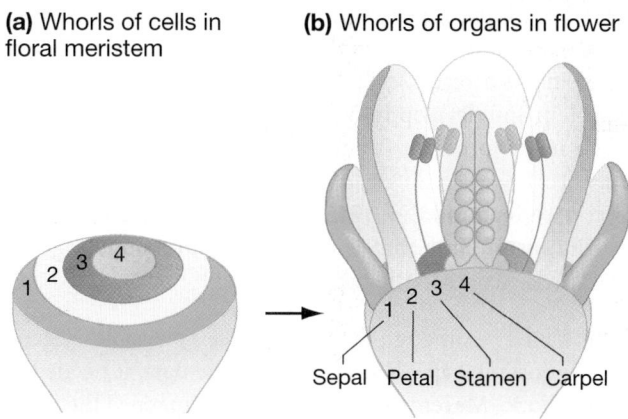

(a) Whorls of cells in floral meristem

(b) Whorls of organs in flower

Sepal Petal Stamen Carpel

FIGURE 24.8 Flowers Are Composed of Four Organs.

production of signals that convert SAMs from vegetative to reproductive development. Through a series of steps, a SAM can produce a **floral meristem;** instead of vegetative structures, such as roots, stems, or leaves, it produces flowers, which contain the plant's reproductive organs. The genes that take part in the regulatory cascade responsible for the maturation of a floral meristem are now well known.

As an *Arabidopsis* flower develops, the floral meristem produces four kinds of organs: **(1)** sepals, **(2)** petals, **(3)** stamens, and **(4)** carpels. Each of these organs is a modified leaf (**FIGURE 24.8**), and all are arranged in a characteristic pattern of whorls within whorls (a whorl is a circular arrangement).

- **Sepals** are located around the outside of the flower and protect the flower as it develops.

- Inside the sepals is a whorl of **petals,** which enclose the male and female reproductive organs.

- The pollen-producing organs, or **stamens,** are located in a whorl inside the petals.

- In the center of the entire structure are egg-producing reproductive organs, or **carpels.** (Ovules most often are located at the base of carpels.)

How does the floral meristem produce these four organs in their characteristic arrangement?

The first hint of an answer came over 100 years ago, when researchers discovered several types of homeotic mutant flowers in popular garden plants. In the mutants, one kind of floral organ was replaced by another. For example, one homeotic mutant had flowers with sepals, petals, another ring of petals, and carpels instead of having sepals, petals, stamens, and carpels. These mutants are similar to *Drosophila* homeotic mutants (see Chapter 22), where individuals have legs or antennae growing in the wrong location in place of the appropriate structure.

Just as an analysis of homeotic mutants in fruit flies triggered a breakthrough in understanding the genetic control of the body plan in animals, an analysis of homeotic flower mutants in *Arabidopsis* triggered a breakthrough in understanding the genetic control of flower structure.

The Genetic Control of Flower Structures

Over 100 years after floral homeotic mutants were first described, Elliot Meyerowitz and colleagues assembled a large collection of *Arabidopsis* homeotic flower-structure mutants. The researchers' goal was to identify and characterize the genes responsible for specifying the four floral organs.

The group found three general classes of mutants that are shown in **FIGURE 24.9**. Some mutants had only carpels and stamens; others had only sepals and carpels; still others had only petals and sepals. The key observation was that each type of mutant lacked the elements found in *two* of the four whorls.

What was going on? The biologists hypothesized that each class of homeotic mutant was caused by a defect in a single gene. They reasoned that if three genes set up the pattern of a flower, the

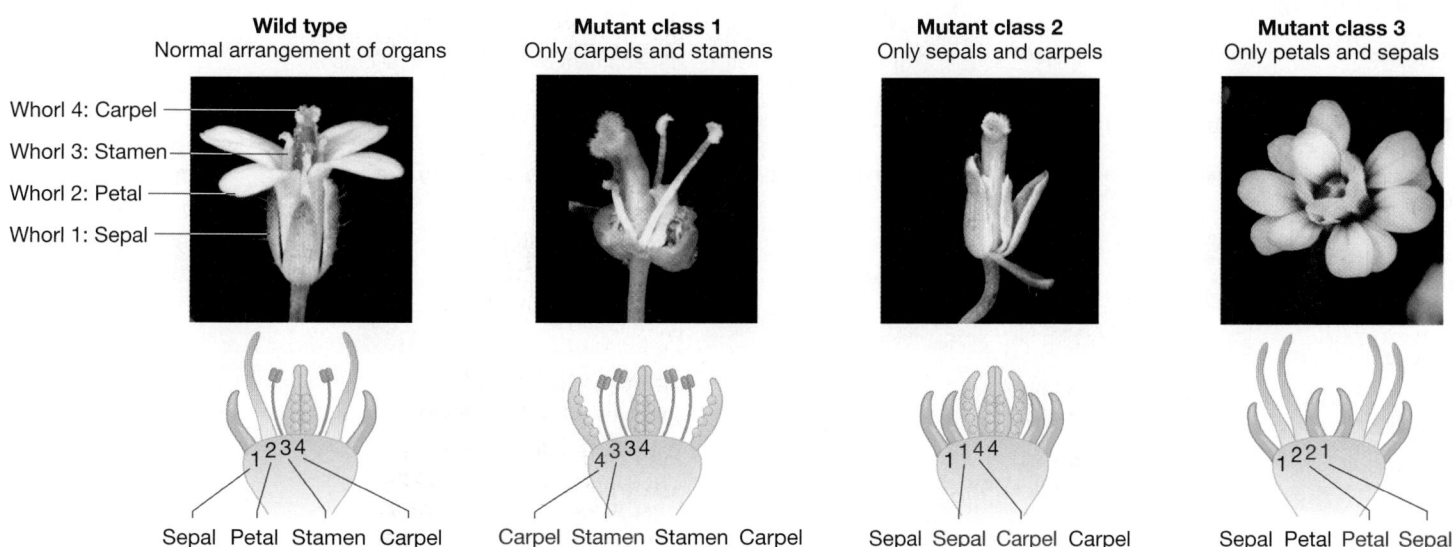

| Wild type
Normal arrangement of organs | Mutant class 1
Only carpels and stamens | Mutant class 2
Only sepals and carpels | Mutant class 3
Only petals and sepals |

Whorl 4: Carpel
Whorl 3: Stamen
Whorl 2: Petal
Whorl 1: Sepal

Sepal Petal Stamen Carpel   Carpel Stamen Stamen Carpel   Sepal Sepal Carpel Carpel   Sepal Petal Petal Sepal

FIGURE 24.9 Homeotic Mutants of *Arabidopsis* Can Have Flower Organs in the Wrong Locations.
Orange labels in the mutants indicate a homeotic transformation (misplaced structure).

mutant phenotypes suggested a hypothesis for how the three gene products interact. Because they referred to the three hypothetical genes as *A*, *B*, and *C*, the hypothesis is called the ABC model.

The ABC Model   Three basic ideas underlie the ABC model (**FIGURE 24.10a**):

- Each of the three genes is expressed in two adjacent whorls.

- Because each gene is expressed in two adjacent whorls, a total of four different combinations of gene products can occur.

- Each of these four combinations of gene products triggers the development of a different floral organ.

Specifically, the Meyerowitz group proposed that (**1**) the A protein alone causes cells to form sepals, (**2**) a combination of A and B proteins sets up the formation of petals, (**3**) B and C combined specify stamens, and (**4**) the C protein alone designates cells as the precursors of carpels.

Does this model explain how the three classes of homeotic mutants occur? The answer is yes, if two additional elements are added to the model:

- The A protein inhibits production of the C protein.

- The C protein inhibits production of the A protein.

With these ideas in mind, patterns of gene expression can be predicted that correspond to the mutant phenotypes. These are shown in **FIGURE 24.10B**.

For example, if the *A* gene is disabled by mutation, then it no longer inhibits the expression of the *C* gene, and all cells produce the C protein. As a result, cells in the outermost whorl express only C protein and develop into carpels, while cells in the whorl just to the inside produce B and C proteins and develop into stamens.

✔ If you understand the ABC model, you should be able to explain why in mutants that lack *C* gene expression, *A* genes become expressed across all whorls.

Testing the Model   Although the ABC model is plausible, elegant, and most important, capable of explaining the data, it needed to be tested. To do so, Meyerowitz and co-workers mapped the genes responsible for the mutant phenotypes and cloned the genes. Once they had isolated the genes, they were able to obtain probes to perform in situ hybridizations (see Chapter 22). The goal was to learn the pattern of expression of the *A*, *B*, and *C* genes and see if that pattern corresponded to the model's predictions.

The mRNAs for the *A*, *B*, and *C* genes showed up in the sets of whorls predicted by the model. The *A* gene is expressed in the outer two whorls, the *B* gene is expressed in the middle two whorls, and the *C* gene is expressed in the inner two whorls. Later work showed that there weren't single *A*, *B*, and *C* genes, but groups of *A*, *B*, and *C* class genes. However, this finding didn't change the outline of the ABC model.

The result that *A*, *B*, and *C* class genes were expressed in the predicted regions supported the ABC model. Just as different

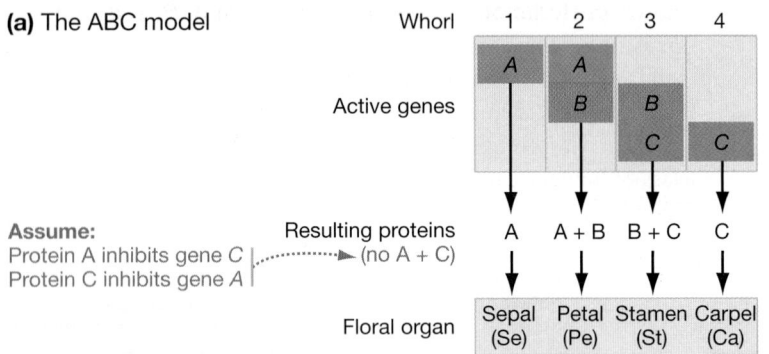

(a) The ABC model

Idea 1: The products of three genes pattern the flower; each gene is expressed in two adjacent whorls.

Idea 2: Four different combinations of proteins occur.

Idea 3: Each protein combination triggers development of a different floral organ.

(b) Predictions of the ABC model

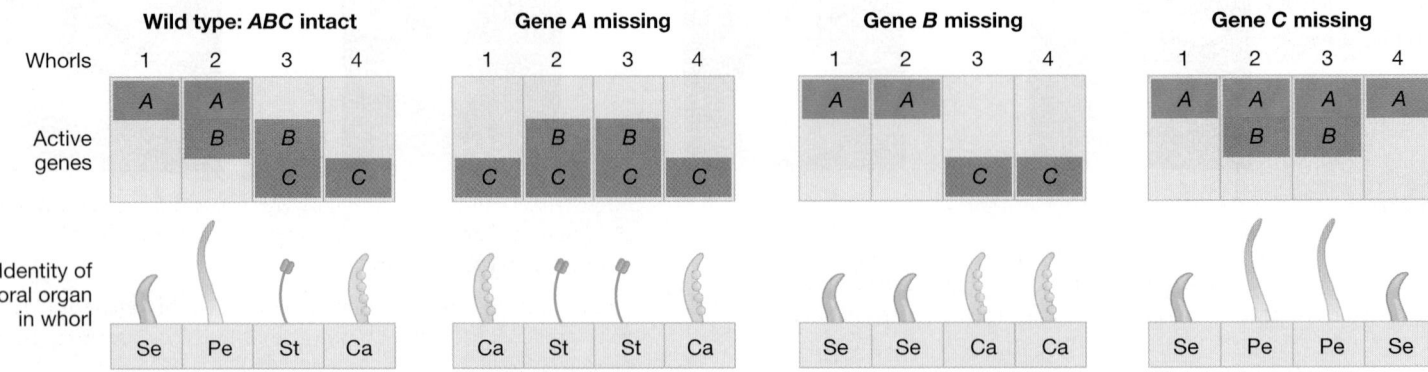

FIGURE 24.10 The ABC Model for Genetic Control of Flower Development. The ABC model is a hypothesis to explain the existence of three types of homeotic mutants in *Arabidopsis* flowers. The regions of gene expression were determined by in situ hybridization.

combinations of *Hox* gene products specify the identity of fly segments, different combinations of floral identity genes specify the parts of a flower.

MADS-Box Genes When Meyerowitz and others analyzed the DNA sequences of the floral organ identity genes, they discovered that all three classes of genes contained a segment called the **MADS box.** The MADS box encodes the MADS domain—a sequence of about 60 amino acids that binds to DNA. Based on this observation, the researchers hypothesized that the floral genes encode transcription factors that bind to regulatory sequences to trigger the expression of genes required for sepal, petal, carpel, and stamen formation.

The researchers found many remarkable similarities between the *MADS-box* genes and the *Hox* genes present in *Drosophila* and other animals.

- *MADS-box* and *Hox* genes encode DNA-binding transcription factors.

- *MADS-box* and *Hox* gene products are both part of regulatory cascades that lead to the specification of structures. *Hox* genes regulate the expression of genes responsible for forming limbs or specific parts of the body; *MADS-box* genes regulate the expression of genes responsible for forming flowers.

- Both sets of genes tell cells where they are in the body and can produce homeotic mutants if they do not work properly.

Biologists are currently working to identify the genes targeted by the ABC proteins, just as they are working to identify the genes controlled by the *Hox* gene products in animals. Eventually, they hope to understand the complete regulatory cascade from initiation of the floral meristem to the expression of proteins that are specific to petals or other floral organs.

Ultimately, the goal is to explore how changes in this cascade could have led to the evolution of the spectacular diversity of flowers observed today. Were mutations in *MADS-box* genes, or the effector genes that they control, responsible for the flamboyant petals of the lady slipper orchid, the bright red sepals of fuchsia, and the elaborate stamens of a sugar maple tree?

Although different genes are involved in plant and animal development, the underlying genetic principles for producing a multicellular body are similar.

Like animals, plants have a genetic toolkit for development. The tools have turned out to be different from those of the animal kit. The differences, however, are not in the types of tools—hammer versus saw—but in their particular forms: nail hammer versus finishing hammer; circular saw versus band saw. For example, plants use MADS-domain transcription factors to specify structures, and animals use Hox transcription factors; plants use graded concentrations of auxin to specify position within the embryo, and (some) animals use graded concentrations of Bicoid. The details are different, but the broad mechanisms are the same.

Why should this be? To return to the opening theme of this chapter, the answer must lie in there being a limited number of ways to build a multicellular organism, coupled with the different evolutionary paths taken in plants and animals to achieve this end. Understanding evolution is the key to understanding biology—and evolution is what you'll explore in the next unit.

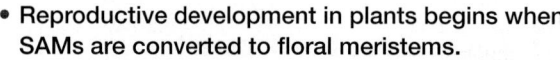

 If you understand that . . .

- Reproductive development in plants begins when SAMs are converted to floral meristems.
- In *Arabidopsis*, development of the four floral organs depends on the expression of regulatory transcription factors encoded by the *A*, *B*, and *C* genes.

✔ **You should be able to . . .**

1. Describe fundamental differences in the generation of plant and animal reproductive cells and structures.

2. Compare and contrast the *MADS-box* genes of plants with the *Hox* genes of animals.

Answers are available in Appendix A.

If you understand . . .

24.1 Embryogenesis

- Embryogenesis is the early stage of development following fertilization. Embryogenesis occurs in the seed and usually includes a period of arrested development that often lasts from weeks to months.

- Body axes are established early in development.

- Three types of embryonic tissues (vascular, ground, epidermis) are formed along the radial axis during embryogenesis.

- Cell migration is not possible in plants. In its absence, tight control of the orientation of the cleavage plane during cell division and differential growth of cells along a particular body axis create the different plant structures.

- A shoot and root meristem arise early in embryogenesis. Meristems contain stem cells that produce all cells for the remainder of the plant's life.

- Master regulatory genes unleash gene-regulatory cascades. These regulatory cascades pattern different regions of the early embryo along the apical–basal and radial axes.

✔ You should be able to predict how the types of patterning mutants obtained by Gerd Jürgens and colleagues (see Figure 24.4) would be different if there were only one master regulatory gene that determined pattern along the apical–basal axis.

24.2 Vegetative Development

- Vegetative development is the development and growth of all plant structures except flowers.
- Vegetative development occurs throughout the plant's life.
- Vegetative development depends on meristems present at the tips of roots and shoots.
- The growth that occurs in vegetative development is finely tuned to the environment, allowing plants to adopt forms that maximize their survival and reproduction.
- Master regulators that control vegetative development are being discovered.

✔ You should be able to describe what an imaginary fly would look like if it underwent the continuous vegetative development found in plants.

24.3 Reproductive Development

- Reproductive development is initiated when a shoot meristem switches from producing cells that will form shoots and leaves

to producing cells that will form flowers, the plant's reproductive structure.

- The switch between vegetative and reproductive development in shoot meristems occurs in response to environmental conditions.
- The ability to switch from vegetative to reproductive development means that plants do not have a devoted germ line—any shoot meristem cell has the potential to produce reproductive cells.
- Once a floral meristem is established, combinations of regulatory transcription factors encoded by A, B, and C classes of genes interact to specify the flower's sepals, petals, stamens, and carpels.

✔ You should be able to predict according to the ABC model the phenotype of a mutant that lacked both the B and C genes.

You should be able to . . .

✔ TEST YOUR KNOWLEDGE
Answers are available in Appendix A

1. From the standpoint of evolution, why are differences in the development of plants and animals expected?

2. Which of the following is *not* a contrast between plant and animal development?
   a. Under certain conditions, plant cells can "de-differentiate" readily.
   b. The fate of a cell is determined in part by its location in the embryo.
   c. Germ cells are not set aside early in plant development.
   d. Plant cells do not move.

3. The MADS box is
   a. part of a *Hox* gene.
   b. a DNA sequence that encodes the DNA-binding portion of a type of transcription factor.
   c. a DNA regulatory sequence for genes involved in flower patterning.
   d. a micro-RNA that controls the timing of flowering.

4. Which of the following does *not* occur during embryogenesis?
   a. formation of the radial axis
   b. production of the suspensor

   c. formation of the cotyledons and hypocotyl
   d. formation of the leaf lateral and proximal–distal axes

5. When does the apical–basal axis first become apparent?
   a. when the epidermal, ground, and vascular tissues form
   b. when the cotyledons, hypocotyl, and root form
   c. when the first cell division produces the apical cell and basal cell
   d. during the globular stage, when the suspensor is complete

6. What evidence suggests that changes in the way the *PHAN* gene is expressed could be partly responsible for evolutionary changes in leaf shape?
   a. If *PHAN* gene expression is manipulated experimentally, individuals produce leaf types found in different species.
   b. Experiments have shown that *PHAN* plays a role in establishing the upper–lower surface axis in leaves.
   c. Sequencing studies and other data have shown that *PHAN* encodes a regulatory transcription factor.
   d. All plant species surveyed to date have a gene homologous to *PHAN*.

7. If auxin acts as a morphogen for determining position along the apical–basal axis, then auxin must be
   a. a transcription factor.
   b. a hormone.
   c. present in a concentration gradient.
   d. synthesized faster than it is degraded.

8. In plants, reproductive tissues may develop late in life, from cells that have undergone mitosis hundreds or thousands of times. How does this differ from animals, and what are the consequences?

9. What qualities of particular meristem cells make them stem cells?

10. In what sense are the epidermal, ground, and vascular tissues produced in the SAMs and RAMs of a 300-year-old oak tree "embryonic"?

11. When in situ hybridization experiments documented where A, B, and C genes were expressed in developing *Arabidopsis* flowers, it was considered strong support for the ABC model. Explain why.

12. Give an example of how each of the fundamental developmental processes—cell proliferation, expansion, interaction, and differentiation—plays a role in plant development.

13. **QUANTITATIVE** Imagine a species that made use of two genes, D and E, to pattern its flowers. Using only these two genes, how many different structures could be specified? In this hypothetical situation—which is not unlike the real situation of the ABC genes in flower patterning—consider that the only thing that's important is whether a gene is on or off (the amount of gene product isn't a factor) and that a default structure can be specified when neither gene is active.

14. Indeterminate growth is the type that occurs when growth is unlimited in duration. In contrast, determinate growth is of limited duration and then stops. Which form of growth is observed in vegetative development and which in reproductive development? Explain your logic.

15. A mutation that causes the B gene to be expressed across the floral meristem (without altering B or C gene expression) would be predicted to lead to

a. petals in the two outer whorls and stamens in the two inner whorls.
b. a pattern of sepals, petals, stamens, and carpels moving from the outer whorl to the center.
c. sepals in the two outer whorls and carpals in the two inner whorls.
d. a pattern of petals, sepals, carpels and stamens moving from the outer whorl to the center.

16. **QUANTITATIVE** Assume that a gene shared by animals and plants mutates at a rate of 1×10^{-6} per cell division. If a germ-line cell in a female mouse undergoes 50 cell divisions before forming an egg, and a cell in an old oak tree undergoes 5000 cell divisions before forming an egg, how much more likely is it to find a mutation in this gene in the oak tree egg relative to the mouse egg?

25 Evolution by Natural Selection

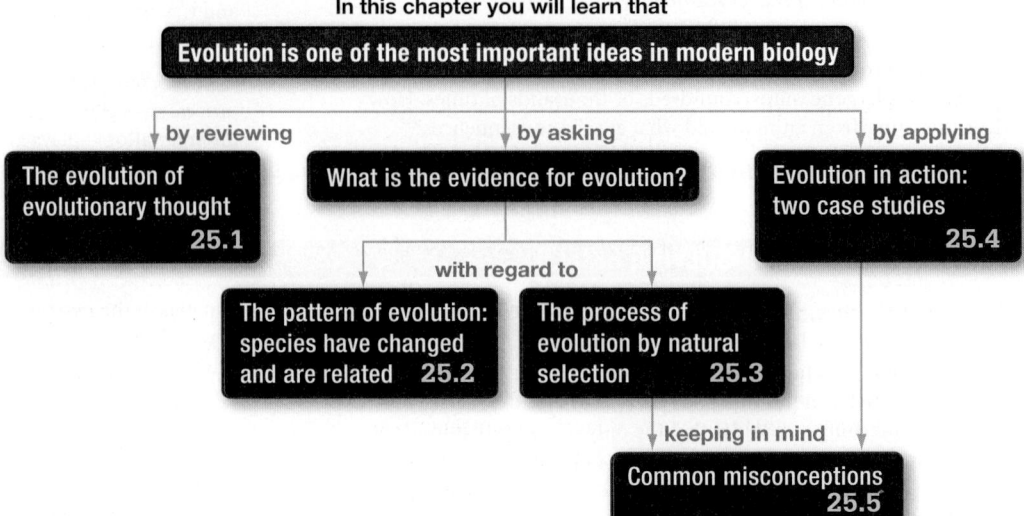

In this chapter you will learn that

Evolution is one of the most important ideas in modern biology

by reviewing → **The evolution of evolutionary thought 25.1**

by asking → **What is the evidence for evolution?**

by applying → **Evolution in action: two case studies 25.4**

with regard to

The pattern of evolution: species have changed and are related 25.2

The process of evolution by natural selection 25.3

keeping in mind

Common misconceptions 25.5

Natural selection acts on individuals in populations such as these sea stars, but only populations evolve. One of Darwin's greatest contributions to science was the introduction of population thinking to the theory of evolution.

This chapter is about one of the great ideas in science. The theory of evolution by natural selection, formulated independently by Charles Darwin and Alfred Russel Wallace, explains how organisms have come to be adapted to environments ranging from arctic tundra to tropical wet forest. It revealed one of the five key attributes of life: Populations of organisms evolve—meaning that they change through time (Chapter 1).

As an example of a revolutionary breakthrough in our understanding of the world, the theory of evolution by natural selection ranks alongside Copernicus's theory of the Sun as the center of our solar system, Newton's laws of motion and theory of gravitation, the germ theory of disease, the theory of plate tectonics, and Einstein's general theory of relativity. These theories are the foundation stones of modern science; all are accepted on the basis of overwhelming evidence.

This chapter is part of the Big Picture. See how on pages 526–527.

✔ When you see this checkmark, stop and test yourself. Answers are available in Appendix A.

Evolution by natural selection is one of the best supported and most important theories in the history of scientific research. But like most scientific breakthroughs, this one did not come easily. When Darwin published his theory in 1859 in a book called *On the Origin of Species by Means of Natural Selection*, it unleashed a firestorm of protest throughout Europe. At that time, the leading explanation for the diversity of organisms was an idea called special creation.

Special creation held that: **(1)** All species are independent, in the sense of being unrelated to each other; **(2)** life on Earth is young—perhaps just 6000 years old; and **(3)** species are immutable, or incapable of change. These beliefs were explained by the instantaneous and independent creation of living organisms by a supernatural being.

Darwin's theory was radically different. How did it differ? Scientific theories usually have two components: a pattern and a process (see Chapter 1):

1. The *pattern component* is a statement that summarizes a series of observations about the natural world. The pattern component is about facts—about how things *are* in nature.

2. The *process component* is a mechanism that produces that pattern or set of observations.

Let's begin with an overview of the evolution of evolutionary thought, and then examine the pattern and process components of the theory of evolution by natural selection.

25.1 The Evolution of Evolutionary Thought

People often use the word revolutionary to describe the theory of evolution by natural selection. Revolutions overturn things—they replace an existing entity with something new and often radically different. A political revolution removes the ruling class or group and replaces it with another. The industrial revolution replaced small shops for manufacturing goods by hand with huge, mechanized assembly lines.

A scientific revolution, in contrast, overturns an existing idea about how nature works and replaces it with another, radically different, idea. The idea that Darwin and Wallace overturned— that species were supernaturally, not naturally, created—had dominated thinking about the nature of organisms in Western civilization for over 2000 years.

Plato and Typological Thinking

The Greek philosopher Plato claimed that every organism was an example of a perfect essence, or type, created by God, and that these types were unchanging. Plato acknowledged that the individual organisms present on Earth might deviate slightly from the perfect type, but he said this deviation was similar to seeing the perfect type in a shadow on a wall. The key to understanding life, in Plato's mind, was to ignore the shadows and focus on understanding each type of unchanging, perfect essence.

Today, philosophers and biologists refer to ideas like this as typological thinking. Typological thinking is based on the idea that species are unchanging types and that variations within species are unimportant or even misleading. Typological thinking also occurs in the Bible's book of Genesis, where God creates each type of organism.

Aristotle and the Great Chain of Being

Not long after Plato developed his ideas, Aristotle ordered the types of organisms known at the time into a linear scheme called the great chain of being, also called the scale of nature (**FIGURE 25.1**). Aristotle proposed that species were organized into a sequence based on increased size and complexity, with humans at the top. He also claimed that the characteristics of species were fixed—they did not change through time.

In the 1700s Aristotle's ideas were still popular in scientific and religious circles. The central claims were that **(1)** species are fixed types, and **(2)** some species are higher—in the sense of being more complex or "better"—than others.

Lamarck and the Idea of Evolution as Change through Time

Typological thinking eventually began to break down. In 1809 the biologist Jean-Baptiste de Lamarck proposed a formal theory of **evolution**—that species are not static but change through time. However, the pattern component of Lamarck's theory was initially based on the great chain of being.

When he started his work on evolution, Lamarck claimed that simple organisms originate at the base of the chain by

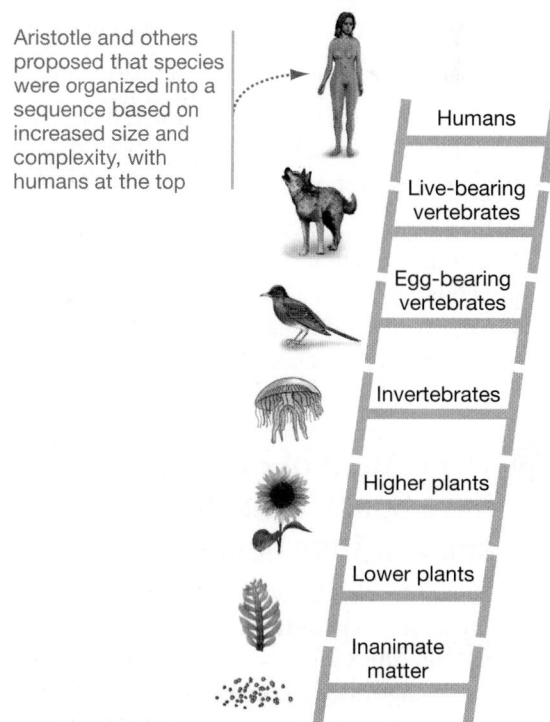

FIGURE 25.1 The Great Chain of Being, or Scale of Nature.

Aristotle and others proposed that species were organized into a sequence based on increased size and complexity, with humans at the top

Humans

Live-bearing vertebrates

Egg-bearing vertebrates

Invertebrates

Higher plants

Lower plants

Inanimate matter

spontaneous generation (see Chapter 1) and then evolve by moving up the chain over time. Thus, Lamarckian evolution is progressive in the sense of always producing larger and more complex, or "better," species. To capture this point, biologists often say that Lamarck turned the ladder of life into an escalator.

Lamarck also contended that species change through time via the inheritance of acquired characters. The idea here is that as an individual develops in response to challenges posed by the environment, its phenotype changes, and it passes on these phenotypic changes to offspring. A classic Lamarckian scenario is that giraffes develop long necks as they stretch to reach leaves high in treetops, and they then produce offspring with elongated necks.

Darwin and Wallace and Evolution by Natural Selection

As his thinking matured, Lamarck eventually abandoned his linear and progressive view of life. Darwin and Wallace concurred. But more important, they emphasized that the process responsible for change through time—evolution—occurs because traits vary among the individuals in a population, and because individuals with certain traits leave more offspring than others do. A **population** consists of individuals of the same species that are living in the same area at the same time.

Darwin and Wallace's proposal was a radical break from the typological thinking that had dominated scientific thought since Plato. Darwin claimed that instead of being unimportant or an illusion, variation among individuals in a population was the key to understanding the nature of species. Biologists refer to this view as **population thinking.**

The theory of evolution by natural selection was revolutionary for several reasons:

1. It overturned the idea that species are static and unchanging.

2. It replaced typological thinking with population thinking.

3. It was scientific. It proposed a mechanism that could account for change through time and made predictions that could be tested through observation and experimentation.

Plato and his followers emphasized the existence of fixed types; evolution by natural selection is all about change and diversity. Now the questions are: What evidence backs the claim that species are not fixed types? What data convince biologists that the theory of evolution by natural selection is correct?

25.2 The Pattern of Evolution: Have Species Changed, and Are They Related?

In *On the Origin of Species*, Darwin repeatedly described evolution as **descent with modification.** He meant that species that lived in the past are the ancestors of the species existing today, and that species change through time.

This view was a radical departure from the independently created and immutable species embodied in Plato's work and in the idea of special creation. In essence, the pattern component of the theory of evolution by natural selection makes two predictions about the nature of species:

1. Species change through time.

2. Species are related by common ancestry.

Let's consider the evidence for each of these predictions in turn.

Evidence for Change through Time

When Darwin began his work, biologists and geologists had just begun to assemble and interpret the fossil record. A **fossil** is any trace of an organism that lived in the past. These traces range from bones and branches to shells, tracks or impressions, and dung. The **fossil record** consists of all the fossils that have been found on Earth and described in the scientific literature.

Why did data in the fossil record support the hypothesis that species have changed through time? And what data from **extant species**—those living today—support the claim that they are modified forms of ancestral species?

The Vastness of Geologic Time Initially, fossils were organized according to their *relative* ages based on a series of principles derived from observations about rock formation. **Sedimentary rocks,** for example, form from sand or mud or other materials deposited at locations such as beaches or river mouths. Sedimentary rocks, along with rocks derived from volcanic ash or lava, are known to form in layers—younger layers are deposited on top of older layers.

Researchers used this information to place fossils in a younger-to-older sequence, based on the fossils' relative position in layers of sedimentary rock (**FIGURE 25.2**). As the scientists observed similarities in rocks and fossils at different sites, they began to create a **geologic time scale:** a sequence of named intervals called eons, eras, and periods that represented the major events in Earth history (see Chapter 28). They also realized that vast amounts of time were required to form the thick layers of sedimentary rock that they were studying, because erosion and deposition of sediments are such slow processes.

This was an important insight. The geologic record indicated that the Earth was much, much older than the 6000 years claimed by proponents of special creation.

After the discovery of radioactivity in the late 1800s, researchers realized that radioactive decay—the steady rate at which unstable or "parent" atoms are converted into more stable "daughter" atoms—furnished a way to assign *absolute* ages, in years, to the relative ages in the geologic time scale.

Radiometric dating is based on three pieces of information:

1. Observed decay rates of parent to daughter atoms

2. The ratio of parent to daughter atoms present in newly formed rocks—such as the amount of uranium atoms versus lead atoms when uranium-containing molten rock first cools (Uranium decays to form lead.)

3. The ratio of parent to daughter atoms present in a particular rock sample

Younger rock layers   **Younger fossils**

Tracks from a mammal-like reptile

Fern

Trilobite

Older rock layers   **Older fossils**

FIGURE 25.2 Sedimentary Rocks Reveal the Vastness of Geologic Time. The relative ages of sedimentary rocks are used to determine the relative ages of fossil organisms because younger layers are deposited on top of older ones. The deepest rock layer in the Grand Canyon is over a billion years old, and the top layer is 270 million years old.

Combining information from these two ratios with information on the decay rate allows researchers to estimate how long ago a rock formed. According to data from radiometric dating, Earth is about 4.6 billion years old, and the earliest signs of life appear in rocks that formed 3.4–3.8 billion years ago.

Data from relative and absolute dating techniques agree: Life on Earth is ancient. A great deal of time has gone by for change to occur.

Extinction Changes the Species Present over Time In the early nineteenth century, researchers began discovering fossil bones, leaves, and shells that were unlike structures from any known animal or plant. At first, many scientists insisted that living examples of these species would be found in unexplored regions of the globe. But as research continued and the number and diversity of fossil collections grew, the argument became less and less plausible.

The issue was finally settled in 1812 when Baron Georges Cuvier published a detailed analysis of an **extinct species**—that is, a species that no longer exists—called the Irish "elk." Scientists accepted the fact of extinction because this gigantic deer was judged to be too large to have escaped discovery and too

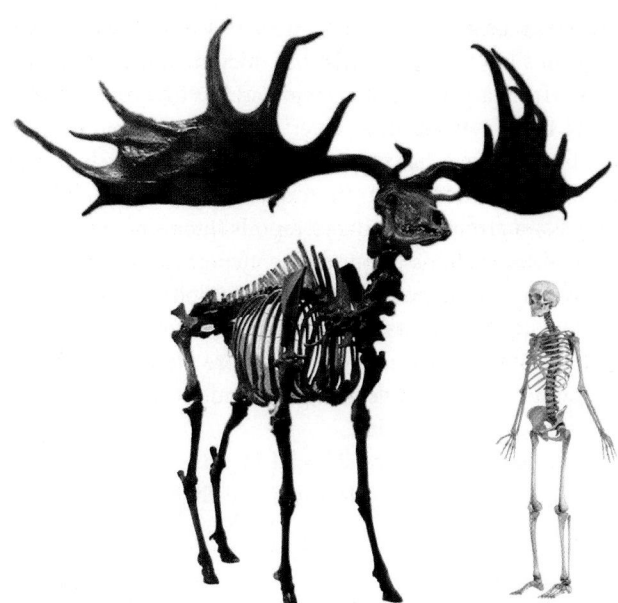

FIGURE 25.3 Evidence of Extinction. The skeleton of the Irish "elk" dwarfs a human. Scientists agreed that the deer was too large and unique to be overlooked if it were alive; it must have gone extinct.

distinctive to be classified as a large-bodied population of an existing species (**FIGURE 25.3**).

Advocates of special creation argued that fossil species were victims of the flood at the time of Noah. Darwin, in contrast, interpreted extinct forms as evidence that species are not static, immutable entities, unchanged since the moment of special creation. His reasoning was that if species have gone extinct, then the array of species living on Earth has changed through time.

Recent analyses of the fossil record suggest that over 99 percent of all the species that have ever lived are now extinct. The data also indicate that species have gone extinct continuously throughout Earth's history—not just in one or even a few catastrophic events.

Transitional Features Link Older and Younger Species Long before Darwin published his theory, researchers reported striking resemblances between the fossils found in the rocks underlying certain regions and the living species found in the same geographic areas. The pattern was so widespread that it became known as the "law of succession." The general observation was that extinct species in the fossil record were succeeded, in the same region, by similar species.

Early in the nineteenth century, the pattern was simply reported and not interpreted. But later, Darwin pointed out that it provided strong evidence in favor of the hypothesis that species had changed through time. His idea was that the extinct forms and living forms were related—that they represented ancestors and descendants.

As the fossil record expanded, researchers discovered species with characteristics that broadened the scope of the law of succession. A **transitional feature** is a trait in a fossil species that is intermediate between those of ancestral (older) and derived (younger) species. For example, intensive work over the past several decades has yielded fossils that document a gradual change over time from aquatic animals that had fins to terrestrial animals that had limbs (**FIGURE 25.4**). Over a period of about 25 million years, the fins of species similar to today's lungfish changed into limbs similar to those found in today's amphibians, reptiles, and mammals—a group called the tetrapods (literally, "four-footed").

These observations support the hypothesis that an ancestral lungfish-like species first used stout, lobed fins to navigate in shallow aquatic habitats. Then they moved onto land, where their descendants became more and more like today's tetrapods in appearance and lifestyle. Lungfish and tetrapod species have clearly changed through time.

Similar sequences of transitional features document changes that led to the evolution of feathers and flight in birds; stomata and vascular tissue in plants; upright posture, flattened faces, and large brains in humans; jaws in vertebrates (animals with backbones); the loss of limbs in snakes; and other traits. Data like these are consistent with predictions from the theory of evolution: If the traits observed in more recent species evolved from traits in more ancient species, then transitional forms are expected to occur in the appropriate time sequence.

The fossil record provides compelling evidence that species have evolved. What data from extant forms support the hypothesis that the characteristics of species change through time?

Vestigial Traits Are Evidence of Change through Time Darwin was the first to provide a widely accepted interpretation of vestigial traits. A **vestigial trait** is a reduced or incompletely developed structure that has no function, or reduced function, but is clearly similar to functioning organs or structures in closely related species.

Biologists have documented thousands of examples of vestigial traits.

- Some whales and snakes have tiny hip and leg bones that do not help them swim or slither.

- Ostriches and kiwis have reduced wings and cannot fly.

- Eyeless, blind cave-dwelling fish have eye sockets.

- Even though marsupial mammals give birth to live young, an eggshell forms briefly early in their development; in some species, newborns have a nonfunctioning "egg tooth" similar to those used by birds and reptiles to break open their shells.

- Monkeys and many other primates have long tails; but our coccyx, illustrated in **FIGURE 25.5**, is too small to help us maintain balance or grab tree limbs for support.

- Many mammals, including primates, are able to erect their hair when they are cold or excited. This behavior manifests itself as goose bumps in humans, but goose bumps are largely ineffective in warming us or signalling our emotional state.

FIGURE 25.4 Transitional Features during the Evolution of the Tetrapod Limb. Fossil species similar to today's lungfish and tetrapods have fin and limb bones that are transitional features. *Eusthenopteron* was aquatic; *Tulerpeton* was probably semiaquatic (mya = million years ago).

✔**QUESTION** How would observations of transitional features be explained under special creation?

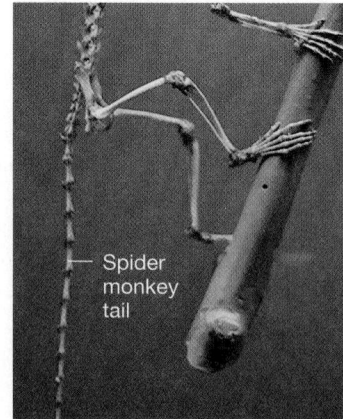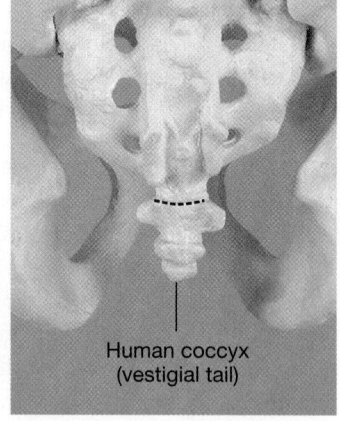

FIGURE 25.5 Vestigial Traits Are Reduced Versions of Traits in Other Species. The tailbone is a human trait that has reduced function. It is no longer useful for balance and locomotion.

✔**QUESTION** How would observations of vestigial traits be explained if evolution occurred via inheritance of acquired characters?

The existence of vestigial traits is inconsistent with the idea of special creation, which maintains that species were perfectly designed by a supernatural being and that the characteristics of species are static. Instead, vestigial traits are evidence that the characteristics of species have changed over time.

Current Examples of Change through Time Biologists have documented hundreds of contemporary populations that are changing in response to changes in their environment. Bacteria have evolved resistance to drugs; insects have evolved resistance to pesticides; weedy plants have evolved resistance to herbicides; the timing of bird migrations, the emergence of insects, and the blooming of flowering plants have evolved in response to climate change. Section 25.4 provides a detailed analysis of research on two examples of evolution in action.

To summarize, change through time continues and can be measured directly. Evidence from the fossil record and living species indicates that life is ancient, that species have changed through the course of Earth's history, and that species continue to change. The take-home message is that species are dynamic—not static, unchanging, and fixed types, as claimed by Plato, Aristotle, and advocates of special creation.

Evidence of Descent from a Common Ancestor

Data from the fossil record and contemporary species refute the hypothesis that species are immutable. What about the claim that species were created independently—meaning that they are unrelated to each other?

Similar Species Are Found in the Same Geographic Area Charles Darwin began to realize that species are related by common ancestry during a five-year voyage he took aboard the English naval ship HMS *Beagle*. While fulfilling its mission to explore and map the coast of South America, the *Beagle* spent a few weeks in the Galápagos Islands off the coast of present-day Ecuador. Darwin had taken over the role of ship's naturalist and, as the first scientist to study the area, gathered extensive collections of the plants and animals found in these islands. Most famous among the birds he collected were the Galápagos finches (featured in Section 25.4) and the Galápagos mockingbirds, pictured in **FIGURE 25.6a**.

Several years after Darwin returned to England, a colleague pointed out that the mockingbirds collected on different islands were distinct species, based on differences in coloration and beak size and shape. This struck Darwin as remarkable. Why would species that inhabit neighboring islands be so similar, yet clearly

(a) Pattern: Although the Galápagos mockingbirds are extremely similar, distinct species are found on different islands.

Nesomimus parvulus

Nesomimus trifasciatus

Nesomimus melanotis

Nesomimus macdonaldi

(b) Recent data support Darwin's hypothesis that the Galápagos mockingbirds share a common ancestor.

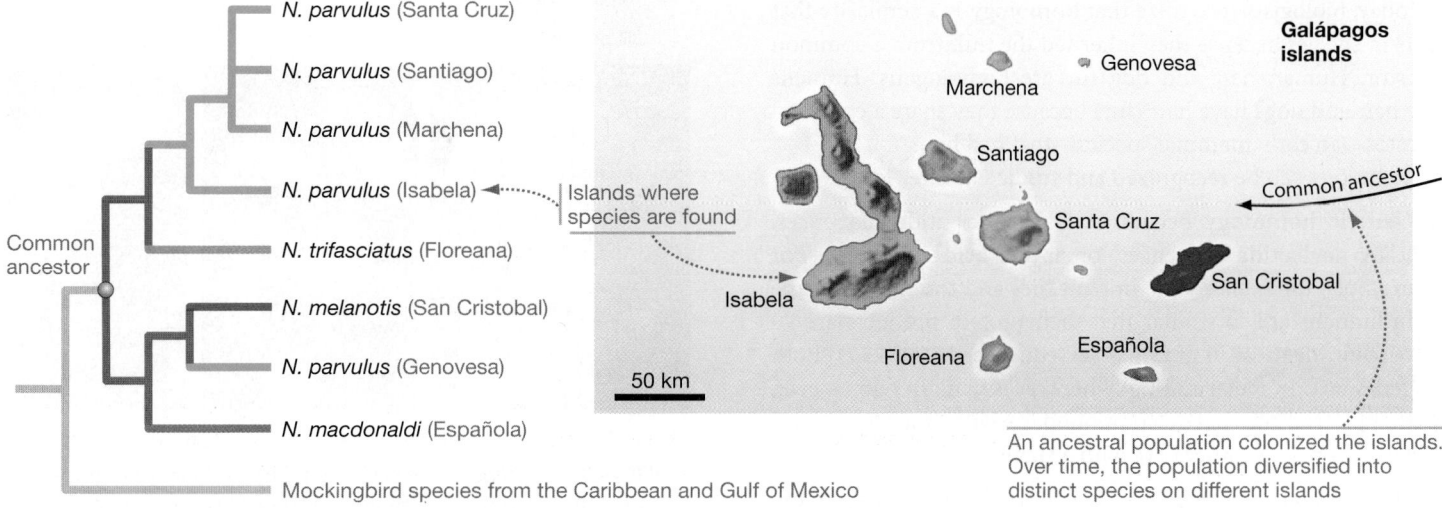

FIGURE 25.6 **Close Relationships among Island Forms Argue for Shared Ancestry.**

Gene:	Amino acid sequence (single-letter abbreviations):
Aniridia (Human)	LQRNRTSFTQEQIEALEKEFERTHYPDVFARERLAAKIDLPEARIQVWFSNRRAKWRREE
eyeless (Fruit fly)	LQRNRTSFTNDQIDSLEKEFERTHYPDVFARERLAGKIGLPEARIQVWFSNRRAKWRREE

FIGURE 25.7 Genetic Homology: Genes from Different Species May Be Similar in DNA Sequence or Other Attributes. Amino acid sequences from a portion of the *Aniridia* gene product found in humans are 90 percent identical to those found in the *Drosophila eyeless* gene product. (For a key to the single-letter abbreviations used for the amino acids, see Figure 3.2.)

distinct? This turns out to be a widespread pattern: In island groups across the globe, it is routine to find similar but distinct species on neighboring islands.

Darwin realized that this pattern—puzzling when examined as a product of special creation—made perfect sense when interpreted in the context of evolution, or descent with modification. The mockingbirds were similar, he proposed, because they had descended from the same common ancestor. That is, instead of being created independently, mockingbird populations that colonized different islands had changed through time and formed new species (**FIGURE 25.6b**).

Recent analyses of DNA sequences in these mockingbirds support Darwin's hypothesis. Researchers have used the DNA sequence comparisons to place the mockingbirds on a **phylogenetic tree**—a branching diagram that describes the ancestor–descendant relationships among species or other taxa (see Chapter 28). As Figure 25.6b shows, the Galápagos mockingbirds are each others' closest living relatives. As Darwin predicted, they share a single common ancestor. (For help with reading phylogenetic trees, see **BioSkills 7** in Appendix B.)

Similar Species Share Homologies Translated literally, homology means "the study of likeness." When biologists first began to study the anatomy of humans and other vertebrates, they were struck by the remarkable similarity of their skeletons, muscles, and organs. But because the biologists who did these early studies were advocates of special creation, they could not explain why striking similarities existed among certain organisms but not others.

Today, biologists recognize that **homology** is a similarity that exists in species because they inherited the trait from a common ancestor. Human hair and dog fur are homologous. Humans have hair and dogs have hair (fur) because they share a common ancestor—an early mammal species—that had hair.

Homology can be recognized and studied at three levels:

1. **Genetic homology** occurs in DNA nucleotide sequences, RNA nucleotide sequences, or amino acid sequences. For example, the *eyeless* gene in fruit flies and the *Aniridia* gene in humans are so similar that their protein products are 90 percent identical in amino acid sequence (**FIGURE 25.7**). Both genes act in determining where eyes will develop—even though fruit flies have a compound eye with many lenses and humans have a camera eye with a single lens.

2. **Developmental homology** is recognized in embryos. For example, early chick, human, and cat embryos have tails and

structures called gill pouches (**FIGURE 25.8**). Later, gill pouches are lost in all three species and tails are lost in humans. But in fish, the embryonic gill pouches stay intact and give rise to functioning gills in adults. To explain this observation, biologists hypothesize that gill pouches and tails exist in chicks, humans, and cats because they existed in the fishlike species that was the common ancestor of today's vertebrates. Embryonic gill pouches are a vestigial trait in chicks, humans, and cats; embryonic tails are a vestigial trait in humans.

3. **Structural homology** is a similarity in adult **morphology,** or form. A classic example is the common structural plan observed in the limbs of vertebrates (**FIGURE 25.9**). In Darwin's own words, "What could be more curious than that the hand of a man, formed for grasping, that of a mole for digging, the leg of the horse, the paddle of the porpoise, and the wing of the bat, should all be constructed on the same pattern, and should include the same bones, in the same relative positions?" An engineer would never use the same underlying structure to design a grasping tool, a digging implement, a walking device, a propeller, and a wing. Instead, the structural homology exists because mammals evolved from the lungfish-like ancestor in Figure 25.4, which had the same general arrangement of bones in its fins.

The three levels of homology interact. Genetic homologies cause the developmental homologies observed in embryos, which then lead to the structural homologies recognized in adults. Perhaps the most fundamental of all homologies is the genetic code.

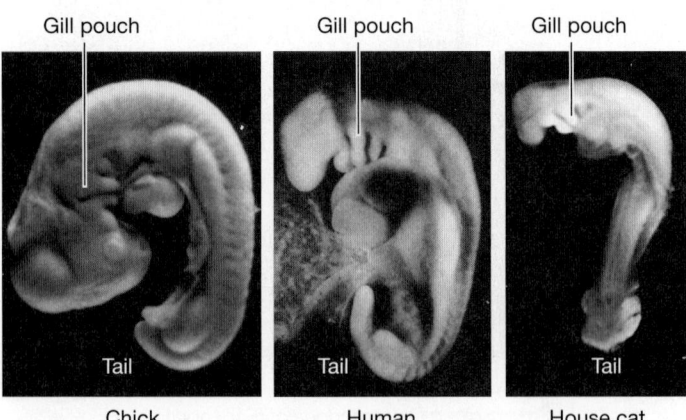

FIGURE 25.8 Developmental Homology: Structures That Appear Early in Development Are Similar. The early embryonic stages of a chick, a human, and a cat show a strong resemblance.

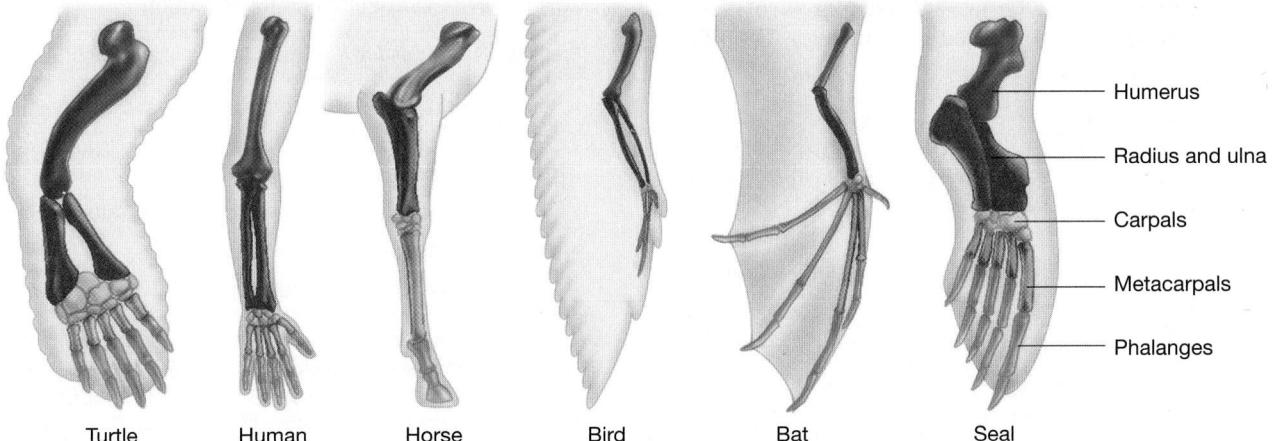

FIGURE 25.9 Structural Homology: Limbs with Different Functions Have the Same Underlying Structure. Even though their function varies, all vertebrate limbs are modifications of the same number and arrangement of bones. (These limbs are not drawn to scale.)

With a few minor exceptions, all organisms use the same rules for transferring the information coded in DNA into proteins (see Chapter 16).

In some cases, hypotheses about homology can be tested experimentally. For example, researchers (1) isolated a mouse gene that was thought to be homologous to the fruit fly *eyeless* gene, (2) inserted the mouse gene into fruit fly embryos, (3) stimulated expression of the foreign gene in locations that normally give rise to appendages, and (4) observed formation of eyes on legs and antennae (**FIGURE 25.10**). The function of the inserted gene was identical to that of *eyeless*. This result was strong evidence that the fruit fly and mouse genes are homologous, as predicted from their sequence similarity.

Homology is a key concept in contemporary biology:

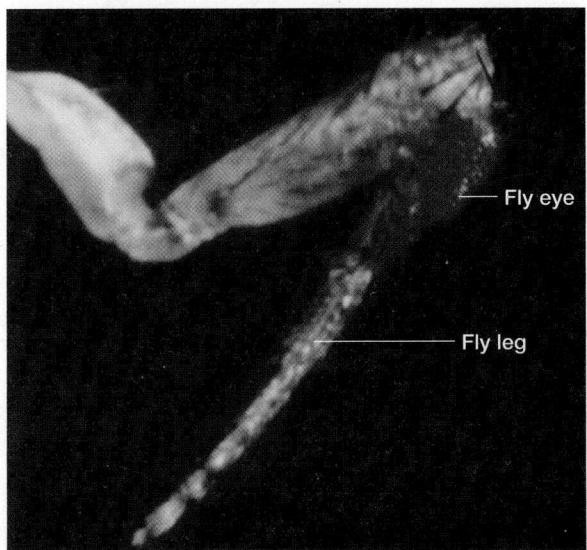

FIGURE 25.10 Evidence for Homology: A Mouse Gene Expressed in Fruit Flies. As an embryo, this fruit fly received a mouse gene that signals where eyes should form. A fruit fly eye formed in the location where the mouse gene was expressed.

- Chemicals that cause cancer in humans can often be identified by testing their effects on mutation rates in bacteria, yeast, zebrafish, mice, and other model organisms because the molecular machinery responsible for copying and repairing DNA is homologous in all organisms (see Chapter 15).

- Drugs intended for human use can be tested on mice or rabbits if the molecules targeted by the drugs are homologous.

- Unknown sequences in the human, rice, or other genomes can be identified if they are homologous to known sequences in yeast, fruit flies, or other well-studied model organisms (see Chapter 21 and **BioSkills 13** in Appendix B).

The theory of evolution by natural selection predicts that homologies will occur. If species were created independently of one another, as special creation claims, these types of similarities would not occur.

Current Examples of Descent from a Common Ancestor Biologists have documented dozens of contemporary populations that are undergoing speciation—a process that results in one species splitting into two or more descendant species. Some populations have served as particularly well-studied examples of speciation in action (see Chapter 27). In most cases, the identity of the ancestral species and the descendant species is known—meaning that biologists have established a direct link between ancestral and descendant species. In addition, the reason for the splitting event is usually known.

The contemporary examples of new species being formed are powerful evidence that species living today are the descendants of species that lived in the past. They support the claim that all organisms are related by descent from a common ancestor.

Evolution's "Internal Consistency"— The Importance of Independent Data Sets

Biologists draw upon data from several sources to challenge the hypothesis that species are immutable and were created independently. The data support the idea that species have descended, with modification, from a common ancestor. **TABLE 25.1** summarizes this evidence.

Perhaps the most powerful evidence for any scientific theory, including evolution by natural selection, is what scientists call internal consistency. This is the observation that data from independent sources agree in supporting predictions made by a theory.

As an example, consider the evolution of whales and dolphins—a group called the cetaceans.

- The fossil record contains a series of species that are clearly identified as cetaceans based on the unusual ear bones found only in this group. Some of the species have the long legs and compact bodies typical of mammals that live primarily on land; some are limbless and have the streamlined bodies typical of aquatic mammals; some have intermediate features.

- A phylogeny of the fossil cetaceans, estimated on the basis of similarities and differences in morphological traits other than limbs and overall body shape, indicates that a gradual transition occurred between terrestrial forms and aquatic, whale-like forms (**FIGURE 25.11**).

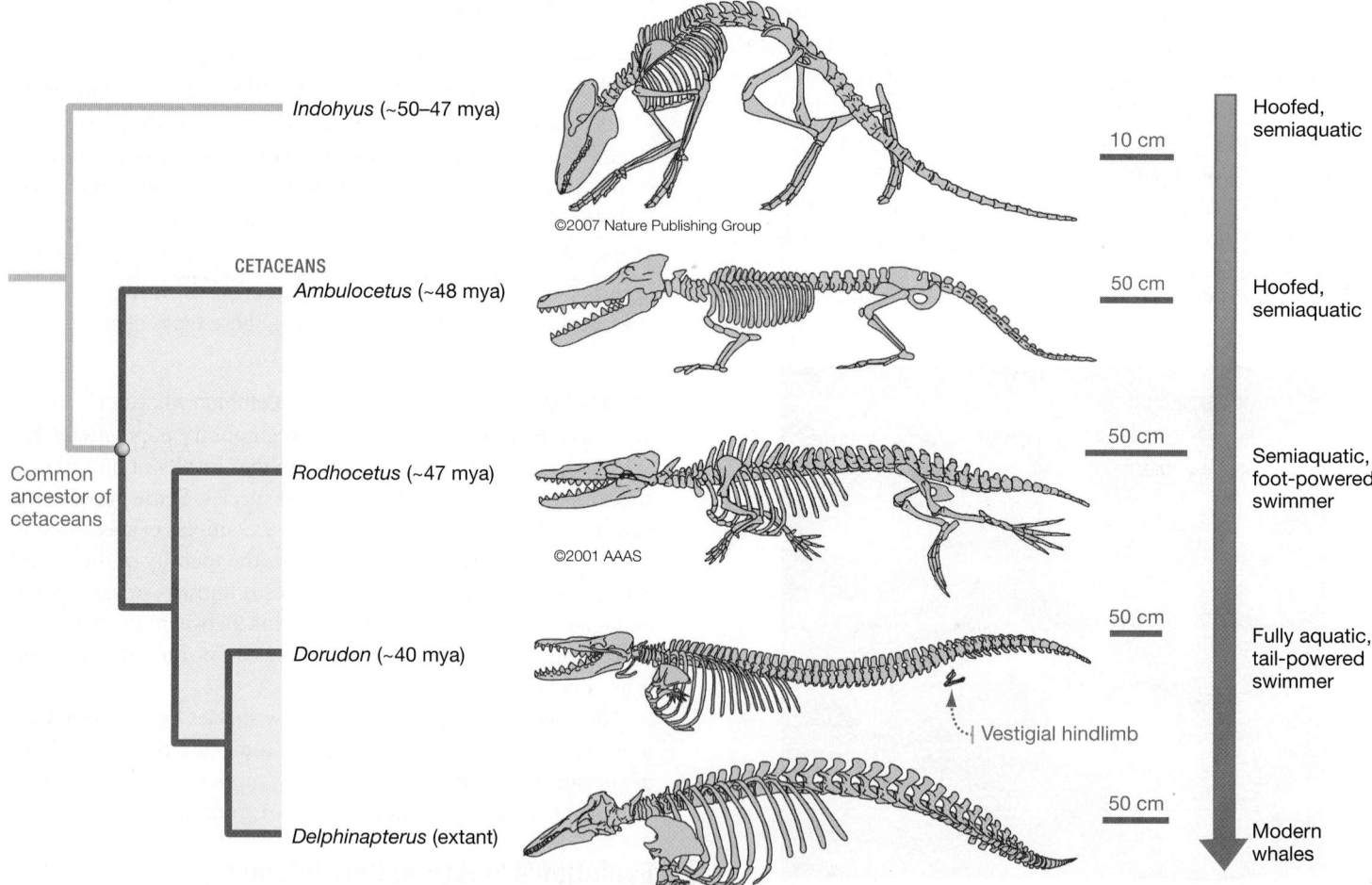

FIGURE 25.11 Data on Evolution from Independent Sources Are Consistent. This phylogeny of fossil cetaceans is consistent with data from relative dating, absolute dating, and phylogenies estimated from molecular traits in living species—all agree that whales evolved from terrestrial ancestors that were related to today's hippos.

- Relative dating, based on the positions of sedimentary rocks where the fossils were found, agrees with the order of species indicated in the phylogeny.

- Absolute dating, based on analyses of radioactive atoms in rocks in or near the layers where the fossils were found, also agrees with the order of species indicated in the phylogeny.

- A phylogeny of living whales and dolphins, estimated from similarities and differences in DNA sequences, indicates that hippos—which spend much of their time in shallow water—are the closest living relative of cetaceans. This observation supports the hypothesis that cetaceans and hippos shared a common ancestor that was semiaquatic.

- Some whales have vestigial hip and limb bones as adults, and some dolphin embryos have vestigial hindlimb buds—outgrowths where legs form in other mammals.

The general message here is that many independent lines of evidence converge on the same conclusion: Whales gradually evolved from a terrestrial ancestor about 50 million years ago.

As you evaluate the evidence supporting the pattern component of the theory of evolution, though, it's important to recognize that no single observation or experiment instantly "proved" the fact of evolution and swept aside belief in special creation. Rather, data from many different sources are much more consistent with evolution than with special creation. Descent with modification is a successful and powerful scientific theory because it explains observations—such as vestigial traits and the close relationships among species on neighboring islands—that special creation does not.

What about the process component of the theory of evolution by natural selection? If the limbs of bats and humans were not created independently and recently, how did they come to be?

check your understanding

 If you understand that . . .
- Species are not static, but change through time.
- Species are related by common ancestry.

 ✔ **You should be able to . . .**

1. Determine what kind of evidence would support the hypothesis that birds evolved from dinosaurs.

2. Explain why the DNA sequences of chimpanzees and humans are about 96 percent similar.

Answers are available in Appendix A.

25.3 The Process of Evolution: How Does Natural Selection Work?

Darwin's greatest contribution did not lie in recognizing the fact of evolution. Lamarck and other researchers had proposed evolution long before Darwin began his work. Instead, Darwin's crucial insight lay in recognizing a process, called natural selection, that could explain the pattern of descent with modification.

Darwin's Inspiration

How did Darwin arrive at his insight? In part, he turned to pigeon breeding—a model system that would be easier to study and manipulate than populations in the wild. Pigeon breeding was popular in England at the time, offering a wealth of experience for Darwin to tap into. Also, pigeons could be maintained easily, and in Darwin's words, "the diversity of the breeds is something astonishing" (see examples in **FIGURE 25.12**).

Darwin crossbred pigeons and observed how characteristics were passed on to offspring. He could choose certain individuals with desirable traits to reproduce, thus manipulating the composition of the population by a process called **artificial selection.** It was clear to Darwin and other breeders that the diverse varieties were all descended from the wild rock pigeons.

Another influence on Darwin was the fortuitous publication of a book by Thomas Robert Malthus, *An Essay on the Principle of Population*, which inspired a great deal of discussion in England at the time. Malthus's studies of human populations in England and elsewhere led him to a startling conclusion: Since many more individuals are born than can survive, a "struggle for existence" occurs as people compete for food and places to live.

Darwin combined his observations of artificial selection with this notion of "struggle for existence" in natural populations, which he knew—from his extensive studies—contained variation. From this synthesis arose his concept of natural selection. Although both Darwin and Wallace arrived at the same idea, Darwin's name is more closely associated with the concept of natural selection because he thought of it first and provided extensive evidence for it in *On the Origin of Species*.

FIGURE 25.12 Diversity of Pigeon Breeds in Captivity. Darwin used the breeding of pigeons as a model system to study how the characteristics of populations can change over time.

Darwin's Four Postulates

Darwin broke the process of evolution by natural selection into four simple postulates (criteria) that form a logical sequence:

1. The individual organisms that make up a population vary in the traits they possess, such as their size and shape.

2. Some of the trait differences are heritable, meaning that they are passed on to offspring. For example, tall parents tend to have tall offspring.

3. In each generation, many more offspring are produced than can possibly survive. Thus, only some individuals in the population survive long enough to produce offspring, and among the individuals that produce offspring, some will produce more than others.

4. The subset of individuals that survive best and produce the most offspring is not a random sample of the population. Instead, individuals with certain heritable traits are more likely to survive and reproduce. **Natural selection** occurs when individuals with certain characteristics produce more offspring than do individuals without those characteristics. The individuals are selected naturally—meaning, by the environment.

Because the selected traits are passed on to offspring, the frequency of the selected traits increases from one generation to the next. We now know that traits are determined by alleles, particular versions of genes (see Chapter 14). Thus, the outcome of evolution by natural selection is a change in allele frequencies in a population over time.

In studying these criteria, you should realize that variation among individuals in a population is essential if evolution is to occur. Darwin had to introduce population thinking into biology because it is populations that change over time. To come up with these postulates and understand their consequences, Darwin had to think in a revolutionary way.

Today, biologists usually condense Darwin's four postulates into a two-part statement that communicates the essence of evolution by natural selection more forcefully: Evolution by natural selection occurs when (1) heritable variation leads to (2) differential reproductive success.

The Biological Definitions of Fitness, Adaptation, and Selection

To explain the process of natural selection, Darwin referred to successful individuals as "more fit" than other individuals. In doing so, he gave the word fitness a definition different from its everyday English usage. Biological **fitness** is the ability of an individual to produce surviving, fertile offspring relative to that ability in other individuals in the population.

Note that fitness is a measurable quantity. When researchers study a population in the lab or in the field, they can estimate the relative fitness of individuals by counting the number of surviving offspring each individual produces and comparing the data.

The concept of fitness, in turn, provides a compact way of formally defining adaptation. The biological meaning of adaptation, like the biological meaning of fitness, is different from its normal English usage. In biology, an **adaptation** is a heritable trait that increases the fitness of an individual in a particular environment relative to individuals lacking the trait. Adaptations increase fitness—the ability to produce viable, fertile offspring. You can see the Big Picture of how adaptation and fitness relate to natural selection on pages 526–527.

Lastly, the term selection has a commonsense meaning in the context of artificial selection. Breeders *choose* which characteristics they want to keep or get rid of in their plant and animal breeds. However, the term selection has a very different meaning in the biological context of natural selection. Here, it refers to a passive process—differential reproduction as a result of heritable variation—not a purposeful choice.

25.4 Evolution in Action: Recent Research on Natural Selection

The theory of evolution by natural selection is testable. If the theory is correct, biologists should be able to test the validity of each of Darwin's postulates—documenting heritable variation and differential reproductive success in a wide array of natural populations.

This section summarizes two examples in which evolution by natural selection is being observed in nature. Literally hundreds of other case studies are available, involving a wide variety of traits and organisms. To begin, let's explore the evolution of drug resistance, one of the great challenges facing today's biomedical researchers and physicians.

Case Study 1: How Did *Mycobacterium tuberculosis* Become Resistant to Antibiotics?

Mycobacterium tuberculosis, the bacterium that causes **tuberculosis**, or TB, has long been a scourge of humankind. It usually infects the lungs and causes fever, coughing, sweats, weight loss, and often death. In Europe and the United States, TB was once as great a public health issue as cancer is now. It receded in importance during the early 1900s, though, for two reasons:

1. Advances in nutrition made people better able to fight off most *M. tuberculosis* infections quickly.

2. The development of antibiotics allowed physicians to stop even advanced infections.

In the late 1980s, however, rates of *M. tuberculosis* infection surged in many countries, and in 1993 the World Health Organization declared TB a global health emergency. Physicians were particularly alarmed because the strains of *M. tuberculosis* responsible for the increase were largely or completely resistant to antibiotics that were once extremely effective.

How and why did the evolution of drug resistance occur? The case of a single patient—a young man who lived in Baltimore—illustrates what is happening all over the world.

A Patient History The story begins when the individual was admitted to the hospital with fever and coughing. Chest X-rays, followed by bacterial cultures of fluid coughed up from the lungs, showed that he had an active TB infection. He was given several antibiotics for 6 weeks, followed by twice-weekly doses of the antibiotic rifampin for an additional 33 weeks. Ten months after therapy started, bacterial cultures from the patient's chest fluid indicated no *M. tuberculosis* cells. His chest X-rays were also normal. The antibiotics seemed to have cleared the infection.

Just two months after the TB tests proved normal, however, the young man was readmitted to the hospital with a fever, severe cough, and labored breathing. Despite being treated with a variety of antibiotics, including rifampin, he died of respiratory failure 10 days later. Samples of material from his lungs showed that *M. tuberculosis* was again growing actively there. But this time the bacterial cells were completely resistant to rifampin.

Drug-resistant *M. tuberculosis* cells had killed this patient. Where did they come from? Could a strain that was resistant to antibiotic treatment have evolved *within* him? To answer this question, a research team analyzed DNA from the drug-resistant strain and compared it with stored DNA from *M. tuberculosis* cells that had been isolated a year earlier from the same patient. After examining extensive stretches from each genome, the biologists were able to find only one difference: a point mutation in a gene called *rpoB*.

A Mutation in a Bacterial Gene Confers Resistance The *rpoB* gene codes for a component of RNA polymerase. This enzyme transcribes DNA to mRNA, so it is essential to the survival and reproduction of bacterial cells (see Chapter 17). In this case, the point mutation in the *rpoB* gene changed a cytosine to a thymine, forming a new allele for the *rpoB* gene (see Chapter 16). This missense mutation caused a change in the amino acid sequence of the RNA polymerase (from a serine to a leucine at the 153rd amino acid)—and a change in its shape.

This shape change proved critical. Rifampin, the antibiotic that was being used to treat the patient, works by binding to the RNA polymerase of *M. tuberculosis* and interfering with transcription. Bacterial cells with the C → T mutation continue to produce offspring efficiently even in the presence of the drug because the drug cannot bind efficiently to the mutant RNA polymerase.

These results suggest the steps that led to this patient's death (**FIGURE 25.13**).

1. By chance, one or a few of the bacterial cells present in the patient, before drug therapy started, happened to have the *rpoB* allele with the C → T mutation. Under normal conditions, mutant forms of RNA polymerase do not work as well as the more common form, so cells with the C → T mutation would not produce many offspring and would stay at low frequency—even while the overall population grew to the point of inducing symptoms that sent the young man to the hospital.

2. Therapy with rifampin began. In response, cells in the population with normal RNA polymerase began to grow much more slowly or to die outright. As a result, the overall bacterial population declined in size so drastically that the patient appeared to be cured—his symptoms began to disappear.

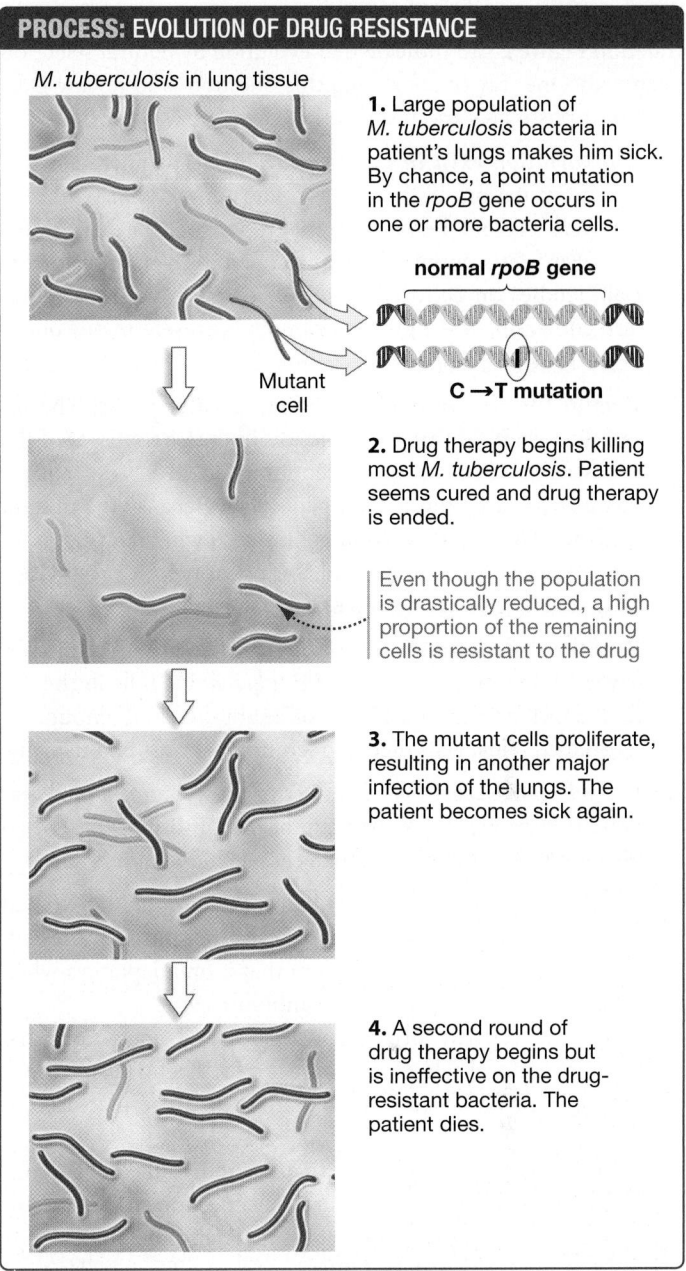

PROCESS: EVOLUTION OF DRUG RESISTANCE

M. tuberculosis in lung tissue

1. Large population of *M. tuberculosis* bacteria in patient's lungs makes him sick. By chance, a point mutation in the *rpoB* gene occurs in one or more bacteria cells.

normal *rpoB* gene

Mutant cell

C → T mutation

2. Drug therapy begins killing most *M. tuberculosis*. Patient seems cured and drug therapy is ended.

Even though the population is drastically reduced, a high proportion of the remaining cells is resistant to the drug

3. The mutant cells proliferate, resulting in another major infection of the lungs. The patient becomes sick again.

4. A second round of drug therapy begins but is ineffective on the drug-resistant bacteria. The patient dies.

FIGURE 25.13 Alleles That Confer Drug Resistance Increase in Frequency When Drugs Are Used.

3. Cells with the C → T mutation continued to increase in number after therapy ended. Eventually the *M. tuberculosis* population regained its former abundance, and the patient's symptoms reappeared.

4. Drug-resistant *M. tuberculosis* cells now dominated the population, so the second round of rifampin therapy was futile.

✔ If you understand these concepts, you should be able to explain: (1) Why the relapse in step 3 occurred, and (2) whether a family member or health-care worker who got TB from this patient at step 3 or step 4 would respond to drug therapy.

Testing Darwin's Postulates   Does the sequence of events illustrated in Figure 25.13 indicate that evolution by natural selection occurred? One way of answering this question is to review Darwin's four postulates and test whether each one was verified:

1. *Did variation exist in the population?*   The answer is yes. Due to mutation, both resistant and nonresistant strains of TB were present before administration of the drug. Most *M. tuberculosis* populations, in fact, exhibit variation for the trait; studies on cultured *M. tuberculosis* show that a mutation conferring resistance to rifampin is present in one out of every 10^7 to 10^8 cells.

2. *Was this variation heritable?*   The answer is yes. The researchers showed that the variation in the phenotypes of the two strains—from drug susceptibility to drug resistance—was due to variation in their genotypes. Because the mutant *rpoB* gene is passed on to daughter cells when a *Mycobacterium* replicates, the allele and the phenotype it produces—drug resistance—are passed on to offspring.

3. *Was there variation in reproductive success?*   The answer is yes. Only a tiny fraction of *M. tuberculosis* cells in the patient survived the first round of antibiotics long enough to reproduce. Most cells died and left no or almost no offspring.

4. *Did selection occur?*   The answer is yes. When rifampin was present, certain cells—those with the drug-resistant allele—had higher reproductive success than cells with the normal allele.

M. tuberculosis individuals with the mutant *rpoB* allele had higher fitness in an environment where rifampin was present. The mutant allele produces a protein that is an adaptation when the cell's environment contains the antibiotic.

This study verified all four postulates and confirmed that evolution by natural selection had occurred. The *M. tuberculosis* population evolved because the mutant *rpoB* allele increased in frequency.

It is critical to note, however, that the individual cells themselves did not evolve. When natural selection occurred, the individual bacterial cells did not change through time; they simply survived or died, or produced more or fewer offspring. This is a fundamentally important point: Natural selection acts on individuals, because individuals experience differential reproductive success. But only populations evolve. Allele frequencies change in populations, not in individuals. Understanding evolution by natural selection requires population thinking.

Drug Resistance: A Widespread Problem   The events reviewed for this single patient have occurred many times in other patients. Recent surveys indicate that drug-resistant strains now account for about 10 percent of the *M. tuberculosis*–causing infections throughout the world.

Unfortunately, the emergence of drug resistance in TB is far from unusual. Resistance to a wide variety of insecticides, fungicides, antibiotics, antiviral drugs, and herbicides has evolved in hundreds of insects, fungi, bacteria, viruses, and plants. In every case, evolution has occurred because individuals with the

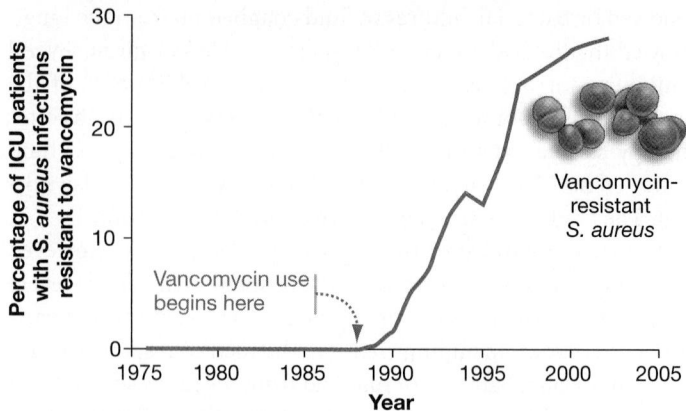

FIGURE 25.14 Trends in Infections Due to Antibiotic-resistant Bacteria. The line indicates changes in the percentage of *S. aureus* infections, acquired in hospitals, that are resistant to the antibiotic vancomycin.

DATA: Centers for Disease Control, 2004.

heritable ability to resist some chemical compound were present in the original population. As the susceptible individuals die from the pesticide, herbicide, or drug, the alleles that confer resistance increase in frequency.

To drive home the prevalence of evolution in response to drugs and other human-induced changes in the environment, consider the data in **FIGURE 25.14**. The graph shows changes through time in the percentage of infections, in intensive care units in the United States, caused by strains of the bacterium *Staphylococcus aureus* that are resistant to the antibiotic vancomycin. Most of these *S. aureus* cells are also resistant to methicillin and other antibiotics—a phenomenon known as multidrug resistance. In some cases, physicians have no effective antibiotics available to treat these infections. ✔ If you understand antibiotic resistance, you should be able to explain why the overprescription of antibiotics by doctors, or the overuse of everyday soaps and cleaners laced with antibiotics, can be a health risk.

Case Study 2: Why Are Beak Size, Beak Shape, and Body Size Changing in Galápagos Finches?

Can biologists study evolution in response to natural environmental change—when humans are not involved? The answer is yes. As an example, consider research led by Peter and Rosemary Grant. For over four decades, these biologists have been investigating changes in beak size, beak shape, and body size that have occurred in finches native to the Galápagos Islands.

Because the island of Daphne Major of the Galápagos is so small—about the size of 80 football fields—the Grants' team has been able to catch, weigh, and measure all the medium ground finches in the island's population (**FIGURE 25.15**) and mark each one with a unique combination of colored leg bands. The medium ground finch makes its living by eating seeds. Finches crack seeds with their beaks.

Early studies of the finch population established that beak size and shape and body size vary among individuals, and that beak

Medium ground finch
(*Geospiza fortis*)

Daphne Major

FIGURE 25.15 Studying Evolution in Action on the Galápagos.

morphology and body size are heritable. Stated another way, parents with particularly deep beaks tend to have offspring with deep beaks. Large parents also tend to have large offspring. Beak size and shape and body size are traits with heritable variation.

Selection during Drought Conditions Not long after the team began to study the finch population, a dramatic selection event occurred. In the annual wet season of 1977, Daphne Major received just 24 millimeters (mm) of rain instead of the 130 mm that normally falls. During the drought, few plants were able to produce seeds, and 84 percent (about 660 individuals) of the medium ground finch population disappeared.

Two observations support the hypothesis that most or all of these individuals died of starvation:

1. The researchers found 38 dead birds, and all were emaciated.

2. None of the missing individuals were spotted on nearby islands, and none reappeared once the drought had ended and food supplies returned to normal.

The research team realized that the die-off was a **natural experiment**. Instead of comparing groups created by direct manipulation under controlled conditions, natural experiments allow researchers to compare treatment groups created by an unplanned change in conditions. In this case, the Grants' team could test whether natural selection occurred by comparing the population before and after the drought.

Were the survivors different from nonsurvivors? The histograms in **FIGURE 25.16** show the distribution of beak sizes in the population before and after the drought. Notice the different scales of the *y*-axes of the two graphs. (For more on how histograms are constructed, see **BioSkills 3** in Appendix B.) On average, survivors tended to have much deeper beaks than did the birds that died.

RESEARCH

QUESTION: Did natural selection on ground finches occur when the environment changed?

HYPOTHESIS: Beak characteristics changed in response to a drought.

NULL HYPOTHESIS: No changes in beak characteristics occurred in response to a drought.

EXPERIMENTAL SETUP:

Weigh and measure all birds in the population before and after the drought.

PREDICTION:

PREDICTION OF NULL HYPOTHESIS:

RESULTS:

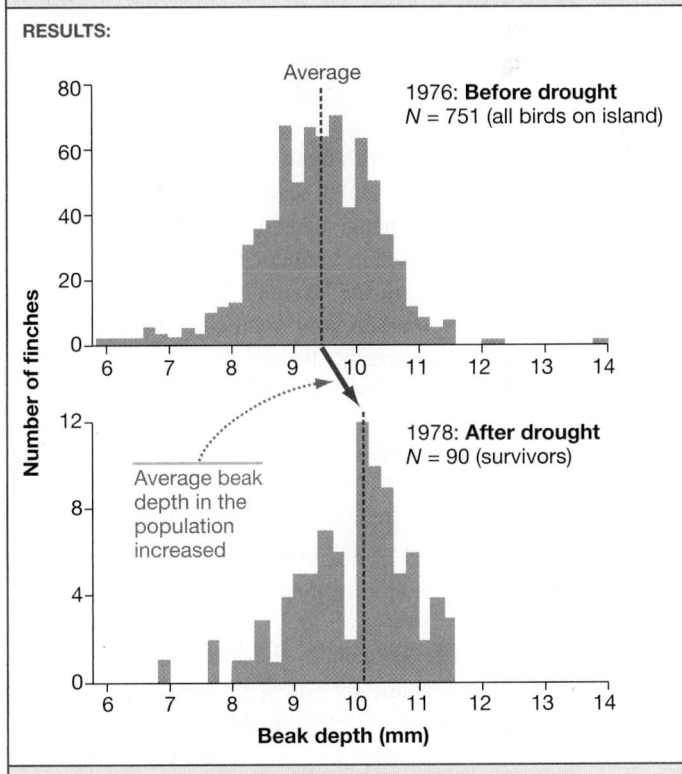

CONCLUSION: Natural selection occurred. The characteristics of the population have changed.

FIGURE 25.16 A Natural Experiment: Changes in a Medium Ground Finch Population in Response to a Change in the Environment (a Drought). The results show the distribution of beak depth in the population of medium ground finches on Daphne Major before and after the drought of 1977. *N* is the population size.

SOURCE: Boag, P. T., and P. R. Grant. 1981. Intense natural selection in a population of Darwin's finches (Geospizinae) in the Galápagos. *Science* 214: 82–85.

✓**EXERCISE** Fill in the predictions made by the two hypotheses.

Why were deeper beaks adaptive? At the peak of the drought, most seed sources were absent and the tough fruits of a plant called *Tribulus cistoides* served as the finches' primary food source. These fruits are so difficult to crack that finches ignore them in years when food supplies are normal. The Grants hypothesized that individuals with particularly large and deep beaks were more likely to crack these fruits efficiently enough to survive.

At this point, the Grants had shown that natural selection led to an increase in average beak depth in the population. When breeding resumed in 1978, the offspring produced had beaks that were half a millimeter deeper, on average, than those in the population that existed before the drought. This result confirmed that evolution had occurred.

In only one generation, natural selection led to a measurable change in the characteristics of the population. Alleles that led to the development of deep beaks had increased in frequency in the population. Large, deep beaks were an adaptation for cracking large fruits and seeds.

Continued Changes in the Environment, Continued Selection, Continued Evolution In 1983, the environment on the Galápagos Islands changed again. Over a seven-month period, a total of 1359 mm of rain fell. Plant growth was luxuriant, and finches fed primarily on the small, soft seeds that were being produced in abundance. During this interval, small individuals with small, pointed beaks had exceptionally high reproductive success—meaning that they had higher fitness than those with large, deep beaks—because they were better able to harvest the small seeds. As a result, the characteristics of the population changed again. Alleles associated with small, pointed beaks increased in frequency.

Over subsequent decades, the Grants have documented continued evolution in response to continued changes in the environment. **FIGURE 25.17** documents changes that have occurred in average body size, beak size, and beak shape over 35 years. From 1972 to 2006, average body size got smaller and average beak size initially increased and then declined. In addition, finch beaks got much pointier.

Long-term studies like this are proving to be powerful because they have succeeded in documenting natural selection in response to changes in the environment.

Which Genes Are under Selection? Characteristics like body size, beak size, and beak shape are polygenic, meaning that many genes—each one exerting a relatively small effect—influence the trait (see Chapter 14). Because many genes are involved, it can be difficult for researchers to know exactly which alleles are changing in frequency when polygenic traits evolve.

To explore which of the medium ground finch's genes might be under selection, researchers in Clifford Tabin's lab began studying beak development in an array of Galápagos finch species. More specifically, they looked for variation in the pattern of expression of cell–cell signals that had already been identified as important in the development of chicken beaks. The hope was that homologous genes might affect beak development in finches.

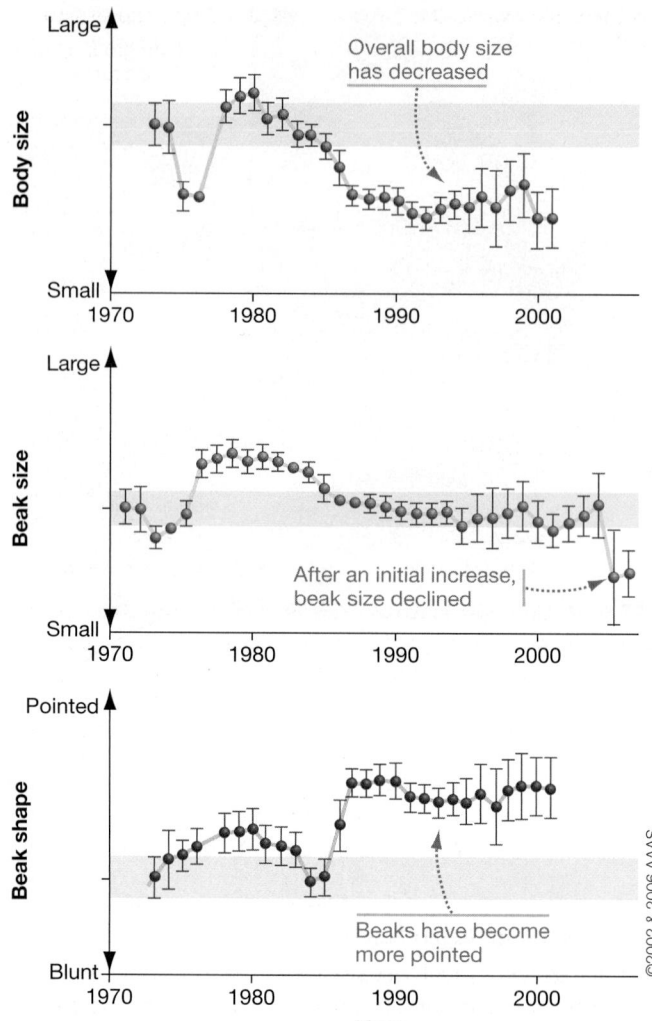

FIGURE 25.17 Body Size, Beak Size, and Beak Shape in Finches Changed over a 35-Year Interval.

DATA: Grant, P. R., and B. R. Grant. 2002. *Science* 296: 707–711; Grant, P. R., and B. R. Grant. 2006. *Science* 313: 224–226.

The researchers struck pay dirt when they carried out in situ hybridization (a technique featured in Chapter 22) showing where a cell–cell signal gene called *Bmp4* is expressed.

- There is a strong correlation between the amount of *Bmp4* expression when beaks are developing in young Galápagos finches and the width and depth of adult beaks (**FIGURE 25.18**).

- When the researchers experimentally increased *Bmp4* expression in young chickens, they found that beaks got wider and deeper than normal.

Similar experiments suggest that variation in alleles for a molecule called calmodulin, which is involved in calcium signaling during development, affects beak length. Based on these data, biologists suspect that alleles associated with *Bmp4* and calmodulin expression may be under selection in the population of medium ground finches that the Grants are studying. If so, the research community will have made a direct connection between natural selection on phenotypes and evolutionary change in genotypes.

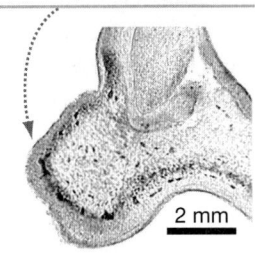

Lower *Bmp4* expression
(dark area) in embryo's beak

Higher *Bmp4* expression
(dark area) in embryo's beak

2 mm

Shallow adult beak

Geospiza fortis

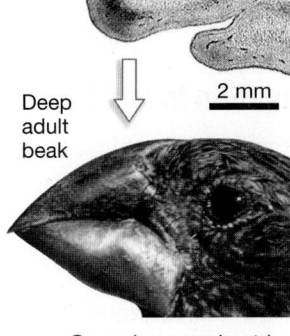

2 mm

Deep adult beak

Geospiza magnirostris

FIGURE 25.18 Changes in *Bmp4* Expression Change Beak Depth and Width. These micrographs are in situ hybridizations (see Chapter 22) showing the location and extent of *Bmp4* expression in young *Geospiza fortis* and *G. magnirostris*. In these and four other species that were investigated, the amount of Bmp4 protein produced correlates with the depth and width of the adult beak.

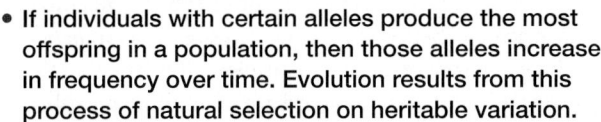

check your understanding

(C)(Y)(U)

If you understand that . . .

- If individuals with certain alleles produce the most offspring in a population, then those alleles increase in frequency over time. Evolution results from this process of natural selection on heritable variation.

✔ **You should be able to . . .**

1. List Darwin's four postulates in your own words and indicate which are related to heritable variation and which are related to differential reproductive success.

2. Explain how data on beak size and shape of Galápagos finch populations provide examples of heritable variation and differential reproductive success.

Answers are available in Appendix A.

25.5 Common Misconceptions about Natural Selection and Adaptation

Evolution by natural selection is a simple process—just the logical outcome of some straightforward postulates. Ironically, it can be extremely difficult to understand.

Research has verified that evolution by natural selection is often misunderstood. To help clarify how the process works, let's consider four of the most common types of misconceptions about natural selection, summarized in **TABLE 25.2** (on page 460).

Selection Acts on Individuals, but Evolutionary Change Occurs in Populations

Perhaps the most important point to clarify about natural selection is that during the process, individuals do not change—only the population does. During the drought, the beaks of individual finches did not become deeper. Rather, the average beak depth in the population increased over time because deep-beaked individuals produced more offspring than shallow-beaked individuals did. Natural selection acted on individuals, but the evolutionary change occurred in the characteristics of the population.

In the same way, individual *M. tuberculosis* cells did not change when rifampin was introduced to their environment. Each of these bacterial cells had the same RNA polymerase alleles throughout its life. But because the mutant allele increased in frequency in the population over time, the average characteristics of the bacterial population changed.

Natural Selection Is Not "Lamarckian" Inheritance There is a sharp contrast between evolution by natural selection and evolution by the inheritance of acquired characters—the hypothesis promoted by Jean-Baptiste de Lamarck. If you recall, Lamarck proposed that **(1)** individuals change in response to challenges posed by the environment (such as giraffes stretching their necks to reach leaves high in the treetops), and **(2)** the changed traits are then passed on to offspring. The key claim is that the important evolutionary changes occur in individuals.

In contrast, Darwin realized that individuals do not change when they are selected. Instead, they simply produce more or less offspring than other individuals do. When this happens, alleles found in the selected individuals become more or less frequent in the population.

Darwin was correct: There is no mechanism that makes it possible for natural selection to change the nature of an allele inside an individual. An individual's heritable characteristics don't change when natural selection occurs. Natural selection just sorts existing variants—it doesn't change them.

Acclimatization Is *Not* Adaptation The issue of change in individuals is tricky because individuals often *do* change in response to changes in the environment. For example, if you were to travel to the Tibetan Plateau in Asia, your body would experience oxygen deprivation due to the low partial pressure of oxygen at high elevations (see Figure 45.2). As a result, your body would produce more of the oxygen-carrying pigment hemoglobin and more hemoglobin-carrying red blood cells. Your body does not normally produce more red blood cells than it needs, because viscous (thick) blood can cause a disease—chronic mountain sickness—that can lead to heart failure.

The increase in red blood cells is an example of what biologists call **acclimatization**—a change in an individual's phenotype that occurs in response to a change in natural environmental conditions. (When this process occurs in study organisms in a laboratory, it is called **acclimation**.) Phenotypic changes due to acclimatization are not passed on to offspring, because no alleles have changed. As a result, acclimatization does not cause evolution.

Misconception	Example
"Evolutionary change occurs in organisms" CORRECTION: • Natural selection just sorts existing variants in organisms; it doesn't change them • Evolutionary change occurs only in populations • Acclimatization ≠ adaptation Selection does not cause neck length to increase in individual giraffes, only in populations	
"Adaptations occur because organisms want or need them" CORRECTION: • Mutation, the source of new alleles, occurs by chance • Evolution is not goal directed or progressive • There is no such thing as a higher or lower organism Tapeworms are not "lower" than their human hosts, just adapted to a different environment	
"Organisms sacrifice themselves for the good of the species" CORRECTION: • Individuals with alleles that cause self-sacrificing behavior die and do not produce offspring, so these alleles are eliminated from the population Lemmings do not jump off cliffs into the sea to save the species	
"Evolution perfects organisms" CORRECTION: • Some traits are nonadaptive • Some traits cannot be optimized due to fitness trade-offs • Some traits are limited by genetic or historical constraints Finch beaks cannot be both deep and narrow, due to genetic constraints	

In contrast, populations that have lived in Tibet for many generations are adapted to this environment through genetic changes. Among native Tibetans, for example, an allele that increases the ability of hemoglobin to hold oxygen has increased to high frequency. In populations that do not live at high elevations, this allele is rare or nonexistent. ✔ **If you understand this concept, you should be able to explain the difference between the biological definition of adaptation and its use in everyday English.**

Evolution Is Not Goal Directed

It is tempting to think that evolution by natural selection is goal directed. For example, you might hear a fellow student say that Tibetans "needed" the new hemoglobin allele so that they could survive at high altitudes, or that *M. tuberculosis* cells "wanted" or "needed" the mutant, drug-resistant allele so that they could survive and continue to reproduce in an environment that included rifampin. This purposeful change does not happen. The mutations that created the mutant alleles in both examples occurred randomly, due to errors in DNA synthesis, and they just happened to be advantageous when the environments changed.

Stated another way, mutations do not occur to solve problems. Mutations just happen. Every mutation is equally likely to occur in every environment. There is no mechanism that enables the environment to direct which mistakes DNA polymerase makes when copying genes. Adaptations do not occur because organisms want or need them.

Evolution Is Not Progressive   It is often appealing to think that evolution by natural selection is progressive—meaning organisms have gotten "better" over time. (In this context, *better* usually means bigger, stronger, or more complex.) It is true that the groups appearing later in the fossil record are often more morphologically complex than closely related groups that appeared earlier. Flowering plants are considered more complex than mosses, and most biologists would agree that the morphology of mammals is more complex than that of the first vertebrates in the fossil record. But there is nothing predetermined or absolute about this tendency.

In fact, complex traits are routinely lost or simplified over time as a result of evolution by natural selection. You've already analyzed evidence documenting limb loss in snakes (Chapter 22) and whales (this chapter). Populations that become parasitic are particularly prone to loss of complex traits. For example, the tapeworm parasites of humans and other mammals have lost their sophisticated digestive tracts and mouths as a result of natural selection—tapeworms simply absorb nutrients directly from their environment.

There Is No Such Thing as a Higher or Lower Organism   The nonprogressive nature of evolution by natural selection contrasts sharply with Lamarck's conception of the evolutionary process, in which organisms progress over time to higher and higher levels on a chain of being (see Figure 25.1).

Under Aristotle's and Lamarck's hypothesis, it is sensible to refer to "higher" and "lower" organisms. But under evolution by natural selection, there is no such thing as a higher or lower organism (**FIGURE 25.19**). Mosses may be a more ancient group than flowering plants, but neither group is higher or lower than the other. Mosses simply have a different suite of adaptations than do flowering plants, so they thrive in different types of environments. A human is no higher than its tapeworm parasite; each is well adapted to its environment.

Organisms Do Not Act for the Good of the Species

Consider the widely circulated story that rodents called lemmings sacrifice themselves for the good of their species. The story claims that when lemming populations are high, overgrazing is so extensive that the entire species is threatened with starvation and extinction. In response, some individuals throw themselves into the sea and drown. This lowers the overall population size and allows the vegetation to recover enough to save the species. Even though individuals suffer, the good-of-the-species hypothesis maintains that the behavior evolved because the group benefits.

The lemming suicide story is false. Although lemmings do disperse from areas of high population density to find habitats with higher food availability, they do not throw themselves into the sea.

To understand why this type of self-sacrificing behavior does not occur, suppose that certain alleles predispose lemmings to sacrifice themselves for others. But consider what happens if alleles exist that prevent this type of behavior—what biologists call

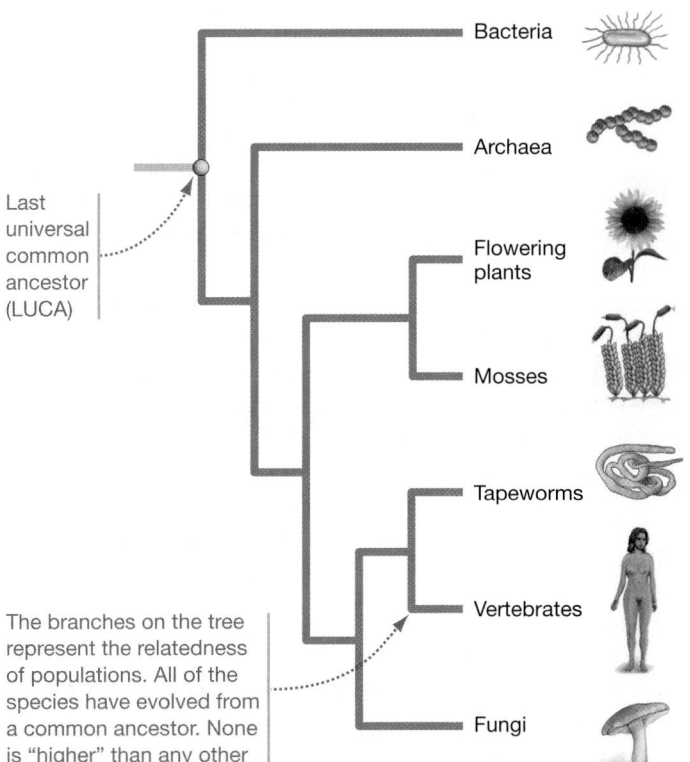

FIGURE 25.19 Evolution Produces a Tree of Life, Not a Progressive Ladder of Life. Under evolution by natural selection, species are related by common ancestry and all have evolved through time. (Not all branches of the tree of life are shown.)

a "selfish" allele. Individuals with self-sacrificing alleles die and do not produce offspring. But individuals with selfish, cheater alleles survive and produce offspring. As a result, selfish alleles increase in frequency while self-sacrificing alleles decrease in frequency. Thus, it is not possible for self-sacrificing alleles to evolve by natural selection.

There Are Constraints on Natural Selection

Although organisms are often exquisitely adapted to their environment, adaptation is far from perfect. A long list of circumstances limits the effectiveness of natural selection; only a few of the most important are discussed here.

Nonadaptive Traits Vestigial traits such as the human coccyx (tailbone) and goose bumps do not increase the fitness of individuals with those traits. The structures are not adaptive. They exist because they were present in the ancestral population.

Vestigial traits are not the only types of structures with no or reduced function. Some adult traits exist as holdovers from structures that appear early in development. For example, human males have nipples despite the absence of mammary glands. Nipples exist in men because they form in the human embryo before sex hormones begin directing the development of male organs instead of female organs.

Perhaps the best example of nonadaptive traits involves evolutionary changes in DNA sequences. A mutation may change a base in the third position of a codon without changing the amino acid sequence of the protein encoded by that gene. Changes such as these are said to be silent. They occur due to the redundancy of the genetic code (see Chapter 16). Silent changes in DNA sequences are extremely common. But because they don't change the phenotype, they can't be acted on by natural selection and are not adaptive.

Genetic Constraints The Grants' team analyzed data on the characteristics of finches that survived the 1977 drought in the Galápagos, and the team made an interesting observation: Although individuals with deep beaks survived better than individuals with shallow beaks, birds with particularly narrow beaks survived better than individuals with wider beaks.

This observation made sense because finches crack *Tribulus* fruits by twisting them. Narrow beaks concentrate the twisting force more efficiently than wider beaks, so they are especially useful for cracking the fruits. But narrower beaks did not evolve in the population.

To explain why, the biologists noted that parents with deep beaks tend to have offspring with beaks that are both deep and wide. This is a common pattern. Many alleles that affect body size have an effect on all aspects of size—not just one structure or dimension. As a result, selection for increased beak depth overrode selection for narrow beaks, even though a deep and narrow beak would have been more advantageous.

The general point here is that selection was not able to optimize all aspects of a trait due to **genetic correlation.** Genetic correlations occur because of pleiotropy, in which a single allele affects multiple traits (see Chapter 14). In this case, selection on alleles for one trait (increased beak depth) caused a correlated, though suboptimal, increase in another trait (beak width).

Genetic correlations are not the only genetic constraint on adaptation. Lack of genetic variation is also important. Consider that salamanders have the ability to regrow severed limbs. Some eels and sharks can sense electric fields. Birds can see ultraviolet light. Even though these traits would possibly confer increased reproductive success in humans, they do not exist—because humans lack the requisite genes.

Fitness Trade-offs In everyday English, the term trade-off refers to a compromise between competing goals. It is difficult to design a car that is both large and fuel efficient, a bicycle that is both rugged and light, or a plane that is both fast and maneuverable.

In nature, selection occurs in the context of fitness trade-offs. A **fitness trade-off** is a compromise between traits, in terms of how those traits perform in the environment. During the drought in the Galápagos, for example, medium ground finches with large bodies had an advantage because they were able to chase off smaller birds from the few remaining sources of seeds. But individuals with large bodies require large amounts of food to maintain their mass; they also tend to be slower and less nimble than smaller individuals. When food is scarce, large individuals are more prone to starvation. Even if large size is advantageous in an environment, there is always counteracting selection that prevents individuals from getting even bigger.

Biologists have documented trade-offs between the size of eggs or seeds that an individual makes and the number of offspring it can produce, between rapid growth and long life span, and between bright coloration and tendency to attract predators.

The message of this research is simple: Because selection acts on many traits at once, every adaptation is a compromise.

Historical Constraints In addition to being constrained by genetic correlations, lack of genetic variation, and fitness trade-offs, adaptations are constrained by history. The reason is simple: All traits have evolved from previously existing traits.

Natural selection acts on structures that originally had a very different function. For example, the tiny incus, malleus, and stapes bones found in your middle ear evolved from bones that were part of the jaw and jaw support in the ancestors of mammals. These bones now function in transmitting and amplifying sound from your outer ear to your inner ear. Biologists routinely interpret these bones as adaptations that improve your ability to hear airborne sounds. But are the bones a "perfect" solution to the problem of transmitting sound from the outside of the ear to the inside? The answer is no. They are the best solution possible, given an important historical constraint. Other vertebrates have different structures involved in transmitting sound to the ear. In at least some cases, those structures may be more efficient than our incus, malleus, and stapes.

To summarize, not all traits are adaptive, and even adaptive traits are constrained by genetic and historical factors. In addition, natural selection is not the only process that causes evolutionary change. Three other processes—genetic drift, gene flow, and mutation—change allele frequencies over time (see Chapter 26). Compared with natural selection, these processes have very different consequences. You can see the Big Picture of how natural selection relates to other evolutionary processes on pages 526–527.

CHAPTER 25 REVIEW

 For media, go to MasteringBiology

If you understand . . .

25.1 The Evolution of Evolutionary Thought

- Plato, Aristotle, and the Bible's book of Genesis consider species as unchanging types. This view is called typological thinking.

- Lamarck proposed a theory of evolution—that species are not static but change through time. He proposed that evolution occurs by the inheritance of acquired characteristics.

- Darwin and Wallace proposed that evolution occurs by natural selection. This was the beginning of population thinking, whereby variation among individuals is the key to understanding evolution.

✔ You should be able to compare and contrast typological thinking and population thinking.

25.2 The Pattern of Evolution: Have Species Changed, and Are They Related?

- Data on (1) the age of the Earth and the fact of extinction; (2) the resemblance of modern to fossil forms; (3) transitional features in fossils; (4) the presence of vestigial traits; and (5) change in contemporary populations show that species change through time.

- Data on (1) the geographic proximity of closely related species; (2) the existence of structural, developmental, and genetic homologies; and (3) the contemporary formation of new species support the consensus that species are related by common ancestry.

- Evidence for evolution is internally consistent, meaning that data from several independent sources are mutually reinforcing.

✔ You should be able to cite examples in support of the statement that species have changed through time and are related by common ancestry.

25.4 The Process of Evolution: How Does Natural Selection Work?

- Darwin developed four postulates that outline the process of evolution by natural selection. These postulates can be summarized by the following statement: Heritable variation leads to differential reproductive success.

- Alleles or traits that increase the reproductive success of an individual are said to increase the individual's fitness. A trait that leads to higher fitness, relative to individuals without the trait, is an adaptation. If a particular allele increases fitness and leads to adaptation, the allele will increase in frequency in the population.

✔ You should be able to explain the difference between the biological and everyday English definitions of fitness.

25.4 Evolution in Action: Recent Research on Natural Selection

- Selection by drugs on the TB bacterium and changes in the size and shape of finch beaks in the Galápagos as a result of seed availability are well-studied examples of evolution by natural selection.

- Both examples demonstrate that evolution can be observed and measured. Evolution by natural selection has been confirmed by a wide variety of studies and has long been considered to be the central organizing principle of biology.

✔ You should be able to predict how changes in *Mycobacterium tuberculosis* populations would be explained under special creation and under evolution by inheritance of acquired characters.

25.5 Common Misconceptions about Natural Selection and Adaptation

- Natural selection acts on individuals, but evolutionary change occurs in populations. Nonheritable changes that occur in individuals are not adaptations and do not result in evolution.

- Evolution is not goal directed and does not lead to perfection. There is no such thing as a higher or lower organism.

- Organisms do not act for the good of the species.

- Not all traits are adaptive, and even adaptive traits are limited by genetic and historical constraints.

✔ You should be able to discuss how adaptations such as the large brains of *Homo sapiens* and the ability of falcons to fly very fast are constrained.

(MB) MasteringBiology

1. **MasteringBiology Assignments**

   Tutorials and Activities Darwin and the Galápagos Islands; Experimental Inquiry: Did Natural Selection of Ground Finches Occur When the Environment Changed?; Natural Selection for Antibiotic Resistance; The Voyage of the *Beagle*: Darwin's Trip Around the World; Reconstructing Forelimbs

   Questions Reading Quizzes, Blue-Thread Questions, Test Bank

2. **eText** Read your book online, search, take notes, highlight text, and more.

3. **The Study Area** Practice Test, Cumulative Test, BioFlix® 3-D Animations, Videos, Activities, Audio Glossary, Word Study Tools, Art

You should be able to . . .

✔ TEST YOUR KNOWLEDGE

Answers are available in Appendix A

1. How can biological fitness be estimated?
   a. Document how long different individuals in a population survive.
   b. Count the number of healthy, fertile offspring produced by different individuals in a population.
   c. Determine which individuals are strongest.
   d. Determine which phenotype is the most common one in a given population.

2. True or false? Some traits are considered vestigial because they existed long ago. *false*

3. What is an adaptation?
   a. a trait that improves the fitness of its bearer, compared with individuals without the trait
   b. a trait that changes in response to environmental influences within the individual's lifetime
   c. an ancestral trait—one that was modified to form the trait observed today
   d. the ability to produce offspring

4. Why does the presence of extinct forms and transitional features in the fossil record support the pattern component of the theory of evolution by natural selection?
   a. It supports the hypothesis that individuals change over time.
   b. It supports the hypothesis that weaker species are eliminated by natural selection.
   c. It supports the hypothesis that species evolve to become more complex and better adapted over time.
   d. It supports the hypothesis that species change over time.

5. Traits that are derived from a common ancestor, like the bones of human arms and bird wings, are said to be _homologous_

6. According to data presented in this chapter, which of the following statements is correct?
   a. When individuals change in response to challenges from the environment, their altered traits are passed on to offspring.
   b. Species are created independently of each other and do not change over time.
   c. Populations—not individuals—change when natural selection occurs.
   d. The traits of populations become more perfect over time.

7. Some biologists summarize evolution by natural selection with the phrase "mutation proposes, selection disposes." Mutation is a process that creates heritable variation. Explain what the phrase means.

8. Explain how artificial selection differs from natural selection.

9. Why don't the biggest and strongest individuals in a population always produce the most offspring?
   a. The biggest and strongest individuals always have higher fitness.
   b. In some environments, being big and strong lowers fitness.
   c. Sometimes the biggest and strongest individuals may choose to have fewer offspring.
   d. Sometimes the number of offspring is not related to fitness.

10. **QUANTITATIVE** The graphs in Figure 25.16 show that the average beak depth of medium ground finches increased after the drought.

However, more finches had deep beaks before the drought than after. Explain this seeming contradiction by calculating the percent of finches that survived the drought.

11. Review the section on the evolution of drug resistance in *Mycobacterium tuberculosis*.
    - What evidence do researchers have that a drug-resistant strain evolved in the patient analyzed in their study, and wasn't instead transmitted from another infected individual?
    - If the antibiotic rifampin were banned, would the mutant *rpoB* gene have lower or higher fitness in the new environment? Would strains carrying the mutation continue to increase in frequency in *M. tuberculosis* populations?

12. Describe how Darwin's four postulates would apply to a population of rabbits sharing a meadow with foxes.

13. Scientists have observed white deer mice living on coastal beaches in Florida and brown deer mice living in nearby forests. Compare and contrast how the theory of evolution by natural selection, special creation, and evolution by inheritance of acquired characters might explain this observation.

14. The average height of humans in industrialized nations has increased steadily for the past 100 years. This trait has clearly changed over time. Most physicians and human geneticists hypothesize that the change is due to better nutrition and a reduced incidence of disease. Has human height evolved?
    a. Yes, because average height has changed over time.
    b. No, because changes in height due to nutrition and reduced incidence of disease are not heritable.
    c. Yes, because height is a heritable trait.
    d. No, because height is not a heritable trait.

15. Scientists hypothesize that humans and chimpanzees diverged from a common ancestor that lived in Africa about 6–7 million years ago. What evidence would support this hypothesis?

16. The geneticist James Crow wrote that successful scientific theories have the following characteristics: (1) They explain otherwise puzzling observations; (2) they provide connections between otherwise disparate observations; (3) they make predictions that can be tested; and (4) they are heuristic, meaning that they open up new avenues of theory and experimentation. Crow added two other elements of scientific theories that he considered important on a personal, emotional level: (5) They should be elegant, in the sense of being simple and powerful; and (6) they should have an element of surprise. How well does the theory of evolution by natural selection fulfill these six criteria?

26 Evolutionary Processes

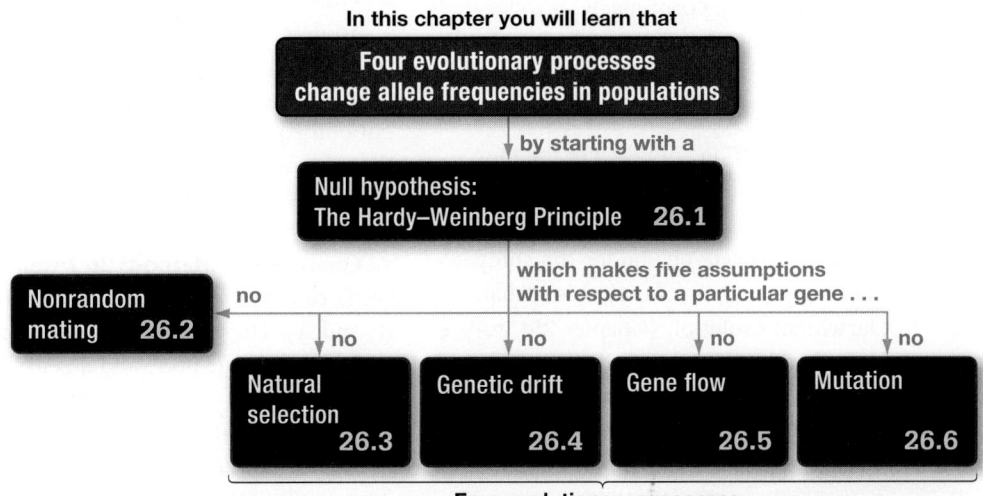

In this chapter you will learn that

Four evolutionary processes change allele frequencies in populations

by starting with a

Null hypothesis: The Hardy–Weinberg Principle   26.1

which makes five assumptions with respect to a particular gene . . .

Nonrandom mating   26.2   no

no   no   no   no

| **Natural selection** 26.3 | **Genetic drift** 26.4 | **Gene flow** 26.5 | **Mutation** 26.6 |

Four evolutionary processes

This albino sea turtle carries rare alleles that prevent pigment formation. The frequency of these alleles changes in the sea turtle population over time due to the evolutionary processes discussed in this chapter.

Since evolution is one of the five key attributes of life, understanding evolutionary processes is essential to understanding living things. How did the diversity of organisms around us (including us) come to be?

It turns out that natural selection (see Chapter 25) is only one of four processes that can shift allele frequencies in populations over time, causing evolution:

This chapter is part of the Big Picture. See how on pages 526–527.

1. *Natural selection* increases the frequency of certain alleles—the ones that contribute to reproductive success in a particular environment. Natural selection is the only one of the four processes that leads to adaptation.

2. *Genetic drift* causes allele frequencies to change randomly. In some cases, drift may cause alleles that decrease fitness to increase in frequency.

✔ When you see this checkmark, stop and test yourself. Answers are available in Appendix A.

3. **Gene flow** occurs when individuals leave one population, join another, and breed. Allele frequencies may change when gene flow occurs, because arriving individuals introduce alleles to their new population and departing individuals remove alleles from their old population.

4. **Mutation** modifies allele frequencies by continually introducing new alleles. The alleles created by mutation may be beneficial or deleterious (detrimental) or have no effect on fitness.

This chapter has two fundamental messages: Natural selection is not the only agent responsible for evolution, and each of the four evolutionary processes has different consequences for genetic variation and fitness.

The first few decades of the 1900s were pivotal for the study of evolutionary processes. Biologists began to apply Mendelian genetics (Chapter 14) to Darwinian evolution (Chapter 25), resulting in an era known as the Modern Synthesis. Evolutionary biologists, mathematicians, and geneticists collaborated to make huge leaps in *quantifying* evolution. One product of this era was a model called the **Hardy–Weinberg principle**, which serves as a mathematical null hypothesis for the study of evolutionary processes. Let's examine the logic of this model and then consider each of the four evolutionary processes in turn.

26.1 Analyzing Change in Allele Frequencies: The Hardy–Weinberg Principle

In 1908, a British mathematician, G. H. Hardy, and a German doctor, Wilhelm Weinberg, each published a major result independently. At the time, it was commonly believed that changes in allele frequency occur simply as a result of sexual reproduction—meiosis followed by the random fusion of gametes (eggs and sperm) to form offspring. Some biologists claimed that dominant alleles inevitably increase in frequency when gametes combine at random. Others predicted that two alleles of the same gene inevitably reach a frequency of 0.5.

To test these hypotheses, Hardy and Weinberg analyzed the frequencies of alleles when many individuals in a population mate and produce offspring. Instead of thinking about the consequences of a mating between two parents with a specific pair of genotypes (as you did with Punnett squares in Chapter 14), Hardy and Weinberg wanted to know what happened in an entire population when *all* of the individuals—and thus all possible genotypes—bred.

Like Darwin, Hardy and Weinberg were engaged in population thinking. A **population** is a group of individuals from the same species that live in the same area at the same time and can interbreed—and that vary in the traits they possess (Chapter 25). w/ trait variation.

The Gene Pool Concept

To analyze the consequences of matings among all of the individuals in a population, Hardy and Weinberg invented a novel approach: They imagined that all of the alleles from all the gametes produced in each generation go into a single group called the **gene pool** and then combine at random to form offspring. Something similar to this happens in species like clams and sea stars, which release their gametes into the water where the gametes mix randomly before combining to form zygotes.

To determine which genotypes would be present in the next generation and in what frequencies, Hardy and Weinberg calculated what happens when two gametes are plucked at random out of the gene pool, many times, and each of these gamete pairs then combines to form offspring. These calculations predict the genotypes of the offspring that would be produced, as well as the frequency of each genotype.

Quantitative Methods 26.1 walks through Hardy's and Weinberg's calculations by focusing on just one gene with two alleles, A_1 and A_2. The letter p is used to symbolize the frequency of A_1 alleles in the gene pool, and q is used to symbolize the frequency of A_2 alleles in the same gene pool. **FIGURE 26.1** illustrates the same calculations a little differently. The figure uses a Punnett square to predict the outcome of random mating—meaning, random combinations of all gametes in a population. (Recall that in Chapter 14 you used Punnett squares to predict the outcome of a mating between two individuals.) The outcome of this analysis is the same as in Quantitative Methods 26.1.

The resulting Hardy–Weinberg principle makes two fundamental claims:

1. If the frequencies of alleles A_1 and A_2 in a population are given by p and q, then the frequencies of genotypes A_1A_1,

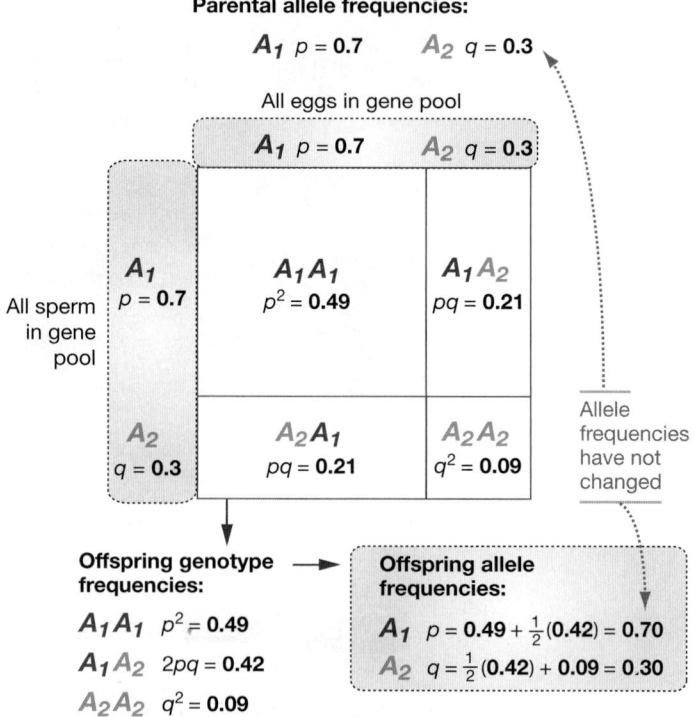

Parental allele frequencies:

A_1 $p = 0.7$ A_2 $q = 0.3$

All eggs in gene pool

A_1 $p = 0.7$ A_2 $q = 0.3$

All sperm in gene pool

	A_1 $p = 0.7$	A_2 $q = 0.3$
A_1 $p = 0.7$	A_1A_1 $p^2 = 0.49$	A_1A_2 $pq = 0.21$
A_2 $q = 0.3$	A_2A_1 $pq = 0.21$	A_2A_2 $q^2 = 0.09$

Allele frequencies have not changed

Offspring genotype frequencies:

A_1A_1 $p^2 = 0.49$
A_1A_2 $2pq = 0.42$
A_2A_2 $q^2 = 0.09$

Offspring allele frequencies:

A_1 $p = 0.49 + \frac{1}{2}(0.42) = 0.70$
A_2 $q = \frac{1}{2}(0.42) + 0.09 = 0.30$

FIGURE 26.1 A Punnett Square Illustrates the Hardy–Weinberg Principle.

Hardy and Weinberg began by analyzing the simplest situation possible—where just two alleles of a particular gene exist in a population. Let's call these alleles A_1 and A_2. The letter p is used to symbolize the frequency of A_1 alleles in the gene pool and q is used to symbolize the frequency of A_2 alleles in the same gene pool. Because there are only two alleles, the two frequencies must add up to 1:

$$p + q = 1$$

Now follow the steps in **FIGURE 26.2**:

Parental Allele Frequencies

Although p and q can have any value between 0 and 1, let's suppose that the initial frequency of A_1 is 0.7 and that of A_2 is 0.3. In this gene pool, 70 percent of the gametes carry A_1 and 30 percent carry A_2.

Offspring Genotype Frequencies

Each time a gamete is involved in forming an offspring, there is a 70 percent chance that it carries A_1 and a 30 percent chance that it carries A_2. In general, there is a p chance that it carries A_1 and a q chance that it carries A_2.

Because only two alleles are present, three genotypes are possible in the offspring generation: A_1A_1, A_1A_2, and A_2A_2. What will the frequency of these three genotypes be? According to the logic of Hardy's and Weinberg's result:

- Frequency of the A_1A_1 genotype: p^2
- Frequency of the A_1A_2 genotype: $2pq$
- Frequency of the A_2A_2 genotype: q^2

The genotype frequencies in the offspring generation must add up to 1, which means that:

$$p^2 + 2pq + q^2 = 1$$

In this numerical example:

$$0.49 + 0.42 + 0.09 = 1$$

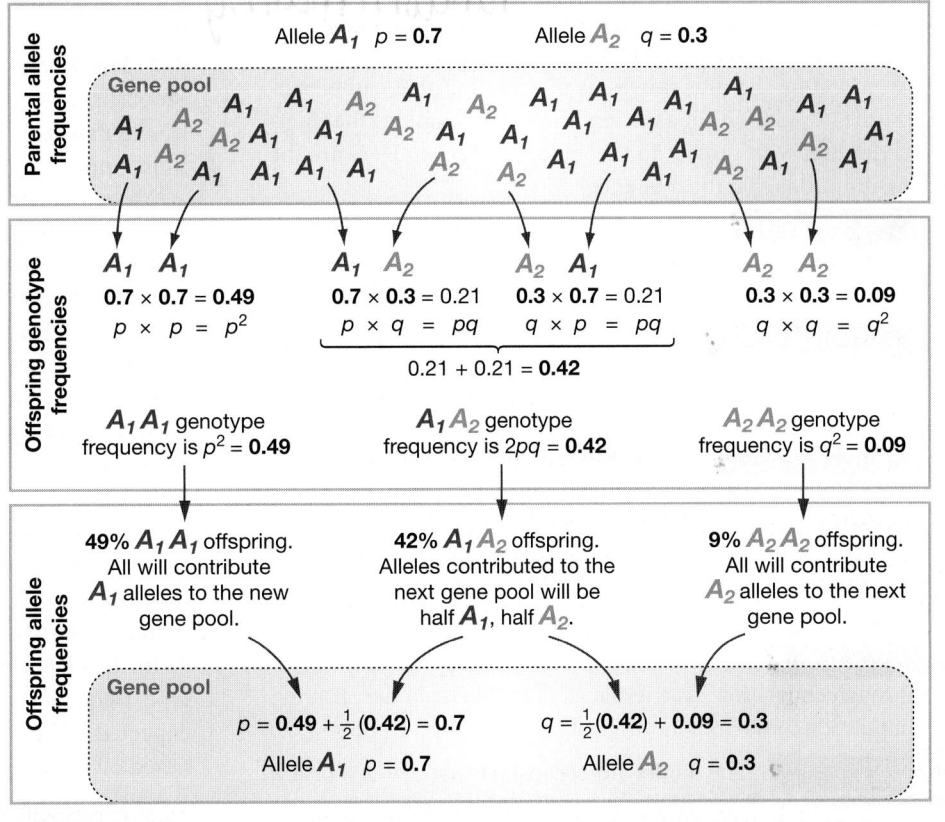

FIGURE 26.2 Deriving the Hardy–Weinberg Principle. To understand the logic behind calculating the frequency of A_1A_2 genotypes, see **BioSkills 5** in Appendix B.

Offspring Allele Frequencies

The easiest way to calculate the allele frequencies in the offspring is to imagine that they form gametes that go into a gene pool. All of the gametes from A_1A_1 individuals carry A_1, so 49 percent (p^2) of the gametes in the gene pool will carry A_1. But half of the gametes from A_1A_2 will also carry A_1, so an additional ½(0.42) = 0.21 (this is ½ × $2pq$ = pq) gametes in the gene pool will carry A_1, for a total of 0.49 + 0.21 = 0.70 or $p^2 + pq = p(p + q) = p$. Use the same logic to figure out the frequency of A_2.

In this example, the frequency of allele A_1 in the offspring generation is still 0.7 and the frequency of allele A_2 is still 0.3. Thus, the frequency of allele A_1 is still p and the frequency of allele A_2 is still q.

No change in allele frequency occurred. Even if A_1 is dominant to A_2, it does not increase in frequency. And there is no trend toward both alleles reaching a frequency of 0.5. This result is called the Hardy–Weinberg principle.

✔ **QUANTITATIVE** If you understand these calculations, you should be able to: (1) calculate the frequencies of the three offspring genotypes if $p = 0.6$ and $q = 0.4$ in the parental gene pool; (2) use your answer to calculate the allele frequencies in the offspring gene pool; (3) determine whether evolution occurred.

A_1A_2, and A_2A_2 will be given by p^2, $2pq$, and q^2, respectively, for generation after generation. That is:

Allele frequencies: $p + q = 1$

Genotype frequencies: $p^2 + 2pq + q^2 = 1$

2. When alleles are transmitted via meiosis and random combinations of gametes, their frequencies do not change over time. For evolution to occur, some other factor or factors must come into play.

What are these other factors?

The Hardy–Weinberg Model Makes Important Assumptions

random mating

The Hardy–Weinberg model is based on important assumptions about how populations and alleles behave. These assumptions helped to define the four processes of evolution that can be acting on the population. In addition, the model assumes that mating is random with respect to the gene in question. Here are the five assumptions that must be met:

1. **Random mating** This condition was enforced by picking gametes from the gene pool at random. Individuals were not allowed to *choose* a mate.

2. **No natural selection** The model assumed that *all* members of the parental generation survived and contributed equal numbers of gametes to the gene pool, no matter what their genotype.

3. **No genetic drift (*random allele frequency changes*)** The model assumed that alleles were picked in their exact frequencies p and q, and not at some different values caused by chance—that is, the model behaved as though the population was infinitely large. For example, allele A_1 did not "get lucky" and get drawn more than 70 percent of the time.

4. **No gene flow** No new alleles were added by immigration or lost through emigration. As a result, all of the alleles in the offspring population came from the original population's gene pool.

5. **No mutation** The model didn't consider that new A_1s or A_2s or other, new alleles might be introduced into the gene pool.

The Hardy–Weinberg principle tells us what to expect if selection, genetic drift, gene flow, and mutation are not affecting a gene, *and* if mating is random with respect to that gene. Under these conditions, the genotypes A_1A_1, A_1A_2, and A_2A_2 should be in the Hardy–Weinberg proportions p^2, $2pq$, and q^2, respectively, and no evolution will occur.

How Does the Hardy–Weinberg Principle Serve as a Null Hypothesis?

Recall that a null hypothesis predicts there are no differences among the treatment groups in an experiment (see Chapter 1). Biologists often want to test whether nonrandom mating is occurring, natural selection is acting on a particular gene, or one of the other evolutionary processes is at work. In addressing questions like these, the Hardy–Weinberg principle functions as a null hypothesis.

Given a set of allele frequencies, the Hardy–Weinberg principle predicts what genotype frequencies will occur when mating is random with respect to that gene, and when natural selection, genetic drift, gene flow, and mutation are not affecting the gene. If biologists observe genotype frequencies that do not conform to the Hardy–Weinberg prediction, it means that something interesting is going on: Either nonrandom mating is occurring (which changes genotype frequencies but not allele frequencies), or allele frequencies are changing for some reason. Further research is then needed to determine which of the five Hardy–Weinberg conditions is being violated.

Let's consider two examples to illustrate how the Hardy–Weinberg principle is used as a null hypothesis: MN blood types and *HLA* genes, both in humans.

Case Study 1: Are MN Blood Types in Humans in Hardy–Weinberg Proportions?

One of the first genes that geneticists could analyze in natural populations was the MN blood group of humans. Most human populations have two alleles, designated *M* and *N*, at this gene.

Because the *MN* gene codes for a protein found on the surface of red blood cells—the *M* allele codes for the M version, and the *N* allele codes for the N version—researchers could determine whether individuals are *MM*, *MN*, or *NN* by treating blood samples with antibodies to each protein (this technique was first introduced in Chapter 11). The *M* and *N* alleles are codominant—meaning that heterozygotes have both M and N versions of the protein on their red blood cells.

To estimate the frequency of each genotype in a population, geneticists obtain data from a large number of individuals and then divide the number of individuals with each genotype by the total number of individuals in the sample.

TABLE 26.1 shows *MN* genotype frequencies for populations from throughout the world and illustrates how observed genotype frequencies are compared with the genotype frequencies expected if the Hardy–Weinberg principle holds. The analysis is based on the following steps:

Step 1 Estimate genotype frequencies by observation—in this case, by testing many blood samples for the *M* and *N* alleles. These frequencies are given in the rows labeled "observed" in Table 26.1.

Step 2 Calculate observed allele frequencies from the observed genotype frequencies. In this case, the frequency of the *M* allele is the frequency of *MM* homozygotes plus half the frequency of *MN* heterozygotes; the frequency of the *N* allele is the frequency of *NN* homozygotes plus half the frequency of *MN* heterozygotes. (You can review the logic behind this calculation in Quantitative Methods 26.1.)

Step 3 Use the observed allele frequencies to calculate the genotypes expected according to the Hardy–Weinberg principle. Under the null hypothesis of random mating and no evolution, the expected genotype frequencies are $p^2 : 2pq : q^2$.

Step 4 Compare the observed and expected values. Researchers use statistical tests to determine whether the differences between the observed and expected genotype frequencies are small enough to be due to chance or large enough to reject the null hypothesis of no evolution and random mating.

Although using statistical testing is beyond the scope of this text (see **BioSkills 4** in Appendix B for a brief introduction to the topic), you should be able to inspect the numbers and comment on them. In these populations, for example, the observed and expected *MN* genotype frequencies are almost identical. (A statistical test shows that the small differences observed are probably due to chance.) For every population surveyed, genotypes

TABLE 26.1 The MN Blood Group of Humans: Observed and Expected Genotype Frequencies

The expected genotype frequencies are calculated from the observed allele frequencies, using the Hardy–Weinberg principle.

Population and Location	Data Type	Genotype Frequencies			Allele Frequencies	
		MM	MN	NN	M	N
Inuit (Greenland)	Observed	0.835	0.156	0.009	0.913	0.087
	Expected	0.834	0.159	0.008		
Native Americans (U.S.)	Observed	0.600	0.351	0.049	0.776	0.224
	Expected	0.602	0.348	0.050		
Caucasians (U.S.)	Observed	0.292	0.494	0.213	0.540	0.460
	Expected	0.290	0.497	0.212		
Aborigines (Australia)	Observed	0.025	0.304	0.672	0.176	0.824
	Expected	0.031	0.290	0.679		
Ainu (Japan)	Observed — Step ❶ →	0.179	0.502	0.319	Step ❷ →	
	Expected ← Step ❹ ↕				← Step ❸	

DATA: W. C. Boyd. 1950. Boston: Little, Brown and Company.

✔ **EXERCISE** Fill in the values for observed allele frequencies and expected genotype frequencies for the Ainu people of Japan.

at the *MN* gene are in Hardy–Weinberg proportions. As a result, biologists conclude that the assumptions of the Hardy–Weinberg model are valid for this locus.

The results imply that when these data were collected, the *M* and *N* alleles in these populations were not being affected by the four evolutionary processes and that mating was random with respect to this gene—meaning that humans were not choosing mates based on their *MN* genotype.

Before moving on, however, it is important to note that a study such as this does not mean that the *MN* gene has never been subject to nonrandom mating, or has never been under selection or genetic drift. Even if selection has been very strong for many generations, one generation of random mating and no evolution will result in genotype frequencies that conform to Hardy–Weinberg expectations.

Case Study 2: Are *HLA* Genes in Humans in Hardy–Weinberg Equilibrium?
Geneticist Therese Markow and colleagues collected data on the genotypes of 122 individuals from the Havasupai tribe native to Arizona. These biologists were studying two genes that are important in the functioning of the human immune system. More specifically, the genes that they analyzed code for proteins that help immune system cells recognize and destroy invading bacteria and viruses.

Previous work had shown that different alleles exist at both the *HLA-A* and *HLA-B* genes, and that the alleles at each gene code for proteins that recognize proteins from slightly different disease-causing organisms. Like the *M* and *N* alleles, *HLA* alleles are codominant.

As a result, the research group hypothesized that individuals who are heterozygous at one or both of these genes may have a strong fitness advantage. The logic is that heterozygous people possess a wider variety of HLA proteins, so their immune

systems can recognize and destroy more types of bacteria and viruses. They should be healthier and have more offspring than homozygous people do.

To test this hypothesis, Markow and her colleagues used their data on observed genotype frequencies to determine the frequency of each allele present. When they used these allele frequencies to calculate the expected number of each genotype according to the Hardy–Weinberg principle, they found the observed and expected values reported in **TABLE 26.2**.

When you inspect these data, notice that there are many more heterozygotes and many fewer homozygotes than expected under Hardy–Weinberg conditions. Statistical tests show it is extremely unlikely that the difference between the observed and expected numbers could occur purely by chance.

These results supported the team's prediction and indicated that one of the assumptions behind the Hardy–Weinberg

TABLE 26.2 *HLA* Genes of Humans: Observed and Expected Genotypes

The expected numbers of homozygous and heterozygous genotypes are calculated from observed allele frequencies, according to the Hardy–Weinberg principle.

Gene	Data Type	Genotype Counts ($n = 122$)	
		Homozygotes	Heterozygotes
HLA-A	Observed	38	84
	Expected	48	74
HLA-B	Observed	21	101
	Expected	30	92

DATA: Markow, T., P. H. Hedrick, K. Zuerlein, et al. 1993. *Journal of Human Genetics* 53: 943–952, Table 3.

principle was being violated. But which assumption? The researchers argued that mutation, gene flow, and drift are negligible in this case and offered two explanations for their data:

1. ***Mating may not be random with respect to the HLA genotype.*** Specifically, people may subconsciously prefer mates with *HLA* genotypes unlike their own and thus produce an excess of heterozygous offspring. This hypothesis is plausible. For example, experiments have shown that college students can distinguish each others' genotypes at genes related to *HLA* on the basis of body odor. Individuals in this study were more attracted to the smell of genotypes unlike their own. If this is true among the Havasupai, then nonrandom mating would lead to an excess of heterozygotes compared with the proportion expected under the Hardy–Weinberg principle.

2. ***Heterozygous individuals may have higher fitness.*** This hypothesis is supported by data collected by a different research team, who studied the Hutterite people living in South Dakota. In the Hutterite population, married women who have the same *HLA*-related alleles as their husbands have more trouble getting pregnant and experience higher rates of spontaneous abortion than do women with *HLA*-related alleles different from those of their husbands. The data suggest that homozygous fetuses have lower fitness than do fetuses heterozygous at these genes. If this were true among the Havasupai, selection would lead to an excess of heterozygotes relative to Hardy–Weinberg expectations.

Which explanation is correct? It is possible that both are. Using the Hardy–Weinberg principle as a null hypothesis allowed biologists to detect an interesting pattern in a natural population. Research continues on the question of why the pattern exists.

Now let's consider each of the processes that can violate the Hardy–Weinberg assumptions—and thus influence evolution.

check your understanding

If you understand that . . .

- The Hardy–Weinberg principle functions as a null hypothesis when researchers test whether nonrandom mating or evolution is occurring at a particular gene.

✔ **QUANTITATIVE You should be able to . . .**

Analyze whether a gene suspected of causing hypertension in humans is in Hardy–Weinberg proportions. In one study, the observed genotype frequencies were A_1A_1 0.574; A_1A_2 0.339; A_2A_2 0.087. (Note: The sample size in this study was so large that a difference of 3 percent or more in any of the observed versus expected frequencies indicated a statistically significant difference—meaning, a difference that was not due to chance.)

Answers are available in Appendix A.

26.2 Nonrandom Mating

In the Hardy–Weinberg model, gametes were picked from the gene pool at random and paired to create offspring genotypes. In nature, however, matings between individuals may not be random with respect to the gene in question. Even in species like clams that simply broadcast their gametes into the surrounding water, gametes from individuals that live close to each other are more likely to combine than gametes from individuals that live farther apart.

The most intensively studied form of nonrandom mating is called **inbreeding,** the mating between relatives. Since relatives share a recent common ancestor, individuals that inbreed are likely to share alleles they inherited from their common ancestor.

How Does Inbreeding Affect Allele Frequencies and Genotype Frequencies?

To understand how inbreeding affects populations, let's follow the fate of alleles and genotypes when inbreeding occurs. As before, focus on a single locus with two alleles, A_1 and A_2, and suppose that these alleles initially have equal frequencies of 0.5.

Now imagine that the gametes produced by individuals in the population don't go into a gene pool. Instead, individuals self-fertilize. Many flowering plants, for example, contain both male and female organs and routinely self-pollinate. Self-fertilization, or selfing, is the most extreme form of inbreeding.

As **FIGURE 26.3a** shows, homozygous parents that self-fertilize produce all homozygous offspring. Heterozygous parents, in contrast, produce homozygous and heterozygous offspring in a 1 : 2 : 1 ratio.

FIGURE 26.3b shows the outcome for the population as a whole. In this figure, the width of the boxes represents the frequency of the three genotypes, which start out at the Hardy–Weinberg ratio of $p^2 : 2pq : q^2$. Notice that the homozygous proportion of the population increases with each generation, while the heterozygous proportion is halved. At the end of the four generations illustrated, heterozygotes are rare. The same outcomes occur more slowly with less extreme forms of inbreeding.

This simple exercise demonstrates two fundamental points about inbreeding:

1. **Inbreeding increases homozygosity.** In essence, inbreeding takes alleles from heterozygotes and puts them into homozygotes.

2. **Inbreeding itself does not cause evolution, because allele frequencies do not change in the population as a whole.**

Nonrandom mating changes only genotype frequencies, not allele frequencies, so is not an evolutionary process itself. ✔ If you understand this concept, you should be able to predict how observed genotype frequencies and allele frequencies should differ from those expected under the Hardy–Weinberg principle when inbreeding is occurring.

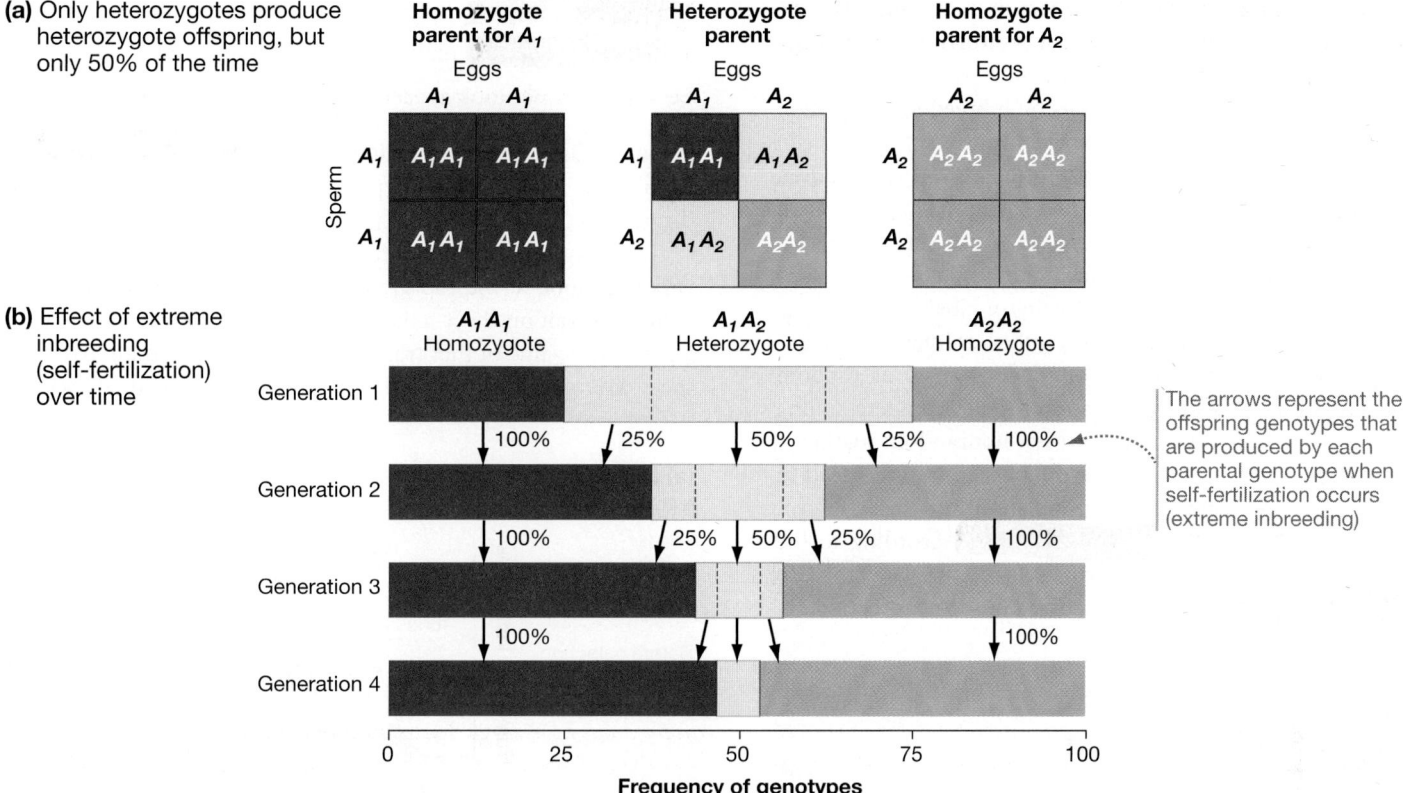

(a) Only heterozygotes produce heterozygote offspring, but only 50% of the time

Homozygote parent for A₁
Eggs
A_1 A_1

Heterozygote parent
Eggs
A_1 A_2

Homozygote parent for A₂
Eggs
A_2 A_2

Sperm

(b) Effect of extreme inbreeding (self-fertilization) over time

A_1A_1 Homozygote | A_1A_2 Heterozygote | A_2A_2 Homozygote

Generation 1

100% | 25% | 50% | 25% | 100%

Generation 2

100% | 25% | 50% | 25% | 100%

Generation 3

100% | 100%

Generation 4

0   25   50   75   100
Frequency of genotypes

The arrows represent the offspring genotypes that are produced by each parental genotype when self-fertilization occurs (extreme inbreeding)

FIGURE 26.3 Inbreeding Increases Homozygosity and Decreases Heterozygosity. (a) Heterozygous parents produce homozygous and heterozygous offspring in a 1:2:1 ratio. **(b)** The width of the boxes corresponds to the frequency of each genotype.

How Does Inbreeding Influence Evolution?

The trickiest point to grasp about inbreeding is that even though it does not cause evolution directly—because it does not change allele frequencies—it can speed the rate of evolutionary change. More specifically, it increases the rate at which natural selection eliminates recessive **deleterious** alleles—alleles that lower fitness—from a population.

Consider **inbreeding depression**—the decline in average fitness that takes place when homozygosity increases and heterozygosity decreases in a population. Inbreeding depression results from two causes:

1. **Many recessive alleles represent loss-of-function mutations.** Because these alleles are usually rare, there are normally very few homozygous recessive individuals in a population. Instead, most loss-of-function alleles exist in heterozygous individuals. The alleles have little or no effect when they occur in heterozygotes, because one normal allele usually produces enough functional protein to support a normal phenotype. But inbreeding increases the frequency of homozygous recessive individuals. Loss-of-function mutations are usually deleterious or even lethal when they are homozygous—they are quickly eliminated by selection.

2. **Many genes—especially those involved in fighting disease—are under intense selection for heterozygote advantage, a selection process that favors genetic diversity.** If an individual is homozygous at these genes, then fitness declines.

The upshot here is that the offspring of inbred matings are expected to have lower fitness than the offspring of outcrossed matings. This prediction has been verified in a wide variety of species. **FIGURE 26.4** shows results from recent efforts to reduce illness in a small population of endangered Florida panthers by introducing females from Texas. Note that the two sets of data points compare the fitnesses of offspring from non-inbred

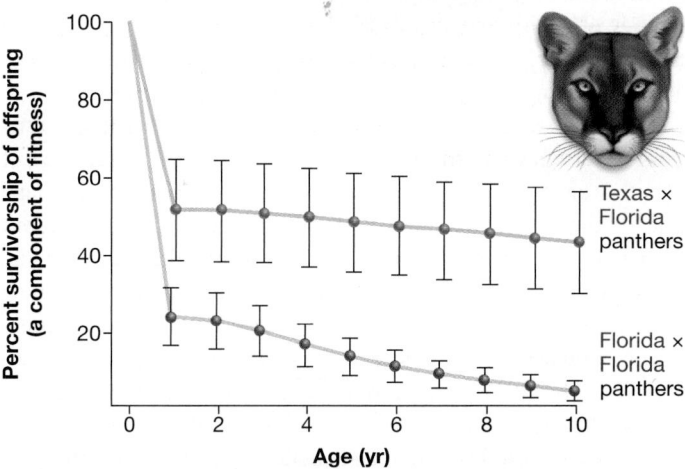

FIGURE 26.4 Inbreeding Depression Occurs in Florida Panthers. Inbreeding depression is the fitness difference between non-inbred and inbred individuals.

DATA: Johnson, W. E., E. P. Onorato, M. E. Roelke, et al. 2010. *Science* 329: 1641–1645.

Texas × Florida matings and inbred Florida × Florida matings. Inbreeding depression is represented by the vertical distance between data points at each age.

Several studies show similar results on inbreeding depression in human populations around the world. Children of first-cousin marriages consistently have a higher mortality rate than children of marriages between nonrelatives. Because inbreeding has such deleterious consequences in humans, it is not surprising that many human societies have laws forbidding marriages between individuals who are related as first cousins or closer.

In insects, vertebrates, and many other animals where inbreeding is uncommon, females often don't mate at random but actively choose certain males, and/or males compete among themselves to secure mates. This form of nonrandom mating is fundamentally different from inbreeding because it *does* lead to changes in allele frequencies in the population, and thus is a form of natural selection—called **sexual selection.** According to the Hardy–Weinberg assumptions, sexual selection violates the "no natural selection" hypothesis, as discussed in the next section.

26.3 Natural Selection

Evolution by natural selection occurs when heritable variation leads to differential success in survival and reproduction (see Chapter 25). Stated another way, natural selection occurs when individuals with certain phenotypes produce more surviving offspring than individuals with other phenotypes do. If certain alleles are associated with the favored phenotypes, they increase in frequency while other alleles decrease in frequency. The result is evolution—a violation of the assumptions of the Hardy–Weinberg model.

How Does Selection Affect Genetic Variation?

When biologists analyze the consequences of selection, they often focus on **genetic variation**—the number and relative frequency of alleles that are present in a particular population. The reason is simple: Lack of genetic variation in a population is usually a bad thing.

To understand why this is so, recall that selection can occur only if heritable variation exists in a population (Chapter 25). If genetic variation is low and the environment changes—perhaps due to the emergence of a new disease-causing virus, a rapid change in climate, or a reduction in the availability of a particular food source—no alleles that have high fitness under the new conditions are likely to be present. As a result, the average fitness of the population will decline. If the environmental change is severe enough, the population may even be faced with extinction.

Natural selection occurs in a wide variety of patterns, or modes, each with different consequences to genetic variation:

- *Directional selection* changes the average value of a trait.
- *Stabilizing selection* reduces variation in a trait.

- *Disruptive selection* increases variation in a trait.
- *Balancing selection* maintains variation in a trait.

Let's take a closer look at each of these four modes in turn.

Mode 1: Directional Selection When **directional selection** occurs, the average phenotype of a population changes in one direction.

To get a sense of how directional selection affects genetic variation, look at the top graph in **FIGURE 26.5a**, which plots the value of a trait on the *x*-axis and the number of individuals with a particular value of that trait on the *y*-axis. (In a graph like this, the *y*-axis could also plot the frequency of individuals with a

(a) Directional selection changes the average value of a trait.

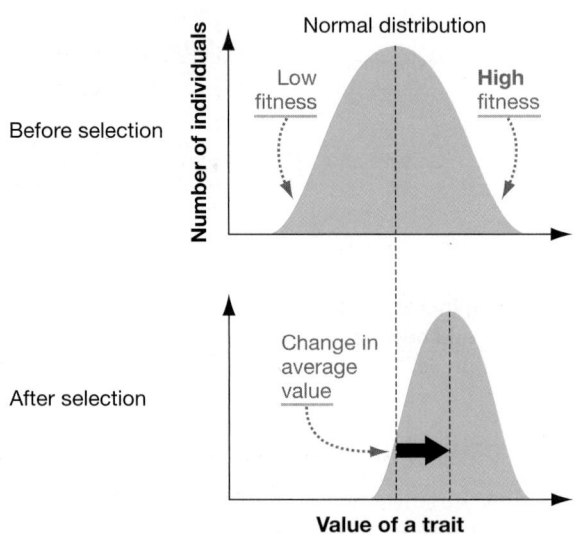

(b) For example, directional selection caused average body size to increase in a cliff swallow population.

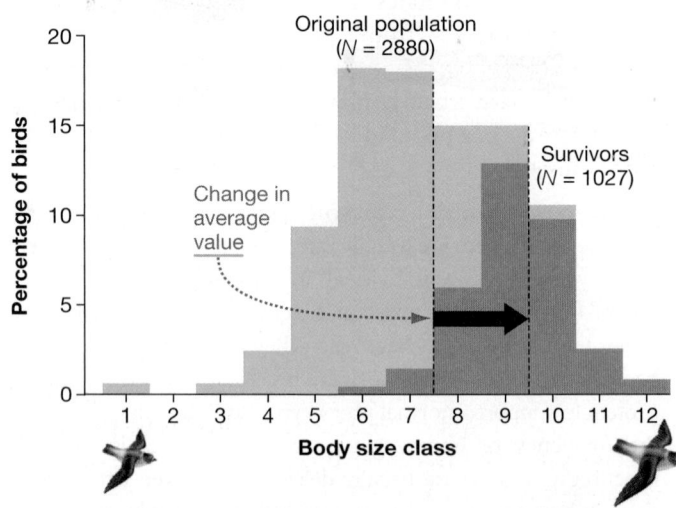

FIGURE 26.5 Directional Selection. When directional selection acts on traits that have a normal distribution, individuals at one end of the distribution experience poor reproductive success.

DATA: Brown, C. R., and M. B. Brown. 1998. *Evolution* 52: 1461–1475.

particular trait value—see **BioSkills 3** in Appendix B.) Note that the trait in question is determined by multiple genes, so it has a bell-shaped, normal distribution in this hypothetical population (Chapter 14).

[handwritten: scale of 0–1]

The second graph in the figure shows what happens when directional selection acts on this trait. Directional selection tends to reduce the genetic diversity of populations. If directional selection continues over time, the favored alleles will eventually approach a frequency of 1.0 while disadvantageous alleles will approach a frequency of 0.0, a clear violation of the assumptions of the Hardy–Weinberg model. Alleles that reach a frequency of 1.0 are said to be fixed; those that reach a frequency of 0.0 are said to be lost. When disadvantageous alleles decline in frequency, **purifying selection** is said to occur.

Now consider data from real organisms in the wild. In 1996 a population of cliff swallows native to the Great Plains of North America endured a six-day period of exceptionally cold, rainy weather. Cliff swallows feed by catching mosquitoes and other insects in flight. Insects disappeared during this cold snap, however, and the biologists recovered the bodies of 1853 swallows that died of starvation.

As soon as the weather improved, the researchers caught and measured the body size of 1027 survivors from the same population. As the histograms in **FIGURE 26.5b** show, survivors were much larger on average than the birds that died.

Directional selection, favoring large body size, had occurred. To explain this observation, the investigators suggested that larger birds survived because they had larger fat stores and did not get as cold as the smaller birds. As a result, the larger birds were less likely to die of exposure to cold and more likely to avoid starvation until the weather warmed up and insects were again available.

If variation in swallow body size was heritable, and if this type of directional selection continued, then alleles that contribute to small body size would quickly decline in frequency in the cliff swallow population.

As it turned out in this case, directional selection was short-lived because the severe weather event was temporary. In other cases, directional selection can be more persistent, such as the directional selection that occurs during global climate change.

Mode 2: Stabilizing Selection When cliff swallows were exposed to cold weather, selection greatly reduced one extreme in the range of phenotypes and resulted in a directional change in the average characteristics of the population. But selection can also reduce both extremes in a population, as illustrated in **FIGURE 26.6a**. This mode of selection is called **stabilizing selection.** It has two important consequences: There is no change in the average value of a trait over time, and genetic variation in the population is reduced.

FIGURE 26.6b shows a classical data set in humans that illustrates stabilizing selection. Biologists who analyzed birth weights and mortality in 13,730 babies born in British hospitals in the 1940s found that babies of average size (slightly over 7 pounds)

survived best. Mortality was high for very small babies and very large babies. This is persuasive evidence that birth weight was under strong stabilizing selection in this population. Alleles associated with high birth weight or low birth weight were subject to purifying selection, and alleles associated with intermediate birth weight increased in frequency.

[handwritten: advantageous]

[handwritten: can cause seperate ↓ species to form]

Mode 3: Disruptive Selection   Disruptive selection has the opposite effect of stabilizing selection. Instead of favoring phenotypes near the average value and eliminating extreme phenotypes, it eliminates phenotypes near the average value and favors

(a) Stabilizing selection reduces the amount of variation in a trait.

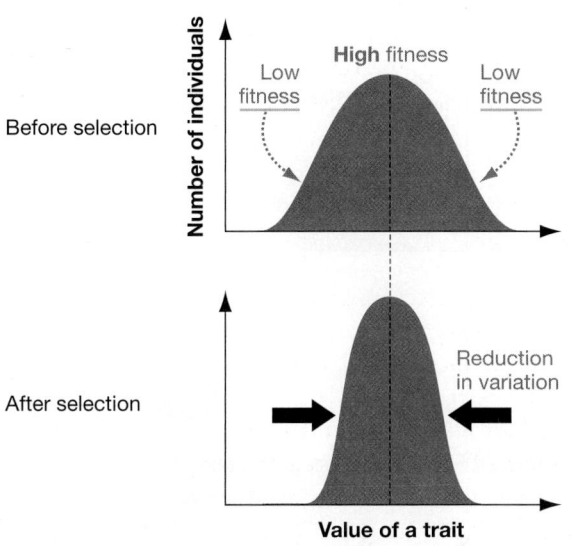

(b) For example, very small and very large babies are the most likely to die, leaving a narrower distribution of birthweights.

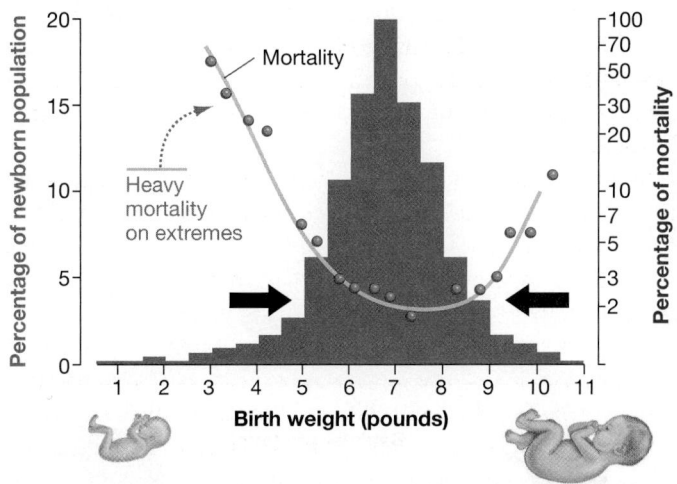

FIGURE 26.6 Stabilizing Selection. When stabilizing selection acts on normally distributed traits, individuals with extreme phenotypes experience poor reproductive success.

DATA: Karn, M. N., H. Lang-Brown, J. J. MacKenzie, et al. 1951. *Annals of Eugenics* 15: 306–322.

extreme phenotypes (**FIGURE 26.7a**). When disruptive selection occurs, the overall amount of genetic variation in the population is maintained.

Recent research has shown that disruptive selection is responsible for the striking distribution of bills of black-bellied seedcrackers (**FIGURE 26.7b**). The data plotted in the graph show that individuals with either short or very long beaks survive best and that birds with intermediate phenotypes are at a disadvantage.

In this case, the agent that causes natural selection is food. At a study site in south-central Cameroon, West Africa, a researcher found that only two sizes of seed are available to the seedcrackers: large and small. Birds with small beaks crack and eat small

seeds efficiently. Birds with large beaks handle large seeds efficiently. But birds with intermediate beaks have trouble with both, so alleles associated with medium-sized beaks are subject to purifying selection. Disruptive selection maintains high overall variation in this population.

Disruptive selection is important because it sometimes plays a part in speciation, or the formation of new species. If small-beaked seedcrackers began mating with other small-beaked individuals, their offspring would tend to be small-beaked and would feed on small seeds. Similarly, if large-beaked individuals chose only other large-beaked individuals as mates, they would tend to produce large-beaked offspring that would feed on large seeds.

In this way, selection would result in two distinct populations. Under some conditions, the populations may eventually form two new species. (See Chapter 27 for a detailed discussion of the process of species formation, which is based on disruptive selection and other mechanisms.)

Mode 4: Balancing Selection

Directional selection, stabilizing selection, and disruptive selection describe how natural selection can act on one polygenic trait in a single generation or episode. Another mode of selection, called **balancing selection,** occurs when no single allele has a distinct advantage. Instead, there is a balance among several alleles in terms of their fitness and frequency. Balancing selection occurs when:

1. Heterozygous individuals have higher fitness than homozygous individuals do, a pattern called **heterozygote advantage**. This pattern of selection is one explanation for the data in Table 26.2, where heterozygotes for the *HLA* genes have a fitness advantage compared to homozygotes. The consequence of this pattern is that genetic variation is maintained in populations.

2. The environment varies over time or in different geographic areas occupied by a population—meaning that certain alleles are favored by selection at different times or in different places. As a result, overall genetic variation in the population is maintained or increased.

3. Certain alleles are favored when they are rare, but not when they are common—a pattern known as **frequency-dependent selection.** For example, rare alleles responsible for coloration in guppies are favored because predators learn to recognize common color patterns. Alleles for common colors get eliminated; alleles for rare colors increase in frequency. As a result, overall genetic variation in the population is maintained or increased.

TABLE 26.3 summarizes the four modes of natural selection. No matter how natural selection occurs, though, its most fundamental attribute is the same: It increases fitness and leads to adaptation. Natural selection results in allele frequencies that deviate from those predicted by the Hardy–Weinberg principle because selection favors certain alleles over others. ✔ If you understand this concept, you should be able to predict how genotype frequencies differ from Hardy–Weinberg proportions under directional selection.

(a) Disruptive selection increases the amount of variation in a trait.

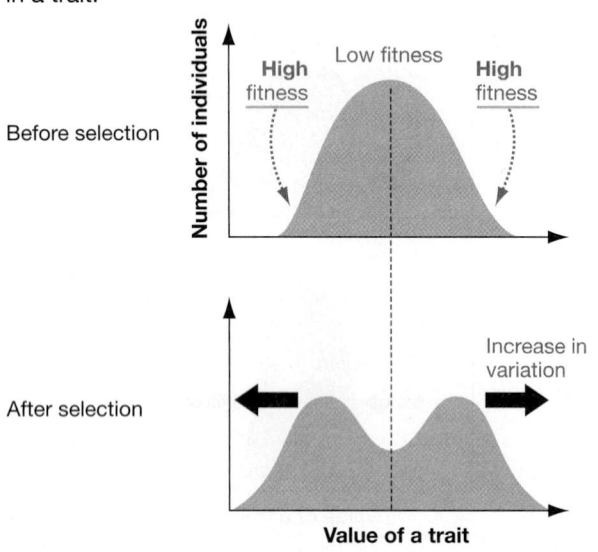

(b) For example, only juvenile black-bellied seedcrackers that had short or extremely long beaks survived long enough to breed.

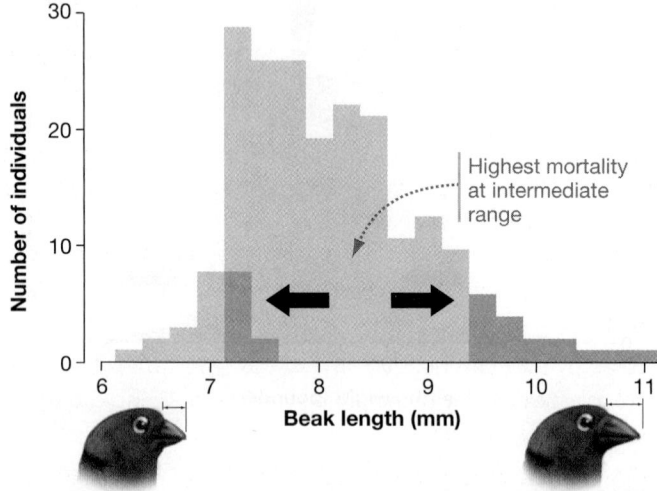

FIGURE 26.7 Disruptive Selection. When disruptive selection occurs on traits with a normal distribution, individuals with extreme phenotypes experience high reproductive success.
DATA: Smith, T. B. 1987. *Nature* 329: 717–719.

SUMMARY TABLE 26.3 Modes of Selection

Mode of Selection	Effect on Phenotype	Effect on Genetic Variation
Directional selection	Favors one extreme phenotype, causing the average phenotype in the population to change in one direction.	Genetic variation is reduced.
Stabilizing selection	Favors phenotypes near the middle of the range of phenotypic variation, maintaining average phenotype.	Genetic variation is reduced.
Disruptive selection	Favors extreme phenotypes at both ends of the range of phenotypic variation.	Genetic variation is increased.
Balancing selection	No single phenotype is favored in all populations of a species at all times.	Genetic variation is maintained.

Sexual Selection

Darwin was the first to recognize that selection based on success in courtship is a mechanism of evolutionary change. Although he introduced the idea of sexual selection in *On the Origin of Species* in 1859, it was not until 12 years later that he detailed his views on sexual selection in *The Descent of Man, and Selection in Relation to Sex.*

Darwin was initially perplexed by seemingly nonadaptive traits such as the extravagantly long and iridescent trains of peacocks. Why was it adaptive for the peacock to spend energy producing a structure that would increase the peacock's vulnerability to predation? Darwin concluded that the peacock's train is adaptive because it helps the peacock to attract mates, thereby increasing its reproductive success.

The peacock example demonstrates mate choice, sometimes referred to as *intersexual selection*—the selection of an individual of one gender for mating by an individual of the other gender. Darwin also recognized that individuals sometimes compete with one another to obtain mates. This form of selection *within* a gender is referred to as *intrasexual selection.*

Theory: The Fundamental Asymmetry of Sex Why would the extravagant trains of peacocks occur in the males but not the peahens? A. J. Bateman addressed this question in the 1940s, followed by Robert Trivers three decades later. The Bateman–Trivers theory contains an observed pattern and a hypothesized process:

- *Pattern* Sexual selection usually acts on males much more strongly than on females. Traits that attract members of the opposite sex are much more highly elaborated in males.

- *Process* The energetic cost of creating a large egg is enormous, whereas a sperm contains few energetic resources. That is, "eggs are expensive, but sperm are cheap."

In most species, females invest much more in their offspring than do males. This phenomenon is called the fundamental asymmetry of sex. It is characteristic of almost all sexual species and has two important consequences:

1. Because eggs are large and energetically expensive, females produce relatively few young over the course of a lifetime. A female's fitness is limited not by her ability to find a mate but primarily by her ability to gain the resources needed to produce more eggs and healthier young.

2. Sperm are so energetically inexpensive to produce that a male can father an almost limitless number of offspring. Thus, a male's fitness is limited not by the ability to acquire the resources needed to produce sperm but by the number of females he can mate with.

The Bateman–Trivers theory of sexual selection makes strong predictions:

- If females invest a great deal in each offspring, then they should protect that investment by being choosy about their mates. Conversely, if males invest little in each offspring, then they should be willing to mate with almost any female.

- If there are an equal number of males and females in the population, and if males are trying to mate with any female possible, then males will compete with each other for mates.

- If male fitness is limited by access to mates, then any allele that increases a male's attractiveness to females or success in male–male competition should increase rapidly in the population, violating the assumptions of Hardy–Weinberg. Thus, sexual selection should act more strongly on males than on females.

Do data from experimental or observational studies agree with these predictions? Let's consider each of them in turn.

Female Choice for "Good Alleles" If females are choosy about which males they mate with, what criteria do females use to make their choice? Recent experiments have shown that in several bird species, females prefer to mate with males that are well fed and in good health. These experiments were motivated by three key observations:

1. In many bird species, the existence of colorful feathers or a colorful beak is due to the presence of the red and yellow pigments called carotenoids.

2. Carotenoids protect tissues and stimulate the immune system to fight disease more effectively.

3. Animals usually cannot synthesize their own carotenoids, but plants can. To obtain carotenoids, animals have to eat carotenoid-rich plant tissues.

These observations suggest that the healthiest and best-nourished birds in a population have the most colorful beaks and feathers. Sick birds have dull coloration because they are using all of their carotenoids to stimulate their immune system. Poorly fed birds have dull coloration because they have few carotenoids available. By choosing a colorful male as the father of her offspring, a female is likely to have offspring with alleles that will help the offspring fight disease effectively and feed efficiently.

To test the hypothesis that females prefer colorful males, a team of researchers experimented with zebra finches (**FIGURE 26.8a**). They identified pairs of brothers and randomly assigned one brother to the treatment group and one brother to the control group. They fed the treatment group a diet that was heavily supplemented with carotenoids, and they fed the control group a diet that was similar in every way except for the additional carotenoids.

As predicted, the males eating the carotenoid-supplemented diet developed more colorful beaks than did the males fed the carotenoid-poor diet. When given a choice of perching with either of the two brothers, 9 out of the 10 females tested preferred the more-colorful male (**FIGURE 26.8b**). These results support the hypothesis that females of this species are choosy about their mates and that they prefer to mate with healthy, well-fed males.

Enough experiments have been done on other bird species to support a general conclusion: Colorful beaks and feathers, along with songs and dances and other types of courtship displays, carry the message "I'm healthy and well fed because I have good alleles. Mate with me."

Female Choice for Paternal Care

Choosing "good alleles" is not the entire story in sexual selection via female choice. In many species, females prefer to mate with males that care for young or that provide the resources required to produce eggs.

Brown kiwi females make an enormous initial investment in their offspring—their eggs routinely represent over 15 percent of the mother's total body weight (**FIGURE 26.9**)—but choose to mate with males that take over all of the incubation and other care of the offspring. It is common to find that female fish prefer to mate with males that protect a nest site and care for the eggs until they hatch. In humans and many species of birds, males provide food, protection, and other resources for rearing young.

To summarize, females may choose mates on the basis of (1) physical characteristics that signal male genetic quality, (2) resources or parental care provided by males, or (3) both. In some species, however, females do not have the luxury of choosing a male. Instead, competition among males is the primary cause of sexual selection.

Male–Male Competition

As an example of research on how males compete for mates, consider data from a long-term study of a northern elephant seal population breeding on Año Nuevo Island, off the coast of California.

Elephant seals feed mostly on marine fish, squid, and octopus, and spend most of the year in the water. But when females are ready to mate and give birth, they haul themselves out of the

(a) Male zebra finch

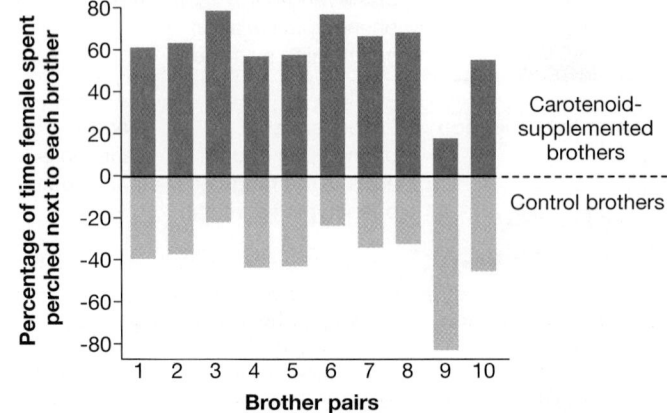

(b) Female choice for colorful beaks

FIGURE 26.8 Female Zebra Finches Prefer Males with Colorful Beaks.
DATA: Blount, J. D., N. B. Metcalf, T. R. Birkhead, et al. 2003. *Science* 300: 125–127.

water onto land. Females prefer to give birth on islands, where newborn pups are protected from terrestrial and marine predators. Because elephant seals have flippers that are ill-suited for walking, females can haul themselves out of the water only on the few beaches that have gentle slopes. As a result, large numbers of females congregate in tiny areas to breed.

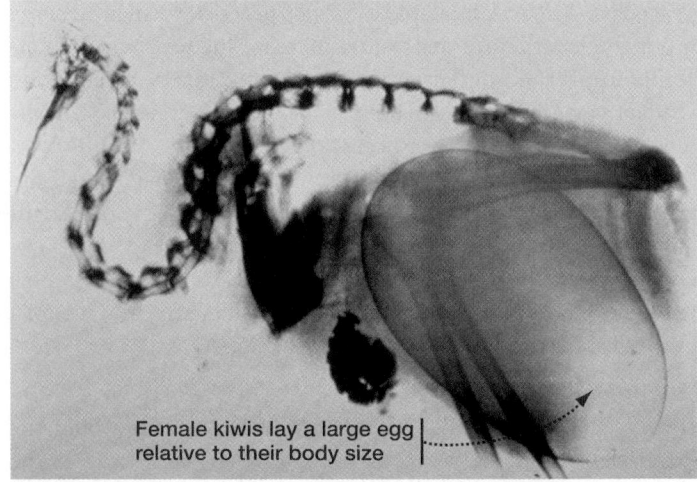

Female kiwis lay a large egg relative to their body size

FIGURE 26.9 In Many Species, Females Make a Large Investment in Each Offspring. X-ray of a female kiwi, ready to lay an egg.

Male elephant seals establish territories on breeding beaches by fighting (**FIGURE 26.10a**). A territory is an area that is actively defended and that provides exclusive or semi-exclusive use by the owner. Males that win battles with other males monopolize matings with the females residing in their territories. Males that lose battles are relegated to territories with fewer females or are excluded from the beach. Fights are essentially slugging contests and are usually won by the larger male. The males stand face to face, bite each other, and land blows with their heads.

Based on these observations, it is not surprising that male northern elephant seals frequently weigh 2700 kg (5940 lbs) and are over four times more massive, on average, than females. The logic here runs as follows:

- Males that dominate beaches with large congregations of females father large numbers of offspring. Males that lose fights father few or no offspring.

- The alleles of dominant males rapidly increase in frequency in the population.

- If the ability to win fights and produce offspring is determined primarily by body size, then alleles for large body size have a significant fitness advantage, leading to the evolution of large male size. The fitness advantage is due to sexual selection, and the consequence is directional selection on large body size.

FIGURE 26.10b provides evidence for intense sexual selection in males. Biologists have marked many of the individuals in the seal population on Año Nuevo to track the lifetime reproductive success of a large number of individuals. The x-axis indicates fitness, plotted as the number of offspring produced over a lifetime. The y-axis indicates the percentage of males in the population that achieved each category of offspring production (0, 1–10, and so on). As the data show, in this population a few males father a large number of offspring, while most males father few or none.

Among females, variation in reproductive success is also high, but it is much lower than in males (**FIGURE 26.10c**). In this species, most sexual selection appears to be driven more by male–male competition than by female choice.

Sexual Dimorphism Results from Sexual Selection   In elephant seals and most other animals studied, most females that survive to adulthood get a mate. In contrast, many males do not. Because sexual selection tends to be much more intense in males than females, males tend to have many more traits that function only in courtship or male–male competition. Stated another way, sexually selected traits often differ sharply between the sexes.

Sexual dimorphism (literally, "two-forms") refers to any trait that differs between males and females. Such traits range from weapons that males use to fight over females, such as antlers and

---

(a) Males compete for the opportunity to mate with females.

(b) Variation in reproductive success is high in males.

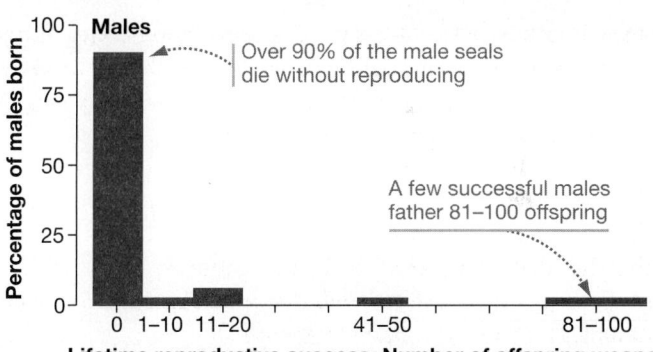

(c) Variation in reproductive success is relatively low in females.

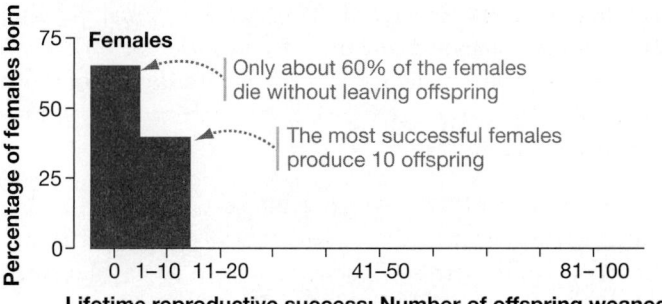

FIGURE 26.10 Intense Sexual Selection in Male Elephant Seals. The histograms show that variation in lifetime reproductive success is higher in **(b)** male northern elephant seals than it is in **(c)** females.

DATA: Le Boeuf, B. J., and R. S. Peterson. 1969. *Science* 163: 91–93.

horns (**FIGURE 26.11a**), to the elaborate ornamentation and behavior used in courtship displays (**FIGURE 26.11b**). Humans are sexually dimorphic in size, distribution of body hair, and many other traits.

✓ If you understand sexual selection, you should be able to (1) define the fundamental asymmetry of sex, and (2) explain why males are usually the sex with exaggerated traits used in courtship.

Take-Home Messages Female choice and male–male competition illustrate how selection can favor certain phenotypes in a population. The adaptive alleles responsible for these phenotypes increase in frequency over time. Thus, evolution occurs and the assumptions of the Hardy–Weinberg model are violated.

Sexual selection is just one type of natural selection. All the other types, sometimes referred to collectively as **ecological (or**

(a) Red deer males use weaponry to compete for mates.

(b) Male raggiana birds of paradise display for females.

FIGURE 26.11 Sexually Selected Traits Are Used to Compete for Mates. Males often have exaggerated traits that they use in fighting or courtship. In many species, females lack these traits.

environmental) selection, favor traits that enable organisms to do things other than obtain mates—such as survive in their physical and biological environments. Different agents of natural selection can act on organisms simultaneously, sometimes favoring the same traits, other times resulting in fitness trade-offs (Chapter 25). For example, sexual selection may favor long trains in peacocks, while ecological selection may favor shorter trains that make the peacocks less vulnerable to predators. The relative importance of different agents of selection can change over time and in space.

Natural selection is the only evolutionary process that results in adaptation, but it is not the only evolutionary process that violates the Hardy–Weinberg assumptions. Let's consider what happens when random changes in allele frequency occur.

26.4 Genetic Drift

Genetic drift is defined as any change in allele frequencies in a population that is due to chance. The process is aptly named, because it causes allele frequencies to drift up and down randomly over time. When drift occurs, allele frequencies change due to blind luck—what is known in statistics as **sampling error.** Sampling error occurs when the allele frequencies of a chosen subset of a population (the sample) are different from those in the total population, by chance. Drift occurs in every population, in every generation, but especially in small populations.

Simulation Studies of Genetic Drift

To understand why genetic drift occurs, imagine a couple marooned on a deserted island. Suppose that at gene A, the wife's genotype is $A_T A_H$ and the husband is also $A_T A_H$. In this population, the two alleles are each at a frequency of 0.5.

Now suppose that the couple produce five children over their lifetime. Half of the eggs produced by the wife carry allele A_T and half carry allele A_H. Likewise, half of the sperm produced by the husband carry allele A_T and half carry allele A_H. To simulate which sperm and which egg happen to combine to produce each of the five offspring, you can flip a coin for each sperm and each egg, where tails represents allele A_T and heads represents allele A_H.

The following coin flips were done by a pair of students in a recent biology class:

	Sperm	Egg	Genotype
First offspring	A_H	A_H	$A_H A_H$
Second offspring	A_T	A_T	$A_T A_T$
Third offspring	A_T	A_H	$A_H A_T$
Fourth offspring	A_H	A_H	$A_H A_H$
Fifth offspring	A_T	A_H	$A_H A_T$

When the parents die, there are a total of 10 alleles in the population. But note that the allele frequencies have changed. In this generation, six of the 10 alleles (60 percent) are A_H; four of the 10 alleles (40 percent) are A_T. Evolution—a change in allele frequencies in a population—occurred due to genetic drift.

Instead of each allele being sampled in exactly its original frequency when offspring formed, as the Hardy–Weinberg principle assumes, a chance sampling error occurred. Allele A_H "got lucky," and allele A_T was "unlucky."

Computer Simulations   **FIGURE 26.12** shows what happens when a computer simulates the same process of random combinations in gametes over time. The program that generated the graphs combines the alleles in a gene pool at random to create an offspring generation, calculates the allele frequencies in the offspring generation, and uses those allele frequencies to create a new gene pool.

In this example, the process was continued for 100 generations. The x-axis on each graph plots time in generations; the y-axis plots the frequency of one of the two alleles present at the A gene in a hypothetical population.

The top graph shows eight replicates of this process with a population size of 4; the bottom graph shows eight replicates with a population of 400. Notice (**1**) the striking differences

between the effects of drift in the small versus large population and (**2**) the consequences for genetic variation when alleles drift to fixation or loss.

Given enough time, drift can be an important factor even in large populations. For example, alleles containing silent mutations, usually in the third position of a codon, do not change the gene product (Chapter 16). As a result, most of these alleles have little or no effect on the phenotype. Yet these alleles routinely drift to high frequency or even fixation over time.

✔ If you understand genetic drift, you should be able to examine the MN blood group genotype frequencies in Table 26.1 and describe how drift could explain differences in genotype frequencies among populations. Note that there are no data indicating a selective advantage for different MN genotypes in different environments.

Key Points About Genetic Drift   These simulated matings illustrate three important points about genetic drift:

1. **Genetic drift is random with respect to fitness.** The changes in allele frequency that it produces are not adaptive.

2. **Genetic drift is most pronounced in small populations.** In the computer simulation, allele frequencies changed much less in the large population than in the small population. And if the couple on the deserted island had produced 50 children instead of five, allele frequencies in the next generation almost certainly would have been much closer to 0.5.

3. **Over time, genetic drift can lead to the random loss or fixation of alleles.** In the computer simulation with a population of 4, it took at most 20 generations for one allele to be fixed or lost. When random loss or fixation occurs, genetic variation in the population declines.

The importance of drift in small populations is a particular concern for conservation biologists because many populations are being drastically reduced in size by habitat destruction and other human activities. Small populations that occupy nature reserves or zoos are particularly susceptible to genetic drift. If drift leads to a loss of genetic diversity, it could darken the already bleak outlook for some endangered species.

Experimental Studies of Genetic Drift

Research on genetic drift began in the 1930s and 1940s with theoretical work that used mathematical models to predict the effect of genetic drift on allele frequencies and genetic variation. In the mid-1950s, Warwick Kerr and Sewall Wright did an experiment to determine how drift works in practice.

Kerr and Wright started with a large laboratory population of fruit flies that contained a **genetic marker**—a specific allele that causes a distinctive phenotype. In this case, the marker was the morphology of bristles. Fruit flies have bristles on their bodies that can be either straight or bent. This difference in bristle phenotype depends on a single gene. Kerr and Wright's lab population contained just two alleles—normal (straight) and "forked" (bent), designated as A_N and A_F respectively. Since the trait is

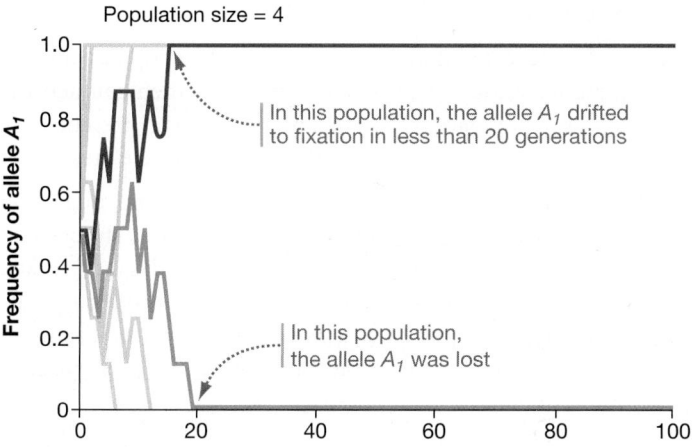

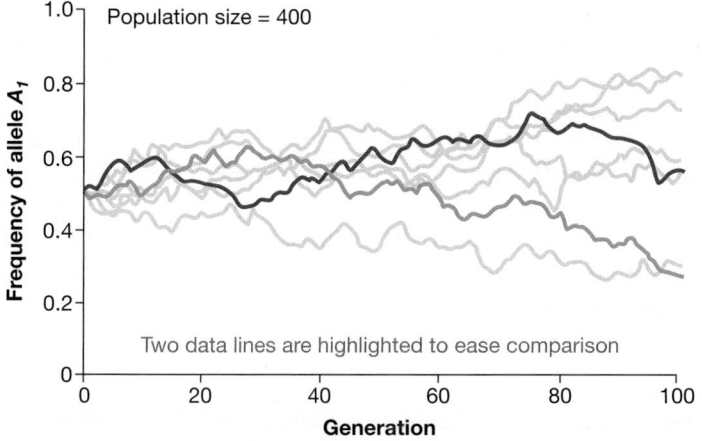

FIGURE 26.12 Genetic Drift Is More Pronounced in Small Populations than in Large Populations.

✔**EXERCISE** Draw a graph predicting what these simulation results would look like for a population size of 4000.

QUESTION: Does genetic drift in lab populations work as predicted by mathematical models?

HYPOTHESIS: Genetic drift causes alleles to be fixed or lost over time.

NULL HYPOTHESIS: Allele frequencies do not change; they stay in Hardy–Weinberg proportions.

EXPERIMENTAL SETUP:

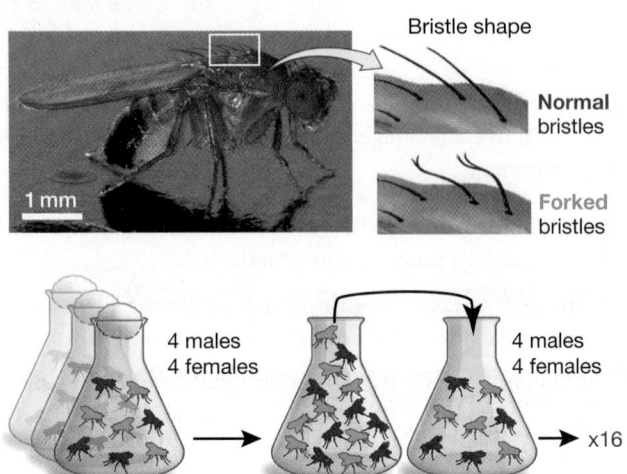

Bristle shape

Normal bristles

Forked bristles

1 mm

4 males 4 females

4 males 4 females

x16

96 x Generation 1   Generation 2

1. Set up 96 cages of fruit flies with starting frequencies of both alleles = 0.5. Allow flies to breed.

2. From the F₁ offspring, randomly choose four males and four females. Allow them to breed.

3. After 16 generations, count the number of populations with either or both of the alleles.

PREDICTION: The number of fruit fly populations with fixed or lost alleles will increase over time. The number of populations with both alleles will decrease.

PREDICTION OF NULL HYPOTHESIS: Both alleles will remain present in all fruit fly populations for all generations.

RESULTS:

Genetic drift reduced allelic diversity in most populations:

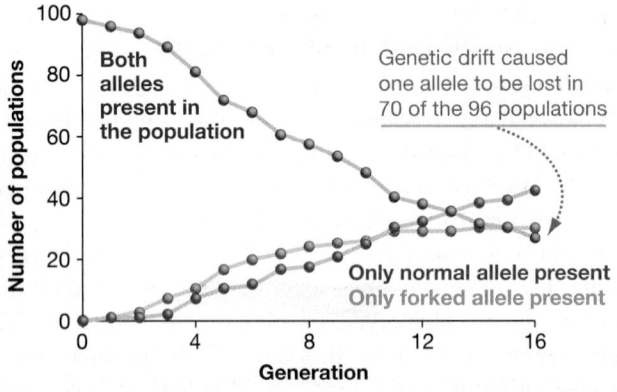

Both alleles present in the population

Genetic drift caused one allele to be lost in 70 of the 96 populations

Only normal allele present
Only forked allele present

Number of populations

Generation

CONCLUSION: Genetic drift occurs in laboratory populations as predicted by mathematical models.

---

sex-linked, males have only one allele (see Chapter 14). In females, the forked allele is recessive to the normal allele, but Kerr and Wright were able to distinguish homozygous females ($A_N A_N$) from heterozygotes ($A_N A_F$).

Kerr and Wright studied drift in these alleles as shown in the experimental setup in **FIGURE 26.13**:

Step 1 The researchers set up 96 small populations in their lab, each consisting of four adult females and four adult males of the fruit fly *Drosophila melanogaster*. They chose individual flies to begin these experimental populations so that the frequency of the normal and forked alleles in each of the 96 starting populations was 0.5. The two alleles do not affect the fitness of flies in the lab environment, so Kerr and Wright could be confident that if changes in the frequency of normal and forked phenotypes occurred, they would not be due to natural selection.

Step 2 After these first-generation adults bred, Kerr and Wright reared their offspring. In the offspring (F₁) generation, they randomly chose four males and four females—meaning that they simply grabbed individuals without caring whether their bristles were normal or forked—from each of the 96 offspring populations and allowed them to breed and produce the next generation.

Step 3 They repeated this procedure until all 96 populations had undergone a total of 16 generations. They then counted the number of populations that had both alleles still present, only the normal allele present, or only the forked allele present.

During the entire course of the experiment, no migration from one population to another occurred. Previous studies had shown that mutations from normal to forked bristles (and forked to normal) are rare. Thus, the only evolutionary process operating during the experiment was genetic drift. It was as if random accidents claimed the lives of all but eight individuals in each generation, so that only eight bred.

Their result? After 16 generations, both alleles were still present in only 26 of the 96 populations—significantly less than the full 96 expected by Hardy–Weinberg proportions. Forked bristles were found on all of the individuals in 29 of the experimental populations. Due to drift, the forked allele had been fixed in these 29 populations and the normal allele had been lost. In 41 other populations, however, the opposite was true: All individuals had normal bristles. In these populations, the forked allele had been lost due to chance.

The message of the study is startling: In 73 percent of the experimental populations (70 out of the 96), genetic drift had reduced allelic diversity at this gene to zero.

FIGURE 26.13 An Experiment on the Effects of Genetic Drift in Small Populations.

SOURCE: Kerr, W. E., and S. Wright. 1954. Experimental studies of the distribution of gene frequencies in very small populations of *Drosophila melanogaster*: I. Forked. *Evolution* 8: 172–177.

✔**QUESTION** Why do you think the researchers decided to start each generation with only eight individuals?

As predicted, genetic drift decreased genetic variation within populations and increased genetic differences between populations. Is drift important in natural populations as well?

What Causes Genetic Drift in Natural Populations?

The random sampling process that takes place during fertilization occurs in every population in every generation in every species that reproduces sexually. Similarly, accidents that remove individuals at random occur in every population in every generation.

It is important to realize, though, that because drift is caused by sampling error, it can occur by *any* process or event that involves sampling—not just the sampling of gametes that occurs during fertilization or the loss of unlucky individuals due to accidents. Let's consider two special cases of genetic drift, called founder effects and bottlenecks.

Founder Effects on the Green Iguanas of Anguilla When a group of individuals immigrates to a new geographic area and establishes a new population, a founder event is said to occur. If the new population is small enough, the allele frequencies in the new population are almost guaranteed to be different from those in the source population—due to sampling error. A change in allele frequencies that occurs when a new population is established is called a **founder effect** (**FIGURE 26.14a**).

In 1995, fishermen on the island of Anguilla in the Caribbean witnessed a founder event involving green iguanas. A few weeks after two major hurricanes swept through the region, a large raft composed of downed logs tangled with other debris floated onto a beach on Anguilla. The fishermen noticed green iguanas on the raft and several onshore. Because green iguanas had not previously been found on Anguilla, the fishermen notified biologists. The researchers were able to document that at least 15 individuals had arrived; two years later, they were able to confirm that at least some of the individuals were breeding. A new population had formed.

During this founder event, it is extremely unlikely that allele frequencies in the new Anguilla population of green iguanas exactly matched those of the source population, which was thought to be on the islands of Guadeloupe.

Colonization events like these have been the major source of populations that occupy islands all over the world, as well as island-like habitats such as mountain meadows, caves, and ponds. Each time a founder event occurs, a founder effect is likely to accompany it, changing allele frequencies through genetic drift.

Genetic Bottleneck on Pingelap Atoll If a large population experiences a sudden reduction in size, a population bottleneck is said to occur. The term comes from the metaphor of a few individuals of a population fitting through the narrow neck of a bottle, by chance. Disease outbreaks, natural catastrophes such as floods or fires or storms, or other events can cause population bottlenecks.

Genetic bottlenecks follow population bottlenecks, just as founder effects follow founder events. A **genetic bottleneck**

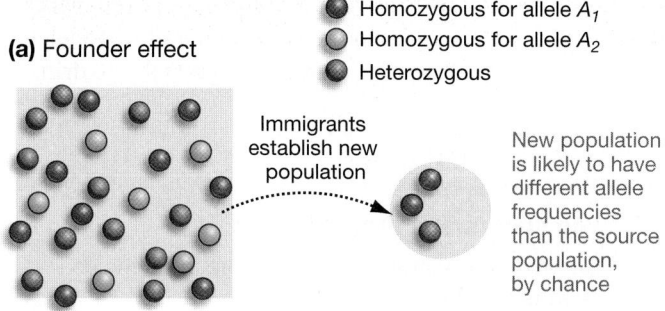

(a) Founder effect

- ● Homozygous for allele A_1
- ○ Homozygous for allele A_2
- ● Heterozygous

Immigrants establish new population

New population is likely to have different allele frequencies than the source population, by chance

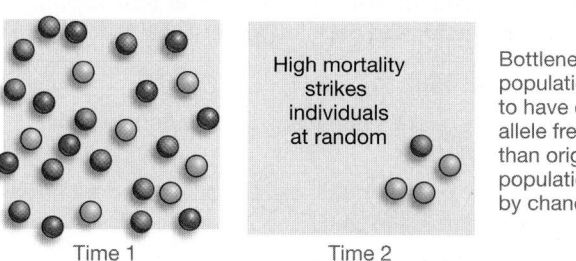

(b) Genetic bottleneck

High mortality strikes individuals at random

Bottlenecked population is likely to have different allele frequencies than original population, by chance

Time 1   Time 2

FIGURE 26.14 Two Causes of Genetic Drift in Natural Populations. The smaller the new population, the higher the likelihood that genetic drift will result in differences in allele frequencies as well as in loss of alleles.

✔**QUANTITATIVE** The original population in **(a)** consists of 9 A_1A_1, 11 A_1A_2, and 7 A_2A_2 individuals (this hypothetical population is very small for simplicity). Compare the frequencies of the A_1 allele in the original and new populations.

is a sudden reduction in the number of alleles in a population (**FIGURE 26.14b**). Genetic drift occurs during genetic bottlenecks and causes a change in allele frequencies.

For an example from a natural population, consider the humans who occupy Pingelap Atoll in the South Pacific. On this island, only about 20 people out of a population of several thousand managed to survive the effects of a typhoon and a subsequent famine that occurred around 1775. The survivors apparently included at least one individual who carried a loss-of-function allele at a gene called *CNGB3*, which codes for a protein involved in color vision.

The *CNGB3* allele is recessive, and when it is homozygous it causes a serious vision deficit called achromatopsia. People with this condition have poor vision and are either totally or almost totally color-blind. The condition is extremely rare in most populations, and the frequency of the *CNGB3* allele is estimated to be under 1.0 percent.

In the population that survived the Pingelap Atoll disaster, however, the loss-of-function allele was at a frequency of about 1/40, or 2.5 percent. If the allele was at the typical frequency of 1.0 percent or less before the population bottleneck, then a large frequency change occurred during the bottleneck, due to drift.

In today's population on Pingelap Atoll, over 1 in 20 people are afflicted with achromatopsia, and the allele that causes the affliction is at a frequency of well over 20 percent. Because the loss-of-function allele is extremely unlikely to be favored by directional selection or heterozygote advantage, researchers hypothesize that the frequency of the allele in this small population has continued to increase over the past 230 years due to drift.

check your understanding

If you understand that . . .
- Genetic drift occurs whenever allele frequencies change due to chance.
- Drift violates the Hardy–Weinberg assumptions and occurs during many different types of events, including random fusion of gametes at fertilization, founder events, and population bottlenecks.

You should be able to . . .
1. Explain why genetic drift leads to a random loss or fixation of alleles.
2. Explain why genetic drift is particularly important as an evolutionary force in small populations.

Answers are available in Appendix A.

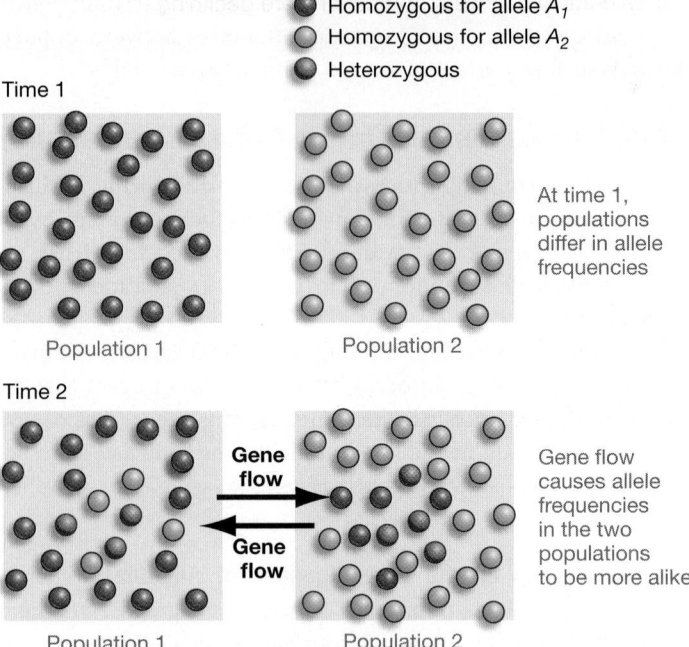

FIGURE 26.15 Gene Flow Makes Allele Frequencies More Similar between Populations. Gene flow can occur in one or both directions. If gene flow continues and no selection occurs, allele frequencies will eventually be identical.

26.5 Gene Flow

When an individual leaves one population, joins another, and breeds, **gene flow** is said to occur—the movement of alleles between populations. Note that while allele flow might be a more apt description, gene flow is the term that evolutionary biologists traditionally use. Also note that while organisms can *emigrate* from a source population or *immigrate* to a new population, the movement of their alleles is called gene flow.

As an evolutionary process, gene flow usually has one outcome: It equalizes allele frequencies between the source population and the recipient population. When alleles move from one population to another, the populations tend to become more alike. To capture this point, biologists say that gene flow homogenizes allele frequencies among populations (**FIGURE 26.15**).

Measuring Gene Flow between Populations

The presence or absence of gene flow has particularly important implications for the conservation of threatened and endangered species. Numerous studies have documented the decline of gene flow between wild populations that have been isolated from one another—for example, by habitat fragmentation. Many studies have also documented the effects of gene flow between wild populations and captive populations.

For example, the captive breeding of fish is increasing around the world as wild populations are being depleted. In some cases, wild fish are used to start captive populations, which are then kept isolated in "farms" until they go to market—except when they escape. In other cases, captive-bred fish are purposefully released into the wild in an effort to supplement the size of wild

populations. What are the effects of gene flow between captive-bred populations and wild populations?

A team of biologists recently studied steelhead trout (*Oncorhynchus mykiss*) in the Hood River of Oregon to answer this question (**FIGURE 26.16a**). Some of the trout in the Hood River are wild, while others were raised in a hatchery and released to supplement the diminishing wild population. The researchers were able to catch, sample, and release all of the steelhead trout in the Hood River each year for three years by trapping them in a dam near the mouth of the river as the fish traveled upstream to their wintering streams.

The researchers extracted DNA from tissue samples and calculated genotypes using eight microsatellite alleles (see Chapter 21) to determine whether the parents of each fish were wild or captive-bred. The team then compared the reproductive fitness—the number of surviving adult offspring—of the trout in three groups:

1. Individuals with two wild parents (Wild × Wild)
2. Individuals with one wild parent and one captive-bred parent (Wild × Captive)
3. Individuals with two captive-bred parents (Captive × Captive)

The graph in **FIGURE 26.16b** shows the fitness results for female steelhead trout. Individuals with one or two captive-bred parents are assigned a value from 0 to 1 relative to the fitness of trout with two wild parents (fitness = 1). Individuals with one captive-bred parent have 16 percent lower fitness than wild trout on average, while individuals with two captive-bred parents have significantly lower fitness, 38 percent lower than wild-bred trout,

(a) Wild steelhead trout populations are declining

(b) Captive-bred trout reduce the fitness of wild populations

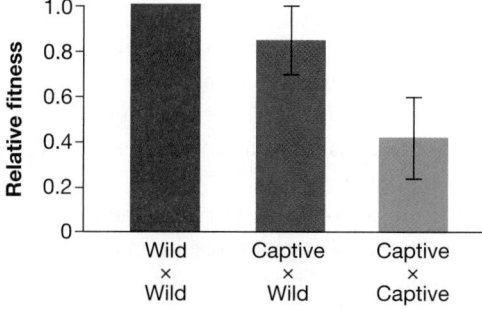

FIGURE 26.16 Gene Flow Reduces Fitness in a Population of Steelhead Trout. In this study, the fitness of female steelhead trout with one or two captive-bred parents is shown relative to females with two wild parents. Data for males are comparable (not shown).

DATA: Araki, H., B. Cooper, and M. S. Blouin. 2009. *Biology Letters* 5(5): 621–624.

on average. The researchers observed similar results for males (not shown).

These data demonstrate that gene flow is occurring from the captive-bred population to the wild population and is causing a reduction in fitness in that population. Related research on farmed Atlantic salmon in Ireland demonstrated that farmed salmon produce smaller eggs, spawn at different times, and have different predator-avoidance behaviors than wild salmon. When farmed salmon were introduced into wild populations, the reproductive success of the wild population decreased.

The conclusion of these studies on steelhead trout and Atlantic salmon is that gene flow between captive-bred and wild populations reduces fitness in the wild populations. As a result, efforts to augment wild populations with captive-bred fish may ultimately result in a *decline* in population size—contrary to the intentions of fisheries biologists.

Gene Flow Is Random with Respect to Fitness

It is not true, though, that gene flow always reduces fitness in the receiving population. If a population has lost alleles due to genetic drift, then the arrival of new alleles via gene flow should increase genetic diversity. If increased genetic diversity results in better

resistance to infections by bacteria or viruses or other parasites, for example, gene flow would increase the average fitness of individuals. This was the result when gene flow was used to aid an endangered population of the Florida panther (see Figures 26.4 and 57.14).

Gene flow is random with respect to fitness—the arrival or departure of alleles can increase or decrease average fitness, depending on the situation. But in every case, movement of alleles between populations tends to reduce their genetic differences.

This latter generalization is particularly important in our own species right now. Large numbers of people from Africa, the Middle East, Mexico, Central America, and Asia are immigrating to the countries of the European Union and the United States. Because individuals from different cultural and ethnic groups are intermarrying and having offspring, allele frequencies are becoming more similar across human populations.

26.6 Mutation

To appreciate the role of mutation as an evolutionary force, let's return to one of the central questions that biologists ask about an evolutionary process: How does it affect genetic variation in a population? Recall that:

- Natural selection often favors certain alleles and leads to a decrease in overall genetic variation.

- Genetic drift tends to decrease genetic diversity over time, as alleles are randomly lost or fixed.

- Gene flow increases genetic diversity in a recipient population if new alleles arrive with immigrating individuals. But gene flow may decrease genetic variation in the source population if alleles leave with emigrating individuals.

If most of the evolutionary processes lead to a loss of genetic diversity over time, what restores it? In sexually reproducing organisms, independent assortment and recombination (crossing over) are major sources of genetic diversity (see Chapter 13). But these processes create new combinations of existing alleles. Where do entirely new alleles come from? The answer to this question is **mutation**.

Mutations can occur in a number of ways (Chapters 16 and 21): *∆ in protein.*

- *Point mutations* If a change in nucleotide sequence occurs in a stretch of DNA that codes for a protein, the new allele may result in a polypeptide with a novel amino acid sequence. If the mutation occurs in a stretch of DNA that codes for regulatory RNA, the new allele may result in a change in regulation of the expression of other alleles.

- *Chromosome-level mutations* One consequence of chromosome mutation is gene duplication, which increases the number of copies of a gene. If duplicated genes diversify via point mutations, they can lose their function or create new alleles.

- *Lateral gene transfer (also known as horizontal gene transfer)* New studies suggest that the transfer of genes from one species to another, rather than from parent to offspring, might be a more important source of genetic variation than previously realized (Chapter 21).

Because errors and chromosome damage are inevitable, mutation constantly introduces new alleles into populations in every generation. Mutation is an evolutionary process that increases genetic diversity in populations.

Even though it consistently leads to an increase in genetic diversity in a population, mutation is random with respect to the affected allele's impact on the fitness of the individual. Changes in the makeup of chromosomes or in specific DNA sequences do not occur in ways that tend to increase fitness or decrease fitness. Mutation just happens. *—random + uncontrollable*

But because most organisms are well adapted to their current habitat, random changes in genes usually result in products that do not work as well as the alleles that currently exist. Stated another way, most mutations in sequences that code for a functional protein or RNA result in deleterious alleles, which lower fitness. Deleterious alleles tend to be eliminated by purifying selection.

On rare occasions, however, mutation in these types of sequences produces a **beneficial** allele—an allele that allows individuals to produce more offspring. Beneficial alleles should increase in frequency in the population due to natural selection.

Because mutation produces new alleles, it can in principle change the frequencies of alleles through time. But does mutation *alone* occur often enough to make it an important factor in changing allele frequencies? The short answer is no.

Mutation as an Evolutionary Process
↳ several mutations must occur to make a significant diunless in simple things like bacteria or archaea

To understand why mutation is not a significant mechanism of evolutionary change by itself, consider that the highest mutation rates that have been recorded at individual genes in humans are on the order of 1 mutation in every 10,000 gametes produced by an individual. This rate means that for every 10,000 alleles produced, on average one will have a mutation at the gene in question.

When two gametes combine to form an offspring, then, at most about 1 in every 5000 offspring will carry a mutation at a particular gene. Now suppose that 195,000 humans live in a population, that 5000 offspring are born one year, and that at the end of that year, the population numbers 200,000. Humans are diploid, so in a population this size there are, in total, 400,000 copies of each gene. Only one of them is a new allele created by mutation, however. Over the course of a year, the allele frequency change introduced by mutation is 1/400,000, or 0.0000025 (2.5×10^{-6}). At this rate, it would take 4000 years for mutation to produce a change in allele frequency of 1 percent.

These calculations support the conclusion that mutation does little on its own to change allele frequencies. Although mutation can be a significant evolutionary process in bacteria and archaea, which have extremely short generation times, mutation in eukaryotes rarely causes a change from the genotype frequencies expected under the Hardy–Weinberg principle. As an evolutionary process, mutation is slow compared with selection, genetic drift, and gene flow.

However, mutation can have a very large effect on evolution when *combined* with genetic drift, gene flow, and selection. Let's consider two examples.

Experimental Studies of Mutation

Consider a lab experiment designed by Richard Lenski and colleagues to evaluate the role that mutation plays over many generations.

Experimental Evolution   Lenski's group focused on *Escherichia coli*, a bacterium that is a common resident of the human intestine. To begin, they set up a large series of populations, each founded with a single cell, and allowed them to replicate for over four years—about 10,000 generations. The strain of *E. coli* used in the experiment is completely asexual and reproduces by cell division. Thus, mutation was the only source of genetic variation in these populations.

The biologists saved a sample of cells from each population at regular intervals during the experiment and stored them in a freezer. Because frozen *E. coli* cells resume growth when they are thawed, the frozen cells serve as an archive of individuals from different generations.

Were cells from the older and newer generations of each population different? Lenski's group used competition experiments to address this question. They grew cells from two different generations on the same plate and compared their growth rates. The populations of cells that were more numerous had grown the fastest, meaning that they were better adapted to the experimental environment.

In this way the researchers could measure the fitness of descendant populations relative to ancestral populations. Relative fitness values greater than 1 meant that recent-generation cells outnumbered older-generation cells following the competition.

Fitness Increased in Fits and Starts   The data from the competition experiments are graphed in **FIGURE 26.17**. Notice that relative fitness increased dramatically—almost 30 percent—over time. But notice also that fitness increased in fits and starts. This pattern is emphasized by the solid line on the graph, which represents a mathematical function fitted to the data points.

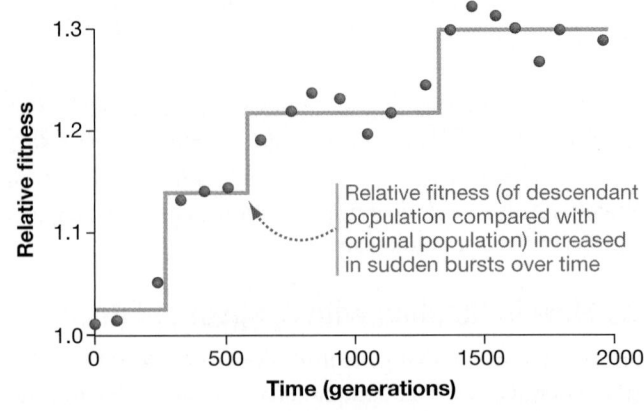

FIGURE 26.17 Evidence of Mutation in Lab Populations of *E. coli*.

DATA: Elena, S. F., V. S. Cooper, and F. E. Lenski. 1996. *Science* 272: 1802–1804.

What caused this stair-step pattern? Lenski's group hypothesized that genetic drift was relatively unimportant in this experiment because population sizes were so large. Instead, they proposed that each jump was caused by a novel mutation that conferred a fitness benefit under selection. Their interpretation was that cells that happened to have the beneficial mutation grew rapidly and came to dominate the population.

After a beneficial mutation occurred, the fitness of the population stabilized—sometimes for hundreds of generations—until another random but beneficial mutation occurred and produced another jump in fitness. These results demonstrate the combined effects of mutation and natural selection.

Studies of Mutation in Natural Populations

Lenski's experiment demonstrates the effect of cumulative mutations in laboratory populations. Now consider a recent study showing how several forms of mutation have combined with natural selection to create color variations, or polymorphisms, in pea aphids.

Lateral Gene Transfer Pea aphids (*Acyrthosiphon pisum*) are small insects that feed on plant sap. They occur in two colors in the wild, red and green, and both colors coexist within populations (**FIGURE 26.18a**). The two phenotypes are maintained in the population due to frequency-dependent selection—ladybird beetles are more likely to prey on red aphids, whereas parasitoid wasps are more likely to lay their eggs in green aphids. The balancing selection on both colors preserves genetic variation in the populations.

The color of aphids is determined by carotenoid pigments, the same group of pigments that give zebra finches their bright orange beaks. Finches and other animals get their carotenoid pigments from the food they eat, mostly from plants.

But there is a twist. Pea aphids do *not* acquire their carotenoids from the plant sap they eat. So where do the pigments come from? Researchers recently examined several bacteria associated with pea aphids and concluded that the carotenoids did not originate from these symbionts either. Instead, other researchers found that the aphids are generating their *own* carotenoids—the first animals ever discovered to have this ability.

If the ancestors of aphids did not have the genes that code for the necessary enzymes for carotenoid biosynthesis, how did the pea aphids obtain this pathway? The answer is by mutation, more specifically by lateral gene transfer—from the genome of a fungal symbiont to the genome of a recent ancestor of the pea aphids. This result was surprising—clear evidence of lateral gene transfer from one eukaryote to another.

Gene Duplication Lateral gene transfer is only part of the mutation story in pea aphids. Carotenoids are a family of molecules with similar pigment and a number of intermediate forms. Two types of enzymes are primarily responsible for the biosynthesis of these molecules in both pea aphids and their fungal ancestor.

A comparison of aphid and fungal genome sequences suggests that after the fungal genes for these two types of enzymes

(a) Red–green color polymorphism in pea aphids

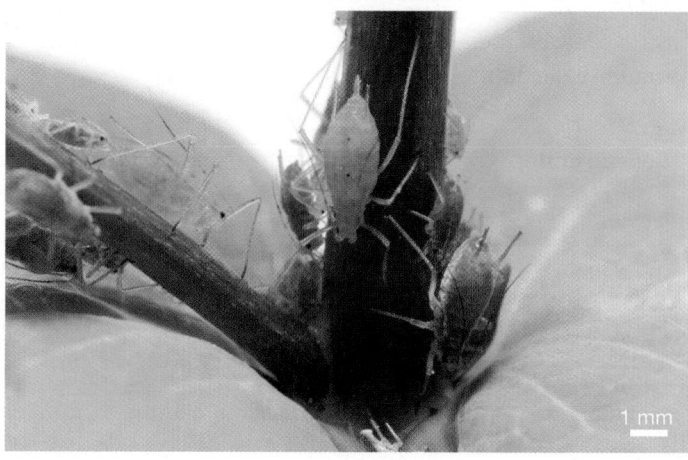

1 mm

(b) Origin of genes for carotenoid synthesis enzymes in aphids

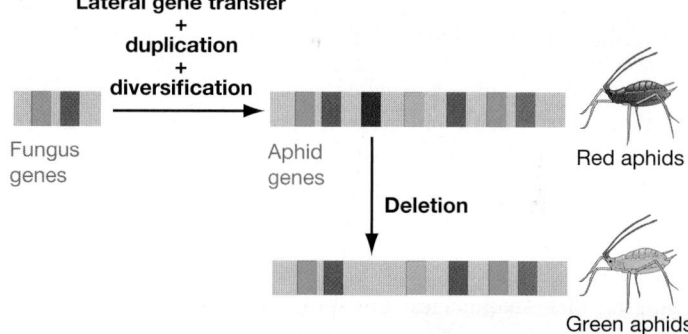

Lateral gene transfer
+
duplication
+
diversification

Fungus genes → Aphid genes

Deletion

Red aphids

Green aphids

FIGURE 26.18 Evidence of Mutation in Pea Aphids.

were transferred to aphids, they underwent further mutations—duplications, sequence diversification by point mutations, and deletion. As a result, red pea aphids have the enzymes necessary to synthesize yellow, green, and red carotenoid pigments, while green pea aphids have the enzymes necessary to synthesize only the yellow and green ones (**FIGURE 26.18b**). The deletion of a gene in the green aphid genome illustrates an important point: Sometimes a loss-of-function mutation can be adaptive. In this case, being green helps aphids to avoid predation by ladybird beetles.

Take-Home Messages

The research on mutations in *E. coli* populations in the lab and the origin of color polymorphism in pea aphids reinforces three fundamental messages about the role of mutation in evolution.

1. Mutation is the ultimate source of genetic variation. Crossing over and independent assortment shuffle existing alleles into new combinations, but only mutation creates new alleles. Mutations just happen—organisms cannot create mutations because they "want" or "need" them.

2. If mutation did not occur, evolution would eventually stop. Recall that natural selection and genetic drift tend to eliminate alleles. Without mutation, eventually there would be no variation for selection and drift to act on.

	Definition/Description	Effect on Genetic Variation	Effect on Average Fitness
Natural selection	Certain alleles are favored	Can maintain, increase, or reduce genetic variation	Can produce adaptation, increasing fitness
Genetic drift	Random changes in allele frequencies; most important in small populations	Tends to reduce genetic variation via loss or fixation of alleles	Random with respect to fitness; usually reduces average fitness
Gene flow	Movement of alleles between populations; reduces differences between populations	May increase genetic variation by introducing new alleles; may decrease it by removing alleles	Random with respect to fitness; may increase or decrease average fitness by introducing high- or low-fitness alleles
Mutation	Production of new alleles	Increases genetic variation by introducing new alleles	Random with respect to fitness; most mutations in coding sequences lower fitness

3. Mutation *alone* is usually inconsequential in changing allele frequencies at a particular gene. However, when considered across the genome and when combined with natural selection, genetic drift, and gene flow, mutation becomes an important evolutionary process.

TABLE 26.4 summarizes these points and similar conclusions about the four evolutionary processes. Each of the four evolutionary processes has different consequences for allele frequencies, all of them violating Hardy–Weinberg predictions. These processes interact to create new species—that is, new branches on the tree of life (Chapter 27). The ultimate result is biological diversity (Chapter 28 and Unit 6). You can see how nonrandom mating and the four evolutionary processes fit into the Big Picture of Evolution on pages 526–527.

CHAPTER 26 REVIEW

For media, go to MasteringBiology

If you understand . . .

26.1 Analyzing Change in Allele Frequencies: The Hardy–Weinberg Principle

- The Hardy–Weinberg principle played an important role in the synthesis of Mendelian genetics and Darwinian evolution.

- The Hardy–Weinberg principle can serve as a null hypothesis in evolutionary studies because it predicts what genotype and allele frequencies are expected to be if mating is random with respect to the gene in question and if none of the four evolutionary processes is operating on that gene.

✔ QUANTITATIVE You should be able to write out the Hardy–Weinberg equations for allele and genotype frequencies and calculate the genotype frequencies expected if the frequency of A_1 is 0.2.

26.2 Nonrandom Mating

- Nonrandom mating changes only genotype frequencies, not allele frequencies, so is not an evolutionary process itself.

- Inbreeding, or mating among relatives, is a form of nonrandom mating. It leads to an increase in homozygosity and a decrease in heterozygosity.

- Inbreeding can accelerate natural selection and can cause inbreeding depression.

✔ You should be able to predict how extensive inbreeding during the 1700s and 1800s affected the royal families of Europe.

26.3 Natural Selection

- Natural selection is the only evolutionary process that produces adaptation.

- Directional selection favors phenotypes at one end of a distribution. It decreases allelic diversity in populations.

- Stabilizing selection eliminates phenotypes with extreme characteristics. It decreases allelic diversity in populations.

- Disruptive selection favors extreme phenotypes and thus maintains genetic variation in populations. Disruptive selection sometimes leads to the formation of new species.

- Balancing selection occurs when no single phenotype is favored; there is a balance among alleles in terms of fitness and frequency. Balancing selection preserves genetic variation.

- Sexual selection is a type of natural selection that leads to the evolution of traits that help individuals attract mates. It usually has a stronger effect on males than on females.

✔ You should be able to explain why natural selection violates the Hardy–Weinberg principle.

26.4 Genetic Drift

- Genetic drift causes random changes in allele frequencies.
- Genetic drift is particularly important in small populations, and it tends to reduce overall genetic diversity.
- Genetic drift can result from random fusion of gametes at fertilization, founder events, and population bottlenecks.

✔ You should be able to suggest how genetic drift could be important to the management of endangered species.

26.5 Gene Flow

- Gene flow equalizes allele frequencies between populations.
- Gene flow can introduce alleles from one population to another when individuals migrate among populations.
- The introduced alleles may have a beneficial, neutral, or deleterious effect.

✔ You should be able to suggest how gene flow could be important to the management of endangered species.

26.6 Mutation

- Mutation is the only evolutionary process that creates new alleles. They may be beneficial, neutral, or deleterious.
- Mutation occurs too infrequently to be a major cause of change in allele frequency alone, but it is important when combined with natural selection, genetic drift, and gene flow.

✔ You should be able to suggest how mutation could be important to the management of endangered species.

(MB) MasteringBiology

1. **MasteringBiology Assignments**

   Tutorials and Activities Causes of Evolutionary Change, Hardy–Weinberg Principle, Mechanisms of Evolution, Three Modes of Natural Selection

   Questions Reading Quizzes, Blue-Thread Questions, Test Bank

2. **eText** Read your book online, search, take notes, highlight text, and more.

3. **The Study Area** Practice Test, Cumulative Test, BioFlix® 3-D Animations, Videos, Activities, Audio Glossary, Word Study Tools, Art

You should be able to . . .

✔ TEST YOUR KNOWLEDGE

1. In what sense is the Hardy–Weinberg principle a null hypothesis?

2. Why isn't inbreeding considered an evolutionary process?
   a. It does not change genotype frequencies.
   b. It does not change allele frequencies.
   c. It does not occur often enough to be important in evolution.
   d. It does not violate the assumptions of the Hardy–Weinberg principle.

3. Why is genetic drift aptly named?
   a. It causes allele frequencies to drift up or down randomly.
   b. It occurs when alleles from one population drift into another.
   c. It occurs when mutations drift into a genome.
   d. It occurs when populations drift into new habitats.

4. What does it mean when an allele reaches "fixation"?
   a. It is eliminated from the population.
   b. It has a frequency of 1.0.
   c. It is dominant to all other alleles.
   d. It is adaptive.

5. True or false? Gene flow can either increase or decrease the average fitness of a population. true

6. Mutation is the ultimate source of genetic variability. Why is this statement correct?
   a. DNA polymerase (the enzyme that copies DNA) is remarkably accurate.
   b. "Mutation proposes and selection disposes."
   c. Mutation is the only source of new alleles.
   d. Mutation occurs in response to natural selection. It generates the alleles that are required for a population to adapt to a particular habitat.

✔ TEST YOUR UNDERSTANDING

7. **QUANTITATIVE** In a population of 2500, how many babies would you expect to have cystic fibrosis, a homozygous recessive condition, if the frequency of the dominant allele is 0.9 and the population is at Hardy–Weinberg equilibrium?
   a. $0.9 \times 2500 = 2025$
   b. $2 \times 0.9 \times 0.1 \times 2500 = 800$
   c. $0.9 \times 0.1 \times 2500 = 400$
   d. $0.1 \times 0.1 \times 2500 = 25$

8. Suggest why inbreeding could cause recessive deleterious alleles for cystic fibrosis to be "purged" from a population.

9. Determine what is incorrect in the following statement: Red aphids mutated their genes so that they could be green and avoid predation by ladybird beetles.

10. Why does sexual selection often lead to sexual dimorphism?

11. Consider an allele that increases reproductive success in elephant seal males versus an allele that increases reproductive success in females. Which allele will increase in frequency faster, and why?

12. Draw a small concept map (**BioSkills 15**) showing how selection, genetic drift, gene flow, and mutation relate to genetic variation.

13. QUANTITATIVE In humans, albinism is caused by loss-of-function mutations in genes involved in the synthesis of melanin, the dark pigment in skin. Only people homozygous for a loss-of-function allele (genotype *aa*) have the albino phenotype. In Americans of northern European ancestry, albino individuals are present at a frequency of about 1 in 10,000 (or 0.0001). Assuming that genotypes are in Hardy–Weinberg proportions, what is the frequency of Caucasians in the United States who carry an allele for albinism?

14. A group of researchers presented artificial calls of the male cricket *Teleogryllus commodus* to female crickets in the lab to measure selection for male calls. They used artificial calls so that they could vary properties such as length of the pause between calls and number of trills in each call. When positive selection was plotted as a function of call characteristics, the shapes of the selection curves started low, peaked in the middle, and ended low. What kind of selection is occurring?
a. sexual selection
b. balancing selection
c. sexual selection and stabilizing selection
d. sexual selection and balancing selection

15. Suppose you were studying several species of human. In one, males never lifted a finger to help females raise children. In another, males provided just as much parental care as females except for actually carrying the baby during pregnancy. How does the fundamental asymmetry of sex compare in the two species? How would you expect sexual dimorphism to compare between the two species?

16. You are a conservation biologist charged with creating a recovery plan for an endangered species of turtle. The turtle's habitat has been fragmented by suburbanization and highway construction into small, isolated, but protected areas. Some evidence indicates that certain turtle populations are adapted to typical freshwater marshes, whereas others are adapted to acidic wetlands or salty habitats. Further, some turtle populations number less than 25 breeding adults, making genetic drift and inbreeding a major concern. In creating a recovery plan, the tools at your disposal are captive breeding, the capture and transfer of adults to create gene flow, or the creation of habitat corridors between wetlands to make migration possible. How would you use gene flow to help this species?

27 Speciation

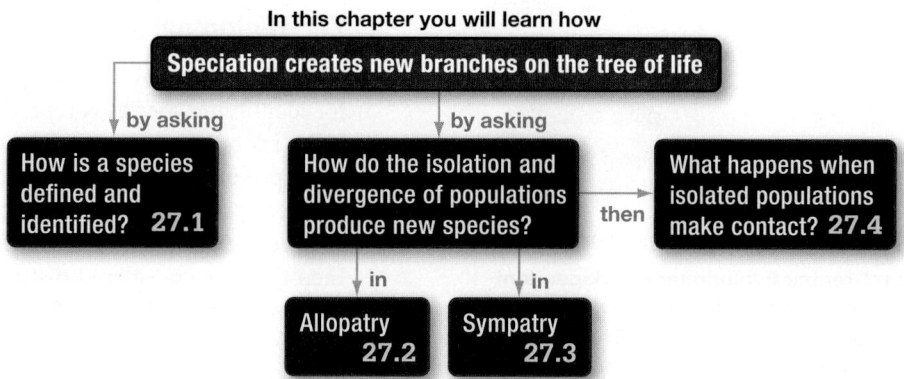

In this chapter you will learn how

Speciation creates new branches on the tree of life

by asking

How is a species defined and identified? 27.1

by asking

How do the isolation and divergence of populations produce new species?

then

What happens when isolated populations make contact? 27.4

in

Allopatry 27.2

in

Sympatry 27.3

New sunflower species have formed recently in southwestern North America. This chapter explains how.

Although Darwin called his masterwork *On the Origin of Species by Means of Natural Selection*, he actually had little to say about how new species arise. Instead, his data and analyses focused on the process of natural selection and the changes that occur *within* populations over time.

Recall that there are four evolutionary processes—natural selection, genetic drift, mutation, and gene flow—and that gene flow makes allele frequencies more similar among populations (see Chapter 26). If gene flow ends, allele frequencies in isolated populations are free to diverge—meaning that the populations begin to evolve independently of each other. For example, when a new mutation creates an allele that changes the phenotype of individuals in one population, there is no longer any way for that allele to appear in the other population. If mutation, selection, and genetic drift cause isolated populations to diverge sufficiently, distinct types, or species, form—that is, the process of speciation takes place.

This chapter is part of the Big Picture. See how on pages 526–527.

✔ When you see this checkmark, stop and test yourself. Answers are available in Appendix A.

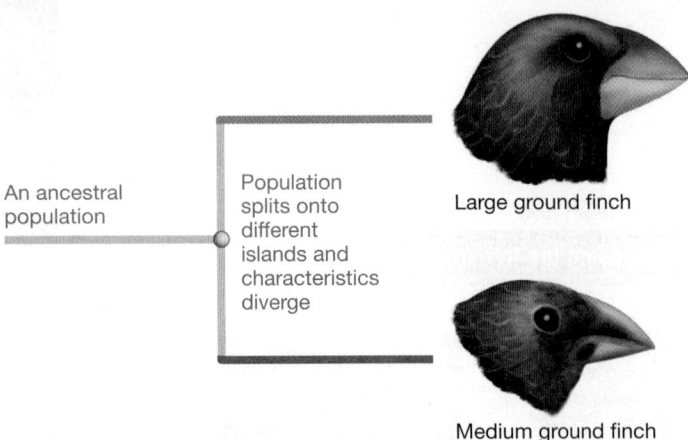

An ancestral population

Population splits onto different islands and characteristics diverge

Large ground finch

Medium ground finch

FIGURE 27.1 Speciation Creates Evolutionarily Independent Populations. The large ground finch and medium ground finch are derived from the same ancestral population. This ancestral population split into two populations isolated by lack of gene flow. Because the populations began evolving independently, over time they acquired the distinctive characteristics observed today.

In essence, speciation results from *genetic isolation* and *genetic divergence*. Genetic isolation results from lack of gene flow, and genetic divergence occurs because selection, genetic drift, and mutation proceed independently in the isolated populations.

Speciation is a splitting event that creates two or more distinct species from a single ancestral species (**FIGURE 27.1**). When speciation is complete, a new branch has been added to the tree of life. Thus, the study of speciation is critical to understanding how life evolved and has many practical implications for the preservation of biodiversity on Earth (see Chapter 57). You can see how speciation fits into the Big Picture of Evolution on pages 526–527.

Let's begin by considering the *pattern* component of studying speciation—how can we define and identify species? Then let's explore the *process* of speciation—how do new species form?

27.1 How Are Species Defined and Identified?

If your friend were to tell you that she was planning to study polar bears and grizzly bears during her summer research project, you would immediately recognize these animals as distinct species. Polar bears and grizzly bears are distinguishable from one another in appearance, behavior, habitat use, and other traits. But what if your friend was going to compare elephants that live in the forests and savannas of Africa? Are these elephants the same species or two different species?

Evolutionary biologists have been wrangling with the definition of species for decades—how can you reliably distinguish two or more species of bears, elephants, or bacteria in the field or fossil record? Although there is no single, universal answer, scientists do agree there is a distinction between the *definition* of species in general and the criteria that can be used to *identify* species in particular cases.

In general, a **species** is defined as an evolutionarily independent population or group of populations. Of the many criteria that can be used to identify species, the three most common ones are (1) the biological species concept, (2) the morphospecies concept, and (3) the phylogenetic species concept.

The Biological Species Concept

According to the **biological species concept,** the main criterion for identifying species is reproductive isolation. This is a logical yardstick because no gene flow occurs between populations that are reproductively isolated from each other. Specifically, if two different populations do not interbreed in nature, or if they fail to produce viable and fertile offspring when matings take place, then they are considered distinct species. The influential evolutionary biologist Ernst Mayr strongly promoted the biological species concept because the criterion of reproductive isolation enables clear evidence of evolutionary independence.

Reproductive isolation can result from a wide variety of events and processes. To organize the various mechanisms that stop gene flow between populations, biologists distinguish between:

- **prezygotic** (literally, "before-zygote") **isolation,** which prevents individuals of different species from mating; and

- **postzygotic** ("after-zygote") **isolation,** in which the offspring of matings between members of different species do not survive or reproduce.

TABLE 27.1 outlines some of the more important mechanisms of prezygotic and postzygotic isolation.

Although the biological species concept has a strong theoretical foundation, it has disadvantages. The criterion of reproductive isolation cannot be evaluated in fossils or in species that reproduce asexually. What's more, the concept is difficult to apply when closely related populations do not happen to overlap with each other geographically. In this case, biologists are left to guess whether interbreeding and gene flow would occur if the populations happened to come into contact.

Further, reproductive isolation can sometimes be a complex gradient rather than an all-or-nothing scenario. For example, if a male lion and female tiger produce a rare "liger" that survives and reproduces in captivity (the ranges of these cats do not overlap in nature), are lions and tigers necessarily the same species?

The Morphospecies Concept

How do biologists identify species when the criterion of reproductive isolation cannot be applied? Under the **morphospecies** ("form-species") **concept,** researchers identify evolutionarily independent lineages by differences in size, shape, or other morphological features. The logic behind the morphospecies concept is that distinguishing features are most likely to arise if populations are independent and isolated from gene flow.

The morphospecies concept is compelling simply because it is so widely applicable. It is a useful criterion when biologists have no data on the extent of gene flow, and it is equally applicable to sexual, asexual, or fossil species. Its disadvantages

TABLE 27.1 Mechanisms of Reproductive Isolation

	Process	Example
Prezygotic Isolation		
Temporal	Populations are isolated because they breed at different times.	Bishop pines and Monterey pines release their pollen at different times of the year.
Habitat	Populations are isolated because they breed in different habitats.	Parasites that begin to exploit new host species are isolated from their original population.
Behavioral	Populations do not interbreed because their courtship displays differ.	To attract male fireflies, female fireflies give a species-specific sequence of flashes.
Gametic barrier	Matings fail because eggs and sperm are incompatible.	In sea urchins, a protein called bindin allows sperm to penetrate eggs. Differences in the amino acid sequence of bindin cause matings to fail between closely related populations.
Mechanical	Matings fail because male and female reproductive structures are incompatible.	In alpine skypilots (a flowering plant), the length of the floral tube varies. Bees can pollinate in populations with short tubes, but only hummingbirds can pollinate in populations with long tubes.
Postzygotic Isolation		
Hybrid viability	Hybrid offspring do not develop normally and die as embryos.	When ring-necked doves mate with rock doves, less than 6 percent of eggs hatch.
Hybrid sterility	Hybrid offspring mature but are sterile as adults.	Eastern meadowlarks and western meadowlarks are almost identical morphologically, but their hybrid offspring are usually infertile.

are that **(1)** it may lead to the naming of two or more species when there is only one **polymorphic species** with differing phenotypes, such as the spotted and black morphs of jaguars; **(2)** it cannot identify **cryptic species,** which differ in traits other than morphology, such as the meadowlarks in Table 27.1; and **(3)** the morphological features used to distinguish species are subjective. In the worst case, different researchers working on the same populations disagree on the characters that distinguish species, such as when paleontologists disagree on the identity of an extinct human species based on measurements of fossil bone fragments.

The Phylogenetic Species Concept

The **phylogenetic species concept** identifies species based on the evolutionary history of populations. The reasoning behind the phylogenetic species concept begins with Darwin's theory that all species are related by common ancestry. In modern terms, all species form a monophyletic ("one-tribe") group—the tree of life.

A **monophyletic group,** also called a **clade** or **lineage,** consists of an ancestral population, all of its descendants, and *only* those descendants. On any given evolutionary tree, there are many monophyletic groups (**FIGURE 27.2a**).

(a) Monophyletic groups

━ **Monophyletic group**: an ancestral population and all descendants

▪┣ **Synapomorphy**: trait unique to a monophyletic group

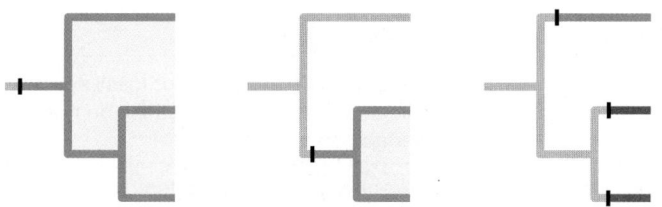

(b) Phylogenetic species: smallest monophyletic groups

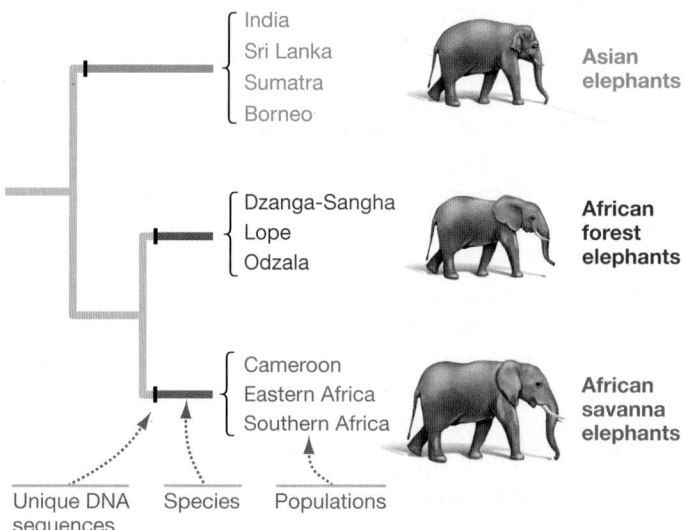

India
Sri Lanka
Sumatra
Borneo
Asian elephants

Dzanga-Sangha
Lope
Odzala
African forest elephants

Cameroon
Eastern Africa
Southern Africa
African savanna elephants

Unique DNA sequences   Species   Populations

FIGURE 27.2 The Phylogenetic Species Concept Is Based on Monophyletic Groups. (a) Monophyletic groups can be mapped on phylogenetic trees. **(b)** Although there are several isolated populations of elephants on two continents, the phylogenetic species concept identifies no more or less than three species of elephants. This tree was created by comparing DNA sequences. Thus, the synapomorphies in this case are DNA sequences unique to each species.

Monophyletic groups, in turn, are identified by traits called synapomorphies ("unique-forms"). A **synapomorphy** is a trait that is found in certain groups of organisms and their common ancestor, but is missing in more distant ancestors. Fur and lactation, for example, are synapomorphies that identify mammals as a monophyletic group.

Synapomorphies are homologous traits that can be identified at the genetic, developmental, or structural level. In many cases, researchers use DNA sequence data to identify synapomorphies and estimate phylogenetic trees. (Chapter 28 explores the data and logic that biologists use to reconstruct phylogenies, and **BioSkills 7** in Appendix B reviews how to read phylogenetic trees.)

Under the phylogenetic species concept, species are defined as the smallest monophyletic groups on the tree of life. Phylogenetic species are made up of populations that share one or more synapomorphies.

The tree in **FIGURE 27.2b**, for example, is based on DNA sequence data from an array of elephant populations. Each of the populations is not distinctive enough to represent separate branches. The phylogenetic species on the tree are labeled Asian elephants, African forest elephants, and African savanna elephants. These are the smallest monophyletic groups on the tree.

The names at some of the tips on this tree (Cameroon, Eastern Africa, Southern Africa, etc.) represent populations within species. These populations may be separated geographically, but their characteristics are so similar that they do not form independent branches on the tree. They are simply part of the same monophyletic group containing other populations.

The phylogenetic species concept has two distinct advantages: **(1)** It can be applied to any population (fossil, asexual, or sexual), and **(2)** it is logical because different species have different synapomorphies only if they are isolated from gene flow and have evolved independently. The approach has a distinct disadvantage, however: Carefully estimated phylogenies are available only for a tiny (though growing) subset of populations on the tree of life.

Critics of the phylogenetic species concept also point out that it would probably lead to recognition of many more species than either the morphospecies or biological species concepts.

Proponents counter that, far from being a disadvantage, the recognition of increased numbers of species might better reflect the extent of life's diversity.

The three major species concepts are summarized in **TABLE 27.2**. In practice, researchers use a combination of species concepts to identify evolutionarily independent populations in nature. Conflicts have occurred, however, as in the case of the dusky seaside sparrow.

Species Definitions in Action: The Case of the Dusky Seaside Sparrow

Seaside sparrows live in salt marshes along the Atlantic and Gulf Coasts of the United States. The scientific name of this species is *Ammodramus maritimus*. Recall that scientific names consist of a genus name followed by a species name (Chapter 1).

Using the morphospecies and biological species concepts, researchers had traditionally named six seaside sparrow "subspecies" (**FIGURE 27.3a**). **Subspecies** are populations that live in discrete geographic areas and have distinguishing features, such as coloration or calls, but are not considered distinct enough to be called separate species. For example, *A. m. nigrescens* (*nigrescens* indicates the subspecies name) lives on the Atlantic Coast of Florida, while *A. m. peninsulae* lives on the Gulf Coast of Florida. These subspecies can interbreed if the geographical barriers to their isolation are removed.

By the late 1960s, biologists began to be concerned about the future of some seaside sparrow subspecies because their habitats were increasingly threatened by agriculture and ocean-front housing. Further, since the seaside sparrow subspecies are physically isolated from one another, and because young seaside sparrows tend to breed near where they hatched, little or no gene flow occurs among populations.

A subspecies called the dusky seaside sparrow (*A. m. nigrescens*) was in particular trouble; by 1980 only six individuals from this population remained. All were males. At this point government and private conservation agencies sprang into action under the Endangered Species Act, a law whose goal is to prevent the extinction of species and, as in this case, subspecies.

SUMMARY TABLE 27.2 **Three Most Common Species Concepts**

	Criterion for Identifying Populations as Species	Advantages	Disadvantages
Biological species	Reproductive isolation between populations (they don't breed and don't produce viable, fertile offspring)	Reproductive isolation = evolutionary independence	Not applicable to asexual or fossil species; difficult to assess if populations do not overlap geographically.
Morphospecies	Morphologically distinct populations	Widely applicable	Subjective (researchers often disagree about how much or what kinds of morphological distinction indicate speciation); misidentifies polymorphic species; misses cryptic species.
Phylogenetic species	Smallest monophyletic group on phylogenetic tree	Widely applicable; based on testable criteria	Relatively few well-estimated phylogenies are currently available.

(a) Each subspecies of seaside sparrow has a restricted range.

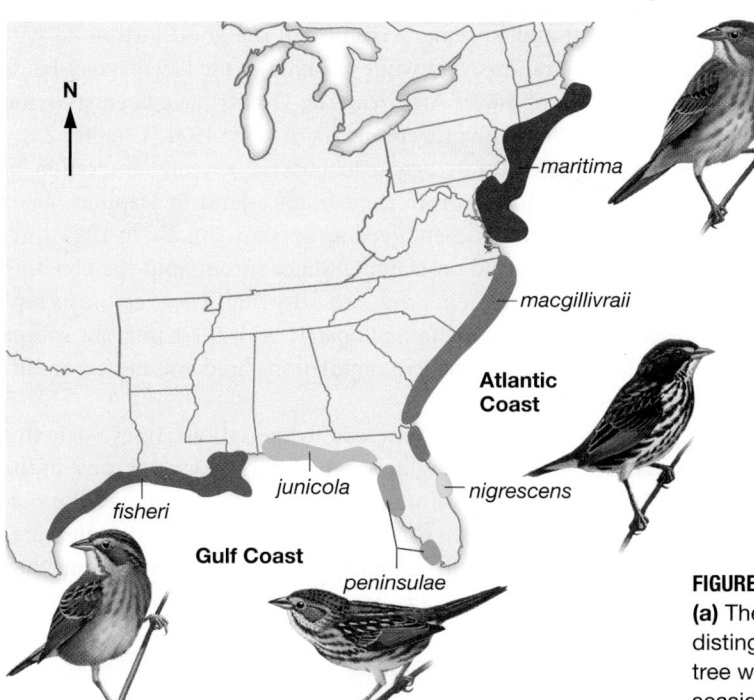

(b) The six subspecies form two monophyletic groups when DNA sequences are compared.

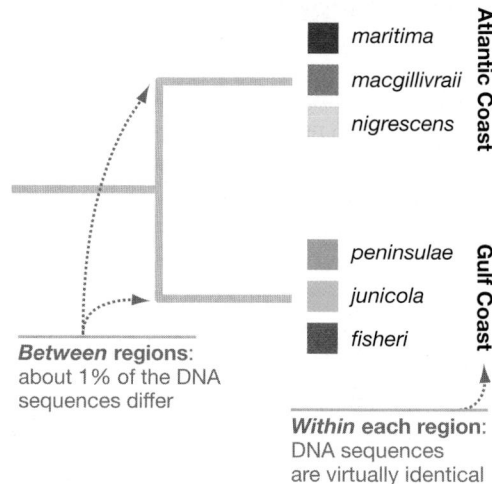

Between regions: about 1% of the DNA sequences differ

Within each region: DNA sequences are virtually identical

FIGURE 27.3 Seaside Sparrows Form Two Monophyletic Groups.
(a) The subspecies of seaside sparrows named on this map are distinguished by their distinctive coloration and songs. **(b)** This phylogenetic tree was constructed by comparing DNA sequences. The tree shows that seaside sparrows represent two distinct monophyletic groups, one native to the Atlantic Coast and the other native to the Gulf Coast.

To launch the rescue program, the remaining male dusky seaside sparrows were taken into captivity and bred with females from the nearby subspecies, *A. m. peninsulae*. Officials planned to use these hybrid offspring as breeding stock for a reintroduction program. The goal was to preserve as much genetic diversity as possible by reestablishing a healthy population of dusky-like birds.

The plan was thrown into turmoil, however, when a different group of biologists estimated the phylogeny of the seaside sparrows by comparing gene sequences. The tree in **FIGURE 27.3b** shows that seaside sparrows represent just two distinct monophyletic groups: one native to the Atlantic Coast and the other native to the Gulf Coast. Under the phylogenetic species concept, the dusky sparrow is part of the same monophyletic group that includes the other Atlantic Coast sparrows.

Officials had unwittingly crossed the dusky males, *A. m. nigrescens* from the Atlantic Coast, with females from the Gulf Coast lineage—a population that may have been geographically isolated from dusky sparrows for more than 250,000 years. Because the goal of the conservation effort was to preserve existing genetic diversity, this was the wrong population to use.

The researchers who did the phylogenetic analysis maintained that the biological and morphospecies concepts had misled a well-intentioned conservation program. Under the phylogenetic species concept, they claimed that officials should have allowed the dusky sparrow to go extinct and then concentrated their efforts on preserving one or more populations from each coast. In this way, the two monophyletic groups of sparrows—and the most genetic diversity—would be preserved.

Under the morphospecies concept, however, officials did the right thing by attempting to preserve all subspecies. They argue that dusky seaside sparrows had distinctive heritable traits, like coloration and songs, that are better "watered down" by crossbreeding than lost completely.

When it comes to conservation, life-and-death decisions like these are crucial. Now our task is to consider a fundamental question about process: How do isolation and divergence produce the event called speciation?

check your understanding

If you understand that . . .

C Y U

- Species are evolutionarily independent because no gene flow occurs between them and other species.
- Biologists use an array of criteria to identify evolutionarily independent groups.

✔ **You should be able to . . .**

1. Explain why the criteria invoked by the biological species, morphospecies, and phylogenetic species concepts allow biologists to identify evolutionarily independent groups.

2. Describe the disadvantages of the biological species, morphospecies, and phylogenetic species concepts.

Answers are available in Appendix A.

27.2 Isolation and Divergence in Allopatry

Speciation begins when gene flow between populations is reduced or eliminated, causing genetic isolation. Genetic isolation happens routinely when populations become geographically separated. Populations that are geographically separated are said to be in **allopatry** ("different-homeland"). Thus, speciation that begins with geographic isolation is known as **allopatric speciation.**

Geographic isolation, in turn, occurs in one of two ways: dispersal or **vicariance,** the physical splitting of a habitat. As **FIGURE 27.4a** illustrates, a population can disperse to a new habitat, colonize it, and found a new population. Alternatively, a new geographic barrier can split a population into two or more subgroups that are physically isolated from each other (**FIGURE 27.4b**).

The case studies that follow address two questions: First, how do dispersal and vicariance events occur? Answering this question takes us into the field of **biogeography**—the study of how species and populations are distributed geographically (see Chapter 52). And second, once populations are physically isolated, how do mutation, genetic drift, and selection produce divergence?

Allopatric Speciation by Dispersal

Peter Grant and Rosemary Grant had the good fortune to witness a colonization event while working in the Galápagos Islands off the coast of South America. The Grants have been studying finches on the island of Daphne Major since 1971 (Chapter 25).

In 1973 the Grants began to observe a yearly migration of large ground finches from their home island to Daphne Major for a few months between breeding seasons. Finally, in 1982, thirteen colonists stayed on Daphne Major throughout the breeding season, and five of them produced offspring. These colonists represented a new population, allopatric with their migrant source population, because the two populations bred and nested on different islands.

Could this dispersal event lead to speciation? To evaluate this question, the Grants caught, weighed, and measured most of the parents and offspring produced on Daphne Major over the succeeding 12 years. When they compared these data with measurements of large ground finches in the migrant population, they discovered that the average beak size in the colonist population was much larger.

Two evolutionary processes could be responsible for the change in beak size:

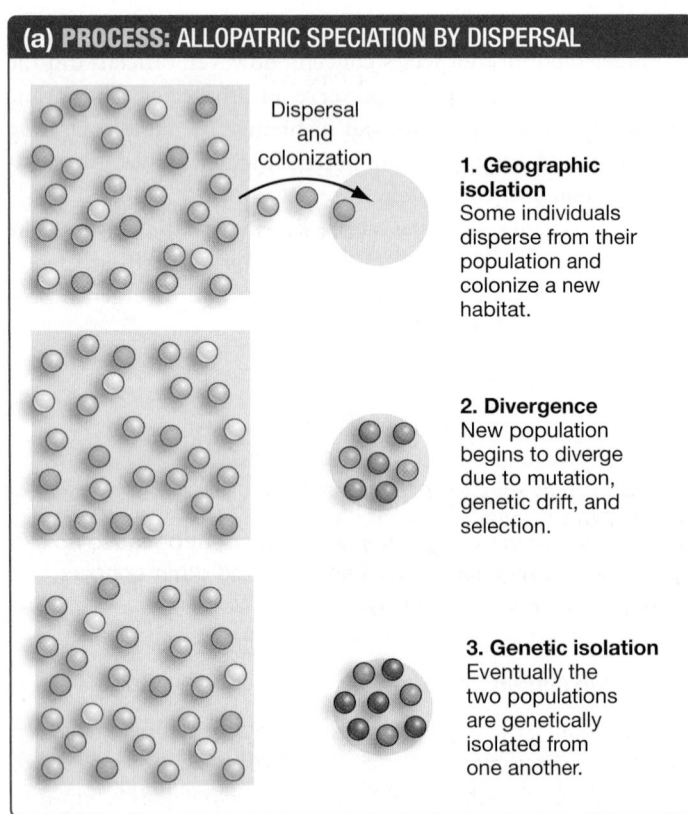

(a) PROCESS: ALLOPATRIC SPECIATION BY DISPERSAL

Dispersal and colonization

1. Geographic isolation
Some individuals disperse from their population and colonize a new habitat.

2. Divergence
New population begins to diverge due to mutation, genetic drift, and selection.

3. Genetic isolation
Eventually the two populations are genetically isolated from one another.

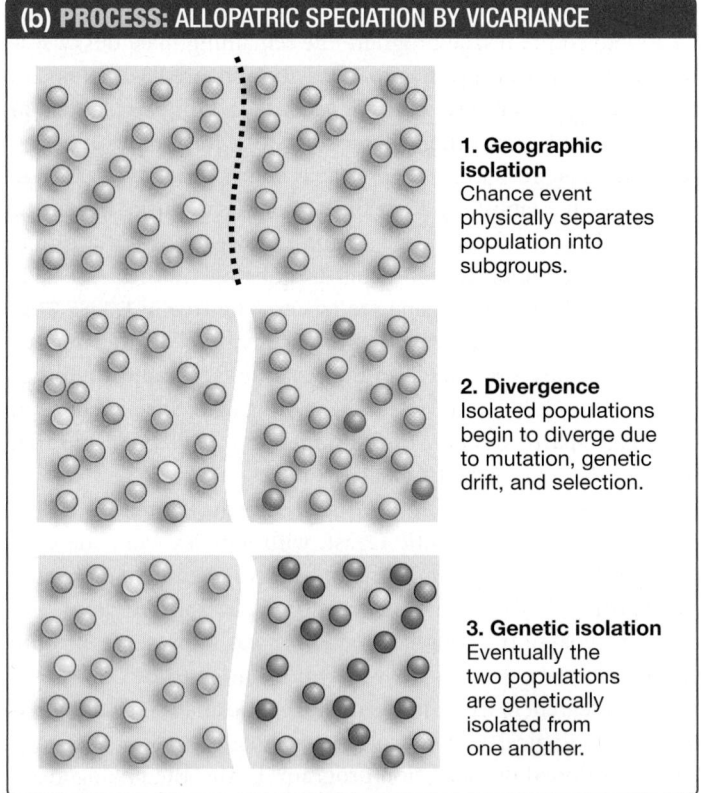

(b) PROCESS: ALLOPATRIC SPECIATION BY VICARIANCE

1. Geographic isolation
Chance event physically separates population into subgroups.

2. Divergence
Isolated populations begin to diverge due to mutation, genetic drift, and selection.

3. Genetic isolation
Eventually the two populations are genetically isolated from one another.

FIGURE 27.4 Allopatric Speciation Begins via Dispersal or Vicariance. (a) When dispersal occurs, colonists establish a new population in a novel location. **(b)** In vicariance, a widespread population becomes fragmented into isolated subgroups. Different colors represent genetic variation.

✔**QUESTION** A continuous salt marsh once existed across what is now central Florida, but it was disrupted by geologic events. How does this discovery support the phylogenetic species concept in seaside sparrows (see Figure 27.3)?

1. Genetic drift produced a colonizing population that happened to have particularly large beaks relative to the migrant population. This is an example of the founder effect (Chapter 26).

2. Natural selection in the new environment could favor alleles associated with large beaks.

The Grants concluded from their detailed observations that both genetic drift and natural selection were at play.

The new population of large ground finches is not yet considered a separate species from the migrant population, because some gene flow continues to occur—such as when a new migrant joins the colonist population. Given enough time, however, the populations may continue to diverge. Dispersal and colonization, followed by genetic drift and natural selection, is thought to be responsible for speciation in Galápagos finches and many other island groups.

Allopatric Speciation by Vicariance

If a new physical barrier such as a mountain range or river splits the geographic range of a species, vicariance has taken place. Such changes are commonplace on Earth due to continental drift (Chapter 28), climate fluctuations, and many other factors (see Chapter 52). For example, geologists estimate that the Isthmus of Panama closed about 3 million years ago, forming a land bridge between North and South America. This vicariance event split the Caribbean and Pacific populations of many marine species, including snapping shrimp (**FIGURE 27.5a**).

A phylogenetic tree of the snapping shrimp shows that many of the species found on either side of the isthmus are **sister species**—meaning that they are each other's closest relative (**FIGURE 27.5b**). This is exactly the pattern predicted if the populations of many species were split in two: the populations on either side of the isthmus subsequently diverged to form distinct species.

To summarize, geographic isolation of populations via dispersal or vicariance produces genetic isolation due to the interruption of gene flow—the first requirement of speciation. When genetic isolation is accompanied by genetic divergence due to mutation, selection, and genetic drift, speciation results.

27.3 Isolation and Divergence in Sympatry

When populations or species live in the same geographic area, or at least close enough to one another to make interbreeding possible, biologists say that they live in **sympatry** ("together-homeland"). Traditionally, evolutionary biologists such as Ernst Mayr predicted that speciation could not occur among sympatric populations because gene flow would easily overwhelm any differences among populations created by genetic drift and natural selection.

But recently, a number of studies have countered this prediction. **Sympatric speciation**—speciation that occurs even

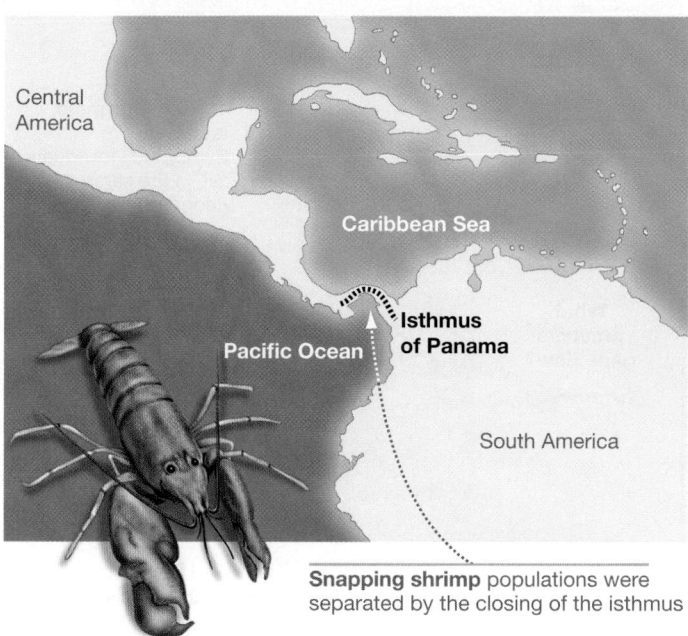

(a) Vicariance event: The closing of the Isthmus of Panama

Central America

Caribbean Sea

Pacific Ocean | Isthmus of Panama

South America

Snapping shrimp populations were separated by the closing of the isthmus

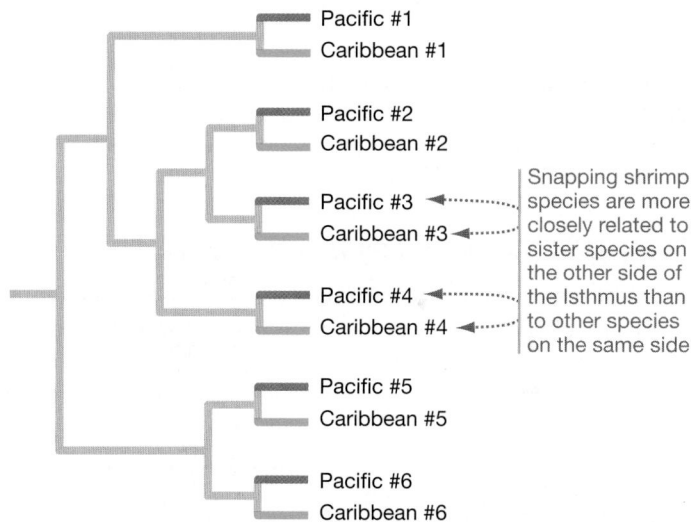

(b) Result: Pairs of sister species straddling the isthmus

Pacific #1
Caribbean #1

Pacific #2
Caribbean #2

Pacific #3
Caribbean #3

Pacific #4
Caribbean #4

Pacific #5
Caribbean #5

Pacific #6
Caribbean #6

Snapping shrimp species are more closely related to sister species on the other side of the Isthmus than to other species on the same side

FIGURE 27.5 Evidence for Speciation by Vicariance in Snapping Shrimp. (a) The Isthmus of Panama separated the Caribbean Sea from the Pacific Ocean about 3 million years ago. **(b)** The most logical interpretation of this phylogenetic tree is that 6 species of snapping shrimp split into 12 species when the Isthmus of Panama formed.

though populations live within the same geographical area (see **FIGURE 27.6** on page 496)—can indeed occur, and serves as a source of new branches on the tree of life. How does it work?

Two types of events can initiate the process of sympatric speciation: **(1)** external events, such as disruptive selection for extreme phenotypes based on different ecological niches (see Chapter 26 for a review of disruptive selection), and **(2)** internal events, such as chromosomal mutations. Let's consider each one in turn.

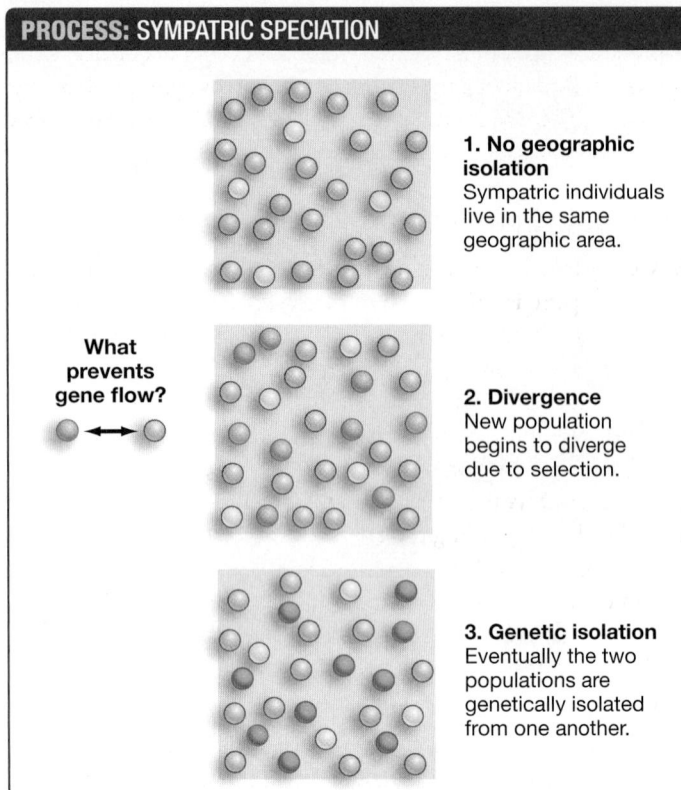

PROCESS: SYMPATRIC SPECIATION

1. No geographic isolation
Sympatric individuals live in the same geographic area.

What prevents gene flow?

2. Divergence
New population begins to diverge due to selection.

3. Genetic isolation
Eventually the two populations are genetically isolated from one another.

FIGURE 27.6 Sympatric Speciation Has Long Perplexed Researchers. Recent studies have documented several mechanisms that reduce gene flow and result in sympatric speciation. Different colors represent genetic variation.

Sympatric Speciation by Disruptive Selection

Biologists use the term **niche** (pronounced *nitch*) to describe the range of ecological resources that a species can use and the range of conditions that it can tolerate (see Chapter 55). A key realization in the study of sympatric speciation was that even though sympatric populations are not geographically isolated, they may become reproductively isolated by adapting to different ecological niches via disruptive selection.

For example, apple maggot flies rely on apples to complete their life cycle—apples are an important part of their niche. Male and female apple maggot flies usually court and mate on apple fruits. The female then lays a fertilized egg inside the fruit, which will be the food source for the growing larva. After the fruit drops off the tree, the larva burrows into the ground and pupates—meaning that it secretes a protective case and undergoes metamorphosis (see Chapter 33). The new adult then emerges the following spring, starting the cycle anew.

Apple trees were introduced to North America from Europe less than 300 years ago, however. Where did apple maggot flies come from? Did they arrive with the apple trees?

Evolutionary biologist Jeffrey Feder and colleagues have been researching this mystery in a long-term study of apple maggot flies and their closest relatives. Phylogenetic trees, estimated from synapomorphies in DNA sequence data, indicate that apple maggot flies are very closely related to hawthorn maggot flies,

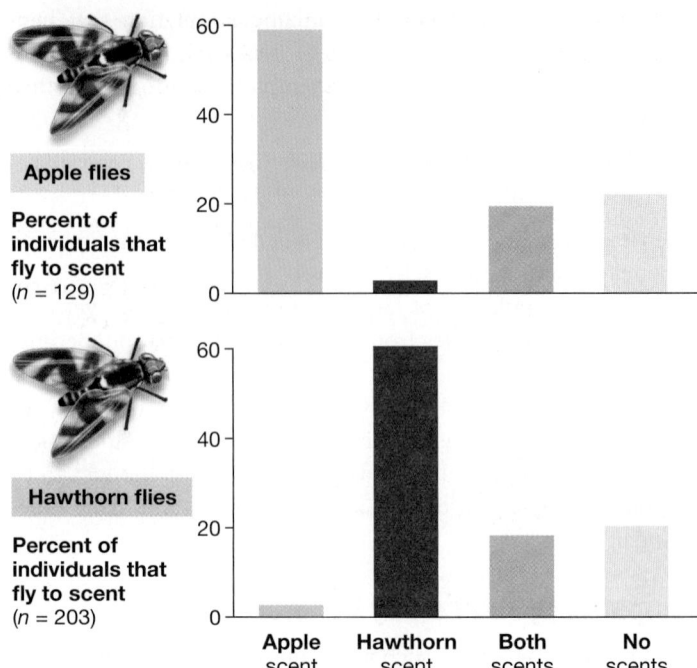

Apple flies

Percent of individuals that fly to scent
(*n* = 129)

Hawthorn flies

Percent of individuals that fly to scent
(*n* = 203)

Apple scent | Hawthorn scent | Both scents | No scents

FIGURE 27.7 Disruptive Selection on Fruit Preference in Flies. Each fly was tested with four types of scent, one at a time, in a lab setting.

DATA: Dambroski, H. R., C. Linn Jr., S. H. Berlocher, et al. 2005. *Evolution* 59: 1953–1964.

✔**QUESTION** Apple maggot flies mate on their food plants. How is this behavior relevant to the population's divergence from hawthorn flies?

which are native to North America. These data suggest that apple maggot flies originated from hawthorn maggot flies following the introduction of apples. Hawthorn flies lay their eggs in hawthorn fruits and apple flies lay their eggs in apples, even though hawthorn trees and apple trees often grow almost side by side. Do the apple flies and hawthorn flies interbreed?

By following marked individuals in the field, Feder's team determined that only about 6 percent of the matings observed are between apple flies and hawthorn flies. The data in **FIGURE 27.7** show why. The bars on the graphs indicate the percentage of apple flies (top) or hawthorn flies (bottom) that land on a surface containing scents from apple, hawthorn, both apple and hawthorn, or neither apple nor hawthorn, in laboratory tests. Notice that

- Apple flies respond most strongly to apple scents; hawthorn flies respond most strongly to hawthorn scents.

- In both types of flies, there is no difference in the response to a mix of both scents and no scent at all.

- Apple flies *avoid* hawthorn scent, and hawthorn flies *avoid* apple scent.

Other experiments have established that (**1**) a fly's ability to discriminate scents has a genetic basis—meaning that apple flies and hawthorn flies have different alleles associated with attraction to fruit; (**2**) specific odor receptor cells are responsible for the difference in scent response; and (**3**) hybrid individuals do not orient to fruit scents as well as their parents.

The upshot is that although apple flies and hawthorn flies live in the same geographic area, prezygotic reproductive isolation is occurring as a result of natural selection for adaptations to two different niches. Apple flies mate on apples (avoiding hawthorn fruits) and hawthorn flies mate on hawthorn fruits (avoiding apples). Hybrid flies have lower fitness due to their reduced success in finding fruits and thus mates, an indication that disruptive selection is occurring (see Chapter 26).

Although they are not yet separate species on the basis of the biological, morphological, or phylogenetic species concepts, apple flies and hawthorn flies are diverging. They are currently in the process of becoming distinct species.

Follow-up studies have shown that the genetic divergence of apple flies and hawthorn flies is not limited to a few alleles associated with attraction to fruit. The divergence of their genomes is widespread—remarkable, given the short time since they began to diverge.

✔ If you understand the sympatric speciation process occurring in these flies, you should be able to hypothesize why natural selection would favor divergence based on the observation that apple fruits drop about 3–4 weeks earlier in the fall than hawthorn fruits.

Although the apple maggot fly's story might seem localized and specific, the events may be common. Biologists currently estimate that over 3 million insect species exist. Most of these species are associated with specific host plants, and thus they occupy specific niches. Based on the data from apple maggot flies, it is reasonable to hypothesize that switching host plants has been a major trigger for speciation throughout the course of insect evolution. Further, speciation itself can trigger more speciation. Researchers have described a species of parasitic wasp that specializes on apple maggot flies.

Sympatric Speciation by Polyploidization

Based on the theory and data reviewed thus far, it is clear that gene flow, genetic drift, and natural selection play important roles in speciation. Can the fourth evolutionary process—mutation—influence speciation as well?

Even though mutation is the ultimate source of genetic variation in populations, it is an inefficient mechanism of evolutionary change on its own—the rate of mutation tends to be low (Chapter 26). Natural selection and genetic drift are usually responsible for amplifying the effects of mutations.

One particular type of mutation, though, turns out to be extremely important in speciation—especially in plants. **Polyploidy** occurs when an error in meiosis or mitosis results in a doubling of the chromosome number—a massive mutation. There are two types of polyploids:

1. **Autopolyploid** ("same-many-form") individuals are produced when a mutation results in a doubling of chromosome number and the chromosomes all come from the same species (**FIGURE 27.8a**).

(a) Formation of **autopolyploid**

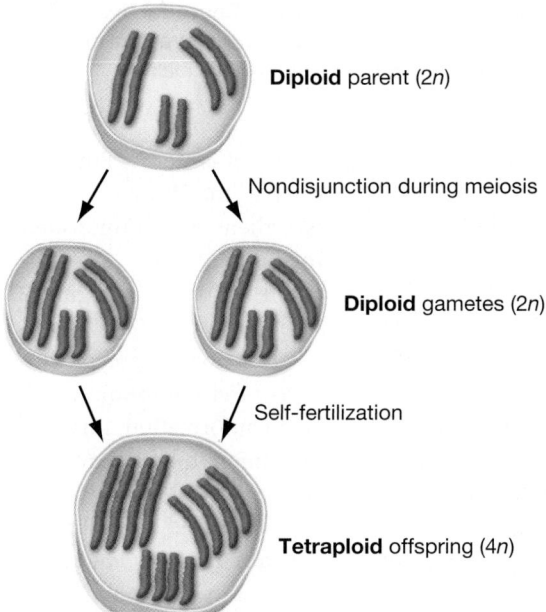

Diploid parent (2n)

Nondisjunction during meiosis

Diploid gametes (2n)

Self-fertilization

Tetraploid offspring (4n)

FIGURE 27.8 Autopolyploids and Allopolyploids. (a) The key event leading to autopolyploidy is nondisjunction during meiosis, resulting in diploid gametes rather than haploid gametes. **(b)** Allopolyploidy can occur by different methods; one example is shown. The key in this case is an error during mitosis that leads to the duplication of chromosomes, enabling the pairing of homologs. In both cases, the polyploids can produce gametes that can self-fertilize.

(b) Formation of **allopolyploid**

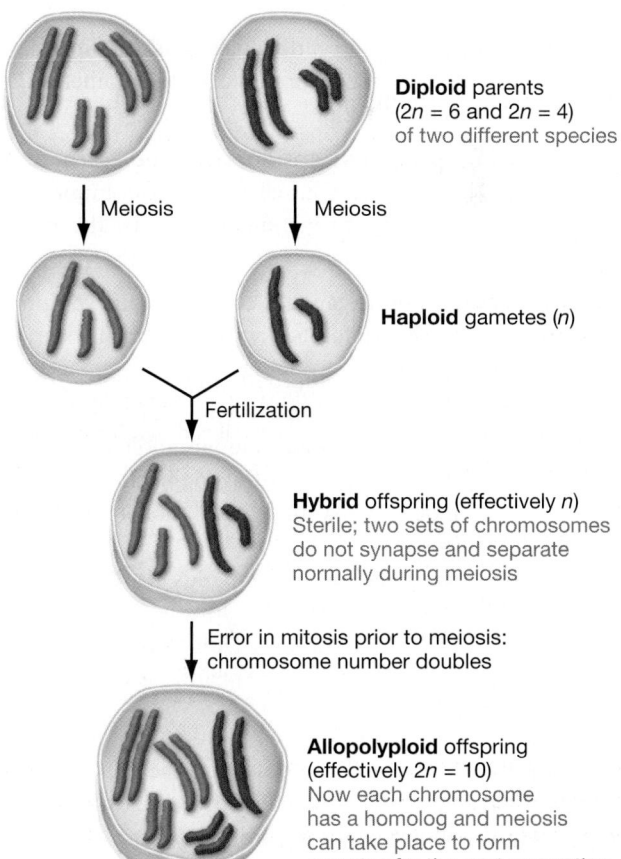

Diploid parents (2n = 6 and 2n = 4) of two different species

Meiosis   Meiosis

Haploid gametes (n)

Fertilization

Hybrid offspring (effectively n) Sterile; two sets of chromosomes do not synapse and separate normally during meiosis

Error in mitosis prior to meiosis: chromosome number doubles

Allopolyploid offspring (effectively 2n = 10) Now each chromosome has a homolog and meiosis can take place to form gametes for the next generation

2. **Allopolyploid** ("different-many-form") individuals are created when parents that belong to different species mate and produce an offspring with two different sets of chromosomes (**FIGURE 27.8b**).

How can these two types of polyploidy lead to speciation? Let's consider a specific example of each.

Autopolyploidy Autopolyploidy is thought to be much less common than allopolyploidy. One example is the maidenhair fern, native to woodlands across North America.

Biologists initially set out to do a routine survey of allelic diversity in a population of these ferns. By chance, they stumbled upon individuals that could produce offspring that were tetraploid (4n) rather than diploid—that is, polyploid mutants within a normal population.

To follow up on the observation, researchers studied the parents of the mutant individuals. The parents turned out to have a defect in meiosis that caused nondisjunction of chromosomes (introduced in Chapter 13). Instead of producing normal, haploid cells as a result of meiosis, the mutant individuals produced diploid cells. These diploid cells eventually led to the production of diploid gametes (see Figure 27.8a). Because maidenhair ferns can self-fertilize, the diploid gametes could combine to form tetraploid offspring.

How could this event lead to speciation? The tetraploid offspring can self-fertilize or mate with their tetraploid parent or each other, but not with other diploids in the population. Consider what happens when that tetraploid individual mates with a diploid individual (**FIGURE 27.9**).

- During meiosis, tetraploid individuals produce diploid gametes, and diploid individuals produce haploid gametes. These gametes unite to form a triploid (3n) zygote.

- Even if this offspring develops normally and reaches sexual maturity, its three homologous chromosomes cannot synapse and separate correctly during meiosis. Thus, the chromosomes are distributed to daughter cells unevenly.

- Because almost all of its gametes contain an uneven number of chromosomes, the triploid individual is virtually sterile.

Polyploid individuals are reproductively isolated from the original diploid population and thus evolutionarily independent, because breeding between diploids and tetraploids generally results in sterile offspring. Thus, according to the biological species concept, speciation occurs in a single generation—instantaneous in evolutionary terms.

✔ If you understand how autopolyploidy works, you should be able to create a scenario explaining how the process gave rise to a tetraploid grape with extra-large fruit, from a diploid population with smaller fruit. (You've probably seen both types of fruit in the supermarket.)

Allopolyploidy As Figure 27.8b illustrates, new polyploid species may be created when two species hybridize. The drawing shows a hybrid offspring forming from a mating between two diploid species. Because the offspring has chromosomes that do

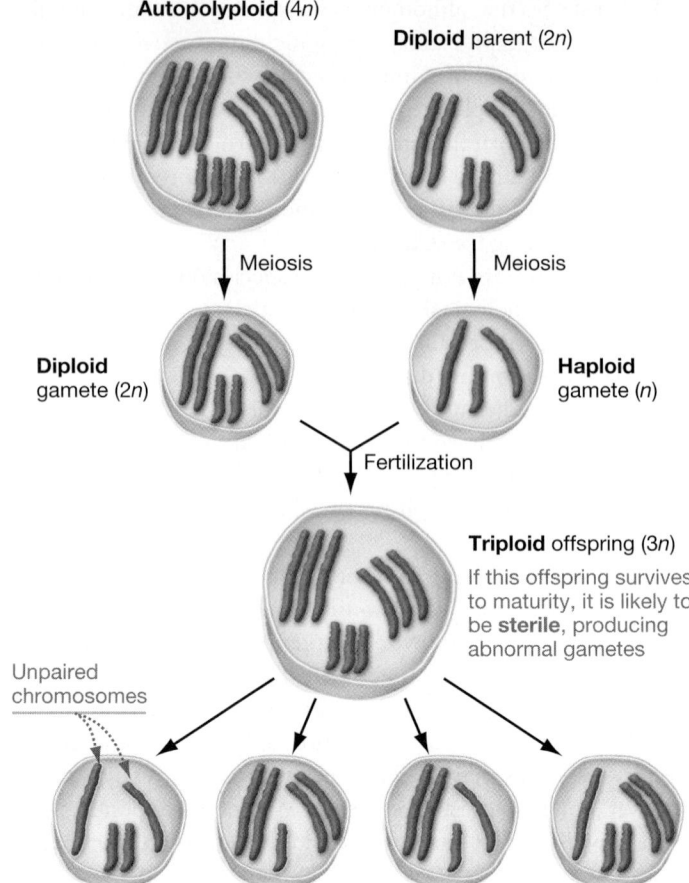

FIGURE 27.9 Polyploidy Can Lead to Reproductive Isolation. The mating diagrammed here illustrates why tetraploid individuals are reproductively isolated from diploid individuals.

not pair normally during meiosis, it is sterile. But if a mutation occurs during mitosis that doubles the chromosome number in this individual before meiosis, then each chromosome gains a homolog and meiosis can proceed normally.

Notice that the allopolyploid offspring in Figure 27.8b appears diploid in the sense that it has two copies of each chromosome (2n = 10). However, it is functionally polyploid because it has two copies each of *two sets* of chromosomes (indicated by blue and red). In terms of genetic information, this offspring has four copies of genes that are common to both parent species, instead of just two.

Exactly this chain of events occurred after three European species of weedy plants in the genus *Tragopogon* were introduced to western North America in the early 1900s. In 1950 a botanist described the first of two new allopolyploid species (**FIGURE 27.10**). Based on an analysis of chromosomes, both were clearly the descendants of matings between the introduced diploids.

Follow-up work by plant geneticists Pam and Doug Soltis has shown that at least one of the new allopolyploid species is expanding its geographic range and outcompeting its parental species. Further, this polyploidization event has occurred multiple times between the same parent species. The Soltises raise an

Tragopogon dubius (2n = 12) *Tragopogon porrifolius* (2n = 12)

Diploid species introduced to North America

Allopolyploid species This polyploidization event occurred multiple times

Tragopogon mirus (2n = 24)

FIGURE 27.10 Allopolyploids Can Form New Species.

check your understanding

If you understand that . . .

- Speciation occurs when populations become isolated genetically and then diverge due to selection, genetic drift, or mutation.

✔ **You should be able to . . .**

Give an example of an event that can lead to the genetic isolation of populations, and explain why selection, drift, and mutation would cause the populations to diverge.

Answers are available in Appendix A.

interesting question: Should the resulting polyploid populations be considered the same or different daughter species?

✔ If you understand how allopolyploidy works, you should be able to create a scenario explaining how a cross between a tetraploid population called Emmer wheat and a wild, diploid wheat gave rise to the hexaploid bread wheat grown throughout the world today.

Why Is Speciation by Polyploidy so Common in Plants? The claim that speciation by polyploidization has been particularly important in plants is backed by the observation that many diploid species have close relatives that are polyploid. Why have polyploids been so successful? The Soltises have addressed this question by examining the genetic implications of polyploidy:

1. Polyploids have higher levels of heterozygosity than their diploid relatives.

2. Polyploids can tolerate higher levels of self-fertilization because they are not as affected by inbreeding depression as their diploid relatives. (See Chapter 26 to review inbreeding.)

3. Genes on duplicated chromosomes can diverge independently, increasing genetic variation in the population.

Genetic variation is a prerequisite for evolution (Chapter 26). Thus, the high genetic diversity of polyploids has enabled a rapid diversification of plant species.

To summarize, speciation by polyploidization is driven by chromosome-level mutations and occurs in sympatry. Compared to the gradual process of speciation by geographic isolation or by disruptive selection in sympatry, speciation by polyploidization is virtually instantaneous. It is fast, sympatric, and common.

27.4 What Happens When Isolated Populations Come into Contact?

Suppose two populations that have been isolated come into contact again. If divergence has taken place and if divergence has affected when, where, or how individuals in the populations mate, then interbreeding is unlikely to take place. In cases such as this, prezygotic isolation exists. When it does, mating between the populations is rare, gene flow is minimal, and the populations continue to diverge.

But what if prezygotic isolation does not exist, and the populations begin interbreeding? The simplest outcome is that the populations fuse over time, as gene flow erases any distinctions between them. Several other possibilities exist, however. Let's explore three of them: reinforcement, hybrid zones, and speciation by hybridization.

Reinforcement

If two populations have diverged extensively and are distinct genetically, it is reasonable to expect that their hybrid offspring will have lower fitness than their parents. This phenomenon has been observed in many cases, such as in the apple maggot fly example. If two populations are well adapted to different habitats, then hybrid offspring will not be well adapted to either habitat. If the two populations have diverged enough genetically, hybrid offspring also may fail to develop normally or may be infertile (see postzygotic isolation in Table 27.1).

When postzygotic isolation occurs, there should be strong selection against interbreeding because hybrid offspring represent a wasted effort on the part of parents—especially for females, due to their typically higher investment in offspring. Individuals that do not interbreed, because of different courtship ritual or pollination system or other form of prezygotic isolation, should be favored because they produce more viable offspring.

Natural selection for traits that isolate populations in this way is called **reinforcement.** The name is descriptive because the selected traits reinforce differences that evolved while the populations were isolated from one another.

Some of the best data on reinforcement come from laboratory studies of closely related fruit fly species in the genus *Drosophila*. Evolutionary biologists Jerry Coyne and Allen Orr analyzed a large series of experiments that tested whether members of closely related fly species are willing to mate with one another. They found an interesting pattern:

- If closely related species are sympatric—meaning that they live in the same area—individuals from the two species are seldom willing to mate with one another.

- If the species are allopatric—meaning that they live in different areas—then individuals from the two species are often willing to mate with one another.

The pattern is logical because natural selection can act to reduce mating between species only if their ranges overlap. Thus, it is reasonable to find that sympatric species exhibit prezygotic isolation but that allopatric species do not. There is a long-standing debate, however, over just how important reinforcement is in other species.

Hybrid Zones

Hybrid offspring are not always dysfunctional. In some cases they are capable of mating and producing viable offspring that have features that are intermediate between those of the two parental populations. When this is the case, hybrid zones can form. A **hybrid zone** is a geographic area where interbreeding occurs and hybrid offspring are common.

Depending on the fitness of hybrid offspring and the extent of breeding between parental species, hybrid zones can be narrow or wide, and long or short lived. As an example of how researchers analyze the dynamics of hybrid zones, let's consider recent work on two bird species.

Townsend's warblers and hermit warblers live in the coniferous forests of North America's Pacific Northwest. In western Washington State, where their ranges overlap, the two species hybridize extensively. As the drawings in **FIGURE 27.11** show, hybrid offspring have characteristics that are intermediate relative to the two parental species.

To explore the dynamics of this hybrid zone, ornithologist Sievert Rohwer teamed up with population geneticist Eldredge Bermingham and others to examine gene sequences in the mitochondrial DNA (mtDNA) of a large number of Townsend's, hermit, and hybrid warblers collected from forests throughout the region. The team found that each of the parental species has certain species-specific mtDNA sequences. This result allowed the researchers to infer how hybridization was occurring.

To grasp the reasoning here, it is critical to realize that mtDNA is maternally inherited in most animals and plants. If a hybrid individual has Townsend's mtDNA, its mother had to be a Townsend's warbler while its father had to be a hermit warbler. In this way, identifying mtDNA types allowed the research team to infer whether Townsend's females were mating with hermit males, or vice versa, or both.

Their data presented a clear pattern: Most hybrids form when Townsend's males mate with hermit warbler females.

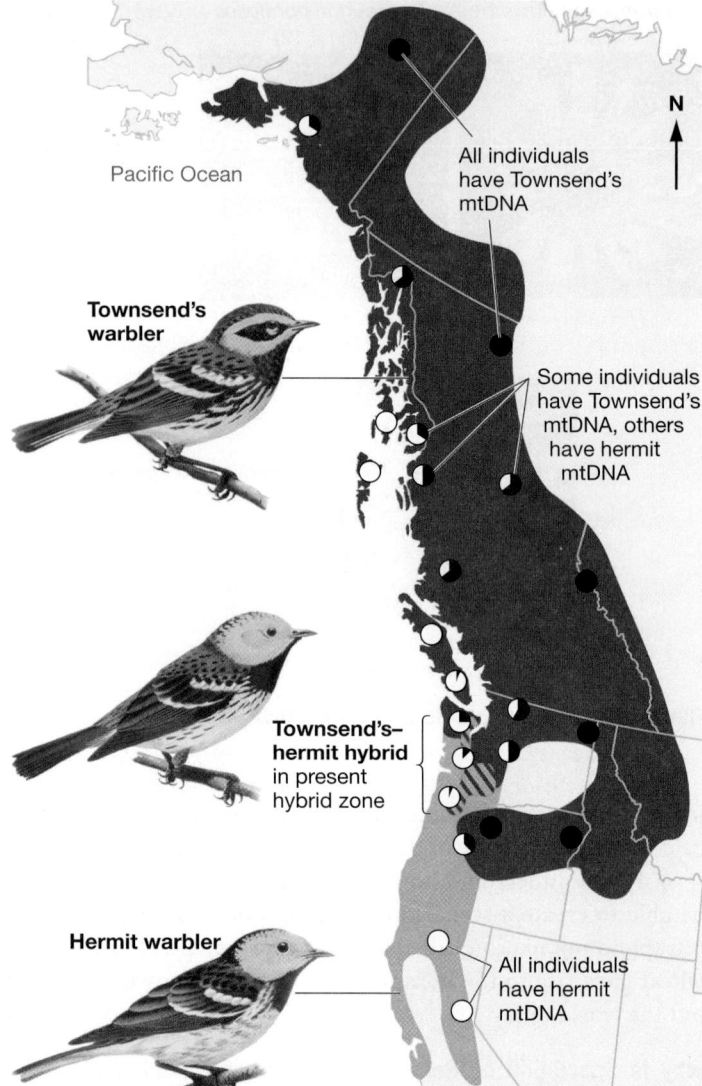

FIGURE 27.11 Analyzing a Hybrid Zone. When Townsend's warblers and hermit warblers hybridize, the offspring have intermediate characteristics. The current ranges of Townsend's warblers (in red) and hermit warblers (in orange) are shown in the map. The small pie charts show the percentage of individuals with Townsend's warbler mtDNA (in black) and hermit warbler mtDNA (in white).

DATA: Rohwer, S., E. Bermingham, and C. Wood. 2001. *Evolution* 55: 405–422.

Rohwer followed up on this result with experiments showing that Townsend's males are extremely aggressive in establishing territories and that they readily attack hermit warbler males. The data suggest that Townsend's males invade hermit territories, drive off the hermit males, and mate with hermit females.

The team also found something completely unexpected. When they analyzed the distribution of mtDNA types along the Pacific Coast and in the northern Rocky Mountains, they discovered that many Townsend's warblers actually had hermit mtDNA. Figure 27.11 shows that in some regions—such as the larger islands off the coast of British Columbia—*all* of the warblers had hermit mtDNA (indicated on the map by white pie charts), even though they looked like full-blooded Townsend's warblers (indicated on the map by the red distribution).

How could this be? To explain the result, the team hypothesized that hermit warblers were once found as far north as Alaska and that Townsend's warblers have gradually taken over their range. Their logic is that repeated mating with Townsend's warblers over time made the hybrid offspring look more and more like Townsend's, even while maternally inherited mtDNA kept the genetic record of the original hybridization event intact.

If this hypothesis is correct, then the hybrid zone should continue moving south. If it does so, hermit warblers may eventually become extinct. In some cases, however, hybridization does not lead to extinction but rather leads to the opposite—the creation of new species.

New Species through Hybridization

If two species interbreed and produce hybrid offspring that can not only survive and reproduce but also possess a unique combination of traits that happen to be adaptive in their particular environment, a new species may result. How would you identify such a species in nature?

Evolutionary biologist Loren Rieseberg and colleagues recently examined the relationships of three sunflower species native to the Southwest: *Helianthus annuus* and *H. petiolaris* are known to hybridize in regions where their ranges overlap, and *H. anomalus* resembles these hybrids (see photo at the start of this chapter). How are these species related?

DNA sequencing studies have shown that some gene regions in *H. anomalus* are remarkably similar to those found in *H. annuus*, while other *H. anomalus* gene sequences are almost identical to those found in *H. petiolaris*. Based on these data, Rieseberg hypothesized that *H. anomalus* originated in hybridization between *H. annuus* and *H. petiolaris*. All three species have the same number of chromosomes, so neither allopolyploidy nor autopolyploidy was involved. Instead, the chromosomes of *H. annuus* and *H. petiolaris* must be similar enough that they can synapse and undergo normal meiosis in hybrid offspring.

The specific hypothesis here is that *H. annuus* and *H. petiolaris* were isolated and diverged as separate species, and later began interbreeding. The hybrid offspring created a third, new species that had unique combinations of alleles from each parental species and therefore different characteristics. This hypothesis is supported by the observation that *H. anomalus* grows in much drier habitats than either of the parental species—suggesting that a unique combination of alleles allowed *H. anomalus* to thrive in dry habitats.

Rieseberg and his colleagues set out to test the hybridization hypothesis by trying to re-create the speciation event experimentally (**FIGURE 27.12**).

Step 1 They mated individuals from the two parental species and raised the offspring in a greenhouse.

Step 2 When these hybrid individuals were mature, the researchers either mated the plants to other hybrid individuals or "backcrossed" them to individuals from one of the parental species.

QUESTION: Can new species arise by hybridization between existing species?

HYPOTHESIS: *Helianthus anomalus* originated by hybridization between *H. annuus* and *H. petiolaris*.

NULL HYPOTHESIS: *Helianthus anomalus* did not originate by hybridization between *H. annuus* and *H. petiolaris*.

EXPERIMENTAL SETUP:

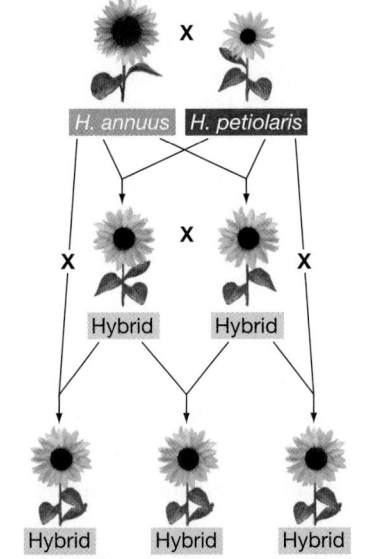

1. Mate *H. annuus* and *H. petiolaris* and raise offspring.

2. Mate F_1 hybrids or backcross F_1s to parental species; raise offspring.

3. Repeat for four more generations.

PREDICTION: Experimental hybrids will have the same mix of *H. annuus* and *H. petiolaris* genes as natural *H. anomalus*.

PREDICTION OF NULL HYPOTHESIS: Experimental hybrids will not have the same mix of *H. annuus* and *H. petiolaris* genes as natural *H. anomalus*.

RESULTS:
DNA comparison of a chromosomal region called S:

(Only colored portions of S region were analyzed)

Experimental hybrid (cross of *H. annuus* and *H. petiolaris*)

Alignment

H. anomalus (naturally occurring species)

Gene region similar to *H. annuus* Gene region similar to *H. petiolaris*

CONCLUSION: New species may arise via hybridization between existing species.

FIGURE 27.12 Experimental Evidence That New Species Can Originate in Hybridization Events.

SOURCE: Rieseberg, L. H., B. Sinervo, C. R. Linder, et al. 1996. Role of gene interactions in hybrid speciation: Evidence from ancient and experimental hybrids. *Science* 272: 741–745.

✔ **QUESTION** Why it is valid to use experiments with living organisms—like this one—to infer what happened during historical speciation events?

	Process	Example
Fusion of the populations	The two populations freely interbreed.	Occurs whenever populations of the same species come into contact.
Reinforcement of divergence	If hybrid offspring have low fitness, natural selection favors the evolution of traits that prevent interbreeding between the populations.	Appears to be common in fruit fly species that occupy the same geographic areas.
Hybrid zone formation	Hybridization occurs in a well-defined geographic area. This area may move over time or be stable.	Many stable hybrid zones have been described; the hybrid zone between hermit and Townsend's warblers appears to have moved over time.
Extinction of one population	If one population or species is a better competitor for shared resources, then the poorer competitor may be driven to extinction.	Townsend's warblers may be driving hermit warblers to extinction.
Creation of new species	If the combination of genes in hybrid offspring allows them to occupy distinct habitats or use novel resources, they may form a new species.	Hybridization between sunflowers gave rise to a new species with unique characteristics.

Step 3 This breeding program continued for four more generations before the experiment ended. Ultimately, the experimental lines were backcrossed twice, and they were mated to other hybrid offspring three times.

The goal of these crosses was to simulate matings that might have occurred naturally.

The experimental hybrids looked like the natural hybrid species, but did they resemble them genetically? To answer this question, Rieseberg's team constructed genetic maps of each population, using a large series of genetic markers (see Chapter 20). Because each parental population had many unique markers in its genomes, the research team hoped to identify which genes found in the experimental hybrids came from which parental species.

Some of their data are diagrammed in the "Results" section of Figure 27.12. The bottom bar in the illustration represents a region called S in the genome of the naturally occurring species *Helianthus anomalus*. As the legend indicates, this region contains three sections of sequences that are also found in *H. petiolaris* (indicated with the color red) and two that are also found in *H. annuus* (indicated with the color orange). The top bar shows the composition of this same region in the genome of the experimental hybrid lines.

The key observation here is that the genetic composition of the synthesized hybrids matches that of the naturally occurring hybrid species. In effect, the researchers succeeded in re-creating a speciation event. Their results provide strong support for the hybridization hypothesis for the origin of *H. anomalus*. ✔ If you understand this experiment, you should be able to suggest one result that would have caused the researchers to reject the hybridization hypothesis.

Secondary contact of two populations can produce a dynamic range of possible outcomes: fusion of the populations, reinforcement of divergence, founding of stable hybrid zones, extinction of one population, or the creation of new species. **TABLE 27.3** summarizes the outcomes of secondary contact.

The study of speciation provides an essential link between the processes of evolution (Chapters 25 and 26) and the tree of life (Chapter 28 and Unit 6). Research on speciation is accelerating, in part due to availability of new genetic and phylogenetic tools, and in part due to the study of human impacts on biodiversity (see Chapter 57). Although human activities usually result in the extinction of species, they can also result in the creation of new species. Humans are changing the shape of the tree of life.

If you understand . . .

27.1 How Are Species Defined and Identified?

- A species is defined as an evolutionarily independent population or group of populations. Researchers use several criteria to identify whether populations represent distinct species.

- The biological species concept uses reproductive isolation as a criterion to identify species.

- The morphospecies concept identifies species using distinctive morphological traits.

- The phylogenetic species concept identifies species as the smallest monophyletic groups on the tree of life.

✔ You should be able to explain whether human populations would be considered separate species under the biological species, morphospecies, and phylogenetic species concepts.

27.2 Isolation and Divergence in Allopatry

- Speciation is a splitting event in which one lineage gives rise to two or more independent descendant lineages.

- Speciation occurs when populations of the same species become genetically isolated by lack of gene flow and then diverge from each other due to selection, genetic drift, and mutation.

- Allopatric speciation occurs when populations diverge in geographic isolation.

- Geographic isolation occurs through dispersal, when small groups of individuals colonize a new habitat, or through vicariance, when a large, continuous population becomes fragmented into isolated habitats.

✔ You should be able to design an experiment that would, given enough time, result in the production of two species from a single ancestral population.

27.3 Isolation and Divergence in Sympatry

- Sympatric speciation occurs when populations diverge genetically despite living in the same geographic area.

- Sympatric speciation can occur when disruptive selection favors individuals that breed in different ecological niches.

- Mutations that produce polyploidy can trigger rapid speciation in sympatry because they lead to reproductive isolation between diploid and tetraploid populations.

✔ You should be able to evaluate whether your experiment on speciation (see previous exercise under 27.2) represents a case of dispersal, vicariance, different habitat use, or polyploidy.

27.4 What Happens When Isolated Populations Come into Contact?

- If gene flow occurs, populations that have diverged may fuse into a single species.

- If prezygotic isolation exists, populations that come back into contact will probably continue to diverge.

- Secondary contact can lead to reinforcement—the evolution of mechanisms that prevent hybridization.

- Gene flow between different species can lead to the formation of hybrid zones that move over time or are stable.

- In some cases, hybridization between species can create new species with unique combinations of traits.

✔ You should be able to predict how a hybrid zone will change over time when hybrid offspring have higher fitness than the parental populations.

(MB) **MasteringBiology**

1. **MasteringBiology Assignments**

   Tutorials and Activities Allopatric Speciation, Defining Species, Polyploid Plants, Speciation by Changes in Ploidy

   Questions Reading Quizzes, Blue-Thread Questions, Test Bank

2. **eText** Read your book online, search, take notes, highlight text, and more.

3. **The Study Area** Practice Test, Cumulative Test, BioFlix® 3-D Animations, Videos, Activities, Audio Glossary, Word Study Tools, Art

You should be able to . . .

1. What distinguishes a morphospecies?
   a. It has distinctive characteristics, such as size, shape, or coloration.
   b. It represents a distinct twig in a phylogeny of populations.
   c. It is reproductively isolated from other species.
   d. It is a fossil from a distinct time in Earth history.

2. Which of the following describes vicariance?
   a. Small populations coalesce into one large population.
   b. A population is fragmented into isolated subpopulations.
   c. Individuals colonize a novel habitat.
   d. Individuals disperse and found a new population.

3. Why are genetic isolation and genetic divergence occurring in apple maggot flies, even though populations occupy the same geographic area?
   a. Different populations feed and mate on different types of fruit.
   b. One population is tetraploid; others are diploid.
   c. The introduction of a nonnative host plant caused vicariance.
   d. Responses to scents have changed due to disruptive selection.

4. The biological species concept can be applied only to which of the following groups?
   a. bird species living today
   b. dinosaurs and bird species living today
   c. dinosaurs
   d. bacteria

5. Why is "reinforcement" an appropriate name for the concept that natural selection should favor divergence and genetic isolation if populations experience postzygotic isolation?
   a. Selection should reinforce high fitness for hybrid offspring.
   b. Selection should reinforce the fact that the diverging populations are "good species" under the morphological species concept.
   c. Hybrid offspring do not develop at all or are sterile when mature because selection reinforces only the success of purebred species.
   d. The selected traits reinforce differences that evolved while the populations were isolated from one another.

6. True or False? Gene flow increases the divergence of populations.

7. Which studies in this chapter represent direct observation of speciation? Which are indirect studies of historical speciation events?

8. When the ranges of two different species meet, a stable "hybrid zone" occupied by hybrid individuals may form. How is this possible?
   a. Hybrid individuals may have intermediate characteristics that are advantageous in a given region.
   b. Hybrid individuals are always allopolyploid and are thus unable to mate with either of the original species.
   c. Hybrid individuals may have reduced fitness and thus be strongly selected against.
   d. One species has a selective advantage, so as hybridization continues, the other species will go extinct.

9. In the case of the seaside sparrow, how did the species identified by the biological species concept, the morphospecies concept, and the phylogenetic species concept conflict?

10. Explain how isolation and divergence are occurring in apple maggot flies. Of the four evolutionary processes (mutation, gene flow, genetic drift, and selection), which two are most important in causing this event?

11. A large amount of gene flow is now occurring among human populations due to intermarriage among people from different ethnic groups and regions of the world. Is this phenomenon increasing or decreasing racial differences in our species? Explain.

12. **QUANTITATIVE** If one species ($2n = 10$) crosses with another species ($2n = 18$), producing an allopolyploid offspring, what is the ploidy of the offspring?
   a. $2n = 10$
   b. $2n = 18$
   c. $2n = 18 + 10 = 28$
   d. $4n = 36 + 20 = 56$

13. Unlike animal gametes, plant reproductive cells do not differentiate until late in life. Why are plants much more likely than animals to produce diploid gametes and polyploid offspring?
   a. Cells that do not divide for long periods accumulate mutations.
   b. Cells that differentiate later in life undergo many rounds of mitosis, so are more likely to accumulate mutations.
   c. Cells that differentiate later in life undergo many rounds of meiosis, so are more likely to accumulate mutations.
   d. Undifferentiated cells are more prone to mutations than differentiated cells.

14. Sexual selection is a type of natural selection that favors individuals with traits that increase their ability to obtain mates, such as female choice for bright orange beaks in zebra finches. Propose a scenario where sexual selection could contribute to divergence in sympatric speciation.

15. A friend says that apple maggot flies prefer apple fruit scents because they need to, in order to survive. Another agrees and adds that the flies acquire the ability to distinguish the apple scents by spending time on the fruit, and that's why their offspring prefer apples. What's wrong with these statements?

16. All over the world, natural habitats are being fragmented into tiny islands as suburbs, ranches, and farms expand, and roads are built to connect them. Explain why this fragmentation process could lead to extinction. Then explain how it could lead to speciation.

28 Phylogenies and the History of Life

In this chapter you will learn that

Vast amounts of change have occurred in the 3.6-billion-year history of life

by asking ↓ ↓ by asking

How can we study the history of life?   **How does the diversity of life undergo big changes?**

↓ via   ↓ via   ↓ via   ↓ via

Phylogenetic trees	The fossil record	Adaptive radiation	Mass extinction
28.1	28.2	28.3	28.4

A fossilized trilobite. The last trilobites disappeared during a mass extinction event analyzed in Section 28.4.

This chapter is about time and change. More specifically, it's about vast amounts of time and profound change in organisms. Both of these topics can be difficult for humans to grasp. Our lifetimes are measured in decades, and our knowledge of history is usually measured in centuries or millennia. But this chapter analyzes events that occurred over millions and even billions of years.

It takes practice to get comfortable analyzing the profound changes that occur in organisms over deep time. To help you get started, the chapter begins by introducing the two major analytical tools that biologists use to reconstruct the history of life: phylogenetic trees and the fossil record. The remaining two sections explore evolutionary episodes called adaptive radiations and mass extinctions, which can cause major changes to the tree of life. The Big Picture of Evolution on pages 526–527 shows how the tree of life relates to speciation (Chapter 27) and the four evolutionary processes (Chapters 25 and 26).

This chapter is part of the Big Picture. See how on pages 526–527.

✔ When you see this checkmark, stop and test yourself. Answers are available in Appendix A.

28.1 Tools for Studying History: Phylogenetic Trees

The evolutionary history of a group of organisms is called its **phylogeny.** A **phylogenetic tree** is a graphical summary of this history, showing the ancestor–descendant relationships among populations, species, or higher taxa, and clarifying who is related to whom. The **tree of life** is the most universal of all phylogenetic trees, depicting the evolutionary relationships among all living organisms on Earth (see Chapter 1). Phylogenetic trees have revolutionized the study of evolution and many related fields. For example:

- Phylogenetic trees can be used in taxonomy to define species (Chapter 27).

- Phylogenetic trees provide evidence for the endosymbiosis theory—that mitochondria and chloroplasts originated as free-living bacteria (Chapter 30).

- Phylogenetic trees can be used to examine how the HIV virus jumped from other primates to humans (Chapter 36).

- Phylogenetic trees can aid in identifying species that are a priority for conservation (Chapter 57).

Phylogenetic trees are extremely effective at summarizing data and testing hypotheses. However, they are unusual diagrams—it can take practice to interpret them correctly.

Take a moment to review the parts of a phylogenetic tree in **TABLE 28.1.** Notice that the tree in this table, like all the trees used in this text, is rooted, which means that the **branches** on the left are more ancient, or ancestral, while the ones on the right are most recent, or derived. Notice also that the **nodes** (forks) represent *hypothetical* common ancestors—the taxa themselves are always located on the branch **tips,** never within the tree. That is, none of the taxa are presumed to be ancestors of others, even if some of the taxa are extinct.

Let's practice reading phylogenetic trees by focusing on how biologists go about building them.

How Do Biologists Estimate Phylogenies?

Like any other pattern or measurement in nature, from the average height of a person in a particular human population to the speed of a passing flock of geese, the historical relationships among species cannot be known with absolute certainty. In addition, humans have not been around to observe most evolutionary events directly. Thus, the relationships depicted in phylogenetic trees must be *estimated* from the best available data.

Creating the Data Matrix The first step in inferring evolutionary relationships is to decide which populations, species, or higher taxa to compare and which characteristics to use—usually morphological or genetic characteristics, or both. For example, to reconstruct relationships among fossil species of humans (there were many; see Chapter 35), scientists analyze aspects of tooth, jaw, and skull structure, since this is often the only

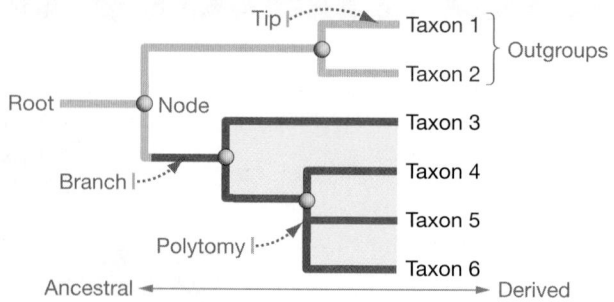

SUMMARY TABLE 28.1   **Anatomy of a Phylogenetic Tree**

Graphical Symbols and Definitions

Branch Represents a population through time

Node (fork) A point within the tree where a branch splits into two or more branches

Outgroup A taxon that diverged prior to the taxa that are the focus of the study; helps to root the tree

Root The most ancestral branch in the tree

Polytomy A node that depicts an ancestral branch dividing into three or more (rather than two) descendant branches; usually indicates that insufficient data were available to resolve which taxa are more closely related

Tip (terminal node) Endpoint of a branch; represents a living or extinct group of genes, species, families, phyla, or other taxa

sort of evidence available. To reconstruct relationships among contemporary human populations, investigators usually compare the sequences of bases in a particular gene.

Consider the simple case of estimating a phylogenetic tree for four closely related species using a short DNA sequence as the character set, where a **character** is any genetic, morphological, physiological, or behavioral characteristic to be studied. The data matrix for this group is shown in **FIGURE 28.1a.** The sequence consists of six characters (nucleotides 1, 2, 3, 4, 5, and 6), each with four possible character states: A, T, C, and G.

Notice that a fifth species is included in the matrix as an **outgroup**—a species that is closely related to the group being studied, but not part of it. Outgroups are used to establish the polarity of each trait—that is, whether a character state is ancestral or derived. An **ancestral trait** is a characteristic that existed in an ancestor; a **derived trait** is one that is a modified form of the ancestral trait, found in a descendant. Derived traits originate via mutation, selection, and genetic drift (see Chapter 26).

It's important to recognize that ancestral and derived traits are relative. If you are comparing mammals with the fossil forms called mammal-like reptiles, then fur and lactation are derived traits. But if you are comparing whales and humans, then fur and lactation are ancestral traits. In the example in Figure 28.1a, the ancestral DNA sequence is estimated to be AAA GCT based on the outgroup; any nucleotides that differ from those in this sequence are considered derived. Since outgroup lineages can

(a) Data matrix

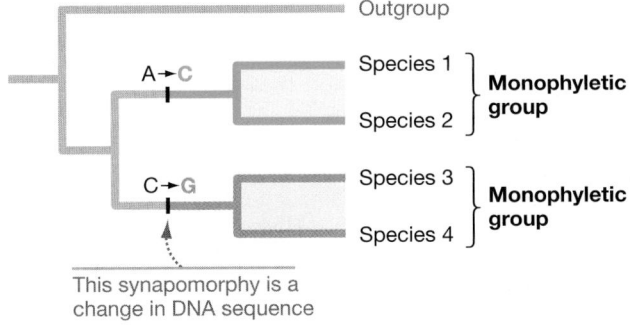

DNA nucleotide sequence

1	2	3	4	5	6	
A	A	A	G	C	T	Outgroup
A	A	C	G	C	T	Species 1
A	A	C	G	C	T	Species 2
A	A	A	G	G	T	Species 3
A	A	A	G	G	T	Species 4

(b) Phylogenetic tree inferred from the data

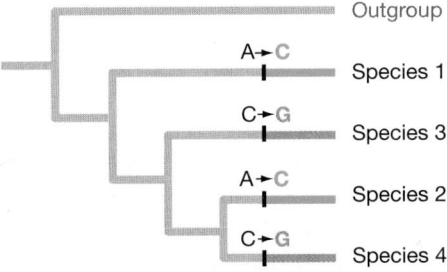

Outgroup

A→C

Species 1 ⎫ Monophyletic
Species 2 ⎭ group

C→G

Species 3 ⎫ Monophyletic
Species 4 ⎭ group

This synapomorphy is a change in DNA sequence

(c) One of many alternate trees

Outgroup

A→C   Species 1

C→G   Species 3

A→C   Species 2

C→G   Species 4

FIGURE 28.1 Constructing a Phylogenetic Tree. Cladistic analysis **(a)** begins with a data matrix and **(b)** builds trees based on the principle that closely related species are likely to share derived traits, called synapomorphies. **(c)** The tree that best fits the data must be selected from many alternate trees (only one shown here).

✔ **EXERCISE** Draw another alternate tree for the same data set.

evolve too, multiple outgroups are often used to estimate ancestral character states (for simplicity, only one is used here).

Using the Data Matrix to Estimate a Tree
The most commonly used method of inferring phylogenetic trees, called cladistics, was introduced by German biologist Willi Hennig in the 1960s. The **cladistic approach** is based on the principle that relationships among species can be reconstructed by identifying shared derived characters—synapomorphies (introduced in Chapter 27).

A **synapomorphy** is a trait found in two or more taxa that is present in their most recent common ancestor but is missing in more distant ancestors. Synapomorphies allow biologists to recognize monophyletic groups—also called **clades** or **lineages.** A **monophyletic group** is an evolutionary unit that includes an ancestral population and all of its descendants, but no others.

SUMMARY TABLE 28.2   Mapping Traits and Groups on Trees

Graphical Symbol	Definition
	Homology Similarity in organisms due to common ancestry (trait **A** is homologous among the red branches)
	Monophyletic group An evolutionary unit that includes an ancestral population and all of its descendants but no others (also called a lineage, or clade)
Synapomorphy	**Synapomorphy** A shared, *derived* trait (trait **A** occurs *only* in the red branches)
	Homoplasy Similarity in organisms due to reasons other than common ancestry (trait **B** arose twice independently)
	Polyphyletic group An unnatural group (shown in purple) that does not include the most recent common ancestor (see segmented animals in Chapter 33)
Loss of trait	**Paraphyletic group** A group (shown in blue) that includes an ancestral population and some of its descendents, but not all (see dicots in Chapter 31, and fish and reptiles in Chapter 35)

See **TABLE 28.2** for a comparison of monophyletic groups and two types of non-monophyletic groups.

In the case of the character set in Figure 28.1a, species 1 and 2 can be recognized as a monophyletic group based on their shared, derived trait, a C in the third position of the DNA sequence. Similarly, species 3 and 4 can be recognized as a different monophyletic group based on their shared, derived trait, a G in the fifth position of the sequence. This logic is summarized by the tree in **FIGURE 28.1b**.

In this example, it is relatively easy to infer a tree that represents the data set in a logical way. In common practice, there are several possible complications to resolve. For example:

- Traits can be similar in two species not because those traits were present in a common ancestor, but because similar traits evolved independently in two different lineages. For example, the A → C changes in DNA sequence at nucleotide 3 may have arisen separately in species 1 and 2 rather than once in their common ancestor (**FIGURE 28.1c**).

- Sometimes a reversal in a change occurs, such as an A → C transition in a nucleotide in one branch followed by a C → A change in the same nucleotide in a subsequent branch, thus creating the appearance that no change occurred.

- Sometimes the species forms a monophyletic group one way according to one trait in the matrix and a different way according to a different trait in the matrix (as would occur if species 1 and 3 had a unique base change at position 1).

- The example in Figure 28.1 shows a small number of species with a small number of traits. The larger the number of species and traits added to the matrix, the greater the number of possible trees and character combinations. There are 15 possible trees for a phylogeny of 4 taxa but over 100,000 trees for a phylogeny of 8 taxa.

How can researchers identify the tree with the best fit to the data? And how can they avoid an erroneous conclusion? Biologists often invoke the logical principle of **parsimony** to address these questions. Under parsimony, the most likely explanation or pattern is the one that implies the least amount of change.

To implement a parsimony analysis, biologists use computer programs that compare the branching patterns that are theoretically possible in a phylogenetic tree and count the number of changes in DNA sequences required to produce each pattern. For example, the tree in Figure 28.1b requires two changes in base sequence; the tree in Figure 28.1c requires four.

Since independent evolution of traits should be rare compared with similarity due to shared descent, the tree that implies the fewest overall evolutionary changes is hypothesized to be the one that most accurately reflects what really happened during evolution. If the branching pattern in Figure 28.1b is the most parsimonious of all possible trees, then biologists conclude that it is the most likely representation of actual phylogeny based on the data in hand.

There are methods other than parsimony to find the tree with the best fit to the data, each guided by different assumptions. However, all the methods are similar in serving as "filters" to identify optimal trees out of a pool of hundreds, thousands, or millions of possible trees. Some researchers use multiple methods to gain more confidence in their results.

It's important to recognize that the trees created using cladistic analysis (called cladograms) focus on branching *patterns*—the branch *lengths* themselves are arbitrary. Some analyses produce trees whose branch lengths represent genetic distance or time since divergence. You can identify these trees by the presence of scale bars (see **TABLE 28.3**). (If you feel that you need more practice interpreting phylogenetic trees, you can turn to **BioSkills 7** in Appendix B at any time.)

How Can Biologists Distinguish Homology from Homoplasy?

The cladistic approach is a powerful method of estimating phylogenies because it is based on one of the central concepts in evolutionary biology—homology (see Table 28.2). **Homology** (literally, "same-source") occurs when traits are similar due to shared ancestry; in contrast, **homoplasy** ("same-form") occurs when traits are similar for reasons other than common ancestry.

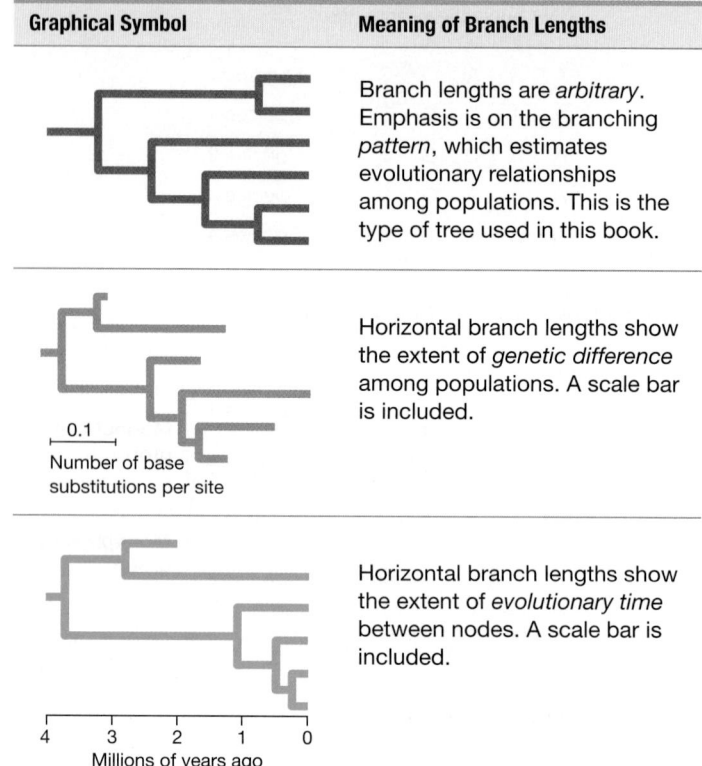

SUMMARY TABLE 28.3 Branch Lengths in Phylogenetic Trees

Graphical Symbol	Meaning of Branch Lengths
	Branch lengths are *arbitrary*. Emphasis is on the branching *pattern*, which estimates evolutionary relationships among populations. This is the type of tree used in this book.
0.1 Number of base substitutions per site	Horizontal branch lengths show the extent of *genetic difference* among populations. A scale bar is included.
4 3 2 1 0 Millions of years ago	Horizontal branch lengths show the extent of *evolutionary time* between nodes. A scale bar is included.

A common cause of homoplasy is convergent evolution. **Convergent evolution** occurs when natural selection favors similar solutions to the problems posed by a similar way of making a living in different species. Since species have many characteristics, some may be homologous with characteristics in other species, while others may be convergent. Let's consider how to distinguish homology and homoplasy.

Are *Hox* Genes in Fruit Flies and Humans Homologous or Convergent? Consider the *Hox* genes of insects and vertebrates (introduced in Chapter 22). Even though insects and vertebrates last shared a common ancestor some 600–700 million years ago, biologists argue that their *Hox* genes are derived from the same ancestral sequences—that is, they are homologous. Several lines of evidence support this hypothesis:

- Groups of *Hox* genes are organized on chromosomes in a similar way. **FIGURE 28.2** shows that *Hox* genes in both insects and vertebrates are found in gene clusters, and similar genes are often found in the same order on the chromosome. Genes with these characteristics are called gene families (Chapter 21). The organization of the gene families is nearly identical among insect and vertebrate species.

- All *Hox* genes share a 180-base-pair sequence called the homeobox. The portion of a protein encoded by the homeobox (introduced in Chapter 22) is almost identical in insects and vertebrates and has a similar function. It binds to DNA and regulates the expression of other genes.

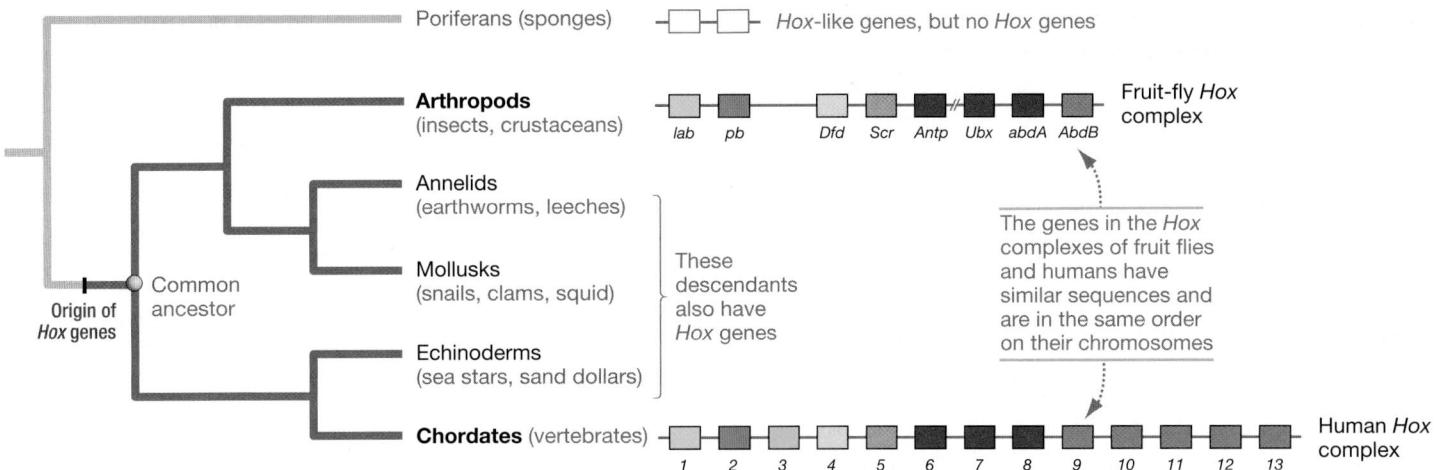

FIGURE 28.2 Homology: Similarities Are Inherited from a Common Ancestor. All animals except sponges have *Hox* complexes that are similar to those illustrated for fruit flies and humans. (Human *Hox* genes occur in four clusters but are summarized here in one for simplicity.)

- The products of the *Hox* genes have similar functions (see Chapter 22): identifying the locations of cells in embryos. They are also expressed in similar patterns in time and space.

Further, if traits found in distantly related lineages are similar due to common ancestry, then similar traits should be found in many intervening lineages on the tree of life—because all of the species in question inherited the trait from the same common ancestor. Indeed, all other lineages of animals that branched off between insects and mammals have similar *Hox* genes. Sponges serve as an outgroup to the clade since they appear to have split off before *Hox* genes arose.

The internal consistency of the genetic, developmental, and phylogenetic data provide strong evidence that *Hox* genes in fruit flies and humans are homologous.

Are Streamlined Bodies in Dolphins and Ichthyosaurs Homologous or Convergent? The extinct aquatic reptiles called ichthyosaurs were strikingly similar to modern dolphins (**FIGURE 28.3**). Both are large marine predators with streamlined bodies and large dorsal fins. Are these characteristics homologous in dolphins and ichthyosaurs?

Figure 28.3 shows a phylogenetic tree based on skeletal characters, such as the number and placement of openings in the skull (since ichthyosaurs are known only from their fossilized skeletons). This analysis shows that dolphins occur within the mammal clade, whereas ichthyosaurs are most closely related to lizards. Notice that the sister groups to dolphins and ichthyosaurs—that is, the most closely related branches on the tree—do not possess streamlined bodies, flippers, and fins.

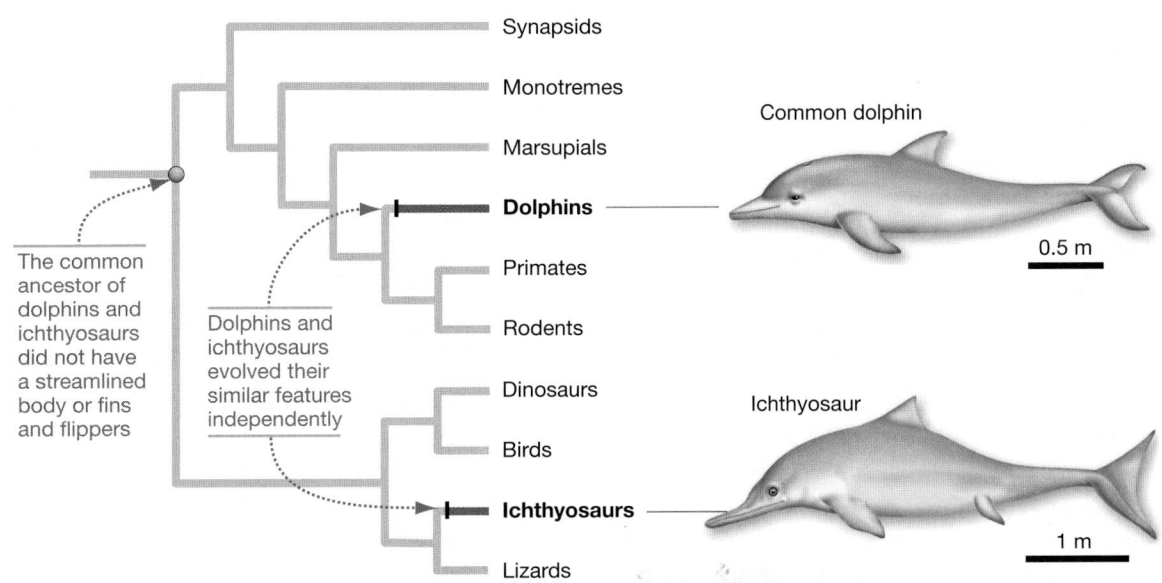

FIGURE 28.3 Homoplasy: Traits Are Similar but Not Inherited from a Common Ancestor. Dolphins and ichthyosaurs look similar but are not closely related.

Based on these data, it is logical to argue that the similarities between ichthyosaurs and dolphins result from convergent evolution. The convergent traits do not occur in the common ancestor of the similar species. Streamlined bodies, flippers, and dorsal fins are adaptations that help *any* aquatic species—whether it is a reptile or a mammal—chase down fish in open water.

Whale Evolution: A Case Study

Let's apply what you've learned about phylogenetic trees and the concept of homology to an interesting evolutionary case study: the origin of a unique clade of mammals, the whales. Scientists since Darwin have puzzled over the rightful place of whales (including dolphins) on the tree of life. Data from the fossil record support the hypothesis that whales evolved from terrestrial mammals. More specifically, hippos, which are semiaquatic, appear to be the closest living relative of today's whales. How can phylogenetic trees be used to test this claim?

Data Set 1: A Phylogeny Based on Morphological Traits Hippos, along with cows, deer, pigs, and camels, are artiodactyls. Members of this mammal group have hooves and an even number of toes. They also share another feature: the unusual pulley shape of an ankle bone called the astragalus. Along with having a unique foot morphology, the shape of the astragalus is a synapomorphy that identifies the artiodactyls as a monophyletic group. Members of the outgroup, perissodactyls (including horses and rhinos), do not possess a pulley-shaped astragalus.

These data support the tree shown in **FIGURE 28.4a**. Notice that whales do not have an astragalus (indeed, no ankles at all) and are shown as an outgroup to the artiodactyls on this tree based on other morphological characteristics. It is logical to map the gain of the pulley-shaped astragalus in the ancestral population, marked by a black bar in Figure 28.4a, because all descendants of that ancestor have the trait, but members of the outgroup do not.

Data Set 2: A Phylogeny Based on DNA Sequence Data When researchers began comparing DNA sequences of artiodactyls and other species of mammals, however, the data showed that whales share many similarities with hippos. These results supported the tree shown in **FIGURE 28.4b**.

Here, it is still logical to map the evolution of the pulley-shaped astragalus at the same ancestor as in Figure 28.4a. But to recognize that the trait had to have been lost in an ancestor of today's whales—which do not have a pulley-shaped astragalus—the tree contains another black bar mapping the trait loss.

The tree supported by the DNA data conflicts with the tree supported by morphological data because it implies that the pulley-shaped astragalus evolved in artiodactyls and then was lost during whale evolution. These two changes (a gain *and* a loss of the astragalus) are less parsimonious than just one change (a gain only).

Data Set 3: Transposable Elements The conflict between the two phylogenies was resolved when researchers analyzed the distribution of the parasitic DNA sequences called **SINEs (short interspersed nuclear elements),** which occasionally insert themselves into the genomes of mammals. (SINEs are transposable elements, similar to the LINEs introduced in Chapter 21.) SINEs provide convincing evidence of relationships because the chance of the same SINE inserting itself in exactly the same place in the genomes of two species is astronomically small.

As the data in **FIGURE 28.4c** show, whales and hippos share several types of SINEs (4, 5, 6, and 7) that are not found in other groups. Other SINEs are present in some artiodactyls but not in others. Camels have no SINEs at all.

To explain these data, biologists hypothesize that no SINEs were present in the population that is ancestral to all of the species in the study. Then, after the branching event that led to the split between the camels and all the other artiodactyls, different SINEs inserted themselves into the genomes of descendant populations.

If this hypothesis is correct, then the presence of a particular SINE represents a derived character. Because whales and hippos share four of these unique derived characters, it is logical to conclude that these animals are closely related.

✔ If you understand this concept, you should be able to explain why SINEs numbered 4–7 are synapomorphies that identify whales and hippos as a monophyletic group, and why the similarity in these SINEs is unlikely to represent homoplasy.

Conclusion: Whales Are Closely Related to Hippos If a phylogenetic tree were estimated based on the SINE data, it would have the same branching pattern as the one shown in Figure 28.4b. (To convince yourself, try the exercise in the figure caption). Thus, most biologists accept the phylogeny shown in Figure 28.4b as the most accurate estimate of whale evolutionary history. According to this phylogeny, whales are artiodactyls and share a relatively recent common ancestor with hippos. This observation inspired the hypothesis that both whales and dolphins are descended from a population of artiodactyls that spent most of their time feeding in shallow water, much as hippos do today.

In 2001, the discovery of fossil artiodactyls supported this hypothesis in spectacular fashion. These fossil species were clearly related to whales—they have an unusual ear bone found only in whales—and yet had a pulley-shaped astragalus.

The combination of DNA sequence data and data from the fossil record has clarified how a particularly interesting group of mammals evolved. What else do fossils have to say?

check your understanding

 If you understand that . . .
- Phylogenies can be estimated by finding synapomorphies that identify monophyletic groups.

 ✔ **You should be able to . . .**

 Determine whether the following traits represent homology or homoplasy: hair in humans and whales; extensive hair loss in humans and whales; limbs in humans and whales; social behavior in humans and certain whales (e.g., dolphins).

Answers are available in Appendix A.

(a) Data set 1 (morphological traits):
Whales diverged before the origin of artiodactyls.

Perissodactyls
(horses and rhinos)

Whale

ARTIODACTYLS
Camel

Peccary

Pig

Gain of pulley-shaped
astragalus

Hippo

Astragalus
(ankle bone)

Deer

Cow

(b) Data set 2 (DNA sequences):
Whales and hippos share a common ancestor.

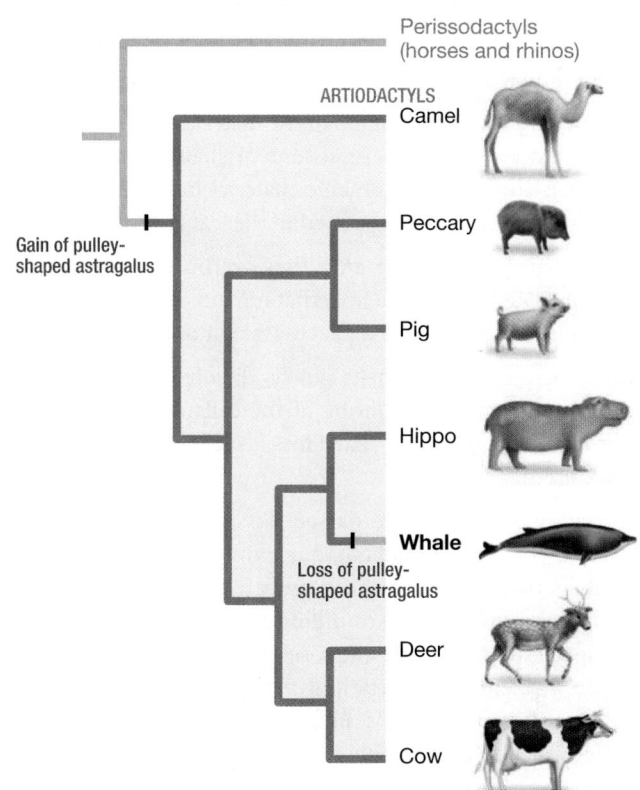

Perissodactyls
(horses and rhinos)

ARTIODACTYLS
Camel

Peccary

Pig

Gain of pulley-
shaped astragalus

Hippo

Whale

Loss of pulley-
shaped astragalus

Deer

Cow

(c) Data set 3 (presence and absence of SINEs): Supports the close relationship between whales and hippos.

Locus	1	2	3	4	5	6	7	8	9	10	11	12	13	14	15	16	17	18	19	20
Cow	0	0	0	0	0	0	0	1	1	1	1	1	1	1	1	1	1	1	0	0
Deer	0	0	0	0	0	0	0	1	?	1	1	1	1	1	1	?	1	1	0	0
Whale	1	1	1	1	1	1	1	0	?	1	0	1	1	0	0	0	?	1	0	0
Hippo	0	?	0	1	1	1	1	0	1	1	0	1	1	0	0	0	?	1	0	0
Pig	0	0	0	?	0	0	0	0	?	0	0	0	?	?	0	0	0	1	1	1
Peccary	?	?	?	?	?	?	?	?	?	?	?	?	?	?	?	?	?	?	1	1
Camel	0	0	0	0	0	0	0	0	0	0	0	0	0	0	0	0	0	0	0	0

1 = SINE present
0 = SINE absent
? = still undetermined

Whales and hippos share four *unique* SINEs (4, 5, 6, and 7)

FIGURE 28.4 Where Do Whales Belong on the Tree of Life?

DATA for (c): Nikaido, M., A. P. Rooney, and N. Okada. 1999. *Proceedings of the National Academy of Sciences, USA* 96: 10261–10266.

✔**EXERCISE** Map the origin of the following groups of SINEs on the tree in part (b): Group (4, 5, 6, 7) identifies hippos and whales as part of a monophyletic group; (8, 11, 14, 15, 17) identifies deer and cows as part of a monophyletic group; (10, 12, 13) identifies hippos, whales, deer, and cows as part of a monophyletic group. What two species does SINEs group (19, 20) identify as part of a monophyletic group?

28.2 Tools for Studying History: The Fossil Record

Phylogenetic analyses are powerful ways to infer the order in which events occurred during evolution and to understand how particular groups of species are related. But only the fossil record provides *direct* evidence about what organisms that lived in the past looked like, where they lived, and when they existed.

A **fossil** is a piece of physical evidence from an organism that lived in the past. The **fossil record** is the total collection of fossils that have been found throughout the world. The fossil record is housed in thousands of private and public collections.

Let's review how fossils form, analyze the strengths and weaknesses of the fossil record, and then summarize major events that have taken place in life's approximately 3.5-billion-year history.

How Do Fossils Form?

Most of the processes that form fossils begin when part or all of an organism is buried in ash, sand, mud, or some other type of sediment. For example, imagine that leaves, pollen, seeds, branches, or entire trees fall onto a muddy swamp, where they are buried by soil and debris before they decay. Once burial occurs, several things can happen.

- If decomposition does not occur, the organic remains can be preserved intact—like the fossil pollen in **FIGURE 28.5a**.

- If sediments accumulate on top of the material and become cemented into rocks such as mudstone or shale, the sediments' weight can compress the organic material below into a thin, carbonaceous film. This happened to the leaf in **FIGURE 28.5b**.

- If the remains decompose *after* they are buried—as did the branch in **FIGURE 28.5c**—the hole that remains can fill with dissolved minerals and create an accurate **cast** of the remains.

- If the remains rot extremely slowly, dissolved minerals can gradually infiltrate the interior of the cells and harden into stone, forming a permineralized fossil, such as petrified wood (**FIGURE 28.5d**).

After many centuries have passed, fossils can be exposed at the surface by erosion, a road cut, quarrying, or other processes. If researchers find a fossil, they can prepare it for study by painstakingly clearing away the surrounding rock.

If the species represented is new, researchers describe its morphology in a scientific publication, name the species, estimate the fossil's age based on dates assigned to nearby rock layers, and add the specimen to a collection so that it is available for study by other researchers. It is now part of the fossil record.

This scenario is based on conditions that are ideal for fossilization: The trees fell into an environment where decomposition was slow and burial was rapid. In most habitats the opposite situation occurs—decomposition is rapid and burial is slow. In reality, then, fossilization is an extremely rare event.

To appreciate this point, consider that there are 10 specimens of the first bird-like dinosaur to appear in the fossil record, *Archaeopteryx*. If you accept an estimate that crow-sized birds native to wetland habitats would have a population size of around 10,000 and a life span of 10 years, and if you accept the current estimate that the species existed for about 2 million years, then you can calculate that about 2 billion *Archaeopteryx* lived. But as far as researchers currently know, only 1 out of every 200,000,000 individuals were fossilized and discovered. For this species, the odds

of becoming part of the fossil record were almost 40 times worse than your odds are of winning the grand prize in a state lottery.

Limitations of the Fossil Record

Before looking at how the fossil record is used to answer questions about the history of life, it is essential to review the nature of this archive and recognize several features.

Habitat Bias   Because burial in sediments is so crucial to fossilization, there is a strong habitat bias in the database. Organisms that live in areas where sediments are actively being deposited—including beaches, mudflats, and swamps—are much more likely to form fossils than are organisms that live in other habitats.

Within these habitats, burrowing organisms such as clams are already underground—pre-buried—at death and are therefore much more likely to fossilize. Organisms that live aboveground in dry forests, grasslands, and deserts are much less likely to fossilize.

Taxonomic and Tissue Bias   Slow decay is almost always essential to fossilization, so organisms with hard parts such as bones or shells are most likely to leave fossil evidence. This requirement introduces a strong taxonomic bias into the record. Clams, snails, and other organisms with hard parts have a much higher tendency to be preserved than do worms.

A similar bias exists for tissues within organisms. For instance, pollen grains are encased in a tough outer coat that resists decay, so they fossilize much more readily than do flowers. Shark teeth are abundant in the fossil record, but shark skeletal elements, which are made of cartilage, are almost nonexistent.

Temporal Bias   Recent fossils are much more common than ancient fossils. This causes a temporal bias in the fossil record. The reason is straightforward—the older a fossil is, the longer it has been exposed to potentially destructive forces.

Older fossils usually occur in sedimentary rock layers deep below newer layers (see Chapter 25). These fossils are more vulnerable to crushing, heating, melting, and distortion by various chemical and physical processes. Older fossils are also more likely to be subducted further into the Earth's interior by the sliding and collision of Earth's tectonic plates. The fossils that survive these destructive forces will be discovered by scientists only if the

(a) Intact fossil (pollen)   **(b)** Compression fossil (leaf)   **(c)** Cast fossil (bark)   **(d)** Permineralized fossil (tree trunk)

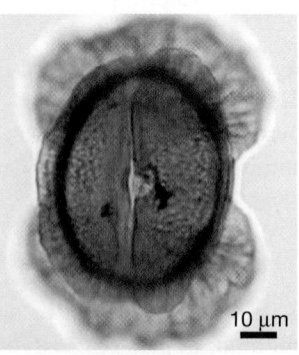   

10 μm

FIGURE 28.5 Fossils Are Formed in Several Ways. Different preservation processes give rise to different types of fossils.

deep layers of sedimentary rock they occupy are exposed to the surface by erosion (such as in the Grand Canyon) or by uplift.

Abundance Bias The fossil record has an abundance bias; it is weighted toward common species. Organisms that are abundant, widespread, and/or present on Earth for long periods of time leave evidence much more often than do species that are rare, local, or ephemeral.

To summarize, the fossil database represents a highly nonrandom sample of the past. **Paleontologists**—scientists who study fossils—recognize that they are limited to asking questions about tiny and scattered segments on the tree of life. ✔ If you understand the biases of the fossil record, you should be able to predict at least two ways that paleontologists could be misled by their observations of fossils.

And yet, as this chapter shows, the fossil record is a scientific treasure trove. Analyzing fossils is the only way scientists have of examining the physical appearance of extinct forms and inferring how they lived. The fossil record is like an ancient library, filled with volumes that give us glimpses of what life was like millions of years before humans appeared.

Life's Time Line

It is very difficult to date the origin of life precisely. Our best estimate is that the Earth started to form about 4.6 billion years ago, and that life began before 3.5 billion years ago—probably in the warm, hydrothermal vents at the bottom of the ocean (see Chapter 2). To organize the tremendous sweep of time between then and now, researchers divide Earth history into segments called eons, eras, and periods.

Originally, geologists used distinctive rock formations or fossilized organisms to identify the boundaries between named time intervals. Later, researchers were able to use radiometric dating to assign absolute dates—expressed as years before the present—to events in the fossil record. Radiometric dating is based on the well-studied decay rates of certain radioactive isotopes (see Chapter 25). By dating rocks near fossils, researchers can also assign an absolute age to many of the species in the fossil record.

To summarize the history of life, researchers create time lines that record key "evolutionary firsts"—the appearance of important new lineages or innovations. It's important to recognize, though, that the times assigned to these first appearances underestimate when lineages appeared and events occurred. The reason is simple: A particular species or lineage can exist for millions of years before leaving fossil evidence. The fossil record and efforts to date fossils are constantly improving, so time lines are always a work in progress.

Precambrian The interval between the formation of Earth about 4.6 billion years ago and the appearance of most animal groups about 542 million years ago (abbreviated mya) is called the **Precambrian** (**FIGURE 28.6**); it is divided into the Hadean, Archaean, and Proterozoic eons.

FIGURE 28.6 Major Events of the Precambrian. Life, photosynthesis, and the oxygen atmosphere all originated in the Precambrian.

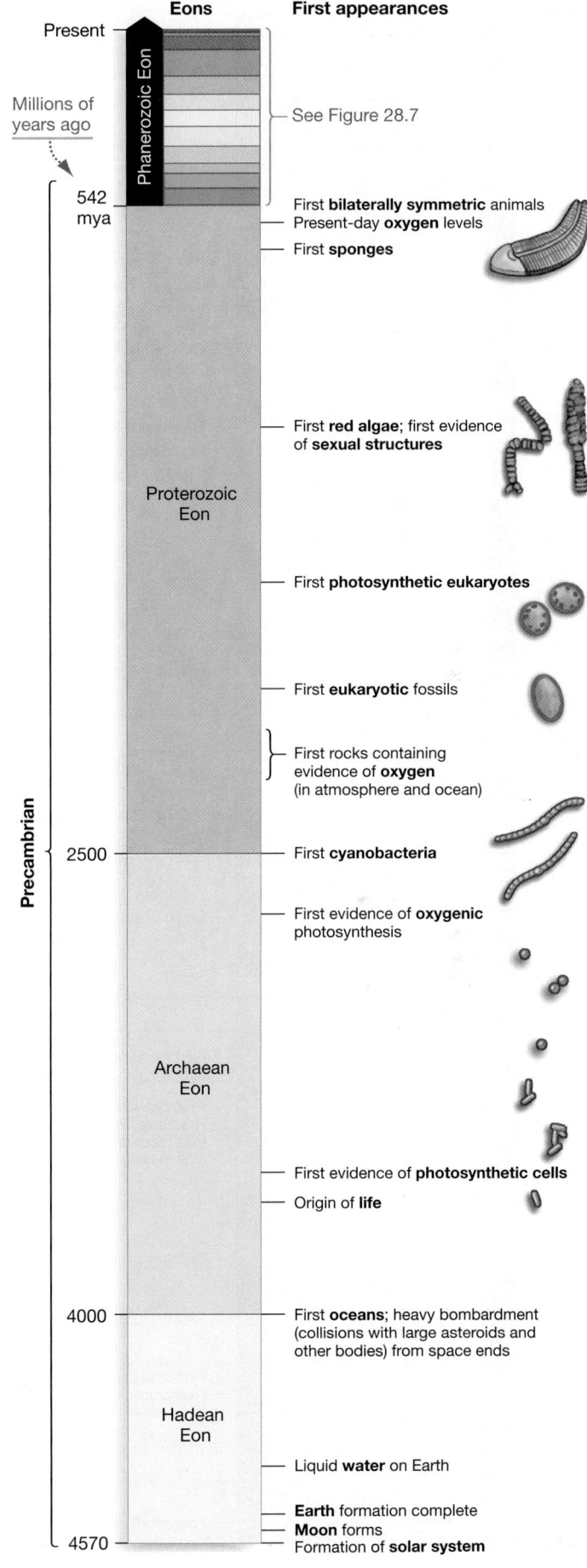

Eons — First appearances

Present

Millions of years ago

542 mya

Phanerozoic Eon

See Figure 28.7

First **bilaterally symmetric** animals
Present-day **oxygen** levels
First **sponges**

First **red algae**; first evidence of **sexual structures**

Proterozoic Eon

First **photosynthetic eukaryotes**

First **eukaryotic** fossils

First rocks containing evidence of **oxygen** (in atmosphere and ocean)

2500

Precambrian

First **cyanobacteria**

First evidence of **oxygenic** photosynthesis

Archaean Eon

First evidence of **photosynthetic cells**

Origin of **life**

4000

First **oceans**; heavy bombardment (collisions with large asteroids and other bodies) from space ends

Hadean Eon

Liquid **water** on Earth

Earth formation complete
Moon forms
Formation of **solar system**

4570

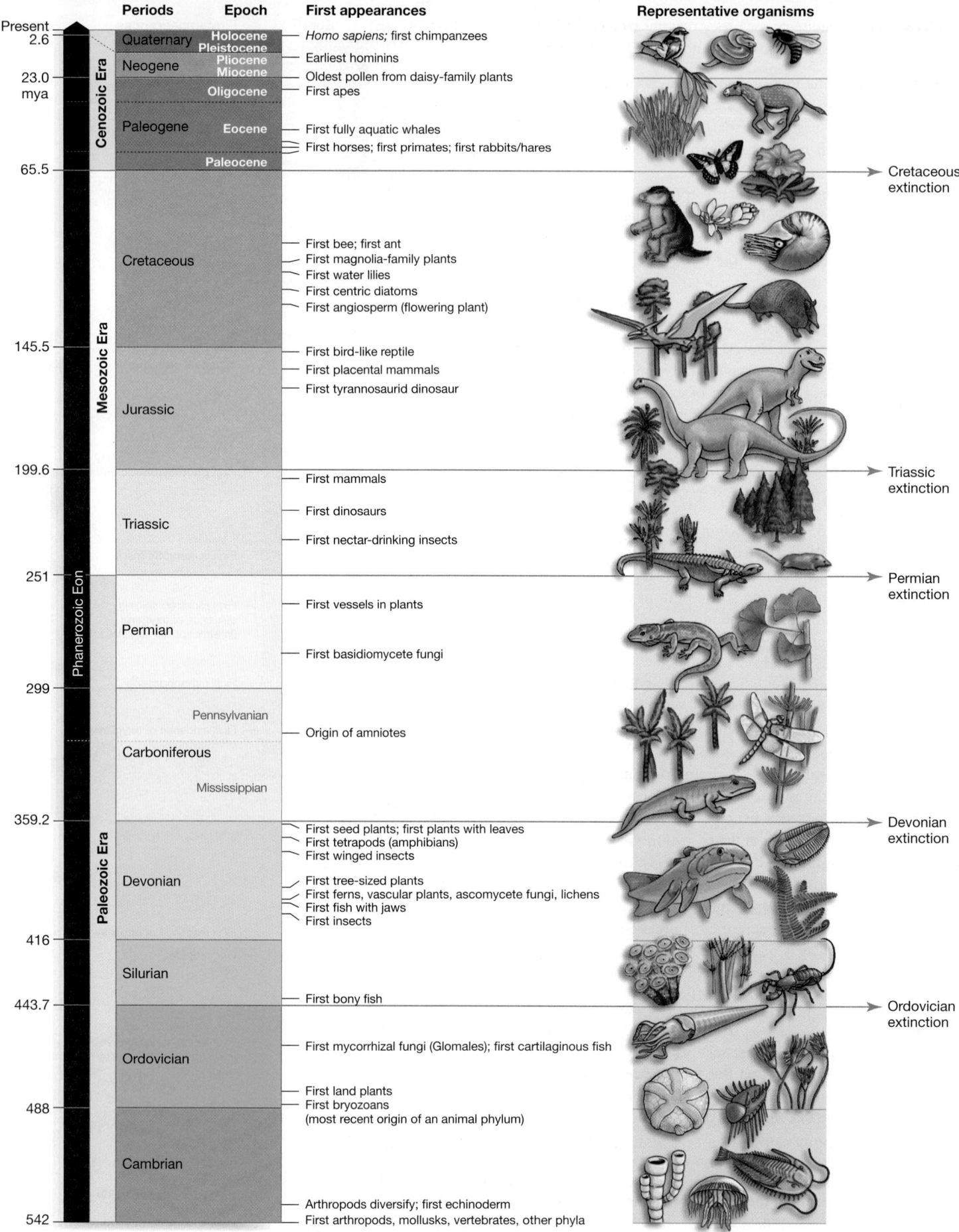

Periods	Epoch	First appearances	Representative organisms

Present
2.6 — Quaternary — Holocene / Pleistocene — *Homo sapiens;* first chimpanzees

Neogene — Pliocene / Miocene — Earliest hominins

23.0 mya — Oldest pollen from daisy-family plants

Oligocene — First apes

Paleogene — Eocene — First fully aquatic whales

First horses; first primates; first rabbits/hares

65.5 — Paleocene — → Cretaceous extinction

Cretaceous — First bee; first ant
First magnolia-family plants
First water lilies
First centric diatoms
First angiosperm (flowering plant)

145.5

Jurassic — First bird-like reptile
First placental mammals
First tyrannosaurid dinosaur

199.6 — → Triassic extinction

Triassic — First mammals
First dinosaurs
First nectar-drinking insects

251 — → Permian extinction

Permian — First vessels in plants

First basidiomycete fungi

299

Pennsylvanian — Origin of amniotes

Carboniferous

Mississippian

359.2 — → Devonian extinction

Devonian — First seed plants; first plants with leaves
First tetrapods (amphibians)
First winged insects
First tree-sized plants
First ferns, vascular plants, ascomycete fungi, lichens
First fish with jaws
First insects

416

Silurian

First bony fish

443.7 — → Ordovician extinction

Ordovician — First mycorrhizal fungi (Glomales); first cartilaginous fish

First land plants
First bryozoans
(most recent origin of an animal phylum)

488

Cambrian

Arthropods diversify; first echinoderm
First arthropods, mollusks, vertebrates, other phyla

542

Cenozoic Era — Mesozoic Era — Paleozoic Era — Phanerozoic Eon

FIGURE 28.7 Major Events of the Phanerozoic Eon (at left). The Phanerozoic began with the initial diversification of animals, continued with the evolution and early diversification of land plants and fungi, and includes the subsequent movement of animals to land. A total of five mass extinctions occurred during the eon.

Here are the important things to know about the Precambrian:

- Life was exclusively unicellular for most of Earth's history.
- Oxygen was virtually absent from the oceans and atmosphere for almost 2 billion years after the origin of life. Photosynthetic bacteria were responsible for the creation of the oxygen atmosphere (see Chapter 29).

Phanerozoic Eon The interval between 542 mya and the present is called the Phanerozoic eon and is divided into three eras (**FIGURE 28.7**). Each of these eras is further divided into intervals called periods, which are themselves divided into epochs.

1. The **Paleozoic** ("ancient life") **era** begins with the appearance of most major animal lineages and ends with the obliteration of almost all multicellular life-forms at the end of the Permian period. The Paleozoic includes the origin and initial diversification of the animals, land plants, and fungi, as well as the appearance of land animals.

2. The **Mesozoic** ("middle life") **era** begins with the end-Permian extinction events and ends with the extinction of the dinosaurs (except birds) and other groups at the boundary between the Cretaceous period and Paleogene period. In terrestrial environments of the Mesozoic, gymnosperms were the most dominant plants and dinosaurs were the most dominant vertebrates.

3. The **Cenozoic** ("recent life") **era** is divided into the Paleogene, Neogene, and Quaternary periods. On land, angiosperms were the most dominant plants and mammals were the largest vertebrates.

At the top of Figure 28.7 you can see that the periods of the Cenozoic are divided into epochs, and that we are presently in the Holocene epoch. However, many scientists have called for the delineation of a new epoch—the Anthropocene ("new human epoch")—to reflect the dramatic physical, chemical, and biological changes that humans are causing on Earth, especially since the industrial revolution. Much discussion will take place before any official change is made.

Changes in the Oceans and Continents The changes in environments and life-forms that are recorded in Figures 28.6 and 28.7 took place in the context of radical changes in climate and in the extent and location of Earth's oceans and continents. Earth's crust is broken into enormous plates that are in constant motion, driven by heat rising from the planet's core.

FIGURE 28.8 summarizes the most recent data on how the positions of the continents and oceans have changed through time,

FIGURE 28.8 Continental Positions during the Phanerozoic (at right). Dramatic changes in the extent and position of the continents took place during the Phanerozoic. These changes affected the total amount of land area, the relative amounts of land in the tropics versus northern latitudes, and the nature of ocean currents.

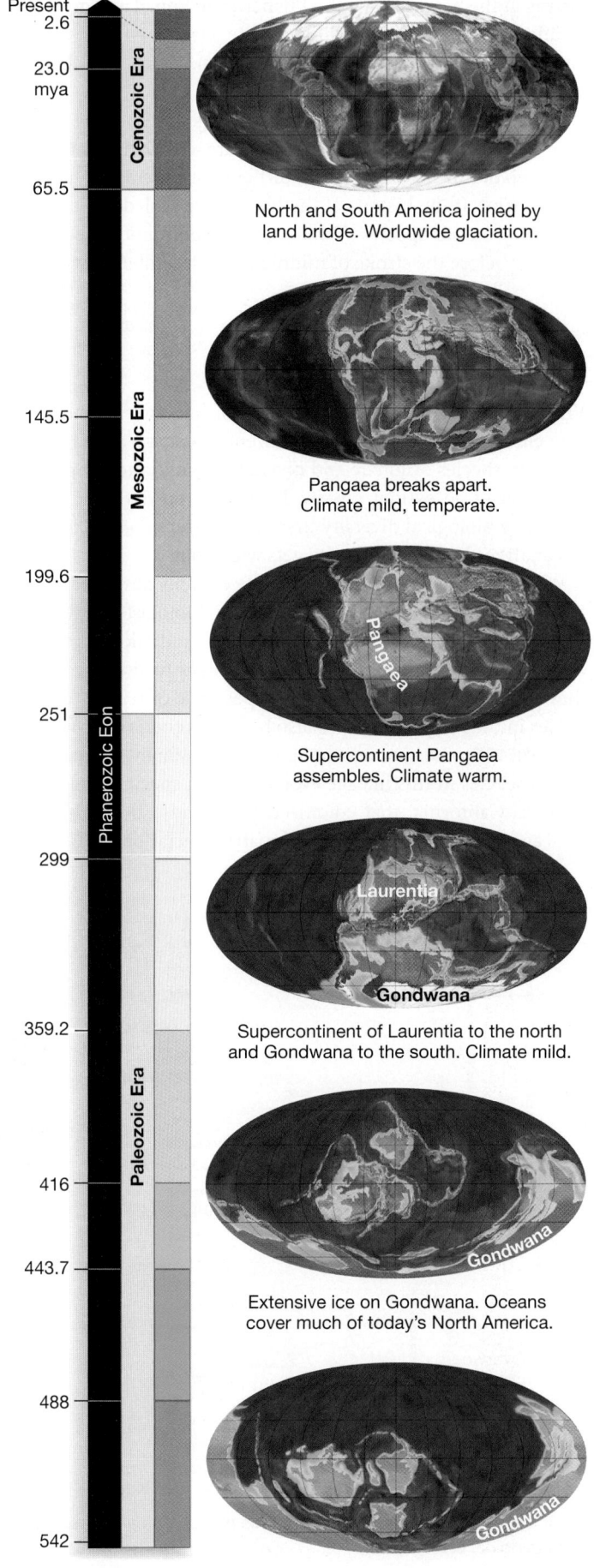

North and South America joined by land bridge. Worldwide glaciation.

Pangaea breaks apart. Climate mild, temperate.

Supercontinent Pangaea assembles. Climate warm.

Supercontinent of Laurentia to the north and Gondwana to the south. Climate mild.

Extensive ice on Gondwana. Oceans cover much of today's North America.

starting at the beginning of the Phanerozoic Eon. The figure also includes notes on the prevailing climate. Over the past 542 million years alone, terrestrial and marine landforms and climates have changed radically.

Taken together, the fundamental message in time-line data is a story of constant change. The changes are well documented, but the sweep of time involved is still difficult for the human mind to comprehend. If the history of the Earth were scaled to fit in one calendar year, our species, *Homo sapiens*, would not appear until one hour before the stroke of midnight on December 31st.

28.3 Adaptive Radiation

When biologists consider the history of life, two of the most compelling types of evolutionary events to study are (1) periods when species originate and diversify rapidly, and (2) periods when species go extinct rapidly. Let's focus on dramatic events that create biological diversity first; the chapter's concluding section analyzes how that diversity gets wiped out.

When a single lineage rapidly produces *many* descendant species with a wide range of adaptive forms, biologists say that an **adaptive radiation** has occurred. Adaptive radiations are noteworthy because they cause dramatic changes to the tree of life. They can be observed as a sudden appearance of related, diverse species in the fossil record and also by phylogenetic analysis.

FIGURE 28.9 provides an example: the Hawaiian silverswords. The 30 species in this lineage evolved from a species of tarweed, native to California, that colonized the islands about 5 million years ago. This is an example of allopatric speciation by dispersal

(see Chapter 27). Compared to speciation rates in other groups, this rate is extremely rapid. Further, today's silverswords vary from mosslike mats to vines and rosettes. They live in habitats ranging from lush rain forests to austere lava flows.

The Hawaiian silverswords fulfill the three hallmarks of an adaptive radiation: (1) they are a monophyletic group, (2) they speciated rapidly, and (3) they diversified ecologically—meaning, in terms of the resources they use and the habitats they occupy.

Biologists use the term niche to describe the range of resources that a species can use and the range of conditions that it can tolerate (see Chapter 27). Silverswords occupy a wide array of niches.

Why Do Adaptive Radiations Occur?

Adaptive radiations are a major pattern in the history of life. But why do some lineages diversify rapidly, while others do not?

Two general mechanisms can trigger adaptive radiations: new resources, and new ways to exploit resources. Let's consider each in turn.

Ecological Opportunity Ecological opportunity—meaning the availability of new or novel types of resources—has driven a wide array of adaptive radiations. For example, biologists explain the diversification of silverswords by hypothesizing that few other flowering plant species were present on the Hawaiian Islands 5 million years ago. With few competitors, the descendants of the colonizing tarweed were able to grow in a wide range of habitats. Over time—by the processes of mutation, genetic drift, and natural selection (discussed in Chapter 26)—some became specialized for growth in different niches.

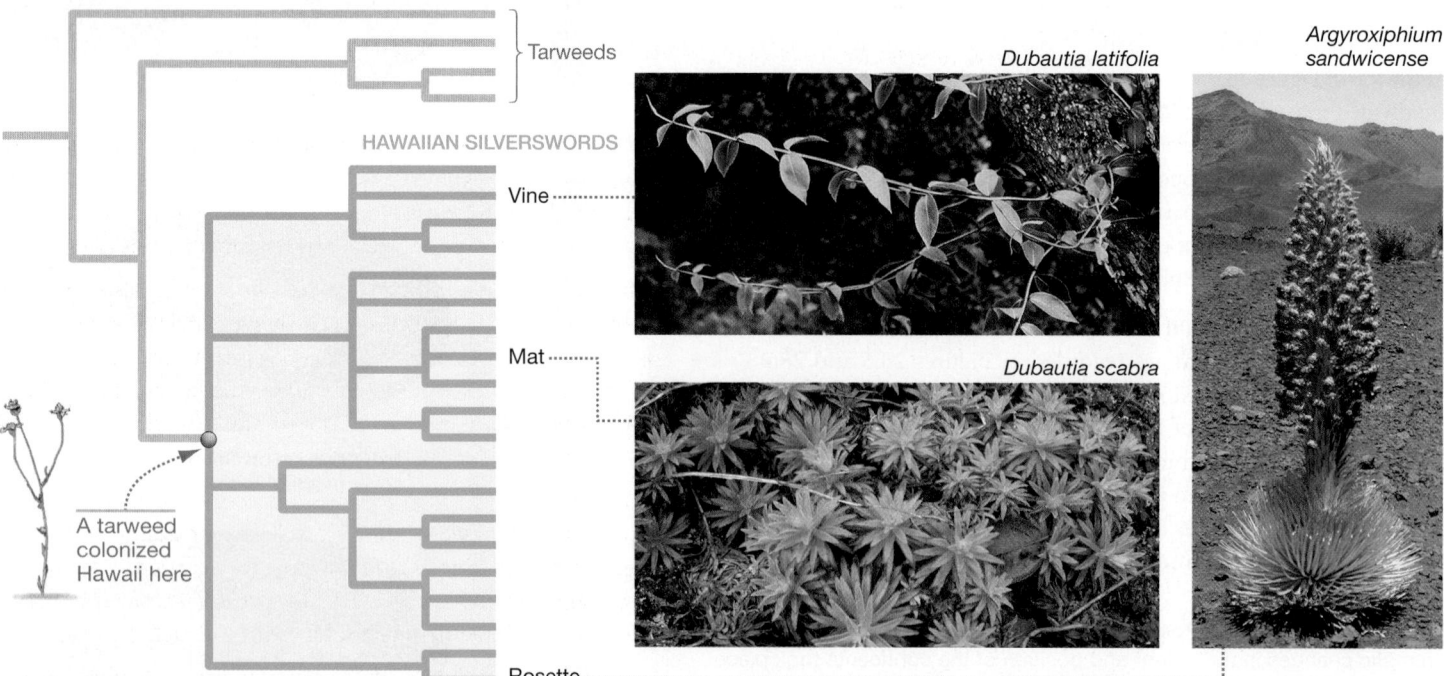

FIGURE 28.9 An Adaptive Radiation. This tree shows a subset of the Hawaiian silverswords, illustrating the extent of morphological divergence.

(a) Short-legged lizard species spend most of their time on the twigs of trees and bushes.

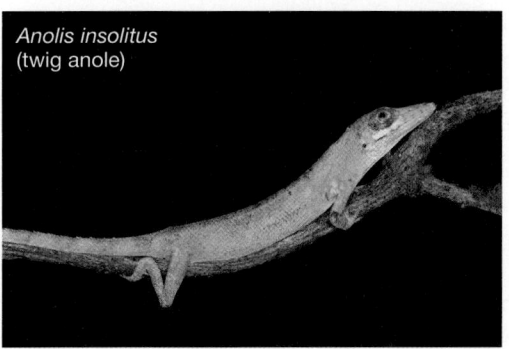

Anolis insolitus
(twig anole)

(b) Long-legged lizard species live on tree trunks and the ground.

Anolis cybotes
(trunk/ground anole)

FIGURE 28.10 Adaptive Radiations of *Anolis* Lizards. (a, b) Species of *Anolis* lizards vary in leg length and tail length. **(c)** Evolutionary relationships among lizard species on the islands of Hispaniola and Jamaica. The initial colonist species was different on these islands, but in terms of how they look and where they live, a similar suite of four species evolved.

(c) The same adaptive radiation of *Anolis* has occurred on different islands, starting from different types of colonists.

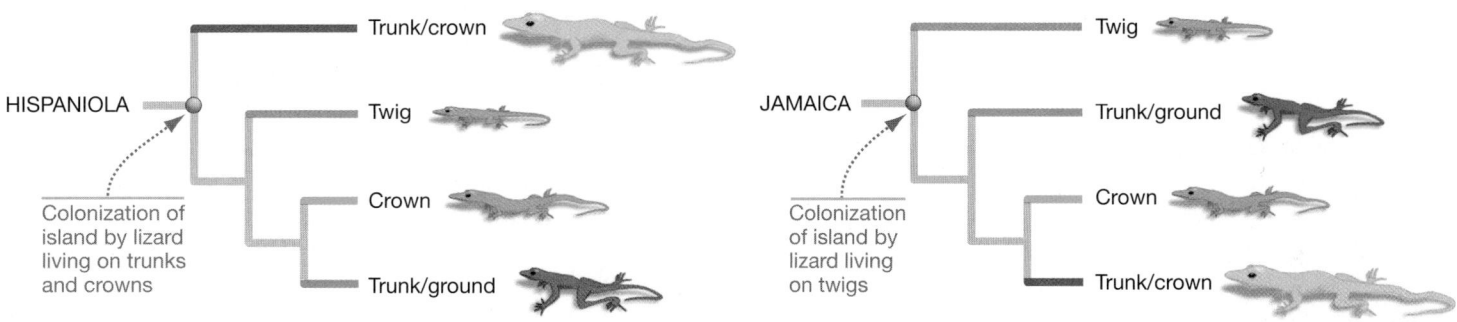

The same type of ecological opportunity was hypothesized to explain the adaptive radiation of *Anolis* lizards on islands in the Caribbean. The lineage includes 150 species. They thrive in a wide array of habitats and have diverse body sizes and shapes. And in most cases, a lizard's size and shape are correlated with the habitat it occupies. For example:

- Species that live on tree twigs have short legs and tails that allow them to move efficiently on narrow surfaces (**FIGURE 28.10a**).

- Species that spend most of their time clinging to broad tree trunks or running along the ground have long legs and tails, making them fast and agile on broad surfaces (**FIGURE 28.10b**).

Most islands in the Caribbean have a distinct suite of lizard species. And in most cases, each island has a species that lives only in the twigs, on the ground, or in other distinctive habitats. The classical explanation for this pattern was that a mini-radiation occurred on each island: An original colonizing group encountered no competitors, and diversified in a way that led to efficient use of the available resources by a group of descendant species.

To test this hypothesis rigorously, Jonathan Losos and his colleagues estimated the phylogeny of *Anolis* from DNA sequence data. The results supported a key claim of the ecological opportunity hypothesis: The lizards on each island were monophyletic.

The critical data, though, are shown in **FIGURE 28.10c**. These phylogenetic trees, for species found on two different islands, are typical. The key observation is that the original colonist on each island was specialized for a different niche. The initial species on Hispaniola lived on the trunks and crowns of trees, while the original colonist on Jamaica occupied twigs.

From different evolutionary starting points, then, an adaptive radiation filled the same niches on both islands. This is exactly what the ecological opportunity hypothesis predicts. In *Anolis* lizards in the Caribbean, a series of mini-radiations were triggered when colonists arrived on an island that had food, space, and other resources, but lacked competitors. ✔ If you understand this concept, you should be able to generate a hypothesis to explain why tarweeds in California and *Anolis* lizards on the mainland are not particularly species-rich or ecologically diverse.

Morphological Innovation   The evolution of a key morphological trait—one that allowed descendants to live in new areas, exploit new sources of food, or move in new ways—triggered many of the important diversification events in the history of life. For example:

- *Flowers* are a unique reproductive structure that helped trigger the diversification of angiosperms (flowering plants). Because flowers are particularly efficient at attracting pollinators, the evolution of the flower made angiosperms more efficient in reproduction. Today angiosperms are far and away the most species-rich lineage of land plants. Over 250,000 species are known.

- *Feathers and wings* gave some dinosaurs the ability to fly. Feathers originally evolved for display or insulation; later they were used in gliding and in powered flight. Today the lineage called birds contains about 10,000 species, whose representatives live in virtually every habitat on the planet.

In each case, the evolution of new morphological features is hypothesized to have supported rapid speciation and ecological divergence.

The Cambrian Explosion

Almost all life-forms were unicellular for almost 3 *billion* years after the origin of life. The exceptions were several lineages of small multicellular algae, which show up in the fossil record about 1 billion years ago. Then, the first animals—early sponges—appear, perhaps as early as 635 million years ago.

In a relatively short time, creatures with shells, exoskeletons, internal skeletons, legs, heads, tails, eyes, antennae, jaw-like mandibles, segmented bodies, muscles, and brains had evolved. It was arguably the most spectacular period of evolutionary change in the history of life. Because much of it occurred during the Cambrian period, this adaptive radiation of animals is called the **Cambrian explosion.**

All organisms were small for 3 billion years; then life got big, in just 50 million years—1/60th of the total time life had existed. Before asking how this happened, let's explore the fossilized evidence for what happened.

The Doushantuo, Ediacaran, and Early Cambrian Fossils
The Cambrian explosion is documented by three major fossil assemblages that record the state of animal life near the beginning of the Cambrian. As **FIGURE 28.11a** indicates, the species collected from each of these intervals are referred to respectively as the Doushantuo fossils (named after the Doushantuo formation in China), Ediacaran fossils (named after the Ediacara Hills, Australia), and Cambrian fossils (from the Burgess Shale in Canada, the Chengjiang region of China, and several other sites around the world).

Each fossil assemblage records a distinctive **fauna**—or collection of animal species:

- *Doushantuo microfossils* include tiny sponges and corals, less than 1 mm across. These creatures—the first animals on Earth—probably made their living by filtering organic debris from the water.

- *Ediacaran faunas* include sponges, jellyfish, and comb jellies as well as fossilized burrows, tracks, and other traces from unidentified animals. None of these organisms have shells, limbs, heads, or feeding appendages (**FIGURE 28.11b, bottom**). Ediacaran animals likely filtered or absorbed organic material as they burrowed in sediments, sat immobile on the seafloor, or floated in the water.

- *Cambrian faunas* are among the most sensational additions ever made to the fossil record. Sponges, jellyfish, and comb jellies are abundant in these rocks, but so are several worm-like creatures and arthropods (**FIGURE 28.11b, top**), in addition to mollusks, echinoderms, and virtually all the other major animal lineages (see Chapter 33). This list includes our own lineage—the chordates—represented by early vertebrate fossils with streamlined, fishlike bodies and a skull made of cartilage (**FIGURE 28.12**). A tremendous increase in the size and morphological complexity of animals occurred, accompanied by diversification in how they made a living. Species in these faunas swam, burrowed, walked, ran, slithered, clung, or floated; there were predators, scavengers, filter feeders, and grazers. The diversification filled many of the ecological niches found in marine habitats today.

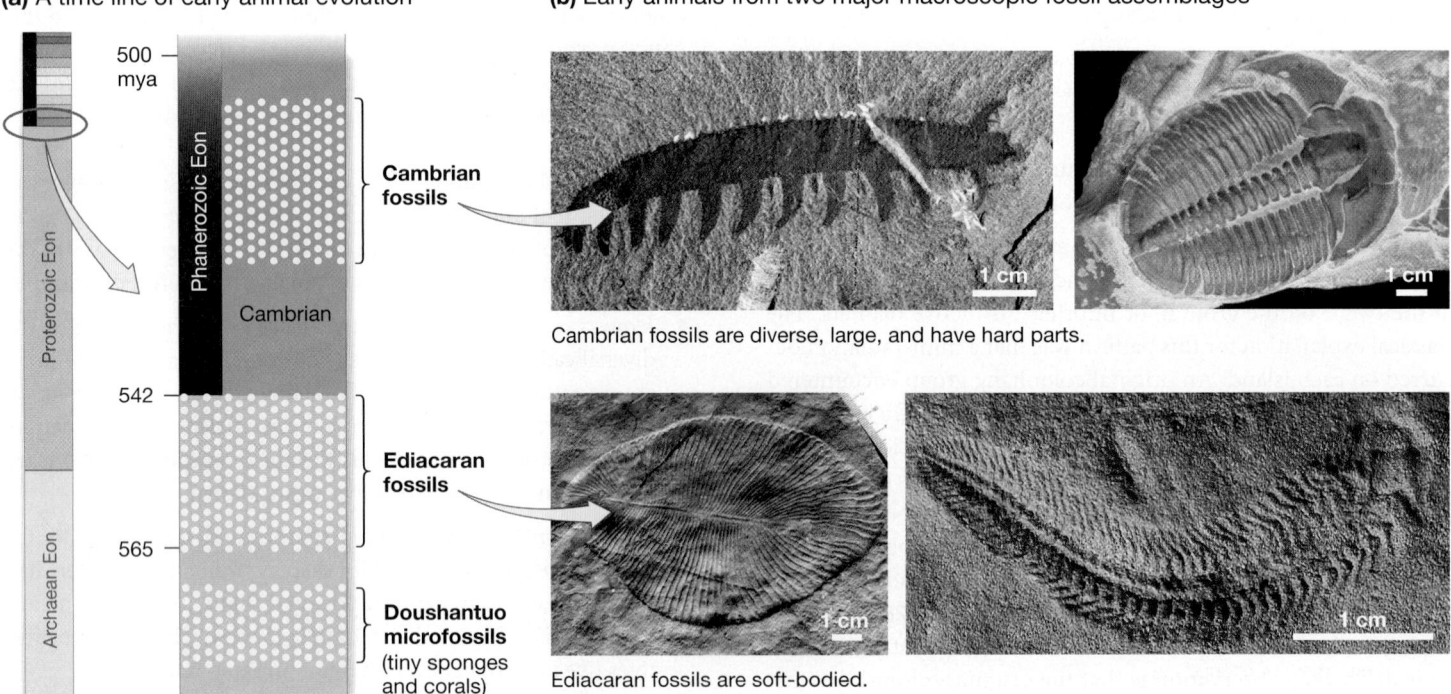

(a) A time line of early animal evolution

(b) Early animals from two major macroscopic fossil assemblages

Cambrian fossils are diverse, large, and have hard parts.

Ediacaran fossils are soft-bodied.

FIGURE 28.11 Fossils Document the Cambrian Explosion. The origin of animals and their diversification during the Cambrian explosion is documented by three major fossil assemblages, two of them represented by macroscopic fossils.

FIGURE 28.12 Life in the Cambrian. This artist's interpretation of animal diversity during the Cambrian is based on fossils found in the Chengjiang formation of China. Chengjiang is famous for its unique fossils of fishlike early vertebrates.

What Triggered the Cambrian Explosion? The Doushantuo, Ediacaran, and Cambrian faunas document what happened during the Cambrian explosion. Now the question is: *How* did all this speciation, morphological change, and ecological diversification come about?

To answer this question, biologists point to an array of data sets and hypotheses:

- *Higher oxygen levels* By analyzing the composition of rocks formed during the Proterozoic, geologists have established

that oxygen levels gradually rose in the atmosphere and ocean. Increased oxygen levels make aerobic respiration more efficient (see Chapter 9); increased aerobic respiration is required to support larger bodies and more active movements. Some biologists suggest that oxygen levels reached a critical threshold at the start of the Cambrian explosion, making the evolution of big, mobile animals possible.

- *The evolution of predation* Before the Cambrian explosion, animals made their living by eating organic material that settled on the seafloor or filtering cells and debris from the water. But Cambrian fossils include shelled animals with holes in the shells—evidence that a predator drilled through and ate the animal inside. When predators evolved, they exerted selection pressure for shells, hard exoskeletons, rapid movement, and other adaptations for prey to defend themselves. This selection would drive morphological divergence.

- *New niches beget more new niches* The vast majority of Doushantuo and Ediacaran animals lived on the ocean floor—in what biologists call benthic habitats. Once animals could move off this substrate, they could exploit algae and other resources that were available above the ocean floor. The presence of animals at an array of depths created selection pressure for the evolution of species that could eat them. In this way, the ability to exploit new niches created new niches, driving speciation and ecological diversification.

- *New genes, new bodies* *Hox* genes play a key role in organizing the development of the body in diverse animals by signaling where cells are in the embryo (Chapter 22). **FIGURE 28.13** summarizes the *Hox* genes found in some major animal

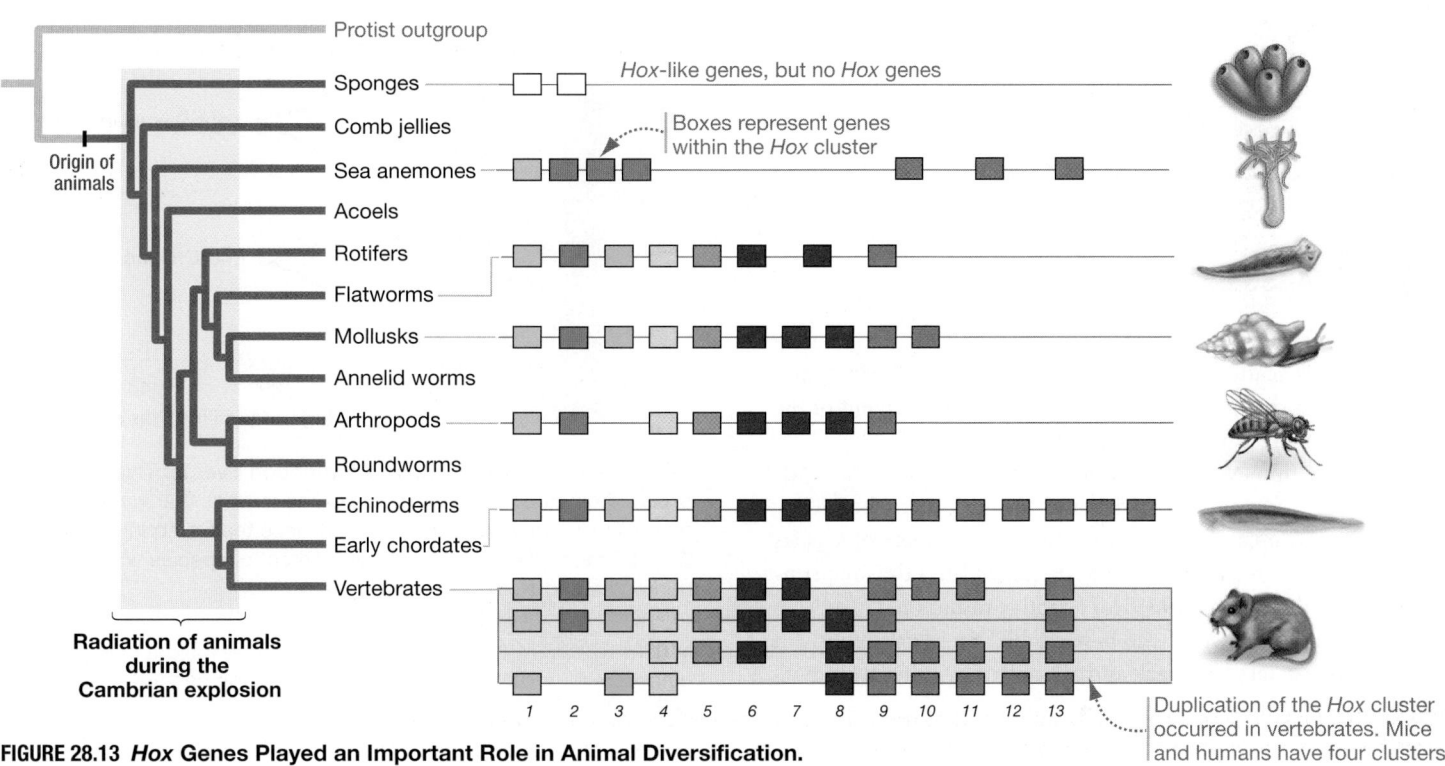

FIGURE 28.13 *Hox* Genes Played an Important Role in Animal Diversification.

groups. The key observation is that the earliest animals in the fossil record had few or no *Hox* genes; most groups that appeared later have more. The idea here is that gene duplication and diversification increased the number of *Hox* genes in animals and made it possible for larger, more complex bodies to evolve.

It's important to recognize that most or all of these hypotheses could be correct. They are not mutually exclusive. If increased oxygen levels made larger bodies and more rapid movement possible, then animals could move into new habitats off the seafloor and become large enough to eat smaller animals. Selection would favor individuals with mutations in *Hox* genes and other DNA sequences that make the development of a large, complex body possible.

About 100 million years after the Cambrian explosion, a similar adaptive radiation occurred after the first plants—the descendants of green algae—adapted to life on land. In the span of about 56 million years, an array of growth forms and most major lineages of plants appear in the fossil record. This "Devonian explosion" resulted from the morphological innovations that allowed green plants to thrive on land, coupled with the ecological opportunity provided by an environment awash in sunlight (Chapter 31).

Adaptive radiations occur when species originate and diversify rapidly. Now let's consider how dramatic events can *wipe out* diversity.

check your understanding

If you understand that . . .

- Adaptive radiations are triggered by ecological opportunity and morphological innovation.

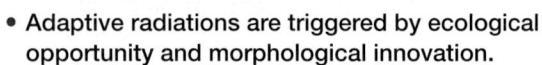

✔ **You should be able to . . .**

Explain the role of ecological opportunity and morphological innovation in the Cambrian explosion.

Answers are available in Appendix A.

28.4 Mass Extinction

A **mass extinction** refers to the rapid extinction of a large number of lineages scattered throughout the tree of life. More specifically, a mass extinction occurs when at least 60 percent of the species present are wiped out within 1 million years.

Mass extinction events are evolutionary hurricanes. They buffet the tree of life, snapping twigs and breaking branches. They are catastrophic episodes that wipe out huge numbers of species and lineages in a short time, giving the tree of life a drastic pruning. They are the polar opposite of adaptive radiation.

Before analyzing two of the best-studied mass extinctions in the fossil record, let's step back and ask how they differ from normal extinction events.

How Do Mass Extinctions Differ from Background Extinctions?

Background extinction refers to the lower, average rate of extinction observed when a mass extinction is not occurring. Although there is no hard-and-fast rule for distinguishing between background and mass extinction rates, paleontologists traditionally recognize and study five historic mass extinction events. **FIGURE 28.14**, for example, plots the percentage of plant and animal lineages called families that died out during each stage in the geologic time scale since the Cambrian explosion. Five spikes in the graph—denoting a large number of extinctions within a short time—are drawn in red. These are referred to as "The Big Five."

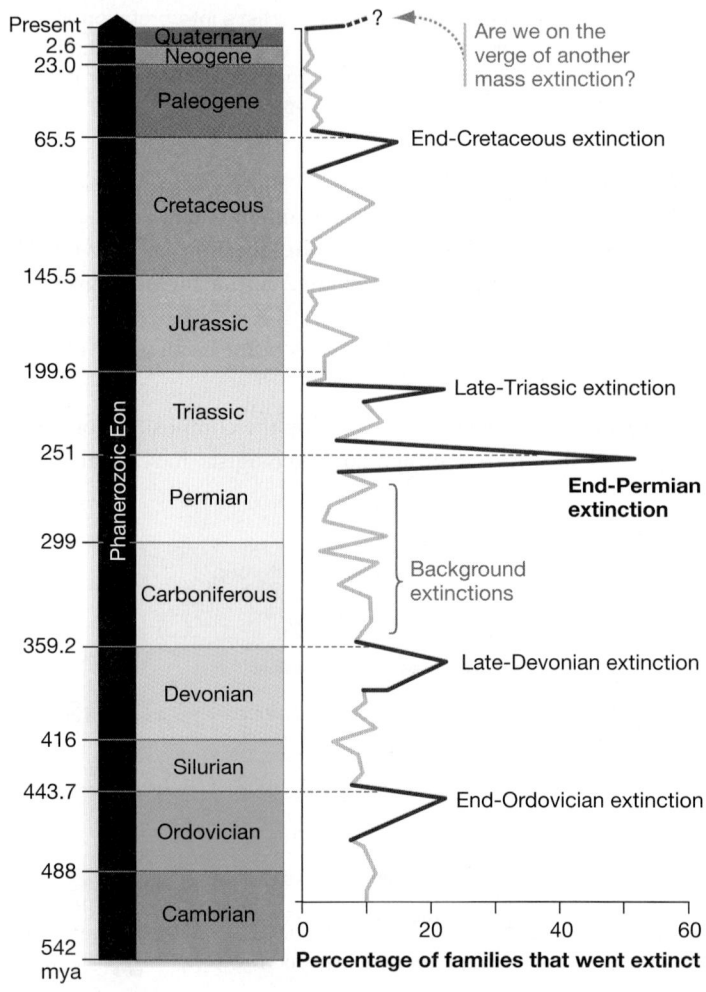

FIGURE 28.14 The Big Five Mass Extinction Events. This graph shows the percentage of lineages called families that went extinct over each interval in the fossil record since the Cambrian explosion. Over 50 percent of families and 90 percent of species went extinct during the end-Permian extinction.

DATA: Benton, M. J., 1995. *Science* 268: 52–58.

✔ **QUANTITATIVE** Which extinction event ended the era of the dinosaurs 65 million years ago? About what percentage of families went extinct?

Biologists are interested in distinguishing between background and mass extinctions because these events have contrasting causes and effects.

- *Background extinctions* are thought to occur when normal environmental change, emerging disease, predation pressure, or competition with other species reduces certain populations to zero.

- *Mass extinctions* result from extraordinary, sudden, and temporary changes in the environment. During a mass extinction, species do not die out because individuals are poorly adapted to normal or gradually changing environmental conditions. Rather, species die out from exposure to exceptionally harsh, short-term conditions—such as huge volcanic eruptions or catastrophic sea-level changes.

To drive these points home, consider a mass extinction event that nearly uprooted the tree of life entirely. The end-Permian extinction, which occurred about 251 million years ago, came close to ending multicellular life on Earth.

The End-Permian Extinction

The end-Permian has been called the Mother of Mass Extinctions. To appreciate the scale of the event, imagine that while walking along a seashore, you identified 100 different species of algae and animals living on the beach and in tide pools and the shallow water offshore. Now imagine that you snapped your fingers, and 90 of those species disappeared forever. Only 10 species are left. An area that was teeming with diverse forms of life would look barren.

This is what happened, all over the world, during the end-Permian extinction. The event was a catastrophe of almost unimaginable proportions. On a personal level, it would be like nine of your ten best friends dying.

Although biologists have long appreciated the scale of the end-Permian extinction, research on its causes is ongoing. Consider the following:

- Flood basalts are outpourings of molten rock that flow across Earth's surface. The largest flood basalts on Earth, called the Siberian traps, occurred during the end-Permian. They added enormous quantities of heat, CO_2, and sulfur dioxide to the atmosphere.

- High levels of atmospheric CO_2 trapped heat, causing intense global warming, the stratification of the oceans, and reaction of deep-ocean sulfur dioxide with water to form sulfuric acid, which is toxic to most organisms.

- Rocks that formed during the interval indicate that the oceans became completely or largely anoxic—meaning that they lacked oxygen. These conditions are fatal to organisms that rely on aerobic respiration.

- There is convincing evidence that sea level dropped dramatically during the extinction event, reducing the amount of habitat available for marine organisms.

- Terrestrial animals may have been restricted to small patches of low-elevation habitats, due to low oxygen concentrations and high CO_2 levels in the atmosphere.

In short, both marine and terrestrial environments deteriorated dramatically for organisms that depend on oxygen to live. A prominent researcher has captured this point by naming the suite of killing mechanisms the "world went to hell hypothesis."

What biologists don't understand is *why* the environment changed so radically, and so quickly. For example, no one has yet been able to establish a convincing connection between the eruption of the Siberian traps and the other, more global, environmental changes that occurred.

The cause of the end-Permian extinction may be the most important unsolved question in research on the history of life, aside from the origin of life itself (see Chapter 2). The cause of the dinosaur's demise, in contrast, is settled.

What Killed the Dinosaurs?

The end-Cretaceous extinction of 65 million years ago is as satisfying a murder mystery as you could hope for, but the butler didn't do it. The **impact hypothesis** for the extinction of the dinosaurs, first put forth in the early 1970s by father-and-son team Luis and Walter Alvarez, proposed that an asteroid struck Earth and snuffed out an estimated 60–80 percent of the multicellular species alive.

Evidence for the Impact Hypothesis   The impact hypothesis was intensely controversial at first. As researchers set out to test its predictions, however, support began to grow (**TABLE 28.4**):

- Worldwide, sedimentary rocks that formed at the Cretaceous–Paleogene (K–P) boundary were found to contain extremely high quantities of the element iridium. Iridium is vanishingly rare in Earth rocks, but it is an abundant component of asteroids and meteorites.

- Shocked quartz and microtektites are minerals that are found only at documented meteorite impact sites. In Haiti and an array of other locations, both shocked quartz and microtektites have been discovered in abundance in rock layers dated to 65 million years ago.

- A crater larger than New Jersey was found just off the northwest coast of Mexico's Yucatán peninsula. Microtektites are abundant in sediments from the crater's walls, and the crater dates to the K–P boundary.

Taken together, these data provided conclusive evidence in favor of the impact hypothesis. Researchers agree that the mystery is solved.

Nature of the Impact   The shape of the Yucatán crater and the distribution of shocked quartz and microtektites dated to 65 mya indicates that the asteroid hit Earth at an angle and splashed material over much of southeastern North America. Based on

Type of Evidence	Explanation
	**Spike of iridium in 65-million-year-old rocks worldwide** Iridium is very rare on Earth but abundant in asteroids and meteors
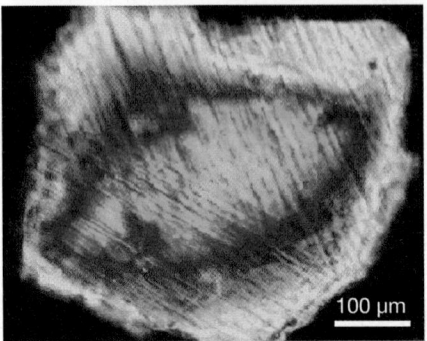	**Spike of shocked quartz in 65-million-year-old rocks worldwide** Shocked quartz forms only under extremely high pressures created from events like impacts (not from volcanoes)
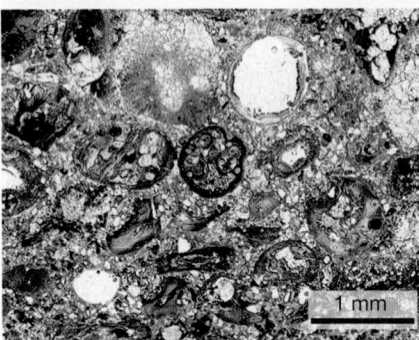	**Spike of micro-tektites in 65-million-year-old rocks in Gulf of Mexico** These glass droplets form only under extremely high temperatures and pressures as rock is splashed, molten, from an impact site
	**Huge, 65-million-year-old crater discovered off coast of Mexico** The crater is 180 kilometers in diameter (absolutely massive), so an impact of this size would have far-reaching effects

currently available data, astronomers and paleontologists estimate that the asteroid was about 10 km across.

To get a sense of the event's scale, consider that Mt. Everest is about 10 km above sea level and that planes cruise at an altitude of about 10 km. Imagine Earth being hit by a rock the size of Mt. Everest, or a rock that would fill the space between you and a jet in the sky.

According to both computer models and geologic data, the consequences of the K–P asteroid strike were nothing short of devastating.

- A tremendous fireball of hot gas would have spread from the impact site; large soot and ash deposits in sediments dated to 65 million years ago testify to catastrophic wildfires, worldwide.

- The largest tsunami in the last 3.5 billion years would have disrupted ocean sediments and circulation patterns.

- The impact site itself is underlain by a sulfate-containing rock called anhydrite. The SO_4^{2-} released by the impact would have reacted with water in the atmosphere to form sulfuric acid, triggering extensive acid rain.

- Massive quantities of dust, ash, and soot would have blocked the Sun for long periods, leading to rapid global cooling and a crash in plant productivity.

Selectivity of the Extinctions The asteroid impact did not kill indiscriminately. Perhaps by chance, certain lineages escaped virtually unscathed while others vanished. Among vertebrates, for example, most dinosaurs (except birds), pterosaurs (flying reptiles), and all of the large-bodied marine reptiles (mosasaurs and plesiosaurs) expired; birds, mammals, crocodilians, amphibians, and turtles survived.

Why? Answering this question has sparked intense debate. For years the leading hypothesis was that the K–P extinction event was size selective. The logic here was that the extended darkness and cold would affect large organisms disproportionately, because they require more food than do small organisms. But extensive data on the survival and extinction of marine clams and snails have shown no hint of size selectivity, and small-bodied dinosaurs (except for birds) perished along with large-bodied forms.

One hypothesis currently being tested is that organisms that were capable of inactivity for long periods—by hibernating or resting as long-lived seeds or spores—were able to survive the catastrophe. Also, recent data show selectivity according to habitat; organisms in the open oceans suffered higher extinction rates than those in coastal waters. The reason for selectivity of the extinctions is still unsolved.

Recovery from the Extinction After the K–P extinction, fern fronds and fern spores dominate the plant fossil record from North America and Australia. These data suggest that extensive stands of ferns replaced diverse assemblages of cone-bearing and flowering plants after the impact. The fundamental message here is that terrestrial ecosystems around the world were radically simplified by the extinction event. In marine environments, some invertebrate groups do not exhibit typical levels of species diversity in the fossil record until 4–8 million years past the K–P boundary. Recovery was slow.

The organisms present in the Paleogene were markedly different from those of the preceding period. The lineage called

Mammalia had begun to diversify before the K–P extinction, during a period of rapid diversification of flowering plants. But early mammals were mostly rat-sized predators and scavengers in the heyday of the dinosaurs. It wasn't until after the dinosaurs became extinct that each of the major lineages of mammals diversified into the orders we recognize today—from bats to bears, and pigs to primates—and average body size increased exponentially. Why? A major branch on the tree of life had disappeared. With competitors removed from many ecological niches, mammals flourished.

The change in the terrestrial vertebrate fauna was not due to a competitive superiority conferred by adaptations such as fur and lactation. Rather, it was due to a chance event: a once-in-a-billion-years collision with a massive rock from outer space.

The Sixth Mass Extinction?

Are mass extinctions events of the distant past? Not likely. Many scientists now propose that life on Earth is on the verge of the sixth mass extinction, precipitated by human impacts such as habitat loss, pollution, overfishing, invasive species, and climate change. Could human activities really cause such dire consequences as massive as those caused by an asteroid impact?

There is mounting evidence that the answer to this question is yes. The exponential increase of human populations, combined with our hunger for resources, has put the diversity of life as we know it at risk.

Unit 6 explores the diversity of life itself. Then Unit 9 examines how human actions are rapidly pruning branches from the tree of life during your lifetime.

CHAPTER 28 REVIEW

 For media, go to MasteringBiology

If you understand . . .

28.1 Tools for Studying History: Phylogenetic Trees

- Phylogenetic trees show the evolutionary history of a group of organisms—who is most closely related to whom. They have revolutionized the study of evolution and many related fields.

- Phylogenetic trees are usually estimated by analyzing shared, derived characters (synapomorphies) that identify monophyletic groups.

- Researchers often use the principle of parsimony to decide which of the many possible trees is most likely to reflect actual evolutionary history. Parsimony assumes that the most likely explanation or pattern is the one that implies the least amount of change.

- Homology occurs when traits are shared due to common ancestry. Homoplasy occurs when traits are similar due to reasons other than common ancestry, such as convergent evolution.

✔ You should be able to explain why upright posture and bipedalism (walking on two legs) is a synapomorphy that defines a monophyletic group called hominins, which includes humans.

28.2 Tools for Studying History: The Fossil Record

- The fossil record is the only direct source of data about what extinct organisms looked like and where they lived.

- The fossil record is biased: Common, recent, and abundant species that burrow and that have hard parts are most likely to be present in the record.

- During the Precambrian, which started with the formation of the Earth 4.6 billion years ago and ended with the origin of most animals about 542 million years ago, life was almost exclusively unicellular.

- Oxygen was virtually absent from the oceans and atmosphere for almost 2 billion years after the origin of life. The oxygen atmosphere was generated by photosynthetic bacteria.

- During the Phanerozoic, which began about 542 million years ago and continues to the present, animals originated and diversified first in the ocean and then on land. Land plants and fungi also originated and diversified.

- If the history of the Earth were scaled to fit in one calendar year, our species, *Homo sapiens*, would not appear until one hour before the stroke of midnight on December 31st.

✔ You should be able to explain why some paleontologists use snail fossils rather than other types of fossils to estimate how biodiversity has changed over time.

28.3 Adaptive Radiation

- Adaptive radiations occur when a single lineage rapidly produces many descendant species with a wide range of adaptive forms.

- Adaptive radiations can be triggered by ecological opportunity, such as the availability of unoccupied niches on an island, or morphological innovations, such as flowers and wings.

- The Cambrian explosion was one of the most spectacular adaptive radiations of all time, producing virtually all of the major animal lineages existing today. During this event, the size and morphological complexity of animals increased tremendously, and animals diversified in how they made a living.

✔ You should be able to explain why the evolution of animals that could swim free of benthic habitats—such as fishlike early vertebrates—created an ecological opportunity.

28.4 Mass Extinction

- Mass extinctions are short-term environmental catastrophes that eliminate most of the species alive. Mass extinctions have altered the course of evolutionary history at least five times.

- Mass extinctions such as the end-Permian and end-Cretaceous prune the tree of life more or less randomly and have marked the end of several branches within prominent lineages and the subsequent rise of new branches.

- After the devastation of a mass extinction, the fossil record indicates that it can take 10–15 million years for ecosystems to recover their former levels of diversity.

- A mass extinction may presently be under way due to the drastic effects of humans on the Earth in recent history.

✔ You should be able to evaluate this statement: Environmental changes caused by humans are eliminating species that are poorly adapted to their natural environment.

MB MasteringBiology

1. MasteringBiology Assignments

Tutorials and Activities Adaptive Radiation, Constructing Phylogenetic Trees, Fossil Record, History of Life, Overview of Macroevolution, Scrolling Geologic Record

Questions Reading Quizzes, Blue-Thread Questions, Test Bank

2. eText Read your book online, search, take notes, highlight text, and more.

3. The Study Area Practice Test, Cumulative Test, BioFlix® 3-D Animations, Videos, Activities, Audio Glossary, Word Study Tools, Art

You should be able to . . .

✔ TEST YOUR KNOWLEDGE

Answers are available in Appendix A

1. Choose the best definition of a fossil.
   a. any trace of an organism that has been converted into rock
   b. a bone, tooth, shell, or other hard part of an organism that has been preserved
   c. any trace of an organism that lived in the past
   d. the process that leads to preservation of any body part from an organism that lived in the past

2. What is the difference between a branch and a node on a phylogenetic tree?
   a. A branch is a population through time; a node is where a population splits into two independent populations.
   b. A branch represents a common ancestor; a node is a species.
   c. A branch is a lineage; a node is any named taxonomic group.
   d. A branch is a population through time; a node is the common ancestor of all species or lineages present on a particular tree.

3. Which of the following best characterizes an adaptive radiation?
   a. Speciation occurs extremely rapidly, and descendant populations occupy a large geographic area.

   b. A single lineage diversifies rapidly, and descendant populations occupy many habitats and ecological roles.
   c. Natural selection is particularly intense, because disruptive selection occurs.
   d. Species recover after a mass extinction.

4. True or false? The dinosaurs (other than birds) went extinct during the asteroid impact at the K–P boundary because they were poorly adapted to their ecological niches.

5. Which of the following is an example of homoplasy?
   a. hair in humans and fur in mice
   b. astragalus ankle bones in hippos and deer
   c. *Hox* genes in humans and flies
   d. streamlined bodies in dolphins and ichthyosaurs

6. True or false? Monophyletic groups are identified by shared, derived traits.

✔ TEST YOUR UNDERSTANDING

Answers are available in Appendix A

7. The text claims that the fossil record is biased in several ways. What are these biases? If the database is biased, is it still an effective tool to use in studying the diversification of life? Why or why not?

8. The initial diversification of animals took place over some 50 million years, mostly during the Cambrian period. Propose an analogy from everyday life to illustrate the significance of this event.

9. Why is parsimony a useful tool for assessing which phylogenetic tree is most accurate? Why was parsimony misleading in the case of the astragalus during the evolution of artiodactyls?

10. Give an example of an adaptive radiation that occurred after a colonization event and another that occurred after a key innovation. In each case, provide a hypothesis to explain why the adaptive radiation occurred.

11. **QUANTITATIVE** The end-Permian extinction was devastating. If each student in your graduating class represented a different species during the Permian and there were 1880 students in your graduating class, about how many students would make it to graduation?

12. **BioSkills 7** in Appendix B recommends a "one-snip test" to identify monophyletic groups—meaning that if you cut any branch on a tree, everything that "falls off" is a monophyletic group. Why is this valid?
   a. Monophyletic groups are nested on a tree—meaning that they are hierarchical.
   b. Monophyletic groups can also be called clades or lineages.
   c. Species are the smallest monophyletic groups on the tree of life.
   d. One snip gets an ancestor and all of its descendants.

13. Suppose that the dying wish of a famous eccentric was that his remains be fossilized. His family has come to you for expert advice. What method would you recommend as most likely to maximize the chances that his wish will be fulfilled?

   a. Place the corpse in an environment where decomposition is rapid (such as a forest).

   b. Place the corpse in an environment where decomposition is slow (such as a swamp or bog).

   c. Bury only the bones, since soft parts don't fossilize well.

   d. Place the corpse in an environment where plenty of oxygen is available.

14. Use the data matrix below to draft at least three alternate phylogenetic trees. Label the outgroup. Indicate which tree is most parsimonious, and explain why. How could you change your analysis to increase your confidence that the tree represents true evolutionary relationships?

Presence or Absence of *Hox* Genes*

	Hox1	*Hox2*	*Hox3*	*Hox4*	*Hox12*
Mouse	1	1	1	1	1
Snail	1	1	1	1	0
Fly	1	1	1	1	0
Sea anemone	1	1	0	0	0
Sponge	0	0	0	0	0

*1 = gene present; 0 = gene absent.

15. Some researchers contend that the end-Permian extinction event was caused by an impact with a large extraterrestrial object. Describe three forms of evidence that you would like to see before you accept this hypothesis.

16. One of the "triggers" proposed for the Cambrian explosion is a dramatic rise in oxygen concentrations that occurred in the oceans about 800 mya. (Review material in Chapter 9 on how oxygen compares with other atoms or compounds as an electron acceptor during cellular respiration; look near the end of Section 9.6, where aerobic and anaerobic respiration are compared.) Make a small concept map outlining the logic behind the "oxygen-trigger" hypothesis for the Cambrian explosion. (See **BioSkills 15** in Appendix B if you need help making concept maps.)

Geneticist and evolutionary biologist Theodosius Dobzhansky said that "Nothing in biology makes sense except in the light of evolution." Use this concept map to study how ideas introduced in Unit 5 fit together.

The key is to connect the four evolutionary processes that work at the level of populations—natural selection, genetic drift, mutation, and gene flow—to processes, events, and outcomes at higher levels of organization: speciation, adaptive radiation, mass extinction, and the tree of life.

It's all about changes in allele frequencies. Over time, small changes that occur between populations lead to large changes that distinguish major lineages on the tree of life.

Note that each box in the concept map indicates the chapter and section where you can go for review. Also, be sure to do the blue exercises in the Check Your Understanding box below.

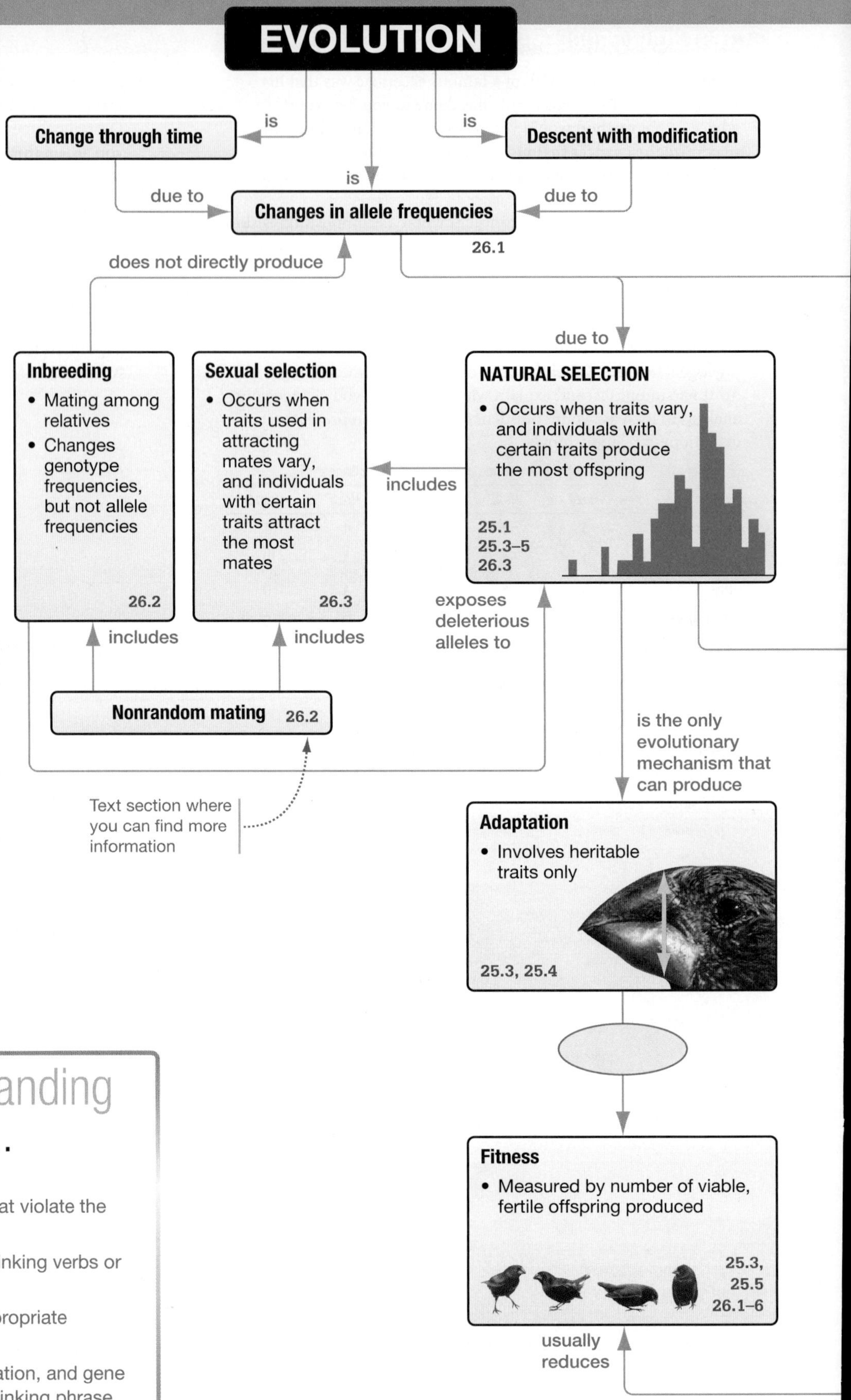

The Big Picture (vertical sidebar text)

EVOLUTION

Change through time ←is— EVOLUTION —is→ **Descent with modification**

Change through time —due to→ **Changes in allele frequencies** ←due to— Descent with modification

EVOLUTION —is→ **Changes in allele frequencies** 26.1

Changes in allele frequencies —does not directly produce→

Changes in allele frequencies —due to→ NATURAL SELECTION

Inbreeding
- Mating among relatives
- Changes genotype frequencies, but not allele frequencies

26.2

Sexual selection
- Occurs when traits used in attracting mates vary, and individuals with certain traits attract the most mates

26.3

NATURAL SELECTION
- Occurs when traits vary, and individuals with certain traits produce the most offspring

25.1
25.3–5
26.3

Natural selection —includes→ Sexual selection

Inbreeding —exposes deleterious alleles to→ Natural selection

Nonrandom mating 26.2

Nonrandom mating —includes→ Inbreeding

Nonrandom mating —includes→ Sexual selection

Text section where you can find more information

Adaptation
- Involves heritable traits only

25.3, 25.4

Natural selection —is the only evolutionary mechanism that can produce→ Adaptation

Fitness
- Measured by number of viable, fertile offspring produced

25.3, 25.5, 26.1–6

Fitness —usually reduces→

check your understanding

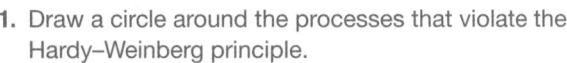

If you understand the big picture . . .
✔ You should be able to . . .

1. Draw a circle around the processes that violate the Hardy–Weinberg principle.

2. Fill in the blue ovals with appropriate linking verbs or phrases.

3. Add a box for "Fossil record" with appropriate connections.

4. Draw arrows linking genetic drift, mutation, and gene flow to the appropriate box using the linking phrase "is random with respect to."

Answers are available in Appendix A.

due to ▼

GENETIC DRIFT
- Changes in allele frequencies due entirely to chance
- Especially important in small populations

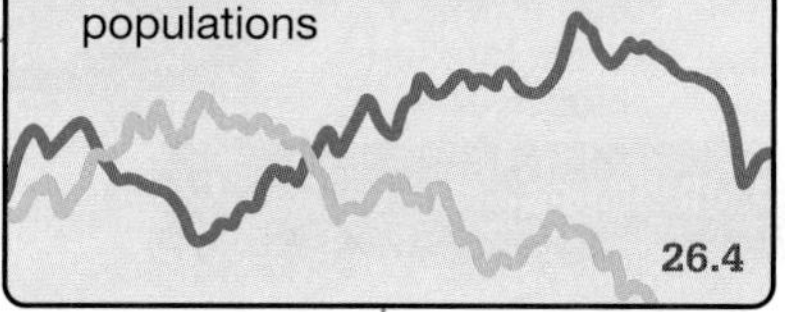

26.4

due to ▼

MUTATION
- Random changes in DNA
- Creates new alleles
- Occurs in every individual in every generation, at low frequency

AT | TAT | TGT | CTA | GTA | CCC | CA

16.4, 26.6

due to ▼

GENE FLOW
- Occurs when individuals move between populations
- Homogenizes allele frequencies between populations

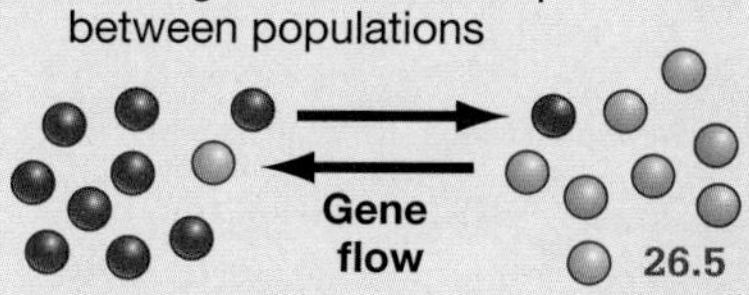

Gene flow

26.5

due to lack of ▲

produces divergence required for

produces divergence required for

provides raw material for

Speciation
Results from:
1. Genetic isolation, followed by
2. Genetic divergence

27.2–4

creates new branches on →

form smallest possible tips on →

The Tree of Life
- Describes the evolutionary relationships among all species

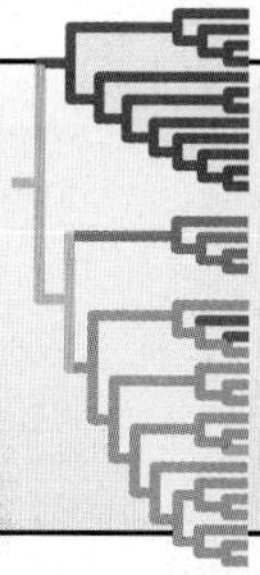

1.3, 28.1

forms new ▼

Species
Evolutionarily independent units in nature, identified by:
1. Reproductive isolation, and/or
2. Phylogenetic analysis, and/or
3. Morphological differences

27.1

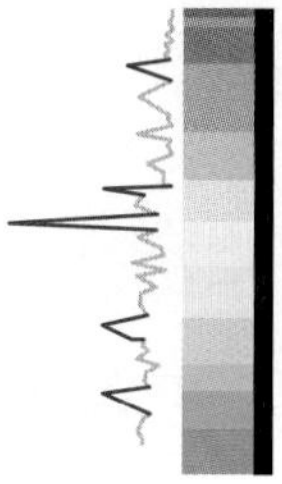

"prune" ▲

Mass extinctions
- 60% of species are lost in less than 1 million years
- 5 events in the past 542 million years. A sixth mass extinction is now underway

28.4

with ▼

Synapomorphies
- Shared, derived traits that are unique to a single lineage
- Arise in a common ancestor

27.1
28.1

that may be →

Key innovations
- Traits that allow species to exploit resources in a new way or use new habitats

28.4

may result in →

may occur after ▲

Adaptive radiations
- Rapid and extensive speciation in a single lineage
- Dramatic divergence in morphology or behavior (species use a wide array of resources/habitats)

28.3

29 Bacteria and Archaea

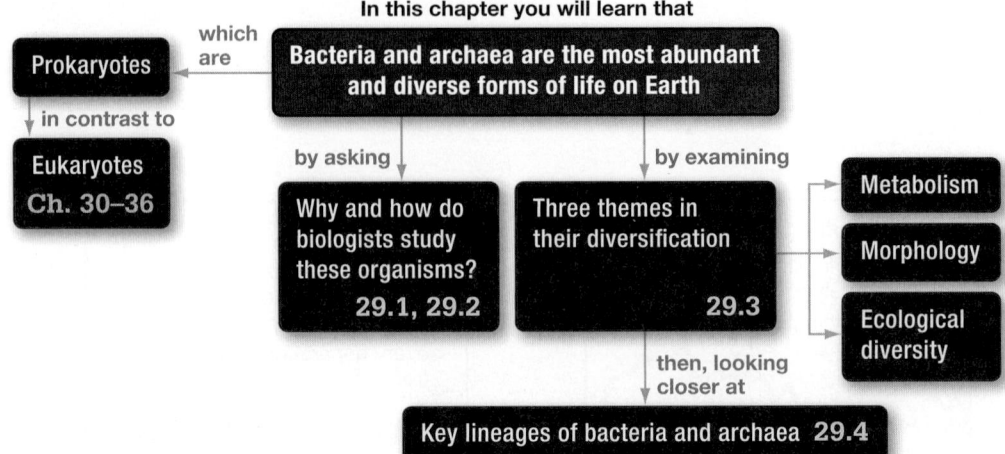

In this chapter you will learn that

Bacteria and archaea are the most abundant and diverse forms of life on Earth

which are → **Prokaryotes**

in contrast to

Eukaryotes Ch. 30–36

by asking → **Why and how do biologists study these organisms? 29.1, 29.2**

by examining → **Three themes in their diversification 29.3**

→ **Metabolism**

→ **Morphology**

→ **Ecological diversity**

then, looking closer at → **Key lineages of bacteria and archaea 29.4**

Although this hot spring looks devoid of life, it is actually teeming with billions of bacterial and archaeal cells.

Bacteria and Archaea (usually pronounced *ar-KEE-ah*) form two of the three largest branches on the tree of life (**FIGURE 29.1**). The third major branch, or domain, consists of eukaryotes and is called the Eukarya. Virtually all members of the Bacteria and Archaea domains are unicellular, and all are prokaryotic—meaning that they lack a membrane-bound nucleus.

Although their relatively simple morphology makes bacteria and archaea appear similar to the untrained eye, in some ways they are strikingly different at the molecular level (**TABLE 29.1**). Organisms in the Bacteria and Archaea domains are distinguished by the types of molecules that make up their plasma membranes and cell walls:

- **Bacteria** have a unique compound called peptidoglycan in their cell walls (see Chapter 5).

✔ When you see this checkmark, stop and test yourself. Answers are available in Appendix A.

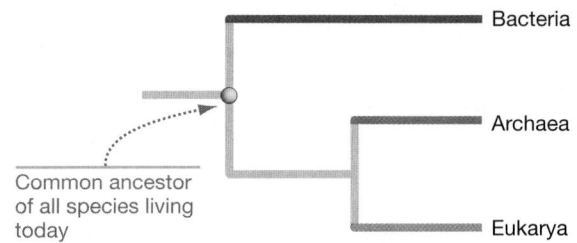

FIGURE 29.1 Bacteria, Archaea, and Eukarya Are the Three Domains of Life. Archaea are more closely related to eukaryotes than they are to bacteria.

✔**QUESTION** Was the common ancestor of all species living today prokaryotic or eukaryotic? Explain your reasoning.

- **Archaea** have unique phospholipids in their plasma membranes—the hydrocarbon tails of the phospholipids are made from isoprene (see Chapter 6).

If you were unicellular, bacteria and archaea would look as different to you as mammals and fish.

In addition, the machinery that bacteria and archaea use to process genetic information is strikingly different. More specifically, the DNA polymerases, RNA polymerases, transcription-initiation proteins, and ribosomes found in Archaea and Eukarya are distinct from those found in Bacteria and similar to each other. These differences have practical consequences: Antibiotics that poison bacterial ribosomes do not affect the ribosomes of archaea or eukaryotes. If all ribosomes were identical, these antibiotics would kill you along with the bacterial species that was supposed to be targeted.

As an introductory biology student, you may know less about bacteria and archaea than about any other group on the tree of life. Before taking this course, it's likely that you'd never even heard of the Archaea. You might be surprised to find out that biologists suspect that the first eukaryotic cells may have formed as a result of a symbiotic event between these two groups of prokaryotes. Notice in Table 29.1 that cells in the domain Eukarya share some features with both Bacteria and Archaea. Studying these shared features has helped biologists understand the early evolution of life on Earth.

This chapter's goal is to convince you that even though bacteria and archaea are tiny, they have an enormous impact on you and the planet in general. By the time you finish reading the chapter, you should understand why a researcher summed up the bacteria and archaea by claiming, "They run this joint."

29.1 Why Do Biologists Study Bacteria and Archaea?

Biologists study bacteria and archaea for the same reasons they study any organisms. First, they are intrinsically fascinating. Discoveries such as finding bacterial cells living a kilometer underground or in 95°C hot springs keep biologists awake at night, staring at the ceiling. They can't wait to get into the lab in the morning and figure out how those cells stay alive.

SUMMARY TABLE 29.1 **Characteristics of Bacteria, Archaea, and Eukarya**

	Bacteria	Archaea	Eukarya
DNA enclosed by a nuclear envelope? (see Chapter 7)	No	No	Yes
Circular chromosome present?	Yes (but linear in some species)	Yes	No (linear)
Organelles enclosed by membranes present? (see Chapter 7)	No	No	Yes
Rotating flagella present? (see Chapter 7)	Yes	Yes	No (flagella and cilia undulate)
Multicellular species?	No (with some exceptions)	No	Yes
Plasma membrane lipids composed of glycerol bonded to unbranched fatty acids by ester linkages? (see Chapter 6)	Yes	No (branched lipids bonded by ether linkages)	Yes
Cell walls, when present, contain peptidoglycan? (see Chapter 5)	Yes	No	No
RNA polymerase composed of >10 subunits?	No (only 5 subunits)	Yes	Yes
Translation initiated with methionine? (see Chapter 17)	No (initiated with N-formylmethionine; f-met)	Yes	Yes

✔**EXERCISE** Using the data in this table, add labeled marks to Figure 29.1 indicating where the following traits evolved: peptidoglycan in cell wall, archaeal-type plasma membrane, archaeal- and eukaryotic-type translation initiation, and nuclear envelope.

Second, there are practical benefits to understanding the species that share the planet with us. Understanding bacteria and archaea is particularly important—in terms of both understanding life on Earth and improving human health and welfare.

Biological Impact

The lineages in the domains Bacteria and Archaea are ancient, diverse, abundant, and ubiquitous. The oldest fossils of any type found to date are 3.5-billion-year-old carbon-rich deposits derived from bacteria. Because eukaryotes do not appear in the fossil record until 1.75 billion years ago, biologists infer that prokaryotes were the only form of life on Earth for at least 1.7 billion years.

Just how many bacteria and archaea are alive today? Although a mere 5000 species have been formally named and described to date, it is virtually certain that millions exist. Consider that over 1000 species of prokaryotes are living in your large intestine right now, and another 700 species are living in your mouth. Well-known microbiologist Norman Pace points out that there may be tens of millions of different insect species but notes, "If we squeeze out any one of these insects and examine its contents under the microscope, we find hundreds or thousands of distinct microbial species." Most of these **microbes** (microscopic organisms) are bacteria or archaea. Virtually all are unnamed and undescribed. If you want to discover and name new species, then study bacteria or archaea.

Abundance Besides recognizing how diverse bacteria and archaea are in terms of numbers of species, it's critical to appreciate their abundance.

- The approximately 10^{13} (10 trillion) cells in your body are outnumbered ten to one by the bacterial and archaeal cells living on and in you. You are a walking, talking habitat—one that is teeming with bacteria and archaea.

- A mere teaspoon of good-quality soil contains *billions* of microbial cells, most of which are bacteria and archaea.

- In sheer numbers, species in a lineage of marine archaea may be the most successful organisms on Earth. Biologists routinely find these cells at concentrations of over 10,000 individuals per milliliter in most of the world's oceans. At these concentrations, one liter of seawater contains a population equivalent to that of the largest human cities. Yet this lineage was not described until the early 1990s.

- Recent research has found enormous numbers of bacterial and especially archaeal cells in rocks and sediments as much as 1600 meters underneath the world's oceans. Although recently discovered, the bacteria and archaea living under the ocean may make up 10 percent of the world's total mass of living material.

- Biologists estimate the total number of individual bacteria and archaea alive today at over 5×10^{30}. If they were lined up end to end, these cells would make a chain longer than the Milky Way galaxy. They contain 50 percent of all the carbon and 90 percent of all the nitrogen and phosphorus found in organisms.

In terms of the total volume of living material on our planet, bacteria and archaea are dominant life-forms.

Habitat Diversity Bacteria and archaea are found almost everywhere. They live in environments as unusual as oxygen-free mud, hot springs, and salt flats. In seawater they are found from the surface to depths of 10,000 meters (m), and at temperatures over 120°C (well above water's boiling point) near submarine volcanoes.

Although there are far more prokaryotes than eukaryotes, much more is known about eukaryotic diversity than about prokaryotic diversity. Researchers who study prokaryotic diversity are exploring one of the most wide-open frontiers in all of science. So little is known about the extent of these domains that recent collecting expeditions have turned up entirely new **phyla** (singular: **phylum**). These are names given to major lineages within each domain. To a biologist, this achievement is equivalent to the sudden discovery of a new group of eukaryotes as distinctive as flowering plants or animals with backbones.

The physical world has been explored and mapped, and many of the larger plants and animals are named. But in **microbiology**—the study of organisms that can be seen only with the aid of a microscope—this is an age of exploration and discovery.

Some Microbes Thrive in Extreme Environments

Bacteria or archaea that live in high-salt, high-temperature, low-temperature, or high-pressure habitats are **extremophiles** ("extreme-lovers"). Studying them has been extraordinarily fruitful for understanding the tree of life, developing industrial applications, and exploring the structure and function of enzymes.

As an example of these habitats, consider hydrothermal vents at the bottom of the ocean, where water as hot as 300°C emerges and mixes with 4°C seawater. At locations like these, archaea are abundant forms of life.

Researchers recently discovered an archaeon that grows so close to these hydrothermal vents that its surroundings are at 121°C—a record for life at high temperature. This organism can live and grow in water that is heated past its boiling point (100°C) and at pressures that would instantly destroy a human. Since high temperature breaks non-covalent bonds holding macromolecules together, extreme heat usually denatures proteins, makes membranes leaky, and separates the strands of the DNA double helix. Biologists are intrigued by how these unusual archaeal cells can thrive under such extreme conditions.

Other discovered bacteria and archaea can grow

- at a pH less than 1.0;

- at temperatures of 0°C under Antarctic ice;

- in water that is virtually saturated with salt (**FIGURE 29.2**).

Extremophiles have become a hot area of research. The genomes of a wide array of extremophiles have been sequenced, and expeditions regularly seek to characterize new species. Why?

- *Origin of life* Based on models of conditions that prevailed early in Earth's history, it appears likely that the first forms of life lived at high temperature and pressure in environments

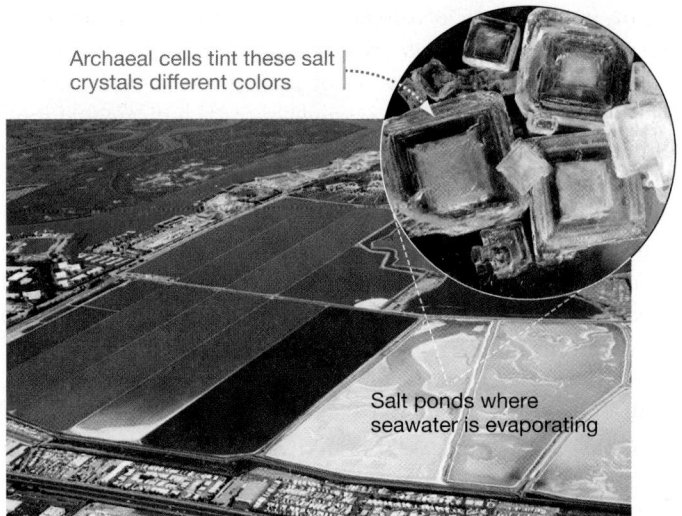

Archaeal cells tint these salt crystals different colors

Salt ponds where seawater is evaporating

FIGURE 29.2 Some Archaeal Cells Live in High-Salt Habitats. The evaporating water is colored red by these photosynthetic cells.

that lacked oxygen—conditions that humans would call extreme. Thus, understanding extremophiles may help explain how life on Earth began.

- **Extraterrestrial life?** In a similar vein, many astrobiologists ("space-biologists") use extremophiles as model organisms in the search for extraterrestrial life. The idea is that if bacteria and archaea can thrive in extreme habitats on Earth, cells might possibly be found in similar environments on other planets or moons of planets.

- **Commercial applications** Because enzymes that function at extreme temperatures and pressures are useful in many industrial processes, extremophiles are of commercial interest as well. Recall that *Taq* polymerase—a DNA polymerase that is stable up to 95°C—is used to run the polymerase chain reaction (PCR) in research and commercial settings (see Chapter 20). This enzyme was isolated from a bacterium called *Thermus aquaticus* ("hot water"), which was discovered in a hot spring in Yellowstone National Park.

Bacteria and archaea may be small, but they thrive in an amazing range of conditions.

Medical Importance

The first paper documenting that an archaeon was associated with a human disease—a dental condition called periodontitis—was published in 2004. But biologists have been studying disease-causing bacteria for over a century.

Of the thousands of bacterial species living in and on your body, only a tiny fraction can disrupt normal body functions enough to cause illness. Bacteria that cause disease are said to be **pathogenic** (literally, "disease-producing"). Pathogenic bacteria have been responsible for some of the most devastating epidemics in human history.

TABLE 29.2 lists some of the bacteria that cause illness in humans. Here are the important things to note:

- Pathogenic forms come from several different lineages in the domain Bacteria.

TABLE 29.2 Some Bacteria That Cause Illness in Humans

Lineage	Species	Tissues Affected	Disease
Firmicutes	*Clostridium tetani*	Wounds, nervous system	Tetanus
	Staphylococcus aureus	Skin, urogenital canal	Acne, boils, impetigo, toxic shock syndrome
	Streptococcus pneumoniae	Respiratory tract	Pneumonia
	Streptococcus pyogenes	Respiratory tract	Strep throat, scarlet fever
Spirochaetes	*Borrelia burgdorferi*	Skin and nerves	Lyme disease
	Treponema pallidum	Urogenital canal	Syphilis
Actinobacteria	*Mycobacterium tuberculosis*	Respiratory tract	Tuberculosis
	Mycobacterium leprae	Skin and nerves	Leprosy
	Propionibacterium acnes	Skin	Acne
Chlamydiales	*Chlamydia trachomatis*	Urogenital canal	Genital tract infection
ε-Proteobacteria	*Helicobacter pylori*	Stomach	Ulcer
β-Proteobacteria	*Neisseria gonorrhoeae*	Urogenital canal	Gonorrhea
γ-Proteobacteria	*Haemophilus influenzae*	Ear canal, nervous system	Ear infections, meningitis
	Pseudomonas aeruginosa	Urogenital canal, eyes, ear canal, lungs	Infections of eye, ear, urinary tract, lungs
	Salmonella enterica	Gastrointestinal tract	Food poisoning
	Yersinia pestis	Lymph and blood	Plague

- Pathogenic bacteria tend to affect tissues at the entry points to the body, such as wounds or pores in the skin, the respiratory and gastrointestinal tracts, and the urogenital canal.

Koch's Postulates Robert Koch was the first person to establish a link between a particular species of bacterium and a specific disease. When Koch began his work on the nature of disease in the late 1800s, microscopists had confirmed the existence of the particle-like organisms people now call bacteria, and Louis Pasteur had shown that bacteria and other microorganisms are responsible for spoiling milk, wine, broth, and other foods. Koch hypothesized that bacteria might also be responsible for causing infectious diseases, which spread by being passed from an infected individual to an uninfected individual.

Koch set out to test this hypothesis by identifying the organism that causes anthrax. Anthrax is a disease of cattle and other grazing mammals that can result in fatal blood poisoning. The disease also occurs infrequently in humans and mice.

To establish a causative link between a specific microbe and a specific disease, Koch proposed that four criteria had to be met:

1. The microbe must be present in individuals suffering from the disease and absent from healthy individuals. By careful microscopy, Koch was able to show that the bacterium *Bacillus anthracis* was always present in the blood of cattle suffering from anthrax, but absent from healthy individuals.

2. The organism must be isolated and grown in a pure culture away from the host organism. Koch was able to grow pure colonies of *B. anthracis* in glass dishes on a nutrient medium, using gelatin as a substrate.

3. If organisms from the pure culture are injected into a healthy experimental animal, the disease symptoms should appear. Koch demonstrated this crucial causative link in mice injected with *B. anthracis*. The symptoms of anthrax infection appeared and then the infected mice died.

4. The organism should be isolated from the diseased experimental animal, again grown in pure culture, and demonstrated by its size, shape, and color to be the same as the original organism. Koch did this by purifying *B. anthracis* from the blood of diseased experimental mice.

These criteria, now called **Koch's postulates,** are still used to confirm a causative link between new diseases and a suspected infectious agent. Microbiologists now recognize that many bacteria cannot be grown in culture, so they use other means of detection for those organisms.

The Germ Theory Koch's experimental results were the first test of the **germ theory of disease.**

- The pattern component of this theory is that certain diseases are infectious—meaning that they can be passed from person to person.

- The process responsible for this pattern is the transmission and growth of certain bacteria and viruses.

Viruses are acellular particles that parasitize cells (see Chapter 36).

The germ theory of disease laid the foundation for modern medicine. Initially its greatest impact was on sanitation—efforts to prevent transmission of pathogenic bacteria. During the American Civil War, for example, records indicate that more soldiers died of bacterial infections contracted from drinking water contaminated with human feces than from wounds in battle. Also during that conflict it was common for surgeons to sharpen their scalpels on their shoe leather, after walking in horse manure.

Fortunately, improvements in sanitation and nutrition have caused dramatic reductions in mortality rates due to infectious diseases in the industrialized countries.

What Makes Some Bacterial Cells Pathogenic? Virulence, or the ability to cause disease, is a heritable trait that varies among individuals in a population. Most *Escherichia coli*, for example, are harmless inhabitants of the gastrointestinal tract of humans and other mammals. But some *E. coli* cells cause potentially fatal food poisoning.

What makes some cells of the same species pathogenic, while others are harmless? Biologists have answered this question for *E. coli* by sequencing the entire genome of a harmless lab strain and the pathogenic strain called O157:H7, which is harmful to humans. The genome of the pathogenic strain is slightly larger because it has acquired virulence genes, including one coding for a protein **toxin.** After entering a host cell, this toxin binds to ribosomes and inhibits protein synthesis, killing the host cells. Because of key differences between the ribosomes of bacteria and eukaryotic cells, only host-cell protein synthesis is blocked by the toxin. Cells lining the blood vessels near the host's intestinal epithelium are most affected by the toxin, and the resulting damage leads to bloody diarrhea and possible death. If sanitation is poor, the pathogenic bacteria are likely to infect many new hosts.

Similar types of studies are identifying the genes responsible for virulence in a wide array of pathogenic bacteria.

The Past, Present, and Future of Antibiotics Antibiotics are molecules that kill bacteria or stop them from growing. They are produced naturally by a wide array of soil-dwelling bacteria and fungi. In these environments, antibiotics are hypothesized to help cells reduce competition for nutrients and other resources.

The discovery of antibiotics in 1928, their development over subsequent decades, and widespread use starting in the late 1940s gave physicians effective tools to combat many bacterial infections.

Unfortunately, extensive use of antibiotics in the late twentieth century in clinics and animal feed led to the evolution of drug-resistant strains of pathogenic bacteria (see Chapter 25). One study found that there are now soil-dwelling bacteria in natural environments that—far from being killed by antibiotics—actually use them as food.

Coping with antibiotic resistance in pathogenic bacteria has become a great challenge of modern medicine. Some researchers even claim that humans may be entering the "post-antibiotic era" in medicine.

New research indicates that bacteria have another advantage: They usually grow as **biofilms,** dense bacterial colonies enmeshed

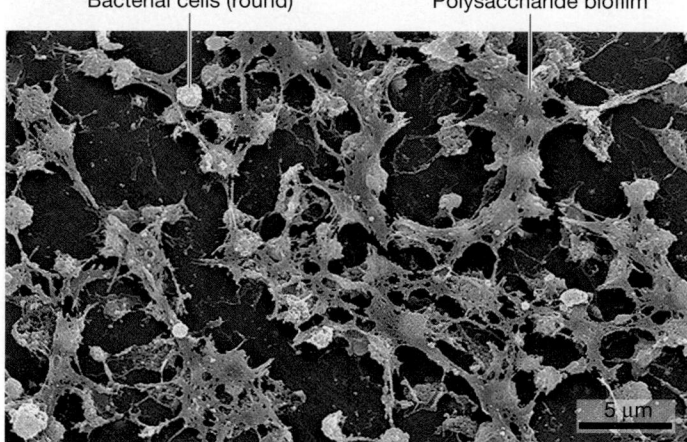

Bacterial cells (round)   Polysaccharide biofilm

5 μm

FIGURE 29.3 Biofilm Growing Inside a Catheter. This micrograph shows *Staphylococcus aureus* cells growing inside a catheter—a tube inserted into a body so that fluids can be withdrawn or injected. Bacterial communities that secrete polysaccharides and adhere to surfaces are sometimes more resistant to antibiotics.

FIGURE 29.4 Bacteria and Archaea Can Play a Role in Cleaning Up Pollution. These clean-up workers are spraying nitrogenous nutrients to encourage bacterial and archaeal growth following the 1989 *Exxon Valdez* oil spill in Alaska.

in a polysaccharide-rich matrix that helps shield the bacteria from antibiotics. Antibiotic-resistant biofilms on medical devices such as catheters are a growing problem in hospitals (**FIGURE 29.3**).

Role in Bioremediation

Bacteria are often in the news because of their dire medical effects. However, only a tiny proportion of bacteria and archaea actually cause disease in humans or other organisms. In the vast majority of cases, bacteria and archaea either have no direct impact on humans or are beneficial. For example, microbes play an important role in wastewater treatment efforts, and researchers are using bacteria and archaea to clean up sites polluted with organic solvents—an effort called **bioremediation.**

Throughout the industrialized world, some of the most serious pollutants in soils, rivers, and ponds consist of organic compounds that were originally used as solvents or fuels but leaked or were spilled into the environment. Most of these compounds are highly hydrophobic. Because they do not dissolve in water, they tend to accumulate in sediments. If the compounds are subsequently ingested by burrowing worms or clams or other organisms, they can be passed along to fish, insects, humans, birds, and other species.

At moderate to high concentrations, these pollutants are toxic to eukaryotes. Petroleum from oil spills and compounds that contain ring structures and chlorine atoms, such as the family of compounds called dioxins, are particularly notorious because of their toxicity to humans.

Fortunately, naturally existing populations of bacteria and archaea can grow in spills and degrade the toxins. This growth can be enhanced using two complementary bioremediation strategies:

1. *Fertilizing contaminated sites to encourage the growth of existing bacteria and archaea that degrade toxic compounds.* After several oil spills, researchers added nitrogen to affected sites as a fertilizer (**FIGURE 29.4**). Dramatic increases occurred in the growth of bacteria and archaea that use hydrocarbons in cellular respiration, probably because the cells used the added nitrogen to synthesize enzymes and other key compounds. In at least some cases, the fertilized shorelines cleaned up much faster than unfertilized sites.

2. *"Seeding," or adding, specific species of bacteria and archaea to contaminated sites.* Seeding shows promise of alleviating pollution in some situations. For example, researchers have recently discovered bacteria that are able to render certain chlorinated, ring-containing compounds harmless. Instead of being poisoned by the pollutants, these bacteria use the chlorinated compounds as electron acceptors during cellular respiration. In at least some cases, the by-product is dechlorinated and nontoxic to humans and other eukaryotes.

To follow up on these discoveries, researchers are now growing the bacteria in quantity, to test the hypothesis that seeding can speed the rate of decomposition in contaminated sediments. Initial reports suggest that seeding may help clean up at least some polluted sites.

29.2 How Do Biologists Study Bacteria and Archaea?

Biologists' understanding of the domains Bacteria and Archaea is advancing more rapidly right now than at any time during the past 100 years—and perhaps faster than our understanding of any other lineages on the tree of life.

As an introduction to the domains Bacteria and Archaea, let's examine a few of the techniques that biologists use to answer questions about them. Some of these research strategies have been used since bacteria were first discovered; some were invented less than 10 years ago.

Using Enrichment Cultures

Which species of bacteria and archaea are present at a particular location, and what do they use as food? To answer questions like these, biologists rely heavily on their ability to culture organisms in the lab. Of the 5000 species of bacteria and archaea described to date, almost all were discovered when they were isolated from natural habitats and grown under controlled conditions in the laboratory.

One classical technique for isolating new types of bacteria and archaea is called **enrichment culture.** Enrichment cultures are based on establishing a specified set of growing conditions—temperature, lighting, substrate, types of available food, and so on. Cells that thrive under the specified conditions increase in numbers enough to be isolated and studied in detail.

To appreciate how this strategy works in practice, consider research on bacteria that live deep below Earth's surface. One study began with samples of rock and fluid from drilling operations in Virginia and Colorado. The samples came from sedimentary rocks at depths ranging from 860 to 2800 meters below the surface, where temperatures are between 42°C and 85°C. The questions posed in the study were simple: Is anything alive down there? If so, what do the organisms use to fuel cellular respiration?

The research team hypothesized that if organisms were living deep below the surface of the Earth, the cells might use hydrogen molecules (H_2) as an electron donor and the ferric ion (Fe^{3+}) as an electron acceptor (**FIGURE 29.5**). Recall that most eukaryotes use sugars as electron donors and use oxygen as an electron acceptor during cellular respiration (see Chapter 9). Fe^{3+} is the oxidized form of iron, and it is abundant in the rocks the biologists collected from great depths. It exists at great depths below the surface in the form of ferric oxyhydroxide, $Fe(OH)_3$. The researchers predicted that if an organism in the samples reduced the ferric ions during cellular respiration, then a black, oxidized, and magnetic mineral called magnetite (Fe_3O_4) would start appearing in the cultures as a by-product of cellular respiration.

What did their enrichment cultures produce? In some cultures, a black compound began to appear within a week. A variety of tests confirmed that the black substance was indeed magnetite. As the "Results" section of Figure 29.5 shows, microscopy revealed the organisms themselves—previously undiscovered bacteria. Because they grow only when incubated at 45°C–75°C, these organisms are considered **thermophiles** ("heat-lovers"). The discovery was spectacular—it was one of the first studies demonstrating that Earth's crust is teeming with organisms to depths of over a mile below the surface. Enrichment culture continues to be a productive way to isolate and characterize new species of bacteria and archaea.

Using Metagenomics

Researchers estimate that of all the bacteria and archaea living today, less than 1 percent have been grown in culture. To augment research based on enrichment cultures, researchers are

---

RESEARCH

QUESTION: Can bacteria live a mile below Earth's surface?

HYPOTHESIS: Bacteria are capable of cellular respiration deep below Earth's surface by using H_2 as an electron donor and Fe^{3+} as an electron acceptor.

NULL HYPOTHESIS: Bacteria from this environment are not capable of using H_2 as an electron donor and Fe^{3+} as an electron acceptor.

EXPERIMENTAL SETUP:

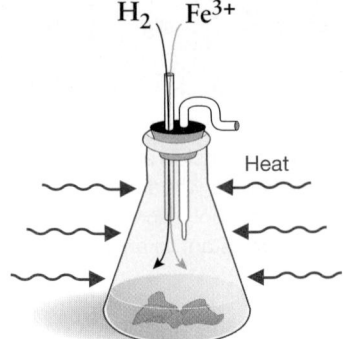

H_2 Fe^{3+}

Heat

Rock and fluid samples

1. Prepare enrichment culture abundant in H_2 and Fe^{3+}; raise temperatures above 45°C.

2. Add rock and fluid samples extracted from drilling operations at depths of about 1000 m below Earth's surface.

PREDICTION: Black, magnetic grains of magnetite (Fe_3O_4) will accumulate because Fe^{3+} is reduced by growing cells and shed as a waste product. Cells will be visible.

PREDICTION OF NULL HYPOTHESIS: No magnetite will appear. No cells will grow.

RESULTS: Cells are visible, and magnetite is detectable.

1 µm

CONCLUSION: At least one bacterial species that can live deep below Earth's surface grew in this enrichment culture. Different culture conditions might result in the enrichment of different species present in the same sample.

FIGURE 29.5 Enrichment Cultures Isolate Large Populations of Cells That Grow under Specific Conditions.

✓**QUESTION** Suppose no organisms had grown in this culture. Explain why the lack of growth would be either strong or weak evidence on the question of whether organisms live a mile below the Earth's surface.

SOURCE: Liu, S. V., J. Zhou, C. Zhang, et al. 1997. Thermophilic Fe(III)-reducing bacteria from the deep subsurface: the evolutionary implications. *Science* 277: 1106–1109.

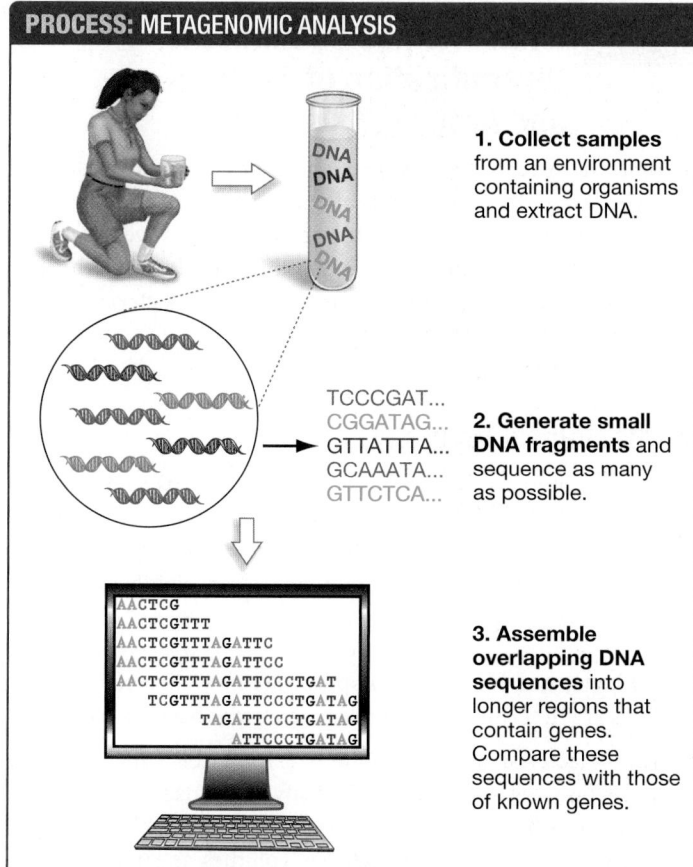

PROCESS: METAGENOMIC ANALYSIS

1. Collect samples from an environment containing organisms and extract DNA.

TCCCGAT...
CGGATAG...
GTTATTTA...
GCAAATA...
GTTCTCA...

2. Generate small DNA fragments and sequence as many as possible.

```
AACTCG
AACTCGTTT
AACTCGTTTAGATTC
AACTCGTTTAGATTCC
AACTCGTTTAGATTCCCTGAT
    TCGTTTAGATTCCCTGATAG
       TAGATTCCCTGATAG
          ATTCCCTGATAG
```

3. Assemble overlapping DNA sequences into longer regions that contain genes. Compare these sequences with those of known genes.

FIGURE 29.6 Metagenomics Allows Researchers to Identify Species That Have Never Been Seen. Metagenomic analysis is used to generate DNA sequences from an environmental sample. That information can then be used to identify novel species and investigate biological processes.

employing a technique called **metagenomics** or **environmental sequencing** (see Chapter 21). Metagenomics is employed in part to document the presence of bacteria and archaea in an environmental sample that cannot be grown in culture. It is based on extracting and sequencing much of the DNA from a sample and then identifying species and biochemical pathways by comparing the DNA sequences with those of known genes. **FIGURE 29.6** outlines the steps performed in a metagenomics study.

Metagenomic analysis allows biologists to rapidly identify and characterize organisms that have never been seen. The technique has revealed huge new branches on the tree of life and produced revolutionary data on the habitats where bacteria and archaea are found.

In one recent study biologists extracted DNA from 125 human fecal samples and generated over 500 billion base pairs of sequence, over 150 times more than the entire human genome. The results they obtained are fascinating:

- In total, the samples contained about 1000 different species of bacteria. Some species were found in most of the samples while others were found in only a few of the humans sampled;

- the vast majority of bacterial species identified were from three phyla: Bacteroidetes, Firmicutes, and Actinobacteria;

- the identified bacterial genes that were shared by all of the human subjects suggest that bacteria play important roles in human physiology, including digestion of complex carbohydrates and synthesis of essential amino acids and vitamins.

Results like these make it clear that humans harbor a diverse ecosystem of symbiotic bacteria. Some bacteria may make us sick, but we depend on many others to stay healthy.

In combination with **direct sequencing**—a technique based on isolating and sequencing a specific gene from organisms found in a particular habitat—metagenomics is revolutionizing biologists' understanding of bacterial and archaeal diversity.

Evaluating Molecular Phylogenies

To put data from enrichment culture and metagenomic studies into context, biologists depend on the accurate placement of species on phylogenetic trees. Recall that phylogenetic trees illustrate the evolutionary relationships among species and lineages (see Chapter 1, Chapter 28, and **BioSkills 7** in Appendix B). They are a pictorial summary of which species are more closely or distantly related to others.

Some of the most useful phylogenetic trees for the Bacteria and the Archaea have been based on studies of the RNA molecules found in the small subunit of ribosomes, or what biologists call 16S and 18S RNA. (See Chapter 17 for more information on the structure and function of ribosomes.) In the late 1960s Carl Woese and colleagues began a massive effort to determine and compare the base sequences of 16S and 18S RNA molecules from a wide array of species. The result of their analysis was the **tree of life,** illustrated in Figure 29.1.

check your understanding

If you understand that . . .

- Enrichment cultures isolate cells that grow in response to specific conditions. They create an abundant sample of bacteria that thrive under particular conditions, allowing further study.
- Metagenomics is based on isolating DNA from samples taken directly from the environment, generating random DNA fragments for sequencing, and then analyzing the DNA sequences to identify the organisms and genes present.

✔ **You should be able to . . .**

1. Design an enrichment culture that would isolate species that could be used to clean up oil spills.
2. Outline a study designed to identify the bacterial and archaeal species present in a soil sample near the biology building on your campus.

Answers are available in Appendix A.

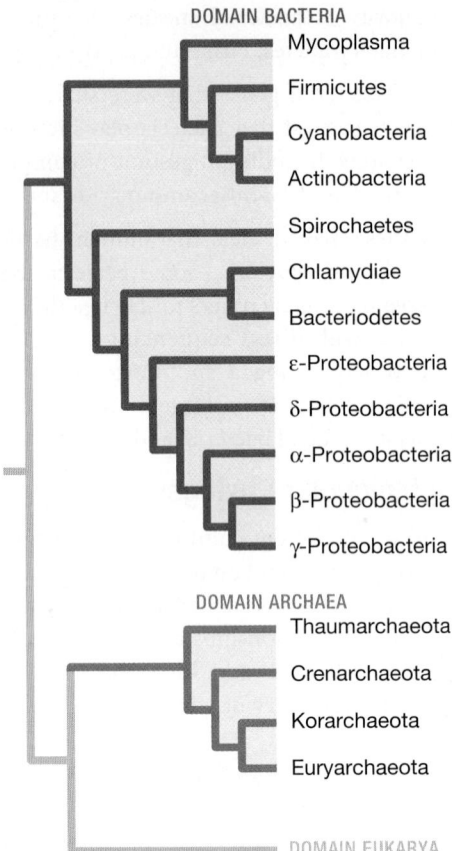

DOMAIN BACTERIA
- Mycoplasma
- Firmicutes
- Cyanobacteria
- Actinobacteria
- Spirochaetes
- Chlamydiae
- Bacteriodetes
- ε-Proteobacteria
- δ-Proteobacteria
- α-Proteobacteria
- β-Proteobacteria
- γ-Proteobacteria

DOMAIN ARCHAEA
- Thaumarchaeota
- Crenarchaeota
- Korarchaeota
- Euryarchaeota

DOMAIN EUKARYA

FIGURE 29.7 Phylogeny of Some Major Lineages in Bacteria and Archaea.

Woese's tree is now considered a classic result. Before its publication, biologists thought that the major division among organisms was between prokaryotes and eukaryotes. But based on data from ribosomal RNA molecules, the major divisions of life-forms are actually the Bacteria, Archaea, and Eukarya. Tracing the early evolutionary history of these domains is extremely difficult since the events distinguishing the lineages took place so long ago. In addition, lateral gene transfer (described in Chapter 21) has blurred the boundaries of the domains. However, recent studies suggest that Eukarya may have formed from a symbiosis between an archaeal cell and a bacterial cell. Traits of both lineages are found within Eukarya (see Chapter 30).

More recent analyses of morphological and molecular characteristics have succeeded in identifying a large series of monophyletic groups within the domains. Recall that a **monophyletic group** consists of an ancestral population and all of its descendants (see Chapter 28). Monophyletic groups can also be called clades or lineages.

The phylogenetic tree in **FIGURE 29.7** summarizes recent results but is still considered highly provisional. Work on molecular phylogenies continues at a brisk pace. Section 29.4 explores some of these lineages in detail, but for now let's turn to the question of how all this diversification took place.

29.3 What Themes Occur in the Diversification of Bacteria and Archaea?

At first, the diversity of bacteria and archaea can seem almost overwhelming. To make sense of the variation among lineages and species, biologists focus on two themes in diversification: morphology and metabolism. Regarding metabolism, the key question is which molecules are used as food. Bacteria and archaea are capable of living in a wide array of environments because they vary in cell structure and in how they make a living.

Morphological Diversity

Because we humans are so large, it is hard for us to appreciate the morphological diversity that exists among bacteria and archaea. To us, they all look small and similar. But at the scale of a bacterium or archaean, different species are wildly diverse in morphology.

Size, Shape, and Motility To appreciate how diverse these organisms are in terms of morphology, consider bacteria alone:

- *Size* Bacterial cells range in size from the smallest of all free-living cells—bacteria called mycoplasmas with volumes as small as 0.15 μm^3—to the largest bacterium known, *Thiomargarita namibiensis*, which has volumes as large as 200 × 10^6 μm^3. Over a billion *Mycoplasma* cells could fit inside an individual *T. namibiensis* (**FIGURE 29.8a**).

- *Shape* Bacterial cells range in shape from filaments, spheres, rods, and chains to spirals (**FIGURE 29.8b**).

- *Motility* Many bacterial cells are motile; their swimming movements are powered by rotating flagella. Depending on the direction of rotation, cells either swim ahead or tumble, which allows them to change direction. Gliding movement, which enables cells to creep along a surface, also occurs in several groups, though the molecular mechanism responsible for this form of motility is still unknown (**FIGURE 29.8c**).

Cell-Wall Composition For single-celled organisms, the composition of the plasma membrane and cell wall are particularly important. The introduction to this chapter highlights the dramatic differences between the plasma membranes and cell walls of bacteria versus archaea.

Within bacteria having cell walls, biologists distinguish two general types of wall using a dyeing system called the **Gram stain.** As **FIGURE 29.9a** shows, Gram-positive cells look purple but Gram-negative cells look pink.

At the molecular level, most cells that are **Gram-positive** have a plasma membrane surrounded by a cell wall with extensive peptidoglycan (**FIGURE 29.9b**). You might recall that peptidoglycan is a complex substance composed of carbohydrate strands that are cross-linked by short chains of amino acids (see Chapter 5). Most cells that are **Gram-negative,** in contrast, have a plasma

(a) Size varies.

Most bacteria are about 1 μm in diameter, but some are much larger.

Smallest (*Mycoplasma mycoides*)

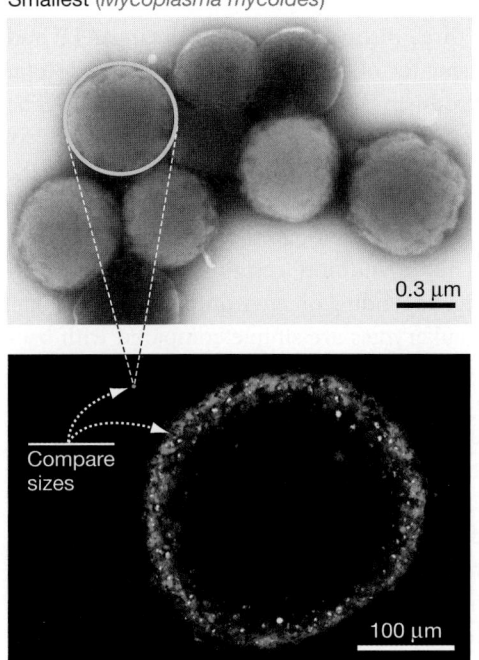

0.3 μm

Compare sizes

100 μm

Largest (*Thiomargarita namibiensis*)

(b) Shape varies...

... from rods to spheres to spirals. In some species, cells adhere to form chains.

Rods, chains of spheres (compost bacteria)

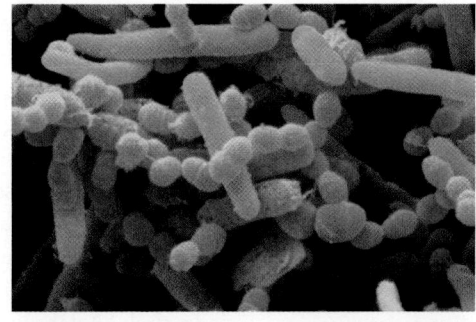

Spirals (*Campylobacter jejuni*)

(c) Motility varies.

Some bacteria are nonmotile, but swimming and gliding are common.

Swimming (*Pseudomonas aeruginosa*)

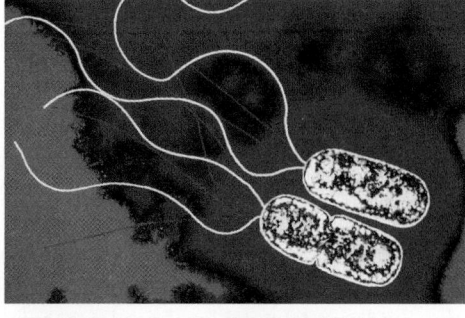

Gliding (*Oscillatoria limosa*)

FIGURE 29.8 Morphological Diversity among Bacteria Is Extensive. Some of the cells in these micrographs have been colorized to make them more visible.

membrane surrounded by a cell wall that has two components—a thin gelatinous layer containing peptidoglycan and an outer phospholipid bilayer (**FIGURE 29.9c**).

Analyzing cell cultures with the Gram stain can be an important preliminary step in treating bacterial infections. Because they contain so much peptidoglycan, Gram-positive cells may respond to treatment by penicillin-like drugs that disrupt peptidoglycan synthesis. Gram-negative cells, in contrast, are more

likely to be affected by erythromycin or other types of drugs that poison bacterial ribosomes.

To summarize, members of the Bacteria and the Archaea are remarkably diverse in their overall size, shape, and motility as well as in the composition of their cell walls and plasma membranes. But when asked to name the innovations that were most responsible for the diversification of these two domains, biologists do not point to their morphological diversity. Instead, they

(a) Gram-positive cells stain more than Gram-negative cells.

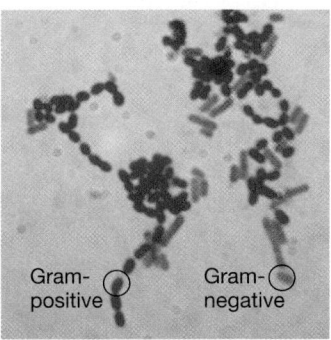

Gram-positive   Gram-negative

(b) Gram-positive cell wall

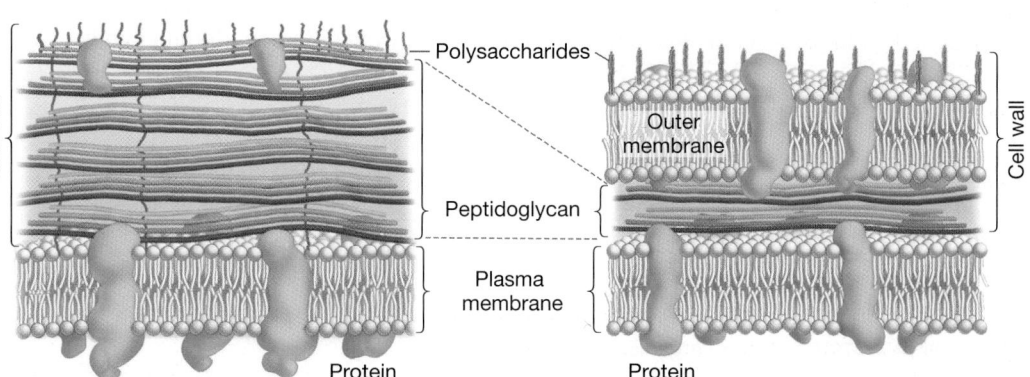

Cell wall

(c) Gram-negative cell wall

Polysaccharides

Outer membrane

Peptidoglycan

Plasma membrane

Protein   Protein

Cell wall

FIGURE 29.9 Gram Staining Distinguishes Two Types of Cell Walls in Bacteria. Cells with extensive peptidoglycan retain a large amount of stain and look purple; others retain little stain and look pink, as can be seen in part (a).

point to metabolic diversity—variation in the chemical reactions that go on inside these cells.

Metabolic Diversity

The most important thing to remember about bacteria and archaea is how diverse they are in the types of compounds they can use as food. Bacteria and archaea are the masters of metabolism. Taken together, they can subsist on almost anything—from hydrogen molecules to crude oil. Bacteria and archaea look small and relatively simple to us in their morphology, but their biochemical capabilities are dazzling.

Just how varied are bacteria and archaea when it comes to making a living? To appreciate the answer, recall that all organisms have two fundamental nutritional needs—acquiring chemical energy in the form of adenosine triphosphate (ATP) and obtaining molecules with carbon–carbon bonds that can be used as building blocks for the synthesis of fatty acids, proteins, DNA, RNA, and other large, complex compounds required by the cell.

Bacteria and archaea produce ATP in three ways:

1. **Phototrophs** ("light-feeders") use light energy to excite electrons. ATP is produced by photophosphorylation (see Chapter 10).

2. **Chemoorganotrophs** oxidize organic molecules with high potential energy, such as sugars. ATP may be produced by cellular respiration—with sugars serving as electron donors—or via fermentation pathways (see Chapter 9).

3. **Chemolithotrophs** ("rock-feeders") oxidize inorganic molecules with high potential energy, such as ammonia (NH_3) or methane (CH_4). ATP is produced by cellular respiration, and inorganic compounds serve as the electron donor.

Bacteria and archaea fulfill their second nutritional need—obtaining building-block compounds with carbon–carbon bonds—in two ways:

1. By synthesizing their own compounds from simple starting materials such as CO_2 and CH_4. Organisms that manufacture their own building-block compounds are termed **autotrophs** ("self-feeders").

2. By absorbing ready-to-use organic compounds from their environment. Organisms that acquire building-block compounds from other organisms are called **heterotrophs** ("other-feeders").

Because there are three distinct ways of producing ATP and two general mechanisms for obtaining carbon, there are a total of six methods for producing ATP and obtaining carbon. The names that biologists use for organisms that employ these six "feeding strategies" are given in **TABLE 29.3**.

Of the six possible ways of producing ATP and obtaining carbon, just two are observed among eukaryotes. But bacteria and archaea do them all. In addition, certain species can switch among modes of living, depending on environmental conditions. In their metabolism, eukaryotes are simple compared with bacteria and archaea. ✔ If you understand the essence of metabolic diversity in bacteria and archaea, you should be able to match the six example species described in **TABLE 29.4** to the appropriate category in Table 29.3.

What makes this remarkable diversity possible? Bacteria and archaea have evolved dozens of variations on the basic processes of respiration and photosynthesis (see Chapters 9 and 10). They use compounds with high potential energy to produce ATP via cellular respiration (electron transport chains) or fermentation, they use light to produce high-energy electrons, and they reduce carbon from CO_2 or other sources to produce sugars or other building-block molecules with carbon–carbon bonds.

The story of bacteria and archaea can be boiled down to two sentences: The basic chemistry required for photosynthesis, cellular respiration, and fermentation originated in these lineages. Then the evolution of variations on each of these processes allowed prokaryotes to diversify into millions of species that occupy diverse habitats. Let's take a closer look.

Producing ATP Through Cellular Respiration: Variation in Electron Donors and Acceptors

Millions of bacterial, archaeal, and eukaryotic species—including animals and some plants—are chemoorganotrophs. These organisms obtain the energy required to make ATP by breaking down organic compounds such as sugars, starch, or fatty acids.

SUMMARY TABLE 29.3 **Six General Methods for Obtaining Energy and Carbon–Carbon Bonds**

		Source of C–C Bonds (for synthesis of complex organic compounds)	
		Autotrophs: self-synthesized from CO_2, CH_4, or other simple molecules	Heterotrophs: from molecules produced by other organisms
Source of Energy (for synthesis of ATP)	**Phototrophs:** from sunlight	photoautotrophs	photoheterotrophs
	Chemoorganotrophs: from organic molecules	chemoorganoautotrophs	chemoorganoheterotrophs
	Chemolithotrophs: from inorganic molecules	chemolitho[auto]trophs	chemolithotrophic heterotrophs

TABLE 29.4 Six Examples of Metabolic Diversity

Example	How ATP Is Produced	How Building-Block Molecules Are Synthesized
Cyanobacteria	via oxygenic photosynthesis	from CO_2 via the Calvin cycle
Clostridium aceticum	via fermentation of glucose	from CO_2 via reactions called the acetyl-CoA pathway
Ammonia-oxidizing archaea (e.g., *Nitrosopumilus* sp.)	via cellular respiration, using ammonia (NH_3) as an electron donor	from CO_2 via the Calvin cycle
Helicobacteria	via anoxygenic photosynthesis	absorb carbon-containing building-block molecules from the environment
Escherichia coli	via fermentation of organic compounds or cellular respiration, using organic compounds as electron donors	absorb carbon-containing building-block molecules from the environment
Beggiatoa	via cellular respiration, using hydrogen sulfide (H_2S) as an electron donor	absorb carbon-containing building-block molecules from the environment

Cellular enzymes can strip electrons from organic molecules that have high potential energy and then transfer these high-energy electrons to the electron carriers NADH and $FADH_2$ (see Chapter 9). These compounds feed electrons to an electron transport chain (ETC), where electrons are stepped down from a high-energy state to a low-energy state (**FIGURE 29.10a**). In eukaryotic cells the ETC is located in the highly folded inner mitochondrial membrane. In bacteria and archaea, this membrane is the plasma membrane.

The energy that is released allows components of the ETC to generate a proton gradient across the plasma membrane (**FIGURE 29.10b**). The resulting flow of protons back through the enzyme ATP synthase results in the production of ATP, via the process called chemiosmosis.

The essence of this process, called **cellular respiration,** is that a molecule with high potential energy serves as an original electron donor and is oxidized, while a molecule with low potential energy serves as a final electron acceptor and becomes reduced. The potential energy difference between the electron donor and electron acceptor is eventually transformed into chemical energy in the form of ATP or is used for other processes (see Chapter 9).

(a) Model of Electron Transport Chain (ETC)

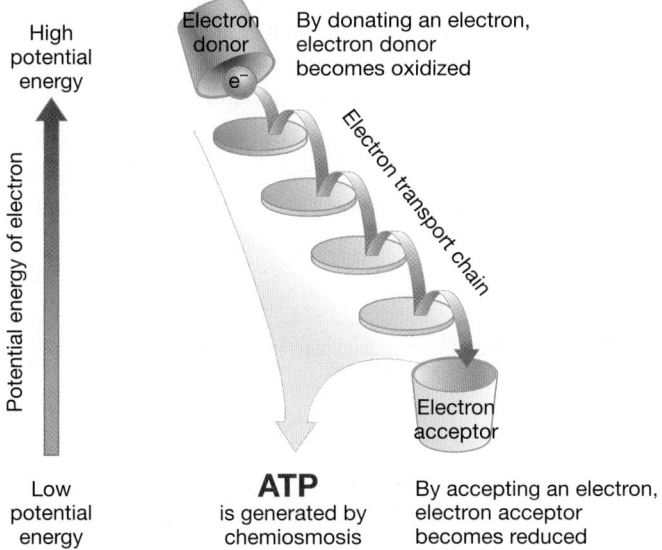

(b) ETC generates proton gradient across plasma membrane.

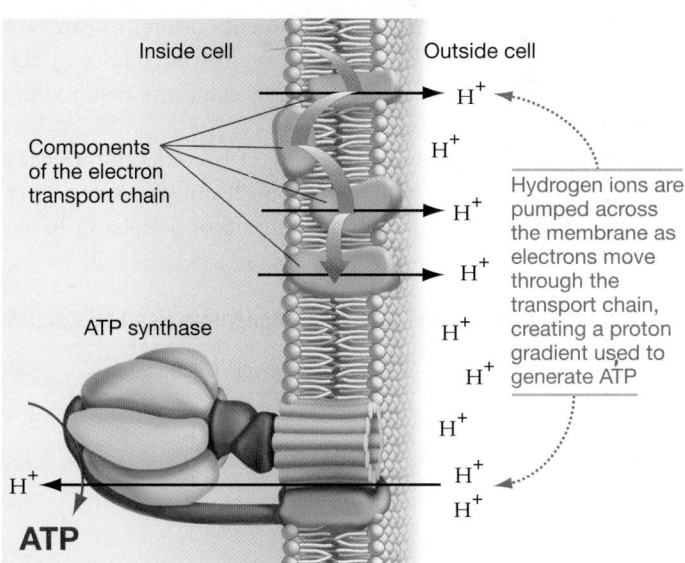

FIGURE 29.10 Cellular Respiration Is Based on Electron Transport Chains. Protons may diffuse away from the cell but a gradient will still form.

✔**EXERCISE** In part (a), add the chemical formula for a specific electron donor, electron acceptor, and reduced by-product for a species of bacteria or archaea. Then write in the electron donor, electron acceptor, and reduced by-product observed in humans.

Most eukaryotes carry out aerobic respiration:

- Organic compounds with high potential energy—often glucose—serve as the original electron donor. When cellular respiration is complete, glucose is completely oxidized to CO_2, which is given off as a by-product.

- Oxygen is the final electron acceptor, and water is also produced as a by-product.

Many bacteria and archaea also rely on these molecules.

It is common, however, to find bacteria and archaea that employ an electron donor other than sugars and an electron acceptor other than oxygen during cellular respiration. These species produce by-products other than carbon dioxide and water (**TABLE 29.5**):

- Molecules with high potential energy serve as electron donors. The substances used as electron donors range from hydrogen molecules (H_2) and hydrogen sulfide (H_2S) to ammonia (NH_3) and methane (CH_4).

- Compounds with relatively low potential energy—including sulfate (SO_4^{2-}), nitrate (NO_3^-), carbon dioxide (CO_2), or ferric ions (Fe^{3+})—act as electron acceptors.

It is only a slight exaggeration to claim that researchers have found bacterial and archaeal species that can use almost any compound with relatively high potential energy as an electron donor and almost any compound with relatively low potential energy as an electron acceptor.

Because the electron donors and electron acceptors used by bacteria and archaea are so diverse, one of the first questions biologists ask about a species is whether it undergoes cellular respiration—and if so, how. The best way to answer this question is through the enrichment culture technique introduced in Section 29.2. Recall that in an enrichment culture, researchers supply specific electron donors and electron acceptors in the medium and try to isolate cells that can use those compounds to support growth.

The remarkable metabolic diversity of bacteria and archaea explains why they play such a key role in cleaning up some types of pollution. Species that use organic solvents or petroleum-based fuels as electron donors or electron acceptors may excrete waste products that are less toxic than the original compounds.

Producing ATP via Fermentation: Variation in Substrates One strategy for making ATP that does not involve electron transport chains is called **fermentation** (see Chapter 9). In fermentation, no outside electron acceptor is used.

Because fermentation is a much less efficient way to make ATP compared with cellular respiration, in many species it occurs as an alternative metabolic strategy when no electron acceptors are available to make cellular respiration possible. In other species, fermentation does not occur at all. But in many bacteria and archaea, fermentation is the only way that cells make ATP.

Although some eukaryotic organisms can ferment glucose to ethanol or lactic acid (see Chapter 9), some bacteria and archaea are capable of using other organic compounds as the starting point for fermentation. Bacteria and archaea that produce ATP via fermentation are still classified as organotrophs, but they are much more diverse in the substrates used. For example:

- The bacterium *Clostridium aceticum* can ferment ethanol, acetate, and fatty acids as well as glucose.

- Other species of *Clostridium* ferment complex carbohydrates (including cellulose or starch), proteins, purines, or amino acids. Species that ferment amino acids produce by-products with names such as cadaverine and putrescine. These molecules are responsible for the odor of rotting flesh.

- Other bacteria can ferment lactose, a prominent component of milk. In some species this fermentation has two end products: propionic acid and CO_2. Propionic acid is responsible for the taste of Swiss cheese; the CO_2 produced during fermentation creates the holes in cheese.

- Many bacterial species in the human digestive tract ferment complex carbohydrates in our diet. The human cells then absorb the by-products and extract even more energy from them using O_2 as the final electron acceptor.

The diversity of enzymatic pathways observed in bacterial and archaeal fermentations extends the metabolic repertoire of these organisms. The diversity of substrates that are fermented also

TABLE 29.5 **Some Electron Donors and Acceptors Used by Bacteria and Archaea**

Electron Donor	By-Product from Electron Donor	Electron Acceptor	By-Product from Electron Acceptor	Category*
Sugars	CO_2	O_2	H_2O	Organotrophs
H_2 or organic compounds	H_2O or CO	SO_4^{2-}	H_2S or S^{2-}	Sulfate reducers
H_2	H_2O	CO_2	CH_4	Methanogens
CH_4	CO_2	O_2	H_2O	Methanotrophs
H_2S or S^{2-}	SO_4^{2-}	O_2	H_2O	Sulfur bacteria
Organic compounds	CO_2	Fe^{3+}	Fe^{2+}	Iron reducers
NH_3	NO_2^-	O_2	H_2O	Ammonia oxidizers
Organic compounds	CO_2	NO_3^-	N_2O, NO, or N_2	Nitrate reducers

*The name biologists use to identify species that use a particular metabolic strategy.

supports the claim that as a group, bacteria and archaea can use virtually any molecule with relatively high potential energy as a source of high-energy electrons for producing ATP.

Producing ATP via Photosynthesis: Variation in Electron Sources and Pigments Instead of using molecules as a source of high-energy electrons, phototrophs pursue a radically different strategy: **photosynthesis.** Among bacteria and archaea, photosynthesis can happen in three different ways:

1. Light activates a pigment called bacteriorhodopsin, which uses the absorbed energy to transport protons across the plasma membrane and out of the cell. The resulting flow of protons back into the cell drives the synthesis of ATP via chemiosmosis (see Chapter 9).

2. A recently discovered bacterium that lives near hydrothermal vents on the ocean floor performs photosynthesis not by absorbing light, but by absorbing geothermal radiation.

3. Pigments that absorb light raise electrons to high-energy states. As these electrons are stepped down to lower energy states by electron transport chains, the energy released is used to generate ATP.

An important feature of this last mode of photosynthesis is that the process requires a source of electrons (see Chapter 10). Recall that in cyanobacteria and plants, the required electrons come from water. When these organisms "split" water molecules apart to obtain electrons, they generate oxygen as a by-product. Species that use water as a source of electrons for photosynthesis are said to complete **oxygenic** ("oxygen-producing") photosynthesis.

In contrast, many phototrophic bacteria use a molecule other than water as the source of electrons. In some cases, the electron donor is hydrogen sulfide (H_2S); a few species can use the ion known as ferrous iron (Fe^{2+}). Instead of producing oxygen as a by-product of photosynthesis, these cells produce elemental sulfur (S) or the ferric ion (Fe^{3+}). This type of photosynthesis is said to be **anoxygenic** ("no oxygen-producing").

The photosynthetic pigments found in plants are chlorophylls *a* and *b* (see Chapter 10). Cyanobacteria have these two pigments. But researchers have isolated seven additional chlorophylls from different lineages of bacterial phototrophs. Each lineage has one or more of these distinctive chlorophylls, and each type of chlorophyll absorbs light best at a different wavelength. ✔ If you understand that different photosynthetic bacteria contain different kinds of light-absorbing pigments, you should be able to explain how several different photosynthetic species can live in the same habitat without competing for light.

Obtaining Building-Block Compounds: Variation in Pathways for Fixing Carbon In addition to acquiring energy, organisms must obtain building-block molecules that contain carbon–carbon bonds. Organisms use two mechanisms to procure usable carbon—either making their own or getting it from other organisms (see Chapters 9 and 10). Autotrophs make their own building-block compounds; heterotrophs consume them.

In many autotrophs, including cyanobacteria and plants, the enzymes of the Calvin cycle transform carbon dioxide (CO_2) into organic molecules that can be used in synthesizing cell material. The carbon atom in CO_2 is reduced during the process and is said to be "fixed." Animals and fungi, in contrast, obtain carbon from living plants or animals, or by absorbing the organic compounds released as dead tissues decay.

Bacteria and archaea pursue these same two strategies. Some interesting twists occur among bacterial and archaeal autotrophs, however. Not all of them use the Calvin cycle to make building-block molecules, and not all start with CO_2 as a source of carbon atoms. For example, consider these biochemical pathways:

- Some proteobacteria are called **methanotrophs** ("methane-eaters") because they use methane (CH_4) as their carbon source. (They also use CH_4 as an electron donor in cellular respiration.) Methanotrophs process CH_4 into more complex organic compounds via one of two enzymatic pathways, depending on the species.

- Some bacteria can use carbon monoxide (CO) or methanol (CH_3OH) as a starting material.

These observations drive home an important message from this chapter: Compared with eukaryotes, the metabolic capabilities of bacteria and archaea are remarkably complex and diverse.

Ecological Diversity and Global Impacts

The metabolic diversity observed among bacteria and archaea explains why these organisms can thrive in such a wide array of habitats.

- The array of electron donors, electron acceptors, and fermentation substrates exploited by bacteria and archaea allows the heterotrophic species to live just about anywhere.

- The evolution of three distinct types of photosynthesis—based on bacteriorhodopsin, geothermal energy, or pigments that donate high-energy electrons to ETCs—extends the types of habitats that can support phototrophs.

The complex chemistry that these cells carry out, combined with their numerical abundance, has made them potent forces for global change throughout Earth's history. Bacteria and archaea have altered the chemical composition of the oceans, atmosphere, and terrestrial environments for billions of years. They continue to do so today.

The Oxygen Revolution Today, oxygen represents almost 21 percent of the molecules in Earth's atmosphere. But researchers who study the composition of the atmosphere are virtually certain that no free molecular oxygen (O_2) existed for the first 2.3 billion years of Earth's existence. This conclusion is based on two observations:

1. There was no plausible source of oxygen at the time the planet formed.

2. Chemical analysis of the oldest Earth rocks suggests that they formed in the absence of atmospheric oxygen.

FIGURE 29.11 Cyanobacteria Were the First Organisms to Perform Oxygenic Photosynthesis.

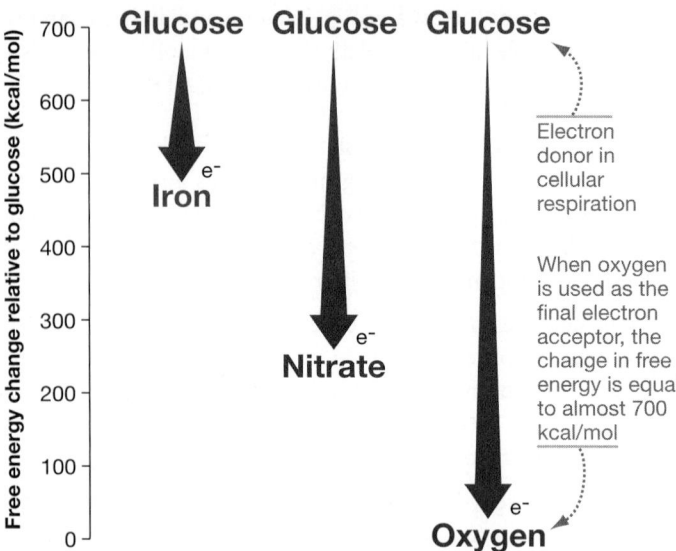

FIGURE 29.12 Cellular Respiration Can Produce More Energy When Oxygen Is the Final Electron Acceptor. More potential energy in glucose can be released when oxygen is the final acceptor compared to other molecules or ions.

DATA: Wilson, D. F., M. Erecińska, and P. L. Dutton. 1974. *Annual Review of Biophysics and Bioengineering 3*: 203–230; Tables 1 and 3.

✔**QUESTION** Which organisms grow faster—those using aerobic respiration or those using anaerobic respiration? Explain your reasoning.

Early in Earth's history, the atmosphere was dominated by nitrogen and carbon dioxide. Where did the oxygen we breathe come from? The answer is cyanobacteria.

Cyanobacteria is a lineage of photosynthetic bacteria. According to the fossil record, species of cyanobacteria first became numerous in the oceans about 2.7–2.5 billion years ago. Their appearance was momentous because cyanobacteria were the first organisms to perform oxygenic ("oxygen-producing") photosynthesis (**FIGURE 29.11**).

The fossil record and geologic record indicate that oxygen concentrations in the oceans and atmosphere began to increase 2.3–2.1 billion years ago. Once oxygen was common in the oceans, cells could begin to use it as the final electron acceptor during cellular respiration. **Aerobic** respiration was now a possibility. Before that, organisms had to use compounds other than oxygen as a final electron acceptor—only **anaerobic** respiration was possible. (Aerobic and anaerobic respiration are introduced in Chapter 9.)

The evolution of aerobic respiration was a crucial event in the history of life. Because oxygen is extremely electronegative, it is an efficient electron acceptor. Much more energy is released as electrons move through electron transport chains with oxygen as the ultimate acceptor than is released with other substances as the electron acceptor.

To drive this point home, study the graph in **FIGURE 29.12**. Notice that the vertical axis plots free energy changes; the graph shows the energy released when glucose is oxidized with iron, nitrate, or oxygen as the final electron acceptor. Once oxygen was available, then, cells could produce much more ATP for each electron donated by NADH or FADH$_2$. As a result, the rate of energy production rose dramatically.

To summarize, data indicate that cyanobacteria were responsible for a fundamental change in Earth's atmosphere—a high concentration of oxygen. Never before, or since, have organisms done so much to alter the nature of our planet.

Nitrogen Fixation and the Nitrogen Cycle In many environments, fertilizing forests or grasslands with nitrogen results in increased growth. Researchers infer from these results that plant growth is often limited by the availability of nitrogen.

Organisms must have nitrogen to synthesize proteins and nucleic acids. Although molecular nitrogen (N$_2$) is extremely abundant in the atmosphere, most organisms cannot use it because of the strong triple bond linking the nitrogen atoms. To incorporate nitrogen atoms into amino acids and nucleotides, all eukaryotes and many bacteria and archaea have to obtain N in a form such as ammonia (NH$_3$) or nitrate (NO$_3^-$).

Certain bacteria and archaea are the only species that are capable of converting molecular nitrogen to ammonia. The steps in the process, called **nitrogen fixation,** are highly endergonic reduction-oxidation (redox) reactions (see Chapter 9). The key enzyme that catalyzes the reaction—nitrogenase—is found only in selected bacterial and archaeal lineages. Many of these organisms are free living, but some form important relationships with plants:

- Some species of cyanobacteria live in association with a water fern that grows in rice paddies and helps fertilize the plants.

- In terrestrial environments, nitrogen-fixing bacteria live in close association with plants—often taking up residence in special root structures called nodules (see Chapter 39).

Why is nitrogenase not found in all organisms? The answer lies in an interesting property of the enzyme. When exposed to O$_2$, nitrogenase is irreversibly poisoned and is degraded. The only organisms with the nitrogenase gene are those that live in anaerobic habitats or are able to protect the enzyme from O$_2$.

Nitrogen fixation is only the beginning of the story, however. A quick glance back at Table 29.5 should convince you that bacteria and archaea use a wide array of nitrogen-containing

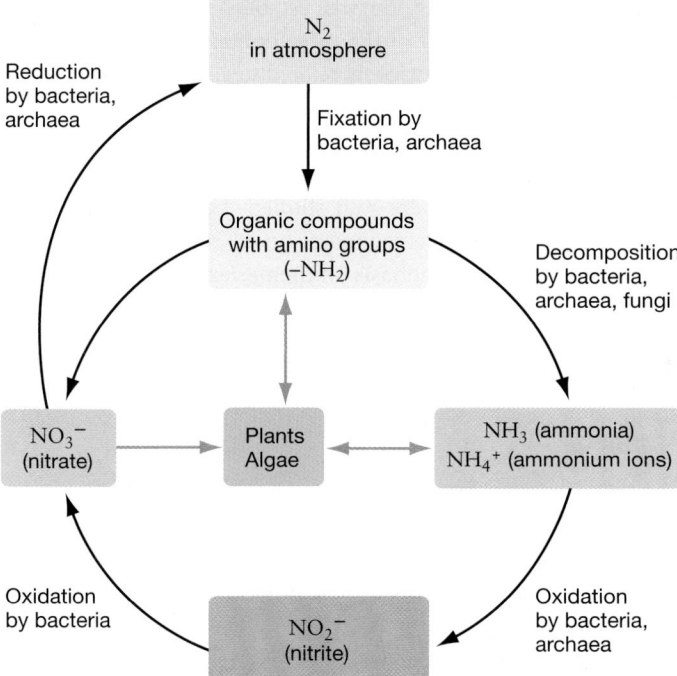

FIGURE 29.13 Bacteria and Archaea Drive the Movement of Nitrogen Atoms through Ecosystems. Nitrogen atoms cycle in different molecular forms.

✔**EXERCISE** Add arrows and labels to indicate that animals ingest amino groups from plants or other animals and release amino groups or ammonia.

compounds as electron donors and electron acceptors during cellular respiration.

To understand why this is important, consider that the nitrite (NO_2^-) produced by some bacteria as a by-product of respiration does not build up in the environment. Instead, other species of bacteria and archaea use it as an electron donor, and it is oxidized to molecular nitrate (NO_3^-). Nitrate, in turn, is reduced to molecular nitrogen (N_2) by yet another suite of bacterial and archaeal species. In this way, bacteria and archaea are responsible for driving the movement of nitrogen atoms through ecosystems around the globe in a process called the **nitrogen cycle** (**FIGURE 29.13**).

Similar types of interactions occur with molecules that contain phosphorus, sulfur, and carbon. In this way, bacteria and archaea play a key role in the cycling of nitrogen and other nutrients. ✔ If you understand the role of bacteria and archaea in the nitrogen cycle, you should be able to provide a plausible explanation of what the composition of the atmosphere and what the nitrogen cycle might be like if bacteria and archaea did not exist.

Nitrate Pollution Most crop plants—including corn, rice, and wheat—do not live in association with nitrogen-fixing bacteria. To increase yields of these crops, farmers use fertilizers that are high in nitrogen. In parts of the world, massive additions of nitrogen in the form of ammonia are causing serious pollution problems.

FIGURE 29.14 shows why. When ammonia is added to a cornfield—in midwestern North America, for example—much of

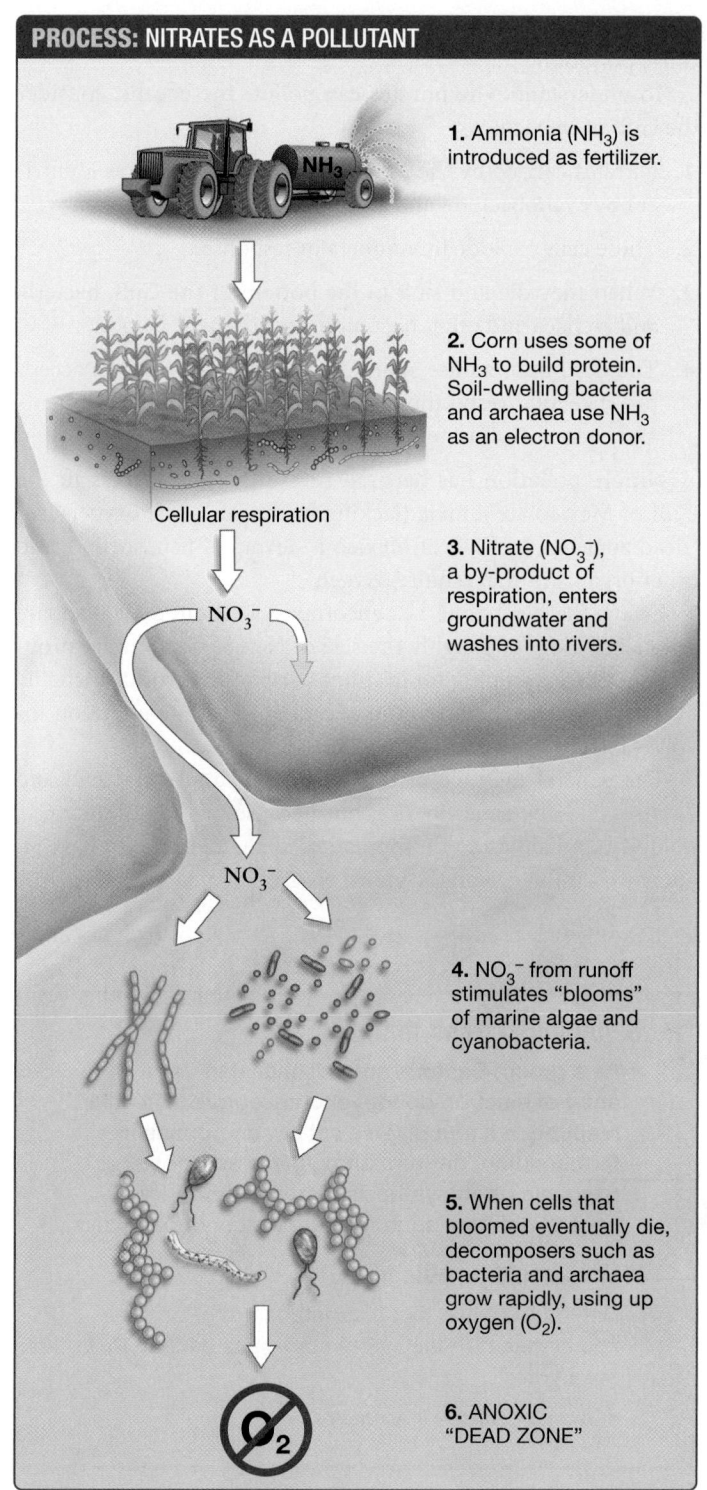

PROCESS: NITRATES AS A POLLUTANT

1. Ammonia (NH_3) is introduced as fertilizer.

2. Corn uses some of NH_3 to build protein. Soil-dwelling bacteria and archaea use NH_3 as an electron donor.

Cellular respiration

3. Nitrate (NO_3^-), a by-product of respiration, enters groundwater and washes into rivers.

4. NO_3^- from runoff stimulates "blooms" of marine algae and cyanobacteria.

5. When cells that bloomed eventually die, decomposers such as bacteria and archaea grow rapidly, using up oxygen (O_2).

6. ANOXIC "DEAD ZONE"

FIGURE 29.14 Nitrates Act as a Pollutant in Aquatic Ecosystems.

it never reaches the growing corn plants. Instead, bacteria and archaea in the soil use a significant fraction of the ammonia as food. Microbes that use ammonia as an electron donor to fuel cellular respiration release nitrite (NO_2^-) as a waste product. Other microbes use nitrite as an electron donor and release nitrate (NO_3^-). Nitrate molecules are extremely soluble in water and tend to be washed out of soils into groundwater or streams.

From there they eventually reach the ocean, where they can cause pollution.

To understand why nitrates can pollute the oceans, consider the Gulf of Mexico:

1. Nitrates carried by the Mississippi River are used as a nutrient by cyanobacteria and algae that live in the Gulf.

2. These cells explode in numbers in response.

3. When they die and sink to the bottom of the Gulf, bacteria and archaea and other decomposers use them as food.

4. The decomposers use so much oxygen as an electron acceptor in cellular respiration that oxygen levels in the sediments and even in Gulf waters decline.

Nitrate pollution has been so severe that large areas in the Gulf of Mexico are anoxic (lacking in oxygen). The oxygen-free "dead zone" in the Gulf of Mexico is devoid of fish, shrimp, and other organisms that require oxygen.

Lately, the dead zone has encompassed about 22,000 square kilometers (km^2)—roughly the size of New Jersey. Similar problem spots are cropping up in other parts of the world. Virtually every link in the chain of events leading to nitrate pollution involves bacteria and archaea.

The general message of this section is simple: Bacteria and Archaea may be small in size, but because of their abundance, ubiquity, and ability to do sophisticated chemistry, they have an enormous influence on the global environment.

check your understanding

If you understand that . . .

• As a group, Bacteria and Archaea can use a wide array of electron donors and acceptors in cellular respiration and a diverse set of compounds in fermentation, perform anoxygenic as well as oxygenic photosynthesis, and fix carbon from several different sources via a variety of pathways.

✔ **You should be able to . . .**

Defend the claim that the metabolism of bacteria and archaea is much more sophisticated than that of eukaryotes.

Answers are available in Appendix A.

29.4 Key Lineages of Bacteria and Archaea

In the decades since the phylogenetic tree identifying the three domains of life was first published, dozens of studies have confirmed the result. It is now well established that all organisms alive today belong to one of the three domains, and that archaea and eukaryotes are more closely related to each other than either group is to bacteria.

Although the relationships among the major lineages within Bacteria and Archaea are still uncertain in some cases, many of the lineages themselves are well studied. Let's survey the attributes of species from selected major lineages within the Bacteria and Archaea, with an emphasis on themes explored earlier in the chapter: their morphological and metabolic diversity, their impacts on humans, and their importance to other species and to the environment.

Bacteria

The name *bacteria* comes from the Greek root *bacter*, meaning "rod" or "staff." The name was inspired by the first bacteria to be seen under a microscope, which were rod shaped. But as the following descriptions indicate, bacterial cells come in a wide variety of shapes.

Biologists who study bacterial diversity currently recognize at least 21 lineages, or phyla, within the domain. Some of these lineages were recognized by distinctive morphological characteristics and others by phylogenetic analyses of gene sequence data. The lineages reviewed here are just a sampling of bacterial diversity.

● Bacteria > Firmicutes

● Bacteria > Cyanobacteria

● Bacteria > Actinobacteria

● Bacteria > Spirochaetes (Spirochetes)

● Bacteria > Chlamydiae

● Bacteria > Proteobacteria

Archaea

The name *archaea* comes from the Greek root *archae*, for "ancient." The name was inspired by the hypothesis that this is a particularly ancient group, which turned out to be incorrect. Also incorrect was the initial hypothesis that archaeans are restricted to hot springs, salt ponds, and other extreme habitats. Archaea live in virtually every habitat known.

Recent phylogenies based on DNA sequence data indicate that the domain is composed of at least four major phyla, called the Thaumarchaeota, Crenarchaeota, Euryarchaeota, and Korarchaeota. The Korarchaeota are known only from direct sequencing studies. They have never been grown in culture, and almost nothing is known about them.

● Archaea > Thaumarchaeota

● Archaea > Crenarchaeota

● Archaea > Euryarchaeota

Bacteria > Firmicutes

The Firmicutes have also been called "low-GC Gram positives" because their cell walls react positively to the Gram stain and because their DNA contains a relatively low percentage of guanine and cytosine (G and C). In some species, G and C represent less than 25 percent of the bases present. There are over 1100 species. ✔ You should be able to mark the origin of the Gram-positive cell wall on Figure 29.7.

Morphological diversity Most are rod shaped or spherical. Some of the spherical species form chains or tetrads (groups of four cells). A few form a durable resting stage called a spore. One subgroup synthesizes a cell wall made of cellulose.

Metabolic diversity Some species can fix nitrogen; some perform anoxygenic photosynthesis. Others make all of their ATP via various fermentation pathways; still others perform cellular respiration, using hydrogen gas (H_2) as an electron donor.

Human and ecological impacts Recent metagenomic studies have shown that members of this lineage are extremely common in the human gut. Species in this group also cause a variety of diseases, including anthrax, botulism, tetanus, walking pneumonia, boils, gangrene, and strep throat. *Bacillus thuringiensis* produces a toxin that is one of the most important insecticides currently used in farming. Species in the lactic acid bacteria group are used to ferment milk products into yogurt or cheese (**FIGURE 29.15**).

Lactobacillus bulgaricus (rods) and *Streptococcus thermophilus*

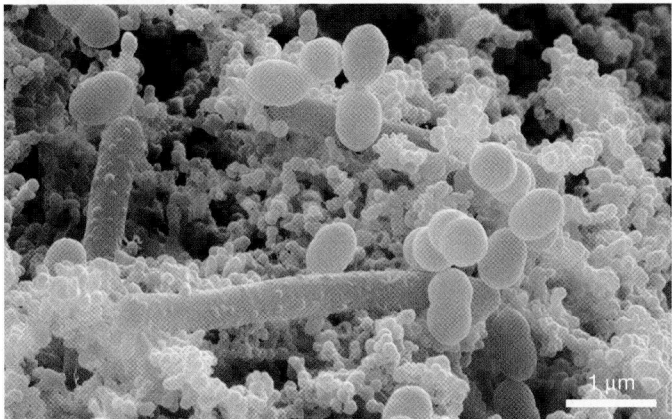

FIGURE 29.15 Firmicutes in Yogurt. (The cells in this scanning electron micrograph have been colorized.)

Bacteria > Cyanobacteria

The cyanobacteria were formerly known as the "blue-green algae"—even though algae are eukaryotes. Only about 80 species of cyanobacteria have been described to date, but they are among the most abundant organisms on Earth. In terms of total mass, cyanobacteria dominate the surface waters in many marine and freshwater environments.

Morphological diversity Cyanobacteria may be found as independent cells, in chains that form filaments (**FIGURE 29.16**), or in the loose aggregations of individual cells called colonies. The shape of *Nostoc* colonies varies from flat sheets to ball-like clusters of cells.

Metabolic diversity All perform oxygenic photosynthesis; many can also fix nitrogen. Because cyanobacteria can synthesize virtually every molecule they need, they can be grown in culture media that contain only CO_2, N_2, H_2O, and a few mineral nutrients. Some species associate with fungi, forming lichens, while others form associations with protists, sponges, or plants. In each case the cyanobacterium provides some form of nutritional benefit to the host. ✔ You should be able to mark the origin of oxygenic photosynthesis on Figure 29.7.

Human and ecological impacts If cyanobacteria are present in high numbers, their waste products can make drinking water smell bad. Some species release molecules called microcystins that are toxic to plants and animals. Cyanobacteria were responsible for the origin of the oxygen atmosphere on Earth. Today they still produce much of the oxygen and nitrogen and many of the organic compounds that feed other organisms in freshwater and marine environments.

Nostoc species

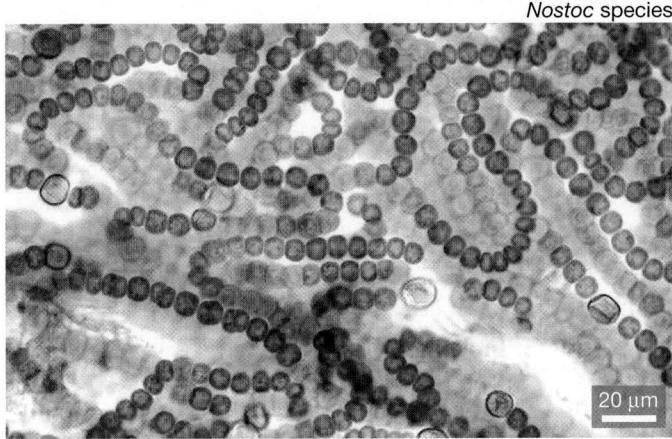

FIGURE 29.16 Cyanobacteria Contain Chlorophyll and Are Green.

Actinobacteria are sometimes called the "high-GC Gram positives" because (1) their cell-wall material appears purple when treated with the Gram stain—meaning that they have a peptidoglycan-rich cell wall and lack an outer membrane—and (2) their DNA contains a relatively high percentage of guanine and cytosine. In some species, G and C represent over 75 percent of the bases present. Over 1100 species have been described to date (**FIGURE 29.17**). ✔ You should be able to mark the origin of this high-GC genome on Figure 29.7.

Morphological diversity Cell shape varies from rods to filaments. Many of the soil-dwelling species are found as chains of cells that form extensive branching filaments called **mycelia.** Because of their morphology they were initially misclassified as fungi, and the incorrect name Actinomyces persists.

Metabolic diversity Many are heterotrophs that use an array of organic compounds as electron donors and oxygen as an electron acceptor. There are several parasitic species that get most of their nutrition from host organisms.

Human and ecological impacts Two serious human diseases, tuberculosis and leprosy, are caused by parasitic *Mycobacterium* species. Over 500 distinct antibiotics have been isolated from species in the genus *Streptomyces*; 60 of these—including streptomycin, neomycin, tetracycline, and erythromycin—are now actively prescribed to treat diseases in humans or domestic livestock. One actinobacterium species is critical to the manufacture of Swiss cheese. Species in the genus *Streptomyces* and *Arthrobacter* are abundant in soil and are vital as decomposers of dead plant and animal material. Some species live in association with plant roots and fix nitrogen; others can break down toxins such as herbicides, nicotine, and caffeine.

Streptomyces griseus

5 µm

FIGURE 29.17 A *Streptomyces* Species That Produces the Antibiotic Streptomycin.

Based on numbers of species, the spirochetes are one of the smaller bacterial phyla: Only 13 genera and a total of 62 species have been described to date.

Morphological diversity Spirochetes are distinguished by their unique corkscrew shape and flagella (**FIGURE 29.18**). Instead of extending into the water surrounding the cell, spirochete flagella are contained within a structure called the outer sheath, which surrounds the cell. When these flagella beat, the cell lashes back and forth and swims forward. ✔ You should be able to mark the origin of the spirochete flagellum on Figure 29.7.

Metabolic diversity Most spirochetes manufacture ATP via fermentation. The substrate used in fermentation varies among species and may consist of sugars, amino acids, starch, or the pectin found in plant cell walls. A spirochete that lives only in the hindgut of termites can fix nitrogen.

Human and ecological impacts The sexually transmitted disease syphilis is caused by a spirochete. Syphilis is thought to have been brought by European explorers to the Western hemisphere, where it was responsible for killing tens of millions of native people. Lyme disease, also caused by a spirochete, is transmitted to humans by deer ticks. Spirochetes are extremely common in freshwater and marine habitats; many live only under anaerobic conditions.

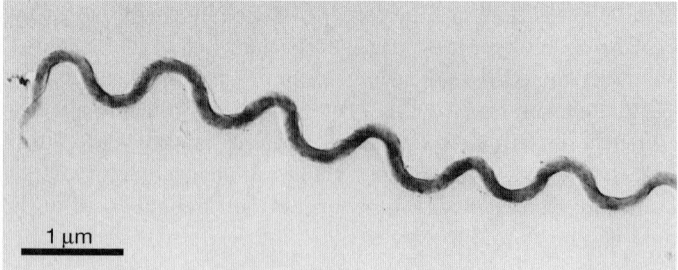

Treponema pallidum

1 µm

FIGURE 29.18 Spirochetes Are Corkscrew-Shaped Cells Inside an Outer Sheath.

In terms of numbers of species living today, Chlamydiae may be the smallest of all major bacterial lineages. Although the group is highly distinct phylogenetically, only 13 species are known. All are Gram-negative.

Morphological diversity Chlamydiae are spherical. They are tiny, even by bacterial standards.

Metabolic diversity All known species in this phylum live as parasites *inside* host cells and are termed **endosymbionts** ("inside-together-living"). Chlamydiae acquire almost all of their nutrition from their hosts. In **FIGURE 29.19**, the chlamydiae have been colored red; the animal cells that they live in are colored brown. ✔ You should be able to mark the origin of the endosymbiotic lifestyle in this lineage on Figure 29.7. (The endosymbiotic lifestyle has also arisen in other bacterial lineages, independently of Chlamydiae.)

Human and ecological impacts *Chlamydia trachomatis* infections are the most common cause of blindness in humans. When the same organism is transmitted from person to person via sexual intercourse, it can cause serious urogenital tract infections. If untreated in women, this disease can lead to ectopic pregnancy, premature births, and infertility. One species causes epidemics of a pneumonia-like disease in birds.

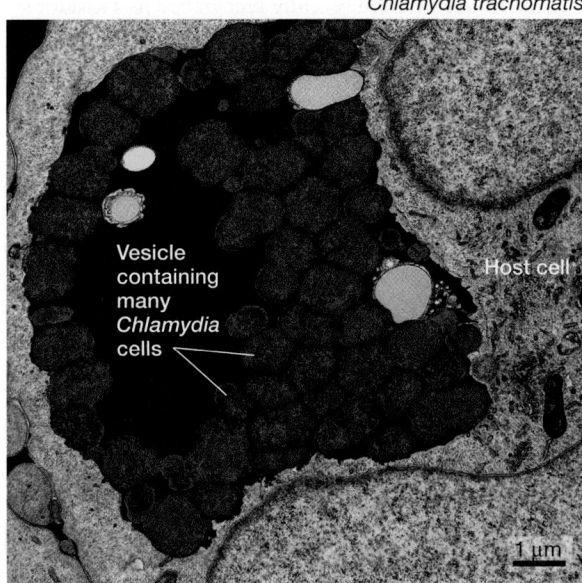

Chlamydia trachomatis

Vesicle containing many *Chlamydia* cells

Host cell

1 µm

FIGURE 29.19 Chlamydiae Live Only Inside Animal Cells.

The approximately 1200 species of proteobacteria form five major subgroups, designated by the Greek letters α (alpha), β (beta), γ (gamma), δ (delta), and ε (epsilon). Because they are so diverse in their morphology and metabolism, the lineage is named after the Greek god Proteus, who could assume many shapes.

Morphological diversity Proteobacterial cells can be rods, spheres, or spirals. Some form stalks (**FIGURE 29.20a**). Some are motile. In one group, cells may move together to form colonies, which then transform into the specialized cell aggregate shown in **FIGURE 29.20b**. This structure is known as a **fruiting body.** At their tips, the fruiting bodies produce cells that are surrounded by a durable coating. These spores sit until conditions improve, and then they resume growth.

Metabolic diversity Proteobacteria make a living in virtually every way known to bacteria—except that none perform oxygenic photosynthesis. Various species may perform cellular respiration by using organic compounds, nitrite, methane, hydrogen gas, sulfur, or ammonia as electron donors and oxygen, sulfate, or sulfur as electron acceptors. Some perform anoxygenic photosynthesis.

Human and ecological impacts *Escherichia coli* may be the best studied of all organisms and is a key species in biotechnology (see Chapter 20, and **BioSkills 13** in Appendix B). Pathogenic proteobacteria cause Legionnaire's disease, cholera, food poisoning, plague, dysentery, gonorrhea, Rocky Mountain spotted fever, typhus, ulcers, and diarrhea. *Wolbachia* infections are common in insects and are often transmitted from mothers to offspring

(a) Stalked bacterium

Caulobacter crescentus

Stalk

1 µm

(b) Fruiting bodies

Chondromyces crocatus

Spores

50 µm

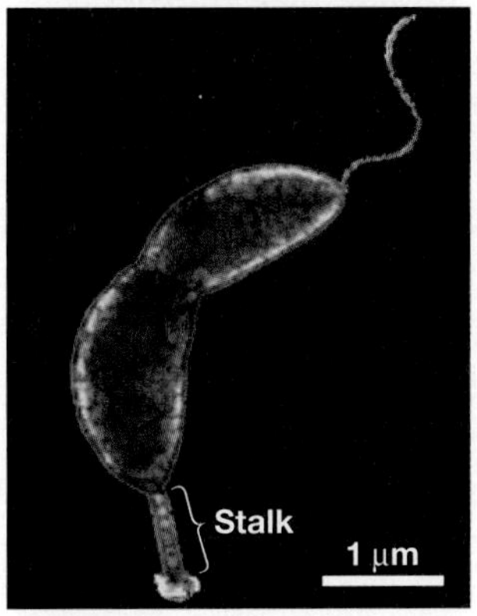

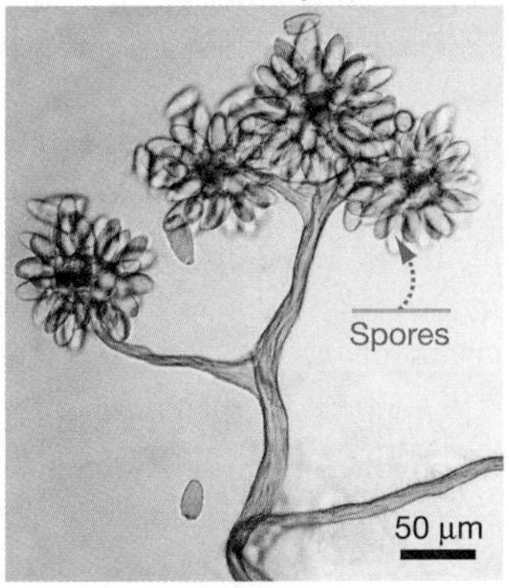

FIGURE 29.20 Some Proteobacteria Grow on Stalks or Form Fruiting Bodies. The stalked bacterium (a) has been colorized.

via eggs. Biologists use *Agrobacterium* cells to transfer new genes into crop plants. Certain acid-loving species of proteobacteria are used in the production of vinegars. Species in the genus *Rhizobium* (α-proteobacteria) live in association with plant roots and fix nitrogen. A group in the δ-proteobacteria, the bdellovibrios, are predators—they drill into other proteobacterial cells and digest them. Because some species use nitrogen-containing compounds as electron acceptors, proteobacteria are critical players in the cycling of nitrogen atoms through terrestrial and aquatic ecosystems.

The Thaumarchaeota were recently recognized as a monophyletic, ancient lineage of archaea. Members of this phylum are extremely abundant in oceans, estuaries, and terrestrial soils. Unlike the extremophiles, species in this lineage are considered mesophilic because they grow best at moderate temperatures.

Morphological diversity Only a few members of this group have been observed, and all consist of rod-shaped cells that are less than 1 micrometer (μm) in length, smaller than typical prokaryotes. One species, *Nitrosopumilus maritimus* (**FIGURE 29.21**), is so abundant it is estimated to constitute possibly 25 percent of the total prokaryotic cell biomass in open oceans.

Metabolic diversity Members of this phylum are called ammonia oxidizers because they use ammonia as a source of electrons and generate nitrite as a by-product. They use the energy from ammonia oxidation to fix CO_2.

Human and ecological impacts Because of their abundance, these organisms are thought to play a major role in Earth's nitrogen and carbon cycles (see Chapter 56). Their presence in deep ocean waters that lack reduced carbon and sunlight may help explain the productivity of these habitats. The species *Cenarchaeum symbiosum* lives as an endosymbiont inside a marine sponge.

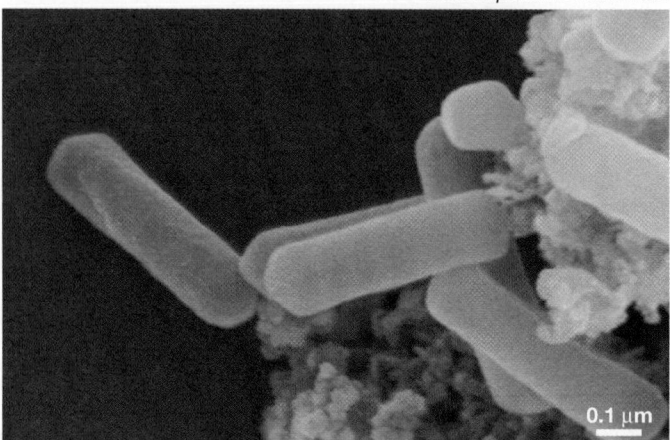
Nitrosopumilus maritimus

0.1 μm

FIGURE 29.21 This Species of Thaumarchaeota Has Rod-Shaped Cells and Is Extremely Abundant in the Open Ocean.

✔**QUANTITATIVE** Use the scale bar in the figure to measure the length of a *N. maritimus* cell. If these cells were placed end to end, how many of them would fit along a meter stick?

The Crenarchaeota got their name because they are considered similar to the oldest archaeans; the word root *cren–* refers to a source or fount. Biologists have named only 37 species so far, but they are virtually certain that thousands are yet to be discovered.

Morphological diversity Crenarchaeota cells can be shaped like filaments, rods, discs, or spheres. One species that lives in extremely hot habitats has a tough cell wall consisting solely of glycoprotein.

Metabolic diversity Depending on the species, cellular respiration can involve organic compounds, sulfur, hydrogen gas, or Fe^{2+} ions as electron donors and oxygen, nitrate, sulfate, sulfur, carbon dioxide, or Fe^{3+} ions as electron acceptors. Some species make ATP exclusively through fermentation pathways.

Human and ecological impacts Crenarchaeota have yet to be used in the manufacture of commercial products. In certain extremely hot, high-pressure, cold, or acidic environments, crenarchaeota may be the only life-form present (**FIGURE 29.22**). Acid-loving species thrive in habitats with pH 1–5; some species are found in ocean sediments at depths ranging from 2500 to 4000 m below the surface.

Sulfolobus species

0.5 μm

FIGURE 29.22 Some Crenarchaeota Live in Sulfur-Rich Hot Springs. The cells in the micrograph have been colorized to make them more visible.

The Euryarchaeota are aptly named, because the word root *eury–* means "broad." Members of this phylum live in every conceivable habitat. Some species are adapted to high-salt habitats with pH 11.5—almost as basic as household ammonia. Other species are adapted to acidic conditions with a pH as low as 0. Species in the genus *Methanopyrus* live near hot springs called black smokers that are 2000 m (over 1 mile) below sea level (**FIGURE 29.23**).

Morphological diversity Euryarchaeota cells can be spherical, filamentous, rod shaped, disc shaped, or spiral. Rod-shaped cells may be short or long or arranged in chains. Spherical cells can be found in ball-like aggregations. Some species have several flagella. Some species lack a cell wall; others have a cell wall composed entirely of glycoproteins.

Metabolic diversity The group includes a variety of methane-producing species. These **methanogens** can use up to 11 different organic compounds as electron acceptors during cellular respiration; all produce CH_4 as a by-product of respiration. In other species of Euryarchaeota, cellular respiration is based on hydrogen gas or Fe^{2+} ions as electron donors and nitrate or sulfate as electron acceptors. Species that live in high-salt environments can use the molecule retinal—which is responsible for light reception in your eyes—to capture light energy and perform photosynthesis.

Human and ecological impacts Species in the genus *Ferroplasma* live in piles of waste rock near abandoned mines. As a by-product of metabolism, they produce acids that drain into streams and pollute them. Methanogens live in the soils of swamps and the guts of mammals (including yours). They are responsible for adding about 2 billion tons of methane to the atmosphere each year. A methanogen in this phylum was also recently implicated in gum disease.

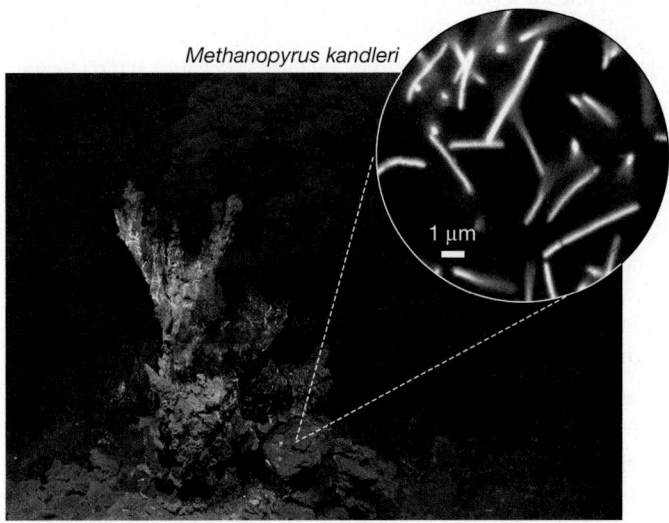

Methanopyrus kandleri

1 μm

FIGURE 29.23 Some Euryarchaeota Cells Live in the Chimneys of "Black Smokers" on the Seafloor.

CHAPTER 29 REVIEW

For media, go to MasteringBiology (MB)

If you understand . . .

29.1 Why Do Biologists Study Bacteria and Archaea?

- Bacteria and archaea are the most abundant organisms on Earth and are found in every habitat that has been sampled.

- Bacteria and archaea are very small, prokaryotic cells, and most are unicellular.

- Bacteria and archaea can be distinguished by their different kinds of membrane lipids and cell walls as well as by their different transcription machinery.

- Bacteria play many beneficial roles in animal digestion, bioremediation, and the production of antibiotics.

- Bacteria cause some of the most dangerous human diseases, including plague, syphilis, botulism, cholera, and tuberculosis.

✔ You should be able to explain the difference between a bacterium, an archaeon, and a eukaryote.

29.2 How Do Biologists Study Bacteria and Archaea?

- Enrichment cultures are used to grow large numbers of bacterial or archaeal cells that thrive under specified conditions.

- Using metagenomic analysis, biologists can study bacteria and archaea that cannot be cultured by extracting DNA directly from an environment and then sequencing and characterizing DNA fragments. Information obtained is used to identify biochemical processes and novel organisms that are then placed on the tree of life.

✔ You should be able to explain how metagenomic analysis might be used to reveal whether bacteria carry out nitrogen fixation in the gut of an insect.

29.3 What Themes Occur in the Diversification of Bacteria and Archaea?

- Metabolic diversity and complexity are the hallmarks of the bacteria and archaea, just as morphological diversity and complexity are the hallmarks of the eukaryotes.

- Among bacteria and archaea, a wide array of inorganic or organic compounds with high potential energy may serve as electron donors in cellular respiration, and a wide variety of inorganic or organic molecules with low potential energy may serve as electron acceptors. Dozens of distinct organic compounds are fermented.

- Photosynthesis is widespread in bacteria. In cyanobacteria, water is used as a source of electrons and oxygen gas is generated as a by-product. But in other species, the electron excited by photon capture comes from a source other than water, and no oxygen is produced.

- To acquire building-block molecules containing carbon–carbon bonds, some bacteria and archaea species use the enzymes of the Calvin cycle to reduce CO_2. But several other biochemical pathways found in bacteria and archaea can also reduce simple organic compounds to sugars or carbohydrates.

- Because of their metabolic diversity, bacteria and archaea play a large role in carbon and nitrogen cycling and alter the global atmosphere, oceans, and terrestrial environments.

- Nitrogen-fixing species provide nitrogen in forms that can be used by many other species, including plants and animals.

✔ You should be able to explain why species that release H_2S as a by-product and that use H_2S as an electron donor often live side by side.

29.4 Key Lineages of Bacteria and Archaea

- Prokaryotes can be divided into two lineages, the Bacteria and the Archaea, based on a wide variety of morphological, biochemical, and molecular characters.

- Bacteria are divided into 21 major lineages including organisms that play major roles in ecosystems as primary producers, decomposers, and parasites.

- Archaea are divided into four major lineages and were thought to exist only in extreme environments; they are now recognized to be widespread.

MB) MasteringBiology

1. MasteringBiology Assignments

Tutorials and Activities Classification of Prokaryotes, Diversity in Bacteria, Tree of Life, Water Pollution from Nitrates

Questions Reading Quizzes, Blue-Thread Questions, Test Bank

2. eText Read your book online, search, take notes, highlight text, and more.

3. The Study Area Practice Test, Cumulative Test, BioFlix® 3-D Animations, Videos, Activities, Audio Glossary, Word Study Tools, Art

You should be able to . . .

✔ TEST YOUR KNOWLEDGE

Answers are available in Appendix A

1. How do the molecules that function as electron donors and those that function as electron acceptors differ?
   a. Electron donors are almost always organic molecules; electron acceptors are always inorganic.
   b. Electron donors are almost always inorganic molecules; electron acceptors are always organic.
   c. Electron donors have relatively high potential energy; electron acceptors have relatively low potential energy.
   d. Electron donors have relatively low potential energy; electron acceptors have relatively high potential energy.

2. What do some photosynthetic bacteria use as a source of electrons instead of water?
   a. oxygen (O_2)
   b. hydrogen sulfide (H_2S)
   c. organic compounds (e.g., CH_3COO^-)
   d. nitrate (NO_3^-)

3. What is distinctive about the chlorophylls found in different photosynthetic bacteria?
   a. their membranes
   b. their role in acquiring energy
   c. their role in carbon fixation
   d. their absorption spectra

4. What are organisms called that use inorganic compounds as electron donors in cellular respiration?
   a. phototrophs
   b. heterotrophs
   c. organotrophs
   d. lithotrophs

5. True or False. Certain aerobic bacteria in the presence of oxygen can convert nitrogen gas to ammonia.

6. Unlike plant cell walls that contain cellulose, bacterial cell walls are composed of _____.

7. What has metagenomic analysis allowed researchers to do for the first time?
   a. sample organisms from an environment and grow them under defined conditions in the lab
   b. isolate organisms from an environment and sequence their entire genome
   c. study organisms that cannot be cultured (grown in the lab)
   d. identify important morphological differences among species

8. Biologists often use the term energy source as a synonym for "electron donor." Why?

9. The text claims that the tremendous ecological diversity of bacteria and archaea is possible because of their impressive metabolic diversity. Do you agree with this statement? Why or why not?

10. Would you predict that disease-causing bacteria, such as those listed in Table 29.2, obtain energy from light, organic molecules, or inorganic molecules? Explain your answer.

11. The text claims that the evolution of an oxygen atmosphere paved the way for increasingly efficient cellular respiration and higher growth rates in organisms. Explain.

12. From what we know about the evolutionary relationship between the three largest domains, as depicted in Figure 29.1, explain the statement, "Prokaryotes are a paraphyletic group."

13. When using Koch's postulates, which of the following is an essential requirement for the suspected pathogen?
    a. It is present in all organisms with the disease.
    b. It can be cultured on an agar plate.
    c. It is pathogenic on a wide variety of organisms.
    d. It can reproduce sexually within the host.

14. The researchers who observed that magnetite was produced by bacterial cultures from the deep subsurface carried out a follow-up experiment. These biologists treated some of the cultures with a drug that poisons the enzymes involved in electron transport chains. In cultures where the drug was present, no more magnetite was produced. Does this result support or undermine their hypothesis that the bacteria in the cultures perform cellular respiration? Explain your reasoning.

15. *Streptococcus mutans* obtains energy by oxidizing sucrose. This bacterium is abundant in the mouths of Western European and North American children and is a prominent cause of cavities. The organism is virtually absent in children from East Africa, where tooth decay is rare. Propose a hypothesis to explain this observation. Outline the design of a study that would test your hypothesis.

16. Suppose that you've been hired by a firm interested in using bacteria to clean up organic solvents found in toxic waste dumps. Your new employer is particularly interested in finding cells that are capable of breaking a molecule called benzene into less toxic compounds. Where would you go to look for bacteria that can metabolize benzene as an energy or carbon source? How would you design an enrichment culture capable of isolating benzene-metabolizing species?

30 Protists

In this chapter you will learn that

Protists are a large and diverse group that tell us about the evolution of life on Earth

by asking ↓

How and why do biologists study these organisms?
30.1, 30.2

by examining ↓

Major themes in their diversification
30.3

→ **Innovations in morphology: Endosymbiosis theory and evolution of multicellularity**

→ **Innovations in feeding, movement, and reproduction**

then, looking closer at ↓

Key lineages of protists
30.4

The brown alga (giant kelp) shown here lives attached to rocks in ocean waters. This algae and other species featured in this chapter are particularly abundant in the world's oceans.

This chapter introduces the third domain on the tree of life: the **Eukarya.** Eukaryotes range from single-celled organisms that are the size of bacteria to sequoia trees and blue whales. The largest and most morphologically complex organisms on the tree of life—algae, plants, fungi, and animals—are eukaryotes.

Although eukaryotes are astonishingly diverse, they share fundamental features that distinguish them from bacteria and archaea:

● Compared to bacteria and archaea, most eukaryotic cells are large, have many more organelles, and have a much more extensive system of structural proteins called the cytoskeleton.

✔ When you see this checkmark, stop and test yourself. Answers are available in Appendix A.

- The nuclear envelope is a **synapomorphy** that defines the Eukarya. Recall that a synapomorphy is a shared, derived trait that distinguishes major monophyletic groups (see Chapter 28).

- Multicellularity is rare in bacteria and unknown in archaea, but has evolved multiple times in eukaryotes.

- Bacteria and archaea reproduce asexually by fission; many eukaryotes reproduce asexually via mitosis and cell division.

- Many eukaryotes undergo meiosis and reproduce sexually.

One of this chapter's fundamental goals is to explore how these morphological innovations—features like organelles and the nuclear envelope—evolved. Another goal is to analyze how morphological innovations allowed eukaryotes to pursue novel ways of performing basic life tasks such as feeding, moving, and reproducing.

In introducing the Eukarya, this chapter focuses on a diverse collection of lineages known as the protists. The term **protist** refers to all eukaryotes that are not land plants, fungi, or animals. Protist lineages are colored orange in **FIGURE 30.1**.

Protists do not make up a monophyletic group. Instead, they refer to a **paraphyletic group**—they represent some, but not all, of the descendants of a single common ancestor. No synapomorphies define the protists. Protists have no trait that is found only in protists and in no other organisms.

By definition, then, the protists are a diverse lot. The common feature among protists is that they tend to live in environments where they are surrounded by water most of the time (**FIGURE 30.2**). Most plants, fungi, and animals are terrestrial, but protists are found in wet soils and aquatic habitats or inside the bodies of other organisms—including, perhaps, you.

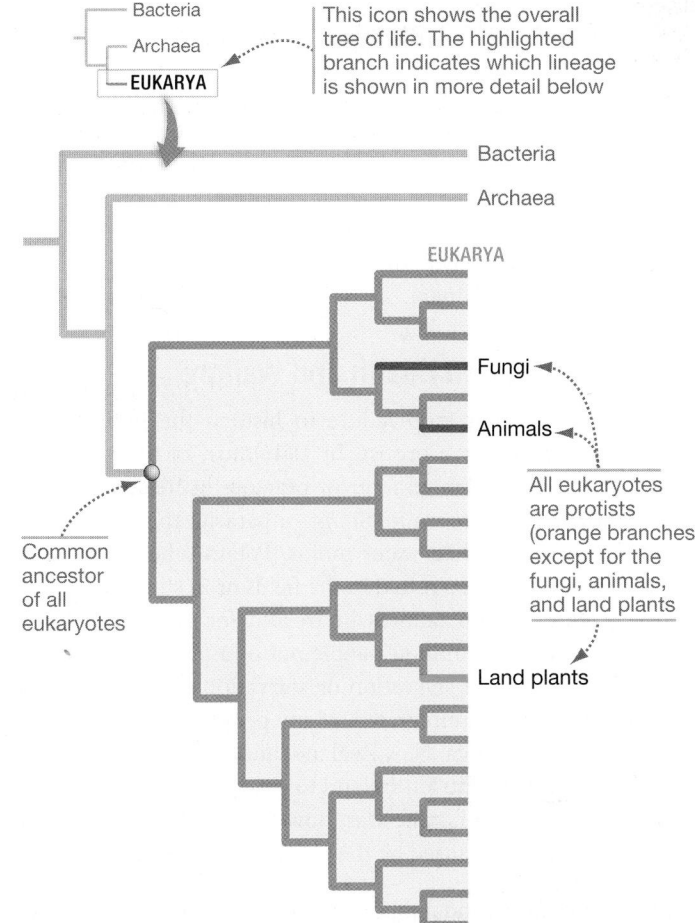

FIGURE 30.1 Protists Are Paraphyletic. The protists include some, but not all, descendants of a single common ancestor.

Open ocean: Surface waters teem with microscopic protists, such as these diatoms.

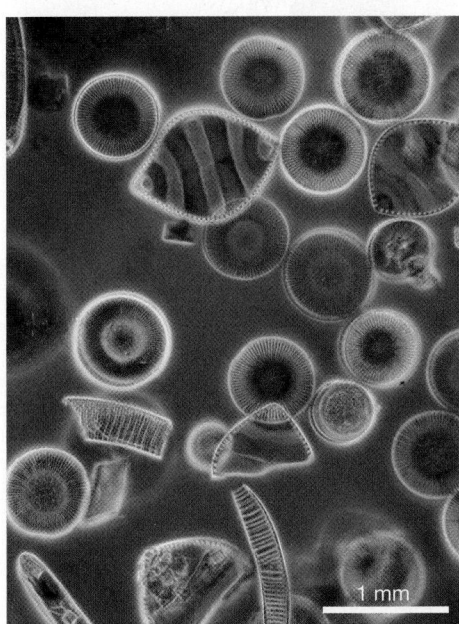

1 mm

Shallow coastal waters: Gigantic protists, such as these kelp, form underwater forests.

Intertidal habitats: Protists such as these sea palms are particularly abundant in tidal habitats.

FIGURE 30.2 Protists Are Particularly Abundant in Aquatic Environments.

30.1 Why Do Biologists Study Protists?

Biologists study protists for three reasons, in addition to their intrinsic interest: (1) they are important medically, (2) they are important ecologically, and (3) they are critical to understanding the evolution of plants, fungi, and animals. The rest of the chapter focuses on why protists are interesting in their own right and how they evolved. Let's first consider their impact on human health and the environment.

Impacts on Human Health and Welfare

The most spectacular crop failure in history, the Irish potato famine, was caused by a protist. In 1845 most of the 3 million acres that had been planted to grow potatoes in Ireland became infested with *Phytophthora infestans*—a parasite that belongs to a group of protists called water molds. Potato tubers that were infected with *P. infestans* rotted in the fields or in storage.

As a result of crop failures in Ireland for two consecutive years, an estimated 1 million people out of a population of less than 9 million died of starvation or starvation-related illnesses. Several million others emigrated. Many people of Irish heritage living in North America, New Zealand, and Australia trace their ancestry to relatives who left Ireland to evade the famine. As devastating as the potato famine was, however, it does not begin to approach the misery caused by the protist *Plasmodium*.

Malaria Physicians and public health officials point to three major infectious diseases that are currently afflicting large numbers of people worldwide: tuberculosis, HIV, and malaria. Tuberculosis is caused by a bacterium (see Chapter 25); HIV is caused by a virus (see Chapter 36). Malaria is caused by a protist—specifically, by apicomplexan species in the eukaryotic lineage Alveolata.

Malaria ranks as one of the world's worst infectious diseases. In India alone, over 30 million people each year suffer from debilitating fevers caused by malaria. At least 300 million people worldwide are sickened by it each year, and nearly 1 million die from the disease annually. The toll is equivalent to eight 747s, loaded with passengers, crashing every day. Most of the dead are children of preschool age.

Five species of the protist *Plasmodium* are capable of parasitizing humans. Infections start when *Plasmodium* cells enter a person's bloodstream during a mosquito bite. As **FIGURE 30.3** shows, *Plasmodium* initially infects liver cells; later, some of the *Plasmodium* cells change into a distinctive cell type that infects the host's red blood cells. The *Plasmodium* cells multiply asexually inside the host cells, which are killed as parasite cells exit to infect additional liver cells or red blood cells. Large numbers of *Plasmodium* cells in a human host's blood increase the chance that a mosquito's blood meal will contain some of them.

After blood from an infected human is transferred to a mosquito during a bite, *Plasmodium* cells differentiate to form gametes. Inside the mosquito, gametes fuse to form a diploid zygote, which undergoes meiosis. Sexual reproduction in the mosquito is beneficial to the parasite because it generates genetic diversity to help evade the host's immune system. The haploid cells that result from meiosis can infect a human when the mosquito bites again.

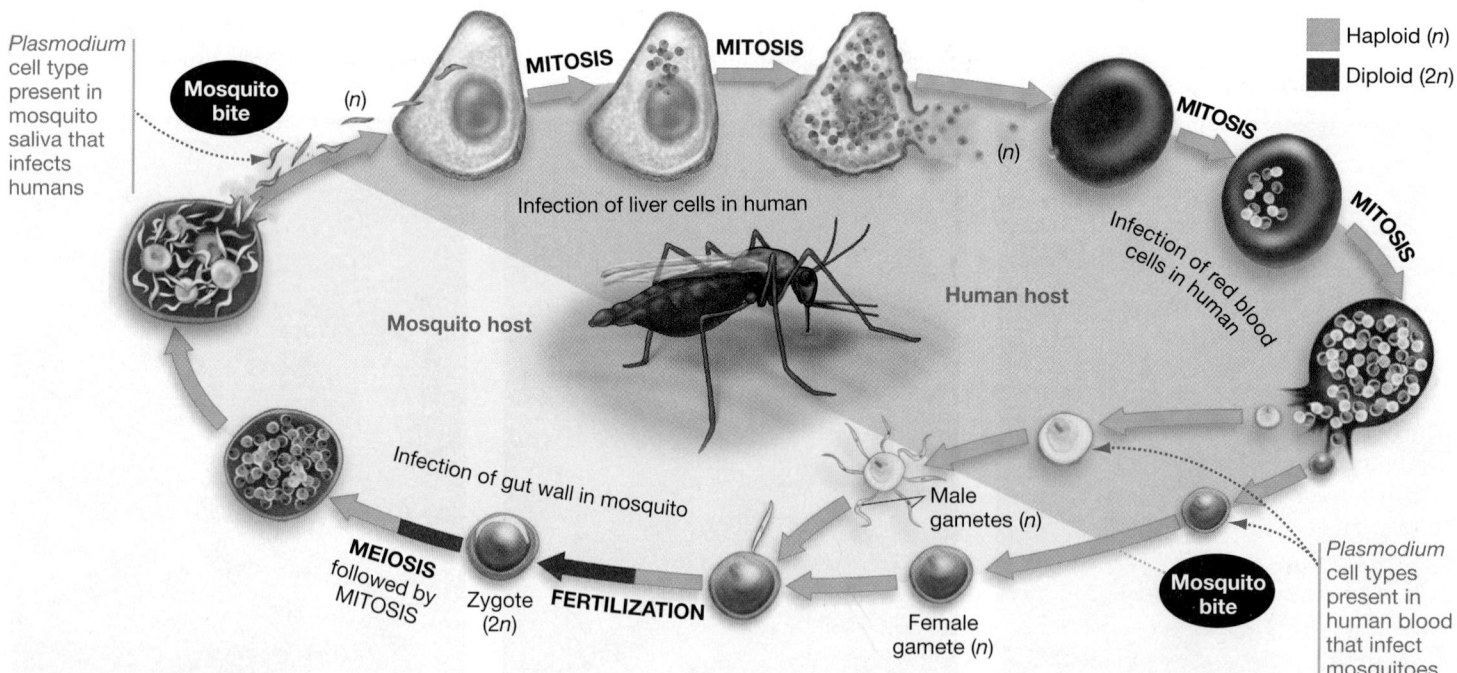

FIGURE 30.3 *Plasmodium* Lives in Mosquitoes and in Humans, Where It Causes Malaria. Over the course of its life cycle, *Plasmodium falciparum* alternates between a mosquito host where sexual reproduction takes place, and a human host where asexual reproduction takes place. In humans, it infects and kills liver cells and red blood cells, contributing to anemia and high fever. In mosquitoes, the protist lives in the gut and salivary glands.

TABLE 30.1 Some Human Health Problems Caused by Protists

Species	Disease
Five species of *Plasmodium,* primarily *P. falciparum* and *P. vivax*	Malaria has the potential to affect 40 percent of the world's total population.
Toxoplasma gondii	Toxoplasmosis may cause eye and brain damage in infants and in AIDS patients.
Many species of dinoflagellates	Toxins released during harmful algal blooms accumulate in clams and mussels and poison people if eaten.
Many species of *Giardia*	Diarrhea due to giardiasis (beaver fever) can last for several weeks.
Trichomonas vaginalis	Trichomoniasis is a reproductive tract infection and one of the most common sexually transmitted diseases. About 2 million young women are infected in the United States each year; some of them become infertile.
Several species of *Leishmania*	Leishmaniasis can cause skin sores or affect internal organs—particularly the spleen and liver.
Trypanosoma gambiense and *T. rhodesiense*	Trypanosomiasis ("sleeping sickness") is a potentially fatal disease transmitted through bites from tsetse flies. Occurs in Africa.
Trypanosoma cruzi	Chagas disease affects 16–18 million people and causes 50,000 deaths annually, primarily in South and Central America.
Entamoeba histolytica	Amoebic dysentery results from severe infections.
Phytophthora infestans	An outbreak of this protist wiped out potato crops in Ireland in 1845–1847, causing famine.

Although *Plasmodium* is arguably the best studied of all protists, researchers have still not been able to devise effective and sustainable measures to control it.

- *Plasmodium* has evolved resistance to some of the drugs used to control its growth in infected people.

- Efforts to develop a vaccine against *Plasmodium* have been fruitless to date, in part because the parasite evolves so quickly. Flu virus and HIV pose similar problems (see Chapter 51).

- Natural selection has favored mosquito strains that are resistant to the insecticides that have been sprayed in their breeding habitats in attempts to control malaria's spread.

Unfortunately, malaria is not the only important human disease caused by protists. **TABLE 30.1** lists some protists that cause human suffering and economic losses.

Harmful Algal Blooms When a unicellular species experiences rapid population growth and reaches high densities in an aquatic environment, it is said to "bloom." Unfortunately, a handful of the many protist species involved in blooms can be harmful.

Harmful algal blooms are usually due to photosynthetic protists called dinoflagellates. Certain dinoflagellates synthesize toxins to protect themselves from predation by small animals called copepods. Because these organisms sometimes have high concentrations of accessory pigments called xanthophylls, their blooms can sometimes discolor seawater (**FIGURE 30.4**).

Algal blooms can be harmful to people because clams and other shellfish filter photosynthetic protists out of the water as food. During a bloom, high levels of toxins can build up in the flesh of these shellfish. Typically, the shellfish themselves are not harmed. But if a person eats contaminated shellfish, several types of poisoning can result.

Paralytic shellfish poisoning, for example, occurs when people eat shellfish that have fed heavily on protists that synthesize poisons called saxitoxins. Saxitoxins block ion channels that have to open for electrical signals to travel through nerve cells

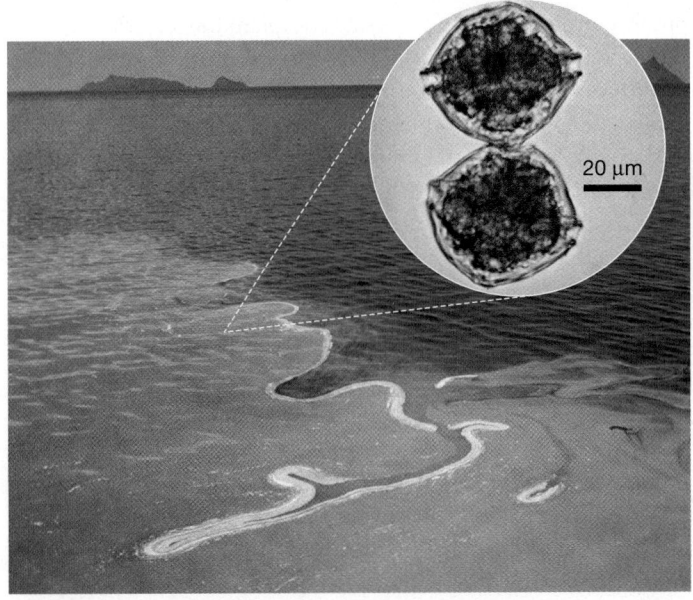

FIGURE 30.4 Harmful Algal Blooms Are Caused by Dinoflagellates.

(see Chapter 46). In humans, high dosages of saxitoxins cause unpleasant symptoms such as prickling sensations in the mouth or even life-threatening symptoms such as muscle weakness and paralysis.

But like the bacteria and archaea, most species of protists are not harmful. Let's have a look at some protists that are essential to healthy ecosystems.

Ecological Importance of Protists

As a whole, the protists represent just 10 percent of the total number of named eukaryote species. Although the number of named species of protists is relatively low, their abundance is extraordinarily high. The number of individual protists found in some habitats is astonishing.

- One milliliter of pond water can contain well over 500 single-celled protists that swim with the aid of flagella.
- Under certain conditions, dinoflagellates can reach concentrations of 60 thousand cells per milliliter of seawater.

Why is this important?

Protists Play a Key Role in Aquatic Food Chains   Photosynthetic protists take in carbon dioxide from the atmosphere and reduce, or "fix," it to form sugars or other organic compounds with high potential energy (see Chapter 10). Photosynthesis transforms some of the energy in sunlight into chemical energy that organisms can use to grow and produce offspring.

Species that produce chemical energy in this way are called **primary producers.** Diatoms, in the Stramenopila lineage, are photosynthetic protists that rank among the leading primary producers in the oceans simply because they are so abundant. Primary production in the ocean represents almost half of the total carbon dioxide that is fixed on Earth.

Diatoms and other organisms that drift in the open oceans or lakes are called **plankton.** The sugars and other organic compounds produced by photosynthetic plankton are the basis of food chains in freshwater and marine environments.

A **food chain** describes nutritional relationships among organisms, and thus how chemical energy flows within ecosystems. In this case, photosynthetic protists and other primary producers are eaten by primary consumers, many of which are protists. Primary consumers are eaten by fish, shellfish, and other secondary consumers, which in turn are eaten by tertiary consumers—whales, squid, and large fish (such as tuna). Many of the species at the base of food chains in aquatic environments are protists.

Could Protists Help Limit Global Climate Change?   Carbon dioxide levels in the atmosphere are increasing rapidly because humans are burning fossil fuels and forests (see Chapter 56). Carbon dioxide traps heat that is radiating from Earth, so high CO_2 levels in the atmosphere contribute to the rise in temperatures associated with global climate change—an issue that many observers consider today's most pressing environmental problem.

The carbon atoms in carbon dioxide molecules move to organisms on land or in the oceans and then back to the atmosphere, in what researchers call the **global carbon cycle.** To restrict global climate change, researchers are trying to figure out ways to decrease carbon dioxide concentrations in the atmosphere and increase the amount of carbon stored in terrestrial and marine environments.

To understand how this might be done, consider the marine carbon cycle diagrammed in **FIGURE 30.5**. The cycle starts when CO_2 from the atmosphere dissolves in water and is taken up by primary producers—photosynthetic plankton called phytoplankton—and converted to organic matter. The plankton are eaten by primary consumers, die and are consumed by decomposers or scavengers, or die and sink to the bottom of the ocean. There they may enter one of two long-lived repositories:

1. *Sedimentary rocks*   Several lineages of protists have shells made of calcium carbonate ($CaCO_3$). When these shells rain down from the ocean surface and settle in layers at the bottom, the deposits that result are compacted by the weight of the water and by sediments accumulating above them. Eventually the deposits turn into rock. The limestone used to build the pyramids of Egypt consists largely of protist shells.

2. *Petroleum*   Although the process of petroleum (oil) formation is not well understood, it begins with accumulations of dead bacteria, archaea, and protists at the bottom of the ocean.

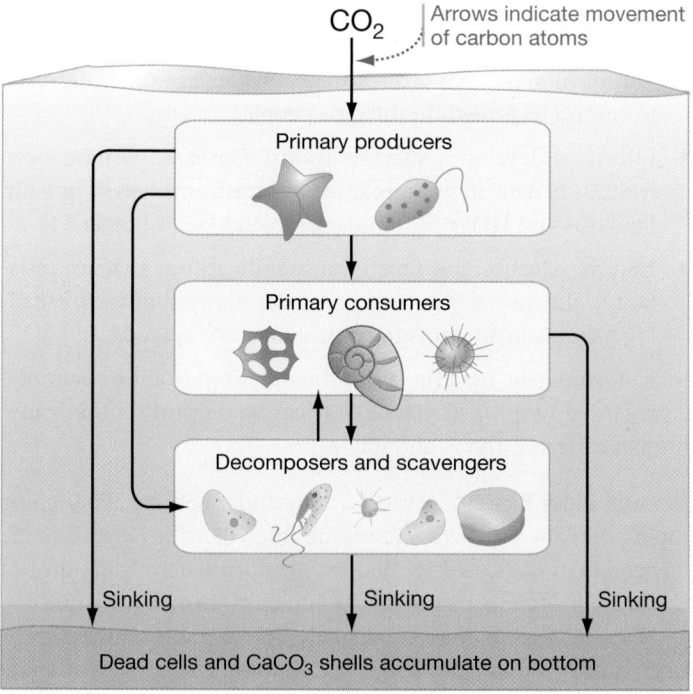

FIGURE 30.5 Protists Play a Key Role in the Marine Carbon Cycle. At the surface, carbon atoms tend to shuttle quickly among organisms. But if carbon atoms sink to the bottom of the ocean in the form of shells or dead cells, they may be locked up for long periods in carbon sinks. (The bottom of the ocean may be miles below the surface.)

Recent experiments have shown that the carbon cycle speeds up when habitats in the middle of the ocean are fertilized with iron. Iron is a critical component of the electron transport chains responsible for photosynthesis and respiration, but it is in particularly short supply in the open ocean. After iron is added to ocean waters, it is not uncommon to see populations of protists and other primary producers increase by a factor of 10.

Some researchers hypothesize that when these blooms occur, the amount of carbon that rains down into carbon sinks in the form of shells and dead cells may increase. If so, then fertilizing the ocean to promote blooms might be an effective way to reduce CO_2 concentrations in the atmosphere.

The effectiveness of iron fertilization is hotly debated, however. Fertilizing the ocean with iron might lead to large accumulations of dead organic matter and the formation of anaerobic dead zones (see Chapter 29). But if further research shows that iron fertilization is safe and effective, it could be added to the list of approaches to limit global climate change.

check your understanding

If you understand that . . .

- Malaria is caused by a protist that lives in mosquitoes and in humans in different parts of its life cycle.
- Harmful algal blooms are caused by protists that produce a toxin as a defense against predation.
- Protists are key primary producers in aquatic environments. As a result, they play a key role in the global carbon cycle.

✔ You should be able to . . .

1. Explain why public health workers are promoting the use of insecticide-treated sleeping nets as a way of reducing malaria.

2. Make a flowchart showing the chain of events that would start with massive iron fertilization and end with large deposits of carbon-containing compounds on the ocean floor.

Answers are available in Appendix A.

30.2 How Do Biologists Study Protists?

Although biologists have made great strides in understanding pathogenic protists and the role that protists play in the global carbon cycle, they have found it extremely difficult to gain any sort of solid insight into how the group as a whole diversified over time. The problem is that the eukaryotic lineages split over a billion years ago and diverged so much that it is not easy to find any overall patterns in the evolution of the group.

Recently, researchers have made progress in understanding protist diversity by combining data on the morphology of key groups with phylogenetic analyses of DNA sequences. However, despite the progress, significant questions remain about how the Eukarya diversified. Let's analyze how this work is being done, beginning with classical results on the morphological traits that distinguish major eukaryote groups.

Microscopy: Studying Cell Structure

Using light microscopy (see **BioSkills 11** in Appendix B), biologists were able to identify and name many of the protist species known today. For example, the early microscopist Anton van Leeuwenhoek identified the parasite *Giardia intestinalis* by examining samples of his own feces.

When transmission electron microscopes became available, a major breakthrough in understanding protist diversity occurred: Detailed studies revealed that protists could be grouped according to overall cell structure, according to organelles with distinctive features, or both.

For example, both light and electron microscopy confirmed that the species that caused the Irish potato famine has reproductive cells with an unusual type of **flagellum.** Flagella are organelles that project from the cell and whip back and forth to produce swimming movements (see Chapter 7). In reproductive cells of this species, one of the two flagella present has tiny, hollow, hairlike projections. Biologists noted that kelp and other species of brown algae also have cells with this type of flagellum.

To make sense of these results, researchers interpreted these types of distinctive morphological features as synapomorphies. Species that have a flagellum with hollow, hairlike projections became known as stramenopiles (literally, "straw-hairs"); the hairs typically have branches at the tip (**FIGURE 30.6**).

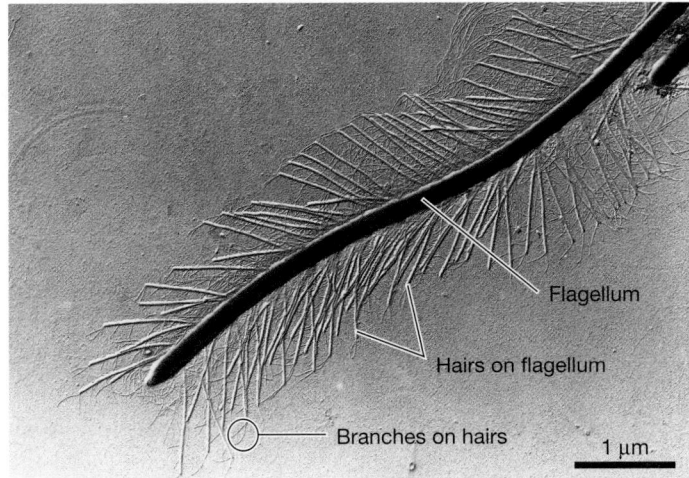

FIGURE 30.6 Species in the Lineage Called Stramenopila Have a Distinctive Flagellum. The unusual, hollow "hairs" that decorate the flagella of stramenopiles often have three branches at the tip.

Lineage	Distinguishing Morphological Features (synapomorphies)
■ Amoebozoa	Cells lack cell walls. When portions of the cell extend outward to move the cell, they form large lobes.
■ Opisthokonta	Reproductive cells have a single flagellum at their base. The cristae inside mitochondria are flat, not tube-shaped as in other eukaryotes. (This lineage includes protists as well as the fungi and the animals. Fungi and animals are discussed separately (see Chapters 32 through 35).
■ Excavata	Most cells have a pronounced "feeding groove" where prey or organic debris is ingested. Most species lack typical mitochondria, although genes derived from mitochondria are found in the nucleus.
■ Plantae	Cells have chloroplasts with a double membrane.
■ Rhizaria	Cells lack cell walls, although some produce an elaborate shell-like covering. When portions of the cell extend outward to move the cell, they are slender in shape.
■ Alveolata	Cells have sac-like structures called alveoli that form a continuous layer just under the plasma membrane. Alveoli are thought to provide support.
■ Stramenopila	If flagella are present, cells usually have two—one of which is covered with hairlike projections.

In recognizing the Stramenopila lineage, investigators hypothesized that because an ancestor had evolved a distinctive flagellum, all or most of its descendants also had this trait. The qualifier *most* is important, because it is not unusual for certain subgroups within a lineage to lose particular traits over the course of evolution, much as humans are gradually losing fur and tailbones.

Eventually, seven major groups of eukaryotes came to be identified on the basis of diagnostic morphological characteristics. These groups and the synapomorphies that identify them are listed in TABLE 30.2. Notice that in almost every case, the synapomorphies that define eukaryotic lineages are changes in structures that protect or support the cell or that influence the organism's ability to move or feed. The plants, fungi, and animals represent subgroups within two of the seven major eukaryotic lineages (see Chapters 31 through 35).

Evaluating Molecular Phylogenies

When researchers began using DNA sequence data to estimate the evolutionary relationships among eukaryotes, the analyses suggested that the seven groups identified on the basis of distinctive morphological characteristics were indeed monophyletic. This was important support for the hypothesis that the distinctive morphological features were shared, derived characters that existed in a common ancestor of each lineage.

The phylogenetic tree in FIGURE 30.7 is the current best estimate of the eukaryotes' evolutionary history. As you read this tree, note that:

- The Amoebozoa and the Opisthokonta—which include fungi and animals—form a monophyletic group called the Unikonta.

- The other five major eukaryotic lineages form a monophyletic group called the Bikonta.

Understanding where the root or base of the tree of Eukarya lies has been more problematic. Finding it will help researchers understand what the common ancestor looked like and how eukaryotic lineages evolved over time. One hypothesis suggested that the first eukaryotic split was between unikonts and bikonts—meaning, between eukaryotes that have one flagellum and those with two flagella. This hypothesis is tentative and controversial, however, given the data analyzed to date. Researchers continue to work on the issue of placing the root of the Eukarya.

Discovering New Lineages via Direct Sequencing

The effort to refine the phylogeny of the Eukarya is ongoing. But of all the research frontiers in eukaryotic diversity, the most exciting may be the one based on the technique called direct sequencing.

Direct sequencing is based on sampling soil or water, analyzing the DNA sequence of specific genes in the sample, and using the data to place the organisms in the sample on a phylogenetic tree. Direct sequencing and metagenomics have led to the discovery of previously unknown but major lineages of Archaea (see Chapter 29). To the amazement of biologists all over the world, the same thing happened when researchers used direct sequencing to survey eukaryotes.

The first direct sequencing studies that focused on eukaryotes were published in 2001. One study sampled organisms at depths from 250 to 3000 meters (m) below the surface in waters off Antarctica; another focused on cells at depths of 75 m in the Pacific Ocean, near the equator. Both studies detected a wide array of species that were new to science.

Investigators who examined the samples microscopically were astonished to find that many of the newly discovered eukaryotes were tiny—from 0.2 micrometers (μm) to 5 μm in diameter. Subsequent work has confirmed the existence of many,

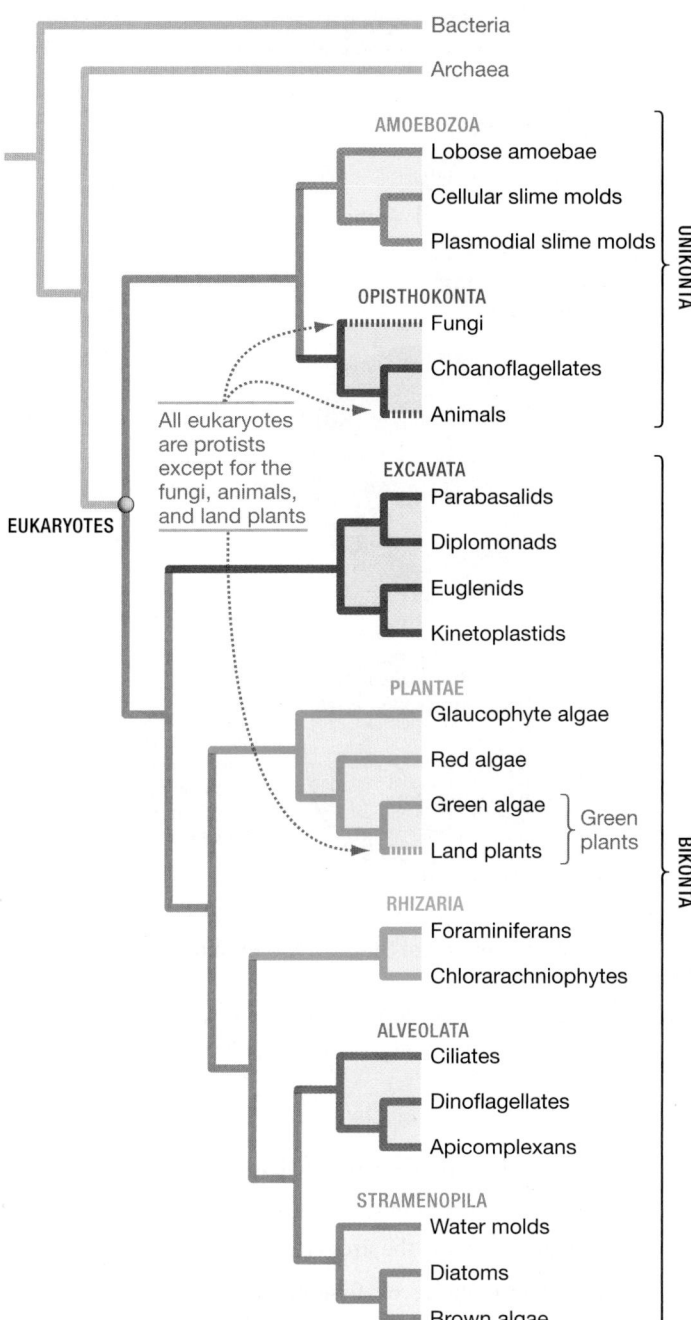

FIGURE 30.7 Phylogenetic Analyses Have Identified Seven Major Lineages of Eukaryotes. This tree shows selected subgroups from the seven major lineages discussed in this chapter. Kinetoplastids are not discussed in detail elsewhere in the chapter, but they include the pathogens *Trypanosoma* and *Leishmania* described in Table 30.1.

diverse species of protists that are less than 0.2 μm in diameter. These cells overlap in size with bacteria, which typically range from 0.5 μm to 2 μm in diameter.

Eukaryotic cells are much more variable in size than previously imagined. A whole new world of tiny protists has only recently been discovered.

check your understanding

If you understand that . . .

- According to the most recent analyses, the domain Eukarya comprises seven major lineages. Members of each lineage have distinctive aspects of cell structure.
- Direct sequencing has allowed investigators to recognize large numbers of previously undescribed eukaryotes, some of which are extremely small.

✔ **You should be able to . . .**

1. Explain why opisthokonts, alveolates, and stramenopiles got their names. (The root *opistho* refers to the back of a cell; the root *kont* refers to a flagellum.)
2. Explain why direct sequencing studies allow researchers to characterize species that have never been seen before.

Answers are available in Appendix A.

30.3 What Themes Occur in the Diversification of Protists?

The protists range in size from bacteria-sized single cells to giant kelp. They live in habitats from the open oceans to the guts of termites. They are almost bewildering in their morphological and ecological diversity. Because they are a paraphyletic group, they do not share derived characteristics that set them apart from all other lineages on the tree of life.

One general theme that can help tie protists together is their amazing diversity. Once an important new innovation arose in protists, it triggered the evolution of species that live in a wide array of habitats and make a living in diverse ways.

What Morphological Innovations Evolved in Protists?

Virtually all bacteria and all archaea are unicellular. Given the distribution of multicellularity in eukaryotes, it is logical to conclude that the first eukaryote was also a single-celled organism.

Further, all eukaryotes alive today have (1) either mitochondria or genes that are normally found in mitochondria, (2) a nucleus and endomembrane system, and (3) a cytoskeleton. Based on the distribution of cell walls in living eukaryotes, it is likely that the first eukaryotes lacked this feature.

Using these observations, biologists hypothesize that the earliest eukaryotes were probably single-celled organisms with mitochondria, a nucleus and endomembrane system, and a cytoskeleton, but no cell wall.

It is also likely that the first eukaryotic cells swam using a novel type of flagellum. Eukaryotic flagella are completely different structures from bacterial flagella and evolved independently.

The eukaryotic flagellum is made up of microtubules, and dynein is the major motor protein. An undulating motion occurs as dynein molecules walk down microtubules (see Chapter 7). The flagella of bacteria and archaea, in contrast, are composed primarily of a protein called flagellin. Instead of undulating, prokaryotic flagella rotate to produce movement.

✔ If you understand the synapomorphies that identify the eukaryotes as a monophyletic group, you should be able to map the origin of the nuclear envelope and the eukaryotic flagellum on Figure 30.7. Once you've done that, let's consider how several of these key new morphological features arose and influenced the subsequent diversification of protists, beginning with one of the traits that define the Eukarya.

Endosymbiosis and the Origin of the Mitochondrion

Mitochondria are organelles that generate adenosine triphosphate (ATP) using pyruvate as an electron donor and oxygen as the ultimate electron acceptor (see Chapter 9).

In 1981 evolutionary biologist Lynn Margulis expanded on a radical hypothesis—first proposed in the nineteenth century—to explain the origin of mitochondria. The **endosymbiosis theory** proposes that mitochondria originated when a bacterial cell took up residence inside another cell about 2 billion years ago.

The theory's name was inspired by the Greek word roots *endo*, *sym*, and *bio* (literally, "inside-together-living"). **Symbiosis** is said to occur when individuals of two different species live in physical contact; **endosymbiosis** occurs when an organism of one species lives inside the cells of an organism of another species.

How and when this momentous event occurred is hotly debated. Some scientists think a primitive amoeba-like eukaryote engulfed a bacterium and simply failed to digest it in its lysosome, allowing it to take up residence. Evidence that was used to support this theory includes the similar feeding habits of today's amoeba, and the existence of some protist species that seem to lack mitochondria.

Recent evidence, however, indicates that all protists originally had mitochondria, and some lost them. A new idea gaining support is that the first eukaryote may have been formed as a result of an endosymbiosis between two prokaryotes—an archaeal host and a bacterium (**FIGURE 30.8**). After this chance union occurred, cells developed nuclei and became much larger. Both of these changes appear to have been triggered by the bacterial invader.

The relationship between the archaeal host and the engulfed bacterial cell was presumed to be stable because a mutual advantage existed between them: The host supplied the bacterium with protection and carbon compounds from its prey, while the bacterium produced much more ATP than the host cell could synthesize on its own.

When Margulis first began promoting the endosymbiosis theory, it met with a storm of criticism—largely because it seemed slightly preposterous. But gradually biologists began to examine it rigorously. For example, endosymbiotic relationships between protists and bacteria exist today. Among the α-proteobacteria alone (see Chapter 29), three major groups are found *only* inside eukaryotic cells. In many cases, the bacterial cells are transmitted to offspring in eggs or sperm and are required for survival.

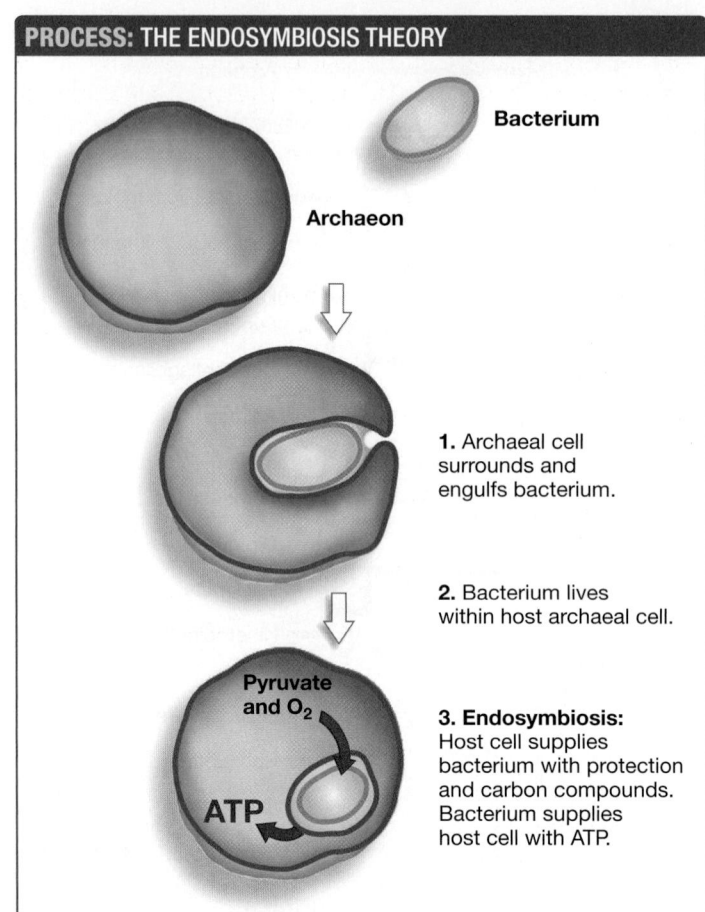

PROCESS: THE ENDOSYMBIOSIS THEORY

Bacterium

Archaeon

1. Archaeal cell surrounds and engulfs bacterium.

2. Bacterium lives within host archaeal cell.

Pyruvate and O_2

ATP

3. Endosymbiosis: Host cell supplies bacterium with protection and carbon compounds. Bacterium supplies host cell with ATP.

FIGURE 30.8 Proposed Initial Steps in the Evolution of the Mitochondrion.

✔**QUESTION** According to this hypothesis, how many membranes should surround a mitochondrion? Explain your logic.

Several observations about the structure of mitochondria are also consistent with the endosymbiosis theory:

- Mitochondria are about the size of an average α-proteobacterium.

- Mitochondria replicate by fission, as do bacterial cells. The duplication of mitochondria takes place independently of division by the host cell. When eukaryotic cells divide, each daughter cell receives some of the many mitochondria present.

- Mitochondria have their own ribosomes and manufacture some of their own proteins. Mitochondrial ribosomes closely resemble bacterial ribosomes in size and composition and are poisoned by antibiotics such as streptomycin that inhibit bacterial, but not eukaryotic, ribosomes.

- Mitochondria have double membranes, consistent with the engulfing mechanism of origin illustrated in Figure 30.8.

- Mitochondria have their own genomes, which are organized as circular molecules—much like a bacterial chromosome. Mitochondrial genes code for a few of the proteins needed to conduct electron transport and RNAs needed to translate the mitochondrial genome.

Although these data are impressive, they are only consistent with the endosymbiosis theory. Stated another way, they do not exclude other explanations. This is a general principle in science: Evidence is considered strong when it cannot be explained by reasonable alternative hypotheses. In this case, the key was to find data that tested predictions made by Margulis's idea against predictions made by an alternative theory: that mitochondria evolved within eukaryotic cells, separately from bacteria.

A breakthrough occurred when researchers realized that according to the "within-eukaryotes" theory, the genes found in mitochondria are derived from nuclear genes in ancestral eukaryotes. Margulis's theory, in contrast, proposed that the genes found in mitochondria were bacterial in origin.

These predictions were tested by studies on the phylogenetic relationships of mitochondrial genes. Specifically, researchers compared gene sequences isolated from eukaryotic mitochondrial DNA with sequences of similar genes isolated from eukaryotic nuclear DNA and with DNA from several species of bacteria. Exactly as the endosymbiosis theory predicted, mitochondrial gene sequences are much more closely related to the sequences from the α-proteobacteria than to sequences from the nuclear DNA of eukaryotes.

The result diagrammed in **FIGURE 30.9** was considered overwhelming evidence that the mitochondrial genome came from an α-proteobacterium rather than from a eukaryote. The endosymbiosis theory was the only reasonable explanation for the data.

The results were a stunning vindication of a theory that had once been intensely controversial. Mitochondria evolved via endosymbiosis between an α-proteobacterium and an archaeal host.

One intriguing feature of mitochondrial genomes is that they typically encode less than 50 genes, whereas the genomes of their bacterial cousins encode about 1500 genes. Most of the genes from the endosymbiotic bacterium moved into the nuclear genome in what was one of the most spectacular **lateral gene transfer** events in the history of life (see Chapter 21). Some unnecessary genes were lost, but many still encode proteins that are synthesized by cytosolic ribosomes and imported into mitochondria.

✔ If you understand the endosymbiosis theory and the evidence for it, you should be able to describe how the chloroplast—the organelle that is the site of photosynthesis in eukaryotes—could arise via endosymbiosis.

Now let's turn to another distinguishing feature of eukaryotic cells—the nucleus, with its unique double membrane called the nuclear envelope. The origin of the nuclear envelope is also a subject of debate.

The Nuclear Envelope The leading hypothesis to explain the origin of the nuclear envelope is based on infoldings of the plasma membrane. As the drawings in **FIGURE 30.10** show, a stepwise process could give rise to small infoldings that were elaborated

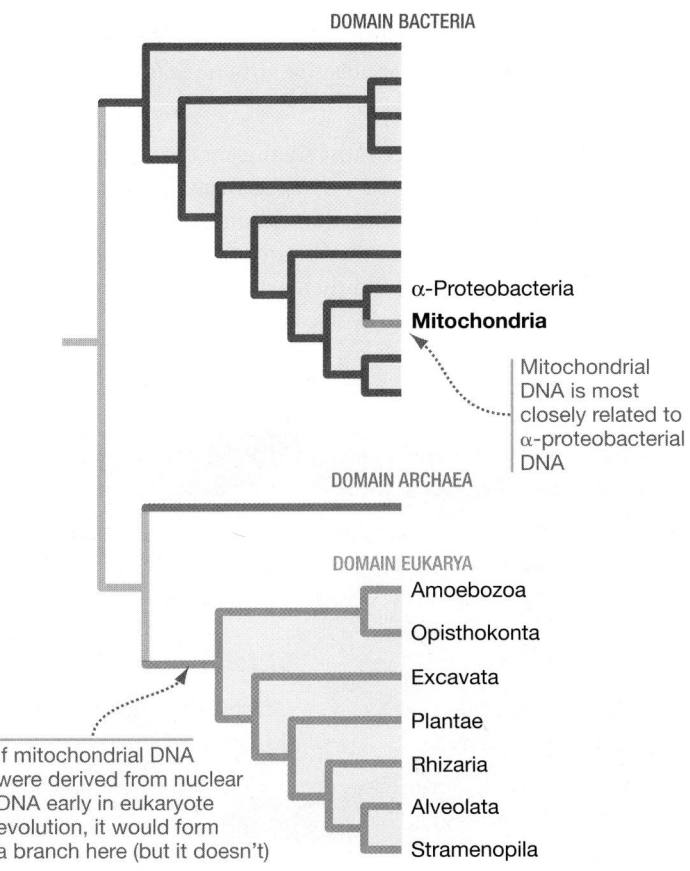

FIGURE 30.9 Phylogenetic Data Support the Endosymbiosis Theory.

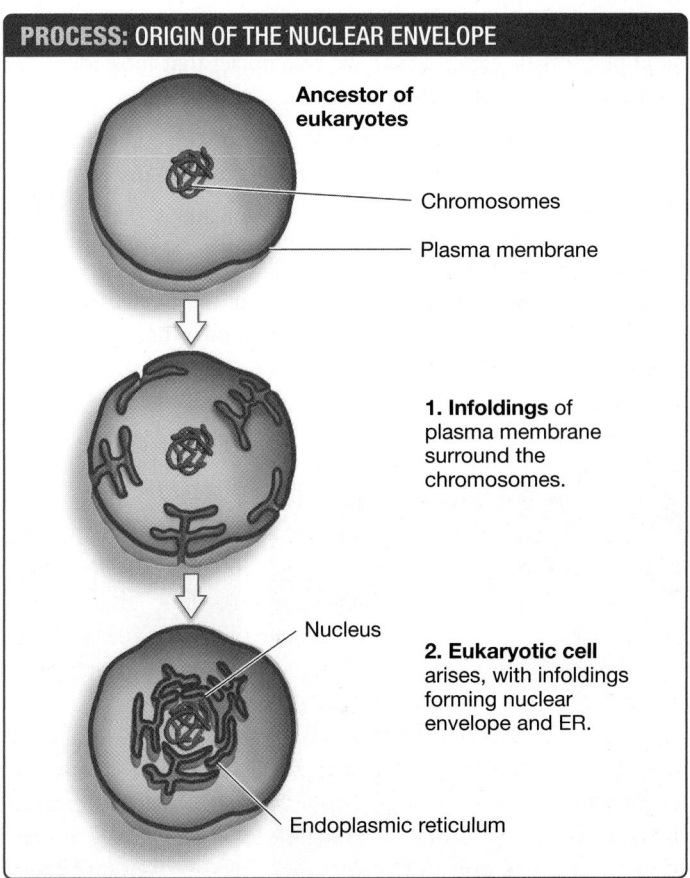

FIGURE 30.10 A Hypothesis for the Origin of the Nuclear Envelope. Infoldings of the plasma membrane, analogous to those shown here, have been observed in bacteria living today.

by mutation and natural selection over time, and the infolding could eventually become detached from the plasma membrane. Note that the infoldings would have given rise to the nuclear envelope and the endoplasmic reticulum (ER) together.

Two lines of evidence support this hypothesis: Infoldings of the plasma membrane occur in some bacteria living today, and the nuclear envelope and ER of today's eukaryotes are continuous (see Chapter 7). ✔ **If you understand the infolding hypothesis, you should be able to explain why these observations support it.**

According to current thinking, the evolution of the nuclear envelope was advantageous because it separated transcription and translation. Recall that RNA transcripts are processed inside the nucleus but translated outside the nucleus (see Chapter 17). In bacteria and archaea, transcription and translation occur together.

With a simple nuclear envelope in place, alternative splicing and other forms of RNA processing could occur—giving the early eukaryotes a novel way to control gene expression (see Chapter 19). The take-home message here is that an important morphological innovation gave the early eukaryotes a new way to manage and process genetic information.

Once a nucleus had evolved, it underwent diversification. In some cases, unique types of nuclei are associated with the founding of important lineages of protists.

- Ciliates (Alveolata) have a diploid micronucleus that is involved only in reproduction, and a polyploid macronucleus where transcription occurs.

- Diplomonads (Excavata) have two nuclei that look identical; it is not known how they interact.

- In foraminifera (Rhizaria), red algae (Plantae), and plasmodial slime molds (Amoebozoa), certain cells contain many nuclei.

- Dinoflagellates (Alveolata) have chromosomes that lack histones and attach to the nuclear envelope.

In each case, the distinctive structure of the nucleus is a synapomorphy that allows biologists to recognize these lineages as distinct monophyletic groups.

Structures for Support and Protection   Many protists have cell walls outside their plasma membrane; others have hard external structures called a **shell;** others have rigid structures inside the plasma membrane. In many cases, these novel structures represent synapomorphies that identify monophyletic groups among protists. For example:

- Diatoms (Stramenopila) are surrounded by a glass-like, silicon-dioxide cell wall (**FIGURE 30.11a**). The cell wall is made up of two pieces that fit together in a box-and-lid arrangement, like the petri plates you may have seen in lab.

- Dinoflagellates (Alveolata) have a cell wall made up of cellulose plates (**FIGURE 30.11b**).

- Within the foraminiferans (Rhizaria), some subgroups secrete an intricate, chambered calcium carbonate shell (**FIGURE 30.11c**).

- Members of other foraminiferan subgroups, and some amoebae, cover themselves with tiny pebbles.

- The parabasalids (Excavata) have a unique internal support rod, consisting of cross-linked microtubules that run the length of the cell.

- The euglenids (Excavata) have a collection of protein strips located just under the plasma membrane. The strips are supported by microtubules and stiffen the cell.

- All alveolates (Alveolata) have distinctive sac-like structures called alveoli, located just under the plasma membrane, that help stiffen the cell.

In many cases, the diversification of protists has been associated with the evolution of innovative structures for support and protection.

Multicellularity   One of the most significant changes in the history of life on Earth occurred when organisms containing more than one cell evolved. The mutations leading to **multicellularity** probably first caused cells to simply stick together after cell division. Selection pressures could then act on these larger, colonial organisms, allowing them to evolve and diversify.

(a) Diatom

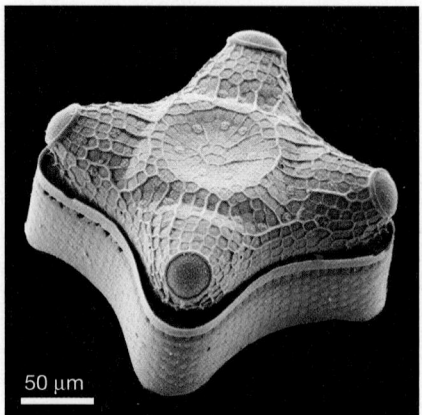

50 µm

Glassy cell wall made of silicon dioxide

(b) Dinoflagellate

10 µm

Tough plates in cell wall made of cellulose

(c) Foraminiferan

50 µm

Chalky, chambered shell made of calcium carbonate

FIGURE 30.11 Hard Outer Coverings in Protists Vary in Composition.

Eventually cells became specialized for different functions. In the simplest multicellular species, certain cells are specialized for producing or obtaining food while other cells are specialized for reproduction. The key point about multicellularity is that not all cells express the same genes.

A few species of bacteria are capable of aggregating and forming structures called fruiting bodies (see Chapter 29). Because cells in the fruiting bodies of these bacteria differentiate into specialized stalk cells and spore-forming cells, they are considered multicellular. But the vast majority of multicellular species are members of the Eukarya.

Multicellularity arose independently in a wide array of eukaryotic lineages, including the green plants, fungi, animals, brown algae, slime molds, and red algae.

To summarize, an array of novel morphological traits played a key role as protists diversified: the mitochondrion, the nucleus and endomembrane system, structures for protection and support, and multicellularity. Evolutionary innovations allowed protists to build and manage the eukaryotic cell in new ways.

Once a new type of eukaryotic cell or multicellular individual existed, subsequent diversification was often triggered by novel ways of finding food, moving, or reproducing. Let's consider each of these life processes in turn.

How Do Protists Obtain Food?

Bacteria and archaea can use a wide array of molecules as electron donors and electron acceptors during cellular respiration (see Chapter 29). Some get these molecules by absorbing them directly from the environment. Other bacteria don't absorb their nutrition—instead, they make their own food via photosynthesis. Many groups of protists are similar to bacteria in the way they find food: They perform photosynthesis or absorb their food directly from the environment.

But one of the most important stories in the diversification of protists was the evolution of a novel method for finding food.

Many protists ingest their food—they eat bacteria, archaea, or even other protists whole. This process is called **phagocytosis** (see Chapter 7). When phagocytosis occurs, an individual takes in packets of food much larger than individual molecules. Thus, protists feed by either (1) ingesting packets of food, (2) absorbing organic molecules directly from the environment, or (3) performing photosynthesis.

Some protists ingest food as well as performing photosynthesis—meaning that they use a combination of feeding strategies. It's also important to recognize that all three lifestyles—ingestive, absorptive, and photosynthetic—can occur within a single lineage.

To drive this last point home, consider the monophyletic group called the alveolates. This lineage has three major subgroups—dinoflagellates, apicomplexans, and ciliates. About half of the dinoflagellates are photosynthetic, but many others are parasitic and absorb nutrients from their hosts. Apicomplexans are parasitic. Ciliates include many species that ingest prey, but some ciliates live in the guts of cattle or the gills of fish. Other ciliate species make a living by holding algae or other types of photosynthetic endosymbionts inside their cells.

The punch line? Within each of the seven major lineages of eukaryotes, different methods for feeding helped trigger diversification.

Ingestive Feeding Ingestive lifestyles are based on eating live or dead organisms or on scavenging loose bits of organic debris. Protists such as the cellular slime mold *Dictyostelium discoideum* are large enough to engulf bacteria and archaea; many protists are large enough to surround and ingest other protists or even microscopic animals.

Feeding by phagocytosis is possible in protists that lack a cell wall. A flexible membrane and dynamic cytoskeleton give these species the ability to surround and "swallow" prey using long, fingerlike projections called **pseudopodia** ("false-feet") (**FIGURE 30.12a**). Phagocytosis was a prerequisite for the endosymbiosis event that led to chloroplasts.

(a) Pseudopodia engulf food.

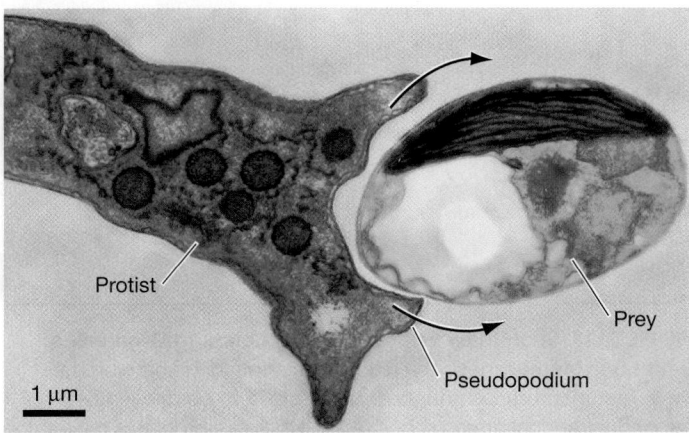

(b) Ciliary currents sweep food into gullet.

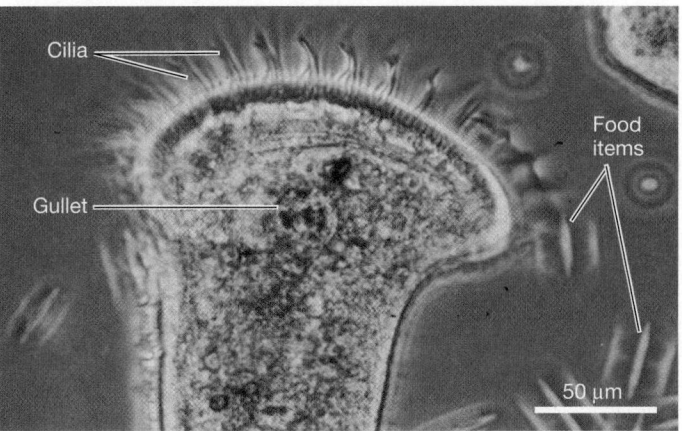

FIGURE 30.12 Ingestive Feeding. Methods of prey capture vary among ingestive protists. **(a)** Some predators engulf prey with pseudopodia; **(b)** other predators sweep them into their gullets with water currents set up by the beating of cilia. Note that the cells in part (a) have been colorized.

Although many ingestive feeders actively hunt down prey and engulf them, others do not. Instead of taking themselves to food, these species attach themselves to a surface. Protists that feed in this way have cilia that surround the mouth and beat in a coordinated way. The motion creates water currents that sweep food particles into the cell (**FIGURE 30.12b**).

Absorptive Feeding Absorptive feeding occurs when nutrients are taken up directly from the environment, across the plasma membrane, usually through transport proteins. Absorptive feeding is common among protists.

Some protists that live by absorptive feeding are **decomposers,** meaning that they feed on dead organic matter, or **detritus.** But many of the protists that absorb their nutrition directly from the environment live inside other organisms. If an absorptive species damages its host, that species is called a **parasite.**

Photosynthesis—Endosymbiosis and the Origin of Chloroplasts

You might recall that photosystems I and II evolved in bacteria, and that both photosystems occur in cyanobacteria (see Chapter 10). None of the basic machinery required for photosynthesis evolved in eukaryotes. Instead, they "stole" it—via endosymbiosis.

The endosymbiosis theory contends that the eukaryotic chloroplast originated when a protist engulfed a cyanobacterium. Once inside the protist, the photosynthetic bacterium provided its eukaryotic host with oxygen and glucose in exchange for protection and access to light. If the endosymbiosis theory is correct, today's chloroplasts trace their ancestry to cyanobacteria.

All of the photosynthetic eukaryotes have both chloroplasts and mitochondria. The evidence for an endosymbiotic origin for the chloroplast is persuasive.

- Chloroplasts have the same list of bacteria-like characteristics presented earlier for mitochondria.

- Many examples of endosymbiotic cyanobacteria are living inside the cells of protists or animals today.

- Chloroplasts contain a circular DNA molecule containing genes that are extremely similar to cyanobacterial genes.

- The photosynthetic organelle of one group of protists, the glaucophyte algae, has an outer layer containing the same constituent (peptidoglycan) found in the cell walls of cyanobacteria.

Like mitochondria, the chloroplast genome is very small compared with genomes of living cyanobacteria; most of the original genes were either lost or transferred to the nucleus.

✔If you understand the evidence for the endosymbiotic origin of the chloroplast, you should be able to add a label indicating the location of chloroplast genes on the phylogenetic tree in Figure 30.7.

Photosynthesis—Primary versus Secondary Endosymbiosis

Which eukaryote originally obtained a photosynthetic organelle? Because all species in the Plantae have chloroplasts with two membranes, biologists infer that the original, or primary,

endosymbiosis occurred in these species' common ancestor. That species eventually gave rise to all subgroups in the Plantae lineage—the glaucophyte algae, red algae, and green plants (green algae and land plants).

But chloroplasts also occur in four of the other major lineages of protists—the Excavata, Rhizaria, Alveolata, and Stramenopila. In these species, the chloroplast is surrounded by more than two membranes—usually four.

To explain this observation, researchers hypothesize that the ancestors of these groups acquired their chloroplasts by ingesting photosynthetic protists that already had chloroplasts. This process, called secondary endosymbiosis, occurs when an organism engulfs a photosynthetic eukaryotic cell and retains its chloroplasts as intracellular symbionts. Secondary endosymbiosis is illustrated in **FIGURE 30.13**.

FIGURE 30.14 shows where primary and secondary endosymbiosis occurred on the phylogenetic tree of eukaryotes. Once

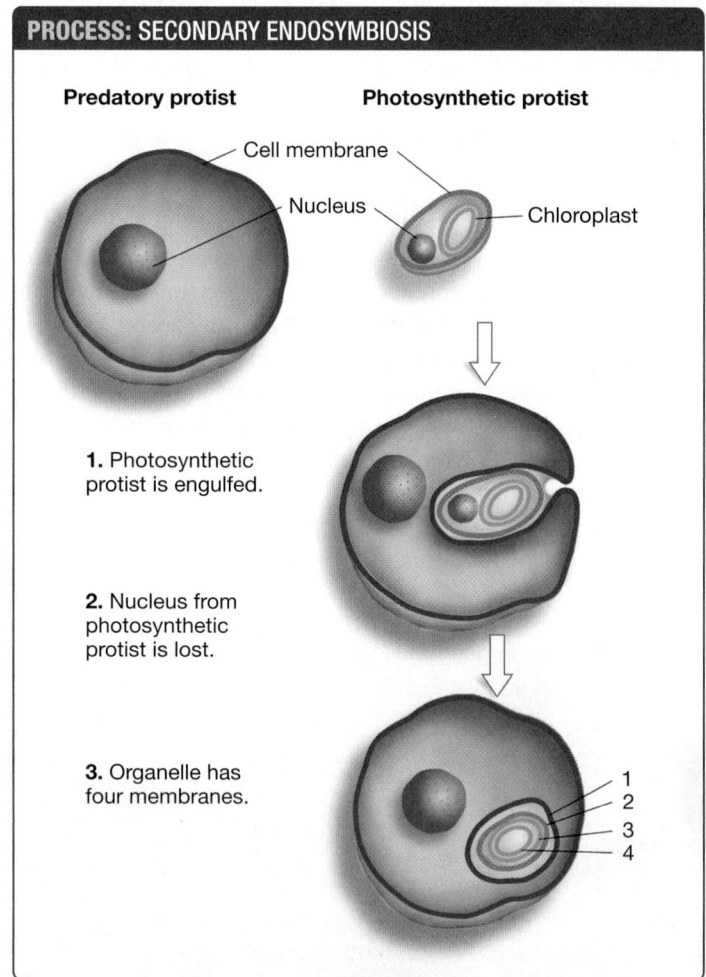

PROCESS: SECONDARY ENDOSYMBIOSIS

Predatory protist   Photosynthetic protist

Cell membrane
Nucleus
Chloroplast

1. Photosynthetic protist is engulfed.

2. Nucleus from photosynthetic protist is lost.

3. Organelle has four membranes.

1
2
3
4

FIGURE 30.13 Secondary Endosymbiosis Leads to Organelles with Four Membranes. The chloroplasts found in some protists have four membranes and are hypothesized to be derived by secondary endosymbiosis. In species where chloroplasts have three membranes, biologists hypothesize that secondary endosymbiosis was followed by the loss of one membrane.

protists obtained the chloroplast, it was "swapped around" to new lineages via secondary endosymbiosis.

✔ If you understand endosymbiosis, you should be able to assess and explain why the primary and secondary endosymbiosis events introduced in this chapter represent the most massive lateral gene transfers in the history of life, in terms of the number of genes moved at once. (To review lateral gene transfer, see Chapter 21.)

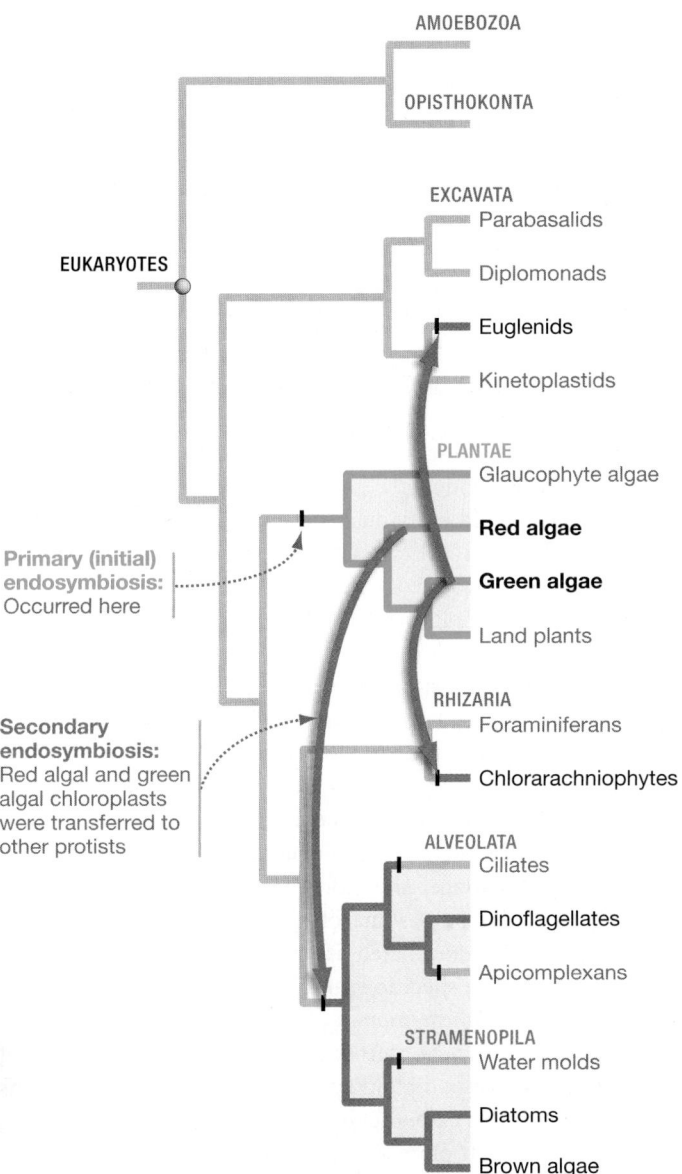

FIGURE 30.14 Photosynthesis Arose in Protists by Primary Endosymbiosis and Then Spread among Lineages via Secondary Endosymbiosis. Biochemical similarities link the chloroplasts found in alveolates and stramenopiles with red algae and the chloroplasts found in euglenids and chlorarachniophytes with green algae. Notice that among the alveolates and stramenopiles, only the dinoflagellates, diatoms, and brown algae have photosynthetic species. In ciliates, apicomplexans, and water molds, the chloroplast has been lost or has changed function.

How Do Protists Move?

Many protists actively move to find food. Predators such as the slime mold *Dictyostelium discoideum* crawl over a substrate in search of prey. Most of the unicellular, photosynthetic species are capable of swimming to sunny locations, though others drift passively in water currents. How are these crawling and swimming movements possible?

Amoeboid motion is a sliding movement observed in some protists. In the classic mode illustrated in **FIGURE 30.15**, pseudopodia stream forward over a substrate, and the rest of the cytoplasm, organelles, and plasma membrane follow. The motion requires ATP and involves interactions between proteins called actin and myosin inside the cytoplasm. The mechanism is related to muscle movement in animals (see Chapter 48). But at the level of the whole cell, the precise sequence of events during amoeboid movement is still uncertain. The issue is attracting researchers' attention, because key immune system cells in humans use amoeboid motion as they hunt down and destroy disease-causing agents.

The other major mode of locomotion in protists involves flagella or cilia (**FIGURE 30.16**, see page 566). Recall that flagella and cilia both consist of nine sets of doublet (paired) microtubules arranged around two central, single microtubules (see Chapter 7). Flagella are long and are usually found alone or in pairs; cilia are short and usually occur in large numbers on any one cell.

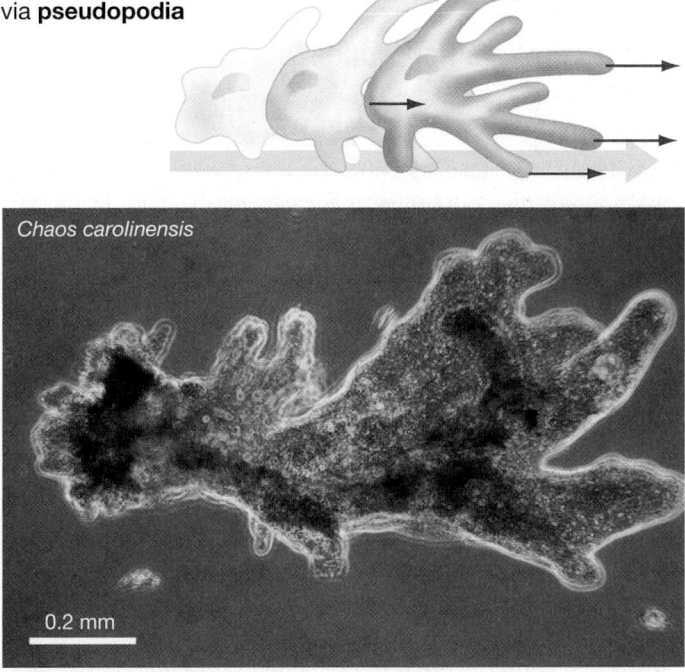

FIGURE 30.15 Amoeboid Motion Is Possible in Species That Lack Cell Walls. In amoeboid motion, long pseudopodia stream out from the cell. The rest of the cytoplasm, organelles, and external membrane follow.

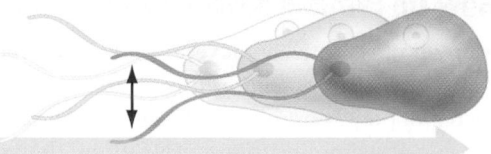

(a) Swimming via flagella

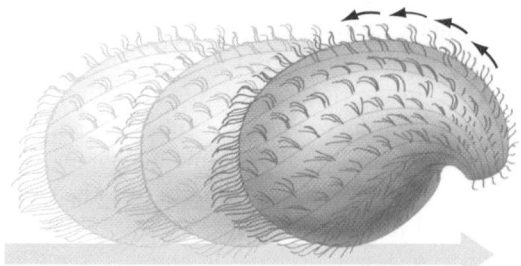

(b) Swimming via cilia

FIGURE 30.16 Many Protists Swim Using Flagella or Cilia.
(a) Flagella are long and few in number. **(b)** Cilia are short and numerous. In many cases they are used in swimming.

Even closely related protists can use radically different forms of locomotion. For example, within the Alveolata, the ciliates swim by beating their cilia, the dinoflagellates swim by whipping their flagella, and mature apicomplexan cells move by amoeboid motion (though their gametes swim via flagella). It is also common to find protists that do not exhibit active movement, but instead float passively in water currents.

Movement is yet another example of the extensive diversification that occurred within each of the seven major monophyletic groups of eukaryotes.

How Do Protists Reproduce?

Sexual reproduction originated in protists. Sexual reproduction can best be understood in contrast to asexual reproduction (see Chapter 13). The key issues are the type of nuclear division involved and the consequences for genetic diversity in the offspring produced.

- Asexual reproduction is based on mitosis and cell division in eukaryotic organisms and on fission in bacteria and archaea. It results in daughter cells that are genetically identical to the parent.

- Sexual reproduction is based on meiosis and fusion of gametes. It results in daughter cells that are genetically different from their parents and from each other.

Most protists undergo asexual reproduction routinely. But sexual reproduction occurs only intermittently in many protists—often at one particular time of year, or when individuals are stressed by overcrowding or scarce food supplies.

The evolution of sexual reproduction ranks among the most significant evolutionary innovations observed in eukaryotes.

Sexual versus Asexual Reproduction The leading hypothesis to explain why meiosis evolved states that genetically variable offspring may be able to thrive if the environment changes. For

example, offspring with genotypes different from those of their parents may be better able to withstand attacks by parasites that successfully attacked their parents (see Chapter 13). This is a key point because many types of parasites, including bacteria and viruses, have short generation times and evolve quickly.

Because the genotypes and phenotypes of parasites are constantly changing, natural selection favors host individuals with new genotypes. The idea is that novel genotypes in offspring generated by meiosis may contain combinations of alleles that allow hosts to withstand attack by new strains of parasites. In short, many biologists view sexual reproduction as an adaptation to fight disease.

✔ If you understand the changing-environment hypothesis for the evolution of sex, you should be able to evaluate and explain whether it is consistent with the observation that many protists undergo meiosis when food is scarce or the population density is high.

Life Cycles—Haploid versus Diploid Dominated

A life cycle describes the sequence of events that occur as individuals grow, mature, and reproduce. The evolution of meiosis introduced a new event in the life cycle of protist species; what's more, it created a distinction between haploid and diploid phases in the life of an individual.

Recall that diploid individuals have two of each type of chromosome inside each cell (see Chapter 13). Haploid individuals have just one of each type of chromosome inside each cell. When meiosis occurs in diploid cells, it results in the production of haploid cells.

The life cycle of most bacteria and archaea is extremely simple: A cell divides, feeds, grows, and divides again. Bacteria and archaea are always haploid.

In contrast, virtually every aspect of a life cycle is variable among protists—whether meiosis occurs, whether asexual reproduction occurs, and whether the haploid or the diploid phase of the life cycle is the longer and more prominent phase.

FIGURE 30.17 illustrates some of this variation. Figure 30.17a depicts the haploid-dominated life cycle observed in many unicellular protists. The specific example given here is the dinoflagellate *Gyrodinium uncatenum*.

To analyze a life cycle, start with **fertilization**—the fusion of two gametes to form a diploid zygote. Then trace what happens to the zygote. In this case, the diploid zygote undergoes meiosis. The haploid products of meiosis then grow into mature cells that eventually undergo asexual reproduction or produce gametes by mitosis and cell division.

Now contrast the dinoflagellate life cycle in Figure 30.17a with the diploid-dominated life cycle in Figure 30.17b. The specific organism shown here is the diatom *Thalassiosira punctigera*. Notice that after fertilization, the diploid zygote develops into a sexually mature, diploid adult cell that can reproduce asexually by dividing mitotically. Meiosis occurs in the adult and results in the formation of haploid gametes, which then fuse to form a diploid zygote. The important contrasts with the

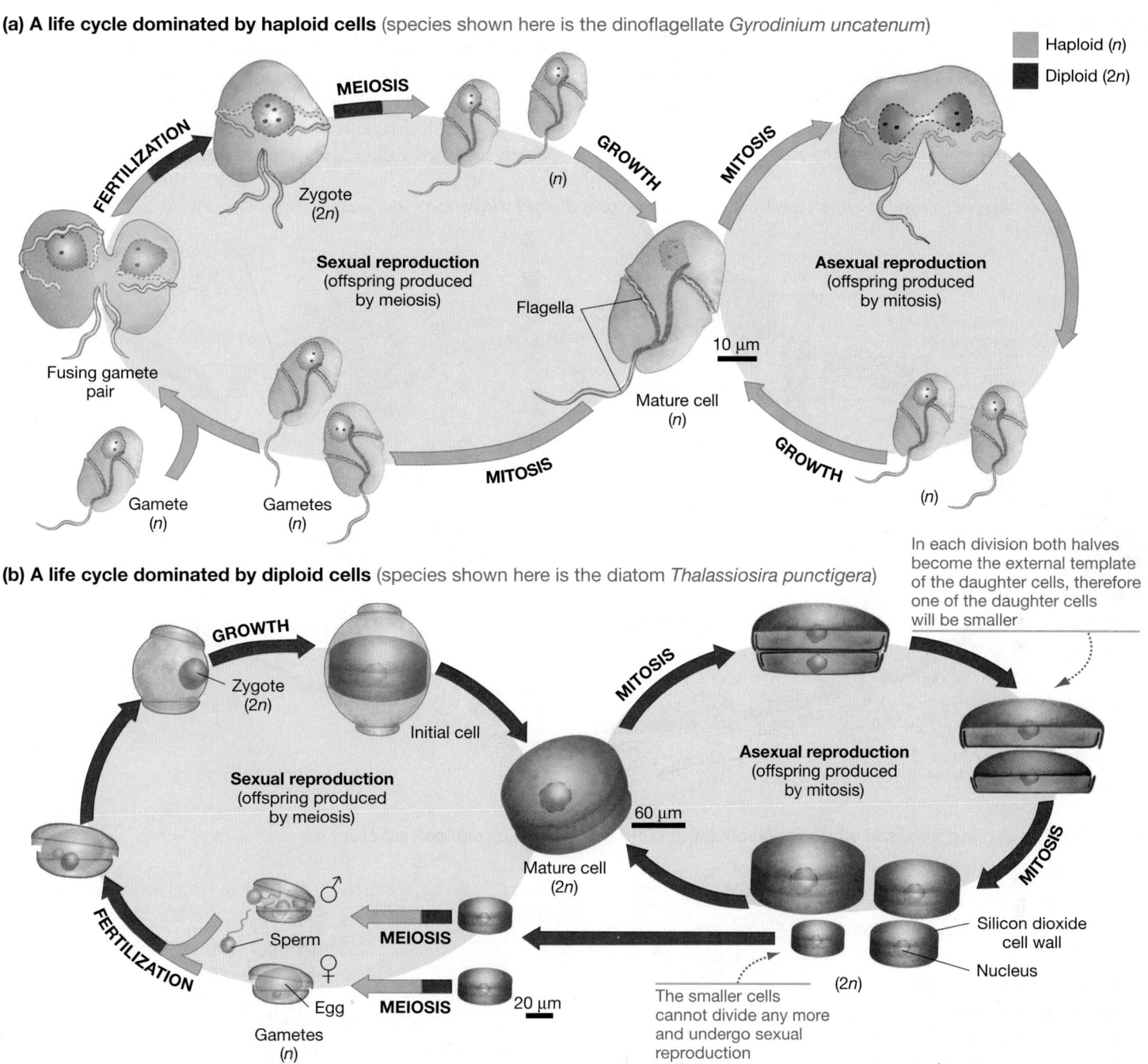

(a) A life cycle dominated by haploid cells (species shown here is the dinoflagellate *Gyrodinium uncatenum*)

Haploid (n)
Diploid (2n)

FERTILIZATION

MEIOSIS

GROWTH

MITOSIS

Zygote
(2n)

(n)

Sexual reproduction
(offspring produced
by meiosis)

Asexual reproduction
(offspring produced
by mitosis)

Fusing gamete
pair

Flagella

10 μm

Mature cell
(n)

MITOSIS

GROWTH

(n)

Gamete
(n)

Gametes
(n)

(b) A life cycle dominated by diploid cells (species shown here is the diatom *Thalassiosira punctigera*)

In each division both halves
become the external template
of the daughter cells, therefore
one of the daughter cells
will be smaller

GROWTH

MITOSIS

Zygote
(2n)

Initial cell

MITOSIS

Sexual reproduction
(offspring produced
by meiosis)

Asexual reproduction
(offspring produced
by mitosis)

60 μm

Mature cell
(2n)

Silicon dioxide
cell wall

♂

Sperm

MEIOSIS

Nucleus

(2n)

FERTILIZATION

♀

Egg

MEIOSIS

20 μm

The smaller cells
cannot divide any more
and undergo sexual
reproduction

Gametes
(n)

FIGURE 30.17 Life Cycles Vary Widely among Unicellular Protists. Many unicellular protists can reproduce by both sexual reproduction and asexual reproduction. The cell may be **(a)** haploid for most of its life or **(b)** diploid for most of its life.

haploid-dominated life cycle are that **(1)** meiosis occurs in the adult cell rather than in the zygote, and **(2)** gametes are the only haploid cells in the life cycle.

Life Cycles—Alternation of Generations In contrast to the relatively simple life cycles of single-celled protists shown in Figure 30.17, many multicellular protists have a multicellular haploid form in one phase of their life cycle alternating with a multicellular diploid form in another phase. This alternation of multicellular haploid and diploid forms is known as **alternation of generations.**

- The multicellular haploid form is called a **gametophyte,** because specialized cells in this individual produce gametes by mitosis and cell division.

- The multicellular diploid form is called a **sporophyte** because it has specialized cells that undergo meiosis to produce haploid cells called spores.

- A **spore** is a single haploid cell that divides mitotically to form a multicellular, haploid gametophyte. A gamete is also a single haploid cell, but its role is to fuse with another gamete to form a zygote.

When alternation of generations occurs, a diploid sporophyte undergoes meiosis and cell division to produce haploid spores. Each spore then divides by mitosis to form a haploid, multicellular gametophyte. The haploid gametes produced mitotically by gametophytes then fuse to form a diploid zygote, which grows into the diploid, multicellular sporophyte.

Gametophytes and sporophytes may be identical in appearance, as in the brown alga called *Ectocarpus siliculosus* (**FIGURE 30.18a**). In many cases, however, the gametophyte and sporophyte look different, as in the brown algae in the genus *Laminaria* (**FIGURE 30.18b**). Among the protists, alternation of generations evolved independently in brown algae, red algae, and other groups.

(a) Alternation of generations in which multicellular haploid and diploid forms look identical (*Ectocarpus siliculosus*)

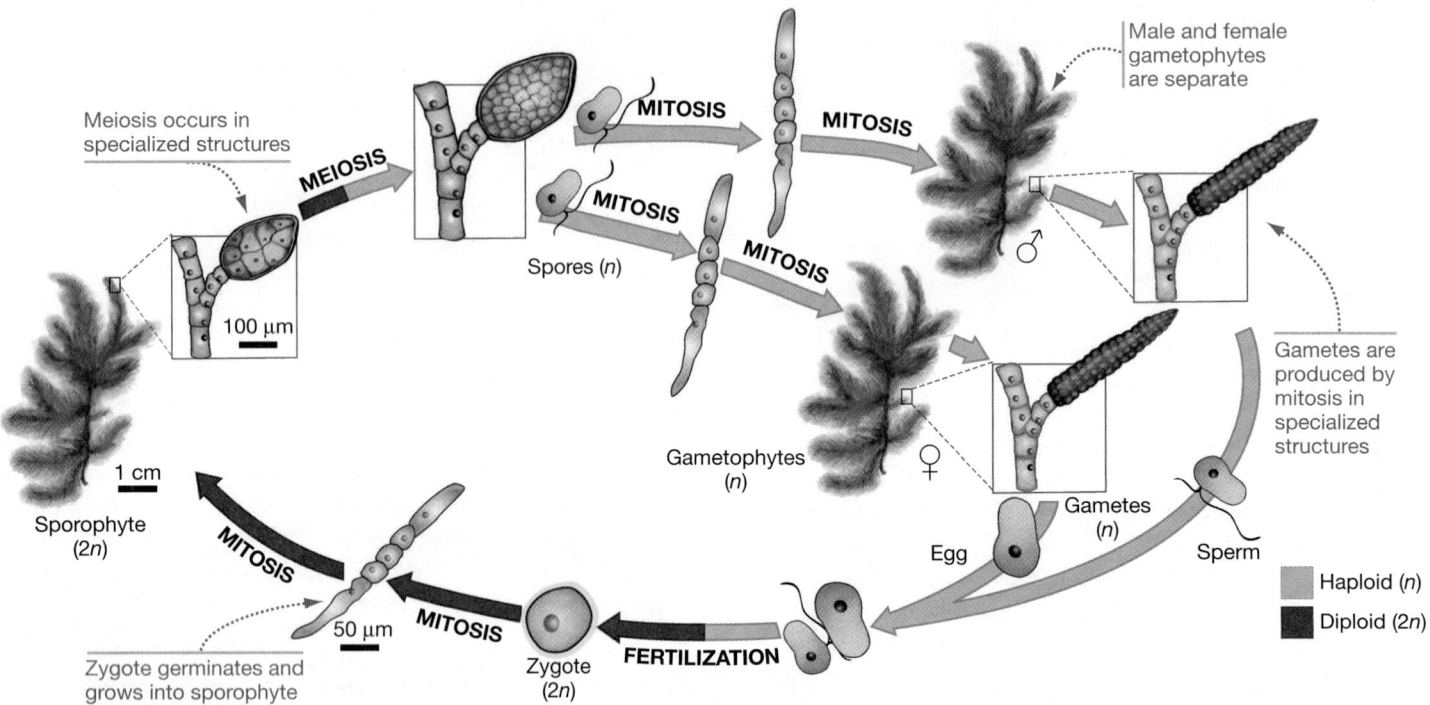

(b) Alternation of generations in which multicellular haploid and diploid forms look different (*Laminaria* sp.)

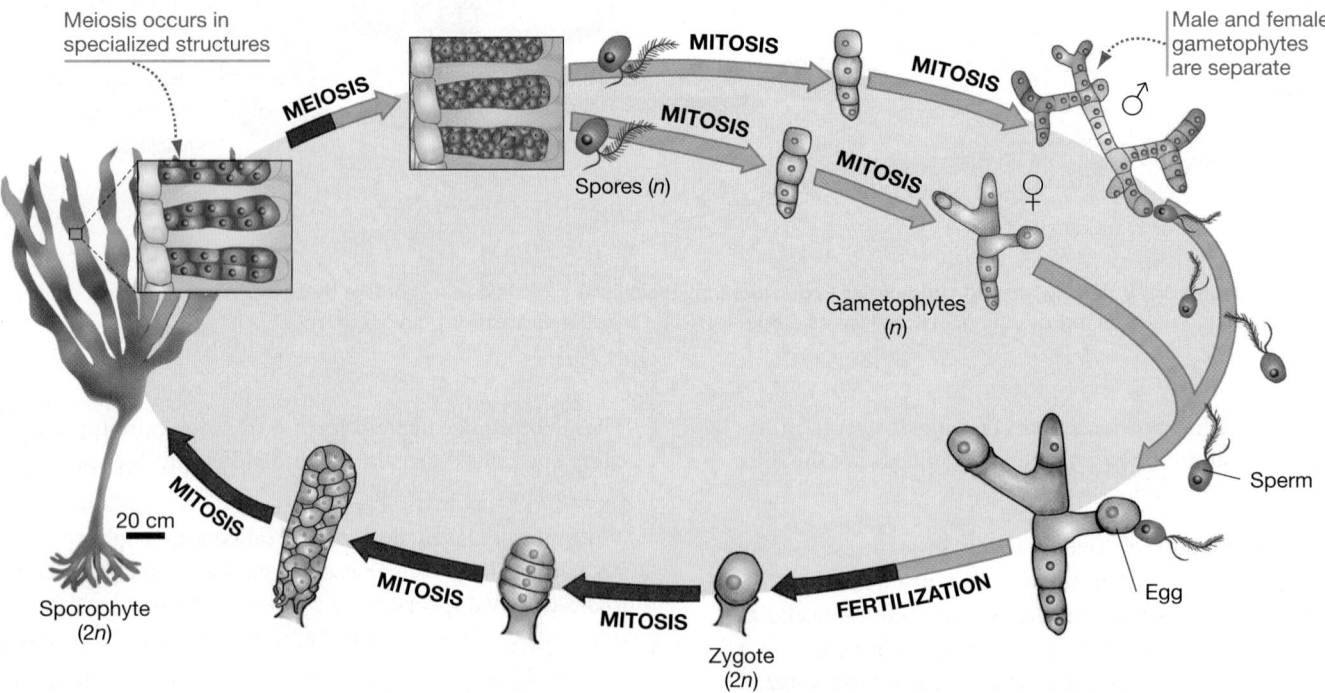

FIGURE 30.18 Alternation of Generations Occurs in Many Multicellular Protists. Compared here, two species of brown algae.

To understand alternation of generations, some students find it helpful to memorize this sentence: Sporophytes produce spores by meiosis; gametophytes produce gametes by mitosis.

✔ If you understand how life cycles vary among multicellular protists, you should be able to (1) define the terms alternation of generations, gametophyte, sporophyte, spore, zygote, and gamete; and (2) diagram a life cycle where alternation of generations occurs, without looking at Figure 30.18.

Why does so much variation occur in the types of life cycles observed among protists? The answer is not known. Variation in life cycles is a major theme in the diversification of protists and is discussed further in the chapter on green algae and land plants (see Chapter 31). Explaining why that variation exists remains a topic for future research.

30.4 Key Lineages of Protists

Each of the seven major Eukarya lineages has at least one distinctive morphological characteristic. But once an ancestor evolved a distinctive cell structure, its descendants diversified into a wide array of lifestyles. For example, parasitic species evolved independently in all seven major lineages. Photosynthetic species exist in most of the seven, and multicellularity evolved independently in at least four. Similar statements could be made about the evolution of life cycles and modes of locomotion.

In effect, each of the seven lineages represents a similar radiation of species into a wide array of lifestyles. Let's take a more detailed look at some representative taxa from six of the seven major lineages, starting with an overview of each one. The seventh major lineage—the opisthokonts—is featured in the chapters on fungi and the animals (see Chapters 32–35). The lineage called green plants is analyzed separately (see Chapter 31).

Amoebozoa

Species in the Amoebozoa lack cell walls and take in food by engulfing it. They move via amoeboid motion and produce large, lobe-like pseudopodia. They are abundant in freshwater habitats and in wet soils; some are parasites of humans and other animals.

Major subgroups in the lineage are lobose amoebae, cellular slime molds, and plasmodial slime molds. The cellular slime mold *Dictyostelium discoideum* is described in detail in **BioSkills 13** in Appendix B. ✔ You should be able to mark the origin of this lineage's amoeboid form on Figure 30.7 and explain whether it evolved independently of the amoeboid form in Rhizaria.

● Amoebozoa > Plasmodial Slime Molds

Excavata

The unicellular species that form the Excavata are named for the morphological feature that distinguishes them—an "excavated" feeding groove found on one side of the cell. Because some lack recognizable mitochondria, excavates were once thought to trace their ancestry to eukaryotes that existed before the origin of mitochondria. But researchers have found that excavates either have

(1) genes in their nuclear genomes that are of mitochondrial origin, or (2) unusual organelles that appear to be vestigial mitochondria. These observations support the hypothesis that the ancestors of excavates had mitochondria, but that these organelles were lost or reduced over time in this lineage. ✔ You should be able to mark the loss of the mitochondrion on Figure 30.7.

● Excavata > Parabasalids, Diplomonads, and Euglenids

Plantae

Biologists are beginning to use the name **Plantae** to refer to the monophyletic group that includes glaucophyte algae, red algae, green algae, and land plants.

The glaucophyte algae are unicellular or colonial. They live as plankton or attached to substrates in freshwater environments—particularly in bogs or swamps. Some glaucophyte species have flagella or produce flagellated spores, but sexual reproduction has never been observed in the group. The chloroplasts of glaucophytes have a distinct bright blue-green color.

All subgroups within Plantae are descended from a common ancestor that engulfed a cyanobacterium, beginning the endosymbiosis that led to the evolution of the chloroplast—their distinguishing morphological feature. This initial endosymbiosis probably occurred in an ancestor of today's glaucophyte algae. To support this hypothesis, biologists point to several important similarities between cyanobacterial cells and the chloroplasts that are found in the glaucophytes.

- Both have cell walls that contain peptidoglycan and that can be disrupted by the antibacterial compounds lysozyme and penicillin.

- The glaucophyte chloroplast also has a membrane outside its wall that is similar to the membrane found in Gram-negative bacteria (see Chapter 29).

Consistent with these observations, phylogenetic analyses of DNA sequence data place the glaucophytes as the sister group to all other members of the Plantae. ✔ You should be able to mark the origin of the plant chloroplast on Figure 30.7.

Green algae and land plants are described later (see Chapter 31); here we consider just the red algae in detail.

● Plantae > Red Algae

Rhizaria

The rhizarians are single-celled amoebae that lack cell walls, though some species produce elaborate shell-like coverings. They move by amoeboid motion and produce long, slender pseudopodia. Over 11 major subgroups in this lineage have been identified and named, including the planktonic organisms called actinopods—which synthesize glassy, silicon-rich skeletons—and the chlorarachniophytes, which obtained a chloroplast via secondary endosymbiosis and are photosynthetic. The best-studied and most abundant group is the foraminiferans. ✔ You should be able to mark the origin of the rhizarian amoeboid form on Figure 30.7.

● Rhizaria > Foraminiferans

Alveolata

Alveolates are distinguished by small sacs, called alveoli, that are located just under their plasma membranes. Although all members of this lineage are unicellular, the ciliates, the dinoflagellates, and the apicomplexans are remarkably diverse in morphology and lifestyle. ✔ **You should be able to mark the origin of alveoli on Figure 30.7.**

- Alveolata > Ciliates, Dinoflagellates, and Apicomplexans

Stramenopila (Heterokonta)

Stramenopiles are sometimes called heterokonts, which translates as "different hairs." At some stage of their life cycle, all stramenopiles have flagella that are covered with distinctive hollow "hairs." The structure of these flagella is unique to the stramenopiles (Figure 30.6). ✔ **You should be able to mark the origin of the "hairy" flagellum on Figure 30.7.** The lineage includes a large number of unicellular forms, although the brown algae are multicellular and include the world's tallest marine organisms, the kelp.

- Stramenopila > Water Molds, Diatoms, and Brown Algae

Amoebozoa > Plasmodial Slime Molds

The plasmodial slime molds got their name because individuals form a large, weblike structure that consists of a single cell containing thousands of diploid nuclei (**FIGURE 30.19**). Like water molds, the plasmodial slime molds were once considered fungi on the basis of their general morphological similarity (see Chapter 32).

Morphology The huge "supercell" form, with many nuclei in a single cell, occurs in few protists other than plasmodial slime molds. ✔ **You should be able to mark the origin of the supercell on Figure 30.7.**

Feeding and locomotion Plasmodial slime molds feed on microorganisms associated with decaying vegetation and move by amoeboid motion or cytoplasmic streaming.

Reproduction Some evidence suggests that when food becomes scarce, part of the amoeba forms a stalk topped by a ball-like structure in which nuclei undergo meiosis and form spores. The spores are then dispersed to new habitats by the wind or small animals. After spores germinate to form amoebae, two amoebae fuse to form a diploid cell that begins to feed and eventually grows into a supercell.

Human and ecological impacts Like cellular slime molds, plasmodial slime molds influence nutrient cycling by feeding on microorganisms. Some species are important model organisms for the study of cell biology and the origin of multicellularity.

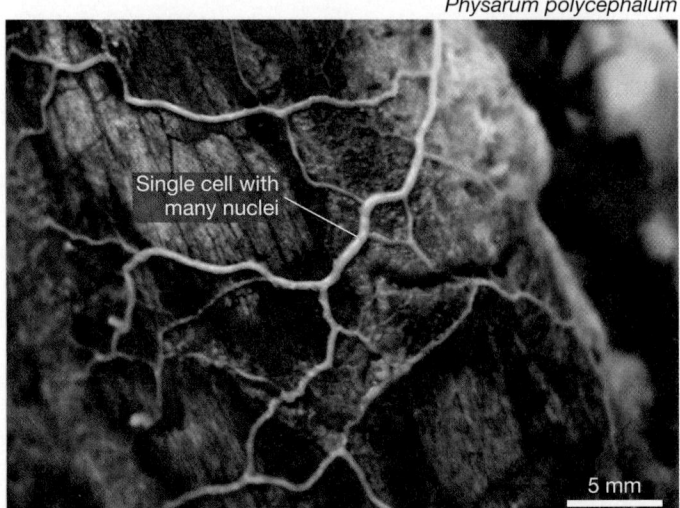

Physarum polycephalum

Single cell with many nuclei

5 mm

FIGURE 30.19 Plasmodial Slime Molds Are Important Decomposers in Forests.

The lineage Excavata includes free-living and symbiotic species, and some that are parasites. All are single-celled, and many lack functional mitochondria.

All of the 300 known species of parabasalids live inside animals; some live only in the guts of termites, where they aid in the digestion of cellulose. The relationship between termites and parabasalids is considered mutualistic, because both parties benefit from the symbiosis.

About 100 species of diplomonads have been named. Many live in the guts of animal species without causing harm to their host; other species live in stagnant-water habitats.

There are about 1000 known species of euglenids. Although most live in freshwater, a few are found in marine habitats.

Morphology Parabasalid cells have a distinctive rod of cross-linked microtubules that runs the length of the cell. The rod is attached to the basal bodies where a cluster of flagella arise (**FIGURE 30.20**). Diplomonad cells have two nuclei, which resemble eyes when viewed under the microscope (**FIGURE 30.21**). Each nucleus is associated with four flagella, for a total of eight flagella per cell. Euglenid cells have a unique system of interlocking proteins just inside the plasma membrane that stiffen and support the cell. Some euglenid cells have a light-sensitive "eyespot" and use flagella to swim toward light (**FIGURE 30.22**).

Feeding and locomotion Some euglenids have chloroplasts and perform photosynthesis, and some diplomonads are parasitic, but most other members of the Excavata feed by engulfing bacteria, archaea, and organic matter. Most swim using their flagella. ✔ You should be able to explain how the observation of ingestive feeding in euglenids relates to the hypothesis that this lineage gained chloroplasts via secondary endosymbiosis.

Giardia intestinalis

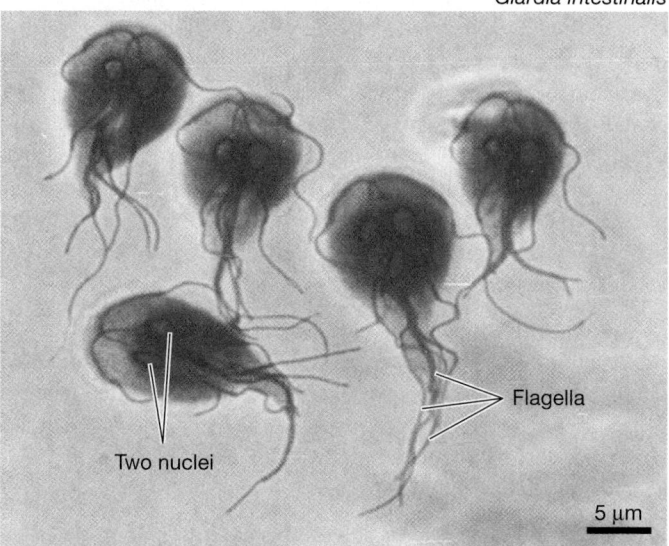

FIGURE 30.21 *Giardia* **Causes Intestinal Infections in Humans.**

Reproduction Nearly all members of the Excavata are known to reproduce only asexually. Sexual reproduction has been observed in just a few species of parabasalids.

Human and ecological impacts Infections of *Trichomonas*, a parabasalid, can sometimes cause reproductive tract problems in humans. However, members of this genus may also live in the gut or mouth of humans without causing harm. The diplomonad *Giardia intestinalis* is a common intestinal parasite in humans and causes giardiasis, or beaver fever. Euglenids are important components of freshwater plankton and food chains.

Trichomonas vaginalis

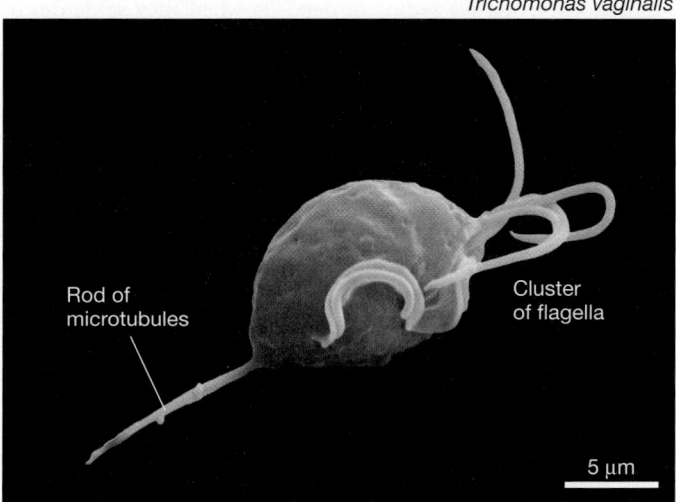

FIGURE 30.20 *Trichomonas* **Causes the Sexually Transmitted Disease Trichomoniasis.** The cell in this micrograph has been colorized.

Euglena velata

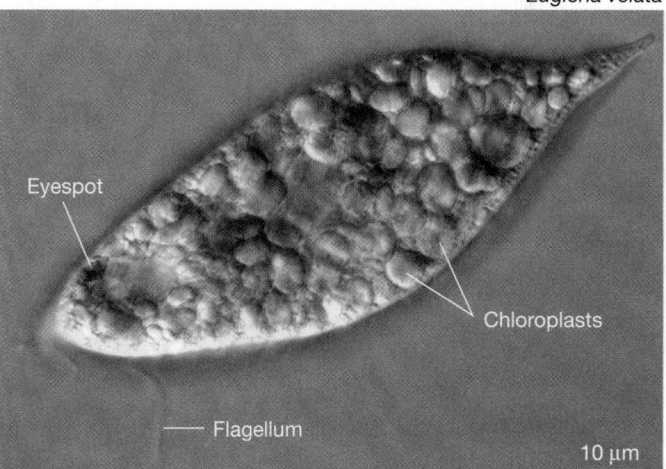

FIGURE 30.22 **Euglenids Are Common in Ponds and Lakes.**

The 6000 species of red algae live primarily in marine habitats. One species lives over 200 m below the surface; another is the only eukaryote capable of living in acidic hot springs. Although their color varies, many species are red because their chloroplasts contain large amounts of the accessory pigment phycoerythrin, which absorbs strongly in the blue and green portions of the visible spectrum. Because blue light penetrates water better than other wavelengths, red algae are able to live at considerable depth in the oceans. ✔ You should be able to mark the origin of high phycoerythrin concentrations on Figure 30.7.

Morphology Red algae cells have walls that are composed of cellulose and other polymers. A few species are unicellular, but most are multicellular. Many of the multicellular species are filamentous; others grow as thin, hard crusts on rocks or coral. Some species have erect, leaf-like structures called thalli (**FIGURE 30.23**).

Feeding and locomotion The vast majority of red algae are photosynthetic, though a few parasitic species have been identified. Red algae are the only type of algae that lack flagella.

Reproduction Asexual reproduction occurs through production of spores by mitosis. Alternation of generations is common, but the types of life cycles observed in red algae are extremely variable.

Human and ecological impacts On coral reefs, some red algae become encrusted with calcium carbonate. These species contribute to reef building and help stabilize the entire reef structure. Cultivation of *Porphyra* (or nori) for sushi and other foods is a billion-dollar-per-year industry in East Asia. Microbiological agar is also derived from the cell walls of various species of red algae.

Mesophyllum lichenoides

1 cm

FIGURE 30.23 Red Algae Adopt an Array of Growth Forms.

Rhizaria > Foraminiferans

Foraminiferans, or forams, got their name from the Latin *foramen*, meaning "hole." Forams produce shells that have holes (see **FIGURE 30.24**) through which pseudopodia emerge. ✔ You should be able to mark the origin of the foram shell on Figure 30.7. Fossil shells of foraminifera are abundant in marine sediments—a continuous record of fossilized forams dates back 530 million years. A foram species was recently found living in sediments at a depth of 11,000 m below sea level. They are abundant in marine plankton as well as bottom habitats.

Morphology Foraminiferan cells generally have multiple nuclei. The shells of forams are usually made of organic material stiffened with calcium carbonate ($CaCO_3$), and most species have several chambers. One species known from fossils was 12 cm long, but most species are much smaller. The size and shape of the shell are traits that distinguish foram species from each other.

Feeding and locomotion Like other rhizarians, forams feed by extending their pseudopodia and using them to capture and engulf bacterial and archaeal cells or bits of organic debris, which are digested in food vacuoles. Some species have symbiotic algae that perform photosynthesis and contribute sugars to their host. Forams do not move actively but simply float in the water.

Reproduction Members of the foraminifera reproduce asexually by mitosis. When meiosis occurs, the resulting gametes are released into the open water, where pairs fuse to form new diploid individuals.

Human and ecological impacts The shells of dead forams commonly form extensive sediment deposits when they settle out of the water, producing layers that eventually solidify into chalk, limestone, or marble. Geologists use the presence of certain foram species to date rocks—particularly during petroleum exploration.

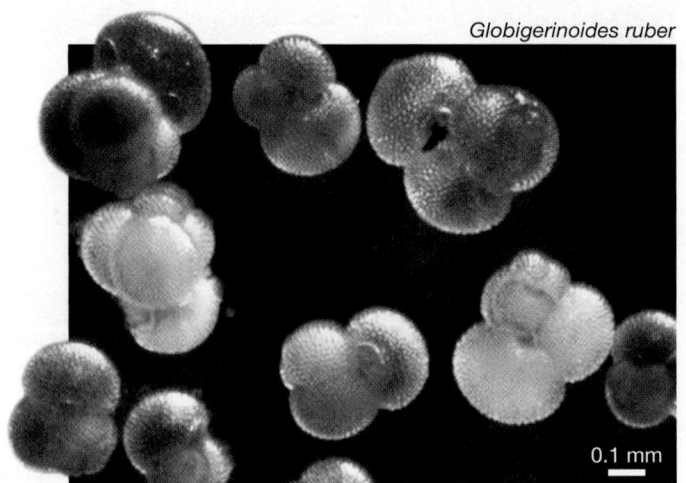

Globigerinoides ruber

0.1 mm

FIGURE 30.24 Forams Are Shelled Amoebae.

Members of the lineage Alveolata comprise three subgroups of single-celled species that are diverse in morphology and nutrition. All alveolates contain flattened, membrane-bound vesicles called alveoli that support the plasma membrane.

Ciliates are named for the cilia that cover them. Some 12,000 species are known from freshwater, marine, and wet soil environments. Most of the 4000 known species of dinoflagellates are marine or freshwater plankton. All of the 5000 known species of apicomplexans are parasitic.

Morphology Ciliate cells have two distinctive nuclei: a polyploid macronucleus that is actively transcribed and a diploid micronucleus that is involved only in gene exchange between individuals (both are stained dark red in **FIGURE 30.25**). ✔ You should be able to mark the origin of the macronucleus/micronucleus structure on Figure 30.7.

Each dinoflagellate species has a distinct shape, in some species maintained by cellulose plates. Apicomplexan cells lack cilia or flagella.

Feeding and locomotion Ciliates may be filter feeders, predators, or parasites. They use cilia to swim and have a mouth area where food is ingested. About half of the dinoflagellates are photosynthetic. They are distinguished by the arrangement of their two flagella, which allows them to swim in a spinning motion: One flagellum projects out from the cell while the other runs around the cell. Some dinoflagellate species are capable of **bioluminescence,** meaning they emit light via an enzyme-catalyzed reaction (**FIGURE 30.26**). Being parasitic, all apicomplexans absorb nutrition directly from their host. Some species can move by amoeboid motion.

Reproduction Ciliates undergo an unusual type of gene exchange called conjugation, in which micronuclei are passed between individuals. Both asexual and sexual reproduction occur in dinoflagellates. In some apicomplexan species, the life cycle involves two distinct hosts, and cells must be transmitted from one host to the next.

Human and ecological impacts Ciliates are abundant marine plankton and are also common in the digestive tracts of mammalian grazers, where they feed on plant matter and help the host animal digest it. Photosynthetic dinoflagellates are important primary producers in marine ecosystems. A few species are responsible for harmful algal blooms (see Figure 30.4). Apicomplexan species in the genus *Plasmodium* cause malaria. *Toxoplasma* is an important pathogen in people infected with HIV (**FIGURE 30.27**).

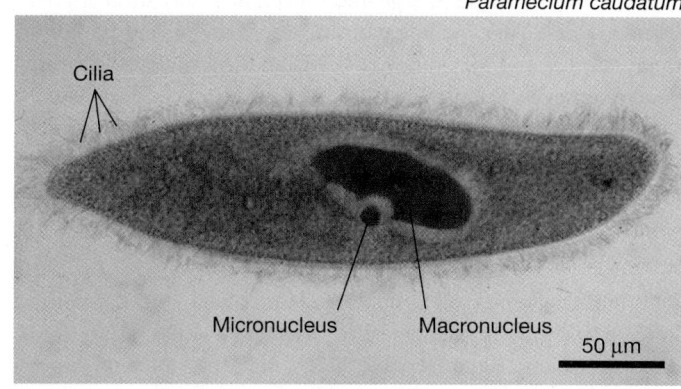

Paramecium caudatum

Cilia

Micronucleus   Macronucleus

50 μm

FIGURE 30.25 Ciliates Are Abundant in Freshwater Plankton.

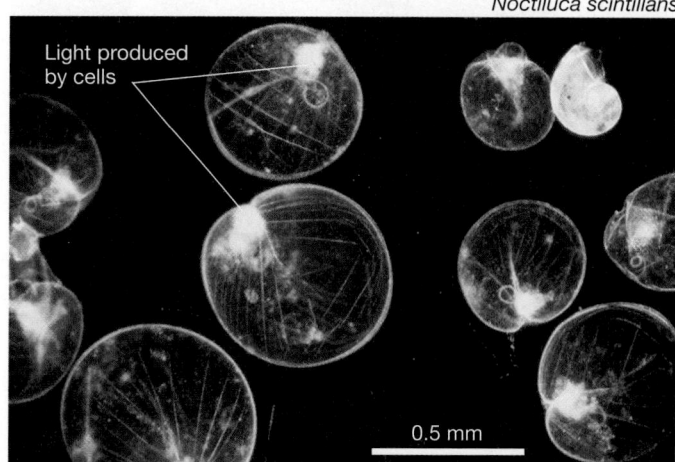

Noctiluca scintillans

Light produced by cells

0.5 mm

FIGURE 30.26 Some Dinoflagellate Species Are Bioluminescent.

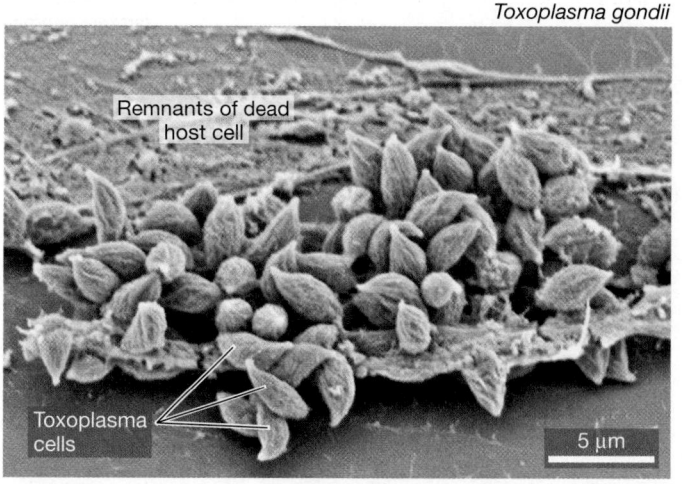

Toxoplasma gondii

Remnants of dead host cell

Toxoplasma cells

5 μm

FIGURE 30.27 *Toxoplasma* Causes Infections in AIDS Patients.

The stramenopiles are a large and morphologically diverse group of eukaryotes. They include parasitic, saprophytic, and photosynthetic species, and they range in size from tiny, single-celled members to some of the largest multicellular organisms in the oceans.

Several morphological features are shared among the subgroups of stramenopiles, including flagella that are covered with tiny, hollow projections (see Figure 30.6). Photosynthetic diatoms and brown algae contain chloroplasts that originated from a secondary endosymbiosis involving a red alga (see Figure 30.13).

Morphology Some water molds are unicellular and some form long, branching filaments called **hyphae** (**FIGURE 30.28**). Most diatoms are unicellular (**FIGURE 30.29**), often in a box-and-lid arrangement (see Figure 30.11a). All species of brown algae are multicellular; the body typically consists of leaflike blades, a stalk known as a stipe, and a rootlike holdfast (**FIGURE 30.30**). ✔ You **should be able to mark the origin of brown algal multicellularity on Figure 30.7.**

Feeding and locomotion Most species of water molds feed on decaying organic material; a few are parasitic. Mature individuals are **sessile**—that is, permanently fixed to a substrate. Both diatoms and brown algae are photosynthetic, but diatoms are planktonic whereas brown algae are sessile.

Reproduction Many species of stramenopiles exhibit diploid-dominant life cycles. Several produce spores via asexual or sexual reproduction. Water mold spores form in special structures, like those shown in Figure 30.28. Most species of brown algae exhibit alternation of generations.

Human and ecological impacts Water molds are extremely important decomposers in freshwater ecosystems. Parasitic forms that affect humans indirectly include the organism that caused the Irish potato famine. Other water mold species are responsible for epidemic diseases of trees, including the diebacks currently occurring in oaks.

Diatoms are considered the most important primary producer in fresh and salt water. Their glassy cell walls settle into massive accumulations that are mined and sold as diatomaceous earth, which is used in filtering applications and as an ingredient in polishes, paint, cosmetics, and other products.

In many coastal areas with cool water, brown algae form kelp forests that are important habitats for a wide variety of animals.

Phytophthora infestans

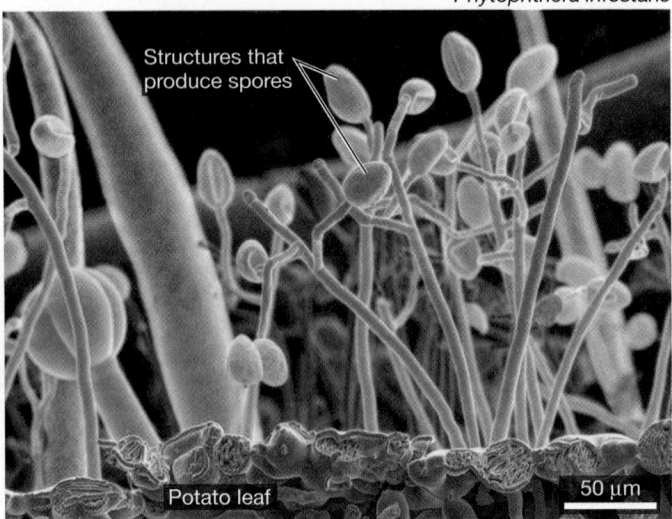

FIGURE 30.28 *Phytophthora Infestans* **Infects Potatoes.** This is a colorized scanning electron micrograph.

Isthmia nervosa

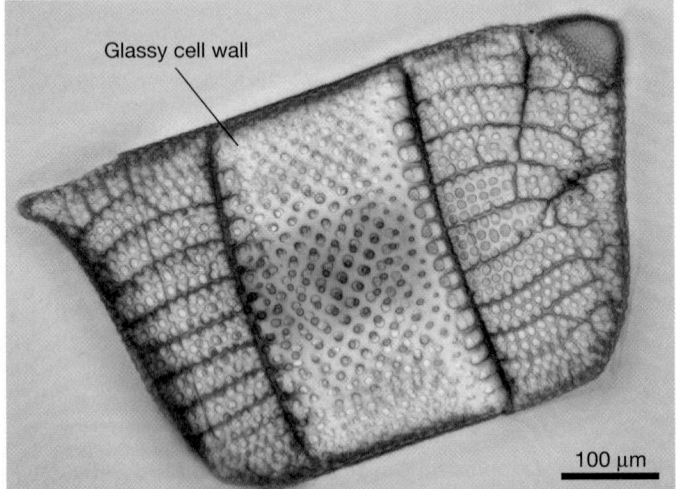

FIGURE 30.29 Diatoms Have Glass-Like Cell Walls.

Durvillaea potatorum

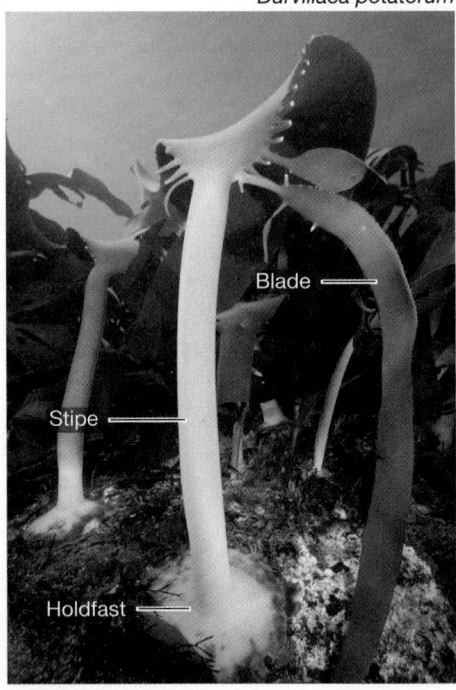

FIGURE 30.30 Many Brown Algae Have a Holdfast, Stalk (Stipe), and Leaflike Blades.

If you understand . . .

30.1 Why Do Biologists Study Protists?

- Protists include all eukaryotes except the land plants, fungi, and animals.

- Protists are often tremendously abundant in marine and freshwater plankton and other habitats. Protists provide food for many organisms in aquatic ecosystems and fix so much carbon that they have a large impact on the global carbon budget.

- Parasitic protists cause several important diseases in humans, including malaria.

- Toxin-producing protists that grow to high densities can result in a harmful algal bloom.

✔ You should be able to explain what is "primary" about primary production by photosynthetic protists.

30.2 How Do Biologists Study Protists?

- Microscopic investigation of protists led to the understanding that many are single celled. Unlike bacteria, they contain mitochondria, a nucleus and endomembrane system, and a cytoskeleton.

- Direct sequencing has revealed important new groups of protists.

- Phylogenies based on molecular data confirmed that there are seven major lineages of protists and that the group is paraphyletic.

✔ You should be able to describe the major differences between eukaryotic cells and prokaryotic cells.

30.3 What Themes Occur in the Diversification of Protists?

- The first eukaryote probably resulted from an ancient endosymbiosis that occurred when a prokaryotic host cell (likely an archaeal cell) engulfed an α-proteobacterium. This bacterium evolved into today's mitochondria.

- The nuclear envelope probably evolved to spatially separate the processes of transcription and translation.

- Multicellularity evolved in several different protist groups independently.

- Protists vary widely in the way they obtain food. They exhibit predatory, parasitic, or photosynthetic lifestyles, which evolved in many groups independently.

- The evolution of ingestive feeding was important for two reasons: (1) It allowed eukaryotes to obtain resources in a new way—by eating bacteria, archaea, and other eukaryotes; and (2) it made endosymbiosis and the evolution of mitochondria and chloroplasts possible.

- The chloroplast's size and its circular chromosome, ribosomes, and double membrane are consistent with the hypothesis that this organelle originated as an endosymbiotic cyanobacterium.

- After primary endosymbiosis occurred, chloroplasts were "passed around" to new lineages of protists via secondary endosymbiosis.

- Protists vary widely in the way they reproduce. They undergo cell division based on mitosis and reproduce asexually. Many protists also undergo meiosis and sexual reproduction at some phase in their life cycle.

- Alternation of generations is common in multicellular protist species—meaning the same species has separate haploid and diploid forms. When alternation of generations occurs, haploid gametophytes produce gametes by mitosis; diploid sporophytes produce spores by meiosis.

✔ You should be able to explain the original source of (1) the two membranes in a plant chloroplast and (2) the four membranes in the chloroplast-derived organelle of *Plasmodium*.

30.4 Key Lineages of Protists

- Protists are a highly diverse group of eukaryotic species organized into seven lineages.

- Protist lineages are defined by DNA sequence evidence and morphological traits such as the presence of chloroplasts or unique flagella.

- A wide range of traits including parasitism, autotrophy, and multicellularity evolved independently in several different lineages of protists.

 MasteringBiology

1. **MasteringBiology Assignments**

   Tutorials and Activities Alternation of Generations in a Protist; Life Cycles of Protists; Tentative Phylogeny of Eukaryotes

   Questions Reading Quizzes, Blue-Thread Questions, Test Bank

2. **eText** Read your book online, search, take notes, highlight text, and more.

3. **The Study Area** Practice Test, Cumulative Test, BioFlix® 3-D Animations, Videos, Activities, Audio Glossary, Word Study Tools, Art

You should be able to . . .

1. Why are protists considered paraphyletic?
   a. They include many extinct forms, including lineages that no longer have any living representatives.
   b. They include some but not all descendants of their most recent common ancestor.
   c. They represent all of the descendants of a single common ancestor.
   d. Not all protists have all of the synapomorphies that define the Eukarya, such as a nucleus.

2. What material is *not* used by protists to manufacture hard outer coverings?
   a. cellulose
   b. lignin
   c. glass-like compounds that contain silicon
   d. mineral-like compounds such as calcium carbonate ($CaCO_3$)

3. What does amoeboid motion result from?
   a. interactions among actin, myosin, and ATP
   b. coordinated beats of cilia
   c. the whiplike action of flagella
   d. action by the mitotic spindle, similar to what happens during mitosis and meiosis

4. According to the endosymbiosis theory, what type of organism is the original ancestor of the chloroplast?
   a. a photosynthetic archaean
   b. a cyanobacterium
   c. an algal-like, primitive photosynthetic eukaryote
   d. a modified mitochondrion

5. True or False: All protists are unicellular.

6. The most important primary producers in marine ecosystems are _____.

7. Which of the following is true of the parasite that causes malaria in humans?
   a. It undergoes meiosis immediately following zygote formation.
   b. It is a complex prokaryote.
   c. It makes gametes by meiosis.
   d. It has a diploid-dominant life cycle.

8. Explain the logic behind the claim that the nuclear envelope is a synapomorphy that defines eukaryotes as a monophyletic group.

9. Consider the endosymbiosis theory for the origin of the mitochondrion. What did each partner provide the other, and what did each receive in return?

10. Why was finding a close relationship between mitochondrial DNA and bacterial DNA considered particularly strong evidence in favor of the endosymbiosis theory?

11. The text claims that the evolutionary history of protists can be understood as a series of morphological innovations that established seven distinct lineages, each of which subsequently diversified based on innovative ways of feeding, moving, and reproducing. Explain how the Alveolata support this claim.

12. Multicellularity is defined in part by the presence of distinctive cell types. At the cellular level, what does this criterion imply?
    a. Individual cells must be extremely large.
    b. The organism must be able to reproduce sexually.
    c. Cells must be able to move.
    d. Different cell types express different genes.

13. Consider the following:
    - *Plasmodium* has an unusual organelle called an apicoplast. Recent research has shown that apicoplasts are derived from chloroplasts via secondary endosymbiosis and have a large number of genes related to chloroplast DNA.
    - Glyphosate is one of the most widely used herbicides. It works by poisoning an enzyme located in chloroplasts.
    - Biologists are testing the hypothesis that glyphosate could be used as an antimalarial drug in humans.

    How are these observations connected?

14. Suppose a friend says that we don't need to worry about the rising temperatures associated with global climate change. Her claim is that increased temperatures will make planktonic algae grow faster and that carbon dioxide (CO_2) will be removed from the atmosphere faster. According to her, this carbon will be buried at the bottom of the ocean in calcium carbonate shells. As a result, the amount of carbon dioxide in the atmosphere will decrease and global warming will decline. Comment.

15. Biologists are beginning to draw a distinction between "species trees" and "gene trees." A species tree is a phylogeny that describes the actual evolutionary history of a lineage. A gene tree, in contrast, describes the evolutionary history of one particular gene, such as a gene required for the synthesis of chlorophyll *a*. In some cases, species trees and gene trees don't agree with each other. For example, the species tree for green algae indicates that their closest relatives are protists and plants. But the gene tree based on chlorophyll *a* from green algae suggests that this gene's closest relative is a bacterium, not a protist. What's going on? Why do these types of conflicts exist?

31 Green Algae and Land Plants

In this chapter you will learn that

Adaptations allowed Green Algae and Land Plants to colonize and dominate the land

asking

Why and how do biologists study these organisms?
31.1, 31.2

examining

Major themes in their diversification
31.3

looking more closely at

Key lineages of green algae and land plants
31.4

focusing on

The transition to life on land

- Controlling water loss
- Surviving intense sunlight
- Growing upright in air
- Reproducing without water
- Using animals to carry pollen and seeds

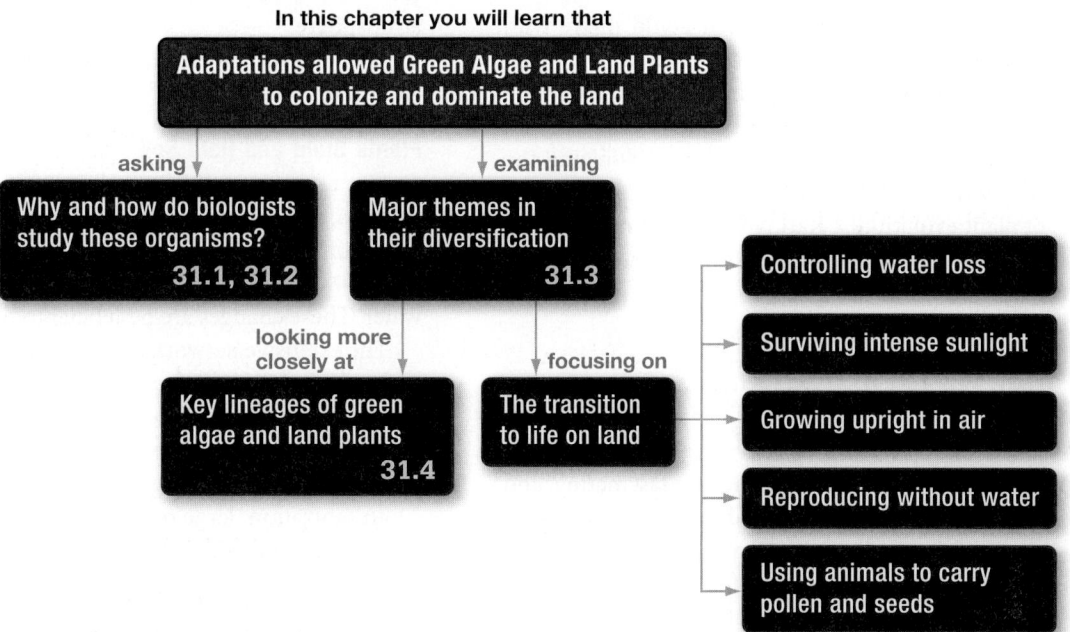

Earth has been called the Blue Planet, but it would be just as accurate to call it the Green Planet.

n terms of their total mass and their importance to other organisms, the green plants dominate terrestrial and freshwater habitats. When you walk through a forest or meadow, you are surrounded by green plants. If you look at pond or lake water under a microscope, green plants are everywhere.

The green plants comprise two major types of organisms: the green algae and the land plants. Green algae are important photosynthetic organisms in aquatic habitats—particularly lakes, ponds, and other freshwater settings. Land plants are the key photosynthesizers in terrestrial environments.

✓ When you see this checkmark, stop and test yourself. Answers are available in Appendix A.

Although green algae have traditionally been considered protists, it is logical to study them along with land plants for two reasons: (1) They are the closest living relatives to land plants and form a monophyletic group with them, and (2) the transition from aquatic to terrestrial life occurred when land plants evolved from green algae.

Land plants were the first organisms that could thrive with their tissues completely exposed to the air instead of being partially or completely submerged. Before they evolved, likely the only life on the continents consisted of bacteria, archaea, and single-celled protists that thrive in wet soils. By colonizing the continents, plants transformed the nature of life on Earth. It was, in the words of plant evolutionist Karl Niklas, "One of the greatest adaptive events in the history of life." Land plants made the Earth green.

31.1 Why Do Biologists Study the Green Algae and Land Plants?

Biologists study the green plants because they are fascinating, and because we couldn't live without them. Along with most other animals and fungi, humans are almost completely dependent on land plants for food. People rely on plants for other necessities of life as well—oxygen, fuel, the fibers used to make clothing and other products, building materials, and pharmaceuticals. But we also rely on land plants for important intangible value such as the aesthetic appeal of landscaping and bouquets. To drive this point home, consider that the sale of cut flowers generates over $1 billion each year in the United States alone.

Based on these observations, it is not surprising that agriculture, forestry, and horticulture are among the most important endeavors supported by biological science. Tens of thousands of biologists are employed in research designed to better understand plants, to increase their productivity, and to create new ways of using them for the benefit of people. Research programs also focus on two types of land plants that cause problems for people: weeds that decrease the productivity of crop plants, and newly introduced species that invade and then degrade natural areas.

Plants Provide Ecosystem Services

An **ecosystem** consists of all the organisms in a particular area, along with physical components of the environment such as the atmosphere, precipitation, surface water, sunlight, soil, and nutrients. Green algae and land plants provide **ecosystem services** because they enhance the life-supporting attributes of the atmosphere, surface water, soil, and other physical components of an ecosystem. Green plants alter the environment in ways that benefit many other organisms.

Plants Produce Oxygen Recall that plants perform oxygenic (literally, "oxygen-producing") photosynthesis (see Chapter 10). In this process, electrons that are removed from water molecules are used to reduce carbon dioxide (CO_2). In the process of stripping electrons from water, plants release oxygen molecules (O_2) as a by-product.

Oxygenic photosynthesis evolved in cyanobacteria and was responsible for the origin of an oxygen-rich atmosphere (see Chapter 29). The evolutionary success of plants continued this trend because plants add huge amounts of oxygen to the atmosphere.

Without the green plants, we and other aerobic organisms would be in danger of suffocating for lack of oxygen.

Plants Build and Hold Soil Fallen leaves, roots, and stems that are not eaten when they are alive provide food for worms, fungi, bacteria, archaea, protists, and other decomposers in the soil. These organisms add organic matter to the soil, which changes the soil structure and the ability of soils to hold nutrients and water. These changes are beneficial to other organisms.

The extensive network of fine roots produced by trees, grasses, and other land plants helps hold soil particles in place. And by taking up nutrients in the soil, plants prevent the nutrients from being blown or washed away.

When areas are devegetated by grazing, farming, logging, or suburbanization, large quantities of soil and nutrients are lost to erosion by wind and water (**FIGURE 31.1**).

Plants Hold Water and Moderate Climate Plant tissues take up and retain water. Intact forests, prairies, and wetlands also prevent rain from quickly running off a landscape: Plant leaves soften the physical impact of rainfall on soil, and plant organic matter builds the soil's water-holding capacity.

By providing shade, plants reduce temperatures beneath them and increase relative humidity. They also reduce the impact of winds that dry out landscapes.

When plants are removed from landscapes to make way for farms or suburbs, habitats become much drier and are subject to more extreme temperature swings.

Plants are Primary Producers Land plants are the dominant primary producers in terrestrial ecosystems. Primary producers convert energy in sunlight into chemical energy (see Chapter 30). The

FIGURE 31.1 Devegetation Leads to Erosion.

sugars that land plants produce by photosynthesis support virtually all of the other organisms present in terrestrial habitats.

Plants are eaten by herbivores ("plant-eaters"), which range in size from insects to elephants. These consumers are eaten by carnivores ("meat-eaters"), ranging in size from the tiniest spiders to polar bears. Humans are an example of omnivores ("all-eaters"), organisms that eat both plants and animals. Omnivores feed at several different levels in the food chain. For example, people consume plants, herbivores such as chicken and cattle, and carnivores such as salmon and tuna.

Because of their role in primary production, land plants are the key to the carbon cycle on the continents. Plants take CO_2 from the atmosphere and reduce it to make sugars. Although green algae and land plants also produce a great deal of CO_2 as a result of cellular respiration, they fix much more CO_2 than they release.

The loss of plant-rich prairies and forests due to suburbanization or conversion to agriculture has contributed to increased concentrations of CO_2 in the atmosphere. Higher carbon dioxide levels, in turn, are partly responsible for the rising temperatures associated with global climate change (see Chapter 56).

Plants Provide Humans with Food, Fuel, Fiber, Building Materials, and Medicines

It is difficult to overstate the importance of plant research to the well-being of human societies. Plants provide our food supply as well as a significant percentage of the fuel, fibers, building materials, and medicines that we use.

Food Agricultural research began with the initial domestication of crop plants, which occurred independently at several locations around the world starting about 12,000 years ago (**FIGURE 31.2a**). By actively selecting individuals with the largest and most nutritious seeds, leaves, or other plant parts year after year, our ancestors gradually changed the characteristics of certain wild species. This process is called **artificial selection** (see Chapter 25).

FIGURE 31.2b compares kernel size in modern maize and its wild ancestor—a grass called teosinte that is native to Central America. Artificial selection has been responsible for dramatic increases in the protein content of maize kernels over the past 100 years (see data in Chapter 1). A current focus in agricultural research is the improvement of crop varieties through genetic engineering (see Chapter 20).

Fuel For perhaps 100,000 years, wood burning was the primary source of energy used by all humans. Wood has now been replaced in the industrialized countries by fossil fuels, including coal and petroleum or natural gas. Each of these forms of energy, however, is derived from plants that either grew during our lifetime or when dinosaurs roamed the earth.

Starting in the mid-1800s, people in England, Germany, and the United States began to mine coal deposits that originally formed during the Carboniferous period some 359–299 million years ago. The coal fueled blast furnaces that smelted vast quantities of iron ore into steel and powered the steam engines that sent trains

(a) Plants were domesticated at an array of locations.

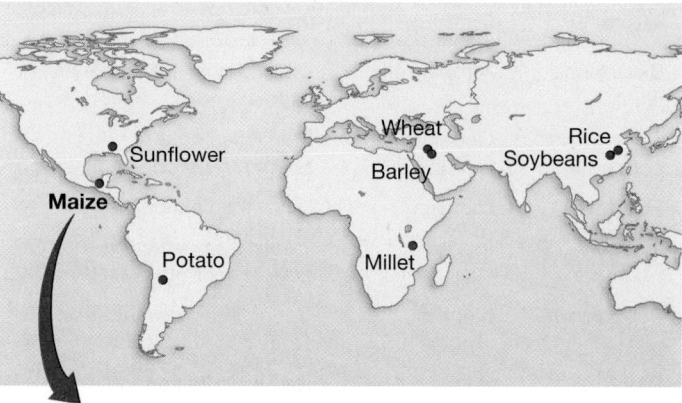

(b) Artificial selection changes the traits of domesticated species.

FIGURE 31.2 Crop Plants Are Derived from Wild Species via Artificial Selection. (a) Crop plants have originated on virtually every continent. **(b)** In some cases, artificial selection has changed domesticated species so radically that they bear little resemblance to their wild ancestors.

streaking across Europe and North America. It is no exaggeration to claim that the organic compounds synthesized by plants more than 300 million years ago fueled the industrial revolution.

Coal still supplies about 20 percent of the energy used in Japan and Western Europe and 80 percent of the energy used in China. Current research is focused on finding cleaner and more efficient ways of mining and burning coal.

Fiber and Building Materials Although nylon and polyester derived from petroleum are increasingly important in manufacturing, cotton and other types of plant fibers are still important sources of raw material for clothing, rope, and household articles like towels and bedding.

Woody plants also provide lumber for houses and furniture. Relative to its density, wood is a stiffer and stronger building material than concrete, cast iron, aluminum alloys, or steel.

Woody plants also provide most of the fibers used in papermaking. The cellulose fibers obtained from trees or bamboo and then used in paper manufacturing are stronger under tension (pulling) than nylon, silk, chitin, collagen, tendon, or bone—even though cellulose is 25 percent less dense. One line

TABLE 31.1 Some Drugs Derived from Land Plants

Source	Compound	Use
Belladonna plant	Atropine	Dilating pupils during eye exams
Opium poppy	Codeine, morphine	Pain relief, cough suppressant (codeine)
Foxglove	Digitalin	Heart medication
Ipecac	Ipecac	Treating amoebic dysentery, poison control
Peppermint	Menthol	Cough suppressant, relief of stuffy nose
Papaya	Papain	Reducing inflammation, treating wounds
Quinine tree	Quinine	Malaria prevention
	Quinidine	Heart medication
Aspen, willow trees	Salicin	Pain relief (aspirin)
Wild yams	Steroids	Precursor compounds for manufacture of birth control pills and cortisone (to treat inflammation)
Pacific yew	Taxol	Treating ovarian cancer
Curare vine	Tubocurarine	Muscle relaxant used in surgery
Rosy periwinkle	Vinblastine, vincristine	Treating leukemia (cancer of blood)

of research is focused on bioengineering to reduce lignin content of certain woody species, so fewer pollutants are generated in extracting lignin during the papermaking process.

Pharmaceuticals In both traditional and modern medicine, plants are a key source of drugs. **TABLE 31.1** lists some of the more familiar medicines derived from land plants; overall, an estimated 25 percent of the prescriptions written in the United States each year include at least one molecule derived from plants.

In most cases, plants synthesize these compounds in order to repel insects, deer, or other types of herbivores. For example, experiments have confirmed that morphine, cocaine, nicotine, caffeine, and other toxic compounds found in plants are effective deterrents to herbivores. Researchers continue to isolate and test new plant compounds for medicinal use in humans and domesticated animals.

31.2 How Do Biologists Study Green Algae and Land Plants?

Given the importance of plants to the planet in general and humans in particular, it is not surprising that knowing as much as possible about plants, including how they evolved, is a key component of biological science. To understand how green plants

originated and diversified, biologists analyze (**1**) morphological traits, (**2**) the fossil record, and (**3**) phylogenetic trees estimated from similarities and differences in DNA sequences from homologous genes and whole genomes.

These approaches are complementary and have produced a remarkably clear picture of how land plants evolved from green algae and then diversified. Let's consider each of these research strategies.

Analyzing Morphological Traits

The green algae include species that are unicellular, colonial, or multicellular and that live in marine, freshwater, or moist terrestrial habitats. The vast majority are aquatic. Although some land plants also live in ponds or lakes or rivers, the vast majority live on land.

Similarities Between Green Algae and Land Plants The green algae have long been hypothesized to be closely related to land plants, because key traits are similar in the two groups.

- Their chloroplasts contain the photosynthetic pigments chlorophyll *a* and *b* and the accessory pigment β-carotene.

- They have similar arrangements of the internal, membrane-bound sacs called thylakoids (see Chapter 10).

- Their cell walls, sperm, and peroxisomes are similar in structure and composition. (Recall that peroxisomes are organelles in which specialized oxidation reactions take place; see Chapter 7.)

- Their chloroplasts synthesize starch as a storage product.

Of all the green algal groups, the two most similar to land plants are the Coleochaetophyceae (coleochaetes) and Charophyceae (stoneworts; **FIGURE 31.3**). Because the species that make up these groups are multicellular and live in ponds and other types of freshwater environments, biologists hypothesize that land plants evolved from multicellular green algae that lived in freshwater habitats.

Major Morphological Differences among Land Plants Based on morphology, the land plants were traditionally clustered into three broad categories:

1. **Nonvascular plants** (also called bryophytes) lack **vascular tissue**—specialized groups of cells that conduct water and

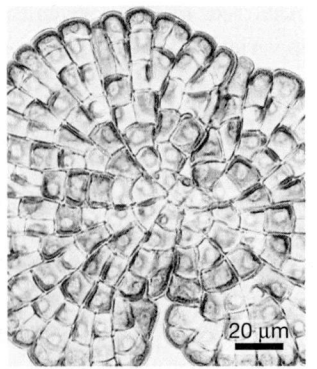

Coleochaetophyceae (coleochaetes)   Charophyceae (stoneworts)

FIGURE 31.3 Most Green Algae Are Aquatic. Examples of species from the green algal lineages closely related to the land plants.

(a) Nonvascular plants

Do not have vascular tissue to conduct water and provide support (for example, mosses)

(b) Seedless vascular plants

Have vascular tissue but do not make seeds (for example, ferns)

(c) Seed plants

Have vascular tissue and make seeds (for example, flowering plants, or angiosperms)

FIGURE 31.4 Morphological Diversity in Land Plants.

nutrients from one part of the plant body to another. Mosses are one group of nonvascular plants (**FIGURE 31.4a**).

2. Seedless vascular plants have vascular tissue but do not make seeds. Instead, they make microscopic spores that are carried by wind to new habitats. Ferns are an example of seedless vascular plants (**FIGURE 31.4b**).

3. Seed plants have vascular tissue. A **seed** consists of an embryo and a store of nutritive tissue, surrounded by a tough protective layer. The flowering plants, or **angiosperms** (literally, "encased seeds"), are a group of seed plants (**FIGURE 31.4c**).

How are nonvascular plants, seedless vascular plants, and seed plants related to each other, and to green algae?

Using the Fossil Record

The first green plants in the fossil record are green algae, found in rocks from 700 to 725 million years old. The first land plant fossils are found in rocks that are about 475 million years old. Because green algae appear long before land plants, the data support the hypothesis that land plants are derived from green algae.

At roughly the same time that green algae appeared and began to diversify, the oceans and atmosphere were starting to become

oxygen rich—as never before in Earth's history. Based on this time correlation, it is reasonable to hypothesize that the evolution of green algae contributed to the rise of oxygen levels on Earth. The origin of the oxygen atmosphere occurred not long before the appearance of animals in the fossil record and may have played a role in their origin and early diversification.

The fossil record of the land plants themselves is massive. In an attempt to organize and synthesize the database, **FIGURE 31.5** breaks it into five time intervals—each encompassing a major event in the diversification of land plants.

Origin of Land Plants The oldest interval begins about 475 million years ago (mya), spans some 60 million years, and documents the origin of the group.

Most of the fossils dating from this period are microscopic. They consist of the reproductive cells called spores and sheets of a waxy coating called cuticle. Several observations support the hypothesis that these fossils came from green plants that were growing on land.

1. Cuticle is a watertight barrier that coats the aboveground parts of today's land plants and helps them resist drying.

2. The fossilized spores are surrounded by a sheetlike coating. Under the electron microscope, the material appears almost

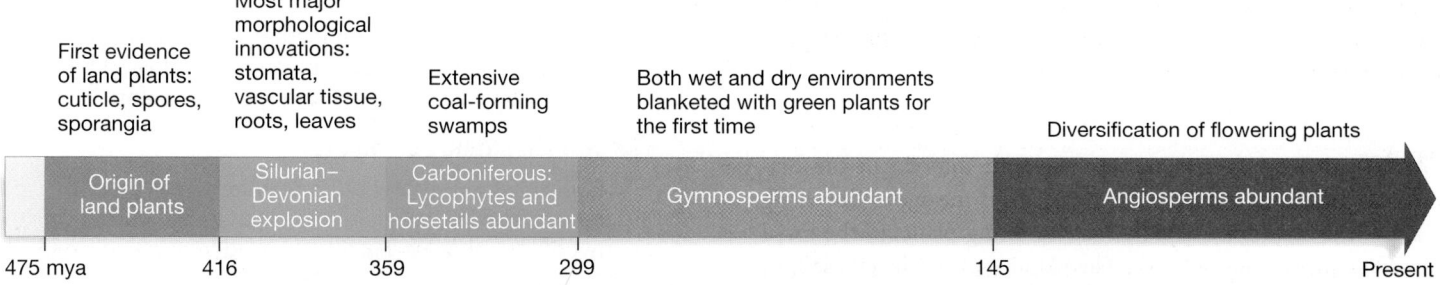

FIGURE 31.5 The Fossil Record of Land Plants Can Be Broken into Five Major Intervals. Note that the first insects are found in the fossil record at about 400 mya, the first terrestrial vertebrate animals at 365 mya, and the first mammals at 190 mya.

identical in structure to a waxy material called **sporopollenin**. Sporopollenin also encases spores and pollen from modern land plants and helps them resist drying.

3. Fossilized spores that are 475 million years old have recently been found in association with spore-producing structures, called **sporangia** (singular: **sporangium**), that are similar to sporangia observed in some nonvascular plants.

Silurian–Devonian Explosion The second major interval in the fossil record of land plants is called the "Silurian–Devonian explosion." In rocks dated 416–359 mya, biologists find fossils from most of the major plant lineages. Virtually all of the adaptations that allow plants to occupy dry, terrestrial habitats are present, including water-conducting tissue and roots.

According to the fossil record, plants colonized the land in conjunction with fungi that grew in a mutually beneficial symbiosis. The fungi grew belowground and helped provide land plants with nutrients from the soil; in return, the plants provided the fungi with sugars and other products of photosynthesis (see Chapters 32 and 39).

The Carboniferous Period The third interval in the fossil history of plants spans the aptly named Carboniferous period. In sediments dated from about 359 to 299 mya, biologists find extensive deposits of coal. Coal is a carbon-rich rock packed with fossil spores, branches, leaves, and tree trunks.

Most of these fossils are derived from seedless vascular plants in the Lycophyta (the lycophytes or club mosses), Equisetophyta (horsetails), and Pteridophyta (ferns). Although the only living lycophytes and horsetails are small, during the Carboniferous these groups were species rich and included a wide array of tree-sized forms.

Because coal formation is thought to start only in the presence of water, the Carboniferous fossils indicate the presence of extensive forested swamps.

Diversification of Gymnosperms The fourth interval in land plant history is characterized by seed plants called **gymnosperms** ("naked-seeds"). Five major groups of gymnosperms are living today: Ginkgophyta (ginkgoes), Cycadophyta (cycads), Cupressophyta (redwoods, junipers, and yews), Pinophyta (pines, spruces, and firs), and Gnetophyta (gnetophytes).

Because gymnosperms grow readily in dry habitats, biologists infer that both wet and dry environments on the continents became blanketed with green plants for the first time during this interval. Gymnosperms are prominent in the fossil record from 299 mya to 145 mya.

Diversification of Angiosperms The fifth interval in the history of land plants is still under way. This is the age of flowering plants—the angiosperms. The first flowering plants in the fossil record appear about 150 mya. The plants that produced the first flowers are the ancestors of today's grasses, orchids, daisies, oaks, maples, and roses.

According to the fossil record, then, the green algae appear first, followed by the nonvascular plants, seedless vascular plants, and seed plants. Organisms that appear late in the fossil record are often less dependent on moist habitats than are groups that appear earlier. For example, the sperm cells of mosses and ferns swim to accomplish fertilization, while gymnosperms and angiosperms produce pollen grains that are transported via wind or insects and that then produce sperm.

These observations support the hypotheses that green plants evolved from green algae and that land plants evolved to colonize dry habitats.

To test the validity of these observations, biologists analyze data sets that are independent of the fossil record. Foremost among these are DNA sequences used to infer phylogenetic trees. Does the phylogeny of land plants confirm or contradict the patterns in the fossil record?

Evaluating Molecular Phylogenies

The phylogenetic tree in **FIGURE 31.6** is a recent version of results emerging from laboratories around the world. The black bars across some branches show when key innovations occurred.

Each of the following bullets states a key observation about this tree, followed by an interpretation—often, a hypothesis that is supported by the observation.

- The green plants are monophyletic.
  Interpretation: A single common ancestor gave rise to all of the green algae and land plants.

- Green algae are paraphyletic.
  Interpretation: The green algae include some but not all of the descendants of a single common ancestor.

- Charophyceae are the closest living relative to land plants.
  Interpretation: Land plants evolved from a multicellular green alga that lived in freshwater habitats.

check your understanding

If you understand that . . .
- Biologists use the fossil record and phylogenetic analyses to study how green plants diversified.
- The data analyzed to date support the hypotheses that green plants are monophyletic and that land plants evolved from multicellular green algae that inhabited freshwater.
- The fossil record and molecular analyses agree that the nonvascular plants evolved first, followed by the seedless vascular plants and the seed plants.

✔ **You should be able to . . .**

Explain why (1) morphological data, (2) the fossil record, and (3) molecular phylogenies all support the hypothesis that land plants evolved from green algae.

Answers are available in Appendix A.

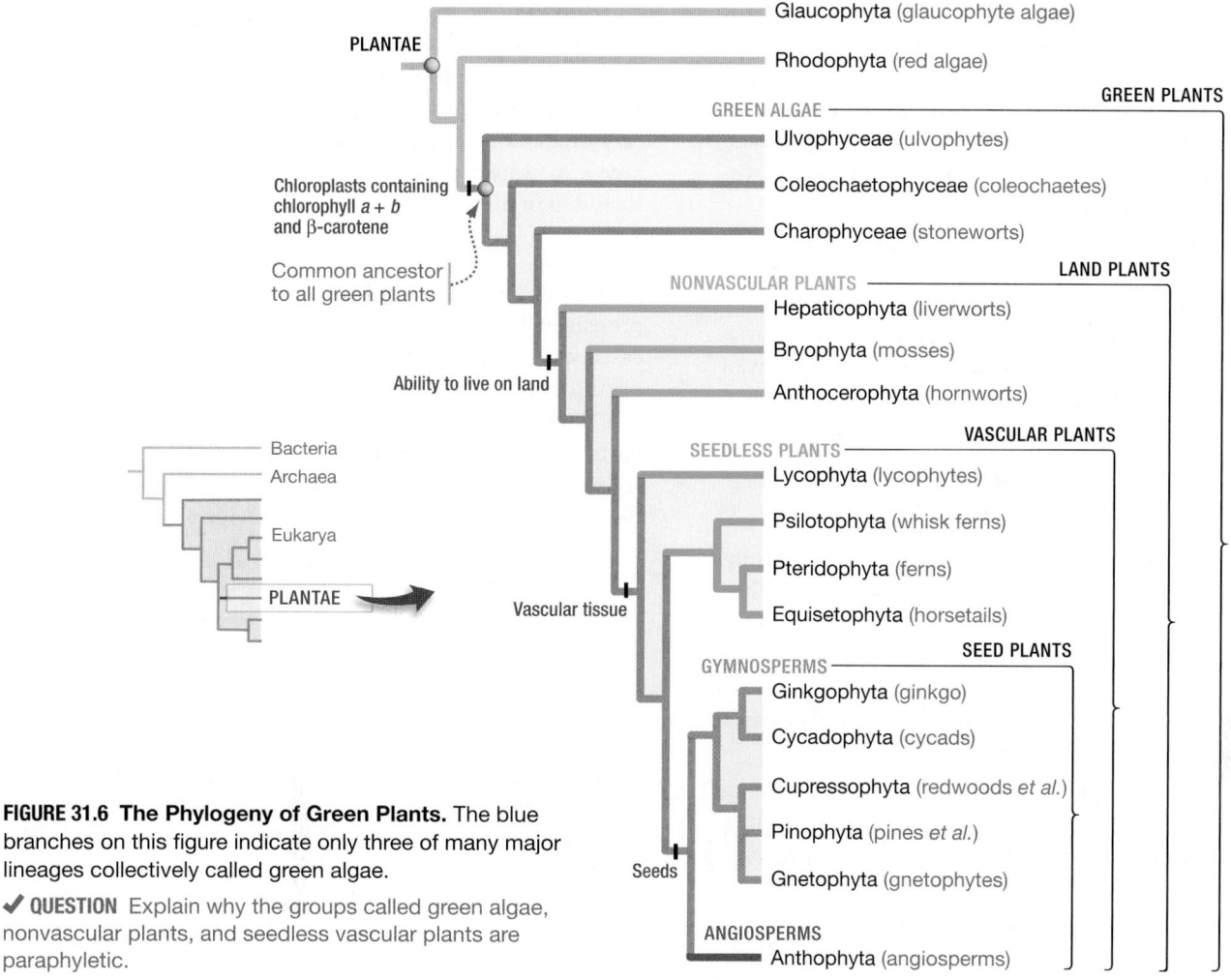

FIGURE 31.6 The Phylogeny of Green Plants. The blue branches on this figure indicate only three of many major lineages collectively called green algae.

✔ **QUESTION** Explain why the groups called green algae, nonvascular plants, and seedless vascular plants are paraphyletic.

• Land plants are monophyletic.
Interpretation: There was only one successful transition from freshwater environments to land.

• The nonvascular plants are the earliest-branching groups among land plants.
Interpretation: The nonvascular plants are the most ancient living group of land plants.

• The nonvascular plants are paraphyletic.
Interpretation: The nonvascular plants include some but not all descendants of a single common ancestor.

• The seedless vascular plants are paraphyletic, but the vascular plants as a whole are monophyletic.
Interpretation: Vascular tissue evolved once.

• The seed plants—the gymnosperms plus angiosperms—are monophyletic.
Interpretation: The seed evolved once.

• The gymnosperms are a monophyletic group, as are the angiosperms.

Interpretation: Among seed plants, there was a major divergence in how seeds develop—either "naked" (in gymnosperms) or protected inside a capsule (in angiosperms).

Although the tree in Figure 31.6 will undoubtedly change and improve as additional data accumulate, biologists are confident about its most fundamental message: The fossil record and the phylogenetic tree based on DNA sequence data agree on the order in which groups appeared. Land plant evolution began with nonvascular plants, proceeded to seedless vascular plants, and continued with the evolution of seed plants.

31.3 What Themes Occur in the Diversification of Land Plants?

Land plants have evolved from algae that grew on the muddy shores of ponds 475 million years ago to organisms that enrich the soil, produce much of the oxygen you breathe and most of the food you eat, and serve as symbols of health, love, and beauty. How did this happen?

Answering this question begins with recognizing the most striking trend in the phylogeny and fossil record of green plants: The most ancient groups in the lineage are dependent on wet habitats, while the more recently evolved groups can live and reproduce in dry—or even desert—conditions. The story of land plants is the story of adaptations that allowed photosynthetic organisms to move from aquatic to terrestrial environments.

Let's first consider adaptations that allowed plants to grow in dry conditions without drying out and dying, and then analyze the evolution of traits that allowed plants to reproduce efficiently on land. This section closes with a brief look at the radiation of flowering plants, which are the most important plants in many of today's terrestrial environments.

The Transition to Land, I: How Did Plants Adapt to Dry Conditions with Intense Sunlight?

For aquatic green algae, terrestrial environments are deadly. Compared with a habitat in which the entire organism is bathed in water, in terrestrial environments only a portion, if any, of the plant's tissues are in direct contact with water. Tissues that are exposed to air tend to dry out and die.

Once green plants made the transition to survive out of water, though, growth on land offered a bonanza of resources.

- *Light* The water in ponds, lakes, and oceans absorbs and reflects light. As a result, the amount of light available to drive photosynthesis is drastically reduced even a meter or two below the water's surface.

- *Carbon dioxide* CO_2—the most important molecule required by photosynthetic organisms—is more abundant in the atmosphere and diffuses more readily there than it does in water.

Natural selection favored early land plants with adaptations that solved the drying problem. These adaptations arose in three steps: (1) preventing water loss, which kept cells from drying out and dying; (2) providing protection from harmful ultraviolet (UV) radiation; and (3) moving water from tissues with direct access to water to tissues without direct access. Let's examine each step in turn.

Preventing Water Loss: Cuticle and Stomata Section 31.2 points out that sheets of the waxy substance called cuticle are present early in the fossil record of land plants, along with encased spores. This observation is significant because the presence of cuticle in fossils is a diagnostic indicator of land plants.

Cuticle is a watertight sealant that covers the aboveground parts of plants and gives them the ability to survive in dry environments (**FIGURE 31.7a**). If biologists had to point to one innovation that made the transition to land possible, it would be the random mutations that led to the production of cuticle.

Covering surfaces with wax creates a problem, however, regarding the exchange of gases across those surfaces. Plants need to take in carbon dioxide (CO_2) from the atmosphere in order to perform photosynthesis. But cuticle is almost as impervious to CO_2 as it is to water.

Most plants solve this problem with a structure called a **stoma** ("mouth"; plural: **stomata**). A stoma consists of an opening surrounded by specialized **guard cells** (**FIGURE 31.7b**). The opening, called a **pore,** opens or closes as the guard cells change shape. When guard cells lose water, they become flaccid or "limp" and the stomata close. Pores are normally closed at night to limit water loss from the plant when CO_2 uptake is not needed. But when

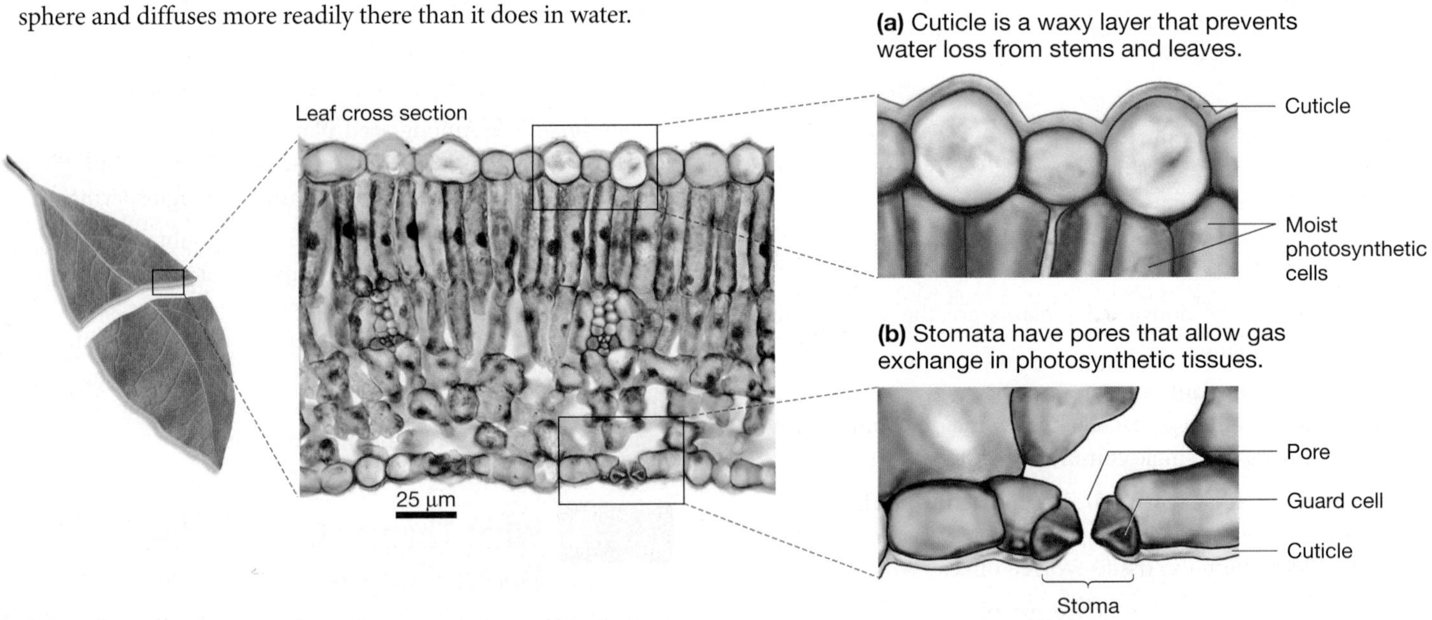

(a) Cuticle is a waxy layer that prevents water loss from stems and leaves.

Leaf cross section

Cuticle

Moist photosynthetic cells

25 μm

(b) Stomata have pores that allow gas exchange in photosynthetic tissues.

Pore

Guard cell

Cuticle

Stoma

FIGURE 31.7 Cuticle and Stomata Are the Most Fundamental Plant Adaptations to Life on Land. In these micrographs, leaf cells have been stained blue to make their structure more visible. **(a)** The interior of plant leaves and stems is extremely moist; cuticle (stained red) prevents water from evaporating away. **(b)** Stomata, which are opened and closed by guard cells, create pores to allow CO_2 to diffuse into the interior of leaves and stems where cells are actively photosynthesizing. Stomata are often on the underside of leaves.

guard cells absorb water and then become turgid or "taut," they open the pore. Open stomata allow CO_2 to diffuse into the interior of leaves and stems where cells are actively photosynthesizing. (The mechanism behind guard-cell movement is explored in Chapter 40.)

Stomata are present in all land plants except the liverworts, which have pores but no guard cells. These data suggest that the earliest land plants evolved pores that allowed gas exchange to occur at breaks in the cuticle-covered surface. Later, the evolution of guard cells gave land plants the ability to regulate gas exchange—and control water loss—by opening and closing their pores.

Providing Protection From UV Irradiation

Life out of water gave land plants a distinct advantage. Because they were exposed to higher light intensities, the plants could carry out photosynthesis faster. However, out of water they were also exposed to the UV rays of the sun. UV light is known to damage DNA by causing thymine dimers (see Chapter 15). Water absorbs UV light, so algae did not face this problem to the same extent.

Researchers think that many species of algae probably colonized the wet soil near their home ponds, but only some were able to survive the harsh sunlight. The plants that survived were those that by chance made compounds that absorb UV light. Most plants today accumulate UV-absorbing compounds, called flavonoids, to protect their DNA from damage. These flavonoids are plant pigments that function as a sunscreen for leaves and stems. Humans use similar compounds to manufacture sunscreen.

The first plants had a cuticle to keep from drying out, and UV-absorbing compounds to act as sunscreen, so they could grow on the water-saturated soils of lake or pond edges. The next challenge? Defying gravity.

The Importance of Upright Growth

Multicellular green algae can grow erect because they float. They float because the density of their cells is similar to water's density. But outside of water, the body of a multicellular green alga collapses. The water that fills its cells is 1000 times denser than air. Although the cell walls of green algae are strengthened by the presence of cellulose, their bodies lack the structural support to withstand the force of gravity and to keep an individual erect in air.

Based on these observations, biologists hypothesize that the first land plants were small or had a low, sprawling growth habit. Besides lacking rigidity, the early land plants would have had to obtain water through pores or through cells that lacked cuticle—meaning they would have had to grow in a way that kept many or most of their tissues in direct contact with moist soil. If this hypothesis is correct, then competition for space and light would have become intense soon after the first plants began growing on land.

In a terrestrial environment, individuals that can grow erect have much better access to sunlight than individuals that are incapable of growing erect. But two problems have to be overcome for a plant to grow erect: (1) transporting water from tissues that are in contact with wet soil to tissues that are in contact with dry air, against the force of gravity; and (2) becoming rigid enough to avoid falling over in response to gravity and wind. As it turns out, vascular tissue solved both problems.

The Origin of Vascular Tissue

Paul Kenrick and Peter Crane explored the origin of water-conducting cells and erect growth in plants by examining the extraordinary fossils found in a rock formation in Scotland called the Rhynie Chert. These rocks formed about 400 million years ago and contain some of the first large plant specimens in the fossil record—as opposed to the microscopic spores and cuticle found in older rocks.

The Rhynie Chert contains numerous plants that fossilized in an upright position. This indicates that many or most of the Rhynie plants grew erect. How did they stay vertical?

By examining fossils with the electron microscope, Kenrick and Crane established that species from the Rhynie Chert contained elongated cells that were organized into tissues along the length of the plant. Based on these data, the biologists hypothesized that the elongated cells were part of water-conducting tissue and that water could move from the base of the plants upward to erect portions through these specialized water-conducting cells.

- Some of the fossilized water-conducting cells had simple, cellulose-containing cell walls like the water-conducting cells found in today's mosses (**FIGURE 31.8a**, see page 586).

- Some of the water-conducting cells present in the early fossils had cell walls with thickened rings containing a molecule called lignin (**FIGURE 31.8b**).

Lignin is a complex polymer built from six-carbon rings. It is extraordinarily strong for its weight and is particularly effective in resisting compressing forces such as gravity.

These observations inspired the following hypothesis: The evolution of lignified cell walls gave stem tissues the strength to remain erect in the face of wind and gravity. Today, the presence of lignin in the cell walls of water-conducting cells is considered the defining feature of vascular tissue. The evolution of vascular tissue allowed early plants to support erect stems and transport water from roots to aboveground tissues. (See Chapter 38 for additional details on the structure and function of plant vascular tissue.)

Elaboration of Vascular Tissue: Tracheids and Vessels

Once simple water-conducting tissues evolved, evolution by natural selection favored more complex tissues that were more efficient in providing support and transport.

In fossils that are about 380 million years old, biologists find the advanced water-conducting cells called tracheids. **Tracheids** are long, thin, tapering cells that have

- a thickened, lignin-containing **secondary cell wall** in addition to a cellulose-based **primary cell wall;** and

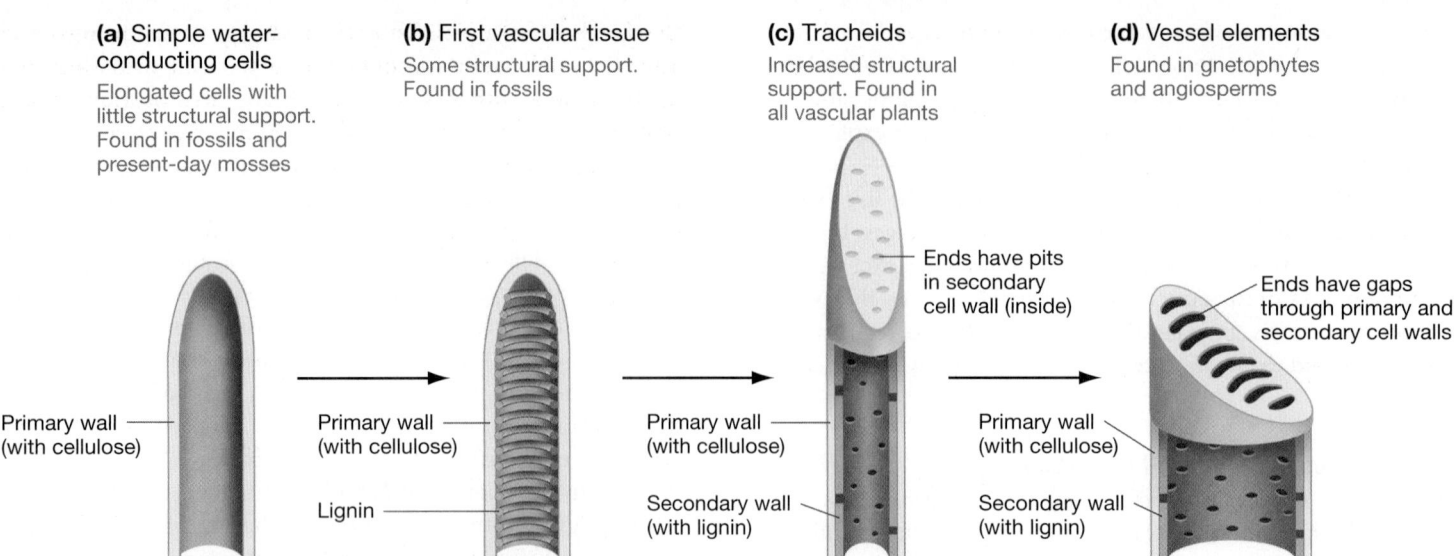

(a) Simple water-conducting cells
Elongated cells with little structural support. Found in fossils and present-day mosses

(b) First vascular tissue
Some structural support. Found in fossils

(c) Tracheids
Increased structural support. Found in all vascular plants

(d) Vessel elements
Found in gnetophytes and angiosperms

Ends have pits in secondary cell wall (inside)

Ends have gaps through primary and secondary cell walls

Primary wall (with cellulose)

Primary wall (with cellulose)

Lignin

Primary wall (with cellulose)

Secondary wall (with lignin)

Primary wall (with cellulose)

Secondary wall (with lignin)

FIGURE 31.8 Evolutionary Sequence Observed in Water-Conducting Cells. According to the fossil record and the phylogeny of green plants, water-conducting cells became stronger over time due to the evolution of lignin and secondary cell walls. Efficient water transport was maintained through pits where the secondary cell wall is missing, or gaps where both the secondary and primary cell wall are absent.

✔ **QUESTION** Biologists claim that vessels are more efficient than tracheids at transporting water, in part because vessels are shorter and wider than tracheids. Why does this claim make sense?

- pits in the sides and ends of the cell where the secondary cell wall is absent, where water can flow efficiently from one tracheid to the next (**FIGURE 31.8c**).

The secondary cell wall gave tracheids the ability to provide better structural support, but water could still move easily through the cells because of the pits. Today, all vascular plants contain tracheids.

In fossils dated to 250–270 million years ago, biologists have documented the most advanced type of water-conducting cells observed in plants. **Vessel elements** are shorter and wider than tracheids, and their upper and lower ends have gaps where both the primary and secondary cell wall are missing. The width of vessels and the presence of open gaps reduce resistance and make water movement more efficient (**FIGURE 31.8d**). In vascular tissue, vessel elements are lined up end to end to form a continuous pipelike structure.

In the stems and branches of some vascular plant species, tracheids or a combination of tracheids and vessels form the extremely strong support material called **wood.** (The anatomy of wood is explained in detail in Chapter 37.) The ability of plants to make lignified vascular tissues allowed them to grow tall and transport water from soil all the way to the top—a remarkable engineering feat.

Mapping Evolutionary Changes on the Phylogenetic Tree

Cuticle, stomata, and vascular tissue were key adaptations that allowed early plants to colonize and thrive on land. **FIGURE 31.9** summarizes how land plants adapted to dry conditions by mapping on the phylogenetic tree where major innovations occurred as the group diversified.

As you study the tree in Figure 31.9, note that fundamentally important adaptations to dry conditions—such as cuticle, pores, stomata, vascular tissue, and tracheids—evolved just once. But convergent evolution also occurred (see Chapter 28). When convergence occurs, similar traits evolve independently in two distinct lineages. Vessels evolved independently in gnetophytes and angiosperms.

The evolution of cuticle, stomata, and vascular tissue made it possible for plants to avoid drying out and to grow upright, while moving water from the base of the plant to its apex. Plants gained adaptations that allowed them not only to survive on land, but also to flourish. Now the question is, how did they reproduce?

The Transition to Land, II: How Do Plants Reproduce in Dry Conditions?

Life cycles of sexually reproducing eukaryotes, including plants, serve several different functions:

- increase genetic variability as a result of meiosis and fertilization;

- increase the number of individuals; and

- disperse individuals to new habitats.

Each of these functions is necessary if a species is to survive. To understand the role of each stage of a life cycle and how life cycles evolve, researchers ask a series of key questions. Is a species mobile or sessile? Does it live in water or on land?

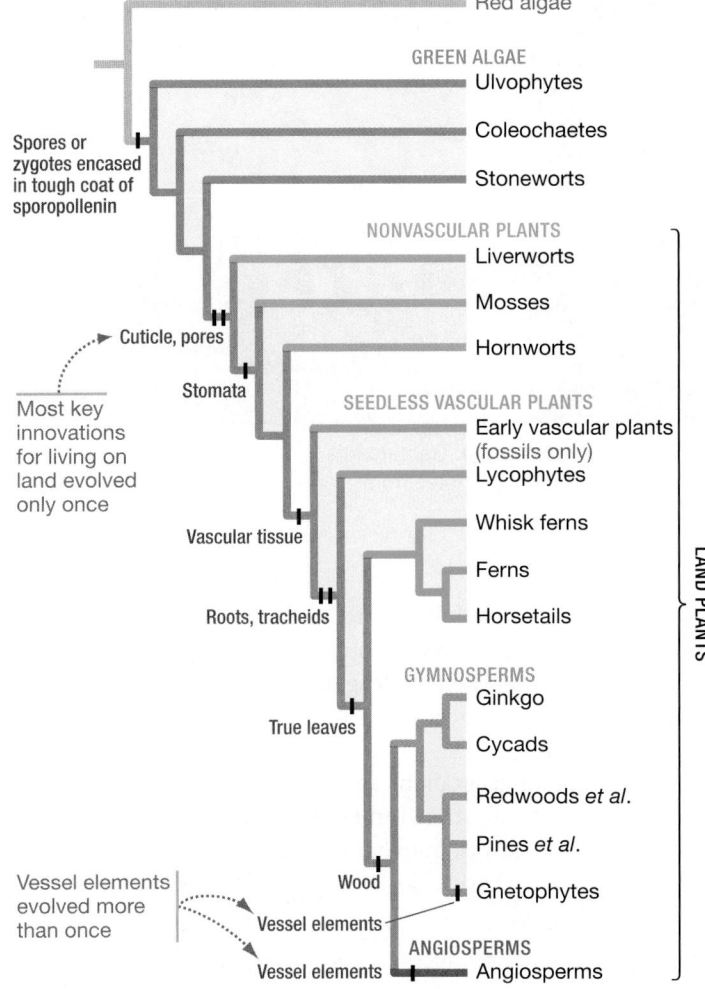

Red algae

GREEN ALGAE

Ulvophytes

Coleochaetes

Stoneworts

Spores or zygotes encased in tough coat of sporopollenin

NONVASCULAR PLANTS

Liverworts

Mosses

Hornworts

Cuticle, pores

Stomata

Most key innovations for living on land evolved only once

SEEDLESS VASCULAR PLANTS

Early vascular plants (fossils only)

Lycophytes

Whisk ferns

Vascular tissue

Ferns

Horsetails

Roots, tracheids

GYMNOSPERMS

Ginkgo

True leaves

Cycads

Redwoods *et al.*

Pines *et al.*

Wood

Gnetophytes

Vessel elements evolved more than once

Vessel elements

ANGIOSPERMS

Vessel elements

Angiosperms

LAND PLANTS

FIGURE 31.9 A Series of Evolutionary Innovations Allowed Plants to Adapt to Life on Land.

✔ **QUESTION** Explain the logic that biologists use to map the location of the innovations on phylogenetic trees.

In some organisms, all three functions are accomplished at one stage of the life cycle. For example, after meiosis coral polyps release many gametes into the water, where they fuse and disperse. Even though adult corals are sessile animals, they use water currents to move their genetically variable gametes and larvae. Terrestrial animals are mobile, so adults can easily disperse and join other adults for mating.

How do green plants reproduce? Keep in mind two important points—the impact of the transition from water to land on the dispersal stage of the life cycle, and the fact that land plants are sessile. Each of these factors has profoundly affected how life cycles of land plants evolved.

Section 31.2 introduces one of the key adaptations for reproducing on land: spores that resist drying because they are encased in a tough coat of sporopollenin. Sporopollenin-like compounds are found in the walls of some green algal zygotes, and thick-walled, sporopollenin-rich spores appear early in

the fossil record of land plants. Based on these observations, biologists infer that sporopollenin-encased spores were one of the innovations that made the initial colonization of land possible.

Two other innovations that occurred early in land plant evolution were instrumental for efficient reproduction in a dry environment: (1) Gametes were produced in complex, multicellular structures; and (2) the embryo was retained on the parent (mother) plant and was nourished by it.

Producing Gametes in Protected Structures   The fossilized gametophytes of early land plants contain specialized reproductive organs called **gametangia** (singular: **gametangium**). Although members of the green algae group Charophyceae (stoneworts) also develop gametangia, the gametangia found in land plants are larger and more complex.

The evolution of an elaborate gametangium was important because it protected gametes from drying and from mechanical damage. Gametangia are present in all land plants living today except angiosperms, where structures inside the flower perform the same functions.

In both the Charophyceae and the land plants, individuals produce distinctive male and female gametangia (**FIGURE 31.10**, see page 588).

- The sperm-producing gametangium is called an **antheridium** (plural: **antheridia**).

- The egg-producing gametangium is called an **archegonium** (plural: **archegonia**).

In terms of their function, antheridia and archegonia are analogous to the testes and ovaries of animals.

Retaining Offspring: Land Plant Embryos Are Nourished by Their Parent   The second innovation that occurred early in land plant evolution involved the eggs that formed inside archegonia. Instead of shedding their eggs into the water or soil, land plants retain them.

Eggs are also retained in the green algal lineages that are most closely related to land plants: In Charophyceae and other closely related groups, sperm swim to the egg, fertilization occurs, and the resulting zygote stays attached to the parent. Either before or after fertilization, the egg or zygote receives nutrients from the mother plant. But the parent plant dies each autumn as the temperature drops. The zygote remains on the dead parental tissue, settles to the bottom of the lake or pond, and overwinters. In spring, meiosis occurs, and the resulting spores develop into haploid adult plants.

In land plants, the zygote is also retained on the parent plant after fertilization. But in contrast to the zygotes of most green algae, the zygotes of all land plants begin to develop on the living parent plant, forming a multicellular embryo that remains attached to the parent and can be nourished by it. This is important because land plant embryos do not have to manufacture

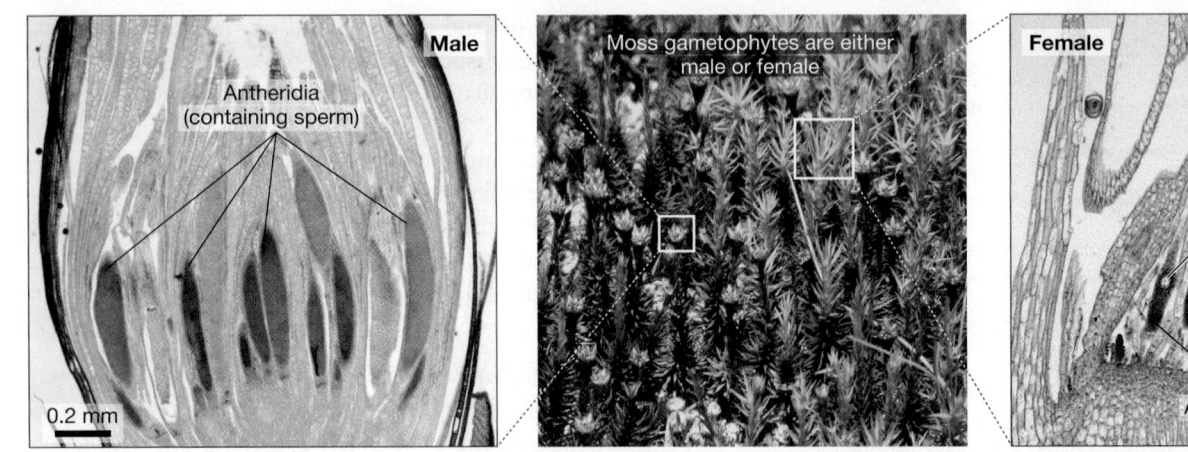

FIGURE 31.10 In All Land Plant Groups but Angiosperms, Gametes Are Produced in Gametangia. Gametangia such as in this moss (*Polytrichium*) are complex, multicellular structures that protect developing gametes from drying and mechanical damage.

their own food early in life. Instead, they receive most or all of their nutrients from the parent plant.

The retention of the embryo was such a key event in land plant evolution that the formal name of the group is **Embryophyta**—literally, the "embryo-plants." The retention of the fertilized egg in **embryophytes** is analogous to pregnancy in mammals, where offspring are retained by the mother and nourished through the initial stages of growth. Thick-walled spores, elaborate gametangia, and the embryophyte condition weren't the only key innovations associated with reproducing on land, though. In addition, all land plants undergo the phenomenon known as **alternation of generations** (see Chapter 30).

Alternation of Generations When alternation of generations occurs, individuals represent a multicellular haploid phase or a multicellular diploid phase. The multicellular haploid stage is called the **gametophyte;** the multicellular diploid stage is called the **sporophyte**. The two phases of the life cycle are connected by distinct types of reproductive cells—gametes and spores.

Although alternation of generations is observed in a wide array of eukaryotic lineages and in some groups of green algae, it does not occur in the algal groups most closely related to land plants. In the coleochaetes and stoneworts, the multicellular form is haploid. Only the zygote is diploid. As **FIGURE 31.11** shows, the zygote undergoes meiosis to form haploid spores. After dispersing with the aid of flagella, the spores begin dividing by mitosis and eventually grow into adult, haploid individuals. You might recall that this haploid-dominant life cycle is common in protists (see Chapter 30).

These data suggest that alternation of generations originated in land plants independently of its evolution in other groups of eukaryotes, and that it originated early in their history—soon after they evolved from green algae. Why was this strategy successful?

Multicellular sporophytes can make many spores and easily disperse them through wind currents. Male gametes

produced by multicellular gametophytes swim through moist soil or water droplets on the surface of the gametophyte. As you'll see below, when male gametes no longer needed to swim, there was no need for a free-living gametophyte stage in the life cycle.

Alternation of generations always involves the same basic sequence of events, illustrated in **FIGURE 31.12**. To review how this type of life cycle works, put your finger on the sporophyte in the figure and trace the cycle clockwise to find the following five key events in a land plant:

1. The sporophyte produces spores by meiosis. Spores are haploid.

2. Spores germinate and divide by mitosis to develop into multicellular, haploid gametophytes.

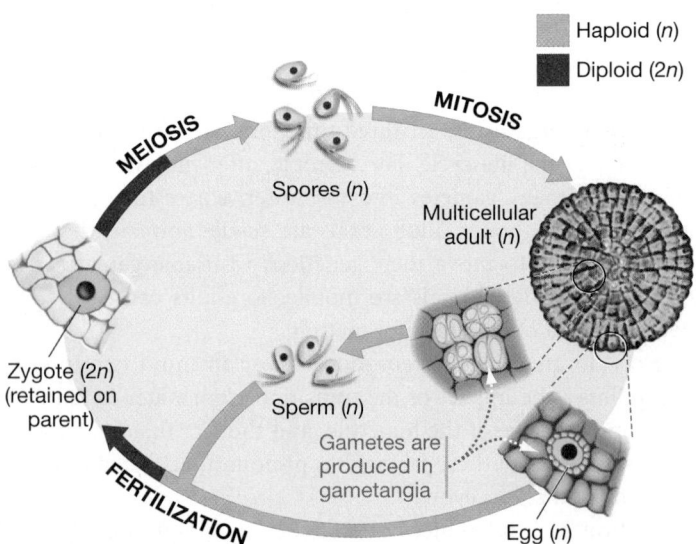

FIGURE 31.11 In Green Algae That Are Closely Related to Land Plants, Only the Zygote Is Diploid. The coleochaetes and stoneworts do not have alternation of generations. The multicellular stage is haploid.

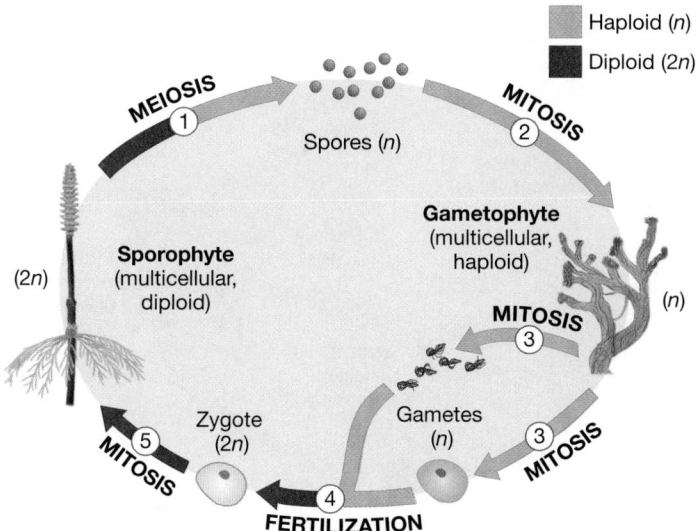

FIGURE 31.12 **All Land Plants Undergo Alternation of Generations.** Alternation of generations always involves the same sequence of five events.

3. Gametophytes produce gametes by mitosis. Both the gametophyte and the gametes are haploid, but gametophytes are multicellular while gametes are unicellular.

4. Two gametes unite during fertilization to form a diploid zygote.

5. The zygote divides by mitosis and develops into a multicellular, diploid sporophyte.

Once you've traced the cycle successfully, take a moment to compare and contrast zygotes and spores.

- Zygotes and spores are both single cells that divide by mitosis to form a multicellular individual. Zygotes develop into sporophytes; spores develop into gametophytes.

- Zygotes are diploid and spores are haploid.

- Zygotes result from the fusion of two haploid cells, such as a sperm and an egg, but spores are not formed by the fusion of gametes.

- Spores are produced by meiosis inside structures called sporangia; gametes are produced by mitosis inside gametangia.

Alternation of generations can be a difficult topic to master, for two reasons: (**1**) It is unfamiliar because it does not occur in humans or other animals, and (**2**) gamete formation results from mitosis—not meiosis, as it does in animals (see Chapter 13).

✔If you understand the basic principles of alternation of generations, you should be able to describe this type of life cycle and explain one hypothesis for why this life cycle might have evolved in land plants.

The Gametophyte-Dominant to Sporophyte-Dominant Trend in Life Cycles
The five steps illustrated in Figure 31.12 occur in all species with alternation of generations. But in land plants, the relationship between the gametophyte and sporophyte is highly variable.

In nonvascular plants such as mosses, the sporophyte is small and short lived and is largely dependent on the gametophyte for nutrition (**FIGURE 31.13**). When you see leafy-looking mosses growing on a tree trunk or on rocks, you are looking at gametophytes. Because the gametophyte is long lived and produces most of the food required by the individual, it is considered the dominant part of the life cycle.

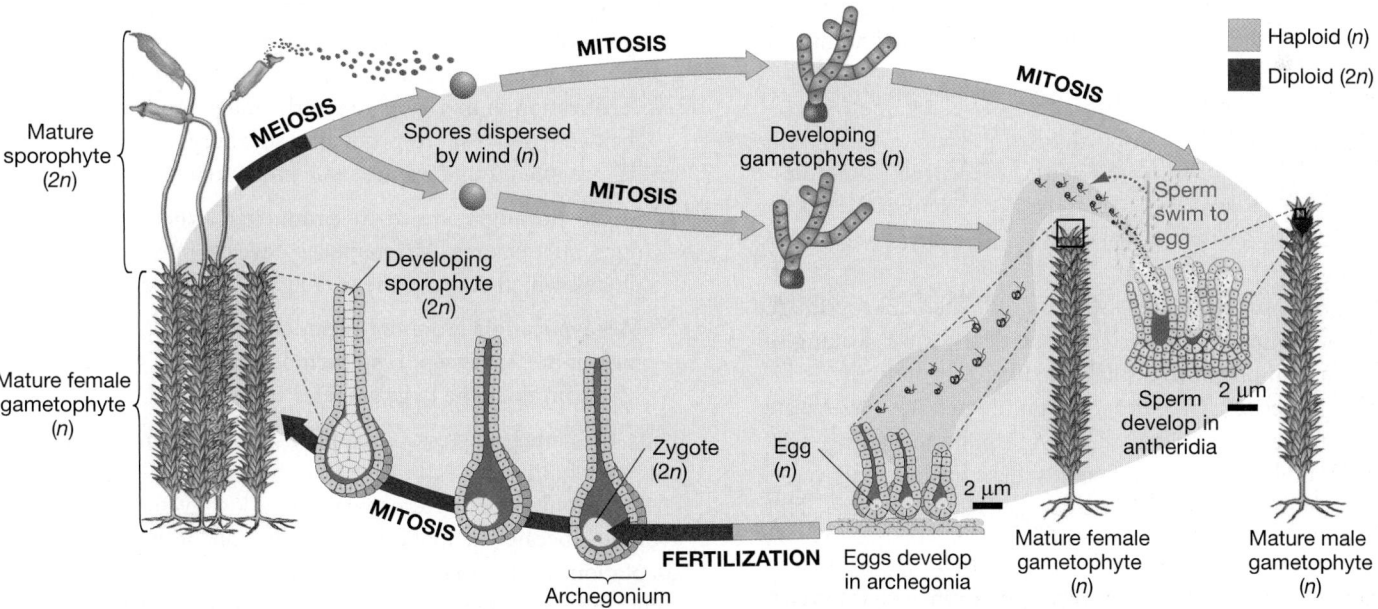

FIGURE 31.13 **Gametophyte-Dominated Life Cycles Evolved Early.** Like today's mosses, the earliest land plants in the fossil record have gametophytes that are much larger and longer lived than the sporophyte. The sporophyte depends on the gametophyte for nutrition.

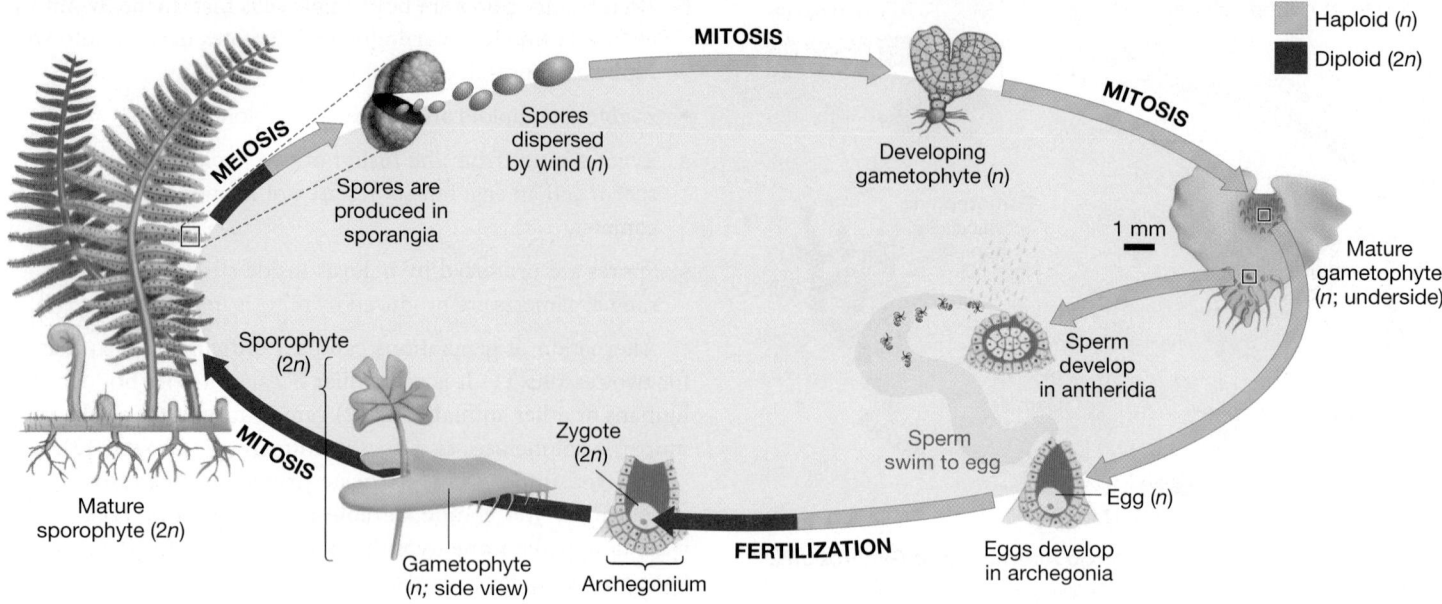

Haploid (n)
Diploid (2n)

MITOSIS

Spores dispersed by wind (*n*)

Spores are produced in sporangia

Developing gametophyte (*n*)

MITOSIS

MEIOSIS

1 mm

Mature gametophyte (*n*; underside)

Sporophyte (2*n*)

Sperm develop in antheridia

MITOSIS

Zygote (2*n*)

Sperm swim to egg

Egg (*n*)

Mature sporophyte (2*n*)

Gametophyte (*n*; side view)

Archegonium

FERTILIZATION

Eggs develop in archegonia

FIGURE 31.14 Sporophyte-Dominated Life Cycles Evolved Later. In lineages that evolved later, such as ferns, the sporophyte is much larger and longer lived than the gametophyte. However, when young, the sporophyte also depends on the gametophyte for nutrition.

✔ **QUESTION** How can you tell that alternation of generations occurs in mosses (Figure 31.13) and ferns (Figure 31.14)?

In contrast, in ferns and other vascular plants the sporophyte is much larger and longer lived than the gametophyte (**FIGURE 31.14**). The ferns you see growing in gardens or forests are sporophytes. You'd have to hunt on your hands and knees to find their free-living gametophytes, which are typically just a few millimeters in diameter. As you'll learn later in the chapter, the gametophytes of gymnosperms and angiosperms are even smaller—they are microscopic—and are retained within the sporophyte. Ferns and other vascular plants are said to have a sporophyte-dominant life cycle.

✔ If you understand the difference between gametophyte-dominant and sporophyte-dominant life cycles, you should be able to examine the photos of hornworts (a nonvascular plant) and horsetails (a seedless vascular plant) in **FIGURE 31.15** and identify which is the gametophyte and which is the sporophyte.

The transition from gametophyte-dominated life cycles to sporophyte-dominated life cycles is one of the most striking of all trends in land plant evolution. To explain why it occurred, biologists hypothesize that sporophyte-dominated life cycles were advantageous because diploid cells can respond to varying environmental conditions more efficiently than haploid cells can—particularly if the individual is heterozygous at many genes. This idea has yet to be tested rigorously, however.

Heterospory In addition to sporophyte-dominated life cycles, another important innovation found in seed plants is called **heterospory**—the production of two distinct types of spores by different structures.

All of the nonvascular plants and most of the seedless vascular plants are homosporous. **Homospory** is the production of a single type of spore. (Among the seedless vascular plants, some lycophytes and a few ferns are heterosporous.) Homosporous species produce spores that develop into bisexual gametophytes that produce both eggs and sperm (**FIGURE 31.16a**). If these gametophytes are isolated in nature, they can self-fertilize and produce offspring. If two bisexual gametophytes are close enough for the exchange of sperm, outcrossing is favored because it increases genetic variation (see Chapter 13). The mechanisms that prevent self-fertilization in this case are poorly understood.

The two types of spore-producing structures in heterosporous species are often found on the same individual (**FIGURE 31.16b**).

1. **Microsporangia** are spore-producing structures that produce microspores. **Microspores** develop into male gametophytes, which produce the small gametes called sperm.

2. **Megasporangia** are spore-producing structures that produce megaspores. **Megaspores** develop into female gametophytes, which produce the large gametes called eggs.

Thus, the gametophytes of seed plants are either male or female, but never both.

The evolution of heterospory was a key event in land plant evolution because it made possible one of the most important adaptations for life in dry environments—pollen.

Pollen The nonvascular plants and the seedless vascular plants have sperm that swim to the egg to perform fertilization. For a sperm cell to swim, there has to be a continuous sheet of water

(a) Hornwort gametophytes and sporophytes

(b) Horsetail gametophyte and sporophyte

FIGURE 31.15 The Reduction of the Gametophyte Is One of the Strongest Trends in Land Plant Evolution. (a) A hornwort with spike-like sporophytes emerging from the gametophyte. **(b)** A horsetail species, both as a tiny, microscopic gametophyte and as a large, macroscopic sporophyte.

between the male and female gametophytes, or a raindrop has to splash sperm onto a female gametophyte.

In species that live in dry environments, these conditions are rare. The land plants made their final break with their aquatic origins and were able to reproduce efficiently in dry habitats when a structure evolved that could move their gametes without the aid of water.

In heterosporous seed plants, the microspore germinates to form a tiny male gametophyte that is surrounded by a tough coat of sporopollenin, resulting in a **pollen grain.**

(a) Nonvascular plants and most seedless vascular plants are **homosporous.**

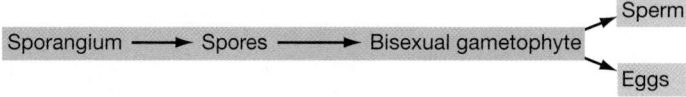

(b) Seed plants are **heterosporous.**

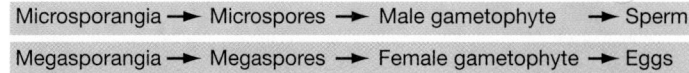

FIGURE 31.16 Unlike Homosporous Plants, Heterosporous Plants Produce Male and Female Spores That Are Morphologically Distinct.

Pollen grains can be exposed to the air for long periods of time without dying from dehydration. They are also tiny enough to be carried to female gametophytes by wind or animals. Upon landing near the egg, the male gametophyte releases the sperm cells that accomplish fertilization.

When pollen evolved, then, heterosporous plants lost their dependence on water to accomplish fertilization. Instead of swimming to the egg as a naked sperm cell, their tiny gametophytes took to the skies.

Seeds The evolution of large gametangia protected the eggs and sperm of land plants from drying. Embryo retention allowed offspring to be nourished directly by their parent, and pollen enabled fertilization to occur in the absence of water.

Retaining embryos has a downside, however: In ferns and horsetails, sporophytes have to live in the same place as their parent gametophyte. Seed plants overcome this limitation. Their embryos are portable and can disperse to new locations. The dispersal stage of the life cycle thus shifted from the haploid spore to the young diploid sporophyte.

A seed is a structure that, like a bird's egg, includes an embryo and a food supply surrounded by a tough coat that allows for effective dispersal of embryos (**FIGURE 31.17**). Spores are an effective dispersal stage in nonvascular plants and seedless vascular plants, but they lack the stored nutrients found in seeds.

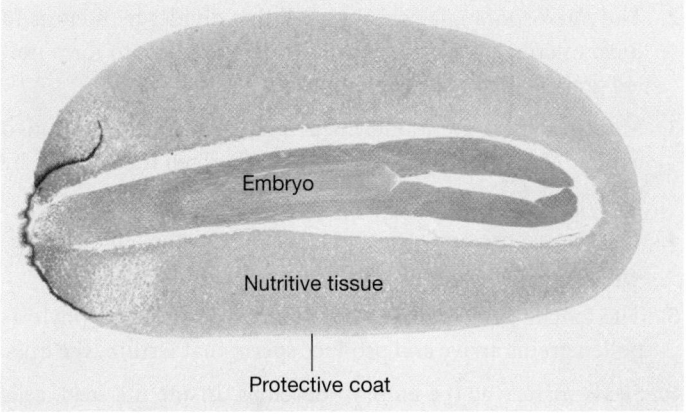

FIGURE 31.17 Seeds Contain an Embryo and a Food Supply and Can Be Dispersed. (This *Pinus* (pine) specimen has been stained.)

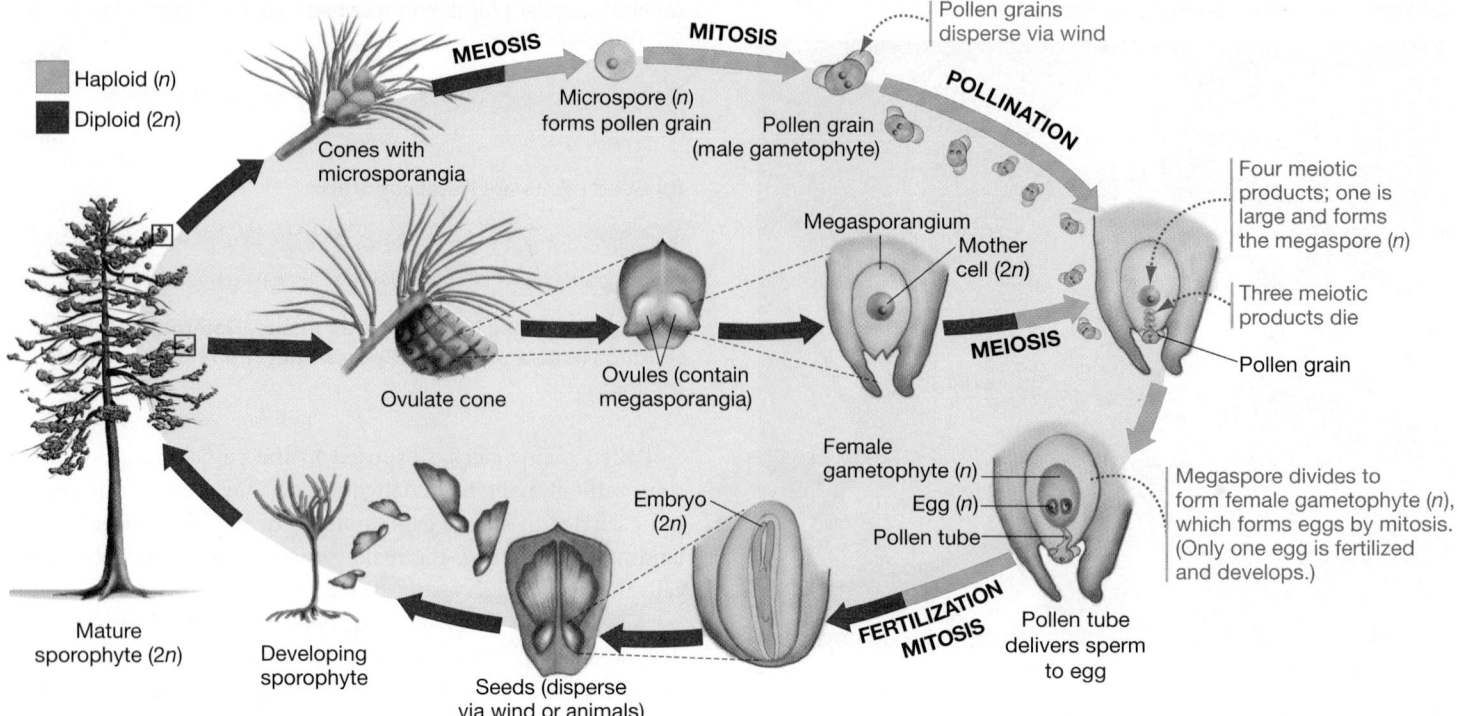

Haploid (*n*)
Diploid (*2n*)

MEIOSIS MITOSIS

Pollen grains disperse via wind

Cones with microsporangia

Microspore (*n*) forms pollen grain

Pollen grain (male gametophyte)

POLLINATION

Megasporangium
Mother cell (*2n*)

Four meiotic products; one is large and forms the megaspore (*n*)

Three meiotic products die

Pollen grain

Ovulate cone

Ovules (contain megasporangia)

MEIOSIS

Female gametophyte (*n*)
Egg (*n*)
Pollen tube

Megaspore divides to form female gametophyte (*n*), which forms eggs by mitosis. (Only one egg is fertilized and develops.)

Embryo (*2n*)

Mature sporophyte (*2n*)

Developing sporophyte

Seeds (disperse via wind or animals)

FERTILIZATION MITOSIS

Pollen tube delivers sperm to egg

FIGURE 31.18 Heterospory in Gymnosperms: Microspores Produce Pollen Grains; Megaspores Produce Female Gametophytes.

✔ **QUESTION** Compare the life cycle of the pine tree in Figure 31.18 with that of the fern pictured in Figure 31.14. Is the gymnosperm gametophyte larger than, smaller than, or about the same size as a fern gametophyte? Compared with ferns, is the gymnosperm gametophyte more or less dependent on the sporophyte for nutrition?

The evolution of heterospory, pollen, and seeds triggered a dramatic radiation of seed plants starting about 299 million years ago. To make sure that you understand these key processes and structures, study the life cycle of the pine tree pictured in **FIGURE 31.18**.

1. Starting with the sporophyte on the left, note that this and many other gymnosperm species have separate structures, called cones, where microsporangia and megasporangia develop. In this case, the two types of spores associated with heterospory develop in separate cones.

2. The microsporangia contain a cell that divides by meiosis to form microspores, which then divide by mitosis to form pollen grains—tiny male gametophytes.

3. Megasporangia are found inside protective structures called ovules. Megasporangia contain a mother cell that divides by meiosis to form a megaspore.

4. The megaspore undergoes mitosis to form the female gametophyte, which contains egg cells.

5. The female gametophyte stays attached to the sporophyte as pollen grains arrive and produce sperm that fertilize the eggs.

6. Seeds mature as the embryo develops. Inside the seed, cells derived from the female gametophyte become packed with nutrients provided by the sporophyte.

When the seed disperses and germinates, the cycle of life begins anew.

Flowers Flowering plants, or angiosperms, are the most diverse land plants living today. About 250,000 species have been described, and more are discovered each year. Their success in terms of geographical distribution, number of individuals, and number of species revolves around a reproductive organ—the **flower.**

Most flowers contain two key reproductive structures: the stamens and carpels illustrated on the left-hand side of **FIGURE 31.19**. Stamens and carpels are responsible for heterospory.

• A **stamen** includes a structure called an anther, where microsporangia develop. Meiosis occurs inside the microsporangia, forming microspores. Microspores then divide by mitosis to form pollen grains.

• A **carpel** contains a protective structure called an **ovary,** where the ovules are found.

The presence of enclosed ovules inspired the name angiosperm ("encased-seed") as opposed to gymnosperm ("naked-seed").

As in gymnosperms, ovules contain the megasporangia. A cell inside the megasporangium divides by meiosis to form the megaspore, which then divides by mitosis to form the female

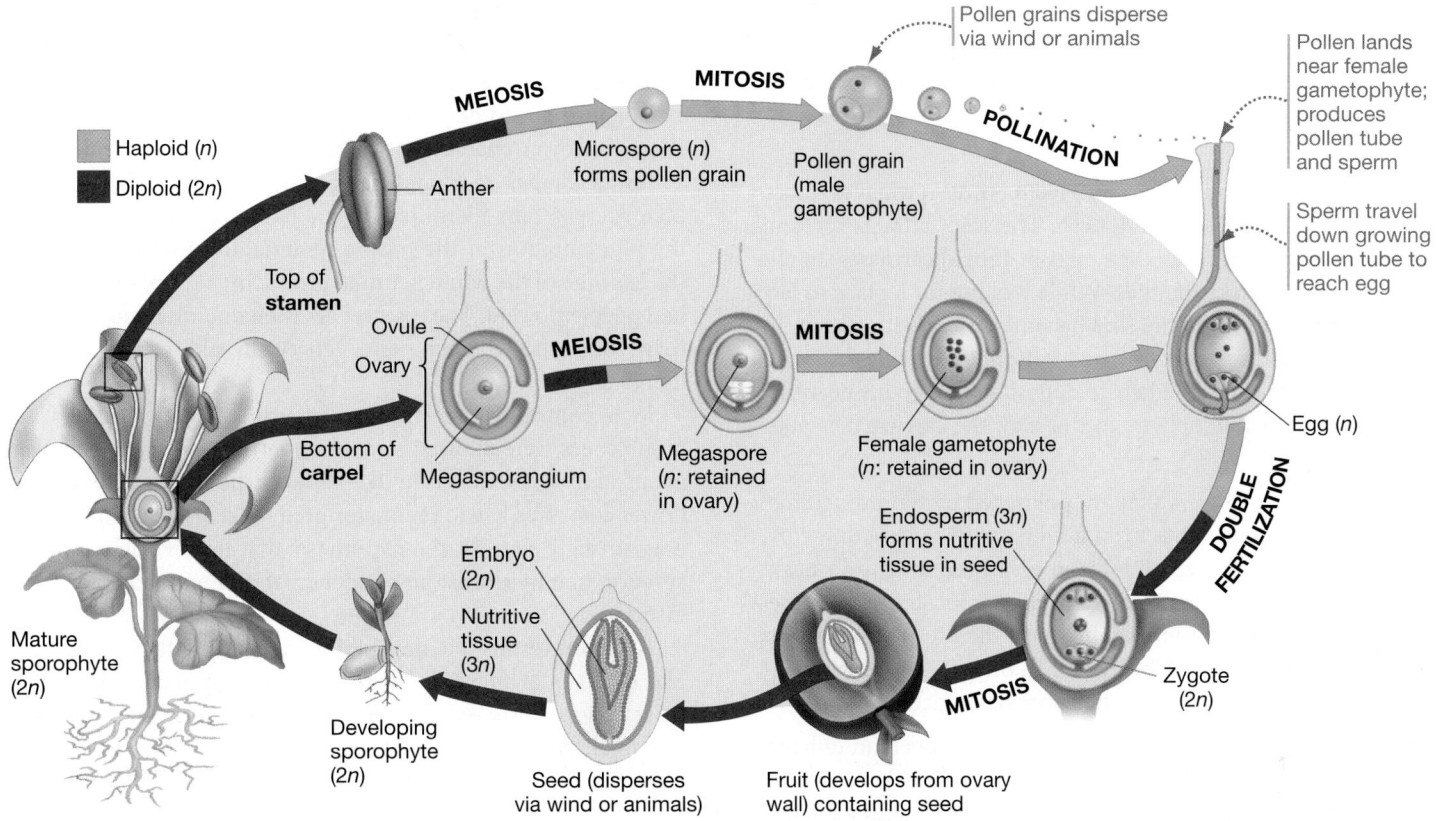

Haploid (*n*)
Diploid (2*n*)

MEIOSIS
MITOSIS

Pollen grains disperse via wind or animals

POLLINATION

Pollen lands near female gametophyte; produces pollen tube and sperm

Sperm travel down growing pollen tube to reach egg

Anther
Microspore (*n*) forms pollen grain
Pollen grain (male gametophyte)

Top of **stamen**

Ovule
Ovary
MEIOSIS
MITOSIS

Bottom of **carpel**
Megasporangium
Megaspore (*n*: retained in ovary)
Female gametophyte (*n*: retained in ovary)
Egg (*n*)

Endosperm (3*n*) forms nutritive tissue in seed

DOUBLE FERTILIZATION

Embryo (2*n*)
Nutritive tissue (3*n*)

Mature sporophyte (2*n*)

Developing sporophyte (2*n*)
Seed (disperses via wind or animals)
MITOSIS
Fruit (develops from ovary wall) containing seed
Zygote (2*n*)

FIGURE 31.19 Heterospory in Angiosperms: Flowers Contain Microspores and Megaspores.

✔ **QUESTION** Gymnosperm pollen grains typically contain from 4 to 40 cells; mature angiosperm pollen grains contain three cells. Gymnosperm female gametophytes typically contain hundreds of cells; angiosperm female gametophytes typically contain seven. Do these observations conflict with the trend of reduced gametophytes during land plant evolution, or are they consistent with it? Explain your logic.

gametophyte. After a pollen grain lands on a carpel and produces sperm, fertilization takes place, as shown on the right-hand side of Figure 31.19.

Angiosperm fertilization is unique, however, because it involves *two* sperm cells. One sperm fuses with the egg to form the diploid (2*n*) zygote, while a second sperm fuses with two nuclei in the female gametophyte to form a triploid (3*n*) nutritive tissue called **endosperm.** The involvement of two sperm nuclei is called **double fertilization.**

The evolution of the flower, then, is an elaboration of heterospory. The key innovation was the evolution of the ovary, which helps protect female gametophytes from insects and other predators.

Double fertilization is another striking innovation associated with the flower, but its adaptive significance is still not well understood. Explaining the significance of double fertilization is a major challenge for biologists interested in understanding how land plants diversified.

Pollination by Insects and Other Animals The story of the flower doesn't end with the ovary. Once stamens and carpels evolved, they became enclosed by modified leaves called **sepals** and **petals.** The

four structures then diversified to produce a fantastic array of sizes, shapes, and colors—from red roses to blue violets. Specialized cells inside flowers also began producing a wide range of scents.

To explain these observations, biologists hypothesize that flowers are adaptations to increase the probability that an animal will perform **pollination**—the transfer of pollen from one individual's stamen to another individual's carpel. Instead of leaving pollination to an undirected agent such as wind, the hypothesis is that natural selection favored structures that reward an animal—usually an insect—for carrying pollen directly from one flower to another.

Under the directed-pollination hypothesis, natural selection has favored flower colors and shapes and scents that are successful in attracting particular types of pollinators. A pollinator is an animal that disperses pollen. Pollinators are attracted to flowers because flowers provide the animals with food in the form of protein-rich pollen or a sugar-rich fluid known as **nectar.** In this way, the relationship between flowering plants and their pollinators is mutually beneficial. The pollinator gets food; the plant gets sex (fertilization).

What evidence supports the hypothesis that flowers vary in size, structure, scent, and color in order to attract specific pollinators?

The first type of evidence on this question is correlational in nature. In general, the characteristics of a flower correlate closely with the characteristics of its pollinator. A few examples of these so-called **pollination syndromes** will help drive this point home:

- *Scent* The carrion flower in **FIGURE 31.20a** produces molecules that smell like rotting flesh. The scent attracts carrion flies, which normally lay their eggs in animal carcasses. In effect, the plant tricks the flies. While looking for a place to lay their eggs on a flower, the flies become dusted with pollen. If the flies are already carrying pollen from a visit to a different carrion flower, they are likely to deposit pollen grains near the plant's female gametophyte. In this way, the carrion flies pollinate the plant.

- *Flower shape* Flowers that are pollinated by hummingbirds typically have petals that form a long, tubelike structure corresponding to the size and shape of a hummingbird's beak (**FIGURE 31.20b**). Nectar-producing cells are located at the base of the tube. When hummingbirds visit the flower, they insert their beaks and harvest the nectar. In the process, they transfer pollen grains attached to their throats or faces.

- *Flower color* Hummingbird-pollinated flowers tend to have red petals (**FIGURE 31.20b**), while bee-pollinated flowers tend to be purple or yellow (**FIGURE 31.20c**). Hummingbirds are attracted to red; bees have excellent vision in the purple and ultraviolet end of the spectrum.

The directed-pollination hypothesis also has strong experimental support. Consider, for example, recent work on South American petunias. Field observations led researchers to conclude that a white-flowered species (*Petunia axillaris*) was primarily pollinated at night by hawk moths, while a purple-flowered species (*P. integrifolia*) was pollinated during the day by bumblebees. (Note that white is highly visible at night, and purple is the color most easily seen by bees.)

Was flower color the key to flower choice by these insects? This is difficult to determine because of the many other differences in flower characteristics between two plant species, such as flower shape, odor, and nectar composition. In this experiment, researchers identified a gene named *AN2* responsible for the pigmentation in the purple petunia and introduced it into the genome of the white petunia, producing new plants that now had purple flowers. The researchers then documented that other than color, the wild-type and genetically altered flowers were identical.

In a controlled, greenhouse experiment, hawk moths and bumblebees were allowed to choose between the two *P. axillaris* plants, and as predicted by the directed-pollination hypothesis, flower color was key to pollinator preference (**FIGURE 31.21**). These data strongly support the hypothesis that flower color is an adaptation that increases the frequency of pollination by particular insects.

Based on results like this, biologists contend that the spectacular diversity of angiosperms resulted, at least in part, from natural selection exerted by the equally spectacular diversity of insect, mammal, and bird pollinators. Mutations that change flower color can change gene flow and start speciation (see Chapter 27).

Fruits The evolution of the ovary was an important event in land plant diversification, but not only because it protected the female gametophytes of angiosperms. It also made the evolution of fruit possible.

A **fruit** is a structure that is derived from the ovary and encloses one or more seeds (**FIGURE 31.22a**). Tissues derived from the ovary are often nutritious and brightly colored (**FIGURE 31.22b**). Animals eat these types of fruits, digest the nutritious tissue

(a) Carrion flowers: Smell like rotting flesh and attract carrion flies

(b) Hummingbird-pollinated flowers: Red, long tubes with nectar at the base

(c) Bee-pollinated flowers: Often bright purple

FIGURE 31.20 Flowers with Different Fragrances, Shapes, and Colors Attract Different Pollinators.

QUESTION: Does flower color influence pollinator preference?

HYPOTHESIS: Hawk moths prefer white petunias while bumblebees prefer purple, even when the flowers are otherwise identical.

NULL HYPOTHESIS: Neither insect has a clear preference for white- or purple-flowered petunias.

EXPERIMENTAL SETUP:

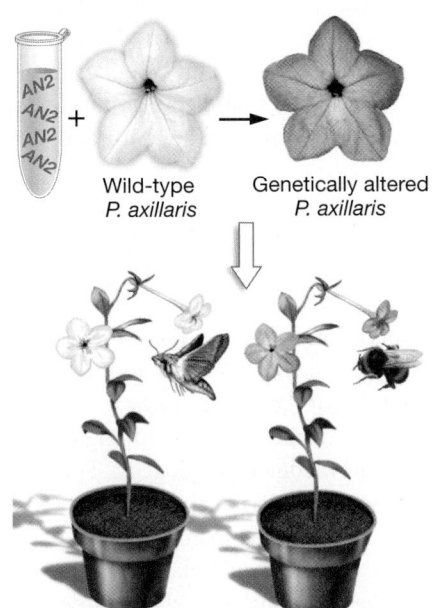

Wild-type
P. axillaris

Genetically altered
P. axillaris

1. Isolate *AN2* gene for purple flower color from a related species (*P. integrifolia*) and insert it into *P. axillaris*.

2. Grow wild-type (white) and genetically altered (purple) *P. axillaris* in controlled greenhouse conditions.

3. Count number of visits by hawk moths and bumblebees to white and purple flowers.

PREDICTION: Hawk moths will visit white-flowered *P. axillaris* more frequently, while bumblebees will visit purple-flowered *P. axillaris* more frequently.

PREDICTION OF NULL HYPOTHESIS: Both insects will visit white and purple flowers equally.

RESULTS:

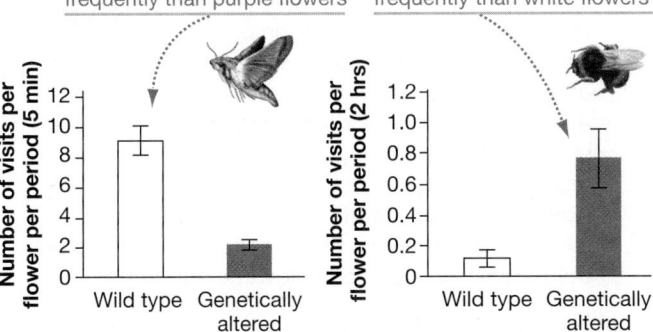

Hawk moths visited white flowers 4 times more frequently than purple flowers

Bumblebees visited purple flowers 7 times more frequently than white flowers

CONCLUSION: Petunia flower preference among bumblebees and hawk moths is significantly influenced by flower color.

FIGURE 31.21 The Adaptive Significance of Flower Color.

SOURCE: Hoballah, M. E., T. Gübitz, J. Stuurman, et al. 2007. Single gene-mediated shift in pollinator attraction in *Petunia. The Plant Cell* 19: 779–790.

✔ **QUESTION** Why did the researchers measure pollination in a greenhouse? Why did they plot flower visits on different scales?

(a) Fruits are derived from ovaries and contain seeds.

Wall of ovary

Seed

(b) Many fruits are dispersed by animals.

FIGURE 31.22 Fruits Are Derived from Ovaries Found in Angiosperms. (a) A pea pod is one of the simplest types of fruit. **(b)** The ovary wall often becomes thick, fleshy, and nutritious enough to attract animals that disperse the seeds inside.

around the seeds, and disperse seeds in their feces. In other cases, the tissues derived from the ovary help fruits disperse via wind or water. The evolution of flowers made efficient pollination possible; the evolution of fruits made efficient seed dispersal possible.

The list of adaptations that allow land plants to reproduce in dry environments is impressive; **FIGURE 31.23** (see page 596) summarizes them. Once land plants had vascular tissue and could grow efficiently in dry habitats, the story of their diversification revolved around traits that allowed sperm cells to reach eggs efficiently and helped seeds disperse to new locations.

The Angiosperm Radiation

For the past 125 million years, land plant diversification has really been about angiosperms. The Anthophyta or angiosperms represent one of the great adaptive radiations in the history of

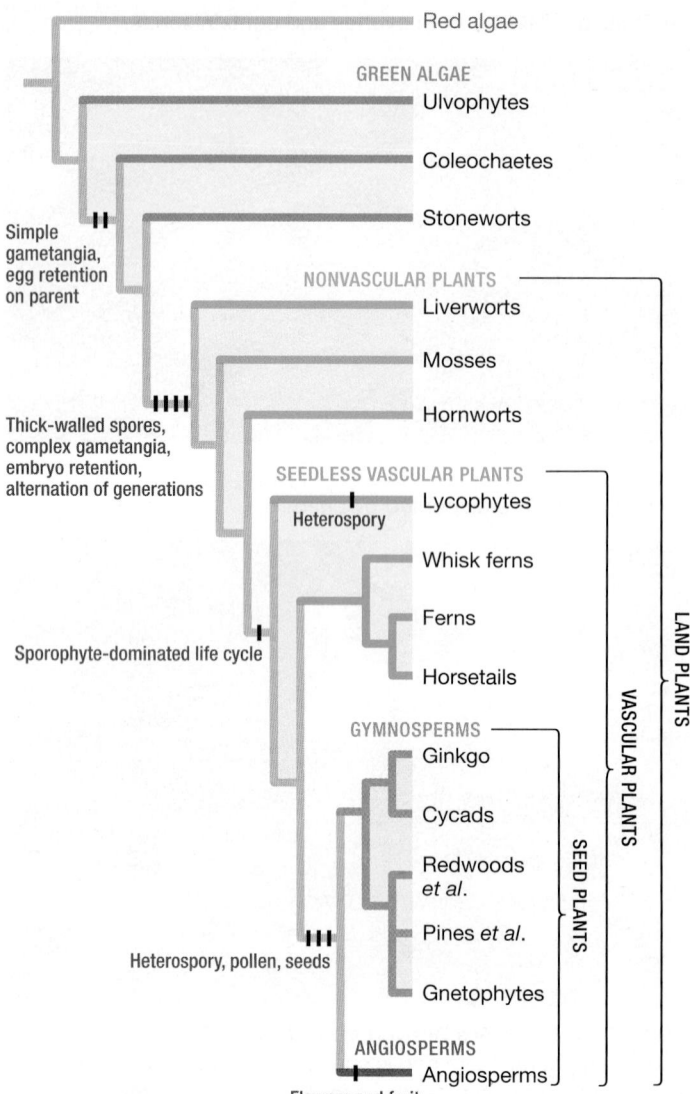

FIGURE 31.23 Evolutionary Innovations Allowed Plants to Reproduce Efficiently on Land.

✔ **EXERCISE** Redwoods, pines, gnetophytes, and angiosperms are the only land plants that do not have flagellated sperm that swim to the egg (at least a short distance). Mark the loss of flagellated sperm on Figure 31.6.

life. An **adaptive radiation** occurs when a single lineage produces a large number of descendant species that are adapted to a wide variety of habitats (see Chapter 28).

The diversification of angiosperms is associated with three key adaptations: (**1**) water-conducting vessels, (**2**) flowers, and (**3**) fruits. In combination, these traits allow angiosperms to transport water, pollen, and seeds efficiently. Based on these observations, it is not surprising that most land plants living today are angiosperms.

On the basis of morphological traits, the 250,000 species of angiosperms identified to date have traditionally been classified into two major groups: the monocotyledons, or **monocots,** and the dicotyledons, or **dicots.** Some familiar monocots include the grasses (such as corn and wheat), orchids, palms, and lilies;

familiar dicots include beans, roses, buttercups, daisies, oaks, and maples.

The names of the two groups were inspired by differences in a structure called the cotyledon. A **cotyledon** ("seed-leaf") stores nutrients and supplies them to the developing embryonic plant. As **FIGURE 31.24** shows, monocots have a single cotyledon (hence the "mono") while dicots have two cotyledons (hence the "di"). The figure also highlights other major morphological differences observed in monocots and dicots, including the arrangement of vascular tissue and leaf veins and the characteristics of flowers.

It would be misleading, however, to think that all species of flowering plants fall into one of these two groups—either monocots or dicots. Recent work has shown that dicots do not form a monophyletic group consisting of a common ancestor and all of its descendants.

To drive this point home, consider the phylogeny illustrated in **FIGURE 31.25**. These relationships were estimated by comparing the sequences of several genes that are shared by all angiosperms. Notice that species with dicot-like characters are found in multiple lineages of the angiosperm phylogenetic tree. Based on this analysis, biologists have concluded that although monocots are monophyletic, dicots are not. Dicots are paraphyletic.

Biologists have adjusted the names assigned to angiosperm lineages to reflect this new knowledge of phylogeny. The most important of these changes was identifying the **eudicots** ("true dicots") as a monophyletic lineage that includes most of the plants once considered dicots. One flowering plant that is neither a monocot nor a eudicot is the magnolia tree. Plant systematists continue to work toward understanding relationships throughout the angiosperm phylogenetic tree; there will undoubtedly be more name changes as knowledge grows.

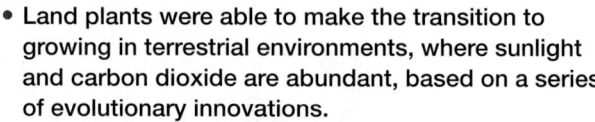

check your understanding

If you understand that . . .

- Land plants were able to make the transition to growing in terrestrial environments, where sunlight and carbon dioxide are abundant, based on a series of evolutionary innovations.
- Adaptations for growing on land included cuticle, stomata, UV-absorbing compounds, and vascular tissue.
- Adaptations for effective reproduction on land included gametangia, the retention of embryos on the parent, pollen, seeds, flowers, and fruits.

✔ **You should be able to . . .**

1. Explain why the evolution of cuticle, UV-absorbing compounds, and vascular tissue was important in survival or reproduction.
2. On Figure 31.6, map where the origin of pollen, flowers, and fruits occurred.

Answers are available in Appendix A.

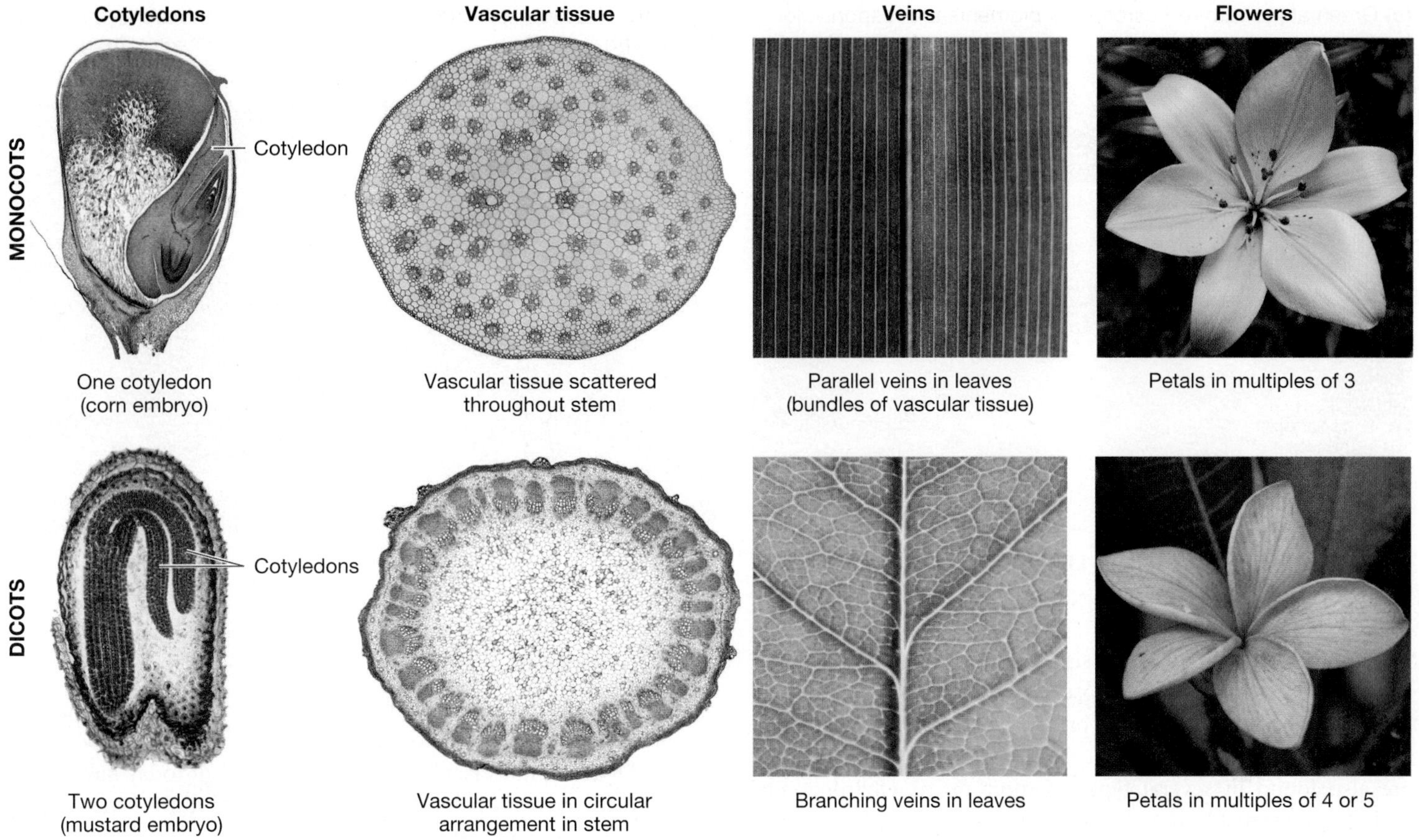

MONOCOTS

Cotyledons

One cotyledon
(corn embryo)

Cotyledon

Vascular tissue

Vascular tissue scattered
throughout stem

Veins

Parallel veins in leaves
(bundles of vascular tissue)

Flowers

Petals in multiples of 3

DICOTS

Two cotyledons
(mustard embryo)

Cotyledons

Vascular tissue in circular
arrangement in stem

Branching veins in leaves

Petals in multiples of 4 or 5

FIGURE 31.24 Four Morphological Differences between Monocots and Dicots. (The stem cross sections, showing vascular tissue and cotyledons, have been stained.)

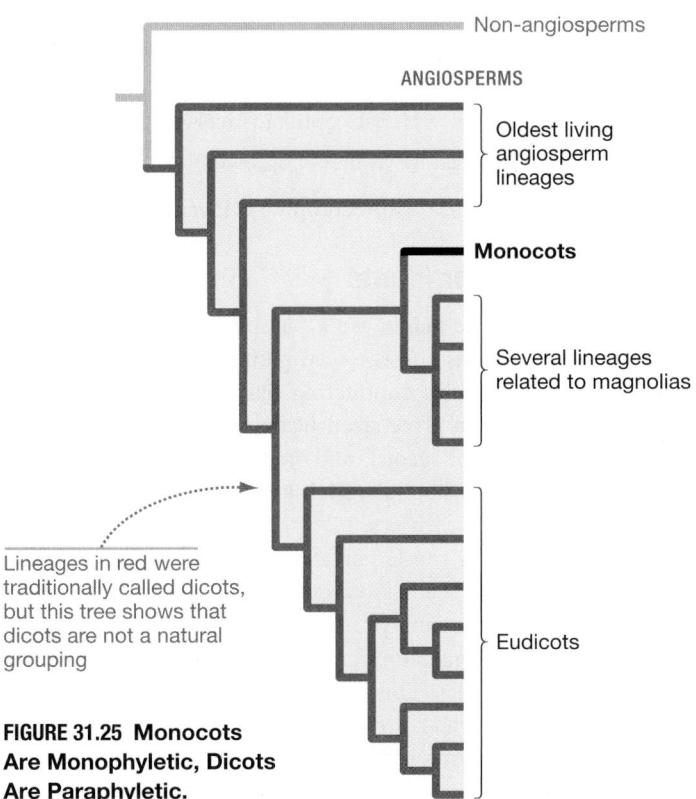

Non-angiosperms

ANGIOSPERMS

Oldest living angiosperm lineages

Monocots

Several lineages related to magnolias

Lineages in red were traditionally called dicots, but this tree shows that dicots are not a natural grouping

Eudicots

FIGURE 31.25 Monocots Are Monophyletic, Dicots Are Paraphyletic.

31.4 Key Lineages of Green Algae and Land Plants

The evolution of cuticle, pores, stomata, UV-absorbing compounds, and water-conducting tissues allowed green plants to grow on land, where resources for photosynthesis are abundant. Once the green plants were on land, the evolution of gametangia, retained embryos, pollen, seeds, and flowers enabled them to reproduce efficiently even in dry environments. The adaptations reviewed in Section 31.3 allowed the land plants to make the most important water-to-land transition in the history of life.

To explore green plant diversity in more detail, let's take a closer look at some major groups of green algae and land plants. The first part of this section considers broad groupings of lineages; the "index cards" that follow focus on particular monophyletic groups.

Green Algae

The **green algae** are a paraphyletic group that totals about 7000 species. Their bright green chloroplasts are similar to those found in land plants. Specifically, green algal chloroplasts have a double membrane and contain chlorophylls *a* and *b*. And like land plants, green algae synthesize starch in the chloroplast as a

(a) Green algae with red carotenoid pigments are responsible for pink snow.

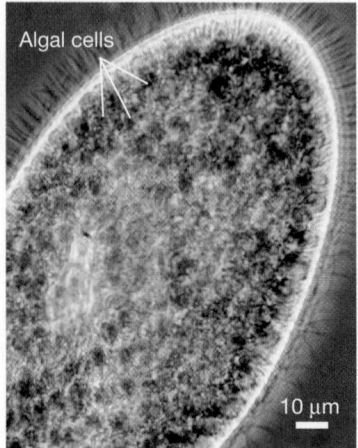

(b) Many unicellular protists harbor green algae.

Algal cells

10 µm

(c) Most lichens are an association between fungi and microscopic green algae.

1 cm

FIGURE 31.26 Some Green Algae Live in Unusual Environments.

storage product of photosynthesis. They also have a cell wall that is composed primarily of cellulose.

Green algae are important primary producers in nearshore ocean environments and in all types of freshwater habitats. They are also found in several types of more exotic environments, including snowfields at high elevations, pack ice, and ice floes. These habitats are often splashed with bright colors due to large concentrations of unicellular green algae (**FIGURE 31.26a**). Although these cells live at near-freezing temperatures, they make all their own food via photosynthesis.

In addition, green algae live in close association with an array of other organisms.

- Unicellular green algae are common endosymbionts in planktonic protists that live in lakes and ponds (**FIGURE 31.26b**). The association is considered mutually beneficial: The algae supply the protists with food; the protists provide protection to the algae.

- **Lichens** are stable associations between green algae and fungi, or between cyanobacteria and fungi, and are often found in terrestrial environments that lack soil, such as tree bark or bare rock (**FIGURE 31.26c**). The algae or cyanobacteria in a lichen are protected from drying by the fungus; in return, they provide sugars produced by photosynthesis.

Of the 17,000 species of lichens described to date, about 85 percent involve green algae. The green algae that are involved are unicellular or grow in long filaments. (Lichens are explored in more detail in Chapter 32.)

Green algae are a large and fascinating group of organisms. In this final section of the chapter, let's take a closer look at just three of the many lineages.

- Green Algae > Ulvophyceae (Ulvophytes)
- Green Algae > Coleochaetophyceae (Coleochaetes)
- Green Algae > Charophyceae (Stoneworts)

Nonvascular Plants

The initial lineages to branch off the phylogeny of living land plants are sometimes known as nonvascular plants.

All of the nonvascular plant species present today have a low, sprawling growth habit. In fact, it is unusual to find species that are more than 5 to 10 centimeters tall.

Individuals are anchored to soil, rocks, or tree bark by structures called **rhizoids.** Although simple water-conducting cells and tissues are found in some mosses, nonvascular plants lack vascular tissue with lignin-reinforced cell walls.

All nonvascular plants have flagellated sperm that swim to eggs through raindrops or small puddles on the plant surface. Spores are dispersed by wind.

- Nonvascular Plants > Hepaticophyta (Liverworts)
- Nonvascular Plants > Bryophyta (Mosses)
- Nonvascular Plants > Anthocerophyta (Hornworts)

Seedless Vascular Plants

The seedless vascular plants are a paraphyletic group between the nonvascular plants and the seed plants. All species of seedless vascular plants have conducting tissues with cells that are reinforced with lignin. Tree-sized lycophytes and horsetails are abundant in the fossil record, and tree ferns are still common inhabitants of certain habitats, such as in New Zealand and on mountain slopes in the tropics.

The sporophyte is the larger and longer-lived phase of the life cycle in all of the seedless vascular plants. The gametophyte is physically independent of the sporophyte, however. Eggs are retained on the gametophyte, and sperm swim to the egg with the aid of flagella. Thus, seedless vascular plants depend on the presence of water for reproduction—they need enough water to form a continuous layer that "connects" gametophytes and allows sperm to swim to eggs. Sporophytes develop on the gametophyte and are nourished by the gametophyte when they are small.

- Seedless Vascular Plants > Lycophyta (Lycophytes, or Club Mosses)
- Seedless Vascular Plants > Psilotophyta (Whisk Ferns)
- Seedless Vascular Plants > Pteridophyta (Ferns)
- Seedless Vascular Plants > Equisetophyta (or Sphenotophyta) (Horsetails)

Seed Plants

The seed plants are a monophyletic group consisting of the gymnosperms—ginkgo, cycads, redwoods, pines, and gnetophytes—and the angiosperms. The group is defined by two key synapomorphies, traits common to a lineage and found in no other lineages: the production of seeds and the production of pollen grains.

1. Seeds are a specialized structure for dispersing embryonic sporophytes to new locations. Seeds are the mature form of a fertilized ovule, the female reproductive structure that encloses the female gametophyte and egg cell.

2. Pollen grains are tiny, sperm-producing gametophytes that are easily dispersed through air as opposed to water.

Seed plants are found in virtually every type of habitat, and they adopt every growth habit known in land plants. Their forms range from mosslike mats to shrubs and vines to 100-meter-tall trees. Seed plants can be **annual** (have a single growing season) or **perennial** (live for many years), with life spans ranging from a few weeks to almost five thousand years. (See Chapters 37 through 41 to further explore the structure and function of seed plants.)

- Seed Plants > Gymnosperms > Ginkgophyta (Ginkgoes)
- Seed Plants > Gymnosperms > Cycadophyta (Cycads)
- Seed Plants > Gymnosperms > Cupressophyta (Redwoods, Junipers, Yews)
- Seed Plants > Gymnosperms > Pinophyta (Pines, Spruces, Firs)
- Seed Plants > Gymnosperms > Gnetophyta (Gnetophytes)
- Seed Plants > Anthophyta (Angiosperms)

Green Algae > Ulvophyceae (Ulvophytes)

The Ulvophyceae are a monophyletic group composed of several diverse and important subgroups, with a total of about 4000 species. Members of this lineage range from unicellular to multicellular.

Many of the large green algae in habitats along ocean coastlines are members of the Ulvophyceae. *Ulva*, the sea lettuce (**FIGURE 31.27**), is a representative marine species. But freshwater lakes and streams also contain large numbers of unicellular or small multicellular species that are free-floating plankton.

Reproduction Most ulvophytes reproduce both asexually and sexually. Asexual reproduction often involves production of spores that swim with the aid of flagella. Sexual reproduction usually results in production of a resting stage—a cell that is dormant in winter. In many species the gametes are not called eggs and sperm, because they are the same size and shape. In most species gametes are shed into the water, so fertilization takes place away from the parent plants.

Life cycle Many unicellular forms are diploid only as zygotes. Alternation of generations occurs in multicellular species. When alternation of generations occurs, gametophytes and sporophytes may look identical or different.

Human and ecological impacts Ulvophyceae are important primary producers in freshwater environments and in coastal areas of the oceans.

Ulva lactuca

5 cm

FIGURE 31.27 Green Algae Are Important Primary Producers in Aquatic Environments.

This group includes 19 species. Most coleochaetes are barely visible to the unaided eye and grow as flat sheets of cells (**FIGURE 31.28**). They are considered multicellular because they have specialized photosynthetic and reproductive cells and because they contain **plasmodesmata** that connect adjacent cells in plants (see Chapter 11).

The coleochaetes are strictly freshwater algae. They grow attached to aquatic plants such as water lilies and cattails or over submerged rocks in lakes and ponds. When they grow near beaches, they are often exposed to air when water levels drop in late summer.

Reproduction Asexual reproduction is common in coleochaetes and involves production of flagellated spores. During sexual reproduction, eggs are retained on the parent and are nourished after fertilization with the aid of transfer cells—a situation similar to that observed in land plants. In some species certain individuals are male and produce only sperm, while other individuals are female and produce only eggs.

Life cycle Alternation of generations does not occur. Multicellular individuals are haploid; the only diploid stage in the life cycle is the zygote.

Human and ecological impacts Because they are closely related to land plants, coleochaetes are studied intensively by researchers interested in how land plants made the water-to-land transition.

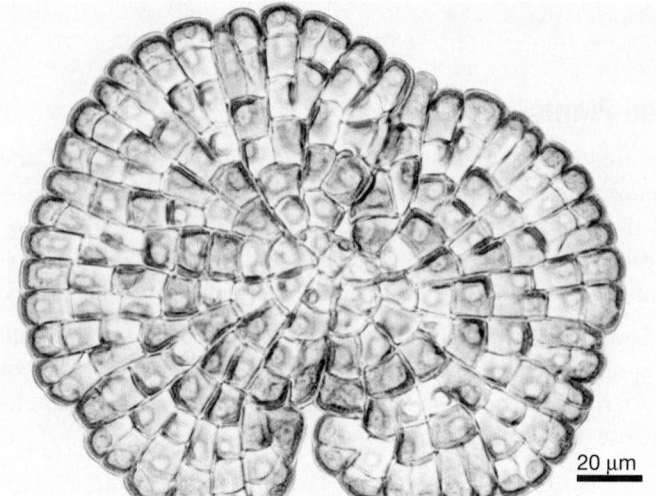

Coleochaete orbicularis

20 μm

FIGURE 31.28 Coleochaetes Are Thin Sheets of Cells.

This group includes several hundred species. They are collectively known as stoneworts, because they commonly accumulate crusts of calcium carbonate ($CaCO_3$) over their surfaces. Like the coleochaetes, they have plasmodesmata and are multicellular. ✔ You should be able to mark the origin of plasmodesmata on Figure 31.6. (They do not occur in ulvophytes.) Some species of stonewort can be a meter or more in length.

The stoneworts are freshwater algae. Certain species are specialized for growing in relatively deep waters, though most live in shallow water near lake beaches or pond edges.

Reproduction Sexual reproduction is common and involves production of prominent, multicellular gametangia similar to those observed in early land plants. In stoneworts, as in coleochaetes, the eggs are retained on the parent plant, which supplies eggs with nutrients before fertilization. ✔ You should be able to mark the origin of egg retention on Figure 31.6. (Egg retention does not occur in ulvophytes.)

Life cycle Alternation of generations does not occur. Multicellular individuals are haploid; the only diploid stage in the life cycle is the zygote.

Human and ecological impacts Some species form extensive beds in lake bottoms or ponds and provide food for ducks and geese as well as food and shelter for fish (**FIGURE 31.29**). They are a good indicator that water is not polluted.

Chara species

5 cm

FIGURE 31.29 Stoneworts Can Form Beds on Lake Bottoms.

Liverworts got their name because some species native to Europe have liver-shaped leaves. According to the medieval *Doctrine of Signatures*, God indicated how certain plants should be used by giving them a distinctive appearance. Thus, liverwort teas were hypothesized to be beneficial for liver ailments. (They are not.) About 6500 species are known. They are commonly found growing on damp forest floors or riverbanks, often in dense mats (**FIGURE 31.30**), or on the trunks or branches of tropical trees.

Adaptations to land Liverworts are covered with cuticle. Some species have pores that allow gas exchange; in species that lack pores, the cuticle is thin.

Reproduction Asexual reproduction occurs when fragments of a plant are broken off and begin growing independently. Some species also produce small structures called **gemmae** asexually, during the gametophyte phase. Mature gemmae are knocked off the parent plant by rain and grow into independent gametophytes. During sexual reproduction, sperm and eggs are produced in gametangia.

Life cycle The gametophyte is the largest and longest-lived phase in the life cycle. Sporophytes are small, grow directly from the gametophyte, and depend on the gametophyte for nutrition. Spores are shed from the sporophyte and are carried away by wind or rain.

Human and ecological impacts When liverworts grow on bare rock or tree bark, their dead and decaying body parts contribute to the initial stages of soil formation. The liverwort *Marchantia polymorpha*, shown in Figure 31.30, is an important model organism in plant biology.

Marchantia polymorpha

1 cm

FIGURE 31.30 Liverworts Thrive in Moist Habitats.

Over 12,000 species of mosses have been named and described to date, and more are being discovered every year—particularly in the tropics.

Although mosses are common in moist forests, they can also be found in more extreme environments, such as deserts and windy, treeless habitats in the Arctic, Antarctic, or on mountaintops. In these severe conditions, mosses are able to thrive because their bodies can become extremely dry without dying. When the weather makes photosynthesis difficult, individuals dry out and become dormant, or inactive. Then when rains arrive or temperatures warm, the plants rehydrate and begin photosynthesis and reproduction.

Adaptations to land One subgroup of mosses contains simple conducting tissues consisting of cells that are specialized for the transport of water or food. But because these cells do not have walls that are reinforced by lignin, they are not considered true vascular tissue. ✔ **You should be able to mark the origin of the simple water-conducting cells and tissues in this moss subgroup on Figure 31.6.** Because they lack true vascular tissue, most mosses are not able to grow much taller than a few centimeters.

Reproduction Asexual reproduction often occurs by fragmentation, meaning that pieces of gametophytes that are broken off by wind or a passing animal can begin growing independently. In many species, sexual reproduction cannot involve self-fertilization because the sexes are separate—meaning that an individual plant produces only eggs, in archegonia, or only sperm, in antheridia. A typical sporophyte produces up to 50 million tiny spores. Spores are usually distributed by wind.

Life cycle The moss life cycle is similar to that of liverworts and hornworts: The sporophyte is retained on the much larger and longer-lived gametophyte and gets most of its nutrition from the gametophyte.

Human and ecological impacts Species in the genus *Sphagnum* are often the most abundant plant in wet habitats of northern environments (**FIGURE 31.31a**, page 602). Because *Sphagnum*-rich environments account for 1 percent of Earth's total land area, equivalent to half the area of the United States, *Sphagnum* species are among the most abundant plants in the world. *Sphagnum*-rich

(Continued on next page)

(a) *Sphagnum* moss is abundant in northern wet habitats.

(b) Semi-decayed *Sphagnum* moss forms peat.

5 mm

FIGURE 31.31 *Sphagnum* **Mosses Are among the Most Abundant Plants in the World.**

habitats are waterlogged, nitrogen poor, and often anaerobic, however, so the decomposition of dead mosses and other plants is slow. As a result, large deposits of semi-decayed organic matter, known as **peat,** accumulate. Researchers estimate that the world's peatlands store about 400 billion metric tons of carbon. If peatlands begin to burn or decay rapidly due to rising temperatures associated with global climate change, the CO_2 released will exacerbate the warming trend (see Chapter 56).

Peat is harvested as a traditional heating and cooking fuel in some countries (**FIGURE 31.31b**). It is also widely used as a soil additive in gardening, because *Sphagnum* can absorb up to 20 times its dry weight in water. This high water-holding capacity is due to the presence of large numbers of dead cells in the leaves of these mosses, which readily fill with water via pores in their walls.

Nonvascular Plants > Anthocerophyta (Hornworts)

Hornworts got their name because their sporophytes have a horn-like appearance (**FIGURE 31.32**) and because "wort" is the Anglo-Saxon word for plant. About 100 species have been described to date. Hornworts can be found worldwide but are more common in tropical regions in damp soil or on the bark of trees.

Adaptations to land In most species of hornworts, sporophytes have stomata that function much like those of vascular plants. Sporophytes have cell layers that change shape when exposed to low humidity, facilitating the release of spores.

Reproduction Depending on the species, gametophytes may contain only egg-producing archegonia or only sperm-producing antheridia, or both. Stated another way, individuals of some species are either female or male, while in other species each individual has both types of reproductive organs.

Life cycle The gametophyte is the longest-lived phase in the life cycle. Although sporophytes grow directly from the gametophyte, they are green because their cells contain chloroplasts. Sporophytes manufacture some of their own food but also get nutrition from the gametophyte. Spores disperse from the parent plant via wind or rain.

Human and ecological impacts Some species harbor symbiotic cyanobacteria that fix nitrogen.

Anthoceros sp.

2 cm

FIGURE 31.32 Hornworts Have Horn-Shaped Sporophytes.

Although the fossil record documents lycophytes that were 2 m wide and 40 m tall, the 1000 species of lycophytes living today are all small in stature (**FIGURE 31.33**). Most lycophytes live on the forest floor or on the branches or trunks of tropical trees. Because of their appearance, they are often called ground pines or club mosses—even though they are neither pines nor mosses.

Adaptations to land Lycophytes are the most ancient land plant lineage with **roots**—a belowground system of tissues that anchors the plant and is responsible for absorbing water and mineral nutrients. Roots differ from the rhizoids observed in nonvascular plants, because roots contain vascular tissue and thus are capable of conducting water and nutrients to the upper reaches of the plant. Unusual leaves called microphylls, which extend from the stems, are a synapomorphy found in lycophytes. ✔ You should be able to mark the origin of microphylls on Figure 31.6.

Reproduction Asexual reproduction can occur by fragmentation or gemmae. During sexual reproduction, spores of some species give rise to bisexual gametophytes—meaning that each gametophyte produces both eggs and sperm. Self-fertilization is extremely rare, however. In the genera *Selaginella* and *Isoetes*, in contrast, heterospory occurs and gametophytes are either male or female.

Life cycle The gametophytes of some species live entirely underground and get their nutrition from symbiotic fungi. In certain species, gametophytes live 6 to 15 years and give rise to a large number of sporophytes over time.

Human and ecological impacts Tree-sized lycophytes were abundant in the coal-forming forests of the Carboniferous period. In coal-fired power plants today, electricity is being generated by burning fossilized lycophyte and fern tissues. Spores of some members are highly flammable. Before the invention of electric flashbulbs, these spores were used as flash powder in early photography.

Lycopodium species

FIGURE 31.33 Lycophytes Living Today Are Small in Stature.

Only two genera of whisk ferns are living today, and there are perhaps six distinct species. Whisk ferns are restricted to tropical regions and have no fossil record. They are extremely simple morphologically; their aboveground parts consist of branching stems that have tiny, scale-like outgrowths instead of leaves (**FIGURE 31.34**).

Because they lack roots and leaves, the whisk ferns were considered an evolutionary "throwback"—a group that retained traits found in the earliest vascular plants. However, molecular phylogenies support an alternative hypothesis: The morphological simplicity of whisk ferns is a derived trait—meaning that complex structures have been lost in this lineage.

Adaptations to land Whisk ferns lack roots. Some species gain most of their nutrition from fungi that grow in association with the whisk ferns' underground stems, called **rhizomes.** Other species grow in rock crevasses or as **epiphytes** ("upon-plants"), meaning that they grow on the branches of other plants—in this case, on

Psilotum nudum

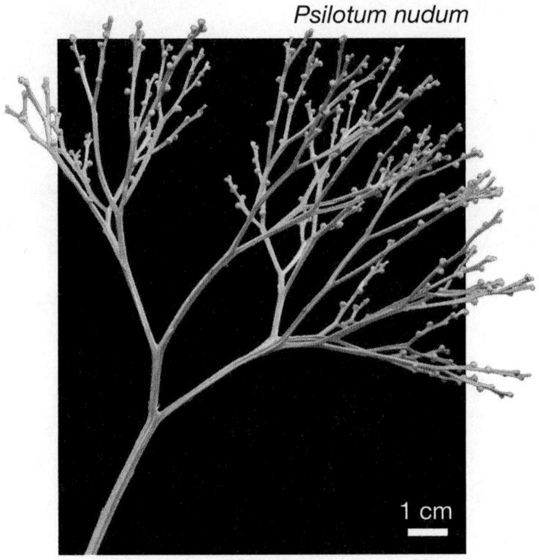

FIGURE 31.34 Psilotophytes Are Extremely Simple Morphologically.

the branches of tree ferns. ✔ After reviewing where leaves and roots originated during land plant evolution (see Figure 31.9), you should be able to mark the loss of leaves and roots in whisk ferns on Figure 31.6.

Reproduction Asexual reproduction occurs via the extension of rhizomes and the production of new aboveground stems. When spores mature, they are dispersed by wind and germinate into gametophytes that contain both archegonia and antheridia.

Life cycle Sporophytes may be up to 30 cm tall, but gametophytes are less than 2 mm long and live under the soil surface. Gametophytes absorb nutrients directly from the soil or from symbiotic fungi. Fertilization takes place inside the archegonium, and the sporophyte develops directly on the gametophyte.

Human and ecological impacts Some whisk fern species are popular landscaping plants, particularly in Japan. The same species can be a serious pest in greenhouses.

With 12,000 species, ferns are by far the most species-rich group of seedless vascular plants. They are particularly abundant in the tropics. About a third of the tropical species are epiphytes, usually growing on the trunks or branches of trees. Species that can grow epiphytically live high above the forest floor, where competition for light is reduced, without making wood and growing tall themselves. The growth habits of ferns are highly variable among species, however, and ferns range in size from rosettes the size of your smallest fingernail to 20-meter-tall trees.

Adaptations to land Ferns are the only seedless vascular plants that have large, well-developed leaves—commonly called **fronds.** Young fronds are coiled and because of their shape are called "fiddleheads"; they uncoil as they develop (**FIGURE 31.35a**). Leaves give the plant a large surface area, allowing it to capture sunlight for photosynthesis efficiently.

Reproduction In a few species, gametophytes reproduce asexually via production of gemmae. Typically, species that can reproduce via gemmae never produce gametes or sporophytes. In most species, however, sexual reproduction is the norm. Ferns are homosporous, but gametophytes develop as males or are bisexual, depending on light intensity, proximity to other gametophytes, and other environmental conditions. In bisexual gametophytes, maturation of male and female gametes is staggered to minimize self-fertilization.

Life cycle Although fern gametophytes contain chloroplasts and are photosynthetic, the sporophyte is typically the larger and longer-lived phase of the life cycle. In mature sporophytes, sporangia are usually found in clusters called **sori** on the undersides of leaves (**FIGURE 31.35b**). The structure of the sporangia is a distinctive feature of ferns. When mature, the side wall of a sporangium ruptures and a thickened row of cells along the other side of the sporangium shrinks, pulling the open sporangium back. When the humidity is low the sporangium springs back, catapulting the spores into the dry air currents. ✔ You should be able to mark the evolution of the distinctive fern sporangium on Figure 31.6.

Human and ecological impacts In many parts of the world, people gather the fiddleheads of ferns in spring as food. Ferns are also widely used as ornamental plants in landscaping.

(a) Fern fronds develop as coiled "fiddleheads."

Sadleria cyatheoides

2 cm

(b) Fern sporangia are often in clusters (sori) on the underside of fronds.

Polypodium vulgare

Sori

5 mm

FIGURE 31.35 Fern Fronds Uncoil as They Develop and Form Clusters of Sporangia (Sori) When They Mature.

Although horsetails are prominent in the fossil record of land plants, just 15 species are known today. All 15 are in the genus *Equisetum*. Translated literally, *Equisetum* means "horse-bristle." Both the scientific name and the common name, horsetail, come from the brushy appearance of the stems and branches in some species (**FIGURE 31.36a**). Horsetails may be locally abundant in wet habitats such as stream banks or marsh edges.

Adaptations to land Horsetails have an adaptation that allows them to flourish in waterlogged, oxygen-poor soils: Their stems are hollow, so oxygen readily diffuses down the stem to reach roots that cannot obtain oxygen from the surrounding soil. Horsetails are also distinguished by having whorled leaves and branches. Horsetail spores have appendages, called elaters, that change position with changes in humidity and cause spores to fall over wet habitats where gametophytes are more likely to survive (**FIGURE 31.36b**). ✔ You should be able to mark the origin of the distinctive spore elaters on Figure 31.6.

Reproduction Asexual reproduction is common in sporophytes and occurs via fragmentation or the extension of rhizomes. From these rhizomes, two types of erect, specialized stems may grow—stems that contain tiny leaves and chloroplast-rich branches and that are specialized for photosynthesis, or stems that bear clusters of sporangia and produce huge numbers of spores by meiosis.

Life cycle Gametophytes perform photosynthesis but are small and short lived. They normally produce both antheridia and archegonia, but in most cases the sperm-producing structure matures first. This pattern is thought to be an adaptation that minimizes self-fertilization and maximizes cross-fertilization.

Human and ecological impacts Horsetail stems are rich in silica granules. The glass-like deposits not only strengthen the stem but also made these plants useful for scouring pots and pans before the invention of other scrubbing tools—hence these plants are often called "scouring rushes."

(a) Fertile shoots bear conelike clusters of sporangia.

(b) As spores inside sporangia dry out, their elaters unfurl, tossing out the spores.

FIGURE 31.36 Horsetails Have a Distinctive Brushy Appearance and Unusual Spores.

Seed Plants > Gymnosperms > Ginkgophyta (Ginkgoes)

Although ginkgoes have an extensive fossil record, just one species is alive today. Leaves from the ginkgo, or maidenhair, tree are virtually identical in size and shape to those observed in fossil ginkgoes that are 150 million years old (**FIGURE 31.37**).

(a) Fossil ginkgo
Ginkgo huttoni

(b) Living ginkgo
Ginkgo biloba

2 cm

5 cm

FIGURE 31.37 The Ginkgo Tree Is a "Living Fossil."

Adaptations to land Unlike most gymnosperms, the ginkgo is **deciduous**—meaning that it loses its leaves each autumn. This adaptation allows plants to be dormant during the winter, when photosynthesis and growth are difficult.

Reproduction and life cycle Sexes are separate—individuals are either male or female. Pollen is transported by wind. Sperm have flagella, however. Once pollen grains land near the female gametophyte and mature, the sperm cells leave the pollen grain and swim to the egg cells.

Human and ecological impacts Although today's ginkgo trees are native to southeast China, they are planted widely as an ornamental. They are especially popular in urban areas because they are tolerant of air pollution. Male trees are usually planted because the seeds produced by female trees contain butyric acid, making them smell like rancid butter. However, in some countries, the inside of the seed is eaten as a delicacy.

Seed Plants > Gymnosperms > Cycadophyta (Cycads)

The cycads are so similar in overall appearance to palm trees, which are angiosperms, that cycads are sometimes called "sago palms." Although cycads were extremely abundant when dinosaurs were present on Earth 150–65 million years ago, only about 140 species are living today. Most are found in the tropics (**FIGURE 31.38**).

Adaptations to land Cycads do not make wood but are supported by stiff stems. They are unique among gymnosperms in having compound leaves—meaning that each leaf is divided into many smaller leaflets. ✔ You should be able to mark the origin of the distinctive cycad leaf on Figure 31.6.

Reproduction and life cycle In cycads each sporophyte individual bears either microsporangia or megasporangia, but not both. Like other seed plants, cycads are heterosporous. Pollen is carried by insects (usually beetles or weevils) or, in some species, wind. Cycad seeds are large and often brightly colored. The colors attract birds and mammals, both of which eat and disperse the seeds.

Human and ecological impacts Cycads harbor large numbers of symbiotic cyanobacteria in specialized, aboveground root structures. The cyanobacteria are photosynthetic and fix nitrogen. The nitrogen acts as an important nutrient for nearby plants as well as for the cycads themselves. Cycads are popular landscaping plants in some parts of the world.

Encephalartos transvenosus

10 cm

FIGURE 31.38 Cycads Resemble Palms but Are Not Closely Related to Them.

Seed Plants > Gymnosperms > Cupressophyta (Redwoods, Junipers, Yews)

The species in this lineage vary in growth form from sprawling juniper shrubs to the world's largest plants. Redwood trees growing along the Pacific Coast of North America can reach heights of over 115 m (375 ft) and have trunk diameters of over 9.5 m (30 ft). These species were recently confirmed as a monophyletic group, independent of the Pinophyta, which also reproduce via cone-bearing structures.

Adaptations to land All of the species in this lineage are trees or large shrubs. Most have narrow leaves, which in many cases are arranged in overlapping scales (**FIGURE 31.39**). Narrow leaves have a small amount of surface area, which is not optimal for capturing sunlight and performing photosynthesis. But because the small surface area reduces water loss from leaves, many species in this lineage thrive in dry habitats or in cold environments where water is often frozen.

Thuja plicata

5 cm

FIGURE 31.39 Some Species in the Cupressophyta Have Scale-like Leaves.

Reproduction and life cycle The species in this group are wind pollinated. As in all seed plants, the female gametophyte is retained on the parent. Thus, fertilization and seed development take place in the female cone. As in gymnosperms, seeds do not form inside an encapsulated structure. Depending on the species, the seeds are dispersed by wind or by seed-eating birds or mammals.

Human and ecological impacts Redwoods, red cedar, white cedar, and yellow cedar have wood that is highly resistant to decay and thus prized for making furniture, decks, and house siding, or for other applications where wood is exposed to the weather. Yew wood is often preferred for making traditional archery bows, and the berry-like cones of juniper are used to flavor gin. The anticancer drug taxol was originally found in Pacific yew trees.

Seed Plants > Gymnosperms > Pinophyta (Pines, Spruces, Firs)

The gymnosperms include two major lineages of cone-bearing species: the pines and allies discussed here, and the group already described that includes redwoods, junipers, and yews. Species in both lineages have a reproductive structure called the cone, which produces microsporangia or megasporangia (**FIGURE 31.40**).

Pinophyta include the familiar pines, spruces, firs, Douglas fir, tamaracks, and true cedars. These are among the largest and most abundant trees on the planet, as well as some of the most long lived. One of the bristlecone pines native to southwestern North America is at least 4750 years old.

Adaptations to land Pinophyta have needle-like leaves whose small surface area allows them to thrive in habitats where water is scarce. Pines are common on sandy soils that have poor water-holding capacity, and spruces and firs are common in cold environments where water is often frozen. All of the living species make wood as a support structure.

Reproduction and life cycle Sexes are separate, and pollen is transferred to female cones by the wind. Female cones take two years to mature and are usually found high in trees. Male cones are usually found lower in the tree—possibly to reduce self-pollination from falling pollen.

Human and ecological impacts In terms of biomass, pines, spruces, firs, and other species in this group dominate forests that grow at high latitudes and high elevations as well as at sandy sites in warmer regions. Their seeds are key food sources for a variety of birds, squirrels, and mice, and their wood is the basis of the building products and paper industries in many parts of the world. The paper in this book was made from species in this group.

(a) Cones that produce microsporangia and pollen

Picea abies

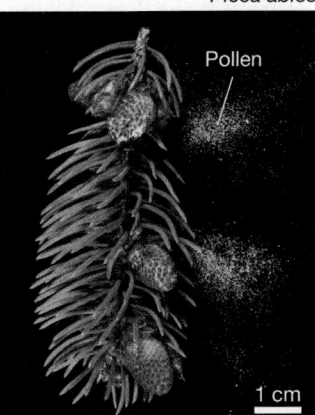

Pollen

1 cm

(b) Cones that produce megasporangia and eggs

Picea abies

5 cm

FIGURE 31.40 Pollen-Bearing Cones Produce Microsporangia; Ovulate Cones Produce Megasporangia.

The gnetophytes comprise about 70 species in three genera: *Gnetum* comprises vines and trees from the tropics; *Ephedra* is made up of desert-dwelling shrubs, including what may be the most familiar gnetophyte—the shrub called Mormon tea, which is common in the deserts of southwestern North America; *Welwitschia* contains a single species that probably qualifies as the world's most bizarre plant (**FIGURE 31.41**), *W. mirabilis*, which is native to the deserts of southwest Africa. Although it has large belowground structures, the aboveground part consists of just two strap-like leaves, which grow continuously from the base and die at the tips. The leaves also split lengthwise as they grow and age.

Adaptations to land Gnetophytes have vessel elements in addition to tracheids. All of the living species make wood as a support structure.

Reproduction and life cycle The microsporangia and megasporangia are arranged in clusters at the end of stalks, similar to the way flowers are clustered in some angiosperms. Pollen is transferred by the wind or by insects. Double fertilization occurs in two of the three genera. As in other gymnosperms, seeds do not form inside an encapsulated structure.

Human and ecological impacts The drug ephedrine was originally isolated from species of *Ephedra* that are native to northern China and Mongolia. Ephedrine is used in the treatment of hay fever, colds, and asthma.

Welwitschia mirabilis

0.5 m

FIGURE 31.41 *Welwitschia* **Is an Unusual Plant.**

The flowering plants, or angiosperms, are far and away the most species-rich lineage of land plants. Over 250,000 species have already been described. They range in size from *Lemna gibba*—a floating, aquatic species that is less than half a millimeter wide—to massive oak trees. Angiosperms thrive in desert to freshwater to rain forest environments and are found in virtually every habitat except the deep oceans. They are the most common and abundant plants in most terrestrial environments.

The defining adaptation of angiosperms is the flower. Flowers are reproductive structures that hold either pollen-producing microsporangia or the megasporangia that produce megaspores and eggs, or both. Nectar-producing cells are often present at the base of the flower, and the color of petals helps attract insects, birds, or bats that carry pollen from one flower to another (**FIGURE 31.42a**). Some angiosperms are pollinated by wind, however. Wind-pollinated flowers lack both colorful petals and nectar-producing cells (**FIGURE 31.42b**).

Adaptations to land In addition to flowers, angiosperms evolved water-conducting vessels, the conducting cells that make water transport particularly efficient. Most angiosperms contain both tracheids and vessels.

Reproduction and life cycle Unlike gymnosperms, angiosperms have a carpel, a structure within the flower that contains an ovary. The ovary encloses one or more ovules, each of which encloses a female gametophyte. In most cases, male gametophytes are carried to female gametophytes by animal pollinators that are inadvertently dusted with pollen as they visit flowers to find food. Depending on the angiosperm species, self-fertilization may be common or absent. When the egg produced by the female gametophyte is fertilized, the ovule develops into a seed. When the ovary matures, it forms a fruit that contains the seed or seeds.

Human and ecological impacts It is almost impossible to overstate the importance of angiosperms to humans and other organisms. In most terrestrial habitats today, angiosperms supply the food that supports virtually every other species. For example, many insects eat flowering plants. Historically, the diversification of angiosperms correlated closely with the diversification of insects, which are by far the most species-rich lineage on the tree of life. It is not unusual for a single tropical tree to support dozens or even hundreds of insect species. Angiosperm seeds and fruits have also supplied the staple foods of virtually every human culture that has ever existed.

(a) Animal-pollinated flower (this species produces both pollen and eggs in the same flower)

Ornithogalum dubium

0.5 cm

(b) Wind-pollinated flower (this species has separate male and female flowers)

Acer negundo

Male flower   1 cm

Acer negundo

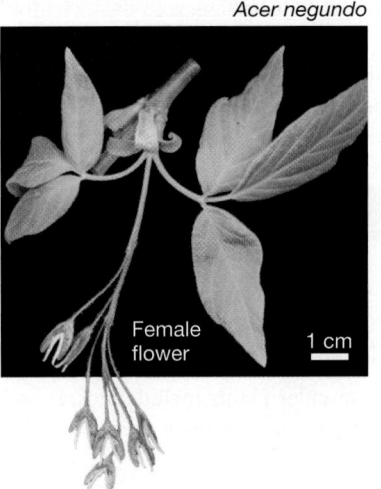

Female flower   1 cm

FIGURE 31.42 Wind-Pollinated Flowers Lack the Colorful Petals and Nectar Found in Most Animal-Pollinated Species.

CHAPTER 31   REVIEW

For media, go to MasteringBiology

If you understand . . .

31.1 Why Do Biologists Study the Green Algae and Land Plants?

- Plants are important primary producers and a source of oxygen, and they hold soil and water in place.

- Plants improve the quality of the environment for other organisms.

- Humans depend on plants for food, fiber, pharmaceuticals, building materials, and fuel.

✔ You should be able to predict how the ongoing and massive loss of plant species and plant communities will affect soils around the world.

31.2 How Do Biologists Study Green Algae and Land Plants?

- Scientists analyze morphological traits, examine the fossil record, and evaluate molecular phylogenies to study and compare green algae and land plants.

- Morphological differences suggest that land plants evolved from multicellular, freshwater green algae from the Charophyceae.

- The fossil record reveals that the first, simple land plants evolved about 475 million years ago and then diversified into plants with vascular systems, seeds, and flowers.

✔ You should be able to describe the synapomorphies that define all green algae and land plants.

31.3 What Themes Occur in the Diversification of Land Plants?

- The evolution of cuticle allowed plant tissues to be exposed to air without dying.

- UV-absorbing compounds allowed the first land plants to tolerate the harsh UV rays from the sun.

- The evolution of stomata, breaks in the cuticle consisting of pores controlled by guard cells, allowed plants to maximize gas exchange and minimize water loss.

- Vascular tissue conducts water and has secondary cell walls reinforced with lignin that provides structural support allowing plants to grow upright. This adaptation reduces competition for light.

- Tracheids are water-conducting cells found in all vascular plants; in addition, angiosperms and gnetophytes have water-conducting cells called vessels.

- All land plants are embryophytes, meaning that eggs and embryos are retained on the parent plant. Consequently, the developing embryo can be nourished by its mother.

- All land plants have alternation of generations. Over the course of land plant evolution, the gametophyte phase became reduced in terms of size and life span, and the sporophyte phase became more prominent. Heterospory also became more common.

- The evolution of pollen was an important breakthrough in the history of life, because sperm no longer needed to swim to the egg—tiny male gametophytes could be transported through the air via wind or animals.

- Seed plant embryos are dispersed from the parent plant to a new location, encased in a protective housing, and supplied with a store of nutrients.

✔ You should be able to discuss (compare and contrast) the advantages and disadvantages of having spores or seeds serve as the dispersal stage in a plant life cycle.

31.4 Key Lineages of Green Algae and Land Plants

- Green algae include single-celled and multicellular freshwater species that share numerous characteristics with land plants. Single-celled species are symbionts with protists, fungi, and animals.

- Nonvascular plants include mosses and liverworts that are small and have a gametophyte-dominant life cycle.

- Seedless vascular plants include lycopods and ferns that have small, free-living gametophytes and large, longer-lived sporophytes. Dispersal is carried out using spores.

- Seed plants include gymnosperms and angiosperms that dominate the world's floras. Animals are often used to disperse seeds and, in angiosperms, to transport pollen between flowers.

 MasteringBiology

1. MasteringBiology Assignments

Tutorials and Activities Fern Life Cycle; Gymnosperms; Highlights of Plant Phylogeny; Moss Life Cycle; Nonvascular Plants; Pine Life Cycle; Plant Evolution and the Phylogenetic Tree; Terrestrial Adaptations of Plants

Questions Reading Quizzes, Blue-Thread Questions, Test Bank

2. eText Read your book online, search, take notes, highlight text, and more.

3. The Study Area Practice Test, Cumulative Test, BioFlix® 3-D Animations, Videos, Activities, Audio Glossary, Word Study Tools, Art

You should be able to . . .

✔ TEST YOUR KNOWLEDGE
Answers are available in Appendix A

1. Which of the following groups is definitely monophyletic?
   a. nonvascular plants
   b. green algae
   c. green plants
   d. seedless vascular plants

2. The appearance of cuticle and stomata correlated with what event in the evolution of land plants?
   a. the first erect growth forms
   b. the first woody tissues
   c. growth on land
   d. the evolution of the first water-conducting tissues

3. What is a pollen grain?
   a. male gametophyte
   b. female gametophyte
   c. male sporophyte
   d. sperm

4. What do seeds contain?
   a. male gametophyte and nutritive tissue
   b. female gametophyte and nutritive tissue
   c. embryo and nutritive tissue
   d. mature sporophyte and nutritive tissue

5. Land plants may have reproductive structures that (a) protect gametes as they develop, (b) allow sperm to be transported in the absence of water, (c) provide stored nutrients and a protective coat so that offspring can be dispersed by wind away from the parent plant, and (d) provide nutritious tissue around seeds that facilitates dispersal by animals. Name each of these four structures, and state which land plant group or groups have each structure.

6. What does it mean to say that a life cycle is gametophyte dominant versus sporophyte dominant?

✔ TEST YOUR UNDERSTANDING
Answers are available in Appendix A

7. In the life cycle of an angiosperm, the egg
   a. forms from meiosis of cells inside the seed.
   b. must be fertilized by a sperm that swims to it with its flagella.
   c. is one of the cells formed by mitosis in the ovule.
   d. is part of a mobile female gametophyte.

8. Soils, water, and the atmosphere are major components of the abiotic (nonliving) environment. Describe how green plants affect the abiotic environment in ways that are advantageous to humans.

9. The evolution of cuticle presented land plants with a challenge that threatened their ability to live on land. Describe this challenge, and explain why stomata represent a solution. Compare and contrast stomata with the pores found in liverworts.

10. Why was the evolution of lignin-reinforced cell walls significant?

11. Explain the difference between homosporous and heterosporous plants. Where are the microsporangium and megasporangium found in a tulip? What happens to the spores that are produced by these structures?

12. Describe the advantage that flowering plants gain by using animals to transfer pollen from one individual to another.

13. Which of the following statements is *not* true?
   a. Green algae in the lineage called Charophyceae are the closest living relatives of land plants.
   b. The nonvascular plants form a monophyletic group.
   c. The horsetails and the ferns form a distinct lineage. They have vascular tissue but reproduce via spores, not seeds.
   d. According to the fossil record and phylogenetic analyses, angiosperms evolved before the gymnosperms. Angiosperms are the only land plants with vessels.

14. Angiosperms such as grasses, oaks, and maples are wind pollinated. The ancestors of these subgroups were probably pollinated by insects, however. As an adaptive advantage, why might a species "revert" to wind pollination? (Hint: Think about the costs and benefits of being pollinated by insects versus wind.) Why is it logical to observe that wind-pollinated species usually grow in dense stands containing many individuals of the same species? Why is it logical to observe that in wind-pollinated deciduous trees, flowers form very early in spring—before leaves form?

15. Vessel elements transport water much more efficiently than tracheids, but they are much more susceptible than tracheids to being blocked by air bubbles. Suggest a hypothesis to explain why the vascular tissue of angiosperms consists of a combination of vessel elements and tracheids.

16. You have been hired as a field assistant for a researcher interested in the evolution of flower characteristics in orchids. Design an experiment to determine whether color, size, shape, scent, or amount of nectar is the most important factor in attracting pollinators to a particular species. Assume that you can change any flower's color with a dye and that you can remove petals or nectar stores, add particular scents, add nectar by injection, or switch parts among species by cutting and gluing.

32 Fungi

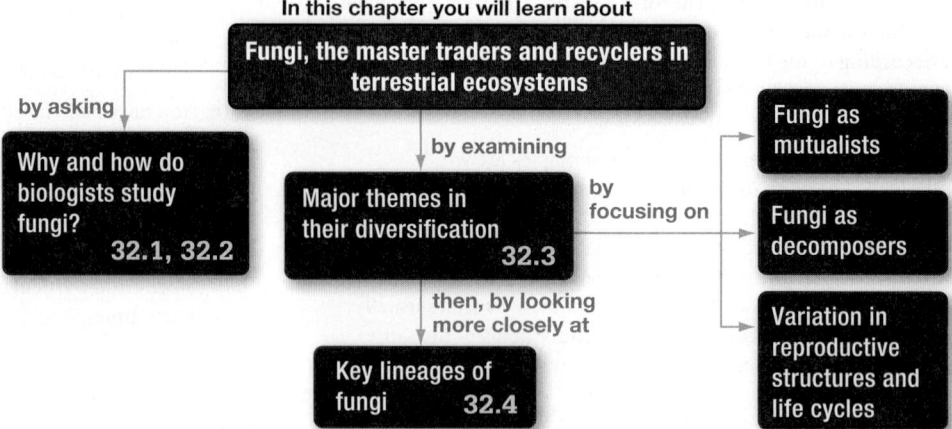

In this chapter you will learn about

Fungi, the master traders and recyclers in terrestrial ecosystems

by asking

Why and how do biologists study fungi?
    32.1, 32.2

by examining

Major themes in their diversification
    32.3

by focusing on

then, by looking more closely at

Key lineages of fungi   **32.4**

Fungi as mutualists

Fungi as decomposers

Variation in reproductive structures and life cycles

Chicken-of-the-woods (*Laetiporus sulphureus*) is a common edible mushroom that tastes somewhat like chicken. Most of its body consists of microscopic filaments that grow into dead trees, decomposing the wood.

Fungi are eukaryotes that grow as single cells or as large, branching networks of multicellular filaments. Familiar fungi include the mushrooms you've encountered in the woods or a grocery store, the molds and mildews in your home, the organism that causes athlete's foot, and the yeasts used in baking and brewing.

Along with the land plants and the animals, the **fungi** are one of three major lineages of large, multicellular eukaryotes that occupy terrestrial environments. When it comes to making a living, the species in these three groups use radically different strategies. Land plants make their own food through photosynthesis. Animals and fungi both feed on plants, protists, fungi, or each other by releasing digestive enzymes and absorbing small molecules, but while many animals carry out digestion in a dedicated digestive tract, fungi release their digestive enzymes to the external environment.

Fungi that absorb nutrients from dead organisms are the world's most important decomposers. Although a few types of organisms are capable of digesting the cellulose in plant cell walls, certain fungi and a handful of bacterial species are the only organisms capable of digesting both the lignin and

✔ When you see this checkmark, stop and test yourself. Answers are available in Appendix A.

cellulose that make up wood. Without fungi, Earth's surface would be piled so high with dead tree trunks and branches that there would be almost no room for animals to move or plants to grow.

Other fungi specialize in absorbing nutrients from living organisms. When fungi absorb these nutrients without providing any benefit in return, they lower the fitness of their host organism and act as parasites. If you've ever had athlete's foot or a vaginal yeast infection, you've hosted a parasitic fungus.

Most of the fungi that live in association with other organisms benefit their hosts, however. In these cases, fungi are not parasites but **mutualists.**

- The roots of most land plants are colonized by an array of mutualistic fungi that provide water and key nutrients such as nitrogen and phosphorus in exchange for sugars that are synthesized by the host plant. The soil around you is alive with an enormous network of fungi that are fertilizing the plants you see aboveground.

- Fungi living inside the shoots of certain plants help ward off herbivores by making toxic compounds.

- Many insects harbor single-celled fungi in their guts that aid their digestion.

- Some insects grow gardens of fungi that they feed with pieces of leaves. The insects then feed off the fungi, which they maintain and continue to cultivate.

In short, fungi are the master traders and recyclers in terrestrial ecosystems. Some fungi release nutrients from dead plants and animals; others obtain nutrients that they then transfer to living plants and animals. Because they recycle key elements such as carbon, nitrogen, and phosphorus and because they transfer key nutrients to plants and animals, fungi have a profound influence on ecosystem productivity and biodiversity. In terms of nutrient cycling, fungi make the world go around.

32.1 Why Do Biologists Study Fungi?

Given their importance to life on land and their intricate relationships to other organisms, it's no surprise that fungi are fascinating to biologists. But there are important practical reasons as well for humans to study fungi. They nourish the plants that nourish us. They affect global climate change, because they are critical to the carbon cycle on land. Unfortunately, many fungi can cause debilitating diseases in humans and crop plants. Let's take a closer look at some of the ways that fungi affect human health and welfare.

Fungi Have Important Economic and Ecological Impacts

In humans, parasitic fungi cause athlete's foot, vaginitis, diaper rash, ringworm, pneumonia, and thrush, among other miseries. But even though these maladies can be serious, in reality no more than approximately 200 species of fungi—out of the hundreds of thousands of existing species—regularly cause illness

in humans. Compared with the frequency of diseases caused by bacteria, viruses, and protists, the incidence of fungal infections in humans is low.

It would be easy to argue, in fact, that fungi have done more to promote human health than degrade it. The first antibiotic that was widely used, penicillin, was isolated from a fungus, and soil-dwelling fungi continue to be the source of many of the most important antibiotics prescribed against bacterial infections.

The major destructive impact that fungi have on people is through the food supply. Fungi known as rusts, smuts, mildews, wilts, and blights cause annual crop losses computed in the billions of dollars. These fungi are particularly troublesome in wheat, corn, barley, and other grain crops (**FIGURE 32.1a**). Saprophytic fungi are also responsible for enormous losses due to spoilage—particularly for fruit and vegetable growers (**FIGURE 32.1b**).

In nature, epidemics caused by fungi have killed 4 billion chestnut trees and tens of millions of American elm trees in North America. The fungal species responsible for these epidemics were accidentally imported on species of chestnut and elm native to other regions of the world. When the fungi arrived in North America and began growing in chestnuts and elms native to North America, the results were catastrophic. The local chestnut and elm populations had virtually no genetic resistance to the pathogens and quickly succumbed. Living chestnut root systems continue to sprout shoots, but these trees rarely live long enough to reproduce.

The chestnut and elm epidemics radically altered the composition of upland and floodplain forests in the eastern United States. Before these fungal epidemics occurred, chestnuts and elms dominated these habitats.

(a) Parasitic fungi infect corn and other crop plants.   (b) Saprophytic fungi rot fruits and vegetables.

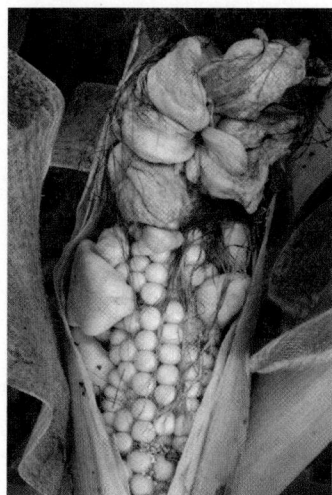 

FIGURE 32.1 Fungi Cause Problems with Crop Production and Storage. (a) A wide variety of grain crops are parasitized by fungi. Corn smut is a serious disease in sweet corn, although in Mexico the smut fungus is eaten as a delicacy. **(b)** Fungi decompose fruits and vegetables as well as leaves and tree trunks.

Fungi also have important positive impacts on the human food supply:

- Mushrooms are consumed in many cultures; in the industrialized nations they are used in sauces, salads, and pizza.

- The yeast *Saccharomyces cerevisiae* was domesticated thousands of years ago; today it and other fungi are essential to the manufacture of bread, soy sauce, cheese, beer, wine, whiskey, and other products. In most cases, domesticated fungi are used in conditions where the cells grow via fermentation, creating ethanol and by-products like the CO_2 that causes bread to rise and beer and champagne to fizz.

- Chocolate is made from the seeds of *Theobroma cacao*, but it is edible only after the seeds are fermented by several species of fungi.

- Enzymes derived from fungi are used to improve the characteristics of foods ranging from fruit juice and candy to meat.

Fungi Provide Nutrients for Land Plants

Fungi that live in close association with plant roots are said to be **mycorrhizal** (literally, "fungal-root"; see **FIGURE 32.2a**). When biologists first discovered how extensive these fungi–plant associations are, they asked an obvious question: Does plant growth suffer if mycorrhizal fungi are absent?

FIGURE 32.2b shows a result typical of many experiments. In this case, seedlings were grown in the presence and absence of the mycorrhizal fungi normally found on their roots. The photographs document that this species grows three to four times faster in the presence of its normal fungal associates than it does without them.

For farmers, foresters, and ranchers, the presence of normal mycorrhizal fungi can mean the difference between profit and loss. Fungi are critical to the productivity of forests, croplands, and rangelands.

Fungi Accelerate the Carbon Cycle on Land

Fungi that make their living by digesting dead plant material are called **saprophytes.** To understand why saprophytic fungi play a key role in today's terrestrial environments, recall that cells in the vascular tissues of land plants have secondary cell walls containing both lignin and cellulose (see Chapter 31). Wood forms when stems grow in girth by adding layers of lignin-rich vascular tissue.

When trees die, certain fungi are the organisms that break down wood into sugars and other small organic compounds (**FIGURE 32.3**). Fungi use these molecules as food, as do many microorganisms. In addition, when fungi die or are eaten, the molecules are passed along to a wide array of other organisms.

Fungi play an important role in cycling carbon atoms through terrestrial ecosystems. Note that there are two basic components of the **carbon cycle** on land:

1. The fixation of carbon by land plants—meaning that carbon in atmospheric CO_2 is reduced to form sugar, which is then used to synthesize cellulose, lignin, and other complex organic compounds in the bodies of plants.

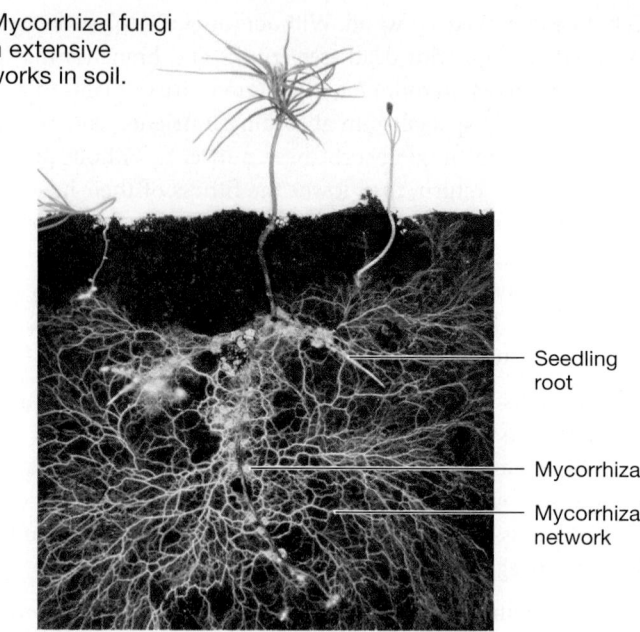

(a) Mycorrhizal fungi form extensive networks in soil.

Seedling root

Mycorrhiza

Mycorrhizal network

(b) Mycorrhizal fungi increase plant growth.

With mycorrhizal fungi   Without

FIGURE 32.2 Plants Grow Better in the Presence of Mycorrhizal Fungi. (a) Root system of a larch tree seedling; the mycelia from a mycorrhizal fungus are visible. **(b)** Typical experimental results when plants are grown with and without their normal mycorrhizal fungi. (Fungi are not visible in the photo.)

2. The release of CO_2 from nearly all organisms as the result of cellular respiration—meaning the oxidation of glucose and production of the ATP that sustains life.

The fundamental point is that, for many carbon atoms, saprophytic fungi connect the two parts of the carbon cycle.

If fungi had not evolved the ability to digest lignin and cellulose soon after land plants evolved the ability to make these compounds, carbon atoms would have been sequestered in wood for millennia instead of being rapidly recycled into glucose molecules and CO_2. Terrestrial environments would be radically different than they are today, and probably much less productive. On land, fungi make the carbon cycle turn much more rapidly than it would without fungi.

FIGURE 32.3 Fungi Speed Up the Carbon Cycle as They Break Down Dead Trees in Terrestrial Ecosystems.

To summarize, biologists study fungi because they affect a wide range of species in nature, including humans. What tools are helping researchers understand the diversity of fungi?

32.2 How Do Biologists Study Fungi?

About 110,000 species of fungi have been described and named to date, and several hundreds more are discovered each year. But the fungi are so poorly studied that the known species are widely regarded as a tiny fraction of the actual total.

To predict the actual number of fungal species alive today, David Hawksworth looked at the ratio of vascular plant species to fungal species in the British Isles—the area where the two groups are the most thoroughly studied. According to Hawksworth's analysis, these islands have an average of six species of fungus for every species of vascular plant. If this ratio holds worldwide, then the estimated total of 275,000 vascular plant species implies that there are perhaps 1.5 million species of fungi.

Although this estimate sounds large, recent data on fungal diversity suggest that it may be an underestimate. Consider what researchers found when they analyzed fungi growing on Barro Colorado Island, Panama: Living on the healthy leaves of just two tropical tree species were a total of 418 distinct morphospecies of fungi. (Recall from Chapter 27 that morphospecies are distinguished from each other by some aspect of morphology.) Because over 310 species of trees and shrubs grow on Barro Colorado, the data suggest that tens of thousands of fungi may be

(a) Single-celled fungi are called yeasts.

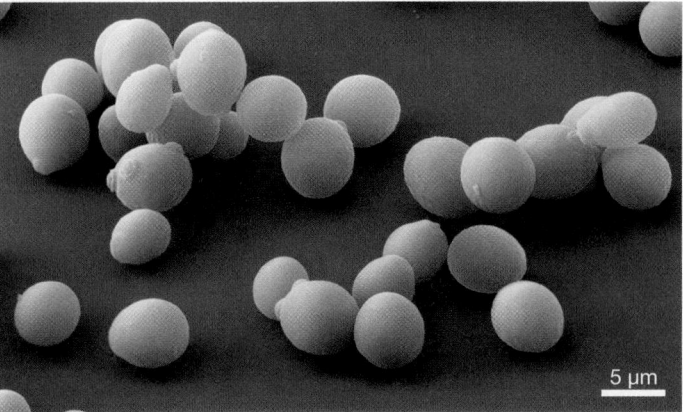

5 µm

(b) Multicellular fungi have weblike bodies called mycelia.

20 mm

FIGURE 32.4 Fungi Have Just Two Growth Forms. Fungi grow **(a)** as single-celled yeasts and/or **(b)** as multicellular mycelia made up of long, thin, highly branched filaments. Both of these scanning electron micrographs are colorized.

native to this island alone. If further work on fungal diversity in the tropics supports these conclusions, there may turn out to be many millions of fungal species.

This viewpoint of fungal diversity was reinforced by an analysis of the fungi living in the guts of beetles. In this study, researchers isolated fungi from 27 species of beetles and characterized them by observing various morphological and physiological traits. In addition, the direct sequencing approach was used to analyze the gene that codes for the RNA molecule in the small subunit of fungal ribosomes (see Chapter 30). The data showed that over 650 fungal species were living in the beetle guts, 200 of which had never been described. Biologists are only beginning to realize the extent of species diversity in fungi.

Let's consider how biologists are working to make sense of all this diversity, beginning with an overview of fungal morphology.

Analyzing Morphological Traits

Compared with animals and land plants, fungi have simple bodies. Only two growth forms occur among them: **(1)** single-celled forms called **yeasts** (**FIGURE 32.4a**), and **(2)** multicellular, filamentous structures called **mycelia** (singular: **mycelium; FIGURE 32.4b**). Many species of fungus grow either as yeasts or as a mycelium, but some regularly adopt both growth forms.

The fossil record for land plants and animals has been extremely useful in understanding their early evolution (see Chapters 25 and 31). In contrast, fossils of fungi are very rare, so biologists rely on morphological and molecular data to explore their past. Because most fungi form mycelia and because this body type is so fundamental to the absorptive mode of life, most studies of fungal morphology have focused on them.

The Nature of the Fungal Mycelium

If food sources are plentiful, mycelia can be long lived and grow to be extremely large. Researchers discovered a mycelium of the fungus *Armillaria* growing across 2100 acres (8.8 km²) in Oregon. This is an area substantially larger than most college campuses. The biologists estimated the individual's weight at hundreds of tons and its age at thousands of years, making it one of the largest and oldest organisms known.

Although most mycelia are much smaller and shorter lived than the individual in Oregon, all mycelia are dynamic. Mycelia constantly grow in the direction of food sources and die back in areas where food is running out. The body shape of a fungus can change almost continuously throughout its life.

The Nature of Hyphae

The filaments within a mycelium are called **hyphae** (singular: **hypha**). As **FIGURE 32.5a** shows, hyphae are long, narrow filaments that branch frequently.

In most terrestrial fungi, each filament is divided into cells by cross-walls called **septa** (singular: **septum; FIGURE 32.5b**). Septa do not close off the cells along hyphae completely. Instead, gaps called pores enable a wide variety of materials, even nuclei and other organelles, to flow from one compartment to the next.

Some fungal lineages have hyphae that are **coenocytic** ("common-celled"; pronounced *see-no-SIT-ick*)—meaning that they are not divided into separate cells and thus lack septa (**FIGURE 32.5c**). Coenocytic fungi have hundreds or thousands of nuclei scattered throughout the mycelium. In effect, they are a single, gigantic, multinucleate cell.

Fungi lack complex, long-distance transport systems like those found in plants and animals. One advantage of most fungi having pores in septa is that nutrients can move rapidly from regions of uptake to regions of mycelial growth. Because nutrients and some organelles can flow through the entire mycelium—at least to a degree—the fungal mycelium is intermediate between a multicellular land plant or animal and an enormous single-celled organism.

Mycelia Have a Large Surface Area

It's important to appreciate just how thin hyphae are. Plant root tips are typically about 1 mm in diameter, but fungal hyphae are typically less than 10 μm in diameter, or 1/100th the size of a root tip. Fungal mycelia can penetrate tiny fissures in soil and absorb nutrients that are inaccessible to plant roots.

Perhaps the most important aspect of mycelia and hyphae, however, is their shape. Because mycelia are composed of complex, branching networks of extremely thin hyphae, fungi have the highest surface-area-to-volume ratios observed in multicellular organisms and are therefore the best at absorption.

(a) Both the reproductive structure and mycelium are composed of hyphae.

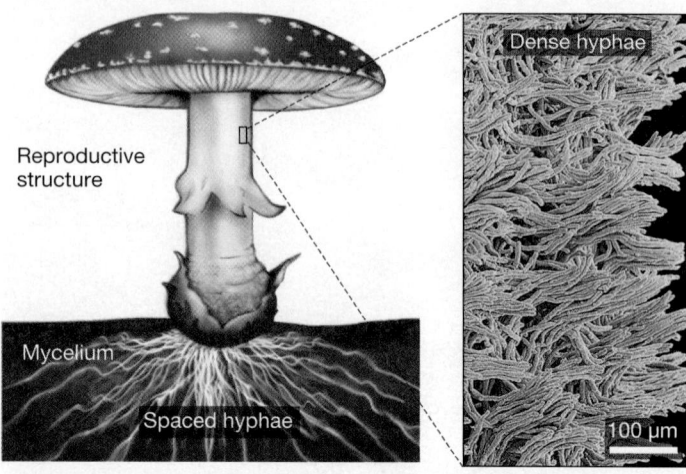

(b) Most hyphae are broken into compartments by septa.

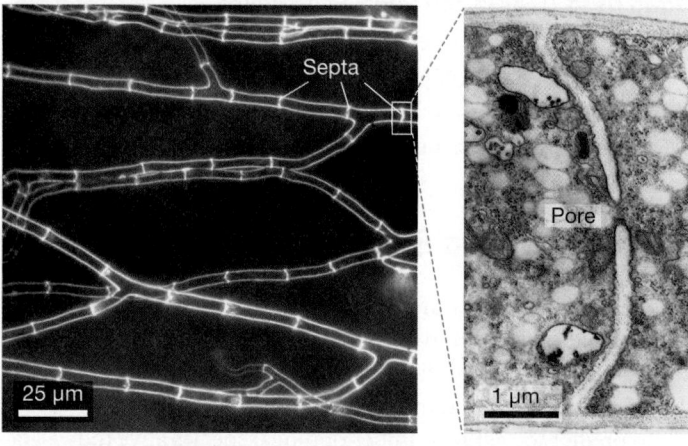

(c) Coenocytic hyphae consist of multinucleate cells.

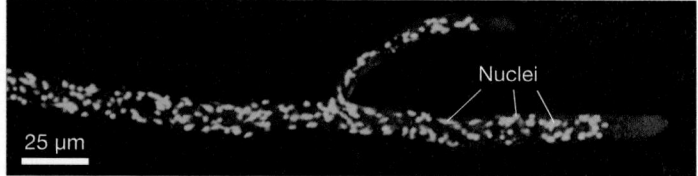

FIGURE 32.5 Multicellular Fungi Have Unusual Bodies. (a) The feeding structure of a fungus is a mycelium, which is made up of hyphae. In some species, hyphae come together to form multicellular structures such as mushrooms, brackets, or morels that emerge from the ground. **(b)** Hyphae of some terrestrial fungi are divided into cell-like compartments by partitions called septa that contain pores. As a result, the cytoplasm of different compartments is continuous. **(c)** Hyphae of some coenocytic fungi are composed of giant, multinucleate cells.

To drive this point home, consider that the hyphae found in any fist-sized ball of rich soil typically have a surface area equivalent to half a page of this book. Because of their large surface area, fungi are extremely efficient at absorbing nutrients.

The extraordinarily high surface area in a mycelium has a downside, however. The amount of water that evaporates from an organism is a function of its surface area—meaning that these organisms are prone to drying out. As a result, fungi are most abundant in moist habitats.

When conditions dry, the fungal mycelium may die back partially or completely. Reproductive cells called **spores** that are produced by sexual or asexual reproduction are resistant to drying, however. As a result, spores can endure dry periods and then germinate to form a new mycelium when conditions improve. Mycelial growth is dynamic, changing with moisture availability or food supply.

Reproductive Structures   Mycelia are an adaptation that supports external digestion and the absorptive lifestyle of fungi. Fungi also produce reproductive organs, which are thick, fleshy structures.

Mushrooms, puffballs, and other dense, multicellular structures that arise from mycelia do not absorb food. Instead, they function in reproduction. Typically they are the only part of a fungus that is exposed to the atmosphere, where drying is a problem. The mass of hyphal filaments on the inside of mushrooms is protected from drying by the densely packed hyphae forming the surface.

In many fungi, including some entire lineages, sexual reproduction has never been observed. In those lineages that do reproduce sexually, important morphological differences among lineages are seen. Most fungal species that undergo sexual reproduction produce one of four types of distinctive reproductive structures.

1. *Swimming gametes and spores*   In certain species that live primarily in water or wet soils, the gametes produced during sexual reproduction have flagella, as do the spores produced during asexual reproduction (**FIGURE 32.6a**). These are the only motile cells known in fungi. Species with swimming gametes are traditionally known as chytrids (pronounced *KYE-trids*).

2. *Zygosporangia*   In some species, haploid hyphae from two individuals meet and become joined, like oxen with a yoke, as shown in **FIGURE 32.6b**. Cells from yoked hyphae fuse to form a distinctive spore-producing structure called a **zygosporangium** (plural: **zygosporangia**; the Greek root *zygo* means to be yoked [see **BioSkills 2** in Appendix B]). Species with a zygosporangium are traditionally known as zygomycetes.

3. *Basidia*   Mushrooms, brackets, and puffballs form specialized club-like cells called **basidia** (singular: **basidium**; "little-club") at the ends of hyphae, each producing four spores (**FIGURE 32.6c**). Species with basidia are traditionally called basidiomycetes or "club fungi."

4. *Asci*   Cups, morels, and some other types of fungi form reproductive structures, specialized sac-like cells called **asci** (singular: **ascus**) at the ends of hyphae, on the surfaces of the fungi, each producing eight spores (**FIGURE 32.6d**). Species with asci are traditionally known as ascomycetes or "sac fungi."

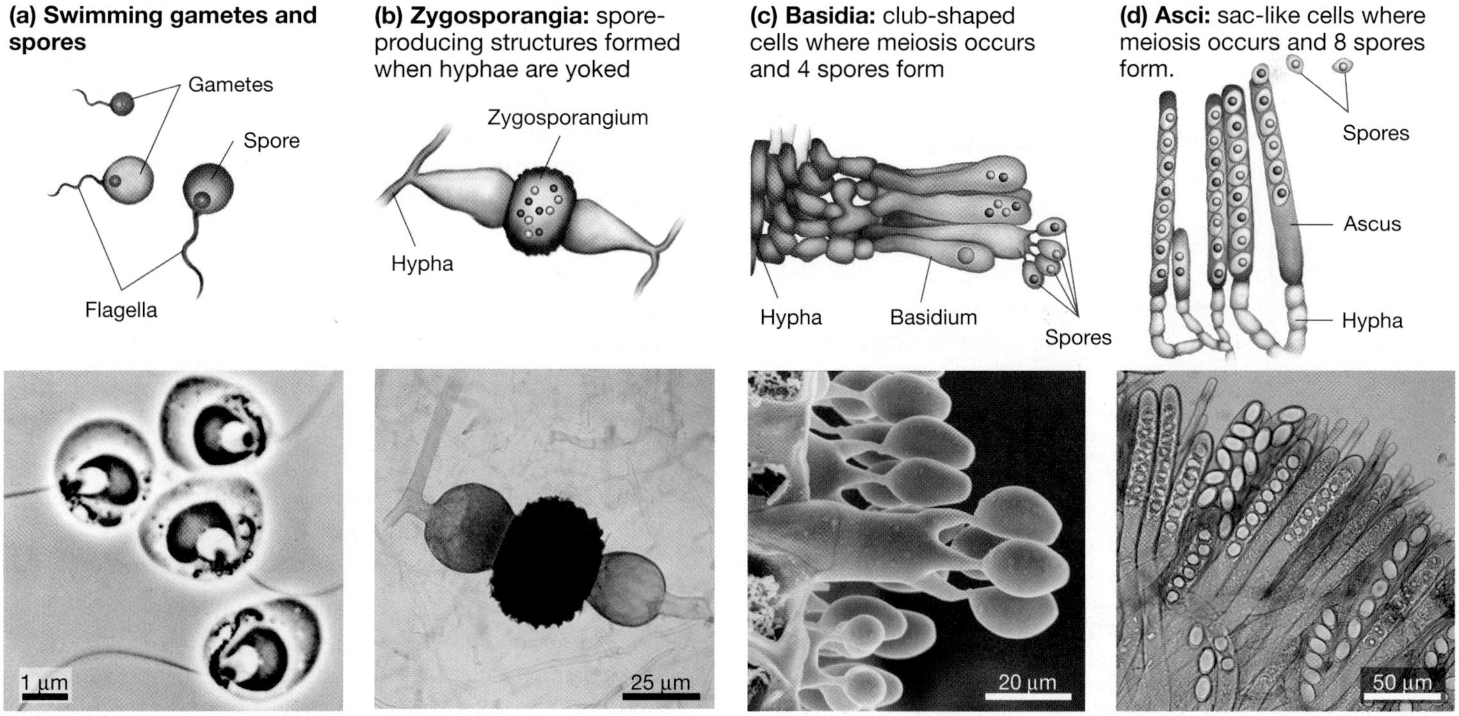

(a) Swimming gametes and spores

Gametes

Spore

Flagella

1 μm

(b) Zygosporangia: spore-producing structures formed when hyphae are yoked

Zygosporangium

Hypha

25 μm

(c) Basidia: club-shaped cells where meiosis occurs and 4 spores form

Hypha   Basidium

Spores

20 μm

(d) Asci: sac-like cells where meiosis occurs and 8 spores form.

Spores

Ascus

Hypha

50 μm

FIGURE 32.6 Four Types of Reproductive Structures Are Observed in Fungi. The dots in the illustrations represent nuclei.

In sum, morphological studies allowed biologists to describe and interpret the mycelial growth habit as an adaptation that makes external digestion and nutrient absorption extremely efficient. Careful analyses of morphological features also allowed researchers to identify four major types of sexual reproductive structures.

Now the question is, within the Fungi, do species that produce swimming gametes and spores, zygosporangia, basidia, and asci each form monophyletic groups—meaning that these distinctive reproductive structures evolved just once? Also, which eukaryotes are most closely related to the fungi?

Evaluating Molecular Phylogenies

Researchers have sequenced and analyzed an array of genes and genomes to establish where fungi fit on the tree of life. In fact, the first complete genome sequence of a eukaryotic species was that of a fungus, the model organism baker's yeast, *Saccharomyces cerevisiae*. The phylogenetic position of fungi as a whole is now well established; the position of lineages within fungi is still the subject of intense research.

Fungi Are Closely Related to Animals   **FIGURE 32.7** shows that fungi are much more closely related to animals than they are to land plants.

The close evolutionary relationship between fungi and animals explains why fungal infections in humans are much more difficult to treat than bacterial infections. Fungi and humans shared a common ancestor relatively recently. As a result, their

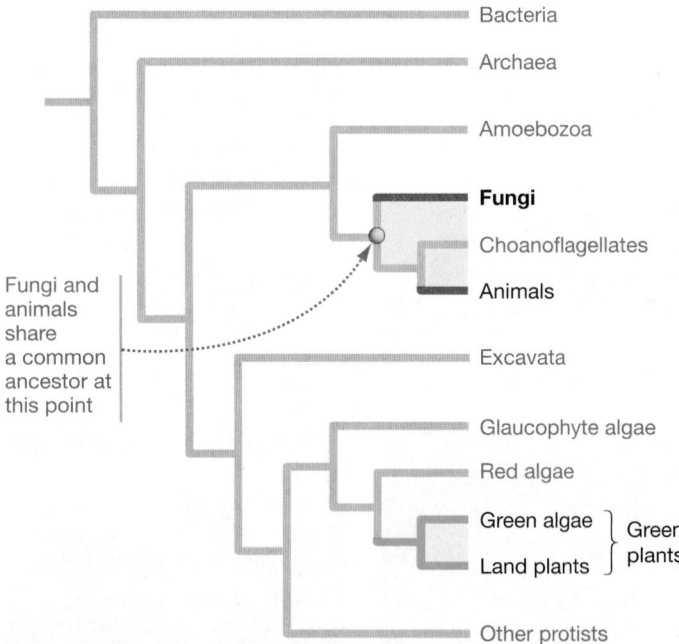

FIGURE 32.7 Fungi Are More Closely Related to Animals than to Land Plants. Phylogenetic tree showing the evolutionary relationships among the green plants, animals, fungi, and some groups of protists. (Choanoflagellates are solitary or colonial protists found in freshwater; see Chapter 33).

enzymes and cell components are similar in structure and function. Drugs that disrupt fungal enzymes and cells are also likely to damage humans.

In addition to DNA sequence data, three key morphological traits link animals and fungi:

1. Most animals and fungi synthesize the tough structural material called chitin (see Chapter 5). Chitin is a prominent component of the cell walls of fungi.

2. The flagella that develop in chytrid spores and gametes are similar to those observed in animals: As in animals, the flagella in chytrids are single, are located at the back of reproductive cells, and move in a whiplike manner.

3. Both animals and fungi store food by synthesizing the polysaccharide glycogen. (Green plants, in contrast, synthesize starch as their storage product.)

What Are the Relationships Among the Major Fungal Groups?

To understand the relationships among species with swimming gametes and spores, zygosporangia, basidia, and asci, biologists have sequenced a series of genes from an array of fungal species and used the data to estimate the phylogeny of the group. The results, shown in **FIGURE 32.8**, support several important conclusions:

- The single-celled, parasitic eukaryotes called microsporidians are actually fungi.
  Interpretation: They are not a distantly related sister group to fungi, as initially thought. This point is important. Researchers are now testing the hypothesis that **fungicides**—molecules that are lethal to fungi—can cure microsporidian infections in bee colonies, silkworm colonies, and AIDS patients.

- The chytrids and zygomycetes are poorly resolved. The actual order of branching events among lineages with these reproductive structures is still not known (in Figure 32.8, they are collapsed into a **polytomy**; see Chapter 28 and **BioSkills 7** in Appendix B).
  Interpretation: Swimming gametes and the zygosporangium evolved more than once. Or, both structures were present in a common ancestor but then were lost in certain lineages.

- An important group called the Glomeromycota is monophyletic.
  Interpretation: The adaptations that allow these species to live in association with plant roots as mycorrhizae (AMF, discussed in Section 32.3) evolved once.

- The basidiomycetes are monophyletic—they form a lineage called Basidiomycota, or club fungi.
  Interpretation: The basidium evolved once.

- The ascomycetes are monophyletic—they form a lineage called Ascomycota, or sac fungi.
  Interpretation: The ascus evolved once.

- Together, the Basidiomycota and Ascomycota form a monophyletic group.
  Interpretation: Because basidiomycetes and ascomycetes both form septate hyphae and large "fruiting" structures, this growth habit evolved once.

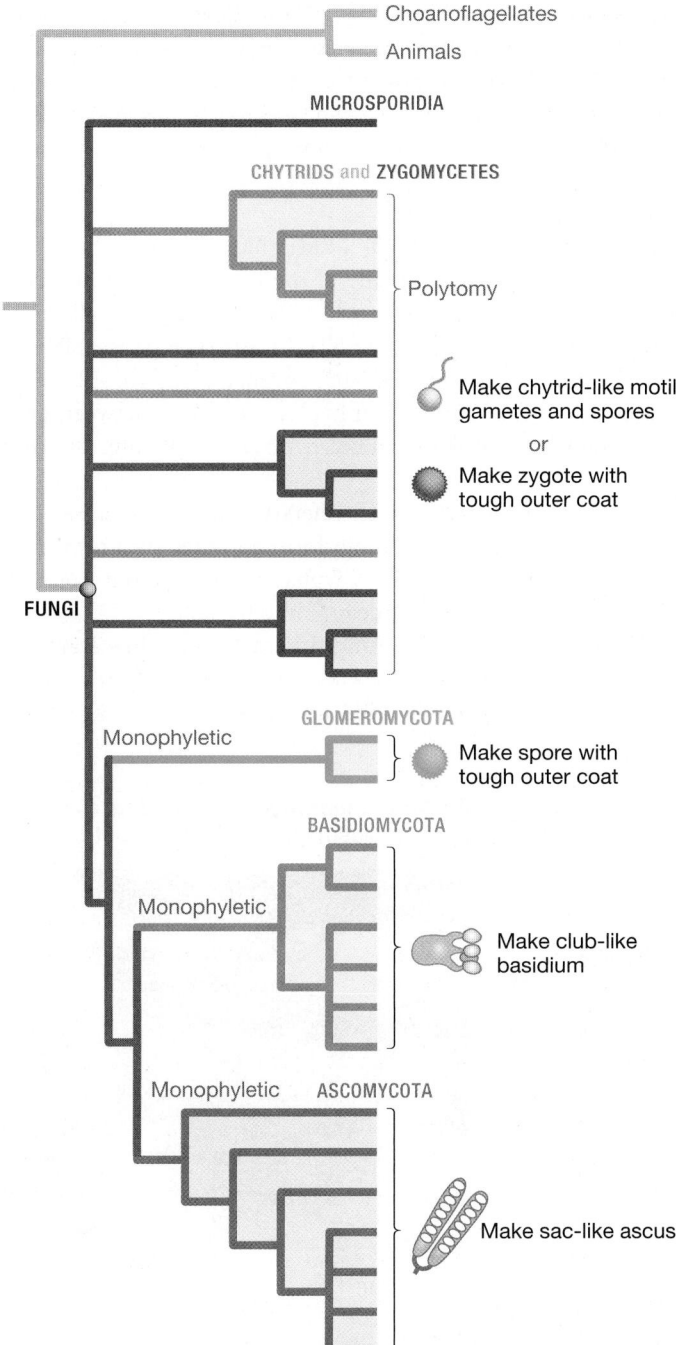

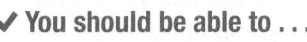

FIGURE 32.8 Phylogeny of the Fungi. A phylogenetic tree based on analyses of DNA sequence data. The icons represent the types of sexual reproductive structures observed in each major lineage.

- The sister group to fungi comprises animals plus protists called choanoflagellates.

  Interpretation: Because choanoflagellates and the most ancient groups of animals are aquatic, and because chytrids are aquatic, it is reasonable to hypothesize that the earliest fungi were aquatic and that the switch to terrestrial life occurred early in the evolution of the Fungi.

  Although recently advances have been made in understanding the evolutionary history of fungi, the phylogenetic tree in

Figure 32.8 is still a work in progress. For example, it is not yet clear where microsporidians are placed relative to several lineages of chytrids and zygomycetes. (Microsporidians lack both swimming gametes and yoked hyphae.) Future work should clarify exactly how fungi diversified.

check your understanding

C Y U

If you understand that . . .
- The bodies of fungi are either single-celled yeasts or multicellular mycelia.
- During sexual reproduction, different groups of fungi produce distinct reproductive structures.

✔ **You should be able to . . .**

1. Explain why mycelia are interpreted as an adaptation to an absorptive lifestyle.
2. Identify the four types of sexual reproductive structures observed in fungi.

Answers are available in Appendix A.

32.3 What Themes Occur in the Diversification of Fungi?

Why are there so many different species of fungi? This question is particularly puzzling given that fungi share a common attribute: They all make their living by absorbing food directly from their surroundings. In contrast to the diversity of food-getting strategies observed in bacteria, archaea, and protists, all fungi make their living in the same basic way. In this respect, fungi are like plants—virtually all of which make their own food via photosynthesis.

The diversification of land plants was driven not by novel ways of obtaining food, but by adaptations that allowed plants to grow and reproduce in a diverse array of terrestrial habitats (see Chapter 31). Recall that associations with animal pollinators were especially important in the diversification of flowering plants. What drove the diversification of fungi? The answer is the evolution of novel methods for absorbing nutrients from a diverse array of food sources.

This section introduces a few of the ways that fungi go about absorbing nutrients from different food sources, as well as how they produce offspring. Let's explore the diversity of ways that fungi do what they do.

Fungi Participate in Several Types of Symbioses

The first plants in the fossil record are closely associated with fungal fossils; the ability to absorb nutrients from fungi may have been crucial in the early evolution of land plants (see Chapter 31). Close associations between land plants and fungi continue today. Researchers estimate that 90 percent of land plants live in physical contact with fungi. Stated another way, fungi and land plants often have a **symbiotic** ("together-living") relationship.

Although some species of fungi live in association with an array of different land plant species, some documented fungi–plant associations are specific. It is not uncommon for one fungal species to live in only a particular type of tissue, in one plant species.

Scientists categorize these symbiotic relationships as **mutualistic** if they benefit both species, **parasitic** if one species benefits at the expense of the other, or **commensal** if one species benefits while the other is unaffected. However, it is sometimes difficult to know if a species really benefits from the relationship.

Not long after associations between fungi and the roots of land plants were discovered, researchers found that two types of plant–mycorrhizal interactions are particularly common, involving **ectomycorrhizal fungi (EMF)** and **arbuscular mycorrhizal fungi (AMF).** The two major types of mycorrhizae have distinctive morphologies, geographic distributions, and functions.

- EMF are usually species from the Basidiomycota, though some ascomycetes participate.

- AMF include species from the Glomeromycota.

Mycorrhizae aren't the only type of symbiotic fungi found in plants, however. Researchers have also become interested in fungi that live in close association with the aboveground tissues of land plants—their leaves and stems. Fungi that live in the aboveground parts of plants are called **endophytic** ("inside-plants").

Recent research has shown that endophytic fungi are much more common and diverse than previously suspected. Further, data indicate that at least some species of endophytes are mutualistic.

Ectomycorrhizal Fungi (EMF)   EMF like the one shown in **FIGURE 32.9a** are found on many of the tree species in the temperate regions of the world, where warm summers alternate with cold winters (see Chapter 52). In this type of association, hyphae form a dense network that covers a plant's root tips. As Figure 32.9a shows, individual hyphae penetrate between cells in the outer layer of the root, but hyphae do not enter the root cells.

The Greek root *ecto*, which refers to "outer," describes this association accurately: The fungi form an outer sheath on root tips that is often 0.1 mm thick. Hyphae also extend out from the sheath-like portion of the mycelium into the soil.

How do these trees and fungi interact, once they start living together? In the habitats where EMF are abundant, nitrogen atoms tend to remain tied up in dead tissues, in amino

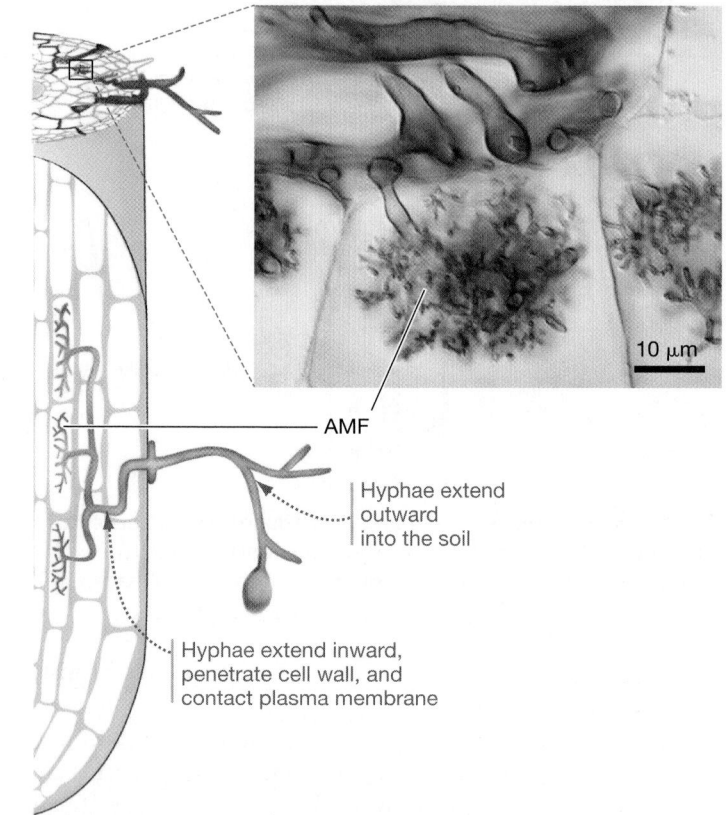

(a) Ectomycorrhizal fungi (EMF) form sheaths around roots and penetrate between root cells.

20 μm

EMF

Hyphae extend outward into the soil

Hyphae extend inward, between cells

Hyphae form a dense, continuous sheath around root

(b) Arbuscular mycorrhizal fungi (AMF) contact plasma membranes of root cells.

10 μm

AMF

Hyphae extend outward into the soil

Hyphae extend inward, penetrate cell wall, and contact plasma membrane

FIGURE 32.9 Mutualistic Fungi Interact with the Roots of Plants in Two Distinct Ways. (a) Ectomycorrhizal fungi (EMF) form a dense network around the roots of plants. Their hyphae penetrate the intercellular spaces of the root but do not enter the root cells. **(b)** The hyphae of arbuscular mycorrhizal fungi (AMF) penetrate the walls of root cells, where they branch into bushy structures or balloon-like vesicles that are in close contact with the root cell's plasma membrane.

acids and nucleic acids, instead of being available in the soil. The hyphae of EMF penetrate decaying material and release enzymes called peptidases that cleave the peptide bonds between amino acids in dead tissues. The amino acids released by this reaction are absorbed by the hyphae and transported to spaces between the root cells of trees, where they can be absorbed by the plant.

EMF are also able to acquire phosphate ions that are bound to soil particles and transfer the ions to host plants. In return, the fungi receive sugars and other complex carbon compounds from the tree.

Researchers have found that when birch tree seedlings are grown with and without their normal EMF in pots filled with forest soil, only the seedlings with EMF are able to acquire significant quantities of nitrogen and phosphorus. Inspired by such data, one biologist has referred to ectomycorrhiza as the "dominant nutrient-gathering organs in most temperate forest ecosystems."

The hyphae of EMF are like an army of miners that discover, excavate, and deliver precious nuggets of nitrogen to trees. The productivity of the world's most important commercial forests depends on EMF.

Arbuscular Mycorrhizal Fungi (AMF)   In contrast to the hyphae of EMF, the hyphae of arbuscular mycorrhizal fungi (AMF) grow *into* the cells of root tissue. The name *arbuscular* ("little-tree") was inspired by the bushy, highly branched hyphae, shown in **FIGURE 32.9b**, that form between the cell walls and the plasma membrane of root cells. AMF are also called **endomycorrhizal fungi,** because they penetrate the interior of root cell walls.

The key point is that the hyphae of AMF penetrate the cell wall and make direct contact with the plasma membrane of root cells. The highly branched hyphae inside the plant cell wall are thought to be an adaptation that increases the surface area available for exchange of molecules between the fungus and its host. However, AMF do not form a tight sheath around roots, as do EMF. Instead, they form a pipeline extending from inside plant roots into the soil well beyond the root.

AMF are found in a whopping 80 percent of all land plant species. They are particularly common in grasslands and in tropical forests. They are also widespread in temperate climates.

What do AMF do? Plant tissues decompose quickly in the grasslands and tropical forests where AMF flourish because the growing season is long and warm. As a result, nitrogen is often readily available to plants. Phosphorus is usually in short supply, though, because it tends to leach out of soils that experience high rainfall.

To explore the nature of fungi–plant symbioses in detail, researchers have used isotopes as tracers for specific elements (**FIGURE 32.10**, see page 622). For example, to test the hypothesis that fungi obtain food in the form of carbon-containing compounds from their plant associates, biologists have introduced radioactively labeled carbon dioxide into the air surrounding plants that do or do not contain symbiotic fungi. The labeled CO_2 molecules are incorporated into the sugars produced during photosynthesis, and the location of the radioactive atoms can then be followed over time by means of a device that detects radioactivity. If plants feed their fungal symbionts, then labeled carbon compounds should be transferred from the plant to the fungi.

To test the hypothesis that plants are receiving nutrients from their symbiotic fungi in return for sugars, researchers have added radioactive phosphorus atoms or the heavy isotope of nitrogen (^{15}N) to potted plants that do or do not contain symbiotic fungi. If fungi facilitate the transfer of nutrients from soil to plants, then plants grown in the presence of their symbiotic fungi should receive much more of the radioactive phosphorus or heavy nitrogen than do plants grown in the absence of fungi.

Experiments with isotopes used as tracers have shown that sugars and other carbon-containing compounds produced by plants via photosynthesis are transferred to their fungal symbionts. In some cases, as much as 20 percent of the sugars produced by a plant end up in their symbiotic fungi. In exchange, the symbiotic fungi facilitate the transfer of phosphorus or nitrogen—or both—from soil to the plant.

Because phosphorus and nitrogen are in extremely short supply in most environments, the nutrients supplied by symbiotic fungi are critical to the success of the plant. In this way, studies with isotopes have supported the hypothesis that most relationships between fungi and land plants are mutually beneficial.

AMF are also extremely important in soil formation. The cell walls of their hyphae contain large quantities of a glycoprotein called **glomalin.** When the cells die, the glomalin enriches the organic matter in soil and helps bind organic compounds to sand or clay particles. Some recent estimates suggest that over 25 percent of all of the organic matter in soil consists of glomalin.

Are Endophytes Mutualists?   Although endophytic fungi were unknown before the 1940s, they are turning out to be both extremely common and highly diverse.

- Biologists in Brazil are examining tree leaves for the presence of fungi; each time they do, they are discovering several new species of endophytes.

- Recall from Section 32.2 that a study in Panama found hundreds of fungal species living in the leaves of just two tree species. These species are endophytes.

Recent research has shown that some endophytes increase the drought tolerance of their host plants. Endophytes found in some grasses also produce compounds that benefit plants. The compounds deter or even kill herbivores. In exchange for these benefits, endophytes absorb sugars from the plant.

Based on these results, biologists have concluded that the relationship between endophytes and grasses is mutualistic. Similar types of anti-herbivore compounds have recently been documented in an endophyte that lives in morning glories.

In other types of plants, however, researchers have not been able to document benefits for the plant host. The current consensus is that at least some endophytic fungi may be commensals—meaning the fungi and the plants simply coexist with no observable effect, either deleterious or beneficial, on the host plant.

QUESTION: Are mycorrhizal fungi mutualistic?

HYPOTHESIS: Host plants provide mycorrhizal fungi with sugars and other photosynthetic products. Mycorrhizal fungi provide host plants with phosphorus and/or nitrogen from the soil.

NULL HYPOTHESIS: No exchange of food or nutrients occurs between plants and mycorrhizal fungi. The relationship is not mutualistic.

EXPERIMENTAL SETUP:

Labeled carbon treatment:
Labeled CO_2
added to air around plant

Mycorrhizal
fungi present

Labeled carbon control:
Labeled CO_2
added to air around plant

No fungi
present

Labeled P or N treatment:

Fungi

Labeled P or N added to soil

Labeled P or N control:

No fungi

Labeled P or N added to soil

PREDICTION FOR LABELED CARBON: A large percentage of the labeled carbon taken up by the plant will be transferred to mycorrhizal fungi. In the control, little labeled carbon will be present in the soil surrounding the roots.

PREDICTION OF NULL HYPOTHESIS, LABELED CARBON: There will be no difference in the localization of carbon in the two treatments.

PREDICTION FOR LABELED P OR N: A large percentage of the labeled P or N taken up by the fungi will be transferred to the plant. In the control, little or no labeled P or N will be taken up by the plant.

PREDICTION OF NULL HYPOTHESIS, LABELED P OR N: There will be no difference between amounts of labeled P or N found in plant in presence or absence of fungi.

RESULTS:

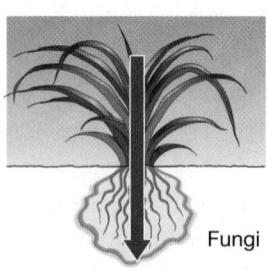

Fungi

Up to 20% of labeled CO_2
taken up by plant is transferred
to mycorrhizal fungi

No
fungi

Little to no labeled carbon
is found in soil

Large amounts of labeled P or N
are found in host plant

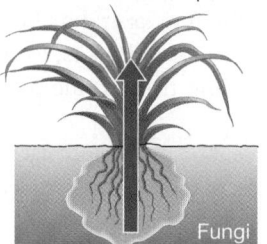

Fungi

Little labeled P or N
is found in host plant

No
fungi

CONCLUSION: The relationship between plants and mycorrhizal fungi is mutualistic. Plants provide mycorrhizal fungi with carbohydrates. Mycorrhizal fungi supply host plants with nutrients.

FIGURE 32.10 Experimental Evidence that Mycorrhizal Fungi and Plants Are Mutualistic. Sugars flow from plants to mycorrhizal fungi; key nutrients flow from mycorrhizal fungi to plants.

SOURCES: Bücking, H., and W. Heyser. 2001. Microautoradiographic localization of phosphate and carbohydrates in mycorrhizal roots of *Populus tremula* × *Populus alba* and the implications for transfer processes in ectomycorrhizal associations. *Tree Physiology* 21: 101–107.

✔ **QUESTION** What do these "labeled nutrient" experiments tell you that you didn't already know from experiments like Figure 32.2, where plants are grown with and without mycorrhizae?

FIGURE 32.11 Leaf-cutter Ants Engage in a Symbiosis with Fungi in Underground Nests. The ants provide the fungus with leaf pieces on which to grow and then the ants feed on the fungus. Of the two partners, only the fungus can digest cellulose.

These results support the general realization that most plants are covered with fungi—from the ends of their branches to the tips of their roots. Throughout their lives, many or even most plants are involved in several distinct types of mutualistic relationships with fungi.

Mutualisms with Other Species Do fungi take up residence with species other than land plants? As you'll recall, the answer is yes.

- **Lichens** are a mutualistic partnership usually between a species of ascomycete and either a cyanobacterium or an alga. The nature of this relationship is explored in Section 32.4.

- Many plant-eating insects harbor diverse arrays of yeast species in their guts, where they may aid digestion or detoxification of plant compounds.

- Some ant species actively farm fungi inside their colonies by inoculating leaf pieces with symbiotic fungi. They fertilize and "weed" their fungal gardens and then harvest parts of the fungi for food (**FIGURE 32.11**).

The many unique properties of fungi have made them useful partners in a wide array of symbiotic relationships.

What Adaptations Make Fungi Such Effective Decomposers?

The saprophytic fungi are master recyclers. Although bacteria and archaea are also important decomposers in terrestrial environments, some fungi and a few bacterial species are the only organisms that can digest wood completely. Given enough time, fungi can turn even the hardest, most massive trees into soft soils.

How do fungi do it? You've already been introduced to two key adaptations:

- The large surface area of a mycelium makes nutrient absorption exceptionally efficient.

- Saprophytic fungi can grow toward the dead tissues that supply their food.

What other adaptations help fungi decompose plant tissues?

Extracellular Digestion Large molecules such as starch, lignin, cellulose, proteins, and RNA cannot diffuse across plasma membranes. Only sugars, amino acids, and other small molecules can enter the cytoplasm of cells through membrane carrier proteins or transporters (see Chapter 6). As a result, fungi, like most animals, have to digest their food before they can absorb it.

Fungi perform **extracellular digestion**—digestion that takes place outside the organism. The simple compounds that result from enzymatic action are then absorbed by the hyphae. Most animals also perform extracellular digestion, but it takes place in a digestive tract (see Chapter 44).

As an example of how this process occurs in fungi, consider the enzymes responsible for digesting lignin and cellulose—the two most abundant organic molecules on Earth.

- Lignin is an extremely strong, complex polymer built from monomers that contain six-carbon rings. Recall that most lignin is found in the secondary cell walls of plant vascular tissues, where it furnishes structural support (see Chapter 31).

- Cellulose is a polymer of glucose and is found in the primary and secondary cell walls of all plant cells.

Some basidiomycetes can degrade lignin and digest cellulose completely—to CO_2 and H_2O. Let's take a closer look at how they do it.

Lignin Degradation Biologists have been keenly interested in understanding how basidiomycetes digest lignin. Paper manufacturers are also interested in this process because they need safe, efficient ways to degrade lignin in order to make soft, absorbent paper products.

To find out how lignin-digesting fungi do it, biologists began analyzing the proteins that these species secrete into extracellular space. After purifying these molecules, the investigators tested each protein for the ability to degrade lignin. Using this approach, investigators from two labs independently discovered an enzyme called lignin peroxidase.

Lignin peroxidase catalyzes the removal of a single electron from an atom in the ring structures of lignin. This oxidation step creates a free radical—an atom with an unpaired electron (see Chapter 2). This is an extremely unstable electron configuration, and it leads to a series of uncontrolled and unpredictable reactions that split the polymer into smaller units.

Biologists have referred to this mechanism of lignin degradation as enzymatic combustion. The phrase is apt: The uncontrolled oxidation reactions triggered by lignin peroxidase are analogous to the uncontrolled oxidation reactions that occur when gasoline burns in a car engine.

The unpredictable nature of the reaction is remarkable, because virtually all of the other reactions catalyzed by enzymes are extremely predictable. The lack of predictability here makes sense, however. Unlike proteins, nucleic acids, and most other polymers with a regular and predictable structure, lignin is extremely heterogeneous. Over 10 types of covalent linkages are routinely found between the monomers that make up lignin. But once lignin peroxidase has created a free radical in the ring structure, any of these linkages can be broken.

However, fungi cannot grow with lignin as their sole source of food. The 6-carbon rings in lignin are extremely difficult to metabolize, and as a result they accumulate in soil. If wood-rotting fungi don't use lignin as food, why do they produce enzymes to digest it? The answer is simple. In wood, lignin forms a dense matrix around long strands of cellulose. Degrading the lignin matrix gives hyphae access to huge supplies of energy-rich cellulose. Saprophytic fungi are like miners. But instead of seeking out rare, gem-like nitrogen or phosphorus atoms as do EMF and AMF, the saprophytes use lignin peroxidase to blast away enormous lignin molecules, exposing rich veins of cellulose that can fuel growth and reproduction.

Cellulose Digestion

Once lignin peroxidase has softened wood by stripping away its lignin matrix, the long strands of cellulose that remain can be attacked by enzymes called cellulases.

Like lignin peroxidase, cellulases are secreted into the extracellular environment by fungi. But unlike the uncontrolled chain reaction catalyzed by lignin peroxidase, degradation of cellulose by cellulases is extremely predictable.

Biologists have purified seven different cellulases from the fungus *Trichoderma reesei*. Two of these enzymes catalyze a critical early step in digestion—they cleave long strands of cellulose into a disaccharide called cellobiose. The other cellulases are equally specific and also catalyze hydrolysis reactions. In combination, the suite of seven enzymes in *T. reesei* transforms long strands of cellulose into a simple monomer—glucose—that the fungus can absorb and use as a source of food.

Variation in Reproduction

Recall from Section 32.2 that fungi may produce swimming gametes and spores, yoked hyphae in which nuclei from different individuals fuse to form a zygosporangium inside a protective structure, or specialized spore-producing cells called basidia and asci. As fungi diversified, in addition to adapting to a variety of food sources, they evolved an array of ways to reproduce.

Spores as Key Reproductive Cells

The spore is the most fundamental reproductive cell in fungi. Spores are the dispersal stage in the fungal life cycle and are produced during both asexual and sexual reproduction. (Recall from Chapter 13 that asexual reproduction is based on mitosis, while sexual reproduction is based on meiosis.) Fungi produce spores in such prodigious quantities that it is not unusual for them to outnumber pollen grains in air samples.

If a spore falls on a food source and is able to germinate, a mycelium begins to form. As the fungus expands, hyphae grow in the direction in which food is most abundant. But if food begins to run out, mycelia respond by making spores, which are dispersed by wind or animals.

Why would mycelia reproduce when food is low? The leading hypothesis to answer this question is that spore production allows starving mycelia to disperse offspring to new habitats where more food might be available. Thus, spore production is favored by natural selection when individuals are under nutritional stress.

Multiple Mating Types

In some fungal lineages, hyphae come in many different mating types. Instead of having morphologically distinct males and females that produce sperm and eggs, hyphae of different mating types appear identical.

Hyphae of the same mating type will not combine during sexual reproduction. When zygomycete hyphae grow close to each other, for example, they will not fuse unless the individuals have different alleles of one or more genes involved in mating. If chemical messengers released by two hyphae indicate that they are of different mating types, then fusion and zygosporangium formation follows.

In this way, mating types function as sexes. Instead of having just two sexes, a single fungal species may have tens of thousands of sexes. The basidiomycete *Schizophyllum commune*, for example, is estimated to have 28,000 mating types.

Why so many? The leading hypothesis to explain the existence of mating types is that it helps generate genetic diversity in offspring by increasing the probability of outcrossing. Genetic diversity, in turn, is known to be advantageous in fighting off infections and responding to changes in the environment (see Chapter 13).

How Does Fertilization Occur?

Compared with green plants, protists, and animals, fertilization in fungi has important unique features:

- Only some chytrids produce gametes, which are not considered eggs and sperm. Both male and female gametes are motile, and female gametes are only slightly larger than male gametes.

- In all other lineages of fungi, fertilization occurs in two distinct steps: (1) fusion of hyphae and (2) fusion of nuclei from the fused hyphae. These two steps can be separated by long time spans and even long distances.

In many fungi, the process of sexual reproduction begins when hyphae from two different mating types fuse to form a hybrid cell. When the cytoplasm of two individuals fuses in this way, **plasmogamy** (pronounced *plaz-MAH-ga-mee*) is said to occur (**FIGURE 32.12**). Often, plasmogamy does not immediately lead to nuclear fusion. Rather, the genetically distinct, haploid nuclei persist in the same cell or mycelium. When two or more genetically distinct nuclei exist within a single mycelium, it is **heterokaryotic** ("different kernel"). Most hyphae in heterokaryotic mycelia are **dikaryotic** ("two kernel")—they are divided by septa, and each cell contains two nuclei, one from each mating type.

The distinct nuclei in dikaryotic and heterokaryotic mycelia function independently, even though gene expression must be coordinated for growth and development to occur. For

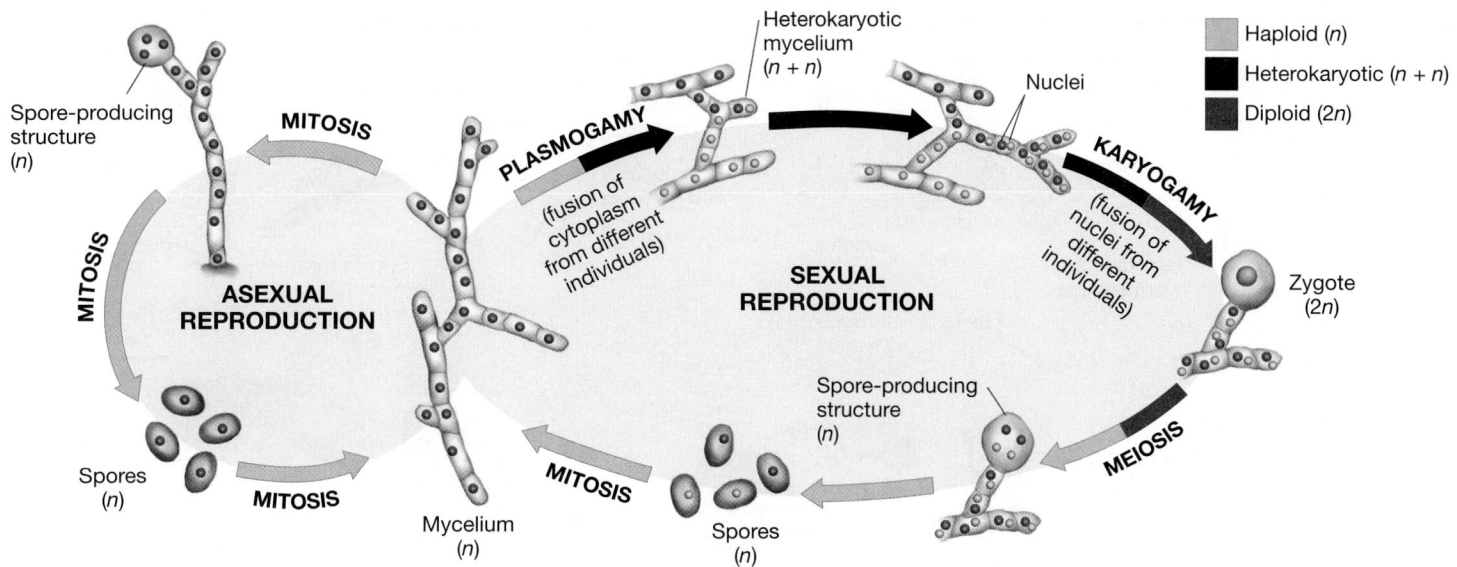

FIGURE 32.12 Fungi Have Unusual Life Cycles. A generalized fungal life cycle, showing both asexual and sexual reproduction.

✔ **QUESTION** Fungi spend most of their lives feeding. Which is the longest-lived component of this life cycle?

example, in septate fungi the two types of nuclei divide as the hyphae expand, so each compartment that is divided by a septum contains one of each of the two types of nuclei. Biologists are currently investigating how the activities of the two nuclei are coordinated.

In a dikaryotic or heterokaryotic mycelium, one or more pairs of unlike nuclei may eventually fuse to form a diploid zygote. The fusion of nuclei is called **karyogamy** (pronounced *ka-ree-AH-ga-mee*). The nuclei that are produced by karyogamy then divide by meiosis to form haploid spores.

✔ **If you understand the relationship between plasmogamy and karyogamy, you should be able to compare and contrast these events with the life cycles of other eukaryotes. For example, which human cells undergo plasmogamy and karyogamy? Are any human cells heterokaryotic?**

Asexual Reproduction As the left side of Figure 32.12 indicates, fungal species can reproduce asexually as well as sexually. In fact, large numbers of ascomycetes have never been observed to reproduce sexually.

During asexual reproduction, spore-forming structures are produced by a haploid mycelium, and spores are generated by mitosis. As a result, offspring are clones—meaning that they are genetically identical to their parent.

Producing microscopic spores that are carried by wind was an adaptation to life on land that allowed terrestrial fungi to disperse widely and to colonize new habitats.

Four Major Types of Life Cycles

Among the sexually reproducing species of fungi, the presence of a heterokaryotic stage and the morphology of the spore-producing structure vary. Morphologies of reproductive structure were introduced in Section 32.2. Let's now take a closer

look at the sexual reproduction portion of each of the four major types of life cycles that have been observed in fungi.

Chytrid Life Cycle The chytrids are the only type of fungi with species that exhibit alternation of generations. **FIGURE 32.13a** (see page 626) shows how alternation of generations occurs in the well-studied genus *Allomyces*. Here are the key stages in its life cycle:

1. Haploid adults form gametangia, in which male and female swimming gametes are produced by mitosis.

2. Gametes from the same individual or different individuals fuse to form a diploid zygote.

3. The zygote grows into a diploid sporophyte.

The life cycle continues when meiosis occurs in the sporophytic mycelium, inside a structure called a sporangium. The haploid spores produced by meiosis disperse by swimming.

Zygomycete Life Cycle In zygomycetes, sexual reproduction starts when hyphae from different mating types fuse, as shown in **FIGURE 32.13b**. Plasmogamy forms a spore-forming structure called a zygosporangium that develops a tough, resistant coat. Inside the zygosporangium, nuclei from the mating partners fuse—meaning that karyogamy occurs.

The zygosporangium can persist if conditions become too cold or dry to support growth. When temperature and moisture conditions are again favorable, however, meiosis occurs and the zygosporangium germinates and forms a sporangium. The meiotic products within the sporangium produce haploid spores.

When spores are released and germinate, they grow into new mycelia. These mycelia can reproduce asexually by making sporangia whose haploid spores are produced by mitosis and dispersed by wind.

(a) Chytrids include the only fungi in which alternation of generations occurs. (Species: *Allomyces macrogynus*)

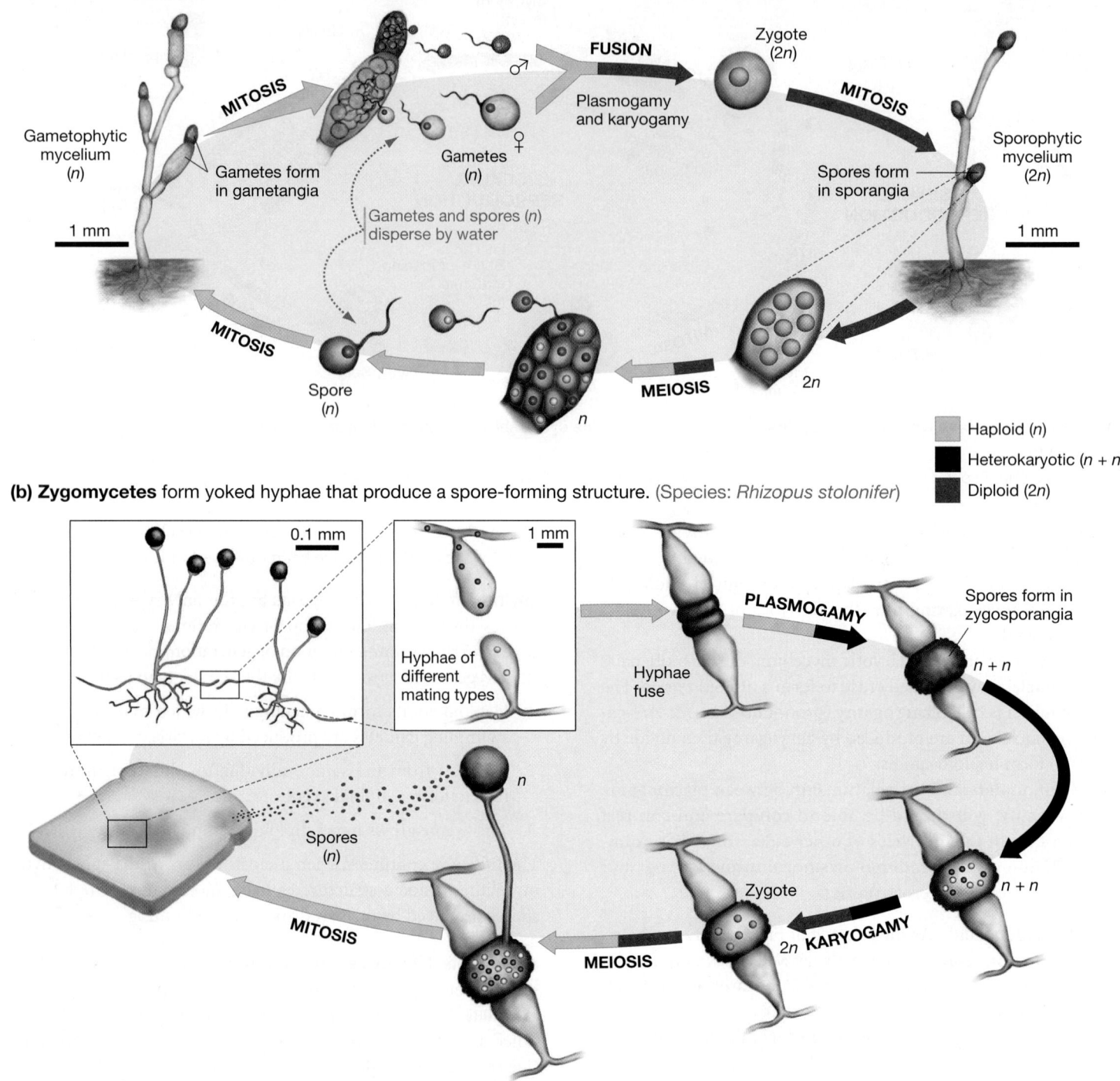

FUSION

Zygote (2n)

MITOSIS

Plasmogamy and karyogamy

Gametophytic mycelium (n)

MITOSIS

Gametes form in gametangia

Gametes (n)

♂ ♀

Sporophytic mycelium (2n)

Spores form in sporangia

1 mm

Gametes and spores (n) disperse by water

1 mm

MITOSIS

Spore (n)

MEIOSIS

2n

n

Haploid (n)

Heterokaryotic (n + n)

Diploid (2n)

(b) Zygomycetes form yoked hyphae that produce a spore-forming structure. (Species: *Rhizopus stolonifer*)

0.1 mm

1 mm

Hyphae of different mating types

Hyphae fuse

PLASMOGAMY

Spores form in zygosporangia

n + n

n + n

Spores (n)

n

Zygote

2n KARYOGAMY

MITOSIS

MEIOSIS

FIGURE 32.13 Variation in Sexual Reproduction in Fungi. The sexual part of the life cycle in four major groups of fungi: **(a)** chytrids, **(b)** zygomycetes, **(c)** Basidiomycota, and **(d)** Ascomycota.

✔ **QUESTION** Asexual reproduction, as shown in Figure 32.12, is extremely common in zygomycetes and ascomycetes. What is the difference between mycelia produced via asexual versus sexual reproduction?

Basidiomycota Life Cycle Mushrooms, bracket fungi, and puff-balls are sexual reproductive structures produced by members of the Basidiomycota (**FIGURE 32.13c**). Even though their size, shape, and color vary enormously from species to species, all basidio-mycete reproductive structures originate from the dikaryotic hy-phae of mated individuals.

Inside a mushroom or bracket fungus or puffball, the club-like, spore-producing cells called basidia form at the ends of dikaryotic hyphae. Karyogamy occurs within the basidia. The diploid nucleus that results undergoes meiosis, yielding four haploid spores. The spores are dispersed by wind and then ger-minate to form a new hyphae.

(c) Basidiomycota have reproductive structures with many spore-producing basidia. (Species: *Amanita muscaria*)

(d) Ascomycota have reproductive structures with many spore-producing asci. (Species: *Cookeina speciosa*)

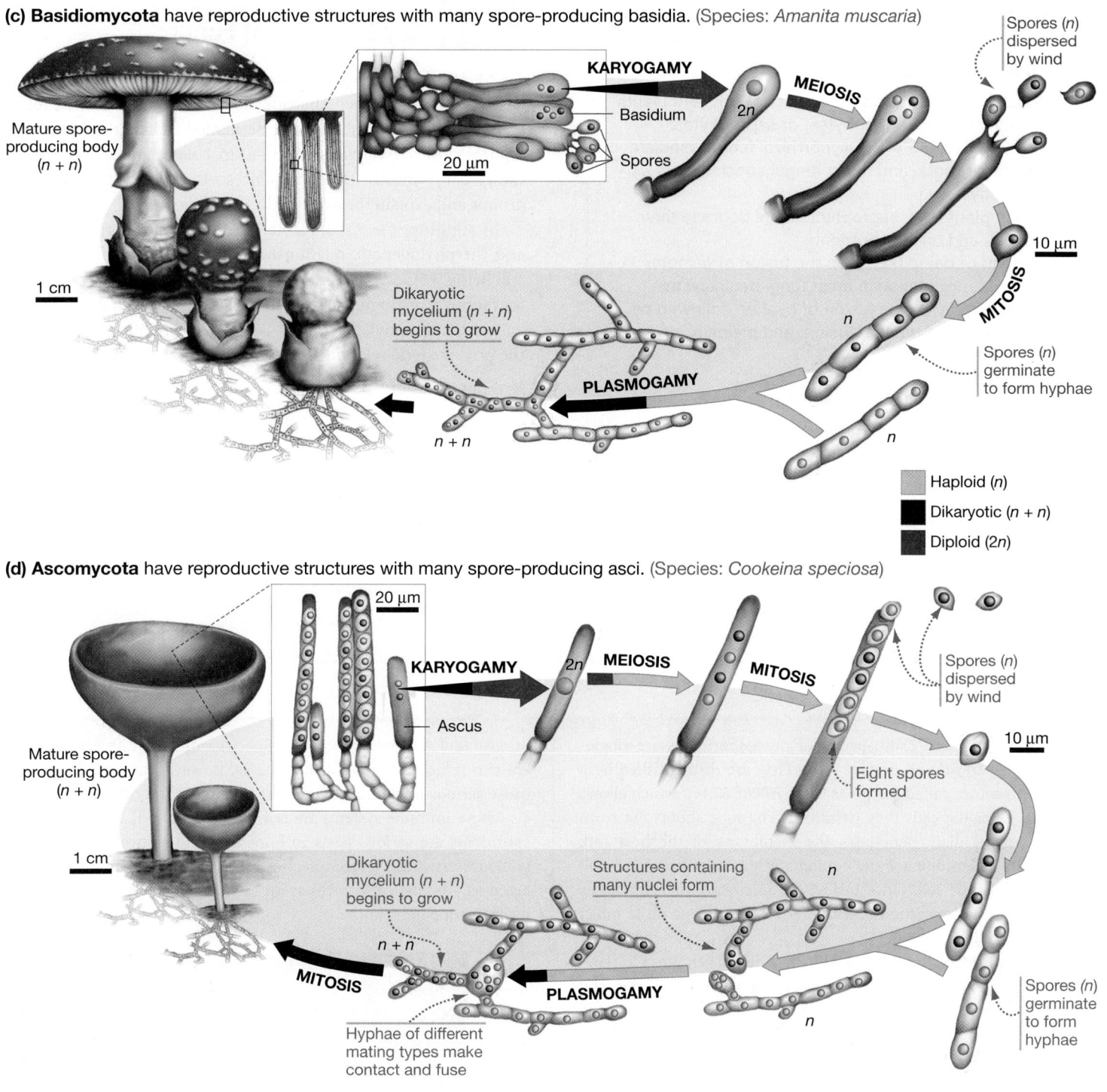

Spores are eventually ejected from the end of the basidia and are dispersed by the wind. It is not unusual for a single puffball or mushroom to produce a billion spores.

Ascomycota Life Cycle The mature sexual reproductive structure in Ascomycota is produced by a dikaryotic hypha.

As **FIGURE 32.13d** illustrates, the process usually begins when hyphae or specialized structures from different mating types fuse, forming a heterokaryotic cell containing many independent nuclei. A short dikaryotic hypha with cells containing two nuclei, one from each parent, emerges and eventually grows into a complex reproductive structure whose hyphae have the sac-like, spore-producing structures called asci at their tips.

After karyogamy occurs inside each ascus, meiosis takes place and eight haploid spores are produced. When the ascus matures, the spores inside are forcibly ejected.

check your understanding

If you understand that . . .

- Most fungi live in close association with land plants as mycorrhizae, endophytes, or saprophytes.
- When plants are alive, mycorrhizal fungi associate with their roots, and many fungal species grow as endophytes.
- When plants die, saprophytic fungi degrade their tissues and release nutrients.
- Instead of being based on the fusion of gametes, sexual reproduction in most fungi begins with plasmogamy, or the fusion of hyphae, followed by karyogamy, or nuclear fusion, and meiosis.

✔ **You should be able to . . .**

1. Describe evidence that mutualism occurs in EMF.
2. Compare and contrast what happens inside a zygosporangium, basidium, and ascus.

Answers are available in Appendix A.

32.4 Key Lineages of Fungi

Based on the current hypothesized phylogeny for the Fungi, it is clear that neither chytrids nor zygomycetes form monophyletic groups and should not be named as single lineages (see Figure 32.8). Researchers continue to collect and analyze additional data—mostly DNA sequences—to identify monophyletic groups and explain the early evolution of fungi.

In addition, the Ascomycota lineage is exceptionally diverse, and the phylogenetic relationships among major subgroups are still being worked out. As a result, the discussion of "key lineages" that follows distinguishes two lifestyles observed in Ascomycota, and not monophyletic groups within the phylum. Research on the phylogenetic relationships within fungi continues.

- Fungi > Microsporidia
- Fungi > Chytrids
- Fungi > Zygomycetes
- Fungi > Glomeromycota
- Fungi > Basidiomycota (Club Fungi)
- Fungi > Ascomycota > Lichen-Formers
- Fungi > Ascomycota > Non-Lichen-Formers

Fungi > Microsporidia

All of the estimated 1500 species of microsporidians are single-celled and parasitic on animal cells. They are distinguished by a unique structure called the polar tube (**FIGURE 32.14**), which allows them to enter the cells they parasitize. The tube shoots out from the microsporidian, penetrates the membrane of the host cell, and acts as a conduit for the contents of the microsporidian cell to enter the host cell. Once inside, the microsporidian replicates and produces a generation of daughter cells, which go on to infect other host cells.

Microsporidians have a dramatically reduced genome. One species has the smallest genome known among eukaryotes—the total number of base pairs present is over 1000 times smaller than the human genome. Unlike other fungi, microsporidians lack functioning mitochondria. As in other fungi, the polysaccharide chitin is a prominent component of their cell wall ✔ **You should be able to mark the origin of the polar tube and the loss of mitochondria on Figure 32.8.**

Absorptive lifestyle Most microsporidians parasitize insects or fish. Because they enter the interior of host cells, they are called intracellular parasites. Their small genome size and lack of functional mitochondria are both adaptations to this lifestyle.

Life cycle Variation in life cycles is extensive. Some species appear to reproduce only asexually. Others produce several different types of sexual or asexual spores, and some must successfully infect several different host species to complete their life cycle.

Human and ecological impacts Species from eight different genera can infect humans. In most cases, however, microsporidians cause serious infections only in AIDS patients and other individuals whose immune systems are not functioning well. Some microsporidians are serious pests in honeybee and silkworm colonies, while others cause severe infections in grasshoppers and are marketed as a biological control agent to reduce crop damage caused by grasshoppers.

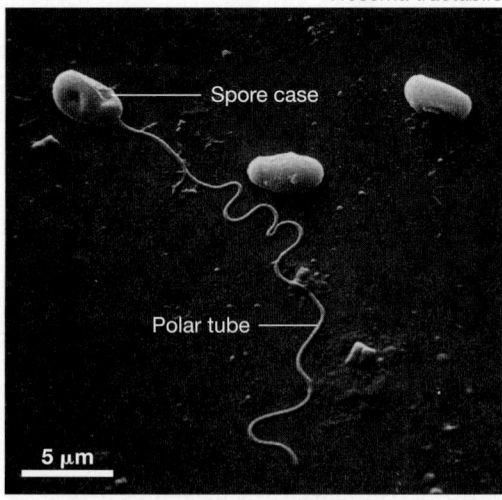

Nosema tractabile

Spore case

Polar tube

5 μm

FIGURE 32.14 Microsporidia Infect Other Cells via a Polar Tube.

Chytrids are largely aquatic and are common in freshwater environments. Some species live in wet soils, and a few have been found in desert soils that are wet only during a rainy season. Species found in dry soils have tough spores that endure harsh conditions. Spores from chytrids have been shown to germinate after a resting period of as long as 31 years. Members of this group are the only fungi that produce motile cells. In each case, the motile gametes use flagella to swim.

Absorptive lifestyle Many species of chytrids have enzymes that allow them to digest cellulose. As a result, these species are important decomposers of plant material in wet soils, ponds, and lakes (**FIGURE 32.15**). However, many of the freshwater species are parasitic, and parasitic chytrids sometimes cause disease epidemics in algae or aquatic insects (including mosquitoes). Other species parasitize mosses, ferns, or flowering plants. Mutualistic chytrids are among the most important of the many organisms living in the guts of deer, cows, elk, and other mammalian herbivores, because the chytrids help these animals digest their food. Chytrids secrete cellulases that degrade the cell walls of grasses and other plants ingested by the animal, releasing sugars that are used by both the chytrids and their hosts.

Life cycle During asexual reproduction, most chytrid species produce spores that swim to new habitats via a flagellum. A few species reproduce sexually as well as asexually and exhibit alternation of generations. (Recall from Section 32.3 that these chytrids are the only fungal species to do so.) In species that reproduce sexually, plasmogamy and karyogamy may occur in a variety of ways: through fusion of hyphae, gamete-forming structures, or gametes.

Human and ecological impacts In addition to the species that serve as decomposers and mutualists, many chytrids are parasites. Biologists are investigating the possibility of using a chytrid that parasitizes mosquito larvae as a biological control agent; potato tubers are sometimes invaded and spoiled by a parasitic chytrid that causes black wart disease. The parasitic chytrid *Batrachochytrium dendrobatidis* is largely responsible for catastrophic declines occurring recently in amphibian populations all over the world.

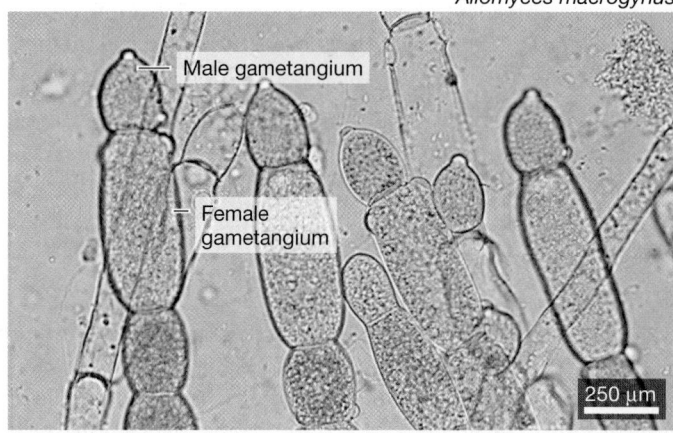

Allomyces macrogynus

FIGURE 32.15 **A Chytrid That is a Common Inhabitant of Soils and Ponds Worldwide.**

The zygomycetes ("yoked-fungi") are primarily soil-dwellers. Their hyphae yoke together and fuse during sexual reproduction and then form a durable, thick-walled zygosporangium.

Absorptive lifestyle Many members of zygomycete lineages are saprophytes and live in plant debris. Some parasitize other fungi, however, or are important parasites of insects and spiders.

Life cycle Asexual reproduction is extremely common. Ball-like sporangia are produced at the tips of hyphae that form stalks, and mitosis results in the production of spores that are dispersed by wind. During sexual reproduction, fusion of hyphae occurs only between individuals of different mating types.

Human and ecological impacts The common bread mold *Rhizopus stolonifer* is a frequent household pest—probably the zygomycete that is most familiar to you. Saprophytic and parasitic members of these lineages are responsible for rotting strawberries and other fruits and vegetables and causing large losses in these food industries. The black bread mold pictured in **FIGURE 32.16** is a familiar example of a saprophytic zygomycete. Some species of *Mucor* are used in the production of steroids for medical use. Species of *Rhizopus* and *Mucor* are used in the commercial production of organic acids, pigments, alcohols, and fermented foods.

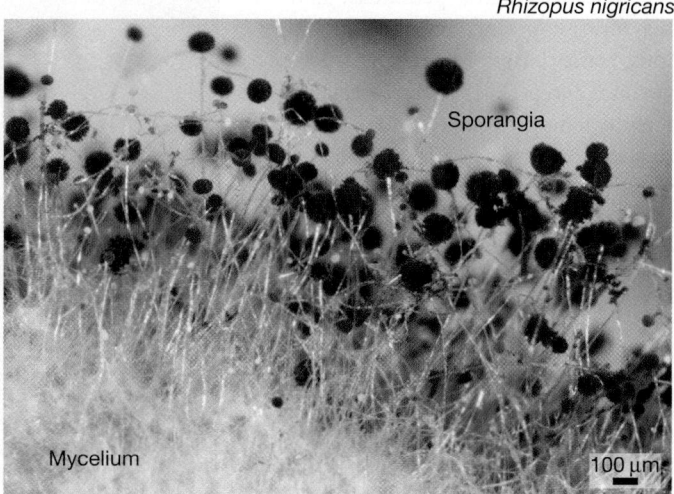

Rhizopus nigricans

FIGURE 32.16 **Black Bread Molds Are Zygomycetes.**

Recent phylogenetic analyses have shown that the arbuscular mycorrhizal fungi (AMF) form a distinct lineage, indicating that Glomeromycota is a major monophyletic group. ✔You should be able to mark the origin of the AMF association on Figure 32.8.

Absorptive lifestyle Recall from Section 32.3 that AMF absorb phosphorus-containing ions or molecules in the soil and transfer them, along with other nutrients and water, into the roots of most of the plants living in grasslands and tropical forests (**FIGURE 32.17**). In exchange, the host plant provides the symbiotic fungi with sugars and other organic compounds.

Life cycle Most species form large spores underground. Glomeromycetes are difficult to grow in the laboratory, so their life cycle is not well known. No one has yet discovered a sexual phase in these species.

Human and ecological impacts Because grasslands and tropical forests occupy large areas of terrestrial habitats on Earth, the AMF are enormously important to both human and natural economies.

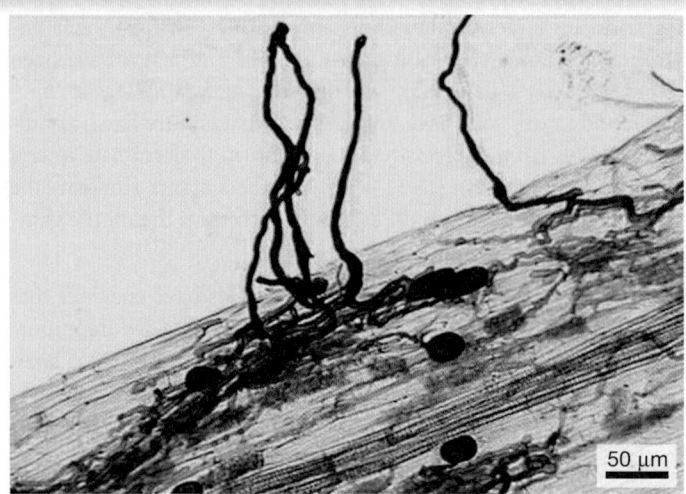

50 μm

FIGURE 32.17 AMF Penetrate the Walls of Plant Root Cells.

Although most basidiomycetes form mycelia and produce multicellular reproductive structures, some species have the unicellular growth form. The group is named for basidia, the club-like cells where meiosis and spore formation occur. ✔You should be able to mark the origin of the basidium on Figure 32.8. About 31,000 species of basidiomycetes have already been described, and more are discovered each year.

Absorptive lifestyle Basidiomycetes are important saprophytes. Along with a few soil-dwelling bacteria, some members of this lineage are the only organisms capable of synthesizing lignin peroxidase and are therefore important in wood decomposition. Some basidiomycetes are ectomycorrhizal fungi (EMF) that associate with trees in temperate forests. One subgroup consists entirely of parasitic forms called rusts, including species that cause serious infections in many crop plants. The plant parasites called smut fungi specialize in infecting grasses; a few infect other fungi. Thus, as in other lineages of fungi, the entire array of lifestyles—saprophytic, mutualistic, and parasitic—is found within Basidiomycota.

Life cycle Asexual reproduction through production of spores is common in the Basidiomycota, although not as prevalent as in other groups of fungi. Asexual reproduction also occurs through growth and fragmentation of mycelia in the soil or in rotting wood, resulting in genetically identical individuals that are physically independent. During sexual reproduction, all basidiomycetes—even unicellular ones—produce basidia. In the largest subgroup in this lineage, basidia form in large, aboveground reproductive structures called mushrooms, brackets, earthstars, or puffballs (**FIGURE 32.18**). Members of the Basidiomycota and Ascomycota are the only fungi that have heterokaryotic (dikaryotic) mycelia. ✔You should be able to mark the origin of the dikaryotic mycelia on Figure 32.8.

Human and ecological impacts EMF are enormously important in forestry. The temperate forests where these fungi are particularly abundant provide most of the hardwoods and softwoods used in building construction, furniture-making, and papermaking. Throughout the world, mushrooms are cultivated or collected from the wild as a source of food. The white button, crimini, and portabella mushrooms you may have seen in grocery stores are all varieties of the same species, *Agaricus bisporus*. Some of the toxins found in poisonous mushrooms are used in biological research; others are hallucinogens.

Astraeus hygrometricus

1 cm

FIGURE 32.18 Puffballs Can Produce Billions of Spores.

About half of the ascomycetes grow in mutualistic association with cyanobacteria and/or single-celled green algae, forming the structures called lichens. Over 15,000 different lichens have been described to date; in most, the fungus involved is an ascomycete (although a few basidiomycetes participate as well). To name a lichen, biologists use the species name assigned to the fungus that participates in the association. Most lichen-formers are found in a single monophyletic group within the Ascomycota. ✔Based on this observation, you should be able to mark the primary origin of the lichen-forming habit on Figure 32.8.

Absorptive lifestyle The fungus in lichens appears to protect the photosynthetic bacterial or algal cells. The fungal hyphae form a dense protective layer that shields the photosynthetic species and reduces water loss. In return, the cyanobacterium or alga provides carbohydrates that the fungus uses as a source of carbon and energy. However, the hyphae of some lichen-forming fungi have been observed to invade algal cells and kill them. This observation suggests a partially parasitic relationship in at least some lichens. The nature of lichen-forming associations is the subject of ongoing research.

Life cycle As **FIGURE 32.19a** shows, many lichens reproduce asexually via the production of small "mini-lichen" structures called soredia that contain both symbionts. Soredia disperse to a new location via wind or water and then develop into a new, mature individual via the growth of the algal or bacterial and fungal symbionts. In addition, the fungal partner may form asci. Spores that are shed from asci germinate to form a small mycelium. If the growing hyphae encounter the appropriate algal or bacterial cells, a new lichen can form.

Human and ecological impacts In abundance and diversity, lichens dominate the Arctic and Antarctic tundras and are extremely common in boreal forests (see Chapter 52). They are the major food of caribou in the winter as well as the most prevalent colonizers of bare rock surfaces throughout the world (**FIGURE 32.19b**). Rock-dwelling lichens are significant because they produce acids that break down the rock surface as they grow—launching the first step in soil formation. Lichens are particularly sensitive to air pollution and are often used as bioindicators of air quality.

(a) Cross section of a lichen, showing three layers

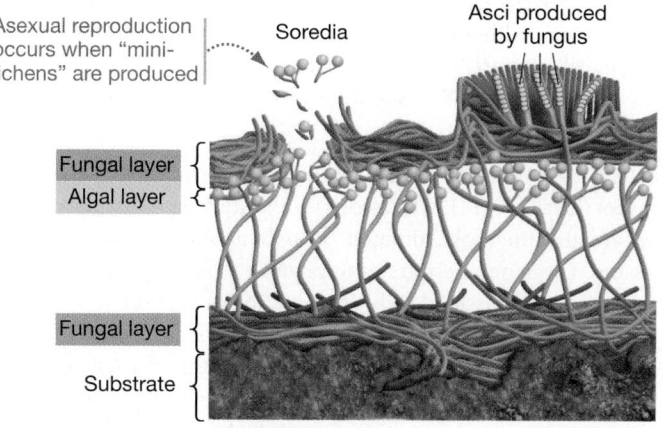

(b) Top view of lichen on a rock

FIGURE 32.19 Lichens Are Associations between a Fungus and a Cyanobacterium or Green Alga. (a) In a lichen, cyanobacteria or green algae live within a dense network of fungal hyphae. **(b)** Lichens often colonize surfaces, such as tree bark or bare rock, where other organisms are rare.

The ascomycetes that do not form lichens are found in virtually every terrestrial habitat as well as some freshwater and marine environments. Although most ascomycetes form mycelia, many are single-celled yeasts. The ascus is a distinguishing characteristic of the Ascomycota. ✔You should be able to mark the origin of the ascus on Figure 32.8.

Absorptive lifestyle Many ascomycetes form mutualistic ecto-mycorrhizal fungi (EMF) associations with tree roots. Ascomycetes are also the most common endophytic fungi on aboveground tissues. Large numbers are saprophytic and are abundant in forest floors and in grassland soils. Parasitic forms are common as well. As in other fungal lineages, the entire array of absorptive lifestyles found in fungi has evolved independently within the Ascomycota.

In addition, about 65 species in one Ascomycota lineage are predatory—primarily on amoebae and other unicellular protists. Some are large enough to capture the microscopic animals called roundworms, however. The predatory ascomycetes trap their prey by means of sticky substances on their cell walls, or they catch prey in snares consisting of looped hyphae (**FIGURE 32.20**). Once a prey individual is captured, hyphae invade its body, digest it, and absorb the nutrients that are released. A predatory ascomycete has been found in fossilized material that is 100 million years old. ✔You should be able to mark the origin of predation on Figure 32.8.

Life cycle The aboveground, asci-bearing reproductive structures of these fungi may be a cup or saucer shape (**FIGURE 32.21**). Spores are also routinely produced asexually, at the ends of specialized hyphae.

Human and ecological impacts Some saprophytic ascomycetes can grow on jet fuel or paint and are used to help clean up contaminated sites. *Penicillium* is an important source of antibiotics, and *Aspergillus* produces citric acid used to flavor soft drinks and candy. Truffles are so highly prized that they can fetch $2000 per pound, and the multibillion-dollar brewing, baking, and wine-making industries would collapse without the yeast *Saccharomyces cerevisiae*. A few parasitic ascomycetes cause infections in humans and other animals. In land plants, parasites from this group cause diseases including Dutch elm disease and chestnut blight.

Arthrobotrys anchonia

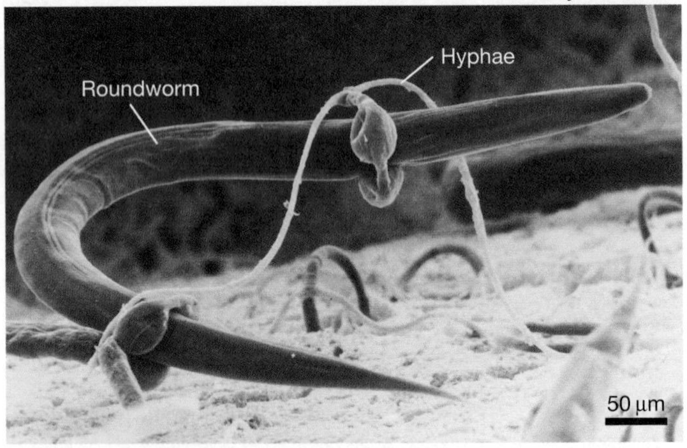

Roundworm
Hyphae
50 μm

FIGURE 32.20 Some Ascomycetes Are Predatory.

Cookeina speciosa

1 cm

FIGURE 32.21 Ascomycete Reproductive Structures are Often Cup Shaped.

If you understand . . .

32.1   Why Do Biologists Study Fungi?

- Many fungi species live in close association with land plants. The roots of most plants are infected by mutualistic fungi that provide the plant with nitrogen, phosphorus, and water in exchange for sugars and other products of plant photosynthesis.

- Parasitic fungi are responsible for devastating blights in crops and other plants.

- Once plants die, saprophytic fungi degrade the lignin and cellulose in wood and use nutrients from decaying plant material.

- Because they free up carbon atoms that would otherwise be locked up in wood, fungi speed up the carbon cycle in terrestrial ecosystems.

- Fungi are used widely in the food processing, cheese-making, baking, and brewing industries.

✔ You should be able to predict the results of using fungicides to experimentally exclude fungi from 10-m × 10-m plots in a forest or grassland near your campus.

32.2   How Do Biologists Study Fungi?

- Morphological analyses of fungi have revealed adaptations that make fungi exceptionally effective at absorbing nutrients from the environment.

- Researchers have determined that fungi exhibit two growth forms—single-celled "yeasts" or mycelia composed of long, filamentous hyphae. Mycelia provide fungi with extremely high surface-area-to-volume ratios. Some fungal bodies are extremely large.

- Septa divide hyphae into cells, but pores in the septa facilitate transport of nutrients and organelles through the hyphae.

- Scientists have determined from morphological studies that fungi possess four known, distinctive types of reproductive structures: (1) chytrids produce motile gametes; (2) zygomycetes make tough zygosporangia; (3) members of the Basidiomycota have club-like, spore-forming structures; and (4) members of the Ascomycota have sac-like, spore-forming structures.

- Recent analyses of DNA sequence data have revealed that fungi are more closely related to animals than they are to plants, and that the Basidiomycota, Ascomycota, and Glomeromycota each form monophyletic groups. Researchers are still trying to understand how the various groups of chytrids and zygomycetes are related to each other and to the other fungal lineages.

✔ You should be able to explain why fungal infections are more difficult to treat than bacterial infections.

32.3   What Themes Occur in the Diversification of Fungi?

- All fungi make their living by absorbing nutrients from living or dead organisms.

- Many fungi live in leaves and stems; in some grasses and other species, these endophytic fungi secrete toxins that discourage herbivores.

- Extracellular digestion, in which enzymes are secreted into food sources, enables fungi to break down extremely large molecules before absorbing them.

- Lignin decomposes through a series of uncontrolled oxidation reactions triggered by the enzyme lignin peroxidase.

- Cellulose digestion occurs in a carefully regulated series of steps, each catalyzed by a specific cellulase.

- Many species of fungi have never been observed to reproduce sexually. Most species can produce haploid spores either sexually or asexually, however.

- Sexual reproduction usually starts when hyphae from genetically different mating types fuse—an event called plasmogamy. If the fusion of nuclei, or karyogamy, does not occur immediately, a heterokaryotic mycelium forms, which in some species is dikaryotic. Heterokaryotic cells may eventually produce spore-forming structures where karyogamy and meiosis take place.

✔ You should be able to explain why most fungi don't need to have gametes to accomplish sexual reproduction.

32.4   Key Lineages of Fungi

- Microsporidia are parasitic, single-celled fungi with reduced genomes and an unusual polar tube used to infect host cells.

- Chytrids are usually aquatic with motile, flagellated spores. Some are terrestrial decomposers and some are parasites.

- Zygomycetes are largely saprophytic on dead plant tissue. During sexual reproduction, they form zygosporangia after hyphae from different mating types are yoked.

- The Glomeromycota is a monophyletic group of soil fungi that form arbuscular mycorrhizae with most land plants.

- The Basidiomycota is a monophyletic group of terrestrial fungi that form four airborne spores on each club-shaped basidium within mushrooms, brackets, and puffballs. Some can degrade wood. Others form ectomycorrhizae with certain trees.

- Ascomycota is a monophyletic group of terrestrial fungi that form eight airborne spores in each ascus. Some form lichens in association with green algae or cyanobacteria.

✔ You should be able to describe some of the characteristics of each major lineage of fungi.

You should be able to . . .

✓ TEST YOUR KNOWLEDGE

Answers are available in Appendix A

1. The mycelial growth habit leads to a body with a high surface-area-to-volume ratio. Why is this important?
   a. Mycelia have a large surface area for absorption.
   b. The hyphae that make up mycelia are long, thin tubes.
   c. Most hyphae are broken up into compartments by walls called septa, although some exist as single, gigantic cells.
   d. Hyphae can infiltrate living or dead tissues.

2. What is plasmogamy?
   a. production of gametes, after fusion of hyphae from different individuals
   b. exchange of nutrients between symbiotic fungi and hosts
   c. fertilization—the fusion of cytoplasm and nuclei from different individuals
   d. fusion of the cytoplasm from different mating types, without nuclear fusion

3. The hyphae of arbuscular mycorrhizal fungi (AMF) form bushy structures after making contact with the plasma membrane of a root cell. Why?
   a. They anchor the fungus inside the root, so the association is more permanent.
   b. They increase the surface area available for the transfer of nutrients.
   c. They produce toxins that protect the plant cells against herbivores.
   d. They break down cellulose and lignin in the plant cell wall.

4. What does it mean to say that a hypha is dikaryotic?
   a. Two nuclei fuse during sexual reproduction to form a zygote.
   b. Two independent nuclei, derived from different individuals, are present in each cell.
   c. The nucleus is diploid or polyploid—not haploid.
   d. It is extremely highly branched, which increases its surface area and thus absorptive capacity.

5. Explain how pores in septa of certain fungi allow mycelia to become larger.

6. Symbiotic partnerships in which one partner benefits and the other is unaffected are called _____.

✓ TEST YOUR UNDERSTANDING

Answers are available in Appendix A

7. The Greek root *ecto* means "outer." Why are ectomycorrhizal fungi, or EMF, aptly named?
   a. Their hyphae form tree-like branching structures inside plant cell walls.
   b. They are mutualistic.
   c. Their hyphae form dense mats that envelop roots but do not penetrate the cell walls.
   d. They transfer nitrogen from outside their plant hosts to the interior.

8. Explain why fungi that degrade dead plant materials are important to the global carbon cycle. Do you accept the text's statement that, without these fungi, "Terrestrial environments would be radically different than they are today and probably much less productive"? Why or why not?

9. Lignin and cellulose provide rigidity to the cell walls of plants. But in most fungi, chitin performs this role. Why is it logical that most fungi don't have lignin or cellulose in their cell walls?

10. Biologists claim that EMF and AMF species are better than plants at acquiring nutrients because they have more surface area and because they are more effective at acquiring phosphorus (P) and/or nitrogen (N). Compare and contrast the surface area of mutualistic fungi and plant roots. Explain why fungi are particularly efficient at acquiring P and N, compared to plants.

11. Using information from Chapters 3 through 6, list two key macromolecules found in plants that contain phosphorus. Because phosphorus is often limiting to plant growth, explain how certain fungi are necessary for rapid plant growth.

12. Compare and contrast the way that fungi degrade lignin with the way that they digest cellulose.

13. Lawns are sometimes fertilized with nitrate that can be washed into neighboring woodlots by rain. If the trees in that woodlot are associated with EMF, what effect might the excess nitrate have on the fungi and/or trees?
    a. Fungal growth may be stimulated, causing the trees to transport more sugars to the fungi.
    b. The fungi may secrete more peptidases to break down the nitrate.
    c. The trees may take up the nitrate directly into their roots and rely less on the fungi.
    d. The excess nitrate may stimulate the fungi to transport more sugar to the trees.

14. The box on chytrids in Section 32.4 mentions that they may be responsible for massive die-offs currently occurring in amphibians. Review Koch's postulates; then design a study showing how you would use Koch's postulates to test the hypothesis that chytrid infections are responsible for the frog deaths (see Chapter 29).

15. Experiments indicate that cellulase genes are transcribed and translated together. If cells are selected to be extremely efficient at digesting cellulose, is this result logical? Would you predict that the gene that codes for lignin peroxidase is transcribed along with the cellulase genes? How would you test your prediction?

16. Many mushrooms are extremely colorful. Fungi do not see, so it is unlikely that colorful mushrooms are communicating with one another. One hypothesis is that the colors serve as a warning to animals that eat mushrooms, much like the bright yellow and black stripes on wasps. Design an experiment capable of testing this hypothesis.

33 An Introduction to Animals

In this chapter you will learn that

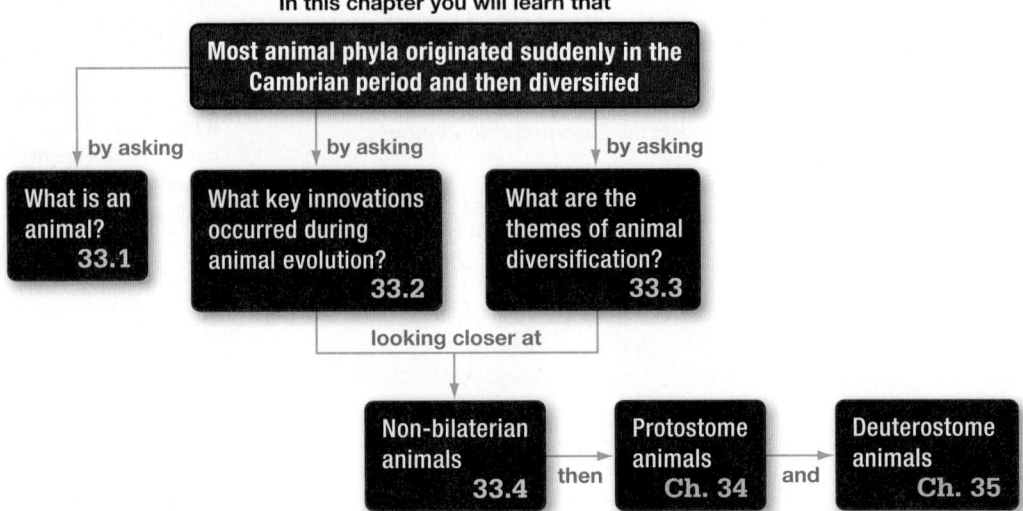

Most animal phyla originated suddenly in the Cambrian period and then diversified

by asking → What is an animal? **33.1**

by asking → What key innovations occurred during animal evolution? **33.2**

by asking → What are the themes of animal diversification? **33.3**

looking closer at ↓

Non-bilaterian animals **33.4** → then → Protostome animals **Ch. 34** → and → Deuterostome animals **Ch. 35**

Jellyfish such as this hydromedusa are among the most ancient of all animals—they appear in the fossil record over 560 million years ago.

You're an animal, and you have plenty of company—animals are an extraordinarily diverse and species-rich lineage on the tree of life.

The radiation of animals began around 550 million years ago during an event called the **Cambrian explosion** (see Chapter 28). After a slow start, diverse animals suddenly appeared in the fossil record with shells, exoskeletons, internal skeletons, legs, heads, tails, eyes, antennae, jaw-like mandibles, segmented bodies, muscles, and brains. The Cambrian explosion was arguably the most spectacular period of evolutionary change in the history of life.

Today, estimates suggest that there are between 8 million and 50 million animal species alive on our planet, some strikingly similar to their Cambrian ancestors, others quite different. These species range in size and complexity from tiny sponges, which attach to a substrate and contain just a few

✔ When you see this checkmark, stop and test yourself. Answers are available in Appendix A.

cell types, to blue whales, which migrate thousands of kilometers each year in search of food and contain trillions of cells, dozens of distinct tissues, an elaborate skeleton, and highly sophisticated sensory and nervous systems.

Only about 1.4 million animal species have been described and given scientific names to date. As scientists race to discover and describe more animal species, the impact of humans on the planet—ranging from habitat loss to pollution, overharvesting, spread of invasive species, and climate change—is causing the modern rate of animal extinction to accelerate. Some species may go extinct before they are ever described (see Chapter 57).

To analyze the almost overwhelming number and diversity of species in the animal lineage, this chapter presents a broad overview of how these species diversified, along with more detailed information on the characteristics of the lineages that appear first in the fossil record. (See Chapters 34 and 35 to focus on the most species-rich groups.) Let's start by considering what characteristics define animals.

33.1 What Is an Animal?

The ancestors to animals were single-celled protists (Chapter 30). **FIGURE 33.1** reviews how animals fit in the tree of life. Animals occur in a clade, or lineage, called Opisthokonta, along with fungi and the single-celled protists called choanoflagellates. Although scientists cannot be sure exactly what the protist ancestor to all animals looked like, it was likely very similar to today's choanoflagellates.

What characteristics define the animal lineage? Although you may be able to recognize most animals as animals when you see them, some animals look more like plants, fungi, or even protists.

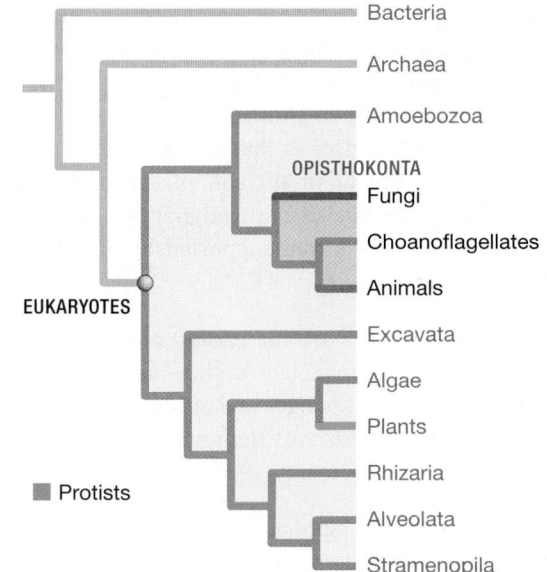

FIGURE 33.1 Choanoflagellates are the Closest Living Relatives to Animals.

Animals are eukaryotes that share key traits:

- *All animals are multicellular*, with cells that lack cell walls but have an extensive **extracellular matrix** (ECM; see Chapter 11). The ECM includes proteins specialized for cell–cell adhesion and communication.

- *All animals are heterotrophs*, meaning that they obtain the carbon compounds they need from other organisms. Most ingest ("eat") their food, rather than absorbing it across their exterior body surfaces.

- *All animals move* under their own power at some point in their life cycle.

- *All animals other than sponges have* (**1**) *neurons* (nerve cells) that transmit electrical signals to other cells; and (**2**) *muscle cells* that can change the shape of the body by contracting.

Multicellular fungi and animals are similar in that they are both multicellular heterotrophs that digest (break down) and absorb nutrients. However, animals are the only multicellular heterotrophs on the tree of life that usually ingest their food first, before they digest it. As a result, digestion in animals typically occurs within the space, or lumen, of the digestive tract rather than in the open exterior, as occurs with fungi.

In most animals, neurons connect to each other, forming a nervous system, and some neurons connect to muscle cells—which may contract in response to electrical signals from neurons. Muscles and neurons are adaptations that allow a large, multicellular body to move efficiently.

What is produced by the combination of multicellularity, heterotrophy, and efficient movement? Eating machines—animals are the largest predators, herbivores, and detritivores on Earth. They are key consumers in virtually every ecosystem, from the deep oceans and tropical forests to alpine ice fields.

Biologists currently recognize 30–35 **phyla**, or major lineages, of animals—the exact number is constantly being debated and revised as additional information comes to light. **TABLE 33.1** (see page 638) lists 29 phyla that are included in most published analyses. How can we make sense of all this diversity? Let's first take a closer look at the key events that occurred during the origin of major animal lineages. Then let's examine the key themes of diversification that occurred *within* lineages.

33.2 What Key Innovations Occurred during the Evolution of Animals?

Biologists who study the evolution of animals consider three types of data:

1. *Fossils* are important because they provide the only direct evidence of what ancient animals looked like, when they existed, and where they lived. However, the fossil record does not represent all animals equally. Fossils are more likely to occur for animals that were abundant, had hard parts, lived in areas where sedimentation was occurring, and/or lived recently (Chapter 28).

TABLE 33.1 An Overview of Major Animal Phyla

Group and Phylum	Common Name or Example Taxa	Number of Described Species
Non-bilaterian Groups		
Porifera	Sponges	8500
Placozoa	Placozoans	1
Ctenophora	Comb jellies	190
Cnidaria	Jellyfish, corals, anemones, hydroids, sea fans	11,500
Acoela	Acoelomate worms	350
Protostomes (lacks typical protostome development)		
Chaetognatha	Arrow worms	120
Protostomes: Lophotrochozoa		
Rotifera	Rotifers	2100
Platyhelminthes	Flatworms	20,000
Nemertea	Ribbon worms	1200
Gastrotricha	Gastrotrichs	400
Acanthocephala	Acanthocephalans	1150
Entoprocta	Entoprocts, kamptozoans	170
Gnathostomulida	Gnathostomulids	100
Annelida	Segmented worms	16,800
Mollusca	Mollusks (clams, snails, octopuses)	85,000
Phoronida	Horseshoe worms	10
Bryozoa	Bryozoa, ectoprocts, moss animals	5700
Brachiopoda	Brachiopods (lamp shells)	550
Protostomes: Ecdysozoa		
Nematoda	Roundworms	25,000
Kinorhyncha	Kinorhynchs	130
Nematomorpha	Hair worms	330
Priapula	Priapulans	16
Onychophora	Velvet worms	165
Tardigrada	Water bears	1045
Arthropoda	Arthropods (spiders, insects, crustaceans)	1,160,000
Deuterostomes		
Echinodermata	Echinoderms (sea stars, sea urchins, sea cucumbers)	7000
Xenoturbellida	Xenoturbillidans	2
Hemichordata	Acorn worms	108
Chordata	Chordates: tunicates, lancelets, sharks, bony fish, amphibians, reptiles (including birds), mammals	65,000

2. *Comparative morphology* provides information about which embryonic, larval, or adult morphological characteristics are common among groups of animals and which are unique to individual lineages (synapomorphies). These data can be used to define the fundamental architecture, or **body plan,** of each animal lineage. In a phylogenetic context, these data can be used to infer which characteristics arose first during the evolution of animals, and which animal groups are more closely related.

3. *Comparative genomics* provides information about the relative similarity of genes or whole genomes of diverse organisms (see Chapter 21). This relatively new source of data is providing dramatic insights into phylogenetic relationships and evolutionary history.

Sometimes these various sources of data on animal evolution suggest the same sequence of evolutionary events. Sometimes they don't. Consider one of the most influential papers on the phylogeny of animals, published in 1997. Using sequences from a gene that codes for RNA in the small subunit of the ribosome, Anna Marie Aguinaldo and colleagues estimated the phylogeny of species from 14 animal phyla. The results were revolutionary, suggesting a different pattern of relationships among animal lineages than had been accepted for years based on morphological data.

The phylogenetic tree in **FIGURE 33.2** is an updated version of the 1997 result, based on further studies of many gene and protein sequences. Some important morphological innovations are mapped onto the tree. Let's explore them by starting at the root of the tree and working toward the branch tips. (See **BioSkills 7** in Appendix B to review how to interpret phylogenetic trees.)

Origin of Multicellularity

Put your finger on the root of the animal tree in Figure 33.2, where the single-celled choanoflagellates outgroup meets the most basal animal branches. What happened at that node?

The first important insight is that animals are a monophyletic group, meaning that all animals share a common ancestor. Data from fossils, comparative morphology, and comparative genomics agree on a single origin of animals.

The second point to note is that sponges (phylum Porifera) include the two most basal, or ancient, lineages of animals. That is, multicellularity appears to have originated in a sponge-like animal. This hypothesis is well supported by diverse evidence. Let's take a closer look.

Fossil Evidence Sponges are the earliest animals to appear in the fossil record, more than 600 million years ago. The early presence of multicellular sponges and absence of fossils of other multicellular organisms are consistent with the basal position of sponges on the phylogeny.

Morphological Evidence Sponges share several key characteristics with the choanoflagellate outgroup:

- Both choanoflagellates and sponges are **sessile,** meaning that adults live permanently attached to a substrate.

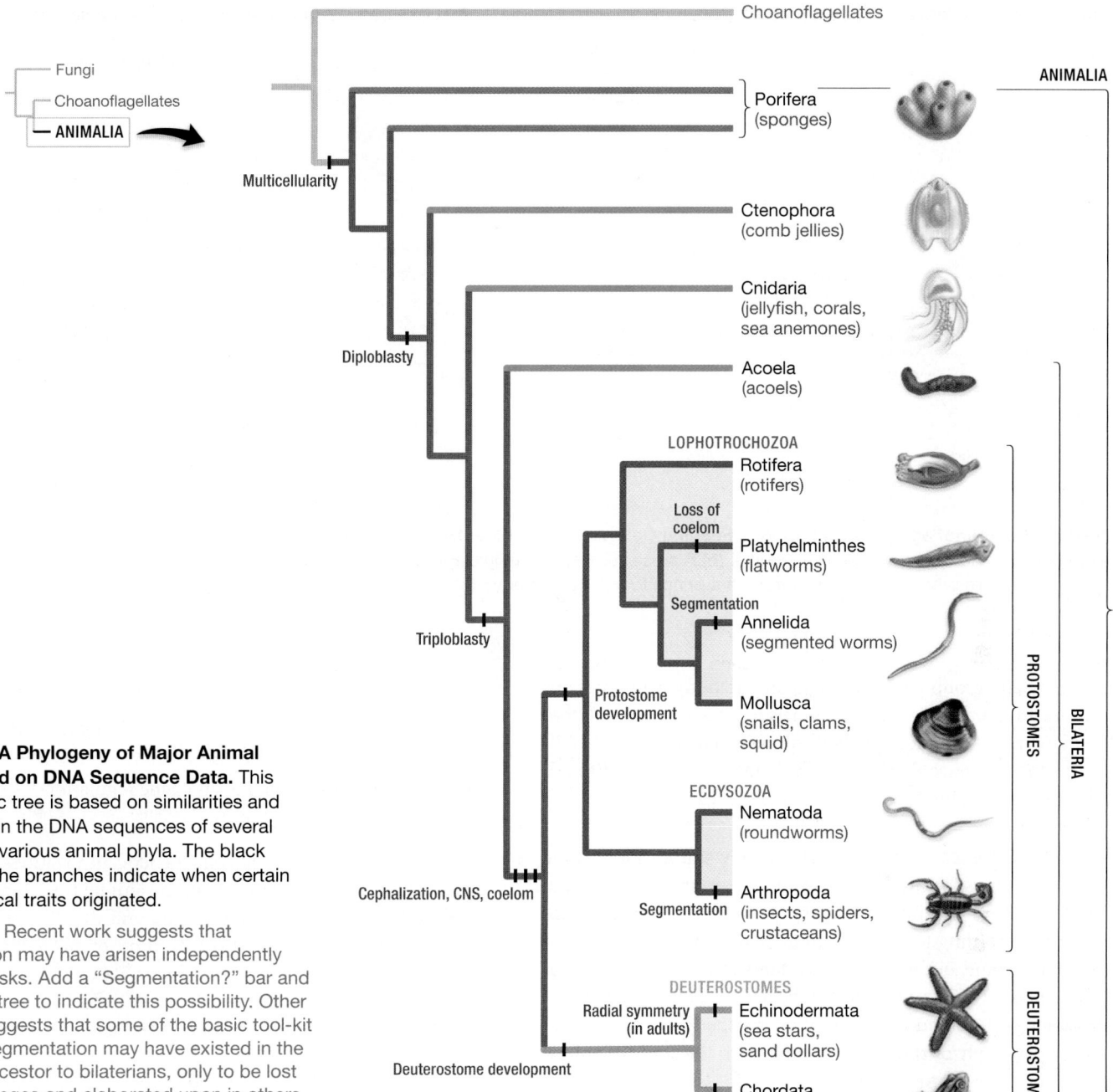

FIGURE 33.2 A Phylogeny of Major Animal Phyla Based on DNA Sequence Data. This phylogenetic tree is based on similarities and differences in the DNA sequences of several genes from various animal phyla. The black bars along the branches indicate when certain morphological traits originated.

✔ **EXERCISE** Recent work suggests that segmentation may have arisen independently within mollusks. Add a "Segmentation?" bar and label to the tree to indicate this possibility. Other research suggests that some of the basic tool-kit genes for segmentation may have existed in the common ancestor to bilaterians, only to be lost in some lineages and elaborated upon in others. Add a "Tool-kit genes for segmentation?" bar and label to the tree to indicate this possibility.

- Both feed in a similar way, using cells with nearly identical morphology. As **FIGURE 33.3** (see page 640) shows, the beating of flagella creates water currents that bring organic debris toward the feeding cells of choanoflagellates and sponges. Sponge feeding cells are called **choanocytes.** Within these choanocytes, food particles are trapped and ingested.

As you can see in the figure, choanoflagellates sometimes form **colonies**—groups of individuals that are attached to each other. Some biologists once considered sponges to be colonies of single-celled protists due to the ability of sponge cells to re-aggregate after being dissociated. But sponges contain many specialized cell types that are dependent on each other, some of which occur in organized layers surrounded by extracellular matrix (ECM). Recent research has shown that some sponges have true **epithelium**—a layer of tightly joined cells that covers the interior and/or exterior surface of the animal. Epithelium is essential to animal form and function (Chapter 42).

Sponges are diverse in size, shape, and composition, distinguished in part by the type of **spicules** they produce—stiff spikes of silica or calcium carbonate ($CaCO_3$) that, along with collagen fibers, provide structural support to the ECM. Despite this morphological diversity, many biologists have described sponges as a

(a) Choanoflagellates are sessile protists; some are colonial.

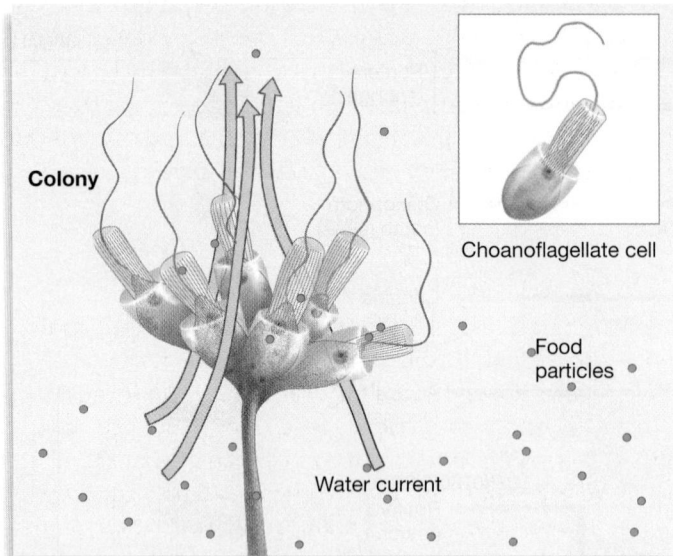

(b) Sponges are multicellular, sessile animals.

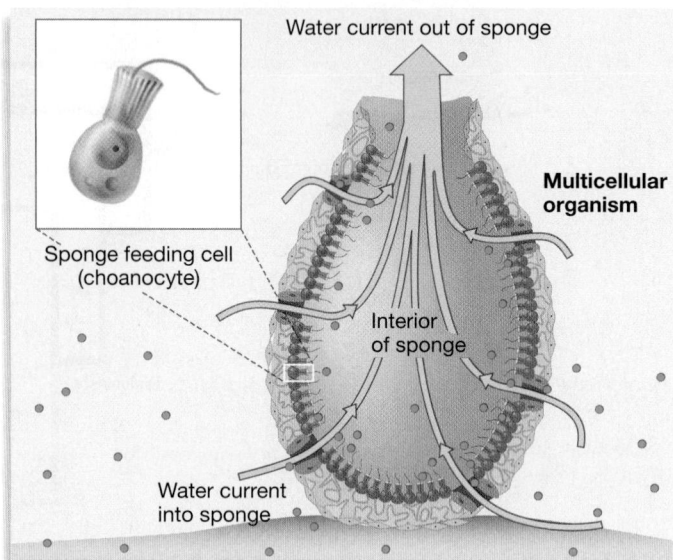

FIGURE 33.3 Choanoflagellates and Sponge Feeding Cells Are Almost Identical in Structure and Function. **(a)** Choanoflagellates are suspension feeders. **(b)** A cross section of a simple sponge as it suspension feeds. The beating of flagella produces a water current that brings food into the body of the sponge, where it can be ingested by feeding cells.

monophyletic group based on their unique body plan—systems of tubes and pores that create channels for water currents. Other biologists reject the monophyly of sponges. The recent availability of gene sequence data for diverse sponges has helped to advance the debate, but there is still no consensus.

Molecular Evidence   Comparative genomic studies support the hypothesis that sponges are the most basal group of animals, but some reject the hypothesis that the phylum Porifera is monophyletic. According to these studies, sponges are paraphyletic—containing some, but not all, descendants of a common ancestor (see Table 28.2). This concept is indicated on Figure 33.2 by two blue lines, instead of one, representing the sponges.

The paraphyly of sponges is important to the study of animal origins because it provides extra support for the hypothesis that the common ancestor of all animals was sponge-like. If sponges were a monophyletic group, it would be more plausible that their distinguishing characteristics could have evolved later rather than in the common ancestor to all animals (the same way that the most recent common ancestor of humans and chimpanzees was not necessarily chimp-like).

What were the other qualities of the animal ancestor? Comparative genomic studies suggest something surprising. Despite the relative morphological simplicity of sponges, they possess a remarkably complex developmental **tool kit** of genes (Chapter 22). This tool kit contains the genes necessary for all the basic molecular processes required by animals:

- Specialization of cell types
- Regulation of cell cycling and growth

- Adhesion among cells, and between cells and extracellular matrix
- Recognition of self and nonself, thus innate immunity
- Developmental signaling and gene regulation
- Programmed cell death

Thus, a series of important genetic innovations appears to have occurred at the very root of the animal tree along with multicellularity. Subsequent duplication and diversification of these genes contributed to the diversification of animal lineages. ✔If you understand this, you should be able to explain why cancer, which typically involves the mutation of genes in the basic animal genetic tool kit, is thought to be as ancient as multicellularity itself.

Origin of Embryonic Tissue Layers

Sponges possess the basic genetic tool kit for cell–cell adhesion and cell–ECM adhesion, and some sponges even have rudimentary epithelium. However, sponges do not have complex **tissues,** groups of similar cells that are organized into tightly integrated structural and functional units. Animals other than sponges are typically divided into two major groups based on the number of embryonic tissue layers they have. Animals whose embryos have two types of tissue are called **diploblasts** (literally, "two-buds"); animals whose embryos have three types of tissue are called **triploblasts** ("three-buds").

The embryonic tissues are organized in layers, called **germ layers** (Chapter 23). In diploblasts these germ layers are called **ectoderm** and **endoderm** (**FIGURE 33.4**). The Greek roots *ecto* and

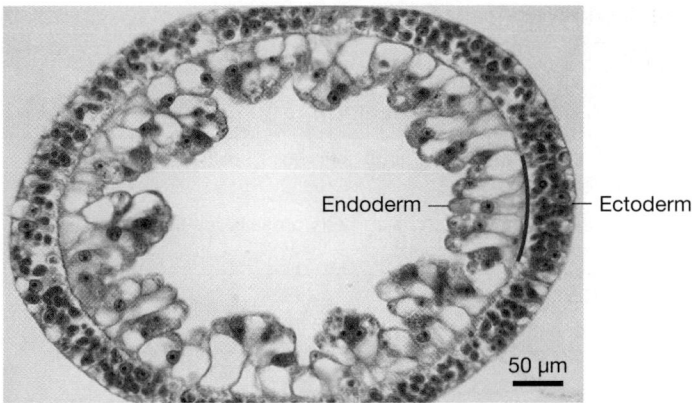

FIGURE 33.4 **Diploblastic Animals Have Bodies Built from Ectoderm and Endoderm.** This is a cross section through the tube-shaped portion of a hydra's body (phylum Cnidaria). The cells have been stained to make them more visible.

endo refer to outer and inner, respectively; the root *derm* means "skin." (See **BioSkills 2** in Appendix B to review common Latin and Greek roots.) In most cases the outer and inner "skins" of diploblast embryos are connected by a gelatinous material that may contain some cells. In triploblasts, however, there is a germ layer called **mesoderm** between the ectoderm and endoderm. (The Greek root *meso* refers to "middle.")

The embryonic tissues found in animals develop into distinct adult tissues, organs, and organ systems. In triploblasts:

- *Ectoderm* gives rise to skin and the nervous system.
- *Endoderm* gives rise to the lining of the digestive tract.
- *Mesoderm* gives rise to the circulatory system, muscle, and internal structures such as bone and most organs.

In general, then, ectoderm produces the covering of the animal and endoderm generates the digestive tract. Mesoderm gives rise to the tissues in between. The same pattern holds in diploblasts, except that (**1**) muscle is simpler in organization and is derived from ectoderm, and (**2**) reproductive tissues are derived from endoderm.

Look again at the animal phylogeny in Figure 33.2. Traditionally, two groups of animals have been recognized as diploblasts: the Ctenophora (pronounced *ten-AH-for-ah*), or comb jellies, and the Cnidaria (pronounced *ni-DARE-ee-uh*), which include the jellyfish, corals, sea pens, hydra, and anemones. Recent data suggest that some cnidarians have true mesoderm and are triploblasts, complicating the question of exactly when the triploblast body plan evolved. However, the take-home message for the animal phylogeny is still the same—the relatively simple diploblast body plan of ctenophores and cnidarians appears to have originated from ancestral animals later than the sponge lineages but before the other major groups of animals, which are all triploblasts.

The evolution of mesoderm was important because it gave rise to the first complex muscle tissue used in movement. Sponges lack muscle and are generally sessile, or nonmoving, as adults. Ctenophores have muscle cells that can change the body's shape, but both larvae and adults swim using cilia. Adult cnidarians have muscle cells that can extend and retract the body, such as in sea anemones, and can enable swimming by jet propulsion in jellyfish; but these movements are relatively simple compared to the complex movements performed by many triploblasts.

Origin of Bilateral Symmetry, Cephalization, and the Nervous System

Body symmetry is a key morphological aspect of an animal's body plan. A body is said to be symmetrical if it can be divided by a plane such that the resulting pieces are nearly identical.

Ctenophores, many cnidarians, and some sponges have **radial symmetry** ("spoke symmetry")—meaning that they have at least two planes of symmetry. For example, almost any plane sectioned through the center of the hydra in **FIGURE 33.5a** produces two identical halves. Radial symmetry evolved independently in the phylum Echinodermata (pronounced *ee-KINE-oh-der-ma-ta*)—a group including species such as sea stars, sea urchins, feather stars, and brittle stars.

Organisms with **bilateral symmetry** ("two-sided symmetry") in contrast, have one plane of symmetry and tend to have a long, narrow body. The annelid worm in **FIGURE 33.5b**, for example, has only one plane of symmetry—running lengthwise down its middle.

Based on the topology of branches on the animal phylogeny in Figure 33.2, radial symmetry appears to have arisen earlier in the evolution of animals than bilateral symmetry. Bilateral symmetry occurs in all triploblastic lineages. Where on the tree did bilateral symmetry originate? The answer is more complex than it would seem at first glance. Let's explore why.

(a) Radial symmetry   **(b)** Bilateral symmetry

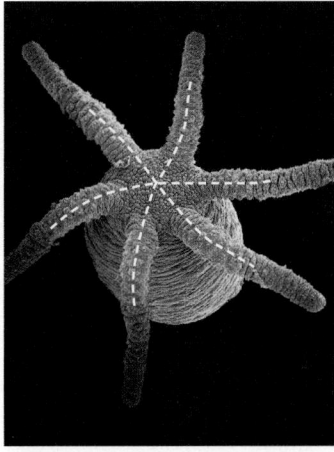 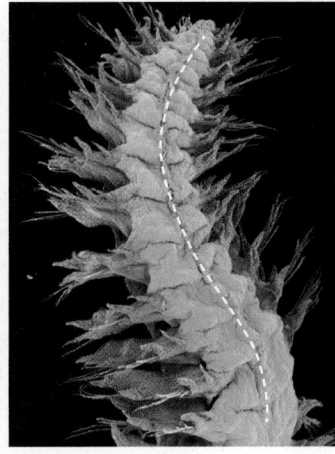

FIGURE 33.5 **Body Symmetry.** These colorized SEMs of **(a)** hydra (looking down on oral surface; phylum Cnidaria) and **(b)** a polychaete worm (phylum Annelida) illustrate the difference between radial and bilateral symmetry.

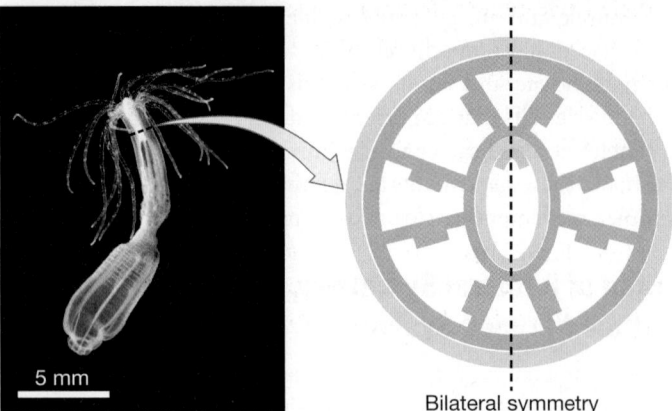

FIGURE 33.6 **Morphological Evidence for Bilateral Symmetry in a Sea Anemone.** Some corals and sea anemones have long been known to have bilateral symmetry despite the outward appearance of radial symmetry. *Nematostella* is a small sea anemone that burrows in soft marine sediments.

Homology or Convergent Evolution?

Nearly all cnidarians appear radially symmetric at first glance, but a closer inspection of their internal morphology reveals bilateral symmetry in some species, especially in sea anemones (**FIGURE 33.6**). Recall that homology is defined as similarity in traits due to inheritance from a common ancestor (Chapter 25). Is the bilateral symmetry in sea anemones homologous to bilateral symmetry in triploblastic animals, or is it an example of convergent evolution?

Developmental biologists John Finnerty, Mark Martindale, and colleagues used developmental regulatory genes as a tool to address this evolutionary question. In triploblastic, bilaterally symmetric animals, called **bilaterians** for short, bilateral symmetry is achieved by the combination of anterior–posterior ("head–tail") axis formation and dorsal–ventral ("back–belly") axis formation. *Hox* genes are important in the development of the anterior–posterior axis (see Chapter 22), and *decapentaplegic* (*dpp*) genes are important in the development of the dorsal–ventral axis.

Finnerty and Martindale predicted that if bilateral symmetry in the sea anemone *Nematostella vectensis* (an increasingly important model organism in biology) is homologous to bilateral symmetry in bilaterians, the pattern of *Hox* and *dpp* gene expression in the sea anemone would be similar to that found in bilaterians. If convergent evolution had occurred, either *Hox* and *dpp* genes would not be expressed in *Nematostella* or their expression patterns would be unrelated to those in bilaterians.

FIGURE 33.7 shows Finnerty and Martindale's results. During *Nematostella* development, *Hox* genes are expressed along the anterior–posterior axis and *dpp* is expressed asymmetrically about the dorsal–ventral axis, a pattern of gene expression that is parallel to that observed in bilaterians. This evidence supports the hypothesis that bilateral symmetry in this sea anemone is homologous to bilateral symmetry in triploblastic animals— meaning that parts of the genetic tool kit that determine bilateral symmetry arose before the evolutionary split of the cnidarian and bilaterian lineages.

RESEARCH

QUESTION: Is bilateral symmetry in sea anemones homologous to bilateral symmetry in bilaterians?

HYPOTHESIS: Bilateral symmetry is homologous in sea anemones and bilaterians.

NULL HYPOTHESIS: Bilateral symmetry arose independently in sea anemones and bilaterians.

EXPERIMENTAL SETUP:

1. Stain gene products (proteins) of *Hox* gene in developing *Nematostella* embryos and larvae to reveal location of expression.

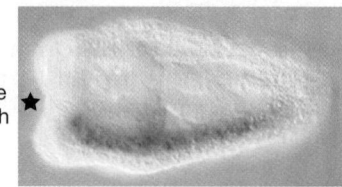

Future mouth ★   Larva + **Hox stain**

2. Repeat for other *Hox* and *dpp* gene products.

PREDICTION: The pattern of expression of *Hox* and *dpp* genes in *Nematostella* will be similar to that found in bilaterians.

PREDICTION OF NULL HYPOTHESIS: The pattern of expression of *Hox* and *dpp* genes in *Nematostella* will be unrelated to that found in bilaterians.

RESULTS: Schematic summary of gene expression patterns (longitudinal section):

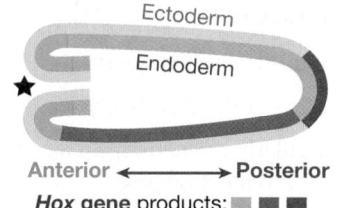

Hox genes are expressed sequentially along the anterior–posterior axis— *as in bilaterians*

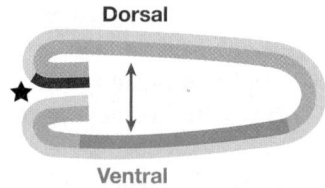

dpp genes are expressed asymmetrically along the dorsal–ventral axis— *as in bilaterians*

CONCLUSION: Bilateral symmetry is homologous in *Nematostella* and bilaterians.

FIGURE 33.7 **Genetic Evidence that Bilateral Symmetry Originated in a Common Ancestor of Cnidaria and Bilateria.**

SOURCE: Finnerty, J. R., K. Pang, P. Burton, et al. 2004. Origins of bilateral symmetry: *Hox* and *dpp* expression in a sea anemone. *Science* 304: 1335–1337.

✔ **QUESTION** What results would have supported the null hypothesis?

However, *Hox* and *dpp* gene expression in *Nematostella* is not identical to that in bilaterians—and the bilateral symmetry in *Nematostella* is not as extensive as in bilaterians. The take-home message is that the entire genetic tool kit for bilateral symmetry did not evolve all at once in the early Cambrian before the diversification of bilaterians. Some components of the tool kit evolved earlier—in the ancestor to cnidarians and bilaterians—and others evolved later, after the split of these two lineages.

Origin of the Nervous System Over 99 percent of modern animals are bilaterally symmetric. Is this body plan a "key innovation" that led to the diversification of the bilaterians during the Cambrian explosion? One hypothesis is that the evolution of the nervous system and the evolution of the head are tightly linked to bilateral symmetry and that together, these characteristics contributed to the radiation of bilaterians.

How are symmetry and nervous systems related? The function of neurons and nervous systems is to transmit and process information in the form of electrical signals (see Chapter 46).

- Sponges generally lack both nerve cells and symmetry.

- Ctenophores and cnidarians have nerve cells that are organized into a diffuse arrangement called a **nerve net** (**FIGURE 33.8a**). These generally radially symmetric animals either float in water or live attached to a substrate. Radially symmetric organisms are more likely to encounter prey and other aspects of the environment in any direction. As a result, a diffuse nerve net can receive and send signals efficiently.

- All other animals have a **central nervous system,** or CNS. In a CNS, some neurons are clustered into one or more large tracts or cords that project throughout the body; others are clustered into masses called **ganglia** (**FIGURE 33.8b**). Most of the bilaterally symmetric animals living today move through

their environment. Bilaterally symmetric organisms tend to encounter prey and other aspects of the environment at the leading end. As a result, it is advantageous to the animal to have many neurons concentrated at that end, with nerve tracts that carry information down the length of the body.

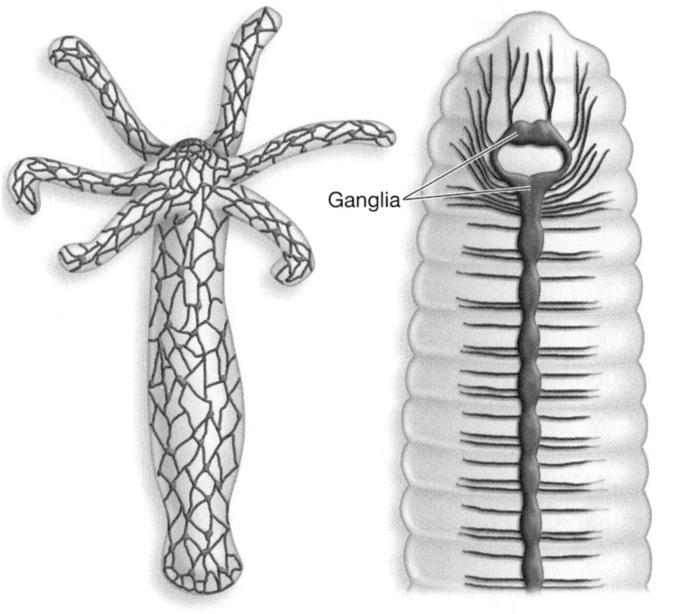

It is intuitive that the evolution of the CNS would coincide with **cephalization:** the evolution of a head, or anterior region, where structures for feeding, sensing the environment, and processing information are concentrated. The large mass of neurons that is located in the head, and that is responsible for processing information to and from the body, is called the cerebral ganglion or **brain.**

To explain the pervasiveness of bilateral symmetry, biologists point out that locating and capturing food is particularly efficient when movement is directed by a distinctive head region and powered by the rest of the body. In combination with the origin of mesoderm, a bilaterally symmetric body plan enabled rapid, directed movement and hunting. Lineages with a triploblastic, bilaterally symmetric, cephalized body had the potential to diversify into an array of formidable eating and moving machines.

Notice in Figure 33.2 that the acoels (phylum Acoela), tiny worms that tend to live among sand grains, are bilaterally symmetric and triploblastic but have a nerve net rather than a CNS—and no head region. Although the position of the acoel branch on the tree is hotly debated, its current position supports the hypothesis that the CNS and cephalization were key to the subsequent radiation of bilaterians.

Acoel worms are named for their lack of a coelom. What is a coelom? What role did it play in the diversification of bilaterians?

Origin of the Coelom

What would you expect an animal body to look like that is triploblastic (formed from ectoderm, mesoderm, and endoderm), bilaterally symmetric, elongated, and cephalized? The basic bilaterian body shape is a *tube within a tube*. The inner tube is the individual's gut with a mouth at one end and an anus at the other, and the outer tube forms the nervous system and skin (**FIGURE 33.9**). The mesoderm in between forms muscles and organs.

(a) Nerve net:
diffuse neurons in hydra

(b) Central nervous system:
clustered neurons in earthworm

Ganglia

FIGURE 33.8 Associations between Body Symmetry and the Nervous System. (a) Radially symmetric animals, like this hydra, have a nerve net. **(b)** Bilaterally symmetric animals, like this earthworm, have a central nervous system.

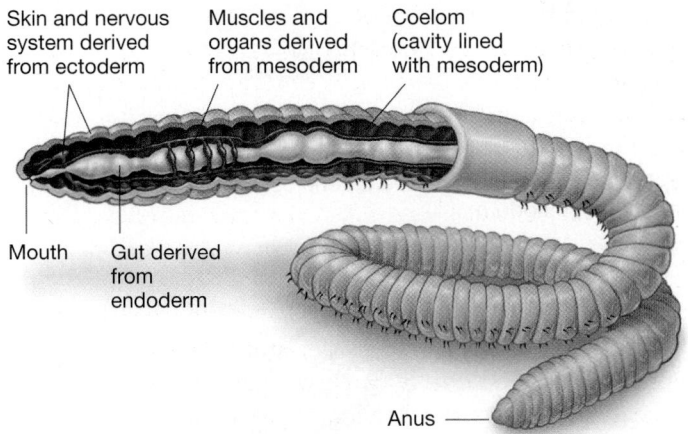

Skin and nervous system derived from ectoderm

Muscles and organs derived from mesoderm

Coelom (cavity lined with mesoderm)

Mouth

Gut derived from endoderm

Anus

FIGURE 33.9 The Tube-within-a-Tube Body Plan Is Common in Animals.

The "tube-within-a-tube" body plan can pose a potential biomechanical and physiological challenge if the inner tube is attached to the outer tube via the mesoderm. One bypass to this physical constraint is a fluid-filled cavity between the inner and outer tubes, called a **coelom** (pronounced *SEE-lum*), as shown in the earthworm in Figure 33.9. The coelom provides a space for the circulation of oxygen and nutrients. It also enables the internal organs to move independently of each other.

Phylogenetic evidence suggests that the coelom arose in the common ancestor of protostomes and deuterostomes (Figure 33.2). The bilaterians that possess a coelom that is completely lined in mesoderm are known as true **coelomates** (**FIGURE 33.10a**). The bilaterians that subsequently lost their coelom, such as the flatworms, are called **acoelomates** ("no-cavity-form"; see **FIGURE 33.10b**). The bilaterians that retained a coelom but lost the mesodermal lining in parts of their coelom, such as nematodes (roundworms), are known as **pseudocoelomates** ("false-cavity-form"; **FIGURE 33.10c**).

The coelom was a critically important innovation during animal evolution, in part because an enclosed, fluid-filled chamber can act as an efficient **hydrostatic skeleton.** Soft-bodied animals with hydrostatic skeletons can move effectively even if they do not have fins or limbs.

(a) Coelomates have an enclosed body cavity completely lined with mesoderm.

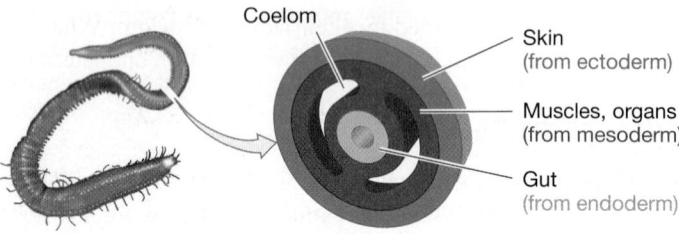

Coelom

Skin (from ectoderm)

Muscles, organs (from mesoderm)

Gut (from endoderm)

(b) Acoelomates have no enclosed body cavity.

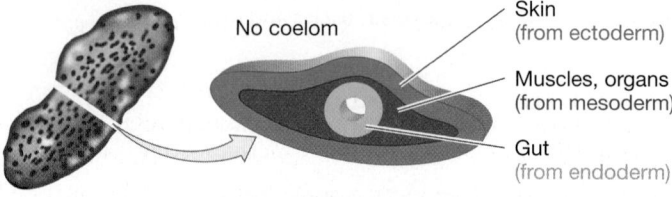

No coelom

Skin (from ectoderm)

Muscles, organs (from mesoderm)

Gut (from endoderm)

(c) Pseudocoelomates have an enclosed body cavity partially lined with mesoderm.

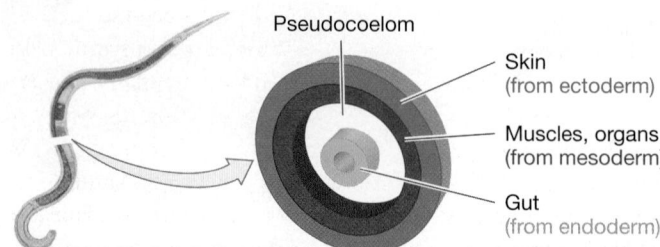

Pseudocoelom

Skin (from ectoderm)

Muscles, organs (from mesoderm)

Gut (from endoderm)

FIGURE 33.10 Animals May or May Not Have a Body Cavity.

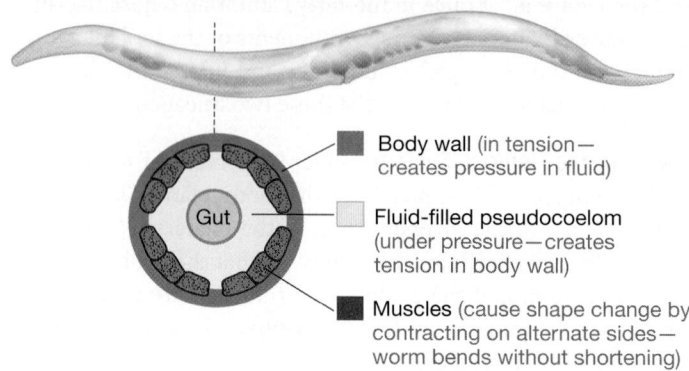

Body wall (in tension— creates pressure in fluid)

Fluid-filled pseudocoelom (under pressure—creates tension in body wall)

Muscles (cause shape change by contracting on alternate sides— worm bends without shortening)

Gut

FIGURE 33.11 Hydrostatic Skeletons Allow Limbless Animals to Move.

For example, movement in nematodes is possible because the fluid inside the pseudocoelom stretches the body wall—much like water stretches the wall of a water balloon—putting the wall under tension (**FIGURE 33.11**). In turn, this tension exerts pressure on the fluid inside the coelom. When certain muscles in the body wall contract against the pressurized fluid, the force is transmitted through the fluid, changing the body's shape.

Coordinated contractions and relaxations of different muscles in the body wall enable various shape changes and forms of locomotion (see Chapter 48). The evolution of the coelom improved the ability of bilaterians to move in search of food.

Origin of Protostomes and Deuterostomes

Turn back to Figure 33.2 and put your finger on the node at the origin of the Bilateria, the monophyletic lineage of bilaterians. Now move your finger one node to the right, past the divergence of the Acoels. The common ancestor that existed at this node during the Cambrian was a bilaterally symmetric triploblast with a CNS, cephalization, and a coelom. This ancestor gave rise to a remarkable radiation of diverse animal lineages, only some of which are included in the tree.

What major groups occur within the Bilaterian coelomates? Despite the staggering diversity of adult animals, common themes during embryonic development have enabled biologists to recognize two major subgroups:

1. **protostomes,** in which the mouth develops before the anus, and blocks of mesoderm hollow out to form the coelom; and

2. **deuterostomes,** in which the anus develops before the mouth, and pockets of mesoderm pinch off to form the coelom.

Translated literally, protostome means "first-mouth" and deuterostome means "second-mouth."

To understand the contrasts in how protostome and deuterostome embryos develop, recall that the three embryonic germ layers form during a process called gastrulation (Chapter 23), which begins when cells move from the outside into the center of the embryo. These cell movements create a pore that opens to the outside (**FIGURE 33.12a**). In protostomes, this pore becomes the mouth. The other end of the gut, the anus, forms later. In

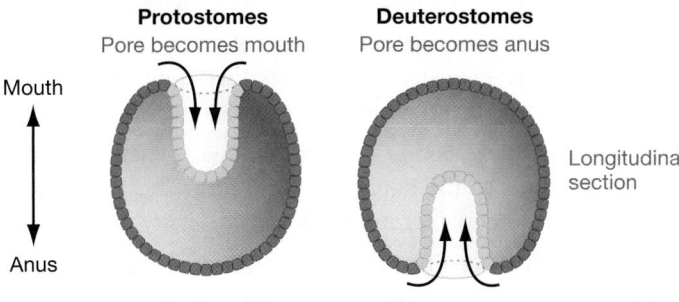

(a) Gastrulation (formation of gut and embryonic germ layers)

Protostomes
Pore becomes mouth

Deuterostomes
Pore becomes anus

Mouth

Anus

Longitudinal section

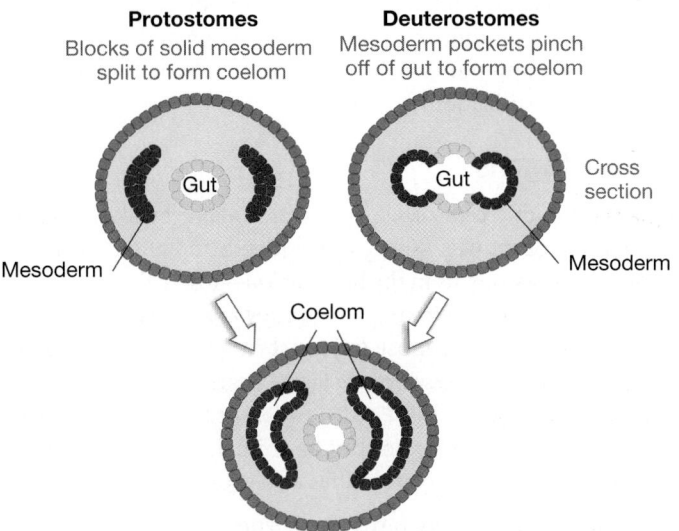

(b) Formation of coelom (body cavity lined with mesoderm)

Protostomes
Blocks of solid mesoderm split to form coelom

Deuterostomes
Mesoderm pockets pinch off of gut to form coelom

Gut

Gut

Cross section

Mesoderm

Mesoderm

Coelom

FIGURE 33.12 In Protostomes and Deuterostomes, Events in Early Development Differ. The differences between protostomes and deuterostomes show that there is more than one way to build a bilaterally symmetric, coelomate body plan.

deuterostomes, however, this initial pore becomes the anus; the mouth forms later.

The other developmental difference between the groups arises as gastrulation proceeds and the coelom begins to form. Protostome embryos have two blocks of mesoderm beside the gut. As the left side of **FIGURE 33.12b** indicates, their coelom begins to form when cavities open within each of the two blocks of mesoderm. In contrast, deuterostome embryos have layers of mesodermal cells located on either side of the gut. As the right side of Figure 33.12b shows, their coelom begins to form when these layers bulge out and pinch off to create fluid-filled pockets lined with mesoderm.

In essence, the protostome and deuterostome patterns of development represent two distinct ways of achieving the same end—the construction of a bilaterally symmetric body that contains a cavity lined with mesoderm. The functional or adaptive significance of the differences—if any—is not known.

Two major groups occur within the protostomes: **(1)** the Lophotrochozoa (pronounced *low-foe-tro-ko-ZOH-ah*) includes the mollusks, annelids, flatworms, and rotifers; and **(2)** the

Ecdysozoa (pronounced *eck-die-so-ZOH-ah*) includes the arthropods and the nematodes. **Lophotrochozoans** grow continuously when conditions are good. **Ecdysozoans** grow by shedding their external skeletons or outer coverings and expanding their bodies.

Species in the phylum Platyhelminthes (flatworms) are protostomes but lack a coelom—meaning that the coelom was lost during flatworm evolution. The coelom has also been reduced in an array of lineages including arthropods—which have limbs and thus rely less on hydrostatic skeletons for movement.

The vast majority of animal species are protostomes (see Chapter 34). However, the largest animals and most voracious predators tend to be deuterostomes (see Chapter 35).

Origin of Segmentation

Segmentation is defined by the presence of repeated body structures. A segmented backbone is one of the defining characteristics of **vertebrates,** a monophyletic lineage within the Chordata that includes fish, reptiles (including birds), amphibians, and mammals. Of the animals that are not vertebrates, called **invertebrates** (a paraphyletic group), segmentation occurs in annelids (earthworms and other segmented worms) and arthropods (insects, spiders, and crustaceans)—both very prolific and diverse lineages. Recent evidence suggests that segmentation also occurs to some extent in mollusks (snails, clams, squid).

Before molecular data were available, biologists grouped segmented animals like annelids and arthropods together in the same clade. Then Aguinaldo's landmark 1997 revision of animal relationships overturned this view, suggesting that segmentation arose independently in these groups (see Figure 33.2).

Is segmentation in different phyla an example of convergent evolution? Applying the same logic that Finnerty and Martindale used in their study of bilateral symmetry in sea anemones, several researchers are investigating whether the genes that regulate segmentation in different phyla are homologous. For example, one recent study found that one of the same gene regulatory pathways that regulate segmentation in arthropods, called the hedgehog pathway, is also important in regulating the segmentation in annelid worms.

The conclusion so far is that some of the tool-kit genes for segmentation may have arisen early in animal evolution, meaning that they are homologous in different phyla. However, these genes would subsequently have been lost in some lineages and co-opted in different ways and elaborated upon in others, contributing to the diversity of bilaterian body plans we see today. Research continues to test this hypothesis.

Why have organisms with segmented bodies been so successful in terms of diversity? One leading hypothesis is that segmentation enables specialization. Small changes in the expression of certain tool-kit genes along the length of a body can result in novel numbers, shapes, and sizes of body segments and appendages. *Hox* genes are a good example (see Chapters 22 and 34). Natural selection can then favor variations that have adaptive specializations for life in certain aquatic or terrestrial environments—leading to diversification.

In summing up this tour of the animal phylogeny, it's important to note that although biologists are increasingly confident that most or all of these inferences about animal relationships are correct, the phylogeny of animals is still a work in progress. As more taxa are sampled, larger numbers of characters are analyzed, and methods of phylogenetic analysis are refined, the resolution of the tree of life will continue to improve. Stay tuned.

check your understanding

If you understand that . . .

- The origin of body plans in the major lineages of animals was marked by changes in four fundamental features: the number of embryonic tissues present; the evolution of nervous systems and a bilaterally symmetric, cephalized body; the evolution of a body cavity; and protostome versus deuterostome patterns of development.

✔ **You should be able to . . .**

1. Describe whether your body has a tube-within-a-tube design. Justify your answer.
2. Explain why cephalization was important.

Answers are available in Appendix A.

33.3 What Themes Occur in the Diversification of Animals?

One of the great unresolved mysteries in the study of animal diversity is this: Why did almost all of the phylum-level body plans evolve all at once during the Cambrian, rather than more gradually over time? Animals diversified tremendously *within* phyla after the Cambrian, but no major new lineages arose during the last 500 million years.

One mechanistic hypothesis is that gene regulatory networks are responsible for limiting the origin of new body plans. These networks determine patterns of animal development and thus lay down the body plan. The scientists studying these networks have observed that once the networks are established, they are very resistant to change—that is, changes tend to be fatal. However, a multitude of "switches" and "plug-ins" can be added to networks, providing a mechanism for diversification within lineages after the body plans are established.

As research continues on the origin of body plans, biologists' understanding of the subsequent diversification within lineages is also steadily improving. Several variables have been hypothesized to play a role in animal diversification (Chapter 28):

- *Higher oxygen levels* An increase in oxygen levels may have made the evolution of big, mobile animals possible due to the efficiency of aerobic respiration.

- *The evolution of predation* The earliest animals were sessile and ate organic material. When predators evolved, they exerted selection pressure on other animals for shells, skeletons, rapid movement, and other adaptations for escaping capture.

- *New niches beget more new niches* As animals diversified, they themselves created new niches that could support yet more ecological diversification.

- *New genes, new bodies* As the animal genetic tool kit evolved—such as the duplication and diversification of *Hox* genes—the potential for morphological diversity increased.

It is likely that a combination of all of these variables was important to the diversification of animals. The result? Animals with the same body plan diversified in their strategies for finding food, ingesting food, and reproducing in different ecological niches. Conversely, animals with different body plans have independently evolved similar strategies for feeding and reproducing when they occupy similar ecological niches. What are the major themes of diversification?

Sensory Organs

The evolution of a cephalized body was a major breakthrough in the evolution of animals. Along with a mouth and brain, a concentration of sensory organs in the head region—where the animal initially encounters the environment—is a key aspect of cephalization.

Certain senses are almost universal in animals (see Chapter 47). The common senses include sight, hearing, taste/smell, and touch (see examples in **TABLE 33.2**). At least some ability to sense temperature is also common. But as animals diversified, a wide array of specialized sensory abilities also evolved. For example:

- *Magnetic field* Many birds, sea turtles, sea slugs, and other animals can detect Earth's magnetic field and use it as an aid in navigation.

- *Electric field* Some aquatic predators, such as sharks, are so sensitive to electric fields that they can detect electrical activity in the muscles of passing prey.

- *Barometric pressure* Some birds can sense changes in air pressure, which may aid them in avoiding storms.

Variation in sensory abilities is important: It allows different species of animals to collect information about a wide array of environments, including the presence of food, predators, and mates.

Feeding

To organize the diversity of ways that animals find food, biologists distinguish *what* individuals eat from *how* they eat. Animals within a lineage often pursue different food sources and feeding strategies when they occupy different niches. For example, sea cucumbers and sea stars both have the same echinoderm body plan, but they have different food sources and feeding strategies—the sea cucumber mops up detritus from the seafloor, and the sea star pries open clams and mussels and devours them.

Conversely, animals from *different* lineages often pursue the *same* food sources and feeding strategies when they occupy similar niches. For example, burrowing sea cucumbers and some burrowing polychaete worms (phylum Annelida and

Sense	Example
Sight *Stimulus:* light Flies use their compound eyes to find food and mates, and to escape predators	
Hearing *Stimulus:* sound Bats use their sense of hearing to find prey and to avoid obstacles in the dark	
Taste/smell *Stimulus:* molecules Some male moths have elaborate antennae to detect pheromones in the air	
Touch *Stimulus:* contact, pressure Sea anemones detect and capture prey using their sense of touch	

Other senses	Examples
Stimulus: temperature (thermal energy)	Pit vipers
Stimulus: magnetic field	Sea turtles
Stimulus: electric field	Sharks
Stimulus: barometric pressure	Birds
Stimulus: gravity	Squid

Echinodermata, respectively) have different body plans but use a similar strategy of consuming detritus on the seafloor.

What Animals Eat: Diversification of Ecological Roles   Animals can be classified as (1) **detritivores** that feed on dead organic matter, (2) **herbivores** that feed on plants or algae, or (3) **carnivores** that feed on animals (see **TABLE 33.3**). **Omnivores** eat both plants and animals. These ecological roles have important implications for whole ecosystems because animal feeding moves both energy and nutrients through food webs (Chapter 56).

Another way to categorize the diverse ecological impacts of animal consumption is to consider the impacts on the organisms that animals consume. **Predators** kill and consume all or most of their prey. Predators are usually larger than their prey and kill them quickly using an array of mouthparts and hunting strategies.

SUMMARY TABLE 33.3   **Diversification of Ecological Roles**

Ecological Role	Example
Detritivores Feed on dead organic matter Millipedes feed on decaying leaves	
Herbivores Feed on plants or algae Pandas eat vast amounts of bamboo	
Carnivores Feed on animals Owls sit and wait for prey	
Omnivores Feed on both plants and animals	*Example:* Humans

Herbivores usually consume plant tissue without killing the whole organism. However, seed predators are an exception because seeds contain an entire—although embryonic—plant.

Like herbivores, **parasites** often harvest nutrients from certain parts of their hosts, but parasites are usually much smaller than their victims. **Endoparasites** live inside their hosts and usually have simple, wormlike bodies. Tapeworms, of the phylum Platyhelminthes (flatworms), are endoparasites with no digestive system. Instead of a mouth, they have hooks or other structures on their head that attach to their host's intestinal wall. Instead of digesting food themselves, they absorb nutrients directly from their surroundings. **Ectoparasites** live on the outside of their hosts. Ectoparasites usually have limbs or mouthparts that allow them to grasp the host, and mouthparts that allow them to pierce their host's skin and suck the nutrient-rich fluids inside. Aphids and ticks are ectoparasites.

Whether animals consume their victims completely or partially, these consumers can drastically affect the fitness of the consumed organisms. Thus, animals are an important agent of natural selection—influencing the evolution of the species they eat (see Chapter 55).

How Animals Feed: Four General Strategies

Animal mouthparts vary, and the structure of an animal's mouthparts correlates closely with its method of feeding. Keep this in mind as you review the four general tactics that animals use to obtain food, summarized in **TABLE 33.4**.

Suspension feeders, also known as **filter feeders,** employ a wide array of structures to trap suspended particles—usually detritus or plankton, small organisms that drift in the currents. Sponges, like the individual illustrated in Figure 33.3, are suspension feeders. So are clams and mussels, which pump water through their bodies and trap suspended food on their feathery gills—structures that also function in gas exchange. Baleen whales suspension feed by gulping water, squeezing it out between the horny baleen plates that line their mouths, and trapping shrimp-like organisms called krill inside. Barnacles capture particles with their feathery, jointed legs.

Because particles float in water much more readily than in air, suspension feeding is particularly common in aquatic environments. Many suspension feeders are sessile.

Many **deposit feeders** digest organic matter in the sediments; their food consists of sand- or mud-dwelling bacteria, archaea, protists, and fungi, along with detritus that settles on the surface of the sediments. Earthworms, for example, are annelids that swallow soil as well as leaves and other detritus on the surface of the soil.

The seafloor is also rich in organic matter, which rains down from the surface and collects in food-rich deposits. These deposits are exploited by a wide array of segmented worms (annelids) as well as sea cucumbers.

Unlike suspension feeders, which are diverse in size and shape and use various trapping or filtering systems, deposit feeders tend to be similar in appearance. They usually have simple mouthparts, and their body shape is wormlike. Like suspension feeders, however, deposit feeders occur in a wide variety of lineages.

SUMMARY TABLE 33.4　**Diversification of Feeding Strategies**

Strategy	Example
Suspension feeders (filter feeders) Capture food by filtering out or concentrating particles floating in water or drifting through the air Barnacles use specialized legs to capture plankton	
Deposit feeders Ingest organic material that has been deposited within a substrate or on its surface Sea cucumbers use feeding tentacles to mop up detritus from the seafloor	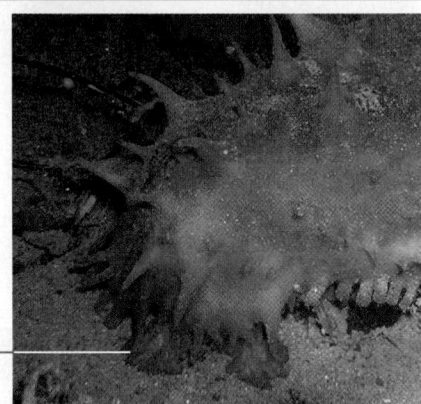
Fluid feeders Suck or mop up liquids like nectar, plant sap, blood, or fruit juice Butterflies and moths drink nectar through their extensible, hollow proboscis	
Mass feeders Take chunks of food into their mouths Lions bite chunks of meat off of prey carcasses	

Fluid feeders range from butterflies and moths that feed on nectar with a straw-like proboscis to vampire bats that feed on blood. Fluid feeders are found in a wide array of lineages and often have mouthparts that allow them to pierce seeds, stems, skin, or other structures in order to withdraw the fluids inside.

Mass feeders ingest chunks of food. The structure of their mouthparts correlates with the type of food pieces that they harvest and ingest. Lions, for example, have razor-like teeth for tearing tough flesh into pieces small enough to be swallowed. Snails use a rasp-like structure called a radula to scrape bits of plant tissue or animal flesh into their mouths.

Movement

Animal locomotion provides an array of important functions: finding food, finding mates, escaping from predators, and dispersing to new habitats. Animals move in highly variable ways; they burrow, slither, swim, fly, crawl, walk, or run—mostly powered by muscle (Chapter 48).

Limbs are a prominent feature of species in many phyla. They develop as outpockets of the body wall and tend to be perpendicular to the anterior–posterior and dorsal–ventral body axes. Limbs can take a variety of forms, from the lobe-like limbs of onychophorans (velvet worms) to the jointed limbs of arthropods and vertebrates to the more flexible tube feet of echinoderms and tentacles of mollusks (**TABLE 33.5**).

Traditionally, biologists have hypothesized that the major types of jointed and unjointed animal limbs were not homologous—that is, not descended from a common ancestor. Because the structure of animal appendages is so diverse, it was logical to maintain that at least some appendages evolved independently of each other. As a result, biologists predicted that completely different genes are responsible for each major type of appendage.

Recent results have challenged this view, however. The experiments in question involve a gene called *Distal-less*, or *Dll*, that was originally discovered in fruit flies. (*Distal* means "away from the body.") The *Dll* protein product seems to deliver a simple message as a fruit fly embryo develops: "Grow appendage out this way."

Biologists in Sean Carroll's lab set out to test the hypothesis that *Dll* might be involved in the initial phase of limb or appendage formation in other animals. The team used a fluorescent marker that sticks to the *Dll* gene product to locate tissues where the gene is expressed. When they introduced the fluorescent marker into embryos from annelids, arthropods, echinoderms, and other phyla, they found that it bound to *Dll* in all of them. More important, the highest concentrations of *Dll* gene products were found in cells that form appendages—even in phyla with wormlike bodies that have extremely simple appendages (**FIGURE 33.13**; see page 650). Other experiments have shown that *Dll* is also involved in limb formation in vertebrates.

Based on these findings, biologists are concluding that all animal appendages have some degree of genetic homology and that they are all derived from appendages that were present in a common ancestor. The idea is that simple appendages evolved early in the history of the Bilateria and that, subsequently, evolution by natural selection produced the diversity of limbs observed today.

SUMMARY TABLE 33.5 **Diversification of Limbs**

Type of Limb	Example
Lobe-like limbs Onychophorans (velvet worms) use lobe-like limbs to crawl	
Jointed limbs The jointed limbs of arthropods (such as this crab) and vertebrates are used for many modes of feeding and locomotion	
Parapodia Polychaete worms use bristled parapodia to crawl and swim	
Tube feet Echinoderms like this sea star use tube feet to crawl	
Tentacles Octopuses use muscular tentacles to crawl and grab prey	

Arthropod	Onychophoran	Annelid

Developing legs

Developing segments

FIGURE 33.13 Genetic Evidence that All Animal Appendages Are Homologous. In diverse organisms, *Dll* gene products (stained green and dark brown) are localized in areas of the embryos where appendages are forming. These data suggest that the appendages are related by common ancestry.

Reproduction

An animal may be efficient at moving and eating, but if it does not reproduce, the alleles responsible for its effective locomotion and feeding will not increase in frequency in the population. Evolution by natural selection occurs when individuals with certain alleles produce more surviving offspring than other individuals do (Chapter 25). Organisms live to reproduce.

Given the array of habitats and lifestyles of animals, it's not surprising that they vary tremendously in how they reproduce (**TABLE 33.6**). A few examples here will help illustrate the diversity of animal reproduction (see Chapter 50 for details).

Asexual or Sexual Reproduction?   In the phylum Rotifera, an entire lineage called the bdelloids (pronounced *DELL-oyds*) reproduces only asexually, through mitosis, by producing diploid eggs that can mature into adults without being fertilized (a process called parthenogenesis). Even certain fish, lizard, and snail species have never been observed to undergo sexual reproduction.

However, sexual reproduction via meiosis and fusion of gametes is much more common than asexual reproduction in animals; and some animals, like coral polyps, can reproduce either way depending on environmental circumstances. Asexual reproduction tends to be more efficient than sexual reproduction, but sexual reproduction leads to greater genetic diversity, which can be favored by natural selection in variable or unfavorable environments (Chapter 13).

Where Does Fertilization Occur?   When sexual reproduction does occur, fertilization may be internal, usually within the body of the female, or external, in the environment.

When internal fertilization takes place, a male typically inserts a sperm-transfer organ into the body of a female. In some animal species, males produce sperm in packets, which females then pick up and insert into their own bodies. But in seahorses, females insert eggs into the male's body, where they are fertilized. (The male is pregnant for a time and then gives birth to live young.)

SUMMARY TABLE 33.6   Diversification of Reproduction

Reproductive Strategy	Example

Asexual reproduction

Polyps within a coral colony are genetically identical clones. They are produced by fission (splitting) or budding

Sexual reproduction
External fertilization

Sea cucumbers reproduce sexually by releasing eggs and sperm into the open water, where fertilization occurs

Internal fertilization

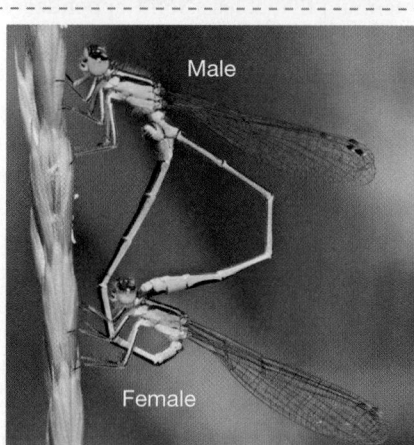

Male

Female

When damselflies copulate, the male holds the female behind her head with claspers and the female retrieves sperm from the male's sperm-transfer organ with the tip of her abdomen

Modes of embryonic development following internal fertilization	Examples
Viviparous species nourish embryos internally and give birth to live young	Most mammals
Oviparous species deposit fertilized eggs; embryos are nourished by yolk	Most insects, birds
Ovoviviparous species retain eggs internally and then give birth to live young	Guppies, garter snakes

External fertilization is extremely common in aquatic species. Females lay eggs onto a substrate or release them into open water. Males shed sperm, which swim, on or near the eggs or into open water.

Where do Embryos Develop? Following internal fertilization, eggs and embryos may be retained in the female's body during development, or fertilized eggs may be laid outside of the body to develop independently of the mother:

- **Viviparous** ("live-bearing") species such as humans and most other mammals nourish embryos inside the body and give birth to live young.

- **Oviparous** ("egg-bearing") species such as chickens and crickets deposit fertilized eggs. The embryos within are nourished by yolk.

- **Ovoviviparous** ("egg-live-bearing") species such as guppies and garter snakes retain eggs inside their body during early development, but the growing embryos are nourished by yolk inside the egg and not by nutrients transferred directly from the mother, as in viviparous species. Ovoviviparous females then give birth to well-developed young.

Most mammals and a few species of sea stars, velvet worms, fish (including sharks), amphibians, and lizards are viviparous; some snails, insects, reptiles, and fishes are ovoviviparous. But the vast majority of animals are oviparous.

Life Cycles

Reproduction is just one component of the diverse life cycles of animals. The vast majority of sexually reproducing animals have diploid-dominant life cycles because the haploid stages—the gametes—are relatively short lived and single celled. But even this general pattern has interesting exceptions, such as haploid males in honeybee colonies. How else do animal life cycles vary?

Perhaps the most spectacular innovation in animal life cycles involves the phenomenon known as **metamorphosis** ("change-form")—a drastic change from one developmental stage to another. The young of some animals, such as zebras, look similar to adults when they are born. Their development is said to be *direct*. However, other animals, such as sea urchins, undergo a dramatic transformation—a metamorphosis—during their life cycle and are said to undergo *indirect* development (**FIGURE 33.14**).

During indirect development, embryogenesis produces **larvae** (singular: **larva**) that look radically different from adults, live in different habitats, and eat different foods. In the case of the sea urchin in Figure 33.14, the larva is bilaterally symmetric and planktonic, and it suspension feeds on single-celled algae; but the adult is radially symmetric and sessile, and it scrapes macroalgae off of rocks.

Through the process of metamorphosis, larvae transform into **juveniles;** they look like adults and live in the same habitats and eat the same foods as adults, but they are still sexually immature. It is only after a period of growth and maturation that juveniles become **adults,** the reproductive stage in the life cycle.

Metamorphosis is extremely common in marine animals as well as in insects (insect metamorphosis is described in Chapter 34)

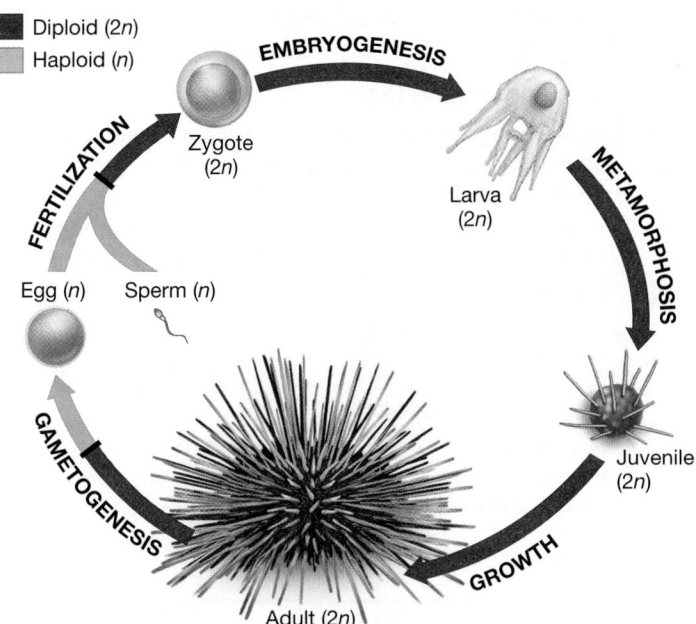

FIGURE 33.14 Many Animal Life Cycles Include Metamorphosis. Most sea urchins have a planktonic, feeding larva that is very different from the sessile adult. Drawings are not to scale.

✔ **EXERCISE** A few sea urchins produce especially large, yolk-rich eggs that develop directly into juveniles and adults, bypassing the feeding larval stage. Circle the part of the life cycle that would be different in this case, and pose a hypothesis to explain why or under what conditions this difference might be adaptive.

and many amphibians. Why would natural selection favor such an indirect path to adulthood? One hypothesis emphasizes dispersal. In marine species that have limited or no movement as adults—such as sponges, corals, sea anemones, clams, and sea urchins—larvae function as a dispersal stage. They are a little like the seeds of many land plants—a life stage that allows individuals to move to new habitats.

Another hypothesis for the occurrence of metamorphosis emphasizes feeding efficiency. Because larvae and adults feed on different foods in different ways, they do not compete with each other and can specialize to take advantage of available food sources. For example, the caterpillar larvae of butterflies can feast on nutritious leaves to acquire the energy needed for growth, while adult butterflies feast on flower nectar to acquire energy needed for finding mates and reproducing.

The diversification of life cycles occurred not only among species in diverse lineages but also within species—such as in organisms that can opt between asexual and sexual reproduction or direct and indirect development, depending on environmental circumstances. The diversification of life cycles, as well as that of sensory structures, ecological roles, feeding strategies, limbs, and reproductive strategies, represent the consequences of evolutionary processes played out in ecological contexts over millions of years.

Let's now examine how the general themes of diversification apply to individual phyla, starting with the most ancient lineages first and then moving on to the extremely diverse protostomes (Chapter 34) and deuterostomes (Chapter 35).

If you understand that . . .

- Most of the animal body plans originated during the Cambrian and subsequently diversified.
- Themes of diversification include evolution of different sensory organs, ecological roles, feeding strategies, type of limbs, type of reproduction, and life cycles.

✔ **You should be able to . . .**

1. Explain how the mouthparts of deposit feeders and mass feeders are expected to differ.
2. Suggest why external fertilization is particularly common in aquatic environments, and internal fertilization is particularly common in terrestrial environments.

Answers are available in Appendix A.

33.4 Key Lineages of Animals: Non-Bilaterian Groups

The fossil record indicates that the phyla Porifera (sponges), Ctenophora (comb jellies), and Cnidaria (jellyfish and others) are the most ancient of all major animal lineages. The phylogeny of animals presented in this chapter is consistent with the fossil record—the first sponges, ctenophores, and cnidarians appear at the base of the tree (Figure 33.2). As you consider each group in more detail, try to pick out which characteristics are unique to the group, and which are common to organisms living in similar environments.

- Porifera (Sponges)
- Ctenophora (Comb Jellies)
- Cnidaria (Jellyfish, Corals, Anemones, Hydroids)

Porifera (Sponges)

About 8500 species of sponges have been described to date. Although a few freshwater species are known, most are marine. All sponges are **benthic,** meaning that they live at the bottom of aquatic environments. Sponges are particularly common in rocky, shallow-water habitats of the world's oceans. One sponge species native to the Caribbean can grow to heights of 2 meters (m).

Sponges have commercial and medical value to humans. The dried bodies of certain sponge species are able to hold large amounts of water and thus are prized for use in bathing and washing. In addition, researchers are increasingly interested in the array of toxins that sponges produce to defend themselves against predators and bacterial parasites—possibly for use in cancer chemotherapy.

Feeding Most sponges are suspension feeders. Their cells beat in a coordinated way to produce a water current that flows through small pores in the outer body wall, into chambers inside the body, and out through a single larger opening (**FIGURE 33.15**). As water passes by feeding cells, organic debris and bacteria, archaea, and small protists are filtered out of the current and then digested. Some deep-sea sponges are predators, however—they capture small crustaceans on hooks that project from the body.

Movement Most adult sponges are sessile, though a few species are reported to move at rates of up to 4 millimeters (mm) per day. Most species produce larvae that swim with the aid of cilia, so they are capable of dispersal. Recent research has confirmed that at least one species can contract its body to expel waste products.

Reproduction Asexual reproduction occurs in a variety of ways, depending on the species. Some sponge cells are totipotent, meaning that when a fragment breaks off, small groups of adult cells have the capacity to develop into complete new adult organisms. Although individuals of most species produce both eggs and sperm, self-fertilization is rare because individuals release their male and female gametes at different times.

Ectyoplasia ferox

35 cm

FIGURE 33.15 Sponges Are Sessile Suspension Feeders.

Ctenophora (Comb Jellies)

Ctenophores are transparent, ciliated, gelatinous diploblasts that live in marine habitats (**FIGURE 33.16**). Although a few species live on the ocean floor, most are planktonic. Only about 190 species have been described to date, but some are abundant enough to represent a significant fraction of the total planktonic biomass. Accidental introductions of the ctenophore *Mnemiopsis leidyi*, which preys on fish larvae, have devastated fish production in the Black Sea and Caspian Sea. If you have gone swimming in an ocean or bay, you may have bumped up against ctenophores without realizing it since they can be abundant yet nearly invisible in the water. (Don't worry, they are harmless.) Some comb jellies can also be seen at night when they emit pulses of faint phosphorescent light.

Feeding Ctenophores are predators, but lack the stinging cells that are characteristic of Cnidarians. Feeding occurs in several ways, depending on the species. Some comb jellies have long tentacles covered with cells that release an adhesive when they contact prey. These tentacles are periodically wiped across the ctenophore's mouth so that captured prey can be ingested. In other species, prey can stick to mucus on the body and be swept toward the mouth by cilia. Still other species ingest large prey whole.

Movement Adults move via the beating of cilia, which occur in comblike plates. The plates form rows that run the length of the body. Ctenophores are the largest adult animals known to use cilia for locomotion. ✔You should be able to indicate the origin of adult swimming via rows of coordinated cilia on Figure 33.2.

Reproduction Most species have both male and female organs and routinely self-fertilize, though fertilization is external. Larvae are free swimming.

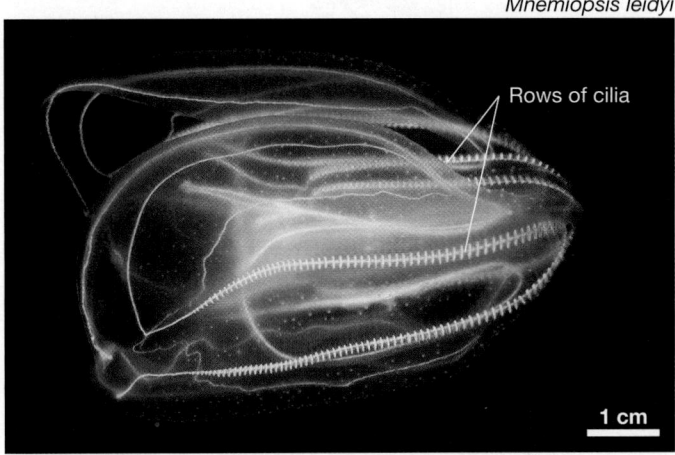
Mnemiopsis leidyi

Rows of cilia

1 cm

FIGURE 33.16 Ctenophores Are Planktonic Predators.

Cnidaria (Jellyfish, Corals, Anemones, Hydroids)

Although a few species of Cnidaria inhabit freshwater, the vast majority of the 11,500 species are marine. They are found in all of the world's oceans, occupying habitats from the surface to the substrate, and are important predators. The phylum comprises four main lineages, as shown in **FIGURE 33.17**.

Many cnidarians are radially symmetric diploblasts consisting of ectoderm and endoderm layers that sandwich gelatinous material known as **mesoglea,** which contains a few scattered ectodermal cells. Recent research indicates that at least some jellyfish are triploblastic and have bilaterally symmetric larvae or are bilaterally symmetric as adults. Cnidarians have a gastrovascular cavity instead of a flow-through gut—meaning there is only one opening to the environment for both ingestion and elimination of wastes (see Chapter 44).

Jellyfish have a life cycle that includes both a sessile **polyp** form that reproduces asexually and a free-floating **medusa** that reproduces sexually, as shown in the case of a hydrozoan in **FIGURE 33.18** (see page 654). Anemones and coral, however, exist only as polyps—never as medusae. Reef-building corals secrete outer skeletons of calcium carbonate that create the physical structure of a coral reef—one of the world's most productive habitats (see Chapter 56).

Feeding A key synapomorphy of the cnidarians is a specialized cell called a **cnidocyte,** which is used in prey capture. When cnidocytes sense a fish or other type of prey, the cells forcibly eject a barbed, spear-like structure, which may contain toxins. The barbs hold the prey, and the toxins subdue it until it can be brought

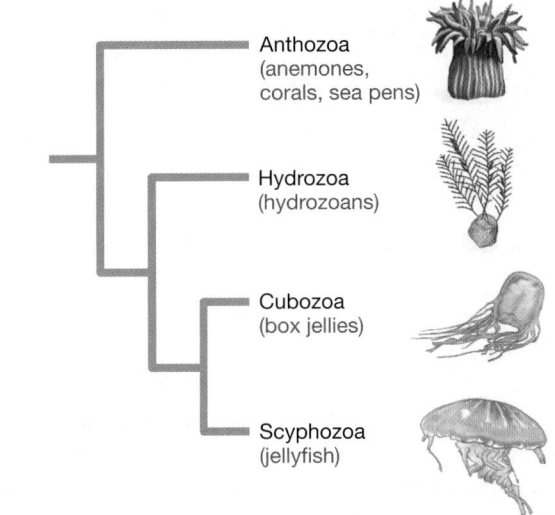

Anthozoa (anemones, corals, sea pens)

Hydrozoa (hydrozoans)

Cubozoa (box jellies)

Scyphozoa (jellyfish)

FIGURE 33.17 There are Four Main Lineages within the Cnidaria.

(Continued on next page)

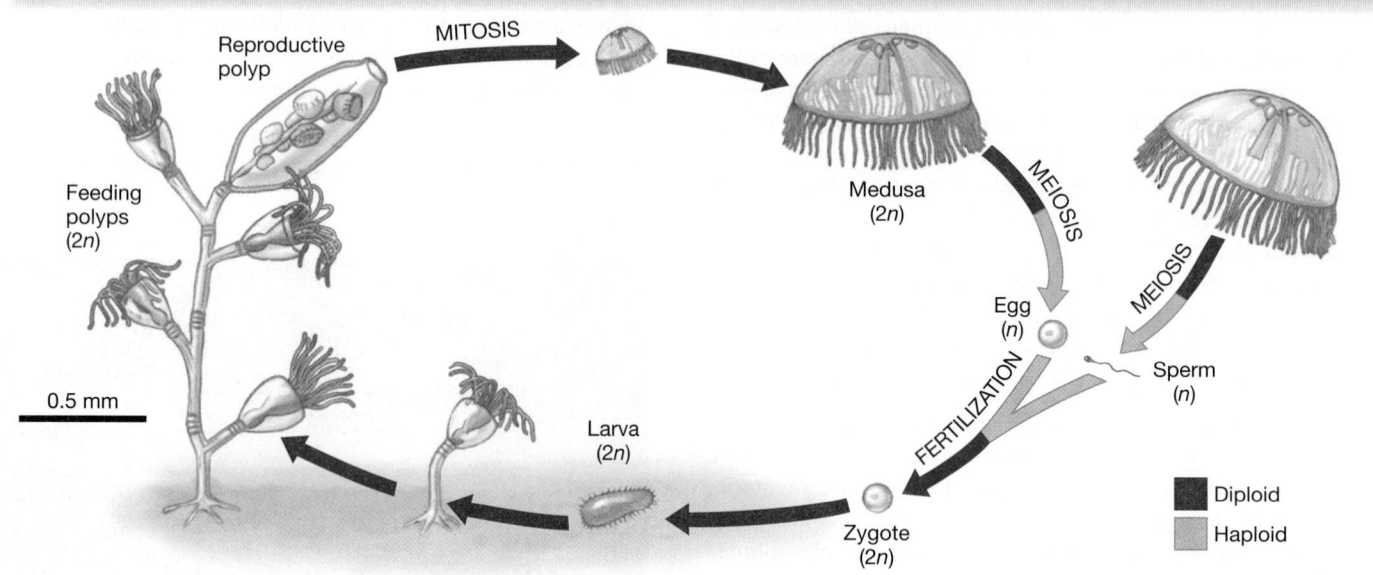

FIGURE 33.18 Cnidarian Life Cycles May Include a Polyp and Medusa Form. This is an example of a hydrozoan called *Obelia*. Colonies are often found attached to seaweed, shells, or rocks and contain hundreds of polyps.

to the mouth and ingested. Cnidocytes are commonly located near the mouths of cnidarians or on elongated structures called tentacles. In a very small number of species, cnidarian toxins can be deadly to humans as well as to prey organisms; in Australia, twice as many people die each year from stings by box jellyfish as from shark attacks (**FIGURE 33.19**). ✔You should be able to indicate the origin of cnidocytes on Figure 33.2.

Besides capturing prey actively, most species of coral, some jellyfish, and many anemones host photosynthetic dinoflagellates. The relationship is mutually beneficial, because the protists supply the cnidarian host with food in exchange for protection.

Movement Both polyps and medusae are capable of movement. In polyps, the gut cavity acts as a hydrostatic skeleton that works with the aid of muscle-like cells to contract or extend the body. Many polyps can also creep along a substrate, using muscle-like cells at their bases. Medusae can contract the muscle-like cells in the rims of their bell-shaped body rhythmically, enabling paddling motions as well as jet propulsion—the forcible flow of water in the opposite direction of movement. Cnidarian larvae swim by means of cilia.

Reproduction Polyps may produce new individuals asexually by **(1)** budding, in which a new organism grows out from the body wall of an existing individual; **(2)** fission, in which an existing adult splits lengthwise to form two individuals; or **(3)** fragmentation, in which parts of an adult regenerate missing pieces to form a complete individual. During sexual reproduction, gametes are usually released from the mouth of a polyp or medusa, and fertilization takes place in the open water. Eggs hatch into larvae that become part of the plankton before settling and developing into polyps.

Chironex fleckeri

FIGURE 33.19 Some Jellyfish Have Long Tentacles, Packed with Cnidocytes That Sting Prey.

If you understand . . .

33.1 What Is an Animal?

- Animals are multicellular, heterotrophic eukaryotes that lack cell walls and ingest their prey.

- All animals move under their own power at some point in their life cycle. All animals other than sponges have specialized nerve and muscle cells that enable complex movements.

- Animals are eating machines, the largest consumers on Earth.

- Animals comprise 30–35 phyla and may number 8 million or more species.

✔ You should be able to predict how the total amount of plant material in a forest would change if all animals were excluded for one year.

33.2 What Key Events Occurred during the Evolution of Animals?

- Animals are a monophyletic group that originated from a protist ancestor that was probably similar to choanoflagellates.

- The basic genetic tool kit for multicellular animals originated at the very root of the animal lineage.

- Sponges lack highly organized, complex tissues other than rudimentary epithelium.

- Most ctenophores and cnidarians have radial symmetry, two embryonic germ layers, and neurons organized into a nerve net. However, some of the regulatory genes responsible for bilateral symmetry arose before the split of cnidarians and bilaterians.

- Most animal species are bilaterians. They have bilateral symmetry, three embryonic tissues, and a coelom—features that gave rise to a tube-within-a-tube body plan.

- Most bilaterians also have a centralized nervous system and cephalized bodies—meaning that a distinctive head region contains the mouth, brain, and sensory organs.

- The bilaterian tube-within-a-tube design is built in one of two fundamental ways—via the protostome or deuterostome patterns of embryonic development.

- Segmentation is a key innovation in annelids, arthropods, and vertebrates.

✔ You should be able to draw a bilaterian that lacks limbs, and label the gut, outer body wall, muscle layers, head region, mouth, anus, brain, and major nerve tracts.

33.3 What Themes Occur in the Diversification of Animals?

- Sensory abilities and sensory structures vary among species and correlate with their habitats and feeding methods.

- Animals fill three different ecological roles depending on their source of food: they are detritivores, herbivores, or carnivores—or omnivores, a combination.

- Animals capture food in four ways: Suspension or filter feeders filter organic material or small organisms from water; deposit feeders swallow sediments; fluid feeders lap or suck up liquids; mass feeders harvest chunks of tissue.

- Although the types of appendages used in animal locomotion range from simple lobe-like limbs to complex jointed legs, the genes that indicate where appendages develop may be homologous.

- Asexual reproduction occurs in many species, but sexual reproduction is predominant. Fertilization may be external or internal, and embryos may complete development inside or outside the mother.

- Development may be direct or indirect. During indirect development, a dramatic morphological transformation takes place from one life stage to another, called metamorphosis.

✔ You should be able to select a phylum and give examples of how the themes of diversification apply within that phylum.

33.4 Key Lineages of Animals: Non-Bilaterian Groups

- Sponges are diverse, benthic suspension feeders.

- Ctenophores are planktonic predators, the largest animals to use cilia for locomotion.

- Cnidarians are diverse and abundant in habitats around the world. The cnidocyte, or stinging cell, is a synapomorphy for this group.

✔ You should be able to recognize sponges, ctenophores, or cnidarians based on their morphology.

(MB) MasteringBiology

1. MasteringBiology Assignments

Tutorials and Activities Animal Phylogenetic Tree; Animal Body Plans; Architecture of Animals

Questions Reading Quizzes, Blue-Thread Questions, Test Bank

2. eText Read your book online, search, take notes, highlight text, and more.

3. The Study Area Practice Test, Cumulative Test, BioFlix® 3-D Animations, Videos, Activities, Audio Glossary, Word Study Tools, Art

You should be able to . . .

1. What synapomorphy (shared, derived trait) distinguishes animals as a monophyletic group, distinct from choanoflagellates?
   a. multicellularity
   b. coloniality
   c. heterotrophy
   d. no cell walls

2. True or false? All coelomates are triploblasts.

3. In _____, the mouth develops before the anus; but in _____, the anus develops before the mouth.

4. Why do some researchers maintain that the limbs of all animals are homologous?
   a. Homologous genes, such as *Dll*, are involved in their development.
   b. Their structure—particularly the number and arrangement of elements inside the limb—is the same.
   c. They all function in the same way—in locomotion.
   d. Animal appendages are too complex to have evolved more than once.

5. In a tube-within-a-tube body plan, what is the interior tube?
   a. ectoderm
   b. mesoderm
   c. the coelom
   d. endoderm

6. Which of the following characteristics does not apply to cnidarians?
   a. usually diploblastic
   b. possess a gastrovascular cavity with one opening
   c. undergo metamorphosis
   d. have a central nervous system

7. Explain the difference between a diploblast and a triploblast. How was the evolution of mesoderm associated with the evolution of a coelom?

8. Why is it ecologically significant that animals are heterotrophic *and* multicellular?

9. QUANTITATIVE To estimate the relative abundance of the major phyla, calculate how many named species of arthropods, mollusks, and nematode worms exist per named species of chordate (see Table 33.1). Do you think these calculations are likely to be underestimates, or overestimates? Why?

10. Explain how an animal mother nourishes an embryo in oviparous species versus viviparous species.

11. Specific genes are known to play an important role in the segmentation of arthropods. What approach could you use to test the hypothesis that segmentation in vertebrates (phylum Chordata) is homologous to segmentation in arthropods (phylum Arthropoda)?

12. Which of the following is an example of homology (similarity due to common ancestry)?
    a. suspension feeding in sponges and clams
    b. ectoparasite lifestyle in aphids and ticks
    c. cnidocytes (stinging cells) in jellyfish and sea anemones
    d. metamorphosis in sea urchins and insects

13. Suppose that a gene originally identified in nematodes (roundworms) is found to be homologous with a gene that can cause developmental abnormalities in humans. Would it be possible to study this same gene in fruit flies? Explain.

14. Some sea anemones can produce large colonies by reproducing asexually, but they can also produce planktonic larvae by reproducing sexually. Predict the circumstances under which each mode of reproduction would be favored.

15. Radial symmetry evolved in echinoderms (sea stars and sea urchins) from bilaterally symmetric ancestors. Predict whether the echinoderm nervous system is centralized or diffuse. Explain your logic.

16. You have a unique opportunity to explore a pristine coral reef in the South Pacific. You come across a small organism that you do not recognize and that is not in your field guide. Which of the following suites of characteristics would be most helpful in identifying the organism's phylum?
    a. sessile, green, 2 cm tall, abundant
    b. radial symmetry, suspension feeder
    c. nonsegmented, sessile, changes shape
    d. radial symmetry, sessile, two tissue layers

34 Protostome Animals

In this chapter you will learn that

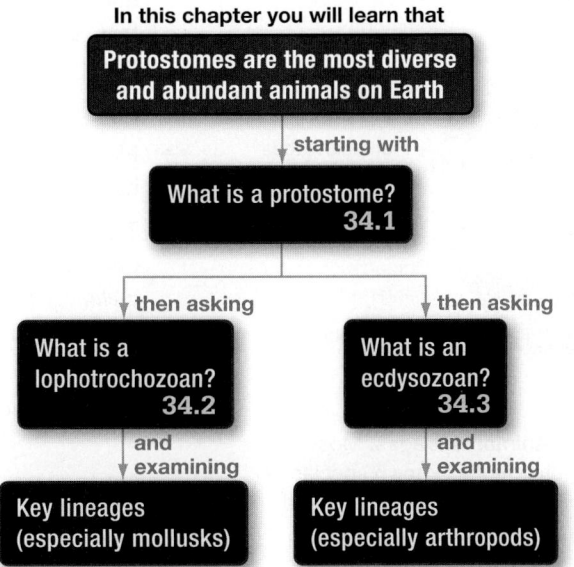

Protostomes are the most diverse and abundant animals on Earth

starting with

What is a protostome? 34.1

then asking

What is a lophotrochozoan? 34.2

then asking

What is an ecdysozoan? 34.3

and examining

Key lineages (especially mollusks)

and examining

Key lineages (especially arthropods)

In numbers of individuals and species richness, protostomes are the most abundant and diverse of all animals—and the insects are by far the most abundant and diverse of the protostomes. The insects shown here come from all over the world.

Protostome animals are spectacularly diverse and abundant. To put this in perspective, consider that of the more than 30 animal phyla known to exist, at least 22 are protostomes. Table 33.1 summarized the major protostome phyla. You will likely recognize many of the more familiar ones:

- *Arthropoda*, including the insects, crustaceans (shrimp, lobsters, crabs), chelicerates (spiders and mites), and myriapods (millipedes, centipedes)

- *Mollusca*, including the snails, clams, octopuses, and squid

- *Annelida*, including the segmented worms such as the earthworms, polychaete worms, and leeches

✔ When you see this checkmark, stop and test yourself. Answers are available in Appendix A.

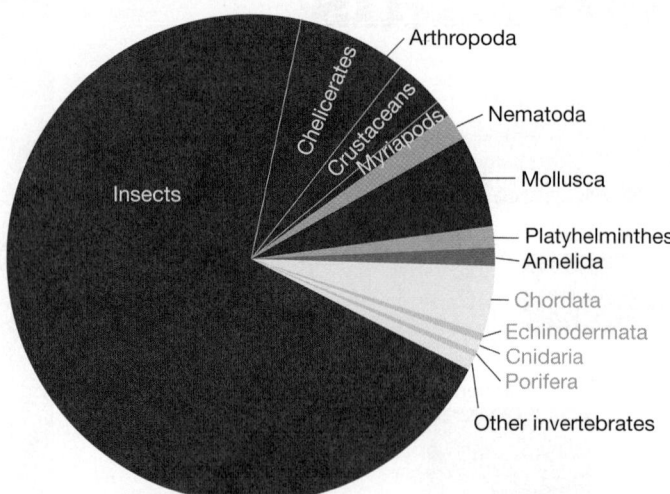

FIGURE 34.1 The Relative Diversity of Animal Lineages. Most animals are protostomes. About 70 percent of all known species of animals on Earth are insects, most of them beetles. (Humans and other vertebrates are deuterostomes, in the phylum Chordata.)

DATA: Chapman, A. D. 2009. *Numbers of Living Species in Australia and the World,* 2nd ed. Canberra: Australian Biological Resources Study.

Some protostome phyla have only a few named species, such as the phylum Priapula, which contains 16 named species. On the other end of the spectrum, over 85,000 mollusks have been named thus far, and about 1.2 million arthropod species, of which about a million are insects. Scientists estimate that the actual number of arthropod species may be over 10 million.

The pie chart in **FIGURE 34.1** shows the relative numbers of named species in various animal phyla. Notice that the wedges representing non-protostome phyla, labeled Chordata, Cnidaria, Echinodermata, and Porifera, are small. Almost all animal species are protostomes.

Certain protostomes are not only diverse (in numbers of species), but also extremely abundant (in numbers of individuals). A typical acre of pasture in England is home to almost 18 *million* individual beetles and up to 9 *billion* roundworms. The world population of ants is estimated to be 1 million billion (1,000,000,000,000,000) individuals.

Given the diversity and abundance of protostomes, you should not be surprised that they fill very important ecological roles. As a group, protostomes live in virtually every aquatic and terrestrial habitat in the world and can be detritivores, herbivores, or carnivores. If one of biology's most fundamental goals is to understand the diversity of life on Earth, then protostomes—particularly the mollusks and arthropods—demand our attention.

The ecological importance of protostomes extends to the health and welfare of humans in profound and diverse ways. For example:

- Protostomes, especially shellfish (clams, oysters, scallops, crabs, shrimp, lobsters), are a major direct source of food for humans, which makes them economically important.

- Humans also rely on protostomes indirectly for their ecosystem services. For example, many farmers rely on protostomes like earthworms to prepare the soil, and on bees and other insects to pollinate their crops.

- Some insects have a massive negative impact on food production by damaging crops, while other insects protect crops from pests.

- Some protostomes produce valuable materials such as silk and pearls.

- Many protostomes play an important role in causing or transmitting human diseases, and are common parasites.

- The protostomes include two of the most important model organisms in all of biological science: the fruit fly *Drosophila melanogaster* and the roundworm *Caenorhabditis elegans* (see **BioSkills 13** in Appendix B).

Let's take a closer look at what defines the protostomes and then explore the diversity within the major phyla.

34.1 What Is a Protostome?

There are two major groups of bilaterally symmetric, triploblastic animals: the protostomes and deuterostomes (Chapter 33). The **protostomes** are distinguished from the deuterostomes by two developmental characteristics:

1. During gastrulation, the initial pore that forms in the embryo becomes the mouth—protostome translates to "first-mouth." (In deuterostomes the pore develops into the anus, and the mouth develops later.)

2. If a **coelom** (a body cavity) forms later in development, it forms from cavities that arise within blocks of mesodermal cells (rather than as mesoderm pockets pinching off the gut, as in deuterostomes).

Phylogenetic studies have long supported the hypothesis that protostomes are a monophyletic group, meaning the protostome developmental sequence arose just once. But biologists have found it difficult to sort out the relationships among protostome lineages. During the 500 million years of evolution that followed the diversification of protostomes during the Cambrian explosion, some morphological characteristics evolved independently in different lineages, resulting in convergent evolution. Other characteristics thought to be synapomorphies—shared, derived traits within a lineage—were lost in some groups. As a result, morphological data have led to many alternative hypotheses about phylogenetic relationships.

Recent analyses of DNA sequence data have rejected many previous hypotheses of relationships based on morphological data. As the phylogenetic tree in **FIGURE 34.2** shows, the recent data support two major subgroups within the protostomes. The Lophotrochozoa (pronounced *low-foe-tro-ko-ZOH-ah*) include mollusks and annelid worms. The Ecdysozoa (pronounced *eck-die-so-ZOH-ah*) include the arthropods.

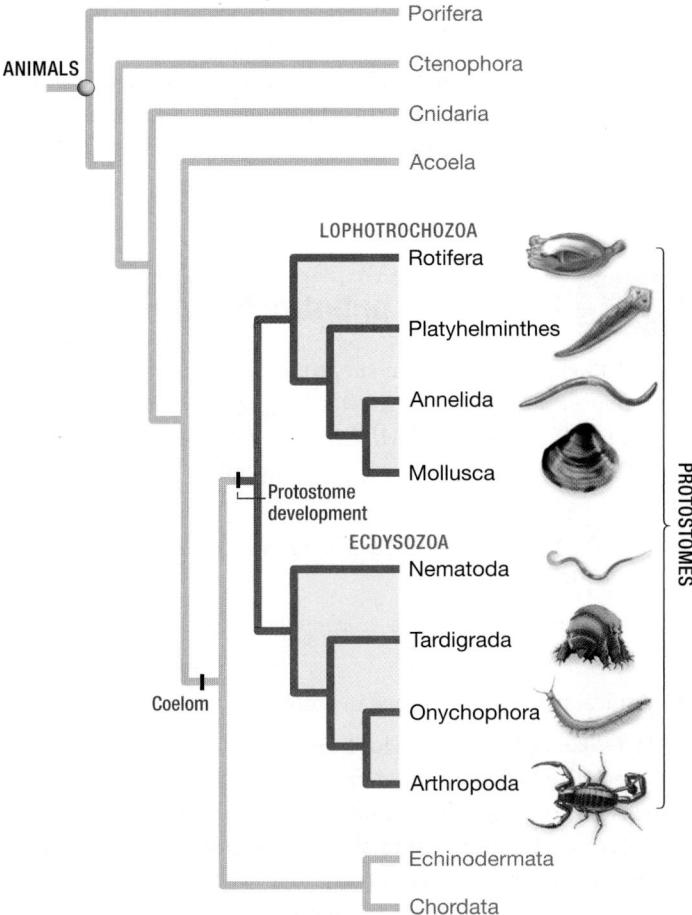

FIGURE 34.2 Protostomes Are a Monophyletic Group Comprising Two Major Lineages. There are 22 phyla of protostomes, but the eight major phyla shown account for about 99.5 percent of the known species.

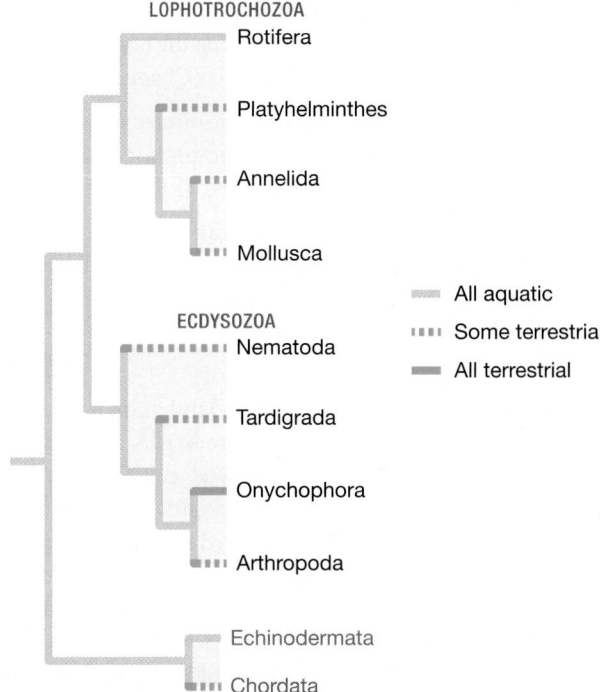

FIGURE 34.3 The Water-to-Land Transition Occurred Several Times in the Protostomes. The transition occurred only once in the deuterostomes, in phylum Chordata.

✔ **QUESTION** According to this tree, terrestrial living arose seven different times independently in protostomes. Why would you not instead conclude that terrestrial living was ancestral and the aquatic lifestyle arose seven times independently?

Before considering each of these groups in more detail, let's examine the key themes in the evolutionary diversification of the protostomes. Recall that all animal lineages evolved diverse ways of sensing the environment, feeding, moving, reproducing, and developing (see Chapter 33). In addition, diversification in protostomes was triggered by the transition from aquatic to terrestrial habitats, and the modular design of many protostome body plans.

The Water-to-Land Transition

Fossils from the Burgess Shale in Canada, Chengjiang in China, and other sites show that protostome lineages originated in marine environments—that is, in the ocean. Today, protostomes are the most abundant animals in the world's oceans as well as in virtually every freshwater and terrestrial habitat. Like land plants and fungi, protostomes made the transition from aquatic to terrestrial environments.

To help put this achievement in perspective, note that green plants made the move from freshwater to land just once (Chapter 31). It is not yet clear whether fungi moved from aquatic habitats to land once or several times (Chapter 32). Only one lineage among deuterostomes moved onto land (Chapter 35).

But protostomes made a water-to-land transition multiple times as they diversified (**FIGURE 34.3**).

Water-to-land transitions are important because they open up entirely new habitats and new types of resources to exploit. Based on this reasoning, biologists propose that the ability to live in terrestrial environments was a key event in the diversification of several protostome phyla.

Several adaptations were required for protostomes to thrive on land, such as the ability to:

1. exchange gases,

2. avoid drying out, and

3. hold up their bodies under their own weight.

Recall that an **adaptation** is a trait that increases the fitness (reproductive success) of individuals in a particular environment.

Land animals exchange gases with the atmosphere readily as long as they have a large, moist surface area that is exposed to the air (Chapter 45). The bigger challenge is to prevent the gas-exchange surface and other parts of the body from drying out. How do terrestrial protostomes solve this problem?

• Roundworms, earthworms, and other terrestrial protostomes live in humid soils or other moist environments and exchange gases across their body surface. They have a high surface-area-to-volume ratio, which increases the efficiency of gas exchange.

- Some terrestrial arthropods and mollusks have gills or other respiratory structures located *inside* the body, minimizing water loss, such as tracheae in insects (Chapter 43).

- Insects evolved a waxy layer to minimize water loss from the body surface, with openings to respiratory passages that can be closed if the environment dries.

- Insect eggs have a thick membrane that keeps moisture in, and the eggs of slugs and snails have a thin calcium carbonate shell that helps retain water. Desiccation-resistant eggs evolved repeatedly in populations that made the transition to life on land.

Unlike land plants and fungi, many land animals can also move to moister habitats if the area they are in gets too dry.

What are some of the mechanical constraints of living on land? Think for a moment of the weightless feeling you experience in a swimming pool. The upward buoyant force provided by the water counteracts the downward pull on your body mass that is exerted by gravity. Without this buoyant force, animals on land require greater structural support to hold their bodies up and to move. Further, as an animal doubles in length (assuming no change in shape), its weight increases by a factor of 8—so larger organisms feel the effect of gravity more than smaller organisms (Chapter 48). This mechanical constraint has limited the size of many protostomes, including insects, on land.

Modular Body Plans

Morphological and physiological diversification has a genetic basis. Until recent decades, biologists assumed that very different genetic instructions were required to create very different organisms. This reasoning seems logical but turns out to be false. Consider that:

- Multicellular animals have a common tool kit of genes that establish the animal body plan during development. *Hox* genes have been featured in several chapters because they are an important part of this tool kit (see Chapters 22 and 33).

- The genetic tool kit can direct the development of dramatically different types and numbers of structures when the genes are expressed at different times and places during development.

The upshot is that the diversification of animal body plans can occur not just by the generation of new genes over time, but also—and especially—by the changing of the expression pattern of existing genes, in particular regulatory genes. Sean Carroll and colleagues demonstrated this principle clearly when they showed that the *Distal-less* gene is essential to the formation of different types of limbs in diverse lineages, ranging from annelids and arthropods to chordates (see Chapter 33).

When comparing diverse organisms within a lineage, genetically based *modularity* in body plans is evident, meaning that a small set of elements can be reused and rearranged to produce a large variety of outcomes. For example, the change in expression patterns of regulatory genes can account for both the dramatic

increase in the number of vertebrae (back bones) in snakes as well as the loss of limbs in this group (Chapter 22). Thus, a change in expression of preexisting genes generated a dramatically different body plan from that of a shorter, limbed ancestor. Keep the concept of modularity in mind as you read on about the diversity of protostomes.

34.2 What Is a Lophotrochozoan?

The 13 phyla of protostome subgroup Lophotrochozoa include the mollusks, annelids, and flatworms (phylum Platyhelminthes; pronounced *plah-tee-hell-MIN-theez*). Lophotrochozoans are named for two of the three unique morphological traits that define the lineage:

1. a feeding structure called a *lophophore*, which is found in three phyla;

2. a type of larva called a *trochophore*, which is common to many of the phyla; and

3. a spiral pattern of cleavage in embryos.

As **FIGURE 34.4a** shows, a **lophophore** (literally, "tuft-bearer") is a specialized structure that rings the mouth and functions in suspension feeding. Lophophores are found in bryozoans (moss animals), brachiopods (lamp shells), and phoronids (horseshoe worms), together called the lophophorates.

Trochophores are a type of larvae common to marine mollusks, marine annelids, and several other phyla in the Lophotrochozoa. As **FIGURE 34.4b** shows, a **trochophore** (literally, "wheel-bearer") larva has a ring of cilia around its middle. These cilia allow swimming and, in some species, sweep food particles into the mouth.

Trochophore larvae, like other larvae, occur in animals that undergo indirect development—where larvae often look radically different from adults, live in different habitats, and eat different foods (Chapter 33). The process of metamorphosis from larvae to adults in marine species is hypothesized to be an adaptation that allows larvae of sessile or slow-moving adults to disperse to new habitats by floating or swimming in the plankton.

Larvae are not unique to lophotrochozoans—even the earliest known animals, the sponges, had larvae. However, recent analyses suggest that the trochophore larva originated early in the evolution of lophotrochozoans and then evolved into different larval types later in some groups.

Recent phylogenetic analyses of lophotrochozoans have inspired the hypothesis that the spiral pattern of cleavage in embryos is a synapomorphy for this monophyletic group. When cells divide at oblique angles to each other during early embryogenesis, a spiraling pattern of cells results in the blastula (**FIGURE 34.5**). This is in contrast to other patterns of cleavage, such as radial cleavage, where cells divide at right angles to each other. Although spiral cleavage has been highly conserved in some lophotrochozoan phyla, it has been modified or lost in others. ✔ **You should be able to indicate the origin of spiral cleavage on Figure 34.2.**

(a) Lophophores function in suspension feeding in adults.

(b) Trochophore larvae swim and may feed.

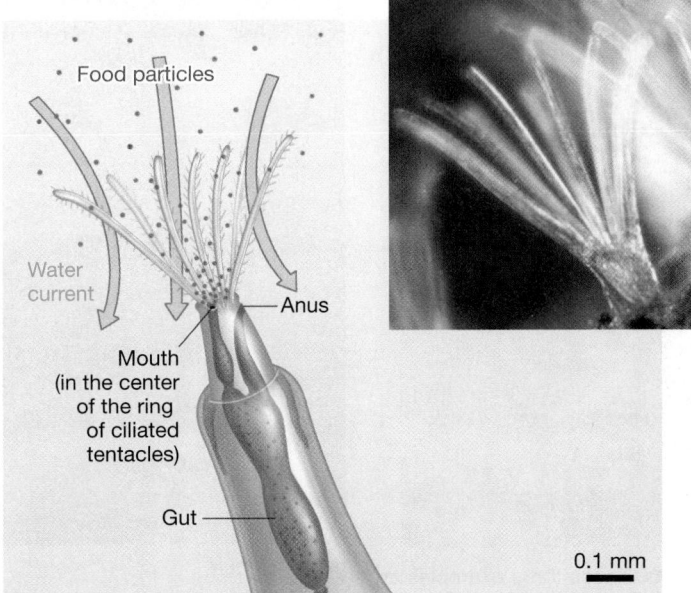
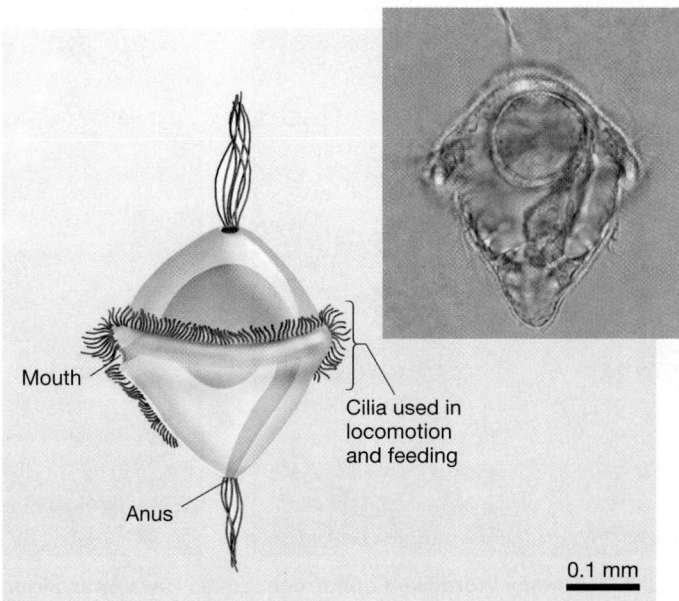

Food particles

Water current

Anus

Mouth (in the center of the ring of ciliated tentacles)

Gut

0.1 mm

Mouth

Cilia used in locomotion and feeding

Anus

0.1 mm

FIGURE 34.4 Lophotrochozoans Have Distinctive Traits. (a) Three phyla of lophotrochozoans, including the bryozoan shown here, have the feeding structure called a lophophore. **(b)** Many phyla of lophotrochozoans have the type of larva called a trochophore.

The implications of spiral cleavage to adult diversity are becoming evident as modern molecular techniques are applied to studies of comparative embryology. For example, spiral cleavage is clockwise in some snails but counterclockwise in others, resulting in adults with right-handed and left-handed whorls, respectively. It turns out that the cleavage pattern determines the downstream expression of *Nodal* and *Pitx*—tool-kit genes that establish the asymmetry of snails. *Nodal* and *Pitx* are also important in the left-right asymmetries of vertebrates, illustrating how homologous tool-kit genes can have similar but different effects in different lineages.

(a) Spiral cleavage is unique to lophotrochozoans.

(b) Radial cleavage, for comparison

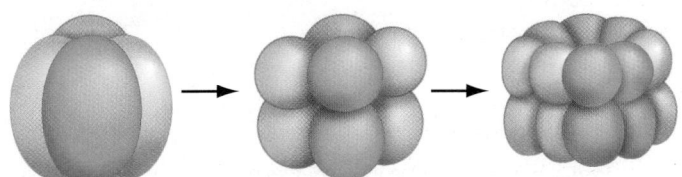

FIGURE 34.5 Spiral Cleavage is a Synapomorphy for the Lophotrochozoans. This pattern of early cell divisions, here contrasted with radial cleavage, has been modified or lost in some lineages. (Some of the cells are stained to help compare the patterns.)

In sum, if you were to observe a lophophore, a trochophore larva, or spiral cleavage in an unfamiliar animal, you could be confident that the animal belongs to the Lophotrochozoa. However, all lophotrochozoans do not possess all three of these characteristics.

Let's briefly consider the diversity of wormlike lophotrochozoans and then look at the most diverse and abundant phylum, the mollusks.

Wormlike Lophotrochozoans

Most protostome phyla include species that have long, thin, tube-like bodies that lack limbs. That is, they are **worms** with a basic tube-within-a-tube design (Chapter 33). The outside tube is the skin, which is derived from ectoderm; the inside tube is the gut, which is derived from endoderm. Muscles and organs derived from mesoderm are located between the two tubes.

Since the coelom evolved in the common ancestor of protostomes and deuterostomes (see Figure 34.2), many lophotrochozoans have a well-developed coelom—not only allowing space for fluids to circulate among organs but also serving as part of a hydrostatic skeleton for movement. The hydrostatic skeleton enables worms to crawl, burrow, and even swim. Specific movements depend on factors such as the organization of muscles within the body, whether the worms are segmented or unsegmented, and the presence or absence of bristles or other structures on the body surface.

As with spiral cleavage and the trochophore larva, however, the coelom was reduced or lost in several lineages. For example, the flatworms—which are mostly small and thin or parasitic—lack a coelom entirely.

(a) Mouthparts for suspension feeding

Feathery mouthparts capture plankton

(b) Mouthparts for deposit feeding

Long, sticky tentacles reach across substrate and deliver organic particles to mouth (worm body remains hidden)

(c) Mouthparts for mass feeding

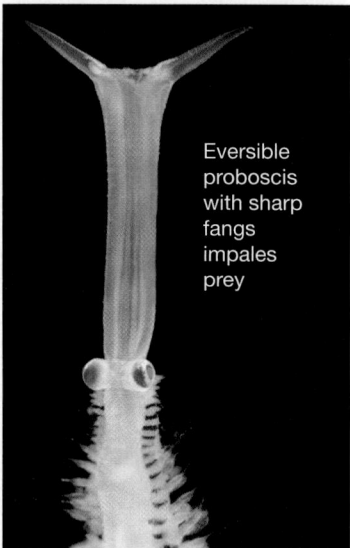

Eversible proboscis with sharp fangs impales prey

FIGURE 34.6 Many Wormlike Lophotrochozoans Have Specialized Mouthparts. These three examples are polychaete worms, phylum Annelida.

✔ **EXERCISE** It is common to find diverse worms in the same habitat. Suggest a hypothesis to explain why this is possible.

Wormlike lophotrochozoans include suspension, deposit, liquid, and mass feeders (see Chapter 33). Exploiting a diversity of foods is possible in part because worms have such a wide variety of mouthparts for capturing and processing food.

The lophophore feeding structure that is used by lophophorates, such as phoronid worms, is an example of one type of specialized mouthpart (see Figure 34.4a). Let's consider several other examples from the phylum Annelida, the segmented worms.

Segmented worms are a diverse group, especially the marine polychaetes (**FIGURE 34.6**). For example, they may live in tubes and extend their feathery tentacles into the water to suspension feed, burrow in the sediments and reach out with their spaghetti-like tentacles to deposit feed, or spear prey with sharp fangs on the tip of a tubular, extensible feeding appendage called a **proboscis.**

While the diversity of wormlike lophotrochozoans is impressive, you can see from the pie wedges in Figure 34.1 that they do not quite compare to the mollusks in terms of diversity.

What Is a Mollusk?

Mollusks exhibit the lophotrochozoan mode of growth and development. But unlike the lophotrochozoans as a whole, the mollusks can be identified by distinctive characteristics. Mollusks have a specialized body plan based on three major components (**FIGURE 34.7**):

1. the **foot,** a large muscle located at the base of the animal and usually used in movement;

2. the **visceral mass,** the region containing most of the main internal organs and the external gill; and

3. the **mantle,** an outgrowth of the body wall that covers the visceral mass, forming an enclosure called the mantle cavity. In many species, the mantle secretes a hard calcium carbonate shell.

✔You should be able to indicate the origin of the molluscan body plan on Figure 34.2.

Once the molluscan body plan evolved, subsequent diversification was largely driven by adaptations that allowed the mollusks to move, feed, or reproduce in novel ways. The result was a dramatic radiation of mollusks into several lineages, including **bivalves** (clams and mussels), **gastropods** (slugs and snails), **chitons** (mollusks with dorsal shells made of plates), and **cephalopods** (squid and octopuses).

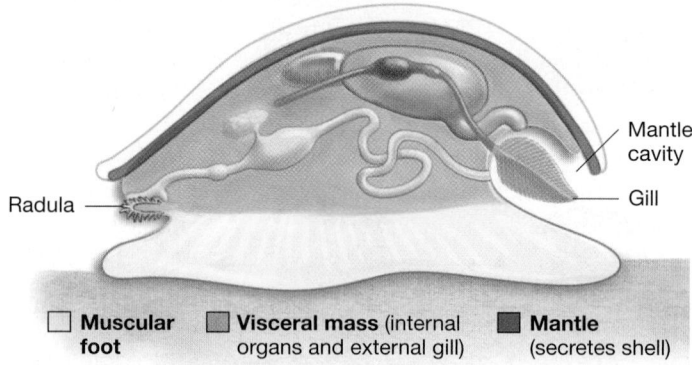

Radula

Mantle cavity

Gill

☐ **Muscular foot**   ☐ **Visceral mass** (internal organs and external gill)   ■ **Mantle** (secretes shell)

FIGURE 34.7 Mollusks Have a Specialized Body Plan. This diagram of a generalized mollusk shows that the mollusk body plan is based on a foot, a visceral mass, and a mantle. Gills are located inside a cavity created by the mantle.

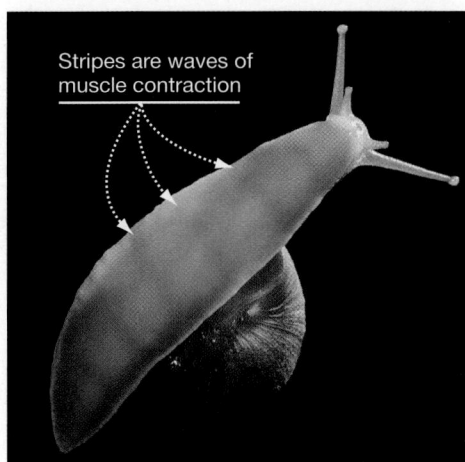

FIGURE 34.8 **How Snails Walk on One Foot.** Waves of muscle contractions, moving forward or backward along the length of the foot, allow mollusks like this snail to creep along a substrate.

Rasp-like **radula** scrapes off pieces of food

FIGURE 34.9 **The Radula Is Unique to Mollusks.** The size and shape of the radular teeth is highly correlated with the type of food eaten by the species.

The Foot Is a Muscular Hydrostat Snails and chitons have a large, muscular foot at the base of the body that—like your tongue—works as a type of hydrostatic skeleton called a muscular hydrostat. How can an animal move with just one foot? Waves of muscle contractions sweep backward or forward along the length of the foot, allowing individuals to crawl along a surface (**FIGURE 34.8**). Biologists in the field of biomechanics have measured the mechanical properties of snail mucus. They observed that it has the remarkable ability to increase traction for the parts of the foot that are not moving, while lubricating motion for the parts that are advancing.

In bivalves such as clams, the foot is modified as a digging appendage. The foot of cephalopods like squid and octopus is modified to form tentacles for crawling and grasping.

The Visceral Mass Separates Internal Organs from the Hydrostatic Skeleton The visceral mass is the space in all mollusks where organs and surrounding fluids are located. This might not seem like much of an innovation, but compare this body plan to that of worms, where the hydrostatic skeleton and internal organs are inseparable. The separation of the foot and visceral mass in mollusks may have enabled diversification of both.

The coelom itself is highly reduced in most mollusks, functioning mostly in reproduction and excretion of wastes. The organs occupy a different type of body cavity called a **hemocoel** ("blood-hollow"; pronounced *HEE-mah-seal*), where body fluids bathe the organs directly in an open circulatory system (see Chapter 45). A hemocoel is different from a coelom because it is not lined in mesoderm.

At the anterior end of the visceral mass, the mouth has a unique molluscan feeding structure called a **radula**, which functions like a rasp or file. The mollusk moves the radula back and forth over the food source, causing the many sharp plates to scrape material so that it can be ingested (**FIGURE 34.9**). Since it is highly unlikely that such a unique structure would evolve more than once, the radula probably evolved early in molluscan

evolution. It was later lost in the bivalves, which acquire food by suspension feeding.

The Mantle Secretes a Shell In many species the mantle secretes a shell made of calcium carbonate. Some mollusk species have a shell consisting of one, two, or eight parts, or valves; others have no shell at all. The ability to secrete a protective shell may have been an important adaptation during the Cambrian and beyond as the number of predators increased. Note, however, that shells made of calcium carbonate are heavy. Thus, there is a trade-off between protection and mobility. As you might predict, mollusks with thick shells are constrained to aquatic habitats where buoyant forces help support the load. The largest and most agile of all aquatic mollusks, cephalopods such as squid and octopus, have highly reduced shells or none at all. Terrestrial mollusks have thin shells or none at all.

The mantle was also modified in many ways to serve diverse functions other than secreting shells. For example, cephalopods have a mantle lined with muscle. When the cavity surrounded by the mantle fills with water and the mantle muscles contract, a stream of water is forced out of a tube called a **siphon.** The force of the water propels the squid in a form of locomotion called jet propulsion (**FIGURE 34.10**, see page 664). The mantle of bivalves such as clams is modified to form two siphons to control incurrent and excurrent water flow over their gills.

Modular Mollusks How did the novel body plan and lifestyle of mollusks such as squid evolve? Mark Martindale and his colleagues tried to address this question by looking at the role of *Hox* genes in the organization of the squid body plan. They found that:

- Squid have nine *Hox* genes that are homologous to those found in other bilaterians (recall that *Hox* genes are important in establishing the anterior–posterior axis of animals).

- Although *Hox* genes are usually organized in a collinear fashion (see Chapter 22), they were co-opted in a new pattern in

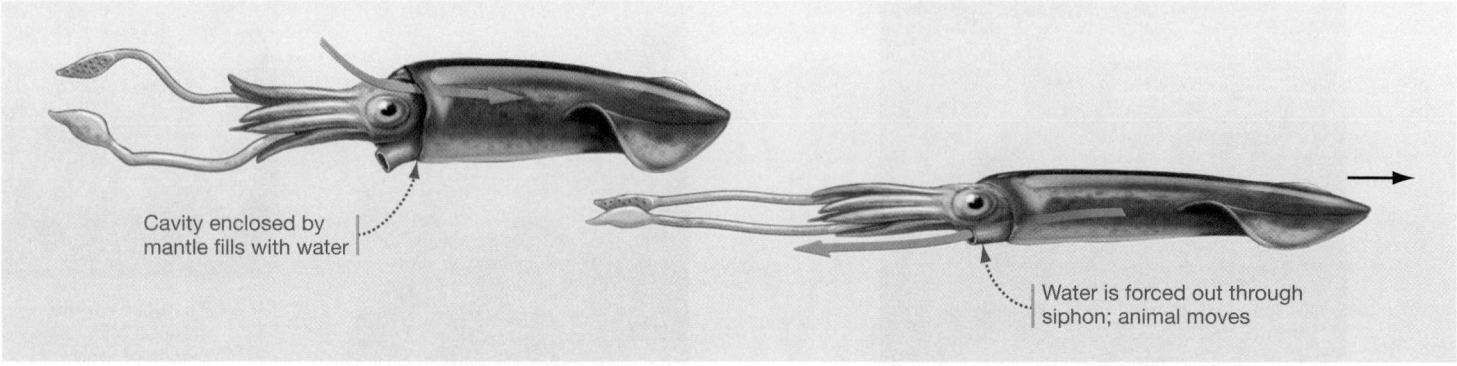

FIGURE 34.10 Jet Propulsion in Mollusks. In jet propulsion, muscular contractions of the mantle force the water out through a movable siphon.

Cavity enclosed by mantle fills with water

Water is forced out through siphon; animal moves

✔ **EXERCISE** Circle the part of the squid that is homologous to the snail foot.

squid, suggesting a mechanism by which a change in the regulation of tool-kit genes could result in novel body structures, such as the modification of the molluscan foot into a crown of muscular tentacles.

The take-home message is that mollusks have a modular design. That is, while all mollusks have a foot, visceral mass, and mantle, the sizes, shapes, and functions of these structures have diversified in ways that enable drastically different methods of feeding, moving, and reproducing.

Let's take a closer look at some of the diversity within the main lophotrochozoan phyla.

check your understanding

If you understand that . . .

• Flatworms, annelids, and mollusks are all diverse phyla within the Lophotrochozoa, and all three phyla made the water-to-land transition.
• The mollusk body plan includes a muscular foot, a visceral mass, and a mantle.

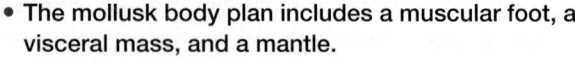 **You should be able to . . .**

1. Predict three ways you would expect a snail that lives on land to differ from a snail that lives in the ocean.
2. Design an experiment to test whether the tentacles of an octopus are homologous to the foot of a snail.

Answers are available in Appendix A.

Key Lineages: Lophotrochozoans

The phyla within Lophotrochozoa are united by shared mechanisms of early development and growth, but they are highly diverse in terms of their resulting morphology. You might not even recognize some of them as having a bilaterally symmetrical body plan.

Let's take a closer look at five key lophotrochozoan phyla: **(1)** Bryozoa, **(2)** Rotifera, **(3)** Platyhelminthes, **(4)** Annelida, and **(5)** Mollusca. Because the Mollusca is so species rich and diverse, the detailed descriptions in this section consider the four most familiar lineages of mollusks separately:

● Lophotrochozoans > Bryozoa (Bryozoans)

● Lophotrochozoans > Rotifera (Rotifers)

● Lophotrochozoans > Platyhelminthes (Flatworms)

● Lophotrochozoans > Annelida (Segmented Worms)

● Lophotrochozoans > Mollusca > Bivalvia (Clams, Mussels, Scallops, Oysters)

● Lophotrochozoans > Mollusca > Gastropoda (Snails, Slugs, Nudibranchs)

● Lophotrochozoans > Mollusca > Polyplacophora (Chitons)

● Lophotrochozoans > Mollusca > Cephalopoda (Nautilus, Cuttlefish, Squid, Octopuses)

Bryozoans, sometimes also called ectoprocts or moss animals, are tiny lophophorate animals that live in colonies that look more like a sponge or coral than a bilaterian (**FIGURE 34.11**). The oldest bryozoan fossils date back to the late Cambrian, giving this group the distinction of being the most recent animal phylum to appear in the fossil record. Today, bryozoans live mostly in marine environments, especially in the tropics, but also in some freshwater environments. So far, 5700 species have been described. The colonies range in size, from millimeters to meters, and in shape, from fans to encrusting sheets.

Feeding Bryozoan individuals, or zooids, suspension feed on plankton using a lophophore (see Figure 34.4).

Movement Most bryozoans are sessile. Only the (non-trochophore) larvae are capable of dispersing to new habitats.

Reproduction Bryozoans can increase the size of their colony by asexual reproduction, by budding. They can also reproduce sexually—zooids are hermaphrodites, some having both male and female reproductive organs at the same time, other serving first as males then as females.

Triphyllozoan species

1 cm

FIGURE 34.11 A Bryozoan Colony. You can see what an individual animal looks like in Figure 34.4.

The 2100 **rotifer** species that have been identified thus far live in damp soils as well as marine and freshwater environments. They are important components of the plankton in freshwater and in brackish (slightly salty) waters. Rotifers have a coelom, and most are less than 1 millimeter (mm) long. Although rotifers do not have a lophophore or a trochophore larval stage, their mode of growth and DNA sequences identify them as a member of the lophotrochozoan lineage.

Feeding Rotifers have a cluster of cilia called a **corona** at their anterior end (**FIGURE 34.12**). In many species, the beating of the cilia in the corona makes suspension feeding possible by creating a current that sweeps microscopic food particles into the mouth. The corona is a synapomorphy of this group. ✔ You should be able to indicate the origin of the corona on Figure 34.2.

Movement Although a few species of rotifers are sessile, most swim via the beating of cilia in the corona.

Reproduction Most reproduction is sexual, but females of some rotifer species can reproduce asexually by a process called **parthenogenesis**—the female produces unfertilized eggs by mitosis; the eggs then hatch into new, asexually produced individuals. In a group of rotifers called the bdelloids, only females have been found and all reproduction appears to be by parthenogenesis. Whether reproduction is sexual or asexual, rotifer development is direct, meaning that fertilized eggs hatch and grow into adults without going through metamorphosis.

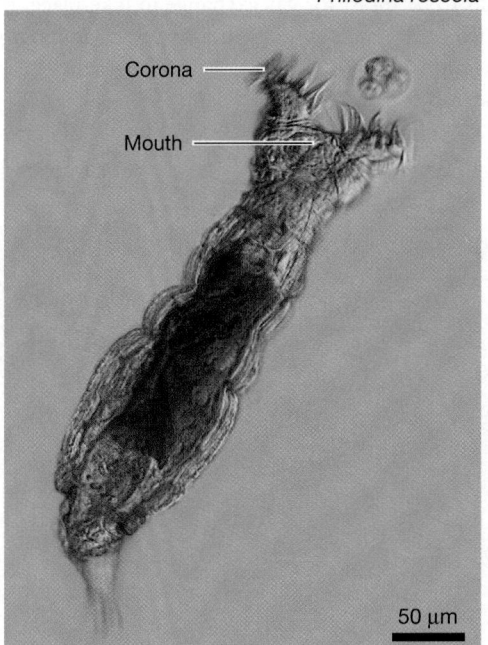

Philodina roseola

Corona

Mouth

50 µm

FIGURE 34.12 Rotifers Are Tiny, Aquatic Suspension Feeders.

The flatworms are a large and diverse phylum. More than 20,000 species have been described in four major lineages:

- *Turbellaria* are free-living flatworms. Although a few turbellarian species are terrestrial, most live on the substrates of freshwater or marine environments. Some marine flatworms are brightly colored and several centimeters long (**FIGURE 34.13a**). However, the freshwater planarians common to biology labs, such as *Dugesia*, are drab and only a few millimeters long.

- *Cestoda* are endoparasitic tapeworms that parasitize fish, mammals, or other vertebrates (**FIGURE 34.13b**).

- *Trematoda* are endoparasitic or ectoparasitic flukes that parasitize vertebrates, arthropods, annelids, and mollusks (**FIGURE 34.13c**). In humans, a fluke is responsible for schistosomiasis—a serious public health issue in many tropical and subtropical nations.

- *Monogenea* are ectoparasites that parasitize fish.

Flatworms are named for the broad, flattened shape of their bodies. (The Greek roots *platy* and *helminth* mean "flat-worm.") Species in the Platyhelminthes are unsegmented and lack a coelom. They also lack structures that are specialized for gas exchange. Further, they do not have blood vessels for circulating oxygen and nutrients to their cells. Based on these observations, biologists interpret the flattened bodies of these animals as an adaptation that gives flatworms an extremely high surface-area-to-volume ratio. This flat body plan allows nutrients and gases to diffuse efficiently to all of the cells inside the animal. The downside is that the body surface has to be moist for gas exchange to take place. Flatworms are restricted to aquatic and moist terrestrial environments. ✔ You should be able to indicate the origin of the flattened, acoelomate body plan on Figure 34.2.

Feeding The Platyhelminthes lack a lophophore and have a digestive tract that is "blind"—meaning it has only one opening for ingestion of food and elimination of wastes.

- Most turbellarians are hunters that prey on protists or small animals; others scavenge dead animals.

- Tapeworms are strictly parasitic and feed on nutrients provided by hosts. They do not have a mouth or a digestive tract, and they obtain nutrients by diffusion across their body wall.

- Flukes are parasites and feed by gulping host tissues and fluids through a mouth.

Movement In general, flatworms do not move much. Some turbellarians can swim by undulating their bodies, and most can creep along substrates with the aid of cilia on their ventral surface. Adult cestodes have hooked attachment structures at their anterior end that permanently attach them to the interior of their host.

Reproduction Turbellarians can reproduce asexually by splitting themselves in half. If they are fragmented as a result of a predator attack, the body parts can regenerate into new individuals. Most turbellarians contain both male and female organs and reproduce sexually by aligning with another individual and engaging in mutual and simultaneous fertilization.

In flukes and tapeworms, individuals reproduce sexually and either cross-fertilize or self-fertilize. The reproductive systems and life cycles of flukes and tapeworms are extremely complex, however. In many cases the life cycle involves two or even three distinct host species; sexual reproduction occurs in the **definitive host,** and asexual reproduction occurs in one or more **intermediate hosts.** For example, humans are the definitive host of the blood fluke *Schistosoma mansoni*, and snails are the intermediate host.

(a) Turbellarians are free living.

Pseudoceros ferrugineus

1 cm

(b) Cestodes are endoparasitic.

Taenia species

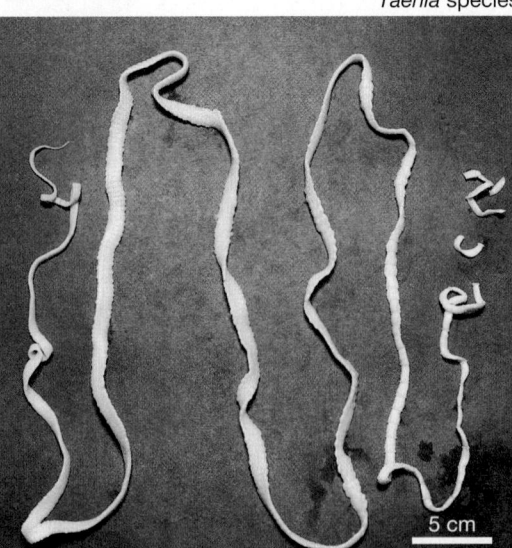

5 cm

(c) Trematodes are endoparasitic.

Dicrocoelium dendriticum

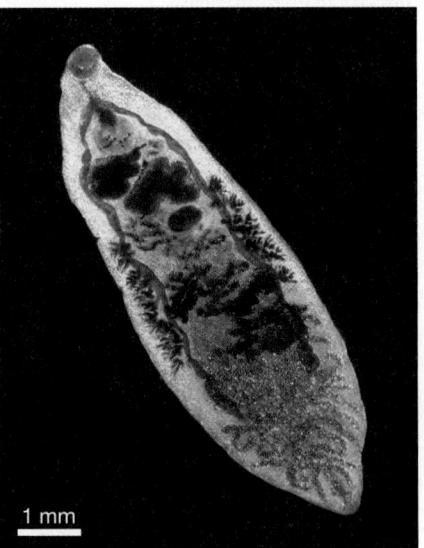

1 mm

FIGURE 34.13 Flatworms Have Simple, Flattened Bodies.

Most **annelids** have a segmented body and a coelom. ✔You should be able to indicate the origin of annelid segmentation on Figure 34.2. The 16,800 described species were traditionally divided into groups called Polychaeta (*pol-ee-KEE-ta*) and Clitellata. Recent analyses have supported Clitellata as a monophyletic group, composed of Hirudinea (leeches), and Oligochaeta (*oh-LIG-oh-keet-a;* earthworms and other oligochaetes). However, the Polychaeta are paraphyletic—that is, including some but not all members of a clade (**FIGURE 34.14**). Further, unsegmented worms called Sipunculida and Echiura, which were traditionally thought to be independent phyla, are actually groups within Annelida—they lost their segmentation.

The common ancestor of the annelids had a key synapomorphy in addition to segmentation: numerous, bristle-like extensions called **chaetae** that extend from appendages called **parapodia.** As the name suggests, polychaetes ("many-bristles"; **FIGURE 34.15a**) have more chaetae than oligochaetes ("few-bristles"; **FIGURE 34.15b**). However, highly mobile polychaetes have larger parapodia and more chaetae than more sedentary polychaetes, which have reduced parapodia and chaetae adapted for burrowing in sediments or living in tubes.

Feeding The polychaete lineages have a wide variety of methods for feeding, including suspension feeding, deposit feeding, and mass feeding (Figure 34.6). Virtually all oligochaetes, in contrast, make their living by deposit feeding in soils. The tunnels that they make are critically important in aerating soil, and their feces contribute large amounts of organic matter. About half of the leeches are ectoparasites that attach themselves to fish or other hosts and suck blood and other body fluids (**FIGURE 34.15c**). The nonparasitic leech species are predators or scavengers.

Movement Polychaetes and oligochaetes crawl or burrow with the aid of their hydrostatic skeletons; the parapodia of polychaetes also act as paddles or tiny feet that aid in movement. Many polychaetes and leeches are good swimmers.

Reproduction Asexual reproduction occurs in some polychaetes and oligochaetes by transverse fission or fragmentation—meaning

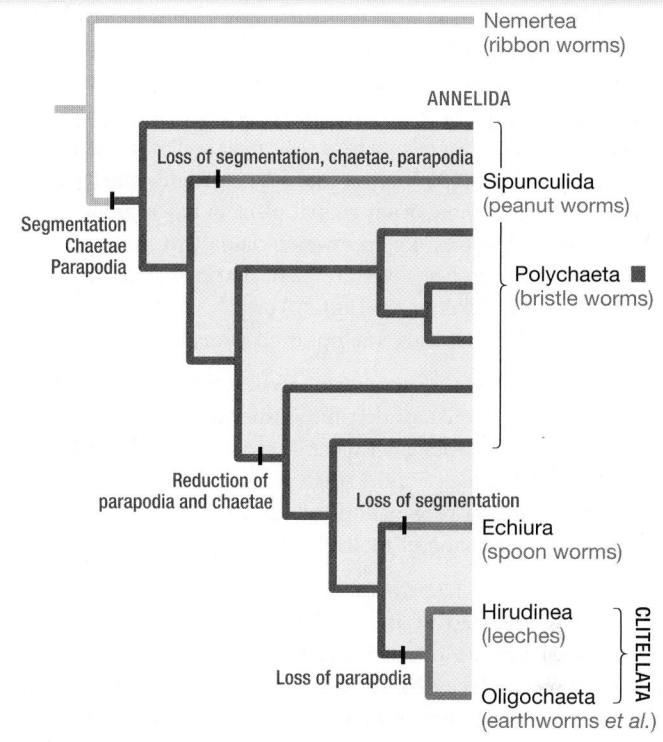

FIGURE 34.14 Phylogeny of Major Groups of Annelids.

that body parts can regenerate a complete individual. Sexual reproduction in polychaetes may begin with internal or external fertilization, depending on the species. Most of the polychaete lineages have separate sexes and usually release their eggs directly into the water. Some species produce eggs that hatch into trochophore larvae, then go on to develop into other larval forms. In oligochaetes and leeches, individuals produce both sperm and eggs and engage in mutual, internal cross-fertilization. Eggs are enclosed in a mucus-rich, cocoon-like structure; after fertilization, offspring develop directly into miniature versions of their parents.

(a) Polychaetes have many bristle-like chaetae.

Hermodice carunculata

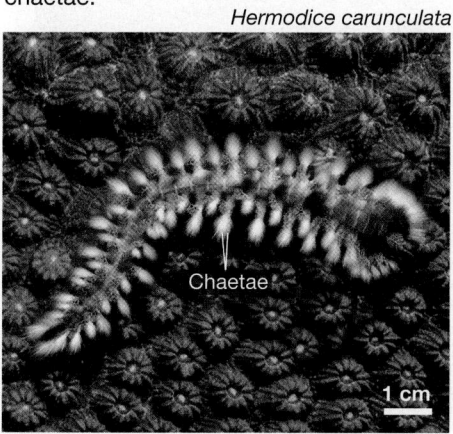

Chaetae

1 cm

(b) Oligochaetes have few or no chaetae.

Chaetogaster species

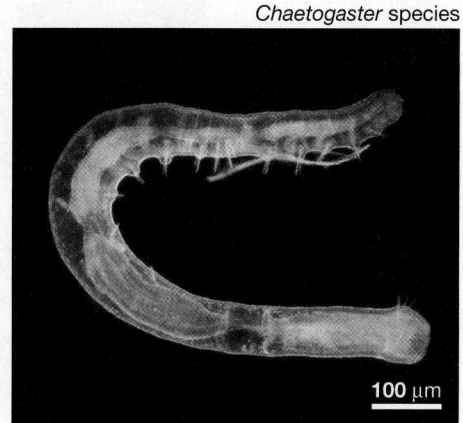

100 μm

(c) Some **leeches** are ectoparasites.

Haemadipsa species

1 cm

FIGURE 34.15 Annelids Are Segmented Worms.

The bivalves are so named because they have a shell with two separate parts, or valves, made of calcium carbonate. The protective shell is hinged, and closes with the aid of muscles attached to the two valves (**FIGURE 34.16**). Because most bivalves live on or under the ocean floor and because they are covered by a hard shell, their bodies are often buried in sediment after death. Bivalves thus fossilize readily and have the most extensive fossil record of any animal, plant, or fungal group.

The smallest bivalves are freshwater clams that are less than 2 mm long; the largest bivalves are giant marine clams that may weigh more than 200 kilograms (kg). All bivalves can sense gravity, touch, and certain chemicals; scallops even have eyes.

Feeding Most bivalves are suspension feeders that take in any type of small animal or protist, or detritus. Suspension feeding is based on a flow of water across gas-exchange structures called **gills** located within the mantle cavity. A single oyster can filter the organic matter from as much as 60 gallons of seawater each day. Bivalves are the only major group of mollusks that lack the feeding structure called a radula.

Movement Clams burrow into sediments with the aid of their muscular foot but are otherwise sedentary. Scallops are able to swim along the surface of soft substrates by clapping their shells together and forcing water out, a form of jet propulsion. Mussels and oysters are sessile, attached to rocks or other solid surfaces.

Chlamys pallium

1 cm

FIGURE 34.16 Bivalves Have Shells with Two Parts.

Reproduction Most bivalves reproduce sexually. Eggs and sperm are shed into the water, and fertilized eggs develop into trochophore larvae. Trochophores then metamorphose into a distinct type of larva called a **veliger,** which continues to feed and swim before settling to the substrate and metamorphosing into an adult form that secretes a shell.

The gastropods ("belly-feet") are named for the large, muscular foot on their ventral side. Marine and terrestrial snails have shells (**FIGURE 34.17**), into which they can retract when they are attacked or when their tissues begin to dry out. Land slugs and nudibranchs (pronounced *NEW-da-branks*; also called sea slugs) lost their shells independently; these gastropods often contain toxins or foul-tasting chemicals to protect them from being eaten. About 70,000 species of gastropods are known, ranging in size from microscopic to almost a meter long. They are the only group of mollusks that made the water-to-land transition.

Feeding The gastropod radula is diverse, varying in size and shape among species (see Figure 34.9). Although most gastropods are herbivores or detritivores, specialized types of teeth allow some gastropods to act as predators. Species called drills, for example, use their radula to bore a hole in the shells of other mollusks.

Movement Waves of contractions along the length of the foot allow gastropods to move by creeping (see Figure 34.8). Sea butterflies are gastropods with a reduced or absent shell but a large, winglike foot that flaps and powers swimming movements.

Reproduction Most reproduction is sexual, but females of some gastropod species can reproduce asexually by producing eggs parthenogenetically. Sexual reproduction in some gastropods begins with internal fertilization. Most marine gastropods produce a veliger larva that may disperse up to several hundred kilometers from the parent. But in some marine species and all terrestrial forms, larvae are not free living. Instead, offspring remain in an egg case while passing through several larval stages and then hatch as miniature versions of the adults.

Maxacteon flammea

5 mm

FIGURE 34.17 Gastropods Have a One-Part Shell or Lack Shells.

Lophotrochozoans > Mollusca > Polyplacophora (Chitons)

The name *Polyplacophora* ("many-plate-bearing") is apt because chitons (pronounced *KITE-uns*) have eight calcium carbonate plates, or valves, along their dorsal surface (**FIGURE 34.18**). The plates form a protective shell. The approximately 1000 species of chitons are exclusively marine, and are found worldwide; the largest reaches 33 centimeters (cm). They are usually found on rocky surfaces in the intertidal zone, where rocks are periodically exposed to air at low tides.

Feeding Chitons have a radula that they use to scrape algae and other organic matter off rocks.

Movement Chitons move by creeping on their broad, muscular foot—as gastropods do.

Reproduction In most chiton species, the sexes are separate and fertilization is external. In some species, however, sperm that are shed into the water enter the female's mantle cavity and fertilize eggs there. Depending on the species, eggs may be enclosed in a membrane and released, or they may be retained until hatching and early development are complete. Most species have trochophore larvae.

Tonicella lineata

1 cm

FIGURE 34.18 Chitons Have Eight Shell Plates.

Lophotrochozoans > Mollusca > Cephalopoda (Nautilus, Cuttlefish, Squid, Octopuses)

The cephalopods ("head-feet") have a well-developed "head" that actually consists of the visceral mass, and a foot that is modified to form **tentacles:** long, thin, muscular extensions that aid in movement and prey capture (**FIGURE 34.19**). Except for the nautilus with its chambered shell, the cephalopod species living today have either highly reduced shells or none at all. Most have large brains and image-forming eyes with sophisticated lenses. They also include the largest of all invertebrates living today, the giant squid, which can be more than 10 m long, including the tentacles. The skin of cephalopods contains sophisticated chromatophore ("color-bearing") cells that enable the rapid color changes used for camouflage and communication.

Feeding Cephalopods are highly intelligent predators that hunt by sight and use their tentacles to capture prey—usually fish or crustaceans. They have a radula as well as a structure called a **beak,** which can exert powerful biting forces. Some cuttlefish and octopuses also inject poisons into their prey to subdue them.

Movement Cephalopods can swim by moving their fins to "fly" through the water, or by jet propulsion using the mantle cavity and siphon (see Figure 34.10). Squid are built for speed and hunt small fish by chasing them down. Octopuses, in contrast, crawl along the substrate using their tentacles. They chase down crabs or other crustaceans, or they pry mussels or clams from the substrate and then use their beaks to crush the exoskeletons of their prey. Most cephalopods are able to escape their own predators by releasing a jet of "ink" that disorients their predators while they slip away.

Reproduction Cephalopods have separate sexes, and some species have elaborate courtship rituals that involve color changes and interaction of tentacles. When a male is accepted by a female, he deposits sperm that are encased in a structure called a **spermatophore.** The spermatophore is transferred to the female, and fertilization is internal. Females lay eggs. When they hatch, juveniles develop directly into adults—that is, there is no larval stage.

Octopus bimaculatus

10 cm

FIGURE 34.19 Cephalopods Have Highly Modified Bodies.

34.3 What Is an Ecdysozoan?

Lophotrochozoans grow continuously and incrementally (**FIGURE 34.20a**)—the ancestral manner of growth for animals. In contrast, the Ecdysozoa are defined by a clear synapomorphy: All ecdysozoans grow intermittently by **molting**—that is, by shedding an exoskeleton or external covering.

The Greek root *ecdysis*, which means "to slip out or escape," is appropriate because during a molt, an individual sheds its outer layer, or **cuticle**—called an **exoskeleton** if it is hard—and slips out of it (**FIGURE 34.20b**). Once the old covering is gone, the body is pumped up with fluid to cause it to expand, and a larger cuticle or exoskeleton forms. As ecdysozoans grow and mature, they undergo a succession of molts, sometimes undergoing dramatic morphological transformations along the way.

A stiff body covering is advantageous because it provides an effective structure for muscle attachment and affords protection. However, the trade-off is that during molting, the soft bodies of ecdysozoans are exposed and vulnerable to predators. When a crab or other crustacean has shed an old exoskeleton, its new exoskeleton takes several hours to harden. During this interval, molting individuals hide and do not feed or move about. Experiments have shown that it is much easier for predators to attack and subdue individuals that are not protected by an exoskeleton during the intermolt period.

A hormone called ecdysone is important in regulating the molting cycle (Chapter 49). Is ecdysone a key innovation for the ecdysozoans? No—studies show that this steroid hormone serves a variety of functions in many other animals. Thus, it appears that ecdysone was co-opted in the ecdysozoans to serve a novel function.

The most prominent of the seven ecdysozoan phyla are the roundworms (Nematoda) and arthropods (Arthropoda). Here, let's consider the implications of having an exoskeleton for the most diverse phylum of animals on Earth, the arthropods.

What Is an Arthropod?

In terms of duration in the fossil record, species diversity, and abundance of individuals, arthropods are the most important phylum within the Ecdysozoa. Arthropods appear in the fossil record over 520 million years ago and have long been the most abundant animals observed in both marine and terrestrial environments. Well over a million living species have been described, and biologists estimate that millions or perhaps even tens of millions of arthropod species have yet to be discovered. In terms of species diversity, arthropods are easily the most successful lineage of eukaryotes.

The Arthropod Body Plan In addition to their bilateral symmetry and other protostome characteristics, **arthropods** (literally "joint-foot") are defined by three key features:

1. a *segmented body*, which is organized into prominent regions, or **tagmata,** such as the head, thorax, and abdomen in the grasshopper (**FIGURE 34.21**);

2. an *exoskeleton* made primarily of the polysaccharide chitin (see Chapter 5), strengthened by calcium carbonate ($CaCO_3$) in crustaceans; and

3. *jointed appendages,* which enable the rigid bodies to move.

Instead of being based on muscle contraction within a hydrostatic skeleton, most arthropod locomotion is based on muscles

(a) Lophotrochozoans grow incrementally.

(b) Ecdysozoans grow by molting.

Growth bands

FIGURE 34.20 Lophotrochozoans and Ecdysozoans Differ in Their Mechanism of Growth. (a) Lophotrochozoans do not molt. The growth bands on this clam show periods of slow and rapid incremental growth. **(b)** Ecdysozoans such as this cicada must crawl out of their old exoskeletons and grow new, larger ones.

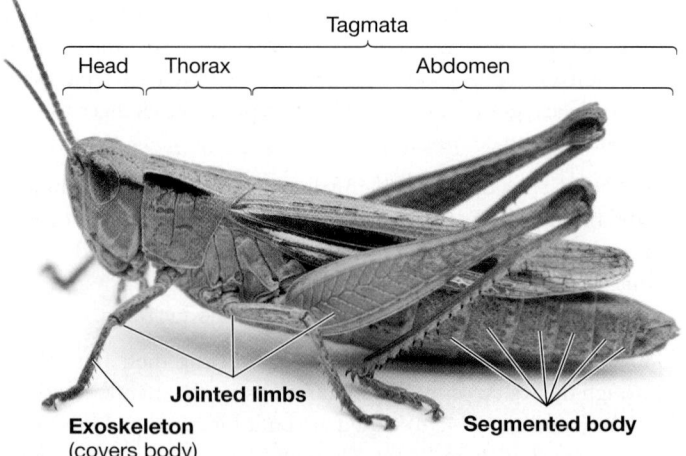

FIGURE 34.21 Arthropods Have a Specialized Body Plan. Arthropods have segmented bodies and jointed limbs, which enable these animals to move despite their hard outer covering, the exoskeleton. Different arthropods have a different number of tagmata, or body regions; insects have three.

that apply force against rigid levers—the exoskeleton—to move legs or wings at the joints. This skeletal architecture enables very rapid and precise movements (Chapter 48).

Arthropods have a highly reduced coelom but, like mollusks, possess an extensive body cavity called a hemocoel. The hemocoel enables movements via a hydrostatic skeleton in caterpillars and other types of arthropod larvae.

The jointed appendages of arthropods are organized in pairs along the segmented body. They have an array of functions: exchanging gases, sensing the environment, feeding, and locomotion via swimming, walking, running, jumping, or flying (Chapter 33). Arthropod appendages provide the ability to sense stimuli and make sophisticated movements in response. ✔ **You should be able to indicate the origin of jointed appendages on Figure 34.2.**

Why Has the Arthropod Body Plan Been So Successful? You have already learned that mollusk bodies are modular in evolutionary terms. The modularity of arthropod segments, however, takes the prize for evolutionary potential.

Studies on the role of tool-kit genes—such as the *Hox* genes—during development of different arthropods have contributed enormously to our understanding of the process of arthropod diversification. These studies show that small changes in the timing and location of expression of certain tool-kit genes along the length of an arthropod can result in novel shapes and sizes of body segments and appendages.

To understand the evolutionary potential of modularity, consider a hypothetical ancestral arthropod with many uniform body segments, each with a pair of uniform appendages. The redundancy of this body plan can provide a fertile starting place for diversification because some of the segments can continue to function the old way, say for crawling, while other segments can become co-opted for new functions. For example, the appendages at the head end could become specialized as antennae or claws without compromising the ability of the animal to crawl.

Thus, genetic opportunity in the form of variation in the expression of tool-kit genes can combine with ecological opportunity by the process of natural selection to result in the diversification of arthropod body segments and appendages. **FIGURE 34.22** shows an example of how diverse, functionally specialized appendages can result in an animal with a "serialized" body plan, a reflection of its modular structure. The crayfish limbs are said to have serial homology.

Origin of the Wing The process of co-opting old parts for new functions is key to the origin of one of the most remarkable adaptations in the history of life—the insect wing. There are more species of animals on the planet today with insect wings than without. According to data in the fossil record, insects were also the first animals on Earth that had wings and could fly. How did wings evolve?

Some biologists have proposed that wings arose as outgrowths from the thorax, independent from the legs. Others have argued that wings arose from gill-like projections on the branched legs of a wingless ancestor, essentially co-opting the gill-like structures

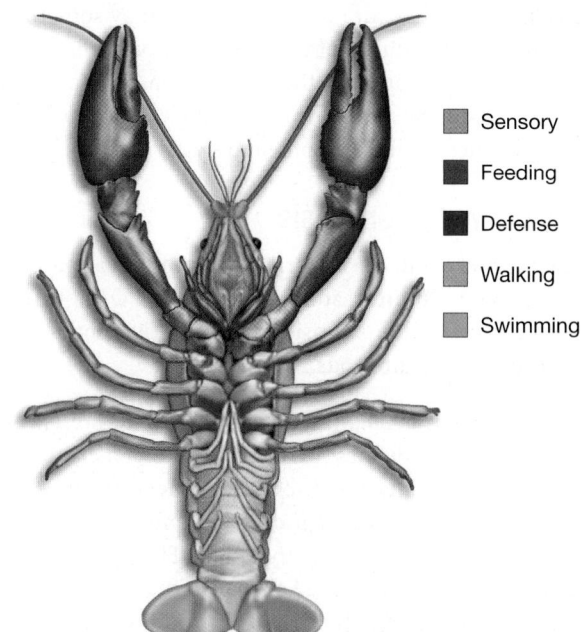

Sensory
Feeding
Defense
Walking
Swimming

FIGURE 34.22 Arthropods Have a Modular Body Plan. This ventral view of a crayfish illustrates how arthropod segments and appendages can be specialized for different functions.

(**FIGURE 34.23**). Insects have unbranched appendages, but arthropods such as crustaceans have branched appendages.

Michalis Averof and Stephen Cohen attempted to settle the wing debate using developmental genetics. They probed the developing limbs of brine shrimp (a member of an ancient lineage of crustaceans) for the presence of tool-kit genes *nubbin* and *apterous*. These genes are known to be important in wing development in insects. The results were striking. *Nubbin* and *apterous* are expressed in the gill-like lobe of the crustacean limb, but nowhere else. These results provide strong support for the origin of insect wings from the gill-like branches of ancestral crustacean limbs.

(a) Independent origin hypothesis

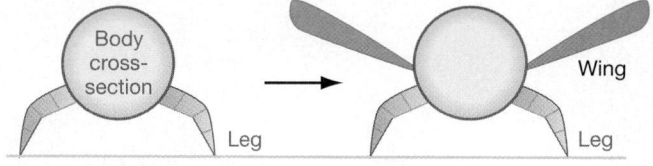

(b) Gill co-option hypothesis

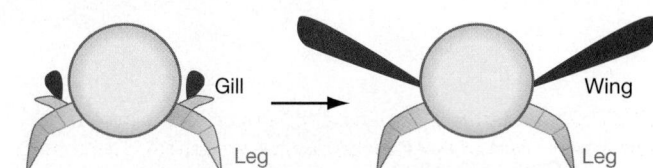

FIGURE 34.23 Alternative Hypotheses for the Origin of Wings in Insects.

Co-opting one structure to form another may sometimes require thousands or millions of years of evolution within a lineage. It is all the more remarkable, then, that dramatic shape changes can also occur during the lifetime of a single individual. Natural selection operates at all developmental stages in a life cycle.

Arthropod Metamorphosis

Many arthropods undergo a morphological transformation from a larva to a very different adult form. Others do not. Let's use insects as a case study to consider the adaptive implications of metamorphosis in arthropods.

Two Types of Insect Metamorphosis   In insects, the presence or absence of a larval stage defines two distinct types of metamorphosis.

In **hemimetabolous** ("half-change") **metamorphosis**—also called **incomplete metamorphosis**—juveniles called nymphs look like miniature versions of the adult. The aphid nymphs in **FIGURE 34.24a**, for example, shed their external skeletons several times and grow—gradually changing from wingless, sexually immature nymphs to sexually mature adults, some of which can fly. But throughout their life, aphids live in the same habitats and feed on the same food source in the same way: They suck sap.

In **holometabolous** ("whole change") **metamorphosis**—also called **complete metamorphosis**—there is a distinct larval stage. As an example, consider the life cycle of the mosquito, illustrated in **FIGURE 34.24b**:

1. Newly hatched mosquitoes live in quiet bodies of freshwater, where they suspension feed on bacteria, algae, and detritus.

2. When a larva has grown sufficiently, the individual stops feeding and moving and secretes a protective case. The individual is now known as a **pupa** (plural: pupae). During pupation, the pupa's body is completely remodeled into a new, adult form.

3. The adult mosquito flies and feeds—females take blood meals from mammals and males drink nectar from flowers.

What Is the Adaptive Significance of Metamorphosis?   In insects, holometabolous metamorphosis is 10 times more common than hemimetabolous metamorphosis. How can this be explained?

Since most adult insects are mobile, using a larval stage for dispersal does not seem to be a sufficient explanation. The leading hypothesis is based on efficiency in feeding. Because juveniles and adults from holometabolous species feed on different materials in different ways and sometimes even in different habitats, they do not compete with each other.

An alternative hypothesis is based on the advantages of functional specialization. In many moths and butterflies, larvae are specialized for feeding, whereas adults are specialized for mating and feed rarely, if ever. Larvae are largely sessile, whereas adults are highly mobile. If specialization leads to higher efficiency in feeding and reproduction and thus higher fitness, then holometabolous metamorphosis would be advantageous. These hypotheses are not mutually exclusive, however—meaning that both could be correct, depending on the species being considered.

Let's explore the diversity of feeding, moving, and reproducing habits of ecdysozoans in more depth by examining the key lineages separately.

Key Lineages: Ecdysozoans

Seven phyla are currently recognized in the Ecdysozoa (see Table 33.1). Second in diversity to Arthropoda, the phylum Nematoda, or roundworms, is especially noteworthy due to its large number of described species and the mind-boggling abundance of individuals. Of the smaller phyla, Onychophora (*on-ee-KOFF-er-uh*) and Tardigrada are of special interest due to their close relationship to arthropods. They are similar to arthropods

(a) Aphid:
Hemimetabolous metamorphosis
(Incomplete metamorphosis)

Nymphs look like miniature versions of adults and eat the same foods.

(b) Mosquito:
Holometabolous metamorphosis
(Complete metamorphosis)

Larvae look substantially different from adults and eat different foods.

FIGURE 34.24 Insect Metamorphosis Can Be Incomplete or Complete.

(a) Onychophorans are called velvet worms.

Peripatus species

0.5 cm

(b) Tardigrades are called water bears.

Echiniscus species

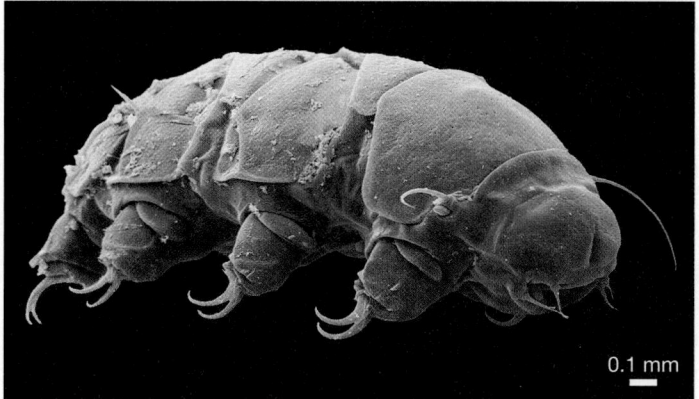

0.1 mm

FIGURE 34.25 Onychophorans and Tardigrades Have Lobe-Shaped Limbs with Claws. The scanning electron micrograph in part (b) has been colorized.

in having a segmented body and limbs. Unlike arthropods, they do not have an exoskeleton and their limbs are not jointed.

The onychophorans, or velvet worms, are small, caterpillar-like organisms that live in moist leaf litter and prey on small invertebrates. Onychophorans have lobe-shaped appendages and segmented bodies with a hemocoel (**FIGURE 34.25a**).

The tardigrades, or water bears, are microscopic animals that live in diverse marine, freshwater, and terrestrial environments, such as in the film of water that covers plants in moist habitats. Tardigrades have a reduced coelom but a prominent hemocoel, and they walk on their clawed, lobe-shaped legs (**FIGURE 34.25b**). Most feed by sucking fluids from plants or animals; others are detritivores, feeding on dead organic matter.

Although the phylogeny of arthropods is still being worked out, most data agree that the phylum as a whole is monophyletic. However, the relationships of lineages within the phylum are currently undergoing a major revision. Insects and myriapods have long been considered sister groups within the arthropods, due in part to their unbranched appendages. However, recent molecular phylogenetic studies—as well as comparisons of new morphological data, developmental data, and evidence in the fossil record—provide strong support for the placement of the insect clade *within* the crustacean lineage, making the crustaceans paraphyletic. It appears that crustaceans successfully invaded land as insects.

In the following exploration of ecdysozoan diversity, insects are described in a separate box, following tradition. However, the revised relationship between insects and crustaceans is then illustrated in the Crustacea box:

- Ecdysozoans > Nematoda (Roundworms)
- Ecdysozoans > Arthropoda > Myriapoda (Millipedes, Centipedes)
- Ecdysozoans > Arthropoda > Insecta (Insects)
- Ecdysozoans > Arthropoda > Crustacea (Shrimp, Lobsters, Crabs, Barnacles, Isopods, Copepods)
- Ecdysozoans > Arthropoda > Chelicerata (Sea Spiders, Horseshoe Crabs, Daddy Longlegs, Mites, Ticks, Scorpions, Spiders)

Species in the phylum Nematoda are commonly called **roundworms** or **nematodes.** Roundworms are unsegmented worms with a pseudocoelom, a tube-within-a-tube body plan, no appendages, and a thick, elastic cuticle that must be molted during growth. Although some nematodes can grow to lengths of several meters, most species are tiny—much less than 1 mm long. They lack specialized systems for exchanging gases and for circulating nutrients. Instead, gas exchange occurs across the body wall, and once ingested, nutrients move by simple diffusion from the gut to other parts of the body.

The nematode *Caenorhabditis elegans* is one of the most thoroughly studied model organisms in biology (see **BioSkills 13** in Appendix B). Species that parasitize humans have also been studied intensively. Pinworms, for example, infect about 40 million people in the United States alone, and *Onchocerca volvulus* causes an eye disease that infects around 20 million people in Africa and Latin America. Advanced infections by the roundworm *Wuchereria bancrofti* result in the blockage of lymphatic vessels, causing fluid accumulation and massive swelling—the condition known as elephantiasis. The potentially fatal disease trichinosis is caused by species in the genus *Trichinella*. The parasites are passed from infected animals to humans by ingesting raw or undercooked meat. An array of symptoms results, including discomfort caused by *Trichinella* larvae encasing themselves in cysts within muscle or other types of tissue (**FIGURE 34.26a**).

A recent phylogenetic analysis of the nematodes has produced an interesting insight: Convergent morphological evolution appears to be extensive. Animal parasitism and plant parasitism both arose several times in different lineages of nematodes. Thus, some nematodes that look very similar and fill a very similar ecological role may be distantly related.

Parasites represent a relatively small fraction of the nematode species that have been described to date, however. The free-living nematodes are important ecologically because they are found in virtually every habitat known (**FIGURE 34.26b**).

While around 25,000 species of roundworms have been named, the actual total may be over a million. Nematodes are also fabulously abundant. Biologists have found 90,000 roundworms in a single rotting apple and have estimated that rich farm soils contain up to 9 billion roundworms per acre. Although they are not the most species-rich animal group, they are among the most abundant. The simple nematode body plan has been extraordinarily successful.

Feeding Roundworms feed on a wide variety of materials, including bacteria, fungi, plant roots, small protists or animals, or detritus. In most cases, the structure of their mouthparts is specialized in a way that increases the efficiency of feeding on a particular type of organism or material.

Movement Roundworms move with the aid of their hydrostatic skeleton. They have no muscles that can change the diameter of the body—their body musculature consists solely of longitudinal muscles that bend the stiff body laterally. ✔ You should be able to indicate the loss of circumferential muscles on Figure 34.2. Most roundworms live in soil or inside a host, so when contractions of their longitudinal muscles cause them to undulate, the movements are resisted by a stiff substrate. As a result, the worm pushes off the substrate, and the body advances.

Reproduction Sexes are separate in most nematode species (but not in *C. elegans*, which is mostly hermaphroditic). Sexual reproduction begins with internal fertilization and culminates in egg laying and the direct development of offspring. Individuals go through a series of four molts over the course of their lifetime.

(a) Parasitic larvae encysted in muscle tissue

Trichinella spiralis

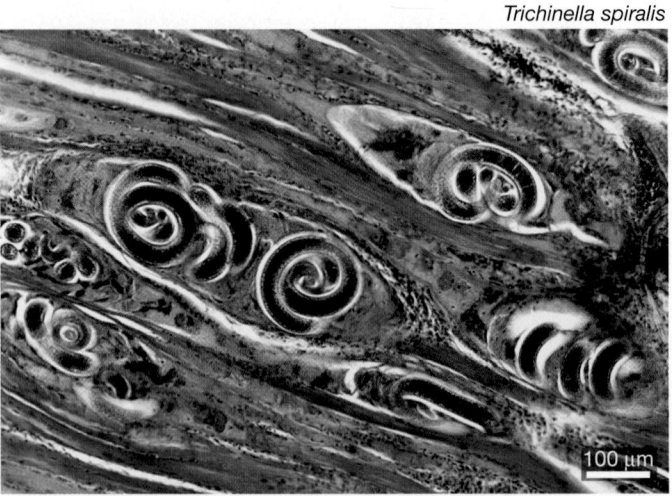

100 μm

(b) Most nematodes are free living.

Caenorhabditis elegans

0.1 mm

FIGURE 34.26 Some Nematodes Are Parasites, but Most Are Free Living.

Ecdysozoans > Arthropoda > Myriapoda (Millipedes, Centipedes)

The **myriapods** have relatively simple bodies, with a head region and a long trunk featuring a series of segments, each bearing one or two pairs of legs (**FIGURE 34.27**). If eyes are present, they consist of a few to many simple structures clustered on the sides of the head. The 16,000 species that have been described to date inhabit terrestrial environments all over the world.

Feeding Millipedes and centipedes have mouthparts that can bite and chew. These organisms live in rotting logs and other types of dead plant material that litters the ground in forests and grasslands. Millipedes are detritivores. Centipedes, in contrast, use a pair of poison-containing fangs just behind the mouth to hunt an array of insects. Large centipedes, with lengths up to 30 cm, can inject enough poison to debilitate a human.

Movement Myriapods walk or run on their many legs; a few species burrow. Some millipedes have over 190 trunk sections, each with two pairs of legs, for a total of over 750 legs. Centipedes usually have fewer than 30 segments, with one pair of legs per segment.

Reproduction Myriapod sexes are separate, and fertilization is internal. Males deposit sperm in packets that are picked up by the female or transferred to her by the male. After females lay eggs in the environment, the eggs hatch into juveniles that develop into adults via a series of molts.

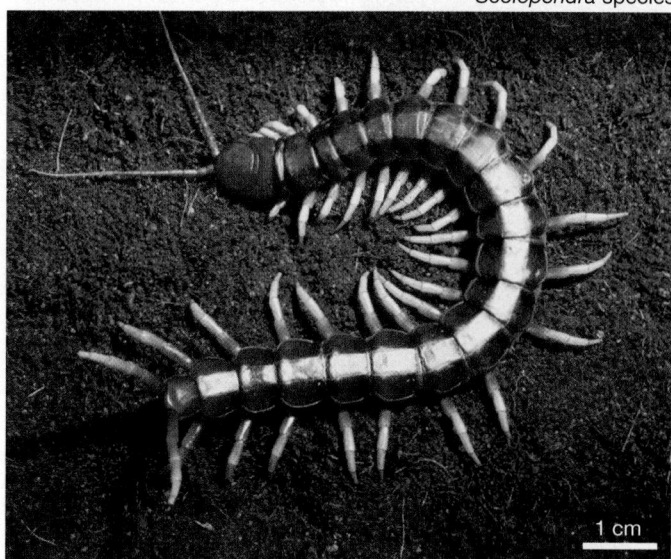

Scolopendra species

1 cm

FIGURE 34.27 Myriapods Have One or Two Pairs of Legs on Each Body Segment.

Ecdysozoans > Arthropoda > Insecta (Insects)

About one million species of insects have been named thus far, but it is certain that many more exist. In terms of species diversity and numbers of individuals, insects dominate terrestrial environments. In addition, the larvae of some species are common in freshwater streams, ponds, and lakes. **TABLE 34.1** (see page 676) describes seven of the most prominent orders of insects. The fruit fly *Drosophila melanogaster,* in the order Diptera, is one of the most thoroughly studied model organisms in biology (see **BioSkills 13** in Appendix B).

Insects are distinguished by having three tagmata—the head, **thorax,** and **abdomen** (see Figure 34.21). Three pairs of walking legs are mounted on the lateral surface of the thorax. Most species have one or two pairs of wings, mounted on the dorsal (back) side of the thorax. Typically the head contains four mouthpart structures, a pair of slender **antennae** that are used to touch and smell, and a pair of compound eyes. A **compound eye** contains many lenses, each associated with a light-sensing, columnar structure (see Chapter 47). (Human and cephalopod eyes are **simple eyes,** meaning they have just one lens.)

Feeding Most insect species have four mouthpart structures (labrum, paired mandibles, paired maxillae, and labium), which have diversified in structure and function in response to natural selection. As a result, insects are able to feed in every conceivable manner and on almost every type of food source available on land. In species with complete (holometabolous) metamorphosis, larvae have wormlike bodies; most tunnel their way through food sources, though some feed by biting leaves. Adults, however, are predators or parasites on plant or animal tissues. Thus, diversification in food sources occurs even within species. Because so many insects make their living by feeding on plant tissues or fluids, the diversification of insects was closely correlated with the diversification of land plants.

Movement Insects use their legs to walk, run, jump, or swim, or they use their wings to fly. When insects walk or run, the sequence of movements usually results in three of their six legs maintaining contact with the ground at all times.

Reproduction Insect sexes are separate, and sexual reproduction is the norm. Mating usually takes place through direct copulation; the male inserts a sperm-transfer organ into the female. Most females lay eggs, but in a few species eggs are retained until hatching. Many species are also capable of reproducing asexually through parthenogenesis, the production of unfertilized eggs via mitosis. Complete metamorphosis is far more common than incomplete metamorphosis.

TABLE 34.1 **Prominent Orders of Insects**

	Order	Description
	Coleoptera ("sheath-winged") Beetles 360,000 known species	**Key traits:** Hardened forewings, called elytra, protect the membranous hindwings that power flight and act as stabilizers. **Feeding:** Adults are important predators and scavengers. Larvae are called grubs and often have chewing mouthparts. **Life Cycle:** All have complete metamorphosis. **Notes:** The most species-rich lineage on the tree of life—about 25 percent of all described species are beetles.
	Lepidoptera ("scale-winged") Butterflies, moths 180,000 known species	**Key traits:** Wings are covered with tiny, often colorful scales. The forewings and hindwings often hook together and move as a unit. **Feeding:** Larvae usually have chewing mouthparts and either bore through food material or bite leaves; adults often feed on nectar. **Life Cycle:** All have complete metamorphosis. **Notes:** Some species migrate long distances.
	Diptera ("two-winged") Flies (including mosquitoes, gnats, midges) 150,000 known species	**Key traits:** Reduced hindwings, called halteres, act as stabilizers during flight. **Feeding:** Adults are usually liquid feeders; many larvae are maggots, which may be parasitic or feed on rotting material. **Life Cycle:** All have complete metamorphosis. **Notes:** The first flies in fossil record appear 240 million years ago.
	Hymenoptera ("membrane-winged") Ants, bees, wasps 115,000 known species	**Key traits:** Membranous forewings and hindwings lock together. **Feeding:** Most ants feed on plant material; most bees feed on nectar; wasps are predatory and often have parasitic larvae. **Life Cycle:** All have complete metamorphosis. Males are haploid (they hatch from unfertilized eggs) and females are diploid (they hatch from fertilized eggs). Larvae are cared for by adults. **Notes:** Most species live in colonies, and many are eusocial.
	Hemiptera ("half-winged") Bugs (including leaf hoppers, aphids, cicadas, scale insects) 85,000 known species	**Key traits:** Some have a thickened forewing with a membranous tip. Mouthparts are modified for piercing and sucking. **Feeding:** Most suck plant juices, but some are predatory. **Life Cycle:** All have incomplete metamorphosis. **Notes:** Called "half-winged" because the forewings of some bugs are hardened near the bases but membranous near the tips.
	**Orthoptera** ("straight-winged") Grasshoppers, crickets 25,000 known species	**Key traits:** Large, muscular hind legs enable jumping. **Feeding:** Most have chewing mouthparts and are leaf eaters. **Life Cycle:** All have incomplete metamorphosis. **Notes:** In many species, males give distinctive songs to attract mates. They produce sound by rubbing their wings against their legs or against each other, a process called stridulation.
	**Odonata** ("toothed") Dragonflies, damselflies 6,500 known species	**Key traits:** Four membranous wings and long, slender abdomens. **Feeding:** Larvae and adults are predatory, often on flies. The "odon" refers to strong, toothlike structures on their mandibles. **Life Cycle:** All have complete metamorphosis: Nymphs are aquatic, adults are terrestrial. **Notes:** They hunt by sight and have reduced antennae (used for touch and smell) but enormous eyes with up to 28,000 lenses.

The 47,000 species of **crustaceans** that have been identified to date live primarily in marine and freshwater environments; they are the most diverse arthropods of the sea. A few species of crab and some isopods (known as sowbugs and pillbugs) are terrestrial, however, and recent phylogenetic evidence suggests that the insects represent a highly successful branch of terrestrial crustaceans (**FIGURE 34.28**). (For the purpose of describing diversity here, insects are covered separately on the previous page.)

Crustaceans—especially the planktonic copepods—are common in surface waters around the world, where they are important consumers. Crustaceans are also important grazers and predators in shallow-water benthic environments.

The segmented body of most crustaceans is divided into two distinct regions, or tagmata: (1) the cephalothorax, which combines the head and thorax, and (2) the abdomen. Many crustaceans have a **carapace**—a platelike section of their exoskeleton that covers and protects the cephalothorax (**FIGURE 34.29**). They are the only type of arthropod with two pairs of antennae, and have sophisticated, compound eyes—usually mounted on stalks. Crustacean appendages are usually branched (insect appendages are unbranched). Other than trilobites—a major arthropod lineage that did not survive the end-Permian mass extinction, 250 million years ago (Chapter 28)—crustaceans were the first arthropods to appear in the fossil record. The tiny ostracods (seed shrimp) have a particularly rich fossil record dating back to the Cambrian.

Feeding Most crustaceans have 4–6 pairs of mouthparts that are derived from jointed appendages, and as a group they use every type of feeding strategy known. Barnacles and many shrimp are suspension feeders that use feathery structures located on head or body appendages to capture prey. Crabs and lobsters are active hunters, herbivores, and scavengers. Typically they have a pair of mouthparts called **mandibles** that bite or chew. Individuals capture and hold their food source with claws or other feeding appendages and then use their mandibles to shred the food into small bits that can be ingested.

Movement Crustaceans typically have many pairs of limbs with diverse forms and functions, including claws, paddle-shaped forms used in swimming, feathery structures used in capturing food that is suspended in water, and slender legs that make sophisticated walking or running movements possible (see Figure 34.22). Barnacles are one of the few types of sessile crustaceans. Adult barnacles cement their heads to a hard substrate, secrete a protective shell of calcium carbonate, and use their feathery legs to capture food particles and transfer them to the mouth.

Reproduction Most individual crustaceans are either male or female, and sexual reproduction is the norm. Fertilization is usually internal, and eggs are typically retained by the female until they hatch. Most crustaceans pass through several distinct larval stages; many species include a larval stage called a **nauplius,** which is usually planktonic. A nauplius has a single eye and appendages that develop into the two pairs of antennae, the mouthparts, and walking/swimming legs of the adult.

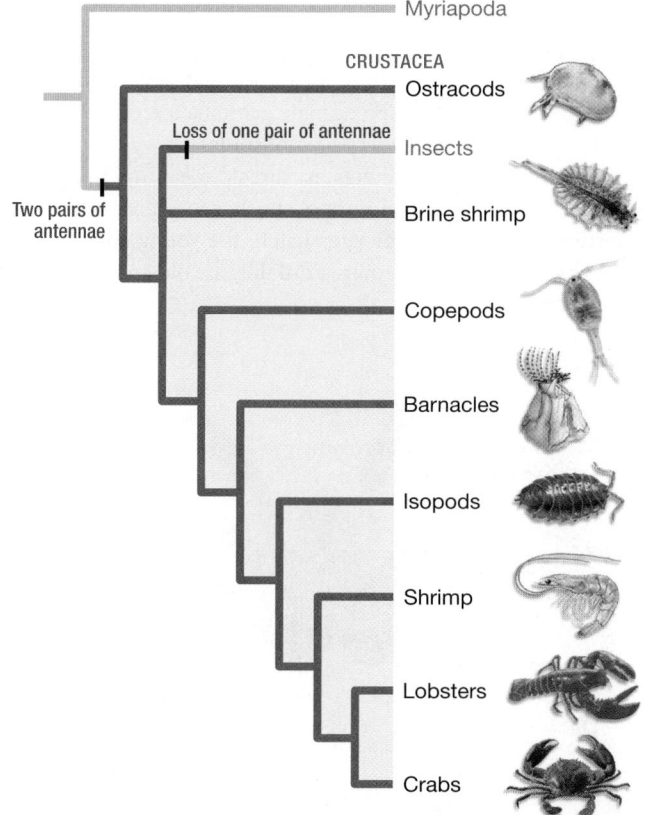

FIGURE 34.28 Phylogeny of the Major Groups of Crustaceans. This phylogeny illustrates the current hypothesis that insects are a terrestrial clade within the crustaceans, overturning traditional classification (which is nonetheless still in widespread use due to its long history).

Cherax quadricarinatus

FIGURE 34.29 Crustaceans Have Two Tagmata: A Cephalothorax and an Abdomen. The carapace protects the cephalothorax (literally "head-chest").

The most prominent lineage within the Chelicerata is the Arachnida, including spiders, scorpions, ticks, mites and daddy longlegs (**FIGURE 34.30**). Most of the 100,000 species of chelicerates are terrestrial, but the horseshoe crabs and sea spiders are marine.

The chelicerate body consists of 2 tagmata, the cephalothorax (prosoma) and abdomen (opisthosoma), and 6 pairs of appendages (**FIGURE 34.31**). The cephalothorax lacks antennae for sensing touch or odor but usually contains eyes. The group is named for a pair of appendages called **chelicerae,** found near the mouth. Depending on the species, the chelicerae are used in feeding, defense, copulation, movement, or sensory reception. In addition to 4 pairs of legs and 1 pair of chelicerae, chelicerates also have a pair of pedipalps, which can be used for diverse functions.

Feeding Spiders, scorpions, and daddy longlegs capture insects or other prey. Although some of these species are active hunters that rove around in search of prey, most spiders are sit-and-wait predators that either pounce on prey that wander near, or that create sticky webs to capture prey. In this case, the spider senses the vibrations of the struggling prey on the web, pounces on it, and administers a toxic bite. Regardless of the mode of capture, spiders generally consume their meals in soup form—they secrete digestive enzymes into the prey to enable digestion to begin externally, then drink up the nutritious liquid. Often all that remains of the prey is an empty exoskeleton.

Mites and ticks are ectoparasitic and use their piercing mouthparts to feed on host animals or, in the case of dust mites, on their hosts' dead skin. Horseshoe crabs eat a variety of animals as well as algae and detritus. Most scorpions feed on insects; the largest scorpion species occasionally eat snakes and lizards. Although the venom of scorpion stings is an effective defense against predators, it is primarily used to kill or paralyze prey.

Movement Like other arthropods, chelicerates move with the aid of muscles attached to an exoskeleton. They walk or crawl on their four pairs of jointed walking legs; some species are also capable of jumping. Horseshoe crabs and some other marine forms can swim slowly. Newly hatched spiders spin long, silken threads that serve as balloons and that can carry them on the wind more than 400 kilometers from the point of hatching.

Reproduction Sexual reproduction is the rule in chelicerates, and fertilization is internal in most groups. Courtship displays are extensive in many chelicerates and may include both visual displays and the release of chemical odorants. In spiders, males may present a dead insect as a gift that the female eats as the pair mates; in some species, the male himself is eaten as sperm is being transferred—the longer the meal lasts, the more successful the sperm transfer. Males use pedipalps to transfer sperm to females. These organs fit into the female reproductive tract in a "lock-and-key" fashion. Differences in the size and shape of male genitalia are often the only way to identify closely related species of spider.

Development in chelicerates is direct, meaning that larval forms and metamorphosis do not occur. In scorpions, females retain fertilized eggs. After the eggs hatch, the young climb on the mother's back. They remain there until they are old enough to hunt for themselves.

FIGURE 34.30 Phylogeny of the Major Groups of Chelicerates.

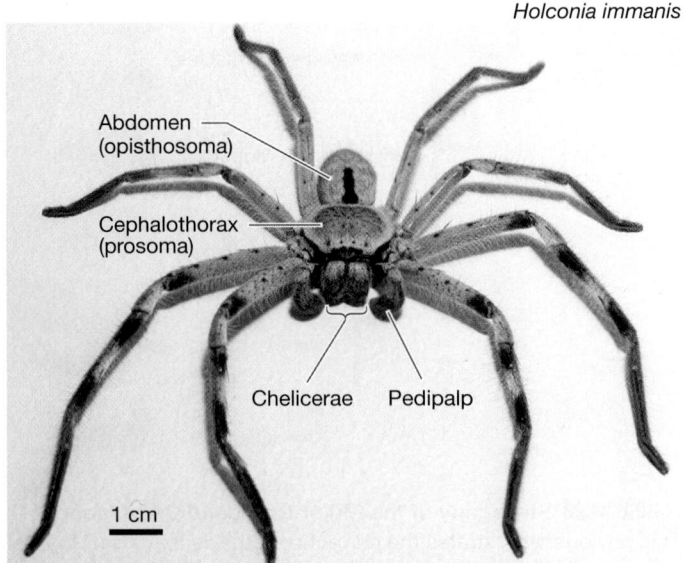

Holconia immanis

FIGURE 34.31 Chelicerates Have Chelicerae and Two Tagmata.

If you understand . . .

34.1 What Is a Protostome?

- Molecular evidence supports the hypothesis that protostomes are a monophyletic group divided into two major subgroups: the Lophotrochozoa and the Ecdysozoa.

- All protostomes are bilaterally symmetric and triploblastic, but the coelom has been reduced or lost in several phyla.

- Protostomes diversified in their methods of sensing their environments, feeding, moving, reproducing, and developing.

- The transition to living on land occurred several times independently during the evolution of protostomes, opening new habitats and driving diversification. The transition was facilitated by watertight shells and exoskeletons, the evolution of specialized structures to minimize water loss and manage gas exchange, and the ability to move to moist locations.

- Many protostome body plans are highly modular due to the underlying expression patterns of developmental tool-kit genes.

✔ You should be able to explain how both ecological opportunity and genetic opportunity were important in the diversification of protostomes.

34.2 What Is a Lophotrochozoan?

- Lophotrochozoans grow incrementally. Some phyla have characteristic feeding structures called lophophores, and many have a distinctive type of larvae called trochophores. Spiral cleavage is a synapomorphy for the group but was modified or lost in some phyla.

- Mollusks, annelids, and flatworms are the most diverse phyla within the Lophotrochozoa. All three phyla made the water-to-land transition.

- The mollusk body plan includes a muscular foot, a visceral mass, and a mantle. These components diversified in many ways, enabling mollusks to occupy most aquatic, and some terrestrial, habitats.

✔ You should be able to explain why the absence of a trait known to be found in a clade, such as a trochophore larva, spiral cleavage, or a coelom, does not necessarily exclude a species from that clade.

34.3 What Is an Ecdysozoan?

- Species in the Ecdysozoa grow by molting—meaning that they shed their old external covering and grow a new, larger one.

- Arthropods and nematodes are the most diverse phyla within the Ecdysozoa. A majority of animals on Earth are arthropods.

- The arthropod body plan is segmented and protected by an exoskeleton made primarily of chitin. Appendages are jointed. The diversification of arthropod body plans is based on variation in the number and functional specialization of segments.

- Complete (holometabolous) metamorphosis enables larvae and adults to specialize, creating opportunities for diversification.

- Crustaceans are the most diverse group of arthropods in the sea, and insects are the most diverse group on land. (Recent data suggest that insects are a terrestrial subgroup of Crustacea.)

✔ You should be able to compare modularity in arthropod body plans with modularity in vertebrates.

(MB) MasteringBiology

1. **MasteringBiology Assignments**

   Tutorials and Activities Characteristics of Invertebrates; Ecdysozoans; Protostome Diversity

   Questions Reading Quizzes, Blue-Thread Questions, Test Bank

2. **eText** Read your book online, search, take notes, highlight text, and more.

3. **The Study Area** Practice Test, Cumulative Test, BioFlix® 3-D Animations, Videos, Activities, Audio Glossary, Word Study Tools, Art

You should be able to . . .

✔ TEST YOUR KNOWLEDGE

Answers are available in Appendix A

1. **True or false?** The transition from water to land reduced protostome diversity.

2. What is a lophophore?
   a. a specialized filter-feeding structure
   b. the single opening in species with a blind gut
   c. a distinctive type of larva with a band of cilia
   d. a synapomorphy that defines lophotrochozoans

3. What is the function of the arthropod exoskeleton?
   a. Because hard parts fossilize more readily than do soft tissues, the presence of an exoskeleton has given arthropods a good fossil record.
   b. It has no well-established function. (Trilobites had an exoskeleton, and they went extinct.)
   c. It provides protection and functions in locomotion.
   d. It makes growth by molting possible.

4. Why is it logical that Platyhelminthes have flattened bodies?
  a. They have simple bodies and evolved early in the diversification of protostomes.
  b. They lack a coelom, so their body cannot form a rounded tube-within-a-tube design.
  c. A flat body provides a large surface area for gas exchange, which compensates for their lack of gas-exchange organs.
  d. They are sit-and-wait predators that hide from passing prey by flattening themselves against the substrate.

5. One trait that is shared by the Lophotrochozoa and Ecdysozoa is_____.

6. Which protostome phylum is distinguished by having body segments organized into tagmata?
  a. Mollusca
  b. Arthropoda
  c. Annelida
  d. Nematoda

✓ TEST YOUR UNDERSTANDING

Answers are available in Appendix A

7. QUANTITATIVE Put the following phyla in order from least diverse (in terms of number of named species) to most diverse: Annelida, Mollusca, Arthropoda, Rotifera.
  a. Annelida, Rotifera, Mollusca, Arthropoda
  b. Mollusca, Rotifera, Annelida, Arthropoda
  c. Rotifera, Annelida, Arthropoda, Mollusca
  d. Rotifera, Annelida, Mollusca, Arthropoda

8. Predict why annelids and arthropods were thought to be closely related, before the phylogenetic analyses in the late 1990s.

9. Pose a hypothesis to explain why a larval stage is adaptive in clams. How could you test your hypothesis?

10. Use your understanding of protostome evolution to predict two adaptations for terrestrial living that occurred in land plants.

11. Pose a hypothesis to explain why the evolution of the wing was such an important event in the evolution of insects.

12. Propose a scenario wherein hemimetabolous metamorphosis would be adaptive.

✓ TEST YOUR PROBLEM-SOLVING SKILLS

Answers are available in Appendix A

13. Recall that the phylum Platyhelminthes includes three major groups: the free-living turbellarians, the parasitic cestodes, and the parasitic trematodes. Because the parasitic forms are so simple morphologically, researchers suggest that they are derived from more complex, free-living forms. Draw a phylogeny of Platyhelminthes that would support the hypotheses that ancestral flatworms were morphologically complex, free-living organisms and that parasitism is a derived condition.

14. Brachiopoda is a phylum within the Lophotrochozoa. Even though they are not closely related to mollusks, brachiopods look and act like bivalve mollusks (clams or mussels). Specifically, brachiopods suspension feed, live inside calcium carbonate shells with two valves that hinge together in some species, and attach to rocks or other hard surfaces on the ocean floor. How is it possible for brachiopods and bivalves to be so similar if they did not share a recent common ancestor?

15. The Mollusca includes a group of about 370 wormlike species called the Aplacophora. Although aplacophorans do not have a shell, their cuticle secretes calcium carbonate scales or spines. They lack a well-developed foot, and some species have a simple radula. They are mostly carnivores and detritivores in the deep sea. The reduced forms of several molluscan synapomorphies suggest that they are ancestral to other mollusks—that is, the earliest group to branch off in the molluscan phylogeny. Evaluate this hypothesis. How could you test it?

16. Examine the figure below summarizing the expression of *Ubx-AbdA Hox* genes (colored red) in different body segments during early development in four crustaceans. The blue appendages represent a novel type of feeding appendage called a maxilliped. Which statement best summarizes the data?

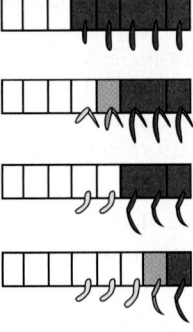

  a. Maxillipeds were lost in segments where *Ubx-AbdA* genes were expressed.
  b. Maxillipeds arose in segments where *Ubx-AbdA* genes were no longer expressed.
  c. Maxillipeds arose in segments where *Ubx-AbdA* genes were expressed.
  d. The presence of maxillipeds does not correlate with the pattern of *Ubx-AbdA* gene expression.

35 Deuterostome Animals

In this chapter you will learn that

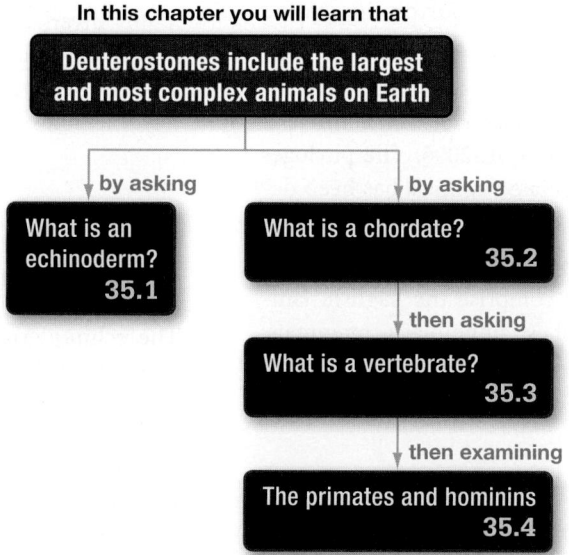

Deuterostomes include the largest and most complex animals on Earth

↓ by asking

What is an echinoderm?
35.1

↓ by asking

What is a chordate?
35.2

↓ then asking

What is a vertebrate?
35.3

↓ then examining

The primates and hominins
35.4

In most habitats the "top predators"—animals that prey on other animals and aren't preyed upon themselves—are deuterostomes.

The **deuterostomes** include the largest-bodied and some of the most morphologically complex of all animals. They range from the sea stars that cling to dock pilings to the fish that dart in and out of coral reefs to the wildebeests that migrate across the Serengeti Plains of East Africa.

The deuterostomes were initially grouped together because they all undergo early embryonic development in a similar way. When a humpback whale, sea urchin, or human is just beginning to grow, the gut starts developing from posterior to anterior—the anus forms first and the mouth second. A coelom, if present, develops from outpocketings of mesoderm (see Chapter 33).

✔ When you see this checkmark, stop and test yourself. Answers are available in Appendix A.

More recently, phylogenies based on molecular evidence have confirmed the monophyly of the deuterostomes. Today, most biologists recognize four deuterostome phyla: the Echinodermata, Hemichordata, Xenoturbellida, and Chordata (**FIGURE 35.1**):

- The echinoderms include the sea stars and sea urchins.

- The hemichordates, or "acorn worms," burrow in marine sands or muds and make their living by deposit feeding or suspension feeding.

- A lone genus with two wormlike species, called *Xenoturbella,* was recognized as a distinct phylum in 2006. The phylogenetic placement of these small, ciliated animals has been debated by scientists for years.

- The chordates include the **vertebrates,** or animals with backbones. The vertebrates, in turn, comprise the hagfish, lampreys, sharks and rays, bony fishes, amphibians, mammals, and reptiles (including birds).

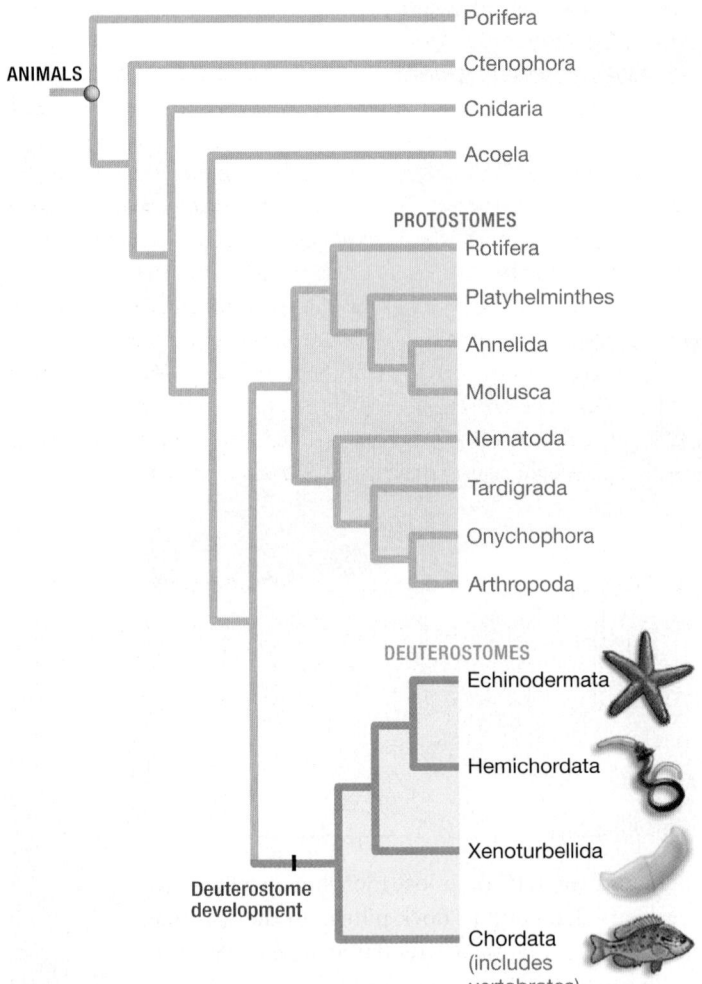

FIGURE 35.1 There Are Four Phyla of Deuterostomes. Vertebrates are in the phylum Chordata.

✔ **QUESTION** Are invertebrates monophyletic or paraphyletic? (Hint: See Table 28.2 if you need help.)

Animals that are not vertebrates are collectively known as **invertebrates.** Over 95 percent of the known animal species are invertebrates—but the vertebrates dominate the deuterostomes.

Although deuterostomes share important features of embryonic development, their adult body plans and their feeding methods, modes of locomotion, and reproductive strategies are highly diverse. This chapter explores deuterostome diversity by introducing the echinoderms and chordates and then taking a more in-depth look at the vertebrates and our own closest ancestors: the hominins.

35.1 What Is an Echinoderm?

The echinoderms (literally, "spiny-skins") were named for the spines or spikes observed in many species. Echinodermata is a large and diverse phylum but is exclusively marine.

In numbers of species and range of habitats occupied, echinoderms are the second-most successful lineage of deuterostomes—next to the vertebrates. To date, biologists have cataloged about 7000 species of echinoderms.

Echinoderms are also abundant. In some deepwater environments, species in this phylum represent 95 percent of the total mass of organisms.

How are these animals put together?

The Echinoderm Body Plan

All deuterostomes are considered bilaterians, because they evolved from an ancestor that was bilaterally symmetric (see Chapter 33). Echinoderm larvae are also bilaterally symmetric (**FIGURE 35.2a**). But a remarkable event occurred early in the evolution of **echinoderms:** the origin of five-sided radial symmetry, called pentaradial symmetry, in adults (**FIGURE 35.2b**).

Radially symmetric animals do not have heads (Chapter 33). As a result, they tend to interact with the environment in all directions at once instead of facing the environment in one direction. Adult echinoderms that are capable of movement tend to move equally well in all directions.

A second noteworthy feature of the echinoderm body is its **endoskeleton:** a hard structure, located just inside a thin layer of epidermal tissue, which provides protection and support (shown without skin in Figure 35.2b). As an individual is developing, cells secrete plates of calcium carbonate inside the skin. Depending on the species, the plates may remain independent and result in a flexible structure or fuse into a rigid case. If the plates do not fuse, they are connected by an unusual tissue that can be reversibly stiff or flexible—depending on conditions.

Another remarkable event in echinoderm evolution was the origin of a unique morphological feature: a series of branching, fluid-filled tubes and chambers called the **water vascular system.** In some groups, one of the tubes is open to the exterior where it meets the body wall, so seawater can flow into and out

(a) Echinoderm larvae are bilaterally symmetric.

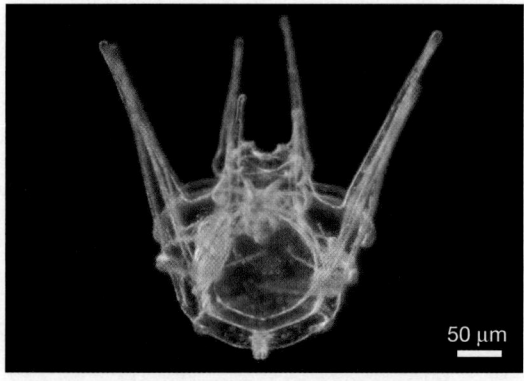

50 μm

(b) Adult echinoderms are radially symmetric.

1 cm

FIGURE 35.2 Body Symmetry Differs in Adult and Larval Echinoderms. (a) Bilaterally symmetric sea urchin larva. **(b)** Radially symmetric adult endoskeleton.

of the system. Inside, fluids move via the beating of cilia that line the interior of the tubes and chambers.

FIGURE 35.3a highlights a particularly important part of the water vascular system called tube feet. **Tube feet** are elongated, fluid-filled appendages, each consisting of a balloon-like *ampulla*

on the inside of the body and a tube-like *podium* (literally, "foot"; plural *podia*) projecting on the outside (**FIGURE 35.3b**). The water vascular system forms a specialized hydrostatic skeleton that operates the tube feet. As tube feet extend and contract in a coordinated way along the base of an echinoderm, they alternately grab and release the substrate or food source, enabling movement and feeding. Tube feet can also aid in respiration by increasing the surface area of soft tissue exposed to the aquatic environment.

Radial symmetry in adults, an endoskeleton of calcium carbonate, and the water vascular system are all synapomorphies—traits that identify echinoderms as a monophyletic group. (Turn to Chapter 28 and **BioSkills 7** in Appendix B to review how to read phylogenetic trees.)

How Do Echinoderms Feed?

Depending on the lineage and species in question, echinoderms make their living by mass feeding on algae or other animals, suspension feeding, or deposit feeding. In most cases, an echinoderm's tube feet play a key role in obtaining food.

Many sea stars, for example, prey on bivalves. Clams and mussels respond to sea star attacks by contracting muscles that close the two halves of their shell tight. But by clamping onto each half of the shell with their tube feet, making their endoskeleton rigid, and pulling, sea stars are often able to pry the shell apart a few millimeters (**FIGURE 35.4a**, see page 684). Once a gap exists, the sea star extrudes its stomach from its body and forces the stomach through the opening in the bivalve's shell. Upon contacting the visceral mass of the bivalve, the stomach of the sea star secretes digestive enzymes. It then begins to absorb the small molecules released by enzyme action. Eventually, only the shell of the prey remains.

Tube feet are also used in suspension feeding (**FIGURE 35.4b**). In most cases, tube feet are extended out into the water. When food particles contact them, the tube feet flick the food down to cilia, which sweep the particles toward the animal's mouth.

(a) Echinoderms have a water vascular system.

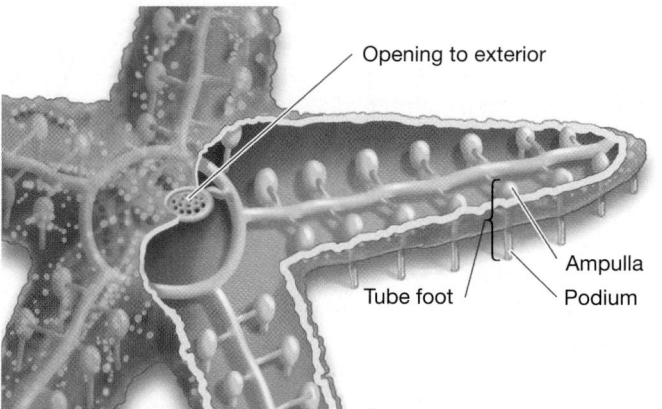

Opening to exterior

Ampulla

Tube foot
Podium

(b) Tube feet project from the underside of the body.

Tube feet

FIGURE 35.3 Echinoderms Have a Water Vascular System. (a) The water vascular system is a series of tubes and chambers that radiates throughout the body, forming a sophisticated hydrostatic skeleton. **(b)** Tube feet aid in movement because they extend from the body and can grab and release the substrate.

(a) Tube feet enable sea stars to get a grip on their prey.

(b) Tube feet enable feather stars to suspension feed.

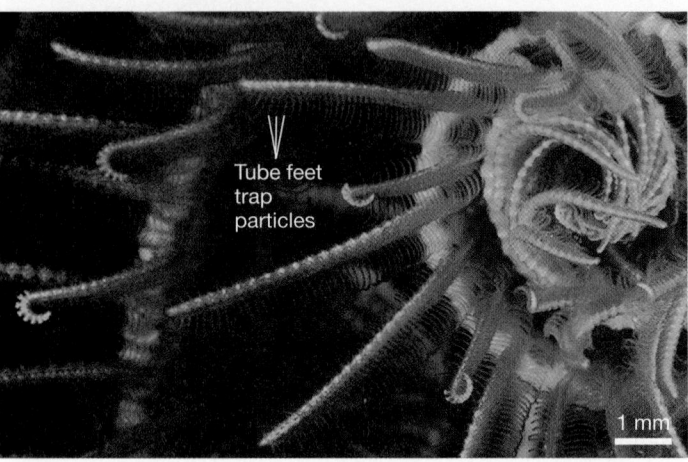

Tube feet
trap
particles

1 mm

FIGURE 35.4 Echinoderms Use Their Tube Feet in Feeding.

In deposit feeders, tube feet secrete mucus that is used to sop up food material on the substrate. The tube feet then roll the food-laden mucus into a ball and move it to the mouth.

Key Lineages: The Echinoderms

The echinoderms living today make up five major lineages: **(1)** feather stars and sea lilies, **(2)** sea stars, **(3)** brittle stars and basket stars, **(4)** sea urchins and sand dollars, and **(5)** sea cucumbers (**FIGURE 35.5**). ✔ You should be able to indicate the origin of pentaradial symmetry in adults, the water vascular system, and the echinoderm endoskeleton on Figure 35.5.

- Most feather stars and sea lilies are sessile suspension feeders.

- Brittle stars and basket stars have five or more long arms that radiate out from a small central disk. They use these arms to suspension feed, deposit feed, or capture small prey animals.

- Sea cucumbers are sausage-shaped animals that suspension feed or deposit feed with the aid of modified tube feet called tentacles that are arranged in a whorl around their mouths.

Sea stars and the sea urchins and sand dollars are described in detail in the two boxes that follow.

- Echinodermata > Asteroidea (Sea Stars)
- Echinodermata > Echinoidea (Sea Urchins and Sand Dollars)

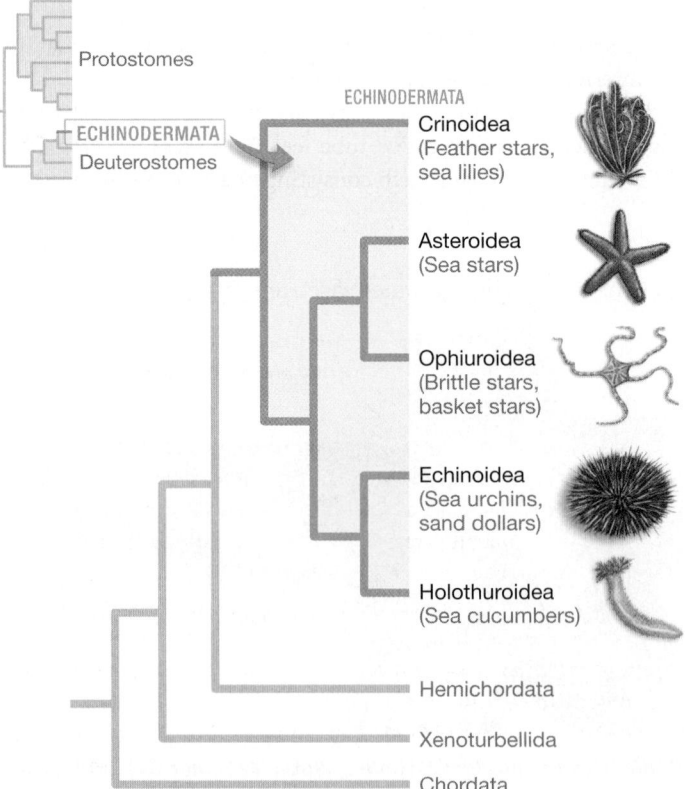

FIGURE 35.5 There Are Five Major Lineages of Echinoderms.

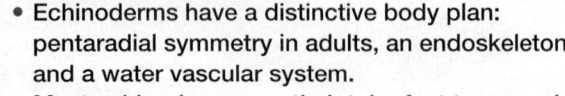

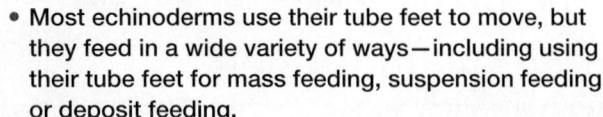

If you understand that . . .

- Echinoderms have a distinctive body plan: pentaradial symmetry in adults, an endoskeleton, and a water vascular system.
- Most echinoderms use their tube feet to move, but they feed in a wide variety of ways—including using their tube feet for mass feeding, suspension feeding, or deposit feeding.

✔ **You should be able to . . .**

Explain how you would determine whether an unidentified animal you encounter along the seashore is an echinoderm.

Answers are available in Appendix A.

The 1900 known species of sea stars have bodies with five or more long arms—in some species, up to 40—radiating from a central region that contains the mouth, stomach, and anus (**FIGURE 35.6**). The sea star's arms are not set off from the central region by clear, joint-like articulations, in contrast to their sister group, the brittle stars. ✔ You should be able to indicate the origin of arms, and the origin of arms that are continuous with the central region, as in sea stars, on Figure 35.5. When fully grown, sea stars can range in size from less than 1 cm in diameter to 1 m across. They live on hard or soft substrates along the coasts of all the world's oceans. Although the spines that are characteristic of some echinoderms are reduced to knobs on the surface of most sea stars, the crown-of-thorns star and a few other species have prominent, upright, movable spines.

Feeding Sea stars are predators or scavengers. Some species pull bivalves apart with their tube feet and feed by everting their stomach into the prey's visceral mass. In some cases, sea star predation on bivalves opens up habitat for other organisms in the community—increasing biodiversity (see Chapter 55). Sponges, barnacles, and snails are also common prey. The crown-of-thorns sea star specializes in feeding on corals and is native to the Indian Ocean and western Pacific Ocean. Its population has skyrocketed recently—possibly because people are harvesting the sea star's major predator, a large snail called the triton, for its shell. Large crown-of-thorns star populations have led to the destruction of large areas of coral reef.

Movement Sea stars crawl with the aid of their tube feet.

Reproduction Sexual reproduction predominates in sea stars, and sexes are separate. At least one sea star arm in each individual is filled with reproductive organs that produce massive amounts of gametes—millions of eggs per female in some species. Some species care for their offspring by holding fertilized eggs on their body until the eggs hatch. Larvae are long-lived and may spend weeks or months in the plankton, possibly dispersing thousands of kilometers. Most sea stars are capable of regenerating arms that are lost in predator attacks or storms. Some species can reproduce asexually by dividing the body in two; each of the two individuals then regenerates the missing half.

Crossaster papposus

5 cm

FIGURE 35.6 Sea Stars May Have Many Arms.

About 1000 species of echinoids are living today; most are sea urchins or sand dollars. Sea urchins have globe-shaped bodies and long spines and crawl along substrates. Sand dollars are flattened and disk-shaped, have short spines, and burrow in soft sediments. ✔ You should be able to indicate the origin of globular or disk-shaped bodies on Figure 35.5.

Feeding Most types of sea urchins are herbivores. In some areas of the world, urchins are extremely important grazers on kelp and other types of algae. In cases when humans have removed predators such as sea otters, and urchin populations have grown in response, the urchins' grazing can destroy kelp forests. The urchins chew away a structure called a holdfast, which anchors kelp to the substrate; the kelp then float away and die. Most echinoids have a unique, jaw-like feeding structure in their mouths, called an Aristotle's lantern, that is made up of five calcium carbonate teeth attached to muscles (**FIGURE 35.7a**, see page 686). In many species, this apparatus can extend and retract as the animal feeds, allowing it to reach more food material. Sand dollars, in contrast, are suspension feeders. The beating of cilia on their spines creates tiny water currents that bring organic material toward the body. The food particles are then trapped in mucus and transported in long strings to the mouth. In this way, sand dollars harvest nutritious detritus found in sand or in other soft substrates (**FIGURE 35.7b**).

Movement Using their spines, sea urchins crawl and sand dollars burrow. Some species also use their tube feet to aid movement, or just to hold on to the substrate to prevent being washed away by waves and currents.

Reproduction Sexual reproduction predominates in sea urchins and sand dollars (see Figure 33.14). Fertilization is external, and sexes are separate. As in sea stars, larvae are long-lived and may disperse long distances. The larvae of some species undergo asexual reproduction by dividing in two or budding off new individuals, before they undergo metamorphosis.

(Continued on next page)

(a) Sea urchins graze with their teeth.

Lytechinus variegatus

Teeth at center
of underside

5 mm

(b) Sand dollars suspension feed.

Echinarachnius parma

1 cm

FIGURE 35.7 Sea Urchins and Sand Dollars Have Different Feeding Strategies.

35.2 What Is a Chordate?

You are a chordate, yet chordates are so diverse that you would probably find it difficult to identify familiar features in some of your fellow chordates—such as the sea squirts, whose small, sessile, globular bodies hardly resemble animals at all. Which characteristics are shared within this phylum?

All **chordates** are defined by the presence of four morphological features at some stage in their life cycles:

1. openings into the throat called **pharyngeal gill slits;**
2. a **dorsal hollow nerve cord,** which runs the length of the body, composed of projections from neurons;
3. a stiff and supportive but flexible rod called a **notochord,** which runs the length of the body; and
4. a muscular, post-anal tail—meaning a tail that contains muscle and extends past the anus.

Electrical signals carried by the dorsal hollow nerve cord coordinate muscle movement, and the notochord stiffens the muscular tail. Together, these traits create a "torpedo"—a long, streamlined animal that can swim rapidly. Fossil evidence suggests that chordates diversified early during the Cambrian radiation of animals, about 550 million years ago.

Three Chordate "Subphyla"

The phylum Chordata is made up of three major lineages: the **(1)** cephalochordates, **(2)** urochordates, and **(3)** vertebrates. Let's examine the four chordate morphological features in each group.

The Cephalochordates Cephalochordates are also called lancelets or amphioxus; they are small, mobile, torpedo-shaped animals with a "fish-like" appearance, and they make their living by suspension feeding (**FIGURE 35.8a**). Adult cephalochordates live in ocean-bottom habitats, where they burrow in sand and suspension feed with the aid of their pharyngeal gill slits.

The dorsal hollow nerve cord in cephalochordates runs parallel to a notochord. Because the notochord stiffens the cephalochordates' bodies, muscle contractions on either side result in fishlike movement when the animals swim during dispersal or mating.

The Urochordates Urochordates are also called tunicates or sea squirts. As **FIGURE 35.8b** shows, pharyngeal gill slits are present in both larvae and adults; these structures function in both feeding and gas exchange.

The dorsal hollow nerve cord, notochord, and tail are present in larvae and in the sexually mature forms of species that move as adults. As in cephalochordates, these features combine to enable swimming movements. As larvae swim or float in the upper water layers of the ocean, they drift to new habitats where food might be more abundant. As a result, larvae function as a dispersal stage in the life cycle.

The Vertebrates Vertebrates (**FIGURE 35.8c**) include the hagfish, lampreys, sharks and rays, several lineages of bony fishes, amphibians, mammals, and reptiles (including birds). In vertebrates, the dorsal hollow nerve cord is the familiar spinal cord—a bundle of nerve cells that runs from the brain to the posterior of the body. Electrical signals from the spinal cord control movements and the function of organs (see Chapter 46).

Structures called pharyngeal pouches are homologous to pharyngeal gill slits and are present in all vertebrate embryos.

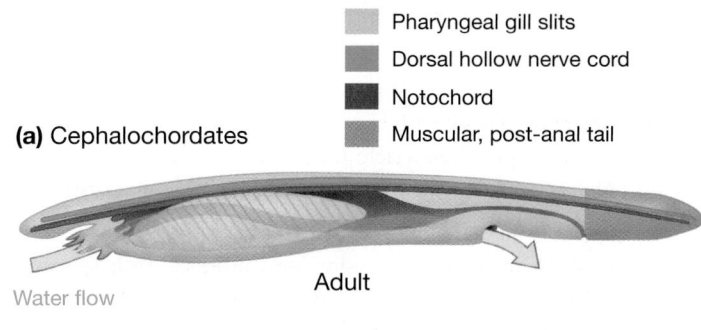

(a) Cephalochordates

Pharyngeal gill slits
Dorsal hollow nerve cord
Notochord
Muscular, post-anal tail

Water flow

Adult

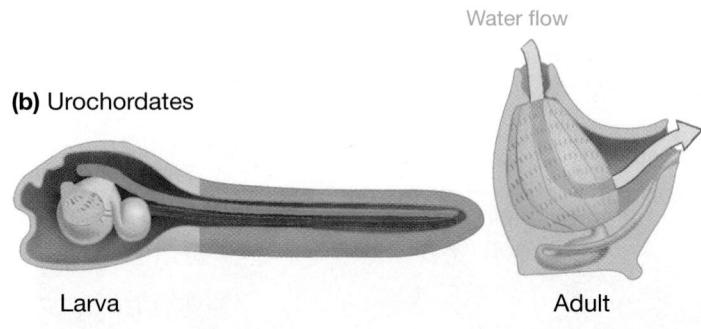

(b) Urochordates

Water flow

Larva

Adult

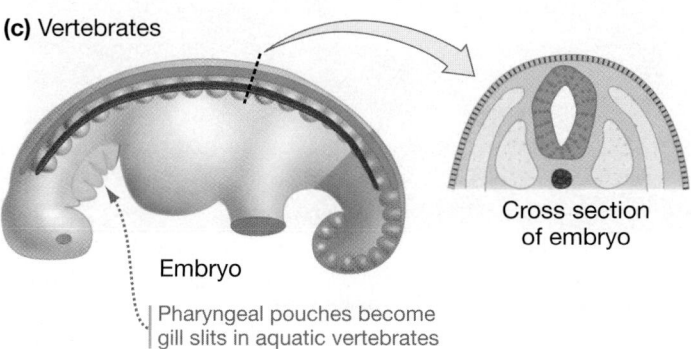

(c) Vertebrates

Cross section of embryo

Embryo

Pharyngeal pouches become gill slits in aquatic vertebrates

FIGURE 35.8 Four Features Distinguish the Three Lineages of Chordates.

In aquatic species, the creases between pouches open into gill slits and develop into part of the main gas-exchange organ—the **gills.** In terrestrial species, however, gill slits do not develop after the pharyngeal pouches form. In these species, the pharyngeal pouches are a vestigial trait (see Chapter 25).

A notochord develops in all vertebrate embryos. It no longer functions in body support and movement, however, for these roles are filled by the vertebral column. Instead, the notochord helps organize the body plan early in development. The notochord secretes proteins that help induce the formation of somites—segmented blocks of tissue that form along the length of the body (Chapter 23). Although the notochord itself disappears, cells in the somites later differentiate into the vertebrae, ribs, and skeletal muscles of the back, body wall, and limbs. In this way, the notochord is instrumental in the development of the trait that gave vertebrates their name.

Key Lineages: The Invertebrate Chordates

There are about 24 species of lancelets, 3000 species of tunicates, and over 62,000 vertebrates. Because they are so species rich and diverse, vertebrate chordates rate their own section in this chapter (Section 35.3). Here let's focus on the tunicates and lancelets. These lineages are sometimes referred to as the invertebrate chordates—meaning, members of the Chordata that lack vertebrae. They are of great interest to evolutionary biologists, who study the morphology, development, and genetics of these animals to better understand the evolutionary history of our own subphylum, the vertebrates.

- Chordata > Cephalochordata (Lancelets)
- Chordata > Urochordata (Tunicates)

Chordata > Cephalochordata (Lancelets)

About two dozen species of cephalochordates have been described to date, all of them found in marine sands. Lancelets—also called amphioxus—have several characteristics that are intermediate between invertebrates and vertebrates. Chief among these is a notochord that is retained in adults, where it functions as an endoskeleton.

Feeding Adult cephalochordates feed by burrowing in sediment until only their heads are sticking out into the water (**FIGURE 35.9**). They take water in through their mouths and trap food particles with the aid of mucus on their pharyngeal gill slits.

Movement Adults have large blocks of muscle arranged in a series along the length of the notochord. Lancelets are efficient swimmers because the flexible, rod-like notochord stiffens the body, making it wriggle when the blocks of muscle contract.

Reproduction Asexual reproduction is unknown, and individuals are either male or female. Gametes are released into the environment, and fertilization is external.

Branchiostoma lanceolatum

1 cm

FIGURE 35.9 Lancelets Look Like Fish but Are Not Vertebrates.

The urochordates are also called tunicates; they comprise two major sub-lineages: the sea squirts (or ascidians) and salps. All of the species described to date live in the ocean. Sea squirts live on the ocean floor (**FIGURE 35.10a**); salps live in open water (**FIGURE 35.10b**). The synapomorphies that define the urochordates include an external coat of polysaccharide, called a tunic, that covers and supports the body; a U-shaped gut; and two body openings, called siphons.

Feeding Adult urochordates use their pharyngeal gill slits to suspension feed. A mucus sheet on the inside of the pharynx traps particles present in the water that enters one siphon, passes through the slits, and leaves through the other siphon.

Movement Larvae swim with the aid of the notochord, which functions as a simple endoskeleton. Larvae are a dispersal stage and do not feed. Adult sea squirts are sessile while salps drift in currents.

Reproduction In most species, individuals produce both sperm and eggs. In some species, both sperm and eggs are shed into the water and fertilization is external; in other species, sperm are released into the water but eggs are retained, so that fertilization and early development are internal. Asexual reproduction by budding is also common in some groups, forming large sessile colonies in some sea squirts, and long chains in salps.

(a) Sea squirt

Rhopalaea crassa

1 cm

(b) Salp

Salpa thompsoni

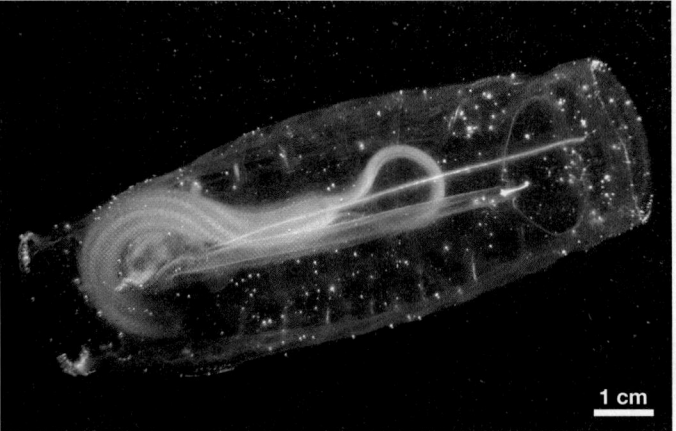

1 cm

FIGURE 35.10 Sea Squirts and Salps Live in Different Habitats.

35.3 What Is a Vertebrate?

The vertebrates are a monophyletic group distinguished by two synapomorphies:

- a column of cartilaginous or bony structures called **vertebrae,** which form along the dorsal side of the body; and

- a **cranium,** a bony, cartilaginous, or fibrous case that encloses the brain. (A skull consists of the cranium in jawless vertebrates, and the cranium and mandible—or jaw—in jawed vertebrates.)

The vertebral column is important because it protects the spinal cord. The cranium is important because it protects the brain along with sensory organs such as eyes. Together, the vertebrae and cranium protect the central nervous system (CNS) and key sensory structures (see Chapter 33).

The vertebrate brain develops as an outgrowth of the most anterior end of the dorsal hollow nerve cord, and it is important to the vertebrate lifestyle. Vertebrates are active predators and herbivores that must sense stimuli in their environments and make

rapid, directed movements in response. More complex brains enable more complex interactions with the environment.

In early vertebrates, the brain was divided into three regions with important sensory functions:

1. The **forebrain** housed the sense of smell;

2. the **midbrain** was associated with vision; and

3. the **hindbrain** was responsible for balance and, in some species, hearing.

All vertebrates living today have retained these three regions, but the structures and functions of the regions have evolved over time. For example, in the jawed vertebrates, or **gnathostomes** (pronounced *NATH-oh-stomes*), the hindbrain consists of enlarged regions called the **cerebellum** and **medulla oblongata.** Part of the forebrain also became elaborated into a large structure called the **cerebrum,** especially in birds and mammals. (The structure and functions of the vertebrate brain are analyzed in detail in Chapter 46.) The evolution of a large, three-part brain, protected by a hard cranium, was a key innovation in vertebrate evolution.

An Overview of Vertebrate Evolution

Vertebrates have been the focus of intense research for well over 300 years, in part because they are large and conspicuous and in part because they include humans. All this effort has paid off in an increasingly thorough understanding of how the vertebrates diversified. Let's consider what the data from the fossil record have to say about key events in vertebrate evolution and then examine the current best estimate of the phylogenetic relationships among vertebrates.

The Vertebrate Fossil Record   Vertebrate fossils are present in the Chengjiang formation of China, dated to the Cambrian explosion. The detailed vertebrate fossils show that the earliest members of this lineage lived in the ocean about 540 million years ago; they had streamlined, fishlike bodies and a cranium made of cartilage (see an artist's interpretation of an early vertebrate species called *Haikouichthys* in Figure 28.12). **Cartilage** is a strong but flexible tissue—found in your ear and the tip of your nose—that consists of scattered cells in a gel-like matrix of polysaccharides and protein fibers (Chapter 42).

A cartilaginous endoskeleton is a basic vertebrate feature; only in the bony fishes and their descendants, the **tetrapods** (literally, "four-footed"), does the skeleton become composed of bone in the adult. **Bone** is a tissue consisting of cells and blood vessels encased in a matrix made primarily of calcium phosphate, along with a small amount of protein fibers. In species with bony skeletons, growing bones are cartilaginous first and become bony only later in their development.

Following the appearance of vertebrates, the fossil record documents a series of key innovations that occurred as this lineage diversified (**FIGURE 35.11**):

- *Bony exoskeleton*   Fossil vertebrates from the early part of the Ordovician period, about 480 million years ago, are the first fossils to have bone. When bone first evolved, it did not occur in the endoskeleton ("inside-skeleton"). Instead, bone was deposited in scalelike plates that formed an **exoskeleton** ("outside-skeleton") (compare with arthropod exoskeletons in Chapter 34). Most of your skull is made up of this type of dermal ("skin") bone—making your skull similar to the "head shields" in Ordovician fishes. Based on the fossils' overall morphology, biologists infer that these animals swam with the aid of a notochord and that they breathed and fed by gulping water and filtering it through their pharyngeal gill slits. Presumably, the bony plates helped provide protection from predators.

- *Jaws*   The first fishes with jaws show up early in the Silurian, about 440 million years ago. The evolution of jaws was significant because it improved the ability of fish to use suction feeding to capture prey, and also to bite, meaning that vertebrates were no longer limited to suspension feeding or deposit feeding. Instead, they could make a living as herbivores or predators. Soon after, jawbones with teeth appear in the fossil record. With jaws and teeth, vertebrates became armed and dangerous. The fossil record shows that a spectacular

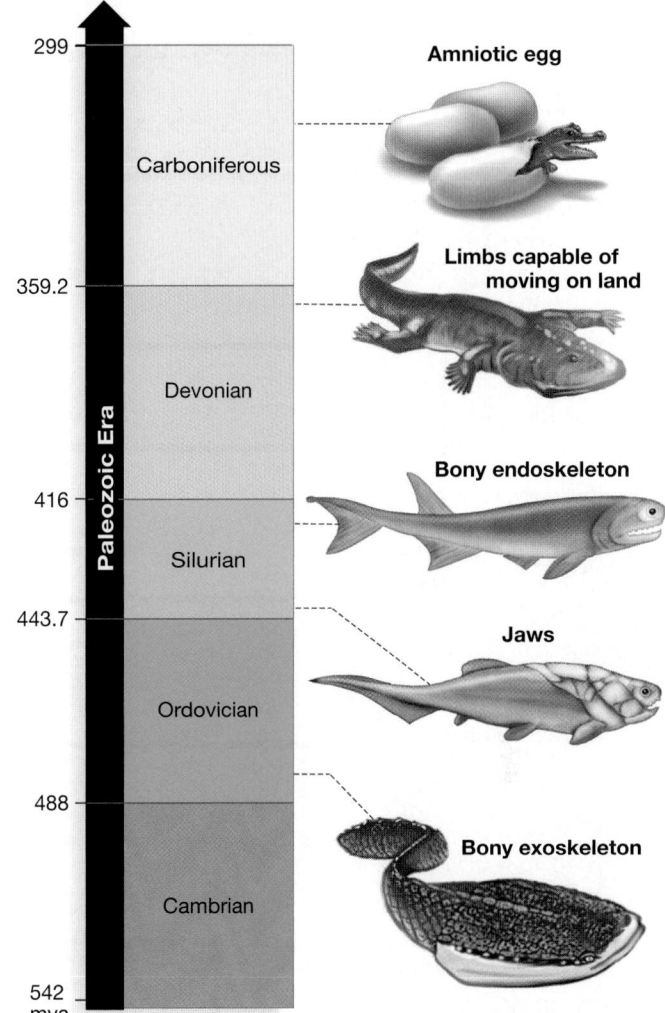

FIGURE 35.11 Timeline of the Early Vertebrate Fossil Record.

radiation of jawed fishes followed, filling marine and freshwater habitats.

- *Bony endoskeleton*   In some lineages of fishes from late in the Silurian period, the cartilaginous endoskeleton began to be stiffened by the deposition of bone. Unlike the dermal bone that had evolved earlier, the bony endoskeleton functioned to support movement—rapid swimming.

- *Limbs capable of moving on land*   The next great event in the evolution of vertebrates was the transition to living on land. The first animals that had limbs and were capable of moving on land are dated to about 365 million years ago, late in the Devonian. These were the first of the tetrapods—animals with four limbs.

- *Amniotic egg*   About 20 million years after the appearance of tetrapods in the fossil record, the first amniotes are present. The Amniota is a lineage of vertebrates that includes all tetrapods other than amphibians. The **amniotes** are named for a signature synapomorphy and adaptation: the amniotic egg. An **amniotic egg** is an egg that has membranes

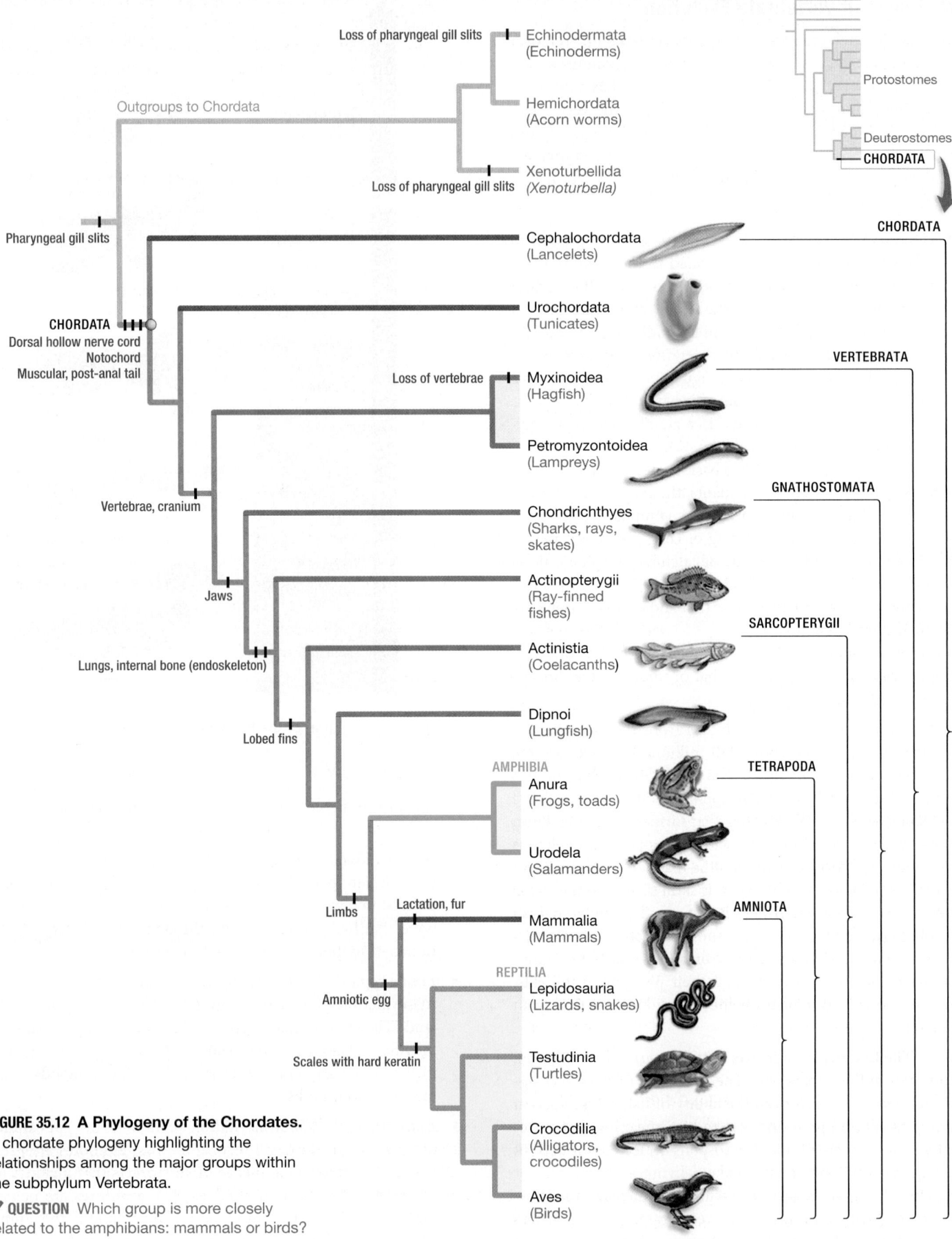

Loss of pharyngeal gill slits — Echinodermata (Echinoderms)

Outgroups to Chordata — Hemichordata (Acorn worms)

Loss of pharyngeal gill slits — Xenoturbellida (*Xenoturbella*)

Pharyngeal gill slits

CHORDATA
Dorsal hollow nerve cord
Notochord
Muscular, post-anal tail

Cephalochordata (Lancelets)

Urochordata (Tunicates)

Loss of vertebrae — Myxinoidea (Hagfish)

Petromyzontoidea (Lampreys)

Vertebrae, cranium

Chondrichthyes (Sharks, rays, skates)

Jaws

Actinopterygii (Ray-finned fishes)

Lungs, internal bone (endoskeleton)

Actinistia (Coelacanths)

Dipnoi (Lungfish)

Lobed fins

AMPHIBIA
Anura (Frogs, toads)

Urodela (Salamanders)

Limbs

Lactation, fur

Mammalia (Mammals)

REPTILIA
Amniotic egg

Lepidosauria (Lizards, snakes)

Scales with hard keratin

Testudinia (Turtles)

Crocodilia (Alligators, crocodiles)

Aves (Birds)

Protostomes

Deuterostomes

CHORDATA

CHORDATA

VERTEBRATA

GNATHOSTOMATA

SARCOPTERYGII

TETRAPODA

AMNIOTA

FIGURE 35.12 A Phylogeny of the Chordates.
A chordate phylogeny highlighting the relationships among the major groups within the subphylum Vertebrata.

✔ **QUESTION** Which group is more closely related to the amphibians: mammals or birds?

surrounding a food supply, a water supply, and a waste repository. These membranes provided support and extra surface area for gas exchange, allowing amniotes to produce larger, better-developed young.

To summarize, the fossil record indicates that vertebrates evolved through a series of major steps, beginning about 540 million years ago with vertebrates whose endoskeleton consisted of a notochord. The earliest vertebrates gave rise to fishes with bony exoskeletons. These fishes gave rise to the jawed vertebrates, including the cartilaginous fishes (sharks and rays) and several lineages of fishes with bony endoskeletons. After the tetrapods emerged and amphibians resembling salamanders began to live on land, the evolution of the amniotic egg paved the way for the diversification of amniotes—specifically, reptiles and the animals that gave rise to mammals. Do phylogenetic trees estimated from analyses of DNA sequence data agree or conflict with these conclusions?

Evaluating Molecular Phylogenies   **FIGURE 35.12** provides a phylogenetic tree that summarizes the relationships among living chordates, especially vertebrates, based on DNA and other molecular sequence data. The labeled bars on the tree indicate where major morphological innovations occurred as lineages diversified. Although the phylogeny of deuterostomes continues to be a topic of intense research, researchers are increasingly confident that the relationships described in Figure 35.12 are accurate. There are a few important take-home messages from the tree:

- *Outgroup to vertebrates*   According to the most recent molecular data, the urochordates are the closest living relatives of the vertebrates.

- *Living outgroup to gnathostomes*   The phylogenetic relationships among hagfish, lampreys, and the gnathostomes have been hotly debated. Two hundred years ago, hagfish and lampreys were grouped together as the class Cyclostomata ("round-mouthed"), based on their unique mouthparts. Then, starting in the 1970s, morphologists began to take a closer look and placed hagfish and lampreys in two independent lineages with lampreys as the sister group to gnathostomes. But molecular phylogenies have not supported this view, instead leaning toward the Cyclostomata grouping. So which is correct? Analysis of a new source of data—evolutionarily conserved microRNAs (Chapter 19)—has decided in favor of the Cyclostomata hypothesis. This result is important because it emphasizes that the ancestor to gnathostomes was probably not lamprey-like. Rather, some vertebrate characteristics appear to have been *lost* in the hagfish/lamprey lineages, especially in hagfish.

- *Fossil outgroup to gnathostomes*   The diverse lineages of extinct armored fishes are not included in the tree of living vertebrates. If pictured, they would occur as outgroups to the gnathostomes—more closely related to jawed vertebrates than are hagfish and lampreys.

- *"Fish"*   No monophyletic group includes all of the living fishlike lineages. Instead, "fishy" organisms are a series of

independent monophyletic groups (the living ones are indicated by blue branches in Figure 35.12). Taken together, they form what biologists call a **grade**—a sequence of lineages that are paraphyletic (that include some, but not all, of the descendants of a common ancestor). ✔ If you understand the difference between a paraphyletic and monophyletic group, you should be able to draw what Figure 35.12 would look like if there actually were a monophyletic group called "The Fishes." (See Table 28.2 if you need help.)

- *Mammals*   Mammals and reptiles are sister groups and thus are more closely related to each other than either is to the amphibians.

- *Birds*   Birds occur within the monophyletic group called reptiles. That is, birds *are* reptiles.

To understand what happened during the diversification of vertebrates, let's explore some of the major morphological innovations involved in feeding, movement, and reproduction in more detail. (For an overview of these themes, see Chapter 33.)

Key Innovations in Vertebrates

Among vertebrates, the most species-rich and ecologically diverse lineages are the ray-finned fishes and the tetrapods. Ray-finned fishes occupy habitats ranging from deepwater environments, which are perpetually dark, to shallow ponds that dry up each year. Tetrapods include the large herbivores and predators in terrestrial environments all over the world. About the same number of species of ray-finned fishes and tetrapods are living today (**FIGURE 35.13**).

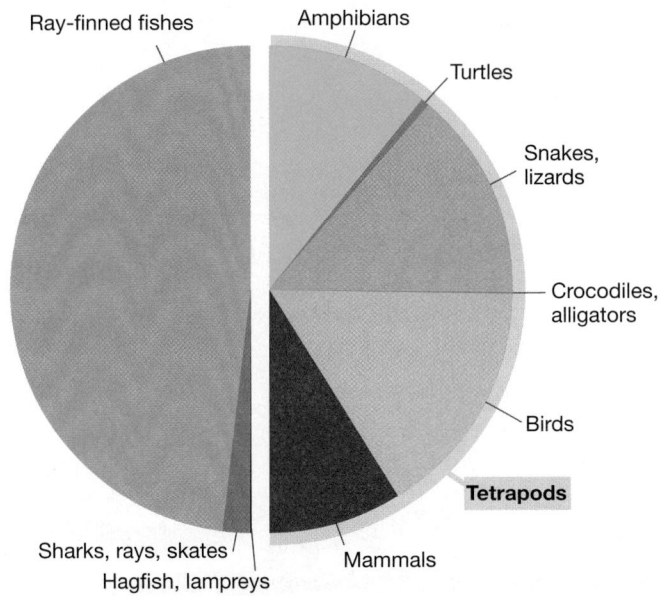

FIGURE 35.13 Relative Species Abundance among Vertebrates. Of the 63,000 species of described vertebrates, about half are ray-finned fishes and about half are tetrapods.

DATA: American Museum of Natural History; birdlife.org; fishbase.org; IUCN red list 2010; Reptile-database.org.

The Vertebrate Jaw The most ancient vertebrates in the fossil record had relatively simple mouthparts that lacked jaws. Until jaws evolved, vertebrates were not able to harvest food by biting.

The leading hypothesis for the origin of the jaw proposes that natural selection acted on developmental regulatory genes that determine the morphology of **gill arches,** which are curved regions of tissue between the gills. The jawless vertebrates have bars of cartilage that stiffen these gill arches. The gill-arch hypothesis proposes that mutation and natural selection increased the size of the most anterior arch and modified its orientation slightly, producing the first working jaw (**FIGURE 35.14**). Four lines of evidence support this hypothesis:

1. Both gill arches and jaws consist of flattened bars of bony or cartilaginous tissue that hinge and bend forward.

2. During development, the same population of cells gives rise to the muscles that move jaws and the muscles that move gill arches.

3. Unlike most other parts of the vertebrate skeleton, both jaws and gill arches are derived from specialized cells in the embryo called neural crest cells.

4. The expression patterns of key developmental regulatory genes, including *Hox* and *Dlx*, are similar in jaws and gill arches.

(a) Jawless vertebrate

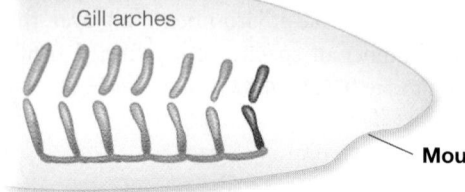

(b) Intermediate form (basal gnathostomes)

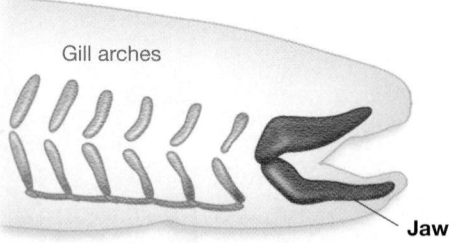

(c) Fossil shark

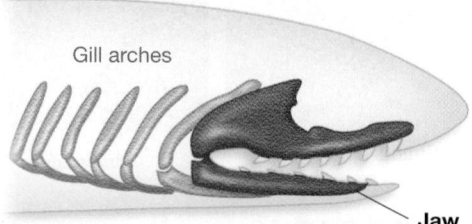

FIGURE 35.14 The Gill-Arch Hypothesis for the Evolution of the Jaw. (a) Gill arches support the gills in jawless vertebrates. **(b)** This intermediate form has yet to be found in the fossil record. **(c)** Fossil sharks that appeared later had more elaborate jaws.

Once jaws originated, changes in size and shape enabled a rapid diversification of feeding strategies from suction feeding to biting. Other modifications to the jaw also occurred:

- In most ray-finned fishes, the jaw is protrusible—meaning it can be extended to nip or bite at food.

- Several species-rich lineages of ray-finned fishes have a second specialized jaw called a **pharyngeal** ("throat") **jaw,** consisting of modified gill arches. It is located in the back of the throat and makes food processing particularly efficient. (For more on the structure and function of pharyngeal jaws, see Chapter 44.)

To summarize, the radiation of ray-finned fishes was triggered largely by the evolution and diversification of the jaw, by modifications that made it possible to protrude the jaw, and by the origin of the pharyngeal jaw. The story of tetrapods is different, however. Although jaw structure varies somewhat among tetrapod groups, the adaptation that triggered their initial diversification involved the ability to move and get to food, not to bite it and process it.

The Tetrapod Limb To understand how tetrapods made the water-to-land transition, consider the morphology and behavior of their closest living relatives, the lungfish (**FIGURE 35.15**). Most living species of lungfishes inhabit shallow, oxygen-poor water. To supplement the oxygen taken in by their gills, they have lungs and breathe air. (Note that although lungs originated before the divergence of ray-finned fishes and the Sarcopterygii, they serve primarily as a swim bladder to maintain buoyancy in ray-finned fishes.) Some lungfishes also have fleshy fins supported by bones and are capable of walking short distances along watery mudflats or the bottoms of ponds. In addition, some species can survive extended droughts by burrowing in mud.

Lungfish have a series of adaptations that allow them to survive on land for short periods. Could a lungfish-like organism have evolved into the first land-dwelling vertebrate with limbs?

Fossils provide strong links between the limbs of the ancestors of today's lungfish and the earliest land-dwelling vertebrates. **FIGURE 35.16** shows a phylogeny of the species involved. Note that during the Devonian, the fossil record documents a series of species that indicate a gradual transition from a lobe-like fin to a limb that could support walking on land. The fossil record

FIGURE 35.15 Lungfish Have Limb-like Fins. Some lungfishes can walk or crawl short distances on their fleshy fins.

is actually more complete than shown here—the figure is just a sample of the tetrapod and tetrapod-like species known from this interval.

The color coding in Figure 35.16 emphasizes that each fin or limb has a single bony element that is proximal (closest to the body; in blue) and then two bones that are distal (farther from the body) and arranged side by side (these are shown in red), followed by a series of distal elements (in orange). Because the structures are similar, and because no other animal groups have limb bones in this arrangement, the evidence for homology is strong.

The hypothesis that tetrapod limbs evolved from fish fins has also been supported by molecular genetic evidence. Recent work has shown that several regulatory proteins involved in the development of cartilaginous fish fins and ray-finned fish fins and the upper parts of mammal limbs are homologous. For example, the proteins produced by several different *Hox* genes are found at the same times during development and in the same locations in fins and limbs.

These data suggest that these appendages are patterned by the same genes. As a result, the data support the hypothesis that tetrapod limbs evolved from fins. Once the tetrapod limb evolved, natural selection elaborated it into diverse structures that are used for running, gliding, crawling, burrowing, swimming, and flying.

Once tetrapods made the water-to-land transition, they had to survive a number of new terrestrial challenges other than locomotion, such as keeping their eggs from drying out.

The Amniotic Egg Amphibians (frogs and toads, salamanders, and limbless species called caecilians) keep their eggs from drying out by laying their jelly-coated eggs in water. In contrast, reptiles (including birds) and the egg-laying mammals produce an amniotic egg, which has a protective covering that reduces the rate of drying significantly. Species that produce amniotic eggs lay them outside of water.

Amphibian eggs are surrounded by a single membrane, whereas amniotic eggs have four. As **FIGURE 35.17** shows, the outermost membrane of an amniotic egg encloses a supply of water in a protein-rich solution called **albumen.** Albumen cushions the developing embryo and provides nutrients. The three inner membranes surround the embryo itself, the yolk provided by the mother, and the waste from the embryo. The additional membranes are thought to be advantageous because they:

1. provide mechanical support—an important consideration outside of a buoyant aquatic environment; and

2. increase the surface area available for exchange of gases and other materials. Efficient diffusion of molecules is important because it allows females to lay larger eggs that hatch into larger, more independent young.

In addition, amniotic eggs are surrounded by a shell. This layer is leathery in lizards and snakes. It is stiffened by some calcium carbonate deposits in turtles and crocodiles, and by extensive

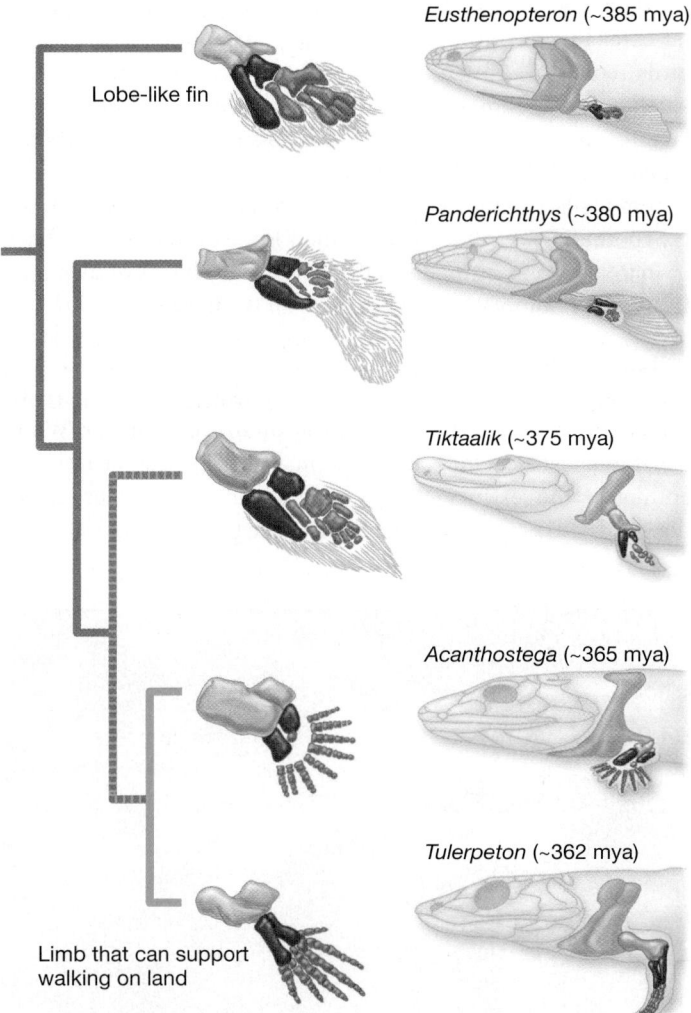

FIGURE 35.16 Evidence for a Fin-to-Limb Transition. The fossil record documents a series of species that indicate a gradual transition from a lobe-like fin to a limb that could support walking on land.

✔ **QUESTION** This phylogeny was estimated from traits other than the morphology of the limb. Why is this important in addressing how limbs evolved?

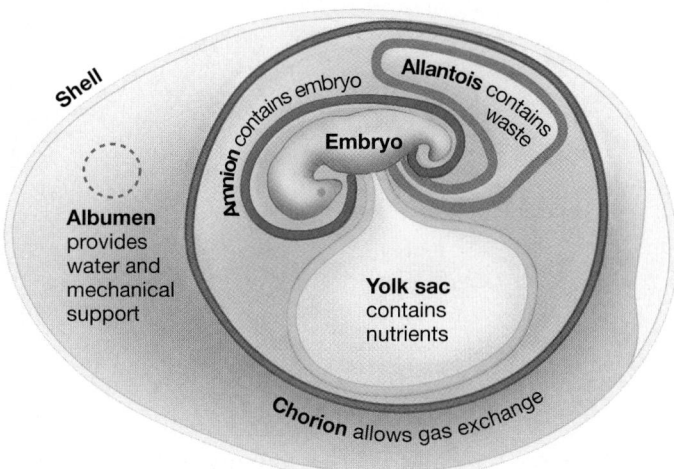

FIGURE 35.17 An Amniotic Egg. Amniotic eggs have four membrane-bound sacs.

calcium carbonate deposits in birds. Lizards, snakes, turtles, and crocodiles often bury their eggs in moist soils, but bird eggs are more watertight and are laid in nests that are exposed to air.

During the evolution of mammals, however, a second major innovation in reproduction occurred—one that eliminated the need for any type of egg laying.

The Placenta Egg-laying animals are said to be **oviparous,** while species that give birth are termed **viviparous** (Chapter 33). In many viviparous animals, females produce an egg that contains a nutrient-rich yolk. Instead of laying the egg, however, the mother retains it inside her body. In these **ovoviviparous** species, the developing offspring depends on the resources in the egg yolk.

In caecilians, some lizards, and most mammals, however, the eggs that females produce have little yolk. After fertilization occurs and the egg is retained, a combination of maternal and embryonic tissues produces a placenta within the uterus or oviduct. The embryo's contributions are the allantois and chorion—membranes that are involved in gas exchange in an amniotic egg. In the placenta, tissues derived from the allantois and chorion are also involved in the diffusion of gases, nutrients, and wastes.

The **placenta** is an organ that is rich in blood vessels and that facilitates a flow of oxygen and nutrients from the mother to the developing offspring, as well as the removal of nitrogenous wastes and carbon dioxide from the offspring (**FIGURE 35.18**). After a development period called **gestation,** the embryo emerges from the mother's body.

Why did viviparity and the placenta evolve? The leading hypotheses are that retaining the embryo inside the body has several advantages:

1. Offspring develop at a more constant, favorable temperature.

2. Offspring are protected.

3. Offspring are portable—mothers are not tied to a nest.

In effect, a placenta and viviparity are mechanisms for increasing the level of **parental care:** behavior by a parent that improves the ability of its offspring to survive. What other types of parental care evolved in vertebrates?

Parental Care In species that provide extensive parental care—even if only in the form of exceptionally large amounts of yolk in eggs—mothers participate in a fitness trade-off (see Chapter 25). They can produce fewer offspring, but the offspring that they do produce have a higher probability of survival.

In animals, mechanisms of parental care include incubating eggs to keep them warm during early development, keeping young warm and dry, supplying young with food, and protecting them from danger. In some insect and frog species, mothers carry eggs or newly hatched young on their bodies. In fishes, parents commonly guard eggs during development and fan them with oxygen-rich water.

Mammals and birds provide particularly extensive parental care. In both groups, the mother (and often the father, in birds) continues to feed and care for individuals after birth or hatching—sometimes for many years. Female mammals also **lactate**—meaning that they produce milk and use it to feed their offspring after birth (**FIGURE 35.19**).

Among large animals, the evolution of extensive parental care is hypothesized to be a major reason for the evolutionary success of mammals and birds. In both lineages, mothers produce relatively small numbers of large, high-quality offspring.

Wings and Flight Following the transition from water to land, the origin of the amniotic egg, and the elaboration of parental care, amniotes made another big transition—into the air. Wings and flight evolved independently in three lineages of tetrapod amniotes: the extinct flying reptiles called pterosaurs (pronounced *TARE-oh-sors*), the bats, and the birds.

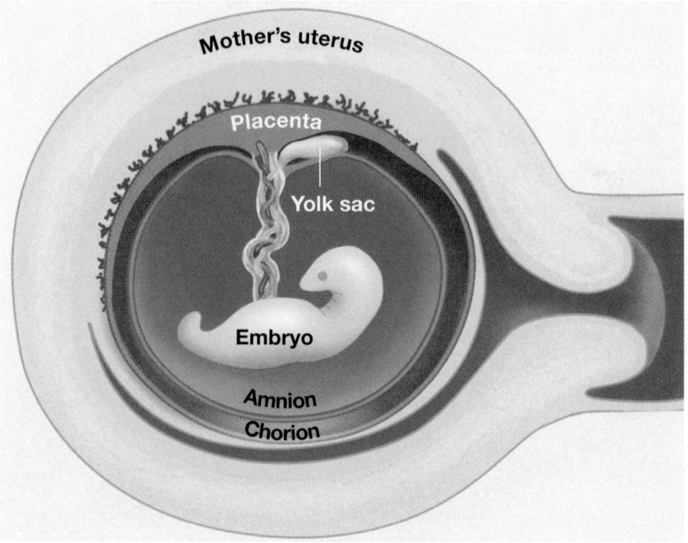

FIGURE 35.18 A Placenta Nourishes a Fetus Internally.

✔ **QUESTION** Compare the relative size of the yolk sac here with that in the amniotic egg (Figure 35.17).

FIGURE 35.19 Parental Care Is Extensive in Mammals. Female mammals, such as this timber wolf, produce highly nutritious milk to nourish their young.

How did flight evolve? The best data sets on this question involve feathered flight in birds. Since the early 2000s, paleontologists have discovered a spectacular series of feathered dinosaur fossils. The newly discovered species and exceptionally well-preserved feather specimens address key questions about the evolution of birds, feathers, and flight:

- **Did birds evolve from dinosaurs?** On the basis of skeletal characteristics, all of these recently discovered fossil species clearly show that birds are part of the monophyletic group called dinosaurs. That is—birds *are* dinosaurs.

- **How did feathers evolve?** The early fossils support a model of feathers evolving in a series of steps, beginning with simple projections from the skin (popularly called "dinofuzz") and culminating in the complex, branched structures familiar today. Although diverse, feathers observed in many non-avian dinosaurs were not strong enough to support gliding or flight. They may have been used for insulation and/or mating displays—recent analysis of pigments in fossil feathers have shown striking color patterns in some dinosaurs (**FIGURE 35.20**).

- **Did birds begin flying from the ground up or from the trees down?** More specifically, did flight evolve with running species that began to jump and glide or make short flights, with the aid of feathers to provide lift? Or did flight evolve from tree-dwelling species that used feathers to glide from tree to tree, much as flying squirrels do today? This question is still unresolved. The observation that many early feathered dinosaurs had "flight" feathers on their legs adds an intriguing piece to the puzzle (see Figure 35.20).

Once dinosaurs evolved feathers and took to the air, the fossil record shows that a series of adaptations made powered, flapping flight increasingly efficient (**FIGURE 35.21**).

- Most dinosaurs have a flat sternum ("breastbone"), but the bird sternum has a projection called the keel, which provides a large surface area to which flight muscles attach.

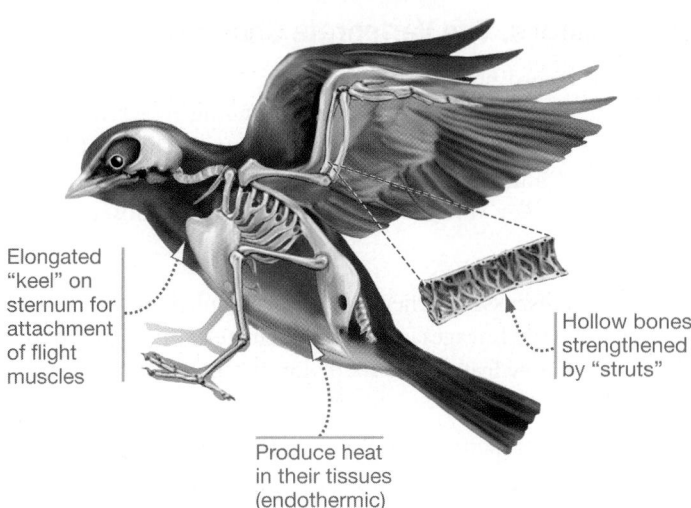

Elongated "keel" on sternum for attachment of flight muscles

Hollow bones strengthened by "struts"

Produce heat in their tissues (endothermic)

FIGURE 35.21 In Addition to Feathers, Birds Have Several Features That Enable Flight.

- Birds are extraordinarily light for their size, primarily because they have a drastically reduced number of bones and because their larger bones are thin walled and hollow—though strengthened by bony "struts."

- Birds are capable of sustained activity year-round because they are **endotherms** ("inside-heated")—they maintain a high body temperature by producing heat in their tissues.

From dinosaurs that jumped or glided from tree to tree, birds have evolved into extraordinary flying machines.

To summarize, the evolution of the jaw gave tetrapods the potential to capture and process a wide array of foods. With limbs, they could move efficiently on land in search of food. The amniotic egg and placenta further enabled vertebrates to diversify into terrestrial niches, and parental care increased the survivorship of their offspring. Finally, some tetrapods were able to defy the pull of gravity, expanding their realm into the air.

check your understanding

If you understand that . . .
- Vertebrates have a distinctive body plan: bilateral symmetry with vertebrae that protect a spinal cord and a cranium that protects the brain.
- An array of key innovations occurred during the evolution of vertebrates: jaws, the tetrapod limb, the amniotic egg, the placenta, extensive parental care, and flight.

✔ You should be able to . . .
1. Explain the adaptive significance of each key innovation.
2. Compare adaptations in protostomes (Chapter 34) and deuterostomes for living on land.

Answers are available in Appendix A.

FIGURE 35.20 Feathers Evolved in Dinosaurs. This artist's depiction of what *Anchiornis huxleyi,* a feathered dinosaur, might have looked like in life incorporates recent data on fossil feather colors. The colors may have been important in mating displays. Unlike modern birds, many feathered dinosaurs had large feathers on their legs.

Key Lineages: The Vertebrate Chordates

The fossil record provides an increasingly clear picture of early vertebrate evolution; intensive and continuing research has explored the origins of key vertebrate innovations such as the jaw, the tetrapod limb, and the amniotic egg. Now that you've been introduced to key vertebrate features, let's take a closer look at the key lineages.

"Fish"   As discussed earlier, fish are a paraphyletic group consisting of diverse lineages, many of which are extinct. Four of the boxes that follow feature the six major lineages of fishes that are living today:

- Myxinoidea (Hagfish) and Petromyzontoidea (Lampreys)
- Chondrichthyes (Sharks, Rays, Skates)
- Actinopterygii (Ray-Finned Fishes)
- Actinistia (Coelacanths) and Dipnoi (Lungfish)

Amphibia   Unlike fish, the **amphibians** form a monophyletic group, the Amphibia. Most amphibians lay their eggs in water and undergo metamorphosis from an aquatic larva to a terrestrial or semiterrestrial adult. The three lineages of amphibians are discussed together in one box:

- Amphibia (Frogs and Toads, Salamanders, Caecilians)

Mammalia   **Mammals** are a monophyletic group, today comprising three major lineages:

- Mammalia > Monotremata (Platypuses, Echidnas)
- Mammalia > Marsupiala (Marsupials)
- Mammalia > Eutheria (Placental Mammals)

The relationships among these groups are shown in **FIGURE 35.22**. Monotremes lay eggs; marsupials have a poorly developed placenta

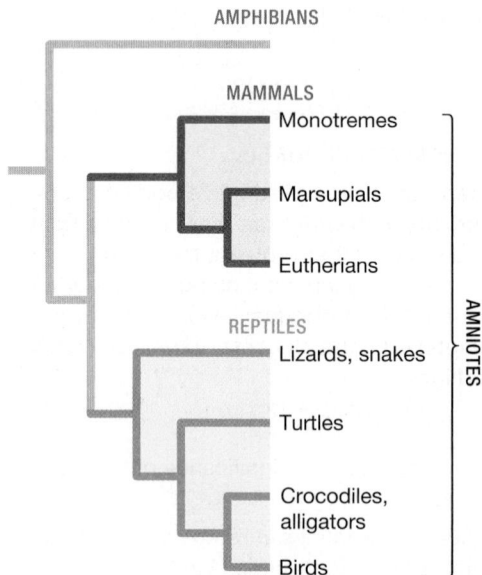

FIGURE 35.22 Mammals and Reptiles Are Monophyletic Groups.

but well-developed pouch for rearing offspring; eutherians have a well-developed placenta and extended pregnancy.

Mammals are endotherms that maintain high body temperatures by oxidizing large amounts of food and generating large amounts of heat. Instead of insulating themselves with feathers, though, mammals retain heat because the body surface is covered with layers of hair or fur. Endothermy evolved independently in mammals and birds. In both groups, endothermy is thought to be an adaptation that enables individuals to maintain high levels of activity—particularly at night or during cold weather.

In addition to being endothermic and having fur, mammals have **mammary glands**—unique structures that make lactation possible. The evolution of mammary glands gave mammals the ability to provide their young with particularly extensive parental care. Mammals are also the only vertebrates with cheek muscles and lips—traits that make suckling milk possible—and the only vertebrates that have a lower jaw formed from a single bone. ✔ **You should be able to mark the origin of mammary glands and fur on Figure 35.22.**

Mammals evolved when dinosaurs and other reptiles were the dominant large herbivores and predators in terrestrial and aquatic environments. The earliest mammals in the fossil record appear about 195 million years ago; most were small animals that were probably active only at night. Many of the 5500 species of mammals living today have good nocturnal vision and a strong sense of smell, as their ancestors presumably did.

The adaptive radiation that gave rise to today's diversity of mammalian orders did not take place until after the dinosaurs went extinct. Long after the dinosaurs were gone, the mammals diversified into lineages that included large herbivores and large predators—ecological roles that had once been filled by dinosaurs and the ocean-dwelling, extinct reptiles called ichthyosaurs and plesiosaurs.

Reptilia   The **reptiles** are a monophyletic group and represent the second major living lineage of amniotes besides mammals (Figure 35.22). One major diagnostic feature distinguishing the reptilian and mammalian lineages is the number and placement of openings in the side of the skull. These skull openings are important: Jaw muscles required for biting and chewing pass through them and attach to bones on the upper part of the skull.

Reptiles have several adaptations for life on land. Their skin is made watertight by a layer of scales made of the protein keratin. Reptiles breathe air through well-developed lungs and lay shelled, amniotic eggs. In snakes and lizards, the egg has a leathery shell; in other reptiles, the shell includes some calcium carbonate.

In many reptiles, the sex of an individual is determined by the environment it experiences during early development. In certain species, for example, high temperatures produce mostly males while low temperatures produce mostly females.

The reptiles include the dinosaurs, ichthyosaurs, plesiosaurs, pterosaurs (flying reptiles), and other extinct lineages that flourished from about 250 million years ago until the mass extinction at the end of the Cretaceous period, 65 million years ago. Today the Reptilia are represented by four major lineages:

- Reptilia > Lepidosauria (Lizards, Snakes)
- Reptilia > Testudinia (Turtles)
- Reptilia > Crocodilia (Crocodiles, Alligators)
- Reptilia > Aves (Birds)

Except for birds, almost all of the reptiles living today are **ectotherms** ("outside-heated")—meaning that individuals do not use internally generated heat to regulate their body temperature. It would be a mistake, however, to conclude that reptiles other than birds do not regulate their body temperature closely. Reptiles bask in sunlight, seek shade, and perform other behaviors to keep their body temperature at an appropriate level.

Chordata > Vertebrata > Myxinoidea (Hagfish) and Petromyzontoidea (Lampreys)

Hagfish and lampreys are the only living jawless vertebrates, comprising about 120 species. Their relationship to jawed vertebrates has been hotly debated; however, the most recent microRNA evidence suggests that the two groups are more closely related to each other than either is to the jawed vertebrates. These data indicated that hagfish and lampreys, collectively the cyclostomes ("round mouths"), may have *lost* many vertebrate characters over time.

Hagfish and lampreys have long, slender bodies and are aquatic (**FIGURE 35.23**). Hagfish are strictly marine; lampreys include both marine and freshwater forms. Most species are less than a meter long when fully grown. Adult hagfish lack any sort of vertebral column, but adult lampreys have small pieces of cartilage along the length of their dorsal hollow nerve cord. Both hagfish and lampreys have brains protected by a cranium, as do all vertebrates.

Feeding Hagfish are scavengers and predators. They deposit feed on the carcasses of dead fish and whales, and some are thought to burrow through ooze at the bottom of the ocean, feeding on polychaete worms and other buried prey. Lampreys, in contrast, are ectoparasites. They attach to fish or other hosts by suction and then use spines in their mouth and tongue to rasp a hole in the side of their victim. Once the wound is open, they suck blood and other body fluids.

Movement Hagfish and lampreys have a well-developed notochord and swim by making undulating movements. Hagfish are particularly flexible, sometimes forming "knots" with their bodies to rid themselves of the copious slime they create as a defense mechanism when disturbed. Lampreys can move themselves upstream, against the flow of water, by attaching their suckers to rocks and looping the rest of their body forward, like an inchworm. Both groups have tail fins, and lamprey have a dorsal (back) fin; but neither group has the paired lateral (side) appendages found in other vertebrates.

Reproduction Virtually nothing is known about hagfish mating or embryonic development. Some lampreys live in freshwater; others are **anadromous**—meaning they spend their adult life in the ocean but swim up streams to breed. Fertilization is external, and adults die after breeding once. Lamprey eggs hatch into larvae that look and act like lancelets (cephalochordates). The larvae burrow into sediments and suspension feed for several years before metamorphosing into free-swimming adults.

(a) Hagfish

Eptatretus stoutii

Petromyzon marinus

(b) Lamprey

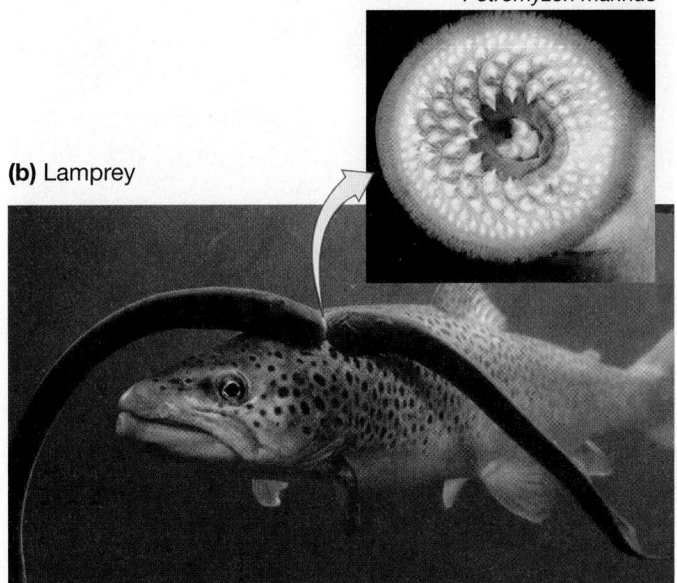

FIGURE 35.23 Hagfish and Lampreys Are Jawless Vertebrates.

Chordata > Vertebrata > Chondrichthyes (Sharks, Rays, Skates)

This lineage is distinguished by a specialized type of reinforced cartilaginous skeleton (*chondrus* is the Greek word for cartilage). Along with jawed vertebrates, or gnathostomes, sharks, rays, and skates also have jaws. They, like the other fish-like gnathostomes, also have paired lateral fins. Paired fins were an important evolutionary innovation because they stabilize the body during rapid swimming—keeping it from pitching up or down, yawing to one side or the other, or rolling. ✔ **You should be able to indicate the origin of paired fins and the chondrichthyan form of cartilage on Figure 35.12.**

Most of the 1200 species of sharks, rays, and skates are marine, though a few species live in freshwater. Sharks have streamlined, torpedo-shaped bodies and an asymmetrical tail—the dorsal portion is longer than the ventral portion (**FIGURE 35.24a**). In contrast, the dorsal–ventral plane of the body in rays and skates is strongly flattened (**FIGURE 35.24b**).

Feeding A few species of sharks and rays suspension feed on plankton, but most species in this lineage are predators. Skates and rays hunt along the ocean floor for snails, clams, and other benthic animals; electric rays capture their prey by stunning them with electric discharges of up to 200 volts. Most sharks, in contrast, are active hunters that chase down prey in open water and bite them. The larger species of sharks feed on large fish or marine mammals. In many marine ecosystems, sharks are referred to as the "top predator" because they are at the top of the food chain—nothing eats them. Yet the largest of all sharks, the whale shark, is a suspension feeder. Whale sharks filter plankton out of water as it passes over their gills.

Movement Rays and skates swim by flapping their greatly enlarged pectoral fins. Sharks swim by undulating their bodies from side to side and beating their large tails.

Reproduction Sharks use internal fertilization, and fertilized eggs may be shed into the water or retained until the young are hatched and well developed. In some viviparous species, embryos are attached to the mother by specialized tissues in a placenta, where the exchange of gases, nutrients, and wastes takes place. (A placenta evolved in certain shark lineages independently of its evolution elsewhere on the tree of life.) Skates are oviparous, but rays are viviparous.

(a) Sharks are torpedo shaped.

Prionace glauca

(b) Skates and rays are flat.

Dasyatis americana

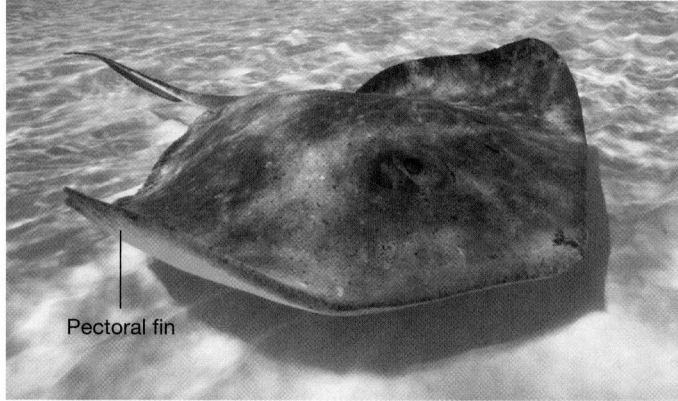

Pectoral fin

FIGURE 35.24 Sharks and Rays Have Cartilaginous Skeletons.

Chordata > Vertebrata > Actinopterygii (Ray-Finned Fishes)

Actinopterygii (pronounced *ack-tin-op-teh-RIJ-ee-i*) means "ray-finned." These **ray-finned fishes** have fins that are supported by long, bony rods arranged in a ray pattern, in addition to a skeleton made of bone (**FIGURE 35.25**). Their bodies are covered with interlocking scales that provide a stiff but flexible covering, and many have a gas-filled **swim bladder** that evolved from the lungs found in the earliest lineages of fish. The evolution of the swim bladder was an important innovation because it allowed ray-finned fishes to avoid sinking. Most tissues are heavier than water, so the bodies of aquatic organisms tend to sink. Sharks and rays, for example, have to swim to avoid sinking. But in ray-finned fishes, gas is added to the bladder when a fish swims down; gas is removed when the fish swims up. In this way, ray-finned fishes maintain neutral buoyancy in water of various depths and thus various pressures. ✔ **You should be able to indicate the origin of rayed fins and the swim bladder on Figure 35.12.**

The actinopterygians are the most successful vertebrate lineage based on number of species, duration in the fossil record, and extent of habitats occupied. About 30,500 species of ray-finned fishes are known, including the smallest known vertebrate—a species whose adult females average less than 8 mm in length.

(Continued on next page)

Holocentrus rufus

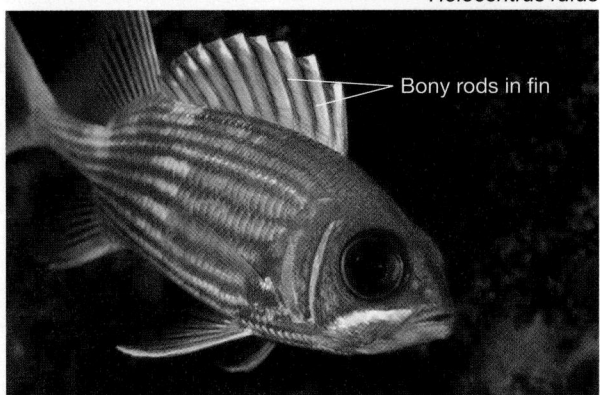

Bony rods in fin

FIGURE 35.25 In Ray-Finned Fishes, Fins Are Supported by Long, Bony Rods.

The most important major lineage of ray-finned fishes is the Teleostei. About 96 percent of all living fish species, including familiar groups like the tuna, trout, cod, and goldfish, are teleosts. The teleosts underwent an adaptive radiation about the same time that mammals did.

Feeding Teleosts can suck food toward their mouths, grasp it with their protrusible jaws, and then process it with teeth on their jaws and with pharyngeal jaws in their throat. The size and shape of the mouth, the jaw teeth, and the pharyngeal jaw teeth all correlate with the type of food consumed. For example, most predatory teleosts have long, spear-shaped jaws armed with spiky teeth, as well as bladelike teeth on their pharyngeal jaws. Besides being major predators, ray-finned fishes are the most important large herbivores in both marine and freshwater environments.

Movement Like other fish, ray-finned fishes swim by alternately contracting muscles on the left and right sides of their bodies from head to tail, resulting in rapid, side-to-side undulations. Their bodies are streamlined to reduce drag in water. Teleosts have a flexible, symmetrical tail, which reduces the need to use their pectoral (side) fins as steering and stabilizing devices during rapid swimming.

Reproduction Most ray-finned fish species rely on external fertilization and are oviparous; some species have internal fertilization with external development; still others have internal fertilization and are viviparous. Although it is common for fish eggs to be released in the water and left to develop on their own, parental care occurs in some species. Parents may carry fertilized eggs on their fins, in their mouth, or in specialized pouches to guard them until the eggs hatch. In other cases the eggs are laid into a nest that is actively guarded and cared for. In freshwater teleosts, offspring develop directly; but marine species have larvae that undergo metamorphosis into a very different juvenile form, which then grows into an adult.

 Chordata > Vertebrata > Actinistia (Coelacanths) and Dipnoi (Lungfish)

Although coelacanths (pronounced *SEEL-uh-kanths*) and lungfish represent independent lineages, they are sometimes grouped together and called **lobe-finned fishes.** Lobe-finned fishes are common and diverse in the fossil record in the Devonian period, about 400 million years ago, but only eight species are living today. They are important, however, because they represent a crucial evolutionary link between the fishes and the tetrapods. Early fishes in the fossil record have fins supported by stiff structures that extend from a base of bone. In ray-finned fishes, the bony elements are reduced. But in lobefins the bony elements extend down the fin, and they branch—similar to the bony elements in the limbs of tetrapods (**FIGURE 35.26**). ✔ You should be able to indicate the origin of extensive fin bones on Figure 35.12.

Coelacanths are marine and occupy habitats 150–700 m below the surface. In contrast, lungfish live in shallow, freshwater ponds and rivers (see Figure 35.15). As their name implies, lungfish have lungs and breathe air when oxygen levels in their habitats drop. Some species burrow in mud and enter a quiescent, sleeplike state when their habitat dries up during each year's dry season.

Feeding Coelacanths prey on fish. Lungfish are omnivorous ("all-eating"), meaning that they eat algae and plant material as well as animals.

Movement Coelacanths and lungfish swim by undulating their bodies; some lungfish can also use their fins to walk along pond bottoms.

Reproduction Sexual reproduction is the rule; fertilization is internal in coelacanths and external in lungfish. Coelacanths are ovoviviparous; lungfish lay eggs. Lungfish eggs hatch into larvae that resemble juvenile salamanders.

Latimeria menadoensis

Fleshy lobes supported by bones

FIGURE 35.26 Coelacanths Are Lobe-Finned Fishes.

Chordata > Vertebrata > Amphibia (Frogs and Toads, Salamanders, Caecilians)

Amphibians are found throughout the world and occupy ponds, lakes, or moist terrestrial environments. Translated literally, their name means "both-sides-living." The name is appropriate because adults of most amphibians feed on land but lay their eggs in water (**FIGURE 35.27**). In many amphibians, gas exchange occurs exclusively or in part across their moist, mucus-covered skin. ✔ You should be able to indicate the origin of "skin-breathing" on Figure 35.12. The 6800 named species of amphibians living today form three distinct clades: (**1**) frogs and toads, (**2**) salamanders, and (**3**) caecilians (pronounced *suh-SILL-ee-uns*). Caecilians lack limbs, and so resemble worms or snakes. They burrow underground in wet tropical regions.

Feeding Adult amphibians are carnivores. Most frogs and toads are sit-and-wait predators that use their long, extensible tongues to capture passing prey. Salamanders also have an extensible tongue, which some species use in feeding. Terrestrial caecilians prey on earthworms and other soil-dwelling animals; aquatic forms eat vertebrates and small fish.

Movement Most amphibians have four well-developed limbs. In water, frogs and toads move by kicking their hind legs to swim; on land, they kick their hind legs out to jump or hop. Salamanders walk on land; in water they undulate their bodies to swim. Caecilians lack limbs; terrestrial forms burrow in moist soils.

Reproduction Reproduction is sexual, breeding occurs in water, and larvae undergo a dramatic metamorphosis into land-dwelling adults. For example, the fishlike tadpoles of frogs and toads develop limbs, and their gills are replaced with lungs. Frogs have external fertilization, but salamanders and caecilians have internal fertilization. Frogs and most salamanders are oviparous, but many caecilians are viviparous.

Rana temporaria

Eggs

FIGURE 35.27 Amphibians Are the Most Ancient Tetrapods.

Chordata > Vertebrata > Mammalia > Monotremata (Platypuses, Echidnas)

The **monotremes** are the most basal lineage of mammals living today and are found in nature only in Australia. Monotremes lay eggs and have metabolic rates that are lower than those of other mammals. Five species exist: one species of platypus (**FIGURE 35.28a**) and four species of echidnas (**FIGURE 35.28b**).

Feeding Monotremes have a leathery beak or bill. The platypus feeds on insect larvae, mollusks, and other small animals in streams. Echidnas feed on ants, termites, and earthworms.

Movement Platypuses swim with the aid of their webbed feet and walk when on land. Echidnas walk on their four legs.

Reproduction Platypuses lay their eggs in a burrow, while echidnas keep their eggs in a pouch on their belly. Young monotremes hatch quickly, and the mother must continue keeping them warm and dry for another four months. Like other mammals, monotremes produce milk and nurse their young. They lack well-defined nipples, however, and instead secrete milk from glands in the skin.

(a) Platypus

Ornithorhynchus anatinus

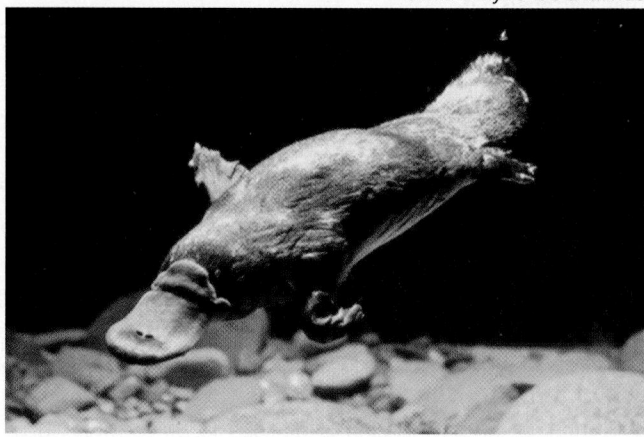

(b) Echidna

Tachyglossus aculeatus

FIGURE 35.28 Platypuses and Echidnas Lay Eggs.

Chordata > Vertebrata > Mammalia > Marsupiala (Marsupials)

The 350 known species of **marsupials** live in the Australian region and the Americas and include the familiar opossums (**FIGURE 35.29**), kangaroos, wallabies, and koala.

Feeding Marsupials are herbivores, omnivores, or carnivores. In many cases, convergent evolution has resulted in marsupials that are similar to placental species in morphology and way of life. For example, a recently extinct marsupial called the Tasmanian wolf was a long-legged hunter similar to the wolves of North America and northern Eurasia. A species of marsupial native to Australia specializes in eating ants and looks and acts much like the South American anteater, which is not a marsupial.

Movement Marsupials move by crawling, gliding, walking, running, or hopping.

Reproduction Although females have a placenta that nourishes embryos during development, the young are born after a short embryonic period and are poorly developed. They crawl from the opening of the female's reproductive tract to her nipples, located within a pouch, where they suck milk. They stay attached to their mother until they grow large enough to move independently. ✔ You should be able to indicate the origin of the placenta and viviparity—traits that are also found in eutherian mammals—on Figure 35.22.

Didelphis virginiana

FIGURE 35.29 Marsupials Give Birth after a Short Embryonic Period. Opossums are the only marsupials living in North America.

Chordata > Vertebrata > Mammalia > Eutheria (Placental Mammals)

The 5100 described species of **placental mammals,** or **eutherians,** are distributed worldwide. They are far and away the most species-rich and morphologically diverse group of mammals.

Biologists group placental mammals into 29 lineages called orders. The six most species-rich orders are the rodents (rats, mice, squirrels; 2255 species), bats (1150 species), insectivores (hedgehogs, moles, shrews; 450 species), artiodactyls (pigs, hippos, whales, deer, sheep, cattle, horses; 329 species), carnivores (dogs, bears, cats, weasels, seals; 285 species), and primates (lemurs, monkeys, apes, humans; 414 species).

Feeding The size and structure of the teeth correlate closely with the diet of placental mammals. Herbivores have large, flat teeth for crushing leaves and other coarse plant material; predators have sharp teeth that are efficient at biting and tearing flesh. The structure of the digestive tract also correlates with the placental mammals' diet. In some plant eaters, for example, the stomach hosts unicellular organisms that digest cellulose and other complex polysaccharides.

Movement In placental mammals, the structure of the limb correlates closely with the type of movement performed. Eutherians fly, glide, run, walk, swim, burrow, or swing from trees. Hindlimbs are reduced or lost in aquatic groups such as whales and dolphins, which swim by undulating their bodies.

Reproduction Eutherians have internal fertilization and are viviparous. An extensive placenta develops from a combination of maternal and fetal tissues; and at birth, young are much better developed than in marsupials—some are able to walk or run minutes after emerging from the mother. ✔ You should be able to indicate the origin of delayed birth (extended development before birth) on Figure 35.22. All eutherians feed their offspring milk until the young have grown large enough to process solid food. A prolonged period of parental care, extending beyond the nursing stage, is common as offspring learn how to escape predators and find food on their own. Some eutherians are highly social (**FIGURE 35.30**).

Suricata suricatta

FIGURE 35.30 Eutherians Are Familiar and Diverse.

Most lizards and snakes are small reptiles with elongated bodies and scaly skin. As these animals grow, they can shed the keratinized surface layer of their skin—all in one piece in snakes, and in flakes in lizards. ✔ You should be able to indicate the origin of scaly skin on Figure 35.22. Most lizards have well-developed jointed legs, but snakes are limbless (**FIGURE 35.31**). The hypothesis that snakes evolved from limbed ancestors is supported by the presence of vestigial hip and leg bones in boas and pythons. About 8900 species of lizards and snakes are alive now.

Feeding Small lizards prey on insects. Although most of the larger lizard species are herbivores, the spectacular 3-meter-long monitor lizard from the island of Komodo is a predator that can eat deer. Snakes are carnivores; some subdue their prey by injecting poison through modified teeth called fangs. Snakes prey primarily on small mammals, amphibians, and invertebrates, which they swallow whole—usually headfirst.

Movement Lizards crawl, climb, or run on their four limbs. Some lizards, such as the water dragons, use their powerful tails for swimming. Snakes and lizards that are limbless burrow through soil, crawl over the ground, or climb trees by undulating their bodies. One genus of snakes in Asia has several adaptations that enable them to glide down from treetops.

Morelia viridis

FIGURE 35.31 Snakes Are Limbless Predators.

Reproduction Although most lizards and snakes lay eggs, many are ovoviviparous; viviparity has also evolved numerous times in lizards. Most species reproduce sexually, but asexual reproduction, via the production of eggs by mitosis (parthenogenesis), is known to occur in six groups of lizards (including the Komodo dragon) and one snake lineage.

The 320 known species of turtles inhabit freshwater, marine, and terrestrial environments throughout the world. The testudines are distinguished by a shell composed of bony plates that fuse to the vertebrae and ribs (**FIGURE 35.32**). The shell functions in protection from predators—when threatened, most turtles can withdraw their head and legs into it. ✔ You should be able to indicate the origin of the turtle shell on Figure 35.22. The turtles' skulls are highly modified versions of the skulls of other reptiles. Turtles lack teeth, but their jawbone and lower skull form a bony beak. They range in size from the 2-m-long, 900-kg leatherback sea turtle to the 8-cm-long, 140-g speckled padloper tortoise. Tortoises are known for their longevity—some individuals are known to have lived well over 150 years.

Feeding Turtles are either carnivorous—feeding on whatever animals they can capture and swallow—or herbivorous. They may also scavenge dead material. Most marine turtles are carnivorous. Leatherback turtles, for example, feed primarily on jellyfish, and they are only mildly affected by the jellyfish's stinging cnidocytes (see Chapter 33). In contrast, species in the lineage of terrestrial turtles called the tortoises are plant eaters.

Testudo hermanni

FIGURE 35.32 Turtles Have a Shell Consisting of Bony Plates.

Movement Turtles swim, walk, or burrow. Some fully aquatic species have feet that are modified to function as flippers.

Reproduction All turtles are oviparous. Other than digging a nest prior to depositing eggs, parental care is lacking. Some aquatic species navigate long distances to their natal beaches with pinpoint accuracy, to lay eggs.

Chordata > Vertebrata > Reptilia > Crocodilia (Crocodiles, Alligators)

About 24 living species of crocodiles and alligators are known. Most are tropical and live in freshwater or marine environments. They have eyes located on the top of their heads, nostrils located at the top of their long snouts, and other adaptations that allow them to sit semi-submerged in water for long periods of time, breathing air and visually monitoring activity around them (**FIGURE 35.33**).

Feeding Crocodilians are predators. Like other toothed vertebrates, except mammals, their jaws are filled with conical teeth that are continually replaced as they fall out during feeding. Their usual method of killing small prey is by biting through the body wall. Large prey are usually subdued by drowning. Crocodilians eat amphibians, turtles, fish, birds, and mammals; one of the common hunting strategies of crocodiles is to leap out of the water to snatch unwary prey that have come to drink.

Movement Crocodiles and alligators walk on land. In water they swim with the aid of their large, muscular tails.

Reproduction Although crocodilians are oviparous, parental care is extensive. Eggs are laid in earth-covered nests or mounds of vegetation that are guarded by the parents. When young inside the eggs begin to vocalize, parents dig them out and often carry the newly hatched young inside their mouths to nearby water.

Alligator mississippiensis

FIGURE 35.33 Alligators Are Adapted for Aquatic Life.

Crocodilian young can hunt when newly hatched but stay near their mother for up to three years. ✔ You should be able to indicate the origin of extensive parental care in crocodilians—which also occurs in birds (and other dinosaurs)—on Figure 35.22.

Chordata > Vertebrata > Reptilia > Aves (Birds)

The fossil record provides conclusive evidence that birds descended from a lineage of dinosaurs that had **feathers.** Feathers provide insulation, are used for display, and furnish the lift, power, and steering required for flight (**FIGURE 35.34**). Birds have many other adaptations that make flight possible, including large breast muscles used to flap the wings. Bird bodies are lightweight because they have a reduced number of bones and organs and because their hollow bones are filled with air sacs linked to the lungs. Instead of teeth, birds have a horny beak. They are endotherms, meaning that they have a high metabolic rate and use the heat produced, along with the insulation provided by feathers, to maintain a constant body temperature. The 10,000 bird species alive today occupy virtually every habitat, including the open ocean. ✔ You should be able to indicate the origin of feathers, endothermy, and flight on Figure 35.22.

Feeding Plant-eating birds usually feed on nectar or seeds. Some birds are omnivores, although many are predators that capture insects, small mammals, fish, other birds, lizards, mollusks, or crustaceans. The shape of a bird's beak correlates with its diet.

Movement Almost all bird species can fly, although flightlessness has evolved repeatedly in certain groups. The size and shape of birds' wings correlate closely with the type of flying they do. Birds that glide or hover have long, thin wings; species that specialize in explosive takeoffs and short flights have short, stocky wings. Many

seabirds are efficient swimmers, using their webbed feet to paddle or flapping their wings to "fly" underwater. Some ground-dwelling birds such as ostrich can run long distances at high speed.

Reproduction Birds are oviparous but provide extensive parental care. In most species, one or both parents build a nest and incubate the eggs. After the eggs hatch, parents feed offspring until they are large enough to fly and find food on their own.

Eudocimus albus

FIGURE 35.34 Birds Are Descendants of Feathered Dinosaurs.

35.4 The Primates and Hominins

Although humans occupy a tiny twig within the mammal branch on the tree of life, there has been a tremendous amount of research on human origins. This section introduces the lineage of mammals called the Primates, the fossil record of human ancestors, and data on the relationships among human populations living today.

The Primates

The lineage known as Primates consists of two main groups: prosimians and anthropoids.

- The **prosimians** ("before-monkeys") typically consist of the lemurs, found in Madagascar, and the lorises, pottos, and tarsiers of Africa and south Asia. However, recent phylogenetic analyses indicate that tarsiers are more closely related to the anthropoids than to the other prosimians, making the prosimians a paraphyletic group (**FIGURE 35.35a**). Most prosimians living today are relatively small in size, reside in trees, and are active at night.

(a) Prosimians

Lemur catta

(b) Anthropoids

Gorilla gorilla

FIGURE 35.35 There Are Two Main Lineages of Primates.

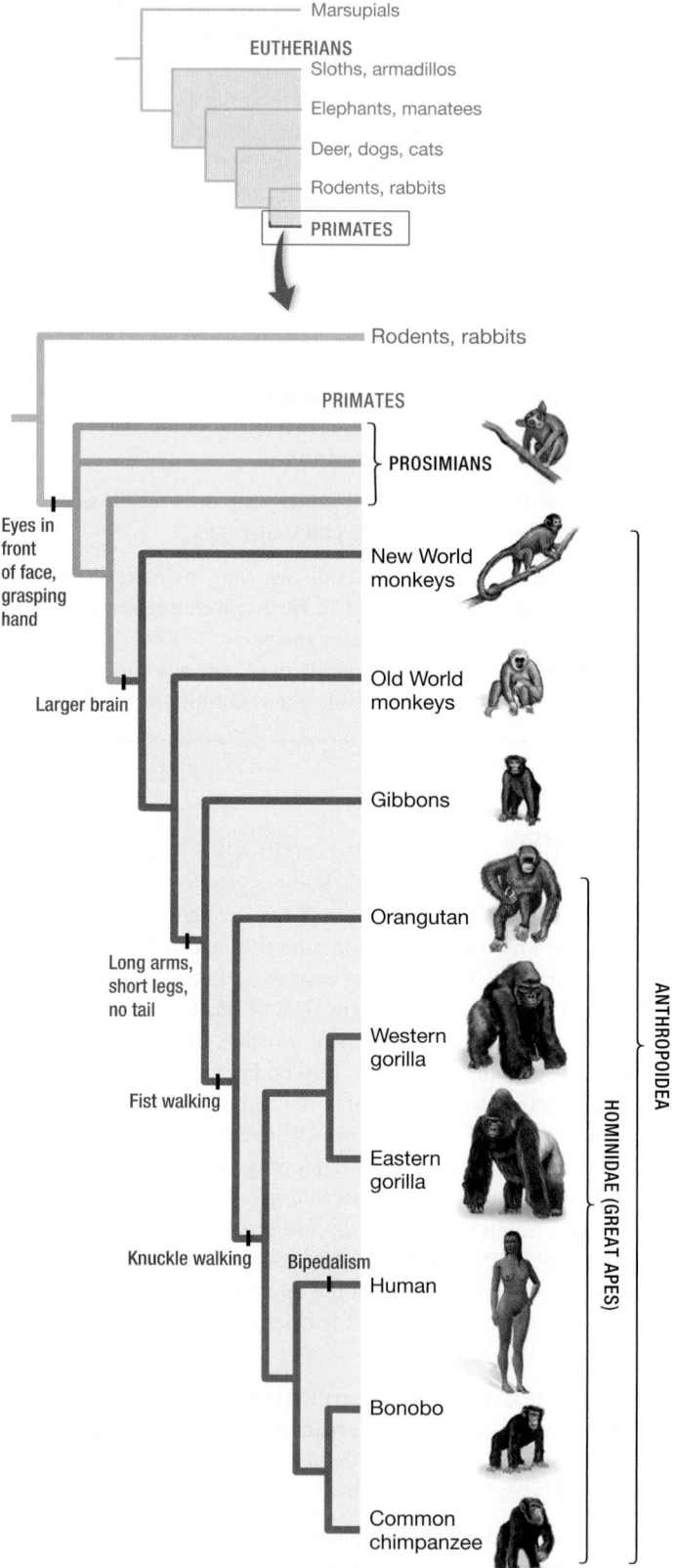

FIGURE 35.36 A Phylogeny of the Primates. Phylogenetic tree estimated from extensive DNA sequence data. According to the fossil record, humans and chimps shared a common ancestor 6 to 7 million years ago.

- The Anthropoidea or **anthropoids** ("human-like") include the New World monkeys found in Central and South America, the Old World monkeys that live in Africa and tropical regions of Asia, the gibbons of the Asian tropics, and the Hominidae, or **great apes**—orangutans, gorillas, chimpanzees, and humans (**FIGURE 35.35b**).

The phylogenetic tree in **FIGURE 35.36** shows the evolutionary relationships among these groups.

What Makes a Primate a Primate? **Primates** tend to have hands and feet that are efficient at grasping, flattened nails instead of claws on the fingers and toes, brains that are large relative to overall body size, color vision, complex social behavior, and extensive parental care of offspring.

Along with other mammal groups that live in trees or make their living by hunting, primates have eyes located on the front of the face. Eyes that look forward provide better depth perception than do eyes on the sides of the face. The hypothesis here is that good depth perception is important in species that run or swing through trees and/or attack prey.

What Makes a Great Ape a Great Ape? The great apes are also called **hominids.** Compared with most types of primate, the hominids are relatively large bodied and have long arms, short legs, and no tail.

Although all of the great ape species except for the orangutans live primarily on the ground, they have distinct ways of walking.

- When orangutans come to the ground, they occasionally walk with their knuckles pressed to the ground. More commonly though, they fist walk—that is, they walk with the backs of their hands pressed to the ground.

- Gorillas and chimps only knuckle walk. They also occasionally rise up on two legs—usually in the context of displaying aggression.

- Humans are the only living great ape that is fully **bipedal** ("two-footed")—meaning they walk upright on two legs. Fossil footprints in Tanzania provide direct evidence that bipedalism occurred at least 3.6 million years ago.

Bipedalism is the synapomorphy that defines the hominins. The **hominins** are a monophyletic group comprising *Homo sapiens* and more than twenty extinct, bipedal relatives.

Fossil Humans

From extensive comparisons of DNA sequence data, it is clear that humans are most closely related to the common chimpanzees and bonobos, and that our next nearest living relatives are the gorillas. According to the fossil record, the common ancestor of chimps and humans lived in Africa 6 to 7 million years ago.

The fossil record of hominins, though not nearly as complete as investigators would like, is rapidly improving. More than 20 species have been named to date, and new fossils that inform the debate over the ancestry of humans are discovered every year.

Although naming the hominin species and interpreting their characteristics remain intensely controversial, most researchers agree that the hominins can be organized into four major groups that appeared after the recently characterized *Ardipithecus ramidus*—the oldest hominin known to date. **TABLE 35.1** summarizes key data for

TABLE 35.1 Characteristics of Selected Hominins

Genus	Species	Location of Fossils	Estimated Average Braincase Volume (cm^3)	Estimated Average Body Size (kg)	Associated with Stone Tools?
■ *Ardipithecus*	A. ramidus	Africa	325	40	no
■ *Australopithecus*	A. afarensis	Africa	460	38	no
	A. africanus	Africa	465	34	no
■ *Paranthropus*	P. boisei	Africa	480	41	no?
■ Early *Homo*	H. habilis	Africa	610	33	**yes**
	H. ergaster	Africa	760	64	**yes**
	H. erectus	Africa, Asia	1000	58	**yes**
■ Recent *Homo*	H. heidelbergensis	Africa, Europe	1200	71	**yes**
	H. neanderthalensis	Middle East, Europe, Asia	1430	72	**yes**
	H. floresiensis	Flores (Indonesia)	380	28	**yes**
	H. sapiens	Middle East, Europe, Asia	1480	64	**yes**

Robson, S. L., and B. Wood. 2008. Hominin life history: Reconstruction and evolution. *Journal of Anatomy* 212: 394–425.

✔ **QUANTITATIVE** How much smaller is the average body size of *H. floresiensis* compared to the average *H. sapiens*? Express your answer as a percentage. How much smaller is the braincase volume of *H. floresiensis* compared to *H. sapiens*? What do these numbers suggest?

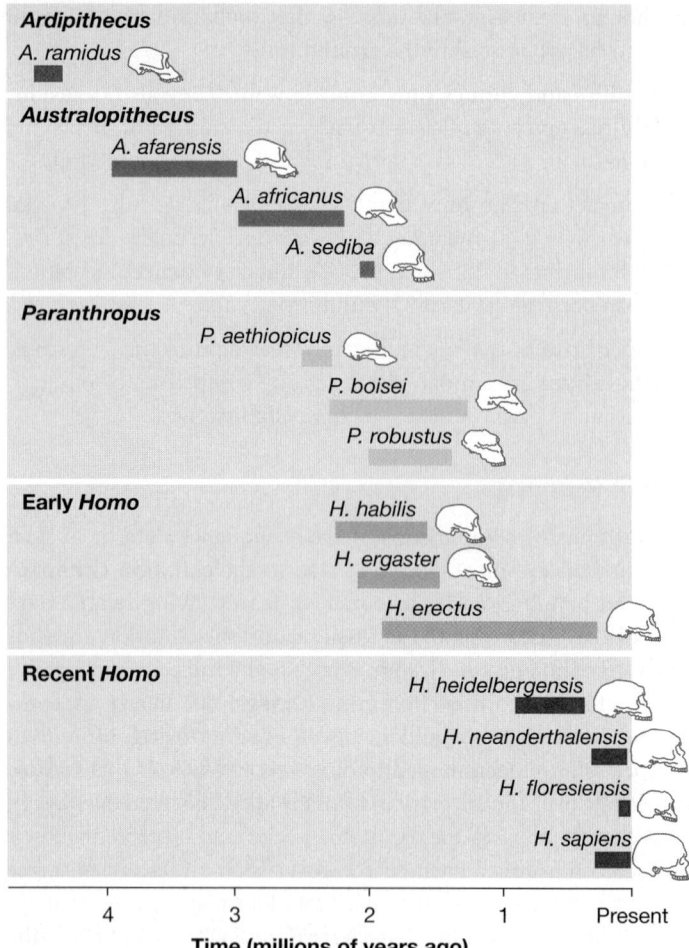

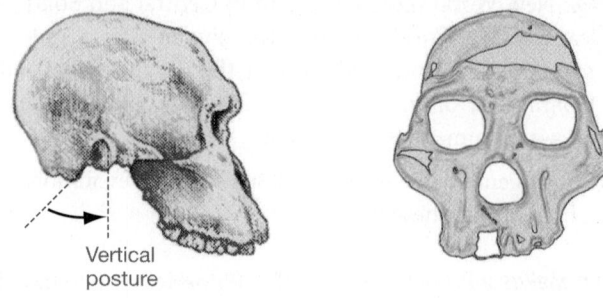

(a) Gracile australopithecines (*Australopithecus africanus*)

Vertical posture

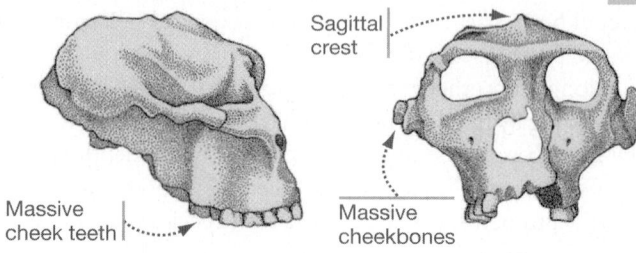

(b) Robust australopithecines (*Paranthropus robustus*)

Sagittal crest

Massive cheek teeth

Massive cheekbones

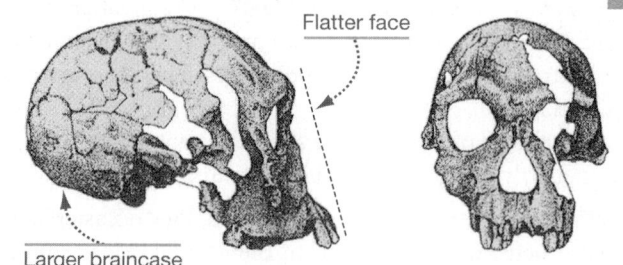

(c) Early *Homo* (*Homo habilis*)

Flatter face

Larger braincase

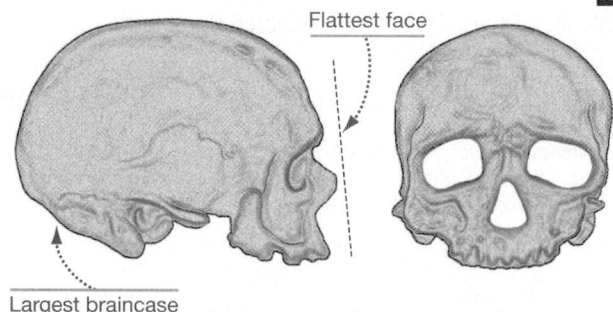

(d) Recent *Homo* (*Homo sapiens*)

Flattest face

Largest braincase

FIGURE 35.37 A Timeline of Hominin Evolution. The colored bars indicate the first appearance and last appearance in the fossil record of each species. (Not all named species are shown.)

DATA: Lieberman, D.E. 2009. *Nature* 459: 41–42.

✔ **QUANTITATIVE** Of the species of hominin shown, how many existed 1.8 million years ago (mya) and 100,000 years ago?

FIGURE 35.38 African Hominins Comprise Four Major Groups. The color key links these fossils with Figure 35.37 and Table 35.1.

✔ **QUESTION** The skulls are arranged as they appear in the fossil record, from most ancient (a) to most recent (d). How did the forehead and brow ridge of hominins change through time?

selected species within the four groups. **FIGURE 35.37** provides the time range of each of these species in the fossil record.

Gracile Australopithecines The small hominins called gracile australopithecines are referred to as *gracile*, or "slender," because they had a slight build. Adult males were about 1.5 meters tall, and their estimated weight was about 36 kg.

The gracile australopithecines are placed in the genus *Australopithecus* ("southern ape"). The name was inspired by the earliest specimens, which came from South Africa. The most recent *Australopithecus* to be named, *A. sediba* in 2010, is also from South Africa.

Several lines of evidence support the hypothesis that the gracile australopithecines were bipedal. The shape of the australopithecine knee and hip are consistent with bipedal locomotion, and the hole in the back of their skulls where the spinal cord connects to the brain is oriented downward (**FIGURE 35.38a**), just as it is in our species, *Homo sapiens*. In chimps, gorillas, and other vertebrates that walk on four feet, this hole is oriented backward.

Robust Australopithecines The robust australopithecines are grouped in the genus *Paranthropus* ("beside-human"). The name *Paranthropus* was inspired by the hypothesis that the three known species are a monophyletic group that was a side branch during human evolution—an independent lineage that went extinct.

Like the gracile australopithecines, these robust australopithecines were bipedal. They were much stockier than the gracile forms, however—about the same height but an estimated 8–10 kilograms (20 pounds) heavier on average. In addition, their skulls were much broader and more robust.

All three *Paranthropus* species had massive cheek teeth and jaws, very large cheekbones, and a sagittal crest—a flange of bone at the top of the skull (**FIGURE 35.38b**). Because muscles that work the jaw attach to the sagittal crest and cheekbones, researchers conclude that these organisms had tremendous biting power—inspiring the nickname "nutcracker man" for one species. Recent analysis of tooth enamel revealed that the biting power of *Paranthropus* was probably used to crush large quantities of tough grasses.

Early *Homo* Species in the genus *Homo* are called **humans.** As **FIGURE 35.38c** shows, species in this genus have flatter and narrower faces, smaller jaws and teeth, and larger braincases than the earlier hominins do. (The **braincase** is the portion of the skull that encloses the brain.)

The appearance of early members of the genus *Homo* in the fossil record coincides closely with the appearance of tools made of worked stone—most of which are interpreted as handheld choppers or knives. Although the fossil record does not exclude the possibility that *Paranthropus* made tools, many researchers favor the hypothesis that extensive toolmaking was a diagnostic trait of early *Homo*.

Recent *Homo* The recent species of *Homo* date from 1.2 million years ago to the present. As **FIGURE 35.38d** shows, these species have even flatter faces, smaller teeth, and larger braincases than the early *Homo* species do. The 30,000-year-old fossil in the figure, for example, is from a population of *Homo sapiens* (our species) called the **Cro-Magnons.**

The Cro-Magnons were accomplished painters and sculptors who buried their dead in carefully prepared graves. There is also evidence that another species, the **Neanderthal** people (*Homo neanderthalensis*) made art and buried their dead in a ceremonial fashion.

Perhaps the most striking recent *Homo*, though, is *H. floresiensis*. This new species was discovered in 2003 on the island of Flores in Indonesia—also home to a species of dwarfed elephants. *H. floresiensis* consisted of individuals that had braincases smaller than those of gracile australopithecines and that were about a meter tall, inspiring the nickname Hobbit. While some researchers propose that the fossils represent a population of *H. sapiens* stunted by disease, most support full species status. Fossil finds suggest that *H. floresiensis* inhabited Flores from about 100,000 to as recently as 12,000 years ago.

What Can Be Deduced from the Hominin Fossil Record? Although researchers do not have a solid understanding of the phylogenetic relationships among the hominin species, several points are clear from the available fossil data.

1. The shared, derived character that defines the hominins is bipedalism.

2. Several species from the lineage were present simultaneously during most of hominin evolution.

3. Compared with the gracile and robust australopithecines and the other great apes, species in the genus *Homo* have extremely large brains relative to their overall body size.

Why did humans evolve such gigantic brains? The leading hypothesis on this question is that early *Homo* began using symbolic spoken language and tools. The logic here is that increased language use and toolmaking triggered natural selection for the capacity to reason and communicate, which required a larger brain.

To summarize, *Homo sapiens* is the sole survivor of an adaptive radiation of hominins that began in Africa 6–7 million years ago. Why all but one species went extinct is still a mystery, though climate change and competition for food and space may have played a part.

The Out-of-Africa Hypothesis

If hominins originated in Africa and then spread around the world, where did *H. sapiens* originate?

The leading hypothesis for the evolution of *H. sapiens* is called the **out-of-Africa hypothesis.** It contends that *H. sapiens* evolved its distinctive traits in Africa and then dispersed throughout the world. Further, it claims that *H. sapiens* evolved independently of the earlier European and Asian species of *Homo*—meaning there was no interbreeding between *H. sapiens* and Neanderthals, *H. erectus*, or *H. floresiensis*. Do the data support this hypothesis?

Fossil Evidence The first *H. sapiens* fossils appear in East African rocks that date to about 195,000 years ago. For some 130,000 years thereafter, the fossil record indicates that our species occupied Africa while *H. neanderthalensis* resided in Europe and the Middle East. Some evidence suggests that *H. erectus* may still have been present in Asia at that time.

In rocks dated between 60,000 and 30,000 years ago, however, *H. sapiens* fossils are found throughout Europe, Asia, Africa, and Australia. *H. erectus* had disappeared by this time, and *H. neanderthalensis* went extinct after coexisting with *H. sapiens* in Europe for thousands of years.

Taken together, the fossil evidence provides strong support for the African origin of *H. sapiens* and subsequent migration, but it cannot resolve the question of whether *H. sapiens* ever interbred with Neanderthals during their coexistence.

Molecular Evidence Some researchers have taken a molecular approach to understanding the history of our species, providing several major insights.

For example, Luigi Cavali-Sforza and others have created phylogenetic trees based on genetic comparisons of diverse human populations living today. These data agree with the pattern in the fossil record. As **FIGURE 35.39** (see page 708) shows, the first lineages to branch off led to descendant populations that live in Africa today. Later branches gave rise to lineages in central Asia, Europe, East Asia, Polynesia, and the Americas. Based on this observation, it is logical to infer that the ancestral population of modern humans lived in Africa.

Did *H. sapiens* leave Africa in one wave? Sequence data from modern Aboriginal Australians support the hypothesis that

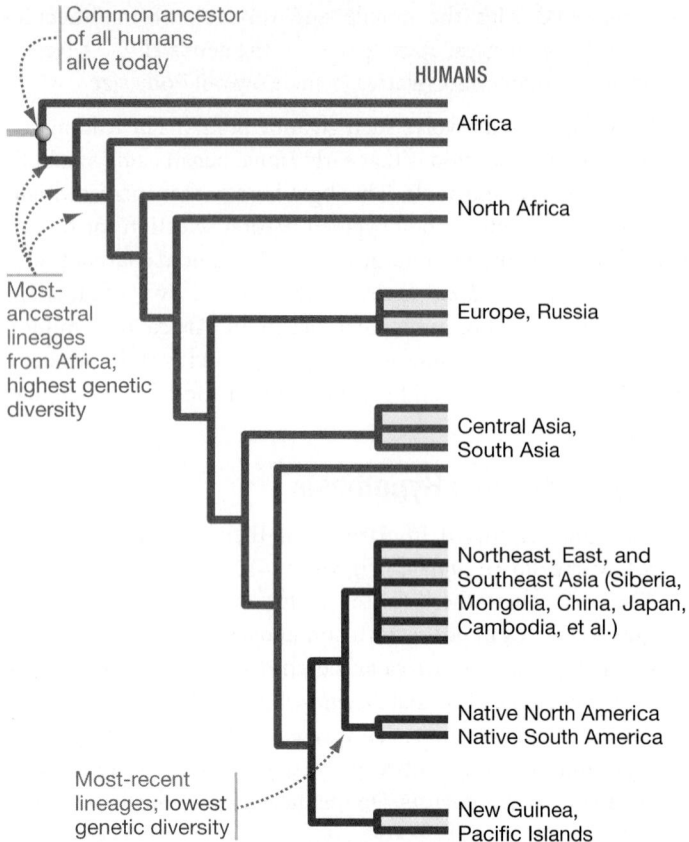

Common ancestor of all humans alive today

HUMANS

Africa

North Africa

Europe, Russia

Central Asia, South Asia

Northeast, East, and Southeast Asia (Siberia, Mongolia, China, Japan, Cambodia, et al.)

Native North America
Native South America

New Guinea, Pacific Islands

Most-ancestral lineages from Africa; highest genetic diversity

Most-recent lineages; lowest genetic diversity

FIGURE 35.39 Phylogeny of Human Populations Living Today. The phylogeny was estimated from DNA sequence data. As predicted by the out-of-Africa hypothesis, the lineages with the highest genetic diversity and deepest history occur in Africa.

the human migration out of Africa occurred in more than one wave—the first wave moved east and south to Australia, and the second wave moved to Europe and the rest of mainland Asia. While the exact migration routes of *H. sapiens* are hotly debated, **FIGURE 35.40** provides a general summary.

What about the question of interbreeding? There are now two lines of evidence that *H. sapiens* interbred with other species as they migrated out of Africa. First, Svante Pääbo and his colleagues compared gene sequences of extinct Neanderthals and diverse modern humans. They made the remarkable observation that 1 to 4 percent of the genome of indigenous Europeans and Asians—but not Africans—is derived from Neanderthals. These data suggest that *H. sapiens* and Neanderthals interbred in the Middle East after *H. sapiens* left Africa, but before migrating through Europe and Asia.

Second, sequence data from modern Asians and Aboriginal Australians suggest that modern humans interbred with a recently discovered descendant population of *H. erectus*, called the Denisovans, in Asia during their first migration out of Africa. As a result, Aboriginal Australians and their island neighbors—descendants of the first migration—inherited about 5 percent of their nuclear DNA from Denisovans.

In summary, the Neanderthal and Denisovan data reject the strict interpretation of the out-of-Africa hypothesis since a small amount of gene flow occurred between *H. sapiens* and Neanderthals as well as between *H. sapiens* and the Denisovans. However, other fossil and molecular evidence supports the general concept of the out-of-Africa hypothesis. That is, the overall data support an out-of-Africa hypothesis with "leakage."

The study of human evolution is an exciting and dynamic field. And humans continue to evolve.

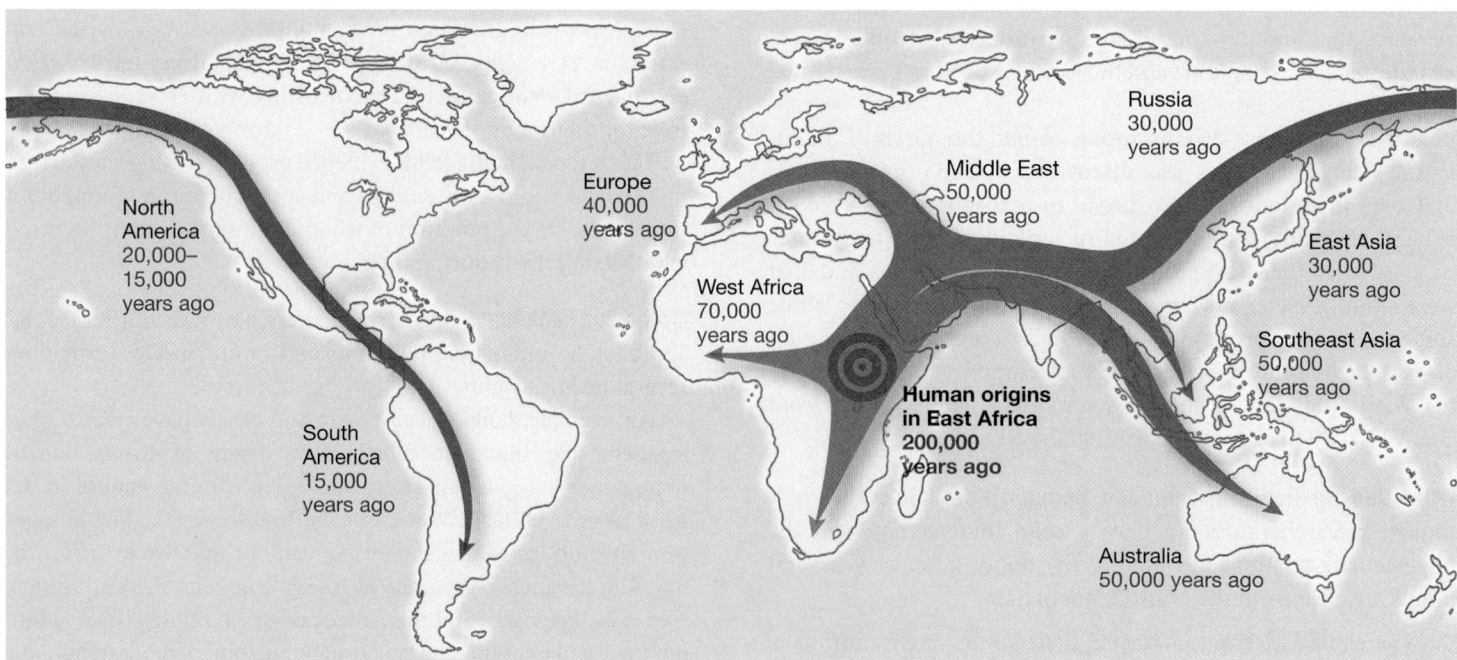

FIGURE 35.40 *Homo sapiens* Originated in Africa and Spread throughout the World. The migration out of Africa probably occurred in two waves—first east and south to Australia, and later to Europe and the rest of Asia and the Americas.

If you understand . . .

35.1 What Is an Echinoderm?

- Echinoderms are among the most important predators and herbivores in marine environments. They include the sea stars, sea urchins, sea cucumbers, and feather stars.

- Echinoderm larvae are bilaterally symmetric but most undergo a metamorphosis into radially symmetric adults.

- Echinoderms have a calcium carbonate endoskeleton.

- The echinoderm water vascular system is composed of fluid-filled tubes and chambers called tube feet. Tube feet may be used for movement, feeding, or respiration.

✔ You should be able to predict what happens when (1) sea stars that prey on mussels and clams are excluded from experimental plots, and (2) sea urchins that graze on kelp are excluded from experimental plots.

35.2 What Is a Chordate?

- Chordates are distinguished by the presence of pharyngeal gill slits, a dorsal hollow nerve cord, a notochord, and a muscular tail that extends past the anus.

- Chordates include cephalochordates, urochordates, and vertebrates.

✔ You should be able to make a simple cartoon of a generic chordate showing its defining features.

35.3 What Is a Vertebrate?

- Vertebrates are distinguished by the presence of a cranium and vertebrae. In some groups of vertebrates, the body plan features an extensive endoskeleton composed of bone.

- The origin of jaws in ray-finned fishes enabled these fishes and their descendants to bite food and process it with teeth.

- The origin of limbs in tetrapods enabled walking and running on land, and flying. Tetrapods include the amphibians, mammals, and reptiles (which include birds).

- The evolution of the amniotic egg allowed tetrapods to lay large eggs on land.

- Parental care was an important adaptation in some groups of ray-finned fishes and tetrapods—particularly mammals.

- Vertebrates are the most important large-bodied predators and herbivores in both marine and terrestrial environments.

✔ You should be able to explain why the tetrapod limb could have evolved in an animal that was aquatic.

35.4 The Primates and Hominins

- Humans are a tiny twig on the tree of life. Chimpanzees and humans diverged from a common ancestor that lived in Africa 6–7 million years ago.

- The fossil record of the past 4.5 million years contains many distinct species of hominins. *Homo sapiens* is the sole surviving representative of an adaptive radiation.

- For most of human evolution, several species were present in Africa or Europe at the same time. Some lineages went extinct without leaving descendant populations.

- The phylogeny of living humans, based on comparisons of DNA sequences, agrees with evidence in the fossil record that *H. sapiens* originated in Africa and later spread throughout Asia, Europe, and the Americas.

✔ You should be able to describe evidence supporting the hypothesis that *Homo sapiens* originated in Africa.

MasteringBiology

1. **MasteringBiology Assignments**

   Tutorials and Activities Allometric Growth; Characteristics of Chordates; Chordates; Deuterostome Diversity; Human Evolution; Primate Diversity

   Questions Reading Quizzes, Blue-Thread Questions, Test Bank

2. **eText** Read your book online, search, take notes, highlight text, and more.

3. **The Study Area** Practice Test, Cumulative Test, BioFlix® 3-D Animations, Videos, Activities, Audio Glossary, Word Study Tools, Art

You should be able to . . .

1. True or false? Both echinoderms and vertebrates have endoskeletons.

2. What trait(s) define(s) the vertebrates?
   a. vertebrae and a cranium
   b. jaws and a spinal cord
   c. endoskeleton constructed of bone
   d. endoskeleton constructed of reinforced cartilage

3. Why are the pharyngeal jaws of many ray-finned fishes important?
   a. They allow the main jaw to be protrusible (extendable).
   b. They make it possible for individuals to suck food toward their mouths.
   c. They give rise to teeth that are found on the main jawbones.
   d. They help process food.

4. The two major lineages that make up the living Amniota are reptiles and _____.

5. In chordates, what is the relationship between the dorsal hollow nerve cord and the spinal cord?
   a. The dorsal hollow nerve cord is found only in embryos; the spinal cord is found in adults.
   b. The dorsal hollow nerve cord stiffens the body and functions in movement; the spinal cord relays electrical signals.
   c. The dorsal hollow nerve cord is found in early fossils of chordates; living species have a spinal cord.
   d. The spinal cord is a type of dorsal hollow nerve cord.

6. Most species of hominins are known only from Africa. Which species have been found in other parts of the world as well?
   a. early *Homo*—*H. habilis* and *H. ergaster*
   b. *H. erectus*, *H. neanderthalensis*, and *H. floresiensis*
   c. gracile australopithecines
   d. robust australopithecines

7. Mammals and birds are both endothermic. Which statement is true?
   a. They both inherited this trait from a common ancestor.
   b. Birds are more closely related to mammals than to reptiles.
   c. Endothermy arose independently in these two lineages.
   d. All reptiles are ectothermic.

8. Explain why each of the four key synapomorphies that distinguish chordates is important in feeding or movement. Why is it possible to say that a certain animal is an "invertebrate chordate"?

9. The cells that make up jaws and gill arches are derived from the same population of embryonic cells. Why does this observation support the hypothesis that jaws evolved from gill arches in fish?

10. Explain how the evolution of the placenta and lactation in mammals improved the probability that mammalian offspring would survive, compared to offspring of species without parental care.

11. The text claims that *Homo sapiens* is the sole survivor of an adaptive radiation that took place over the past 6–7 million years. Do you agree with this statement? Why or why not?

12. Describe a phylogeny of modern humans that would reject the out-of-Africa hypothesis.

13. There is some evidence that pharyngeal gill slits occur in certain species of echinoderms that appear early in the fossil record. If confirmed, what do these data suggest?
    a. Echinoderms are chordates.
    b. Pharyngeal gill slits were present in the earliest echinoderms and lost later.
    c. Some lineages of echinoderms are more closely related to chordates than others.
    d. Pharyngeal gill slits should not be used as a trait in phylogenetic analysis.

14. Animals in the species *Xenoturbella* have extremely simple, wormlike bodies. For example, they have a blind gut (only one opening) and no brain. Propose a hypothesis to explain how this simple body plan evolved in this lineage.

15. In arthropods such as crustaceans and insects, changes in the expression patterns of tool-kit genes enabled a dramatic diversification of their segmented appendages and bodies. Design an experiment to test whether this concept applies to vertebrates.

16. **QUANTITATIVE** Genetic diversity in living human populations is highest in Africa and decreases as a function of distance from Africa. Draw a graph to show this result. Add a label where you would expect to find a data point for the Yanomamö tribe of South America. Why is it important to use indigenous people for this study?

36 Viruses

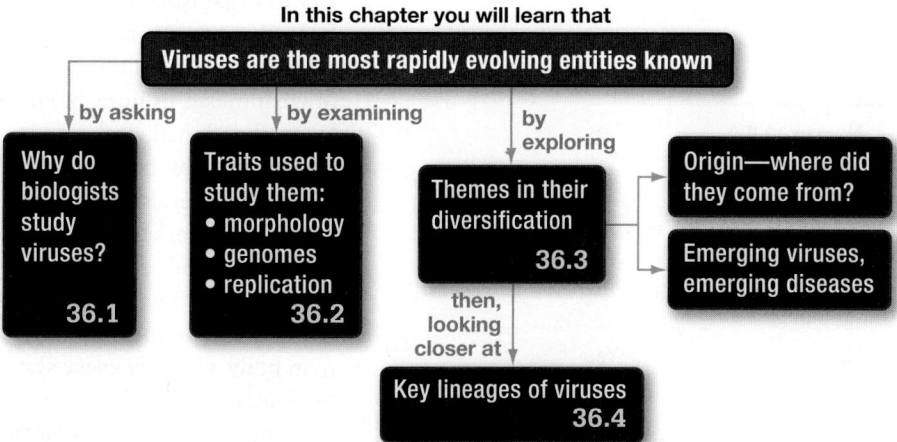

In this chapter you will learn that

Viruses are the most rapidly evolving entities known

by asking

Why do biologists study viruses?

36.1

by examining

Traits used to study them:
- morphology
- genomes
- replication

36.2

by exploring

Themes in their diversification

36.3

Origin—where did they come from?

Emerging viruses, emerging diseases

then, looking closer at

Key lineages of viruses
36.4

Photomicrograph created by treating seawater with a fluorescing compound that binds to nucleic acids. The smallest, most abundant dots are viruses. The larger, numerous spots are bacteria and archaea. The largest splotches are protists.

f you have ever been laid low by a high fever, cough, scratchy throat, body ache, and debilitating lack of energy, you may have wondered what hit you. What hit you was probably a **virus:** an obligate, intracellular parasite.

Why obligate? Viral replication is *completely* dependent on host cells. Why intracellular? Viruses must enter a host cell for replication to occur. Why parasite? Viruses reproduce at the expense of their host cells.

Viruses can also be defined by what they are not:

- They are not cells and are not made up of cells, so they are not considered organisms.

- They cannot manufacture their own ATP or amino acids or nucleotides, and they cannot produce proteins on their own.

Viruses enter a **host cell,** take over its biosynthetic machinery, and use that machinery to manufacture a new generation of viruses. Outside of host cells, viruses simply exist.

✔ When you see this checkmark, stop and test yourself. Answers are available in Appendix A.

Characteristics	Viruses	Organisms
Hereditary material	DNA or RNA; can be single stranded or double stranded	DNA; always double stranded
Plasma membrane present?	No	Yes
Can carry out transcription independently?	No—even if a viral polymerase is present, transcription of viral genomes requires use of nucleotides provided by host cell	Yes
Can carry out translation independently?	No	Yes
Metabolic capabilities	Virtually none	Extensive—synthesis of ATP, reduced carbon compounds, vitamins, lipids, nucleic acids, etc.

When you have the flu, influenza viruses enter the cells that line your respiratory tract and use the machinery inside to make copies of themselves. Every time you cough or sneeze, you eject millions or billions of their offspring into the environment. If one of those infectious particles is lucky enough to be breathed in by another person, it may enter his or her respiratory tract cells and start a new infection.

Viruses are not organisms. Most biologists would argue that viruses are not alive, because they are dependent on their host cell to satisfy the five attributes of life (see Chapter 1). Yet viruses have a genome, they are superbly adapted to exploit the metabolic capabilities of their host cells, and they evolve. **TABLE 36.1** summarizes some characteristics of viruses and how they differ from organisms.

The diversity and abundance of viruses almost defy description. Each type of virus infects a specific unicellular species or cell type in a multicellular species, and nearly all organisms examined thus far are parasitized by at least one kind of virus. The surface waters of the world's oceans teem with bacteria and archaea, yet viruses outnumber them in this habitat by a factor of 10 to 1. If you leaned over a boat and filled a wine bottle with seawater, it would contain about 10 billion virus particles—close to one and a half times the world's population of humans.

36.1 Why Do Biologists Study Viruses?

Any study of life's diversity would be incomplete unless it included a look at the acellular parasites that contribute to that diversity. Viruses have directly participated in organismal diversity by introducing foreign genes into cellular genomes. They can pick up cellular genes and shuttle them from one organism to another. In this way, viruses can promote **lateral gene transfer** (see Chapter 21).

In addition, viruses also contribute their own genetic material to organisms. Researchers estimate that 5–8 percent of the human genome consists of remnants of viral genomes from past infections. Some of these viral genes have even evolved to be part of what makes us human. For example, a protein that is encoded by an abandoned viral gene is necessary for proper development and function of the human placenta.

Viruses are also important from a practical standpoint. To health-care workers, agronomists, and foresters, these parasites are a persistent—and sometimes catastrophic—source of misery and economic loss. Much of the research on viruses is motivated by the desire to minimize the damage they can cause. In the human body, virtually every system, tissue, and cell can be infected by at least one kind of virus, and each of these viruses often infects more than one site (**FIGURE 36.1**).

Recent Viral Epidemics in Humans

Physicians and researchers use the term **epidemic** (literally, "upon-people") to describe a disease that rapidly affects a large number of individuals over a widening area.

Viruses have caused the most devastating epidemics in recent human history. During the eighteenth and nineteenth centuries, it was not unusual for Native American tribes to lose 90 percent of their members over the course of a few years to measles, smallpox, and other viral diseases spread by contact with European settlers. To appreciate the impact of these epidemics, think of 10 close friends and relatives—then remove nine.

An epidemic that is worldwide in scope is called a **pandemic.** The influenza outbreak of 1918–1919, called Spanish flu, qualifies as the most devastating pandemic recorded to date. The strain of influenza virus that emerged in 1918 infected people worldwide and was particularly **virulent**—meaning it tended to cause severe disease. The viral outbreak occurred just as World War I was drawing to a close and killed far more people in eight months than did the five-year conflict itself.

FIGURE 36.2 illustrates the impact of this pandemic in relation to the life expectancy in the United States during the twentieth century. Worldwide, the Spanish flu is thought to have killed up to 50 million people. An estimated 20 million died in the first 8 months of the pandemic.

Current Viral Pandemics in Humans: AIDS

In terms of the total number of people affected, the measles and smallpox epidemics and the 1918 influenza pandemic are almost certain to be surpassed by the incidence of **acquired immune deficiency syndrome (AIDS).** AIDS is an affliction caused by the **human immunodeficiency virus (HIV).**

HIV is now the one of the most intensively studied of all viruses. Since the early 1980s, governments and private corporations from around the world have spent hundreds of millions of dollars on

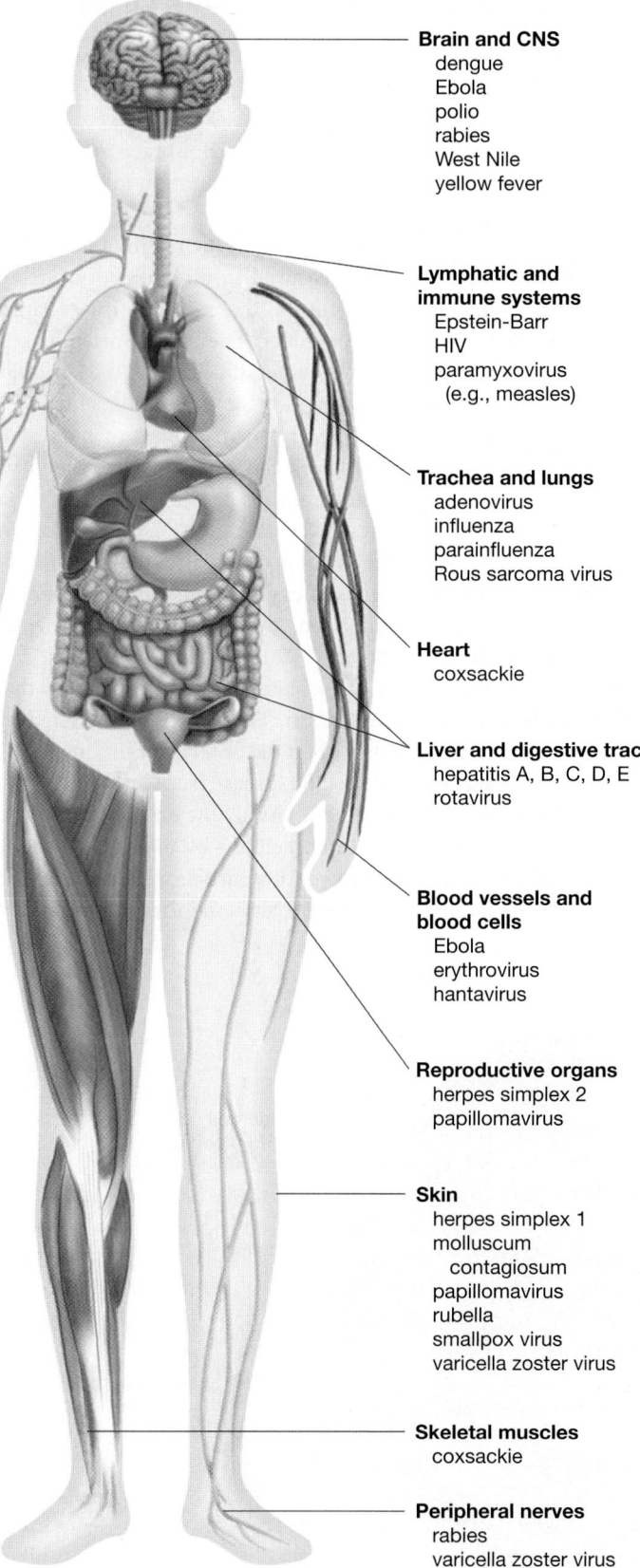

Brain and CNS
dengue
Ebola
polio
rabies
West Nile
yellow fever

Lymphatic and immune systems
Epstein-Barr
HIV
paramyxovirus
 (e.g., measles)

Trachea and lungs
adenovirus
influenza
parainfluenza
Rous sarcoma virus

Heart
coxsackie

Liver and digestive tract
hepatitis A, B, C, D, E
rotavirus

Blood vessels and blood cells
Ebola
erythrovirus
hantavirus

Reproductive organs
herpes simplex 2
papillomavirus

Skin
herpes simplex 1
molluscum
 contagiosum
papillomavirus
rubella
smallpox virus
varicella zoster virus

Skeletal muscles
coxsackie

Peripheral nerves
rabies
varicella zoster virus

FIGURE 36.1 Human Organs and Systems That Are Parasitized by Viruses. Viruses shown in this figure may infect more than one tissue, but for simplicity, usually only one tissue is shown.

(a) Deadly impact of the 1918 influenza pandemic

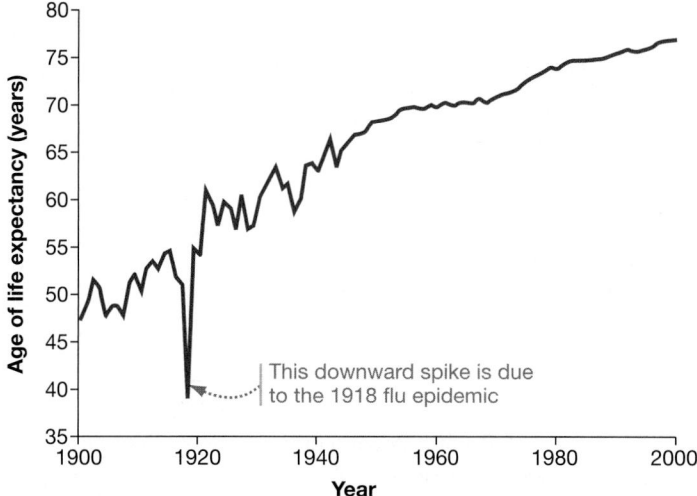

This downward spike is due to the 1918 flu epidemic

(b) Emergency hospital at the start of the 1918 pandemic

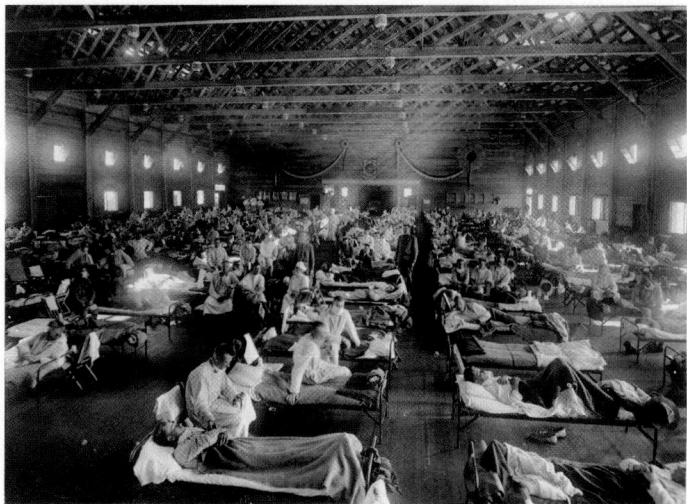

FIGURE 36.2 Life Expectancy of Humans over the 20th Century in the United States. (a) This graph illustrates the dramatic effect of the 1918 influenza pandemic in an otherwise promising trend. **(b)** The first recorded cases of the 1918 Spanish flu were reported at this U.S. Army emergency hospital (Camp Funston, Kansas).

DATA: Arias, E. 2010. United States life tables, 2006. *National Vital Statistics Reports,* 58 (21): 1–40, Table 10. Hyattsville, MD: National Center for Health Statistics.

HIV research. Given this virus's current and projected impact on human populations around the globe, the investment is justified.

How Does HIV Cause Disease?   Like other viruses, HIV parasitizes specific types of cells—most notably, the helper T cells of the **immune system,** which is the body's defense system against disease. Helper T cells are crucial to the immune system's response to invading pathogens (see Chapter 51).

If an HIV particle succeeds in infecting a helper T cell and reproduces inside, the cell will die as hundreds of new progeny particles are released and infect more cells. Although the body continually replaces helper T cells, the number produced does not keep pace with the number being destroyed by HIV. As

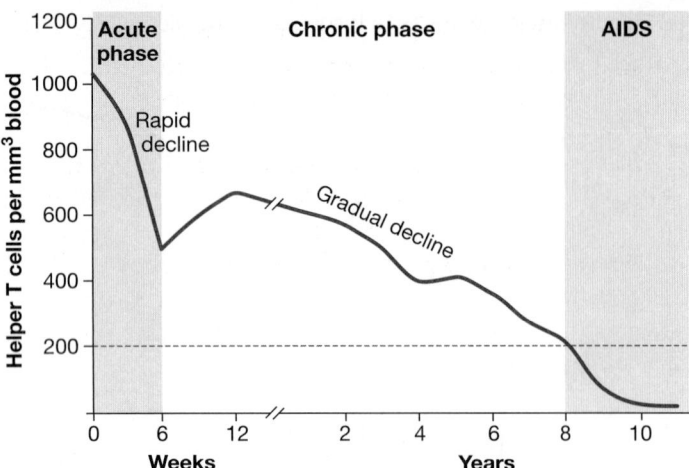

FIGURE 36.3 T-Cell Counts Decline during an HIV Infection.
Graph of changes in the number of helper T cells that are present in the bloodstream over time, based on data from a typical patient infected with HIV. The acute phase may be associated with symptoms such as fever. Few or no disease symptoms occur in the chronic phase. AIDS typically occurs when T-cell counts dip below 200 per mm³ of blood.

DATA: Pantaleo, G., and Fauci, A. S. 1996. *Annual Review of Microbiology* 50: 825-854.

a result, the total number of helper T cells in the bloodstream gradually declines as an HIV infection proceeds (**FIGURE 36.3**).

As the T-cell count drops, the immune system's responses to invading pathogens become less and less effective. Eventually, too few helper T cells are left to fight off pathogens efficiently, allowing bacteria, fungi, protists, or other viruses to multiply unchecked. In almost all cases, one or more of these "opportunistic" infections proves fatal. HIV kills people indirectly—by making them susceptible to diseases that normally do not arise in those with functioning immune systems.

What Is the Scope of the AIDS Pandemic? Researchers with the United Nations AIDS program estimate that AIDS has already killed almost 30 million people worldwide. HIV infection rates have been highest in east and central Africa, where one of the greatest public health crises in history is now occurring. In Botswana, for example, blood-testing programs have confirmed that over 20 percent of the population is HIV positive. Although there may be a lag of as much as 8 to 12 years between the initial infection and the onset of illness, virtually all people who become infected with the virus will die of AIDS.

Currently, the UN estimates the total number of HIV-infected people worldwide at about 34 million. An additional 2.7 million people are infected each year. Most infectious diseases afflict the very young and the very old, but because HIV is primarily a sexually transmitted disease, young adults are most likely to contract the virus and die.

People who become infected with HIV in their late teens or twenties die of AIDS in their twenties or thirties. Tens of millions of people are being lost in the prime of their lives. Physicians, politicians, educators, and aid workers all use the same word to describe the epidemic's impact: staggering.

36.2 How Do Biologists Study Viruses?

Many researchers who study viruses focus on two goals: (**1**) developing vaccines that help hosts fight off disease if they become infected and (**2**) developing antiviral drugs that prevent a virus from replicating efficiently inside the host. Both types of research begin with attempts to isolate the virus in question to learn more about its structure and mode of replication.

Isolating viruses takes researchers into the realm of nanobiology, in which structures are measured in billionths of a meter. (One nanometer, abbreviated nm, is 10^{-9} meter.) Viruses range from about 20 to 300 nm in diameter. They are dwarfed by eukaryotic cells and even by most bacterial cells (**FIGURE 36.4**). Millions of viruses can fit on the period at the end of this sentence.

If virus-infected cells can be grown in culture or harvested from a host individual, researchers can usually isolate the virus by passing solutions of infected cultures or patient samples (e.g., blood, saliva, feces) through a filter. The filters used to study viruses have pores that are large enough for viruses to pass through but are too small to admit cells. If exposing susceptible host cells to the filtrate results in infection, the hypothesis that a virus caused the disease is supported.

In this way, researchers can isolate a virus and confirm that it is the causative agent of infection. These steps are inspired by Koch's postulates, which established the criteria for linking a specific infectious agent with a specific disease (see Chapter 29).

Once biologists have isolated a virus, how do they study and characterize it? The answer is threefold—by analyzing (**1**) the structure of the extracellular infectious particle, referred to as the **virion;** (**2**) the nature of the genetic material that is transmitted

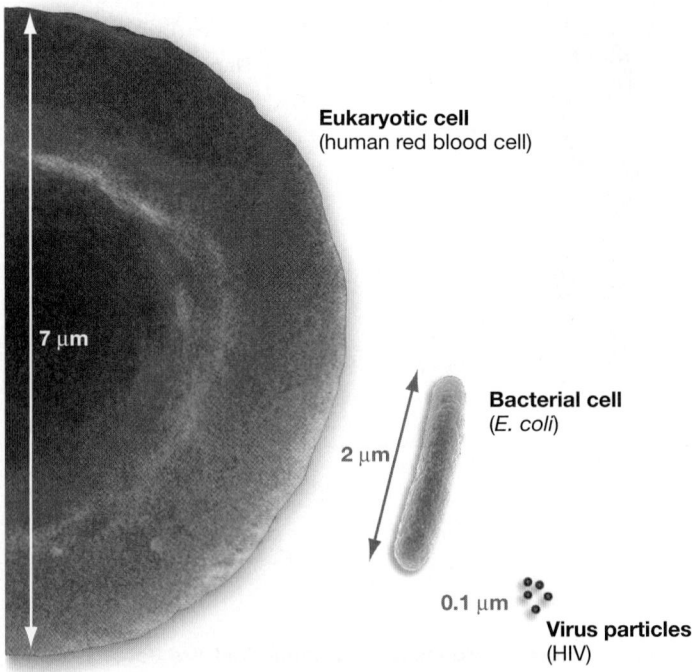

Eukaryotic cell
(human red blood cell)

7 μm

Bacterial cell
(*E. coli*)

2 μm

0.1 μm

Virus particles
(HIV)

FIGURE 36.4 Viruses Are Tiny.

by virions from one host to another; and **(3)** variations in how viruses replicate. Let's begin with morphological traits.

Analyzing Morphological Traits

To see the virion of a particular virus, researchers usually rely on transmission electron microscopy (see **BioSkills 11** in Appendix B). Only the very largest viruses, such as the smallpox virus, are visible with a light microscope. Electron microscopy has revealed that viruses come in a wide variety of shapes, and many viruses can be identified by shape alone.

In terms of overall structure, most viruses fall into two general categories: **(1)** those that are enclosed by just a shell of protein called a **capsid** and **(2)** those enclosed by both a capsid and a membrane-like **envelope.**

Most viruses produce virions with capsid shapes that are either helical or icosahedral (an icosahedron is a polyhedron with 20 triangular faces) (**FIGURE 36.5a** and **b**). Some viruses, like bacteriophage T4 and the smallpox virus, however, have more complex capsid shapes (**FIGURE 36.5c** and **d**). A virion's capsid serves two functions: it protects the genome while outside the host and is also able to release the genome when infecting a new cell.

Nonenveloped viruses, often called "naked" viruses, use only the capsid shell to protect their genetic material. The naked icosahedral virion illustrated in Figure 36.5b is an adenovirus. You undoubtedly have adenoviruses on your tonsils or in other parts of your upper respiratory passages right now.

Enveloped viruses also have genetic material inside a capsid, or bound to capsid proteins, but the capsid is further enclosed in one or more membranes like the smallpox virus shown in Figure 36.5d. The envelope consists of viral proteins embedded in a phospholipid bilayer derived from a membrane found in a host cell—specifically, the host cell in which the virion was manufactured. Later sections of this chapter will detail how viruses obtain their envelope from an infected host cell.

Analyzing the Genetic Material

In addition to morphology, viruses can be categorized based on the nature of their genome. This is not true for cells. DNA is the hereditary material in all cells, and information flows from DNA to mRNA to proteins (Chapter 16). Although all cells follow this pattern, which is called the central dogma of molecular biology, many viruses break it.

This conclusion traces back to work done in the 1950s, when Heinz Fraenkel-Conrat and colleagues were able to separate the protein and nucleic acid components of a plant virus known as the tobacco mosaic virus, or TMV. Surprisingly, the nucleic acid

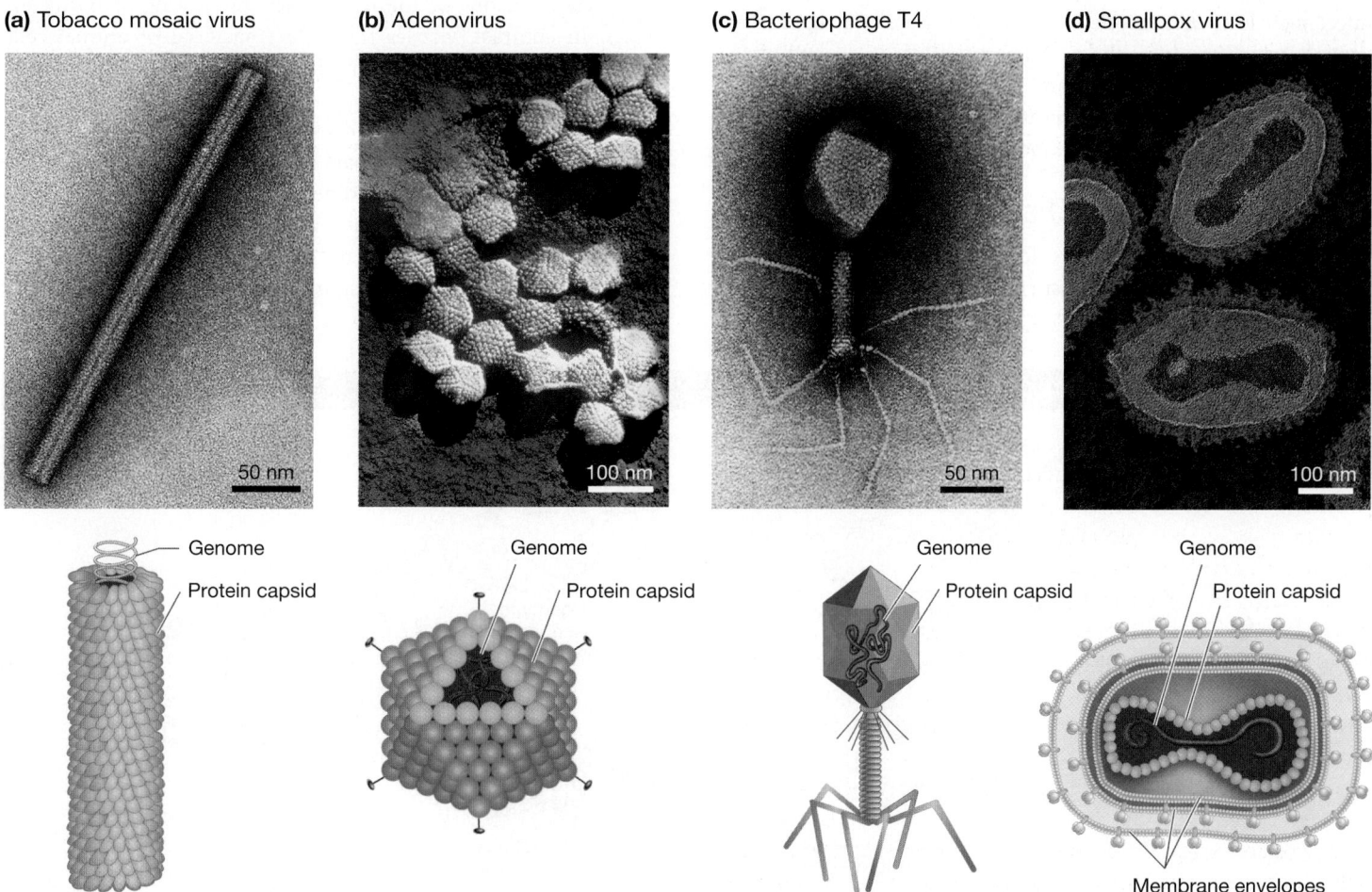

(a) Tobacco mosaic virus **(b)** Adenovirus **(c)** Bacteriophage T4 **(d)** Smallpox virus

50 nm   100 nm   50 nm   100 nm

Genome — Protein capsid   Genome — Protein capsid   Genome — Protein capsid   Genome — Protein capsid   Membrane envelopes

FIGURE 36.5 Viruses Vary in Size and Shape, and May Be "Naked" or Enveloped.

portion of this virus consisted of RNA, and the RNA of TMV, by itself, could infect plant tissues and cause disease. This was a confusing result because it showed that the RNA—not DNA—functions as TMV's genetic material.

Subsequent research revealed an amazing diversity of viral genome types. In some groups of viruses, such as the agents that cause measles and flu, the genome consists of RNA. In others, such as the viruses that cause herpes and smallpox, the genome is DNA. Viral genomes may be linear or circular and may consist of a single molecule or be broken up into several different segments.

Further, the RNA and DNA genomes of viruses can be either single stranded or double stranded. The single-stranded RNA genomes can also be classified as "positive sense" or "negative sense" or "ambisense."

- In a **positive-sense** RNA virus, the genome contains the same sequences as the mRNA required to produce viral proteins.

- In a **negative-sense** RNA virus, the base sequences in the genome are complementary to those in viral mRNAs.

- In an **ambisense** RNA virus, the genome has at least one strand that contains two regions: one is positive sense and the other negative sense.

The nature of viruses has been understood only since the 1940s, but they have been the focus of intense research ever since. Although likely millions of types of virus exist, they all appear to infect their host cells in one of two general ways: via **replicative growth,** which produces the next generation of virions and often kills the host cell, or in a dormant manner that suspends production of virions and allows the virus to coexist with the host for a period of time. Let's look first at the more immediately deadly mode.

Analyzing the Phases of Replicative Growth

Six phases are common to replicative growth in virtually all viruses: **(1)** attachment to a host cell and entry into the cytosol;

(2) transcription of the viral genome and production of viral proteins; **(3)** replication of the viral genome; **(4)** assembly of a new generation of virions; **(5)** exit from the infected cell; and **(6)** transmission to a new host.

FIGURE 36.6 shows the replicative growth of a bacteriophage. A **bacteriophage** (literally, "bacteria-eater") is a virus that infects bacterial cells. This cycle is referred to as the **lytic cycle,** since it ends with lysis (destruction) of the cell. One thing should stand out as you examine this process: in a single replicative cycle, one virus particle can produce many progeny.

FIGURE 36.7 shows how a bacteriophage's tremendous capacity for replication results in nonlinear growth that proceeds in a stepwise manner. This growth curve is very different from what is observed in the host cells, where each cell produces only two progeny in each generation.

Each type of virus has a particular way of entering a host cell and completing the subsequent phases of the cycle. Let's take a closer look at each phase.

How Do Viruses Enter a Cell? The replicative cycle of a virus begins when a free virion enters a target cell. This is no simple task. All cells are protected by a plasma membrane, and many also have a cell wall. How do viruses breach these defenses, insert themselves into the cytosol inside, and begin an infection?

Most plant viruses are inserted directly into the host-cell cytosol via abrasions or the mouthparts of sucking or biting insects. In contrast, viruses that infect bacterial or animal cells must first attach to a specific molecule on the cell wall or plasma membrane.

After a bacteriophage attaches to its host, it uses an enzyme called lysozyme to degrade part of the cell wall and expose the plasma membrane. The genome of the bacteriophage is then transferred to the cytosol in a process referred to as uncoating, which varies depending on the structure of the virus. The T4 bacteriophage shown in Figure 36.6, for example, uncoats its

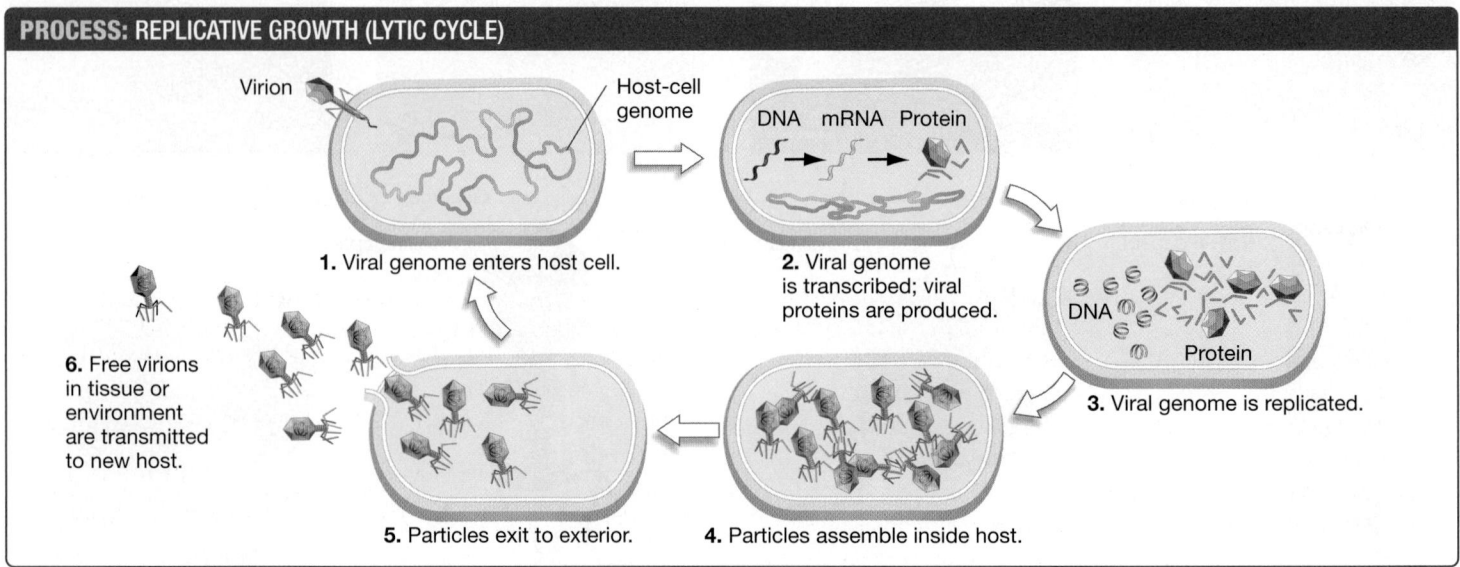

PROCESS: REPLICATIVE GROWTH (LYTIC CYCLE)

Virion — Host-cell genome

1. Viral genome enters host cell.

DNA mRNA Protein

2. Viral genome is transcribed; viral proteins are produced.

DNA — Protein

3. Viral genome is replicated.

4. Particles assemble inside host.

5. Particles exit to exterior.

6. Free virions in tissue or environment are transmitted to new host.

FIGURE 36.6 The Viral Replicative Cycle. Many viruses follow the same general cycle as this bacteriophage model.

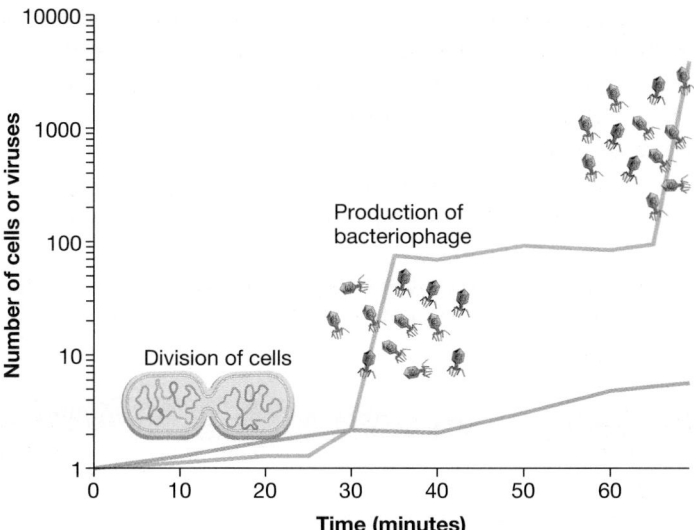

FIGURE 36.7 Growth Patterns Differ between Viruses and Their Host. Cells reproduce in an exponential manner, as indicated by a linear pattern when plotted on a logarithmic scale. The production of virions is not exponential; instead, reproduction of bacteriophage like T4 exhibits a stepwise pattern.

DATA: Courtesy of the Undergraduate Biotechnology Laboratory, California Polytechnic State University, San Luis Obispo.

✔**QUANTITATIVE** Based on the information and trends in this graph, predict the number of extracellular virions that would be produced after 90 minutes if all of those released by 40 minutes were allowed to infect new host cells. Compare this to the number of cells you would predict to be present after 90 minutes.

genome by injecting it through a hollow needle that is stabbed into the bacterial membrane.

Viruses that attack animal cells must first attach to one or more specific molecules in the host cell's plasma membrane. These molecules, often referred to as virus receptors, are typically either membrane proteins or the carbohydrates attached to glycoproteins or glycolipids.

To appreciate how investigators identify virus receptors, consider research on HIV. In 1981—right at the start of the AIDS epidemic—biomedical researchers realized that people with AIDS had few or no T cells possessing a particular membrane protein called **CD4**. These cells, symbolized as CD4$^+$, are called helper T cells because of their key supportive role in the immune response (see Chapter 51).

Two research groups hypothesized that CD4 serves as the receptor for HIV attachment. To test this hypothesis, the researchers systematically blocked different membrane proteins on helper T cells and determined which membrane proteins are involved in HIV attachment. They used **antibodies** to specifically bind to host membrane proteins, including several that bound to CD4 (see **BioSkills 9** in Appendix B to review the activity of antibodies).

The key assumption in their experiments was that antibodies directed against the virus receptor would prevent HIV from entering and infecting the cell. After completing their work, both research teams reached exactly the same result: Antibodies to CD4 inhibited HIV from infecting host cells (**FIGURE 36.8**).

RESEARCH

QUESTION: Does CD4 protein function as the receptor HIV uses to enter host cells?

HYPOTHESIS: CD4 is the membrane protein HIV uses to enter cells.

NULL HYPOTHESIS: CD4 is not the membrane protein HIV uses to enter cells.

EXPERIMENTAL SETUP:

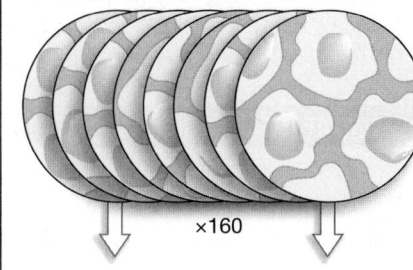

×160

1. **Take 160 identical samples of helper T cells** from a large population of T cells growing in culture.

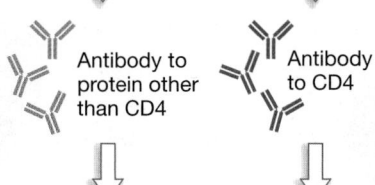

Antibody to protein other than CD4

Antibody to CD4

2. **Add a different antibody to each sample of cells**—each antibody will "block" a specific membrane protein.

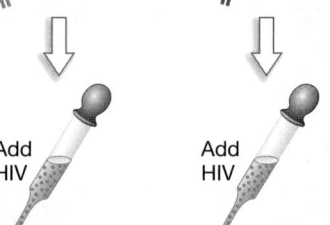

Add HIV

Add HIV

3. **Add a constant number of HIV virions** to all samples. Incubate cultures under conditions optimal for virus entry.

PREDICTION: HIV will not infect cells with antibody to CD4 but will infect other cells.

PREDICTION OF NULL HYPOTHESIS: HIV will infect cells with antibody to CD4.

RESULTS:

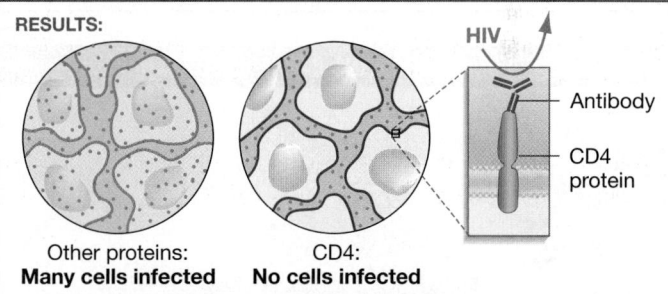

HIV

Antibody

CD4 protein

Other proteins:
Many cells infected

CD4:
No cells infected

CONCLUSION: HIV uses CD4 as the receptor to enter helper T cells. Thus, only cells with free, unbound CD4 on their surface can be infected by HIV.

FIGURE 36.8 Experiments Confirmed that CD4 Is the Receptor Used by HIV to Enter Host Cells. In this experiment, the antibodies added to each culture bound to a specific protein found on the surface of helper T cells. Antibody binding blocked the membrane protein, so the protein could not be used by HIV to gain entry to the cells.

SOURCE: Dalgleish, A. G., P. C. Beverley, P. R. Clapham, et al. 1984. The CD4 (T4) antigen is an essential component of the receptor for the AIDS retrovirus. *Nature* 312: 763–767. Also Klatzmann, D., E. Champagne, S. Chamaret, et al. 1984. T-lymphocyte T4 molecule behaves as the receptor for human retrovirus LAV. *Nature* 312: 767–768.

✔ **QUESTION** Does this experiment show that CD4 is the only membrane protein required for HIV entry? Explain why or why not.

Although CD4 is necessary for HIV attachment, uncoating the viral genome will not occur unless the virion also binds to a second membrane protein, called a **co-receptor.** In most individuals, membrane proteins called CXCR4 and CCR5 function as co-receptors. Once the virion binds to both CD4 and a co-receptor, the lipid bilayers of the virion's envelope and the plasma membrane of the T cell fuse (**FIGURE 36.9a**). When fusion occurs, HIV has breached the cell boundary. The viral capsid then enters the cytosol and disassembles to release the genomic RNAs, and infection proceeds.

Another common uncoating process is shown in **FIGURE 36.9b**, where the virion is first internalized via endocytosis (see Chapter 7). After a virion, in this case influenza, attaches to a component on the cell surface, it is pulled into the cell in the form of a vesicle called an endosome. When the endosome acidifies—a normal part of endocytosis—the structure of the viral attachment proteins change in a way that promotes fusion of the viral envelope and the endosomal membrane.

Endocytosis is also commonly used by naked viruses to enter animal cells. As with enveloped viruses, acidification of the

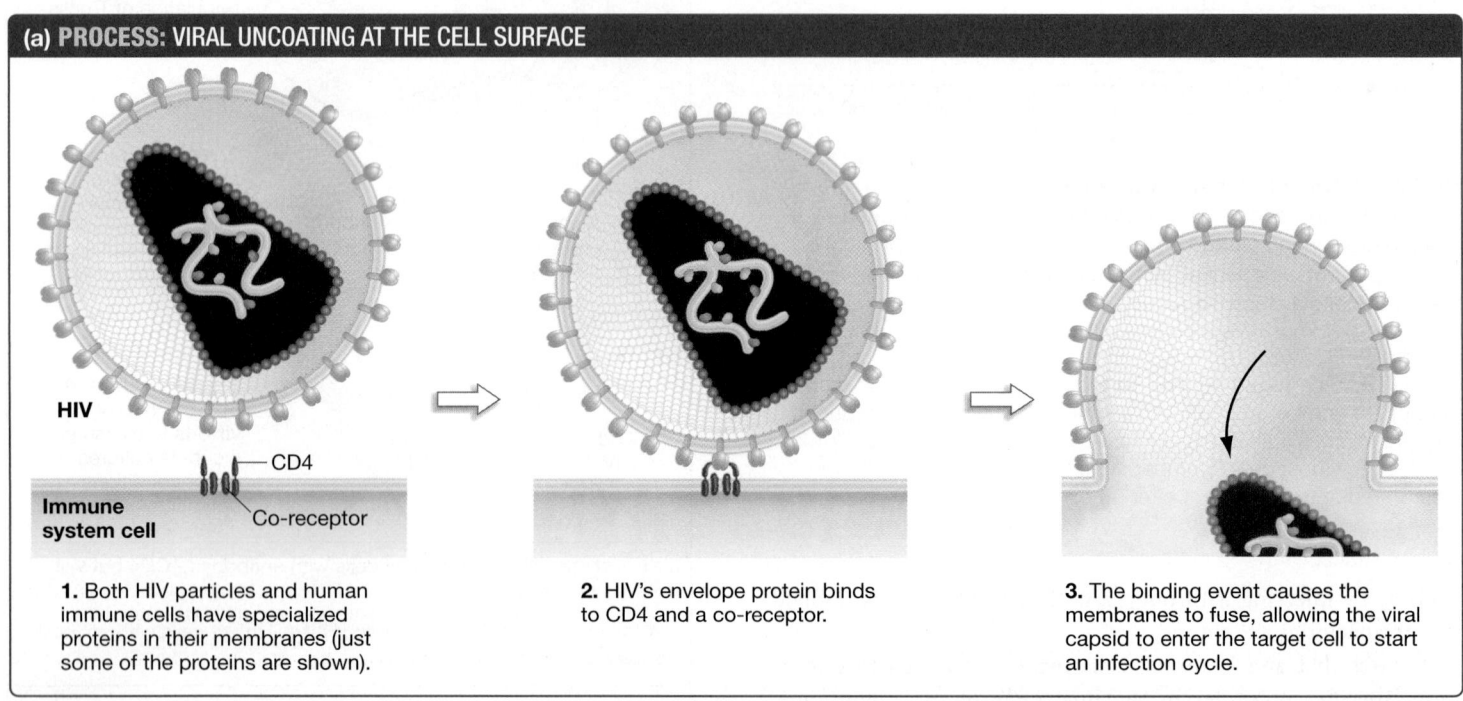

(a) PROCESS: VIRAL UNCOATING AT THE CELL SURFACE

HIV

CD4

Immune system cell

Co-receptor

1. Both HIV particles and human immune cells have specialized proteins in their membranes (just some of the proteins are shown).

2. HIV's envelope protein binds to CD4 and a co-receptor.

3. The binding event causes the membranes to fuse, allowing the viral capsid to enter the target cell to start an infection cycle.

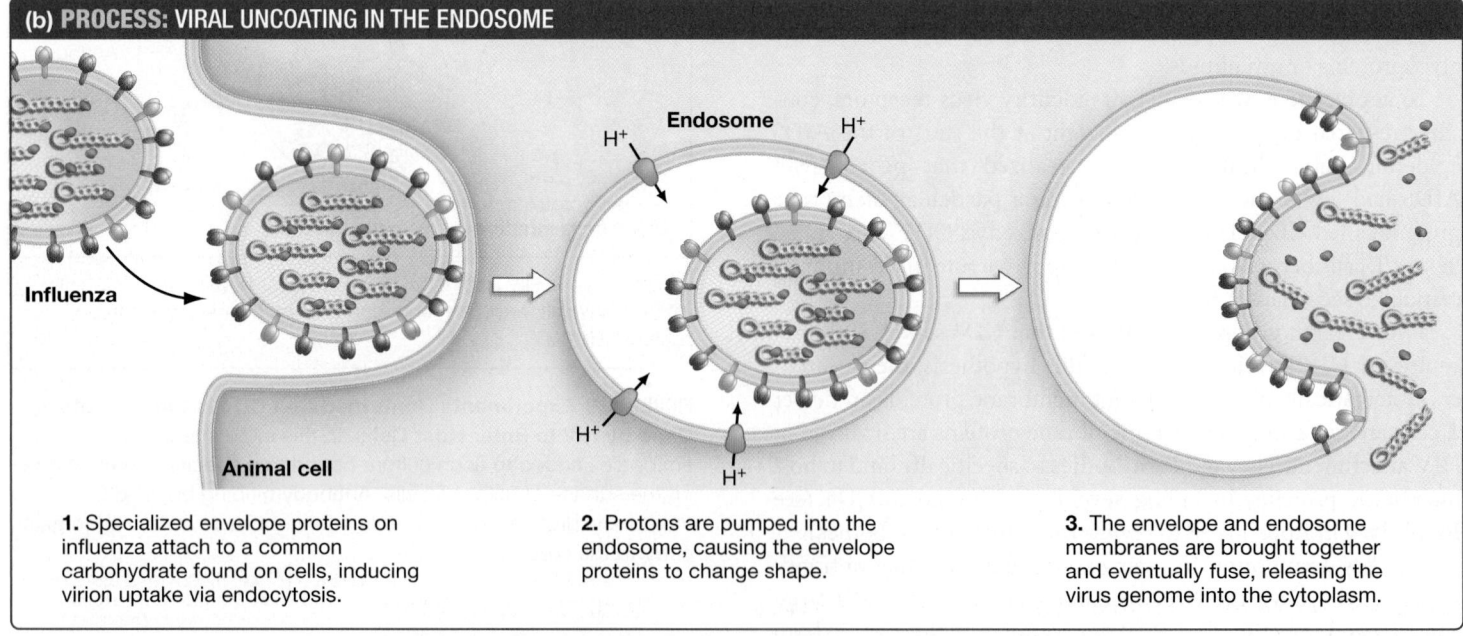

(b) PROCESS: VIRAL UNCOATING IN THE ENDOSOME

Endosome

H+   H+

Influenza

H+

H+

Animal cell

1. Specialized envelope proteins on influenza attach to a common carbohydrate found on cells, inducing virion uptake via endocytosis.

2. Protons are pumped into the endosome, causing the envelope proteins to change shape.

3. The envelope and endosome membranes are brought together and eventually fuse, releasing the virus genome into the cytoplasm.

FIGURE 36.9 Enveloped Viruses Bind to Host Membranes and Uncoat via Fusion.

endosome serves as a signal for uncoating the genetic material. In this case, low pH alters the structure of the capsid proteins, which disrupts the endosomal membrane so the viral genetic material may enter the cytosol.

The discovery of virus receptors and requirements for uncoating has inspired a search for compounds that block these early events of the replicative cycle. As a result, drugs have been designed to block attachment and uncoating of HIV by targeting the HIV envelope protein and host CCR5 co-receptor. Drugs that interfere with either viral infection or replication of viruses are called **antivirals.**

How Do Viruses Produce Proteins?

Production of viral proteins begins soon after the virus uncoats its genome. Depending on the virus, transcription of the viral genome may be accomplished by either host or viral RNA polymerases. In all viruses, however, translation of viral transcripts is entirely dependent on the host. Viruses lack the ribosomes, amino acids, ATP, and most of the other biosynthetic machinery required for translating their own mRNAs into proteins.

Viral proteins are produced and processed in one of two ways, depending on whether the proteins end up in the envelope of a virion or in a capsid.

Viral mRNAs that code for envelope proteins are translated as if they were mRNAs for the cell's own membrane proteins. They are translated by ribosomes attached to the rough endoplasmic reticulum (rough ER), where carbohydrates are added to the protein (see Chapter 7). Depending on the virus, the proteins then may be transported to the Golgi apparatus for further processing. In some viruses, such as HIV, the finished glycoproteins are transported to the plasma membrane, where they are assembled into new virions.

A different route is taken by mRNAs that code for proteins that make up the capsid of a virion. These mRNAs are translated by free ribosomes in the cytosol, just as if they were cellular mRNAs for cytosolic proteins.

In some viruses, long polypeptide sequences called polyproteins are cut into individual proteins by viral enzymes called **proteases.** These enzymes cleave viral polyproteins at specific locations—a critical step in the production of finished viral proteins. In HIV, for example, a polyprotein must be cleaved before it can assemble into new viral capsids.

The discovery that HIV produces a protease triggered research for drugs that would inhibit the enzyme. This effort got a huge boost when researchers identified the three-dimensional structure of HIV's protease, using X-ray crystallographic techniques (see **BioSkills 11** in Appendix B). The enzyme has an opening in its interior where the active site is located (**FIGURE 36.10a**). Viral polyproteins fit into the opening and are cleaved at the active site.

Based on these data, researchers immediately began searching for molecules that could fit into the opening and prevent protease from functioning by blocking the active site (**FIGURE 36.10b**). Several HIV protease inhibitors are currently being used to interfere with viral replication.

How Do Viruses Copy Their Genomes?

In addition to transcription and translation, viruses must also copy their genetic material to make a new generation of virions. Viruses depend on the host cell for nucleotide monomers. In addition, some DNA viruses also depend on the host-cell DNA polymerase machinery to replicate their genomes.

In viruses that have an RNA genome, however, the virus must supply its own enzyme to make copies of its genome. Viral enzymes called **RNA replicases** function as *RNA-dependent* RNA polymerases. In other words, RNA replicases synthesize RNA *from an RNA template*, using ribonucleotides provided by the host cell. For example, the RNA replicases of positive-sense single-stranded RNA viruses first convert the genome into double-stranded RNA and then produce multiple positive-sense copies from the negative-sense complementary strand.

(a) HIV's protease enzyme

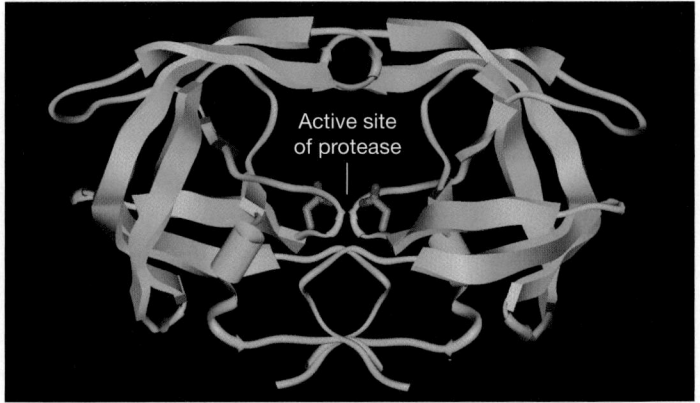

(b) Could a drug block the active site?

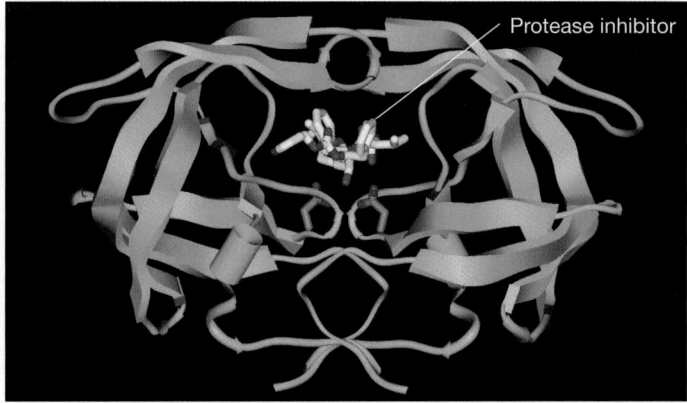

FIGURE 36.10 The Three-Dimensional Structure of HIV's Protease. (a) Ribbon diagram depicting the three-dimensional shape of HIV's protease enzyme. **(b)** Once protease's structure was solved, researchers began synthesizing compounds that were predicted to fit into the active site and prevent the enzyme from working.

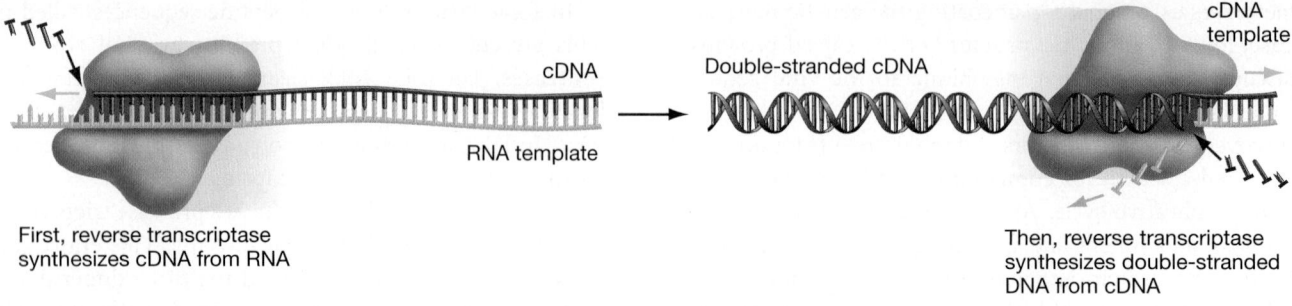

First, reverse transcriptase
synthesizes cDNA from RNA

Then, reverse transcriptase
synthesizes double-stranded
DNA from cDNA

FIGURE 36.11 Reverse Transcriptase Catalyzes Synthesis of a Double-Stranded DNA from an RNA Template. The DNA produced by reverse transcriptase is called a cDNA because its base sequence is complementary to the RNA template.

But not all RNA viruses replicate using RNA replicases. In certain RNA viruses, the genome is first transcribed from RNA to DNA by a viral enzyme called **reverse transcriptase.** This enzyme is an unusual DNA polymerase—one that can make DNA from either an RNA or a DNA template. It first makes **complementary DNA,** or **cDNA,** from a single-stranded RNA template (see Chapter 20). Reverse transcriptase then removes the RNA strand and catalyzes the synthesis of a second, complementary DNA strand, resulting in a double-stranded DNA (**FIGURE 36.11**). The DNA copy is then inserted into the host genome and used as a template for host cellular machinery—including an RNA polymerase—to produce viral mRNAs and genomic RNAs.

Viruses that reverse-transcribe their genome in this way are called **retroviruses** ("backward viruses"). The name is apt because the initial flow of genetic information in this type of virus is RNA → DNA—the opposite of the central dogma (see Chapter 16). HIV is an example of a reverse-transcribing RNA virus.

The first antiviral drugs that were developed to combat HIV act by inhibiting reverse transcriptase. Logically enough, drugs of this type are called reverse transcriptase inhibitors.

How Are New Virions Assembled? Once viral proteins have been produced and the viral genome has been replicated, assembly of a new generation of virions can take place.

During assembly, viral genomes are packaged into capsids. Some viruses also include copies of non-capsid proteins, often enzymes like polymerases, inside the capsid. In many cases, the details of the assembly process are not well understood. Some viruses assemble the capsid first and then use motor proteins to pull the viral genome inside, while others assemble the capsid around the genomic material.

Enveloped viruses use the host endomembrane system (see Chapter 7) to transport their envelope proteins to the appropriate membrane for assembly. For example, virions of HIV and influenza assemble at the host cell's surface and acquire their envelope from its plasma membrane. Other enveloped viruses may assemble at the surface of internal membranes, such as the rough ER or Golgi apparatus.

In most cases, self-assembly occurs—though some viruses produce proteins that provide scaffolding where new virions are put together. To date, researchers have yet to develop drugs that inhibit the assembly process during a viral infection.

How Do Progeny Virions Exit an Infected Cell? Most viruses leave a host cell in one of two ways: by budding from cellular membranes or by bursting out of the cell. In general, enveloped viruses bud; nonenveloped viruses burst.

Viruses that bud from one of the host cell's membranes take some of that membrane with them. As a result, their envelope includes host-cell phospholipids along with envelope proteins encoded by the viral genome. In **FIGURE 36.12a**, HIV is shown budding from the plasma membrane. After the budding step, the HIV capsid is further processed by its protease, giving the virion its characteristic cone-like appearance.

Viruses that bud through internal membranes, such as those of the rough ER or Golgi apparatus, are secreted from the cell by being escorted through the endomembrane system (see Chapter 7). Regardless of the membrane used for producing enveloped viruses, exit from the host cell does not require the death of the host.

In contrast, most nonenveloped viruses release their virions from the cell by lysing it—commonly referred to as the burst. For example, bacteriophages produce lytic enzymes that break down the cell wall of the host. Because the cell exerts pressure on the wall, the cell will explode when the wall is damaged—dispersing a new generation of virions into the environment. **FIGURE 36.12b** shows T4 bacteriophages bursting from an infected *E. coli* cell.

How Are Virions Transmitted to New Hosts? Once the replicative cycle is complete, dozens to several hundred newly assembled virions are in the extracellular space. What happens next?

If the host cell is part of a multicellular organism, the new generation of virions may be transported through the body—often via the bloodstream or lymphatic system. In vertebrates, antibodies produced by the immune system may bind to the virions and mark them for destruction. But if a virion contacts an appropriate host cell before it encounters antibodies, then a new replicative cycle will be started.

The long-term success of the virus, however, is dependent on its ability to be transmitted through the environment from one organism to another. For example, when people cough, sneeze,

(a) Budding of enveloped viruses

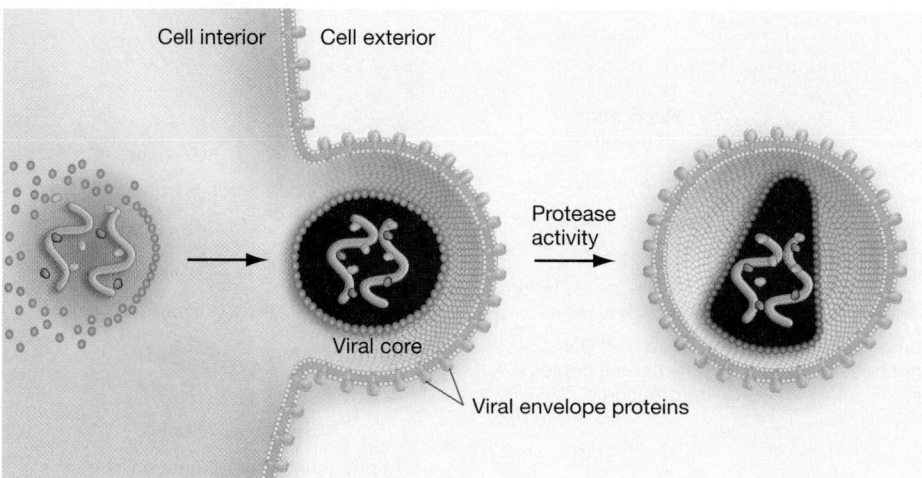

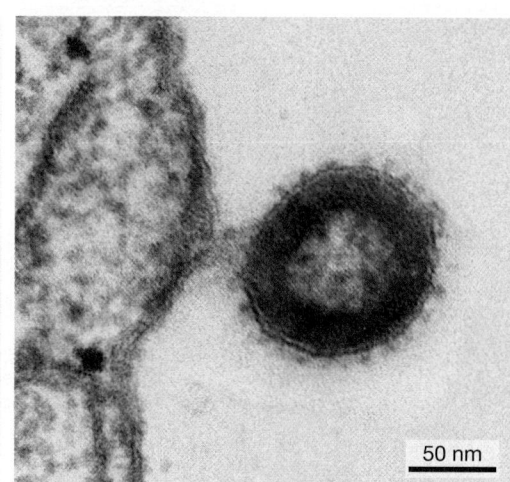

50 nm

(b) Bursting of nonenveloped viruses

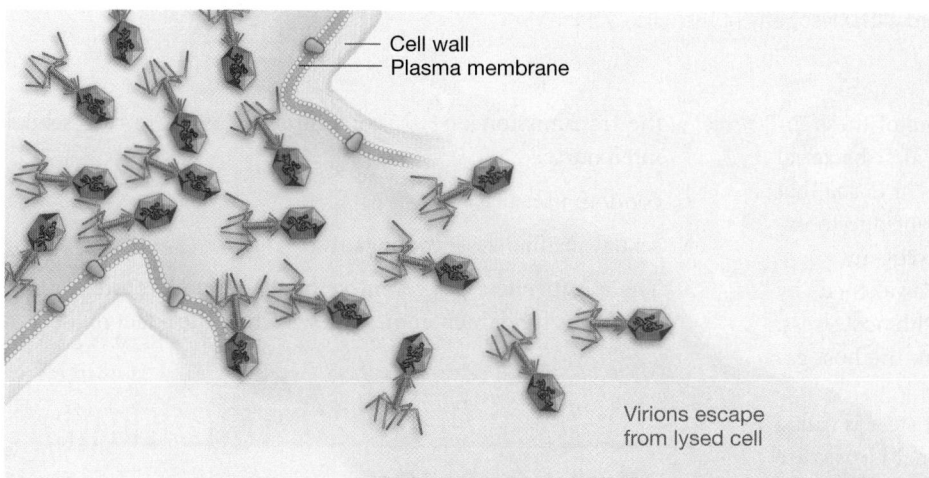

Cell wall
Plasma membrane

Virions escape
from lysed cell

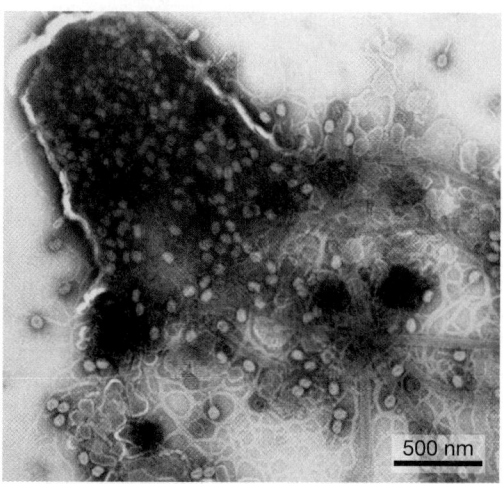

500 nm

FIGURE 36.12 Viruses Leave Infected Cells by Budding or Bursting. (a) In enveloped viruses such as HIV, virions are formed as they bud from membranes of the host cell. **(b)** Naked viruses normally exit the host cell by bursting the cell, as in T4 bacteriophage. Bursting kills the host cell; budding does not require cell death to occur.

✔ **QUESTION** Propose a hypothesis to explain why cells often die after extensive budding of enveloped viruses.

wipe a runny nose, or defecate, they help rid their body of viruses and bacteria. But they also expel the pathogens into the environment, sometimes directly onto a new host.

From the virus's point of view, a new host represents an unexploited habitat brimming with resources in the form of target cells. The situation is analogous to that of a multicellular animal dispersing to a new habitat and colonizing it. The alleles carried by these successful colonists increase in frequency in the total population. In this way, natural selection favors alleles that allow viruses to do two things: **(1)** replicate within a host and **(2)** be transmitted to new hosts.

✔ If you understand how different types of viruses produce a new generation of virions, you should be able to compare and contrast the replicative cycles of bacteriophage T4 and HIV.

Analyzing How Viruses Coexist with Host Cells

All viruses undergo replicative growth, but some can arrest the replicative cycle and enter a dormant state. Note that only certain types of viruses are capable of switching between replicative growth and dormant coexistence with the host cell. In bacteriophages, this alternate type of infection is called **lysogeny** (**FIGURE 36.13**, page 722).

The onset of lysogeny is triggered by molecular cues in the host that push the virus out of the replicative cycle. Instead of actively transcribing and replicating the viral DNA, it becomes incorporated into the host's chromosome and the expression of most of the viral genes is shut down.

Although no virions are produced during lysogeny, the host's DNA polymerase replicates the viral DNA each time the cell divides. Copies of the viral genome are passed on to daughter cells

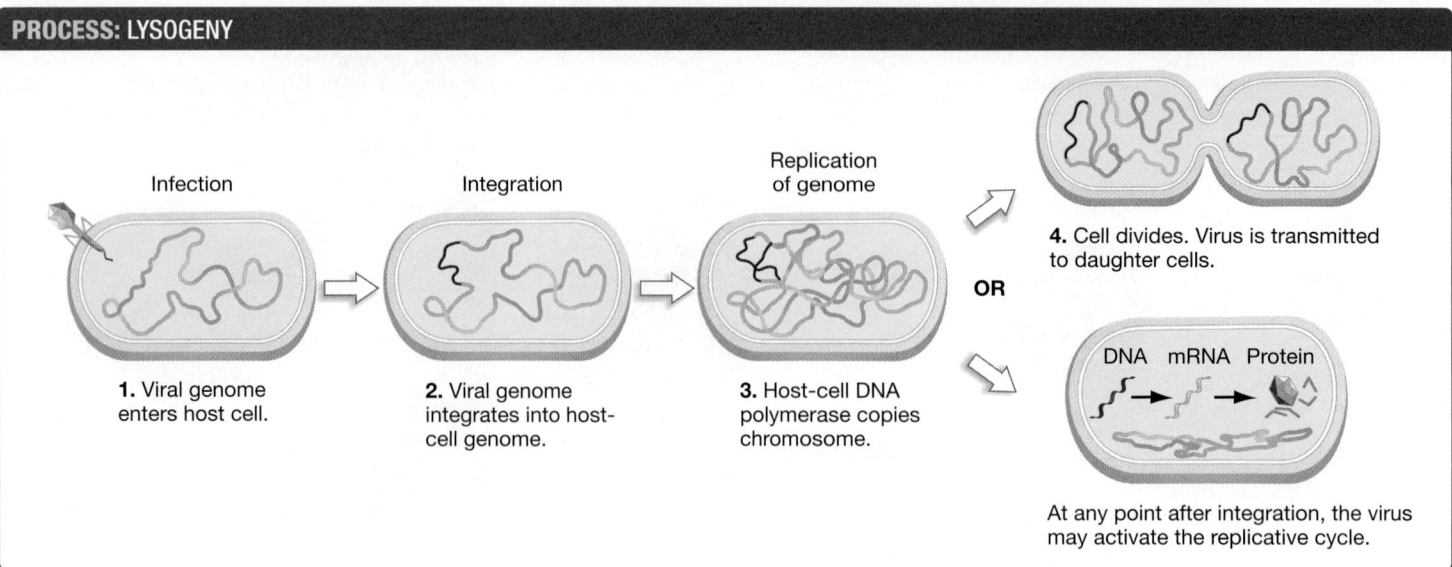

Infection

Integration

Replication of genome

OR

4. Cell divides. Virus is transmitted to daughter cells.

1. Viral genome enters host cell.

2. Viral genome integrates into host-cell genome.

3. Host-cell DNA polymerase copies chromosome.

DNA   mRNA   Protein

At any point after integration, the virus may activate the replicative cycle.

FIGURE 36.13 Some Bacteriophages Can Switch between the Lytic Cycle and Lysogeny. Molecular cues from the host can direct certain bacteriophages toward either lysogeny or the lytic cycle.

just as if it were one of the host's own genes. Some of these introduced genes have been known to significantly alter bacterial hosts. For example, many of the pathogenic strains of *E. coli* that have recently emerged in humans are more virulent due to expression of phage toxins that were introduced via lysogeny.

An integrated bacteriophage genome may be "awakened" by different cues from the host cell, often associated with host stress. Activation leads to excision of the viral DNA from the host genome and reentry into the replicative growth cycle.

In viruses that infect animal cells, the dormant state is called **latency.** As with bacteriophages, only certain animal viruses are capable of arresting their replicative growth cycle. Depending on the virus involved, the genome may or may not be integrated into the host genome.

HIV has a well-characterized latency period that occurs at the start of many new infections. After reverse transcriptase makes a double-stranded DNA version of the viral genome, the viral DNA is inserted into the host-cell genomic DNA. But integration alone is not enough to trigger expression of the viral genes. The helper T cell host must first be activated by the immune response to drive expression of the HIV genome. Until this occurs, HIV remains hidden and silent.

Current antivirals are ineffective against an HIV infection when HIV genes aren't being expressed. For this reason, reducing the likelihood of **transmission**—the spread of pathogens from one individual to another—is presently the most effective way to combat a viral disease like AIDS.

HIV is transmitted from person to person via body fluids such as blood, semen, or vaginal secretions. Faced with decades of disappointing results in drug and vaccine development, public health officials are aggressively promoting preventive medicine, through

* aggressive treatment of venereal diseases—because the lesions caused by chlamydia, genital warts, and gonorrhea encourage

the transmission of HIV-contaminated blood during sexual intercourse;

* condom use; and

* sexual abstinence or monogamy.

The effectiveness of preventive medicine underscores one of this chapter's fundamental messages: Viruses are a fact of life.

check your understanding

If you understand that . . .

* Viruses replicate using the energy, substrates, and protein synthesis machinery of their host cells to produce a new generation of virions.
* After infecting a cell, the manner in which viruses proceed through the steps of the replicative cycle will vary depending on the type of virus.

✔ **You should be able to . . .**

Evaluate the claim that the viral replicative cycle more closely resembles the mass production of automobiles than the reproduction of cells.

Answers are available in Appendix A.

36.3 What Themes Occur in the Diversification of Viruses?

The tree of life will never be free of viruses. Mutation and natural selection guarantee that viral genomes will continually adapt to the defenses offered by their hosts, whether those defenses are devised by an immune system or by biomedical researchers.

Two other points are critical to recognize about viral diversity: (1) Biologists do not have a solid understanding of how new viruses originate, but (2) it is certain that viruses will continue to diversify. Because most viral polymerases have high error rates and because viruses lack error repair enzymes, mutation rates are extremely high. Many viruses change constantly—giving them the potential to evolve rapidly.

Where Did Viruses Come From?

No one knows how viruses originated. To address this question, biologists are currently considering three hypotheses to explain where viruses came from.

Origin in Plasmids and Transposable Elements? Like viruses, plasmids and transposable elements are acellular, mobile genetic elements that replicate with the aid of a host cell (see Chapters 20 and 21). Certain viruses are actually indistinguishable from plasmids except for one feature: They encode proteins that form a capsid and allow the genes to exist outside of a cell.

Some biologists hypothesize that simple viruses represent "escaped gene sets." This hypothesis proposes that mobile genetic elements are descended from clusters of genes that physically escaped from prokaryotic or eukaryotic chromosomes long ago.

According to this hypothesis, the escaped gene sets took on a mobile, parasitic existence because they happened to encode the information needed to replicate themselves at the expense of the genomes that once held them. In the case of viruses, the hypothesis is that the escaped genes included the instructions for making a protein capsid and possibly envelope proteins. According to the escaped-genes hypothesis, it is possible that each of the distinct types of viruses represent distinct "escape events."

To support the escaped-genes hypothesis, researchers would need to discover a virus that had so recently derived from intact prokaryotic or eukaryotic genes that the viral genomes still strongly resembled the DNA sequence of those genes.

Origin in Symbiotic Bacteria? Some researchers contend that DNA viruses with large genomes trace their ancestry back to free-living bacteria that once took up residence inside eukaryotic cells. The idea is that these organisms degenerated into viruses by gradually losing the genes required to synthesize ribosomes, ATP, nucleotides, amino acids, and other compounds.

Although this idea sounds speculative, it cannot be dismissed lightly. For example, there is evidence that the mitochondria and chloroplasts of eukaryotic cells originated as intracellular symbionts from ancestors that were independent, free-living cells (see Chapter 30). Investigators contend that, instead of evolving into intracellular symbionts that aid their host cell, DNA viruses became parasites capable of replication and transmission from one host to another.

To support the degeneration hypothesis, researchers have pointed to mimivirus, the largest known virus that infects certain protists. In addition to its large size, the genome of this virus was found to contain some of the genes involved in protein synthesis—genes that are common to cells, but had not previously been observed in viruses. It is still not clear, though, whether these genes are remnants from an ancestral cell or were acquired from a host cell that was infected relatively recently. The discovery of a cell that possesses a genome similar to that of the mimivirus would support the degeneration hypothesis for the origin of this virus.

Origin at the Origin of Life? Recently, some researchers have started to discuss a third alternative to explain the origin of viruses: that they trace their ancestry back to the first, RNA-based forms of life on Earth. If this hypothesis is correct, then the RNA genomes of some viruses are descended from genes found in early inhabitants of the RNA world (see Chapter 4). Some researchers have even suggested that retrovirus-like parasites may have been responsible for transforming the genetic material of cells from RNA to DNA.

To support this hypothesis, advocates point to the ubiquity of viruses—which suggests that they have been evolving along with organisms since life began. The RNA world-origin hypothesis also addresses a problem that weakens the other two hypotheses: Several proteins that are commonly expressed in many viruses are not expressed in any cell examined thus far. If viruses originated at the time of the origin of the cell, then the genes coding for these proteins may have come from an RNA-world pool of genes instead of a cell.

Currently, there is no one widely accepted view of where viruses came from. Because viruses are so diverse, all three hypotheses may be valid.

Emerging Viruses, Emerging Diseases

Although it is not known how the various types of viruses originated, it is certain that viruses will continue to diversify. With alarming regularity, the front pages of newspapers carry accounts of deadly viruses that are infecting humans for the first time.

HIV is an example of a virus responsible for an **emerging disease:** a new illness that suddenly affects significant numbers of individuals in a host population. Another, and perhaps the most infamous of such outbreaks occurred in 1995, when the Ebola virus infected 200 individuals in the Democratic Republic of Congo and had a fatality rate of 80 percent. In both cases, the causative agents were considered emerging viruses because they had switched from their traditional host species to a new host—humans.

Some Emerging Viruses Arise from Genome Reassortment Each year the World Health Organization worries about a new influenza pandemic, one that might resemble the devastating outbreak of 1918. The small changes that arise from the influenza virus's error-prone RNA replicase are primarily responsible for the need for yearly vaccine updates. But influenza can also acquire alleles that are entirely new to the strain. (A virus **strain** consists of populations that have similar characteristics and is the lowest, or most specific, level of taxonomy for viruses.) How do new strains of influenza originate?

Influenza has a single-stranded RNA genome that consists of eight segments, most of which encode only one protein. If two viruses infect the same cell, the replicated genomic segments

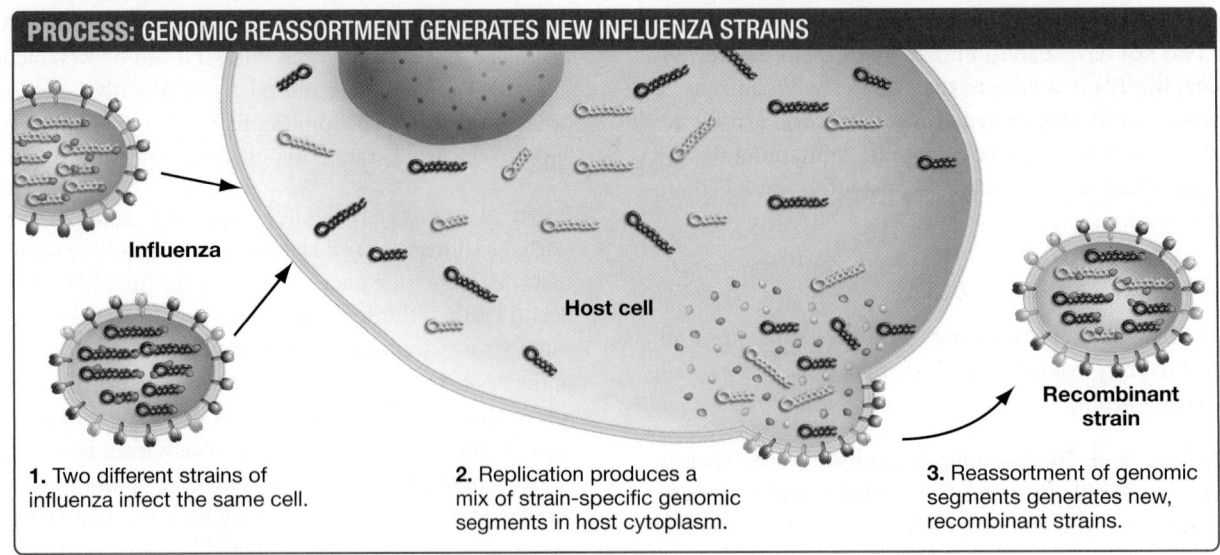

PROCESS: GENOMIC REASSORTMENT GENERATES NEW INFLUENZA STRAINS

Influenza

Host cell

Recombinant strain

1. Two different strains of influenza infect the same cell.

2. Replication produces a mix of strain-specific genomic segments in host cytoplasm.

3. Reassortment of genomic segments generates new, recombinant strains.

FIGURE 36.14 Influenza Can Generate New Strains via Genomic Reassortment.

are randomly shuffled in the cytosol. Thus when the progeny are assembled, many have segments from each parent virus (**FIGURE 36.14**).

The ability to randomly mix genomic segments becomes particularly significant when two different strains infect the same cell. In addition to humans, different strains of influenza can infect other animals, such as birds and pigs. While the avian strains do not efficiently infect humans, they can infect pigs—and so can human viral strains. The last point is key— pigs can serve as mixing vessels to produce new recombinant strains. Pandemic strains are thought to emerge from this type of reassortment.

Using Phylogenetic Trees to Understand Emerging Viruses How do researchers know that an emerging virus has "jumped" to a new host? The answer is to analyze the evolutionary history of the virus in question and then estimate a phylogenetic tree that includes its close relatives.

HIV, for example, belongs to a group of viruses called the lentiviruses, which infect a wide range of mammals including house cats, horses, goats, and primates. (*Lenti* is a Latin root that means "slow"; here it refers to the long period observed between the start of an infection by these viruses and the onset of the diseases they cause.) Consider the conclusions that can be drawn from the phylogenetic tree of HIV, shown in **FIGURE 36.15**.

- *There are immunodeficiency viruses.* Many of HIV's closest relatives parasitize cells that are part of the immune system. Several of them cause diseases with symptoms reminiscent of AIDS. These viruses infect monkeys and chimpanzees and are called simian immunodeficiency viruses (SIVs).

- *There are two HIVs.* There are two distinct types of human immunodeficiency viruses, called HIV-1 and HIV-2. Although both can cause AIDS, HIV-1 is far more virulent and is the better studied of the two.

HIV-1's closest known relatives are immunodeficiency viruses isolated from chimpanzees that live in central Africa. In contrast, HIV-2's closest relatives are immunodeficiency viruses that parasitize monkeys called sooty mangabeys.

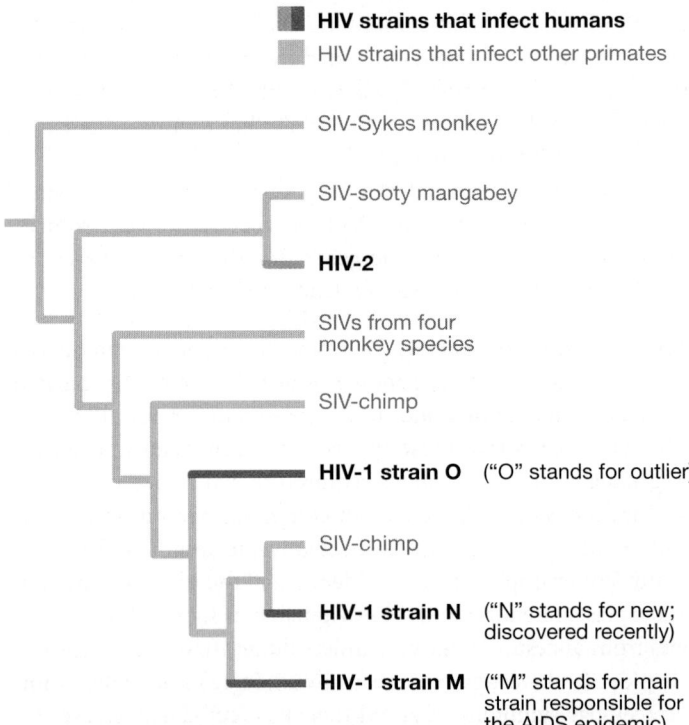

HIV strains that infect humans

HIV strains that infect other primates

SIV-Sykes monkey

SIV-sooty mangabey

HIV-2

SIVs from four monkey species

SIV-chimp

HIV-1 strain O ("O" stands for outlier)

SIV-chimp

HIV-1 strain N ("N" stands for new; discovered recently)

HIV-1 strain M ("M" stands for main strain responsible for the AIDS epidemic)

FIGURE 36.15 Phylogeny of HIV Strains and Types. Phylogenetic tree showing the evolutionary relationships among some of the immunodeficiency viruses that infect primates—including chimpanzees, humans, and several species of monkeys.

✔**EXERCISE** On the appropriate branches, indicate where an SIV jumped to humans (draw and label bars across the appropriate branches).

In central Africa, where HIV-1 infection rates first reached epidemic proportions, contact between chimpanzees and humans is extensive. Chimps are hunted for food and kept as pets. Similarly, sooty mangabeys are hunted and kept as pets in western Africa, where HIV-2 infection rates are highest.

- *Multiple "jumps" have occurred.* Several strains of HIV-1 exist. The most important are called O for outlier (meaning, the most distant group relative to other strains), N for new, and M for main. Each of these strains likely represents an independent origin of HIV-1 from a chimp SIV strain.

The last point is particularly important. The existence of distinct strains suggests that HIV-1 has jumped from chimps to humans several times. It may do so again in the future.

Responding to a Virus Outbreak Physicians become alarmed when they see a large number of patients with identical, and unusual, disease symptoms in the same geographic area over a short period of time. The physicians report these cases to public health officials, who take on two urgent tasks: (1) identifying the agent that is causing the new illness and (2) determining how the disease is being transmitted.

Several strategies can be used to identify the pathogen responsible for an emerging disease. In 1993, an outbreak of an unknown pulmonary syndrome rocked the U.S. Southwest, killing half of the individuals who came down with the disease. Officials recognized strong similarities between symptoms in these cases and symptoms caused by a hantavirus native to northeast Asia, called Hantaan virus. The Hantaan virus rarely causes disease in North America, but it is known to infect rodents.

To determine whether a Hantaan-like virus was responsible for the U.S. outbreak, researchers began capturing mice in the homes and workplaces of afflicted people. About a third of the captured rodents tested positive for the presence of a Hantaan-like virus. Genome sequencing studies confirmed that the virus was a previously undescribed type of hantavirus, and it was aptly named the *sin nombre* virus (Spanish for "no-name virus"). Further, the sequences found in the mice matched those found in infected patients. Based on these results, officials were confident that this rodent-borne hantavirus was causing the wave of human infections.

The next step in the research program, identifying how the agent is being transmitted, is equally critical and takes old-fashioned detective work. By interviewing patients about their activities, researchers called epidemiologists decide how each patient could have acquired the virus. Was there a particular environment or source of food or water shared by those who were infected? Was the illness showing up in health-care workers who were in contact with infected individuals, implying that it was being transmitted from human to human?

If the virus identified as the cause of an illness can be transmitted efficiently from person to person, then the outbreak has the potential to become an epidemic. To date, the sin nombre virus has not been transmitted efficiently from person to person in the United States—only inefficiently from mice to people.

check your understanding

If you understand that . . .

- The origin of viruses is unknown, but three hypotheses have been developed to address it.
- Emerging viruses arise either from genomic reassortment between different strains of virus or via switching host species.

✓ **You should be able to . . .**

1. Compare and contrast the three hypotheses for the origin of viruses.
2. Predict what type of a mutation in an avian flu virus would make it more dangerous to humans, and explain how this change might occur.

Answers are available in Appendix A.

36.4 Key Lineages of Viruses

Because scientists are almost certain that viruses originated more than once throughout the history of life, there is no such thing as the phylogeny of all viruses. Stated another way, unlike the organisms discussed in previous chapters, viruses have no single phylogenetic tree that represents their evolutionary history. Instead, researchers most often focus on comparing base sequences in the genetic material of small, closely related groups of viruses to reconstruct the phylogenies of particular lineages, exemplified by the tree in Figure 36.15.

To organize the genomic diversity of viruses, researchers group them into seven general categories based on the nature of their genetic material and how they replicate. David Baltimore first proposed this approach in 1971; thus, it is called the Baltimore classification. **FIGURE 36.16** (page 726) summarizes the seven classes.

Notice that each of the Baltimore classes has its own unique strategy for transcription and replication of the genetic material. (See Section 36.2 to review how viruses copy their genomes.) All these strategies, however, share the need to produce mRNA, which is then translated by the host cell to make viral proteins.

Within the Baltimore classification system, researchers further distinguish viruses according to (1) virion morphology, (2) nature of the host species, and (3) the type of disease resulting from infection. Using these criteria, biologists identify a total of about 70 virus families.

Although the phylogenetic relationships among viruses is not as clear as what is observed in organisms, those within families are grouped into distinct genera for convenience. Within genera, biologists identify and name types of virus, such as HIV, the measles virus, and smallpox. Within each of these viral types, populations with distinct characteristics may be identified and named as strains.

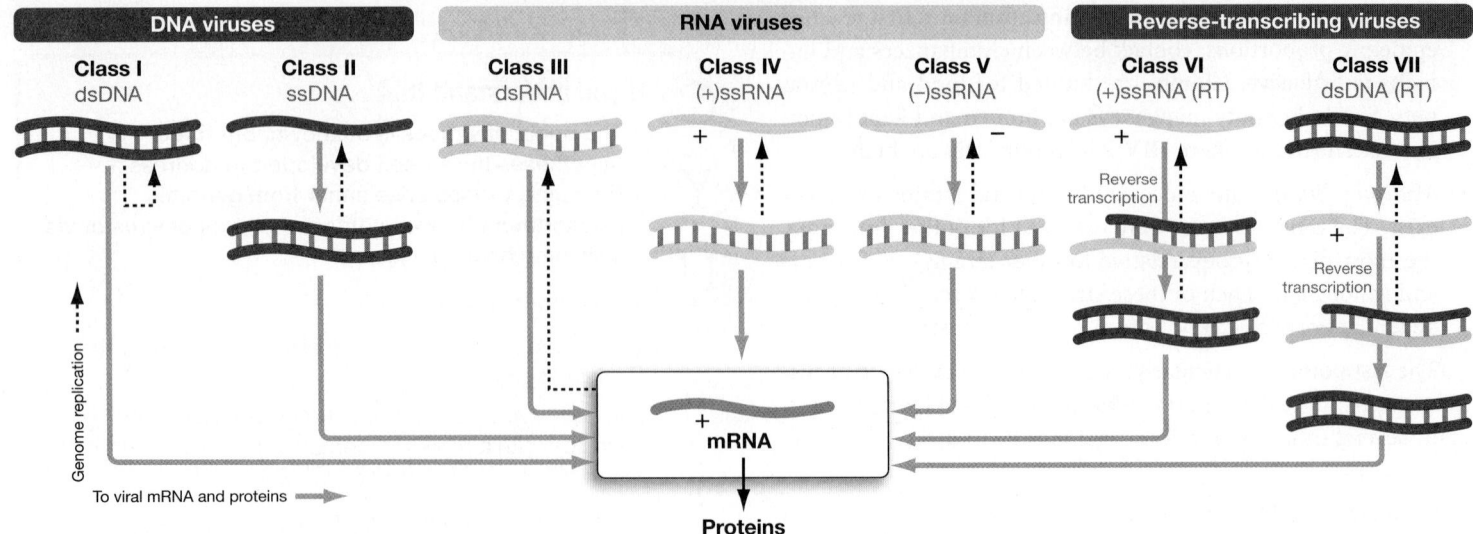

FIGURE 36.16 The Seven Different Strategies for Expression of Viral Genetic Material Converge on the Translation of mRNA. Notice that, although classes I and IV have the same types of genomes as classes VII and VI, they differ in the ways their genomes are replicated.

To get a sense of viral diversity, let's survey some of the most prevalent groups that can be distinguished based on the nature of their genetic material and mode of replication.

- Double-Stranded DNA (dsDNA) Viruses
- Double-Stranded RNA (dsRNA) Viruses
- Positive-Sense Single-Stranded RNA ([+]ssRNA) Viruses
- Negative-Sense Single-Stranded RNA ([−]ssRNA) Viruses
- RNA Reverse-Transcribing Viruses (Retroviruses)

Double-Stranded DNA (dsDNA) Viruses

The double-stranded DNA viruses are a large group, composed of some 21 families and 65 genera. Smallpox is perhaps the most familiar of these viruses (**FIGURE 36.17**). Although smallpox had been responsible for millions of deaths throughout human history, it was eradicated by worldwide vaccination programs. In 1977, smallpox was declared to be extinct in the wild; the only remaining samples of the virus are stored securely in research labs.

Genetic material As their name implies, the genes of these viruses consist of a single molecule of double-stranded DNA. The molecule may be linear or circular.

Host species These viruses parasitize hosts throughout the tree of life, with the notable exception of land plants. They include bacteriophage families called the T-series (T1–T7) and λ, which infect *E. coli*. Other common animal viruses with double-stranded DNA genomes include the human papilloma virus (HPV), herpesviruses, and adenoviruses.

Replicative cycle In most double-stranded DNA viruses that infect eukaryotes, viral genes have to enter the nucleus to be replicated. Often these viruses will replicate their genomes only during S phase, when the host cell's chromosomes are being replicated. Viruses like HPV are able to induce the cells to enter S phase despite the absence of the normal growth signals. This ability may, in fact, be responsible for the link between HPV infections and cervical cancer in women.

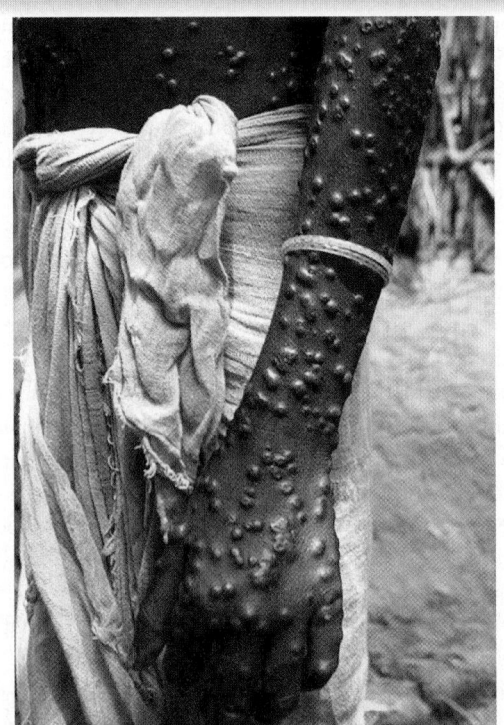

FIGURE 36.17 In Humans, Smallpox Can Cause Disease with Easily Recognizable Symptoms.

Double-Stranded RNA (dsRNA) Viruses

There are 7 families of double-stranded RNA viruses and a total of 22 genera. Most of the viruses in this group are nonenveloped.

Genetic material In some families, virions have a genome consisting of 10–12 double-stranded RNA molecules; in other families, the genome is composed of just 1–3 RNA molecules.

Host species A wide variety of organisms, including fungi, land plants, insects, vertebrates, and bacteria, are hosts for viruses with double-stranded RNA genomes. Particularly prominent are viruses that cause disease in rice, corn, sugarcane, and other crops. The bluetongue virus, a double-stranded RNA virus that is transmitted through midges and other biting insects, causes an often fatal disease that has significantly affected the livestock industry (**FIGURE 36.18**). In humans, reovirus and rotavirus infections are the leading cause of infant diarrhea and are responsible for over 110 million cases and a devastating 440,000 infant deaths each year.

Replicative cycle Once inside the cytosol of a host cell, the double-stranded genome of these viruses serves as a template for the synthesis of mRNAs, which are then translated into viral proteins. Some of the proteins form the capsids for a new generation of virions. The genome is copied by a viral enzyme that converts the mRNA transcripts into double-stranded RNA within the newly formed capsids.

FIGURE 36.18 dsRNA Viruses Parasitize a Wide Array of Organisms, Such as Livestock Infected by the Bluetongue Virus.

Positive-Sense Single-Stranded RNA ([+]ssRNA) Viruses

This is the largest group of viruses known. It has 21 families and 81 genera that include both enveloped and nonenveloped morphologies.

Genetic material The sequence of bases in a positive-sense single-stranded RNA virus is the same as that of a viral mRNA. Stated another way, the genome does not need to be transcribed in order for proteins to be produced. Depending on the species, the genome consists of 1–3 double-stranded RNA molecules.

Host species Most of the commercially important plant viruses belong to this group. Because they kill groups of cells in the host plant and turn patches of leaf or stem white, they are often named mottle viruses, spotted viruses, chlorotic (meaning, lacking chlorophyll) viruses, or mosaic viruses. **FIGURE 36.19** shows a healthy cowpea leaf and a cowpea leaf that has been attacked by a positive-sense single-stranded RNA virus. This group of viruses is also known to infect bacteria, fungi, and animals. Viruses within this group that infect humans include the rhinovirus (common cold), polio, SARS CoV, West Nile virus, and hepatitis A, C, and E.

Replicative cycle When the genome of these viruses enters a host cell, the positive-sense single-stranded RNA is immediately translated into a protein, often a polyprotein that is cleaved to generate

functionally distinct proteins. These proteins include proteases that process the polypeptides and polymerases that copy the genome and produce mRNA for translation.

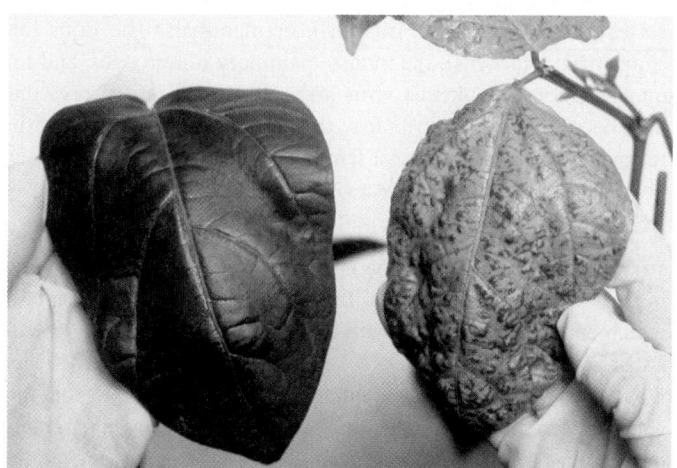

Healthy leaf   Leaf infected with virus

FIGURE 36.19 (+)ssRNA Viruses, Such as the Cowpea Mosaic Virus, Cause Important Plant Diseases.

Negative-Sense Single-Stranded RNA ([−]ssRNA) Viruses

This group has 7 families and 30 genera. Most members of this group are enveloped, but some negative-sense single-stranded RNA viruses lack an envelope.

Genetic material The sequence of bases in a negative-sense single-stranded RNA virus is opposite in polarity to the sequence in a viral mRNA. Stated another way, the virus genome is complementary to the viral mRNA. Depending on the family, the genome may consist of one single-stranded RNA molecule or up to eight separate RNA molecules.

Host species A wide variety of plants and animals are parasitized by viruses that have negative-sense single-stranded RNA genomes. If you have ever suffered from the flu, the mumps, or the measles, then you are painfully familiar with these viruses (**FIGURE 36.20**). The Ebola, sin nombre, and rabies viruses also belong to this group.

Replicative cycle When the genome of a negative-sense single-stranded RNA virus enters a host cell, a viral RNA-dependent RNA polymerase must accompany it to make new viral RNAs. Some of these transcripts serve as mRNA to be translated into viral proteins, while others function as templates for replicating the viral genome.

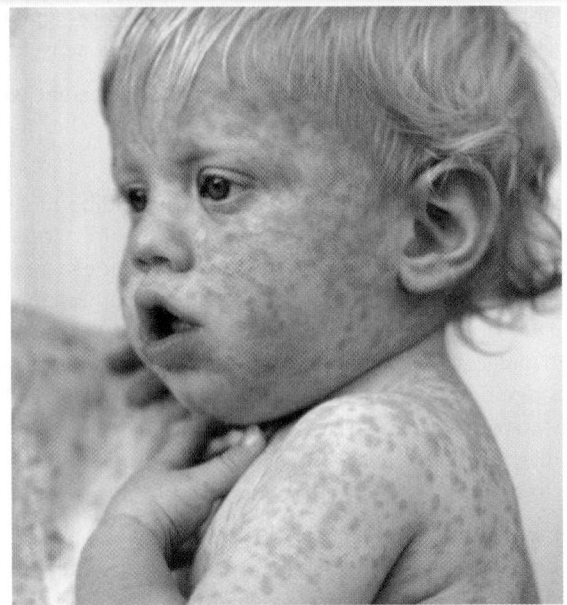

FIGURE 36.20 (−)ssRNA Viruses, Such as the Measles Virus, Cause Some Common Childhood Diseases.

RNA Reverse-Transcribing Viruses (Retroviruses)

The genomes of the RNA reverse-transcribing viruses are composed of positive-sense single-stranded RNA. This group has only one family, called the retroviruses, with 7 genera.

Genetic material Virions have two copies of a positive-sense single-stranded genome.

Host species Species in this group are known to parasitize only vertebrates—specifically birds, fish, or mammals. The Rous sarcoma virus (chickens), the mouse mammary tumor virus, and the murine (mouse) leukemia virus are well-studied retroviruses that have been shown to contribute to the development of cancer. Of the retroviruses, HIV is the most familiar and, in terms of the human population, by far the deadliest virus in this group (**FIGURE 36.21**).

Replicative cycle Although retroviruses have positive-sense single-stranded RNA, their genome is not translated immediately after it enters the host-cell cytosol. Instead, viral reverse transcriptase, which enters the cell along with the genome, catalyzes the synthesis of a double-stranded DNA version of the viral genome. This DNA is transported into the nucleus and integrated into the host genome by a viral protein called integrase. The virus may remain dormant for a period until induced by host-cell signals, such as the activation of HIV via the immune response. Activation starts up the replicative cycle. Genes are transcribed to mRNA to begin the production of a new generation of virions.

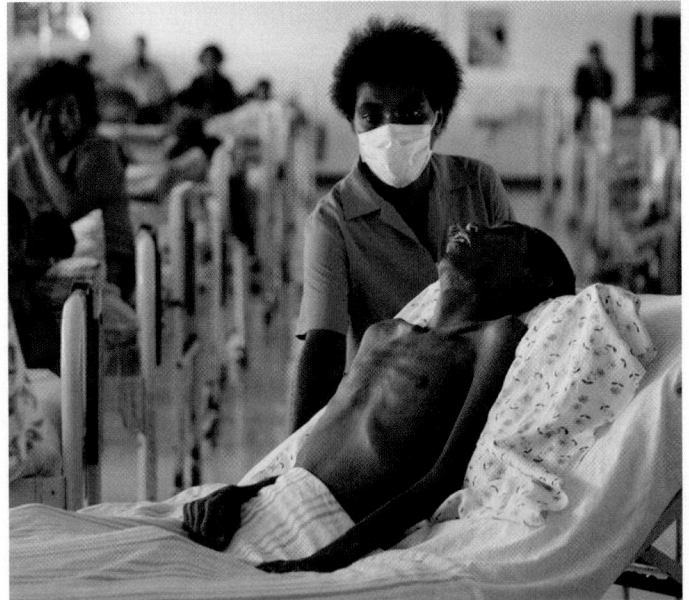

FIGURE 36.21 Some Retroviruses, Such as HIV, Severely Impair the Physiology of the Host.

If you understand . . .

36.1 Why Do Biologists Study Viruses?

- Viruses cause illness and death in organisms from all three domains of life.

- Viruses are specialists—different types of viruses infect particular species and types of cells.

✓ You should be able to explain how HIV causes AIDS and why it takes close to eight years before the clinical signs appear.

36.2 How Do Biologists Study Viruses?

- Most viruses have a capsid made of protein that is either icosahedral or helical in shape. Some viruses are covered by a host-derived membrane called an envelope.

- Viral genomes exhibit considerable diversity. The genetic material may consist of one or more molecules of DNA or RNA that is either double stranded or single stranded.

- Viral entry often depends on specific interactions between viral proteins and molecules on the host-cell surface, called virus receptors. Attachment is followed by uncoating, which is when the viral genome is released into the cytosol.

- The protein-production and genome-replication phases of the replicative cycle depend on biosynthetic machinery, chemical energy, and substrates provided by the host cell.

- Enveloped viruses exit a cell by budding; naked viruses often exit a cell by lysis—killing the host cell in the process.

- Some viruses may enter a dormant phase, when they do not produce virions, but instead coexist with the host cell and transmit genetic material to daughter cells when the host divides.

✓ You should be able to explain how three of the six phases of a viral replicative cycle can be stopped by antivirals, using HIV as an example.

36.3 What Themes Occur in the Diversification of Viruses?

- Three hypotheses have been developed to explain the origin of viruses: (1) They are escaped gene sets from cells; (2) they are the products of degenerate cellular parasites; (3) they coevolved with cells from the RNA world.

- Viruses continue to evolve. Factors influencing rates of viral evolution are errors during genome replication and genomic reassortment.

- Many viruses currently infecting humans are examples of emerging diseases—that is, diseases caused by viruses transmitted from other host species.

✓ You should be able to apply what you know of Darwin's four postulates on natural selection (see Chapter 25) to explain how it is possible for viruses to evolve even though they are not alive.

36.4 Key Lineages of Viruses

- In addition to morphology and the nature of the genetic material, viruses also vary in how the genome is replicated and transcribed to produce mRNA.

✓ You should be able to compare and contrast the transcription and replication of the RNA genomes of class IV and class VI viruses.

(MB) **MasteringBiology**

1. MasteringBiology Assignments

Tutorials and Activities HIV Replicative Cycle; HIV Reproductive Cycle; Phage Lysogenic and Lytic Cycles; Phage Lytic Cycle; Retrovirus (HIV) Reproductive Cycle; Simplified Viral Reproductive Cycle; Viral Replication

Questions Reading Quizzes, Blue-Thread Questions, Test Bank

2. eText Read your book online, search, take notes, highlight text, and more.

3. The Study Area Practice Test, Cumulative Test, BioFlix® 3-D Animations, Videos, Activities, Audio Glossary, Word Study Tools, Art

You should be able to . . .

1. What do host cells provide for viruses?
   a. nucleotides and amino acids
   b. ribosomes
   c. ATP
   d. all of the above

2. How do viruses that infect animals enter an animal's cells?
   a. The viruses pass through a wound.
   b. The viruses bind to a membrane component.
   c. The viruses puncture the cell wall.
   d. The viruses lyse the cell.

3. What does reverse transcriptase do?

4. When do most enveloped virions acquire their envelope?
   a. during entry into the host cell
   b. during budding from the host cell
   c. as they burst from the host cell
   d. as they integrate into the host cell's chromosome

5. In the viral replicative cycle, what reaction do viral proteases catalyze?
   a. polymerization of amino acids into peptides
   b. folding of long peptide chains into functional proteins
   c. cutting of polyprotein chains into functional proteins
   d. assembly of virions

6. What features distinguish the seven major categories of viruses?

7. The outer surface of a virion consists of either a membrane-like envelope or a protein capsid. How does the outer surface correlate with a virus's mode of exiting a host cell, and why?

8. Compare and contrast the bacteriophage lytic cycle versus lysogeny by addressing (1) replication of the viral genome, (2) production of virions, and (3) effect on the host cell.

9. Propose a reason why the development of antiviral drugs is more difficult than developing the antibiotics used to treat bacterial infections.

10. How does the diversity of viral genome types support the hypothesis that viruses originated several times independently?

11. What types of data convinced researchers that HIV originated when a simian immunodeficiency virus "jumped" to humans?

12. Of the viruses highlighted in Section 36.4, predict which of the following would be able to make viral proteins by injecting nothing more than its genome into a suitable host cell.
    a. cowpea mosaic ([+]ssRNA) virus
    b. bluetongue (dsRNA) virus
    c. measles ([−]ssRNA) virus
    d. human immunodeficiency (RNA reverse-transcribing) virus

13. Suppose you could isolate a virus that parasitizes the pathogen *Staphylococcus aureus*—a bacterium that causes acne, boils, and a variety of other afflictions in humans. How could you test whether this virus might serve as a safe and effective treatment for a staph infection?

14. If you were in charge of the government's budget devoted to stemming the AIDS epidemic, would you devote most of the resources to drug development or preventive medicine? Defend your answer.

15. Latency is a key adaptation in some animal viruses and likely evolved to prolong the infection by avoiding a strong and effective immune response. Which of the following types of viruses would you expect to require periods of latency?
    a. viruses that have large genomes and require a long time for replication
    b. viruses that require a long time for transmission to new host organisms
    c. viruses that require a long time for assembling into complex structures
    d. viruses that infect cells of the immune system

16. Consider these two contrasting definitions of life: (1) An entity is alive if it is capable of replicating itself via the directed chemical transformation of its environment; (2) An entity is alive if it is an integrated system for the storage, maintenance, replication, and use of genetic information. According to these definitions, are viruses alive? Explain.

37 **Plant Form and Function**

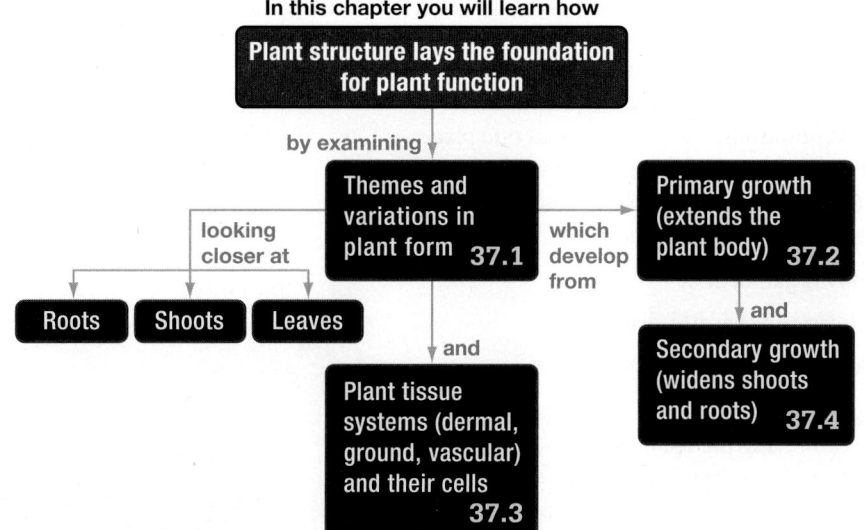

In this chapter you will learn how

Plant structure lays the foundation for plant function

by examining

Themes and variations in plant form 37.1

looking closer at

Roots   Shoots   Leaves

which develop from

Primary growth (extends the plant body) 37.2

and

Secondary growth (widens shoots and roots) 37.4

and

Plant tissue systems (dermal, ground, vascular) and their cells 37.3

All plants are able to harvest diffuse resources and concentrate them in cells and tissues, but their forms and strategies are diverse. These baobab trees in Madagascar may live to be hundreds of years old despite drought conditions, in part by storing water in their enormous trunks.

This chapter is part of the Big Picture. See how on pages 840–841.

Photosynthetic plants carry out the most remarkable biochemistry of any terrestrial organisms. Using the energy in sunlight and the simplest of starting materials—carbon dioxide, water, and ions containing nitrogen, phosphorus, potassium, and other key atoms—plants synthesize thousands of different carbohydrates, proteins, nucleic acids, and lipids. They use these compounds to build bodies that may live for thousands of years.

This feat is even more impressive when you consider that the simple starting materials that plants need to grow are tiny and diffuse—carbon dioxide molecules, water molecules, nitrate ions, and other resources are usually found at low concentrations over a large area. To gather the raw materials required for their sophisticated biosynthetic machinery, a plant's roots and shoots grow outward, extending the individual into the soil and atmosphere.

✔ When you see this checkmark, stop and test yourself. Answers are available in Appendix A.

In essence, a plant's body harvests diffuse resources and concentrates them in cells and tissues. The structure of its body is dynamic, because most plants exhibit **indeterminate growth**; that is, they grow throughout their lives. A 4750-year-old bristlecone pine has roots and shoots that are still growing. In response to favorable conditions, a plant sends shoots and roots in the most promising directions, seeking light and the simple compounds it requires.

The contrast between the plant and animal way of life is striking. Most animals move around, eat concentrated sources of food, and avoid stressful conditions. But plants stay in one place, extend their roots and shoots to harvest diffuse resources, make their own food, and cope with stress where they stand.

This chapter focuses on three fundamental questions:

1. How is the plant body organized?

2. Why are plants so diverse in size and shape?

3. How do plants grow throughout their lives?

Instead of surveying the entire catalog of land plants, though, the focus here is on the flowering plants, or angiosperms. Recall that angiosperms are the most recent major group of plants to appear in the fossil record (see Chapter 31). With about 250,000 species described, they are by far the most abundant, species rich, and geographically widespread plant group today. It is also difficult to overstate their economic and medical importance to humans. Most of the food we eat and many of the medicines we use are derived from angiosperms.

If you look outside or down the produce aisle of a grocery store, you will see a wide diversity of plants and plant products. By the time you finish this chapter, you'll understand how these plant bodies are put together and how they grow. Exploring questions about the anatomy of flowering plants is vital to understanding the world at large as well as the other chapters in this unit. (You can review the importance of plant form and function in the Big Picture of plant and animal physiology on pages 840–841.)

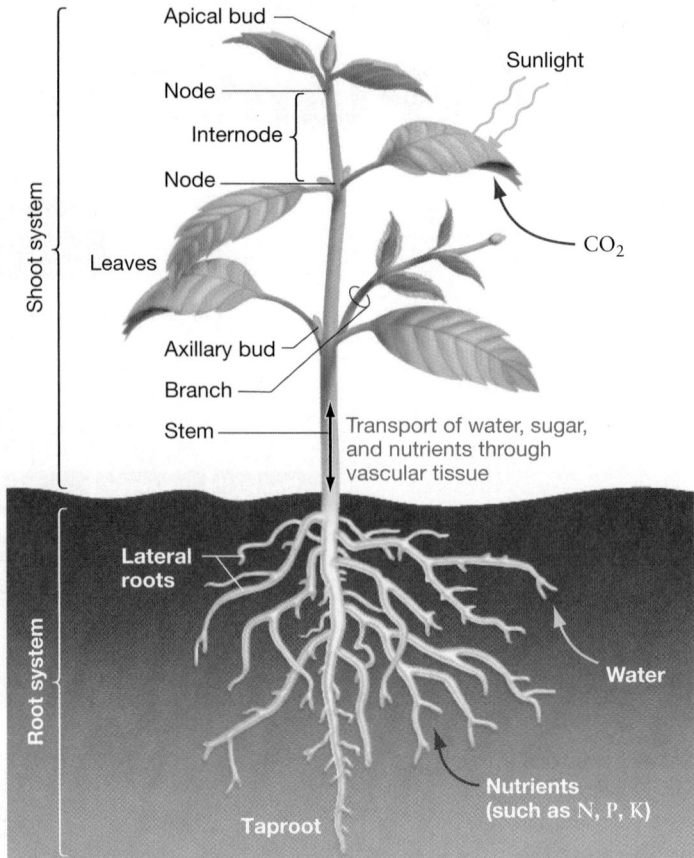

FIGURE 37.1 **Typical Root and Shoot Systems Acquire and Transport Resources.** Shoot systems are specialized for harvesting light and CO_2. Root systems absorb water and key nutrients such as nitrogen (N), phosphorus (P), and potassium (K).

✔QUESTION Suppose that this plant's growth was limited by access to light and nutrients, and that the plant had a new deposit of nutrient-rich soil to the left and much more sunlight suddenly available to the right. Explain what you would expect this individual to look like in one month.

37.1 Plant Form: Themes with Many Variations

Plants—along with algae, cyanobacteria, and a variety of protists—obtain energy and carbon to grow and reproduce (see Chapter 10). Plants use light energy to synthesize carbohydrates using carbon dioxide from the air and water from the soil.

For photosynthesis to occur efficiently, plants need large amounts of light and carbon dioxide, along with water as an electron source. Plants also need large amounts of water to fill their cells and maintain them at normal volume and pressure.

To synthesize nucleic acids, enzymes, phospholipids, and the other molecules needed to build and maintain cells, plants must obtain nitrogen (N), phosphorus (P), potassium (K), magnesium (Mg), and a host of other nutrients. Most of these key elements exist in nature as ions that dissolve in water found in soil.

FIGURE 37.1 labels the major structures in the two basic systems that plants use to acquire the resources they need for photosynthesis. A belowground portion called the **root system** anchors the plant and takes in water and nutrients from the soil; an aboveground portion called the **shoot system** harvests light and carbon dioxide from the atmosphere to produce sugars. Both systems grow throughout the life of the individual, allowing the plant to increase in size, acquire resources, and reproduce.

In most plants, vascular tissue connects the root and shoot systems. Through vascular tissue, water and nutrients are transported from roots to shoots; sugars can be transported in both directions. As you learn about how the plant form has evolved to serve necessary functions, compare and contrast what you learn with how animals have adapted to new environmental challenges. In many ways the two groups are remarkably similar.

The Importance of Surface Area/Volume Relationships

Before exploring the nature of root and shoot systems in more detail, it's important to recognize a key structural relationship that is critical to their function.

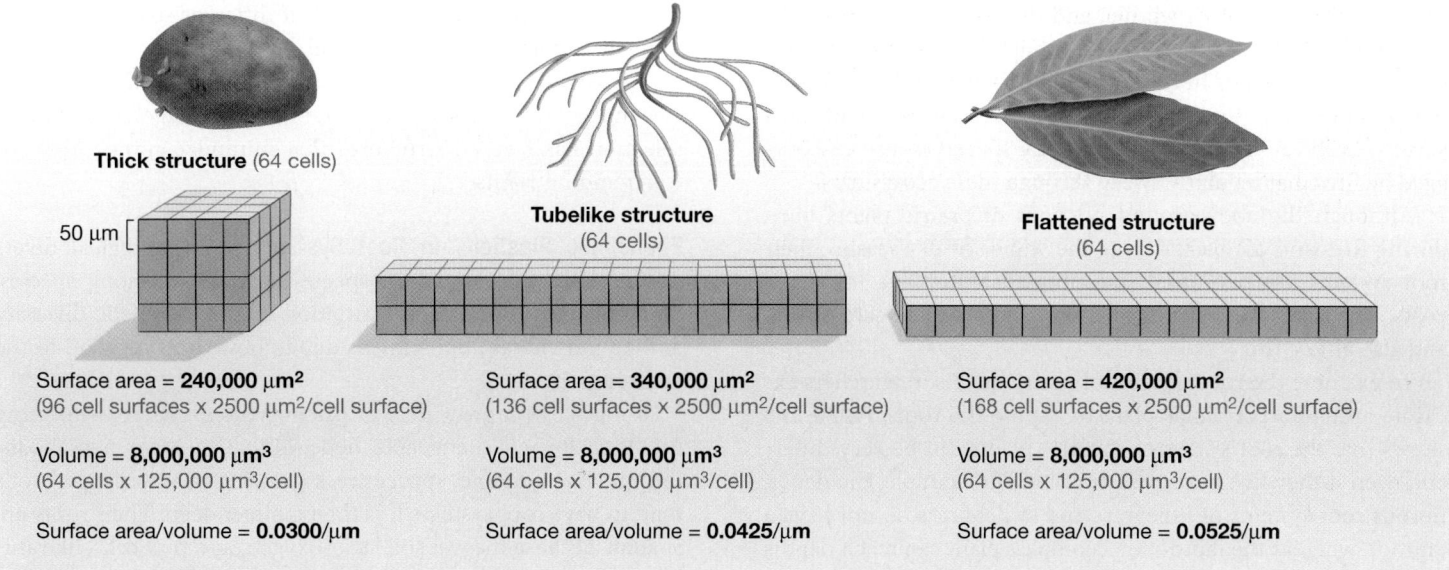

FIGURE 37.2 The Morphology of Roots and Leaves Gives Them a High Surface-Area-to-Volume Ratio. In this example, the "thick structure" represents a tree trunk or potato-like storage organ; the "tubelike structure" represents a root; the "flattened structure" represents a leaf. Note that each schematic structure has the same number of cells and the same total volume—but a very different surface area.

Root and shoot systems both function in absorption—of water and key ions, or of light, respectively. Absorption takes place across a surface. But the cells that use the absorbed molecules and light occupy a volume. Thus, a plant body is more efficient as an absorption-and-synthesis machine when it has a large surface area relative to its volume.

FIGURE 37.2 illustrates this point. In this example, the cells in a plant are represented by cubes; the side of each cell is 50 μm long. Thus, each face of a cell has a surface area of $50 \times 50 = 2500$ μm²; each cell has a volume of $50 \times 50 \times 50 = 125,000$ μm³. Follow the calculations in the figure and note that:

- If 64 cells are arranged in a cube, the surface area/volume relationship is 0.0300/μm.

- If 64 cells are arranged in a long tube, the surface area/volume relationship is 0.0425/μm.

- If 64 cells are arranged in a flat sheet, the surface area/volume relationship is 0.0525/μm.

This simple exercise has an important punch line: Tubes and sheets have much more surface area relative to their volume than cubes. It's no surprise, then, that the absorptive regions of a root system are tubelike, and the absorptive regions of a shoot system are the flattened structures called leaves. Storage tissues such as tubers and seeds have a low surface-area-to-volume ratio because they are not involved in absorption.

The Root System

Many root systems have a vertical **taproot**, as well as numerous **lateral roots** that run more or less horizontally. The root system anchors the plant in soil, absorbs water and ions from the soil, conducts water and selected ions to the shoot, obtains energy in the form of sugar from the shoot, and stores material produced in the shoot for later use.

Root systems can be impressive in extent. For example, in 1937 botanist Howard Dittmer grew a winter rye plant in a container full of soil for four months. He then unearthed the plant and meticulously measured its roots. The root system of this single individual contained more than 14 billion identifiable structures with a combined length of over 11,000 km—over one-fourth of Earth's circumference! The total surface area of the root system was almost 640 m², or about the size of 1.5 basketball courts. A root system like this is clearly adapted to absorbing diffuse resources located underground.

Other studies have shown that (1) the roots of trees routinely extend beyond the width of their aboveground canopy, and (2) it is not unusual for a plant's root system to represent over 80 percent of its total mass. Many plants devote a great deal of energy and resources to the growth of their root systems.

Although most root systems contain the same general structures, the root systems are diverse. This diversity can be analyzed on three levels:

1. morphological diversity among species;

2. phenotypic plasticity, or changes in the structure of an individual's root system in response to the environment; and

3. modified roots that are specialized for unusual functions.

Let's consider each level in turn.

Morphological Diversity in Root Systems As an example of the range of morphological diversity observed in the root systems of angiosperms, consider prairie plants.

Prairies are grassland ecosystems found in areas of the world such as central North America, the Serengeti Plain of East Africa, the Pampas region of Argentina, and the steppes of central Asia (see Chapter 52). Rain is abundant enough in these areas to support a lush growth of **herbaceous plants**—seed plants that lack woody tissue. Rain is scarce enough to exclude trees and most shrubs, however. The growth of woody species is also discouraged by fires that regularly sweep through these ecosystems.

Although the aboveground portions of prairie plants burn during fires and die back during the winter or dry season, their root systems are **perennial,** meaning that they live for many years. The root system sends up a new shoot system each spring and also after a fire.

To examine the root systems of prairie plants, researchers excavate around a particular plant to expose the roots. **FIGURE 37.3** shows that the root systems of prairie plants can be very different, even if they live next to each other. For example, the dense, fibrous root systems of junegrass and switchgrass do not have a taproot, whereas the taproot of a compass plant can reach depths of over 4.5 m.

Diversity in root system structure has important consequences for competition between species. The tips of the smallest roots are where most water and nutrient absorption take place,

but in prairies, the small root tips of different species are often found at different depths in the soil.

To explain the diversity of root systems observed among species that grow in the same habitat, biologists suggest that natural selection has favored structures that minimize competition for water and nutrients.

Phenotypic Plasticity in Root Systems Morphological diversity in roots occurs within species as well as among species. Some of the within-species variation is due to genetic diversity among individuals, but some is due to how roots respond to the environment.

Roots show a great deal of **phenotypic plasticity**—meaning that their form is changeable, depending on environmental conditions. For example, spruce trees growing in waterlogged soils tend to have root systems less than a meter deep. Their roots are shallow because the wet soil lacks oxygen, and root cells, like animal cells, suffocate in the anoxic conditions. The same tree growing in drier soil would develop a root system several meters deep.

The key point is that even genetically identical individuals will have very different root systems if they grow in different environments.

Phenotypic plasticity is particularly important in plants because they grow throughout their lives, and sometimes their environment changes. Root systems that grow into nutrient-rich septic fields or sewer pipes that leak human waste are a prime example. Roots actively grow into areas of soil where resources are abundant; roots stop growing or die back in areas where resources are used up or lacking.

Modified Roots The taproots and fibrous roots illustrated earlier do not begin to exhaust the types of roots found among plants (**TABLE 37.1**). For example, some roots are **adventitious**—meaning they develop from an unusual source, the shoot system instead of the root system.

- In ivy, anchor roots—adventitious roots that grow from nodes in the shoot system—help anchor individuals to brick walls or other structures.

- The prop roots of corn are adventitious roots that help brace individuals in windy weather.

Even roots that are part of the root system can have specialized functions—meaning that they do things other than absorbing water and nutrients and anchoring the shoot system.

The pneumatophores of mangroves in the genus *Avicennia* are specialized lateral roots that function in gas exchange. These mangroves grow in submerged habitats where fine silt is deposited, cutting off the oxygen supply to their roots. Their root cells do not suffocate, however, because oxygen from the atmosphere can diffuse into the root system through the pneumatophores. These roots grow upward—not downward—in response to gravity.

The thick taproot of some biennial plants, such as carrots, stores starch during the first of the plants' two growing seasons. Carrot roots, the ones we eat, are normally harvested at this point. If carrots were not harvested, the plants would resume

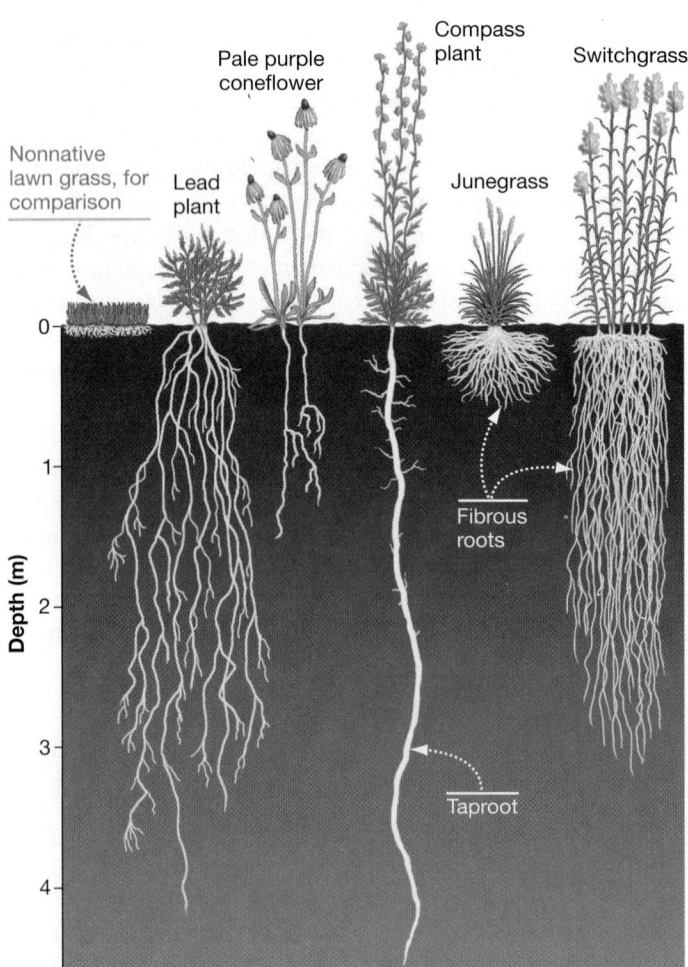

FIGURE 37.3 Plants Have Diverse Root Systems. The roots of prairie plants that live side by side can be very different.

✔**QUESTION** Why are lawns not drought tolerant?

Modified Root	Example
Anchor roots These adventitious roots anchor stems to walls and to other plants	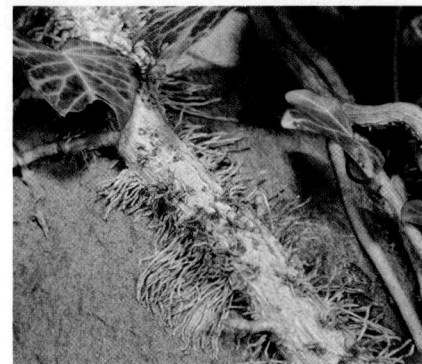
Prop roots These adventitious roots stabilize the stem	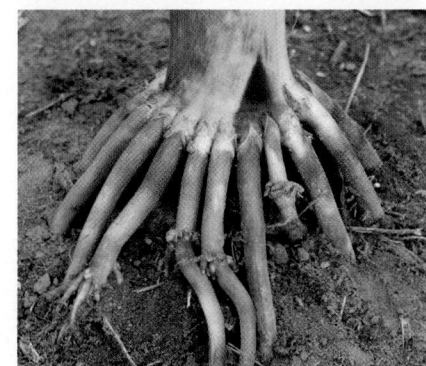
Pneumatophores Pneumatophores allow gas exchange between roots and atmosphere	

✔**QUESTION** Why do root cells need oxygen?

growth after the winter and use that stored energy during a second summer to "go to seed," producing a large flowering shoot. Starch in the taproot makes the energy-intensive reproductive process possible.

The Shoot System

As Figure 37.1 indicates, the shoot system has an array of important anatomical features.

- The shoot system consists of one or more **stems,** which are vertical aboveground structures.
- A stem consists of **nodes,** where leaves are attached, and **internodes,** or segments between nodes.

- A **leaf** is an appendage that projects from a stem laterally. Leaves usually function as photosynthetic organs.
- The nodes where leaves attach to the stem are also the site of **axillary** (or **lateral**) **buds,** which form just above the site of leaf attachment.
- If conditions are appropriate, an axillary bud may grow into a **branch**—a lateral extension of the shoot system.
- The tip of each stem and branch contains an **apical bud,** where growth occurs that extends the length of the stem or branch.
- If conditions are appropriate, apical or axillary buds may develop into flowers or other reproductive structures.

In essence, the shoot system is a repeating series of nodes, internodes, leaves, and apical and axillary buds. As plants grow, the number of nodes, internodes, and leaves increases. After an initial period of growth, however, an internode does not increase much in length over time.

As with root systems, diversity in shoots can be analyzed on three levels: morphological diversity among species, phenotypic plasticity within individuals, and modified shoots with specialized functions.

Morphological Diversity in Shoot Systems   The shoot systems of plants, essentially the visible part of the plant, range in size from species like the tiny (<5 mm diameter) duckweed that you may have seen growing on the surface of stagnant ponds to redwood trees that reach heights of over 100 m (300 ft) and giant sequoias with trunks that weigh 2.6 million kg (over 5.7 million lb)—about the same as 10 diesel locomotives.

The shape of the shoot system also varies a great deal among species. For example, the manner in which new branches are added as the shoot system grows affects the shape of the individual and its ability to compete for light. As **FIGURE 37.4** shows, a plant growing with wide branching angles and short internodes has a very different shape from that of a plant with narrow branching angles and long internodes.

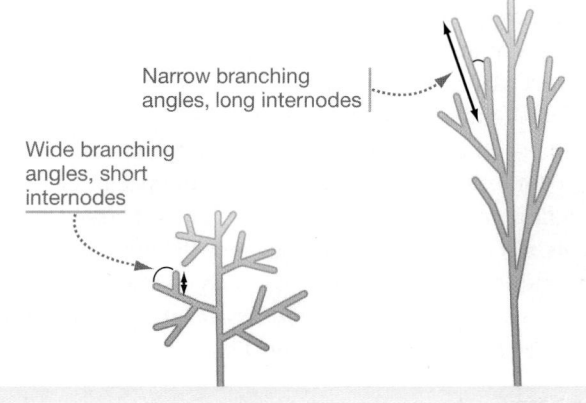

Narrow branching angles, long internodes

Wide branching angles, short internodes

FIGURE 37.4 Plant Form Can Vary as a Function of Branch Angle and Internode Length.

Variation in the size and shape of the shoot system is important: It allows plants of different species to harvest light at different locations and thus minimize competition. It also allows plants to thrive in a wide array of habitats.

As an example of how the shape of a shoot system varies among species in different environments, consider the silversword plants native to Hawaii. You might recall that all of the silverswords are believed to be descended from the same ancestor—a species of tarweed that arrived in Hawaii from the west coast of North America about 5 million years ago (see Chapter 28).

Silverswords represent an adaptive radiation: a lineage that rapidly split into many species occupying a wide array of habitats. Their shoot systems are particularly diverse in size, shape, and growth habit (see Figure 28.9). Some silverswords grow low to the ground in dense mats; some form bunched rosettes of leaves; some are vines; others are woody shrubs or even small- to medium-sized trees.

Biologists interpret this diversity of shoot systems as a suite of adaptations for harvesting light and carbon dioxide in different environments.

- In lush habitats, where competition for light is intense, woody individuals that grow tall are favored by natural selection.

- In dry, windblown habitats, individuals with short stems or rosettes thrive because they require less water than taller individuals, and they don't blow over.

The adaptive radiation of silverswords has been based in part on diversification in their shoot systems.

Phenotypic Plasticity in Shoot Systems   The size and shape of an individual's shoot system can vary dramatically based on variation in growing conditions: temperature, exposure to wind, and availability of water, nutrients, and light.

This conclusion was driven home in an experiment conducted by Jens Clausen and colleagues in the late 1930s. These biologists transplanted several species of herbaceous plants between sites along an elevational gradient: from sea level to alpine habitats. In each case, the transplanted individuals were propagated from cuttings—meaning that they were genetically identical to individuals growing at the other locations. As the "Results" section in **FIGURE 37.5** shows, the overall size and shape of the shoot system varied markedly among locations.

Because a plant's shoot system continues to grow over the course of its lifetime, it can respond to changes in environmental conditions just like the root system can. Experiments on phototropism, for example, established that shoot systems can bend toward light if an individual is shaded on one side (see Chapter 40). Plants also undergo differential growth, producing more branches and leaves in regions of the body that are exposed to the highest light levels. A plant's shoot system grows in directions that maximize its chances of capturing light.

Modified Stems   As in root systems, the stems of many plants have modified structures to serve various functions (**TABLE 37.2**).

- Many desert cacti have highly modified stems. Instead of functioning primarily to support leaves, cactus stems often

---

RESEARCH

QUESTION: How much does a plant's growth form depend on its environment?

HYPOTHESIS: (No explicit hypothesis—the goal of this experiment was to explore the interaction between genetic makeup versus environmental influence on size and shape.)

EXPERIMENTAL SETUP:

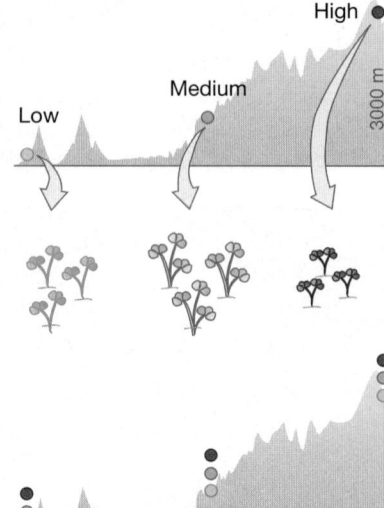

1. **Take cuttings** from individuals of *Potentilla glandulosa* growing in low-, medium-, and high-elevation habitats in the Sierra Nevada mountains.

2. **Propagate** genetically identical individuals.

3. **Transplant** individuals from each source population into each habitat (low, medium, and high elevation). Allow to grow and observe mature plants.

PREDICTION: (No explicit predictions.)

RESULTS:

Examples of mature plants observed:

"High" plant grown at low elevation

"High" plant grown at medium elevation

"High" plant grown at high elevation

CONCLUSION: Environmental conditions have a profound influence on body size and shape (genetically identical plants look different at each site). BUT, genetic makeup also has a large influence on plant morphology. (Though not pictured here, plants from each source population look different, even when grown in the same habitat.)

FIGURE 37.5 Experimental Evidence for Phenotypic Plasticity in Shoot Systems.

SOURCE: Clausen, J., D. D. Keck, and W. M. Hiesey. 1945. Experimental studies on the nature of species. II. Plant evolution through amphiploidy and autoploidy, with examples from the Madiinae. Publication No. 564. Washington, DC: Carnegie Institute of Washington.

✔ **QUESTION** Why was it important for the researchers to propagate the individuals from cuttings?

SUMMARY TABLE 37.2 Form and Function of Modified Stems

Modified Stem	Example
Water-storage structures Cactus stems (shown in cross section) store water	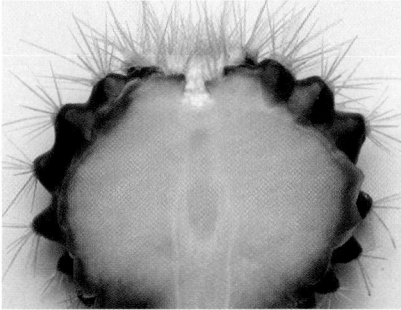
Stolons Strawberry stolons produce new individuals at nodes aboveground	 Stolon
Rhizomes Rhizomes produce new individuals at nodes belowground, and store carbohydrates	 Rhizome
Tubers Tubers such as potatoes store carbohydrates such as starch	
Thorns Thorns provide protection from herbivores	

enlarge into water-storage organs. Water accounts for up to 98 percent of the weight of a cactus stem. These cactus stems also contain the plant's photosynthetic tissue.

- **Stolons** are modified stems that grow horizontally along the soil surface, producing adventitious roots and leaves at each node. Because new plants form at these nodes, stolons function in asexual reproduction (see Chapter 13).

- Like stolons, **rhizomes** are stems that grow horizontally instead of vertically. They produce new plants at nodes and thus participate in asexual reproduction. But while stolons grow aboveground, rhizomes spread belowground. One quaking aspen plant in Utah was found to have 47,000 stems rising from a network of rhizomes, making it one of the largest organisms known. What we see as individual aspen trees are actually stems that are connected underground.

- **Tubers** are underground, swollen rhizomes that function as carbohydrate-storage organs. The eyes of a potato—a typical tuber—are nodes in the stem where new branches may arise.

- **Thorns** are modified stems that help protect the plant from attacks by large **herbivores,** or plant eaters, such as deer, giraffe, or cattle.

The Leaf

In most plant species, the vast majority of photosynthesis occurs in the part of the shoot system we call leaves. The total area of leaf produced by a single plant can be enormous—a single tree can have hundreds of thousands of leaves with a total surface area equivalent to that of a football field. All of this area is available for absorbing photons and supporting photosynthesis.

A simple leaf (**FIGURE 37.6a** on page 738) is composed of just two major structures: an expanded portion called the **blade** and a stalk called the **petiole.** But leaves exhibit many variations on the central theme of a flattened structure specialized for performing photosynthesis.

Morphological Diversity in Leaves Glance outside or stroll through a garden, and you'll find many types of simple leaves with an easily recognizable blade and petiole. (Grass leaves will stump you, though, because they lack petioles entirely.) You will also find compound leaves that have blades divided into a series of leaflets (**FIGURE 37.6b**). You may even encounter doubly compound leaves, which have leaflets that are again divided (**FIGURE 37.6c**).

Not all leaf blades are thin with a large surface area, however. For example, plants that thrive in deserts and in cold, dry habitats tend to have needle-shaped leaves (**FIGURE 37.6d**). The leading hypothesis to explain this pattern is based on two observations: (1) Water is often in short supply in these environments because it is scarce in deserts, or frozen and thus unavailable in cold habitats; and (2) leaves with large surface areas lose large amounts of water through an evaporative process called **transpiration** (discussed in Chapter 38).

Thus, needlelike leaves are interpreted as adaptations that minimize transpiration in water-scarce habitats. Small, narrow

(a) Simple leaves have a petiole and a single blade.

Blade

Petiole

(b) Compound leaves have blades divided into leaflets.

(c) Doubly compound leaves are large yet rarely damaged by wind or rain.

(d) Needlelike leaves are characteristic of species adapted to very cold or hot climates.

FIGURE 37.6 Leaves Vary in Size and Shape. The structures in parts **(a)** through **(c)** represent single leaves; part **(d)** shows two leaves.

✔**QUESTION** For capturing photons, what is the advantage of having a leaf with a large surface area? In terms of wind damage and water loss, what is the disadvantage of having a leaf with a large surface area?

leaves are also much less susceptible to wind damage than are large, broad leaves.

The arrangement of leaves on a stem can vary as much as leaf shape. For example, leaves can be:

- arranged to alternate on either side of the stem (**FIGURE 37.7a**);
- paired opposite each other on the stem (**FIGURE 37.7b**);
- arranged in a whorl (**FIGURE 37.7c**);
- found in a compact basal arrangement where internodes are extremely short—leading to the rosette growth form (**FIGURE 37.7d**).

Leaf shape and the arrangement of leaves on a stem are usually determined genetically, so these characters are often used for plant identification.

Phenotypic Plasticity in Leaves Even though leaves do not grow continuously, they exhibit phenotypic plasticity just like root systems and stems do.

Leaves from the same individual that grow in sun versus shade serve as a prominent example of phenotypic plasticity in leaf morphology. As the oak tree leaves in **FIGURE 37.8** show:

- *Sun leaves* are thicker and have a relatively small surface area, which reduces water loss in areas of the body where light is abundant.
- *Shade leaves* are relatively thin and broad, providing a high surface area that maximizes absorption of rare photons.

Water loss is less of a problem for shade leaves because temperatures are cooler in shade than in bright sun.

(a) Alternate leaves

(b) Opposite leaves

(c) Whorled leaves

(d) Rosette

FIGURE 37.7 The Arrangement of Leaves on Stems Varies.

Grown in shade | Grown in sun

FIGURE 37.8 Phenotypic Plasticity in Leaves. These leaves came from the same tree.

Modified Leaves   Not all leaves function primarily in photosynthesis; some perform other roles (**TABLE 37.3**).

- Onion bulbs consist of thickened leaf bases that store nutrients, separated by highly condensed internodes.

- The thick leaves of plants called succulents, such as aloe vera, store water.

- The tendrils that enable garden peas and other vines to climb are modified leaflets or leaves.

- The bright red leaves of poinsettias attract pollinators to the tiny yellow flowers that they surround.

 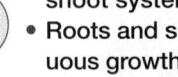

check your understanding

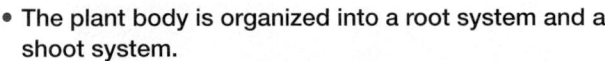 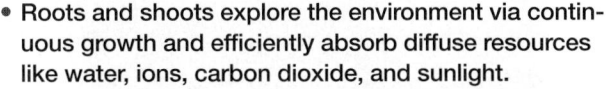

If you understand that . . .

- The plant body is organized into a root system and a shoot system.
- Roots and shoots explore the environment via continuous growth and efficiently absorb diffuse resources like water, ions, carbon dioxide, and sunlight.
- Roots and stems may also function to anchor the plant, store water, produce offspring asexually, provide protection, or store carbohydrates.
- Leaves vary among species and within individuals, and they may be modified to store food or water, capture insects, make climbing possible, attract pollinators, or protect the stem.

✔ You should be able to . . .

1. Diagram a generalized version of the angiosperm body, labeling each major part.

2. Provide two examples each of root systems, stems, and leaves that differ in structure and/or function from the generalized body shown in Figure 37.1.

Answers are available in Appendix A.

SUMMARY TABLE 37.3   **Form and Function of Modified Leaves**

Modified Leaves	Example
Bulbs Onion leaves store food	 Stem / Leaves
Succulent leaves Aloe vera leaves store water	
Tendrils Pea tendrils aid in climbing	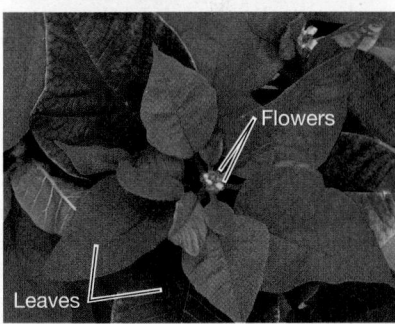
Floral mimics Red poinsettia leaves attract pollinators	Flowers Leaves
Traps Pitcher plant leaves trap entering insects, which are discouraged from flying out by the "hood," then are digested	Hood Digestive enzymes or bacteria

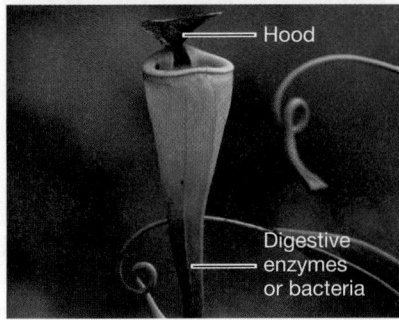

- The tubelike leaves of the pitcher plant trap insects. When insects enter, they feed on the plant's nectar and appear to become dizzy. Eventually they fall into the bottom of the tube and drown in water that has accumulated. The insects are then digested by bacteria or enzymes secreted by the plant, and the released nutrients are taken up by epidermal cells.

- Cactus **spines** are modified leaves that protect the stem (see top image in Table 37.2).

The variability of plant roots, stems, and leaves is impressive. Diversity, plasticity, and dynamism are recurring themes in the study of plant anatomy.

37.2 Primary Growth Extends the Plant Body

Plants grow continuously because they have many **meristems**—populations of undifferentiated cells that retain the ability to undergo mitosis and produce new cells. When meristematic cells divide, some of the daughter cells remain in the meristem, allowing the meristem to persist. Other cells, though, undergo differentiation. You might recall that differentiation is a developmental process that produces a specialized cell—one that expresses only certain genes and has a distinctive structure and function (see Chapter 22).

Apical meristems are located at the tip of each root and shoot. As cells in apical meristems divide, enlarge, and differentiate, root and shoot tips extend the plant body outward, allowing it to explore new space.

This process is **primary growth,** which is common to all plants. The major consequence of primary growth is to increase the length of the root and shoot systems. Cells that are derived from apical meristems form the primary plant body.

To understand how primary growth occurs, let's look at the overall organization of the primary plant body and then delve into a detailed look at its tissues and cells.

How Do Apical Meristems Produce the Primary Plant Body?

Whether located in the root or the shoot, apical meristems give rise to three distinct populations of **primary meristem** cells: protoderm, ground meristem, and procambium. These cells are partially differentiated but retain the character of meristematic cells because they keep dividing.

The three types of primary meristematic cells are important because they give rise to three major tissue systems that extend throughout the plant body. A **tissue** is a group of cells that functions as a unit.

FIGURE 37.9 indicates where the apical meristems and the primary meristems—protoderm, ground meristem, and procambium—are found in shoots and roots.

- **Protoderm** gives rise to the **dermal** (literally, "skin") **tissue system.** The dermal tissue system, or **epidermis,** is a single layer of cells that covers the plant body and protects it.

- **Ground meristem** gives rise to the **ground tissue system,** which makes up the bulk of the plant body and is responsible for photosynthesis and storage.

- **Procambium** gives rise to the **vascular tissue system,** which provides support and transports water, nutrients, and photosynthetic products between the root system and shoot system.

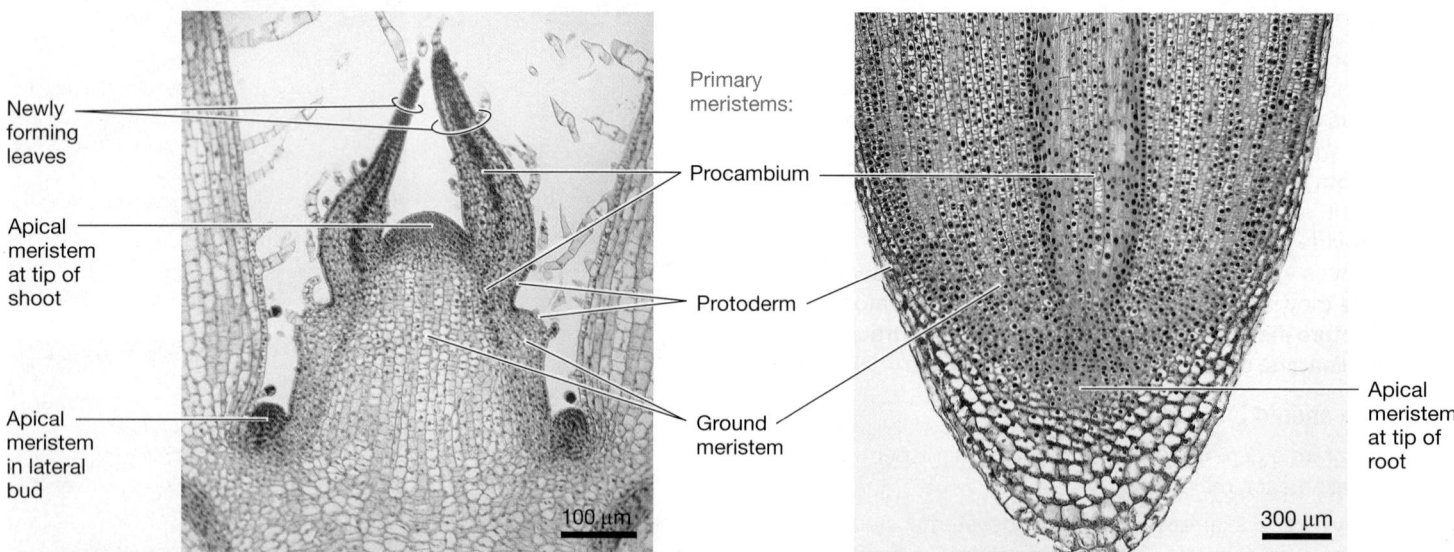

(a) Apical and primary meristems in a shoot

Newly forming leaves

Apical meristem at tip of shoot

Apical meristem in lateral bud

100 μm

(b) Apical and primary meristems in a root

Primary meristems:

Procambium

Protoderm

Ground meristem

Apical meristem at tip of root

300 μm

FIGURE 37.9 The Structure of Apical and Primary Meristems in the Shoot and Root. Apical meristems consist of small, similar-looking cells that divide when conditions are favorable. Three types of primary meristem cells—protoderm, ground meristem, and procambium—are derived from the apical meristem and consist of partially differentiated cells that can still divide.

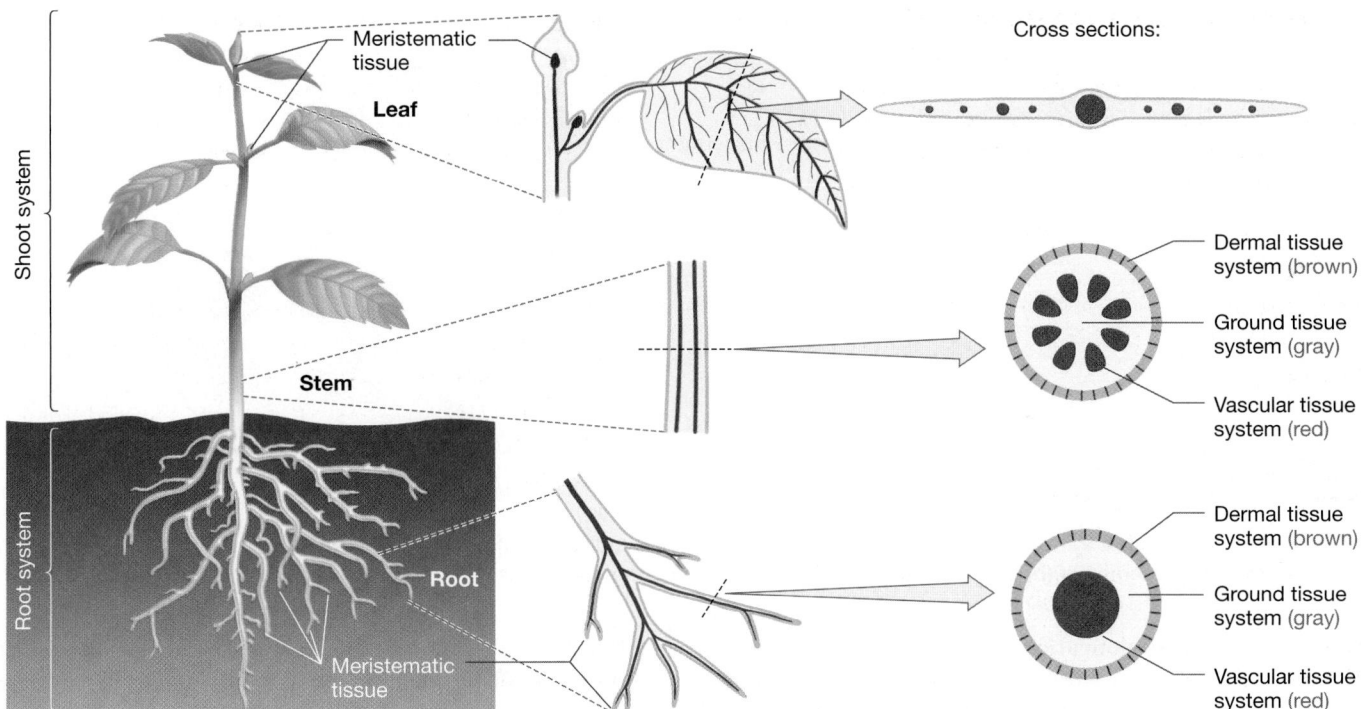

FIGURE 37.10 The Primary Plant Body Comprises the Dermal, Ground, and Vascular Tissue Systems. The dermal, ground, and vascular tissue systems arise from the protoderm, ground meristem, and procambium, respectively, as illustrated in Figure 37.9.

Vascular tissue runs through ground tissue, so the cells that make up ground tissue are usually close to cells that conduct the water and nutrients they need. **FIGURE 37.10** shows how the dermal, ground, and vascular tissues are distributed in the plant body. In contrast to these tissues, apical meristematic cells are localized at the tips of shoots and roots.

The key point to remember is that the dermal, ground, and vascular tissue systems are originally derived from cells in primary meristems, which originated from apical meristems. Thus, they represent the primary plant body.

How Is the Primary Root System Organized?

Roots have several features that allow them to grow into new regions of the soil, so they can furnish cells throughout the plant body with water and key nutrients.

As **FIGURE 37.11** shows, a group of cells called the **root cap** protects the root apical meristem. Cells produced by the meristem constantly replenish the cap, which regularly loses cells.

In addition to protecting the root tip, root cap cells are important in sensing gravity and determining the direction of growth. They also synthesize and secrete a slimy, polysaccharide-rich substance that helps lubricate the root tip, reducing friction and protecting the meristem as it is pushed through the soil.

Three distinct populations of cells exist behind the root cap:

1. The **zone of cellular division** contains the apical meristem, where cells are actively dividing, along with the protoderm, ground meristem, and procambium, where additional cell division occurs.

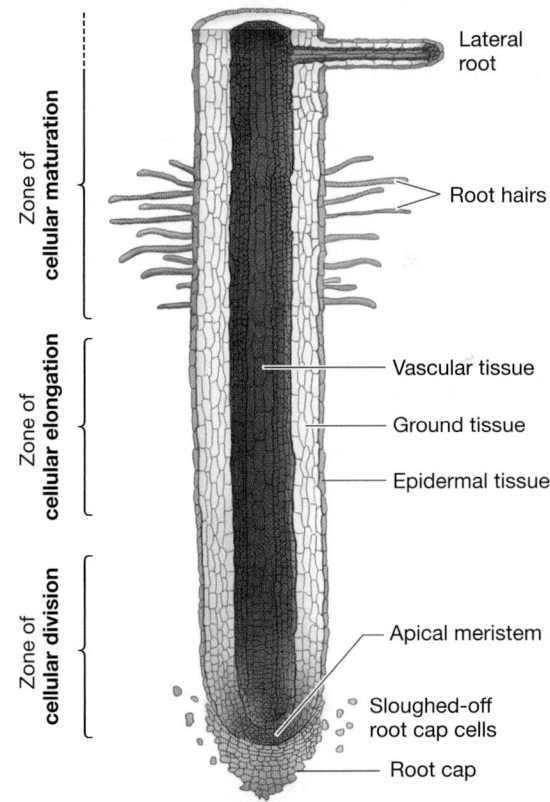

FIGURE 37.11 Roots Extend into the Soil via Growth of Apical Meristems and Cell Elongation. This is a longitudinal (lengthwise) section. The zone of cellular maturation is actually much larger than can be shown here. Most absorption of water and nutrients occurs at root hairs.

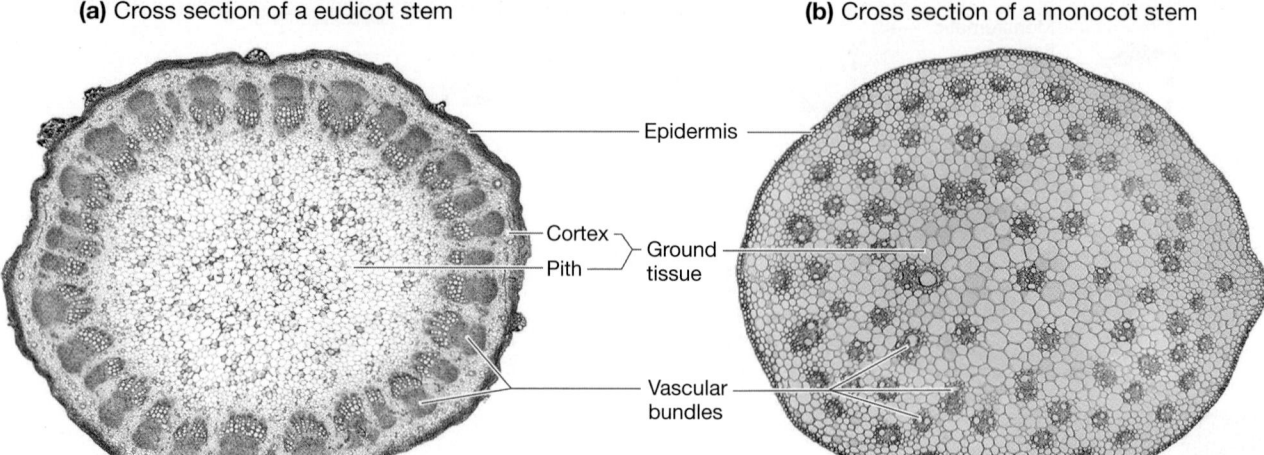

(a) Cross section of a eudicot stem

(b) Cross section of a monocot stem

Epidermis

Cortex
Pith

Ground
tissue

Vascular
bundles

1 mm

0.5 mm

FIGURE 37.12 Stems Contain a Variety of Cell and Tissue Types. Vascular bundles are **(a)** arranged in a ring near the perimeter of eudicot stems but **(b)** scattered throughout the ground tissue in monocots.

2. The **zone of cellular elongation** is made up of cells that are recently derived from the primary meristematic tissues and that increase in length.

3. The **zone of cellular maturation** is where older cells complete their differentiation into dermal, vascular, and ground tissues.

The zone of cellular elongation is the region most responsible for the growth of roots through the soil. The cells in this region increase in length by taking up water. Their expansion provides the force that pushes the root cap and apical meristem through the soil. When environmental conditions are good, roots can extend by as much as 4 centimeters per day.

The zone of cellular maturation is the most important root segment in terms of water and nutrient absorption. In this region, epidermal cells produce outgrowths called **root hairs,** which greatly increase the surface area of the dermal tissue. Root hairs furnish the actual sites of water and nutrient absorption. The rest of the root system provides structural support for the root hairs, conducts water and ions to the shoot, stores the products of photosynthesis, and anchors the plant in the soil. Uptake of water and nutrients by root hairs is vital to plants (portions of Chapters 38 and 39 focus on how these processes occur).

The zone of cellular maturation is also where lateral roots begin to grow. In contrast to lateral branches in the shoot, which arise from meristems in axillary, or lateral buds (see Figure 37.1), lateral roots arise from cells within a ring of cells surrounding the vascular tissue and then erupt through the surrounding ground tissue.

How Is the Primary Shoot System Organized?

If you visit a garden regularly, you can imagine how the growing root tips push through the soil and expand to form complex networks underground. But even a casual observer can watch the growth of shoot systems over time, as the tips of stems extend and branch and as new leaves form and expand.

Just behind each shoot apical meristem, the primary meristematic cells give rise to dermal, ground, and vascular tissues.

FIGURE 37.12a shows how these tissues are arranged in the stem of a sunflower plant when it matures. Note that the vascular tissues are grouped into **vascular bundles,** which form strands running the length of the stem.

In sunflowers and other eudicots, the vascular bundles are arranged in a ring near the stem's perimeter (see Chapter 31). The ground tissue that the vascular tissue runs through is divided into two major regions: **pith,** the ground tissue that is inside the vascular bundles, and **cortex,** the ground tissue that is outside the vascular bundles.

The arrangement of the vascular bundles and ground tissue is dramatically different in the stems of monocots such as grasses, lilies, and orchids, however. As **FIGURE 37.12b** shows, vascular bundles in monocot stems are normally scattered throughout the ground tissue. (Review other comparisons of monocots and eudicots in Chapter 31.)

Now let's drill down a bit deeper, and look at the composition of the dermal, ground, and vascular tissue systems. Each of these tissue systems is made up of an array of distinct cell and tissue types. What do they look like, and what are their functions?

37.3 Cells and Tissues of the Primary Plant Body

Recall that all eukaryotic cells, including those of plants and animals, share most of their key characteristics: chromosomes enclosed in a nuclear envelope, a plasma membrane studded with proteins that regulate the passage of materials in and out, mitochondria that produce ATP by oxidizing sugars, and an array of other organelles that synthesize or degrade key molecules (see Chapter 7).

In addition, plant cells have several features that are absent in animal cells (**FIGURE 37.13a**):

1. All plant cells are surrounded by a stiff, cellulose-rich **cell wall** that supports the cell and defines its shape.

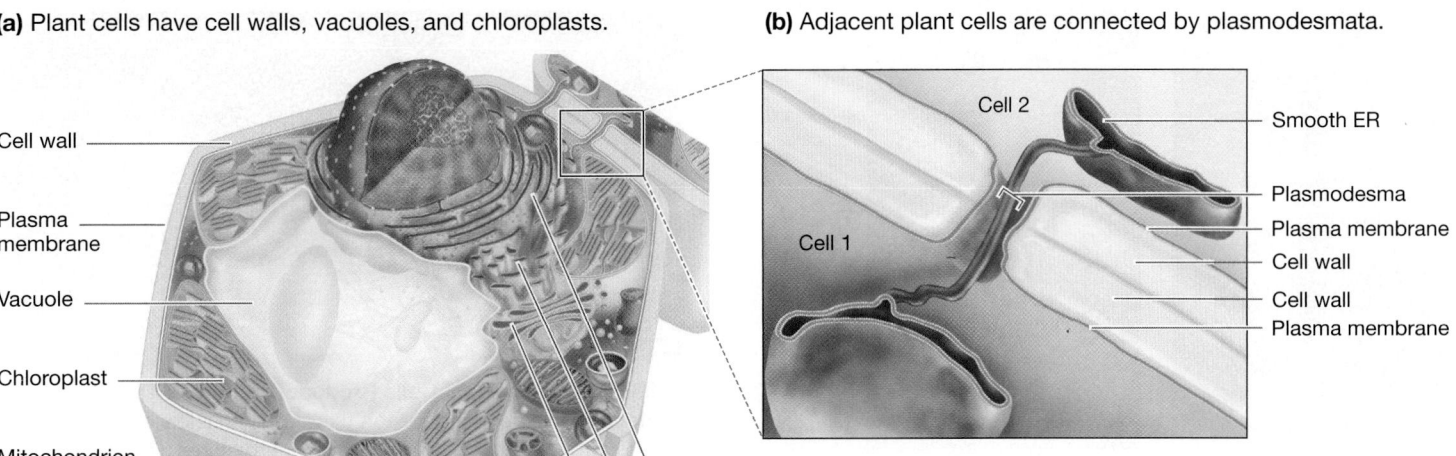

(a) Plant cells have cell walls, vacuoles, and chloroplasts.

Cell wall

Plasma membrane

Vacuole

Chloroplast

Mitochondrion

Rough endoplasmic reticulum

Smooth endoplasmic reticulum

Golgi apparatus

(b) Adjacent plant cells are connected by plasmodesmata.

Cell 2

Cell 1

Smooth ER

Plasmodesma

Plasma membrane

Cell wall

Cell wall

Plasma membrane

FIGURE 37.13 A Generalized Plant Cell. Plant vacuoles are similar to animal lysosomes; however, the cell wall, chloroplasts, and plasmodesmata are unique to plants. Unlike the extracellular matrix of animals, the plant cell wall is rigid.

2. The cytoplasm of adjacent plant cells is often connected via **plasmodesmata** (singular: **plasmodesma**) (see Chapter 11). Plasmodesmata consist of cytoplasm and segments of smooth endoplasmic reticulum (smooth ER) that run through tiny, membrane-lined gaps in the cell wall (**FIGURE 37.13b**).

3. Plant cells often contain several types of organelles that are not found in animals—specifically chloroplasts and a large, membrane-bound organelle called a vacuole, which fills most of the cell's volume.

   Chloroplasts are the site of photosynthesis (see Chapter 10). Non-photosynthetic cells found in roots, seeds, flower petals, and other locations may have organelles that are related to chloroplasts but are specialized for storing pigments, starch, oils, or proteins.

   Vacuoles, which contain an aqueous solution called **cell sap,** store wastes and in some cases also digest wastes, as do animal lysosomes. In addition, plant vacuoles store water and nutrients. They may also hold pigments that provide color, or poisons that deter herbivores.

   Another important distinction between plant cells and animal cells is that plant cells do not change position once they form. Some animal cells migrate within the body either early in the development of an individual or as mature (differentiated) cells.

   Let's examine the cells and tissues of the three primary plant tissue systems in turn.

The Dermal Tissue System

Dermal tissue is the interface between the organism and the external environment. Its primary function in shoots is to protect the plant body—from water loss, disease-causing agents, and herbivores. In roots the dermal tissue includes root hairs, and it functions primarily in water and nutrient absorption.

Most tissues are made up of several different cell types, each with a distinct structure and function. Let's consider the cell types found in dermal tissue.

Epidermal Cells Protect the Surface   Most of the cells in the dermal tissue system are epidermal cells, which in shoots are flattened and usually lack chloroplasts.

Epidermal cells in the shoot system fulfill their protective role in part by secreting the **cuticle**—a waxy layer that forms a continuous sheet on the surface (see Chapter 31). Waxes are lipids and are therefore highly hydrophobic. As a result, the presence of cuticle on stems and leaves drastically reduces the amount of water that these structures lose by evaporation.

From a human perspective, the water-repellent properties of cuticle make it a valuable ingredient in polishes and lipsticks. The carnauba wax used in car and floor polishes, for example, is secreted by epidermal cells in the leaves of carnauba palms native to Brazil.

Besides minimizing water loss, cuticle forms a barrier to protect the plant from viruses, bacteria, and the spores or growing hyphae of parasitic fungi. In this way, the plant epidermis forms the first line of defense against disease-causing agents, or **pathogens.**

Waxes found in the cuticle can also be detrimental to the plant, however, by reducing gas exchange. This can be a serious problem because photosynthesis depends on the free flow of carbon dioxide to photosynthetic cells. The problem is solved by specialized structures in dermal tissue, called stomata.

Stomata Regulate Gas Exchange and Water Loss   Most land plants have structures called **stomata** (singular: **stoma**) that allow carbon dioxide to enter photosynthetically active tissues.

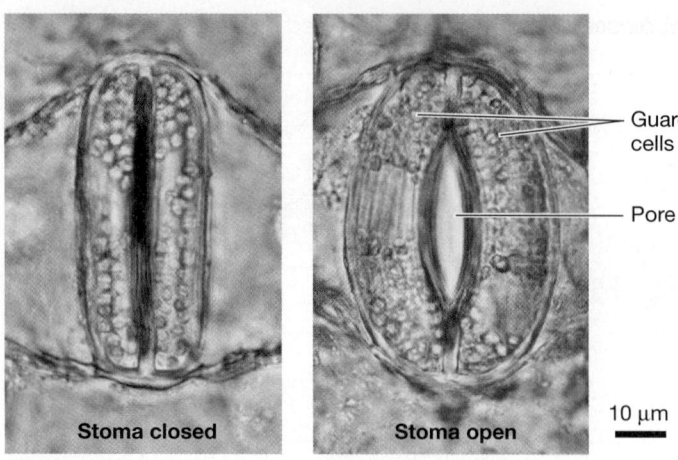

Guard
cells

Pore

Stoma closed | Stoma open

10 µm

FIGURE 37.14 A Stoma's Guard Cells Regulate the Opening of a Pore.

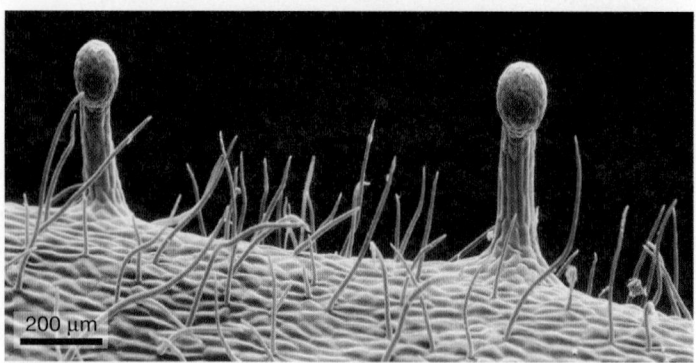

FIGURE 37.15 Epidermal Cells Produce Trichomes That Provide Protection. Some trichomes on this leaf are hairlike extensions of a single cell, while others are multicellular structures. (The structures that are colored orange here hold toxins.)

A stoma consists of two specialized **guard cells,** which change shape to open or close an opening in the epidermis known as a **pore** (**FIGURE 37.14**).

When stomata are open, CO_2, O_2, water vapor, and other gases can move between the atmosphere and the interior of the plant by diffusion. Stomata open when CO_2 is needed inside the leaf. When stomata are open, water vapor diffuses from the moist interior of the leaf to the atmosphere. Stomata close when CO_2 is not needed. This prevents large amounts of water from being lost by transpiration. In a later chapter, we explore the molecular mechanisms responsible for stomatal opening and closing (see Chapter 40).

Trichomes Perform an Array of Functions   In addition to minimizing water loss and regulating gas exchange in shoots, cells in dermal tissue may protect the individual from the damaging effects of intense sunlight and attacks by herbivores.

Trichomes are hairlike appendages made up of specialized epidermal cells. They are found in shoot systems and come in a wide variety of shapes, sizes, and abundances.

Depending on the species, trichomes may (**1**) keep the leaf surface cool by reflecting sunlight; (**2**) reduce water loss by forming a dense mat that limits transpiration; (**3**) provide barbs, or store toxic compounds that thwart herbivores (**FIGURE 37.15**); or (**4**) trap and digest insects.

The Ground Tissue System

Most photosynthesis, as well as most carbohydrate storage, takes place in ground tissue. Cells in ground tissue are also responsible for most of the synthesis and storage of specialized products such as colorful pigments, the chemical signals called hormones, and toxins required for defense.

If the primary function of the dermal tissue system is protection, the ground tissue is all about producing and storing valuable molecules.

Ground tissue is made up of three distinct tissue types: parenchyma (pronounced *pa-REN-ki-ma*), collenchyma (*ko-LEN-ki-ma*), and sclerenchyma (*skle-REN-ki-ma*).

Parenchyma Are "Workhorse" Cells   **Parenchyma cells** have relatively thin cell walls and are the most abundant and versatile plant cells. Groups of parenchyma cells form parenchyma tissue.

The parenchyma tissue in leaves consists of parenchyma cells filled with chloroplasts, and it is the primary site of photosynthesis (**FIGURE 37.16a**). But in other organs, parenchyma cells store starch granules (**FIGURE 37.16b**). When you eat a potato or an apple, you are ingesting primarily parenchyma cells in ground tissue.

Many parenchyma cells are **totipotent,** meaning they retain the capacity to divide and develop into a complete, mature plant. The totipotency of parenchyma cells is important in healing wounds and in reproducing asexually via stolons or rhizomes. In each case, parenchyma cells may begin to divide, grow, and differentiate to form new roots and shoots.

The totipotency of parenchyma cells also allows gardeners to clone plants by making cuttings. For example, if you cut a piece of stem from a coleus plant and place it in water, parenchyma cells will divide to produce a mass of undifferentiated cells called a **callus.** Roots then develop from the callus (**FIGURE 37.17**), and the new individual can be planted in soil.

Bananas, seedless grapes, and several other commercially important plants cannot undergo sexual reproduction to produce seeds. Instead, they are propagated entirely by cuttings.

Collenchyma Cells Function Primarily in Shoot Support   **Collenchyma cells** have cell walls that are thicker in some areas than others, and their overall shape is longer and thinner than that of parenchyma cells. Groups of collenchyma cells form collenchyma tissue that supports the plant body.

Even when collenchyma cells are mature, their cell walls retain the ability to stretch and elongate. As a result, collenchyma cells can continue to lengthen as they provide structural support to the growing regions of shoots.

Collenchyma tissue is often found just under the epidermis of stems, especially outside vascular bundles. The ability of collenchyma cells to stretch allows stems to flex in the wind without tearing or breaking. The "strings" you may have peeled from a

(a) In leaves: photosynthesis and gas exchange

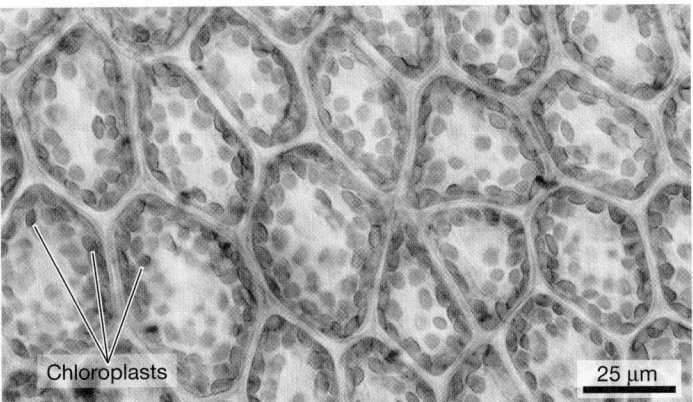

Chloroplasts

25 µm

(b) In roots: carbohydrate storage

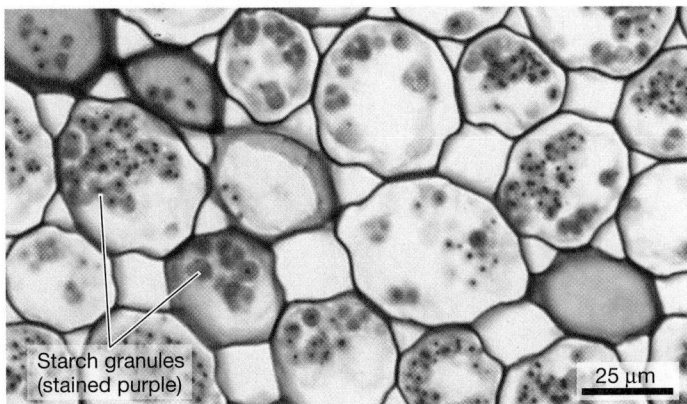

Starch granules (stained purple)

25 µm

FIGURE 37.16 Parenchyma Cells Perform a Wide Array of Tasks.

✔**EXERCISE** Give an example of a gene that is likely to be expressed in these leaf cells, but not in the root cells.

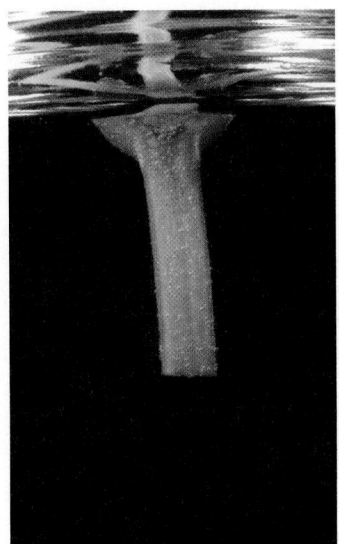

FIGURE 37.17 Parenchyma Cells in Cut Stems Can Form Adventitious Roots. Parenchyma cells in a cut coleus stem (left) divide to form a mass of undifferentiated cells called a callus, which then sprouts roots (right).

stalk of celery or rhubarb—which is actually the petiole of a celery or rhubarb leaf—include many strands of collenchyma cells (**FIGURE 37.18**).

Sclerenchyma: Two Types of Specialized Support Cells The cells that are classified as **sclerenchyma** produce a thick **secondary cell wall** in addition to the relatively thin **primary cell wall** found in all cells. Unlike a primary cell wall, the secondary cell wall contains the tough, rigid compound **lignin** in addition to **cellulose** (see Chapter 31).

Collenchyma cells can support actively growing parts of the plant because they have an expandable primary cell wall. In contrast, the nonexpandable secondary cell wall of sclerenchyma cells specializes them for supporting stems and other structures after growth has ceased. Another key difference between collenchyma and sclerenchyma is that sclerenchyma cells are usually dead at maturity—meaning they contain no cytoplasm.

Ground tissue typically contains two types of sclerenchyma cells: fibers and sclereids.

- **Fibers** are extremely elongated. The fiber cells from ramie plants, a species of Asian nettle, can be over half a meter long.

(a) Cross section of celery stalk   **(b)** Close-up of "string," in cross section   **(c)** Collenchyma cells, in cross section

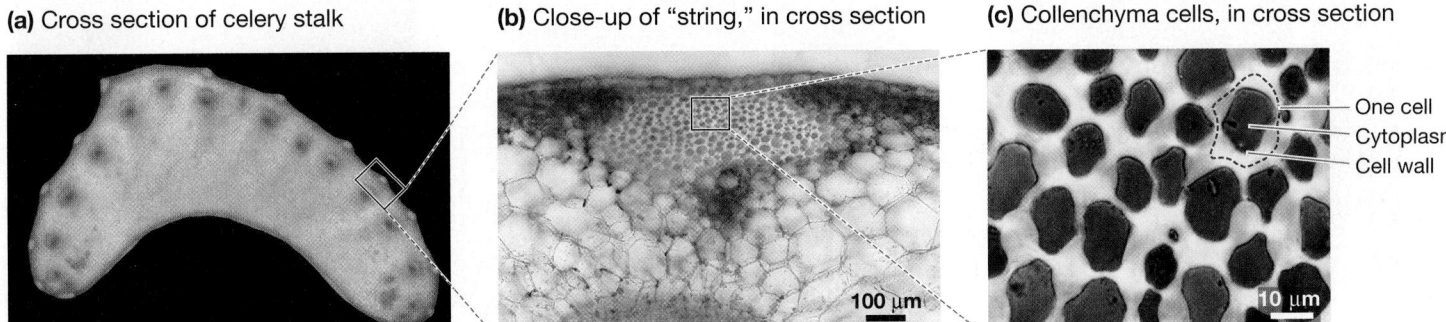

100 µm

One cell
Cytoplasm
Cell wall

10 µm

FIGURE 37.18 Collenchyma Cells Support Growing Tissues. A celery stalk is actually a petiole; the strands you can peel from it are columns of collenchyma cells.

(a) Fibers

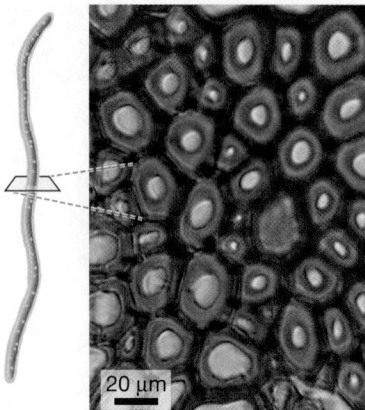

20 µm

(b) Sclereids

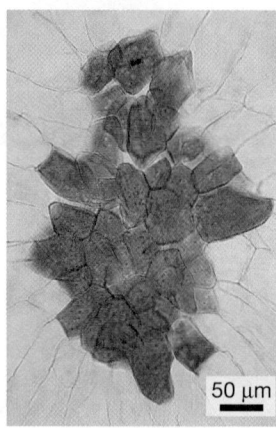

50 µm

FIGURE 37.19 Sclerenchyma Cells Support Mature Tissues.
(a) Fibers and **(b)** sclereids (stained pink) have thickened secondary cell walls. These cells provide support or protection for tissues that are no longer growing.

Fiber cells are important in the manufacture of paper, hemp or jute ropes, or linen and other fabrics (**FIGURE 37.19a**).

- **Sclereids** are relatively short, have variable shapes, and often function in protection. The tough coats of seeds and the thick shells of nuts are composed of sclereids; these cells are also responsible for the gritty texture of pears (**FIGURE 37.19b**).

The Vascular Tissue System

The vascular tissue system functions in support and in long-distance transport of water and dissolved nutrients. It also moves the products of photosynthesis that are made and stored in ground tissue.

Plant tissues that consist of a single cell type are called simple tissues; tissues that contain several types of cells are termed complex tissues. The vascular tissue system is made up of two complex tissues, xylem and phloem.

- **Xylem** (pronounced *ZYE-lem*) conducts water and dissolved ions in one direction: from the root system to the shoot system.

- **Phloem** (*FLO-em*) conducts sugar, amino acids, chemical signals, and other substances in two directions: from roots to shoots and from shoots to roots.

Xylem Structure   Two important cell types in xylem tissue are tracheids and vessel elements.

- In all vascular plants, xylem contains water-conducting cells called **tracheids** (*TRAY-kee-ids*).

- In angiosperms and species in the group Gnetophyta, xylem also contains conducting cells called **vessel elements** (see Chapter 31).

Tracheids and vessel elements have thick, lignin-containing secondary cell walls that are often deposited in ringlike or spiral patterns. Both tracheids and vessel elements are dead at maturity. As a result, they have no membranes and are filled with the fluids that they conduct instead of with cytoplasm.

Tracheids are long, slender cells with tapered ends (**FIGURE 37.20a**). The sides and ends of tracheids have **pits,** which are gaps in the secondary cell wall where only the primary cell wall is present. Because the cell is dead, pits have no plasma membrane spanning the opening. When water is moving up a plant through tracheids, it moves from cell to cell both vertically and laterally through pits, because that is where resistance to flow is lowest.

Vessel elements, in contrast, are shorter and wider than tracheids (**FIGURE 37.20b**). In addition to having pits, vessel elements have **perforations**—openings that lack both primary and secondary cell walls. In some species, the ends of vessel elements

(a) Tracheids are long, tapered, and have pits.

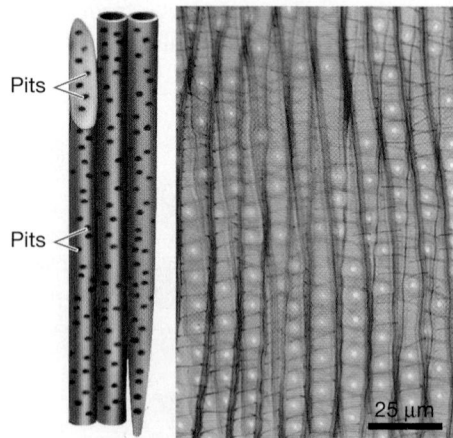

Pits

Pits

25 µm

(b) Vessel elements are short and wide and have perforations as well as pits.

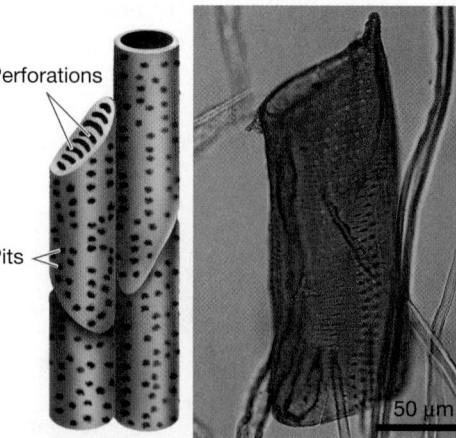

Perforations

Pits

50 µm

(c) Tracheids and vessel elements are found together in vascular tissue.

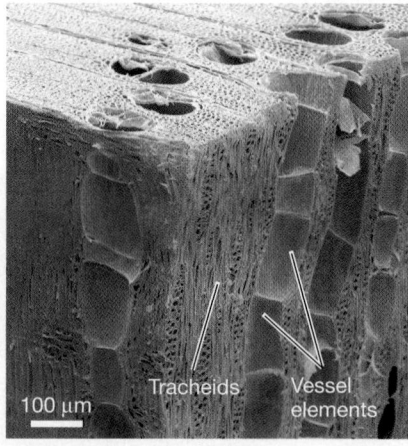

100 µm

Tracheids   Vessel elements

FIGURE 37.20 Xylem May Contain Two Types of Water-Conducting Cells. (a) Tracheids are long and thin compared to **(b)** vessel elements, which are much shorter and wider. **(c)** Both types of water-conducting cells are found in the vascular tissue of angiosperms. The image in (a) is a light micrograph of stained tissue; the images in (b) and (c) are colorized scanning electron micrographs.

lack any cell wall at all, and stacked cells form open pipes called vessels. Vessel elements conduct water more efficiently than do tracheids, because their width and perforations offer less resistance to flow.

In angiosperms, tracheids and vessel elements are found adjacent to each other (**FIGURE 37.20c**). Xylem also contains sclerenchyma cells (fibers) and some parenchyma cells that transport materials laterally in the stem—not vertically.

Phloem Structure   In most vascular plants, phloem is made up primarily of two specialized types of parenchyma cells: sieve-tube elements and companion cells. Both are alive at maturity, lack lignified secondary cell walls, and arise from division of a common precursor cell.

- **Sieve-tube elements** are long, thin cells that have perforated ends called **sieve plates** (**FIGURE 37.21**). They are responsible for transporting sugars and other nutrients.

- **Companion cells** are not conducting cells, but instead provide materials to maintain the cytoplasm and plasma membrane of sieve-tube elements.

Interestingly, sieve-tube elements lack nuclei and most other organelles, but they are directly connected to adjacent companion cells by means of numerous plasmodesmata (see Figure 37.13b). Companion cells contain most of the organelles normally found in a plant cell and support the function of sieve cells.

Companion cells are also involved in loading and unloading carbohydrates and other nutrients into and out of sieve-tube elements (see Chapter 38 for more detail). Phloem also contains sclerenchyma tissue (fibers and sclereids), which provides mechanical support.

TABLE 37.4 (page 748) summarizes the major tissues and cell types found in the dermal, ground, and vascular tissue systems. Once you've mastered the structure and function of the primary plant body, you're ready to consider the next level of complexity: secondary growth.

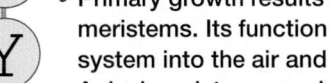

check your understanding

C Y U

If you understand that . . .

- Primary growth results from cell division in apical meristems. Its function is to extend the shoot system into the air and the root system into the soil.
- Apical meristems produce three types of primary meristematic cells: protoderm, ground meristem, and procambium. The dermal, ground, and vascular tissue systems that arise from these meristematic cells extend throughout the individual and make up the primary plant body.
- The dermal system protects the individual and controls the exchange of gases and nutrients with the environment; the ground system produces and stores the carbohydrates that make life possible; the vascular system moves those molecules from place to place and holds the plant up. Each of these systems consists of an array of distinctive cell and tissue types.

✔ **You should be able to . . .**

1. Explain the relationship between an apical meristem and the three primary meristematic tissues.

2. Describe the structure and function of epidermal cells, parenchyma cells, tracheids, and sieve-tube elements.

Answers are available in Appendix A.

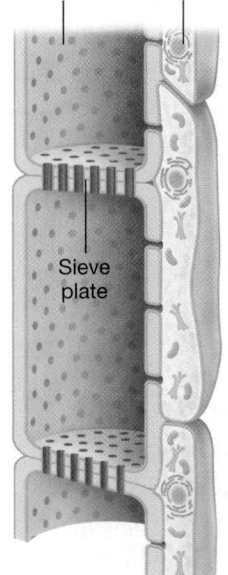

Sieve-tube element (few organelles)   Companion cell (many organelles)

Sieve plate

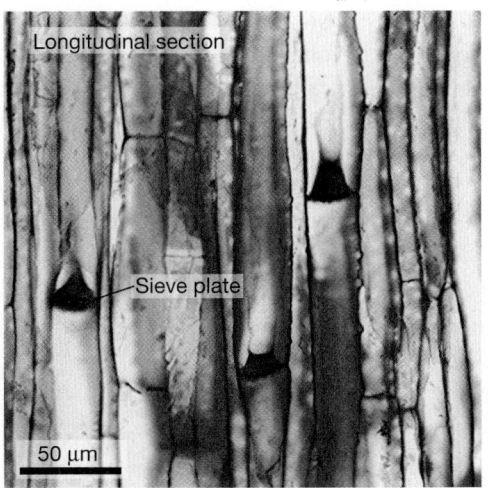

Longitudinal section

Sieve plate

50 μm

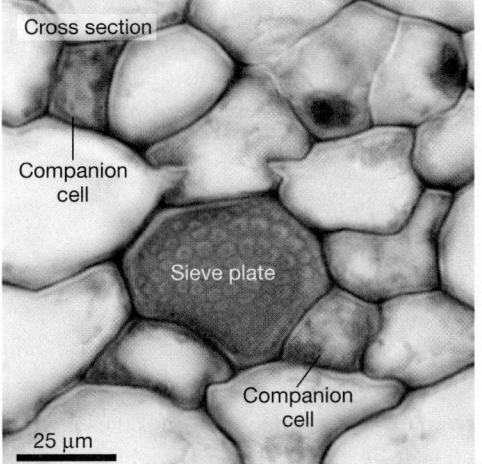

Cross section

Companion cell

Sieve plate

Companion cell

25 μm

FIGURE 37.21 Phloem Consists of Sieve-Tube Elements and Companion Cells. Sieve-tube elements conduct sucrose throughout the body; companion cells support sieve-tube elements.

Tissues Present	Description of Tissue	Function
Dermal Tissue System (arises from protoderm)		
Epidermal	Complex tissue consisting of epidermal cells, guard cells, trichome cells, and root hair cells	Shoots: Protection, gas exchange Roots: Protection, water and nutrient absorption
Ground Tissue System (arises from ground meristem)		
Parenchyma	Simple tissue consisting of parenchyma cells	Synthesis and storage of sugars and other compounds
Collenchyma	Simple tissue consisting of collenchyma cells	Support (flexible cell walls)
Sclerenchyma	Simple tissues consisting of sclerenchyma cells: sclereids or fibers	Support and protection (rigid cell walls; dead at maturity)
Vascular Tissue System (arises from procambium)		
Xylem	Complex tissue consisting of tracheids, vessels, and parenchyma cells and sclerenchyma cells (fibers)	Transport of water and ions; structural support (tracheids, vessels and fibers are dead at maturity)
Phloem	Complex tissue consisting of parenchyma cells (sieve-tube elements, companion cells) and sclerenchyma cells (fibers)	Transport of sugars, amino acids, hormones, etc.; support (fibers are dead at maturity)

37.4 Secondary Growth Widens Shoots and Roots

Primary growth increases the length of roots and shoots; its major function is to extend the reach of the root and shoot system and thus increase a plant's ability to absorb photons and acquire carbon dioxide, water, and ions.

In some plants, **secondary growth** increases the width of the plant body. Its major function is to increase the amount of conducting tissue available and provide the structural support required for extensive primary growth. Recall that the evolution of conducting tissues allowed the vascular plants to grow taller to compete for light (see Chapter 31). Without the support provided by secondary growth, however, roots would not be massive enough to anchor large shoot systems, and long stems would fall over or break. Only with secondary growth could vascular plants become the giants seen today.

What Is a Cambium?

Secondary growth produces **wood** and occurs in species that have a cambium in addition to apical meristems. A **cambium** (plural: cambia) is a special type of meristem (also called a lateral meristem) that differs from an apical meristem in two ways:

1. A cambium forms a cylinder that runs the length of a root or stem and is made up of a single layer of meristematic cells. In contrast, apical meristems are localized at root tips and shoot tips and are dome shaped.

2. In a cambium, cells divide in a way that increases the width of roots and shoots (**FIGURE 37.22a**). Cells in an apical meristem divide in a way that extends the root and shoot tips.

As **FIGURE 37.22b** shows, there are two distinct types of cambium in plants that undergo secondary growth.

- A cylinder of meristematic cells called the **vascular cambium** is located between the secondary xylem and secondary phloem, inside the stem and root.

- Another cylinder of meristematic cells called the **cork cambium** is located near the perimeter of the stem and root.

One other observation is critical to understanding how cambia work: The cork cambium produces new cells primarily to the outside. Vascular cambium, in contrast, generates new layers of cells both to the inside and outside. The new cells formed to the inside push all of the other cells toward the outside, causing an increase in girth.

TABLE 37.5 summarizes the major mature tissue types and cell types that arise from each type of meristem. Now let's consider how these meristems work.

What Does Vascular Cambium Produce?

Vascular cambium in roots and stems produces both phloem and xylem (see Figure 37.22b). New cells that are produced to the outside of the vascular cambium differentiate into secondary phloem; new cells produced to the inside differentiate into secondary xylem. (In contrast, the procambium at each apical meristem produces *primary* phloem and *primary* xylem.)

(a) Cross section of a 3-year-old *Tilia* stem

(b) Vascular cambium and cork cambium produce wood and bark.

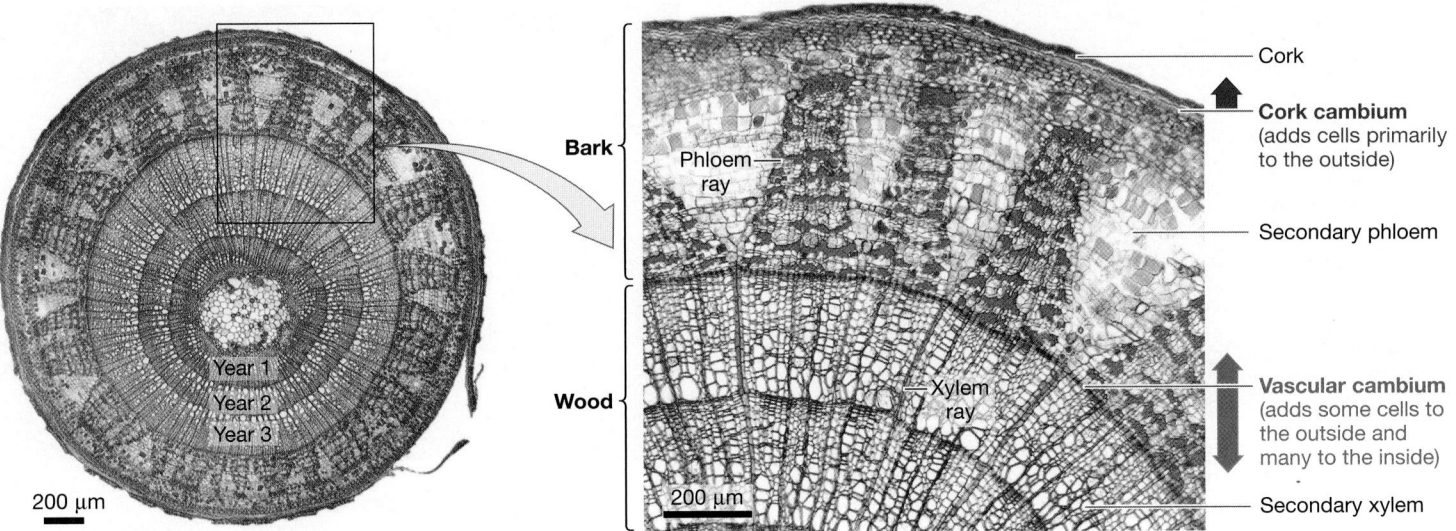

FIGURE 37.22 Vascular Cambium Is Responsible for Secondary Growth. Tree trunks contain two types of cylindrical cambia: cork cambium and vascular cambium. Bark consists primarily of the cork cells produced by the cork cambium and secondary phloem. Wood consists of secondary xylem and parenchyma cells produced by vascular cambium.

Secondary phloem and secondary xylem cells are not always produced simultaneously. In most cases, the vascular cambium produces many more secondary xylem cells than secondary phloem cells.

Primary phloem and xylem are produced by all vascular plants, but secondary phloem and xylem are found only in some lycophytes, all gymnosperms, and certain angiosperms.

Structurally, primary and secondary phloem and primary and secondary xylem are complex tissues, made up of more than one cell type. Functionally, primary and secondary phloem are similar, as are primary and secondary xylem.

- Primary and secondary phloem function in sugar transport.

- Primary and secondary xylem function in water transport and structural support.

Secondary xylem makes up the structural material called wood, while secondary phloem makes up the inner part of a tree's bark, which we describe later.

Besides producing conducting cells such as sieve-tube elements, tracheids, and vessel elements, the vascular cambium also produces sclerenchyma cells (fibers) for additional strength, along with some parenchyma cells. The parenchyma cells are formed in rows that radiate laterally across the xylem and phloem. These rows of cells are called **rays** (see Figure 37.22b), and they form a living conduit through which water and nutrients are transported laterally across the stem.

It's important to realize, though, that the results of cell division in vascular cambium are highly asymmetrical. As the

SUMMARY TABLE 37.5 Components of Secondary Growth

Meristematic Tissue	Mature Tissue	Direction of Growth	Mature Cell Composition	Mature Tissue Function
Cork Cambium	Cork	Produced to the outside	Cork cells	Protection
Vascular Cambium	Secondary phloem	Produced to the outside	Parenchyma cells (sieve-tube elements, companion cells) and sclerenchyma cells (fibers)	Transport of sugars, amino acids, hormones, etc.
	Secondary xylem*	Produced to the inside	Tracheids, vessels, parenchyma cells (arranged in rays) and sclerenchyma cells (fibers)	Transport of water and ions; structural support

*Secondary xylem is also called wood.

FIGURE 37.23 Lenticels in the Bark of Paperbark Cherry (*Prunus serrula*). Lenticels are natural breaks in cork tissue through which oxygen can enter a stem.

vascular cambium grows, all of the secondary xylem is retained and accumulates, but the primary xylem in the center of the stem eventually clogs and may rot away. In addition, the outermost secondary phloem becomes fragmented and compressed as the stem increases in diameter. As a result, mature woody roots and stems are dominated by secondary xylem, or wood.

✔ If you understand this concept, you should be able to explain why a nail hammered into a tree at eye level doesn't move up as the tree grows.

What Does Cork Cambium Produce?

The cork cambium's primary role is to produce **cork cells** to the outside (Figure 37.22b). Together with the secondary phloem, the cork cambium and cork cells make up the tissue called **bark.**

Bark provides a particularly tough barrier in species whose cork cells secrete a strong secondary cell wall containing lignin. Bark also helps prevent water loss because cork cells produce a layer of wax and other molecules inside their cell walls, making them impermeable to water and gases. Gas exchange can still occur between the atmosphere and living tissues inside the stem, though—through small, spongy openings in the bark called **lenticels** (**FIGURE 37.23**).

Cork cells die when they mature. As a stem continues to widen, the cork layer often cracks and flakes, and the outer layers might even slough off the tree.

Bark is important because it protects the woody stem from damage and pathogens as it increases in girth. As a woody stem or root matures, the epidermal tissue produced by the apical meristem during primary growth is replaced by the bark, which takes over the role of preventing water loss and protecting the stem and root from pathogens and herbivores. In some species, exceptionally thick bark can even protect the shoot system from fire damage. Redwood trees, for example, are adapted to fire-prone habitats and can have bark that is 20 cm (12 in.) thick.

(a) Heartwood and sapwood have different functions.

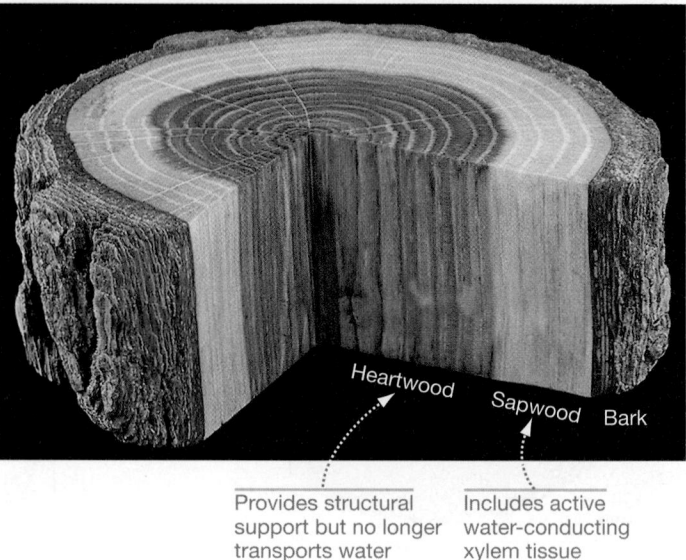

Heartwood — Provides structural support but no longer transports water

Sapwood — Includes active water-conducting xylem tissue

Bark

(b) Growth rings result from seasonal variation in cell size.

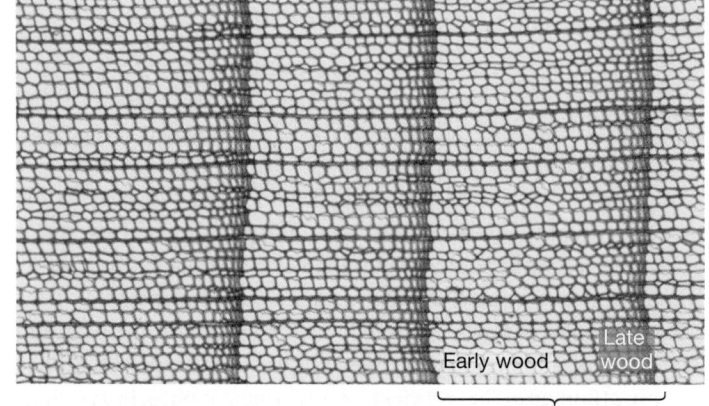

Early wood   Late wood

One growth ring

(c) Patterns in growth rings can tell a tree's history.

Thick growth rings before onset of acid rain

Thin growth rings after onset of acid rain

Bark

FIGURE 37.24 Anatomy of a Tree Trunk. (a) Unstained section of wood. **(b)** Section of wood, stained to show individual cells. **(c)** Unstained section through a fir tree from Germany's Black Forest.

✔**EXERCISE** In part (c), pick a thick growth ring and label the early wood and late wood.

The Structure of a Tree Trunk

Trees are perennial plants that live for many years. As a tree matures and grows in width, the innermost xylem layers stop transporting water—only the xylem from the most recent years actually transports fluid.

Heartwood and Sapwood Xylem that no longer transports begins accumulating protective compounds secreted by other tissues. These compounds form resins, gums, and other complex mixtures. The deposition of these molecules causes the oldest portions of secondary xylem to become darker than the younger portions.

The darker-colored, inner xylem region is called **heart-wood,** while the lighter-colored, outer xylem is called **sapwood** (**FIGURE 37.24a**). If you look closely at wood furniture or flooring, you may see a color difference between sapwood and heart-wood in some boards.

Annual Growth Rings Another important phenomenon occurs in environments where the vascular cambium stops growing for a portion of each year. This period of **dormancy** occurs during the winter in cold climates and during the dry season in tropical habitats.

When the vascular cambium resumes growth in the spring or at the start of the rainy season, it produces large, relatively thin-walled cells, called early wood. As the growing season nears its end, conditions tend to dry out or become cooler; the secondary xylem cells that are produced at this time tend to be smaller, thicker walled, and darker and are called late wood. Thus, when growth is seasonal, regions of large, thin-walled cells alternate with layers of small, thick-walled cells, resulting in annual growth rings in the secondary xylem (**FIGURE 37.24b**).

Analyzing patterns in tree growth rings is an important field of study in biology. Because trees grow faster when moisture and nutrients are plentiful, wide tree rings are reliable indicators of wet years. In contrast, narrow rings signal drought years—or in the case of the fir tree shown in **FIGURE 37.24c**, years when abundant acid rain, due to air pollution, reduced growth.

By studying the growth rings in fossil trees and extremely old living trees, biologists can often assemble a continuous record that dates back thousands of years. In doing so, they gain a better understanding of climate changes that occurred in the past. With continued research, researchers also hope to predict how forests might respond to the rising temperatures associated with the global climate change that is currently under way.

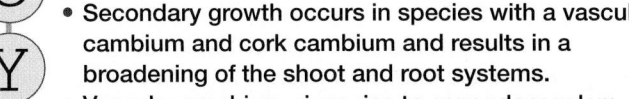
check your understanding

If you understand that . . .

- Secondary growth occurs in species with a vascular cambium and cork cambium and results in a broadening of the shoot and root systems.
- Vascular cambium gives rise to secondary xylem and secondary phloem tissues. Cork cambium gives rise to the protective tissue called bark.

✔ **You should be able to . . .**

1. Describe what the rings in a cross section of a tree trunk would look like if the individual is heavily shaded on one side but exposed to sunlight on the other side.

2. Draw from memory a three-year-old woody stem in cross section, labeling secondary xylem and phloem, vascular and cork cambium, and cork. Add arrows showing the direction of growth in each meristem.

Answers are available in Appendix A.

CHAPTER 37 REVIEW

For media, go to MasteringBiology MB

If you understand . . .

37.1 Plant Form: Themes with Many Variations

- The root and shoot systems of plants are specialized for harvesting the light, water, and nutrients required for performing photosynthesis. Structures involved in absorption have a high surface-area-to-volume ratio.

- Roots extract water and nutrients such as nitrogen, phosphorus, and potassium from the soil.

- The shoot system consists of all aboveground portions of the plant, including stems and leaves.

- Leaves carry out photosynthesis by capturing light and carbon dioxide from the atmosphere and usually consist of a flattened blade that extends from a petiole.

- The overall morphology of root and shoot systems varies widely among plant species, allowing individuals to reduce competition for resources.

- Roots and shoots may be modified to perform a variety of other functions, however, including nutrient storage, water storage, protection, and asexual reproduction.

- Because roots and shoots grow throughout life, a plant is able to respond appropriately to changes in environmental conditions.

✔ You should be able to explain why phenotypic plasticity in roots and shoots is expected to be more important (1) in environments where conditions are variable versus stable, and (2) in long-lived versus short-lived species.

37.2 Primary Growth Extends the Plant Body

- Each apical meristem gives rise to three primary meristematic tissues: protoderm, ground meristem, and procambium.

- The primary meristematic tissues give rise to the dermal, ground, and vascular tissue systems, which extend throughout the plant body.

- Behind the root apical meristem, or the zone of cellular division, cells become longer in the zone of cellular elongation and acquire specialized functions in the zone of cellular maturation.

✔ You should be able to explain why most of the water and nutrient absorption by roots takes place in the zone of cellular maturation.

37.3 Cells and Tissues of the Primary Plant Body

- The dermal tissue system is usually one cell layer thick and plays a role in protection and water absorption by roots. In shoots the epidermis synthesizes the cuticle, which aids in water conservation.

- Stomata are formed by pairs of epidermal guard cells that open and close to control CO_2 uptake and water loss by leaves.

- The ground tissue system performs photosynthesis and stores carbohydrates and other compounds.

- The vascular tissue system transports materials throughout the plant. Within the vascular system, xylem tissue transports water and dissolved ions up the plant; phloem tissue transports sugars up and down.

- Ground tissue contains (1) parenchyma cells, which function in material synthesis, transport and storage; (2) collenchyma cells, which provide structural support for growing regions; and (3) sclerenchyma cells, fibers and sclereid cells, that strengthen regions of the body that have stopped growing.

✔ You should be able to predict the results of an experiment where a drug was used to (1) poison an apical meristem in the shoot and an apical meristem in the root, and (2) selectively poison protoderm cells in a shoot apical meristem.

37.4 Secondary Growth Widens Shoots and Roots

- In some plant species, shoots and roots are widened by cylindrical meristems called vascular cambium that produce secondary xylem and secondary phloem, and cork cambium that produces cork.

- Cork cells produce waxes that protect stems from water loss.

- Wood consists of secondary xylem, while bark consists of all tissue outside of the vascular cambium.

✔ You should be able to predict the results of an experiment where a drug was used to (1) slow the growth of the vascular cambium but not the cork cambium, and (2) slow the growth of both the vascular and cork cambia on one side of a tree trunk.

(MB) **MasteringBiology**

1. **MasteringBiology Assignments**

Tutorials and Activities Plant Growth; Primary and Secondary Growth; Primary and Secondary Growth in Plants; Root, Stem, and Leaf Sections

Questions Reading Quizzes, Blue-Thread Questions, Test Bank

2. **eText** Read your book online, search, take notes, highlight text, and more.

3. **The Study Area** Practice Test, Cumulative Test, BioFlix® 3-D Animations, Videos, Activities, Audio Glossary, Word Study Tools, Art

---

You should be able to . . .

✔ TEST YOUR KNOWLEDGE

Answers are available in Appendix A

1. Which of the following functions is not performed by parenchyma cells?
   a. synthesizing key molecules
   b. providing mechanical support
   c. performing photosynthesis
   d. storing nutrients

2. What is a sieve-tube element?
   a. the sugar-conducting cell found in phloem
   b. the widened, perforation-containing, water-conducting cell found only in angiosperms
   c. the nutrient- and water-absorbing cell found in root hairs
   d. the nucleated and organelle-rich support cell found in phloem

3. Which statement best characterizes primary growth?
   a. It does not occur in roots, only in shoots.
   b. It leads to the development of cork.

   c. It produces the dermal, ground, and vascular tissues.
   d. It produces rings of xylem and phloem tissue as well as rings of cork tissue.

4. Which statement best characterizes secondary growth?
   a. It results from divisions of the vascular and cork cambium cells.
   b. It increases the length of the plant stem.
   c. It results from divisions in the apical meristem cells.
   d. It often produces phloem cells to the inside and xylem cells to the outside of the vascular cambium.

5. To illustrate the concept that a plant structure can be modified for many different functions, give an example of a leaf and a stem that protect individuals against large herbivores.

6. Compare and contrast the structure and function of collenchyma and sclerenchyma cells.

7. How do tracheids differ from vessel elements, in addition to their overall shape?
   a. Tracheids are stacked end to end to form continuous, open columns.
   b. In tracheids, water flows from cell to cell primarily through gaps in the secondary cell wall called pits.
   c. Tracheids are dead at maturity.
   d. Tracheids have secondary cell walls reinforced with lignin.

8. Describe the general function of the shoot system and the general function of the root system. Which tissues are continuous throughout these two systems? Suggest a hypothesis to explain why the shoot and root systems of different species are so variable in size and shape.

9. Explain why continuous growth enhances the phenomenon known as phenotypic plasticity.

10. What does cuticle do? What do stomata do? Predict how the thickness of cuticle and the number of stomata differ in plants from wet habitats versus dry habitats.

11. Compare and contrast the roles of parenchyma cells in the ground tissue system versus the vascular tissue system.

12. Describe how the vascular cambium produces secondary xylem and secondary phloem.

13. Which one of the following tissues is most likely to be the source of nutrition for insects that kill trees?
    a. primary xylem
    b. secondary xylem
    c. vascular cambium
    d. cork

14. Identify the structure you are consuming when you eat each of the following vegetables: asparagus, Brussels sprouts, celery, spinach, carrot, potato.

15. Why do mahogany trees, and other trees that grow in tropical rain forests, lack growth rings?

16. Trees can be killed by girdling—meaning the removal of bark and vascular cambium in a ring all the way around the tree. Explain why.

38 Water and Sugar Transport in Plants

In this chapter you will learn how

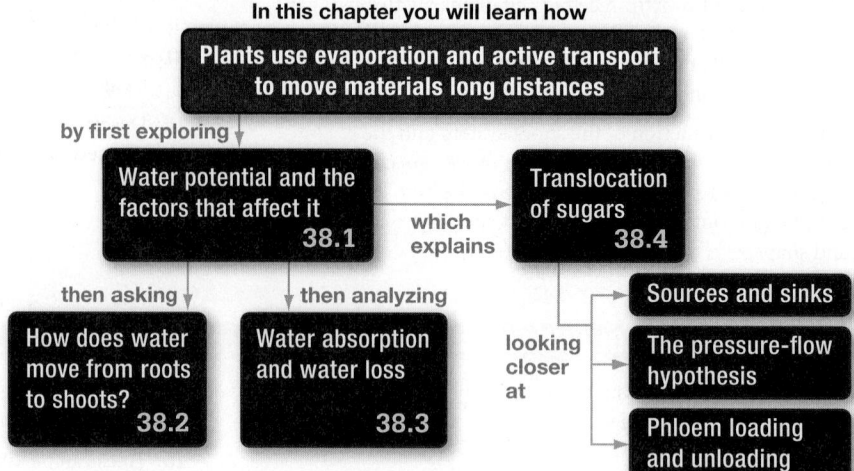

Plants use evaporation and active transport to move materials long distances

by first exploring

Water potential and the factors that affect it
38.1

which explains

Translocation of sugars
38.4

then asking

How does water move from roots to shoots?
38.2

then analyzing

Water absorption and water loss
38.3

looking closer at

Sources and sinks

The pressure-flow hypothesis

Phloem loading and unloading

This chapter explores how plants move water from their roots to their leaves and how they transport sugars to all of their tissues—sometimes over great distances.

This chapter is part of the Big Picture. See how on pages 840–841.

On a hot summer day, a large deciduous tree can lose enough water to fill three 55-gallon drums. To understand why, recall that the surfaces of leaves are dotted with stomata (see Chapter 37), which in most plants open during the day so gas exchange can occur between the atmosphere and the cells inside the leaf. This exchange is crucial. For photosynthesis to take place, leaf cells must acquire carbon dioxide (CO_2).

There's a catch, however. While stomata are open, the moist interior of the leaf is exposed to the dry atmosphere. As a result, large quantities of water evaporate from the leaf. If the lost water is not replaced with water absorbed by roots, plant cells will dry out and die.

Evaporation from leaves is a challenge for plants, but it is also beneficial. Because water enters a plant through its roots and exits through its leaves, there is a regular flow of water from roots to shoots. Minerals that are absorbed by roots are carried to the leaves by this flow. Without this

✔ When you see this checkmark, stop and test yourself. Answers are available in Appendix A.

movement of water, shoots would be unable to receive the minerals they need to grow.

In addition, under hot conditions, evaporation cools the plant, just like sweating cools your body. Heavy rates of evaporation can lower leaf temperatures by as much as $10-15°C$.

But the source of water can be a long distance from where it is needed. In the case of certain redwood trees, the leaves that lose water may be 100 meters (m) from the root hairs that absorb water. How do plants transport water against the force of gravity—in some cases, the length of a football field? And how do plants move the sugar they produce from active photosynthetic sites to sites where sugar is stored or needed for growth? These questions are the heart and soul of this chapter. Answering them is a fundamental part of understanding how plants work. (You can review plant form and function in the Big Picture of Plant and Animal Form and Function, on pages 840–841.)

38.1 Water Potential and Water Movement

Loss of water via evaporation from the aerial parts of a plant is called **transpiration.** Transpiration occurs whenever two conditions are met: (**1**) Stomata are open, and (**2**) the air surrounding leaves is drier than the air inside leaves.

The first condition is usually met during the day, when photosynthesis takes place. The second condition occurs whenever atmospheric humidity is less than 100 percent.

One of the most astonishing observations in biology is that water moves from roots to leaves passively—that is, with no expenditure of ATP. Plants do not need a heart muscle to pump water from roots to shoots. Even in 100-m-tall redwood trees, water flows passively from the root system to the shoot system. This movement occurs because of differences in the potential energy of water.

What Is Water Potential?

Recall that potential energy is stored energy (see Chapter 2). Changes in potential energy are associated with changes in position, such as the position of a molecule or an electron.

Biologists use the term **water potential** to indicate the potential energy that water has in a particular environment compared with the potential energy of pure water at room temperature and atmospheric pressure. Under these conditions, pure water has a water potential of 0.

Water potential is symbolized by the Greek letter ψ (psi, pronounced sigh). Differences in water potential determine the direction that water moves. Water always flows from areas of higher water potential to areas of lower water potential. Several factors can contribute to differences in water potential.

What Factors Affect Water Potential?

To understand how water moves from cell to cell in a plant, first consider the cell in the beaker on the left in **FIGURE 38.1a**. Notice

that it is suspended in a **solution**—a homogenous, liquid mixture containing several substances. In this case, the solution consists of water and dissolved substances, or **solutes.**

In the beaker on the left, the solute concentrations in the cell and in the surrounding solution are the same. Such a solution is said to be **isotonic** to the cell (see Chapter 6). When two solutions are isotonic, there is no net movement of water between them.

The Role of Solute Potential What happens when the cell is transferred to the beaker on the right in Figure 38.1a? This beaker contains pure water, which has no solutes. As a result, the solution surrounding the cell is strongly **hypotonic** relative to the cell.

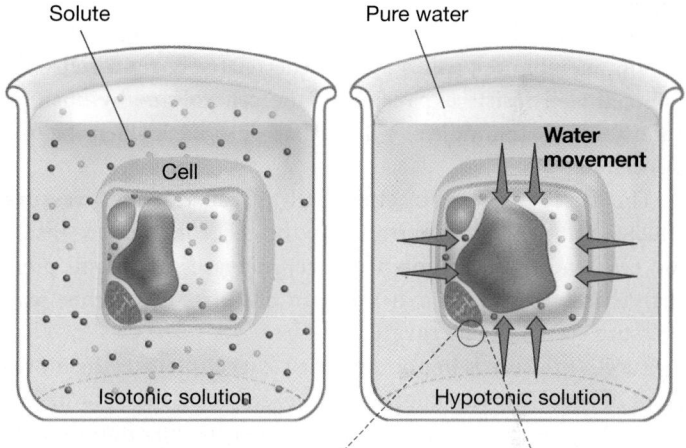

(a) Solute potential is the tendency of water to move by osmosis.

Solute potential inside cell and in surrounding solution is the same. No net movement of water.

Cell is placed in pure water. The cell's solute potential is low relative to its surroundings. Water moves into cell via osmosis.

Solute

Pure water

Cell

Water movement

Isotonic solution

Hypotonic solution

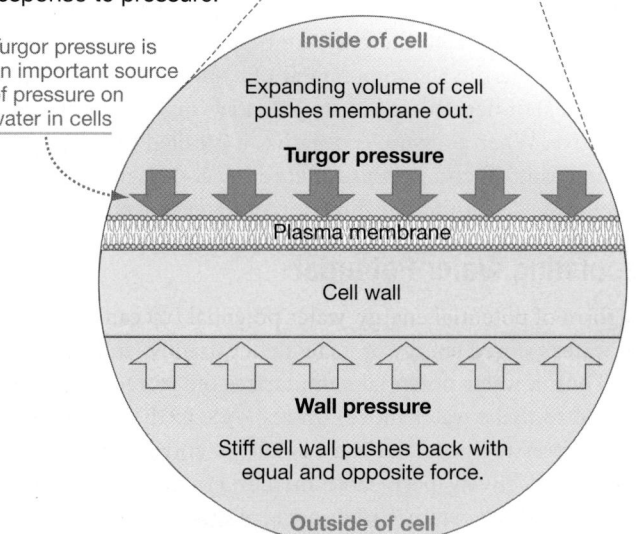

(b) Pressure potential is the tendency of water to move in response to pressure.

Turgor pressure is an important source of pressure on water in cells

Inside of cell

Expanding volume of cell pushes membrane out.

Turgor pressure

Plasma membrane

Cell wall

Wall pressure

Stiff cell wall pushes back with equal and opposite force.

Outside of cell

FIGURE 38.1 Water Potential Has Two Major Components: Solute Potential and Pressure Potential.

The concentration of the solution is important because when solutions are separated by a selectively permeable membrane, such as the plasma membrane of a cell, water passes through the membrane from regions of low solute concentration to regions of high solute concentration. This movement of water across membranes, in response to differences in solute concentration, is called **osmosis** (see Chapter 6).

The tendency for water to move in response to differences in solute concentrations is called the **solute potential (ψ_S).** The presence of solutes can only decrease a solution's solute potential below that of pure water.

The solute potential of a solution is defined by its total solute concentration relative to pure water. If water contains a high concentration of solutes, then it has a low solute potential compared with pure water. Water moves down its potential gradient toward the region of higher solute concentration.

The Role of Pressure Potential

When an animal cell is placed in a hypotonic solution and water enters the cell via osmosis, the volume of the cell increases until the cell bursts. This does not happen to plant cells, however.

If a plant cell swells in response to incoming water, its plasma membrane pushes against the relatively rigid cell wall. The cell wall resists expansion of the cell volume by pushing back, much as the walls of a basketball push back when the ball is inflated.

The force exerted by the wall is called **wall pressure** (**FIGURE 38.1b**). As water moves into the cell, the pressure inside the cell, known as **turgor pressure,** increases until wall pressure is induced. Cells that are firm and that experience wall pressure are said to be **turgid.**

Turgor pressure is important because it counteracts the movement of water due to osmosis. In the example on the right in Figure 38.1a, the solute potential favors water moving into the cell. However, the rigid cell wall limits the amount of water that can enter the cell.

Pressure potential (ψ_P) refers to any kind of physical pressure on water. Inside a cell, the pressure potential consists of turgor pressure. While the solute potential of a solution can be only 0 (pure water) or negative, pressure potential can be either positive or negative. When pressure is negative, it is called tension.

When osmosis and pressure affect a cell at the same time, how do biologists determine the direction of water movement?

Calculating Water Potential

As a form of potential energy, water potential (ψ) can be thought of as water's stored energy or its tendency to move to a new position. Thus, a water potential summarizes the stored energy that will tend to make water move—in response to the combined effects of a pressure potential and a solute potential.

When selectively permeable membranes are present, water tends to move by osmosis from areas of high solute potential to areas of low solute potential. When no membranes are present to stop it, water moves from areas of high pressure potential to areas of low pressure potential.

If we ignore the effects of gravity, water potential is defined by the following equation:

$$\psi = \psi_P + \psi_S$$

In words, the potential energy of water in a particular location is the sum of the pressure potential and the solute potential that it experiences.

Assigning Units of Pressure and Signs   Water potential is measured in units called **megapascals (MPa, 10^6 Pa).** A **pascal (Pa)** is a unit of measurement commonly applied to pressures—force per unit area. A car tire is inflated to about 0.2 MPa, and the water pressure in home plumbing is usually 0.2 to 0.3 MPa.

Solute potentials (ψ_S) are always negative because they are measured relative to the solute potential of pure water. Because pure water contains no dissolved substances, it has a solute potential of 0 megapascal (MPa). And because there are always some solutes inside a cell, the water inside always has a solute potential lower than that of pure water. Therefore pure water will always tend to move *into* the cell. Increasing the concentration of solutes in a cell lowers its solute potential even more—making it more negative.

In contrast to the solute potential in cells, the pressure potential (ψ_P) from turgor pressure is positive inside *living* cells. But as you will see later, the pressure potential in dead cells can sometimes be negative. Let's first consider how solute and pressure potentials combine to drive water movement into and out of living cells.

Water Movement in the Absence of Pressure   In the U-shaped tube on the left side of **FIGURE 38.2a**, two solutions are separated by a selectively permeable membrane. The system is open to the atmosphere and thus is not under additional pressure, meaning $\psi_P = 0$ MPa.

Note that the left side of the tube contains pure water, which has a ψ_S of 0 MPa. The ψ_S for the solution on the right side of the membrane is −1.0 MPa. Because water potential is higher on the left side of the tube than on the right side, water moves from left to right.

The right side of Figure 38.2a models the same situation with a cell that is initially **flaccid**—meaning it has no turgor pressure and thus a pressure potential of 0. Note that the cell has been placed in a solution of pure water. Because the cell has low solute potential (−1.0 MPa) and the pure water has a higher water potential than the cell, water enters the cell via osmosis.

✔ QUANTITATIVE   If you understand the concept of solute potential, you should be able to explain (1) how you would change the solute potential of the pure water on the left of Figure 38.2a to make it lower than the solute potential in the solution to the right, and (2) what the consequences for water movement in the tube would be.

Water Movement in the Presence of a Solute Potential and Pressure Potential   On the left side of **FIGURE 38.2b**, the concentrations in the U-shaped tube are the same as in Figure 38.2a, but the solution on the right side of the tube experiences pressure exerted by a plunger. If the force on the plunger produces a

(a) Solute potentials differ.

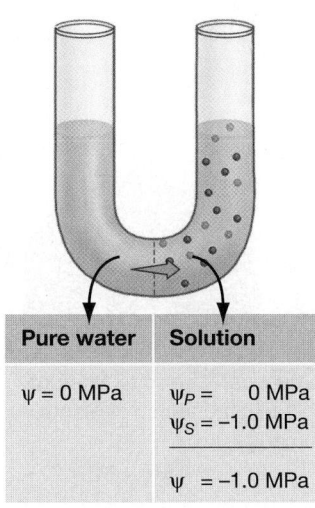

Pure water	Solution
$\psi = 0$ MPa	$\psi_P = 0$ MPa
	$\psi_S = -1.0$ MPa
	$\psi = -1.0$ MPa

Water moves left to right—from area with higher water potential to area with lower water potential

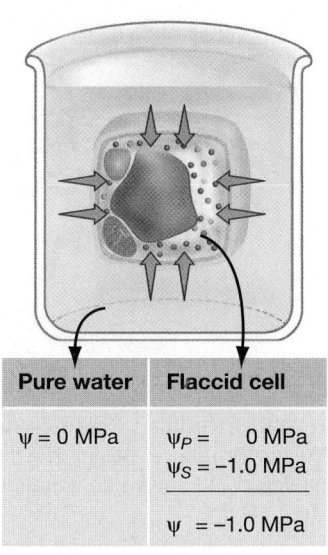

Pure water	Flaccid cell
$\psi = 0$ MPa	$\psi_P = 0$ MPa
	$\psi_S = -1.0$ MPa
	$\psi = -1.0$ MPa

Water moves into cell—from area with higher water potential to area with lower water potential

(b) Solute and pressure potentials differ.

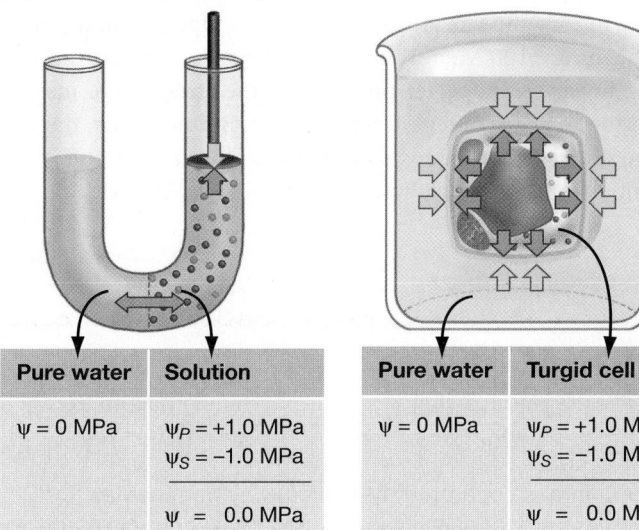

Pure water	Solution
$\psi = 0$ MPa	$\psi_P = +1.0$ MPa
	$\psi_S = -1.0$ MPa
	$\psi = 0.0$ MPa

Water potentials are equal— no net movement

Pure water	Turgid cell
$\psi = 0$ MPa	$\psi_P = +1.0$ MPa
	$\psi_S = -1.0$ MPa
	$\psi = 0.0$ MPa

Water potentials are equal— no net movement

FIGURE 38.2 Solute Potential and Pressure Potential Interact.

✔**QUANTITATIVE** In the left side of part (a), does the solute potential on the right side of the tube increase, decrease, or stay the same when water flows from left to right by osmosis?

pressure potential of 1.0 MPa on the right side, and if the ψ_S for the solution on the right side of the membrane is still -1.0 MPa, then the water potential of the right side is -1.0 MPa + 1.0 MPa = 0.

In this case, the water potential on both sides of the membrane is equal and there will be no net movement of water. If the force on the plunger is greater than 1.0 MPa, the solution on the right side would have a *higher* water potential and water would flow from right to left.

The right side of Figure 38.2b models this situation in a living cell. In this case, the incoming water creates turgor pressure. When the positive turgor pressure ($+1.0$ MPa) plus the cell's negative solute potential (-1.0 MPa) equals 0 MPa—the water potential of pure water—the system reaches equilibrium. At equilibrium, there is no additional net movement of water. In this way, turgor pressure acts like the plunger in Figure 38.2b.

✔ **QUANTITATIVE** If you understand the concept of pressure potential, you should be able to explain how you would use the plunger in Figure 38.2b to create a negative pressure instead of a positive pressure.

Water Potentials in Soils, Plants, and the Atmosphere

The water contained within leaf, stem, or root tissues has a pressure potential and a solute potential, just like the water inside a cell does. Likewise, both the soil surrounding the root system and the air around the shoot system have a water potential.

Water Potential in Soils In moist soil, the water that fills crevices between soil particles usually contains relatively few solutes and normally is under little pressure. As a result, its water potential tends to be high relative to the water potential found in a plant's roots, which is higher in solutes.

There are important exceptions to this rule, however.

- *Salty soils* The soils near ocean coastlines have water potentials as low as -4MPa or less due to high solute concentrations. This is much lower than the water potential typically found inside plant roots.

- *Dry soils* When soils dry, water no longer flows freely in the spaces between soil particles. All of the remaining water adheres tightly to soil particles, creating a tension that lowers the water potential of soil water.

When the water potential in soil drops, water is less likely to move from soil into plants. If soil water potential is low enough, water may even move from plants to the soil. This situation would be deadly to plants.

This is an enormously important issue for world agriculture. When soils are irrigated to boost crop yields, much of the water evaporates. The solutes in the irrigation water are left behind and tend to collect in the first few inches below the surface. Over time, then, irrigated soils tend to become salty. In some parts of the world, formerly productive soils have become so salty that they are now abandoned as cropland.

How Do Plants That Are Adapted to Salty or Dry Habitats Cope? Salt-adapted species often respond to low water potentials in soil by accumulating solutes in their root cells, which lowers their solute potential. These plants have enzymes that increase the concentration of certain organic molecules in the cytoplasm. As a result, they can keep the water potential of their tissues even lower than the water potential of salty soils.

Species that are adapted to dry sites cope by tolerating low solute potentials. For example, let's look at the changes in solute

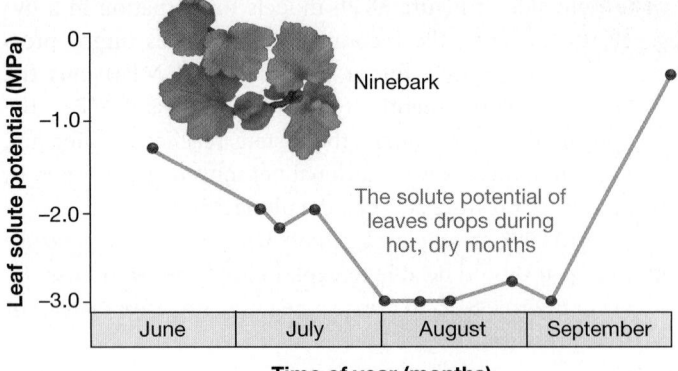

FIGURE 38.3 Plants with Low Solute Potentials Can Grow in Dry Soils. As soils dry, their water potential declines. But if a plant's solute potential also drops, it can maintain a water-potential gradient that continues to bring water into the plant.

DATA: Cline, R. G., and W. S. Campbell. 1976. *Ecology* 57: 367–373.

FIGURE 38.4 Wilting Occurs When Water Loss Leads to Loss of Turgor Pressure. Wilting is a life-threatening condition in plants, analogous to severe dehydration in humans.

potential that were recorded in tissue from a shrub species called ninebark. In **FIGURE 38.3** each data point represents the solute potential recorded on a particular day; the lines between the data points are drawn simply to make the trend clear.

Ninebarks thrive on dry sites in the Rocky Mountains of western North America. Notice that the solute potential of tissue is relatively high in June, at the start of the growing season. In the Rockies, June tends to be rainy and cool. July and August are progressively hotter and drier, however. As the graph indicates, the solute potential in tissue drops dramatically during this period.

This is a key observation. As the summer progresses and water potential in soil drops, ninebark shrubs are able to keep acquiring water and grow because the solute potentials of their tissues can drop to stay below the soil water potential.

Plants that are adapted to wetter sites cannot tolerate such low solute potentials in their tissues. When conditions get hot and dry, they have to close their stomata and stop photosynthesis—meaning that their growth will stop or slow down dramatically compared to dry-adapted plants.

If plants keep their stomata open and leaf cells lose water faster than the water is replaced, the positive turgor pressure normally inside living cells drops to 0. If the cells do not regain turgor quickly, they are at risk of dehydration and death as cells shrivel like grapes drying into raisins. When an entire tissue loses turgor pressure, it will **wilt** (**FIGURE 38.4**).

Turgor pressure is required for growth to occur; without it, cells cannot expand once cell division is complete. In addition, turgor pressure provides structural support—which is why turgid plants do not wilt. Unless corrected, extensive wilting may lead to the death of the tissue and, eventually, the plant.

Water Potential in Air In the atmosphere, water exists as a vapor with a solute potential of 0 MPa. The pressure exerted by water vapor in the atmosphere depends on temperature and humidity. The lower the pressure potential, the faster the rate of evaporation from liquid water to the atmosphere.

- When air is dry, there are few water molecules present and the pressure they exert is low, increasing the rate of evaporation.

- When air is warm, water molecules move farther apart and also exert lower pressure.

Warm, dry air has an extremely low water potential, often approaching −100 MPa. When the weather is cool and rainy or foggy, however, the water potential of the atmosphere may be equal to the water potential inside a leaf. But normally, the water potential of the atmosphere is much lower than the water potential inside a leaf, so water evaporates quickly.

check your understanding

If you understand that . . .

- Water moves along a water-potential gradient, from areas of high water potential to areas of low water potential.

- In areas separated by a selectively permeable membrane, part of water's potential energy is made up of its solute potential—its tendency to move via osmosis.

- Water also has a pressure potential. In living plant cells, for example, the cell wall can exert pressure on water and affect its pressure potential.

- Soils, plant tissues, and the atmosphere all have a water potential made up of a solute potential and a pressure potential.

✔ **You should be able to . . .**

1. Explain why the water potential of a cell in equilibrium with pure water can be zero even though it contains solutes.

2. Compare and contrast the solute potential of an irrigated agricultural field before and after a long, hot day of evaporation.

Answers are available in Appendix A.

Low water potential
Atmosphere ψ: –90 MPa
(Changes with humidity;
usually very low)

Leaf ψ: –0.8 MPa
(Depends on transpiration rate;
low when stomata are open)

Root ψ: –0.6 MPa
(Medium–high)

Soil ψ: –0.3 MPa
(High if moist;
low if extremely dry)

High water potential

FIGURE 38.5 A Water-Potential Gradient Exists between Soil, Plants, and Atmosphere. Water moves from regions of high water potential to regions of low water potential.

In most cases, water potential is highest in soil, lower in roots, lower yet in leaves, and lowest in the atmosphere. This situation sets up a **water-potential gradient** that causes water to move up through the plant. To move *up* a plant, water moves *down* the water-potential gradient that exists between the soil, its tissues, and the atmosphere. When it does so, it replaces the water lost through transpiration (**FIGURE 38.5**).

38.2 How Does Water Move from Roots to Shoots?

Suppose that you are caring for the wilted plant in Figure 38.4. If you add water to the soil, the water potential of the soil increases and water will move into the plant along a water-potential gradient.

Water moves into root cells by osmosis, but how does it move up to the shoots, against gravity? Biologists have tested three major hypotheses for how the water could be transported to shoots:

1. Root pressure—a pressure potential that develops in roots—could drive water up against the force of gravity.

2. Capillary action could draw water up the cells of xylem.

3. Cohesion-tension, a force generated in leaves by transpiration, could pull water up from roots.

Note that all three hypotheses could be correct—they are not mutually exclusive.

Let's begin by considering how water and solutes move from the soil across the root and into the **vascular tissue,** which contains xylem and phloem. Then we can analyze each of the three hypotheses for how water and solutes finish the trip up the xylem to the shoot system.

Movement of Water and Solutes into the Root

To understand how water enters a root, consider the cross section through a young buttercup root shown in **FIGURE 38.6** (see also Chapter 37). Starting at the outside of the root and working inward, notice that several distinct tissues are present:

*   The **epidermis** (literally, "outside skin") is a single layer of cells. In addition to protecting the root, some epidermal cells produce **root hairs,** which greatly increase the total surface area of the root.

*   The **cortex** consists of ground tissue—usually parenchyma cells—and stores carbohydrates.

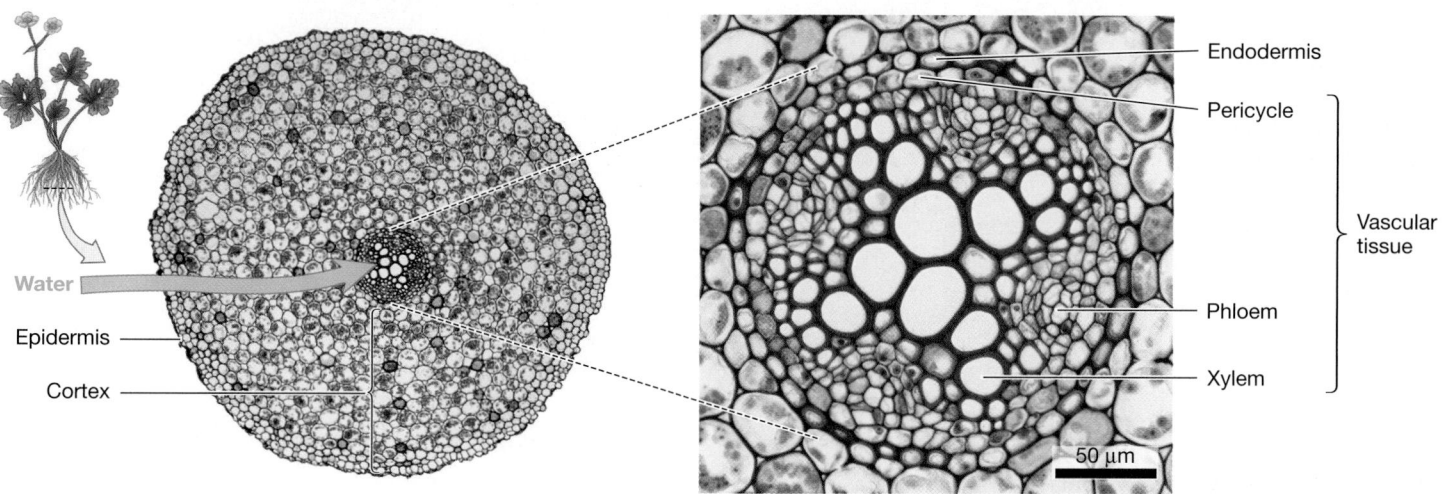

FIGURE 38.6 In Roots, Water Has to Travel through Several Tissue Layers to Reach Vascular Tissue. This cross section through a buttercup root shows the anatomy that is typical of roots in eudicots.

- The **endodermis** ("inside skin") is a cylindrical layer of cells that forms a boundary between the cortex and the vascular tissue. The function of the endodermis is to control ion uptake and prevent ion leakage from the vascular tissue.

- The **pericycle** ("around-circle") is a layer of cells that forms the outer boundary of the vascular tissue. The pericycle can become meristematic and produce lateral roots.

- Conducting cells of the vascular tissue function in transport between roots and shoots, and are located in the center of roots in buttercups and other eudicots. Notice that, in these plants, phloem is situated between each of four arms formed by xylem, which is arranged in a cross-shaped pattern.

Three Routes through Root Cortex to Xylem   When water enters a root along a water-potential gradient, it does so through root hairs. As water is absorbed, it moves through the root cortex toward the xylem through three distinct routes (**FIGURE 38.7**).

1. The *transmembrane route* is based on flow through aquaporin proteins—water channels located in the plasma membranes of root cells (see Chapter 6). Some water may also diffuse directly across plasma membranes.

2. The *apoplastic route* is outside the plasma membranes. The **apoplast** consists of cell walls, which are porous, and the spaces that exist between cells. The name apoplastic is apt because movement takes place outside the plasma membrane and cell interior. Water moving along the apoplastic route *must* eventually pass into the cytoplasm of endodermal cells before entering xylem.

3. The *symplastic route* is inside the plasma membranes. The **symplast** consists of the cytosol and the continuous connections through cells that exist via plasmodesmata (see Chapter 11).

In essence, water can flow through tissues by crossing membranes via aquaporins, or by passing around cells via the apoplast, or by passing through the symplast via plasmodesmata.

The Role of the Casparian Strip   The situation changes when water reaches the endodermis. Endodermal cells are tightly packed and secrete a narrow band of wax called the **Casparian strip.** This layer is composed primarily of a compound called **suberin,** which forms a water-repellent cylinder at the endodermis.

The Casparian strip blocks the apoplastic route by preventing water from moving through the walls of endodermal cells and

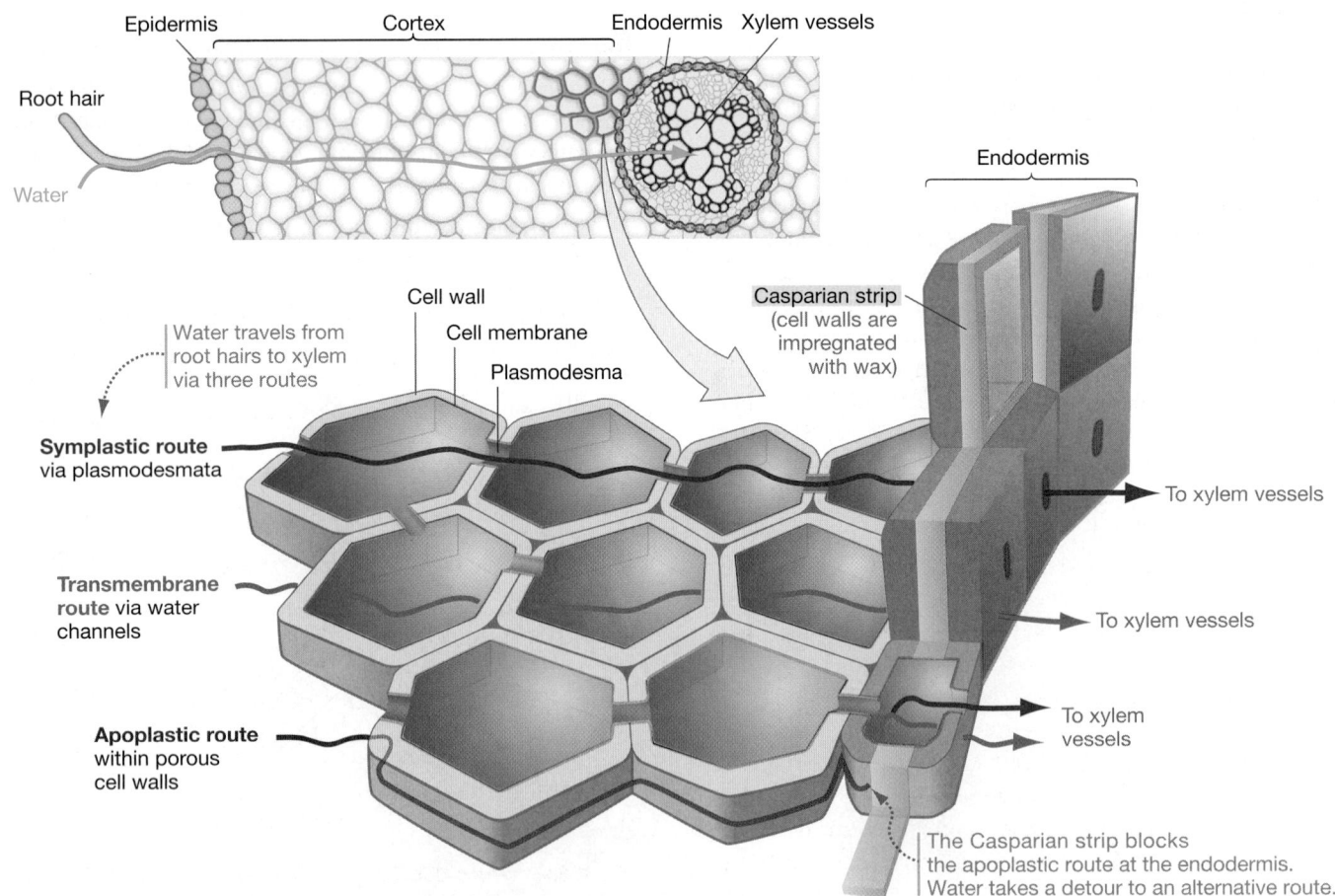

FIGURE 38.7 How Water Travels from Root Hair to Xylem.

✔**EXERCISE** Add dots representing aquaporins in the transmembrane route.

proceeding to the vascular tissue (Figure 38.7). The Casparian strip does not affect water and ions that move through the symplastic route.

The Casparian strip is important because it means that for water and solutes to reach vascular tissue, they have to move into the cytoplasm of an endodermal cell. Endodermal cells, in turn, act as filters.

By forcing water and ions to cross at least two membranes on their way from the soil to the xylem, one entering the symplast outside the Casparian strip and one leaving the symplast inside the Casparian strip, plants can use specific channel and carrier proteins (see Chapter 6) to control what moves to the shoots. Endodermal cells allow ions such as potassium (K^+) that are needed by the plant to pass through to the vascular tissue. In contrast, these cells can prevent the passage of ions such as sodium (Na^+) or heavy metals that are not needed or may be harmful.

✔ If you understand this concept, you should be able to predict what would happen if a plant had a mutation that prevented synthesis of suberin and formation of the Casparian strip.

Water Movement via Root Pressure

Have you ever walked barefoot through grass early in the morning and gotten your feet wet? Where does that water come from? Let's explore this phenomenon.

Movement of ions and water into the root xylem is responsible for the process known as **root pressure.** Recall that root pressure is one of three hypothesized mechanisms for moving water up xylem, from root to shoot.

The Casparian strip in endodermal cells is essential for root pressure to develop. Without an apoplastic barrier between the xylem and the environment, ions and water would simply leak out of roots.

Stomata normally close during the night, when photosynthesis is not occurring and CO_2 is not needed. Their closure minimizes water loss and slows the movement of water through plants. But roots often continue to accumulate ions that their epidermal cells acquire from the soil as nutrients, and these nutrients are actively pumped into the xylem. The influx of ions lowers the water potential of xylem below the water potential in the surrounding cells.

As water flows into xylem from other root cells in response to the solute gradient, a positive pressure is generated at night that forces fluid up the xylem. More water moves up xylem and into leaves than is being transpired from the leaves.

In low-growing plants, enough water can move to force water droplets out of the leaves, a phenomenon known as **guttation.** Some of the water that gets your feet wet early in the morning spent part of the night moving through plants (**FIGURE 38.8**).

At one time, positive root pressure formed the basis of a leading hypothesis to explain how water moves from roots to leaves in trees. However, research showed that over long distances, such as the height of a tree, the force of root pressure is not enough to overcome the force of gravity on the water inside xylem.

FIGURE 38.8 Root Pressure Causes Guttation. When ions accumulate in the xylem of roots at night, enough water enters the xylem via osmosis to force water up and out of low-growing leaves.

In addition, researchers demonstrated that cut stems, which have no contact with the root system, are still able to transport water to leaves. Biologists concluded that there must be some other mechanism involved in the long-distance transport of water. What is it?

Water Movement via Capillary Action

Researchers have also evaluated a hypothesis based on the phenomenon of **capillarity** (also called capillary action), or movement of water up a narrow tube. When a thin glass tube is placed upright in a pan of water, water creeps up the tube (**FIGURE 38.9**). The movement occurs in response to three forces: (**1**) adhesion, (**2**) cohesion, and (**3**) surface tension. Let's consider each force in turn.

Adhesion is a molecular attraction among unlike molecules. In this case, water interacts with a solid substrate—such as the glass walls of a capillary tube or the cell walls of tracheids or vessel elements—through hydrogen bonding. As water molecules

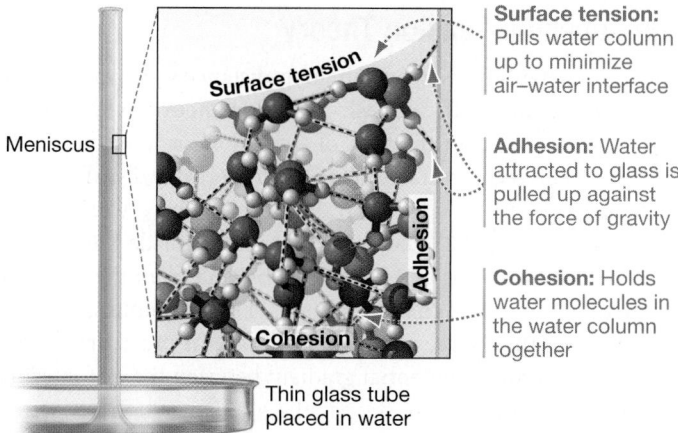

FIGURE 38.9 Water Can Rise in Xylem via Capillarity. Capillary action occurs through a combination of three forces, all of which are generated by hydrogen bonding and together are acting against gravity.

bond to each other and adhere to the side of the tube, they are pulled upward.

Cohesion is a molecular attraction among like molecules, such as the hydrogen bonding that occurs among molecules in water. Because water molecules cohere, the upward pull by adhesion is transmitted to the rest of the water column. The water column rises against the pull of gravity.

The effects of adhesion, cohesion, and gravity are responsible for the formation of a concave surface boundary called a **meniscus** (plural: **menisci**). A meniscus forms at most air–water interfaces—including those found in narrow tubes. Menisci form because adhesion and cohesion pull water molecules up along the sides of the tube, while gravity pulls the water surface down in the middle.

Surface tension is a force that exists among water molecules at an air–water interface. In the body of a water column, all the water molecules present are surrounded by other water molecules and form hydrogen bonds in all directions. Water molecules at the surface, however, can form hydrogen bonds only with the water molecules beside and below them. As a result, surface molecules share stronger attractive forces. This enhanced attraction results in tension that minimizes the total surface area. When a meniscus forms, surface tension exerts a pull against gravity.

Capillary action results when adhesion creates an upward pull at the water–container interface, surface tension creates an upward pull all across the surface, and cohesion transmits both forces to the water below. All three forces counteract the effect of gravity, and the result is capillarity.

Like root pressure, capillarity can transport water only a limited distance. Capillary action moves water along the surfaces of mosses and other low-growing, non-vascular plants; but it can raise the water in the xylem of a vertical stem only about 1 m.

Thus, root pressure and capillary action cannot explain how water moves from soil to the top of a redwood tree—an organism that can grow 5 to 6 stories higher than the Statue of Liberty. How then does it actually happen?

The Cohesion-Tension Theory

The leading hypothesis to explain long-distance water movement in vascular plants is the **cohesion-tension theory,** which states that water is pulled to the tops of trees along a water-potential gradient, via forces generated by transpiration at leaf surfaces.

To understand how the cohesion-tension force works, start with step 1 in **FIGURE 38.10**. Notice that spaces inside the leaf are filled with moist air as a result of evaporation from the surfaces of surrounding cells. When a stoma opens, this humid air is exposed to the atmosphere, which in most cases is much drier. This creates a steep water-potential gradient between the leaf interior and its surroundings. The steeper the gradient, the faster water vapor diffuses out through the stomata.

Step 2 shows that as water is lost from the leaf to the atmosphere, the humidity of the gas-filled space inside the leaf drops. In response, more water evaporates from the walls of the parenchyma cells. At the microscopic level, cell walls are not perfectly smooth, so menisci form at the air–water interface. As more water evaporates, the menisci become steeper and the total area of the air–water interface increases.

In 1894 Henry Dixon and John Joly hypothesized that the formation of steep menisci produces a force capable of pulling water up from the roots, dozens or hundreds of meters into the air. Is this really possible?

The Role of Surface Tension in Water Transport The key concept in the cohesion-tension theory is that the negative force or pull (tension) generated at the air–water interface is transmitted through the water outside of leaf cells (step 3 in Figure 38.10), to the water in xylem (step 4), to the water in the vascular tissue of roots (step 5), and finally to the water in the soil (step 6).

The transmission of pulling force from the leaf surface to the root is possible because (**1**) there are continuous columns of water throughout the plant, and (**2**) all of the water molecules present hydrogen bonds to one another (cohesion).

Note that the plant does not expend energy to create the pulling force. The force is generated by energy from the Sun, which drives evaporation from the leaf surface. Water transport is solar powered.

In effect, the cohesion-tension theory of water movement states that, because of the hydrogen bonding between water molecules, water is pulled up through xylem in continuous columns.

Creating a Water Potential Gradient You can also think about the cohesion-tension theory in terms of water potentials. Note that because tracheids and vessels are dead at maturity, the water in xylem does not cross plasma membranes. As a result, water does not move between cells by osmosis. In xylem, water movement is driven entirely by differences in pressure potential. The water in a column of xylem cells moves by **bulk flow**—a mass movement of molecules along a pressure gradient.

The pulling force generated at menisci in leaf cell walls lowers the pressure potential of water in leaves. Even though the tension created at each meniscus is relatively small, there are so many menisci in the leaves of an entire plant that the tension created by summing many small pulling forces is remarkable. It creates a water-potential gradient between leaves and roots that is steep enough to overcome the force of gravity and pull water up long distances.

To appreciate just how great the forces involved are, think of the vessel elements or tracheids in xylem as groups of drinking straws. When you use a straw, the vacuum that you create causes liquid to rise. Sucking on a drinking straw creates a pressure difference of about -0.1 MPa, which can draw water up a maximum of about 10 m. In contrast, the negative pressure exerted by the menisci in leaves can be as low as -2.0 MPa. The tension in xylem tissue may be 10 times the amount of pressure on a fully inflated car tire. (But note that the pressure on a car tire is positive, not negative.) The force is enough to draw water up 100 m.

The Importance of Secondary Cell Walls If you suck on a drinking straw hard enough, the pressure gradient between the

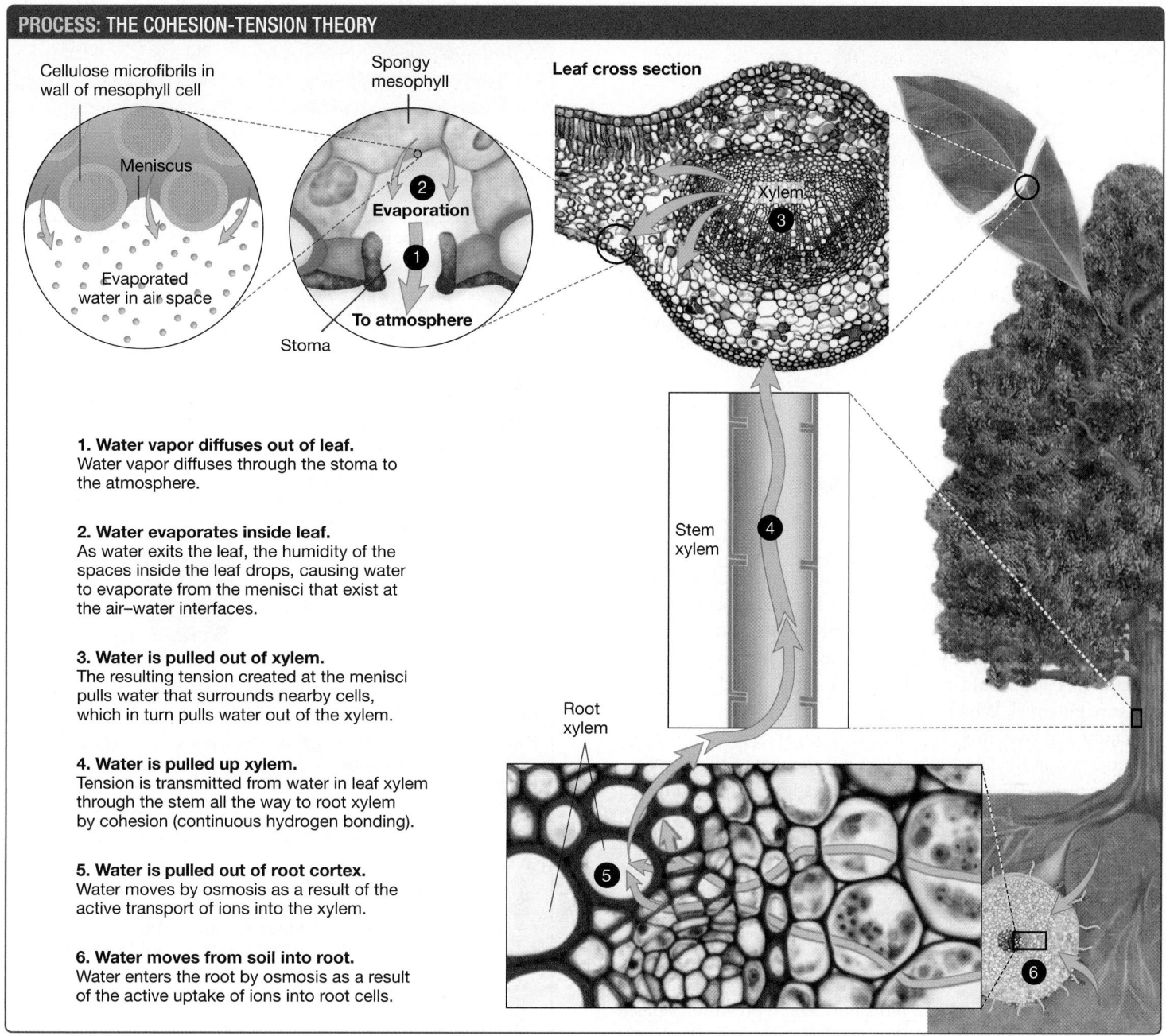

PROCESS: THE COHESION-TENSION THEORY

Cellulose microfibrils in wall of mesophyll cell

Meniscus

Evaporated water in air space

Spongy mesophyll

Evaporation

To atmosphere

Stoma

Leaf cross section

Xylem

Stem xylem

Root xylem

1. Water vapor diffuses out of leaf.
Water vapor diffuses through the stoma to the atmosphere.

2. Water evaporates inside leaf.
As water exits the leaf, the humidity of the spaces inside the leaf drops, causing water to evaporate from the menisci that exist at the air–water interfaces.

3. Water is pulled out of xylem.
The resulting tension created at the menisci pulls water that surrounds nearby cells, which in turn pulls water out of the xylem.

4. Water is pulled up xylem.
Tension is transmitted from water in leaf xylem through the stem all the way to root xylem by cohesion (continuous hydrogen bonding).

5. Water is pulled out of root cortex.
Water moves by osmosis as a result of the active transport of ions into the xylem.

6. Water moves from soil into root.
Water enters the root by osmosis as a result of the active uptake of ions into root cells.

FIGURE 38.10 Transpiration Creates Tension That Is Transmitted from Leaves to Roots.

inside of the straw and the atmosphere can overcome the stiffness of the straw and cause it to collapse. How can vascular tissue withstand negative pressures as low as −2.0 MPa without collapsing?

The answer is found in a key adaptation: the secondary thickenings characteristic of the cell walls in tracheids and vessel elements. The cells in vascular tissue have walls that are reinforced with tough lignin molecules (see also Chapter 31).

The evolution of lignified secondary cell walls was an important event in the evolution of land plants because it allowed vascular tissue to withstand extremely negative pressures. The result? Tall trees.

What Evidence Do Biologists Have for the Cohesion-Tension Theory? If the cohesion-tension theory is correct, the water present in xylem should experience a strong pulling force. A simple experiment supports this prediction.

If you find a leaf that is actively transpiring and cut its petiole, the watery fluid in the xylem, or xylem sap, withdraws from the edge toward the inside of the leaf (**FIGURE 38.11**, see page 764). According to the cohesion-tension theory, this observation is due to a transpirational pull at the air–water interface in parenchyma cells of leaves.

Although the observation that xylem sap is under tension was important, the cohesion-tension theory remained controversial.

Petiole

When you cut a petiole, watery fluid (xylem sap) withdraws from the cut surface

Pull

Pull

FIGURE 38.11 Xylem Sap in Cut Stems "Snaps Back."

For example, several studies failed to document rapid changes in xylem pressure when temperature and humidity—and thus the severity of the water-potential gradient—were changed experimentally. Advocates of the theory blamed the negative results on instruments that could not document small, rapid changes in pressure potential. Who was right?

Chunfang Wei, Melvin Tyree, and Ernst Steudle answered the question in the late 1990s with an instrument called a xylem pressure probe (**FIGURE 38.12**). A xylem pressure probe includes an oil-filled glass tube that can be inserted directly into the xylem of a leaf. The oil transmits changes in pressure within the xylem to a gauge, allowing researchers to record changes in xylem pressure instantly and directly.

To test the cohesion-tension theory, the researchers altered xylem pressure by raising light levels to alter transpiration rates in the leaves of corn plants. Their results?

- The graph in Figure 38.12 plots how xylem pressure changed as the researchers changed light levels over a 7-minute period. Note that as light intensity increased, the xylem pressure probe documented increased tension, or pull—negative pressure.

- In addition, higher light levels reduced the weight of the entire plant—suggesting that higher transpiration rates caused water loss.

Both observations are consistent with predictions that follow from the cohesion-tension theory.

These experiments convinced most biologists that increased transpiration leads to increased tension on xylem sap. Rising tension, in turn, lowers the water potential of leaves and exerts a pull on water in the roots, where the water potential is high. On the basis of these and other results, most biologists now accept the cohesion-tension theory.

RESEARCH

QUESTION: Do direct measurements of pressure in xylem tissue support the cohesion-tension theory?

HYPOTHESIS: Increasing transpiration by raising light intensity will lower xylem pressure in leaves.

NULL HYPOTHESIS: Increasing light intensity will not affect xylem pressure in leaves.

EXPERIMENTAL SETUP:

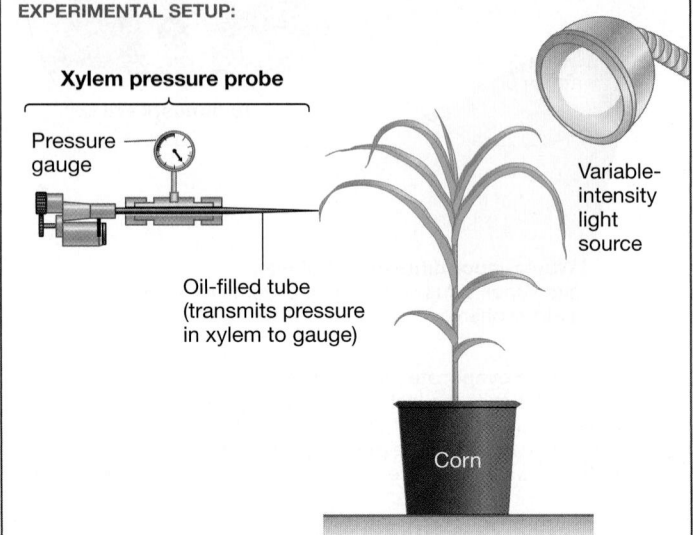

Xylem pressure probe

Pressure gauge

Oil-filled tube (transmits pressure in xylem to gauge)

Variable-intensity light source

Corn

PREDICTION: Xylem pressure will decrease as transpiration increases with higher light levels.

PREDICTION OF NULL HYPOTHESIS: Xylem pressure will not be affected by light intensity.

RESULTS:

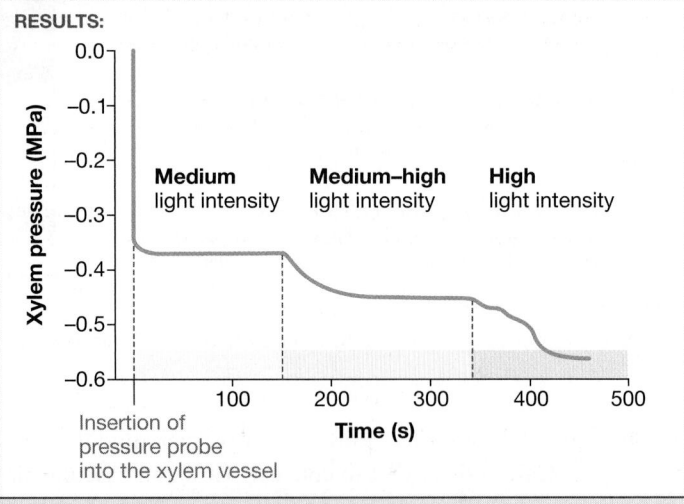

Medium light intensity

Medium–high light intensity

High light intensity

Insertion of pressure probe into the xylem vessel

CONCLUSION: Xylem pressure decreases when light intensity increases. The data support the cohesion-tension theory.

FIGURE 38.12 Measuring Changes in Pressure inside Xylem.

SOURCE: Wei, C., M. T. Tyree, and E. Steudle. 1999. Direct measurement of xylem pressure in leaves of intact maize plants. A test of the cohesion-tension theory taking hydraulic architecture into consideration. *Plant Physiology* 121: 1191–1205.

✓**QUANTITATIVE** Suppose the researchers had chosen to plot changes in the water-potential gradient between the xylem in roots and leaves on the *y*-axis, during the same experiment. What would the graph look like?

If you understand that . . .

- Water can move a short distance in xylem via root pressure or capillarity.
- Long-distance transport of water depends on movement along a steep water-potential gradient. This gradient is created primarily by the negative pressure potential of water in leaves, due to surface tension that develops in response to transpiration.

✔ **You should be able to . . .**

1. Explain how transpiration at the surface of a meniscus near a stoma creates a pull on the water in leaves and xylem.

2. Predict what happens to the meniscus when each of the following occurs: A nearby stoma closes, a rain shower starts, and weather changes as dry air blows in.

Answers are available in Appendix A.

38.3 Water Absorption and Water Loss

One of the most important features of the cohesion-tension theory is that it does not require plants to expend energy to lift the water that the roots take in by osmosis. To absorb water, roots must expend energy to accumulate the ions that maintain a solute potential that is lower than the soil. Once the water with dissolved nutrients is in the root xylem, the Sun furnishes the energy required to pull the solution from roots to shoots—not ATP supplied by the plant.

1. Energy from the Sun heats water molecules at the air–water interface inside leaves enough to break the hydrogen bonds between them and cause transpiration.

2. Rapid transpiration creates deep menisci in the walls of leaf cells, causing tension that lowers the water potential of leaves.

3. Hydrogen bonding between water molecules transmits this tension down to water molecules in the roots.

Xylem acts as a passive conduit—a set of narrow pipes that allows water to move from a region of high water potential (the roots) to a region of low water potential (the leaves). Water flows from roots to shoots as long as the water-potential gradient—from root to leaf to atmosphere—is intact.

When soils begin to dry, however, it becomes difficult for plants to replace water being lost via transpiration. If water is not replaced fast enough, the solute potentials of leaves drop and leaves and branches begin to wilt. In response, stomata may close down partially or completely to reduce transpiration rates and conserve water. However, closing stomata affects the ability of plants to carry on photosynthesis, because CO_2 acquisition slows or stops.

The balance between conserving water and maximizing photosynthesis is termed the photosynthesis-transpiration compromise.

This compromise is particularly delicate for species that grow on dry sites. How do they cope?

Limiting Water Loss

Plants that thrive in dry sites have several adaptations that help them slow transpiration and limit water loss. Consider the oleander plant, which is native to the dry shrub-grassland habitats of southern Eurasia. The micrograph in **FIGURE 38.13** shows a cross section through an oleander leaf, which has the following special features:

- A particularly thick cuticle covers the upper surface of oleander leaves. This waxy layer minimizes water loss from cells that are directly exposed to sunlight. In general, species that are adapted to dry soils have much thicker cuticles than do species adapted to wet soils.

- The stomata of oleanders are located on the undersides of their leaves, inside deep pits in the epidermis. Hairlike extensions of epidermal cells called **trichomes** (see Section 37.3) shield these pits from the atmosphere. The leading hypothesis to explain these traits is that they slow the loss of water vapor from stomata to the dry air surrounding the leaf by creating a layer of still air surrounding the stomata.

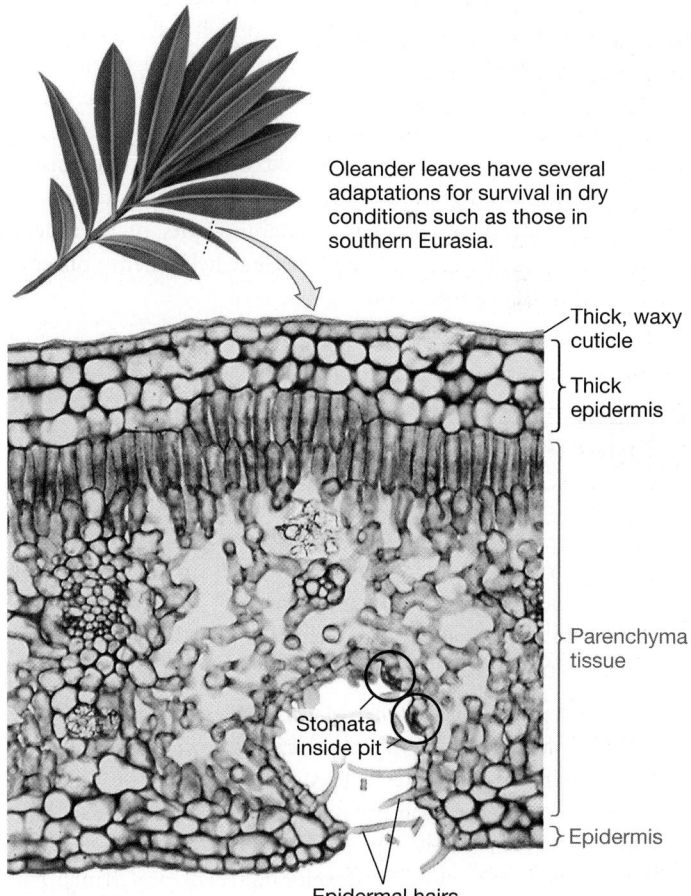

Oleander leaves have several adaptations for survival in dry conditions such as those in southern Eurasia.

Thick, waxy cuticle

Thick epidermis

Parenchyma tissue

Stomata inside pit

Epidermis

Epidermal hairs

FIGURE 38.13 Species That Are Adapted to Dry Habitats Have Modified Leaf Structures.

Other species have other adaptations for limiting water loss. For example, recall that many species adapted to water-limited habitats—either cold environments where water is often frozen or deserts where rainfall is rare—have needlelike leaves (see Chapter 37). Long, thin leaf shapes minimize the surface area exposed to sunlight and thus minimize transpiration. Plants adapted to wetter habitats, in contrast, tend to have broad leaves with a large surface area.

Obtaining Carbon Dioxide under Water Stress

Many of the species that thrive in deserts and other hot, dry habitats can continue photosynthesizing even when soil moisture content is low. Recall that two novel biochemical pathways, **crassulacean acid metabolism (CAM)** and **C_4 photosynthesis,** allow plants to increase CO_2 concentrations in their leaves and conserve water (see Chapter 10).

CAM plants open their stomata at night and store the CO_2 that diffuses into their tissues by adding the carbon dioxide molecules to organic acids. When sunlight is available during the day and photosynthesis begins, the CO_2 molecules are released from the organic acids and converted to sugar by **rubisco**—the enzyme, found in all green plants, that initiates the Calvin cycle. In this way, CAM plants can photosynthesize and grow even with their stomata closed during the day.

C_4 plants minimize the extent to which their stomata open because they use CO_2 so efficiently. Mesophyll cells in C_4 plants take up CO_2 and add it to organic acids. The CO_2 is then transferred to specialized cells called **bundle-sheath cells** (see Chapter 10), where the Calvin cycle operates. In effect, the C_4 cycle is a mechanism for concentrating carbon dioxide in cells deep inside the leaf, so stomata do not have to be wide open continuously.

Like the cuticle and stomata-containing pits of oleanders, CAM and C_4 plants have adaptations that help them conserve water by limiting transpiration.

38.4 Translocation of Sugars

Translocation is the movement of sugars by bulk flow throughout a plant—specifically, from sources to sinks. In vascular plants, a **source** is a tissue where sugar enters the phloem; a **sink** is a tissue where sugar exits the phloem.

Where do sources and sinks occur in a plant? The answer often depends on the time of year.

- *During the growing season* Mature leaves and stems that are actively photosynthesizing produce sugar in excess of their own needs. These tissues act as sources. Sugar moves from leaves and stems to a variety of sinks, where sugar use is high and production is low. Apical meristems, lateral meristems, developing leaves, flowers, developing seeds and fruits, and storage cells in roots all act as sinks (**FIGURE 38.14**).

- *Early in the growing season* When a plant resumes growth after the winter or the dry season, sugars move from storage

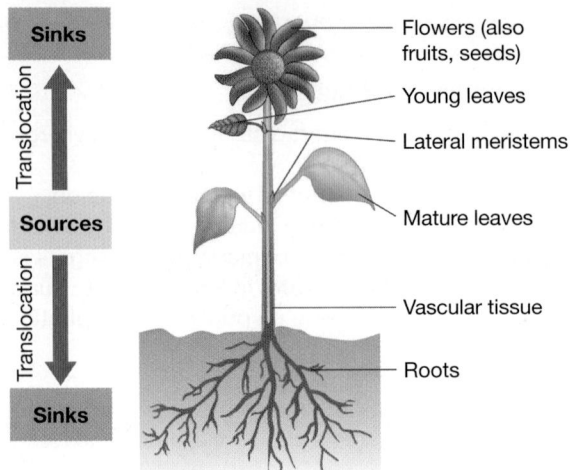

FIGURE 38.14 Sugars Move from Sources to Sinks.

areas to growing areas. Storage cells in roots and seeds act as sources; developing leaves act as sinks.

Tracing Connections between Sources and Sinks

To explore the relationship between sources and sinks in more detail, consider research on sugar-beet plants that were exposed to carbon dioxide molecules containing the radioactive isotope ^{14}C. The goal was to track where carbon atoms moved after they were incorporated into sugars via photosynthesis.

The location of ^{14}C atoms inside a plant can be documented in two ways: (1) by measuring the amount of radioactive CO_2 emanating from different tissues, or (2) by laying plant parts on X-ray film and allowing the radioactivity to expose and blacken the film.

Many researchers have enclosed individual leaves of intact plants in a bag and introduced a known amount of radioactive CO_2 for a short period of time. One typical experiment documented that mature leaves retained just over 9 percent of the labeled carbon. In contrast, growing leaves retained 67 percent. These data are consistent with the prediction that fully expanded leaves act as sources of sugar, while actively growing leaves and roots act as sinks.

In similar experiments, researchers have exposed all the leaves on a growing plant to labeled carbon. One such experiment found that over 16 percent of the total carbon was translocated to root tissue within 3 hours. This result is consistent with the prediction that, during the growing season, roots also act as sinks.

Similar experiments support two generalizations.

1. Sugars can be translocated rapidly—typically 50–100 centimeters per hour (cm/hr).

2. There is a strong correspondence between the physical locations of sources and sinks.

The second point is particularly interesting. For example, mature leaves that act as sources send sugar to tissues on the same side of the plant (**FIGURE 38.15a**). In addition, experiments with tall herbaceous plants show that leaves on the upper part of the stem send sugar to apical meristems, but leaves on the lower part of the plant send sugar to the roots (**FIGURE 38.15b**).

Why would leaves send sugar to tissues on a certain side or part of the body? The answer hinges on understanding the structure of phloem.

The Anatomy of Phloem

Two specialized parenchyma cell types make up phloem: **sieve-tube elements** and **companion cells** (see Chapter 37). Unlike the tracheids and vessel elements that make up most of the xylem, sieve-tube elements and companion cells are alive at maturity.

Recall that, in most plants, sieve-tube elements lack nuclei and most other organelles. They are connected to one another, end to end, by perforated **sieve plates** (**FIGURE 38.16**). The pores create a direct connection between the cytoplasms of adjacent cells.

(a) Source leaves send sugar to the same side of the plant.

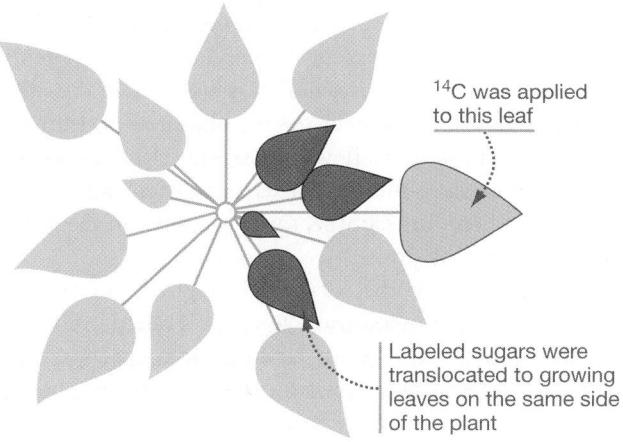

¹⁴C was applied to this leaf

Labeled sugars were translocated to growing leaves on the same side of the plant

(b) Source leaves send sugar to tissues on the same end of the plant.

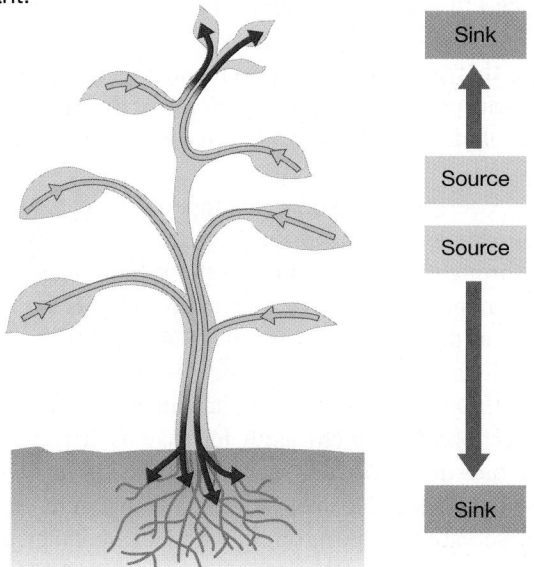

Sink

Source

Source

Sink

FIGURE 38.15 Sources Supply Sinks on the Same Side and Same End of the Body.

Companion cells, in contrast, have nuclei and a rich assortment of ribosomes, mitochondria, and other organelles. Companion cells function as "support staff" for sieve-tube elements.

You might also recall that the phloem in primary and secondary vascular tissue is continuous throughout the plant—meaning that there is a direct anatomical connection between the phloem from the tips of shoots to the tips of roots (see Chapter 37). The sieve-tube elements in phloem represent a continuous system for transporting sugar throughout the plant body.

Each vascular bundle runs the length of stems and roots, and certain bundles extend into specific branches, leaves, and lateral roots. Phloem sap does not move from one vascular bundle to another—instead, each bundle is independent.

Based on these results, the physical relationships observed between sources and sinks in herbaceous plants are logical. For example, the phloem in the leaves on one side of a plant connects directly with the phloem of branches, stems, and roots on the same side of the individual, through a specific set of vascular bundles.

The phloem sap that flows through vascular tissue is often dominated by the disaccharide **sucrose**—table sugar. Phloem sap can contain small amounts of minerals, amino acids, mRNAs, hormones, and other compounds as well. How does this solution move? What mechanism is responsible for translocating sugars from sources to sinks?

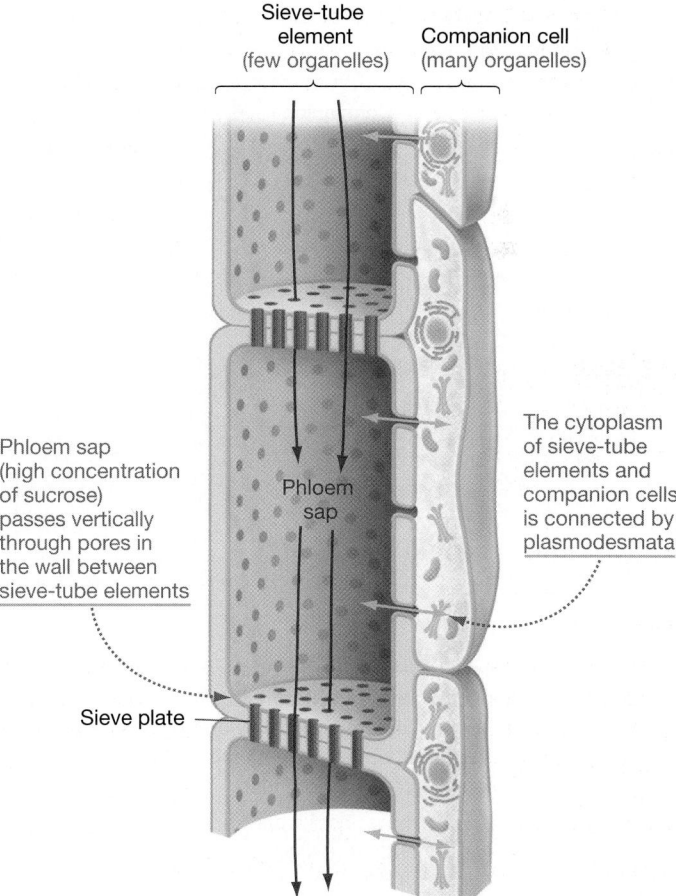

Sieve-tube element (few organelles)

Companion cell (many organelles)

Phloem sap (high concentration of sucrose) passes vertically through pores in the wall between sieve-tube elements

Phloem sap

The cytoplasm of sieve-tube elements and companion cells is connected by plasmodesmata

Sieve plate

FIGURE 38.16 Sieve-Tube Elements Are Connected by Pores.

The Pressure-Flow Hypothesis

In 1926 Ernst Münch proposed the **pressure-flow hypothesis,** which states that events at source tissues and sink tissues create a pressure potential gradient in phloem (**FIGURE 38.17**). The water in phloem sap moves down this pressure gradient, and sugar molecules are carried along by bulk flow.

Like the cohesion-tension theory for water transport in xylem, the pressure-flow hypothesis is based on movement along a water-potential gradient created by changes in pressure potential. Unlike the cohesion-tension model, however, transpiration does not provide the driving force to move phloem sap. Instead, differences between turgor pressure in the phloem near source tissues and turgor pressure in the phloem near sink tissues generate

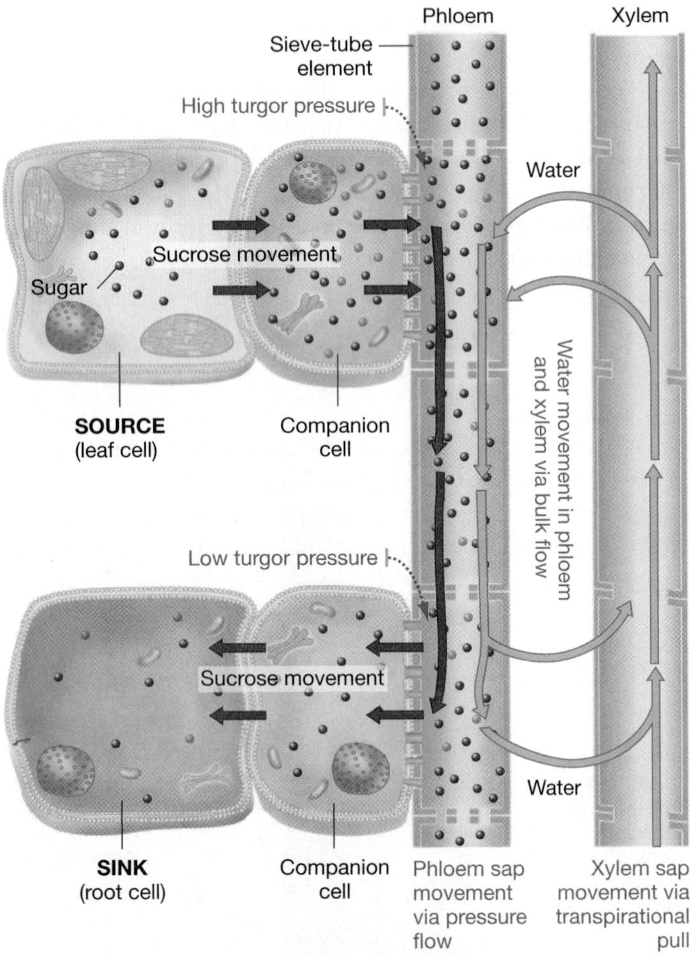

FIGURE 38.17 The Pressure-Flow Hypothesis: High Turgor Pressure Near Sources Causes Phloem Sap to Flow to Sinks. The pressure-flow hypothesis predicts that water cycles between xylem and phloem, and that water movement in phloem, is a response to a gradient in pressure potential.

✔ **EXERCISE** This diagram shows how water moves between xylem and phloem in the middle of the growing season, when leaves are sources and roots are sinks. Add new arrows, in new colors, to indicate the direction of water and phloem sap flow in spring, when roots act as sources and leaves act as sinks.

the necessary force. Creating these differences in turgor pressure usually requires an expenditure of ATP.

Creating High Pressure Near Sources and Low Pressure Near Sinks

To understand how Münch's model works, start with the source cell at the upper left in Figure 38.17. The small red arrows reflect Münch's proposal that sucrose moves from source cells into companion cells and from there into sieve-tube elements.

Because of this phloem loading, the phloem sap in the source tissue has a high concentration of sucrose. Compared to the water in the adjacent xylem cells in a vascular bundle, the phloem sap has a very low solute potential.

As the blue arrows in the upper part of the diagram show, water moves along a water-potential gradient—flowing passively from xylem across the selectively permeable plasma membrane of sieve-tube elements. In response, turgor pressure begins to build in the sieve-tube elements in the source region.

What is happening at the sink? Münch proposed that cells in the sink (bottom left in Figure 38.17) remove sucrose from the phloem sap by passive or active transport. As a result of this phloem unloading—a loss of solutes—the water potential in sieve-tube elements increases until it is higher than the water potential in adjacent xylem cells. As the blue arrows at the bottom of the figure show, water flows across the selectively permeable membranes of sieve-tube elements into xylem along a water-potential gradient. In response, turgor pressure in the sieve-tube elements in the sink drops.

The net result of these events is high turgor pressure in phloem at a source tissue and low turgor pressure in phloem at the sink, created by the loading and unloading of sugars, respectively. This difference in pressure potential drives phloem sap from source to sink via bulk flow. There is a one-way flow of sucrose and a continuous loop of water movement. Water returns to the source tissue via the xylem.

Testing the Pressure Flow Model

The pressure-flow hypothesis is logical, given the anatomy of vascular tissue and the principles that govern water movement. But has any experimental work supported the theory? Some of the best tests have relied on aphids—small insects that make their living ingesting phloem sap.

Aphids insert a syringe-like mouthpart, called a stylet, into sieve-tube elements. The pressure on the fluid in these cells forces it through the stylet and into the aphid's digestive tract. Excess water and sucrose that the aphid does not need is excreted out its anus as droplets of "honeydew" (**FIGURE 38.18**).

If the aphids are then severed from their stylets, sap continues to flow out through the stylets. This phenomenon allows researchers to collect phloem sap efficiently for analysis. It also confirms that the aphids do not actively suck the fluid. As predicted by the pressure-flow model, phloem is indeed under pressure.

This observation supports one of the fundamental predictions of the pressure-flow hypothesis. Now the question is, How does sucrose enter and leave phloem in a way that sets up the water-potential gradient?

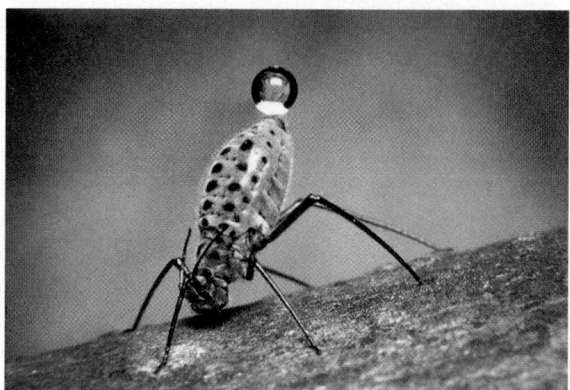

FIGURE 38.18 **Aphids Feed on Phloem Sap.** The tip of this aphid's mouthpart (the stylet) is in a sieve-tube element within the plant stem. The droplet emerging from the aphid's anus is honeydew, which consists of sugary phloem sap.

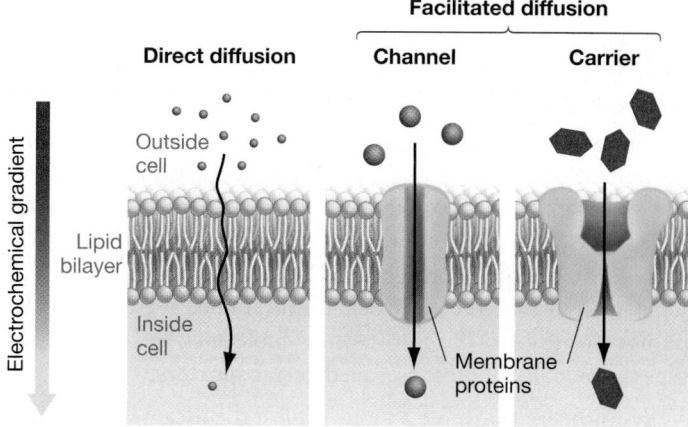

FIGURE 38.19 **Passive Transport Is Based on Diffusion.** In passive transport, ions or molecules diffuse across membranes—meaning they follow their electrochemical gradient. The movement can occur directly through the phospholipid bilayer or be facilitated by a channel or carrier protein.

Phloem Loading

In contrast to the cohesion-tension model of water movement in xylem, pressure flow often requires that plants expend energy to set up a water-potential gradient in phloem. To establish a high-pressure potential in sieve-tube elements near source cells, large amounts of sugar have to be transported into the phloem sap—enough to raise the solute concentration of sieve-tube elements. This requirement is illustrated in Figure 38.17, top left.

In some cases, loading is active—it requires an expenditure of ATP and some sort of membrane transport system. But when sucrose concentrations in source cells are extremely high, movement of sucrose into sieve-tube elements can also occur via passive diffusion through plasmodesmata.

Phloem unloading at sinks can also be active or passive, depending on the tissue. When sugar is unloaded against its concentration gradient, an expenditure of ATP and a second membrane transport mechanism is required. How do phloem loading and unloading occur? What specific membrane proteins are involved?

To answer these questions, let's start by reviewing how transport proteins make it possible for sugars and other large or charged substances to cross a phospholipid bilayer.

How Are Sucrose and Other Solutes Transported across Membranes?
Passive transport (**FIGURE 38.19**) occurs when ions or molecules move across a plasma membrane by diffusion—that is, along their electrochemical gradient. The adjective passive is appropriate because no expenditure of energy is required for the movement to occur.

Recall that small, uncharged molecules diffuse across phospholipid bilayers rapidly (Figure 38.19, left). But ions and many large molecules diffuse across phospholipid bilayers slowly if at all, even when their movement is favored by a strong electrochemical gradient. To diffuse rapidly, they must avoid direct contact with the phospholipid bilayer by passing through a membrane protein.

Two types of membrane protein—channels and carriers—facilitate the passive diffusion of specific ions or molecules (see Chapter 6).

- **Channel proteins** form pores that selectively admit certain ions (Figure 38.19, middle).

- **Carrier proteins** work like enzymes, undergoing a conformational change that transports a bound molecule across the lipid bilayer (Figure 38.19, right).

Channels and carriers are responsible for **facilitated diffusion**.

Active transport (**FIGURE 38.20**) occurs when ions or molecules move across a plasma membrane against their electrochemical gradient. The adjective active is appropriate because cells must expend energy in the form of ATP to move solutes in an energetically unfavorable direction.

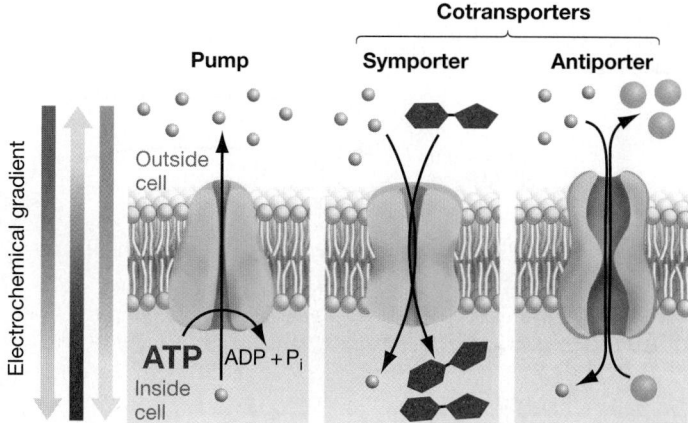

FIGURE 38.20 **Active Transport Moves Ions or Molecules against an Electrochemical Gradient.** All forms of active transport require an expenditure of ATP. Pumps use ATP directly; cotransport proteins use ATP indirectly. Cotransport depends on a previous expenditure of ATP by a pump.

Active transport always involves membrane proteins. **Pumps** are proteins that change shape when they bind ATP or a phosphate group from ATP. As they move, pumps transport ions or molecules against an electrochemical gradient.

Pumps establish an electrochemical gradient that favors the movement of an ion or molecule across the plasma membrane. For example, the pump on the left side of Figure 38.20 has established an electrochemical gradient for bringing the ions or molecules symbolized by the light gray balls into the cell.

In many cases, the electrochemical gradients established by pumps are used to transport other molecules or ions by two types of membrane proteins called **cotransporters.**

- **Symporters** transport solutes *against* a concentration gradient, using the energy released when a different solute moves in the same direction *along* its electrochemical gradient. The red molecules in the middle of Figure 38.20 are moving through a symporter.

- **Antiporters** work in a similar way, except that the solute being transported against its concentration gradient moves in the direction *opposite* that of the solute moving down its concentration gradient. The orange ions in Figure 38.20, right, are moving through an antiporter.

When solutes move through cotransport proteins, **secondary active transport** occurs.

Active transport, secondary active transport, and passive transport are all involved in moving sugars around plants. Let's look first at events at source tissues, where the active transport of sucrose into sieve-tube elements results in a high-pressure potential. The chapter concludes with a look at how the same molecules are unloaded to maintain low-pressure potentials at sinks.

How Are Sugars Concentrated in Sieve-Tube Elements at Sources?

Because sucrose may be more highly concentrated in companion cells than in photosynthetic cells where it is produced,

researchers hypothesized that sucrose transport from source cells into companion cells may be active. Another key observation—that strong pH differences exist between the interior and exterior of phloem cells—suggested that sucrose might enter companion cells with protons, by secondary active transport.

FIGURE 38.21 explains the logic behind this hypothesis. Note this key claim: A membrane protein in companion cells hydrolyzes ATP and uses the energy that is released to transport protons (H^+) across the membrane to the exterior of the cell. Proteins like these are called **proton pumps,** or more formally, **H^+-ATPases.**

Proton pumps establish a large difference in charge and in hydrogen ion concentration on the two sides of the membrane. The resulting electrochemical gradient favors the entry of protons into the cell.

The right side of the figure shows the second key claim: A symporter acts as a conduit for protons and sucrose to enter the cell together. With this symporter, protons move along their electrochemical gradient; sucrose moves against its concentration gradient.

If phloem loading depends on the activity of a proton–sucrose symporter, a mutation in the gene encoding this protein should result in a plant that cannot export sucrose from leaves to various sink tissues. What would this mutant plant look like?

Michael Sussman's lab used the model plant *Arabidopsis thaliana* (see **BioSkills 13** in Appendix B) to isolate a mutant lacking a gene called *SUC2*. This gene was known to be expressed in vascular tissue and to encode a proton–sucrose symporter.

Mutant plants grew much more slowly than wild-type plants, indicating that something was indeed wrong with the mutant (**FIGURE 38.22a**). When the researchers looked at leaves under a microscope, they found that chloroplasts from mutant plants accumulated huge amounts of starch, an indication that they could not efficiently export sugar to the rest of the plant (**FIGURE 38.22b**). To test this hypothesis, the researchers then applied radioactive sucrose to a single source leaf on mutant and wild-type plants and measured where it went 5 hours later.

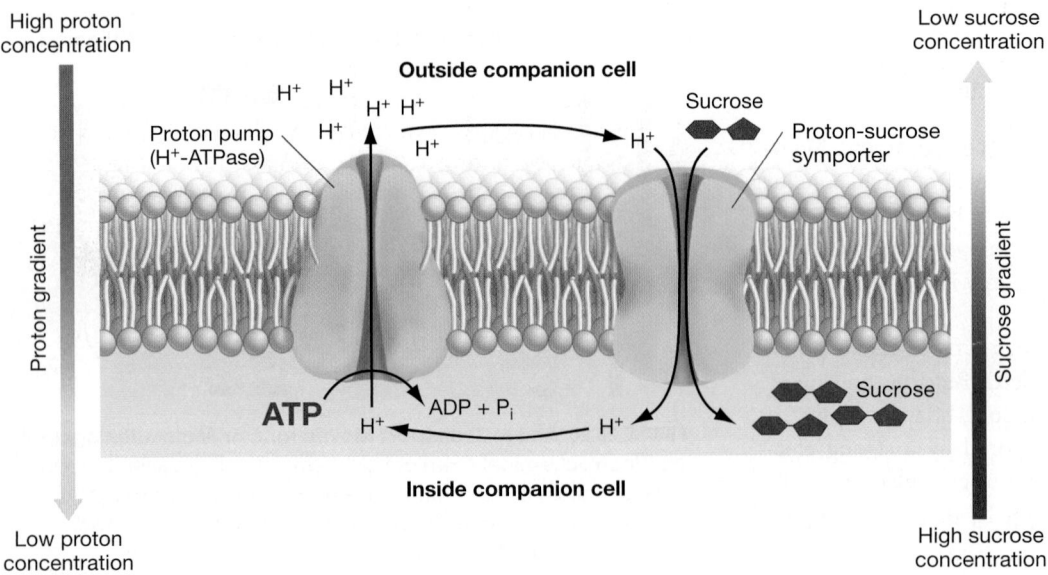

FIGURE 38.21 A Model for Cotransport of Protons and Sucrose. According to the model of cotransport, a proton pump hydrolyzes ATP to move hydrogen ions to the exterior of the cell. The resulting high concentration of H^+ outside the cell establishes an electrochemical gradient that allows the transport of sucrose into the cell against its concentration gradient.

(a) 21-day-old wild-type versus mutant plants

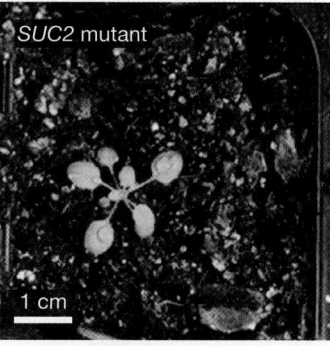

(b) Starch accumulation in wild type versus mutant

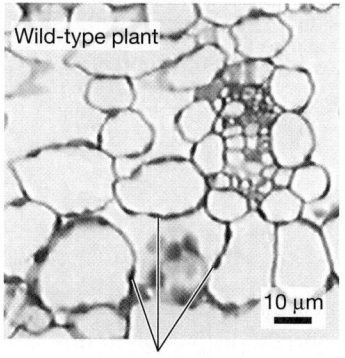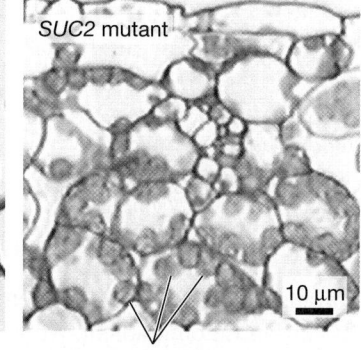

Normal chloroplasts   Starch-filled chloroplasts

(c) Sucrose movement in wild type versus mutant

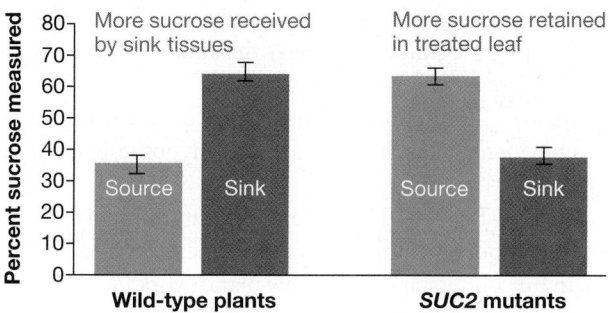

FIGURE 38.22 The Role of a Proton–Sucrose Symporter in Long-Distance Transport in Plants.

DATA: Gottwald, J. R., P. J. Krysan, J. C. Young, et al. 2000. *Proceedings of the National Academy of Sciences* 97: 13979–13984.

✔**QUESTION** Why did the researchers measure radioactivity after only 5 hours, and not wait a longer period such as 24 hours?

As **FIGURE 38.22c** shows, mutant leaves that were supplied with radioactive sucrose retained much more of the sucrose than wild-type leaves. Also, sink tissues in the mutant plants received much less radioactive sucrose than wild-type sink tissues. Without the proton–sucrose symporter in companion cells, mutant plants were impaired in their ability to load sucrose into phloem cells for long-distance transport and the excess sugar accumulated in leaves as starch.

There is now considerable genetic and biochemical evidence to support the following model for phloem loading:

1. Proton pumps in the membranes of companion cells create a strong electrochemical gradient that favors a flow of protons into companion cells.

2. A symporter in the membranes of companion cells uses the proton gradient to bring sucrose into companion cells from the surrounding cell walls and intercellular spaces around source cells.

3. Once inside companion cells, sucrose travels into sieve-tube elements via plasmodesmata.

Although work on the mechanism of phloem loading in *Arabidopsis* and other species continues, most researchers are convinced that proton pumps and proton–sucrose symporters play a key role. Now, once sucrose has been loaded into sieve-tube elements near sources and follows a water-potential gradient to sinks, how is it unloaded?

Phloem Unloading

The membrane proteins that are involved in transporting sucrose molecules out of the phloem, and the mechanism of movement, vary among different types of sinks within the same plant. Mechanisms for phloem unloading also vary among different species of plants.

To appreciate this diversity, consider how sucrose is unloaded in the phloem of sugar beets—a crop grown for the storage tissues in its root, which is a major source of the granulated and powdered sucrose sold in grocery stores.

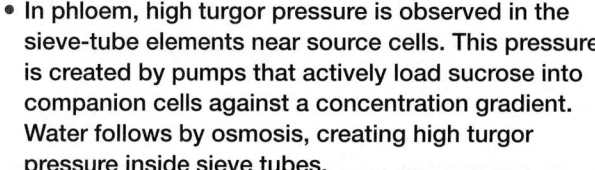

If you understand that . . .

- Phloem sap moves from areas of high water potential to areas of low water potential.

- In phloem, high turgor pressure is observed in the sieve-tube elements near source cells. This pressure is created by pumps that actively load sucrose into companion cells against a concentration gradient. Water follows by osmosis, creating high turgor pressure inside sieve tubes.

- At sinks, turgor pressure is much lower than it is at sources because storage cells or growing tissues remove sucrose from phloem sap. The removal of solutes in phloem causes water to leave phloem and enter xylem.

✔ **You should be able to . . .**

1. Explain why flow of phloem sap in a stem often changes direction between early spring and midsummer.

2. Explain the adaptive value of sieve cells having no nuclei and few other organelles.

Answers are available in Appendix A.

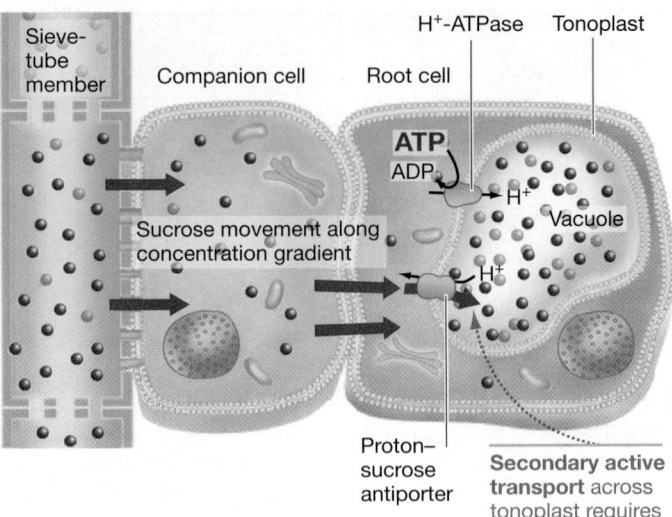

(a) Phloem unloading into growing leaves of sugar beets

Sieve-tube member Companion cell Developing leaf cell

Sucrose movement along concentration gradient

Sucrose is used for metabolism and biosynthesis

Passive transport

(b) Phloem unloading into roots of sugar beets

Sieve-tube member Companion cell Root cell H⁺-ATPase Tonoplast

ATP
ADP
H^+
Vacuole
H^+

Sucrose movement along concentration gradient

Proton–sucrose antiporter

Secondary active transport across tonoplast requires a proton gradient

FIGURE 38.23 Phloem Unloading Occurs when Sucrose Is Moved Out of Phloem Tissue in Sinks. The mechanism of phloem unloading can vary from sink to sink, such as in **(a)** young leaves and **(b)** roots of the same plant.

In sugar beets, sucrose is unloaded along a concentration gradient into an important sink: young, growing leaves. The passive transport occurs because sucrose is rapidly used up inside the cells to provide energy for ATP synthesis and carbon for the synthesis of cellulose, proteins, nucleic acids, and phospholipids needed by the growing cells (**FIGURE 38.23a**).

In the roots of the same plant, however, an entirely different mechanism is responsible for unloading sucrose. Root cells in this species have a large vacuole that stores sucrose. The membrane surrounding this organelle is called the **tonoplast.** It contains two proteins that work together to accumulate sucrose in the vacuole, much like the phloem loading process described earlier. A different proton pump hydrolyzes ATP and uses the energy released to transport protons into the vacuole, against its concentration gradient (**FIGURE 38.23b**). A proton–sucrose

cotransporter then uses the proton gradient to move sucrose up its concentration gradient. An important distinction in this case is that the cotransporter is an antiporter, while in source tissues it is a symporter.

The active transport of sucrose into the vacuole allows sucrose to move passively from phloem into the storage cells, keeping the water potential of the phloem sap near the sink low—as required by the Münch pressure-flow model.

To summarize, more than seven decades of research provide convincing evidence that the pressure-flow hypothesis is fundamentally correct.

The cohesion-tension theory for water movement and the pressure-flow model for phloem sap movement represent major advances in our understanding of how plants work.

CHAPTER 38   REVIEW

For media, go to MasteringBiology

If you understand . . .

38.1 Water Potential and Water Movement

- Plants lose water as an inevitable consequence of exchanging gases with the atmosphere. The flow of water from soil to air via plant tissues transports minerals to the shoot and follows a water-potential gradient.

- Water potential (ψ) is a measure of the tendency of water to move down its potential energy gradient and is expressed as megapascals (MPa).

- In plants, water potential has two components: (**1**) a solute potential, formed by the concentration of solutes in a cell or tissue; and (**2**) a pressure potential, provided by the cell wall and other factors. The water potential of a cell, tissue, or plant is the sum of its solute potential and pressure potential.

- When selectively permeable membranes are present, water moves by osmosis from areas of high solute potential to areas of low solute potential.

- When no membranes are present, water moves by bulk flow from areas of high pressure to areas of low pressure, independently of differences in solute potential.

✔ You should be able to explain why water in the xylem of a root moves up through the shoot in response only to a pressure gradient—not a solute gradient.

38.2 How Does Water Move from Roots to Shoots?

- According to the cohesion-tension theory, water is pulled in one continuous column from the soil to roots to shoots, against the force of gravity, by the surface tension caused by transpiration from leaves.

- Surface tension occurs at menisci that form as water evaporates from the walls of leaf cells, and it is transmitted downward via hydrogen bonding between water molecules. In this way, the energy in sunlight is responsible for the movement of water from roots to shoots.

✔ You should be able to suggest a hypothesis for why the pressure potential of water in a leaf, and thus the rate of water movement up a stem, change when the Sun goes behind a cloud.

38.3 Water Absorption and Water Loss

- Plants that occupy dry habitats have traits that limit the amount of water they lose to transpiration. These include narrow, needle-like leaves or stomata that are located in pits on the undersides of their leaves.

- CAM and C_4 photosynthetic cycles are adaptations in some plants that limit water loss in dry habitats.

✔ You should be able to explain how CAM plants avoid losing very much water despite living in deserts.

38.4 Translocation of Sugars

- Translocation is the movement of sucrose and other products through the plant.

- According to the Münch pressure-flow model, sugars move from sources to sinks via bulk flow along a pressure gradient that develops in phloem.

- A pressure gradient is generated by the active transport of sugars into sieve-tube elements in source tissues, coupled with the transport of sucrose out of sieve-tube elements at sink tissues. Water moves osmotically from xylem into sieve-tube elements near sources and cycles back to xylem near sinks.

✔ You should be able to explain why the pressure-flow model would not work if xylem and phloem were not bundled together.

(MB) **MasteringBiology**

1. **MasteringBiology Assignments**

   Tutorials and Activities Solute Transport in Plants; Translocation of Phloem Sap; Transport of Xylem Sap; Water Transport in Plants: The Transpiration-Cohesion-Tension Mechanism; Water Transport in Plants: Transpiration

   Questions Reading Quizzes, Blue-Thread Questions, Test Bank

2. **eText** Read your book online, search, take notes, highlight text, and more.

3. **The Study Area** Practice Test, Cumulative Test, BioFlix® 3-D Animations, Videos, Activities, Audio Glossary, Word Study Tools, Art

You should be able to . . .

✔ TEST YOUR KNOWLEDGE

Answers are available in Appendix A

1. Under what conditions does the rate of transpiration increase?
   a. when the temperature of a leaf decreases
   b. when stomata close at night
   c. during rainstorms, when atmospheric pressure is low
   d. when the weather changes and air becomes drier

2. The cells of a certain plant species can accumulate solutes to create very low solute potentials. Which of the following statements is correct?
   a. The plant's transpiration rates will tend to be extremely low.
   b. The plant can compete for water effectively and live in dry soils.
   c. The plant will grow most effectively in soils that are saturated with water year round.
   d. The plant's leaves will wilt easily.

3. What forces are responsible for capillarity?
   a. adhesion of water molecules to the sides of xylem cells, cohesion of water molecules to each other, and surface tension
   b. surface tension created by transpiration and cohesion of water molecules in a continuous flow from leaf to root

   c. high solute potentials created by the entry of ions during the night, when transpiration rates are low, followed by an influx of water
   d. gravity and wall pressure (from the sides of xylem cells)

4. What is a proton pump?
   a. a membrane protein that transports sucrose against a concentration gradient
   b. a membrane protein that transports protons against an electrochemical gradient
   c. a membrane protein that transports protons *along* an electrochemical gradient and sucrose *against* a concentration gradient
   d. any membrane protein that acts as a channel—meaning it does not consume ATP

5. The Casparian strip is located in what type of root cells?

6. Proteins in sieve-tube elements must be encoded by nuclei in what cells?

7. Why is the transport of phloem sap considered an active process?
   a. The manufacture of sucrose via photosynthesis is driven by the energy in sunlight.
   b. Transpiration is driven by the energy in sunlight.
   c. ATP is used to transport sucrose into companion cells near sources, against a concentration gradient.
   d. In spring, phloem sap moves against the force of gravity.

8. Draw a plant cell in pure water. Add dots to indicate solutes inside the cell. Now add dots to indicate an increase in solute potential inside the cell. Add an arrow showing the direction of water movement in response. Add arrows showing the direction of wall pressure and turgor pressure in response to water movement. Repeat the same exercise; but this time, add solutes to the solution outside the cell at a concentration that is greater than the inside of the cell.

9. Compare and contrast the forces involved in transporting water in xylem via root pressure, capillarity, and transpiration. Which of these mechanisms is/are passive?

10. Why are "cohesion-tension" and "pressure-flow" sensible names for the hypotheses analyzed in this chapter?

11. How does the movement of solutes through cotransport proteins result in phloem loading?

12. A seed is a sink when it is forming inside the parent plant. When is it a source?

13. A mutant plant lacking the ability to pump protons out of leaf companion cells will be unable to do which of the following?
   a. transpire from leaves
   b. load sucrose into sieve cells
   c. carry out photosynthesis
   d. transport water through the xylem

14. Suppose that plants over 1 meter tall had to expend energy to transport water from their roots to their leaves. What would be the consequences in terms of growth rates and overall height?

15. When young trees are transplanted to a new site, it takes several weeks or months for their root systems to grow and establish a high capacity to take up water. If a heat wave occurs during this period, the trees are likely to die—but not of starvation or loss of turgor. What kills them?

16. A recent paper indicates that the aquaporins in plasma membranes of cells throughout a plant close in response to drought stress. How does the closing of these channels help plant cells maintain turgor?

39 **Plant Nutrition**

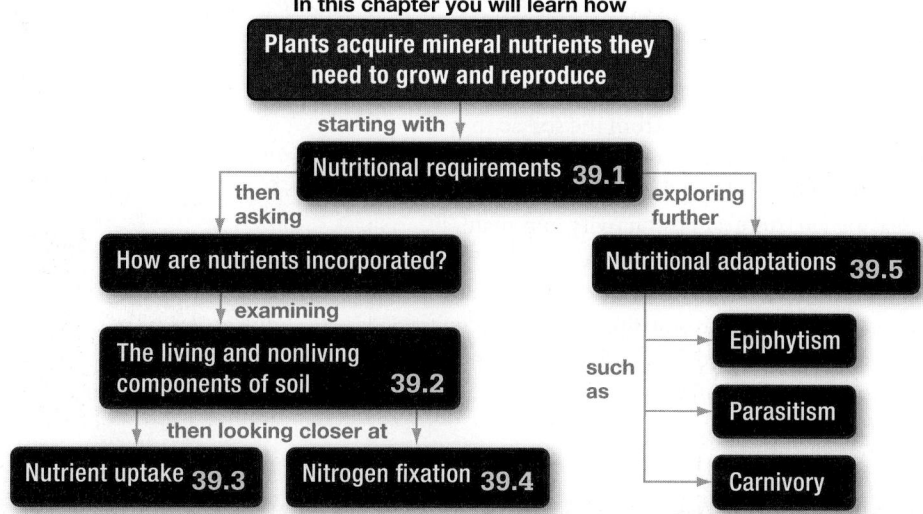

In this chapter you will learn how

Plants acquire mineral nutrients they need to grow and reproduce

starting with ↓

Nutritional requirements 39.1

then asking ↓

exploring further ↓

How are nutrients incorporated?

Nutritional adaptations 39.5

examining ↓

The living and nonliving components of soil 39.2

such as

→ **Epiphytism**

then looking closer at ↓

→ **Parasitism**

Nutrient uptake 39.3   **Nitrogen fixation** 39.4

→ **Carnivory**

In most plants, roots obtain the water and key nutrients required for individuals to survive and thrive.

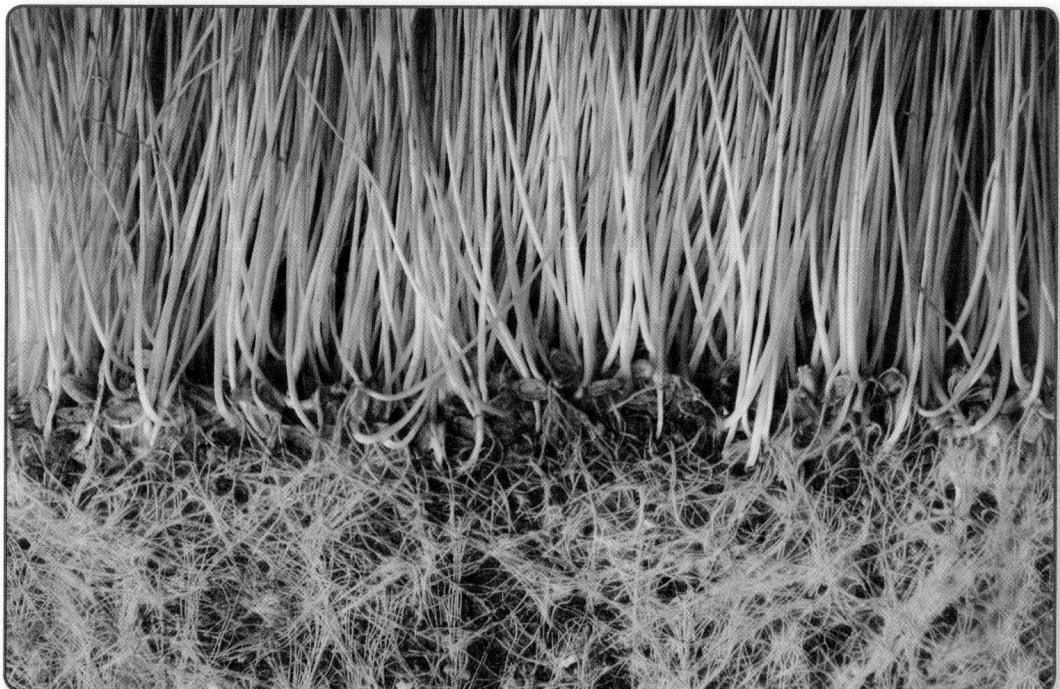

The most urgent tasks facing any organism are to acquire (**1**) carbon-containing molecules that will be used as cellular building blocks and (**2**) the chemical energy required to make ATP. Plants acquire both of these by producing sugar through the process of photosynthesis.

Yet plants cannot live on sugar alone. Besides making the carbohydrates they need, plants synthesize all of their own nucleic acids, amino acids, enzymes, chlorophylls, enzyme cofactors, and other molecules necessary to grow and reproduce. Plants do some of the world's most impressive synthetic organic chemistry.

A plant's ability to perform sophisticated reactions depends on its capacity to harvest as raw materials a wide variety of elements in the form of simple ions and molecules. In addition to carbon dioxide and water, plants have to obtain nitrogen, phosphorus, potassium, sulfur, magnesium, and other elements. Most of these nutrients exist in soil, the majority as ions that are dissolved in soil

This chapter is part of the Big Picture. See how on pages 840–841.

✔ When you see this checkmark, stop and test yourself. Answers are available in Appendix A.

water at low—sometimes extremely low—concentrations. Once these ions are inside roots, the one-way flow of water up the xylem carries the nutrients throughout the plant body.

The plant body is an efficient machine for harvesting these diffuse resources and concentrating them in cells and tissues. You have been introduced to how the organization and growth of the plant root and shoot systems make resource acquisition possible (see Chapter 37) and how water and nutrients are transported throughout the plant body (see Chapter 38). This chapter concentrates on how plants take up elements from the soil so that these nutrients can be transported to the cells that need them.

Questions about nutrition are fundamental to understanding how plants work, increasing agricultural productivity and maintaining the productivity of forests that supply lumber and fuel—as well as mitigate temperature increases associated with global climate change (see Chapter 56) by transforming atmospheric CO_2 into wood. Let's begin by analyzing the basic nutritional needs of plants—the equivalent of the minimum daily requirements in humans.

39.1 Nutritional Requirements of Plants

What do plants need to live? In the early 1600s, Jean-Baptiste van Helmont performed a classic experiment designed to answer this question. Van Helmont wanted to know where the mass of a growing plant comes from, and he used a willow tree as a study organism.

As **FIGURE 39.1** shows, he began by placing 90 kilograms (kg) of soil in a pot with a 2-kg willow sapling. He allowed the plant to grow for five years, adding only water. At the end of the experiment, he weighed the willow and the soil. The willow had gained about 74 kg while the soil had lost only 60 grams (g).

Where had the additional 74 kg of tree come from? Because he was not aware that gases have mass, van Helmont hypothesized that the new plant material came from water. He also ignored the loss of 60 g in the soil, chalking it up to measurement error.

As it turned out, van Helmont's measurements were not the problem—his conclusions were. Most of the added mass of the tree came from carbon dioxide in the atmosphere. The 60 g removed from the soil contained vital elements—the nutrients that are the focus of this chapter. What are they?

About half the elements in the periodic table—more than 60—can be found in the tissues of one or more plant species. The question that biologists, farmers, and foresters ask is, Which of these elements are essential for growth and reproduction in most species, and in what quantities?

Which Nutrients Are Essential?

Biologists define an **essential nutrient** as an element or compound that is required for normal growth and reproduction—meaning that the plant cannot complete its life cycle without this nutrient. Essential nutrients cannot be synthesized by the organism.

Researchers test whether a nutrient is essential by denying a specific element to plants and documenting what happens to the plants over time. For most vascular plants, 17 elements are

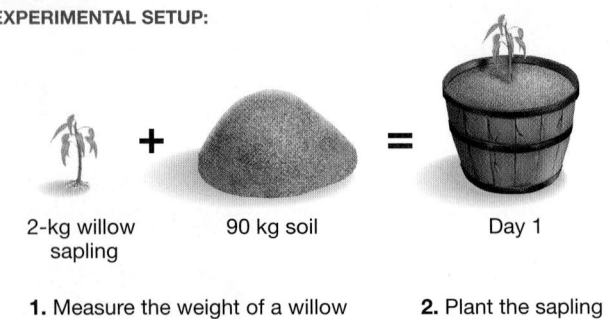
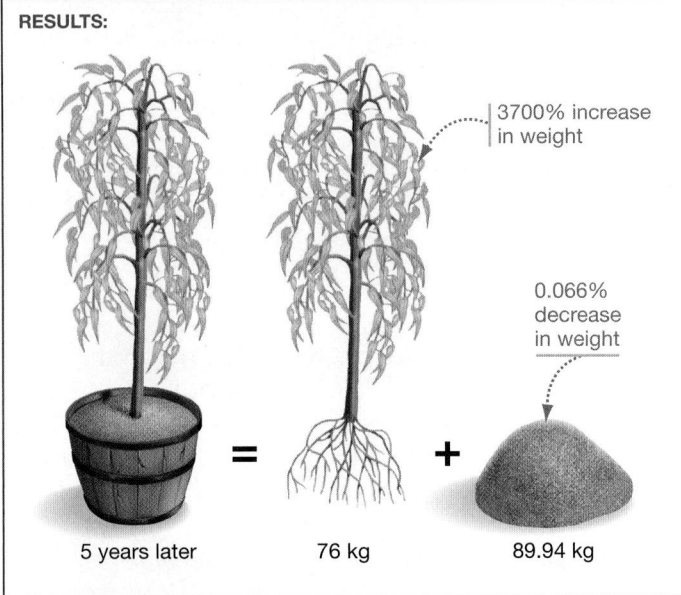
FIGURE 39.1 An Early Experiment on the Role of Soil in Plant Nutrition.

✔ **QUESTION** Many nonbiologists think that most of a plant's mass comes from soil or water. Describe an experiment that would convince someone that most of a plant's mass comes from CO_2.

essential. Just three of these—carbon, hydrogen, and oxygen—typically make up about 96 percent of the dry weight of a plant. The remaining 14 elements are sometimes called mineral nutrients, because they originate in soil.

Although different classification schemes for the essential elements have been proposed, the most common is based on distinguishing nutrients that are obtained from water or air versus soil and then dividing soil nutrients into macronutrients and micronutrients based on their abundance in plants (**TABLE 39.1**, see page 778).

Macronutrients Certain elements in the soil are required by plants in relatively large quantities; these are called **macronutrients.** Some of these are major components of nucleic acids, proteins, and phospholipids, all of which are plentiful in plants.

Among the macronutrients, nitrogen (N), phosphorus (P), and potassium (K) are particularly important because they often act as **limiting nutrients,** meaning their availability limits plant growth. If N, P, and/or K are added in appropriate quantities to soil as fertilizer, plant growth usually increases. This observation explains why the leading ingredients in virtually every commercial fertilizer are N, P, and K.

Micronutrients In contrast to macronutrients, **micronutrients** are required in small quantities. When plant tissues are dried and analyzed, micronutrients are typically present in only trace amounts. Instead of acting as components of macromolecules, micronutrients usually function as cofactors for specific enzymes—substances that are required for normal enzyme function (see Chapter 3).

It's important not to underestimate the importance of micronutrients, even though only tiny amounts are needed. For example, a typical plant contains just one molybdenum atom for every 60 million hydrogen atoms in its body, not including water. Yet plants die without molybdenum, because it functions as a cofactor for several enzymes involved in nitrogen processing. What happens to plants when other essential nutrients are missing?

What Happens When Key Nutrients Are in Short Supply?

In some cases, biologists can examine a plant that is growing poorly and diagnose a nutrient deficiency (**FIGURE 39.2**). For example, if older leaves are in poor condition, the problem is probably due to lack of N, P, K, or magnesium. These elements are mobile—meaning they are readily transported from older leaves

to younger leaves when they are in short supply—so older leaves deteriorate first when these elements are scarce. Immobile nutrients like iron or calcium, in contrast, stay tied up in older leaves. When they are in short supply, younger leaves are the first to show deficiency symptoms. Deficiencies in the various nutrients produce symptoms specific to each nutrient (see Table 39.1).

In large part, the ability to diagnose nutrient deficiencies is based on studies involving hydroponic growth systems. **Hydroponic growth** takes place in liquid cultures, without soil, so researchers can precisely control the availability of each nutrient.

Consider an experiment on copper deficiency in tomatoes (**FIGURE 39.3**). Researchers grew seedlings in two types of treatments. One treatment consisted of flasks containing water and

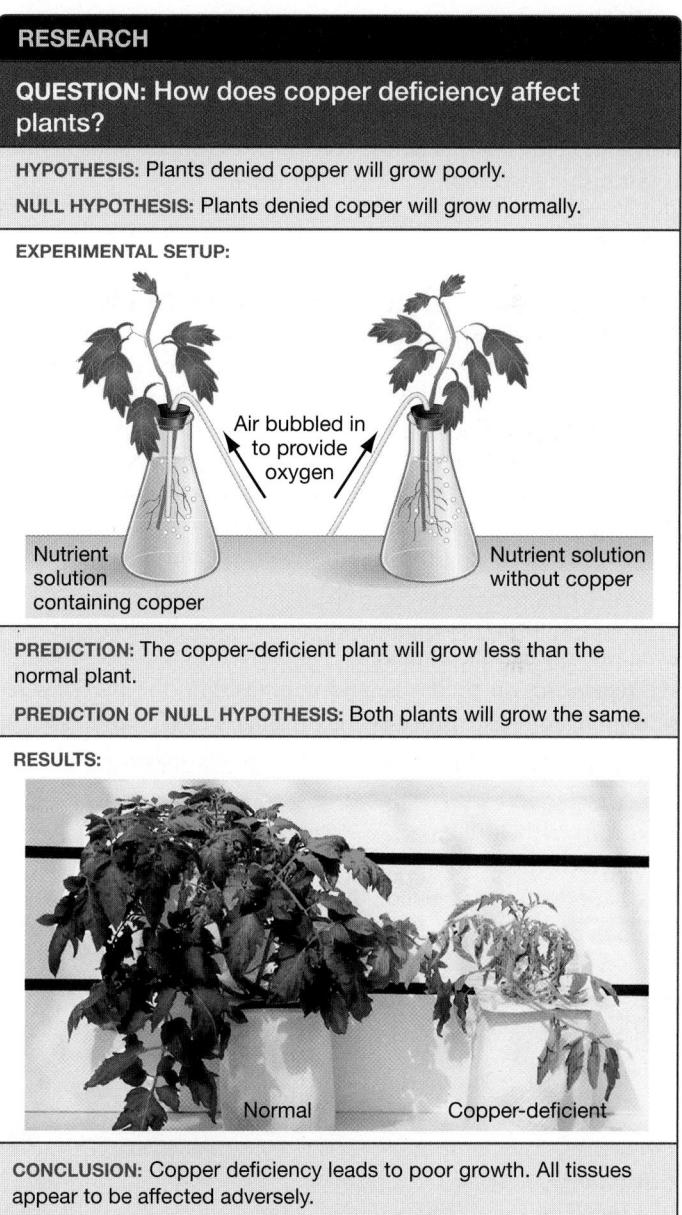

RESEARCH

QUESTION: How does copper deficiency affect plants?

HYPOTHESIS: Plants denied copper will grow poorly.
NULL HYPOTHESIS: Plants denied copper will grow normally.

EXPERIMENTAL SETUP:

Air bubbled in to provide oxygen

Nutrient solution containing copper

Nutrient solution without copper

PREDICTION: The copper-deficient plant will grow less than the normal plant.

PREDICTION OF NULL HYPOTHESIS: Both plants will grow the same.

RESULTS:

Normal   Copper-deficient

CONCLUSION: Copper deficiency leads to poor growth. All tissues appear to be affected adversely.

FIGURE 39.3 Hydroponics Is Used to Study Nutrient Deficiencies.
SOURCE: Arnon, D. I., and P. R. Stout. 1939. *Plant Physiology* 14: 371–375.

✔**QUESTION** What problem would arise if this experiment had been done in soil with and without added copper?

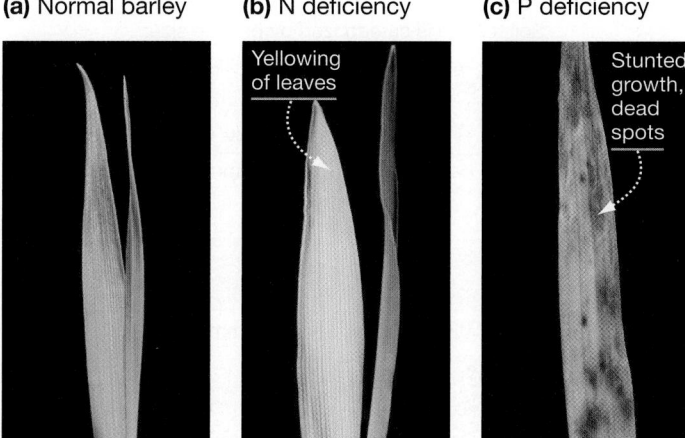

(a) Normal barley   **(b)** N deficiency   **(c)** P deficiency

Yellowing of leaves

Stunted growth, dead spots

FIGURE 39.2 Nutrient Deficiencies Can Have Distinctive Symptoms.

all the essential nutrients in the relative concentrations that are optimal for tomato growth. The second treatment was identical, except that the nutrient solution lacked copper.

As Figure 39.3 shows, copper-deprived individuals have stunted shoots, unnaturally light-colored foliage, and curled leaves. Given copper's role as a cofactor or component of several enzymes involved in redox reactions required for ATP production, it is understandable that all tissues in the plant were severely affected. And because copper is a micronutrient, it is reasonable to expect that a relatively small amount would cure the deficiency. In line with this prediction, the researchers found that the symptoms were prevented if the plants were cultured in a solution containing just 0.002 milligrams per liter (mg/L) of copper. Analogous studies have been done on the other essential nutrients.

For farmers, foresters, and plant ecologists, understanding which nutrients are essential, and why, is basic to understanding why certain plants thrive and others fail. Now, where do these nutrients come from? The answer—soil—is simple. But soil itself is astonishingly complex.

39.2 Soil: A Dynamic Mixture of Living and Nonliving Components

The process of soil building begins with solid rock. As **FIGURE 39.4** shows, **weathering**—the forces applied by rain, running water, temperature changes, and wind—continually breaks tiny pieces off large rocks. The weathering process is accelerated if small cracks develop in the rock. If plant roots grow into the crack, they expand as they grow, widen the crack, and break off small flakes or pebbles. A similar effect occurs in high latitudes or at high elevations when water enters the cracks, freezes in winter, expands, and breaks off pieces.

Depending on their size and composition, the particles resulting from these processes are called gravel, sand, silt, or clay. These rock fragments are the first ingredient in soil. As organisms occupy the substrate, they add dead cells and tissues and feces. This decaying organic matter is called **humus** (pronounced *HEW-muss*).

With time, soil eventually becomes a complex and dynamic mixture of inorganic particles, organic particles, and living organisms. In a single gram of soil there are normally dozens of small animals and plant roots; hundreds of thousands of protists, fungi, and microscopic animals; and hundreds of millions of bacteria and archaea (**FIGURE 39.5**).

Both the parent rock that contributes inorganic soil components and the organisms and organic matter that occupy soils vary from one site to another. **Texture**—the proportions of gravel, sand, silt, and clay—and other soil qualities vary as well. Soil texture is important for several reasons:

- Texture of a soil affects the ability of roots to penetrate and obtain water and nutrients, as well as to anchor and support the plant body. For example, soil that is dominated by clay-sized particles tends to compact and resist root penetration.

- Texture affects a soil's ability to hold water and make it available to plants. Water tends to adhere to clay and silt particles but runs through sand and gravel.

- A soil's texture and water content dictate the availability of oxygen. Like other eukaryotes, plants have to take in oxygen to use as an electron acceptor during cellular respiration. The oxygen used by plant root cells is found in air pockets among soil particles. This explains why overwatering a plant is just as detrimental as underwatering it: Overwatering drowns a plant's roots.

The best soils for plants, called loams, contain roughly equal amounts of sand, silt, and clay, along with a high proportion of humus.

Loams that have good texture and large amounts of organic matter can take thousands of years to develop through the weathering of rocks and the continual addition of humus. Unfortunately, it can take just a few years of abuse by humans for loams to blow or wash away.

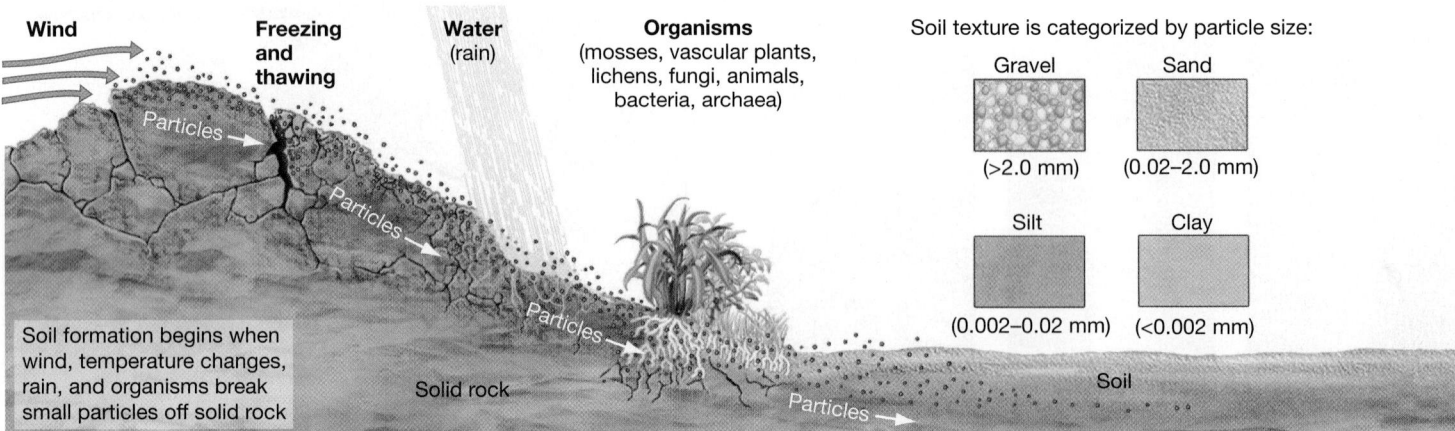

FIGURE 39.4 Soil Formation Begins with Weathering of Rock.

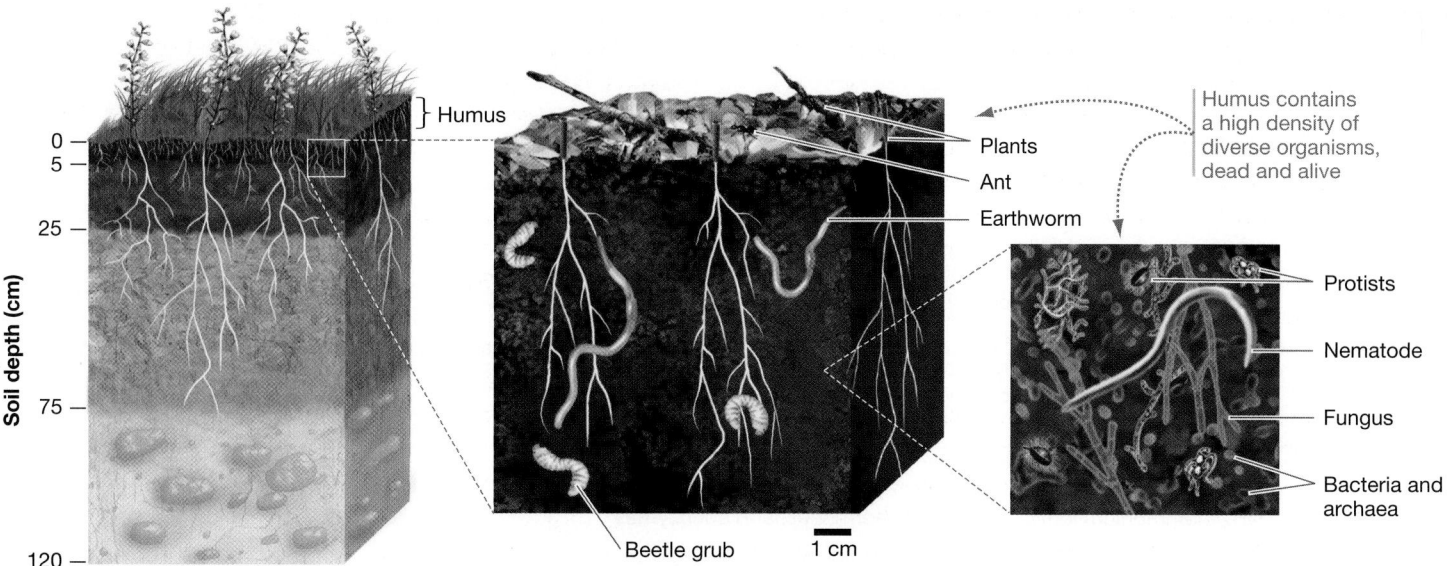

FIGURE 39.5 Mature Soils Are a Complex Mixture of Organic and Inorganic Components. In addition to mineral particles, soil contains humus—organic material derived from dead organisms—and a wide array of living organisms.

The Importance of Soil Conservation

Soil erosion occurs when soil is carried away from a site by wind or water. Soil erosion occurs naturally, such as when rivers cut away at their banks and carry material downstream. In most natural environments, though, the rate of soil formation exceeds the rate of soil erosion, so soils build up over time. Unfortunately, the situation can change dramatically when humans exploit an area.

When plant cover is removed for forestry, farming, or suburbanization, stems and leaves can't lessen the force of wind and rain, and plant roots no longer hold soil particles in place. The results can be devastating. At some locations in the United States,

8–10 centimeters (cm) of topsoil were blown away during the Dust Bowl of the 1930s, when drought and poor farming practices left thousands of hectares of soil unprotected (**FIGURE 39.6a**).

U.S. soil erosion rates have declined dramatically since then. Still, researchers estimate that almost 30 percent of all croplands in the United States are eroding too fast to maintain their long-term productivity. Deforestation is also exposing forest soils, contributing to disasters such as the mudslides and flooding that occurred in the Dominican Republic and Haiti in 2004, killing close to 5000 people. **FIGURE 39.6b** provides an aerial view of devastating mudslides that occurred in 2007 in the U.S. state of Washington, after recent deforestation. Worldwide, an estimated 36 billion tons of soil are lost to erosion every year.

(a) Wind erosion in the United States, 1930s

(b) Mudslides caused by deforestation

FIGURE 39.6 Soil Erosion Can Have Devastating Consequences.

TABLE 39.1 Essential Nutrients

Element	Form Available to Plants	Functions	Average % Dry Weight*	Deficiency Symptoms
Obtained from Water or Air: H_2O or CO_2				
Oxygen	O_2, H_2O	Electron acceptor in cellular respiration; major component of organic compounds	45	Usually affects roots: cells suffocate, leading to root rot and wilting
Carbon	CO_2	Substrate for photosynthesis; major component of organic compounds	45	Slow growth (starvation)
Hydrogen	H_2O	Major component of organic compounds; functions in electrical balance and establishment of electrochemical gradients	6	Slow growth due to cell death (desiccation)
Obtained from Soil: Macronutrients				
Nitrogen	NO_3^- (nitrate) NH_4^+ (ammonium ion)	Component of proteins, nucleic acids, ATP, chlorophyll, hormones, and coenzymes	1.5	Failure to thrive; chlorosis (yellowing of older leaves)
Potassium	K^+	Necessary for osmotic adjustment in cells; required for synthesis of organic molecules; cofactor for some enzymes	1.0	Chlorosis at margins of leaves or in mottled pattern; weak stems; short internodes
Calcium	Ca^{2+}	Regulatory functions; role in cell wall structure; stabilizes membranes; involved in signal transduction; enzyme cofactor	0.5	Necrosis (small spots of dead cells) in meristems; deformation of young leaves; stunted, highly branched root system
Magnesium	Mg^{2+}	Chlorophyll component; activates many enzymes	0.2	Chlorosis between leaf veins; premature leaf drop
Phosphorus	$H_2PO_4^-$ (dihydrogen phosphate ion) HPO_4^{2-} (hydrogen phosphate ion)	Component of ATP, nucleic acids, phospholipids, and several coenzymes	0.2	Stunted growth in young plants; dark green leaves with necrosis
Sulfur	SO_4^{2-} (sulfate ion)	Component of proteins containing methionine and cysteine; electron transport proteins and coenzymes	0.1	Stunted growth; chlorosis
Obtained from Soil: Micronutrients				
Chlorine	Cl^- (chloride ion)	Needed for water-splitting step of photosynthesis; functions in water balance and electrical balance	0.01	Wilting at leaf tips; general chlorosis and necrosis of leaves or development of bronze color
Iron	Fe^{3-} (ferric ion) Fe^{2-} (ferrous ion)	Necessary for chlorophyll synthesis; component of cytochromes and ferredoxin; enzyme cofactor	0.01	Chlorosis between veins of young leaves
Manganese	Mn^{2+}	Involved in photosynthetic O_2 evolution; enzyme activator; important in electron transfer	0.005	Chlorosis between leaf veins and small necrotic spots
Zinc	Zn^{2+}	Involved in synthesis of the plant hormone auxin, maintenance of ribosome structure, enzyme activation	0.002	Small internodes; stunted and distorted ("puckered") leaves
Boron	$H_2BO_3^-$ (borate ion)	Strengthens cell walls; required for pollen tube growth and normal membrane function	0.002	Black necrosis in young leaves and buds
Copper	Cu^+ (cuprous ion) Cu^{2+} (cupric ion)	Cofactor of some enzymes; present in lignin of xylem	0.0006	Light-green leaves with necrotic spots; twisted and malformed leaves
Nickel	Ni^{2+}	Cofactor for enzyme functioning in nitrogen metabolism	[no data]	Necrosis at leaf tips
Molybdenum	MoO_4^{2-} (molybdate ion)	Cofactor in nitrogen reduction; essential for nitrogen fixation	0.00001	Chlorosis between veins; necrosis of older leaves

*These percentages were obtained by drying vascular plants and documenting what proportion of the waterless mass consists of various elements.

If soil is managed carefully, however, it can be a renewable resource. Techniques that maintain long-term soil quality and productivity are the basis of **sustainable agriculture** and sustainable forestry. Farmers can reduce soil loss dramatically by

- planting rows of trees as windbreaks;

- using techniques that minimize the amount of plowing and tilling needed to control weeds; and

- planting crops in strips that follow the contour of hillsides.

Farmers can also maintain soil quality by adding organic material in the form of manures and, soon after a harvest, planting cover crops that are plowed in and allowed to decompose.

What Factors Affect Nutrient Availability?

The elements required for plant growth are found in the soil not as atoms, but as **ions.** The second column of Table 39.1 lists the ion forms of some essential nutrients available to plants. Notice that some of these nutrients are available as elemental ions, such as K^+ or Cl^-, while others exist as molecular ions, such as HPO_4^{2-} or NO_3^-.

Anions and Cations Behave Differently   The ions present in soil tend to behave in one of two ways, depending on their charge (**FIGURE 39.7**). Anions—ions with negative charges—usually dissolve in soil water, because they interact with water molecules via hydrogen bonding. Phosphate ions, an exception to this rule, tend to form insoluble complexes with iron, aluminum, calcium, or other positively charged cations.

Because they exist as solutes, negatively charged anions are readily available to plants for absorption. They are also easily washed out of the soil by rain, however. The loss of nutrients via the movement of water through soil is called **leaching.**

Cations—ions with positive charges—dissolve in soil water but are not as immediately available as anions. In solution, cations interact with the negative charges found on two types of soil particles: (**1**) organic matter that is rich in negatively charged organic acids; and (**2**) the surfaces of the tiny, sheetlike particles called clay, which are rich in mineral anions (see Figure 39.7).

Organic soils that contain clay tend to retain nutrients, because few positively charged cations leach away and because these soils hold water (and thus anions) better than sandy soils do. The presence of clay makes cations more difficult for plants to extract and use because the cations are tightly bound to the clay.

The Role of Soil pH   In addition to soil texture, other factors can influence the availability of essential elements. Perhaps the most important of these is soil pH.

Recall that the pH scale indicates the relative concentration of hydrogen ions in a solution, and that it ranges from 0 to 14 (see Chapter 2). Soils with low pH have a relatively high concentration of hydrogen ions and are considered acidic. Soils with a high pH contain relatively few hydrogen ions and are termed basic or alkaline.

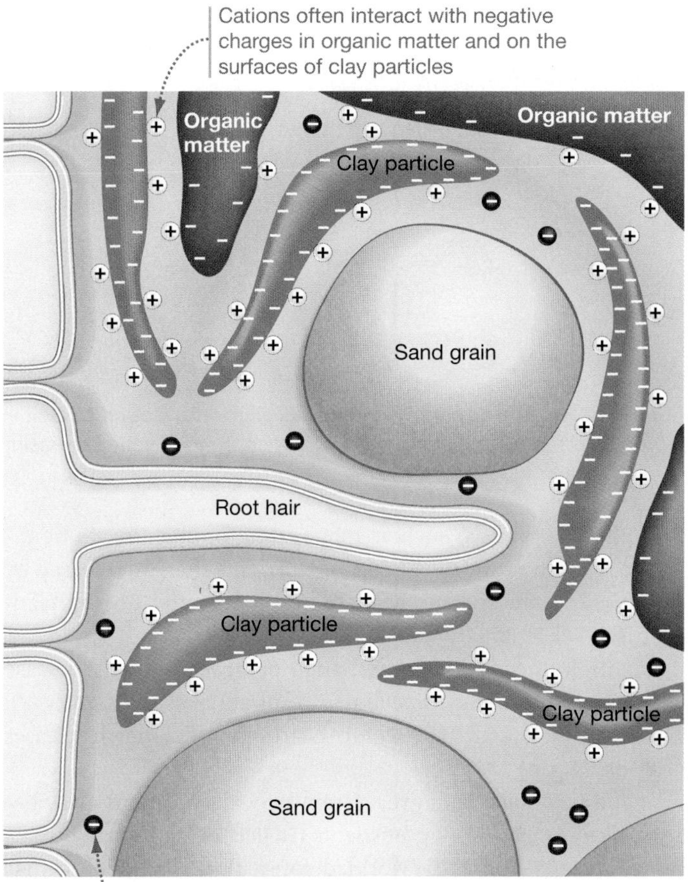

Cations often interact with negative charges in organic matter and on the surfaces of clay particles

Anions usually dissolve in soil water; they are readily available for absorption by root hairs

FIGURE 39.7 Cations Tend to Bind to Clay Particles and Organic Matter; Anions Stay in Solution.

Acidic soils are found in regions such as conifer forests, where the decomposition of organic matter produces carbonic acid, phosphoric acid, or nitric acid. Alkaline soils, in contrast, are common in regions where limestone ($CaCO_3$) is abundant. When limestone reacts with water, the calcium ions that are released take the place of protons that cling to soil particles. The protons then react with CO_3^{2-} to form bicarbonate ions (HCO_3^-), lowering the hydrogen ion concentration of the soil and raising its pH.

Most plants thrive in soils with a relatively neutral pH—between 6 and 7. But some plants, such as blueberries, grow best in acidic soils (pH around 3 to 5); others such as lavender, rosemary, and fennel, thrive on alkaline soils (pH above 8).

Cation Exchange   Soil pH affects the availability of plant nutrients in a number of ways. For example, the presence of protons in soil water can cause the release of cations that are bound to soil particles. The process responsible is called **cation exchange.** Cation exchange occurs when protons or other soluble cations bind to negative charges on soil particles and

(a) Cation exchange releases nutrients...

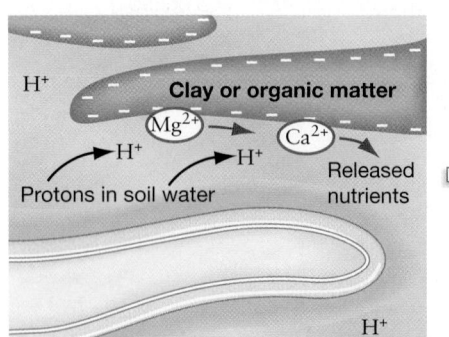

(b) ... which are absorbed by roots...

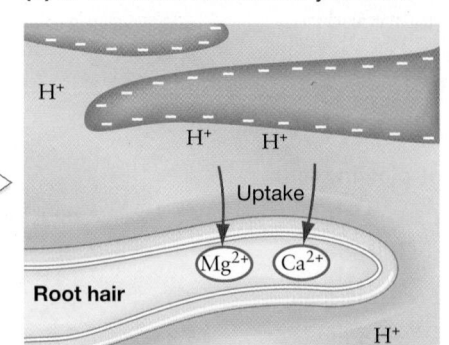

or

(c) ... or leached in heavy rains.

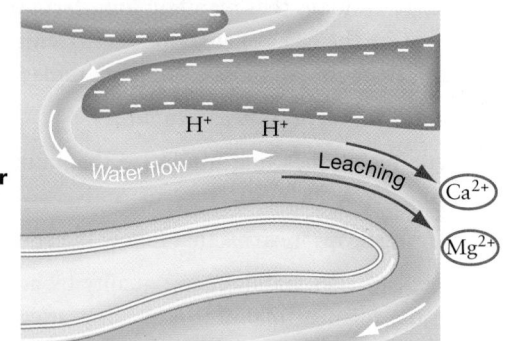

FIGURE 39.8 Cation Exchange Releases Nutrients Bound to Soil Particles. When cation exchange occurs, a proton binds to negative charges on clay or organic matter, releasing bound cations.

cause bound cations, such as magnesium or calcium, to be released from the soil (**FIGURE 39.8a**). Nutrients that are released by cation exchange become available for uptake by nearby plant roots (**FIGURE 39.8b**).

Plants influence cation exchange because root cells release CO_2 as a by-product of cellular respiration (see Chapter 9). The CO_2 reacts with H_2O to form carbonic acid, which releases protons.

If soil is too acidic, however, rain may leach cations away before the roots can take up nutrients (**FIGURE 39.8c**). In many tropical rain forests around the world, deforestation for agriculture has resulted in acidic, nutrient-poor soils that are not productive.

To summarize, anions stay in solution in soil water. They are readily available to plants but may wash away easily. Positive ions, in contrast, tend to bind to soil particles but can be released by cation exchange.

TABLE 39.2 details how the presence of sand, clay, and organic matter affects ion availability and other soil properties, including water availability. Based on these properties, it is not surprising that the most productive soils are loams that are a mixture of sand, clay, and organic matter.

Nutrients are found at extremely low concentrations in soil but at high concentrations in plant cells. How are plants able to bring N, P, K, and other key elements into their bodies against a concentration gradient?

39.3 Nutrient Uptake

In most species of plants, the root system is the site of nutrient uptake. You have been introduced to the general features of root anatomy (see Chapter 37) and the role of roots in water uptake (see Chapter 38). Now we need to explore the detailed anatomy of roots and analyze events that occur in the plasma membranes of their epidermal and cortex cells.

Most nutrient uptake occurs just above the growing root tip, in the region called the **zone of maturation.** Recall that epidermal cells in this part of the root have extensions called **root hairs** (**FIGURE 39.9**). Root hairs dramatically increase the surface area available for nutrient and water absorption. For example, the root system of a single annual rye plant can have a total surface area the size of a basketball court. Most of this area—60 percent—is found in an estimated 10^{10} root hairs.

Root hairs are so numerous and so efficient at absorbing nutrients from soil that, over time, they create a "zone of nutrient depletion" in soil immediately surrounding them. The creation of this mined-out region is why continued root growth is vital to a plant's health. Because roots continue to grow throughout a plant's life, the zone of maturation is continually entering new and potentially nutrient-rich areas of soil.

If an epidermal cell encounters soil where nutrients are available, what happens? How do ions enter?

TABLE 39.2 Effects of Soil Composition on Soil Properties

Composition	Water Availability	Nutrient Availability	Oxygen Availability	Root Penetration Ability
Sand	Low: water drains through	Low: poor capacity for cation exchange; anions leach out	High: many air-containing spaces	High: does not pack tight
Clay	High: water clings to charged surface	High: large capacity for cation exchange; anions remain in solution	Low: few air-containing spaces	Low: packs tight
Organic matter	High: water clings to charged surface	High: source of nutrients; large capacity for cation exchange; anions remain in solution	High: many air-containing spaces	High: does not pack tight

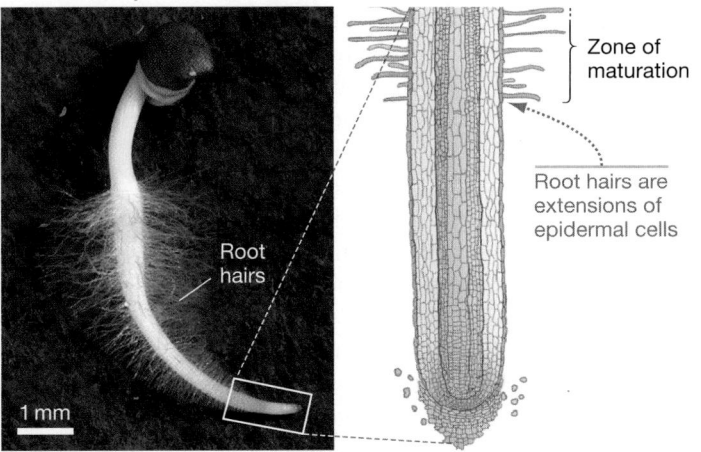

FIGURE 39.9 Root Hairs Increase the Surface Area Available for Nutrient Absorption.

Mechanisms of Nutrient Uptake

Ions, small molecules, and even large molecules can pass through the plant cell wall freely. The plasma membrane, in contrast, is highly selective. Recall that the plasma membrane is a fluid, sheetlike structure consisting of a phospholipid bilayer studded with proteins (see Chapter 6). Because the interior of the phospholipid bilayer is nonpolar and hydrophobic, it resists the passage of ions—even ions that are required by the plant. Some membrane proteins span the bilayer, however, and allow specific ions to cross the membrane.

Because root hairs have such a large surface area, they contain large numbers of membrane proteins that bring nutrients into the cytosol of root cells. These membrane proteins cannot import the ions that the plant needs completely on their own, however. They often function in tandem with another protein, a proton pump.

Establishing a Proton Gradient Plants harvest diffuse nutrients and concentrate them in their tissues. In an epidermal cell, ions such as potassium (K^+), hydrogen phosphate (HPO_4^{2-}), and nitrate (NO_3^-) are many times more concentrated than they are in the soil water outside the cell. For these and other ions to enter the cell, they have to cross the plasma membrane against a strong concentration gradient. How is this possible?

As **FIGURE 39.10a** shows, the answer hinges on **proton pumps,** or H^+-ATPases. These proteins are found in the plasma membranes of root epidermal and cortex cells, and they are similar to the pumps that make it possible for companion cells to load sucrose into phloem against a strong concentration gradient. Recall that when a phosphate group from ATP binds to the pumps, they change conformation in a way that allows them to transport protons to the exterior of the cell (see Chapter 38).

The activity of proton pumps leads to a strong excess of protons on the exterior of the plasma membrane relative to the interior. In the case of root cells, this differential results in a strong pH gradient favoring the movement of protons back into the cell. In addition, the outside of the membrane becomes positively charged relative to the inside. Stated another way, there is a separation of charge—a **voltage**—across the membrane.

Proton pumps are found in all types of plant cells, and all plant cell membranes carry a voltage. Because voltage is a form of potential energy, the charges that are separated by the membrane create a **membrane potential,** or difference in electrical charge across a cell membrane. How is the membrane potential used to transport ions?

Using a Proton Gradient to Import Cations In the membranes of root cells, the electrical gradient established by proton pumps, which favors the entry of positive ions, is strong enough to overcome the pH gradient, which opposes the entry of these cations. In essence, plant cells are batteries that are charged up to attract nutritionally necessary cations.

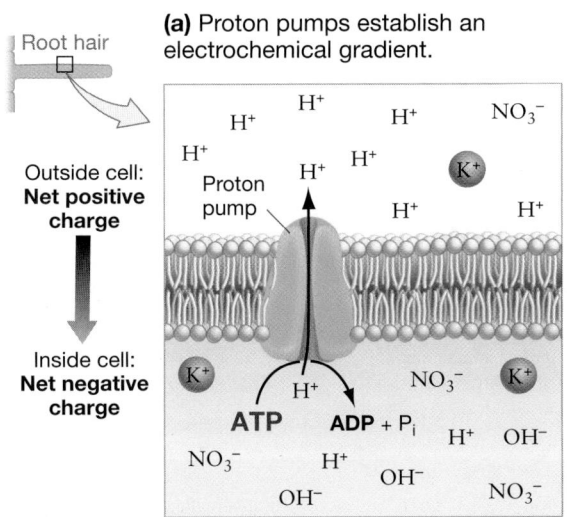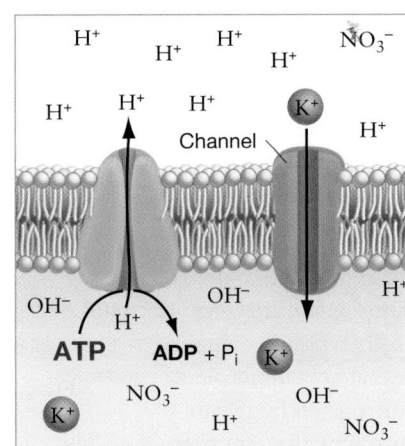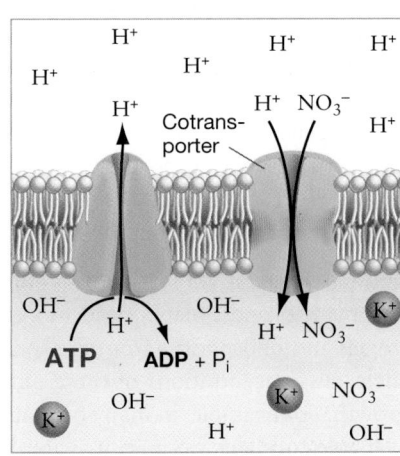

(a) Proton pumps establish an electrochemical gradient.

(b) Cations enter root hairs via channels.

(c) Anions enter root hairs via cotransporters.

FIGURE 39.10 Ions Enter Roots along Electrochemical Gradients Created by Proton Pumps.

✔**EXERCISE** In part (b), add an arrow labeled "Proton gradient." In part (c), add an arrow labeled "Electrical gradient" and indicate its positive–negative polarity.

To drive this point home, consider potassium cations. Potassium is an essential nutrient in plants because it is required as a cofactor by over 40 enzymes. In addition, it is found at relatively high concentrations inside plant cells and plays a key role in bringing water into cells via osmosis and maintaining normal turgor pressure.

Researchers who added radioactive potassium ions to the solution outside a root cell and followed their movement found that K^+ does indeed accumulate in cells due to an electrochemical gradient (**FIGURE 39.10b**). In follow-up experiments, biologists were able to isolate the membrane protein responsible for the uptake of K^+ and sequence the gene that encodes the protein.

✔ If you understand this concept, you should be able to explain what happens to cation uptake if (1) H^+-ATPases fail, and (2) a molecule blocks a cation channel.

Cations like K^+ enter root hairs through membrane proteins along an electrochemical gradient, but how is it possible for anions to enter? The negatively charged interior of the cell should repel these ions.

Using a Proton Gradient to Import Anions   **FIGURE 39.10c** shows how anions such as NO_3^- are able to enter root hairs against their electrochemical gradient. The key is that anions enter through membrane transport proteins called cotransporters (see Chapter 38), which transport two solutes at once. In this case, the cotransporter is a symporter: It brings NO_3^- into the cell along with a proton, and both solutes move in the same direction.

Here's the key observation: So much energy is released when a proton enters the cell along its electrochemical gradient that nitrate, phosphate ions, or other anions can be cotransported against their electrochemical gradients.

To summarize, the electrochemical gradient set up by proton pumps makes it possible for plant roots to absorb key cations and anions via ion channels and symporters, respectively. ✔ If you understand this concept, you should be able to explain what happens to anion uptake if (1) H^+-ATPases fail, and (2) a molecule blocks an anion symporter.

Once ions have crossed the plasma membrane of an epidermal or cortex cell in the root, they continue to move via the symplast and through the endodermis toward the central vascular tissue. Recall that the endodermis is the site of the Casparian strip, which blocks the movement of ions and water through the apoplast (see Chapter 38). Once inside the endodermal layer, ions are actively transported out of cells, where they accumulate in the xylem sap. The ions are then transported passively by bulk flow to tissues throughout the plant.

Nutrient Transfer via Mycorrhizal Fungi   To synthesize proteins and nucleic acids, plants need to extract large quantities of nitrogen and phosphorus from the soil. But soils typically contain such low concentrations of these nutrients that they limit plant growth. For example, in many habitats soil nitrogen is largely unavailable to plants because it is found in proteins that are part of the organic matter. How do plants absorb enough N and P to satisfy their nutritional needs? In most cases, they rely on partnerships with fungi that live in close association with their roots.

You might recall that fungi and plant roots that live in association are called **mycorrhizae** (literally, "fungus-root") (see Chapter 32). Mycorrhizal fungi and plants are **symbiotic** ("living-together"), meaning that they live in physical contact with each other. Biologists estimate that more than 80 percent of all vascular plant species associate with mycorrhizal fungi.

In some habitats, plants receive large quantities of nitrogen from mycorrhizal fungi. The fungal symbionts in these associations are particularly efficient at digesting macromolecules (proteins and nucleic acids) in decaying organic material and absorbing the amino acids (a source of nitrogen) and phosphate ($H_2PO_4^-$) that are released. As supported by experimental evidence described earlier (see Chapter 32), the mycorrhizal fungi then transport the nutrients from soil to plant roots (**FIGURE 39.11**). In exchange for these nutrients, the plant symbionts transfer sugars and other photosynthetic products to the fungi. Because there is a reciprocal exchange of nutrients, the symbiotic relationship is considered **mutualistic,** or mutually beneficial.

Fungi are particularly efficient at acquiring the nutrients required by plants, for two reasons:

1. Networks of filamentous hyphae increase the surface area available for absorbing nutrients by up to 700 percent.

2. Fungi can acquire nutrients from macromolecules in the soil that are unavailable to non-mycorrhizal plants.

Most plant species grow slowly and are overwhelmed by competitors if denied their mycorrhizal associates.

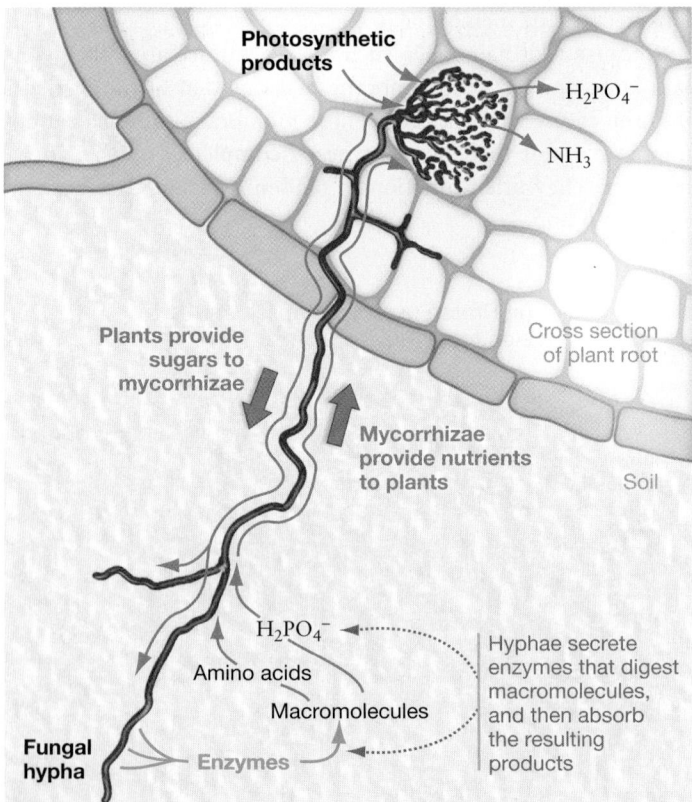

FIGURE 39.11 Most Mycorrhizal Fungi and Plants Are Mutualists. The fungus illustrated here provides nutrients to its host plant in exchange for photosynthetic products.

Mechanisms of Ion Exclusion

Plants have sophisticated systems for absorbing nutrients using membrane transporters and energy from the proton gradients they generate. Not all ion uptake is beneficial, however.

Certain types of natural soils—as well as soils that have been contaminated by waste products from mining or smelting operations—contain enough cadmium, zinc, nickel, lead, or other metals to poison enzymes in most plants. Sodium is also detrimental at high concentrations: Too much Na^+ inside cells can disrupt enzyme function, and too much Na^+ in extracellular spaces can create a solute potential that pulls enough water out of cells to result in a loss of turgor.

Sodium poisoning is a key issue in an array of environments, including (1) ocean coastlines, (2) habitats near roads that are treated with salt to melt ice and snow, and (3) irrigated farmlands. When soils are irrigated, solutes are left behind when water evaporates from the surface—leading to salt buildup over time.

How do plants exclude ions that are detrimental?

Passive Exclusion Ions move across roots following the same routes that water follows: the apoplastic, transmembrane, and symplastic pathways (see Chapter 38). Many ions do not enter the root symplast, simply because epidermal and cortex cells lack the requisite membrane transporters. If cells lack the membrane protein required for a certain ion to enter the cell, the ion won't enter. This is a form of passive exclusion.

Ions that move across the root cortex via the apoplastic pathway may never make it to xylem. To understand why, recall that the Casparian strip forces all solutes that travel through the root apoplast to cross the plasma membrane of endodermal cells before moving into the xylem (see Chapter 38). Ions that cannot enter endodermal cells via membrane transporters are excluded from entering the rest of the plant.

By having membrane transporters only for the ions that it requires, a plant can control which ions enter the symplast. Ions that reach the endodermis via the apoplastic pathway are blocked by the presence of the Casparian strip, so endodermal cells act as a selective filter—preventing unnecessary ions from reaching the xylem. The Casparian strip also qualifies as a mechanism for the passive exclusion of metals, sodium, or other ions (**FIGURE 39.12**).

The number of ion transporters present in a root is also important in determining the toxicity to certain ions. For example, salt-tolerant strains of corn transport much less salt into their roots than salt-intolerant strains—possibly because of genetic variation in the number of sodium channels found in root cells. Among species, rice is notoriously sensitive to salt buildup. Barley, in contrast, is relatively tolerant to high salt concentrations in soil.

Active Exclusion by Metallothioneins Plants also have mechanisms for coping with toxins once they are inside their cells. This is important because even essential nutrients are toxic at high enough concentrations.

Copper channels, for example, admit the small amounts of copper ions required for normal cell function. But plants that grow on soils near copper-mining operations experience large concentration gradients that favor an influx of this nutrient, so a surplus is likely to build up inside the plant body. How do plants deal with copper and other types of excess nutrients before they poison key enzymes?

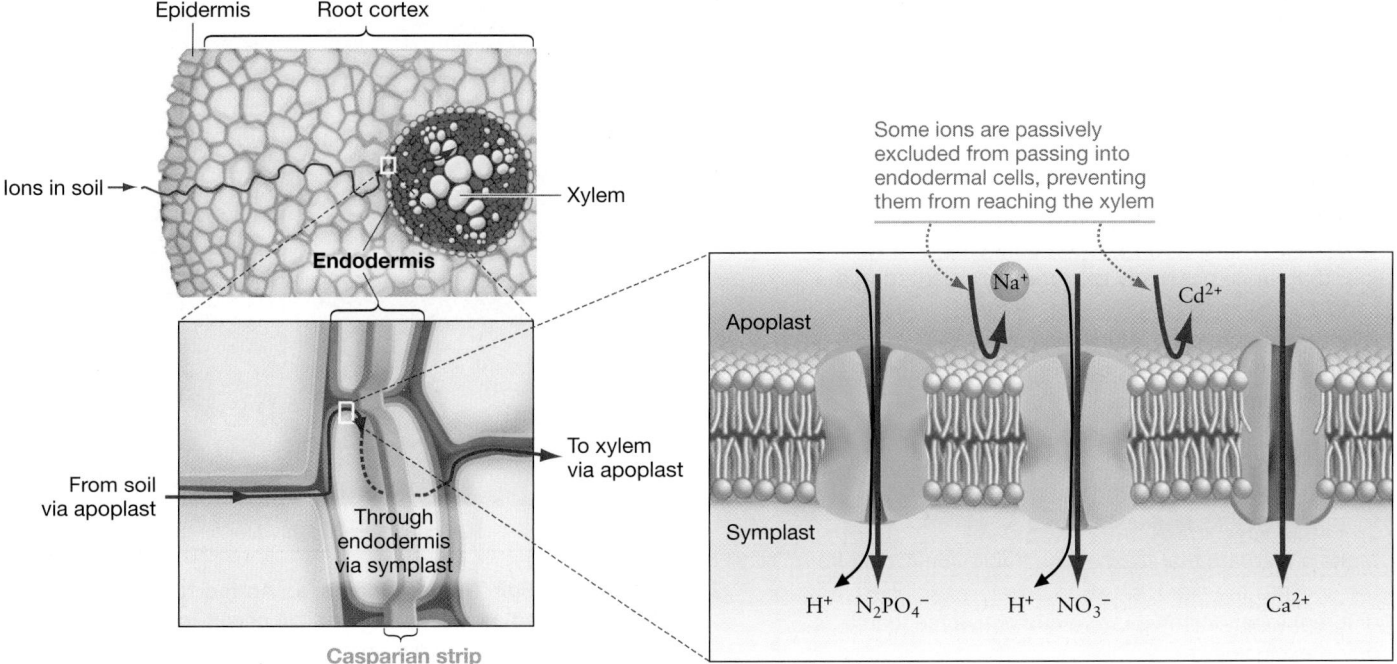

FIGURE 39.12 Passive Exclusion Occurs at the Plasma Membrane. Certain ions are excluded from cells because it is difficult for them to cross the plasma membrane.

One mechanism for coping with toxic concentrations of metals involves small proteins called **metallothioneins.** Metallothioneins bind to metal ions and prevent them from acting as a poison. Producing metallothionein proteins requires an expenditure of energy and represents a form of active exclusion.

Genes for metallothioneins have been found in a wide variety of organisms—bacteria, fungi, and animals as well as plants. Recent research on the mustard-family species *Arabidopsis thaliana* has shown that individuals from populations with a high tolerance for copper produce many more metallothionein proteins than do individuals from populations with a low tolerance for copper. (For more on the importance of *Arabidopsis thaliana* as a model organism, see **BioSkills 13**, Appendix B.)

Active Exclusion by Antiporters A second mechanism for actively neutralizing specific toxins involves transport proteins located in the **tonoplast**—the membrane surrounding the large, central vacuole. Proteins in the tonoplast membrane allow plants to actively remove toxic substances from the cytosol and store them in the vacuole.

Perhaps the best-studied example involves proteins that move sodium ions from the cytosol into vacuoles, where they cannot poison enzymes (**FIGURE 39.13a**). H^+-ATPases in the tonoplast move protons into the vacuole, creating an electrochemical

check your understanding

(C)(Y)(U)

If you understand that . . .

- Plants absorb most of the nutrients they need via membrane proteins in root hairs.
- Proteins in the plasma membranes of root-hair, cortex, and endodermal cells pump protons out of the cells, creating an electrochemical gradient that favors the entry of selected ions via ion channels or cotransporters.
- In many species, mycorrhizal fungi are important for bringing ions that contain nitrogen or phosphorus atoms into the root.
- If no membrane proteins in the epidermal, cortex, or endodermal cells of a root admit a certain type of ion, that ion is passively excluded from entering the root.
- Plants can actively exclude toxic ions by pumping them into vacuoles or binding them to metallothionein proteins.

✔ You should be able to . . .

1. Explain how plants generate and use electrical power to import cations and anions.
2. Make a diagram that traces a cadmium ion (Cd^{2+}) from the soil into the plant, showing how it is excluded at different locations by passive and active mechanisms of ion exclusion.

Answers are available in Appendix A.

(a) In the tonoplast, antiporters send H^+ out of, and Na^+ into the vacuole.

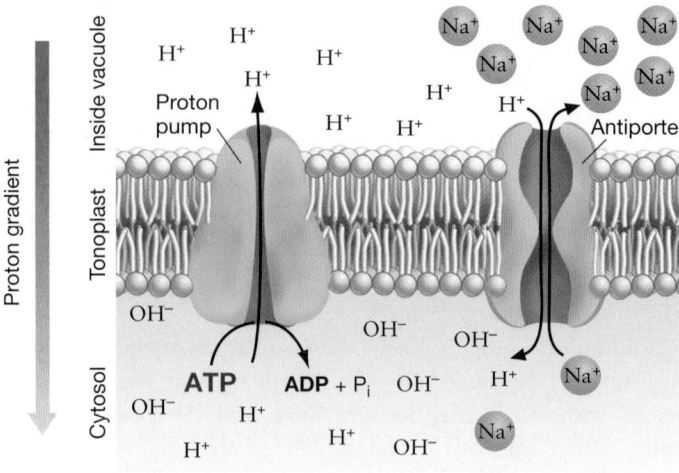

(b) Plants with additional H^+/Na^+ antiporters tolerate salt.

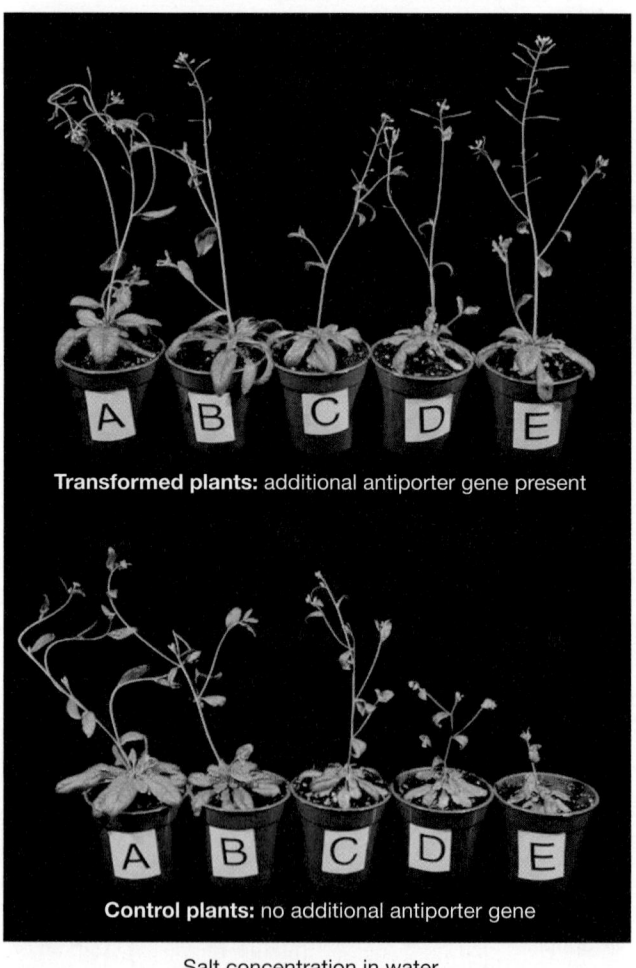

Transformed plants: additional antiporter gene present

Control plants: no additional antiporter gene

Salt concentration in water
Low ⟵——————————————⟶ High

FIGURE 39.13 In Salt-Tolerant Plants, an Antiporter Concentrates Sodium in Vacuoles. (a) In the tonoplast of salt-tolerant species, proton pumps establish a proton gradient. This gradient allows antiporters to concentrate sodium ions inside the vacuole, where they cannot poison enzymes in the cytoplasm. **(b)** *Arabidopsis* plants that were transformed with the gene for the antiporter protein grow well even when watered with salty water.

gradient that favors the movement of H^+ out of the vacuole. A transport protein that functions as an **antiporter** then uses this gradient to conduct protons *out* of the vacuole, move them down their electrochemical gradient, and bring sodium ions *into* the vacuole—against the sodium concentration gradient.

Recall that an antiporter is a cotransporter involved in moving the solutes in opposite directions instead of the same direction—as solutes do through symporters (see Chapter 38).

Knowledge of this H^+/Na^+ antiporter has created a great deal of excitement among biologists, because it offers a way to alter crop plants genetically so that they can grow well in salty soils created by poor irrigation practices. As an example, consider an experiment in which normal *Arabidopsis* plants and individuals transformed with extra copies of the *Arabidopsis* gene for the H^+/Na^+ antiporter were exposed to different levels of NaCl. The genetically transformed plants accumulated more sodium in their vacuoles, and as **FIGURE 39.13b** shows, they were able to grow efficiently even when exposed to high concentrations of salt.

39.4 Nitrogen Fixation

Nitrogen gas (N_2) makes up 80 percent of the atmosphere. Unfortunately, plants and other eukaryotes cannot use nitrogen in this form. Nitrogen gas is unreactive because it takes a great deal of energy to break the triple bond between the two nitrogen atoms.

To synthesize amino acids, nucleic acids, and other nitrogen-containing compounds, plants normally absorb nitrogen in forms such as ammonium (NH_4^+) or nitrate ions. But these ions are in short supply in many soils, meaning that plant growth is often limited by the availability of usable nitrogen. Most crop plants grow much faster when they receive a nitrogen-containing fertilizer.

Nitrogen-based fertilizers have drawbacks, however. Fertilizer production is extremely energy intensive, and in many parts of the world, ammonia and other nitrogen-based fertilizers are too expensive for farmers. In more affluent regions, these fertilizers are used so extensively that they are causing serious pollution problems (see Chapter 29). For these and other reasons, there is intense interest in understanding the molecular basis of a phenomenon called biological nitrogen fixation.

The Role of Symbiotic Bacteria

Among all the organisms on the tree of life, only a few species of bacteria and archaea are able to absorb N_2 from the atmosphere and convert it to ammonia (NH_3), nitrites (NO_2), or nitrates. This process is called **nitrogen fixation.**

Nitrogen fixation requires a series of specialized enzymes and cofactors, including a large multi-enzyme complex called nitrogenase. The process is extremely energy demanding. An expenditure of 8 high-energy electrons and 16 ATP molecules is required for nitrogenase to reduce one molecule of N_2 to two molecules of NH_3. The process can be summarized as follows:

$$N_2 + 8\,e^- + 8\,H^+ + 16\,ATP \longrightarrow 2\,NH_3 + H_2 + 16\,ADP + 16\,P_i$$

In some cases, bacterial cells that are capable of nitrogen fixation take up residence *inside* plant root cells. Although several different bacteria and plant hosts can be involved, the best-studied nitrogen-fixing bacteria are members of the genus *Rhizobium* that associate with plants in the **legume** family, which includes peas, soybeans, clover, and alfalfa. Members of the bacterial genus *Rhizobium* and closely related species are often called **rhizobia.**

Because of their ability to fix nitrogen, legumes are extremely important agricultural crops. A legume crop is often planted after harvesting a grain crop such as corn or wheat, so it can add nitrogen to the soil. This type of crop rotation results in "free" nitrogen fertilizer that benefits the next grain crop. Because legumes are rich in nitrogen, they are an important part of the human diet in many parts of the world.

As **FIGURE 39.14** shows, the infected root cells of legumes form distinctive structures called **nodules,** where nitrogen-fixing rhizobia are found. The nodules are pink because they contain an iron-containing molecule called leghemoglobin (short for legume hemoglobin). **Leghemoglobin** is related to the hemoglobin that carries oxygen in your blood. Like hemoglobin, leghemoglobin binds oxygen.

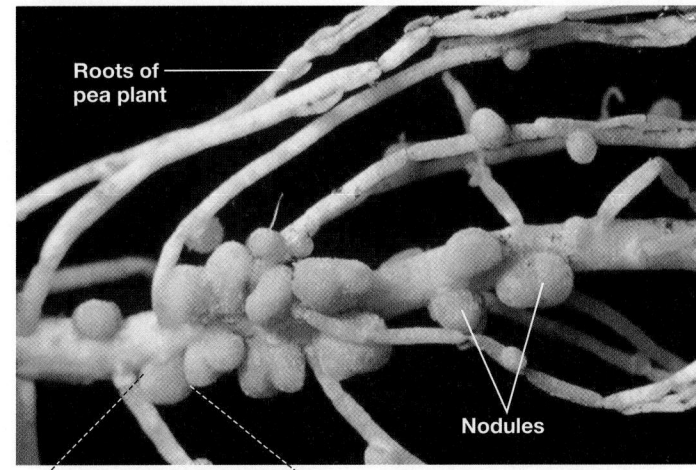

Roots of pea plant

Nodules

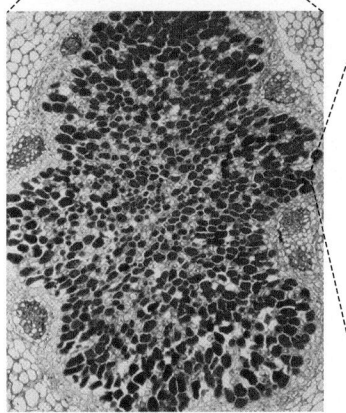

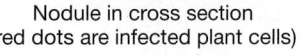
Nodule in cross section (red dots are infected plant cells)

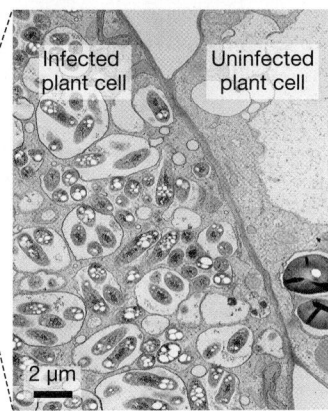
Infected plant cell

Uninfected plant cell

2 μm

Close-up of rhizobia inside vesicles, inside infected plant cell

FIGURE 39.14 In Some Plants, Roots Form Nodules Where Nitrogen-Fixing Bacteria Live.

Leghemoglobin is important because nitrogenase—the enzyme complex responsible for nitrogen fixation—is poisoned by the presence of oxygen. Inside root nodules, oxygen molecules bind to leghemoglobin instead of binding to nitrogenase. In this way, leghemoglobin keeps levels of free oxygen low enough for nitrogenase to function while maintaining a source of oxygen for electron transport to occur. Intense cellular respiration and ATP synthesis are necessary to supply nitrogenase with the energy it requires.

If roots allowed bacterial pathogens to enter, the plant would possibly die from the subsequent infection. How do roots distinguish beneficial bacteria from pathogens and allow only the rhizobia to infect the roots?

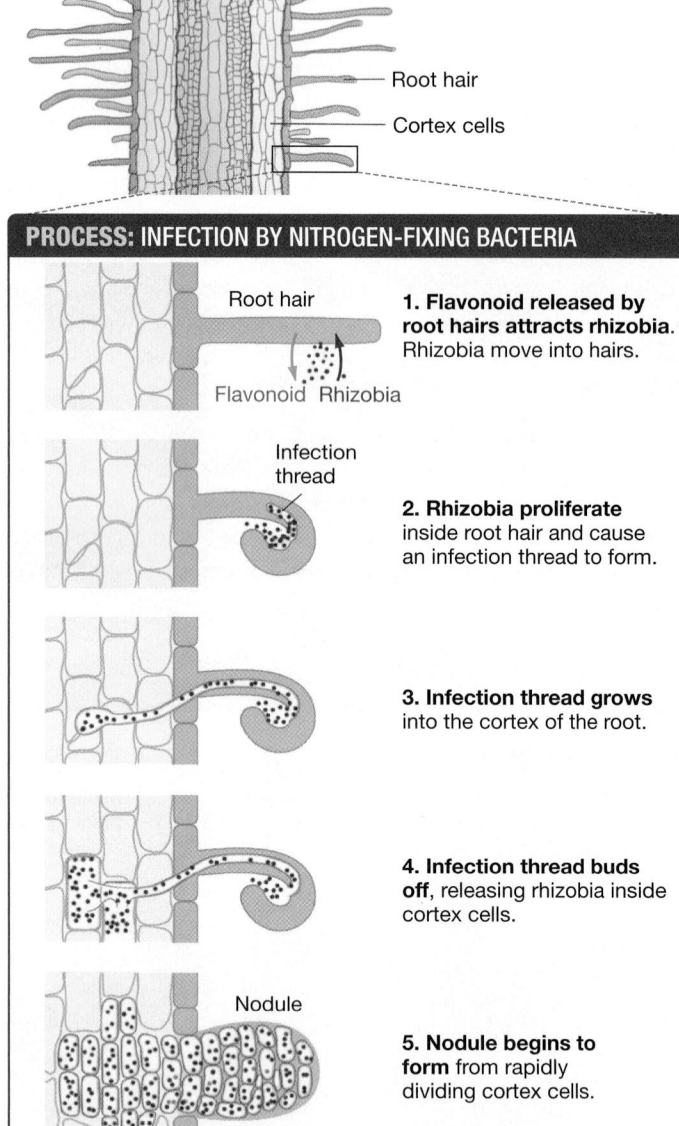

Root

Root hair

Cortex cells

PROCESS: INFECTION BY NITROGEN-FIXING BACTERIA

Root hair

Flavonoid Rhizobia

1. Flavonoid released by root hairs attracts rhizobia. Rhizobia move into hairs.

Infection thread

2. Rhizobia proliferate inside root hair and cause an infection thread to form.

3. Infection thread grows into the cortex of the root.

4. Infection thread buds off, releasing rhizobia inside cortex cells.

Nodule

5. Nodule begins to form from rapidly dividing cortex cells.

FIGURE 39.15 Infection by Nitrogen-Fixing Bacteria Is a Multistep Process. After rhizobia bind to a root hair, they enter the cytoplasm in a membrane-lined infection thread and travel into the root cortex, where they enter cortex cells.

How Do Nitrogen-Fixing Bacteria Infect Plant Roots?

Like mycorrhizal fungi and their host plants, legumes and rhizobia have a mutualistic relationship. The nitrogen-fixing bacteria provide the plant with ammonia, while the legume provides the bacteria with carbohydrates and protection. The association is costly for the host plant, however. For the bacteria to synthesize the enormous amounts of high-energy electrons and ATP required to manufacture ammonia, plants have to supply them with large quantities of sugar.

Given the cost of the plant–bacterium interaction and its importance to both species, it's not surprising that the infection process is carefully regulated.

When a legume seed germinates, its roots do not contain a population of rhizobia. Instead, the roots make contact with bacterial cells existing in the soil, the rhizobia infect root cells, and the root cells grow into a nodule.

The first event in infection is a recognition step that occurs between a legume-family plant and its symbiotic bacterium. First, young roots release compounds called flavonoids. When rhizobia contact the flavonoids, the bacteria respond by producing sugar-containing molecules called **Nod factors** (for *nod*ule formation). Nod factors, in turn, bind to signaling proteins on the membrane surface of root hairs.

The recognition step is specific to the species involved. Each legume species produces a different flavonoid that acts as a recognition signal, and each rhizobium species responds with one or more unique Nod factors. When researchers have switched recognition signals or Nod factors between species, the recognition step fails.

When Nod factors bind to the root-hair surface, they set off a chain of events that leads to dramatic morphological changes in the host legume, as **FIGURE 39.15** shows.

1. The bacterial cells contact the root-hair surface.

2. Rhizobia begin to multiply and move, through an invagination of the root-hair plasma membrane called an **infection thread.**

check your understanding

C

Y

U

If you understand that . . .

• Certain species of bacteria infect plant root cells and fix nitrogen, which is then made available to the plant.

✔ **You should be able to . . .**

Generate hypotheses addressing two additional questions about nitrogen fixation:

1. Why don't all plants have symbiotic bacteria that fix nitrogen?

2. Why is the interaction that establishes the infection by nitrogen-fixing bacteria so complex?

Answers are available in Appendix A.

3. The infection thread extends into the root, invading the cortex.

4. Portions of the infection thread bud off, forming membrane-bound clusters within cortex cells, each filled with rhizobia.

5. The infected cortex cells divide, forming root nodules.

The interaction between rhizobia and legumes is one of the most complex and best studied of all mutualisms.

39.5 Nutritional Adaptations of Plants

Based on the data available so far, over 95 percent of vascular plants take up nutrients from soil. And over 80 percent of all plants supplement their "diet" with nutrients acquired from mycorrhizal fungi, while a small but significant fraction associate with nitrogen-fixing bacteria. Perhaps 99 percent of all living plant species make their own sugar through the process of photosynthesis.

What about the small number of plant species that don't follow these rules? Some appear to live on air, some parasitize other plants, and others catch insects and digest them.

Epiphytic Plants

Species from a diverse array of plant lineages do not absorb nutrients from soil. In fact, these species never even make contact with soil. As **FIGURE 39.16a** shows, they often grow on the trunks or branches of trees. For this reason they are called **epiphytes** ("upon-plants").

In northern forests, it is common for mosses and ferns to grow epiphytically on tree trunks. The so-called Spanish moss that hangs from oak trees in the southeastern United States is another familiar epiphyte—although this species is actually not a moss but a bromeliad, a relative of the pineapple. In the tropics and subtropics, one-third of all ferns grow as epiphytes, and

there are thousands of species of epiphytic orchids, bromeliads, and lycophytes.

Epiphytes absorb most of the water and nutrients they need from rainwater, dust, and particles that collect in their tissues or in the crevices of bark. As **FIGURE 39.16b** shows, some epiphytic bromeliads have leaves that grow in rosettes and form "tanks" that collect water and organic debris. In such cases, nutrients are actually absorbed through the leaves themselves.

Parasitic Plants

Parasites are organisms that live in close physical contact with individuals from another species and that lower the fitness of those individuals, usually by obtaining water or nutrients from the host. Biologists estimate that about 3000 species of angiosperms are parasitic—less than 1 percent of the plant species that have been studied and named to date.

Some plant parasites are non-photosynthetic and obtain all of their nutrition by tapping into the vascular tissue of the host individual. But most parasitic plants make their own sugars through photosynthesis and tap the xylem of other species for water and essential nutrients.

For example, the mistletoe (**FIGURE 39.17**, see page 790) is green and photosynthetic but has structures called **haustoria** (singular: **haustorium**) that penetrate a host's xylem and extract water and ions that the mistletoe uses as nutrients. In the forests of western North America, mistletoe infection is a serious cause of economic loss.

Carnivorous Plants

Carnivorous plants trap insects and other animals, kill them, and then digest the prey to absorb its nutrients. Carnivorous species make their own carbohydrates via photosynthesis but use carnivory to supplement the nitrogen available in the environment. Most are found in bogs or other habitats where nitrogen is scarce or unavailable.

(a) Epiphytes grow on trees.

(b) Water-holding "tanks" formed by leaves of an epiphyte

Tanks

FIGURE 39.16 Epiphytes Are Adapted to Grow in the Absence of Soil. The environments where many epiphytes grow—tree trunks and branches—are dry and nutrient poor.

Mistletoe

Host tree

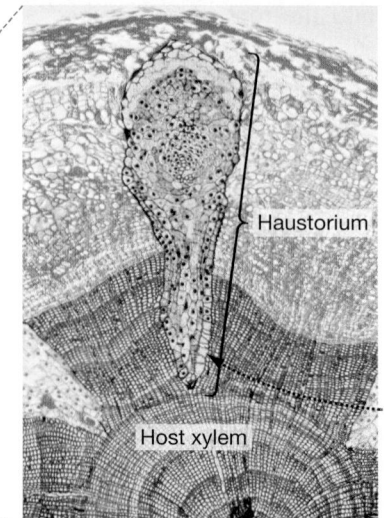

Haustorium

Host xylem

FIGURE 39.17 Some Plant Parasites Tap into the Vascular Tissue of Their Hosts.

✓**QUESTION** The bromeliads in Figure 39.16 and mistletoe both grow on other plants. How do they differ in their relationship to the trees that support them?

Mistletoe haustoria penetrate host xylem and extract water and ions

Modified Leaves Form an Array of Trapping Mechanisms Carnivorous plants use modified leaves or roots to trap insects. For example, the leaves of pitcher plants form tubes (see Table 37.3). Prey are enticed to the tube by an attractive odor and then have a difficult time climbing back out. Eventually they fall into the pool of water below, where they drown. Enzymes released by the plant digest the prey, and the plant absorbs the nutrients obtained from the insects' dead bodies.

Sundews have an alternative hunting strategy: They have modified leaves that function like flypaper. The leaf surfaces develop hairs that exude a sticky substance and trap insects (**FIGURE 39.18**). Glands near the sundew's trap release enzymes that digest the prey. The leaf then absorbs the nutrients that are released.

In the Venus flytrap, modified leaves trap insects mechanically. Sensory hairs protrude from the epidermis of each leaf. When an insect lands on the trap and bumps two or three of these hairs, the hair cells produce an electrical signal and the leaf responds by snapping shut (see Chapter 40). If the prey continues to stimulate hairs, the leaf secretes enzymes that digest the prey.

Costs and Benefits of Carnivory Initially, research on carnivorous plants focused on documenting that meat-eating was adaptive. Early experiments in natural environments supported this hypothesis by showing that, compared to individuals of the same species that are not fed fruit flies, carnivorous plants that were fed fruit flies grew faster and flowered more often.

Follow-up work explored the costs as well as the benefits of carnivory. For example, several teams of researchers showed that when carnivorous plants are provided with nitrogen-based fertilizers, they produce fewer of their specialized insect-trapping leaves and a higher proportion of leaves that function primarily in photosynthesis. These results suggest carnivory is a trait that shows phenotypic plasticity (see Chapter 37): Plants increase investment in prey-capture devices when nitrogen is rare, but they decrease investment in prey-capturing structures when nitrogen is readily available.

Currently, some research groups are doing "evo-devo" research (see Chapter 22) on carnivory. They are working to identify the novel alleles that made the evolution of carnivory possible in certain groups. Meat-eating may be relatively rare in plants, but it has inspired interesting new directions for research.

Trapped insect

FIGURE 39.18 Sundews Have Modified Leaves with Sticky Surfaces That Catch Insects.

If you understand . . .

39.1 Nutritional Requirements of Plants

- About 96 percent of the dry weight of a typical plant consists of carbon, hydrogen, and oxygen. Plants obtain these elements by absorbing carbon dioxide from the atmosphere and water from the soil.

- The other 4 percent of the plant body consists of a complex suite of elements, acquired from soil, that are required for normal growth and reproduction. These essential elements are usually absorbed in the form of ions.

- Nutrients are usually classified as macronutrients, needed in large supply, and micronutrients, needed in only trace amounts.

- When plants are lacking in an essential nutrient, they usually develop predictable deficiency symptoms.

✔ You should be able to design a field experiment that would determine whether nitrogen is a limiting nutrient in a habitat near your campus.

39.2 Soil: A Dynamic Mixture of Living and Nonliving Components

- Soil is a complex and dynamic mixture of inorganic particles such as clay and sand, organic matter, and organisms. Soil provides plants with oxygen, water, and nutrients as well as a physical substrate for anchoring and supporting the plant body.

- Plants absorb nutrients in the form of ions. Anions in soil water are normally available to roots, but cations adhere to soil particles and are made available only after cation exchange.

✔ You should be able to explain why loam is an ideal soil for most plants.

39.3 Nutrient Uptake

- Nutrients are present at low concentrations in the soil surrounding roots. Plants import nutrients against a concentration gradient by pumping protons into the extracellular space.

- A large excess of protons outside the root hair creates a strong electrochemical gradient favoring the entry of positively charged ions through membrane transporters.

- The proton gradient in root-hair membranes also allows negatively charged ions to cross the plasma membrane via symporters.

- In addition to acquiring nutrients by establishing a proton gradient, many plants obtain nitrogen or phosphorus through a mutualistic relationship with mycorrhizal fungi in exchange for carbohydrates derived from photosynthesis.

- Passive and active systems also exist for excluding certain ions. This is important because all nutrients are toxic at high concentrations.

✔ You should be able to predict the consequences of mutations that allow root cells to establish a membrane voltage that is much more negative than normal, without a large increase in energy expenditure.

39.4 Nitrogen Fixation

- Certain plants, including those in the legume family, are capable of symbiosis with nitrogen-fixing bacteria. The bacteria colonize the interior of root cells after exchanging species-specific signals.

- Nitrogen-fixing bacteria receive protection and sugar from the host plant, in exchange for producing ammonia, a useable form of nitrogen for the plant, from nitrogen gas.

✔ You should be able to design an experiment, using radioactive carbon and the heavy isotope of nitrogen ($^{15}N_2$), that would test whether the rhizobia–pea plant interaction is mutualistic.

39.5 Nutritional Adaptations of Plants

- Epiphytes grow on the trunks and branches of other plants and absorb mineral nutrients from rainwater, dust, or decaying matter trapped in the bark of their host plant.

- Some parasitic plants produce their own carbohydrates through photosynthesis but obtain water and nutrients by penetrating the xylem of host plants.

- Carnivorous plants have evolved mechanisms for trapping and digesting insects and other animals. The primary benefit of carnivory is to obtain nitrogen in low-nitrogen environments.

✔ You should be able to explain why epiphytes in tropical rain forests and desert plants often have similar adaptations to minimize water loss

(MB) MasteringBiology

1. **MasteringBiology Assignments**

   Tutorials and Activities Global Soil Degradation; How Plants Obtain Minerals from Soil; Nitrogen Nutrition in Plants; Soil Formation and Nutrient Uptake

   Questions Reading Quizzes, Blue-Thread Questions, Test Bank

2. **eText** Read your book online, search, take notes, highlight text, and more.

3. **The Study Area** Practice Test, Cumulative Test, BioFlix® 3-D Animations, Videos, Activities, Audio Glossary, Word Study Tools, Art

You should be able to . . .

1. Which of the following characteristics defines an element as essential for a particular species?
   a. It has to be added as fertilizer to achieve maximum seed production.
   b. If it is missing, a plant cannot grow or reproduce normally.
   c. If it is present at high concentration, plant growth increases.
   d. If it is absent, other nutrients may be substituted for it.

2. Why is the presence of clay particles important in soil?
   a. They provide macronutrients—particularly nitrogen, phosphorus, and potassium.
   b. They bind metal ions, which would be toxic if absorbed by plants.
   c. They allow water to percolate through the soil, making oxygen-rich air pockets available.
   d. The negative charges on clay bind to positively charged ions and prevent them from leaching.

3. Where does most nutrient uptake occur in roots?
   a. at the root cap, where root tissue first encounters soil away from the zone of nutrient depletion
   b. at the Casparian strip, where ions must enter the symplast before entering xylem cells
   c. in the symplastic and apoplastic pathways
   d. in root hairs, in the zone of maturation

4. Why are proton pumps in root-hair plasma membranes important?
   a. They pump protons into cells, generating a membrane potential (voltage).
   b. They allow toxins to be concentrated in vacuoles, so the toxins do not poison enzymes in the cytoplasm.
   c. They set up an electrochemical gradient that makes it possible for roots to absorb cations and anions.
   d. They set up the membrane voltage required for action potentials to occur.

5. Some plants obtain nitrogen from symbiotic fungi or bacteria. What were the original sources of N that the fungi and bacteria obtained?

6. Before the formation of a root nodule, which partner secretes a Nod factor, the plant or the bacterium?

7. Why is the relationship between most mycorrhizal fungi and their host plants considered mutualistic?
   a. They live in close physical association.
   b. Both species benefit from the association.
   c. The host plant cannot live without the mycorrhizae.
   d. The mycorrhizae cannot live without the host plant.

8. A farmer is concerned that her corn crop is suffering from iron deficiency. Design an experiment that uses hydroponic cultures to identify (1) the symptoms of iron deficiency in corn and (2) the minimum amount of iron required to support normal growth rates.

9. Explain why carnivorous and parasitic plants are most common in nutrient-poor habitats.

10. Would you expect plant productivity to be higher in sandy soils or in soils containing both sand and clay? Why?

11. Why is it important for plants to exclude certain ions? Summarize the difference between active and passive exclusion mechanisms.

12. Why is it logical for biologists to (1) distinguish nutrients that are obtained from the atmosphere or water (C, H, and O) from mineral nutrients obtained from soil, and (2) distinguish macronutrients from micronutrients?

13. **QUANTITATIVE** There is a conflict between van Helmont's data on willow tree growth and the data on essential nutrients listed in Table 39.1. According to the table, nutrients other than C, H, and O should make up about 4 percent of a willow tree's weight. Most or all of these nutrients should come from soil. But van Helmont claimed that the soil in his experiment lost just 60 g, while the tree gained 74,000 g. Calculate the percentage of added weight contributed by soil. State at least one hypothesis to explain the conflict between expected and observed results. How would you test this hypothesis?

14. Acid rain occurs when sulfur oxides and nitrous oxides released by cars or factories react with water vapor, forming sulfuric and nitric acids, respectively. In areas affected by acid rain, cations can quickly leach out of the soil. Explain why.

15. Design an experiment using the drug vanadate, which poisons proton pumps, to test the hypothesis that phosphate ions enter cells via an H^+/HPO_4^{2-} cotransporter.

16. Alder trees have nitrogen-fixing bacteria associated with their roots and have a much higher proportion of nitrogen in their tissues than do spruce trees that grow in the same habitat. Which species should decay faster after death: alder or spruce? Explain your logic.

40 Plant Sensory Systems, Signals, and Responses

In this chapter you will learn how

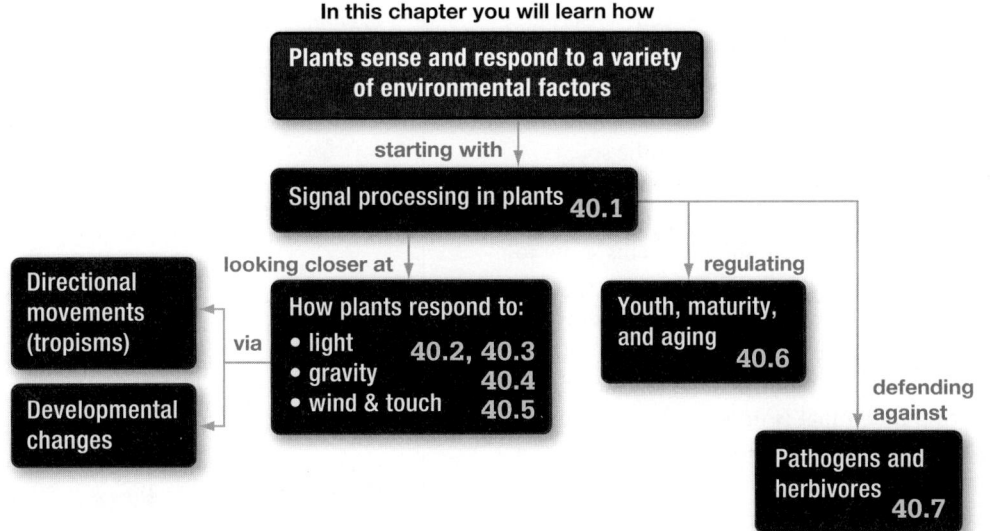

Plants sense and respond to a variety of environmental factors

starting with ↓

Signal processing in plants 40.1

looking closer at ↓   *regulating* ↓

Directional movements (tropisms)

via

How plants respond to:
- light   40.2, 40.3
- gravity   40.4
- wind & touch   40.5

Youth, maturity, and aging 40.6

Developmental changes

defending against

Pathogens and herbivores 40.7

Plants have sophisticated information processing systems. These radish seedlings sense the presence of light and are bending toward it.

This chapter is part of the Big Picture. See how on pages 840–841.

Imagine standing in place for several hundred years, like an oak tree. Each spring you produce flowers. All summer you absorb light from the Sun, carbon dioxide from the atmosphere, and water and nutrients from the soil. Water, ions, and sugars flow up and down your vascular tissue. For six months or more your body grows upward, outward, and downward. In fall, thousands of your offspring drop to the ground as acorns. Then as winter approaches, your metabolism slows and you stop growing. Like a hibernating animal, you spend the long, hard months of winter in a state of suspended animation, before resuming growth the following spring.

✔ When you see this checkmark, stop and test yourself. Answers are available in Appendix A.

To stay alive, an oak needs to gather information about its environment. It has to sense the season of the year, the time of day, the pull of gravity, the force of wind, and attacks by enemies. It needs to sense when its leaves are being shaded and which leaves are receiving the wavelengths of light that are required to support photosynthesis.

Plants may not have eyes or ears, but they can sense light, gravity, pressure, and wounds. They have the equivalent of a sense of smell, because they can perceive certain airborne molecules. It can even be argued that they have a sense of taste, because their roots sense the presence of nutrients in the soil.

In addition to gathering information about the conditions around them, plants have to respond in an appropriate way. They do not jump or swim or run, but their shoot systems grow toward light or end up shorter and stockier in response to wind. In response to gravity, shoots grow up and roots grow down. In response to touch, the modified leaves of a Venus flytrap shut fast enough to catch flying insects. If a plant is being attacked or if it senses that a neighboring individual is under attack, it may lace its tissues with toxic compounds or mobilize other defenses.

The message of this chapter is simple: Plants have sophisticated systems for collecting information about their environment and responding in ways that maximize their chances of surviving, thriving, and producing offspring. The ability to gather information and respond to it is one of the five fundamental attributes of life (see Chapter 1).

40.1 Information Processing in Plants

Every environment is full of information. But, like other organisms, plants monitor only aspects of the environment that matter to them—those that affect their ability to stay alive and produce offspring.

FIGURE 40.1 summarizes the three-step process by which plants gather, process, and respond to the information they monitor:

1. Receptor cells receive an external signal and change it to an intracellular signal.

2. Receptor cells send a signal to cells in other parts of the body that can respond to the information.

3. Responder cells receive this signal, transduce it to an intracellular signal, and change their activity in a way that produces an appropriate response.

Let's briefly consider how each of these steps works, then explore some of the details of how plants sense and respond to light, gravity, and other signals from the environment.

How Do Cells Receive and Transduce an External Signal?

When you text a friend, the signal travels from you to their receiver via airwaves. When the message arrives, the receiving cell

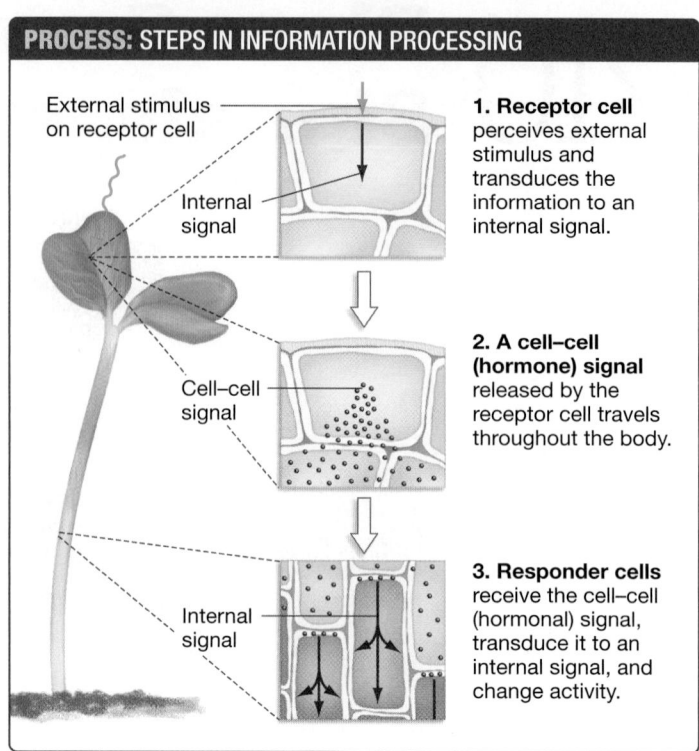

PROCESS: STEPS IN INFORMATION PROCESSING

External stimulus on receptor cell

Internal signal

1. Receptor cell perceives external stimulus and transduces the information to an internal signal.

Cell–cell signal

2. A cell–cell (hormone) signal released by the receptor cell travels throughout the body.

Internal signal

3. Responder cells receive the cell–cell (hormonal) signal, transduce it to an internal signal, and change activity.

FIGURE 40.1 In Many Cases, Cell–Cell Signals Link a Stimulus and a Response.

phone changes the information in the airwaves into electrical signals and then into words that your friend can understand.

Plants work in much the same way. When light strikes a sensory cell, for example, the information that it carries has to be changed into a form that is meaningful to that cell. Signals from the environment are usually detected by proteins specialized for that function. Receptor proteins change shape in response to an environmental stimulus, such as being struck by a particular wavelength of light, having pressure applied, or binding to a particular type of molecule.

When a receptor changes shape in response to a stimulus, the information changes form—from an external signal to an intracellular signal. This process is called **signal transduction** (**FIGURE 40.2**); the verb transduce means "to convert energy from one form to another." Once information has been transduced to an intracellular form, it travels down what biologists call a signal transduction pathway.

You might recall that there are two basic types of signal transduction pathways in all organisms: phosphorylation cascades and second messengers (see Chapter 11). Both begin with a receptor protein in the plasma membrane.

- **Phosphorylation cascades** are triggered when the change in the receptor protein's shape leads to the transfer of a phosphate group (PO_4^{3-}) from ATP to the receptor or an associated protein. Phosphorylation activates proteins involved in signal transduction cascades, causing them to phosphorylate and activate a different set of proteins, which in turn catalyze the phosphorylation and activation of still other proteins, and so on.

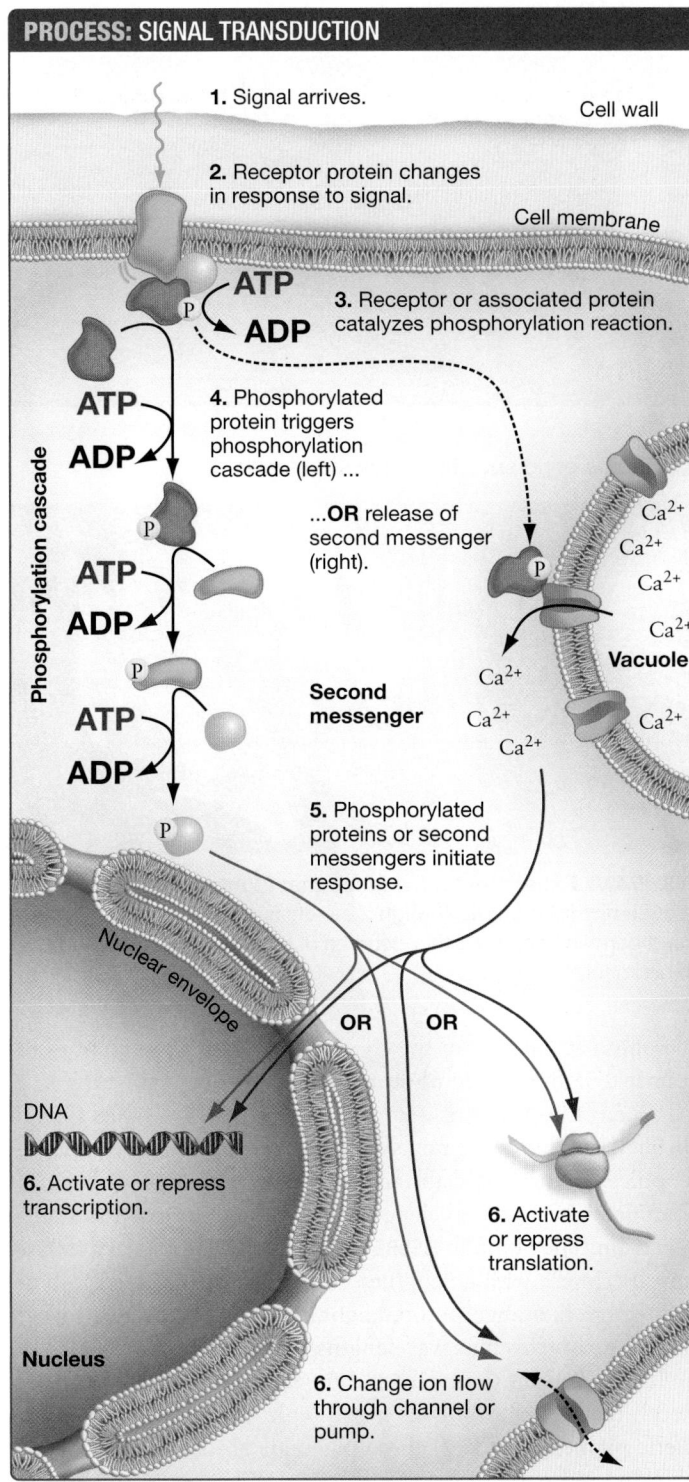

PROCESS: SIGNAL TRANSDUCTION

1. Signal arrives.

Cell wall

2. Receptor protein changes in response to signal.

Cell membrane

ATP

ADP

P

3. Receptor or associated protein catalyzes phosphorylation reaction.

Phosphorylation cascade

ATP

ADP

P

ATP

ADP

P

4. Phosphorylated protein triggers phosphorylation cascade (left) ...

...**OR** release of second messenger (right).

P

Ca^{2+}
Ca^{2+}
Ca^{2+}
Ca^{2+}

Vacuole

Ca^{2+}

Second messenger

Ca^{2+}

Ca^{2+}

Ca^{2+}

ATP

ADP

P

5. Phosphorylated proteins or second messengers initiate response.

Nuclear envelope

OR   **OR**

DNA

6. Activate or repress transcription.

6. Activate or repress translation.

Nucleus

6. Change ion flow through channel or pump.

FIGURE 40.2 Signal Transduction Changes an External Signal to an Internal Signal. After an environmental or cell–cell signal is transduced to an intracellular signal, it triggers a change in the cell's activity.

- **Second messengers** are produced when receptor proteins trigger the production of intracellular signals or their release from storage areas. Calcium ions (Ca^{2+}) stored in the vacuole, ER, or cell wall are one of several ions or molecules that function as the second messenger in plants.

In some cases, phosphorylation cascades and second messengers interact.

Signal transduction primes the receptor cell for action. In many cases, though, the cells that receive information from the environment are located in a part of the body that is distant from the cells that need to respond to the information. How does information from activated receptor cells get to responder cells?

In most cases, the answer involves a **hormone**—an organic compound that is produced in small amounts in one part of a plant and transported to target cells, where it causes a physiological response. Signal transduction in a receptor cell often results in the release of a hormone that carries information to responder cells.

How Are Cell–Cell Signals Transmitted?

Because plants perceive such a wide array of stimuli, they have a wide array of hormones coursing through their bodies. These signals may be transmitted from cell to cell by specialized transport proteins in cell membranes, in xylem sap or phloem sap, or by simple diffusion from the originating cell. In most cases the signals act on target tissues throughout the body.

Plant cells routinely receive information from several different hormones at the same time, so it is common for different types of hormones to interact with each other and modulate the cell's response.

How Do Cells Respond to Cell–Cell Signals?

Cells are exposed to a constant stream of hormones. Many of these signals have little to no effect on what is happening inside a given cell. The reason is simple: Hormones can elicit a response only if a cell has an appropriate receptor.

If a receptor exists on or in the cell, an arriving hormone binds to it. When the receptor changes shape in response, the effect is like a knock on the door of your room. The signal—often received at the cell's periphery—is rapidly transduced to a series of phosphorylated proteins or the release of a second messenger inside the cell. The result may be a dramatic change in the cell's activity.

To understand why these binding events can have such significant effects on cell activity, look again at Figure 40.2. Activation of a signal transduction cascade results in the production of many phosphorylated proteins, or the release of many second messengers. These events amplify the original signal many times. In this way, low concentrations of plant hormones can have a large impact on target cells. Hormones are small molecules in low concentrations, but because the information they carry is amplified during signal transduction, they produce big results.

As Figure 40.2 shows, the response to hormone binding can include

- activation of membrane transport proteins (see Chapter 6), which produces a change in the membrane's electrical potential or the cell wall's pH; or

- changes in gene expression—via alterations in transcription activators or repressors (see Chapter 19) or the translation machinery—that result in new suites of proteins or RNAs in the cell.

The response to the signal is important. When cells respond to a hormone, the change in their activity helps the plant cope with the environmental change sensed by the receptor cell.

check your understanding

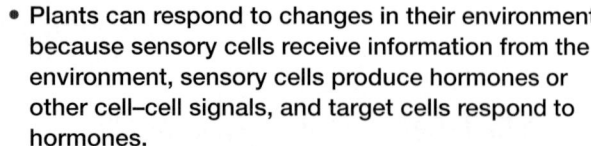

C
Y
U

If you understand that . . .

- Plants can respond to changes in their environment because sensory cells receive information from the environment, sensory cells produce hormones or other cell–cell signals, and target cells respond to hormones.
- Information processing in sensory cells and the cells that respond to cell–cell signals involves three steps:

  Step 1 A receptor molecule changes in response to the signal.

  Step 2 A signal transduction pathway transforms the hormonal signal into an intracellular signal.

  Step 3 The intracellular signal triggers a response—changes in the transcription of target genes, activity of specific enzymes or transport proteins, or other changes in cell activity.

✔ **You should be able to . . .**

1. Explain why only certain cells in the plant body—not all—respond to an environmental signal, and why only certain cells respond to a hormone.

2. Explain how cell–cell signals are amplified inside a target cell.

Answers are available in Appendix A.

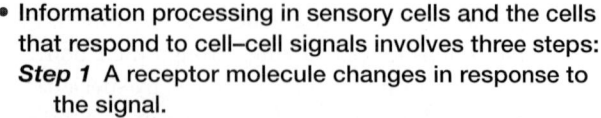

40.2 Blue Light: The Phototropic Response

Most of the general principles of plant communication emerged from studies of how plants respond to light. For example, consider the claim that plants are highly selective about the information they process. Light is made up of a wide array of wavelengths (see Chapter 10), but plants sense and respond to only a few.

This conclusion traces back to experiments that Charles Darwin and his son Francis performed in 1881 with coleoptiles of a monocot called reed canary grass. A **coleoptile** is a modified leaf that forms a sheath protecting the emerging shoots of young grasses.

The Darwins germinated seeds in the dark, placed the young, straight coleoptiles next to a light source, and noted that they grew toward the light—a response, due to differential cell elongation, called bending (**FIGURE 40.3a**). You have probably seen the same response in houseplants that are near a window (see

(a) Shoots bend toward full-spectrum light.

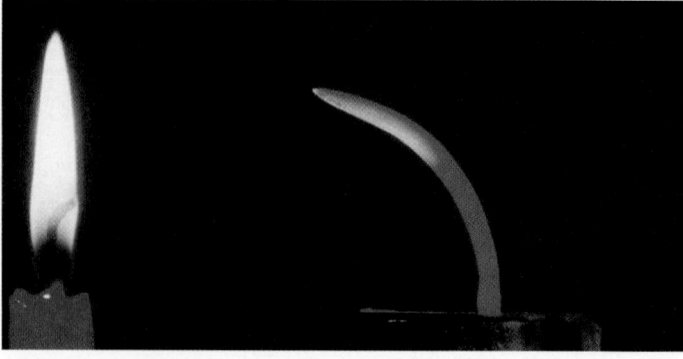

(b) Shoots bend specifically toward blue light.

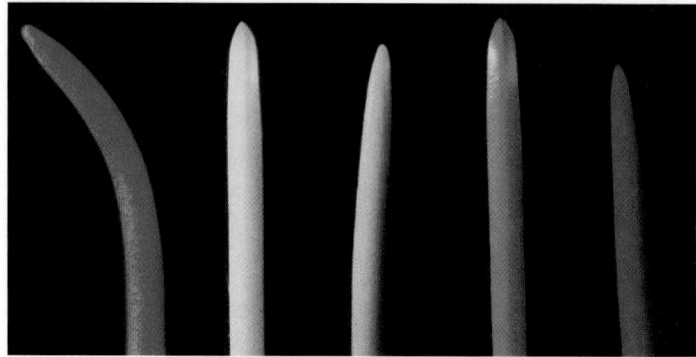

FIGURE 40.3 Experimental Evidence that Plants Sense Specific Wavelengths of Light. Though a candle is shown here, the Darwins used paraffin lamps or other sources of artificial light in their original experiments.

the photo at the start of the chapter). Directed movement in response to light is called **phototropism** (literally, "light-turn").

When the Darwins exposed coleoptiles to light filtered through a solution of potassium dichromate, however, the coleoptiles did not bend toward the light. Potassium dichromate solutions filter out wavelengths in the blue part of the visible spectrum. The photo in **FIGURE 40.3b** shows the results of a follow-up experiment with coleoptiles that have been exposed to blue, yellow, green, orange, and red light. Bending occurs only toward light that contains blue wavelengths.

It is important to understand the specificity of this response. Recall that (**1**) chlorophylls *a* and *b* are the primary photosynthetic pigments, and (**2**) these pigments absorb strongly in the blue and red parts of the spectrum (see Chapter 10). Plants exhibit a phototropic response if blue wavelengths are available, but show no response if blue wavelengths are not present. Plants move toward blue light, which is important in photosynthesis.

Let's look at what happens when blue light strikes a receptor cell, and follow the signal to examine the cells that undergo differential growth in response.

Phototropins as Blue-Light Receptors

Although biologists knew that the blue-light receptor must be a **pigment**—a molecule that absorbs certain wavelengths of light—it took decades to find it. A key breakthrough came in the early 1990s, when researchers found a membrane protein in the tips of emerging shoots that gains a phosphate group in response to blue light. Researchers hypothesized that the membrane protein becomes activated when it is phosphorylated in response to blue light, and that the activated protein then triggers the phototropic response.

Subsequent work succeeded in isolating the gene that codes for the membrane protein. The gene, named *PHOT1*, was found by analyzing mutant *Arabidopsis thaliana* individuals that do *not* show a phototropic response to blue light. Further experiments by Shin-ichiro Inoue and colleagues showed that *PHOT1* codes for a blue-light receptor. An important question remained, however. Is the PHOT1 protein activated by phosphorylation, and does this trigger bending of the plant?

As a first step in answering this question, Inoue and colleagues identified an amino acid in the PHOT1 protein that becomes phosphorylated after plants are exposed to blue light. During this analysis, they constructed a mutated *PHOT1* gene, called *phot1-m*, which coded for an altered protein with an amino acid that cannot be phosphorylated. This altered protein provided the researchers with a way to test whether phosphorylation of the PHOT1 protein was required for a phototropic response. **FIGURE 40.4** shows their experimental setup.

The key to the experiment was to compare bending in response to blue light in wild-type *Arabidopsis* with the response in mutant plants that lacked a PHOT1 protein. When Inoue and colleagues introduced the *phot1-m* gene into a mutant plant, the altered protein failed to rescue phototropism. However, when the normal *PHOT1* gene was introduced into a mutant plant, phototropism was restored.

The current consensus is that *PHOT1* encodes a blue-light receptor in plants. A phototropic response is initiated when this receptor protein is phosphorylated.

Recent research indicates that *Arabidopsis* has a second blue-light receptor related to PHOT1, called PHOT2. Collectively, **photoreceptors** that detect blue light and initiate phototropic responses are known as **phototropins.**

Given its importance to photosynthesis, it's not surprising that blue light triggers an array of responses in addition to bending. The phototropins, for example, trigger signal transduction cascades that result in at least two other responses.

1. Chloroplast movements inside leaf cells. These movements put chloroplasts in positions to optimize light absorption. In high light the chloroplasts move to the sides of cells to shade each other while in low light they spread out.

2. Opening of stomata. As a result, carbon dioxide can diffuse into cells as blue light triggers photosynthesis.

In addition to the phototropins, several other kinds of blue-light receptors have been identified in plants that also help control the opening of stomata, as well as stem elongation and flower production. Phototropism, however, ranks as the best studied of all blue-light responses. How is the signal from PHOT proteins transmitted to the cells that respond by bending?

RESEARCH

QUESTION: Is PHOT1 activated by phosphorylation?

HYPOTHESIS: Phototropism in *Arabidopsis* is triggered by phosphorylation of PHOT1 protein.

NULL HYPOTHESIS: Phototropism in *Arabidopsis* is not triggered by phosphorylation of PHOT1 protein.

EXPERIMENTAL SETUP:

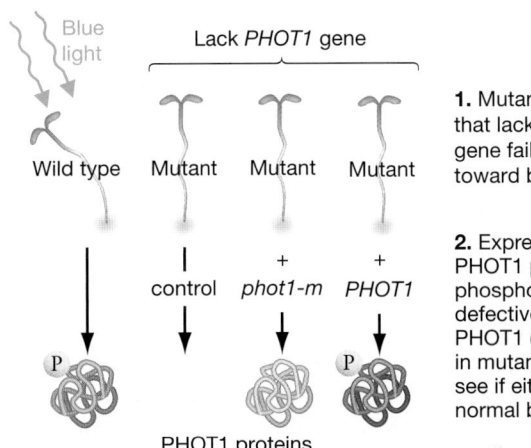

1. Mutant plants that lack the *PHOT1* gene fail to bend toward blue light.

2. Express normal PHOT1 protein and a phosphorylation-defective form of PHOT1 (phot1-m) in mutant plants to see if either rescues normal bending.

PREDICTION: Mutant plants expressing the normal *PHOT1* gene will bend toward blue light, but mutant plants expressing the defective form of *PHOT1* (*phot1-m*) will not bend toward blue light.

PREDICTION OF NULL HYPOTHESIS: Mutant plants expressing either the normal or the defective form of the *PHOT1* gene will both bend toward blue light.

RESULTS:

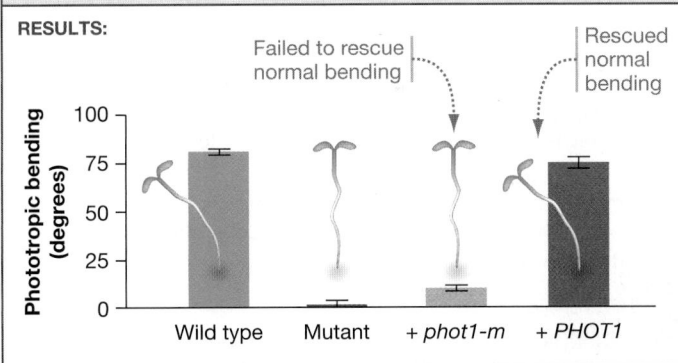

CONCLUSION: Phosphorylation of PHOT1 triggers phototropic bending in response to blue light.

FIGURE 40.4 The Role of PHOT1 Phosphorylation in Phototropic Bending: An Experimental Test.

SOURCE: Inoue, S., T. Kinoshita, M. Matsumoto, et al. 2008. Blue light-induced autophosphorylation of phototropin is a primary step for signaling. *Proceedings of the National Academy of Sciences* 105: 5626–5631.

✔ **QUESTION** Why did the researchers put the mutant and normal PHOT1s into PHOT1-deficient plants and not wild-type plants?

Auxin as the Phototropic Hormone

Long before the phototropins were identified, biologists knew that receptor cells responded to blue light by releasing a hormone. The Darwins established this result when they followed

QUESTION: Where is light sensed to initiate phototropism in grass seedlings?

HYPOTHESIS: Light is sensed at the tip of a coleoptile.

NULL HYPOTHESIS: Light is not sensed at the tip of the coleoptile.

EXPERIMENTAL SETUP AND RESULTS:

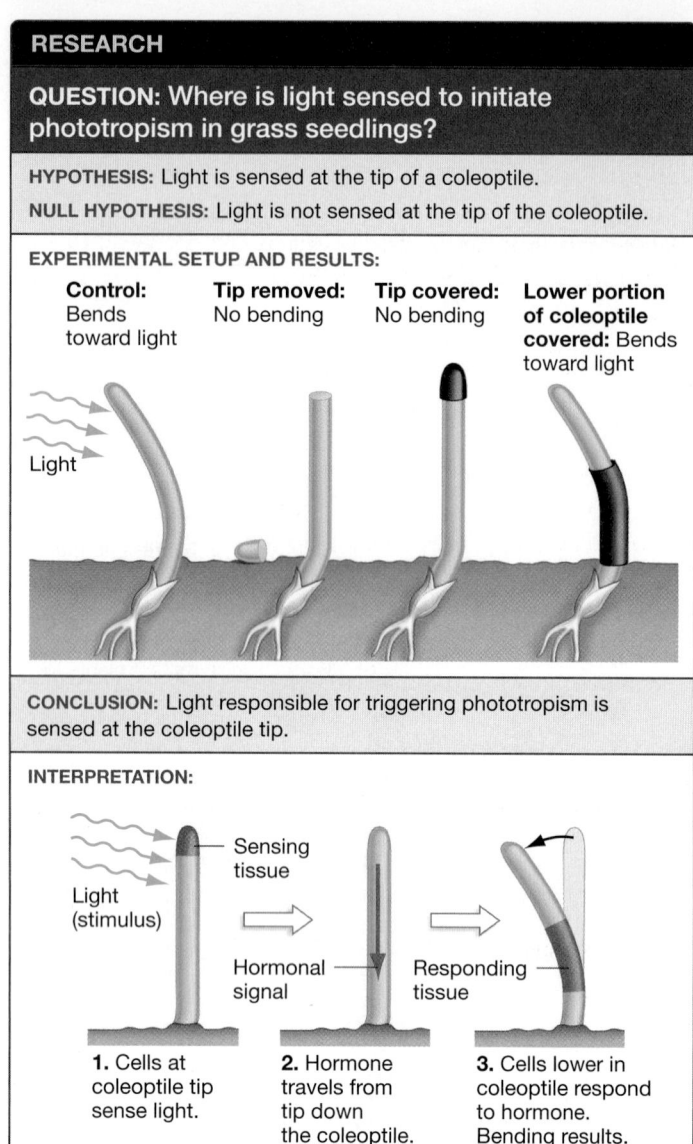

Control: Bends toward light

Tip removed: No bending

Tip covered: No bending

Lower portion of coleoptile covered: Bends toward light

Light

CONCLUSION: Light responsible for triggering phototropism is sensed at the coleoptile tip.

INTERPRETATION:

Light (stimulus) — Sensing tissue

Hormonal signal

Responding tissue

1. Cells at coleoptile tip sense light.

2. Hormone travels from tip down the coleoptile.

3. Cells lower in coleoptile respond to hormone. Bending results.

FIGURE 40.5 The Sensory and Response Cells Involved in Phototropism Are Not the Same.

SOURCE: Darwin, C., and F. Darwin. 1897. *The Power of Movement in Plants.* New York: D. Appleton & Co.

✔**QUESTION** A critic could argue that this experiment lacked appropriate controls for the treatments labeled "Tip removed" and "Tip covered." Suggest better controls for these treatments than the unmanipulated individual at the far left.

up on their initial experiments. **FIGURE 40.5** shows their experimental setup and results.

- If they removed the tips of coleoptiles, they found that the decapitated seedlings stopped bending toward the light.

- If they covered the tips of coleoptiles with opaque material, the seedlings did not bend toward light.

- If they put opaque collars below the tips, in the area where bending occurs, the seedlings bent toward light normally.

These data provided evidence that the blue-light sensors were located in the tips of the coleoptiles, and not in the part of the coleoptile that bends.

How did the sensory cells in the tip communicate with the cells that actually elongate? The Darwins proposed that phototropism depends on "some matter in the upper part which is acted on by light, and which transmits its effects to the lower part." Their hypothesis was that a substance produced at the tip of the coleoptile acts as a signal and is transported to the area of bending. This was the first explicit hypothesis stating that hormones—signaling molecules that can act at a distance—must exist.

The hormone hypothesis was not tested rigorously until 1913, when Peter Boysen-Jensen

1. cut the tips off young oat coleoptiles;

2. put a tip and a porous block of the gelatinous compound called agar on some of the decapitated coleoptiles; and

3. put a tip and a piece of mica (a nonporous mineral) on the rest of the decapitated coleoptiles.

As **FIGURE 40.6a** shows, only the coleoptiles treated with the porous agar block showed normal phototropism. Based on this observation, Boysen-Jensen concluded that the phototropic signal was indeed a chemical and that it could diffuse. He also inferred that the molecule was water soluble, because the agar that he used was a water-based gelatin.

Twelve years later, Frits Went extended these results. He placed the decapitated tips of oat coleoptiles on agar blocks, with the goal of collecting the hypothesized hormone controlling phototropism. Then he did something clever: He placed agar blocks that had been exposed to oat tips off-center on the decapitated coleoptiles of other individuals (**FIGURE 40.6b**). He also did the same with agar blocks that had *not* been exposed to oat tips, as a control (not shown).

Even though the coleoptiles were kept in the dark during the entire experiment, they responded by bending if their agar block had been exposed to oat tips. In this way, Went succeeded in producing the phototropic response without the stimulus of light.

Because it promotes cell elongation in the shoot, Went named the hormone **auxin** (from the Greek *auxein,* "to increase"). Auxin was the first plant hormone ever discovered.

Isolating and Characterizing Auxin After years of effort, researchers in two laboratories independently succeeded in isolating and characterizing auxin. The hormone turned out to be indoleacetic acid, or IAA. It was difficult to identify because it is present in such low concentrations. Depending on the species and tissue involved, IAA concentrations range from 3 to 500 nanograms per gram of tissue. (The prefix *nano–* refers to billionths.)

Like the other plant hormones introduced in this chapter, auxin is a small molecule with a relatively simple structure. It is present in quantities so small that its concentration is difficult to measure. Yet its impact is huge. Auxin can bend stems—producing, in some cases, tree trunks that are permanently bowed.

(a) The phototropic signal is a chemical.

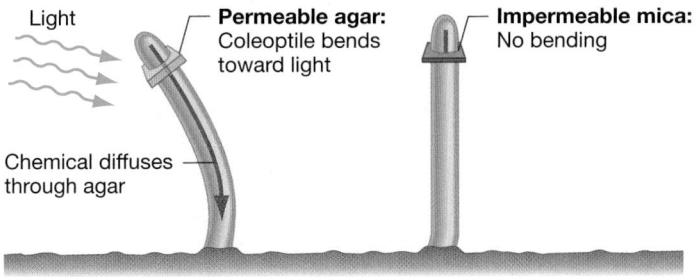

Light

Permeable agar: Coleoptile bends toward light

Impermeable mica: No bending

Chemical diffuses through agar

(b) The hormone can cause bending in darkness.

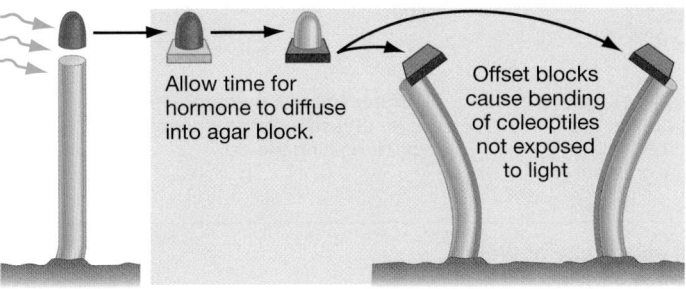

Allow time for hormone to diffuse into agar block.

Offset blocks cause bending of coleoptiles not exposed to light

(c) The hormone causes bending by elongating cells.

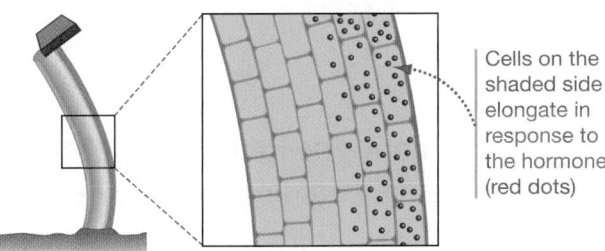

Cells on the shaded side elongate in response to the hormone (red dots)

FIGURE 40.6 Experimental Evidence Supports the Hormone Hypothesis for Phototropism. (a) Coleoptiles bend in response to light if substances from the tip are allowed to move downward. **(b)** If bending can take place in darkness, then light is not directly required for the response. Only the hormone is required. **(c)** During the phototropic response, bending occurs because cells on the shaded side of the coleoptile elongate.

The Cholodny–Went Hypothesis Went's experiments were a breakthrough in research on information processing: They confirmed the hormone hypothesis and led to the discovery and characterization of IAA. But Went's experiments also inspired an important hypothesis for *how* the hormone produces the bending response. Working independently, both N. O. Cholodny and Went proposed that phototropism results from an asymmetric distribution of auxin.

The Cholodny–Went hypothesis makes three predictions:

1. Auxin produced in the tips of coleoptiles moves from the lighted side of the tip to the shaded side.

2. Auxin is then transported straight down one side of the coleoptile.

3. The asymmetric distribution of auxin causes cells on the shaded side of the coleoptile to elongate more than cells on the illuminated side (**FIGURE 40.6c**).

Bending results. In essence, bending results from differential elongation.

To test the Cholodny–Went model, Winslow Briggs used thin sheets of mica inserted into coleoptile tips to block the movement of auxin. As predicted by the model, auxin concentrations were the same on the lighted and shaded sides of divided tips. If a path for auxin movement was maintained, more of the auxin accumulated on the shaded side of the tip.

The Cholodny–Went asymmetric distribution model was correct. It explained how auxin leads to asymmetric cell elongation, and thus the bending response called phototropism. Biologists now know that the stems of eudicots and the coleoptiles of monocots respond to phototropic signals in essentially the same way.

The Cell Elongation Response How do cells in the stem respond to auxin? Experiments in corn plants and *Arabidopsis* suggest that auxin-binding proteins are found in stem and leaf cells. Researchers proposed that once auxin binds to these receptors, the signal transduction cascade that follows increases the activity of membrane H^+-ATPases, or proton pumps, in the plasma membrane.

Recall that **proton pumps** use the energy in ATP to drive protons out of the cell against an electrochemical gradient (see Chapter 38). Because the pH of the cell wall decreases when

check your understanding

C
Y
U

If you understand that . . .

The chain of events involved in phototropism can be summarized as follows:

• When phototropins in coleoptile-tip cells absorb blue light, auxin is redistributed to the shaded side of the tip.

• Auxin is transported down the shoot and binds to receptors in target cells.

• These cells elongate when activated receptors lead to the activation of proton pumps, acidification of the cell wall, and activation of expansin proteins.

✔ **You should be able to . . .**

1. Assuming that you have a large supply of purified auxin, state how you would manipulate a large bed of roses so their stems bend toward the east.

2. Based on the information in this section, predict the phenotype of a mutant plant in which the auxin-binding protein in stems is always in the active state, even in the absence of auxin.

Answers are available in Appendix A.

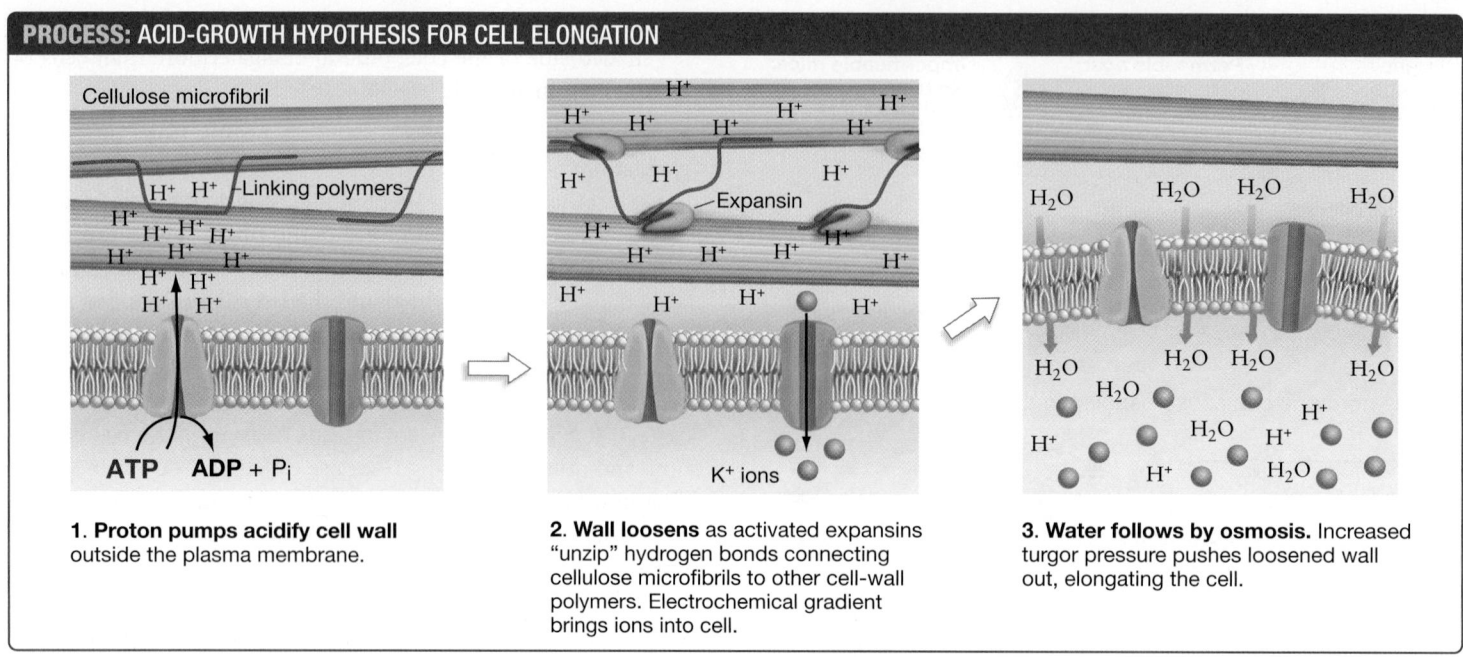

1. Proton pumps acidify cell wall outside the plasma membrane.

2. Wall loosens as activated expansins "unzip" hydrogen bonds connecting cellulose microfibrils to other cell-wall polymers. Electrochemical gradient brings ions into cell.

3. Water follows by osmosis. Increased turgor pressure pushes loosened wall out, elongating the cell.

FIGURE 40.7 The Acid-Growth Hypothesis Requires Expansion of Cell Wall and Intake of Water. When activated proton pumps lower the pH outside the cell membrane in response to auxin; a series of events lead to elongation of the cell.

H^+-ATPases are active, the idea that these pumps are responsible for cell elongation became known as the **acid-growth hypothesis.**

To understand the rationale behind the acid-growth hypothesis, it's important to realize that two things have to happen for a plant cell to get larger:

1. The cell wall has to expand to create a larger volume.

2. Water has to enter the cell and generate turgor pressure on the cell wall to make an increase in volume possible.

As **FIGURE 40.7** shows, both processes are triggered by pumping protons into the cell wall.

When proton pumping lowers the pH of the wall to 4.5, cell-wall proteins called **expansins** are activated. Expansins "unzip" the hydrogen bonds that form between cellulose microfibrils and other polymers in the cell wall, loosening the structure.

As protons are pumped out of the cell, an electrochemical gradient is established. The inside of the membrane becomes much more negative than the outside, favoring the entry of potassium (K^+) or other positively charged ions. As the concentration of solutes increases inside the cell, water follows via osmosis.

This is a key point: Cells don't move water directly. Instead, they create an osmotic gradient that favors water movement. Pumping protons *out* of a cell is a way to bring water *into* the cell. The incoming water increases turgor pressure, which pushes out the loosened cell wall. The cell gets bigger.

The upshot? When cell walls on one side of a stem are acidified in response to a signal from auxin, bending results. Auxin's role in phototropism is an exquisite example of information processing in plants.

Given the importance of light to plants, it shouldn't be surprising that they have photoreceptors that detect other wavelengths of light. We will explore one of these photoreceptors in the next section.

40.3 Red and Far-Red Light: Germination, Stem Elongation, and Flowering

Plants are sensitive to wavelengths in the red and far-red portions of the visible spectrum, as well as to blue light. This sensitivity is interesting because red wavelengths (about 660 to 700 nm) and far-red wavelengths (over 710 nm) signal very different things to a plant.

*   Red light drives photosynthesis, just as blue light does.

*   Far-red wavelengths are not absorbed strongly by photosynthetic pigments, so they tend to pass through leaves. As a result, far-red wavelengths are prominent in light that is filtered through tree leaves before it reaches the forest floor. Far-red light indicates shade.

The Red/Far-Red "Switch"

The first hint that plants monitor red and far-red light emerged from studies by H. A. Borthwick and colleagues in the early 1950s on how lettuce seeds germinate. By exposing lettuce seeds to various wavelengths of light and plotting their frequency of germination, researchers discovered that germination rates peak when seeds receive red light (about 660 nm).

This observation made sense, because lettuce thrives best when it grows in bright sunlight. But the stimulatory effect of just 1 minute of red light disappeared if seeds were then exposed to 4 minutes of far-red light.

TABLE 40.1 How Do Red Light and Far-Red Light Affect the Germination of Lettuce Seeds?

Biologists exposed moistened lettuce seeds to flashes of light containing one of two wavelengths: red or far-red (FR). After exposure to light, the seeds were held in the dark for several days.

Light Exposure	Germination (%)
None (control)	9
Red	98
Red → FR	54
Red → FR → Red	100
Red → FR → Red → FR	43
Red → FR → Red → FR → Red	99
Red → FR → Red → FR → Red → FR	54
Red → FR → Red → FR → Red → FR → Red	98

SOURCE: H. A. Borthwick, S. B. Hendricks, M. W. Parker, et al. 1952. A reversible photoreaction controlling seed germination. *PNAS* 38: 662–666, Table 1.

✔ **QUANTITATIVE** According to the data above, what is the average germination rate of lettuce seeds that were last exposed to red light? To far-red light? How would these values compare with the germination rate of seeds that are buried underground and receive no light at all?

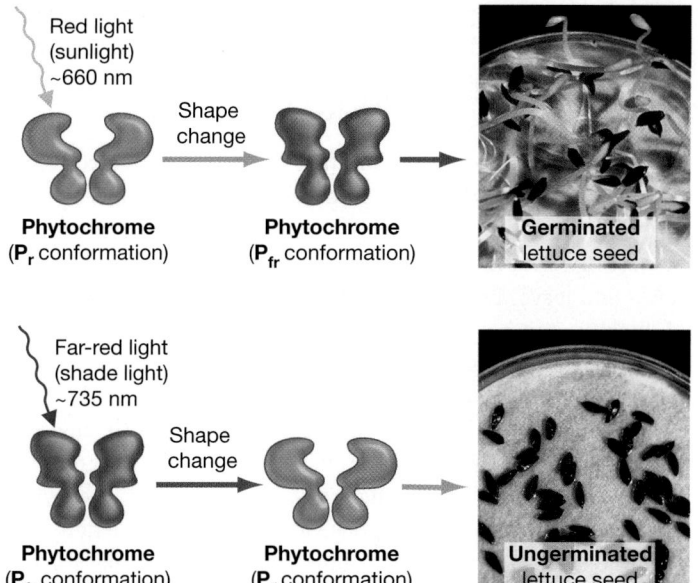

FIGURE 40.8 The Photoreversibility Hypothesis for Phytochrome Behavior. According to the photoreversibility hypothesis, phytochrome switches between the P_r conformation and the P_{fr} conformation when it absorbs red or far-red light, respectively. Only the P_{fr} form promotes germination.

This observation also made sense because far-red light indicates that the seeds are shaded. Wavelengths near 735 nm inhibit germination the most effectively.

Follow-up experiments showed that red and far-red light act like an on-off switch for lettuce seed germination (**TABLE 40.1**). Red light promotes lettuce germination; far red inhibits it.

The key observation, though, is that the last wavelength sensed by the seed determines whether germination occurs at a high rate. How could this happen?

Phytochromes as Red/Far-Red Receptors

To interpret the red/far-red switch in seed germination, biologists hypothesized that the same pigment absorbs both wavelengths. Further, they suggested that the pigment exists in two shapes, or conformations: One shape absorbs red light, and one shape absorbs far-red light.

The idea was that switching behavior, or **photoreversibility,** occurs because light absorption makes the photoreceptor pigment change shape, like a light switch moving up or down in response to touch. Each conformation would be responsible for a different response.

Biologists called the hypothesized pigment **phytochrome** ("plant-color"). Phytochrome was thought to be a specialized light receptor, different from any of the pigments involved in absorbing light during photosynthesis or phototropism.

FIGURE 40.8 illustrates the photoreversibility hypothesis. One conformation of phytochrome, called P_r (phytochrome red), absorbs red light. Another conformation of the same molecule,

called P_{fr} (phytochrome far-red), absorbs far-red light. According to the photoreversibility hypothesis, each conformation switches to the other when it absorbs its specific wavelength.

How Were Phytochromes Isolated?

Young corn and bean plants lengthen their stems in response to light deprivation or exposure to excessive far-red light. If you have grown these seeds indoors, you have seen this response to far-red light firsthand. The shoots get long and spindly.

Corn and beans normally grow in open sunlight. If they are grown indoors under incandescent light, which is deficient in red wavelengths as compared with far-red wavelengths, the plants react as though they are being shaded—they attempt to grow high enough to reach full sunlight. This behavior suggests that young corn and bean plants have a receptor protein for far-red light.

To follow up on this observation, researchers purified proteins from corn coleoptiles and succeeded in isolating one that was photoreversible. Specifically, when the protein was placed in solution and exposed to alternating red and far-red light, the color of the solution switched back and forth between blue and blue-green. The color switches supported the hypothesis that the same protein absorbed red as well as far-red light. Phytochrome was found.

Follow-up work showed that a region of the phytochrome molecule changes shape in response to red and far-red light. The shape changes cause phytochrome to take on or lose a phosphate group, just as the shape changes triggered by blue light result in phosphorylation of the phototropins. Conformational changes in proteins are frequently associated with a change in phosphorylation.

The photoreversibility hypothesis explains the red/far-red switch seen in seed germination and stem elongation. But can it also explain other plant responses? Let's consider another key event in plant growth—flowering—and the signals that trigger it.

Signals That Promote Flowering

Anatomically, a flower is a modified shoot that develops from a compressed stem and highly modified leaves. In essence, flower formation begins when an apical meristem stops making energy-harvesting stems and leaves and begins to produce the modified stems and leaves that make up flowers (see Chapter 24). Instead of making more food through photosynthesis, a sporophyte commits to investing energy in sexual reproduction—by producing gametophytes that will produce gametes.

When does this happen? Early experiments on the environmental signals that promote flowering focused on the number of hours of light and dark during a day. As it turns out, the switch that triggers flowering in response to changes in day length also shows red/far-red photoreversibility. Phytochrome plays a key role in a phenomenon called photoperiodism.

Photoperiodism is any response by an organism that is based on photoperiod—the relative lengths of day and night. In plants, the ability to measure photoperiod is important because it allows individuals to respond to seasonal changes in climate—for example, to flower when pollinators are available and when resources for producing seeds are abundant. Let's take a closer look.

Responding to Changes in Photoperiod Experiments on photoperiodism in plants have shown that, with respect to flowering, plants fall into three main categories:

1. **Long-day plants** bloom in midsummer, when days are longest and nights shortest. Radishes, lettuce, spinach, corn, irises, and other long-day plants flower only when days are longer than a certain length—usually between 10 and 16 hours, depending on the species.

2. **Short-day plants** bloom in spring, late summer, or fall. Asters, chrysanthemums, poinsettias, and other short-day plants flower only if days are shorter than a certain species-specific length.

3. **Day-neutral plants** flower without regard to photoperiod. Day length has no effect on flowering in roses, snapdragons, dandelions, tomatoes, cucumbers, and many **weeds**—plants that are adapted to grow in soils that have been disturbed enough to remove or damage existing vegetation.

How do plants sense changes in day length and initiate the chain of events that leads to flowering? Early experiments suggested that phytochrome is involved.

FIGURE 40.9 summarizes some of the key experiments. Researchers showed that interrupting the night period with a light flash changed the flowering response. In short-day plants, for example, flowering was inhibited if the dark period was interrupted by red light—wavelengths around 660 nm. But a subsequent flash of far-red light—wavelengths around 735 nm—erased the effect. In some way, the phytochrome switch linked changes in day length to flowering.

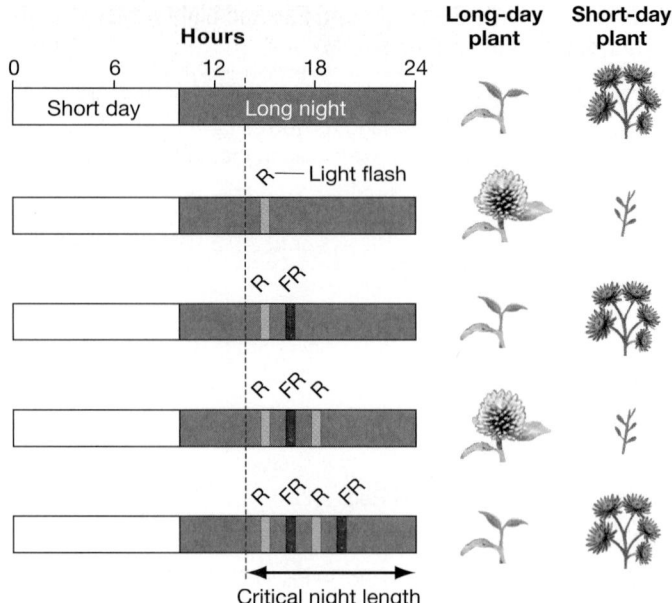

FIGURE 40.9 Flashes of Red Light and Far-Red Light Switch the Photoperiod Response On and Off. If flashes of red light (R) and far-red light (FR) are alternated during the night, the plant's flowering response correlates with the last light it experiences. A flash of red light turns a long night into a short night; a flash of far-red light restores the perception of a long night.

More recent research has shown that phytochrome's effect is closely tied to the molecular mechanisms of timekeeping in plants. Plants have a clock that is reset each morning. Clock protein levels rise during the day and trigger expression of a gene called *CONSTANS* (*CO*). The CO protein is a transcription factor that affects the production of a flowering hormone. When phytochrome is activated by light, it stabilizes CO—so that CO accumulates in cells.

- In long-day plants, high levels of CO stimulate production of the flowering hormone.

- In short-day plants, high levels of CO *inhibit* production of the flowering hormone.

Many questions remain, however, and research is continuing. In the meantime, it is clear that a complex interaction takes place between clock proteins, phytochrome activation, and the production of the signal that initiates flowering. What is that signal?

Discovery of the Flowering Hormone Since the 1930s, biologists have known that exposing even one leaf on a plant to the conditions necessary to induce flowering may result in the whole plant flowering. This result suggests that the signal to flower comes from leaves and travels to the apical meristem.

Grafting experiments, in which an organ from one individual is physically attached to a different individual, support this result. If you provide an experimental plant with the appropriate flower-triggering night length and then cut off a leaf or stem and graft it onto a second, experimental plant that has never been exposed

QUESTION: Can signals from different plants induce flowering?

HYPOTHESIS: Signals from grafted leaves can induce flowering.

NULL HYPOTHESIS: Signals from grafted leaves cannot induce flowering.

EXPERIMENTAL SETUP:

Leaf exposed to **short-night** photoperiod

Graft

Day-neutral plant exposed to **long-night** photoperiod

PREDICTION: Plant will flower.

PREDICTION OF NULL HYPOTHESIS: Plant will not flower.

RESULTS:

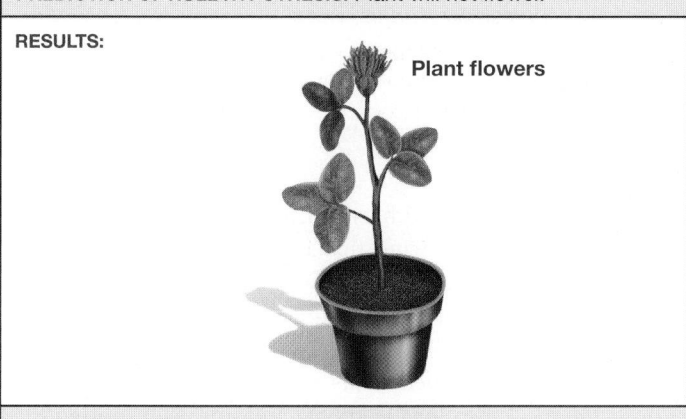

Plant flowers

CONCLUSION: A signal from a grafted leaf can induce flowering.

FIGURE 40.10 Experimental Support for the Hypothesis that a Hormone for Flowering Exists.

SOURCE: Lang, A., M. K. Chailakhyan, and I. A. Frolova. 1977. Promotion and inhibition of flower formation in a day-neutral plant in grafts with a short-day plant and a long-day plant. *Proceedings of the National Academy of Sciences, USA* 74: 2412–2416.

✔ **QUESTION** For this experiment to support the claim that a flowering hormone exists, a control treatment needs to be carried out. What is this control treatment? If the hypothesis is correct, what is the predicted outcome of the control treatment?

to the correct photoperiod for flowering, the experimental plant will flower (**FIGURE 40.10**).

Like the experiments on phototropism with agar blocks (Figure 40.6), this result supported the hypothesis that some substance produced in the transplanted leaf must travel up the recipient plant. This substance would then trip a developmental switch in the apical meristem, causing the change from vegetative growth to flowering. Biologists were so convinced that flowering must be induced by a hormone that they named it **florigen,** even though the actual hormone had not been discovered.

Almost 80 years later, researchers finally found the florigen molecule. This work began by focusing on the *FLOWERING LOCUS T (FT)* gene in *Arabidopsis*, which is known to promote flowering when activated. When researchers exposed leaves to short nights, they found that the gene was expressed in the leaf vascular tissue. Subsequently, some research indicates that the protein product of the *FT* gene is transported from leaves to the shoot apical meristem, and that the protein's presence triggers the activation of genes required for converting the stem to a flower.

Details are emerging on how the information present in the P_r/P_{fr} switch is translated into action that affects germination, stem elongation, flowering, timekeeping, and other responses to red and far-red light.

Given their reliance on photosynthesis for both energy and reduced carbon compounds, it isn't surprising that plants are highly tuned to their light environment. They also need to tell which end is up. Next, let's explore how they sense gravity.

check your understanding

If you understand that . . .

- Plants respond to red and far-red light via phytochromes, which change shape and activity when they absorb red or far-red light.
- Red light acts as a sunlight or day indicator, while far-red light acts as a shade or night indicator.
- Flowering occurs at specific times of the year and is controlled by night length through phytochrome signals.

✔ **You should be able to . . .**

1. Explain why it is adaptive for red light to trigger germination in lettuce seeds while far-red light inhibits it.

2. Explain why it is adaptive for changes in night length to affect the tendency to flower in some species.

Answers are available in Appendix A.

40.4 Gravity: The Gravitropic Response

The wavelengths, quantity, and direction of light that a plant receives change with the season, weather, time of day, and shading by other plants. However, gravity is constant and unidirectional. Light means food; gravity provides information about how the plant should orient itself in space.

Shoots usually respond to gravity by growing in an upward direction; roots usually respond by growing downward or laterally. How do plants sense gravity, so that it can be used as a signal to orient the body?

In 1881 Charles and Francis Darwin published one of the first experimental results on **gravitropism** ("gravity-turn")—the ability to move in response to gravity. Recall that the ends of root tips are covered by a protective tissue called the **root cap** (see

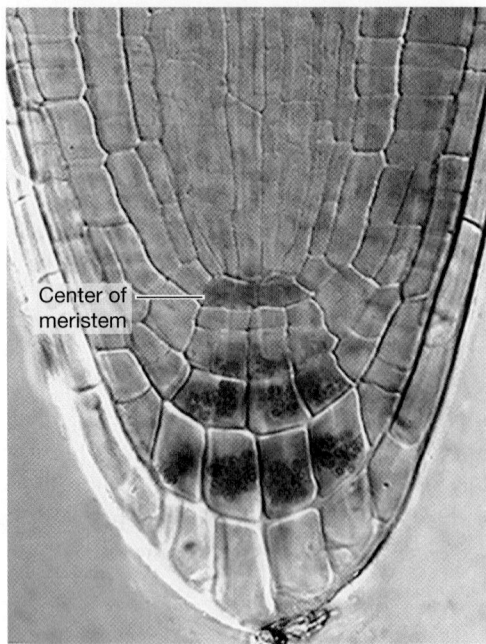

FIGURE 40.11 Gravity Sensing Occurs in the Root Cap. The root cap is a protective structure. Gravity-sensing cells in the center of the cap contain starch granules (stained dark purple).

Chapter 37). The Darwins found that roots stop responding to gravity if the root caps are removed. This observation suggested that gravity sensing occurs somewhere inside the root cap.

Recently biologists demonstrated precisely which cells are involved in gravity sensing in *Arabidopsis* roots. By killing tiny blocks of cells with laser beams, researchers showed that cells located at the center of the root cap are the most important for regulating the gravitropic response. Importantly, these cells always contain starch (**FIGURE 40.11**). Cells in the root cap respond to gravity and initiate gravitropism. Does starch play a role in sensing gravity?

The Statolith Hypothesis

The leading explanation for how plants sense gravity—the **statolith hypothesis**—is based on two interconnected ideas:

1. **Amyloplasts** are organelles that contain starch granules; starch is denser than water. These organelles are pulled to the bottom of root cap cells by the force of gravity (**FIGURE 40.12**).

2. The position of the amyloplasts activates sensory proteins located in the plasma membrane. These sensory proteins initiate the gravitropic response.

The statolith hypothesis was inspired by animals that use dense particles to sense gravity. Lobsters, for example, take up grains of sand that become positioned in specialized gravity-sensing organs in their antennae. The grains of sand are called **statoliths** ("place-stones"). When the animal tilts or flips over, the statoliths move in response to gravity. Inside the organ, the sand grain ends up pushing against a sensory cell. When this cell is activated, it indicates that the animal is no longer upright.

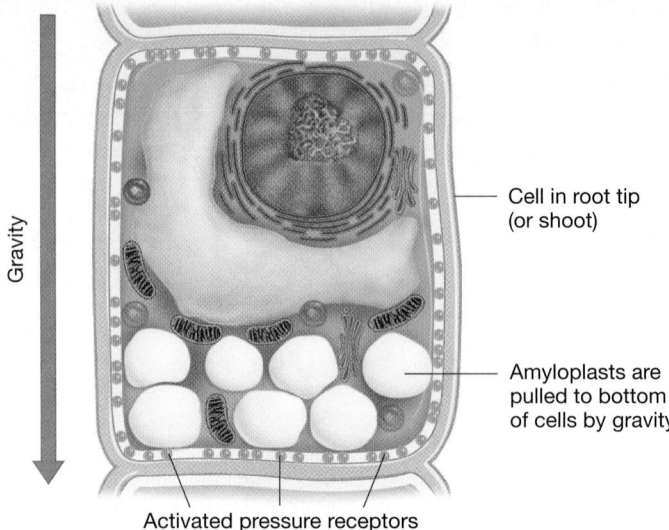

FIGURE 40.12 The Statolith Hypothesis States that Amyloplasts Stimulate Sensory Cells. Amyloplasts are filled with starch. They are dense, so they sink in response to gravity. The statolith hypothesis predicts that pressure receptors in the plasma membrane become activated as a result.

✔**EXERCISE** Assuming that the receptor is a protein, suggest a hypothesis to explain how signal transduction occurs in response to pressure from a statolith.

According to the statolith hypothesis, the same thing happens in root cap cells. If the wind tips a plant over, for example, the amyloplasts settle onto the new "lower" cell walls. The weight activates receptors, which signal that the root no longer faces in the correct direction.

Although recent experiments strongly support the statolith hypothesis, the search for the gravity receptor itself continues. In contrast, the second and third steps of information processing in response to gravity—the production of a cell–cell signal and the response of target cells—are much better understood.

Auxin as the Gravitropic Signal

Root cap cells that sense changes in the direction of gravitational pull respond by changing the distribution of auxin in the root tip. **FIGURE 40.13** illustrates this chain of events.

Step 1 Under normal conditions, auxin flows down the middle of the root, then toward the perimeter, and finally away from the root cap.

Step 2 If the root is tipped, sensory receptors trigger changes in the position of auxin transport proteins that redistribute auxin.

Step 3 Auxin is redistributed: The lower portion of the root receives increased concentrations of auxin; the upper portion receives lower concentrations.

Step 4 Because high auxin concentrations inhibit growth in roots, the differences in auxin concentrations trigger differential growth. Cells in the lower portion of the root elongate more slowly compared with cells in the upper portion. The result is bending.

Note that the way root cells respond to auxin redistribution during the gravitropic response is opposite to the way that cells in the stem respond during phototropism. In stems, high concentrations of auxin lead to *increased* cell elongation and bending. In roots, high concentrations of auxin lead to *decreased* cell division and elongation. Roots bend as cells on the upper side of the zones of cellular division and elongation (see Chapter 37) continue to grow.

check your understanding

If you understand that . . .

- Cells in root caps sense gravity via pressure that amyloplasts exert on receptors.
- Changes in gravity-sensing cells result in a redistribution of auxin and changes in the growth rate of root tips.

✔ **You should be able to . . .**

1. Explain the similarities and differences that exist between auxin redistribution in shoots and roots and the responses called phototropism and gravitropism.
2. Explain why auxin could be considered a gravitropic hormone.

Answers are available in Appendix A.

40.5 How Do Plants Respond to Wind and Touch?

Light and gravity are not the only physical forces that plants sense and respond to. Plants also react to mechanical stresses such as wind and touch.

Let's analyze two of the best-studied responses to pushing or pulling forces: growth responses that lead to exceptionally stout, stiff stems, and movement responses that make a Venus flytrap snap shut.

Changes in Growth Patterns

When plants are buffeted by wind, receptor cells transduce the mechanical force into an internal signal in the form of phosphorylated proteins or second messengers. Although the exact receptor and transduction pathway are not known, studies in *Arabidopsis thaliana* have shown that transcription of a large suite of genes is turned up or down in response to touch or other mechanical stimuli that mimic the effect of wind. Some of the new protein products act to stiffen cell walls, resulting in plants that are shorter and stockier than plants that do not experience repeated vibrations or touching.

FIGURE 40.14 shows what this response looks like in *Arabidopsis*. In this experiment, plants were either touched lightly 2 times

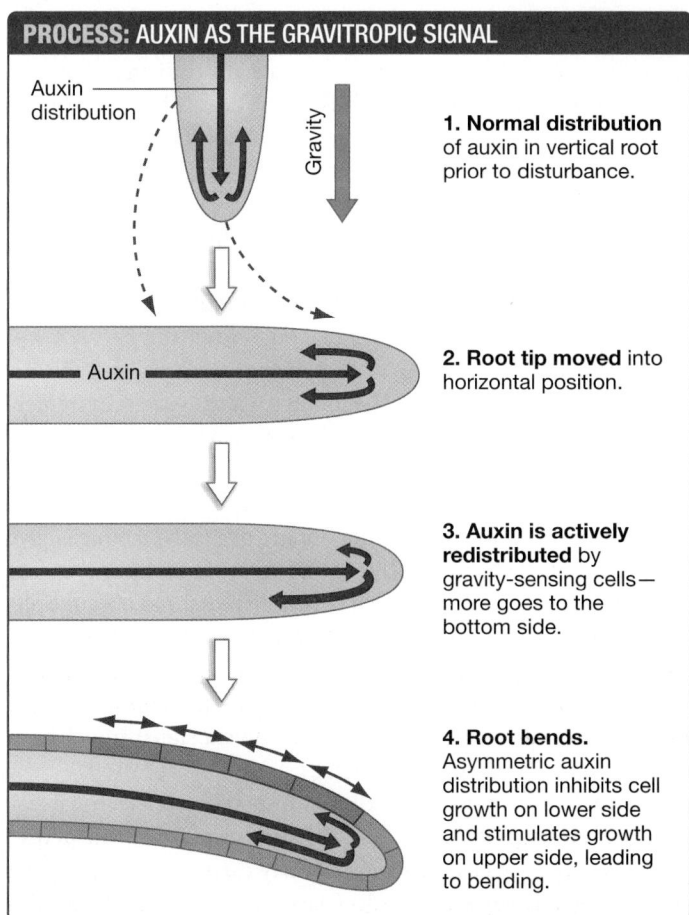

PROCESS: AUXIN AS THE GRAVITROPIC SIGNAL

Auxin distribution

Gravity

1. Normal distribution of auxin in vertical root prior to disturbance.

Auxin

2. Root tip moved into horizontal position.

3. Auxin is actively redistributed by gravity-sensing cells—more goes to the bottom side.

4. Root bends. Asymmetric auxin distribution inhibits cell growth on lower side and stimulates growth on upper side, leading to bending.

FIGURE 40.13 The Auxin Redistribution Hypothesis for Gravitropism. This sequence of events might begin when a growing root tip hits a rock and is displaced horizontally or when a plant is tipped in a windstorm and partially uprooted.

per day or left untouched. The flowing stalks of the touched plants were much shorter.

In response to wind, then, plants change their growth patterns in ways that make them more likely to withstand the force, stay upright, and live long enough to produce flowers and fruit.

FIGURE 40.14 Plant Growth Changes in Response to Wind or Touch. The *Arabidopsis* plants on the right were touched several times daily, while plants on the left were untouched.

Movement Responses

In some cases, plants respond to touch by moving. This response, **thigmotropism** ("touch-bending"), can be fast. For example, species that grow by climbing up objects or other plants may have modified leaves or stems that form long, thin structures called tendrils. When a tendril makes contact with an object, it responds by wrapping itself around the item as fast as one or more times per hour (**FIGURE 40.15**). Plants that climb in this way are able to grow taller toward light without needing to invest considerable resources in making wood.

Movement is even faster in "touch-sensitive" plants. A Venus flytrap, for example, closes fast enough to catch insects. Extremely rapid movements like this are possible when a touch-receptor cell transduces the mechanical signal to an electrical signal.

To understand how plants use electrical signaling, recall that proton pumps in the plasma membrane of most plant cells give them a negative charge relative to the exterior environment (see Chapter 38). The separation of charges creates a membrane voltage, or **membrane potential.** If a receptor protein responds to touch by allowing ions to flow across the membrane—which changes the amount of charge on either side—then the membrane potential changes. In this way, the mechanical signal (touch) can be transduced to an electrical signal.

To travel from a sensory cell to a response cell, electrical signals are propagated in a characteristic form called an **action potential.** In animals, action potentials carry signals along the plasma membrane of nerve cells (see Chapter 46). In a Venus flytrap, action potentials generated by sensory hairs inside the trap race across the leaf at a rate of about 10 cm/sec.

How do plant action potentials move across tissues, when those tissues lack nerve cells? Recall that most plant cells are connected by membrane-lined **plasmodesmata** (see Chapter 37). Action potentials move more slowly in plants than in animals, but they can still carry information long distances in a hurry.

When the action potentials in a Venus flytrap reach cells on the outer surface of the trap, the cells change shape and push the

FIGURE 40.15 Thigmotropism Is Movement in Response to Touch. Portions of grape plants called tendrils wind around support structures after contacting them. The attachment provided by the tendrils allows plants to climb.

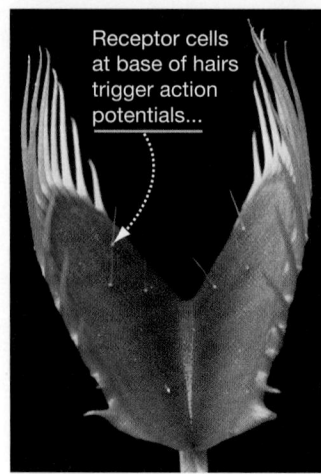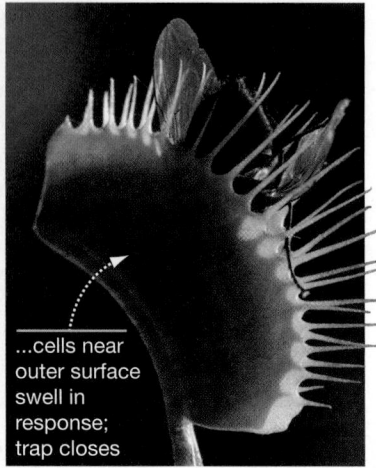

Receptor cells at base of hairs trigger action potentials...

...cells near outer surface swell in response; trap closes

FIGURE 40.16 Venus Flytraps Close in Response to Action Potentials Generated by Sensory Hairs.

trap shut (**FIGURE 40.16**). The response is rapid enough to resemble the way an animal's muscle contracts in response to an action potential.

Next, let's explore how various signaling molecules control a plant's growth throughout its life cycle.

40.6 Youth, Maturity, and Aging: The Growth Responses

Plants grow throughout their lives, from the time they germinate until the time they die. As the plant body grows, it matures into an efficient machine for absorbing sunlight, water, nutrients, and other diffuse resources. But growth is not constant. It speeds up when water, light, and nutrients are abundant and slows or stops when conditions are poor, or when it is time for leaves to drop or fruits to ripen.

Controlling growth in response to changes in age or environmental conditions is one of the most important aspects of information processing in plants. Hormones play key roles in regulating growth.

To explore how plants grow in response to changing conditions, let's consider six of the best-studied hormones involved in growth responses, which are appropriately referred to as plant growth regulators. We'll start with auxin—the molecule responsible for phototropism and gravitropism.

Auxin and Apical Dominance

When **apical dominance** occurs, growth is restricted to the main stems; the lateral buds in the axils of each leaf remain dormant. But if the apical bud dies or is eaten by an herbivore, the dormancy of the lateral buds is broken and lateral branches begin to grow. **FIGURE 40.17** shows this phenomenon in action.

What caused the change? Because auxin is produced in shoot tips, researchers suspected that it might have a role in apical dominance as well as phototropism. This hypothesis was

(a) Apical meristem intact

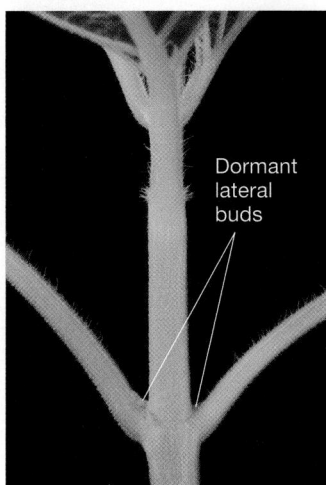

Dormant lateral buds

(b) Apical meristem cut off

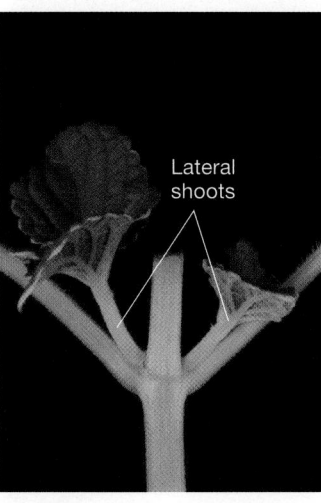

Lateral shoots

FIGURE 40.17 When Apical Dominance Occurs, Growth of Lateral Buds Is Suppressed. (a) The stem of a coleus plant is shown still intact. **(b)** The same plant, several weeks after the stem was cut. The lateral shoots will orient themselves vertically.

confirmed when it was shown that apical dominance could be sustained by applying auxin to a shoot's cut surface after its tip had been removed.

Auxin's role in apical dominance suggests that tip cells send a constant stream of information down to other organs and tissues. A stop in the signal means that apical growth has been interrupted. In response, lateral branches sprout and begin to take over for the main shoot. Now the question is, How does this signal move?

Polar Transport of Auxin   Auxin transport is **polar,** or unidirectional. If radioactively labeled auxin is added to the top end of a stem segment, the hormone emerges from the basal end. But if labeled auxin is added to the basal end of the segment, it is not transported toward the top. Auxin is the only plant hormone known to be transported through individual cells in one direction only.

Studies with labeled auxin have also shown that the hormone is transported all the way down the stem through the root, via parenchyma cells in the ground tissue and vascular tissue. Auxin enters the apical end of cells via a specialized membrane protein, diffuses to the other end of the cell, and then is transported out by carrier proteins located only in the basal portion of the plasma membrane. Labeled auxin moves from cell to cell at about 10 cm/hr—approximately 10 times more slowly than substances traveling in the phloem or xylem.

Because enzymes break down some auxin molecules as they travel down the long axis of the plant, polar transport sets up a strong gradient in auxin concentration. Auxin concentrations are much higher in shoots than they are in roots.

What Is Auxin's Overall Role?   Auxin clearly plays a key role in controlling growth via apical dominance, phototropism, and gravitropism. But this chemical messenger has other important effects as well:

- Auxin produced by seeds within the fruit promotes fruit development.

- Falling auxin concentrations are involved in the **abscission,** or shedding of leaves and fruits, associated with the genetically programmed aging process called senescence.

- The presence of auxin in growing roots and shoots is essential not only for the proper differentiation of xylem and phloem cells in vascular tissue but also for the development of vascular cambium (see Chapter 37).

- Auxin stimulates the development of adventitious roots in tissue cultures and cuttings.

Auxin has so many different effects on plants that it has been difficult for biologists to understand its overall role. Recently, several investigators have proposed that auxin's overall function is to signal where cells are in the plant body. The idea is that auxin concentration identifies where a cell is located relative to the root–shoot axis; a high concentration would inform the cell that it was near the tip of the shoot.

If conditions relating to the root–shoot axis change—for instance, a windstorm tips the plant or a deer eats the shoot apex—changes in auxin concentration effectively signal how the individual's tissues should respond. Phototropism, gravitropism, apical dominance, and the production of adventitious roots are all ways of coping with changes along the root–shoot axis of the plant.

Cytokinins and Cell Division

Cytokinins are a group of plant hormones that promote cell division. (*Cyto* is the Greek root for cell; *kinin* refers to kinesis, meaning movement or division.)

The Discovery of Cytokinins   When biologists were first attempting to grow plant cells and embryos in culture, they found that coconut milk, which stores nutrients used by growing coconut embryos, promoted cell division. This was the first hint that certain molecules can promote cell division in plants. Later experiments showed that molecules derived from the nitrogenous base adenine also stimulate the growth of cells in culture.

Eventually, naturally occurring adenine derivatives that stimulate growth were discovered in corn and apples, and were named cytokinins. Zeatin, the cytokinin found in the most species, is derived from adenine.

Cytokinins are synthesized in root tips, young fruits, seeds, growing buds, and other developing organs. But most of the zeatin and other cytokinins that are active in plants are synthesized in the apical meristems of roots and transported up into the shoot system via the xylem. Biologists still add cytokinins to plant cells growing in culture, to stimulate cell division (see **BioSkills 12** in Appendix B).

How Do Cytokinins Promote Cell Division?   After years of searching, biologists have now isolated and characterized a group of closely related proteins that act as cytokinin receptors. When cytokinins bind to these receptors in the plasma membranes of target cells, the receptors activate genes that regulate cell division.

Recent research on cytokinins has explored whether they affect molecules that regulate the cell cycle, including the cyclins and the cyclin-dependent kinases (Cdks; see Chapter 12). Recall that activated cyclins and Cdks allow cells to progress through checkpoints in the cell cycle and continue dividing.

To assess whether cytokinins affect cell-cycle genes, researchers grew *Arabidopsis* cells in culture so the nutrients and other molecules available could be carefully controlled. The biologists starved the cells of cytokinins for a day and then added the hormones again to half of the cells. When they assessed the level of mRNA from a cyclin gene called *CycD3*, they documented significant increases in the cells that were treated with cytokinins compared with the level in untreated cells.

This is strong evidence that cytokinins regulate growth by activating genes that keep the cell cycle going. In the absence of cytokinins, cells arrest at the G_2 checkpoint in the cell cycle and stop dividing (**FIGURE 40.18**).

Gibberellins and ABA: Growth and Dormancy

In high latitudes and at high elevations, most seeds and mature plants start growing in spring. Conditions for growth are good at that time of year, because temperatures are warming and soil moisture levels are usually at their peak. Seedlings and mature plants continue to grow throughout the summer and early fall if moisture and nutrients are still available.

During drought conditions, however, growth stops. Growth also stops in the embryos inside seeds. Embryos begin to develop as seeds mature, but cease this initial growth and remain dormant throughout the cold winter months. **Dormancy** is a temporary state of reduced metabolic activity.

Which signals initiate growth in response to changing environmental conditions, and which signals stop it? Two hormones provide the answer. **Gibberellins,** a large family of closely related compounds, stimulate growth in plants. **Abscisic acid,** commonly abbreviated **ABA,** inhibits growth. In at least some cases, the two hormones interact like start and stop signals.

The Discovery of Gibberellins

Over 100 years ago, Japanese farmers noticed that some of their rice seedlings grew exceptionally quickly but fell over before they could be harvested. Biologists found that the plants with this "foolish seedling disease" were infected with the fungus *Gibberella fujikuroi*.

Researchers confirmed a causal connection between *Gibberella* infection and rapid stem elongation when they treated rice seedlings with an extract from the fungus. As predicted, the treated seedlings produced abnormally long shoots.

The active component in the extract was eventually isolated and named gibberellic acid (GA), which is a gibberellin. Follow-up research showed that rice plants produce their own gibberellin but respond to applications of additional hormone by elongating their stems. In effect, the infected rice seedlings were suffering from a gibberellin overdose. Fungi that manipulate their host

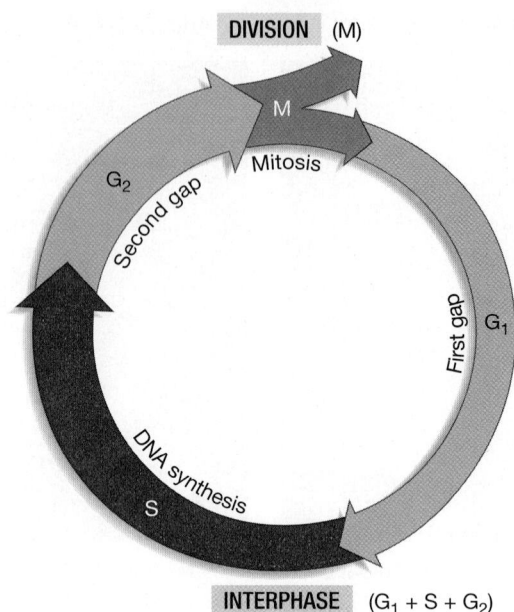

FIGURE 40.18 Cytokinins Affect the Cell Cycle.

plants in this way release their spores higher off the ground, so they are likely to blow farther away from their parent.

Gibberellins are found in a wide array of fungi and plants. Most plant species produce several different gibberellins that are active hormones.

Even though gibberellins have dramatic effects on growth, they are present in vanishingly small concentrations. In growing stems and leaves, active forms of gibberellin may be present in concentrations of about 10 nanograms per gram of tissue.

Defective Gibberellin Genes Cause Dwarfing To find the genes that are responsible for producing gibberellins, biologists analyzed mutant plants with abnormal stem length (**FIGURE 40.19**). Recall that Gregor Mendel analyzed the transmission of two alleles at a single gene that affected stem height in garden peas (see Chapter 14). One allele was associated with tall stems; the other was associated with dwarfed growth. The tall allele was dominant to the dwarf allele.

The gene responsible for the stem-length differences in garden peas is known as *Le* (for *le*ngth). Early work on dwarf mutants showed that they attain normal height if they are treated with the gibberellin called GA_1. This observation suggested that dwarf peas can respond to gibberellins normally—meaning that the problem is not with a hormone receptor.

In follow-up experiments, biologists treated dwarf peas with a radioactively labeled molecule used in the synthesis of GA_1. These plants did not produce radioactively labeled GA_1, even though plants with the normal allele did. Based on these results, researchers became convinced that the *Le* locus encodes an enzyme involved in GA synthesis.

Investigators recently identified the gene at the *Le* locus and confirmed that in dwarf plants, it contains a mutation that renders the enzyme inactive. Over 100 years after Mendel did his

FIGURE 40.19 Dwarfed Individuals May Have Mutations That Affect Gibberellins. Dwarfed plants have much shorter stems than do normal individuals of the same age. By analyzing dwarfed individuals, biologists were able to identify genes involved in gibberellin synthesis.

✓**QUESTION** If dwarfed individuals receive the same amount of sun and thus perform as much photosynthesis as taller individuals, they often produce more flowers and seeds. Why?

experiments, biologists finally understood the molecular basis of the dwarfing phenotype he studied.

In stems, gibberellins appear to promote both cell elongation and rates of cell division. But it is well established that auxin also promotes cell elongation, and that cytokinins also promote cell division. Research continues on how GAs, cytokinins, and auxin interact on the molecular level to control plant growth and development.

Gibberellins and ABA Interact during Seed Dormancy and Germination Many plants produce seeds that have to undergo a period of drying or a period of cold, wet conditions before they are able to germinate. A requirement for drying ensures that mature seeds will not sprout on the parent plant; an obligatory cold period prevents seeds from germinating just before the onset of winter, when young seedlings would be more likely to perish. Some seeds also have to receive a dose of red light, which indicates that they are in a sunny location.

In essence, then, seeds have an "off" setting that discourages germination and an "on" setting that initiates growth. The appropriate state for the on/off switch is determined by environmental cues such as temperature, moisture, and light.

By applying hormones to seeds, researchers learned that in many plants, ABA is the signal that inhibits seed germination and gibberellins are the signal that triggers germination. To understand how these messengers interact, researchers have studied a specific event: the production of an enzyme called α-amylase in germinating oat or barley seeds.

α-Amylase acts as a digestive enzyme that breaks the bonds between the sugar subunits of starch. (Your saliva contains an amylase that acts on the starch in food. In humans and other mammals, this enzyme initiates carbohydrate digestion in the mouth.)

FIGURE 40.20 shows that, during the germination of a barley seedling, α-amylase is released from a tissue called the aleurone layer. The enzyme diffuses into the starch-filled storage tissue in the seed called endosperm, which is dead when seeds germinate. Digestion of starch releases sugars that can be used by the growing embryo. Adding GA to the aleurone layer increases the production and release of α-amylase; adding ABA to that layer decreases α-amylase levels.

Research on the molecular interaction between GA and ABA carries several important messages:

- A cell's response to a hormone often occurs because specific genes are turned on or off.

- Hormones don't act on genes directly. Instead, a receptor on the surface of a cell or in the cytosol receives the message and

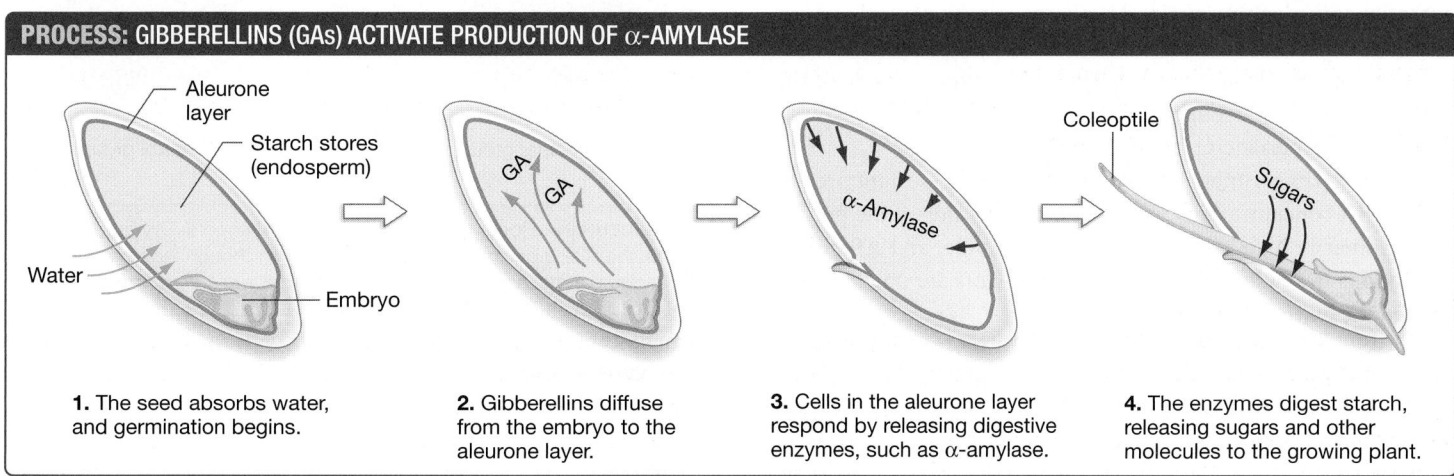

PROCESS: GIBBERELLINS (GAs) ACTIVATE PRODUCTION OF α-AMYLASE

1. The seed absorbs water, and germination begins.

2. Gibberellins diffuse from the embryo to the aleurone layer.

3. Cells in the aleurone layer respond by releasing digestive enzymes, such as α-amylase.

4. The enzymes digest starch, releasing sugars and other molecules to the growing plant.

FIGURE 40.20 The Molecular Mechanism of Gibberellin Action in Seed Germination.

responds by initiating a signal transduction cascade, which activates specific gene regulatory proteins—the activators and repressors of transcription (see Chapter 19).

- Different hormones interact at the molecular level because they induce different gene regulatory proteins, which increase or decrease expression of key genes.

The logic runs as follows: Different hormones trigger the production of different regulatory transcription factors. Hormone concentration affects the amount of each transcription factor produced. Changes in transcription factors are responsible for changes in gene expression.

GA's role in seed germination has important commercial applications. For example, in the production of beer, brewers routinely use gibberellins in the malting process—the conversion of starches stored in barley seeds to sugars. The sugars that are released in response to GA treatment support fermentation by yeast and provide flavor and alcohol in the finished beer.

ABA Closes Guard Cells in Stomata One of the major challenges faced by land plants is replacing water that is lost to the atmosphere when **stomata** are open (see Chapter 38). Because stomata open in response to blue light, they allow gas exchange to occur while the plant is receiving the wavelengths of light used in photosynthesis.

However, if plant roots are unable to obtain enough water to replace what is lost from the leaves, stomata close. Closing stomata is an adaptive response when roots cannot find adequate water, because continued transpiration would lead to wilting and potential tissue damage.

Early work on the mechanism of stomatal closing suggested that ABA is involved. For example, applying ABA to the exterior of stomata causes them to close within seconds.

To explore the hypothesis that a hormone regulates stomatal closing, researchers performed the experiments summarized in **FIGURE 40.21**. The fundamental idea was to grow plants whose roots had been divided. Only one side of the experimental plants was watered, while both sides of the control plants were watered. During this treatment, investigators documented that the water potential of the leaves remained the same in both control and experimental plants. Yet the stomata of experimental plants began to close. This result suggested that roots from the dry side of the pot were signaling drought stress, even though the leaves were not actually experiencing a water shortage.

Follow-up experiments have supported two important predictions: ABA concentrations in roots on the dry side of the pot are extraordinarily high relative to the watered side, and ABA concentrations in the leaves of experimental plants are much higher than in the leaves of control plants.

These results suggest that ABA from roots is transported to leaves and that it serves as an early warning of drought stress. In doing so, ABA overrides the signal from the blue-light photoreceptors introduced earlier in this chapter and causes stomata to close to conserve water.

RESEARCH

QUESTION: Can roots communicate with shoots?

HYPOTHESIS: Roots that are dry can signal shoots to close stomata.

NULL HYPOTHESIS: Roots cannot communicate with the shoot.

EXPERIMENTAL SETUP:

1. Divide roots of many plants into two sides.

2. In experimental group, water one side.

3. In control group, water both sides.

4. In both groups, measure water potential of leaves and observe stomata.

PREDICTION: Stomata in experimental plants will close; stomata in control plants will stay open.

PREDICTION OF NULL HYPOTHESIS: Stomata in both experimental and control plants will stay open.

RESULTS: No difference between experimental and control plants in water potential of leaves.

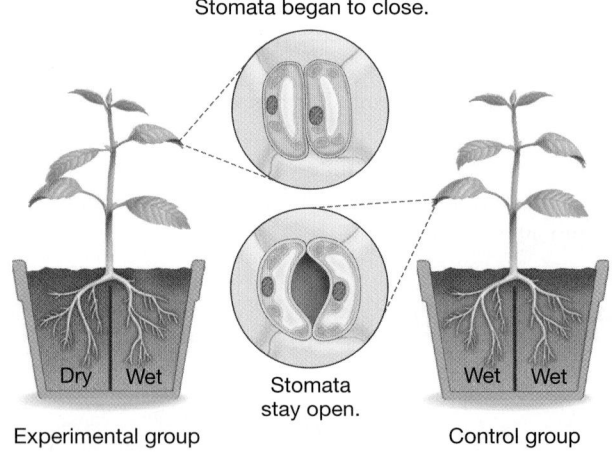

Stomata began to close.

Stomata stay open.

Experimental group   Control group

CONCLUSION: Roots can communicate with shoots. Dry roots signal the shoot and cause stomata to close, even though leaves are receiving sufficient water (from roots on the wet side of the plant).

FIGURE 40.21 Experimental Evidence that Roots Produce an "It's Too Dry" Signal.

SOURCE: Blackman, P. G., and W. J. Davies. 1985. Root to shoot communication in maize plants of the effects of soil drying. *Journal of Experimental Botany* 36: 39–48.

✔**QUESTION** Why was it important to show that the water potential of leaves was the same in the two treatments?

The Molecular Mechanism of Guard-Cell Closure To understand how stomata open and close, recall that a stoma consists of two **guard cells** (see Chapter 37). When the vacuoles of guard cells are filled with water, the cells are turgid. The shape of turgid guard cells results in an open pore between the two cells, which allows gas exchange between the atmosphere and the interior of the leaf. But when vacuoles lose water and guard cells become flaccid and lack turgor, cell shape changes in a way that closes the pore and stops both gas exchange and loss of water via transpiration.

Based on these observations, pursuing the question of how stomata open and close is the same as asking how guard cells become turgid or flaccid—meaning, how water flows into or out of vacuoles. Activation of PHOT by blue light leads to water entry and stomatal opening; activation of ABA receptors leads to water exit and stomatal closing.

Remember that cells do not transport water directly. Instead, they change ion concentrations, creating osmotic gradients that result in water movement.

Considering this process, it shouldn't be surprising to learn that guard-cell opening and closing is based on changes in the activity of H^+-ATPases and ion transporters in the plasma membrane.

When photoreceptors are stimulated by blue light, large numbers of protons are pumped out of each guard cell. As **FIGURE 40.22a** shows, increased H^+-ATPase activity creates a strong electrochemical gradient that brings potassium and chloride ions into

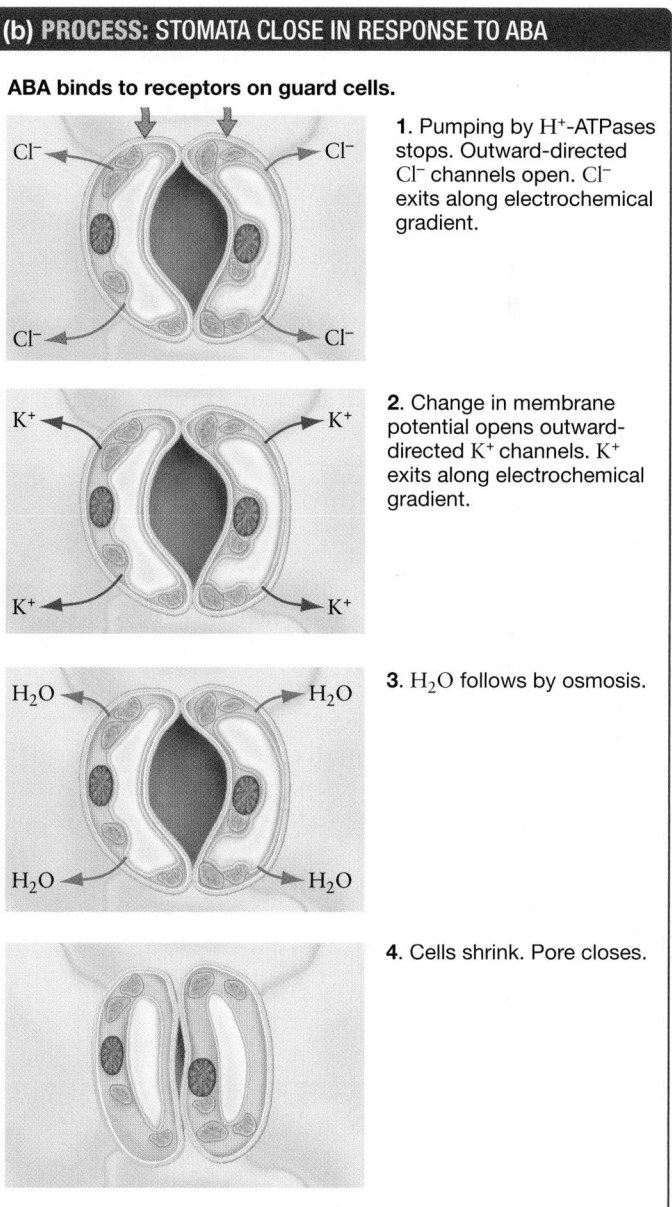

(a) PROCESS: STOMATA OPEN IN RESPONSE TO BLUE LIGHT

Blue light strikes photoreceptor.

H^+ H^+
H^+ H^+

1. Pumping by H^+-ATPases increases. Protons leave guard cells.

K^+ K^+
Cl^- Cl^-
H^+ H^+

2. K^+ and Cl^- enter cells along electrochemical gradients via inward-directed K^+ channels and H^+/Cl^- cotransporters.

H_2O H_2O
H_2O H_2O

3. H_2O follows by osmosis.

4. Cells swell. Pore opens.

(b) PROCESS: STOMATA CLOSE IN RESPONSE TO ABA

ABA binds to receptors on guard cells.

Cl^- Cl^-
Cl^- Cl^-

1. Pumping by H^+-ATPases stops. Outward-directed Cl^- channels open. Cl^- exits along electrochemical gradient.

K^+ K^+
K^+ K^+

2. Change in membrane potential opens outward-directed K^+ channels. K^+ exits along electrochemical gradient.

H_2O H_2O
H_2O H_2O

3. H_2O follows by osmosis.

4. Cells shrink. Pore closes.

FIGURE 40.22 Changes in Ion Flows Are Responsible for Opening and Closing Stomata. (a) Activated blue-light receptors trigger ion flows into guard cells. The cells swell when water follows by osmosis. **(b)** Activated ABA receptors trigger ion flows out of guard cells. The cells shrink when water follows by osmosis.

the interior of the guard cells. Water follows the incoming ions via osmosis. The upshot? The cells swell and pores open.

But as **FIGURE 40.22b** shows, guard cells respond to ABA in a very different way. When ABA reaches the guard cells, two things happen:

1. H^+-ATPases are inhibited.

2. Channels that allow chloride and other anions to leave along their electrochemical gradients are opened.

When the anions leave guard cells, the change in membrane potential causes outward-directed potassium channels to open. Large amounts of K^+ leave the cells, and water follows by osmosis. The result is a loss of turgor and closing of the pore.

Whether it acts on guard cells or seeds, ABA fulfills a general role in plants as a dormancy or "no-growth" signal. In many cases, its action depends on input from other hormones and photoreceptors. To survive and reproduce successfully, plants have to integrate information from a variety of sources.

Brassinosteroids and Body Size

When you went through puberty, you experienced a growth spurt triggered by surges in steroid hormones called testosterone and estradiol. In plants, growth spurts are triggered by surges in similar steroid hormones called **brassinosteroids.**

The name brassinosteroid was inspired by two observations:

1. The hormones were initially discovered in 1979 in *Brassica napus*—a crop plant that is the source of the canola oil you may use in cooking.

2. They are **steroids**—part of a family of lipid-soluble compounds (see Chapter 6).

Brassinosteroids promote growth and are a key regulator of overall body size in plants. In the model organism *Arabidopsis thaliana*, for example, mutant individuals that cannot synthesize brassinosteroids are extremely dwarfed (see **BioSkills 13**, see Appendix B).

Recent research has highlighted a fascinating difference between the brassinosteroids and the steroid hormones found in animals. Steroid hormones in animals act by entering cells, binding to receptors in the cytosol, and forming a hormone-receptor complex that enters the nucleus (see Chapter 49). This complex then binds to DNA and directly changes gene expression. Because steroids are lipid soluble, researchers were not surprised to find that testosterone and estradiol—animal hormones—cross the plasma membrane before binding to a receptor.

Brassinosteroids, in contrast, are more hydrophilic than animal hormones and never enter the cell. Instead, they bind to receptors on the plasma membrane and activate signal transduction events—probably phosphorylation cascades—that lead to changes in gene expression.

The genes for the synthesis of brassinosteroids appear to be homologous with the genes required for steroid hormone synthesis in animals, meaning that they are derived from a similar gene in the common ancestor of plants and animals. So how do brassinosteroids interact with auxin, cytokinins, and gibberellins to regulate growth and body size? These are questions for future research.

Ethylene and Senescence

Senescence is a regulated process of aging and eventual death of an entire organism or organs such as fruits and leaves. Like most aspects of plant growth and development, senescence is regulated by complex interactions between several hormones in response to changes in temperature, light, and other factors.

The hormone most strongly associated with senescence is **ethylene.** Like other plant hormones, ethylene is simple in structure and active at small concentrations. Unlike other plant hormones, however, ethylene is a gas at normal temperatures.

Ethylene is synthesized from the amino acid methionine and is involved in the following senescence events:

1. fruit ripening, which eventually leads to the aging and rotting of fruit;

2. flowers fading; and

3. leaf abscission—meaning their detachment and fall.

In addition, ethylene influences plant growth, and it is a stress hormone induced by drought and other conditions. Ethylene regulates a surprisingly large range of physiological responses.

The Discovery of Ethylene Ethylene was initially discovered in ancient China, when fruit growers noticed that burning incense in closed rooms made pears ripen faster. Westerners made a similar observation in the late 1800s, when gas street lamps came into wide use in cities and the plants growing near leaky gas lines dropped their leaves prematurely.

Researchers showed that ethylene in lamp gas was the molecule responsible for the leaf loss; ethylene is likewise present in incense smoke. In the 1930s ethylene was also found in the gases that are released by ripe apples.

Subsequently, biologists documented sharp spikes in ethylene production during fruit ripening in tomatoes, bananas, and certain other species in addition to apples. Follow-up research on these species showed that ethylene induces (1) the production of some of the enzymes required for the ripening process, and (2) an increase in cellular respiration, which furnishes ATP.

Ethylene and Fruit Ripening During ripening, stored starch is converted to sugar, enhancing sweetness; protective toxins are removed or destroyed; cell walls are degraded, softening the fruit; chlorophyll is broken down; and pigments and aromas that signal ripeness are produced. Biologists interpret fruit ripening as an adaptation that enhances its attractiveness to birds, mammals, and other animals that disperse seeds to new locations.

Today, fruit sellers manipulate ethylene levels to control fruit ripening. For example, they treat green bananas with ethylene after the bananas have been shipped, to stimulate ripening (**FIGURE 40.23**). Conversely, apples are stored in warehouses with high concentrations of CO_2 and low concentrations of O_2, which inhibits ethylene production in the fruit. Apples stored under these conditions can be sold long after their original harvest date, when untreated fruits have rotted.

+ Ethylene

FIGURE 40.23 Ethylene Speeds Ripening and Other Aspects of Senescence. These bananas are identical, except that the bunch on the right was exposed to the plant hormone ethylene.

Ethylene and Leaf Abscission Ethylene's effects on leaf senescence and leaf abscission involve complex interactions with auxin and cytokinins. In addition to being synthesized in apical meristems, auxin is produced in healthy leaves. It is then transported from the leaf to the stem through the petiole. In response to age or to changes in ambient temperature or day length, leaves sometimes produce less auxin (**FIGURE 40.24**). As a result, cells in a region of the leaf petiole called the **abscission zone** become more sensitive to ethylene in the tissue.

Increased ethylene sensitivity activates enzymes that weaken the walls of cells in the abscission zone. At the same time, chlorophyll in the leaf is broken down, and nutrients are withdrawn and stored in parenchyma cells in the stem. Eventually the cell walls at the base of the petiole degrade enough that the leaf falls.

Applications of cytokinins, in contrast, reverse these effects and dramatically extend the life span of leaves. As a result, ethylene and cytokinins are thought to have opposite effects on at least some of the processes involved in senescence.

An Overview of Plant Growth Regulators

Understanding how different signals interact is an exciting frontier in research on plant hormones that are growth regulators. Knowledge of the environmental signals that change hormone levels and the genes that they regulate is rapidly improving. In contrast, a great deal remains to be learned about how the information is carried by signal transduction pathways, how those pathways interact, and how plant cells respond.

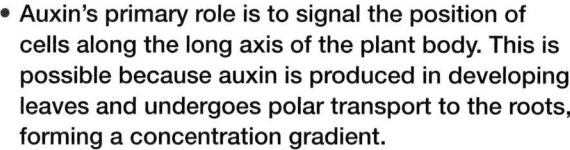
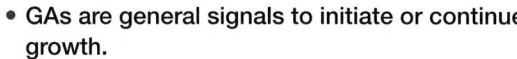

check your understanding

C Y U

If you understand that . . .

- Auxin's primary role is to signal the position of cells along the long axis of the plant body. This is possible because auxin is produced in developing leaves and undergoes polar transport to the roots, forming a concentration gradient.
- GAs are general signals to initiate or continue growth.
- ABA is a general signal to stop growth or remain dormant.
- Ethylene is a signal that controls senescence.
- Brassinosteroids promote large body size.
- GAs and ABA interact at the molecular level to control seed germination and dormancy.
- Signals from blue light and ABA interact to control the opening and closing of stomata.

✔ **You should be able to . . .**

1. Predict the effects of watering a large number of daisy plants, each in a pot in a greenhouse, with water that contains either ABA or GAs.
2. Write a hypothesis to explain the adage "one rotten apple spoils the whole bushel."

Answers are available in Appendix A.

PROCESS: LEAF SENESCENCE AND ABSCISSION

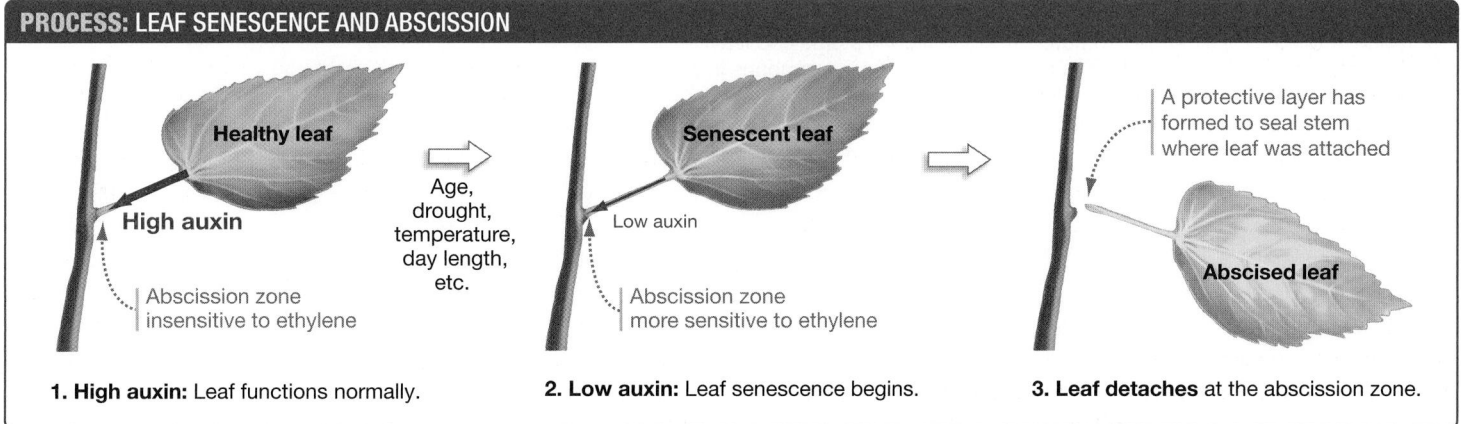

1. High auxin: Leaf functions normally.   **2. Low auxin:** Leaf senescence begins.   **3. Leaf detaches** at the abscission zone.

FIGURE 40.24 Leaves Drop in Response to Signals from Auxin and Ethylene. Older leaves produce much less auxin than young leaves do, leading to an increase in ethylene sensitivity and leaf abscission.

TABLE 40.2 provides notes on the structure and function of hormones discussed in this section. As you study this table, two key observations should emerge:

1. It is common for a single hormone to affect many different target tissues. This means that there can be an array of responses to the same cell–cell signal. To interpret this pattern, biologists point out that hormones may carry a common message to a variety of tissues and organs. Auxin can define the root–shoot axis of the body; gibberellins trigger stem growth; cytokinins promote cell division; ABA slows or prevents growth; ethylene signals senescence; and brassinosteroids increase overall mass.

2. In most cases, several different hormones can affect the same response. Stated another way, hormones do not work independently—they interact with each other. To make sense of this pattern, biologists point out that individual hormones tend to be produced in response to an environmental cue at a certain location, such as water availability at root tips. Many environmental cues may be changing at the same time, however. For plants to respond appropriately, they need to integrate information from various environmental cues perceived at various locations in the body.

Cross-talk is the interaction between the signal transduction cascades triggered by different hormones (see Chapter 11). The

SUMMARY TABLE 40.2 **Plant Growth Regulators**

Hormone	Notes	Chemical Structure
Auxin	• Involved in cell elongation and apical dominance • Promotes differentiation of xylem and phloem • Helps to define long axis of body (phototropism and gravitropism responses) • First plant hormone isolated and characterized • Produced in shoot apical meristems and young leaves	
Cytokinins	• Promote cell division in the presence of auxin • Promote chloroplast development and break lateral bud dormancy • Delay senescence (aging) • Produced in root apical meristems, many other tissues	
Gibberellins (GAs)	• Promote stem growth via both cell elongation and division • Promote seed germination • Produced in apical meristems, immature seeds, and anthers (pollen-producing organs)	
Abscisic Acid (ABA)	• Inhibits bud growth and seed germination • Induces closure of stomata in response to water stress • Acts as a stress hormone analogous to cortisol in humans • Produced in almost all cells	
Brassinosteroids	• Promote cell elongation in stems and leaves • Structurally related to steroid hormones in animals • Act on receptors at cell surface • Produced in almost all tissues	
Ethylene	• Exists in gas form • Involved in fruit ripening • Induces senescence of fruits, flowers and leaves • Produced in all organs when plants are under stress	

key insight to remember is that signaling systems form communication networks. Cross-talk is the molecular mechanism responsible for integrating information from many sensory cells and signals.

The complex interactions among hormones involved in the growth response have a purpose: allowing individuals to survive and thrive long enough to reproduce. The same can be said for the hormones involved in protecting plants from danger.

40.7 Pathogens and Herbivores: The Defense Responses

Plants cannot run away from danger. Instead, they have to stand and fight.

Like humans and other animals, plants are constantly threatened by an array of disease-causing viruses, bacteria, and parasitic fungi. In addition, plant roots are susceptible to attacks by nematodes—the soil-dwelling roundworms (see Chapter 34).

Disease-causing agents are termed **pathogens.** If plants were not able to sense attacks by pathogens and respond to them quickly and effectively, the landscape would be littered with dead and dying vegetation.

The waxy cuticle that covers epidermal cells is an effective barrier to most viruses, bacteria, fungi, and other disease-causing agents, and the structures called thorns, spines, and trichomes help protect leaves and stems from damage by herbivores (see Chapter 37).

In addition, many plants lace their tissues with **secondary metabolites**—molecules that are closely related to compounds in key synthetic pathways but are not found in all plants. Some secondary metabolites function in plant defense by poisoning herbivores.

- The flavorful oils in peppermint, lemon, basil, and sage have insect-repellent properties.

- The pitch that oozes from pines and firs contains a molecule called pinene, which is toxic to bark beetles.

- The pyrethroids produced by *Chrysanthemum* plants are a common ingredient in commercial insecticides.

- Molecules called tannins are found in a wide array of plant species; when they are ingested by animals, tannins bind to digestive enzymes and make the herbivores sick.

- Compounds like opium, caffeine, cocaine, nicotine, and tetrahydrocannabinol (THC) disrupt the nervous systems of plant-eating insects and vertebrates.

Although these defenses are effective, they are also expensive to produce in terms of the ATP and materials invested. It is not surprising that plants often produce defenses or increase their existing defenses only in direct response to attacks by pathogens or herbivores.

Responses to attacks in both plants and animals are called **inducible defenses,** because they are induced by the presence of a threat. Let's first consider how plants sense and respond to viruses and other pathogens and then explore what they do when attacked by insects and other herbivores.

How Do Plants Sense and Respond to Pathogens?

Like animals, plants have remarkably swift and diverse defense mechanisms to resist attacks by viruses, bacteria, and fungi. Plants use a wide variety of sensory proteins to detect invading pathogens and use signaling pathways to carry out their defense. Successful pathogens, in turn, have evolved elaborate mechanisms to evade these defenses.

An Evolutionary Arms Race Most plant pathogens are very specific, infecting only one or a few host species. This is in part due to the long history of coevolution of the pathogen and its host (see Chapter 55).

If a pathogen evolves a mechanism to exploit its host more effectively, the host may evolve new abilities to detect the pathogen and defend itself from attack. In response, the pathogen may then evolve an alternate strategy to evade detection by the host.

The complex interactions that have resulted from this back-and-forth "arms race" have proved challenging for biologists to understand. The human food supply is estimated to be reduced 50 percent worldwide because of plant disease. In the mid-1800s, Ireland experienced a potato famine in which millions of people faced starvation because of a potato pathogen. Pathogens can also be devastating to native species. The once-dominant American chestnut tree was nearly wiped out by a fungal pathogen in the early 1900s.

To cause disease, plant pathogens must first get past the surface of a host plant and begin growing in the apoplast, outside the plasma membranes of cells. Most pathogens enter through stomata or wounds. Once inside the plant, some pathogens kill host cells and feed on the cell debris. More commonly, a pathogen feeds on nutrients from a host cell while it is still alive. How do the plant cells respond to an attack in progress?

The Hypersensitive Response Plant cells contain a variety of receptor proteins that can bind to molecules derived from pathogen cells. When contact is made, these 'surveillance' proteins signal to the plant cell that a pathogen is present, which triggers a cascade of signals that lead to

1. stomatal closure, preventing more pathogen cells from entering the plant;

2. production of toxin molecules targeted to the pathogen;

3. reinforcement of the neighboring plant cell walls to limit movement of the pathogen; and

4. rapid suicide of cells in the infected region.

These events take place within hours of the start of an infection and are termed the **hypersensitive response (HR).** The HR results in small brown spots in leaves—those cells dead by

FIGURE 40.25 The Hypersensitive Response Protects Plants from Pathogens. This red maple leaf is riddled with evidence of past infections.

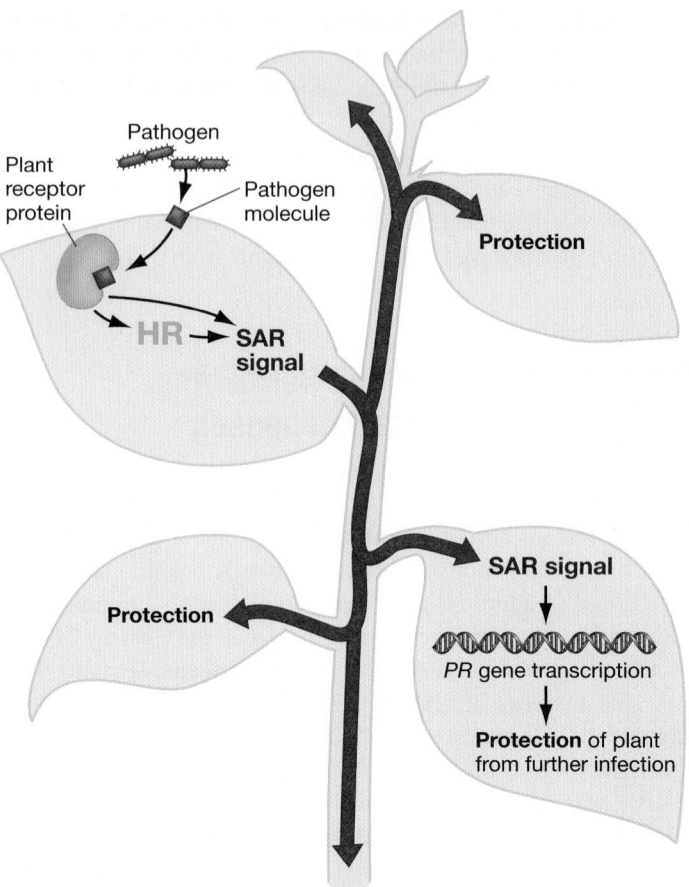

FIGURE 40.26 The Hypersensitive Response Produces a Signal That Induces Systemic Acquired Resistance (SAR). This diagram summarizes the current consensus on how the HR and SAR interact.

suicide—that accumulate over the growing season (**FIGURE 40.25**). By the end of the growing season, it is rare to find leaves lacking these brown spots. Plants that can detect an infection quickly and induce an HR remain healthy and show no further signs of disease. The HR is similar to the cell-mediated immune response in mammals, which leads to the death of infected cells (see Chapter 51).

✔ If you understand this concept, you should be able to explain the adaptive value of the hypersensitive response despite the death of plant cells.

In the 1990s biologists discovered that a mutant of *Pseudomonas*, a bacterial pathogen, could infect plant cells if it was artificially introduced inside leaves, but the pathogen was unable to cause disease if applied to the outside of leaves. This mutant could not produce a small molecule called coronatine, but the role coronatine played in infection was not clear.

Plants normally respond to *Pseudomonas* by closing their stomata within an hour of detecting the pathogen, effectively slamming the door on future invasions. The normal bacteria, however, are able to interfere with this defense by causing the stomata to re-open several hours later, allowing more bacteria to enter the leaf. In 2006, Maeli Melotto and coworkers discovered that coronatine is the molecule that causes the stomata to re-open. Without the ability to make coronatine, the mutant bacteria couldn't re-open the stomata and were at a clear disadvantage.

Plant defenses are based on good communication not only from cell to cell, as in the HR, but also throughout the plant and even to neighboring plants.

An Alarm Hormone Extends the HR Once the HR is under way in a localized area of infection, a hormone produced at the infection site travels throughout the body and triggers a slower

and more widespread set of events called **systemic acquired resistance (SAR).** Over the course of several days, SAR primes cells throughout the root or shoot system for resistance to assault by a pathogen—even cells that have not been directly exposed to the disease-causing agent.

FIGURE 40.26 illustrates how the HR and SAR are thought to work together. In addition to triggering the HR, detection of a pathogen leads to the release of a hormone that initiates SAR. This signal acts globally as well as locally—that is, at the point of infection—and results in the expression of a large suite of genes called the pathogenesis-related (*PR*) genes.

When biologists set out to locate the hormone responsible for SAR, they found that levels of **methyl salicylate (MeSA)**—a molecule derived from salicylic acid—increase dramatically after tissues are infected with a pathogen. Follow-up work revealed these findings:

- Phloem sap leaving infected sites has elevated levels of MeSA.
- Treatments that reduce MeSA reduce or abolish SAR.
- Adding MeSA to the lower leaves of tobacco plants leads to SAR in the upper, untreated leaves.

Biologists are working to understand these complex interactions in order to genetically engineer crop plants that can resist diseases. Despite these interventions, however, pathogens continue to evolve, in a never-ending arms race.

How Do Plants Sense and Respond to Herbivore Attack?

Over a million species of insects have been discovered and named so far. Most of them make their living by eating leaves, stems, phloem sap, seeds, roots, or pollen. Plants have effective induced defenses in response to pathogens like viruses, bacteria, and fungi. But can they ramp up defenses in response to insect attack?

The Role of Proteinase Inhibitors When researchers started studying why some plant tissues are more palatable and digestible than others, biochemists discovered that many seeds and some storage organs, such as potato tubers, contain proteins called **proteinase inhibitors.**

Proteinase inhibitors block the enzymes—found in the mouths and stomachs of animals—that are responsible for digesting proteins. When an insect or a mammalian herbivore ingests a large dose of a proteinase inhibitor, the herbivore gets sick. As a result, herbivores learn to detect proteinase inhibitors by taste and avoid plant tissues containing high concentrations of these molecules.

Although many plant tissues contain proteinase inhibitors in low concentrations, biologists wanted to test the hypothesis that these proteins might also be part of an induced defense by the plant. To evaluate this idea, researchers allowed herbivorous beetles to attack one leaf on each of several potato plants.

- In liquid extracted from the other leaves on the attacked plants, proteinase inhibitor concentrations averaged 336 µg per mL of leaf juice.

- In control plants, where no insect damage had occurred, proteinase inhibitor levels averaged just 103 µg per mL.

This result supported the hypothesis that a hormone produced by wounded cells travels to undamaged tissues and induces the production of proteinase inhibitors.

The Discovery of Systemin Biologists isolated the wound-response hormone by purifying the compounds found in tomato leaves and testing them for the ability to induce the synthesis of proteinase inhibitors. The hormone that is active in tomato plants and closely related species turned out to be **systemin,** a polypeptide just 18 amino acids long. Systemin was the first peptide hormone ever described in plants.

Researchers who labeled copies of systemin with a radioactive carbon atom, injected the hormone into plants, and then monitored its location confirmed that systemin moves from damaged tissues to undamaged tissues.

Currently, work on the production of systemin and proteinase inhibitors focuses on determining each step in the signal

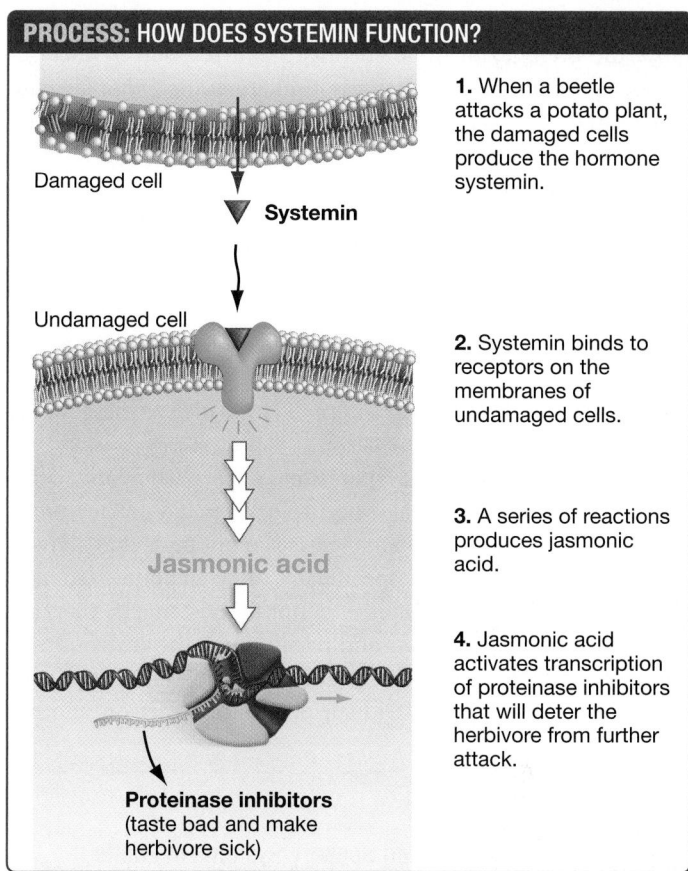

PROCESS: HOW DOES SYSTEMIN FUNCTION?

Damaged cell

Systemin

Undamaged cell

Jasmonic acid

Proteinase inhibitors (taste bad and make herbivore sick)

1. When a beetle attacks a potato plant, the damaged cells produce the hormone systemin.

2. Systemin binds to receptors on the membranes of undamaged cells.

3. A series of reactions produces jasmonic acid.

4. Jasmonic acid activates transcription of proteinase inhibitors that will deter the herbivore from further attack.

FIGURE 40.27 Signals from Insect-Damaged Cells Prepare Other Cells for Attack. Systemin, a hormone produced by herbivore-damaged cells, initiates a protective response in undamaged cells.

transduction pathway that alerts undamaged cells to danger (**FIGURE 40.27**). The data have yielded the following results:

1. Systemin is released from damaged cells.

2. Systemin travels through the body via phloem and binds to membrane receptors on target cells.

3. The activated receptor triggers a series of chemical reactions that eventually synthesize a molecule called jasmonic acid.

4. Jasmonic acid activates the production of at least 15 new gene products, including proteinase inhibitors.

In this way, plants build potent concentrations of insecticides in tissues that are in imminent danger of attack.

"Talking Trees": Responses from Nearby Plants When an insect starts munching on a leaf, volatile compounds evaporate from the leaf's surface and travel through the air. In the 1980s, researchers began to suspect that plants growing near other plants under herbivore attack "eavesdrop" on these volatiles. In response, they increase their own defenses—even though they've yet to be attacked.

The "talking trees" hypothesis has received extensive support, primarily through experiments based on exposing plants to volatiles in the absence of any insects. Even if the volatiles come from an entirely different species, research has shown that plants can sense the presence of these chemicals and respond by increasing the production of proteinase inhibitors and other defense compounds.

In some cases, plants that are under attack even call for help from other organisms.

Pheromones Released from Plant Wounds Recruit Help from Wasps Caterpillars and other herbivorous insects have enemies of their own—often wasps that lay their eggs in the insects' bodies. When wasp eggs hatch inside a caterpillar, the wasp larvae begin eating the host from the inside out. An organism that is free living as an adult but parasitic as a larva, such as these wasps, is called a **parasitoid** (**FIGURE 40.28**).

Biologists observed that parasitoids are particularly common when insect outbreaks occur in croplands. They wondered whether wounded plants release compounds that actively recruit parasitoids. More specifically, they hypothesized that plants produce **pheromones**—chemical messengers that are synthesized by an individual and released into the environment, and that elicit a response from a different individual. Hormones act on cells inside an individual; pheromones act on another individual.

Parasitized caterpillar

Wasp larvae emerging from caterpillar

FIGURE 40.28 Parasitoids Kill Herbivores. When this plant was first attacked by this caterpillar, it produced pheromones that attracted a female wasp. The wasp laid her eggs in the caterpillar. As the wasp larvae grew and developed into pupae, they devoured the caterpillar.

SUMMARY TABLE 40.3 Selected Sensory Systems in Plants

Stimulus	Receptor	Signal Transduction	Response	Adaptive Significance
Blue light	PHOT1 and other phototropins in stem and leaves	PHOT1 autophosphorylates; remainder of signal transduction systems unknown	Phototropism occurs; also involved in stomatal opening	Stems grow toward light with wavelengths needed for photosynthesis
Red light	Phytochrome in seeds and elsewhere	Phytochrome changes to P_{fr} form and activates responses	Seed germinates	Sunlight triggers germination
Far-red light	Phytochrome in stem and elsewhere	P_r moves into nuclei and induces genetic responses	Stems lengthen	Species that require full sunlight attempt to escape shade from leaves
Gravity	Proteins located in plasma membrane	Details unknown	Cells on opposite side of root or shoot elongate; tissue curves	Roots grow down; shoots grow up
Touch or wind	Stretch receptors; location unknown	Details unknown, but result is transcription activation in target genes	Stems grow shorter and thicker; in some cases vining (twining) growth	Individual is more resistant to damage; plants grow toward light
Touch	Receptor hair cell in Venus flytrap	Electrical changes in receptor cell's plasma membrane trigger action potentials	Target cells change shape; trap shuts	Plant can capture prey
Pathogens	Receptor protein products	Details unknown	Hypersensitive response (HR); death of infected cells	Pathogens starve, so infection is slowed or stopped
Herbivores	Unknown; activated in response to molecule from herbivore	Details unknown	Insecticide production; signals to parasitoids	Herbivores are sickened or killed

To explore this idea, researchers analyzed compounds that were released from corn seedlings during attacks by caterpillars. The insect-damaged leaves produced 11 molecules that were not produced by undamaged leaves. These compounds were not released by leaves that had been cut with scissors or crushed with a tool; only insect damage triggered their production.

To follow up on this result, the investigators put female wasps in an arena that contained leaves damaged by insects and leaves that had suffered only mechanical damage. In more than two-thirds of the tests that were performed, the wasps preferred to fly toward the insect-damaged leaves.

These results support the hypothesis that plants produce wasp attractants in response to attack by caterpillars. Biologists are increasingly convinced that plants can produce pheromones that recruit help in the form of egg-laden wasps.

From the research that has been done on plant sensory systems, it is abundantly clear that plants don't just sit there. These organisms may be stationary, but they constantly monitor and respond to a wide array of information about their environment (**TABLE 40.3**). Gaining a better understanding of phototropism, gravitropism, response to disease, and other aspects of plant behavior forms an exciting frontier in biology.

CHAPTER 40   REVIEW

For media, go to MasteringBiology

If you understand . . .

40.1   Information Processing in Plants

- In most cases, information processing in plants starts when a receptor protein changes shape in response to a stimulus.
- When signal transduction occurs, an external signal is changed into an intracellular signal.
- In receptor cells, signal transduction culminates in the production of hormones that are transported throughout the plant body.
- If a hormone binds to a receptor on a target cell, signal transduction occurs and culminates in changes in gene expression, altered translation rates, or changes in the activity of specific membrane pumps, channels, or ion carriers.

✔ You should be able to explain the analogy between how a plant sensory cell works and the following events: A person sees a barn on fire and calls the rural fire department; the dispatcher rings a siren; in response, members of the volunteer fire department race to the station to get the pumper truck.

40.2   Blue Light: The Phototropic Response

- Phototropins are blue-light receptors that become phosphory-lated and then initiate signal transduction pathways.

- In the phototropic response, cells near the tips of coleoptiles sense changes in blue light and respond by altering the distribution of the hormone auxin. Cells on one side of the coleoptile elongate in response to auxin much more than do cells on the other side of the coleoptile. In this way, plants bend toward sunlight.

✔ You should be able to predict whether the phototropic response differs in plants that require high-light conditions versus plants that thrive best in low-light conditions.

40.3   Red and Far-Red Light: Germination, Stem Elongation, and Flowering

- Phytochrome is a red/far-red switch that allows plants to detect shade from other plants and regulates a wide variety of plant growth responses including germination, stem elongation, and flowering.
- The phytochrome protein changes shape when it absorbs red light. The same protein changes to an alternate shape when it absorbs far-red light.

✔ You should be able to state a hypothesis to explain why animals do not have phytochromes (a red/far-red light switch).

40.4 Gravity: The Gravitropic Response

- Cells in the center of root caps contain amyloplasts with starch granules that function as statoliths and cause roots to grow down in response to gravity.

- Redistribution of auxin in response to gravity results in downward growth of roots and upward growth of shoots.

✓ You should be able to explain why starch granules make good gravity-sensing bodies.

40.5 How Do Plants Respond to Wind and Touch?

- Some plants respond to touch, including wind, by growing more slowly and becoming stocky.

- Thigmotropism is a direction response to touch, such as tendril coiling and closing of Venus flytraps.

✓ You should be able to describe the adaptive significance of the response to wind.

40.6 Youth, Maturity, and Aging: The Growth Responses

- Several key hormones regulate plant growth.

- Auxin establishes and maintains the long axis of the plant body, playing a key role in phototropism and gravitropism and maintaining apical dominance. All of these responses rely on the polar transport of auxin, which establishes a gradient in auxin from the plant's shoot tips to its root tips.

- Cytokinins promote cell division by regulating the cell cycle, and they delay senescence.

- Gibberellins (GAs) signal that conditions for growth are good and promote the initiation or continuation of growth and development.

- Abscisic acid (ABA) signals that environmental conditions are bad by suppressing growth, enforcing dormancy, and closing stomata.

- Regulation of dormancy and growth by ABA and by GAs are examples of how hormones interact—allowing plants to integrate information from several different stimuli and respond appropriately.

- Brassinosteroids are steroid hormones that regulate growth.

- Ethylene is a gaseous hormone that triggers senescence and fruit ripening, and it plays a role in leaf abscission.

✓ You should be able to suggest a hypothesis explaining why some hormones are not transported in a single direction.

40.7 Pathogens and Herbivores: The Defense Responses

- Plants have receptor proteins that detect pathogen infection and trigger the hypersensitive response, which limits the spread of disease.

- Herbivores trigger plant responses that lead to the synthesis of molecules that make plants less palatable.

- Localized infection or herbivory stimulates long-distance signaling and defense responses throughout the plant.

- Volatile signals triggered by herbivory attract parasitoids that attack the herbivores.

✓ You should be able to explain why cell suicide is beneficial to a plant.

(MB) MasteringBiology

1. **MasteringBiology Assignments**

   Tutorials and Activities Experimental Inquiry: What Effect Does Auxin Have on Coleoptile Growth?; Flowering Lab Leaf Abscission; Plant Defenses; Plant Hormones; Plant Responses to Light; Sensing Light

   Questions Reading Quizzes, Blue-Thread Questions, Test Bank

2. **eText** Read your book online, search, take notes, highlight text, and more.

3. **The Study Area** Practice Test, Cumulative Test, BioFlix® 3-D Animations, Videos, Activities, Audio Glossary, Word Study Tools, Art

You should be able to . . .

✓ TEST YOUR KNOWLEDGE

Answers are available in Appendix A

1. Which of the following statements about phytochrome is *not* correct?
   a. It is photoreversible.
   b. Its function was understood long before the protein itself was isolated.
   c. The P_{fr} form activates the responses to red light.
   d. It is involved in guard-cell opening.

2. If a plant is touched repeatedly over the course of many days or if it experiences long-term exposure to wind, what happens?
   a. Growth of the root system is accelerated, making the plant more stable.
   b. Large-scale changes in gene expression occur, resulting in shoots that are short and stout.
   c. Electrical signals cause leaves to fold up, avoiding damage.
   d. Continued mechanical stimulation indicates that the individual is threatened with destruction, so it initiates flowering in an attempt to reproduce before it dies.

3. Which of the following statements about hormones is correct?
   a. They tend to be large molecules.
   b. They exert their effects only on the same cells that produce them.
   c. They can exert strong effects only when they are present in high concentrations.
   d. They trigger a response by binding to receptors in target cells.

4. What evidence suggests that ABA from roots can signal guard cells to close?
   a. If roots are given sufficient water, guard cells close anyway.
   b. If roots are dry, guard cells begin to close—even though leaves are not experiencing water stress.
   c. Applying ABA on guard cells directly causes them to close.
   d. If roots are dry, ABA concentrations in leaf cells drop dramatically.

5. What is the rapid death of plant cells in response to a pathogen attack called?

6. Unlike animal steroid hormone receptors, where are brassinosteroid receptors located?

✔ TEST YOUR UNDERSTANDING   Answers are available in Appendix A

7. Why was it logical to predict that amyloplasts function as statoliths?
   a. They are dense and settle to the bottom of gravity-sensing cells.
   b. They are present only in gravity-sensing cells.
   c. They make a direct physical connection with membrane proteins that have been shown to be the gravity-receptor molecule.
   d. Their density changes in response to gravity.

8. Phytochromes can be considered "shade detectors," while phototropins such as PHOT1 can be considered "sunlight detectors." Explain why these characterizations are valid.

9. What does transduce mean? Compare and contrast two types of transduction processes described for a sensory or a defense process.

10. A plant's response to a given hormone depends on the cells or tissues that receive the signal, the plant's developmental stage or age, the concentration of the hormone, and the concentration of other plant hormones that are present. Provide examples that support each of these claims.

11. Discuss the general role that ethylene serves in plants, and predict how produce dealers deal with benefits or costs of the effects of ethylene.

12. Suppose that a mutant plant is unable to make methyl salicylate. Explain why it is not likely to survive in the wild.

✔ TEST YOUR PROBLEM-SOLVING SKILLS   Answers are available in Appendix A

13. To explore how hormones function, researchers have begun to transform plants with particular genes. In one experiment, a gene involved in cytokinin synthesis was introduced into tobacco plants. Which one of the following results would be expected?
    a. Individuals produced more lateral branches.
    b. Stems grew extremely tall and slender.
    c. Roots were incapable of responding to gravity.
    d. Stomata were closed most of the time.

14. In general, small seeds that have few food reserves must be exposed to red light before they will germinate. (Lettuce is an example.) In contrast, large seeds that have substantial food reserves typically do not depend on red light as a stimulus to trigger germination. State a hypothesis to explain these observations.

15. In many species native to tropical wet forests, seeds do not undergo a period of dormancy. Instead, they germinate immediately. Make a prediction about the role of ABA in these seeds. How would you test your prediction?

16. Researchers have shown that transforming plants with genes encoding resistance proteins makes these plants resistant to more pathogens. How might the pathogens evolve when exposed to these genetically altered plants?

41 Plant Reproduction

In this chapter you will learn that

Reproduction is key to the success of flowering plants

starting with ↓

Reproductive systems: unifying principles 41.1

looking closer at ↓

Reproductive structures 41.2 → in → **Flowering plants**

involved in ↓

Pollination and fertilization 41.3 → using → **Animals and wind to move pollen from flower to flower**

resulting in ↓

The seed (and embryo) 41.4 → which uses → **Animals, wind, and water to be dispersed**

This chapter focuses on the structure and function of plant reproductive structures like those of this flower.

t would be difficult to overemphasize the importance of the reproductive organs and processes that are analyzed in this chapter—for plants, for biologists, and for you.

- For plants, every structure in the body and every physiological process—from water transport to photosynthesis—exist for one reason: to maximize the chances that the individual will produce off-spring. Reproduction is the unconscious goal of everything that an organism does (see Chapter 1).

- For biologists, plant reproduction is not only fundamental to understanding how plants work but also the basis for major industries. Agriculture, horticulture, forestry, biotechnology, and eco-logical restoration draw extensively on what biologists know about plant reproduction.

This chapter is part of the Big Picture. See how on pages 840–841.

✔ When you see this checkmark, stop and test yourself. Answers are available in Appendix A.

- For you, plant reproduction for the most part means food. Human diets are based on consuming plant reproductive structures—primarily the seeds and fruits derived from flowers. A **flower** is a reproductive structure that produces gametes, attracts gametes from other individuals, nourishes embryos, and develops seeds and fruits. **Seeds** consist of an embryo and nutrient stores surrounded by a protective coat. **Fruits** develop from the flower's seed-producing organ and contain seeds.

This chapter's analysis of plant reproduction focuses on angiosperms, for three reasons: (**1**) Angiosperms represent over 85 percent of the land plants described to date; (**2**) virtually every important domesticated plant is an angiosperm; and (**3**) aspects of reproduction in other land plant lineages were described previously (see Chapter 31). By the end of this chapter, you'll appreciate the practical aspects of flowers as well as their beauty.

41.1 An Introduction to Plant Reproduction

Plant reproductive structures and processes vary among species. Consider just one aspect of reproductive organs—size. Flowers vary from microscopic to the size of a small child; seeds and fruits range from dustlike particles to coconuts.

Fortunately for students of plant biology, several basic principles unify this diversity of reproductive systems. Let's begin by defining sex.

Sexual Reproduction

Most plants reproduce sexually. **Sexual reproduction** is based on meiosis and fertilization, and it results in offspring that are genetically unlike each other and unlike their parents (see Chapter 13). In plants, meiosis and fertilization occur in alternate phases of a life cycle.

To review briefly, **meiosis** is a type of nuclear division that results in four daughter cells, each of which has half the number of chromosomes present in the parent cell. **Fertilization** is the fusion of haploid cells termed **gametes.** The result of fertilization is the production of a single, diploid cell called the **zygote,** which will develop into a multicellular individual. Male gametes, or **sperm,** are small cells that contribute genetic information in the form of DNA but few or no nutrients to the offspring. Female gametes, or **eggs,** also contain DNA. But the female parent contributes nutrient stores as well. The important point is that although both sperm and egg contribute DNA, the male and female parents contribute vastly different amounts of other resources to the offspring.

Meiosis and fertilization result in offspring that are genetically unlike the parents. Sexual reproduction produces genetically diverse offspring that can potentially be more successful than their parents at warding off attacks from viruses, bacteria, and other pathogens (see Chapter 13).

The advantages of sexual reproduction are common to all eukaryotes that undergo meiosis. When and where meiosis occurs, however, is highly variable. Let's take a closer look.

The Land Plant Life Cycle

In most animals, meiosis leads directly to the formation of gametes. In plants, the situation is much different.

Land plants are characterized by a life cycle with two distinct multicellular forms—one diploid and one haploid. An individual in the diploid phase of the life cycle is called a **sporophyte,** while an individual in the haploid phase of the life cycle is called a **gametophyte.**

This type of life cycle, called **alternation of generations,** has evolved independently in various protists and land plant groups (see Chapters 30 and 31).

What Is Alternation of Generations? Key features of alternation of generations are reviewed in **FIGURE 41.1.** When alternation of generations occurs, meiosis does not lead directly to the formation of gametes, as it does in humans and other animals. Instead, it leads to the production of haploid cells called spores. A **spore** is a cell that grows directly into an adult individual. Several structures and processes are common to all land plant life cycles:

1. Meiosis occurs in sporophytes and results in the production of haploid spores. Unlike zygotes, spores are not produced by the fusion of two cells. Unlike gametes, spores produce an adult without fusing with another cell. Meiosis and spore production occur inside structures called **sporangia.**

2. Spores undergo mitosis and divide to form multicellular, haploid gametophytes.

3. Gametophytes produce gametes using mitosis.

4. Fertilization occurs when two gametes fuse to form a diploid zygote.

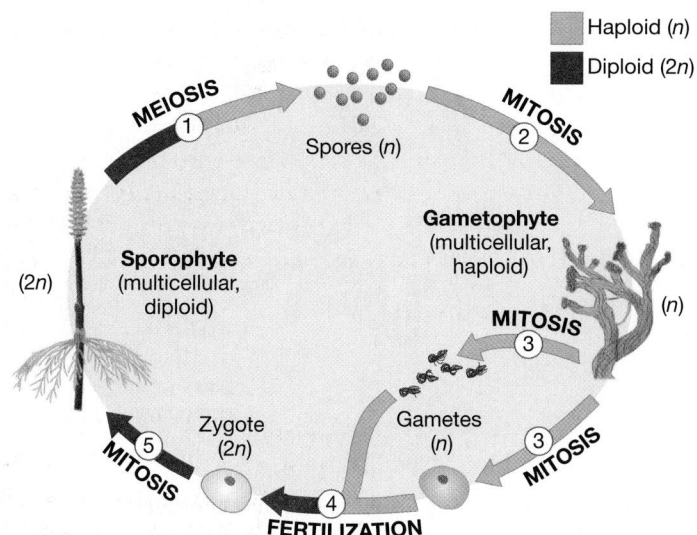

FIGURE 41.1 All Plants Undergo Alternation of Generations.

(a) Mosses: Gametophytes are large and long lived; the sporophyte is small, short lived and nutritionally dependent on the gametophyte.

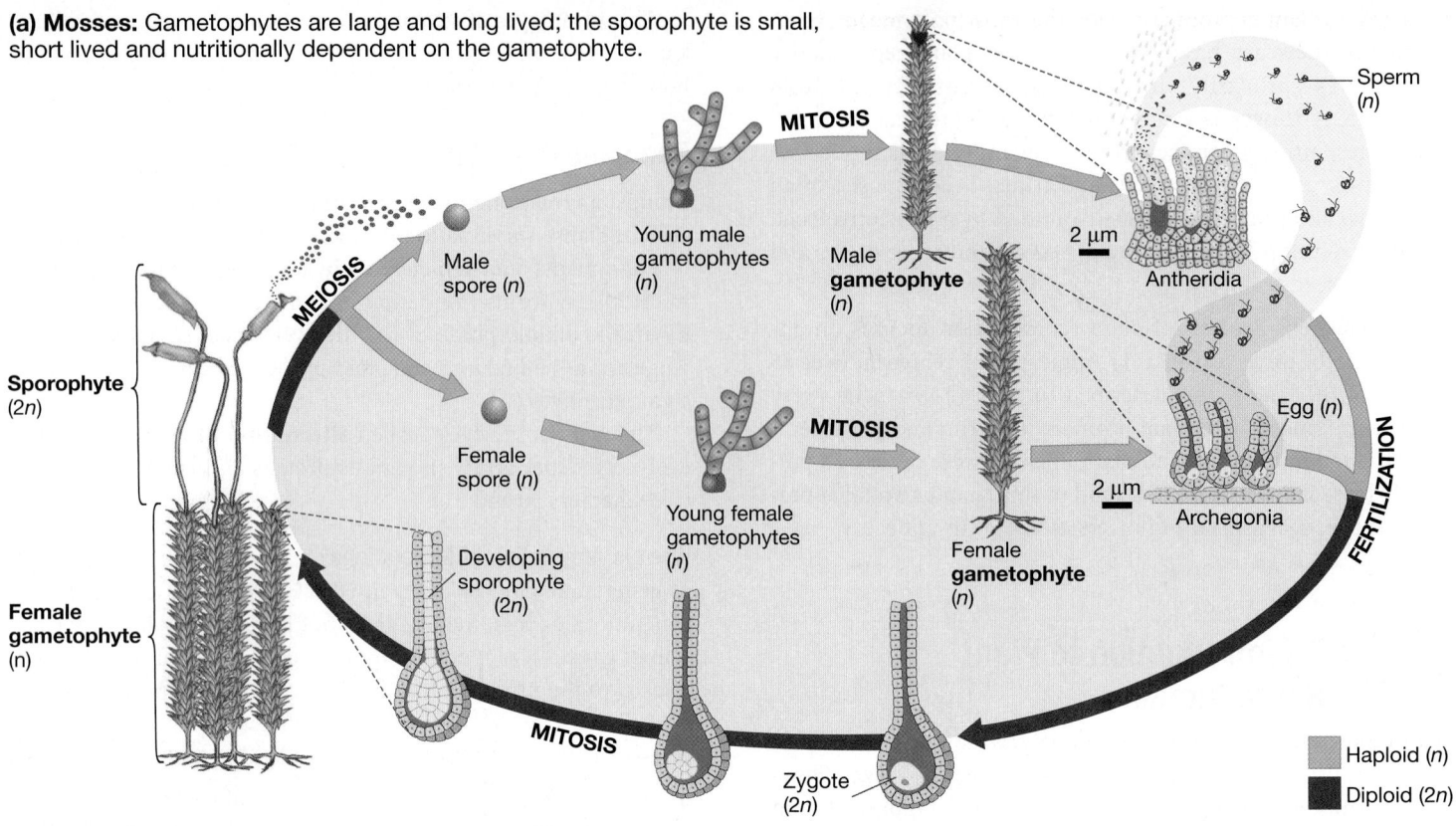

(b) Angiosperms: Sporophyte is large and long lived; gametophytes are small (microscopic), short lived, and nutritionally dependent on the sporophyte.

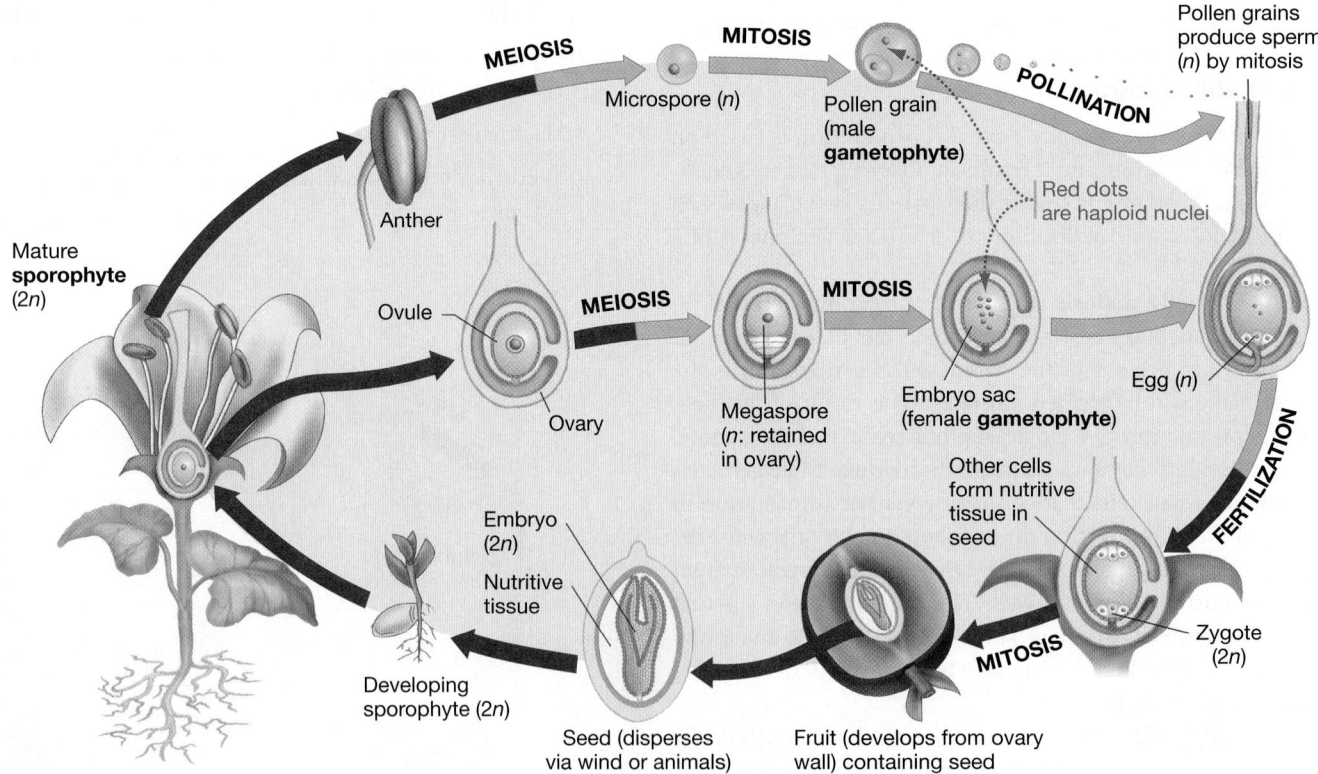

FIGURE 41.2 There Is Wide Variation in Plant Life Cycles. (a) Mosses, **(b)** Angiosperms.

5. The zygote undergoes mitosis and grows to form the multi-cellular sporophyte.

A good way to keep these terms straight is to remember that sporophyte means spore-plant, while gametophyte means gamete-plant. Sporophytes produce spores by meiosis. Gametophytes produce gametes by mitosis.

Variation in Plant Life Cycles Land plant species vary a great deal in how large and long lived the gametophyte and sporophyte are relative to each other (see Chapter 31). The examples in **FIGURE 41.2** represent the extremes of the variation in life cycles observed within land plants.

- Figure 41.2a diagrams the life cycle of a moss. Notice that the largest stage in the moss life cycle is the gametophyte. The sporophyte is dependent on the gametophyte for nutrition.

- Figure 41.2b diagrams the life cycle of an angiosperm. Notice that the male and female gametophytes are microscopic, physically separate, and completely dependent on the sporophyte for nutrition.

✔ If you understand the angiosperm life cycle, you should be able to (1) identify the male spore and female spore in Figure 41.2b, and (2) provide evidence to support the statement that in angiosperms, female gametophytes never leave their parent plant.

The non-vascular plants (sometimes called bryophytes), which include mosses, are the sister group to all other land plants, and they are found early in the fossil record. Angiosperms, in contrast, are the most recent lineage of land plants to appear in the fossil record and the most derived branch on the evolutionary tree. Based on these observations, biologists have concluded that over the course of land plant evolution, gametophytes became reduced while sporophytes became more conspicuous.

Why this trend occurred is still unknown. The adaptive significance of the gametophyte and sporophyte phases remains a challenge for biologists interested in plant reproduction.

Asexual Reproduction

Asexual reproduction does not involve fertilization and results in the production of **clones**—genetically identical copies of the parent plant.

Some plants extend their life indefinitely by asexual reproduction. The oldest of all known plants is a ring of creosote bushes in the Mojave Desert of California. The bushes comprise a clone that originated from a parent plant estimated to have germinated 12,000 years ago.

Although all asexual reproduction is based on mitosis, a wide array of mechanisms exists.

- **FIGURE 41.3a** shows grass shoots and roots emerging from nodes on underground horizontal stems called **rhizomes.** If the individuals emerging from the nodes become separated from the parent plant, they represent asexually produced offspring.

- The gladiolus plant in **FIGURE 41.3b** has propagated itself via modified stems called **corms,** which grow under the surface of the soil.

- The kalanchoe in **FIGURE 41.3c** produces "plantlets" (small plants) from meristematic tissue located along the margins of its leaves. When the plantlets mature, they drop off the parent plant and grow into independent individuals.

- In dandelion and certain other species, mature seeds can form without fertilization occurring. This phenomenon, known as **apomixis,** results in seeds that are genetically identical to the parent.

The key characteristic of asexual reproduction is efficiency. If an herbivore or a disease wipes out the plants surrounding a grass plant, the grass can quickly send out horizontal stems. Its asexually produced offspring are likely to fill the unoccupied space before seeds from competitors can establish themselves and grow. The parent plant can also nourish these progeny as they become established.

(a) Rhizome

(b) Corm

(c) Plantlets

FIGURE 41.3 The Mechanisms of Asexual Reproduction Are Diverse. These are just three of many mechanisms of asexual reproduction in plants.

Although asexual reproduction is extremely common in plants, it does have a downside. The most important is that a fungus or other disease-causing agent that infects an individual plant will probably succeed in infecting the plant's cloned offspring as well, even if they are no longer physically connected.

This hypothesis is based on the observation that plants fight disease with a wide variety of molecules (see Chapter 40). Because sexually produced offspring are genetically unlike their parents, these offspring have unique combinations of disease-fighting molecules and may be able to resist infections that devastate their parents.

This is an important point in agriculture and horticulture because asexually propagated apples, bananas, and other crops are more susceptible to epidemics than are sexually propagated species. How does sexual reproduction occur?

41.2 Reproductive Structures

Each major group of plants, from mosses to angiosperms, has a characteristic variation on the theme of alternation of generations, as well as characteristic male and female reproductive structures (see Chapter 31). Here, though, the focus is on the flower.

The General Structure of the Flower

Structurally, all flowers are variations on a theme. They are made up of four basic organs that are essentially modified leaves: (1) sepals, (2) petals, (3) stamens, and (4) one or more carpels. These organs are attached to a compressed portion of stem called the receptacle (**FIGURE 41.4a**).

Not all four organs are present in all flowers, however; and as **FIGURE 41.4b** shows, the coloration, size, and shape of these four components are fabulously diverse. Let's consider each of the four parts, in turn.

Sepals Form an Outer, Protective Whorl Sepals are leaflike structures that make up the outermost parts of a flower. Sepals are usually green and photosynthetic, and they are relatively thick compared with other parts of the flower.

Because they are attached to the receptacle in a circle or whorled arrangement, sepals enclose the flower bud as it develops and grows—protecting young buds from damage by insects or disease-causing agents. The entire group of sepals in the flower is called the **calyx.**

Petals Furnish a Visual Advertisement Like sepals, **petals** are arranged around the receptacle in a whorl. Often brightly colored and scented, petals function to advertise the flower to bees, flies, hummingbirds, and other pollinators.

In some cases, the color of the petals correlates with the visual abilities of particular animals. Bees, for example, respond strongly to wavelengths in the blue and purple regions of the light spectrum, as well as yellow (they don't see red well). Flowers

(a) Basic parts of a flower

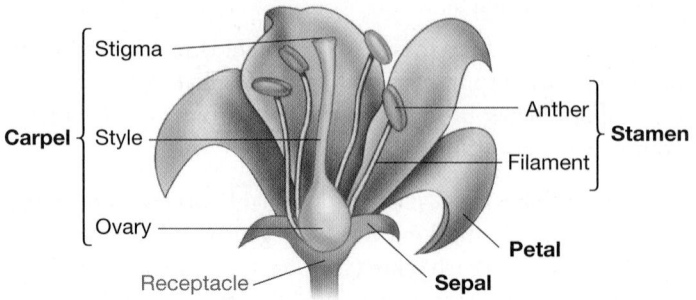

(b) Examples of flower diversity

FIGURE 41.4 The Basic Structures in Flowers Are Highly Variable. **(a)** Flowers comprise sepals, petals, stamens, and carpels. **(b)** The four parts vary among species.

FIGURE 41.5 Insects See in the Ultraviolet Range. (a) The inflorescence (flower cluster) of a black-eyed Susan, seen by the unaided human eye. **(b)** The same structure, photographed with a camera that records ultraviolet wavelengths that are visible to bees but invisible to humans.

that attract bees, in turn, often have yellow, blue, or purple petals with ultraviolet patches.

The ultraviolet regions of petals in "bee flowers" frequently highlight the center of the flower (**FIGURE 41.5**), where the stamens and carpels are located. Why? In these flowers, the base of the petals contains a gland called a **nectary.** The nectary produces the sugar-rich fluid **nectar,** which is harvested by many of the animals that visit flowers, along with pollen. In the process of collecting pollen or nectar, the visiting animal usually deposits pollen from a different plant on the female parts—accomplishing pollination.

The entire group of petals in a flower is called the **corolla.** In some species, the petals within the corolla vary in size, shape, and function:

- Flattened petals may provide a landing pad for flying insects.
- Elongated, tubelike petals frequently have a nectary at their base that can be reached only by animals with a long beak or tongue-like proboscis.
- Some petals protect the reproductive organs located inside the corolla.
- Specialized cells in some petals synthesize and release molecules that provide a scent attractive to certain species of pollinating animals.

In contrast, wind-pollinated angiosperms such as oaks, birches, pecans, and grasses have flowers that have small petals or no petals at all, and they lack nectaries. These species do not invest in structures that aren't required for pollination.

Stamens Produce Pollen   **Stamens** are reproductive structures that produce male gametophytes—also known as pollen grains. The male gametophytes, in turn, produce sperm.

Each stamen consists of two components:

1. a slender stalk termed the **filament,** and
2. the pollen-producing organs called **anthers** (see Figure 41.4a).

The anther is the business end of the stamen—where meiosis and pollen formation take place. The filament holds the stamen in a place where wind, insects, hummingbirds, bats, or other agents can make contact with the pollen grains produced in the anther.

Carpels Produce Ovules   The fourth reproductive structure is the **carpel,** which produces female gametophytes. A carpel consists of three regions:

1. The **stigma** is a sticky tip that receives pollen.
2. The **style** is a slender stalk.
3. The **ovary** is an enlarged structure at the base of the carpel (see Figure 41.4a).

Inside the ovary, female gametophytes are produced in structures called **ovules.** An ovary may contain more than one ovule. When the female gametophytes that are produced inside ovules mature, they produce eggs.

The "Sex" of Flowers Varies   In most angiosperm species, stamens and carpels are produced on the same individual. Flowers that contain both stamens and carpels are referred to as **perfect.**

Flowers can also be **imperfect,** however, meaning they contain either stamens *or* carpels, but not both. Imperfect flowers that contain only stamens can be considered "male" flowers. Similarly, imperfect flowers that contain only carpels can be considered "female" flowers.[1]

In some cases, separate stamen- or carpel-producing flowers occur on the same individual. Species like these, including the corn plants illustrated in **FIGURE 41.6a** (see page 828), are **monoecious** (literally, "one-house"). In corn, the tassel is a collection of stamen-producing "male" flowers, and the ear contains a group of carpel-producing "female" flowers.

Species that are monoecious or that have perfect flowers are like hermaphroditic species of animals, which contain both male and female reproductive organs and produce sperm and eggs (see Chapter 50).

In contrast, some species with imperfect flowers are **dioecious** ("two-houses")—meaning that each individual plant produces either stamen-bearing flowers only, and could be considered "male," or carpel-bearing flowers only and could be considered "female." *Cannabis sativa* is a dioecious species (**FIGURE 41.6b**). Dioecious plants are like animal species with individuals that have either male or female reproductive organs—not both.

How Are Female Gametophytes Produced?

What purposes do the three parts of the carpel serve? The function of the stigma and style will become clear in Section 41.3; for now, let's concentrate on what happens inside the ovary.

---

[1]Technically, flowers are not referred to as male and female. Instead, they are staminate or carpellate. Staminate flowers produce stamens, which produce pollen grains, which produce male gametes (sperm). Carpellate flowers produce carpels, which contain ovaries. Female gametophytes develop inside ovaries and produce female gametes (eggs). For convenience, though, the text will sometimes refer to male and female flowers and reproductive structures.

(a) Corn is monoecious.

Male
flowers

Female
flowers

(b) *Cannabis* is dioecious.

Male
flowers

Female
flowers

FIGURE 41.6 "Male" and "Female" Flowers Can Occur on the Same Individual or on Different Individuals. (a) The tassels of corn contain male flowers; ears contain female flowers. **(b)** In *Cannabis sativa*, male and female flowers are found on different individuals.

FIGURE 41.7 illustrates a longitudinal section showing the inside of a typical angiosperm ovary. Notice that it contains one or more ovules. Each ovule contains a structure called the megasporangium, which contains a diploid cell called the megasporocyte. (The use of "mega" is appropriate, because these structures are much larger than their counterparts in the stamen.) The megasporangium is comparable to spore-producing organs found in other plants, such as the sporangia found on the back of fern leaves.

When you study Figure 41.7, note four important points:

1. The megasporocyte divides by meiosis.

2. Four haploid nuclei called **megaspores** result from meiosis, but three degenerate. No one is sure how or why this happens.

3. The surviving megaspore divides by mitosis to produce a structure with haploid nuclei. This is the female gametophyte—usually known as the **embryo sac.**

4. The haploid nuclei segregate to different positions in the embryo sac, and cell walls form around some of them.

In the carpel, then, a diploid megasporocyte divides by meiosis to form a megaspore, which then divides by mitosis to form the female gametophyte. Female gametophytes are encased in an ovary, are retained in the flower, and produce an egg.

In many angiosperms, the embryo sac contains eight haploid nuclei and seven cells. Typically, two **polar nuclei** stay together within one central cell—the largest cell in the ovule. The number of polar nuclei varies among species, however.

The egg cell is located at one end of the female gametophyte, near an opening in the ovule called the **micropyle** ("little-gate"). The micropyle is where a sperm nucleus will enter the ovule before fertilization.

How Are Male Gametophytes Produced?

FIGURE 41.8 provides a detailed look at the stamen and the steps that occur in the production of male gametophytes. Recall that a stamen consists of two major parts: an anther and a filament.

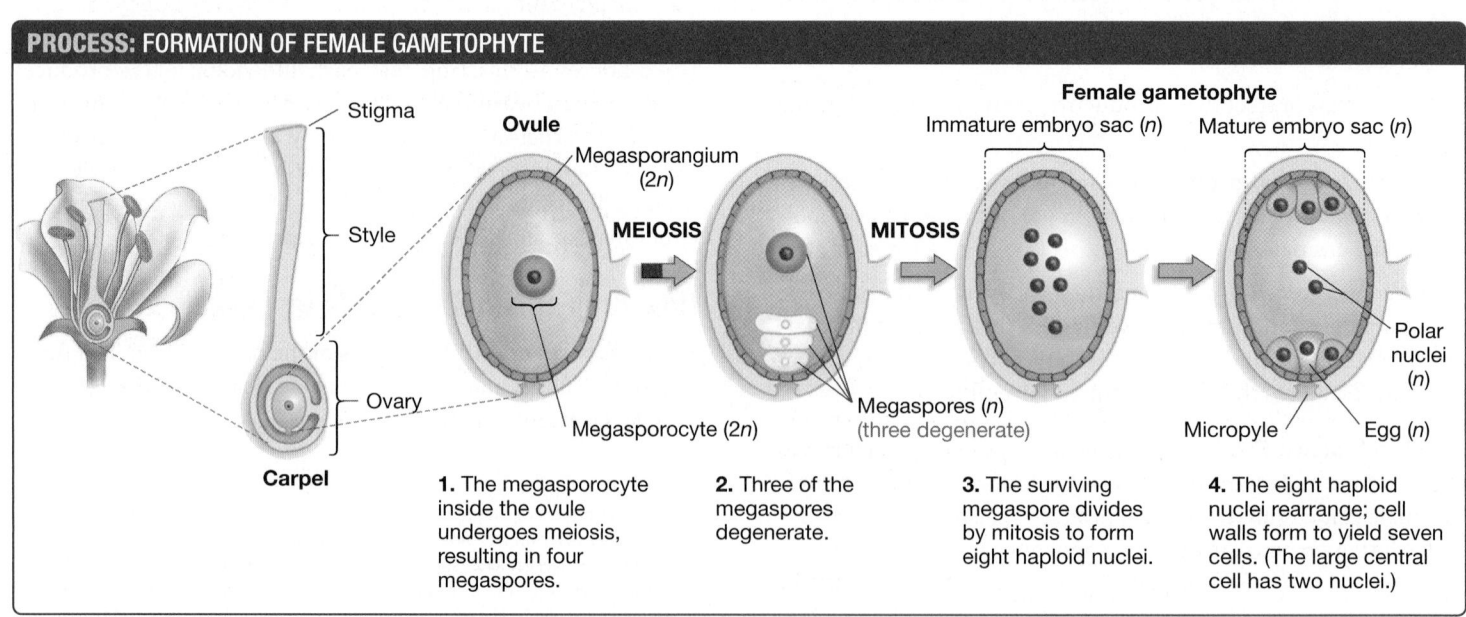

PROCESS: FORMATION OF FEMALE GAMETOPHYTE

Stigma

Style

Ovary

Carpel

Ovule

Megasporangium (2n)

MEIOSIS

Megasporocyte (2n)

Megaspores (n) (three degenerate)

MITOSIS

Female gametophyte

Immature embryo sac (n)

Mature embryo sac (n)

Polar nuclei (n)

Micropyle

Egg (n)

1. The megasporocyte inside the ovule undergoes meiosis, resulting in four megaspores.

2. Three of the megaspores degenerate.

3. The surviving megaspore divides by mitosis to form eight haploid nuclei.

4. The eight haploid nuclei rearrange; cell walls form to yield seven cells. (The large central cell has two nuclei.)

FIGURE 41.7 In Angiosperms, Megaspores Produce Female Gametophytes.

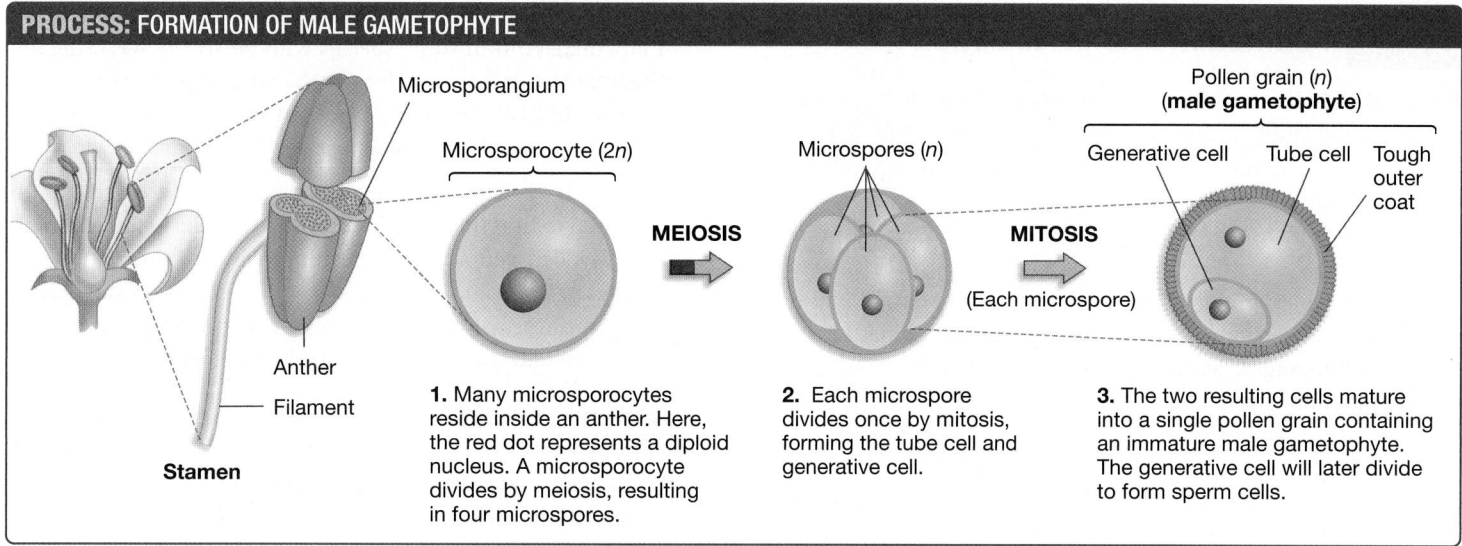

1. Many microsporocytes reside inside an anther. Here, the red dot represents a diploid nucleus. A microsporocyte divides by meiosis, resulting in four microspores.

2. Each microspore divides once by mitosis, forming the tube cell and generative cell.

3. The two resulting cells mature into a single pollen grain containing an immature male gametophyte. The generative cell will later divide to form sperm cells.

FIGURE 41.8 In Angiosperms, Microspores Produce Male Gametophytes.

✔ **QUESTION** Define a gametophyte. Why do pollen grains conform to the definition of a gametophyte?

Inside the anther, structures known as microsporangia contain diploid cells called microsporocytes.

When you study Figure 41.8, note three important points:

1. Microsporocytes undergo meiosis.

2. Each haploid cell that results is a **microspore.** Normally, all of the microspores survive. Microspores divide by mitosis.

3. The two nuclei that result from mitotic division in a microspore form a haploid, immature male gametophyte, also known as the **pollen grain.**

In the anther, then, a diploid microsporocyte divides by meiosis to form microspores, which then divide by mitosis to form male gametophytes. Male gametophytes are dispersed from the flower and eventually produce sperm.

At the immature stage—before it has produced sperm—the male gametophyte consists of two cells: a small generative cell enclosed within a larger tube cell. The male gametophyte is considered mature when the haploid generative cell undergoes mitosis and produces two sperm cells.

In some species, this maturation step occurs while pollen is still in the anther. In other species, maturation and sperm production don't occur until after the pollen grain lands on a stigma and begins to grow. The reasons for this difference are not understood.

The wall of a pollen grain develops a tough outer coat that includes the watertight compound called **sporopollenin** (introduced in Chapter 31). This coat protects the male gametophyte when the pollen is released from the parent plant into the environment. Depending on the species, pollen grains may be dispersed by an animal, the wind, or water currents.

And now we're finally ready to explore the moment of truth: How does a pollen grain get to the mature carpel of the same species, where an egg cell is waiting?

check your understanding

C Y U

If you understand that . . .

- In angiosperm sporophytes, flowers produce megaspores and microspores that develop into female and male gametophytes, respectively.
- The female reproductive structures called carpels contain ovaries. Ovaries enclose structures called ovules. Female gametophytes are produced inside ovules.
- Formation of a female gametophyte begins when a diploid megasporocyte inside an ovule undergoes meiosis. The product of meiosis is a haploid megaspore. The megaspore divides by mitosis to form the female gametophyte—including the egg and polar nuclei.
- Male gametophytes are produced inside reproductive structures called anthers.
- Formation of the male gametophyte begins when a diploid microsporocyte undergoes meiosis to form haploid microspores. Microspores divide by mitosis to form the male gametophyte—including a tube cell and a generative cell that will divide by mitosis to form two sperm cells.

✔ **You should be able to . . .**

1. Compare and contrast a megasporocyte and a microsporocyte.

2. Compare and contrast a female gametophyte and a male gametophyte.

Answers are available in Appendix A.

41.3 Pollination and Fertilization

Pollination is the transfer of pollen grains from an anther to a stigma; fertilization occurs when a sperm and an egg actually unite to form a diploid zygote. The two events are separated in space and time.

Pollination is not restricted to angiosperms. Gymnosperms (introduced in Chapter 31) also package their male gametophytes into pollen grains. This section will focus on pollination and fertilization in flowering plants. Managing pollination and fertilization in angiosperms is a critical challenge for fruit growers and plant breeders.

In addition, angiosperms' pollination and fertilization systems are thought to be key to their evolutionary success. What aspects of pollination and fertilization allowed flowering plants to become so successful in terms of their numbers of species?

Pollination

Pollen can fall on the stigma of the same individual or the stigma of a different individual. **Self-fertilization,** or selfing, occurs when a sperm and an egg from the same individual combine to produce an offspring. In most cases, though, plants **outcross**—meaning that sperm and eggs from different individuals combine to form an offspring. Outcrossing is the result of **cross-pollination**—when pollen is carried from the anther of one individual to the stigma of a different individual.

Selfing versus Outcrossing: Costs and Benefits Selfing and outcrossing each have advantages and disadvantages. The primary advantage of selfing is that successful pollination is virtually assured—it doesn't depend on agents other than the plant itself.

Biologists have documented the benefit of pollination assurance by hand-pollinating plants that normally outcross. In most cases, the hand-pollinated individuals produce far more seed than do individuals that are pollinated naturally.

Other things being equal, self-fertilization should result in the production of many more seeds than outcrossing. Other things are not equal, however. Selfing has a distinct disadvantage: Even though it still involves meiosis, selfed offspring are usually much less diverse genetically than outcrossed offspring are (see Chapter 13). In some cases, selfed offspring may also suffer from inbreeding depression (see Chapter 26).

Although outcrossing is riskier in terms of the chances that pollination will occur, it results in genetically diverse offspring that may be much more successful at warding off attacks from viruses, bacteria, and other pathogens (see Chapter 13).

Outcrossing is much more common than selfing. In many cases, plants have elaborate mechanisms to prevent selfing:

- *Temporal avoidance* In some species that have perfect flowers, male and female gametophytes mature at different times. Thus, selfing cannot occur.

- *Spatial avoidance* Selfing isn't possible in dioecious species and may be rare in monoecious species, unless pollinators transfer pollen between different-"sexed" flowers on the same individual. And in some species with perfect flowers, the anthers and stigma are so far apart that self-pollination is extremely unlikely—if pollen falls inside the flower, there is almost no chance that it will land on the stigma.

- *Molecular matching* In species that normally outcross, it is common to find that the interaction of proteins on the surfaces of the pollen and stigma determines whether pollination will be successful. Pollination is blocked if proteins on the pollen-grain surface match proteins on the stigma—indicating that the pollen and stigma are from the same individual.

Pollination Syndromes Cross-pollination can be accomplished in various ways: Pollen can be carried from flower to flower by physical agents such as wind or water, or by organisms such as insects, birds, or bats.

Animals visit flowers to eat pollen grains, harvest nectar, or both. As an animal feeds from a flower, pollen grains adhere to its body incidentally. When the same individual visits another flower of the same species to feed, some of these grains are deposited on a stigma of the second flower.

In most cases, animal pollination is an example of **mutualism:** a mutually beneficial relationship between two species. Pollinators usually benefit by receiving food; flowering plants gain by having their male gametophytes transferred to a different individual so that outcrossing takes place.

Pollination syndromes are suites of flower characters that are associated with certain types of pollinators. For example, many bird-pollinated flowers tend to be red and unscented, and they open during the day when birds are active. In contrast, moths and bats are usually active at night. If they feed on nectar or pollen, the flowers they visit tend to be white—and thus more visible in low light, have a strong scent, and open at night.

Structures associated with pollination syndromes are thought to be adaptations: traits that increase the fitness of individuals in a particular environment. In this case, flowers and pollinators have adaptations that increase pollination frequency and feeding efficiency, respectively. To capture this point, biologists say that **coevolution** has occurred (see Chapter 55).

One of the most famous examples of coevolution in pollination involves a species called Darwin's orchid (**FIGURE 41.9**), which is native to Madagascar. When first discovered by Western scientists, the orchid attracted a great deal of attention because it has a "spur" that can be as much as 28 cm (11 in) long, and there is a nectary at its base. Charles Darwin hypothesized that it must be pollinated by a moth with a tongue-like proboscis as long as the spur. The idea seemed preposterous at the time, but 40 years later, a hawkmoth species with a proboscis that averages 24 cm in length was discovered pollinating the orchid in the wild.

Why Did Pollination Evolve? In mosses, ferns, and other groups that do not form pollen, sperm have flagella and swim to the egg through droplets of water, or are transferred on water droplets that cling to the legs of tiny soil insects, such as springtails and mites. In conifers and most other gymnosperms, wind transmits

Flower · Pollinator

Extremely long spur with nectary at the base

Extremely long proboscis

FIGURE 41.9 Pollinators Can Sometimes Be Inferred from Flower Structure and Color. Charles Darwin hypothesized that this white orchid flower with an extremely long spur must be pollinated by a moth with a tongue-like proboscis that was long enough to reach the nectar at the end. He was right.

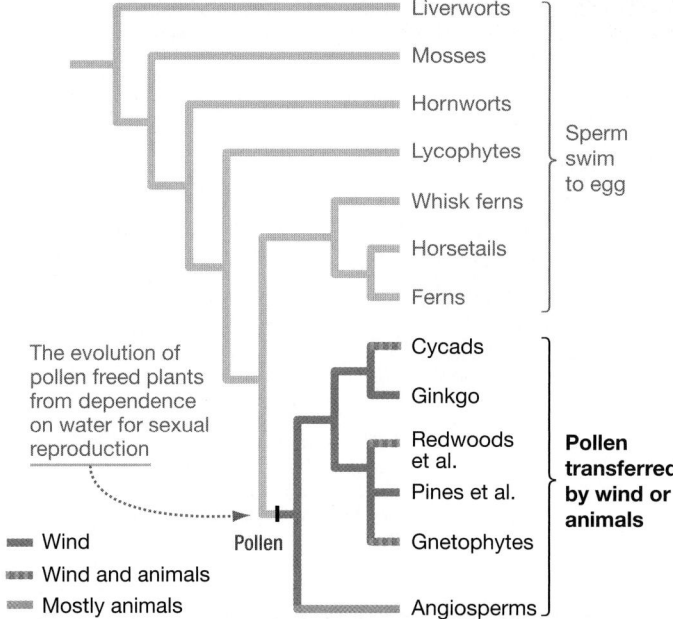

Liverworts
Mosses
Hornworts
Lycophytes
Whisk ferns
Horsetails
Ferns

Sperm swim to egg

The evolution of pollen freed plants from dependence on water for sexual reproduction

Cycads
Ginkgo
Redwoods et al.
Pines et al.
Gnetophytes
Angiosperms

Pollen transferred by wind or animals

Wind
Pollen

━━ Wind
━━ Wind and animals
━━ Mostly animals

FIGURE 41.10 Pollen Is a Relatively Recent Innovation in Land Plant Evolution. Evolutionary relationships among the major groups of plants. In the lineages colored gray, sperm have flagella and swim to the egg. In the lineages colored blue or orange, sperm are produced by pollen grains.

pollen from male to female. In some of the other groups that produce pollen, such as the cycads, gnetophytes, and angiosperms, many species are pollinated by animals—particularly by insects.

When these observations are mapped onto the phylogenetic tree shown in **FIGURE 41.10**, two important patterns emerge.

1. Pollination evolved late in land plant evolution. Mosses and other groups that do not form pollen appear first in the fossil record of land plants. Conifers and other groups that are strictly or primarily wind pollinated evolved later but before angiosperms.

2. Seed plants do not need water for sexual reproduction to occur. As a result, the evolution of pollen allowed these species to be much less dependent on wet habitats. Along with the evolution of the seed—highlighted in Section 41.4—pollen paved the way for the colonization of drier environments.

In addition, it's critical to realize that pollination became a much more precise process when plants began to recruit animals to act as pollinators. Wind-borne pollen grains have a low probability of landing successfully on a flower stigma. Animal-borne pollen, in contrast, is much more likely to be successfully transferred to flowers of the same species.

Wind-pollinated species invest in making large numbers of pollen grains; animal-pollinated species make fewer pollen grains but invest in structures that attract and reward animals.

In effect, plants "pay" nectar- and pollen-eating animals to work for them. Wind is free, but animals are more precise. Animal pollination is an important adaptation because it makes sexual reproduction much more efficient.

Does Pollination by Animals Encourage Speciation? In addition to affecting the fitness of individual plants, does pollination by animals make the formation of entirely new species more likely?

To answer this question, consider the following situation. A biologist has documented that two populations of a mountain-dwelling species called the alpine skypilot have flowers with different characteristics.

- Alpine skypilots that grow in forested habitats at or below timberline have small flowers with short stalks and an aroma described as "skunky."

- Individuals that grow in the tundra habitats above timberline have large flowers with long stalks and a sweet smell.

These differences are interesting, because different insects pollinate the two populations. Small flies are abundant at slightly lower elevations, are attracted to skunky odors, and pollinate the timberline individuals; large bumblebees are abundant at higher elevations, are attracted to sweet odors, and pollinate the tundra flowers.

Experiments have shown that bumblebees prefer to pollinate big flowers—probably because larger flowers can support their larger mass. Because flies and bumblebees prefer to visit different types of flowers, gene flow between the two skypilot populations is low and they are evolving distinct characteristics. The two populations may be on their way to becoming different species.

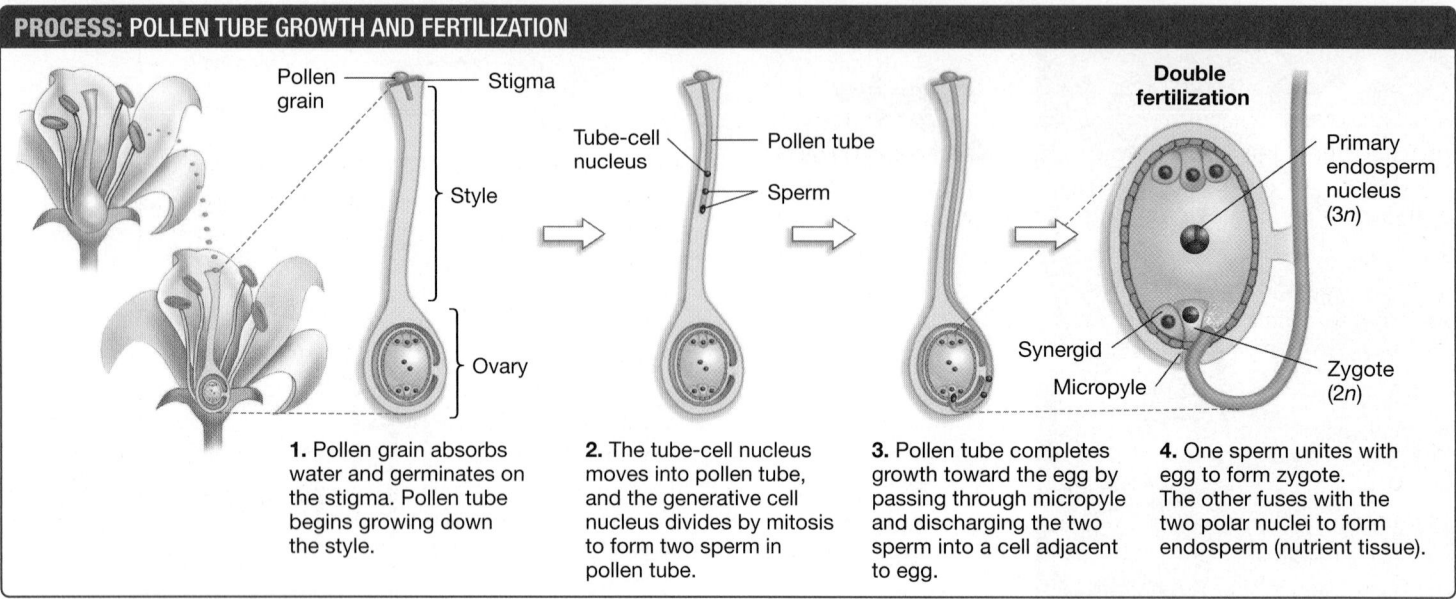

1. Pollen grain absorbs water and germinates on the stigma. Pollen tube begins growing down the style.

2. The tube-cell nucleus moves into pollen tube, and the generative cell nucleus divides by mitosis to form two sperm in pollen tube.

3. Pollen tube completes growth toward the egg by passing through micropyle and discharging the two sperm into a cell adjacent to egg.

4. One sperm unites with egg to form zygote. The other fuses with the two polar nuclei to form endosperm (nutrient tissue).

FIGURE 41.11 Double Fertilization Produces a Zygote and an Endosperm Nucleus. When the pollen tube reaches the female gametophyte, one sperm nucleus fertilizes the egg while the second fuses with the polar nuclei.

The message here is that evolutionary changes in the size or food-finding habits of a pollinator affect the angiosperm populations they pollinate. In return, changes in flower size and shape affect the insects pollinating that population. If a small population of Darwin's orchids evolved longer spurs, for example, the hawkmoth pollinators that lived in that area would be under intense selection that favored the evolution of a longer proboscis.

Because mutation continuously introduces variations in traits, insect and angiosperm populations frequently change, diverge, and form new species. Changes in pollination can trigger the evolution of new species. It is no surprise that insects and angiosperms are exceptionally species-rich groups.

It is clear that pollination was a crucial innovation during plant evolution. Now let's get down to mechanics. What happens once a pollen grain is deposited on a stigma?

Fertilization

FIGURE 41.11 walks you through the steps in fertilization.

Step 1 After landing on the stigma of a mature flower from the same species, a pollen grain absorbs water and germinates. **Germination** is a resumption of growth and development.

Step 2 When the male gametophyte germinates, a long tubular cell called a **pollen tube** grows through the stigma and down the length of the style. The direction of growth is affected by chemical attractants, called LUREs, which are small proteins released by synergids. Synergids are haploid cells in the female gametophyte that lie close to the egg. In the species illustrated in the figure, the tube-cell nucleus and the generative cell travel down the length of the tube, and the generative cell divides to form two sperm.

Step 3 When the pollen tube reaches the micropyle of the ovule, it grows through it and enters the interior of the female gametophyte.

Step 4 Fertilization—the fusion of sperm and egg to form a diploid zygote—occurs. In most plant groups, fertilization is straightforward—sperm and egg simply combine, and a diploid nucleus is formed. In angiosperms, however, an unusual event called **double fertilization** takes place. One sperm nucleus unites with the egg nucleus to form the zygote. The other sperm nucleus moves through the female gametophyte and fuses with the polar nuclei in the central cell. In most cases, two polar nuclei are present and a large triploid ($3n$) cell forms.

check your understanding

If you understand that . . .

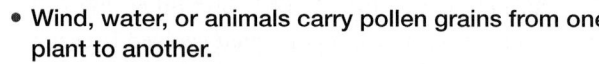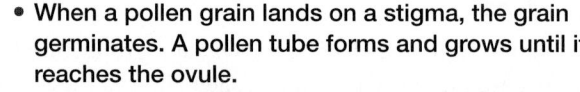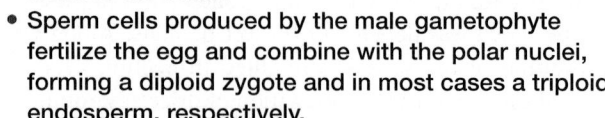

- Wind, water, or animals carry pollen grains from one plant to another.
- When a pollen grain lands on a stigma, the grain germinates. A pollen tube forms and grows until it reaches the ovule.
- Sperm cells produced by the male gametophyte fertilize the egg and combine with the polar nuclei, forming a diploid zygote and in most cases a triploid endosperm, respectively.

✔ **You should be able to . . .**

1. Explain why insects increase their fitness by visiting flowers and why flowers increase their fitness by rewarding insects.

2. Describe the function of the cells produced by double fertilization.

Answers are available in Appendix A.

✔ If you understand double fertilization, you should be able to identify in a female gametophyte, immediately after fertilization occurs, which cells are haploid, diploid, and triploid.

The triploid nucleus resulting from this second fertilization undergoes mitosis and cytokinesis to form the **endosperm** ("inside-seed") tissue. In most species, endosperm is triploid and functions to store nutrients. Endosperm cells are loaded with starch or oils (lipids) plus proteins and other nutrients that the embryo will need after it germinates. But before germination can occur, seeds must develop and be dispersed.

41.4 The Seed

Fertilization triggers the development of a young, diploid sporophyte. In angiosperms, the first stage in the sporophyte's life is the maturation of the seed.

As a seed matures, the embryo and endosperm develop inside the ovule and become surrounded by a covering called a **seed coat.** At the same time, the ovary around the ovule develops into a fruit, which encloses and protects the seed (or seeds, if a single ovary contains multiple ovules). In addition to providing protection, fruits often aid in dispersing seeds away from the parent plant. The mature seed consists of an embryo, a food supply—originating with endosperm—and a seed coat. In most cases, mature seeds leave the parent plant encased in a fruit.

Along with pollen, the evolution of the seed was a crucial innovation as land plants diversified. Because seeds contain stored nutrients, they allow offspring to be much more successful in colonizing habitats that are crowded with competitors than are offspring produced from spores, which are single cells. As a young plant emerges from the seed, it can subsist on stored nutrients until it is well enough established to absorb water from the soil and feed itself via photosynthesis.

Let's analyze how seeds work, beginning with a closer look at how the embryo develops.

Embryogenesis

Recall that **embryogenesis** is the process by which a single-celled zygote becomes a multicellular embryo (see Chapter 24). As **FIGURE 41.12** shows, embryogenesis in angiosperms can be analyzed as a four-step process.

Step 1 The zygote divides to form the two daughter cells.

Step 2 The lower daughter cell, or basal cell, divides to form a cell that forms part of the root tip and a cell that produces a row of cells. This row of cells, called the suspensor, provides a route for nutrient transfer from the parent plant to the developing embryo. The upper daughter cell, or apical cell, is the parent of almost all the cells in the embryo. It divides to form a mass of cells.

Step 3 Cells within the mass differentiate into groups conforming to one of the three adult tissue types (see Chapter 37). The exterior layer of embryonic cells, the **protoderm,** is the progenitor of the adult dermal tissue, or epidermis. The **ground meristem**—the cells just within the protoderm—gives rise to the ground tissue found in adults. And the **procambium** is a group of cells in the core of the embryo that becomes the vascular tissue.

Step 4 As the embryo continues to develop, the long axis of the plant begins to emerge and several important structures take shape: **(1) cotyledons** ("seed-leaves"), which absorb nutrients from the endosperm and supply them to the rest of the embryo, **(2)** the **hypocotyl** ("under-cotyledon"), which is the embryonic stem, and **(3)** the **radicle,** or embryonic root. Some embryos also have an **epicotyl** ("above-cotyledon"), which is a portion of the embryonic stem that extends above the cotyledons.

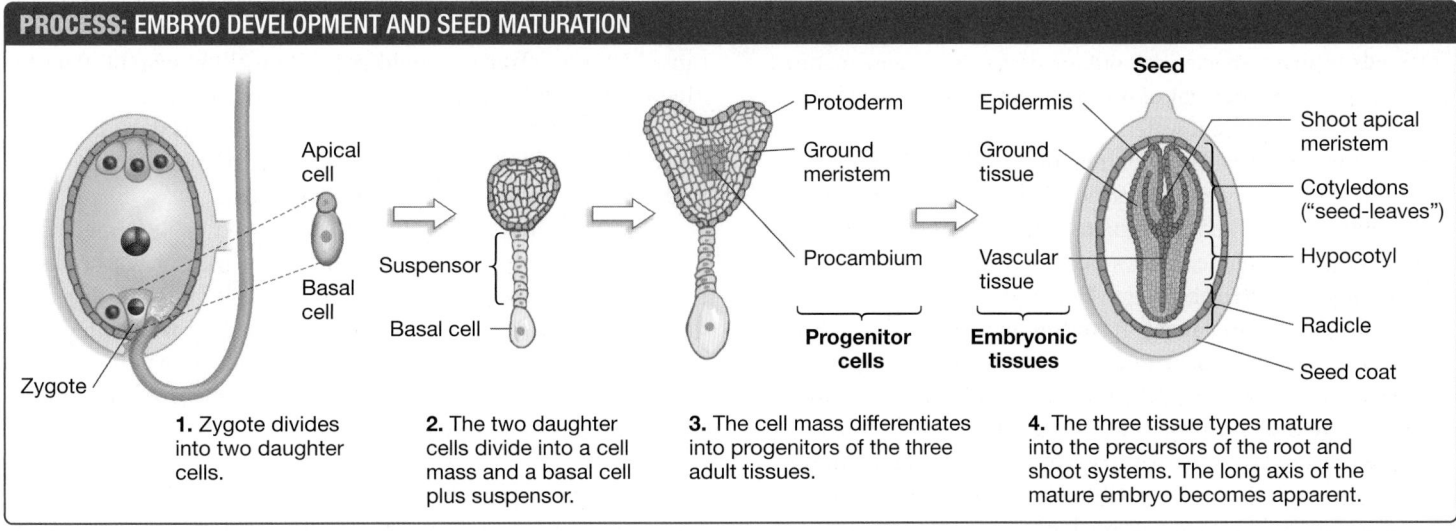

PROCESS: EMBRYO DEVELOPMENT AND SEED MATURATION

1. Zygote divides into two daughter cells.

2. The two daughter cells divide into a cell mass and a basal cell plus suspensor.

3. The cell mass differentiates into progenitors of the three adult tissues.

4. The three tissue types mature into the precursors of the root and shoot systems. The long axis of the mature embryo becomes apparent.

FIGURE 41.12 Embryonic Tissues and Structures Develop Inside Seeds. The embryo inside a seed has cotyledons and the beginnings of root and shoot systems. The embryonic epidermis and the ground and vascular tissues are organized in distinct layers.

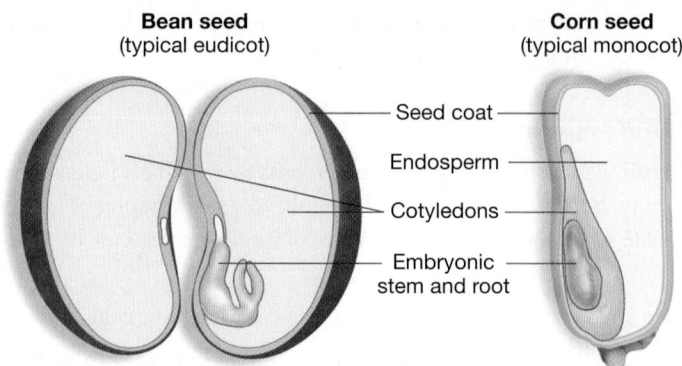

Bean seed
(typical eudicot)

Corn seed
(typical monocot)

Seed coat
Endosperm
Cotyledons
Embryonic
stem and root

FIGURE 41.13 Seeds Contain an Embryo and a Food Supply Surrounded by a Tough Coat. In beans (left), the nutrients in the endosperm have been absorbed by the cotyledons and stored. In corn (right), the endosperm is intact.

✔ If you understand the basic steps in embryonic development, you should be able to compare and contrast a cotyledon with a radicle in terms of their tissue composition and the progenitor cells they originated from.

Recall that one prominent lineage of angiosperms—the monocots—have just one cotyledon, whereas eudicots have two (see Chapter 31). In most eudicots, the cotyledons take up the nutrients that were initially stored in the endosperm and store them again. In these species, there is no endosperm left by the time the seed matures—instead, the cotyledons function as the nutrient storage organ in mature seeds. **FIGURE 41.13** compares the seed structure in beans and corn—a representative eudicot and monocot, respectively.

By the time a seed matures, then, the three major embryonic tissue types have developed. The precursors of the root and shoot systems, along with the seed leaves, have formed. Once these events are accomplished, the seed tissues dry and the embryo becomes quiescent—meaning it stops growing and waits.

The Role of Drying in Seed Maturation

The seeds of many species dry out, or desiccate, as they mature. Water makes up 90 percent of normal plant cells, but dried seeds contain just 5–20 percent water.

Loss of water is an adaptation that prevents seeds from germinating until after they are dispersed. The dry condition of seeds ensures that they will not germinate until water is available. This is adaptive in temperate species because water is crucial to the survival of germinated seedlings. Dry seeds are also less susceptible than wet seeds to damage from freezing.

How do the cellular structures in the embryo and endosperm survive the drying process? When researchers reduce the amount of water surrounding isolated membranes or isolated proteins to the levels observed in extremely dry seeds, the membranes disintegrate and the proteins denature. Clearly, something is happening at the molecular level in seeds to keep these cell components intact.

Researchers established that one of these "somethings" involves sugars. As water leaves the seed during drying, sugars

become concentrated and maintain the integrity of plasma membranes and proteins. If drying is extreme, the sugars form an extremely viscous liquid that contains little if any water. Substances such as this are considered vitrified, or glass-like (glass is a liquid solution with the viscosity of a solid). Biologists propose that this glassy state helps maintain the integrity of plasma membranes and proteins in seeds that experience extremely dry conditions. When seeds imbibe water, the glassy sugars dissolve and germination proceeds.

Drying is only one part of the seed maturation process, however. Equally important is the development of tissues surrounding the seed itself. In many cases, these tissues are required for the seed to be dispersed from the parent plant.

Fruit Development and Seed Dispersal

Fertilization in angiosperms initiates the development of the fruit as well as the seed and embryo.

Fruit Structure   Fruits come in three basic types (**FIGURE 41.14**).

- **Simple fruits** like the cherry develop from a single flower that contains a single carpel or several carpels that are fused together. This is the most common type of fruit.
- **Aggregate fruits** like the blackberry also develop from a single flower, but one that contains many separate carpels.
- **Multiple fruits** like the pineapple develop from many flowers and thus many carpels.

As a fruit matures, the walls of the ovary thicken to form the **pericarp,** the part of the fruit that surrounds and protects the seed or seeds (**FIGURE 41.15**). Fruits can be dry when they are mature, as in nuts, or fleshy, as in cherries and tomatoes. Note that fruits are formed from tissues derived from the mother—not the embryo.

Fruit Function   Fruits have two functions: They protect seeds from physical damage and seed predators, and they frequently aid in seed dispersal. Dispersal is important to the fitness of the young sporophyte. This is especially true in long-lived species, in which the offspring would compete with the parent plant for light, water, and nutrients if there were no dispersal.

Fruits sometimes split open and release seeds to be dispersed directly. In many cases, however, seeds are dispersed to new locations while they are still enclosed in the fruit.

Dry fruits may simply fall to the ground or be dispersed by wind, propulsion, or animals. Some dry fruits have hooks or barbs that adhere to passing animals, while nuts are dispersed by seed predators. Fruits that are dispersed by wind often have external structures to catch the breeze and extend the distance they travel; the fruits of dandelions and maple trees are familiar examples. Fruits that float, such as coconuts, can disperse seeds in water.

Some plants actually disperse dry fruits via propulsion. The sandbox tree, for example, produces a seed pod that shrinks as it dries. Eventually the pod splits apart violently, spraying seeds in all directions with so much force that the plant is sometimes

(a) Simple fruit (e.g., cherry):
Develops from a single flower with one carpel or fused carpels

One carpel

(b) Aggregate fruit (e.g., blackberry):
Develops from a single flower with many separate carpels

Many carpels

(c) Multiple fruit (e.g., pineapple):
Develops from many flowers with many carpels

Many flowers

FIGURE 41.14 Three Major Types of Fruits. The structure of a fruit depends on the number of carpels found in each flower and whether or not the carpels are fused.

✔**QUESTION** When fruits ripen, their color changes in a way that makes them more conspicuous to fruit eaters. State a hypothesis to explain why this color change might increase the fitness of an individual.

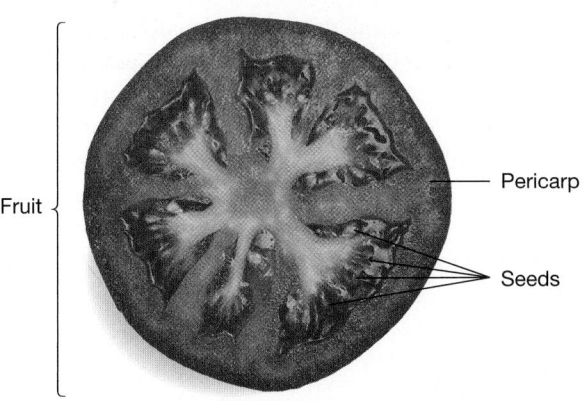

Fruit

Pericarp

Seeds

FIGURE 41.15 As Fruits Mature, the Ovary Wall Develops into a Pericarp That Surrounds the Seed or Seeds. A fruit consists of a pericarp and the enclosed seeds.

called the dynamite tree. The bursting seed pod sounds like a pistol shot, and seeds can be scattered as much as 40 m away from the parent plant. Similarly, the dwarf mistletoe fruit fills with sugars as it matures. Enough water follows via osmosis to make the fruit explode and shoot seeds as far as 5 m.

Animals are the most common dispersal agent for fleshy fruits. Just as plants have evolved to use animals for pollination, they have also evolved to use specific types of animals for seed dispersal. In cases like this, seed dispersal is an example of mutualism. The plant provides a fruit rich in sugars and other nutrients; in return, the animal carries the fruit to a new location and excretes the seeds along with a supply of fertilizer.

Mammals and birds are the most common seed dispersers. Mammals are often active at night and use their well-developed sense of smell to locate fruits. As you might imagine, fruits of mammal-dispersed seeds are usually dull colored and fragrant. In contrast, birds are usually active in the day and see well. Fruits of bird-dispersed seeds are usually brightly colored.

Some animals are seed predators that compete with dispersers. For example, birds usually swallow fruit whole and disperse seeds over long distances. Fruit-eating mice, on the other hand, chew their food, often killing the seeds inside. How can plants discourage seed predators such as mice from eating their fruits? The answer is one that might be familiar to you. Some plants such as chili peppers lace their fruits with a spicy-hot repellent called capsaicin. But does it work?

An experiment by ecologists Joshua Tewksbury and Gary Nabhan in 2001 tested the hypothesis that capsaicin promoted dispersal by the curve-billed thrasher and deterred predation by cactus mice (**FIGURE 41.16**, see page 836). Each animal was offered three kinds of fruit: hackberries, fruits from a strain of chilies that can't synthesize capsaicin (non-pungent chilies), and pungent chilies that have lots of capsaicin. All three fruits looked similar and had equivalent nutritional value. For each animal tested, the researchers recorded the percentage of each fruit eaten during a specific time interval. They then calculated the average amount of fruit that was eaten by five test individuals from each species.

Curve-billed thrashers ate all three fruits equally, but the mice ate fewer non-pungent chilies and avoided the capsaicin-laced fruits entirely. In a follow-up experiment, the researchers fed non-pungent chilies to each kind of animal. When the seeds had passed through the animal's digestive tract and were excreted, the researchers collected and planted the seeds. About 60 percent of the seeds that passed through the birds germinated, but none of those eaten by the mice germinated. Capsaicin appears to be an effective deterrent to seed predation.

Seed Dormancy

Once they have dispersed from the parent plant, seeds may not germinate for a period of time. This condition is known as **dormancy.**

Dormancy is usually a feature of seeds from species that inhabit seasonal environments, where for extended periods of time conditions may be too cold or dry for seedlings to thrive. Based on this observation, dormancy is interpreted as an adaptation that allows seeds to remain viable until conditions improve.

Consistent with this hypothesis, dormancy is rare or nonexistent in seeds produced by plants that inhabit tropical wet forests or other areas where conditions are suitable for germination year-round.

What molecular mechanisms are responsible for dormancy?

How Do Hormones Regulate Dormancy? The answer to this question is not well understood, but the hormone abscisic acid (ABA) plays important roles in seed development and dormancy (see Chapter 40). Mutants of some plants that cannot either make or respond to ABA exhibit a property called vivipary, in which the seeds germinate on the parent plant as soon as they are mature. In many species ABA triggers the accumulation of storage compounds, desiccation tolerance, and the prevention of germination.

After a seed has been dispersed from the parent plant, it may remain dormant in the soil for years or even centuries before it will germinate. The current record for longevity is a lotus seed that germinated after being submerged in an ancient lakebed in China for 1450 years.

What finally causes dormant seeds to germinate? Thanks to the wide range of habitats in which plants are adapted, germination depends on a wide variety of internal and external factors.

How Dormancy Is Broken? The coats of some seeds are thick enough to prevent water and oxygen from physically reaching the embryo. For germination to occur, these seed coats must be disrupted, or **scarified.**

Crop seeds that require scarification are placed in large, revolving drums with pieces of sandpaper that abrade and scarify the seeds. When planted, these scarified seeds germinate quickly and uniformly. In nature, impermeable seed coats protect nutritious seeds from being killed by soil bacteria and fungi, and they often prevent seeds from germinating for many years.

To germinate, most seeds must experience particular environmental conditions. Species native to high latitudes or high

---

RESEARCH

QUESTION: Does the presence of capsaicin in chilies deter some predators but not others?

HYPOTHESIS: Capsaicin deters cactus mice (seed predators) but not birds (seed dispersers).

NULL HYPOTHESIS: Cactus mice and birds respond to capsaicin in the same way.

EXPERIMENTAL SETUP:

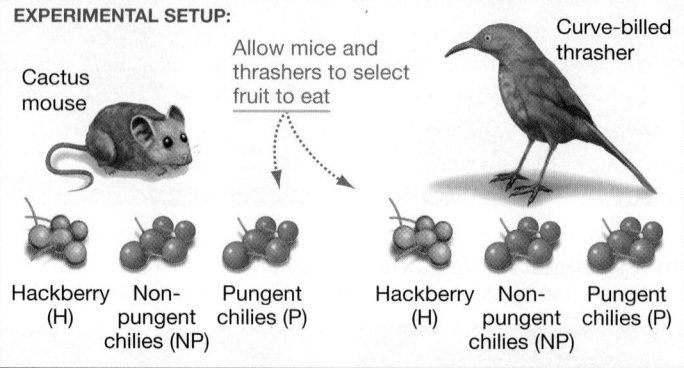

PREDICTION: Both will eat hackberry, but only thrashers will eat pungent chilies.

PREDICTION OF NULL HYPOTHESIS: No difference between thrashers and mice in fruit consumed.

RESULTS:

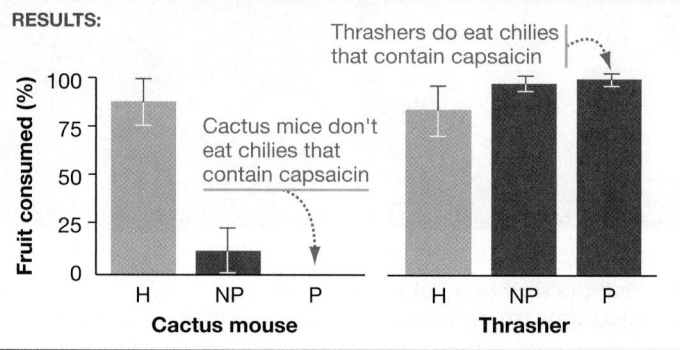

CONCLUSION: The presence of capsaicin deters cactus mice but not thrashers.

FIGURE 41.16 Experimental Support for the Hypothesis that Capsaicin Deters Seed Predators but Not Seed Dispersers.
SOURCE: Tewksbury, J. J., and G. P. Nabhan. 2001. Directed deterrence by capsaicin in chilies. *Nature* 412: 403–404.

✔ **QUESTION** Why were hackberries used in this experiment?

---

elevations often produce seeds that must undergo cool, wet conditions before they will germinate. Without this level of control, germination before the ensuing winter would likely result in death of the seedling. Recent studies indicate that germination is regulated by two hormones, gibberellin and ABA, and that the levels of these hormones in seeds are affected by temperature (see Chapter 40). After experiencing a winter, gibberellin levels increase and germination begins.

Because small seeds have few nutrient reserves in their cotyledons or endosperm, many small-seeded species need to germinate near the soil surface, where individuals are exposed to light and can feed themselves via photosynthesis. Lettuce seeds and other small seeds must be exposed to red light before they will break dormancy and germinate (see Chapter 40). Red light is an important environmental cue, because wavelengths in the red portion of the light spectrum are absorbed by chlorophyll. Red light and blue light indicate that sunlight is abundant and that the seedling will not be shaded by other plants.

Finally, many of the seeds produced by species native to habitats where wildfires are frequent, such as the California chaparral and South African fynbos, have an unusual chemical requirement to break dormancy: They must be exposed to fire or smoke before they will germinate. In fact, the commercial food product "liquid smoke" induces germination in these seeds as well as actual smoke does. In fire-prone habitats, it is advantageous for seeds to germinate after fire has cleared away existing vegetation.

The message here is that dormancy can be broken in response to a wide variety of environmental cues. In general, the cue that triggers germination is a reliable signal that conditions for seedling growth are favorable for a particular species in a particular environment.

Seed Germination

Even if specific environmental signals are required to break dormancy, seeds do not germinate without water. Water uptake is the first event in germination. Once the seed coat allows water penetration, water enters by moving along a steep water-potential gradient, because the seed is so dry.

Water uptake in a typical angiosperm seed has three distinct phases:

Phase 1 Germination begins with a rapid influx of water. Oxygen consumption and protein synthesis in the seed increase dramatically, but no new messenger RNAs are transcribed. Based on these observations, biologists have concluded that some of the key early events in germination are driven by mRNAs that are stored in the seed before maturation.

Phase 2 The second phase is an extended period during which water uptake stops. New mRNAs are transcribed and translated into protein products. Mitochondria also begin to multiply. In effect, seeds take up enough water in phase 1 to hydrate their existing proteins and membranes and then begin to manufacture the proteins and mitochondria needed to support growth.

Phase 3 Water uptake resumes as growth begins. This renewed phase of water uptake enables cells to develop enough turgor pressure to enlarge. Eventually, the seedling bursts from the seed coat.

FIGURE 41.17 shows what happens as eudicot and monocot embryos emerge from the seed. The radicle emerges first, and it subsequently develops into the mature root system. This is important because the seedling must have a source of water in order to grow. Initially, leaves are less important because of the nutrients stored in the seed.

In eudicots, the shoot system with its cotyledons usually emerges shortly after the radicle appears. In monocots such as corn, the emerging shoot is protected by the coleoptile. Note that in eudicots, the emerging stem has a hook shape. Like the coleoptile of monocots, this trait is thought to protect the apical meristem from damage as the shoot works its way upward through rough soil particles.

The next major event in the seedling's life occurs when either the cotyledons or the earliest leaves produced by the growing seedling commence photosynthesis. The seedling is said to be established when the young plant no longer relies on food reserves in its endosperm or cotyledons; instead, it receives all of its nourishment from its own photosynthetic products. With this, a new generation is under way.

(a) Beans are eudicots with cotyledons that emerge aboveground.

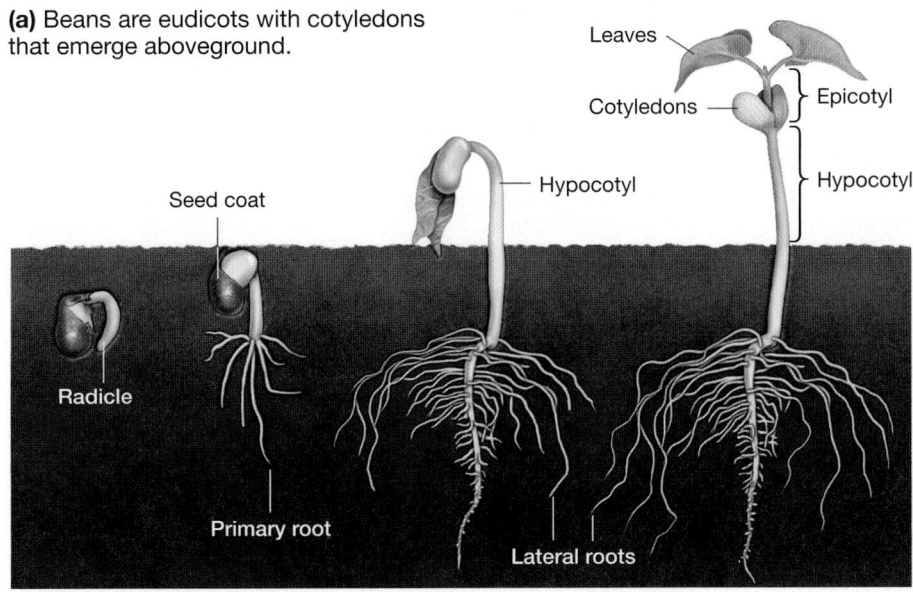

(b) Corn is a monocot with a cotyledon that remains belowground.

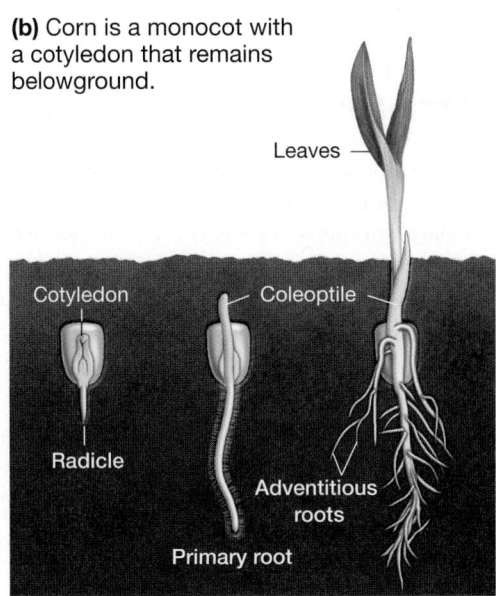

FIGURE 41.17 The Germination Sequence Varies among Species.

✔**QUESTION** In which of these species are cotyledons photosynthetic?

If you understand . . .

41.1 An Introduction to Plant Reproduction

- Plants undergo alternation of generations, in which a diploid sporophyte phase alternates with a haploid gametophyte phase. Sporophytes produce spores by meiosis. Gametophytes produce gametes by mitosis.

- The relative size and life span of the gametophyte and sporophyte phases vary a great deal among plant groups.

- In the most basal groups of land plants, gametophytes are larger and longer-lived than sporophytes, and sporophytes depend on gametophytes for nutrition.

- In angiosperms, or flowering plants, sporophytes are the large and long-lived phase where photosynthesis takes place; gametophytes consist of just a few cells.

✔ You should be able to state the ploidy of a sporophyte, gametophyte, spore, gamete, and zygote.

41.2 Reproductive Structures

- In angiosperms, male and female gametophytes are microscopic and are produced inside flowers. Flowers are made up of sepals, petals, stamens, and one or more carpels.

- The lower part of the carpel, the ovary, contains one to many ovules. Within the ovule, a megasporocyte undergoes meiosis, producing a megaspore that develops into the female gametophyte.

- In the anthers of stamens, microsporocytes undergo meiosis. The resulting microspores develop into male gametophytes, which are enclosed in pollen grains.

✔ You should be able to describe what would happen if all meiotic products in megasporangia went on to form female gametophytes.

41.3 Pollination and Fertilization

- Pollination occurs when pollen grains are transported to the stigma of the carpel. In most cases, the structure of a flower correlates with the morphology and behavior of its pollinator.

- If allowed to germinate on the stigma, a pollen grain sends a long pollen tube down the style. Two sperm travel down the pollen tube and enter the female gametophyte.

- In double fertilization, one sperm fuses with the egg to form a zygote, while the other fuses with polar nuclei within the female gametophyte. The fusion of sperm and polar nuclei produces endosperm—nutritive tissue that in most species is triploid.

✔ You should be able to explain why the endosperm of a seed is called a triploid tissue and how it compares with the zygote.

41.4 The Seed

- Seeds contain an embryo and a food supply surrounded by a coat.

- The development of an angiosperm embryo includes the formation of dermal tissue (epidermis), ground tissue, and vascular tissue layers and the development of the radicle, hypocotyl, and cotyledons.

- As an embryo develops, endosperm cells divide to form a nutrient-rich tissue. In addition, cells along the outside of the ovules form a protective seed coat, and the ovary develops into a fruit.

- In many cases, the mature fruit contains structures that help disperse the mature seed via wind, water, propulsion, or animals.

- Many seeds do not germinate immediately but instead experience a period of dormancy.

- A wide variety of conditions, ranging from scarification to exposure to red light, may break seed dormancy. In many cases, the event that triggers germination ensures that the seed germinates when environmental conditions are favorable.

- Germination begins when the seed takes up water and mRNAs already present in the seed are translated. It ends when the radicle breaks the seed coat and begins to penetrate the soil.

✔ You should be able to explain why it is adaptive for many seeds from plants native to high latitudes to require a cold period before germination.

(MB) **MasteringBiology**

1. **MasteringBiology Assignments**

   Tutorials and Activities Angiosperm Life Cycle; Angiosperm Life Cycle; Fruit Structure and Development; Reproduction in Flowering Plants; Seed and Fruit Development

   Questions Reading Quizzes, Blue-Thread Questions, Test Bank

2. **eText** Read your book online, search, take notes, highlight text, and more.

3. **The Study Area** Practice Test, Cumulative Test, BioFlix® 3-D Animations, Videos, Activities, Audio Glossary, Word Study Tools, Art

You should be able to . . .

1. What is the major evolutionary trend in land plant life cycles?
   a. Instead of being approximately the same size and shape, gametophytes and sporophytes began to look different.
   b. Sporophytes became larger and long lived while gametophytes became drastically reduced.
   c. In lineages that evolved more recently, such as angiosperms, spores are no longer produced.
   d. Sporophytes began to rely on gametophytes for all of their nutritional needs.

2. What happens when double fertilization occurs?
   a. Two zygotes are formed, but only one survives.
   b. Two sperm fertilize the egg, forming a triploid zygote.
   c. One sperm fertilizes the egg, while another sperm fuses with the polar nuclei.
   d. One sperm fertilizes the egg, while two other sperm fuse with a polar nucleus.

3. What is a fruit?
   a. an expanded ovary that includes a seed or seeds
   b. a structure consisting of an embryo and a food supply surrounded by a tough coat
   c. a female gametophyte
   d. a male gametophyte

4. Why is the interaction between angiosperms and pollinators considered mutualistic?
   a. New species can form if mutant flowers attract new types of pollinators.
   b. Flowers may have an array of traits, including corolla shape, color, scent, and the presence of nectar, to attract a specific type of pollinator.
   c. Wind pollination is much "cheaper," but animal pollination is much more precise.
   d. Angiosperms get their pollen dispersed, while pollinators get food.

5. In the angiosperm life cycle, which cells undergo meiosis? Which cells are spores? Which structures are gametophytes?

6. What events take place during pollination and fertilization?

7. Why is the emergence of the radicle an important first step in germination?
   a. Its hook helps protect the shoot that emerges later.
   b. It carries out photosynthesis to supply the embryo with food.
   c. It is important for establishing a supply of water to the growing embryo.
   d. It is necessary to break the seed coat.

8. In terms of maximizing reproductive success, what is the advantage of asexual reproduction? What is the disadvantage?

9. How do the structure and function of sepals and petals differ? How would you expect these structures to differ in species that are pollinated by wind versus bumblebees?

10. What are the advantages and disadvantages of self-fertilization versus those of outcrossing?

11. Consider a carpel, an ovary, and an ovule. Which is responsible for producing the female gametophyte, and which produces the pericarp of a fruit? Are these structures part of the sporophyte, the gametophyte, or a combination of the two?

12. What is the relationship between the endosperm of corn and the cotyledons of beans?

13. What happens when outcrossing occurs?
    a. Inbred offspring are produced.
    b. The same flower can be visited by many different types of pollinators—not a single specialist species.
    c. Gametes from different individuals fuse to form a zygote.
    d. Gametes from the same individual fuse.

14. **QUANTITATIVE** Because wind is less efficient than animals at moving pollen from plant to plant, wind-pollinated plants generate far more pollen than animal-pollinated plants. How much of that pollen is wasted? Consider a typical field-corn plant that makes one ear. Ears contain about 600 kernels, each of which is fertilized by an individual pollen grain. Male flowers clustered in the tassel at the top of the plant release about 10 million pollen grains per plant. Using these numbers, what percentage of corn pollen is wasted?

15. Some flowering plants "cheat" their pollinators because they offer no food reward. Likewise, certain pollinators cheat plants by removing nectar from flowers without picking up pollen. (In some cases, they do so by chewing through the petals that hold the store of nectar.) Speculate on the types of mutations that might modify insect behavior and/or plant structure in a way that limits cheating and enforces mutualism.

16. Consider the following fruits: an acorn, a cherry, a burr, and a dandelion seed. Based on the structure of each of these fruits, predict how the seed is dispersed. Design a study that would estimate the average distance that each type of seed is dispersed from the parent plant.

The Big Picture

Plants and animals are diverse lineages of multicellular eukaryotes. They are different in important ways. Each lineage evolved independently from a different single-celled protist—plants with the ability to make their own food by photosynthesis, and animals reliant on obtaining energy from other organisms. Further, plants are sessile, while most animals are capable of complex movements and locomotion.

Yet despite these differences, plants and animals face many of the same challenges to survive and reproduce in water and on land. Use this concept map to explore some of their similarities and differences in form and function.

Note that each box in the concept map indicates the chapters where you can go for more information. Also, be sure to do the blue exercises in the Check Your Understanding box below.

Plant and Animal FORM AND FUNCTION

Evolutionary processes 25 & 26 *including*

are the product of

occurring in

Changing ecological contexts Unit 9

because plants and animals have

create

Abiotic environment
- Temperature
- Light
- Water availability
- pH, salinity, dissolved gases, nutrients
- Habitat structure 52

Biotic environment
- Parental care
- Competitors
- Predators, prey
- Parasites, hosts
- Mutualists 54, 55

Key FUNCTIONS to survive and reproduce

correlate with

including

REGULATION OF WATER AND IONS

NUTRITION

PLANTS

REGULATION OF WATER AND IONS
- Water moves along a water potential gradient
- Turgor pressure provides structural support
- Terrestrial plants lose water evaporatively by transpiration, regulated by stomata and waxy surfaces 38

NUTRITION
- Autotrophic; make their own food by photosynthesis
- Take in sunlight, CO_2, and minerals
- Obtain ions from soil, symbiotic fungi, or bacteria 39

ANIMALS

- Osmotic stress varies in marine, freshwater, and terrestrial habitats
- Urinary system maintains homeostasis of water and electrolytes while managing excretion of nitrogenous wastes
- Terrestrial animals limit evaporative loss 42

- Heterotrophic; must eat food to acquire energy
- Take in carbohydrates, fats, proteins, vitamins, and minerals
- Obtain nutrients by ingestion, digestion, and absorption 43

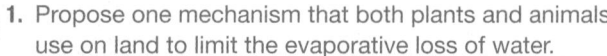

check your understanding

C Y U

If you understand the big picture . . .
✔ You should be able to . . .

1. Propose one mechanism that both plants and animals use on land to limit the evaporative loss of water.

2. Give an example of a method that both plants and animals use to protect their eggs and sperm from drying out.

3. Explain one constraint of large body size for both plants and animals.

4. Explain where cellular respiration fits into this map.

Answers are available in Appendix A.

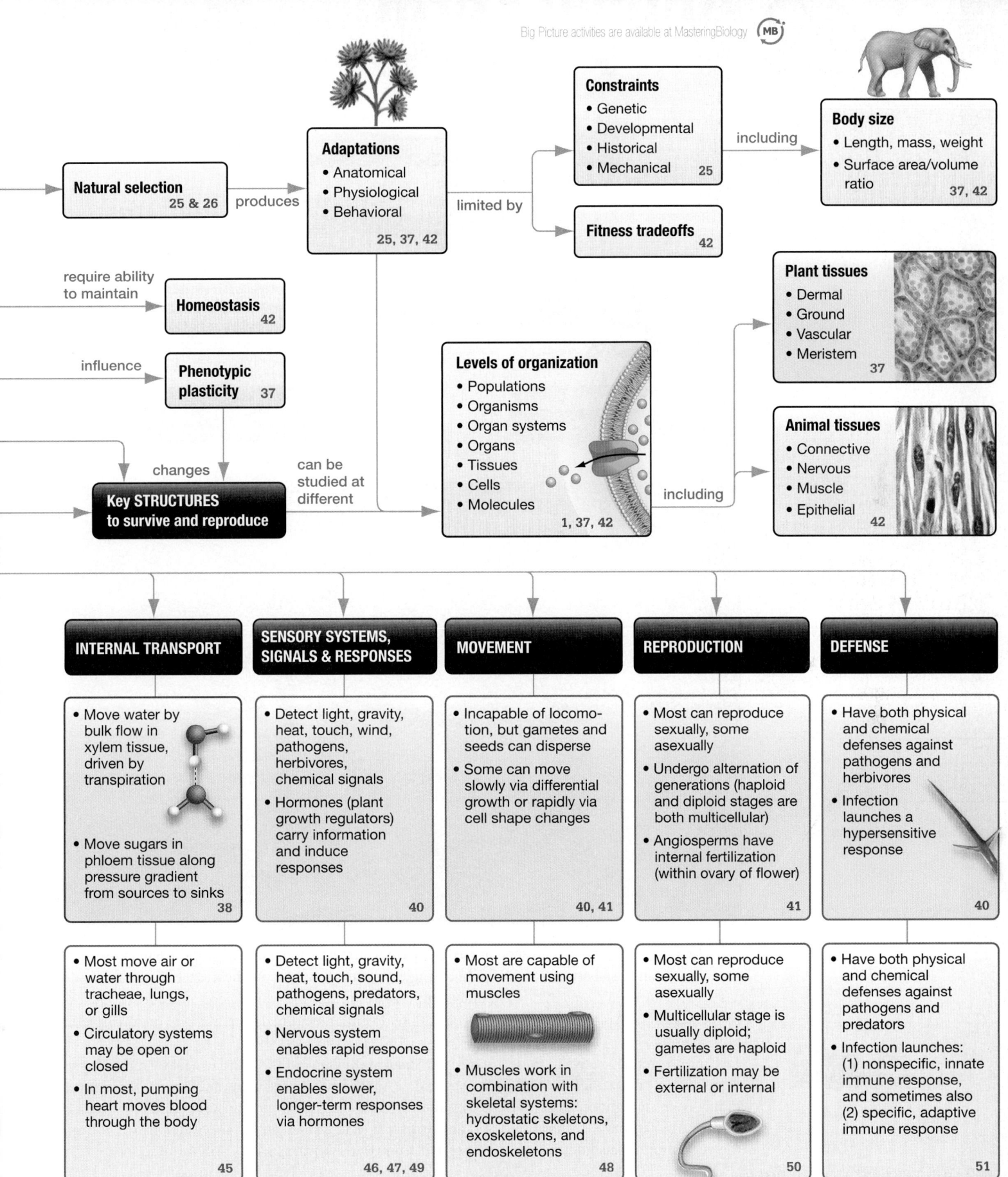

Big Picture activities are available at MasteringBiology (MB)

Adaptations
- Anatomical
- Physiological
- Behavioral

25, 37, 42

Natural selection

25 & 26

produces →

limited by →

Constraints
- Genetic
- Developmental
- Historical
- Mechanical 25

including →

Body size
- Length, mass, weight
- Surface area/volume ratio

37, 42

Fitness tradeoffs 42

require ability to maintain →

Homeostasis 42

influence →

Phenotypic plasticity 37

changes →

Key STRUCTURES to survive and reproduce

can be studied at different →

Levels of organization
- Populations
- Organisms
- Organ systems
- Organs
- Tissues
- Cells
- Molecules

1, 37, 42

including →

Plant tissues
- Dermal
- Ground
- Vascular
- Meristem

37

Animal tissues
- Connective
- Nervous
- Muscle
- Epithelial

42

INTERNAL TRANSPORT

- Move water by bulk flow in xylem tissue, driven by transpiration
- Move sugars in phloem tissue along pressure gradient from sources to sinks

38

- Most move air or water through tracheae, lungs, or gills
- Circulatory systems may be open or closed
- In most, pumping heart moves blood through the body

45

SENSORY SYSTEMS, SIGNALS & RESPONSES

- Detect light, gravity, heat, touch, wind, pathogens, herbivores, chemical signals
- Hormones (plant growth regulators) carry information and induce responses

40

- Detect light, gravity, heat, touch, sound, pathogens, predators, chemical signals
- Nervous system enables rapid response
- Endocrine system enables slower, longer-term responses via hormones

46, 47, 49

MOVEMENT

- Incapable of locomotion, but gametes and seeds can disperse
- Some can move slowly via differential growth or rapidly via cell shape changes

40, 41

- Most are capable of movement using muscles
- Muscles work in combination with skeletal systems: hydrostatic skeletons, exoskeletons, and endoskeletons

48

REPRODUCTION

- Most can reproduce sexually, some asexually
- Undergo alternation of generations (haploid and diploid stages are both multicellular)
- Angiosperms have internal fertilization (within ovary of flower)

41

- Most can reproduce sexually, some asexually
- Multicellular stage is usually diploid; gametes are haploid
- Fertilization may be external or internal

50

DEFENSE

- Have both physical and chemical defenses against pathogens and herbivores
- Infection launches a hypersensitive response

40

- Have both physical and chemical defenses against pathogens and predators
- Infection launches: (1) nonspecific, innate immune response, and sometimes also (2) specific, adaptive immune response

51

42 Animal Form and Function

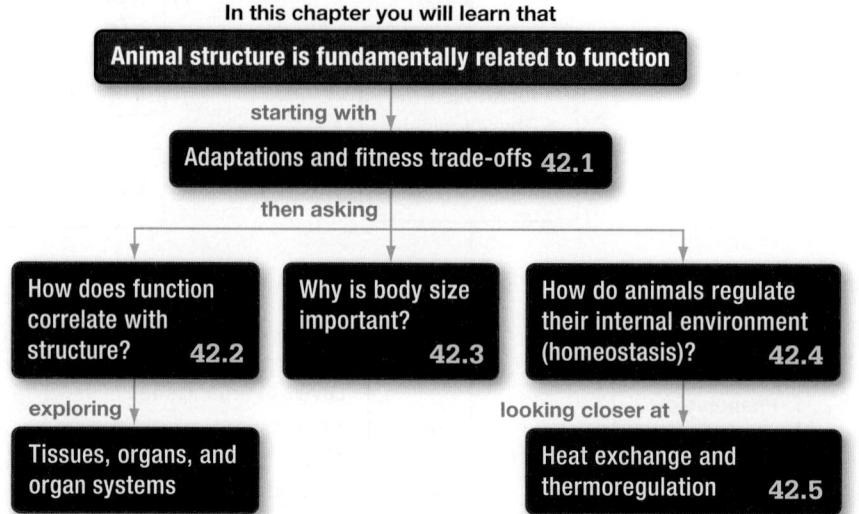

In this chapter you will learn that

Animal structure is fundamentally related to function

starting with ↓

Adaptations and fitness trade-offs 42.1

then asking ↓

| How does function correlate with structure? **42.2** | Why is body size important? **42.3** | How do animals regulate their internal environment (homeostasis)? **42.4** |

exploring ↓

Tissues, organs, and organ systems

looking closer at ↓

Heat exchange and thermoregulation 42.5

Oryx are adapted to desert life. They have an exceptional ability to withstand heat and can acquire most of their water from the food they eat.

This chapter is part of the Big Picture. See how on pages 840–841.

The Sahara and Arabian deserts are extreme environments. In some parts of the Sahara, several years can pass between rainfalls. In a single day, temperatures in these deserts can fluctuate between −0.5°C and 37.5°C; midday temperatures in summer can exceed 50°C. Yet few places in the Sahara or Arabian deserts are devoid of life. Even large animals, such as the oryx, thrive in both regions.

How do animals native to these areas cope? Small animals avoid the midday heat by retreating to a cool burrow deep underground or the shade of a small shrub. Large mammals have a harder time hiding from sunlight but possess traits that allow them to keep cool and conserve water.

The Arabian oryx, for example, seldom drinks. It gets 86 percent of the water it needs from the vegetation it eats and the other 14 percent from water synthesized as a by-product of cellular

✔ When you see this checkmark, stop and test yourself. Answers are available in Appendix A.

respiration—what biologists call **metabolic water.** To conserve what water they have, Arabian oryx produce extremely concentrated urine and exceptionally dry fecal pellets.

In addition, oryx do not sweat to cool off. Instead, their body temperature rises from a normal 37°C to just over 40°C as temperatures increase during the day. The excess body heat is released during the cool desert nights, when their body temperature returns to normal.

In contrast to oryx, humans may die of dehydration when denied water for just three days. And for a human, having a body temperature of 40°C is a life-threatening situation—equivalent to running a fever of 104°F. How do oryx do it? What aspects of their anatomy and physiology allow them to thrive in such an extreme environment?

Anatomy is the study of an organism's physical structure. **Physiology** is the study of how the physical structures in an organism function. The anatomy and physiology of an oryx are clearly different from those of a human—or those of a shark, frog, bird, fruit fly, or crab, for that matter. This chapter is an introduction to the study of anatomy and physiology in animals. You can review the importance of animal form and function in the Big Picture on pages 840–841.

42.1 Form, Function, and Adaptation

Biologists who study animal anatomy and physiology are studying **adaptations**—heritable traits that allow individuals to survive and reproduce in a certain environment better than individuals that lack those traits (see Chapter 25).

Recall that adaptation results from evolution by natural selection. Natural selection, in turn, occurs whenever individuals with certain alleles leave more offspring that survive to reproductive age than do individuals with different alleles. Because of this difference in reproductive success, the frequency of the selected alleles increases from one generation to the next.

Oryx with alleles that allow them to extract more water from their feces survive better and produce more offspring than do oryx with alleles that allow water to be lost in feces. The ability to produce extremely dry fecal pellets is an adaptation that helps oryx thrive in water-short environments.

The Role of Fitness Trade-Offs

Adaptations increase fitness—the ability to produce viable offspring. But no adaptation is "perfect." Instead, adaptations are limited by which alleles are present in a population and by the nature of the traits that already exist—because all adaptations derive from preexisting traits.

The human spine, for example, is a highly modified form of the vertebral column in ancestors that walked on all fours (see Chapter 35). The modifications in the human spine can be considered adaptations to support our upright posture, but they are far from perfect—most adults experience back pain at some point during their lives. The evolution of the human spine has been constrained by the nature of the ancestral trait and by a lack of alleles that would improve its structure and function.

The most important constraint on adaptation, though, may be **trade-offs**—inescapable compromises between traits. For example, it takes a lot of energy to produce offspring through the process of reproduction and also to mount an immune response during an infection. Animals sometimes do not have enough energy to satisfy both needs. In these cases, a trade-off emerges: The animal may devote more energy to reproduction at the expense of strong immune function, or vice versa, or both traits might be negatively affected.

How do biologists study trade-offs in animal physiology? Let's consider experimental work on trade-offs in crickets. During mating, a male cricket deposits a **spermatophore** on the female's genital opening (**FIGURE 42.1a**). This spermatophore consists of a packet of sperm surrounded by a large, gelatinous mass. After mating, the female begins to eat the gelatinous mass (**FIGURE 42.1b**). The sperm packet remains behind and the sperm slowly begin to enter her reproductive tract.

The longer it takes for the female to eat the mass, the more sperm are transferred to her, increasing the number of eggs fertilized by the male. It might seem advantageous for the male to make as large a spermatophore as possible. However, there is a

(a) Decorated crickets mating

(b) The female eats the gelatinous mass.

FIGURE 42.1 Mating in Crickets Involves Transfer of a Spermatophore from the Male to the Female.

QUESTION: Is there a trade-off between reproductive and immune function in male crickets?

HYPOTHESIS: Male crickets need to make an energy trade-off between reproductive function and immune function.

NULL HYPOTHESIS: No energy trade-off between reproductive function and immune function is required.

EXPERIMENTAL SETUP:

1. Remove spermatophores from male crickets:

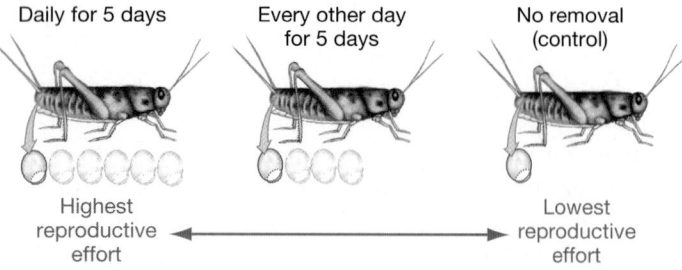

Daily for 5 days   Every other day for 5 days   No removal (control)

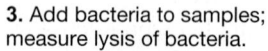

Highest reproductive effort ◄────────────► Lowest reproductive effort

2. Draw hemolymph samples from the three sets of crickets.

3. Add bacteria to samples; measure lysis of bacteria.

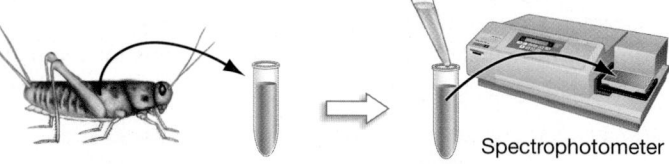

Spectrophotometer

EXPERIMENTAL SETUP:

1. Inject male crickets:

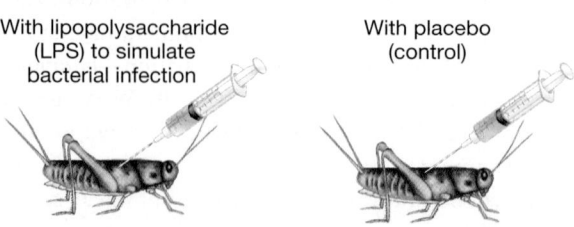

With lipopolysaccharide (LPS) to simulate bacterial infection   With placebo (control)

2. Remove spermatophores and measure size of gelatinous mass.

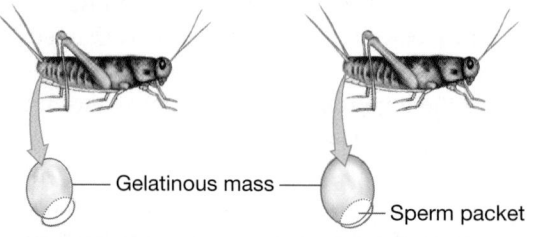

Gelatinous mass —   — Sperm packet

PREDICTION: Hemolymph from males forced to produce more spermatophores will exhibit lower lytic activity than controls.

PREDICTION OF NULL HYPOTHESIS: There will be no difference in lytic activity between treated males and control males.

PREDICTION: Spermatophores from LPS-injected males will have smaller gelatinous masses than those from control males.

PREDICTION OF NULL HYPOTHESIS: There will be no difference in gelatinous mass size between treated and control males.

RESULTS:

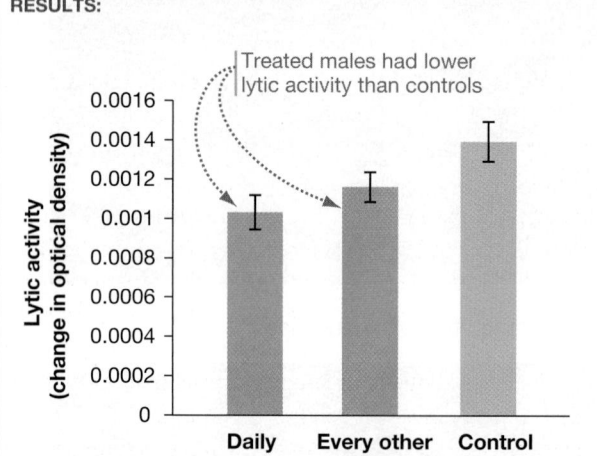

Treated males had lower lytic activity than controls

RESULTS:

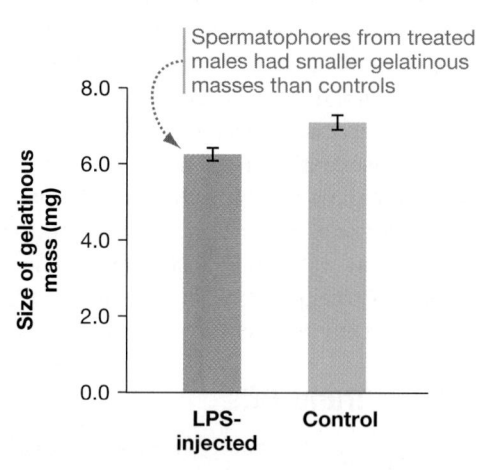

Spermatophores from treated males had smaller gelatinous masses than controls

CONCLUSION: Male crickets must make an energy trade-off to support increased reproductive effort.

FIGURE 42.2 Trade-offs Between Reproduction and Immune Function in Male Crickets. This study showed that males with experimentally increased investment in reproduction had a lower ability to kill bacteria than control males, and that males with experimentally stimulated immune function had lower reproductive function than control males.

SOURCE: Kerr, A. M., S. N. Gershman, and S. K. Sakaluk. 2010. Experimentally induced spermatophore production and immune responses reveal a trade-off in crickets. *Behavioral Ecology* 21: 647–654.

✔**QUESTION** If the null hypothesis were supported by the results, what would each graph look like?

cost involved. Spermatophores weigh up to 6 percent of a male cricket's body mass, and it takes males more than three hours to make a new one. During the time males spend making new spermatophores, they lose other mating opportunities.

Losing out on new mates while making spermatophores may not be the only cost a male cricket accrues. When an insect is infected with a pathogen such as a bacterium, it mounts an immune response. Activation of the insect defense system is energetically expensive. Is there a trade-off in male crickets between producing spermatophores and mounting an immune response?

To answer this question, biologists carried out an experiment using a powerful method—a reciprocal design. They altered the energetic investment crickets made into each of the systems in question—reproduction and immune function—and observed the effect on the other system (**FIGURE 42.2**).

First, the biologists removed spermatophores from male crickets, causing them to make new ones, daily or every other day. The biologists then took a sample of the crickets' hemolymph, or blood, and added bacteria to find out whether the hemolymph was able to destroy the pathogen. As bacteria are lysed, or ruptured, the opaque hemolymph becomes clear. Lytic activity was measured using a spectrophotometer to calculate the change in the opacity, or optical density, of the hemolymph over time.

Compared to control crickets, which did not have their spermatophores removed, the experimental crickets had lower lytic activity (fewer bacteria destroyed). Even more striking, the crickets that had to make new spermatophores daily had lower lytic activity than those that had to make new spermatophores only every other day. This suggested that increased investment into making spermatophores was traded off against investment into immune function.

Next, the biologists injected crickets with a component of the cell walls of bacteria called lipopolysaccharide (LPS), which caused the crickets to mount an immune response. What was the effect on spermatophore size? The result was dramatic. Male crickets injected with LPS produced smaller spermatophores than control crickets. When the researchers looked more closely, they realized the size of the sperm packet remained largely unchanged, but the size of the gelatinous mass decreased in injected crickets. This suggests that mounting an immune response may reduce male crickets' reproductive success. How? By causing males to make smaller gelatinous masses, which decreases the time females spend eating them, reducing sperm transfer.

Taken together, these two sets of experiments provide very strong evidence that there is a trade-off between reproductive and immune function in crickets.

Trade-offs, such as the compromise between energetic investment and the competing demands of reproductive and immune function, are common in nature. Desert animals that sweat to cool off are threatened with dehydration. An eagle's beak is superbly adapted for tearing meat but not for weaving nesting materials together. In studying animal anatomy and physiology, biologists study compromise and constraint as well as adaptation.

Adaptation and Acclimatization

In biology, adaptation refers to a genetic change in a population in response to natural selection exerted by the environment. Short-term, reversible responses to environmental fluctuations are referred to as **acclimatization.** Acclimatization is a phenotypic change in an individual in response to short-term changes in the natural environment. Acclimation is similar to acclimatization, but acclimation refers to changes that occur in organisms in a laboratory setting.

If you moved to Tibet, your body would acclimatize to high elevation by making more of the oxygen-carrying pigment hemoglobin and more hemoglobin-carrying red blood cells. But populations that have lived in Tibet for many generations are adapted to this environment through genetic changes. Among native Tibetans, for example, an allele that increases the ability of hemoglobin to hold oxygen has increased to high frequency. In populations of Tibetans that do not live at high elevations, this allele is rare or nonexistent.

The ability to acclimatize is itself an adaptation. Light-skinned humans, for example, vary in the ability to tan in response to sunlight. Some individuals tan easily—they have alleles that allow them to acclimatize efficiently to environments with intense sunlight—while others do not. In this and many other cases, the ability to acclimatize is a genetically variable trait that can respond to natural selection.

42.2 Tissues, Organs, and Systems: How Does Structure Correlate with Function?

If a structure found in an animal is adaptive—meaning that it helps the individual survive and produce viable offspring—it is common to observe that the structure's size, shape, or composition correlates closely with its function.

For example, recall that biologists have documented extensive changes in beak size and shape in medium ground finches from the Galápagos Islands (see Chapter 25). Such changes are due to natural selection. Individuals with deep beaks are better able to crack the large fruits that predominate during drought years, while individuals with small beaks are better able to harvest the small seeds that predominate during wet years.

As **FIGURE 42.3** (see page 846) shows, a strong correlation between diet and beak structure is also found *among* species of Galápagos finch. Species with large, cone-shaped beaks eat large seeds; species with small, cone-shaped beaks eat small seeds. Species with long, tweezer-like beaks pick insects off tree trunks or other surfaces.

The mechanism responsible for these structure–function correlations is straightforward: If a mutant allele alters the size or shape of a structure in a way that makes it function more efficiently, individuals who have that allele will produce more offspring than will other individuals. As a result, the allele will increase in frequency in the population over time.

Species of Galápagos finch		Food source
Geospiza fuliginosa		Small seeds
Geospiza fortis		Medium seeds
Geospiza magnirostris		Large seeds
Certhidea olivacea		Insects, nectar

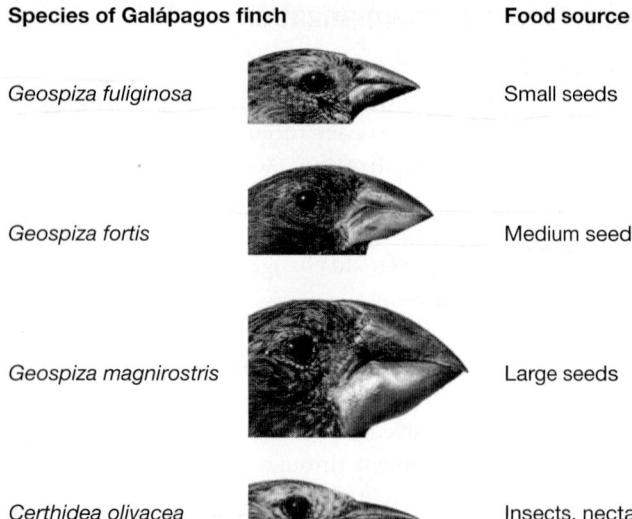

FIGURE 42.3 In Animal Anatomy and Physiology, Form Often Correlates with Function.

Structure–Function Relationships at the Molecular and Cellular Levels

Correlations between form and function start at the molecular level. For example, earlier chapters emphasized that the shape of proteins correlates with their role as enzymes, structural components of the cell, or transporters. The membrane proteins called channels form pores that allow specific ions or molecules to pass in or out of cells (see Chapter 6). The ends and interior of a channel are hydrophilic, which allows the protein to interact with the surrounding solution or the interior of the cell, while the perimeter is hydrophobic—allowing the protein to interact with the lipid bilayer. The protein's structure fits its function.

Similar correlations between structure and function occur at the level of the cell. In fact, it is possible to predict a cell's specialized function by examining its internal structure. Cells that manufacture and secrete hormones or digestive enzymes are packed with rough endoplasmic reticulum (ER) and Golgi apparatuses; cells that store energy are dominated by large fat droplets; cells that ingest and destroy invading bacteria have many lysosomes.

The overall shape of a cell can also correlate with its function. For example, cells that are responsible for transporting materials into or out of the body often have extremely large areas of plasma membrane. As a result, they have room to accommodate the thousands of membrane channels, transporters, and pumps required for extensive transport.

Tissues Are Groups of Cells That Function as a Unit

Animals are **multicellular,** meaning that their bodies contain distinct types of cells that are specialized for different functions. Frequently, animal cells that are similar in structure and function are physically attached to each other and form a tissue. A **tissue** is a group of cells that function as a unit.

SUMMARY TABLE 42.1 Connective Tissues

Type	Example
Loose *soft* extracellular matrix; provides padding	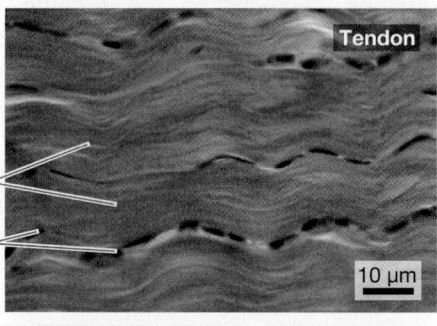
Dense *fibrous* extracellular matrix; provides connections	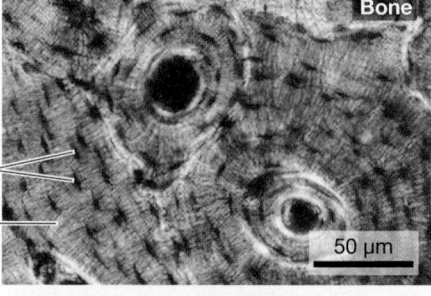
Supporting *firm* extracellular matrix; functions in structural support and protection	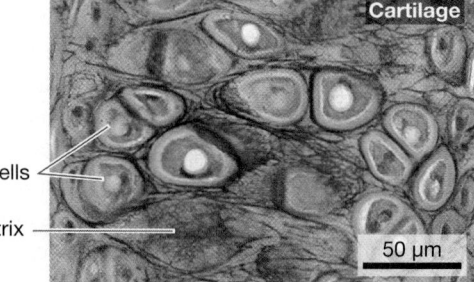
Fluid *liquid* extracellular matrix; functions in transport	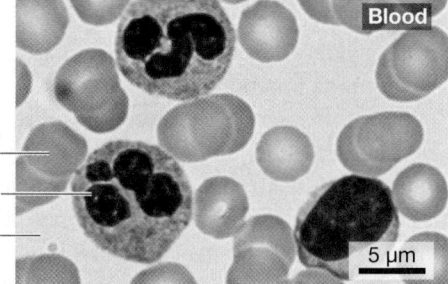

The embryonic tissues called ectoderm, mesoderm, and endoderm are found in most animals (see Chapter 23 and Chapter 33). As an individual develops, the embryonic tissues give rise to four adult tissue types: (**1**) connective tissue, (**2**) nervous tissue, (**3**) muscle tissue, and (**4**) epithelial tissue. In each case, the structure of the tissue correlates closely with its function. Let's consider each type in turn.

Connective Tissue   **Connective tissue** consists of cells that are loosely arranged in a liquid, jellylike, or solid matrix. The matrix comprises extracellular fibers and other materials, and it is secreted by the connective tissue cells themselves (**TABLE 42.1**). Each type of connective tissue secretes a distinct type of extracellular matrix. The nature of the matrix determines the nature of the connective tissue.

- **Loose connective tissue** contains an array of fibrous proteins in a soft matrix; it serves as a packing material between organs or padding under the skin. Reticular connective tissue—in lymphoid organs such as the spleen and bone marrow—and adipose tissue, or fat tissue, are loose connective tissues made up of cells suspended in a matrix of fibers and fluid.

- **Dense connective tissue** is found in the tendons and ligaments that connect muscles, bones, and organs. As Table 42.1 shows, the matrix in tendons and ligaments is dominated by tough collagen fibers (introduced in Chapter 11).

- **Supporting connective tissue** has a firm extracellular matrix. **Bone** and **cartilage** are connective tissues that provide structural support for the vertebrate body as well as protective enclosures for the brain and other components of the nervous system.

- **Fluid connective tissue** consists of cells surrounded by a liquid extracellular matrix. **Blood,** which transports materials throughout the vertebrate body, contains a variety of cell types and has a specialized extracellular matrix called plasma (see Chapter 45).

Nervous Tissue   **Nervous tissue** consists of nerve cells, which are also called **neurons,** and several types of supporting cells. Neurons transmit electrical signals, which are produced by changes in the permeability of the cell's plasma membrane to ions (see Chapter 46). Supporting cells have many functions, including regulating ion concentrations in the space surrounding neurons, supplying neurons with nutrients, or serving as scaffolding or support for neurons.

Although they vary widely in shape, all neurons have projections that contact other cells. As **FIGURE 42.4** shows, most neurons have two distinct types of projections from the cell body, where the nucleus is located: (**1**) highly branched, relatively short processes called **dendrites,** and (**2**) a relatively long structure called an **axon.** Dendrites contact other cells and transmit electrical signals from them to the cell body; the axon carries electrical signals from the cell body to other cells.

Muscle Tissue   **Muscle tissue** was a key innovation in the evolution of animals—like nervous tissue, it appears in no other

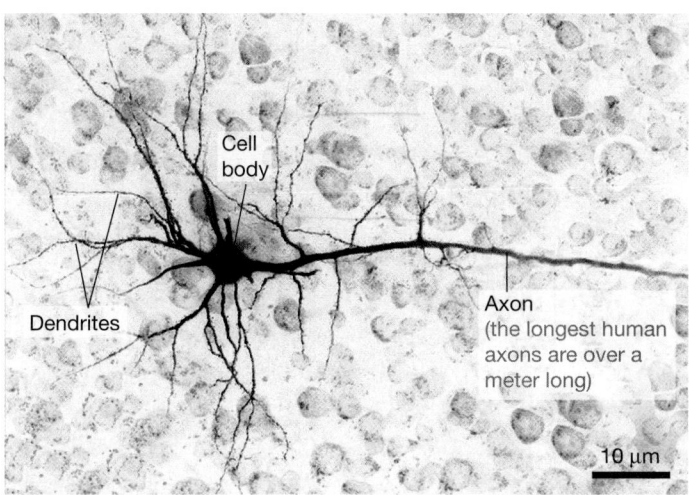

FIGURE 42.4 Neurons Transmit Electrical Signals. In a neuron, information is transmitted from dendrites to the cell body to the axon.

✔**QUESTION** Based on the structure of a neuron, does it provide signals to specific cells and tissues, or does it broadcast signals widely throughout the body?

lineage on the tree of life (see Chapter 33). Some of the functions of muscle include movement of the body, pumping of the heart, and mixing of food in the gastrointestinal tract. There are three types of muscle tissue (**FIGURE 42.5**); you, along with other vertebrates, have all three.

1. **Skeletal muscle** attaches to the bones of the skeleton and exerts a force on them when it contracts. Skeletal muscle is responsible for most body movements. In addition, it encircles the openings of the digestive and urinary tracts and controls swallowing, defecation, and urination.

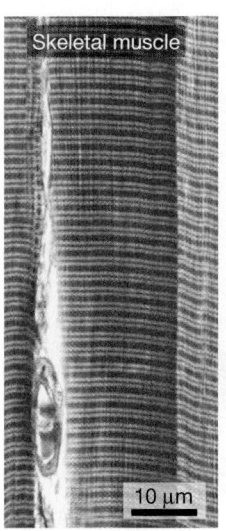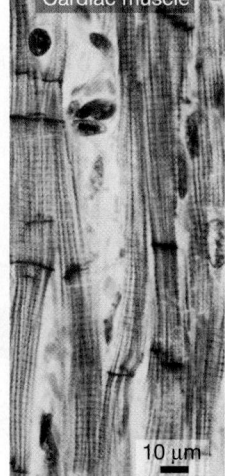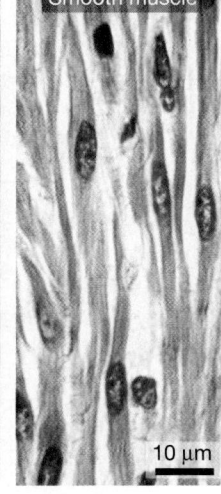

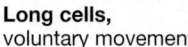

| Long cells, | Branched cells, | Tapered cells, |
| voluntary movement | involuntary movement | involuntary movement |

FIGURE 42.5 Muscle Tissues Comprise Cells That Contract. The three types of muscle tissue have distinctive structures and functions.

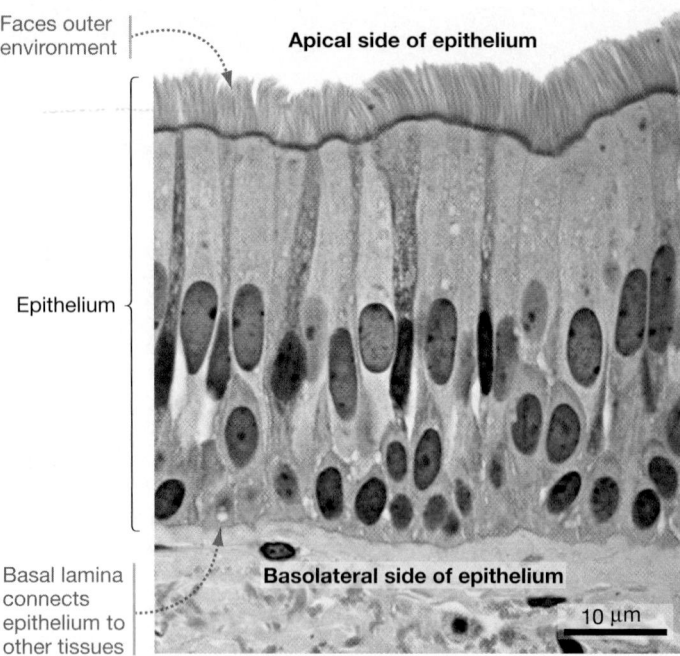

Faces outer environment

Apical side of epithelium

Epithelium

Basal lamina connects epithelium to other tissues

Basolateral side of epithelium

10 μm

FIGURE 42.6 Epithelial Cells Provide Protection and Regulate which Materials Pass across Body Surfaces.

2. **Cardiac muscle** makes up the walls of the heart and is responsible for pumping blood throughout the body.

3. **Smooth muscle** cells, which are tapered at each end, form a muscle tissue that lines the walls of the digestive tract and the blood vessels.

Muscle tissue and movement are explored in detail in Chapter 48.

Epithelial Tissues **Epithelial tissues** are also called **epithelia** (singular: **epithelium**). Epithelium covers the outside of the body, lines the surfaces of organs, and forms glands. An **organ** is a structure that serves a specialized function and consists of several tissues; a **gland** is an organ that secretes specific molecules or solutions such as hormones or digestive enzymes.

Epithelia form the interface between the interior of an organ or body and the exterior. In addition to providing protection, epithelial tissues are gatekeepers. Water, nutrients, and other substances are transported, often selectively, across epithelia.

Because the primary function of epithelium is to act as a barrier and protective layer, it's not surprising to observe that epithelial cells typically form layers of closely packed cells (**FIGURE 42.6**).

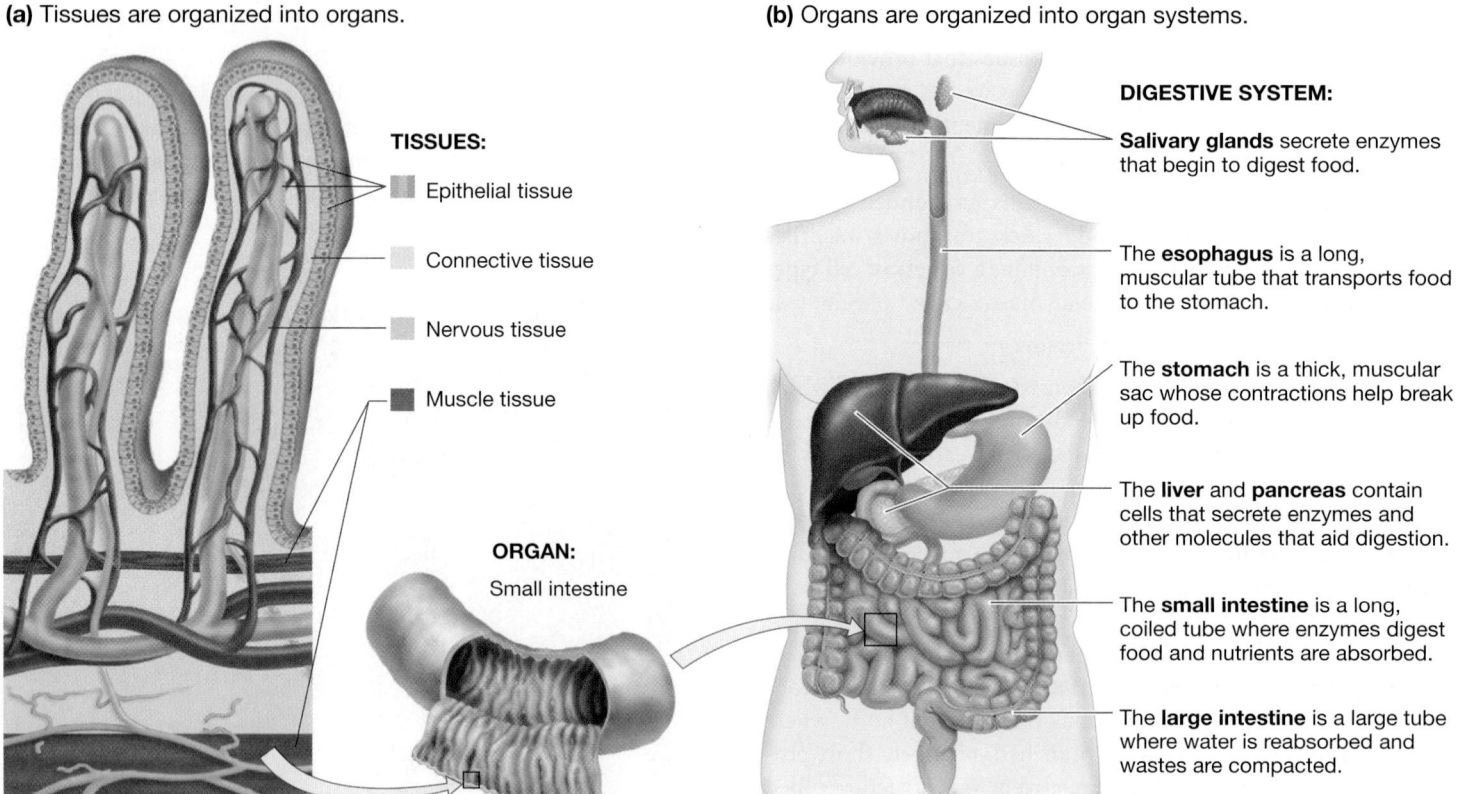

(a) Tissues are organized into organs.

TISSUES:

Epithelial tissue

Connective tissue

Nervous tissue

Muscle tissue

ORGAN:

Small intestine

(b) Organs are organized into organ systems.

DIGESTIVE SYSTEM:

Salivary glands secrete enzymes that begin to digest food.

The **esophagus** is a long, muscular tube that transports food to the stomach.

The **stomach** is a thick, muscular sac whose contractions help break up food.

The **liver** and **pancreas** contain cells that secrete enzymes and other molecules that aid digestion.

The **small intestine** is a long, coiled tube where enzymes digest food and nutrients are absorbed.

The **large intestine** is a large tube where water is reabsorbed and wastes are compacted.

FIGURE 42.7 Organs Are Composed of Tissues; Organ Systems Are Made Up of Organs. (a) The human small intestine is an organ composed of all four major tissue types. **(b)** The human digestive system is essentially one long tube divided into chambers where food is processed and nutrients are absorbed. The salivary glands, liver, and pancreas are organs that secrete specific enzymes or compounds into the tube.

In many cases, adjacent epithelial cells are joined by structures that hold them tightly together, such as tight junctions and desmosomes (introduced in Chapter 11).

Epithelial tissue has polarity, or sidedness. An epithelium has an **apical** side, which faces away from other tissues and toward the environment, and a **basolateral** side, which faces the interior of the animal and connects to connective tissues. This connection is made by a layer of fibers called the **basal lamina.**

The apical and basolateral sides of an epithelium have distinct structures and functions. Epithelial cells, for example, line the surface of your trachea, or windpipe. The apical side of these cells secretes mucus and is covered with cilia that help sweep away dust, bacteria, and viruses. The basolateral side lacks these features but is cemented to the basal lamina.

Epithelial cells have short life spans. The cells that line your esophagus—the tube connecting your mouth and stomach—live for 2 to 3 days, while the cells that line your large intestine live for a maximum of 6 days. Muscle cells and neurons, in contrast, normally live as long as the individual animal does. Epithelial cells are short lived because they are exposed to harsh environments, where they are likely to be killed or scraped away.

The tissue as a whole does not wear away, however, because it includes cells that actively undergo mitosis and cytokinesis—producing new epithelial cells to replace those lost on the side that faces the environment.

Organs and Organ Systems

Cells with similar functions are organized into tissues, and tissues are organized into specialized structures called organs. Recall that an organ is a structure that serves a specialized function and consists of several types of tissues. The small intestine, for example, consists of muscle, nervous, connective, and epithelial tissues (**FIGURE 42.7a**).

An **organ system** consists of groups of tissues and organs that work together to perform one or more functions. Using the digestive system as an example, **FIGURE 42.7b** illustrates how the structure of organs correlates with their function and how the components of an organ system work together in an integrated fashion.

Because an animal's body contains molecules, cells, tissues, organs, and organ systems, biologists who study animal anatomy and physiology must work at various levels of organization to understand how that body operates.

FIGURE 42.8 illustrates these levels of organization, using the human nervous system as an example. Because the structure and function of each component in the body are integrated with those of other components, and because each level of organization is integrated with other levels of organization, the organism as a whole is greater than the sum of its parts. In other words, an organism is more than just a collection of individual systems, and each system is more than just a collection of individual cells or tissues or even organs.

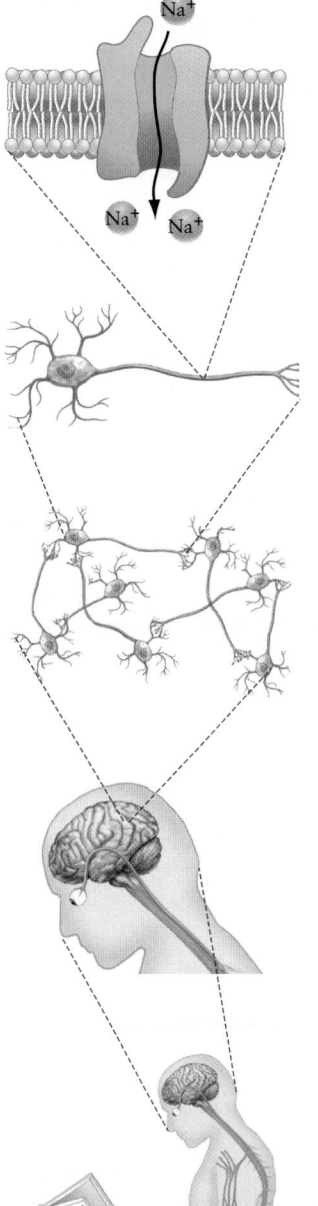

Atomic and molecular levels:
Membrane protein in neurons regulates flow of ions.

Cellular level:
Electrical signal travels down length of neuron.

Tissue level:
Electrical signals travel from cell to cell in nervous tissue.

Organ level:
Nervous tissue and connective tissue in brain aid in sight, smell, memory, and thought.

Organ system level:
Brain and nerves send signals throughout the body to control breathing, digestion, movement, and other functions.

Organism level:
Nervous system coordinates the functions of other systems to support life.

FIGURE 42.8 Biologists Study Anatomy and Physiology at Many Levels. The levels of organization within an organism are not independent of each other. Instead, they are tightly integrated.

In effect, each subsequent chapter in this unit focuses on a different organ system found in animals, beginning with the excretory system and ending with the immune system. Each of these systems can be interpreted as a suite of adaptations and trade-offs. Each system accomplishes a specific task required for survival and reproduction, and each works in conjunction with other systems.

Before delving into the various systems, however, it's essential to examine general phenomena that affect all systems in animals. Let's start by looking at how body size affects animal physiology.

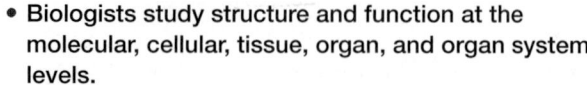

check your understanding

If you understand that . . .

- Biologists study structure and function at the molecular, cellular, tissue, organ, and organ system levels.
- Events at each level of organization in an individual interact to form an integrated whole that responds to the environment in appropriate ways.

✔ **You should be able to . . .**

Describe and compare the structure and function of the four major types of animal tissues.

Answers are available in Appendix A.

42.3 How Does Body Size Affect Animal Physiology?

Animals are living machines, made up of molecules, cells, tissues, organs, and organ systems that have changed over time in response to natural selection.

The laws of physics affect the anatomy and physiology of a living machine. The force of gravity, for example, limits how large an animal can be and still move efficiently. Or consider the forces exerted by the medium in which animals live. Because water is much denser than air, it is harder for animals to move through water. As a result, fish and aquatic mammals have much more streamlined bodies than terrestrial animals do.

Physical laws clearly affect body size. Just as clearly, body size has pervasive effects on how animals function. Large animals need more food than small animals do. Large animals also produce more waste, take longer to mature, reproduce more slowly, and tend to live longer. Conversely, small animals are more susceptible to damage from cold and dehydration than large animals are, because they lose heat and water faster. Juveniles and adults of the same species face different challenges simply because their body sizes are different.

Why is body size such an important factor in how animals work? How do biologists study the consequences of size? Let's consider each question in turn.

Surface Area/Volume Relationships: Theory

From microscopic roundworms to gigantic blue whales, animals span an incredible range of body masses—a total of 12 orders of magnitude. Many of the challenges posed by increasing size are based on this fundamental relationship between surface area and volume.

You learned previously that the relationship between the surface area and volume of the roots and shoots of plants affects water and light absorption by the plant (see Chapter 37). Similarly, surface area is important in animals because oxygen and nutrients such as glucose must diffuse into the animal's body, and waste products such as urea and carbon dioxide must diffuse out. The rate at which these and other molecules and ions diffuse depends in part on the amount of surface area available for diffusion. In contrast, the rate at which nutrients are used and heat and waste products are produced depends on the volume of the animal.

The contrast between processes that depend on surface area and those that depend on volume is important for a simple reason. As an animal gets larger, its volume increases much faster than its surface area does.

Reviewing a little basic geometry will convince you why this is so. As **FIGURE 42.9a** shows:

- The surface area of a cube increases as a function of its linear dimension *squared*. Because a cube has six sides, the surface area of a cube of length l is $6l^2$ (six times the area of any one side).

- The volume of the same structure increases as a function of its linear dimension *cubed*. Hence, the volume (or mass) of a cube of length l is l^3.

Area has two dimensions; volume has three. In general:

$$\text{Surface area} \propto (\text{length})^2$$
$$\text{Volume (or mass)} \propto (\text{length})^3$$
$$\text{Surface area} \propto (\text{volume})^{2/3}$$

(The symbol $\propto$ means "is proportional to.")

FIGURE 42.9b graphs the consequences of these relationships. The *x*-axis plots the length of a side in a cube; the *y*-axis plots the cube's volume (orange line) or surface area (yellow line). As a cube gets bigger, its surface area increases much more slowly than does its volume (or mass).

The same general relationship holds for cells, tissues, organs, and organ systems. Larger cells, for example, have lower surface area/volume ratios than smaller cells. Quantities that are based on volume, such as body mass, increase disproportionately fast with increases in linear dimensions.

✔ If you understand the relationship between surface area and volume, you should be able to predict which of the following has the higher surface area/volume ratio: a newborn or an adult human.

How does the relationship between surface area and volume affect animal form and function?

(a) What are the surface area and volume of each cube?

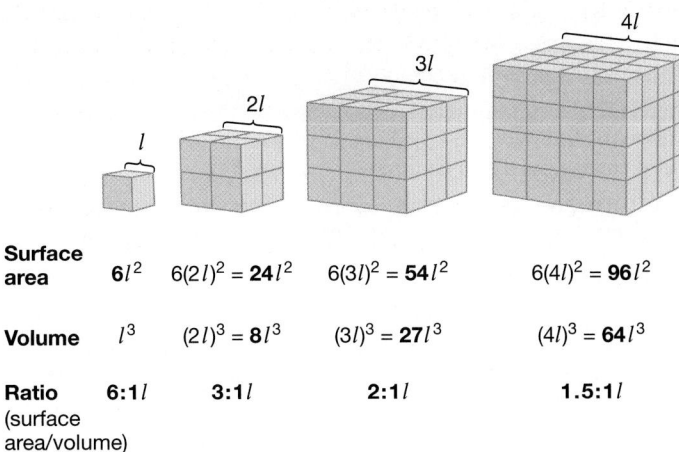

Surface area	$6l^2$	$6(2l)^2 = 24l^2$	$6(3l)^2 = 54l^2$	$6(4l)^2 = 96l^2$
Volume	l^3	$(2l)^3 = 8l^3$	$(3l)^3 = 27l^3$	$(4l)^3 = 64l^3$
Ratio (surface area/volume)	6:1l	3:1l	2:1l	1.5:1l

(b) Surface area and volume of a cube versus length of a side

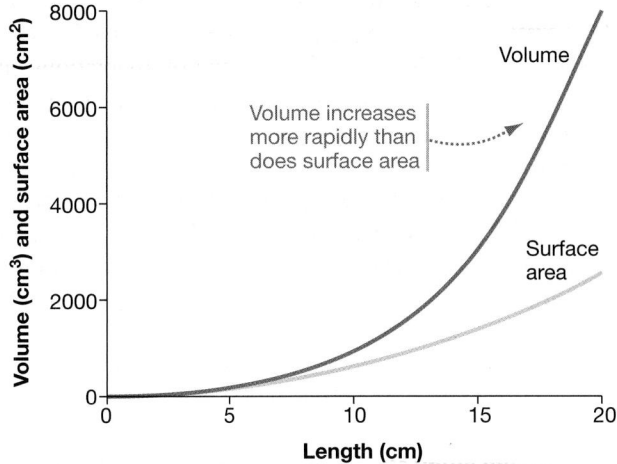

FIGURE 42.9 Surface Area and Volume Change as a Function of Overall Size. (a) The surface area of an object increases as the square of the length (l). The volume increases as the cube of the length. **(b)** Volume increases much more rapidly than does surface area as linear dimensions increase.

Surface Area/Volume Relationships: Data

As an example of how surface area/volume relationships affect an animal's physiology, consider the metabolic rate of mammals. **Metabolic rate** is the overall rate of energy consumption by an individual. Because consumption and production of energy in mammals depend largely on aerobic respiration, metabolic rate is often measured in terms of oxygen consumption, and it is typically reported in units of milliliters of O_2 consumed per hour.

Because it is so much larger, an elephant consumes a great deal more oxygen per hour than a mouse does. But what is going on at the levels of cells and tissues in these animals?

Comparing Mice and Elephants To compare metabolic rates in different species, biologists divide metabolic rate by overall mass and report a mass-specific metabolic rate in units of milliliters of oxygen per gram per hour (mL O_2/g/hr). This mass-specific metabolic rate gives the rate of oxygen consumption per gram of tissue.

Because an individual's metabolic rate varies dramatically with its activity, the accepted convention is to report the **basal metabolic rate (BMR)**—the rate at which an animal consumes oxygen while at rest, with an empty stomach, under normal temperature and moisture conditions.

FIGURE 42.10 plots per-gram or "mass-specific" BMR as a function of average body mass. Notice that the x-axis on the graph is logarithmic, to make it easier to compare very small species with very large ones. For help with logarithms, see **BioSkills 6** in Appendix B.

The graph's take-home message? On a per-gram basis, small animals have higher BMRs than do large animals. An elephant has more mass than a mouse, but a gram of elephant tissue consumes much less energy than a gram of mouse tissue does.

The leading hypothesis to explain this pattern is based on surface area/volume ratios. Many aspects of metabolism—including

oxygen consumption, food digestion, delivery of nutrients to tissues, and removal of wastes and excess heat—depend on exchange across surfaces. As an organism's size increases, its mass-specific metabolic rate must decrease. Otherwise the surface area available for exchange of materials would fail to keep up with the metabolic demands generated by the organism's enzymes.

Changes during Development A king salmon weighs a few milligrams or less at hatching but grows into an adult weighing 50 kg or more—a millionfold increase in body mass. To explore the consequences of this change, biologists have studied how gas

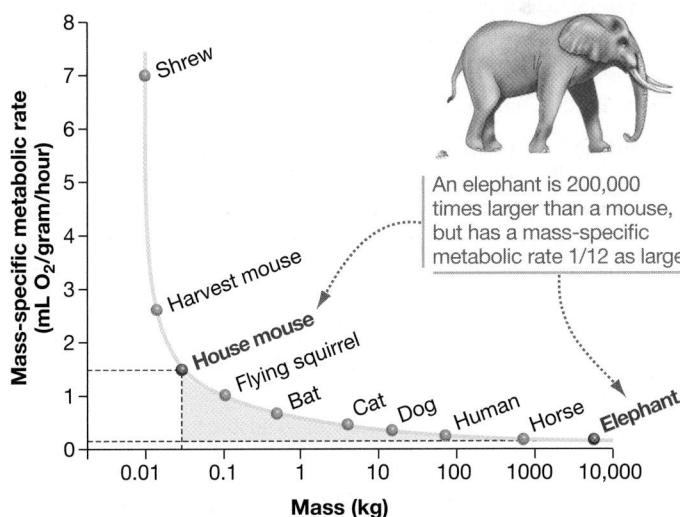

FIGURE 42.10 Small Animals Have Higher Relative Metabolic Rates than Large Animals Do. Overall body mass, plotted on a logarithmic scale, versus metabolic rate per gram of tissue.

✔**QUESTION** Which mammal has to eat more to support each gram of its tissue: a puppy or an adult dog?

exchange—uptake of oxygen and removal of carbon dioxide—occurs in newly hatched Atlantic salmon.

Like most fish species, young salmon have rudimentary gills but also exchange gases across their skin. In aquatic animals, **gills** are organs that allow the exchange of gases and dissolved substances between the animals' blood and the surrounding water.

To document the amount of gas exchange that occurs in the gills versus the skin, researchers inserted the heads of individual larval and juvenile salmon through a pinhole in a soft rubber membrane and then recorded the rate of oxygen uptake on either side of the membrane. As the "Experimental Setup" section of **FIGURE 42.11** indicates, the gills were responsible for oxygen uptake on one side of the membrane; skin was responsible for oxygen uptake on the other side of the membrane.

The graph in the figure's "Results" section plots the percentage of total oxygen uptake that took place across the skin (green line) versus gills (purple line), as a function of body mass. Each data point represents the recordings from an individual fish.

Note that newly hatched larvae take up most of the oxygen they need by diffusion across the body surface. As an individual grows, however, its skin surface area decreases in relation to its volume. To avoid suffocation, individuals switch from skin-breathing to gill-breathing at a body mass of about 0.1 grams. What makes gills so effective in oxygen uptake?

Adaptations That Increase Surface Area

If the function of a cell or tissue depends on diffusion, it usually has a shape that increases its surface area relative to its volume. Flattening, folding, and branching are effective ways for structures to have a high surface area/volume ratio:

- *Flattening* Fish have **gill lamellae** (**FIGURE 42.12a**)—thin sheets of epithelial cells that provide the gill with an extremely high surface area relative to its volume. Because the surface available is so large, gases are able to diffuse across the gills rapidly enough to keep up with the growth in the volume of a developing fish.

- *Folding* In portions of the digestive tract where nutrients are transported into the body, the surface of the structure is folded. Extending from these folds are narrow projections called **villi** (**FIGURE 42.12b**). Together, the folds and villi make an extensive surface area available. Folded surfaces are common in diffusion-dependent organs.

- *Branching* The highly branched network shown in **FIGURE 42.12c** is a system of small, thin-walled blood vessels called **capillaries.** Capillaries have a high surface area available for gases, nutrients, and waste products to diffuse into and out of blood; branching increases the surface available in each square centimeter of tissue. In general, highly branched structures increase the surface area available for diffusion.

The amount of surface area created by flattening, folding, and branching can be impressive. The highly branched capillaries in a human have a total surface area of up to 1000 m²; extensive folding gives a total surface area of about 140 m² in your lungs and

QUESTION: Newly hatched salmon can breathe through their skin and through their gills. Which predominates?

HYPOTHESIS: The relative amount of gas exchange across gills and skin changes as a salmon grows.

NULL HYPOTHESIS: The relative amount of gas exchange across gills and skin does not change as a salmon grows.

EXPERIMENTAL SETUP:

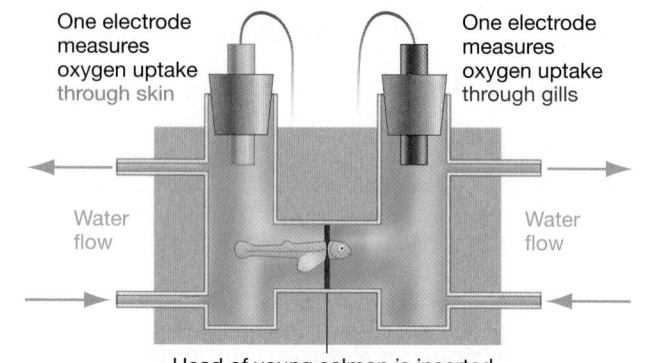

One electrode measures oxygen uptake through skin

One electrode measures oxygen uptake through gills

Water flow

Water flow

Head of young salmon is inserted through a small hole in rubber membrane

PREDICTION: Juveniles will exchange a higher percentage of gas across gills and a lower percentage of gas across skin than larvae.

PREDICTION OF NULL HYPOTHESIS: Juveniles and larvae will exchange the same percentage of gas across gills and skin.

RESULTS:

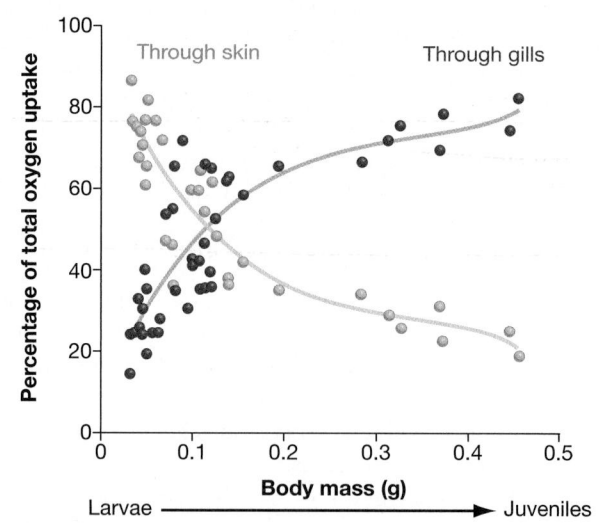

CONCLUSION: Breathing changes from skin to gills as larvae grow. Interpretation: Gills provide larger surface area relative to increasing volume of body.

FIGURE 42.11 How Do Young Salmon Breathe?

SOURCE: Wells, P. R., and A. W. Pinder. 1996. The respiratory development of Atlantic salmon. *Journal of Experimental Biology* 199: 2737–2744.

✓**QUESTION** Suppose the experimenters had measured oxygen uptake on either side of the apparatus in the absence of a fish. What would the results be?

(a) Flattening: fish gill lamellae

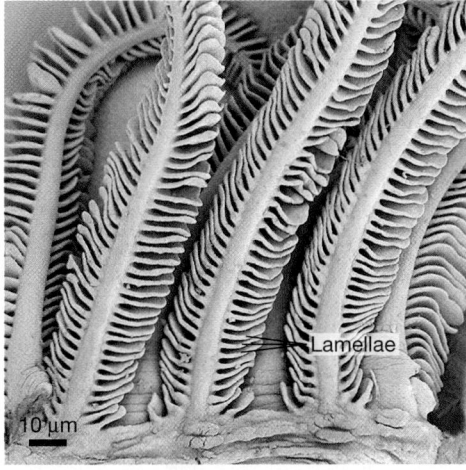

Lamellae

10 µm

(b) Folding: intestinal folds and villi

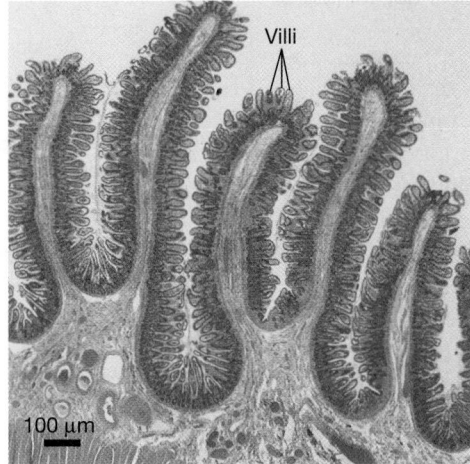

Villi

100 µm

(c) Branching: capillaries

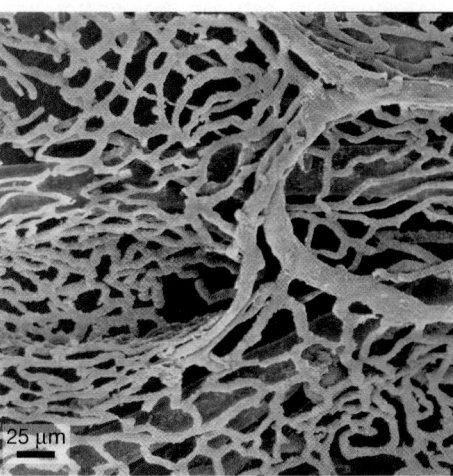

25 µm

FIGURE 42.12 Certain Structures Have High Surface Area/Volume Ratios. The micrograph in part (c) has been colorized to highlight capillaries, colored pink.

250 m^2 in your small intestine. For comparison, a doubles tennis court has a surface area of 261 m^2.

Surface area/volume relationships have a pervasive influence on the structure and function of animals. They will be an issue in almost every chapter in this unit.

check your understanding

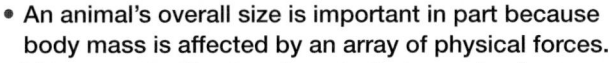
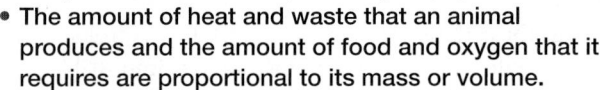
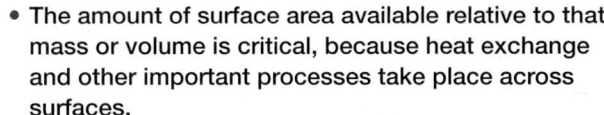

C Y U

If you understand that . . .
- An animal's overall size is important in part because body mass is affected by an array of physical forces.
- The amount of heat and waste that an animal produces and the amount of food and oxygen that it requires are proportional to its mass or volume.
- The amount of surface area available relative to that mass or volume is critical, because heat exchange and other important processes take place across surfaces.

✔ You should be able to . . .
1. Explain why large animals have a relatively low surface area/volume ratio.
2. Explain why a squirrel needs to eat more energy per unit body mass than a bear.

Answers are available in Appendix A.

42.4 Homeostasis

Adaptation and surface area/volume ratios are important themes in the analysis of animal form and function. So is homeostasis.

Homeostasis (literally, "alike-standing") is defined as stability in the chemical and physical conditions within an animal's cells, tissues, and organs. Although conditions may vary as an animal's environment changes, internal chemical and physical states are usually kept within a tolerable range.

Homeostasis: General Principles

Many of the structures and processes observed in animals can be interpreted as mechanisms for maintaining homeostasis with respect to some quantity, such as pH, temperature, or calcium ion concentration. Let's review some important general ideas about homeostasis and then analyze how homeostasis can be maintained in the face of environmental fluctuations.

Two Approaches to Achieving Homeostasis Constancy of physiological state can be achieved by two processes: (1) conformation or (2) regulation.

1. *Conformation* The body temperature of Antarctic rock cod closely matches that of the surrounding seawater, which is typically −1.9°C (seawater is still liquid at this temperature because of the high concentration of solutes). The rock cod does not actively regulate its body temperature to match that of seawater. Instead, its body temperature remains constant because it conforms to the temperature of its constant surroundings. If rock cod were placed into a different environment, for example warmer water, they would not maintain a constant body temperature of −1.9°C. Because these animals are adapted to such constant environmental conditions, their physiological systems cannot function properly outside of these conditions.

2. *Regulation* Regulatory homeostasis is based on mechanisms that adjust the internal state to keep it within limits that can be tolerated, no matter what the external conditions. For example, a dog maintains a body temperature of about 38°C whether it's cold or hot outside. If the dog's body temperature rises, it might pant to cool off and maintain homeostasis. If body temperature falls, the dog might shiver to bring its temperature back up to the target value. There

are, however, limits to regulatory homeostasis. If the external temperature gets too cold or hot, animals cannot maintain homeostasis and will die.

The Role of Epithelium Because epithelium is the interface between the internal and external environments, it plays a key role in achieving homeostasis. Epithelium is responsible for forming an internal environment that can be dramatically different from the external environment such that physical and chemical conditions inside an animal can be maintained at relatively constant levels.

As subsequent chapters will show, many epithelial cells are studded with membrane proteins that regulate the transport of ions, water, nutrients, and wastes. No molecule can enter or leave the body without crossing an epithelium. Homeostasis is possible because epithelia control this exchange.

Why Is Homeostasis Important? Much of the answer to this question is based on enzyme function. Recall that enzymes are proteins that catalyze chemical reactions within cells (see Chapter 4). Temperature, pH, and other physical and chemical conditions have a dramatic effect on the structure and function of enzymes. Most enzymes function best under a fairly narrow range of conditions.

Other processes depend on homeostasis, too. Temperature changes affect membrane permeability and how quickly solutes diffuse. To take an extreme case, the expansion of water as it freezes can rip cells apart if tissues are allowed to drop much below 0°C. Conversely, extremely high temperatures can denature proteins—meaning that they lose their tertiary structure and cease to function.

When homeostasis occurs, conditions inside the body allow molecules, cells, tissues, organs, and organ systems to function at an optimal level. However, occasional departures from homeostasis can represent important adaptations. For example, a fever is a response to an infection by a pathogen. This increase in body temperature can help fight off the pathogen.

The Role of Regulation and Feedback

To achieve homeostasis, most animals have regulatory systems that monitor internal conditions such as temperature, blood pressure, blood pH, and blood glucose. If one of these variables changes, a homeostatic system acts quickly to modify it. Like the thermostat in a home heating system, each of these systems has a **set point**—a normal or target range of values for the controlled variable.

Animals have a set point for blood pH, blood oxygen and nutrient concentrations, and other parameters. In most mammals, the set point for body temperature is somewhere between 35°C and 39°C. How does an individual maintain its tissues at the set point despite changes in activity and the environment?

A homeostatic system consists of three general components: a sensor, an integrator, and an effector. **FIGURE 42.13** shows how these components interact:

1. A **sensor** is a structure that senses some aspect of the external or internal environment.

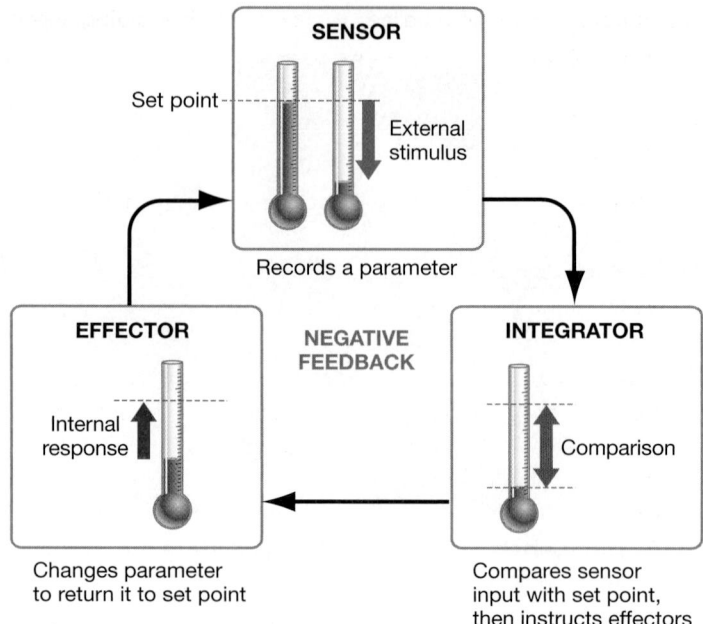

FIGURE 42.13 Animals Achieve Homeostasis through Negative Feedback. Many animals use homeostatic systems similar to this one to maintain a preferred range of hydration, blood pH, blood pressure, calcium ion concentration, body temperature, and so on.

2. An **integrator** evaluates the incoming sensory information and "decides" whether a response is necessary to achieve homeostasis. (The word decides is in quotation marks because the decision is not a conscious one.)

3. An **effector** is any structure that helps restore the desired internal condition.

Without these three elements, homeostatic control systems can't maintain a desired set point; homeostasis is impossible.

Homeostatic systems are based on negative feedback. When **negative feedback** occurs, effectors reduce or oppose the change in internal conditions. For example, a rise in blood pH triggers effectors that act to reduce that rise. Blood pH returns to the set point in response to this negative feedback.

Homeostatic systems are a key aspect of one of the five attributes of life (see Chapter 1): acquiring information from the environment and responding to it. Subsequent chapters in this unit explore how animals use sensor–integrator–effector systems to achieve homeostasis with respect to the solute concentrations of their cells and tissues, their oxygen supply, and nutrient availability. In this chapter, let's focus on how different animals achieve homeostasis with respect to body temperature.

42.5 How Do Animals Regulate Body Temperature?

All animals exchange heat with their environment. Heat flows "downhill," from regions of higher temperature to regions of lower temperature. If an individual is warmer than its surroundings, it will lose heat; if it is cooler than its environment, it will gain heat.

How does heat exchange occur?

Mechanisms of Heat Exchange

As **FIGURE 42.14** shows, animals exchange heat with the environment in four ways: conduction, convection, radiation, and evaporation.

1. **Conduction** is the direct transfer of heat between two physical bodies that are in contact with each other. For instance, when a turtle sits on a warm rock, heat is transferred from the rock to its body. The rate at which conduction occurs depends on the surface area of transfer, the steepness of the temperature difference between the two bodies, and how well each body conducts heat.

2. **Convection** is a special case of conduction. During conduction, heat is transferred between two solids; but during convection, heat is exchanged between a solid and a moving liquid or gas. For example, the heat loss that occurs when wind blows on your skin is due to convection. As the speed of the air or water flow increases, so does the rate of heat transfer.

3. **Radiation** is the transfer of heat between two bodies that are not in direct physical contact. All objects, including animals, radiate energy as a function of their temperature. The Sun radiates heat; so does your body, but to a much lesser degree.

4. **Evaporation** is the phase change that occurs when liquid water becomes a gas. Conduction, convection, and radiation can cause heat gain or loss, but evaporation leads only to heat loss. The turtle in the photograph is losing heat as water evaporates off its shell and skin. Because of the extensive hydrogen bonding in liquid water, a large amount of energy is needed

FIGURE 42.14 Four Methods of Heat Exchange. The arrows indicate the direction of heat exchange.

to heat water and produce evaporation (see Chapter 2). If you get overheated on a summer day, splashing water on your skin and sweating will allow you to use evaporative heat loss to cool your body. Conversely, getting wet on a cold day can be deadly. The water on your skin absorbs so much heat from your body that your temperature may drop dangerously.

Heat exchange is critical in animal physiology because individuals that get too hot or too cold may die. Overheating can cause enzymes and other proteins to denature and cease functioning. It may also lead to excessive water loss and dehydration. A sharp drop in body temperature, in contrast, can slow enzyme function and energy production. In humans, both heat stroke and hypothermia ("under-heating") are life-threatening conditions.

Although most organisms conform to the environmental temperature, some animals can regulate body temperature to keep it in an optimal range. Let's take a closer look.

Variation in Thermoregulation

The ability of animals to **thermoregulate,** or control body temperature, varies widely. Two ways to organize this variation are by examining (**1**) how animals obtain heat, and (**2**) whether body temperature is held constant.

An **endotherm** ("inner-heat") produces adequate heat to warm its own tissues, while an **ectotherm** ("outer-heat") relies principally on heat gained from the environment. Endotherms and ectotherms represent two extremes along a continuum of heat sources. Many animals fall somewhere in between.

There are also two extremes on a continuum describing whether animals hold their body temperature constant: **Homeotherms** ("alike-heat") keep their body temperature constant, while **heterotherms** ("different-heat") allow their body temperature to rise or fall depending on environmental conditions.

Humans, along with most birds and most other mammals, are strictly endothermic homeotherms. These species produce their own heat and maintain a constant body temperature. In contrast, most freshwater and terrestrial invertebrates, fishes, amphibians, and reptiles are ectothermic heterotherms whose body temperatures change throughout the day and seasonally. But many animal species lie somewhere between these extremes:

- Some desert-adapted mammals, such as the oryx featured in the introduction to this chapter, allow their body temperature to rise during the hotter part of the day—meaning they are somewhat heterothermic.

- Small mammals that inhabit cold climates lose heat rapidly because their surface area is large relative to their volume. To survive when temperatures are cold, species such as dormice reduce their metabolic rate and allow their body temperature to drop, a form of heterothermy. This condition is called **torpor.** In animals in which torpor persists for weeks or months, it is called **hibernation.**

- Naked mole rats are mammals, but lack insulation because they have no fur. They live in underground tunnels and allow their body temperature to rise and fall with burrow

(a) A hornet preys on a honeybee.

1 mm

(b) A swarm of bees surrounds a hornet ...

(c) ... forming a hot defensive ball.

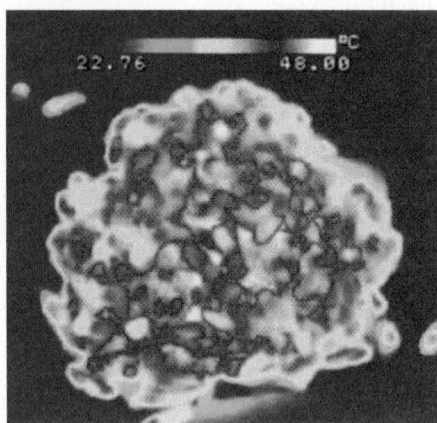

22.76   48.00

FIGURE 42.15 Honeybees Use Heat to Kill Predators.

temperatures. They are heterothermic and intermediate between ectotherms and endotherms.

- Japanese honeybees exhibit heterothermy when defending their hives from predatory hornets. The honeybees swarm an invading hornet and contract their flight muscles repeatedly to collectively produce heat endothermically (**FIGURE 42.15**). The temperature within the swarm rises to 47°C (117°F), killing the hornet but not the honeybees, which can tolerate temperatures up to 50°C (122°F).

Even in a homeothermic endotherm such as a mammal or bird, body temperature can vary widely in different body regions. When a Canada goose is standing on ice, its feet may be at a temperature of just 9°C, even though its body core is at 35°C. Similar variations exist in tuna and mackerel. These fish are ectotherms but generate heat to warm certain sections of their bodies, such as their eyes or swimming muscles.

Endothermy and Ectothermy: A Closer Look

Endotherms can warm themselves because their basal metabolic rates are extremely high—the heat given off by the high rate of chemical reactions is enough to warm the body. Mammals and birds retain this heat because they have elaborate insulating structures such as feathers or fur.

Ectotherms can also generate heat as a by-product of metabolism. The amount of heat they generate is small compared with the amount generated by endotherms, however, because ectotherms have relatively low metabolic rates. The most important sources of heat gain in ectotherms are radiation and conduction: Ectotherms bask in sunlight or lie on warm rocks or soil.

Endothermy and ectothermy are best understood as contrasting adaptive strategies. Because endotherms maintain a high body temperature at all times, they can be active in winter and at night. Their high metabolic rates also allow them to sustain high levels of aerobic activities, such as running or flying.

These abilities come at a cost, however: To fuel their high metabolic rates, endotherms have to obtain large quantities of energy-rich food. The energy used to produce heat is then unavailable for other energy-demanding processes, such as reproduction and growth.

In contrast, ectotherms are able to thrive with much lower intakes of food. And because they are not oxidizing food to provide heat, they can use a greater proportion of their total energy intake to support reproduction.

What's the downside of ectothermy? Chemical reaction rates are temperature dependent, so muscle activity and digestion slow dramatically as the body temperature of an ectotherm drops. As a result, ectotherms are more vulnerable to predation in cold weather and in general are less successful than endotherms at inhabiting cold environments or remaining active in cool nighttime temperatures.

In short, each suite of adaptations has advantages and disadvantages. Like all adaptations, endothermy and ectothermy involve trade-offs.

Temperature Homeostasis in Endotherms

Thermoregulation is an important aspect of homeostasis in many animals. **FIGURE 42.16** illustrates how the sensor–integrator–effector components of mammals function in thermoregulation.

Temperature receptors located throughout the body constantly monitor information about body temperature. For example, temperature receptors in the skin sense cooling or heating, and they respond by altering the pattern of electrical signals that they send to adjacent neurons. Receptors in the brain region called the **hypothalamus** respond in a similar fashion to changes in blood temperature.

The electrical signals that originate with temperature receptors are transmitted to an integrator, also located in the hypothalamus. Current evidence indicates that separate centers in the hypothalamus of the brain sense and integrate increases and decreases in body temperature.

If a mammal is cold, cells in the hypothalamus send signals to effectors that return body temperature to the set point. Signals from the hypothalamus might induce shivering to generate

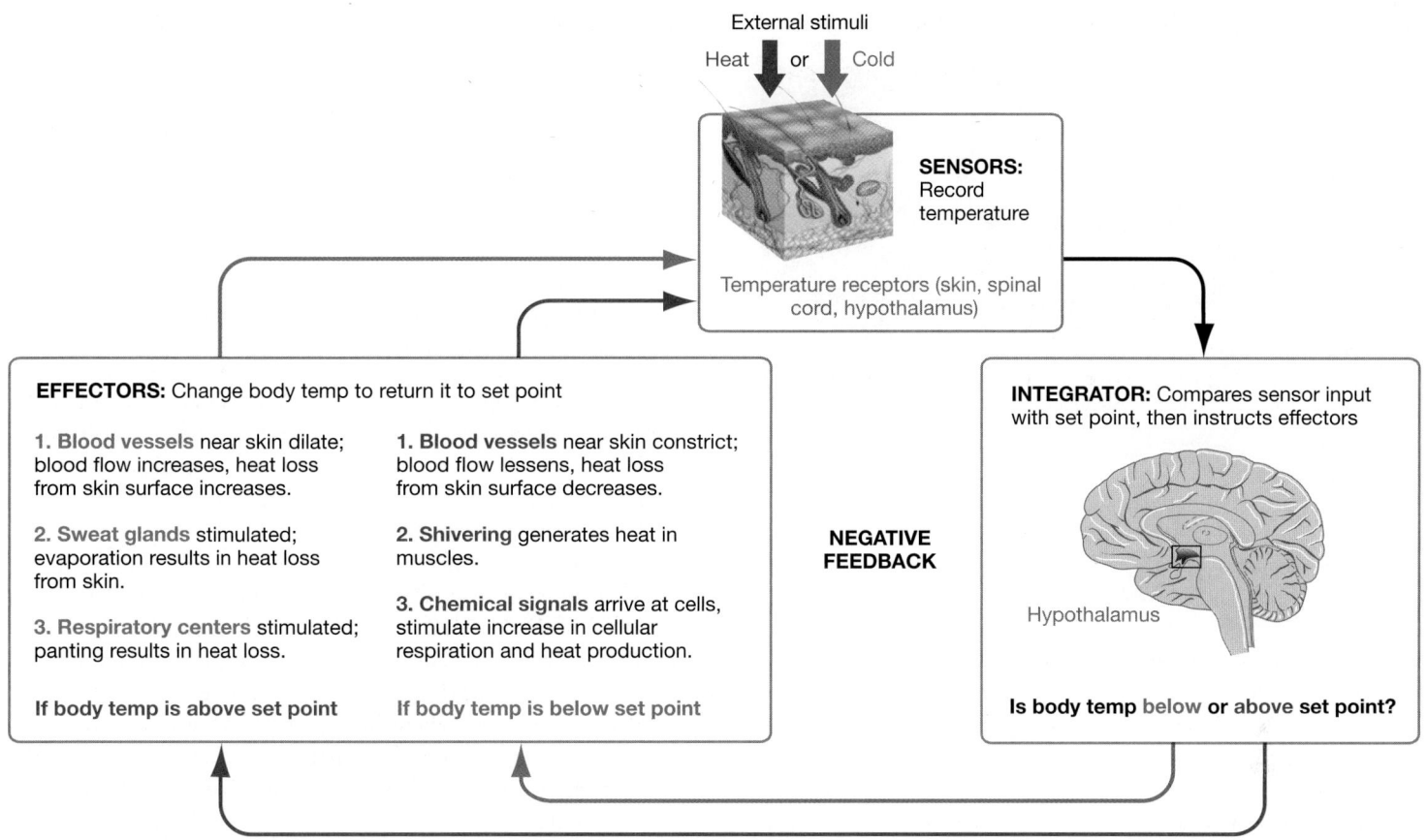

External stimuli

Heat or Cold

SENSORS: Record temperature

Temperature receptors (skin, spinal cord, hypothalamus)

EFFECTORS: Change body temp to return it to set point

1. **Blood vessels** near skin dilate; blood flow increases, heat loss from skin surface increases.

2. **Sweat glands** stimulated; evaporation results in heat loss from skin.

3. **Respiratory centers** stimulated; panting results in heat loss.

If body temp is above set point

1. **Blood vessels** near skin constrict; blood flow lessens, heat loss from skin surface decreases.

2. **Shivering** generates heat in muscles.

3. **Chemical signals** arrive at cells, stimulate increase in cellular respiration and heat production.

If body temp is below set point

NEGATIVE FEEDBACK

INTEGRATOR: Compares sensor input with set point, then instructs effectors

Hypothalamus

Is body temp below or above set point?

FIGURE 42.16 Mammals Regulate Temperature through Negative Feedback. In mammals, a set point for temperature is maintained by a complex negative feedback system that includes integrators in the anterior and posterior hypothalamus and sensors located throughout the body. The set point varies among species, from 30°C in monotremes to over 39°C in rabbits.

warmth, and fluffing of fur or feathers to improve insulation and retain heat. Signals from the same or nearby cells can also result in the release of blood-borne chemical signals that increase the rate at which cellular respiration takes place throughout the body—generating more body heat.

But if the same individual is too hot, an integrator in the hypothalamus sends signals that initiate sweating or panting—responses that cool the body. Other signals can induce behavioral changes that slow heat production, such as seeking shade or a cool burrow and then resting. Within cells, temperature spikes that are dramatic enough to denature proteins may activate **heat-shock proteins** (introduced in Chapter 3). Heat-shock proteins speed the refolding of proteins—a key step in the recovery process.

In response to either cooling or heating, behavioral and physiological responses move the body temperature back toward the set point via negative feedback. Figure 42.16 makes several points about the effectors that maintain homeostasis:

- It is common to observe redundancy in feedback systems—there are usually several ways to change a parameter.

- Feedback systems usually work in "antagonistic pairs": One set of responses increases a parameter while a corresponding set of responses decreases it.

- Input from sensors and integrators is constant, so feedback systems are constantly making fine adjustments relative to the set point.

Countercurrent Heat Exchangers

Homeothermic endotherms such as birds and mammals have sophisticated systems for thermoregulation. One of their most impressive adaptations allows them to minimize heat loss from limbs. Heat loss is a particularly important problem for mammals that live in aquatic environments. If you've ever gone swimming in cold water, you can appreciate the problem faced by seals, otters, and whales. Water is such an effective conductor of heat that aquatic organisms lose metabolic heat rapidly. To conserve heat, otters have dense, water-repellent fur that maintains a layer of trapped air next to the skin. Seals and whales are insulated by thick layers of fatty blubber. In addition, some marine mammals have specially arranged blood vessels that minimize heat loss.

An Example: Gray Whale Tongues Gray whales have a feature that minimizes heat loss from their tongue, which is exposed to cold water during feeding. The tongue contains bundles of arteries and veins. Each bundle contains an artery that carries warm, oxygenated blood from the body core. The artery is encircled by

(a) Tongue of gray whale

Cross section of blood
vessels of tongue

Arteries are surrounded
by veins in each bundle

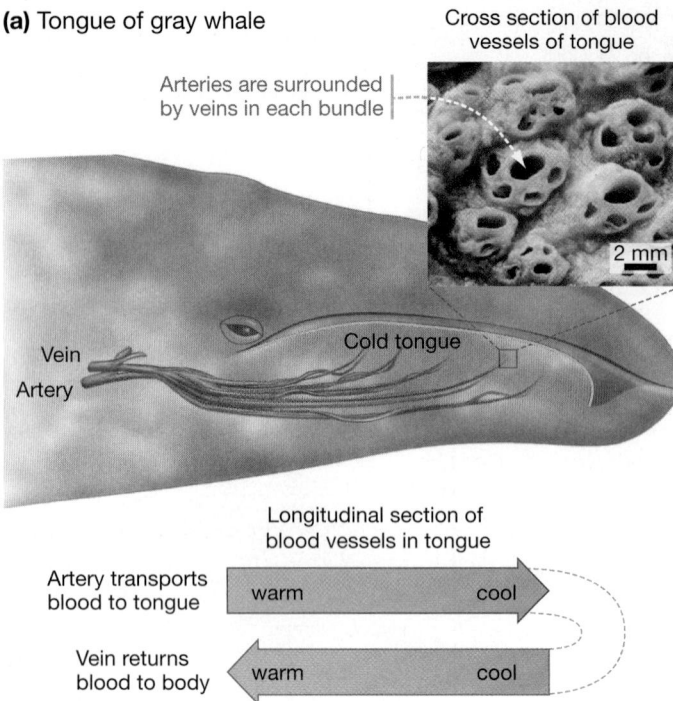

2 mm

Vein
Artery

Cold tongue

Longitudinal section of
blood vessels in tongue

Artery transports
blood to tongue

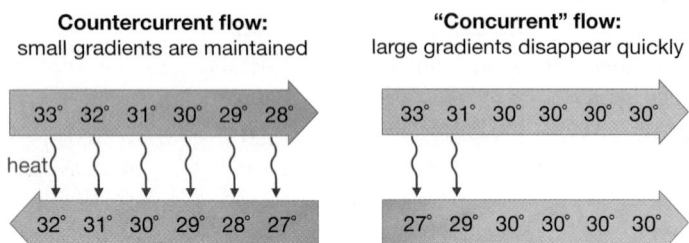

warm   cool

Vein returns
blood to body

warm   cool

(b) Contrasting countercurrent with "concurrent" heat exchange

Countercurrent flow:
small gradients are maintained

"Concurrent" flow:
large gradients disappear quickly

33° 32° 31° 30° 29° 28°   33° 31° 30° 30° 30° 30°

 heat

32° 31° 30° 29° 28° 27°   27° 29° 30° 30° 30° 30°

FIGURE 42.17 Countercurrent Exchangers Conserve Heat.
(a) Bundles of arteries and veins in whale tongues form heat
exchangers that minimize heat loss from the tongue to the cold
ocean water during feeding. **(b)** Countercurrent arrangements are
much more efficient than "concurrent" arrangements. The data
given here are hypothetical.

smaller veins, which transport cool blood from the tongue surface back toward the body core (**FIGURE 42.17a**).

The key is that the two types of blood vessel are arranged in
an antiparallel fashion. Vessels with warm blood traveling out to
the tongue are in close contact with vessels containing cool blood
traveling back to the body.

This type of arrangement, with fluids flowing through adjacent pipes in opposite directions, is called a **countercurrent exchanger.** The "exchanger" part of the name is apt because in a
case like the whale's tongue, heat is exchanged between the warm
blood in the artery and the cool blood in the veins.

To see how the countercurrent exchange system works, study
the longitudinal section in **FIGURE 42.17b**. Note that the fluid that

enters the countercurrent heat exchanger is initially warm but
steadily transfers heat to the adjacent, cooler fluid flowing in
the opposite direction. There is a warmer-to-cooler gradient
between the two currents at every point along the length of the
countercurrent exchanger.

If the two solutions ran in the same direction, as in the right
side of Figure 42.17b, the gradient between the two solutions
would disappear quickly as the source current cooled and the
recipient current heated. Countercurrent exchangers, including
heat exchangers, are efficient because they maintain a gradient
between the two fluids along their entire length.

A Multiplier Effect   A second key point about countercurrent
heat exchangers concerns the temperature differential between
the solutions. Although the differential is small at any point
along the pipes, there is a large temperature differential from one
end of each pipe to the other.

In effect, small differences in heat along the length of the exchanger sum up to create a large overall temperature gradient
from beginning to end. The longer the system, the greater the
overall differential will be. To highlight this property, the systems
are sometimes called countercurrent multipliers.

Similar heat-conserving arrangements of arteries and veins
are found in the flippers of whales and dolphins, and in the
legs of many mammals and birds that live in cold terrestrial
environments.

Countercurrent exchangers are just one of many sophisticated
adaptations you'll encounter in this unit—structures and systems
that allow animals to thrive in a wide array of environments.

check your understanding

If you understand that . . .

C
Y
U

- Two major aspects of temperature regulation
  vary among animal species: the amount of heat
  generated by the animal's own tissues and the
  degree to which body temperature varies over time.
- Negative feedback allows endothermic
  homeotherms to maintain homeostasis with respect
  to body temperature.
- Countercurrent heat exchangers are efficient
  adaptations for minimizing heat loss from
  extremities.

✓ **You should be able to . . .**

1. Discuss the advantages and disadvantages of
   endothermy and ectothermy.
2. Predict how the temperature of a whale's tongue
   would change if it employed concurrent rather than
   countercurrent flow of blood in its arteries and veins.
   Explain.

Answers are available in Appendix A.

If you understand . . .

42.1 Form, Function, and Adaptation

- Animal structures and their functions represent adaptations, which are heritable traits that improve survival and reproduction in a certain environment.

- Adaptations involve trade-offs, or inescapable compromises between traits.

- Acclimatization is a reversible response to the environment that improves physiological function in that environment.

✓ You should be able to provide examples of an adaptation and acclimatization of a mammal to a cold environment.

42.2 Tissues, Organs, and Systems: How Does Structure Correlate with Function?

- Cells with a common function are grouped together into four general types of tissue: connective tissue, nervous tissue, muscle tissue, and epithelial tissue.

- Epithelium forms the interface between the animal's external and internal environments.

- Organs are structures that are composed of two or more tissues that together perform specific tasks.

- Organ systems comprise organs that work together in an integrated fashion to perform one or more functions.

✓ You should be able to predict the effect of a drug that loosens the tight junctions between epithelial cells.

42.3 How Does Body Size Affect Animal Physiology?

- Large animals have smaller surface area/volume ratios than small animals. As animals grow, the volume increases more rapidly than the surface area.

- Large animals have low mass-specific metabolic rates, because they have a relatively small surface area for exchanging the oxygen and nutrients required to support metabolism, the wastes produced by metabolism, and heat.

- The relatively high surface area of small animals means that they lose heat extremely rapidly.

✓ You should be able to explain why it would be impossible for a gorilla the size of King Kong to have fur. (In answering this question, explain how the surface area/volume ratio of a normal-sized gorilla would compare to Kong's; then relate this to the role of surface area and volume in heat generation and transfer and the function of fur.)

42.4 Homeostasis

- Homeostasis refers to relatively constant physical and chemical conditions inside the body.

- Homeostasis can be achieved by conformation to a constant environment or by regulation in a fluctuating environment.

- Animals have a set point, or target value, for blood pH, tissue oxygen concentration, nutrient concentrations, temperature, and other parameters.

- When a condition in the body is not at its set point, negative feedback occurs. Responses to negative feedback return conditions to the set point and result in homeostasis.

✓ You should be able to explain how, due to acclimatization as well as adaptation, the homeostatic system for maintaining body temperature in a species of mammal might change as global temperatures rise.

42.5 How Do Animals Regulate Body Temperature?

- Animals vary from endothermic to ectothermic and from homeothermic to heterothermic.

- Most mammals have a set point for body temperature. If an individual overheats, it will pant or sweat and seek a cool environment; if an individual is cold, it will shiver, bask in sunlight, or fluff its fur.

- Countercurrent heat exchangers work by placing warm and cool fluids next to each other and running in opposite directions.

✓ The dinosaur *Apatosaurus* is one of the largest terrestrial animals that has ever lived—over 20 m in length and weighing over 20 metric tons. Do you believe that *Apatosaurus* was homeothermic or heterothermic? Explain.

MB MasteringBiology

1. **MasteringBiology Assignments**

   Tutorials and Activities Connective Tissue; Epithelial Tissue; Homeostasis; Muscle Tissue; Nervous Tissue; Overview of Animal Tissues; Surface Area/Volume Relationships; Thermoregulation

   Questions Reading Quizzes, Blue-Thread Questions, Test Bank

2. **eText** Read your book online, search, take notes, highlight text, and more.

3. **The Study Area** Practice Test, Cumulative Test, BioFlix® 3-D Animations, Videos, Activities, Audio Glossary, Word Study Tools, Art

You should be able to . . .

1. True or False: The increase in red blood cell count in tourists visiting Tibet is an example of acclimatization.

2. _____ tissues form the interface between the inside of an animal's body and the environment.

3. As an animal gets larger, which of the following occurs?
a. Its surface area grows more rapidly than its volume.
b. Its volume grows more rapidly than its surface area.
c. Its volume and surface area increase in perfect proportion to each other.
d. Its volume increases, but its total surface area decreases.

4. Which of the following best describes the set point in a homeostatic system?
a. the cells that collect and transmit information about the state of the system
b. the cells that receive information about the state of the system and that direct changes to the system
c. the various components that produce appropriate changes in the system
d. the target or "normal" value of the parameter in question

5. What does it mean to say that an animal is a heterothermic endotherm?
a. Its body temperature varies, but most of its heat is produced by its own tissues.
b. Its body temperature varies because it gains most of its heat from sources outside its body.
c. Its body temperature does not vary, because most of its heat is produced by its own tissues.
d. Its body temperature does not vary, even though it gains most of its heat from sources outside its body.

6. Which of the following is an advantage that ectotherms have over endotherms of the same size?
a. They require much less food.
b. They are less vulnerable to predation events during cold weather.
c. They can remain active in cold weather or at nighttime—when temperatures cool.
d. They have higher metabolic rates and grow much more quickly.

7. Why is epithelium a particularly important tissue in achieving homeostasis with respect to temperature?

8. The metabolic rate of a frog in summer (at 35°C) is about eight times higher than in winter (at 5°C). Compare and contrast the individual's ability to move, exchange gases, and digest food at the two temperatures. During which season will the frog require more food energy, and why?

9. Consider the following:
- absorptive sections of digestive tract
- capillaries
- beaks of Galápagos finches
- fish gills

In each case, how does the structure relate to the function?

10. Why is the surface area/volume ratio different in a small sphere versus a large sphere? If materials diffuse into and out of each sphere, in which of the two will diffusion occur more efficiently relative to the volume? Explain your answer.

11. QUANTITATIVE Consider three cubes with lengths of 1 cm, 5 cm, and 10 cm. Calculate the surface area and the volume of each cube, and plot them on a graph with length on the x-axis and surface area and volume on the y-axis. Explain how the graph shows the relationship between size and surface area/volume ratio.

12. Which of the following is an example of negative feedback? Explain why each is or is not negative feedback.
a. Uterine contractions during childbirth stimulate release of a hormone that stimulates more uterine contractions.
b. A viral infection stimulates the hypothalamus to increase the set point for body temperature.
c. High blood sugar concentrations stimulate the release of the hormone insulin, which stimulates the sugars to move from the blood into cells.
d. The body temperature of the Arabian oryx rises during the hottest parts of the day.

13. When food is scarce, bigger *Geospiza fortis* (medium ground finches) win contests over seeds. Biologists have documented that there is strong natural selection in favor of large body size under these conditions, and that the finch population evolves in response. If so, why aren't *G. fortis* a lot bigger than they are?

14. What data would you need to collect in order to document that adaptation is occurring in a particular population of lizards, as the warming associated with global climate change intensifies?

15. An engineer has to design a system for dissipating heat from a new type of car engine that runs particularly hot. Recall that heat is gained and lost as a function of surface area. Suggest ideas to consider that are inspired by biological structures with exceptionally high surface area/volume ratios.

16. Suppose a friend of yours is trying to decide whether to buy a pet turtle or a pet mouse. In making the decision, all he cares about is how much it will cost to feed the animal. Your friend says he can't decide because the two animals weigh the same and their food costs the same per pound. As a biologist, what's your advice?
a. Buy the turtle, because it is an ectotherm and will therefore consume less food than the mouse.
b. Buy the turtle, because it is an endotherm and will therefore consume less food than the mouse.
c. Buy the mouse, because it is an ectotherm and will therefore consume less food than the turtle.
d. Buy the mouse, because it is an endotherm and will therefore consume less food than the turtle.

43 Water and Electrolyte Balance in Animals

In this chapter you will learn that

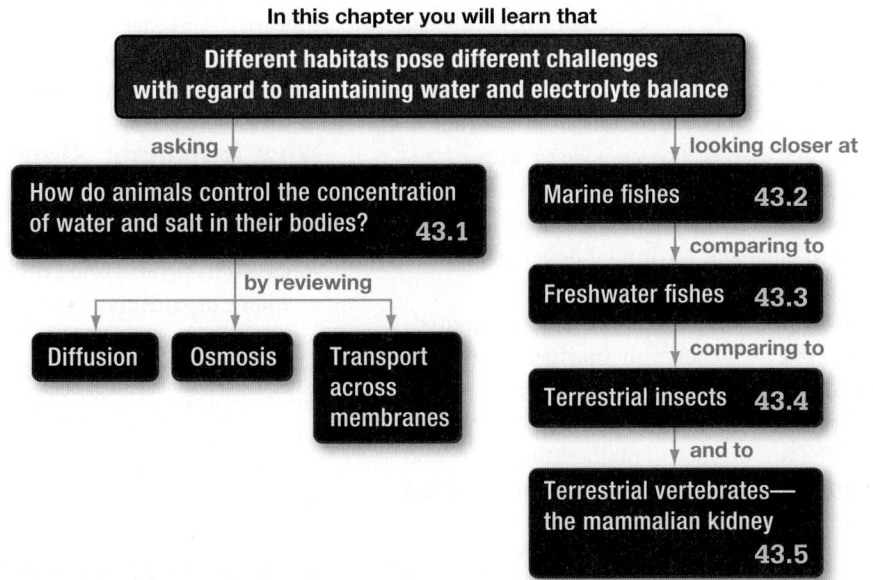

Different habitats pose different challenges
with regard to maintaining water and electrolyte balance

asking

How do animals control the concentration
of water and salt in their bodies? **43.1**

by reviewing

Diffusion

Osmosis

Transport across membranes

looking closer at

Marine fishes **43.2**

comparing to

Freshwater fishes **43.3**

comparing to

Terrestrial insects **43.4**

and to

Terrestrial vertebrates—the mammalian kidney **43.5**

Terrestrial animals lose water every time they breathe, defecate, and urinate. For many animals, drinking is an important way to gain water and achieve homeostasis. This chapter explores how terrestrial and aquatic animals maintain water balance.

This chapter is part of the Big Picture. See how on pages 840–841.

The chemical reactions that make life possible occur in an aqueous solution. If the balance of water and dissolved substances in the solution is disturbed, those chemical reactions—and life itself—may stop. Humans can stay alive for weeks without eating but can survive just three days without drinking water. If hurricanes introduce enough freshwater to the ocean shore to disrupt normal salt concentrations, marine animals die. Maintaining water balance is a matter of life or death.

An animal achieves water balance when its intake of water equals its loss of water. Water balance is an important element in homeostasis—the ability to keep cells and tissues in constant and favorable conditions.

✔ When you see this checkmark, stop and test yourself. Answers are available in Appendix A.

Water balance is intimately associated with sustaining a balanced concentration of electrolytes throughout the body. An **electrolyte** is a compound that dissociates into ions when dissolved in water. Electrolytes got their name because they conduct electrical current.

In many animals, the most abundant electrolytes are sodium (Na^+), chloride (Cl^-), potassium (K^+), and calcium (Ca^{2+}). Cells require precise concentrations of these ions to function normally. In humans, electrolyte imbalances can lead to muscle spasms, confusion, irregular heart rhythms, fatigue, paralysis, or even death.

Water and electrolyte balance is also associated with excretion. Animals produce urine to excrete wastes. Urine contains water, so excretion of urine inevitably leads to water loss. The amount of water lost in the urine of an animal depends both on its hydration state and on the type of wastes it produces.

This chapter is focused on a single question: How do animals maintain water and electrolyte balance in marine, freshwater, and terrestrial environments? Answering it will introduce you to some of the most complex and important homeostatic systems known. You can review the importance of maintaining water and electrolyte balance in the Big Picture of Plant and Animal Form and Function, on pages 840–841.

43.1 Osmoregulation and Excretion

In organisms, electrolytes and water move by two processes—diffusion and osmosis (introduced in Chapter 6).

- **Diffusion** is the net movement of substances from regions of higher concentration to regions of lower concentration.

- **Osmosis** is a special case of diffusion. It is the net movement of water from regions of higher water concentration to regions of lower water concentration, across a selectively permeable membrane.

A membrane that exhibits **selective permeability,** such as a phospholipid bilayer, is a membrane that some solutes can cross more easily than other solutes can. When either diffusion or osmosis occurs, net movement of ions and molecules occurs along their **concentration gradient.**

FIGURE 43.1a illustrates how dissolved substances, or **solutes,** move down their concentration gradients via diffusion across a selectively permeable membrane. When the solutes are randomly distributed throughout the solutions on both sides of the membrane, equilibrium is established. Molecules continue to move back and forth across the membrane at equilibrium, but at equal rates.

As **FIGURE 43.1b** shows, water can also move down its concentration gradient (see Chapter 6). The concentration of dissolved substances in a solution, measured in osmoles[1] per liter, is the solution's **osmolarity**. Because solutes interact with water molecules, an increase in solute concentration reduces the amount of water available for osmosis, effectively lowering water concentration. If dissolved substances are separated by a selectively permeable membrane and the solutes cannot cross that membrane, water moves from areas of lower osmolarity—that is, where solute concentrations are lower and water concentration is higher—to areas of higher osmolarity—where solute concentrations are higher and water concentration is lower.

Now let's examine how osmosis and diffusion affect water and ion balance living in different environments.

What Is Osmotic Stress?

Osmotic stress occurs when the concentration of dissolved substances in a cell or tissue is abnormal. It means that water and solute concentrations are different from their set point.

Many organisms respond to osmotic stress by osmoregulating, just as they respond to heat or cold stress by thermoregulating (see Chapter 42). **Osmoregulation** is the process by which living organisms control the concentration of water and electrolytes in their bodies.

Not all animals encounter osmotic stress. For many marine invertebrates such as sponges, jellyfish, and flatworms, achieving homeostasis with respect to water and electrolyte balance is straightforward. Seawater is a fairly constant ionic and osmotic environment, and it nearly matches the normal electrolyte concentrations found within these animals. These species are **osmoconformers.**

Seawater is **isosmotic** with respect to tissue in these species. Stated another way, solute concentrations inside and outside these animals are equal. Because the tissues of osmoconforming marine invertebrates are isosmotic to seawater, diffusion and osmosis don't alter water and electrolyte balance and induce osmotic stress.

Osmotic Stress in Seawater, in Freshwater, and on Land

In contrast to most marine invertebrates, marine bony fishes are **osmoregulators.** These fishes actively regulate osmolarity inside their bodies to achieve homeostasis.

By osmoregulating, marine bony fishes keep the osmolarity of their tissues lower than seawater. The difference in osmolarity is most important in the gills, which are organs involved in gas exchange. For gas exchange to occur with the environment, the epithelial cells on the surfaces of the gills must be in direct contact with seawater. But because there is a large difference in the concentration of solutes between the inside of each cell and the seawater outside, water tends to flow by osmosis out of the gill epithelium (**FIGURE 43.2**).

Seawater is **hyperosmotic** in comparison to the tissues of marine fishes—the solution inside the body contains fewer solutes than does the solution outside. If the water that marine fishes lose across their gills is not replaced, the cells will shrivel and die. These species face a trade-off between gas exchange and water and electrolyte balance.

Marine fishes replace the lost water by drinking large quantities of seawater. Drinking brings in excess electrolytes, however.

---

[1]The unit osmole is similar to a mole except that it takes into account molecules that dissociate in solution. For example, because NaCl dissociates into Na^+ and Cl^- in solution, adding 1 mole of NaCl is equivalent to adding 2 osmoles.

(a) PROCESS: DIFFUSION

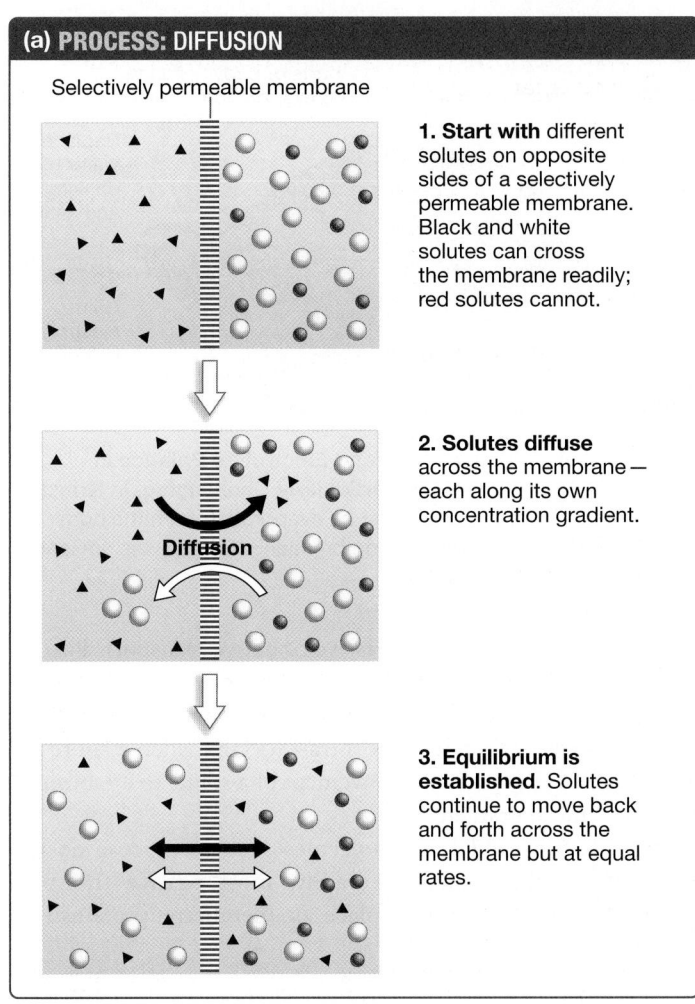

Selectively permeable membrane

1. Start with different solutes on opposite sides of a selectively permeable membrane. Black and white solutes can cross the membrane readily; red solutes cannot.

Diffusion

2. Solutes diffuse across the membrane—each along its own concentration gradient.

3. Equilibrium is established. Solutes continue to move back and forth across the membrane but at equal rates.

(b) PROCESS: OSMOSIS

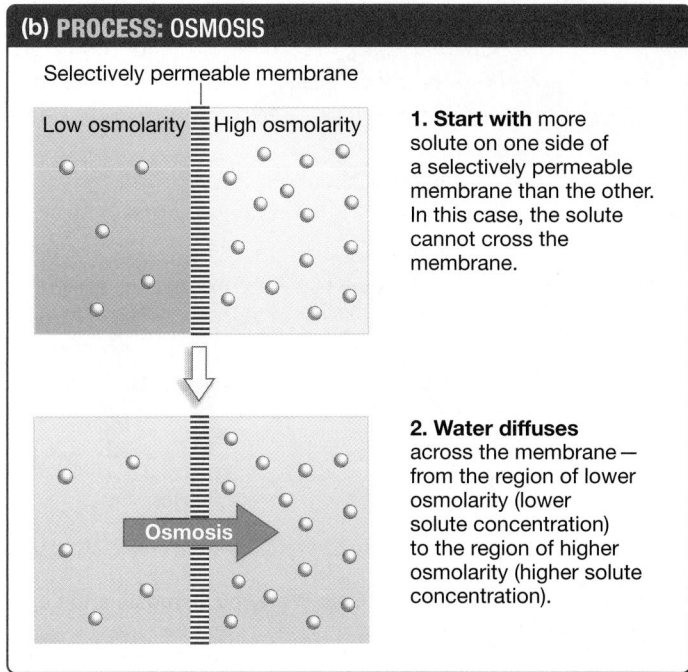

Selectively permeable membrane

Low osmolarity | High osmolarity

1. Start with more solute on one side of a selectively permeable membrane than the other. In this case, the solute cannot cross the membrane.

Osmosis

2. Water diffuses across the membrane—from the region of lower osmolarity (lower solute concentration) to the region of higher osmolarity (higher solute concentration).

FIGURE 43.1 Solutes Move Down a Concentration Gradient via Diffusion; Water Moves Down a Concentration Gradient via Osmosis. (a) Diffusion occurs any time a solute is at higher concentration in one location than another. **(b)** Osmosis is a special case of diffusion involving the movement of water across a selectively permeable membrane.

✔ **QUESTION** Why doesn't the presence of red molecules on the right side of part (a) affect the movement of the black and white molecules?

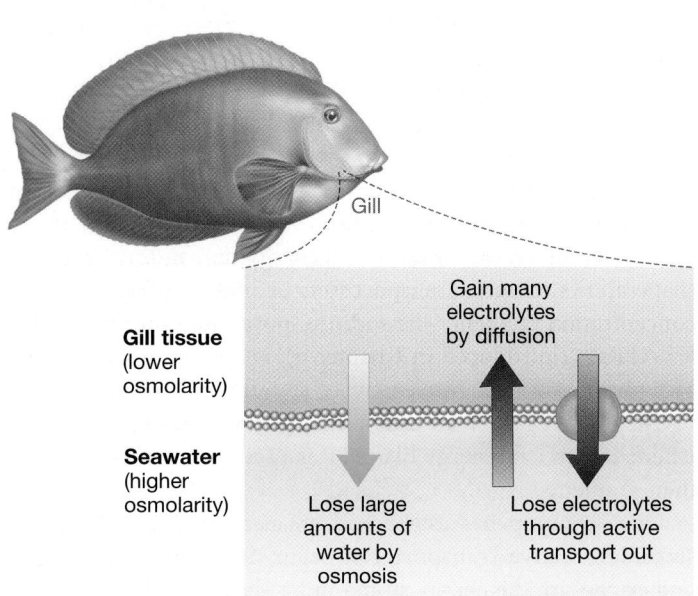

Gill tissue (lower osmolarity)

Seawater (higher osmolarity)

Gain many electrolytes by diffusion

Lose large amounts of water by osmosis

Lose electrolytes through active transport out

FIGURE 43.2 Marine Fishes Lose Water by Osmosis and Gain Electrolytes by Diffusion.

Electrolyte balance is thrown even further out of whack because ions and other solutes diffuse into the gill epithelium, following a concentration gradient from seawater to tissues.

To rid themselves of these excess electrolytes, marine fishes have to actively pump ions out of their bodies and back into seawater, using membrane proteins found in the gill epithelium.

Freshwater fishes osmoregulate in an environment dramatically different from the ocean. Marine fishes are under osmotic stress because they lose water and gain solutes; freshwater fishes are under osmotic stress because they gain water and lose solutes.

Why? In the gills of freshwater fishes, epithelial cells contain more solutes than the solution outside. The freshwater is **hyposmotic** relative to the fishes' tissues. As a result, the fishes gain water via osmosis across the gill epithelium (**FIGURE 43.3**, see page 864), which puts the body under osmotic stress. Just as in marine fishes, there is a trade-off between gas exchange and osmoregulation.

If a freshwater fish does not get rid of incoming water, its cells will burst and the individual will die. To achieve homeostasis and survive, freshwater fishes excrete large amounts of water in their urine and do not drink.

In addition to gaining water, freshwater fishes undergo osmotic stress because ions and other solutes tend to diffuse out of the gill epithelium into the environment, along a concentration

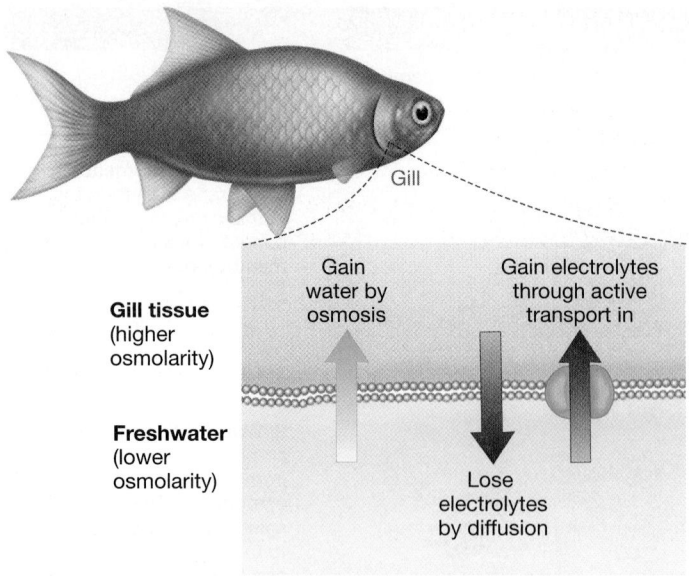

FIGURE 43.3 Freshwater Fishes Gain Water by Osmosis and Lose Electrolytes by Diffusion.

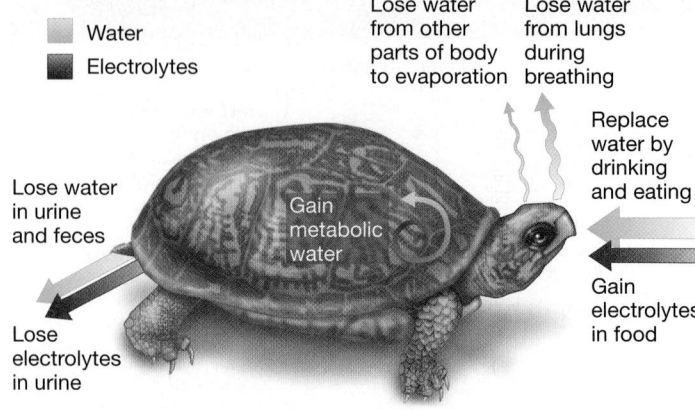

FIGURE 43.4 Maintaining Water and Electrolyte Balance in Terrestrial Environments Is Particularly Challenging. In terrestrial environments, animals lose water by evaporation from the body surface and as water vapor during breathing. Electrolytes are lost primarily in the urine and feces.

gradient. Freshwater animals must replace electrolytes that are lost by obtaining them in food or by actively transporting them from the surrounding water—usually across the gills.

What about land animals? In terms of water balance, terrestrial environments are similar to the ocean. Land animals constantly lose water to the environment, just like many marine animals do. In this case, however, the process involved is not osmosis but evaporation (see Chapter 42).

The epithelial cells that line a turtle's lung and a fruit fly's gas exchange structures have a moist surface, in order to protect the integrity of their plasma membranes and to facilitate diffusion of gases across respiratory epithelia. Because the atmosphere is almost always drier than the wet gas exchange surface, terrestrial animals lose water by evaporation. Once again, there is a trade-off between breathing and balance of water and electrolytes.

Water balance is further complicated because animals lose water in the form of urine, and some terrestrial species lose additional water when they sweat or pant to lower their body temperature (**FIGURE 43.4**). The lost water has to be replaced by drinking, ingesting water in food, or gaining metabolic water—the H_2O produced during cellular respiration (see Chapter 9). The relative importance of each of these methods of replacing water depends on the animal. For example, desert animals like the Arabian oryx (see Chapter 42) do not have access to drinking water throughout much of the year, so they rely more on ingested and metabolic water.

How Do Electrolytes and Water Move across Cell Membranes?

What molecular mechanisms allow animals to cope with the diverse challenges they face in maintaining water and electrolyte balance? You might recall from earlier chapters that solutes move across membranes by passive or active transport. **Passive transport** is driven by diffusion along an electrochemical gradient and does not require an expenditure of energy in the form of ATP. **Active transport,** in contrast, occurs when a source of energy like ATP powers the movement of a solute to establish an electrochemical gradient.

Because ions and large molecules such as glucose do not cross phospholipid bilayers readily, most passive transport and all active transport take place via membrane proteins (see Chapter 6).

- In many cases, passive transport occurs through **channels**—proteins that form a pore, or opening, that selectively admits specific ions or molecules (**FIGURE 43.5a**).

- Passive transport also occurs via **carrier proteins,** which are transmembrane proteins that bind a specific ion or molecule and transport it across the membrane by undergoing a conformational change.

- When solutes move from an area of high to low concentration via channels or carrier proteins, **facilitated diffusion** is said to occur.

Active transport is based on membrane proteins called pumps, which change conformation when they bind ATP or are phosphorylated (**FIGURE 43.5b**). This energy-demanding change in shape allows them to transport ions or molecules against their concentration gradient. The **sodium–potassium pump,** or Na$^+$/K$^+$-ATPase (introduced in Chapter 6), is a very important type of pump in animals. Movement of sodium and potassium ions by Na$^+$/K$^+$-ATPase is an example of **primary active transport,** where a source of energy like ATP is used to move ions against their gradients.

Once a pump has established an electrochemical gradient, **secondary active transport** can occur. Specifically, a **cotransporter** can use the energy stored in an electrochemical gradient across a cell membrane to transport a different solute *against* its

(a) Passive transport is based on diffusion down a gradient.

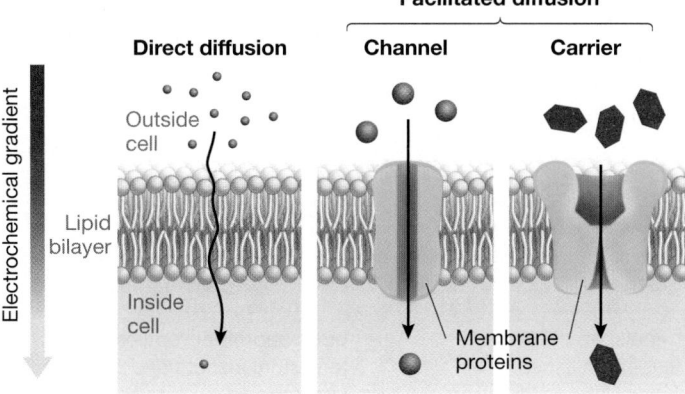

(b) Active transport requires an energy source to move molecules against a gradient.

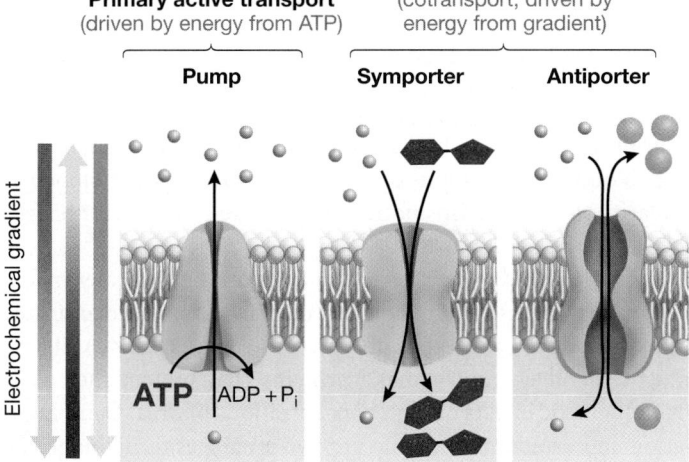

FIGURE 43.5 Mechanisms of Passive and Active Transport.
(a) In passive transport, solutes diffuse along their electrochemical gradient either directly or through a membrane protein.
(b) In primary active transport, a pump uses ATP to move solutes against their electrochemical gradients. In secondary active transport (cotransport), energy stored in an electrochemical gradient for one molecule is used to move another molecule against its gradient.

electrochemical gradient. A cotransporter that moves solutes in the same direction is called a **symporter;** a cotransporter that moves solutes in opposite directions is called an **antiporter.**

How does water move? To date, there are no known mechanisms for actively transporting water across plasma membranes. Instead, cells use pumps to transport ions and set up an osmotic gradient; water then follows by osmosis—often through the specialized membrane channels called **aquaporins** (see Chapter 6). In essence, cells move water by moving solutes.

The body excretes excess solutes, along with waste products, using a urinary system. Because solutes and wastes often must be dissolved in water to be excreted, water balance is fundamentally related to excretion.

Types of Nitrogenous Wastes: Impact on Water Balance

Animal cells contain amino acids and nucleic acids that are used to synthesize proteins, RNA, and DNA. Excess amino acids and nucleic acids can be broken down in catabolic reactions that result in the production of **ammonia** (NH_3). Ammonia is toxic to cells, because at high concentrations it raises the pH of intracellular and extracellular fluids enough to poison enzymes.

Because ammonia must be dissolved in water, excretion of ammonia inevitably leads to water loss. However, the amount of water lost during excretion depends on the method of ammonia excretion.

Forms of Nitrogenous Waste Vary among Species
How do animals get rid of ammonia safely and efficiently? Different species solve the problem in different ways (see **TABLE 43.1**, page 866).

- In freshwater fishes, ammonia is diluted to low concentration and excreted in watery urine.

- In freshwater and marine fishes, ammonia diffuses across the gills into the surrounding water along a concentration gradient.

- In mammals (including humans) and adult amphibians, enzyme-catalyzed reactions convert ammonia to a much less toxic compound called **urea,** which is excreted in urine.

- In terrestrial arthropods, birds, and other reptiles, reactions convert ammonia to **uric acid,** the white, paste-like substance that you have probably seen in bird feces.

Uric acid is a particularly interesting form of nitrogenous waste. Compared with urea and ammonia, uric acid is extremely insoluble in water—which explains why it is so difficult to wash bird droppings off a car. As a result, birds, snakes, lizards, and terrestrial arthropods can get rid of excess nitrogen while losing little water.

Why Do Nitrogenous Wastes Vary among Species?
The type of nitrogenous waste produced by an animal correlates with its lineage—its evolutionary history. For example, mammals excrete urea while reptiles (including birds) and insects excrete uric acid (Table 43.1).

Evolutionary history is not the entire story, however. Waste production is also related to the habitat that a species occupies, and thus the amount of osmotic stress it endures.

- Terrestrial birds conserve water by excreting about 90 percent of their nitrogenous waste as uric acid and only 3–4 percent as ammonia, but aquatic birds such as ducks excrete just 50 percent of their excess nitrogen as uric acid and 30 percent as ammonia.

- Tadpoles are aquatic and excrete ammonia, but many adult frogs and toads are terrestrial and excrete urea.

- Production of urea and uric acid is particularly common in animals—such as reptiles—that live in dry habitats.

Attribute	Ammonia	Urea	Uric Acid
Solubility in water (moles/liter)	high	medium	very low
Water loss (amount required for excretion of waste)	high	medium	very low
Energy cost (amount of ATP required)	low	high	high
Toxicity	high	medium	low
Groups where it is the primary waste	bony fishes, aquatic invertebrates	mammals, amphibians, cartilaginous fishes	birds and other reptiles, most terrestrial insects and spiders (arthropods)
Method of synthesis	product of breakdown of amino acids and nucleic acids	synthesized in liver, starting with amino groups from amino acids	synthesis starts with amino acids and nucleic acids
Method of excretion	in urine, and diffuses across gills	in urine (mammals); diffuses across gills (sharks)	with feces

To make sense of these observations, biologists point out that there is a fitness trade-off between the energetic cost of excreting each type of waste and the benefit of conserving water. Ammonia excretion requires a large water loss but little energy expenditure, because the molecule isn't processed by enzymes. Uric acid excretion, in contrast, requires almost no loss of water but a sophisticated series of enzyme-catalyzed, energy-demanding reactions. Different trade-offs are favored in different environments.

Now that you have a basic understanding of the osmoregulatory and excretory challenges facing animals, let's take a closer look at how marine, freshwater, and terrestrial animals maintain water and electrolyte homeostasis.

43.2 Water and Electrolyte Balance in Marine Fishes

Marine fishes experience severe osmotic stress because they live in water with a very high osmolarity. Since their divergence over 400 million years ago, marine bony and cartilaginous fishes have evolved distinct strategies for dealing with osmotic stress.

Osmoconformation versus Osmoregulation in Marine Fishes

Marine bony fishes are osmoregulators. Recall that they maintain a lower blood osmolarity than seawater by drinking to replace water lost via osmosis, and by actively transporting electrolytes out of the body. This process comes with a significant energetic cost.

Cartilaginous fishes, such as sharks, have a different strategy. They are **osmoconformers**—their blood osmolarity nearly matches that of the sea water. However, the composition of their blood is quite different from seawater. Sharks have low

concentrations of ions in their blood, but they maintain relatively high blood concentrations of urea. This increases their blood osmolarity so that it is isosmotic with seawater. The result? Sharks lose less water by osmosis.

If sharks avoid water loss to the surrounding seawater by osmoconforming, then why don't all marine fishes osmoconform? The answer lies in a biological trade-off (see Chapter 42). Sharks must expend energy to make proteins that protect their cells from the toxic effects of high concentrations of urea. Osmoregulation and osmoconformation are two strategies for living in the ocean, and each has its own costs and benefits.

Even though they are osmoconformers, sharks still maintain relatively low concentrations of salt (NaCl) in their blood. Sharks need to excrete salt because sodium and chloride ions diffuse into their gill cells from seawater, down the concentration gradients for the ions. Research on the molecular mechanism of salt excretion in sharks revealed two key points:

1. The mechanism is general. The salt-excreting system discovered in sharks is found in a wide array of species, including *Homo sapiens*. It is functioning in your kidneys, right now.

2. The mechanism represents a critically important concept in physiology. Plant and animal cells use active transport to set up a strong electrochemical gradient for one ion—typically Na^+ in animals and H^+ in plants. The sodium or proton gradient is then used to transport a wide array of other substances, without further expenditure of energy.

This is fundamental stuff. Let's dig in.

How Do Sharks Excrete Salt?

Research on salt excretion in sharks focused on an organ called the **rectal gland,** which secretes a concentrated salt solution. To determine how this gland works, researchers studied it in vitro—meaning outside the shark's body, in a controlled laboratory environment. The basic approach was to dissect rectal glands, immerse them in a solution with a defined composition and osmolarity, and analyze the fluid that the rectal gland produced in response.

Early experiments showed that normal salt excretion occurred only if the solution in the rectal gland contained ATP. This result supported the hypothesis that salt excretion involves active transport. Ions can be concentrated only if they are actively transported against a concentration gradient. The question was, How does concentration occur?

The Role of Na$^+$/K$^+$-ATPase An energy-demanding mechanism for salt excretion implies that a protein in the plasma membrane of epithelial cells is actively pumping Na$^+$, Cl$^-$, or both. The best-characterized candidate was the Na$^+$/K$^+$-ATPase.

To test the hypothesis that the sodium–potassium pump is involved in salt excretion by sharks, biologists used a plant defense compound called **ouabain** (pronounced *WAA-bane*). This molecule is toxic to animals because it binds to Na$^+$/K$^+$-ATPase and prevents it from functioning.

Just as predicted, rectal glands that are treated with ouabain stop producing a concentrated salt solution. This was strong evidence that Na$^+$/K$^+$-ATPase is essential for salt excretion.

A Molecular Model for Salt Excretion Subsequent work has shown that salt excretion in sharks is a multistep process, summarized in **FIGURE 43.6**.

1. Na$^+$/K$^+$-ATPase pumps sodium ions out of epithelial cells across the basolateral surface, into the extracellular fluid. The pump creates an electrochemical gradient favoring the diffusion of Na$^+$ into the cell. This "master gradient" allows the cell to transport other ions without an additional expenditure of energy.

2. Na$^+$, Cl$^-$, and K$^+$ all enter the cell by secondary active transport through a cotransporter, powered by the Na$^+$ master gradient. Note that this master gradient causes Na$^+$ to diffuse into the cell *along* its electrochemical gradient, allowing Cl$^-$ and K$^+$ to move into the cell *against* their electrochemical gradients. Movement of Cl$^-$ and K$^+$ is possible only because these ions move through a cotransporter with Na$^+$.

3. As Cl$^-$ builds up inside the cell, Cl$^-$ diffuses down its concentration gradient into the lumen of the gland through a chloride channel located in the apical membrane.

4. Following their electrochemical gradient, Na$^+$ ions diffuse into the lumen of the gland through spaces between the cells.

In many animals, epithelial cells that transport sodium and chloride ions contain the same combination of membrane proteins found in the shark rectal gland. These species include

- marine birds and reptiles that drink seawater and excrete NaCl via salt glands in their nostrils;
- marine fishes that excrete salt from their gills (see Figure 43.2);
- mammals that transport salt in their kidneys.

Research on the shark rectal gland also produced an unforeseen benefit for biomedical research. Several years after the shark chloride channel was characterized, investigators identified a human protein called cystic fibrosis transmembrane regulator

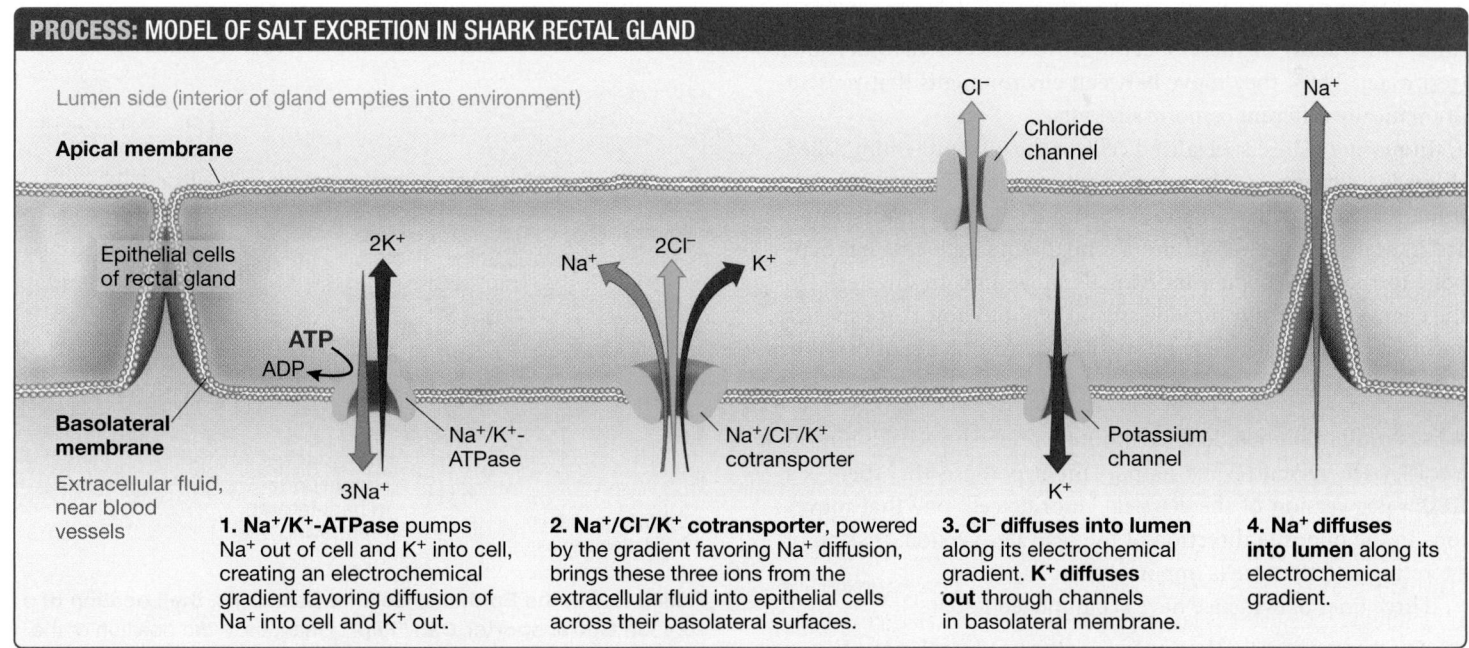

PROCESS: MODEL OF SALT EXCRETION IN SHARK RECTAL GLAND

FIGURE 43.6 The Shark Rectal Gland Rids the Body of Excess Salt.

✔**QUESTION** Which of these membrane proteins are involved in (1) primary active transport, (2) secondary active transport, or (3) passive transport?

(CFTR). Cystic fibrosis is the most common genetic disease in populations of northern European extraction (see Chapter 6). Although the disease was known to be associated with defects in the CFTR protein, no one knew what the molecule did.

When investigators realized that the amino acid sequence of CFTR is 80 percent identical to that of the shark chloride channel, it was their first hint that CFTR is involved in Cl^- transport. Subsequent studies supported the hypothesis that cystic fibrosis results from a defect in a chloride channel. With this result, studies on water and electrolyte balance in sharks shed light on an important human disease.

43.3 Water and Electrolyte Balance in Freshwater Fishes

Research on the shark rectal gland and the gills of marine bony fishes has uncovered the molecular mechanisms of salt balance in marine fishes. How do freshwater fishes achieve homeostasis with respect to electrolytes?

How Do Freshwater Fishes Osmoregulate?

Recall from Figure 43.3 that freshwater fishes have to cope with an osmotic stress that is opposite the challenge facing marine fishes. Freshwater fishes lose electrolytes across their gill epithelium by diffusion across a concentration gradient. To maintain homeostasis, they have to actively transport ions back into the body across the gill epithelium. How do they do this?

Salmon and Sea Bass as Model Systems To understand how freshwater fishes gain electrolytes, researchers have focused on sea bass and several species of salmon. Over the course of a lifetime, individuals of these species move between seawater and freshwater. Thus, they move between environments that present dramatically different osmotic stressors.

In marine fishes, specialized cells in the gill epithelium called chloride cells move salt using the combination of membrane proteins illustrated in Figure 43.6. When sea bass and salmon are in seawater, these cells are abundant and active. What happens to these cells when individuals move into freshwater? Do the changes that occur provide any insight into how these species acclimatize to the new environment and avoid dying of osmotic stress?

A Freshwater Chloride Cell? Although research is continuing at a brisk pace, recent results support the hypothesis that there is a freshwater version of the classical chloride cell: one that moves ions in the opposite direction of the seawater version. Instead of excreting salt, these cells import it.

Three lines of evidence have accumulated to date:

1. *Osmoregulatory cells may be in different locations.* Young salmon taken from freshwater versus seawater have chloride cells in different locations on the gills. Adult salmon

from freshwater versus seawater show the same pattern, and similar changes have been observed in other fish species that switch between freshwater and seawater habitats. The pattern suggests that when the nature of osmotic stress changes, the nature of the gill epithelium changes. Specifically, active pumping of ions takes place in a different population of cells in seawater versus freshwater.

2. *Different forms of Na^+/K^+-ATPase may be activated.* The salmon genome contains genes for several different forms of the Na^+/K^+-ATPase. There is now strong evidence that different forms are activated when individuals are in seawater versus freshwater.

3. *The orientation of key transport proteins "flips."* In sea bass, researchers have been able to stain epithelial cells to determine the location of the cotransporter illustrated in Figure 43.6—the one that brings Na^+, Cl^-, and K^+ into the cell. When individuals are in seawater, the protein is located in the basolateral side of chloride cells. But when individuals are in freshwater, the protein is located in the apical side (**FIGURE 43.7**).

Taken together, the data suggest that freshwater fishes have a freshwater version of the chloride cell, with different forms of Na^+/K^+-ATPase and transporters that result in the import rather than export of ions. Identifying the mechanisms of electrolyte uptake in freshwater fishes is an important challenge for researchers who want to know how aquatic organisms cope with osmotic stress.

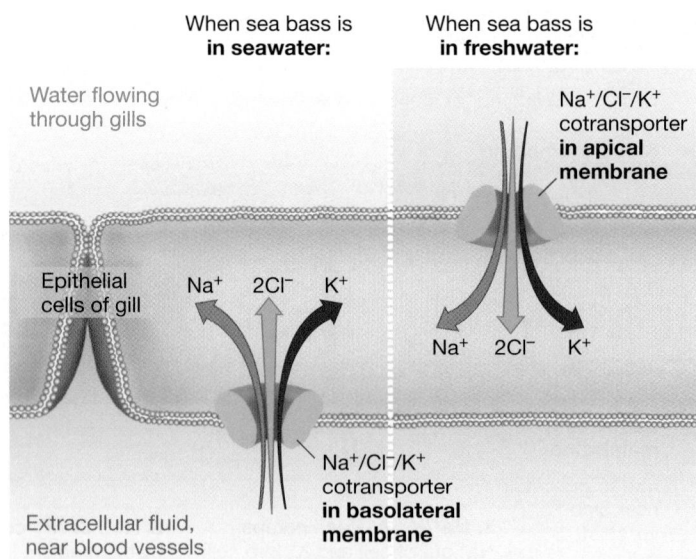

FIGURE 43.7 In the Epithelial Cells of Bass Gills, the Location of a Key Ion Cotransporter Can "Flip." Changes in the position of the $Na^+/Cl^-/K^+$ cotransporter help sea bass deal with osmotic stress in both seawater and freshwater environments. (Other membrane channels and pumps are present but not shown here.)

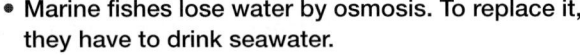
If you understand that . . .

- Marine fishes lose water by osmosis. To replace it, they have to drink seawater.
- Marine fishes have to rid themselves of salt. They gain salt when they drink seawater or when sodium and chloride ions diffuse into their cells along a concentration gradient.
- Freshwater fishes gain water by osmosis. They have to rid themselves of excess water by urinating.
- Freshwater fishes lose electrolytes to the surrounding water by diffusion. They gain electrolytes in their food and by active transport from the surrounding water.

✔ You should be able to . . .

Predict what happens to the osmolarity of a freshwater fish's tissues when epithelial cells in its gills are treated with ouabain—a molecule that poisons Na^+/K^+-ATPase.

Answers are available in Appendix A.

43.4 Water and Electrolyte Balance in Terrestrial Insects

By studying extreme situations or unusual organisms, biologists can often gain insight into how organisms cope with more moderate environments. In studies on the molecular mechanisms of water and electrolyte balance in terrestrial insects, the most valuable model organisms have been the desert locust and a common household pest called the flour beetle. (You may have seen the larvae of flour beetles, called mealworms, in bags of flour that were not shut tightly enough to keep adults from entering and breeding.)

Desert locusts and flour beetles live in environments where osmotic stress is severe. These insects rarely, if ever, drink—simply because little or no water is available in the habitats they occupy.

How do they maintain water and electrolyte balance? The answer has two parts: They minimize water loss from their body surface, and they carefully regulate the amount of water and electrolytes that they excrete in their urine and feces. Let's look at each issue in turn.

How Do Insects Minimize Water Loss from the Body Surface?

Terrestrial animals breathe by exposing an extremely thin layer of epithelium to the atmosphere (see Chapter 45). Oxygen diffuses into this epithelium, and carbon dioxide diffuses out. But water constantly leaks across the thin respiratory surface and is lost to the atmosphere via evaporation.

Evaporation from the body surface itself is another threat—a particular challenge to insects, because they are small. Small organisms have a high surface area/volume ratio (see Chapter 42). Insects have a relatively large surface area from which to lose water but a small volume in which to retain it.

How do desert locusts, flour beetles, and other insects minimize water loss during gas exchange? In these species, gas exchange occurs across the membranes of epithelial cells that line the **tracheae,** an extensive system of tubes. The insect tracheal system connects with the atmosphere at openings called **spiracles** (**FIGURE 43.8a**). Muscles just inside each spiracle open or close the pore, much as guard cells open or close the pores in plant leaves and stems.

When investigators manipulated insects called *Rhodnius* so that their spiracles stayed open and placed the animals in a dry environment, the insects died within three days. These data support the hypothesis that the ability to close spiracles is an important adaptation for minimizing water loss during respiration. If an insect is under osmotic stress, it may be able to close its spiracles and wait until conditions improve before resuming activity.

FIGURE 43.8b shows how insects minimize evaporation from the surface of their bodies. This diagram is a cross-sectional view of the exoskeleton of an insect, which consists of a tough,

(a) Spiracles can be closed to minimize water loss from tracheae.

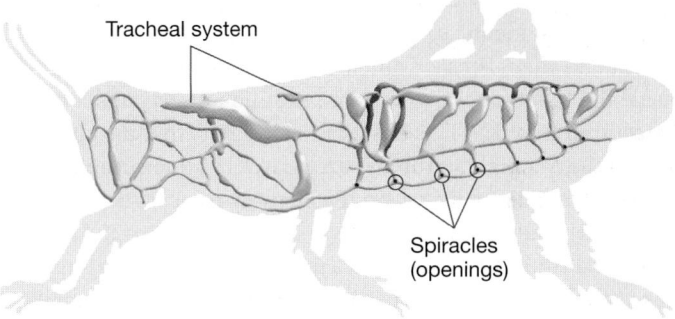

(b) Except at spiracles, the insect body is covered with wax.

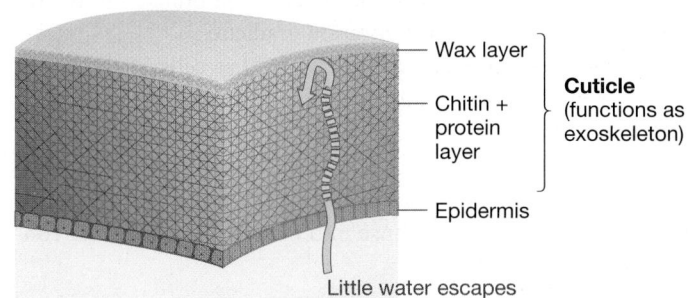

FIGURE 43.8 In Desert Locusts, Adaptations Limit Water Loss during Respiration and from the Body Surface.

✔ **QUESTION** In desert grasshoppers, spiracles are located in a row along the underside of the abdomen. Why would this location help minimize water loss?

nitrogen-containing polysaccharide called chitin and layers of protein, with a layer of wax on the outside. This combination of chitin, protein, and wax is known as the **cuticle.**

Recall that waxes, a type of lipid, are highly hydrophobic and thus highly impermeable to water (see Chapter 6). Researchers who removed the wax from insect exoskeletons have confirmed that the rate of water loss from the body surface increases sharply. Based on this observation, the wax layer is interpreted as an adaptation that minimizes evaporative water loss.

For insects, minimizing water loss is only half the battle in avoiding osmotic stress. To maintain homeostasis, insects must also carefully regulate the composition of a blood-like fluid called **hemolymph.** Hemolymph is pumped by the heart and transports electrolytes, nutrients, and waste products.

To maintain water and electrolyte balance, insects rely on excretory organs called **Malpighian tubules,** and on their hindgut—the posterior portion of their digestive tract (**FIGURE 43.9a**).

Filtrate Forms in the Malpighian Tubules As Figure 43.9a shows, Malpighian tubules have a large surface area, are in direct contact with the hemolymph, and empty into the hindgut. The Malpighian tubules are responsible for forming a **filtrate** from the hemolymph. This "pre-urine" then passes into the hindgut, where it is processed and modified before excretion.

To explore how the Malpighian tubules work, biologists collected filtrate from the lumen of Malpighian tubules in mealworms and compared its composition with that of hemolymph from the same individuals. The two solutions were roughly isosmotic, but not identical.

How do they differ? Compared to hemolymph, the solution inside the tubules has a lower concentration of Na^+ and a higher concentration of K^+. This observation suggests that the epithelial cells in the Malpighian tubules are relatively impermeable to sodium ions but contain a pump that actively transports potassium ions into the tubules.

To test this hypothesis, researchers dissected the tubules, rinsed them, and bathed their interior and exterior in a solution containing a known concentration of potassium ions. When they measured the concentration of electrolytes on either side of the membrane over time, they found that K^+ accumulated in the tubule lumen, against its concentration gradient.

This result supported the hypothesis that cells in the membranes of Malpighian tubules contain pumps that actively transport potassium ions into the lumen of the organ. Subsequent work has shown that a high concentration of potassium ions brings water into the tubules by osmosis. Other electrolytes and nitrogenous wastes then diffuse into the filtrate along their concentration gradients.

The Hindgut: Selective Reabsorption of Electrolytes and Water
The filtrate that accumulates inside the Malpighian tubules flows into the hindgut, where it joins material emerging from the digestive tract. If an insect is osmotically stressed due to a shortage of electrolytes and water, electrolytes and water from the filtrate are reabsorbed in the hindgut and returned to the hemolymph,

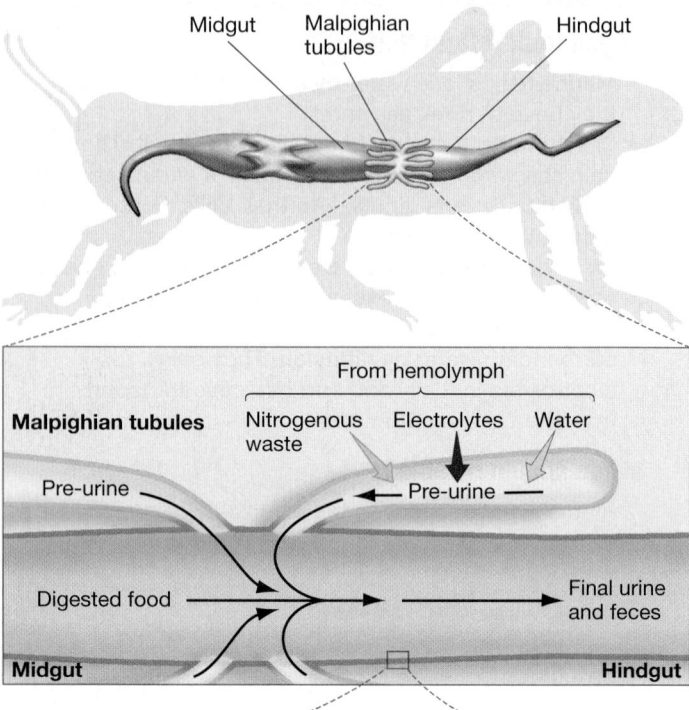

(a) Malpighian tubules produce an isosmotic pre-urine.

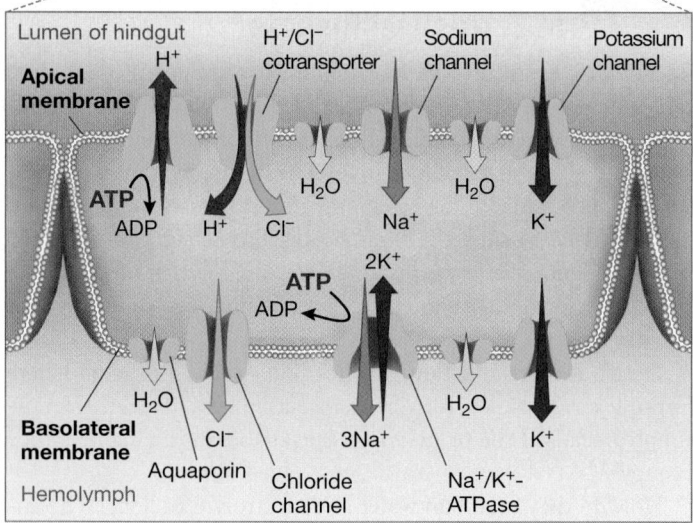

(b) Under osmotic stress, the hindgut reabsorbs electrolytes and water to form a hyperosmotic urine.

FIGURE 43.9 In Insects, Urine Forms in the Malpighian Tubules and Hindgut. (a) The isosmotic filtrate that forms in the Malpighian tubules empties into the hindgut. **(b)** In the hindgut, the primary driving forces for reabsorption of electrolytes and water are H^+ pumps in the apical membrane and Na^+/K^+-ATPases in the basolateral membrane.

while the uric acid remains in the hindgut. Reabsorption results in formation of hyperosmotic final urine, conservation of water, and efficient elimination of nitrogenous wastes.

In desert locusts, flour beetles, and other species in extremely dry environments, 80 to 95 percent of the water in the filtrate is reabsorbed and kept inside the body. The ability to recover this water allows these insects to live in dry habitats such as deserts and flour bins. How does reabsorption happen?

The mechanism involves a series of specific membrane pumps and channels, not unlike the system found in the chloride cells of fishes. To study it, researchers positioned the hindgut epithelium from a desert locust as a sheet dividing two solutions. They manipulated electrolyte concentrations on either side of the hindgut wall and measured changes in the solutions over time.

For example, when investigators removed K^+ and Na^+ from the solution on the lumen side of the organ, water reabsorption stopped. These data established that the hindgut's ability to recover water from urine depends on ion movement: The epithelial cells in the hindgut transport ions out of the filtrate and into the hemolymph. Water follows by osmosis, forming concentrated urine.

How do the ions move? By poisoning the experimental membranes with ouabain, biologists confirmed that Na^+/K^+-ATPase is involved in moving ions out of the lumen and into the hemolymph. Ouabain-treated membranes continued to transport Cl^-, however. Years of experiments on the locust hindgut resulted in the model in **FIGURE 43.9b**:

- H^+ is pumped into the hindgut lumen, creating an electrical gradient that favors movement of K^+ into the cells through potassium channels. Water follows via osmosis.

- H^+ diffuses along its gradient from the lumen into the cell through a H^+/Cl^- cotransporter, bringing Cl^- into the cell against its gradient.

- In the basolateral membrane, Na^+/K^+-ATPase sets up electrochemical and osmotic gradients that favor diffusion of Na^+ from the hindgut lumen into the epithelial cell, and diffusion of Cl^- and K^+ from the epithelial cell into the hemolymph. Water moves into the hemolymph by osmosis.

Regulating Water and Electrolyte Balance: An Overview Several general principles that have emerged from studies of insect excretion turn out to be relevant to vertebrate systems as well:

- Water moves only by osmosis—it is not pumped directly. Water moves between cells or body compartments via osmotic gradients that are set up by the active transport of ions.

- The formation of the filtrate is not particularly selective. Most of the molecules present in the hemolymph are also present in the Malpighian tubules.

- In contrast to filtrate formation, reabsorption is highly selective. The protein pumps and channels involved in reabsorption are highly specific for certain ions and molecules. Waste products do not pass through the hindgut membrane. Instead, they remain in the hindgut and are eliminated along with the feces. Only valuable ions and molecules are reabsorbed.

- In contrast to filtrate formation, reabsorption is tightly regulated. The membrane pumps and channels involved in reabsorption are activated and deactivated in response to osmotic stress. If an insect is dehydrated, virtually all of the water in the filtrate is reabsorbed. But if it has plenty to drink, reabsorption

does not occur and the urine is watery and hyposmotic to the individual's hemolymph. The system is dynamic and allows precise control over water and electrolyte balance.

Given the success of insects in terms of the numbers of species and individuals and the array of habitats they occupy, it is clear that their systems for maintaining water and electrolyte balance are remarkably effective.

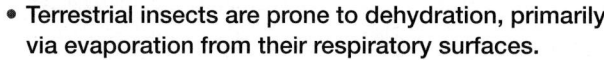

check your understanding

C Y U

If you understand that . . .

- Terrestrial insects are prone to dehydration, primarily via evaporation from their respiratory surfaces.
- Terrestrial insects have a cuticle, respiratory system, and excretory system that are adapted to conserve water.
- In terrestrial insects, urine production begins with formation of a filtrate that is isosmotic with hemolymph, followed by selective and tightly regulated reabsorption of ions, nutrients, and water.

✓ **You should be able to . . .**

Explain how the following traits are involved in water retention:

1. Excretion of ammonia in the form of uric acid.
2. Selective reabsorption of electrolytes in the hindgut.

Answers are available in Appendix A.

43.5 Water and Electrolyte Balance in Terrestrial Vertebrates

With respect to water loss, terrestrial vertebrates face the same hazards that terrestrial insects do. Crocodiles, turtles, lizards, frogs, birds, and mammals lose water from their body surfaces, from the surface of their lungs every time they breathe, and in their urine. Electrolytes are also lost in urine and, in some species, in sweat. To replace the water they lose, most terrestrial vertebrates drink. They also ingest electrolytes in food.

In land-dwelling vertebrates, osmoregulation occurs primarily through events that take place in the key organ of the urinary system, the **kidney.** The kidney is responsible for water and electrolyte balance, as well as the excretion of nitrogenous wastes. In terms of function, it is analogous to the Malpighian tubules and hindgut of insects. Let's take a closer look at the kidney in mammals.

The Structure of the Mammalian Kidney

Mammalian kidneys occur in pairs and tend to be bean shaped. A large blood vessel called the renal artery brings blood that contains nitrogenous wastes into the organ; the renal vein is a large blood vessel that carries "clean" blood away.

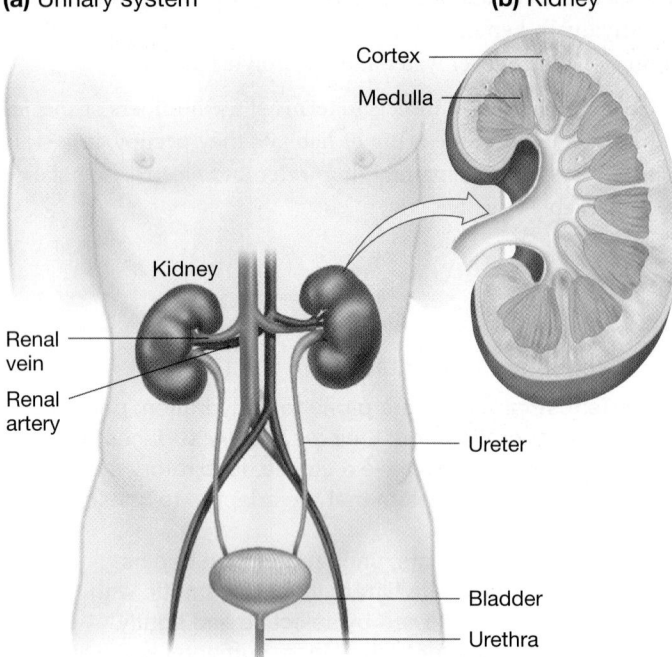

(a) Urinary system   **(b)** Kidney

FIGURE 43.10 **Anatomy of the Human Urinary System and Kidney.**
(a) In mammals, kidneys are paired and are located near the spinal
column. **(b)** The kidney has an outer region called the cortex and an
inner area called the medulla.

The urine that forms in the kidney is transported via a long tube
called the **ureter** to a storage organ, the **bladder.** From the bladder,
urine is transported to the body surface through the **urethra** and
then excreted. In most vertebrates, the kidneys are located near the
dorsal (back) side of the body. **FIGURE 43.10a** summarizes the parts
of the urinary system.

Most of the kidney's mass is made up of small structures called
nephrons. The **nephron** is the basic functional unit of the kidney.
The work involved in maintaining water and electrolyte balance
occurs in the nephron.

Most of the approximately 1 million nephrons in a human
kidney are located in the outer region of the organ, or **cortex**
(**FIGURE 43.10b**). But some nephrons extend from the cortex into
the kidney's inner region, or **medulla.**

The Function of the Mammalian Kidney: An Overview

The nephron shares important functional characteristics with
the insect excretory system:

- Water cannot be transported actively—it moves only by osmosis.

- To move water, cells in the kidney set up strong osmotic
  gradients.

- By regulating these gradients and specific channel proteins,
  kidney cells exert precise control over loss or retention of water and electrolytes.

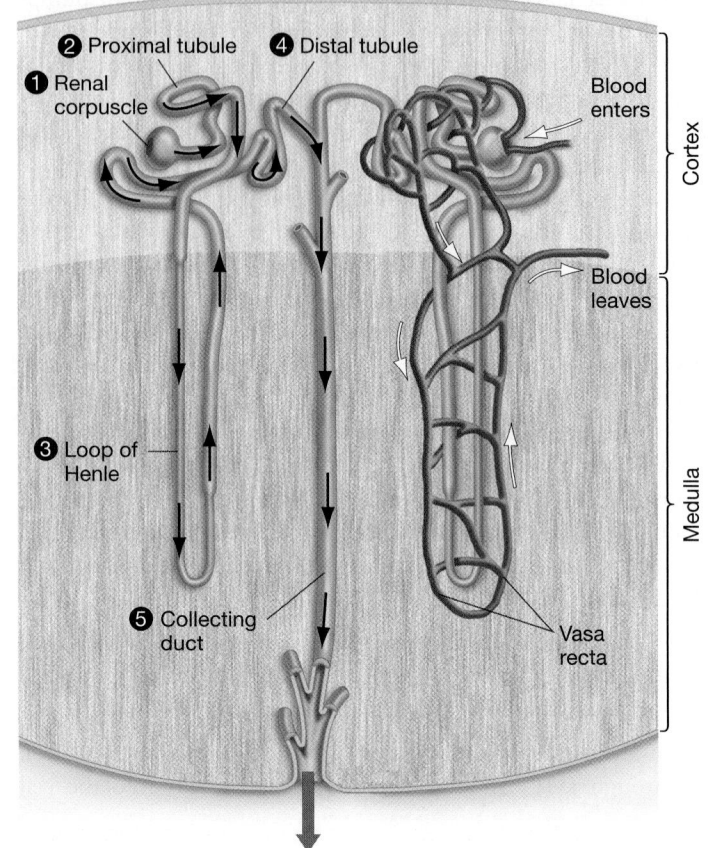

(a) The structure of the
nephron and collecting duct

(b) Blood vessels serve
each nephron.

Final urine to ureter

FIGURE 43.11 **A Nephron Has Four Major Parts, Empties into a
Collecting Duct, and Is Served by Blood Vessels.** Urine formation
begins in the renal corpuscle and ends in the collecting duct.

FIGURE 43.11a provides a detailed view of the nephron. Note
that it has four major regions and is closely associated with a tube
called the collecting duct. **FIGURE 43.11b** illustrates a second key
point about the anatomy of the nephron: It is served by blood
vessels that wrap around each of its four regions.

The four major nephron regions and the collecting duct each
have a distinct function:

1. The *renal corpuscle* filters blood, forming a filtrate or "pre-
   urine" consisting of ions, nutrients, wastes, and water.

2. The *proximal tubule* has epithelial cells that reabsorb nutrients, valuable ions, and water from the filtrate into the
   bloodstream.

3. The *loop of Henle* establishes a strong osmotic gradient in the
   interstitial fluid surrounding the loop, with osmolarity increasing as the loop descends.

4. The *distal tubule* reabsorbs ions and water in a regulated
   manner—one that helps maintain water and electrolyte balance according to the body's needs.

5. The *collecting duct* may reabsorb more water to maintain homeostasis. In addition, urea leaves the base of the collecting

duct and contributes to the osmotic gradient in the interstitial fluid set up by the loop of Henle.

The blood vessels associated with the nephron play a key role as well: They bring "dirty" blood into the nephron and then take away the molecules and ions that are reabsorbed from the initial filtrate.

Now let's delve into the details. The sections that follow trace the flow of material through each component of the nephron and out of the collecting duct.

Filtration: The Renal Corpuscle

In terrestrial vertebrates, urine formation begins in the **renal corpuscle** (literally, "kidney-little-body"). Notice in Figure 43.11a that the nephron consists of a tube that is closed at one end and open at the other. The closed end is the beginning of the nephron; the open end is the terminus of the collecting duct.

As **FIGURE 43.12a** shows, the closed end of the nephron forms a capsule that encloses a cluster of tiny blood vessels, or capillaries. These vessels bring blood to the nephron from the renal artery. Collectively, the cluster of capillaries is called the **glomerulus** ("ball of yarn"). The region of the nephron that surrounds the glomerulus is named **Bowman's capsule.** Together, the glomerulus and Bowman's capsule make up the renal corpuscle.

FIGURE 43.12b illustrates a key feature of the glomerular capillaries: They have large pores, or openings. In addition, they are surrounded by unusual cells whose membranes fold into a series of slits and ridges.

The structure of the renal corpuscle allows it to function as a **filtration** device. Water and small solutes from the blood pass through the pores and slits into the Bowman's capsule. Filtration is based on size: Proteins, cells, and other large components of blood do not fit through the pores and do not enter the nephron. They remain in the blood instead.

Stated another way, urine formation starts with a size-selective filtration step—with blood pressure supplying the force required to perform filtration. In vertebrates, blood is under higher pressure than the surrounding tissues because it is pumped by the heart through a closed system of vessels. This pressure is enough to force water and small solutes through the pores in the glomerulus, so the renal corpuscle strains large volumes of fluid without expending energy in the form of ATP.

✔ If you understand this concept, you should be able to describe the contents of blood on one side of the filter (glomerulus) and the contents of the filtrate on the other side.

It is critical to note two additional facts about the filtration step in urine formation:

1. The renal corpuscles of a human kidney are capable of producing about 180 liters of filtrate per day. This is an impressive volume—think of 180 one-liter bottles of soft drink arranged on a supermarket shelf.

2. About 99 percent of the filtrate is reabsorbed—only a tiny fraction of the original volume is actually excreted as urine.

Filtering large volumes from the blood allows wastes to be removed effectively; pairing this process with selective

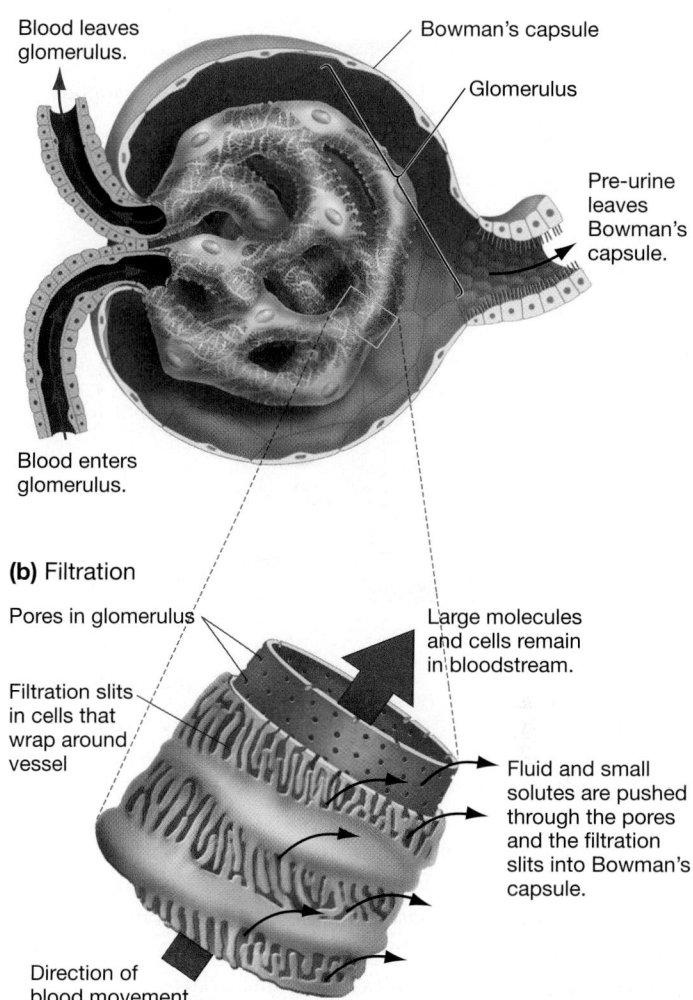

(a) Anatomy of the renal corpuscle

Blood leaves glomerulus.

Bowman's capsule

Glomerulus

Pre-urine leaves Bowman's capsule.

Blood enters glomerulus.

(b) Filtration

Pores in glomerulus

Large molecules and cells remain in bloodstream.

Filtration slits in cells that wrap around vessel

Fluid and small solutes are pushed through the pores and the filtration slits into Bowman's capsule.

Direction of blood movement

FIGURE 43.12 Urine Formation Begins When Blood Is Filtered in the Renal Corpuscle. (a) The renal corpuscle consists of Bowman's capsule and the glomerulus. **(b)** The capillaries in the glomerulus have pores and are surrounded by cells that have filtration slits. Blood pressure forces water and small molecules out of the capillaries, through the slits, and into Bowman's capsule.

reabsorption allows waste excretion to occur with a minimum of water and nutrient loss.

Reabsorption: The Proximal Tubule

Where does filtrate reabsorption begin? Filtrate leaves Bowman's capsule and enters a convoluted structure called the **proximal tubule.** The filtrate inside this tubule contains water and small solutes such as urea, glucose, amino acids, vitamins, and electrolytes. Some of these molecules are waste products; others are valuable nutrients.

Active Transport Occurs in Epithelial Cells The epithelial cells of the proximal tubule have a prominent series of small projections, called **microvilli** ("little shaggy hairs"), facing the lumen. The microvilli greatly expand the surface area of this epithelium.

(a) Microvilli expand surface area of lumen of proximal tubule.

(b) Model of selective reabsorption in proximal tubules

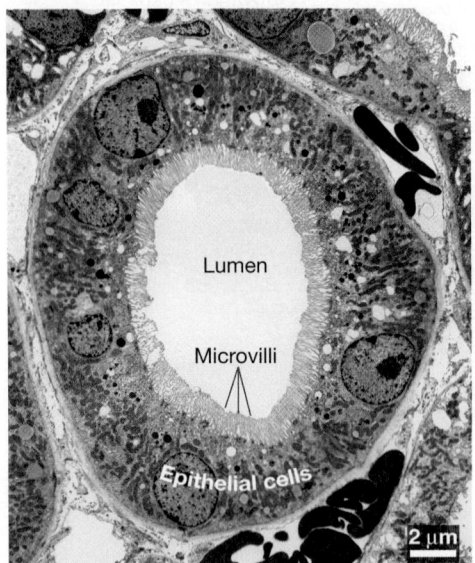

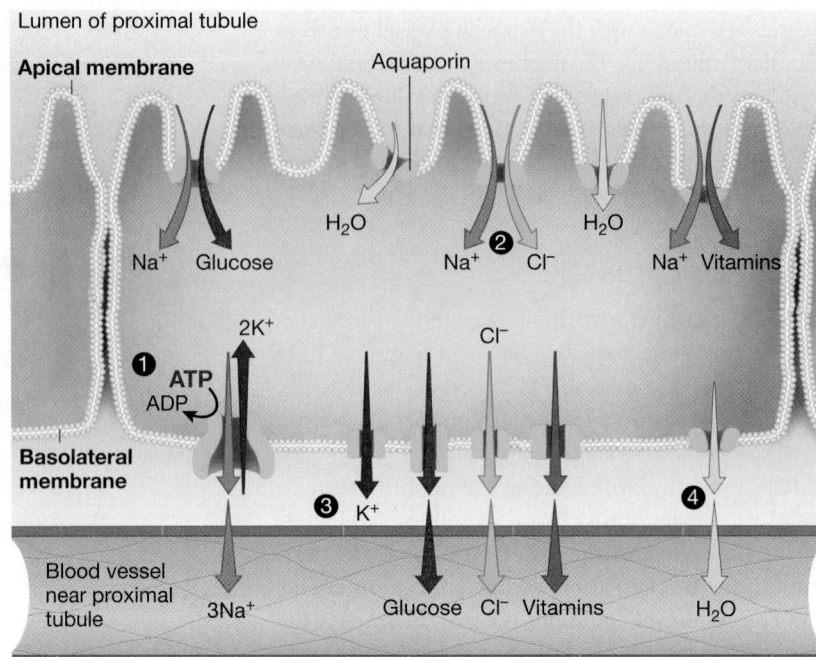

FIGURE 43.13 Water, Electrolytes, and Nutrients Are Reabsorbed in the Proximal Tubule.

A large surface area provides space for membrane proteins that act as pumps, channels, and cotransporters (**FIGURE 43.13a**).

Epithelial cells in the proximal tubule are also packed with mitochondria, which suggests that ATP-demanding active transport is occurring. Based on these observations, biologists hypothesized that the proximal tubule functions in the active transport of selected molecules out of the filtrate.

By injecting solutions of known composition into isolated rabbit and rat proximal tubules in the presence or absence of ATP, researchers confirmed that selected electrolytes and nutrients are actively reabsorbed from the filtrate that enters the tubules.

When solutes move from the proximal tubule into epithelial cells, and then into the bloodstream, water follows along the osmotic gradient. In this way, valuable solutes and water are reabsorbed and returned to the bloodstream.

Ion and Water Movement Is Driven by a "Master Gradient"

FIGURE 43.13b summarizes the current model of the molecular mechanisms involved in selective reabsorption in the proximal tubule:

1. Na^+/K^+-ATPase in the basolateral membranes removes Na^+ from the interior of the cell. The active transport of sodium ions out of the cell creates a gradient favoring the entry of Na^+ from the lumen.

2. In the apical membrane adjacent to the lumen, Na^+-dependent cotransporters use this gradient to remove valuable ions and nutrients selectively from the filtrate. The movement of Na^+ into the cell, *along* its concentration gradient, provides the means for moving other solutes *against* a concentration gradient.

3. The solutes that move into the cell diffuse across the basolateral membrane into the interstitial fluid and then nearby blood vessels.

4. Water follows the movement of ions from the proximal tubule into the cell and then out of the cell and into blood vessels. Recall that water moves by osmosis across the membranes of these epithelial cells through membrane proteins called aquaporins.

Almost all of the nutrients, along with about two-thirds of the NaCl and water that is originally filtered by the renal corpuscle, are reabsorbed in the proximal tubule. The osmolarity of the tubular fluid is unchanged despite this huge change in volume, however, because water reabsorption is proportional to solute reabsorption.

In effect, then, the cells that line the proximal tubule act as a recycling center. The filtration step in the renal corpuscle is based on size; the reabsorption step in the proximal tubule selectively retrieves small substances that are valuable. The pumps and cotransporters in the proximal tubule reabsorb nutrients, electrolytes, and water but leave wastes. As the filtrate flows into the loop of Henle, it has a relatively high concentration of waste molecules and a relatively low concentration of nutrients.

Creating an Osmotic Gradient: The Loop of Henle

In mammals, the fluid that emerges from the proximal tubule enters the **loop of Henle**—named for Jacob Henle, who described it in the early 1860s. In most nephrons, the loop is short and barely enters the medulla. But in about 20 percent of the nephrons present in a human kidney, the loop is long and plunges from the cortex of the kidney deep into the medulla.

In 1942 Werner Kuhn offered a hypothesis, inspired by countercurrent heat exchangers, to explain what the loop of Henle does. Recall that a countercurrent heat exchanger is a system in which two adjacent fluids flow through pipes in opposite directions (see Chapter 42). Kuhn proposed that the loop of Henle is a countercurrent exchanger and multiplier. It doesn't exchange heat, however. Instead, it sets up an osmotic gradient.

Specifically, Kuhn proposed that the osmolarity of the filtrate in the loop of Henle is low in the cortex and high in the medulla. Further, Kuhn maintained that the osmolarity in the interstitial fluid surrounding the loop mirrors the gradient inside the loop. This is a key point.

Testing Kuhn's Hypothesis A series of papers published during the 1950s and early 1960s supplied important experimental support for the countercurrent exchange model. **FIGURE 43.14** reproduces two particularly important data sets, obtained by comparing the osmolarity of kidney tissue slices. In both graphs, the *x*-axis shows the location in the kidney, from cortex to medulla. The *y*-axis indicates osmolarity, measured either as the percentage of the maximum observed or as solute concentration.

- Figure 43.14a shows data on the osmolarity of fluid inside the loop of Henle. The vertical lines represent the range of values observed at a particular location. As predicted by Kuhn's model, a strong gradient in osmolarity exists from the cortex to the medulla.

- Figure 43.14b shows that the concentrations of Na^+, Cl^-, and urea also increase sharply from the cortex to the medulla in the interstitial fluid outside the loop of Henle.

These results suggested that the solutes responsible for the gradient outside the loop are Na^+, Cl^-, and urea. The change in concentration of urea turned out to be particularly important.

How Is the Osmotic Gradient Established? The loop of Henle has three distinct regions: the descending limb, the thin ascending limb, and the thick ascending limb (**FIGURE 43.15a**, see page 876). The thin and thick ascending limbs differ in the thickness of their walls. Do the three regions also differ in their permeability to water and solutes?

It took over 15 years of experiments performed in laboratories around the world to formulate a definitive answer to that question. Researchers punctured the loop of Henle of rodents with a micropipette, analyzed the composition of the fluid inside, and compared it with the nephron's final product—urine.

In the ascending limb of the loop of Henle, Na^+ and Cl^- constituted at least 60 percent of the solutes; urea constituted about 10 percent. But in the distal tubule, urea was the major solute. These data suggested that Na^+ and Cl^-, but not urea, were being removed somewhere in the ascending limb. How?

Na^+ and Cl^- were also present at high concentrations in the tissue surrounding the thick ascending limb, so researchers hypothesized that sodium might be actively pumped out of this portion of the nephron. The hypothesis was that the active transport of Na^+ out of the filtrate in the thick ascending limb would create an electrical gradient that would also favor the loss of Cl^-.

(a) Fluid inside the loop of Henle

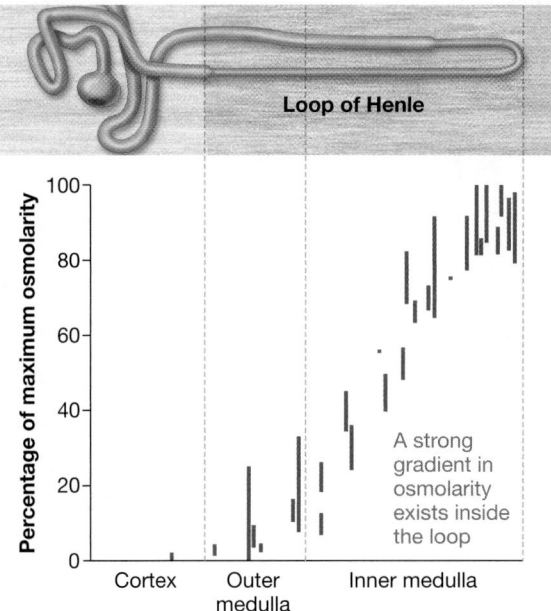

(b) Interstitial fluid outside the loop of Henle

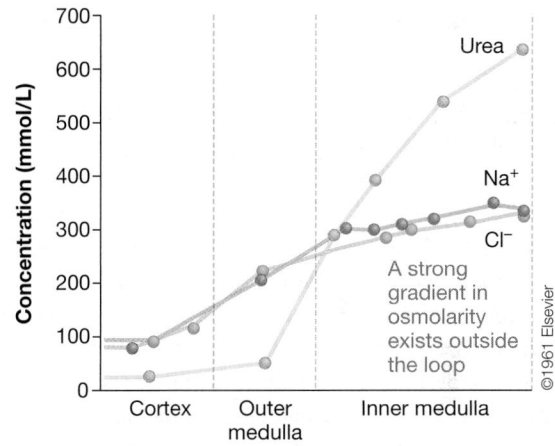

©1961 Elsevier

FIGURE 43.14 Data Confirm the Existence of a Strong Osmotic Gradient Both Inside and Outside the Loop of Henle. As the nephron plunges into the inner medulla, the concentration of dissolved solutes increases both inside **(a)** and outside **(b)** the loop of Henle.

DATA: K. J. Ullrich, K. Kramer, and J.W. Boyer. 1961. *Progress in Cardiovascular Diseases* 3: 395–431.

Follow-up experiments using ouabain and other poisons supported the hypothesis that sodium ions are actively transported out of the solution inside the thick ascending limb, and chloride ions follow along an electrochemical gradient. The epithelial cells responsible for salt excretion are configured almost exactly like the epithelium of the shark rectal gland (see Figure 43.6).

What is happening in the descending limb and the thin ascending limb of the loop of Henle? By injecting solutions of known concentration into the nephrons of rabbits, biologists documented that the descending limb is highly permeable to water but almost completely impermeable to solutes. The thin ascending limb of the loop, in contrast, is highly permeable to Na^+

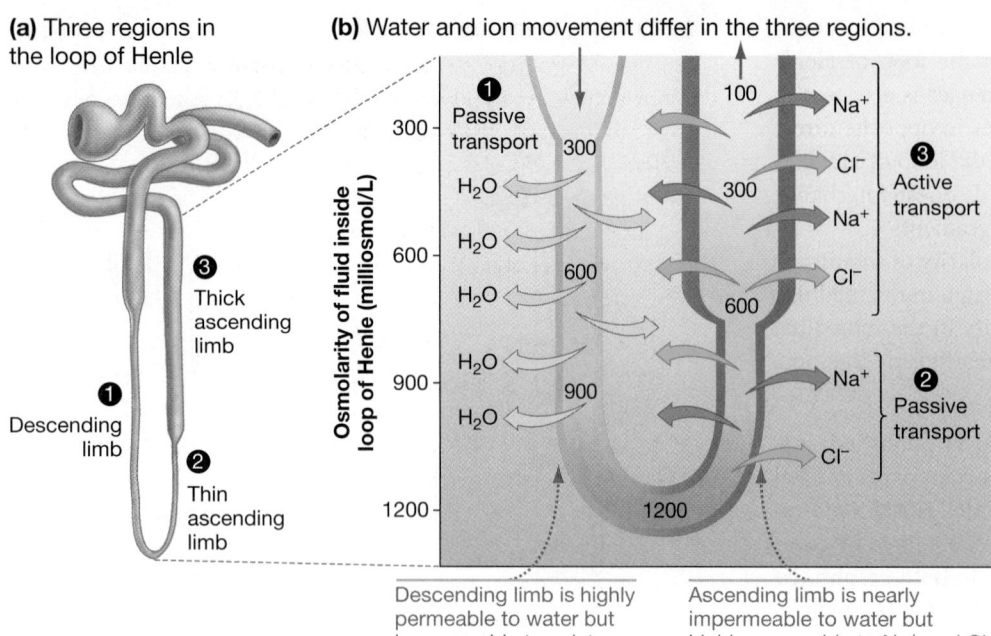

(a) Three regions in the loop of Henle

(b) Water and ion movement differ in the three regions.

❸ Thick ascending limb

❶ Descending limb

❷ Thin ascending limb

Descending limb is highly permeable to water but impermeable to solutes

Ascending limb is nearly impermeable to water but highly permeable to Na⁺ and Cl⁻

FIGURE 43.15 The Loop of Henle Maintains an Osmotic Gradient because Water Leaves the Descending Limb and Salt Leaves the Ascending Limb. The values inside the loop in part (b) represent the osmolarity of the filtrate.

and Cl⁻, moderately permeable to urea, and almost completely impermeable to water.

A Comprehensive View of the Loop of Henle   All of the observations just summarized came together in 1972 when two papers, published independently, proposed the same comprehensive model for how the loop of Henle works. To understand this model, follow the events in **FIGURE 43.15b:**

1. As fluid flows down the descending limb, the fluid inside the loop loses water to the interstitial fluid surrounding the nephron. This movement of water is passive—it does not require an expenditure of ATP. The water follows an osmotic gradient created by the ascending limb. At the bottom of the loop—in the inner medulla—the fluids inside and outside the nephron have high osmolarity. The filtrate does not continue to lose water, though, because the membrane in the ascending limb is nearly impermeable to water.

2. The fluid inside the nephron loses Na⁺ and Cl⁻ in the thin ascending limb. The ions move passively, along their concentration gradients.

3. Toward the cortex, the osmolarity of the surrounding interstitial fluid is low. Additional Na⁺ and Cl⁻ ions are actively transported out of the nephron in the thick ascending limb.

The countercurrent flow of fluid, combined with changes in permeability to water and in the types of channels and pumps that are active in the epithelium of the nephron, creates a self-reinforcing system. The presence of an osmotic gradient stimulates water and ion flows that in turn maintain an osmotic gradient.

Here's how it works: Movement of NaCl from the ascending limb into surrounding tissue increases the osmolarity of the fluid outside the descending limb, which results in a flow of water out of the water-permeable walls of the descending limb, via osmosis. This loss of water in the descending limb increases the osmolarity

in the fluid entering the ascending limb. The high concentration of salt in the fluid at the base of the ascending limb triggers a passive flow of ions out—reinforcing the osmotic gradient.

✔ If you understand this concept, you should be able to predict what happens to the osmotic gradient when the drug furosemide inhibits membrane proteins that pump sodium and chloride ions out of the thick ascending limb. Specifically, how does this drug affect (1) water reabsorption in the descending limb, (2) the osmolarity of the filtrate in the bottom of the loop of Henle, and (3) reabsorption of salt in the thin ascending limb?

The Vasa Recta Removes Water and Solutes That Leave the Loop of Henle   What happens to the water and salt that move out of the loop from the filtrate into the interstitial fluid? They quickly diffuse into the **vasa recta,** a network of blood vessels that runs along the loop and eventually joins up with small veins in the kidney. As a result, the water and electrolytes that are reabsorbed are returned to the bloodstream instead of being excreted in urine (**FIGURE 43.16**).

The removal of water that leaves the descending limb is particularly important. If it were not drawn off into the bloodstream, it would dilute the concentrated fluid outside the loop of Henle and quickly destroy the osmotic gradient.

The Collecting Duct Leaks Urea   Urea is the solute that is most responsible for the steep osmotic gradient in the space surrounding the nephron. Urea is at high concentration in the interstitial fluid in the inner medulla and low concentration in the outer medulla. This gradient exists because the innermost section of the collecting duct is permeable to urea.

Although the system created by the nephron, vasa recta, and collecting duct may seem complex, its outcome is simple: the creation and maintenance of a strong osmotic gradient with the minimum possible expenditure of energy.

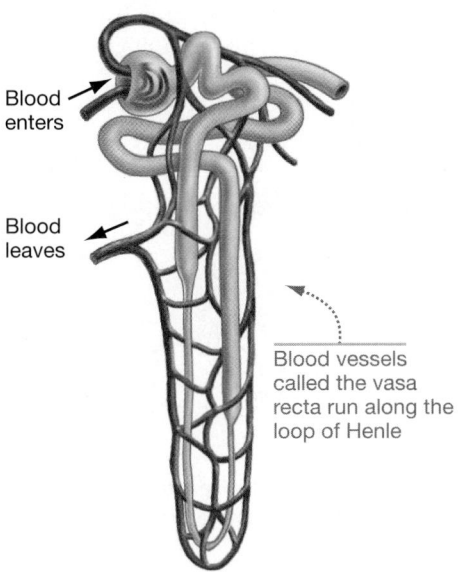

FIGURE 43.16 Blood Supply to the Loop of Henle. Water and solutes reabsorbed from the loop of Henle move into blood vessels (vasa recta).

Blood enters

Blood leaves

Blood vessels called the vasa recta run along the loop of Henle

Regulating Water and Electrolyte Balance: The Distal Tubule and Collecting Duct

The first three steps in urine formation—filtration, reabsorption, and establishment of an osmotic gradient—result in a fluid that is slightly hyposmotic to blood. Once the filtrate has passed through the loop of Henle, the major solutes that it contains are urea and other wastes along with a low concentration of ions.

The filtrate that enters the distal tubule is always dilute. In contrast, the urine that leaves the collecting duct is dilute when the individual is well hydrated but concentrated when the individual is dehydrated. How is this possible?

Urine Formation Is under Hormonal Control The answer is based on two observations about the **distal tubule** and **collecting duct:** Their activity is **(1)** highly regulated, and **(2)** altered in response to osmotic stress. The amount of Na^+, Cl^-, and water that is reabsorbed in the distal tubule and in the collecting duct varies with the animal's hydration.

Changes in the distal tubule and collecting duct are controlled by **hormones**—signaling molecules (introduced in Chapter 11 and explored further in Chapter 49). Specifically:

- If Na^+ levels in the blood are low, the adrenal glands release the hormone **aldosterone,** which leads to activation of sodium–potassium pumps and reabsorption of Na^+ in the distal tubule. Water follows by osmosis. Aldosterone saves sodium and water.

- If an individual is dehydrated, the brain releases **antidiuretic hormone (ADH).** (The term diuresis refers to increased urine production, so antidiuresis means inhibition of urine production.) ADH is also referred to as vasopressin or arginine vasopressin. ADH saves water.

How Does ADH Work? Epithelial cells of the collecting duct are joined by tight junctions (see Chapter 11), making them impermeable to water and solutes. ADH has two important effects on the permeability of epithelial cells in the collecting duct:

1. ADH triggers the insertion of aquaporins into the apical membrane. As a result, cells become much more permeable to water and large amounts of water are reabsorbed.

2. ADH increases permeability to urea, which is reabsorbed into the surrounding fluid. This helps create a concentration gradient favoring water reabsorption from the filtrate.

As **FIGURE 43.17a** shows, water leaves the collecting duct passively—following the concentration gradient maintained by

(a) ADH present: Collecting duct is highly permeable to water.

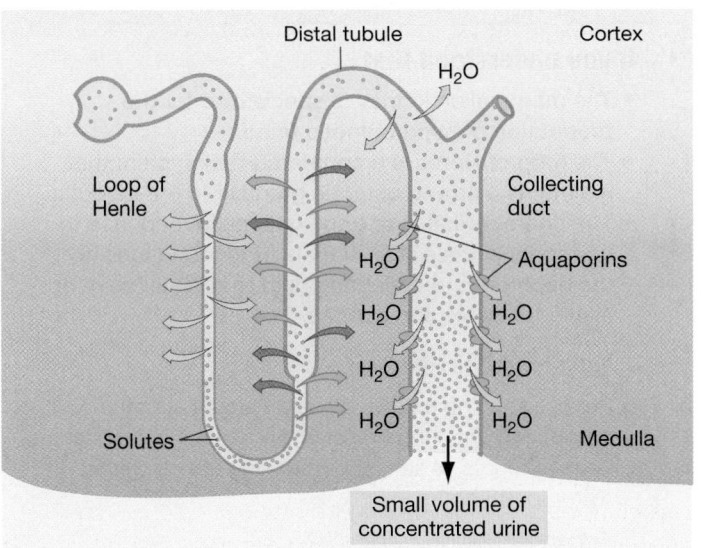

(b) No ADH present: Collecting duct is not permeable to water.

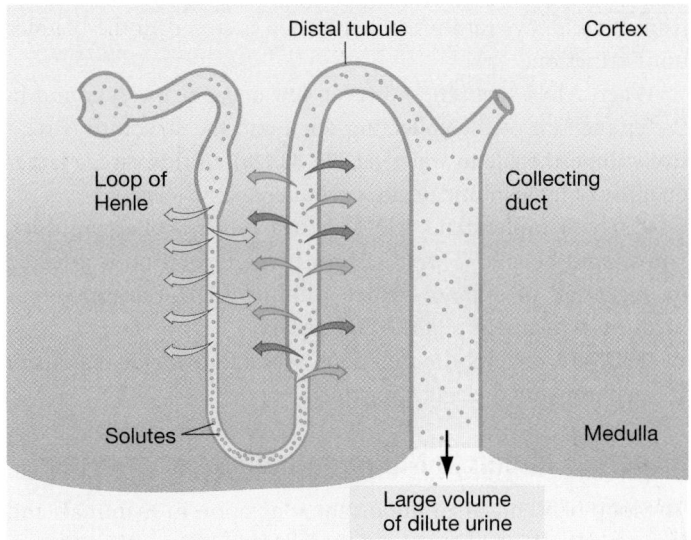

FIGURE 43.17 ADH Regulates Water Reabsorption by the Collecting Duct.

Structure	Function
Renal Corpuscle (Bowman's capsule and glomerulus)	Size-selective filtration: forms filtrate from blood (water and other small substances enter nephron)
Proximal Tubule	Reabsorbs electrolytes (active transport), nutrients, water
Loop of Henle	Maintains osmotic gradient in interstitial fluid from outer to inner medulla
• Descending limb	• Permeable to water (passive transport out of filtrate)
• Thin ascending limb	• Permeable to Na^+, Cl^- (passive transport out of filtrate)
• Thick ascending limb	• Active transport of Na^+, Cl^- out of filtrate
Distal Tubule	With aldosterone: Reabsorbs Na^+. Without aldosterone: Does not reabsorb Na^+.
Collecting Duct	Regulates water retention and loss
• Main portion	• With ADH: Water leaves filtrate; produces small volume of urine that is hyperosmotic to blood
	Without ADH: Water stays in filtrate; produces large volume of urine that is hyposmotic to blood
• Innermost portion	• Urea leaks out by passive transport to establish and/or maintain high osmolarity of inner medulla

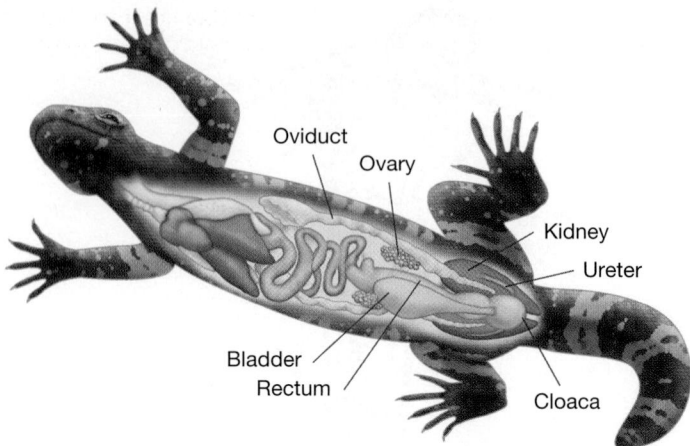

FIGURE 43.18 The Cloaca Is a Cavity into Which the Urinary, Gastrointestinal, and Reproductive Tracts Empty.

fishes, amphibians, and non-avian reptiles lack loops of Henle, and their kidneys are therefore unable to produce concentrated urine.

Many fishes and amphibians do not need to produce concentrated urine. But conserving water is important in reptiles, especially those inhabiting deserts. Recall that reptiles produce nitrogenous wastes in the form of uric acid, which is excreted with very little water and is therefore hyperosmotic to their body tissues. If reptiles' kidneys produce isosmotic urine, how does it become hyperosmotic?

In most reptiles, the ureters empty isosmotic urine into the **cloaca,** a cavity into which the urinary, gastrointestinal, and reproductive tracts all empty (**FIGURE 43.18**). Reptiles are able to absorb water from urine across the wall of the cloaca into the bloodstream. Eventually, a semisolid uric acid paste is excreted along with the feces.

the loop of Henle. When ADH is present, water is conserved and urine is strongly hyperosmotic relative to blood. The collecting duct is the final place in which the composition of the filtrate can be altered. Urine exiting the collecting ducts moves from the kidneys into ureters and then is stored in the bladder until urination.

When ADH is absent, however, few aquaporins are found in the epithelium of the collecting duct, and the structure is relatively impermeable to water (**FIGURE 43.17b**). In this case, a larger quantity of hyposmotic urine is produced.

✔ If you understand ADH's effect on the collecting duct, you should be able to predict how urine formation is affected by ingestion of ethanol, which inhibits ADH release, versus nicotine, which stimulates ADH release.

TABLE 43.2 reviews the functions of the four major regions of the nephron and the collecting duct.

Urine Formation in Nonmammalian Vertebrates

The loop of Henle is an important adaptation in mammals and some birds. Water loss is reduced because these animals can produce urine that is hyperosmotic to their blood. In contrast,

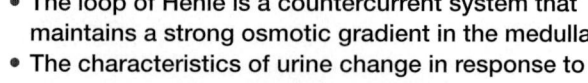

If you understand that . . .

- The mammalian kidney is specialized for the production of hyperosmotic urine.
- The loop of Henle is a countercurrent system that maintains a strong osmotic gradient in the medulla.
- The characteristics of urine change in response to hormonal signals. The signals trigger changes in the nephron and collecting duct that either save or eliminate water and solutes.

✔ You should be able to . . .

Predict how the following events would affect urine production: drinking massive amounts of water, eating large amounts of salt, and refraining from drinking water for 48 hours.

Answers are available in Appendix A.

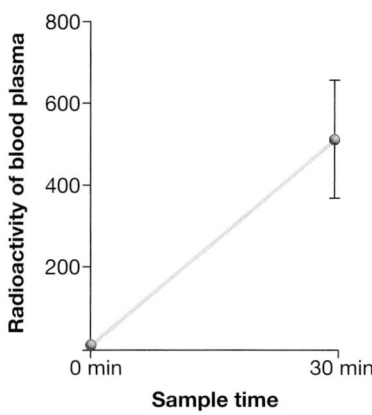

(a) Change following injection of radioactive water into bladder

(b) Change following injection of pure water into bladder or stomach

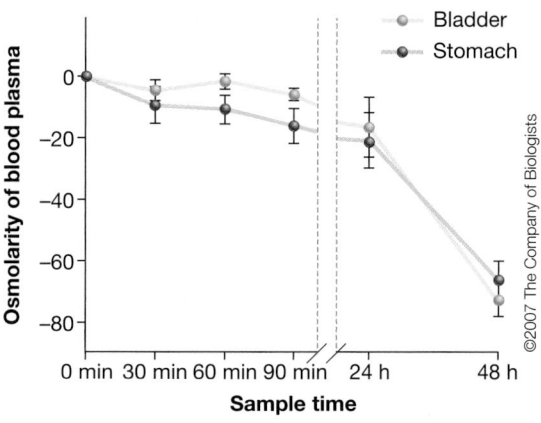

FIGURE 43.19 Gila Monsters Can Reabsorb Water from Their Bladders.
(a) Radioactively labeled water passes from bladder into bloodstream. **(b)** Pure water reduces blood osmolarity at a similar rate in dehydrated lizards whether added to stomach or to bladder.

DATA: Davis, J. R., and D. F. DeNardo. 2007. *The Journal of Experimental Biology* 210: 1472–1480.

✔**QUESTION** What physiological trade-offs might be imposed by carrying a large amount of water in the bladder?

©2007 The Company of Biologists

Some reptiles also have a bladder that collects the isosmotic urine from the ureters and stores it before emptying it into the cloaca. When water is available, these reptiles drink a lot, and their bladders fill up with dilute urine. Researchers hypothesized that the bladder acts as a "canteen" to store water during times when water is unavailable.

To test this hypothesis, the investigators injected the bladders of dehydrated Gila monsters—a large desert lizard—with radioactively labeled water and drew blood samples over time. The radioactivity of the lizards' blood increased within a half hour, which meant that the radioactively labeled water from the bladder was being absorbed into the bloodstream (**FIGURE 43.19a**).

The researchers also examined whether water in the bladder can rehydrate lizards to the same extent as drinking water.

Working with two groups of dehydrated Gila monsters, they injected water into the bladders of one group, and—to mimic drinking—they injected water into the stomachs of the second group. They found that the osmolarity of the plasma decreased at the same rate in both groups (**FIGURE 43.19b**).

So, the bladder can indeed act as a "canteen," allowing lizards to carry a water supply that they can access when water is scarce. As water is reabsorbed from the bladder, the urine becomes more and more concentrated, but the osmolarity of the blood remains low.

Whether considering the mammalian nephron or the Gila monster bladder, the vertebrate urinary system is remarkably effective in regulating water and electrolyte balance and achieving homeostasis.

CHAPTER 43   REVIEW

For media, go to MasteringBiology

If you understand . . .

43.1 Osmoregulation and Excretion

- Solutes move across membranes via passive transport, facilitated diffusion, or active transport. Water moves across membranes by osmosis.

- In most animals, epithelial cells that selectively transport water and electrolytes are responsible for homeostasis.

- The mechanisms involved in regulating water and electrolyte balance vary widely among animal groups because different habitats present different types of osmotic stress.

- The type of nitrogenous waste excreted by an animal is affected by its phylogeny and its habitat type. Most fishes excrete ammonia; mammals and most adult amphibians excrete urea; and insects, reptiles, and birds excrete uric acid.

✔ You should be able to explain how an electrochemical gradient for sodium ions makes it possible for other ions and water to move across a membrane passively.

43.2 Water and Electrolyte Balance in Marine Fishes

- Seawater is strongly hyperosmotic in relation to the tissues of marine bony fishes, so the fishes tend to lose water by osmosis and gain electrolytes by diffusion.

- Marine bony fishes are osmoregulators, whereas cartilaginous fishes such as sharks are osmoconformers.

- Epithelial cells in the shark rectal gland and in the gills of marine bony fishes excrete excess salt using sodium–potassium pumps and sodium–potassium–chloride cotransporters located in the basolateral membrane.

- Salt-excreting cells similar to those found in the shark rectal gland occur in the gills of marine fishes, the salt glands of marine birds and other reptiles, and the kidneys of mammals.

✔ You should be able to compare and contrast the mechanisms by which cartilaginous and bony fishes achieve water and electrolyte balance in the ocean.

43.3 Water and Electrolyte Balance in Freshwater Fishes

- Freshwater is strongly hyposmotic in relation to the tissues of freshwater fishes, so the fishes tend to gain water by osmosis and lose electrolytes by diffusion.

- Epithelial cells in the gills of freshwater fishes import ions using sodium–potassium pumps (Na^+/K^+-ATPases) located in the basolateral membrane and sodium–potassium–chloride cotransporters located in the apical membrane.

✔ You should be able to describe the evidence for the hypothesis that new chloride cells form when salmon migrate from the ocean to freshwater.

43.4 Water and Electrolyte Balance in Terrestrial Insects

- A waxy coating on the insect exoskeleton limits evaporative water loss. The openings to insect respiratory organs close when osmotic stress is severe.

- The Malpighian tubules of insects form a filtrate that is isosmotic with the hemolymph. If pumps in the epithelium of the hindgut are activated, then electrolytes and water are reabsorbed from the filtrate and returned to the hemolymph.

- Insects can form hyperosmotic urine that minimizes water loss during the excretion of nitrogenous wastes.

✔ You should be able to predict what type of nitrogenous waste is excreted by aquatic insects.

43.5 Water and Electrolyte Balance in Terrestrial Vertebrates

- Nephrons in the vertebrate kidney form a filtrate in the renal corpuscle and then reabsorb valuable nutrients, electrolytes, and water in the proximal tubule.

- A solution containing urea and electrolytes flows through the loop of Henle of mammalian kidneys, where changes in the permeability of epithelial cells to water and salt—along with active transport of salt—create a steep osmotic gradient.

- If antidiuretic hormone (ADH) increases the water permeability of the collecting duct, water is reabsorbed along the osmotic gradient and hyperosmotic urine is produced.

- The nephrons of fishes, amphibians, and reptiles do not have loops of Henle and therefore cannot produce urine that is hyperosmotic to the body fluids. Some of these vertebrates are able to reabsorb water from the cloaca or bladder to produce hyperosmotic urine.

✔ You should be able to explain why desert-dwelling kangaroo rats, which are under extreme osmotic stress due to lack of drinking water, have extremely long loops of Henle relative to their body size.

(MB) **MasteringBiology**

1. MasteringBiology Assignments

Tutorials and Activities Control of Water Reabsorption; Kidney Structure and Function; Mammalian Kidney; Nephron Function; Structure of the Human Excretory System

Questions Reading Quizzes, Blue-Thread Questions, Test Bank

2. eText Read your book online, search, take notes, highlight text, and more.

3. The Study Area Practice Test, Cumulative Test, BioFlix® 3-D Animations, Videos, Activities, Audio Glossary, Word Study Tools, Art

You should be able to . . .

✔ TEST YOUR KNOWLEDGE

Answers are available in Appendix A

1. True or False: Sodium–potassium ATPase pumps move three sodium molecules out of a cell in exchange for two potassium ions, establishing an electrochemical gradient.

2. Which one of the following statements is true of fishes that live in freshwater?
   a. The environment is isosmotic with respect to their tissues. As a result, they do not require a specialized organ to maintain water and electrolyte balance.
   b. They lose water to their environment primarily through the gills. They replace this water by drinking.
   c. Water enters epithelial cells in their gills via osmosis. Electrolytes leave the same cells via diffusion.
   d. They have specialized cells that actively pump Na^+ and Cl^- from blood into epithelial cells, so the ions can be excreted.

3. Which of the following organisms would lose the most water by osmosis across its gills?
   a. marine bony fish
   b. shark
   c. freshwater bony fish
   d. freshwater invertebrate

4. Which of the following gives the correct sequence of the regions of the nephron with respect to fluid flow?
   a. distal tubule → ascending limb of loop of Henle → descending limb of loop of Henle → proximal tubule → collecting duct
   b. ascending limb of loop of Henle → descending limb of loop of Henle → proximal tubule → distal tubule → collecting duct
   c. proximal tubule → ascending limb of loop of Henle → descending limb of loop of Henle → distal tubule → collecting duct
   d. proximal tubule → descending limb of loop of Henle → ascending limb of loop of Henle → distal tubule → collecting duct

5. What effect does antidiuretic hormone (ADH) have on the nephron?
   a. It increases water permeability of the descending limb of the loop of Henle.
   b. It decreases water permeability of the descending limb of the loop of Henle.
   c. It increases water permeability of the collecting duct.
   d. It decreases water permeability of the collecting duct.

6. Fill in the blank: In Gila monsters, the organ in which water from urine is reabsorbed into the bloodstream is the _____.

✔ TEST YOUR UNDERSTANDING   Answers are available in Appendix A

7. Which of the following organisms would you expect to have the highest concentrations of proteins that protect cells from the toxic effects of ammonia?
   a. shark
   b. freshwater bony fish
   c. terrestrial reptile
   d. mammal

8. Compare and contrast the types of nitrogenous wastes observed in animals. Identify which compound can be excreted with a minimum of water, which is most toxic, and which waste product is found in fishes, in mammals, and in insects. Which type of nitrogenous waste would you expect to find produced by embryos inside eggs laid on land?

9. The chloride cells of fish gills are sometimes called mitochondria-rich cells due to their high density of mitochondria. How does high mitochondrial density relate to the functional role of chloride cells? Would you expect other epithelial cells involved in ion transport to contain large numbers of mitochondria? Explain.

10. This chapter introduced a number of features that help terrestrial animals reduce water loss, including the layer of wax found on insect exoskeletons, the ability of insects to close spiracles, the excretion of nitrogenous wastes as insoluble uric acid, and long loops of Henle in the mammalian kidney. Predict how each of these traits differs in animals that live in very humid versus very dry habitats. How would you test your predictions?

11. Explain why mammals would not be able to produce concentrated urine if they lacked loops of Henle.

12. Scientists have noted that marine invertebrates tend to be osmoconformers, while freshwater invertebrates tend to be osmoregulators. Suggest an explanation for this phenomenon.

✔ TEST YOUR PROBLEM-SOLVING SKILLS   Answers are available in Appendix A

13. Biologists recently have been able to produce mice that lack functioning genes for aquaporins. How does their urine compare to that of individuals with normal aquaporins?
    a. lower volume and lower osmolarity
    b. lower volume and higher osmolarity
    c. higher volume and lower osmolarity
    d. higher volume and higher osmolarity

14. Suppose a grazing mammal ingested the leaves of a plant containing the toxin ouabain. What effect would this toxin have on the mammal's urine volume? Explain.

15. You have isolated a segment of a rat nephron and introduced a solution of known composition. When you compare the fluid collected at the end of the segment with the test solution introduced at the beginning, you find that the volume has decreased by 30 percent and the Na^+ concentration has decreased by 30 percent, but the urea concentration has increased by 50 percent. What processes in this segment might account for these changes?

16. **QUANTITATIVE** To test the hypothesis that mussels are osmoconformers, researchers exposed mussels to water of varying osmolarities and then drew hemolymph samples from the mussels. Is the researchers' hypothesis supported by the data, shown below? Explain.

Water Osmolarity	Hemolymph Osmolarity
250	261
500	503
750	746
1000	992

44 Animal Nutrition

In this chapter you will learn how

Animals ingest, digest, and absorb nutrients to survive and thrive

starting with ↓

Essential nutrients 44.1

asking ↓   then asking ↓

How are nutrients incorporated?

How is nutritional homeostasis maintained?

by examining ↓

Mouthparts 44.2

and

Digestive tracts 44.3

by exploring ↓

A classic example of nutrient imbalance: Type 2 diabetes 44.4

A crocodile has just caught a fish. Animals obtain nutrients by ingesting food.

This chapter is part of the Big Picture. See how on pages 840–841.

Animals get the two basic requirements for life—(1) chemical energy for synthesizing ATP and (2) carbon-containing compounds for building complex macromolecules—by ingesting other organisms. In short, they are heterotrophs: They eat to live.

The types of food that are available to different animals vary widely, and food is often in dangerously short supply. From these observations, you might expect animals to have evolved many different means for obtaining food and to be under intense natural selection for making efficient use of the food they have. How do animals get their food, and how do they process it? Which substances in food are used as nutrients, and how do humans and other animals maintain appropriate levels of key nutrients in their bodies?

✔ When you see this checkmark, stop and test yourself. Answers are available in Appendix A.

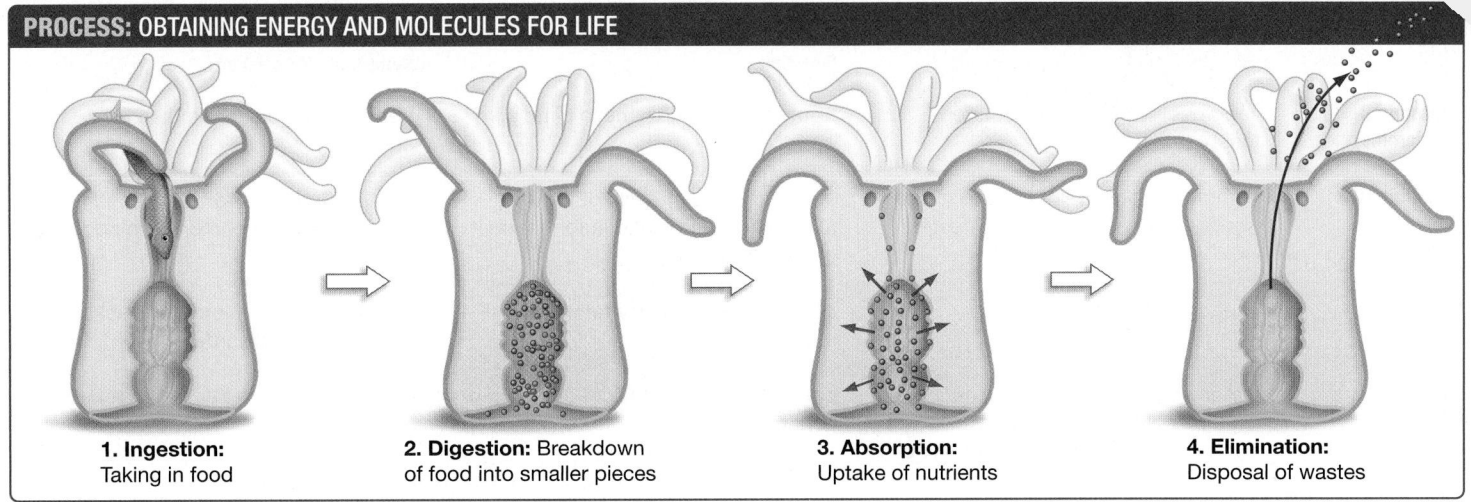

1. Ingestion: Taking in food

2. Digestion: Breakdown of food into smaller pieces

3. Absorption: Uptake of nutrients

4. Elimination: Disposal of wastes

FIGURE 44.1 Animal Nutrition Is a Four-Step Process.

If you're like most people, you've probably given a fair amount of thought to the food you eat but little thought to what happens to that food once it's inside you. As a meal moves through a digestive system, its physical characteristics change dramatically. Large packets of food that enter the mouth are reduced to monomers that can be absorbed into the bloodstream.

Research on feeding and digestion is fundamental to understanding these basic aspects of animal biology, but research on animal nutrition has important practical applications as well. For example, this chapter addresses questions about why several nutrition-related diseases, including diabetes mellitus and obesity, are on the rise in many human populations.

Just as plant nutrition is fundamental to the productivity of agricultural and natural ecosystems, nutrition is a basic component of animal health and welfare. For you or any other animal to stay alive, food must be ingested, digested, and absorbed, and then the wastes must be eliminated (**FIGURE 44.1**). Let's begin with a look at what animals must eat to live.

44.1 Nutritional Requirements

Humans and other animals get the chemical energy and carbon-containing building blocks they need from carbohydrates, proteins, and fats. All these substances are carbon compounds with high potential energy. Previous chapters analyzed the structures of carbohydrates and fats (see Chapter 5 and Chapter 6), and detailed how these compounds are used to synthesize ATP and key macromolecules (see Chapter 9).

A carbohydrate, protein, or fat is an example of a **nutrient**: a substance that an organism needs to remain alive. **Food** is any material that contains nutrients.

The amount of energy provided by foods is measured in kilocalories (on food labels, kilocalories are referred to as Calories).

Because fats and other lipids contain more C—H bonds than do carbohydrates such as starch and sugars, fats provide over twice the energy per gram. Specifically, carbohydrates and proteins provide about 4 kcal/g, and fats provide about 9 kcal/g.

Understanding which nutrients an individual needs, and in what amounts, are basic issues in research on human and animal nutrition. Let's consider what humans need to maintain good health.

In 1943 the Food and Nutrition Board of the U.S. National Academy of Sciences[1] published the first Recommended Dietary Allowances (RDAs). The goal of the RDAs was to specify the amount of each essential nutrient that an individual must ingest to meet the needs of most healthy people. **Essential nutrients** are nutrients that cannot be synthesized and must be obtained from the diet.

In addition to obtaining chemical energy and the "building blocks" for chemical synthesis from carbohydrates and other compounds in food, humans require several other essential nutrients:

- Of the 20 amino acids required to manufacture most proteins, humans can synthesize 12 from simpler building blocks (see Chapter 3). The others are called **essential amino acids** because they must be obtained from food.

- **Vitamins** are organic, or carbon-containing, compounds that are vital for health but are required in only minute amounts. They have a variety of roles; several function as coenzymes in critical reactions (see Chapter 3). **TABLE 44.1** (see page 884) lists a few of the vitamins for which RDAs have been established, notes their functions, and indicates the problems that develop if they are missing in the diet.

---

[1]The National Academy of Sciences is a group of scientists and engineers that advises the U.S. Congress on scientific and technical matters. Its Food and Nutrition Board is made up of biologists who specialize in animal nutrition.

Some Important Vitamins Required by Humans

	Source	Function	Symptoms if Deficient
B₁ (thiamine)	legumes, whole grains, potatoes, peanuts	formation of coenzyme in citric acid cycle	beriberi (fatigue, nerve disorders, anemia)
Vitamin B₁₂	red meat, eggs, dairy products; also synthesized by bacteria in intestine	coenzyme in synthesis of proteins and nucleic acids and in formation of red blood cells	anemia (fatigue and weakness due to low hemoglobin content in blood)
Niacin	meat, whole grains	component of coenzymes NAD$^+$ and NADP$^+$	pellagra (digestive problems, skin lesions, nerve disorders)
Folate	green vegetables, oranges, nuts, legumes, whole grains; also synthesized by bacteria in intestine	coenzyme in nucleic acid and amino acid metabolism	anemia
Vitamin C (ascorbic acid)	citrus fruits, tomatoes, broccoli, cabbage, green peppers	used in collagen synthesis, prevents oxidation of cell components, improves absorption of iron	scurvy (degeneration of teeth and gums)
Vitamin D	fortified milk, egg yolk; also synthesized in skin exposed to sunlight	aids absorption of calcium and phosphorus in small intestine	rickets (bone deformities) in children; bone softening in adults

- **Minerals** are inorganic substances used as components of enzyme cofactors or structural materials (see **TABLE 44.2**). Some, such as calcium and phosphorus, are needed in relatively large quantities. Others, such as iron and magnesium, are required in small or trace amounts.

- **Electrolytes** are mineral ions that influence osmotic balance and are required for normal membrane function. Sodium (Na$^+$), potassium (K$^+$), and chloride (Cl$^-$) are the major ions in the human body.

To obtain nutrients, animals must ingest them, usually via a mouth. The structure of animal mouthparts is therefore often highly specialized to capture specific types of food.

44.2 Capturing Food: The Structure and Function of Mouthparts

Instead of making their own food, as plants do, animals obtain the energy and nutrients they need by feeding on other organisms. Biologists assign animal feeding techniques to one of four strategies (see Chapter 33):

1. **Suspension feeders,** such as sponges and tubeworms, filter small organisms or bits of organic debris from water by means of cilia, mucus-lined "nets," or other structures.

2. **Deposit feeders,** including earthworms and sea cucumbers, swallow sediments and other types of deposited material rich in organic matter.

3. **Fluid feeders** suck or lap up fluids—like blood or nectar.

4. **Mass feeders** are the majority of animals. They seize and manipulate chunks of food by using mouthparts such as jaws and teeth, beaks, or special toxin-injecting organs.

Mouthparts as Adaptations

The types of food that animals harvest range from soupy solutions in decaying carcasses to nuts inside of hard shells. Solutions have to be lapped up; nuts have to be cracked. Given the diversity of food sources that animals exploit, it is not surprising that a wide variety of mouthpart structures have evolved to facilitate capturing and processing food.

Natural selection has closely matched the structure of animal mouthparts to their function in obtaining food. For example,

- Mammals chew their food and swallow distinct packets or boluses. The extinct mammal in **FIGURE 44.2a** illustrates one example of the many tooth shapes that evolved from the relatively simple and uniform teeth in the common ancestor of all mammals. Diversification of tooth shape has allowed mammals to exploit a wide range of foods.

- Complex skull, jawbones, and associated musculature have evolved in snakes. The highly flexible structure that results allows snakes to ingest large prey whole (**FIGURE 44.2b**).

The reason that there is such a close correlation between the structure and function of mouthparts is simple: Natural selection is particularly strong when it comes to food capture, because obtaining nutrients is so fundamental to fitness—the ability to produce offspring.

Let's pursue the correlation between mouthparts and food sources further, by analyzing the structure and function of jaws and teeth in what may be the most diverse lineage within any vertebrate family: the cichlid fishes of Africa.

A Case Study: The Cichlid Jaw

The cichlids that inhabit the Rift Lakes of East Africa are a spectacular example of **adaptive radiation**—the diversification of a

(a) The large canines of saber-toothed cats stabbed and sliced prey.

(b) A highly mobile cranium and jaws allows snakes to swallow large prey whole.

FIGURE 44.2 Mouthpart Structure Correlates with Function.

single ancestral lineage into many species, each of which lives in a different habitat or employs a distinct feeding method (see Chapter 28). Lake Victoria, for example, is home to 300 cichlids that live nowhere else.

In general, each Lake Victoria cichlid species feeds on a different specific item, but as a group they exploit almost every food source in the lake: planktonic organisms, crust-forming algae, leaflike algae, eggs, fish scales, fish fins, whole fish, plants, insects, and snails.

How can a group of closely related species exploit so many different food sources? Many fish species have pharyngeal (throat) jaws located well behind the normal oral (mouth) jaws. Non-cichlids use their pharyngeal jaws to move food down their throats, but cichlids can also use theirs to bite. This is possible

TABLE 44.2 **Major Minerals and Electrolytes Required by Humans**

	Source in Diet	Function	Symptoms if Deficient
Calcium (Ca)	dairy products, green vegetables, legumes	bone and tooth formation, nerve signaling, muscle response	loss of bone mass, slow growth
Chloride (Cl⁻)	table salt or sea salt, vegetables, seafood	fluid balance in cells, protein digestion in stomach (HCl), acid–base balance	weakness, loss of muscle function
Fluorine (F)	fluoridated water, seafood	maintenance of tooth structure	higher frequency of tooth decay
Iodine (I)	iodized salt, algae, seafood	component of the thyroid hormones thyroxine and T_3	goiter (enlarged thyroid gland)
Iron (Fe)	meat, eggs, whole grains, green leafy vegetables, legumes	enzyme cofactor; synthesis of hemoglobin and electron carriers	anemia, weakness
Magnesium (Mg)	whole grains, green leafy vegetables	enzyme cofactor	nerve disorders
Phosphorus (P)	dairy products, meat, grains	bone and tooth formation; synthesis of nucleotides and ATP	weakness, loss of bone
Potassium (K⁺)	dairy products; meat; nuts; fruits; potatoes, legumes, and other vegetables	nerve signaling, muscle response, acid–base balance	weakness, muscle cramps, loss of muscle function
Sodium (Na⁺)	table salt or sea salt, seafood	nerve signaling, muscle response, blood pressure regulation	weakness, muscle cramps, loss of muscle function, nausea, confusion
Sulfur (S)	any source of protein	amino acid synthesis	swollen tissues, degener... of liver, mental retardatio...

FIGURE 44.3 Rift Lake Cichlids Have Two Sets of Biting Jaws. Oral jaws capture food; pharyngeal jaws process it.

because the upper pharyngeal jaw attaches to the skull in cichlids, and because the muscles of their lower pharyngeal jaw allow it to move against the upper jaw (**FIGURE 44.3**).

In addition to acting as a second set of biting jaws that make food processing more efficient, cichlid pharyngeal jaws provide a more specialized set of toothlike structures. These protuberances vary in size and shape among cichlids, correlating with their function, such as crushing snail shells, tearing fish scales, or compacting algae (**FIGURE 44.4**).

These observations are part of a large body of evidence supporting a general pattern in animal evolution: In response to natural selection, mouthparts have diversified to exploit a diversity of food sources. The structures of jaws, teeth, and other mouthparts correlate with their functions in harvesting and processing food.

44.3 How Are Nutrients Digested and Absorbed?

Animals ingest just about every type of food conceivable. **Ingestion** is the process of bringing food into the **digestive tract**—also known as the alimentary (literally, "nourishment") canal or gastrointestinal (GI) tract. The digestive tract is a chamber or tube where **digestion,** the breakdown of food, takes place. Interestingly, food in the digestive tract is technically *outside* the body. Various glands secrete enzymes into the digestive tract that digest food into particles small enough for efficient **absorption**—the uptake of specific ions and molecules across the membrane of the digestive tract.

Digestion is a key process in animals because they, unlike plants, unicellular organisms, and certain parasites, do not acquire nutrients as individual molecules. Instead, they take in packets of food that must be broken down into small pieces. Nutrients must be extracted from the small pieces, and waste materials must be eliminated. How and where does this processing occur?

An Introduction to the Digestive Tract

Digestive tracts come in two general designs:

1. **Incomplete digestive tracts** have a single opening that doubles as the location where food is ingested and wastes are eliminated. The mouth opens into a chamber, called a gastrovascular cavity, where digestion takes place (**FIGURE 44.5**).

2. **Complete digestive tracts** have two openings—they start at the mouth and end at the anus. The interior of this tube communicates directly with the external environment via these openings (**FIGURE 44.6**).

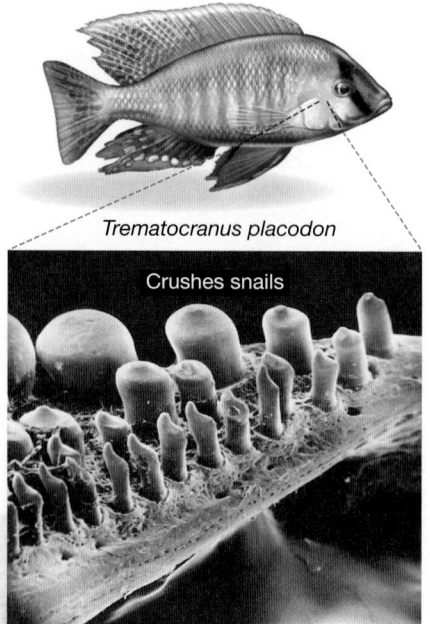

Trematocranus placodon

Crushes snails

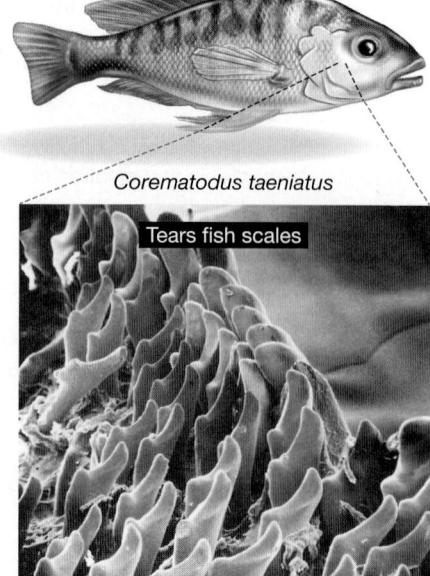

Corematodus taeniatus

Tears fish scales

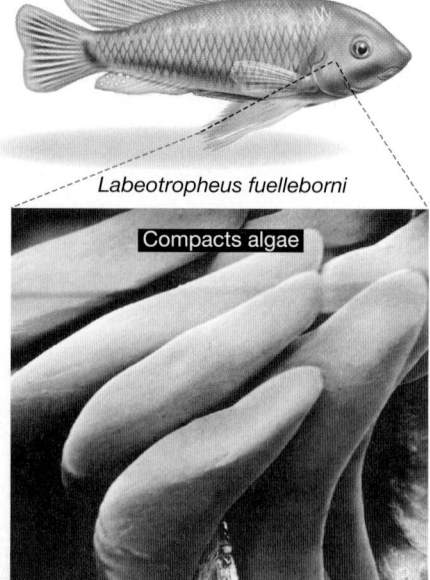

Labeotropheus fuelleborni

Compacts algae

FIGURE 44.4 In Cichlids, the Structure of the Pharyngeal Jaw Correlates with the Type of Food Ingested.

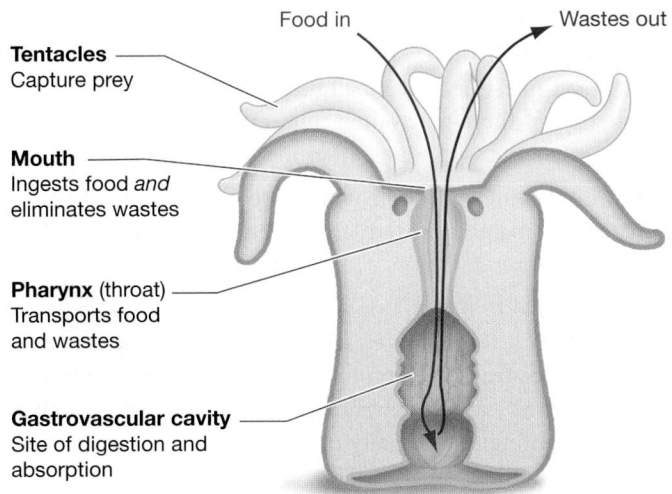

Tentacles
Capture prey

Mouth
Ingests food *and*
eliminates wastes

Pharynx (throat)
Transports food
and wastes

Gastrovascular cavity
Site of digestion and
absorption

Food in   Wastes out

FIGURE 44.5 Anemones Have an Incomplete Digestive Tract.
Anemones use stinging cells located on their tentacles to capture
small fishes, crustaceans, and other prey. Prey are taken into the
mouth and digested in the gastrovascular cavity; then wastes are
excreted through the mouth.

A tubelike digestive tract has three advantages:

1. It allows animals to feed on large pieces of food, expanding
   the range of food sources that can be ingested.

2. Different chemical and physical processes can be confined to
   different compartments within the canal, so that they occur
   independently of each other and in a prescribed sequence.
   The stomach, for example, provides an acidic environment
   for digestion. Ingested material proceeds from there to the
   small intestine, where enzymes are specialized to function in
   a slightly alkaline environment.

3. Because there is a one-way flow of food and wastes, material
   can be ingested and digested without interruption, instead of
   alternating with waste removal as occurs in an incomplete digestive tract.

The digestive tract is only one part of the digestive system,
however. Several vital organs and glands are connected to the digestive tract. These accessory structures contribute digestive enzymes and other products to specific portions of the tract. They
include the salivary glands, liver, gallbladder, and pancreas (see
Figure 44.6).

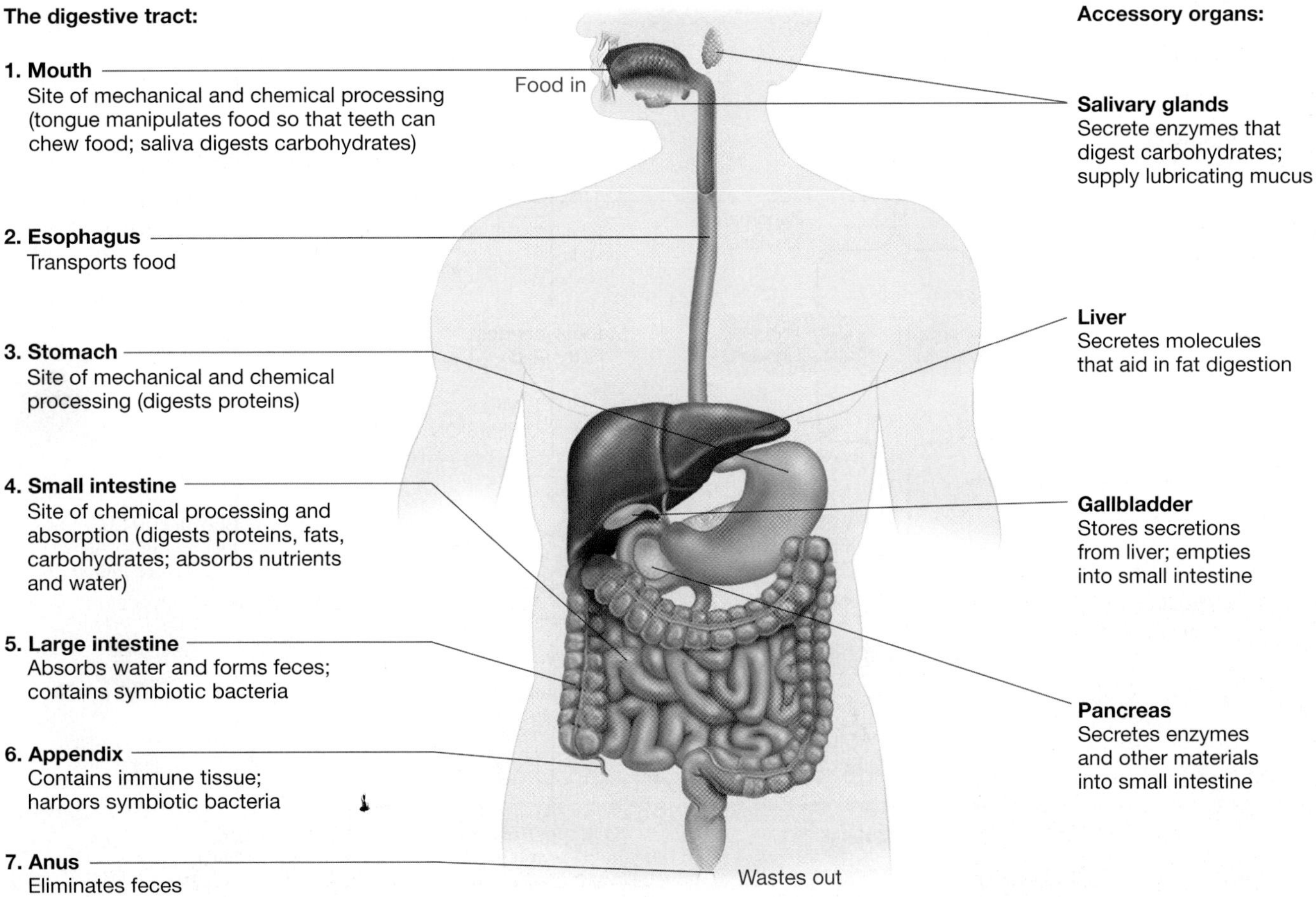

The digestive tract:

1. Mouth
Site of mechanical and chemical processing
(tongue manipulates food so that teeth can
chew food; saliva digests carbohydrates)

2. Esophagus
Transports food

3. Stomach
Site of mechanical and chemical
processing (digests proteins)

4. Small intestine
Site of chemical processing and
absorption (digests proteins, fats,
carbohydrates; absorbs nutrients
and water)

5. Large intestine
Absorbs water and forms feces;
contains symbiotic bacteria

6. Appendix
Contains immune tissue;
harbors symbiotic bacteria

7. Anus
Eliminates feces

Food in

Wastes out

Accessory organs:

Salivary glands
Secrete enzymes that
digest carbohydrates;
supply lubricating mucus

Liver
Secretes molecules
that aid in fat digestion

Gallbladder
Stores secretions
from liver; empties
into small intestine

Pancreas
Secretes enzymes
and other materials
into small intestine

FIGURE 44.6 Humans Have a Complete Digestive Tract. In humans, the digestive tract is a tube that runs from the
mouth to the anus. The salivary glands, liver, gallbladder, and pancreas are not part of the tract itself. Instead, they
secrete material into the tract at specific points.

An Overview of Digestive Processes

Before analyzing the function of each component of the digestive system in detail, let's consider the general changes that happen to food on its way through the digestive tract. In this brief overview and in the detailed discussion that follows, humans will serve as a model species—simply because so much is known about human digestion.

In mammals, digestion begins with the tearing and crushing activity of teeth. Chewing reduces the size of food particles and softens them. Humans augment the mechanical breakdown of food by their use of knives and cooking. In fact, the invention of cutting tools and cooking, which make food easier to chew, is the leading hypothesis to explain why average tooth size has declined steadily over the past several million years of human evolution.

Distinct chemical changes occur as food moves through each compartment in the digestive tract (**FIGURE 44.7**):

- In the mouth, enzymes in the saliva begin the chemical breakdown of carbohydrates.
- Chemical digestion of protein begins in the acidic environment of the stomach.

- Chemical processing of the three major types of macromolecules—carbohydrates, proteins, and lipids—is completed in the small intestine. The small molecules that result from the digestion of these macromolecules are absorbed in the small intestine, along with water, vitamins, and ions.

- In the large intestine, or colon, more water is absorbed. The material remaining in the large intestine is **feces,** which are eventually excreted.

Because digestion is so important to understanding how animal bodies work, let's analyze each step in more detail. As the following sections track food on its journey from the mouth to the anus in humans—a particularly well-studied animal—watch for notes highlighting the diversity of structures found in the digestive tracts of nonhuman animals.

The Mouth and Esophagus

If you hold a cracker in your mouth long enough, it will start to taste sweet. The sensation occurs because an enzyme in your saliva hydrolyzes some of the starch molecules in the cracker to

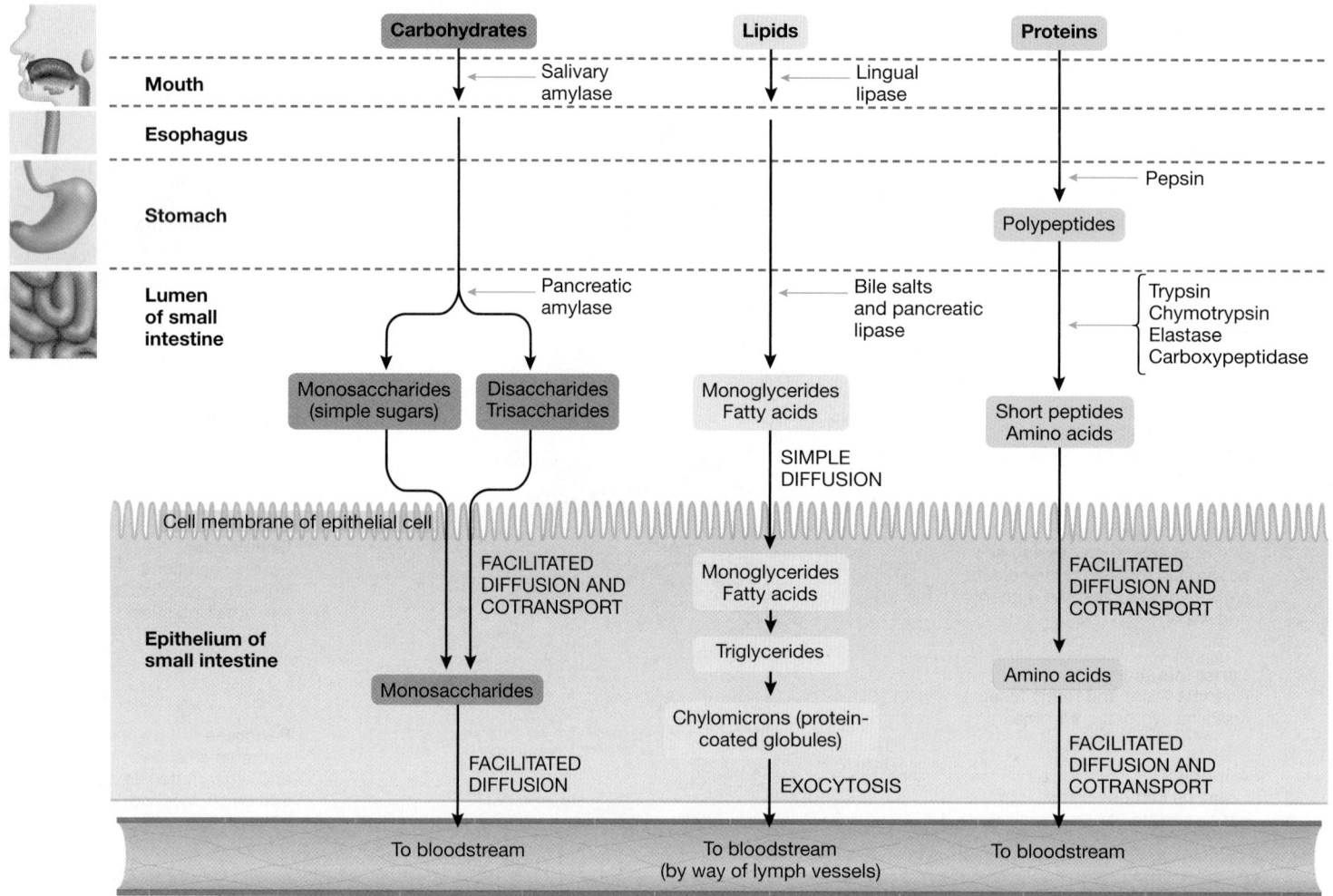

FIGURE 44.7 Carbohydrates, Lipids, and Proteins Are Processed in a Series of Steps. Three key types of macromolecules enter the digestive system (top of diagram). As they proceed through the digestive tract, they are broken apart by various enzymes. Simple sugars, fatty acids, and amino acids then enter epithelial cells in the small intestine and are transported to the bloodstream.

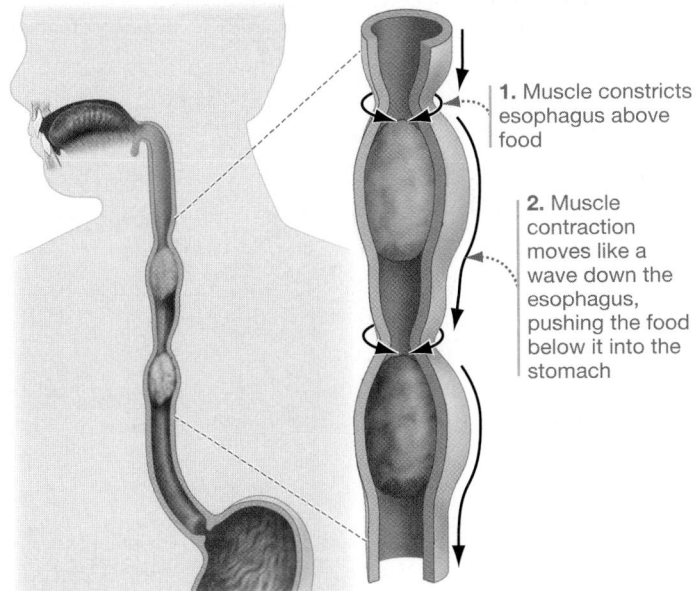

1. Muscle constricts esophagus above food

2. Muscle contraction moves like a wave down the esophagus, pushing the food below it into the stomach

FIGURE 44.8 In the Esophagus, Peristalsis Transports Food to the Stomach. Peristalsis is a wave of muscle contraction that constricts the esophagus behind food and expands it in front of food. The wave of contraction begins at the oral end of the esophagus and propels the food toward the stomach.

maltose. Maltose is a disaccharide that is split in the small intestine to form two glucose monomers.

Starch breakdown was actually the first enzyme-catalyzed reaction ever discovered. In the early 1800s several researchers found that a component of certain plant extracts digested starch; in 1831 the same activity was discovered in human saliva.

Digestion Starts in the Mouth **Salivary amylase,** the enzyme responsible for starch digestion in the mouth is one of the best-studied enzymes. Amylase cleaves bonds to release maltose from starch and glycogen, initiating the digestion of starch.

Cells in the tongue synthesize and secrete another important salivary enzyme, **lingual lipase** (lingual refers to the tongue), which begins the digestion of lipids by breaking triglycerides into diglycerides and fatty acids.

Salivary glands in the mouth not only produce amylase but also release water and glycoproteins called mucins. When mucins contact water, they form the slimy substance called **mucus.** The combination of water and mucus makes food soft and slippery enough to be swallowed.

Peristalsis Moves Material Down the Esophagus Once food is swallowed, it enters a muscular tube called the **esophagus,** which connects the mouth and stomach. A wave of muscular contractions called **peristalsis** propels food down the esophagus. About 6 seconds after being swallowed, food reaches the bottom of the esophagus. Because peristalsis actively moves material along the esophagus, you can swallow even when your mouth is lower than your stomach, such as when you bend over to drink from a stream.

In response to nerve signals, the muscles in the esophagus contract and relax in a coordinated fashion (**FIGURE 44.8**). The resulting wave of muscle contractions propagates down the tube, propelling the food mass ahead of it. These nerve signals are not the result of conscious choice but are a reflex—an automatic reaction to a stimulus—that is stimulated by the act of swallowing.

A Modified Esophagus: The Bird Crop In an array of bird species, the esophagus has a prominent, widened segment called the **crop** where food can be stored and, in some cases, processed.

The structure and function of the crop varies among the bird species that have one. In many groups, the crop is a simple sac that holds food and regulates its flow into the stomach. In these species, the crop is interpreted as an adaptation that allows individuals to eat a large amount in a short time; they then retreat to a safe location while digestion occurs. In addition, some birds store food in their crops and then regurgitate it into the mouths of their young.

The crop has independently evolved into a digestive organ in two leaf-eating species of bird. Leaves are difficult to digest because they contain a large amount of cellulose (see Chapter 5). In the enlarged crop of species called the hoatzin and kakapo, bacteria that are capable of breaking down cellulose perform digestion. The bacterial cells, along with the fatty acids that result from bacterial metabolism, leak out of the crop into the stomach and are used as food by the bird.

The Stomach

Although little if any digestion occurs in the esophagus of most animals, the situation changes dramatically when food reaches the stomach. The **stomach** is a tough, muscular pouch in the digestive tract, bracketed on both ends by ringlike muscles called **sphincters,** which control the passage of material (**FIGURE 44.9**).

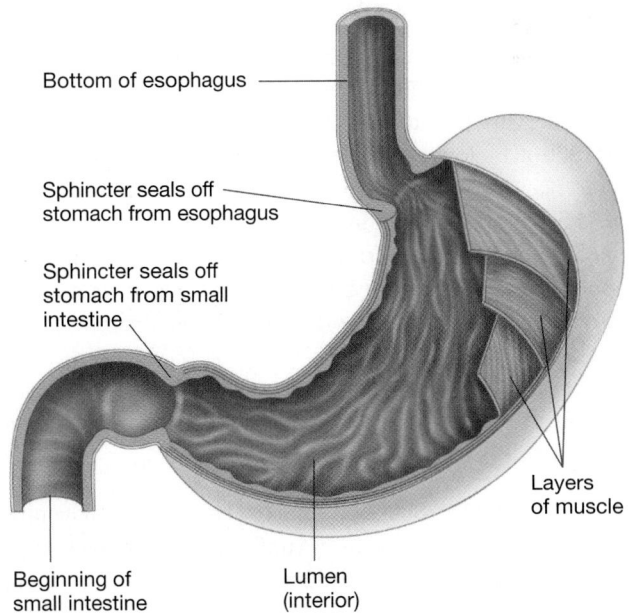

Bottom of esophagus

Sphincter seals off stomach from esophagus

Sphincter seals off stomach from small intestine

Beginning of small intestine

Lumen (interior)

Layers of muscle

FIGURE 44.9 The Stomach Is a Muscular Pocket of the Digestive Tract. The stomach provides an acidic environment for protein digestion. Its muscular contractions mix food and break it into smaller pieces.

When a meal fills the stomach, muscular contractions churn and mix the stomach contents to a uniform consistency and solute concentration. A certain amount of mechanical breakdown of food also results from this churning. The other main function of the stomach is the partial digestion of proteins.

Compared with the mouth or esophagus (or virtually any other tissue), the lumen of the stomach is highly acidic. Early researchers documented this fact by analyzing vomit or material collected by sponges that were tied to strings, swallowed, and pulled back up; chemists confirmed that the predominant acid in the stomach is hydrochloric acid (HCl).

Not long after this discovery, a physician named William Beaumont established that digestion takes place in the stomach. He reached this conclusion through an extraordinary series of experiments on a young man named Alexis St. Martin.

The Stomach as a Site of Protein Digestion

In 1822, when St. Martin was 19 years old, a shotgun accidentally discharged into his abdomen, leaving a series of wounds. Despite repeated attempts, Beaumont was unable to close a hole in St. Martin's stomach. Eventually Beaumont inserted a small tube through the opening; the tube remained in St. Martin's body for the rest of his life. (Today, biologists insert tubes into various parts of the digestive tract of cows or sheep to study how these animals digest different types of feed.)

With the tube in place, Beaumont was able to tie a string onto small pieces of meat or vegetables, insert the food directly into St. Martin's stomach, and draw it out after various intervals. Beaumont also removed liquid from inside the stomach and observed how this gastric (stomach) juice acted on food in vitro. His experiments showed that gastric juice digests food—particularly meat.

Theodor Schwann later purified the enzyme that is responsible for digesting proteins in the stomach and named it **pepsin.** Because it destroys proteins, biologists hypothesized that pepsin must be synthesized and stored in cells while it is in an inactive form—otherwise it would kill the cells that make it.

In 1870 a biologist established through careful microscopy the presence of granules in specialized stomach cells called chief cells. These granules were hypothesized to be a pepsin precursor. Follow-up work confirmed this hypothesis. The precursor compound, which came to be called pepsinogen, is converted to active pepsin by contact with the acidic environment of the stomach.

Secretion of a protein-digesting enzyme in inactive form is important: It prevents destruction of proteins in the cells where the enzyme is synthesized.

Which Cells Produce Stomach Acid?

The acidic environment of the human stomach denatures (unfolds) proteins so that pepsin can digest them efficiently. But where does the acid come from?

Researchers who were studying the anatomy of the stomach wall noticed clusters of distinctive **parietal cells** located in pits in the stomach lining (**FIGURE 44.10a**). An investigator also

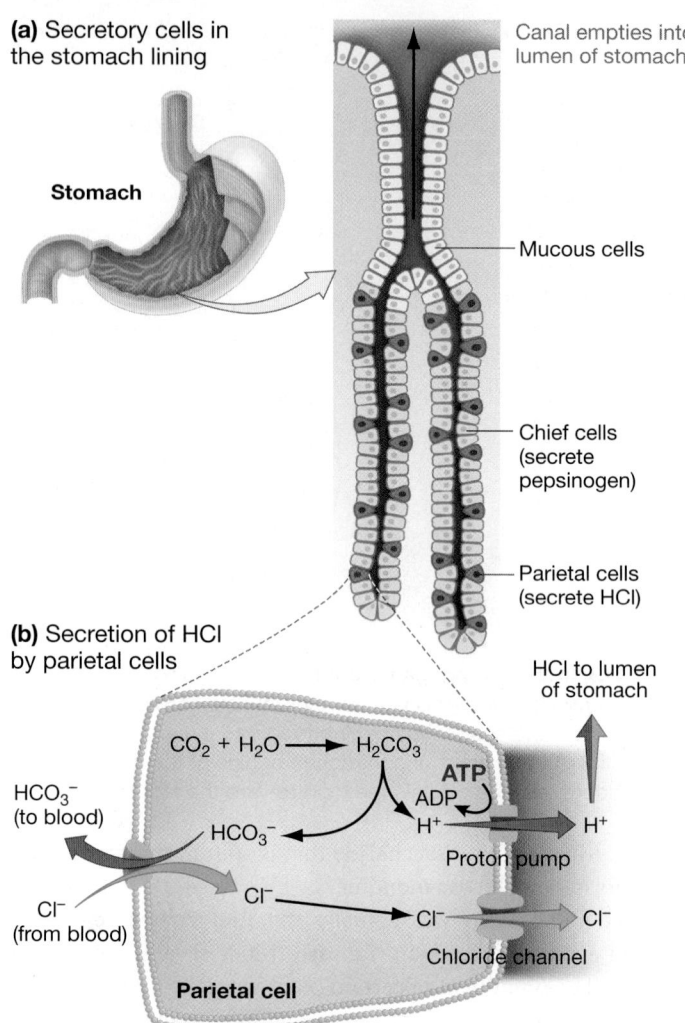

(a) Secretory cells in the stomach lining

Stomach

Canal empties into lumen of stomach

Mucous cells

Chief cells (secrete pepsinogen)

Parietal cells (secrete HCl)

(b) Secretion of HCl by parietal cells

HCl to lumen of stomach

$CO_2 + H_2O \longrightarrow H_2CO_3$

ATP

ADP

HCO_3^- (to blood)

HCO_3^-

H^+

H^+

Proton pump

Cl^- (from blood)

Cl^-

Cl^-

Cl^-

Chloride channel

Parietal cell

FIGURE 44.10 Cells in the Stomach Lining Secrete Mucus, Pepsinogen, and Hydrochloric Acid.

documented that the shape and activity of these cells appeared to vary as the digestion of a meal proceeded. Based on these observations, he inferred that parietal cells are the source of the HCl in gastric juice, which may have a pH as low as 1.5.

Earlier microscopists had shown that another type of cell, called a **mucous cell,** secretes additional mucus that is found in gastric juice. Mucus lines the gastric epithelium and protects the stomach from damage by HCl. To summarize, these anatomical studies showed that the epithelium of the stomach contains several types of secretory cells, each of which is specialized for a particular function.

How Do Parietal Cells Secrete HCl?

The first clue to how parietal cells manufacture HCl emerged in the late 1930s, when a researcher found a high concentration of an enzyme called carbonic anhydrase in parietal cells.

This result was interesting because **carbonic anhydrase** catalyzes the formation of carbonic acid (H_2CO_3) from carbon dioxide and water. In solution, the carbonic acid that is formed immediately dissociates to form a proton (H^+) and the bicarbonate ion (HCO_3^-):

$$CO_2 + H_2O \rightleftharpoons H_2CO_3 \rightleftharpoons H^+ + HCO_3^-$$

A second clue to the formation of HCl came in the 1950s, when transmission electron microscopes allowed researchers to analyze parietal cells at high magnification (see **BioSkills 11** in Appendix B). The micrographs showed that parietal cells are packed with mitochondria. Because mitochondria produce ATP, the structure of parietal cells suggested that they might function in active transport.

Later work confirmed this hypothesis by showing that the protons formed by the dissociation of carbonic acid are actively pumped into the lumen of the stomach. Subsequent studies showed that chloride ions from the blood enter parietal cells in exchange for bicarbonate ions, via a cotransport protein, and then move into the lumen through a chloride channel. **FIGURE 44.10b** diagrams the current model for HCl production.

Ulcers as an Infectious Disease An **ulcer** is an eroded area in an epithelium; it exposes the underlying tissues to damage. Ulcers in the lining of the stomach or in the duodenum—the initial section of the small intestine—can cause intense abdominal pain.

For decades, physicians thought that gastric and duodenal ulcers resulted from the production of excess acid in the stomach. They treated ulcers by prescribing alkaline compounds (bases—see Chapter 2) that neutralized hydrochloric acid in the stomach. In the 1980s, however, scientists discovered that most ulcers are associated with infections from a bacterium called *Helicobacter pylori*. Physicians now routinely prescribe antibiotics to relieve ulcers.

The Ruminant Stomach It is common for animals to have a stomach or stomach-like organ. The structure and function of this organ can vary, however, depending on the nature of the diet. In cattle, sheep, goats, deer, antelope, giraffe, and pronghorn—species that are collectively called **ruminants**—the stomach is specialized for digesting cellulose—not proteins.

Animals do not produce the enzymes required to digest cellulose. Yet cellulose is the main carbohydrate in the leaves, stems, and twigs that ruminants ingest.

Like the hoatzin and kakapo birds described earlier in this chapter, ruminants are able to harvest energy from cellulose thanks to a combination of specialized anatomical structures and symbiotic relationships with bacteria and unicellular protists. The microbes ferment the cellulose to produce food for themselves; meanwhile, other by-products of the fermentation, as well as some of the microbes themselves, are used as food by the ruminant. This is an example of **symbiosis,** in which members of two different species live in close contact with each other.

As **FIGURE 44.11** shows, ruminant mammals have four-chambered stomachs.

1. Food initially enters the largest chamber, the rumen, which serves as a fermentation vat. The rumen is packed with symbiotic bacteria and protists. These organisms have enzymes

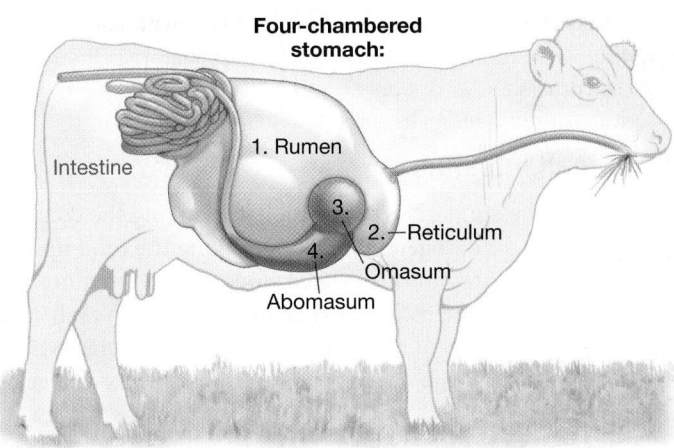

FIGURE 44.11 Ruminant Stomachs Facilitate the Digestion of Cellulose by Symbiotic Organisms. Ruminants obtain many of their nutrients from symbiotic bacteria and protists that live in the rumen and reticulum chambers of the stomach.

called **cellulase** capable of breaking apart the chemical bonds in cellulose, yielding glucose. The rumen is an oxygen-free environment, and the symbiotic organisms produce ATP from this glucose via fermentation, releasing fatty acids as a by-product (see Chapter 9). These fatty acids are absorbed by the ruminant and used as an energy source.

2. The chamber adjacent to the rumen, called the reticulum, is similar in function. After plant material has been partially digested in the rumen and the reticulum, the animal regurgitates portions of that material into its mouth, forming a cud. The ruminant chews that regurgitated material further to enhance mechanical breakdown and then re-swallows it.

3. Processed food that moves out of the rumen enters the third chamber, or omasum, where water and some minerals are absorbed.

4. The final chamber is the abomasum, which contains the ruminant's own digestive enzymes and corresponds to a true stomach.

Most of a ruminant's food consists of (1) fatty acids and other compounds produced as waste products of fermentation reactions in symbiotic organisms, and (2) the symbiotic cells themselves.

The Avian Gizzard The avian gizzard is another prominent type of modified stomach. Birds do not have teeth and cannot chew food into small pieces. Instead, most species swallow sand and small stones that lodge in the gizzard. As this muscular sac contracts, food is pulverized by the grit.

The gizzard is particularly large and strong in bird species that eat coarse foods such as seeds and nuts. The gizzard of a wild turkey, for example, can crack large walnuts.

Like the crop, the gizzard is interpreted as an adaptation that allows birds to ingest food quickly—without needing to chew—and digest it later. Biologists invoke the same hypothesis to explain why ruminants chew cud. The ability to regurgitate material and finish chewing while hiding in a place safe from predators is thought to increase fitness.

The Small Intestine

In humans, the stomach is responsible for mixing the contents of a meal into a homogenous slurry, breaking the food up mechanically, and providing the acid and enzymes required to partially digest proteins. Peristalsis in the stomach wall then moves small amounts of material through the valve created by a sphincter muscle at the base of the stomach and into the small intestine.

The **small intestine** is a long tube that is folded into a compact space within the abdomen. In the small intestine, partially digested food mixes with secretions from the pancreas and the liver and begins a journey of about 6 m (20 ft). When passage through this structure is complete, digestion is finished, and most nutrients—along with large quantities of water—have been absorbed.

Folding and Projections Increase Surface Area The surface area available for nutrient and water absorption in the small intestine is nothing short of remarkable (see Chapter 42). As **FIGURE 44.12** shows, the organ's epithelial tissue is folded and covered with fingerlike projections called **villi** (singular: **villus**). In turn, the cells that line the surface of villi have tiny projections on their apical surfaces called **microvilli** (singular: **microvillus**). Microvilli project into the lumen of the digestive tract.

If the small intestine lacked folds, villi, and microvilli, it would have a surface area of about 3300 cm^2 (3.6 ft^2). Instead, the epithelium covers about 2 million cm^2 (over 2200 ft^2)—an area about the size of a tennis court.

The enormous surface area of the small intestine increases the efficiency of nutrient absorption. And because each villus contains blood vessels and a lymphatic vessel called a **lacteal**, nutrients pass quickly from epithelial cells into the body's transport systems. (The circulatory system and lymphatic system are analyzed in Chapter 45 and Chapter 51, respectively.)

✔ If you understand the importance of surface area in the small intestine, you should be able to explain why surface area is so much higher in this structure than it is in the stomach or esophagus.

To understand how digestion is completed and absorption occurs, let's explore what happens to proteins, lipids, and carbohydrates as they move through this section of the digestive tract—again using humans as a model organism.

Protein Processing by Pancreatic Enzymes The acidic environment of the stomach destroys the secondary and tertiary structures of proteins. In addition, pepsin cleaves the peptide bonds next to certain amino acids, reducing long polypeptides to relatively small chains of amino acids. In the small intestine, protein digestion is completed so that individual amino acids can enter the bloodstream and be transported to cells throughout the body.

How do these final stages of protein digestion occur? By the end of the nineteenth century, it was established that enzymes in the small intestine digest polypeptides to monomers. Later work showed that each of these protein-digesting enzymes, or **proteases,** is specific to certain types or configurations of amino acids in a polypeptide chain. Thus, a suite of proteases is required to completely digest polypeptides to amino acid monomers.

In addition, by 1900 biologists had determined that proteases are synthesized in an inactive form in the **pancreas,** which is connected to the small intestine by the pancreatic duct. Like the production of inactive pepsinogen by chief cells in the stomach,

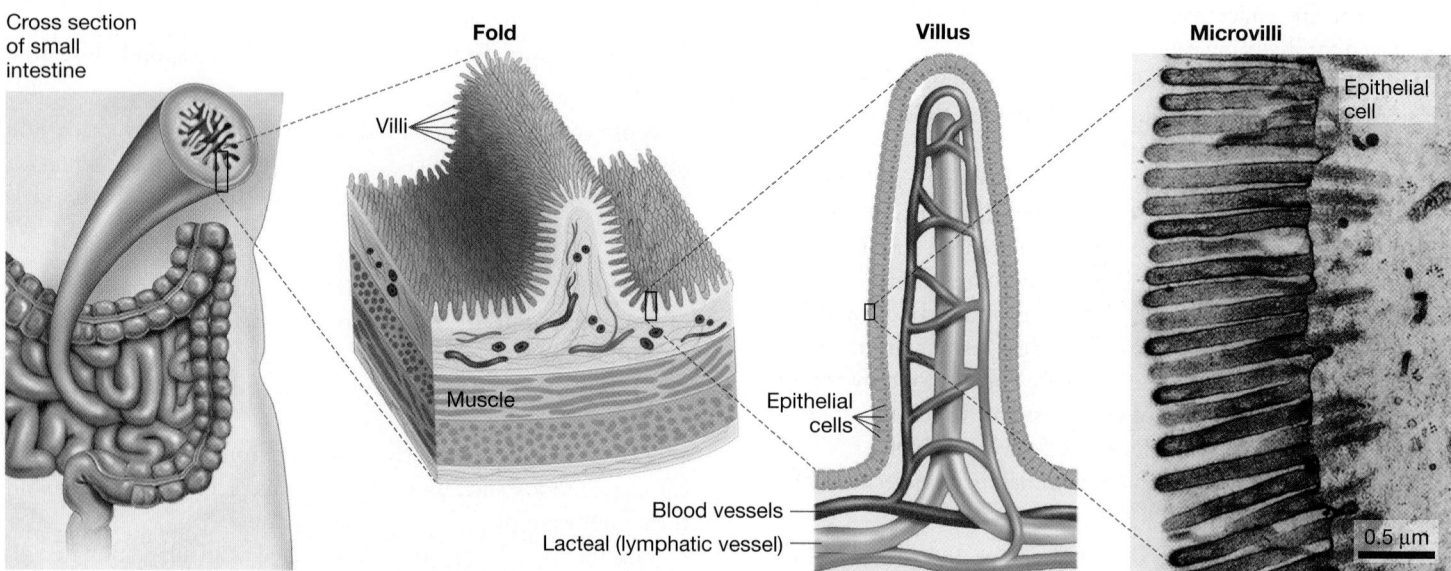

FIGURE 44.12 The Small Intestine Has an Extremely Large Surface Area. The villi that project from folds in the small intestine are covered with microvilli (colorized brown in the micrograph at the far right).

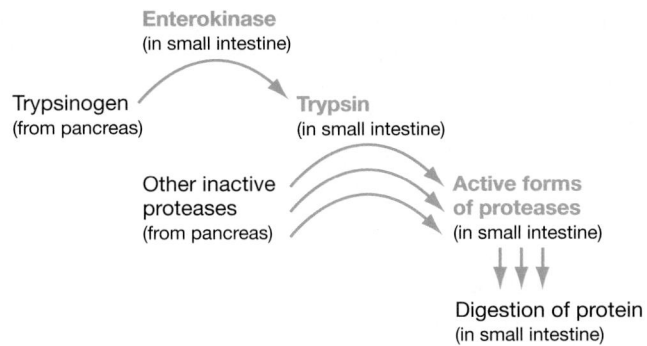

FIGURE 44.13 Enterokinase Triggers an Enzyme-Activation Cascade in the Small Intestine.

the production of digestive enzymes in an inactive conformation prevents pancreatic cells from digesting themselves.

It took decades of work to understand how pancreatic enzymes are activated in the small intestine. In 1900 a researcher showed that contact with juice from the upper part of the small intestine activates pancreatic enzymes. Activation ceased to occur when he heated the intestinal juice, and because he knew that heat denatures enzymes (like all proteins), he hypothesized that the agent responsible for activating the pancreatic enzymes was also an enzyme. He called the unknown enzyme enterokinase.

Decades later, a researcher succeeded in purifying a pancreatic enzyme called trypsinogen and demonstrated that enterokinase activates it in vitro by phosphorylating it, resulting in the active enzyme **trypsin.** Trypsin then triggers the activation of other protein-digesting enzymes, such as chymotrypsin, elastase, and carboxypeptidase (**FIGURE 44.13**). These enzymes are also synthesized by the pancreas and secreted in an inactive form. Once these enzymes are activated in the upper reaches of the small intestine, each begins cleaving specific peptide bonds. Eventually, polypeptides are broken up into amino acid monomers.

What Regulates the Release of Pancreatic Enzymes? Digestive enzymes are needed only when food reaches the small intestine. Based on this simple observation, it was logical to predict that their release would be carefully controlled.

A classic experiment by William Bayliss and Ernest Starling, published in 1902, established how pancreatic enzymes are controlled. Bayliss and Starling began by cutting the nerves that connect to the pancreas and small intestine of a dog. Electrical signaling between the two organs via neurons was now impossible. But when the researchers introduced a weak HCl solution into the upper reaches of the animal's small intestine, to simulate the arrival of material from the stomach, its pancreas secreted enzymes in response.

This observation was startling: The small intestine had successfully signaled the pancreas that food had arrived, even though the nerves connecting the two organs had been cut.

Starling hypothesized that a chemical messenger must be involved, and that the chemical messenger must originate in the small intestine and travel to the pancreas via the blood. He tested this idea by cutting off a small piece of the small intestine,

grinding it up, and injecting the resulting solution into a vein in the animal's neck (see Figure 49.4). Minutes later, the pancreas sharply increased secretion.

Bayliss and Starling had discovered the first **hormone**—a chemical messenger that influences physiological processes at very low concentrations. The molecule they detected, which they called **secretin,** is produced by the small intestine in response to the arrival of food from the stomach.

Follow-up work showed that secretin's primary function is to induce a flow of bicarbonate ions (HCO_3^-) from the pancreas to the small intestine. The bicarbonate is important because it neutralizes the acid arriving from the stomach.

Researchers also discovered a second hormone produced in the small intestine. Called cholecystokinin (pronounced *ko-la-sis-toe-KIN-in*), it induces secretion from the liver and gallbladder as well as the pancreas. **Cholecystokinin** (literally, "bile-bag-mover") stimulates the secretion of digestive enzymes from the pancreas and the secretion of molecules from the gallbladder that aid in processing lipids.

Hormones are involved in stomach function as well. For example, after being stimulated by nerves or the arrival of food, certain stomach cells produce the hormone **gastrin.** In response, parietal cells begin secreting HCl.

How Are Carbohydrates Digested and Transported? In addition to manufacturing protein-digesting enzymes, the pancreas produces nucleases and an amylase that is similar to the salivary enzyme introduced earlier. **Nucleases** digest the RNA and DNA in food; **pancreatic amylase** continues the digestion of carbohydrates that began in the mouth. Carbohydrate digestion ends with the release of monosaccharides such as glucose.

When digestion of proteins and carbohydrates is complete in the small intestine, the resulting slurry is a mixture of nutrients, water, indigestible plant fibers from food, and bacterial cells that live symbiotically in the gut. What molecular mechanisms make it possible for epithelial cells to transport nutrients, like glucose, from the lumen of the small intestine into the bloodstream?

Two general principles apply to carbohydrate and protein absorption: **(1)** It is highly selective, meaning proteins in the plasma membranes of microvilli are responsible for bringing specific nutrients into the cell; and **(2)** it is active, meaning ATP is expended to bring nutrients into the epithelium across a concentration gradient.

Work over the past several decades has demonstrated both these principles. One of the key results grew out of a series of experiments during the 1980s, which established that glucose absorption depends on the presence of an electrochemical gradient favoring an influx of sodium ions into the epithelium. Based on this finding, biologists hypothesized that the apical membranes of epithelial cells must contain a series of cotransporters—membrane proteins that would bring a nutrient molecule into the cell along with sodium ions. To confirm that such a sodium–glucose cotransporter exists, investigators set out to find the gene that codes for the hypothesized membrane protein.

The researchers began by purifying mRNAs from rabbit intestinal cells (**FIGURE 44.14**), which presumably were transcribing the cotransporter genes. Then the team separated the mRNAs by size via gel electrophoresis (see **BioSkills 9** in Appendix B) and injected one of each type of mRNA into a series of frog eggs—cells that do not normally transport glucose. The frog cells translated the rabbit mRNAs into proteins.

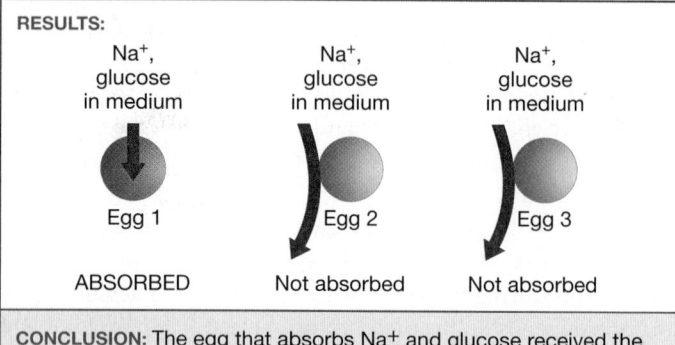

RESEARCH

QUESTION: How is glucose transported into epithelial cells of the small intestine?

HYPOTHESIS: Glucose enters epithelial cells along with sodium ions via a Na$^+$-glucose cotransporter protein.

NULL HYPOTHESIS: Glucose transport does not depend on Na$^+$ transport.

EXPERIMENTAL SETUP:

1. Purify mRNA from intestinal cells.

2. Separate mRNAs by size via gel electrophoresis.

3. Inject individual mRNAs into frog eggs. Test each egg—can it absorb Na$^+$ and glucose?

PREDICTION: An egg will be able to absorb Na$^+$ and glucose, because it received the mRNA that codes for the Na$^+$-glucose cotransporter.

PREDICTION OF NULL HYPOTHESIS: None of the eggs will be able to absorb Na$^+$ and glucose.

RESULTS:

Na$^+$, glucose in medium	Na$^+$, glucose in medium	Na$^+$, glucose in medium
Egg 1	Egg 2	Egg 3
ABSORBED	Not absorbed	Not absorbed

CONCLUSION: The egg that absorbs Na$^+$ and glucose received the mRNA from the Na$^+$-glucose cotransporter gene.

FIGURE 44.14 The Experimental Protocol for Locating the Na$^+$-Glucose Cotransporter Gene.

SOURCE: Wright, E. M. 1993. The intestinal Na$^+$/glucose cotransporter. *Annual Review of Physiology* 55: 575–589.

✔**QUESTION** Why did the researchers inject the RNAs into frog eggs instead of into rabbit epithelial cells?

When tested, one of the experimental eggs was able to import Na$^+$ and glucose in tandem. The biologists inferred that this egg had received the mRNA for the rabbit Na$^+$-glucose cotransporter. The researchers made a DNA copy of the mRNA (using techniques introduced in Chapter 20), analyzed it to determine the sequence of the gene, and from that inferred the amino acid sequence of the membrane protein.

The discovery of the Na$^+$-glucose cotransporter inspired a three-step model for glucose absorption:

1. Na$^+$/K$^+$-ATPase (sodium–potassium pump) in the basolateral membrane of the epithelial cells creates an electrochemical gradient that favors the entry of Na$^+$.

2. Glucose from digested food enters the cell along with sodium via the Na$^+$-glucose cotransporter in the apical membrane.

3. Glucose diffuses into nearby blood vessels through a glucose carrier in the basolateral membrane.

If this configuration of pumps, cotransporters, and carriers sounds familiar, there is a good reason: The same combination of membrane proteins occurs in the proximal tubule of the kidney, where the proteins are responsible for the reabsorption of sodium and glucose from urine (see Chapter 43).

Follow-up work showed that in the small intestine—just as in the proximal tubule—other cotransporters are responsible for the absorption of other monosaccharides and of amino acids, and specific channels and carriers in the basolateral membrane are responsible for the transport of each substance to the blood.

Digesting Lipids: Bile and Transport The pancreatic secretions include digestive enzymes that act on fats, in addition to enzymes that act on proteins and carbohydrates. Like the lingual lipase added to saliva in the mouth, the enzyme **pancreatic lipase** breaks certain bonds present in complex fats and results in the release of fatty acids and other small lipids.

Recall that fats are insoluble in water (see Chapter 6). As a result, they tend to form large globules as they are churned in the stomach. Before pancreatic lipase can act, the large fat globules that emerge from the stomach must be broken up—a process known as **emulsification.**

In the small intestine, emulsification results from the action of small molecules called bile salts. As **FIGURE 44.15** shows, bile salts function like the detergents that researchers use to break up the lipids in plasma membranes (see Chapter 6).

Bile salts are synthesized in the **liver,** an organ that performs an array of functions related to digestion, and secreted in a complex solution called **bile,** which is stored in the **gallbladder.** When bile enters the small intestine, it raises the pH and emulsifies fats. Once fats are broken into small globules, which increases their surface area, they can be attacked by pancreatic lipase and digested. **TABLE 44.3** summarizes the major digestive enzymes.

The monoglycerides and fatty acids released by lipase activity enter small intestine epithelial cells by simple diffusion and are processed into protein-coated globules called chylomicrons, which move by exocytosis into lacteals—see Figure 44.12—near

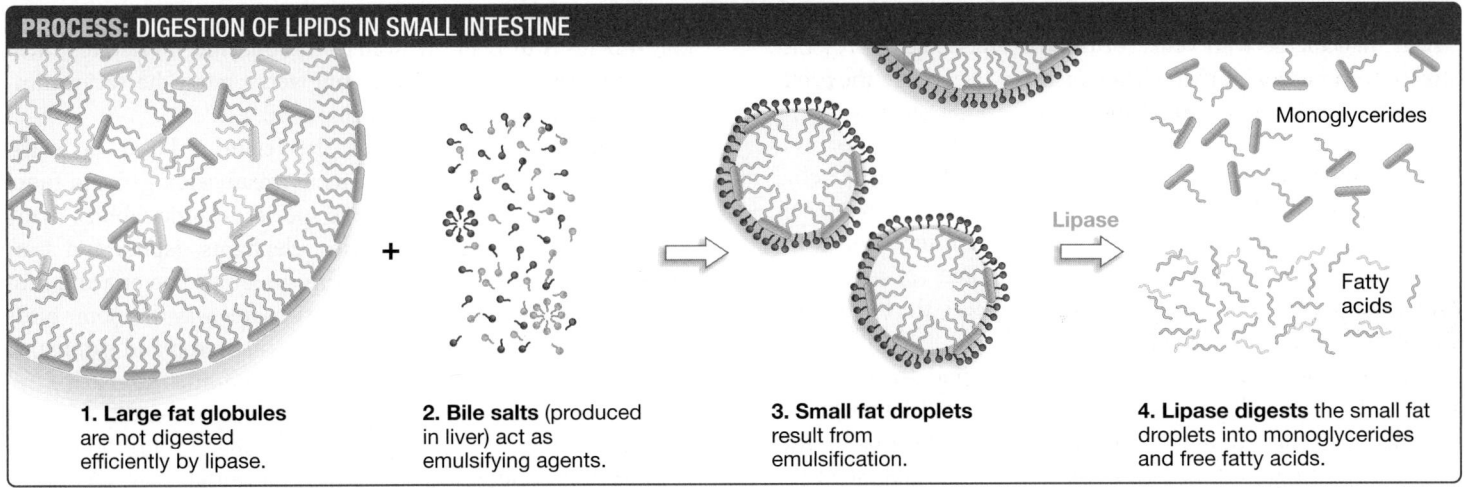

1. **Large fat globules** are not digested efficiently by lipase.

2. **Bile salts** (produced in liver) act as emulsifying agents.

3. **Small fat droplets** result from emulsification.

4. **Lipase digests** the small fat droplets into monoglycerides and free fatty acids.

FIGURE 44.15 Emulsifying Agents and Lipases Digest Lipids in the Small Intestine. Once bile salts break up large fat globules, lipase can digest fats efficiently.

SUMMARY TABLE 44.3 **Digestive Enzymes in Mammals**

Digestion is accomplished by the enzymes listed here, by HCl produced in the stomach in response to the hormone gastrin, and by bile salts from the liver. Bile salts are stored in the gallbladder. They are released in response to the hormone cholecystokinin and emulsify fats in the small intestine.

	Where Synthesized	Regulation	Function
Carboxypeptidase	Pancreas	Released in inactive form in response to cholecystokinin from small intestine; activated by trypsin	In small intestine, breaks peptide bonds in polypeptides—releasing amino acids
Chymotrypsin	Pancreas	Released in inactive form in response to cholecystokinin from small intestine; activated by trypsin	In small intestine, breaks peptide bonds in polypeptides—releasing amino acids
Elastase	Pancreas	Released in inactive form in response to cholecystokinin from small intestine; activated by trypsin	In small intestine, breaks peptide bonds in polypeptides—releasing amino acids
Lingual lipase	Salivary glands	Released in response to taste and smell stimuli	In mouth and stomach, breaks bonds in fats—releasing fatty acids and monoglycerides
Nucleases	Pancreas	Released in response to cholecystokinin from small intestine	In small intestine, break apart nucleic acids—releasing nucleotides
Pancreatic amylase	Pancreas	Released in response to cholecystokinin from small intestine	In small intestine, breaks apart carbohydrates—releasing sugars
Pancreatic lipase	Pancreas	Released in response to cholecystokinin from small intestine	In small intestine, breaks bonds in fats—releasing fatty acids and monoglycerides
Pepsin	Stomach	Released in inactive form (pepsinogen); activated by low pH in stomach lumen	In stomach, breaks peptide bonds between certain amino acids in proteins—releasing polypeptides
Salivary amylase	Salivary glands	Released in response to taste and smell stimuli	In mouth, breaks apart carbohydrates—releasing sugars
Trypsin	Pancreas	Released in inactive form (trypsinogen) in response to cholecystokinin from small intestine; activated by hormone enterokinase from small intestine	In small intestine, breaks specific peptide bonds in polypeptides—releasing amino acids

the epithelial cells. The lacteals merge with larger lymph vessels, which then merge with veins. In this way, fats enter the bloodstream without clogging small blood vessels. Eventually, the products of fat digestion end up in adipose tissue and other tissues.

How Is Water Absorbed?   When solutes from digested material are absorbed into the epithelium of the small intestine, water follows passively by osmosis. This is an important mechanism for (1) absorbing water that has been ingested, and (2) reclaiming liquid that was secreted into the digestive tract in the form of saliva, mucus, and pancreatic fluid.

This mechanism of water absorption inspired an important medical strategy called oral rehydration therapy. If a patient has diarrhea, clinicians frequently prescribe dilute solutions of glucose and electrolytes to be taken orally. When the glucose in the drink is absorbed in the small intestine through sodium-glucose cotransporters, enough water and sodium follow to prevent the life-threatening effects of dehydration. This simple medication saves thousands of lives every year. ✔ **If you understand this concept, you should be able to predict at least two effects of a molecule that selectively blocks the sodium-glucose cotransporter.**

The Large Intestine

By the time digested material reaches the large intestine of a human, a large amount of water (approximately 5 liters per day in total) and virtually all of the available nutrients have been absorbed. The primary function of the **large intestine** is to compact the wastes that remain and to absorb enough water to form feces.

These processes occur in the **colon**—the main section of the large intestine. Feces are held in the **rectum,** which is the final part of the large intestine, until they can be excreted. Although the kidneys are responsible for maintaining water balance, water absorption in the large intestine is important for keeping the body well hydrated.

In addition to compacting wastes and absorbing water, the human colon contains symbiotic microorganisms that digest cellulose. Although fermentation of cellulose is not as important to an omnivorous human as it is to an herbivore like a ruminant, bacteria in the human colon also produce several important nutrients, such as vitamin K.

Aquaporins Play a Key Role in Water Reabsorption   Water is absorbed in the large intestine through aquaporins. Recall that **aquaporins** are proteins that act as water channels in plasma membranes, thus increasing the rate of water movement via osmosis (see Chapter 6). The best-studied of these proteins, called AQP1 for aquaporin 1, is common in the nephrons of the kidney. AQP1 is one of the proteins responsible for water reabsorption from urine along the osmotic gradient (described in Chapter 43). To date, four distinct aquaporins have been found in the large intestines of rats, mice, and humans. Researchers are working to unravel exactly how the activity of these aquaporins is regulated.

Variations in Structure and Function   The size and function of the large intestine varies dramatically among animals. In insects,

the posteriormost portion of the digestive tract, called the hindgut, functions to reabsorb water and ions and excrete uric acid and feces (see Chapter 43). Among vertebrates, the various lineages of fish have no large intestine at all.

In some herbivorous species, the **cecum,** a blind sac at the proximal end of the large intestine, is greatly enlarged and functions in digesting cellulose. These species include rabbits, many rodents, some marsupials, horses, elephants, tapirs, and leaf-eating primates. The cecum contains symbiotic bacteria and protists that ferment cellulose, in much the same way that some birds house cellulose-digesting symbionts in their crop and ruminants house cellulose-digesting symbionts in a modified stomach.

In humans, a narrow pouch called the **appendix** emerges from the cecum. The function of the appendix has long been debated. It has often been described as vestigial—a reduced or incompletely developed trait that is a vestige of evolutionary ancestry (see Chapter 25)—partly because it does not perform any obvious vital function. Indeed, if it becomes inflamed, it can be surgically removed from a patient and cause no ill effects. However, although not vital, the appendix may have important functions: it is imbued with immune cells, and it appears to act as a haven for symbiotic microorganisms that inhabit the colon. For example, serious infections can result in diarrhea that flushes symbiotic bacteria from the colon. The appendix may provide the additional bacteria needed for recolonizing the colon after such an episode.

Rabbits and some other species with a cecum are able to excrete the cecum's contents as pellets, which the animal then reingests and passes through the digestive tract a second time. This particular example of **coprophagy,** or excrement eating, allows the animal to absorb more nutrients from the food.

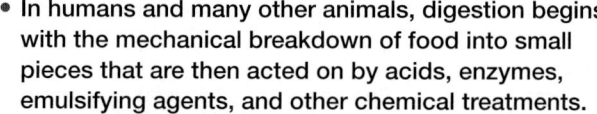
check your understanding

If you understand that . . .
- In humans and many other animals, digestion begins with the mechanical breakdown of food into small pieces that are then acted on by acids, enzymes, emulsifying agents, and other chemical treatments.
- Distinct compartments within a digestive tract have distinct structures and functions.
- Nutrients are absorbed through specific membrane proteins in epithelial cells of the digestive tract.

✔ **You should be able to . . .**
1. Explain how each compartment in the human digestive tract aids the ingestion and digestion of food, absorption of nutrients, and excretion of wastes.
2. Predict the consequences of treating a person with drugs that inhibit the release of bile salts, that inactivate trypsin, or that block the action of the Na^+-glucose cotransporter in the epithelial cells of the small intestine.

Answers are available in Appendix A.

Another striking structural variation occurs in the rectum of amphibians, reptiles, and birds. Their urine flows from the kidneys into an enlarged portion of the large intestine called the **cloaca** (see Chapter 43). Instead of having two orifices for excretion, these species have one.

44.4 Nutritional Homeostasis—Glucose as a Case Study

When digestion is complete, amino acids, fatty acids, ions, and sugars enter the bloodstream and are delivered to the cells that need them. The body uses or stores these nutrients in order to maintain homeostatic levels in the blood and avoid imbalances. Too much of a nutrient or too little can be problematic or even fatal.

The illness **diabetes mellitus** is a classic example of nutrient imbalance. People with diabetes mellitus experience abnormally high levels of glucose in their blood. Over the course of a lifetime, chronically elevated blood glucose can lead to an array of complications, including blindness, impaired circulation, and heart failure. What causes the imbalance? More important, how is glucose homeostasis normally maintained?

The Discovery of Insulin

In 1879, researchers removed the pancreas from a dog and observed that the dog's blood glucose became very high. This experiment suggested that the pancreas secretes a compound needed for removing glucose from the blood.

When other investigators cut up pancreatic tissues and injected extracts into diabetic dogs, however, they observed no response. Eventually, they decided that digestive enzymes in the pancreas were probably destroying the active agent during the extraction process.

Then, in 1921, Frederick Banting and Charles Best conducted a breakthrough experiment:

1. They began by tying off a dog's pancreatic duct—the tube through which digestive enzymes flow to the small intestine. The logic was that blocking the secretion of digestive enzymes might kill the cells that synthesize them.

2. The investigators waited several weeks for the cells near the duct to die and then removed the pancreas.

3. They froze the pancreatic tissue, ground it up, and injected an extract into a diabetic dog.

4. To their delight, the dog's blood sugar levels stabilized, and the dog became more active and healthy looking.

After Banting and Best repeated the experiment and observed the same result, they grew increasingly confident that they had located the source of the molecule responsible for lowering blood glucose levels. The molecule came to be called insulin.

Insulin's Role in Homeostasis

Insulin is a hormone that is secreted by the pancreas when blood glucose levels are high. It travels through the bloodstream and binds to receptors on cells throughout the body (see Chapter 49 for more detail on the structure and function of hormones and other chemical signals).

In response, cells that have insulin receptors increase their rate of glucose uptake and processing. Specifically, insulin stimulates cells in the liver and skeletal muscle to import glucose from the blood and synthesize glycogen from glucose monomers. As a result, glucose levels in the blood decline.

If blood glucose levels fall, as they do when food is lacking, cells in the pancreas secrete a hormone called **glucagon.** In response to glucagon, cells in the liver catabolize glycogen, and also produce glucose via **gluconeogenesis,** the synthesis of glucose from non-carbohydrate compounds. As a result, glucose levels in the blood rise (**FIGURE 44.16**).

Insulin and glucagon interact to form a negative feedback system capable of achieving homeostasis with respect to glucose concentrations in the blood.

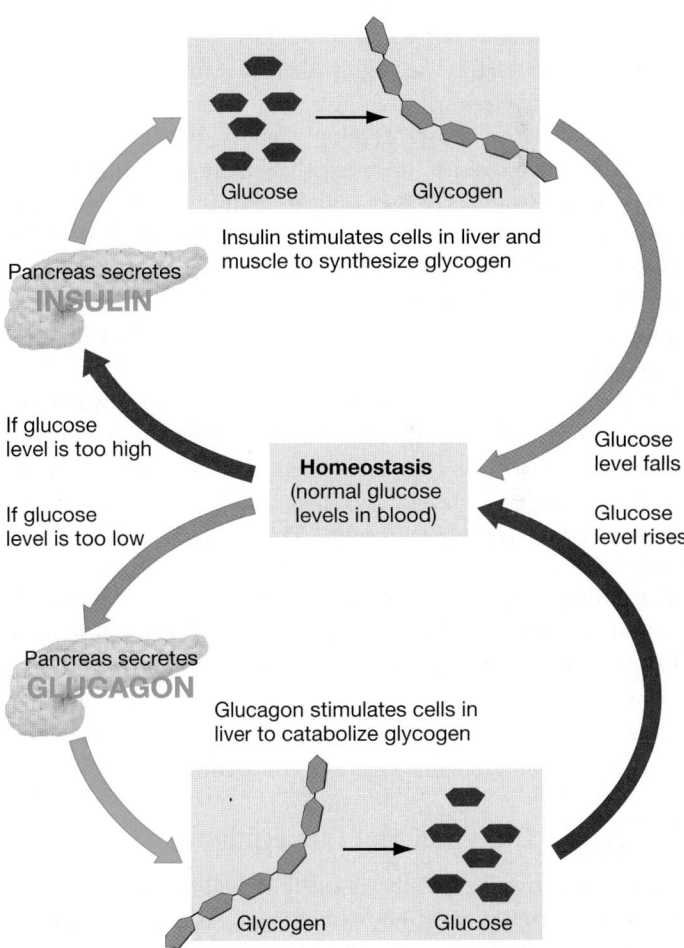

FIGURE 44.16 Insulin and Glucagon Provide Negative Feedback in a Homeostatic System. Both insulin and glucagon are secreted by cells in the pancreas but have opposite effects on blood glucose concentrations.

✔ QUESTION After reading the descriptions of type 1 and type 2 diabetes mellitus, decide which arrow is disrupted in individuals with type 2 diabetes mellitus and which arrow is disrupted in individuals with type 1 diabetes mellitus.

Diabetes Mellitus Has Two Forms

Diabetes mellitus develops in people who (1) do not synthesize sufficient insulin or (2) are resistant to insulin, meaning that insulin does not effectively activate its receptor in target cells. The first condition is called type 1 diabetes mellitus; the second condition is type 2 diabetes mellitus. In both cases, effector cells do not receive the signal that would result in a drop in blood glucose levels.

Glucose imbalance has a direct effect on urine formation. Normally, signals from insulin keep blood glucose levels low enough that all of the glucose can be reabsorbed from the filtrate formed in the kidney. But when blood glucose levels are high, so much glucose enters the nephron that it cannot all be reabsorbed. High glucose concentrations in the filtrate increase its osmolarity and decrease the amount of water reabsorbed from it. More water leaves the body, leading to high urine volume in both types of diabetes mellitus.

The word diabetes means "to run through"; water "runs through" people with diabetes. In addition, the word mellitus means "honeyed (sweet)." Before chemical methods of analyzing urine were available, physicians would taste the patient's urine. If the urine was sweet, it was highly likely that the patient was suffering from diabetes mellitus.

Why do these diseases develop? Type 1 diabetes mellitus is an autoimmune disease, meaning that the body's immune system mistakenly targets its own cells for destruction. In the case of type 1 diabetes mellitus, the insulin-producing cells of the pancreas are destroyed. Type 2 diabetes mellitus occurs when the receptors for insulin no longer function correctly or are reduced in number. The primary risk factor for developing type 2 diabetes mellitus is obesity.

Currently, type 1 diabetes mellitus is treated with insulin injections and careful attention to diet; type 2 diabetes mellitus is managed primarily through prescribed diets, exercise, and monitoring blood glucose levels, as well as taking drugs that increase cellular responsiveness to insulin. The challenge is to achieve homeostasis with respect to blood glucose in the absence of the body's normal regulatory mechanisms.

✔ If you understand the difference between type 1 and type 2 diabetes mellitus, you should be able to explain how, in individuals with each disease and without disease, insulin receptors in the plasma membrane of liver cells interact with insulin and how the liver cells respond to high glucose concentrations in the blood.

The Type 2 Diabetes Mellitus Epidemic

An epidemic of type 2 diabetes mellitus is currently under way in certain human populations. For example, as of 2011, a total of 11 percent of U.S. adults aged 20 and over were diabetic, and this figure jumps to 27 percent in U.S. adults aged 65 and over. About 90% to 95% of cases are type 2 diabetes mellitus. Minority groups such as African Americans and Hispanics are experiencing much higher rates of new cases of type 2 diabetes mellitus than white non-Hispanic Americans.

Because there is a strong association between the prevalence of diabetes mellitus in parents and in children, researchers have long suspected that some individuals have a genetic predisposition for developing the disease. Researchers have so far identified alleles at 38 different genes that predispose individuals to type 2 diabetes mellitus (using techniques introduced in Chapter 20 and Chapter 21).

However, there is also strong evidence that environmental conditions have an important impact on the incidence of type 2 diabetes mellitus. As an example, consider the Pima Indians of North America.

The Pima consist of two main populations—one in southwest Arizona in the United States and one in a remote area of the Sierra Madre mountains of Mexico. The bars graphed in **FIGURE 44.17a** indicate the average percentage of individuals with type 2 diabetes mellitus, by gender, in Mexican people who are

(a) Variation in average incidence of type 2 diabetes mellitus

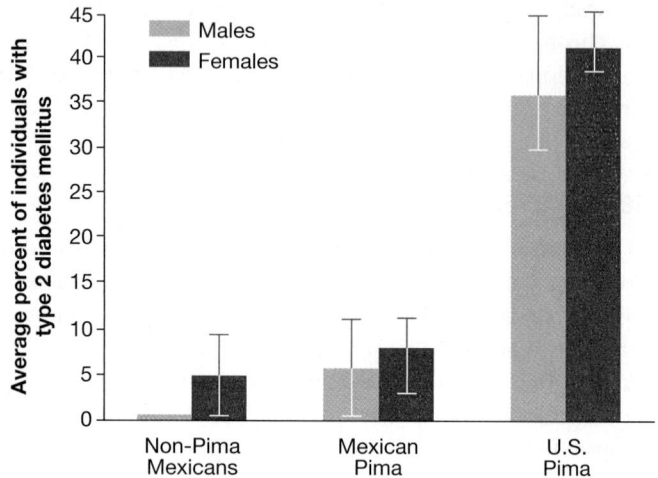

(b) Variation in average body mass index

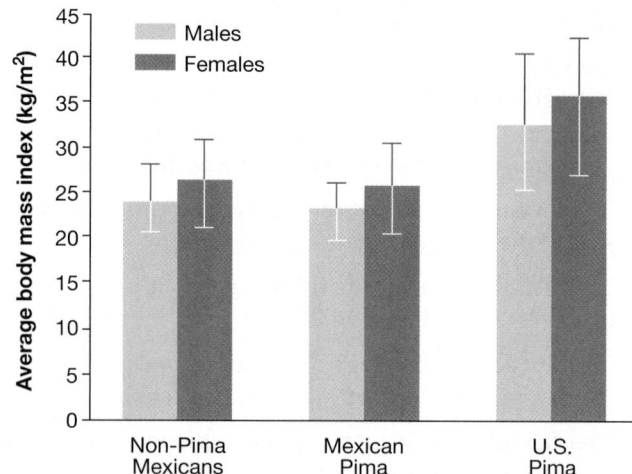

FIGURE 44.17 The Incidence of Type 2 Diabetes Mellitus Is Correlated with Obesity.

DATA: Schultz, L. O., P. H. Bennett, E. Ravussin, et al. 2006. *Diabetes Care* 29: 1866–1871.

✔**QUANTITATIVE** In the graphs above, it is apparent that U.S. Pima have higher rates of type 2 diabetes mellitus and higher body mass indices than the other groups. Does this prove that obesity causes diabetes mellitus? Why or why not?

not members of the Pima and in the two populations of Pima. Although researchers have been unable to find any significant genetic differences between the two groups of Pima, the data indicate there are dramatic differences in the incidence of type 2 diabetes mellitus.

The incidence of type 2 diabetes mellitus is correlated with obesity. As the data graphed in **FIGURE 44.17b** show, Pima from the United States are more likely to be obese than individuals in the other two populations in this study. A person who has a body mass index greater than or equal to 30 is considered obese. The **body mass index** is calculated as weight (in kilograms) divided by height squared (in meters).

The differences in body mass index among the Pima are explained by differences in physical activity and diet. Pima men in Mexico are physically active an average of 33 hours a week, while their counterparts in the United States are active just 12.1 hours per week. Pima women in Mexico average 22 hours of physical activity a week; Pima women in the United States average just 3.1. In addition, Pima in the United States get approximately 40 percent of their calories from fat, compared to 15 percent in the traditional Pima diet.

Although Pima people in the United States have the highest rates of type 2 diabetes mellitus recorded for any people to date, the same trends are occurring in populations all over the world. Nutrition-related diseases have become a major public health concern.

Type 2 diabetes mellitus used to be called adult-onset diabetes because the disease typically appeared in adults. However, that name has been abandoned as more and more young people are diagnosed with the disease. While a multitude of genes can predispose a child or adult to developing diabetes mellitus, the disease is triggered by environmental factors including diet, lack of exercise, and obesity.

In a five-year span in the past decade, obesity rates increased by 10 percent in American children overall and by over 30 percent in children from low-income families and certain racial minority groups. Scientists estimate that one in three children born nowadays will develop diabetes mellitus, and that rates are likely to continue rising. It is clear that diabetes mellitus is a modern-day epidemic.

check your understanding

If you understand that . . .

- Insulin and glucagon interact to regulate glucose concentrations in the blood.
- Diabetes mellitus results from a lack of homeostasis in the concentration of the nutrient glucose in the blood.

✔ **You should be able to . . .**

1. Identify the causes of each type of diabetes mellitus.

2. Prescribe what a person with type 1 diabetes mellitus should do when the blood glucose level is too high and when the blood glucose level is too low.

Answers are available in Appendix A.

CHAPTER 44 REVIEW

For media, go to MasteringBiology

If you understand . . .

44.1 Nutritional Requirements

- The diets of animals include fats, carbohydrates, and proteins that provide energy; vitamins that serve as coenzymes and perform other functions; ions required for water balance and for nerve and muscle function; and elements that are incorporated into molecules synthesized by cells.

- Fats contain the most energy of all nutrients: 9 kcal/g compared to 4 kcal/g for carbohydrates and proteins, making fats an efficient way to store energy in the body.

- ✔ **QUANTITATIVE** You should be able to calculate and compare the caloric content of skim milk and whole milk. Per serving, skim milk contains 12 g carbohydrates, 8 g protein, and no fat; whole milk contains 12 g carbohydrates, 8 g protein, and 8 g fat.

44.2 Capturing Food: The Structure and Function of Mouthparts

- Most animals are mass feeders that obtain food by seizing and manipulating it with mouthparts such as teeth, jaws, or beaks or with special toxin-injecting organs.

- Natural selection has modified mouthparts in different species to act as efficient tools for obtaining particular types of food.

- ✔ You should be able to predict the types of mouthparts and the nature of the digestive tract in a mammal that eats only nectar from flowers.

44.3 How Are Nutrients Digested and Absorbed?

- In most animals, the digestive tract begins at the mouth and ends at the anus.

- In many species, chemical digestion of food also begins in the mouth. In mammals, an enzyme in saliva called salivary amylase begins to hydrolyze the bonds linking glucose monomers in starch, glycogen, and other carbohydrates.

- Once food is swallowed, it is propelled down the esophagus by peristalsis.

- Digestion continues in the stomach. In humans, the stomach is a highly acidic environment that denatures proteins and in which the enzyme pepsin begins the cleavage of peptide bonds that link amino acids.

- Food passes from the stomach into the small intestine, where it is mixed with secretions from the pancreas and liver.

- In the small intestine, carbohydrate digestion is continued by pancreatic amylase; fats are emulsified by bile salts and digested by pancreatic lipase; protein digestion is completed by a suite of pancreatic enzymes.

- Cells that line the small intestine absorb the nutrients released by digestion. In many cases, uptake is driven by an electrochemical gradient established by Na^+/K^+-ATPase that favors a flow of Na^+ into the cell.

- As solutes leave the lumen of the small intestine and enter cells, water follows by osmosis.

- Water reabsorption is completed in the large intestine, where feces form.

- The structure of organs in the digestive tract varies widely among species, in ways that support efficient processing of the food a particular species ingests.

✔ You should be able to compare the optimal pH for amylase and pepsin.

44.4 Nutritional Homeostasis— Glucose as a Case Study

- Diabetes mellitus is a condition in which concentrations of glucose in the blood are chronically too high.

- Type 1 diabetes mellitus is caused by a defect in the production of insulin—a hormone secreted by the pancreas that promotes the uptake of glucose from the blood.

- Type 2 diabetes mellitus is characterized by a failure of cells to respond to insulin.

- The development of type 2 diabetes is correlated with obesity and is reaching epidemic proportions in some populations.

✔ You should be able to explain how each type of diabetes mellitus results in high blood glucose.

You should be able to . . .

✔ TEST YOUR KNOWLEDGE

Answers are available in Appendix A

1. What does secretin stimulate?
   a. secretion of HCO_3^- from the pancreas
   b. secretion of HCl from the stomach epithelium
   c. secretion of digestive enzymes from the pancreas
   d. uptake of glucose from the bloodstream by cells throughout the body

2. What is an incomplete digestive tract?
   a. a digestive tract that is missing a major compartment—for example, fish lack a large intestine
   b. a ruminant-type tract, where food is regurgitated and re-chewed before digestion is completed
   c. a reduced or vestigial tract like that found in animals that absorb nutrition directly across the body wall
   d. a digestive system with a single opening for ingestion and excretion

3. In mammals, how and where are carbohydrates digested?
   a. by lipases in the small intestine
   b. by pepsin and HCl in the stomach
   c. by aquaporins in the large intestine
   d. by amylases in the mouth and small intestine

4. Cellulose is fermented in which of the following structures in rabbits?
   a. small intestine
   b. cecum
   c. abomasum
   d. rumen

5. A hormone that reduces blood glucose is _____, while a hormone that increases blood glucose is _____.

6. True or False: Type 1 diabetes mellitus is more prevalent in the United States than type 2 diabetes mellitus.

7. Explain how the structure of each of the following digestive tract compartments correlates with its function: the bird crop, cow rumen, and elephant large intestine.

8. Why are many digestive enzymes produced in an inactive form and then activated in the lumen of the digestive tract?

9. Explain why oral rehydration therapy with a solution of sodium and glucose is effective.
   a. The sodium and glucose decrease urine output.
   b. The sodium and glucose facilitate water absorption by the small intestine.
   c. The sodium and glucose help kill the intestinal bacteria.
   d. The sodium and glucose make the person thirsty.

10. How does the high type 2 diabetes mellitus rate among Pima Indians in the United States provide evidence for an environmental effect on development of the disease?

11. What do the shapes of human teeth suggest about their function in terms of the types of food humans are adapted to eat?

12. Predict what would happen if a type 1 diabetic took too much insulin.

13. Doctors sometimes prescribe proton pump inhibitors to treat patients with acid reflux. How do these drugs work?

14. An elderly man arrives at the hospital severely dehydrated. The physicians find that there is glucose in his urine and that his blood glucose does not decline much upon administration of an insulin shot. What is he likely to be suffering from?
    a. type 1 diabetes mellitus
    b. type 2 diabetes mellitus
    c. diarrhea
    d. gastric ulcer

15. Scientists who backed the hypothesis that secretion from the pancreas is under nervous control strenuously objected to the conclusions reached in the experiment that led to the discovery of hormonal control. They said the experiment was inconclusive, because of the likelihood that not all the nerves contacting the small intestine had been cut. The biologists who did the experiment replied that even if not all the nerves had been cut, the result was still valid. In your opinion, who is correct? Why?

16. Among vertebrates, the large intestine occurs only in lineages that are primarily terrestrial (amphibians, reptiles, and mammals). Propose a hypothesis to explain this observation.

45 Gas Exchange and Circulation

In this chapter you will learn that

> **Animals have adaptations for gas exchange across body surfaces and circulation within their bodies**

via ↓

> **Respiratory and circulatory systems** **45.1**

comparing ↓

> **O_2 and CO_2 exchange in air vs. water** **45.2**

asking ↓

> **How do different gas exchange organs work?** **45.3**

exploring ↓

> **O_2 and CO_2 transport in blood** **45.4**

asking ↓

> **How do different circulatory systems work?** **45.5**

During intense exercise, animal circulatory systems deliver large amounts of oxygen to tissues and remove large amounts of carbon dioxide. This chapter explores how gas exchange occurs in animals that live in aquatic or terrestrial environments.

This chapter is part of the Big Picture. See how on pages 840–841.

Animal cells are like factories that run 24 hours a day. Inside the plasma membrane, the chemical reactions that sustain life require a steady input of raw materials. Those reactions also produce a steady stream of wastes. Earlier in this unit, you learned how solid and liquid waste materials are removed from the cells and excreted from the body (see Chapter 43). You also examined how nutrients enter the body (see Chapter 44). Let's now turn our attention to two other major questions:

1. How are two of the most important molecules in the economy of the cell—the oxygen (O_2) required for cellular respiration and the carbon dioxide (CO_2) produced by cellular respiration—exchanged with the environment?

✔ When you see this checkmark, stop and test yourself. Answers are available in Appendix A.

2. How are these gases—along with wastes, nutrients, and other types of molecules—transported throughout the body?

Understanding gas exchange and circulation is fundamental to understanding how animals work. If either process fails, the consequences are dire. Consider the maladies that can develop when gas exchange or circulation is disrupted in humans: anemia, pneumonia, tuberculosis, malaria, heart disease, and stroke are just a few examples.

Let's begin with an overview of animal respiratory and circulatory systems and then plunge into the details of how gases are exchanged and transported.

45.1 The Respiratory and Circulatory Systems

When the mitochondria inside animal cells are producing ATP via cellular respiration, they consume oxygen and produce carbon dioxide. To support continued ATP production, cells have to obtain oxygen and expel excess carbon dioxide continuously (see The Big Picture: Energy, pages 198–199).

How does this gas exchange occur between an animal's environment and its mitochondria? In most cases, gas exchange involves the four steps illustrated in **FIGURE 45.1**:

1. **Ventilation,** the movement of air or water through a specialized gas exchange organ, such as lungs or gills.

2. **Gas exchange,** the diffusion of O_2 and CO_2 between air or water and the blood at the respiratory surface.

3. **Circulation,** the transport of dissolved O_2 and CO_2 throughout the body—along with nutrients, wastes, and other types of molecules—via the circulatory system.

4. **Cellular respiration,** the cell's use of O_2 and production of CO_2. In tissues, where cellular respiration has led to low O_2 levels and high CO_2 levels, gas exchange occurs between blood and cells.

Steps 1 and 2 are accomplished by the **respiratory system,** the collection of cells, tissues, and organs responsible for gas exchange between the individual animal and its environment. In essence, a respiratory system consists of structures for conducting air or water to a surface where gas exchange takes place.

In some animals the gas exchange surface is the skin, but in most species it is located in a specialized organ like the lungs of tetrapods, the tracheae of insects, or the gills found in mollusks, arthropods, and fishes. Section 45.3 analyzes the structure and function of gills, tracheae, and lungs in detail.

Step 3 in Figure 45.1 is usually accomplished by a **circulatory system,** which moves O_2, CO_2, and other materials around the body. In many cases, a muscular heart propels a specialized, liquid transport tissue throughout the body via a system of vessels.

Keeping in mind this broad overview of gas exchange and circulatory systems, let's dive into the details by exploring how oxygen and carbon dioxide move between an animal's body and its environment.

45.2 Air and Water as Respiratory Media

Gas exchange between the environment and cells is based on diffusion. Under normal conditions, oxygen concentrations are relatively high in the environment (for example, in air that you inhale or in ocean water) and low in tissues, while carbon dioxide levels are relatively high in tissues and low in the environment. Thus oxygen tends to move from the environment into tissues, and carbon dioxide tends to move from tissues into the environment.

How much oxygen and carbon dioxide are present in the atmosphere versus the ocean? What factors in air and water influence how quickly these gases move by diffusion?

How Do Oxygen and Carbon Dioxide Behave in Air?

The atmosphere is composed primarily of nitrogen (76%) and oxygen (21%) and has trace amounts of argon (0.93%) and CO_2 (0.03%). Nitrogen (N_2) and argon are not important to animals living at sea level and are usually ignored in analyses of gas exchange.

However, the data are actually a little misleading. To understand why, consider that the *percentage* of O_2 in the atmosphere does not vary with elevation. The atmosphere at the top of Mt. Everest is

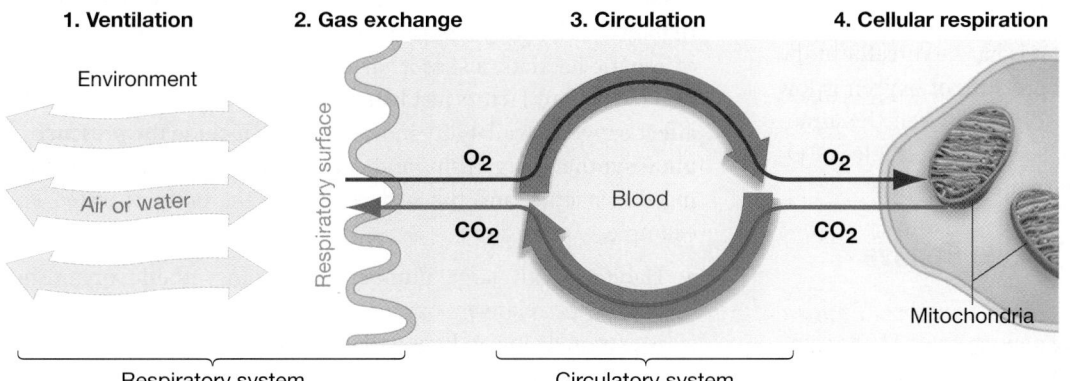

1. Ventilation   **2. Gas exchange**   **3. Circulation**   **4. Cellular respiration**

Environment

Air or water

Respiratory surface

O_2

CO_2

Blood

O_2

CO_2

Mitochondria

Respiratory system   Circulatory system

FIGURE 45.1 Gas Exchange Involves Ventilation, Circulation, and Respiration. In animals, oxygen and carbon dioxide are exchanged across the surface of a lung, a gill, the skin, or some other gas exchange organ. In many species these gases are then transported to and from cells—where gas exchange again takes place—in a fluid tissue such as blood.

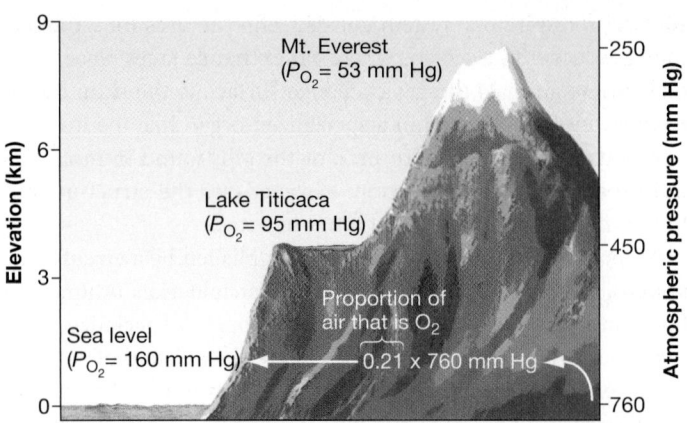

FIGURE 45.2 Oxygen Makes Up 21 Percent of the Atmosphere, but Its Partial Pressure Depends on Altitude. Atmospheric pressure, and thus oxygen partial pressure (P_{O_2}), fall with increasing elevation.

composed of 21 percent oxygen, just as it is at sea level. The key difference is that far fewer molecules of oxygen and other atmospheric gases are present at high elevations than at sea level. Air at the top of Mt. Everest is less dense than air at sea level, so less oxygen is present. To understand how gases move by diffusion, it is important to express their presence in terms of partial pressures instead of percentages. Pressure is a type of force. A **partial pressure** is the pressure of a particular gas in a mixture of gases.

To calculate the partial pressure of a particular gas, multiply the fractional composition (the fraction of air the gas comprises) by the total pressure exerted by the entire mixture (atmospheric pressure). The calculation is valid because the total pressure of a mixture of gases is the sum of the partial pressures of all the individual gases. This relationship is called Dalton's law.

For example, **FIGURE 45.2** shows that the total atmospheric pressure at sea level is 760 mm Hg (millimeters of mercury). If you multiply this value by 0.21, which is the fraction of air that is O_2, you obtain a partial pressure of oxygen, abbreviated P_{O_2}, at sea level of 160 mm Hg. Because the atmospheric pressure is only about 250 mm Hg at the top of Mt. Everest, the P_{O_2} there is only $0.21 \times 250 = 53$ mm Hg.

Oxygen and carbon dioxide diffuse between the environment and cells along their respective partial pressure gradients, just as solutes diffuse along their electrochemical gradients. In both air and water, O_2 and CO_2 move from regions of high partial pressure to regions of low partial pressure. It is hard to breathe at the top of Mt. Everest because the partial pressure of oxygen is low there—meaning that the diffusion gradient between the atmosphere and your lung tissues is small, so fewer molecules of O_2 diffuse into your tissues when you take a breath.

How Do Oxygen and Carbon Dioxide Behave in Water?

To obtain oxygen, water breathers face a much more challenging environment than air breathers do. Aquatic animals live in an environment that contains much less oxygen than the environments inhabited by terrestrial animals. At 15°C, a liter of air can contain up to 209 milliliters (mL) of O_2, while a liter of water may contain a maximum of only 7 mL of O_2. To extract a given amount of oxygen, an aquatic animal has to process 30 times more water than the amount of air a terrestrial animal breathes.

In addition, water is about a thousand times denser than air and flows much less easily. As a result, water breathers have to expend much more energy to ventilate their respiratory surfaces than do air breathers.

What Affects the Amount of Gas in a Solution? Oxygen and carbon dioxide diffuse into water from the atmosphere, but the amount of gas that dissolves in water depends on several factors:

- *Solubility of the gas in water* Oxygen has very low solubility in water. Only 0.003 mL of oxygen dissolves in 100 mL of water for each increase of 1 mm Hg in oxygen partial pressure. Because of this low solubility, blood contains a molecule that binds to oxygen and delivers it to tissues. Without this carrier molecule, the rate of blood flow to tissues would have to increase dramatically to meet oxygen demand.

- *Temperature of the water* As the temperature of water increases, the amount of gas that dissolves in it decreases. Other things being equal, warm-water habitats have much less oxygen available than cold-water habitats do. For a fish, breathing in warm water is comparable to a land-dwelling animal breathing at high elevation.

- *Presence of other solutes* Because seawater has a much higher concentration of solutes than does freshwater, seawater can hold less dissolved gas. At 10°C, up to 8.02 mL of O_2 can be present per liter of freshwater versus only 6.35 mL of O_2 per liter of seawater. As a result, freshwater habitats tend to be more oxygen rich than marine environments.

- *Partial pressure of the gas in contact with the water* Gases move from regions of high partial pressure to regions of low partial pressure. So if the partial pressure in a liquid exceeds that in the adjacent gas, the gas will bubble up out of the liquid. This is what happens when the cap is removed from a bottle of carbonated beverage. The partial pressure of carbon dioxide in the newly opened drink is much higher than it is in the atmosphere.

What Affects the Amount of Oxygen Available in an Aquatic Habitat? The partial pressure of oxygen varies in different types of aquatic habitats, just as it varies with altitude on land. In addition to the four factors just listed, other important considerations affect oxygen's availability in water. These include the presence of photosynthetic organisms and decomposers, the amount of mixing that occurs, and the surface area of the body of water. For example,

- Habitats with large numbers of photosynthetic organisms tend to be relatively oxygen rich. In contrast, oxygen content is extremely low in bogs and other stagnant-water habitats because oxygen is quickly depleted by decomposers that use it as an electron acceptor in cellular respiration.

- Unless currents mix water almost continuously, water near the surface has much higher oxygen content than water near the bottom of the same habitat.

- Shallow ponds and streams tend to be much better oxygenated than deep bodies of water because shallower bodies have a higher ratio of surface area to volume.

- Rapids, waterfalls, and other types of whitewater are the most highly oxygenated of all aquatic environments because a large surface area is exposed to the atmosphere as water splashes over rocks and logs and because air bubbles are incorporated into the water.

Now let's consider the structure and function of ventilatory organs. How do the gills of fishes, the tracheae of insects, and the lungs of mammals cope with the differences between air and water?

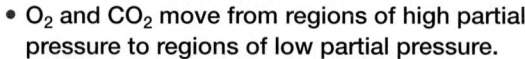 check your understanding

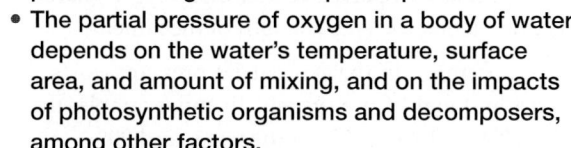

If you understand that . . .

- O_2 and CO_2 move from regions of high partial pressure to regions of low partial pressure.
- The partial pressure of oxygen in a body of water depends on the water's temperature, surface area, and amount of mixing, and on the impacts of photosynthetic organisms and decomposers, among other factors.
- Water breathing is much more difficult than air breathing, in part because the partial pressure of oxygen in water is much lower than its partial pressure in air.

✔ **You should be able to . . .**

1. Explain why oxygen partial pressures are relatively high in mountain streams and relatively low at the ocean bottom.
2. For each of three aquaria, decide whether a large or a small amount of air should be bubbled in. One contains warm-water species, one contains vigorous algal growth, and one contains sedentary animals.

Answers are available in Appendix A.

45.3 Organs of Gas Exchange

Many small animals lack specialized gas exchange organs, such as gills or lungs. Instead, they obtain O_2 and eliminate CO_2 by diffusion across the body surface. This is possible because their size and shape give them an extraordinarily high ratio of surface area to volume (see Chapter 42). In sponges, jellyfish, flatworms, and other species, diffusion across the body surface is rapid enough to fulfill their requirements for taking in O_2 and expelling CO_2.

Most of these animals are restricted to living in wet environments, however. For gas exchange to take place efficiently, the surface where it occurs has to be thin, and thin tissues are prone to water loss. Living in wet or humid environments allows animals to exchange gases across their outer surface while avoiding dehydration.

In contrast, animals that are large or that live in dry habitats need some sort of specialized respiratory organ. Respiratory organs provide a greater surface area for gas exchange—enough to meet the demands of a large body filled with cells. In terrestrial animals, respiratory organs are located inside the body, which helps minimize water loss.

Biologists have long marveled at the efficiency of gills and lungs. To appreciate why, let's examine the physical factors that control diffusion rates and then look at the structure and function of these respiratory organs.

Physical Parameters: The Law of Diffusion

In 1855 Adolf Fick derived an equation regarding diffusion, based on the results of experiments he had performed on the behavior of gases. **Fick's law of diffusion** states that the rate of diffusion of a gas depends on five parameters: the solubility of the gas in the aqueous film lining the gas exchange surface; the temperature; the surface area available for diffusion; the difference in partial pressures of the gas across the gas exchange surface; and the thickness of the barrier to diffusion (**FIGURE 45.3**).

Fick's law identifies traits that allow animals to maximize the rate at which oxygen and carbon dioxide diffuse across surfaces. Specifically, Fick's law states that gases diffuse at the highest rates when three conditions are met:

1. *A* is large, meaning a large area is available for gas exchange. Based on Fick's law, it is not surprising that the respiratory surface in the human lungs would cover about 140 m²—about a quarter of a basketball court—if the epithelium were spread flat.

2. *D* is small, meaning the respiratory surface is extremely thin. In the human lung, this barrier to diffusion is only 0.2 μm thick—about 1/200th of the thickness of this page.

3. $P_2 - P_1$ is large, meaning the partial pressure gradient of the gas across the surface is large. High partial pressure gradients are maintained in part by having an efficient circulatory

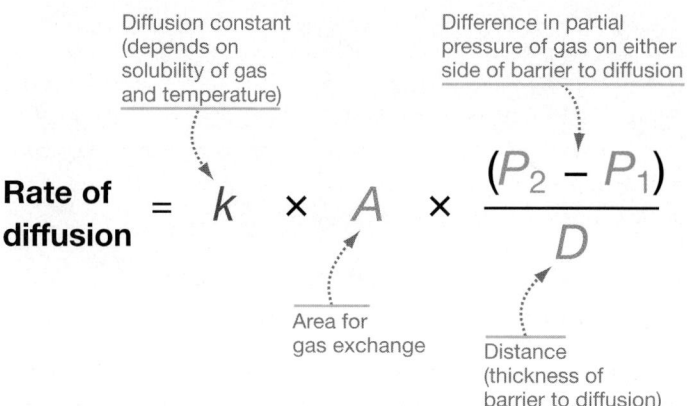

FIGURE 45.3 Fick's Law Describes the Rate of Diffusion.

system in close contact with the gas exchange surface. When blood flows close to the respiratory surface, oxygen is rapidly taken away from the area where inward diffusion is occurring, and carbon dioxide is rapidly brought into the area where outward diffusion is occurring. As a result, $P_2 - P_1$ stays high.

What other aspects of gill and lung structure affect diffusion rate? To answer this question, let's delve into the anatomy of these respiratory organs.

How Do Gills Work?

Gills are outgrowths of the body surface or throat that are used for gas exchange in aquatic animals. Gills are efficient solutions to the problems posed by water breathing, primarily because they present a large surface area for the diffusion of gases across a thin epithelium.

In some species of invertebrates, such as the red tubeworm (**FIGURE 45.4a**), gills project from the body surface and contact the surrounding water directly. In other invertebrate species, such as the crayfish (**FIGURE 45.4b**), gills are located inside the exoskeleton or body wall. If gills are internal, water must be driven over them by cilia, the limbs, or other specialized structures.

In contrast to the diversity of gills found in invertebrates, the gills of bony fishes are all similar in structure. Fish gills are located on both sides of the head and in teleosts (see Chapter 35) consist of four arches, as **FIGURE 45.5** shows.

How Do Fishes Ventilate Their Gills? To move water through their gills so gas exchange can take place, most fishes open and close their mouth and **operculum,** the stiff flap of tissue that covers the gills. The pumping action of the mouth and operculum creates a pressure gradient that moves water over the gills.

In contrast, tuna and other fishes that are particularly fast swimmers force water through their gills by swimming with their mouths open. This process is called ram ventilation.

Regardless of how fish gills are ventilated, water flows in one direction through gills, passing over long, thin structures called **gill filaments** that extend from each gill arch. Each gill filament is composed of hundreds or thousands of **gill lamellae.** Gill lamellae are sheetlike structures, shown in detail at the bottom of Figure 45.5. Note that a bed of small blood vessels called capillaries runs through each lamella.

The Fish Gill Is a Countercurrent System The one-way flow of water through gill lamellae has a profound impact on gill function, for a simple reason: The flow of blood through the capillary bed in each lamella is in the opposite direction to the flow of water. As a result, each lamella functions as a countercurrent exchanger.

Recall that countercurrent exchangers are based on two adjacent fluids flowing in opposite directions (see Chapter 42). **FIGURE 45.6** illustrates why the countercurrent flow is so critical.

Note two key points about the left side of the figure, where water and blood flow in opposite directions:

1. A slight gradient in partial pressure of oxygen (here, 10 percent) exists along the entire length of the lamella.

2. A large difference in oxygen partial pressure exists between the start and end of the system. In this example, the difference is $100\% - 15\% = 85\%$ in the water and $90\% - 5\% = 85\%$ in the blood.

The upshot? Most of the oxygen in the incoming water has diffused into the blood.

Now look at the right side of the figure, where water and blood flow in the same direction.

1. A large gradient in partial pressure of oxygen (here, 100 percent) exists at the start of the system. The gradient in oxygen partial pressures declines rapidly and eventually disappears.

(a) External gills are in direct contact with water.

(b) Internal gills must have water brought to them.

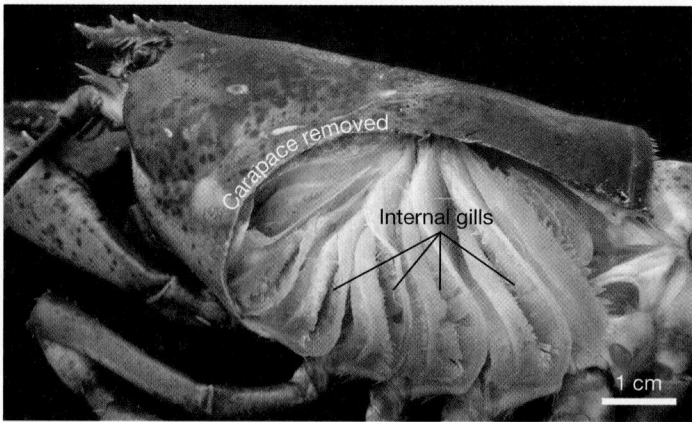

FIGURE 45.4 Gills Can Be External or Internal. (a) Red tubeworms are marine polychaetes with gills that protrude from the body. **(b)** Lobster gills are located inside the main body wall. A portion of the crayfish exoskeleton has been removed to expose the gills.

✔**QUESTION** Predict what the advantages and disadvantages of external versus internal gills might be in terms of fitness.

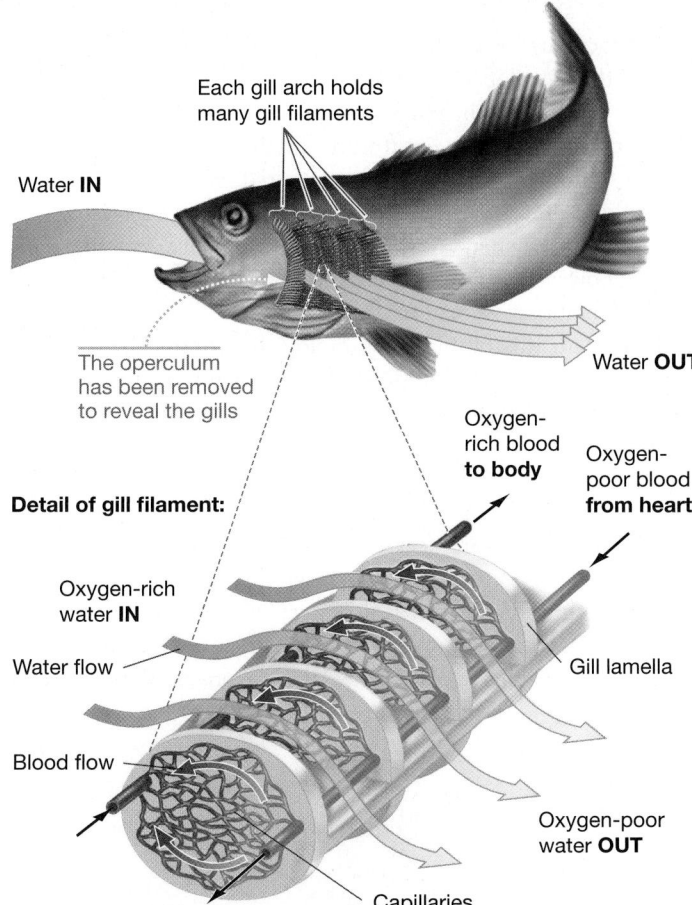

Each gill arch holds
many gill filaments

Water **IN**

The operculum
has been removed
to reveal the gills

Detail of gill filament:

Oxygen-rich
water **IN**

Water flow

Blood flow

Water **OUT**

Oxygen-
rich blood
to body

Oxygen-
poor blood
from heart

Gill lamella

Oxygen-poor
water **OUT**

Capillaries

FIGURE 45.5 Fish Gills Are a Countercurrent Exchange System.
In fish gills, water and blood flow in opposite directions. In the
blood vessels, red represents oxygenated blood, blue represents
deoxygenated blood, and purple represents mixed blood.

2. A relatively small difference in oxygen partial pressure exists
between the start and end of the system. In this example, the
difference is $100\% - 50\% = 50\%$ in the water and $50\% - 0\% =$
50% in the blood.

With this arrangement, only half of the oxygen in the incoming
water has diffused into the blood.

Countercurrent flow makes fish gills extremely efficient at extracting oxygen from water because it ensures that a difference in
the partial pressure of oxygen and carbon dioxide in water versus
blood is maintained over the entire gas exchange surface.

The effect of countercurrent exchange is to maximize the
$P_2 - P_1$ term in Fick's law of diffusion, averaged over the entire
gill surface. Based on this observation, biologists cite countercurrent exchange as another example of how gills are optimized for
efficient gas exchange.

How Do Insect Tracheae Work?

As noted earlier, air and water are dramatically different ventilatory media, because they have different densities, viscosities,
and abilities to hold oxygen and carbon dioxide. In addition, the

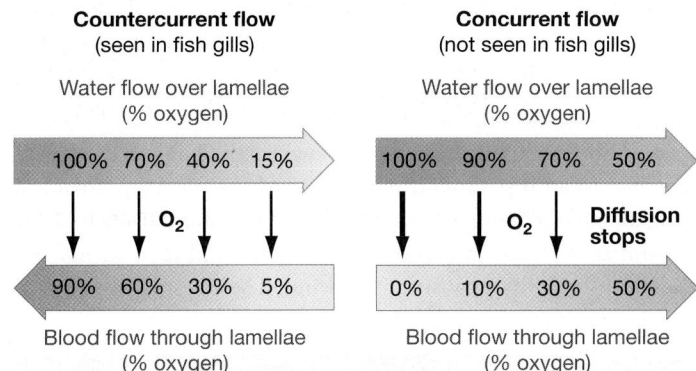

Countercurrent flow
(seen in fish gills)

Concurrent flow
(not seen in fish gills)

Water flow over lamellae
(% oxygen)

Water flow over lamellae
(% oxygen)

100% 70% 40% 15%

100% 90% 70% 50%

O_2

O_2

**Diffusion
stops**

90% 60% 30% 5%

0% 10% 30% 50%

Blood flow through lamellae
(% oxygen)

Blood flow through lamellae
(% oxygen)

**FIGURE 45.6 Countercurrent Exchange Is Much More Efficient
than Concurrent Exchange.** In the countercurrent system of fish
gills, oxygen is transferred along the entire length of the capillaries.

✔ **QUESTION** If you understand this concept, you should be able to
predict what would happen to oxygen transfer from water to blood
if flow were concurrent.

consequences of exposing the gas exchange surface to air versus
water differ.

In aquatic habitats, ventilation tends to disrupt water and
electrolyte balance, and homeostasis must be maintained by an
active osmoregulatory system. Osmosis causes marine animals
to lose water across their gas exchange surface and freshwater
animals to gain water (see Chapter 43). Diffusion tends to cause
marine animals to gain sodium, chloride, and other ions, and
freshwater animals to lose them.

In contrast, breathing leads to a loss of water by evaporation in terrestrial environments. How do terrestrial animals
minimize water loss while maximizing the efficiency of gas
exchange?

To answer this question, consider the tracheal system of insects. Recall that insects have an extensive system of air-filled
tubes called **tracheae** located within the body (see Chapter 43).
These connect to the exterior through openings called **spiracles,**
which can be closed to minimize the loss of water by evaporation. The ends of tracheae are tiny and highly branched. This design allows the tracheal system to transport air close enough to
cells for gas exchange to take place directly across their plasma
membranes.

Air moves from the atmosphere into the tracheae and then
through the tracheae to the tissues in the insect's body. Is simple
diffusion efficient enough to ventilate the system, or is some type
of breathing mechanism involved?

If insects ventilate their tracheal system with simple diffusion,
then according to Fick's law, the distance that the air would have
to diffuse down the tracheae would be very important. In other
words, the rate of diffusion of air down the tracheae would be
limited in part by the length of the tracheae. In effect, only very
small insects have tracheae that are short enough to be efficiently
ventilated by simple diffusion.

Many insects are far too large to deliver gases to and from
their tissues by simple diffusion alone. How do these large insects ventilate their tracheal system?

Consider a study on the sweet potato hawkmoth. Yutaka Komai set out to investigate how the amount of O_2 delivered to the wing muscles changes during flight. To document the partial pressure of oxygen in these muscles, he inserted a needlelike electrode into the flight muscles of a hawkmoth. The electrode was attached to an instrument that measured the partial pressure of oxygen, and the hawkmoth was tethered to a stand (**FIGURE 45.7**).

Komai recorded P_{O_2} as the insect rested and then stimulated the moth to fly by exposing it to wind. The "Results" section of

Figure 45.7 shows how P_{O_2} changed during one such experiment. (The procedure was repeated on several individuals, and the same pattern was observed.)

To read this graph, put your finger on the left-hand end of the orange line and move it toward the right, to where the insect started flying and the line changes to red. Note that P_{O_2} levels in the flight muscles dropped initially. This is not surprising, because flight is an energetically demanding activity. As flying continued, however, P_{O_2} levels recovered until they were nearly as high as when they were at rest.

To explain this observation, biologists propose that the tracheae are alternately compressed and dilated as the muscles around them contract and relax. The muscle contractions and relaxations produce pressure changes that cause the volume of space inside the tracheal system to change (**FIGURE 45.8**). This is a key point: If the volume occupied by a fixed amount of gas expands, the gas pressure decreases. But if the volume that a fixed amount of gas occupies declines, the gas pressure increases.

The volume of the tracheal system increases when muscles relax, causing pressure inside the system to go down and air from the atmosphere to rush in. What happens when muscles

---

HYPOTHESIS: Air moves through the tracheal system faster during physical activity.

NULL HYPOTHESIS: Physical activity does not affect the rate of air movement.

EXPERIMENTAL SETUP:

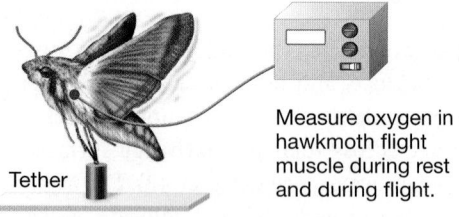

Measure oxygen in hawkmoth flight muscle during rest and during flight.

Tether

PREDICTION: Flying will increase ventilation of tracheal system, causing increase in P_{O_2} in wing muscle.

PREDICTION OF NULL HYPOTHESIS: Flying will not increase ventilation of tracheal system; P_{O_2} in flight muscle will decline steadily during flight.

RESULTS:

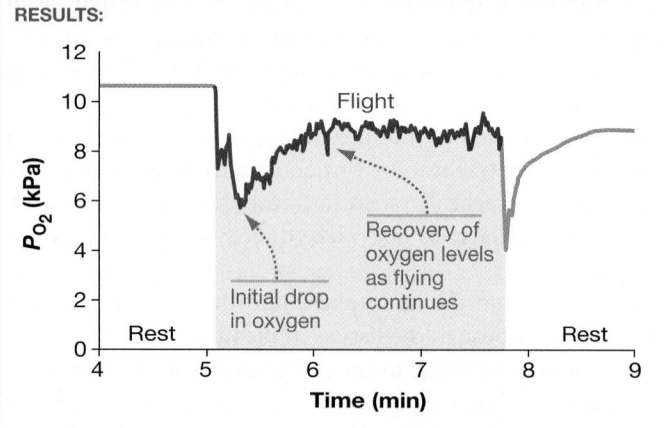

Flight

Initial drop in oxygen

Recovery of oxygen levels as flying continues

Rest

Rest

P_{O_2} (kPa)

Time (min)

CONCLUSION: Muscular contractions may help ventilate the tracheal system in at least some insects.

FIGURE 45.7 Research Suggests the Tracheal System Is Ventilated during Movement. On the graph's y-axis, kPa stands for kilopascal, a unit of pressure (or force per unit area).

SOURCE: Komai, Y. 1998. *The Journal of Experimental Biology* 201: 2359–2366.

✔**QUESTION** Why was it important for the researcher to repeat this experiment with several different individuals?

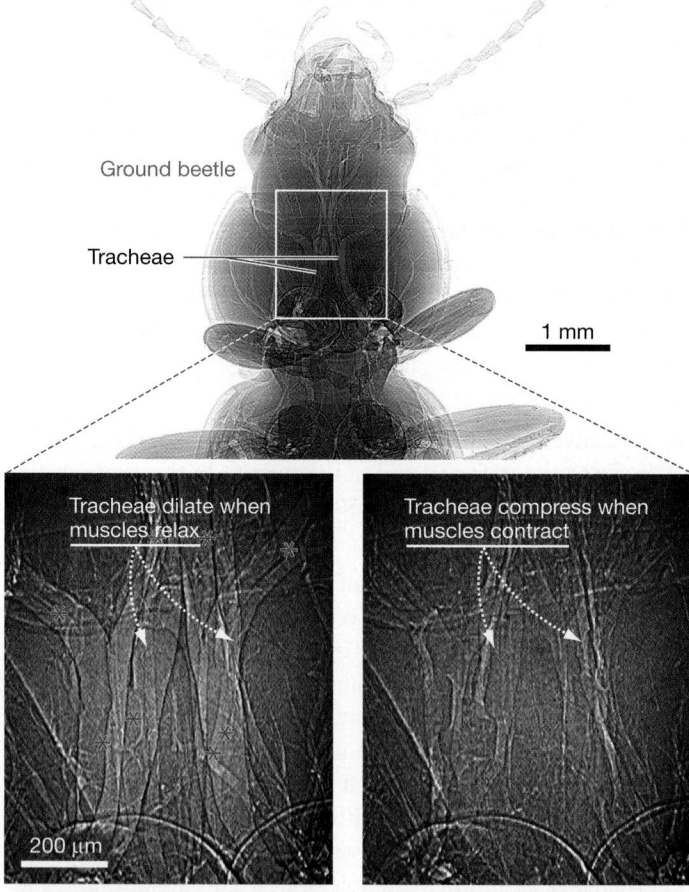

Ground beetle

Tracheae

1 mm

Tracheae dilate when muscles relax

Tracheae compress when muscles contract

200 μm

FIGURE 45.8 Tracheae Are Dilated and Compressed during Ventilation. Muscle relaxation and contraction during flight or other activity alternately dilates and compresses tracheae, causing pressure changes that promote air flow in and out of the tracheae. Asterisks in the left image mark tracheal tubes that are compressed in the right image.

contract? The opposite—the volume of the tracheae decreases, pressure inside the system increases, and gas moves out of the tracheae into the atmosphere.

The action of the abdominal and flight muscles therefore stimulates air flow through the insect tracheal system, causing gases to move more quickly than they would by diffusion alone.

In large insects, the movement of gases is further promoted by larger tracheae diameters. These larger diameters increase another variable in Fick's law, the cross-sectional area for gas exchange. As evidence, researchers have shown that the diameter of the tracheae in large beetles is proportionally much larger than that of small beetles. Disproportionally large tracheae allow effective ventilation even in the largest living beetles, which can reach impressive lengths of over 15 cm (6 inches) and weigh over 50 g (3 ounces), the size of a submarine sandwich!

Why don't beetles get even larger than this? Researchers hypothesize that if they did, the size of their tracheae would be so large that there would not be enough space in their limbs for much else. The beetles would lack sufficient muscles and other tissues to support their huge bodies.

In the Paleozoic era about 300 million years ago (mya), however, giant insects flourished. While the largest dragonflies today are about 15 cm in wingspan, during the Paleozoic they grew to be as large as 70 cm (2.3 feet)! How can this be explained?

One major hypothesis posits that giant insects evolved during the Paleozoic era because the atmospheric oxygen concentration was much higher than it is today (**FIGURE 45.9**). Recall that according to Fick's law, the rate of diffusion of a gas from one point to another is affected by the difference in partial pressures of the gas at the two points. During the latter part of the Paleozoic, the value of $P_2 - P_1$ was increased in the equation shown in Figure 45.3. This may have increased the rate of movement of gases through the tracheal system, permitting efficient ventilation in extremely large insects.

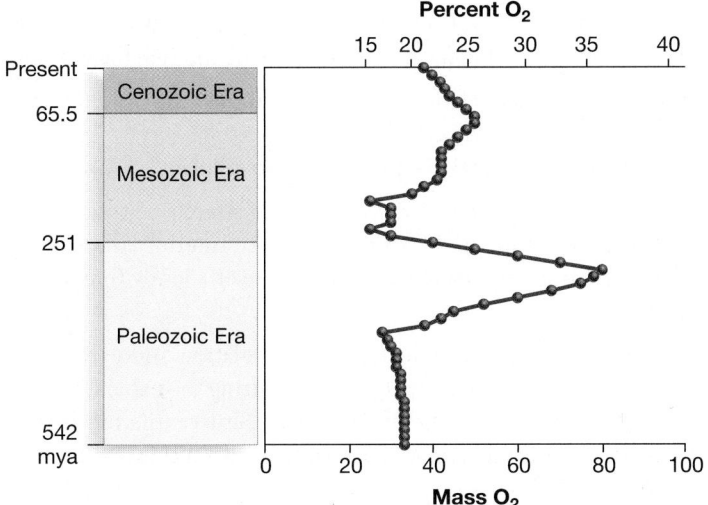

FIGURE 45.9 Atmospheric Oxygen Levels Peaked during the Paleozoic Era.

DATA: Berner, R. A. 1999. *Proceedings of the National Academy of Sciences* 96: 10955–10957.

About 50 to 100 million years later, at the start of the Mesozoic era, atmospheric oxygen concentrations declined precipitously. Giant insects went extinct, likely in part because they could no longer supply their tissues with enough oxygen.

How Do Vertebrate Lungs Work?

In most terrestrial vertebrates, air enters the body through both the nose and mouth. A tube known as the **trachea** (not to be confused with the tracheae of insects) carries the inhaled air to narrower tubes called **bronchi** (singular: **bronchus**). The bronchi branch off into yet narrower tubes, the **bronchioles.** The organs of ventilation, the lungs, enclose the bronchioles and part of the bronchi (**FIGURE 45.10a**, see page 910).

Lungs are internal organs that are used for gas exchange. In addition to being found in terrestrial vertebrates—amphibians, reptiles (including birds), and mammals—they occur in many fishes and certain invertebrates.

Lung Structure and Ventilation Vary among Species   The amount of lung surface area available for gas exchange varies a great deal among species. In frogs and other amphibians, the lung is a simple sac lined with blood vessels. The lungs of mammals, in contrast, are finely divided into tiny sacs called **alveoli** (singular: **alveolus; FIGURE 45.10b**).

Each human lung contains approximately 150 million alveoli, which give mammalian lungs about 40 times more surface area for gas exchange than an equivalent volume of frog lung tissue. As **FIGURE 45.10c** shows, an alveolus provides an interface between air and blood that consists of a thin aqueous film, a layer of epithelial cells, some extracellular matrix (ECM) material, and the wall of a capillary.

In addition to total surface area, the other major feature of lungs that varies among species is mode of ventilation. In the lungs of snails and spiders, air movement takes place by diffusion only. Vertebrates, in contrast, actively ventilate their lungs by pumping air via muscular contractions.

One mechanism for pumping air is **positive pressure ventilation,** used by frogs and related animals. A frog lowers the floor of its throat, increasing the volume there and thus drawing in air from the atmosphere through the nasal passages and into the oral cavity. The animal then closes the nasal passages and contracts its throat muscles. These actions increase the pressure on air in the oral cavity and force it into the lungs.

In effect, frogs push air into their lungs. In contrast, humans and other mammals pull air into their lungs. How does this **negative pressure ventilation** work?

Ventilation of the Human Lung   The pressure inside the human chest cavity is about 5 mm Hg less than atmospheric pressure. This negative pressure surrounding the lung is just enough to keep the lung expanded. If a wound penetrates the chest wall and the pressure differential between the chest cavity and the atmosphere disappears, the lung on the side of the injury will collapse like a deflated balloon.

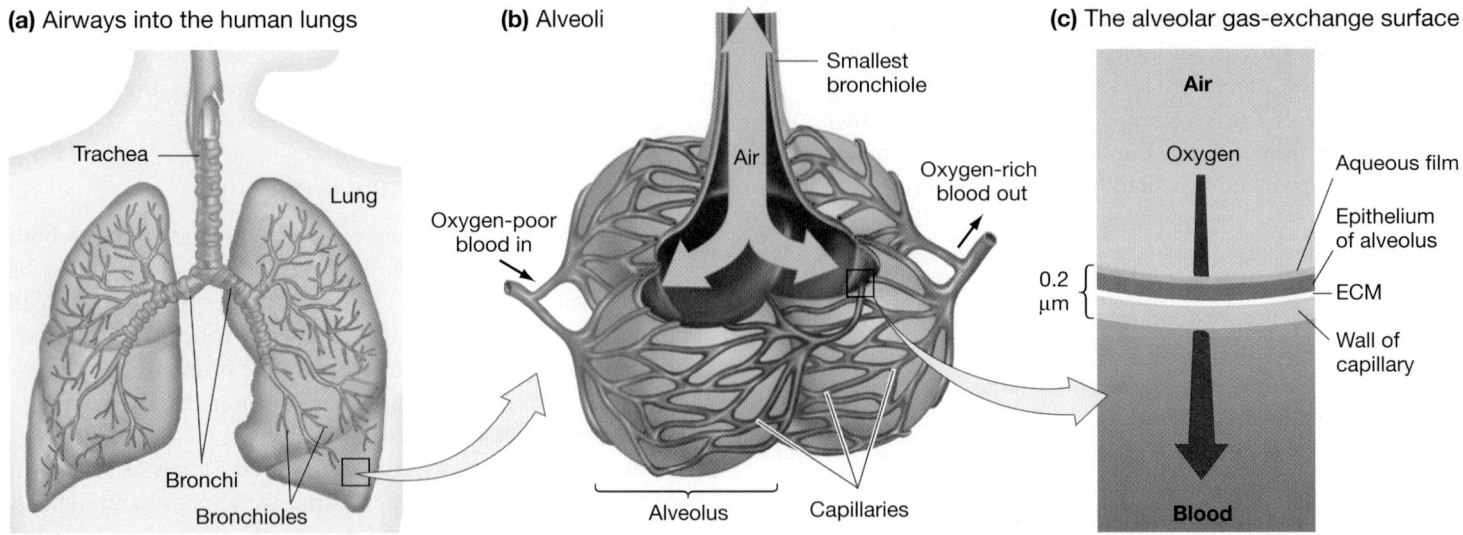

(a) Airways into the human lungs

Trachea

Lung

Bronchi

Bronchioles

(b) Alveoli

Smallest bronchiole

Air

Oxygen-poor blood in

Oxygen-rich blood out

Alveolus

Capillaries

(c) The alveolar gas-exchange surface

Air

Oxygen

Aqueous film

Epithelium of alveolus

ECM

Wall of capillary

0.2 μm

Blood

FIGURE 45.10 Lungs Offer a Thin Membrane with a Large Surface Area for Gas Exchange between Air and Blood. (a) The human respiratory tract branches repeatedly from the largest airway, the trachea, to the smallest, called bronchioles. The system of airways ends in clusters of tiny sacs called alveoli. **(b)** Alveoli are covered with capillary networks and are **(c)** the site of gas exchange.

Humans ventilate their lungs by changing the pressure within their chest cavity between about −5 mm Hg and −8 mm Hg relative to the atmosphere. As **FIGURE 45.11a** shows, inhalation is based on increasing the volume of the chest cavity and thus lowering the pressure. The change is caused by a downward motion of the thin muscular sheet called the **diaphragm** and an outward motion of the ribs. As the pressure surrounding the lungs drops, air flows into the airways along a pressure gradient.

Exhalation, in contrast, is a passive process—the volume of the chest cavity decreases as the diaphragm and rib muscles relax. Because the lung is elastic, it returns automatically to its original, collapsed shape if it is not stretched or compressed. During exercise, though, exhalation is an energy-demanding, active process.

The changes in pressure that occur during negative pressure ventilation are analogous to changing the pressure within a jar, as shown in **FIGURE 45.11b**.

About 450 mL of air moves into and out of the lungs in an average breath. Only about two-thirds of this volume actually participates in gas exchange, however, because 150 mL of the air occupies **dead space**— air passages that are not lined by a respiratory surface. The trachea and bronchi shown in Figure 45.10a, for example, represent dead space.

Breathing is much more efficient during exercise, when the chest cavity undergoes larger changes in volume. When a person is breathing hard, over 2500 mL of air can move with each inhalation–exhalation cycle, but the 150 mL of dead space stays the same.

Ventilation of the Bird Lung Flight is one of the most energy-demanding activities performed by animals. Even so, some birds fly tens of thousands of kilometers during annual migrations.

Even more impressive, geese regularly fly over the top of Mt. Everest. At this elevation, the partial pressure of oxygen is so low that most humans would immediately black out if they were flown there and dropped off. How are birds able to extract enough oxygen from the atmosphere to support long flights and to fly at high elevations?

FIGURE 45.12 provides a diagram of ventilation in birds.

1. During inhalation, air flows through the trachea and enters two large air sacs posterior to the lungs.

2. During exhalation, air leaves the posterior air sacs and enters tiny, branching airways, called parabronchi, in the posterior portion of the lungs.

3. During the next inhalation, air moves into parabronchi in the anterior part of the lungs and on to a system of air sacs anterior to the lungs.

4. During the next exhalation, air moves out of the anterior sacs, through the trachea, and out to the atmosphere. Meanwhile, air from the second inhalation is now flowing through the lungs.

The key conclusion from these observations? Airflow through the avian lung is unidirectional.

Why is the avian respiratory system so efficient?

- Dead space is restricted to the short stretch of trachea between the mouth and the opening of the anterior air sacs. As a result, birds use inhaled air much more efficiently than mammals do.

- Gas exchange occurs during both inhalation and exhalation. In contrast, no gas exchange occurs during the exhalation half of the respiratory cycle in mammals. Bird ventilation resembles the continuous ventilation of fish gills in this respect.

- Blood circulates through the bird lung in capillaries that cross the parabronchi perpendicularly. This crosscurrent pattern is less efficient than the countercurrent circulation of fish gills but far more efficient than the weblike arrangement of capillaries that surrounds mammalian alveoli.

(a) Lungs expand and contract in response to changes in pressure inside the chest cavity.

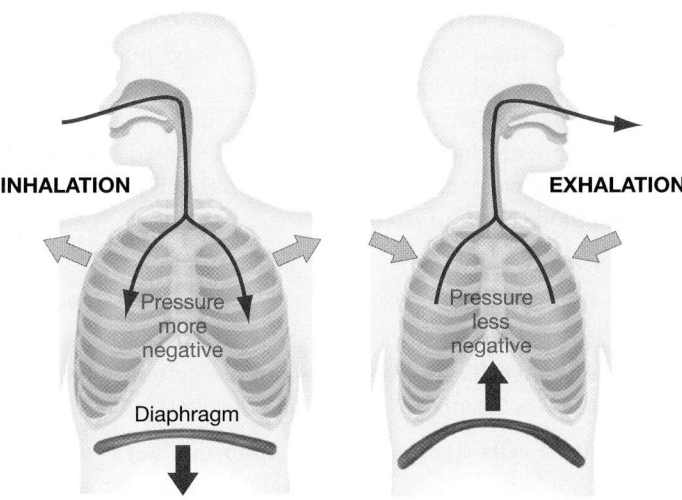

INHALATION   EXHALATION

Pressure more negative

Pressure less negative

Diaphragm

(b) Ventilatory forces can be modeled by a balloon in a jar.

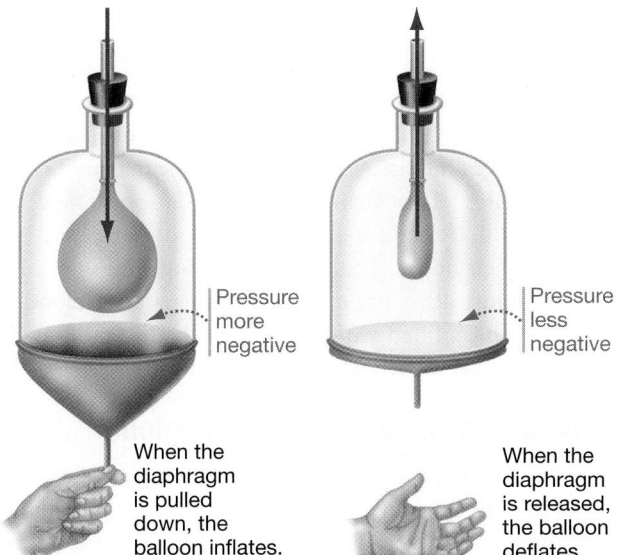

Pressure more negative

Pressure less negative

When the diaphragm is pulled down, the balloon inflates.

When the diaphragm is released, the balloon deflates.

FIGURE 45.11 Changes in the Volume of the Chest Cavity Drive Negative Pressure Ventilation. (a) Inhalation: When the diaphragm and rib muscles contract, the volume of the lung cavity increases, lowering pressure. In response, air flows into the lungs. Exhalation: When the diaphragm and rib muscles relax, the volume of the lung cavity decreases, causing internal pressure to increase. In response, lung volume decreases—due to elasticity of the lungs—and air flows out. **(b)** A model of negative pressure ventilation.

Homeostatic Control of Ventilation

An animal is in trouble if its mechanisms of homeostasis fail to maintain blood oxygenation or to eliminate carbon dioxide. Adequate adenosine triphosphate (ATP) production depends on maintaining the partial pressures of oxygen and carbon dioxide within a narrow range, both during rest and vigorous exercise. How is ventilation controlled to achieve this critical homeostasis?

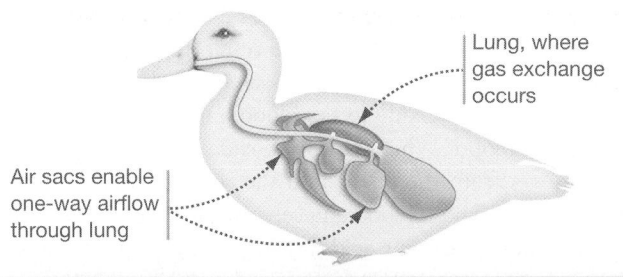

Lung, where gas exchange occurs

Air sacs enable one-way airflow through lung

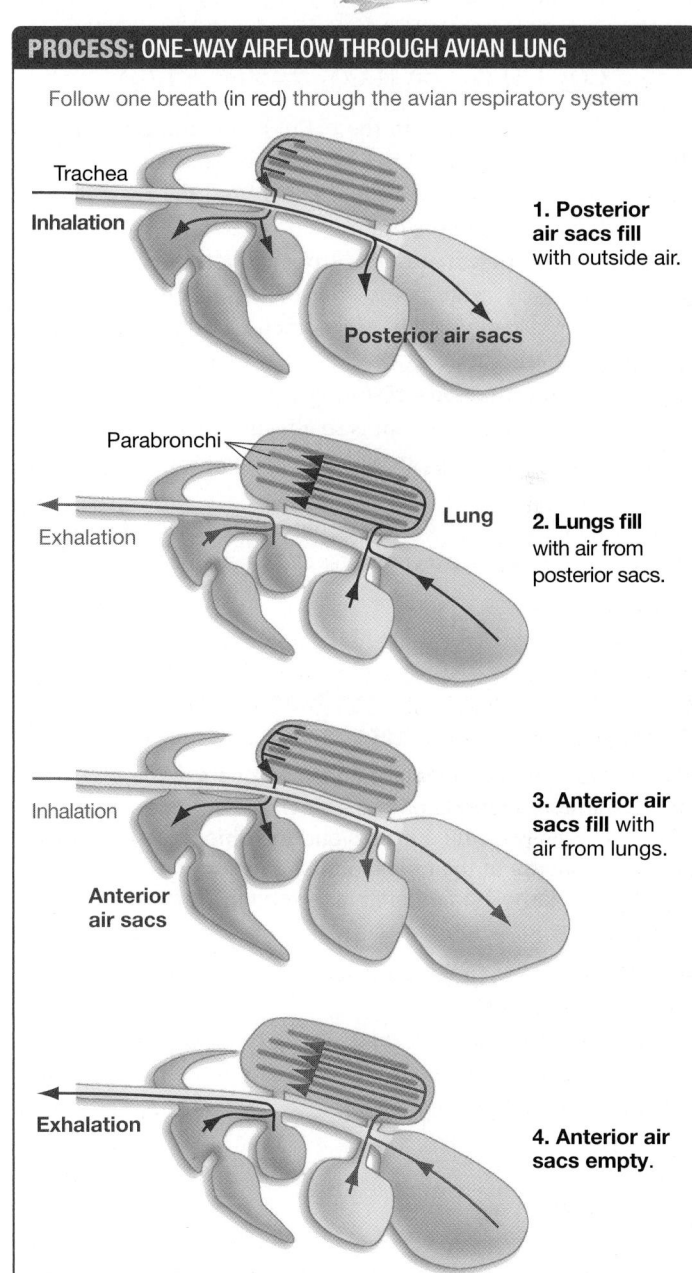

PROCESS: ONE-WAY AIRFLOW THROUGH AVIAN LUNG

Follow one breath (in red) through the avian respiratory system

Trachea

Inhalation

1. Posterior air sacs fill with outside air.

Posterior air sacs

Parabronchi

Exhalation

Lung

2. Lungs fill with air from posterior sacs.

Inhalation

3. Anterior air sacs fill with air from lungs.

Anterior air sacs

Exhalation

4. Anterior air sacs empty.

FIGURE 45.12 Air Flows in One Direction through the Bird Lung, Maximizing Gas Exchange.

When mammals are resting, the rate of breathing is established by the medullary respiratory center, an area at the base of the brain, just above the spinal cord. This center stimulates the rib and diaphragm muscles to contract about 12 to 14 times per minute in humans.

But during exercise, things change. Active muscle tissue takes up more oxygen from the blood. As a result, the partial pressure of oxygen (P_{O_2}) in blood drops. Those same muscles release larger quantities of carbon dioxide to the blood, raising its partial pressure (P_{CO_2}) in blood. When carbon dioxide reaches the brain, it rapidly diffuses from the blood into the cerebrospinal fluid that bathes the brain. In both blood and cerebrospinal fluid, CO_2 reacts with water to form carbonic acid (H_2CO_3), which then dissociates to release a hydrogen ion (H^+) and a bicarbonate ion (HCO_3^-):

$$CO_2 + H_2O \rightleftharpoons H_2CO_3 \rightleftharpoons H^+ + HCO_3^-$$

The result is a slight drop in the pH of blood and cerebrospinal fluid. The change in blood pH is sensed by specialized neurons located near the large arteries that travel from the heart into the neck and to the base of the brain.

Signals from these neurons or from pH detectors in the medullary respiratory center cause the breathing rate to increase. The resulting rise in ventilation rate increases the rate of oxygen delivery to the tissues and the rate at which carbon dioxide is eliminated from the body, restoring P_{O_2} and P_{CO_2} to their resting levels. This control system is so effective that it can maintain stable blood levels of oxygen and carbon dioxide even during intense exercise.

Now let's look more closely at how blood transports oxygen and carbon dioxide between the gas exchange surface and an animal's tissues.

check your understanding

If you understand that . . .

- The rate of diffusion of gases across a respiratory surface depends on the area of the surface, the thickness of the surface, and the difference in partial pressures of the gases across the surface.
- Most large-bodied animals exchange gases via gills, tracheae, or lungs.

✔ You should be able to . . .

1. Identify two features that are common to gills, tracheae, and lungs, as well as one trait that is unique to each.
2. When you hold your breath, what happens to the P_{O_2}, P_{CO_2}, and pH of your blood?

Answers are available in Appendix A.

45.4 How Are Oxygen and Carbon Dioxide Transported in Blood?

Blood is a connective tissue that consists of cells in a watery extracellular matrix. Besides carrying oxygen and carbon dioxide between cells and the lungs, blood transports nutrients from the digestive tract to other tissues in the body, moves waste products to the kidney and liver for processing, conveys hormones from glands to target tissues, delivers immune system cells to sites of infection, and distributes heat throughout the body.

Given the wide variety of functions that blood serves, it is not surprising that it is a complex tissue. In an average human, 50–65 percent of the blood volume is composed of an extracellular matrix called **plasma.** The remainder of the volume comprises cells and cell fragments that are collectively called formed elements.

The formed elements in blood include platelets, several types of white blood cells, and red blood cells.

- **Platelets** are cell fragments that act to minimize blood loss from ruptured blood vessels. They do so by releasing material that helps form the blockages known as clots.
- **White blood cells** are part of the immune system. They fight infections (as Chapter 51 will explain in detail).
- **Red blood cells** transport oxygen from the lungs to tissues throughout the body. They also play a role in transporting carbon dioxide from tissues to the lungs. In humans, red blood cells make up 99.9 percent of the formed elements.

The human body synthesizes new red blood cells at the rate of 2.5 million per second to replace old red blood cells, which die at the same rate. These red blood cells last for about 120 days. Red blood cells, white blood cells, and platelets develop from stem cells located in the tissue inside bone (bone marrow).

Vertebrates other than mammals transport oxygen in red blood cells that retain their nuclei. But in mammals, red blood cells lose their nuclei as they mature, along with their mitochondria and most other organelles. Mammalian red blood cells are essentially bags filled with approximately 280 million copies of the oxygen-carrying molecule **hemoglobin.**

Structure and Function of Hemoglobin

Even though oxygen is not highly soluble in water, it is often found in high concentrations in blood. Blood has a high oxygen-carrying capacity because O_2 readily binds to the hemoglobin molecules in red blood cells.

The evolution of hemoglobin was a key event in the diversification of animals. By increasing the oxygen-carrying capacity of blood, hemoglobin made it possible for cellular respiration rates to increase. High rates of ATP production, in turn, support high rates of growth, movement, digestion, and other activities.

Hemoglobin is a tetramer, meaning that it consists of four polypeptide chains (**FIGURE 45.13**). Each of the four polypeptide chains binds to a nonprotein group called **heme** (represented by black circles in the figure). Each heme molecule, in turn, contains an iron ion (Fe^{2+}) that can bind to an oxygen molecule. As a result, each hemoglobin molecule can bind up to four oxygen molecules. In blood, 98.5 percent of the oxygen is bound to hemoglobin; only 1.5 percent is dissolved in plasma.

What Is Cooperative Binding? Blood leaving the human lungs has a P_{O_2} of about 100 mm Hg, while at rest the muscles and other tissues have a P_{O_2} of about 40 mm Hg. This partial pressure difference creates a diffusion gradient that unloads O_2 from hemoglobin to the tissues.

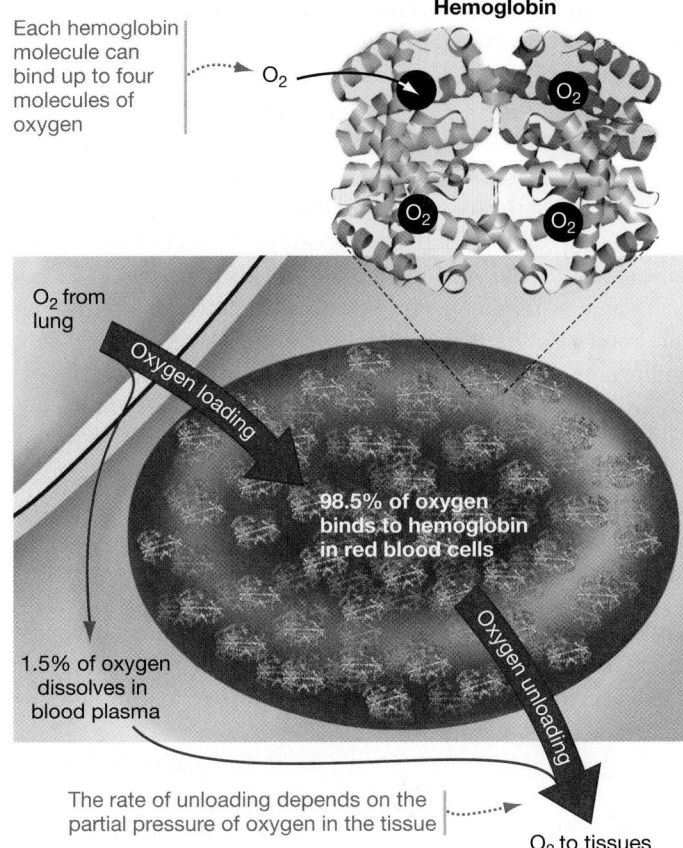

Each hemoglobin molecule can bind up to four molecules of oxygen

Hemoglobin

O_2

O_2

O_2

O_2

O_2 from lung

Oxygen loading

98.5% of oxygen binds to hemoglobin in red blood cells

1.5% of oxygen dissolves in blood plasma

Oxygen unloading

The rate of unloading depends on the partial pressure of oxygen in the tissue

O_2 to tissues

FIGURE 45.13 Hemoglobin Transports Oxygen to Tissues.

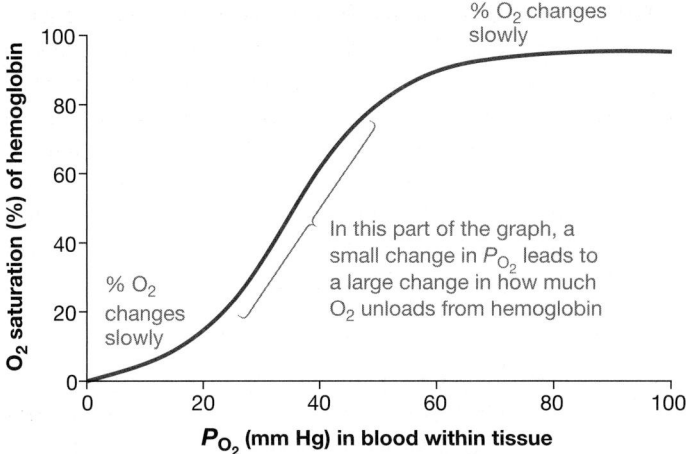

% O_2 changes slowly

In this part of the graph, a small change in P_{O_2} leads to a large change in how much O_2 unloads from hemoglobin

% O_2 changes slowly

P_{O_2} (mm Hg) in blood within tissue

FIGURE 45.14 The Oxygen–Hemoglobin Equilibrium Curve Is Sigmoidal. A sigmoidal curve has three distinct regions.

Researchers who studied the dynamics of O_2 unloading in tissues found the pattern shown in **FIGURE 45.14**. This graph plots the percentage of O_2 saturation of hemoglobin in red blood cells versus the P_{O_2} in the blood within tissues. If saturation of hemoglobin is 100 percent, it means that every possible binding site in hemoglobin contains an oxygen molecule.

The graph in Figure 45.14 is called an **oxygen–hemoglobin equilibrium curve,** or an oxygen dissociation curve or hemoglobin saturation curve. Note that the x-axis plots the partial pressure of oxygen in tissues. In effect, this represents "demand." Oxygen-depleted tissues, where demand for oxygen is high, are on the left-hand part of the horizontal axis; oxygen-rich tissues are toward the right. The y-axis, in contrast, plots the percentage of hemoglobin molecules in blood that are saturated with oxygen—a measure of "supply," or how many oxygen molecules on average are bound to hemoglobin. Each 25 percent change in saturation corresponds to an average of one additional oxygen molecule bound per hemoglobin molecule or one oxygen molecule delivered to tissues.

The most remarkable feature of the oxygen–hemoglobin equilibrium curve is that it is sigmoidal, or S-shaped. The pattern occurs because the binding of each successive oxygen molecule to a subunit of the hemoglobin molecule causes a conformational change in the protein that makes the remaining subunits much more likely to bind oxygen.

This phenomenon is called **cooperative binding.** Conversely, the loss of a bound oxygen molecule changes hemoglobin's conformation in a way that makes the loss of additional oxygen molecules more likely.

Why Is Cooperative Binding Important? To understand why cooperative binding is important, use **FIGURE 45.15a** (see page 914) to figure out what happens to hemoglobin saturation when oxygen demand in tissues changes.

Let's begin with some basic observations. When blood arrives at tissues from the lungs, hemoglobin saturation is close to 100 percent. At rest, tissue P_{O_2} is typically about 40 mm Hg. But during exercise, cells are using so much oxygen in cellular respiration that tissue P_{O_2} drops to about 30 mm Hg. Now,

- Put your finger on the x-axis at 40 mm Hg, trace the dashed line up until it hits the equilibrium curve, and check where this point is on the y-axis. The answer is about 52 percent. This means that when tissues are at rest, hemoglobin unloads about half of its oxygen (about 48 percent) to the tissues.

- Now put your finger on the x-axis at 30 mm Hg, trace the dashed line up, and check hemoglobin saturation at this point on the curve. The answer is about 22 percent. This means that when tissues are exercising, hemoglobin unloads about 78 percent of its oxygen to tissues.

Here's the punch line: In response to a relatively small change in tissue P_{O_2}, there is a relatively large change in the percentage of saturation of hemoglobin. The large change occurs because the equilibrium curve is extremely steep in the range of P_{O_2} values commonly observed in tissues. The curve is steep because of cooperative binding. Cooperative binding is important because it makes hemoglobin exquisitely sensitive to changes in the of P_{O_2} tissues.

✔ If you understand this concept, you should be able to explain the consequences of an oxygen–hemoglobin equilibrium curve with a middle section that is even steeper than the one shown in Figure 45.15a.

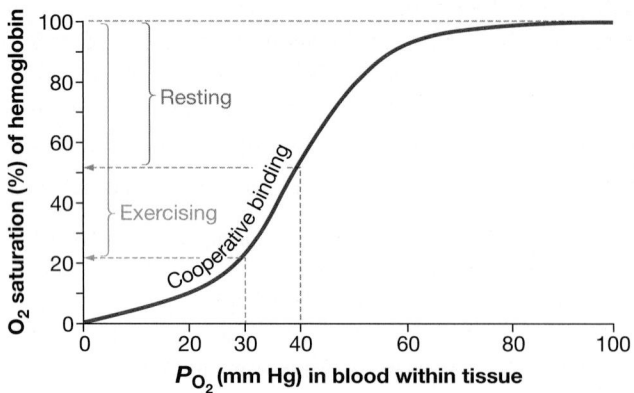

(a) With cooperative binding, large amounts of O_2 are delivered to resting and exercising tissues.

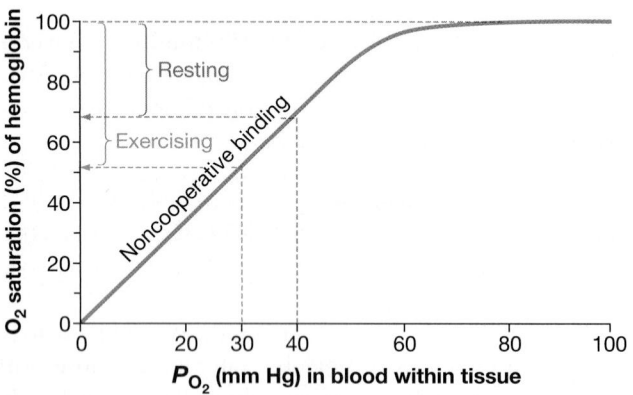

(b) Without cooperative binding, smaller amounts of O_2 would be delivered to resting and exercising tissues.

FIGURE 45.15 Cooperative Binding of O_2 by Hemoglobin Results in Greater O_2 Delivery than Noncooperative Binding. The hemoglobin is almost 100 percent saturated with oxygen until it arrives at tissues.

If cooperative binding did not occur, all four subunits of hemoglobin would load or unload oxygen independently of each other. They would lose or gain oxygen in direct proportion to the partial pressure of oxygen in the blood, until the molecule was either completely saturated or unsaturated with oxygen. There would be a linear relationship between hemoglobin saturation and tissue P_{O_2}—not a sigmoidal one.

FIGURE 45.15b shows noncooperative binding. As the dashed lines on this graph indicate, a relatively small change in oxygen delivery would occur when tissue P_{O_2} changes from its resting level of about 40 mm Hg to 30 mm Hg. Specifically, hemoglobin would unload about 100% − 68% = 32% of its oxygen when tissues are at rest, and about 100% − 52% = 48% of its oxygen during exercise. This difference—about 16 percent—is much less than the 30 percent change observed with cooperative binding.

How Do Temperature and pH Affect Oxygen Unloading from Hemoglobin? Cooperative binding is only part of the story behind oxygen delivery. Hemoglobin—like other proteins—is sensitive to changes in pH and temperature.

As noted above, the temperature and partial pressure of CO_2 rise in active muscle tissue during exercise. The CO_2 produced by exercising muscle reacts with the water in blood to form carbonic acid, which dissociates and releases a hydrogen ion. As a result, the pH of the blood in exercising muscle drops.

Decreases in pH alter hemoglobin's conformation. These shape changes make hemoglobin more likely to unload O_2 at any given value of tissue P_{O_2}. As **FIGURE 45.16** shows, this phenomenon, known as the **Bohr shift,** causes the oxygen–hemoglobin equilibrium curve to shift to the right when pH declines.

The Bohr shift is important because it makes hemoglobin more likely to release oxygen during exercise or other conditions in which P_{CO_2} is high, pH is low, and tissues are under oxygen stress. Increasing temperature has the same result: It shifts the curve to the right, representing a greater unloading of oxygen to tissues at any given P_{O_2}.

Oxygen Delivery by Hemoglobin Is Extremely Efficient To appreciate cooperative binding and the Bohr shift in action, consider an experiment on how the oxygen transport system in rainbow trout responds to sustained exercise.

To begin, biologists had fish swim continuously against a current in a water tunnel. As the researchers increased the speed of the current and thus the swimming speed of the fish, they periodically sampled the O_2 content of arterial and venous blood. Arterial blood is freshly oxygenated and moving away from the gills; venous blood is returning to the gills from the rest of the body.

Not surprisingly, the biologists found that arterial O_2 levels remained fairly constant as swimming speed increased—meaning the gills continued to work efficiently enough to saturate hemoglobin with oxygen.

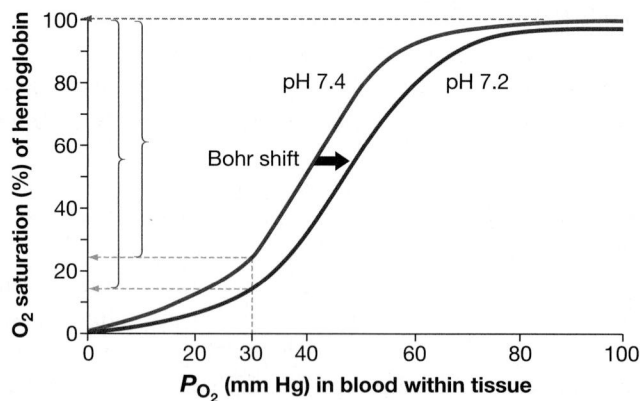

FIGURE 45.16 The Bohr Shift Makes Hemoglobin More Likely to Release Oxygen to Tissues with Low pH. As pH drops, oxygen becomes less likely to stay bound to hemoglobin at all values of tissue P_{O_2}. Exercising tissues have lower pH than resting tissues and thus receive more oxygen from hemoglobin.

✔ **QUANTITATIVE** Estimate how much more oxygen is unloaded from hemoglobin at pH 7.2 than at pH 7.4 when the tissue has a P_{O_2} of 30 mm Hg.

In contrast, the O_2 content of venous blood, which had undergone gas exchange with the tissues, dropped steadily as swimming speed increased. When the fish had reached their maximum sustainable speed, virtually all the oxygen that had been available in the blood had been extracted.

The data show that in hard-working tissues, the combination of increased temperature, lower pH, and lower P_{O_2} caused hemoglobin to become almost completely deoxygenated. Oxygen-delivery systems based on hemoglobin are extremely efficient.

Comparing Hemoglobins Hemoglobin molecules from different individuals or species may vary in ways that affect fitness—the ability to survive and produce viable offspring. As an example, consider the oxygen–hemoglobin equilibrium curves in **FIGURE 45.17**. The curve in dark red is from a pregnant woman; the curve in light red is from a fetus she is carrying.

The hemoglobin found in fetuses is encoded by different genes than adult hemoglobin and has a distinctive structure and function. Specifically, the oxygen–hemoglobin equilibrium curve for fetal hemoglobin is shifted to the left with respect to the curve for adult hemoglobin.

Recall that the rightward shift of the curve means that hemoglobin is more likely to release oxygen. The leftward shift in Figure 45.17 means that fetal hemoglobin is less likely to give up oxygen—it binds oxygen more tightly. Stated another way, fetal hemoglobin has a higher affinity for oxygen than adult hemoglobin does at every P_{O_2}.

This shift is crucial. In the placenta, hemoglobin from the mother's blood gets close to hemoglobin from the fetus's blood. Because fetal hemoglobin has such a high affinity for oxygen, there is a transfer of oxygen from the mother's blood to the fetus's blood. The difference in hemoglobin structure and function ensures an adequate supply of oxygen as the fetus develops.

CO_2 Transport and the Buffering of Blood pH

The carbon dioxide that is produced by cellular respiration in the tissues enters the blood, where it reacts with water to form carbonic acid, which dissociates into bicarbonate and hydrogen ions. Recall that the resulting drop in blood pH stimulates an increase in breathing rate. Rapid exhalation of CO_2 then counteracts the drop in blood pH.

Homeostasis with respect to blood pH is reinforced by an elegant series of events that take place inside red blood cells. Biologists were able to work out what was happening when they discovered large amounts of the enzyme **carbonic anhydrase** in red blood cells.

The Role of Carbonic Anhydrase and Hemoglobin Recall that carbonic anhydrase catalyzes the formation of carbonic acid from carbon dioxide in water (see Chapter 44). Consequently, CO_2 that diffuses into red blood cells is quickly converted to bicarbonate ions and protons. The same reaction occurs in the plasma surrounding red blood cells, although much more slowly in the absence of the enzyme.

Why is the carbonic anhydrase activity in red blood cells so important? The answer has two parts.

1. The protons produced by the enzyme-catalyzed reaction induce the Bohr shift, which makes hemoglobin more likely to release oxygen.

2. The partial pressure of CO_2 in blood drops when carbon dioxide is converted to soluble bicarbonate ions, maintaining a strong partial pressure gradient favoring the entry of CO_2 into red blood cells.

Thus, carbonic anhydrase activity makes CO_2 uptake from tissues more efficient (**FIGURE 45.18**).

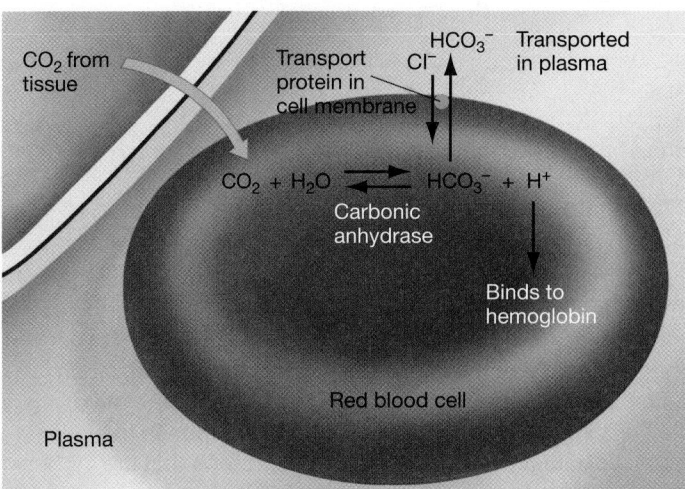

FIGURE 45.18 Carbonic Anhydrase Is Key to CO_2 Transport in Blood. When CO_2 diffuses into red blood cells, carbonic anhydrase quickly converts it to carbonic acid, which dissociates into a bicarbonate ion (HCO_3^-) and a proton (H^+). This reaction maintains the partial pressure gradient favoring the entry of CO_2 into red blood cells. The protons produced by the reaction bind to deoxygenated hemoglobin. Most CO_2 in blood is transported to the lungs in the form of HCO_3^-.

✔ QUESTION This diagram shows the sequence of events in tissues. After reading the rest of Section 45.4, explain what happens when the red blood cell in the diagram reaches the lungs.

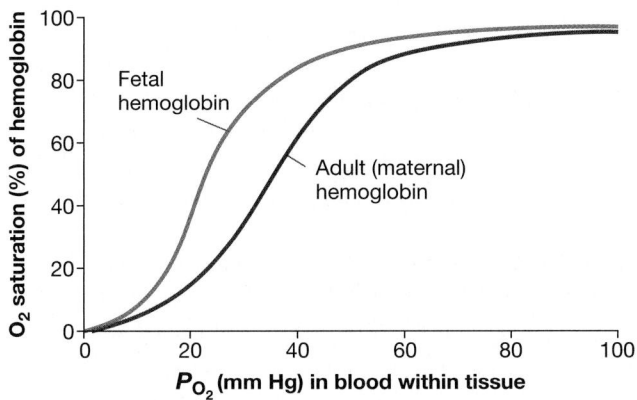

FIGURE 45.17 Fetal Hemoglobin Binds Oxygen More Tightly than Maternal Hemoglobin.

Once bicarbonate ions form in the red blood cell, they are transported into the blood plasma. The outcome is that most CO_2 is transported in blood (specifically in plasma) in the form of HCO_3^-—the bicarbonate ion. In contrast, the protons produced by the reaction stay inside red blood cells.

What ultimately happens to these protons? When hemoglobin is carrying few oxygen molecules, it has a high affinity for protons. As a result, it takes up much of the H^+ that is produced by the dissociation of carbonic acid. The hemoglobin acts as a **buffer**—a compound that minimizes changes in pH.

What Happens When Blood Returns to the Lungs? When deoxygenated blood reaches the alveoli, its environment changes dramatically. In the lungs, a partial pressure gradient favors the diffusion of CO_2 from plasma and red blood cells to the atmosphere within the alveoli. As CO_2 diffuses from the blood into the alveoli, P_{CO_2} in blood declines.

The drop in blood P_{CO_2} reverses the chemical reactions that occurred in tissues:

1. Hydrogen ions (protons) leave their binding sites on hemoglobin.

2. Protons react with bicarbonate to form CO_2.

3. CO_2 diffuses into the alveoli and is exhaled from the lungs.

In the meantime, hemoglobin has picked up O_2. Hemoglobin's affinity for oxygen is high in the alveoli because blood pH rises as P_{CO_2} declines.

When blood leaves the lungs, it has unloaded its carbon dioxide, and its hemoglobin is saturated with oxygen. The cycle begins anew.

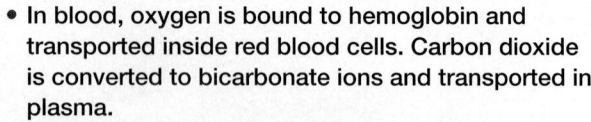
check your understanding

C Y U

If you understand that . . .

- In blood, oxygen is bound to hemoglobin and transported inside red blood cells. Carbon dioxide is converted to bicarbonate ions and transported in plasma.
- Hemoglobin has several properties that make it an effective transport protein, including cooperative binding of oxygen, the Bohr shift response to low pH, and the ability to bind the protons that are generated when carbon dioxide is converted to bicarbonate ions.

✔ **You should be able to . . .**

Predict how the oxygen–hemoglobin equilibrium curves measured in Tibetan people, whose ancestors have lived at high elevations for thousands of years, compare to curves from people whose ancestors have lived at sea level for many generations.

Answers are available in Appendix A.

45.5 Circulation

According to Fick's law, differences in the partial pressure of gases are only part of the story when it comes to understanding diffusion rates. Surface area—A in the equation featured in Figure 45.3—also plays a key role.

Animals without circulatory systems have various ways of maximizing the surface area available for diffusion of gases and other key solutes:

- Animals that are only a few millimeters in size, like rotifers and tardigrades, have a small enough volume that diffusion over their body surface is adequate to keep them alive.

- The flattened bodies of flatworms and tapeworms give these animals a high surface-area-to-volume ratio (see Chapter 42). In these species, too, molecules are exchanged with the environment directly across the outer body surface.

- Diffusion across the body wall also occurs in roundworms, where gas exchange is facilitated by muscular contractions in the body wall. Diffusion is enhanced as roundworms circulate fluids by sloshing them back and forth.

- Jellyfish and corals have a large, highly folded gastrovascular cavity that offers a large surface area for exchange of molecules with the environment.

In larger animals, however, the problem of providing a large enough surface area for diffusion is solved by a circulatory system. A circulatory system carries transport tissues called blood or hemolymph into close contact with every cell in the body. In this case, "close contact" is a distance of about 0.1 mm between the smallest blood vessels and cells within tissues. Diffusion is efficient at this scale.

To explore how circulatory systems work, let's start by distinguishing the two most basic types—open and closed. Open circulatory systems occur in most invertebrates, while all vertebrates have closed systems.

What Is an Open Circulatory System?

In an **open circulatory system,** a fluid connective tissue called hemolymph is actively pumped throughout the body in vessels. The hemolymph is not confined exclusively to the vessels, however. Instead, hemolymph comes into direct contact with tissues. As a result, the molecules being exchanged between hemolymph and tissues do not have to diffuse across the wall of a vessel. Hemolymph transports wastes and nutrients and may also contain oxygen-carrying pigments, some cells, and clotting agents.

FIGURE 45.19 illustrates the open circulatory system in a clam. Note that a muscular organ called the **heart** pumps hemolymph into vessels that empty into an open, fluid-filled space. Hemolymph is returned to vessels and then the heart when the heart relaxes and its internal pressure drops below that in the body cavity. General body movements also move hemolymph to and from the heart.

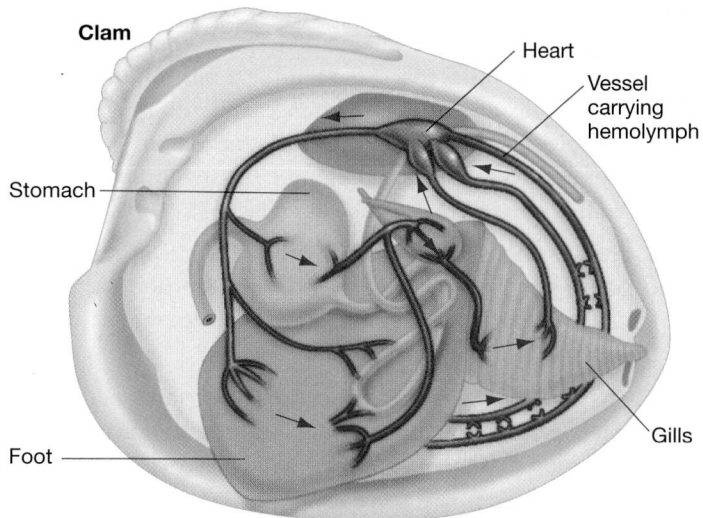

Clam

Heart

Vessel carrying hemolymph

Stomach

Foot

Gills

FIGURE 45.19 Clams Have Open Circulatory Systems. This drawing shows a clam with the top of the shell removed. Red arrows show the direction of hemolymph flow.

Because it moves throughout the volume of the body, hemolymph is under relatively low pressure in open circulatory systems. As a result, hemolymph flow rates may also be low. These features make open circulatory systems most suitable for relatively sedentary organisms, which do not have high oxygen demands.

Insects, with their rapid movements and more active lifestyles, are an exception to this rule. In the open circulatory systems of insects, the limitations imposed by low hemolymph pressure are overcome by their tracheal respiratory system, which delivers oxygen directly to the tissues.

Another characteristic of open circulatory systems, because they lack discrete, continuous vessels, is that the flow of hemolymph cannot be directed toward tissues that have a high oxygen demand and CO_2 buildup. An open circulatory system moves hemolymph throughout an animal's body in much the same way that a ceiling fan moves air throughout a room in a house.

Crustaceans are an important exception to this rule, however. Even though their circulatory system is classified as open, these species have a network of small vessels that can preferentially send hemolymph to tissues with the highest oxygen demands.

What Is a Closed Circulatory System?

In a **closed circulatory system,** blood flows in a continuous circuit through the body, under pressure generated by a heart. Because the blood is confined to vessels, a closed system can generate enough pressure to maintain a high flow rate.

In a closed circulatory system, blood flow can also be directed in a precise way in response to the tissues' needs. For example, blood can be shunted to leg muscles during exercise, to the intestines after a meal, or to regions of the brain engaged in particular mental tasks.

Which Lineages Have Closed Circulatory Systems? Closed circulatory systems are found in vertebrates and a few other lineages where individuals tend to be active. Earthworms and other annelids, for example, have a closed circulatory system and exchange gases with the environment across their thin, moist skin, which has a dense supply of capillaries. As a result, annelids are able to obtain and circulate enough oxygen to support intense muscular activity. Most live as active burrowers and hunters.

A similar situation occurs in squid, octopuses, and other cephalopods that hunt down prey. The closed circulatory system of these mollusks generates high rates of blood flow, which oxygenates their muscles well enough to support rapid movements and a predatory lifestyle.

Closed circulatory systems contain various types of blood vessels, each having a distinct structure and function. Let's review the major types of blood vessels and then consider how the vessels of a closed circulatory system interact with the lymphatic system.

Types of Blood Vessels An enormous amount of tubing is required to distribute blood within "diffusion distance" of every cell in the body. If all the blood vessels in a human body were laid end to end, they would stretch about 100,000 km (over 60,000 miles).

Blood vessels are classified as follows:

- **Arteries** are tough, thick-walled vessels that take blood away from the heart. Small arteries are called **arterioles.**

- **Capillaries** are vessels whose walls are just one cell thick, allowing exchange of gases and other molecules between blood and tissues. Networks of capillaries are called **capillary beds.**

- **Veins** are thin-walled vessels that return blood to the heart. Small veins are called **venules.**

The structure of arteries, capillaries, and veins correlates closely with their functions in a closed circulatory system. For example, the heart ejects blood into a large artery, usually called the **aorta.** All arteries have both muscle fibers and elastic fibers in their walls, but elastic fibers dominate the walls of the aorta. As a result, the aorta can expand when blood enters it under high pressure from the heart.

When a contraction of the heart ends, the diameter of the aorta returns to its resting state. This elastic response propels blood away from the heart and augments the force generated by the heart contraction. Similar types of secondary pumping action occur to some extent in other arteries as well. This feature helps maintain forward blood flow in the period between heart contractions.

The walls of arteries and arterioles have a thick layer composed of smooth muscle fibers. When the muscle fibers relax, the vessel diameter increases, resistance to flow is reduced, and blood flow increases in the tissues served by the vessel. But when these muscle fibers contract, the vessel diameter decreases, increasing resistance to flow and slowing the flow of blood in the vessel. In this way, blood flow to tissues can be carefully regulated by signals from the nervous system.

Capillaries are the smallest blood vessels. Their walls are only one cell layer thick, and they are just wide enough to let red blood cells through one at a time (**FIGURE 45.20a**). The extreme thinness of capillaries and the dense network they form throughout the body make them suitable sites for the exchange of gases, nutrients, and wastes between blood and the other tissues.

In some organs, such as the liver, the walls of capillaries contain many small openings that further diminish the barrier to diffusion between blood and the tissues. Despite their thinness, it is rare for capillaries to rupture, because blood pressure drops dramatically as blood passes through arterioles on its way to capillary beds.

After blood from arteries and arterioles passes through capillaries, veins carry it back to the heart. Because blood is under relatively low pressure by the time it exits the tissues, veins have thinner walls and larger interior diameters than arteries do (**FIGURE 45.20b**).

Blood flow in veins is speeded by muscle activity in the extremities, which compresses large veins. Larger veins also contain one-way **valves,** which are thin flaps of tissue that prevent any backflow of blood.

All veins contain some muscle fibers, which contract in response to signals from the nervous system, decreasing the diameter and overall volume of the vessels. Blood pressure in a closed circulatory system is regulated, in part, by actively adjusting the volume of blood contained within the veins.

What Is Interstitial Fluid? The relatively high operating pressure of closed circulatory systems, combined with the thinness of capillaries, produces a small but steady leakage of fluid from these blood vessels into the surrounding space. The area between cells is called interstitial space; the fluid that fills it is **interstitial fluid.** Blood cells are retained within capillaries, so interstitial fluid resembles plasma in its electrolyte composition.

Why does interstitial fluid build up? In 1896 Ernest Starling proposed that two forces were at work (**FIGURE 45.21**):

1. There is an outward-directed hydrostatic force in capillaries, created by the pressure on blood generated by the heart. This force is analogous to the pressure that drives water through the wall of a leaky garden hose.

2. There is also an inward-directed osmotic force across the capillary walls, created by the higher concentration of solutes in the blood plasma than in the interstitial space.

Starling reasoned that at the end of the capillary nearest to an arteriole, the hydrostatic force (the blood pressure) would exceed the osmotic force. If so, then in that location fluid would move out of the capillary into the interstitial space. But because blood pressure drops as fluid passes through a long, thin tube, Starling proposed that at the venous end of the capillary, the inward-directed osmotic force would exceed the outward-directed hydrostatic force. Thus, the fluid that was lost at the arteriolar end of the capillary would be largely reclaimed at the venous end.

Note the adverb "largely," however—not all interstitial fluid is reabsorbed by capillaries. In Figure 45.21, not all of the fluid entering the interstitial space at the arteriolar end of the capillary

(a) Capillaries are small and extremely thin walled.

(b) Veins and arteries differ in structure.

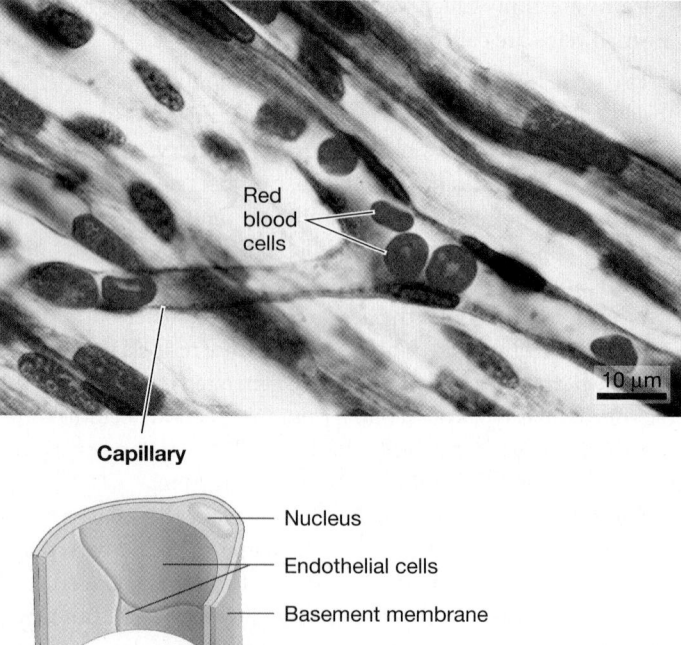

Red blood cells

10 µm

Capillary

Nucleus

Endothelial cells

Basement membrane

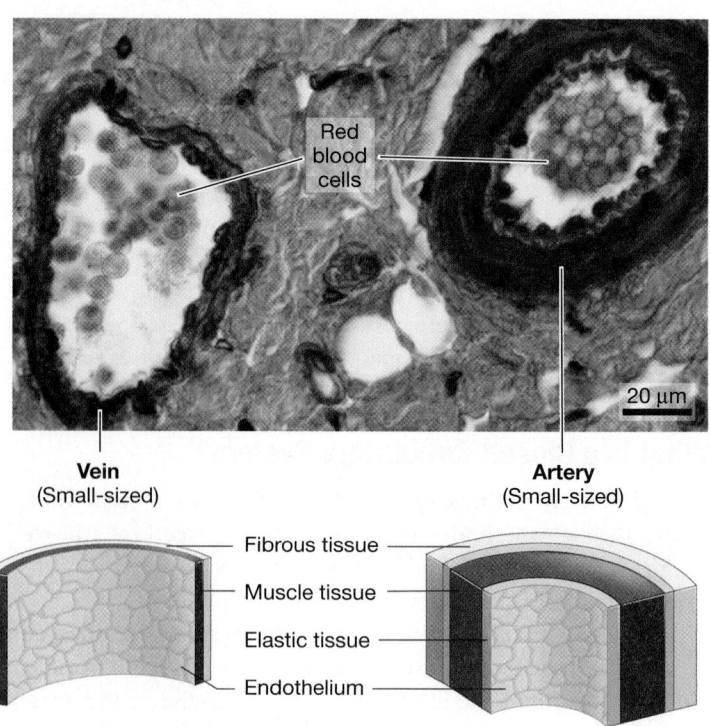

Red blood cells

20 µm

Vein
(Small-sized)

Artery
(Small-sized)

Fibrous tissue

Muscle tissue

Elastic tissue

Endothelium

FIGURE 45.20 The Structures of Capillaries, Veins, and Arteries Reflect Their Different Functions.
Notice the differences in relative wall thickness and overall size.

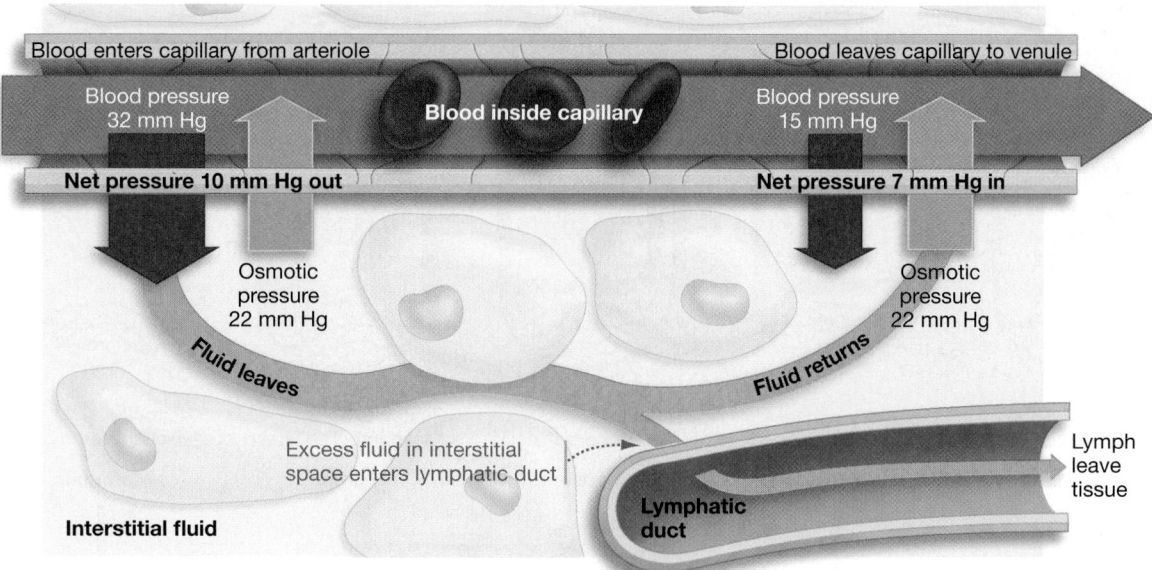

The balance of blood pressure and osmotic forces favors fluid loss from the beginning (inflow end) of capillaries and fluid recovery at the other (outflow) end. Fluid that is not recovered by the capillaries is transported out of the tissue as lymph, which eventually rejoins the blood circulation.

bed has reentered the bloodstream at the venous end—there is a net buildup of interstitial fluid. What happens to this fluid?

The Role of the Lymphatic System   Starling proposed that because interstitial fluid is continually added to the interstitial space, there must be a mechanism for draining the excess fluid. In fact, the fluid is collected in the **lymphatic system:** a collection of thin-walled, branching tubules called lymphatic vessels that permeate all tissues. Interstitial fluid that enters the lymphatic ducts is called **lymph.** Lymphatic vessels join with one another, like the tributaries of a river, to form larger vessels. The largest lymphatic vessels return excess interstitial fluid, in the form of lymph, to the major veins entering the heart.

The importance of the lymphatic system becomes evident when lymphatic vessels are damaged or blocked. For example, a disease called elephantiasis results when the lymphatic vessels in the extremities are blocked by parasitic worms that are transmitted from person to person via mosquito bites. The affected limbs swell dramatically because the lymph cannot be drained, and the skin thickens, cracks, and becomes very painful.

How Does the Heart Work?

In animals with closed circulatory systems, the heart contains at least two chambers: There is at least one thin-walled **atrium** (plural: **atria**), which receives blood, and at least one thick-walled **ventricle,** which generates the force required to propel blood out of the heart and through the circulatory system. Atria are separated from ventricles by atrioventricular valves.

The phylogenetic tree in **FIGURE 45.22** (see page 920) shows the evolutionary relationships among some major vertebrate groups and a simplified sketch of the heart and circulatory system for each lineage. Two points are particularly important to note:

1. The number of atria and ventricles in the heart increased as vertebrates diversified. Fish hearts have one atrium and one ventricle; amphibians, turtles, lizards, and snakes have two atria and one ventricle; crocodilians, birds, and mammals have two atria and two ventricles. It is common to refer to these as two-, three-, and four-chambered hearts, respectively.

2. In fishes, the circulatory system forms a single circuit—one loop services the gills and the body. In other lineages, there are separate circuits to the lungs and to the body.

Why Did Multichambered Hearts and Multiple Circulations Evolve?   To understand these evolutionary patterns, let's first consider the relatively simple heart and circulatory system found in fishes. A single atrium and ventricle and single circuit are adequate in fishes. Even though blood pressure drops as blood passes through the gills—due to the mechanical resistance to flow that occurs in the gills' capillary beds—blood pressure stays high enough to move blood throughout the body. This is largely because fishes live in the neutrally buoyant environment of water, where gravity does not have a large impact on blood flow.

In contrast, gravity has a much larger effect on circulation in land-dwelling vertebrates. Gravity has a particularly antagonistic effect on the flow of blood to elevated portions of the body. To overcome gravity in terrestrial environments, blood must be pumped at high pressure. However, the capillaries and alveoli of the lungs are too thin to withstand high pressures.

The successful solution in terrestrial vertebrates was the evolution of two separate pumping circuits:

1. The **pulmonary circulation** is a lower-pressure circuit to and from the lung.

2. The **systemic circulation** is a higher-pressure circuit to and from the rest of the body.

The pulmonary and systemic circulations are completely separated in the four-chambered hearts of crocodilians, birds, and mammals. On the other hand, the pulmonary and systemic

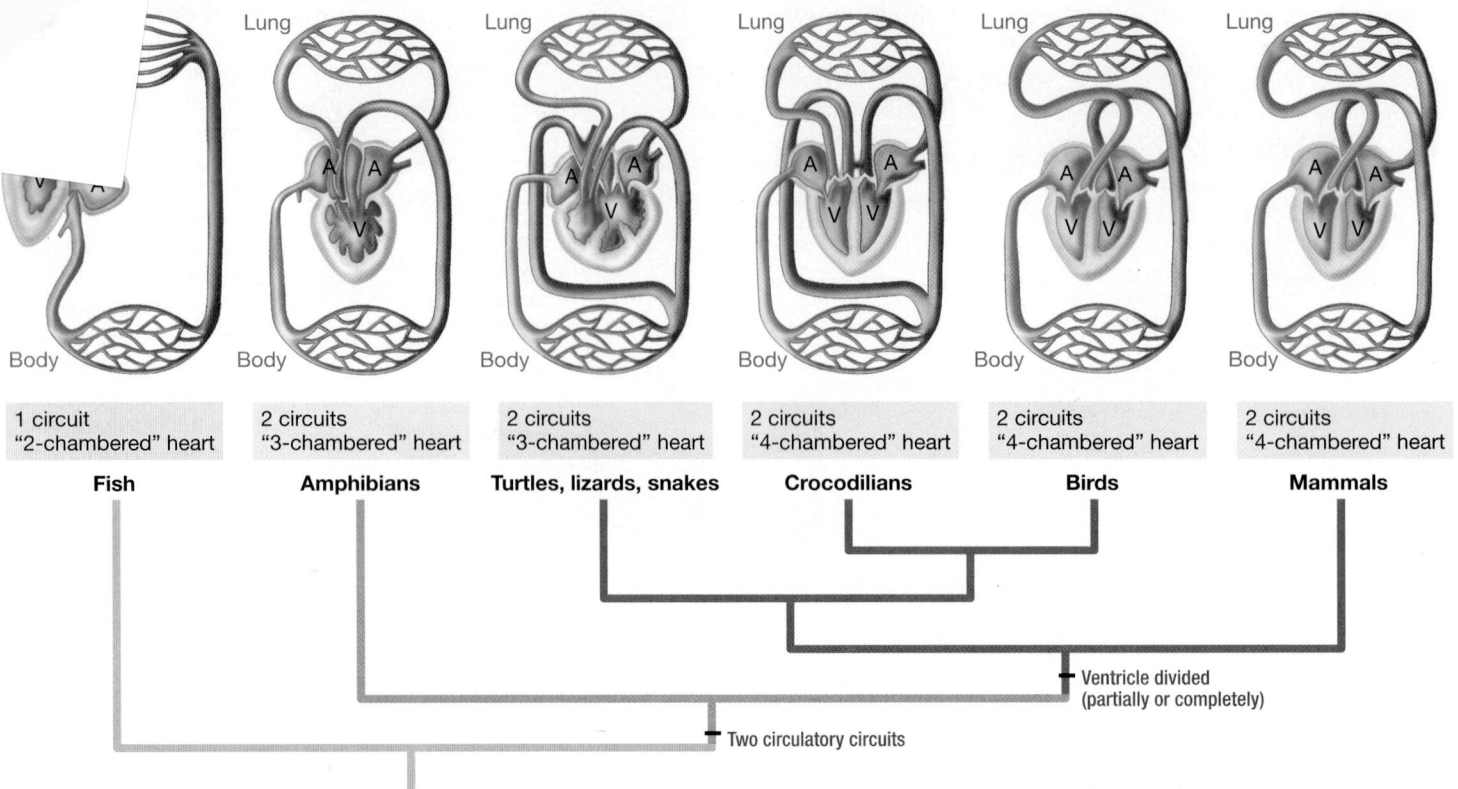

FIGURE 45.22 As Vertebrate Circulatory Systems Evolved, the Number of Atria and Ventricles Increased.
"A" denotes the atria—chambers that receive blood coming into the heart from the body and the gills or lungs.
"V" denotes the ventricles—chambers that pump blood out to the gills or lungs and the body.

circulations are only partially separated in the hearts of amphibians, turtles, lizards, and snakes. In these lineages, blood from the right and left atria may mix in the common ventricle before being expelled from the heart to the lungs or to the body. Turtles, lizards, and snakes, however, have partially divided ventricles that can limit the amount of mixing that occurs there (Figure 45.22).

In addition, turtles, lizards, and snakes have a bypass vessel running from the right side of the ventricle directly into the

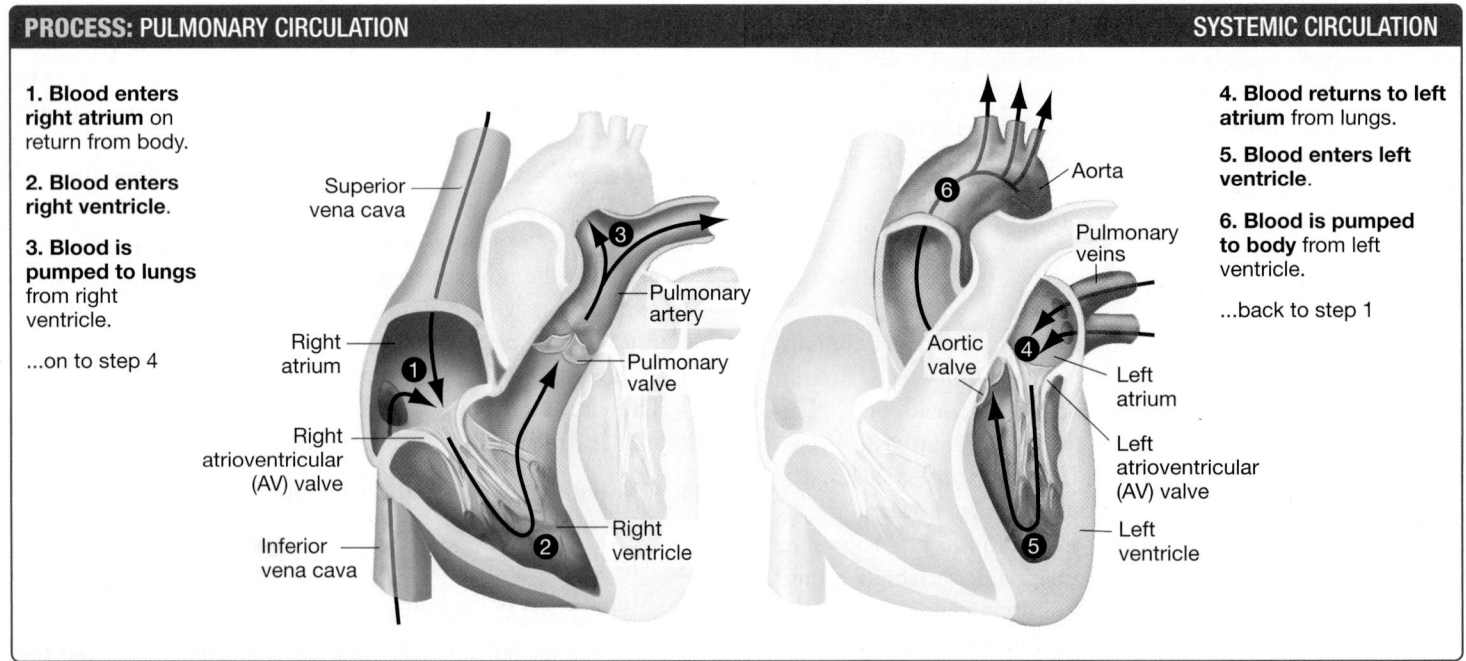

FIGURE 45.23 The Human Heart Maintains Separation of Oxygenated and Deoxygenated Blood. Blood flows through the chambers in the sequence shown.

systemic circulation. This bypass vessel is also observed in the unusual four-chambered hearts of crocodilians. The bypass vessels have an important function: They shunt blood from the pulmonary to the systemic circulation when the animal is underwater and not breathing. The result is a great reduction in blood flow to the lungs at those times.

Unlike reptiles, the hearts of birds and mammals have completely divided ventricles and lack a bypass vessel. This configuration completely separates the pulmonary and systemic circuits.

The Human Heart Your heart is located in the chest cavity, between your lungs, and is roughly the size of your fist. As **FIGURE 45.23** shows, the human circulatory system returns blood from the body to the right atrium of the heart. This blood is low in oxygen, and it arrives via two large veins called the inferior (lower) and superior (upper) **venae cavae** (singular: **vena cava**).

When the muscles that line the right atrium contract, they send deoxygenated blood to the right ventricle. The right ventricle, in turn, contracts and sends blood out to the lungs, via the **pulmonary artery.** In this way, the right ventricle powers the movement of blood through the pulmonary circulation.

Blood flows from atrium to ventricle to artery in only one direction, because one-way valves separate the heart's chambers from each other and from the adjacent arteries. As Figure 45.23 indicates, the valves are flaps, oriented to ensure a one-way flow of blood with little or no backflow. If heart valves are damaged or defective, the resulting backflow can be heard through a stethoscope. The backflow reduces the organ's efficiency and is called a **heart murmur.**

After blood circulates through the capillary beds in the lung's alveoli and becomes oxygenated, it returns to the heart through the **pulmonary veins.** The oxygenated blood enters the left atrium.

When the left atrium contracts, it pushes blood into the left ventricle. The walls of the left ventricle are so thick with muscle cells that their contraction sends oxygenated blood at high pressure through the aorta and into the arteries and capillaries that make up the systemic circulation.

FIGURE 45.24 summarizes the flow pattern through the human circulatory system and the blood gas concentrations at various points in the pulmonary and systemic circulations. Notice that blood vessels are called arteries or veins according to the direction of blood flow relative to the heart, not because of the oxygen content of the blood in them. Thus, the pulmonary artery is called an artery because it takes blood away from the heart, even though this blood is low in oxygen.

In healthy individuals, gas exchange in the lungs and tissues is rapid relative to the rate at which blood flows through capillaries. This situation is beneficial: It ensures that the maximum amount of oxygen is taken up and the maximum amount of carbon dioxide is released.

Similarly, the slow passage of blood through capillaries means that a maximum amount of oxygen is released from blood to tissues and a maximum amount of carbon dioxide is taken up. As

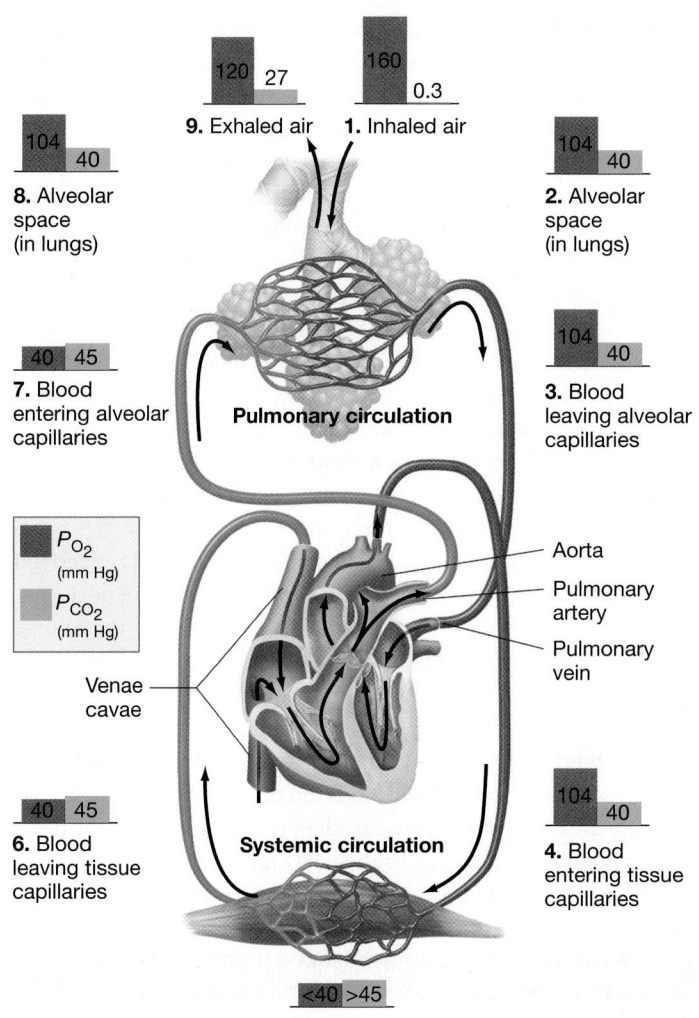

FIGURE 45.24 Partial Pressures of Gases Vary throughout the Human Circulatory System.

✔ **QUESTION** Why are the partial pressures of oxygen and carbon dioxide in exhaled air intermediate in magnitude between the partial pressures in inhaled and alveolar air?

a result, partial pressures for oxygen and carbon dioxide in the tissues and the systemic veins are equal.

Electrical Activation of the Heart Like other muscle cells, cardiac muscle cells contract in response to electrical signals. In invertebrates, the electrical impulses that trigger contraction come directly from the nervous system. But in vertebrates, a group of cells in the heart itself is responsible for generating the initial signal.

The cells that initiate contraction in the vertebrate heart are known as **pacemaker cells.** They are located in a region of the right atrium called the **sinoatrial (SA) node.**

The SA node and the muscle cells of the heart receive input from the nervous system and from chemical messengers carried in the blood. These inputs are important for regulating both the heart rate and the strength of ventricular contraction. In this way, the amount of blood moving through the circulatory system varies in response to electrical signals and hormones. During the "fight or flight" response (introduced in Chapter 49), for example, a chemical signal called epinephrine causes both heart rate and contraction strength to increase—sending more blood through the body in preparation for rapid movement.

As carefully regulated as the heart rate and contraction strength may be, a vertebrate heart will continue to beat even if all nerves supplying it are severed. Why? An electrical impulse that stimulates contraction is generated in the SA node and rapidly conducted throughout the right and left atria. The signal spreads quickly from cell to cell because of a striking property of cardiac muscle cells: They form physical and electrical connections with each other.

All cardiac muscle cells branch to contact several other cardiac muscle cells, join end to end with these neighboring cells (see Figure 42.5), and connect to them by specialized structures called **intercalated discs.** Because these discs contain many gap junctions (cell-to-cell connections described in Chapter 11), electrical signals pass directly from one cardiac muscle cell to the next.

The electrical activation of the heart is reflected in the orange line on the bottom of **FIGURE 45.25**. This line is an **electrocardiogram,** or **EKG**—a recording of the electrical events that occur over the course of a cardiac cycle. An EKG recording is generated by amplifying the overall electrical signal conducted from the heart to the chest wall through the tissues of the body. By inspecting an EKG, physicians can diagnose disturbances of heart rhythm and detect damage to the heart muscle.

The drawings above the graph in Figure 45.25 show where the key electrical events are happening.

1. The SA node generates an electrical signal.

2. The signal from the SA node is quickly propagated to atrial muscle cells. As a result, the atria contract simultaneously and eject blood into the ventricles.

3. As the atria begin to contract, the signal is conducted to an area of the heart called the **atrioventricular (AV) node.** The AV node delays the signal slightly before passing it to the ventricles. The delay allows the ventricles to fill completely with blood from the atria before they contract.

4. After the delay, the electrical impulse is rapidly transmitted through specialized fibers in the muscular wall that separates the ventricles. The impulse spreads through both ventricles, causing them to contract as the atria relax. The ventricles empty efficiently because the signal and the resulting muscular contraction move from the bottom up to the top of each ventricle—toward the arteries that allow blood to exit.

5. The final electrical event occurs as the ventricles relax and their cells recover—restoring their electrical state before contraction.

✔ If you understand these concepts, you should be able to predict how the amount of blood ejected from the ventricles would change if there were no delay at the AV node.

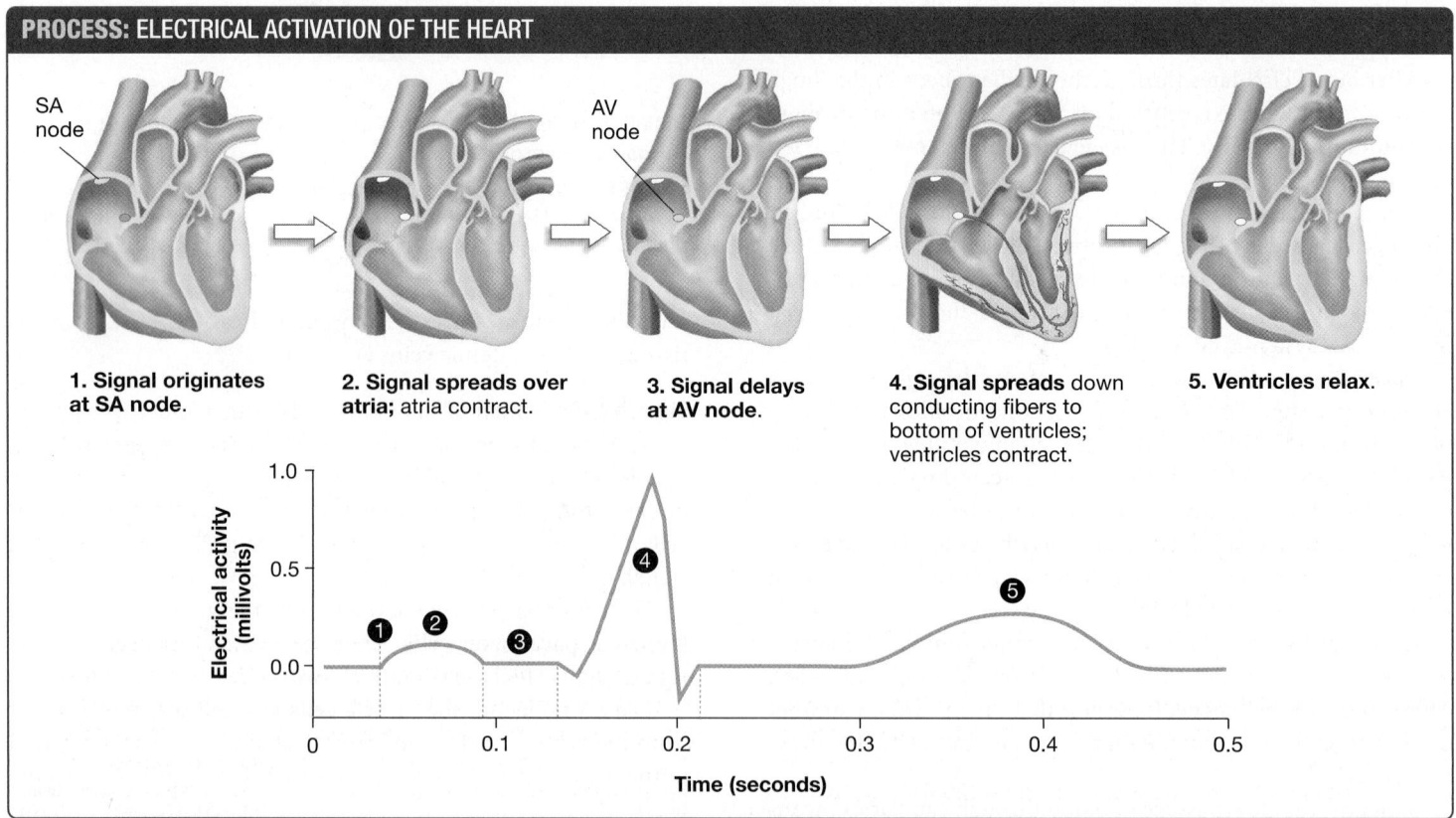

PROCESS: ELECTRICAL ACTIVATION OF THE HEART

1. **Signal originates at SA node.**

2. **Signal spreads over atria;** atria contract.

3. **Signal delays at AV node.**

4. **Signal spreads** down conducting fibers to bottom of ventricles; ventricles contract.

5. **Ventricles relax.**

FIGURE 45.25 Sequential Electrical Activation Leads to Coordinated Contraction of the Heart. The rate and strength of contractions control the pressure in the chambers and arteries.

The Cardiac Cycle The electrical signals originating from the SA node ensure that the atria contract simultaneously, and the delay at the AV node ensures that the atria are relaxed by the time the ventricles contract. The contraction phases of the atria and the ventricles, called **systole,** are therefore closely coordinated with their relaxation phases, or **diastole.**

This sequence of contraction and relaxation is called the **cardiac cycle.** It consists of one diastole and one systole for both atria and ventricles.

Ventricular contraction (ventricular systole) leads to a rapid increase in pressure within both ventricles, as recorded in the dark purple line in **FIGURE 45.26**. Blood is ejected into the pulmonary artery and the aorta when ventricular pressure exceeds the pressure within each respective artery. Blood pressure measured in the systemic arterial circulation at the peak of ventricular ejection into the aorta is called the **systolic blood pressure.** Blood pressure measured just before ventricular ejection is called the **diastolic blood pressure.**

Clinicians report blood pressure measurements in fractional notation, where systolic pressure is the numerator and diastolic pressure is the denominator. People with blood pressures consistently higher than 140/90 mm Hg have high blood pressure, or **hypertension.**

Hypertension is a serious concern to physicians because it can lead to a variety of defects in the heart and circulatory system. Abnormally high blood pressure puts mechanical stress on arteries. If the walls of an artery fail, the individual may experience heart attack, stroke, kidney failure, and burst or damaged vessels.

Patterns in Blood Pressure and Blood Flow

As blood moves through capillaries, blood pressure drops dramatically—as the top graph in **FIGURE 45.27** indicates. The bottom graph on this figure shows why. Trace the line labeled "Total area" on this graph to see that as arteries branch, rebranch, and eventually form networks of capillaries, the total cross-sectional area of blood vessels in the circulatory system increases—even though the size of individual blood vessels decreases as blood moves away from the heart. Note two key points:

1. Blood pressure in the capillaries drops dramatically (on the top graph) because the amount of mechanical resistance to flow is a function of total cross-sectional area of the vessels—meaning that friction losses are high in capillary beds. The pulsing nature of the pressure observed in the larger arteries diminishes as a result of this pressure drop.

2. As the line labeled "Velocity" in the bottom graph indicates, the velocity of blood flow also decreases significantly in capillary beds relative to arteries and veins, because the same

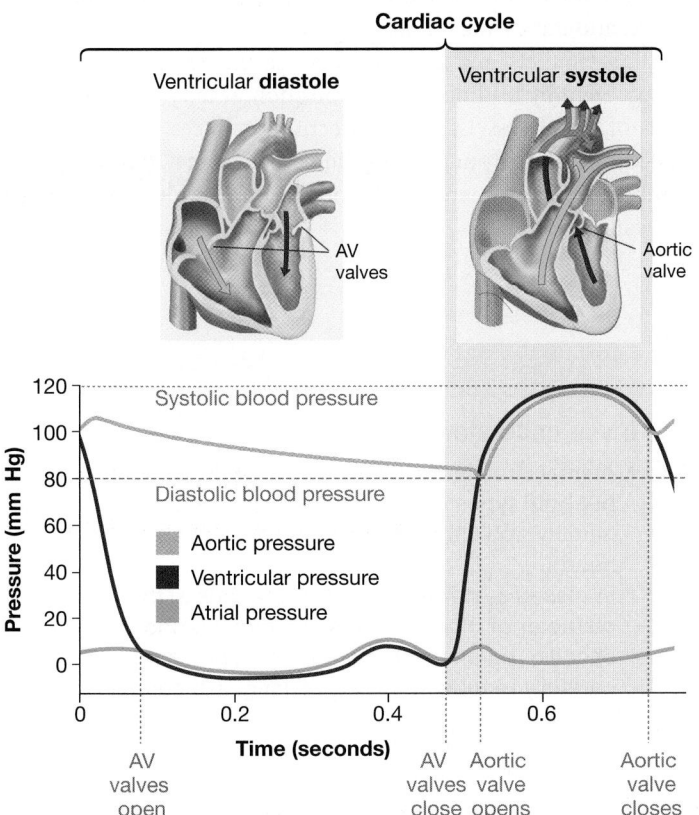

FIGURE 45.26 Blood Pressure Changes during the Cardiac Cycle. These data show the pressures created by the left ventricle and left atrium in the course of a cardiac cycle. In this example, the blood pressure measured in the upper arm would be 120/80 mm Hg. Right ventricle and pulmonary artery pressures would produce a similar pattern except that the blood pressure in the pulmonary artery would be much lower—closer to 25/8 mm Hg.

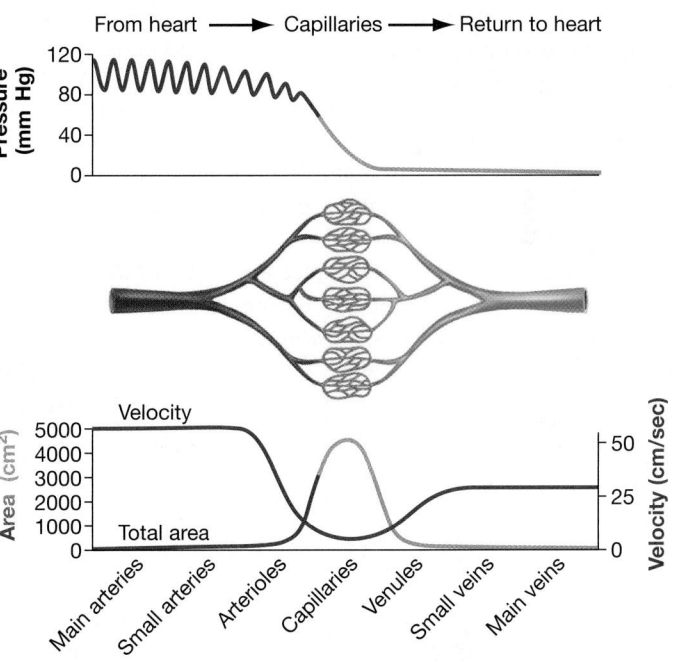

FIGURE 45.27 Blood Pressure Drops Dramatically in the Circulatory System. The top graph shows how blood pressure changes as blood leaves the heart and travels through arteries, capillaries, and veins, as in the branching pattern of the middle diagram. In arteries near the heart, each heartbeat causes fluctuations in blood pressure. These pressure pulses disappear in the capillaries, so blood flows there at a steady speed. The bottom graph plots the total area of blood vessels shown in the diagram, as well as the velocity of blood flow through the vessels.

amount of fluid is passing through a much larger area. Recall that the slow rate of blood flow through capillaries is important: It provides sufficient time for gases, nutrients, and wastes to diffuse between tissues and blood.

Why Is Regulation of Blood Pressure and Blood Flow Important?

The general patterns of blood pressure and blood flow diagrammed in Figure 45.27 don't tell the entire story, however. Blood movement is carefully regulated at an array of points throughout the circulatory system.

Recall that the walls of arterioles are comprised partially of smooth muscle. Contraction or relaxation of this muscle can constrict or allow blood flow to specific regions of the body. This means that the nervous system, along with certain chemical messengers in the circulation, can accurately control blood flow to various tissues by contracting or relaxing the muscle. For example, arterioles in the skin may dilate during exercise, diverting blood flow to the skin to eliminate excess heat. This accounts for the flushed facial appearance induced by vigorous exercise. Regulating blood flow is important for maintaining homeostasis with respect to body temperature.

As another example of how blood pressure and blood flow are regulated, consider what happens if you sit long enough for blood to pool in your lower extremities, under the influence of gravity. If you stand up rapidly, your blood pressure can drop enough to reduce blood flow to the brain and cause dizziness or even a blackout. More serious drops in blood pressure, due to severe dehydration or blood loss, can be fatal. Fortunately, decreases in blood pressure elicit a powerful homeostatic response.

Homeostatic Control of Blood Pressure
Recall that all homeostatic responses involve (1) sensors that detect the change in condition, (2) an integrator that processes information about the change, and (3) effectors that diminish the impact of the change (see Chapter 42).

Specialized pressure-sensing receptors called **baroreceptors** detect changes in blood pressure. Baroreceptors are found in the walls of the heart and the major arteries, both as they leave the heart and as they enter the neck. This distribution is logical, because the head tends to be the most elevated part of terrestrial animals, and preserving blood flow to the brain is the highest priority.

When baroreceptors transmit nerve signals to the brain (the integrator) indicating a serious fall in blood pressure, a rapid, three-component effector response ensues:

1. Cardiac output—the volume of blood leaving the left ventricle—increases. This is due to an increase in heart rate and an increase in stroke volume, which is the amount of blood ejected from the ventricles during each cardiac cycle. (Cardiac output = heart rate $\times$ stroke volume.)

2. Arterioles serving the capillaries of certain tissues constrict to divert blood to more critical organs. (This occurs in tissues like the skin and intestines, which can endure short-term restrictions in their blood supply without damage.)

3. Veins constrict, decreasing their overall volume. Because more than half of the blood in the circulatory system is contained within the veins, constriction of these vessels shifts blood volume toward the heart and arteries to maintain blood pressure and flow to vital organs.

This coordinated response is mediated both by a portion of the nervous system called the sympathetic nervous system (Chapter 46) and by hormones (Chapter 49) produced by the adrenal glands. Sympathetic nerves and the hormones involved in blood pressure regulation deliver their messages directly to (1) the SA node to increase heart rate, (2) the ventricles of the heart to increase stroke volume, and (3) the muscular walls of the arteries and veins to modify their total volume.

Cardiovascular Disease
A healthy circulatory system is obviously critical to your well-being. Indeed, **cardiovascular disease,** which is a group of ailments collectively affecting the heart and blood vessels, is the number one cause of death in humans worldwide.

Many factors contribute to cardiovascular disease, including age, tobacco use, poor diet, obesity, inactivity, and genetics. As people age, their blood vessels harden and lose elasticity—a condition called **arteriosclerosis.** Nicotine in tobacco causes constriction of blood vessels, further reducing the diameter of blood vessels and increasing blood pressure.

High-fat diets and lack of physical activity can compound the problem by leading to the deposition of fatty plaques on the walls of blood vessels, which effectively reduces their diameter. The loss of elasticity and the decline in the diameter of the vessels combine to cause increased blood pressure, which can weaken the walls of arteries.

check your understanding

(C)(Y)(U)

If you understand that . . .

- Animal circulatory systems may be open or closed, but both types of systems circulate blood or hemolymph via pressure generated by one or more hearts.

- In closed systems, regulated changes in the diameter of blood vessels can direct blood to specific regions, and overall blood pressure is carefully regulated through changes in cardiac output.

✔ You should be able to . . .

1. Explain how the mammalian lymphatic system and circulatory system interact.

2. Make a labeled diagram showing how blood circulates through the mammalian heart.

Answers are available in Appendix A.

If the arteries that deliver blood to the heart muscle become completely blocked, a **myocardial infarction,** or heart attack, can occur. In a myocardial infarction, a portion of heart tissue dies within minutes when it is deprived of oxygen. Depending on the location and extent of damage, myocardial infarction can lead to rapid death.

Over 17 million people worldwide died from cardiovascular disease in 2004, and this number is projected to reach 23 million by the year 2030. With the incidence of obesity and diabetes reaching epidemic proportions, these numbers may become even higher. Effectively combating cardiovascular disease will require improved diet, reduced tobacco use, greater amounts of physical activity, and better access to healthcare across the globe.

CHAPTER 45 REVIEW

For media, go to MasteringBiology

If you understand . . .

45.1 The Respiratory and Circulatory Systems

- Animal gas exchange involves ventilation, exchange of gases between the environment and the blood, and exchange of gases between blood and tissues.

- Animal circulation involves transportation of gases, nutrients, wastes, and other substances throughout the body.

✔ You should be able to identify the process that uses oxygen inside animal cells.

45.2 Air and Water as Respiratory Media

- As media for exchanging oxygen and carbon dioxide, air and water are dramatically different.

- Compared with water, air contains much more oxygen and is much less dense and viscous. As a result, terrestrial animals have to process a much smaller volume of air to extract the same amount of O_2, and the amount of work required to do so is less than in aquatic animals.

- Both terrestrial and aquatic animals pay a price for exchanging gases: Land-dwellers lose water to evaporation; freshwater animals lose ions and gain excess water; marine animals gain sodium and chloride and lose water.

✔ You should be able to explain why no water breathers have the extremely high rates of metabolism required for endothermy.

45.3 Organs of Gas Exchange

- The structure of gills, tracheae, lungs, and other gas exchange organs minimizes the cost of ventilation while maximizing the rate at which O_2 and CO_2 diffuse.

- Consistent with predictions made by Fick's law of diffusion, respiratory epithelia tend to be extremely thin and to be folded to increase surface area.

- In fish gills, countercurrent exchange ensures that the differences in O_2 and CO_2 partial pressures between water and blood are high over the entire length of the ventilatory surface.

- Insect tracheae carry air directly to and from tissues.

- In bird lungs, structural adaptations lead to a high ratio of useful ventilatory space to dead space.

- Breathing rate is regulated to keep the carbon dioxide content of the blood stable during both rest and exercise.

✔ You should be able to explain why no large animals have gas exchange occurring only across the skin.

45.4 How Are Oxygen and Carbon Dioxide Transported in Blood?

- The tendency of hemoglobin to give up oxygen varies as a function of the P_{O_2} in surrounding tissue in a sigmoidal fashion. As a result, a relatively small change in tissue P_{O_2} causes a large change in the amount of oxygen released from hemoglobin.

- Oxygen binds less tightly to hemoglobin when pH is low. Because CO_2 tends to react with water to form carbonic acid, the existence of high CO_2 partial pressures in exercising muscle tissues lowers their pH and makes oxygen less likely to stay bound to hemoglobin and more likely to be unloaded into tissues.

- The CO_2 that diffuses into red blood cells from tissues is rapidly converted to carbonic acid by the enzyme carbonic anhydrase. The protons that are released as carbonic acid dissociates bind to deoxygenated hemoglobin. In this way, hemoglobin acts as a buffer that takes protons out of solution and prevents large fluctuations in blood pH.

✔ You should be able to explain why exposure to large doses of carbon monoxide (CO) can lead to suffocation. (The CO found in furnace and engine exhaust and in cigarette smoke binds to the heme groups in hemoglobin 210 times more tightly than does oxygen.)

45.5 Circulation

- In many animals, blood or hemolymph moves through the body via a circulatory system consisting of a pump (heart) and vessels.

- In open circulatory systems, overall pressure is low and tissues are bathed directly in hemolymph.

- In closed circulatory systems, blood is contained in vessels that form a continuous circuit. Containment of blood allows higher pressures and flow rates, as well as the ability to direct blood flow accurately to tissues that need it the most.

- In organisms with a closed circulatory system, a lymphatic system returns excess fluid that leaks from the capillaries back to the circulation.

- In amphibians and some reptiles, blood from the pulmonary and systemic circuits may be mixed in the single ventricle.

- In mammals and birds, a four-chambered heart pumps blood into two circuits, which separately serve the lungs and the rest of the body. Crocodilians have a similar heart with a bypass vessel that can shunt blood from the pulmonary to the systemic circuit.

- The cardiac cycle is controlled by electrical signals that originate in the heart itself.

- Heart rate, cardiac output, and contraction of both arterioles and veins are regulated by chemical signals and by electrical signals from the brain.

- Cardiovascular disease is the leading cause of death in humans.

- ✔ You should be able to predict the characteristics of the circulatory systems of terrestrial animals with high activity rates.

(MB) MasteringBiology

1. **MasteringBiology Assignments**

   Tutorials and Activities Gas Exchange; Gas Exchange in the Lungs and Tissues; Gas Transport in Blood; Human Heart; Human Respiratory System; Mammalian Cardiovascular System Function; Mammalian Cardiovascular System Structure; Path of Blood Flow in Mammals; Transport of Respiratory Gases

   Questions Reading Quizzes, Blue-Thread Questions, Test Bank

2. **eText** Read your book online, search, take notes, highlight text, and more.

3. **The Study Area** Practice Test, Cumulative Test, BioFlix® 3-D Animations, Videos, Activities, Audio Glossary, Word Study Tools, Art

You should be able to . . .

✔ TEST YOUR KNOWLEDGE
Answers are available in Appendix A

1. O_2 will diffuse from blood to tissue faster in response to which of the following conditions?
   a. an increase in the P_{O_2} of the tissue
   b. a decrease in the P_{O_2} of the tissue
   c. an increase in the thickness of the capillary wall
   d. a decrease in the surface area of the capillary

2. Which of the following does blood *not* do?
   a. transport O_2 and CO_2
   b. distribute body heat
   c. produce new red blood cells and other formed elements
   d. buffer against pH changes

3. In insects, what is the adaptive significance of spiracles?
   a. They dilate and constrict during flight or other types of movement, functioning as a "breathing" mechanism.
   b. They open into the body cavity, allowing direct contact between hemolymph and tissues.
   c. They are thin and highly branched, offering a large surface area for gas exchange.
   d. They close off tracheae to minimize water loss.

4. Which of the following is *not* an advantage of breathing air over breathing water?
   a. Air is less dense than water, so it takes less energy to move during ventilation.
   b. Oxygen diffuses faster through air than it does through water.
   c. The oxygen content of air is greater than that of an equal volume of water.
   d. Air breathing leads to high evaporation rates from the respiratory surface.

5. Which of the following promotes oxygen release from hemoglobin?
   a. a decrease in temperature
   b. a decrease in CO_2 levels
   c. a decrease in pH
   d. a decrease in carbonic anhydrase

6. Describe the ways in which an open circulatory system is less efficient than a closed circulatory system.

✔ TEST YOUR UNDERSTANDING
Answers are available in Appendix A

7. Describe the changes in oxygen delivery that occur as a person proceeds from a resting state to intense exercise.

8. Explain how most carbon dioxide is transported in the blood. In humans, why don't intense exercise and rapid production of CO_2 lead to a rapid reduction in blood pH?

9. Why is ventilation in birds considered much more efficient than the respiratory system of humans and other mammals?

10. Explain how each parameter in Fick's law of diffusion is reflected in the structure of the mammalian lung.

11. Carp are fishes that thrive in stagnant-water habitats with low oxygen partial pressures. Compared with the hemoglobin of many other fish species, carp hemoglobin has an extremely high affinity for O_2. Is this trait adaptive? Explain your answer.

12. Frog lungs have lower surface area for gas exchange than mammalian lungs. Which of the following observations explains this?

a. Frog tissue needs more oxygen than mammalian tissue.
b. Frogs breathe more quickly than mammals.
c. Frogs also obtain oxygen via diffusion across the skin.
d. Frog lung tissue has a greater density of capillary beds than mammalian lung tissue.

✓ TEST YOUR PROBLEM-SOLVING SKILLS

Answers are available in Appendix A

13. Predict how the icefish native to the Antarctic are able to transport oxygen and carbon dioxide in their blood even though they lack hemoglobin.

14. Compare and contrast the respiratory and circulatory systems of an insect and a human.

15. Why did separate systemic and pulmonary circulations evolve in species that have the high-pressure circulatory system required for rapid movement of blood?

16. Under certain conditions, cells produce a metabolic byproduct called 2,3-diphosphoglycerate (DPG). Which of the following would support the hypothesis that DPG helps increase oxygen unloading from tissues?

a. DPG shifts the oxygen–hemoglobin equilibrium curve to the left.
b. DPG decreases percentage saturation of hemoglobin with oxygen at a P_{O_2} of 30 mm Hg.
c. DPG increases the pH of the tissues.
d. DPG is produced in greater quantity at low altitude than at high altitude.

46 Animal Nervous Systems

In this chapter you will learn how

Animals use electrical signaling along neurons for internal communication

beginning with

Principles of electrical signaling **46.1**

then asking

How does the action potential work? **46.2**

finally looking at

Synapses—the interface between neurons **46.3**

then exploring

The vertebrate nervous system **46.4**

including

The brain, learning, and memory

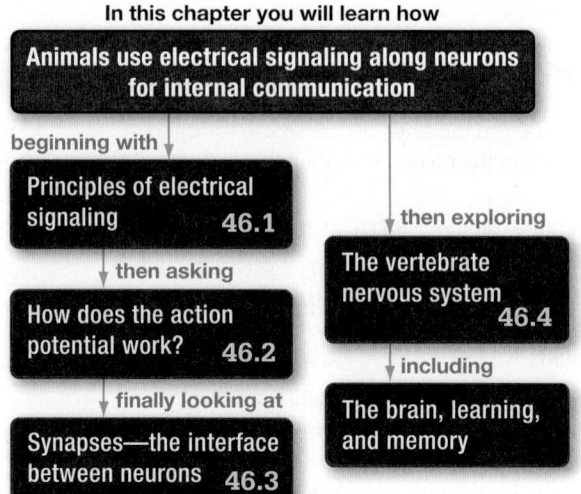

Diffusion spectrum imaging reveals the activity and trajectory of neurons in the brain. Such advances in brain imaging are allowing neurobiologists to study how neurons in the brain communicate with one another.

This chapter is part of the Big Picture. See how on pages 840–841.

Most students and professional biologists are attracted to the study of neurobiology because they want to understand the human brain as well as higher-order processes like consciousness, intelligence, emotion, learning, and memory. But early in the history of neurobiology, researchers realized that the human brain—with its billions of cells—was much too complex to study productively.

Instead biologists did the same thing they did in the early days of studying cells, genetics, and evolution: They started simply. Early research in neurobiology focused on the function of individual nerve cells, or **neurons,** the cells mainly responsible for the working of the brain and the rest of the nervous system. Neurons conduct information in the form of electrical signals from point to point

✔ When you see this checkmark, stop and test yourself. Answers are available in Appendix A.

in the body at speeds of up to 200 m/sec (450 mph). Thus electrical signaling is a crucial aspect of information processing—one of the five attributes of life (introduced in Chapter 1).

Initial research on the electrical properties of single neurons laid a broad foundation for more recent studies of how the human brain works. This chapter proceeds in the same way. Let's begin by focusing on the neurons themselves and how they use electrical signaling to communicate within the body and with each other. By the end of the chapter, you'll be considering how the brain is organized and how phenomena like memory work.

46.1 Principles of Electrical Signaling

The evolution of neurons was a key event in the diversification of animals, along with the evolution of muscles (see Chapter 33). All animals except sponges have neurons and muscle cells. Neurons transmit electrical signals; muscles can respond to signals from neurons by contracting.

Neurons and muscle cells made rapid movement possible. These innovations in cell structure and function enabled animals to evolve into today's flatworms, squid, beetles, sharks, and birds.

Recall that neurons are organized into two basic types of nervous systems:

1. The diffuse arrangement of cells called a **nerve net,** found in cnidarians (jellyfish, hydra, anemones) and ctenophores (comb jellies).

2. A **central nervous system (CNS)** that includes large numbers of neurons aggregated into clusters called ganglia.

In most cases, animals with a CNS have a large cerebral ganglion, or brain, located in their anterior end. You also might recall that this phenomenon—the evolution of a bilaterally symmetric body with structures for information gathering and processing located at the head end—is known as cephalization (see Chapter 33).

Cephalization made animals into efficient eating and moving machines: They face the environment in one direction, and sensory appendages take in information and send it to a nearby brain for processing. After integrating information from an array of sensory cells, the brain sends electrical signals to muscles and other tissues that respond to the sensory stimuli.

Types of Neurons in the Nervous System

The sensory cells that are responsible for gathering information respond to light, sound, touch, or other stimuli. Sensory cells in the animal's skin, eyes, ears, mouth, and nose transmit information about the environment. Sensory cells inside the body monitor conditions that are important in homeostasis, such as blood pH and temperature. In this way, sensory cells monitor conditions both outside and inside the body. Many sensory cells are **sensory neurons** (**FIGURE 46.1**), which carry information to the brain and spinal cord.

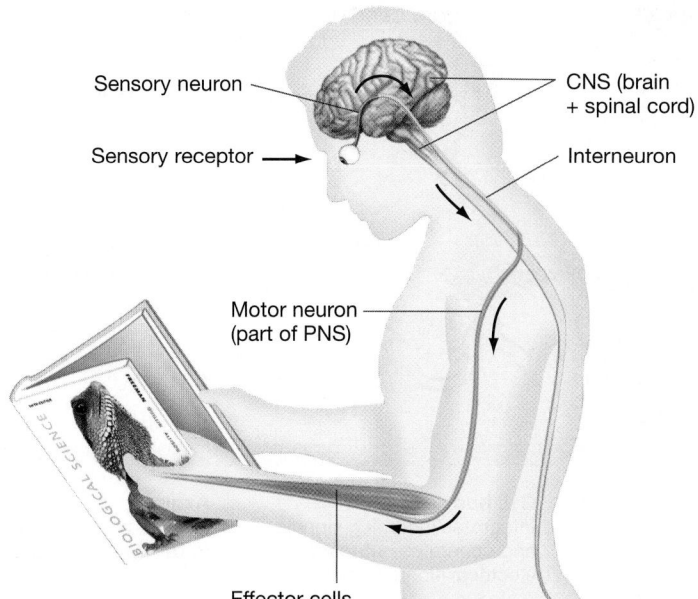

FIGURE 46.1 The Brain Integrates Sensory Information and Sends Signals to Effector Cells. In most cases, sensory neurons send information to the brain, where it is integrated with information from other sources. Once integration is complete, a response is sent to effector cells through motor neurons.

Together, the brain and spinal cord form the CNS of vertebrates. One function of the CNS is to integrate information from many sensory neurons. Cells in the CNS called **interneurons** (literally, "between-neurons"), which pass signals from one neuron to another, perform this integration.

Interneurons also make connections between sensory neurons and **motor neurons,** which are nerve cells that send signals to effector cells in glands or muscles. Motor neurons and sensory neurons are bundled together into long, tough strands of nervous tissue called **nerves.** Recall that effectors are structures that bring about a physiological change in an organism (see Chapter 42).

All neurons and other components of the nervous system that are outside the CNS are considered part of the **peripheral nervous system,** or **PNS.** Section 46.4 describes the structure and function of the PNS in more detail.

Typically, sensory information from receptors in the PNS is sent to the CNS, where it is processed. Then a response is transmitted back to appropriate parts of the body via motor neurons. When you stub your toe, pain receptors in your toe relay sensory information to the brain, which then modifies your movements to avoid further injury. You might hop a bit, for example, and then limp to avoid further pressure on that toe.

The Anatomy of a Neuron

Neurons are difficult to study because they are small, transparent, and morphologically complex. So when Camillo Golgi discovered that some neurons become visible when samples of preserved tissue are treated with a solution containing silver nitrate, his finding was a major advance. The year was 1898.

(a) Information flows from dendrites to the axon.

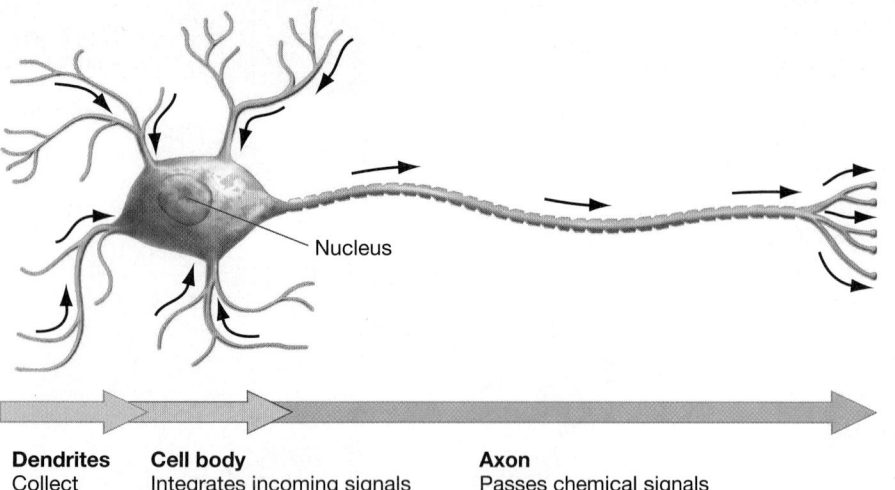

Dendrites
Collect chemical signals

Cell body
Integrates incoming signals and generates outgoing electrical signal to axon

Axon
Passes chemical signals to dendrites of another cell or to an effector cell

(b) Neurons form networks for information flow.

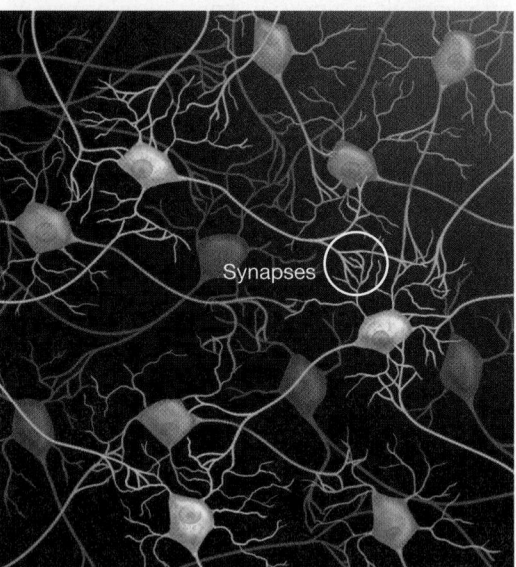

Synapses

FIGURE 46.2 How Does Information Flow in a Neuron? (a) The structure of a generalized neuron. **(b)** Most neurons receive inputs from many different neurons and send projections to several different neurons.

Through the early decades of the twentieth century, the work of Golgi and Santiago Ramón y Cajal revealed several important points about the anatomy of neurons. Most neurons have the same three parts, shown in **FIGURE 46.2a**:

1. A cell body, which contains the nucleus

2. A highly branched group of relatively short projections called dendrites

3. One or more relatively long projections called axons

Dendrites are rarely more than 2 mm long, but axons can be over a meter in length. The number of dendrites and their arrangement vary greatly from cell to cell. Further, many neurons in the brain have only dendrites and lack axons.

A **dendrite** receives a signal from the axons of adjacent cells; a neuron's **axon** sends a signal to the dendrites and cell bodies of other neurons (**FIGURE 46.2b**). In short, dendrites receive signals; axons pass them on. The **cell body,** or **soma,** also receives signals from axons of other cells. Incoming signals are integrated in the dendrites and cell body, and an outgoing signal is sent to the axon.

Ramón y Cajal maintained that the plasma membrane of each neuron is separated from those of adjacent neurons. This hypothesis was confirmed in the 1950s, when images from electron microscopes showed that most neurons are separated from one another at their junctions by tiny spaces called synapses.

Given that neurons transmit electrical signals, how do they produce them?

An Introduction to Membrane Potentials

Ions carry an electric charge. In virtually all cells, the cytoplasm and extracellular fluids adjacent to the plasma membrane contain unequal distributions of ions. As a result, cells are inherently electrical in nature.

A difference of electrical charge between any two points creates an **electrical potential,** or a **voltage.** If the positive and negative charges of the ions on each side of a plasma membrane do not balance each other, the membrane will have an electrical potential.

When an electrical potential exists across a plasma membrane, the separation of charges is called a **membrane potential.** If there is a large separation of charges across the membrane, the membrane potential is large.

It is important to remember that membrane potentials refer only to a separation of charge immediately adjacent to the plasma membrane, on either side of the membrane. Even if there is a large membrane potential, there may be no charge separation slightly farther away from the membrane.

Units and Signs Membrane potentials are measured in units called millivolts. The **volt** is the standard unit of electrical potential, and a **millivolt (mV)** is 1/1000 of a volt. As a comparison, an AA battery that you buy in the store has an electrical potential of 1500 mV between its + and − terminals. In neurons, membrane potentials are typically about 65 to 80 mV.

By convention, membrane potentials are always expressed in terms of inside relative to outside (the outside value is defined as 0). Because there are usually more negatively charged ions and fewer positively charged ions on the inside surface of a membrane relative to its outside surface, membrane potentials are usually negative in sign.

Electrical Potential, Electric Currents, and Electrical Gradients Membrane potentials are a form of potential energy. Recall that potential energy is energy based on the position of matter.

To convince yourself that ions have potential energy when a membrane potential exists, consider what would happen if the membrane were removed. Ions would spontaneously move from the region of like charge to the region of unlike charge—causing a flow of charge. This flow of charge, called an **electric current,** would occur because like charges repel and unlike charges attract.

However, charge is not the only form of potential energy. Ions also have different concentrations across membranes. Therefore, a membrane potential also includes energy stored as the concentration gradients of charged ions on the two sides of the membrane. Recall that the combination of an electrical gradient and a concentration gradient is an **electrochemical gradient** (see Chapter 6).

What do all these facts have to do with neuron function? Neurons use the electrochemical gradient of ions across their membranes to power the signals that allow neurons to communicate with one another and with other tissues.

How Is the Resting Potential Maintained?

When a neuron is not communicating with other cells or tissues, the difference in charge across its membrane is called the **resting potential.** To understand why the resting potential exists, consider the distribution of the various ions and other charged molecules on the two sides of the neuron's plasma membrane, shown in **FIGURE 46.3**:

- The interior side of the membrane has relatively low concentrations of sodium (Na^+) and chloride (Cl^-) ions, a relatively high concentration of potassium ions (K^+), and some organic anions—amino acids and other organic molecules that have dropped a proton and thus carry a negative charge.

- In the extracellular fluid, sodium and chloride ions predominate.

If each type of ion diffused across the membrane in accordance with its concentration gradient, organic anions and K^+ would leave the cell, while Na^+ and Cl^- would enter.

Ions cannot cross phospholipid bilayers readily, however. They cross plasma membranes efficiently in only three ways (see Chapter 6):

1. Flowing along their electrochemical gradient through an **ion channel**—a protein that forms a pore in the membrane and allows a specific ion to diffuse along its gradient

2. Carried, via a membrane cotransporter protein, with another ion that diffuses along its electrochemical gradient

3. Pumped against an electrochemical gradient by a membrane protein that hydrolyzes adenosine triphosphate (ATP)

How does the distribution of ions illustrated in Figure 46.3 come to be? The answer hinges on two types of membrane proteins: an ion channel that allows potassium ions to diffuse along their gradient, and a pump that actively transports sodium and potassium ions against their gradients. Let's consider each in turn.

The K^+ Leak Channel At rest, the plasma membrane of neurons is relatively impermeable to most cations. However, neurons have a relatively high number of potassium channels, called K^+ **leak channels,** that allow K^+ to leak out of the cell. K^+ ions slowly

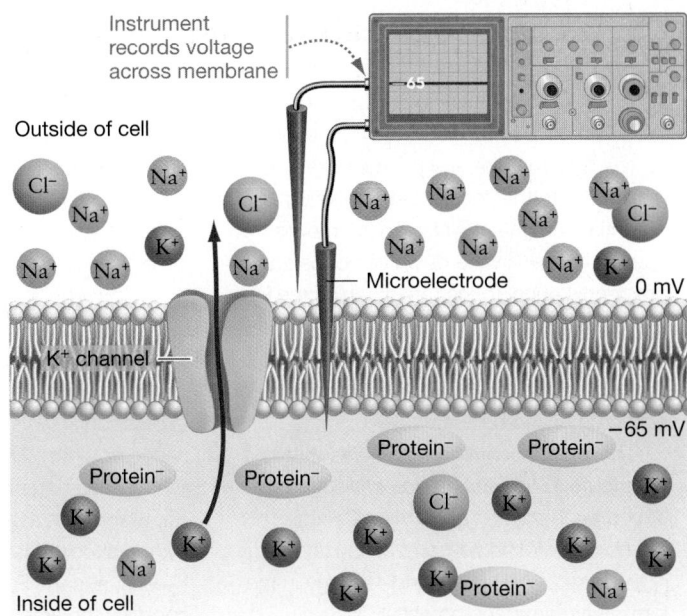

FIGURE 46.3 Neurons Have a Resting Potential. In resting neurons, the membrane is selectively permeable to K^+. As K^+ leaves the cell along its concentration gradient, the inside of the membrane becomes negatively charged relative to the outside. To measure a neuron's membrane potential, researchers insert a microelectrode into the cell and compare that reading with the reading outside the cell.

✓**QUESTION** Will K^+ continue to leave the cell indefinitely? Explain why or why not.

exit the neuron through the leak channels along their electrochemical gradient.

As K^+ moves from the interior of the cell to the exterior through leak channels, the inside of the cell becomes more and more negatively charged relative to the outside. This buildup of negative charge inside the cell begins to attract K^+ and counteract the concentration gradient that had favored the movement of K^+ out.

As a result of the counteracting influences, the membrane reaches a voltage at which equilibrium exists between the concentration gradient that favors movement of K^+ out and the electrical gradient that favors movement of K^+ in. At this voltage, there is no longer a net movement of K^+. This voltage is called the **equilibrium potential** for K^+. **Quantitative Methods 46.1** (see page 932) shows how equilibrium potentials for individual ions are calculated.

Although Cl^- and Na^+ cross the plasma membrane much less readily than does K^+, some movement of these ions also occurs through a small number of leak channels that are selective for each of them. As a result, each type of ion has an equilibrium potential. The membrane as a whole has a membrane potential that combines the effects of the individual ions.

The Role of the Na^+/K^+-ATPase What is responsible for the higher concentration of Na^+ on the outside and K^+ on the inside in the first place? It is the sodium–potassium pump, Na^+/K^+-ATPase, which actively pumps Na^+ out of the cell and K^+ into

An equilibrium potential exists whenever there is a concentration gradient for an ion across a membrane and the membrane is permeable to that ion. At the equilibrium potential, the rate at which an ion moves across the membrane down its concentration gradient is equal to the rate at which the ion moves, in the opposite direction, down its electrical gradient. It's the voltage at which the concentration and electrical gradients acting on an ion balance out.

The ion concentration gradients that contribute to equilibrium potentials are produced by the action of Na$^+$/K$^+$-ATPase. This membrane protein converts chemical energy in the form of ATP to electrical energy in the form of a membrane potential.

To calculate an equilibrium potential, biologists use the Nernst equation—a formula that converts the energy stored in a concentration gradient to the energy stored as an electrical potential. The concentration gradient for an ion is symbolized

as $[ion]_o/[ion]_i$, where $[ion]_o$ and $[ion]_i$ are the concentrations of the ion outside and inside the cell, respectively. The equilibrium potential for that ion is symbolized as E_{ion}. The Nernst equation specifies the equilibrium potential for a given ion as

$$E_{ion} = 2.3 \frac{RT}{zF} \log \frac{[ion]_o}{[ion]_i}$$

In this expression, z is the valence of the ion (for instance, +1 for potassium) and the expression RT/F is known as the thermodynamic potential. The three terms in the thermodynamic potential are the gas constant (R), which acts as a constant of proportionality; the absolute temperature (T), measured in Kelvins; and the Faraday constant (F), which specifies the amount of charge carried by 1 mol of an ion with a valence of +1 or −1.

Note that the Nernst equation is based on the base-10 logarithm of the ion concentration ratio, $[ion]_o/[ion]_i$. Thus, the thermodynamic potential specifies the voltage required to balance a tenfold concentration ratio

across the membrane. (For more on using logarithms, see **BioSkills 6** in Appendix B.)

The Nernst equation applies to only a single ion. For example, suppose that the potassium concentration inside the axon of a squid neuron has been measured as 400 mM, yet the outside concentration for this ion is only 20 mM (**Table 46.1**). Potassium has a charge of +1, so at 20°C the RT/zF part of the equation yields +25 mV. In this case, the equilibrium potential for potassium becomes

$$E_K = 2.3 \times 25\text{ mV} \times \log \frac{20\text{ mM}}{400\text{ mM}}$$

$$E_K = 58\text{ mV} \times \log 0.05 = -75\text{ mV}$$

The minus sign indicates that the interior of the axon is negatively charged with respect to the exterior.

Repeating this process for Na$^+$ ions and Cl$^-$ ions yields the following results for the squid axon:

$$E_{Na} = 58\text{ mV} \times \log \frac{440\text{ mM}}{50\text{ mM}} = +54.8\text{ mV}$$

$$E_{Cl} = -58\text{ mV} \times \log \frac{560\text{ mM}}{51\text{ mM}} = -60\text{ mV}$$

Again, notice that the equilibrium potential given by the Nernst equation is calculated independently for each ion.

✔**QUANTITATIVE** If you understand this concept, you should be able to calculate the equilibrium potential at 20°C for calcium (Ca^{2+}) in a cell where the intracellular concentration is 0.0001 mM and the extracellular concentration is 1 mM.

TABLE 46.1 Concentration of Important Ions across a Squid Neuron's Plasma Membrane at Rest

Ions	Cytoplasm Concentration	Extracellular Concentration	Equilibrium Potential
Na$^+$	50 mM	440 mM	+54.8 mV
K$^+$	400 mM	20 mM	−75 mV
Cl$^-$	51 mM	560 mM	−60 mV
Organic anions	385 mM	—	—

the cell. More specifically, the energy gained by the sodium–potassium pump when it receives a phosphate group from one ATP is used to move three Na$^+$ ions out of the cell and two K$^+$ ions into the cell (**FIGURE 46.4**).

Active transport via Na$^+$/K$^+$-ATPase ensures that eventually the concentration of K$^+$ is much higher on the inside of the plasma membrane than outside, while the concentration of Na$^+$ is lower inside than outside. In addition to setting up concentration gradients of K$^+$ and Na$^+$, the pump establishes an electrical gradient: The outward movement of three positive charges and inward movement of two positive charges each time the ATPase pumps makes the interior of the membrane less positive (more negative) than the outside.

To summarize, the neuron has a negative resting membrane potential because the Na$^+$/K$^+$-ATPase pumps three cations out for only two cations in, and because K$^+$ exits the neuron through

leak channels along its electrochemical gradient. The resting potential represents energy stored as concentration and electrical gradients in various ions. ✔ If you understand this concept, you should be able to predict what would happen to the membrane potential if Na$^+$ or K$^+$ were allowed to diffuse freely across the membrane.

Using Microelectrodes to Measure Membrane Potentials

During the 1930s and 1940s, A. L. Hodgkin and Andrew Huxley focused on what has become a classic model system in the study of electrical signaling: the axons of squid.

Squid live in the ocean and are preyed on by fishes and whales. When a squid is threatened, electrical signals travel down the axons to muscle cells. When these muscles contract,

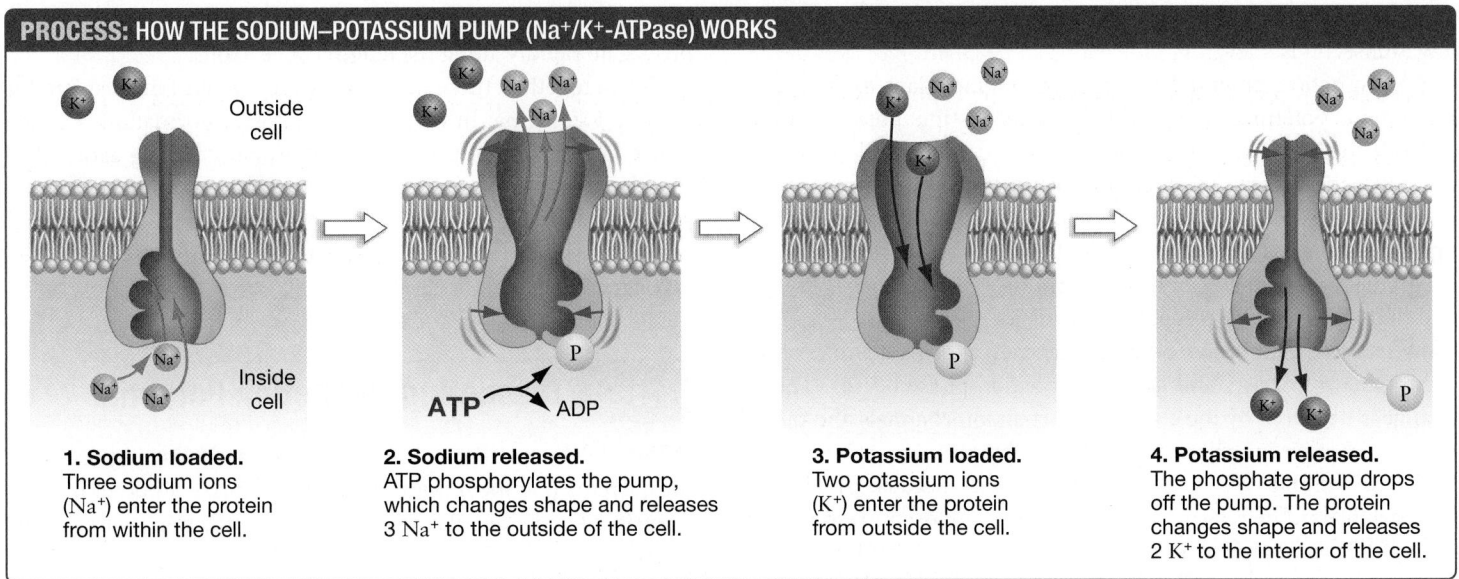

PROCESS: HOW THE SODIUM–POTASSIUM PUMP (Na⁺/K⁺-ATPase) WORKS

Outside cell

Inside cell

ATP → ADP

1. Sodium loaded.
Three sodium ions (Na⁺) enter the protein from within the cell.

2. Sodium released.
ATP phosphorylates the pump, which changes shape and releases 3 Na⁺ to the outside of the cell.

3. Potassium loaded.
Two potassium ions (K⁺) enter the protein from outside the cell.

4. Potassium released.
The phosphate group drops off the pump. The protein changes shape and releases 2 K⁺ to the interior of the cell.

FIGURE 46.4 In Neuron Membranes, Na⁺/K⁺-ATPase Imports Potassium Ions and Exports Sodium Ions. By following radioactive Na⁺ and K⁺, biologists found that the pump transports 3 Na⁺ out of the cell for every 2 K⁺ brought in. Notice that this pump operates via a conformational change.

water is expelled from a cavity in the squid's body. As a result, the squid lurches away from danger by jet propulsion (see Chapter 34). The extremely rapid electrical signal produced in the squid is an adaptation that helps avoid predation.

Hodgkin and Huxley decided to study the squid's axon simply because it is so large. Many of the axons found in humans are a mere 2 μm in diameter, but the squid axon is about 500 μm in diameter—large enough that the researchers could record membrane potentials by inserting a wire down its length.

By measuring the voltage difference between the wire inside the cell and an electrode outside the cell, Hodgkin and Huxley could record the voltage across the plasma membrane and observe how it changed in response to stimuli. Later researchers developed glass microelectrodes that were tiny enough to be inserted into smaller neurons to record membrane voltage.

With their relatively simple early equipment, Hodgkin and Huxley were able to record the axon's resting potential. They documented that it can be disrupted by an event called the action potential when the axon is stimulated.

What Is an Action Potential?

An **action potential** is a rapid, temporary change in a membrane potential. It may qualify as the most important type of electrical signal in cells. When stimulated, neurons mount action potentials that allow them to communicate with other neurons, muscles, or glands.

Although Hodgkin and Huxley initially studied action potentials in the squid giant axon, subsequent work has shown that the action potential has the same general characteristics in all species and in all types of neurons.

A Three-Phase Signal　FIGURE 46.5 shows the form of the action potential that Hodgkin and Huxley recorded from the squid's

giant axon—the signal that allows the squid to jet away from predators. The action potential has three distinct phases:

1. **Depolarization** of the membrane. In its resting state, a membrane is said to be polarized because the charges on the two sides are different. Depolarization means that the membrane becomes less polarized than before. During the depolarization phase, the membrane potential becomes less polarized as it moves from highly negative toward zero and then is briefly positive.

2. A rapid **repolarization,** which changes the membrane potential back to negative.

3. A **hyperpolarization** phase, when the membrane is slightly more negative than the resting potential.

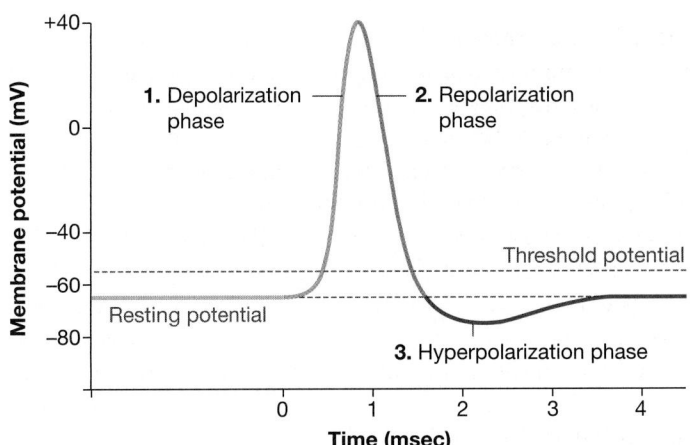

FIGURE 46.5 Action Potentials Have the Same General Shape. An action potential is a stereotyped change in membrane potential— meaning that it occurs the same way every time.

Together, all three phases of an action potential occur within a few milliseconds.

For an action potential to begin in a squid giant axon, the membrane potential must shift from its resting potential of −65 mV to about −55 mV. If the membrane depolarizes less than that, an action potential does not occur. But if this **threshold potential** is reached, certain channels in the axon membrane open and ions rush into the axon, following their electrochemical gradients. The inside of the membrane becomes less negative and then positive with respect to the outside.

When the membrane potential reaches about +40 mV, an abrupt change occurs and the repolarization phase begins. The change is triggered by the closing of certain ion channels and the opening of other ion channels in the membrane.

To summarize, an action potential occurs because specific ion channels in the plasma membrane open or close in response to changes in voltage. An action potential always has the same three-phase form, even though the size of the resting potential, threshold potential, and peak depolarization may vary among species or even among types of neurons in one species.

An "All-Or-None" Signal That Propagates   Hodgkin and Huxley made other important observations about the action potential. Besides being fast and having three distinct phases, it is an all-or-none event.

- There is no such thing as a partial action potential.
- All action potentials for a given neuron are identical in magnitude and duration.
- Action potentials are always propagated down the entire length of the axon.

For example, when an action potential was recorded at a particular point on a squid axon, an action potential that was identical in shape and size would be observed farther down the same axon soon afterward. Neurons are said to have **excitable membranes**

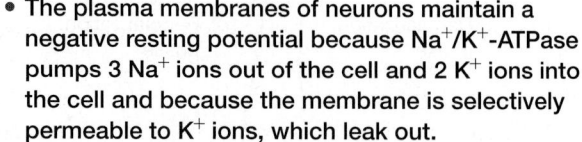

If you understand that . . .

- The plasma membranes of neurons maintain a negative resting potential because Na$^+$/K$^+$-ATPase pumps 3 Na$^+$ ions out of the cell and 2 K$^+$ ions into the cell and because the membrane is selectively permeable to K$^+$ ions, which leak out.
- Action potentials are three-phase, all-or-none signals that propagate down the length of a neuron.

✔ **You should be able to . . .**

1. Predict what would happen to the resting potential of a squid axon if potassium leak channels were blocked.
2. Explain why only the frequency of action potentials—not their size—contains information.

Answers are available in Appendix A.

because they are capable of generating action potentials that propagate rapidly along the length of the axons.

Taken together, these observations suggested a mechanism for electrical signaling. In the nervous system, information is coded in the form of action potentials that propagate along axons. The frequency of action potentials—not their size—is the meaningful signal. In the squid's giant axon, increased frequency of action potentials signals muscles to contract. As a result, the animal escapes from danger.

46.2 Dissecting the Action Potential

Of Na$^+$, Cl$^-$, and K$^+$, which ion or ions are most important in the currents that form the action potential? Are different ions responsible for the depolarization and repolarization phases of the event?

Hodgkin made a crucial start in answering this question when he realized that +40 mV was close to the equilibrium potential for Na$^+$ in the squid giant axon (see **Quantitative Methods 46.1**). If sodium channels opened early in the action potential, then Na$^+$ should flow into the neuron until the membrane potential was about +40 mV. How could this hypothesis be tested?

Distinct Ion Currents Are Responsible for Depolarization and Repolarization

To understand the currents responsible for the action potential, Hodgkin and Huxley recorded electrical activity in squid axons that were bathed in solutions containing different concentrations of ions.

- Removing Na$^+$ from the solution surrounding the axon abolished the production of action potentials.
- When axons were bathed in solutions with various concentrations of Na$^+$, the peak membrane potential paralleled the concentration of Na$^+$. If Na$^+$ concentration outside the cell was high, the peak was high. If Na$^+$ concentration outside the cell was low, the peak was low.

These experiments furnished strong support for the hypothesis that the action potential begins when Na$^+$ flows into the neuron. In other words, sodium ions are responsible for the depolarization phase.

What happens during the repolarization phase? Using radioactive K$^+$, Hodgkin and Huxley showed that there was a strong flow of potassium ions out of the cell during the repolarization phase.

The action potential consists of a strong inward flow of sodium ions followed by a strong outward flow of potassium ions.

✔ If you understand this concept, you should be able to add labels that read, "Sodium channels open—Na$^+$ enters" and "Potassium channels open—K$^+$ leaves" to Figure 46.5.

How Do Voltage-Gated Channels Work?

The action potential depends on **voltage-gated channels**—membrane proteins that open and close in response to changes

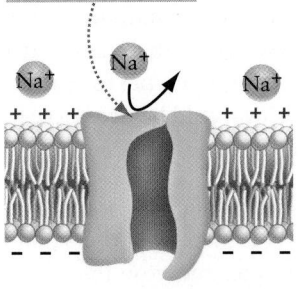 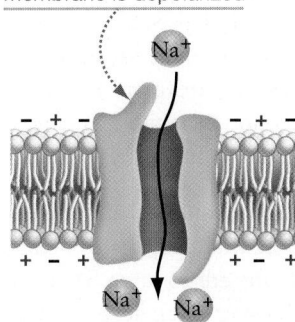

At the resting potential, voltage-gated Na⁺ channels are closed

Conformational changes open voltage-gated channels when the membrane is depolarized

FIGURE 46.6 The Shape of a Voltage-Gated Channel Depends on the Membrane Potential. Changes in the conformation of voltage-gated channels are responsible for changes in a neuron's permeability to ions.

in membrane voltage. The shape of a voltage-gated channel, and thus its ability to admit ions, changes in response to the charges present at the inside of the membrane. **FIGURE 46.6** shows a simple model of how voltage-gated sodium channels change as a function of membrane potential.

Hodgkin and Huxley confirmed that voltage-gated channels exist using a technique called voltage clamping. **Voltage clamping** allows researchers to hold the voltage of a region of an axon at any desired level and record the electrical currents that occur at that voltage. When the researchers held the squid axon at various voltages, different currents resulted. These experiments supported the hypothesis that the behavior of the ion channels depends on voltage.

Patch Clamping and Studies of Single Channels   Studying individual ion channels became possible in the 1980s when Erwin Neher and Bert Sakmann perfected a technique known as **patch clamping**. As **FIGURE 46.7** shows, the researchers touched a membrane with a fine-tipped microelectrode and applied suction to capture a single ion channel within the electrode's tip.

Using this technique, the researchers were able to document for the first time the currents that flowed through individual

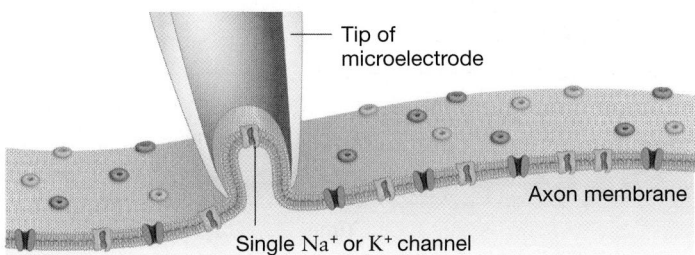

Tip of microelectrode

Axon membrane

Single Na⁺ or K⁺ channel

FIGURE 46.7 Patch Clamping Provided Insights into the Action Potential. Patch clamping makes use of extremely fine-tipped microelectrodes. The goal is to isolate one channel and record from it.

channels, and they showed that different ion channels behave differently:

- Voltage-gated channels are either open or closed.   There is no gradation in channel behavior. This conclusion is based on the shape of the recorded current: Current flow starts and stops instantly, and the size of the current is always the same.

- Sodium channels open quickly after depolarization.   They stay open for about a millisecond, close, and remain inactive for 1 to 2 msec. That explains why the cell can repolarize: Once the sodium channels close, there is a lag before they can open again.

- Potassium channels open with a delay after depolarization. They continue to flip open and closed until the membrane repolarizes. Once the membrane returns to the resting potential, these channels remain closed.

Positive Feedback Occurs during Depolarization   More detailed experiments on Na⁺ channels also explained why the action potential is an all-or-none event. The key observation was that Na⁺ channels become more likely to open as a membrane depolarizes. As a result, an initial depolarization leads to the opening of more Na⁺ channels, which depolarizes the membrane further, which leads to the opening of additional Na⁺ channels.

The opening of Na⁺ channels exemplifies **positive feedback**— meaning that the occurrence of an event makes the same event more likely to recur. When a fuse is lit, for example, the heat generated by the oxidation reaction accelerates the reaction itself, which generates still more heat and leads to additional oxidation reactions and keeps the fuse burning. Positive feedback is rare in organisms: It cannot be employed as a feedback mechanism under many circumstances because it often leads to uncontrolled events. The opening of Na⁺ channels during an action potential is one of the few examples known.

✔ If you understand how voltage-gated sodium and potassium channels work, you should be able to (1) explain how positive feedback occurs in the opening of Na⁺ channels, and (2) predict what would happen if Na⁺ and K⁺ channels were open at the same time.

Using Neurotoxins to Identify Channels and Dissect Currents   In addition to using voltage clamping and patch clamping, researchers have used neurotoxins to explore the dynamics of voltage-gated channels. **Neurotoxins** are poisons that affect neuron function—often resulting in convulsions, paralysis, or unconsciousness. They come from sources as diverse as venomous snakes and foxglove plants.

For example, when biologists treated giant axons from lobsters with tetrodotoxin from puffer fish, they found that the resting potential in treated neurons was normal, but action potentials were abolished. More specifically, the flow of K⁺ out of the cell was normal but the influx of Na⁺ was wiped out. Researchers concluded that tetrodotoxin blocks the voltage-gated Na⁺ channel, probably by binding to a specific site on the channel protein.

How Is the Action Potential Propagated?

To explain how action potentials propagate down an axon, Hodgkin and Huxley suggested the model illustrated in **FIGURE 46.8a**.

Step 1 The influx of Na⁺ at the start of an action potential repels intracellular cations, causing them to spread away from the sodium channels.

Step 2 As positive charges are pushed farther from the initial sodium channels, they depolarize adjacent portions of the membrane.

Step 3 Nearby voltage-gated Na⁺ channels open when the membrane reaches threshold, resulting in an action potential.

In this way, action potentials are continuously regenerated at adjacent areas of the plasma membrane (**FIGURE 46.8b**). The response is all or none because action potentials are always regenerated along the entire length of the axon.

Why don't action potentials propagate back up the axon in the direction of the cell body? To answer this question, recall that once Na⁺ sodium channels have opened and closed, they are less likely to open again for a short period. This is known as the **refractory** state. Action potentials are propagated in one direction only, because "upstream" sodium channels, in the direction of the cell body, are in the refractory state.

The hyperpolarization phase, in which the membrane is more negative than the resting potential, also keeps the charge that spreads upstream from triggering an action potential in that direction, because a much stronger stimulus would be necessary to raise the membrane potential to threshold potential.

Axon Diameter Affects Speed Understanding how the action potential propagates helped researchers explain why the squid's axons are so large. When sodium ions enter the axon interior at the start of an action potential, they repulse intracellular cations, causing them to flow along the inside of the membrane. Cations moving down axons with larger diameters meet less resistance to flow than those moving down narrow axons. As a result, the charge spreads down the membrane more quickly.

The upshot is that the squid's giant axon and other large-diameter neurons transmit action potentials much faster than small axons can. The squid axon's large size is an adaptation that makes particularly rapid signaling possible.

(a) PROCESS: PROPAGATION OF ACTION POTENTIAL

Neuron / Axon

Na⁺ Na⁺ Na⁺ Na⁺ Na⁺ Na⁺ Na⁺ Na⁺

1. Na⁺ enters axon.

2. Charge spreads; membrane "downstream" depolarizes.

Depolarization at next ion channel

3. Downstream voltage-gated channel opens in response to depolarization.

FIGURE 46.8 Action Potentials Propagate because Charge Spreads Down the Membrane. (a) An action potential starts with an inflow of Na⁺. The influx of positive charge attracts negative charges inside the cell and repels positive charges. As a result, cations spread away from the channel where the Na⁺ enters and depolarize nearby regions of the neuron. Voltage-gated Na⁺ channels open in response. **(b)** The action potential is propagated down the axon as a wave of depolarization, but there is no loss of signal because action potentials at one area of the membrane always stimulate action potentials at adjacent areas of the membrane.

(b) Action potential spreads as a wave of depolarization.

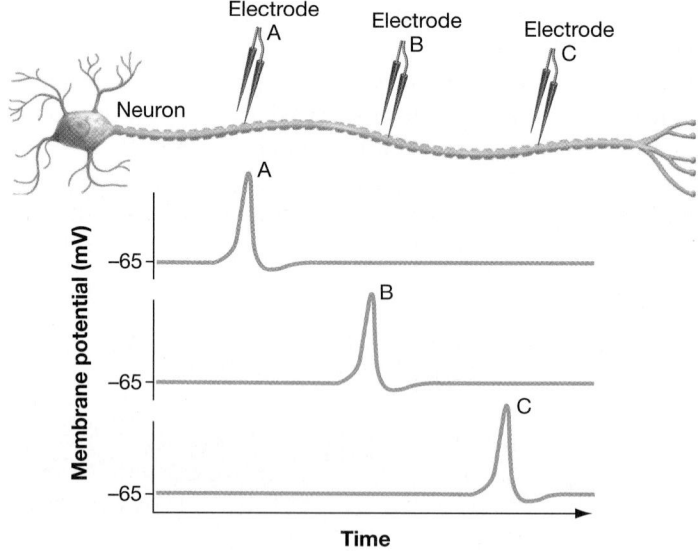

Myelination Affects Speed Relatively few vertebrates have giant axons. Instead, vertebrates—and some invertebrates—have specialized accessory cells whose membranes wrap around the axons of certain neurons and increase the efficiency of action potential propagation.

In the central nervous system, the specialized accessory cells are called **oligodendrocytes.** In the peripheral nervous system, described in Section 46.4, the cells are **Schwann cells** (**FIGURE 46.9a**). Oligodendrocytes and Schwann cells are two of several types of nervous system cells that support neurons. Collectively, these accessory cells are called **glia.**

When oligodendrocytes or Schwann cells wrap around an axon, they form a **myelin sheath,** which acts as a type of electrical insulation. As action potentials spread down an axon, the myelin sheath prevents charge in the form of ions from leaking back out across the plasma membrane of the neuron.

Consequently, the cations moving down the membrane are able to spread until they hit an unmyelinated section of the axon,

called a **node of Ranvier** (**FIGURE 46.9b**). The node has a dense concentration of voltage-gated Na^+ and K^+ channels, so new action potentials can occur at the node.

Electrical signals "jump" from node to node down a myelinated axon much faster than they can move down an unmyelinated axon of the same diameter. In an unmyelinated axon, sodium and potassium channels are found in all locations, and action potentials occur continuously down its length. Myelination is interpreted as an adaptation that makes rapid transmission of electrical signals possible in axons that have a small diameter.

To appreciate the importance of myelination, consider what happens when it is disrupted. If myelin degenerates, the transmission of electrical signals slows considerably. The autoimmune disease **multiple sclerosis (MS)** develops when the immune system mistakenly targets the oligodendrocytes that make up the myelin sheath in the CNS. As damage to myelin increases and electrical signaling is impaired, coordination among neurons is

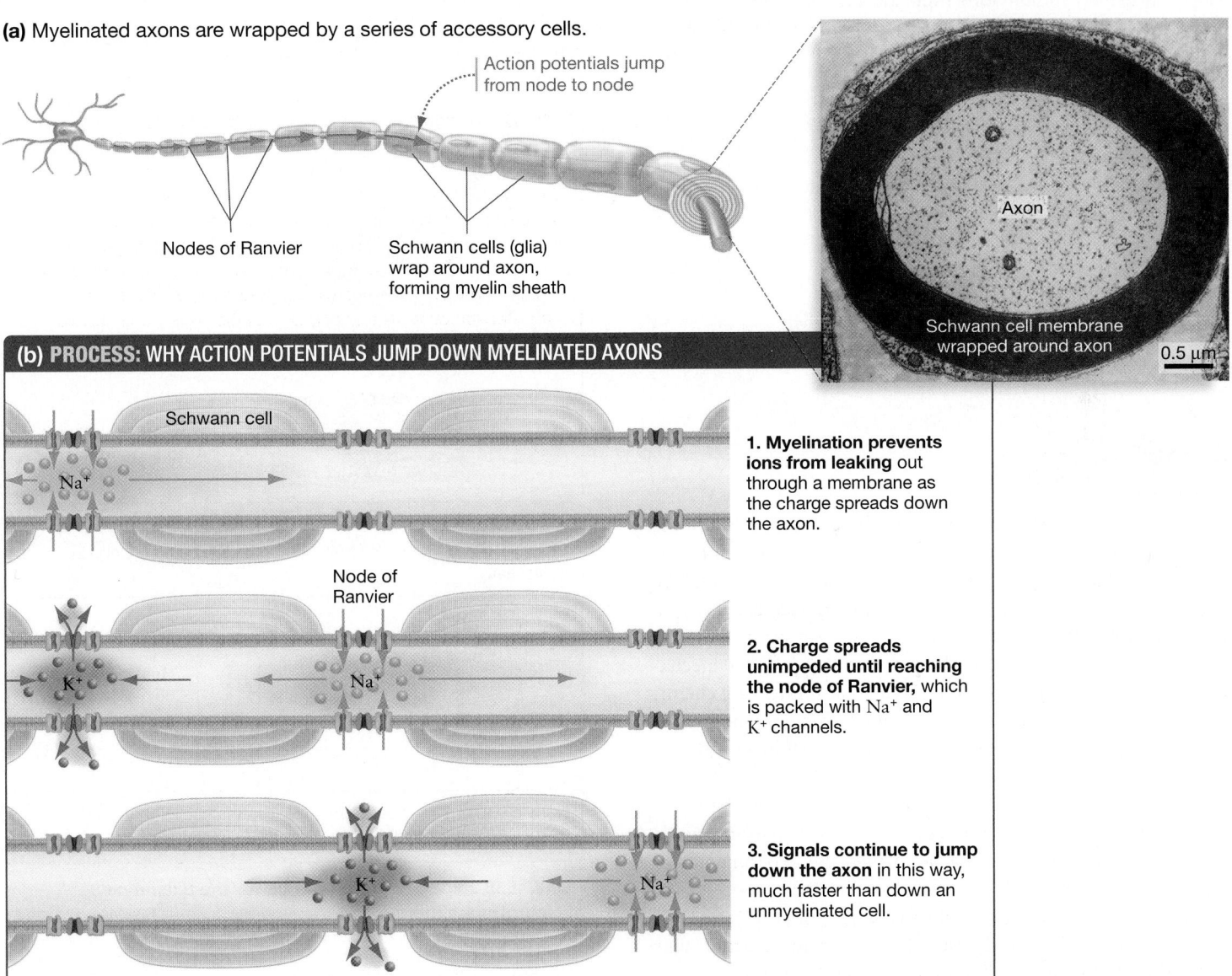

(a) Myelinated axons are wrapped by a series of accessory cells.

Action potentials jump from node to node

Nodes of Ranvier

Schwann cells (glia) wrap around axon, forming myelin sheath

Axon

Schwann cell membrane wrapped around axon

0.5 μm

(b) PROCESS: WHY ACTION POTENTIALS JUMP DOWN MYELINATED AXONS

Schwann cell

Na^+

1. Myelination prevents ions from leaking out through a membrane as the charge spreads down the axon.

Node of Ranvier

K^+ Na^+

2. Charge spreads unimpeded until reaching the node of Ranvier, which is packed with Na^+ and K^+ channels.

K^+ Na^+

3. Signals continue to jump down the axon in this way, much faster than down an unmyelinated cell.

FIGURE 46.9 Action Potentials Propagate Quickly in Myelinated Axons.

affected and muscles weaken. The symptoms of MS are highly variable; in severe cases, the disease progresses and can be crippling.

What happens once an action potential has traveled the length of the axon? In most neurons, the membrane at the end of the axon is adjacent to the membrane of another neuron's dendrite but separated from its surface by a tiny gap. Let's investigate what happens when an action potential arrives at this interface between cells.

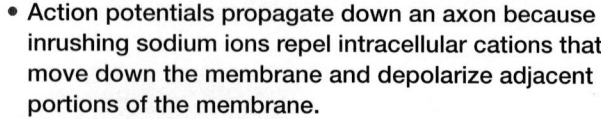

check your understanding

If you understand that . . .

- During an action potential, membrane voltage undergoes rapid changes due to an inflow of sodium ions followed by an outflow of potassium ions.
- Action potentials propagate down an axon because inrushing sodium ions repel intracellular cations that move down the membrane and depolarize adjacent portions of the membrane.

✔ You should be able to . . .

1. Explain why the action potential is an all-or-none phenomenon.
2. Predict what would happen if batrachotoxin, the poison produced by some species of poison dart frogs, were applied to the membrane. (Batrachotoxin binds to voltage-gated sodium channels and maintains them in an open state.)

Answers are available in Appendix A.

46.3 The Synapse

The cytoplasm of most neurons is not directly connected to the cytoplasm of other neurons, yet neurons are able to communicate with one another. Based on this observation, there must be some indirect mechanism that transmits signals from cell to cell, across their plasma membranes.

In the 1920s, Otto Loewi showed that this indirect mechanism involves **neurotransmitters**. Neurotransmitters are chemical messengers that transmit information from one neuron to another neuron, or from a neuron to a target cell in a muscle or gland.

Loewi knew that signals from the vagus nerve slow the heart. To test the hypothesis that the signal from nerve to heart muscle is delivered by a chemical, he performed the experiment diagrammed in **FIGURE 46.10**.

First, Loewi isolated the vagus nerve and heart of a frog. As predicted, the heart rate slowed when he stimulated the vagus nerve electrically. Next, he took the solution that bathed the first heart, applied it to another heart—this time one that was not

RESEARCH

QUESTION: How is information transferred from one neuron to another?

HYPOTHESIS: Molecules called neurotransmitters carry information from one neuron to the next.

NULL HYPOTHESIS: Information is not transferred between neurons in the form of molecules.

EXPERIMENTAL SETUP:

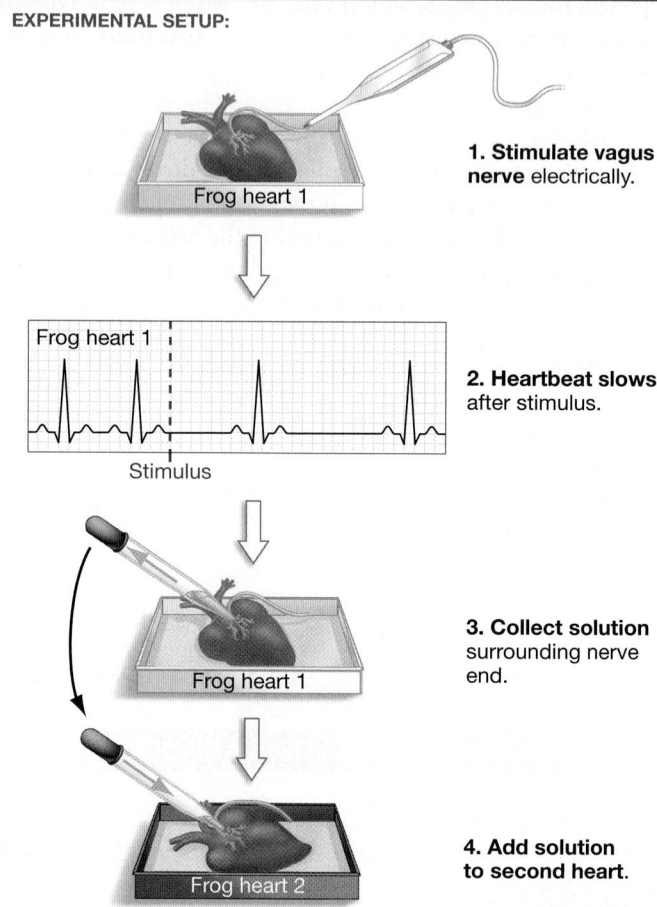

1. Stimulate vagus nerve electrically.

2. Heartbeat slows after stimulus.

3. Collect solution surrounding nerve end.

4. Add solution to second heart.

PREDICTION: The heartbeat will slow.

PREDICTION OF NULL HYPOTHESIS: There will be no change in heartbeat.

RESULTS:

Heartbeat slows after solution is added.

CONCLUSION: The vagus nerve releases molecules that slow heartbeat. Neurotransmitters carry information.

FIGURE 46.10 Experimental Evidence for the Existence of Neurotransmitters.

SOURCE: Loewi, O. 1921. Über humorale Übertragbarkeit der Herznervenwirkung. *Pflügers Archiv European Journal of Physiology* 189: 239–242.

✔ **QUESTION** What would be an appropriate control for this experiment?

accompanied by a vagus nerve—and showed that the second heart rate slowed as well.

This result provided strong evidence for the chemical transmission of electrical signals. The vagus nerve had released a neurotransmitter into the bath.

Synapse Structure and Neurotransmitter Release

When transmission electron microscopy became available in the 1950s, biologists finally understood the physical nature of the interface, or **synapse,** between neurons. As **FIGURE 46.11** shows, (1) the membranes of axons and dendrites are separated by a tiny space, the **synaptic cleft,** and (2) the ends of axons contain numerous sac-like structures, called **synaptic vesicles.** Synaptic vesicles were hypothesized to be storage sites for neurotransmitters.

Anatomical observations such as these, combined with chemical studies of the synapse, led to the model of synaptic transmission illustrated in **FIGURE 46.12**. Notice that the "sending" cell is the **presynaptic neuron** and the "receiving" cell is the **postsynaptic neuron.**

Step 1 An action potential arrives at the end of the axon.

Step 2 The depolarization created by the action potential opens voltage-gated calcium channels located near the synapse, in the presynaptic membrane. The electrochemical gradient for Ca^{2+} results in the inflow of calcium ions through the open channels.

Step 3 In response to the increased calcium concentration inside the axon, synaptic vesicles fuse with the membrane and release neurotransmitters into the gap between the cells, the synaptic cleft. The delivery of neurotransmitters into the cleft is an example of exocytosis (see Chapter 7).

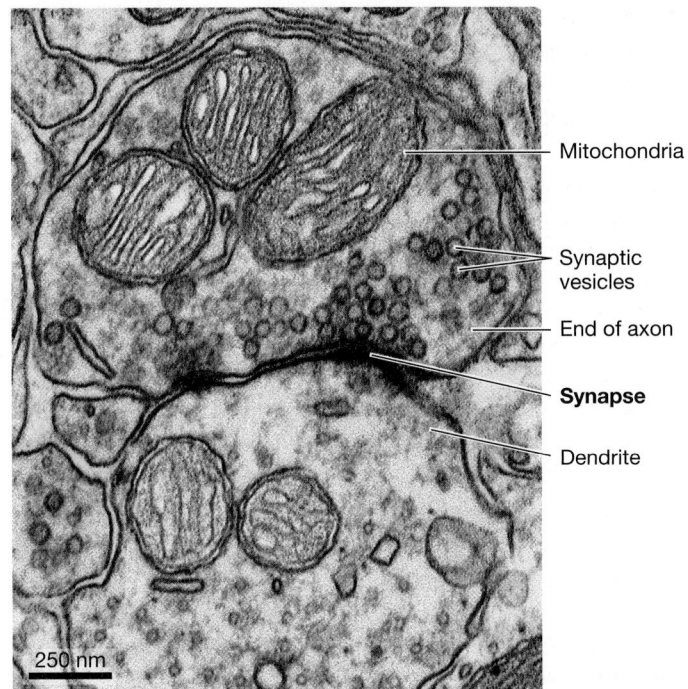

FIGURE 46.11 Synaptic Vesicles Cluster Near Synapses. A cross section of the site where an axon meets a dendrite.

Step 4 Neurotransmitters bind to receptors on the postsynaptic cell, leading to changes in the membrane potential of the postsynaptic cell. The combined effect on membrane potential of many neurotransmitters binding may trigger an action potential in the postsynaptic cell.

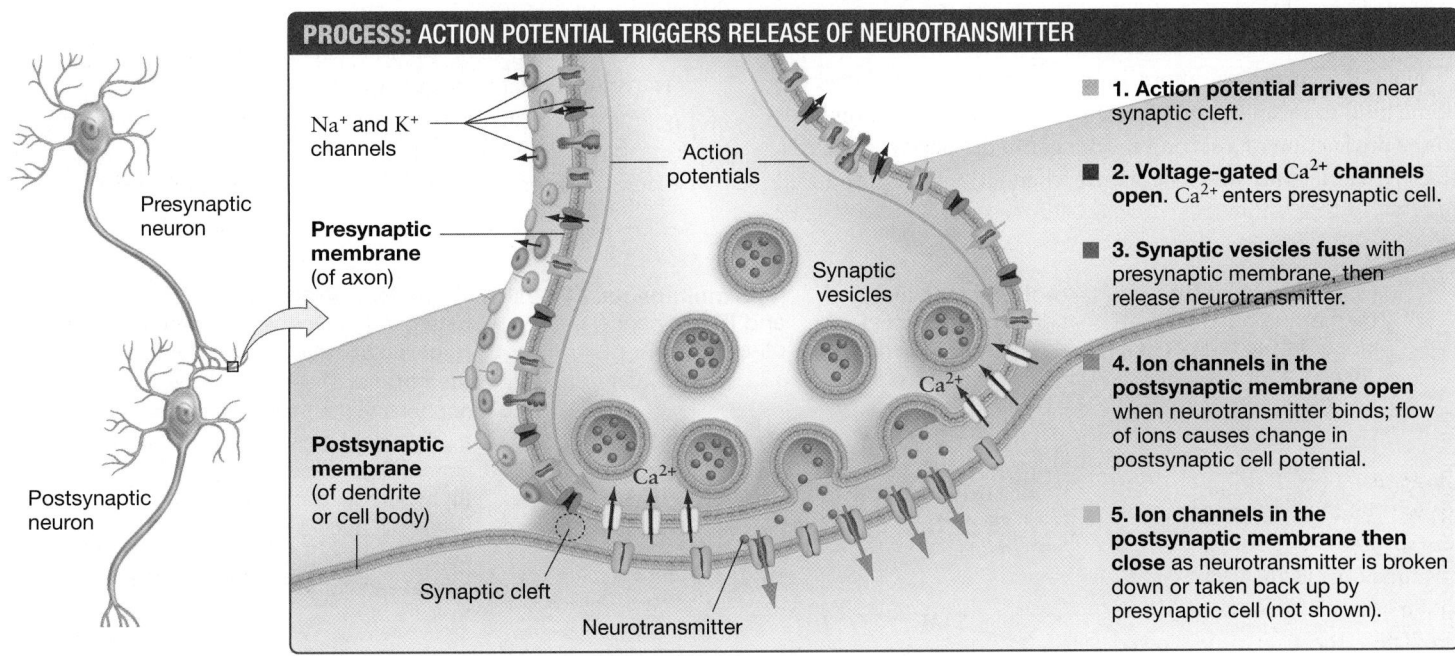

PROCESS: ACTION POTENTIAL TRIGGERS RELEASE OF NEUROTRANSMITTER

1. **Action potential arrives** near synaptic cleft.

2. **Voltage-gated Ca^{2+} channels open.** Ca^{2+} enters presynaptic cell.

3. **Synaptic vesicles fuse** with presynaptic membrane, then release neurotransmitter.

4. **Ion channels in the postsynaptic membrane open** when neurotransmitter binds; flow of ions causes change in postsynaptic cell potential.

5. **Ion channels in the postsynaptic membrane then close** as neurotransmitter is broken down or taken back up by presynaptic cell (not shown).

FIGURE 46.12 Neurons Meet and Transfer Information at Synapses. The sequence of events that occurs when an action potential arrives at a synapse.

Step 5 The response ends as the neurotransmitter is broken down and taken back up by the presynaptic cell.

Is the model correct? Let's begin by analyzing the role of neurotransmitters.

What Do Neurotransmitters Do?

Researchers can look for neurotransmitters by stimulating a neuron, collecting the molecules that are released, and analyzing them chemically. To find the receptor for a particular neurotransmitter, researchers can attach a radioactive atom or other type of label to the neurotransmitter and add it to neurons. Once the labeled transmitter has bound to its receptor, the receptor protein can be isolated and analyzed.

Using techniques such as these, biologists have discovered and characterized a wide array of neurotransmitters and receptors. Some of them are listed in **TABLE 46.2**.

By patch-clamping receptors, biologists confirmed that many neurotransmitters function as ligands. A **ligand** is a molecule that binds to a specific site on a receptor molecule. Many neurotransmitters are ligands that bind to receptors called **ligand-gated channels.** These are channel proteins that open in response to binding by a specific ligand—just as voltage-gated channels open in response to a change in voltage.

When a neurotransmitter binds to a ligand-gated ion channel in the postsynaptic membrane, the channel opens and admits a flow of ions along an electrochemical gradient. In this way, the neurotransmitter's chemical signal is transduced to an electrical signal—a change in the membrane potential of the postsynaptic cell. ✔ If you understand this concept, you should be able to envision a membrane with a resting potential of −65 mV and explain what happens to the membrane potential when a ligand-gated ion channel opens and allows chloride ions to leave the cell. How does this affect the likelihood that the postsynaptic cell will mount an action potential?

Not all neurotransmitters bind to ion channels, however. Some bind to receptors that activate enzymes whose action leads to the production of a second messenger in the postsynaptic cell. Recall that **second messengers** are chemical signals produced

inside a cell in response to a chemical signal that arrives at the cell surface (see Chapter 11).

The second messengers induced by neurotransmitters may trigger changes in enzyme activity, gene transcription, or membrane potential. (Chapter 49 explores the cellular role of second messengers in detail.)

Postsynaptic Potentials

What happens when a neurotransmitter binds to a ligand-gated ion channel in the postsynaptic cell?

Ligand-gated sodium channels on the membranes of dendrites are in particularly high concentration near synapses. When neurotransmitters bind, these channels open to let sodium into the cell, causing depolarization (**FIGURE 46.13a**). In most cases, depolarization makes an action potential in the postsynaptic cell more likely because the membrane potential approaches threshold. Changes in the postsynaptic cell that bring the membrane potential closer to threshold are called **excitatory postsynaptic potentials (EPSPs).**

If the receptor activity at the synapse leads to an outflow of potassium ions or an inflow of chloride ions or other anions in the postsynaptic cell, the postsynaptic membrane hyperpolarizes—making action potentials less likely to occur in the postsynaptic cell (**FIGURE 46.13b**). Changes in the postsynaptic cell that make the membrane potential more negative are called **inhibitory postsynaptic potentials (IPSPs).**

If an EPSP and an IPSP occur at the same time in the same place, they cancel each other out (**FIGURE 46.13c**). Synapses also can be modulatory—meaning that their activity modifies a neuron's response to other EPSPs or IPSPs.

Postsynaptic Potentials Are Graded It is critical to realize that, unlike action potentials, EPSPs and IPSPs are not all-or-none events. Instead, they are graded in size.

The size of an EPSP or IPSP depends on the amount of neurotransmitter that is released at the synapse at a given time. A higher concentration of neurotransmitter in the synaptic cleft leads to a larger EPSP or IPSP. Both types of signal are short lived because neurotransmitters do not bind irreversibly to channels

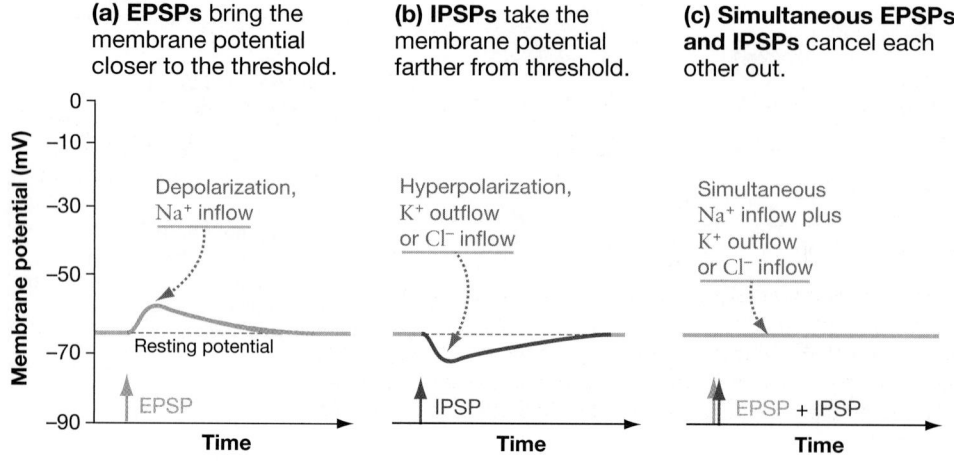

(a) EPSPs bring the membrane potential closer to the threshold.

(b) IPSPs take the membrane potential farther from threshold.

(c) Simultaneous EPSPs and IPSPs cancel each other out.

FIGURE 46.13 Events at the Synapse May Lead to Depolarization or Hyperpolarization of the Postsynaptic Membrane. These recordings show what happens to the membrane potential of a postsynaptic neuron with the arrival of signals that cause **(a)** depolarization, **(b)** hyperpolarization, or **(c)** no change. The last result occurs because simultaneous depolarizing and hyperpolarizing signals cancel each other out.

TABLE 46.2 Categories of Neurotransmitters

Neurotransmitter	Site of Action	Action*	Drugs That Interfere†
Acetylcholine	Neuromuscular junction, some CNS pathways	Excitatory (inhibitory in some parasympathetic neurons)	• Botulism toxin blocks release • Black widow spider venom increases, then eliminates, release • Alpha-bungarotoxin (in some snake venoms) binds to and blocks receptor
Monoamines			
Norepinephrine	Sympathetic neurons, some CNS pathways	Excitatory or inhibitory	• Ritalin (used for attention deficit hyperactivity disorder) increases release • Some antidepressants prevent reuptake
Dopamine	Many CNS pathways	Excitatory or modulatory	• Cocaine prevents reuptake • Amphetamine prevents reuptake
Serotonin	Many CNS pathways	Inhibitory or modulatory	• MDMA (ecstasy) causes increased release
Amino Acids			
Glutamate	Many CNS pathways	Excitatory	• PCP (angel dust) blocks receptor
Gamma-aminobutyric acid (GABA)	Some CNS pathways	Inhibitory	• Ethanol mimics response to GABA
Peptides			
Endorphins, enkephalins, substance P	Sensory pathways (pain)	Excitatory, modulatory, or inhibitory	

*Excitatory neurotransmitters make action potentials more likely in postsynaptic cells; inhibitory neurotransmitters make action potentials less likely; modulatory neurotransmitters modify the response at other synapses.

† Drugs that prevent reuptake of neurotransmitters increase their activity.

in the postsynaptic cell. Instead, they are quickly inactivated by enzymes or taken up by the presynaptic cell and recycled.

If either the amount or life span of neurotransmitters is artificially altered, the normal functioning of neurons is altered. The drugs cocaine and amphetamine, for example, exert their effects by inhibiting the uptake and recycling of particular neurotransmitters (see Table 46.2).

Summation and Threshold How do EPSPs and IPSPs affect the postsynaptic cell? As **FIGURE 46.14a** shows, the dendrites and the cell body of a neuron typically make hundreds or thousands of synapses with other cells. At any instant, the EPSPs and IPSPs that occur at each of these synapses lead to short-lived changes in membrane potential in the dendrites and cell body of the postsynaptic cell.

(a) Most neurons receive information from many other neurons.

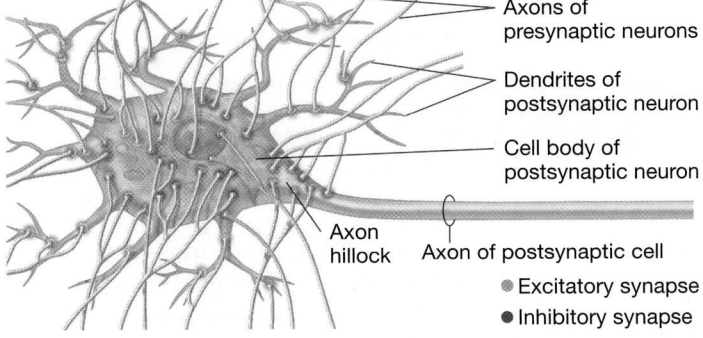

Axons of presynaptic neurons
Dendrites of postsynaptic neuron
Cell body of postsynaptic neuron
Axon hillock
Axon of postsynaptic cell
● Excitatory synapse
● Inhibitory synapse

(b) Postsynaptic potentials sum.

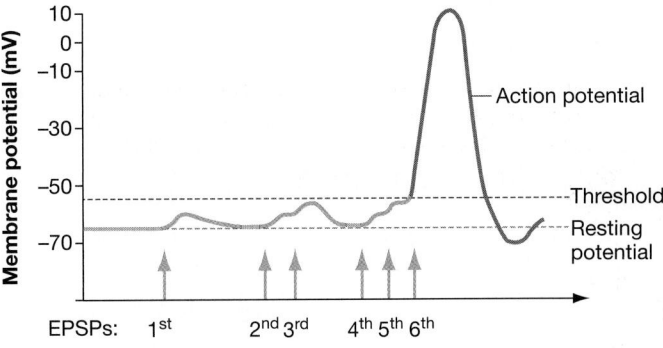

Action potential
Threshold
Resting potential
EPSPs: 1st   2nd 3rd   4th 5th 6th

FIGURE 46.14 Neurons Integrate Information from Many Synapses. (a) The dendrites and cell body of a neuron typically receive signals from hundreds or thousands of other neurons. **(b)** When action potentials arrive close together in time, the postsynaptic potentials sum. Here the first EPSP is insufficient to generate an action potential; two EPSPs arriving close together undergo summation but do not reach threshold; the summing of three EPSPs arriving close together does exceed the threshold. This example is simplified—in reality, hundreds or thousands of IPSPs and EPSPs sum to determine action potential frequency.

If an IPSP and EPSP occur close together in space or time, the changes in membrane potential tend to cancel each other out. But if several EPSPs occur close together in space or time, they sum and make the neuron more likely to reach threshold and fire an action potential (**FIGURE 46.14b**). The additive nature of post-synaptic potentials is termed **summation.**

The sodium channels that trigger action potentials in the postsynaptic cell are located near the start of the axon at a site called the **axon hillock** (see Figure 46.14a). As IPSPs and EPSPs are received and interact throughout the dendrites and cell body, charge spreads to the axon hillock. If the membrane at the axon hillock depolarizes past the threshold potential, enough voltage-gated sodium channels open to trigger positive feedback and an action potential. Once an action potential starts at the axon hillock, it propagates down the axon to the next synapse.

Summation is critically important. Because neurons receive input from many synapses, and because IPSPs and EPSPs sum, information in the form of electrical signals is modified at the synapse before being passed along. An action potential in a pre-synaptic neuron does not always lead to an action potential in a postsynaptic neuron—the response by the postsynaptic cell depends on the information it receives from a wide array of neurons.

check your understanding

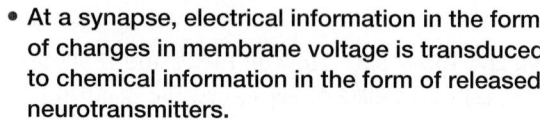

If you understand that . . .

- At a synapse, electrical information in the form of changes in membrane voltage is transduced to chemical information in the form of released neurotransmitters.
- Binding of a neurotransmitter to its receptor causes a change in the membrane potential of the postsynaptic cell.

✓ **You should be able to . . .**

Predict the effect on an EPSP of an increase in the concentration of a synaptic enzyme that breaks down an excitatory neurotransmitter.

Answers are available in Appendix A.

46.4 The Vertebrate Nervous System

The first three sections of this chapter examined electrical signaling at the level of molecules, membranes, and individual cells. This section discusses electrical signaling at the levels of tissues, organs, and systems.

To begin, let's consider the overall anatomy of the vertebrate nervous system, and how researchers explore the function of the most complex organ known: the human brain. The chapter concludes by returning to the molecular level (and to studies of invertebrates) to introduce recent work on learning and memory.

What Does the Peripheral Nervous System Do?

Recall from Section 46.1 that the central nervous system (CNS) is made up of the brain and spinal cord and is concerned primarily with integrating information. The peripheral nervous system (PNS) is made up of neurons outside the CNS.

What functions do the cells of the PNS control? Anatomical and functional studies indicate that the PNS consists of two divisions with distinct functions:

1. The **afferent division** transmits sensory information to the CNS.

2. The **efferent division** carries commands from the CNS to the body.

Neurons in the afferent division monitor conditions inside and outside the body. Once information from afferent neurons has been processed in the CNS, neurons in the efferent division carry signals that allow the body to respond to changed conditions in an appropriate way.

As **FIGURE 46.15** shows, the afferent and efferent divisions are part of a hierarchy of PNS functions. The efferent division is further divided into a **somatic nervous system,** which controls voluntary movements, and an **autonomic nervous system,** which controls internal processes such as digestion and heart rate.

- The somatic nervous system carries out voluntary responses, which are under conscious control. Skeletal muscles serve as the effectors.

- The autonomic nervous system carries out involuntary responses, which are not under conscious control. Smooth muscle, cardiac muscle, and glands serve as the effectors.

Many organs and glands are served by two functionally distinct types of autonomic nerves, summarized in **FIGURE 46.16**: One type inhibits activity, whereas the other promotes it.

1. Nerves in the **parasympathetic nervous system** promote "rest and digest" functions that conserve or restore energy.

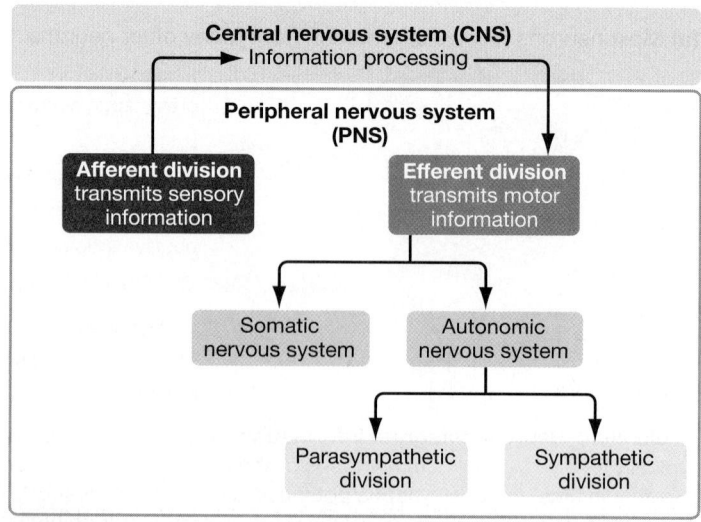

FIGURE 46.15 The Nervous System Has Several Functional Divisions.

For example, the parasympathetic nerves that synapse on the heart slow it down, while those that serve the digestive tract stimulate its activity.

2. Nerves in the **sympathetic nervous system** typically prepare organs for stressful "fight or flight" situations. Sympathetic nerves speed up the heart rate, stimulate the release of glucose from the liver, and inhibit action by digestive organs.

Functional Anatomy of the CNS

Parasympathetic nerves originate at the base of the brain or the base of the spinal cord. Most sympathetic nerves also originate in the spinal cord, but they emerge along the middle of its length. Similarly, most sensory neurons project axons to the spinal cord, and most somatic motor neurons project from the spinal cord.

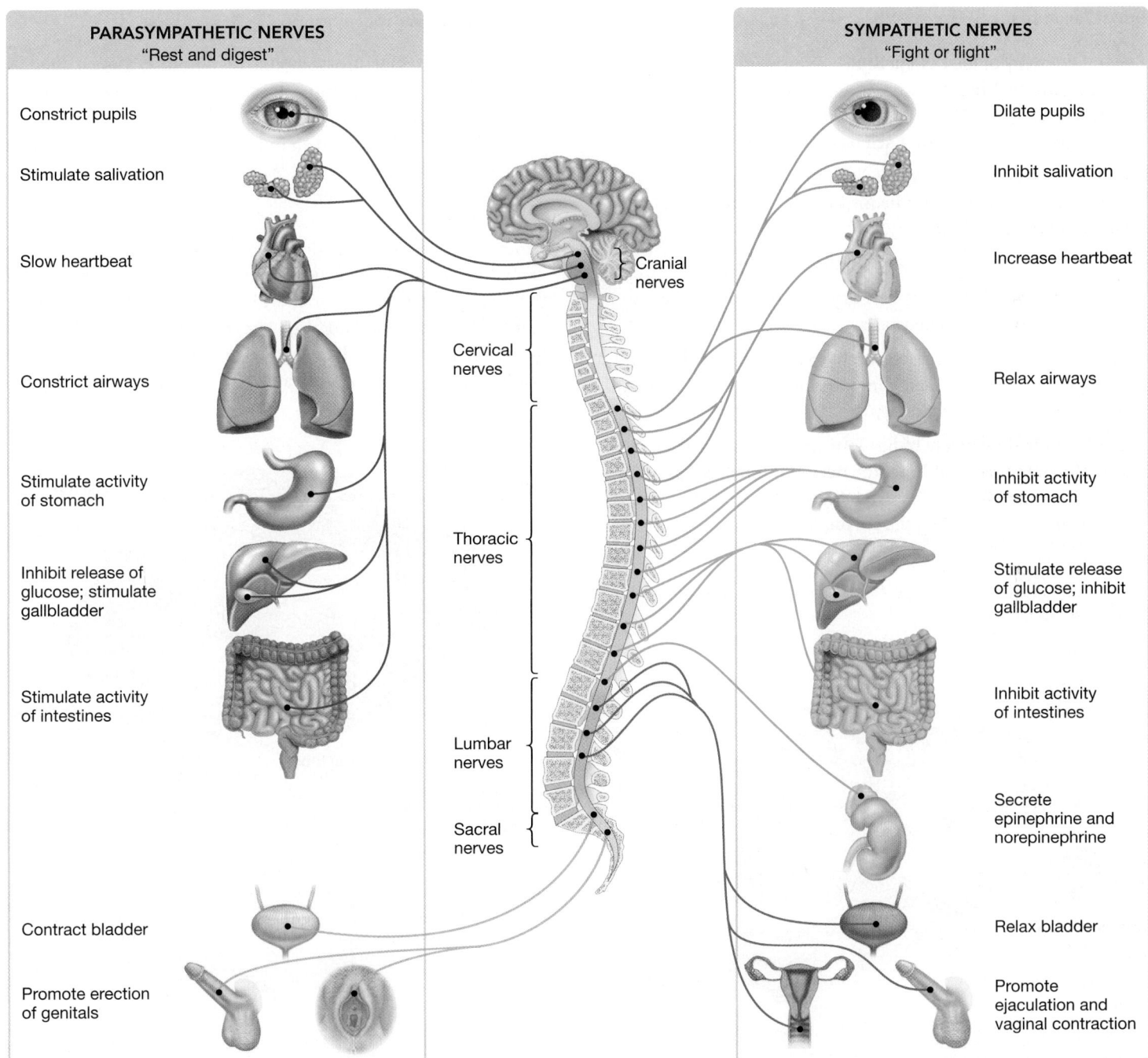

PARASYMPATHETIC NERVES
"Rest and digest"

Constrict pupils

Stimulate salivation

Slow heartbeat

Constrict airways

Stimulate activity of stomach

Inhibit release of glucose; stimulate gallbladder

Stimulate activity of intestines

Contract bladder

Promote erection of genitals

Cranial nerves

Cervical nerves

Thoracic nerves

Lumbar nerves

Sacral nerves

SYMPATHETIC NERVES
"Fight or flight"

Dilate pupils

Inhibit salivation

Increase heartbeat

Relax airways

Inhibit activity of stomach

Stimulate release of glucose; inhibit gallbladder

Inhibit activity of intestines

Secrete epinephrine and norepinephrine

Relax bladder

Promote ejaculation and vaginal contraction

FIGURE 46.16 The Autonomic Nervous System Controls Internal Processes.

✓QUESTION Explain how the responses listed here for pupils, heartbeat, and liver support the rest-and-digest versus fight-or-flight functions.

In effect, then, the spinal cord serves as an information conduit. It collects and transmits information throughout the body. Virtually all the information that travels to or from the spinal cord is sent to the brain for processing. An exception is a spinal **reflex,** in which sensory neurons stimulate interneurons in the spinal cord that then directly stimulate motor neurons. This process allows you to withdraw your hand rapidly from a hot burner before your brain becomes aware of the pain.

What happens once sensory signals arrive at the brain? How are thousands of signals integrated to allow an animal to respond to stimuli? Let's begin our exploration of the brain by delving into its anatomy.

General Anatomy of the Human Brain Nineteenth-century anatomists established that the human brain is made up of the four structures labeled in **FIGURE 46.17**: the cerebrum, cerebellum, diencephalon, and brain stem. Each has a distinct function.

- The **cerebrum** makes up the bulk of the human brain. It is divided into left and right hemispheres and is the seat of conscious thought and memory.

- The **cerebellum** coordinates complex motor patterns.

- The **diencephalon** relays sensory information to the cerebrum and controls homeostasis.

- The **brain stem** connects the brain to the spinal cord. It is the autonomic center for regulating cardiovascular, digestive, and other involuntary functions.

Each cerebral hemisphere has four major areas, or lobes: the **frontal lobe,** the **parietal lobe,** the **occipital lobe,** and the **temporal lobe** (**FIGURE 46.18**). The two hemispheres are connected by a thick band of axons called the **corpus callosum.**

The relative size of the entire brain, and of its component structures, varies greatly among vertebrates (**FIGURE 46.19**). For example, compare the size of the cerebrum in fishes and mammals. In fishes, the cerebrum is quite small and is involved mainly in olfactory sensation. In mammals it is very large and contains regions specialized for memory and reasoning, in addition to the processing of multiple sensory and motor functions. Your own cerebrum is three times bigger than in other species of comparably sized mammals, reflecting its role in the higher brain functions that are so advanced in humans.

What methods do researchers use to explore the function of each area within the cerebrum?

Mapping Functional Areas in the Cerebrum I: Lesion Studies
Early work on brain function studied people with specific mental deficits caused by areas of brain damage, or lesions. Paul Broca, for example, studied an individual who could understand language but could not speak. After the person's death in 1861, Broca examined the patient's brain and discovered a damaged area, or lesion, in the left frontal lobe of the cerebrum. Broca hypothesized that this region is responsible for speech. More generally, he formulated the hypothesis that specific regions of the brain are specialized for coordinating particular functions.

Broca's claim that functions are localized to specific brain areas has been verified through extensive efforts to map the cerebrum. In some cases, advances were made by studying the

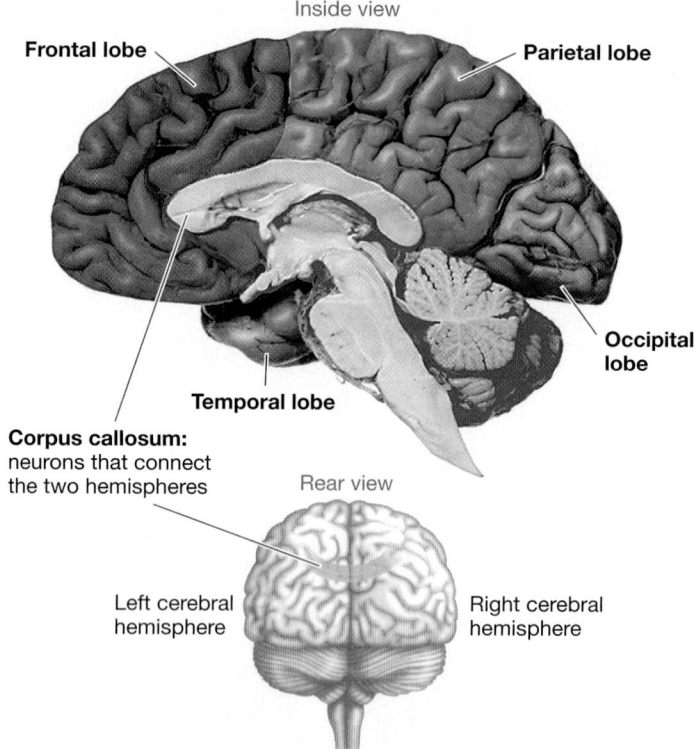

FIGURE 46.18 The Human Cerebrum Has Four Lobes and Two Hemispheres.

✔**EXERCISE** On your own head, point to each of the labeled areas.

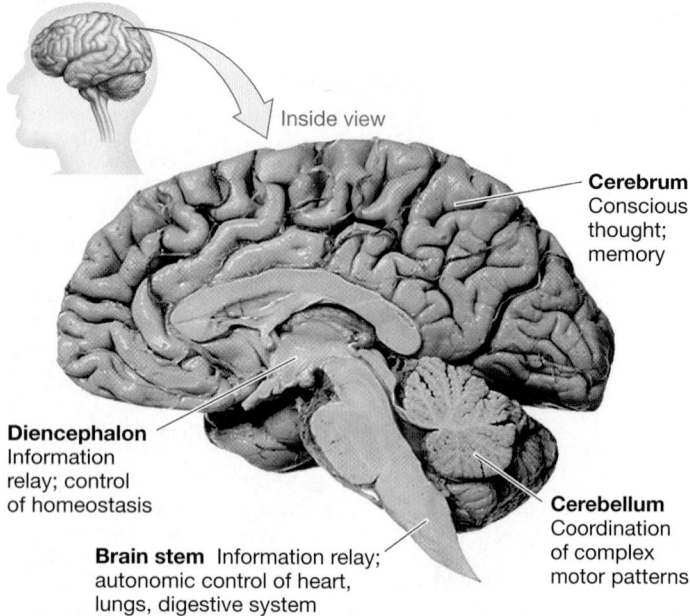

FIGURE 46.17 The Human Brain Contains Four Distinct Structures.

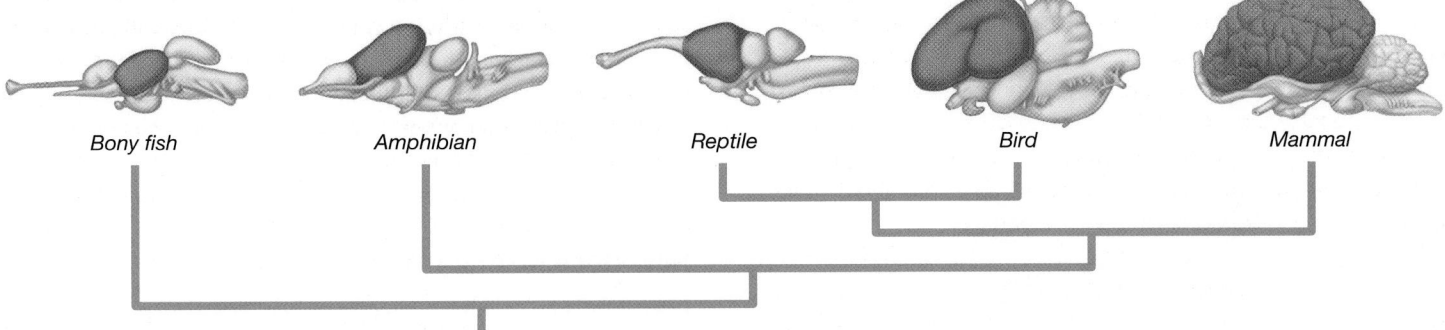

FIGURE 46.19 The Relative Sizes of Brain Regions Differ among Vertebrate Lineages. For example, the cerebrum (in orange) tends to be larger in birds and mammals than in basal vertebrates.

behavior of people known to have had lesions to specific brain regions or to have had specific portions of their brains removed.

In 1848, for example, Phineas Gage was working on a railroad construction site when an iron rod over an inch in diameter was blasted through his skull. The rod entered beneath his left eye, exited through his forehead (**FIGURE 46.20**), and landed some 80 feet away. Miraculously, Gage survived this accident. However, his personality did not fare as well. After the accident, Gage's physician reported him to be "fitful," "irreverent," and "obstinate," a dramatic change from his previous personality. The iron rod had damaged Gage's frontal lobe, providing some of the first clues that this part of the brain plays a role in personality and emotion.

Another fascinating case is that of a 27-year-old man named Henry Gustav Molaison (referred to as "H.M." until his death). In 1953, surgeons treated him for life-threatening seizures by removing a small portion of his temporal lobe and about two thirds of his hippocampus, a structure at the inner edge of the temporal lobe. The man recovered (he died in December 2008, at age 83),

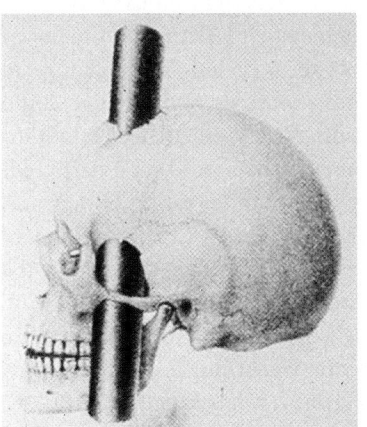 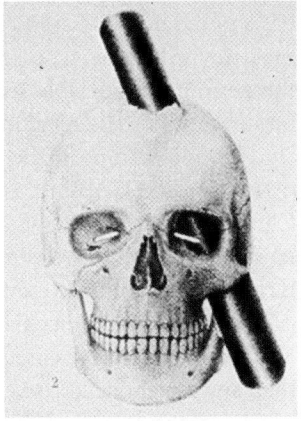

FIGURE 46.20 The Frontal Lobe Injury of Phineas Gage Affected His Personality. In 1848 a railroad worker named Phineas Gage recovered from a brain injury in which an iron rod was blasted through the frontal lobe of his cerebrum. The resulting change in his personality cued scientists that the frontal lobe plays a role in emotion and temperament.

had normal intelligence, and vividly remembered his childhood, but he had no short-term memory. Brenda Milner, who studied this individual for over 40 years, had to introduce herself to him every time they met; he could not even recognize a recent picture of himself. Based on case histories like Molaison's and studies of memory in laboratory animals, a consensus has emerged that several aspects of memory are governed by the hippocampus and interior sections of the temporal lobe.

Mapping Functional Areas in the Cerebrum II: Electrical Stimulation of Conscious Patients While working with severe epileptics (people suffering from seizures), Wilder Penfield pioneered a different approach to studying brain function. These individuals were scheduled to have seizure-prone areas of their brains surgically removed. While the patients were awake and under a local anesthetic, Penfield electrically stimulated portions of their cerebrums. His immediate goal was to map essential areas that should be spared from removal if possible.

When Penfield stimulated specific areas, patients reported sensations or experienced movement in particular regions of the body. From these responses, Penfield was able to map the sensory regions of the cerebrum shown in **FIGURE 46.21** (see page 946), as well as the adjacent motor regions. Brain surgeons still use this technique to map critical areas near tumors and seizure-prone areas.

Perhaps the most striking of Penfield's findings was that, on occasion, patients would respond to stimulation of their temporal lobe by having what appeared to be flashbacks. After one region was stimulated, a woman said, "I hear voices. It is late at night around the carnival somewhere—some sort of traveling circus. . . . I just saw lots of big wagons that they used to haul animals in."

Was this a memory, stored in a small set of neurons that Penfield happened to stimulate? The hypothesis that memories are somehow stored in specialized cells is controversial. However, a recent study provided support for it.

Researchers attached tiny electrodes that recorded the electrical activity of individual neurons in the temporal lobes and hippocampi of study subjects. They then showed each subject a set of images of celebrities, places, and objects. In one subject,

(a) Top view of cerebrum

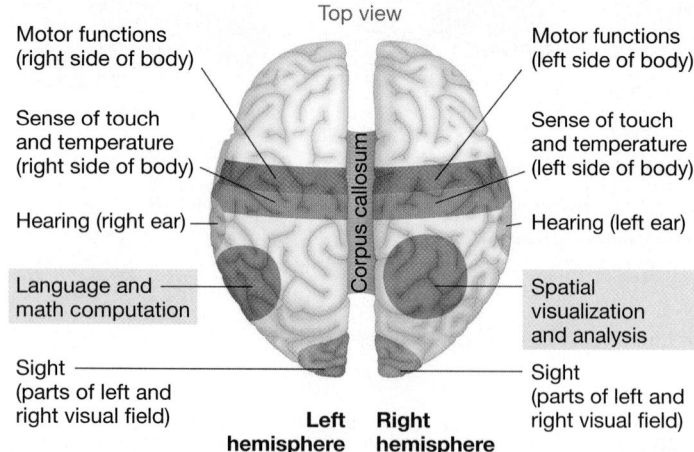

Top view

Motor functions
(right side of body)

Motor functions
(left side of body)

Sense of touch
and temperature
(right side of body)

Sense of touch
and temperature
(left side of body)

Corpus callosum

Hearing (right ear)

Hearing (left ear)

Language and
math computation

Spatial
visualization
and analysis

Sight
(parts of left and
right visual field)

Sight
(parts of left and
right visual field)

**Left
hemisphere**

**Right
hemisphere**

(b) Cross section through area responsible for sense of touch
and of temperature

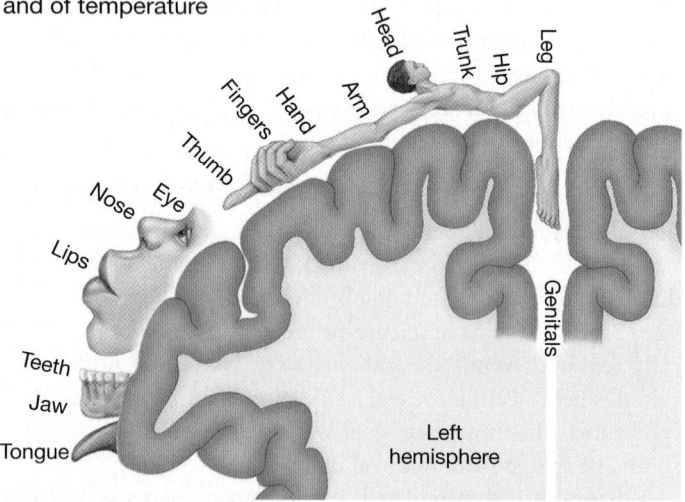

Head

Trunk

Hip

Leg

Fingers

Hand

Arm

Thumb

Nose

Eye

Lips

Teeth

Jaw

Tongue

Genitals

Left
hemisphere

FIGURE 46.21 Specific Brain Areas Have Specific Functions.
(a) Map of the brain, in top view (as if the person is looking at the
top of the page), showing the functions of some major regions.
The map was compiled from studies of people with damaged brain
areas or in whom brain regions had been removed surgically. Note
that the corpus callosum is not actually as wide as shown here.
(b) Researchers mapped the area responsible for the sense of
touch and of temperature by stimulating neurons in the brains of
patients who were awake. The size of the icons corresponds to the
amount of brain area devoted to sensing those parts.

✔**QUESTION** Is there a correlation between the size of the brain area
devoted to sensing a particular body part and the size of that body
part? Explain.

a specific neuron fired when the subject viewed any of several
images of the actress Jennifer Aniston (**FIGURE 46.22**). The neuron
did not fire for other images showing spiders, buildings, other
actresses, or even Aniston with another person. It appeared that,
through experience, at least one of this patient's neurons became
singularly devoted to the concept of Jennifer Aniston.

It is important to note that the researchers were testing only a
tiny subset of the neurons present in the human brain, so there

may be more than one "Jennifer Aniston neuron" in the brain.
But the take-home message is that the formation or retrieval of
memories associated with specific concepts involves specific
neurons. Researchers are still examining exactly how these neurons interact with other parts of the brain during the processes of
learning and remembering.

Have other approaches to studying memory been productive?

How Does Memory Work?

Learning is an enduring change in behavior that results from a
specific experience in an individual's life. **Memory** is the retention of learned information. Learning and memory are thus
closely related and are often studied in tandem. As an introduction to how researchers explore these phenomena, let's first examine work that focuses on neurons and then review research at
the molecular level.

Recording from Single Neurons during Memory Tasks How do
the action potentials generated by a cell change as learning and
memory take place? Researchers have attempted to answer this
question by recording from individual neurons in the temporal
lobes of humans.

Before operating on patients who were still awake and about
to undergo surgery to remove seizure-prone areas of their brains,
physicians placed electrodes in specific brain regions. Then they
projected words or names of objects on a screen and asked the
individuals to read them silently, read them aloud, or remember them and repeat them later. The data showed that individual
neurons in the cerebrum's temporal lobe are relatively quiet while
patients identify objects but extremely active when the individual
remembers the objects and repeats their names aloud.

What do such data mean? Neurons in the temporal lobe are
most active during memory tasks. So how could action potentials from particular cells make memory possible?

Documenting Changes in Synapses Research on the molecular
basis of memory is based on the idea that learning and memory
must involve some type of short-term or long-term change in
the neurons responsible for these processes. This change could
be structural or chemical in nature. Structural changes might include modifications in the number of synapses that a particular
neuron makes, or even the destruction and creation of neurons.
Chemical changes might involve alterations in the amount of
neurotransmitter released at certain synapses or changes in the
number of receptors present in postsynaptic cells.

To explore the molecular basis of learning and memory, Eric
Kandel's group has focused on an animal much easier to study
than any vertebrate, the sea slug *Aplysia californica* (**FIGURE 46.23a**).
Much of their work has explored the reflex diagrammed in
FIGURE 46.23b: When the siphon, a structure on the animal's back,
is touched—for example, by a stream of water—the sea slug
responds by withdrawing its gill. The reflex is produced by a
sensory neuron that is activated by touch and a motor neuron
that projects to a gill muscle. Retracting the gill protects it from
predators.

| 0 action potentials/sec | 8 action potentials/sec | 0 action potentials/sec | 6 action potentials/sec | 0 action potentials/sec |

FIGURE 46.22 Single-Neuron Recording Reveals that Some Neurons in the Brain Recognize Specific Concepts. The graphs below each image show how a single neuron fires in response to images of actress Jennifer Aniston but not to other images.

DATA: Quiroga, R. Q., L. Reddy, G. Kreiman, et al. 2005. *Nature* 435: 1102–1107.

Early work established that this simple reflex is modified by learning. For example, *Aplysia* also withdraw their gills when their tails are given an electrical shock. If shocks to the tail are repeatedly combined with a very light touch to the siphon—too light to normally get a response on its own—an *Aplysia* will learn to withdraw its gills in response to a light siphon touch alone.

Follow-up studies on this reflex showed that the neurons involved in learning release the neurotransmitter **serotonin,** which causes an EPSP in the motor neuron to the gill. Repeated application of serotonin mimics what happens at the synapse during learning, when the neurons fire repeatedly. As **FIGURE 46.24a** (see page 948) shows, experimental application of serotonin leads to higher EPSPs, meaning that the motor neuron is more likely to

generate action potentials. These results suggest that in *Aplysia*, changes in the nature of the synapse form the molecular basis of learning and memory. This change in responsiveness of the synapse is termed **synaptic plasticity.**

Recently Kandel's team replicated these results with sensory and motor neurons growing in culture. **FIGURE 46.24b** shows two *Aplysia* motor neurons on a culture plate. Each motor neuron synapses with the same sensory neuron. To mimic the learning process, the investigators applied serotonin to one synapse five times over a short period. When they stimulated the sensory neuron a day later, they found a huge increase in EPSPs in the motor neuron postsynaptic to that synapse (**FIGURE 46.24c**). The experimental postsynaptic cells had also established additional synapses. The structure and behavior of the experimental

(a) Sea slug *Aplysia californica*

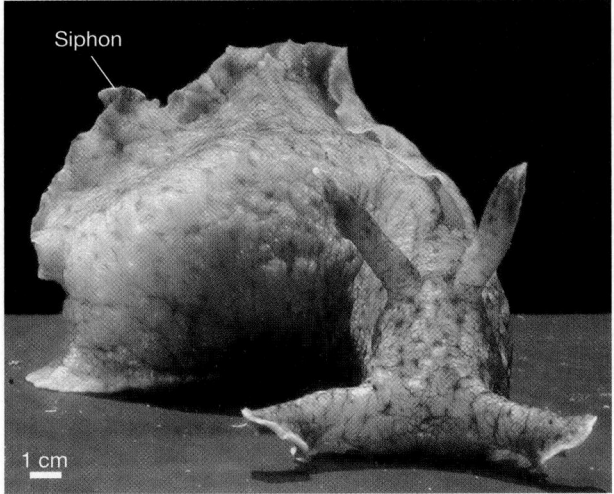

(b) Gill-withdrawal reflex protects the gills during an attack.

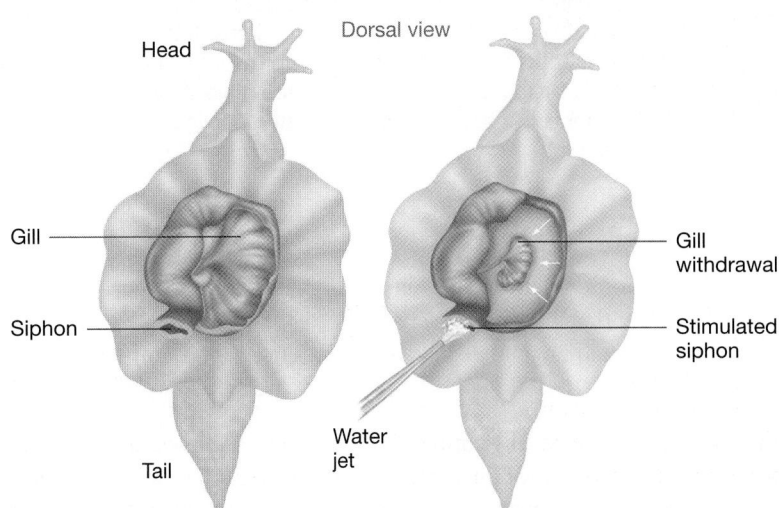

FIGURE 46.23 The Gill-Withdrawal Reflex in *Aplysia* Is a Model System in Learning and Memory.

(a) Postsynaptic potentials changing as a result of learning

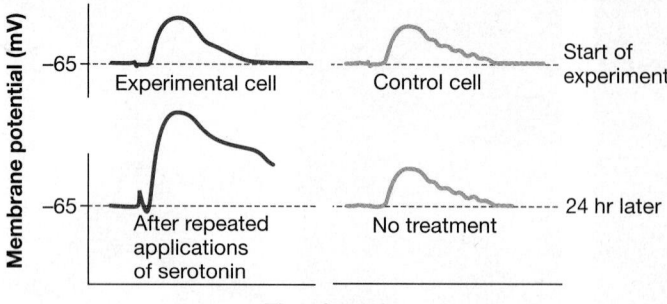

Start of experiment

Experimental cell   Control cell

24 hr later

After repeated applications of serotonin   No treatment

Time (msec)

(b) Applying neurotransmitters to neurons in culture

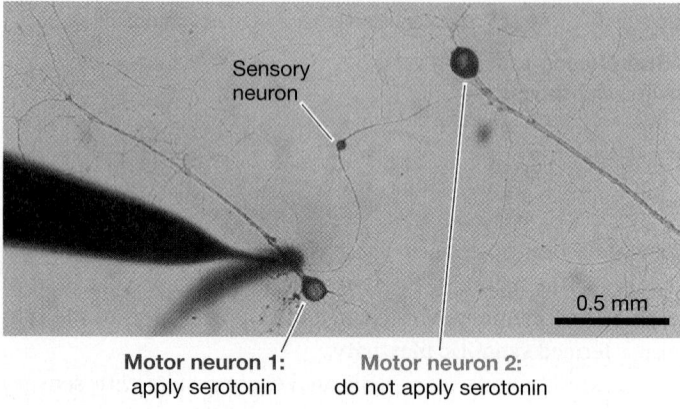

Sensory neuron

0.5 mm

Motor neuron 1: apply serotonin   Motor neuron 2: do not apply serotonin

(c) Repeating *Aplysia* experiment in culture

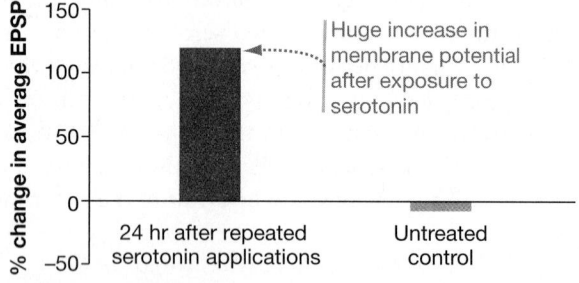

Huge increase in membrane potential after exposure to serotonin

24 hr after repeated serotonin applications   Untreated control

FIGURE 46.24 Learning and Memory Involve Changes in Synapses. (a) Experiment showing that repeated application of serotonin changes the behavior of postsynaptic neurons. **(b)** An in vitro study of interactions between *Aplysia* motor neurons and sensory neurons. **(c)** Histogram depicting the percentage increase in postsynaptic potentials that occurs after repeated application of serotonin, as in part **(a)**.

DATA: Martin, K. C., A. Casadio, H. Zhu, et al. 1997. *Cell* 91: 927–938.

postsynaptic motor neuron had changed, based on its experience. Neurons that had not received the repeated stimulation with serotonin had normal postsynaptic responses and numbers of synapses.

Documenting Changes in Neurons Synapses change over time, but what about the cells of the central nervous system themselves? When early neurobiologists dissected adult cadaver brains, they found that none of the neurons showed signs of mitosis. From these observations they concluded that **neurogenesis**—the formation of new neurons—does not occur in adults.

In 1983, Steven Goldman and Fernando Nottebohm revisited this long-standing theory. They examined whether new neurons can be formed in the brains of songbirds. Songbirds were chosen because in their brains, specific regions that control singing behavior undergo dramatic seasonal changes in size. The researchers hypothesized that these changes result from an increased rate of neurogenesis during the reproductive season.

To test this hypothesis, the researchers injected canaries (**FIGURE 46.25a**) with radiolabeled thymidine, which becomes incorporated into newly synthesized DNA. The researchers later collected the birds' brains, cut them into very thin slices, mounted the slices onto microscope slides, and brought them into contact with photographic film. In this technique, called autoradiography (see **BioSkills 9** in Appendix B), any new cells with radiolabeled thymidine in their DNA would expose the film, thereby indicating the presence and location of cells that were produced after the bird was injected.

The researchers found that adult songbirds are indeed able to produce large numbers of new neurons. Dozens of studies since Goldman and Nottebohm's experiment have confirmed that new neurons are incorporated into the song-control regions of the brain each spring, causing them to grow dramatically in size (**FIGURE 46.25b**). The new neurons promote learning and memory in the song-control system.

Other scientists around the same time and, indeed, even earlier obtained autoradiographic evidence of neurogenesis in adult mammals. However, the sheer number of new cells was much lower in mammals than in birds, so these reports did

check your understanding

 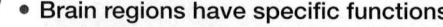

If you understand that . . .

- The CNS and PNS work together to gather information about the external and internal environments, process that information, and signal muscles, glands, and other tissues to make appropriate responses to the information.
- Brain regions have specific functions.
- Changes in synapses and neurogenesis allow animals to learn and create memories.

✔ **You should be able to . . .**

1. Describe the research strategies that allowed biologists to localize particular functions to specific regions in the brain.
2. Identify the key mechanisms behind learning and memory.

Answers are available in Appendix A.

(a) Male canaries (left) sing to attract females (right).

(b) In songbirds, the size of the brain's song-control region increases in spring.

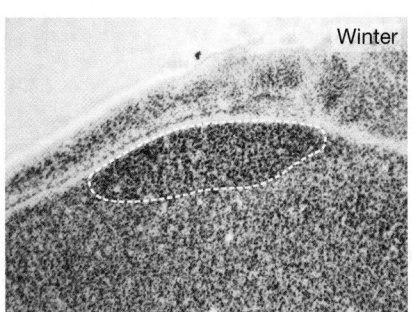
Winter

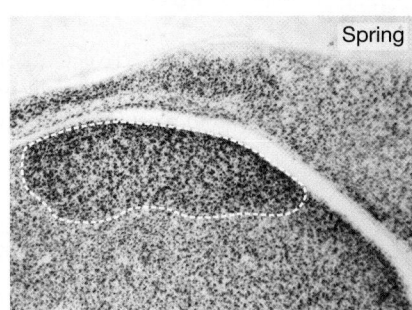
Spring

FIGURE 46.25 Neurogenesis Occurs in the Adult Songbird Brain.

not convince all researchers. Skeptics initially suggested that the labeled cells in mammals might actually be glia rather than neurons.

A breakthrough came in the early 1990s with the development of new techniques that specifically label neurons. The markers used in these techniques bind to proteins that are found only in neurons and not in glial cells. Since that time, adult neurogenesis has been definitively shown to occur in all vertebrates, including humans.

Scientists are only beginning to understand the functional role of these new neurons. Is it the same as that of neurons made during development? Can neurogenesis be used to help heal brain injuries? Some studies have even suggested that chronic stress during childhood (the stress of poverty or bullying, for example) can reduce the ability to make new neurons, thereby negatively affecting learning in adults.

Results like these reinforce a growing consensus that learning, memory, and the control of complex behaviors involve not only molecular and structural changes in synapses but also changes in the number of neurons. Further, most researchers now agree that at least some aspects of long-term memory involve changes in gene expression. Chemical messengers called hormones also cause changes in gene expression in target cells (see Chapter 49). But before investigating how hormones work, let's focus on the electrical signals involved in vision, hearing, taste, and movement—the subject of Chapters 47 and 48.

CHAPTER 46 REVIEW

 For media, go to MasteringBiology (MB)

If you understand . . .

46.1 Principles of Electrical Signaling

- All neurons have a cell body and multiple short dendrites that receive signals from other cells. Most neurons also have a single axon that transmits electrical signals to other neurons or to effector cells in glands or muscles.

- Studies of the squid giant axons established that neurons have a resting potential created by the sodium–potassium pump and potassium leak channels. When Na^+/K^+-ATPase hydrolyzes ATP, it transports 3 Na^+ out of the cell and 2 K^+ in.

✔ You should be able to diagram the plasma membrane of a neuron. Add symbols to show the relative concentrations of Na^+, K^+, and Cl^-. Add labels indicating the role of K^+ leak channels and the Na^+/K^+-ATPase.

46.2 Dissecting the Action Potential

- Studies of the squid axon established that the action potential is a rapid, all-or-none change in membrane potential.

- An action potential begins with an inflow of Na^+ that depolarizes the membrane. An outflow of K^+ follows and repolarizes the membrane.

- Both Na^+ and K^+ flow through voltage-gated channels.

- As Na^+ flows in, it repulses cations that spread from the site of the action potential, causing the nearby membrane to depolarize enough to trigger an action potential in that adjacent portion of the membrane.

- Propagation takes place most rapidly in large axons or myelinated axons.

✔ You should be able to explain why every action potential in a given neuron is identical.

46.3 The Synapse

- When action potentials arrive at a synapse, synaptic vesicles fuse with the axon's membrane and deliver neurotransmitters into the synapse that bind to receptors on the membrane of a postsynaptic cell.

- One class of receptors functions as ligand-gated channels. In response to binding by a neurotransmitter, these channels open and admit ions that depolarize or hyperpolarize the postsynaptic cell's membrane.

- Postsynaptic potentials sum.

- If the membrane at the axon hillock of the postsynaptic neuron depolarizes to a threshold value, an action potential is triggered.

✓ You should be able to explain how summation in the postsynaptic cell relates to the claim that the cell body integrates information from many different synapses.

46.4 The Vertebrate Nervous System

- The CNS consists of the brain and spinal cord (in vertebrates); the PNS consists of all nervous system components outside the CNS.

- In vertebrates, the PNS contains somatic and autonomic components. The somatic PNS signals the effector cells producing movement; the autonomic PNS monitors internal conditions and signals the effector cells that change the activity of tissues and organs.

- Early efforts to map functional regions of the brain were based on analyzing deficits in individuals with brain lesions or on stimulating certain regions of the cerebrum.

- Research has established that learning and memory involve modifications in synapses. After learning takes place, certain neurons release more or less neurotransmitter, or make additional synapses, in response to stimulation.

- Recent research shows that vertebrates are capable of adult neurogenesis, which may be an important component of learning and memory.

✓ You should be able to predict the effects of a brain stem lesion in an animal.

(MB) MasteringBiology

1. MasteringBiology Assignments

Tutorials and Activities Action Potentials, How Neurons Work (1 of 3): Neuron Structure and Resting Potential; How Neurons Work (2 of 3): The Action Potential; How Neurons Work (3 of 3): Conduction of an Action Potential; How Synapses Work (1 of 2): Chemical Synapses; How Synapses Work (2 of 2): Postsynaptic Potentials; Membrane Potentials; Nerve Signals: Action Potentials; Neuron Structure; Signal Transmission at a Chemical Synapse; The Vertebrate Nervous System

Questions Reading Quizzes, Blue-Thread Questions, Test Bank

2. eText Read your book online, search, take notes, highlight text, and more.

3. The Study Area Practice Test, Cumulative Test, BioFlix® 3-D Animations, Videos, Activities, Audio Glossary, Word Study Tools, Art

You should be able to . . .

✓ TEST YOUR KNOWLEDGE   *Answers are available in Appendix A*

1. Which ion leaks across a neuron's membrane to help create the resting potential?
   a. Ca^{2+}
   b. K^+
   c. Na^+
   d. Cl^-

2. Why did the squid axon become a model system for studying electrical signaling in animals?
   a. Its action potentials are particularly large and frequent.
   b. It is the tissue from which researchers initially isolated Na^+/K^+-ATPase.
   c. Squids are abundant and easy to obtain.
   d. It was large enough to support intracellular recording by the first microelectrodes.

3. How does myelination affect the propagation of an action potential?
   a. It speeds propagation by increasing the density of voltage-gated channels.
   b. It speeds propagation by increasing electrochemical gradients favoring Na^+ entry.

   c. It speeds propagation because cations do not leak out of the membrane as they spread down the axon.
   d. It slows down propagation because Na^+ channels exist only at unmyelinated nodes (nodes of Ranvier).

4. In a neuron, what creates the electrochemical gradient favoring the outflow of K^+ when the cell is at rest?
   a. Na^+/K^+-ATPase
   b. voltage-gated K^+ channels
   c. voltage-gated Na^+ channels
   d. ligand-gated Na^+/K^+ channels

5. Which of the following brain regions is responsible for formation of new memories?
   a. brain stem
   b. hippocampus
   c. frontal lobe
   d. cerebellum

6. Draw a graph of an action potential and label the axes. Label the parts of the graph, and explain which ion flow or flows are responsible for each part.

7. Explain why the Na$^+$/K$^+$-ATPase is "electrogenic"—meaning that it creates a voltage across a membrane.

8. Explain the difference between a ligand-gated K$^+$ channel and a voltage-gated K$^+$ channel.

9. Describe the role of summation in postsynaptic cells.

10. Compare and contrast the somatic and autonomic components of the PNS.

11. Predict the effect on heart rate of a drug that activates sympathetic nervous system activity.

12. Why is memory thought to involve changes in particular synapses?
    a. In some systems, an increased release of neurotransmitters occurs after learning takes place.
    b. In some systems, the type of neurotransmitter released at the synapse changes after learning takes place.
    c. When researchers stimulated certain neurons electrically, individuals replayed memories.
    d. People who lack short-term memory have specific deficits in synapses within the brain regions responsible for memory.

13. Explain why drugs that prevent neurotransmitters from being taken back up by the presynaptic cell have dramatic effects on the activity of postsynaptic neurons.

14. Discuss the pros and cons of lesion studies and electrical stimulation of conscious patients in determining the functions of particular brain structures.

15. **QUANTITATIVE** The data points on the graph record the movement of potassium ions (in μSiemens, a unit of conductance) observed when neuron membranes were voltage-clamped at the array of voltages indicated on the *x*-axis. The "Neuron with venom" data were generated by adding black mamba venom to the solution bathing the neuron (black mambas are snakes). Which of these conclusions can you draw from the data?

    a. The venom opens all potassium channels.
    b. The venom blocks all potassium channels.
    c. There is more than one type of potassium channel; at least one is affected by the venom while others are not.
    d. Black mamba venom does not affect potassium ion movement.

16. Studies of the brain show that mature neurons do not undergo mitosis to give rise to new neurons, yet adult neurogenesis occurs in all vertebrates. Suggest a hypothesis as to the source of these new neurons.

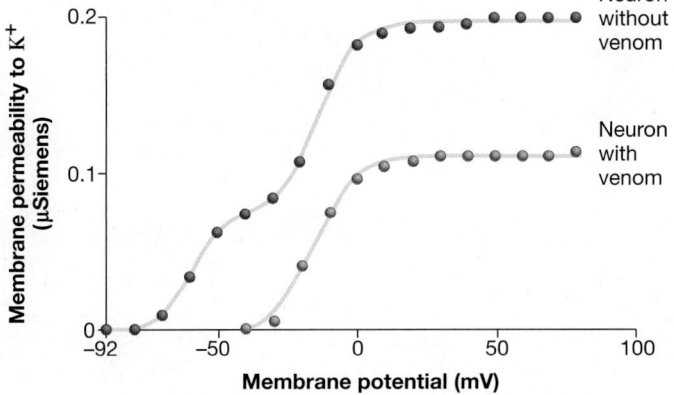

47 Animal Sensory Systems

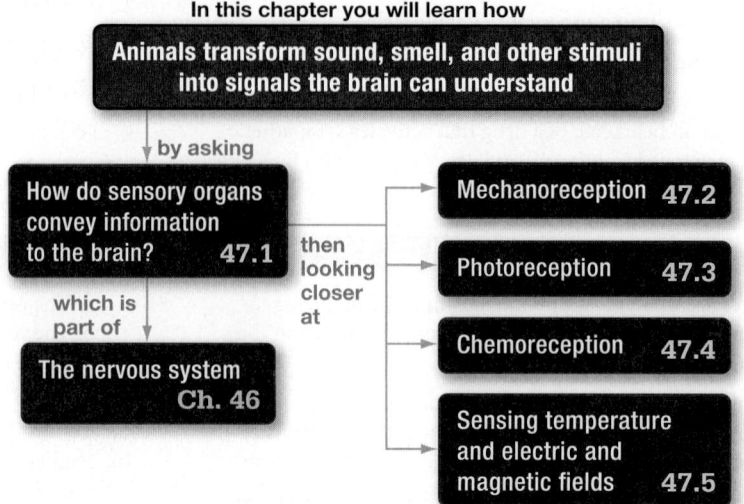

In this chapter you will learn how

Animals transform sound, smell, and other stimuli into signals the brain can understand

by asking

How do sensory organs convey information to the brain? 47.1

which is part of

The nervous system Ch. 46

then looking closer at

Mechanoreception 47.2

Photoreception 47.3

Chemoreception 47.4

Sensing temperature and electric and magnetic fields 47.5

In many species of moth, males have much larger antennae than females do. Receptor cells on the males' feathery antennae detect airborne chemical signals that are produced by sexually mature females. As a result, males can locate females in total darkness.

This chapter is part of the Big Picture. See how on pages 840–841.

Many adult moths are active at night, when it is difficult or impossible to see. Instead of *looking* for a mate under these challenging circumstances, sexually mature female moths release a chemical attractant into the air. Male moths can detect even a single molecule of the attractant, due to receptor cells located on their large, feathery antennae. Guided by an airborne gradient of attractant molecules, a male moth flies unerringly toward a female.

As they patrol in search of these airborne molecules, however, male moths are hunted by bats. Like moths, bats are active almost exclusively at night. Instead of hunting by sight, like a falcon or a cheetah, bats hunt with the aid of sonar: They emit a train of high-pitched sounds as they fly and then listen for echoes that indicate the direction and shape of objects in their path. If the object is a moth, the bat flies toward it, catches the moth in its mouth, and eats it.

✔ When you see this checkmark, stop and test yourself. Answers are available in Appendix A.

But some moth species can hear bat calls. When moths detect sounds from an onrushing bat, they tumble out of the sky in chaotic escape flights.

If you were out at night as these dramas unfolded, at best you might be dimly aware that bats and moths were flying about. Humans cannot smell moth attractants or hear the sounds that most bats emit when flying. It took decades of careful experimentation for biologists to understand how moths and bats sense the world around them and respond to the information they receive.

Sensing and interpreting changes in the environment are fundamental to how animals work. Let's begin with a basic question: How are sounds, smells, and other stimuli transformed into a signal that the brain can understand?

47.1 How Do Sensory Organs Convey Information to the Brain?

As a moth flies through the night, its brain receives streams of signals from an array of sensory organs. Antennae provide information about the concentration of female attractant molecules; ears located on various parts of the body send data on the presence of high-pitched sounds; detectors for balance and gravity transmit signals about the body's orientation in space.

Each type of sensory information is detected by a sensory neuron or by a specialized receptor cell that makes a synapse with a sensory neuron. As **FIGURE 47.1** shows, the moth's nervous system integrates the sensory input—the information from sensory neurons—and responds with motor output, via electrical signals, to specific muscle groups (effectors).

The ability to sense a change in the environment depends on three processes:

1. *Transduction,* the conversion of an external stimulus to an internal signal in the form of action potentials along sensory neurons

2. *Amplification* of the signal

3. *Transmission* of the signal to the central nervous system (CNS)

The first step in the sequence, transduction, requires a sensory receptor cell specialized for converting light, sound, touch, or some other signal into an electrical signal. Sensory receptors are located throughout the body and are categorized by type of stimulus:

- **Mechanoreceptors** respond to distortion caused by pressure.
- **Photoreceptors** respond to particular wavelengths of light.
- **Chemoreceptors** perceive specific molecules.
- **Thermoreceptors** detect changes in temperature.
- **Nociceptors** sense harmful stimuli such as tissue injury.
- **Electroreceptors** detect electric fields.
- **Magnetoreceptors** detect magnetic fields.

With such a broad range of possible sensory detectors, it is no wonder that animals can monitor and respond to a wide array of changes in their environments.

Now, how do sensory cells receive information from the environment and report it to the brain, so an appropriate response can occur?

Sensory Transduction

During the resting state in most sensory cells, the inside of the plasma membrane is more negative than the exterior (see Chapter 46). When ion flows cause the interior to become more positive (less negative), the membrane is **depolarized.** When ions cause the cell interior to become more negative than the resting potential, the membrane is **hyperpolarized.**

Although sensory receptors can detect a remarkable variety of stimuli, they all transduce sensory input—such as light, sounds, touch, and odors—to a change in membrane potential. In this way, different types of information are transduced to a common type of signal—one that can be interpreted by the brain.

If a sensory stimulus induces a large change in a sensory receptor's membrane potential, there is a change in the firing rate of action potentials sent to the brain. The amount of depolarization that occurs in a sound-receptor cell, for example, is proportional to the loudness of the sound. If the depolarization passes

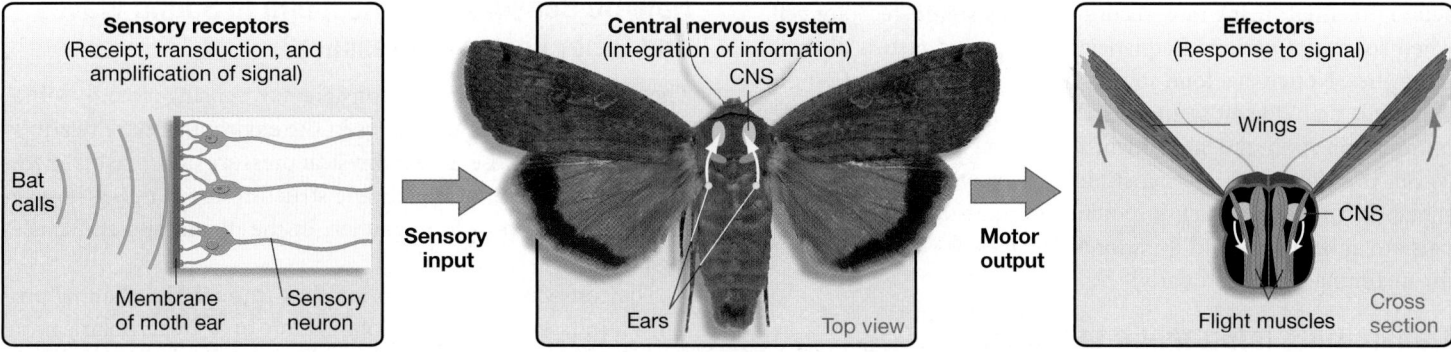

FIGURE 47.1 Sensory Systems, the CNS, and Effectors Are Linked. Sensory neurons relay information about conditions inside and outside an animal to the central nervous system. After integrating information from many sensory neurons, the CNS sends signals to muscles.

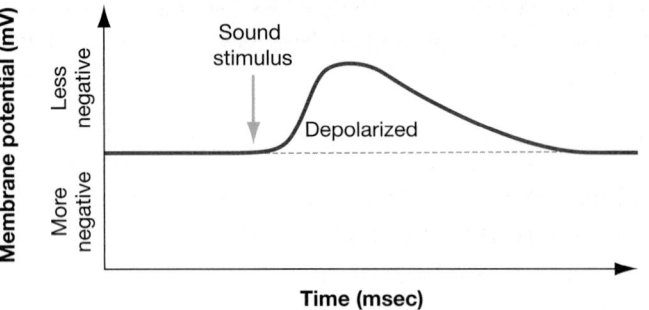

(a) Sound-receptor cells depolarize in response to sound.

Membrane potential (mV)

Less negative / More negative

Sound stimulus

Depolarized

Time (msec)

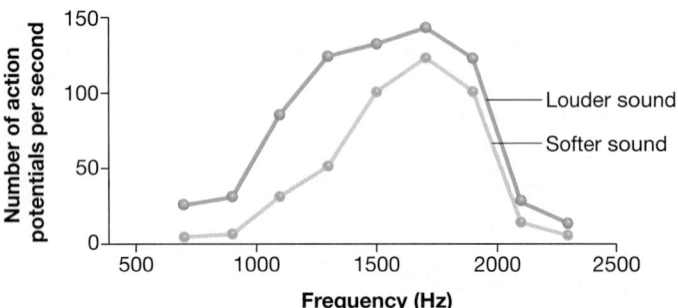

(b) Sound-receptor cells respond more strongly to louder sounds.

Number of action potentials per second

Louder sound

Softer sound

Frequency (Hz)

FIGURE 47.2 Sensory Inputs Change the Membrane Potential of Receptor Cells. (a) In response to sensory stimuli, ions flow across the membranes of receptor cells and either depolarize or hyperpolarize the membrane. **(b)** The rate at which action potentials occur in an auditory neuron transmits information about the nature and intensity of the sensory stimulus. (The frequency of a sound, measured in hertz, determines its pitch.)

DATA: Rose, J. E., J. E. Hind, D. J. Anderson, et al. 1971. *Journal of Neurophysiology* 34: 685–699.

threshold, enough voltage-gated sodium channels open to trigger action potentials that are relayed to the brain.

FIGURE 47.2a shows the membrane potential from a sound-receptor cell. If you put your finger on the red line and trace to the right, you'll notice that when the experimenter played a sound, the sound-receptor cell depolarized for a short time in response. Other sensory cells work in a similar way.

Recall that all action potentials from a given neuron are identical in size and shape (see Chapter 46). **FIGURE 47.2b** graphs the action potential "firing rate" recorded from a sound-receptor cell when sounds at various frequencies were played at two distinct intensities. Notice that loud sounds induce a higher rate of action potentials than do soft sounds. In this way receptor cells provide information about the intensity of a stimulus.

But if all types of external stimuli are converted to electrical signals in the form of action potentials, and if all action potentials are alike in size and duration, how does the brain interpret the incoming signals properly?

Transmitting Information to the Brain

There are two keys to understanding how the brain interprets sensory information. First, receptor cells tend to be highly specific. For example, each receptor cell in a human ear responds best to certain pitches of sound. Some receptors are more sensitive to low-pitched sounds, and others respond best to high-pitched sounds. The pattern of action potentials from a cell contains information about the pitch of the sound that is being received, its intensity, and how long the stimulus lasts.

The second key point: Each type of sensory neuron sends its signal to a specific portion of the brain. Axons from sensory neurons in the human ear project to the temporal lobes at the sides of the brain, but axons from sensory receptors in the eye deliver action potentials to the occipital lobe at the back of the brain. Different regions of the brain are specialized for interpreting different types of stimuli.

Now that the basic principles of sensory reception and transduction have been introduced, let's delve into the details of the major sensory systems.

47.2 Mechanoreception: Sensing Pressure Changes

Animals have a variety of mechanisms for **mechanoreception**—the sensation of pressure changes. Crabs, for example, have a fluid-filled organ that helps them sense the pressure created by gravity. The organ, known as a **statocyst,** is lined with pressure-receptor cells and contains a small calcium-rich particle that normally rests on the bottom of the organ. But if the crab is tipped or flipped over, this particle presses against receptors that are *not* on the bottom of the organ. When the brain receives action potentials from these receptors, it responds by activating muscles that restore the animal to its normal posture.

It's also common for animals to have cells that are responsible for detecting direct physical pressure on skin, as well as pressure-receptor cells that monitor how far muscles or blood vessels are stretched. Animals detect sound waves, which produce pressure in air, with hearing, and some aquatic animals detect pressure waves in water via a lateral line system. These pressure-sensing systems are all based on the same mechanism.

Let's briefly examine the general design of a mechanoreceptor cell and its response to pressure, then investigate the specific structures involved in vertebrate hearing and the lateral line system.

How Do Sensory Cells Respond to Sound Waves and Other Forms of Pressure?

The mechanoreceptors responsible for sensing sound, vibrations, and other pressure waves in the environment are relatively simple. In every case, direct physical pressure on a plasma membrane or distortion of membrane structures by bending changes the conformation of ion channels in the membrane and causes the channels to open or close.

The consequent change in ion flow through the channel proteins results either in a depolarization or a hyperpolarization. This changes the pattern of action potentials in a sensory neuron.

The Structure of Hair Cells In vertebrates, ion channels that respond to pressure are often found in hair cells. **Hair cells** are

(a) Hair cells have many stereocilia and one kinocilium.

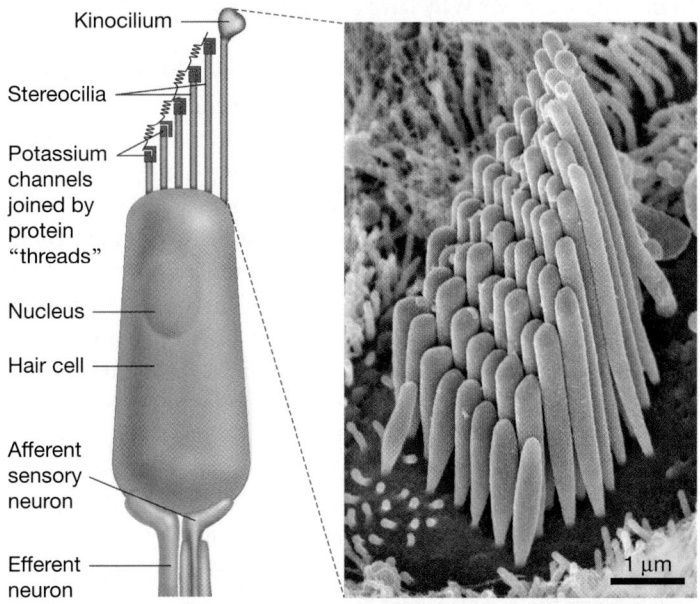

Kinocilium

Stereocilia

Potassium channels joined by protein "threads"

Nucleus

Hair cell

Afferent sensory neuron

Efferent neuron

1 μm

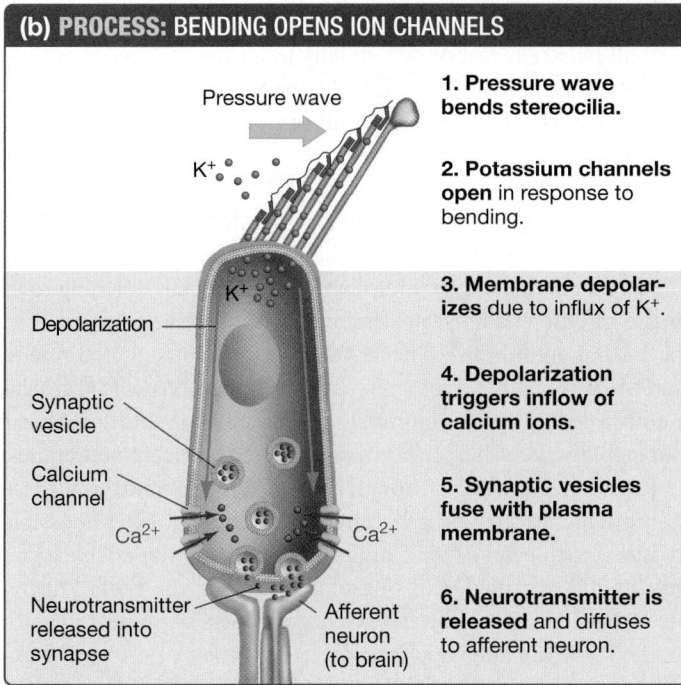

(b) PROCESS: BENDING OPENS ION CHANNELS

Pressure wave

K^+

K^+

Depolarization

Synaptic vesicle

Calcium channel

Ca^{2+} Ca^{2+}

Neurotransmitter released into synapse

Afferent neuron (to brain)

1. **Pressure wave bends stereocilia.**

2. **Potassium channels open** in response to bending.

3. **Membrane depolarizes** due to influx of K^+.

4. **Depolarization triggers inflow of calcium ions.**

5. **Synaptic vesicles fuse with plasma membrane.**

6. **Neurotransmitter is released** and diffuses to afferent neuron.

FIGURE 47.3 Hair Cells Transduce Sound Waves to Electrical Signals.

pressure-receptor cells, illustrated in **FIGURE 47.3a**, named for their stiff outgrowths called **stereocilia** (singular: **stereocilium**). The "hairy-looking" stereocilia are microvilli that are reinforced by actin filaments.

Many hair cells also have a single **kinocilium,** a true cilium that contains a 9 + 2 arrangement of microtubules (introduced in Chapter 7). Hair cells are found in the ears of land-dwelling vertebrates and the lateral line system in many species of fishes and some amphibians.

As Figure 47.3a shows, the stereocilia in hair cells are arranged in order of increasing height; if a kinocilium is present, it is the tallest of all the projections. These structures extend into a fluid-filled chamber.

Signal Transduction in Hair Cells   If stereocilia are bent in the direction of the kinocilium in response to pressure (**FIGURE 47.3b**), the distortion causes potassium ion (K^+) channels in the stereocilia to open. This is the common theme connecting pressure-sensing cells: Bending opens ion channels.

Recall that the opening of K^+ channels usually causes an outflow of K^+ that hyperpolarizes neurons (see Chapter 46). Hair cell plasma membranes respond differently, however, because they are bathed by extracellular fluid with an extraordinarily high K^+ concentration. As a result, the equilibrium potential for K^+ in hair cells is 0 mV instead of the −85 mV in a typical neuron. The resting potential of the hair cell plasma membrane is −70 mV, so when the channels open, K^+ rushes in and causes a depolarization of approximately 20 mV.

In hair cells, depolarization causes an inflow of calcium ions, which triggers an increase in the amount of neurotransmitter released at the synapse between the hair cell and a sensory neuron. The end result is excitation of the postsynaptic cell, meaning that it becomes more likely to fire an action potential to the brain via an afferent sensory neuron. You might recall that afferent neurons are part of the peripheral nervous system and conduct information to the CNS (see Chapter 46).

If sound pressure waves bend stereocilia the other way, however, the K^+ channels close, and the cell hyperpolarizes by 5 mV. Hyperpolarization of the hair cell decreases the amount of neurotransmitter released at the synapse and inhibits the postsynaptic sensory neuron, making it less likely to trigger action potentials.

How can bending affect ion channels? Electron micrographs show that tiny threads connect the tips of stereocilia to each other. One hypothesis contends that when the stereocilia are bent, the threads somehow pull on the ion channels in the wall of the next tallest stereocilium and open them like tiny trapdoors (see Figure 47.3b, step 2). This hypothesis remains to be confirmed, however. Researchers still do not fully understand how the ion channels involved in pressure reception work.

Hearing: The Mammalian Ear

Hearing is the sensation produced by the wavelike changes in air pressure called sound. A sound consists of waves of pressure in air or in water. The number of pressure waves that occur in 1 second is called the **frequency** of the sound, reported in units called hertz (Hz), or cycles per second. When you hear different sound frequencies, you perceive them as different **pitches.** A high-pitched sound may have a frequency in the range of 8000 Hz, whereas a low-pitched sound frequency might be 1000 Hz.

The ear transduces sound waves into action potentials that send information to the brain. To understand how changes in the membrane potential of hair cells result in hearing, let's focus on the human ear as a case study. The human ear has three sections:

the **outer ear, middle ear,** and **inner ear** (**FIGURE 47.4**). A membrane separates each section from the next.

The path of sound through the ear is traced in Figure 47.4. The outer ear, which projects from the head, collects incoming pressure waves and funnels them into a tube known as the ear canal. At the inner end of the ear canal (see the lower part of Figure 47.4), the waves strike the **tympanic membrane,** or eardrum, which separates the outer ear from the middle ear.

The repeated cycles of air compression cause the tympanic membrane to vibrate back and forth with the same frequency as the sound wave. The vibrations are passed to three tiny bones in the middle ear called the **ear ossicles,** which vibrate against one another in response. The last ossicle, the **stapes** (pronounced *STAY-peez*), vibrates against a membrane called the **oval window,** which separates the middle ear from the inner ear. The oval window oscillates in response and generates waves in the fluid inside a chamber known as the **cochlea** (pronounced *KOK-lee-ah*). These pressure waves are sensed by hair cells in the cochlea.

In effect, the ear translates airborne waves into fluid-borne waves. The system seems extraordinarily complex, though, for such a simple result. Why doesn't the outer ear canal lead directly to the oval window? Why have a middle ear at all?

The Middle Ear Amplifies Sounds Biologists began to understand the function of the middle ear when they recognized two key aspects of its structure. First, the size difference between the tympanic membrane and the oval window is important. The tympanic membrane is about 15 times larger than the oval window, causing the amount of vibration induced by sound waves to increase by a factor of 15 when it reaches the oval window. This phenomenon is similar to taking the same amount of force used to bang on a very large door and applying it to a very small door.

In addition, the three ossicles act as levers that further amplify vibrations from the tympanic membrane. The overall effect in mammals is to amplify sound by a factor of 22—meaning that soft sounds are amplified enough to stimulate hair cells in the cochlea. Thus, biologists interpret the mammalian middle ear as an adaptation for increasing sensitivity to sound.

To summarize, the mammalian outer ear transmits sound waves from the environment to the middle ear; the middle ear amplifies these waves enough to stimulate the hair cells within the cochlea of the inner ear.

If all hair cells responded equally to all frequencies of sound, you would be able to perceive only one pitch. Everyone's voice—indeed, every noise—would sound the same. How can hair cells distinguish different frequencies?

The Cochlea Detects the Frequency of Sounds As **FIGURE 47.5a** shows, the cochlea is a coiled tube with a set of internal membranes that divide it into three chambers. Hair cells, forming rows in the middle chamber, are embedded in a tissue that sits atop the **basilar membrane** (**FIGURE 47.5b**). In addition, the hair cells' stereocilia touch yet another, smaller surface called the **tectorial membrane.** (The kinocilium is not present in a mature cochlear hair cell.) In effect, hair cells are sandwiched between membranes.

Researchers struggled for decades to understand how these membranes affect hair cell function. It is virtually impossible to study cochleas in living organisms, because the cochleas are tiny, complex, coiled, and buried deep inside the skull. During the 1920s and 1930s, however, Georg von Békésy pioneered work on the structure and function of these organs by performing experiments on cochleas that he had dissected from fresh human cadavers.

Von Békésy was able to vibrate the oval window and record how the cochlea's internal membranes moved in response. He found that when a pressure wave traveled down the fluid in the upper and lower chambers, the basilar membrane vibrated in response. His key finding, though, was that sounds of different frequencies caused the basilar membrane to vibrate maximally at specific points along its length (**FIGURE 47.6**). When the basilar membrane vibrated in a particular location, the stereocilia of the hair cells there were bent one way and then the other by the tectorial membrane.

Von Békésy also noted that the basilar membrane is stiff near the oval window and flexible at the other end. This is why each

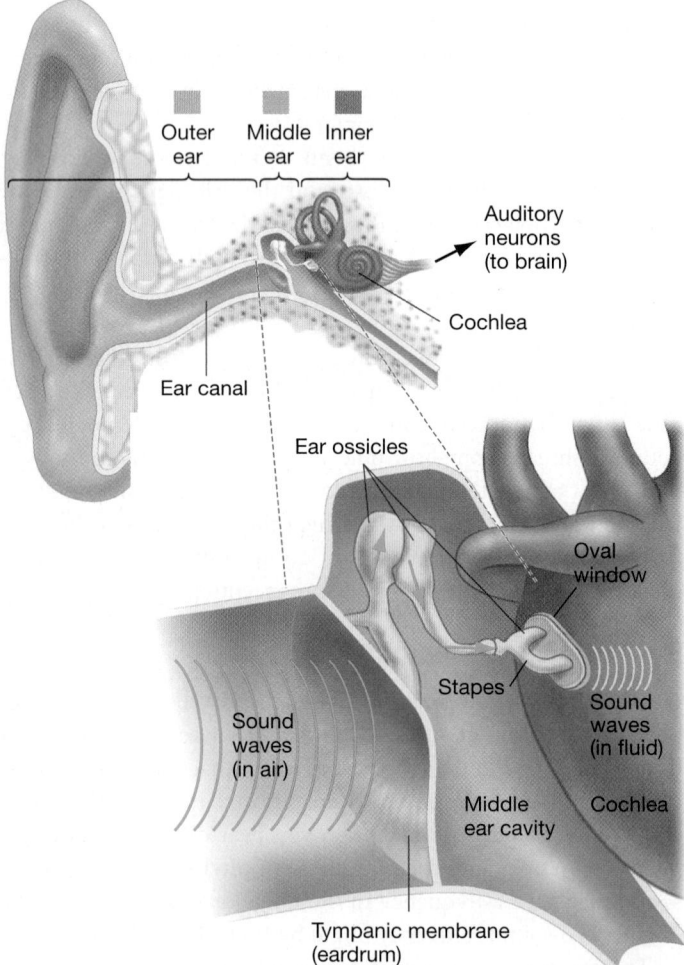

FIGURE 47.4 Mammals Have an Outer Ear, Middle Ear, and Inner Ear. The middle ear starts with the tympanic membrane and ends in the oval window of the cochlea.

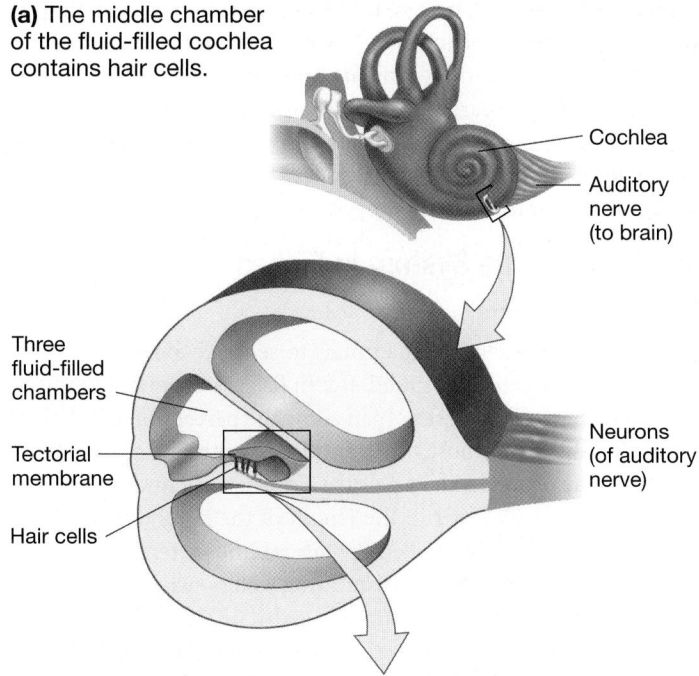

(a) The middle chamber of the fluid-filled cochlea contains hair cells.

Cochlea

Auditory nerve (to brain)

Three fluid-filled chambers

Tectorial membrane

Hair cells

Neurons (of auditory nerve)

(b) Hair cells are sandwiched between membranes.

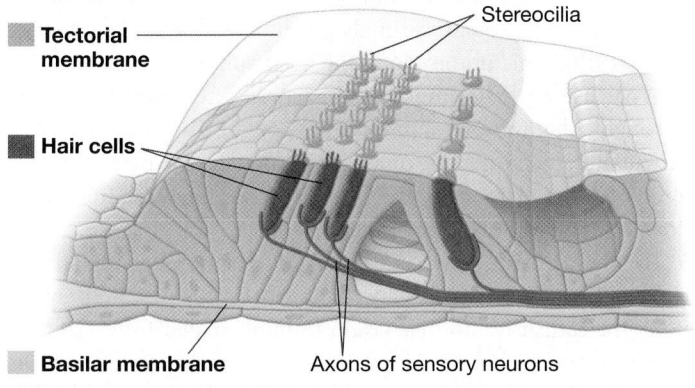

Stereocilia

Tectorial membrane

Hair cells

Basilar membrane

Axons of sensory neurons

FIGURE 47.5 The Human Cochlea Contains Fluid-Filled Chambers Separated by Membranes.

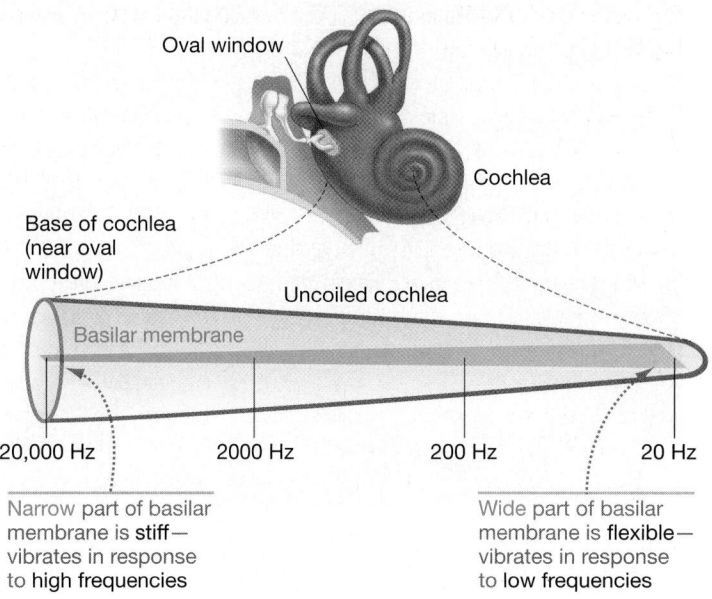

Oval window

Cochlea

Base of cochlea (near oval window)

Uncoiled cochlea

Basilar membrane

20,000 Hz   2000 Hz   200 Hz   20 Hz

Narrow part of basilar membrane is **stiff**—vibrates in response to **high** frequencies

Wide part of basilar membrane is **flexible**—vibrates in response to **low** frequencies

FIGURE 47.6 The Basilar Membrane Varies in Stiffness. Different parts of the basilar membrane respond to different sound frequencies.

segment vibrates in response to a different frequency of sound. Just as a stiff drumhead produces a high-pitched sound and a loose drumhead yields a low-pitched sound, high-frequency sounds cause the stiff part of the basilar membrane to vibrate; low-frequency sounds cause the flexible part to vibrate.

To summarize, certain portions of the basilar membrane vibrate in response to specific frequencies and result in the bending of hair cell stereocilia. In this way, hair cells in a particular place on the membrane respond to sounds of a certain frequency. When the occipital lobe of the cerebrum receives action potentials from neurons associated with specific hair cells, it interprets them as a particular pitch—meaning a specific frequency of sound. The result is the sense called hearing.

Complex sounds contain a wide variety of frequencies and trigger particular combinations of hair cells. Through experience, the brain learns which combinations of frequencies represent music, a fire alarm, or a best friend's voice.

Compared with the hearing of many mammals, human hearing is not particularly acute. Humans can hear sounds between 20 Hz and 20,000 Hz (20,000 Hz is equal to 20 kHz, or kilohertz). But some mammals can sense low-frequency infrasounds that are too low for humans to hear (*infra–* means below or under); others are aware of high-frequency ultrasounds that are above the range of human hearing (*ultra–* means beyond).

Elephants Detect Infrasound When Katherine Payne was observing elephants at a zoo in the mid-1980s, she noticed a subtle throbbing in the air. Payne knew that infrasound can produce such sensations.

To test the hypothesis that the elephants were producing infrasonic vocalizations, Payne returned to the zoo with microphones that could pick up sounds at extremely low frequency. At normal speed, the tape she made was silent. But when she raised the pitch of the sounds by speeding up the tape, she heard a chorus of cow-like noises. The elephants were calling to each other, using low-frequency sounds.

According to follow-up research, elephants have the best infrasonic hearing of any land mammal. Because infrasound can travel exceptionally long distances, biologists hypothesize that infrasonic calls allow wild elephants to coordinate their movements when they are miles apart.

Recent research suggests that in addition to detecting infrasound via their large ears, elephants use their feet to detect a by-product of infrasound: seismic vibrations traveling in the ground.

Bats Detect Ultrasound Ultrasonic hearing in bats was discovered in the late 1930s, when Donald Griffin borrowed the only ultrasonic apparatus then in existence from Robert Galambos,

Bat sonar
Returning sound waves

FIGURE 47.7 Bats Emit Ultrasonic Sound Waves That Bounce Off Surfaces. Returning sound waves are sensed by the bat's inner ear.

a fellow graduate student. Griffin used the machine to demonstrate that flying bats constantly emit ultrasounds. In follow-up experiments, he documented that a bat with cotton in its ears, or with its mouth taped shut, crashed into walls when released in a room. Blindfolded bats, in contrast, never crashed.

Griffin and Galambos concluded that bats use sound echoes (sonar) to navigate. This concept, termed **echolocation,** was an outlandish idea at the time. When Galambos described it at a meeting in 1940, another scientist shook him by the shoulders and said, "You can't really mean that!"

Bats generate high-frequency sound waves with the larynx, or voice box. These waves "bounce" off of surfaces (including those of insects), producing echoes that the bat perceives in its inner ear (**FIGURE 47.7**). Recall that different sections of the basilar membrane are specialized for sensing sound waves of different frequencies. In bats, a huge area of the basilar membrane is specialized for sensing the high-frequency sounds of the returning

echoes. Similarly, the part of the brain used to process sound is unusually large in bats, highlighting the extremely important role of echolocation in their navigation and hunting.

More recent research has shown that dolphins, shrews, and certain other animals besides bats use sonar. In fact, it is likely that at least some of these species perceive shapes with their ears better than they do with their eyes.

The Lateral Line System in Fishes and Amphibians

Hair cells in the ear allow mammals to sense changes in air pressure that are perceived as sound. But in fishes and aquatic amphibians, hair cells in a different organ allow the perception of pressure changes in water. In most fishes and larval amphibians, groups of hair cells are embedded in gel-like domed structures called cupulae inside canals that run the length of the body (**FIGURE 47.8a**), forming a sensory organ called the **lateral line system.**

Pressure changes in the surrounding water—whether resulting from waves, an animal swimming nearby, or some other force—cause changes in the pressure of water moving through the lateral line system (**FIGURE 47.8b**). These changes cause kinocilia and stereocilia on the hair cells to bend, and the distortion leads to a change in action potentials along sensory neurons to the brain. In this way, most aquatic animals get information about pressure changes at specific points along the head and body.

What use do fishes make of the lateral line system? It seems reasonable that the lateral line system could be helpful for identifying mates, locating prey, or avoiding predators. However, fishes could also use vision, smell, or other senses for these functions. How important is the lateral line system?

To answer this question, Kirsten Pohlmann, Jelle Atema, and Thomas Breithaupt studied how nocturnal catfish locate prey. At night, the catfish cannot use vision to hunt smaller fishes, so they must use stimuli that persist in the wake of their prey after it has moved on. The researchers hypothesized that the nocturnal catfish hunt smaller fishes using the lateral line system.

(a) The lateral line system consists of a series of canals running along the head and body.

(b) Water enters the canals through pores and bends kinocilia on hair cells, activating sensory neurons.

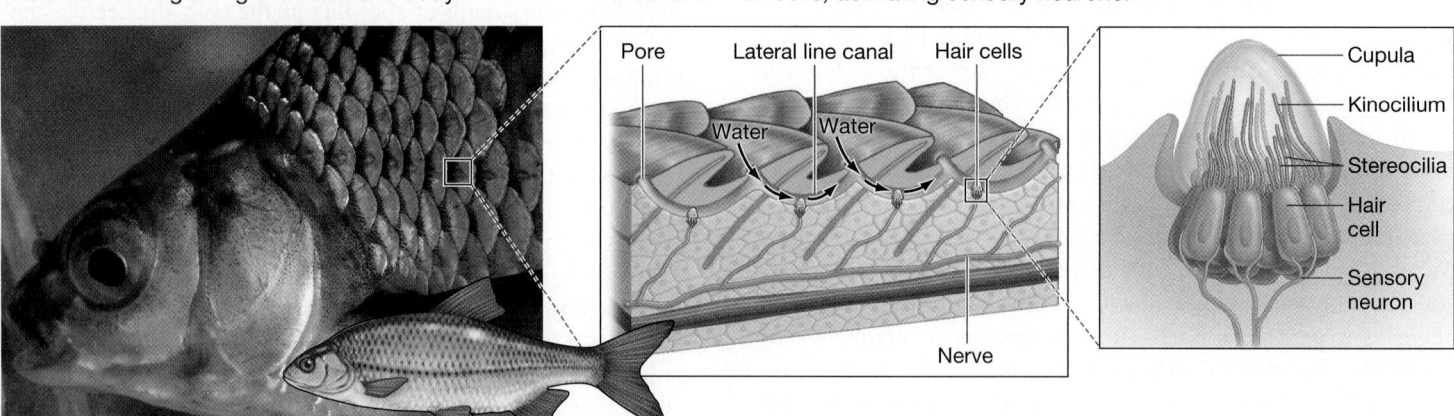

FIGURE 47.8 The Lateral Line System Detects Pressure Waves in Water. (a) Lateral line canals lie just under the fish's epidermis. **(b)** Hair cells within the canals have kinocilia and stereocilia embedded in a gel-like cupula.

To test this hypothesis, the researchers conducted an ablation experiment on the catfish (**FIGURE 47.9**). Ablation is the removal or blocking of a process. In this case, the researchers ablated the lateral line system in one group of catfish, using a chemical that blocks the hair cells from responding to pressure waves. In another group of catfish, they ablated the ability to smell or taste the water by surgically removing the lobes of the brain responsible for these senses.

While the catfish whose smell and taste were ablated took the same amount of time as the non-ablated control group to locate and capture prey, it took the catfish whose lateral line was ablated over twice as long. The researchers concluded that the lateral line system is much more important than other senses in the successful hunting of nocturnal fishes.

RESEARCH

QUESTION: What cues do predatory fish use to detect prey at night?

HYPOTHESIS: Predatory fish detect prey using the lateral line system.

ALTERNATE HYPOTHESIS: Predatory fish detect prey using smell and taste.

NULL HYPOTHESIS: Predatory fish use neither sense to detect prey.

EXPERIMENTAL SETUP:

1. Three experimental groups of catfish (*n*=16 for each group):

Lateral line ablated	Smell and taste ablated	Control fish

2. Acclimate catfish to dark tank at night.

3. Add guppies to tank and record capture success rate.

PREDICTION: Fish with their lateral line ablated will have lower capture success than control fish.

PREDICTION OF ALTERNATE HYPOTHESIS: Fish with smell and taste ablated will have lower capture success than control fish.

PREDICTION OF NULL HYPOTHESIS: There will be no capture success difference among groups.

RESULTS:

Percentage of Guppies Captured

Lateral line ablated	Smell and taste ablated	Control fish
17%	60%	65%

CONCLUSION: Nocturnal predatory catfish use the lateral line system to detect prey.

FIGURE 47.9 The Lateral Line System Is Used for Predation.

✔**QUESTION** If you wanted to conduct a study similar to this one on a diurnal species of catfish, what other sense would you need to control for, and how would you do it?

SOURCE: Pohlmann, K., J. Atema, and T. Breithaupt. 2004. The importance of the lateral line in nocturnal predation of piscivorous catfish. *The Journal of Experimental Biology* 207: 2971–2978.

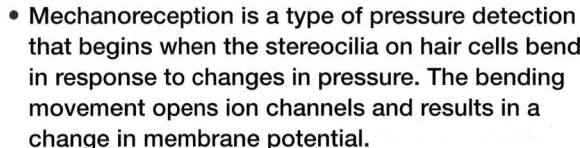

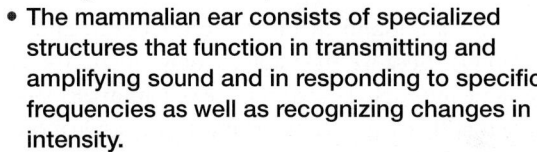

If you understand that . . .

- Mechanoreception is a type of pressure detection that begins when the stereocilia on hair cells bend in response to changes in pressure. The bending movement opens ion channels and results in a change in membrane potential.
- The mammalian ear consists of specialized structures that function in transmitting and amplifying sound and in responding to specific frequencies as well as recognizing changes in intensity.
- The lateral line system in fishes and aquatic amphibians allows them to sense pressure waves in water caused by prey, predators, and potential mates.

✔ You should be able to . . .

1. Predict and explain the effect on hearing in each of the following cases: a punctured eardrum, a mutation that results in dramatically shortened stereocilia, and an age-related loss of flexibility in the basilar membrane.

2. Describe the function of the lateral line system in fishes.

Answers are available in Appendix A.

47.3 Photoreception: Sensing Light

Most animals have a way to sense light. The organs involved in **photoreception** range from simple light-sensitive eyespots in flatworms to the sophisticated, image-forming eyes of vertebrates, cephalopod mollusks, and arthropods.

Variation in the structure of light-sensing organs illustrates an important general principle about the sensory abilities of animals: In most cases, a species' sensory abilities correlate with the environment it lives in and its mode of life—how it finds food and mates. Eyes and other sensory structures are adaptations that allow individuals to thrive in a particular environment. Salamander species that live in meadows and forests have sophisticated eyes; those that live in lightless caves have no functional eyes at all.

Keep this point in mind as you delve into the details of how insects and vertebrates see.

The Insect Eye

Insects have a **compound eye,** composed of hundreds or thousands of light-sensing columns called **ommatidia.** As **FIGURE 47.10** shows, each ommatidium has a lens that focuses light onto a small number of receptor cells—usually four. The receptor cells, in turn, send axons to the brain.

Each ommatidium contributes information about one small piece of the visual field, not unlike the contribution of a single pixel on a computer monitor. Thus, the more ommatidia in a compound eye, the better the resolution—meaning the resolving power, or ability to distinguish objects.

In addition, the presence of many light-sensing columns makes species with compound eyes particularly good at detecting movement. Insects that hunt by sight, such as damselflies and dragonflies, have particularly large numbers of ommatidia.

The Vertebrate Eye

Compound eyes are found in insects, crustaceans, and certain other arthropods. Because they appear only in species that are part of the same monophyletic group (see Chapter 28), researchers conclude that this type of eye structure evolved just once—in an ancestor of today's arthropods.

In contrast, the **simple eye**—a structure with a single lens that focuses incoming light onto a layer of many receptor cells—evolved independently in several widely divergent groups, including annelids, cephalopod mollusks (squid and octopuses), and vertebrates. Let's examine the vertebrate version of the simple eye more closely.

The Structure of the Vertebrate Eye   **FIGURE 47.11a** shows the major structures in a typical vertebrate eye:

- The outermost layer of the structure is a tough rind of white tissue called the sclera. This is the "white of the eye."
- The front of the sclera forms the **cornea,** a transparent sheet of connective tissue.

- The **iris** is a pigmented, round muscle just inside the cornea. The iris can contract or expand to control the amount of light entering the eye.
- The **pupil** is the hole in the center of the iris.
- Light enters the eye through the cornea and passes through the pupil and a curved, clear **lens.**
- Together, the cornea and lens focus incoming light onto the retina in the back of the eye. The **retina** contains a thin layer of cells with photoreceptors and several layers of neurons.

FIGURE 47.11b provides a closer look at the retina, which is attached to the rest of the eye by a single layer of pigmented epithelial cells. From back to front, the retina comprises three distinct cell layers:

1. The photoreceptors, sensory cells that respond to light, are held in place by the pigmented epithelium.

2. Photoreceptors synapse with an intermediate layer of connecting neurons called **bipolar cells.**

3. Bipolar cells in the intermediate neuron layer connect with one another and with neurons called **ganglion cells,** which form the innermost layer of the retina. The axons of the ganglion cells project to the brain via the **optic nerve.**

Note that vertebrate eyes—including yours—have a blind spot because there are no photoreceptor cells where the optic nerve leaves the retina. If light falls in this area, there are no sensory cells available to respond. No signal is sent to the brain, so the stimulus isn't seen.

What Do Rods and Cones Do?   Early anatomists established that the photoreceptors in vertebrate eyes come in two distinct types: small rod-shaped cells and cone-shaped cells, called **rods** and **cones,** respectively (**FIGURE 47.12a**). When technical advances allowed changes in the membrane potentials of these cells to be recorded, it became clear that rods and cones differ in function as well as structure.

(a) Ommatidia are the functional units of insect eyes.

(b) Each ommatidium contains receptor cells that send axons to the CNS.

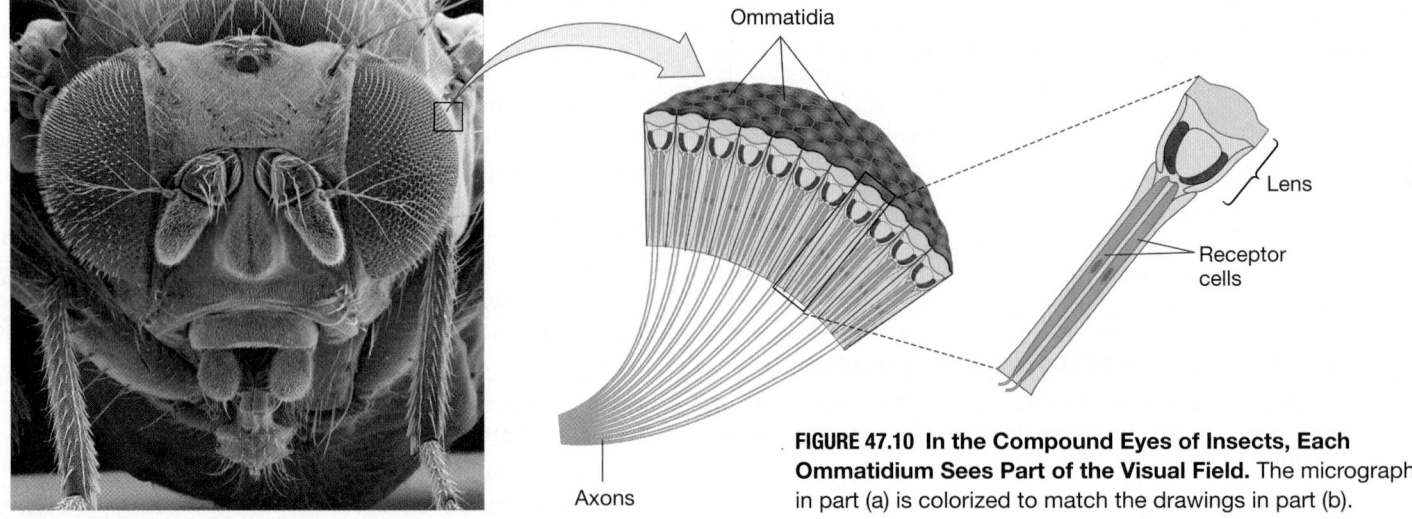

Ommatidia

Lens

Receptor cells

Axons

FIGURE 47.10 In the Compound Eyes of Insects, Each Ommatidium Sees Part of the Visual Field. The micrograph in part (a) is colorized to match the drawings in part (b).

(a) The structure of the vertebrate eye

(b) In the retina, cells are arranged in layers.

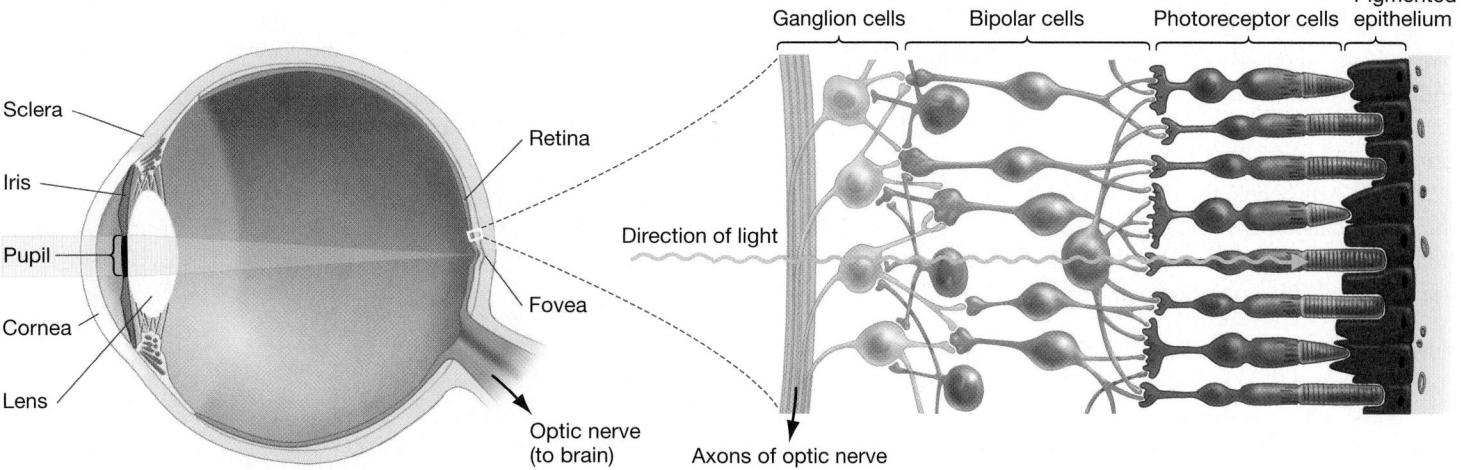

FIGURE 47.11 Simple Eyes Have a Single Lens That Focuses Incoming Light on Receptor Cells. (a) Light passes through the pupil of the eye and is focused onto the retina by the cornea and lens. **(b)** The photoreceptor cells that respond to light are in the "outermost" layer of the retina, furthest from the light source.

(a) Rods and cones contain stacks of membranes.

(b) Rhodopsin is a transmembrane protein complex.

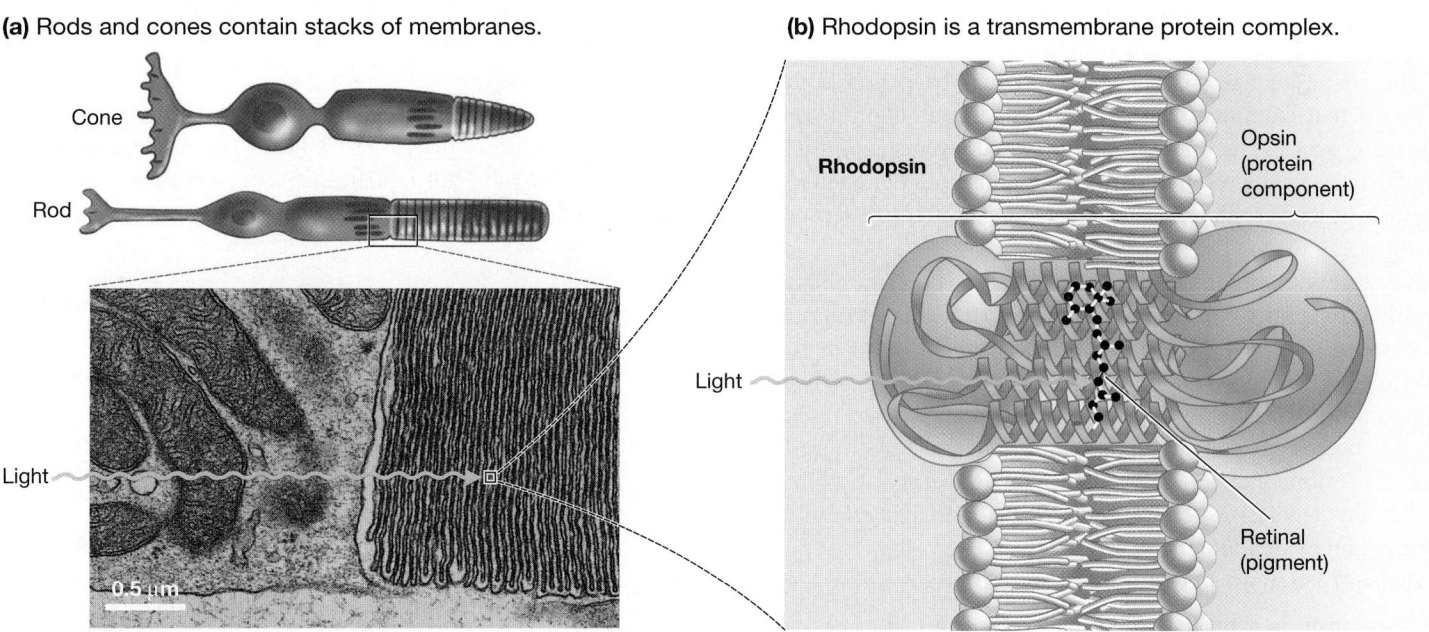

(c) The retinal molecule inside rhodopsin changes shape when retinal absorbs light.

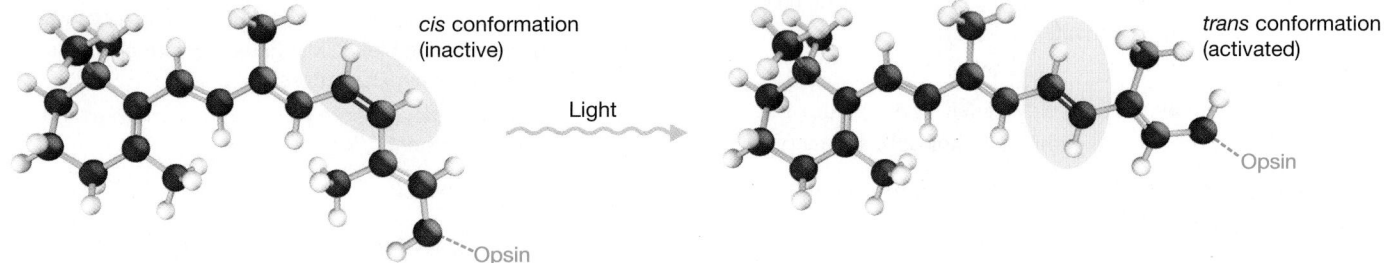

FIGURE 47.12 Rods and Cones Are Packed with Transmembrane Proteins That Contain the Pigment Retinal.
(a) Rods and cones have membranous disks containing thousands of opsin molecules (called rhodopsin in rods).
(b) Each opsin holds one retinal molecule. **(c)** Retinal changes conformation when it absorbs light. In response, opsin also changes shape.

✔**QUESTION** Explain why the change from *cis* to *trans* shown in part (c) would affect opsin's function.

Rods are sensitive to dim light but not to color. Cones, in contrast, are much less sensitive to faint light but are stimulated by different wavelengths (i.e., by colors). These discoveries explained why night vision is largely black and white—at night, the rods do most of the work.

Rods dominate most of the retina, but one small spot in the center of the retina has only cones. This is the **fovea** (see Figure 47.11a). When people focus on an object, their eyes move so that the image falls on the fovea of each eye. Based on these observations, biologists concluded that the high density of cones in the fovea maximizes the resolution of the image.

How Do Rods and Cones Detect Light?

As Figure 47.12a shows, rods and cones have segments that are packed with membrane-rich disks. The membranes contain large quantities of a transmembrane protein called **opsin**. Each opsin molecule is associated with a molecule of the pigment **retinal**. In rod cells, the two-molecule complex is called **rhodopsin** (**FIGURE 47.12b**).

Experiments with isolated retinal, opsin, and rhodopsin molecules confirmed that retinal changes shape when it absorbs light. Specifically, the number-11 carbon in the retinal molecule changes from the *cis* conformation to the *trans* conformation (**FIGURE 47.12c**). Retinal is a light switch.

The shape change that occurs in retinal triggers a series of events that culminate in a different stream of action potentials being sent to the brain. The sequence of events is unusual, though, because the receipt of a light stimulus does not open ion channels or trigger the release of a neurotransmitter to a sensory neuron.

In vertebrates, the molecular basis of vision is a shape change in retinal that closes ion channels and decreases the amount of neurotransmitter being released to the sensory neuron. In rod cells, electrical activity across the membrane, as well as neurotransmitter release, are maximized in the dark. When retinal has not been activated by light, sodium channels in the rod's plasma membrane are open (**FIGURE 47.13a**), and entry of sodium continually depolarizes the rod cells. Exposure to light transmits information by *inhibiting* both processes.

FIGURE 47.13b shows how shutting down happens:

1. Rhodopsin is activated when light causes retinal to change shape from the *cis* to *trans* conformation.

2. Rhodopsin activation causes a membrane-bound molecule called transducin to activate the enzyme phosphodiesterase (PDE).

3. PDE breaks down a nucleotide called cyclic guanosine monophosphate (cGMP) to guanosine monophosphate (GMP).

4. As cGMP levels decline, cGMP-gated sodium channels in the plasma membrane of the rod cell close.

5. When sodium channels close, Na$^+$ entry decreases and the membrane hyperpolarizes.

✔ If you understand this sequence of events, you should be able to explain why cGMP acts as both a second messenger and a ligand in this system, and how transducin compares to G

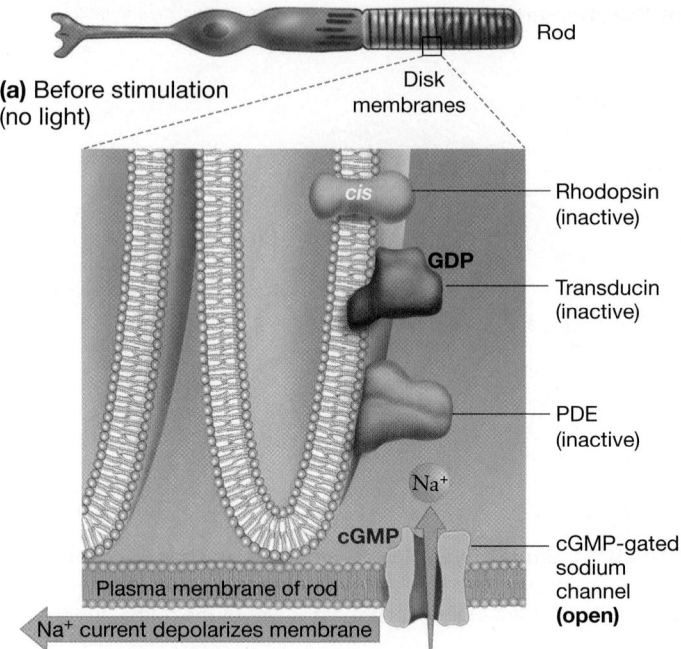

(a) Before stimulation (no light)

Rod

Disk membranes

Rhodopsin (inactive)

GDP

Transducin (inactive)

PDE (inactive)

Na$^+$

cGMP

cGMP-gated sodium channel **(open)**

Plasma membrane of rod

Na$^+$ current depolarizes membrane

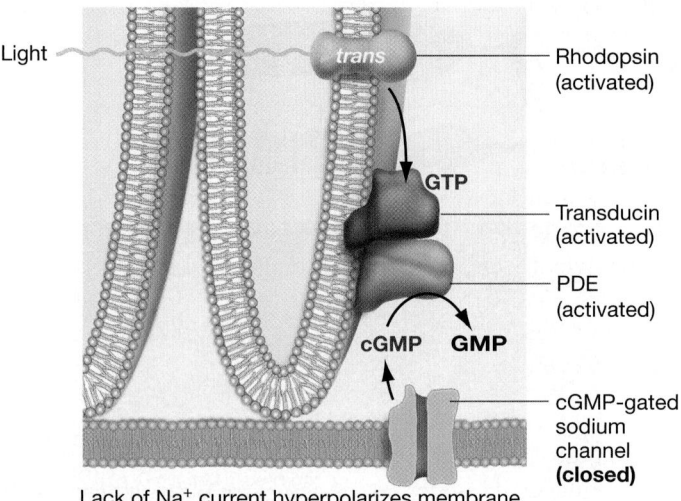

(b) After stimulation (light)

Light

trans

Rhodopsin (activated)

GTP

Transducin (activated)

PDE (activated)

cGMP GMP

cGMP-gated sodium channel **(closed)**

Lack of Na$^+$ current hyperpolarizes membrane

FIGURE 47.13 A Signal-Transduction Pathway Connects Light Absorption to Changes in Membrane Potential. (a) An unstimulated photoreceptor. Notice that sodium ions flow into the cell when light is *not* being received. **(b)** A stimulated photoreceptor. Activation of rhodopsin leads to a reduction in cGMP concentration. With less cGMP available, cGMP-gated sodium channels close, and the membrane hyperpolarizes.

✔**QUESTION** The inflow of sodium ions into a photoreceptor cell is called "the dark current." Why?

proteins (introduced in Chapter 11). You should also be able to explain why shutting Na$^+$ channels results in hyperpolarization.

In response to the ensuing change in membrane potential, smaller quantities of the neurotransmitter glutamate are discharged at the synapse. The decrease in neurotransmitter indicates to the postsynaptic bipolar cell that the rod has absorbed

FIGURE 47.14 People with Red–Green Color Blindness Cannot Distinguish Red from Green. These images show what a person with red–green color blindness would see, compared to a person with normal color vision.

light. As a result, a new pattern of action potentials is sent to the brain, via neurons called ganglion cells. Axons from ganglion cells are bundled into the optic nerve.

This system is exquisitely sensitive: Biologists have recorded a measurable change in the membrane potentials of rod cells in response to a single photon of light. But how do humans and other animals perceive color?

Color Vision: The Puzzle of Dalton's Eye To answer this question, consider the research program initiated by John Dalton[1] in the late eighteenth century. At the age of 26, Dalton realized that he and his brother saw colors differently than other people did. To them, red sealing wax and green laurel leaves appeared to be the same color, and a rainbow exhibited only two hues. Dalton and his brother could not differentiate the colors red and green. This condition is called red–green color blindness (**FIGURE 47.14**).

In a lecture delivered in 1794, Dalton explained his perceptions by hypothesizing that red wavelengths failed to reach his retinas. Further, he hypothesized that because a normal eyeball is filled with clear fluid, and because blue fluids absorb red light, his defective vision resulted from the presence of bluish fluid rather than clear fluid in his eyes.

---

[1]Dalton was an accomplished physicist. He was the first proponent of the atomic theory of matter and formulated Dalton's law on the partial pressures of gases (introduced in Chapter 45). Red–green color blindness is sometimes called daltonism in his honor.

To test this hypothesis, Dalton left instructions that his eyes should be removed after his death and examined to see if the fluid inside was blue. When he died 50 years later, an assistant dutifully removed the eyes from Dalton's corpse and examined them. The fluid inside the eye was not blue at all, however, but slightly yellow—the normal color for an older person. Further, when the back was cut off one eye and colored objects were viewed through the lens, the objects looked perfectly normal. Dalton's hypothesis was incorrect.

Color Vision: Multiple Opsins What, then, caused Dalton's color blindness? The key to answering this question was the discovery that the human retina contains three types of color-sensitive photoreceptors: blue, green, and red cones, named for the colors that they best perceive.

To follow up on this result, biologists analyzed opsin molecules from the three cell types and found that each had a distinct amino acid sequence. The three proteins are now called the blue, green, and red opsins (or S, M, and L, for short, medium, and long wavelengths, respectively). Although retinal is the light-absorbing molecule in all photoreceptor cells, the different opsin molecules cause each type to respond to a different range of wavelengths of light.

Based on these results, biologists hypothesized that the brain distinguishes colors by combining signals initiated by the three classes of opsins. **FIGURE 47.15**, for example, graphs how much light is absorbed across a range of wavelengths by the S, M, and

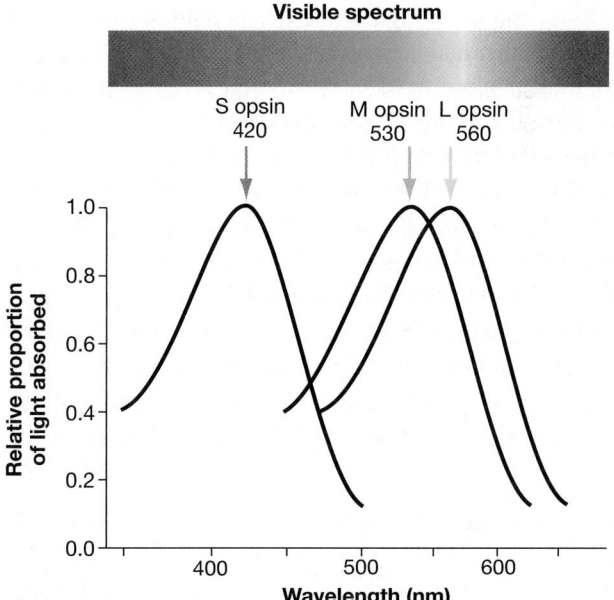

FIGURE 47.15 Color Vision Is Possible because Different Opsins Absorb Different Wavelengths of Light. Each human cone cell contains one of three different types of opsin. Each opsin absorbs a different range of wavelengths.

✔**QUESTION** The retinal molecules in S, M, and L opsins are identical. What is the likely reason that they respond to different wavelengths of light?

L opsins of humans. Notice that a wavelength of 560 nanometers (nm) stimulates L cones strongly, M cones to an intermediate degree, and S cones not at all. In response to the corresponding signals from these cells, the brain perceives the color yellow.

Does this hypothesis explain Dalton's color blindness? According to the data in Figure 47.15, wavelengths from green to red do not stimulate S opsin at all. It is thus unlikely that S opsin is involved in red–green color blindness. Did Dalton fail to distinguish red and green because his M or L cones were defective? Research has shown that red–green color-blind people lack either functional M or L cones, or both. Was the same true of Dalton?

This question was answered in the 1990s, when the genes for the M and L opsins were sequenced. Remarkably, Dalton's eyes had been preserved. Researchers managed to extract DNA from the 150-year-old tissue and analyze his M opsin genes and L opsin genes. They found that Dalton had a normal *L* allele but lacked a functional *M* allele. As a result, he did not have green-sensitive cones. The puzzle of Dalton's color vision was solved.

Red–green color blindness, or the inability to distinguish red and green due to an absence of either M or L cones, is estimated to affect about 5 to 10 percent of men and less than 0.5 percent of women. It is more prevalent in men than in women because it is an X-linked trait (see Chapter 14). In contrast, the inability to distinguish blue from other colors, which would result from absence of the S cone, is an extremely rare, autosomal dominant trait.

Do Other Animals See Color? What about other animals—do they see color the way humans do? The answer is, probably not. Animals that are active at night have relatively few cone cells and many rods, giving them high sensitivity to light but poor color vision. On the other hand, many vertebrate and invertebrate species have four or more types of opsins and probably perceive a world of colors that is much richer than ours.

In general, the types of opsins found in a species correlate with the environment it inhabits and its mode of life. For example,

- A marine fish called the coelacanth (pronounced *SEE-luh-kanth*), which lives in water 200 m deep, has two opsins that respond to the blue region of the spectrum (wavelengths of 478 nm and 485 nm). As a result, coelacanths perceive several distinct hues of blue that we would perceive as a single color. These opsins offered an adaptive advantage to coelacanths because wavelengths in the yellow and red parts of the spectrum do not penetrate well into deep water—only blue light exists in the coelacanth's habitat.

- In humans and other primates that eat fruit, two of the three opsins are sensitive to wavelengths around 550 nm. The presence of these opsins allows individuals to distinguish between the greens, yellows, and reds of unripe and ripe fruits.

- Many animals (for example, some insects and birds) can see ultraviolet light, which has shorter wavelengths than humans can see. This ability can be vital. Certain flowers have ultraviolet patterns that serve as signals for insect pollinators. Also, many birds have strong ultraviolet patterns in their plumage

that are invisible to us but are important criteria used by females of some species for selecting mates.

check your understanding

If you understand that . . .

- In vertebrates, light detection begins when retinal changes shape after absorbing light.
- Stimulation of rhodopsin or the photopsins by light causes conformation changes in retinal, which triggers a series of events resulting in the closing of sodium channels in the membrane of a photoreceptor cell. The resulting change in membrane voltage triggers a change in neurotransmitter release from the receptor cell.

✔ **You should be able to . . .**

1. Explain why retinal can be thought of as a "light switch."

2. Predict the type of vision loss that results from each of the following: detachment of the retina following a car accident, a mutation that knocks out the gene for S opsin, and an age-related clouding of the lens.

Answers are available in Appendix A.

47.4 Chemoreception: Sensing Chemicals

Chemoreception occurs when chemicals bind to chemoreceptors to initiate action potentials in sensory neurons. The sense of taste, called **gustation,** and the sense of smell, called **olfaction,** originate in chemoreceptors. Chemoreceptors detect the presence of particular molecules by undergoing a change in membrane potential when a specific compound is present. In this way, information about the presence of a particular chemical is transduced to an electrical signal in the body.

Until recently, taste and smell were poorly understood in comparison with vision and hearing. It is easy to see why. Eyes and ears respond to relatively simple stimuli—light and sound waves. The tongue and nose, in contrast, respond to thousands of different chemicals. Consequently, taste and smell were difficult to study until techniques became available for identifying how particular molecules bind to certain receptors. But in the past 20 years, research on the chemosenses has exploded.

Taste: Detecting Molecules in the Mouth

The chemoreceptor cells that sense taste are clustered in structures known as **taste buds.** Although humans have taste buds scattered around the mouth and throat, most taste buds are located on the tongue (**FIGURE 47.16**). A taste bud contains about 100 spindle-shaped taste cells, which contain taste receptors and synapse onto sensory neurons.

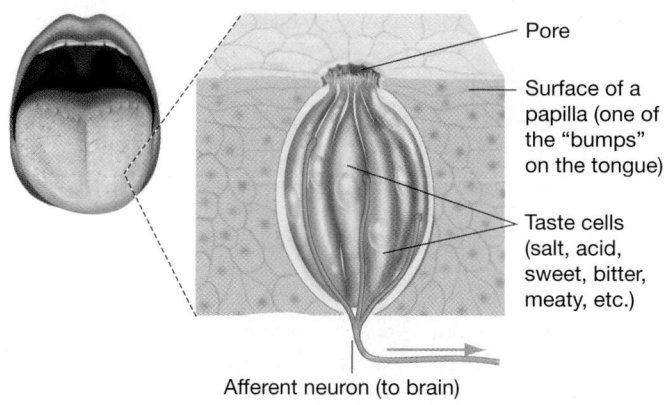

FIGURE 47.16 **Taste Buds Contain Many Types of Chemoreceptors.**

How do these receptors work on a molecular level, and how do they produce the sensation of taste? Early taste research focused on the hypothesis that four "basic tastes" existed: salty, sour, bitter, and sweet.

Salt and Sour   Researchers who analyzed the membrane proteins in taste cells found strong evidence that salt and sour sensations result from the activity of ion channels.

- The sensation of saltiness is due primarily to sodium ions (Na^+) dissolved in food. These ions flow into certain taste cells through open Na^+ channels and depolarize the cells' membranes.

- Sourness is due in part to the presence of protons (H^+), which flow directly into certain taste cells through H^+ channels and depolarize the membrane.

The sour taste of grapefruit and other citrus fruits, for example, results from the release of protons by citric acid. In general, the lower the pH of a food, the more it depolarizes a taste cell's plasma membrane, and the more sour the food tastes.

Compared to salt and sour, the molecular mechanisms responsible for the sensations of bitterness and sweetness have been much more difficult to identify. Researchers have only recently been able to document that certain food molecules actually bind to specific receptors on taste cells to cause bitter and sweet tastes.

Why Do Many Different Foods Taste Bitter?   Bitterness has been difficult to understand because molecules with very different structures, such as orange peel or unsweetened cocoa, are all perceived as bitter. How is this possible?

An answer began to emerge after researchers confirmed that some humans genetically lack the ability to taste certain bitter substances. In 1931, Arthur Fox was synthesizing phenylthiocarbamide (PTC) and accidentally blew some of it into the air. A nearby colleague complained of a bitter taste in his mouth, but Fox could not taste anything. Follow-up research confirmed that the ability to taste PTC is inherited and polymorphic. About 25 percent of Americans cannot sense the molecule.

To find the gene responsible for this trait, biologists compared the distribution of genetic markers observed in "tasters"

and "nontasters." The mapping effort recently narrowed down this gene's location to several candidate chromosomal segments. (Chapter 20 introduced this type of gene hunt.) In one of the regions, researchers found a family of 40 to 80 genes that encode transmembrane receptor proteins.

Follow-up work has documented that each protein in the family binds to a different type of bitter molecule. A taste cell, however, can have many different receptor proteins from this family. As a result, many different molecules can depolarize the same cell and cause the sensation of bitterness.

Why are so many genes devoted to detecting bitterness? Many of the molecules that bind to these receptors are found in toxic plants; most animals react to bitter foods by spitting them out and avoiding them in the future. In essence, bitterness indicates, "This food is dangerous; don't swallow it."

The proliferation of genes responsible for detecting bitterness hints that the trait is extremely adaptive—meaning that individuals with the capacity to detect a wide array of bitter compounds produce more offspring than do individuals that lack this ability.

What Is the Molecular Basis of Sweetness and Other Tastes?   Inspired by progress on bitter receptors, research teams used a similar approach—analyzing mutant mice that could *not* sense sweetness—to understand how the sweet receptor works.

In humans and mice, three closely related membrane receptors are responsible for detecting sweetness as well as glutamate and other amino acids. Glutamate triggers the sensation called **umami,** which is the meaty taste of the molecule monosodium glutamate (MSG). Glutamate is sensed by one particular pair of the three receptors. Sweetness is sensed by a different pair of receptors.

Recent work has solved a long-standing question about sweet sensation—why so many different types of sugars trigger the same sensation. As it turns out, a single receptor has binding sites for multiple types of sweet compounds, meaning that a variety of molecules can stimulate action potentials from the same receptor.

Currently, several teams are studying how the different taste sensations are conveyed to the brain and interpreted. Although taste is beginning to reveal its secrets, the complete story will probably not be known for many years.

Olfaction: Detecting Molecules in the Air

Taste allows animals to assess the quality of their food before swallowing it. Olfaction, in contrast, allows animals to monitor airborne molecules that convey information. Wolves and domestic dogs, for example, can distinguish millions of different airborne molecules at vanishingly small concentrations. The molecules that constitute odor contain information about the movements and activities of prey and other members of their own species.

Odorants Provide Information about the Environment   Airborne molecules that convey information about food or the environment are called **odorants**. When they reach the nose, they diffuse

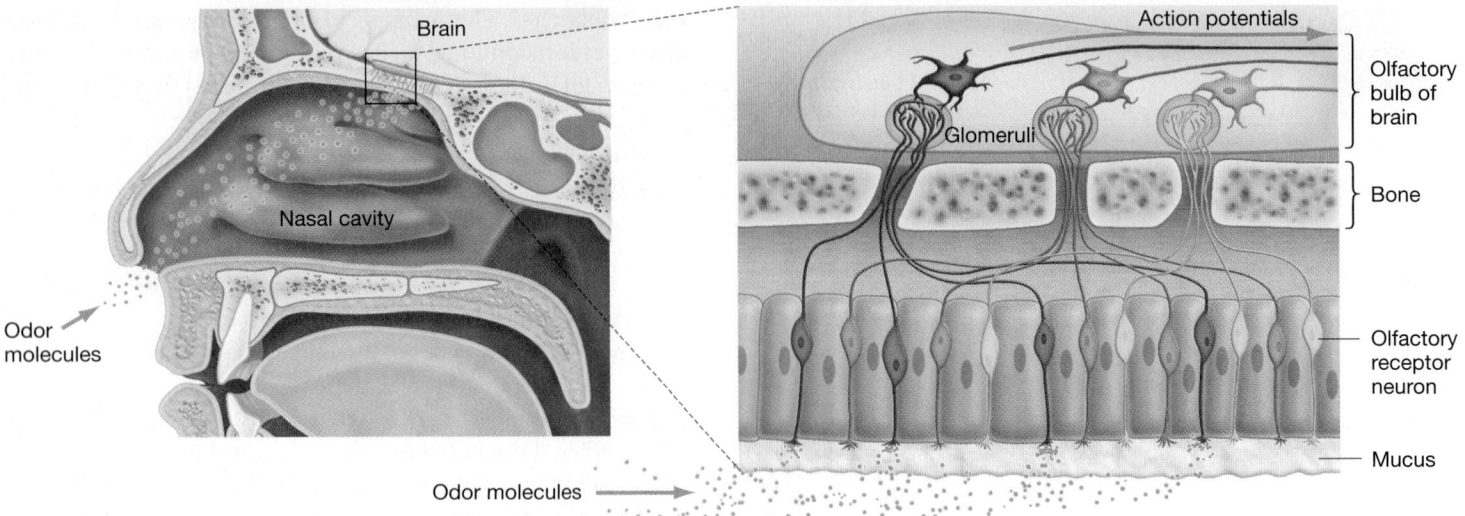

FIGURE 47.17 In Mammals, Chemoreceptor Cells in the Nose Respond to Specific Odorants. Each of the chemosensory neurons in the nose has one type of odor receptor protein on its dendrites. Sensory neurons with the same receptor project to the same glomerulus, or section within the olfactory bulb of the brain.

into a mucus layer in the roof of the nose (**FIGURE 47.17**). There, they activate olfactory neurons by binding to membrane-bound receptor proteins. Axons from these neurons project up to the **olfactory bulb,** the part of the brain where olfactory signals are processed and interpreted.

Understanding the anatomy of the odor-recognition system was a relatively simple task. Understanding how receptor neurons distinguish one molecule from another was much more difficult. Initially, investigators hypothesized that receptors respond to a small set of "basic odors," such as musky, floral, minty, and so on. The idea was that each basic odor would be detected by its own type of receptor, much like the way gustation works.

In 1991, Linda Buck and Richard Axel discovered a gene family in mice that comprises hundreds of distinct coding regions and encodes receptor proteins on the surface of olfactory receptor neurons. Follow-up experiments confirmed that each receptor protein binds to a small set of molecules.

Further work established that most, if not all, vertebrates possess this family of genes. Mammals typically have 800 to 1500 of these genes, meaning mammals could produce at least 800 to 1500 different olfactory receptor proteins. In humans, however, about half of these genes have mutations that probably render them nonfunctional. This observation may explain why the sense of smell is so poor in humans compared with that of most other mammals. But the sense of smell is even worse in whales and dolphins—70% to 80% of their olfactory receptor genes are nonfunctional. These animals rely far more on hearing and vision than on smell to sense their environment.

Buck and Axel's announcement inspired a series of questions. How many different receptors occur in the membrane of each neuron involved in odor reception? How does the brain make sense of the input from so many different receptors?

Recently, scientists determined that each olfactory neuron has only one type of receptor, and neurons with the same type of receptor are linked to distinct regions in the olfactory bulb of the brain. These regions are called **glomeruli** (meaning little balls).

Follow-up work from Axel's lab indicated that each smell recognized by mice is associated with the activation of a different subset of the 2000 glomeruli in the brain. For example, the activation of clumps 130, 256, and 1502 might be perceived as the smell "cinnamon." In essence, then, the sensing of odorants is similar to the visual system's use of three cones to perceive many colors; but odor reception works on a much larger scale.

Interestingly, odorant receptors recently have been identified in locations other than the typical olfactory epithelium—for example, in tissues of the heart and pancreas. Receptors have also been identified in the cell membranes of sperm cells, where they appear to play an important role in guiding the sperm toward the egg. Research on this complex and impressive sense continues at a furious pace.

Pheromones Provide Information about Members of the Same Species Recall from the introduction to this chapter that male silk moths are able to locate female moths from miles away. Males have much larger antennae than females, suggesting that they use these antennae in a sex-specific manner.

These factors led scientists to hypothesize that female moths release a chemical into the environment that binds to chemoreceptors on the males' antennae and acts as an attractant. In 1959, scientists identified this chemical and named it bombykol, after the scientific name of the silk moth (*Bombyx mori*). Bombykol was the first chemically characterized **pheromone,** meaning a chemical that is secreted into the environment and that affects the behavior or physiology of animals of the same species.

Thousands of pheromones have since been identified in invertebrates and vertebrates alike, performing such roles as alerting other members of a beehive to an intruder or signaling a male rodent that a female is ovulating. In insects, pheromones typically bind to receptors on the antennae. In tetrapod vertebrates, pheromone receptors are often localized in the **vomeronasal organ,** a sensory organ in the nasal region but distinct from the olfactory bulb. The vomeronasal organ and the olfactory bulb send signals to different parts of the brain, although some animals also sense odorants with the vomeronasal organ. For example, a snake may follow a scent trail of prey using its vomeronasal organ.

Do humans release pheromones? This is a hotly debated question. In 1971, Martha McClintock demonstrated that the menstrual cycles of women living in close contact with one another become synchronized as a result of a secretion from the women's armpits, but the responsible chemical has yet to be identified. Other studies have suggested that secretions from men's armpits can alter hormone levels in women, but both the amount of the chemical secreted and the reaction to it vary dramatically among men and women. Until scientists more fully understand the mechanisms by which these human "pheromones" achieve their effects, the jury is out.

check your understanding

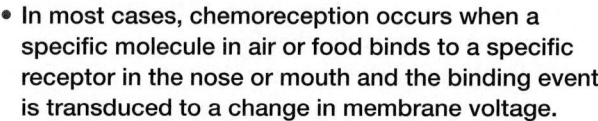

If you understand that . . .

- In most cases, chemoreception occurs when a specific molecule in air or food binds to a specific receptor in the nose or mouth and the binding event is transduced to a change in membrane voltage.
- Odorants bind to chemoreceptors to initiate action potentials to the brain, where action potentials from specific receptors are interpreted as particular smells.
- Pheromones bind to chemoreceptors that send action potentials to brain regions that then affect the behavior or physiology of the animal.

✔ **You should be able to . . .**

1. Discuss why a loss in chemosensory ability occurs when you burn your tongue with extremely hot food.
2. Develop a hypothesis to explain why the vomeronasal organ appears to be vestigial in most primates.

Answers are available in Appendix A.

47.5 Other Sensory Systems

The stimuli and senses discussed so far are the ones you are likely most familiar with. But animals can sense much more than pressure waves, light, and chemicals. All animals can sense temperature and painful stimuli, and some can even perceive electric or magnetic fields. Let's start with temperature and examine each of these other stimuli and senses in turn.

Thermoreception: Sensing Temperature

Recall that many animals thermoregulate to maintain body temperature within an acceptable range (see Chapter 42). Virtually every physiological process, from digestion to metabolic rate, is temperature-dependent, so the ability to sense temperature changes in the environment and respond accordingly is crucial.

Thermoreception Helps Animals Thermoregulate Animals detect heat energy by **thermoreception** and adjust their behaviors or physiological processes, such as shivering and sweating, in response.

Some thermoreceptors are located in the central nervous system. In mammals, the hypothalamus is the brain region that senses departures from homeostatic body temperature and sends signals to effectors to restore homeostasis (see Figure 42.16).

Thermoreceptors also are commonly found on skin and other outer surfaces of animals, so that changes in the temperature of the environment can be sensed. As an example, several types of thermoreceptors have been identified in mammals. Some receptors depolarize in response to cooling, and others depolarize in response to heating. Picking up a cold object stimulates "cold receptors" in your skin, resulting in an increase in the rate of action potentials in sensory neurons that inform your brain that the object is cold.

Interestingly, extreme temperatures are sensed by a different type of receptor, called a nociceptor, that also senses other painful stimuli such as those produced by certain chemicals, excessive pressure, and tissue damage. If you touch a hot stove burner, the pain you feel arises primarily via stimulation of nociceptors.

Pit Vipers Have Extremely Sensitive Thermoreceptors The pit vipers are a group of snakes named after the two temperature-sensitive pits just beneath their nostrils (**FIGURE 47.18a**, see page 968). Inside each pit is a membrane lined with exquisitely sensitive thermoreceptors—a rattlesnake's thermoreceptors can sense changes in temperature as little as 0.003°C.

Pit vipers use these thermoreceptors to sense the heat energy given off by prey, predators, and even possible burrows in which to hide. The brains of these snakes may combine visual and thermal stimuli into a "thermal image" that might look something like an image from an infrared camera (**FIGURE 47.18b**). Even in complete darkness, rattlesnakes can strike prey with deadly precision.

✔ If you understand this, you should be able to predict whether a rattlesnake could strike effectively at its prey with (1) its eyes covered but pits exposed, (2) its eyes and pits covered with cotton cloth, and (3) its eyes and pits covered with an opaque heat-blocking material.

Electroreception: Sensing Electric Fields

All animals give off weak electrical impulses that arise from the activity of their nerves and muscles. Since water is a good conductor of these electrical impulses, many kinds of fishes use **electroreception,** or sensation of electric fields, to locate prey, detect predators, and navigate.

Sharks Use Electroreception to Hunt and Navigate In sharks, some parts of the lateral line system are specialized to detect electrical fields rather than pressure. Tiny pores scattered across a shark's head contain structures called **ampullae of Lorenzini**

(a) Pit vipers have temperature-sensitive pits.

(b) Warm animals emit infrared radiation.

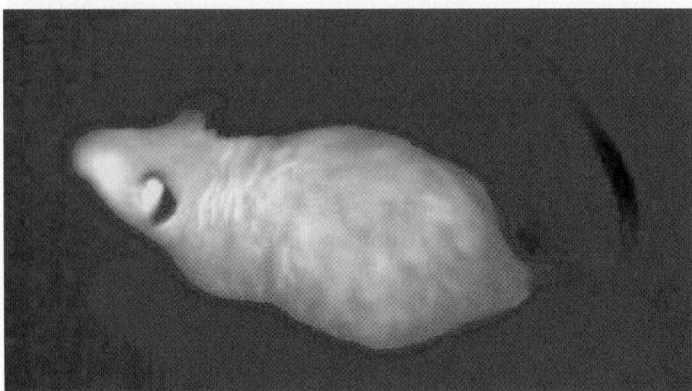

FIGURE 47.18 Pit Vipers Use Thermoreception to Detect Prey. (a) Pits are lined with extremely sensitive thermoreceptors. **(b)** Pit vipers can detect infrared radiation given off by rodents and other prey.

(**FIGURE 47.19**). These ampullae are lined with hair cells that depolarize in response to a change in electrical polarity and send signals to the shark's brain via sensory neurons.

The sensitivity of these ampullae is remarkable—sharks can detect electrical currents as weak as a nanovolt (a billionth of a volt). This ability allows sharks to detect prey that are far away or even buried in the sand at the ocean floor. Combined with their exquisite sense of olfaction—sharks can detect a single drop of blood in a million drops of ocean water—electroreception makes them finely tuned predators.

Although data are scarce, scientists have some evidence to suggest that sharks also use electroreception to navigate. Ocean currents moving through the Earth's magnetic field generate weak electrical currents. Sharks have been observed in the ocean orienting themselves to these fields as well as to artificially created fields in the laboratory.

Electrogenic Fishes Generate Electric Fields Electrogenic fishes have specialized organs near their tails that generate

FIGURE 47.19 Ampullae of Lorenzini on Sharks' Heads Detect Electrical Fields to Help Locate Prey and Navigate.

electric fields stronger than those of regular nerves or muscles. The currents generated by the electric organ move in an arc through the water (**FIGURE 47.20**). Any item located within that arc will disrupt the currents, allowing the fish's electroreceptors to detect it. In this way, electrogenic fishes use their electric organs to locate prey, sense predators, navigate through murky water, and even communicate with other fishes.

Some electrogenic fishes have the ability to produce extremely strong currents that stun or kill their prey. The electric organs of electric eels take up over 80 percent of their body mass and can generate a 500-volt change in electrical potential and 1 ampere of current in the water around them. This amount of current is enough to kill a person swimming in water with an electric eel.

Magnetoreception: Sensing Magnetic Fields

The Earth produces a magnetic field as it rotates on its axis. Just as a compass uses this magnetic field to determine direction, animals may home in on magnetic fields while navigating. **Magnetoreception** has been described in many groups of organisms, including bacteria, fungi, invertebrates, and all vertebrate classes.

In general, studies of the mechanisms by which animals sense the Earth's magnetic fields are in their infancy. However, scientists are confident that at least several distinct mechanisms have evolved. Recall that sharks can sense magnetic fields indirectly via electric fields produced by ocean currents. Terrestrial organisms, in contrast, have the ability to sense magnetic fields directly.

In 1968, German scientists noticed that European robins being kept in the laboratory with no visual cues to the outside began to sit at one end of their cages at the beginning of the migratory season. This behavior led the scientists to hypothesize that birds use magnetoreception to determine direction as they migrate.

Since then, support for this hypothesis has accumulated. Application of artificial magnetic fields changes the robins' orientation. For example, when placed into circular chambers with artificial magnetic fields, birds always orient themselves in relation to the artificial field. Furthermore, disruption of the

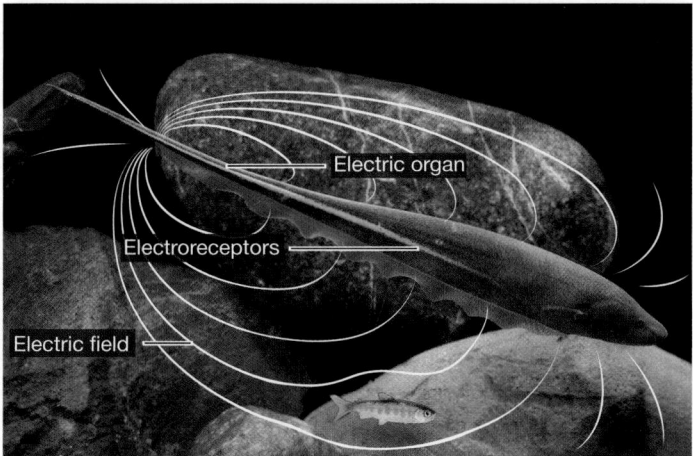

FIGURE 47.20 **Electrogenic Fishes Can Create Strong Electric Fields.** The fish emits an electric current with its electric organ. When a prey object disrupts the electric field, the fish detects its presence using electroreceptors.

magnetic field prevents birds from navigating properly. When scientists fitted homing pigeons with little caps that reversed the polarity of the magnetic field, the pigeons flew in the direction opposite to the one they were trained to fly in.

What enables birds to sense magnetic fields? It is likely a combination of factors. One hypothesis is that deposits of iron inside sensory neurons in the beak play a role in stimulating a response to changes in the magnetic field. In support of this hypothesis, cutting these neurons prevents birds from responding to artificial changes in the magnetic field in the lab. To complicate matters, however, magnetoreception in birds is apparently also dependent on vision. Covering the right eye—but not the left eye—of migrating birds interferes with their ability to navigate using magnetic cues.

This complication highlights a key feature of sensory perception: Animals do not use individual senses in isolation, but rather combine sensations of many types when locating prey, evading predation, or communicating with other animals. Whether it is a homing pigeon using magnetoreception and photoreception to find its way home, or a rattlesnake using a visual and thermal image to strike at a mouse, all animals depend on sensory systems that work together to provide the animal with the information it requires to survive.

check your understanding

 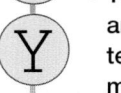

If you understand that . . .

- Thermoreceptors, nociceptors, electroreceptors, and magnetoreceptors convey information about temperature, painful stimuli, electric fields, and magnetic fields, respectively.
- Animals combine input from multiple senses to locate prey, escape from predators, and communicate with other animals.

✔ **You should be able to . . .**

1. Identify one or more habitat types in which electrogenic fishes may have a survival advantage.

2. Propose a mechanism by which a female sea turtle migrates to lay eggs on the beach she was born on decades earlier.

Answers are available in Appendix A.

CHAPTER 47   REVIEW

For media, go to MasteringBiology

If you understand . . .

47.1   How Do Sensory Organs Convey Information to the Brain?

- If the membrane potential of a sensory cell is altered substantially enough in response to a stimulus, the pattern of action potentials that it sends to the brain changes. In this way, sensory stimuli as different as sound and light are transduced to electrical signals.

- The brain is able to distinguish different types of stimuli because axons from different types of sensory neurons project to different regions of the brain.

✔ You should be able to suggest a hypothesis to explain why people who have had limbs amputated experience "phantom pain"—the perception that their missing tissue hurts.

47.2   Mechanoreception: Sensing Pressure Changes

- Pressure receptors detect direct physical stimulation, including stimulation from sound.

- Hair cells, the major sensory detectors in the vertebrate ear, undergo a change in membrane potential in response to bending of their stereocilia.

- Sound waves of a certain frequency cause a certain part of the cochlea's basilar membrane to vibrate. Hair cells at this location stimulate action potentials in sensory neurons in response to the vibration.

- Hair cells in the lateral line systems of fishes and aquatic amphibians are stimulated by pressure changes in the water.

✔ You should be able to identify the key feature common to gravity sensing, hearing, and the lateral line system.

47.3 Photoreception: Sensing Light

- In the vertebrate eye, photoreceptors are located in rods and cones. Although these two cell types differ in structure and function, both contain molecules that consist of retinal paired with opsin (called rhodopsin in rod cells).

- The rhodopsin found in rods is stimulated by even the faintest light.

- Color vision is possible because cones contain opsins that respond to specific wavelengths of light absorbed by retinal.

- Humans distinguish colors based on the pattern of stimulation of three types of opsins found in cones. People who lack one of the functional cone opsins are color blind, meaning they cannot distinguish as many colors as people with all three opsins can.

✔ You should be able to explain how different animal species are able to see different colors.

47.4 Chemoreception: Sensing Chemicals

- Chemoreceptors detect the presence of certain foodborne or airborne molecules.

- Taste buds contain taste cells with membrane proteins that respond to toxins, salt, acid, and other types of molecules in food. Sodium ions and protons enter taste cells via channels and depolarize the membrane directly, producing the sensations of saltiness and acidity. Sugars and some toxic compounds bind to membrane receptors and trigger action potentials that are interpreted by the brain as sweet and bitter flavors, respectively.

- Smell, or olfaction, is used to scan molecules from the outside environment. Airborne chemicals are detected by hundreds of different odor-receptor proteins located in the membranes of receptor cells in the nose.

✔ You should be able to explain why people with nasal congestion complain that food tastes bland, and why the brain can perceive so many different flavors based on inputs from just four or five basic types of taste receptors.

47.5 Other Sensory Systems

- Thermoreceptors respond to changes in temperature.

- Nociceptors respond to painful stimuli, including extreme temperatures, certain chemicals, excessive pressure, and tissue damage.

- Electroreceptors contain modified hair cells that respond to electric fields.

- Magnetoreceptors respond to magnetic fields and are often used in navigation and orientation.

✔ You should be able to explain why the chemical capsaicin, found in chili peppers, feels hot on the tongue.

(MB) **MasteringBiology**

1. **MasteringBiology Assignments**

   Tutorials and Activities Structure and Function of the Eye; Vertebrate Eye

   Questions Reading Quizzes, Blue-Thread Questions, Test Bank

2. **eText** Read your book online, search, take notes, highlight text, and more.

3. **The Study Area** Practice Test, Cumulative Test, BioFlix® 3-D Animations, Videos, Activities, Audio Glossary, Word Study Tools, Art

You should be able to . . .

✔ TEST YOUR KNOWLEDGE

Answers are available in Appendix A

1. What is echolocation?
   a. use of echoes from high-frequency vocalizations to detect objects
   b. use of extremely low-frequency vocalizations to communicate over long distances
   c. vision based on input from many independent lenses, functioning like pixels on a computer screen
   d. variation in the structure of opsin proteins, which allows animals to see different colors

2. In the human ear, how do different hair cells respond to different frequencies of sound?
   a. Waves of pressure move through the fluid in the cochlea.
   b. Hair cells are "sandwiched" between membranes.
   c. Receptors in the stereocilia of each hair cell are different; each receptor protein responds to a certain range of frequencies.
   d. Because the basilar membrane varies in stiffness, it vibrates in certain places in response to certain frequencies.

3. Which of the following comparisons of rods and cones is *false*?
   a. Most human eyes have one type of rod and three types of cones.
   b. Rods are more sensitive to dim light than cones are.
   c. Nocturnal animals have fewer rods than diurnal animals.
   d. Both rods and cones use retinal and opsins to detect light.

4. Which of the following statements about taste is *true*?
   a. Sweetness is a measure of the concentration of hydrogen ions in food.
   b. Sodium ions from foods can directly depolarize certain taste cells.
   c. All bitter-tasting compounds have a similar chemical structure.
   d. Membrane receptors are involved in detecting acids.

5. True or false: A rattlesnake's facial pits detect the presence of prey using photoreception.

6. What type of sensory system do migrating birds use to detect direction?

7. Considering that sounds and odors both trigger changes in the patterns of action potentials from sensory cells, how does the brain perceive which sense is which when the action potentials reach the brain?
   a. The action potentials stimulated by sounds are different in size and shape from those stimulated by an odor.
   b. The axons from different sensory neurons go to different areas of the brain.
   c. Mechanoreception is not consciously perceived by the brain, whereas chemoreception is.
   d. Chemoreception is not consciously perceived by the brain, whereas mechanoreception is.

8. Give three examples of how the sensory abilities of an animal correlate with its habitat or method of finding food and mates.

9. **QUANTITATIVE** Scientists collected data on the date of onset of the menstrual cycles of a group of women who moved into a college dormitory together in the fall. The *y*-axis of the graph shows the average difference (in days) between the onset of a woman's cycle and the average onset date of the rest of the women. Describe how these data provide evidence for the existence of a human pheromone.

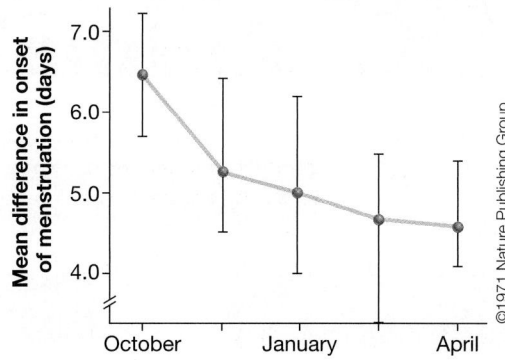

DATA: McClintock, M. K. 1971. *Nature* 229: 244–245, Figure 1.

10. Explain how the bending of stereocilia in a hair cell and binding by a bitter-tasting molecule can both result in an ion channel opening.

11. Scientists generally think that a "good hypothesis" is one that is reasonable and testable and inspires further research into the phenomenon. Using these criteria, was Dalton's hypothesis about color vision a good hypothesis? Was it correct? Explain your answer.

12. Design experiments to test the hypothesis that electric eels are both electrogenic and electroreceptive.

13. Which of the following animals is *least* likely to have a well-developed vomeronasal organ?
    a. fish
    b. salamander
    c. rodent
    d. snake

14. Compare and contrast the lateral line system of fishes with electroreception in sharks.

15. A company is marketing perfumes that it claims are attractive to the opposite sex because they contain "pure reaction-grade human pheromones." Explain why this is unlikely to be truth in advertising.

16. Houseflies have about 800 ommatidia in each of their compound eyes. Dragonflies, in contrast, have up to 10,000 ommatidia per eye. Houseflies feed by lapping up watery material from piles of excrement or rotting carcasses, which they locate by scent. Dragonflies are aerial predators and hunt by sight. How would you test the hypothesis that the large numbers of ommatidia in dragonflies make them more efficient hunters?

48 Animal Movement

In this chapter you will learn that

Muscle movement was a key innovation in the evolution of animals

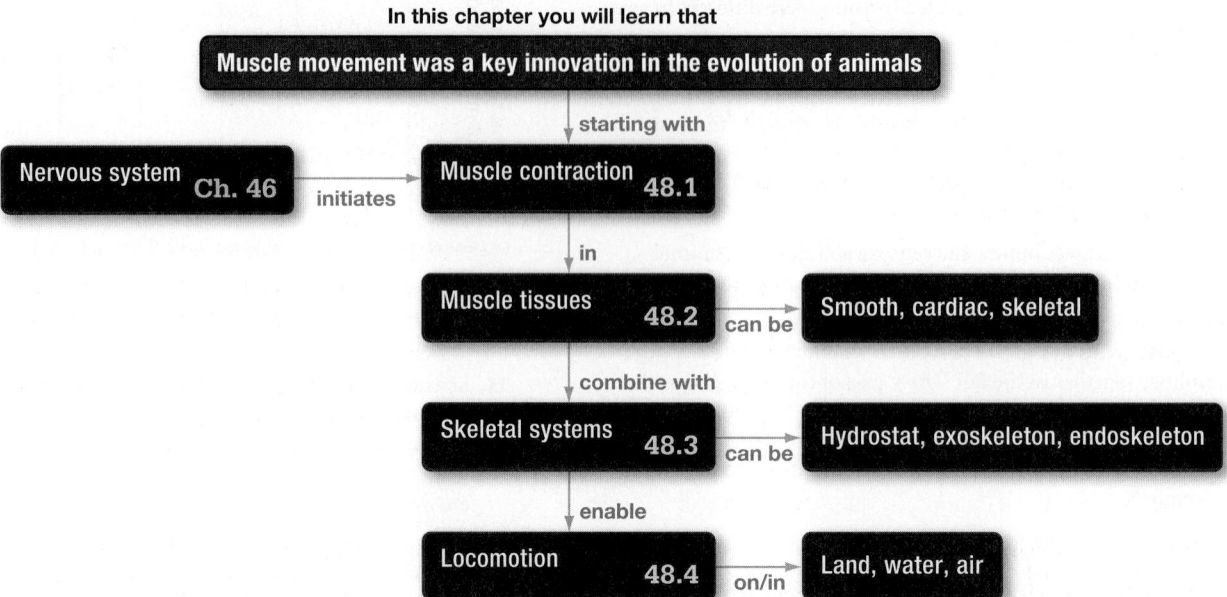

starting with

| Nervous system Ch. 46 | —initiates→ | Muscle contraction 48.1 |

↓ in

| Muscle tissues 48.2 | —can be→ | Smooth, cardiac, skeletal |

↓ combine with

| Skeletal systems 48.3 | —can be→ | Hydrostat, exoskeleton, endoskeleton |

↓ enable

| Locomotion 48.4 | —on/in→ | Land, water, air |

Basilisk lizards are able to run on water—literally. This impressive escape strategy demonstrates the extent to which muscle-generated movements have diversified among animals. For most animals, complex muscle movements make the difference between life and death.

This chapter is part of the Big Picture. See how on pages 840–841.

You may have discovered while studying biology that plants and animals are more similar to each other than they initially appear. For example, plants and animals both require water and nutrients in specific quantities, have highly specialized tissues and complex reproductive structures, and launch defenses against parasites and predators. You can see a comparison of these and other traits in the Big Picture of Plant and Animal Form and Function on pages 840–841.

However, animals possess a quality that clearly distinguishes them from plants and other organisms: movement by virtue of muscle contractions. Muscle-generated movements were a key innovation in animal evolution. Rapid movements, along with sophisticated sensory structures

✔ When you see this checkmark, stop and test yourself. Answers are available in Appendix A.

(Chapter 47) and complex information processing systems (Chapter 46), were vital to animal diversification (Chapter 33), and these attributes made animals efficient eating machines in diverse ecosystems.

Muscles generate movement by exerting force and causing shape changes. Movement falls into two general categories:

1. *Movement of the entire organism relative to its environment.* **Locomotion,** movement of an animal under its own power, enables animals to seek food, water, mates, and shelter as well as avoid predators. Modes of locomotion include undulating, jetting, swimming, walking, running, jumping, gliding, and flying (introduced in Chapter 33).

2. *Movement of one part of the animal relative to other parts (not involved in locomotion).* This type of movement also has important functions—for example, to ventilate gills for gas exchange and to grasp prey. Even sessile organisms like sea anemones and barnacles use complex, muscle-generated movements to survive and reproduce.

How do animals accomplish their spectacular movements? This chapter starts by probing into the mechanism of muscle contraction that serves as the "engine" for most animal movements and then considers how muscle and skeletal systems work together to produce locomotion. The latter discussion introduces a research field called **biomechanics,** in which the principles of physics and engineering are applied to questions about the mechanical structure and function of organisms. Let's jump in.

48.1 How Do Muscles Contract?

The mechanism responsible for the contraction of muscle has fascinated and perplexed scientists for many centuries. Before the advent of microscopes and modern research techniques, scientists could only speculate about what makes muscles contract and relax.

Early Muscle Experiments

In the second century C.E., the Roman physician and philosopher Galen proposed that spirits flowed from nerves into muscles, inflating them and increasing their diameter. This "inflation" hypothesis persisted into the seventeenth century, when the French philosopher René Descartes suggested that nerves carry fluid from the pineal gland—which is in the brain and was considered to be the seat of the soul—to the muscles, making them shorten and swell.

Later that century, Dutch anatomist Jan Swammerdam tested Descartes' inflation hypothesis with a simple yet elegant experiment. He placed a piece of frog muscle into an airtight syringe with the nerve protruding through a small hole in the side and a small drop of water in the tip. He then stimulated the nerve, causing the muscle to contract. If the muscle's volume changed during contraction, the drop of water in the tip of the syringe would move. But it did not. The volume of the muscle remained constant.

Swammerdam's experiment demonstrated an important point: The contraction mechanism is inherent to the muscle itself—muscle is not like a balloon, and the nerve is not like a water hose filling a balloon. This insight was confirmed by Italian scientist Luigi Galvani in the 1790s, when he severed the nerve to a frog's leg muscle and then connected the two sides of the cut with a metal conductor. Contraction occurred. He concluded that the nerve and muscle possess "animal electricity" that can induce contraction.

If the shape of muscle does not change by inflation, what is the mechanism of muscle contraction?

The Sliding-Filament Model

Early microscopists established that the muscle tissue in vertebrate limbs and hearts is composed of slender fibers. A **muscle fiber** is a long, thin muscle cell. Within each of the muscle cells are many threadlike, contractile filaments called **myofibrils.** The myofibrils inside muscle fibers often look striped or striated due to the alternating light–dark units called **sarcomeres,** which repeat along the length of a myofibril (**FIGURE 48.1**).

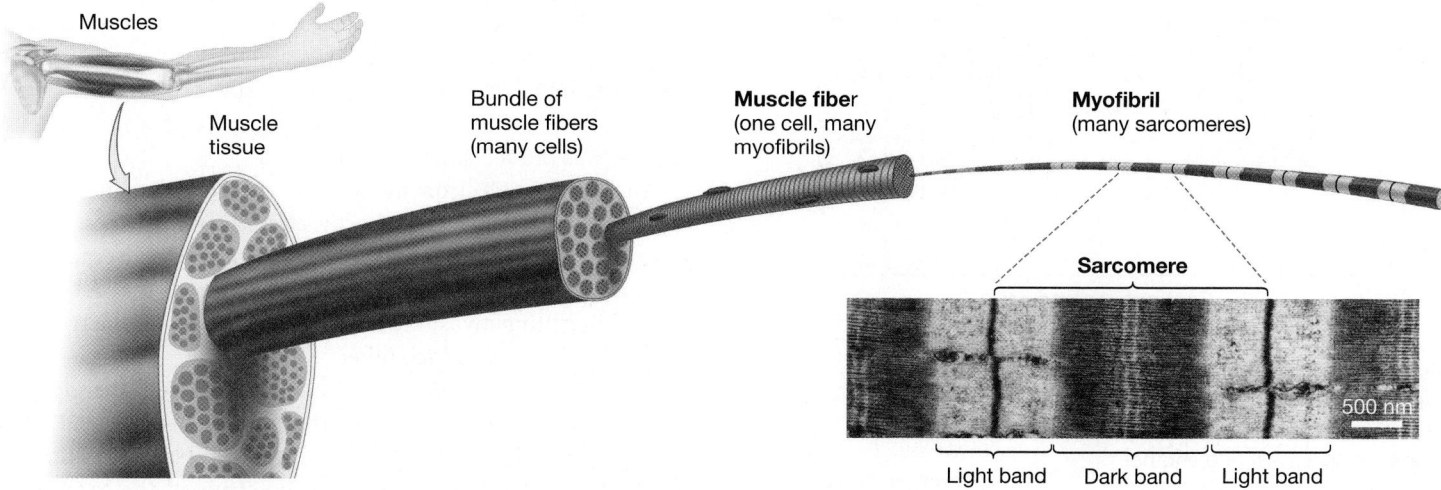

FIGURE 48.1 Muscle Cells Contain Many Myofibrils, Which Contain Many Sarcomeres. Skeletal muscle cells (fibers) have a striped appearance due to repeating sarcomeres, which are units of alternating light–dark bands.

The microscopists observed that sarcomeres shorten as myofibrils contract. Sarcomeres then lengthen when the cell relaxes and an external force stretches the muscle. Based on these observations, it became clear that the question of how muscles contract simplifies to the question of how sarcomeres shorten.

Biologists knew that the answer must involve the two types of protein that had been found in sarcomeres: **actin** and **myosin**. But they did not know the shapes of these molecules or how they were arranged within the sarcomere. Did both types of molecules span the entire length of the sarcomere? Or were the filaments restricted to certain bands within the sarcomere?

In 1952, biologist Hugh Huxley produced electron micrographs of sarcomeres in cross section. He observed that there were two types of filaments, **thin filaments** and **thick filaments**, and that these filaments overlapped in the dark bands but not in the light bands. Huxley and his collaborator Jean Hanson also observed that sarcomeres stripped of their myosin had no dark bands. They concluded that the thick filaments must be composed of myosin, and the thin filaments must be composed of actin.

How did myosin and actin interact to shorten the sarcomere? In 1954, Huxley and Hanson hit on the key insight when they observed how the light and dark bands in sarcomeres changed when the muscles were relaxed versus contracted. Overall, the width of the dark bands did not change during a contraction, but the light bands became narrower.

To explain these observations, Huxley and Hanson proposed the **sliding-filament model** illustrated in **FIGURE 48.2**. The hypothesis was that the filaments slide past one another during a contraction. That is, the sarcomere can shorten with no change in lengths of the thin and thick filaments themselves:

- The distance from point A to point C does not change, and the distance from point B to point D does not change.

- Points A and B move closer to each other during contraction, as do points C and D.

Another pair of researchers, Andrew F. Huxley (no relation to Hugh) and Rolf Niedergerke, published the same result at the same time. Follow-up research has shown that the Huxley–Hanson model is correct in almost every detail.

Structurally, thin filaments are composed of two coiled chains of actin, a common component of the cytoskeleton of eukaryotic cells (see Chapter 7). One end of each thin filament is anchored to a structure called the **Z disc,** which forms the end wall of the sarcomere. The other end of a thin filament is free to interact with thick filaments. Thick filaments are composed of multiple strands of myosin. They span the center of the sarcomere and are free at both ends to interact with thin filaments.

To appreciate how the sliding-filament model works, consider the following analogy: Two large trucks are parked 50 m apart, facing each other. Each has a long rope attached to the front bumper. Six burly weightlifters stand in a line in front of each truck, grab onto the rope, and pull, hand over hand, so that the two trucks roll toward one another. ✔ If you understand the model, you should be able to explain which elements in the analogy represent the Z discs, which are the thin filaments, and which elements are the thick filaments.

How Do Actin and Myosin Interact?

How does this sliding action occur at the molecular level? Early work on the three-dimensional structure of myosin revealed that each myosin molecule is made up of a pair of subunits with "tails" coiled around one another and "heads" bent to the side. Each myosin head can bind to actin, and the head region can catalyze the hydrolysis of ATP into adenosine diphosphate (ADP) and a phosphate ion. These results suggested that myosin—not actin—was the site of active movement.

In addition, electron microscopy revealed that myosin and actin are locked together shortly after an animal dies, and its muscles enter the stiff state known as rigor mortis. Because ATP is unavailable in dead tissue, the data suggested that ATP is involved in getting myosin to release from actin once the two molecules have bound to each other.

Later, Ivan Rayment and colleagues solved the detailed three-dimensional structure of the myosin head (**FIGURE 48.3**). Using X-ray crystallographic techniques (see **BioSkills 11** in Appendix B), Rayment's group determined the location of the actin-binding site and examined how myosin's structure

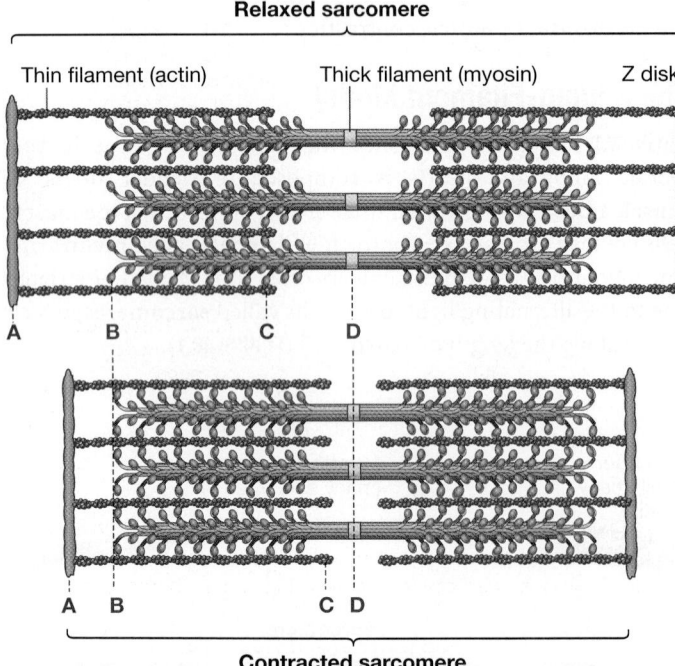

Relaxed sarcomere

Thin filament (actin)   Thick filament (myosin)   Z disk

A   B   C   D

A B   C D

Contracted sarcomere

FIGURE 48.2 The Sliding-Filament Model Explains Important Aspects of Sarcomere Contraction. When a sarcomere contracts, the lengths of the thin filaments (distance from A to C) and thick filaments (distance from B to D) do not change. Rather, the filaments slide past one another.

✔ **QUESTION** According to the model shown here, why is the dark band in a sarcomere (see Figure 48.1) dark and the light band light?

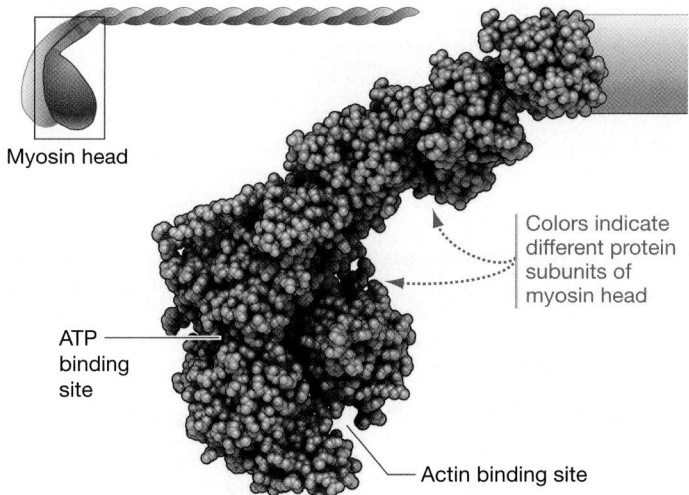

Myosin head

Colors indicate different protein subunits of myosin head

ATP binding site

Actin binding site

FIGURE 48.3 Myosin's "Head" Binds ATP and Actin. Each myosin molecule consists of two subunits with their tails coiled together and their two heads exposed. The heads contain binding sites for ATP and actin; when one of these subunits binds, its myosin head changes shape.

changed when ATP or ADP bound to it. As predicted, the protein's conformation changed significantly when bound to ATP versus ADP.

Based on these data, Rayment and co-workers proposed a four-step model for actin–myosin interaction (**FIGURE 48.4**):

Step 1 ATP binds to the myosin head, causing a conformational change that releases the head from the actin in the thin filament.

Step 2 When ATP is hydrolyzed to ADP and inorganic phosphate, the neck of the myosin straightens and the head pivots. The myosin head then binds to a new actin subunit farther down the thin filament. In this position, the myosin head is "cocked" in its high-energy state, ready for the power stroke.

Step 3 When inorganic phosphate is released, the neck bends back to its original position. This bending, called the power stroke, moves the entire thin filament relative to the thick filament.

Step 4 After ADP is released, the myosin head is ready to bind to another molecule of ATP.

As ATP binding, hydrolysis, and release continue, the two ends of the sarcomere are pulled closer together. (The transition from step 4 to step 1 cannot occur after death. Rigor mortis sets in as ATP supplies run out.)

The same basic ratcheting mechanism between actin and myosin is responsible for the amoeboid movement observed in amoebae and slime molds (see Chapter 30) as well as the streaming of cytoplasm observed in algae and land plants. Actin and myosin have played a critical role in the diversification of eukaryotes because they make movement possible in the absence of cilia and flagella.

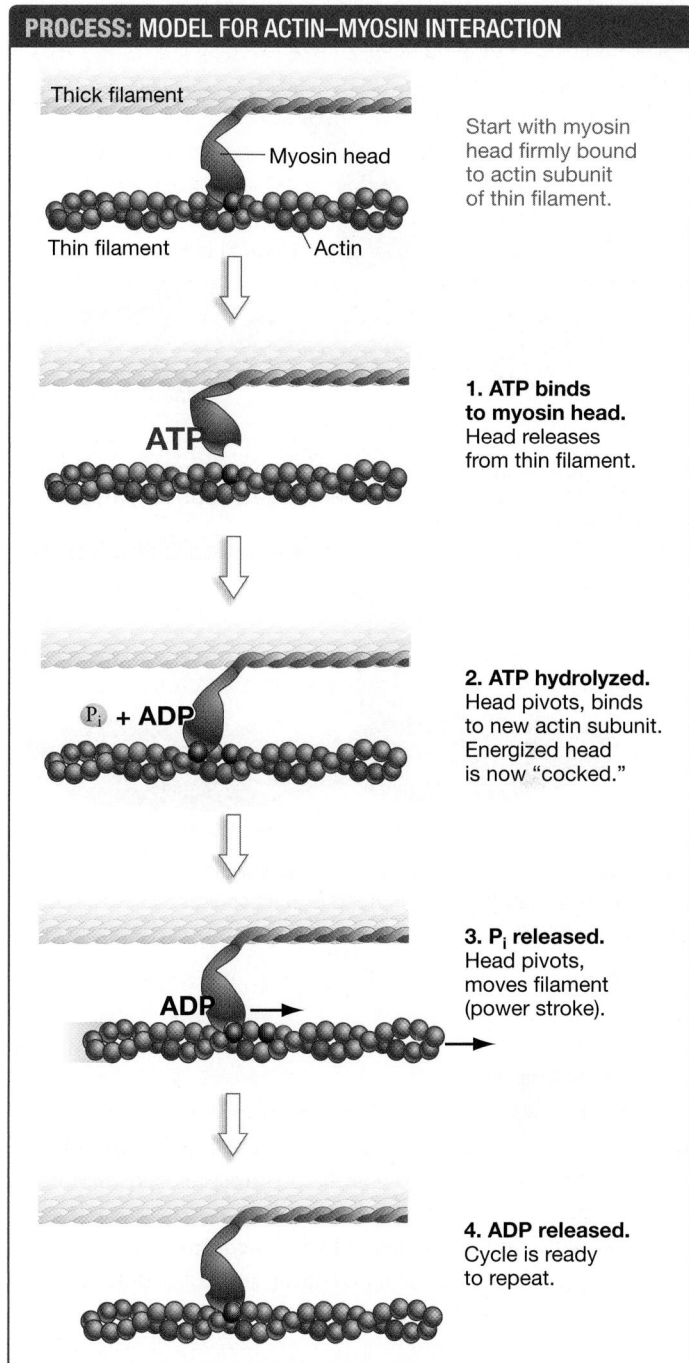

PROCESS: MODEL FOR ACTIN–MYOSIN INTERACTION

Thick filament

Myosin head

Thin filament

Actin

Start with myosin head firmly bound to actin subunit of thin filament.

ATP

1. ATP binds to myosin head. Head releases from thin filament.

P_i + ADP

2. ATP hydrolyzed. Head pivots, binds to new actin subunit. Energized head is now "cocked."

ADP

3. P_i released. Head pivots, moves filament (power stroke).

4. ADP released. Cycle is ready to repeat.

FIGURE 48.4 Myosin and Actin Interact during Muscle Contraction. Summary of the current model of how myosin and actin interact as a sarcomere contracts. The four steps repeat rapidly. (Only one head of the myosin molecule is shown, for simplicity.)

Considering that ATP is almost always available in living muscles, how do muscles ever stop contracting and relax? Besides actin, thin filaments contain two key proteins called **tropomyosin** and **troponin.** Tropomyosin and troponin work together to block the myosin binding sites on actin. The myosin–actin interaction cannot then occur, and thick and thin

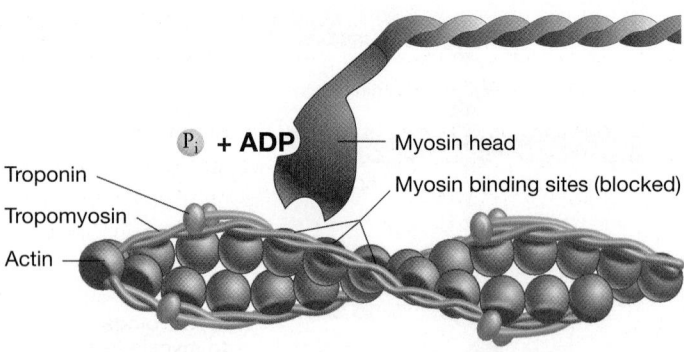

(a) Muscle relaxed: Tropomyosin and troponin work together to block the myosin binding sites on actin.

P_i + **ADP**

Myosin head

Troponin

Myosin binding sites (blocked)

Tropomyosin

Actin

(b) Contraction begins: When a calcium ion binds to troponin, the troponin–tropomyosin complex moves, exposing myosin binding sites.

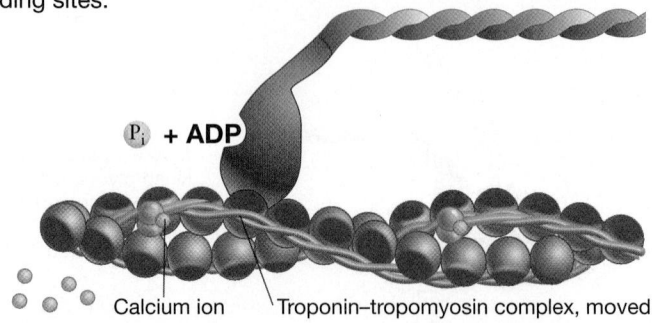

P_i + **ADP**

Calcium ion

Troponin–tropomyosin complex, moved

FIGURE 48.5 Troponin and Tropomyosin Regulate Muscle Activity. Note that the myosin head is in its energized state when a muscle is relaxed.

filaments cannot slide past each other. As a result, the muscle relaxes (**FIGURE 48.5a**).

But when calcium ions bind to troponin, the troponin–tropomyosin complex moves in a way that exposes the myosin binding sites on actin. As **FIGURE 48.5b** shows, myosin then binds and contraction can begin.

How are calcium ions released so that contraction can begin? The process begins with the arrival of an action potential—an electrical signal from a motor neuron.

How Do Neurons Initiate Contraction?

You are probably sitting as you read. If so, contract your calf muscles to point your toes. Your nervous system just played a critical role in controlling the timing of your muscle contractions. First, your central nervous system—your brain—received input from an array of sensory cells in your peripheral nervous system, such as the ones in the retina of your eyes as you read this paragraph. Then your brain integrated this information and triggered action potentials in the motor neurons of the peripheral nervous system of your legs. (See Chapter 46 to review the structure and function of the nervous system.)

FIGURE 48.6 summarizes what happens when an action potential from a motor neuron arrives at a muscle cell:

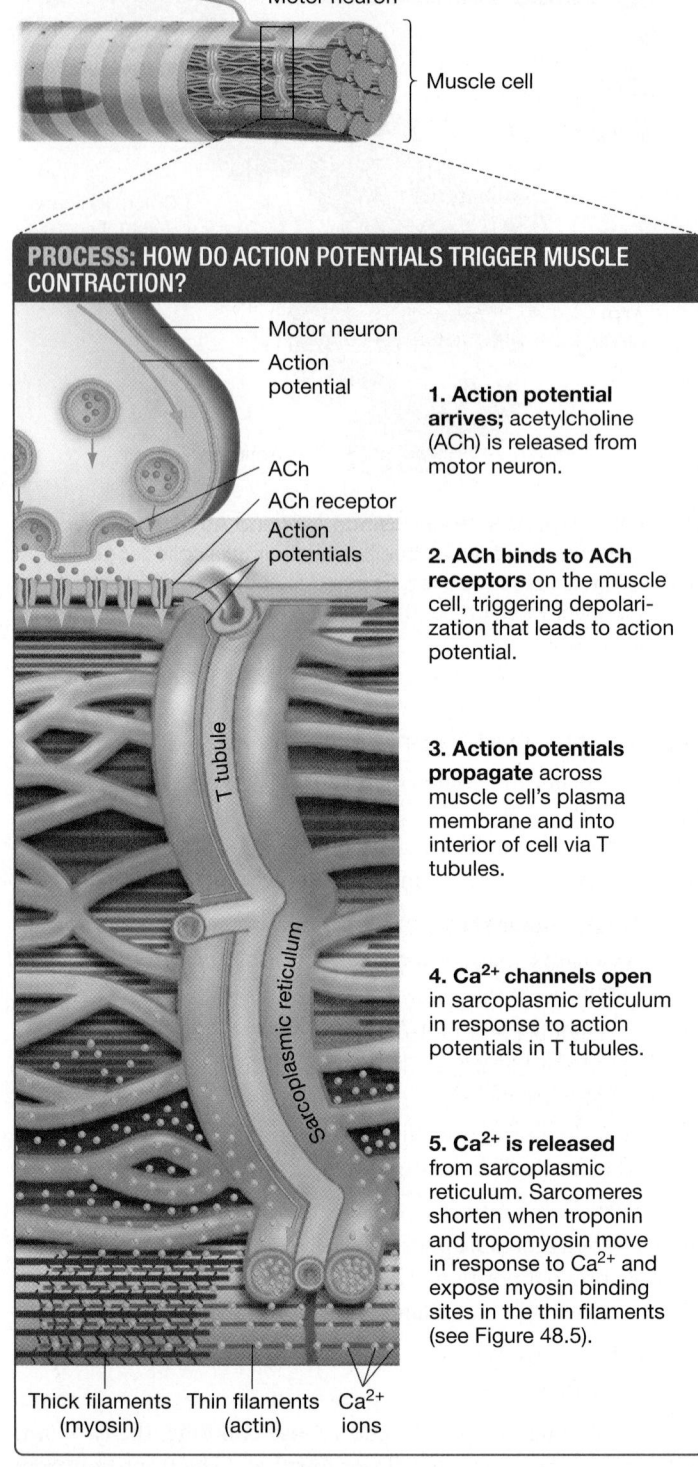

Motor neuron

Muscle cell

PROCESS: HOW DO ACTION POTENTIALS TRIGGER MUSCLE CONTRACTION?

Motor neuron

Action potential

ACh

ACh receptor

Action potentials

T tubule

Sarcoplasmic reticulum

Thick filaments (myosin)   Thin filaments (actin) Ca^{2+} ions

1. Action potential arrives; acetylcholine (ACh) is released from motor neuron.

2. ACh binds to ACh receptors on the muscle cell, triggering depolarization that leads to action potential.

3. Action potentials propagate across muscle cell's plasma membrane and into interior of cell via T tubules.

4. Ca^{2+} channels open in sarcoplasmic reticulum in response to action potentials in T tubules.

5. Ca^{2+} is released from sarcoplasmic reticulum. Sarcomeres shorten when troponin and tropomyosin move in response to Ca^{2+} and expose myosin binding sites in the thin filaments (see Figure 48.5).

FIGURE 48.6 Action Potentials Trigger Ca^{2+} Release. Action potentials at the neuromuscular junction trigger the release of Ca^{2+}, which binds to troponin–tropomyosin and allows myosin to form a cross-bridge with actin.

1. Action potentials trigger the release of the neurotransmitter **acetylcholine** from the motor neuron into the synaptic cleft between the motor neuron and the muscle cell. (See Figure 46.12 on page 939 for a review of the synaptic cleft.)

2. Acetylcholine diffuses across the synaptic cleft and binds to acetylcholine receptors on the plasma membrane of the muscle cell. By recording voltage changes in muscle cells, biologists showed that a membrane depolarization occurs in response to the binding of acetylcholine. If enough acetylcholine is applied to a muscle cell, depolarization triggers action potentials in the fiber itself.

3. The action potentials propagate along the length of the muscle fiber and spread into the interior of the fiber via invaginations of the muscle cell membrane called **T tubules.** (The T stands for transverse, meaning extending across.)

4. T tubules intersect with extensive sheets of smooth endoplasmic reticulum called the **sarcoplasmic reticulum.** When an action potential passes down a T tubule and reaches one of these intersections, a protein in the T-tubule membrane changes conformation and opens calcium channels in the sarcoplasmic reticulum.

5. Calcium causes the myosin binding sites on the actin filaments to be exposed, enabling the contraction to begin.

By this series of events, the interaction of the nervous system and muscle tissue at the neuromuscular junction precisely regulates muscle contractions—and thus complex movement.

check your understanding

If you understand that . . .

- Muscles shorten when thick filaments of myosin slide past thin filaments of actin in a series of binding events mediated by the hydrolysis of ATP.
- Muscle cells shorten in response to action potentials, which trigger the release of calcium ions that enable actin and myosin to interact.

✔ **You should be able to . . .**

1. Describe the sliding-filament model.
2. Predict the effect on muscle function of drugs that have the following actions: increase acetylcholine release at the neuromuscular junction, prevent conformational changes in troponin, and block uptake of calcium ions into the sarcoplasmic reticulum.

Answers are available in Appendix A.

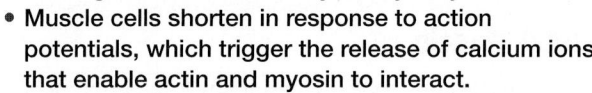 **Muscle Tissues**

How do muscle cells and muscle tissues vary? After years of careful anatomical study, biologists concluded that animals have three classes of muscle tissue: **(1)** smooth muscle, **(2)** cardiac muscle, and **(3)** skeletal muscle. You, and all other vertebrates, have all three.

Smooth, cardiac, and skeletal muscle share several properties. They all contract as described by the sliding-filament model, and they all contract in response to electrical stimulation. However, the three classes of muscle also differ in important ways, summarized in **TABLE 48.1** (see page 978):

- *Voluntary versus involuntary* **Voluntary muscles** can contract in response to conscious thought (and also by unconscious reflexes) and are stimulated by neurons in the somatic division of the peripheral nervous system. **Involuntary muscles** contract only in response to unconscious electrical activity and are stimulated and inhibited by neurons in the autonomic division of the peripheral nervous system. (See Chapter 46 for a review of the somatic and autonomic divisions of the nervous system.)

- *Multinucleate versus uninucleate* Muscle cells may have one or many nuclei depending on the size of the cells.

- *Striated versus unstriated* In some muscle cells, the actin and myosin filaments are aligned in rows forming sarcomeres, giving the cells and tissues a banded appearance; for this reason, it is often called **striated muscle.** Other muscle cells are unstriated.

Let's apply these characteristics to each class of muscle tissue in more detail.

Smooth Muscle

Smooth muscle cells are unbranched, tapered at each end, and often organized into thin sheets. They lack the sarcomeres that are found in skeletal and cardiac muscle; hence, they are unstriated and appear smooth. Smooth muscle cells are relatively small and have a single nucleus.

Smooth muscle is essential to the function of the lungs, blood vessels, digestive system, urinary bladder, and reproductive system. Bronchioles in the lungs have a layer of smooth muscle that controls the size of airways; similarly, smooth muscles in the blood vessels can contract or relax to alter blood-flow patterns and blood pressure (Chapter 45). Layers of smooth muscle in the gastrointestinal tract help mix and move food (Chapter 44), and uterine smooth muscle is responsible for expelling the fetus during birth (see Chapter 50).

Smooth muscle is innervated by autonomic motor neurons, and it is thus involuntary. Acetylcholine from parasympathetic ("rest-and-digest") neurons aids digestion by stimulating contraction of smooth muscle in the stomach and intestine (Chapter 46).

In contrast, sympathetic ("fight-or-flight") neurons release the neurotransmitter **norepinephrine;** and the adrenal glands adjacent to the kidneys release the hormone **epinephrine,** also called **adrenaline** (see Chapter 49). Norepinephrine and epinephrine have the opposite effect to that of acetylcholine: they inhibit contraction of muscle in the stomach and intestine. Different smooth muscles are stimulated or inhibited by autonomic motor neurons depending on the type of neurotransmitter and neuron receptors they have.

Smooth Muscle	Cardiac Muscle	Skeletal Muscle
25 µm	25 µm	25 µm
Location		
Intestines, arteries, other	Heart	Attached to the skeleton
Function		
Move food, help regulate blood pressure, etc.	Pump blood	Move skeleton
Cell characteristics		
	Intercalated discs	Nuclei
Single nucleus	1 or 2 nuclei	Multinucleate
Unstriated	Striated	Striated
Unbranched	Branched; intercalated discs form direct cytoplasmic connection end to end	Unbranched
No sarcomeres	Contains sarcomeres	Contains sarcomeres
Activity is "involuntary," meaning that signal from motor neuron is not required	Activity is "involuntary," meaning that signal from motor neuron is not required	Activity is "voluntary," meaning that signal from somatic motor neuron is required

Cardiac Muscle

Cardiac muscle makes up the walls of the heart and is responsible for pumping blood throughout the body. Unlike smooth muscle, cardiac muscle cells contain sarcomeres and are striated. Further, cardiac muscle cells have a unique branched structure, and they are directly connected end-to-end via specialized regions called intercalated discs. These discs are critical to the flow of electrical signals from cell to cell and thus to the coordination of the heartbeat (Chapter 45).

Like smooth muscle, cardiac muscle is involuntary—it contracts following spontaneous depolarizations. During rest, parasympathetic neurons release acetylcholine onto the heart. This neurotransmitter slows down the rate of depolarization of cardiac cells. The result is a slower heart rate.

During exercise, or when an animal is frightened, stressed, or otherwise stimulated, sympathetic neurons release norepinephrine onto the heart, and the adrenal gland releases epinephrine. Norepinephrine and epinephrine have an opposite effect to that of acetylcholine: they increase heart rate and strengthen the force of cardiac muscle contraction. The result is that more blood is pumped from the heart—an essential component of the fight-or-flight response of the sympathetic nervous system (Chapter 46).

Skeletal Muscle

Skeletal muscle consists of exceptionally long, unbranched muscle fibers. For example, a muscle fiber of a cat may be 0.4 mm wide and 40 mm long—enormous compared to most cells. These large cells result from the fusion of many smaller embryonic cells during development, accounting for the multiple nuclei spread out along the cell. Each muscle fiber is packed with myofibrils, each of which may contain thousands of sarcomeres, giving skeletal muscle its striated appearance.

Skeletal muscle is so named because it usually attaches to the skeleton. When skeletal muscle contracts, it exerts a pulling force on the skeleton, causing it to move—powering the sprint of cheetahs, the flight of hummingbirds, and the pinch of crab claws. In addition, skeletal muscle encircles the openings of the digestive and urinary tracts and controls swallowing, defecation, and urination.

A significant fraction of the body of many animals is composed of skeletal muscle. For example, 63 percent of the body weight of trout is skeletal muscle, and mammals—including humans—of all sizes are 40 to 45 percent muscle. Clearly, skeletal muscle plays an important role in animal biology. Some of the major skeletal muscles in the human body are shown in **FIGURE 48.7**.

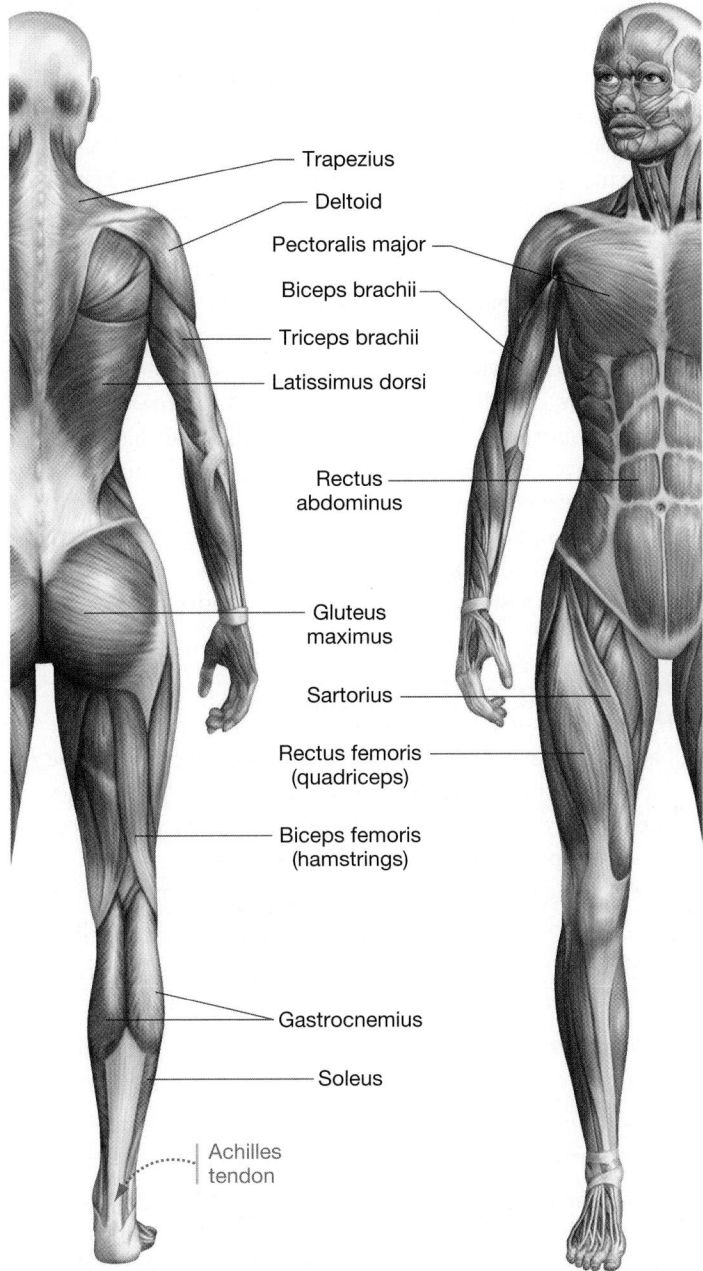

FIGURE 48.7 **Major Skeletal Muscles Make Up a Large Portion of the Human Body.**

Skeletal muscle is distinguished from cardiac and smooth muscle in being voluntary. Skeletal muscle must be stimulated by somatic motor neurons to contract (Chapter 46). If these motor neurons are damaged, as can occur with a spinal cord injury, skeletal muscle cannot contract and becomes paralyzed.

Although all muscles contract as described by the sliding-filament model, not all skeletal-muscle fibers have the same contractile properties. The force output of skeletal muscles depends on (1) the relative proportion of different fiber types, (2) the organization of fibers within the muscle, and (3) how the muscle is used. Let's take a closer look at these sources of variation in muscle performance.

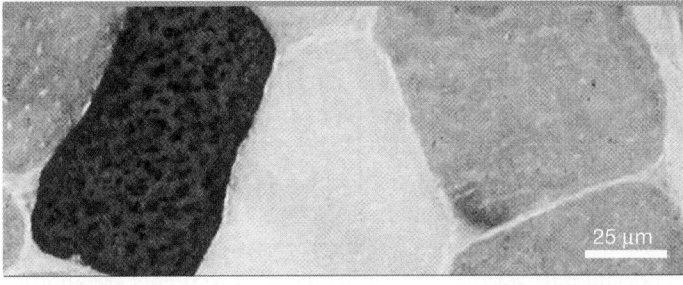

Slow fiber	Fast fiber	Intermediate fiber
Red	White	Pink or red
High myoglobin concentration	Low myoglobin concentration	High myoglobin concentration
Derive most ATP via aerobic respiration (slow oxidative)	Derive most ATP via glycolysis (fast glycolytic)	Derive ATP from glycolysis and aerobic respiration (slow glycolytic)
Many mitochondria	Few mitochondria	Many mitochondria
Slow twitch	Fast twitch	Intermediate twitch
Fatigues slowly	Fatigues quickly	Intermediate fatigue

Skeletal-Fiber Types Skeletal muscle fibers can be divided into general types based on their structural and functional characteristics, summarized in **TABLE 48.2**:

- **Slow muscle fibers** (slow oxidative fibers) appear red because they contain a high concentration of myoglobin, an iron-bearing pigment that carries oxygen (similar to but distinct from hemoglobin in the blood). Slow fibers contract slowly because the myosin hydrolyzes ATP at a slow rate. They also fatigue slowly because they have many mitochondria and can generate steady quantities of ATP using oxidative phosphorylation—that is, aerobic respiration (Chapter 9)—thanks to the plentiful supply of oxygen delivered by myoglobin.

- **Fast muscle fibers** (fast glycolytic fibers) appear white because they have a low myoglobin concentration. They contract rapidly because the myosin hydrolyzes ATP at a rapid rate, but they also fatigue rapidly because their primary source of ATP is glycolysis rather than aerobic respiration.

- **Intermediate muscle fibers** (fast oxidative fibers) appear pink or red. Their contractile properties vary but are intermediate between slow and fast fibers because they derive ATP from both glycolysis and aerobic respiration.

The different fiber types are present in all skeletal muscles, but their relative abundances differ from muscle to muscle. Slow fibers are abundant in muscles specialized for endurance—such as the leg muscles of birds that excel at swimming or walking (the "dark meat" of chicken legs). In humans, the soleus muscle in the back of the calf is an example of a muscle with a high proportion of slow fibers—it helps to keep you upright when you stand.

Fast fibers contract and relax up to three times faster than slow fibers, making them well suited for bursts of activity. The "white meat" of chicken breasts, specialized only for quick bursts of flight to escape predators, are made primarily of fast fibers. The muscles that control your eye movements are another example.

✔️ If you understand muscle fiber types, you should be able to predict whether the breast meat of pigeons (which are capable of prolonged flights) is composed of dark meat or white meat and explain why there is a color difference between these muscle types.

Can humans change their fiber types through training? Experiments have shown that endurance training can increase the density of blood vessels, mitochondria, and myoglobin in muscle fibers, enabling athletes to improve their muscle performance. However, training does not change slow fibers to fast fibers, nor the reverse. The ratio of muscle-fiber types in muscles is heritable—that is, genetically determined.

Skeletal-Fiber Organization   The force of a muscle is proportional to the cross-sectional area of the muscle—the number of sarcomeres lined up side by side exerting a pull in synchrony. By contrast, the length change of the muscle is determined by the length of the muscle fibers—how many sarcomeres are lined up in a row in each fiber. Thus, the arrangement of muscle fibers within a muscle influences the contractile properties of the muscle.

For a given muscle volume, some muscles are organized to maximize length change, because the fibers are parallel to each other in long bands (**FIGURE 48.8**, left)—the longer the chain of

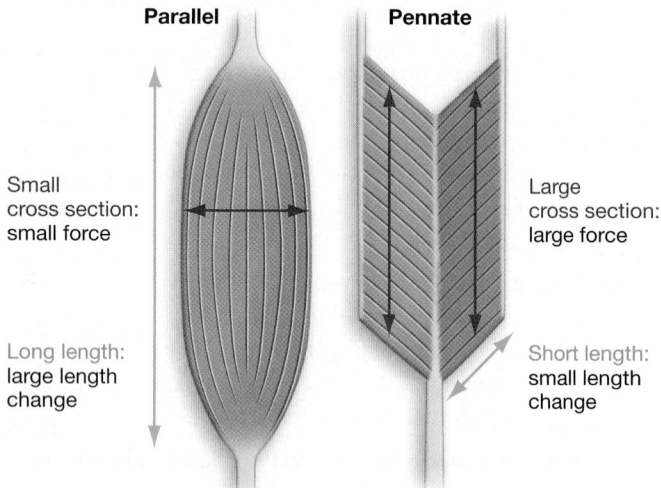

Parallel   **Pennate**

Small cross section: small force

Large cross section: large force

Long length: large length change

Short length: small length change

FIGURE 48.8 Muscle-Fiber Patterns Affect Contractile Properties. Most muscles have fibers arranged in either a parallel or pennate pattern. The orientation of muscle fibers affects the contractile properties of the muscle.

sarcomeres in series in the myofibril, the greater the length change. The sartorius muscle in the human thigh has parallel fibers.

Other muscles are organized to maximize force, because the fibers are in a diagonal, or pennate, pattern (penna means feather; **FIGURE 48.8**, right)—the greater the number of sarcomeres pulling in parallel, the greater the force produced. The gastrocnemius muscle in the human calf is a pennate muscle.

Context of Muscle Contraction   The relative abundances of fiber types and organization of muscle fibers are not sufficient to account for the diversity of muscle contraction properties of skeletal muscles. Muscle tension also varies according to how extended the muscle is when it contracts and how rapidly it is allowed to shorten—if at all. Muscles like the quadriceps in your thigh can even exert a force while they are lengthening, such as when you ease your weight down a step.

These circumstances depend on the interaction between the muscle and the skeleton. Let's take a closer look at how the muscle and skeletal systems interact to produce movement.

48.3 Skeletal Systems

All a muscle can do is pull. How can complex movements be accomplished using an engine that can only pull? Also, muscles are limited in how much they can shorten. How, then, can they cause dramatic shape changes in animal bodies?

Muscle forces and shape changes are transmitted to other parts of the body and to the environment via the skeleton. Skeletal systems perform four main functions:

1. **Protection**   from physical and biological assaults.

2. **Maintenance of body posture**   despite the downward pull of gravity and the vagaries of wind and waves.

3. **Re-extension of shortened muscles**   If no mechanism of re-extension existed, muscles would shorten only once.

4. **Transfer of muscle forces**   to other parts of the body and to the environment, enabling a much greater range of force production and shape change than can be accomplished by muscle alone.

The relative importance of these roles varies among animals according to their lifestyles and environments. For example, natural selection has favored turtles with robust shells in some environments—a skeletal adaptation for protection. In other environments, natural selection has favored highly reduced shells—an adaptation for rapid locomotion. There are trade-offs between protection and mobility (see Chapter 25).

Despite the stunning diversity in animal bodies, virtually all animals can be considered to have one (or more) of three types of skeletal systems:

1. Hydrostatic skeletons use the hydrostatic pressure of enclosed body fluids or soft tissues to support the body.

2. Endoskeletons have rigid structures inside the body.

3. Exoskeletons have rigid structures on the outside of the body.

Let's consider how each skeletal system transmits muscle forces and shape changes.

Hydrostatic Skeletons

Despite their squishy appearance, soft-bodied animals do have skeletons—hydrostatic skeletons. First let's look at how they are built, and then consider how they function.

Structure **Hydrostatic skeletons** ("still-water skeletons"), or hydrostats, are constructed of an extensible body wall in tension surrounding a fluid or deformable tissue under compression. When fluid is under compression, its pressure increases. The pressurized internal fluid, rather than a rigid structure, enables soft-bodied animals to maintain posture, re-extend muscles, and transfer muscle forces to the environment.

Hydrostatic skeletons occur in diverse animals, from sea anemones and jellyfish to mollusks and many types of worms (Chapter 34). Hydrostats also support *parts* of animals, such as the tongues and penises of humans and the tube feet of echinoderms (see Chapter 35).

The structures of hydrostatic skeletons are diverse as well. The body wall of hydrostats may include different numbers and orientations of muscle layers and fiber-reinforced cuticles or connective tissues. The interior may include seawater, coelomic fluid, blood, or soft organs such as intestines. Hydrostatic skeletons composed mostly of muscle, such as tongues and tentacles, are called muscular hydrostats (Chapter 34).

Function How do animals with hydrostatic skeletons move? Consider an earthworm. Its body wall consists of a cuticle reinforced with collagen fibers as well as two layers of muscle—longitudinal muscles, oriented along the length of the animal, and circumferential muscles, oriented in bands around each segment. When the circumferential muscles contract, they make the segments narrower and squeeze the internal coelomic fluid and tissue, thus increasing internal pressure. The pressure pushes outward in all directions, extending the relaxed longitudinal muscles, lengthening the segment (**FIGURE 48.9**).

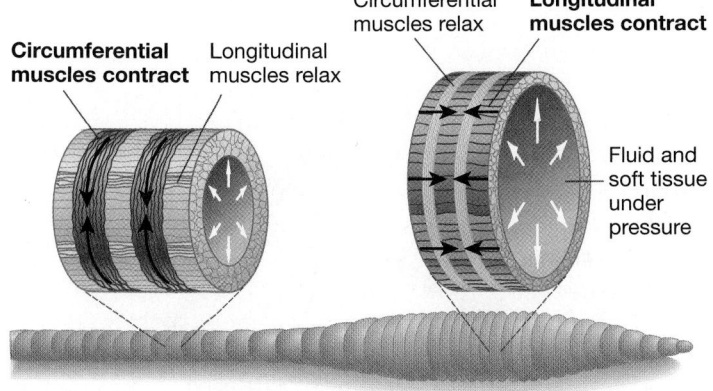

Circumferential muscles contract | Longitudinal muscles relax

Circumferential muscles relax | **Longitudinal muscles contract**

Fluid and soft tissue under pressure

FIGURE 48.9 Antagonistic Muscle Groups Cause Shape Changes in Hydrostatic Skeletons. The pressure of the internal fluid or tissue transmits forces between muscle groups and between muscles and the environment.

When the longitudinal muscles contract and the circumferential muscles relax, the reverse occurs—the segments become wider and shorter, pushing sideways against the soil. Alternating contractions of longitudinal and circumferential muscles pass down the earthworm in waves, called **peristalsis.** In this way, earthworms move forward (or backward) within their underground burrows.

Longitudinal and circumferential muscles in earthworms make up an **antagonistic muscle group,** a group of two or more muscles that re-extend one another via the skeleton.

✔ **If you understand how a hydrostatic skeleton works, you should be able to explain how earthworms can push laterally against the walls of their burrows despite having muscles that can only pull.**

Endoskeletons

Even though parts of you, like your tongue, are supported by a hydrostatic skeleton, your endoskeleton is what keeps you standing up and on the move.

Structure **Endoskeletons** ("inside skeletons") are rigid structures that occur within the body. Even the most ancient of animal lineages, the sponges, secrete spicules—stiff spikes of silica or calcium carbonate ($CaCO_3$)—that provide structural support for the body. In echinoderms, the endoskeleton consists of calcium carbonate plates just beneath the skin—fused into a rigid case in sea urchins, but suspended in a flexible matrix that enables bending of the arms in sea stars.

The vertebrate endoskeleton differs from those of most sponges and echinoderms—and from hydrostatic skeletons—in that it is composed of rigid levers (the bones) separated by joints. Vertebrates change the shapes of their bodies largely by changing the *joint angles* between bones in the limbs and between the limbs and the body, rather than changing the shapes of body segments themselves.

Vertebrate skeletons are composed of four main elements:

1. **Bones** are made up of cells in a hard extracellular matrix of calcium phosphate ($CaPO_4$) with small amounts of calcium carbonate and protein fibers. The adult human body contains 206 bones (**FIGURE 48.10a**, see page 982). The meeting places where adjacent bones interact are called **articulations,** or **joints.** Bones articulate in ways that limit the range of motion, for example enabling a swivel in the shoulder joint but a hinge in the elbow joint (**FIGURE 48.10b**).

2. **Cartilage** is made up of cells scattered in a gelatinous matrix of polysaccharides and protein fibers. Cartilage can be quite rigid, such as in the clam-crushing jaws of some stingrays, or more rubbery, such as in the pads that cushion the joints in your knees and back.

3. **Tendons** are bands of fibrous connective tissue, primarily collagen, that connect skeletal muscles to bones. The ropelike structure of tendons transmits muscle forces to precise locations on the bones, sometimes quite a distance away from the muscle itself.

(a) Bones of the human endoskeleton

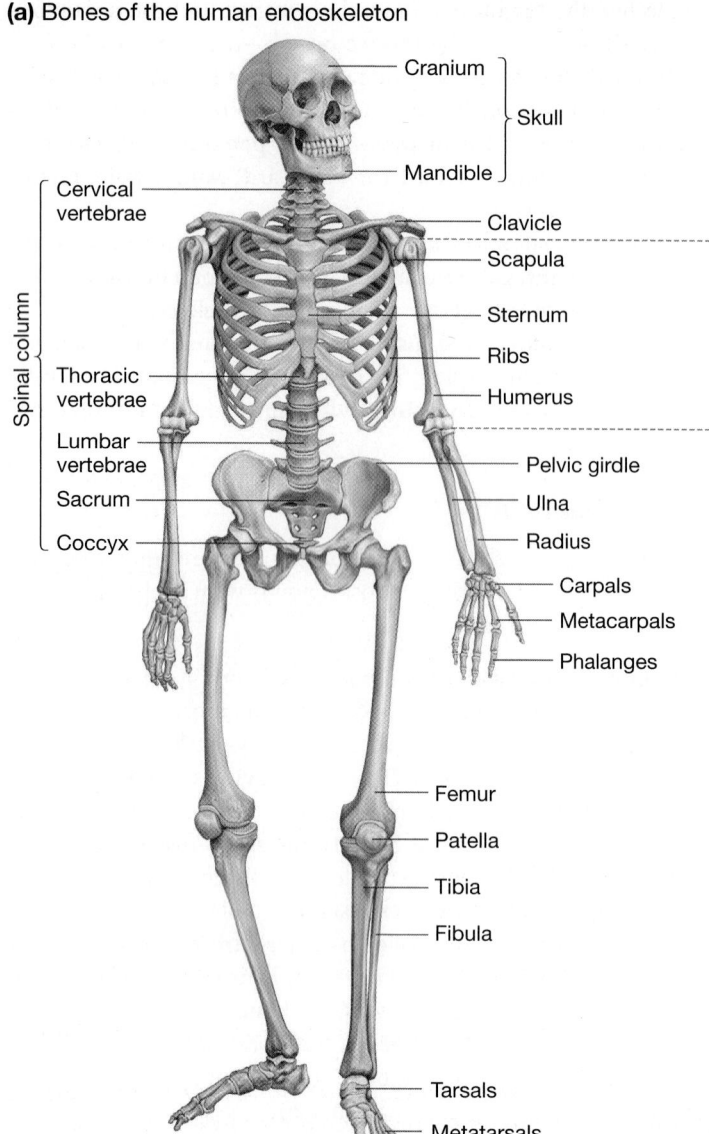

- Cranium
- Skull
- Mandible
- Cervical vertebrae
- Clavicle
- Scapula
- Sternum
- Ribs
- Thoracic vertebrae
- Humerus
- Lumbar vertebrae
- Pelvic girdle
- Sacrum
- Ulna
- Coccyx
- Radius
- Carpals
- Metacarpals
- Phalanges
- Femur
- Patella
- Tibia
- Fibula
- Tarsals
- Metatarsals
- Phalanges

Spinal column

FIGURE 48.10 Bones of the Human Endoskeleton. Since bones are rigid and cannot change shape themselves, they articulate at joints that make specific types of movement possible, such as the swiveling and hinging shown here.

(b) Joints enable movement

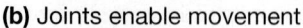
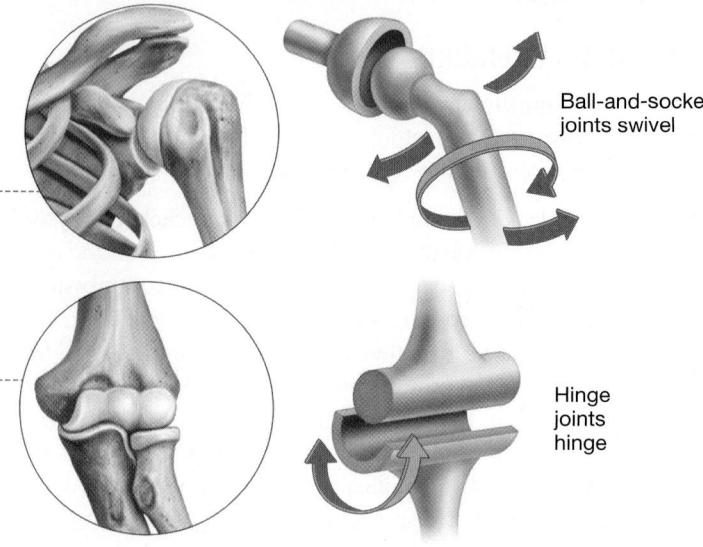

Ball-and-socket joints swivel

Hinge joints hinge

flexors, muscles that pull bones closer together, decreasing the joint angle between them. They swing your lower leg back toward your thigh, reducing the angle of your knee joint. The quadriceps muscles in the front of your thigh are **extensors,** muscles that increase the angle of a joint. They straighten your leg at the knee joint (**FIGURE 48.11**).

The hamstrings and quadriceps muscles accomplish the large swing of the lower leg by inserting into different locations on the tibia (the shin bone). The articulation of the tibia with the femur (the thigh bone) serves as the pivot point for this lever. Along with enabling a shape change, the bone transmits the forces exerted within the thigh muscles to the foot—such as when the extension of your leg enables you to kick a ball.

The role of bones is primarily mechanical, but they also serve several physiological functions. Chief among these is storage of calcium and other minerals. When blood calcium falls to low levels, the bones release calcium to maintain blood–calcium homeostasis. The interior of long bones, called bone marrow, is also

4. **Ligaments** are bands of fibrous connective tissue, primarily collagen, that bind bones to other bones. Ligaments stabilize the joints.

Function   Bones and cartilages are structures that do a good job of resisting compression (pushing) and bending, whereas tendons and ligaments do a good job of resisting tension (pulling). These structures combine with muscle in ways that enable the efficient transmission of muscle forces and shape changes.

Vertebrate skeletons move by means of changes in joint angles controlled by antagonistic muscle groups. For example, consider how your thigh muscles flex and extend your knee joint. The hamstring muscles in the back of your thigh are

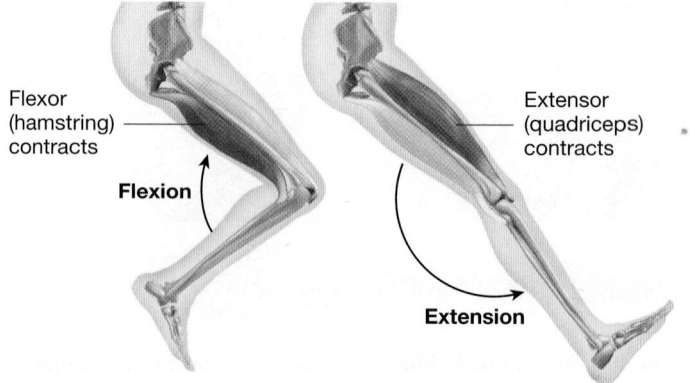

Flexor (hamstring) contracts

Flexion

Extensor (quadriceps) contracts

Extension

FIGURE 48.11 Endoskeletons Move by Contraction and Relaxation of Flexor and Extensor Muscles.

the source of red blood cells needed to carry oxygen in the blood (Chapter 45) and white blood cells needed for the immune system (Chapter 51).

Exoskeletons

The mechanical function of rigid cuticle in exoskeletons is similar in many ways to that of endoskeletons. Exoskeletons occur primarily in arthropod animals, including insects, crustaceans, and arachnids (spiders, ticks, scorpions).

Structure An **exoskeleton** ("outside skeleton") is an exterior skeleton that encloses and protects an animal's body. The origin of the exoskeleton was a key innovation that preceded the spectacular diversification of arthropods. Arthropods, especially insects, are the most diverse and abundant animals on Earth (Chapter 34).

The material composition of exoskeletons varies. Insect exoskeletons consist of a cuticle formed from a composite of proteins and the polysaccharide chitin (Chapter 43). Chitinous ingrowths of the skeleton form **apodemes,** where muscles attach. Crustaceans such as crabs and lobsters have a cuticle that is mineralized with calcium carbonate, making their shells relatively thick and hard—and heavy. Most crustaceans are marine, so their buoyancy in water helps support their weight.

Function Like vertebrates, arthropods have paired flexor–extensor muscles that operate their jointed skeletons, causing movements that are based on changes in joint angles rather than changes in the dimensions of the segments themselves (**FIGURE 48.12**). Unlike vertebrates, however the muscles of arthropods must be packed *within* the skeletal tubes.

One solution to the problem of the interior placement of muscles is that many arthropod muscles have the pennate, or feather-like arrangement of muscle fibers illustrated in Figure 48.8. This arrangement boosts force output by effectively increasing the muscle cross-sectional area but not the muscle width during a contraction.

The disadvantage of pennate muscles is that their length change is small, so they have limited range of motion. Arthropods compensate for this constraint in part by the placement of their apodemes, which can transduce a small shortening of a muscle into a large change in joint angle.

The rigid levers of vertebrate skeletons can grow continuously as the rest of the body grows. But since the rigid exoskeletons of arthropods encase the growing soft tissue like a suit of armor, they must be shed—molted—periodically and replaced with a bigger one. Arthropods are vulnerable to predation during molts because their skeleton is soft and dysfunctional at that time (Chapter 34).

Hydrostatic skeletons, endoskeletons, and exoskeletons all transmit muscle forces and shape changes to other parts of the body and to the environment. While hydrostatic skeletons are the most widespread in terms of the number of animal phyla that contain species with skeletons of this type (virtually all—even arthropods, echinoderms, and vertebrates), the lever-based, segmented, jointed skeletons of arthropods and vertebrates win the prize for overall functional diversity (Chapters 34 and 35).

check your understanding

If you understand that . . .

- Hydrostatic skeletons, endoskeletons, and exoskeletons transmit muscle forces and shape changes to other parts of the animal and to the environment.
- Antagonistic muscles re-extend one another via the skeleton.
- In contrast to hydrostatic skeletons, shape changes in endoskeletons and exoskeletons involve changes in angles between rigid segments rather than in the segments themselves.

✔ **You should be able to . . .**

1. Compare and contrast the structure and function of hydrostatic skeletons, endoskeletons, and exoskeletons.

2. Predict what would happen if neurons simultaneously stimulated the biceps and triceps muscles to contract.

Answers are available in Appendix A.

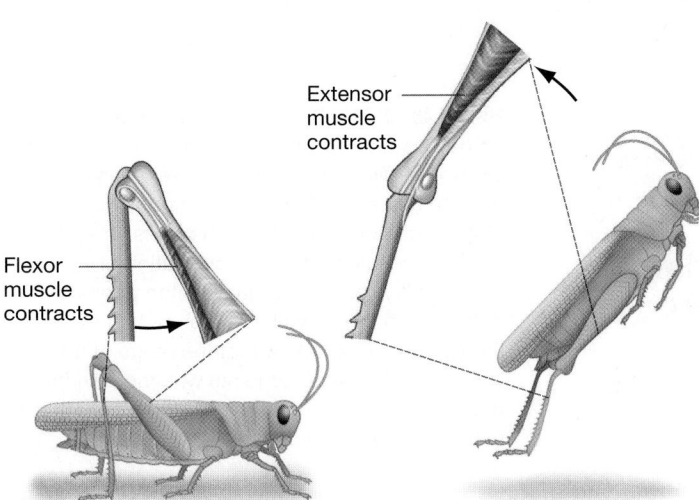

FIGURE 48.12 Exoskeletons Move by Contraction and Relaxation of Flexor and Extensor Muscles.

Extensor muscle contracts

Flexor muscle contracts

48.4 Locomotion

The most spectacular capability conferred on animals by the combination of muscle contractions and skeletal systems is efficient locomotion. Animals locomote to seek food, water, mates, and shelter, and to avoid predators. In the process, some animals migrate thousands of miles per year, and others perform astonishing feats of acrobatics. Locomotion has been shaped by natural selection and has been central to complex ecological relationships since the radiation of animals during the Cambrian Explosion more than 500 million years ago (Chapter 33).

Many of the diverse modes of locomotion are already familiar to you. Here are some examples.

- **On land** Crawling, walking, running, climbing, hopping, jumping, burrowing
- **In water** Undulating, jetting, swimming, rowing
- **In air** Flying, gliding

What variables do biologists analyze to unlock the secrets of animal locomotion?

How Do Biologists Study Locomotion?

Experimental studies on locomotion have increased exponentially in recent years, partly due to the conceptual breakthrough offered by the field of biomechanics—applying the principles of engineering to quantify the mechanics of organisms. Biomechanics studies the physical act of locomotion at different levels, including

- material properties
- structures
- motions
- forces
- energetics
- ecology and evolution

Let's examine each of these levels, starting with materials.

How Are the Material Properties of Tissues Important to Locomotion? Because the active contractile properties of skeletal muscle are central to understanding the generation of forces in locomotion, a great deal of research has been devoted to analyzing the contractile properties of muscle, introduced in Section 48.1. However, the passive material properties of the skeletal elements are also essential to understanding the transmission of forces.

Consider how you move. A pioneer of biomechanics, R. McNeill Alexander, observed that the movement and energy exchange of the center of mass of a person walking is mechanically similar to that of an upside-down pendulum, whereas running is mechanically similar to a bouncing ball. Based on this insight, Alexander hypothesized that the large tendons of the lower legs of terrestrial animals—for example, the Achilles tendon at the back of your ankle (see Figure 48.7)—work as springs when animals run.

To test this hypothesis, Alexander measured the elastic properties of tendon by clamping a piece of tendon in an engineering device that measures the pulling force of the tissue as it is stretched to different lengths and released. The tendon returned 93 percent of the energy invested, losing only 7 percent to heat. This high rate of elastic-energy storage explains how tendons add a spring to the steps of runners, reducing the amount of muscle-generated power that must be generated from step to step.

The importance of the material properties of muscles and skeletal tissues is most obvious when they fail, whether in subtle or catastrophic ways. You may have firsthand experience with the debilitating consequences of broken bones or strained tendons.

How Is Musculoskeletal Structure Adapted for Locomotion? Many biologists begin their study of locomotion by examining the size and shape of the skeletal elements. Careful measurements of skeleton geometries can reveal a great deal about the posture of the organism, range of motion of joints, and skeletal function in general. For example, a team of biologists recently used structural analysis to test the hypothesis that an early tetrapod amphibian, called *Ichthyostega,* could walk on all four limbs like today's salamanders (see **FIGURE 48.13**).

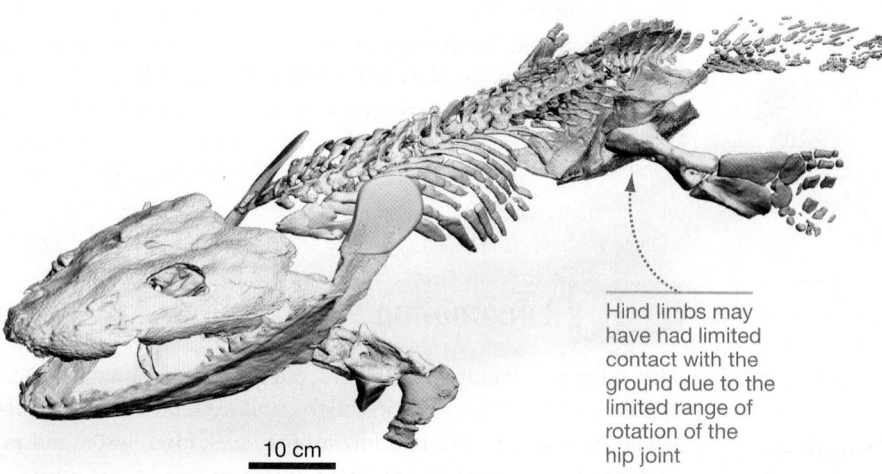

Hind limbs may have had limited contact with the ground due to the limited range of rotation of the hip joint

10 cm

FIGURE 48.13 The Relationship between Structure and Function Can Be Studied Using Computer Models. This computer image of an early tetrapod, *Ichthyostega,* was built using high-resolution scans of fossil bones. The 3D model measures the range of rotation of the major limb joints based on skeleton geometry, determining the overall range of motion of the limbs. The model found that the range of rotation of the shoulder and hip joints was more limited than previously thought, rejecting the hypothesis that these animals could walk on all fours.

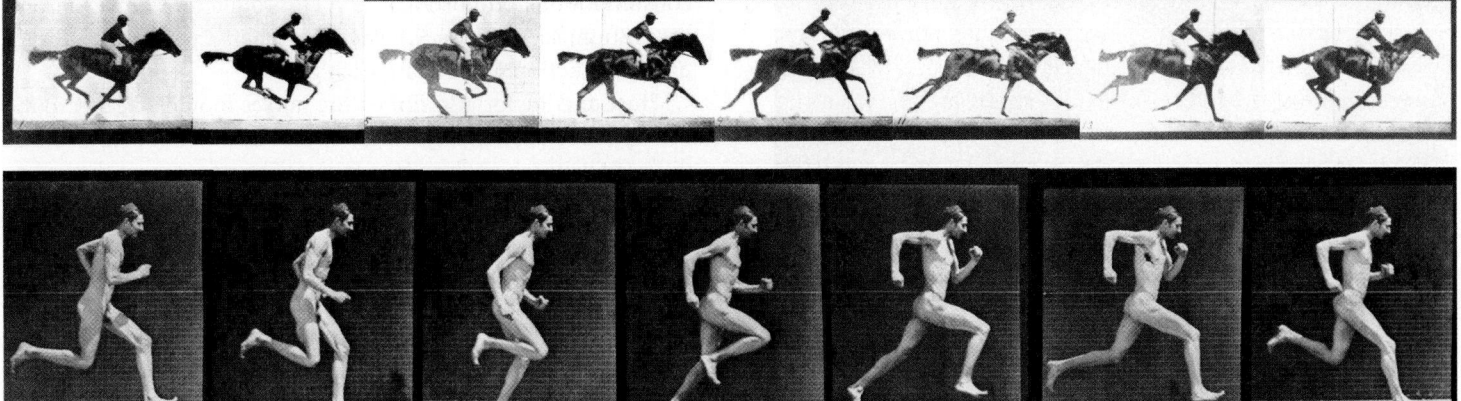

FIGURE 48.14 The Motions of Locomotion Can Be Captured on Film. Eadweard Muybridge shot many photo series of animals during locomotion to enable precise analysis of limb and body motions over time.

Similar mechanical principles can be applied to relate the shapes of wings to flying ability, and the shapes of aquatic animals to their ability to swim through water. Sometimes, however, you just have to watch the action itself to understand how an animal uses its body to locomote.

What Does Locomotion Look Like in Living Animals?

All the motions of different parts of the skeleton, such as the angular rotation of limbs and the pattern of footfall on the ground, together produce locomotion.

Photographer Eadweard Muybridge is famous for his pioneering photo sequences of locomotion. His work with animals reportedly began when he was commissioned by a racehorse owner to settle a wager on whether horses are ever completely airborne during a gallop. His results are shown in the top sequence of **FIGURE 48.14**. High-speed video and digital images of many other animals have since been recorded, providing insights into many forms of locomotion, such as

- The gait of human sprinters (Figure 48.14, bottom)
- The complex wing-beat patterns of hovering bees and hummingbirds
- The upright, bipedal (two-footed) gait of basilisk lizards running across the surface of water (see chapter opening image)
- The footfall pattern that prevents centipedes from tripping on their own legs
- The aerial undulating of snakes that glide down from treetops
- The limb-like use of fins in lungfish "walking"
- Peristalsis in the muscular feet of crawling snails

For most animals, the pattern of movement during locomotion varies with speed. For example, horses walk at slow speeds, trot at intermediate speeds, and gallop at fast speeds—each gait has a distinct pattern of leg motions.

Computers facilitate the analysis of the many images captured in motion studies. The results are themselves insightful, but they also serve as an important stepping-stone to understanding the forces involved in locomotion.

What Forces Are Involved in Locomotion?

If an animal wants to move forward, it must push something backward, as predicted by Newton's third law of motion. Otherwise, the animal could move but not get anywhere, like a person with smooth shoes on slick ice. The types of forces that are important to locomotion vary according to whether the animal is locomoting on land, in water, in air, or some combination of the three.

On land, gravitational forces and inertial forces dominate. The gravitational force experienced by an animal is its weight, which is the product of its mass and the acceleration due to gravity (9.8 m/sec² on Earth). Weight is important on land because most terrestrial animals must hold themselves up to move forward. Inertial forces are proportional to mass and velocity, and they represent resistance of bodies and limbs to acceleration and deceleration. Note in the horse and human photo sequences in Figure 48.14 that the arms and legs must swing back and forth dramatically—an energy-intensive process of acceleration and deceleration.

In water, gravitational forces are less important than on land, due to the counteracting buoyant forces supporting the animal's weight in water. However, aquatic animals must overcome drag, the force that resists forward motion through fluids. Convergent evolution of torpedo-shaped bodies has occurred in diverse aquatic animals, from tuna to dolphins and ichthyosaurs, due to the strong selection for bodies that minimize drag during rapid locomotion (Chapter 28). Aquatic animals that move more slowly face less drag and thus are morphologically more diverse.

Water and air are both fluids, but air is a thousand times less dense than water. Buoyant forces are therefore negligible in air, making gravitational forces very important to fliers—most animals that locomote by flying or gliding have adaptations that make them lightweight. They must produce a force called lift to counteract gravity, and they must also minimize drag. As a result, fast fliers tend to have very streamlined shapes.

FIGURE 48.15 Visualization of Airflow Is Used to Analyze the Forces Involved in the Hovering of a Bat. The lift and drag forces acting on flying and swimming animals can be measured by observing fluid flow around the animal. The arrows represent the velocity of tiny water droplets illuminated by a laser in front of a high-speed camera.

How are forces measured? Scientists can measure ground force—the force with which a terrestrial animal strikes the substratum during a step—by coaxing an animal to walk, run, or hop upon an instrument called a force plate. Quantifying forces in fluids is more nuanced. It often requires indirect measurement by visualizing the airflow or waterflow around the animal, such as the flow of air around the wings of a hovering bat shown in **FIGURE 48.15**.

What Is the Cost of Locomotion?

Animals must spend energy to find food and mates and escape from predators. However, the more energy an animal spends on locomotion, the less it can spend on producing offspring. There is strong selection pressure to minimize the cost of locomotion. How do biologists measure this cost?

To get a sense of variables that determine the cost of locomotion, consider the classic studies by Alexander on the gait transition from walking to running in humans. Alexander discovered that walking is an efficient mode of transport because you exchange potential energy at the top of your stride with kinetic energy midstride. However, the resulting pendulum-like motion of your center of mass is not efficient at higher speeds, when it becomes more cost effective to run using your spring-like tendons and other skeletal tissues to store energy between strides.

Alexander hypothesized that animals locomote using the most energy-efficient gait at each speed. To test this hypothesis in horses, physiologists Dan Hoyt and Richard Taylor trained horses to walk, trot, and gallop at a range of unnatural speeds—for example, trotting at a speed where the horse would normally have preferred to gallop. Hoyt and Taylor fitted each horse with an oxygen mask and ran it on a treadmill, so that they could measure oxygen consumption and speed simultaneously. The rate of oxygen consumption is a measure of energy use. (Oxygen consumption is proportional to ATP production during aerobic respiration; see Chapter 9.) The researchers then plotted energy use as a function of speed for each gait.

Hoyt and Taylor also filmed the horses moving freely around their paddock and measured the speeds at which the horses used different gaits when given free choice. The researchers then compared the lab data to the gait preference in the paddock.

The graph in **FIGURE 48.16** shows Hoyt and Taylor's results. The three curves at the top represent the energy used per distance traveled at the three gaits—the dips in the curves indicate the speeds at which the gaits were most efficient. The bars at the bottom of the graph show which gaits and speeds the horses chose when they were able to locomote freely in the paddock. The data support the hypothesis that the horses use the most energy-efficient gaits at different speeds and avoid intermediate speeds where the cost of locomotion is higher.

Similar studies have been conducted for diverse animals running, swimming, and flying, with similar results. Natural selection favors animals that locomote efficiently because they have more energy available for other vital activities.

Evolution and Ecology: What Is the Context? Some biologists study animal locomotion with a strictly mechanistic focus, investigating proximate causes via "how" questions. Alexander's study of the elastic energy storage of the Achilles tendon falls in this category. However, biologists also examine the ultimate causation of locomotion by asking "why" questions, which require an analysis of evolutionary history and ecological context.

For example, why do dolphins and ichthyosaurs have such similar torpedo-like shapes? Both taxa are fast-swimming predators in their marine environments, yet phylogenetic analysis shows that they are not closely related (see Figure 28.3). Rather, the torpedo-like shape evolved independently in both lineages because it substantially reduces drag during rapid locomotion in water, improving the efficiency of locomotion and thus increasing fitness (the number of viable, fertile offspring produced).

But locomotion cannot be "perfected" over evolutionary time. Natural environments are spatially complex and change over time, so the context for locomotion is not static. Further, there are numerous constraints and fitness trade-offs to different aspects of locomotion. Some trade-offs occur within locomotory systems (such as the trade-off between stability and maneuverability) while others occur between locomotion and other aspects of behavior (such as using appendages both for locomotion and feeding). Overall, a thorough study of locomotion requires an integrative approach with evolutionary and ecological context.

Size Matters

Animals that use muscles to power locomotion span a vast range of sizes—an astonishing 10 orders of magnitude (10,000,000,000), from tiny ants to giant whales. Many of these animals also grow over a large size range during their development. As is true for many other aspects of animal structure and function, size matters.

QUESTION: Do horses minimize the cost of locomotion?

HYPOTHESIS: Horses choose gaits that minimize energy use at different speeds.

NULL HYPOTHESIS: Horses do not choose gaits based on cost of locomotion.

EXPERIMENTAL SETUP:

1. Measure oxygen consumption of horses trained to walk, trot, and gallop at a range of speeds on a treadmill. Calculate energy used per distance travelled at different speeds.

2. Videotape the same horses locomoting freely in the paddock, and measure the gaits and speeds they choose to use naturally.

PREDICTION: For each gait, there is a range of speeds where energy use is minimized. Horses will favor these gaits and speeds.

PREDICTION OF NULL HYPOTHESIS: There will be no correlation between chosen gaits and energy consumption.

RESULTS:

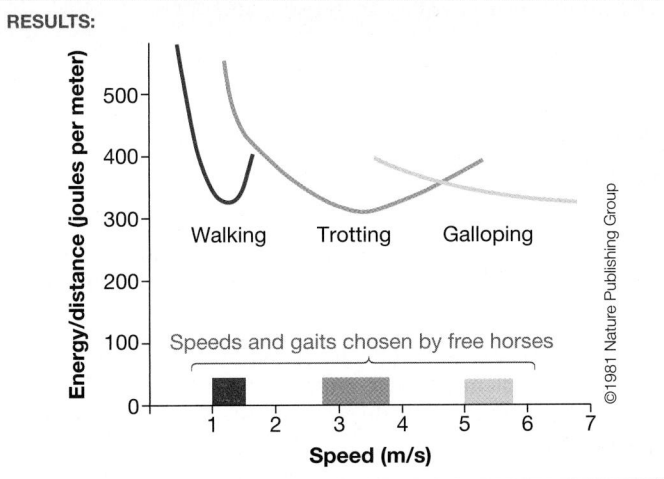

CONCLUSION: Horses choose gaits that minimize energy use at different speeds and avoid speeds with high energy consumption.

FIGURE 48.16 Horses Minimize the Cost of Locomotion by Choosing Appropriate Gaits.

SOURCE: Hoyt, D. F., and C. R. Taylor. 1981. Gait and the energetics of locomotion in horses. *Nature* 292: 239–240.

✔**QUANTITATIVE** Use the graph to estimate the relative energy expense of galloping rather than trotting at 3.5 meters/second (m/s).

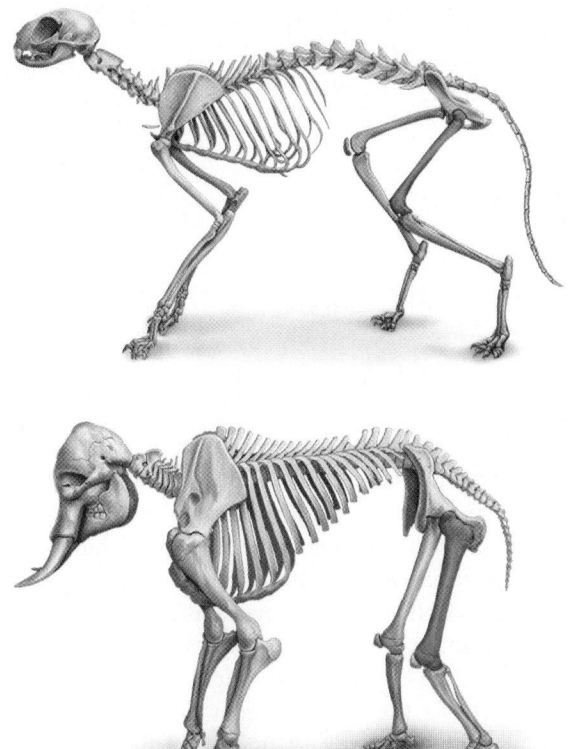

FIGURE 48.17 Size Influences Skeleton Geometry. When scaled to the same size, you can see that the elephant has thicker bones and a more upright stance than a cat does, due to the disproportionate burden of gravity on larger animals. The femurs are highlighted for comparison.

Two organisms may be geometrically similar—that is, they may have exactly the same proportions—but if they are different sizes, they are more different than they seem. To start, the ratio of surface area to volume decreases as the organism gets larger, because surface area is proportional to length squared, while volume is proportional to length cubed (see Chapter 42). This concept has far-reaching implications for physiology. It also has important mechanical implications.

The weight of an animal is proportional to its volume, and the ability of leg bones to support the weight is proportional to their cross-sectional area. Thus, large terrestrial organisms have disproportionately hefty skeletal elements to avoid breaking their legs—something that Galileo observed 400 years ago when he compared skeletons of small and large organisms (**FIGURE 48.17**).

So, in the late 1980s, it was surprising when Andrew Biewener discovered that animals of different sizes maintain similar stresses in their skeletal tissues. How do they do that?

Biewener observed that posture and behavior are important variables. Small animals locomote in a more crouched posture and make great leaps, while large animals like elephants are more straight-legged, so that their skeletons, rather than their muscles, can support their body weight. Larger animals also locomote more gently. If a house cat were enlarged to be the size of an elephant, it would break its own bones when attempting a pounce.

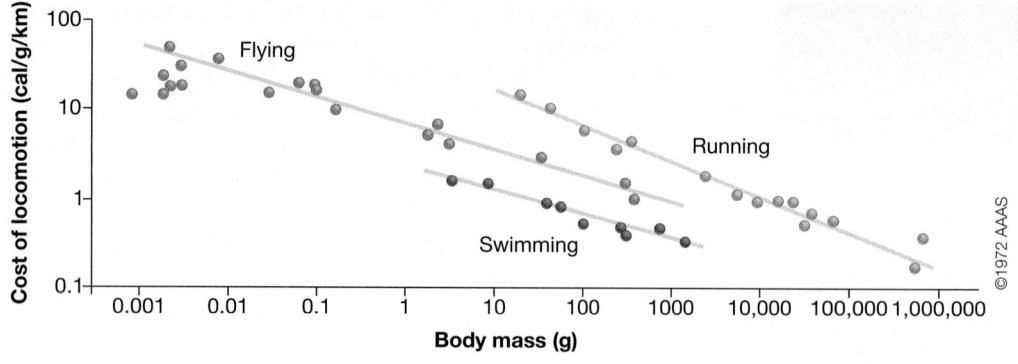

FIGURE 48.18 Cost of Locomotion for Swimming, Running, and Flying Animals Decreases with Mass.

DATA: Schmidt-Nielsen, K. 1972. Locomotion: Energy cost of swimming, flying, and running. *Science* 177: 222–228.

✔**QUANTITATIVE** About how much more costly is it to run than to swim for animals with a body mass of 100 g?

Size is also of paramount importance in determining how organisms locomote through fluids. Fluids have an inherent viscosity, or stickiness, that is of minor concern to animals that are very large and/or fast (thus have high inertia), like dolphins, which can glide through water with little effort. But viscosity is of enormous concern to animals that are very small and/or slow (thus have low inertia), like plankton. If you were the size of a grain of rice and went for a swim in a lake, the water would feel like corn syrup.

Size also affects the cost of locomotion. Physiologist Knut Schmidt-Nielsen, in the early 1970s, was the first to make an energy comparison between different modes of locomotion. To account for widely different body sizes and distances traveled, he normalized all the data to the energy cost in calories per gram of body per kilometer traveled and then plotted the data on a log–log plot (see **BioSkills 6** in Appendix B).

Schmidt-Nielson's data showed that larger organisms use less energy per gram of body tissue per distance traveled than smaller organisms—all the lines slope down to the right (**FIGURE 48.18**). The data also show that for any given body size, running is a more expensive mode of transport than flying, and swimming is the least expensive.

Many other aspects of locomotion can also be plotted as a function of body size. The results reveal general principles that can be used to make predictions for diverse animals, even those that are extinct.

The take-home message? The laws of physics establish very definite constraints on the realm of possible sizes, shapes, and modes of locomotion in animals. Understanding these principles helps illuminate themes and variations among the diversity of animals.

CHAPTER 48 REVIEW

For media, go to MasteringBiology (MB)

If you understand that . . .

48.1 How Do Muscles Contract?

- The muscles of vertebrate limbs are composed of muscle cells, called muscle fibers, that contain threadlike contractile elements called myofibrils, each divided into contractile units called sarcomeres.

- Sarcomeres appear striated, or banded, due to the aligned arrangement of thick filaments (myosin) and thin filaments (actin).

- Sarcomeres shorten when thick filaments of myosin slide past thin filaments of actin in a series of binding events mediated by the hydrolysis of ATP.

- Calcium ions play an essential role in muscle contraction by making the actin in thin filaments available for binding by myosin.

- Acetylcholine released from somatic motor neurons is the neurotransmitter that stimulates contraction of skeletal muscle.

✔ You should be able to predict the primary symptom of botulism, which occurs when a toxin prevents release of acetylcholine from the neuromuscular junction. Explain your answer.

48.2 Muscle Tissues

- Smooth muscle lines bronchioles, blood vessels, the gastrointestinal tract, and certain reproductive organs. Smooth muscle cells are small and unstriated and have a single nucleus. Contractions are involuntary.

- Cardiac muscle occurs in the heart and forces blood through the circulatory system. Cardiac muscle cells are striated, branched, and connected to one another by intercalated discs. Contractions are involuntary.

- Most skeletal muscles are attached to the skeleton and are responsible for voluntary movement of the body. Skeletal muscle cells are long, striated, and multinucleate.

- Skeletal muscle fibers are specialized to contract slowly or quickly and to have a high or low endurance. These properties depend on the concentration of myoglobin present and the use of aerobic respiration and/or glycolysis for the production of ATP.

- Skeletal muscle fibers are organized in a parallel arrangement, which maximizes shortening, or a pennate pattern, which maximizes force production.

✔ You should be able to describe the major structural and functional differences among the three classes of muscle tissue.

48.3 Skeletal Systems

- Hydrostatic skeletons are composed of a body wall in tension surrounding a fluid or soft tissue under compression.

- Endoskeletons are internal skeletons, surrounded by soft tissue. In vertebrates, the jointed endoskeleton is composed of bones, cartilages, tendons, and ligaments.

- Exoskeletons are external skeletons, enclosing soft tissue. In arthropods, the muscles occur within the rigid, jointed cuticle composed of chitin, proteins, and sometimes minerals. Exoskeletons must be shed to enable growth.

- Movement is based on antagonistic muscle groups that act on a skeleton. In vertebrates and arthropods, these muscles are called flexors and extensors. They change the joint angle between rigid skeletal segments—especially in limbs.

✔ You should be able to compare and contrast the types of shape changes that are caused by muscle contractions in the different types of skeletons.

48.4 Locomotion

- Locomotion is movement relative to the environment and requires the transmission of muscle forces to the land, water, or air surrounding the animal.

- Locomotion can be studied at different levels: material properties, structures, motions, forces, and energetics, and within the larger context of evolution and ecology.

- Locomotion on land is usually dominated by gravitational and inertial forces. Swimmers must overcome drag. Fliers must overcome drag and must generate enough lift to overcome gravitational forces.

- Body size is important to the mechanics of locomotion.

✔ You should be able to give general examples of how physical constraints affect the locomotion of animals on land, in water, and in air.

You should be able to . . .

✔ TEST YOUR KNOWLEDGE

Answers are available in Appendix A

1. Which of the following classes of muscle is/are voluntary?
   a. skeletal muscle
   b. cardiac muscle
   c. smooth muscle
   d. all of the above

2. Which of the following is a neurotransmitter that stimulates contraction of skeletal muscle?
   a. norepinephrine
   b. adrenaline
   c. acetylcholine
   d. calcium

3. In muscle cells, myosin molecules continue moving along actin molecules as long as
   a. ATP is present and troponin is not bound to Ca^{2+}.
   b. ADP is present and tropomyosin is released from intracellular stores.

   c. ADP is present and intracellular acetylcholine is high.
   d. ATP is present and intracellular Ca^{2+} is high.

4. True or false: The postural muscles of your legs are composed mostly of slow muscle fibers.

5. Which of the following is critical to the function of most exoskeletons, endoskeletons, and hydrostatic skeletons?
   a. Muscles interact with the skeleton in antagonistic groups.
   b. Muscles attach to each of these types of skeleton via tendons.
   c. Muscles extend joints by pushing skeletal elements.
   d. Segments of the body or limbs are extended when paired muscles relax in unison.

6. True or false: A large animal will experience twice the gravitational forces of a small animal half its size if their geometries are the same.

7. How did data on sarcomere structure inspire the sliding-filament model of muscle contraction? Explain why the observation that muscle cells contain many mitochondria and extensive smooth endoplasmic reticulum turned out to be logical once the molecular mechanism of muscular contraction was understood.

8. If a sprinter began an endurance running program, which of the following would occur?
   a. Some slow fibers would become fast fibers.
   b. Some fast fibers would become slow fibers.
   c. Some fibers would develop a higher mitochondrial density.
   d. Some fibers would develop a lower mitochondrial density.

9. Acetylcholine has very different effects on cardiac muscle and skeletal muscle. Explain.

10. Rigor mortis is the stiffening of a body after death that occurs when myosin binds to actin but cannot unbind. Why does myosin bind, and what prevents it from unbinding?

11. R. McNeill Alexander discovered that the arch of the human foot operates like a spring during running. Predict how a runner's oxygen consumption would change if a runner wore shoes that prevented the arches from changing shape. Explain your reasoning.

12. Explain how the physical constraints of locomotion on land, in water, and in air influence convergent evolution.

13. Predict the effect of ingestion of the toxin atropine on heart rate. Atropine is a naturally occurring compound in many poisonous nightshade plants. It blocks acetylcholine receptors in the heart. Explain your logic.

14. The force exerted by the shortening of a sarcomere depends on the length of the sarcomere when the stimulus is received. Based on your understanding of the sliding-filament model, predict how force would differ if the sarcomere were in a greatly stretched state at the beginning of the contraction versus in an average resting position. Explain your reasoning.

15. **QUANTITATIVE** The speed at which you switch from a walk to a run can be predicted using what is called the Froude number, based on the relative importance of gravitational and inertial forces of your pendulum-like walking gait.

$$\text{Froude number} = \frac{(\text{speed of locomotion})^2}{\text{gravitational acceleration} \times \text{leg length}}$$

Most mammals change from a walk to a run at a Froude number of 0.5. If gravitational acceleration is 9.8 m/s^2, and your leg length is 0.9 m, at what speed are you likely to switch from a walk to a run?
   a. 2.0 m
   b. 1.9 m/s
   c. 2.1 m/s
   d. 2.0 m/s^2

16. If you have seen the film *Jurassic Park*, you can imagine the nightmarish sight of *Tyrannosaurus rex* in high-speed pursuit of you as you flee in a jeep. Evaluate whether the data collected on the biomechanics of locomotion in living species are sufficient to determine how fast a large dinosaur like *T. rex* could run.

49 Chemical Signals in Animals

In this chapter you will learn that

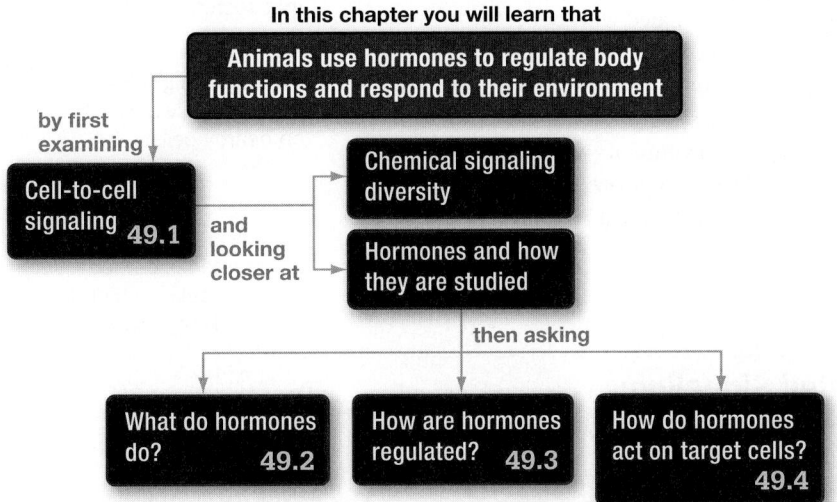

Animals use hormones to regulate body functions and respond to their environment

by first examining

Cell-to-cell signaling 49.1

and looking closer at

Chemical signaling diversity

Hormones and how they are studied

then asking

What do hormones do? 49.2

How are hormones regulated? 49.3

How do hormones act on target cells? 49.4

The spectacular transformation that occurs during insect metamorphosis is triggered by chemical signals called hormones.

This chapter is part of the Big Picture. See how on pages 840–841.

In response to sights, sounds, and other sensory stimuli, an animal's nervous system sends rapid messages, in the form of action potentials, to precise locations in the body. In many cases, these messages result in immediate, temporary responses such as muscle contractions and movement.

In response to changes in external or internal conditions, cells in the central nervous system (CNS) or the endocrine system release certain molecules. These molecules produce longer-term responses in a broad range of tissues and organs.

The **endocrine system** is a collection of organs and cells that secrete chemical signals into the bloodstream. A chemical signal that circulates through body fluids and affects distant target cells is called a **hormone.**

✔ When you see this checkmark, stop and test yourself. Answers are available in Appendix A.

As you read this, a large suite of hormones is coursing through your circulatory system. These molecules are regulating the maturation of sperm or eggs by your reproductive system, changing the composition of the urine forming in your kidneys, and controlling the release of digestive enzymes in your gastrointestinal tract. Earlier in your life, changes in hormone concentrations led to the dramatic physical changes associated with puberty.

The goal of this chapter is to explore how hormones and other types of internal chemical signals work in animals. Together, animal nervous systems and endocrine systems process information about the internal and external environment—a function that is one of the five key attributes of life (see Chapter 1).

Let's begin with an overview of chemical signaling systems and then plunge into analyzing how hormones regulate the activity of target cells.

49.1 Cell-to-Cell Signaling: An Overview

Animal chemical signals are present in extremely low concentrations but can have enormous effects on their target cells. Unlike action potentials, which are electrical impulses that have a short-term effect on a single cell or on a small population of adjacent cells, the messages that chemical signals carry can have a relatively long-lasting effect.

In combination, electrical and chemical signals allow animals to coordinate the activities of cells throughout the body. They are the mechanism responsible for maintaining trillions of cells as an integrated unit called an individual.

Major Categories of Chemical Signals

The chemical signals found in animals have diverse structures and functions. **TABLE 49.1** summarizes how biologists go about organizing the diversity of chemical signals, based on where the molecules originate and where they act.

Notice that the names for most of the categories use the Greek word root *crin*, meaning separated. Its use captures something essential about how chemical signals act: They are released from cells and thus are separated from them.

It's important to note that these five classes of chemical messenger do not coincide with five structurally distinct classes of molecules. For example, the endocrine signals found in a particular organism routinely belong to several families of chemical compounds, ranging from amino acid derivatives to lipids. And a particular family of molecules—say, peptides or the lipids called steroids—may function as endocrine, autocrine, and paracrine signals in the same individual.

Autocrine Signals Act on the Same Cell That Secretes Them
Translated literally, autocrine means same-separated. The name is appropriate because **autocrine** signals affect the same cell that releases them.

Perhaps the best-studied autocrine signals are **cytokines** ("cell-movers"). Most cytokines amplify the response of a cell

Type of Chemical Signal	Source and Target
Autocrine signals act on the same cell that secretes them.	
Paracrine signals diffuse locally and act on nearby cells.	
Endocrine signals are hormones carried between cells by blood or other body fluids.	
Neural signals diffuse a short distance between neurons.	
Neuroendocrine signals (neurohormones) are hormones released from neurons.	

to a stimulus. An example is interleukin 2, which in the course of fighting an infection is synthesized and released by a type of white blood cell called a T cell. Interleukin 2 activates T cells to help eliminate the infection. It also causes the cells to divide repeatedly, producing more activated T cells for host defense.

Paracrine Signals Act on Neighboring Cells   Translated literally, **paracrine** means beside-separated. Paracrine signals diffuse locally and act on target cells near the source cell. Cytokines, for example, may act as paracrine signals as well as autocrine signals, because they can trigger responses by other nearby cells of the immune system (see Chapter 51).

It is common, in fact, to observe that a single chemical messenger can be assigned to more than one category of signal, based on its mode of action. Like some cytokines, the cell–cell signals named **insulin, glucagon,** and **somatostatin** cross categories. These molecules are produced by three distinct populations of cells within a region of the **pancreas** called the islets of Langerhans. The molecules act on nearby pancreatic cells as paracrine signals and ensure a smooth, steady response to

(a) Endocrine pathway

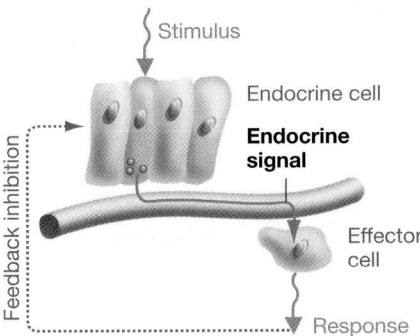

(b) Neuroendocrine pathway

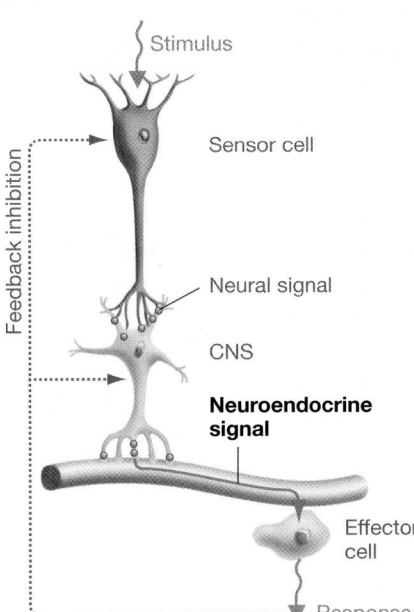

(c) Neuroendocrine-to-endocrine pathway

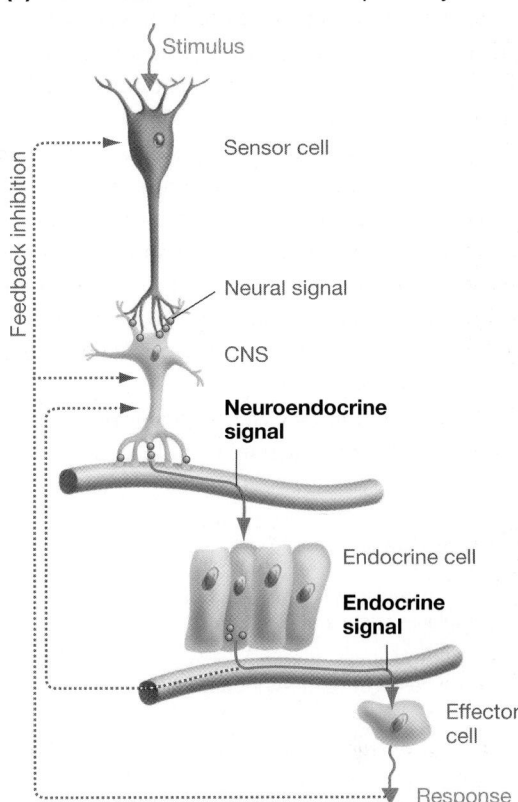

FIGURE 49.1 Hormones Act via Three Pathways and Are Regulated by Negative Feedback.

changing blood-glucose levels. But they also act as hormones, in that they are released into the blood or other body fluids and affect distant cells—in this case, controlling the concentration of glucose in the blood.

Endocrine Signals Are Hormones **Endocrine** ("inside-separated") signals are carried to distant cells by blood or other body fluids. The cells that produce endocrine signals may be organized into discrete organs called **glands** or may be interspersed among the cells of other organs—as are the islets of Langerhans in the pancreas.

Because hormones are well studied and particularly important to understanding how animals work, they serve as the focus of this chapter. Many or most of the principles discussed are relevant to the other categories of animal cell–cell signals as well, however.

Neural Signals Are Neurotransmitters You might recall that when an action potential arrives at a synapse, it triggers the release of neurotransmitters that bind to receptors on the postsynaptic cell and induce a change in membrane potential—altering the tendency for the postsynaptic cell to fire action potentials (see Chapter 46).

Neural signaling can be very fast, because action potentials propagate rapidly and neurotransmitters have to diffuse only a short distance—across the tiny gap between two neurons, called the synaptic cleft. Neural signals also tend to be short-lived, because the signaling molecules are broken down or taken back up by the presynaptic cell.

Neuroendocrine Signals Act at a Distance Even though they are released from neurons, **neuroendocrine** ("nerve-inside-separated") signals are considered hormones and are therefore often called **neurohormones.** They share a key attribute with endocrine signals: They act on distant cells. Unlike neural signals, they do not act at the adjacent synapse.

Antidiuretic hormone (ADH; also called vasopressin) is a particularly well studied neuroendocrine signal. ADH is produced by neurons that originate in the hypothalamus of the brain. But instead of acting as a neural signal, ADH acts on cells in the collecting duct of the kidney to help regulate water excretion (see Chapter 43).

Hormone Signaling Pathways

In plants, sensory cells perceive changes in the environment and broadcast a hormonal signal that triggers an appropriate response from effector cells (see Chapter 40). Some animal hormones are also sent directly from endocrine cells to effector cells, in response to a stimulus (**FIGURE 49.1a**). But frequently, hormonal signaling in animals involves additional steps.

In many cases, information about external or internal conditions is gathered by sensory receptors and then integrated by neurons in the central nervous system (CNS) before the production of a hormonal signal. Neurons in the CNS respond by releasing neuroendocrine signals that act on effector cells directly (**FIGURE 49.1b**) or stimulate cells in the endocrine system, which respond by producing a hormone (**FIGURE 49.1c**).

All three types of signaling pathway—direct from an endocrine cell, direct from the CNS, or CNS-to-endocrine-system—are regulated by **negative feedback,** or **feedback inhibition.** In feedback inhibition, the product of a process inhibits its production. Negative feedback is key to homeostasis (see Chapter 42).

In the endocrine pathway, the hormone produced by the effector cells feeds back on endocrine cells, lowering production of the hormone and down-regulating the response (Figure 49.1a). An effector's response also feeds back to cells that initiate the neuroendocrine and neuroendocrine-to-endocrine pathways. A change in input from these cells then lowers production of the signal and reduces the response (see Figures 49.1b and 49.1c). Neuroendocrine-to-endocrine signaling pathways have an additional layer of regulation, because the endocrine signal usually inhibits production of the neuroendocrine signal (Figure 49.1c).

The take-home message? The nervous system and endocrine system are tightly integrated. Endocrine signals are released in response to electrical signals; in turn, endocrine signals modulate the electrical signals transmitted by the nervous system.

Feedback inhibition in the endocrine system is analogous to temperature control by a heat-sensitive thermostat. If the temperature is too high, the thermostat sends a signal that turns the furnace off; if the temperature is too low, the thermostat sends a signal that turns the furnace on. The result is a constant air temperature. In animal cell–cell signaling, feedback inhibition reduces production or secretion of the hormone, or both.

✔ If you understand this concept, you should be able to explain which parts of the thermostat analogy correspond to the sensory input, CNS, cell-to-cell signal, and effector in a hormone signaling pathway. You should also be able to predict what happens to hormone concentrations when feedback inhibition fails in a hormone signaling pathway.

Hypothalamus

Growth-hormone-releasing hormone: stimulates release of GH from pituitary gland

Corticotropin-releasing hormone (CRH): stimulates release of ACTH from pituitary gland

Thyrotropin-releasing hormone: stimulates release of TSH from thyroid gland

Gonadotropin-releasing hormone (GnRH): stimulates release of FSH and LH from pituitary gland

Antidiuretic hormone (ADH): promotes reabsorption of H_2O by kidneys

Oxytocin: induces labor and milk release from mammary glands in females

■ Polypeptides
■ Amino acid derivatives
■ Steroids

Pineal gland

Melatonin: regulates sleep-wake cycles and seasonal reproduction

Anterior pituitary gland

Growth hormone (GH): stimulates growth factors

Adrenocorticotropic hormone (ACTH): stimulates adrenal glands to secrete glucocorticoids such as cortisol

Thyroid-stimulating hormone (TSH): stimulates thyroid gland to secrete thyroid hormones

Follicle-stimulating hormone (FSH) and luteinizing hormone (LH): stimulate production of gametes and sex steroid hormones

Prolactin (PRL): stimulates mammary gland growth and milk production in females

Parathyroid glands (on dorsal side of thyroid gland)

Parathyroid hormone (PTH): increases blood Ca^{2+}

Thyroid gland

Thyroid hormones, thyroxine (T_4) and triiodothyronine (T_3): increase metabolic rate and heart rate; promote growth

Adrenal glands

Epinephrine: produces many effects related to short-term stress response

Cortisol: produces many effects related to short-term and long-term stress responses

Aldosterone: increases reabsorption of Na^+ by kidneys

Pancreas (islets of Langerhans)

Insulin: decreases blood glucose

Glucagon: increases blood glucose

Ovaries (in females)

Estradiol: regulates development and maintenance of secondary sex characteristics in females; other effects

Progesterone: prepares uterus for pregnancy

Kidneys

Erythropoietin (EPO): stimulates synthesis of red blood cells

Testes (in males)

Testosterone: regulates development and maintenance of secondary sex characteristics in males; other effects

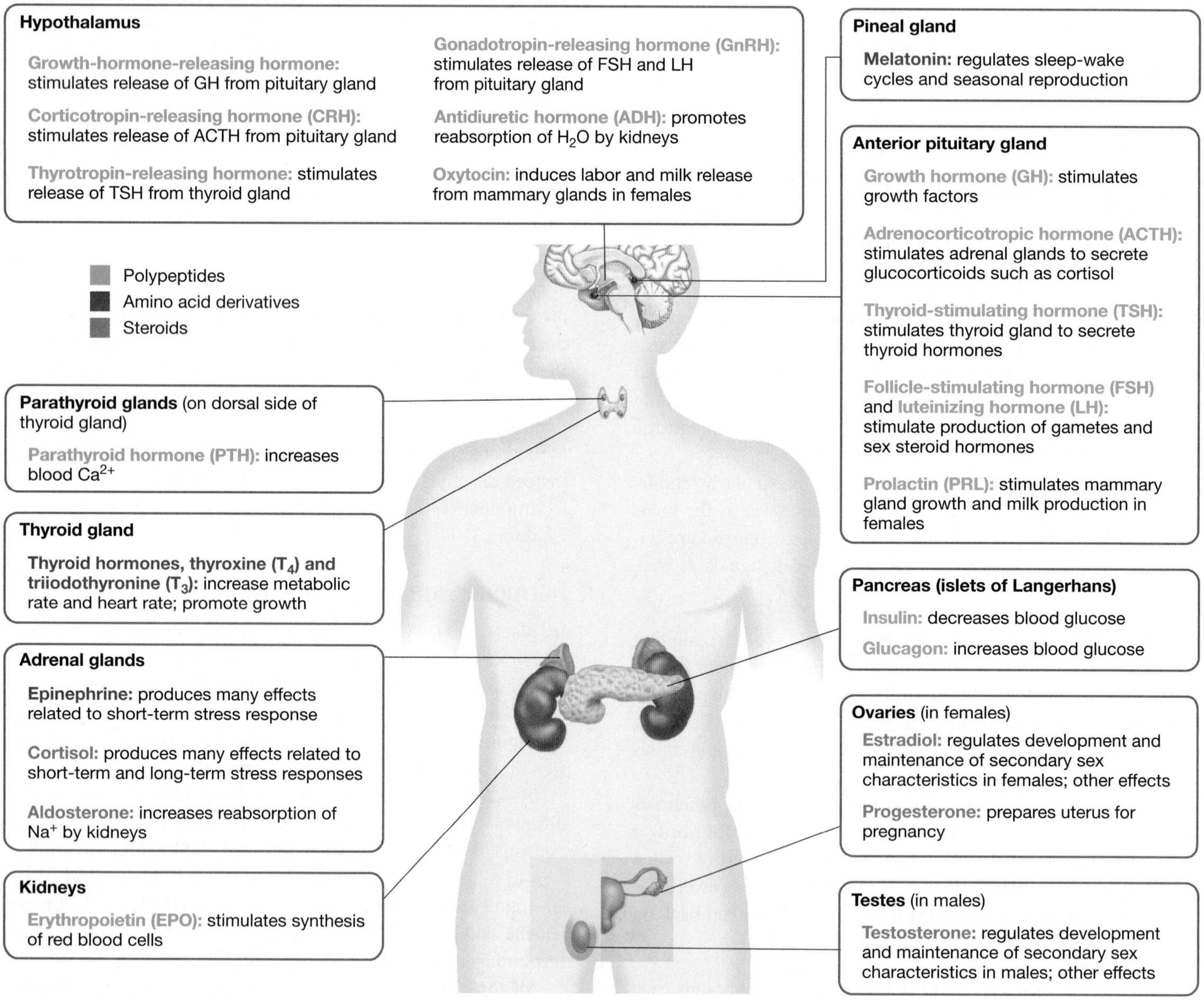

FIGURE 49.2 Humans Possess a Diverse Array of Endocrine Glands and Hormones. This list is only partial. The heart, gastrointestinal tract, adipose (fat) tissue, and many other organs and cells also produce hormones.

What Makes Up the Endocrine System?

The endocrine system is the collection of cells, tissues, and organs responsible for hormone production and secretion. Organs that secrete a hormone into the bloodstream are called **endocrine glands.**

The tissues and organs that make up the endocrine system vary widely among animals. For example, neurons that manufacture and secrete hormones are particularly important in insects, where they regulate molting, metamorphosis, and other processes. Salmon have an unusual gland that secretes a hormone responsible for regulating calcium ion concentration.

Even within one species, the diversity of endocrine system components can be impressive. For example, **FIGURE 49.2** shows major human glands with endocrine functions. As mentioned earlier, hormone-secreting cells are not always organized into discrete glands. In many cases, they are located in other kinds of organs. (The islets of Langerhans were mentioned earlier as an example.)

It's important to note that not all glands in the body are part of the endocrine system. **Exocrine glands,** in contrast to endocrine glands, deliver their secretions through outlets called ducts into a space other than the circulatory system. Most of the digestive glands are either exocrine glands—an example is the salivary glands—or mixed endocrine and exocrine glands, such as the pancreas (see Chapter 44). The exocrine cells of the pancreas secrete digestive enzymes through ducts into the intestine. The endocrine portion of the pancreas consists of cells that secrete insulin and glucagon directly into the bloodstream.

At first glance, the diversity of hormones, glands, and their effects can seem overwhelming, especially considering that

Figure 49.2 represents just a partial catalog for a single species. But the picture can be simplified somewhat by recognizing that most animal hormones belong to one of just three major structural families: polypeptides, amino acid derivatives, or steroids.

Chemical Characteristics of Hormones

FIGURE 49.3 illustrates the three major classes of chemicals that can act as hormones in animals:

1. Polypeptides, which are chains of amino acids linked by peptide bonds (see Chapter 4)

2. Amino acid derivatives

3. Steroids, which are a family of lipids distinguished by a four-ring structure (see Chapter 6)

Secretin, a hormone produced in the small intestine to stimulate the exocrine portion of the pancreas, is a polypeptide; the hormone epinephrine is synthesized in the medulla of the **adrenal glands** from the amino acid tyrosine; and the hormone cortisol is synthesized in the cortex of the adrenal glands from the steroid cholesterol.

Hormone Concentrations Are Low, but Their Effects Are Large
Hormones vary widely in structure but share a common characteristic: They have profound effects on individuals, even though they are present at vanishingly small concentrations. As an example, consider research that led to the discovery of **growth hormone (GH).**

Several researchers noted that rats and other laboratory animals stopped growing when their pituitary glands were removed. Based on this observation, it was widely suspected that

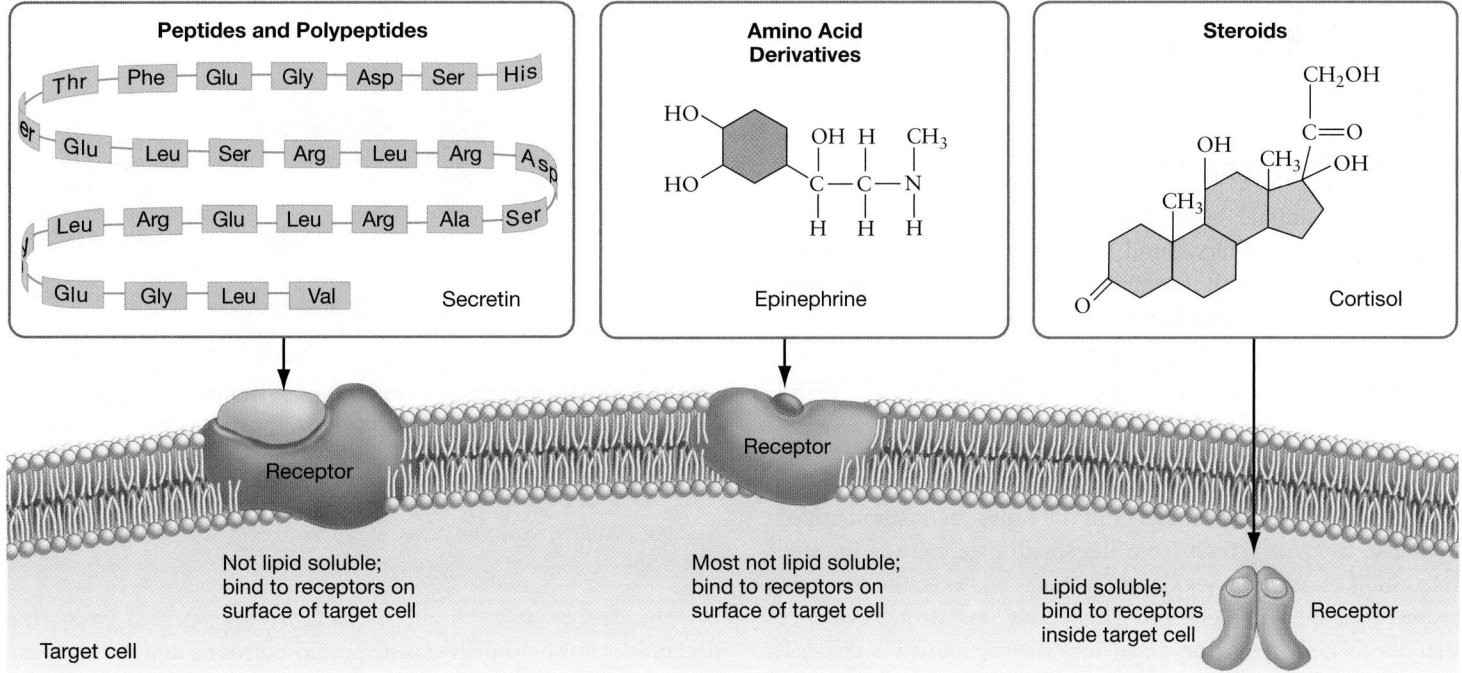

FIGURE 49.3 Most Animal Hormones Belong to One of Three Chemical Families.

the pituitary produces a chemical signal that promotes cell division and other aspects of growth.

To test this hypothesis, a research group purified a polypeptide from cow pituitary glands, injected the polypeptide into lab rats, and documented rapidly accelerated growth. When the researchers injected 0.01 mg of the molecule each day for nine days into rats that lacked pituitary glands, the width of the growth plates in the rats' leg bones increased by 50 percent. The individuals also gained an average of 10 g compared with rats that lacked a pituitary and did not receive the hormone treatment. Stated another way, an additional 0.09 mg of hormone led to a weight gain of 10,000 mg.

Further, 1 kg of cow pituitary tissue yielded a mere 0.04 g of growth hormone. By mass, the hormone makes up just 4 one-thousandths of 1 percent of the cow pituitary.

Only Some Hormones Can Cross Cell Membranes Given that small amounts of polypeptide, amino-acid-derived, and steroid hormones have large effects on the activity of cells, organs, and systems, how do the three types of hormones differ? The major difference is that steroids are lipid soluble, but polypeptides and most amino acid derivatives are not (see Figure 49.3).

Important exceptions to this rule are the **thyroid hormones, triiodothyronine** (also known as T_3) and **thyroxine** (also known as T_4), that are produced by the **thyroid gland.** The thyroid hormones are derived from the amino acid tyrosine but are lipid soluble.

Differences in solubility are important because steroids and thyroid hormones cross plasma membranes much more readily than do other types of hormones. To affect a target cell, all polypeptides and most amino acid derivatives bind to a receptor on the cell surface. Lipid-soluble hormones, in contrast, can diffuse through the plasma membrane and bind to receptors inside the cell.

Before exploring the consequences of this difference in more detail, let's consider how researchers discovered the array of hormones introduced here.

How Do Researchers Identify a Hormone?

Research on animal hormones began in earnest in the early 1900s, when researchers found that adding dilute hydrochloric acid (HCl) to the small intestine, to mimic the arrival of acidic material from the stomach, stimulated the pancreas to secrete compounds that neutralized the acid (see Chapter 44).

How could stimulating the small intestine lead to a response by the pancreas? The working hypothesis was that a chemical traveled from the small intestine to the pancreas by way of the bloodstream, to signal the arrival of acid.

To test this idea, researchers did the experiment summarized in **FIGURE 49.4**. An extract from the small intestine was injected into blood vessels in a dog's neck. A short time later, the pancreas secreted an alkaline solution. This result was strong evidence that the extract from the small intestine contained a cell–cell signaling molecule. The hormone was later purified and named **secretin.**

RESEARCH

QUESTION: How could stimulating the small intestine with acid cause the pancreas to secrete buffers that neutralize the acid?

HYPOTHESIS: A chemical messenger from the small intestine reaches the pancreas via the bloodstream.

NULL HYPOTHESIS: There is no blood-borne chemical messenger from the small intestine that acts on the pancreas.

EXPERIMENTAL SETUP:

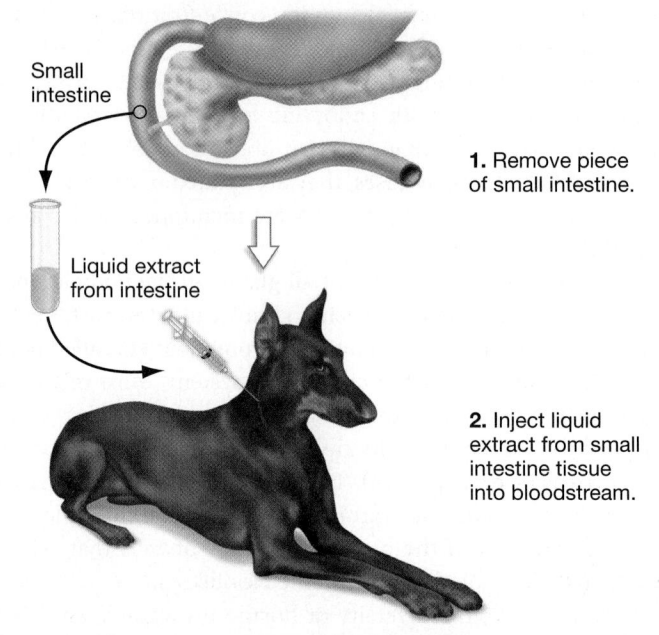

Small intestine

Liquid extract from intestine

1. Remove piece of small intestine.

2. Inject liquid extract from small intestine tissue into bloodstream.

PREDICTION: Liquid extract in bloodstream will signal pancreas to secrete buffer.

PREDICTION OF NULL HYPOTHESIS: Liquid extract in bloodstream will not signal pancreas to secrete buffer.

RESULTS:

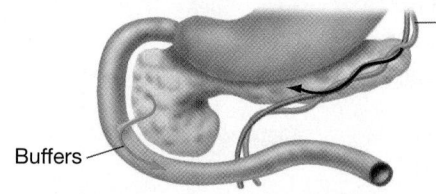

Blood vessels

A short time after extract is injected into bloodstream, pancreas secretes buffer into intestine.

Buffers

CONCLUSION: A hormone from the small intestine (secretin) signals cells in the pancreas via the bloodstream.

FIGURE 49.4 Researchers Studied Digestive Processes to Find Experimental Evidence of a Hormone. Follow-up work succeeded in isolating the molecule secretin.

SOURCE: Bayliss, W. M., and E. H. Starling. 1902. The mechanism of pancreatic secretion. *Journal of Physiology* 28: 325–353.

✔**QUESTION** What would be an appropriate control in this experiment?

In essence, using liquid extracts from suspected endocrine glands is a way to deliver a suspected hormone and evaluate the effect. The opposite strategy has also been productive: observing the consequences of removing specific organs or tissues. For example, removing the adrenal glands of an animal rapidly leads to death due to low blood sodium, low blood sugar, and low blood pressure. But injecting adrenal extracts into the blood of animals lacking adrenal glands corrects these abnormalities. Results like these provided strong evidence that hormones secreted by the adrenal glands help regulate blood-sugar concentrations and blood pressure.

A Breakthrough in Measuring Hormone Levels

Documenting an association between a particular gland or hormone and an effect in the body is just a first step. To understand hormone action, researchers have to figure out how these signals help animals stay alive and produce offspring.

The key to this goal is the ability to quantify the levels of hormones circulating in the bloodstream. Given that hormones are present in such small amounts, this ability eluded researchers for decades.

The breakthrough came in the 1950s when Rosalind Yalow developed the **radioimmunoassay**, a feat for which she later received the Nobel Prize in Medicine. In a radioimmunoassay, the quantity of hormone in a blood sample can be estimated by adding radioactively labeled hormones that compete with the unknown amount of hormone to bind to an antibody (see **BioSkills 9** in Appendix B). This technique revolutionized the field of endocrinology and the treatment of endocrine diseases, allowing the precise concentrations of hormones to be measured from blood samples drawn from animals or patients. Currently, radioimmunoassays are used every day in endocrine research and in medical laboratories. Much of what researchers know about hormone function was learned in studies employing the radioimmunoassay.

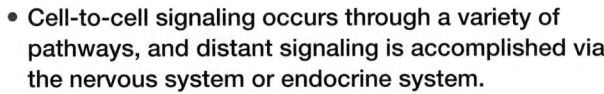

check your understanding

C Y U

If you understand that . . .
- Cell-to-cell signaling occurs through a variety of pathways, and distant signaling is accomplished via the nervous system or endocrine system.
- Three main chemical families of hormones affect cells in different ways.

✔ **You should be able to . . .**
1. Discuss why it is often difficult to differentiate between the functions of the nervous and endocrine systems.
2. Predict the location of thyroid hormone receptors.

Answers are available in Appendix A.

49.2 What Do Hormones Do?

At the beginning of this chapter, you read that hormones are chemical messengers. If so, what do hormones "say"?

A first step in answering this question is to recognize that a single hormone can exert a variety of effects. For example, the thyroid hormones, T_3 and T_4, stimulate metabolism in humans and thus oxygen consumption throughout the body. But they also promote growth, increase heart rate, and stimulate the synthesis of many important macromolecules.

A second step in grasping what hormones do is to recognize that several different hormones may affect the same aspect of physiology. Insulin, glucagon, epinephrine, and cortisol all influence glucose levels in the blood.

Some hormones have extremely diverse effects; in other hormones, the functions appear to overlap. These observations begin to make sense when hormone action is viewed in the context of the whole organism. Hormones coordinate the activities of cells in three arenas: (1) development, growth, and reproduction; (2) response to environmental challenges; and (3) maintenance of homeostasis.

Let's analyze each of these functions in turn.

How Do Hormones Direct Developmental Processes?

In animals, as in plants, hormones play a key role in regulating growth and development. Growth hormones and sex hormones play crucial roles in promoting cell division, increasing overall body size, and promoting sexual differentiation as an individual matures; certain hormones direct the development of particular cells and tissues at critical junctures in an individual's life.

Let's explore two of the most dramatic examples of hormonal control—metamorphosis in amphibians and in insects—and then survey other developmental processes that are affected by hormone action.

T_3's Role in Amphibian Metamorphosis Frogs, toads, and salamanders are called amphibians ("double-lives") because in most species, juveniles live in water while adults live on land. The process of changing from an immature, aquatic larva to a sexually mature, terrestrial frog, toad, or salamander is an example of **metamorphosis** ("change-form") (**FIGURE 49.5**; see page 998).

Two sets of complementary experiments, published in 1912 and 1916, established that frog metamorphosis depends on thyroid hormones. Researchers could induce frog tadpoles to undergo metamorphosis by feeding them ground-up thyroid glands from horses; they could prevent metamorphosis by surgically removing the tadpoles' thyroid glands.

Follow-up work showed that the thyroid hormone triiodothyronine, or T_3, is responsible for many of the changes observed in metamorphosis. In response to a signal from the brain, the pituitary gland secretes thyroid-stimulating hormone, which stimulates the thyroid gland to produce thyroxine (T_4). The T_4 is then converted to T_3 at the target tissues.

In juvenile amphibians, cells respond to increased levels of T_3 by growing and forming new structures, such as legs. Other structures—such as a tadpole's tail—disintegrate or are absorbed. Still other tissues change structure and function. For example,

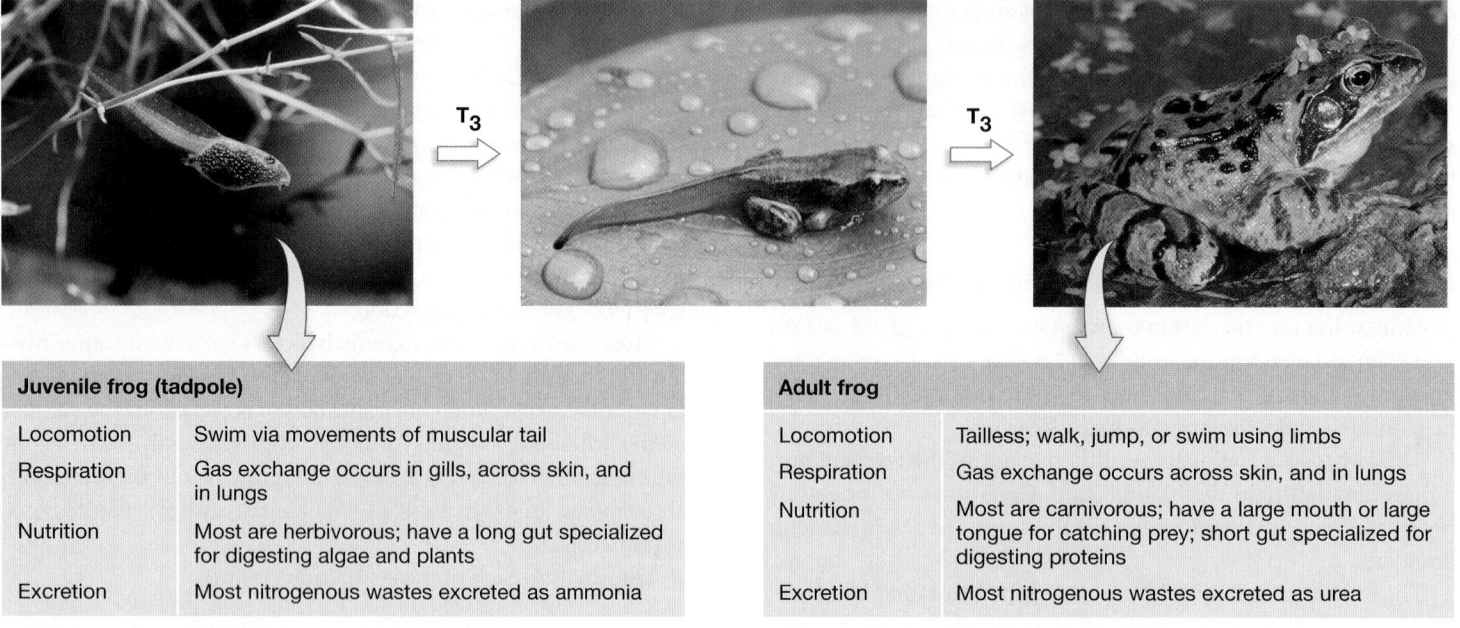

Juvenile frog (tadpole)	
Locomotion	Swim via movements of muscular tail
Respiration	Gas exchange occurs in gills, across skin, and in lungs
Nutrition	Most are herbivorous; have a long gut specialized for digesting algae and plants
Excretion	Most nitrogenous wastes excreted as ammonia

Adult frog	
Locomotion	Tailless; walk, jump, or swim using limbs
Respiration	Gas exchange occurs across skin, and in lungs
Nutrition	Most are carnivorous; have a large mouth or large tongue for catching prey; short gut specialized for digesting proteins
Excretion	Most nitrogenous wastes excreted as urea

FIGURE 49.5 Amphibian Metamorphosis Is a Continuous Process. When metamorphosis begins in a frog, toad, or salamander, the individual stays active and feeding. The continuous and gradual transition from juvenile to adult is mediated by T_3.

changes in existing cells are responsible for the switch from a tadpole's long intestine, specialized for digesting plant material, to an adult's short intestine, specialized for digesting insects and other prey. In the liver, cells respond to T_3 by manufacturing the enzymes required to excrete urea—the nitrogenous waste product released by adults—instead of the ammonia produced by tadpoles.

✔ If you understand the basic principles of hormone action, you should be able to suggest a hypothesis explaining why different frog cells can respond to T_3 in such different ways.

Hormone Interactions Regulate Insect Metamorphosis Some insects show a remarkable type of juvenile-to-adult transition, called holometabolous metamorphosis (see Chapter 33). In species that undergo this process, juveniles are called larvae. As in amphibians, insect larvae look completely different from adults, live in different habitats, and eat different food.

As larvae grow, they undergo a series of molts in which they shed their old exoskeleton, expand their bodies, and produce a new exoskeleton. After a specific number of these juvenile molts, however, they secrete a tough case called a pupal case. Inside the pupal case, specific populations of larval cells give rise to a completely new adult body. The rest of the larval body disintegrates (**FIGURE 49.6**).

In insects, metamorphosis depends on interactions between two hormones. If **juvenile hormone (JH)** is present at a high concentration in the larva, surges of the hormone **ecdysone** induce the growth of a juvenile insect via molting. But if JH levels are low, ecdysone triggers a complete remodeling of the body—metamorphosis—and the transition to adulthood and sexual maturity.

Sexual Development and Activity in Vertebrates In mammals and other vertebrates, long-distance cell-to-cell signals play key roles as embryos develop. Hormones also direct anatomical and physiological changes that occur later in life. Some of the most important of these changes involve the reproductive organs.

FIGURE 49.6 Insect Metamorphosis Occurs during a Resting Stage. When metamorphosis begins in a holometabolous insect, the individual enters a resting stage called the pupa.

Events early in development dictate whether the sex organs, or **gonads,** of a vertebrate embryo become male (**testes**) or female (**ovaries**). This process is called primary sex determination. In mammals, primary sex determination depends on genes on sex chromosomes.

Once testes or ovaries develop, they begin producing different hormones. In human males, the early testes produce two hormones:

1. A steroid hormone called **testosterone** induces early development of the male reproductive tract.

2. A polypeptide hormone called **Müllerian inhibitory substance** inhibits development of the female reproductive tract.

In females, the ovaries produce the steroid hormone **estradiol,** which is in the family of molecules called **estrogens.** Estradiol is required for further development of the female reproductive tract.

Sex hormones also play a key role in the juvenile-to-adult transition. When humans reach early adolescence, for example, surges of sex hormones lead to the physical and emotional changes associated with **puberty.** These developmental changes create the adult phenotype and the ability to produce offspring.

In boys, surges of sex hormones lead to changes that include enlargement of the penis and testes and growth of facial and body hair. In girls, an increased concentration of estradiol leads to the enlargement of breasts, the onset of menstruation, and other changes. In both sexes, a growth surge begins at puberty. This growth is stimulated by growth hormone (GH) produced in the pituitary gland (see Figure 49.2). GH regulates growth factors, which are signals that control the cell cycle (see Chapter 12). Growth originates in the epiphyseal plate, a small piece of cartilage separating the shaft from the end at both extremities of long bones.

Puberty is associated with a growth spurt because the effect of GH on the human skeleton is enhanced by the action of sex hormones, which surge during adolescence. Although GH and sex hormones continue to be produced long after puberty, growth in humans stops when the epiphyseal plates are replaced with bone, and no further growth is possible.

The sex hormones continue to play a key role in adults. In humans, sex hormones are instrumental in regulating sperm production and the menstrual cycle (see Chapter 50). The result of this steady release of sex hormones is that humans can mate year-round. In many animals, reproductive behavior is instead confined to specific times of the year. In these species, environmental cues such as increasing day length, warmth, or the onset of seasonal rains trigger the release of sex hormones (see Chapter 53).

How Photoperiod Affects Sex Hormone Release The increase in day length—or increasing **photoperiod**—during spring is particularly important in stimulating the release of sex hormones in seasonally reproducing mammals, lizards, and birds. The lengthening spring photoperiod is sensed by photoreceptors, which are sensory receptors that respond to light (see Chapter 47). These photoreceptors send signals to the hypothalamus, a brain region that initiates a series of signals directing production of sex hormones.

The location of the photoreceptors depends on the animal. In mammals, photoreceptors in the retinas of the eyes sense light and send signals to the **pineal gland** via a pathway leading through the brain and spinal cord. The pineal gland secretes the hormone **melatonin,** which relays photoperiodic information to the hypothalamus. It also regulates sleep–wake cycles.

Maximal melatonin secretion occurs in the dark, so stimulation of the pineal gland by photoreceptors *reduces* melatonin secretion. Animals therefore experience a daily rhythm in melatonin levels, which are highest at night. As photoperiod increases in spring, there are fewer hours of dark, resulting in lower *overall* levels of melatonin (**FIGURE 49.7**). This decline in melatonin "informs" the hypothalamus to stimulate the testes and ovaries to make sex hormones.

Many diurnal lizard species do not rely solely on retinal photoreceptors. These lizards have a small hole in the top of their skull, covered only by a thin layer of skin; this hole is often called the pineal eye or parietal eye (**FIGURE 49.8**, see page 1000). The pineal gland lies just below this opening and responds to direct light. Interestingly, in birds, neither retinal nor pineal photoreceptors are responsible for relaying photoperiodic information to the hypothalamus. It appears that their photoreceptors are located diffusely throughout the brain.

The common thread among mammals, lizards, birds, and other vertebrates is that information about photoperiod is transduced from a signal detected by photoreceptors into a hormonal signal originating at the hypothalamus. The role of the hypothalamus is discussed in more detail in Section 49.3.

Some Chemicals Can Disrupt Hormone Signaling In 1962, Rachel Carson stunned the world with her book *Silent Spring,* in which she described how commercially produced synthetic chemicals—even at low concentrations—can adversely affect humans and

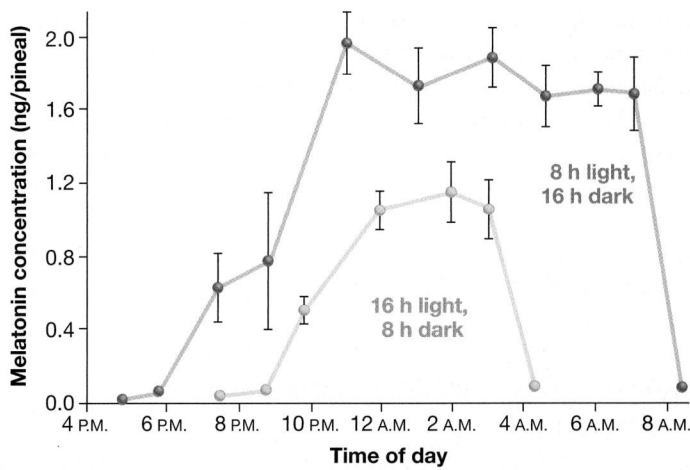

FIGURE 49.7 Melatonin Levels Are Affected by Time of Day and Photoperiod in Seasonal Breeders. Melatonin levels are highest at night, and they are higher overall when photoperiod is short (8 h light, 16 h dark) than when it is long (16 h light, 8 h dark).

DATA: Vanecek, J. 1998. *Physiological Reviews* 78: 687–721.

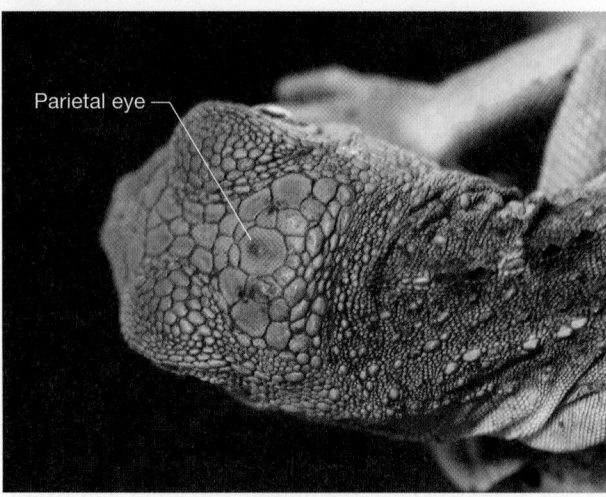
Parietal eye

FIGURE 49.8 Some Diurnal Reptiles Have a Parietal Eye. This tiny hole in the skull allows sunlight to directly stimulate photoreceptors in the pineal gland.

wildlife. Since then, evidence has continued to accumulate that many pesticides, industrial chemicals, and other pollutants interfere with normal endocrine function in many animal species.

In the early 1990s, scientists coined the term **endocrine disruptor** to describe chemicals that interfere with normal hormonal signaling. This interference can happen in a number of ways, including binding of the chemical to hormone receptors and altering the metabolism of hormones, thereby affecting physiological processes and behavior.

Many endocrine disruptors are **xenoestrogens**—foreign chemicals that bind to estrogen receptors and induce estrogen-like effects. For example, exposure to the widely used herbicide atrazine causes feminization and even sex change of male frogs, induces reproductive-tissue abnormalities and reduced spawning in fishes, and is associated with low sperm counts and increased risk of birth defects in humans.

Another well-known xenoestrogen is bisphenol A (BPA), an industrial chemical used in many plastic or metal products, including containers for food and water. Laboratory animals exposed to BPA show abnormal development of reproductive and brain tissues, along with increased risk of some types of cancers.

As research continues into the possible health effects of xenoestrogens and other endocrine disruptors, many states and countries are limiting or banning their use. For example, atrazine is currently banned in the European Union, and BPA is banned in some countries for use in products such as baby bottles.

How Do Hormones Coordinate Responses to Stressors?

When an animal is thrust into a dangerous or unpredictable situation, hormones play a part in both the short-term and long-term responses. Let's explore each of these response categories in turn.

Short-Term Responses to Stress The short-term reaction, called the **fight-or-flight response,** is triggered by the sympathetic nervous system (see Chapter 46). If you were being chased by a grizzly bear, action potentials from your sympathetic nerves would stimulate your adrenal medulla and lead to the release of the hormone **epinephrine,** also known as **adrenaline.** (The Greek word roots *epi* and *nephron* mean "top-kidney"; the Latin word roots *ad* and *renal* also mean "top-kidney.")

To determine how epinephrine affects the body, researchers injected human volunteers with a saline solution—as a control—or epinephrine. Each point graphed in the "Results" section of **FIGURE 49.9** represents data from one of the volunteers; the data in the table are average values from the seven study participants.

The data indicate dramatic increases in an array of physiological processes: concentrations of free fatty acids and glucose in the blood, pulse rate, blood pressure, and oxygen consumption by the brain. In addition, the volunteers reported strong subjective feelings of anxiety and excitement.

Other experiments showed that epinephrine redirects blood away from the skin and digestive system and toward the heart, brain, and muscles. Epinephrine relaxes smooth muscle in blood vessels leading to these tissues—increasing blood delivery.

Taken together, the responses to epinephrine lead to a state of heightened alertness and increased energy use that prepares the body for rapid, intense action such as fighting or fleeing. Epinephrine coordinates the activities of cells in many organs and systems throughout the body to prepare an individual to cope with a life-threatening situation.

Long-Term Responses to Stress If you have ever experienced the fight-or-flight response, you may recall that the state is short lived. Once an epinephrine "rush" wears off, most people feel exhausted and want to rest and eat.

What happens if the stress continues and turns into a long-term condition? In the course of a lifetime, it is not unusual for a person to experience periods of starvation or fasting, prolonged emotional distress, or chronic illness. How do hormones help humans and other animals cope with extended stress?

Early studies of long-term stress in human subjects suggested a role for the hormone **cortisol,** which is produced in the adrenal cortex. Increased levels of cortisol were found in airplane pilots and crew members during long flights, athletes who were training for intense contests, parents of children undergoing treatment for cancer, and college students who were preparing for final exams. Why?

What Does Cortisol Do? In humans, cortisol's primary role is to ensure the continuing availability of glucose for use by the brain during long-term stress. Cortisol manages three main processes that maintain glucose production:

1. Cortisol induces the synthesis of liver enzymes that make glucose from amino acids and other chemical precursors.

2. Cortisol makes adipose tissue—fat tissue—and resting muscles resistant to insulin. Insulin normally stimulates **adipocytes**

QUESTION: How does epinephrine affect the body?

HYPOTHESIS: Epinephrine causes changes involved in the fight-or-flight response.

NULL HYPOTHESIS: Epinephrine is not involved in the fight-or-flight response.

EXPERIMENTAL SETUP:

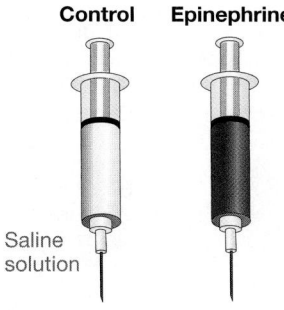

Control   Epinephrine

Saline solution

1. **Inject** human volunteers with saline solution or epinephrine.

2. **Document changes** in fatty-acid and glucose concentrations in blood, pulse rate, blood pressure, and oxygen consumption in brain.

PREDICTION: Epinephrine increases fatty-acid and glucose concentrations in blood, pulse rate, blood pressure, and brain oxygen consumption relative to controls.

PREDICTION OF NULL HYPOTHESIS: No differences in physiological state of individuals based on molecule injected.

RESULTS:

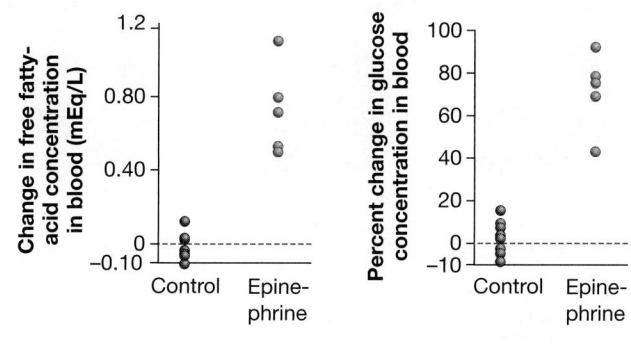

	Control	Epinephrine
Pulse rate (beats/min)	78.3	89.6
Blood pressure (average, mm Hg)	90.9	108.7
O₂ consumption in brain (cc O₂/100 g/min)	3.41	4.16

CONCLUSION: Epinephrine causes an array of changes associated with the fight-or-flight response.

FIGURE 49.9 Epinephrine Prepares the Body for Action.

SOURCES: King, B. D., L. Sokoloff, and R. L. Wechsler. 1952. The effects of *l*-epinephrine and *l*-nor-epinephrine upon cerebral circulation and metabolism in man. *Journal of Clinical Investigation* 31: 273–279. Mueller, P. S., and D. Horwitz. 1962. Plasma free fatty acid and blood glucose responses to analogues of norepinephrine in man. *Journal of Lipid Research* 3: 251–255.

✔ **QUESTION** Why did researchers bother to inject volunteers in the control group with saline? Why inject them with anything?

and resting muscle cells to remove glucose from the bloodstream. But when cortisol makes these cells resistant to insulin, glucose is reserved for use by the brain and exercising muscles.

3. Cortisol promotes the release of fatty acids—the body's major fuel molecules—from adipose tissue, for use by the heart and muscles.

Because of its importance in regulating blood glucose, cortisol is referred to as a **glucocorticoid.**

The long-term stress response comes at a high price, however, as any victim of a serious injury or illness knows. Glucocorticoids make amino acids available for glucose synthesis by promoting the degradation of contractile proteins in muscle. The resulting loss of muscle mass may cause severe weakness. Also, glucocorticoids impair wound healing and suppress immune and inflammatory responses. These processes are costly in terms of energy use, but reducing them makes the body more susceptible to infection.

The overall concept here is that the long-term stress response is a compromise—a fitness trade-off (see Chapter 42). The fuel requirements of the brain are met at the expense of other tissues and organs.

✔ If you understand the effects of chronically elevated cortisol in stressed individuals, you should be able to predict its effect on their ability to heal wounds and fight pathogens.

How Are Hormones Involved in Homeostasis?

Recall that homeostasis is the maintenance of relatively constant physical and chemical conditions inside the body. Homeostatic systems depend on three components (see Figure 42.14):

1. A sensory receptor that monitors conditions relative to a normal value, or set point

2. An integrator that processes information from the sensor

3. Effector cells that return conditions to the set point

In homeostatic systems, messages often travel from integrators to effectors in the form of hormones.

Leptin and Energy Reserves Healthy animals keep energy in reserve for use during periods of decreased food availability. This energy reserve typically takes the form of the lipid triglyceride (see Chapter 6). Triglyceride is an effective energy-storage molecule because large amounts of ATP can be generated when its three fatty-acid subunits are oxidized.

Although some triglyceride is present in muscle cells, most is stored in adipocytes. These cells make up the fat bodies found in insects and other species and the adipose tissue of mammals. Even a lean 70-kg human male stores enough energy in adipose tissue to survive for over 30 days without eating.

Is there a homeostatic system that regulates triglyceride stores? An answer started to emerge in the 1970s, when biologists began studying mutations in mouse genes called *obese* and *diabetic*. Homozygous *obese* (*ob/ob*) and *diabetic* (*db/db*) mice eat large quantities of food and move much less than heterozygous

ob/ob

ob/+

FIGURE 49.10 In Mice, Mutations in the *obese* and *diabetic* Genes Can Cause Obesity. These mice are siblings from the same litter. At the *ob* gene, the lean mouse is heterozygous and the obese mouse is homozygous.

or wild-type siblings do. The homozygotes also become extremely obese (**FIGURE 49.10**).

To test the hypothesis that the *obese* and *diabetic* gene products are involved in cell–cell signaling, researchers turned to an experimental technique called **parabiosis**, which works as follows:

1. Two closely related animals are surgically united by suturing the pelvis, shoulders, and abdominal walls together.

2. The newly created twin animal is allowed to recover from the procedure.

3. Within a short time, capillaries form between the two parabiotic partners. As a result, the two individuals develop a shared circulatory system.

4. The united circulatory system permits the passage of certain hormones and other long-lived molecules between the two partners. The system does not allow passage of molecules that are rapidly metabolized, such as glucose and fatty acids.

5. No new nerves grow between the two animals. As a result, they can influence one another through endocrine signals but not through electrical signals.

When researchers performed parabiosis between *db/db, ob/ob,* and lean mice, the results were striking (**FIGURE 49.11**):

- When a *db/db* animal was joined to either a lean animal or an *ob/ob* animal, the *db/db* mouse continued to eat and grow normally; but its partner stopped eating, lost weight, and eventually died of apparent starvation.

- When *ob/ob* animals were joined to lean animals, the *ob/ob* mice ate less food and gained weight less rapidly than did *ob/ob* mice joined to other *ob/ob* mice.

In addition, when two *db/db* animals were joined, both partners ate and grew as expected—both became obese. There was no difference between the two in survival.

To interpret these results, biologists hypothesized that mice produce a satiation, or "stop-eating," hormone—a negative feedback signal in homeostasis with respect to fat stores. The explanation

RESEARCH

QUESTION: Are the *diabetic* (*db*) and *obese* (*ob*) gene products involved in endocrine function?

HYPOTHESIS: The gene products are involved in hormonal signaling that affects appetite.

NULL HYPOTHESIS: The gene products are not involved in hormonal signaling.

EXPERIMENTAL SETUP:

Perform parabiotic surgery on closely related mice with different genotypes, so blood-borne products will pass between the mice in each pair.

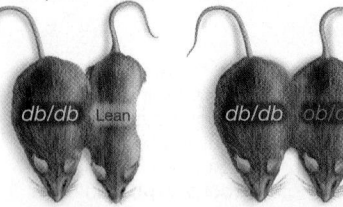 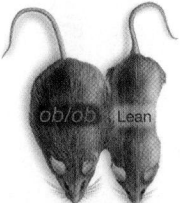

PREDICTION: *db/db* mice parabiosed to either *ob/ob* or lean mice will gain weight while their partners lose weight; *ob/ob* mice parabiosed to a lean mouse will lose weight.

PREDICTION OF NULL HYPOTHESIS: No changes in phenotypes of parabiotic mice will occur.

RESULTS:

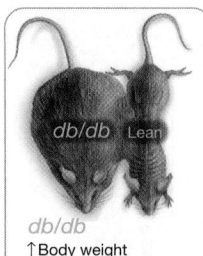 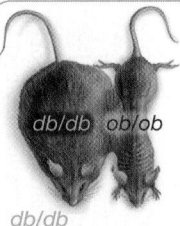

db/db
↑Body weight
↑Adipose tissue mass

Lean
↓ Food intake
↓ Body weight
↓ Adipose tissue mass
(death by starvation)

db/db
↑Body weight
↑Adipose tissue mass

ob/ob
↓ Food intake
↓ Body weight
↓ Adipose tissue mass
(death by starvation)

ob/ob
↓ Food intake
↓ Body weight
↓ Adipose tissue mass

Lean
No change

CONCLUSION: Gene products of *db* and *ob* are involved in hormonal signaling that affects appetite.

FIGURE 49.11 Parabiosis of Genetically Obese and Lean Mice Provides Evidence of a "Satiation Hormone."

SOURCE: Coleman, D. L. 1973. Effects of parabiosis of obese with diabetes and normal mice. *Diabetologia* 9: 294–298.

✔**QUESTION** Suppose a mouse were doubly homozygous *(ob/ob db/db)*. Would its phenotype be the same as or different from singly homozygous individuals? Explain your reasoning.

was that *db/db* mice lack the receptor required for the hormone to affect target cells.

As *db/db* mice got fatter and fatter, they would produce more and more hormone—to no avail. This model explained why the signal from *db/db* mice greatly reduced food intake in their parabiotic partners but had no effect on the *db/db* mouse.

The satiation hormone itself was postulated to be encoded by the *obese* gene. As a result, *ob/ob* mice do not produce the hormone. They respond to the signal if it is available from a normal partner, however.

A key idea here is that the two genotypes produce the same phenotype because they disable different parts of the same hormone-signaling system. For a system to function, the hormone must be present and the target tissues must have receptors for that specific hormone.

This model was confirmed in 1994, when the *obese* gene product, called **leptin,** was shown to be a polypeptide hormone. Leptin is secreted into the blood by adipocytes and interacts with a specific receptor located in many tissues, including areas of the brain known to control feeding behavior.

In mice, leptin injections correct the obesity of *ob/ob* (leptin-deficient) mice but not of *db/db* (leptin-receptor-deficient) individuals. Unfortunately, leptin injections are not helpful for the vast majority of obese humans.

Follow-up work has shown that leptin levels in the blood vary in proportion to total adipose tissue mass. When adipose mass falls below a set point, the leptin level in the blood also falls. The brain senses the decrease in leptin level and generates both an increase in appetite and a decrease in energy expenditure. These responses promote eating and restore energy balance.

When sufficient food intake has occurred to restore triglyceride stores, leptin levels rise. The result is diminished appetite, increased energy expenditure, and the stabilization of adipose tissue mass. This is a striking example of homeostasis achieved by feedback inhibition.

ADH, Aldosterone, and Water and Electrolyte Balance Recall that when an individual is dehydrated, **antidiuretic hormone (ADH)** is released from the pituitary gland (see Chapter 43). ADH increases the permeability of the kidney's collecting ducts to water, causing water to be reabsorbed from urine and saved.

ADH is instrumental in achieving homeostasis with respect to water balance. For example, the ethanol in alcoholic beverages inhibits the release of ADH from the pituitary. People who imbibe large quantities of these beverages produce large quantities of dilute urine. The resulting water loss can lead to dehydration and nausea—symptoms associated with an alcoholic hangover.

When sodium concentration in body fluids is low, **aldosterone** is released from the adrenal cortex (see Chapter 43). Because aldosterone increases reabsorption of sodium ions in the distal tubules of the kidney, it plays a key role in homeostasis with respect to electrolyte concentrations and overall volume of body fluids. Adrenal hormones with this effect are called **mineralocorticoids.**

ADH saves water; aldosterone saves sodium. Together, they are key players in maintaining water and electrolyte balance.

EPO and Oxygen Availability Erythropoietin (EPO) is a crucial element in the homeostatic system for blood oxygen levels. When blood oxygen levels fall, the kidneys and other tissues release EPO, which stimulates the production of red blood cells. The more red blood cells, the higher the oxygen-carrying capacity of blood is. If you moved to a place at high elevation and experienced chronic oxygen deficit, your body would respond by releasing EPO and increasing red blood cell concentration. This explains why many athletes train at high altitudes.

Unfortunately, some endurance athletes have turned to EPO injections as a way to increase the oxygen-carrying capacity of their blood and give themselves a competitive edge. The practice is dangerous, as well as illegal. EPO abuse is thought to be responsible for the collapse and death of several cyclists during races in the mid-1990s.

The increased viscosity of blood in EPO abusers is accentuated during exercise, when blood plasma volume drops due to dehydration. The combination of high blood viscosity and low blood volume can impair blood flow through capillaries, increasing the risk of tissue damage and blood clotting. If clots form in blood vessels that lead to the heart or brain, heart attack or stroke may occur.

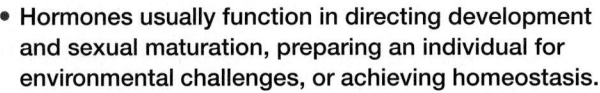

check your understanding

If you understand that . . .
- Hormones usually function in directing development and sexual maturation, preparing an individual for environmental challenges, or achieving homeostasis.
- Endocrine disruptors and stressors can disrupt hormone signaling and homeostasis.

✔ **You should be able to . . .**
1. Compare and contrast the roles of the hormones involved in metamorphosis of amphibians and insects.
2. Explain how the various changes induced by elevated cortisol result in a response to long-term stress.

Answers are available in Appendix A.

49.3 How Is the Production of Hormones Regulated?

Most hormones are released in response to an environmental cue or a message from an integrator in a homeostatic system. Often, the nervous system is closely involved. For example, environmental cues that signal the onset of the breeding season or the presence of a predator are received by sensory receptors and interpreted by the brain. Similarly, integration in most homeostatic systems is done by neurons in the central nervous system (CNS)—the brain and spinal cord (see Chapter 46).

Based on these observations, the short answer to the question posed in the title of this section is simple: In many cases, hormone production is directly or indirectly controlled by the nervous system.

The Hypothalamus and Pituitary Gland

The **pituitary gland,** located at the base of the brain, is directly connected to the brain region called the **hypothalamus.** This physical link between the hypothalamus and pituitary is the basis of the connection between the CNS and the endocrine system.

The pituitary has two distinct segments: the **anterior pituitary** and the **posterior pituitary.** In 1930, a biologist documented the consequences of removing the entire pituitary in laboratory rats: The animals stopped growing, could not maintain a normal body temperature, and suffered atrophy (shrinkage) of their genitals, thyroid glands, and adrenal cortexes. Not surprisingly, their life span shortened dramatically.

These experiments suggested that, in addition to secreting growth hormone, the pituitary secretes hormones that regulate the production of a wide variety of other hormones. As a case study, let's look at the pituitary hormone that acts on the adrenal glands.

Controlling the Release of Glucocorticoids

Early work on rats suggested that a molecule from the pituitary gland affects the adrenal gland. This molecule soon came to be called **adrenocorticotropic hormone,** or **ACTH** (*adreno* refers to the adrenal glands; *cortico* refers to the outer portion, or cortex, of the gland; and *tropic* means affecting the activity of).

ACTH was purified and characterized in 1943. When human volunteers were injected with ACTH, cortisol levels in their blood rose (**FIGURE 49.12a**). This result provided evidence that ACTH is a regulatory hormone. The adrenal cortex secretes glucocorticoids in response to ACTH released from the pituitary.

What regulates ACTH? Biologists from two laboratories independently showed that ACTH is released in response to a molecule produced by the hypothalamus. After years of effort, a different team of researchers succeeded in purifying a peptide—just 41 amino acids long—called **corticotropin-releasing hormone (CRH).** When the hypothalamus releases CRH, it stimulates cells in the anterior pituitary to secrete ACTH into the bloodstream.

Feedback Inhibition by Glucocorticoids

What *stops* glucocorticoid secretion? The key is to recognize that glucocorticoids themselves suppress ACTH production by the pituitary gland. Glucocorticoids accomplish feedback inhibition—they suppress their own production.

When human volunteers were injected with cortisol, ACTH levels in their bloodstream dropped dramatically (**FIGURE 49.12b**). Cortisol also inhibits release of CRH from the hypothalamus. Thus, if glucocorticoid levels become too high, ACTH levels fall. But if glucocorticoid levels become too low, ACTH levels rise and drive a compensatory increase in glucocorticoid production. **FIGURE 49.12c** summarizes the relationships among CRH, ACTH, and the glucocorticoid cortisol.

What happens when feedback inhibition fails? Certain pituitary tumors diminish the ability of cortisol to suppress ACTH production, leading to persistently high blood ACTH and

(a) Results of injecting ACTH into human volunteers

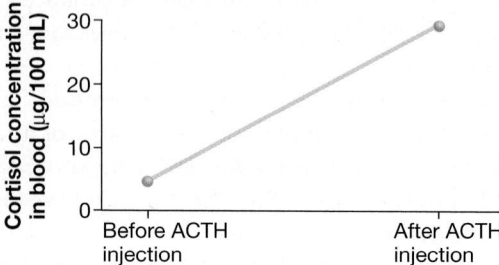

(b) Results of injecting cortisol into human volunteers

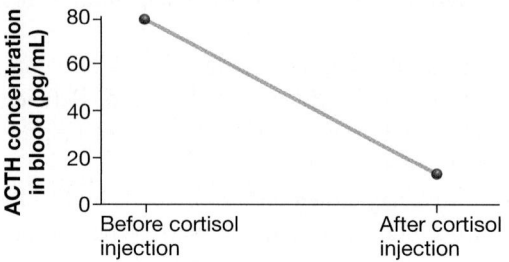

(c) Feedback inhibition by cortisol on ACTH

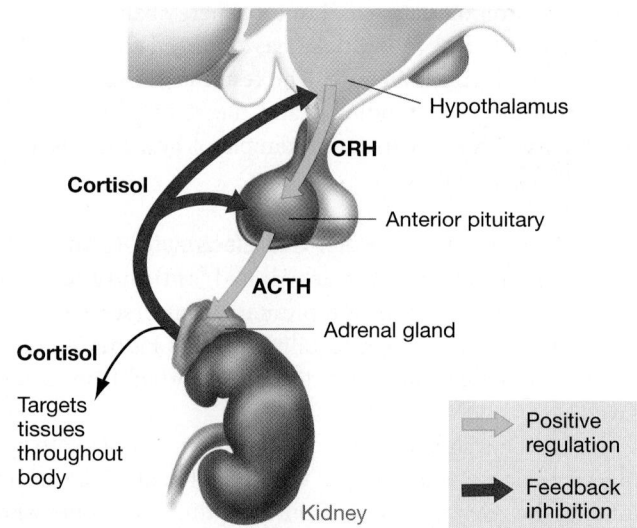

FIGURE 49.12 Cortisol Exerts Feedback Inhibition on ACTH Release. (a) ACTH stimulates release of cortisol. **(b)** Cortisol inhibits ACTH release. **(c)** The interaction between cortisol, ACTH, and CRH is an example of feedback inhibition.

DATA: Upton, G. V., A. Corbin, C. C. Mabry, et al. 1973. *Acta Endocrinologica* 73: 437–443.

✔**EXERCISE** How would you use the data in part (a) to devise a test for adrenal failure in humans?

cortisol levels. The result is **Cushing's disease,** an unrelenting stress response that depletes the body's protein reserves. It is fatal if not treated.

Patterns in Glucocorticoid Release

Under ordinary circumstances, the production of CRH by the hypothalamus displays a daily rhythm and reaches its highest level in the early morning hours. This pattern drives a corresponding daily rhythm in ACTH production and blood cortisol level.

Ordinarily, the morning peak in blood cortisol levels coincides with arousal and initiation of the day's activities—and the effect is saving glucose for use by the brain. As an aside, the unpleasant symptoms of jet lag are due in part to the daily cortisol rhythm being out of synchrony with local time for several days after you arrive in a new time zone.

When the brain processes stimuli that produce pain or anxiety, it initiates a long-term stress response and a sustained increase in CRH production. Increased CRH production causes blood ACTH and cortisol levels to remain much higher throughout the day than they are in the unstressed state.

The Hypothalamic–Pituitary Axis—An Overview The CRH-ACTH-glucocorticoid relationship is just one of many hormone systems based on interactions among the hypothalamus, pituitary, and target glands or cells. The **hypothalamic–pituitary axis** actually forms two anatomically distinct systems (**FIGURE 49.13**). The anterior pituitary develops from cells in an embryo's mouth and throat lining; the posterior pituitary is an extension of the brain.

Two distinct populations of neurons in the hypothalamus influence the posterior versus anterior sections of the pituitary gland. Both types of hypothalamic neurons synthesize and release neurohormones and are called **neurosecretory cells.** The neurosecretory cells release hormones under the control of brain regions responsible for integrating information about the external or internal environment. For example, information about an upcoming exam or athletic contest might trigger action potentials that lead to the release of CRH.

The Posterior Pituitary Even though both parts of the gland contain neurosecretory cells, the anterior pituitary and posterior pituitary function in different ways. As Figure 49.13a indicates, the posterior portion of the pituitary is an extension of the hypothalamus itself.

Neurosecretory cells that project from the hypothalamus produce the hormones ADH and oxytocin, which are then stored

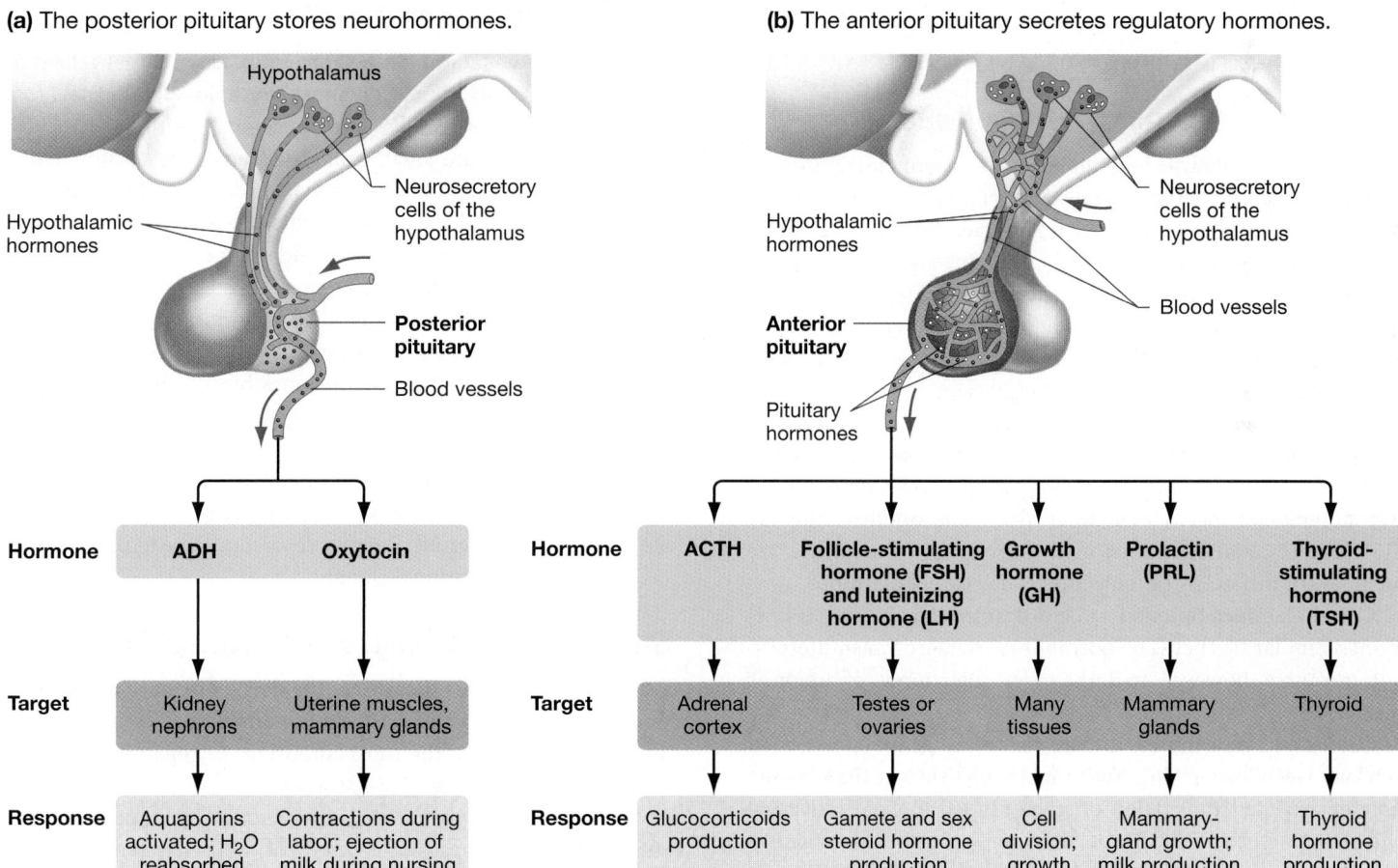

(a) The posterior pituitary stores neurohormones.

(b) The anterior pituitary secretes regulatory hormones.

FIGURE 49.13 **The Hypothalamus and Pituitary Interact Closely. (a)** Developmentally and anatomically, the posterior pituitary is an extension of the hypothalamus. Neurosecretory cells in the hypothalamus extend directly into the posterior pituitary and secrete ADH (vasopressin) and oxytocin. **(b)** The hypothalamus and the anterior pituitary communicate indirectly, via blood vessels. Hormones produced by other populations of neurosecretory cells in the hypothalamus travel in the blood to the anterior pituitary, where they control the release of pituitary hormones.

in the posterior pituitary. From there, ADH and oxytocin are released into the bloodstream. This is an example of the neuro-endocrine pathway of hormone action. Recall that ADH aids in the reabsorption of water by the kidneys. **Oxytocin** helps induce labor and milk release in female mammals.

The Anterior Pituitary   Unlike the situation in the posterior pituitary, the hypothalamus and anterior pituitary are connected indirectly. Neurosecretory cells from the hypothalamus secrete stimulatory or inhibitory hormones into tiny blood vessels to the anterior pituitary. In response, the anterior pituitary alters its secretion of hormones that enter the general bloodstream and act on target tissues or glands. This is an example of the neuroendocrine-to-endocrine pathway of hormone action.

Many of the hormones produced by the anterior pituitary stimulate the production of other hormones. The anterior pituitary hormones include ACTH; **follicle-stimulating hormone (FSH)** and **luteinizing hormone (LH),** which are involved in stimulating the gonads to produce sex hormones and gametes; GH; **prolactin,** which stimulates mammary gland growth and milk production in mammals; and **thyroid-stimulating hormone (TSH),** which triggers the production of thyroid hormones.

Control of Epinephrine by Sympathetic Nerves

When biologists analyze how the nervous system and endocrine system interact to control the release of epinephrine, the distinction between the two systems begins to blur. Section 49.2 introduced how epinephrine acts as an endocrine signal. During the fight-or-flight response, sympathetic nerves trigger the release of epinephrine from the adrenal medulla into the bloodstream. But in addition, some sympathetic nerves release the related molecule **norepinephrine** directly onto target cells.

In effect, the endocrine system broadcasts a messenger by secreting it into the bloodstream whereas the nervous system delivers a chemical messenger directly to particular cells. Epinephrine and norepinephrine, which differ from one another only by the presence of an additional methyl group on epinephrine, are members of the family of molecules called **catecholamines.**

Catecholamines function as neurotransmitters as well as hormones. Similarities between hormones and neurotransmitters do not end there, however. In some cases, their mode of action is similar.

Some neurotransmitters initiate changes in gene expression in neurons (see Chapter 46). You might recall that in the sea slug *Aplysia,* repeated application of serotonin led to gene activation and changes in the behavior of the synapse. By altering synapses in this way, neurotransmitters play a central role in learning and memory.

Similarly, many hormones exert their effects by activating particular genes in target cells. Understanding how gene activation occurs in response to neurotransmitters and hormones is the subject of intense research at laboratories around the world. It is also the subject of Section 49.4.

 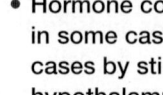
If you understand that . . .
- Hormone concentrations are tightly regulated—in some cases by negative feedback, in other cases by stimulatory or inhibitory signals from the hypothalamus.

✔ **You should be able to . . .**
1. Explain how feedback inhibition occurs in production of ACTH.
2. Discuss the relationship between processing centers in the brain, neurosecretory cells in the hypothalamus, and hormone-secreting cells in the anterior pituitary.

Answers are available in Appendix A.

49.4   How Do Hormones Act on Target Cells?

The key to understanding how hormones act on target cells is to recognize that only some hormones are lipid soluble and cross plasma membranes readily (see Figure 49.3). More specifically, steroid hormones are small lipids that enter cells without difficulty. But the peptide and polypeptide hormones—as well as most amino acid derivatives, with the exception of thyroid hormones—do not cross plasma membranes easily because of their large size and electrical charge.

Differences in the lipid solubility of hormones are important because they relate to the location of hormone receptors in a target cell. Steroid and thyroid hormones bind to intracellular receptors, while most amino acid derivatives and all polypeptides bind to receptors on the surface of the plasma membrane.

To compare these two distinct paths of hormone action, let's consider how estradiol and epinephrine affect target cells. As a steroid and a nonsteroid, they serve as model systems for target cell responses to hormonal signals.

Steroid Hormones Bind to Intracellular Receptors

Estrogens are steroids that direct the development of female secondary sex characteristics in many animal species. In humans and other mammals, the most important estrogen is the molecule estradiol (formally, 17 β-estradiol).

Because of its importance in reproduction by humans and domesticated animals, estradiol's mode of action has been the topic of intense investigation for over 50 years. How do target cells receive the signal carried by estradiol?

Identifying the Estradiol Receptor   In 1966, biologists succeeded in isolating the estradiol receptor in laboratory rats. **FIGURE 49.14** indicates the experimental approach. In essence, the research team introduced labeled hormone molecules into females and

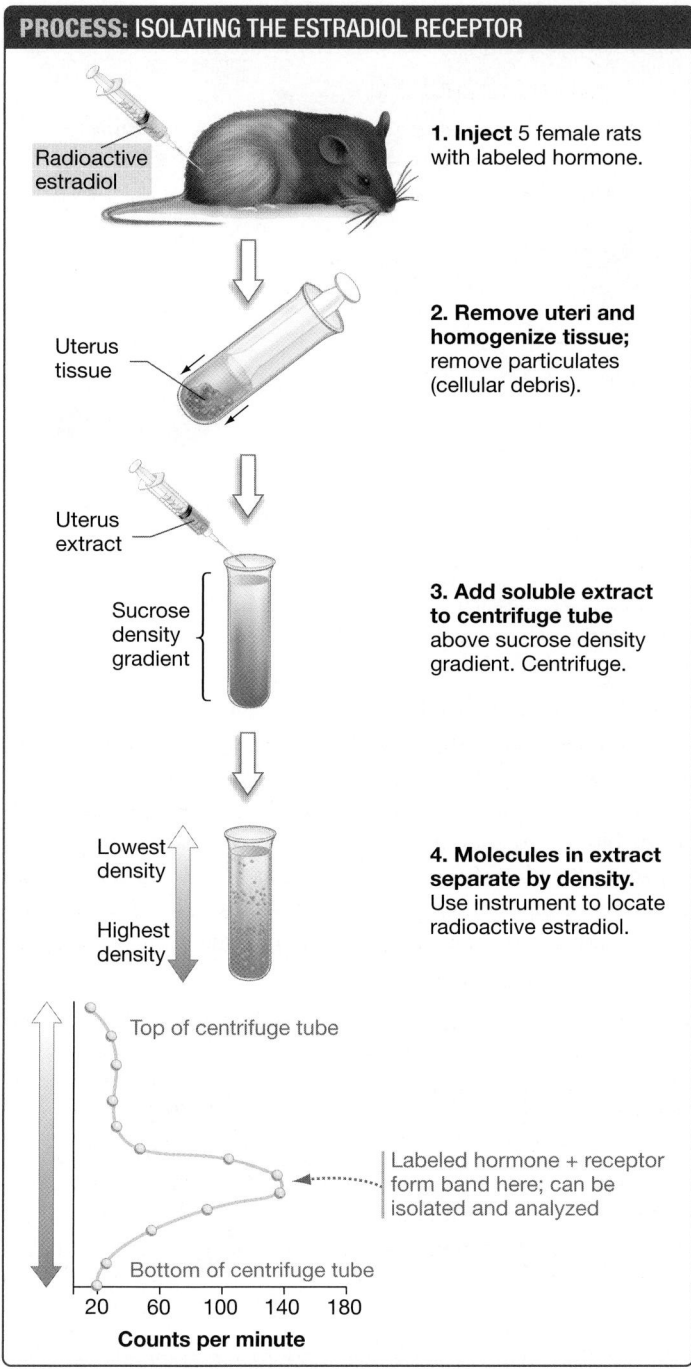

PROCESS: ISOLATING THE ESTRADIOL RECEPTOR

Radioactive estradiol

1. Inject 5 female rats with labeled hormone.

Uterus tissue

2. Remove uteri and homogenize tissue; remove particulates (cellular debris).

Uterus extract

Sucrose density gradient

3. Add soluble extract to centrifuge tube above sucrose density gradient. Centrifuge.

Lowest density

Highest density

4. Molecules in extract separate by density. Use instrument to locate radioactive estradiol.

Top of centrifuge tube

Labeled hormone + receptor form band here; can be isolated and analyzed

Bottom of centrifuge tube

20   60   100   140   180
Counts per minute

FIGURE 49.14 Labeled Hormones Can Be Used to Find Hormone Receptors. If radioactive estradiol binds to its receptor in the uterus, the hormone–receptor complex should form a distinct band of radioactivity when molecules from uterine cells are separated by centrifugation.

DATA: Toft, D., and J. Gorski. 1966. *Proceedings of the National Academy of Sciences, USA* 55: 1574–1581.

then used density-gradient centrifugation—a technique introduced in **BioSkills 10** in Appendix B—to separate molecules in target cells by size.

When centrifugation was complete, the biologists found that the labeled estradiol was concentrated in a narrow band comprised of the labeled hormone bound to its receptor. After purifying the receptor molecule, they found that proteinase enzymes could destroy it. Based on this result, they inferred that the estradiol receptor was a protein.

Follow-up experiments established that the estradiol receptor is primarily located in the nucleus but is not associated with the nuclear envelope. Further, the receptor is found only in estradiol target tissues, including the uterus, hypothalamus, and mammary glands. The latter finding was particularly exciting, because it clarified how hormones act in a tissue-specific way.

This is a crucial point: Hormones are broadcast throughout the body via the bloodstream, but they act only on cells that express the appropriate receptor. Target cells respond to a particular hormone because they contain a receptor for that hormone.

Later, biologists found that the gene for the estradiol receptor is similar to the genes that encode receptors for the glucocorticoids, testosterone, and other steroid hormones. This result suggested that all steroid receptors are descended from an ancestral receptor molecule, and that the binding of any steroid hormone to its receptor affects the target cell via a similar mechanism.

What happens once the hormone–receptor complex is present inside the nucleus of a target cell?

Documenting Changes in Gene Expression   During the 1970s and 1980s, work in several laboratories suggested that estradiol and other steroid hormones affect gene transcription after they bind to their receptors. For example, researchers injected laboratory animals with estradiol or other steroid hormones and documented changes in the mRNAs and proteins produced in target cells. These data showed that steroid hormones can cause dramatic changes in the amount or timing of mRNA production by a large number of genes.

How? The estradiol receptor, like other members of the steroid-hormone receptor family, has two copies of a distinctive DNA-binding region called a zinc finger. DNA-binding domains are sections of a protein that make physical contact with DNA. The presence of zinc fingers in the estradiol receptor suggested that once estradiol had bound to it, the hormone–receptor complex might affect gene expression by binding directly to DNA.

Follow-up work confirmed that steroid hormone–receptor complexes bind to specific sites in DNA called **hormone-response elements.** Hormone-response elements are located just "upstream" (in the 5′ direction) from the start of target genes. Gene expression changes when a regulatory molecule such as a steroid hormone–receptor complex binds to the hormone-response element for that gene.

FIGURE 49.15 (see page 1008) summarizes the current model of how steroid hormones affect target cells. Because each hormone–receptor complex leads to the production of many copies of the gene product, the signal from the hormone is amplified. In this way, a small number of hormone molecules produces a large change in the activity of target cells and tissues.

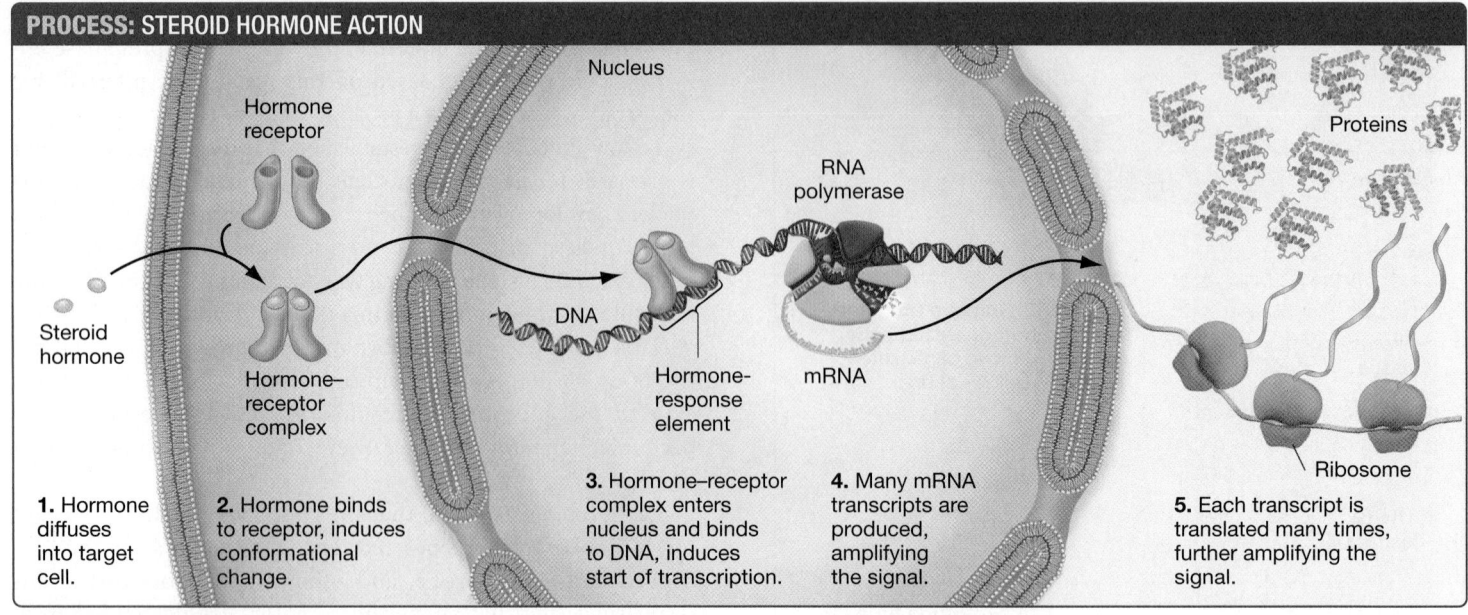

FIGURE 49.15 Steroid Hormones Bind to Receptors Inside Target Cells and Affect Gene Expression.

Hormones That Bind to Cell-Surface Receptors

Most polypeptide hormones and amino-acid-derived hormones are not lipid soluble. For these molecules to affect a cell, they must bind to receptors on the cell surface. Because the messenger never enters the target cell, its message must be transduced—changed into a form that is active inside the cell. This phenomenon is known as **signal transduction** (Chapter 11).

To explore how signal transduction occurs, let's first examine hormone receptors that reside in the plasma membrane and then explore the molecules that process the message inside the cell. In both cases, epinephrine will be used as a model system.

Identifying the Epinephrine Receptor In 1948 a biologist published an exhaustive set of studies on how epinephrine affects dogs, cats, rats, and rabbits. The responses fell into two distinct categories, depending on the tissue being considered. To explain this observation, the researcher suggested that epinephrine binds two distinct types of receptor. He called these hypothetical proteins the alpha receptor and the beta receptor.

Follow-up work with molecules that block epinephrine receptors documented that there are actually two types of alpha receptors and two types of beta receptors. Thus, there are four distinct epinephrine receptors. Each is found in a distinct tissue type, and each induces a different response from the cell.

The discovery of four epinephrine receptors reinforces the concept of tissue specificity observed in experiments on the estradiol receptor. Hormones are transmitted throughout the body, not unlike a cell phone signal that is broadcast through the atmosphere. But their message is received only by cells with the appropriate receptor—just as a cell phone signal is received only by equipment with the appropriate antenna. Since there are four distinct epinephrine receptors, the same hormone can trigger different effects in different cells. What happens once epinephrine binds to one of these receptors?

What Acts as the Second Messenger? Signal transduction occurs when a chemical message binds to a cell-surface receptor and triggers a response inside the cell.

How does a signal from epinephrine increase glucose levels in the blood? To answer this question, biologists focused on the enzyme **phosphorylase**, which catalyzes a reaction that cleaves glucose molecules off glycogen (**FIGURE 49.16a**). Phosphorylase exists in active and inactive forms; the enzyme switches between

(a) Phosphorylase catalyzes the production of glucose from glycogen.

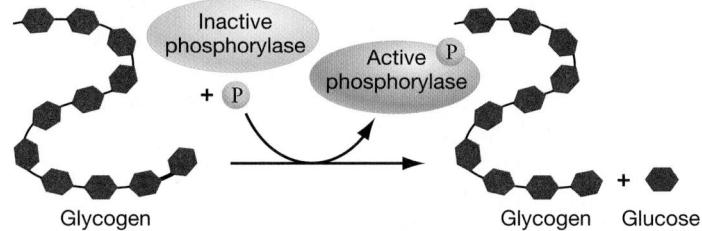

(b) Phosphorylase is activated in response to epinephrine.

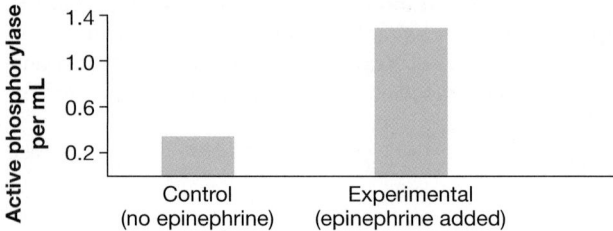

FIGURE 49.16 Epinephrine Activates the Enzyme That Catalyzes the Formation of Glucose from Glycogen. (a) Phosphorylase is activated when an enzyme adds a phosphate group to it. **(b)** When epinephrine is added to cell-free extracts from liver tissue, the amount of activated phosphorylase increases dramatically.

DATA: Rall, T. W., E. W. Sutherland, and J. Berthet. 1957. *Journal of Biological Chemistry* 224: 463–475.

these states when it is phosphorylated or dephosphorylated by another enzyme.

Phosphorylase is present in liver cells—the primary source of blood glucose during the fight-or-flight response. As predicted, when researchers added epinephrine to extracts from homogenized (ground up) liver cells, much larger amounts of phosphorylase were activated relative to cell extracts that did not receive epinephrine (**FIGURE 49.16b**).

This observation suggested there was something in the homogenized cells that activated phosphorylase when epinephrine was present. By purifying components from the liver cell extracts and testing them one by one, researchers eventually found the ingredient that activated phosphorylase: a molecule called cyclic adenosine monophosphate, or **cyclic AMP (cAMP).**

The role of cAMP in epinephrine signaling was confirmed when researchers studied epinephrine's effects on the rat heart. During the fight-or-flight response, heart rate and the contractile force of the heart rise dramatically—increasing cardiac output. Cardiac output is a measure of how much blood the heart pumps per unit time.

The graphs in **FIGURE 49.17** show how cardiac muscle cells respond to epinephrine. In each graph, the *x*-axis plots time, in seconds, after epinephrine is applied.

- *Top* Almost immediately, there is a striking increase in cAMP levels inside the cells.

- *Middle* A few seconds later, the contractile force of the cells increases, peaking about 18 seconds after the hormone's arrival.

- *Bottom* Phosphorylase activity also rises, but peaks later—about 40 seconds after the signal.

To capture the importance of cAMP in triggering these effects, biologists refer to it as a second messenger. Recall that a **second messenger** is a nonprotein signaling molecule that increases in concentration inside a cell in response to a received signal—a molecule that binds at the surface (see Chapter 11).

A Phosphorylation Cascade   How does cAMP transfer the hormonal signal on the cell surface to phosphorylase inside the cell? Follow-up work revealed the mechanism for epinephrine action on liver cells (**FIGURE 49.18**).

When epinephrine binds to its receptor, it activates a G protein (see Chapter 11), which then activates the enzyme adenylyl cyclase. Adenylyl cyclase catalyzes a reaction that converts ATP to cAMP. Next, cAMP initiates a chain of events called a **signal transduction cascade,** by binding to an enzyme called

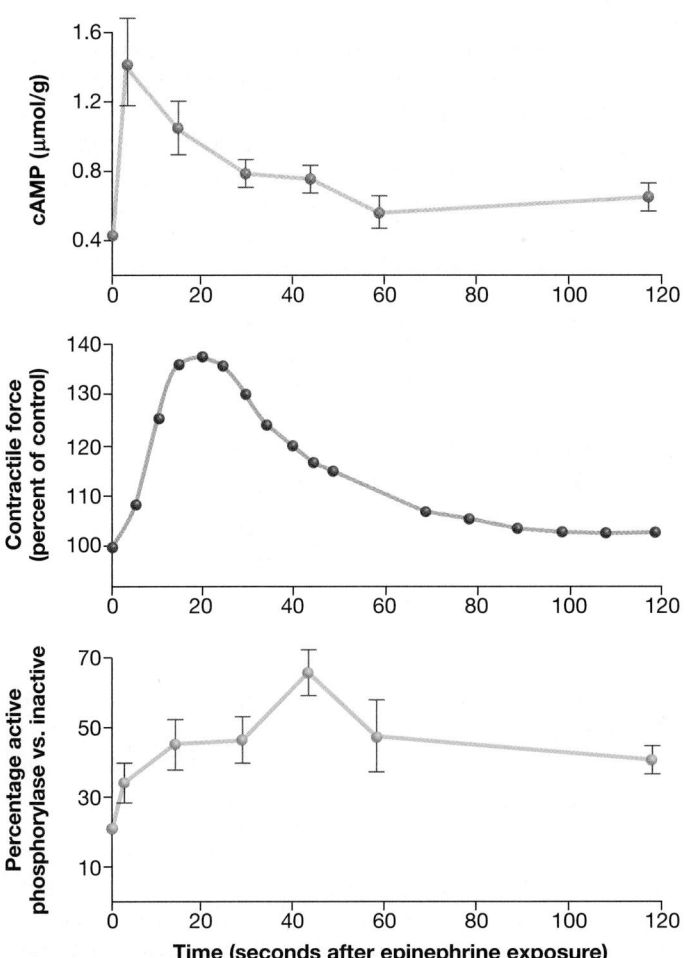

FIGURE 49.17 The Chemistry and Activity of Heart Muscle Cells Change in Response to Epinephrine.

DATA: Robison, G. A., R. W. Butcher, I. Oye, et al. 1965. *Molecular Pharmacology* 1: 168–177.

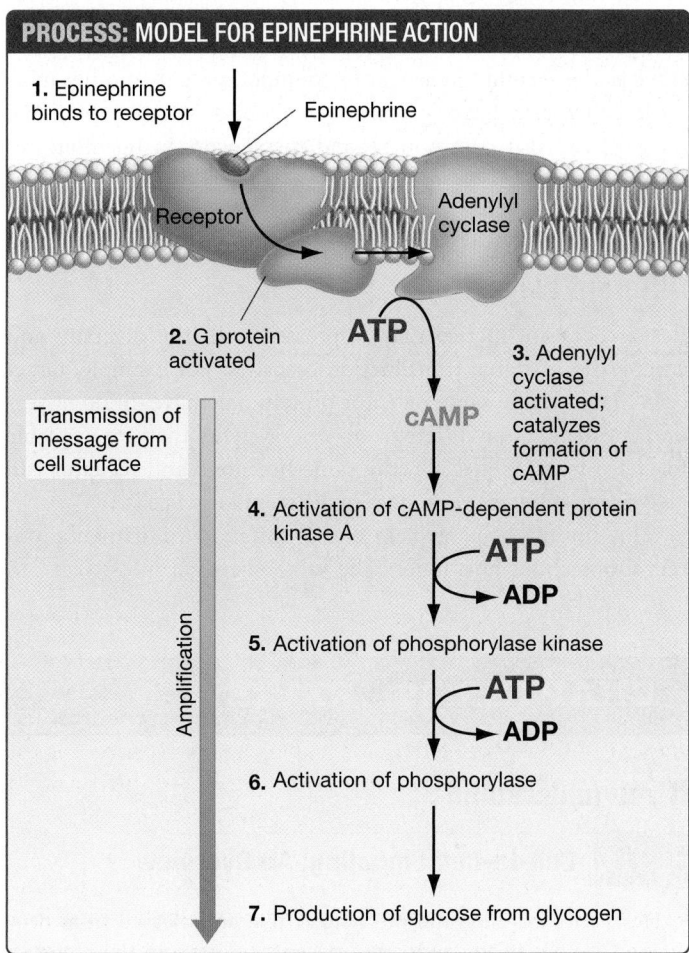

FIGURE 49.18 Epinephrine Triggers a Signal Transduction Cascade. Epinephrine's signal is amplified at each of steps 3–6.

cAMP-dependent protein kinase A. This enzyme responds by phosphorylating the enzyme phosphorylase kinase, which then phosphorylates phosphorylase.

In studying this model, it is crucial to recognize two points:

1. cAMP transmits the signal from the cell surface to the signaling cascade.

2. Together, cAMP production and the subsequent phosphorylation events amplify the original signal from epinephrine.

To appreciate the second point, consider that, in response to stimulation by the hormone–receptor complex, adenylyl cyclase is thought to catalyze the formation of at least 100 molecules of cAMP. In turn, each of these cAMP molecules activates many molecules of cAMP-dependent protein kinase A. Subsequently, each protein kinase molecule activates many molecules of phosphorylase kinase, and so on.

In this way, the binding of just a single molecule of epinephrine may trigger the release of millions or even billions of glucose molecules. Amplification through a signal transduction cascade explains why tiny amounts of hormones can have such huge effects on an individual.

The model in Figure 49.18 was inspired by experiments on the epinephrine receptor called the beta-1 receptor. But other researchers showed that when epinephrine binds to an alpha-1 receptor, a completely different signal transduction event occurs. In this and many other receptor systems, calcium ions (Ca^{2+}) serve as the second messenger in conjunction with another molecule called IP_3. Diacylglycerol (DAG) and 3′, 5′-cyclic GMP (cGMP) are also common second messengers in hormone response systems.

Why Do Different Target Cells Respond in Different Ways?

Researchers are increasingly impressed with the diversity and complexity of signal transduction cascades. For example, target cells that have the same receptor protein may have different second messengers or different enzyme systems that are available for activation. As a result, the same hormone and receptor can give rise to different responses in different target cells.

This finding helps explain one of the most fundamental observations about hormones: The same chemical messenger can trigger different responses in cells from different organs or in cells at different developmental stages. The reason is that the cells contain different receptors, second messengers, amplification steps, protein kinases, enzymes, or transcriptionally active genes.

To summarize this section, steroid and thyroid hormones tend to exert their effects through changes in gene expression. In contrast, polypeptide and most amino-acid-derived hormones activate a specific protein or set of proteins, usually by phosphorylation. Steroid and thyroid hormones activate transcription factors that lead to the production of new proteins; nonsteroid hormones trigger signal transduction cascades that activate existing proteins.

check your understanding

C Y U

If you understand that . . .

- Hormones act on target cells by binding to receptors.
- The response to a hormone is tissue specific, because only certain cells contain receptors for particular hormones.
- In response to binding by lipid-soluble hormones, hormone–receptor complexes bind to DNA and induce changes in gene expression.
- In response to binding by lipid-insoluble hormones, activated cell-surface receptors induce the production of second messengers and the activation of signal transduction cascades that result in the phosphorylation of existing proteins.

✔ **You should be able to . . .**

1. Predict what would happen to the estradiol response if an individual had mutations that changed the DNA sequence of its hormone-response element.

2. Suppose a steroid hormone binds to a cell-surface receptor. Explain how its mode of action would compare to a polypeptide hormone.

Answers are available in Appendix A.

CHAPTER 49 REVIEW

For media, go to MasteringBiology

If you understand . . .

49.1 Cell-to-Cell Signaling: An Overview

- Hormones are chemical messengers that are released from neurons or cells of the endocrine system, circulate in the blood or other body fluids, and trigger a response in target cells containing an appropriate receptor.

- Hormones have a variety of chemical structures. Most animal hormones are polypeptides, amino acid derivatives, or steroids.

✔ You should be able to explain the relationship between electrical signals from the nervous system and chemical signals from the nervous system and endocrine system, as they work in combination to coordinate the body's response to environmental change.

49.2 What Do Hormones Do?

- Together with the nervous system, hormones coordinate the activities of diverse cells and tissues. A single hormone may affect a wide array of cells and tissues and induce a variety of responses.

- Estradiol is an example of a hormone that regulates development and sexual maturation. Estradiol stimulates the formation of female sex characteristics in human embryos and the maturation of these tissues in adolescence.

- Melatonin is a hormone that regulates reproductive physiology in response to seasonal changes in day length.

- Epinephrine and cortisol are examples of hormones that help individuals cope with environmental changes. These hormones activate the short-term and long-term responses to stressors, triggering the fight-or-flight response or inducing changes that conserve glucose for use by the brain.

- Hormones are involved in a wide array of homeostatic interactions. For example, hormones are involved in directing cells that modify the concentrations of water, sodium ions, and other components of the blood and interstitial fluid. Homeostasis with respect to triglyceride stores is also under endocrine control.

✓ You should be able to explain why hormones—instead of electrical signals from the nervous system—are primarily responsible for regulating responses to environmental change, embryonic and sexual development, and homeostasis.

49.3 How Is the Production of Hormones Regulated?

- In many cases, the release of a hormone is regulated by chemical messengers from the anterior pituitary.

- Hormone-secreting cells in the anterior pituitary are regulated by hormones released by the hypothalamus region of the brain.

- The long-term stress response is a well-studied example of hormone regulation. The brain responds to long-term stress by triggering the release of the hypothalamic hormone CRH. CRH activates the release of ACTH by the pituitary gland, which stimulates the production of cortisol by cells in the adrenal cortex. Because cortisol inhibits the production of ACTH and CRH, the chain of events is regulated by feedback inhibition.

✓ You should be able to predict the consequences if negative feedback fails to occur after the release of CRH.

49.4 How Do Hormones Act on Target Cells?

- Animal hormones have two basic modes of action. Steroid and thyroid hormones are lipid soluble, cross plasma membranes readily, and often bind to receptors inside cells. Most polypeptide and amino-acid-derived hormones are not lipid soluble; they bind to receptors located in the membranes of target cells. In both cases, the response to a hormone is tissue specific because only certain cells express certain receptors.

- Most steroid and thyroid hormones act by inducing a change in gene expression.

- Polypeptide and most amino-acid-derived hormones trigger signal transduction cascades that activate one or more target proteins by phosphorylation.

- Although they are produced in tiny concentrations, hormones have large effects because they trigger gene expression or because their message is amplified through a signal transduction cascade.

✓ You should be able to predict whether a cell can respond to more than one hormone at a time, and explain why or why not.

(MB) MasteringBiology

1. **MasteringBiology Assignments**

Tutorials and Activities Coordination of the Endocrine and Nervous Systems; Endocrine System Anatomy; Hormone Actions on Target Cells; Human Endocrine Glands and Hormones; Peptide Hormone Action; Steroid Hormone Action; What Role Do Genes Play in Appetite Regulation?

Questions Reading Quizzes, Blue-Thread Questions, Test Bank

2. **eText** Read your book online, search, take notes, highlight text, and more.

3. **The Study Area** Practice Test, Cumulative Test, BioFlix® 3-D Animations, Videos, Activities, Audio Glossary, Word Study Tools, Art

You should be able to . . .

1. Both epinephrine and cortisol are involved in the body's response to stress. How do the two molecules differ?
   a. Cortisol is an amino acid derivative; epinephrine is a steroid.
   b. Cortisol binds to receptors on the plasma membranes of target cells; epinephrine binds to receptors in the interior of target cells.
   c. Epinephrine mediates the short-term response; cortisol mediates the short- and long-term responses.
   d. Cortisol controls the release of epinephrine from the adrenal glands.

2. How do steroid hormones differ from polypeptide hormones and most amino-acid-derived hormones?
   a. Steroids are lipid soluble and cross plasma membranes readily.
   b. Polypeptide and amino-acid-derived hormones are longer lived in the bloodstream and thus exert greater signal amplification.
   c. Polypeptide hormones are the most structurally complex and induce permanent changes in target cells.
   d. Only polypeptide and steroid hormones bind to receptors in the plasma membrane.

3. Which of the following developmental processes is *not* controlled by hormones?
   a. the initial development of male and female gonads, soon after fertilization
   b. overall growth
   c. molting in insects and other invertebrate animals
   d. metamorphosis in insects and other invertebrate animals

4. What is a hormone-response element?
   a. a receptor for a steroid hormone
   b. a receptor for a polypeptide hormone
   c. a segment of DNA where a hormone–receptor complex binds
   d. an enzyme that is activated in response to hormone binding and produces a second messenger

5. Where are the photoreceptors that sense day length located in mammals? In lizards? In birds?

6. True or False: In hormone systems, feedback inhibition occurs when the presence of a hormone inhibits release of the hormone.

✔ TEST YOUR UNDERSTANDING *Answers are available in Appendix A*

7. Hormones are present in tiny concentrations, yet have large effects on target cells and on the individual as a whole. This is possible because a single hormone molecule
   a. can bind to many receptors at once.
   b. can activate many intracellular enzymes.
   c. can initiate transcription of DNA by binding to cell-surface receptors.
   d. can bind irreversibly to its receptor.

8. Why is the observation that one hormone may bind to more than one type of receptor important?

9. Compare and contrast the structure and function of the anterior and posterior pituitary glands.

10. Compare and contrast the modes of action of steroid and nonsteroid hormones.

11. Design a study to test the hypothesis that the symptoms of jet lag are caused by disruption of normal daily cortisol rhythms.

12. You see an advertisement on television for a leptin supplement that claims to result in dramatic weight loss. Explain why this supplement is unlikely to be effective.

✔ TEST YOUR PROBLEM-SOLVING SKILLS *Answers are available in Appendix A*

13. You are a physician supervising a patient's recovery from the surgical removal of the posterior pituitary. Which of the following symptoms would you expect to have to treat?
    a. amenorrhea (cessation of the menstrual cycle)
    b. cessation of growth
    c. dehydration
    d. high blood-glucose levels

14. Suppose that during a detailed anatomical study of a marine invertebrate, you found a small, previously undescribed structure. How would you test the hypothesis that the structure is a gland that releases one or more hormones?

15. Suppose that a researcher announces the discovery of a hormone that affects the metabolism of fats in lab rats. Preliminary data indicate that the hormone is a polypeptide, about 50 amino acids long. The researcher suggests that the hormone activates transcription of a gene for a protein that stimulates fat breakdown. Explain why this researcher's hypothesis is reasonable or unreasonable.

16. **QUANTITATIVE** Scientists set out to test the hypothesis that atrazine is an endocrine disruptor that feminizes male amphibians. They treated male amphibians with atrazine, then later compared their circulating testosterone concentrations with those of control males and females. The results are shown here. Bars with different letters are significantly different from one another. Was the hypothesis supported? Why or why not?

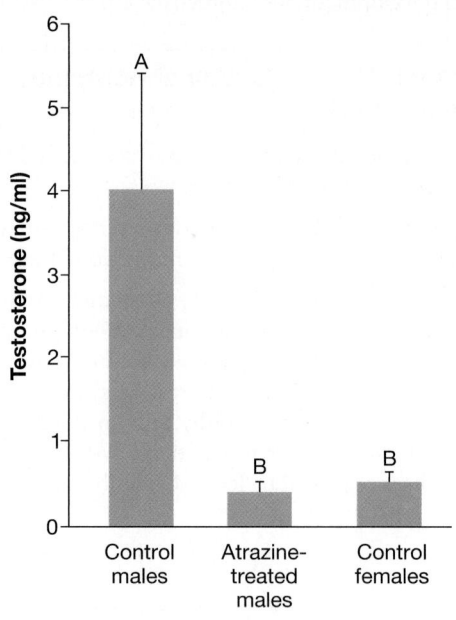

DATA: Hayes, T. B., A. Collins, M. Lee, et al. 2002. Proceedings of the National Academy of Sciences, USA 99: 5476–5480.

50 Animal Reproduction

In this chapter you will learn that

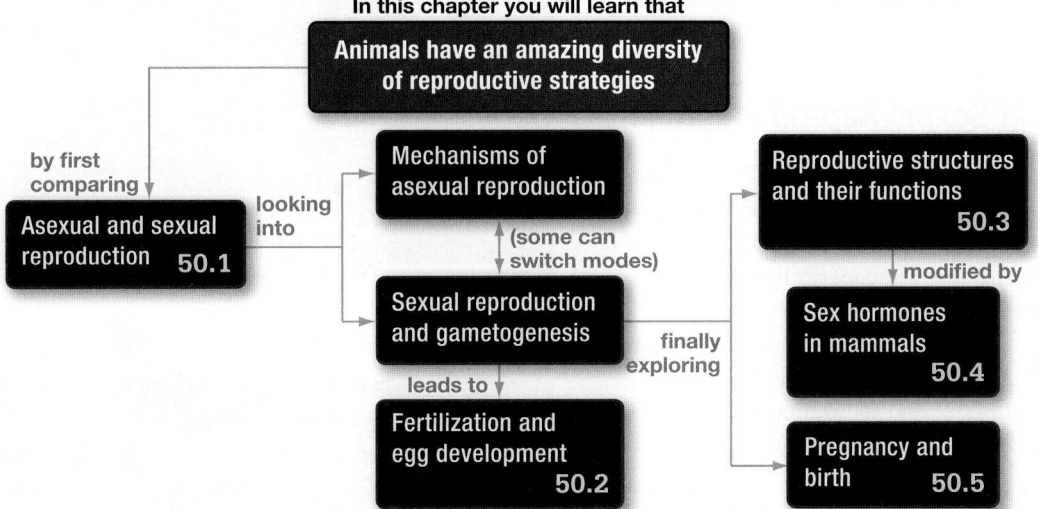

Animals have an amazing diversity of reproductive strategies

by first comparing → Asexual and sexual reproduction **50.1**

looking into →

Mechanisms of asexual reproduction

(some can switch modes)

Sexual reproduction and gametogenesis

leads to ↓

Fertilization and egg development **50.2**

Reproductive structures and their functions **50.3**

modified by ↓

Sex hormones in mammals **50.4**

finally exploring →

Pregnancy and birth **50.5**

The swollen, red rump of this female Hamadryas baboon indicates that she is about to produce an egg. She will probably mate with several males before the egg is fertilized. This chapter discusses how hormones regulate the female reproductive cycle and the consequences of multiple mating.

All the cells, tissues, organs, and systems introduced in Unit 8 exist for one reason: They allow animals to survive long enough and gather enough resources to reproduce. Stated another way, producing offspring is the reason that adaptations exist. Reproduction is the underlying purpose of virtually everything that an animal does. Replication is a fundamental attribute of life (see Chapter 1).

But if evolution by natural selection explains *why* animals reproduce, the goal of this chapter is to explore *how* reproduction occurs. Section 50.1 introduces research on animals that cycle between asexual and sexual modes of reproduction and explains how both modes occur at the cellular level. Section 50.2 surveys the diverse ways in which animals accomplish fertilization once gametes have formed. The final three sections focus on mammalian reproduction, using humans as the primary model organism.

This chapter is part of the Big Picture. See how on pages 840–841.

✔ When you see this checkmark, stop and test yourself. Answers are available in Appendix A.

The chapter includes topics of urgent practical interest, such as human birth control methods and the recent and dramatic declines in the death rate of human mothers during childbirth. Understanding and manipulating animal reproductive systems is an important issue for physicians, veterinarians, farmers, zookeepers, conservation biologists, and many others in biology-related professions.

50.1 Asexual and Sexual Reproduction

Several earlier chapters have explored how asexual and sexual reproduction differ. When reproduction is asexual, it occurs without fusion of gametes. **Asexual reproduction** is usually based on mitosis and results in offspring that are genetically identical to their parent. **Sexual reproduction,** in contrast, is based on meiosis and fusion of gametes. Due to genetic recombination during meiosis and the fusion of haploid gametes—usually from different parents—during fertilization, sexual reproduction results in offspring that are genetically different from each other and from their parents.

Let's first consider how animals engage in asexual reproduction to make genetically identical copies of themselves, then review the mechanisms responsible for sexual reproduction.

How Does Asexual Reproduction Occur?

In thousands of animal species, individuals can **clone** themselves—that is, produce large numbers of identical copies of themselves asexually.

There are three main mechanisms of asexual reproduction:

1. **Budding** An offspring begins to form within or on a parent (**FIGURE 50.1a**). The process is complete when the offspring—a miniature version of the parent—breaks free and begins to grow on its own.

2. **Fission** An individual simply splits into two or more descendants (**FIGURE 50.1b**).

3. **Parthenogenesis** (literally, "virgin-origin") Females produce offspring without fertilization by a male. Usually, parthenogenetic eggs are produced by mitosis, or by meiosis that occurs after chromosome number has doubled, but without crossing over and recombination. In both of these cases, offspring are genetically identical to the mother. Parthenogenesis occurs in a wide diversity of lineages, including certain invertebrates, fishes, lizards, snakes, and birds (**FIGURE 50.1c**).

Many animal species regularly switch between reproducing asexually and reproducing sexually. Why?

Switching Reproductive Modes: A Case History

Daphnia are crustaceans that live in freshwater habitats throughout the world. In a typical year, *Daphnia* produce only diploid, female offspring throughout the spring and summer, via parthenogenesis. The eggs produced by parthenogenesis develop in a structure called a brood pouch (**FIGURE 50.2**) and are released when the female molts her exoskeleton.

(a) Budding in hydra

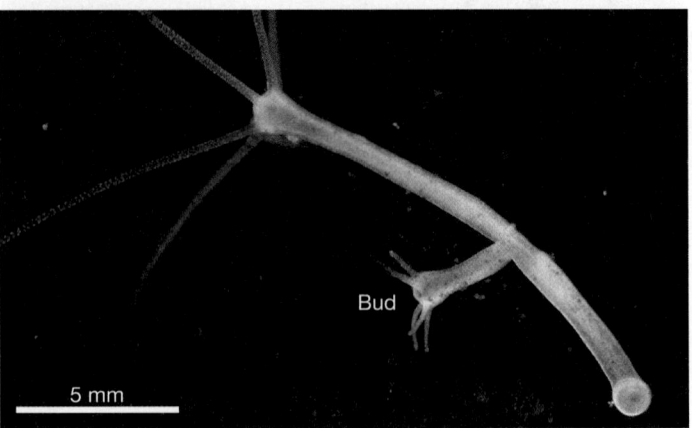

(b) Fission in anemones

(c) Parthenogenesis in lizards

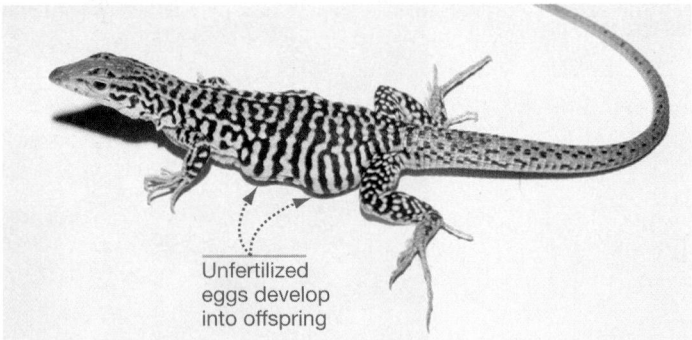

Unfertilized eggs develop into offspring

FIGURE 50.1 Mechanisms of Asexual Reproduction in Animals Are Diverse.

In late summer or early fall, however, many *Daphnia* females begin producing unfertilized, asexually produced eggs, some of which develop into males. Once the males have matured, sexual reproduction ensues: Haploid **sperm,** the male gametes, produced by meiosis in the males, fertilize haploid **eggs,** the female gametes, that females produce via meiosis.

Fertilization is the fusion of sperm and egg. In *Daphnia*, fertilized eggs are released into a durable case that falls to the bottom

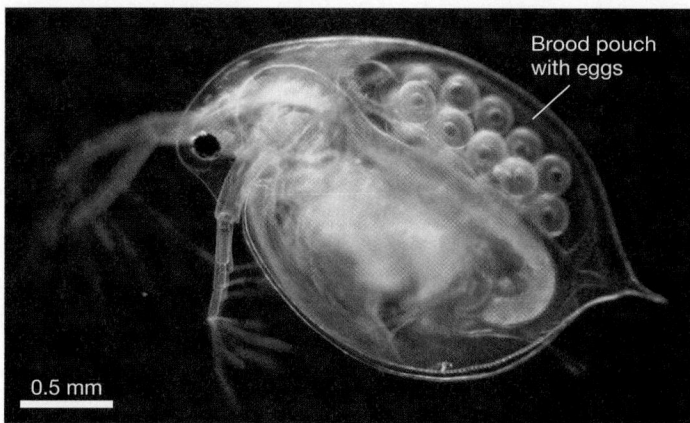

FIGURE 50.2 Female *Daphnia* Can Produce Eggs by Parthenogenesis.

of the pond or lake for the winter. In spring, the sexually produced offspring hatch and the females begin reproducing asexually.

Biologists try to explain observations like this at two levels (see Chapter 53):

1. **Proximate causation** addresses *how* a trait is produced. When researchers identify the genetic, developmental, hormonal, or neural mechanisms responsible for a phenotype, they are working at the proximate level.

2. **Ultimate causation** addresses *why* a trait occurs, in terms of its effect on fitness. Researchers who work at the ultimate level try to understand the evolutionary history of traits.

Let's consider work on proximate aspects of the asexual–sexual switch in *Daphnia* and then consider ultimate causation.

What Environmental Cues Trigger the Switch? For decades, most researchers contended that day length triggered the asexual–sexual switch. The idea was that the shortening days of late summer or fall affected sensors in the brain (see Chapter 49), and these receptors produced electrical or hormonal signals that induced the production of males and haploid eggs.

In 1965, however, biologists showed that high population densities are also a factor. Researchers who brought *Daphnia pulex* populations into the lab and kept day length constant found a strong, positive correlation between population density and the percentage of females reproducing sexually. When you read the graph in **FIGURE 50.3a**, note that the *x*-axis is logarithmic (see **BioSkills 6** in Appendix B). The highest proportion of sexually reproducing offspring is found at high density.

Another group of investigators built on this result by pinpointing the specific aspects of crowding that affected the animals. These biologists brought a closely related species—*D. magna*—into the laboratory and altered day length, the amount of food available to individuals, and the quality of the water the *Daphnia* occupied. To vary water quality, the investigators used either clean water or "crowded" water taken from tanks where *D. magna* were being maintained at high density.

As **FIGURE 50.3b** shows, individuals in the study population switched to sexual reproduction only if they were exposed to poor-quality, "crowded" water, low food availability, *and* short day lengths. In short, *D. magna* needs three different cues from the environment to switch to sexual reproduction. Two of these cues were associated with high population density; the third is associated with the onset of winter.

(a) Sexual reproduction is more common in crowded populations of *Daphnia pulex* than in sparse populations.

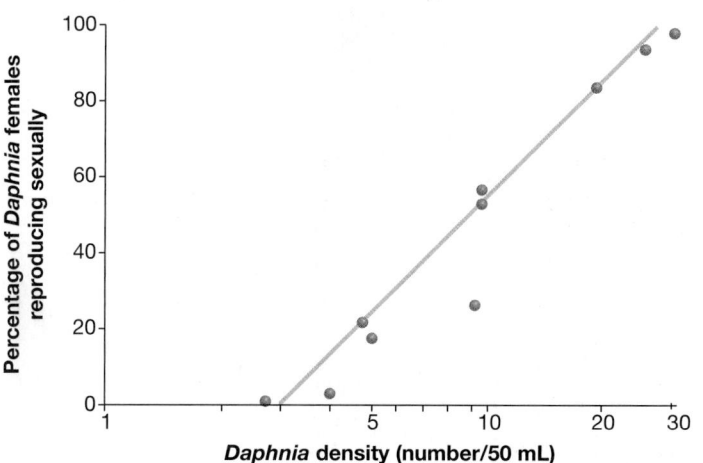

(b) In *Daphnia magna*, which environmental cues trigger the switch to sexual reproduction?

Water quality	Food concentration	Day length	Sexual broods (%)
Clean	Low	Short	0
Crowded	Low	Short	44
Clean	Low	Long	0
Crowded	Low	Long	0
Clean	High	Short	0
Crowded	High	Short	0
Clean	High	Long	0
Crowded	High	Long	0

FIGURE 50.3 In *Daphnia*, Environmental Cues Signal the Switch from Asexual to Sexual Reproduction.
(a) There is a strong, positive correlation between the percentage of females that reproduce sexually and the density of the population (plotted on a log scale here). **(b)** Environmental conditions were varied experimentally for *Daphnia*. "Crowded" water was taken from tanks containing dense populations.

DATA: Kleiven, O. T., P. Larsson, and A. Hoboek. 1992. *Oikos* 65: 197–206.

✔**QUESTION** How would you go about determining which molecule or molecules in "crowded" water serve as a signal that triggers sexual reproduction?

Why Do *Daphnia* Switch between Asexual and Sexual Reproduction? *Daphnia* appear to start sexual reproduction when conditions worsen. Why?

The leading hypothesis in answer to this question points out that sexually produced offspring are genetically diverse. When the environment changes, genetically diverse offspring are likely to include individuals that will survive better and reproduce more than will offspring that are identical to their parents. Parasites evolve rapidly, changing the environment for their hosts. According to the hypothesis, snails with greater genetic variability are likely to adapt better to those changes.

Genetically variable offspring have higher fitness in environments with rapidly evolving parasites, deteriorating physical conditions, or other types of rapid environmental change. Genetic variability increases the likelihood that at least some of the offspring will be able to survive and reproduce in such an environment. ✔ If you understand this concept, you should be able to predict the conditions under which asexual reproduction would be found as the only method of producing offspring.

To date, however, the variable-environment hypothesis has yet to be tested rigorously in *Daphnia*. Research on the adaptive significance of sex in this species and other organisms continues.

Mechanisms of Sexual Reproduction: Gametogenesis

The mitotic cell divisions, meiotic cell divisions, and developmental events that produce male and female gametes, or sperm and eggs, are collectively called **gametogenesis** (see Chapter 23). **Spermatogenesis** is the formation of sperm; **oogenesis** is the formation of eggs (**FIGURE 50.4**).

In the vast majority of animals, gametogenesis occurs in a sex organ, or **gonad.** Male gonads are called **testes;** female gonads are called **ovaries.** Early in development, reproductive cells known as germ cells enter the testes and ovaries and give rise to diploid cells that will undergo gametogenesis.

Spermatogenesis in Mammals   Figure 50.4a summarizes the events that take place during spermatogenesis. Note that in the male gonad, diploid cells called **spermatogonia** (singular: **spermatogonium**) divide by mitosis. Some of the resulting cells continue to function as spermatogonia; others change to form specialized cells that are committed to developing into sperm.

The specialized cells produced by spermatogonia are called **primary spermatocytes.** They undergo meiosis I and produce two **secondary spermatocytes,** which then undergo meiosis II. The result is four haploid cells called **spermatids.**

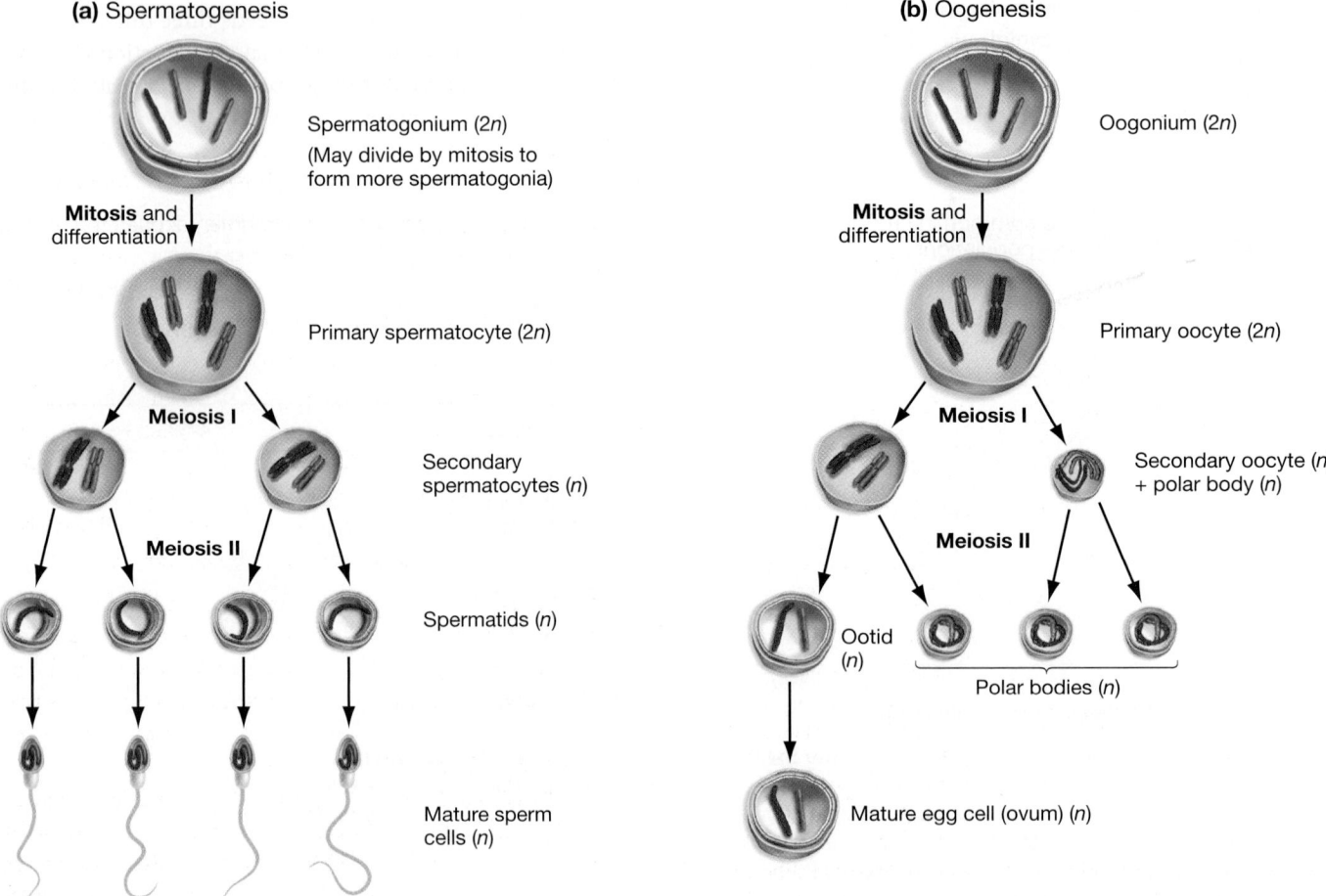

(a) Spermatogenesis

Spermatogonium (2*n*) (May divide by mitosis to form more spermatogonia)

Mitosis and differentiation

Primary spermatocyte (2*n*)

Meiosis I

Secondary spermatocytes (*n*)

Meiosis II

Spermatids (*n*)

Mature sperm cells (*n*)

(b) Oogenesis

Oogonium (2*n*)

Mitosis and differentiation

Primary oocyte (2*n*)

Meiosis I

Secondary oocyte (*n*) + polar body (*n*)

Meiosis II

Ootid (*n*)

Polar bodies (*n*)

Mature egg cell (ovum) (*n*)

FIGURE 50.4 Gametogenesis in Sexually Reproducing Animals Produces Sperm or Eggs via Meiosis.

Each haploid spermatid matures into a sperm—a cell that is specialized for carrying a haploid genome from the male through the female reproductive tract and fertilizing an egg. The production of spermatogonia, primary spermatocytes, and sperm occurs continuously throughout a male's adult life.

Structure and Function of Sperm   As a mammalian sperm cell matures, it acquires the four main compartments shown in **FIGURE 50.5**: the head, neck, midpiece, and tail.

- The head region contains the nucleus and an enzyme-filled structure called the **acrosome.** The enzymes stored in the acrosome allow the sperm to penetrate the barriers surrounding the egg.

- The neck encloses a centriole that will combine with a centriole contributed by the egg to form a centrosome (see Chapter 7). The centrosome is required for spindle formation during mitosis (see Chapter 12).

- The midpiece is packed with mitochondria (see Chapter 9), which produce the ATP required to power movement.

- The tail region consists of a **flagellum**—a long structure, composed of microtubules and surrounded by plasma membrane, that whips back and forth to make swimming possible (see Chapter 7).

Sperm are race cars—stripped down, streamlined, souped-up cells that are specialized for racing other sperm to the egg. Eggs, in comparison, are like semitrailers—bulky, far less mobile storage containers that are packed with valuable merchandise and securely locked.

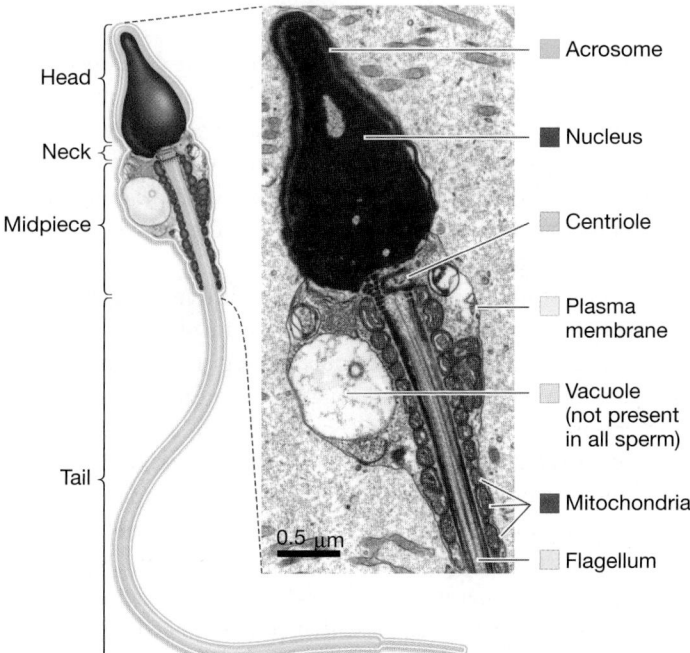

FIGURE 50.5 Mammalian Sperm Are Specialized for Motility and Fusing with an Egg Cell. The morphology of human sperm is typical of many mammal species.

Oogenesis in Mammals   The top of Figure 50.4b highlights an important similarity between spermatogenesis and oogenesis: In the female gonad, diploid cells called **oogonia** (singular: **oogonium**) divide by mitosis. Some of the resulting cells continue to function as oogonia; others change to form specialized cells that are committed to producing an egg. In this respect, oogonia and spermatogonia are similar.

However, subsequent steps in gametogenesis are markedly different in females. The specialized cells produced by an oogonium are called **primary oocytes.** When these cells undergo meiosis, only one of the haploid products, known as a **secondary oocyte,** can eventually mature into an egg. The secondary oocyte is arrested in the final stages of meiosis II until it is fertilized by a sperm; it then completes meiosis to become an ootid, which matures into an **ovum,** or a mature egg cell.

The other cells produced by meiosis in females have a tiny amount of cytoplasm and do not mature into eggs. Because the distribution of cytoplasm is so unequal during each meiotic division in females, the smaller cells are called **polar bodies.** (Recall that polar refers to inequality or opposites.) Polar bodies degenerate shortly after their formation.

In addition, the production of primary oocytes stops early in development in many mammals; in humans, it stops before a female fetus is born. The primary oocytes enter prophase of meiosis I during fetal development but then stop developing for months, years, or decades.

Structure and Function of Eggs   Animal gametes are easy to tell apart: Sperm are small and motile; eggs are large and nonmotile. Eggs are large mainly because they contain the nutrients required for the embryo's early development. The quantity of nutrients present in the egg varies widely among species, however.

- In mammals, embryos start to obtain nutrition through a maternal organ called the **placenta** within a week or two after fertilization. Thus, the egg only has to supply nutrients for early development and is relatively small.

- In species where females lay eggs into the environment, the stores in the egg are the *only* source of nutrients until organs have formed and a larva or juvenile hatches and begins to feed. In these species, the nutrients required for early development are provided by **yolk**—a fat- and protein-rich cytoplasm that is loaded into egg cells as they mature. Yolk may be present as one large mass or as many small granules.

Just outside the plasma membrane of eggs, a fibrous, mat-like sheet of glycoproteins called the **vitelline envelope** forms and surrounds the egg. In many aquatic animals, such as the sea urchin, a large gelatinous matrix known as a jelly layer surrounds the vitelline envelope to further enclose and protect the egg (see **FIGURE 50.6a**, on page 1018).

In the eggs of humans and other mammals, the vitelline envelope structure is unusually thick and is called the **zona pellucida**. This structure is surrounded by a layer of cells, called the corona radiata, that a sperm must penetrate before it can fertilize the oocyte (**FIGURE 50.6b**).

(a) Sea urchin eggs are surrounded by a jelly coat.

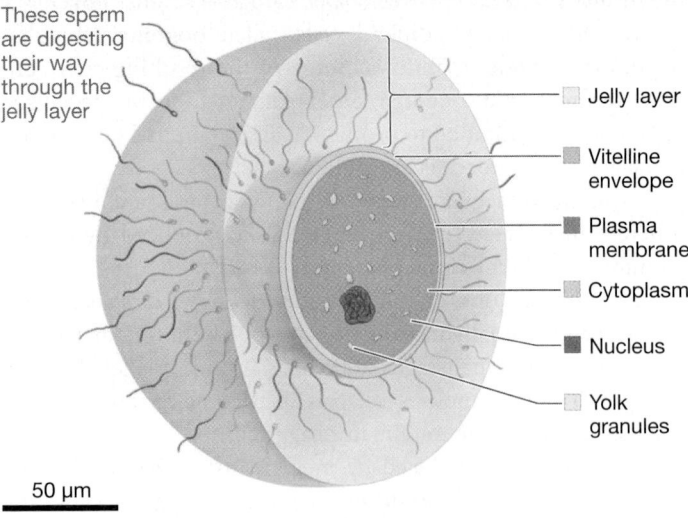

These sperm are digesting their way through the jelly layer

- Jelly layer
- Vitelline envelope
- Plasma membrane
- Cytoplasm
- Nucleus
- Yolk granules

50 μm

(b) Human oocytes are surrounded by a protective layer called the corona radiata.

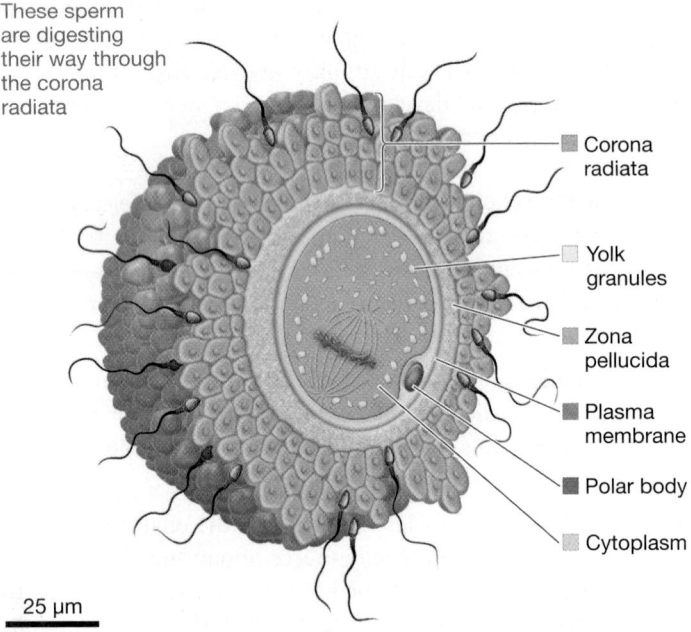

These sperm are digesting their way through the corona radiata

- Corona radiata
- Yolk granules
- Zona pellucida
- Plasma membrane
- Polar body
- Cytoplasm

25 μm

FIGURE 50.6 Eggs Are Surrounded by Protective Structures.

50.2 Fertilization and Egg Development

Fertilization is the joining of a sperm and an egg to form a diploid **zygote.** Recall that during this process, a sperm makes contact with an egg and penetrates the egg's membrane. The molecular mechanisms of sperm–egg binding have been well characterized using the sea urchin as a model organism (see Chapter 23).

The following discussion focuses on the range of fertilization mechanisms observed among animals and the question of what happens to fertilized eggs. In many species, individuals release their gametes into their environment synchronously (at the same time), and external fertilization occurs. In other animals, sperm is deposited into the reproductive tracts of females, and internal fertilization occurs.

External Fertilization

Most animals that rely on external fertilization live in aquatic environments. The correlation between external fertilization and aquatic environments is logical, because gametes and embryos must be protected from drying. If external fertilization occurred in a terrestrial environment, either the gametes or the resulting zygote would likely die of desiccation.

Another observation is that species with external fertilization tend to produce exceptionally large numbers of gametes. For example, a female sea star *Asterias amurensis* typically releases 100,000,000 eggs into the surrounding seawater during spawning. Males release many times that number of sperm. The leading hypothesis to explain this pattern is that the probability of a sperm and egg meeting in an ocean or lake is small unless large numbers of gametes are present. In addition, production of a large number of embryos is beneficial because so many of them fall prey to predators.

Given that sperm and eggs from different individuals must be released into the environment synchronously for external fertilization to work, how is gamete release coordinated? The answer has two parts.

1. Gametogenesis usually occurs in response to environmental cues, such as lengthening days and warmer water temperatures, that indicate a favorable season for breeding.

2. Gametes are released in response to specific cues from individuals of the same species.

In fishes and other aquatic animals with well-developed eyes, external fertilization by spawning is often the culmination of an elaborate courtship ritual between a male and female. In contrast, courtship behavior appears to be much less important—or even absent—in species such as clams, sea urchins, and sea cucumbers. How do animals that cannot see their mates time the release of their gametes?

Recent research indicates that the chemical messengers called **pheromones** (see Chapter 47) might be involved in synchronizing gamete release. For example, biologists maintained two groups of sea cucumbers under natural conditions of light and temperature for 15 months. In one treatment, individuals were isolated in separate tanks; in another, they were kept in tanks containing other sea cucumbers. The researchers found that all the individuals that shared their tanks released gametes during the normal spawning period. In contrast, only about 10 percent of the individuals maintained in isolation released gametes.

Although these data suggest that pheromones might be involved in coordinating external fertilization in sea cucumbers, they are not definitive. Only recently have biologists begun to identify the actual pheromones involved in mating of marine invertebrates. Accumulation of data on pheromones in multiple species will help solidify the case that spawning is coordinated by pheromones.

Internal Fertilization

Internal fertilization occurs in the vast majority of terrestrial animals as well as in a significant number of aquatic animals. Internal fertilization occurs in one of two ways:

1. After copulation, in which males deposit sperm directly into the female reproductive tract with the aid of a copulatory organ, usually called a **penis.**

2. After males package their sperm into a structure called a **spermatophore,** which is then picked up and placed into the female's reproductive tract by the male or female.

In some salamander species, for example, the male places the spermatophore on the ground within its territory. Later the female picks it up with her **cloaca,** a chamber that both the reproductive and excretory systems flow into and that opens to the environment. In this case, the result is internal fertilization without any direct physical contact between the sexes.

What Is Sperm Competition? In terms of understanding animal behavior, perhaps the most important insight about internal fertilization originated with Geoff Parker. In 1970 Parker published experiments, on dung flies, that confirmed the existence of **sperm competition**—competition between sperm from different males to fertilize the eggs of the same female.

Parker's experiments consisted of a series of matings of one female with two males. In each experiment, the two males were selected in such a way that Parker could distinguish between their offspring. The proportion of offspring fathered by each male was not 50:50. Instead, whichever male was last to copulate fathered an average of 85 percent of the offspring produced.

In addition to suggesting that sperm were competing to fertilize eggs, these data indicated that, in this experiment, the second male won. Follow-up research has confirmed that **second-male advantage** is widespread, although not universal, in insects and some other animal groups. How does it occur?

Sperm Competition and Second-Male Advantage To explore why second-male advantage occurs, biologists turned to the fruit fly *Drosophila melanogaster*. A research group introduced a gene into male fruit flies that produced sperm with green fluorescent tails (**FIGURE 50.7**). When a mating by a green-spermed male was followed with a second mating—by a male having normal-colored sperm—many fewer green-tailed sperm were found in the female's sperm-storage area compared with the number observed when no second mating took place.

To interpret this finding, the biologists suggested that the sperm of the second male physically dislodged the first male's gametes from the female sperm-storage area and inserted themselves in their place. The researchers also showed that the fluid that accompanies sperm during fertilization is able to displace stored sperm from competing males. More recent research has shown that in addition to these factors, females are able to eject

RESEARCH

QUESTION: How does the "second-male advantage" occur in sperm competition?

HYPOTHESIS: In sperm-storage areas, sperm from the second male displaces sperm from the first male.

NULL HYPOTHESIS: The mechanism does not involve sperm displacement from sperm-storage areas.

EXPERIMENTAL SETUP:

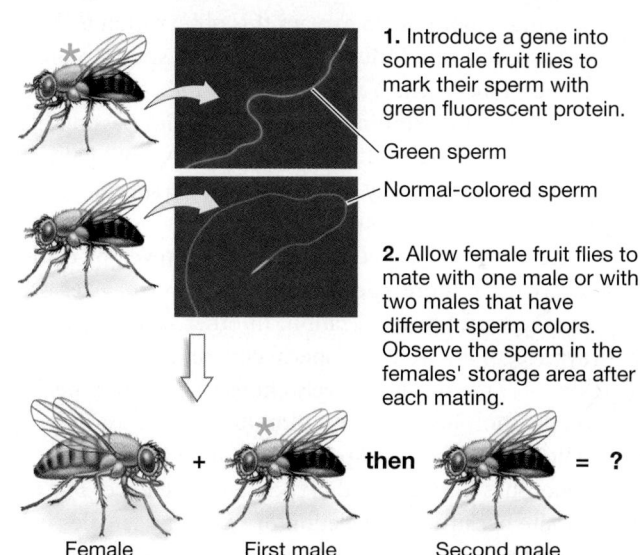

1. Introduce a gene into some male fruit flies to mark their sperm with green fluorescent protein.

Green sperm

Normal-colored sperm

2. Allow female fruit flies to mate with one male or with two males that have different sperm colors. Observe the sperm in the females' storage area after each mating.

Female + First male then Second male = ?

PREDICTION: When females mate twice, little sperm from the first male remains in storage.

PREDICTION OF NULL HYPOTHESIS: When females mate twice, most or all of the sperm deposited by the first male is still present.

RESULTS:

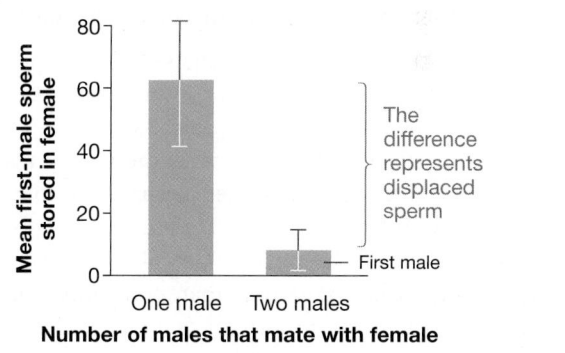

The difference represents displaced sperm

First male

CONCLUSION: If a female mates a second time, most sperm from the first male she mated with disappears.

FIGURE 50.7 Experimental Evidence Supports Second-Male Advantage in *Drosophila*. The graph shows the average number of green sperm stored in females when males with green sperm were the only male to mate or the first of two males to mate.

Price, C. S. C., K. A. Dyer, and J. A. Coyne. 1999. Sperm competition between *Drosophila* males involves both displacement and incapacitation. *Nature* 400: 449–452.

✔**QUESTION** Based on these data, is the claim that sperm from the second male physically displaces sperm from the first male valid?

sperm from their reproductive tracts. Together these mechanisms resulted in the second male's sperm fertilizing most of the eggs laid.

Why Is Testes Size Variable among Species? Research on sperm competition has recently contributed another major finding. In species where females routinely mate with multiple males before laying eggs or giving birth, males have extraordinarily large testes for their size and produce proportionally larger numbers of sperm.

The leading hypothesis to explain this observation is that fertilization is similar to a lottery in which each sperm represents a ticket. The more tickets a male enters in the competition, the higher his chance of "winning" fertilizations and passing his alleles on to the next generation. Males with exceptionally large testes produce exceptionally large numbers of sperm, and they are more likely to win the lottery.

The lottery model has been challenged, however, by evidence that females often store sperm and exert control over which sperm are successful in fertilization. In other words, females do not always accept the results of sperm competition passively.

Females of some species actively choose which male performs the last copulation before fertilization takes place. In other species, females physically eject sperm from undesirable males. This phenomenon has been dubbed cryptic female choice. The name is appropriate because the selection of sperm by females is hidden from males.

Unusual Mating Strategies

Studies on animal sexual reproduction have documented some of the most remarkable behaviors and structures observed in nature. The list that follows is just a sampling, offered simply to illustrate the diversity of mating arrangements in animals.

- *Femmes fatales* When Australian redback spiders mate, the male does a somersault after inserting his penis-like organ. The somersault places his dorsal surface in front of the female's mouthparts. In many cases the female responds by eating the male. (Biologists refer to females who cannibalize males as femmes fatales—a phrase used to describe murderous human females in movies.) Cannibalized males copulate longer and fertilize more eggs than non-cannibalized males, probably because sperm transfer continues until the meal is over.

- *Giant sperm* In the fruit fly *Drosophila bifurca*, males average 1.5 mm in total body length. Their sperm, however, are each 6 cm long. The coiled, bulky sperm fill the female's sperm-storage area and make it impossible for another male's sperm to enter.

- *Infidelity* When researchers assess paternity in bird species that appear to be monogamous, they find that up to 60 percent of nests contain at least one offspring fathered by a male that is not mated to the resident female. In most cases, females actively solicit copulations from males holding nearby territories.

FIGURE 50.8 Snails Stab with Love Darts. Stabbing a mate with a mucus-covered love dart increases reproductive success.

- *Love darts* Many snails and slugs are **hermaphroditic,** meaning an individual has both male and female gonads. During mating, two individuals simultaneously receive and deposit sperm to fertilize each other's eggs. In some species, individuals fire mucus-covered "love darts" from their genitalia into the mating partner (**FIGURE 50.8**). A substance in the mucus increases the chance that sperm will successfully fertilize eggs.

- *Hypodermic insemination* Some bedbugs have hypodermic penises, meaning that they function like a hypodermic needle. Males force the organ through the female's abdominal wall and deposit sperm directly into her body cavity.

Why Do Some Females Lay Eggs While Others Give Birth?

For females, many aspects of fertilization and egg laying vary among species: for example, a female's size or age at first breeding; the number and size of eggs she produces; the number of times she reproduces over the course of a lifetime. Here let's consider a different aspect of variation in animal reproduction: whether eggs are laid outside the body or retained inside the body.

In **oviparous** ("egg-bearing") animals, the embryo completes most of its development inside an egg that is laid into the environment. In some oviparous species, such as sea stars, sea urchins, and most insects, no further care is provided by the parents; the eggs and embryos are left to fend for themselves. Birds, however, incubate their eggs and feed the young after hatching; fish may guard their eggs from predators and fan the clutches to oxygenate them.

In **viviparous** ("live-bearing") species, embryonic development takes place entirely within the mother's body. The embryo attaches to the reproductive tract of the mother and receives nutrition directly from her—via diffusion from her circulatory system. In **ovoviviparous** species, offspring develop inside the mother's body but are nourished by nutrient-rich yolk stored in the egg.

Why does oviparity exist in some groups and viviparity or ovoviviparity in others? Biologists tackled this question by studying the lizard genus *Sceloporus*. Some *Sceloporus* populations are oviparous; others are ovoviviparous.

To understand how oviparity and ovoviviparity evolved, the biologists analyzed a phylogenetic tree—based on molecular and morphological data—of many *Sceloporus* species (**FIGURE 50.9**). Two conclusions should make sense to you (see **BioSkills 7** in Appendix B for help with interpreting phylogenetic trees):

1. Because the basal branches of the tree represent oviparous species, egg laying probably represents the original or ancestral condition.

2. As the red branches on the tree show, ovoviviparity evolved independently in two groups.

Using a similar research strategy, biologists have found that ovoviviparous populations of sea stars have evolved from oviparous populations on several occasions.

Why did natural selection favor these changes between egg laying and live birth? Ovoviviparity or viviparity should evolve when it leads to higher numbers of surviving young. Researchers have long hypothesized that natural selection favoring live birth should be especially strong in cold habitats. Low temperatures slow the development of embryos. Thus, in cold habitats, it might be advantageous for females to retain eggs inside their bodies so that they can thermoregulate to maintain the offspring at a more favorable temperature for development. This hypothesis is supported by the fact that ovoviviparous *Sceloporus* species live in the highlands of the southwestern United States and central Mexico, where temperatures can be cool.

Sceloporus lizards have been a productive group in which to study this question because the trait has evolved multiple times in extremely closely related species—making it easier to find correlations between live birth and other characteristics, such as living in cold habitats. The issue is much more difficult to study in mammals, where viviparity evolved just once. Monotremes (duck-billed platypus and echidna) are oviparous, while marsupials and placental mammals are viviparous (see Chapter 35). It

is still not clear why, and how, viviparity evolved in mammals before the marsupial–placental split. Is the cold-habitats hypothesis relevant? To date, nobody knows.

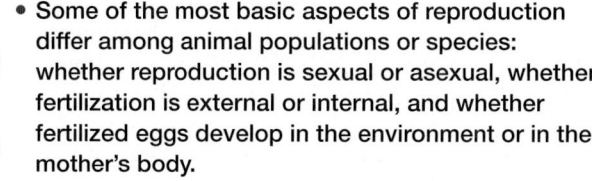

50.3 Reproductive Structures and Their Functions

The first two sections of this chapter considered (1) the broad contrast between asexual and sexual reproduction and (2) general patterns in fertilization and egg care. Now let's explore the mechanics of sexual reproduction in more detail. The first task is to understand the anatomy of the male and female reproductive systems—focusing primarily on mammals.

The Male Reproductive System

In humans, the external anatomy of the male reproductive system consists of the scrotum and the penis. The saclike **scrotum** holds the testes; the penis functions as the organ of copulation necessary for internal fertilization.

Biologists who have compared these structures among animal species have been struck by their variability and complexity. For example, a scrotum occurs only in certain mammal species. Whales, elephants, hedgehogs, moles, and many other mammal groups, along with all other vertebrates, lack the structure entirely and instead have testes that are located well within the abdominal cavity. Among species that do have a scrotum, the structure's size and shape vary from tiny to prominent. In many primate species, the scrotum is brightly colored and appears to function in courtship or other sexual display behaviors.

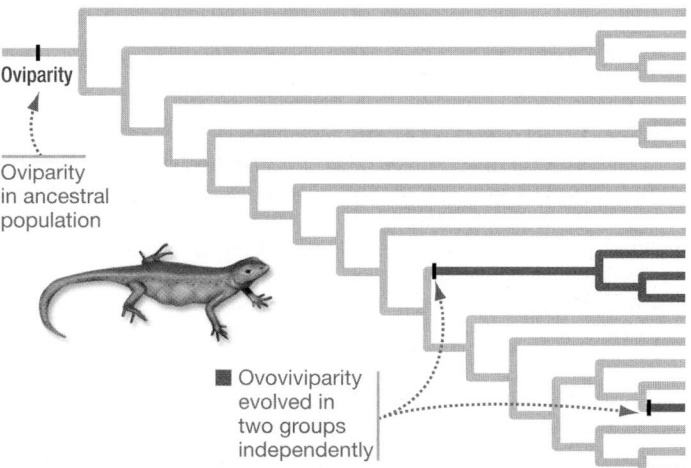

FIGURE 50.9 Ovoviviparity Has Evolved More than Once in *Sceloporus* Lizards. Each twig on this phylogenetic tree represents a *Sceloporus* species or population from central Mexico.

How Does External Anatomy Affect Sperm Competition? Diversity in scrotal morphology pales in comparison with variation in the structure of the penis or other types of male **genitalia,** meaning external copulatory organs. In many groups of insects and spiders, for example, closely related species are morphologically identical except for their distinctive genitalia. Why are these organs so diverse?

The leading hypothesis on this question is that certain genitalia shapes give males an advantage in sperm competition. To test this idea, biologists measured the size of the spines found on the tips of male genitalia in a species of seed beetle (**FIGURE 50.10a**) and then documented which males were most successful at fertilizing eggs when females mated with two males.

The *x*-axis in **FIGURE 50.10b** plots variation in the length of genital spines; the *y*-axis plots the proportion of eggs fertilized by the second male to copulate with a female, during experimental matings. The data points on the graph indicate genital spine length in each second-to-mate male tested. Note that second-to-mate males with the longest genital spines fathered a higher percentage of offspring than did others. The mechanism appears to be the ability of long spines to stick in the female's reproductive tract and prolong copulation time, even if the female is ready for copulation to end.

The upshot? Size matters, in seed beetles. Natural selection that occurs during sperm competition may explain why genitalia are so diverse among insect and spider species.

Internal Anatomy of Human Male Reproductive Organs

Although it consists of many structures, the reproductive system in human males has just three basic functional components. **FIGURE 50.11** shows the relevant structures in side view and front view.

1. *Spermatogenesis and sperm storage* Sperm are produced in the testes and stored nearby in the **epididymis.**

2. *Production of accessory fluids* Complex solutions form in the **seminal vesicles, prostate gland,** and **bulbourethral gland.** These accessory fluids are added to sperm before **ejaculation,** or expulsion from the body. **TABLE 50.1** lists some components of these accessory fluids, which combine with sperm to form **semen.**

3. *Transport and delivery* A **vas deferens** is a muscular tube that transports sperm from the epididymis to the short **ejaculatory duct,** where it is mixed with accessory fluids to form semen. The semen then enters the **urethra,** a longer tube that passes through the penis and services both the reproductive and urinary systems in males. The semen is expelled during ejaculation.

The composition of the accessory fluids varies widely among animals. In many insects, spiders, and vertebrates, molecules in the accessory fluids congeal after they arrive in the female reproductive tract and plug it. Experiments have shown that these copulatory plugs can serve as an effective deterrent to future matings. In some species, though, females or second males actively remove them.

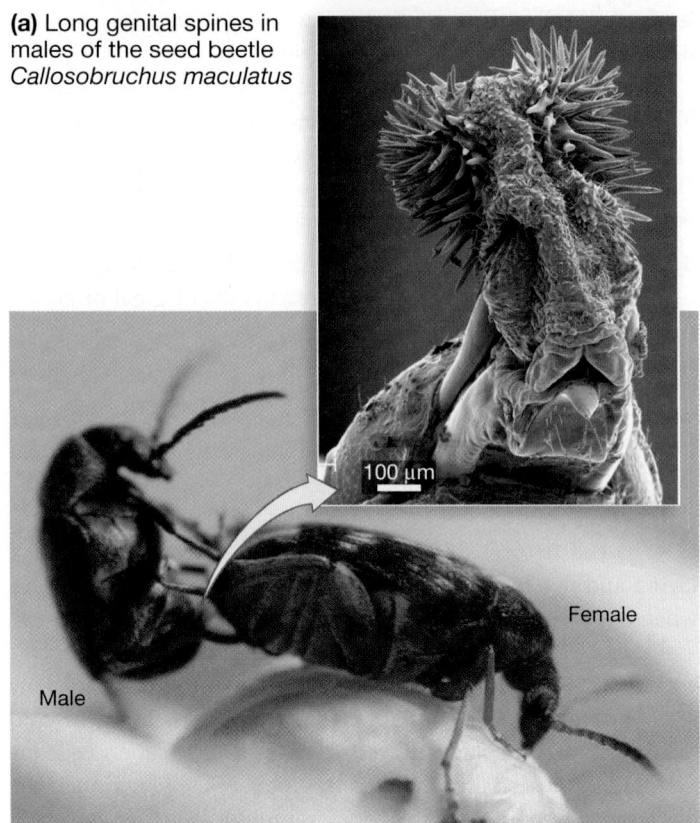

(a) Long genital spines in males of the seed beetle *Callosobruchus maculatus*

100 µm

Female

Male

(b) During sperm competition, males with longer genital spines father more offspring.

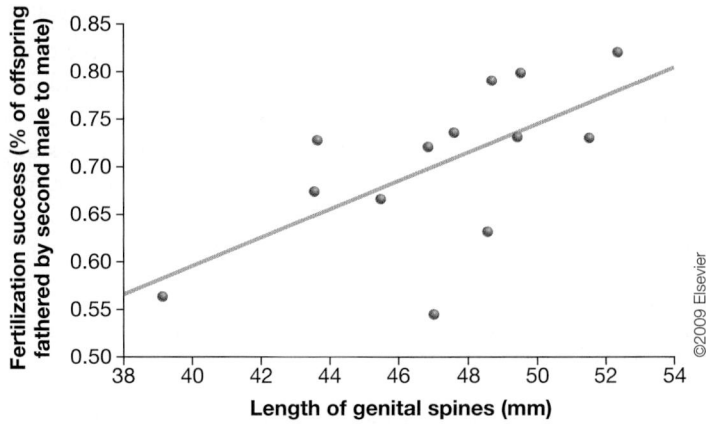

©2009 Elsevier

FIGURE 50.10 In Some Insects and Spiders, Variation of Male Genitalia among Individuals May Affect Reproductive Success. (a) An elaborate male reproductive structure in a seed beetle, used to transfer sperm to the female. The spiky structures at the top of the inset photo are called genital spines. **(b)** Data from experiments on reproductive success during sperm competition.

DATA: Hotzy, C., and G. Arnqvist. 2009. *Current Biology* 19: 404–407.

Another diverse aspect of male internal anatomy is a bone inside the penis called the **baculum.** Some mammal species, including humans, lack this feature. But in rodents, the shape of the baculum is so variable that it can be used to distinguish among species. And in seals, baculum size correlates with mating system—the mating practices observed in a species. For example,

(a) Side view

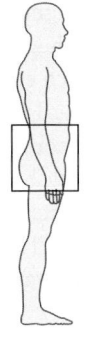

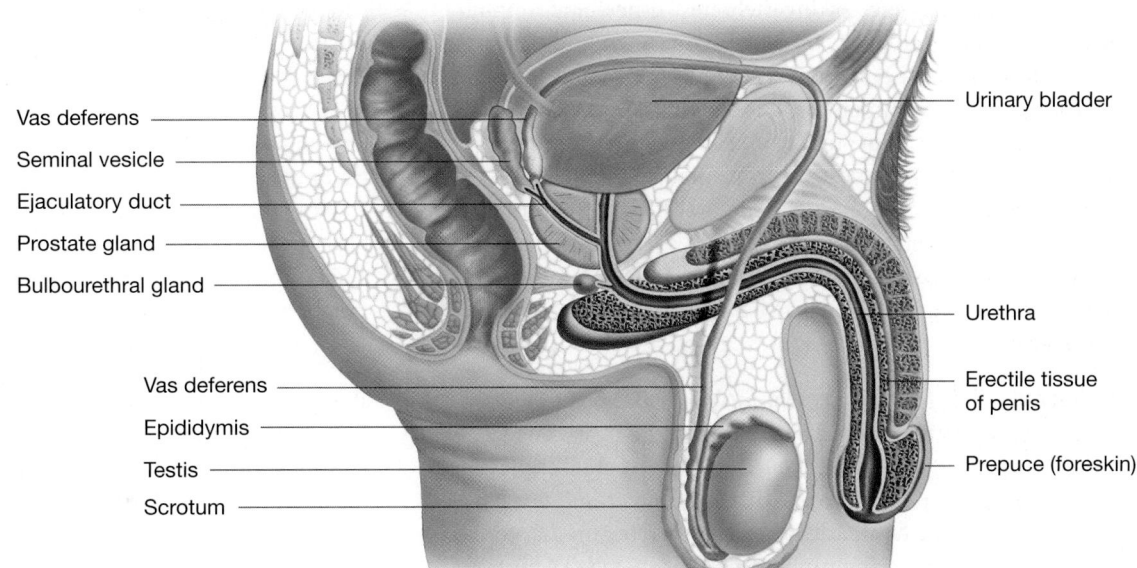

Vas deferens

Seminal vesicle

Ejaculatory duct

Prostate gland

Bulbourethral gland

Urinary bladder

Urethra

Erectile tissue of penis

Prepuce (foreskin)

Vas deferens

Epididymis

Testis

Scrotum

(b) Front view

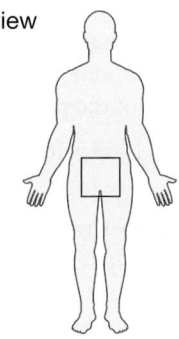

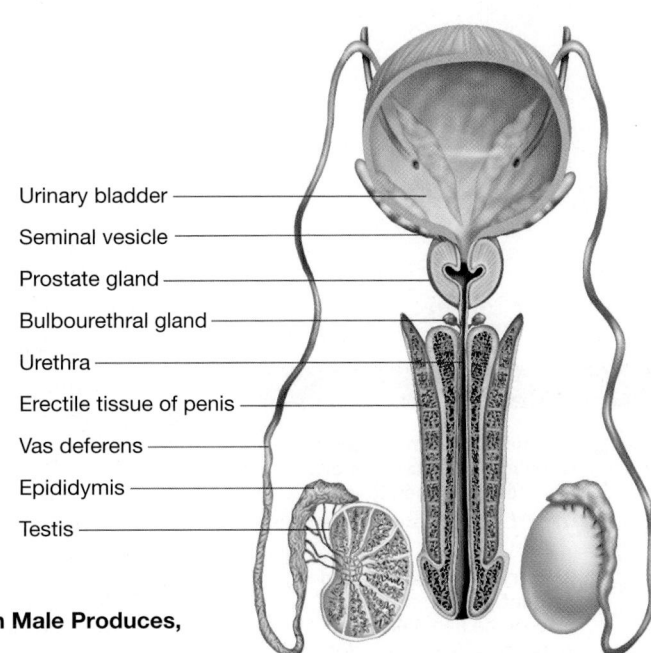

Urinary bladder

Seminal vesicle

Prostate gland

Bulbourethral gland

Urethra

Erectile tissue of penis

Vas deferens

Epididymis

Testis

FIGURE 50.11 The Reproductive Tract in a Human Male Produces, Stores, and Transports Sperm.

SUMMARY TABLE 50.1 Accessory Fluids in Human Semen

Source	Content	Function
Seminal vesicles	Fructose (a sugar)	Source of chemical energy for sperm movement
	Prostaglandins	Stimulate smooth-muscle contractions in uterus
Prostate gland	Antibiotic compound	Prevent urinary tract infections in males
	Citric acid	Nutrient used by sperm
Bulbourethral gland	Alkaline mucus	Lubricate tip of penis; neutralize acids in urethra

in seal species in which females routinely mate with several males before becoming pregnant, males have not only large testes for their size but also a large baculum. The testes and baculum are much smaller in species in which females mate with a single male. In all species where it appears, the baculum helps stiffen the penis during copulation.

The Female Reproductive System

In animals, the most important part of the female reproductive system is the ovary—where meiosis occurs and mature egg cells, or ova, are produced. In the vast majority of species, the mature egg cell is a membrane-bound structure consisting of a haploid nucleus, a full complement of other organelles, and a large supply of nutrients in the form of yolk.

In the gelatinous eggs of invertebrates, fishes, and amphibians, the extensive outer layers and plasma membranes play a key role in binding sperm from the same species and initiating fertilization. Other animals produce membrane-rich amniotic eggs, which evolved as an innovation that allowed tetrapods to lay large eggs that do not desiccate on land (see Chapter 35).

To probe variation in egg structure and female reproductive systems further, let's consider two highly specialized examples, the reproductive systems of female birds and mammals. Birds lay an amniotic egg protected by a hard shell; most mammals are viviparous.

The Reproductive Tract of Female Birds

FIGURE 50.12 diagrams the bird reproductive system. Starting with the ovary, the labels on the drawing indicate the sequence of events that takes place as the egg moves down the reproductive tract. The result is a hard-shelled egg—like the familiar chicken eggs you buy in the store—that can be laid into the environment and incubated. All birds are oviparous.

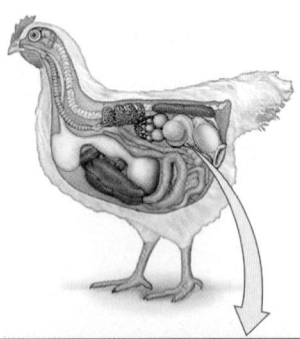

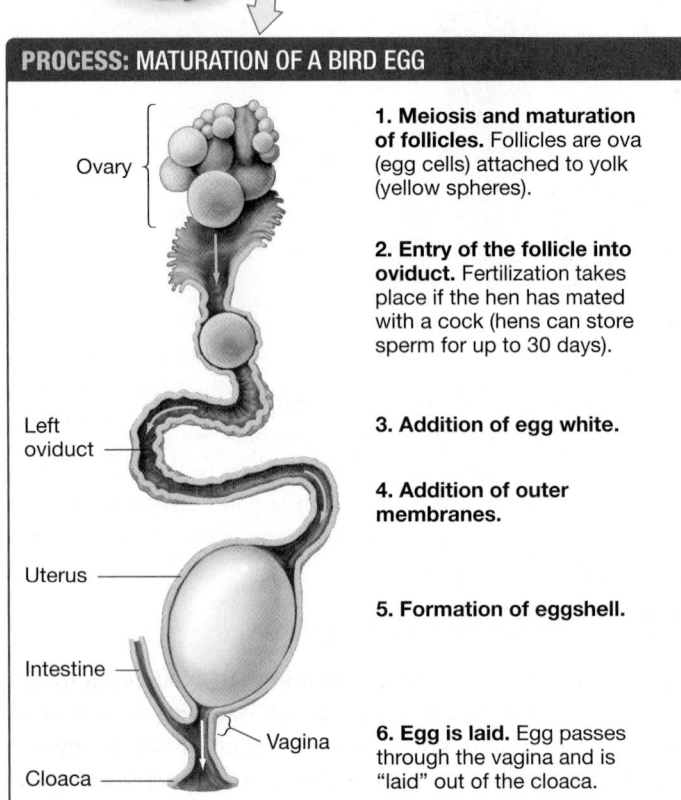

PROCESS: MATURATION OF A BIRD EGG

Ovary

Left oviduct

Uterus

Intestine

Vagina

Cloaca

1. **Meiosis and maturation of follicles.** Follicles are ova (egg cells) attached to yolk (yellow spheres).

2. **Entry of the follicle into oviduct.** Fertilization takes place if the hen has mated with a cock (hens can store sperm for up to 30 days).

3. **Addition of egg white.**

4. **Addition of outer membranes.**

5. **Formation of eggshell.**

6. **Egg is laid.** Egg passes through the vagina and is "laid" out of the cloaca.

FIGURE 50.12 All Birds Are Oviparous.

From the time the egg is released from the ovary until the zygote undergoes mitosis and the embryo begins to develop, a bird egg is a single cell. The ostrich egg, which can be over 15 cm (6 inches) in diameter, contains one of the largest single cells known in animals. The structure stores enough nutrients and water to sustain development until hatching.

Note that although male birds have two testes, females of most bird species have just one functional ovary. The presence of a single working ovary is thought to be an adaptation that reduces weight and makes flight more efficient.

External and Internal Anatomy of Human Females

FIGURE 50.13 shows side and front views of the human female reproductive system. The external anatomy features the **labia minora** (singular: **labium minus**) and the **labia majora** (singular: **labium majus**), the opening of the urethra, and the opening of the vagina.

- The labia are folds of skin that cover the urethral and vaginal openings.
- The **clitoris** is an organ that develops from the same population of embryonic cells that gives rise to the penis in males. It becomes erect during sexual stimulation and is covered with a protective sheath called the *prepuce*, homologous with the prepuce (foreskin) that covers the end of the penis.
- The urethral opening, where urine is expelled, is separate from the reproductive structures.
- The **vagina**, or birth canal, is the chamber where semen is deposited during sexual intercourse and where the baby is delivered during childbirth.

The internal anatomy of the female reproductive system in humans and other mammals is dominated by structures serving the system's two functions:

1. *Production and transport of eggs* Eggs are produced in the paired ovaries. During **ovulation**, a secondary oocyte is expelled from the ovary and enters the **oviduct**, also known as the **fallopian tube**, where fertilization may take place. Fertilized eggs are then transported through the oviduct to the muscular sac called the **uterus.**
2. *Development of offspring* The uterus is where embryonic development takes place. During childbirth, the developed embryo (now called a fetus) passes through an opening in the **cervix**—the bottom part of the uterus—and into the vagina.

Passage of the fetus through the cervix and vagina is the norm in mammals. However, the spotted hyena has evolved a fascinating alternative. In female hyenas, the birth canal is enclosed within an inch-wide, 7-inch-long clitoris that resembles a male's penis (**FIGURE 50.14**). Mating and birthing occur through the clitoris, which often tears during delivery of the Chihuahua-sized cubs.

Although the adaptive significance of the hyena clitoris is unknown, scientists have determined that it results from very high levels of male-typical sex hormones in females. These hormones also cause females to be larger and more muscular than males and socially dominant over them.

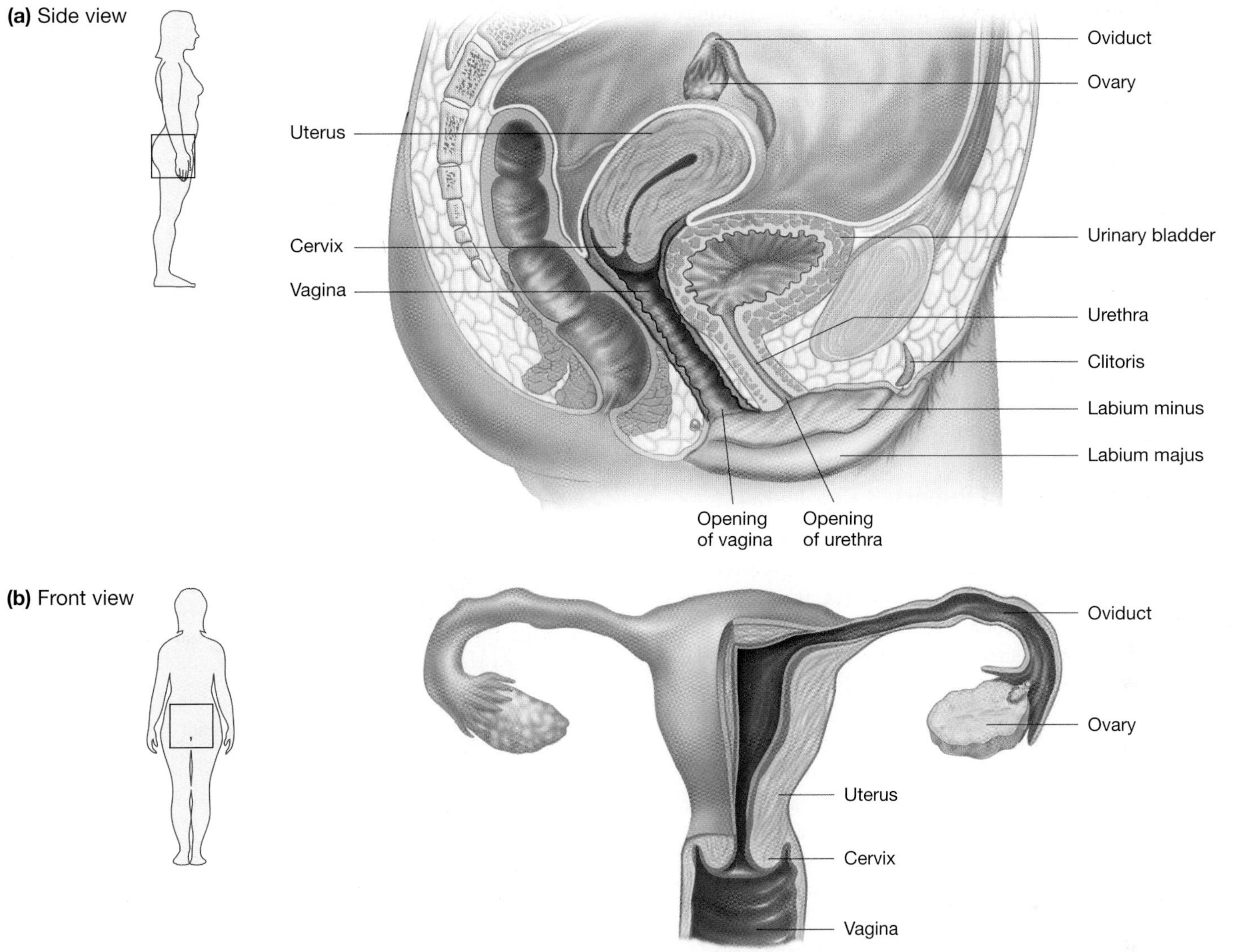

(a) Side view

Uterus

Cervix

Vagina

Oviduct

Ovary

Urinary bladder

Urethra

Clitoris

Labium minus

Labium majus

Opening of vagina   Opening of urethra

(b) Front view

Oviduct

Ovary

Uterus

Cervix

Vagina

FIGURE 50.13 The Reproductive Tract in a Human Female Produces Eggs and Nurtures the Embryo and Fetus.

✔**QUESTION** Generate a hypothesis to explain why male and female gonads are paired in mammals. How would you test your hypothesis?

FIGURE 50.14 Female Spotted Hyenas Have Enlarged Clitori. High levels of testosterone are responsible for the male-like anatomy of female hyenas. They copulate and give birth through the long, narrow clitoris.

Let's take a further look at how sex hormones control reproduction, focusing on a mammal—specifically, humans—as an illustrative example. But it is important to note that the same or very similar hormones control reproduction in all vertebrates, from fishes to birds.

50.4 The Role of Sex Hormones in Mammalian Reproduction

Recall that the sex hormones **testosterone** and **estradiol** are steroids; the latter belongs to a class of hormones known as estrogens (see Chapter 49). Testosterone and estradiol bind to receptors within the cytoplasm or nucleus of target cells. The resulting hormone–receptor complexes bind to DNA and trigger changes in gene expression.

Testosterone and estradiol are classified as gonadal hormones because they are produced in the gonads. Most testosterone is synthesized in specialized cells inside the testes; most estradiol and other estrogens are synthesized in the ovaries. More specifically, the female sex hormones are produced by cells that surround each developing egg. These surrounding cells form a structure called a **follicle** around the egg.

The sex hormones play a key role in three events:

1. Development of the reproductive tract in embryos

2. Maturation of the reproductive tract during the transition from childhood to adulthood

3. Regulation of spermatogenesis and oogenesis in adults

To explore the action of sex hormones, let's take a closer look at their role in the transition from human juvenile to adult.

Which Hormones Control Puberty?

Puberty is the process that leads to sexual maturity in humans. In amphibians, the juvenile-to-adult transition is triggered by the hormone T_3 (triiodothyronine); in insects, the transition occurs in response to ecdysone (see Chapter 49). But in humans, the transition is directed by increased levels of gonadal hormones—testosterone in boys and estradiol in girls.

A group of physicians recently offered dramatic evidence for testosterone's role: a 2-year-old boy whose symptoms included an enlarged penis, the development of pubic hair, and facial acne. Through careful interviewing, the doctors discovered that the child's father was a bodybuilder who was smearing a testosterone cream on his own arms in an attempt to build muscle mass. The child was apparently absorbing enough testosterone from being carried by his father to trigger puberty-like symptoms. All his symptoms except for penile enlargement subsided once his father discontinued the testosterone cream.

What Regulates the Gonadal Hormones?   Researchers suggested that sex hormone production is regulated by the hypothalamic–pituitary axis (see Chapter 49). Recall that chemical signals from the hypothalamus lead to the release of regulatory hormones from the pituitary gland, which then cause the release of hormones from other glands.

Two advances made it possible to test this hypothesis rigorously:

1. Researchers isolated a hormone called **gonadotropin-releasing hormone (GnRH)** from the hypothalamus.

2. Investigators noted that boys and girls who were entering puberty experienced pulses in the concentration of two pituitary hormones, **luteinizing hormone (LH)** and **follicle-stimulating hormone (FSH).**

These observations inspired the hypothesis that GnRH directs LH and FSH pulses, which then trigger increases in testosterone and estradiol (**FIGURE 50.15**).

To test this idea, researchers administered pulses of GnRH to boys and girls who had deficits in their hypothalamus and were experiencing delays in the onset of puberty. As predicted, the

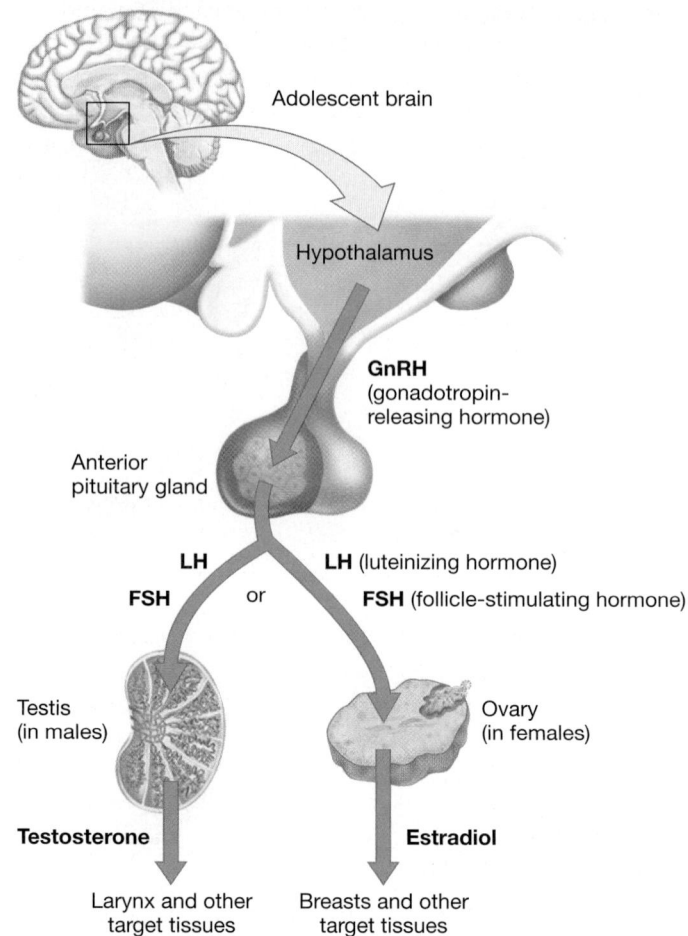

FIGURE 50.15 In Humans, Puberty Is Triggered by Hormones from the Hypothalamus and Pituitary.

✔**QUESTION** How does control of testosterone and estradiol production compare with control of cortisol release by the adrenal gland? (See Figure 49.12c.)

GnRH treatment induced surges in LH and FSH, followed by puberty onset.

What Regulates the Hypothalamic and Pituitary Hormones?   The model in Figure 50.15 raises a question: What triggers GnRH increases at the appropriate age? Although this question remains unanswered, there is some evidence that nutritional state is involved. For example:

• The current average age for the onset of menstruation in females in the United States is slightly over 12 years. This is much earlier than the average age of 17 years during the eighteenth and nineteenth centuries, when the general nutritional state of the population was poorer.

• Among girls living today, individuals with large fat stores tend to enter puberty earlier than do girls who are thin.

• The hormone leptin is produced by fat cells (see Chapter 49), and leptin levels are higher in people with more fat. In mice, leptin accelerates the onset of puberty, suggesting that leptin may be involved in triggering the GnRH surge.

Research continues on how age, nutritional condition, and perhaps other factors interact to promote the release of GnRH.

If you recall the discussion of how the adrenal hormone cortisol is controlled, however, you might suspect that the model of sex-hormone regulation in Figure 50.15 is simplified (see Chapter 49). Many hormones participate in negative feedback—also called feedback inhibition—meaning that the presence of the hormone inhibits the factor that triggers its release.

Do sex hormones participate in negative feedback? The short answer to this question is yes. To appreciate the details, let's investigate hormonal control of the human menstrual cycle.

Which Hormones Control the Menstrual Cycle in Mammals?

FIGURE 50.16 illustrates the sequence of events in the human ovary during the **menstrual cycle,** a monthly reproductive cycle. Although the cycle's length varies among women, 28 days is about average.

In conjunction with changes in the ovary illustrated in the figure, the lining of the uterus undergoes a dramatic thickening and regression. Ultimately, if fertilization and implantation do not occur, part of the uterine lining sloughs off and is expelled through the vagina.

Day 0 in the menstrual cycle is marked by the beginning of **menstruation**—the expulsion of the uterine lining. The remainder of the cycle has two distinct phases:

1. *Follicular phase* A follicle matures during the **follicular phase,** which lasts an average of 14 days. Primary oocytes complete meiosis I during this phase. Ovulation occurs when the follicle is mature and releases its secondary oocyte into the oviduct.

2. *Luteal phase* The **luteal phase** begins with ovulation and averages 14 days in length. Its name was inspired by the formation

and subsequent degeneration of a structure called the **corpus luteum** ("yellowish body") from the ruptured follicle.

The regular occurrence of ovulation throughout the year makes human females extremely unusual among mammals. Although some mammals ovulate multiple times during the year, most ovulate only during a single prescribed breeding season—often in response to environmental cues such as changing photoperiod; less often in response to cues from males.

In addition, only humans and other great apes menstruate. In the vast majority of mammals, the lining of the uterus is reabsorbed if pregnancy does not occur. These females are described as having an **estrous cycle** and are sexually receptive only during estrus—also known as being "in heat."

Whether an estrous or menstrual cycle occurs, the basic sequence of events, with a follicular phase preceding ovulation and a luteal phase following ovulation, is shared among mammals. Hormonal control of the estrous and menstrual cycles is also similar.

How Do Pituitary and Ovarian Hormones Change during a Menstrual Cycle? By monitoring hormone concentrations in the blood or urine of a large number of women over the course of the menstrual cycle, researchers were able to document dramatic changes in the concentrations of estradiol and several other hormones. During each cycle,

- LH and FSH are produced in the anterior pituitary gland in response to GnRH;

- the steroid hormone **progesterone** is produced along with estrogens, including estradiol, in the ovaries.

Several observations jump out of the data in **FIGURE 50.17** (see page 1028):

1. LH levels are fairly constant except for a spike that begins just before ovulation, suggesting that LH might be the trigger for this event.

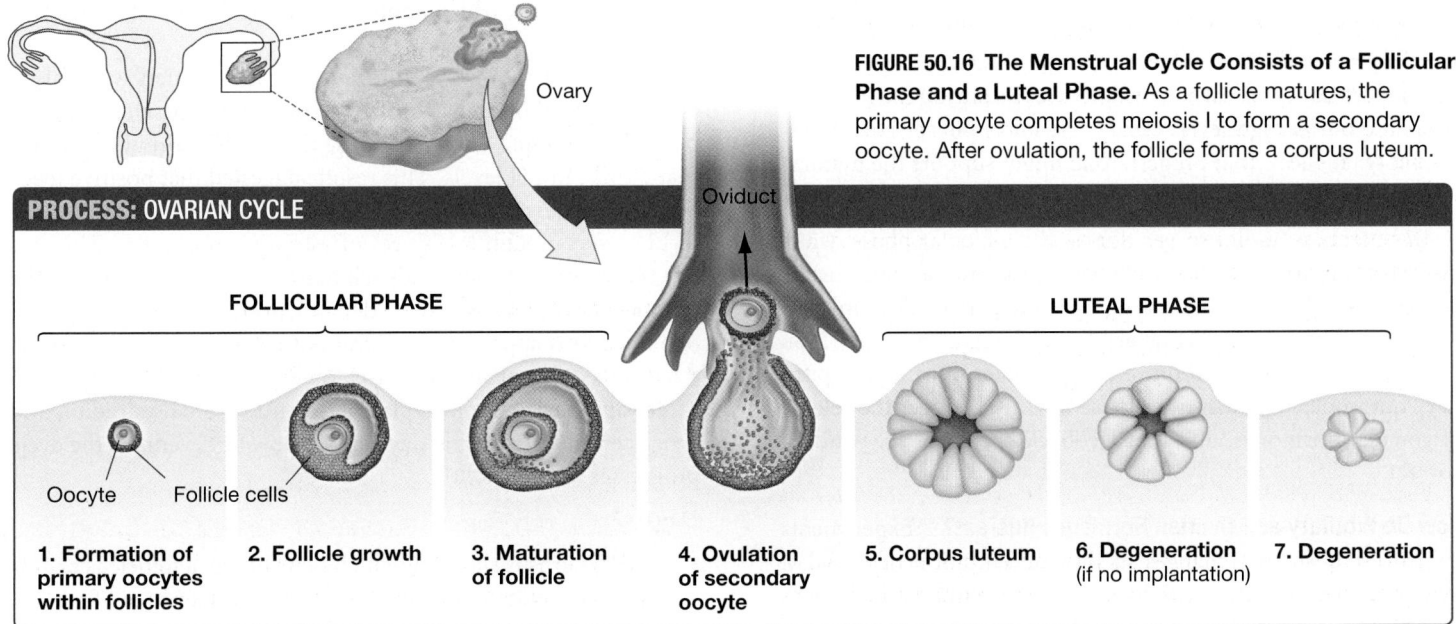

FIGURE 50.16 The Menstrual Cycle Consists of a Follicular Phase and a Luteal Phase. As a follicle matures, the primary oocyte completes meiosis I to form a secondary oocyte. After ovulation, the follicle forms a corpus luteum.

PROCESS: OVARIAN CYCLE

Ovary

Oviduct

FOLLICULAR PHASE

LUTEAL PHASE

Oocyte   Follicle cells

1. Formation of primary oocytes within follicles

2. Follicle growth

3. Maturation of follicle

4. Ovulation of secondary oocyte

5. Corpus luteum

6. Degeneration (if no implantation)

7. Degeneration

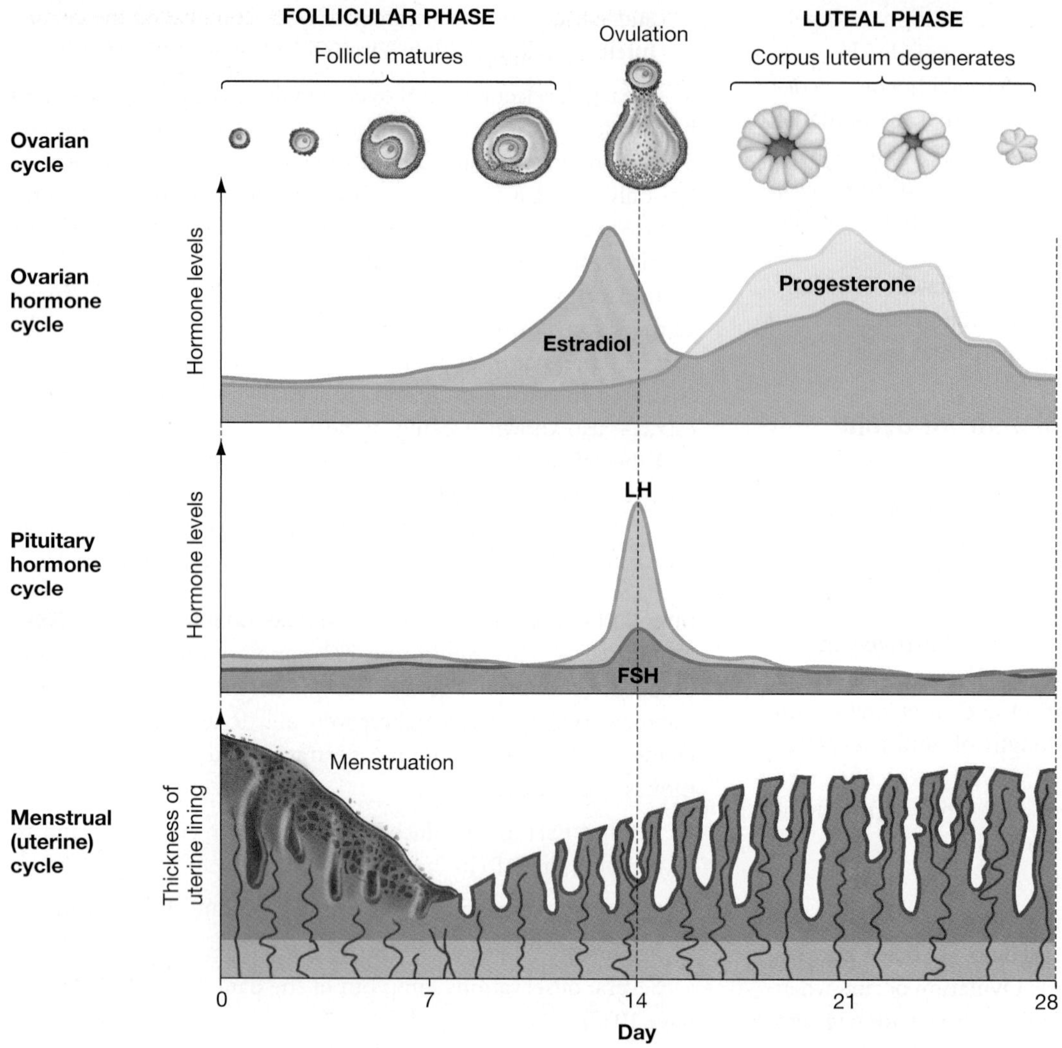

FIGURE 50.17 Hormones Regulate Events in the Human Menstrual Cycle.

2. FSH concentrations are relatively high during the follicular phase and low during the luteal phase, though they also make a small spike before ovulation.

3. Estradiol concentrations change in a complex way, highlighted by a peak late in the follicular phase.

4. Progesterone is present at very low levels during the follicular phase but at high levels during the luteal phase. This observation suggests that progesterone might support the maturation of the thickened uterine lining.

In general, estradiol surges during the follicular phase, while progesterone surges during the luteal phase. Similar patterns are observed in other mammals, ranging from marsupials to mice.

Now the question is: How are these changes regulated? You might hypothesize (based on reading Chapter 49) that the pituitary hormones released in response to GnRH trigger the release of gonadal hormones, and that feedback occurs. If so, you'd be correct.

How Do Pituitary and Ovarian Hormones Interact? Experiments helped establish that changes in the concentration of estradiol and progesterone affect the release of the pituitary hormones

LH and FSH. Researchers worked with three volunteers whose ovaries had been removed because of cancerous growths or other problems. The women were receiving low, maintenance-level doses of estradiol, which appeared to exert negative feedback on LH and FSH.

But when the investigators injected the women with larger doses of estradiol or with progesterone, dramatic changes took place. For example, a large increase in estradiol stimulated a dramatic spike in LH levels. This result suggested that positive feedback was occurring. High levels of estradiol increase the release of LH, even though low doses of estradiol suppress it. This is a key observation: When feedback occurs, estradiol's effect on the pituitary hormones is dosage dependent. In contrast, progesterone injections appeared to inhibit both FSH and LH. Progesterone exerts only negative feedback on the pituitary hormones.

To summarize the interplay between LH, FSH, estradiol, and progesterone, let's start at day 0 and follow key events as the cycle progresses (**FIGURE 50.18**).

Day 0–7

- As the uterus is shedding much of its lining, a follicle is beginning to develop in one ovary under the influence of FSH.

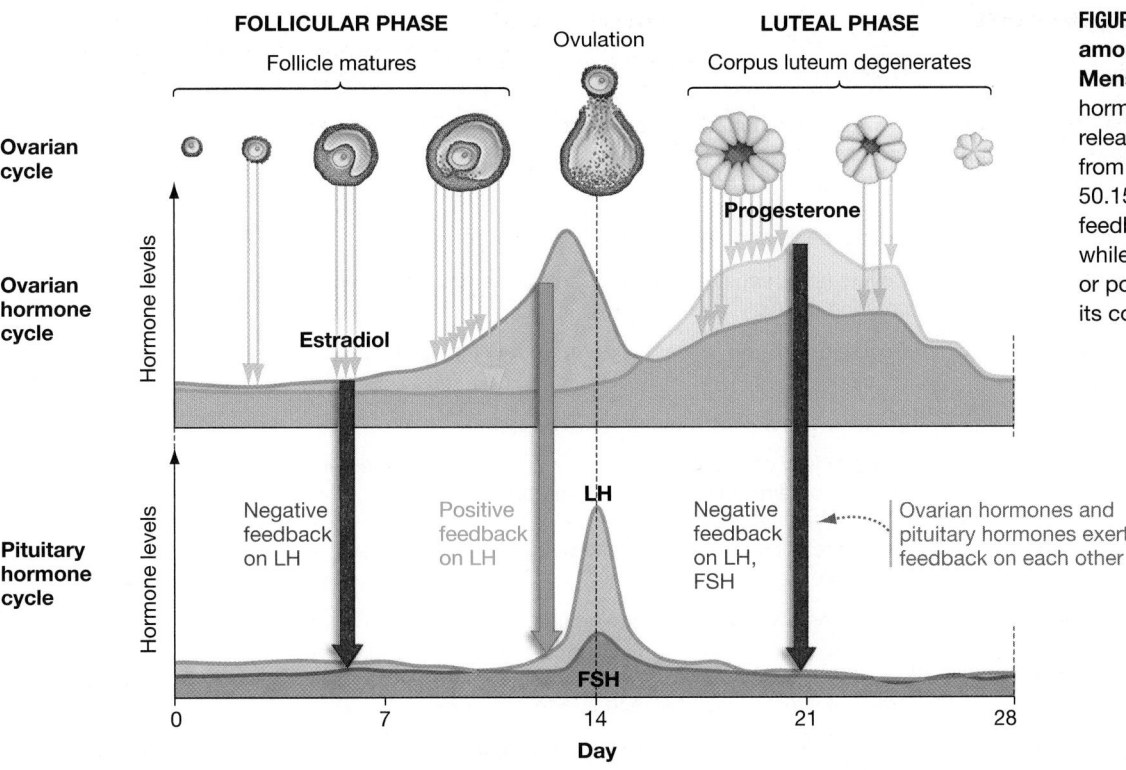

FOLLICULAR PHASE
Follicle matures

Ovulation

LUTEAL PHASE
Corpus luteum degenerates

Ovarian cycle

Ovarian hormone cycle

Progesterone

Estradiol

Pituitary hormone cycle

Hormone levels

Negative feedback on LH

Positive feedback on LH

LH

Negative feedback on LH, FSH

Ovarian hormones and pituitary hormones exert feedback on each other

FSH

0 7 14 21 28

Day

FIGURE 50.18 Complex Interactions among Hormones Regulate the Menstrual Cycle. The pituitary hormones FSH and LH control the release of estradiol and progesterone from reproductive tissues (see Figure 50.15). Progesterone exerts negative feedback on the pituitary hormones, while estradiol exerts either negative or positive feedback, depending on its concentration.

- The follicle produces estradiol and a small amount of progesterone.
- While its levels are still relatively low, estradiol suppresses LH secretion through negative feedback inhibition.

Day 8–14

- As the follicle grows, its production of estradiol gradually increases. The increase in estradiol stimulates mitosis and an increase in cell number in the uterine lining.
- The enlarged follicle produces large quantities of estradiol, which begin to exert positive feedback on LH secretion.
- Positive feedback results in a spike in LH levels, just after estradiol concentrations peak.
- The LH surge triggers ovulation and ends the follicular phase.

✔ If you understand that estradiol's effect on LH changes from negative regulation to positive regulation, you should be able to predict the effect of (1) injecting a female volunteer with high estradiol concentrations early in a cycle, and (2) using a drug to keep estradiol production low throughout the cycle.

Day 15–21

- As the corpus luteum develops from the remains of the ruptured follicle, it secretes large amounts of progesterone and small quantities of estradiol, in response to LH.
- The rise in progesterone lowers production of LH and FSH and activates the thickened uterine lining, creating a spongy tissue with a well-developed blood supply. In this way, progesterone fosters an environment that supports embryonic development if fertilization occurs.

Day 22–28

- If fertilization does not occur, the corpus luteum degenerates.
- Progesterone levels fall as the corpus luteum shrinks.
- When progesterone declines, the thickened lining of the uterus degenerates. This causes the menstrual bleeding that marks the first day of the next cycle.
- GnRH, LH, and FSH are released from the inhibitory control that progesterone exerts.
- LH and FSH levels rise, and a new menstrual cycle begins.

The interplay between ovarian and pituitary hormones is similar in other mammal species that have been studied to date.

Manipulating Hormone Levels to Prevent Pregnancy Data on hormonal control of the menstrual cycle opened new avenues in birth control research. Specifically, researchers have found that manipulating levels of progesterone and estradiol can prevent ovulation and serve as a safe and effective method of **contraception,** or preventing unwanted pregnancies.

Hormone-based contraceptive methods deliver synthetic versions of progesterone or of progesterone and estradiol. These hormones suppress the release of GnRH, FSH, and LH through negative feedback inhibition. Because an LH spike does not occur during the follicular phase of the cycle, the follicle does not mature and ovulation does not occur.

In the United States, birth control pills are the most widely used contraceptive method. Hormone-containing pills are taken for three weeks and then stopped for one week to allow menstruation to occur. Other popular hormonal contraceptives include injections and hormone-secreting patches and implants.

Type	Name	Mode of Action	Percent Effectiveness*
Hormone-based methods	The Pill, the Patch, the Ring, the Shot, the Implant	Provides continuous or cyclical delivery of progesterone, or progesterone plus estradiol.	92 to 99.9[†]
Barrier methods	Condom	Covers penis and prevents sperm from entering uterus.	85
	Female condom	Covers labia, vagina, and cervix and prevents sperm from entering uterus.	79
	Diaphragm	Covers cervix and prevents sperm from entering uterus.	84
	Sponge	Covers cervix and prevents sperm from entering uterus; also contains a molecule that immobilizes sperm.	84
	Spermicide	Foam or jelly covers cervix and prevents sperm from entering uterus; contains a molecule that immobilizes sperm.	71
Behavioral methods	Rhythm method	Couple refrains from vaginal intercourse around time of ovulation.	80
	Withdrawal	Man withdraws penis before ejaculation.	73
Fertilization prevention	Intrauterine device (IUD)	Small T-shaped structure inserted into uterus; induces uterus to produce substances hostile to sperm and eggs.	99
	Emergency contraception	Delivers progesterone or progesterone and estradiol after unprotected vaginal intercourse.	92
Pregnancy termination	Mifepristone	Blocks progesterone receptors so menstruation occurs even after fertilization and implantation.	92

*"Percent Effectiveness" indicates the average percentage of women who do not become pregnant during one year of typical use.

[†]Depends on delivery system used.

But as **TABLE 50.2** indicates, hormone-based methods are just one of several approaches to preventing pregnancy. Other methods work by preventing sperm from contacting the oocyte or by interfering with implantation of the embryo.

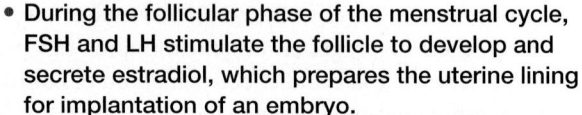

check your understanding

If you understand that . . .

- During the follicular phase of the menstrual cycle, FSH and LH stimulate the follicle to develop and secrete estradiol, which prepares the uterine lining for implantation of an embryo.
- During the luteal phase, progesterone protects the uterine lining. If fertilization does not occur, progesterone drops and the lining is shed.

✔ **You should be able to . . .**

1. Describe the function of FSH, and explain how FSH levels increase at the end of one cycle and the start of the next cycle.
2. Predict the consequences of a drug that inhibits the release of FSH.

Answers are available in Appendix A.

The "Percent Effectiveness" column on the far right of the table indicates the average percentage of women who do not become pregnant during one year of typical use of that method. Percent effectiveness usually increases dramatically if couples are able to use a method "perfectly"—meaning, exactly as specified for optimal effectiveness, during every episode of sexual intercourse.

When sperm and egg do unite successfully, the menstrual cycle is interrupted. The corpus luteum does not degenerate, and progesterone and estradiol levels stay high. Menstruation does not occur. Instead, the mother is now pregnant.

50.5 Pregnancy and Birth in Mammals

Viviparity allows the mother to provide a warm, protected environment for offspring during early development. Oviparous species that guard or incubate their eggs also provide warm, safe surroundings for their young.

Any investment that parents make in an offspring comes at a cost, however: The more a mother invests in each offspring, the lower the number of offspring she can produce.

Pregnancy and **lactation**—providing milk that nourishes offspring after birth—represent some of the most extreme forms of parental care known in animals. And in some mammal species, parental care continues long after lactation ends. Humans, for example, are largely or completely dependent on their parents for protection and nutrition until puberty or young adulthood.

Let's examine how mammals make this investment, starting with marsupials and then turning to the eutherian mammals.

Gestation and Early Development in Marsupials

You might recall that mammals comprise three major monophyletic groups: monotremes, eutherians, and marsupials (see Chapter 35). Although all mammals are endothermic, have fur, and lactate, the three lineages are distinguished by their mode of reproduction:

- Monotremes lay eggs and incubate them until hatching.

- In eutherians, mothers carry offspring internally for relatively long periods of development and nourish them via a placenta.

- In marsupials, the corpus luteum is not maintained as it is in eutherians, and no placenta forms.

In marsupials, young are ejected from the mother's body at the end of the estrous cycle. As a result, they are far less developed than eutherian mammal offspring, which undergo a lengthier **gestation**—the developmental period that takes place inside the mother.

The jaws, gut, lungs, and forelimbs of a newly born marsupial are relatively well developed at birth. As a result, the offspring is able to climb from its mother's vagina to a nipple, which is usually enclosed in a pouch created by a flap of skin (**FIGURE 50.19a**). The offspring clamps onto the nipple and continues to develop, fed by the mother's milk (**FIGURES 50.19b** and **50.19c**).

Even after growing large enough to leave the pouch and begin moving and feeding on its own, offspring will return to the pouch for protection. Marsupial mothers invest a great deal in their offspring, even though a relatively short period of development takes place inside their bodies.

Major Events during Human Pregnancy

Marsupials and eutherians differ sharply in terms of how long the developing embryo is retained inside the mother's body. Let's consider humans as a model organism in eutherian reproduction.

When a secondary oocyte is released from the human ovary, the cell is viable for less than 24 hours. Human sperm, in contrast, remain capable of fertilizing an egg for up to five days. Therefore, sexual intercourse in humans has to occur less than five days before ovulation or immediately after ovulation for pregnancy to result.

Although an ejaculation may contain hundreds of millions of sperm, most die as they travel through the uterus. Only 100 to 300 actually succeed in reaching the oviduct, where fertilization takes place.

You might recall that when sperm and oocyte meet, enzymes released from the head of the sperm create a path through the material surrounding the oocyte membrane. Once the membranes of the oocyte and sperm have fused, the oocyte nucleus completes meiosis II. The two nuclei then unite to form a diploid zygote.

Implantation Smooth-muscle contractions in the oviduct gradually move the zygote toward the uterus. As it travels, the cell begins to divide by mitosis. By the time it reaches the lining of the uterus, the embryo consists of a hollow ball of cells. It then undergoes **implantation**—meaning that it becomes embedded in the thickened, vascularized wall of the uterus. It will stay in the uterus for approximately 270 days (9 months).

Once the embryo is implanted in the uterine lining, its cells begin synthesizing and secreting the hormone human chorionic gonadotropin; this hormone is later produced in larger quantities by the placenta. **Human chorionic gonadotropin (hCG) is a chemical messenger that prevents the corpus luteum from degenerating. When hCG is present, the ovary continues secreting progesterone, and the menstrual cycle is arrested.** Since hCG is excreted in the mother's urine, it is the chemical used to detect pregnancy in pregnancy tests.

The First Trimester Human gestation is divided into 3-month stages called trimesters. Not long after implantation is complete, mass movements of cells result in the formation of the three major embryonic tissues, called ectoderm, endoderm, and mesoderm (see Chapter 23). By 8 weeks into gestation, these tissues have differentiated into the various organs and systems of

(a) Brushtail possum shortly after birth

(b) 1.5 months after birth

(c) 3.5 months after birth

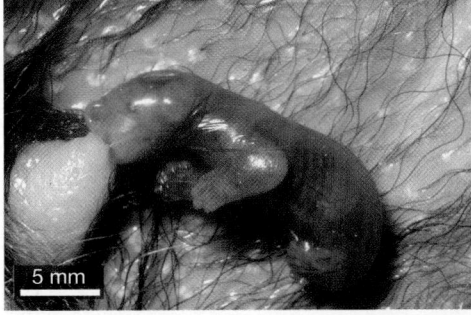
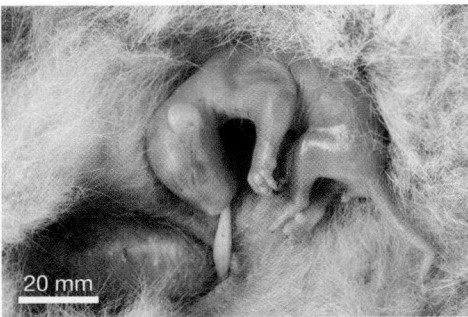

FIGURE 50.19 Marsupials Trade a Long Gestation Period for a Long Lactation Period. Relative to eutherians, marsupial offspring—like this brushtail possum—spend a short time developing in the uterus and a relatively long time being fed milk after birth and being protected in the mother's pouch.

the body. Also by this time, the heart has begun pumping blood through a circulatory system. The embryo at this stage is called a **fetus** (FIGURE 50.20a).

Early in development, the embryonic ectoderm contributes to several important membranes. One of these membranes, the **amnion,** completely surrounds the embryo. The amnion eventually fills with amniotic fluid, which provides the embryo with a protective cushion.

The other key event in the first trimester is the formation of the placenta. This organ, which starts to form on the uterine wall a few weeks after implantation, is composed of tissues from both mother and embryo (see Chapter 35). Because the placenta contains a dense supply of blood vessels from the mother, it provides nutrition for the growing fetus. Arteries transport blood from the circulatory system of the fetus, through the **umbilical cord,** to an extensive capillary bed in the placenta. This capillary bed provides a large surface area for the exchange of gases, nutrients, and wastes between maternal and fetal blood, even though the maternal and fetal blood do not commingle.

The placenta secretes a variety of hormones, including large amounts of progesterone and estrogens. Because these hormones suppress the release of GnRH, LH, and FSH through negative feedback, they prevent the maturation and ovulation of additional follicles. By the end of the first trimester, the placenta is producing more than enough progesterone to replace the amount that had been produced by the corpus luteum, which has degenerated by this time. In essence, the placenta takes over from the corpus luteum in secreting the hormones required to maintain the pregnancy.

✔ If you understand how hormones influence pregnancy, you should be able to explain (1) why women who produce low levels of progesterone from the corpus luteum are prone to miscarriage, and (2) why a pregnancy test administered to a woman in her third trimester would register negative.

The Second and Third Trimesters After the fetal organs and placenta form during the first trimester, the rest of development consists mainly of growth (**FIGURES 50.20b** and **50.20c**). During the last weeks of pregnancy, the brain and lungs undergo particularly dramatic growth and development. If a baby is born prematurely,

intervention may be required to keep the baby alive until the lungs can complete their development.

The machinery and level of hospital care required by premature infants emphasize just how superbly adapted mothers are for nourishing a growing fetus in the uterus. It costs hundreds of thousands of dollars for health care providers to do what mothers do naturally in the last trimester. Let's take a closer look at this critical aspect of pregnancy.

How Does the Mother Nourish the Fetus?

In oviparous and ovoviviparous species, mothers produce relatively large eggs that contain all of the nutrients and fluids that the embryo needs for development until hatching. But in some viviparous species, eggs are relatively small and contain almost no nutrients.

In species such as humans, the developing embryo depends on the mother's body for oxygen, chemical energy in the form of sugars, amino acids and other raw materials for growth, and waste removal. What physiological changes occur in human mothers to accommodate these demands?

Oxygen Exchange between Mother and Fetus During pregnancy, a mother's respiratory and circulatory systems change in ways that increase the efficiency of nutrient transfer and gas exchange with the fetus. For example:

- A woman's total blood volume expands by as much as 50 percent during pregnancy.

- To accommodate the increase in blood volume, maternal blood vessels dilate (widen) and blood pressure drops.

- The mother's heart enlarges and beats faster, increasing her total cardiac output by almost 50 percent.

- The mother's breathing rate and breathing volume increase to accommodate the fetus's demand for oxygen and its production of carbon dioxide.

In addition, important adaptations heighten the efficiency of gas exchange between the mother and the embryo, in the placenta. In many species, such as sheep, maternal and fetal blood

(a) 1st trimester

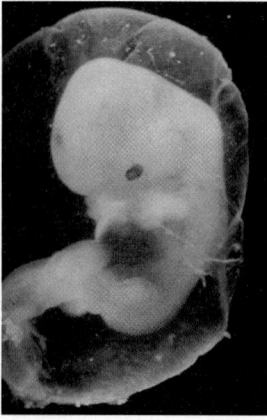

(b) 2nd trimester

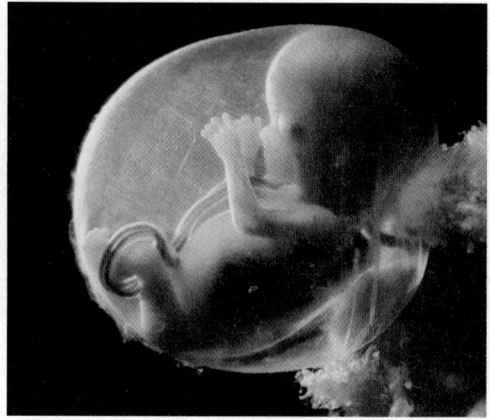

(c) 3rd trimester

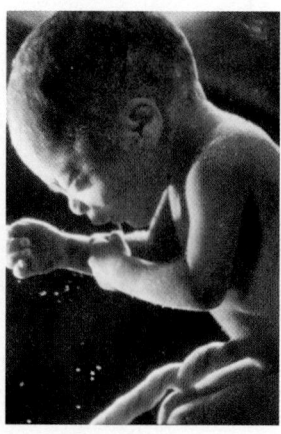

FIGURE 50.20 Development of the Human Fetus Can Be Divided into Three Trimesters.

in the placenta flow in a countercurrent fashion (**FIGURE 50.21a**). Countercurrent flows maintain a concentration gradient that increases the efficiency of diffusion or other types of exchange (see Chapter 42).

Countercurrent flow does not occur in the human placenta. Instead, oxygen exchange between mother and fetus is efficient because of another mechanism. Maternal arteries in humans empty into a space at the junction of the maternal and fetal portions of the placenta. This space is packed with small projections called villi, which contain the fetal blood vessels. Thus, a large surface area from the fetus is bathed with highly oxygenated maternal blood. The fetal villi are analogous to the alveoli of the lungs (see Chapter 45), which provide a large surface area for gas exchange.

FIGURE 50.21b illustrates another adaptation that increases the rate of oxygen delivery to a human fetus. Notice that the axes on this graph give the partial pressure of oxygen in blood versus the percentage of hemoglobin that holds oxygen at that pressure. Recall that this type of graph is an **oxygen–hemoglobin equilibrium curve** (see Chapter 45).

The key message of this graph is that the data for fetal hemoglobin are shifted to the left of the data for adult hemoglobin. This means that the fetus's blood always has a higher affinity for

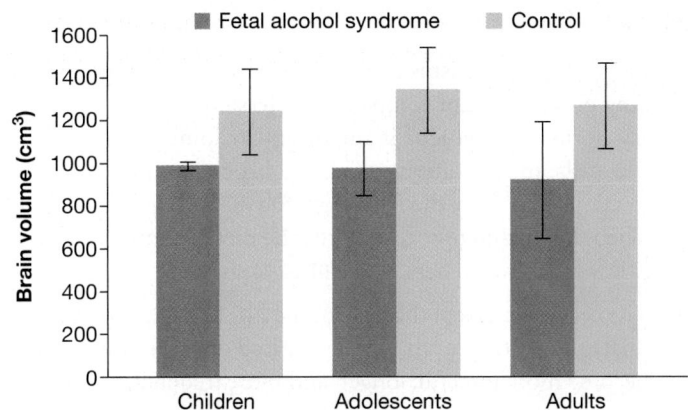

FIGURE 50.22 Patients with Fetal Alcohol Syndrome Have Reduced Brain Size. The effect is most pronounced in young people.

DATA: Swayze, V. W., II, Johnson, V. P., Hanson, J. W., et al. 1997. *Pediatrics* 99: 232–240.

oxygen than does the mother's blood. Even if the partial pressure of oxygen is the same in the mother and fetus, oxygen will move from the mother's blood to the fetus's blood because it is held more tightly by the fetal hemoglobin.

Biologists interpret this pattern as an adaptation. The high oxygen affinity of fetal hemoglobin ensures that the fetus is always able to acquire oxygen from the mother.

Toxic Chemicals Can Be Transferred from Mother to Fetus

Mothers and embryos exchange more than nutrients and wastes—they can also exchange dangerous molecules. As an example, consider the thalidomide tragedy. During the 1950s, hundreds of children were affected by their mothers' consumption of the tranquilizer thalidomide, which was prescribed to treat morning sickness. The molecule diffused into the fetal bloodstream and caused birth defects—often a dramatic shortening of the arms.

Although thalidomide is now banned for use by pregnant women, alcohol use continues to affect newborns. Compared to children of mothers who do not drink alcohol, children of mothers who imbibe ethanol are at high risk for hyperactivity, severe learning disabilities, and depression. Collectively, these symptoms are termed **fetal alcohol syndrome (FAS).**

To investigate why FAS occurs, researchers used a technique called magnetic resonance imaging (MRI) to compare the total brain volumes of patients suffering from fetal alcohol syndrome with those of healthy controls. They found that the brains of children with FAS were over 20 percent smaller than those of healthy children (**FIGURE 50.22**). In adolescents and adults with FAS, brain volume was 28 percent smaller than in healthy people of the same age, suggesting that prenatal exposure to alcohol stunts brain development throughout life.

The MRIs also revealed numerous structural abnormalities in the brain. These abnormalities, along with the reduced brain volume, are thought to be responsible for dramatic reductions in IQ and other measures of intelligence and learning ability. On the basis of results like these, public health officials strongly advise pregnant women against drinking *any* alcohol.

(a) Countercurrent blood flow in the sheep placenta

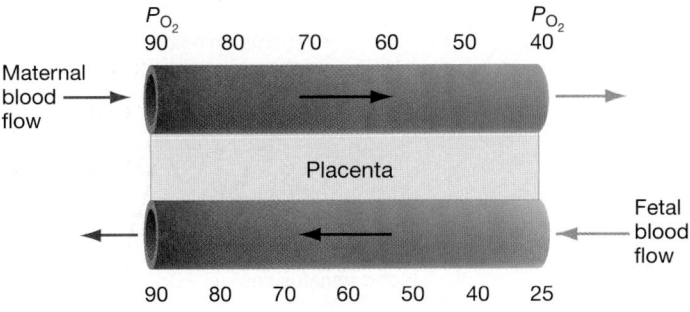

(b) Oxygen–hemoglobin saturation curves in the human fetus and mother

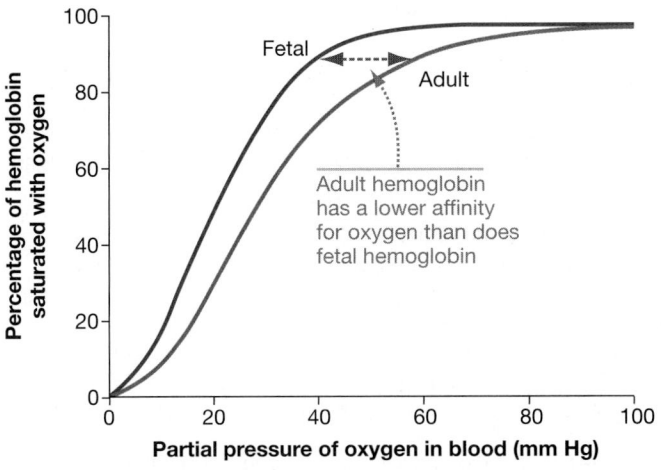

FIGURE 50.21 Mammals Have Adaptations That Increase Delivery of Oxygen to the Fetus.

Birth

Although the mechanisms responsible for initially triggering the birthing process are not completely understood, the posterior pituitary hormone **oxytocin** is important in stimulating smooth-muscle cells in the uterine wall to begin contractions. The contractions that expel the fetus from the uterus constitute **labor.**

When the birthing process starts, the uterus contracts at relatively low frequency. Then, as **FIGURE 50.23** shows,

1. The cervical canal at the base of the uterus begins to open, or dilate. Once the cervix is fully dilated, uterine contractions become more forceful, longer, and more frequent.

2. The fetus is expelled through the cervix and into the vagina.

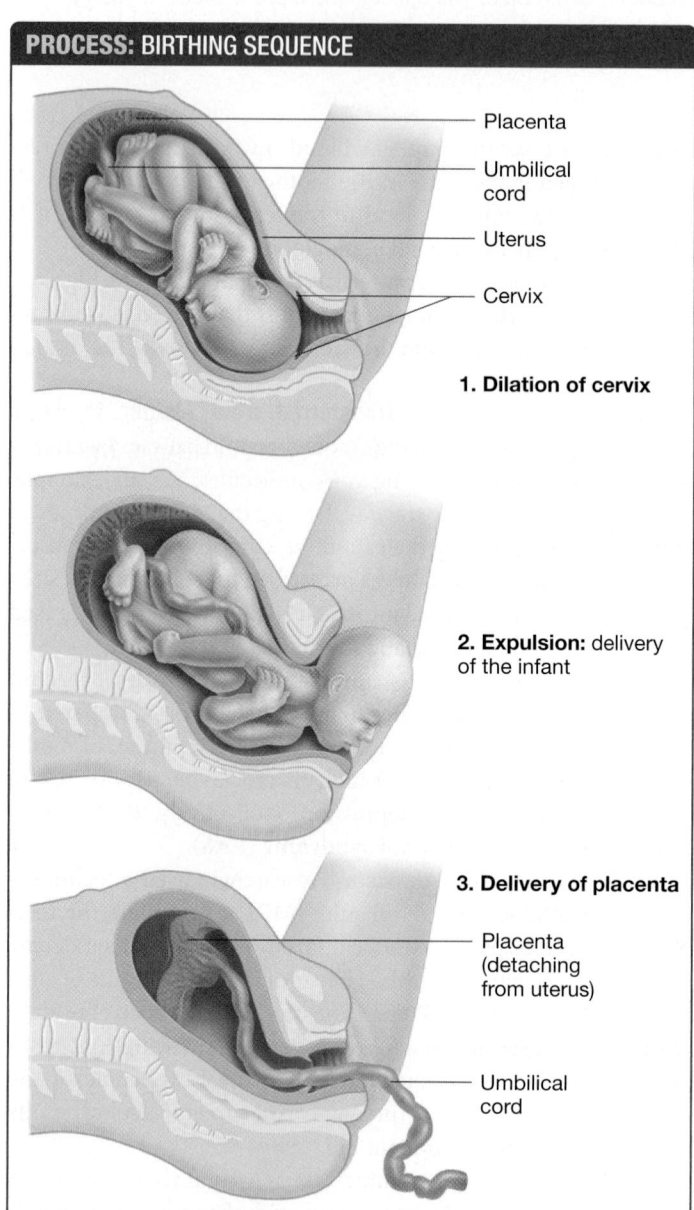

PROCESS: BIRTHING SEQUENCE

Placenta
Umbilical cord
Uterus
Cervix

1. Dilation of cervix

2. Expulsion: delivery of the infant

3. Delivery of placenta

Placenta (detaching from uterus)

Umbilical cord

FIGURE 50.23 Human Birth Occurs in Three Stages.

✔**QUESTION** Human babies are normally born headfirst and face down, as shown. Predict the consequences of a baby positioned face up or feet first in the birth canal.

3. After the baby is delivered, the placenta remains attached to the uterine wall. At this point, caregivers clamp and cut the umbilical cord that connects the child and the placenta. When the mother delivers the placenta and accompanying membranes, birth is complete.

Although this description sounds straightforward, in reality a large number of complications are possible. For example, **FIGURE 50.24** shows the number of Swedish mothers who died, per 100,000 live births, in five-year intervals between 1760 and 1980. In 1760, approximately 1.4 mothers died for every 100 infants successfully delivered. In most cases, the cause of death was blood loss or infection following delivery. Note that the steepest drops in mortality rate occurred with the introduction of hand-washing in the late 1870s.

As a result of sterile techniques, antibiotics, and blood transfusion technology, Sweden's mortality rate has now declined to less than 0.007 percent. Improved nutrition, sanitation, and medical care have also reduced infant mortality rates in many countries.

The huge decline in the rate of death associated with childbirth qualifies as one of the great triumphs of modern medicine. Unfortunately, because many developing nations lack sterile facilities and antibiotics, the mortality of mothers and infants remains high in those countries.

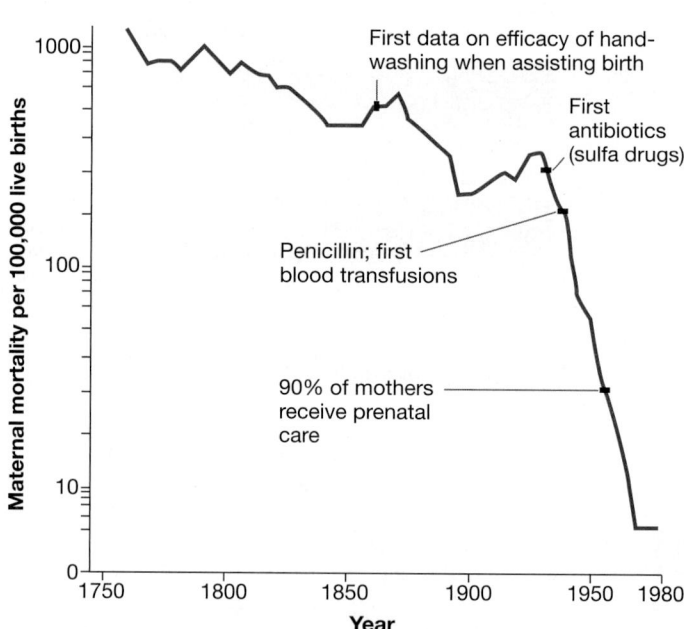

FIGURE 50.24 Maternal Mortality in Childbirth Has Decreased Dramatically. These data are from Sweden; the scale on the vertical axis is logarithmic. *Note:* Data showing that hand-washing dramatically reduced maternal mortality during childbirth were published in 1861, but doctors resisted the idea until well after Pasteur had published the germ theory of disease in 1878.

DATA: Högberg, U., and I. Joelsson. 1985. *Acta Obstetricia et Gynecologica Scandinavica* 64: 583–592.

✔**QUANTITATIVE** If a woman living in Sweden in 1760 had 10 children over the course of her lifetime, what was her overall chance of dying in childbirth?

If you understand . . .

50.1 Asexual and Sexual Reproduction

- Asexual reproduction produces offspring with genetic material from the mother only.

- Sexually produced offspring are genetically unique because of recombination during meiosis and the fusion of haploid gametes from different parents during fertilization.

- Asexual reproduction is favored in constant environments, whereas sexual reproduction is favored in changing environments, or when food availability is low and availability of mates is high.

✔ You should be able to explain why sexually produced offspring would have higher fitness than asexually produced offspring in a population of animals that is highly parasitized.

50.2 Fertilization and Egg Development

- In human males, spermatogenesis is continuous throughout adult life, but in human females, all primary oocytes are formed early in development. Meiosis of oocytes is arrested for long periods of time, and cell division after meiosis is so unequal in females that just one egg—not four—is produced from each primary oocyte.

- Fertilization is external in many aquatic animals but internal in almost all terrestrial species.

- When sperm competition occurs, males have large testes relative to their body size, and the last male to mate usually fathers a disproportionately large number of offspring.

- Once eggs are fertilized, females, depending on their species, may lay the eggs or may retain them and give birth to live offspring. In some groups at least, viviparity may be an adaptation that increases the survival of young in cold habitats.

✔ You should be able to explain why fish that care for their young are expected to produce fewer eggs per year than fish that do not care for their young, and also whether sperm competition is common in fish.

50.3 Reproductive Structures and Their Functions

- In humans, the male reproductive system includes structures specialized for producing and storing sperm, synthesizing other components of semen, or transporting and delivering semen.

- The female reproductive system includes structures specialized for producing eggs, receiving sperm, and nourishing offspring during early development.

✔ You should be able to suggest ways to surgically sterilize human males and females for birth control.

50.4 The Role of Sex Hormones in Mammalian Reproduction

- In mammals, GnRH from the hypothalamus triggers the release of FSH and LH from the pituitary gland. The pituitary hormones regulate the production of the gonadal hormones, testosterone and estradiol, in the testes and ovaries, respectively.

- During the human menstrual cycle, estradiol and progesterone exert feedback (estradiol, positive and negative feedback; progesterone, negative only) on the production of FSH and LH. Interactions between the pituitary and ovarian hormones are responsible for regulating cyclical changes in the ovaries and uterus.

✔ You should be able to predict whether FSH and LH concentrations would increase or decrease following ovariectomy (surgical removal of ovaries).

50.5 Pregnancy and Birth in Mammals

- If fertilization occurs, the developing embryo and placenta secrete the hormone hCG, which arrests the menstrual cycle and allows pregnancy to continue.

- During the first trimester, the embryo becomes implanted in the thickened uterine wall, organs develop, and the nutritive organ called the placenta forms.

- To make rapid growth possible during the second and third trimesters, the mother's heart rate and blood volume increase. Nutrients and gases are exchanged efficiently in the placenta.

✔ You should be able to explain why the placenta is highly vascularized (consists mainly of blood vessels).

MasteringBiology

1. **MasteringBiology Assignments**

   Tutorials and Activities Human Gametogenesis; Human Reproduction; Reproductive System of the Human Female; Reproductive System of the Human Male; Sex Hormones and Mammalian Reproduction

   Questions Reading Quizzes, Blue-Thread Questions, Test Bank

2. **eText** Read your book online, search, take notes, highlight text, and more.

3. **The Study Area** Practice Test, Cumulative Test, BioFlix® 3-D Animations, Videos, Activities, Audio Glossary, Word Study Tools, Art

You should be able to . . .

1. What term describes the mode of asexual reproduction in which offspring develop from unfertilized eggs?
   a. parthenogenesis; **b.** budding; **c.** regeneration; **d.** fission

2. In sperm competition, what is "second-male advantage"?
   a. the observation that when females mate with two males, each male fertilizes the same number of eggs
   b. the observation that when females mate with two males, the second male fertilizes most of the eggs
   c. the observation that females routinely mate with at least two males before laying eggs or becoming pregnant
   d. the observation that accessory fluids prevent matings by second males—for example, by forming copulatory plugs

3. How are the human penis and clitoris similar?
   a. They develop from the same population of embryonic cells.
   b. Both develop during the earliest stages of puberty.
   c. Both contain the urethra.
   d. Both produce accessory fluids required during sexual intercourse.

4. Which of the following hormones would you expect to be more elevated in female hyenas compared to females of other species of mammals?
   a. follicle-stimulating hormone
   b. testosterone
   c. progesterone
   d. oxytocin

5. What two pituitary hormones are involved in regulating the human menstrual cycle?

6. True or false: The corpus luteum is retained upon implantation due to the presence of the hormone human chorionic gonadotropin (hCG).

7. What is the fundamental difference between sexual and asexual reproduction, in terms of the characteristics of offspring?

8. Summarize the experimental evidence that *Daphnia* require three cues to trigger sexual reproduction. Discuss what these cues indicate about the environment. Generate a hypothesis for why sexual reproduction is adaptive for these animals.

9. Contrast spermatogenesis with oogenesis in humans. How do these processes differ with respect to numbers of cells produced, gamete size, and timing of the second meiotic division?

10. Give examples of negative feedback and positive feedback in hormonal control of the human menstrual cycle. Why can high estradiol levels be considered a "readiness" signal from a follicle?

11. In hormone production, how do the follicle and corpus luteum compare?
    a. Both produce primarily estradiol.
    b. Both produce primarily progesterone.
    c. The follicle produces more estradiol than progesterone; the corpus luteum produces more progesterone than estradiol.
    d. The follicle produces mostly progesterone; the corpus luteum produces estradiol and progesterone.

12. Suppose that females of a certain insect species routinely mate with two males and choose their mates based on courtship displays or other traits. Predict whether females would be choosier about the first male or the second male that they mate with, or equally choosy about both males. Explain the logic behind your prediction.

13. Researchers have recently developed methods for cloning mammals. In effect, biologists can now induce asexual reproduction in species that do not normally reproduce asexually. Suppose this practice becomes so widespread in the future that most of the sheep in the world become genetically identical. Discuss some possible consequences of this development.

14. Suppose that you've been given a chance to study several populations of lizards—some of them viviparous, some of them oviparous. Which populations would you expect to produce especially large eggs, relative to their overall body size?

15. When marsupial eggs are developing, a shell membrane appears for a short time and then disintegrates. State a hypothesis to explain this observation.

16. **QUANTITATIVE** Consider the data in the table below, showing the average body mass index (BMI, where higher values represent heavier girls for a given height; see Chapter 44) of pre-pubertal and post-pubertal girls. Which conclusion can you make from the data?

Age	BMI	
	Pre-pubertal	Post-pubertal
11	−0.22	0.75
12	−0.28	0.52
13	−0.56	0.34

DATA: Anderson, S. E., G. E. Dallal, and A. Must. 2003. *Pediatrics* 111: 844–850.

a. At a given age, there are more girls with low BMI than with high BMI.
b. At a given age, girls with high BMI are more likely to hit puberty than girls with low BMI.
c. 11-, 12-, and 13-year-olds are equally likely to hit puberty.
d. There is no relationship between BMI and age at puberty.

51 The Immune System in Animals

In this chapter you will learn how

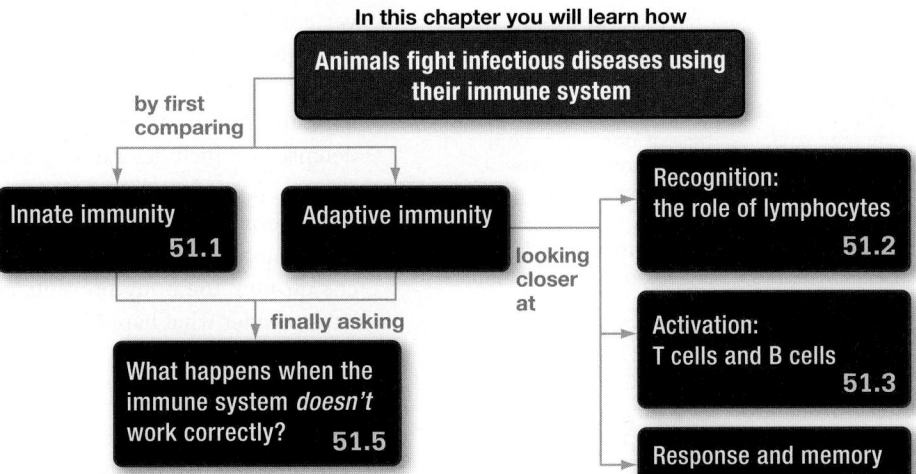

Animals fight infectious diseases using their immune system

by first comparing

Innate immunity
51.1

Adaptive immunity

finally asking

What happens when the immune system *doesn't* work correctly?
51.5

looking closer at

Recognition: the role of lymphocytes
51.2

Activation: T cells and B cells
51.3

Response and memory
51.4

Colorized scanning electron micrograph of human immune system cells attacking *Wuchereria bancrofti*—a parasitic nematode that clogs lymph vessels and causes the dramatic swelling called elephantiasis. This chapter explores how immune system cells are able to recognize parasitic worms, bacteria, and viruses as foreign and eliminate them.

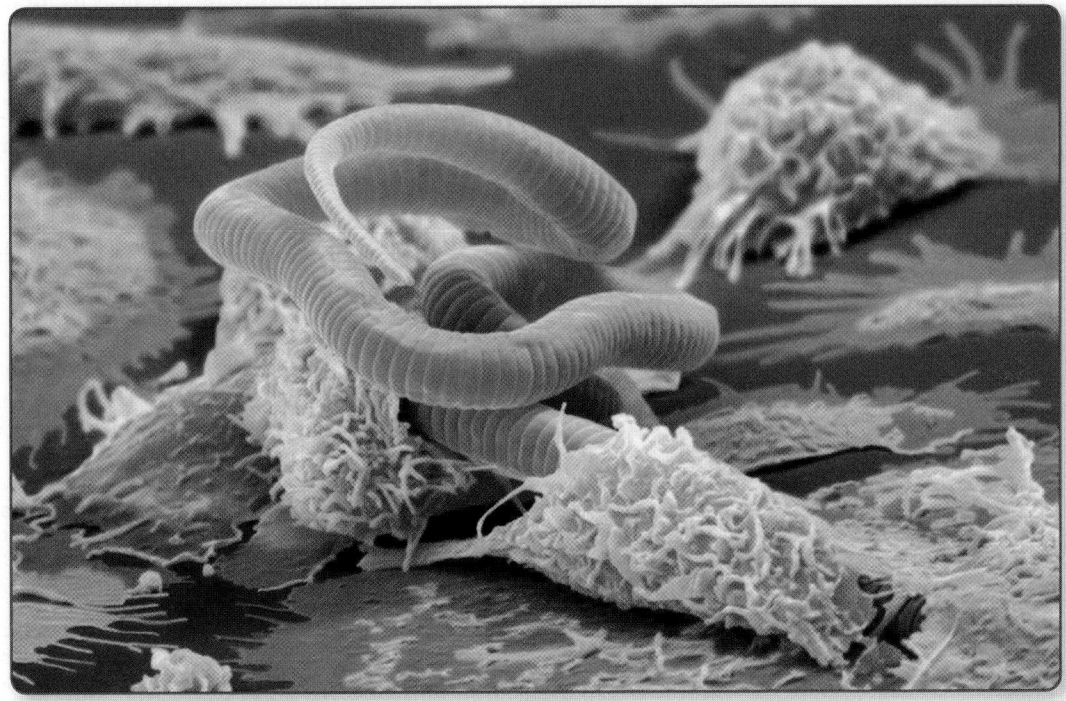

nfectious disease threatens the survival of every living organism. For example, animals alone may be infected by thousands of different disease-causing bacteria, viruses, parasitic worms, fungi, and protists—collectively referred to as **pathogens.** Given the ability of these pathogens to cause illness and death, it is remarkable that so many animals stay healthy for most of their lives.

If an animal contracts a bacterial or viral illness, however, it may eventually recover, even without medical intervention. Frequently, after recovering from an infection, the animal will be immune to (literally, "exempt" from) the disease—that is, it will not contract the same disease in the future. **Immunity** is a resistance to or protection against a disease-causing pathogen.

Thus, the **immune system** is responsible for defending animals against pathogens. The successful function of the immune system involves three key processes. It must **(1)** prevent the entry of

This chapter is part of the Big Picture. See how on pages 840–841.

✔ When you see this checkmark, stop and test yourself. Answers are available in Appendix A.

Innate Immune System	Adaptive Immune System
Occurs in all animals	Occurs only in vertebrates
Has both cell-mediated and secreted components	Has both cell-mediated and secreted components
Rapid response	Slow response
Broadly specific response against types of pathogens	Specific response against pathogen strains
Response does not vary when infections recur (no memory)	Response is more rapid and efficient when infections recur (memory)

potential pathogens; (**2**) detect the presence of a pathogen by distinguishing it from the animal's own body; and (**3**) eliminate the pathogen.

When biologists began analyzing the immune system, they found that certain immune system cells are ready to respond to foreign invaders at all times, while others must be activated first. Cells that are always ready confer **innate immunity.** Cells that must first be selectively activated to tailor their response against a specific pathogen confer **adaptive immunity.**

In combination, the innate and adaptive immune responses form a powerful system for protecting individuals against a formidable and ever-changing array of pathogens. **TABLE 51.1** illustrates the different roles of the innate and adaptive immune systems. This chapter explores the different roles of these two systems and how their cells and secreted cellular products collaborate to battle infections. It then considers the problems that arise when the immune system does not function properly.

51.1 Innate Immunity

Innate immunity is so named because it is inherent in all animals and is ready to go from the moment of birth. In contrast, adaptive immunity occurs in only 1 percent of animals—the vertebrates—and is not fully developed until 6 months after birth. Clearly, the innate immune system on its own has succeeded in protecting a spectacular abundance and diversity of animals, both invertebrates and vertebrates. It is the first line of defense and includes exterior anatomical structures that protect animals from invading pathogens as well as interior detection and response systems.

To launch an investigation into innate immunity, let's first focus on how the body prevents entry by foreign invaders. Then let's consider what happens if some do get in.

Barriers to Entry

The most effective way for animals to avoid getting an infection is to prevent pathogens from entering their bodies in the first place. In humans and many other animals, the most important deterrent to infection is the exterior surface. For example,

- The armored bodies of insects are difficult for pathogens to enter because they are covered with the tough layer called cuticle, along with a layer of wax (see Chapter 43).

- Soft-bodied organisms like slugs, snails, and earthworms are covered with a protective layer of **mucus** (the adjective is mucous), a slimy mix of glycoproteins and water that traps pathogens and sloughs off.

- Human skin has an outer layer of dead cells that are reinforced with tough fibers of the protein keratin.

Besides providing a tough physical barrier, these outer surfaces also ensure a restrictive chemical environment. The oil secreted by skin cells is converted to fatty acids by bacteria that live

Airways (lining of trachea)
Most pathogens are trapped in mucus before they can reach the lungs. Beating cilia sweep pathogens up and out of the airway.

Mucus-secreting cells   Ciliated cells

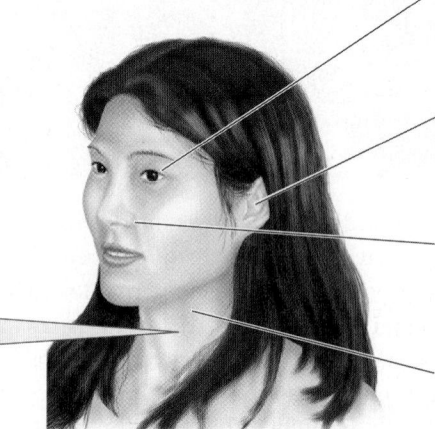

Eyes
Blinking wipes tears across the eye. Tears contain the antibacterial enzyme lysozyme.

Ears
Hairs and earwax trap pathogens in the passageway of the external ear.

Nose
The nasal passages are lined with mucous secretions and hairs that trap pathogens.

Digestive tract
Pathogens are trapped in saliva and mucus, then swallowed. Most are destroyed by the low pH of the stomach.

FIGURE 51.1 The Body Protects Openings. The scanning electron micrograph of mucus-secreting and ciliated cells has been colorized.

harmlessly on the surface. The fatty acids lower the pH of the surface to about 5, creating a dry, acidic environment that prevents the growth of most pathogens.

Unless this protective exterior surface is broken by an injury, the most vulnerable places in animal bodies are the gaps in the surface where the digestive tract, reproductive tract, gas-exchange surfaces, and sensory organs make contact with the environment.

How Are Openings in the Body Protected?

As **FIGURE 51.1** shows, gaps in the outer surface of animal bodies are protected by a complex array of chemical substances that discourage pathogen entry.

The mucus that protects the surface of soft-bodied invertebrates is equally important in protecting the surface openings in vertebrates. For example, many of the pathogens that you breathe in or ingest while eating or drinking stick to the mucus that lines your airways and digestive tract. Pathogens that are stuck in mucus cannot come in contact with the plasma membranes of epithelial tissues. In mammals, many of these pathogens are either coughed out or swallowed and killed in the acidic environment of the stomach. In the respiratory tract, they are swept out of harm's way by the beating of cilia (see Figure 51.1).

Gaps in the body that are not covered with mucous layers — the eyes, for example—are often protected by other types of secretions. Your ears are protected by waxy secretions, and your eyes by tears that contain the enzyme **lysozyme**. Lysozyme acts as an **antibiotic** by digesting bacterial cell walls.

How Do Pathogens Gain Entry?

When preventive measures fail, as they sometimes do, pathogens gain entry to tissues beneath the skin. Flu viruses, for example, have an enzyme on their surface that disrupts the mucous lining of the respiratory tract. When the outer surface of the virus makes contact with a host cell beneath the mucous layer, the virus is able to enter the cell and begin an infection.

Falls, wounds, and other types of physical trauma provide another important mode of entry. When the skin is broken, bacteria and other pathogens gain direct access to the tissues inside.

To viruses, bacteria, and fungi, your body is a tropical paradise, brimming with resources. Given that a single bacterium could give rise to a population of 100 trillion in a day, something must be done, and fast, when the outer defenses of the body are penetrated. What happens then?

The Innate Immune Response

If foreign invaders penetrate the body's protective barrier, particular cells initiate the **innate immune response**—the body's first response to pathogens. As a group, these cells are called white blood cells to distinguish them from red blood cells. More formally, they are known as **leukocytes** ("white-cells").

The leukocytes involved in innate immunity provide an immediate, *generic* response that is directed against the general type of pathogen encountered. The response is considered generic since it is directed against broad groups of pathogens instead of

specific organisms. For example, the innate response is able to distinguish between fungi and bacteria but cannot identify a specific strain within either group.

How Are Pathogens Recognized by the Innate Immune System?

The answer to this question emerged from a breakthrough in the late 1990s, when results from unrelated studies came together. Earlier, German researchers had identified a protein that was important in the early development of fruit flies. When flies were mutated to no longer express this protein, their larvae developed abnormally. The strange appearance of the larvae inspired the researchers to name it the Toll protein (in German, "toll" means amazing).

Jules Hoffmann and his colleagues found that Toll protein also served an important immune function: Adult fruit flies lacking Toll could not defend themselves from fungal infection (see **FIGURE 51.2**). Just a year later, two separate labs identified Toll-like

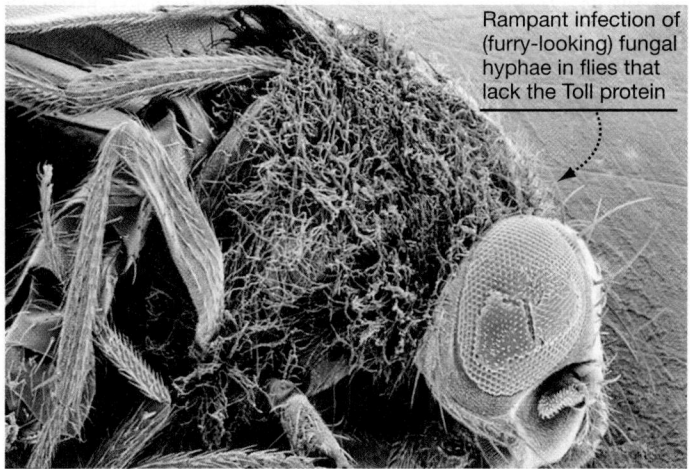

Rampant infection of (furry-looking) fungal hyphae in flies that lack the Toll protein

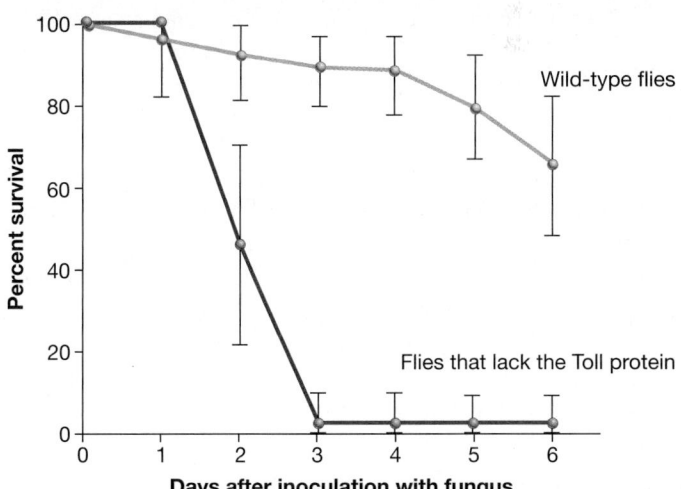

FIGURE 51.2 Toll Protein Is Critical to Immune Response.
Researchers learned this by comparing the survival rate of wild-type flies and mutants unable to express Toll protein after exposure to a fungal pathogen. The error bars represent the confidence interval for each point with $p \leq 0.05$. This means that there is 95 percent confidence that if the experiment were repeated, the value for each time point would fall within the bracketed range.

DATA: Lemaitre, B., et al. 1996. *Cell* 86: 973–983.

proteins in humans, demonstrating their universality. This was an exciting discovery, but the function of these Toll-like proteins was still not understood.

The riddle of Toll-like proteins was solved by Bruce Beutler and his colleagues, who studied how mutant mice reacted to antigens. An **antigen** is any foreign molecule that can initiate an immune system response. Earlier researchers observed that a molecule called lipopolysaccharide (LPS), harvested from the surface of Gram-negative bacteria (see Chapter 29), reliably caused a powerful immune response in animals. However, Beutler had a mutant mouse that did not respond to this antigen—and the mouse had defective Toll-like proteins.

Beutler had discovered that the Toll-like protein functions as a receptor that receives the signal that a pathogen is present, in this case a Gram-negative bacterium. This breakthrough was key to understanding how the innate immune system could become engaged.

Toll-like receptors (TLRs) are a subset of a larger group of proteins called **pattern-recognition receptors,** which serve as sentinels to signal the presence of molecules associated with pathogens. The TLRs have been observed in fungi and plants, suggesting that they arose in a common ancestor of eukaryotes. Eleven TLRs have been identified in humans, each one responding to a different kind of antigen. For example,

- TLR2 binds to zymosan, a common molecule in fungi.

- TLR4 binds to lipopolysaccharide (LPS) from Gram-negative bacteria (the one in Beutler's study).

- TLR5 binds to flagellin from the flagella of motile bacteria.

- TLR7 binds to single-stranded viral RNA.

What all these antigens have in common is that they are ubiquitous within the broad group in which they have been identified (e.g., *all* of the countless Gram-negative bacteria produce LPS), yet they do not occur in the host animal. Thus, these antigens serve as reliable signals of attack for the innate immune system. A mere 11 TLRs in humans is enough to detect virtually any type of invasion.

Pattern-Recognition Receptors Transduce Signals

When pattern-recognition receptors, such as TLRs, on the surface of a leukocyte receive the signal that an invader is present, they trigger a signal cascade within the cell that will have different consequences depending on which TLRs were activated (see Chapter 11 for an introduction to signal transduction). This step is vital to determining and orchestrating the innate immune responses to follow. For example,

- When the Toll protein in healthy flies detects a fungal infection, a signal cascade activates the expression and secretion of antimicrobial peptides, which function as antibiotics to destroy the fungal pathogens.

- When TLR4 in humans is activated by LPS, a signal cascade leads to the expression and secretion of **cytokines** ("cell movers"). Cytokines are a class of diverse molecules that signal other parts of the immune system in various ways, such as

attracting other immune cells to the site of infection or stimulating other immune cells into action.

- When TLR7 in humans detects the single-stranded RNA of a virus, the cell may produce and secrete a specific type of cytokine called interferon, which stimulates neighboring cells to improve their resistance to the imminent viral infection.

✔ **QUANTITATIVE** If you understand the role of pattern-recognition receptors, you should be able to return to Figure 51.2 and predict how the same two groups of flies would respond to a Gram-negative bacterial pathogen. Draw a line on the graph to represent flies lacking the Toll protein after treatment, and explain your reasoning. Assume the survival rate of wild-type flies would be similar to what is shown in the figure.

Once the general signal of an invasion is received by TLRs, the first response is sent out and a cascade of further actions occurs that result in a fully engaged immune response. For example, consider what would happen if you were to trip on the sidewalk while running across campus and scrape your elbow. If this wound resulted in an infection, you would observe redness, swelling, pain, and heat in the infected area. What is responsible for all of these changes?

The Inflammatory Response in Humans The physical signs of an infected injury are a direct consequence of the innate immune response being called into action to contain and eliminate the infection. **FIGURE 51.3** summarizes the major steps in this **inflammatory** ("in-flames") **response,** a multistep, innate immune response observed in an array of animals.

Step 1 A break in the skin allows bacteria to enter the body. If capillaries and other small blood vessels are broken, the wound bleeds.

Step 2 Immediately after the injury, blood components called **platelets** release proteins that form clots and lessen bleeding. Other clotting proteins in the blood form cross-linked structures that help wall off the wound and reduce blood loss.

Step 3 Wounded tissues and leukocytes called **macrophages** secrete **chemokines** ("chemical-movers"), which are signaling molecules that recruit other cells to the site of infection. The localized production of chemokines is important because it forms a gradient that marks a path to the wound site.

Step 4 **Mast cells** release chemical messengers such as **histamine** that constrict blood vessels at the site of the wound—reducing blood flow and thus blood loss. Mast cells also release other signaling molecules that induce blood vessels slightly farther from the wound to dilate and become more permeable.

Step 5 The combination of dilated blood vessels and a chemokine gradient is like a 911 call that provides specific directions to the scene of a fire. Leukocytes called neutrophils move out of dilated blood vessels and migrate to the site of the infection. **Neutrophils** are major players in the innate response. They destroy invading cells by phagocytosis ("cell eating")—meaning that they engulf them. Once inside neutrophils, the invading cells are killed with a complex array of antimicrobial

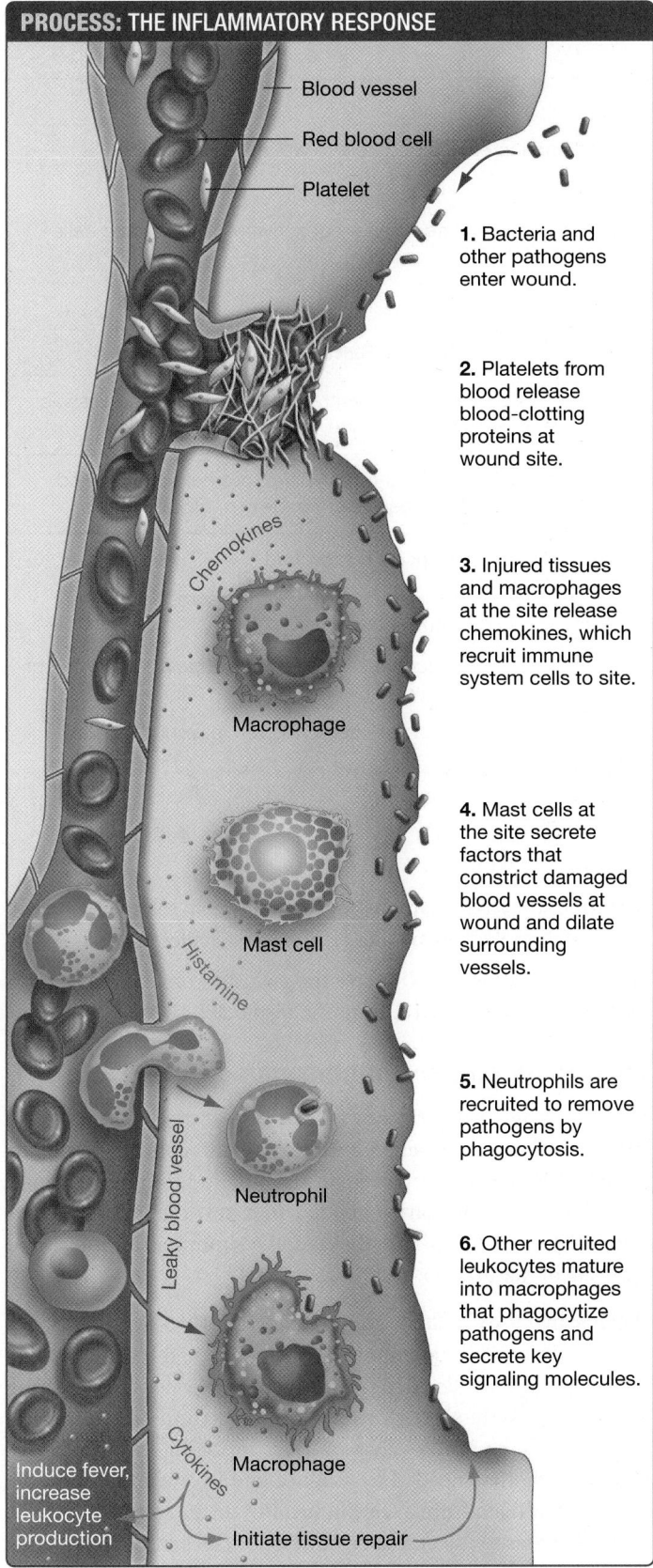

- Blood vessel
- Red blood cell
- Platelet

1. Bacteria and other pathogens enter wound.

2. Platelets from blood release blood-clotting proteins at wound site.

Chemokines

3. Injured tissues and macrophages at the site release chemokines, which recruit immune system cells to site.

Macrophage

4. Mast cells at the site secrete factors that constrict damaged blood vessels at wound and dilate surrounding vessels.

Histamine

Mast cell

Leaky blood vessel

5. Neutrophils are recruited to remove pathogens by phagocytosis.

Neutrophil

6. Other recruited leukocytes mature into macrophages that phagocytize pathogens and secrete key signaling molecules.

Cytokines

Macrophage

Induce fever, increase leukocyte production

Initiate tissue repair

FIGURE 51.3 The Inflammatory Response of Innate Immunity Has Many Elements.

✔**QUESTION** How do these events lead to the classical signs of an infection, which include rubor (reddening), calor (heat), and tumor (swelling), among others?

compounds—including lysozyme, which degrades bacterial cell walls. Cells that perform this activity are collectively referred to as phagocytes.

Step 6 Cells that will mature into macrophages arrive. Besides secreting chemokines, these macrophages produce additional cytokines that have an array of effects. Some of these cytokines stimulate bone marrow to make and release additional leukocytes, induce fever—an elevated body temperature that aids in healing—and activate cells involved in tissue repair and wound healing. Like neutrophils, macrophages are also phagocytes that help clear invaders from the area.

This overview actually simplifies the situation—many other cell types and cell–cell signals are involved in responding to pathogens at a site of infection.

The inflammatory response continues until all foreign material is eliminated and the wound is repaired. **TABLE 51.2** on page 1042 summarizes key cells and molecules involved in the response.

But what happens when the innate immune system of a vertebrate fails to contain and eliminate invading pathogens?

check your understanding

If you understand that . . .

- The innate immune response occurs when macrophages and mast cells that reside in tissues, and neutrophils that circulate in the blood, react in a generic way to signals from invading pathogens.

✔ You should be able to . . .

Describe how the following steps that first-aid workers use when initially treating a wound mimic events in the innate immune response:

1. Applying direct pressure to close blood vessels
2. Cleaning the wound to remove dirt and debris
3. Applying bandages containing compounds that halt blood flow from the wound

Answers are available in Appendix A.

51.2 Adaptive Immunity: Recognition

In addition to the innate immune response, vertebrates have evolved the ability to recognize specific antigens, and can differentiate between different species and even different strains of pathogens. The leukocytes involved can customize their response to the particular invader, so this arm of the immune system is often referred to as the **adaptive immune response.**

Given the array of pathogens that exist, an animal is almost certain to be exposed to an enormous variety of antigens in the course of its lifetime. Is there a limit to how many different antigens its adaptive immune system responds to?

(a) Key Cells

Name	Primary Function
Mast cells	Release signals that constrict blood flow from wound and increase blood flow to wound area
Neutrophils	Kill invading cells via phagocytosis
Macrophages	Release cytokines that recruit other cells to wound site and stimulate a variety of activities; kill invading cells via phagocytosis

(b) Key Signaling Molecules

Name	Produced By	Received By	Message/Function
Histamine	Mast cells	Blood vessels	High concentration constricts blood vessels at wound site, among other activities
Chemokines*	Injured tissues and macrophages in tissues	Neutrophils and macrophages	Mark path to wound; promote dilation and increased permeability of blood vessels
Cytokines other than chemokines	Macrophages	Leukocytes	Mark path to wound
		Bone marrow	Increase production of macrophages and neutrophils
		CNS	Induce fever by raising set point for control of body temperature
		Local tissues	Stimulate cells involved in wound repair

*Note that chemokines are a subset of cytokines.

Research conducted in the early 1920s answered this question. Researchers synthesized organic compounds that do not exist in nature, injected the novel molecules into rabbits, and observed whether they activated the animals' adaptive immune system. To the amazement of the scientists, the rabbits were able to recognize and respond to every novel antigen tested. More specifically, each of the animals produced proteins in their blood called antibodies that specifically bound to the particular antigen that was injected. **Antibodies** are secreted proteins that bind to a specific part of a specific antigen.

The take-home message is that the adaptive immune system can specifically respond to an almost limitless array of antigens. This observation, and subsequent research on antibody production, led to the identification of four key characteristics of the adaptive immune response:

1. *Specificity* Antibodies and other components of the adaptive immune system bind only to specific sites on specific antigens.

2. *Diversity* The adaptive immune response recognizes an almost limitless array of antigens.

3. *Memory* The adaptive immune response can be reactivated quickly if it recognizes antigens from a previous infection.

4. *Self–nonself recognition* Molecules that are produced by the individual do not act as antigens, meaning that the adaptive immune system can distinguish between self and nonself. Nonself molecules are antigens; self molecules are not.

The next question is, What are the cells and organs that are responsible for this adaptive immune response?

An Introduction to Lymphocytes

The leukocytes that carry out the major features of the adaptive immune response are called **lymphocytes.** In contrast to the diversity of leukocytes involved in the innate immune response, lymphocytes are primarily divided into two distinct cell types that differ in their role in the adaptive immune response and their site of maturation.

The Discovery of B Cells and T Cells In 1956 a group of researchers provided an important insight into the adaptive immune system, by accident. The biologists were investigating the immune system's response to the bacterium *Salmonella typhimurium*, a common cause of food poisoning in humans. To do this work, they needed to produce and isolate antibodies to a particular antigen from *S. typhimurium*—an antigen that is toxic. Their plan was to inject a large number of chickens with the antigen and collect the antibodies that the treated animals produced in response.

In addition to injecting many normal chickens, though, the biologists happened to include some chickens that had undergone experimental removal of an organ called the bursa. Six of the nine chickens that lacked a bursa and that were injected with the antigen died. The other three birds without a bursa survived, but failed to produce antibodies to the antigen. In contrast, all of the chickens with the bursa intact survived and produced large quantities of antibodies. To make sense of these results, the researchers proposed that the bursa is critical for antibody production and that antibodies are important in neutralizing the toxic antigens.

Not long after this observation was published, three groups of scientists independently conducted a related experiment. To explore the function of the thymus in mammals, these groups removed the organ from newborn mice. Mice lacking a thymus developed pronounced defects in their immune systems. For example, when pieces of skin from other mice were grafted onto the experimental individuals, their immune systems did not recognize the tissue as foreign. In contrast, individuals with an intact thymus quickly mounted an immune response that killed the foreign skin cells.

The results of these and follow-up experiments showed that lymphocytes from the bursa and thymus have different functions. The two types of lymphocytes became known as bursa-dependent and thymus-dependent lymphocytes, or B cells and T cells, respectively.

- **B cells** produce antibodies. Later work showed that in humans and other species where the bursa organ does not exist, B cells mature in bone marrow.

- **T cells** are involved in graft rejection along with an array of other immune functions, including recognizing and killing host cells that are infected with a virus. T cells mature in the thymus.

The requirement for maturation in a particular organ distinguishes the lymphocytes from other leukocytes that were introduced in Section 51.1. What is the origin of these lymphocytes, and where do they go after they mature?

Where Are Lymphocytes Found? The colored structures in **FIGURE 51.4** mark the major sites in the human body that play a key role in the life of a lymphocyte.

- *Lymphocyte origin* All lymphocytes are produced in **bone marrow**—a tissue that fills the internal cavities in bones.

- *Lymphocyte maturation* B cells mature in the bone marrow in humans and many other animals. T cells mature in the **thymus**—an organ located in the upper part of the chest of vertebrates (just behind the breastbone in humans).

- *Lymphocyte activation* Lymphocytes recognize antigens and become activated in the spleen and lymph nodes. The **spleen** is an organ located near the stomach in the abdominal cavity and is also involved in destroying old blood cells. **Lymph nodes** are small, oval organs that are located all around the body. Lymph nodes filter the lymph passing through them. Recall that **lymph** is a mixture of fluid and lymphocytes (see Chapter 45).

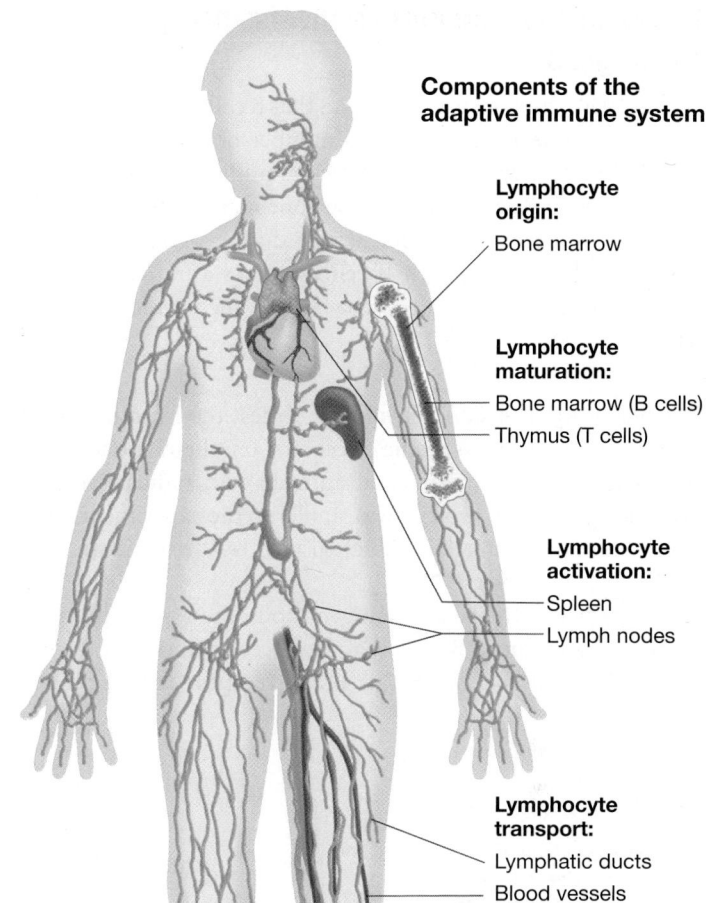

Components of the adaptive immune system

Lymphocyte origin:
Bone marrow

Lymphocyte maturation:
Bone marrow (B cells)
Thymus (T cells)

Lymphocyte activation:
Spleen
Lymph nodes

Lymphocyte transport:
Lymphatic ducts
Blood vessels

FIGURE 51.4 Lymphocytes Are Formed, Activated, and Transported in the Immune System.

The liquid portion of lymph originates in fluid that is forced out of capillaries by blood pressure.

- *Lymphocyte transport* Lymphocytes circulate through the blood and the secondary organs of the immune system—lymph nodes, spleen, and lymphatic ducts. Lymphatic ducts are thin-walled, branching tubules that transport lymph throughout the body and, along with lymph nodes, make up the **lymphatic system.**

Besides being found in bone marrow, the thymus, spleen, lymph nodes, lymphatic ducts, and blood vessels, large numbers of lymphocytes, along with other leukocytes, are associated with skin cells and with epithelial tissues that secrete mucus—primarily in the digestive tract and respiratory tract. Collectively, the immune system cells found in the gut and respiratory organs are called **mucosal-associated lymphoid tissue (MALT).** Leukocytes in the skin and MALT are important because they guard points of entry against pathogens.

Now let's turn to one of the most fundamental questions in immunology: How are B cells and T cells able to recognize so many different antigens? Let's begin with a look at the B-cell and T-cell receptors—the molecules that start the adaptive immune response.

Lymphocytes Recognize a Diverse Array of Antigens

By the 1960s, biologists understood that B cells can produce antibodies to a seemingly limitless number of antigens and that each antibody is specific to a particular antigen. They hypothesized that each B cell formed in the bone marrow has thousands of copies of a receptor on its surface that recognizes only one antigen. What's more, each B cell expresses a unique receptor relative to the other B cells. What would such a receptor look like?

The Discovery of B-Cell Receptors To test the hypothesis that B cells have antigen-specific receptors on their surfaces, researchers injected experimental animals with radioactively labeled antigens. This strategy was similar to the experiments with labeled estradiol that allowed biologists to isolate that hormone's receptor (see Chapter 49).

As predicted, the labeled antigens bound to a protein on the surface of only those B cells that produced antibody to the antigen. Chemical analysis of this **B-cell receptor (BCR)** protein showed that it has the same overall structure as the antibodies produced by the B cells and secreted into the blood.

The BCR protein consists of two distinct polypeptides (**FIGURE 51.5a**). The smaller polypeptide is called the **light chain.** The larger polypeptide is roughly twice the size of the light chain and is called the **heavy chain.** Each BCR has two copies of the light chain and two copies of the heavy chain that are all held together by disulfide bonds. The heavy chain includes a transmembrane domain that anchors the protein in the plasma membrane of the B cell.

The antibodies produced by a B cell are identical in structure to its BCR, except that they lack the transmembrane domains. Instead of being inserted into the plasma membrane, antibodies are secreted from the cell and circulate throughout the body.

Both the BCR and the antibodies produced by B cells belong to a family of proteins called the **immunoglobulins (Ig).** Immunoglobulins are crucial to the adaptive immune response.

TABLE 51.3 shows the five classes of immunoglobulin proteins that act as antibodies. The five types are symbolized IgG, IgD, IgE, IgA, and IgM. Each class is distinguished by unique amino acid sequences in the heavy-chain region, and each has a distinct

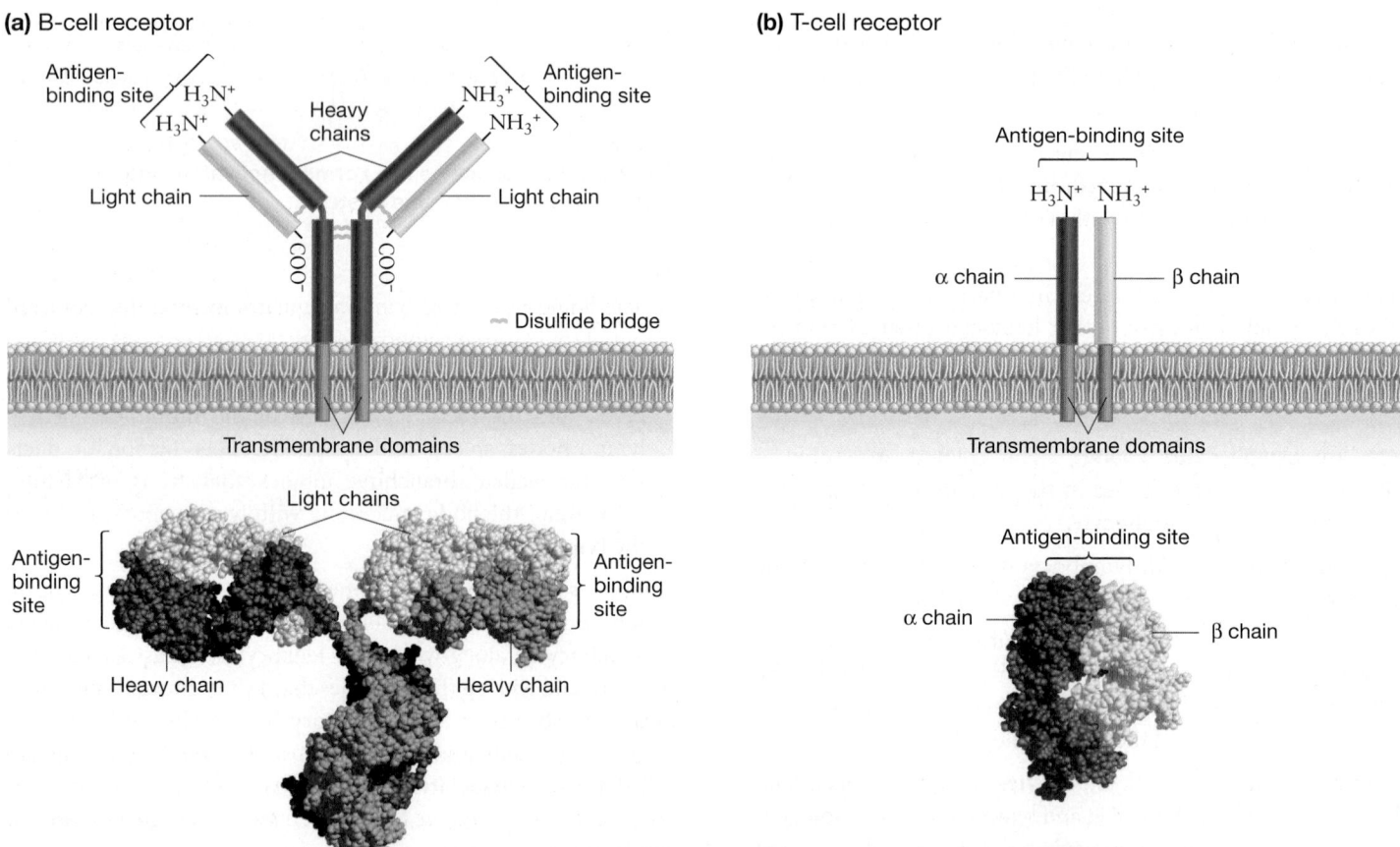

(a) B-cell receptor

Antigen-binding site
H_3N^+
H_3N^+
Heavy chains
NH_3^+
NH_3^+
Antigen-binding site
Light chain
Light chain
COO^-
COO^-
Disulfide bridge
Transmembrane domains

Light chains
Antigen-binding site
Antigen-binding site
Heavy chain
Heavy chain

(b) T-cell receptor

Antigen-binding site
H_3N^+ NH_3^+
α chain
β chain
Transmembrane domains

Antigen-binding site
α chain
β chain

FIGURE 51.5 B-Cell Receptors and T-Cell Receptors Have Transmembrane Domains and Antigen-Binding Sites. (a) Schematic and space-filling models of the B-cell receptor, which is shaped like a Y. **(b)** The shape of the T-cell receptor resembles one "arm" of the Y-shaped B-cell receptor. The space-filling models for each receptor do not show the transmembrane domains that anchor them in the plasma membrane.

TABLE 51.3 Five Classes of Immunoglobulins

Name	Structure (Secreted Form)	Function
IgG	Monomer	The most abundant type of secreted antibody. Circulates in blood and interstitial fluid. Protects against bacteria, viruses, and toxins.
IgD	Monomer	Present on membranes of immature B cells; rarely secreted. Serves as BCR.
IgE	Monomer	Secreted in minute amounts. Involved in response to parasitic worms. Also responsible for hypersensitive reaction that produces allergies.
IgA	Dimer	Most common antibody in breast milk, tears, saliva, and the mucus lining the respiratory and digestive tracts. Prevents bacteria and viruses from attaching to mucous membranes; helps immunize breastfed newborns.
IgM	Pentamer	First type of secreted antibody to appear during an infection. Binds many antigens at once; effective at clumping viruses and bacteria so that they can be killed. Monomeric form also serves as BCR.

function in the immune response. Besides acting as antibodies, IgD and monomeric forms of IgM also serve as BCRs.

The Discovery of T-Cell Receptors It took much longer for researchers to isolate and characterize the **T-cell receptor (TCR).** The technique used to identify BCRs was not useful for the receptors on T cells, suggesting that they cannot bind antigens on their own. It turns out that the T cells require other cells to **(1)** process the antigens and **(2)** present them to the TCRs. This means that for a TCR to recognize an antigen, the foreign molecule must first undergo a complex process called **antigen presentation.**

This is a fundamentally important distinction. B cells bind to antigens directly; T cells bind only to antigens that are displayed by other cells.

Other data showed that the TCR is composed of two protein chains: an alpha (α) chain and a beta (β) chain (**FIGURE 51.5b**). The overall shape of the TCR is similar to one of the two "arms" of an antibody or BCR molecule.

Antibodies and Receptors Bind to Epitopes Antibodies, BCRs, and TCRs do not bind to the entire antigen. Instead, they bind to a selected region of the antigen called an **epitope.**

To understand the relationship between an antigen and an epitope, consider that every bacterium, virus, fungus, and protist is made up of a large number of different molecules. Many of these molecules are antigens, because they would be recognized

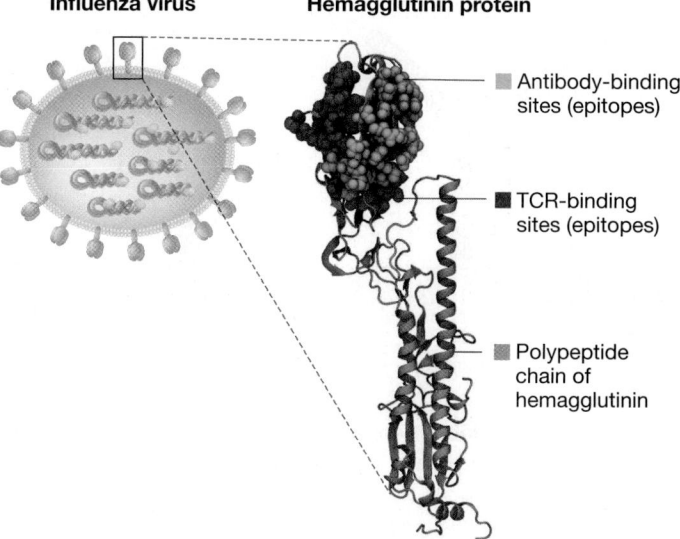

FIGURE 51.6 Most Antigens Have Multiple Epitopes for B-Cell Receptors and T-Cell Receptors. The envelope of the influenza virus includes the protein hemagglutinin. This version of hemagglutinin has several distinct epitopes for BCRs and TCRs. One example of each type is highlighted as a space-filling model.

as being foreign to your cells. In turn, each antigen may have many different epitopes, where binding by antibodies and lymphocyte receptors actually takes place.

FIGURE 51.6 illustrates a protein called hemagglutinin, which is found on the surface of the influenza virus. This antigen has several distinct epitopes, two of which are identified in the figure. Often, the epitopes for BCRs are different from those recognized by TCRs because of how epitopes are presented to the different receptors: BCRs bind to epitopes in the context of the intact antigen, while TCRs bind to epitopes that have been processed and presented by other cells. Each epitope is recognized by a particular antibody, BCR, or TCR. It is not unusual for an antigen to have between 10 and 100 different epitopes.

How do the immunoglobulins recognize specific epitopes? The answer to this question emerged through detailed studies of the heavy and light chains.

What Is the Molecular Basis of Antibody Specificity and Diversity?

In the 1950s, biologists developed an important model system for studying immunoglobulin structure. The model cells were B-cell tumors, or myelomas, that could be grown in laboratory culture.

Each type of myeloma arises from a single cell that produces a single type of antibody. When researchers compared the immunoglobulins produced by different myelomas, they found that the sequence of amino acids in the light chains consisted of two regions: one that is virtually identical among the myelomas, and another that is unique. These light-chain segments have come to be known as the **constant (C) regions** and **variable (V) regions,** respectively. Heavy chains also have a C region and a V region.

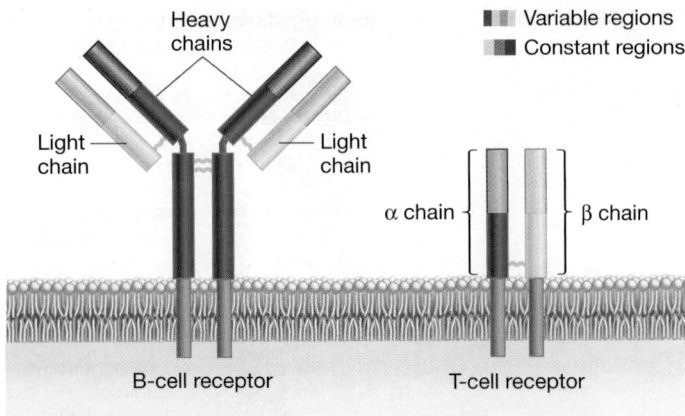

FIGURE 51.7 The Variable Regions of B-Cell Receptors and T-Cell Receptors Face Away from the Plasma Membrane.

FIGURE 51.7 makes two important points:

1. The *V* regions of heavy and light chains in a BCR are adjacent to one another and face away from the plasma membrane.

2. The *V* and *C* regions of a TCR are arranged in a similar manner as the BCR.

The presence of unique amino acid sequences in the *V* regions of every antibody, BCR, and TCR explains why each of these proteins binds to a unique epitope. Your body can respond to an almost limitless number of antigens because the number of different BCRs, antibodies, and TCRs is virtually limitless. How does all this variation come to be?

The Discovery of Gene Recombination In 1965, W. J. Dryer and J. Claude Bennett proposed a fantastic-sounding explanation for how immunoglobulin genes code for so many different variable regions and thus so many different proteins. They hypothesized that as a lymphocyte is maturing, a segment from a variable-region gene is cut and combined with a segment from the constant-region gene. Further, they proposed that this cutting and pasting is done in a different way in each lymphocyte. The result is that each lymphocyte has a novel "*V* + *C* gene" for the light-chain protein.

At the time, this type of genetic recombination in cells had not been observed, and most researchers considered the hypothesis wildly implausible.

Eleven long years passed before the first experimental evidence was obtained to support their hypothesis. In 1976, researchers showed that the amount of DNA in the *V* + *C* region of mature lymphocytes is shorter than it is in immature lymphocytes. This is exactly what the gene-recombination hypothesis predicts.

The flurry of studies inspired by this result showed that the genes for light chains have dozens of different *V* segments, several different joining (*J*) segments, and a single constant (*C*) segment. The heavy-chain gene includes diversity (*D*) segments along with a set of *V* and *J* segments similar to those in the light-chain

gene. The genes that encode the α and β chains of the TCR have a similar arrangement of distinct segments, each with multiple versions.

As a lymphocyte matures, the various gene regions are mixed and matched to produce unique receptors. **FIGURE 51.8** illustrates the steps involved in the production of the BCR light chain. A similar process occurs in the DNA encoding the heavy chain gene. Consider how these recombination events lead to BCR diversity:

1. In the light chain, one of the 40 *V* segments recombines with one of the 5 *J* segments. This step can produce $40 \times 5 = 200$ different light chains.

2. In the heavy chain, any one of the 51 *V* segments, 27 *D* segments, and 6 *J* segments can recombine.

3. The light-chain and heavy-chain rearrangements occur independently, but the final BCR includes the polypeptides from each gene to form the antigen-specific variable region.

✔ **QUANTITATIVE** If you understand how unique genes for the light and heavy chains of the BCR are made, you should be able to calculate the number of different antigen-specific BCRs that could made by this process in (1) a human and (2) a single human B cell.

In addition, gene segments do not always join precisely during DNA recombination. Some variation occurs where the *V* and *D* segments join and where the *D* and *J* segments join. As a result, an estimated 10^{10} to 10^{14} different BCRs can form in a single individual. TCR production is just as diverse.

Gene recombination is the molecular mechanism responsible for the specificity and diversity of the adaptive immune system. This process allows each lymphocyte to produce a unique BCR, antibody, or TCR—enabling it to recognize a unique epitope. The charges and geometry at each surface make the way the receptor and epitope fit together extremely specific.

How Does the Immune System Distinguish Self from Nonself?

If the adaptive immune system has the remarkable ability to generate specific, targeted responses against virtually any substance, what keeps it from reacting against itself? If a receptor responded to a **self molecule**—that is, a molecule belonging to the host—the receptor would trigger an immune response. Because this type of response is extremely rare compared to responses against foreign antigens, biologists hypothesized that there must be some mechanism for eliminating self-reactive B cells and T cells.

To test this hypothesis, researchers injected B cells and T cells possessing anti-self receptors into mice and found that the injected lymphocytes were eliminated. Follow-up work showed that if B cells and T cells maturing in the bone marrow and thymus have anti-self receptors, the cells are likely to be destroyed or permanently inactivated before they leave these organs. The conclusion? Some form of "self-education" occurs during the maturation phase of lymphocyte development.

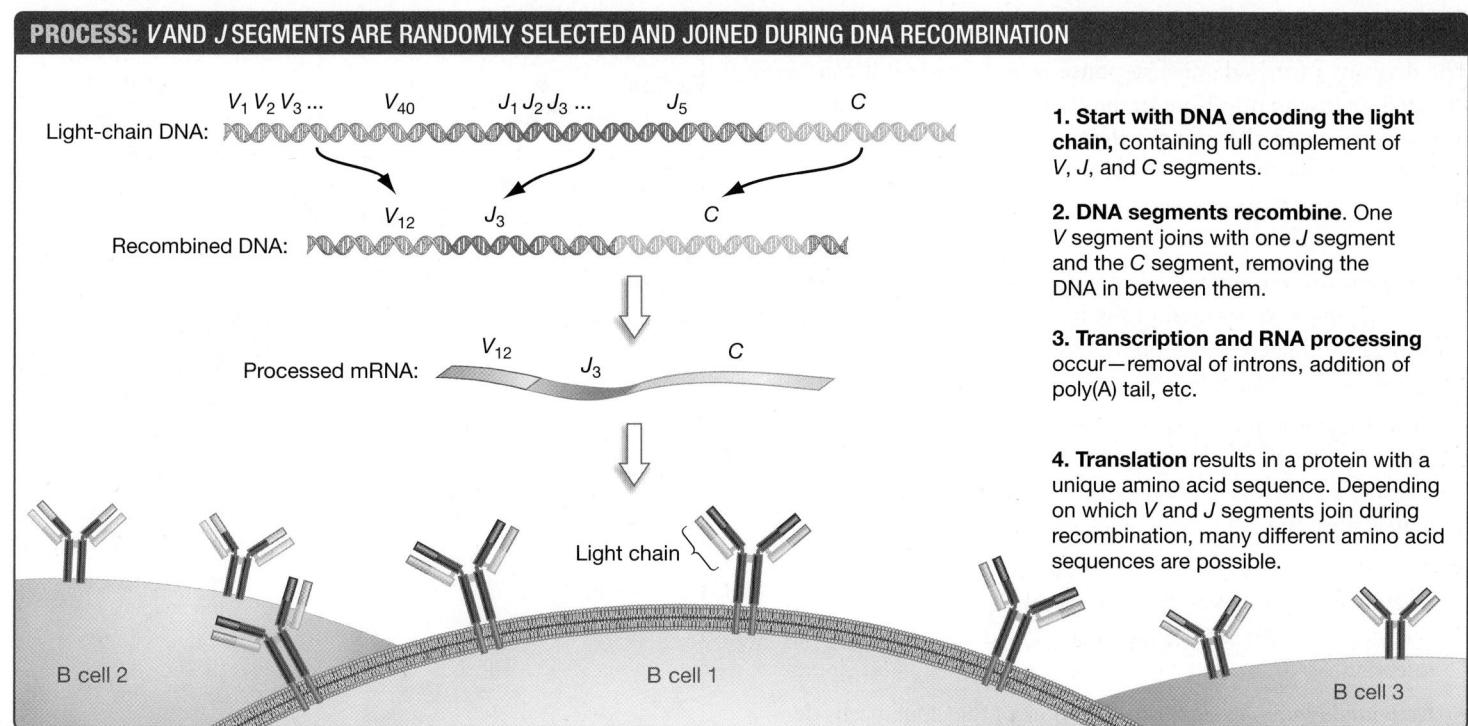

$V_1 V_2 V_3 \ldots$ V_{40} $J_1 J_2 J_3 \ldots$ J_5 C

Light-chain DNA:

V_{12} J_3 C

Recombined DNA:

V_{12} J_3 C

Processed mRNA:

Light chain

B cell 2 B cell 1 B cell 3

1. **Start with DNA encoding the light chain,** containing full complement of *V*, *J*, and *C* segments.

2. **DNA segments recombine.** One *V* segment joins with one *J* segment and the *C* segment, removing the DNA in between them.

3. **Transcription and RNA processing** occur—removal of introns, addition of poly(A) tail, etc.

4. **Translation** results in a protein with a unique amino acid sequence. Depending on which *V* and *J* segments join during recombination, many different amino acid sequences are possible.

FIGURE 51.8 As B Cells Mature, Immunoglobulin-Gene Segments Recombine to Form a Single Gene. In the mature B cell 1, the final light-chain gene consists of the V_{12}, J_3, and *C* segments spliced together; the final heavy-chain gene might consist of the V_{48}, D_{22}, J_1, and *C* segments spliced together (not shown).

Researchers have also identified another type of T cell that acts as a regulatory cell. Regulatory T cells suppress certain parts of the immune system to limit the intensity of normal responses, and may help inhibit any self-reactive cells that slip through the self-education system. Defective or insufficient regulatory T cells may be partly responsible for the development of immune disorder diseases (see Section 51.5).

Even after the self-reactive lymphocytes are inactivated or removed, the structural diversity still present in the receptors of mature B and T cells is sufficient for recognizing virtually any foreign antigen. But this diversity alone is not enough to mount an effective immune response. The enormous repertoire of possible BCRs and TCRs means that only a few cells will express any particular epitope-specific receptor. To engage the invading pathogen successfully, these few lymphocytes must reproduce. How does this selective expansion take place?

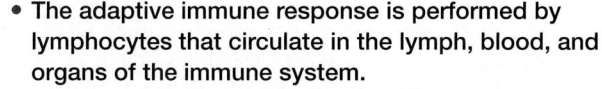

If you understand that . . .

- The adaptive immune response is performed by lymphocytes that circulate in the lymph, blood, and organs of the immune system.
- Lymphocytes can respond to a wide array of antigens because gene segments recombine to produce receptors that bind to unique epitopes on antigens.
- Lymphocytes that express self-reactive receptors are eliminated or inactivated to prevent immune responses against the body's own cells.

✔ You should be able to . . .

1. Cite two key events that must occur during lymphocyte maturation to ensure a safe and effective adaptive immune response.
2. Explain the benefit of assembling the variable region from discrete segments rather than having multiple intact variable regions.

Answers are available in Appendix A.

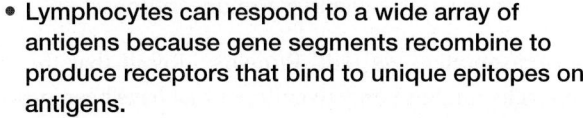

51.3 Adaptive Immunity: Activation

Lymphocytes are normally in a resting, or inactive, state. Over the course of a day, an inactive lymphocyte may hang out in the skin or MALT, enter the lymphatic vessels, migrate through a lymph node, cross over into the blood, pass through the spleen, return to the blood, and so on.

If an inactive lymphocyte does not encounter the epitope that it is programmed to respond to, the cell eventually dies. As it turns out, this is the fate of most of the lymphocytes that originate from the bone marrow. During an infection, however, some of these inactive lymphocytes will go on to mount a massive attack that is targeted against the specific invader.

CHAPTER 51 The Immune System in Animals **1047**

The Clonal Selection Theory

The diversity of the adaptive response is useful only if it can be directed against an infection. In the 1950s, Frank Bernet and colleagues developed the **clonal selection theory** to explain how only the most useful lymphocytes are activated during an infection. Their theory made several key claims about how the adaptive immune system works:

- *Antigens are recognized by receptors on B cells and T cells.* Each lymphocyte formed in the bone marrow or thymus expresses a unique receptor on its surface that binds to a unique epitope in an antigen.

- *Lymphocytes require receptor-epitope binding to become activated.* When the receptor on a lymphocyte binds to an epitope, a switch is thrown in the lymphocyte that may ultimately push the cell from a quiescent to an activated state.

- *Activated lymphocytes are cloned.* An activated lymphocyte divides and makes many identical copies of itself. In this way, specific cells are selected and cloned in response to an infection.

- *Activated lymphocytes endure.* Some of the cloned cells descended from an activated lymphocyte persist long after the pathogen is eliminated. As a result, the cloned cells are able to respond quickly and effectively if the infection recurs.

FIGURE 51.9 summarizes the assertions made by the clonal selection theory, using a B cell model. ✔ If you understand the clonal selection theory, you should be able to explain what "clonal" and "selection" refer to.

The activation of the relevant B cells and T cells, however, is a carefully controlled, stepwise process. The mechanism is reminiscent of the precautions that nations with powerful missiles take to avoid accidental deployment. For the most dangerous weapons, the signal to launch is checked and cross-checked, using a series of codes and signals. In the immune system, the checking and cross-checking occur through protein–protein interactions on the surfaces of cells, and the release and receipt of cytokines and other signaling molecules.

Let's first take a look at how T cells are activated and then examine how they play a role in the full activation of B cells.

T-Cell Activation

As the innate immune response is battling the invaders at the site of infection, leukocytes known as **dendritic** ("tree-like") **cells** are gobbling up antigens and debris via phagocytosis. These dendritic cells are like messengers that collect antigens from the battle scene and then report to the lymph nodes, where they present them to lymphocytes. Antigen presentation is a key event that links the innate and adaptive arms of the immune system; without this transfer of information, the adaptive immune response would not be activated to respond to the infection.

To understand how the activation system works, let's explore how the antigens taken up by dendritic cells are processed and presented to T lymphocytes. T cells are classified as CD4$^+$ or

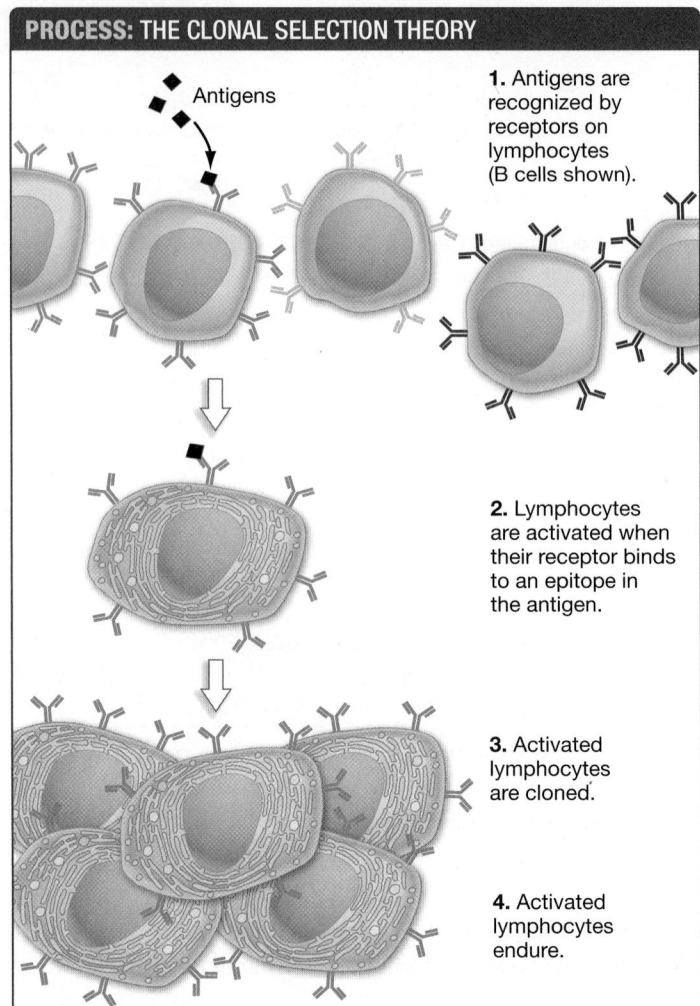

PROCESS: THE CLONAL SELECTION THEORY

1. Antigens are recognized by receptors on lymphocytes (B cells shown).

2. Lymphocytes are activated when their receptor binds to an epitope in the antigen.

3. Activated lymphocytes are cloned.

4. Activated lymphocytes endure.

FIGURE 51.9 The Clonal Selection Theory. The clonal selection theory maintains that certain lymphocytes are "selected" by binding to an antigen and that the selected cells then multiply.

CD8$^+$, based on the presence of key proteins called **CD4** or **CD8** on their plasma membranes. CD4$^+$ T cells and CD8$^+$ T cells have distinct functions in the adaptive immune response.

Antigen Presentation via MHC Proteins Recall that the receptors on T cells can bind only to epitopes that have been processed and presented on the surface of other cells. The surface proteins responsible for presenting these epitopes are called **major histocompatibility (MHC) proteins.** MHC proteins have a groove that binds to small peptide fragments that are typically 8 to 20 amino acids in length (see **FIGURE 51.10**).

MHC proteins come in two types, called **class I** and **class II MHC proteins.** Dendritic cells present peptides in both classes of MHC proteins, but the origins of the peptides differ between the two: class I MHC proteins bind antigens inside the endoplasmic reticulum (ER); class II MHC proteins bind antigens inside endosomes. The proteins that are processed and loaded onto the class I MHC are derived from the cell's cytosol, while those loaded onto the class II MHC are obtained from the external environment.

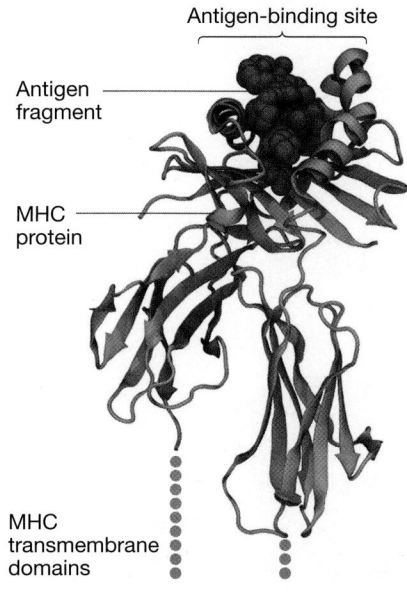

Antigen-binding site

Antigen fragment

MHC protein

MHC transmembrane domains

FIGURE 51.10 MHC Structure Promotes Peptide Binding. A peptide processed from the hemagglutinin antigen fits onto the epitope-binding groove in this class II MHC protein like a hot dog in a bun. The class I MHC has a similar epitope-binding groove.

FIGURE 51.11 shows how dendritic cells process antigens and load the peptides onto the class II MHC molecules (a similar process takes place for class I molecules in the ER).

Step 1 Dendritic cells ingest antigens at a site of infection.

Step 2 The antigen is moved from the phagosome to an endosome, where an enzyme complex breaks the protein into small peptide fragments (see Chapter 7).

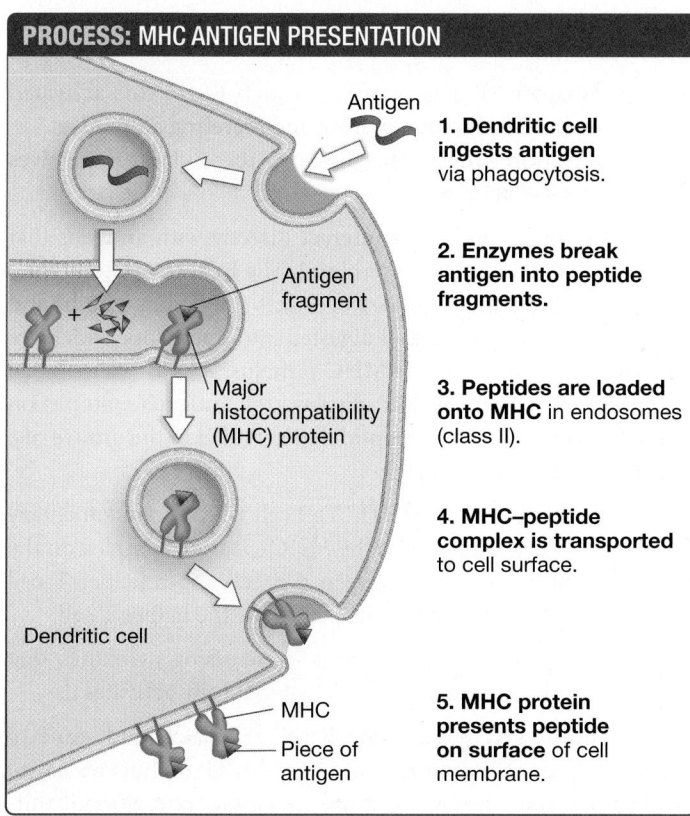

PROCESS: MHC ANTIGEN PRESENTATION

Antigen

1. **Dendritic cell ingests antigen** via phagocytosis.

Antigen fragment

Major histocompatibility (MHC) protein

2. **Enzymes break antigen into peptide fragments.**

3. **Peptides are loaded onto MHC** in endosomes (class II).

4. **MHC–peptide complex is transported** to cell surface.

Dendritic cell

MHC

Piece of antigen

5. **MHC protein presents peptide on surface** of cell membrane.

FIGURE 51.11 Dendritic Cells Present Antigens. Dendritic cells take in antigens, break them into pieces, and present the fragments in the groove of an MHC protein.

Step 3 Some of the peptides are bound to the groove of the class II MHC molecules, which were made in the rough ER and transported to the endosome.

Step 4 The MHC–peptide complexes are exported from the endosomes into vesicles for transport to the cell surface.

Step 5 The vesicles fuse with the plasma membrane, displaying the MHC–peptide complexes on the cell surface.

It's important to note that humans have several genes encoding class I and class II MHC proteins. As a result, you can produce several distinct proteins of each class that vary in the type of peptide that is presented. In addition, the MHC genes are among the most polymorphic of any genes known—meaning that many different alleles exist in the population (see Chapter 14). Because so many distinct alleles exist, a wide array of peptides can be bound and presented—allowing these cells to activate a response to many different pathogens.

How Are T Cells Activated by Antigen-Presenting Cells? FIGURE 51.12 illustrates what happens when T cells recognize the MHC–peptide complexes on a dendritic cell. $CD4^+$ T cells interact with class II MHC–bound epitopes on dendritic cells; $CD8^+$ T cells interact with class I MHC–bound epitopes.

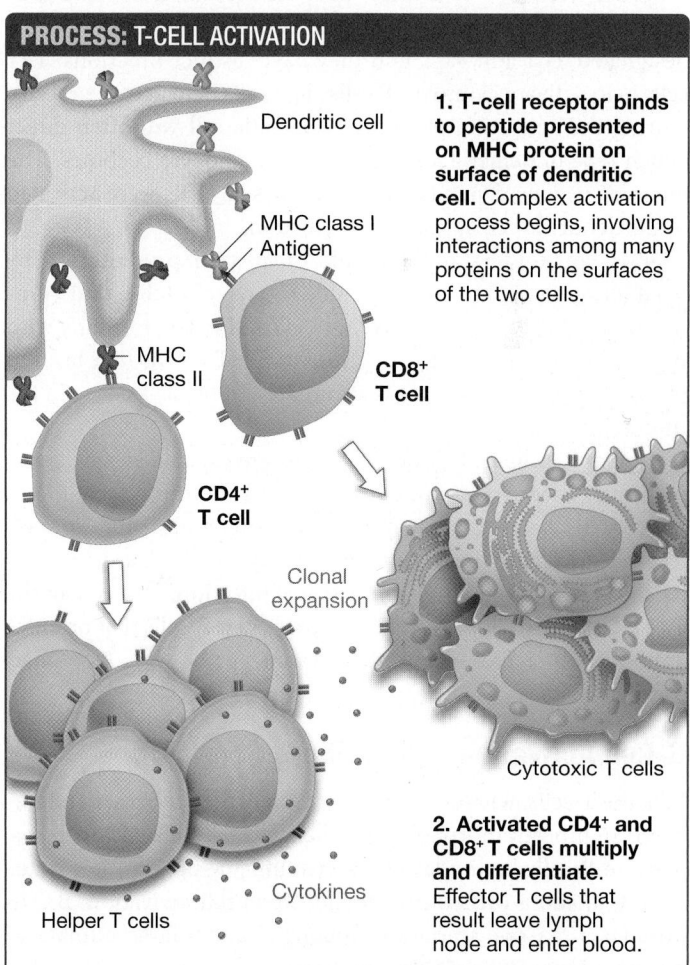

PROCESS: T-CELL ACTIVATION

Dendritic cell

MHC class I Antigen

MHC class II

$CD8^+$ T cell

$CD4^+$ T cell

1. **T-cell receptor binds to peptide presented on MHC protein on surface of dendritic cell.** Complex activation process begins, involving interactions among many proteins on the surfaces of the two cells.

Clonal expansion

Cytotoxic T cells

Cytokines

Helper T cells

2. **Activated $CD4^+$ and $CD8^+$ T cells multiply and differentiate.** Effector T cells that result leave lymph node and enter blood.

FIGURE 51.12 T Cells Are Activated by Interacting with MHC–Peptide Complexes.

To the T cell, the antigen-presenting cell carries the message "I've found antigen—response required." If the TCR binds, the activation process starts in response to interactions with proteins on the dendritic cell. In most cases, full activation of CD8$^+$ T cells also requires interactions with cytokines produced by activated CD4$^+$ T cells.

An activated T cell divides to produce a series of genetically identical daughter cells. This event, called clonal expansion, is a crucial step in the adaptive immune response. It leads to a large population of lymphocytes capable of responding specifically to the antigen that has entered the body. During this clonal expansion, many morphological changes also occur that prepare the cells for their specific functional roles in the immune response.

Cytotoxic T Cells and Helper T Cells When activated CD8$^+$ T cells undergo clonal expansion, the daughter cells develop into **cytotoxic** ("cell-poison") **T cells,** also known as cytotoxic T lymphocytes (CTLs) or killer T cells. The adjectives cytotoxic and killer are appropriate: CD8$^+$ T cells kill cells that are infected with an intracellular pathogen.

In contrast, the daughter cells of activated CD4$^+$ lymphocytes differentiate into **helper T cells.** The adjective helper is also appropriate: Helper T cells assist with the activation of other cells in the immune response. There are two types of helper T cells, designated T$_H$1 and T$_H$2, and they have distinct functions: T$_H$1 cells help activate cytotoxic T cells; T$_H$2 cells help activate B cells. During the activation phase, the dendritic cell will often direct which type of helper T cell the CD4$^+$ lymphocyte becomes. The outcome usually depends on which types of TLRs were activated in the dendritic cell at the site of infection.

✔ If you understand the role of antigen presentation by MHC proteins, you should be able to create a table that summarizes the roles of the class I and class II MHC proteins with regard to the origin of the peptides, type of T cells that bind to them, and the activity that is stimulated in these T cells via the interaction.

Helper T cells and cytotoxic T cells are capable of participating in the adaptive immune response and are thus referred to as **effector** T cells. Following clonal expansion and maturation, effector T cells leave the lymph node through a lymphatic duct, enter the blood, and migrate to the site of infection. One role of the T$_H$2 cells is to identify and activate particular B cells that produce antibodies directed against the invading pathogens. How does this work?

B-Cell Activation and Antibody Secretion

Like the T cells, when B cells are activated, they replicate and undergo significant morphological changes. As **FIGURE 51.13a** shows, inactive B cells have a large nucleus, little cytoplasm, few mitochondria, and a ruffled membrane. Upon full activation, B cells produce a massive amount of rough ER and a large number of mitochondria (**FIGURE 51.13b**).

Recall that many of the proteins synthesized in a cell's rough ER are inserted into the plasma membrane or secreted from the

(a) Inactive lymphocyte

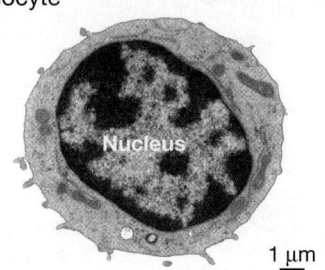

(b) Activated lymphocyte

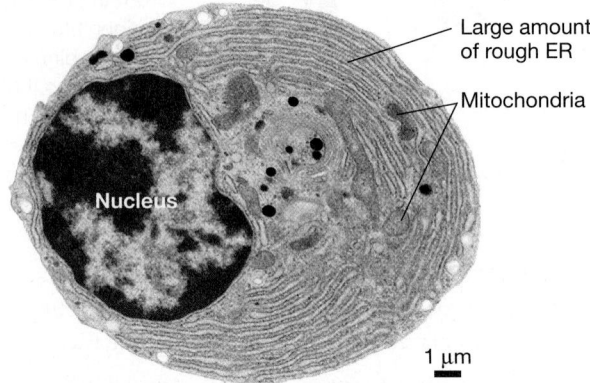

FIGURE 51.13 Lymphocytes Exist in Two States: Inactive and Activated. (a) This inactive B cell has a small amount of cytoplasm and few organelles. **(b)** This activated B cell, called a plasma cell, has extensive rough endoplasmic reticulum (ER) and many mitochondria—suggesting that a great deal of protein synthesis is taking place.

cell (see Chapter 7). The increased rough ER in this activated B cell is required for manufacturing and secreting antibodies.

The activation process that leads to these changes involves several steps (**FIGURE 51.14**):

Step 1 The BCRs on B cells interact directly with antigens that are floating free in lymph or blood. The BCR–antigen interaction results in the first part of B-cell activation. The bound antigen is internalized and digested into fragments, which are then loaded onto class II MHC proteins. As a result, a B cell that encounters its antigen displays the antigen's epitopes on its own surface, and the peptides are cradled in the groove of a class II MHC protein.

Step 2 When an *activated* CD4$^+$ T$_H$2 cell with a complementary receptor arrives, it binds to the MHC–peptide complex on the B cell. The interaction between a B cell and a helper T cell supplies activation signals that stimulate the helper T cell.

Step 3 The helper T cell responds by releasing cytokines that provide a second signal to the B cell that fully activates it.

Step 4 The now fully activated B cell replicates, and some of the daughters differentiate into effector B lymphocytes called **plasma cells.** Plasma cells produce and secrete large quantities of antibodies. Recall that antibodies are identical to B-cell receptors, except that they lack a transmembrane domain and are secreted instead of being found in the plasma membrane.

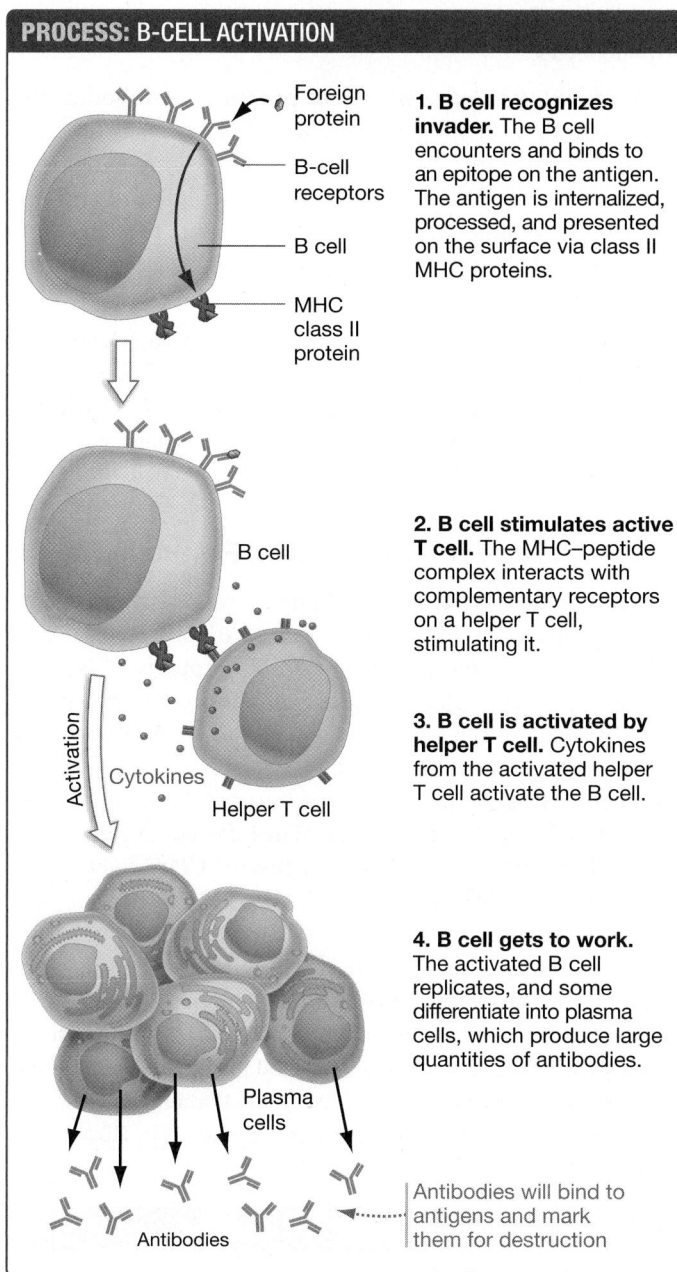

PROCESS: B-CELL ACTIVATION

1. B cell recognizes invader. The B cell encounters and binds to an epitope on the antigen. The antigen is internalized, processed, and presented on the surface via class II MHC proteins.

Foreign protein
B-cell receptors
B cell
MHC class II protein

2. B cell stimulates active T cell. The MHC–peptide complex interacts with complementary receptors on a helper T cell, stimulating it.

B cell
Activation
Cytokines
Helper T cell

3. B cell is activated by helper T cell. Cytokines from the activated helper T cell activate the B cell.

4. B cell gets to work. The activated B cell replicates, and some differentiate into plasma cells, which produce large quantities of antibodies.

Plasma cells
Antibodies

Antibodies will bind to antigens and mark them for destruction

FIGURE 51.14 B Cells Are Activated by Binding to Antigens and Interacting with Helper T Cells.

When B cells are first activated to replicate by binding to the free antigen, they migrate to a specialized area in the lymph node called the germinal center. There, the DNA sequences that code for the variable region of the immunoglobulin gene undergo rapid mutations that modify the variable region of the receptors. B cells with receptors that bind best to the free antigen live and produce daughter cells; those that bind to the antigen less effectively die.

This process of **somatic hypermutation** fine-tunes the immune response. The BCRs that result from somatic hypermutation bind to the antigen more tightly than those formed during maturation in the bone marrow.

Once the effector B cells and T cells have been activated, the adaptive immune response is in full swing. Cytotoxic and helper T cells move into the site of infection, and antibodies specific to the invading pathogen begin to circulate in the blood and lymph. The adaptive immune system has recognized the invaders and initiated its response.

Now let's look at what happens once the secreted antibodies and activated cells are recruited to the site of infection. For pathogens, the results are usually devastating.

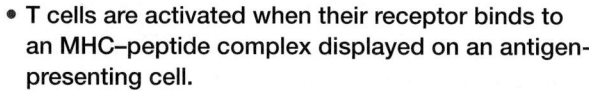

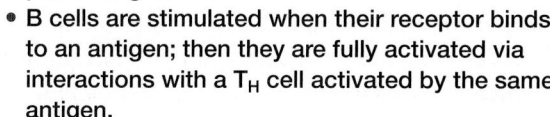

If you understand that . . .

- T cells are activated when their receptor binds to an MHC–peptide complex displayed on an antigen-presenting cell.
- B cells are stimulated when their receptor binds to an antigen; then they are fully activated via interactions with a T_H cell activated by the same antigen.
- Activated lymphocytes undergo dramatic changes in morphology and divide rapidly.

✔ **You should be able to . . .**

Generate a hypothesis to explain the observation that humans who are heterozygous for the genes encoding MHC proteins tend to be healthier than individuals who are homozygous for these genes.

Answers are available in Appendix A.

51.4 Adaptive Immunity: Response and Memory

In combination with the leukocytes involved in the innate immune system, the cells of the adaptive immune system are almost always successful in eliminating threats from bacteria, parasites, fungi, and viruses.

The array of mechanisms used by the adaptive immune system to dispose of foreign invaders are broadly grouped into two responses:

1. The **cell-mediated response** is promoted by T_H1 cells and involves cytotoxic T lymphocytes (activated $CD8^+$ cells) among others. This response primarily takes place via cell–cell contact.

2. The **humoral response** is promoted by T_H2 cells and involves the production of antibodies and other proteins secreted into the blood and lymph. (The Latin root *humor* means fluid.)

To understand how the adaptive immune system actually eliminates pathogens, let's examine how these two responses deal with invaders that remain extracellular versus those that reside within cells.

How Are Extracellular Pathogens Eliminated?

During the innate immune response to an infection, macrophages and dendritic cells at the site phagocytize some of the invaders. In addition to killing the foreign cells, these leukocytes process and present antigens via the MHC class II proteins. Macrophages display the processed epitopes on their surfaces at the site of infection, while dendritic cells move to the lymph nodes to present epitopes of the antigens. If the epitopes are recognized by helper T cells, the now activated T cells move into the site of infection.

If an activated T_H1 cell binds to the class II MHC proteins presented by these antigen-laden macrophages, two things happen. First, the phagocytic activity of the macrophages is enhanced. Second, the T_H1 cells secrete cytokines that recruit additional phagocytic cells to the site—increasing the inflammatory response.

In addition, antibodies from plasma cells begin attaching to extracellular bacteria, fungi, viruses, and other foreign material. These bound antibodies interfere with the infection in four ways (**FIGURE 51.15**):

1. *Opsonization ("preparation for eating")* Antibodies from plasma cells begin coating pathogens at the infection site. Pathogens that are tagged with antibodies are readily destroyed by phagocytes.

2. *Neutralization* Coated pathogens are blocked from interacting with—and thus infecting—host cells. Their participation in the infection is neutralized.

3. *Agglutination ("gluing together")* In many cases, antibodies cause the clumping of antigens, including cells and viruses, via a process called **agglutination.** Each antibody has at least two binding sites (pentameric IgM antibodies have 10), so a single antibody can bind epitopes on foreign cells and cross-link them. Clumped cells and viruses are not

able to infect the cells of the body and are easy targets for phagocytes.

4. *Co-stimulation of complement proteins* Antibodies that are bound to pathogens also stimulate a lethal group of proteins called the **complement system.** Complement proteins circulate in the bloodstream and assemble at antigen–antibody complexes. When complement proteins activate, they punch deadly holes in the plasma membranes of pathogens.

Within a few days, this combination of killing mechanisms—armies of phagocytic cells and complement proteins that home in on antibody-tagged material—usually eliminates all of the extracellular pathogens. But what about those pathogens that reside within the cells of the body?

How Are Intracellular Pathogens Eliminated?

Virtually all nucleated cells in the body express class I MHC proteins on their cell surface. Recall that these MHC proteins display peptides that were processed from cytosolic proteins. This means that if a cell were infected, peptides from the foreign proteins expressed inside the cell would be loaded into some of the MHC proteins and presented to circulating T cells.

Cells that display the antigens of intracellular pathogens are effectively waving a flag that says, "I'm infected. If you destroy me, you'll destroy the infection." As activated CD8+ cells migrate into the area, those that recognize the class I MHC–peptide complex are stimulated to respond to the signal.

After cytotoxic T cells have recognized their target, they form a tight attachment with the cell to direct the secretion of molecules from the T cell to the target cell's surface without disturbing adjacent cells. Some of these secreted products assemble into pores in the target cell's plasma membrane that allow other molecules from the T cell to pass through. As shown in **FIGURE 51.16**,

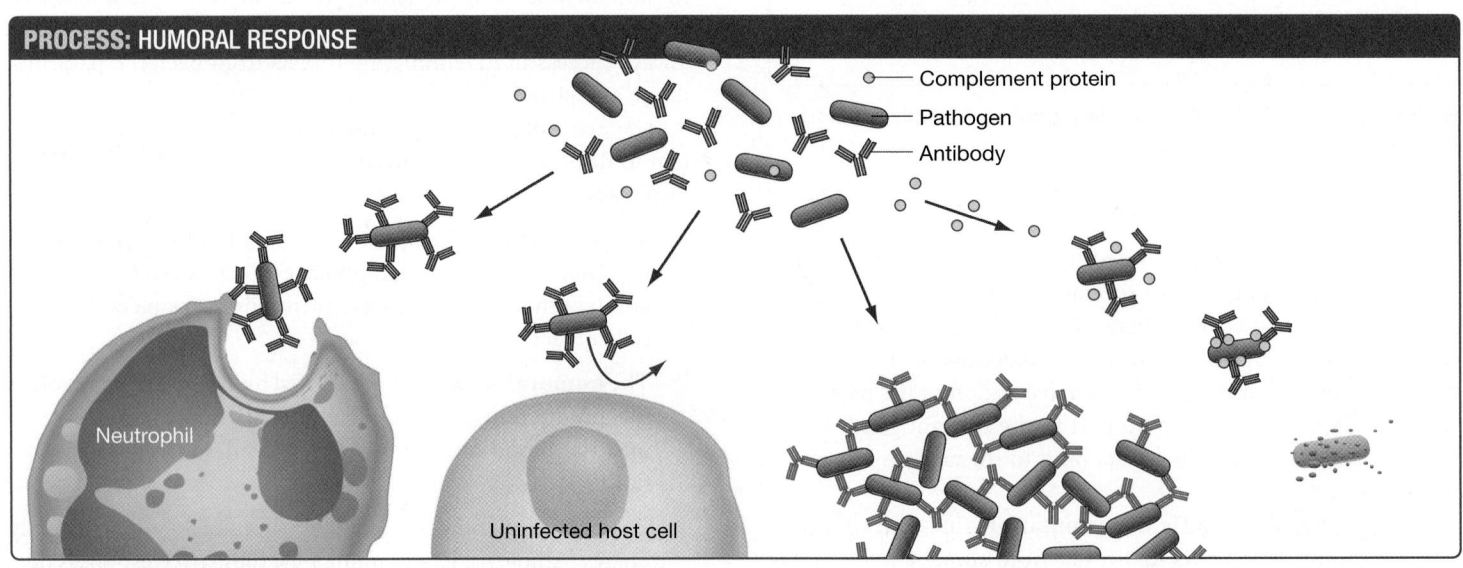

PROCESS: HUMORAL RESPONSE

Complement protein
Pathogen
Antibody

Neutrophil

Uninfected host cell

| Opsonization | Neutralization | Agglutination | Co-stimulation of complement proteins |

FIGURE 51.15 The Humoral Response Eliminates Extracellular Pathogens.

these molecules pass directly into the cytoplasm of the target cell and activate a signaling cascade that causes the cell to self-destruct via **apoptosis** (see Chapter 22). The result of apoptosis is the death and fragmentation of a cell into smaller vesicles, called apoptotic bodies, which are ingested by phagocytes like macrophages.

Once the cytotoxic T cell has delivered this signal, often referred to as the "kiss of death," the T cell releases the dying cell and seeks out another infected cell to kill. Over time, all of the infected cells are eliminated.

If the pathogen is able to replicate only inside host cells, as is the case with viruses, this cell-mediated response limits the spread of the infection by preventing the production of new generations of pathogens.

Why Does the Immune System Reject Foreign Tissues and Organs?

The innate and adaptive immune responses have devastating effects on invading pathogens. Unfortunately, they are equally deadly in response to tissues or organs that are introduced into a patient to heal a wound or cure a disease.

Consider the problems that can arise with blood transfusions. You might recall that certain individuals have red blood cells with membrane glycoproteins called A and B (see Chapter 14). These molecules act as antigens if they are introduced into a person whose own blood cells lack those glycoproteins.

For example, if you have type A blood, it means that your blood cells have the A glycoprotein. If your blood is transfused into a person who lacks the A glycoprotein—meaning someone who has type B or type O blood—the recipient's immune system will recognize the A glycoprotein as an antigen and mount a devastating response against it. For a blood transfusion to be successful, the recipient may have to receive blood that lacks the A and B glycoproteins entirely (type O) or that contains the same "antigens" found in his or her own blood.

Similar problems arise in organ transplants, except that the molecules directing the rejection of these tissues are the class I MHC proteins. The adaptive system mounts a response against the foreign MHC proteins, and the innate response is activated based on the *absence* of "self MHC" signals. If transplanted cells do not display the same class I MHC proteins as the host, components of the innate immune system kill them.

To prevent strong immune reactions against transplanted organs or tissues, physicians seek donors who have MHC proteins that are extremely similar to those of the recipient, as is often seen in siblings. Even with close relatives, however, molecular differences will exist between the donor and recipient. Thus, physicians must also treat the recipient with drugs that suppress the immune response.

Thanks to steady improvements in drug development and in systems for matching MHC types between donors and recipients, the success rate for organ transplants has improved dramatically in recent years.

As the blood transfusion and organ transplant examples show, the immune system rejects foreign cells because they either contain nonself molecules or lack molecules that serve as indicators of self. To your immune system, a mismatched blood transfusion or an organ transplant is indistinguishable from a massive influx of bacteria, viruses, or other foreign invaders.

Responding to Future Infections: Immunological Memory

In addition to producing the cells that implement the humoral and cell-mediated responses, activated B cells and T cells produce specialized daughter cells called memory cells. **Memory cells** do not participate in the initial adaptive response, or **primary immune response.** Instead, they provide a surveillance

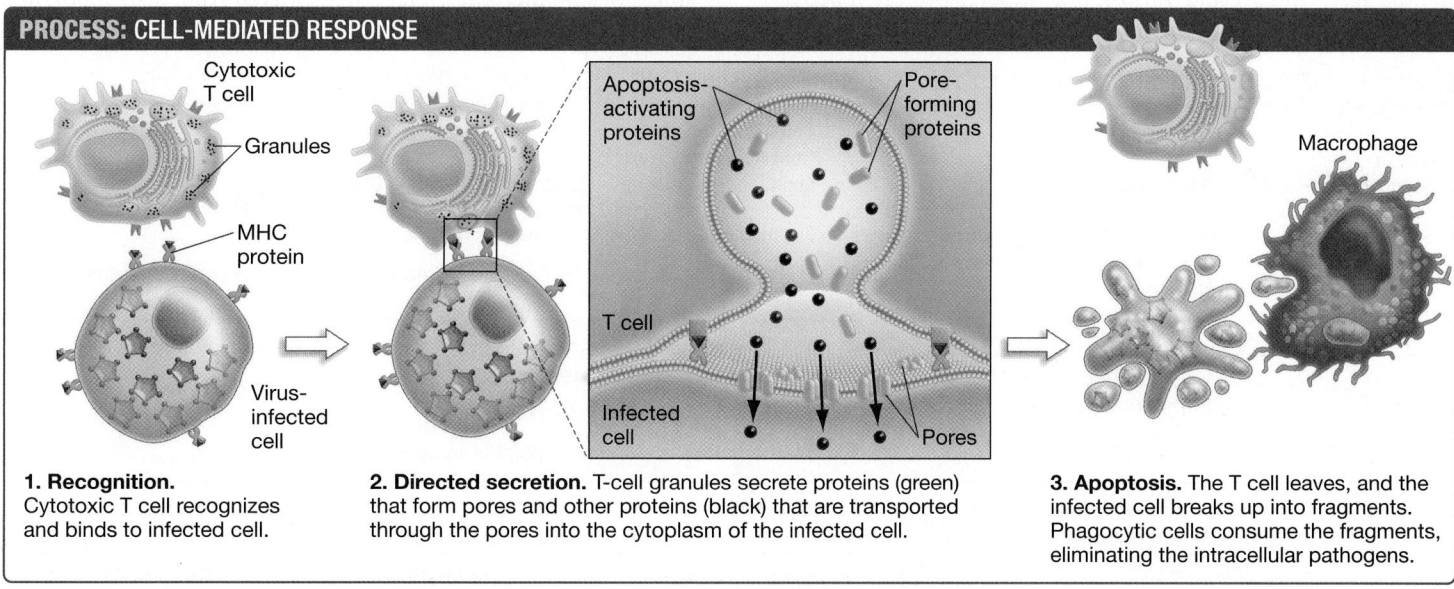

PROCESS: CELL-MEDIATED RESPONSE

Cytotoxic T cell

Granules

MHC protein

Virus-infected cell

Apoptosis-activating proteins

Pore-forming proteins

T cell

Infected cell

Pores

Macrophage

1. Recognition. Cytotoxic T cell recognizes and binds to infected cell.

2. Directed secretion. T-cell granules secrete proteins (green) that form pores and other proteins (black) that are transported through the pores into the cytoplasm of the infected cell.

3. Apoptosis. The T cell leaves, and the infected cell breaks up into fragments. Phagocytic cells consume the fragments, eliminating the intracellular pathogens.

FIGURE 51.16 The Cell-Mediated Response Eliminates Intracellular Pathogens.

service after the original infection has been cleared. Memory cells remain in the spleen and lymph nodes for years or decades, ready to provide a rapid response should an infection with the same antigen recur.

The production of memory cells is a hallmark of the vertebrate immune response. It occurs only to a limited degree, if at all, in invertebrates. How do memory cells protect an individual from future infections?

The Secondary Response Is Strong and Fast   If the same antigen enters the body a second time, memory cells recognize certain epitopes of the antigen and trigger a second adaptive response, or **secondary immune response. FIGURE 51.17** compares the rate of antibody production during the first and second exposures to a virus. The launching of a secondary immune response by means of memory cells is known as **immunological memory.**

The secondary immune response is faster and more efficient than the primary response. It is faster because the presence of an expanded pool of memory T and B cells increases the likelihood that lymphocytes with the correct antigen-specific receptors will find the antigen and activate quickly. It is more efficient because some of the memory B cells pass through another round of somatic hypermutation, the same process that occurred at the start of the primary immune response (see Section 51.3). Memory B cells with receptors that bind best to the antigen's epitope live and produce daughter cells.

To review immunological memory, starting with how B cells and T cells are activated, expanded, and differentiated into the various effector cells, study **TABLE 51.4.**

Vaccination Leads to Immunological Memory   The production and effectiveness of memory cells explain phenomena that have been observed throughout recorded history. For example, when plague struck Athens in 430 B.C., the historian Thucydides noted that only people who had recovered from the illness could nurse the sick, because they did not become ill a second time. In the middle ages, Chinese and Turkish practitioners protected people

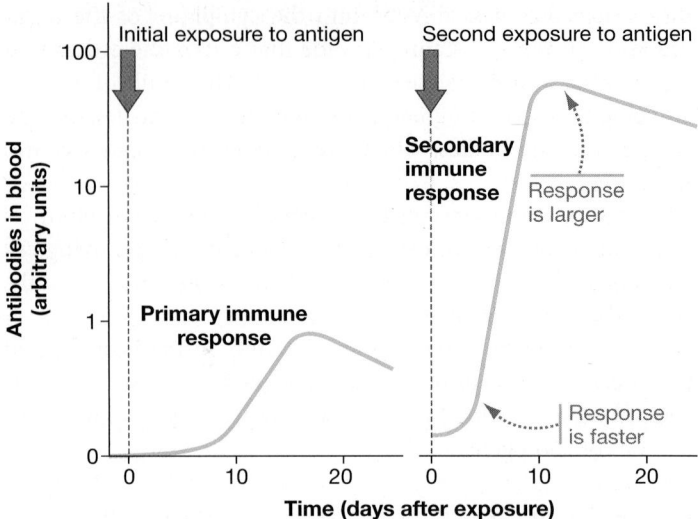

FIGURE 51.17 The Secondary Immune Response Is Faster and Stronger than the Primary Response. The graphs represent results that are commonly observed when biologists inject the same antigen into a mouse at two different times and measure the concentration of antibody in the blood (plotted on logarithmic scale). The change in antibody concentration over time indicates the speed and strength of the response to the antigen.

✔**EXERCISE** Imagine you injected a new antigen at the same time you began the second exposure to the original antigen. Draw a line on the graph to show what you predict would be the response to this new antigen in terms of antibody concentration.

from smallpox by intentionally exposing them to the dried crusts of smallpox pustules taken from infected individuals. These are examples of **immunization**—the conferring of immunity to a particular disease.

In the late 1700s Edward Jenner refined this immunization technique. In Jenner's day, milkmaids were considered pretty because their faces were not pockmarked with scars from smallpox infections. Jenner knew that cows suffered from a smallpox-like

SUMMARY TABLE 51.4 Activation and Function of Adaptive Immune System Cells

Type of Lymphocyte	Method of Activation	Cells That Result from Activation and Clonal Expansion	Function of Resulting Cells
B cell	Receptor binds to free antigen, then class II MHC–peptide complex interacts with TCR of T_H2 cell	Plasma cells	Secrete antibodies
		Memory B cells	Participate in secondary response (secrete antibodies)
CD4$^+$ T cell	Receptor binds to class II MHC–peptide complex on dendritic cell or other antigen-presenting cell	T_H1 helper T cells	Activate cytotoxic T cells; regulate inflammatory response
		T_H2 helper T cells	Activate B cells
		Memory T cells	Participate in secondary response
CD8$^+$ T cell	Receptor binds to class I MHC–peptide complex on dendritic cell or other antigen-presenting cell; T cell is then stimulated by cytokines of T_H1 cells	Cytotoxic T cells	Kill infected host cells that present complementary class I MHC–peptide complex
		Memory T cells	Participate in secondary response

disease called cowpox. He reasoned that milkmaids became immune to smallpox because they had been exposed to cowpox while milking cows.

To test this hypothesis, Jenner inoculated a boy with fluid from a cowpox pustule. Later he inoculated the same child with fluid from a smallpox pustule. That is, he first inoculated the boy with cowpox pathogens and then with smallpox pathogens.

As predicted, the boy did not contract smallpox. Jenner's technique was named **vaccination.** (The Latin root *vacca* means cow.) Vaccination is the introduction of a **vaccine,** a preparation containing antigens from a weakened or altered pathogen, to prime the body's immune system so it fights later infections effectively.

The data in Figure 51.17 explain why vaccination is an effective defense against certain pathogens—it speeds up the body's response to an infection. The antigens used in a vaccine are usually components of a pathogen's exterior, because antibodies can readily attach to them.

Consider the three general types of vaccines against viruses:

1. *Subunit vaccines* consist of isolated viral proteins: familiar examples include vaccines against hepatitis B and influenza.

2. *Inactivated viruses* have been damaged by chemical treatments—often exposure to formaldehyde—or exposure to ultraviolet light. They do not cause infections, but are antigenic. If you have been vaccinated for hepatitis A or polio, you may have received an inactivated virus.

3. *Attenuated viruses* are also called "live" virus vaccines because they consist of complete virus particles. Researchers make these viruses harmless by culturing them on cells from species other than the normal host. In adapting to growth on the atypical cells, the viruses usually lose the ability to grow

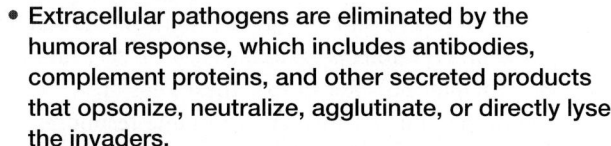

check your understanding

If you understand that . . .

C
Y
U

* Extracellular pathogens are eliminated by the humoral response, which includes antibodies, complement proteins, and other secreted products that opsonize, neutralize, agglutinate, or directly lyse the invaders.
* Intracellular pathogens are eliminated by activated cytotoxic T cells that recognize epitopes displayed on infected cells and kill them before the pathogens inside can replicate.
* Immunological memory provides a rapid, efficient secondary response against future infections with the same pathogen.

✓ **You should be able to . . .**

Compare and contrast inactivated and attenuated vaccines, and predict which would result in a more effective cell-mediated immune response.

Answers are available in Appendix A.

rapidly in their normal host cells. The smallpox and measles vaccines consist of attenuated viruses.

After vaccination, the body mounts a primary immune response that produces memory cells. If a second infection occurs later, these memory cells respond quickly and eliminate the threat before illness appears. Vaccinations function like fire drills or earthquake preparedness exercises—they prepare the immune system for a specific threat.

Unfortunately, viruses such as the human immunodeficiency virus (HIV) mutate so rapidly that traditional vaccines have been ineffective for immunization. Memory cells generated from HIV vaccination are unlikely to recognize the changed epitopes in the rapidly evolving strains. Currently, the only "cure" for HIV infection is prevention.

51.5 What Happens When the Immune System *Doesn't* Work Correctly?

The vertebrate immune system is a marvel of adaptation. A healthy immune system is able to defeat the vast majority of infections without medical intervention and with little impact on the body. The immune system is a formidable threat to pathogens, but if its response is dysfunctional, it can also be a liability to the health and even survival of the animal.

To appreciate this point, let's first examine what happens when the immune system activates inappropriate responses and then look at the consequences when it fails to respond.

Allergies

For some people, walking through a field of ragweed or petting a cat is a prescription for developing a runny nose, itchy eyes, and labored breathing. For other people, being stung by a bee or eating a peanut is a life-threatening experience. These are examples of allergies—immune responses that are overreacting to foreign substances.

An **allergy,** or allergic reaction, is an abnormal response to an antigen. For reasons that are still not clear, certain people produce the IgE class of antibodies in response to specific molecules found in cat dander, nuts, plant pollen, or other products. Molecules that trigger this response are called **allergens** instead of antigens.

Allergic reactions are considered inappropriate responses because the normal role of IgE antibodies is to defend against infections by parasitic worms. Recent research has also shown that in areas where infections with intestinal worms are common, allergies are almost unknown. These data suggest that allergies don't occur if IgEs are expressed normally—that is, in response to worm infections.

The presence of IgE antibodies in an allergic response is important because it triggers a series of events known as the **hypersensitive reaction.** When a susceptible person is first exposed to an allergen, the IgE antibodies that are produced bind to receptors on mast cells and certain other leukocytes. This binding occurs at the constant region of the IgE heavy chains—the "tail" of the antibody. Once this binding event occurs, the cells (and the person) are said to be sensitized.

If the person is later exposed to the same allergen, the variable regions of these IgE antibodies bind to the allergen molecules and send a signal to the sensitized cells to rapidly secrete large quantities of histamine, cytokines, and other compounds. In response to these molecules, blood vessels dilate and become more permeable, smooth-muscle cells contract, and other cells secrete mucus. Common manifestations of these hypersensitive reactions are hay fever, hives, and asthma.

In more severe responses, blood vessels can dilate to the point where blood pressure plummets, oxygen delivery to the brain is reduced dramatically, and the person loses consciousness. In addition, severe smooth-muscle contractions in the digestive system and respiratory system can induce vomiting, diarrhea, and complete constriction of the airway passages. This combination of events, known as anaphylactic shock, is lethal unless the person receives immediate medical attention.

What can be done about allergies? The number one approach is to avoid the allergen. If exposure does occur, drugs called antihistamines can sometimes reduce symptoms by blocking histamine receptors. In some cases, corticosteroids applied to the skin can reduce swelling. Longer-term solutions are still in the experimental stage, however. Developing effective treatments for allergies is an important research frontier in biomedicine.

Although allergic reactions are considered abnormal, they are still directed against substances that are foreign to the body. What happens when the immune system turns against the body's own tissues?

Autoimmune Diseases

An immune response directed against molecules or cells that normally exist in the host is known as **autoimmunity.** Autoimmune reactions often result in disease due to the destruction of the body's own cells and structures.

- Multiple sclerosis (MS) results from the production of cytotoxic T cells that attack the myelin sheath of nerve fibers (see Chapter 46). Because damage to myelin reduces the efficiency of nerve signaling, coordination problems result.

- Rheumatoid arthritis develops when self-reactive antibodies alter the lining of joints, causing painful inflammation.

- Type 1 diabetes mellitus occurs when cytotoxic T cells attack and kill insulin-secreting cells in the pancreas, resulting in a lack of insulin and inability to regulate blood glucose levels (see Chapter 44).

The cause and treatment of immune-mediated diseases are active areas of research. The mechanisms behind these diseases are still poorly understood, but an interesting connection is emerging. Beginning in the late 1960s, several studies have shown that the risk of developing allergies and autoimmune diseases increases in individuals who live in homes with a high level of sanitation.

Results from these studies have led to the development of the **hygiene hypothesis,** which claims that autoimmune and allergic responses arise in individuals who, because of hygienic practices, have experienced less exposure to pathogens and parasites. Recall that allergies are virtually absent in areas where intestinal worms are common. This effect is also seen for autoimmune diseases, which do not involve the IgE-based hypersensitivity reaction.

This correlation between hygiene and immune disorders points to the close ties that have been forged during the coevolution of the immune system and the invaders it defends us against. Low infant mortality and other benefits of sanitary lifestyles appear to come at the cost of inappropriate immune responses.

But, what happens when the immune system fails to respond at all?

Immunodeficiency Diseases

Children who are born with a genetic disorder called severe combined immunodeficiency (SCID) lack a normal immune system and are unable to fight off infections. The results are devastating. Without a functional immune system, the number of potential infectious agents grows to include "opportunistic" pathogens that would normally be incapable of causing disease.

You may recall that one form of SCID, designated SCID-X1, is caused by mutations in a gene on the X chromosome (see Chapter 20). The gene codes for a receptor protein necessary for the development of T cells. In another type of SCID, called Omenn syndrome, there is a genetic defect in one of the enzymes responsible for DNA recombination in maturing lymphocytes. As a result, cells are unable to generate normal T-cell and B-cell receptors.

In both cases, the adaptive immune response is badly impaired. Afflicted individuals suffer debilitating illness from infections that other children fight off easily. If not given a bone-marrow transplant or kept in a completely sterile environment, children with SCID typically will die before they are 2 years old.

Similarly, people who are infected with the **human immunodeficiency virus (HIV)** suffer from a progressive failure of the immune system. HIV infects and kills $CD4^+$ T cells and macrophages (see Chapter 36). Recall that these cells are required for both the humoral and cell-mediated immune responses. As the infection continues, populations of $CD4^+$ T cells gradually decline to a point where the immune system can no longer mount an effective response to infection.

Eventually, HIV-infected people develop **acquired immune deficiency syndrome (AIDS).** They succumb to illnesses that physicians almost never see in people with healthy immune systems.

Over the past three decades, the incidence of HIV infections has skyrocketed throughout the world, devastating the immune systems of millions of people. This, coupled with the marked increase in allergies and autoimmune diseases in the developed world, illustrates the importance of continued research into the many nuances of the immune system.

If you understand . . .

51.1   Innate Immunity

- Animals protect themselves from infection by establishing barriers that prevent the entry of pathogens.

- The innate immune system provides a rapid, generic response against broad classes of pathogens based on the recognition of pathogen-associated molecules by pattern-recognition receptors.

- During the innate immune system's inflammatory response, leukocytes respond to an infection by phagocytizing foreign material and cells and releasing cytokines that stimulate neighboring cells to respond to the infection and recruit more leukocytes into the area.

✔ You should be able to explain the role of Toll-like receptors in the innate immune response.

51.2   Adaptive Immunity: Recognition

- The vertebrate immune system possesses T cells and B cells that recognize specific epitopes on antigens via unique receptors found on the T- and B-cell surfaces.

- The receptor proteins expressed on the surfaces of T cells and B cells are generated through DNA recombination—a rearrangement of gene segments.

- Because every receptor that results from DNA recombination is slightly different, the immune system is able to recognize and respond to an almost limitless array of antigens.

✔ You should be able to predict how the BCR locus in the genomic DNA of immature B cells is different from that in mature B cells.

51.3   Adaptive Immunity: Activation

- The clonal selection theory explains how the most appropriate cells for controlling an infection are selected and expanded to mount an effective immune response.

- T cells are activated when their receptors recognize epitopes displayed in MHC proteins on antigen-presenting cells. Full activation and differentiation of T cells often requires secondary signals from cytokines.

- B cells are activated when their receptors bind to free antigens. B cells differentiate into plasma cells and memory cells only when fully activated by helper T cells.

✔ You should be able to explain how antigen-presenting cells of the innate immune response are required for the differentiation of B cells into plasma cells.

51.4   Adaptive Immunity: Response and Memory

- Extracellular pathogens are eliminated via phagocytosis and soluble components secreted by the humoral response, which include antibodies. Antibodies contribute to the response by opsonizing, neutralizing, or agglutinating pathogens, or by co-stimulating the lytic activity of the complement proteins.

- Intracellular pathogens are eliminated via the cell-mediated response, in which infected host cells are induced to self-destruct by cytotoxic T cells.

- The immune system rejects blood and tissue transplants based on the presence of nonself molecules, often in the form of carbohydrates or MHC proteins, or on a lack of self recognition.

- The production of memory lymphocytes, via a primary infection or vaccination, allows the immune system to respond rapidly and effectively to future infections by the same pathogen.

✔ You should be able to explain how viruses are eliminated by both humoral and cell-mediated immune responses.

51.5   What Happens When the Immune System *Doesn't* Work Correctly?

- Allergies are abnormal, hypersensitive responses that are generated against allergens, which are antigens that are not associated with infection.

- Autoimmune diseases occur when self-reactive lymphocytes elicit an immune response against the body's own tissues.

- Immunodeficiency diseases arise from genetic mutations or viral infections that disrupt the function of key components in the immune system.

✔ You should be able to predict the consequences of producing IgE against a self molecule.

(MB) MasteringBiology

1. MasteringBiology Assignments

Tutorials and Activities Acquired Immunity; Adaptive Immune Response; Immune Responses; Inflammatory Response

Questions Reading Quizzes, Blue-Thread Questions, Test Bank

2. eText Read your book online, search, take notes, highlight text, and more.

3. The Study Area Practice Test, Cumulative Test, BioFlix® 3-D Animations, Videos, Activities, Audio Glossary, Word Study Tools, Art

You should be able to . . .

1. What is the primary difference between the innate and adaptive responses?
   a. The innate response does not distinguish between pathogens, while the adaptive response does.
   b. Only the innate response is triggered by antigens.
   c. The adaptive response generates immunological memory and is more specific than the innate response.
   d. The innate response does not kill cells; the adaptive response does.

2. The overall role of the inflammatory response is to
   a. contain and eliminate foreign cells and material at the site of infection.
   b. increase heat at the site of infection to activate enzymes used in the immune response.
   c. produce antibodies that bind to invading cells and eliminate them.
   d. increase blood flow at the site of the wound to flush out invading pathogen.

3. What is the difference between an epitope and an antigen?
   a. An epitope is any foreign substance; an antigen is a foreign protein.
   b. An epitope is the part of an antigen where an antibody or lymphocyte receptor binds.
   c. An antigen is the part of an epitope where an antibody or lymphocyte receptor binds.
   d. Antigens are recognized by B cells and antibodies; epitopes are recognized by T cells.

4. How do B-cell receptors and antibodies differ?

5. What is one of the differences between $CD4^+$ and $CD8^+$ T cells?
   a. $CD4^+$ cells are immature and $CD8^+$ cells are mature.
   b. $CD4^+$ cells are activated and $CD8^+$ cells are not.
   c. $CD4^+$ cells interact with class II MHC proteins, and $CD8^+$ cells interact with class I MHC proteins.
   d. $CD4^+$ cells activate cell-mediated responses, and $CD8^+$ cells activate humoral responses.

6. What is the hygiene hypothesis?

7. Explain how B cells are fully activated to differentiate into plasma cells.

8. Compare and contrast how B-cell receptors and T-cell receptors recognize antigens.

9. What do vaccines need to contain in order to be effective? Why don't we have vaccines for HIV?

10. Why is clonal selection necessary for the adaptive immune response but not the innate immune response?
    a. The adaptive response uses receptors to recognize pathogens, and the innate response does not.
    b. There is more diversity in the receptors in the adaptive response compared to the innate response.
    c. Cells in the innate response do not require activation, and those in the adaptive response do.
    d. This process is used for targeting pathogens, and the innate response is used only to stop blood flow from the wound.

11. Compare and contrast the interaction between pathogens and the pattern-recognition receptors on leukocytes versus that between antigens and BCRs or TCRs.

12. Explain how DNA recombination leads to the production of almost limitless numbers of different B-cell receptors, T-cell receptors, and antibodies.

13. Which of the following outcomes would be expected if there were a mutation that inhibited somatic hypermutation?
    a. The diversity of pattern-recognition receptors would be significantly lowered.
    b. B and T lymphocytes would not be able to produce receptors that recognize antigens.
    c. The adaptive immune system would not be able to respond to pathogens.
    d. The secondary immune response to a repeat infection would produce the same antibodies as those made in the primary immune response.

14. Predict how self-reactive B cells are identified and eliminated during maturation.

15. It seems astonishing that the immune system can produce antibodies to compounds that were recently synthesized for the first time in the lab. Given that viruses and other pathogens evolve rapidly, why would natural selection favor individuals having genes that made extensive antibody diversity possible?

16. During World War II, physicians discovered that if they removed undamaged skin from a patient and grafted it onto the site of a burn on the same patient, the tissue healed well. But if the grafted tissue came from a different individual, the skin graft was rejected. What is responsible for this difference?

52 An Introduction to Ecology

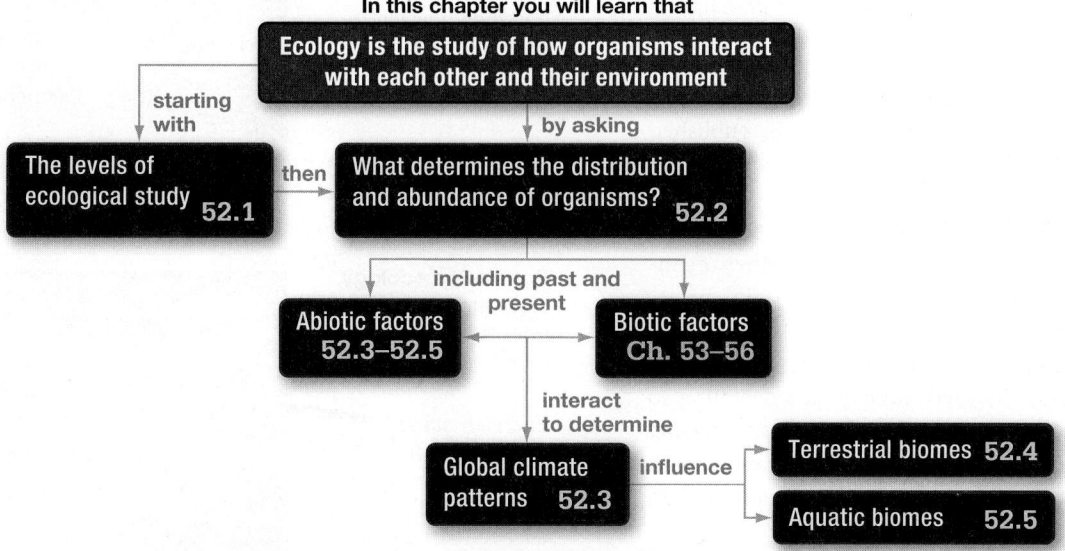

In this chapter you will learn that

Ecology is the study of how organisms interact with each other and their environment

starting with

The levels of ecological study **52.1**

then → *by asking*

What determines the distribution and abundance of organisms? **52.2**

including past and present

Abiotic factors **52.3–52.5** ← → Biotic factors **Ch. 53–56**

interact to determine

Global climate patterns **52.3** — *influence* → Terrestrial biomes **52.4**

Aquatic biomes **52.5**

This heavily oiled bird is being rescued from the Gulf of Mexico following the Deepwater Horizon oil spill in 2010. Increasingly, ecology is the study of how humans are altering the environment in ways that make life difficult for other organisms.

BIG PICTURE

This chapter is part of the Big Picture. See how on pages 1196–1197.

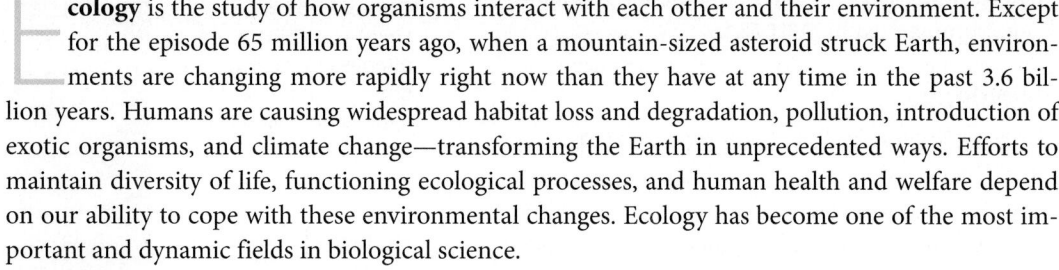

cology is the study of how organisms interact with each other and their environment. Except for the episode 65 million years ago, when a mountain-sized asteroid struck Earth, environments are changing more rapidly right now than they have at any time in the past 3.6 billion years. Humans are causing widespread habitat loss and degradation, pollution, introduction of exotic organisms, and climate change—transforming the Earth in unprecedented ways. Efforts to maintain diversity of life, functioning ecological processes, and human health and welfare depend on our ability to cope with these environmental changes. Ecology has become one of the most important and dynamic fields in biological science.

Ecology is also among the most synthetic and integrative fields in biological science. In this unit, you'll be applying what you've learned about biochemistry, genetics, development, physiology,

✔ When you see this checkmark, stop and test yourself. Answers are available in Appendix A.

evolution, and the diversity of life. Ecologists work in subdisciplines ranging from ecological developmental biology to physiological ecology and conservation genetics.

Of these subdisciplines, the intersection of ecology and evolution has been appreciated the longest. The diversity of life (Unit 6) is the result of the evolutionary processes (Unit 5) as they played out in a changing ecological context over 3.6 billion years. Evolution and ecology are intertwined and codependent. It was no coincidence that the German biologist Ernst Haeckel first defined "oekologie" less than a decade after Charles Darwin published *On the Origin of Species by Means of Natural Selection* in 1859 (see Chapter 25).

Ecology's primary goal is to understand the distribution and abundance of organisms. How do interactions with other organisms and the physical environment dictate why certain species live where they do, and how many individuals live there? Some biologists ask why orangutans are restricted to forests in Borneo and Northern Sumatra and why their numbers are declining so rapidly. Other biologists create mathematical models to predict how quickly the human immunodeficiency virus will increase in India, or how long it will take for the current human population of 7 billion to double.

Distribution and abundance are core questions in today's world. Land-use changes and climate change are having dramatic effects on species' distributions, and expanding human populations are causing the abundance of many species to decline, sometimes to zero. Let's delve in.

52.1 Levels of Ecological Study

To understand why organisms live where they do and in what numbers, biologists consider ecology at several levels of analysis. This is a common strategy in biological science. Cell biologists study how cells work, at different levels of organization—from individual molecules to cellular structures, cells, and multicellular organisms. Physiologists analyze processes at the level of ions and molecules as well as whole cells, tissues and organs, and complete systems. You can review the levels of organization in The Big Picture of Doing Biology (pages 16–17).

In ecology, researchers work at five main levels: (1) organisms, (2) populations, (3) communities, (4) ecosystems, and (5) the biosphere. The levels aren't rigid or exclusive; researchers routinely draw on multiple levels—including lower levels—to explore a particular question. Let's consider how these levels apply to sockeye salmon, summarized in **TABLE 52.1**.

Organismal Ecology

At the organismal level, researchers explore the morphological, physiological, and behavioral adaptations that allow individuals to live in a particular area (see Chapter 53). For example, after spending four or five years feeding and growing in the ocean, sockeye salmon travel hundreds or thousands of kilometers to return to the stream where they hatched. Females create nests in

Level	Sockeye Salmon Example
Organismal ecology How do individuals interact with each other and their physical environment? Salmon migrate from saltwater to freshwater environments to breed	
Population ecology How and why does population size change over space and time? Each female salmon produces thousands of eggs. On average, only a few offspring will survive to return to the same stream to breed	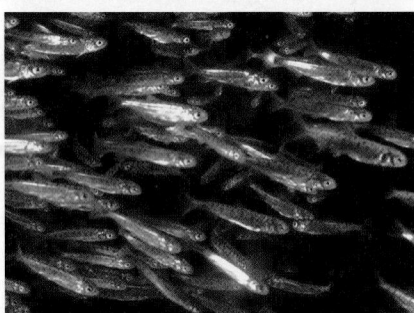
Community ecology How do species interact, and what are the consequences? Salmon are prey as well as predators	
Ecosystem ecology How do energy and nutrients cycle through the local environment? Salmon die and then decompose, releasing nutrients that are used by other organisms	
Global ecology How is the biosphere affected by global changes in nutrient cycling and climate? Worldwide populations of salmon are affected by climate change	

the gravel stream bottom and lay eggs. Nearby males compete for the chance to fertilize eggs as they are laid. When breeding is finished, all the adults die.

Biologists want to know how these individuals interact with their physical surroundings and with other organisms in and around the stream. How do individuals cope with the transition from living in salt water to living in freshwater? Which females get the best nesting sites and lay the most eggs, and which males are most successful in fertilizing eggs?

Population Ecology

A **population** is a group of individuals of the same species that lives in the same area at the same time. When biologists study population ecology, they focus on how the number and distribution of individuals in a population change over time (see Chapter 54).

For example, researchers use mathematical models to predict the future of salmon populations. Many of these populations have declined due to overharvesting and the loss of habitat quality caused by pollution and dam construction. If the factors that affect population size can be described accurately enough, mathematical models can assess the impact of proposed dams, changes in weather patterns, altered harvest levels, or specific types of protection efforts.

Community Ecology

A biological **community** consists of populations of different species that interact with each other within a particular area. Community ecologists ask questions about the nature of the interactions between species and the consequences of those interactions (see Chapter 55). Research might concentrate on predation, parasitism, and competition or explore how communities respond to fires, floods, and other disturbances.

As an example, consider the interactions among salmon and other species in the marine and stream communities where they live. When they are at sea, salmon eat smaller fish and are themselves hunted and eaten by orcas, sea lions, humans, and other mammals; when they return to freshwater to breed, they are preyed on by bears and bald eagles. In both marine and freshwater habitats, salmon are subject to parasitism and disease. They are also heavily affected by disturbances—particularly changes in their food supply and the quality of their breeding streams.

Ecosystem Ecology

Ecosystem ecology is an extension of organismal, population, and community ecology. An **ecosystem** consists of all the organisms in a particular region along with nonliving components. These physical and chemical, or **abiotic** (literally, "not-living"), components include air, water, and the nonliving parts of soil. At the ecosystem level, biologists study how nutrients and energy move among organisms and through the surrounding atmosphere and soil or water (see Chapter 56).

Salmon are interesting to study at the ecosystem level because they link marine and freshwater ecosystems. They harvest nutrients in the ocean and then, when they migrate, die, and decompose, they transport those molecules to streams and streamside forests. In this way, salmon transport chemical energy and nutrients from one habitat to another.

Global Ecology

Biologists define the **biosphere** as a thin zone surrounding the Earth where all life exists—the sum of all terrestrial and aquatic ecosystems. It extends to about 5 km below the land surface and 10 km down in the deepest trenches of the sea to over 10 km up into the atmosphere. This may sound like a thick layer, but it represents only 0.002 percent of the diameter of the Earth—and most of life occupies a much thinner layer, comparable to the layer of varnish on a typical classroom globe.

The young field of global ecology is growing rapidly as ecologists scramble to quantify the effects of human impacts on the biosphere. Humans are modifying landscapes and releasing massive amounts of energy and nutrients to the biosphere, causing changes that transcend individual ecosystems.

Since the biosphere is practically a closed system, the actions of people in one part of the world may alter distant ecosystems, creating difficult education and public policy challenges. For example, salmon are sensitive to changes in water temperature. Who is accountable for the impacts of climate change on salmon?

Conservation Biology Applies All Levels of Ecological Study

The five levels of ecological study are synthesized and applied in conservation biology. **Conservation biology** is the effort to study, preserve, and restore threatened genetic diversity in populations, species diversity in communities, and ecosystem function (see Chapter 57).

Conservation biologists are like physicians. But instead of prescribing treatments for sick people, they prescribe remedies for threatened environments to preserve biodiversity, clean air, pure water, and productive soils.

Each chapter in this unit focuses on a different level of ecological study, ending with the study of biodiversity and conservation. Turn to the Big Picture of Ecology (pages 1196–1197) to visualize how the different levels of ecology interact. In this introductory chapter, let's take a global look at factors that determine the distribution and abundance of organisms. In particular, what abiotic factors set the stage for life on Earth?

52.2 What Determines the Distribution and Abundance of Organisms?

When you look around at the plants and animals in your neighborhood, you can tell right away that you are not located deep in the Amazon rain forest—no scarlet macaws, jaguars, anacondas, blue morpho butterflies, or giant kapok trees out there. Different organisms live in different places. Why?

Understanding why organisms are found where they are is one of ecology's most basic tasks. The study of how organisms are distributed geographically is called **biogeography.** Let's consider how both abiotic and biotic factors determine the distribution and abundance of organisms.

Abiotic Factors

The most fundamental observation about the **range,** or geographic distribution, of organisms is simple: No one species can survive the full array of environmental conditions present on Earth.

Temperature is particularly important in determining ranges because temperature has a big impact on the physiology of organisms, and organisms are limited in their ability to regulate their own temperatures (see Chapter 42). *Thermus aquaticus* can thrive in hot springs with temperatures above 70°C because its enzymes do not denature at near-boiling temperatures. But the same cells would die instantly if they were transplanted to the frigid waters near polar ice—even though vast numbers of other microbes are present. The cold-water species have enzymes that work efficiently at near-freezing temperatures.

No organism thrives in both hot springs and cold ocean water. The reason is fitness trade-offs (introduced in Chapter 25). No enzyme can function well in both extremely high temperatures and extremely low temperatures. Because of fitness trade-offs, organisms tend to be adapted to a limited set of physical conditions, or abiotic factors—for example, to a particular temperature and moisture regime on land, or a particular salinity, water depth, and movement regime in water. Every organism has a specific range of tolerance of abiotic conditions.

Adaptation to a limited set of abiotic conditions is just the start of the story explaining why a particular species lives where it does. For example, the temperature and moisture regimes of southern California, in North America, are almost identical to those found in the Mediterranean region of Europe and northern Africa. But the species present at the sites are almost completely different. Why?

Biotic Factors

The ability of a species to persist in a given area is often limited by **biotic** ("living") factors—meaning interactions with other organisms (see Chapters 54 and 55). As an example, consider the distributions of hermit warblers and Townsend's warblers along the Pacific Coast of North America (introduced in Chapter 27).

- Both hermit warblers and Townsend's warblers live in evergreen forests.

- Experiments have shown that male Townsend's warblers directly attack male hermit warblers and evict them from breeding territories.

- The geographic range of Townsend's warblers has been expanding steadily at the expense of hermit warblers.

These observations support the hypothesis that a biotic factor—competition with another species—is limiting the range of hermit warblers. Competition is not the only biotic factor to affect species range, however.

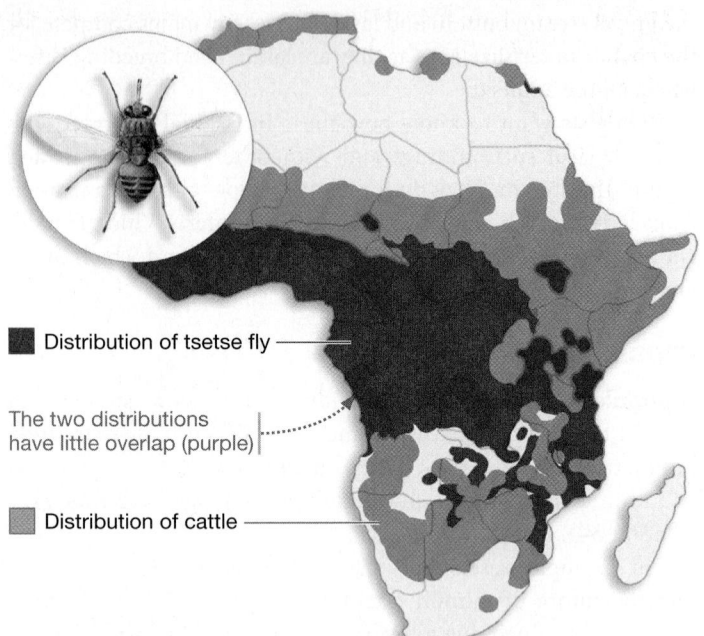

FIGURE 52.1 In Africa, Tsetse Flies Limit the Distribution of Cattle. Tsetse flies carry a disease that is fatal to cattle.
DATA: Food and Agriculture Organization of the United Nations

✓**QUESTION** What do you think would happen to the range of cattle in Africa if the tsetse fly were eradicated?

- Yucca moths lay their eggs only in the flowers of yucca plants. As a result, this species does not exist outside the range of yucca plants.

- In Africa, the range of domestic cattle is limited by the distribution of tsetse flies (**FIGURE 52.1**). Tsetse flies transmit a parasite that causes the disease trypanosomiasis, which is fatal in cattle (and can be fatal in humans unless treated).

- Most of the songbirds native to Hawaii are limited to alpine habitats—above the range of the mosquitoes that transmit avian malaria.

These examples hint at a key insight: The distribution of organisms is determined not only by conditions present today but also by events in the past. Let's take a closer look at the importance of history.

History Matters: Past Abiotic and Biotic Factors Influence Present Patterns

The abiotic and biotic factors that influence the distribution of organisms are dynamic—they are constantly changing, in part because the Earth itself is changing. The landforms and oceans that may appear static now have been in constant flux for the entire history of life—mountains rise and fall, islands and lakes form and disappear, and whole continents crash together and break apart (see Chapter 28). Associated fluctuations in climate cause glaciers and ice caps to form and melt and sea levels to rise and fall.

These events have an important impact on **dispersal,** the movement of individuals from their place of origin to the location where they live and breed as adults. For example, the formation 3 million years ago of the land bridge from North America to South America, today known as the Isthmus of Panama (see Figure 27.5), created opportunities for dispersal from each continent to the other, introducing new biotic interactions on both continents. Meanwhile, this same geological event separated the Pacific Ocean from the Caribbean Sea, creating barriers to dispersal for marine species.

Alfred Russel Wallace was among the first to document this phenomenon. In addition to codiscovering evolution by natural selection (see Chapter 25), Wallace founded the study of biogeography.

The Wallace Line: Barriers to Dispersal   While working in the Malay Archipelago, Wallace realized that the plants and animals native to the more northern and western islands were radically different from the species found on the more southern and eastern islands. Sumatra, to the northwest, has tigers, rhinos, and other species with close relatives in southern Asia. New Guinea, to the southeast, has tree kangaroos and other species with close relatives on Australia. But both islands are tropical wet forests.

This biogeographical demarcation, now known as the **Wallace line,** separates species with Asian and Australian affinities (**FIGURE 52.2**). It exists because a deep trench in the ocean maintained a water barrier to dispersal, even when ocean levels dropped during the most recent glaciations. As a result, landforms on either side of the line remained isolated at a time when most of the other islands became connected.

The Influence of Humans   The events Wallace studied occurred before the dispersal of modern humans around the globe from Africa, beginning at least 70,000 years ago (see Chapter 35). Humans have influenced the distribution of species for thousands of years by hunting, clearing forests, and physically moving organisms around. Humans have transported thousands of plants, birds, insects, and other species across physical barriers to new locations—sometimes purposefully and sometimes by accident.

If an **exotic species**—one that is not native—is introduced into a new area, spreads rapidly, and competes successfully with native species, it is said to be an **invasive species.** In North America alone, dozens of invasive species have had devastating effects on native plants and animals. Examples include kudzu (**FIGURE 52.3**, see page 1064), purple loosestrife, reed canary grass, garlic mustard, Russian thistle (also known as tumbleweed), European starlings, African honeybees, red imported fire ants, Argentine ants, and zebra mussels. Similar lists could be compiled for other regions of the world.

The microorganisms that humans introduce to new areas can also dramatically affect species distributions by causing disease. A microscopic fungus introduced to North America in the 1920s swept across the continent, wiping out most Dutch elm trees in its path in a matter of decades.

Humans continue to influence the distribution of species by affecting both biotic and abiotic factors.

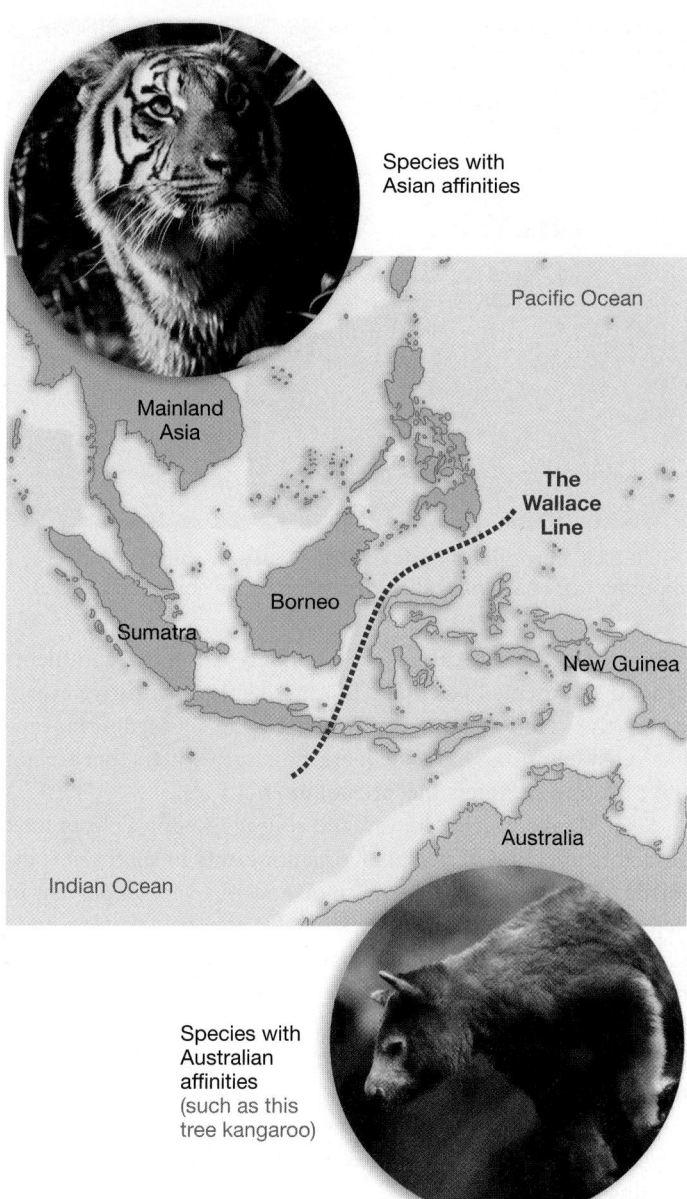

Species with Asian affinities

Pacific Ocean

Mainland Asia

The Wallace Line

Borneo

Sumatra

New Guinea

Australia

Indian Ocean

Species with Australian affinities (such as this tree kangaroo)

FIGURE 52.2 The "Wallace Line" Divides Organisms with Asian versus Australian Affinities.

Biotic and Abiotic Factors Interact

In many cases, biotic and abiotic factors interact to determine the distribution and abundance of organisms. How can this interaction be measured?

Consider the case of the Argentine ant, *Linepithema humile*, native to South America and rapidly expanding around the world. These small ants form unusually large supercolonies because they mix freely with other Argentine ants while fighting aggressively against native ants. This invasion creates a huge pest-control problem in homes, businesses, and farms, and it also has grave ecological implications. For example, horned lizards are in decline in southern California due to the disappearance of the native ants that they are specialized to eat.

FIGURE 52.3 Introduced Species Can Cause Problems. Kudzu, imported from Asia, has taken over parts of the southeastern U.S.

Argentine ants are dispersed by humans to different continents in food shipments like coffee. But these ants don't thrive everywhere they land—they are most successful in Mediterranean-like ecosystems that have adequate moisture, winters that are not too cold, and summers that are not too hot.

Sean Menke, David Holway, and colleagues took a closer look at factors affecting dispersal of Argentine ants in southern California. They observed that soil moisture was very important in determining the species' range. They also observed that the invasive ants tend to wipe out native ants. But does competition with native ants *interact* with abiotic factors in some way, or do abiotic factors trump biotic factors in this invasive species' success?

To find out, the researchers set up 28 study plots along a contact zone between Argentine ants and native ants at the local edge of the Argentine ant's range—an edge that had been stable for over 10 years. What was keeping the invasive ants from extending their range?

The study plots were randomly assigned to four groups varying in soil moisture and competition, as shown in the "Experimental Setup" section of **FIGURE 52.4**. By counting the number of Argentine ants found in ant baits in each plot at the start of the experiment and again after 12 weeks, the researchers could measure the relative importance of soil moisture and competition to the dispersal of the invasive ants.

As the graph in the "Results" section of Figure 52.4 shows, the Argentine ants did not spread into dry study plots (red bars), even when native ants were not present. They did, however, spread into irrigated plots (blue bars), even when native ants were present—but the native ants slowed them down.

The take-home message is that the interaction of biotic and abiotic factors slowed the spread of invasive ants more than the abiotic factors alone. Argentine ants are at a competitive advantage in moist soils, but at a disadvantage compared to native ants in dry soils. This information can help scientists to predict the dispersal pattern of Argentine ants and design strategies to limit the spread of these invaders.

RESEARCH

QUESTION: Do abiotic and biotic factors interact to determine the distribution of a species?

HYPOTHESIS: Abiotic and biotic factors interact to determine the distribution of invasive Argentine ants.

NULL HYPOTHESIS: Abiotic conditions alone determine the distribution of invasive Argentine ants.

EXPERIMENTAL SETUP:

1. Set up 28 plots (10 × 10 m) in the contact zone between Argentine ants and native ants. Measure invasive ant activity in each plot using ant baits.

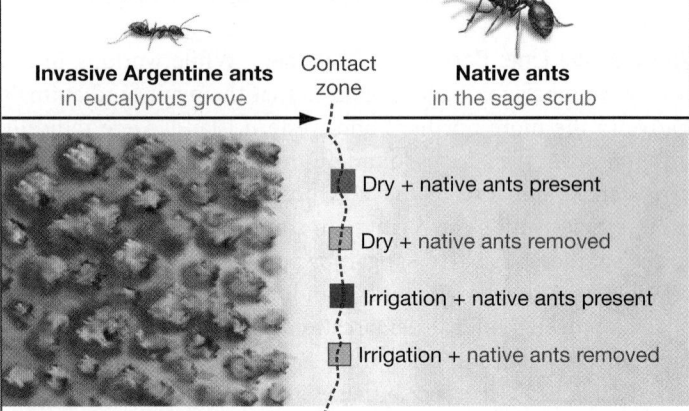

Invasive Argentine ants
in eucalyptus grove

Contact zone

Native ants
in the sage scrub

- Dry + native ants present
- Dry + native ants removed
- Irrigation + native ants present
- Irrigation + native ants removed

2. Vary soil moisture and presence of native ants as shown above.

3. After 12 weeks, remeasure invasive ant activity in each plot.

PREDICTION: Invasive ants will displace native ants only in irrigated plots—more so in plots where native ants were removed.

PREDICTION OF NULL HYPOTHESIS: Invasive ants will displace native ants only in irrigated plots—with no difference in plots with native ants present and native ants removed.

RESULTS:

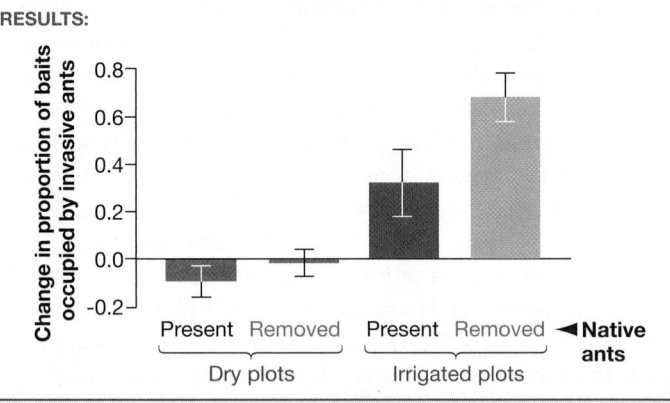

CONCLUSION: Abiotic factors (soil moisture) and biotic factors (competition with native ants) interact in determining the distribution of Argentine ants.

FIGURE 52.4 Experimental Test of the Relative Importance of Biotic and Abiotic Factors in Limiting the Dispersal of a Species. SOURCE: Menke, S. B., et al. 2007. Biotic and abiotic controls of Argentine ant invasion success at local and landscape scales. *Ecology* 88(12): 3164–3173.

✔QUANTITATIVE The bars in the graphs represent the means of several plots. How many replicate plots were used for each treatment? Why were replicates important for this experiment?

The range of every species on Earth is limited by a combination of abiotic and biotic factors that occurred in the past, and that occur in the present. The range of a particular species depends on the capacity to survive climatic conditions, the ability to find food and avoid being eaten, and the ability to disperse.

If a modest amount of irrigation can have a large effect on the dispersal of an ant species, you can imagine how abiotic factors like precipitation can have far-reaching consequences on ecology on a global scale. Let's take a closer look at climate as a major abiotic determinant of the distribution of life on the planet.

52.3 Climate Patterns

Species are adapted to the abiotic conditions present where they live. An array of abiotic factors affect organisms, including the chemistry of water in aquatic habitats, the nature of soils in terrestrial habitats, and **climate**—the prevailing, *long-term* weather conditions found in an area. **Weather** consists of the specific *short-term* atmospheric conditions of temperature, precipitation, sunlight, and wind.

Although your local meteorologist might not always make a good prediction on the night's snowfall, climate occurs in relatively predictable global patterns.

Why Are the Tropics Wet?

One of the most striking climate patterns on Earth involves precipitation. When average annual rainfall is mapped for regions around the globe, it is clear that areas along the equator receive the most moisture. In contrast, locations about 30 degrees latitude north and south of the equator are among the driest on the planet. Why do these patterns occur?

Simple questions often have fascinating answers. It turns out that a major cycle in global air circulation, called a **Hadley cell**, is responsible for making the Amazon River basin wet and the Sahara Desert dry. Hadley cells are named after George Hadley, who in 1735 conceived of the idea of enormous air circulation patterns.

To understand how Hadley cells work, put your finger on the equator in **FIGURE 52.5a** and follow the arrows.

- Air that is heated by the strong sunlight along the equator expands, lowering its pressure and causing it to rise. Note that this warm air holds a great deal of moisture, because warm water molecules tend to stay in vapor form instead of condensing into droplets.

- As air rises above the equator, it radiates heat to space. It also expands into the larger volume of the upper atmosphere, which lowers its density and temperature.

- As rising air cools, its ability to hold water declines. When water vapor cools, water condenses.

The result? High levels of precipitation occur along the equator. The story is just beginning, however.

- As more air is heated along the equator, the cooler, "older" air above Earth's surface is pushed poleward.

- When the air mass has cooled enough, its density increases and it begins to sink.

(a) Circulation cells exist at the equator ...

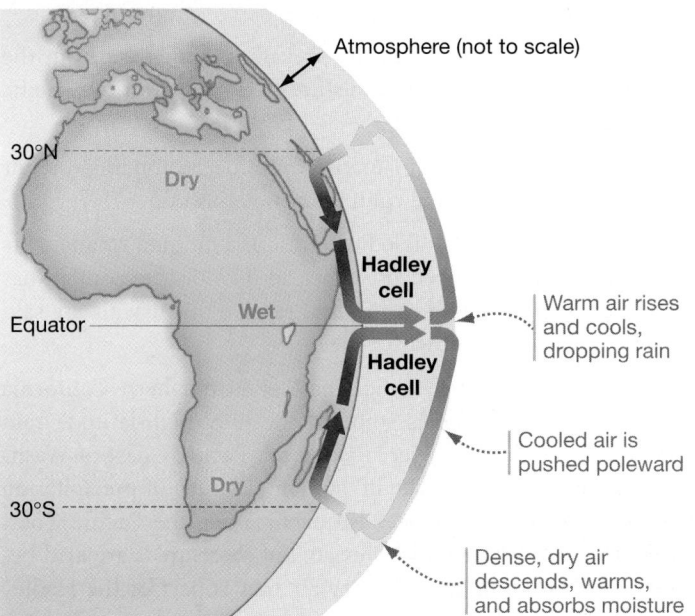

(b) ... and at higher latitudes.

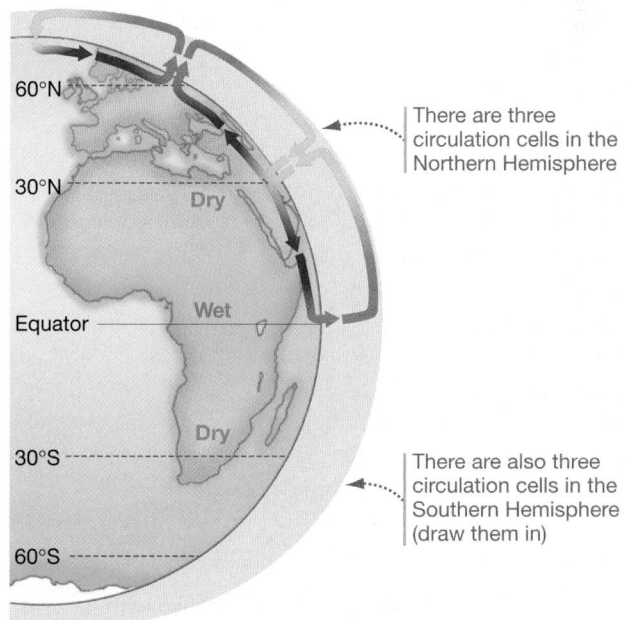

FIGURE 52.5 Global Air Circulation Patterns Affect Rainfall. Hadley cells explain why the tropics are much wetter than regions near 30° latitude north and south. Regions with rising air, such as those at the equator, tend to be wetter than regions with descending air.

- As it sinks, it absorbs more and more solar radiation reflected from Earth's surface and begins to warm while being pushed to the surface by higher-density air above it.

- As the air warms, it also gains water-holding capacity. Thus the air approaching the Earth "holds on" to its water, and little rain occurs where it returns to the surface.

The result? A band of deserts in the vicinity of 30° latitude north and south. These areas are bathed in warm, dry air.

Similar air circulation cells occur between 30° latitude and 60° latitude, and between 60° latitude and the poles, north and south (**FIGURE 52.5b**). ✔ If you understand how Hadley cells work, you should be able to add diagrams to Figure 52.5b showing the three cells that occur in the Southern Hemisphere.

Why Are the Tropics Warm and the Poles Cold?

In general, areas of the world are warm if they receive a large amount of sunlight per unit area; they are cold if they receive a small amount of sunlight per unit area. Over the course of a year, regions at or near the equator receive much more sunlight per unit area—and thus much more energy in the form of heat—than regions that are closer to the poles.

- At the equator, the Sun is often directly overhead. As a result, sunlight strikes the surface at or close to an angle of 90 degrees. At these angles, Earth receives a maximum amount of solar radiation per unit area (**FIGURE 52.6**).

- Because Earth's surface slopes away from the equator, the Sun strikes the surface at lower and lower angles moving toward the poles. When sunlight arrives at a low angle, much less energy is received per unit area.

The pattern of decreasing average temperature with increasing latitude is a result of Earth's spherical shape.

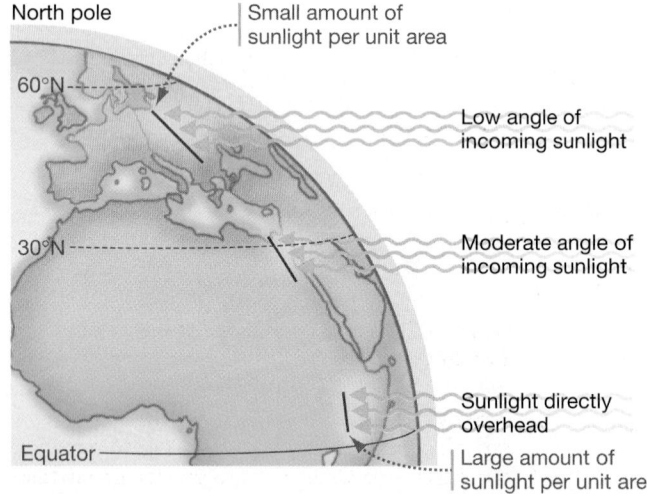

FIGURE 52.6 Solar Radiation per Unit Area Declines with Increasing Latitude. Over the course of a year, the Sun is frequently almost directly overhead at the equator. As a result, equatorial regions receive a large amount of solar radiation per unit area compared to high-latitude regions.

What Causes Seasonality in Weather?

The striking global patterns in temperature and moisture that you've just reviewed are complicated by the phenomenon of seasons. Seasons are defined as regular, annual fluctuations in temperature, precipitation, or both.

As **FIGURE 52.7** shows, seasonality occurs because Earth is tilted on its axis by 23.5 degrees. As a result of this incline, the Northern Hemisphere is tilted toward the Sun in June. In this position, it faces the Sun most directly. In June, the Northern Hemisphere presents its least-acute angle to the Sun and receives the largest amount of solar radiation per unit area.

The Southern Hemisphere, in contrast, is tilted away from the Sun in June. As a result, it presents a steep angle for incoming sunlight and receives its smallest quantity of solar radiation per unit area.

The upshot? In June it is summer in the Northern Hemisphere and winter in the Southern Hemisphere. Conversely, in December it is summer in the Southern Hemisphere and winter in the Northern Hemisphere. In March and September, the equator faces the Sun most directly. During these months, the tropics receive the most solar radiation. If Earth did not tilt on its axis, there would be no seasons.

What Regional Effects Do Mountains and Oceans Have on Climate?

The broad patterns of climate that are dictated by global heating patterns, Hadley cells, and seasonality are modified by regional effects. The most important of these are due to the presence of mountain ranges and proximity to an ocean.

FIGURE 52.8 shows how mountain ranges affect regional climate, using the Cascade Mountain range along the Pacific Coast of North America as an example.

- In this area of the world, the prevailing winds are from the west. These winds bring moisture-laden air from the Pacific Ocean onto the continent.

- As the air masses begin to rise over the mountains, the air cools and releases large volumes of water as rain.

- Once cooled air has passed the crest of a mountain range, the air is relatively dry because much of its moisture content has already been released. Areas that receive this dry air are said to be in a **rain shadow.**

The area along the Pacific Coast from northern California to southeast Alaska hosts some of the only high-latitude rain forests in the world. One area along the Pacific Coast of Washington State gets an average of 340 cm (134 in.) of precipitation each year. Mountain ranges also occur along the Pacific Coast of Southern California and Mexico, but these areas are arid because they are dominated by dry air that is part of the Hadley circulation.

Logically enough, one of the few high-latitude deserts in the world exists to the east of the Cascade Mountains. It is created by a rain shadow and averages less than 25 cm (10 in.) of moisture

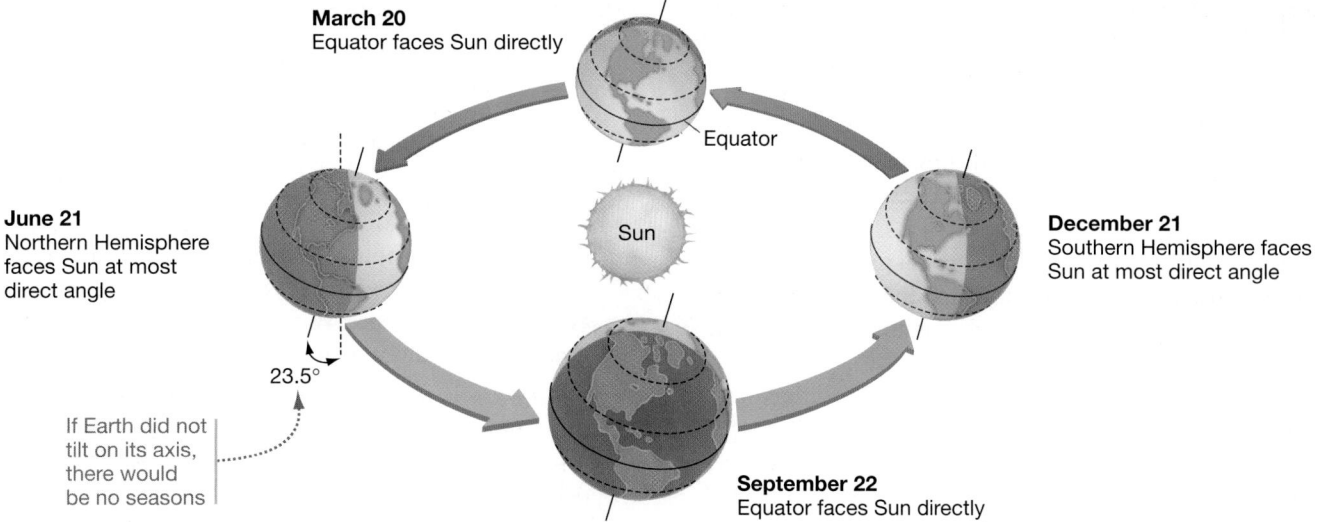

March 20
Equator faces Sun directly

Equator

June 21
Northern Hemisphere
faces Sun at most
direct angle

Sun

December 21
Southern Hemisphere faces
Sun at most direct angle

23.5°

If Earth did not
tilt on its axis,
there would
be no seasons

September 22
Equator faces Sun directly

FIGURE 52.7 Earth's Orbit and Tilt Create Seasons at High Latitudes. Seasons occur due to annual variation in the amount of solar radiation that different parts of the Earth receive.

annually. Similar rain shadows create the Great Basin desert of western North America, the Atacama Desert of South America, and the dry conditions of Asia's Tibetan Plateau.

While the presence of mountain ranges tends to produce extremes in precipitation, the presence of an ocean has a moderating influence on temperature. Water has an extremely high **specific heat,** meaning that it has a large capacity for storing heat energy (Chapter 2). Because water molecules form hydrogen bonds with each other, it takes a great deal of heat energy to boil water or melt ice—in fact, to change the temperature of water at all. The hydrogen bonds have to be broken before the water molecules themselves can begin to absorb heat and move faster, which we measure as increased temperature.

Because water has a high specific heat, it can absorb a great deal of heat from the atmosphere in summer, when the water temperature is cooler than the air temperature. As a result, the ocean moderates summer temperatures on nearby landmasses. This is why the air temperature is cooler near the ocean on a hot summer's day than it is a few miles inland. Similarly, the ocean releases heat to the atmosphere in winter, when the water temperature is warmer than the air temperature. This is why on a calm winter's day the air temperature can be warmer near the

ocean than it is a few miles inland. Because of the large capacity of water for storing heat energy, islands and coastal areas have much more moderate climates than do inland areas.

How do all these climate patterns combine to determine the distribution of organisms on Earth?

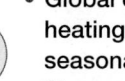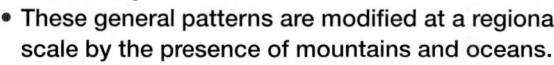

check your understanding

If you understand that . . .

- Global climate patterns are dictated by differential heating of Earth's surface, Hadley cells, and seasonality.
- These general patterns are modified at a regional scale by the presence of mountains and oceans.

✔ **You should be able to . . .**

1. Predict whether the North Pole is rainy.
2. Predict where on the globe the highest risk of skin cancer would occur and explain why.

Answers are available in Appendix A.

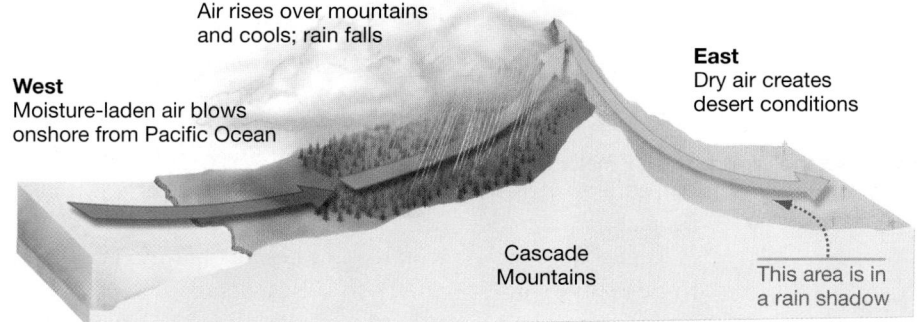

Air rises over mountains
and cools; rain falls

West
Moisture-laden air blows
onshore from Pacific Ocean

East
Dry air creates
desert conditions

Cascade
Mountains

This area is in
a rain shadow

FIGURE 52.8 Mountain Ranges Create Rain Shadows. When moisture-laden air rises over a mountain range, the air cools enough to condense water vapor. As a result, moisture falls to the ground as precipitation. On the other side of the mountain, little water is left to provide precipitation.

52.4 Types of Terrestrial Biomes

If you could walk from the equator in South America to the North Pole, you would notice startling changes in the organisms around you. Lush tropical forests with broad-leaved evergreen trees would give way to seasonally dry forests and then to deserts. The deserts would yield to the vast grasslands of central North America, which terminate at the boreal forests of the subarctic. If you pressed on, you would reach the end of the trees and the beginning of the most northerly community—the arctic tundra.

If you walked from sea level to the top of the Cascade Mountains of British Columbia, you would experience similar types of changes. Wet forests along the coast would give way to drier forests at higher elevations. Near the peaks, trees would thin and eventually yield to alpine tundras filled with hardy, low-growing plants. Broad-leaved evergreen forests, deserts, grasslands, and tundras are **biomes**: regions characterized by distinct abiotic characteristics and dominant types of vegetation.

Natural Biomes

The global distribution of the most common types of naturally occurring terrestrial biomes is shown in **FIGURE 52.9**—ignoring, for the moment, the effects of humans on the landscape. Notice that different ecosystems may belong to the same biome. For example, the Sonoran Desert in North America and the Sahara

Desert in Africa are both part of the same biome—subtropical desert—but each is a unique ecosystem with its own community of indigenous organisms.

Each of the terrestrial biomes found around the world is associated with a distinctive set of abiotic conditions determined largely by climate:

- *Temperature* is critical because the enzymes that make life possible work at optimal efficiency only in a narrow range of temperatures (see Chapter 8). Temperature also affects the availability of moisture: Water freezes at low temperatures and evaporates rapidly at high temperatures.

- *Moisture* is significant because it is required for life, and because terrestrial organisms constantly lose water to the environment through evaporation or transpiration. To stay alive, they must reduce water loss and replace lost water.

- *Sunlight* is essential because it is required for photosynthesis.

- *Wind* is important because it exacerbates the effects of temperature and moisture. Wind increases heat loss due to evaporation and convection, and it increases water loss due to evaporation and transpiration. Wind also has a direct physical impact on organisms such as birds, flying insects, and plants—it pushes them around.

Of the four components of climate, variation in temperature and moisture are far and away the most important in determining plant distribution and abundance. More specifically, the nature of the terrestrial biome that develops in a particular region

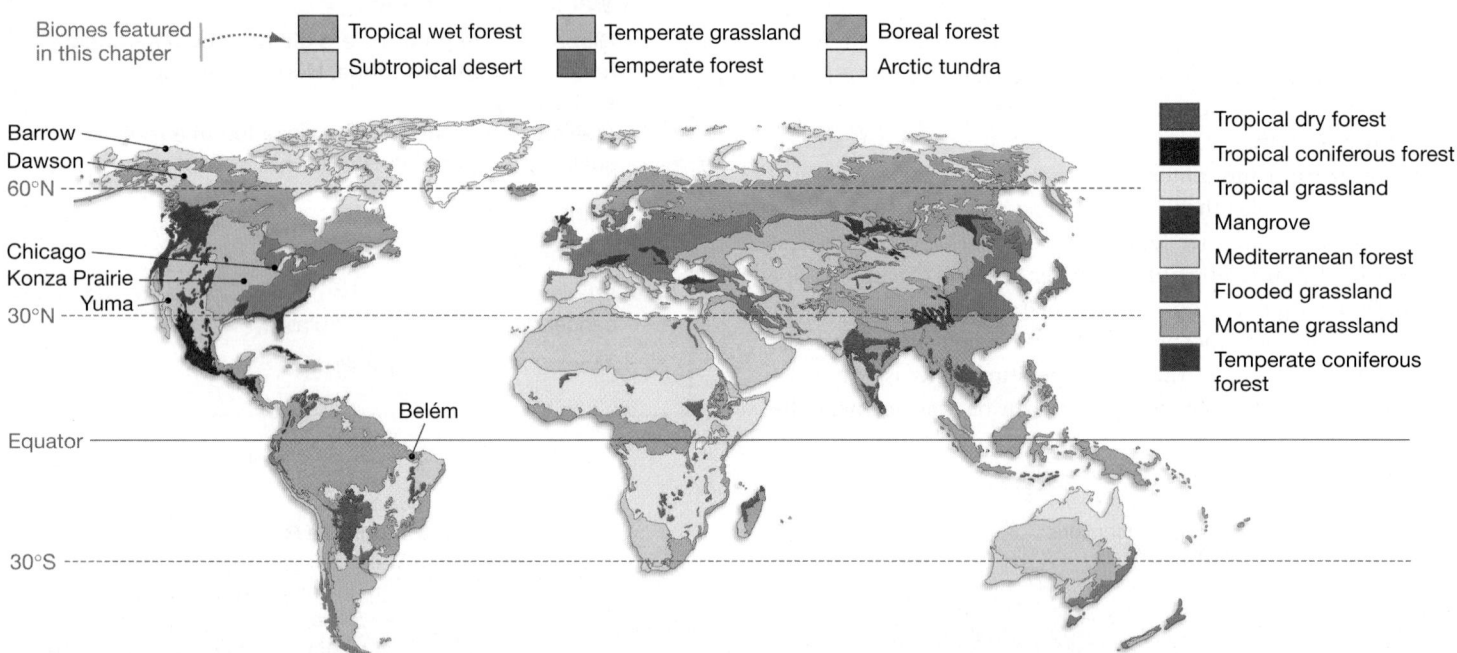

FIGURE 52.9 Distinct Terrestrial Biomes Are Found throughout the World. The six major biomes listed above the map are featured in this chapter, citing data from the labeled locations on the left.

DATA: Olson, D. M., E. Dinerstein, E. D. Wikramanayake, et al. 2001. *BioScience* 51: 933–938.

is governed by (1) average annual temperature and precipitation, and (2) annual variation in temperature and precipitation. Each biome contains species that are adapted to a particular temperature and moisture regime.

Biologists are particularly concerned with how temperature and moisture influence **net primary productivity (NPP).** NPP is defined as the total amount of carbon that is fixed per year minus the amount that is oxidized during cellular respiration. Fixed carbon that is consumed in cellular respiration provides energy for the organism but is not used for growth—that is, production of **biomass** (see Chapter 56).

NPP is crucial because it represents the organic matter that is available as food for other organisms. In terrestrial environments, NPP is often estimated by measuring **aboveground biomass**—the total mass of living plants, excluding roots.

Photosynthesis, plant growth, and NPP are maximized on land when temperatures are warm and conditions are wet. Conversely, photosynthesis cannot occur efficiently at low temperatures or under drought stress. Photosynthetic rates are maximized in warm temperatures, when enzymes work efficiently, and in humid weather, when stomata can remain open and CO_2 is readily available.

To help you understand how temperature and precipitation influence biomes, let's take a detailed look at the six terrestrial biomes that represent Earth's most extensive natural vegetation types (see boxes on pages 1070–1072). In each case, you'll be analyzing temperature and precipitation data from a specific location ranging from the wet tropics to the arctic (labeled in Figure 52.9). Data plotted in red indicate the daily average temperature in each biome, graphed throughout the year; data plotted in purple indicate the average monthly precipitation.

- Terrestrial Biomes > Tropical Wet Forest
- Terrestrial Biomes > Subtropical Deserts
- Terrestrial Biomes > Temperate Grasslands
- Terrestrial Biomes > Temperate Forests
- Terrestrial Biomes > Boreal Forests
- Terrestrial Biomes > Arctic Tundra

Anthropogenic Biomes

The map in Figure 52.9 shows the distribution of terrestrial biomes as they would occur naturally. This map is important because it represents the product of 3.6 billion years of interplay between geology, climate, evolution, and ecology.

Now for the rest of the story. The exponential increase in the human population (Chapter 54) is driving land-use changes and other impacts that are so significant that some researchers have called for a new epoch of history, the Anthropocene (see Chapter 28). Consider these data:

- More than 75 percent of Earth's ice-free land shows evidence of direct alteration by humans—for example, farming, logging, and urban development (*not* including the effects of pollution, invasive species, or climate change).

- The remaining regions, or wildlands, account for just 11 percent of terrestrial net primary productivity.

What does the distribution of biomes look like with human impact factored in? Biogeographer Erle Ellis and colleagues have pioneered efforts to map anthropogenic biomes, or "anthromes" worldwide, and with eye-opening results. **FIGURE 52.10** zooms in on one part of the world, eastern North America, to show a close-up of the distribution of anthropogenic biomes. The remaining wildlands (pale gray-green on the map) are so sparse that they are difficult to locate.

As you can see, humans are directly affecting the distribution of terrestrial ecosystems by physically changing the landscape, one forest and suburb at a time. Humans are also changing the global distribution of ecosystems in a more indirect way—by changing the global climate.

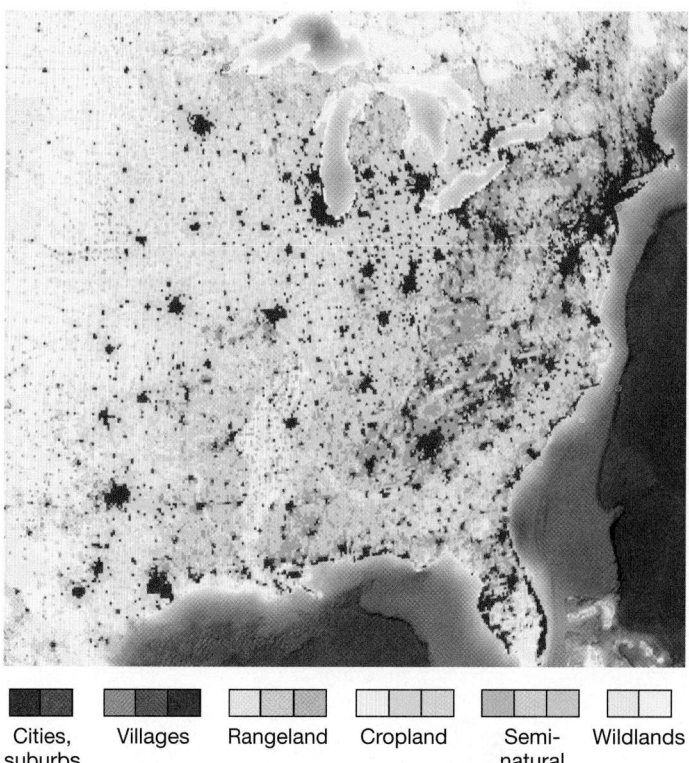

| Cities, suburbs | Villages | Rangeland | Cropland | Semi-natural | Wildlands |

FIGURE 52.10 Anthropogenic Biomes in Eastern North America. Several color subcategories have been consolidated in the key for simplicity. For example, cropland can be irrigated or rain-fed, residential, sparsely populated, or remote.

DATA: Ellis, E. C., K. Klein Goldewijk, S. Siebert, et al. 2010. *Global Ecology and Biogeography* 19: 589–606.

✔ **QUESTION** Where do the remaining wildlands occur?

Tropical wet forests—also called tropical rain forests—are found in equatorial regions around the world. Plants in this biome have broad leaves and are evergreen. Favorable year-round growing conditions produce riotous growth, leading to extremely high productivity, aboveground biomass, and species diversity. It is not unusual to find over 200 tree species in a single study plot. The diversity of plant sizes and growth forms in wet forest communities produces extraordinary structural diversity, presenting a wide array of habitat types for animals.
✔You should be able to explain why vines and epiphytes increase the productivity of tropical wet forests.

Temperature Compared to other biomes, tropical wet forests show almost no seasonal variation in temperature (**FIGURE 52.11**). This is important because temperatures are high enough to support growth throughout the year.

Precipitation Even the driest month of the year (November in Belém, Brazil) receives over 5 cm (2 in.) of rainfall—considerably more than the *annual* rainfall of many deserts.

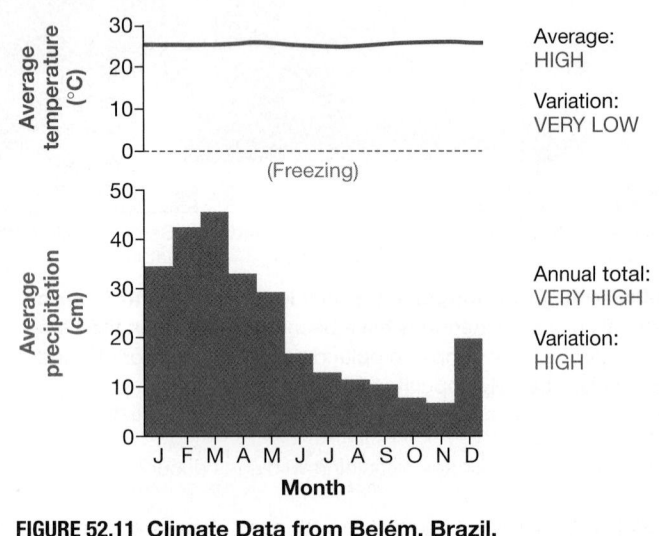

FIGURE 52.11 Climate Data from Belém, Brazil.

Subtropical deserts are found throughout the world at 30 degrees latitude north and south—a pattern that is explained in Section 52.3. Desert species adapt to the extreme temperatures and aridity by growing at a low rate year-round, or breaking dormancy and growing rapidly in response to any rainfall. Cacti can grow year-round because they have no leaves, or small leaves modified as spines; a thick, waxy coating; and the CAM pathway for photosynthesis (see Chapter 10). The average productivity of desert communities is extremely low.
✔You should be able to explain why most desert plant species have a small surface area.

Temperature Mean monthly temperatures in subtropical deserts vary more than in tropical wet climates (**FIGURE 52.12**). Temperatures fall below freezing in some subtropical deserts.

Precipitation The most striking feature of the subtropical desert climate is low precipitation; average annual precipitation in Yuma, Arizona, is just 7.5 cm (3 in.).

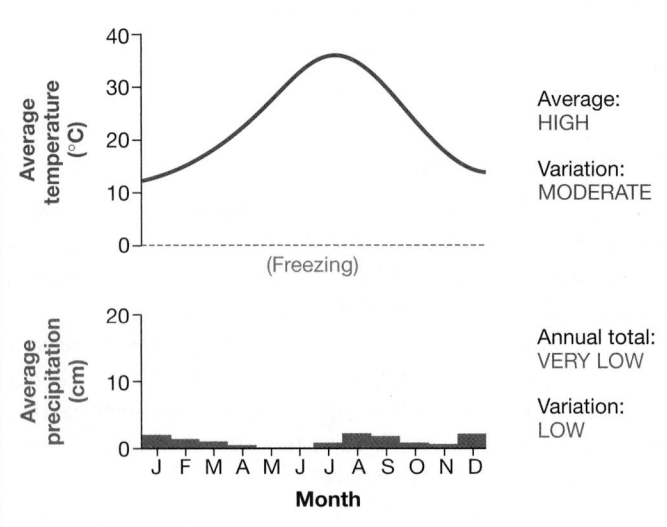

FIGURE 52.12 Climate Data from the Sonoran Desert.

Temperate grasslands are commonly called prairies in North America and steppes in central Eurasia. Grasses are the dominant life-form in temperate grasslands for one of two reasons: (1) Conditions are too dry to enable tree growth, or (2) encroaching trees are burned out by prairie fires. In contrast to deserts, plant life is extremely dense—virtually every square centimeter is filled, not only on the surface, but also underground where root systems are extensive. Although the productivity of temperate grasslands is generally lower than that of forest communities, grassland soils are often highly fertile. ✔You should be able to explain why grasslands of North America and Eurasia are the breadbaskets of those continents.

Temperature In the **temperate** zone, plant growth is possible only in spring, summer, and fall months when moisture and warmth are adequate (**FIGURE 52.13**).

Precipitation Conditions are still slightly too hot and dry to support forests; average annual precipitation in Konza Prairie (Kansas) is 83.4 cm (32.5 in.).

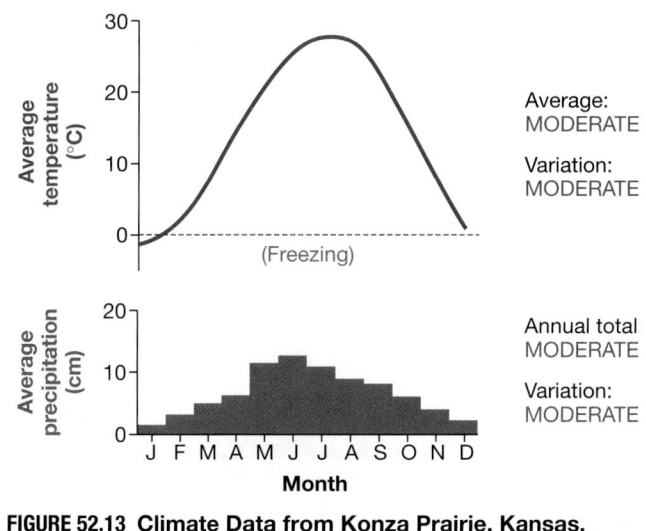

Average:
MODERATE

Variation:
MODERATE

Annual total:
MODERATE

Variation:
MODERATE

FIGURE 52.13 Climate Data from Konza Prairie, Kansas.

Temperate forests in North America and Europe are dominated by deciduous species, which are leafless in winter and grow new leaves each spring and summer. Needle-leaved evergreens are also common. Most temperate forests have productivity levels that are lower than those of tropical forests yet higher than those of deserts or grasslands. The level of diversity is also moderate. Temperate forests have moderate productivity simply because temperatures do not support photosynthesis year-round. ✔You should be able to predict the vegetation found in regions with climate characteristics intermediate between those of grasslands and temperate forests.

Temperature Temperate forests experience moderate fluctuations in mean monthly temperatures, sometimes falling below freezing, as in the Chicago area (**FIGURE 52.14**).

Precipitation Precipitation in temperate forests is moderate, and relatively constant throughout the year compared to grassland climates; average annual precipitation in the Chicago area is 85 cm (34 in.), exceeding 5 cm (2 in.) during most months.

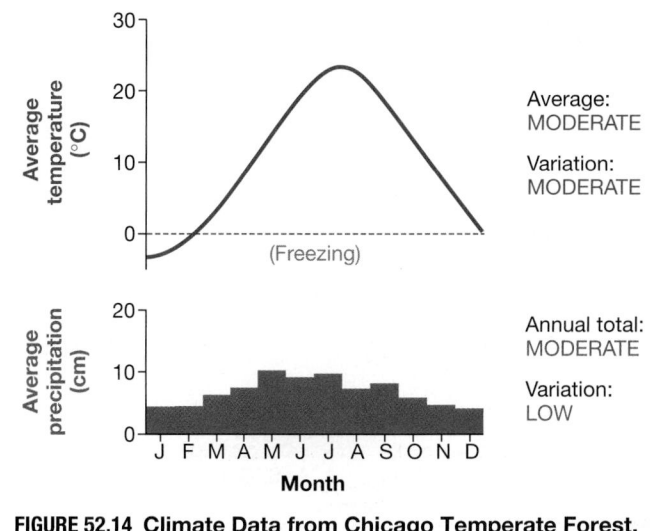

Average:
MODERATE

Variation:
MODERATE

Annual total:
MODERATE

Variation:
LOW

FIGURE 52.14 Climate Data from Chicago Temperate Forest.

Boreal forests, or **taiga,** stretch across most of Canada, Alaska, Russia, and northern Europe and are dominated by highly cold-tolerant conifers, including pines, spruce, fir, and larch trees. Productivity is low, but aboveground biomass is high because slow-growing tree species may be long lived and gradually accumulate large standing biomass. Boreal forests also have exceptionally low species diversity—there are just seven or fewer tree species in Alaska. They also lack most elements of structure seen in tropical or even temperate forests—they often have just a tree layer and a ground layer. ✔You should be able to predict how the global distribution of boreal forests will change in response to global warming.

Temperature Boreal forests have very cold winters and cool, short summers—an annual fluctuation of over 40°C in the Yukon Territory (**FIGURE 52.15**), and up to 70°C in some areas.

Precipitation Annual precipitation is low, but temperatures are so cold that evaporation is minimal; as a result, moisture is usually abundant enough to support tree growth.

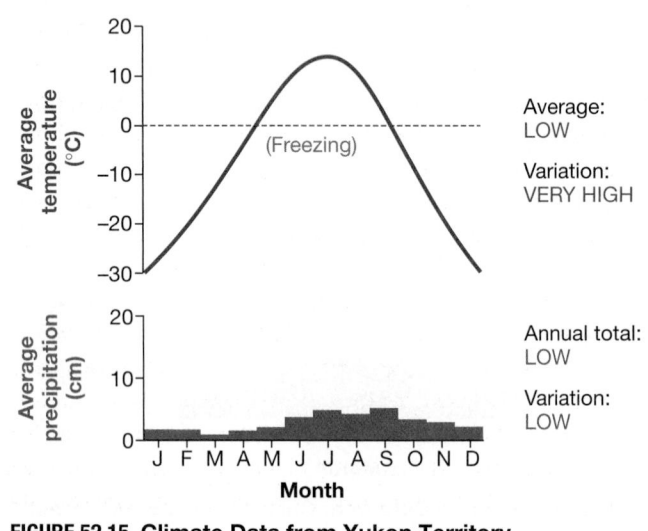

Average:
LOW

Variation:
VERY HIGH

Annual total:
LOW

Variation:
LOW

FIGURE 52.15 Climate Data from Yukon Territory.

Arctic tundra occurs throughout the arctic regions of the Northern Hemisphere. Tundra has low species diversity, low productivity, and low aboveground biomass—and it is treeless. Most tundra soils are in the perennially frozen state known as **permafrost,** which limits both the release and uptake of nutrients. Unlike desert biomes, however, the ground surface in tundra communities is completely covered with plants or lichens. Animal diversity also tends to be low, although insect abundance—particularly of biting flies—can be staggeringly high. ✔You should be able to explain why the types of vegetation in arctic tundra are similar to the types found in alpine tundra, which occurs at high elevation.

Temperature The growing season is very short, 10–12 weeks at most in Barrow, Alaska; for the remainder of the year, temperatures are below freezing (**FIGURE 52.16**).

Precipitation Precipitation on the arctic tundra is extremely low, but many arctic soils are saturated year-round due to extremely low evaporation rates.

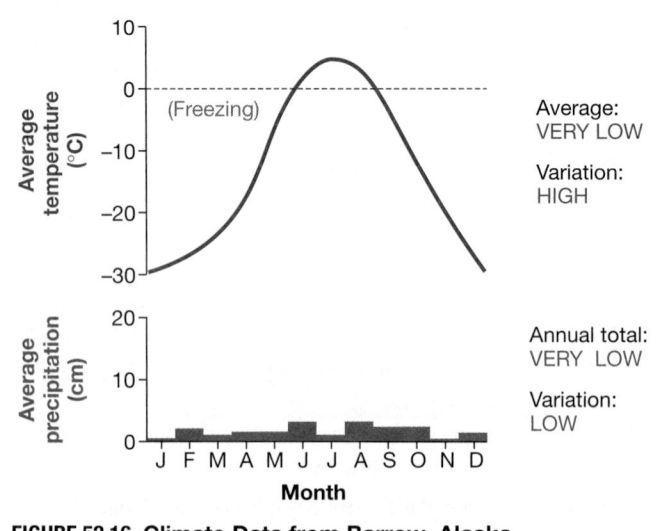

Average:
VERY LOW

Variation:
HIGH

Annual total:
VERY LOW

Variation:
LOW

FIGURE 52.16 Climate Data from Barrow, Alaska.

How Will Global Climate Change Affect Terrestrial Biomes?

It is now well established that human-caused emissions of greenhouse gases like CO_2 are causing climate change (see Chapter 56). Climate has changed frequently and radically throughout the history of life, but recent increases in average temperatures around the globe, combined with a projected increase of up to 5.8°C over the next 100 years, represent one of the most rapid periods of climate change since life began.

How will global climate change affect the overall distribution of life on Earth? Biologists use four tools to make predictions:

1. *Simulation studies* are based on computer models of weather patterns in local regions. By increasing average temperatures in these models, researchers can predict how wind and rainfall patterns, storm frequency and intensity, and other aspects of weather and climate will change. In effect, simulation studies predict how temperature and precipitation profiles, like those in this chapter's "Terrestrial Biome" boxes, will change—and how the distribution of organisms might change as a result.

2. *Observational studies* are based on long-term monitoring at fixed sites around the globe. Researchers are documenting changes in key physical variables—temperature and precipitation—as well as changes in the distribution and abundance of organisms, such as the timing of insect emergence and bird migrations. By extrapolation, they can predict how change may continue at these sites.

3. *Historical studies* examine the relationship between CO_2 levels, climate change, and the distribution and abundance of organisms during events that occurred millions of years ago. For example, researchers have been able to measure the pattern of change in the biodiversity of the Amazonian rain forest as a function of past climate fluctuations. These patterns provide valuable data for models designed to predict future changes.

4. *Experiments* are designed to simulate changed climate conditions and to record responses by the organisms present.

Even though the scientific community did not become alert to the reality of global climate change until the 1980s, a large number of multiyear experiments have already been done. Because arctic regions are projected to experience particularly dramatic changes in temperature and precipitation, many of these experiments have been conducted in the tundra biome.

FIGURE 52.17 shows one example of how these experiments are set up. Transparent, open-topped chambers are placed at random locations around a study site, and the characteristics of the vegetation inside the chambers are recorded annually. The same measurements are made in randomly assigned study plots that lack chambers.

Across experiments, the average annual temperature inside the open-topped chambers increases by 1–3°C. This increase is consistent with conservative projections for how temperatures will change over the next century.

FIGURE 52.17 Experimental Increases in Temperature Change Species Composition in Arctic Tundra. The transparent, open-topped chambers shown here are an effective way to increase average annual temperature. Inside the chambers, species composition changes relative to otherwise similar sites nearby.

✓**QUESTION** Why does a clear chamber like this increase average temperature inside?

In an analysis of data from 11 such study sites scattered throughout the arctic, several strong patterns emerged:

- Overall, species diversity decreases.

- Compared with control plots, grasses and shrubs increase inside the chambers, and mosses and lichens decrease.

These experimental results support simulation and observational studies predicting that arctic tundra environments are giving way to boreal forest. Biomes are changing right before our eyes.

Note that not all areas are expected to transition smoothly and incrementally from one biome to another based on changes in temperature and/or rainfall. There is growing evidence that small perturbations in climate in some areas will cause abrupt changes to biomes. For example, scientists predict that decreased rainfall over the Amazon will cause a rapid transition from forest to grassland due to positive feedback loops in the system. Decreases in forest cover will decrease the rate of rain-cloud formation, which will further decrease the rainfall. Decreased rainfall will increase the prevalence of fires, which will further decrease forest cover. The predicted result is a rapid change—a tipping point—from a forested stable state to a grassland stable state, with dramatic effects on biodiversity.

Other simulation and observational studies have converged on a remarkable conclusion: Increases in average global temperature are increasing variability in temperature and precipitation. Stated another way, global climate change is making climates more extreme. Average temperature and average annual precipitation in many regions may not be changing substantially, even though variation is. For example, thus far the frequency of monsoon rains, cyclones, hurricanes, and other types of storm events is not changing. However, the amount of precipitation and the wind speeds associated with these events are increasing. The severity of droughts is also on the rise.

Both human land-use patterns and climate change cause dramatic alterations in the composition and distribution of terrestrial biomes. These factors also affect aquatic biomes. Let's take a closer look.

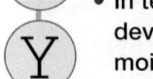
52.5 Types of Aquatic Biomes

About 70 percent of the surface of the Earth is covered in water. What types of biomes occur there? And what are the characteristics that distinguish aquatic biomes?

Let's think back to the sockeye salmon in the beginning of the chapter. Most salmon begin their lives in streams, travel to the ocean to feed and grow, then return to their natal streams to mate and die. Thus salmon migrate between a freshwater aquatic biome (a stream) and a marine aquatic biome (the ocean). The major abiotic factors that distinguish streams from oceans include salinity, water depth, water flow, and nutrient availability. Let's consider each factor in turn.

Salinity

Water is an excellent solvent—one reason why it is so essential to life (see Chapter 2). The water in clouds is nearly pure, but as rainwater percolates through soils and flows across rocks, it picks up solutes. The freshwater of ponds, lakes, and rivers has very low solute concentrations. However, as this freshwater steadily flows into the ocean, it continually adds solutes. Over millions of years, this process has resulted in a relatively high concentration of solutes in seawater.

The proportion of solutes dissolved in water determines its **salinity,** generally measured as the number of grams of solute per kilogram of water—a unitless number described as parts per thousand. The salinity of freshwater varies from about 0.06 to 0.3 parts per thousand, whereas the salinity of the open ocean is fairly constant at 35 parts per thousand.

What makes the ocean "salty"? Solutes with a positive charge—like sodium, potassium, magnesium, and calcium—combine with solutes with a negative charge—like chlorine, bromine, and fluorine—to form salts. The concentration of different salts in the ocean is determined by their solubility in water. For example, calcium carbonate has a low solubility in water, so it precipitates as limestone at the bottom of the ocean. But sodium chloride (familiar to you as table salt) has such a high solubility in water that it makes up about 86 percent of sea salt.

Salinity has dramatic effects on osmosis and water balance in organisms (see Chapters 38 and 43); species have physiological adaptations that allow them to cope with a specific range in salinity. Thus salinity is a major determinant of the distribution of organisms in aquatic biomes.

Water Depth

Water absorbs and scatters light, so the amount and types of wavelengths available to organisms change dramatically as water depth increases. Light availability has a major influence on productivity.

At the water surface, all wavelengths of light are equally available. But as **FIGURE 52.18a** shows, ocean water dramatically removes light in the red region of the visible spectrum, resulting in the familiar blue hue of underwater scenes (**FIGURE 52.18b**). This is important because wavelengths in the red region are unavailable for photosynthesis underwater, despite the high absorption peaks for red wavelengths in key photosynthetic pigments like chlorophyll (see Chapter 10).

The quantity of light available to organisms also diminishes rapidly with increasing depth. In pure seawater, the total amount of light available at a depth of 10 m is less than 40 percent of what it is at the surface; virtually no light reaches depths greater than 40 m (131 ft).

Further, seawater is rarely pure and may contain high concentrations of algae, sediments, and other light-reducing components. The **turbidity,** or cloudiness, of water is an important determinant of light penetration. Turbidity is sometimes caused by natural processes such as erosion of river sediments by floodwaters and the erosion of coastal sediments by wave action. Turbidity is also caused by many human activities, such as runoff from agricultural fields and algae blooms caused by nutrient pollution (see Chapter 56).

Water Flow

Water movement is a critical factor in aquatic ecosystems because it presents a physical challenge. It can literally sweep organisms away.

Organisms that live in fast-flowing streams have to cope with the physical force of the water, which constantly threatens to move them downstream. Stream-dwelling organisms like

(a) Red wavelengths are not available underwater.

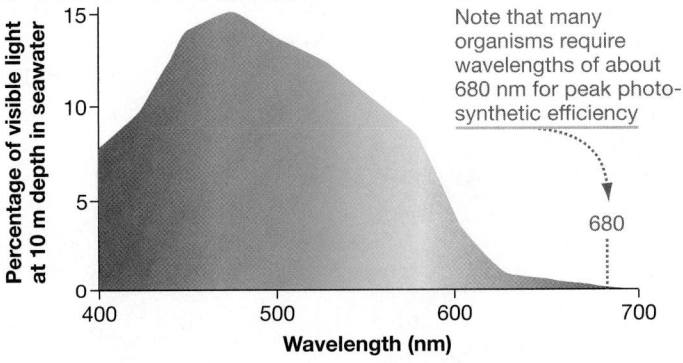

Note that many organisms require wavelengths of about 680 nm for peak photosynthetic efficiency

(b) Blue wavelengths dominate underwater.

FIGURE 52.18 Light Is Limited Underwater.
DATA: Saffo, M. B. 1987. *BioScience* 37: 654–664.

mayfly larvae have streamlined shapes and behavioral adaptations that help them to maintain their position in the stream. Marine organisms that live in intertidal regions are also exposed to water flow—sometimes to violent wave action during storms—and to tidal currents. A growing field of biomechanics (see Chapter 48) studies the mechanical implications of water flow to the physiology, evolution, and ecology of aquatic organisms.

Water flow can also influence non-mechanical aspects of the abiotic environment, including the availability of oxygen, light, and nutrients. For example, the fast-moving water near the source of streams tends to be high in O_2, clear, and low in nutrients. The slower-flowing water downstream tends to be lower in O_2, more turbid, and more nutrient rich.

Nutrient Availability

In many aquatic ecosystems, nutrients such as nitrogen and phosphorus are in short supply. If water is moving, such as in a small salmon stream, nutrients tend to be washed away. If water is still, nutrients tend to fall to the bottom and collect in the form

---

PROCESS: OCEAN UPWELLING

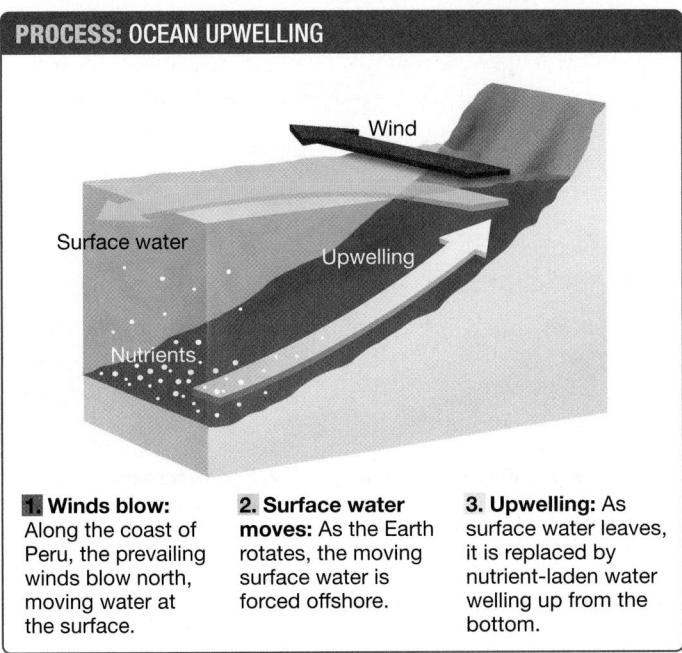

1. Winds blow: Along the coast of Peru, the prevailing winds blow north, moving water at the surface.

2. Surface water moves: As the Earth rotates, the moving surface water is forced offshore.

3. Upwelling: As surface water leaves, it is replaced by nutrient-laden water welling up from the bottom.

FIGURE 52.19 Ocean Upwellings Bring Nutrients to the Surface.

of debris. Nutrient levels are important: The scarcity of nutrients limits growth rates in the photosynthetic organisms that provide food for other species.

Let's consider two important events that affect nutrient availability in lakes and oceans. In both cases, they are mechanisms that bring nutrients from the bottom up to the water surface, where they can nourish the growth of photosynthetic species.

Ocean Upwelling In the oceans, nutrients in the sunlit surface waters are constantly lost in the form of dead organisms that rain down into the depths. In certain coastal regions of the world's oceans, however, nutrients are brought up to the surface by currents that cause upwellings.

FIGURE 52.19 shows how this happens off the coast of Peru. Here the prevailing winds blow along the coastline, pushing surface water to the north. Because the Earth rotates constantly, this wind-driven water current is slowly moved offshore. As the surface water moves away from the coast, it is steadily replaced by water moving up from the ocean bottom.

The upwelling water is nutrient rich. In effect, it recycles nutrients that earlier had fallen to the ocean floor. The nutrients support luxuriant growths of photosynthetic cells, which feed an array of small grazing animals, which in turn feed small fish called anchoveta. When currents are favorable and upwelling is steady, the anchoveta fishery is the most productive in the world.

Lake Turnover Because they were scoured by glaciers as recently as 10,000 years ago, the higher latitudes of the world host many lakes. Each year, these bodies of water undergo remarkable changes known as the spring and fall **turnovers.**

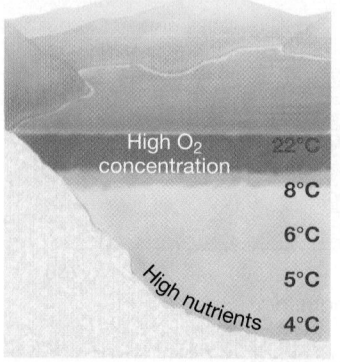
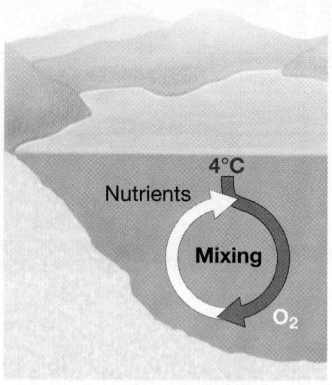

1. Winter stratification: Dense 4°C water at the bottom becomes nutrient rich while colder water near surface becomes oxygenated.

2. Spring turnover: Surface water warms to 4°C and sinks, carrying O_2 down and driving nutrients up.

3. Summer stratification: Dense 4°C water at the bottom becomes nutrient rich while warmer water near surface becomes oxygenated.

4. Fall turnover: Surface water cools to 4°C and sinks, carrying O_2 down and driving nutrients up.

FIGURE 52.20 In Temperate Regions, Lakes Turn Over Each Spring and Fall.

The spring and fall lake turnovers occur in response to changes in air temperature. To see how this happens, examine the temperature profiles in **FIGURE 52.20**.

Note that in winter, water at the surface of a northern lake is locked up in ice at a temperature of 0°C. The water just under the ice is slightly warmer and is relatively oxygen rich because it was exposed to the atmosphere as winter set in. Because water is densest at 4°C, the water at the bottom of the lake is at 4–5°C. The bottom water is also oxygen poor because organisms that decompose falling organic material use up available oxygen in the process.

A gradient in temperature such as this is called a **thermocline** ("heat-slope"). Thermal stratification is said to occur.

The ice begins to melt, however, when spring arrives. The temperature of the water on the surface rises until it reaches 4°C. The water at the surface of the lake is now heavier than the water below it. As a result, the water on the surface sinks. The water at the bottom of the lake is displaced and comes to the surface, completing the spring turnover.

During the spring turnover, water at the bottom of the lake carries sediments and nutrients from the bottom up to the surface. This flush of nutrients triggers a rapid increase in the growth of photosynthetic algae and bacteria; biologists call it the spring bloom. The lake then stratifies again in the summer.

When temperatures cool in the fall, the water at the surface reaches a temperature of 4°C and sinks, displacing water at the bottom and creating the fall turnover. The fall turnover is important because it brings oxygen-rich water from the surface down to the bottom, and because it again brings nutrients from the bottom up to the sunlit surface waters.

Without the spring and fall turnovers, most freshwater nutrients would remain on the bottom of lakes. These aquatic ecosystems would be much less productive as a result.

The boxes that follow summarize the types of aquatic biomes that are classified based on whether they are freshwater or marine as well as on their water depth, water flow, and nutrient availability.

- Aquatic Biomes > Freshwater > Lakes and Ponds
- Aquatic Biomes > Freshwater > Wetlands
- Aquatic Biomes > Freshwater > Streams
- Aquatic Biomes > Freshwater/Marine > Estuaries
- Aquatic Biomes > Marine > Oceans

How Are Aquatic Biomes Affected by Humans?

Just as on land, humans are affecting aquatic biomes worldwide by causing both direct and indirect effects. For example:

Direct Effects   Humans cause physical changes to aquatic ecosystems by filling or draining wetlands, damming streams, and removing water from ponds and rivers for agricultural irrigation. Humans also cause biological changes by transporting organisms—such as the invasive zebra mussel, which has become a scourge in the Great Lakes of North America and other regions of the world—and overexploiting marine resources, especially top predators such as tuna and sharks, causing cascading effects on food webs (see Chapters 56 and 57).

Indirect Effects   Global climate change influences aquatic ecosystems by the same mechanisms discussed for terrestrial ecosystems. In addition, increased atmospheric CO_2 is causing waterways to become more acidic, limiting the ability of organisms like corals to build skeletons, and having other far-reaching effects (see Chapter 56).

The importance of climate and other abiotic factors to the distribution and abundance of organisms was the main focus of this chapter. Next, let's take a closer look at the biotic factors. How do interactions among organisms of the same species and interactions among different species affect the overall distribution and abundance of organisms?

Lakes and ponds are distinguished from each other by size. Ponds are small; lakes are large enough that the water can be mixed by wind and wave action. Most natural lakes and ponds occur in high latitudes—they formed in depressions created by the scouring action of glaciers thousands of years ago.

Water depth Biologists describe the structure of lakes and ponds by naming zones, some of which overlap (**FIGURE 52.21**).

- The **littoral** ("seashore") **zone** consists of the shallow waters along the shore, where flowering plants are rooted.

- The **limnetic** ("lake") **zone** is offshore and comprises water that receives enough light to support photosynthesis.

- The **benthic** ("depths") **zone** occurs at the substrate.

- Regions of the littoral, limnetic, and benthic zones that receive sunlight are part of the **photic zone.**

- Portions of a lake or pond that do not receive sunlight make up the **aphotic zone.**

Water flow and nutrient availability Water movement is driven by wind and temperature. The littoral and limnetic zones are typically much warmer and better oxygenated than the benthic zone, but the benthic zone is relatively nutrient rich because dead and decomposing bodies sink and accumulate there. The movement of nutrients from the benthic to littoral zones is often dependent on seasonal turnovers.

Organisms Cyanobacteria, algae, and other microscopic organisms, collectively called **plankton,** live in the photic zone, as do the fish and small crustaceans—organisms related to shrimp and crabs—that eat them. In shallow parts of the photic zone, rooted plants are common. Animals (invertebrates and fish) that consume dead organic matter, or **detritus,** are particularly abundant in the benthic zone. ✔You should be able to predict which zone of lakes and ponds produces the most grams of organic material per square meter per year, and to explain your logic.

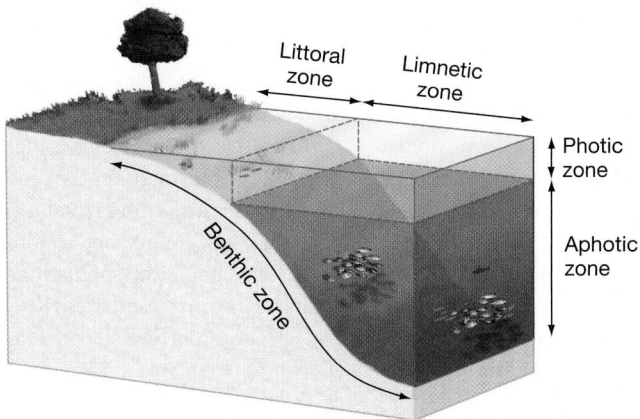

FIGURE 52.21 Lakes Have Distinctive Zones.

Wetlands are shallow-water habitats where the soil is saturated with water for at least part of the year. They are usually distinguished from terrestrial habitats by the presence of "indicator plants" that grow only in saturated soils.

Water depth Wetlands are distinct from lakes and ponds for two reasons: They have only shallow water, and they have **emergent vegetation**—meaning plants that grow above the surface of the water. All or most of the water in wetlands receives sunlight; emergent plants capture sunlight before it strikes the water.

Water flow and nutrient availability Freshwater marshes and swamps are wetland types characterized by a slow but steady flow of water. **Bogs,** in contrast, develop in depressions where water flow is low or nonexistent. If water is stagnant, oxygen is used up during the decomposition of dead organic matter faster than it enters via diffusion from the atmosphere. As a result, bog water is oxygen poor or even anoxic, resulting in slow decomposition rates. Organic acids and other acids build up, lowering the pH of the water. At low pH, nitrogen becomes unavailable to plants. As a result, bogs are nutrient poor. Marshes and swamps, in contrast, are relatively nutrient rich. In terms of total area, bogs are nearly as common as marshes and swamps but occur mostly at northern latitudes.

Organisms The combination of acidity, lack of available nitrogen, and anoxic conditions makes bogs extremely unproductive habitats. Marshes and swamps, in contrast, offer ample supplies of oxygenated water and sunlight, along with nutrients available from rapid decomposition. As a result, they are extraordinarily productive. **Marshes** lack trees and typically feature grasses, reeds, or other nonwoody vegetation; **swamps** are dominated by trees and shrubs. Because their physical environments are so different, there is little overlap in the types of species found in bogs, marshes, and swamps. ✔You should be able to explain why carnivorous plants, which capture and digest insects, are relatively common in bogs but rare in marshes and swamps.

Streams are bodies of water that move constantly in one direction. Creeks are small streams; rivers are large streams.

Water depth Most streams are shallow enough that sunlight reaches the bottom. Availability of sunlight is usually not a limiting factor for organisms, except when turbidity is high.

Water flow and nutrient availability The structure of a typical stream varies along its length. Where it originates at a mountain glacier, lake, or spring, a stream tends to be cold, narrow, and fast. As it descends toward a lake, ocean, or larger river, a stream accepts water from tributaries and becomes larger, warmer, and slower. Oxygen levels tend to be high in fast-moving streams because water droplets are exposed to the atmosphere when moving water splashes over rocks or other obstacles. In contrast, slow-moving streams that lack rapids tend to become relatively oxygen poor. Also, cold water holds more oxygen than warm water does (see Chapter 45). Slow-moving streams tend to be more nutrient rich than fast-moving streams, because decaying matter does not flush away as quickly.

Organisms It is rare to find photosynthetic organisms in small, fast-moving streams; nutrient levels tend to be low, and most of the organic matter present consists of leaves and other materials that fall into the water from outside the stream. Fish, insect larvae, mollusks, and other animals have adaptations that allow them to maintain their positions in the fast-moving portions of streams. As streams widen and slow down, conditions become more favorable for the growth of algae and plants, and the amount of organic matter and nutrients increases. As a result, the same stream often contains completely different types of organisms near its source and near its end, or mouth. ✔You should be able to explain why, in terms of their ability to perform cellular respiration, fish species found in cold, fast-moving streams tend to be much more active than fish species found in warm, slow-moving streams.

Estuaries form where rivers meet the ocean—meaning that freshwater mixes with salt water. In essence, an estuary includes saline marshes (from slightly to highly saline) as well as the body of water that moves in and out of these environments. Salinity varies with **(1)** changes in river flows—it declines when the river floods and increases when the river ebbs—and **(2)** with proximity to the ocean. Salinity has dramatic effects on osmosis and water balance (see Chapters 38 and 43); species that live in estuaries have physiological adaptations that allow them to cope with variations in salinity.

Water depth Most estuaries are shallow enough that sunlight reaches the substrate. Water depth may fluctuate dramatically, however, in response to tides, storms, and floods.

Water flow and nutrient availability Water flow in estuaries fluctuates daily and seasonally due to tides, storms, and floods. The fluctuation is important because it alters salinity, which in turn affects which types of organisms are present. Estuaries are nutrient rich because nutrient-laden sediments are deposited when flowing river water slows as it enters the ocean.

Organisms Because the water is shallow and sunlit, and because nutrients are constantly replenished by incoming river water, estuaries are among the most productive environments on Earth. They serve as a nursery for young fish, which feed on abundant vegetation, benthic invertebrates, and plankton while hiding from predators. Estuaries also tend to have high species diversity—some species may occur throughout the estuary while others occupy specialized zones according to salinity and other abiotic factors. Estuaries are also important feeding grounds for both residential and seasonal bird populations. ✔You should be able to explain why estuaries and freshwater marshes contain few of the same species, even though they are both shallow-water habitats filled with rooted plants.

Oceans form a continuous body of salt water and are remarkably uniform in chemical composition. Regions within an ocean vary markedly in their physical characteristics, however, and have profound effects on the organisms found there.

Water depth Biologists describe the structure of oceans by naming zones (**FIGURE 52.22**).

- The **intertidal** ("between tides") **zone** consists of a rocky, sandy, or muddy beach that is exposed to the air at low tide but submerged at high tide.

- The **neritic zone** extends from the intertidal zone to depths of about 200 m. Its outermost edge is defined by the end of the **continental shelf**—the gently sloping, submerged portion of a continental plate.

- The **oceanic zone** is the "open ocean"—the deepwater region beyond the continental shelf.

- The bottom of the ocean is the **benthic zone.**

- The intertidal and sunlit regions of the neritic, oceanic, and benthic zones make up a **photic zone.**

- Areas that do not receive sunlight are in an **aphotic zone.**

The ocean is indeed very deep (average depth = 3.7 km), but its depth is small compared to its breadth, like a piece of paper.

Water flow and nutrient availability Water movement in the ocean is dominated by different processes at different depths. In the intertidal zone, tides and wave action are the major influences. In the neritic zone, currents that bring nutrient-rich water from the benthic zone of the deep ocean toward shore have a heavy impact. Throughout the ocean, large-scale currents circulate water in the oceanic zone in response to prevailing winds and the Earth's rotation. In general, nutrient availability is dictated by water movement. The neritic and intertidal zones are relatively nutrient rich because they receive nutrients from rivers and upwelling. The oceanic zone, in contrast, constantly loses nutrients due to a steady rain of dead organisms drifting to the benthic zone.

Organisms Each zone in the ocean is populated by distinct species that are adapted to the physical conditions present. Organisms that live in the intertidal zone must be able to withstand physical pounding from waves and desiccation and high temperatures at low tide. Productivity is high, however, due to the availability of sunlight and nutrients contributed by estuaries as well as by currents that sweep in nutrient-laden sediments from offshore areas.

Productivity is also high on the outer edge of the neritic zone, due to nutrients contributed by upwelling at the edge of the continental plate. Almost all of the world's major marine fisheries exploit organisms that live in the neritic zone. In the tropics, shallow portions of the neritic zone may host **coral reefs**—large assemblages of colonial marine corals (related to sea anemones; see Chapter 33) that provide habitat for many other organisms. Mutualistic algae living within the corals are the autotrophs in coral reef ecosystems. Because the water is warm and sunlight penetrates to the ocean floor in these habitats, coral reefs are among the most productive environments in the world (see Chapter 56).

If coral reefs are the rain forests of the ocean, then the oceanic zone is the desert. Sunlight is abundant in the photic zone of the open ocean, but nutrients are extremely scarce. When the photosynthetic organisms and the animals that feed on them die, their bodies drift downward out of the photic zone and are lost. In the open ocean, the benthic zone can be several kilometers below the surface, and there is no mechanism for bringing nutrients back up from the bottom. The aphotic zone of the open ocean is also extremely unproductive because light is absent and photosynthesis is impossible. Most organisms present in the aphotic zone survive on the rain of dead bodies from the photic zone. ✔You should be able to predict whether animals that live in the aphotic zone have functioning eyes.

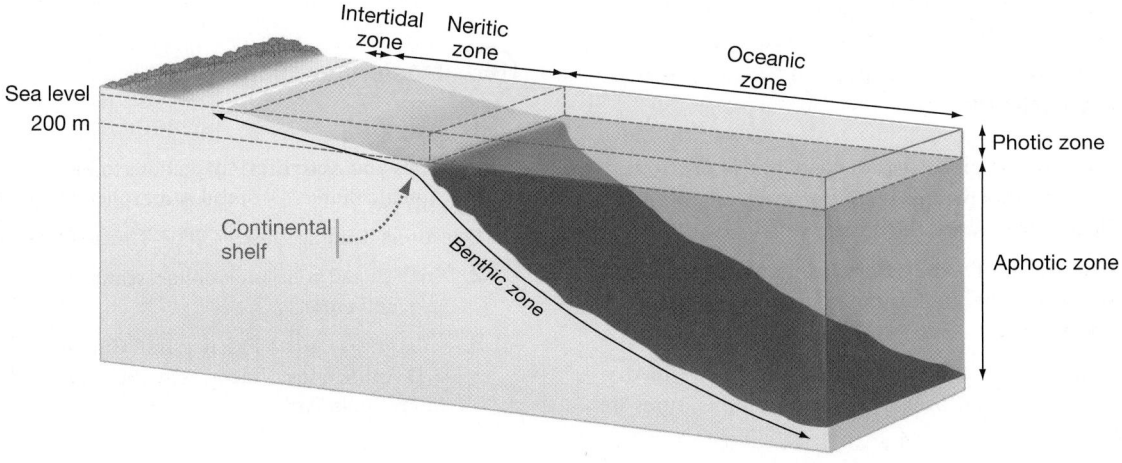

FIGURE 52.22 Oceans Have Distinctive Zones.

If you understand . . .

52.1 Levels of Ecological Study

- Biologists study ecology at five main levels: **(1)** organisms, **(2)** populations, **(3)** communities, **(4)** ecosystems, and **(5)** the biosphere.

- The goal of organismal ecology is to understand how individuals interact with other organisms and their physical surroundings.

- Population ecology focuses on how and why populations grow or decline in time and space.

- Community ecology is the study of how different species interact.

- Ecosystem ecology analyzes interactions between organisms and their abiotic environment—particularly the flow of energy and nutrients.

- Global ecology studies processes that transcend ecosystems, such as the effects of climate change.

- To preserve endangered species and communities, conservation biologists apply analytical methods and results from all levels.

✔ You should be able to explain why these five levels are hierarchical.

52.2 What Determines the Distribution and Abundance of Organisms?

- For each species, a unique combination of abiotic and biotic factors determines where individuals live and the size of populations.

- Understanding historical events, such as the movement of entire continents, is important to interpreting current patterns of species distributions.

- Abiotic and biotic factors often interact to produce a different effect on species distributions than either type of factor would have on its own.

✔ You should be able to predict what would happen to the range of Argentine ants in southern California if dry scrubland is converted to irrigated agricultural fields.

52.3 Climate Patterns

- Climate varies around the globe because sunlight is distributed asymmetrically—equatorial regions receive on average more solar radiation than do regions toward the poles.

- Hadley cells create bands of wet and dry habitats, and the tilt of Earth's axis causes seasonality in the amount of sunlight that non-equatorial regions receive.

- The presence of mountains can create local areas of wet or dry habitats, and proximity to an ocean moderates temperatures in nearby terrestrial habitats.

✔ You should be able to explain how climate would be expected to vary around the Earth if the planet were cylindrical and spun on an axis that was at a right angle to the Sun.

52.4 Types of Terrestrial Biomes

- Terrestrial biomes are defined by climatic regimes and distinct types of terrestrial vegetation.

- The major terrestrial biomes include tropical wet forest, subtropical desert, temperate grassland, temperate forest, boreal forest, and arctic tundra.

- Because photosynthesis is most efficient when temperatures are warm and water supplies are ample, the productivity and degree of seasonality in biomes varies with temperature and moisture.

- Humans are changing the terrestrial landscape, creating new anthropogenic biomes—such as cities, suburbs, farms, and ranchlands—in place of natural biomes.

- Global climate change is causing temperature profiles and precipitation patterns to change in ecosystems around the planet.

✔ You should be able to predict how global climate change may affect terrestrial biomes due to increased transpiration from plants.

52.5 Types of Aquatic Biomes

- Salinity, water depth, water flow, and nutrient availability are key factors determining types of aquatic biomes.

- Aquatic biomes include lakes and ponds, wetlands, streams, estuaries, and oceans.

- Humans are affecting aquatic biomes by physically altering them (such as by damming a river), by changing their chemistry (such as by polluting and causing ocean acidification), and by changing species compositions (such as by introducing invasive species).

✔ You should be able to explain why the open ocean is similar to the desert, and why it is not.

MasteringBiology

1. **MasteringBiology Assignments**

   Tutorials and Activities Adaptations to Biotic and Abiotic Factors; Aquatic Biomes; Tropical Atmospheric Circulation

   Questions Reading Quizzes, Blue-Thread Questions, Test Bank

2. **eText** Read your book online, search, take notes, highlight text, and more.

3. **The Study Area** Practice Test, Cumulative Test, BioFlix® 3-D Animations, Videos, Activities, Audio Glossary, Word Study Tools, Art

You should be able to . . .

1. Name the five main levels of study in ecology.

2. Why are certain exotic species considered "invasive"?
   a. They are found in areas where they are not native.
   b. They were introduced by humans—often accidentally.
   c. They spread aggressively and displace native species.
   d. They benefit from being in a new environment.

3. Where do rain shadows exist?
   a. the part of a mountain that receives prevailing winds and heavy rain
   b. the region beyond a mountain range that receives dry air masses
   c. the region along the equator where precipitation is abundant
   d. the region near 30° N and 30° S latitude that receives hot, dry air masses

4. What is the predominant type of vegetation in a tropical wet forest?

   a. shrubs and bunchgrasses
   b. herbs, grasses, and vines
   c. broad-leaved deciduous trees
   d. broad-leaved evergreen trees

5. What is one expected consequence of global climate change?
   a. Average rainfall will increase.
   b. Average rainfall will decrease.
   c. Variability in rainfall will increase.
   d. We cannot make predictions about future rainfall.

6. Typically, where are oxygen levels highest and nutrient levels lowest in a stream?
   a. near its source
   b. near its mouth, or end
   c. where it flows through a swamp or marsh
   d. where it forms an estuary

7. Write questions about the ecology of humans that are relevant to each of the five levels of ecological study.

8. Explain how biotic and abiotic factors determine the distribution of invasive Argentine ants.

9. Why does the Australian Outback receive so little rainfall?
   a. The Australian Outback is in a rain shadow.
   b. A Hadley cell drops warm, dry air in this region.
   c. A Hadley cell pulls warm, moist air up from this region.
   d. The climate in Australia is moderated by the surrounding ocean.

10. Contrast the productivities of the intertidal, neritic, and oceanic zones of marine environments. Explain why large differences in productivity exist.

11. Compare and contrast the terrestrial biomes found at increasing elevation on a mountain with the biomes found at increasing latitude in Figure 52.9.

12. Compare the distribution of the natural terrestrial biomes of eastern North America with the distribution of anthropogenic biomes.

13. Mars has an even more pronounced tilt than Earth does. Does Mars experience seasons? Why or why not?
    a. Yes, the tilt will cause seasons because the Northern and Southern Hemispheres will be tilted toward the Sun, then away, at different times of the year.
    b. Yes, because the tilt will cause the orbit of Mars to be closer to the Sun during part of the year.
    c. No, because the tilt of Mars does not affect the distance of its orbit from the Sun.
    d. No, because the tilt of Mars does not affect the average amount of solar radiation received.

14. Very similar reptile fossils have been discovered on the east coast of South America and the west coast of Africa. Pose a hypothesis to explain this distribution.

15. Scientists are concerned about the future of species that are adapted to high elevations. Do you think their concern is justified?

16. Scientists predict that global climate change will cause a greater increase in average temperature at higher latitudes than in the tropics. This seems like good news since most of the world's biodiversity occurs in the tropics. However, recent data suggest that a smaller temperature change in the tropics will cause a *larger* physiological response there than the larger temperature change will in organisms in temperate areas. Based on what you have learned in this chapter, pose a hypothesis to explain this result.

53 Behavioral Ecology

In this chapter you will learn that

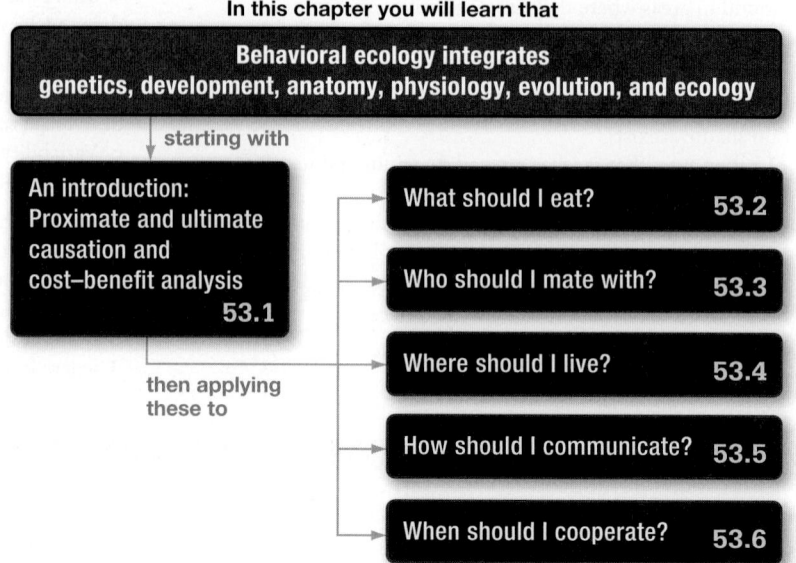

**Behavioral ecology integrates
genetics, development, anatomy, physiology, evolution, and ecology**

starting with

An introduction:
Proximate and ultimate
causation and
cost–benefit analysis
53.1

then applying
these to

What should I eat? **53.2**

Who should I mate with? **53.3**

Where should I live? **53.4**

How should I communicate? **53.5**

When should I cooperate? **53.6**

This male satin bowerbird is decorating his display area with objects of his preferred color: blue. A female will inspect the bower. She may mate with the male if she finds him, and the display, attractive enough.

This chapter is part of the Big Picture. See how on pages 1196–1197.

To biologists, **behavior** is action—specifically, the response to a stimulus. Pond-dwelling bacteria swim toward drops of blood that fall into the water. Time-lapse movies of growing sunflowers document the steady movement of their flowering heads throughout the day, as they turn to face the shifting Sun. Filaments of the fungus *Arthrobotrys* form loops and then release a molecule that attracts roundworms. When a roundworm touches a loop, the fungal filaments swell—ensnaring the worm. The fungus then grows into the worm's body and digests it.

Among animals, action may be frequent and even spectacular. Peregrine falcons reach speeds of up to 320 km/hr (200 mph) when they dive in pursuit of ducks or other flying prey. Hognose snakes

✔ When you see this checkmark, stop and test yourself. Answers are available in Appendix A.

go belly-up and feign death—mouths agape—when harassed by predators. Deep inside a hive, honeybees communicate about food sources by "dancing."

Ecology is the study of how organisms interact with their physical and biological environments; **behavioral ecology** is the study of how organisms respond to particular stimuli from their environments. What does an *Anolis* lizard do when it gets too hot? How does that same individual respond when a member of the same species approaches it? Or when a house cat approaches? You can see how behavior fits into the Big Picture of Ecology on pages 1196–1197.

Although all organisms respond in some way to signals from their environment, most behavioral research is performed on animals, especially vertebrates, arthropods, or mollusks. With their sophisticated nervous systems and skeletal–muscular systems, these animals can sense, process, and respond rapidly to a wide array of environmental stimuli.

53.1 An Introduction to Behavioral Ecology

Behavior is a particularly integrative field in biological science. While the primary focus of behavioral ecology is at the level of the organism (see Chapter 52 and The Big Picture of Doing Biology on pages 16–17), this field integrates levels from molecules up to ecosystems, depending on the question being asked. To understand why animals and other organisms do what they do, researchers have to ask questions about genetics, development, anatomy, physiology, evolution, and ecology.

To make sense of this diversity, it's helpful to recognize that behavioral ecologists ask questions and test hypotheses at two fundamental levels—proximate and ultimate. Let's first explore these levels and then consider different types of behaviors.

Proximate and Ultimate Causation

The eminent evolutionary biologist Ernst Mayr wrote an influential paper on "cause and effect in biology" in 1961. He made a clear distinction between two types of biological causation:

1. **Proximate** (or mechanistic) **causation** explains *how* actions occur in terms of the genetic, neurological, hormonal, and skeletal–muscular mechanisms involved. For example, does a neural or hormonal signal trigger an animal's courtship display? Does the display vary according to diet, photoperiod, temperature, or other factors?

2. **Ultimate** (or evolutionary) **causation** explains *why* actions occur—based on their evolutionary consequences and history. (Behavior is just like any other phenotype in that it can evolve by natural selection.) Is a particular behavior currently adaptive, meaning that it increases an individual's fitness—its ability to produce viable, fertile offspring? If so, how does the behavior help individuals to produce offspring in a particular environment?

It's important to recognize that efforts to explain behavior at the proximate and ultimate levels are complementary. To understand what an organism is doing, biologists want to know how the behavior happens and why.

To illustrate the proximate and ultimate levels of causation, consider the Argentine ant (*Linepithema humile*) (introduced in Chapter 52). These ants originated in Argentina but have been shipped inadvertently with fruits and other cargoes around the world. Argentine ants are the subject of intense research due to their extreme success as an invasive species.

Argentine Ant Behavior In their native South America, Argentine ants live in colonies and defend their feeding areas, or **territories,** by fighting with neighboring Argentine ants. However, in places where they have been introduced, such as southern California, Argentine ants from adjacent colonies are "friendly" to each other—workers and queens move freely from nest to nest, forming huge "supercolonies." But when the Argentine ants from supercolonies meet native ants, they are especially aggressive (**FIGURE 53.1**). What's going on?

Proximate Causes Ants identify each other by smelling waxy hydrocarbon "tags" on each other's exoskeletons. Biochemical research has shown that introduced Argentine ants accept most other members of their species as nest mates because they smell the same as nest mates. That is, they display the same scent tags.

Follow-up research suggests that the introduced ants smell similar because the genes that determine their scent tags are similar; unlike the Argentine ant populations in South America, there is little genetic variation for this trait among colonies in California. Thus the proximate cause for unusually friendly behavior among Argentine ants is the genetically determined scent signal that says "nest mate: don't fight." By contrast, the very unfamiliar scents from native ants signal "foreigner: attack!"

Ultimate Causes Why is genetic diversity in the supercolonies low? Argentine ants appear to have low genetic variation due to a genetic bottleneck that occurred when the first few ants were introduced to California (see Chapter 26 for a review of genetic bottlenecks).

FIGURE 53.1 Tiny Invasive Argentine Ants Attack a Native Ant.

This chance event happened to be adaptive. Since Argentine ants in supercolonies are not fighting to maintain their feeding territories, they have more time and energy available to produce offspring, thus increasing their fitness. The resulting high density of the supercolonies contributes to their invasiveness. A single, tiny Argentine ant is rarely successful in combat against a single, larger native ant. But when Argentine ants recruit help and fight in mobs, they are a powerful force.

Note, however, that low genetic diversity has its dangers (see Chapter 13). If the supercolony becomes exposed to a virulent pathogen or other environmental change, the entire colony may collapse—a possibility that humans are trying to exploit.

This example illustrates how behavioral ecologists work at the interface of many fields and at both proximate and ultimate levels to address *how* and *why* questions about organisms in their environments. Before considering some of the most prominent questions in behavioral ecology, let's take a closer look at types of behavior.

Types of Behavior: An Overview

Behavior encompasses a diverse array of responses to stimuli. To establish a framework to organize this diversity, let's explore both axes of the graph shown in **FIGURE 53.2**.

Behaviors Vary in Their Flexibility The horizontal axis of Figure 53.2 maps how flexible a behavior is in terms of its expression within and between individuals of the same species. This axis forms a continuum ranging from fixed, stereotyped behaviors to highly flexible, conditional behaviors.

Consider what happens to a kangaroo rat that is feeding at night and hears the sound of a rattlesnake's rattle. It jumps back—away from the direction of the stimulus. The jump-back behavior is performed the same way every time. Highly inflexible,

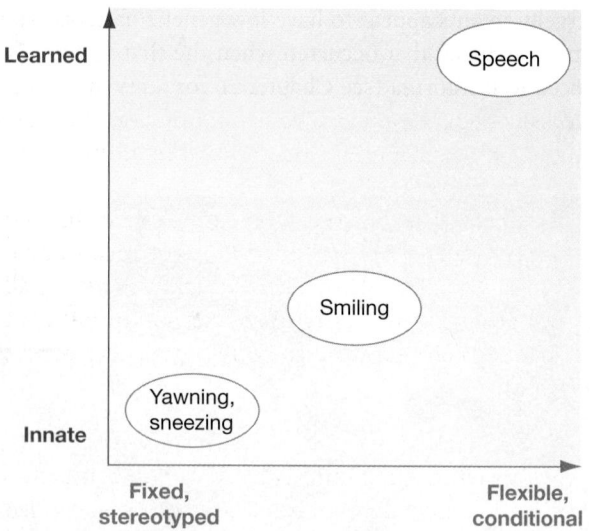

FIGURE 53.2 Behavioral Traits Vary in Flexibility and Dependence on Learning. These two axes represent one way to organize the diversity of behavioral traits. Three human behaviors are mapped as examples.

stereotyped behavior patterns like a kangaroo rat's jump-back behavior are called **fixed action patterns,** or **FAPs.** In humans, yawns and sneezes are familiar FAPs.

At the other end of this axis, behaviors show a great deal of flexibility in response to environmental conditions. For example, human speech is highly flexible and dependent upon factors ranging from inherited sound-making ability to upbringing, education, emotional state, circumstance, and so on. Most behaviors lie between these two extremes of inflexibility and flexibility.

Behaviors Vary in the Extent to Which They Are Learned The vertical axis of Figure 53.2 plots how much learning is involved in the nature of the behavioral response.

Behaviors that require no learning are said to be innate. **Innate behavior** is inherited—meaning that it is passed genetically from parents to offspring. FAPs are examples of innate behaviors because they are "hardwired"—the jump-back behavior of kangaroo rats and your ability to sneeze are not learned.

When the pioneering behavioral researchers Konrad Lorenz, Niko Tinbergen, and Karl von Frisch first began to explore animal behavior experimentally, they focused on these types of inflexible, innate actions—such as the way goose chicks imprint on their mother and follow her around. This research strategy is common in biological science. If you are addressing a new question—whether it's how DNA is transcribed or how ions cross plasma membranes—it's helpful to start by studying the simplest possible system. Once simple systems are understood, investigators can delve into more complex situations and questions. Lorenz, Tinbergen, and von Frisch won a joint Nobel Prize in 1973 for their behavioral work.

At the other end of the continuum from innate behavior is **learning,** an enduring change in behavior that results from a specific experience in an individual's life. Behaviorist B. F. Skinner is famous for his experiments with rats and mice showing that their behaviors are subject to conditioning. For example, a rat can be trained to pull a lever if food is released as a reward.

The examples plotted in Figure 53.2 indicate that there is a general tendency for innate behaviors to be more fixed and for learned behaviors to be more flexible. For example, in humans, yawning is a relatively fixed, innate behavior while speech is flexible and highly dependent upon learning.

Flexible, Learned Behaviors Often Involve Choice Most animals have a range of actions that they can perform in response to a situation. Animals take in information from the environment and, based on that information, make decisions about what to do. Animals make choices.

To link condition-dependent behavior to fitness—and thus the ultimate level of explanation—biologists use a framework called **cost–benefit analysis.** Animals appear to weigh the costs and benefits of responding to a particular situation in various ways. Costs and benefits are measured in terms of their impact on fitness—the ability to produce viable, fertile offspring.

Note the phrase "appear to weigh" in the previous paragraph. The decisions made by nonhuman organisms are not—as far as is known—conscious. An Argentine ant does not identify another ant by smell and then consider the pros and cons of reacting as friend or foe. Instead, evolution has shaped the ant's genome to direct a nervous and endocrine system that is capable of (1) taking in information and (2) directing behavior that is likely to pass that genome on to the next generation.

Five Questions in Behavioral Ecology

The following five sections of this chapter introduce some of the most prominent questions in behavioral ecology. In each case, the question is posed as a decision that an animal has to make. The ensuing discussion is divided into research on how the question has been answered at the proximate level and at the ultimate level.

The goals of this chapter are to explore some key topics in behavioral ecology and illustrate the spirit and synthetic nature of the field. Let's begin with an introduction to research on a resource that every animal has to find in its environment: food.

53.2 What Should I Eat?

When animals seek food, they are **foraging.** In some cases, the food source that a species exploits is extremely limited. Giant pandas, for example, subsist almost entirely on the shoots and leaves of bamboo. But individuals must still make decisions about which bamboo plants to harvest and which parts of a particular plant to eat and which to ignore.

In most cases, animals have a relatively wide range of foods that they exploit over the course of their lifetime. To understand why individuals eat the way they do, let's consider (1) research on genetic variation in foraging behavior and (2) the concept of optimal foraging.

Proximate Causes: Foraging Alleles in *Drosophila melanogaster*

When Marla Sokolowski was working as an undergraduate research assistant, she noticed that some of the fruit fly larvae she was studying tended to move after feeding at a particular location, while others tended to stay put. She reared these "rovers" and "sitters" to adulthood, bred them, and found that the offspring of rovers also tended to be rovers, while offspring of sitters tended to be sitters (**FIGURE 53.3**). Thus, this aspect of behavior is inherited. In fruit flies, variation in larval foraging behavior is at least partly due to genetic variation among individuals.

Using information from the *Drosophila melanogaster* genome sequence (see Chapter 21 and BioSkills 13 in Appendix B) and genetic mapping techniques (Chapter 20), Sokolowski found and cloned a gene associated with the rover–sitter difference. She named it *foraging* (*for*).

Follow-up work showed that the protein product of the *for* gene involved in a signal transduction cascade (see Chapters 11 and 49)—meaning it is involved in the response to a cell–cell

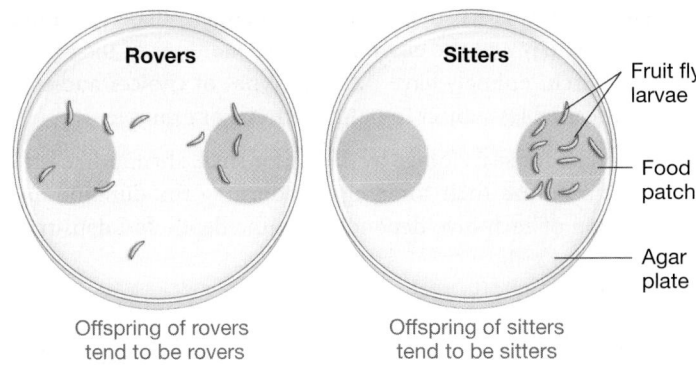

FIGURE 53.3 Foraging Behavior in Fruit Fly Larvae Is Heritable.

signal. Rovers and sitters tend to behave differently when they are foraging because they have different alleles of the *for* gene.

Genes that are homologous to *for* have now been found in honeybees and the roundworm *Caenorhabditis elegans.* Currently, researchers are focused on finding out which signaling pathway uses the *for* gene product and how different alleles change feeding behavior. Presumably, the signaling pathway is associated with feeding or nervous system development. When this aspect of fly feeding is thoroughly understood at the proximate level, biologists will know why variants of the same protein make larvae more likely to move or stay in place after feeding.

Ultimate Causes: Optimal Foraging

By altering population density experimentally and documenting the reproductive success of rovers and sitters, Sokolowski has shown that the rover allele in fruit fly larvae is favored when population density is high and food is in short supply. In contrast, the sitter allele reaches high frequency in low-density populations, where food is abundant. Rovers do better at high density because they are more likely to find unused patches of food; sitters do better at low density because they don't waste energy moving around.

These results are exciting because they link a proximate mechanism with an ultimate outcome. In this case, the presence of certain alleles is responsible for a difference in fitness in specific types of habitats.

In *Drosophila* larvae, foraging is a relatively inflexible and innate behavior—individuals don't "decide" to be rovers or sitters. It is the variability of food supplies that maintains both alleles in the population. But in *Drosophila* adults and in most animal species, foraging behavior is much more flexible and dependent upon learning.

Introduction to Optimal Foraging When biologists set out to study why animals forage in a particular way, they usually start by assuming that individuals make decisions that maximize the amount of usable energy they take in, given the costs of finding and ingesting their food and the risk of being eaten while they're at it. This hypothesis—that animals maximize their feeding efficiency—is called **optimal foraging.**

The specific costs and benefits of foraging depend on the feeding strategy and the environmental conditions of the organism. Different animals have different types of choices and thus different variables subject to optimization. For example:

- Antarctic fur seals forage by diving for small shrimp-like krill. They optimize their foraging by adjusting the duration and location of each dive depending on the depth and density of patches of krill.

- Shorebirds called oystercatchers feed on a variety of prey, including clams, which they have to open to eat. They optimize their foraging by selecting clams of intermediate size. The small clams provide less energy than the large clams, but opening the large clams can take more time and effort, and it can damage the birds' bills.

How do biologists test the optimal foraging hypothesis?

A Test of Optimal Foraging in Desert Gerbils Consider one example from the long-term study of optimal foraging in gerbils. Behavioral ecologists Zvika Abramsky, Michael Rosenzweig, and Aziz Subach posed an interesting question: Do gerbils "weigh" the costs and benefits of foraging? More specifically, if the presence of predators *decreases* the amount of time gerbils spend foraging, and the presence of extra food *increases* foraging time, how much extra food will offset the risk of predation?

To answer this question, Abramsky and his team set up an experiment using two replicate 2-hectare plots in the natural habitat of the gerbil *Gerbillus allenbyi* in the Negev Desert of Israel (**FIGURE 53.4**). Each of these plots had two subplots: a control half and a treatment half, with portals in the fencing to allow gerbils to switch sides. Abramsky started by removing all the gerbils from the plots, setting up seed stations to control the amount of supplementary food that would be available, and setting up sampling stations to measure gerbil track density (number of tracks per given area) during each trial. Earlier research had shown that gerbil track density served as a reliable indicator of the time gerbils spent foraging.

Abramsky and his team then introduced 34 gerbils into each subplot and conducted the following three treatments, replicating their experiments on three different nights:

1. No extra predation risk or seeds in treatment subplots.

2. Increased predation risk—captive barn owls were trained to fly over the treatment subplots three times per hour during the two-hour trials.

3. Increased predation risk plus increased seed availability— varying quantities of supplementary seeds were offered to the treatment subplots during different trials.

What did they find out? The graph at the bottom of Figure 53.4 shows that foraging activity in control and treatment subplots was the same in the absence of owls and extra seed. But when owls were flown over treatment subplots, foraging activity declined dramatically—and some gerbils fled through portals into the control subplots, increasing foraging activity there.

RESEARCH

QUESTION: Do gerbils "weigh" the costs and benefits of foraging?

HYPOTHESIS: Gerbils reconcile the risk of predation and the benefits of extra food availability.

NULL HYPOTHESIS: Foraging activity is independent of predation and food availability.

EXPERIMENTAL SETUP:

1. Start with 34 gerbils in each 1-hectare desert subplot in the Negev Desert of Israel.

2. Use the density of gerbil tracks in sampling stations as an index of foraging activity in control versus treatment subplots where:

Treatment 1	Treatment 2	Treatment 3
No owl fly-overs No extra seeds	Owl flyovers No extra seeds	Owl flyovers Extra seeds of varying quantities

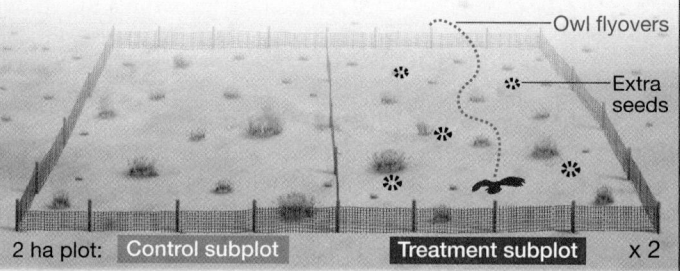

2 ha plot: Control subplot | Treatment subplot x 2

3. Replicate experiment on three different nights.

PREDICTION: A certain amount of added seeds will compensate for the decrease in foraging activity due to predation risk.

PREDICTION OF NULL HYPOTHESIS: Gerbil foraging activity will be independent of the presence of owls, extra seeds, or both.

RESULTS:

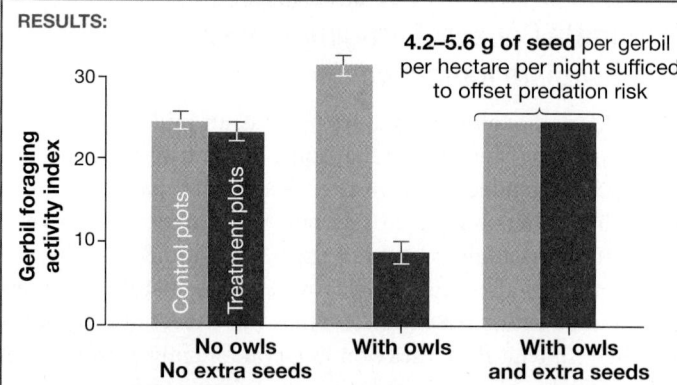

CONCLUSION: Gerbils weigh the relative risk of predation and benefits of extra seeds when they forage.

FIGURE 53.4 Optimal Foraging in Desert Gerbils

SOURCE: Abramsky, Z., M. L. Rosenzweig, and A. Subach. 2002. The costs of apprehensive foraging. *Ecology* 85: 1330–1340.

✔ **QUESTION** Why did the researchers measure foraging activity in the treatment plot before adding owls or seeds?

The researchers also observed a direct relationship between the gerbils' foraging activity and the amount of supplementary seed offered, even with owl flyovers. It turns out that 5 grams of supplementary seeds per individual per hectare per night was sufficient to compensate for the extra risk of predation—that is, to eliminate the difference in foraging between control and treatment subplots. This study was insightful because it enabled predation risk to be measured in energetic terms.

The logic behind optimal foraging is that animals that maximize their feeding efficiency will have more time and energy available for reproduction—thus higher fitness. Animals that forage optimally are expected to be favored by natural selection—thus pass on their heritable behavior patterns to the next generation.

check your understanding

C Y U

If you understand that . . .
- Feeding behavior can be explored at both proximate and ultimate levels.
- Optimal foraging minimizes fitness costs and maximizes fitness benefits.

✔ **You should be able to . . .**

Predict how the results of the gerbil foraging experiment would change if the owls were permitted to *hunt* in the treatment subplots rather than just flying over, and pose a proximate explanation for the change.

Answers are available in Appendix A.

53.3 Who Should I Mate With?

Biologists sometimes claim that almost all animal behavior revolves around food or sex. Section 53.2 focused on food; here the subject is sex. Let's start by considering the proximate mechanisms responsible for triggering sexual activity in lizards and then go on to explore the ultimate question of why animals choose their mates.

Proximate Causes: How Is Sexual Activity Triggered in *Anolis* Lizards?

Anolis carolinensis lives in the woodlands of the southeastern United States. After spending the winter under a log or rock, males emerge in January and establish feeding and breeding territories (**FIGURE 53.5a**). Females become active a month later, and the breeding season begins in April. By May, females are laying an egg every 10–14 days. By the time the breeding season is complete three months later, the eggs produced by a female will total twice her body mass.

FIGURE 53.5b graphs how sexual condition changes throughout the year, by plotting the size of the testes in males and ovaries of

(a) Display behavior of a male *Anolis* lizard

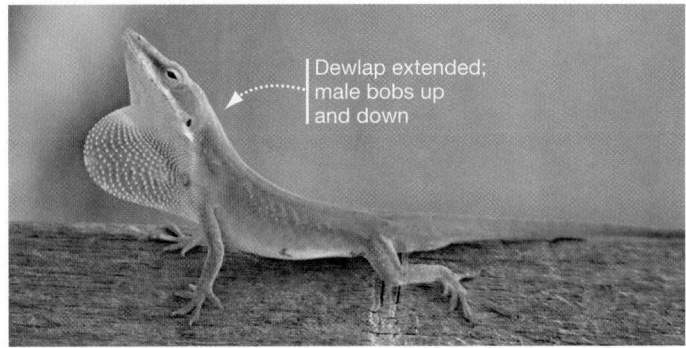

Dewlap extended; male bobs up and down

(b) Changes in sexual organs through the year

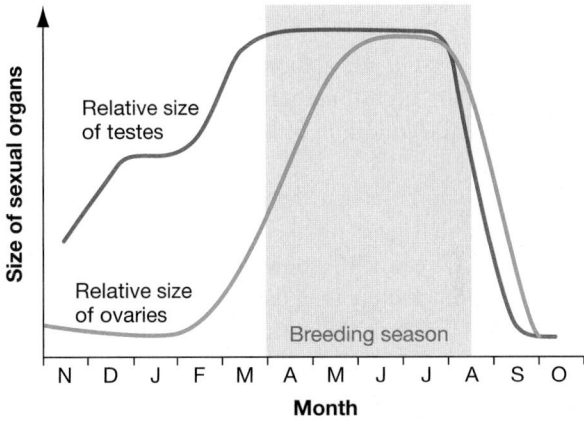

FIGURE 53.5 Sexual Behavior in *Anolis* Lizards Is Seasonal.
DATA: Crews, D. 1975. *Science* 189: 1059–1065.

females—relative to an individual's overall body size. Individuals come into breeding condition in spring, although males experience growth of gonads long before females do. In both sexes, the gonads shrink in size during the fall and winter.

Testosterone and Estradiol   What causes the onset of sexual behavior? The proximate answer is sex hormones—testosterone in males and estradiol in females. Testosterone is produced in the testes of males, and estradiol in the ovaries of females (see Chapter 50). The evidence for the effects of sex hormones is direct. Testosterone injections induce courtship activity in castrated males with no prior courtship activity; estradiol injections induce sexual activity in females whose ovaries have been removed.

But a full proximate explanation must also consider environmental and social cues. What environmental cues trigger the production of sex hormones in early spring? How do male and female *Anolis* lizards synchronize their sexual readiness?

Testing the Effects of Light and Social Stimulation   To answer these questions, behavioral ecologist David Crews brought a large group of sexually inactive adult lizards into the laboratory during the winter and divided them into five treatment groups. The physical environment was exactly the same in all treatments.

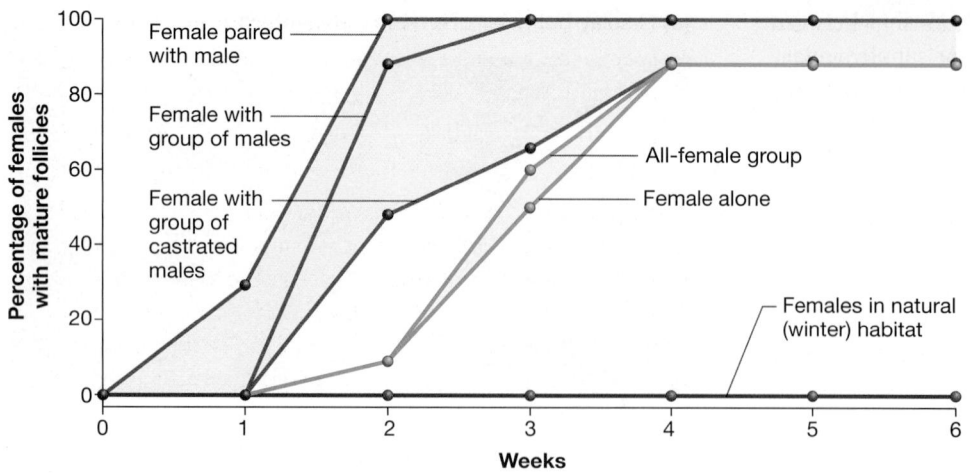

FIGURE 53.6 Exposure to Springlike Conditions and Breeding Males Stimulates Female Lizards to Produce Eggs. The percentage of female lizards with mature follicles, plotted over time. Five treatment groups were exposed to springlike light and temperature in the lab; data are also plotted for females left in a natural (winter) habitat. Each data point represents the average value for a group of 6–10 females.

DATA: Crews, D. 1975. *Science* 189: 1059–1065.

✓**QUESTION** Some critics contend that the females that remained outside do not represent a legitimate control in this experiment. What conditions would represent a better control?

Each lizard received identical food, high "daytime" temperatures, and slightly lower "nighttime" temperatures. Crews also continued to monitor the condition of lizards that remained in natural habitats nearby.

To test the hypothesis that changes in day length signal the arrival of spring and trigger initial changes in sex hormones, Crews exposed the five treatment groups in the laboratory to artificial lighting that simulated the long days and short nights of spring.

To test the hypothesis that social interactions among individuals are responsible for synchronizing sexual behavior, he varied the social setting among the five treatment groups:

1. a single isolated female;

2. a group of females;

3. a single female with a single male;

4. a single female with a group of castrated (nonbreeding) males; and

5. a single female with a group of uncastrated (breeding) males.

Each week, Crews examined the ovaries of females in each treatment. He also monitored the ovaries of females in nearby natural habitats under winter conditions.

As **FIGURE 53.6** shows, the differences in the animals' reproductive systems were dramatic. The data points on this graph represent the average percentage of females with mature ovarian follicles, in each treatment; the points are connected by lines to help you keep track of the different treatments and see the overall patterns. Note two key points:

1. Females that were exposed to springlike conditions began producing eggs; females in the field that were not exposed to springlike conditions did not.

2. Females that were exposed to breeding males began producing eggs much earlier than did the females placed in the other treatment groups.

These results support the hypothesis that two types of stimulation are necessary to produce the hormonal changes that lead to sexual behavior. Females need to experience springlike light and temperatures *and* exposure to breeding males.

Visual Cues from Males Trigger Female Readiness What is it about breeding males that triggers earlier estradiol release and egg production in females?

When males court females, they bob up and down in push-up fashion and extend a brightly colored patch of skin called a dewlap (see Figure 53.5a). To test the hypothesis that this visual stimulation triggers changes in estradiol production, Crews repeated the previous experiment but added a twist: He placed some females with males that had intact dewlaps, and other females with males whose dewlaps had been surgically removed.

The result? Females that were living with dewlap-less males were slow to produce eggs. They were just as slow, in fact, as the females in the first experiment that had been grouped with castrated males. These latter females had not been courted at all.

These data suggest that the dewlap is a key visual signal. The experiments succeeded in identifying the environmental cues—long day length and bobbing dewlaps—that trigger hormone production and the onset of sexual behavior.

Ultimate Causes: Sexual Selection

Why are environmental cues and bobbing dewlaps important to mating lizards? One reason is that lizards that succeed in synchronizing their sexual readiness will likely bear more offspring and have higher fitness than those that do not. *Anolis* lizards are most successful if they reproduce early in the spring when food supplies are increasing but snakes and other predators are not yet hunting to feed their own young. Failing to synchronize their sex or having sex too late would be disastrous for the *Anolis* lizards.

Also, bobbing dewlaps serve as an important signal in mate choice. A specific pattern of selection called **sexual selection** favors individuals possessing traits that increase their ability to obtain mates (introduced in Chapter 26). According to theory and experimental data:

• Females are usually the gender that is pickiest about mate choice because females spend a disproportionate amount of energy on their gametes compared to males.

• Females choose males that contribute good alleles and/or resources to their offspring.

Research on bird species such as zebra finches (see Chapter 26) has shown that individuals with particularly long tails, bright colors, and energetic courtship displays are usually the healthiest and best fed in the population. Because it takes large amounts of energy and resources to produce traits like these, they serve as an honest signal of male quality.

Are bright dewlaps honest signals of male quality? So far, the brightness of lizard dewlaps does not appear correlated with the nutritional health of the lizards. However, lizards with enlarged dewlaps tend to have high bite forces, which are important in fighting with other males to defend their feeding and breeding territories. Males are known to defend their territories fiercely, sometimes with costly consequences. The bobbing dewlap territorial displays serve as honest signals of physical vigor that can help to avert actual confrontation.

There are two types of sexual selection. When females choose males, intersexual ("between-sexes") selection is said to occur. When males compete with one another for mates, intrasexual ("within-sexes") selection is taking place (see Chapter 26). The data on *Anolis* lizards so far suggest that even though female sexual readiness is triggered by male displays, male competition for territories is the main factor determining mate choice.

Data on mate choice in *Anolis* lizards reinforce a central theme of this chapter: Animals usually make decisions in a way that maximizes their fitness.

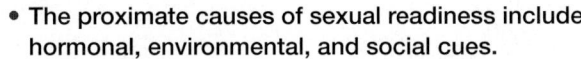
check your understanding

If you understand that . . .

C Y U

- The proximate causes of sexual readiness include hormonal, environmental, and social cues.
- At the ultimate level, animals use physical and behavioral cues to choose mates that will increase their fitness.

✔ **You should be able to . . .**

1. Propose one proximate cause of human sexual readiness that is similar to that in *Anolis* lizards.
2. Propose one ultimate cause of human mate choice that is similar to that in *Anolis* lizards.

Answers are available in Appendix A.

53.4 Where Should I Live?

Most animals select the food they eat and the individuals they mate with. In most cases, they also select the habitat where they live. Biologists have explored an array of questions related to habitat selection:

- Should juveniles disperse from the area where they were raised, or should they stay?

- How large an area should be defended against competitors in species that maintain a territory for breeding or feeding, as in *Anolis* lizards?

- Do individuals have higher fitness when they occupy crowded, high-quality habitats or uncrowded, poor-quality habitats?

Let's take a closer look at one of these "lifestyle" decisions: the proximate and ultimate mechanisms responsible for migration. In ecology, **migration** is defined as the long-distance movement of a population associated with a change of seasons. At the proximate level, how do animals know where to go between one seasonal "home" and another? At the ultimate level, what are the fitness benefits of having more than one home range that could overcome the costs of long-distance migrations?

Proximate Causes: How Do Animals Navigate?

To organize research into how animals find their way during migratory movements, biologists distinguish three categories of navigation:

1. **Piloting** is the use of familiar landmarks.

2. **Compass orientation** is movement that is oriented in a specific direction.

3. **True navigation (map orientation)** is the ability to locate a specific place on Earth's surface.

Imagine that you asked someone to blindfold you, drive you to a remote area 100 miles from home, and drop you off by the side of the road without your cell phone. After removing your blindfold, how would you find your way home? Which type of navigation would you use?

Piloting If you were lucky, you might recognize some landmarks and remember from a previous experience how to return home. The offspring of some migratory birds and mammals navigate in a similar way. They memorize the route when following their parents one or more times—south in the fall and north in the spring—and later are able to pilot the route on their own.

Compass Orientation If you were dropped off on the side of the road and did not recognize any specific landmarks, which way would you start walking? If you had a compass and knew which direction to head, that would be a big help. Once you got closer to home and found a familiar landmark, you could pilot the rest of the way home.

How do animals perform compass orientation? To date, most research on compass orientation has been done on migratory birds, such as European robins. To determine where north is, these animals appear to use the Sun, the stars, and Earth's magnetic field.

The Sun is difficult to use as a compass reference because its position changes during the day. It rises in the east, is due south at noon (in the Northern Hemisphere), and sets in the west. To use the Sun as a compass reference, then, an animal must have an internal clock that defines morning, noon, and evening. Fortunately, most animals have such a clock. The **circadian clock** that exists in organisms maintains a 24-hour rhythm of chemical activity. The clock is set by the light–dark transitions of day and night. It tells individuals enough about the time of day that they can use the Sun's position to find magnetic north.

The situation is actually simpler on clear nights because migratory birds in the Northern Hemisphere can use the North Star to find magnetic north and select a direction for migration. But what if the weather is cloudy?

Under these conditions, migratory birds appear to orient using Earth's magnetic field. There are two main hypotheses for how animals detect magnetism. One hypothesis contends that animals can detect magnetism by their visual system, through a chemical reaction that involves electron transfer among molecules. An alternative hypothesis maintains that individuals have small particles of magnetic iron—the mineral called magnetite—in their bodies, such as in the upper beaks of birds. Changes in the positions of magnetic particles, in response to Earth's magnetic field, could then be detected and provide reliable information for compass orientation. So far, the data suggest that birds may be capable of both mechanisms.

Although research on mechanisms of compass orientation continues, one important point is clear: Birds and other organisms have multiple mechanisms of finding a compass direction. At least some species can use a Sun compass, a star compass, and a magnetic compass. Which system an individual uses depends on the weather and other circumstances.

True Navigation (Map Orientation) If you were dropped off on the side of the road with a compass *and a map* of your location, finding your way home would be easy.

Do animals have the equivalent of map orientation? Neurobiologist and ecologist Kenneth Lohmann and his colleagues tested this question in green sea turtles. Sea turtles have the remarkable ability to travel around vast areas of the ocean and then return to a specific coastal nesting or feeding site with pinpoint accuracy. How do they do it?

The Earth's magnetic field varies around its surface in predictable gradients of both intensity and angle, or inclination. Lohmann's hypothesis was that the turtles use this magnetic field as a source of precise positional information, much like we use latitude and longitude coordinates to locate ourselves on a map.

To test this hypothesis, Lohmann captured green sea turtles at their coastal feeding grounds in Melbourne Beach, Florida, fitted them with soft harnesses attached to a tracking device, and then let them swim one at a time in a circular pool (**FIGURE 53.7**). A coil system around the pool was used to change the magnetic field to replicate the magnetic signature of locations to the north and south of Melbourne Beach.

The data at the bottom of Figure 53.7 show how the turtles responded. Each dot represents the average swimming direction for one turtle. The arrow represents the average for the whole group of turtles. When the magnetic field in the pool simulated a location 337 km north of Melbourne Beach, the turtles swam south, the direction of "home." When the magnetic field in the pool simulated a location 337 km south of Melbourne Beach, the

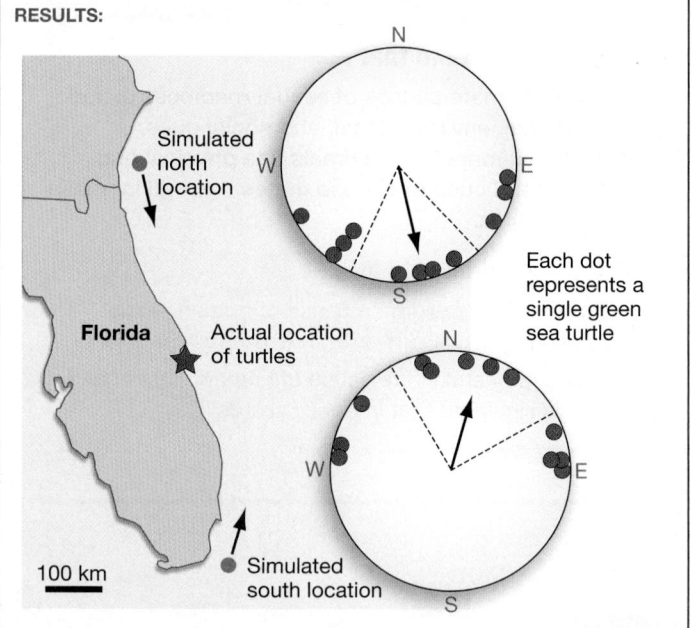
©2004 Nature Publishing Group

FIGURE 53.7 Sea Turtles Navigate Using Map Orientation

SOURCE: Lohmann, K. J., C. M. F. Lohmann, L. M. Ehrhart, et al. 2004. Geomagnetic map used in sea-turtle navigation. *Nature* 428: 909–910

✔ **QUESTION** Sketch what the data for the two sites would have to look like to reject Lohmann's hypothesis.

turtles swam north. Note that the direction of magnetic north did not change in either case, so compass orientation alone could not account for the ability of the turtles to navigate. Rather, the turtles were able to locate themselves on the map—they possess map orientation.

Ultimate Causes: Why Do Animals Migrate?

Why do some animals have more than one home range? Migratory movements can be spectacular in their extent and in the navigation challenges they pose:

- Arctic terns nest during summer in tundra in the far north. One population flies south along the coast of Africa to wintering grounds off Antarctica and then flies back north along the eastern coast of South America (**FIGURE 53.8**). Chicks follow their parents south but make the return trip on their own. The journey totals over 32,000 km (20,000 miles).

- Many of the monarch butterflies native to North America spend the winter in the mountains of central Mexico or southwest California. Tagged individuals are known to have flown over 3000 km (1870 miles) to get to a wintering area. In spring, monarchs begin the return trip north. They do not live to complete the trip, however. Instead, mating takes place along the way, and females lay eggs en route. Although adults die on the journey, offspring continue to head north. After several generational cycles of reproduction and death in the original habitats in northern North America, individuals again migrate south to overwinter. Instead of a single generation making the entire migratory cycle, then, the round trip takes several generations.

- Salmon that hatch in rivers along the Pacific Coast of North America and northern Asia migrate to the ocean when they are a few months to several years old, depending on the species. After spending several years feeding and growing in the North Pacific Ocean, they return to the stream where they hatched. There they mate and die (see Chapter 52).

In most cases, at the ultimate level it is relatively straightforward to generate hypotheses explaining why migration exists.

- Arctic terns feed on fish that are available in different parts of the world at different seasons. Thus, individuals that do not migrate might starve; individuals that do migrate should achieve higher reproductive success.

- Monarch butterflies may achieve higher reproductive success by migrating to wintering areas and new breeding areas than by trying to overwinter in northern North America.

- Salmon eggs and young are safer and thrive better in freshwater habitats than they do in the ocean. But adult salmon can find much more food in saltwater habitats than in freshwater and thus grow to bigger size, which confers a benefit in competition for nest sites and mates.

Testing these hypotheses rigorously is more difficult because researchers must compare the fitness of individuals that do and do not migrate. Recently, however, researchers have documented the evolution of sockeye salmon populations that do not migrate to the ocean—in watersheds where migratory sockeye were introduced by humans. The nonmigratory populations are much smaller bodied than the migratory populations—presumably because less food is available in the freshwater habitats.

The fitness of migrant and nonmigrant sockeye salmon appears to be similar in this case, due to fitness trade-offs (see Chapter 25). The larger-bodied migratory salmon are more successful in competing for nesting sites and mates. However, the migratory salmon also experience higher costs in terms of time, energy, and predation risk.

At the ultimate level, researchers are also struggling to understand how a nonmigratory population can evolve into a migratory one that travels to a completely different part of the world. It's likely that habitat changes occurring in response to global climate change (see Chapter 56) will furnish an example. Changes in the range and arrival times of migratory species have already been documented; it may be only a matter of time before a nonmigratory population begins to migrate, to cope with habitat shifts caused by global climate change.

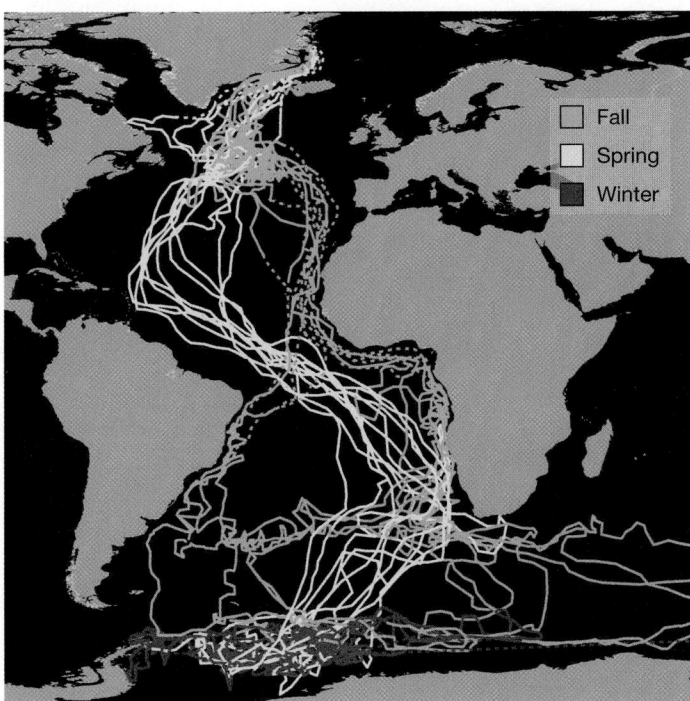

FIGURE 53.8 Arctic Terns Have Spectacular Migrations. The map shows the pathway of 11 arctic terns for a year based on data collected using tiny geolocators.

Legend: Fall / Spring / Winter

53.5 How Should I Communicate?

The song of a bird, the bobbing dewlap of a male lizard, and the sentences in this text all have the same overall goal: communication. In biology, **communication** is defined as any process in which a signal from one individual modifies the behavior of a recipient individual. A **signal** is any information-containing behavior or characteristic. Communication is a crucial component of animal behavior. It creates a stimulus that elicits a response.

By definition, communication is a social process. For communication to occur, it is not enough that a signal is sent; the signal must be received and acted on. How does this process play out in honeybees?

Proximate Causes: How Do Honeybees Communicate?

Honeybees are highly social animals that live in hives. Inside the hive, a queen lays eggs that are cared for by workers. Besides caring for young and building and maintaining the hive, workers obtain food for themselves and other members of the colony by gathering nectar and pollen from flowering plants.

Biologists noticed that bees appear to recruit to food sources, meaning that if a new source is discovered by one or a few individuals, many more bees begin showing up over time. This observation inspired the hypothesis that food-finders have some way of communicating the location of food to others. How do bees "talk" to each other?

The Dance Hypothesis In the 1930s Karl von Frisch began studying bee communication by observing bees that built hives inside the glass-walled chambers he had constructed. He found that if he placed a feeder containing sugar water near one of these observation hives, a few of the workers began moving in a circular pattern on the vertical, interior walls of the hive.

Von Frisch called these movements the "round dance" (**FIGURE 53.9a**). Other workers appeared to follow the progress of the dance, first by touching the displaying bee as it danced and then by flying away from the hive in search of the food source.

To investigate the function of these movements further, von Frisch placed feeders containing sugar water at progressively greater distances from the hive. Using this technique, he was able to get bees to visit feeders at a distance of several kilometers from the hive. By catching bees at the feeders and dabbing them with paint, he could individually mark successful food-finders. Follow-up observations at the hive confirmed that marked foragers (**1**) danced when they returned to the hive, and (**2**) returned to the food source with unmarked bees.

To explain these data, von Frisch proposed that the round dance contained information about the location of food. Because bee hives are completely dark, he hypothesized that workers got information from the dance by touching the dancer and following the dancer's movements.

Communicating Directions and Distances When von Frisch began placing feeders at longer distances from the hive, he found that successful food-finders used a new type of display. He named these movements the "waggle dance," because they combined circular movements like those of the round dance with short, straight runs (**FIGURE 53.9b**). During these runs, the dancer vigorously moved her abdomen from side to side.

These observations supported the hypothesis that both the round dance and waggle dance communicate information about food sources. Recent work has shown that the round and waggle dances are actually the same type of behavior—round dances just have a short waggle phase.

Three key observations allowed von Frisch to push our understanding of bee language further:

1. The orientation of the waggle part of the dance varied.

2. The direction of the waggle run correlated with the direction of the food source from the hive.

3. The length of the straight, "waggling" run was proportional to the distance the foragers had to fly to reach the feeder.

(a) The round dance

(b) The waggle dance

Other bee workers follow the progress of the dance by touching the displaying individual

FIGURE 53.9 Honeybees Perform Two Types of Dances. (a) During the round dance, successful food-finders move in a circle. **(b)** During the waggle dance, successful foragers move in a circle but then make straight runs through the circle. During the straight part of the dance, the dancer waggles her abdomen.

By varying the location of the food source and observing the orientation of the waggle dance given by marked workers, von Frisch was able to confirm that dancing bees were communicating the position of the food relative to the current position of the Sun. For example:

- If food is directly away from the Sun's current position, marked bees do the waggle portion of their dance directly downward (**FIGURE 53.10a**).

- If the food is 90 degrees to the right of the Sun, marked bees waggle 90 degrees to the right of vertical (**FIGURE 53.10b**).

These results are nothing short of astonishing. Honeybees do not have large brains, yet they are capable of symbolic language. What's more, they are able to interpret the angle of the waggle dance performed on a vertical surface and to respond by flying horizontally along the corresponding angle.

Further work has confirmed that the dance language of bees includes several modes of communication: tactile information in the movements themselves, sounds made during the dance, and scents that indicate the nature of the food source.

Ultimate Causes: Why Do Honeybees Communicate the Way They Do?

At the ultimate level, it's straightforward to come up with a hypothesis for why honeybees use a complex method of communication to share information about food sources. If more workers find a rich food source quickly, then the total amount of resources harvested as well as the number of offspring produced by the hive should increase. Thus the frequency of bee colonies with successful communication should increase over time.

In general, why do animals use the modes of communication that they do? One of the most general observations about communication is that the type of signal used by an organism correlates with its habitat. Bees, ants, and termites that live in hives or underground rely on olfactory and tactile communication. Similarly, bats, wolves, and other animals that are active at night communicate via sound or scent. In contrast, animals that are active during the day and that live in open or treeless habitats tend to rely on visual communication.

Sound travels much farther than light in aquatic habitats. Based on this observation, it is logical to observe that humpback whales rely on songs for long-distance communication. In some cases, humpback whale songs can travel hundreds of kilometers. Groups of whales use acoustic communication to keep together during migrations, and individual males sing to attract mates and warn rivals.

Each mode of communication has advantages and disadvantages. Songs and calls can carry information over long distances, but are short lived. Thus they have to be repeated to be effective, and frequent repetition requires a large expenditure of time and energy. In addition, acoustic communication attracts predators. It is no surprise that when a hawk or falcon approaches a marsh inhabited by courting red-winged blackbirds, things get very quiet. Communication systems have been honed by natural selection to maximize their benefits and minimize their costs.

(a) Straight runs down the wall of the hive indicate that food is opposite the direction of the Sun.

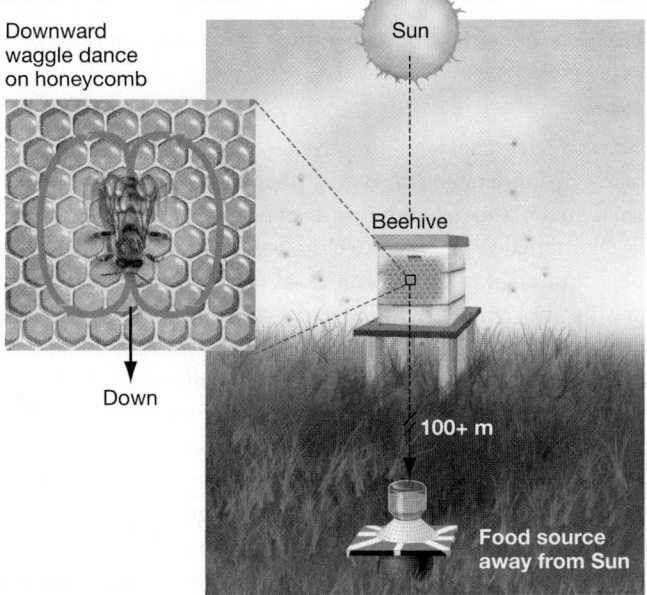

(b) Straight runs to the right indicate that food is 90° to the right of the Sun.

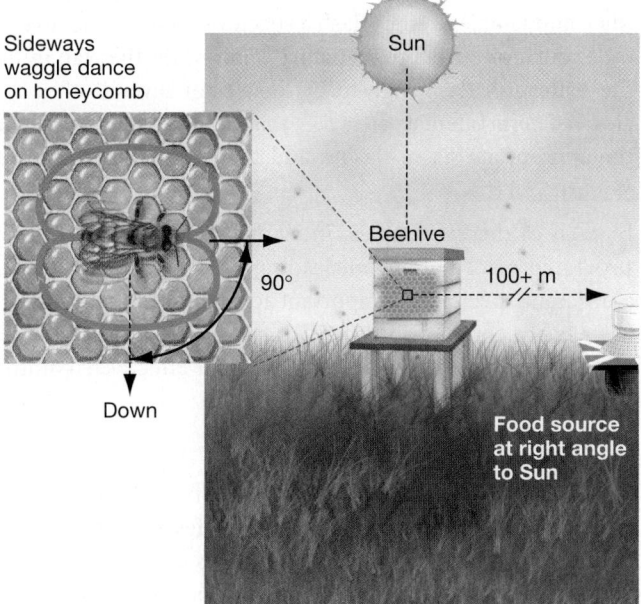

FIGURE 53.10 The Direction of the Waggle Dance Indicates the Location of Food Relative to the Sun.
Waggle dances are done only when the food is at least 100 m from the hive.

✔**EXERCISE** The length of the straight run in the waggle dance indicates the relative distance of the food. Diagram a waggle dance that indicates food in the direction of the Sun and twice as far away as the food sources indicated by the dances drawn here.

When Is Communication Honest or Deceitful?

At the ultimate level, one of the questions that biologists ask about communication concerns the quality of the information. Is the signal reliable?

As far as is known, honeybees are always honest. Stated another way, honeybee dances consistently "tell the truth" about the location of food sources. This observation is logical, because the honeybees that occupy a hive are closely related and cooperate extensively in the rearing of offspring. As a result, it is advantageous for an individual to convey information accurately. If a food-finder provided inaccurate or misleading information to its hivemates, fewer offspring would be reared and the food-finder's fitness would be reduced.

In many cases, however, natural selection has favored the evolution of deceitful communication. Recent research on deceitful communication has highlighted just how complex interactions between signalers and receivers can be. A few examples will help drive this point home.

Deceiving Individuals of Another Species **FIGURE 53.11** illustrates two of the hundreds of examples of deceitful communication that have been documented among members of different species. Many of the best-studied instances involve prey deceiving predators or predators deceiving prey:

- Hognose snakes are famous for their theatrical ability to play dead (Figure 53.11a). When perturbed by a predator, hognose snakes turn belly up, produce a foul odor, hang their tongues out of their gaping mouths, and may even produce a couple of droplets of blood from their mouths. When turned back onto their bellies, they quickly flip onto their backs again and sustain the charade until the coast is clear.

- Male and female fireflies flash a species-specific signal to each other during courtship. Predatory *Photuris* fireflies can mimic the pattern of flashes given by females of several other species. A *Photuris* female attracts a male of different species with the appropriate set of flashes and then attacks and eats him (Figure 53.11b).

In each of these examples, individuals increase their fitness by providing inaccurate or misleading information to members of a different species. It's important to realize, though, that the signaler is not consciously "lying." Instead, natural selection has simply favored certain behavioral traits that effectively communicate deceitful information.

Deceiving Individuals of the Same Species In some cases, natural selection has favored the evolution of traits or actions that deceive members of an organism's own species. Perhaps the best-studied type of deceit in nonhuman animals involves the mating system of bluegill sunfish. Male bluegills set up nesting territories in the shallow water along lake edges, fertilize the eggs laid in their nests, care for the developing embryos by fanning them with oxygen-rich water, and protect newly hatched offspring from large predators.

(a) Hognose snakes play dead to avoid being eaten.

(b) Female *Photuris* fireflies flash the courtship signal of another species and then eat males that respond.

"Femme fatale"

"Victim"

FIGURE 53.11 Deceitful Communication Is Common in Nature.

Some males cheat on this system, however, by mimicking females. To understand how this happens, examine the female, normal male, and female-mimic male in **FIGURE 53.12**. Mimics look like females but have well-developed testes and produce large volumes of sperm. Mimics also act like females during courtship movements with territory-owning males. They even adopt the usual egg-laying posture. When normal females approach the nest and begin courtship, the mimics join in.

The territory-owning male appears to tolerate the mimic, courting the two females at the same time. But when the actual female begins to lay eggs, the mimic responds by releasing sperm—and thus fertilizing some of the eggs—and then darting away. In this way, the mimic fathers offspring but does not help care for them. In addition, mimics do not have to expend time and energy in growing large and making and defending a nest.

Why aren't all males mimics? It turns out that as the number of mimics increases, the reproductive success of mimics declines. This is a common observation in studies of deceitful

FIGURE 53.12 In Bluegill Sunfish, Some Males Look and Act Like Females.

Territorial male

Female

Female-mimic male

communication—cheating works best within a population when it is relatively rare. The logic behind this hypothesis runs as follows: If deceit becomes extremely common, then natural selection will strongly favor individuals that can detect and avoid or punish liars. But if liars are rare, then natural selection will favor individuals that are occasionally fooled but are more commonly rewarded by responding to signals in a normal way. This is an example of frequency-dependent selection (see Chapter 26).

Further, as the number of mimic males increases in sunfish, competition with other mimics for mating opportunities increases and aggression by territory-owning males on mimic males also increases, lowering the reproductive success of the mimics.

check your understanding

If you understand that . . .

- Communication is an exchange of information between individuals.
- In most instances, the mode of communication that animals use maximizes the probability that the information will be transferred efficiently.
- Communication can be honest or deceitful, depending on the nature of the information received.

✔ **You should be able to . . .**

1. Explain the costs and benefits of auditory, olfactory, and visual communication.
2. Predict why natural selection favors individuals that can detect deceitful communication and either avoid "liars" or punish them.

Answers are available in Appendix A.

53.6 When Should I Cooperate?

The types of behavior reviewed thus far all have a key common element: They help individuals respond to environmental stimuli in a way that increases their fitness. There is a type of behavior that appears to contradict this pattern, however: altruism.

Altruism is behavior that has a fitness cost to the individual exhibiting the behavior and a fitness benefit to the recipient of the behavior. It is the formal term for self-sacrificing behavior. Altruism decreases an individual's ability to produce offspring but helps others produce more offspring.

In terms of ultimate causation, altruistic behavior appears to be paradoxical. If certain alleles make an individual more likely to be altruistic, those alleles should be selected against (Chapter 25). Meanwhile, individuals with selfish alleles should be more likely to survive and produce more offspring. Over time, selfish alleles should increase in frequency while self-sacrificing alleles should decrease in frequency. Does altruism actually occur? Let's consider four explanations for apparent altruistic behavior.

Kin Selection

There is no question that self-sacrificing behavior occurs in nature. For example, black-tailed prairie dogs perform a behavior called alarm calling (**FIGURE 53.13**). These burrowing mammals live in large communities, called towns, throughout the Great Plains region of North America. When a badger, coyote, hawk, or other predator approaches a town, some prairie dogs give alarm calls that alert other prairie dogs to run to mounds and scan for the threat.

Giving these calls is risky. In several species of ground squirrels and prairie dogs, researchers have shown that alarm-callers draw attention to themselves by calling and are in much greater danger of being attacked than non-callers are.

FIGURE 53.13 Black-Tailed Prairie Dogs Are Highly Social.
Prairie dogs live with their immediate and extended family within large groups called towns. The individual with the upright posture has spotted an intruder and may give an alarm call.

The coefficient of relatedness, *r*, varies between 0.0 and 1.0. If two individuals have no identical alleles that were inherited from the same ancestor, then their *r* value is 0.0. Because every allele in pairs of identical twins is identical, their coefficient of relatedness is 1.0.

What about other relationships? **FIGURE 53.14a** shows how *r* is calculated between half-siblings. (To review what the boxes, circles, and lines in a pedigree mean, see Figure 14.22.) Half-siblings share one parent. Thus, *r* represents the probability that half-siblings share alleles as a result of inheriting alleles from their common parent. It is critical to realize that in each parent-to-offspring link of descent, the probability of any particular allele being transmitted is 1/2. This is so because meiosis distributes alleles from the parent's diploid genome to their haploid gametes randomly. Thus half the gametes produced by a parent get one of the alleles present at each gene, and half the gametes produced get the other allele. Half-siblings are connected by two such parent-to-offspring links. The overall probability of two half-siblings sharing the same allele by descent is $1/2 \times 1/2 = 1/4$. (To review rules for combining probabilities, see **BioSkills 5** in Appendix B.)

To think about this calculation in another way, focus on the red arrows in Figure 53.14a. The left arrow represents the probability that the mother transmits a particular allele to her son. The right arrow represents the probability that the mother transmits the same allele to her daughter. Both probabilities are 1/2. Thus the probability that the mother transmitted the same allele to both her son and daughter is $1/2 \times 1/2 = 1/4$.

FIGURE 53.14b shows how *r* is calculated between full siblings. The challenge here is to calculate the probability of two individuals sharing the same allele as a result of inheriting it through their mother or through their father. The probability that full siblings share alleles as a result of inheriting them from one parent is 1/4. Thus the probability that full siblings share alleles inherited from either their mother or their father is $1/4 + 1/4 = 1/2$.

(a) What is the probability that half-siblings inherit the same allele from their common parent?

r **between half-siblings:**
$\frac{1}{2} \times \frac{1}{2} = \frac{1}{4}$

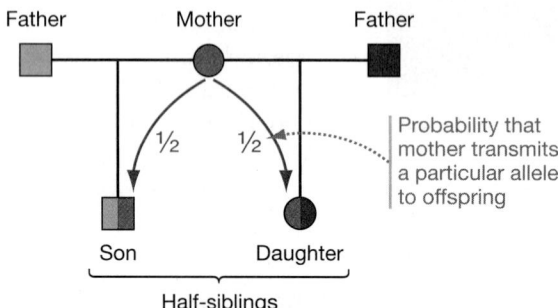

Probability that mother transmits a particular allele to offspring

Half-siblings

(b) What is the probability that full siblings inherit the same allele from their father or their mother?

Probability that they inherit same allele from **father**:
$\frac{1}{2} \times \frac{1}{2} = \frac{1}{4}$

Probability that they inherit same allele from **mother**:
$\frac{1}{2} \times \frac{1}{2} = \frac{1}{4}$

r **between full siblings:**
$\frac{1}{4} + \frac{1}{4} = \frac{1}{2}$

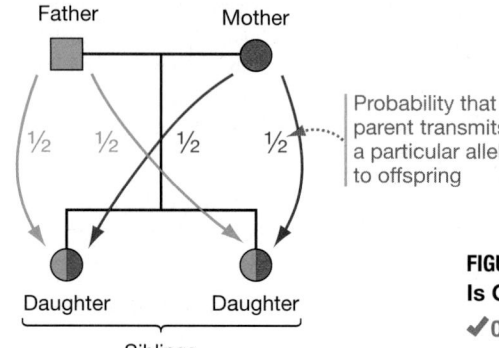

Probability that parent transmits a particular allele to offspring

Siblings

FIGURE 53.14 The Coefficient of Relatedness (*r*) Is Calculated from Information in Pedigrees.

✔ **QUANTITATIVE** What is the *r* between first cousins?

Hamilton's Rule   How can natural selection favor the evolution of self-sacrificing behavior? William D. Hamilton answered this question by creating a mathematical model to assess how an allele that contributes to altruistic behavior could increase in frequency in a population.

To model the fate of altruistic alleles, Hamilton represented the fitness cost of the altruistic act to the actor as *C* and the fitness benefit to the recipient as *B*. Both *C* and *B* are measured in units of offspring produced. His model showed that the allele could spread if

$$Br > C$$

where *r* is the **coefficient of relatedness.** The coefficient of relatedness is a measure of how closely the actor and beneficiary are related. Specifically, *r* measures the fraction of alleles in the actor and beneficiary that are identical by descent—that is, inherited from the same ancestor (**Quantitative Methods 53.1**).

The formulation $Br > C$ is called **Hamilton's rule.** It states that altruistic behavior is most likely when three conditions are met:

1. The fitness benefits are high for the recipient.

2. The altruist and recipient are close relatives.

3. The fitness costs to the altruist are low.

When Hamilton's rule holds, alleles associated with altruistic behavior will be favored by natural selection—because close relatives are very likely to have copies of the altruistic alleles. These alleles will increase in frequency in the population.

Inclusive Fitness   Hamilton's rule is important because it shows that individuals can pass on their alleles to the next generation not only by having their own offspring but also by helping close relatives produce more offspring. To capture this point, biologists refer to direct fitness and indirect fitness.

- Direct fitness is derived from an individual's own offspring. Parents can increase their direct fitness by spending resources to ensure the welfare of their offspring. This behavior is called parental care (Chapter 35).

- Indirect fitness is derived from helping relatives produce more offspring than they could produce on their own.

The combination of direct and indirect fitness components is called **inclusive fitness.**

Biologists use the term **kin selection** to refer to natural selection that acts through benefits to relatives at the expense of the individual. While kin selection decreases the direct fitness of an individual, it increases the indirect fitness, resulting in an overall increase in an individual's inclusive fitness. ✔ If you understand kin selection, you should be able to explain why it is likely to be common in humans.

Testing Hamilton's Rule Does Hamilton's rule work? Do animals really favor relatives when they act altruistically? To test the kin-selection hypothesis, behavioral ecologist John Hoogland studied which of the inhabitants of a black-tailed prairie dog town were most likely to give alarm calls.

Within a large prairie dog town, individuals live in small groups called coteries that share the same underground burrow. Members of each coterie defend a territory inside the town.

By tagging offspring that were born over several generations, Hoogland identified the genetic relationships among individuals in the town. He determined whether each individual had:

- no close genetic relatives in its coterie;

- no offspring in the coterie but at least one sibling, cousin, uncle, aunt, niece, or nephew; or

- at least one offspring or grandoffspring in the coterie.

The kin-selection hypothesis predicts that individuals who do not have close genetic relatives nearby will rarely give an alarm call. To evaluate this prediction, Hoogland recorded the identity of callers during 698 experiments. In these studies, a stuffed badger was dragged through the colony on a sled.

Were prairie dogs with close relatives nearby more likely to call, or did kinship have nothing to do with the probability of alarm calling? The bar charts in **FIGURE 53.15** show the average proportion of times that individuals in each of the three categories called during an experiment. The data indicate that black-tailed prairie dogs are much more likely to call if they live in a coterie that includes close relatives.

This same pattern—of preferentially dispensing help to kin—has been observed in many other species of social mammals and birds. Most cases of self-sacrificing behavior that have been analyzed to date are consistent with Hamilton's rule and are hypothesized to be the result of kin selection.

Manipulation

Note that altruism refers to a type of behavior with specific fitness consequences to an organism, not to the generosity of an organism. Sometimes altruistic behavior is involuntary.

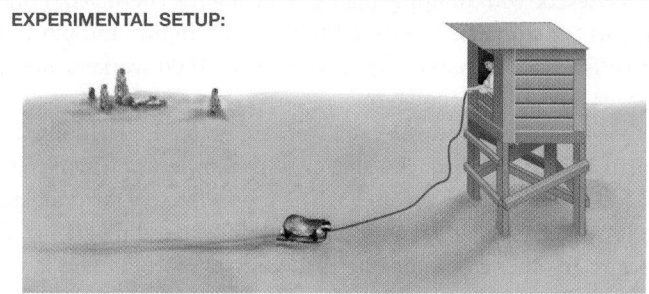

RESEARCH

QUESTION: Do black-tailed prairie dogs prefer to help relatives when they give an alarm call?

HYPOTHESIS: Individuals give an alarm call only when close relatives are near.

NULL HYPOTHESIS: The presence of relatives has no influence on the probability of alarm calling.

EXPERIMENTAL SETUP:

1. Determine relationships among individuals in prairie dog coterie.

2. Drag stuffed badger across territory of coterie.

3. From observation tower, record which members of coterie give an alarm call.

4. Repeat experiment 698 times. Each prairie dog coterie is tested 6–9 times over a 3-year period.

PREDICTION OF KIN-SELECTION HYPOTHESIS: Individuals in coteries that contain a close genetic relative are more likely to give an alarm call than are individuals in coteries that do not contain a close genetic relative.

PREDICTION OF NULL HYPOTHESIS: The presence of relatives in coteries will not influence the probability of alarm calling.

RESULTS:

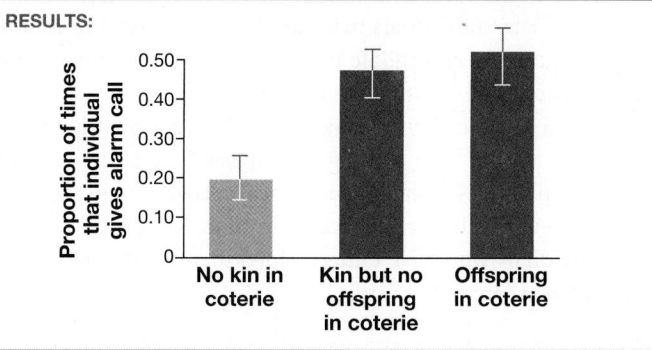

CONCLUSION: Alarm calling usually benefits relatives.

FIGURE 53.15 Experimental Evidence That Black-Tailed Prairie Dogs Are More Likely to Give Alarm Calls if Relatives Are Nearby.

SOURCE: Hoogland, J. L. 1983. Nepotism and alarm calling in the black-tailed prairie dog (Cynomys ludovicianus). *Animal Behavior* 31: 472–479.

✔ **QUESTION** What would be an appropriate control in this experiment, and what hypothesis would it test?

For example, one of the most famous applications of Hamilton's rule is in explaining the complex social structure of organisms such as the bees discussed earlier, as well as ants and wasps. In these societies, workers sacrifice most or all of their direct reproduction to help rear the queen's offspring, a phenomenon called **eusociality.**

Why would workers be so altruistic? In insect societies that have a single, diploid queen who mates once with a *haploid* male, all the workers are highly related, with a coefficient of relatedness of 0.75—much higher than the 0.5 of their daughters. These numbers suggest that it would be more adaptive for workers to raise sisters than to raise their own offspring, supporting Hamilton's rule.

However, the queens in some insect societies mate with diploid males, or with multiple males, reducing the coefficient of relatedness of workers. Inclusive fitness is still important, but it is not sufficient to explain why less than 1 in 1000 workers would lay eggs. What else is going on?

One factor is that workers are sometimes manipulated into altruistic behavior. The queen and/or other workers police the nest and kill any eggs laid by workers. The more effective the policing, the less likely the workers are to continue expending energy to lay eggs. In other cases, the queen suppresses reproduction in the workers via pheromones or other chemical signals. Manipulation, or "enforced altruism," makes sense at the ultimate level because efficient colonies will be favored by natural selection, while rampant "cheating" can cause colonies to collapse.

Reciprocal Altruism

During long-term studies of highly social animals such as horses, elephants, dolphins, lions, chimpanzees, and vampire bats, biologists have observed nonrelatives helping each other. Chimps and other primates, for example, may spend considerable time grooming unrelated members of their social group—cleaning their fur and removing ticks and other parasites from their skin. In vampire bats, individuals that have been successful in finding food are known to regurgitate blood meals to non-kin that have not been successful and that are in danger of starving.

How can self-sacrificing behavior like this evolve if kin selection is not acting? The leading hypothesis to explain altruism among nonrelatives is called **reciprocal altruism:** an exchange of fitness benefits that are separated in time. Reciprocal altruists help individuals who have either helped them in the past or are likely to help them in the future.

Data that have been collected so far support the reciprocal altruism hypothesis in some instances.

- Among vervet monkeys, individuals are most likely to groom unrelated individuals that have groomed or helped them in the past.

- Vampire bats are most likely to donate blood meals to non-kin that have previously shared food with them.

Reciprocal altruism is also widely invoked as an explanation for the helpful and cooperative behavior commonly observed among unrelated humans.

Cooperation and Mutualism

Some behaviors may appear altruistic, but actually may be selfish because they increase, rather than decrease, the fitness of the animal exhibiting the behavior. That is, the benefits of assisting others can exceed the costs.

- African wild dogs cannot hunt large prey alone, but they can succeed by hunting as a pack. The fitness of all the dogs increases when they work together. This is an example of cooperation among members of the same species.

- Female horses that spend more time in grooming relationships with other female horses tend to suffer less harassment from males—and to have more foals.

When cooperation occurs among individuals of different species, mutualism is said to occur. The relationship between cleaner shrimp and their host fish is mutualistic because both the shrimp and the fish benefit when the shrimp clears off—and eats—parasites from the fish's gills and jaws (see Chapter 55).

To summarize, organisms are altruistic when doing so either increases their direct fitness or their inclusive fitness, or when they are coerced. As with feeding, mating, finding a place to live, and communicating, helping others has costs and benefits that must be weighed by the organism. This "decision making" occurs at the proximate level through genetic and physiological mechanisms, and at the ultimate level through evolutionary processes.

If you understand . . .

53.1 An Introduction to Behavioral Ecology

- Proximate causation: Experiments and observations focus on understanding how specific gene products, neuron activity, and hormonal signals cause behavior.

- Ultimate causation: Researchers seek to understand the adaptive significance of behavior, or why it enables individuals to survive and reproduce, thus increasing their fitness.

- Behaviors vary in the extent to which they are inflexible or flexible, and in the extent to which they are innate or learned.

- Animals appear to weigh the costs and benefits of responding to a particular situation in various ways. Costs and benefits are measured in terms of their impact on fitness.

✔ You should be able to explain the proximate and ultimate causes for increased sex drive in human teens.

53.2 What Should I Eat?

- Proximate causation: Fruit fly larvae will be sitters or rovers depending on which alleles they carry for a certain gene.

- Ultimate causation: The sitter and rover alleles are maintained in fruit fly populations because the fitness consequences of the alleles depend on fluctuating environmental circumstances.

- Optimal foraging theory predicts that animals will make foraging decisions that maximize energy gain and minimize energy loss, time spent, and predation risk, among other factors. For example, gerbils balance the benefits of foraging for food with the costs of exposure to owls.

✔ You should be able to suggest what variables you would have to consider in testing the principle of optimal foraging in a web-building spider.

53.3 Who Should I Mate With?

- Proximate causation: The onset of sexual behavior in *Anolis* lizards depends on surges of the sex hormones testosterone and estradiol that are triggered by springlike temperatures *and* courtship displays from males.

- Ultimate causation: Sexual selection favors individuals with traits that increase their ability to obtain mates. Females tend to choose mates that provide good alleles and valuable resources. Males compete for the opportunity to mate with females.

- Sexual selection in *Anolis* lizards is determined more by male–male competition than by female choice.

✔ You should be able to generate a hypothesis for why nest decorations are a reliable signal of male quality in bowerbirds (see chapter-opening photo).

53.4 Where Should I Live?

- Many animals migrate between different habitats rather than staying in one place due to the seasonable availability of food and the location of good nesting and mating places.

- Proximate causation: Animals can find their way by following familiar landmarks (piloting); by getting compass information from the Sun, the stars, and Earth's magnetic field; and/or by using the magnetic field for true navigation (map orientation).

- Ultimate causation: The benefits of migrating to food sources and mating and nesting areas must outweigh the energetic costs and danger of the migration.

✔ You should be able to provide an evolutionary hypothesis for the observation that some bird species do not migrate if people supply food for them in feeders.

53.5 How Should I Communicate?

- Honeybees have a symbolic language—their dancing movements communicate accurate information about the direction and distance of food sources.

- Proximate causation: Animals communicate with touch, sounds, scents, and/or sight depending on the environment.

- Ultimate causation: The ability to communicate effectively increases the fitness of animals.

- Communication can be honest or deceitful. Deceitful communication can be adaptive when it is rare.

✔ You should be able to predict the consequences for deceitful communication in bluegill sunfish if most large, territory-owning males were fished out of a lake.

53.6 When Should I Cooperate?

- Proximate causation: Altruistic behavior can be voluntary or coerced.

- Ultimate causation: Altruistic behavior appears paradoxical because it lowers the direct fitness of an individual.

- When altruistic behavior is directed toward close relatives, alleles that lead to self-sacrificing may increase in frequency due to kin selection (increasing the inclusive fitness of the altruist).

- Some animals that live in close-knit social groups engage in reciprocal altruism—meaning they exchange help over time.

- Other animals cooperate because everyone benefits, such as in cooperative foraging. Cooperation among species is called mutualism.

✔ You should be able to describe the characteristics of a species for which helpful behavior is expected to be common.

(MB) MasteringBiology

1. **MasteringBiology Assignments**

   Tutorials and Activities Animal Behavior and Learning; Homing Behavior in Digger Wasps

   Questions Reading Quizzes, Blue-Thread Questions, Test Bank

2. eText Read your book online, search, take notes, highlight text, and more.

3. The Study Area Practice Test, Cumulative Test, BioFlix® 3-D Animations, Videos, Activities, Audio Glossary, Word Study Tools, Art

You should be able to . . .

1. What do proximate explanations of behavior focus on?
   a. how displays and other types of behavior have changed through time, or evolved
   b. the functional aspect of a behavior, or its "adaptive significance"
   c. genetic, neurological, and hormonal mechanisms of behavior
   d. appropriate experimental methods when studying behavior

2. What do ultimate explanations of behavior focus on?

3. What unit(s) do biologists use when analyzing the costs and benefits of behavior?
   a. proximate versus ultimate
   b. body size (e.g., large versus small)
   c. time
   d. fitness—the ability to survive and produce offspring

4. Which of the following statements about the waggle dance of the honeybee is not correct?
   a. The length of a waggling run is proportional to the distance from the hive to a food source.
   b. Sounds and scents produced by the dancer provide information about the nature of the food source.
   c. The speed of waggling is proportional to the distance from the hive to a food source.
   d. The orientation of the waggling run provides information about the direction of the food from the hive, relative to the Sun's position.

5. Why are biologists convinced that the sex hormone testosterone is required for normal sexual activity in male *Anolis* lizards?
   a. Male *Anolis* lizards with larger testes court females more vigorously than do males with smaller testes.
   b. Testosterone is not found in female *Anolis* lizards.
   c. Male *Anolis* lizards whose gonads had been removed did not develop dewlaps.
   d. Male *Anolis* lizards whose gonads had been removed did not court females.

6. Why does altruism seem paradoxical?
   a. Sometimes altruistic behavior is actually selfish.
   b. Altruism does not actually help others.
   c. Alleles that cause an organism to behave altruistically should be selected against since these alleles should lower the organism's fitness.
   d. Animals behave altruistically to help the species, but sometimes their behavior harms the species.

7. Explain the ultimate causes of honeybee dancing.

8. Is it true that all organisms forage optimally? Why or why not?

9. Why do behaviors like *Anolis* lizard bobbing displays serve as honest signals of male quality?

10. What environmental stimuli cause changes in hormone levels that lead to egg laying in *Anolis* lizards? What biologically relevant information do these stimuli provide?

11. Evaluate this statement: For green sea turtles to return from the open ocean to the beach where they were hatched, they must have both a "map" and a "compass."

12. **QUANTITATIVE** Hamilton's rule states that an altruistic allele could spread in a population if $Br > C$, where B represents the fitness benefit to the recipient, r is the coefficient of relatedness between altruist and recipient, and C represents the fitness cost to the altruist. If $r = 0.5$ between the altruist and the recipient, what would the ratio of costs to benefits have to be for the altruistic allele to spread?
   a. $C/B > 0.5$
   b. $C/B > 0$
   c. $C/B < 0.5$
   b. $C/B < 0$

13. To date, the rover and sitter alleles have been shown to affect foraging behavior only in fruit fly larvae. Design an experiment to test the hypothesis that adults with the rover allele tend to fly farther in search of food sources than adults with the sitter allele.

14. You are invited on a summer internship to study the behavioral ecology of howler monkeys in Costa Rica. Your task will be to follow a troop of howler monkeys during the day and identify what types of fruits and leaves the troop consumes. Which of these general principles will help guide your work? Select all that apply.
    a. condition-dependent behavior
    b. optimal foraging
    c. Hamilton's rule
    d. cost–benefit analysis

15. **QUANTITATIVE** A biologist once remarked that he'd be willing to lay down his life to save two brothers or eight cousins. Explain what he meant.

16. Based on the theory of reciprocal altruism, predict the conditions under which people are expected to donate blood.

54 Population Ecology

In this chapter you will learn that

The number of individuals in populations change over space and time

starting with
an overview of

Distribution and abundance
54.1

looking closer at

Demography
54.2

assists
study of

Population growth
54.3

applies to

including

Human population growth
54.5

affects

threatens

Population dynamics
54.4

informs

Conservation of populations
54.6

This vivid school of crescent-tail bigeyes off the coast of Papua New Guinea demonstrates how individuals associate together in populations. The distribution and abundance of individuals in populations changes over space and time.

f you asked a biologist to name two of today's most pressing global issues, she might say global climate change and extinction of species. If you were asked to name two of the most pressing issues facing your region, you might say traffic and the price of housing. All four problems have a common cause: recent and dramatic increases in the size of the human population.

In 1940, for example, Mexico City had a population of about 1.6 million. The city grew to 5.4 million in 1960, 13.9 million in 1980, and over 21 million in 2010. How much bigger will it get over the next 50 years? And how large will the entire human population be by the time your kids are in college?

This chapter is part of the Big Picture. See how on pages 1196–1197.

✔ When you see this checkmark, stop and test yourself. Answers are available in Appendix A.

A **population** is a group of individuals of the same species that live in the same area at the same time; **population ecology** is the study of how and why the number of individuals in a population changes over time and space. (You can see how population ecology fits into the Big Picture of Ecology on pages 1196–1197).

With the explosion of human populations across the globe, the massive destruction of natural habitats, changes in climate, and the resulting threats to species throughout the tree of life, population ecology has become a vital field in biological science. The mathematical and analytical tools introduced in this chapter help biologists predict changes in population size and design management strategies to save threatened species.

Let's start by considering some of the basic tools that biologists use to study populations and then follow up with examples of how biologists study changes in the population size of humans and other species over time. The chapter concludes by asking how all of these elements fit together in efforts to limit human population growth and conserve biodiversity.

54.1 Distribution and Abundance

If you were interested in studying the population biology of the blue-ringed octopus, Florida strangler fig, or any other species, you might want to start by asking, "Where do they live, and how many are there?" Distribution and abundance are two fundamental concepts in ecology.

Most species identification guides show the **range,** or geographic distribution, of different species. Two types of factors determine range:

- *Abiotic factors*, such as temperature, rainfall, the presence of geographical structures like mountains and oceans, and large-scale ongoing and historical processes such as continental drift (see Chapter 52);

- *Biotic factors*, such as the past and current presence of other species that provide habitat, food, or competition (discussed in detail in Chapter 55).

Ranges are dynamic—in constant flux as abiotic and biotic factors change over time.

The range of a species can also be considered at different scales. For example, the map in **FIGURE 54.1** shows the vast global range of the common lizard, *Lacerta vivipara* (sometimes called *Zootoca vivipara*), from Ireland in western Europe to Japan in eastern Asia. This map indicates that *L. vivipara* can live in a variety of habitats and is cold adapted—it lives farther north than other reptiles, and only at high altitudes at the southern extent of its range.

Lizards are not distributed uniformly within this range, however. The **population density**—the number of individuals per unit area—of *L. vivipara* varies throughout its range. If you zoom in on one region, such as the British Isles, you can observe that *L. vivipara* is remarkably widespread, yet absent from some areas. Zooming in further still, the distribution of *L. vivipara* is

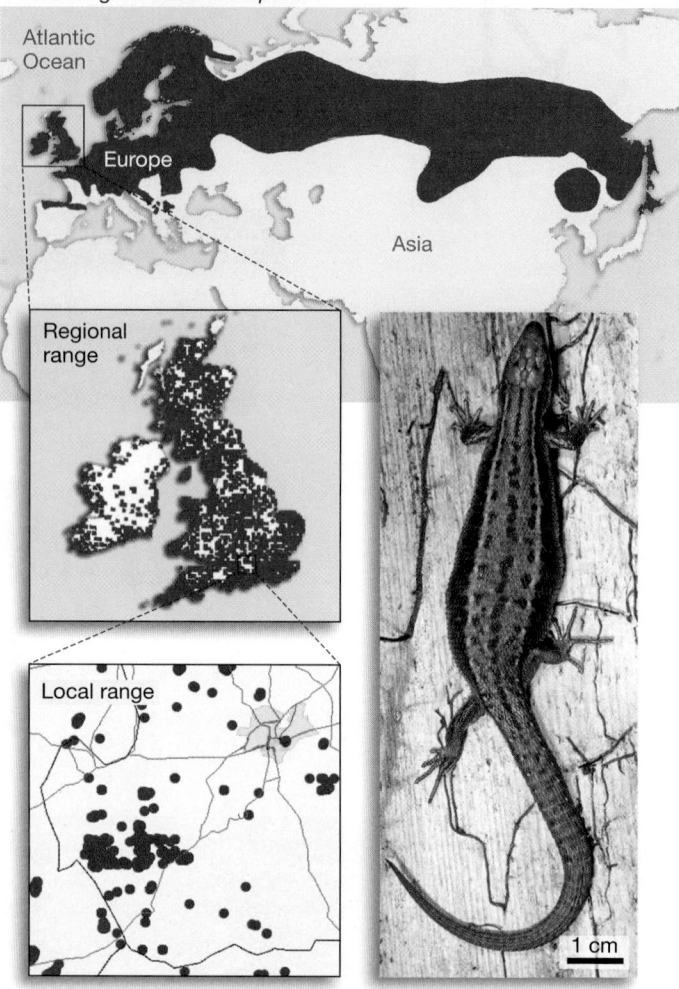

Global range of *Lacerta vivipara*

FIGURE 54.1 Geographic Range of *Lacerta vivipara*.

clumped according to the distribution of suitable habitats, such as meadows and open woodland.

In general, individual organisms can be arranged in different patterns within populations. The arrangement may be

- *random*, if the position of each individual is independent of the others, as may occur when seeds are dispersed by the wind;

- *clumped*, if the quality of the habitat is patchy or the organisms associate in social groups (such as schools of fish); or

- *uniform*, if negative interactions occur among individuals, such as competition for space, water, or other resources.

Within a certain patch of habitat, the distribution of *L. vivipara* individuals may be relatively uniform due to their maintenance of feeding and mating territories (see Chapter 53). However, as with most species, the distribution of lizards is clumped overall due to varying density of individuals in habitat patches of different quality. The particular dispersion pattern of a population has both proximate causes, in terms of physiological and behavioral mechanisms, and ultimate causes, in terms

To begin a mark–recapture study, researchers catch individuals and mark them with leg bands, ear tags, or some other method of identification. After the marked individuals are released, they are allowed to mix with the unmarked animals in the population for a period of time. Then a second trapping effort is conducted, and the percentage of marked individuals that were captured is recorded.

To estimate the total population from these data, researchers make a key assumption: The percentage of marked and recaptured individuals is equal to the percentage of marked individuals in the entire population. This assumption should be valid if individuals are not moving in and out of the study area, and if no bias exists regarding which individuals are caught in each sample attempt. It

is important that individuals do not learn to avoid or seek out traps after being caught once and that they do not change their behavior or die as a result of being marked or trapped.

The relationship between marked and unmarked individuals can be expressed algebraically:

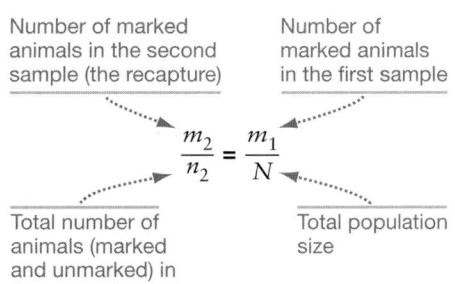

Number of marked animals in the second sample (the recapture)

Number of marked animals in the first sample

$$\frac{m_2}{n_2} = \frac{m_1}{N}$$

Total number of animals (marked and unmarked) in the second sample

Total population size

(Eq. 54.1)

Having measured m_2, n_2, and m_1, the researcher can estimate N.

✔**QUANTITATIVE** If you understand how to estimate population size in a mark–recapture study, you should be able to do so in this example: Suppose researchers marked 255 animals and later were able to trap a total of 162 individuals in the population, of which 78 were marked. What is the estimate for total population size? Solve for N (give the nearest integer) and check your answer.[1] (Answer is given below.)

[1] $N = 530$.

---

of evolutionary adaptations (see Chapter 53 for an overview of proximate and ultimate causation).

If a species range is small, it may consist of a single population of interbreeding individuals. If a range is large, it may consist of many populations—sometimes contiguous, and sometimes isolated in space. If individuals from a species occupy many small patches of habitat, so that they form many independent populations, they are said to represent a **metapopulation**—a population of populations connected by migration. Because humans are reducing large, contiguous areas of forest and grasslands to isolated patches or reserves, more and more species are being forced into a metapopulation structure.

How can the population size, density, or distribution pattern of individuals be determined? It depends on the species. For sedentary (largely immobile) organisms such as trees and barnacles, researchers sample habitats by counting the individuals that occur along lines of known position and length—called *transects*—or inside rectangular plots—called *quadrats*—set up at random locations in the habitat. These counts can be extrapolated to the entire habitat area to estimate the total population size. In addition, they can be compared to later censuses to document trends over time.

In contrast, counting organisms that are mobile and that do not congregate into herds or flocks or schools is much more challenging. For example, the number of lizards inside sample quadrats or along transects changes constantly as individuals move. Further, it can be difficult to track whether a particular individual has already been counted. But if individuals can be captured and then tagged in some way, the total population size of a mobile species can be estimated by using a research approach called mark–recapture (**Quantitative Methods 54.1**).

Note that individuals within populations vary, and populations are not static in time and space. To get a better understanding of

how researchers quantify population size and structure, let's first examine a field of study in ecology called demography.

54.2 Demography

The number of individuals present in a population depends on four processes: birth, death, immigration, and emigration.

- Populations grow as a result of births—here, meaning any form of reproduction—and **immigration,** which occurs when individuals enter a population by moving from another population.

- Populations decline due to deaths and **emigration,** which occurs when individuals leave a population to join another population.

Analyzing birth rates, death rates, immigration rates, and emigration rates is fundamental to **demography**: the study of factors that determine the size and structure of populations through time.

To predict the future of a population, biologists have to know something about its makeup. They need to know the population's **age structure**—how many individuals of each age are alive. They also need to know how likely individuals of different ages are to survive to the following year, how many offspring are produced by females of different ages, and how many individuals of different ages immigrate and emigrate each **generation**—the average time between a mother's first offspring and her daughter's first offspring.

If a population consists primarily of young individuals with a high survival rate and reproductive rate, the population size should increase over time. But if a population comprises chiefly old individuals with low reproductive rates and low survival rates, then it is almost certain to decline over time. Life tables are useful tools for analyzing this type of information.

Life Tables

A **life table** summarizes the probability that an individual will survive and reproduce in any given time interval over the course of its lifetime. Life tables were invented almost 2000 years ago; ancient Romans used them to predict food needs. In modern times, life insurance companies use life tables to predict the likelihood of a person dying at a given age. Biologists use life tables to study the demographics of endangered species.

***Lacerta vivipara*: A Case Study** To understand how researchers use life tables, consider again the lizard *L. vivipara*. As their name suggests—vivipara means "live birth"—most populations are ovoviviparous and give birth to live young (see Chapter 33).

European biologists Henk Strijbosch and Raymond Creemers set out to construct a life table of a low-elevation population of *L. vivipara* in the Netherlands, with the goal of comparing the results to data that other researchers had collected from *L. vivipara* populations in the mountains of Austria and France and in lowland habitats in Britain and Belgium. They wanted to know whether populations that live in different environments vary in basic demographic features.

To begin the study, Strijbosch, Creemers, or their students visited their study site daily during the seven months that these lizards are active during the year. Each day, the research team captured and marked as many individuals as possible.

Because this program of daily monitoring continued for seven years, the biologists were able to document the number of young produced by each female in each year of her life. If a marked individual was not recaptured in a subsequent year, they assumed that it had died sometime during the previous year. The data allowed researchers to calculate the number of individuals that survived each year in each particular age group as well as how many offspring each female produced. What did the numbers reveal?

TABLE 54.1 shows the data Strijbosch and Creemers collected for females. In most cases, biologists focus on females when calculating life-table data, because the number of males present rarely affects population growth. There are almost always enough males present to fertilize all of the females in breeding condition, so growth rates depend entirely on females. Let's walk through Table 54.1 step by step.

Age Class and Number Alive The first column in Table 54.1 lists the age classes of *L. vivipara*. An **age class** is a group of individuals of a specific age—for example, all female lizards between 4 and 5 years old. These data show that *L. vivipara* females have a maximum life span of about seven years.

But not all females live to be this old. The second column in the table shows how many female lizards survived to reach each age class. Suppose 1000 female *L. vivipara* are born in a particular year. These individuals represent a **cohort**—a group of the same age that can be followed through time. How many individuals would survive to age 1, age 2, and so on? On average, 424 would still be alive one year later, and 308 would still be alive two years later.

Survivorship The third column in Table 54.1 converts the data in the second column to proportions. **Survivorship**—a key component of a life table—is defined as the proportion of offspring produced that survive, on average, to a particular age. For example, the survivorship from birth to age 1 was 0.424 in the Netherlands population. Survivorship from birth to age 2 was 0.308.

Life tables do not usually include data bars—the bars in this table, such as the orange bars for survivorship, are intended to help you visualize the patterns in the numbers in the different columns. Since survivorship varies by several orders of magnitude, in this case from 1.000 to 0.002, biologists typically plot the logarithm of the number of survivors versus age. Using a

TABLE 54.1 Life Table for *Lacerta vivipara* Females in the Netherlands

Age Class (x)	Number of Survivors (N_x)	Survivorship (l_x)	×	Age-Specific Fecundity (m_x)	=	Average Births/Year/Original Female $(l_x m_x)$
0	1000	1.000		0.00		0.00
1	424	0.424		0.04		0.02
2	308	0.308		1.47		0.45
3	158	0.158		2.07		0.33
4	57	0.057		2.44		0.14
5	10	0.010		3.25		0.04
6	7	0.007		3.25		0.03
7	2	0.002		3.25		+ 0.01

$R_0 = 1.00 =$ Net reproductive rate

SOURCE: H. Strijbosch and R. C. M. Creemers, 1988. Comparative demography of sympatric populations of *Lacerta vivipara and Lacerta agilis*. *Oecologia* 76: 20–26. With kind permission of Springer Science and Business Media.

(a) Three general types of survivorship curves

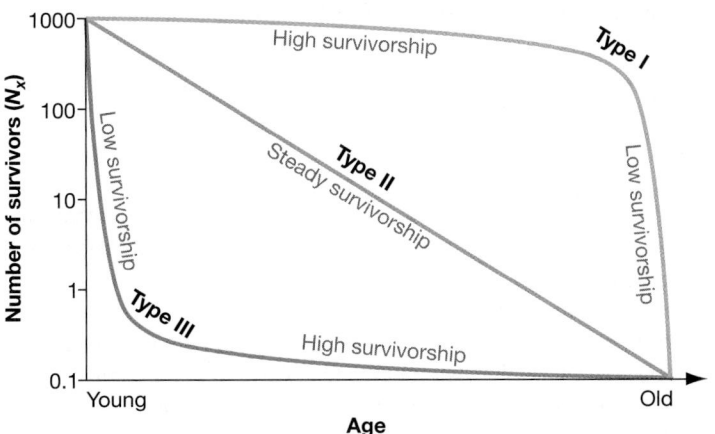

(b) Exercise: Survivorship curve for *Lacerta vivipara*

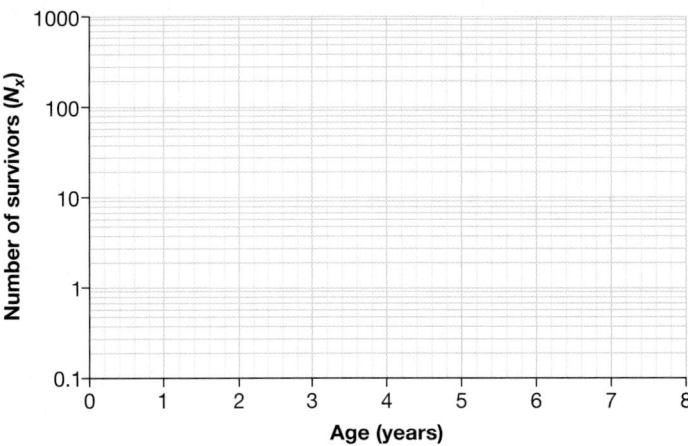

FIGURE 54.2 Survivorship Curves Identify When Mortality Rates Are Low, Steady, or High.

✔QUANTITATIVE Fill in the graph in part (b) with data on survivorship of *Lacerta vivipara* given in Table 54.1. Compare the shape of the curve to the generalized graphs in part (a).

logarithmic scale on the *y*-axis is a matter of convenience—it makes patterns easier to see (see **BioSkills 6** in Appendix B). The resulting graph is called a **survivorship curve.** Studies on a wide variety of species indicate that three general types of survivorship curves exist (**FIGURE 54.2a**).

1. Humans have what biologists call a type I survivorship curve. In this pattern, survivorship throughout life is high—most individuals approach the species' maximum life span.

2. Type II survivorship curves occur in species where individuals have about the same probability of dying in each year of life. Blackbirds and other songbirds have this type of curve.

3. Many plants have Type III curves—a pattern defined by extremely high death rates for seeds and seedlings but high survival rates later in life.

    FIGURE 54.2b provides a graph for you to plot survivorship of *L. vivipara*, using the data reported in Table 54.1. Note the logarithmic scale on the *y*-axis of the graph. What type of survivorship curve does *L. vivipara* have?

Fecundity　The fourth column in Table 54.1 shows the data for **fecundity,** the number of female offspring produced by each female in the population. Because the researchers documented the reproductive output of the same *L. vivipara* lizard females year after year, they were able to calculate a quantity called **age-specific fecundity**: the average number of female offspring produced by a female in each age class. The blue data bars in Table 54.1 illustrate the pattern of fecundity in lizards of different ages: older female lizards have a higher age-specific fecundity than younger females.

Reproductive Rate　If survivorship is higher in young lizards and fecundity is higher in older lizards, what is the outcome? The purple bars in the fifth column of Table 54.1 show the product of survivorship multiplied by fecundity—most of the offspring in the population are produced by females of intermediate age.

The sum of the purple bars is a measure of **net reproductive rate,** which indicates whether the population is increasing or decreasing (as long as immigration and emigration are insignificant). Since this sum is equal to 1, this population of *L. vivipara* is stable. **Quantitative Methods 54.2** (see page 1106) walks through the calculations in the life table in more detail.

How do the data on the Netherlands populations of *L. vivipara* compare with populations in different types of habitats?

The Role of Life History

The life-table data in Table 54.1 are interesting because they contrast with results from other populations of *L. vivipara*. Consider data on fecundity:

- In the Netherlands, almost no 1-year-old female *L. vivipara* reproduce.

- In Brittany, France, 50 percent of 1-year-old female lizards reproduce.

- In the mountains of Austria, females don't begin breeding until they are 4 years old.

The contrasts are equally stark when looking at survivorship. In the Austrian population, most females live much longer than do individuals in either lowland population—in the Netherlands or France.

In Brittany, fecundity is high but survivorship is low; in Austria, fecundity is low but survivorship is high. The population in the Netherlands is intermediate. In this species, key aspects of the life table vary dramatically among populations.

What Are Fitness Trade-Offs?　Why isn't it possible for *L. vivipara* females to have both high fecundity and high survival? The answer is **fitness trade-offs** (see Chapters 25 and 42).

Fitness trade-offs occur because every individual has a restricted amount of time and energy at its disposal—meaning that its resources are limited. If a female lizard devotes a great deal of

If immigration and emigration are not occurring, the data in a life table can be used to calculate a population's growth rate. This approach is logical, because survivorship and fecundity are ways to express death rates and birth rates—the other two factors influencing population size. To see how a population's growth rate can be estimated from life-table data, let's look at each component in a life table more carefully. Recall that life tables usually focus exclusively on females.

Survivorship is symbolized as l_x, where x represents the age class being considered. Survivorship for age class x is calculated by dividing the number of females in that age class (N_x) by the number of females that existed as offspring (N_0):

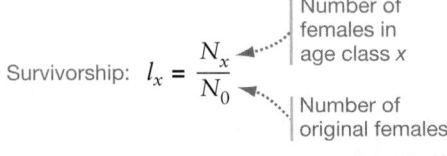

Survivorship: $l_x = \dfrac{N_x}{N_0}$

— Number of females in age class x

— Number of original females

(Eq. 54.2)

Age-specific fecundity is symbolized m_x, where x again represents the age class being considered. Age-specific fecundity is calculated as the total number of female offspring produced by females of a particular age, divided by the total number of females of that age class present. It represents the average number of female offspring produced by a female of age x.

Documenting age-specific survivorship and fecundity allows researchers to calculate the average number of female offspring produced by females during each age of life, as $l_x m_x$. For example, if 30 percent of the females in a population live to be 2 years old, and if 2-year-old females have an average of 2 offspring, then the average number of births at age 2, per female born into the population, is $l_x m_x = 0.30 \times 2 = 0.6$.

Summing $l_x m_x$ values over the entire life span of a female gives the net reproductive rate, R_0, of a population:

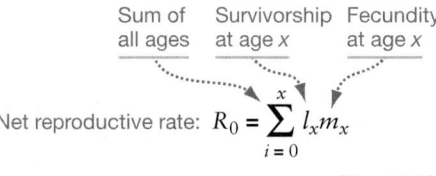

Sum of all ages   Survivorship at age x Fecundity at age x

Net reproductive rate: $R_0 = \displaystyle\sum_{i=0}^{x} l_x m_x$

(Eq. 54.3)

Here the Greek letter sigma (Σ) stands for "sum." The expression on the right-hand side of the equals sign is read, "sum the $l_x m_x$ values from each year of life, starting from birth or hatching ($i = 0$), through age x."

The net reproductive rate represents the growth rate of a population per generation. The logic behind the equation for R_0 is that the growth rate of a population per generation equals the average number of female offspring that each female produces over the course of her lifetime. A female's average lifetime reproduction, in turn, is a function of survivorship and fecundity at each age class. In Table 54.1, R_0 is the sum of the survivorship multiplied by the fecundity values in the right column.

If R_0 is greater than 1, then the population is increasing in size. If R_0 is less than 1, then the population is declining. ✔**QUANTITATIVE** If you understand these concepts, you should be able to use the data in Table 54.1 to determine how many female offspring an average *L. vivipara* female produces over the course of her lifetime,[1] and describe whether the population is growing, stable, or declining.[2]

[1] 1.00; [2] The population is stable.

energy to producing a large number of offspring, she is not able to devote that same energy to her immune system, growth, nutrient stores, or other traits that increase survival. That is, a female can maximize fecundity, maximize survival, or strike a balance between the two.

An organism's **life history** describes how an individual allocates resources to growth, reproduction, and activities or structures that are related to survival. Traits such as survivorship, age-specific fecundity, age at first reproduction, and growth rate are all aspects of an organism's life history. Understanding variation in life history is all about understanding fitness trade-offs.

In almost all cases, biologists find that life history is shaped by natural selection in a way that maximizes an individual's fitness in its environment. In *L. vivipara* populations, biologists contend that females who live a long time but mature late and have few offspring each year have high fitness in cold, high-elevation habitats, such as Austria. In these habitats, females have to reduce their reproductive output and put more energy into traits that increase survival in a harsh environment. In contrast, females who have short lives but mature early and have large numbers of offspring each year do better in warm, low-elevation habitats, such as Brittany.

✔ If you understand the role that fitness trade-offs play in determining life-history patterns, you should be able to (1) predict how survivorship and fecundity should compare in *L. vivipara* in the warmest versus coldest parts of their range, and (2) comment on why females in these populations lay many eggs instead of giving birth to a relatively small number of live young.

Life-History Patterns across Species The life-history data you've reviewed thus far all involve comparisons of populations within a species. But the same patterns exist when many different species are compared.

In general, individuals from species with high fecundity tend to grow quickly, reach sexual maturity at a young age, and produce many small eggs or seeds. The mustard plant *Arabidopsis thaliana*, for example, germinates and grows to sexual maturity in just 4 to 6 weeks. In this species, individuals usually live only a few months but may produce as many as 10,000 tiny seeds. They live fast and die young.

In contrast, individuals from species with high survivorship tend to grow slowly and invest resources in traits that reduce damage from enemies and increase their own ability to compete for water, sunlight, or food. A coconut palm, for example, may take a decade to mature but live 60–70 years and produce offspring each

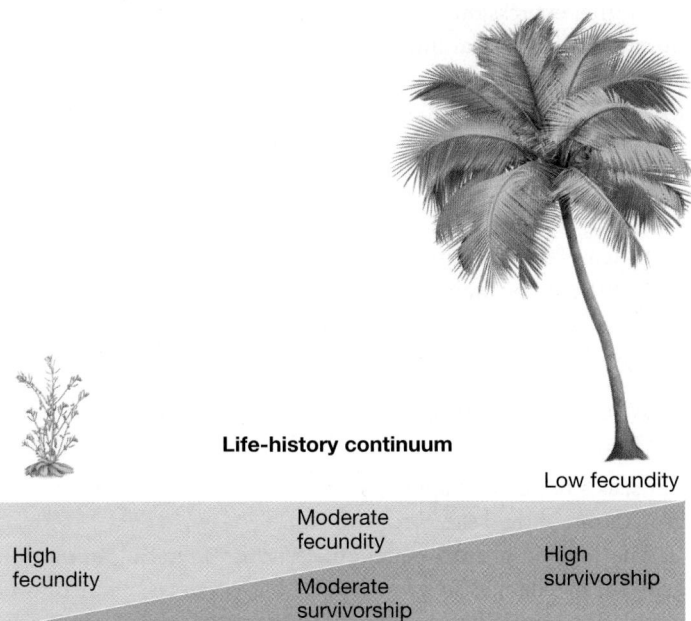

Life-history continuum

Low fecundity

High fecundity

Moderate fecundity

Moderate survivorship

High survivorship

Low survivorship

FIGURE 54.3 Life-History Traits Form a Continuum. Every organism can be placed somewhere on this life-history continuum. The placement of a species is most meaningful when it is considered relative to closely related species.

year. Coconut palms invest resources in making stout stems that allow them to grow tall, a relatively extensive root system, large fruits (coconuts), molecules that make herbivores sick, and enzymes that reduce infections from disease-causing fungi. These traits increase survivorship but decrease fecundity.

An *Arabidopsis thaliana* plant and a coconut palm represent two ends of a broad continuum of life-history characteristics (**FIGURE 54.3**). How does a species' location on this continuum affect its population growth rate?

54.3 Population Growth

Life tables provide a snapshot of a population, linking life-history patterns to population growth. Let's take a closer look at changes in population size over time. For conservation biologists, analyzing and predicting changes in population size is fundamental to managing threatened species.

Recall that four processes affect a population's size: Births and immigration add individuals to the population; deaths and emigration remove them. It follows that a population's overall growth rate is a function of birth rates, death rates, immigration rates, and emigration rates.

Quantifying the Growth Rate

A population's growth rate is the change in the number of individuals in the population (ΔN) per unit time (Δt). Here the Greek symbol Δ (delta) means change.

If no immigration or emigration is occurring, then a population's growth rate is equal to the number of individuals (N) in the population times the difference between the birth rate per individual (b) and death rate per individual (d). The difference between the birth rate and death rate per individual is called the per capita rate of increase and is symbolized r. (Per capita means "for each individual.")

If the per capita birth rate is greater than the per capita death rate, then r is positive and the population is growing. But if the per capita death rate begins to exceed the per capita birth rate, then r becomes negative and the population declines. Within populations, r varies through time. Its value can be positive, negative, or 0.

When conditions are optimal for a particular species—meaning birth rates per individual are as high as possible and death rates per individual are as low as possible—then r reaches a maximal value called the **intrinsic rate of increase**, r_{max}. When this occurs, a population's growth rate is expressed as

$$\frac{\Delta N}{\Delta t} = r_{max}N$$

In species such as *Arabidopsis* and fruit flies, which breed at a young age and produce many offspring each year, r_{max} is high. In contrast, r_{max} is low in species such as giant pandas and coconut palms, which take years to mature and produce few offspring each year. Stated another way, r_{max} is a function of a species' life-history traits.

Each species has a characteristic r_{max} that does not change. But at any specific time, a population has an instantaneous growth rate, or per capita rate of increase, symbolized by r. Whereas r_{max} tells you what the maximum growth rate is, r tells you what it is at a particular time. So r is always less than or equal to r_{max}.

The instantaneous growth rate of a population at a particular time is actually likely to be much lower than r_{max}. A population's r is also likely to be different from r values of other populations of the same species (such as in upland and lowland populations of *L. vivipara*), and to change over time (such as in wet years versus dry years). The instantaneous growth rate is dynamic. However, let's consider the simplest case first, where r is constant.

Exponential Growth

The graph in **FIGURE 54.4** (see page 1108) plots changes in population size, for various values of r, under the condition known as exponential growth. **Exponential population growth** occurs when r does not change over time.

The key point about exponential growth is that the growth rate does not depend on the number of individuals in the population. Biologists say that this type of population growth is **density independent.**

It's important to emphasize that exponential growth adds an increasing number of individuals as the total number of individuals, N, gets larger. As an extreme example, an r of 0.02 per year in a population of 1 billion adds over 20 million individuals per year. The same growth rate in a population of 100 adds just

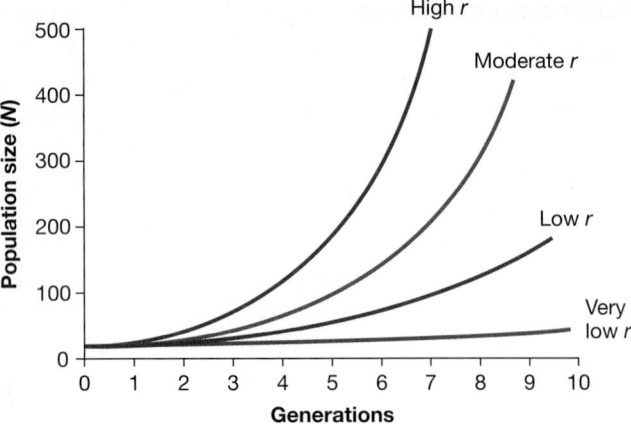

FIGURE 54.4 Exponential Growth Is Independent of Population Size. When the per capita growth rate *r* does not change over time, exponential growth occurs. Population size may increase slowly or rapidly, depending on the size of *r*.

over 2 individuals per year. Even if *r* is constant, the number of individuals added to a population is a function of *N*. The *rate* of increase is the same, but the number of individuals added is not.

In nature, exponential growth is common in two circumstances: **(1)** a few individuals found a new population in a new habitat, or **(2)** a population has been devastated by a storm or some other type of catastrophe and then begins to recover, starting with a few surviving individuals. But it is not possible for exponential growth to continue indefinitely.

If *Arabidopsis* populations grew exponentially for a long period of time, they would eventually fill all available habitat. When population density gets very high, the population's per capita birth rate will decrease and the per capita death rate will increase, causing *r* to decline. Stated another way, growth is often **density dependent.**

Logistic Growth

To analyze what happens when growth is density dependent, biologists use a parameter called carrying capacity. **Carrying capacity, *K*,** is defined as the maximum number of individuals in a population that can be supported in a particular habitat over a sustained period of time.

The carrying capacity of a habitat depends on many factors: food, space, water, soil quality, and resting or nesting sites. Carrying capacity can change from year to year, depending on conditions.

A Logistic Growth Equation If a population of size *N* is below the carrying capacity *K*, then the population should continue to grow. More specifically, a population's growth rate is proportional to $(K - N)/K$:

$$\frac{\Delta N}{\Delta t} = r_{max} N \left[\frac{K - N}{K} \right]$$

This expression is called a logistic growth equation.

In the expression $(K - N)/K$, the numerator defines the number of additional individuals that can be accommodated in a habitat with carrying capacity *K*; dividing $K - N$ by *K* turns this number of individuals into a proportion. Thus $(K - N)/K$ describes the proportion of "unused resources and space" in the habitat. It can be thought of as the environment's resistance to growth.

- When *N* is small, then $(K - N)/K$ is close to 1 and the growth rate should be high. For example: $(1000 - 10)/1000 = 0.99$.

- As *N* gets larger, $(K - N)/K$ gets smaller. For example: $(1000 - 900)/1000 = 0.10$.

- When *N* is at carrying capacity (meaning that $K = N$), then $(K - N)/K$ is equal to 0 and growth stops.

Thus, as a population approaches a habitat's carrying capacity, its growth rate should slow.

The logistic growth equation describes **logistic population growth,** or changes in growth rate that occur as a function of population size. Whereas exponential growth is density independent, logistic growth is density dependent. **Quantitative Methods 54.3** (see pages 1110–1111) explores population growth models in more detail and explains how they relate to the net reproductive rate calculated in the life table.

Graphing Logistic Growth   **FIGURE 54.5a** illustrates density-dependent growth in a hypothetical population. The graph plots changes in population size over time, and has three sections.

1. Initially, growth is exponential—meaning that *r* is constant.

2. With time, *N* increases to the point where competition for resources or other density-dependent factors begins to occur. As a result, the growth rate begins to decline.

3. When the population is at the habitat's carrying capacity, the growth rate is 0—the graph of population size versus time is flat.

This is exactly what happened in an experiment on laboratory populations of ciliates, *Paramecium aurelia* and *P. caudatum* (see Chapter 30). Russian biologist Georgii Gause placed 20 individuals from one of the *Paramecium* species into 5 mL of a solution. He created many replicates of these 5-mL environments for each species separately. He kept conditions as constant as possible by adding the same number of bacterial cells every day for food, washing the solution every second day to remove wastes, and maintaining the pH at 8.0.

In the graphs in **FIGURE 54.5b**, each data point represents population size for one of the species—the average of the replicates in the experiment. Note that both species exhibited logistic growth in this environment. The carrying capacity differed in the two species, however. The maximum density of *P. aurelia* averaged 448 individuals per mL, but that for *P. caudatum* averaged just 128 individuals per mL. (When the two species are grown together, one species gets eliminated; see Chapter 55.)

In both of the *Paramecium* species in this experiment, exponential growth could be sustained for only about five days. What factors cause growth rates to change?

(a) Density dependence: Growth rate slows at high density.

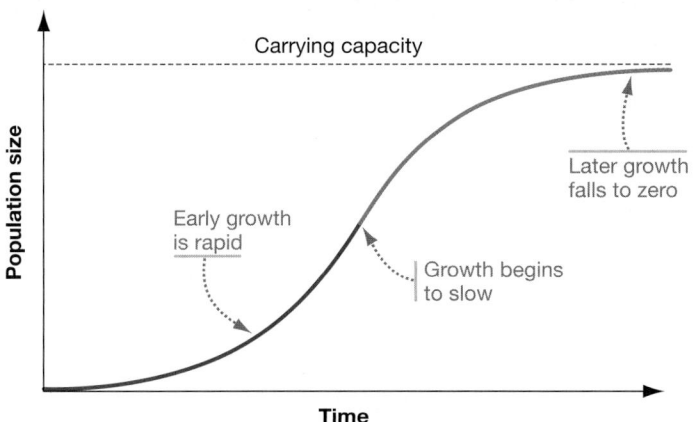

(b) Logistic growth in ciliates

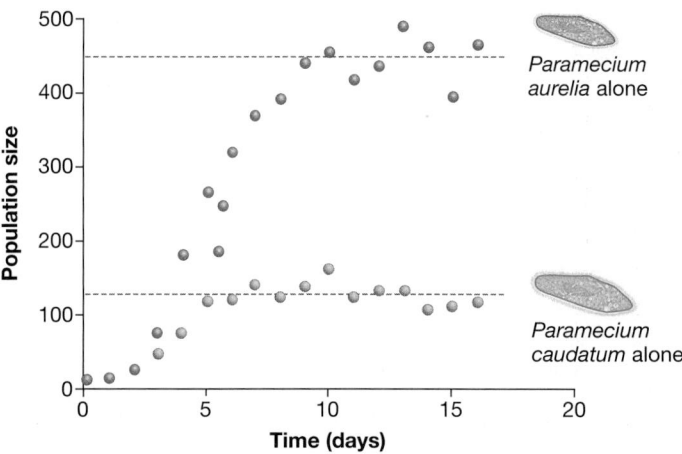

FIGURE 54.5 Logistic Growth Is Dependent on Population Size. (a) A curve illustrating the pattern predicted by the logistic growth equation. This pattern often occurs when a small number of individuals colonize an unoccupied habitat. Initially, *r* is high because competition for resources is low to nonexistent. Carrying capacity depends on the quality of the habitat and can vary over time. **(b)** Data from laboratory experiments with two species of *Paramecium*.

DATA: Gause, G. F. 1934. *The Struggle for Existence*. New York: Hafner Press.

What Limits Growth Rates and Population Sizes?

Population sizes change as a result of two general types of factors:

1. *Density-independent factors* alter birth rates and death rates irrespective of the number of individuals in the population, and they usually involve changes in the abiotic environment—variation in weather patterns, or catastrophic events such as cold snaps, hurricanes, volcanic eruptions, or drought.

2. *Density-dependent factors* change in intensity as a function of population size, and they are usually biotic. When trees crowd each other, they have less water, nutrients, and sunlight at their disposal and make fewer seeds.

A Closer Look at Density Dependence To get a better understanding of how biologists study the density-dependent factors that affect population size, consider the data in **FIGURE 54.6**.

The graph in Figure 54.6a presents results of an experimental study of a coral-reef fish called the bridled goby. Each data point represents the average of eight artificial reefs constructed by ecologists Graham Forrester and Mark Steele from pieces of real coral reefs. The initial density, plotted along the *x*-axis, represents the number of bridled gobies released per area at the start of the experiment. The proportion surviving, plotted on the vertical axis, represents the individuals that were still living at that reef 30 days later. The researchers compared the survivorship of

(a) Survival of gobies declines at high population density.

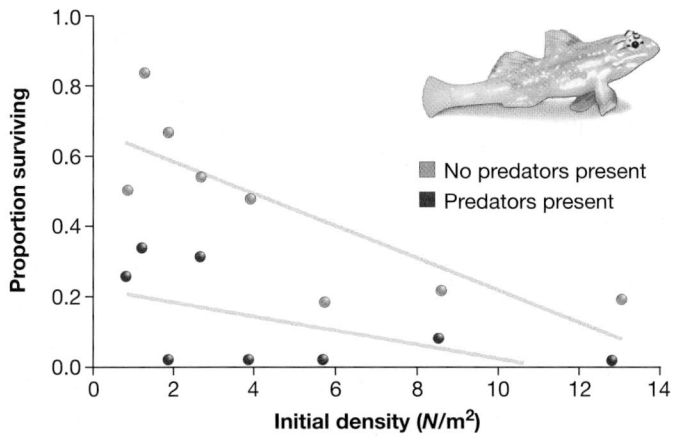

(b) Fecundity of sparrows declines at high population density.

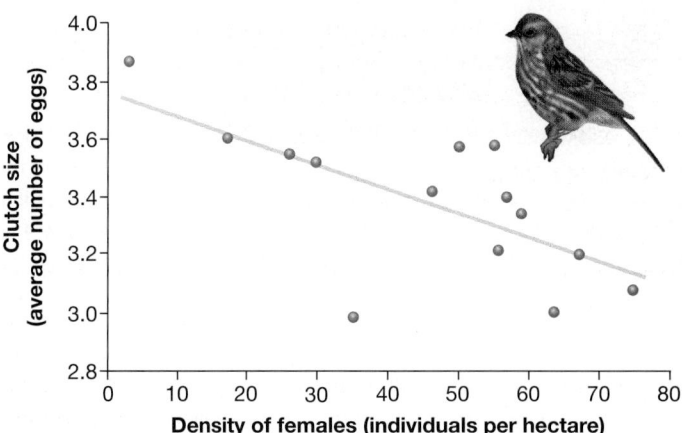

FIGURE 54.6 Density-Dependent Growth Results from Changes in Survivorship and Fecundity.

DATA: Forrester, G. E. and M. A. Steele 2000. *Ecology* 81: 2416–2427; Arcese, P., and J. N. M. Smith. 1988. *Journal of Animal Ecology* 57: 119–136.

✔**QUESTION** When song sparrow populations are at high density and extra food is provided to females experimentally, average clutch size is much higher than expected from the data in part (b). Based on these data, state a hypothesis to explain why average clutch size declines at high density.

To explore how biologists model changes in population size in more detail, let's consider data on whooping cranes, which are large, wetland-breeding birds native to North America (**FIGURE 54.7**). Whooping cranes may live more than 20 years in the wild, but it is common for females to be six or seven years old before they breed for the first time. Female cranes lay just one or two eggs per year and usually rear just one chick. Based on these life-history characteristics, you should understand why the growth rate of the whooping crane populations is extremely low.

Conservation biologists monitor this species closely because hunting and habitat destruction reduced the total number of individuals in the world to about 20 in the mid-1940s. Since then, intensive conservation efforts have resulted in a current total population of about 571. That number includes a group of 279 individuals that breed in Wood Buffalo National Park in the Northwest Territories of Canada.

FIGURE 54.7 Whooping Cranes Have Been the Focus of Intensive Conservation Efforts.

In addition to the Wood Buffalo population, two new populations of cranes have been established by releasing offspring raised in captivity. One of these groups lives near the Atlantic coast of Florida year-round; the other group migrates from breeding areas in northern Wisconsin to wintering areas along Florida's Gulf Coast.

According to the biologists who are managing whooping crane recovery, the species will no longer be endangered when the two newly established populations have at least 25 breeding pairs, and when neither population needs to be supplemented with captive-bred young in order to be self-sustaining. How long will this take?

Discrete Growth

Whooping cranes breed once per year, so the simplest way to express a crane population's growth rate is to compare the number of individuals at the start of one breeding season to the number at the start of the following year's breeding season.

To create a general expression for how populations grow over a discrete time interval, biologists use N to symbolize population size. N_0 is the population size at time zero (the starting point), and N_1 is the population size one breeding interval later. In equation form, the growth rate is given as

$$\text{Finite rate of increase:} \quad \lambda = \frac{N_1}{N_0}$$

Population size one breeding interval later

Population size at time zero

(Eq. 54.4)

The parameter λ (lambda) is called the **finite rate of increase.** (In mathematics, a *finite rate* refers to an observed rate over a given period of time. A *parameter* is a variable or constant term that affects the shape

of a function.) The current population size at Wood Buffalo is 279. Suppose that biologists count 300 cranes on the breeding grounds next year. The population growth rate could be calculated as $290/279 = 1.039 = \lambda$. Stated another way, the population will have grown at the rate of 3.9 percent per year. Rearranging the expression in Equation 54.4 gives

$$N_1 = N_0\lambda \quad \textbf{(Eq. 54.5)}$$

Stated more generally, the size of the population at the end of year t will be given by:

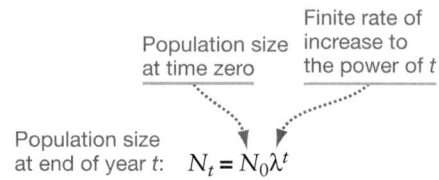

Population size at time zero

Finite rate of increase to the power of t

Population size at end of year t: $\quad N_t = N_0\lambda^t$

(Eq. 54.6)

This equation summarizes how populations grow when breeding takes place seasonally. The size of the population at time t is equal to the starting size, times the finite rate of increase multiplied by itself t times.

In a sense, λ works like the interest rate at a bank. For species that breed once per year, the "interest" on the population is compounded annually. A savings account with a 5 percent annual interest rate increases by a factor of 1.05 per year.

If a population is growing, then its λ is greater than one. The population is stable when λ is 1.0 and declining when λ is less than 1.0.

If a population's age structure is stable, meaning that the proportion of females in each age class is not changing over time, then its finite rate of increase also has a simple relationship to its net reproductive rate (discussed in Quantitative Methods 54.2):

gobies with natural predators present (orange data points), and also with predators excluded from the study areas by cages (red data points). The higher the original density of gobies on the reef, the least likely these fish were to survive. More specifically, a majority of the gobies in the low-density trials survived the study period. In contrast, only about 20 percent of the gobies

in the high-density trials survived. This graph shows a strong density-dependent relationship in *survivorship* both with and without predators.

The graph in Figure 54.6b is from Peter Arcese and James Smith's long-term study of song sparrows on Mandarte Island, British Columbia. Each data point represents a different year.

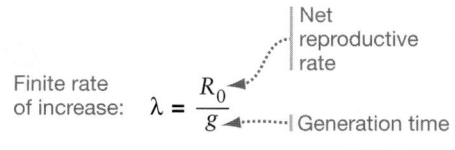

Finite rate of increase: $\lambda = \dfrac{R_0}{g}$

(with labels: Net reproductive rate → R_0; Generation time → g)

(Eq. 54.7)

In essence, dividing the net reproductive rate by generation time transforms it into a discrete rate.

Continuous Growth

The parameters λ and r (the finite rate of increase and the per capita growth rate, respectively) have a simple mathematical relationship. The best way to understand their relationship is to recall that λ expresses a population's growth rate over a discrete interval of time. In contrast, r gives the population's per capita growth rate at any particular instant. This is why r is also called the instantaneous rate of increase. The relationship between the two parameters is given by

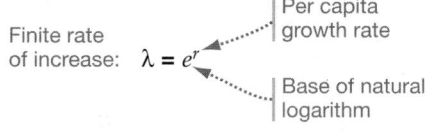

Finite rate of increase: $\lambda = e^r$

(with labels: Per capita growth rate → r; Base of natural logarithm → e)

(Eq. 54.8)

where e is about 2.72. (For help with using logarithms, see **BioSkills 6** in Appendix B. Also, note that the relationship between any finite rate and any instantaneous rate is given by finite rate $= e^{\text{instantaneous rate}}$.)

Substituting Equation 54.8 into Equation 54.6 gives

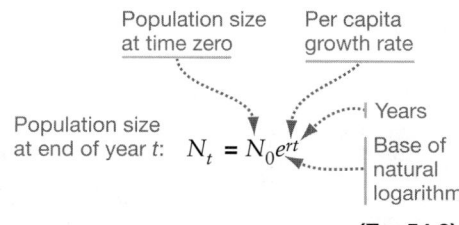

Population size at end of year t: $N_t = N_0 e^{rt}$

(with labels: Population size at time zero → N_0; Per capita growth rate → r; Years → t; Base of natural logarithm → e)

(Eq. 54.9)

This expression summarizes how populations grow when they breed continuously, like humans and bacteria do, instead of at defined intervals. For species that breed continuously, the "interest" on the population is compounded continuously. When the growth rates λ and r are equivalent, however, the differences between discrete and continuous growth are negligible.

Because r represents the growth rate at any given time, and because r and λ are so closely related, biologists routinely calculate r for species that breed seasonally. In the whooping crane example, $\lambda = 1.039$ per year $= e^r$. To solve for r, take the natural logarithm of both sides. (**BioSkills 6** in Appendix B explains how to calculate a natural logarithm using your calculator.) In this case, $r = 0.038$ per year.

The instantaneous rate of increase, r, is also directly related to the net reproductive rate, R_0:

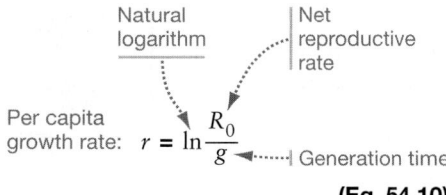

Per capita growth rate: $r = \ln \dfrac{R_0}{g}$

(with labels: Natural logarithm → $\ln$; Net reproductive rate → R_0; Generation time → g)

(Eq. 54.10)

Thus, r can be calculated from life-table data. It is a more useful measure of growth rate than R_0, because r is independent of generation time.

To summarize, biologists have developed several ways of calculating and expressing a population's growth rate:

Finite rate of increase (λ): easiest to understand, but applies only to populations that breed seasonally.

Net reproductive rate (R_0): can be calculated directly from life-table data, but is dependent on generation time.

Instantaneous rate of increase (r): more difficult conceptually, but is the most useful expression for growth rate because it is independent of generation time and is relevant for species that breed either seasonally or continuously.

Applying the Models

To get a better feel for r and for Equation 54.9, consider the following series of questions about whooping cranes. The key to answering these questions is to realize that Equation 54.9 has just four parameters. Given three of these parameters, you can calculate the fourth. ✔**QUANTITATIVE** If you understand this concept, you should be able to solve these three problems:

1. If 20 individuals were alive in 1941 and 571 existed in 2011, what is r? Here $N_t = 571$, $N_0 = 20$, and $t = 70$ years. Substitute these values into Equation 54.9 and solve for r. Then check your answer at the end of this box.[1]

2. The most recent report issued by the biologists working on the Wood Buffalo crane recovery program estimates that that the flock should be able to sustain an r of 0.048 for the foreseeable future. If the flock currently contains 279 individuals, how long will it take that population to double? Here $N_0 = 279$ and $N_t = 2 \times 253 = 558$. In this case, you solve Equation 54.9 for t. Then check your answer.[2]

3. In 2002, a pair of birds in the flock that lives in Florida year-round successfully raised offspring (nine years after the first cranes were introduced there). In 2003, another pair of cranes in this population bred successfully. If the number of breeding pairs had continued to double each year, how long would it have taken to reach the goal of 25 breeding pairs? (Note that $\lambda = 2.0$ if a population is doubling each year.) Solve for r first (Equation 54.8); then, given that $N_0 = 2$ and $N_t = 25$, solve for t and check your answer.[3]

[1] $r = 0.048$; [2] $t = 14$ years; [3] $t = 3.64$ years.

The density of females, plotted along the horizontal axis, is the number of females that bred on the island; the clutch size, plotted on the vertical axis, is the average number of eggs laid by each female. Females at low density had about 0.5 eggs more per clutch, on average, than females at high density. This graph indicates a strong density-dependent relationship in *fecundity*.

Density-dependent changes in survivorship and fecundity cause logistic population growth. In this way, density-dependent factors define a particular habitat's carrying capacity. If gobies get crowded, they die or emigrate. If song sparrows are crowded, the average number of eggs and offspring that they produce declines.

Carrying Capacity Is Not Fixed It's important to recognize that *K* varies among species and populations. *K* varies because for any particular species, some habitats are better than other habitats due to differences in food availability, space, and other density-dependent factors. Stated another way, *K* varies in space. It also varies with time, because conditions are better in some years than in others.

What's more, the same habitat may have a very different carrying capacity for different species—as the experiment with growing paramecia in 5-mL vials showed. The same area will tend to support many more individuals of a small-bodied species than of a large-bodied species, for example, simply because large individuals demand more space and resources.

These simple observations help explain the variation in total population size that exists among species and among populations of the same species.

check your understanding

If you understand that . . .

- Populations grow exponentially unless slowed by density dependence.
- Density-dependent factors that influence population growth rates include competition for resources such as food and sunlight.

✔**You should be able to . . .**

Propose a density-dependent factor that limits growth in *Lacerta vivipara* and design an experiment to test this hypothesis.

Answers are available in Appendix A.

54.4 Population Dynamics

The tools introduced in the previous two sections provide a foundation for exploring how biologists study **population dynamics**—changes in populations through time and space. Let's pull together what you have learned about the spatial distribution of populations, demography, and population growth rates—and bring immigration and emigration into the discussion—by considering one of the most important patterns observed: the "blinking on and off" of fragmented metapopulations.

How Do Metapopulations Change through Time?

Glanville fritillaries are an endangered species of butterfly native to the Åland islands off the coast of Finland (**FIGURE 54.8a**). They are a well-studied example of a metapopulation because they occupy isolated patches of habitat within their geographic range. Recent research on Glanville fritillaries by Ilkka Hanski and colleagues illustrates the consequences of dynamic metapopulation structure for endangered species.

(a) Glanville fritillaries live in patches of meadow in Finland.

(b) A metapopulation is made up of small, isolated populations.

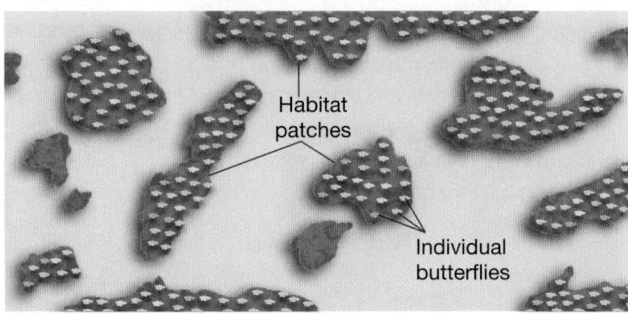

Although some populations go extinct over time...

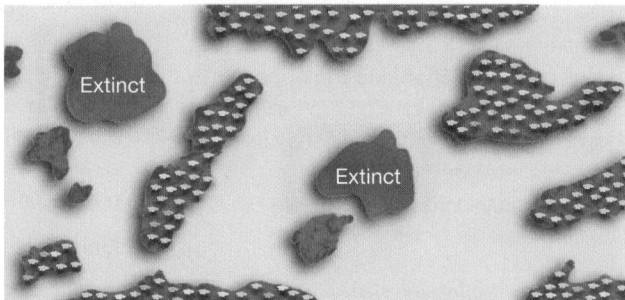

...migration can restore or establish populations.

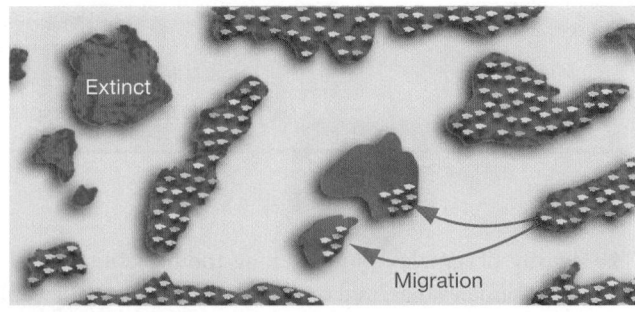

FIGURE 54.8 Metapopulation Dynamics Depend on Extinction and Recolonization. The overall size of a metapopulation stays relatively stable even if populations go extinct. These populations may be restored by migration, or unoccupied habitats might be colonized.

Metapopulations Should Be Dynamic Research on Glanville fritillaries began with a survey of meadow habitats on the islands. Because fritillary caterpillars feed on just two types of host plant, *Plantago lanceolata* and *Veronica spicata*, the team was able to pinpoint potential butterfly habitats.

Hanski's group estimated the fritillary population size within each patch of habitat by counting the number of larval webs in the patch. Of the 1502 meadows that contained the host plants, 536 had Glanville fritillaries. The patches ranged from 6 m² to 3 ha in area. Most had only a single breeding pair of adults and one larval group, but the largest population contained hundreds of pairs of breeding adults and 3450 larvae.

FIGURE 54.8b illustrates how a metapopulation like this is expected to change over the years.

- Given enough time, each population within the larger metapopulation is expected to go extinct. The cause could be a catastrophe, such as a storm; it could also be a disease outbreak or a sudden influx of predators.

- Migration from nearby populations can reestablish populations in empty habitat fragments.

In this way, the balance between extinction and recolonization exists within a metapopulation. Even though populations blink on and off over time, the overall metapopulation is maintained at a stable number of individuals.

An Experimental Test Does this metapopulation model correctly describe the population dynamics of fritillaries? To answer this question, Hanski's group conducted a mark–recapture study. They caught and marked many individuals, released them, and revisited fritillary habitats later to recapture as many individuals as possible (see Quantitative Methods 54.1).

Of the 1731 butterflies that the biologists marked and released, 741 were recaptured over the course of the summer study period. Of the recaptured individuals, 9 percent were found in a previously unoccupied patch. This migration rate is high enough to suggest that patches where a population has gone extinct will eventually be recolonized.

Hanski and co-workers repeated the survey two years after their initial census. Just as the metapopulation model predicted, some populations had gone extinct and others had been created. On average, butterflies were lost from 200 patches each year; 114 unoccupied patches were colonized and newly occupied each year. The overall population size was relatively stable even though constituent populations came and went.

To summarize, the history and future of a metapopulation is driven by the birth and death of populations, just as the dynamics of a single population are driven by the birth and death of individuals. Further, migration is an important source of individuals to recolonize patches of vacated habitats. As the final section in this chapter shows, these dynamics have important implications for conserving biodiversity.

Why Do Some Populations Cycle?

Density-dependent factors can be based on *intraspecific* ("within-species") interactions, such as competition among members of a cohort for food, or *interspecific* ("between-species") interactions, such as predation, parasitism, or competition among species for food (discussed in detail in Chapter 55). Analyzing population cycles has been a particularly productive way to understand how intraspecific and interspecific interactions interact. **FIGURE 54.9** shows a classic case of population cycling that involves two species: snowshoe hare and lynx in northern Canada.

The *x*-axis on this graph plots a 50-year interval; the *y*-axes show changes in hare population density (on the left, with a logarithmic scale) and changes in lynx density (on the right). Data for hares are plotted in tan; data for lynx are plotted in blue. Note three key points:

1. The scales on the *y*-axes are different—there are many more hares than lynx. (The *y*-axis for hares is logarithmic to compress the values so they can be compared more easily to changes in the lynx population.)

2. The hare and lynx populations cycle every 10 years on average.

3. Changes in lynx density lag behind changes in hare density by about two years.

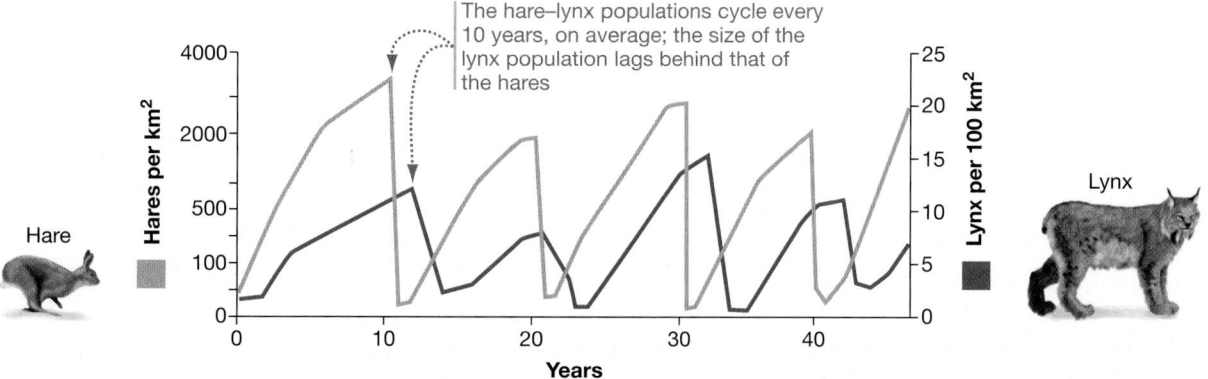

FIGURE 54.9 The Hare and Lynx Population Cycles Are Synchronous, but Lagged.

DATA: MacLulich, D. A. 1937. *University of Toronto Studies Biological Series* 43.

Snowshoe hares are herbivores—they subsist on leaves and stems that grow close to the ground. Lynx are predators, and they subsist mainly on snowshoe hares.

Is It Food or Predation? To explain the hare–lynx cycle, biologists posed two hypotheses based on density-dependent factors:

1. Hares use up all their food when their populations reach high density and starve; in response, lynx also starve.

2. Lynx populations reach high density in response to increases in hare density. At high density, lynx eat so many hares that the prey population crashes.

Stated another way, hares control lynx population size, or lynx control hare population size. The cycle could be controlled by intraspecific competition for food or interspecific interactions.

A Field Experiment To test these hypotheses, Charles Krebs and colleagues set up a series of 1-km^2 study plots in boreal forest habitats that were as identical as possible (**FIGURE 54.10**).

- Three plots were left as unmanipulated controls.
- One plot was ringed with an electrified fence made up of a mesh that excluded lynx but allowed hares to pass freely.
- Two plots received additional food for hares year-round (fertilizer was used to increase plant growth).

RESEARCH

QUESTION: What factors control the hare–lynx population cycle?

HYPOTHESIS: Predation, food availability, or a combination of those two factors controls the hare–lynx cycle.

NULL HYPOTHESIS: The hare–lynx cycle isn't driven by predation, food availability, or a combination of those two factors.

EXPERIMENTAL SETUP:

Document hare population in seven study plots (similar boreal forest habitats, each 1 km^2) from 1987 to 1994 (most of a population cycle).

3 plots: Unmanipulated controls

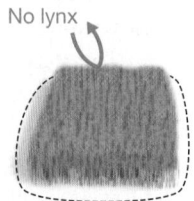

No lynx

1 plot: Electrified fence excludes lynx but allows free access by hares.

Extra food

2 plots: Supply extra food for hares.

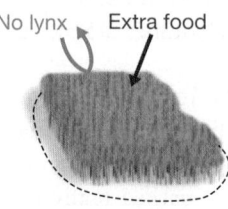

No lynx Extra food

1 plot: Electrified fence excludes lynx but allows free access by hares; supply extra food for hares.

PREDICTION: Hare populations in at least one type of manipulated plot will be higher than the average population in control plots.

PREDICTION OF NULL HYPOTHESIS: Hare populations in all of the plots will be the same.

RESULTS:

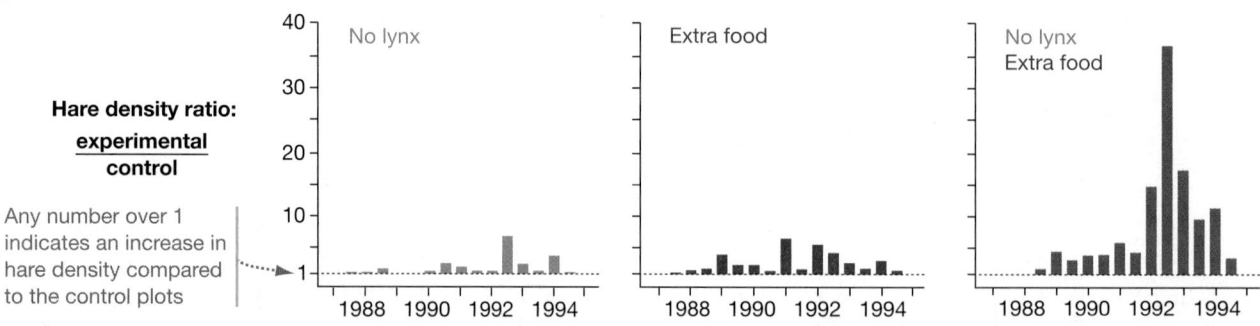

Hare density ratio:

$$\frac{\text{experimental}}{\text{control}}$$

Any number over 1 indicates an increase in hare density compared to the control plots

CONCLUSION: Hare populations are limited by both predation and food availability. When predation and food limitation occur together, they have a greater effect than either factor does independently.

©1995 AAS

FIGURE 54.10 Experimental Evidence that Predation and Food Availability Drive the Hare–Lynx Cycle.

SOURCE: Krebs, C. J., S. Boutin, R. Boonstra, et al. 1995. Impact of food and predation on the snowshoe hare cycle. *Science* 269: 1112–1115.

✔**QUESTION** Why do you think the extra food did not cause a more dramatic increase in hare density at the *beginning* of the population cycle?

- One plot had a predator-exclusion fence and was also supplemented with food for hares year-round.

The biologists then used the mark–recapture method and radio collars to monitor the size of the hare and lynx populations over an 8-year period.

The graphs in Figure 54.10 plot average hare density in the lynx-exclusion, food-addition, and combination plots relative to controls. A value of 1 on the y-axis means that hare density in the experimental and control plots were the same. Each bar represents the average population density over a six-month interval.

The graphs should convince you that plots with predators excluded showed higher hare populations than did control plots. This result supports the hypothesis that predation by lynx reduces hare populations. Plots with supplemental food also had higher populations than controls. But the plot with supplemental food *and* predators excluded had a hare population that became as much as 35 times denser than the population in the control plots.

Note that the effects of the treatments were not even over the course of the study—most of the effects occurred during the decline phase of the cycle, which occurred from 1992 to 1994. As control plots were crashing, the experimental plots retained some or many hares.

These data support the hypothesis that hare populations are limited by availability of food as well as by predation and that food availability and predation intensity interact—meaning the combined effect of food and predation is much larger than their impact in isolation. The leading hypothesis to explain this combined effect is that when hares are at high density, individuals are weakened by nutritional stress and are more susceptible to predation.

Research on population cycles has increased our understanding of density dependence factors in population dynamics. Now let's consider factors that determine dynamics in the human population.

Human Population Growth

As the life-table data presented in Section 54.1 showed, age dramatically affects the probability that an individual will reproduce and survive to the following year. A population's age structure, in turn, dramatically influences the population's growth over time.

Age Structure in Human Populations

To study age structure in our species, researchers use stacks of horizontal bars to plot the number of males and the number of females in each age cohort—often in five-year intervals between birth (0) and 100. The resulting graph is called a population pyramid or age pyramid.

Age Pyramids   In countries where industrial and technological development took place generations ago, like Sweden, survivorship has been high and fecundity low for several decades. An age pyramid like that in **FIGURE 54.11a** results. The most striking pattern in these data is that most of the age classes contain similar numbers of people. The evenness occurs because about the same number of infants are being born each year, and because most of them survive to old age.

In contrast, the age distribution is bottom heavy in less-developed nations like Honduras (**FIGURE 54.11b**). These populations are dominated by the very young, because they are undergoing rapid growth—the number of children being born recently is higher than the number of people existing in older age classes.

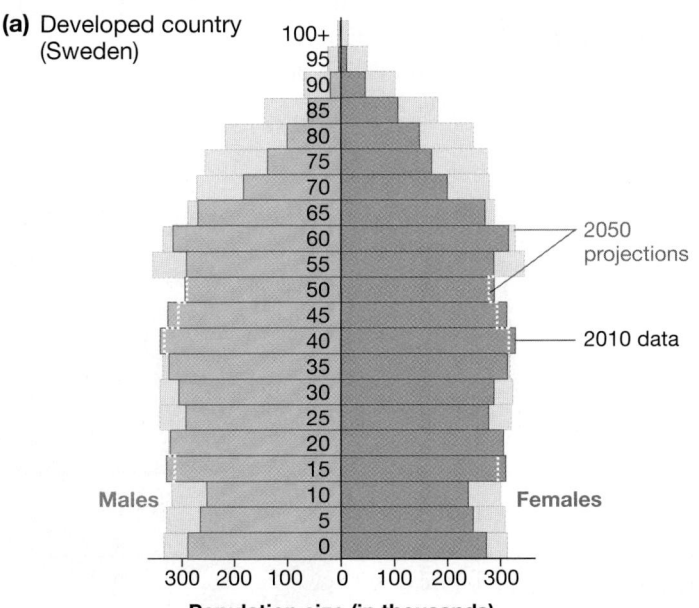

FIGURE 54.11 The Age Structure of Human Populations Varies Dramatically.

DATA: United Nations, *World Population Prospects: The 2010 Revision*.

Analyzing an age pyramid can give biologists important information about a population's history. But studying age distributions can also help researchers predict a population's future.

Look again at Figure 54.11a, and note that the white lines and lighter bars show what the age structure of Sweden is projected to be in 2050. Modest changes in the age distribution will occur because survivorship has increased while fecundity remains the same. But the population is not expected to grow quickly, because only modest numbers of individuals reach reproductive age. The projections highlight a major public policy concern in the industrialized countries: how to care for an increasingly aged population.

Now consider Figure 54.11b again. Thanks to dramatic improvements in health care, most young Hondurans survive to reproductive age. The projected age distribution in 2050 does not continue to flare out, however, because fecundity is declining. The projections illustrate a major public policy concern in less-developed countries: providing education and jobs for an enormous group of young people who will be reaching adulthood during the lifetimes of preceding generations.

Population "Momentum"   The data in Figure 54.11b make another important point: Because of recent and rapid population growth in developing countries, overall population size will increase dramatically in these nations over the course of your lifetime. Honduras will have many more people in 2050 than it does now.

A large part of this increase will be due to increased survivorship. But the number of offspring being born each year is also expected to stay high, even though fecundity is predicted to *decline*. This result seems paradoxical, but it is based on a key observation: These populations now contain so many young women that the overall number of births will stay high, even though the average number of children *per female* is much lower than it was a generation ago.

To capture this point, biologists say that these populations have momentum or inertia. Combined with high survivorship, age structures like these make continued increases in the total human population almost inevitable.

Analyzing Change in the Growth Rate of Human Populations

Fossil and molecular evidence indicates that *Homo sapiens* originated in Africa about 200,000 years ago, then migrated into Asia, Europe, Australia, and eventually North and South America (see Chapter 35). The geographic range of our species now consists of most of the land surface of the planet. How did our population size change over that time?

FIGURE 54.12 plots changes in the human population over the past 2000 years. Although the curve's shape looks superficially similar to exponential growth, the growth rate for humans has—until recently—actually *increased* over time, leading to a steeply rising curve.

It is almost impossible to overemphasize just how dramatically the human population has grown recently. It took all of human history to reach a population size of 1 billion in 1804, and only 123 years to reach 2 billion. Most recently, it took only 12 years to add 1 billion people to the population to reach 7 billion.

Consider that until your grandparents' generation, no individual had lived long enough to see the human population double. But many members of the generation born in the early 1920s have lived to see the population *triple*. And to drive home the impact of continued population doubling, consider this: If you were given a penny on January 1, then $0.02 on January 2, $0.04 on January 3, and so on, you would be handed $10,737,418.00 on January 31.

The astonishing growth rate of the human population has raised a sense of alarm that we will overshoot the carrying capacity of our planet—and this concern is not new. For example, Thomas Robert Malthus's book, *An Essay on the Principle of Population*, was widely influential in the early 1800s. Malthus warned that the population cannot continue to increase unchecked—competition for resources will eventually slow population growth, whether by famine, war, or voluntary reduction of family size. Malthus was an important source of inspiration for Charles Darwin in his formulation of the theory of evolution by natural selection (Chapter 25).

FIGURE 54.12 The Human Population Has Been Growing Rapidly.
Graph based on estimates from historical and census data.
DATA: Vince, G. 2011. *Science* 334: 32–37.

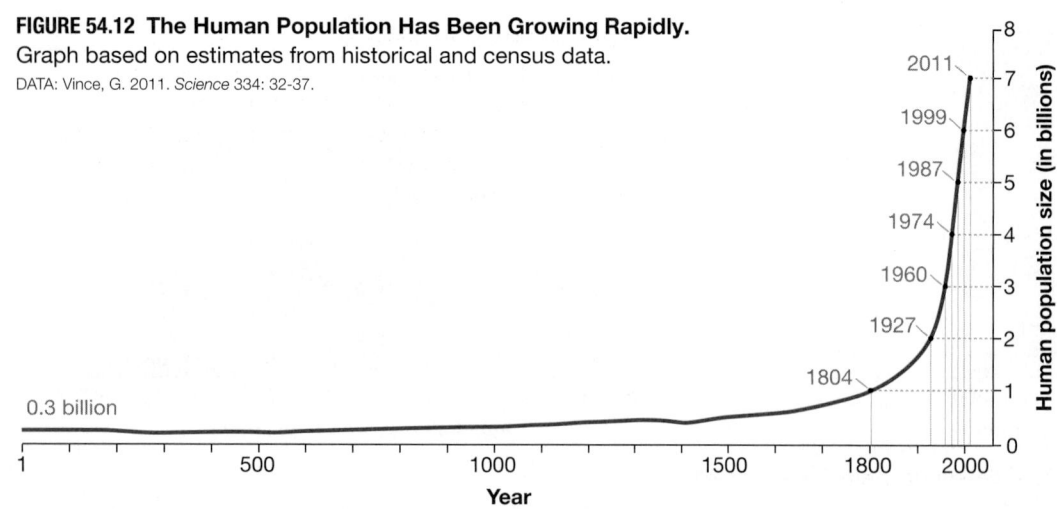

How Large Is the Current Human Population? As this book goes to press, the world population is estimated at over 7 billion. About 77 million additional people—more than double the population of California—are being added each year.

The consequences of recent and current increases in human population are profound (see Chapter 57). Aside from being the primary cause of habitat loss and species extinctions, overpopulation is linked by researchers to declines in living standards, mass movements of people, political instability, and acute shortages of water, fuel, and other basic resources in many parts of the world.

The one encouraging trend in the data is that the growth rate of the human population has already peaked and begun to decline. The highest growth rates occurred between 1965 and 1970, when populations increased at an average of 2.04 percent per year. Between 1990 and 1995, the overall growth rate in human populations averaged 1.46 percent per year; the current growth rate is 1.2 percent annually.

In humans, r may be undergoing the first long-term decline in history. The question is: Will the human population stabilize or decline in time to prevent global—and potentially irreversible—damage?

Will Human Population Size Peak in Your Lifetime? The United Nations Population Division makes regular projections for how human population size will change between now and 2050. For most readers of this book, these projections describe what the world will look like as you reach your early 60s.

The UN projections are based on three scenarios, which hinge on different values for **fertility** rates—the average number of surviving children that each woman has during her lifetime. Currently, the worldwide average fertility rate is 2.5 children per woman. This represents an enormous reduction in fertility from the 1950s, which averaged 5.0 children per woman. **FIGURE 54.13** shows how total population size is expected to change by 2050

if average fertility continues at 2.5 (high), or declines to 2.0 (medium), or 1.6 (low) children per woman.

The middle number in the projections is statistically the most likely scenario. Its value is close to the replacement rate of the population, 2.1. The **replacement rate** is the average fertility required for each woman to produce exactly enough offspring to replace herself and her offspring's father. (It is slightly above 2.0 to account for the mortality of some women before they bear children). When this fertility rate is sustained for a generation, $r = 0$, and there is **zero population growth.**

A glance at Figure 54.13 should convince you that the three scenarios are starkly different. If average fertility around the world stays at its current level, world population will be closing in on 11 billion about the time you might consider retiring—a 50 percent increase over today's population with no signs of peaking. The low-fertility projection, in contrast, predicts that the total human population in 2050 will be just over 8 billion and will have peaked.

The Role of Fertility Rates The UN's population projections make an important point: The future of the human population hinges on fertility rates—on how many children each woman living today decides to have. Those decisions, in turn, depend on a wide array of factors, including how free women are to choose their family size and how much access women have to education.

Why education? When women are allowed to become educated, they tend to delay having children and have a smaller overall family size. Access to education and reliable birth control methods (see Chapter 50), in addition to overall economic development and access to quality health care, will play a large role in determining how world population changes over your lifetime.

To summarize, the human population may be approaching the end of a period of rapid growth that lasted well over 500 years. Changes in fertility rates will be a major factor in determining how quickly population growth rates decline and how large the population eventually becomes.

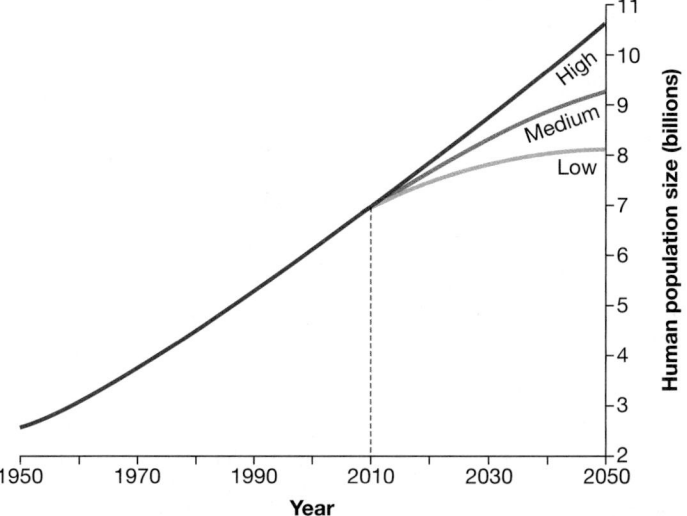

FIGURE 54.13 Projections for Human Population Growth to 2050. The "High," "Medium," and "Low" labels refer to fertility rates.

DATA: United Nations, *World Population Prospects: The 2010 Revision.*

check your understanding

If you understand that . . .

- A population's age structure reflects changes in survivorship and fertility rates over time.
- Age structure affects population dynamics because it dictates how many females of reproductive age are in the population.

✓**You should be able to . . .**

1. Explain why the age structure of human populations differs between developed and developing nations.

2. Explain why changes in fertility rates have such a dramatic impact on projections for the total human population in 2050.

Answers are available in Appendix A.

The human population boom is central to ecology because the growing human population dramatically affects other organisms via habitat destruction, pollution, overharvesting of fish and other resources, and climate change. As the human population rises, the impact on other species increases.

54.6 How Can Population Ecology Help Conserve Biodiversity?

When designing programs to save species threatened with extinction, conservation biologists draw heavily on concepts and techniques from population ecology. As habitat destruction and global climate change push species into decline, the study of population growth rates and population dynamics has taken on an increasingly applied tone.

To illustrate one of the ways that biologists apply the theory and results introduced in this chapter, let's analyze how an understanding of life-table data and geographic structure can help direct conservation action.

Using Life-Table Data

Suppose that you were in charge of reintroducing a population of lizards to a nature reserve, and that earlier research had documented survivorship and fecundity data (**FIGURE 54.14a**).

Your initial plan is to take 1000 newly hatched female lizards from a captive breeding center and release them into the natural habitat, along with enough males for breeding to occur. Is this population likely to become established and grow on its own, or will you need to keep introducing offspring that have been raised in captivity?

Making Population Projections To answer this question, you need to use the life-table data in Figure 54.14a and calculate two values:

1. How many adults survive to each age class each year?

(a) Life table data from previous study

Age (x)	Survivorship (l_x)	Fecundity (m_x)
0 (birth)	----	0.0
1	0.33	3.0
2	0.20	4.0
3	0.04	5.0

■ 1st-year cohort
■ 2nd-year cohort
■ 3rd-year cohort
■ 4th-year cohort

(b) Predictions: Fate of first-generation females

Year	0 (newborns)	1-year-olds	2-year-olds	3-year-olds	Total population size (N)
1st	1000 (just introduced)				1000
2nd	990 (= 330 × 3.0)	330 (= 1000 × 0.33)			1320 (= 990 + 330)
3rd	800 (= 200 × 4.0)		200 (= 1000 × 0.20)		?
4th	200 (= 40 × 5.0)			40 (= 1000 × 0.04)	?

(c) Predictions: Fate of first- and second-generation females

Year	0 (newborns)	1-year-olds	2-year-olds	3-year-olds	Total population size (N)
1st	1000				1000
2nd	990	330			1320
3rd	800 + 981 (981 = 327 × 3.0)	327 (= 990 × 0.33)	200		?
4th	200 + 792 (792 = 198 × 4.0)		198 (= 990 × 0.20)	40	?

(d) Predictions: Fate of first-, second-, and third-generation females

Year	0 (newborns)	1-year-olds	2-year-olds	3-year-olds	Total population size (N)
4th	200 + 792 + ?	?	198	40	?

FIGURE 54.14 Life-Table Data Can Be Used to Project the Future of a Population.

✔QUANTITATIVE Fill in the blanks indicated by the gray question marks. Note that in the bottom row, you must calculate the number of 1-year-olds in the 3rd-year cohort (indicated by blue) and the number of newborns produced by those 1-year-olds (part of the 4th-year cohort, in green).

2. How many offspring are produced by each adult age class—each year over the course of several years?

FIGURE 54.14b starts these calculations for you. The predicted fate of the original 1000 females is indicated in red. Note that in the start of the second year, just 330 of these year-old individuals are expected to survive; only 200 will survive to be 2-year-olds; only 40 will survive to be 3-year-olds.

How many offspring will this first cohort produce? Remember that just 330 of the original 1000 females are expected to remain after one year. As the purple number in Figure 54.14b indicates, these survivors have an average of 3.0 female offspring apiece. Thus, they contribute 990 new female offspring to the population. In the second year, the 200 females that are left from the original cohort have an average of 4.0 female offspring each and contribute 800 juveniles. In their third year, the 40 surviving females average 5.0 offspring and contribute 200 juveniles.

FIGURE 54.14c extends the calculations by predicting what will happen as the offspring of the original females, indicated in purple, begin to breed. ✔QUANTITATIVE By continuing the analysis until the third year, you should be able to predict whether the population of lizards will stay the same, decline, or increase over time. In this way, life-table data can be used to predict the future of populations.

Altering Values for Survivorship and Fecundity

Part of the value of a population projection based on life-table data—like the one begun in Figure 54.14—is that it allows biologists to alter values for survivorship and fecundity at particular ages and assess the consequences. For example, suppose that a predatory snake began preying on juvenile lizards. The survivorship of juveniles would decline, decreasing the number of individuals that survive to reproduce each year—with potentially dire consequences for the population size over time.

Analyses like this allow biologists to determine which aspects of survivorship and fecundity are especially sensitive for particular species. The studies done to date support some general conclusions:

- Whooping cranes, sea turtles, spotted owls, and many other endangered species have high juvenile mortality, low adult mortality, and low fecundity. In these species, the fate of a population is extremely sensitive to increases in adult mortality. Based on this insight, conservation biologists have recently begun an intensive campaign to reduce the loss of adult female sea turtles in fishing nets (**FIGURE 54.15**). Previously, most conservation action had focused on protecting eggs and nesting sites.

- Climate change has enormous implications for the life-history traits of populations. For example, the sex of many reptiles is determined by the temperature of the eggs in the nest. Researchers have observed a change in sex ratio of some reptile species due to the warmer climate, and this finding has significant implications for the growth rates of these populations. Direct measurements of life-history traits of populations in the field are essential to mathematical models used to predict ongoing effects of global climate change.

In some or even most cases, the population projections made from life-table data may be too simplistic to be useful. For example, conservationists may need to expand the basic demographic models to account for occasional disturbances such as fires or storms or disease outbreaks. What about species whose overall population dynamics are dictated by a metapopulation structure?

Preserving Metapopulations

Habitat destruction caused by suburbanization and other human activities leaves small populations isolated in pockets of intact habitat (**FIGURE 54.16**; see Chapter 57). Work on Glanville fritillaries and other species has shown that a small, isolated population—even one within a nature preserve—is unlikely to survive over the long term. What can be done about this?

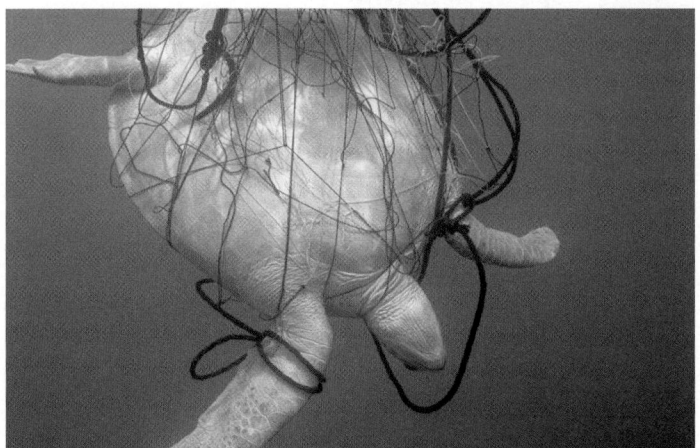

FIGURE 54.15 The mortality of adult sea turtles in fishing nets has a dramatic impact on sea turtle populations. Data from population ecology helps to prioritize conservation action.

FIGURE 54.16 Human activities are turning once intact populations into metapopulations by dividing habitats into fragments.

In Glanville fritillaries, data collected by Hanski's group have identified four attributes of populations that are most likely to persist:

1. They have larger population sizes.

2. They occupy larger geographical ranges.

3. They are closer to neighboring populations (and hence more likely to be colonized).

4. They have higher genetic diversity.

Results like these have important messages for conservation biologists:

- Areas that are being protected for threatened species should be substantial enough in area to maintain large populations that are unlikely to go extinct in the near future. Large populations are more likely than small populations to endure disasters such as storms, and they are much less susceptible to the damaging effects of inbreeding.

- When it is not possible to preserve large tracts of land, an alternative is to establish systems of smaller tracts that are connected by corridors of habitat, so that migration between patches is possible. Constructing "wildlife corridors" is particularly important when patches of appropriate habitat are separated by highways, subdivisions, rivers, or other obstacles to dispersal.

- If the species that is threatened exists as a metapopulation, it is crucial to preserve at least some patches of unoccupied habitat to provide future homes for immigrants. If a population is lost from a preserve, the habitat should continue to be protected so it can be colonized in the future. This can be difficult to do—nonscientists are often skeptical about preserving scarce land that does not (yet) contain the species of interest.

These results are made more complex by climate change. Data on existing populations show that changes in temperature and precipitation are already changing the ranges of species. For some species with large ranges, such as the lizard *Lacerta vivipara* (see Figure 54.1), local extinctions may occur in some populations, such as in the mountains of Spain, while the range expands northward elsewhere. Other species with smaller ranges or the inability to relocate rapidly are likely to go extinct. Population ecology helps quantify past trends in the distribution and abundance of organisms as well as predict future trends, enabling conservation biologists to prioritize their efforts.

CHAPTER 54 REVIEW

For media, go to MasteringBiology

If you understand . . .

54.1 Distribution and Abundance

- The geographic ranges of species vary in size and patchiness and are determined by a combination of abiotic and biotic factors.

- Individuals in a population have a random, clumped, or uniform distribution depending on the availability of resources, dispersal ability, social behavior, and other factors.

- There are many methods for measuring population size, population density, and distribution patterns—the best method to use depends on the behavior of the species.

✓ You should be able to devise a method for using the pellets (feces) of snowshoe hares to estimate population density.

54.2 Demography

- Life tables summarize how likely it is that individuals of each age class in a population will survive and reproduce.

- Three general types of survivorship curves have been observed: (I) high survivorship throughout life, (II) a constant probability of dying at each year of life, and (III) high mortality in juveniles followed by high survivorship in adults.

- There is a trade-off between survival and reproduction due to the limited availability of resources.

- In analyzing life history, there is a continuum between populations or species with high survivorship and low fecundity and populations or species with low survivorship and high fecundity.

✓ You should be able to predict how the data in the life tables constructed in ancient Rome would differ from a life table for present-day Rome.

54.3 Population Growth

- The growth rate of a population can be calculated from life-table data or from the direct observation of changes in population size over time.

- Exponential growth occurs when the per capita growth rate, r, does not change over time.

- During logistic growth, growing populations approach the carrying capacity, K, of their environment and r declines to 0.

- Density dependence in population growth is due to competition for resources, disease, predation, or other factors that increase in intensity when population size is high. At high density, survivorship and fecundity decrease.

✓ QUANTITATIVE You should be able to draw a logistic growth curve, describe how r changes along the length of the curve, and suggest factors responsible for each change in r.

54.4 Population Dynamics

- In some metapopulations, populations may blink on and off over time even though the overall metapopulation size is stable. Migration can reestablish populations in empty habitat fragments.

- Predator–prey interactions and food availability are density-dependent factors that drive the population cycle of the snowshoe hare and other species.

- ✓ Mosquitoes lay their eggs in water and have an aquatic larval stage. You should be able to predict a pattern of mosquito density over time as a function of rain.

54.5 Human Population Growth

- Age structure has a profound impact on population dynamics. A population with few juveniles and many adults past reproductive age, like the human populations of developed nations, may be declining or stable in size. In contrast, a population with a large proportion of juveniles is likely to increase rapidly in size.

- The total human population has been increasing rapidly and is currently over 7 billion.

- Based on various scenarios for average female fertility, the total human population in 2050 is expected to be between 8 and 11 billion. The future of the human population hinges on fertility rates.

- ✓ You should be able to explain why the human population is projected to continue rapid growth, even though fertility rates are declining.

54.6 How Can Population Ecology Help Conserve Biodiversity?

- Data from population ecology studies help biologists design effective management strategies for endangered species and help prevent other species from becoming endangered.

- ✓ You should be able to explain why a small, isolated population is virtually doomed to extinction, but why population size may be stable in a species consisting of a large number of small, isolated populations.

(MB) MasteringBiology

1. MasteringBiology Assignments

Tutorials and Activities Age Pyramids and Population Growth; Analyzing Age-Structure Pyramids; Human Population Growth; Human Population Growth and Regulation; Investigating Survivorship Curves; Modeling Population Growth; Population Ecology; Population Ecology: Logistic Growth, Techniques for Estimating Population Density and Size

Questions Reading Quizzes, Blue-Thread Questions, Test Bank

2. eText Read your book online, search, take notes, highlight text, and more.

3. The Study Area Practice Test, Cumulative Test, BioFlix® 3-D Animations, Videos, Activities, Audio Glossary, Word Study Tools, Art

You should be able to . . .

✓ TEST YOUR KNOWLEDGE
Answers are available in Appendix A

1. What is the defining feature of exponential growth?
   a. The population is growing very quickly.
   b. The growth rate is constant.
   c. The growth rate increases rapidly over time.
   d. The growth rate is very high.

2. List four factors that define population growth.

3. In what populations does exponential growth tend to occur?
   a. populations that colonize new habitats
   b. populations that experience intense competition
   c. populations that experience high rates of predation
   d. populations that have surpassed their carrying capacity

4. True or False? Climate change can influence the population growth of species.

5. If most individuals in a population are young, why is the population likely to grow rapidly in the future?
   a. Death rates will be low.
   b. The population has a skewed age distribution.
   c. Immigration and emigration can be ignored.
   d. Many individuals will begin to reproduce soon.

6. Why have population biologists become particularly interested in the dynamics of metapopulations?
   a. because humans exist as a metapopulation
   b. because whooping cranes exist as a metapopulation
   c. because many populations are becoming restricted to small islands of habitat
   d. because metapopulations explain why populations occupying large, contiguous areas are vulnerable to extinction

✓ TEST YOUR UNDERSTANDING
Answers are available in Appendix A

7. Describe the life-history traits of elephants.
   a. high survivorship, high fecundity
   b. low survivorship, high fecundity
   c. high survivorship, low fecundity
   d. low survivorship, low fecundity

8. Make a table to show what types of traits you would expect to see on the left, middle, and right of the life history continuum (see Figure 54.3). Consider growth habit (herbaceous, shrub, or tree) as well as relative disease- and predator-fighting ability, seed size, seed number, and body size.

9. Offer a hypothesis to explain why humans have undergone near-exponential growth for over 500 years. Why can't exponential growth continue indefinitely? Describe two examples of density-dependent factors that may influence human population growth.

10. Explain why biologists want to maintain (a) "habitat corridors" that connect populations in a metapopulation, and (b) unoccupied habitat that is appropriate for the species in question.

11. Predict which geographic subpopulations of *Lacerta vivipara* may be most vulnerable to the effects of climate change.

12. What is the difference between r (the per capita rate of increase) and r_{max} (the maximum or intrinsic growth rate)?

13. When wild plant and animal populations are logged, fished, or hunted, only the oldest or largest individuals tend to be taken. Many of the commercially important species are long lived and are slow to begin reproducing. If harvesting is not regulated carefully and exploitation is intense, what impact does harvesting have on a population's age structure? How might harvesting affect the population's life table and growth rate?

14. Design a system of nature preserves for an endangered species of beetle whose larvae feed on only one species of sunflower. The sunflowers tend to be found in small patches that are scattered throughout dry grassland habitats.

15. Make a rough sketch of the age distribution in developing versus developed countries, and explain why the shapes of the diagrams are different. How is AIDS, which is a sexually transmitted disease, affecting the age distribution in countries hard hit by the epidemic?

16. In most species the sex ratio is at or near 1.00, meaning that the number of males and females is approximately equal. In China, however, there is a strong preference for male children. According to the 2000 census there, the sex ratio of newborns is almost 1.17, meaning that close to 117 boys are born for every 100 girls. Based on these data, researchers project that China will soon have about 50 million more men than women of marriageable age. Evaluate whether this skewed sex ratio will affect the population growth rate in China.
    a. No, growth rate is not affected by males.
    b. Yes, growth rate will increase.
    c. Yes, growth rate will decrease.
    d. No, the sex ratio affects only fecundity.

55 Community Ecology

In this chapter you will learn that

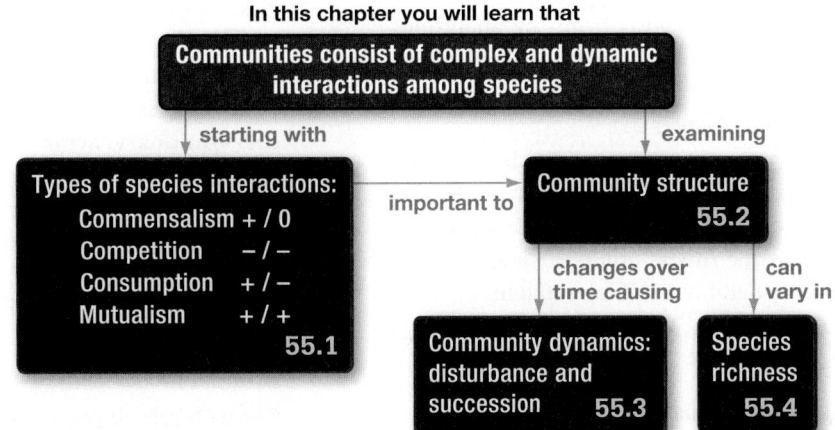

Communities consist of complex and dynamic interactions among species

starting with → ← examining

Types of species interactions:
Commensalism + / 0
Competition – / –
Consumption + / –
Mutualism + / +
55.1

important to →

Community structure
55.2

changes over time causing ↓   ↓ can vary in

Community dynamics: disturbance and succession **55.3**

Species richness **55.4**

This chapter explores how species that live in coral reefs and other communities interact with each other, and how communities change over time.

Populations are dynamic. They grow and decline over time and through space (Chapter 54). However, populations do not exist in isolation—individuals within populations also interact with individuals of other species, forming complex assemblages called communities.

A biological **community** consists of all of the species that interact in a certain area. Ecologists want to know how communities work, and how to manage them in a way that will preserve biodiversity, ecosystem function, and an environment that people want to live in. You can see how community ecology fits into the Big Picture of Ecology on pages 1196–1197.

The goal of this chapter is to explore how species interact with each other and to analyze the structure and dynamics of biological communities—how they develop and change over time. Let's delve in.

This chapter is part of the Big Picture. See how on pages 1196–1197.

✔ When you see this checkmark, stop and test yourself. Answers are available in Appendix A.

55.1 Species Interactions

The species in a community interact constantly. Organisms eat each other, exchange nutrients, compete for resources, and provide habitats for each other. In many cases, a population's fate is tightly linked to the other species that share its habitat.

To study species interactions, biologists begin by analyzing the effects of one species on the fitness of another. Recall that the **fitness** of an individual is defined by its ability to survive and produce viable, fertile offspring (Chapter 25).

Does the relationship between two species provide a fitness benefit to members of one species (a "+" interaction) but hurt members of the other species (a "−" interaction)? Or does the association have no effect on the fitness of a participant (a "0" interaction)? There are four general types of interactions:

1. **Commensalism** occurs when one species benefits but the other species is unaffected (+/0).

2. **Competition** occurs when individuals use the same resources—resulting in lower fitness for both (−/−).

3. **Consumption** (including herbivory, predation, and parasitism) occurs when one organism eats or absorbs nutrients from another. The interaction increases the consumer's fitness but decreases the victim's fitness (+/−).

4. **Mutualism** occurs when two species interact in a way that confers fitness benefits to both (+/+).

Let's consider each type of species interaction in more detail. As you analyze each one, watch for three key themes:

1. *Species interactions can affect the distribution and abundance of a particular species.* Species interactions can affect geographic ranges, such as the effect of disease-carrying tsetse flies on the distribution of cattle in Africa (Chapter 52). Species interactions can also have impacts on population size, such as the effects of lynx on hare populations in northern Canada (Chapter 54).

2. *Species act as agents of natural selection when they interact.* Deer are fast and agile in response to natural selection exerted by their major predators, wolves and cougars. The speed and agility of deer, in turn, promote natural selection that favors wolves and cougars that are fast and that have superior eyesight, senses of smell, and hunting strategies. To capture this point, biologists say that **coevolution** is occurring—a pattern of evolution where two species influence each other's adaptations over time.

As species interact over time, it is common to observe a **coevolutionary arms race**—a repeating cycle of reciprocal adaptation. In human civilizations, an arms race occurs when one nation develops a new weapon, which prompts a rival nation to develop a defensive weapon, which pushes the original nation to manufacture an even more powerful weapon, and so on. In biology, coevolutionary arms races occur between predators and prey, parasites and hosts, and other types of interacting species.

3. *The outcome of interactions among species is dynamic and conditional.* Just as two people may alternately help or compete with each other depending on circumstances, the interactions between different organisms may change over time and space.

Commensalism

Of the four general types of interactions, commensalism (+/0) is the least studied. Commensalism can be challenging to demonstrate, because an absence of fitness benefits or costs ("0") can be difficult to quantify. Commensalism is also very conditional.

An example is the relationship between antbirds and the army ants that they follow in the tropics. As the ants march along the forest floor, they hunt insects and small vertebrates. The antbirds follow and pick off prey species that fly or jump up out of the way of the ants (**FIGURE 55.1**). The antbirds benefit, but the ants are not affected.

However, if antbird attacks start to force other insects into the path of the ants, then both ants and birds benefit, and the

(a) Ants stir up insects while hunting ...

(b) ... antbirds tag along and benefit.

FIGURE 55.1 Commensals Gain a Fitness Advantage but Don't Affect the Species They Depend On. Birds that associate with army ants are commensals. They have no measurable fitness effect on the ants, but gain from the association by capturing insects that try to fly or climb out of the way of the ants.

relationship becomes mutualistic. But if the antbirds begin to steal prey that would otherwise be taken by ants, then the relationship becomes competitive. The outcome of the interaction may depend on the number and types of prey, birds, and ants present and may change over time.

Competition

Competition (−/−) lowers the fitness of the individuals involved for a simple reason: When competitors use resources, those resources are not available to help other individuals survive better and produce more offspring.

Competition that occurs between members of the same species is called **intraspecific** (literally, "within species") **competition.** Intraspecific competition for space, sunlight, food, and other resources intensifies as a population's density increases. As a result, intraspecific competition is a major cause of density-dependent growth (see Chapter 54). **Interspecific** ("between species") **competition** occurs when individuals from different species use the same limiting resources.

Using the Niche Concept to Analyze Interspecific Competition

Early work on interspecific competition focused on the concept of the niche. A **niche** can be thought of as the range of resources that the species is able to use, or the range of conditions it can tolerate. For example, *Anolis* lizards that occupy the crowns of trees on the island of Jamaica occupy a different niche than *Anolis* lizards that live on the ground (Chapter 28).

G. Evelyn Hutchinson proposed that a species' niche could be envisioned by plotting its habitat requirements along a series of axes. **FIGURE 55.2a**, for example, represents one niche axis for a hypothetical species. In this case, the habitat requirement plotted is the size of seeds eaten by members of this population, determined by variables such as mouth or tooth size. Other niche axes could represent other types of foods used or the temperatures, humidity, and other environmental conditions tolerated by the species. In these types of graphs, the *y*-axis represents the number or proportion of resources used, averaged over the population.

Interspecific competition occurs when the niches of two species overlap. The two species plotted in **FIGURE 55.2b**, for example, compete for seeds of intermediate size—assuming that seeds are a limited resource. Individuals that use this "joint" resource are at a disadvantage relative to individuals that use other resources, because they are likely to acquire fewer seeds and have lower fitness as a result. However, if both species experience a similar decrease in fitness due to the overlap of their niches, then **symmetric competition** is said to occur and both species may persist in the area of overlap, even if in reduced numbers.

What Happens when One Species Is a Better Competitor?
When **asymmetric competition** occurs, one species suffers a greater fitness decline than the other species does. The outcome of this interaction depends on the amount of overlap in their niches.

What outcomes are possible? To explore this type of competitive interaction, G. F. Gause conducted a series of experiments with two species of the unicellular pond-dweller *Paramecium.*

(a) One species eats seeds of a certain size range.

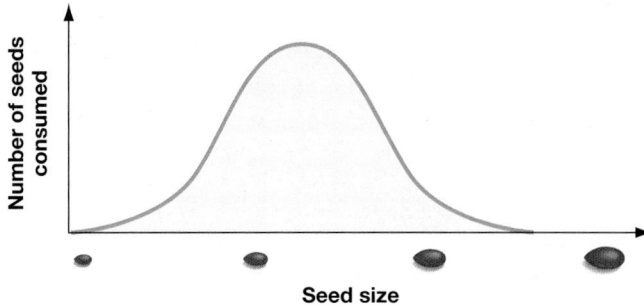

(b) Partial niche overlap: competition for seeds of intermediate size

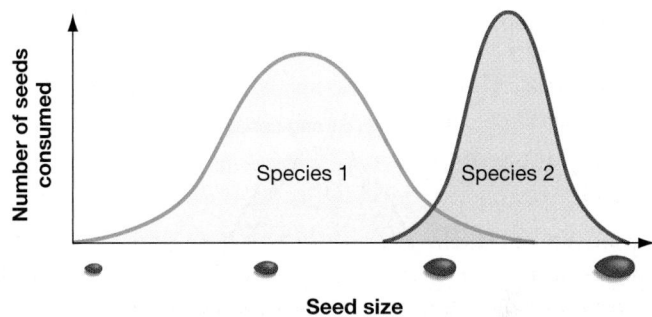

FIGURE 55.2 Niche Overlap Leads to Competition. (a) A graph describing one aspect of a species' fundamental niche, meaning the range of resources that it can use or range of conditions it can tolerate. **(b)** Competition occurs when the niches of different species overlap.

- When *P. caudatum* and *P. aurelia* grew in separate laboratory cultures, both species exhibited logistic growth—the number of individuals increased rapidly at first, then leveled off as the population reached its carrying capacity (see Chapter 54).

- When the two species grew in the same culture together, only the *P. aurelia* population exhibited a logistic growth pattern. *P. caudatum*, in contrast, was driven to extinction (**FIGURE 55.3a**, see page 1126).

The stronger competitor drove out the weaker competitor because the two laboratory populations had completely overlapping niches (**FIGURE 55.3b**). Based on these data, Gause proposed that two species that occupy the same niche cannot coexist, a hypothesis called the **competitive exclusion principle.** Some biologists invoke this principle to explain the disappearance of Neanderthals (*Homo neanderthalensis*) after thousands of years of coexistence with our species, *Homo sapiens* (see Chapter 35).

But what happens if the niches of the two species do not overlap completely? The species that is the weaker competitor should be able to retreat to, or persist in, the area of non-overlap. In cases like this, an important distinction arises.

- A **fundamental niche** is the total theoretical range of environmental conditions that a species can tolerate.

(a) Observation: asymmetric competition

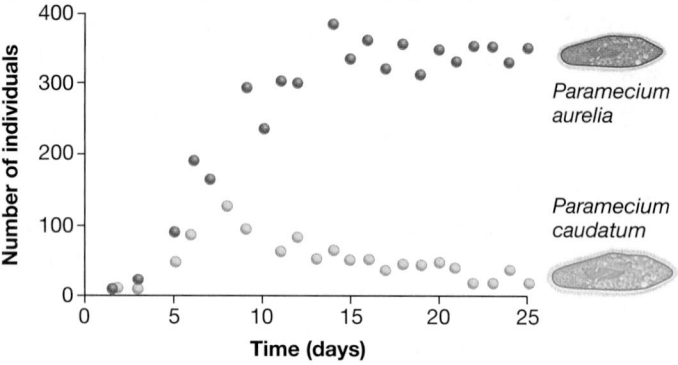

(b) Explanation: competitive exclusion due to complete niche overlap

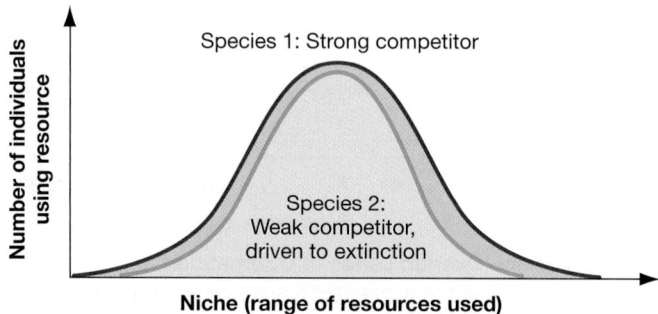

FIGURE 55.3 Competitive Exclusion in *Paramecium*.

DATA: Gause, G. F. 1934. *The Struggle for Existence*. New York: Hafner Press.

- A **realized niche** is the portion of the fundamental niche that a species actually occupies, given limiting factors such as competition with other species (**FIGURE 55.4**).

Let's take a look at an important field experiment that demonstrated the difference between a fundamental and realized niche.

Experimental Studies of Competition in Nature Joseph Connell began a classic study of competition in the late 1950s, after observing an interesting pattern in an intertidal rocky shore in Scotland. He noticed that there were two species of barnacles

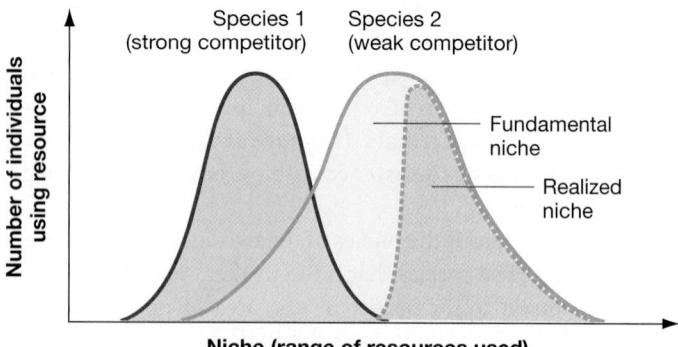

FIGURE 55.4 Fundamental Niches Are Broader than Realized Niches.

with distinctive distributions. The adults of one species, *Chthamalus stellatus*, occurred in an upper intertidal zone. The adults of the other species, *Semibalanus balanoides*, were restricted to a lower intertidal zone. The upper intertidal zone is a more severe environment for barnacles, because it is exposed to the air and intense heat for longer periods at low tide each day.

Although adult barnacles live attached to rocks, their planktonic larvae are mobile. The young of both species were found together in the lower intertidal zone. Why aren't *Chthamalus* adults found there as well?

To answer this question, Connell considered two hypotheses:

1. Adult *Chthamalus* are outcompeted in the lower intertidal zone by *Semibalanus*.

2. Adult *Chthamalus* are absent from the lower intertidal zone because they do not thrive in the physical conditions there.

Connell tested these hypotheses by performing the experiment shown in **FIGURE 55.5**. He began by removing a number of rocks that had been colonized by *Chthamalus* from the upper intertidal zone and transplanting them into the lower intertidal zone. He screwed the rocks into place and allowed *Semibalanus* larvae to colonize them. Once the spring colonization period was over, Connell divided each rock into two sides. In one half, he removed all *Semibalanus* that were in contact with or next to a *Chthamalus*.

This experimental design allowed Connell to document *Chthamalus* survival in the absence of competition with *Semibalanus*, and compare it with survival during competition. This is a common experimental strategy in competition studies: One of the competitors is removed, and the response by the remaining species is observed.

Connell's results supported the competition hypothesis. On the rocks with both species present, *Semibalanus* killed many of the *Chthamalus* by growing against them and lifting them off the substrate. As the orange bar in the "Results" section of Figure 55.5 shows, *Chthamalus* survival was much higher when all of the *Semibalanus* were removed.

The fundamental niches of these two species partially overlap. Under the conditions tested in this experiment, *Semibalanus* is a superior competitor for space, but *Chthamalus* is able to thrive in a smaller realized niche where *Semibalanus* cannot survive. ✔ If you understand this result, you should be able to sketch and label a graph of the fundamental and realized niches of the two barnacle species as a function of depth in the intertidal zone.

Fitness Trade-Offs in Competition Why haven't *Semibalanus* and other successful competitors taken over the world? Biologists answer this question by invoking the concept of **fitness trade-offs**—inevitable compromises in adaptation (see Chapter 25).

The key insight is that the ability to compete for a particular resource—like space on rocks or edible seeds of a certain size—is only one aspect of an organism's niche. If individuals are extremely good at competing for a particular resource, then they are probably not as good at enduring drought conditions, warding off disease, or preventing predation.

QUESTION: Why is the distribution of adult *Chthamalus* restricted to the upper intertidal zone?

HYPOTHESIS: Adult *Chthamalus* are outcompeted in the lower intertidal zone.

NULL HYPOTHESIS: Adult *Chthamalus* do not thrive in the physical conditions of the lower intertidal zone.

EXPERIMENTAL SETUP:

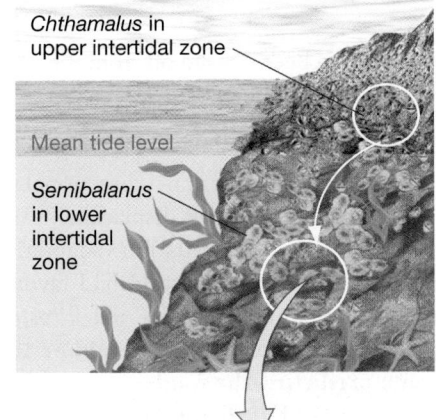

Chthamalus in upper intertidal zone

Mean tide level

Semibalanus in lower intertidal zone

1. Transplant rocks containing young *Chthamalus* to lower intertidal zone.

2. Let *Semibalanus* colonize the rocks.

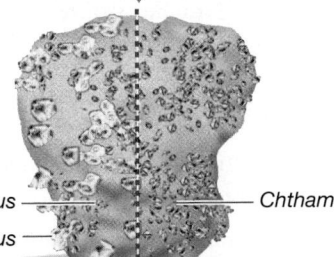

3. Remove *Semibalanus* from half of each rock. Monitor survival of *Chthamalus* on both sides.

Chthamalus + *Semibalanus*

Chthamalus

PREDICTION: *Chthamalus* will survive better in the absence of *Semibalanus*.

PREDICTION OF NULL HYPOTHESIS: *Chthamalus* survival will be low and the same in the presence or absence of *Semibalanus*.

RESULTS:

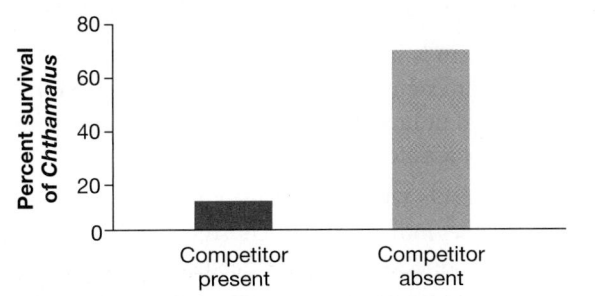

CONCLUSION: *Chthamalus* do not occur in the lower intertidal zone because they are outcompeted by *Semibalanus*.

FIGURE 55.5 Experimental Evidence of Competition.

SOURCE: Connell, J. H. 1961. The influence of interspecific competition and other factors on the distribution of the barnacle *Chthamalus stellatus*. *Ecology* 42: 710–723.

✔ **QUESTION** Why was it important to carry out both treatments on the same rock? Why not use separate rocks?

In the case of *Semibalanus* and *Chthamalus* growing in the intertidal zone, the fitness trade-off is rapid growth and success in competing for space versus the ability to endure the harsh physical conditions of the upper intertidal zone. *Semibalanus* are fast-growing and large; *Chthamalus* grow slowly but can survive long exposures to the air and to intense sun and heat. Neither species can do both things well.

Also, competitive ability is not an inherent quality of species—it depends on environmental conditions. For example, Argentine ants have unremarkable competitive abilities in their native South America, due to the checks and balances provided by thousands of years of species interactions. But when these ants were introduced to California and the Mediterranean in food shipments, they began to attack aggressively and outcompete native species (Chapter 52). The remarkable competitive success of invasive species poses a threat to communities around the world—and conservation biologists are very concerned about it (see Chapter 57).

Mechanisms of Coexistence: Niche Differentiation It's important to realize that because competition is a −/− interaction, there is strong natural selection on both species to avoid it. **FIGURE 55.6** shows the predicted outcome of competition over many generations: An evolutionary change in traits reduces the amount of niche overlap, and thus the amount of competition.

An *evolutionary* change in resource use, caused by competition over generations, is called **niche differentiation** or resource partitioning. The evolutionary change that occurs in species' traits, and that enables species to exploit different resources, is called **character displacement.** Character displacement makes niche differentiation possible.

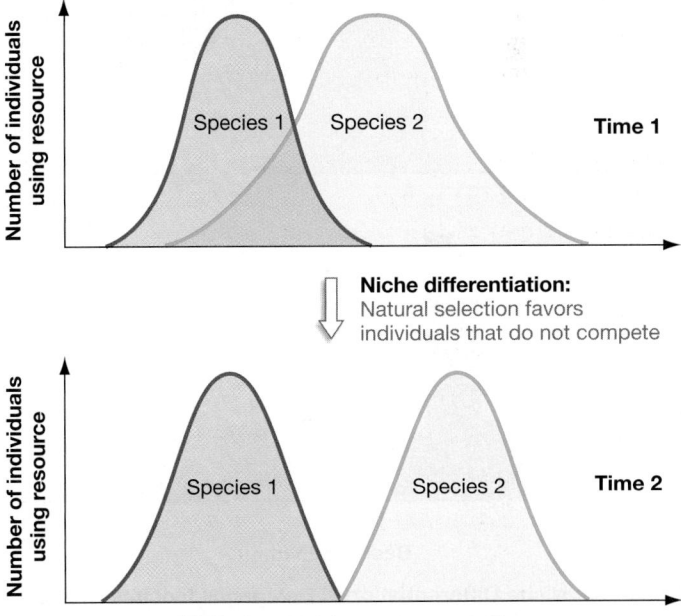

FIGURE 55.6 Competition Can Lead to Niche Differentiation. Over evolutionary time, natural selection favors individuals that do not compete.

One spectacular example of character displacement involves the Galápagos finches, famous for their diversity in beak shapes (Chapter 25). Specializations in beak shape are adaptations for harvesting particular food types, such as small or large seeds.

Beak shape varies not only among species of Galápagos finches but also within species. For example, **FIGURE 55.7** shows the frequency of beak depths (in this case, measured from the top half of the beak only) in a population of medium ground finch, *Geospiza fortis*, on the island of Daphne Major in the 1940s. You can see on the *x*-axis that beak depth, an important measure of beak shape, ranged from 3.5 to 5.6 mm. However, on the island of Santa Cruz at around the same time, *G. fortis* lived alongside the small ground finch, *G. fuliginosa*, and had a broader and higher range of beak depths, from 4.0 to 7.5 mm. From these data you could hypothesize either that competition causes character displacement in *G. fortis*, or that the difference between populations is due to other factors such as predation or food availability.

No data were available from the 1940s to test these hypotheses, but Peter and Rosemary Grant observed character displacement in *G. fortis* on Daphne Major more recently. In addition to recording six beak size measurements for hundreds of medium ground finches over three decades starting in 1970, they also made careful observations of feeding behavior, the amount of rainfall each year, the availability of different types of seeds, and the presence of other species.

The Grants observed that the large ground finch, *G. magnirostris*, established a breeding population on Daphne Major in 1982,

living alongside or "sympatric" with *G. fortis*. True to its name (*magnirostris* translates to "large beak"), this finch has a large beak capable of opening large, hard, drought-tolerant seeds. What effect did competition with *G. magnirostris* have on *G. fortis*?

At first the Grants observed a slow but steady decline in average beak size in *G. fortis* after the arrival of *G. magnirostris*. Then in 2004, during a severe drought when competition with *G. magnirostris* was at its most intense, most finches died of starvation. The Grants observed that

• Only *G. fortis* that could eat extremely small seeds efficiently could survive.

• The larger-beaked *G. fortis* had been chased off from the feeding sites with the large seeds by *G. magnirostris*.

• Overall, the average beak size of *G. fortis* declined sharply.

The Grants had observed evolution in action—directional selection on *G. fortis* beak size causing character displacement (reduced average beak size) and niche differentiation (adaptation for different seed sizes). That is, natural selection had favored *G. fortis* individuals that did not compete with *G. magnirostris*. ✔ If you understand niche differentiation, you should be able to sketch and label a graph predicting the fundamental niches of the barnacles *Semibalanus* and *Chthamalus* many years after Connell's experiment, if niche differentiation were to occur.

Consumption

Consumption is a +/− interaction that occurs when one organism eats another. There are three major types of consumption:

1. **Herbivory** takes place when **herbivores** ("plant-eaters") consume plant tissues. Caterpillars chew leaves; marine iguanas feed on algae.

2. **Parasitism** occurs when a **parasite** consumes relatively small amounts of tissue or nutrients from another individual, called the **host**. Parasitism often occurs over a long period. It is not necessarily fatal, and parasites are usually small relative to their host. An array of worms, ticks, mites, and unicellular protists parasitizes humans. In some cases, though, the term parasitism is used more broadly, to denote use of other resources. For example, birds that lay their eggs in other species' nests and induce the other species to raise the young are called social parasites.

3. **Predation** occurs when a predator kills and consumes all or most of another individual. The consumed individual is called the prey. Woodpeckers eat bark beetles; ladybird beetles devour aphids; orcas eat seals. Although many researchers use predation to refer specifically to **carnivores** ("meat-eaters"), predation can also refer to the consumption of plants, especially seeds (which contain entire plant embryos).

To illustrate how biologists analyze the impact of consumption, let's consider a series of questions about herbivores, parasites, predators, and their victims.

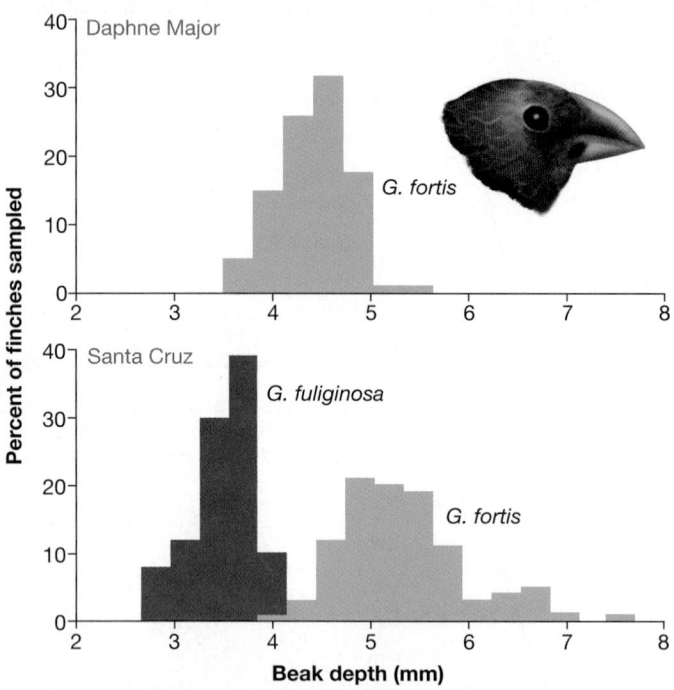

FIGURE 55.7 Niche Differentiation in Galápagos Finches Average beak depths in *Geospiza fortis* were higher on the island with a competitor present.

DATA: Grant, P. 1999. *Ecology and Evolution of Darwin's Finches*. Princeton, NJ: Princeton University Press.

Constitutive Defenses From the standpoint of fitness, consumption is costly for prey and beneficial for consumers. Prey individuals do not passively give up their lives to increase the fitness of their predators, however. Natural selection strongly favors traits that allow individuals to avoid being eaten. Prey species may hide, flee, poison or threaten to poison their predators, school together to confuse their predators, or deploy armor or weapons in defense (**TABLE 55.1**). Traits like these are called **standing or constitutive defenses**, because they are present even in the absence of predators.

The diversity and specificity of constitutive defenses is nothing short of mind-boggling. Consider the discovery made by naturalist Henry Walter Bates as he collected butterflies in the Amazonian rain forest in the mid-1800s. Bates found an astonishing diversity of butterflies in his study area—many more species than in all of Europe—and there was a surprising pattern among the diversity. He noticed pairs of species from different taxonomic groups that were not closely related, yet had very similar wing patterns and occupied the same parts of the forest. One example of this pattern of one species resembling another, called **mimicry,** is shown in **FIGURE 55.8.**

Based on what you have learned about niche differentiation, you might have expected that the traits of species occupying similar niches would become more *different* over time, not more alike. What could explain this seeming paradox in butterflies?

Bates had an insight when he observed that some of the butterflies exuded foul substances when he handled them and that these butterflies were passed over by would-be predators. Concluding that the foul substance thwarted predators, he hypothesized that natural selection favored mimic species that

FIGURE 55.8 Why Do these Butterflies Look So Similar? Henry Walter Bates observed that these two Amazonian butterflies are distantly related, but look more like each other than either does to its closer relatives.

SUMMARY TABLE 55.1 Constitutive Defenses

Type of Defense	Example
Cryptic coloration Cryptic coloration helps prey such as this leaf grasshopper escape detection	
Escape behavior Prey may have adaptations to detect predators and run, fly, jump, or swim away	
Toxins and other defense chemicals Prey may lace their tissues with toxic compounds. Many advertise their toxicity with warning coloration	
Schooling/flocking Predators can become confused by groups of prey	
Defense armor and weapons Prey may have protective shells, sharp spines, or other defense structures	

resembled the unpalatable species, a form of mimicry now known as **Batesian mimicry.** Follow-up experimental studies supported this hypothesis—and recent genetic studies have discovered some of the molecular mechanisms responsible for the shared color patterns.

Bates also noticed that some pairs of similar but distantly related butterflies are *both* unpalatable. What was the fitness advantage of this pattern? Several hypotheses were proposed, but German naturalist Fritz Müller hit on the key idea—that the existence of similar-looking unpalatable prey in the same habitat increases the likelihood that predators will learn to avoid them. Thus, co-mimicry boosts the fitness of both species. This type of mimicry, now known as **Müllerian mimicry,** was subsequently supported by field experiments.

Butterflies are not the only organisms to engage in mimicry. Bates and Müller both observed diverse examples, some of them quite shocking, such as a caterpillar with a close resemblance to a venomous snake. **FIGURE 55.9** provides a comparison of Batesian and Müllerian mimics in the case of stinging and nonstinging insects with the familiar yellow and black warning coloration.

One of the take-home messages from studies of mimicry is that the +/− consumption interaction between two species can cause unexpected consequences to other species in the community. That is, natural selection in the form of predator–prey interactions can result in mutualism, commensalism, and other types of interactions among prey species.

Inducible Defenses Although constitutive defenses can be extremely effective, they can also be expensive due to the energy and resources that must be devoted to producing and maintaining them. Based on this observation, it should not be surprising to learn that many prey species have **inducible defenses:** physical, chemical, or behavioral defensive traits that are induced in the prey in response to the presence of a predator.

Induced defenses decline in prey species if predators leave the habitat. Inducible defenses are efficient energetically, but they are slow—it takes time to produce them.

To see how inducible defenses work, consider research on the blue mussels that live in an estuary along the coast of Maine. Biologists had documented that predation on mussels by crabs was high in an area of the estuary with relatively slow tidal currents (a "low-flow" area) but low in an area of the estuary with relatively rapid tidal currents (a "high-flow" area). Mark Bertness and colleagues hypothesized that if blue mussels possess inducible defenses, then heavily defended prey individuals should occur in the low-flow area, where predation pressure is higher, but not in the high-flow area, where water movement reduces the number of crabs present. This is exactly the trend they observed: Blue mussels had thicker shells in the part of the estuary with many crabs.

The correlation between crab density and mussel shell thickness is not enough to demonstrate that predation pressure induced shell thickening. One alternate hypothesis is that crabs eliminated mussels with thin shells from the low-flow areas. The observed differences could also be due to variation in light, temperature, or other abiotic factors that might affect mussel traits but have nothing to do with predation.

To test the inducible-defense hypothesis more rigorously, Bertness and his team carried out the experiment diagrammed in **FIGURE 55.10**. The tank on the right allowed the researchers to measure shell growth in mussels that were "downstream" from crabs, such that they were exposed to the scent of crabs. As predicted by the inducible-defense hypothesis, the mussels exposed to a crab in this way developed significantly thicker shells than did mussels that were not exposed to the scent of a crab. These results suggest that, even without direct contact, mussels can sense the presence of crabs and increase their investment in defenses.

Are Animal Predators Efficient Enough to Reduce Prey Populations?

Prey are typically smaller than predators, have larger litter or clutch sizes, and tend to begin reproducing at a younger age. As a result, they have a much larger intrinsic growth rate. This quantity, symbolized r_{max}, is the maximum growth rate that a population can achieve under ideal conditions (Chapter 54).

If prey reproduce rapidly and are also well defended, can predators kill enough of them to reduce the prey population significantly?

Data from predator-removal programs—in which human hunters actively kill wolves, cougars, coyotes, or other predators—suggest that the answer is yes. For example, when a wolf control program in Alaska during the 1970s decreased predator abundance, the population of moose, on which wolves prey, tripled. This observation suggests that wolves had reduced this moose population far below the carrying capacity—the number that could be supported by the available space and food (see Chapter 54).

Other types of experiments are consistent with this result. For example, the regular population cycles of snowshoe hares appear

(a) Batesian mimics: look dangerous, are **not dangerous**

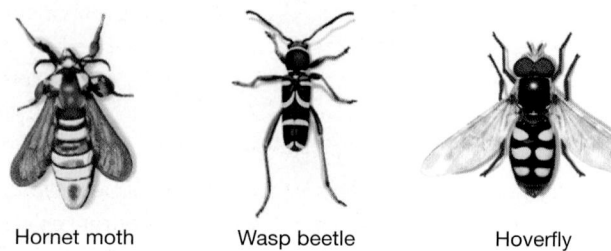

Hornet moth   Wasp beetle   Hoverfly

(b) Müllerian mimics: look dangerous, are **dangerous**

Paper wasp   Bumblebee   Honeybee

FIGURE 55.9 Prey Species Display Two Basic Forms of Mimicry.

QUESTION: Are mussel defenses induced by the presence of crabs?

HYPOTHESIS: Mussels increase investment in defense in the presence of crabs.

NULL HYPOTHESIS: Mussels do not increase investment in defense in the presence of crabs.

EXPERIMENTAL SETUP:

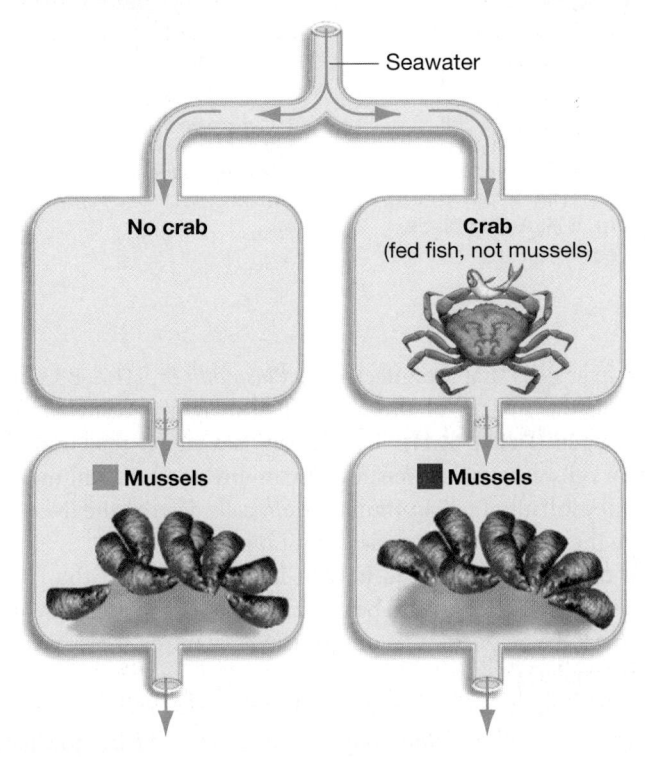

PREDICTION: Mussels downstream of the crab tank will have thicker shells than mussels downstream of the empty tank.

PREDICTION OF NULL HYPOTHESIS: Mussels in the two tanks will have shells of equal thickness.

RESULTS:

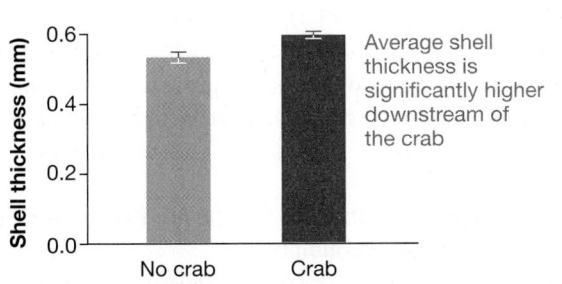

Average shell thickness is significantly higher downstream of the crab

CONCLUSION: Mussels increase investment in defense when they detect crabs. Shell thickness is an inducible defense.

FIGURE 55.10 Experimental Evidence for Inducible Defenses.
SOURCE: Leonard, G. H., M. D. Bertness, and P. O. Yund. 1999. Crab predation, waterborne clues, and inducible defenses in the blue mussel, *Mytilus edulis. Ecology* 80: 1–14.

✔ **QUESTION** Why did the researchers feed the crabs fish instead of mussels?

to be driven, at least in part, by density-dependent increases in predation by lynx (Chapter 54). Taken together, the data indicate that in many instances, predators are efficient enough to reduce prey populations below carrying capacity.

Why Don't Herbivores Eat Everything—Why Is the World Green?

If predators affect the size of populations in prey that can run or fly or swim away, then consumers should have a devastating impact on plants and on mussels, anemones, sponges, and other sessile (nonmoving) animals.

In some cases, this prediction turns out to be correct. For example, consider the results of a recent **meta-analysis**—a study of studies, meaning an analysis of a large number of data sets on a particular question. Biologists who compiled the results of more than 100 studies on herbivory found that the median percentage of mass removed from aquatic algae by herbivores was 79 percent. Herbivores eat the vast majority of algal food available in aquatic biomes. The figure dropped to just 30 percent for aquatic plants, however, and only 18 percent for terrestrial plants.

Why don't herbivores eat more of the food available on land? Stated another way, why is the world green? Biologists routinely consider two hypotheses:

1. *Top-down control hypothesis: Herbivore populations are limited by predation and disease.* The "top-down" name is inspired by the food-chain concept (introduced in Chapter 30 and explored in detail in Chapter 56). In a food chain, herbivores are "on top of" plants.

   The logic of top-down control is simple: Predators and parasites remove herbivores that eat plants. As a result, a great deal of plant material remains uneaten.

   In a recent test of this idea, researchers monitored herbivory on islands created by a 1986 dam project in Venezuela. On some small islands in the new lake, predators disappeared. On the predator-free islands, there are now many more herbivores than on similar sites nearby where predators are present. As predicted by the top-down control hypothesis, a much higher percentage of the total plant material is being eaten on the predator-free islands. For example, predator-free islands had just 25 percent of the small trees found on islands that contained predators.

2. *Bottom-up limitation hypothesis: Plant tissues offer poor nutrition and are well defended.* The "bottom-up" name reflects the position of plants as the base or bottom of a food chain. The idea here is that herbivore numbers are limited because plant tissues offer poor nutrition, or that plant tissues aren't eaten because they are toxic.

   The poor-nutrition aspect of this hypothesis was inspired by the observation that plant tissues have less than 10 percent of the nitrogen found in animal tissues, by weight. If the growth and reproduction of herbivores are limited by the availability of nitrogen, then their populations will be low and the impact of herbivory on plants relatively slight. Herbivores could eat more plant material to gain nitrogen, but at

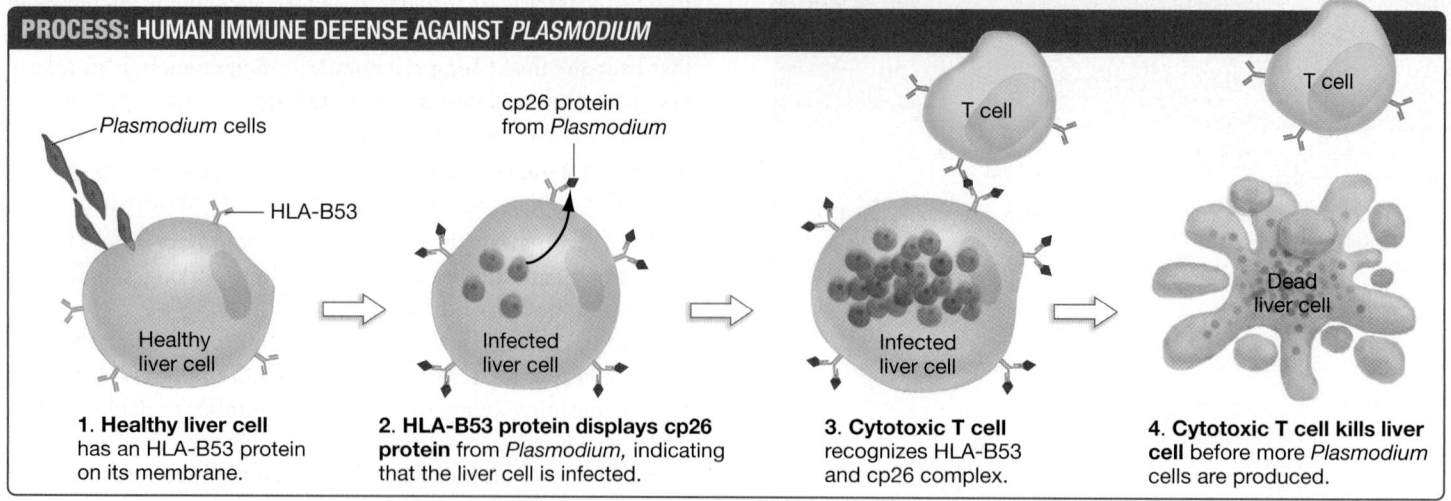

PROCESS: HUMAN IMMUNE DEFENSE AGAINST *PLASMODIUM*

Plasmodium cells

HLA-B53

Healthy
liver cell

cp26 protein
from *Plasmodium*

Infected
liver cell

T cell

Infected
liver cell

T cell

Dead
liver cell

1. Healthy liver cell has an HLA-B53 protein on its membrane.

2. HLA-B53 protein displays cp26 protein from *Plasmodium*, indicating that the liver cell is infected.

3. Cytotoxic T cell recognizes HLA-B53 and cp26 complex.

4. Cytotoxic T cell kills liver cell before more *Plasmodium* cells are produced.

FIGURE 55.11 Interactions between the Human Immune System and *Plasmodium*. If HLA-B53 binds to a particular *Plasmodium* protein, then infected cells are recognized and destroyed.

a cost—they would be exposed to predation and expend energy processing the food. Indeed, when nitrogen concentration in plants is increased experimentally by fertilization, it is not unusual to see significant increases in herbivore growth rates.

What's more, most plant tissues are defended by weapons such as thorns, prickles, or hairs, or by potent poisons such as nicotine, caffeine, and cocaine. The punchline? Both top-down control and bottom-up limitations—such as nitrogen limitation and effective defense—are important factors in limiting the impact of herbivory. The mix of factors that keeps the world green varies from species to species and habitat to habitat.

Adaptation and Arms Races   Over the long term, how do species that interact via consumption affect each other's evolution? When predators and prey or herbivores and plants interact over time, coevolutionary arms races result. Traits that increase feeding efficiency evolve in predators and herbivores. In response, traits that make prey unpalatable or elusive evolve. In counter-response, selection favors consumers with traits that overcome the prey adaptation. And so on. Coevolution is a major driver of evolutionary change.

To see a coevolutionary arms race in action, consider interactions between humans and the most serious of all human parasites—species in the genus *Plasmodium*. *Plasmodium* species are unicellular protists that cause malaria (Chapter 30). Malaria kills at least 1 million people a year, most of them young children. Recent data suggest that humans and *Plasmodium* are locked in a coevolutionary arms race.

In West Africa, for example, there is a strong association between an allele called *HLA-B53* and protection against malaria. Malaria infections start when *Plasmodium* cells enter a person's bloodstream during a mosquito bite. The *Plasmodium* individuals then infect liver cells, where they begin to multiply.

In liver cells that are infected by *Plasmodium*, HLA-B53 proteins on the surface of the liver cell display a parasite protein called cp26 (**FIGURE 55.11**). The display is a signal that immune system cells can read. It means "I'm an infected cell. Kill me before they kill all of us." Immune system cells destroy the liver cell before the parasite cells inside can multiply.

In this way, people who have at least one copy of the *HLA-B53* allele are better able to beat back malarial infections. People who have the *HLA-B53* allele appear to be winning the arms race against malaria.

Follow-up research has shown that the arms race is far from won, however. *Plasmodium* populations in West Africa now have a variety of alleles for the protein recognized by HLA-B53. Some of these variants bind to HLA-B53 and trigger an immune response in the host, but others escape detection (**TABLE 55.2**). To make sense of these observations, researchers suggest that natural selection has favored the evolution of *Plasmodium* strains with weapons that counter HLA-B53.

Can Parasites Manipulate Their Hosts?   To thrive, parasites do not just have to invade tissues and grow while evading defensive responses by their host. They also have to be transmitted to new

TABLE 55.2 **Some *Plasmodium* Strains Are Particularly Effective at Infecting Humans**

Plasmodium Strain	Infection Rate	Interpretation
cp26	Low	HLA-B53 binds to these proteins. Immune response is effective.
cp29	Low	
cp27	High	HLA-B53 does not bind to these proteins. Immune response is not as effective.
cp28	Average	

hosts. To a parasite, an uninfected host represents uncolonized habitat, teeming with resources. What have biologists learned about how parasites are transmitted to new hosts?

To answer this question, consider species of tree-dwelling ants (**FIGURE 55.12a**) that are parasitized by nematodes (roundworms). When nematodes have matured inside a host individual, they lay eggs in the ant's abdomen. The exoskeleton in the area becomes translucent, and the yellow eggs inside make the structure look reddish instead of the normal black color (**FIGURE 55.12b**). Infected ants also hold the abdomen up in a "flagging" posture, making them look like the berries that grow on the trees occupied by the ants (**FIGURE 55.12c**). Experiments have shown that fruit-eating birds are much more likely to pluck the abdomen from infected ants than uninfected ants.

Why is this interesting? To complete their life cycle, the nematodes have to grow inside birds, before being shed in bird feces that are subsequently eaten by ants.

To interpret their observations, biologists suggest that these nematodes not only change the appearance of the ants, but they also manipulate their behavior. The changes make the parasite much more likely to be transmitted to a new host.

Biologists have now cataloged a large number of case studies like this, from an array of parasites and hosts. Parasite manipulation of hosts is another example of extensive coevolution.

(a) Uninfected ant   **(b)** Infected ant

(c) Infected ants resemble berries that are eaten by birds.

FIGURE 55.12 A Parasite That Manipulates Host Behavior. When the tropical ant *Cephalotes atratus* becomes infected with parasitic nematodes, its appearance and behavior change.

✔**EXERCISE** Design an experiment to test the hypothesis that infected ants are more likely than uninfected ants to be eaten by birds.

Using Consumers as Biocontrol Agents Research on the dynamics of predator–prey interactions and other aspects of consumption has paid off in practical benefits: In some cases, biologists have been able to control pests by introducing predators or parasites.

The invasive plant purple loosestrife, for example, had succeeded in taking over large expanses of wetlands in North America. Through careful experimentation, however, biologists were able to find two insect species that feed exclusively on purple loosestrife. By introducing the herbivores into wetlands, researchers have been able to reduce purple loosestrife infestations dramatically.

In agriculture and forestry, the use of predators, herbivores, and parasites as biocontrol agents is a key part of **integrated pest management**: strategies to maximize crop and forest productivity while using a minimum of insecticides or other types of potentially harmful compounds. Integrated pest management is a promising and growing area of research. However, ecologists warn that introducing species to communities to limit one species can have unexpected consequences to others, so it is of utmost importance to proceed with extreme caution.

Mutualism

Mutualisms are +/+ interactions that involve a wide variety of organisms and rewards.

- Bees visit flowers to harvest nectar and pollen. The bees benefit by using nectar and pollen as a food source; flowering plants benefit because in the process of visiting flowers, foraging bees carry pollen from one plant to another and accomplish pollination. Adaptations found in flowering plants increase the efficiency of pollination (Chapters 31 and 41).

- One of the most important of all mutualisms occurs between mycorrhizal fungi and plant roots. For example, experimental evidence indicates that mycorrhizal fungi receive sugars and other carbon-containing compounds in exchange for nitrogen or phosphorus acquired by the plant partner (Chapter 32). The plant feeds the fungus; the fungus fertilizes the plant.

- Mutualism occurs between nitrogen-fixing bacteria and the plant species that house them in their tissues (Chapters 29 and 37). The host plants provide sugars and protection to the bacteria; the bacteria supply ammonia or nitrate in return.

- **FIGURE 55.13a** (see page 1134) shows ants in the genus *Crematogaster*, which live in acacia trees native to Africa. The ants live in bulbs at the base of acacia thorns and feed on small structures that grow from leaf tips. The ants protect the tree by attacking and biting herbivores, and by cutting vegetation from the ground below the host tree.

- **FIGURE 55.13b** illustrates cleaner shrimp in action. These shrimp pick external parasites from the jaws and gills of fish. In this mutualism, one species receives dinner while the other obtains medical attention.

The Role of Natural Selection in Mutualism As the examples you've just reviewed show, the rewards from mutualistic interactions range from transporting gametes to acquiring food, housing,

(a) Mutualism between ants and acacia trees

Entrance
to ant
colony

(b) Mutualism between cleaner shrimp and fish

FIGURE 55.13 Mutualisms Take Many Forms. (a) In certain species of acacia tree, ants in the genus *Crematogaster* live in large bulbs at the base of tree spines and attack herbivores that threaten the tree. The ants eat nutrient-rich tissue produced at the tips of leaves. **(b)** Cleaner shrimp remove and eat parasites that take up residence on the gills of fish.

✔**EXERCISE** Name a possible cost of these mutualistic associations to the acacia tree, the ants, the cleaner shrimp, and the host fish.

medical help, and protection. It is important to note, however, that even though mutualisms benefit both species, the interaction does not involve individuals from different species being altruistic or "nice" to each other.

Expanding on a point that Charles Darwin introduced in 1862, Judith Bronstein described mutualisms as "a kind of reciprocal parasitism; that is, each partner is out to do the best it can by obtaining what it needs from its mutualist at the lowest possible cost to itself." Her point is that the benefits received in a mutualism are a by-product of each individual pursuing its own self-interest by maximizing its fitness—its ability to survive and produce viable offspring.

In this light, it is not surprising that some species "cheat" on mutualistic systems. For example, deceit pollination occurs when certain species of plants produce a showy flower but no nectar reward. Pollinators receive no reward for making a visit and carrying out pollination. Evolutionary studies show that deceit pollinators evolved from ancestral species that did provide a reward. Over time, a +/+ interaction evolved into a +/− interaction.

Mutualisms Are Dynamic A recent experimental study of mutualism provides another good example of the dynamic nature of these interactions. This study focused on ants and treehoppers. Treehoppers are small, herbivorous insects that feed by sucking sugar out of the phloem of plants. Treehoppers excrete the sugary solution called honeydew from their posteriors. The ants, in turn, harvest the honeydew for food.

It is clear that ants benefit from this association. But do the treehoppers? J. Hall Cushman and Thomas Whitham hypothesized that the ants might protect the treehoppers from their major predator, jumping spiders. These spiders feed heavily on juvenile treehoppers.

To test the hypothesis that ants protect treehoppers, Cushman and Whitham studied ant–treehopper interactions over a three-year period. As **FIGURE 55.14** shows, the researchers marked out a 1000-m² study plot. Each year they removed the ants from one group of the treehopper host plants inside the plot but left the others alone to serve as a control. Then they compared the growth and survival of treehoppers on plants with and without ants.

Recall that this is a common research strategy for studying species interactions. To assess the fitness costs or benefits of the interaction, researchers remove one of the participants experimentally and document the effect on the other participant's survival and reproduction, compared with the survival and reproduction of control individuals that experience a normal interaction.

In both the first and third years of the study, the number of treehopper young on host plants increased in the treatment with ants but showed a significant decline in the treatment with no ants. This result supports the hypothesis that treehoppers benefit from the interaction with ants, because the ants protect the treehoppers from predation by jumping spiders.

In the second year of the study, however, Cushman and Whitham found a very different pattern. There was no difference in offspring survival, adult survival, or overall population size between treehopper populations with ants and those without ants.

Why? The researchers were able to answer this question because they also measured the abundance of spiders that prey on treehoppers in each of the three years. Their census data showed that in the second year of the study, spider populations were very low.

Based on these results, Cushman and Whitham concluded that the benefits of the ant–treehopper interaction depend entirely on predator abundance. Treehoppers benefit from their interaction with ants in years when predators are abundant, but are unaffected in years when predators are scarce. If producing honeydew is costly to treehoppers, then the +/+ mutualism changes to a +/− interaction when spiders are rare.

RESEARCH

QUESTION: Is the relationship between ants and treehoppers mutualistic?

HYPOTHESIS: Ants harvest food from treehoppers. In return, they protect treehoppers from jumping spiders.

NULL HYPOTHESIS: Ants harvest food from treehoppers but are not beneficial to treehopper survival.

EXPERIMENTAL SETUP:

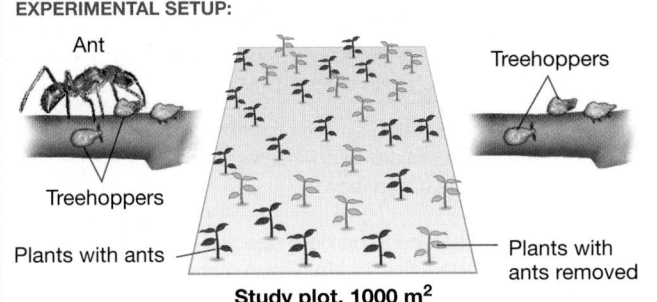

Ant

Treehoppers

Treehoppers

Plants with ants

Plants with ants removed

Study plot, 1000 m²

PREDICTION: Treehopper reproduction will be higher when ants are present than when ants are absent.

PREDICTION OF NULL HYPOTHESIS: There will be no difference in the number of young treehoppers on the plants.

RESULTS:

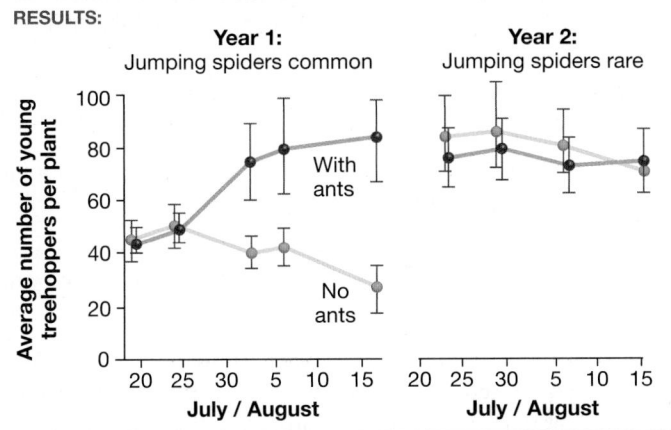

Year 1:
Jumping spiders common

Year 2:
Jumping spiders rare

With ants

No ants

Average number of young treehoppers per plant

July / August

July / August

CONCLUSION: Treehoppers benefit from the interaction with ants only when jumping spiders are common.

FIGURE 55.14 Experimental Evidence of Mutualism. The data points on the graphs represent average values from many study plots.

Cushman, J. H., and T. G. Whitham. 1989. Conditional mutualism in a membracid–ant association: temporal, age-specific, and density-dependent effects. *Ecology* 70: 1040–1047.

✔**QUESTION** The researchers assigned the treatments to different plants at random, and they did many replicates of each treatment. Why was this method important?

Mutualism is like parasitism, competition, and other types of species interactions in an important respect: The outcome of the interaction depends on current conditions. Because the costs and benefits of species interactions are fluid, an interaction between the same two species may range from parasitism to mutualism to competition. **TABLE 55.3** (see page 1136) summarizes the fitness effects, short-term impacts on population size, and long-term evolutionary aspects of species interactions.

If you understand that . . .

- Natural selection favors the evolution of traits that decrease competition, resulting in changes in species' niches over time.
- Consumers may reduce the population size of the species they feed on, and they often exert strong natural selection for effective defense mechanisms.
- Mutualisms benefit the species involved and can lead to highly coevolved associations, such as mycorrhizal fungi and symbiotic nitrogen-fixing bacteria.
- Coevolution is a major driver of evolutionary change.

✔ **You should be able to . . .**

1. Explain why niche differentiation (resource partitioning) does not involve a conscious choice by the individuals involved.

2. Explain what a coevolutionary arms race is, and give an example.

Answers are available in Appendix A.

55.2 Community Structure

Up to this point, the focus has been on interactions that occur between two species. But in a community, each species interacts with dozens or hundreds or even thousands of other species within a defined area. The resulting complexity of interactions is mind-boggling. How can biologists begin to characterize communities? In general, a community structure has four key attributes:

1. the total number of species;

2. the relative abundance of those species;

3. the sum of interactions among all species; and

4. the physical attributes of the community, which may include abiotic factors such as geographic size or altitude gradient, and biotic factors such as the physical structure provided by the primary vegetation type, as in a forest or grassland (see Chapter 52).

The processes that shape a particular community can be quite complex. To get a better sense of these processes, let's consider some of the ways that a community's structure can be either predictable or unpredictable.

How Predictable Are Communities?

The first question that biologists asked about woodlands and deserts and shorelines concerned structure: Do biological communities have a tightly prescribed organization and composition, or are they merely loose assemblages of species?

Type of Interaction	Fitness Effects	Short-Term Impact: Distribution and Abundance	Long-Term Impact: Coevolution
Commensalism	+/0	Population size and range of commensal may depend on population size and distribution of host.	Strong selection on commensal to increase fitness benefits in relationship; no selection on host
Competition	−/−	Reduces population size of both species; if competition is asymmetric, the weaker competitor may become locally extinct, or it may retreat to a part of its niche that does not overlap with the stronger competitor.	Niche differentiation via selection to reduce competition
Consumption	+/−	Impact on prey population depends on predator density, prey density, and effectiveness of defenses.	Strong selection on prey for effective defense; strong selection on consumer for traits that overcome defenses
Mutualism	+/+	Population size and range of both species are dependent on each other.	Strong selection on both species to maximize fitness benefits and minimize fitness costs of relationship

The Clements–Gleason Dichotomy   Beginning with a paper published in 1936, Frederick Clements hypothesized that biological communities are stable and orderly entities with a highly predictable composition. His reasoning was that species interactions are so extensive—and coevolution so important—that communities have become highly integrated and interdependent units in nature. Stated another way, the species within a community cannot live without each other.

In developing this hypothesis, Clements likened the development of a plant community to the development of an individual organism. He argued that communities develop over time by passing through a series of predictable stages dictated by extensive interactions among species. Eventually, this developmental progression culminates in a final stage, known as a **climax community.** The climax community is stable—it does not change over time.

According to Clements, the nature of a climax community is determined by the area's climate. He predicted that if a fire or other disturbance destroys the climax community, it will reconstitute itself by repeating its predictable developmental stages.

Henry Gleason, in contrast, contended that the community found in a particular area is neither stable nor predictable. He claimed that plant and animal communities are ephemeral associations of species that just happen to share similar climatic requirements.

According to Gleason, it is largely a matter of chance whether a similar community develops in the same area after a disturbance occurs. Gleason downplayed the role of biotic factors, such as species interactions, in structuring communities. Instead, chance historical events—for example, which seeds and juvenile animals happened to arrive after a disturbance—were key elements in determining which species are found at a particular location.

Which viewpoint is more accurate? Let's consider experimental and observational data.

An Experimental Test   To explore the predictability of community structure experimentally, ecologists David Jenkins and Arthur Buikema constructed 12 identical ponds (**FIGURE 55.15**). They filled the ponds at the same time with water that contained enough chlorine to kill any preexisting organisms. The Clements and Gleason hypotheses make clear and contrasting predictions:

- If community structure is predictable, then each pond should develop the same community of species once the chlorine has vaporized and made the water habitable.

- If community structure is unpredictable, then each pond should develop a different community.

To test these predictions, the researchers sampled water from the ponds repeatedly for one year. They measured temperature, chemical makeup, and other physical characteristics of the water and recorded the diversity and abundance of each planktonic species by examining the samples under the microscope. Species that make up **plankton** drift in the water and swim little, if at all.

Their results? The rows of colored bars on the graph in Figure 55.15 indicate whether a particular species was found in each of the 12 ponds. Note that the researchers found a total of 61 planktonic species in all of the ponds. However, individual ponds had only 31 to 39 species apiece. This observation is important. Each pond contained just half to two-thirds of the total number of species that lived in the experimental area and that were available for colonization. Several species occurred in most or all 12 of the ponds, but each pond had a unique species assemblage. Why?

To explain their results, Jenkins and Buikema contended that some species are particularly good at dispersing and are likely to colonize all or most of the available habitats. Other species disperse more slowly and tend to reach only one or a few of the

QUESTION: Are communities predictable or unpredictable?

COMMUNITIES-ARE-PREDICTABLE HYPOTHESIS: The group of species present at a particular site is highly predictable.

COMMUNITIES-ARE-UNPREDICTABLE HYPOTHESIS: The group of species present at a particular site is highly unpredictable.

EXPERIMENTAL SETUP:

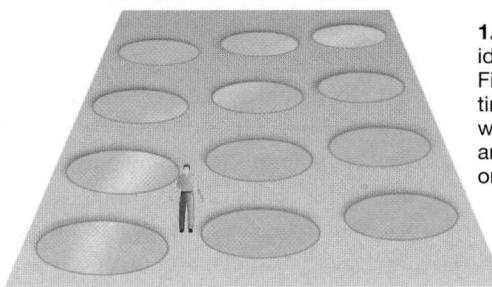

1. Construct 12 identical ponds. Fill at the same time and sterilize water so that there are no preexisting organisms.

2. Examine water samples from each pond. Identify each plankton species present in each sample.

COMMUNITIES-ARE-PREDICTABLE PREDICTION: Identical plankton communities will develop in all 12 ponds.

COMMUNITIES-ARE-UNPREDICTABLE PREDICTION: Different plankton communities will develop in different ponds.

RESULTS (AFTER 1 YEAR):

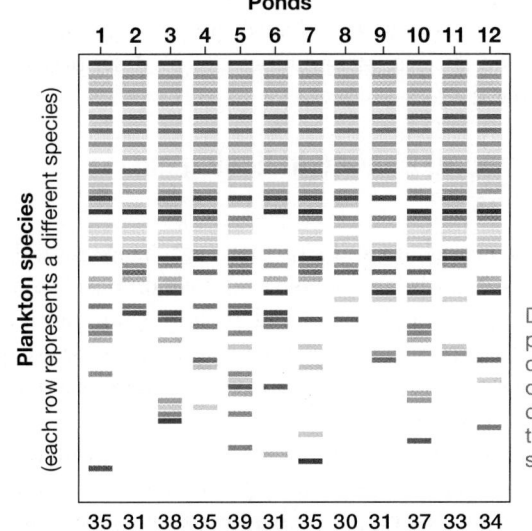

Ponds

Plankton species (each row represents a different species)

Different ponds contain different combinations of species

Total species in each pond

CONCLUSION: Although about half of the species present appear in all or most ponds, each pond has a unique composition. Both hypotheses are partially correct.

FIGURE 55.15 Experimental Evidence that Identical Communities Do Not Develop in Identical Habitats.

SOURCE: Jenkins, D. G., and A. L. Buikema. 1998. Do similar communities develop in similar sites? A test with zooplankton structure and function. *Ecological Monographs* 68: 421–443.

✔ **QUESTION** At least 20 species are found in most or all of the ponds. Does this observation suggest that community composition is predictable or unpredictable?

---

available habitats. Further, the investigators proposed that the arrival of certain competitors or predators early in the colonization process greatly affects which species are able to invade successfully later. As a result, the specific details of community assembly and composition are somewhat contingent and difficult to predict. At least to some degree, communities are a product of chance and history.

Mapping Current and Past Species' Distributions If communities are predictable assemblages, then the ranges of species that make up a particular community should be congruent—meaning that the same group of species should almost always be found living together.

When biologists began documenting the ranges of tree species along elevational gradients, however, they found that species came and went independently of each other. As you go up a mountain, for example, you might find white oak trees growing with hickories and chestnuts. As you continue to gain elevation, chestnuts might disappear, but white oaks and hickories remain. Farther upslope, hickories might drop out of the species mix as white pines and red pines start to appear.

Data on the historical composition of plant communities supported these observations. Studies of fossil pollen documented that the distribution of plant species and communities at specific locations throughout North America has changed radically since the end of the last ice age about 11,000 years ago.

An important pattern emerged from these data: Species do not come and go in the fossil record in tightly integrated units. Instead, the ranges of individual species tend to change independently of one another.

The overall message of research on community structure is that Clements's position was too extreme and that Gleason's view may be closer to being correct. Although both biotic interactions and climate are important in determining which species exist at a certain site, chance and history also play a large role.

How Do Keystone Species Structure Communities?

Even though communities are not predictable assemblages dictated by obligatory species interactions, biotic interactions are still critically important. And not all species are equal. The presence of a particular plant, for example, can have a huge effect on community structure, such as the presence of redwood trees in a redwood forest. Removal of these structural species by harvesting, disease, competition with an invasive species, climate change, or other factors can cause dramatic changes to community structure.

The presence of certain consumers can also have an enormous impact on the species present. In some cases, the structure of an entire community can change dramatically if a single species of predator or herbivore is removed from a community or added to it.

As an example of this research, consider an experiment that Robert Paine conducted in intertidal habitats of the Pacific Northwest of North America. In this environment, the sea star

(a) Predator: *Pisaster ochraceous*

(b) Prey: *Mytilus californianus*

(c) Effect of keystone predator on species richness

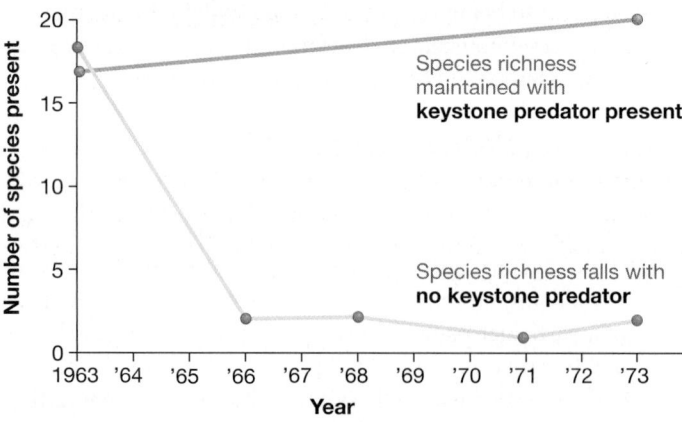

FIGURE 55.16 Keystone Predation Alters Community Structure in a Rocky Intertidal Habitat. Each data point in part (c) represents the average number of species present in several study plots.

DATA: Paine, R. T. 1966. *American Naturalist* 100: 65–75.

Pisaster ochraceous is an important predator (**FIGURE 55.16a**). Its preferred food is the California mussel *Mytilus californianus*.

When Paine removed *P. ochraceous* from experimental areas, what had been diverse communities of algae and invertebrates became overgrown with solid stands of mussels (**FIGURE 55.16b**).

M. californianus is a dominant competitor, but its populations had been held in check by sea star predation.

The graph in **FIGURE 55.16c** shows the number of species present in experimental plots where *P. ochraceous* was present or excluded. When the predator was absent, the species richness of the habitat changed radically.

To capture the effect that a predator such as *P. ochraceous* can have on a community, Paine coined the term keystone species. A **keystone species** has a much greater impact on the distribution and abundance of the surrounding species than its abundance and total biomass would suggest.

An important take-home message of Paine's study is that sea stars are not just affecting their prey. They are also indirectly affecting many other species that interact with their prey. When considering the effects of a keystone species or any other species on the community, it is helpful for biologists to think in terms of the **food web,** the network of exchanges of energy and nutrients among producers, consumers, and decomposers in an ecosystem (see Chapter 56). Food webs focus only on feeding interactions, not all types of species interactions, but are central to the structure of communities.

55.3 Community Dynamics

Once biologists had a basic understanding of how species interact and how communities are structured, they turned to questions about how communities change through time. Like cells, individuals, and species, communities can be described in one word: dynamic.

Disturbance and Change in Ecological Communities

Community composition and structure may change radically in response to changes in abiotic and biotic conditions. Biologists have become particularly interested in how communities respond to disturbance. A **disturbance** is any strong, short-lived disruption to a community that changes the distribution of living and/or nonliving resources.

Forest fires, windstorms, floods, the fall of a large canopy tree, disease epidemics, and short-term explosions in herbivore numbers all qualify as disturbances. These events are important because they alter light levels, nutrients, unoccupied space, or some other aspect of resource availability.

Biologists have come to realize that the impact of disturbance is a function of three factors: (**1**) the type of disturbance, (**2**) its frequency, and (**3**) its severity—for example, the speed and duration of a flood, or the intensity of heat during a fire. Biologists have also come to realize that most communities experience a characteristic type of disturbance. In most cases, disturbances occur with a predictable frequency and severity.

To capture this point, biologists refer to a community's **disturbance regime.** For example, fires kill all or most of the existing trees in a boreal forest every 100 to 300 years, on average.

FIGURE 55.17 Giant Sequoia after a Fire.

(a) Fire scars in the growth rings

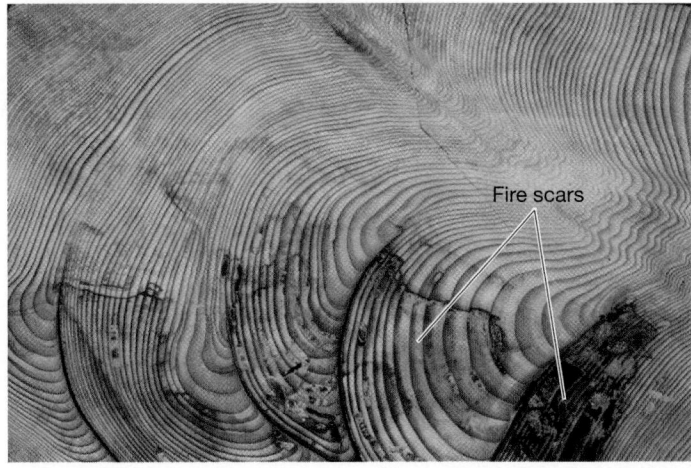

Fire scars

(b) Reconstructing history from fire scars

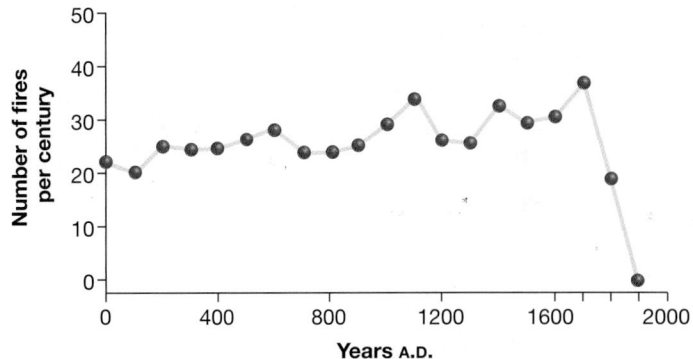

FIGURE 55.18 History of Disturbance in a Fire-Prone Community. Because trees form one ring (light band/dark band) every year, researchers can count the rings to determine how often fires have occurred during the last 2000 years in giant sequoia groves.
DATA: Swetnam, T. W. 1993. *Science* 262: 885–889.

In contrast, many small-scale tree falls, usually caused by windstorms, occur in temperate and tropical forests every few years.

To appreciate why biologists are so interested in understanding disturbance regimes, consider a recent study on the fire history of giant sequoia groves in California (**FIGURE 55.17**). Giant sequoias grow in small, isolated groves on the western side of the Sierra Nevada range. Individuals live more than a thousand years, and many have been scarred repeatedly by fires.

Thomas Swetnam, a specialist on tree-ring research, obtained samples of cross sections through the bases of 90 giant sequoias in five different groves. As **FIGURE 55.18a** shows, the cross sections contained many growth rings that had been scarred by fire. To determine the date of each disturbance, he counted tree rings back from the present.

The graph in **FIGURE 55.18b** plots the average number of fires that occurred each century in the groves, from the year 0 C.E. to 1900. Swetnam found that in most of the groves, 20–40 fires had occurred each century. The data indicated that each tree had been burned an average of 64 times.

This study established that fires had been extremely frequent in these forests for most of recorded history. Because not enough time would pass for large amounts of fuel to accumulate between fires, they were probably of low severity. However, recent human intervention has reduced the frequency of forest fires—with unintended and unfortunate consequences. When fires do occur, the buildup of fuel causes massive conflagrations that are often devastating to the forest community.

Partly because of this work, the biologists responsible for managing sequoia groves now set controlled fires or let low-intensity natural fires burn instead of suppressing them immediately. To maintain communities in good condition, biologists have to ensure that the normal disturbance regime occurs. Otherwise, community composition changes dramatically.

Similarly, studies of disturbance regimes along the Colorado River in southwestern North America inspired land managers to begin releasing a huge pulse of water from the reservoirs behind dams, at intervals. The floods that have resulted were designed to mimic the natural disturbance regime. According to follow-up studies, the artificial floods appear to have benefited the plant and animal communities downstream by depositing nutrients and creating additional sandbar habitat.

Succession: The Development of Communities after Disturbance

Severe disturbances remove all or most of the organisms from an area. The recovery that follows is called **succession.** The name was inspired by the observation that species with certain types of life history strategies tend to succeed others over time.

- **Primary succession** occurs when a disturbance removes the soil and its organisms as well as organisms that live above the surface. Glaciers, floods, volcanic eruptions, and landslides often initiate primary succession.

- **Secondary succession** occurs when a disturbance removes some or all of the organisms from an area but leaves the soil intact. Fire and logging are examples of disturbances that initiate secondary succession.

How does recovery proceed after a severe disturbance? Imagine a landscape before primary succession begins, such as the area surrounding the Kilauea volcano in Hawaii. There are no animals, no plants, *and no soil.* The process of soil building begins with solid rock, which weathers over time to produce gravel, sand, silt, and clay (Chapter 39). As organisms occupy the substrate, they add dead cells and tissues and feces. With time, soil eventually becomes a complex and dynamic mixture of inorganic particles, organic particles, and living organisms that enable plants to take root and add structure to the landscape.

During secondary succession, soil is already present. **FIGURE 55.19** shows a typical sequence of plant communities that develops as succession proceeds. Early successional communities are dominated by species that are short lived, small in stature, and disperse their seeds over long distances. Late successional or climax communities are dominated by species that tend to be long lived, large, and good competitors for resources such as light and nutrients.

The particular sequence of species that appears over time is called a successional pathway. What determines the pattern and rate of species replacement during succession at a certain time and place? To answer this question, biologists focus on three factors:

1. the particular traits of the species involved;

2. how the species interact; and

3. historical and environmental circumstances, such as the size of the area involved and weather conditions.

Let's consider each factor in turn.

The Role of Species Traits   Species traits, such as dispersal capability and the ability to withstand extreme dryness, are particularly important early in succession. As common sense would predict, recently disturbed sites tend to be colonized by plants and animals with good dispersal ability. When these organisms arrive, however, they often have to endure harsh environmental conditions.

Pioneering species tend to have "weedy" life histories. A **weed** is a plant that is adapted for growth in disturbed soils. Because soil is often disturbed when biomass is removed, weeds tend to thrive at the start of secondary succession.

Early successional species devote most of their energy to reproduction and little to competitive ability. They are at the "high fecundity, low survivorship" end of the life-history continuum (Chapter 54).

More specifically, pioneering species have small seeds, rapid growth, and a short life span, and they begin reproducing at an early age. As a result, they have a high reproductive rate (high r_{max}—see Chapter 54). But in addition, they can tolerate severe abiotic conditions such as high light levels, poor nutrient availability, and desiccation.

The Role of Species Interactions   Once colonization is under way, the course of succession tends to depend less on how species

PROCESS: SECONDARY SUCCESSION

1. Pioneering species
Weedy species become established in disturbed soils.

2. Early successional community
Weedy species are replaced by longer-lived herbaceous species.

3. Mid-successional community
Shrubs and short-lived trees begin to invade.

4. Climax community
Long-lived tree species mature.

FIGURE 55.19 Secondary Succession in Midlatitude Temperate Forests. Succession leads to the development of a temperate forest from a disturbed state (in this case, an abandoned agricultural field).

cope with aspects of the abiotic environment and more on how they interact with other species. This change occurs because plants that grow early in succession change abiotic conditions in a way that makes the conditions less severe.

For example, because plants provide shade, they reduce temperatures and increase humidity. Their dead bodies also add organic material and nutrients to the soil. During succession, existing species can have one of three effects on subsequent species:

1. **Facilitation** takes place when the presence of an early-arriving species makes conditions more favorable for the arrival of certain later species, such as by providing shade or nutrients.

2. **Tolerance** means that existing species do not affect the probability that subsequent species will become established.

3. **Inhibition** occurs when the presence of one species inhibits the establishment or regrowth of another. For example, a plant species that requires high light levels to germinate may be inhibited late in succession by the presence of mature trees that prevent sunlight from reaching the forest floor. Alternately, an established species may produce a chemical that inhibits the growth of other species.

The Role of Chance and History   In addition to species traits and species interactions, the pattern and rate of succession depend on their historical and environmental context. For example, the communities that developed after forest fires disturbed Yellowstone National Park in 1988 depended on the size of the burned patch and how hot the fire had been at that location.

Succession is also affected by the particular weather or climate conditions that occur during the process. Variation in weather and climate causes different successional pathways to occur in the same place at different times.

Analyzing species traits, species interactions, and the historical–environmental context provides a useful structure for understanding why particular successional pathways occur. To see this theoretical framework in action, let's examine data on the course of succession that has occurred in Glacier Bay, Alaska.

A Case Study: Glacier Bay, Alaska   An extraordinarily rapid and extensive glacial recession is taking place at Glacier Bay (**FIGURE 55.20**). In just 200 years, glaciers that once filled the bay have retreated approximately 100 km, exposing tracts of barren glacial sediments to colonization. Because of this event, Glacier Bay has become an important site for studying succession.

Figure 55.20 shows the plant communities found in the area.

- Locations that have been ice free for 20 years or less do not have a continuous plant cover. Instead, they host scattered individuals of willow and a small shrub called *Dryas*.

- Areas that have been deglaciated for about 100 years are inhabited by dense thickets of a shrub called Sitka alder and scattered Sitka spruce trees.

- The oldest sites, ranging from 150 to 200 years of soil exposure, are dense forests of Sitka spruce and western hemlock.

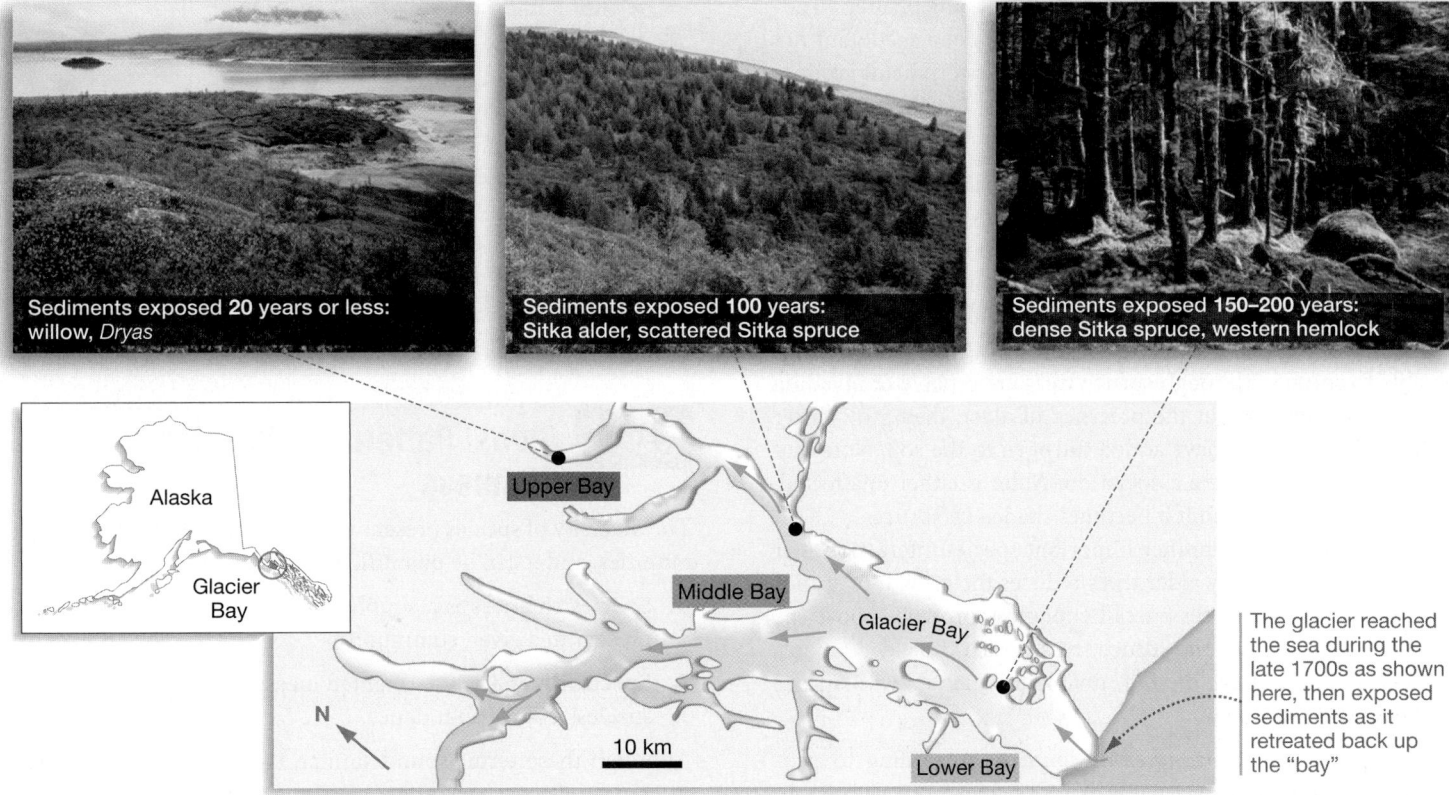

Sediments exposed **20 years or less**: willow, *Dryas*

Sediments exposed **100 years**: Sitka alder, scattered Sitka spruce

Sediments exposed **150–200 years**: dense Sitka spruce, western hemlock

Alaska

Glacier Bay

Upper Bay

Middle Bay

Glacier Bay

Lower Bay

N

10 km

The glacier reached the sea during the late 1700s as shown here, then exposed sediments as it retreated back up the "bay"

FIGURE 55.20 Successional Stages in Glacier Bay.

These observations inspired a hypothesis for the pattern of succession in Glacier Bay: With time, the youngest communities of *Dryas* and willow succeed to alder thickets, which subsequently become dense spruce–hemlock forests. The key claim is that the entire bay is following a single successional pathway.

Research has challenged this hypothesis, however. Biologists who reconstructed the history of each community by studying tree rings found that at least two distinct successional pathways have occurred:

1. In the lower part of the bay, soon after the ice retreated, Sitka spruce began growing and quickly formed dense forests. Western hemlock arrived after spruce and is now common.

2. At middle-aged sites in the upper part of the bay, alder thickets were dominant for several decades, and spruce is just beginning to become common. These forests will probably never be as dense as the ones in the lower bay, however, and there is no sign that western hemlock has begun to establish itself.

In contrast, the youngest sites in the uppermost part of the bay may be following a third pathway. Alder thickets became dominant fairly early, but spruce trees are scarce. Instead, cottonwood trees are abundant.

These data raise an important question: How do species traits, species interactions, and dispersal patterns interact to generate the observed successional pathways?

- *Species traits* Western hemlock is abundant at older sites but largely absent from young ones. This pattern is logical for this species, because its seeds germinate and grow only in soils containing a substantial amount of organic matter. In addition, young hemlock trees can tolerate deep shade but not bright sunlight. Because of these traits, Western hemlock does not tolerate early successional conditions, but thrives later in succession.

- *Species interactions* Sitka alder facilitates the growth of Sitka spruce. The facilitating effect occurs because symbiotic bacteria that live inside nodules on the roots of alder convert atmospheric nitrogen (N_2) to nitrogen-containing molecules that alder use to build proteins and nucleic acids. When alder leaves fall and decay or roots die, the nitrogen becomes available to spruce. Although spruce trees are capable of invading and growing without the presence of alder, they grow faster when alder stands have added nitrogen to the soil. Note that spruce gains from the association. Alder is either unaffected, or negatively affected if it becomes shaded by spruce.

  Competition is another important species interaction. For example, shading by alder trees reduces the growth of spruce trees until spruce trees are tall enough to protrude above the alder thicket. Once the spruce trees breach the alder canopy, however, alder trees die out, unable to compete with spruce trees for light.

- *Historical and environmental context* According to geological evidence, glacial ice was more than 1100 m thick in the upper part of Glacier Bay during the mid-1700s. Because

forests grow to an elevation of only 700 m or 800 m in this part of Alaska, the glacier eliminated all of the existing forests. But in the lower part of the bay, the ice was substantially thinner. As a result, some forests remained on the mountain slopes beyond the ice, even at the height of glaciation.

The difference is crucial. In the lower part of the bay, nearby forests provided a ready source of Sitka spruce and western hemlock seed. But in the upper part of the bay, no forests remained to contribute seed. As a result, two dramatically different successional pathways were set in motion.

To summarize, successional pathways are determined by the adaptations that certain species have to their abiotic environment, interactions among species, and the history of the site. Species traits and species interactions tend to make succession predictable; history and chance events contribute a degree of unpredictability to succession.

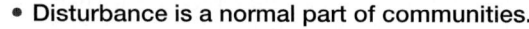

check your understanding

If you understand that . . .

- Disturbance is a normal part of communities.
- The impact of a disturbance depends on its type, frequency, and severity.
- After a disturbance occurs, a succession of species and communities replace those that were lost.
- The exact sequence of species observed is a function of their traits, their interactions, and the history of the site.

✔ **You should be able to . . .**

1. Explain how early successional species alter the environment in ways that make growing conditions more difficult for themselves.

2. Explain how the presence or absence of a species like alder, where nitrogen fixation occurs, might alter the course of succession.

Answers are available in Appendix A.

55.4 Global Patterns in Species Richness

The diversity of species present is a key feature of biological communities, and it can be quantified in two ways.

1. **Species richness** is a simple count of how many species are present in a given community.

2. **Species diversity** is a weighted measure that incorporates the species' relative abundance.

Although these terms sound similar, they can be very different in communities where a few species dominate. In such a case, species richness might be high but species diversity low (see

Consider the composition of the three hypothetical communities shown in **FIGURE 55.21**. These data can be used to compare two measures of diversity, *species richness* and *species diversity*. Species richness is simply the number of species found in a community. In this case, communities 1 and 2 have equal species richness, and community 3 is lower in richness by one species. It is important to note, however, that communities 2 and 3 have similar relative abundances of each species, or what biologists call high *evenness*. Community 1, in contrast, is highly uneven. Fifty-five percent of the individuals in community 1 belong to species A, and other species are relatively rare. An uneven community has lower effective diversity than its species richness would indicate.

To take evenness into account, other diversity indices have been developed. A simple example is the Shannon index of species diversity, given by the following equation:

$$H' = -\sum_{i=1}^{n} p_i \ln p_i$$

In this equation, p_i is the proportion of individuals in the community that belong to species i. The index is summed over all the species in the study (n).

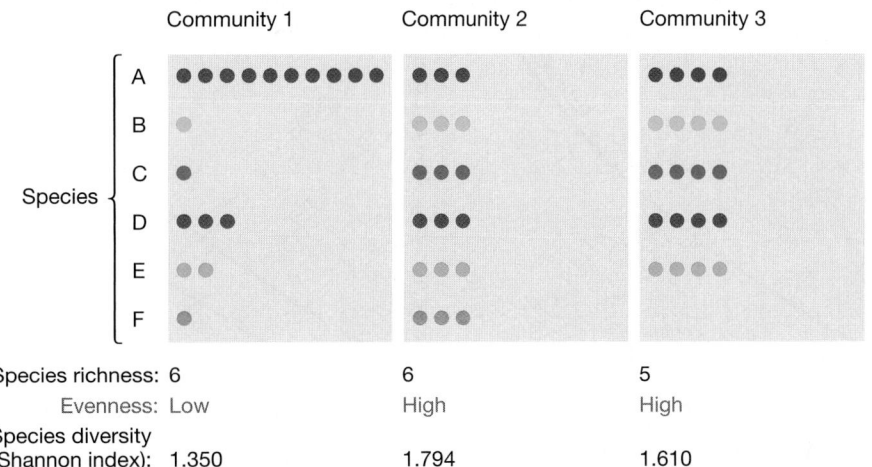

FIGURE 55.21 **Species Diversity Can Be Quantified.**

	Community 1	Community 2	Community 3
Species richness:	6	6	5
Evenness:	Low	High	High
Species diversity (Shannon index):	1.350	1.794	1.610

The Shannon index of species diversity for the hypothetical communities shown in Figure 55.21 was calculated by **(1)** computing the proportion of individuals in each community that belong to each species, **(2)** taking the natural logarithm of each of these proportions (see **BioSkills 6** in Appendix B for a review of logarithms), **(3)** multiplying each natural logarithm times the proportion for each species, and **(4)** summing the total across the six species in the community and multiplying by −1. ✔ **QUANTITATIVE If you understand the equation, you should be able to do these calculations and get the same result given in Figure 55.21.** Notice that while communities 1 and 2 have the same species richness, community 2 has higher diversity because of its greater evenness. Community 3 has lower species richness than community 1 but higher diversity.

Quantitative Methods 55.1. At scales above that of relatively small study plots, however, it is rare to have data on relative abundance. As a result, species richness is the most popular measure, and biologists sometimes use the terms species richness and species diversity interchangeably.

To introduce research on richness and diversity, let's focus on a simple question: Why are some communities more species rich than others?

Predicting Species Richness: The Theory of Island Biogeography

When researchers first began counting how many species are present in various areas, a strong pattern emerged: Larger patches of habitat contain more species than do smaller patches of habitat. The observation is logical because large areas should contain more types of niches and thus support higher numbers of species.

Early work on species richness highlighted another pattern, however—one that was harder to explain. Islands in the ocean have smaller numbers of species than do areas of the same size on continents. Why?

As it turned out, answering this question has had far-reaching implications in ecology and conservation biology. This situation often happens in biological science: Research on a seemingly esoteric question—often done out of the sheer joy of discovery—turns out to have important implications for completely unrelated issues, or even applications to practical problems that had previously been intractable.

The Role of Immigration and Extinction Robert MacArthur and Edward O. Wilson tackled the question of why islands in the ocean have smaller numbers of species than do areas of the same size on continents. They assumed that speciation occurs so slowly that the number of species on an island is a product of just two processes: immigration and extinction, which should vary with the number of species already present.

- Immigration rates should decline as the number of species on the island increases, because individuals that arrive are more likely to be members of a species that is already there. In addition, competition should prevent new species from becoming established when many species are already present on an island—meaning that immigration should seldom result in establishment of a new population.

(a) Species richness depends on the number of existing species.

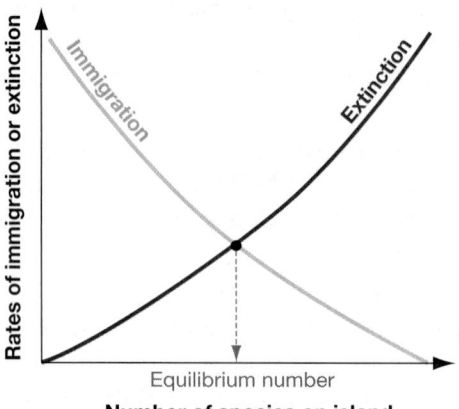

(b) Species richness depends on island size.

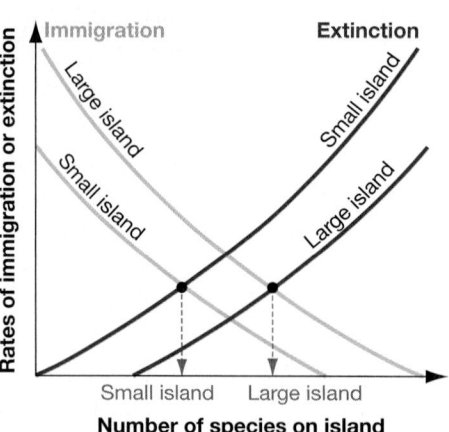

(c) Species richness depends on remoteness of the island.

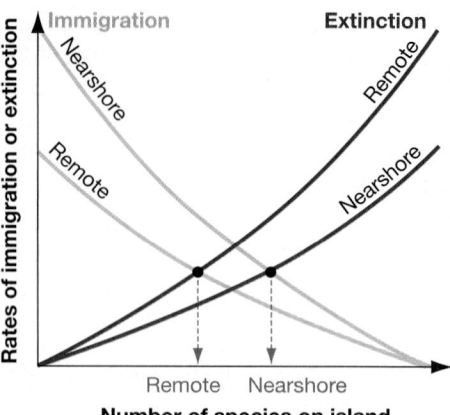

FIGURE 55.22 Theoretical Models Predicting Species Richness as a Function of Island Characteristics. The models illustrate how two key factors affecting species richness—immigration and extinction—depend on the number of existing species, island size, and island remoteness.

✔**QUESTION** Suppose that the mainland habitats closest to an island were wiped out by suburbanization. How would this affect the curves in part (c)?

- Extinction rates should increase as species richness increases, because niche overlap and competition for resources will be more intense.

The curves in **FIGURE 55.22a** illustrate these two predictions. Note that when the two processes interact, the result is an equilibrium—a balance between the arrival of new species and the extinction of existing ones. If species richness changes due to a hurricane or fire or some other disturbance, then continued immigration and extinction should restore the equilibrium value.

The Role of Island Size and Isolation MacArthur and Wilson also realized that immigration and extinction rates should vary as a function of island size and how far the island is from a continent or other source of immigrants.

- Immigration rates should be higher on large islands that are close to mainlands, because immigrants are more likely to find large islands that are close to shore than small ones that are far from shore.

- Extinction rates should be highest on small islands that are far from shore, because fewer resources are available to support large populations, and because fewer individuals arrive to keep the population going.

Their theory makes two predictions: Species richness should be higher on (**1**) larger islands than smaller islands (**FIGURE 55.22b**), and (**2**) nearshore islands versus remote islands (**FIGURE 55.22c**).

Applying the Theory The MacArthur–Wilson model, called the theory of island biogeography, is important for several reasons:

- It is relevant to a wide variety of island-like habitats such as alpine meadows, lakes and ponds, and caves. It is also relevant to national parks and other reserves that are surrounded by human development, forming islands of habitat.

- It is relevant to species that have a metapopulation structure (Chapter 54). Immigration and extinction rates in subpopulations should follow the dynamics predicted by the theory.

- It made specific predictions that could be tested. For example, researchers have measured species richness on tiny islands, removed all of the species present, and then measured whether the same number of species recolonized the island and reached an equilibrium number. Predictions about immigration and extinction rates have also been measured by observing the same islands over time.

- It can help inform decisions about the design of natural preserves. In general, the most species-rich reserves should be ones that are (**1**) relatively large, and (**2**) located close to other relatively large habitat areas (Chapters 54 and 57).

Global Patterns in Species Richness

Biologists have long understood that large habitat areas tend to be species rich, and the theory of island biogeography has been successful in framing thinking about how species richness should vary among island-like habitats. But researchers have had a much more difficult time explaining what may be the most striking pattern in species richness.

The Latitudinal Gradient Biologists who began cataloging the flora and fauna of the tropics in the mid-1800s quickly

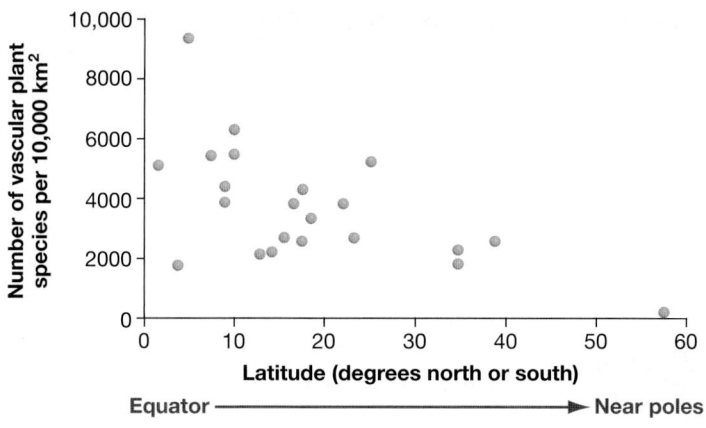

FIGURE 55.23 Species Richness of Vascular Plants Has a Strong Latitudinal Gradient.

DATA: Reid, W. V., and K. R. Miller. 1989. *Keeping Options Alive: The Scientific Basis for Conserving Biodiversity*. Washington, DC: World Resources Institute.

recognized that these habitats contain many more species than do temperate or subarctic environments. Data compiled in the intervening years have confirmed the existence of a strong latitudinal gradient in species diversity—for communities as a whole as well as for many taxonomic groups.

In birds, mammals, fish, reptiles, many aquatic and terrestrial invertebrates, and trees, species diversity declines as latitude increases. **FIGURE 55.23** graphs data on the number of vascular plant species present from the equator toward the poles. This pattern is widespread, although not universal—a number of marine groups, as well as shorebirds, show a *positive* relationship between latitude and diversity. Why does a latitudinal gradient often occur?

To explain why species diversity might decline with increasing latitude, biologists have to consider two fundamental principles:

1. The causal mechanism must be abiotic, because latitude is a physical phenomenon produced by Earth's shape (see Chapter 52). It must be a physical factor that varies predictably with latitude and that could produce changes in species diversity.

2. The species diversity of a particular area is the sum of four processes: speciation, extinction, immigration (colonization), and emigration (dispersal). Thus, the latitudinal gradient must be caused by an abiotic factor that affects the rate of speciation, extinction, immigration, or emigration in a way that would lead to more species in the tropics and fewer near the poles.

Over 30 hypotheses have been proposed to explain the latitudinal gradient. Let's consider three of the most prominent ideas.

The High-Productivity Hypothesis High productivity in the tropics could promote high diversity by increasing speciation rates and decreasing extinction rates. (Recall from Chapter 52 that productivity is the total amount of photosynthesis per unit area per year.) The logic here is that increased biomass production supports more herbivores and thus more predators,

parasites, and scavengers. If higher population sizes lead to more intense competition, then speciation rates should increase as niche differentiation occurs within populations of herbivores, predators, parasites, and scavengers.

The evidence for the high-productivity hypothesis is still inconclusive.

- The hypothesis is supported by global patterns in productivity (see Chapter 56). For example, productivity is extremely high in biomes such as tropical rain forests and coral reefs, which have high species diversity.

- It is not supported by experimental addition of fertilizer to aquatic or terrestrial communities. In most of these studies, productivity increases but species diversity decreases. However, critics contend that the temporal and spatial scale of these experiments is too small to be relevant to global patterns.

- It is not supported in some highly productive habitats, such as estuaries (see Chapter 52), where species diversity can be low.

The Energy Hypothesis The energy hypothesis is an extension of the high-productivity hypothesis. Recognizing that some highly productive habitats are species poor, investigators proposed that the key factor can't be productivity alone.

The energy hypothesis claims that high temperatures increase species diversity by increasing both productivity and the likelihood that organisms can tolerate the physical conditions in a region. The idea is that high energy inputs—rather than high nutrient inputs—are the key to high speciation rates and low extinction rates.

Although the energy hypothesis is relatively new, some data support it. For example, biologists who analyzed data on gastropods and other marine invertebrates have documented a strong correlation between the temperature of marine waters and species diversity.

The Area and Age Hypothesis Temperate and arctic latitudes were repeatedly scoured by ice sheets over the last 2 million years, but tropical regions were not. In addition, the land area of the tropics has historically been larger than the land area at high latitudes. Thus, tropical regions have had more time and more space for speciation to occur than other regions have.

Recent data suggest, however, that tropical forests were dramatically reduced in size by widespread drying trends during the ice ages. Existing tropical forests may be much younger than originally thought. If so, then the contrast in the age of northern and southern habitats may not be enough to explain the dramatic difference in species diversity.

What can you conclude from all this? The current consensus is that each of the factors discussed here may influence the species diversity of communities. It is tempting to look for single causes to explain important patterns such as the latitudinal gradient. But in this case, it is probably more realistic to say there are multiple causes.

If you understand . . .

55.1 Species Interactions

- Two interacting species may each experience a net fitness cost (−) or benefit (+) from the interaction. These costs and benefits depend on the conditions that prevail at a particular time and place and may change through time.

- Commensalism (+/0) occurs when one species benefits but the other is unaffected.

- Competition (−/−) occurs when the niches of two species overlap—meaning they use the same resources. Competition may result in the complete exclusion of one species (the weaker competitor). It may also result in niche differentiation, in which competing species evolve traits that allow them to exploit different resources.

- Consumption (+/−) occurs when consumers eat part or all of other individuals, which resist through constitutive defenses or inducible defenses.

- Predators are efficient enough to reduce the size of many prey populations.

- Levels of herbivory are relatively low in terrestrial ecosystems, because predation and disease limit herbivore populations, plants provide little nitrogen, and many plants have defenses.

- Parasites are consumers that generally spend all or part of their life cycle in or on their host (or hosts). They usually have traits that allow them to escape host defenses and even manipulate host behavior to increase their likelihood of transmission to a new host. In turn, hosts have evolved counter-adaptations that help fight off parasites.

- Mutualism (+/+) provides individuals with food, shelter, transport of gametes, defense against predators, or other services. For each species involved, the costs and benefits of mutualism may vary over time and from place to place.

- Interactions among species have two main outcomes: (1) They affect the distribution and abundance of the interacting species, and (2) they are agents of natural selection and thus affect the evolution of the interacting species, by a process called coevolution.

✔ You should be able to give an example of how a mutualistic relationship can evolve into a parasitic one.

55.2 Community Structure

- The attributes of community structure include the number of species present, their relative abundance, the sum of their interactions, and the physical attributes of the community as determined by a combination of abiotic and biotic factors.

- Historical and experimental evidence supports the view that communities are dynamic rather than static, and that their composition is neither entirely predictable nor stable over time.

- The assemblage of species found in a community is primarily a function of climate and chance historical events.

- In some cases, the structure of an entire community can change dramatically if a single species is removed from it or added to it.

✔ You should be able to predict the effects of an invasive species on community structure.

55.3 Community Dynamics

- Each community has a characteristic disturbance regime—meaning a type, severity, and frequency of disturbance that it experiences.

- Three types of factors influence the pattern of succession that occurs after a disturbance: (1) A species' physiological traits influence when it can successfully join a community; (2) interactions among species influence if and when a species can become established; and (3) the historical and environmental context of the site affects which species are present.

✔ You should be able to explain why climate makes the three successional pathways documented in Glacier Bay similar, and why chance historical events make them different.

55.4 Global Patterns in Species Richness

- Species richness is higher in large islands near continents than in small, isolated islands, due to differences in immigration and extinction.

- The theory of island biogeography is relevant to a wide variety of isolated habitats.

- Species richness is higher in the tropics and lower toward the poles, but the mechanism responsible for this pattern is still hotly debated.

✔ You should be able to propose how the theory of island biogeography could be applied by conservation biologists to species richness in a national park.

MasteringBiology

1. **MasteringBiology Assignments**

   Tutorials and Activities Experimental Inquiry: Can a Species Niche Be Influenced by Interspecific Competition?; Exploring Island Biogeography; Interspecific Interactions; Life Cycle of a Malaria Parasite; Primary Succession; Species Area Effect and Island Biogeography; Succession

   Questions Reading Quizzes, Blue-Thread Questions, Test Bank

2. **eText** Read your book online, search, take notes, highlight text, and more.

3. **The Study Area** Practice Test, Cumulative Test, BioFlix® 3-D Animations, Videos, Activities, Audio Glossary, Word Study Tools, Art

You should be able to . . .

1. True or false? Species act as agents of natural selection when they interact.

2. What is niche differentiation?
   a. the evolution of traits that reduce niche overlap and competition
   b. interactions that allow species to occupy their fundamental niche
   c. the degree to which the niches of two species overlap
   d. the claim that species with the same niche cannot coexist

3. The relationship between ants and treehoppers in the presence of spiders is an example of
   a. commensalism.
   b. competition.
   c. consumption.
   d. mutualism.

4. What is one advantage of inducible defenses?
   a. They are always present; thus, an individual is always able to defend itself.
   b. They make it impossible for a consumer to launch surprise attacks.
   c. They result from a coevolutionary arms race.
   d. They make efficient use of resources, because they are produced only when needed.

5. Pioneer species tend to have high _____ and lower survivorship.

6. Which of these factors is *not* generally correlated with species diversity?
   a. latitude
   b. productivity
   c. longitude
   d. island size

7. The text claims that species interactions are conditional and dynamic. Do you agree with this statement? Why or why not? Cite specific examples to support your answer.

8. Why does the phrase "coevolutionary arms race" appropriately characterize the long-term effects of species interactions?
   a. Both plants and animals have evolved weapons for defense that are so effective that many plants are not eaten, and predators cannot reduce prey populations to extinction.
   b. Adaptations that give one species a fitness advantage in an interaction are likely to be countered by adaptations in the other species that eliminate this advantage.
   c. In all species interactions except for mutualism, at least one species loses (suffers decreased fitness).
   d. Even mutualistic interactions can become parasitic if conditions change. As a result, interacting species are always "at war."

9. Biologists have tested the hypotheses that communities are highly predictable versus highly unpredictable. State the predictions that these hypotheses make with respect to (a) changes in the distribution of the species in a particular community over time, and (b) the communities that should develop at sites where abiotic conditions are identical. Which hypothesis appears to be more accurate?

10. What is a disturbance? Consider the role of fire in a forest. Compare and contrast the consequences of high-frequency versus low-frequency fire, and high versus low severity of fire.

11. Summarize the life-history attributes of early successional species. Why are these attributes considered adaptations?

12. Explain why high productivity should lead to increased species richness in habitats such as tropical rain forests and coral reefs.

13. Some insects harvest nectar by chewing through the wall of the structure that holds the nectar. As a result, they obtain a nectar reward, but pollination does not occur. Suppose that you observed a certain bee species obtaining nectar in this way from a particular orchid species. Over time, how might you expect the characteristics of the orchid population to change in response to this bee behavior?

14. You are a walking, talking community that includes trillions of bacterial and archaeal cells. Your gut, in particular, contains a complex microbial community. In some circumstances, the use of probiotics, which stimulate the rapid growth of bacteria that are mutalistic or commensal with humans, can eliminate the need to use antibiotics, which can wipe out helpful bacteria along with

harmful bacteria. The use of probiotics is an example of which process?
   a. succession
   b. competitive exclusion
   c. parasitism
   d. niche differentiation

15. Suppose that a two-acre lawn on your college's campus is allowed to undergo succession. Describe how species traits, species interactions, and the site's history might affect the community that develops.

16. Design an experiment to test the hypothesis that increasing species richness increases a community's productivity.

56 Ecosystems and Global Ecology

In this chapter you will learn how

Ecosystems are connected by the flow of energy and nutrients

by asking → How does energy flow through ecosystems? **56.1**

considering

by asking → How do nutrients cycle through ecosystems? **56.2**

including a look at ↓ Global patterns in productivity

Human activities

including a look at ↓ Global biogeochemical cycles

← affect   affect →

↓ cause

Global climate change **56.3**

affect   affect

The reservoir at the Harbeespoort Dam in South Africa is notorious for its spectacular algae blooms caused by nutrient pollution. This chapter explores how humans are altering ecosystems all over the planet.

The data in this chapter carry a simple, but important, message: Some of the most urgent problems facing humans are rooted in ecosystems and global ecology. Acid rain, nitrate pollution, declining freshwater supplies, soil erosion, and global climate change are just a few of the large-scale issues that ecologists are documenting.

The chapter has a simple theme, as well: In nature, everything is connected. This is because the components of an **ecosystem**—the species present in a region, along with abiotic components such as the soil, climate, water, and atmosphere—are linked by the flow of energy and the cycling of **nutrients** needed to sustain life, such as carbon and nitrogen. Further, ecosystems are themselves linked by the global exchange of energy and nutrients throughout the **biosphere,** the thin zone of life surrounding the Earth.

This chapter is part of the Big Picture. See how on pages 1196–1197.

✔ When you see this checkmark, stop and test yourself. Answers are available in Appendix A.

Humans are removing species, changing landscapes, and adding massive amounts of energy and nutrients to ecosystems all over the planet. The biological, chemical, and geological consequences have been so dramatic that scientists have proposed naming a new epoch in the geological timescale—the Anthropocene ("new human epoch") (Chapter 28).

Let's begin our analysis of large-scale ecology by considering the basics of how energy flows through the components of an ecosystem. Then, let's examine the cycling of nutrients. Finally, you will see how the flow of energy and the cycling of nutrients are central to understanding global climate change. You can see how ecosystems fit into the Big Picture of Ecology on pages 1196–1197.

56.1 How Does Energy Flow through Ecosystems?

One of the defining characteristics of life is the use of energy (Chapter 1). Energy enters ecosystems via the primary producers. A **primary producer** is an **autotroph** (literally, "self-feeder")—an organism that can synthesize its own food from inorganic sources. In most ecosystems, primary producers use solar energy to manufacture food via photosynthesis (see Chapter 10). But in deep-sea hydrothermal vents and iron-rich rocks deep below Earth's surface, primary producers use the chemical energy contained in inorganic compounds such as methane (CH_4) or hydrogen sulfide (H_2S) to make food (see Chapter 29).

Primary producers do not *create* energy. Rather, they are *transforming* the energy in sunlight or inorganic compounds into the chemical energy stored in sugars. Recall that the first law of thermodynamics states that energy cannot be created or destroyed, but only transferred and transformed (Chapter 8).

The total amount of chemical energy produced in a given area and time period is called **gross primary productivity (GPP)**. The primary producers use this chemical energy in two ways:

1. *Cellular Respiration* It takes energy just to stay alive. Primary producers have to use chemical energy to fuel cellular respiration (or in anaerobic microbes, fermentation) to produce the energy currency, ATP (Chapter 9). ATP, in turn, is used to fuel a diverse array of metabolic processes.

2. *Growth and Reproduction* Chemical energy that isn't used for cellular respiration can be applied to growth and reproduction. Energy that is invested by primary producers in building new tissue or offspring is called **net primary productivity (NPP)**.

Because cellular respiration, growth, and reproduction are not perfectly efficient (second law of thermodynamics), some energy is lost during these processes as molecular waste and heat without being used. Linking these concepts together:

$$NPP = GPP - R$$

(where R = energy used in cellular respiration, or lost)

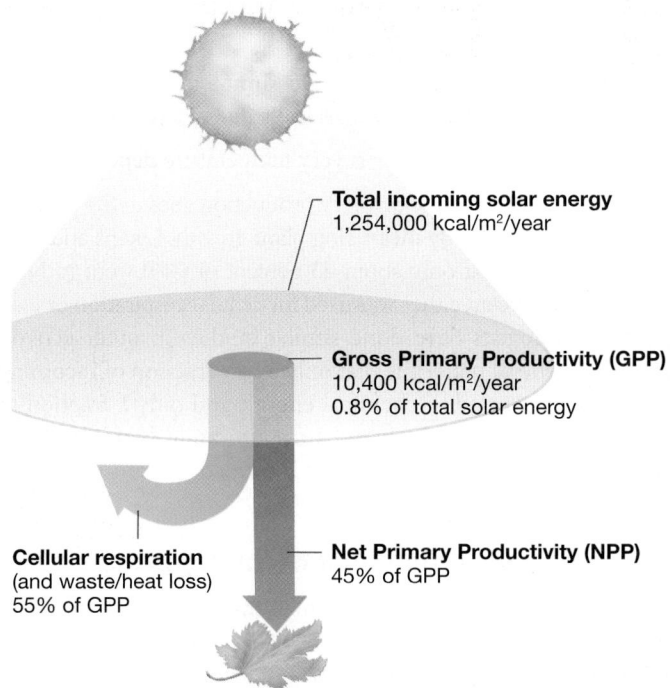

FIGURE 56.1 Measuring the Energy That Enters Ecosystems. These data are from the Hubbard Brook Experimental Forest.

NPP represents the total amount of chemical potential energy that is stored in organic material, or **biomass.** NPP is critically important because it is the amount of energy available via primary producers to the other organisms of an ecosystem.

How Efficient Are Autotrophs at Capturing Solar Energy?

How much of the energy that enters an ecosystem is captured as NPP? To answer this question, let's look at data from the world's most intensively studied ecosystem. Since the early 1960s, ecologists Gene Likens, Herbert Bormann, and colleagues have been studying how energy and nutrients flow through a temperate forest ecosystem in the northeastern United States: the Hubbard Brook Experimental Forest in New Hampshire.

During one year, from June 1, 1969, to May 31, 1970, 1,254,000 kcal of solar radiation per square meter (m^2) entered this ecosystem in the form of sunlight. If this amount of energy were available in the form of electricity, it would easily power two 150-watt lightbulbs per square meter of land, burning continuously for one year.

By documenting rates of photosynthesis in a large sample of forest plants, Likens and Bormann calculated that the plants used 10,400 kcal/m^2 of this energy in photosynthesis—a mere 0.8 percent of the incoming sunlight (**FIGURE 56.1**). Why so inefficient? For comparison, the best solar panels produced today have an efficiency of about 20 percent in normal outdoor conditions.

The answer includes several components.

- Even when conditions are ideal, the pigments that drive photosynthesis can absorb only a fraction of the light wavelengths available and thus a fraction of the total energy received.

- Plants in temperate biomes have drastically reduced photosynthetic rates in winter.

- If conditions get dry during the summer, stomata close to conserve water. Photosynthesis stalls due to a lack of CO_2.

- The efficiency of enzymes is very temperature dependent.

How much of the gross primary production goes to the production of new biomass? By measuring plant growth, Likens and Bormann estimated that only about 45 percent of GPP went to NPP. The other 55 percent either was used for cellular respiration or lost.

Other biologists have done similar studies in other ecosystems. The general pattern is that only a tiny fraction of incoming sunlight is converted to chemical energy, and only a fraction of this gross primary productivity is used to build biomass. What happens to the new biomass once it is formed?

What Happens to the Biomass of Autotrophs?

Graduate student Raymond Lindeman made a key insight about the fate of autotroph biomass in the 1940s, based on his research at Cedar Creek Bog in Minnesota. Lindeman proposed the *energy flow model* for ecosystems, whereby energy could flow from autotrophs through other organisms in the form of biomass.

- **Consumers** eat living organisms. **Primary consumers** eat primary producers; **secondary consumers** eat primary consumers; **tertiary consumers** eat secondary consumers, and so on. You are a primary consumer when you eat a salad, a secondary consumer when you eat beef, and a tertiary consumer when you eat tuna.

- **Decomposers,** or **detritivores,** obtain energy by feeding on the remains of other organisms or waste products. Dead animals and dead plant tissues ("plant litter") are collectively known as **detritus.** Many fungi and bacteria are decomposers, but humans rarely act as decomposers.

Energy flows when one organism eats another. However, a key take-home message is that energy constantly dissipates as it flows through ecosystems and is ultimately "lost" as heat, whereas nutrients continue to cycle (**FIGURE 56.2**). Consider that all the *energy* stored in your tissues right now ultimately will be released into the atmosphere, but all the carbon, nitrogen, and other *nutrients* present in your tissues will persist in other organisms and the abiotic environment for millennia.

Fifteen years after Lindeman's model was published, Howard Odum carried the concept of energy flow to the next level by quantifying energy flow in an aquatic ecosystem. Odum categorized organisms according to their roles. Organisms that obtained energy from the same type of source were said to occupy the same **trophic** ("feeding") **level.**

Trophic Structure: Food Chains and Food Webs Biologists often begin with a simple model and add complexity. A **food chain** focuses on one possible pathway of energy flow among trophic levels in an ecosystem. **FIGURE 56.3** shows two examples of food chains: a **grazing food chain** and a **decomposer food chain.**

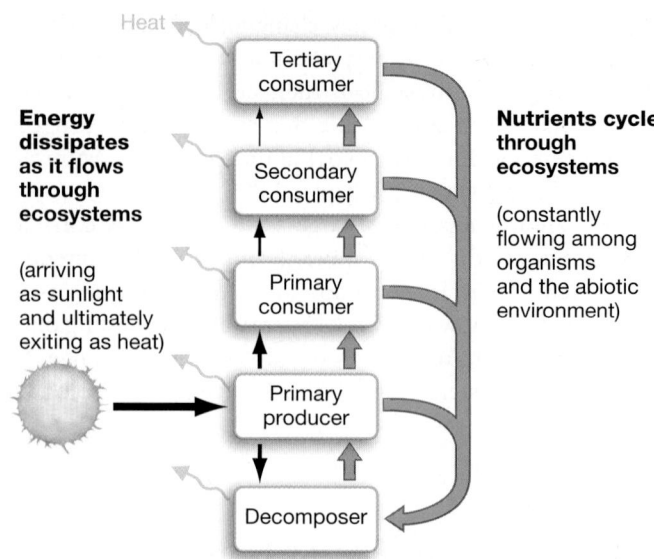

FIGURE 56.2 Energy Eventually Dissipates, but Nutrients Cycle.

At the second trophic level, the decomposer food chain has a **primary decomposer,** which feeds on plant detritus, rather than a primary consumer. Note several key points:

- At higher trophic levels, the grazing and decomposer food chains often merge. For example, the robin in the figure functions as

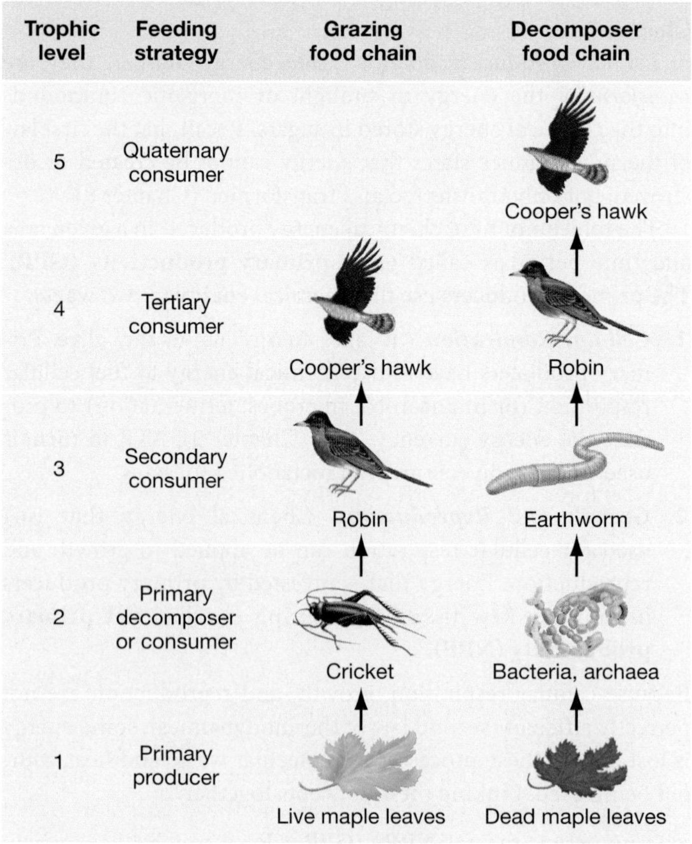

Trophic level	Feeding strategy	Grazing food chain	Decomposer food chain
5	Quaternary consumer		Cooper's hawk
4	Tertiary consumer	Cooper's hawk	Robin
3	Secondary consumer	Robin	Earthworm
2	Primary decomposer or consumer	Cricket	Bacteria, archaea
1	Primary producer	Live maple leaves	Dead maple leaves

FIGURE 56.3 Trophic Levels Identify Steps in Energy Transfer.

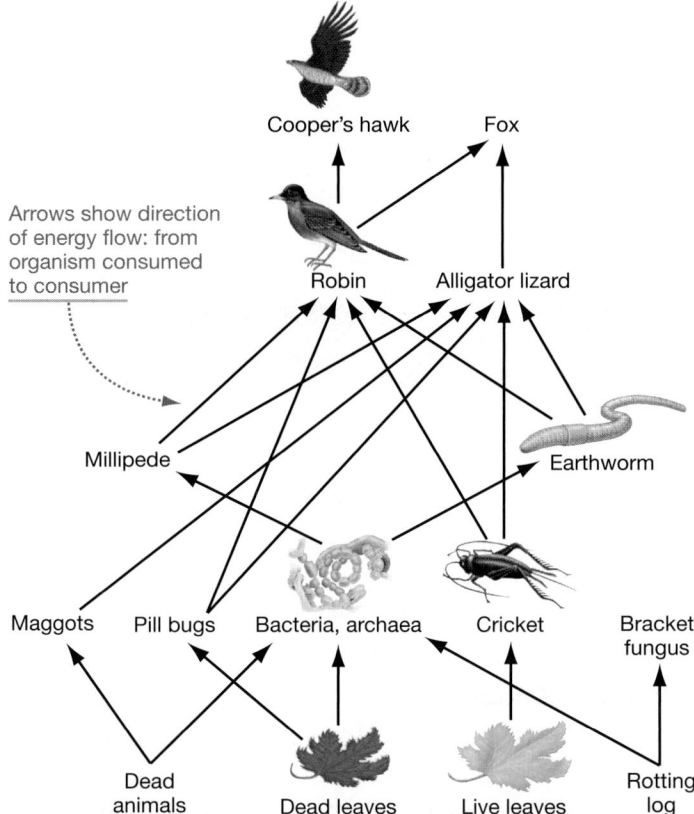

Arrows show direction of energy flow: from organism consumed to consumer

Cooper's hawk   Fox

Robin   Alligator lizard

Millipede   Earthworm

Maggots   Pill bugs   Bacteria, archaea   Cricket   Bracket fungus

Dead animals   Dead leaves   Live leaves   Rotting log

FIGURE 56.4 Food Webs Describe Trophic Relationships in an Ecosystem. Food webs offer a more comprehensive analysis of feeding relationships than food chains do (compare to Figure 56.3). This food web shows only a fraction of the total feeding relationships and species of a temperate deciduous forest.

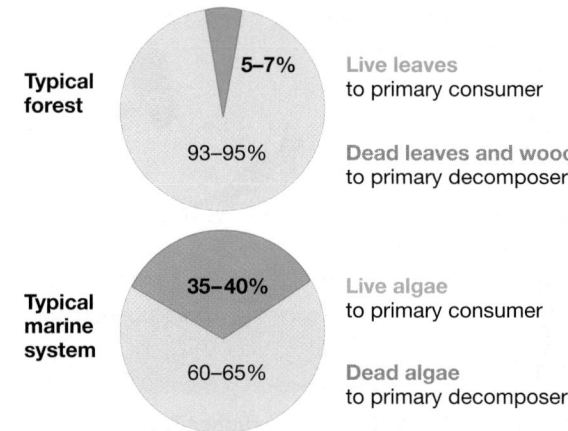

Typical forest

5–7% Live leaves to primary consumer

93–95% Dead leaves and wood to primary decomposer

Typical marine system

35–40% Live algae to primary consumer

60–65% Dead algae to primary decomposer

FIGURE 56.5 How Much Biomass Gets Eaten Dead versus Alive?

a secondary consumer in the grazing food chain and a tertiary consumer in the decomposer food chain.

- It is common for consumers—such as humans and robins—to feed at multiple trophic levels.

Because most consumers feed on a wide array of organisms and at multiple trophic levels, food chains are embedded in more complex **food webs.** The food web shown in **FIGURE 56.4** is a more detailed description of the trophic relationships among the organisms in Figure 56.3. A complete food web would include *all* the organisms interacting in this ecosystem, along with estimates of the amount of energy transferred at each link. Food webs are a way to summarize energy flows and to document the complex trophic interactions that occur in ecosystems.

Energy Flow to Grazers versus Decomposers   Figures 56.3 and 56.4 show that the primary producers—maple leaves in this case—are a source of energy whether alive or dead. How much biomass gets eaten alive versus dead?

The answer to this question varies among habitats. In forest ecosystems, the percentage of NPP consumed alive can be as low as 5–7 percent, because indigestible wood isn't transferred to other organisms until it decays. But in marine habitats, where algae account for most of the primary production, the percentage

of NPP that gets consumed alive is much higher—often 35–40 percent (**FIGURE 56.5**). This is an important point, for two reasons:

1. If you want to understand energy flow in a forest, you have to understand decomposers. Decomposition is less of a factor in marine ecosystems, where primary consumers are key.

2. The decomposer food chain is "leaky." At Hubbard Brook, large amounts of energy leave the forest ecosystem in the form of detritus that washes into streams. In these streams, photosynthesis by aquatic algae and plants introduced only about 10 kcal/m^2/yr versus 6039 kcal/m^2/yr that washed in from the surrounding forest. In marine ecosystems, energy exits in the form of dead bodies that rain down to the ocean floor.

The "dead versus alive" percentage can vary from season to season or even week to week according to ambient temperatures, light levels, nutrient levels, and species interactions.

Energy Transfer between Trophic Levels

All ecosystems share a characteristic pattern: The total biomass produced each year declines from lower trophic levels up to the higher levels. To understand why, let's go back to the Hubbard Brook Experimental Forest.

Chipmunks are small rodents that are seed predators at Hubbard Brook. On average, these primary consumers harvest 31 kcal/m^2 of energy each year. Of that total, how much is used to create biomass? Careful measurements revealed that, on average:

- 17.7 percent is unused—meaning that it is not digested and absorbed. Instead, it is excreted.

- 80.7 percent is used for cellular respiration.

- 1.6 percent goes into the production of new chipmunk tissue by growth and reproduction.

The same pattern would be seen in a secondary consumer—a Cooper's hawk, for example, that eats chipmunks and other primary consumers. Consider the energy that a hawk expends to catch prey. This energy cannot be used to make hawk biomass.

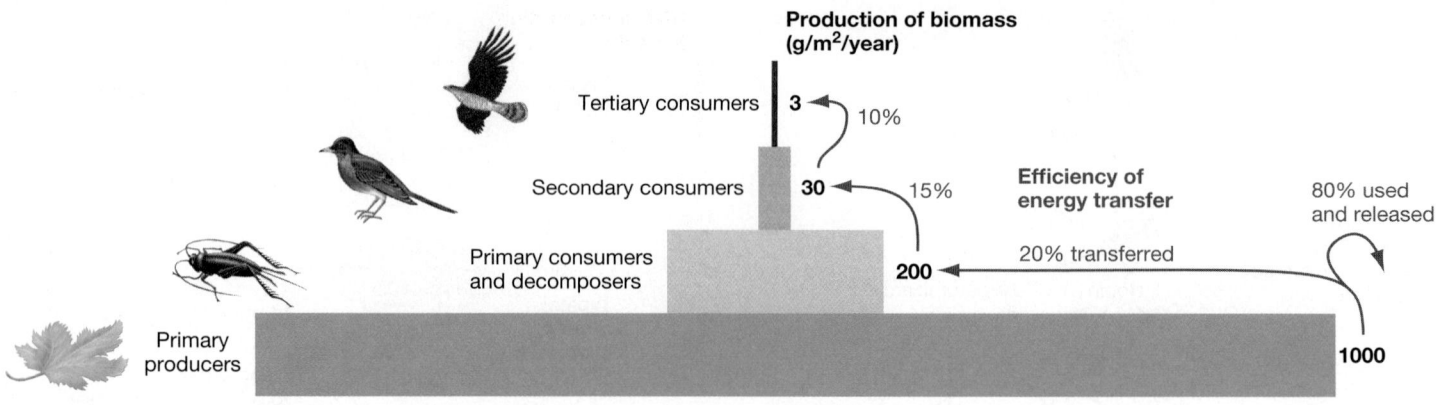

FIGURE 56.6 Productivity Declines at Higher Trophic Levels. This pattern is called the pyramid of productivity.

The general point is simple: Because most of the total energy consumed is used for cellular respiration rather than growth and reproduction, the amount of biomass produced at the second trophic level must be less than the amount at the first trophic level; productivity at the third trophic level must be less than that at the second.

The Pyramid of Productivity The data in **FIGURE 56.6** depict the biomass produced at trophic levels 1–4 in a temperate deciduous forest. A separate horizontal bar represents each trophic level. Note that the graph reports two numbers:

1. Productivity is a rate, measured here in units of grams of biomass produced by growth and reproduction per square meter of area each year ($g/m^2/year$).

2. Efficiency is a ratio and thus dimensionless—the units cancel out. (It is reported here as a percentage.) Efficiency is the fraction of biomass transferred from one trophic level to the next.

Biomass production at each trophic level varies widely among ecosystems. But as a rough rule of thumb, the efficiency of biomass transfer from one trophic level to the next is only about 10 percent. ✔ QUANTITATIVE If you understand this concept, you should be able to explain why it takes about 10 times more energy to grow a kilogram of beef than a kilogram of wheat.

Efficiency Varies It's important to recognize that the "10 percent rule" in ecological efficiency masks a great deal of interesting variation. For example:

• Large mammals are more efficient at producing biomass than small mammals, because large animals have a smaller surface-area-to-volume ratio and lose less heat (see Chapter 42).

• Biomass production is much more efficient in ectotherms than in endotherms. Ectotherms rely principally on heat gained from the environment and do not oxidize sugars to keep warm (see Chapter 42). As a result, they devote much less energy to cellular respiration than endotherms do. Chipmunks devote just 1.5 percent of their energy intake to growth; lizards and other ectotherms may devote 20–40 percent.

✔ If you understand this concept, you should be able to explain why it may be more efficient to eat lobster than chicken, even though the two animals feed at the same trophic level.

Biomagnification

The pyramid of productivity helps explain how some pollutants become concentrated in food webs. Some types of molecules do not break down quickly in the environment and are neither digested by organisms nor excreted efficiently. These include heavy metals like mercury, emitted mostly from coal-fired power plants, and organic compounds called POPs—persistent organic pollutants—which are used for herbicides, pesticides, and other industrial applications. Mercury and POPs undergo **biomagnification**—meaning that they increase in concentration at higher levels in a food chain. Why?

Biomagnification begins when persistent atoms or molecules are taken up from air or water by primary producers. A very large quantity of pollutants can be taken up by primary producers—even if the concentration of the pollutant in the environment is low—because of the very large biomass of primary producers at the bottom of the productivity pyramid (Figure 56.6).

Remember that organisms at each trophic level eat about 10 g of tissue to make 1 g of their own tissue. When the primary consumers consume the biomass of the primary producers, they take up all the pollutants but concentrate them in their smaller biomass—because little or none of the pollutant is metabolized or excreted. As a result, the pollutant becomes more concentrated as it moves up each trophic level (**FIGURE 56.7**). The greater the number of trophic levels, the higher the consequences to the top consumers. In many cases, concentrations get high enough in secondary and tertiary consumers to poison them.

One POP, called DDT (dichlorodiphenyltrichloroethane), was made famous by ecologist Rachel Carson in her 1962 book, *Silent Spring*. Carson was alarmed that many birds were dying in areas that were sprayed with DDT to control mosquitoes—the pesticide caused birds to lay thin-shelled eggs that became crushed in the nest. Carson's resulting campaign for responsible chemical use spurred the beginning of the environmental

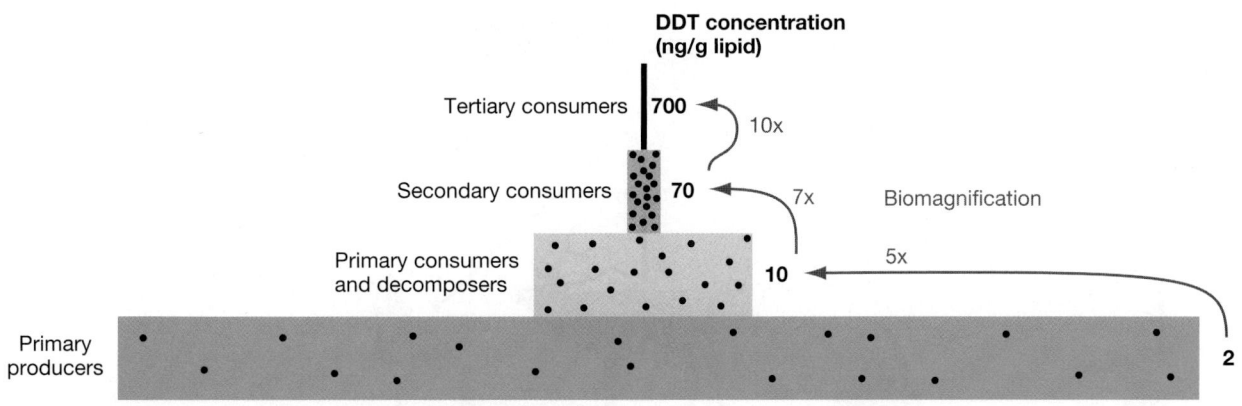

FIGURE 56.7 Biomagnification Can Lead to High Concentrations of Toxic Substances at Higher Trophic Levels. DDT is a pesticide once commonly used to control mosquitoes and other insects.

movement. DDT was banned in the United States in 1972. Populations of top predators like ospreys and bald eagles are back on the rise.

Humans are top predators too, and are vulnerable to biomagnification. ✔ **If you understand biomagnification, you should be able to explain how you could get mercury poisoning from eating tuna every day, but not from eating the same quantity of sardines every day.**

Top-Down Control and Trophic Cascades

Top predators are not abundant compared to other organisms in the food web (as indicated by the tiny biomass at the peak of the productivity pyramid in Figure 56.6), and usually don't get eaten by other consumers. Are top predators important?

One way to answer this question is to remove top predators from an ecosystem and see what happens. Robert Paine used this approach in his study of the keystone species, *Pisaster ochraceous* (Chapter 55). When Paine removed this sea star from a rocky intertidal habitat on the Pacific Coast, mussels took over—and dramatically changed which other species were present in the community. When a consumer limits a prey population in this way, biologists say that **top-down control** is occurring.

It is rarely practical to remove a top predator from an ecosystem, but sometimes events present an opportunity for a natural experiment. That's exactly what happened in the Greater Yellowstone ecosystem of western North America, where wolves were hunted out in the early 1900s but reintroduced in 1995.

What effect did the absence of wolves have on the Greater Yellowstone ecosystem? During the 1990s, ecologists William Ripple and Robert Beschta had been trying to determine why aspen trees had been declining for most of the last century. Then they made an important connection—the number of aspen trees started to decline at the same time that the wolves had disappeared. Ripple and Beschta hypothesized that there was an indirect link between aspen trees and wolves.

How could they test their hypothesis? One way was to see if the number of aspen trees would begin to increase as the reintroduced wolf population increased. Although the Yellowstone

wolf population has grown to only about 150 individuals since reintroduction, their presence has already led to an increase in aspen—and to many other changes in the food web (**FIGURE 56.8**, see page 1154).

- In Yellowstone, wolves primarily feed on elk. As elk numbers declined and individuals changed their feeding habits due to fear of being in the open, populations of favorite elk foods like aspen, cottonwood, and willow have increased.

- The changes in willow and other plant species triggered an increase in beavers, which compete with elk for these plants.

- Beavers dam streams, forming ponds that provide habitat for frogs, turtles, and an array of other species, whose numbers have increased.

- Wolves do not tolerate coyotes. As coyotes declined, mouse populations increased. The increase in mouse populations led to larger populations of hawks that prey on mice.

When changes in top-down control cause conspicuous effects two or three links away in a food web, biologists say that a **trophic cascade** has occurred. Trophic cascades have been observed in many ecosystems. Atlantic cod and other large predators, for example, have virtually disappeared from large parts of the Atlantic Ocean due to overfishing. Shrimp and crab—species that cod used to eat—have increased in response; the primary producers that shrimp and crab eat have decreased.

The message from these studies is clear: The removal of top predators from ecosystems can have widespread, and sometimes unexpected, effects on the abundance of many species in a community. The underlying cause is energy flow—one of the factors that links components in an ecosystem.

Global Patterns in Productivity

Top predators are not the only players that can profoundly affect the abundance of species in a community. The productivity of primary producers is also very important, and it can have a "bottom-up" effect. Let's consider two big-picture questions about NPP: Where is most of it produced, and what limits its production to those areas?

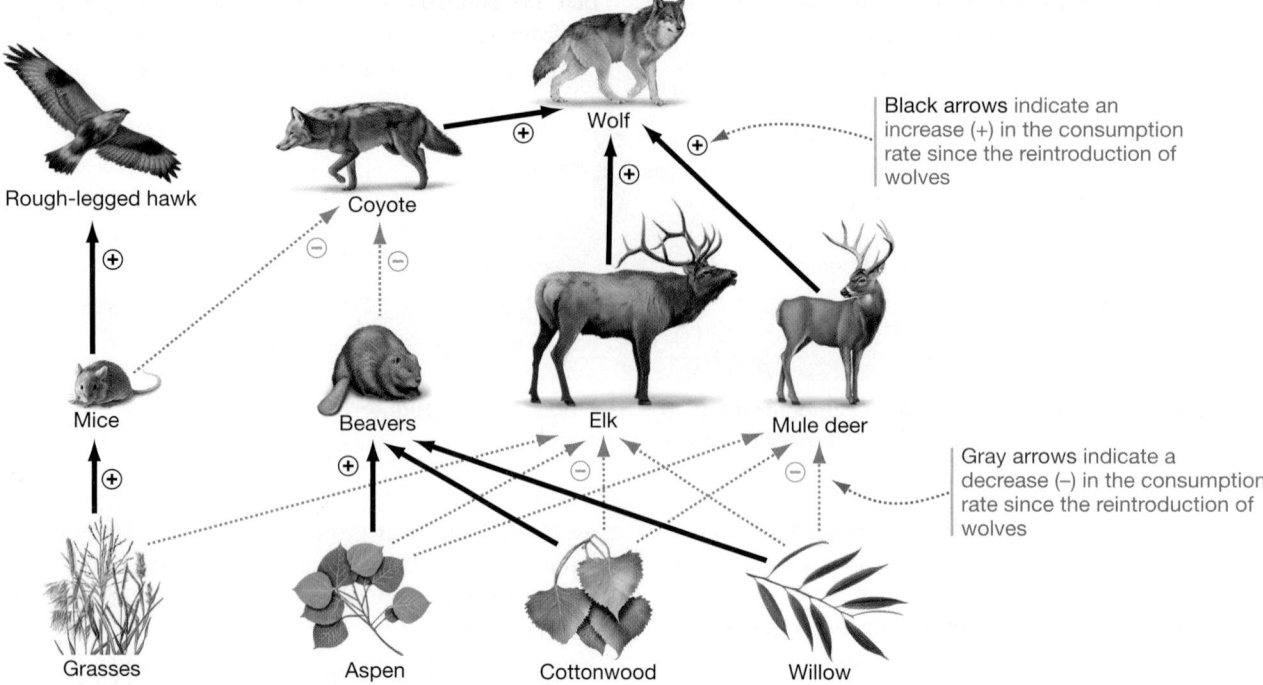

FIGURE 56.8 The Reintroduction of Wolves to Yellowstone Caused a Trophic Cascade. This food web shows only a few of the hundreds of species affected by the reintroduction of wolves.

Keep in mind that productivity is limited by any factor that limits the rate of photosynthesis. The top candidates are temperature and the availabilities of water, sunlight, and nutrients (see Chapter 10 and Unit 7). Let's consider how different limiting factors prevail in different environments.

Is Productivity Higher on the Land or in the Sea? FIGURE 56.9 summarizes data on NPP from around the globe. Note that the highest NPP is indicated in red, yellow, and green, and that these colors occur only on land. In general, NPP on land is much higher than it is in the oceans. Why?

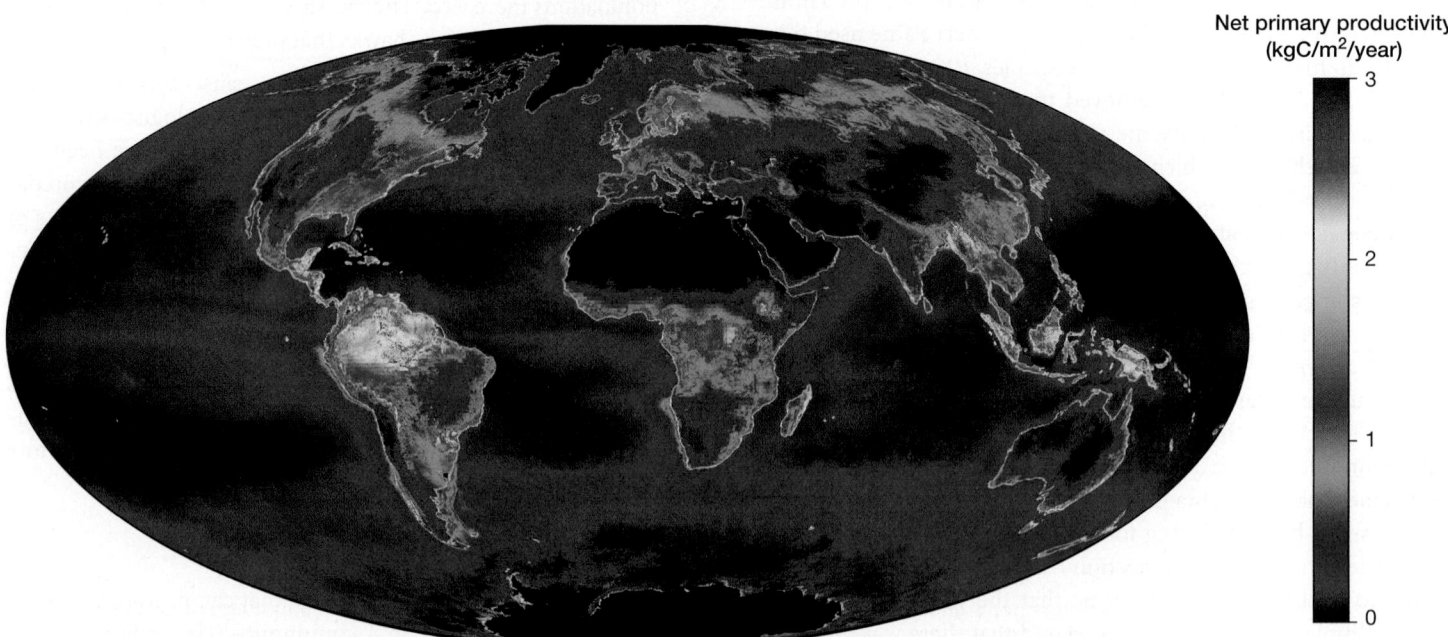

FIGURE 56.9 Net Primary Productivity Varies among Geographic Regions. The terrestrial ecosystems with the highest primary productivity are found in the tropics; tundra and deserts have the lowest. The highest productivity in the oceans occurs in nutrient-rich coastal areas.

DATA: http://earthobservatory.nasa.gov.

The leading hypothesis to explain this pattern is simply that much more light is available to drive photosynthesis on land than in marine environments. Water quickly absorbs many of the light wavelengths that are used in photosynthesis (see Chapter 52). This simple observation also underlines the importance of the evolution of land plants (see Chapter 31). Before green plants invaded the land, NPP in terrestrial environments may have been as low as it is in today's oceans.

Which Terrestrial Ecosystems Are Most Productive? The terrestrial ecosystems with highest productivity are located in the wet tropics. Except in the world's major deserts (such as the large black area representing the Sahara Desert in northern Africa), NPP on land declines from the equator toward the poles. Why? The availability of sunlight and average temperature decreases with latitude. The low productivity in deserts is an exception attributed to the low annual rainfall caused by air circulation patterns (see Chapter 52).

You might recall that the equator-to-poles gradient in NPP is thought to be important in explaining the equator-to-poles gradient in species richness that occurs in most types of plants and animals (see Chapter 55). Areas of high productivity tend to be "hotspots" for high species richness (see Chapter 57).

The global productivity of terrestrial ecosystems is limited by a combination of temperature and availability of water and sunlight. At a local level, however, NPP on land is also limited by nutrient availability. When biologists fertilize a terrestrial ecosystem with nitrogen or phosphorus, NPP almost always increases. Recent research has also shown that NPP in temperate and boreal forests is increasing due to nitrogen that is blowing in from agricultural fields hundreds or thousands of miles away, where it was originally applied as a fertilizer.

Which Marine Ecosystems Are Most Productive? Productivity patterns in marine ecosystems contrast strongly with those of terrestrial environments. Marine productivity is highest along coastlines, and it can be as high near the poles as in the tropics. Why? The shallow water along coasts receives more nutrients:

- Rivers carry nutrients from terrestrial ecosystems and deposit them on the coasts.

- Nearshore ocean currents bring nutrients from the cold, deep water of the oceanic zone up to the surface, a process called upwelling (Chapter 52).

The coastal areas colored deep blue in Figure 56.9 include most of the world's most important fisheries.

In contrast, the oceanic zones, colored black in Figure 56.9, have extremely low NPP. They lack both river inputs and upwelling. In addition, nutrients in organisms near the surface of the open ocean—where sunlight is abundant—constantly fall to dark, deeper waters in the form of dead cells, where they are no longer available to organisms near the surface. Typically, a square meter (m^2) of open ocean produces a maximum of 35 g (1.2 oz) of organic matter each year. In terms of productivity per m^2, the open ocean is a desert.

Which Biomes Are Most Productive? **FIGURE 56.10** presents NPP data a different way—organized by biome instead of by geography

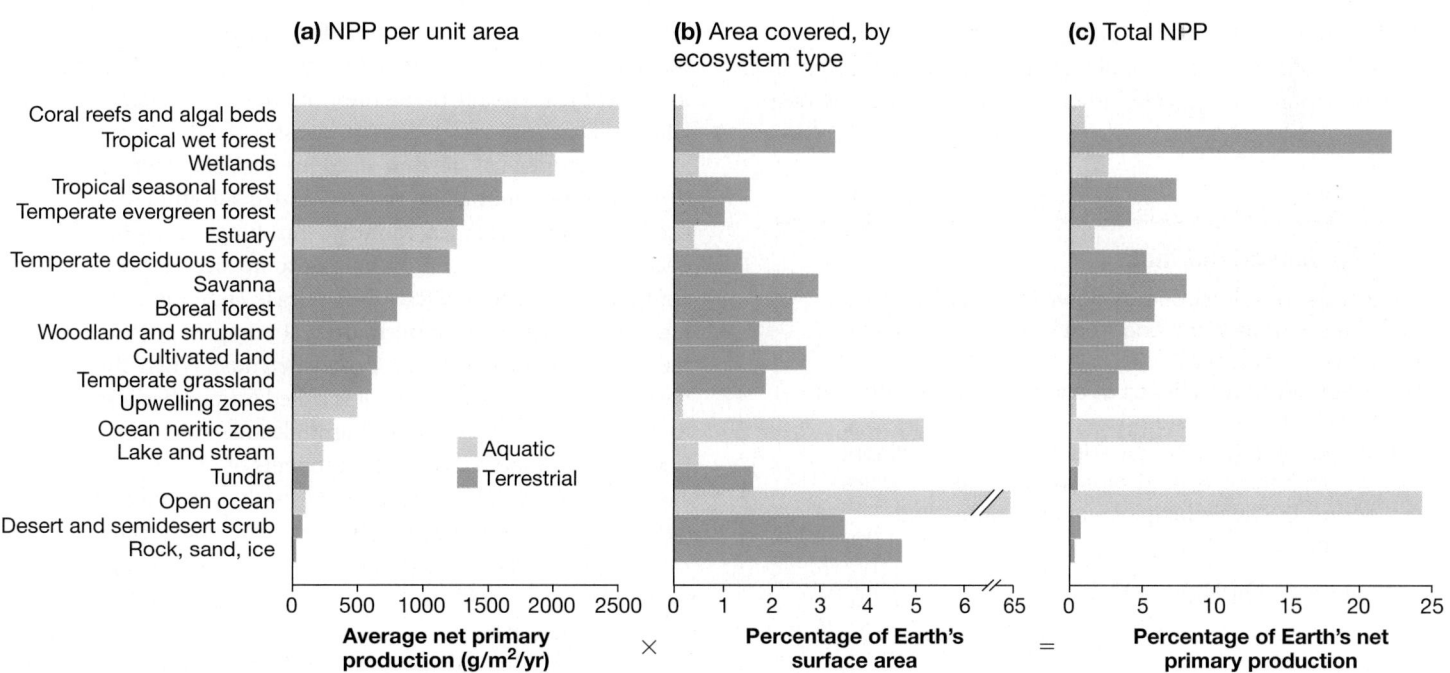

(a) NPP per unit area

(b) Area covered, by ecosystem type

(c) Total NPP

Coral reefs and algal beds
Tropical wet forest
Wetlands
Tropical seasonal forest
Temperate evergreen forest
Estuary
Temperate deciduous forest
Savanna
Boreal forest
Woodland and shrubland
Cultivated land
Temperate grassland
Upwelling zones
Ocean neritic zone
Lake and stream
Tundra
Open ocean
Desert and semidesert scrub
Rock, sand, ice

Aquatic
Terrestrial

Average net primary production (g/m²/yr)

Percentage of Earth's surface area

Percentage of Earth's net primary production

FIGURE 56.10 Net Primary Productivity Varies among Biomes.

DATA: Leith, H., and R. H. Whittaker (eds.). 1975. *Primary Productivity in the Biosphere*. Ecological Studies and Synthesis Vol. 14. New York: Springer-Verlag.

✔**QUANTITATIVE** Estimate the average NPP per unit area for the open ocean and the total area covered by the open ocean. How do these numbers explain the results in column (c)?

(see Chapter 52). Figure 56.10a provides data on average NPP per square meter per year for each biome; Figure 56.10b documents the total area that is covered by each type of ecosystem; and Figure 56.10c presents the percentage of the world's total productivity—the result of multiplying the data in part (a) by the data in part (b). In each case, data from different types of biomes are plotted as horizontal bars. Note these observations:

- Tropical wet forests and tropical seasonal forests (which have a dry season) cover less than 5 percent of Earth's surface but together account for over 30 percent of total NPP.

- Among aquatic ecosystems, the most productive habitats are algal beds and coral reefs, wetlands, and estuaries.

- Although NPP per square meter is extremely low in the open ocean, this biome is so extensive that its total NPP is high.

One of the reasons biologists are so concerned about the destruction of tropical rain forests and coral reefs is that these are the world's most productive habitats.

The "Human Appropriation" Of the total NPP available on our planet, how much are humans using? A research group led by Helmut Haberl answered this question recently in terms of *potential* NPP—meaning, NPP that would exist without human activity. Using data from a wide array of sources, Haberl's team estimated that humans are currently preventing or appropriating 24 percent of potential NPP. Of this percentage,

- 53 percent is directly harvested and used (timber, food, etc.);

- 40 percent is prevented through land-use changes (such as parking lots and other forms of development); and

- 7 percent is destroyed due to human-induced fires.

This analysis is astonishing. One species—our own—is appropriating almost a quarter of the planet's biomass. Given increases in human population projected to occur over your lifetime (see

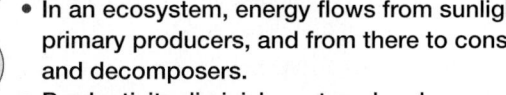
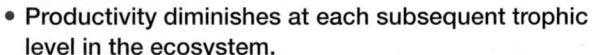

check your understanding

If you understand that . . .

- In an ecosystem, energy flows from sunlight to primary producers, and from there to consumers and decomposers.
- Productivity diminishes at each subsequent trophic level in the ecosystem.
- In most ecosystems, NPP is limited by conditions that limit the rate of photosynthesis: temperature and/or the availability of sunlight, water, and nutrients.

✔**You should be able to . . .**

1. Explain why net productivity diminishes with increasing trophic levels.
2. Predict the effect of global warming on NPP.

Answers are available in Appendix A.

Chapter 54), the demand for NPP is likely to reach the physical limit of what can be appropriated.

Humans are affecting the flow of energy through ecosystems as well as the cycling of nutrients within and among ecosystems. Let's take a closer look at nutrient cycling.

56.2 How Do Nutrients Cycle through Ecosystems?

Energy is transferred when one organism eats another. But tissues also contain carbon (C), nitrogen (N), phosphorus (P), sulfur (S), calcium (Ca), and other elements that act as nutrients and are essential for normal metabolism, growth, and reproduction. The transfer of both energy and nutrients links the biotic and abiotic components of ecosystems.

Atoms are constantly reused as they move through trophic levels, but they also spend time suspended in air, dissolved in water, or held in soil. The path an element takes as it moves from abiotic systems through producers, consumers, and decomposers and back again is referred to as its **biogeochemical** ("life-Earth-chemical") **cycle.** Because humans are disturbing biogeochemical cycles on a massive scale, research on the topic is exploding.

Nutrient Cycling within Ecosystems

To understand the basic features of a biogeochemical cycle, study the generalized terrestrial nutrient cycle in **FIGURE 56.11**. Start with the "Uptake" arrow on the bottom left, which indicates that nutrients such as nitrogen and phosphorus are taken up from the soil by plants and assimilated into plant tissue; carbon enters primary producers as carbon dioxide from the atmosphere. Note that if the plant tissue is eaten, the nutrients pass to the consumer food web; if the plant tissue dies, the nutrients enter the region labeled "Detritus," which contains wastes and dead organisms.

What happens as detritus decomposes? Nutrients that reside in plant litter, animal excretions, and dead animal bodies are used by bacteria, archaea, roundworms, fungi, and other primary decomposers. These microscopic decomposers and the carbon-containing compounds that they release combine to form what biologists call **soil organic matter**, a complex mixture of partially and completely decomposed detritus. When the detritus becomes completely decayed, it is called **humus** because it is rich in a family of carbon-containing molecules called humic acids. (Chapter 39 described other components of the soil.)

Eventually, decomposition converts the nutrients in soil organic matter to an inorganic form. For example, cellular respiration by soil-dwelling bacteria and archaea converts the nitrogen present in amino acids to ammonium (NH_4^+) or nitrate (NO_3^-) ions, and the carbon to carbon dioxide. Once decomposition is accomplished, the nutrients are available for uptake by plants, completing the cycle of nutrient flow through ecosystems.

What Factors Control the Rate of Nutrient Cycling? Of the many links in a nutrient cycle, the decomposition of detritus most

TERRESTRIAL NUTRIENT CYCLE

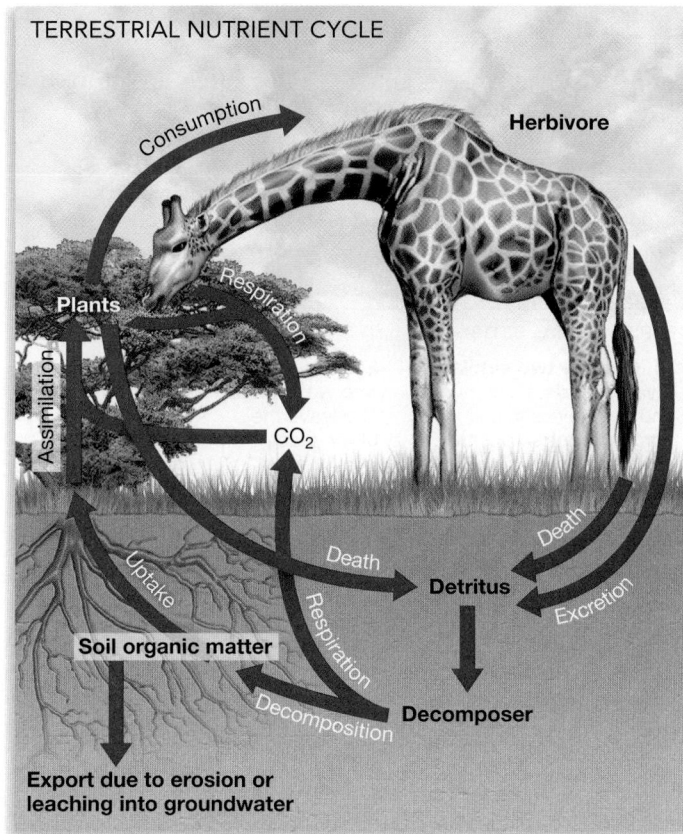

FIGURE 56.11 Generalized Terrestrial Nutrient Cycle. Nutrients cycle from organism to organism in an ecosystem via assimilation, consumption, and decomposition. Nutrients are exported from ecosystems when water or organisms leave the area.

(a) Boreal forest: Accumulation of detritus and organic matter

(b) Tropical wet forest: Almost no organic accumulation

FIGURE 56.12 Temperature and Moisture Affect Decomposition Rates.

often limits the overall rate at which nutrients move through an ecosystem. Until decomposition occurs, nutrients stay tied up in intact tissues. The decomposition rate, in turn, is influenced by three types of factors:

1. abiotic conditions such as oxygen availability, temperature, and precipitation;

2. the quality of the detritus as a nutrient source for the fungi, bacteria, and archaea that accomplish decomposition; and

3. the abundance and diversity of detritivores present.

To appreciate the importance of abiotic conditions on the decomposition rate, consider the difference in detritus accumulation between boreal forests and tropical wet forests—two moist biomes that differ in temperature. Boreal forests are cool, so the rate of decomposition is slow. Tropical wet forests are warm all year long, so the rate of decomposition is high. **FIGURE 56.12** illustrates typical soils from boreal forests and tropical wet forests. Notice that the uppermost part of the soil in a boreal forest consists of a thick layer of partially decomposed detritus and organic matter. In a tropical forest, this layer is virtually absent.

The contrast occurs because the cold conditions in boreal forests limit the metabolic rates of decomposers. As a result, decomposition fails to keep up with the input of detritus, and organic matter accumulates. In the tropics, conditions are so favorable for fungi, bacteria, and archaea that decomposition keeps pace with inputs. Nutrients cycle slowly through boreal forests but rapidly through wet tropical forests. As a result, most of the biomass in a tropical forest is living. ✔ **If you understand this concept, you should be able to explain why farms created by clearing tropical rain forests tend to perform poorly.**

The quality of detritus also exerts a powerful influence on the growth of decomposers, and thus the decomposition rate. Decomposition is inhibited if detritus is **(1)** low in nitrogen, or **(2)** high in lignin. When nitrogen is scarce, such as in acidic bogs, cells have a difficult time making additional proteins and nucleic acids. Lignin—an important constituent of wood—is extremely difficult to digest (see Chapter 32). The presence of lignin is a major reason why wood takes much longer to decompose than leaves do. Only basidiomycete fungi and a few bacteria have the enzymes required to completely degrade lignin.

The presence of oxygen is also critically important. If soils or water become depleted of oxygen, such as in some wetlands, decomposition slows dramatically. Fungi can perform cellular respiration only by using oxygen as the final electron acceptor (see Chapter 9). And even though many bacteria and archaea can thrive in the absence of oxygen and continue decomposition, the growth rates of anaerobic species are a tiny fraction of rates in aerobic bacteria and archaea (to understand why, see Chapter 29).

Sources of Local Nutrient Export and Import Although nutrients are constantly cycled within individual ecosystems, they can also be lost from one ecosystem and exported to another.

- If an herbivore eats a plant and moves out of the ecosystem before excreting the nutrients or dying, the nutrients are exported.

- Nutrients leave ecosystems when flowing water or wind removes particles or inorganic ions and deposits them somewhere else.

Several of the major impacts that humans have on ecosystems—such as farming, logging, burning, and soil erosion—accelerate nutrient export. Farming and logging remove biomass in the form of plants; burning releases nutrients to the atmosphere; soil erosion moves nutrients that are bound to soil particles or dissolved in water. Further, these processes are often interrelated, as when logging leads to soil erosion and runoff.

For an ecosystem to function normally, nutrients that are exported must be replaced by imports. There are three major mechanisms to replace nutrients:

1. Ions that act as nutrients are released as rocks weather (see Chapter 39).

2. Nutrients can blow in on soil particles or arrive dissolved in streams.

3. Nitrogen—a particularly important nutrient in maintaining NPP—is added when nitrogen-fixing bacteria (introduced in Chapter 39) convert molecular nitrogen (N_2) in the atmosphere to usable nitrogen in ammonium or nitrate ions.

In ecosystems that humans manage, nutrients usually have to be actively replaced to maintain productivity. People accomplish this by adding fertilizers or planting nitrogen-fixing crops.

An Experimental Study   The ideas presented thus far are logical and provide a strong conceptual foundation for thinking about nutrient cycles. But to explore nutrient movement in more depth, researchers have turned to experiments.

One of the first large-scale experiments on nutrient cycling took place at the Hubbard Brook Experimental Forest, introduced earlier in the chapter. A group of researchers led by Likens and Bormann began by choosing two similar **watersheds**—areas drained by a single stream—for study. Before starting the experiment, they documented that 90 percent of the nutrients in the watersheds was in soil organic matter; 9.5 percent was in plant biomass. The nutrients in the soil were either dissolved in water or attached to small particles of sand or clay (see Chapter 39).

To begin the experimental treatment, the researchers cut all vegetation, including the trees, in one of the two watersheds (**FIGURE 56.13**). In each of the following three years, the clear-cut watershed was treated with an herbicide to prevent vegetation from regrowing. As a result, the experimental watershed was devegetated. Researchers left the other watershed untreated to serve as a control.

The graph in Figure 56.13 documents the amount of total dissolved substances that washed out of the streams in each watershed over the course of the four years of devegetation. Losses from the devegetated site were typically over 10 times as high as they were from the control site.

Why? Plant roots were no longer able to hold soil particles in place, so some of the soil itself washed away. Further, plant roots

RESEARCH

QUESTION: How does the presence of vegetation affect the rate of nutrient export in a temperate-forest ecosystem?

HYPOTHESIS: Presence of vegetation lowers the rate of nutrient export because it increases soil stability and recycling of nutrients.

NULL HYPOTHESIS: Presence of vegetation has no effect on the rate of nutrient export.

EXPERIMENTAL SETUP:

1. Choose two similar watersheds. Document nutrient levels in soil organic matter, plants, and streams.

2. Devegetate one watershed, and leave the other intact.

3. Monitor the amount of dissolved substances in streams.

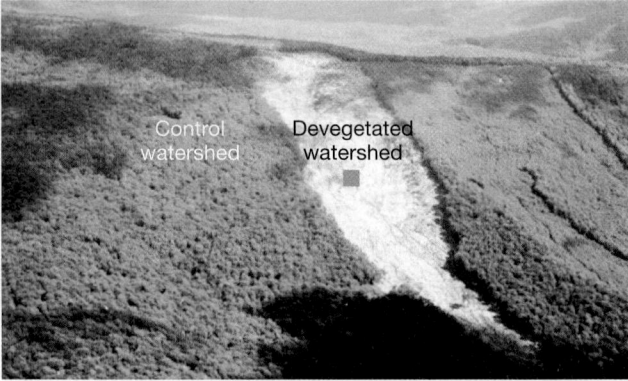

PREDICTION: Amount of dissolved substances in stream in devegetated watershed will be much higher than amount of dissolved substances in stream in control watershed.

PREDICTION OF NULL HYPOTHESIS: No difference will be observed in amount of dissolved substances in the two streams.

RESULTS:

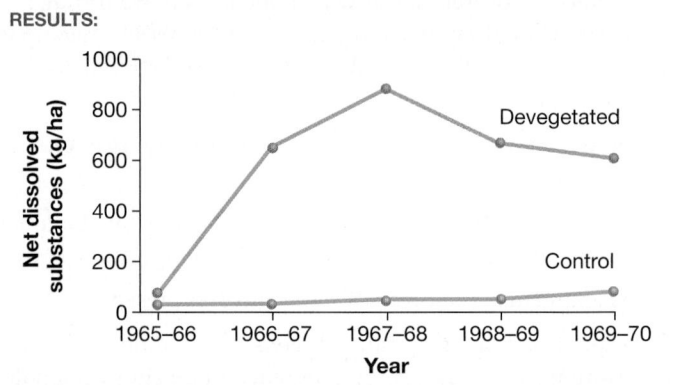

CONCLUSION: Presence of vegetation limits nutrient export. Removing vegetation leads to large increases in nutrient export.

FIGURE 56.13 Experimental Evidence that Deforestation Increases the Rate of Nutrient Export from Ecosystems.

SOURCE: Likens, G. E., F. H. Bormann, N. M. Johnson, et al. 1970. Effects of forest cutting and herbicide treatment on nutrient budgets in the Hubbard Brook watershed-ecosystem. *Ecological Monographs* 40: 23–47.

✔ **QUESTION** Propose a hypothesis to explain why the level of nutrient export from the experimental site appears to peak and then decline.

no longer took up and recycled nutrients that were dissolved in the remaining soil water. Instead, massive quantities of nitrate and other nutrients washed out of the soil and were exported from the ecosystem in runoff water.

Long-term devegetation of this type has occurred in formerly forested areas of the Middle East, North Africa, and elsewhere, due to intensive farming and grazing. Nutrients that used to occur within those soils—and the soils themselves—were exported. Today, the productivity of these regions is a tiny fraction of what it once was. Once soil is lost, it is difficult to regain—soil takes a very long time to form (Chapter 39).

Global Biogeochemical Cycles

When nutrients leave one ecosystem, they enter another. In this way, the movement of ions and molecules among ecosystems links local biogeochemical cycles into one massive global system. As an introduction to how these global biogeochemical cycles work, let's consider the global water, nitrogen, and carbon cycles. These cycles have recently been heavily modified by human activities—with serious ecological consequences.

To understand how biogeochemical cycling works on a global scale, researchers focus on three fundamental questions:

1. What are the nature and size of the reservoirs—areas or "compartments" where elements are stored for a period of time? In the case of carbon, the biomass of living organisms is an important reservoir, as are sediments and soils. Another significant carbon reservoir is buried in the form of coal and oil.

2. How fast does the element move between reservoirs, and what processes are responsible for moving elements from one reservoir to another? The global photosynthetic rate, for example, measures the rate of carbon flow from carbon dioxide (CO_2) in the atmosphere to living biomass. When fossil fuels burn, carbon that was buried in coal or petroleum moves into the atmosphere as CO_2.

3. How does one biogeochemical cycle interact with another biogeochemical cycle? For example, researchers are trying to understand how changes in the nitrogen cycle affect the carbon cycle.

The Global Water Cycle   A simplified version of the **global water cycle** appears in **FIGURE 56.14**. The diagram shows the estimated amount of water that moves between major reservoirs over the course of a year, powered by energy from the sun.

To analyze this cycle, begin at the left with evaporation of water out of the ocean, and the subsequent precipitation of water back into the ocean. For the marine component of the cycle, evaporation exceeds precipitation—meaning that, over the oceans, there is a net gain of water to the atmosphere.

When this water vapor moves over the continents, it is joined by a small amount of water that evaporates from lakes and streams and a large volume of water that is transpired (lost) by plants (see Chapter 38). The total volume of water in the atmosphere over land is balanced by the amount of rain and other forms of precipitation that falls on the continents.

The cycle is completed by the water that moves from the land back to the oceans via streams and **groundwater**—water that is found underground. Overall, there is no net gain or loss of water.

Note that 97.5 percent of the Earth's water is salt water (in the ocean). Most of the remaining 2.5 percent of freshwater is locked up in the ice caps and glaciers, leaving about 0.5 percent available for drinking in lakes and rivers, and in groundwater. Most

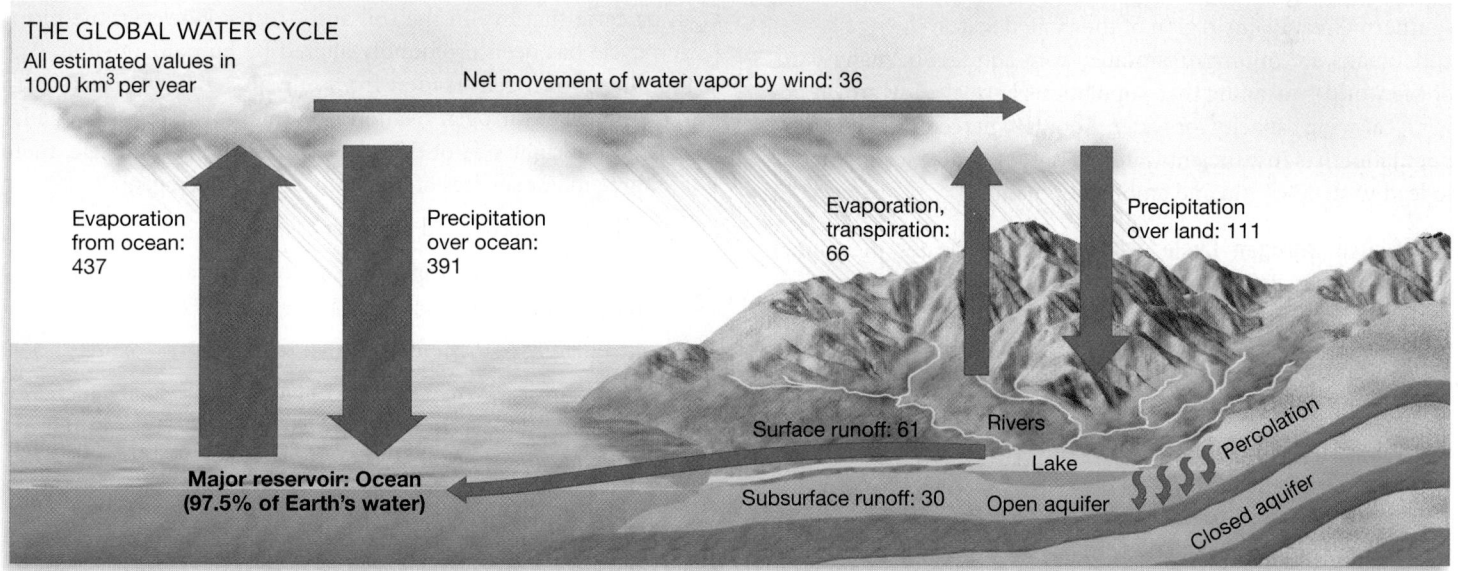

THE GLOBAL WATER CYCLE
All estimated values in 1000 km³ per year

Net movement of water vapor by wind: 36

Evaporation from ocean: 437

Precipitation over ocean: 391

Evaporation, transpiration: 66

Precipitation over land: 111

Major reservoir: Ocean (97.5% of Earth's water)

Surface runoff: 61

Rivers

Lake

Percolation

Subsurface runoff: 30

Open aquifer

Closed aquifer

FIGURE 56.14 The Global Water Cycle.

DATA: Oki, T., and S. Kanae. 2006. *Science* 313: 1068–1072.

✔**QUESTION** Predict how the amount of water evaporated from the ocean is changing in response to global warming. Discuss one possible consequence of this change.

groundwater is stored in **aquifers**—layers of porous rock, sand, or gravel that are saturated with water. Aquifers are said to be closed (contained) when nonporous rock layers overlie them, and open (uncontained) when they can be replenished by water percolating down from the surface (Figure 56.14). Closed aquifers may take thousands of years to recharge.

What effects are humans having on the water cycle? Perhaps the simplest and most direct impacts concern rates of groundwater depletion.

- Asphalt and concrete surfaces reduce the amount of water that percolates from the surface to deep soil and rock layers.

- When grasslands and forests are converted to agricultural fields, root systems that hold water are lost. More water runs off into streams and less percolates into groundwater.

- Irrigated agriculture, along with industrial and household use, is removing massive amounts of water from groundwater storage and bringing it to the surface. The Ogallala–High Plains Aquifer is the largest underground reservoir of freshwater in the United States, underlying 430,000 square kilometers (176,700 square miles) of the Midwest from South Dakota to Texas, and supplying 30 percent of the irrigation water for the country. Billions of gallons are withdrawn each day. The water in this aquifer is estimated to be 10,000 years old, and it is being depleted much faster than it is being recharged (**FIGURE 56.15**), raising concerns about the future supply. Most of the aquifer is closed; only the areas shown in blue have recharged.

The **water table**—the level where soil is saturated with stored water—is dropping on every continent. For example, between 1986 and 2006, the water table north of Beijing, China, experienced a drop of 61 m (200 ft). Throughout India, water tables are falling at a rate of between 1 m and 3 m per year. Similar rates are being documented in Yemen, parts of Mexico, and states in the southern Great Plains region of the United States.

Humans are mining freshwater from aquifers in many parts of the world—meaning that populations have already grown beyond carrying capacity for water. About 40 percent of the world's population has insufficient water resources. Water scarcity tends to lead to strife within and among populations.

The Global Nitrogen Cycle   **FIGURE 56.16** illustrates the **global nitrogen cycle.** To follow the cycle, start at the top in the atmosphere. This represents the main reservoir for nitrogen in the form of nitrogen gas (N_2)—a biologically inert molecule that comprises about 78 percent of the air you breathe. Nitrogen is usable by organisms only when it is reduced, or "fixed," meaning that it is converted from unusable atmospheric N_2 to usable ammonium (NH_4^+) or nitrate ions (NO_3^-; see Chapter 39).

The fixation of nitrogen is usually a slow process, so nitrogen often limits primary productivity in terrestrial and aquatic ecosystems. Reduced nitrogen builds up slowly in soils and water over time and then cycles relatively rapidly among soil, microorganisms, fungi, plants, and animals within both terrestrial and aquatic ecosystems. Ultimately, bacteria digest nitrogen-containing molecules and return N_2 to the atmosphere, completing the cycle.

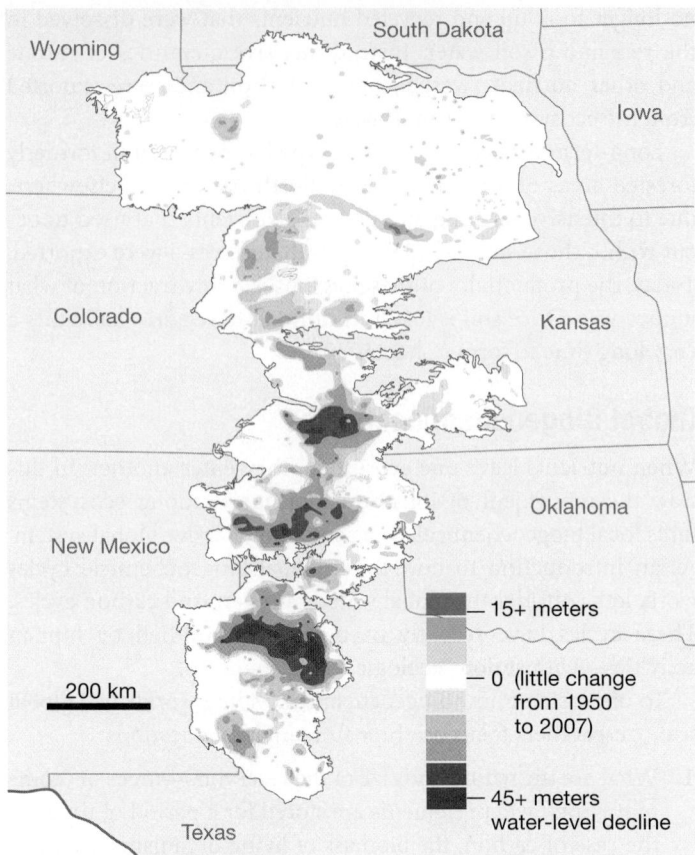

FIGURE 56.15 Declining Water Levels in the Ogallala–High Plains Aquifer from 1950 to 2007.
DATA: U.S. Geological Survey.

Nitrogen fixation occurs naturally due to lightning-driven reactions in the atmosphere and from enzyme-catalyzed reactions in bacteria that live in the soil and oceans. However, the nitrogen cycle has been profoundly altered by human activities. The amount of nitrogen fixation from human sources almost equals the amount of nitrogen fixation from natural sources. As the labels on the right side of the graph in **FIGURE 56.17** indicate, there are three major sources of this human-fixed nitrogen:

1. industrially produced fertilizers;

2. the cultivation of crops, such as soybeans and peas that harbor nitrogen-fixing bacteria; and

3. the burning of fossil fuels, which releases nitric oxide.

Adding nitrogen to terrestrial ecosystems usually increases productivity. But overfertilization with nitrogen has important consequences.

- Nitrogen-laced runoff from agricultural fields causes algae blooms in aquatic ecosystems, such as the lake shown in **FIGURE 56.18**. When the algae die and are decomposed by microbes, oxygen is depleted from the water, causing the formation of oxygen-free "dead zones" (see Chapter 29).

- When nitrogen is added to experimental plots in grasslands in midwestern North America, a few competitively dominant

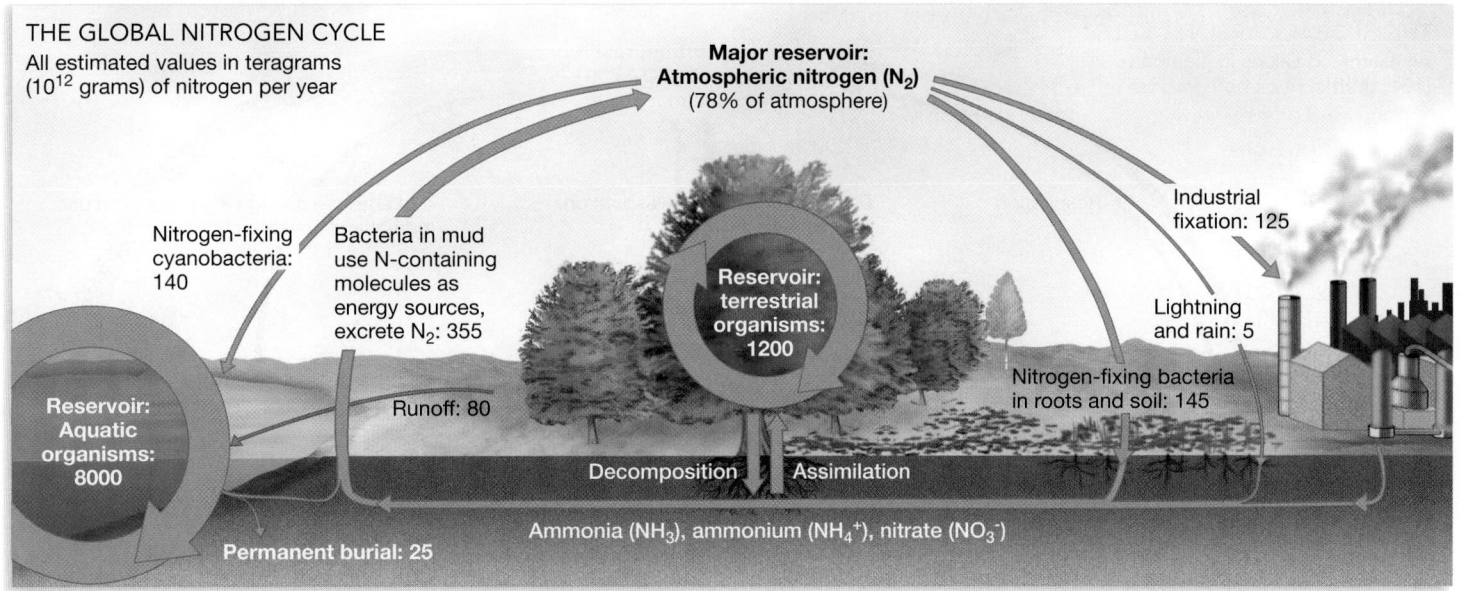

THE GLOBAL NITROGEN CYCLE

All estimated values in teragrams (10^{12} grams) of nitrogen per year

**Major reservoir:
Atmospheric nitrogen (N$_2$)**
(78% of atmosphere)

Nitrogen-fixing cyanobacteria: 140

Bacteria in mud use N-containing molecules as energy sources, excrete N$_2$: 355

Reservoir: terrestrial organisms: 1200

Industrial fixation: 125

Lightning and rain: 5

Nitrogen-fixing bacteria in roots and soil: 145

Reservoir: Aquatic organisms: 8000

Runoff: 80

Decomposition   Assimilation

Ammonia (NH$_3$), ammonium (NH$_4^+$), nitrate (NO$_3^-$)

Permanent burial: 25

FIGURE 56.16 The Global Nitrogen Cycle. Nitrogen enters ecosystems as ammonia or nitrate via fixation from atmospheric nitrogen. It is exported in runoff and as nitrogen gas given off by bacteria that use nitrogen-containing compounds as an electron acceptor.

DATA: Gruber, N., and J. N. Galloway. 2008. *Nature* 451: 293–296.

species grow rapidly and displace other species that do not respond to nitrogen inputs as strongly. Productivity increases, but species diversity decreases (Chapter 57).

The Global Carbon Cycle The **global carbon cycle** documents the movement of carbon among terrestrial ecosystems, the oceans, and the atmosphere (**FIGURE 56.19**, see page 1162). Of these three reservoirs, the ocean is by far the largest. The atmospheric reservoir is also important despite its relatively small size, because carbon moves into and out of it rapidly—through organisms.

In both terrestrial and aquatic ecosystems, photosynthesis is responsible for taking carbon out of the atmosphere and incorporating it into tissue. Cellular respiration, in contrast, releases carbon that has been incorporated into living organisms to the atmosphere, in the form of carbon dioxide.

How have humans changed the carbon cycle? For thousands of years, humans have altered the global carbon cycle by deforesting land for agriculture and settlements—releasing the carbon stored in trees to the atmosphere.

But the human influence on the carbon cycle increased sharply during the industrial revolution. Burning fossil fuels moves massive amounts of carbon from an inactive geological reservoir, in the form of petroleum or coal (the remains of ancient organisms), to an active reservoir—the atmosphere. When you burn gasoline, you are releasing carbon atoms that have been locked up in petroleum reservoirs for hundreds of millions of years.

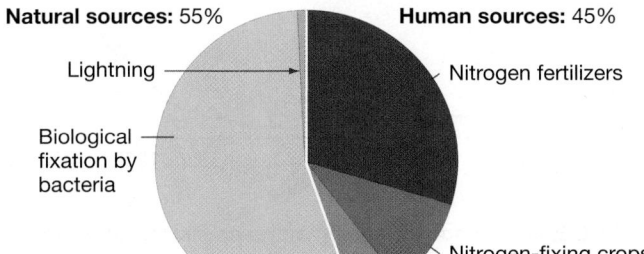

Sources of nitrogen fixation

Natural sources: 55%   **Human sources: 45%**

Lightning

Biological fixation by bacteria

Nitrogen fertilizers

Nitrogen-fixing crops

Burning of fossil fuels

FIGURE 56.17 Humans Are Adding Nitrogen to Ecosystems.

DATA: Canfield, D. E., A. N. Glazer, and P. G. Falkowski. 2010. *Science* 330: 192–196.

FIGURE 56.18 Nutrient Runoff Causes Harmful Algae Blooms.
Blooms, in turn, can lead to oxygen-starved dead zones.

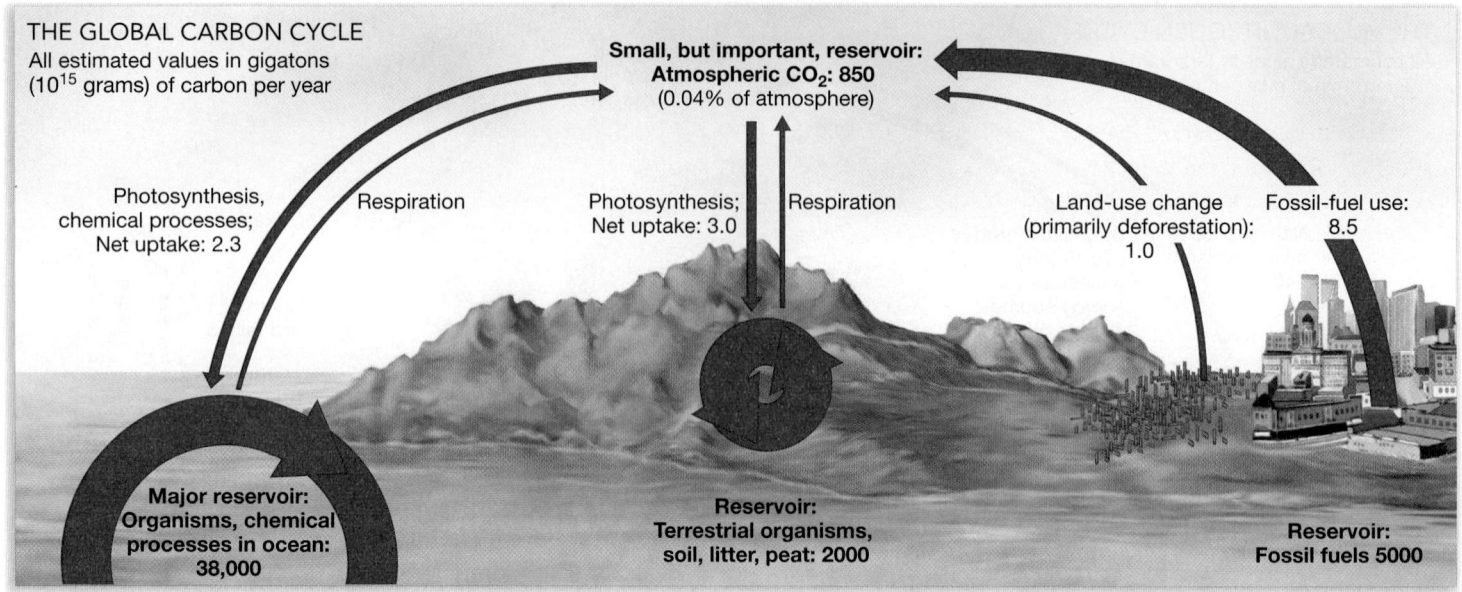

THE GLOBAL CARBON CYCLE
All estimated values in gigatons (10^{15} grams) of carbon per year

Small, but important, reservoir:
Atmospheric CO$_2$: 850
(0.04% of atmosphere)

Photosynthesis, chemical processes; Net uptake: 2.3

Respiration

Photosynthesis; Net uptake: 3.0

Respiration

Land-use change (primarily deforestation): 1.0

Fossil-fuel use: 8.5

Major reservoir:
Organisms, chemical processes in ocean: 38,000

Reservoir:
Terrestrial organisms, soil, litter, peat: 2000

Reservoir:
Fossil fuels 5000

FIGURE 56.19 The Global Carbon Cycle. The arrows indicate how carbon moves into and out of ecosystems. Photosynthesis and cellular respiration move carbon in and out of the atmosphere in the form of CO$_2$. The burning of fossil fuels moves carbon from an inactive geological reservoir, in the form of petroleum or coal, to an active reservoir, the atmosphere.

DATA: Hönisch, B., A. Ridgwell, D. N. Schmidt, et al. 2012. *Science* 335: 1058–1063.

FIGURE 56.20a highlights the dramatic increase in carbon released from fossil-fuel burning over the past century; **FIGURE 56.20b** shows corresponding changes in CO$_2$ concentrations. Note that CO$_2$ levels in the atmosphere have now risen to 392 parts per million (ppm), far above the 280 ppm that was typical before 1860. Analysis of CO$_2$ levels in air bubbles trapped in Antarctic ice reveals that CO$_2$ fluctuated within a range of about 170 to 300 ppm for the extent of the record—800,000 years. This means that current CO$_2$ levels are 30 percent higher than the highest level measured over the last 800,000 years.

Why is the increase in atmospheric CO$_2$ so important? Because it changes the climate.

check your understanding

If you understand that . . .

• Biogeochemical cycles trace the movement of water and elements within and among ecosystems.
• Biologists analyze three aspects of these cycles: (1) the nature and size of reservoirs where elements are stored or captured, (2) the processes responsible for moving elements among reservoirs, and (3) interactions among cycles.

✔**You should be able to . . .**

Compare how human activities are altering the global water, nitrogen, and carbon cycles.

Answers are available in Appendix A.

(a) Increases in fossil-fuel use

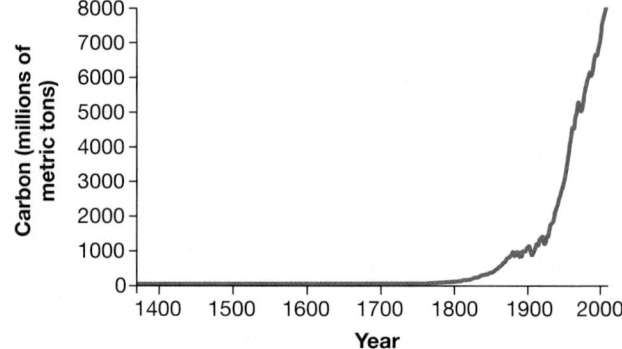

(b) Global changes in atmospheric CO$_2$ over time

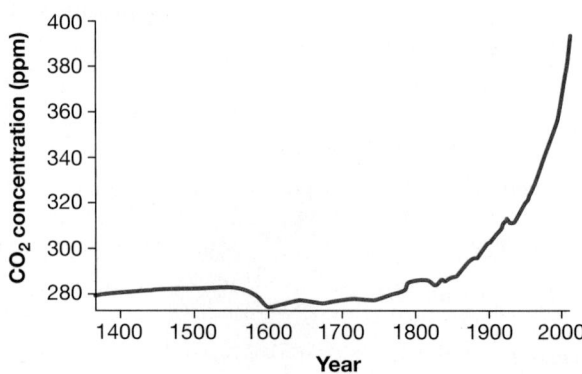

FIGURE 56.20 Humans Are Causing Increases in Atmospheric Carbon Dioxide. (a) Rates of carbon flow from fossil-fuel burning have increased as human populations have increased. **(b)** For centuries, average CO$_2$ concentrations in the atmosphere were about 280 parts per million (ppm).

DATA: Carbon Dioxide Information Analysis Center.

56.3 Global Climate Change

If you are in your late teens or early twenties as you read this text, you are part of a generation that will experience the most dramatic episode of environmental change in human history. The epic event has three main interacting drivers: changes in land use (Chapters 52 and 57), the massive loss of species (see Chapter 57), and global climate change, also called global warming.

Global warming and global climate change refer to the same phenomenon, but with different emphases:

- **Global warming** refers to the increase in the *average* temperature of the planet. The average temperature is very important in determining physical, chemical, and biological processes.

- **Global climate change** refers to the sum of all the changes in local temperature and precipitation patterns that result from global warming—including the frequency and intensity of storms, droughts, and other events. The effects of global warming vary depending on geography, local weather patterns, ocean currents, and other factors. But averaged over the entire planet and over time, the Earth is already much warmer than it was just a few decades ago and is projected to get even warmer.

Why is this warming such a problem? Let's explore the underlying cause of global climate change and then consider the consequences.

What Is the Cause of Global Climate Change?

Climate change is normal. Throughout Earth's history, the average temperature of the atmosphere and local weather patterns have fluctuated. Scientists today are not alarmed because there is change; they are alarmed because the *rate* of change is unprecedented, and because it is caused by human activities. What is going on?

The Greenhouse Effect By the 1950s, scientists were aware that humans were releasing tremendous amounts of CO_2 into the atmosphere. The common reassuring argument at the time was that the vast ocean would be able to absorb this excess CO_2.

However, Roger Revelle, a pioneer in oceanography, and his colleague Hans Suess reported in 1957 that the complex chemistry of the ocean limits how much CO_2 it can absorb. Their data supported a stunning hypothesis posed by the eminent chemist and physicist Svante Arrhenius half a century earlier: The atmospheric CO_2 would rise, which would lead to global warming. Why? Carbon dioxide is a **greenhouse gas**: It traps heat that has been radiated from Earth and keeps it from being lost to space, similar to the way the glass of a greenhouse traps heat (**FIGURE 56.21**).

How could the global warming hypothesis be tested? At the time of Revelle and Suess's research, there were no reliable records of baseline atmospheric CO_2 levels—local CO_2 measurements varied widely due to fluctuating emissions from cities and other variables. So Revelle hired a young geochemist, Charles David

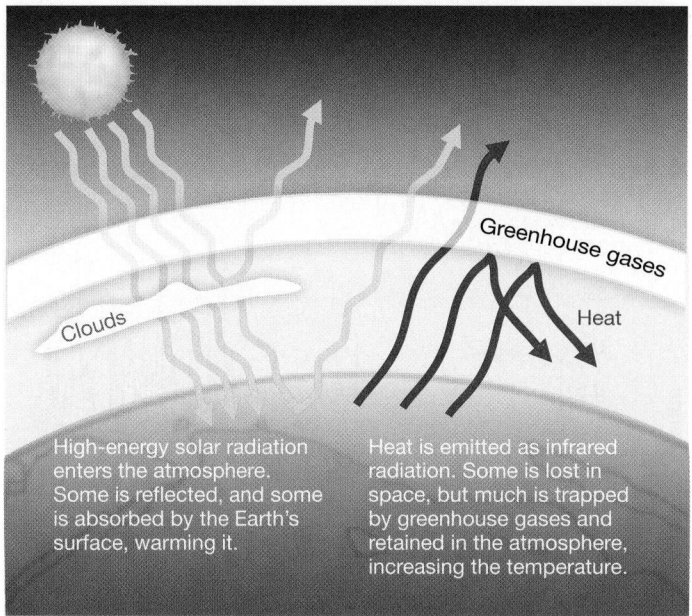

FIGURE 56.21 The Greenhouse Effect. Greenhouse gases in the atmosphere have an effect that is similar to glass in a greenhouse—high-energy radiation in the form of light can easily pass through the gases, but low-energy infrared radiation gets trapped by them.

(Within figure:)
High-energy solar radiation enters the atmosphere. Some is reflected, and some is absorbed by the Earth's surface, warming it.

Heat is emitted as infrared radiation. Some is lost in space, but much is trapped by greenhouse gases and retained in the atmosphere, increasing the temperature.

Clouds

Greenhouse gases

Heat

Keeling, to gather precise measurements of CO_2 concentrations in remote, pristine locations, one of which was the Mauna Loa Observatory in Hawaii. By 1961, Keeling was able to confirm that global CO_2 was rising. Despite many obstacles, he persevered in collecting invaluable records for more than 40 years.

FIGURE 56.22 (see page 1164) shows CO_2 data from Mauna Loa over the last three decades, along with temperature and solar energy data.

- There is no indication that the rate of increase of CO_2 is slowing, despite an array of regulations and incentives to reduce CO_2 emissions.

- The temperature fluctuates, but there is a clear increase in average temperature.

- Even though the amount of solar energy cycles regularly, there is no net increase over time. Thus, fluctuations in solar energy are not sufficient to account for the marked increase in temperature, as one alternative hypothesis had predicted.

Carbon dioxide is only one of several important greenhouse gases in the atmosphere, including methane (CH_4) and nitrous oxide (N_2O). All these gases collectively increase the atmosphere's heat-trapping potential—causing global warming.

Why Is the Climate Changing So Rapidly? The rate of increase of atmospheric CO_2 and temperature are alarming. Why so fast? There are two main reasons. First, the human population has exploded in size. The planet now has over 7 billion people, a 700 percent increase over a mere two centuries—a blink of an eye in Earth's history (see Chapter 54). But an equally important part of the answer involves resource use. The average per capita (per

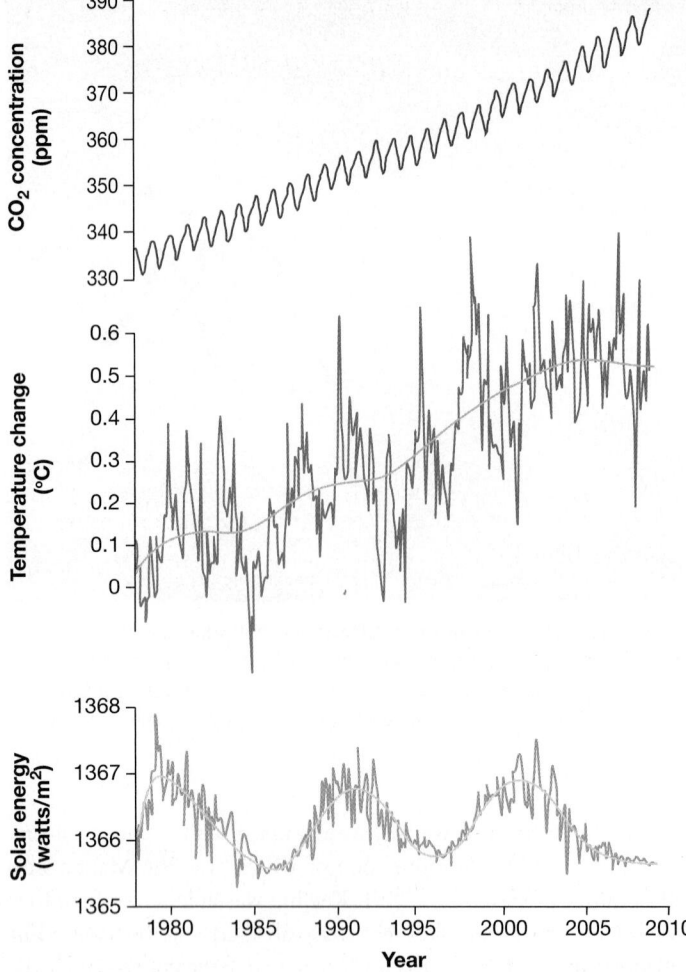

FIGURE 56.22 Recent Changes in Atmospheric CO$_2$, Temperature, and Solar Radiation.

DATA: NOAA; Karl, T. R., J. M. Melillo, T. C. Peterson, and S. J. Hassol (eds.). 2009. *Global Climate Change Impacts in the United States*. Cambridge University Press.

✔ **QUESTION** Notice the annual cycles in CO$_2$. Why are atmospheric CO$_2$ concentrations low in the Northern Hemisphere in summer and high in winter?

person) use of fossil fuels is skyrocketing—especially in industrialized countries.

The data in **FIGURE 56.23** illustrate the consequences of population size and resource use. By multiplying the total population size of a country (left graph) by the quantity of CO$_2$ emissions per person in that country (center graph), the total emissions per country can be calculated (right graph). For example, even though the average person in China causes only a quarter of the CO$_2$ emissions that an average person from the United States does, the total emissions from China surpass all others due to its enormous population size.

Conversely, although the United States represents less than 5 percent of the world population, Americans produce an astonishing one-fifth of the global CO$_2$ emissions due to their heavy use of fossil fuels. Looking at the data in another way, the population of the United States was twice that of Nigeria in 2008, but

the average American produced 22 times as much CO$_2$ as the average Nigerian that year.

The bottom line is that the recent exponential increase of the human population, plus the steep increase in per capita fossil fuel use—as well as land-use changes such as deforestation—have caused a steep increase in the emission of CO$_2$ and other greenhouse gasses. These changes, in turn, have caused the recent and dramatic increase in the average temperatures and climate variability around the globe. The rapid rate of climate change is ultimately rooted in human behavior.

How Much Will the Climate Change?

In 1988 an international group composed of hundreds of scientists, called the Intergovernmental Panel on Climate Change (IPCC), was formed to evaluate the consequences of well-documented increases in CO$_2$ and other heat-trapping gases to climate. The group has since produced a series of reports summarizing the state of scientific knowledge on the issue. How has climate already changed, and what changes can we expect in the future?

Temperature The latest IPCC report concludes that the average global temperature increased 0.74°C (1.3°F) during the 20th century, which was double the increase of the 19th century. More recently, four independent records of temperature summarized by NASA concur that the last decade has been the warmest in recorded history.

Predicting the future state of a system as complex and variable as Earth's climate is extremely difficult. Global climate models are the primary tool that scientists use to make these projections. A global climate model is based on a large series of equations that describe how the concentrations of various gases in the atmosphere, solar radiation, transpiration rates, and other parameters interact to affect climate.

The models currently being used suggest that average global temperature will undergo additional increases of 1.1–6.4°C (2.0–11.5°F) by the year 2100. The low number is based on models that assume no further increase in greenhouse gas emissions over present levels; the high number is based on models that assume continued intensive use of fossil fuels and increases in greenhouse gas emissions. To get a sense of what these numbers mean, consider the data in **FIGURE 56.24**. Under a higher-emissions scenario, parts of the southern United States that currently experience about 60 days per year of temperatures above 90°F are expected to experience 150 or more of such days by the end of the century.

Predicting how temperature will change in the future is particularly difficult because ecosystems respond to warming in ways that increase or decrease the concentrations of CO$_2$ and other greenhouse gases and that affect other drivers of global climate, such as the size of the polar ice caps. In essence, warming has a domino effect on other factors that may either further accelerate or reduce warming.

Documenting Positive and Negative Feedback Positive feedback in climate change occurs when changes due to global warming

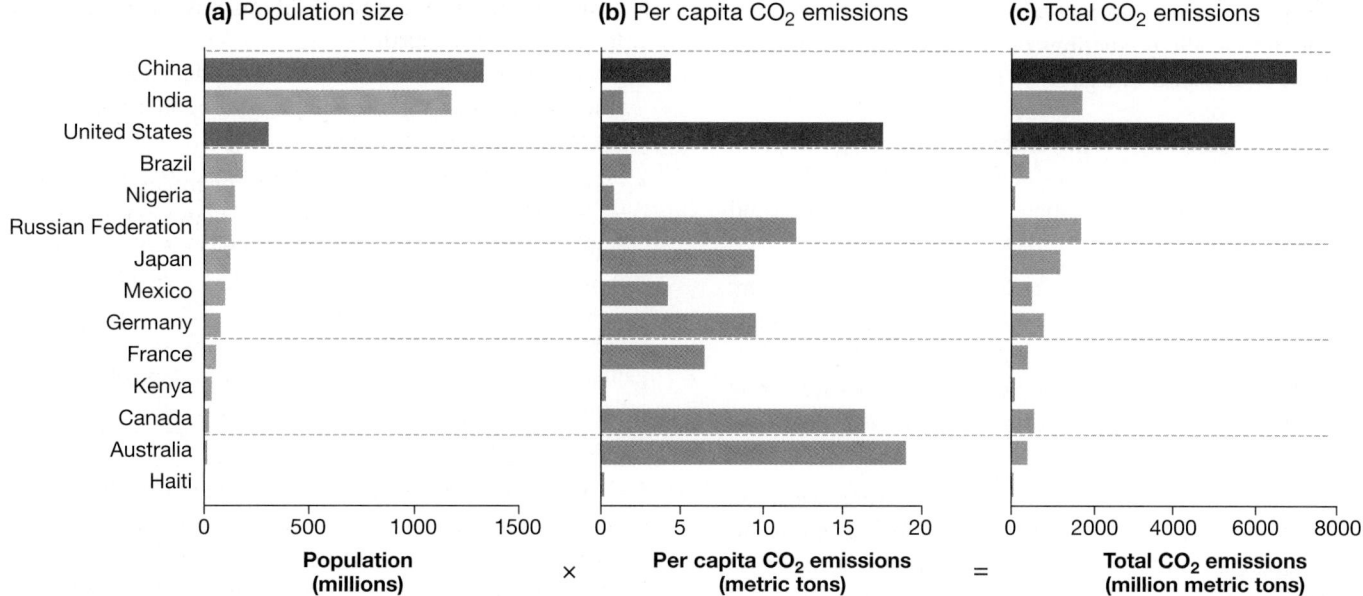

(a) Population size **(b) Per capita CO₂ emissions** **(c) Total CO₂ emissions**

China
India
United States
Brazil
Nigeria
Russian Federation
Japan
Mexico
Germany
France
Kenya
Canada
Australia
Haiti

Population (millions) × **Per capita CO₂ emissions (metric tons)** = **Total CO₂ emissions (million metric tons)**

FIGURE 56.23 Total Emissions Depend on Population Size and Resource Use.
These data are from 2008. The bars for China and the United States are emphasized for comparison.

DATA: Carbon Dioxide Information Analysis Center (http://cdiac.ornl.gov).

✔**QUANTITATIVE** Estimate what the total emissions in China will be if the average release of CO_2 per person doubles due to increased use of fossil fuels.

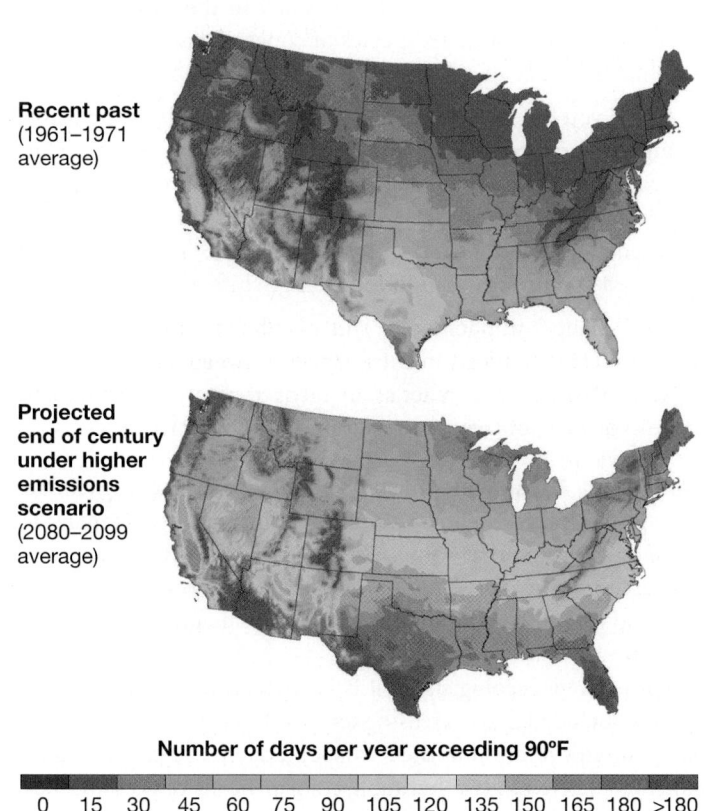

Recent past
(1961–1971 average)

Projected end of century under higher emissions scenario
(2080–2099 average)

Number of days per year exceeding 90°F

0   15   30   45   60   75   90   105   120   135   150   165   180   >180

FIGURE 56.24 More Extreme Temperatures and Continued High Emissions Are Expected in the Future.

DATA: Karl, T. R., J. M. Melillo, and T.C. Peterson (eds.). 2009. *Global Climate Change Impacts in the United States*. New York: Cambridge University Press.

result in the further acceleration of warming. Many types of positive feedback have been documented. For example:

- Warmer and drier climate conditions have increased the frequency of forest fires in the Rocky Mountains of North America as well as in the boreal forests of Canada. Forest fires release CO_2, which leads to even more warming.

- Traditionally, tundra sequesters carbon in the form of soil organic matter, because decomposition rates are extremely low. During a series of warm summers, however, researchers found that decomposition rates in tundra increased sufficiently to release carbon from stored soil organic matter and to produce a net flow of carbon to the atmosphere.

- Other studies of soil microbes observed that elevated CO_2 levels triggered release of CH_4 and N_2O, greenhouse gases that are less abundant than CO_2 but have a more potent effect on warming.

- Melting of the polar ice caps exposes more open water to the sun. The darker open water absorbs more solar radiation than ice, increasing the rate of warming there.

Negative feedback in climate change occurs when changes due to global warming result in increased uptake and sequestration of CO_2 and other greenhouse gases—meaning that global warming should be reduced. For example, recent experiments have shown that the growth rates of many plant species, including some agricultural crops (but not others), increase in direct response to increasing atmospheric CO_2, thus acting to reduce atmospheric CO_2.

Will ecosystem responses increase or decrease global warming? While the specific predictions of different models vary, the overall trend is the same—it's going to keep getting hotter.

Climate Factors Other than Temperature   Increasing temperatures affect other aspects of climate, and elevated levels of greenhouse gases themselves affect climate, with various negative and positive feedback interactions.

- Temperatures affect the water cycle, changing where, when, and how much water falls as precipitation. Average precipitation has increased about 5 percent in the United States during the past 50 years. Models predict that northern areas may become wetter in the future, but the Southwest is likely to become drier.

- Temperature and precipitation are becoming more variable, and the severity of extreme weather events such as monsoon rains, tornados, and hurricanes is increasing, in part because warmer air can hold more moisture (Chapter 52).

- Warmer temperatures have caused the sea level to rise due to expansion of the oceans and the melting of glaciers and polar ice caps.

- The melting of the ice caps affects other aspects of climate such as ocean currents and cloud cover.

   Climate is an abiotic component of ecosystems. How does climate change affect organisms?

Consequences to Organisms

Even though global temperatures have risen only slightly in comparison with projections for the next 50–100 years, biologists have already documented dramatic impacts on organisms. The main types of consequences of global climate change to organisms are summarized in **TABLE 56.1**.

Phenology Shifts   The timing of seasonal events, or **phenology,** is changing in many biomes. Spring arrives in the United States 10 days to two weeks earlier than it did just 20 years ago. As a result, many biological activities associated with spring occur earlier. Recent studies have confirmed that migratory bird species are arriving on their breeding grounds in the northeastern United States an average of 13 days earlier than in previous decades, and fruit trees are ripening as much as 18 days earlier.

   Note, however, that different species in an ecosystem do not necessarily synchronize changes in their timing. Mismatches can result, disrupting trophic interactions and other types of species interactions (Chapter 55). For example:

- Insectivorous birds like great tits (*Parus major*) rely on the abundance of caterpillars to feed their hatchlings. However, a 20-year study has shown that the peak caterpillar biomass now occurs earlier due to the warmer spring, but the reproductive timing of great tits has not changed. Thus, there is a mismatch that has resulted in fewer and lower-weight great tit fledglings.

- Other researchers observed interactions in a community of plankton in the North Sea for 40 years. Plankton at different trophic levels responded to climate change differently over this period, causing mismatches in trophic relationships and thus a restructuring of the food web.

- Mismatches in interactions due to phenology are not always trophic. The mutualistic interaction between flowers and pollinators is disrupted when the timing of flowering and the emergence of insect pollinators becomes asynchronous.

Geographic Range Shifts   The geographic ranges of many organisms are shifting as their "climate envelopes" of suitable habitat move toward the poles, toward higher altitudes, or toward areas of adequate precipitation.

- Until recently, the range of mountain pine beetles was limited in North America by freezing temperatures. Milder temperatures have triggered massive infestations of beetles in areas where they had not been abundant before (Table 56.1). Researchers fear that an ongoing expansion of these voracious beetles northward and eastward in Canada will threaten the boreal forest ecosystem.

- Copepods are small crustaceans that serve as key primary consumers in marine plankton. Copepods have been studied intensively because they are important commercially: They are eaten by fish that are in turn eaten by other fish that are eaten by humans. A 40-year study in the North Atlantic showed that a southern species of copepod has moved north, while the range of cold-water species has declined.

- When you buy a package of seeds or a potted plant, the planting instructions often indicate which hardiness zones are appropriate for its growth. The U.S. Department of Agriculture released an updated map of hardiness zones in 2012 based on recent temperature and precipitation data. The zones have shifted northward—signifying a range shift of many species.

As with changes in phenology, shifts in the geographic distribution of species can lead to mismatches between species and resources. Mismatches can act as an agent of natural selection and cause evolution, or they can result in the extinction of species.

Evolutionary Adaptation   In some populations, changing climate and species interactions are already causing allele frequencies to change—meaning that some species are evolving in response to global climate change. There are clearly both evolutionary winners and losers, depending on the ecological context and adaptive capacity of the organisms.

   For example, ecologists Paul Brakefield and Peter de Jong recently concluded that increasing temperatures in the Netherlands have caused evolution in the two-spot ladybird beetle, *Adalia bipunctata*. There are two heritable color patterns of these beetles: black with red spots and red with black spots (Table 56.1). The frequency of the black morph has declined from 60 percent to 20 in one area in just 30 years. It turns out that the black morphs have lower fitness under warmer conditions because they absorb more solar energy and overheat.

How predictable are evolutionary adaptations to climate change? Luc de Meester and colleagues conducted artificial selection experiments with the water flea, *Daphnia magna*, at different temperatures to try to measure its adaptive capacity. They observed that isolated clones of *Daphnia* evolved in different ways than *Daphnia* living in the presence of competitors, predators, and parasites. Thus, predictive models must consider the complexity of species interactions when making predictions about adaptive capacity—a formidable task.

If changes occur too quickly relative to the life span of the organism, local extinction is likely to occur, rather than adaptation.

Extinctions Biologists are concerned that rapid changes in climate, coupled with other stressors such as land-use changes, will continue to cause high extinction rates (discussed in detail in Chapter 57).

- Behavioral ecologist Barry Sinervo and his team of colleagues found that 12 percent of local lizard populations have gone extinct in Mexico since 1975 because the temperature is changing too rapidly for the lizard populations to adapt. The team used physiological models to discover the mechanism—since lizards were unable to regulate their body temperatures in the heat, they stayed out of the sun during the day, limiting their ability to search for food and thus lowering their fitness.

- As the climate warms, alpine tundra habitats are disappearing. Some plants and animals that have restricted geographic ranges and grow on mountaintops are "running out of habitat" and going extinct.

- Researchers are concerned that habitats may change so rapidly that long-lived trees with limited dispersal will not be able to keep up—their ranges will not be able to change fast enough.

- Polar bears live on sea ice in the Arctic. As sea ice melts, the polar bears are losing their habitat (Table 56.1). If present trends continue, polar bears may be extinct in the wild by 2100.

- Amphibians are in rapid decline globally due to numerous threats—40 percent of amphibian species face imminent extinction. Local changes to temperature and precipitation not only shrink or eliminate the already narrow ranges of many amphibians but also create conditions that favor the chytrid fungus that is causing widespread amphibian mortality.

- Coral reefs are extremely productive in part because corals house symbiotic algae that perform photosynthesis. Warmer water temperatures cause some reef-building corals to expel their photosynthetic algae. When this "coral bleaching" continues due to sustained exposure to warm water, corals begin to die of starvation.

Effects of Ocean Acidification Although the ocean's capacity to absorb CO_2 is not limitless, it is still vast—about one-third of global CO_2 emissions per year is absorbed by the oceans. This absorption has complex consequences for organisms.

When atmospheric CO_2 dissolves, it reacts with water to form bicarbonate (H_2CO_3), which then drops a proton to form

SUMMARY TABLE 56.1 Effects of Climate Change on Organisms

Type of Effect	Example
Phenology Shifts The change in the timing of seasonal events such as migration, flowering, hatching, and feeding can result in mismatches between species; such as flower and pollinator.	
Geographic Range Shifts The geographic redistribution of species to more favorable climate envelopes can change ecological interactions. Mountain pine beetles are spreading north and east, devastating pine forests.	
Evolutionary Adaptation Change in allele frequencies can increase fitness in new conditions. The red morph of the ladybird beetle is more adaptive than heat-absorbing black morph in warmer environments.	
Extinction Species that are not able to survive and/or reproduce under new physical or biological conditions go extinct. Polar bears rely on sea ice, which is melting.	
Acidification The ocean becomes more acidic as it absorbs CO_2, causing a wide range of consequences, such as the slowing of skeletal growth in corals, and coral bleaching.	

carbonic acid (HCO_3^-). Increased CO_2 levels in the atmosphere are increasing the rate of this reaction, and the protons that are released are lowering the pH of the oceans. ✔ **If you understand this concept, you should be able to predict what happens to the pH of a glass of water when you blow bubbles into it with a straw.** The acidity of seawater has climbed by 30 percent over the past 150 years—it's now more acidic than it has been for 20 million years. How does acidity affect organisms?

Research is showing that increasing acidity has devastating effects on diverse processes, ranging from the ability of larval fish to smell predators to the ability of sperm to swim. But by far the greatest amount of research has gone into understanding the effects of acidity on skeleton growth. Increased acidity reduces the availability of calcium carbonate ions to organisms that build calcium carbonate skeletons—such as corals, mollusks, and echinoderms. The more acidic the water, the slower the skeletons form. In some conditions, skeletons can dissolve altogether.

Coral reefs are the subject of a tremendous amount of research because they are among the most productive and species-rich ecosystems in the world. Reefs are large, undersea structures built by colonies of corals that secrete calcium carbonate skeletons. A reduction in skeleton growth due to acidity, changes in temperature, or a combination of the two (in addition to bleaching caused by these factors) threatens the ability of corals to create habitat and food for other organisms.

Consequences to Ecosystems

The combination of local changes in climate and corresponding effects on organisms result in changes to whole ecosystems. Understanding the complex impacts of climate change on ecosystems—and predicting future changes—are among the most challenging and important tasks of ecologists today.

As the first two sections of this chapter emphasized, NPP is the basis of both energy flow and nutrient cycling in ecosystems. Thus, ecologists are monitoring effects of climate change on NPP closely.

How Is NPP Changing on Land? A study from 1982 to 1999 showed an overall increase in global terrestrial productivity. However, the most recent data set, from 2000 to 2009—the warmest decade on record—revealed a different trend. The green areas of **FIGURE 56.25** indicate increased productivity, while the blue areas indicate decreased productivity. Overall, the global terrestrial NPP declined.

This result does not seem intuitive at first, because an increase in atmospheric CO_2 would be expected to increase rates of photosynthesis. The researchers of this study, Maosheng Zhao and Steven Running, point out that large-scale droughts in the Southern Hemisphere have decreased NPP in some of the world's most productive forests, counteracting the overall increase in NPP in the Northern Hemisphere.

Other studies have supported this conclusion. For example, a drying trend in the Amazon rain forest, combined with clear-cutting and burning, threatens to turn one of the world's largest carbon sinks into a carbon source. These data illustrate the complex interaction of effects of CO_2, temperature, and precipitation on productivity.

How Is NPP Changing in the Oceans? As on land, there is considerable variation in local productivity in the oceans. For example, productivity has dropped dramatically in large areas of the pelagic zone, but it has increased in a number of other regions. Overall, scientists agree that there is a general correlation between surface water temperatures and decreased productivity—NPP is decreasing in the oceans. What's going on?

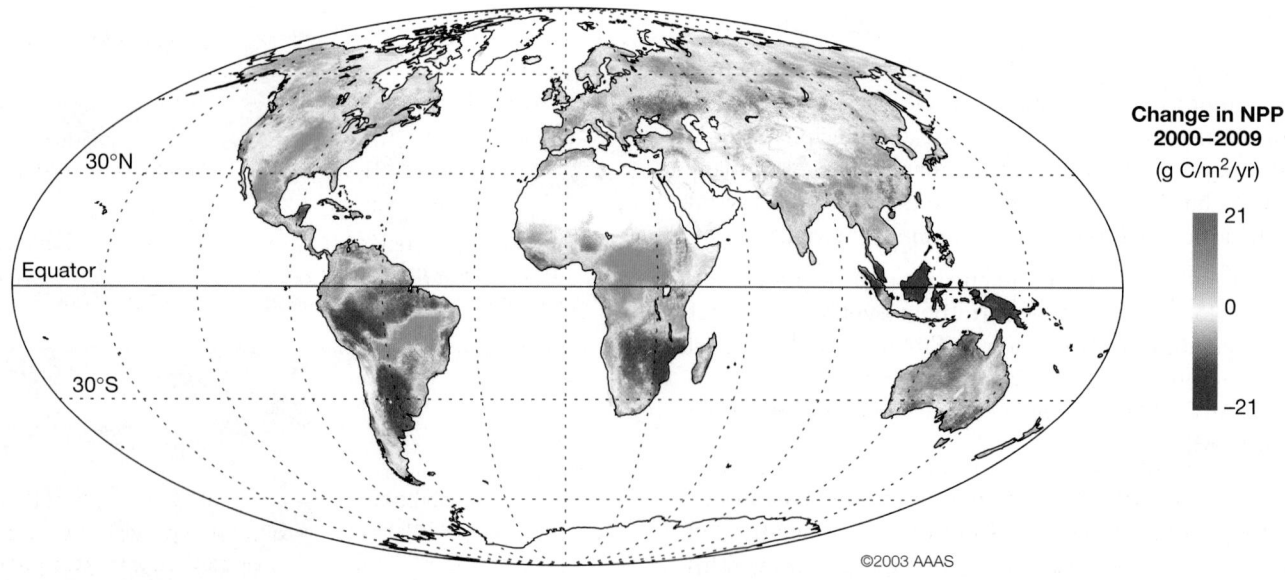

©2003 AAAS

FIGURE 56.25 Recent Changes in Terrestrial NPP. The data shown represent the average annual change in terrestrial NPP from 2000 to 2009.

DATA: Zhao, M., and S. W. Running. 2010. *Science* 329: 940–943.

One of the leading explanations for the drop in marine productivity hinges on changes in water density. To understand the logic behind this hypothesis, recall that lakes can become stratified—that is, have layers of different densities (see Chapter 52). Stratification occurs because water reaches its highest density at 4°C and becomes much less dense at higher temperatures.

Seawater also becomes stratified. This layering is important because the 4°C water in the benthic zone is nutrient rich due to the rain of decomposing organic material from the surface. When warmer surface water becomes lighter, water currents are much less likely to be strong enough to overcome the density difference and bring nutrient-rich, 4°C water all the way up to the surface, where nutrients can spur the growth of photosynthetic bacteria and algae (**FIGURE 56.26**). In this way, the warming global climate is causing surface waters to become more nutrient poor.

Local and Global Consequences   What are the consequences of these unprecedented global changes in NPP? There are two main take-home messages:

1. Changes in NPP cause feedback on the rate and pattern of climate change. If terrestrial and marine NPP continue to decline, the rate of absorption of CO_2 will decline, exacerbating climate change with far-reaching effects to humans and other organisms.

2. Changes in NPP combine with other factors, such as changes in precipitation patterns and ocean acidification, to alter food web dynamics—and thus ecosystem function. Maintaining

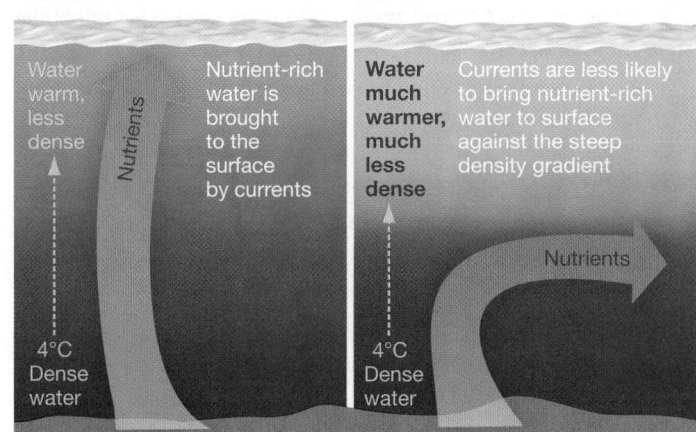

FIGURE 56.26 Temperature Stratification in the Ocean May Affect Productivity.

ecosystem function is critical to the preservation of biodiversity and the ecosystem services that humans rely upon for survival (see Chapter 57).

This chapter began with the theme that everything in nature is connected. As you've seen, humans are very much a part of these connections. Understanding how humans fit into the big picture is key to taking actions to improve ecosystem health and slow global climate change.

CHAPTER 56 REVIEW

For media, go to MasteringBiology

If you understand . . .

56.1 How Does Energy Flow through Ecosystems?

- Net primary productivity (NPP) is the foundation of ecosystems because it represents the energy, in the form of biomass, available for consumption by other organisms. Humans are reducing and appropriating a large percentage of NPP.

- The feeding relationships among species in a particular ecosystem are described by a food web. Energy is constantly dissipated as it flows from producers to consumers and decomposers in a food web.

- Organisms that acquire energy from the same type of source are said to occupy the same trophic level. Because energy transfer from one trophic level to the next is inefficient, ecosystems have a pyramid of productivity: Biomass production is highest at the lowest trophic level and lower at each higher trophic level.

- Persistent toxins can be magnified in concentration as they are passed up trophic levels in a food web.

- Changes in the species composition of a food web can cause trophic cascades.

- Among terrestrial ecosystems, NPP is highest in tropical wet forests and tropical dry forests. The productivity of terrestrial ecosystems is limited by warmth and moisture.

- Among aquatic ecosystems, productivity is highest in coral reefs, wetlands, and estuaries. Nutrient availability is the key constraint in aquatic ecosystems.

- Although productivity is extremely low in the oceanic zone, the area covered by this ecosystem is so extensive that it accounts for the highest percentage of Earth's overall productivity.

✓ You should be able to explain why it is more energy efficient to eat a pound of tofu (bean curd) than a pound of hamburger.

56.2 How Do Nutrients Cycle through Ecosystems?

- Unlike energy that enters an ecosystem and is ultimately dissipated as heat, nutrients such as carbon, nitrogen, phosphorus, and sulfur move indefinitely through ecosystems in biogeochemical cycles.

- To analyze nutrient cycles, biologists focus on the nature of the reservoirs where elements reside and the processes that move elements between reservoirs.

- The rate of nutrient cycling is strongly affected by the rate of decomposition of detritus.

- Nutrients "leak" from one ecosystem to another when biomass is exported. Experiments have shown that vegetation serves an important role in preventing nutrient export by reducing runoff.

- Humans are harvesting freshwater from reservoirs faster than it can be replenished by natural processes.

- Human activities have doubled the amount of usable nitrogen that enters ecosystems each year, leading to increased productivity and to pollution as well as to loss of biodiversity.

- The burning of fossil fuels by humans releases massive amounts of carbon from an inactive geological reservoir to the atmosphere. Current CO_2 levels are 30 percent higher than the highest level measured over the last 800,000 years.

✔ You should be able to describe conditions under which decomposition rates in terrestrial and marine environments could be extremely low—creating a carbon sink due to a buildup of organic matter.

56.3 Global Climate Change

- Average global temperatures are increasing rapidly because land-use changes and burning of fossil fuels have increased the flow of carbon in the form of CO_2 into the atmosphere. CO_2 is one of the greenhouse gases that trap heat in the atmosphere.

- Both positive and negative feedback are occurring in response to global warming.

- Average temperature is increasing, but local temperatures and precipitation vary. The variability of temperature and rainfall is also increasing. Storm events are becoming more severe.

- Biologists have documented an array of consequences of climate change to organisms, including changes to phenology, geographic ranges, adaptations, extinctions, and acidification.

- Recent data show that the warming associated with climate change is causing a decrease in average NPP both on land and in the oceans, influencing both climate and ecosystem function.

✔ You should be able to propose two mechanisms that humans could adopt to reduce global warming, and explain the logic behind each.

(MB) MasteringBiology

1. **MasteringBiology Assignments**

   Tutorials and Activities Animal Food Production Efficiency and Food Policy; Atmospheric CO_2 and Temperature Changes; Carbon Cycle; Energy Flow and Chemical Cycling; Energy Flow through Ecosystems; What Factors Influence the Loss of Nutrients from a Forest Ecosystem? Food Webs; Global Carbon Cycle; Global Freshwater Resources; Greenhouse Effect; Municipal Solid Waste Trends in the U.S.; Nitrogen Cycle; Prospects for Renewable Energy; Pyramids of Production

   Questions Reading Quizzes, Blue-Thread Questions, Test Bank

2. **eText** Read your book online, search, take notes, highlight text, and more.

3. **The Study Area** Practice Test, Cumulative Test, BioFlix® 3-D Animations, Videos, Activities, Audio Glossary, Word Study Tools, Art

You should be able to . . .

✔ TEST YOUR KNOWLEDGE

Answers are available in Appendix A

1. True or false? An ecosystem comprises a community and its abiotic environment.

2. Which of the following ecosystems would you expect to have the highest primary production?
   a. subtropical desert
   b. temperate grassland
   c. boreal forest
   d. tropical dry forest

3. Most of the net primary productivity that is consumed is used for what purpose?
   a. respiration by primary consumers
   b. respiration by secondary consumers
   c. growth by primary consumers
   d. growth by secondary consumers

4. What is biomagnification?
   a. biological imaging that provides images or data at fine spatial scales

   b. conversion of an energy input (e.g., sunlight) to chemical energy in the form of sugars, by primary producers
   c. accumulation of certain molecules at high concentration at upper trophic levels
   d. increases in nutrients that lead to pollution

5. Which of the following is normally the longest-lived reservoir for carbon?
   a. atmospheric CO_2
   b. marine plankton (primary producers *and* consumers)
   c. petroleum
   d. wood

6. Devegetation has what effect on ecosystem dynamics?
   a. It increases belowground biomass.
   b. It increases nutrient export.
   c. It increases rates of groundwater recharge (penetration of precipitation to the water table).
   d. It increases the pool of soil organic matter.

7. **QUANTITATIVE** If the GPP of a grassland is 5000 kcal/m^2/year and 55 percent is used up by cellular respiration, what is the NPP?
   a. 2250 kcal/m^2/year
   b. 2750 kcal/m^2/year
   c. 5000 kcal/m^2/year
   d. Not enough information is provided.

8. Predict why the large-scale removal of sharks from the oceans has caused some coral reefs to become overgrown with algae, even though sharks don't eat algae.

9. Explain why decomposition rates in a field in Nebraska would differ from the decomposition rates in a field in the Amazon. How do decomposers regulate nutrient availability in an ecosystem?

10. If you were to select a study organism to measure the effects of climate change on trophic interactions in food webs, would corals be a good choice? Explain your reasoning.

11. Use forests as an example to contrast positive and negative feedbacks on global warming.

12. Why are the open oceans nutrient poor? Why are coastal areas and intertidal habitats relatively nutrient rich?

13. Suppose you had a small set of experimental ponds at your disposal and an array of pond-dwelling algae, plants, and animals. How could you use a radioactive isotope of nitrogen to study the effects of fertilizer runoff in these experimental ecosystems?

14. Suppose that record snows blanket your campus this winter. Your friend says this is proof that global warming isn't really occurring. What is the flaw in your friend's logic?
    a. The average temperature of the Earth is not actually increasing.
    b. Global warming refers to temperatures, but snow is a type of precipitation.
    c. While the average global temperature is increasing, local temperatures and precipitation will vary.
    d. He is confusing global warming and global climate change.

15. Suppose that herbivores were removed from an area within a temperate deciduous forest ecosystem (e.g., by fencing to exclude herbivorous mammals and by spraying insecticides to kill herbivorous insects). Predict what would happen to the rate of nitrogen cycling. Explain the logic behind your prediction.

16. During the Carboniferous period, rates of decomposition slowed even though plant growth was extensive (probably due to the formation of vast, oxygen-poor swamp habitats). As a result, large amounts of biomass accumulated in terrestrial environments (much of this biomass is now coal). The fossil record indicates that atmospheric oxygen increased, atmospheric carbon dioxide decreased, and global temperatures dropped. Explain why.

57 Biodiversity and Conservation Biology

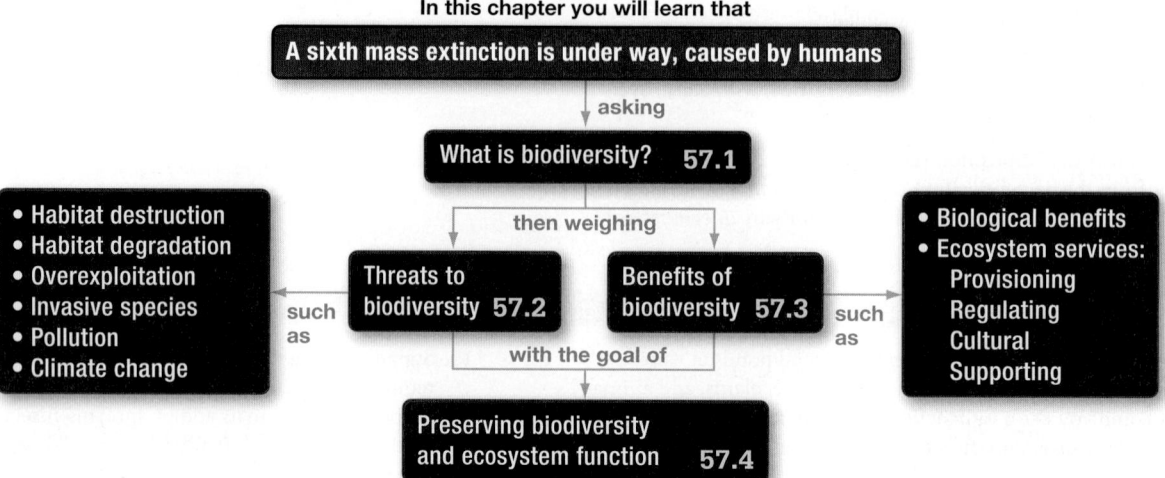

In this chapter you will learn that

A sixth mass extinction is under way, caused by humans

asking

What is biodiversity? 57.1

then weighing

- Habitat destruction
- Habitat degradation
- Overexploitation
- Invasive species
- Pollution
- Climate change

such as

Threats to biodiversity 57.2

Benefits of biodiversity 57.3

such as

- Biological benefits
- Ecosystem services:
  Provisioning
  Regulating
  Cultural
  Supporting

with the goal of

Preserving biodiversity and ecosystem function 57.4

The island of Borneo contains some of the most spectacular rain forest in the world, yet it is being cleared at an alarming rate to make way for oil palm plantations. This action threatens both biodiversity and ecosystem function.

This chapter is part of the Big Picture. See how on pages 1196–1197.

When conservation biologist Willie Smits gazed out upon an area near Samboja on the Southeast Asian island of Borneo in 2002, he saw a devastated wasteland:

- almost total extinction of plant and animal life;
- depleted soil fertility and failing crops;
- unusually severe climate: droughts, floods, fires; and
- a destitute human population: 50 percent unemployed, high crime rate, poor nutrition, low life expectancy.

Things were not always this way. Borneo is famous for having some of the most luxuriant rain forests on the planet—15,000 species of plants, 222 species of mammals, and hundreds of birds,

✔ When you see this checkmark, stop and test yourself. Answers are available in Appendix A.

amphibians, and fish. Six thousand of these species occur nowhere else on Earth. What happened in Samboja?

Starting in the 1950s, loggers started cutting down the valuable hardwood trees. Following the eventual clear-cut, the remnants were burned to make room for oil palm plantations to supply the world's demand for cooking oil, products such as lipstick, and biodiesel. This transition to agriculture has been pervasive in Borneo. In just 20 years, Borneo has lost as much forest as would fit in Florida (**FIGURE 57.1a**).

The orangutans have been one major casualty of this land transformation, which is how Smits became interested in Samboja (**FIGURE 57.1b**). For these apes, already threatened by poaching for the pet trade, the massive loss of habitat was the last straw. Smits realized that if he wanted to save the orangutans, he would have to save the rain forest.

The scenario at Samboja is not unique. As the human population soars (Chapter 54) and the available resources plummet, more and more wild areas on Earth are being tamed, often with species losses and unintended consequences for ecosystems and humans. The goal of this chapter is to size up the problem and then explore

(a) Rapid deforestation of Borneo

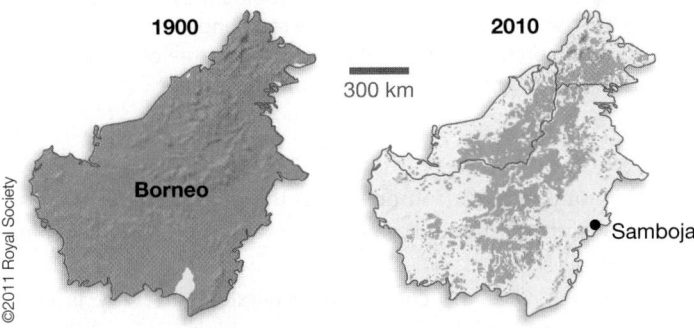

(b) Orangutans are one casualty of deforestation

FIGURE 57.1 The Deforestation of Borneo (a) Borneo is a large island in Southeast Asia, once completely forested. **(b)** The clearing of lowland rain forest for oil palm plantations has had a devastating effect on species like the orangutan. Conservationist Willie Smits was inspired by the plight of these animals to work toward a solution.

positive steps that can enable a happier outcome. Let's begin by defining a concept that is key to the discussion—biodiversity.

57.1 What Is Biodiversity?

In essence, **biodiversity** means biological diversity. But this concept is more complex than it might seem at first. Perhaps the simplest way to think about biodiversity is in terms of the tree of life—the phylogenetic tree of all organisms (introduced in Chapter 1 and explored in detail in Chapters 29 through 36). If biologists are eventually able to estimate the complete tree of life (using the techniques reviewed in Chapter 28), the branches would represent all of the lineages of organisms living today; the tips would represent all of the species.

When biodiversity increases, branches and tips are added, and the tree of life gets fuller and bushier. When extinctions occur, tips and perhaps branches are removed; the tree of life becomes thinner and sparser (see Chapter 28 and The Big Picture of Evolution on pages 526–527).

Coordinated, multinational efforts are under way to estimate the tree of life—particularly major lineages such as the land plants and vertebrates—and to understand how the tree has changed over time. But even though the tree of life is a compelling way to document biodiversity, it does not tell the entire story.

Biodiversity Can Be Measured and Analyzed at Several Levels

To get a complete understanding of the diversity of life, biologists recognize and analyze biodiversity at the genetic, species, and ecosystem levels. Let's consider each level in turn.

Genetic Diversity **Genetic diversity** is the total genetic information contained within all individuals of a population, species, or group of species. It is measured as the number and relative frequency of all genes (and their alleles) present in a group.

At the genetic level, biodiversity is everywhere around you, from the variety of apples at the grocery store to the variation in singing ability of sparrows in your backyard. Distinct populations of the same species often contain unique alleles that are maintained by genetic drift or have been favored by natural selection as adaptations to local conditions (see Chapter 26).

Recently, efforts to catalog genetic diversity have been advanced by two technical breakthroughs (see Chapter 21):

1. the ability to sequence the entire genome of multiple members of the same species, and

2. the research protocol called environmental sequencing.

In environmental sequencing, researchers take a sample from the soil or water present in a habitat and sequence all or most of the genes present. It's a way to document genetic diversity even if the species present are unknown.

Genetic diversity is important because it represents the adaptive capacity of a population or higher taxonomic group—the ability of that group to persist over time despite changes in the environment.

Species Diversity   The diversity of species present is a key feature of biological communities. It is usually quantified in two ways:

1. **Species richness** is a simple count of how many species are present in a given community.

2. **Species diversity** is a weighted measure that incorporates the species' relative abundance.

The key difference is in evenness. In communities that are dominated by a few species, species richness might be high but species diversity low (see Chapter 55). Be aware that since relative abundance data are often not available, biologists sometimes use the terms interchangeably.

Ecologists traditionally measure species diversity by spending hours at their field sites employing transects, traps, quadrats, mark-and-recapture techniques, and other methods to quantify the species present (see Chapter 54). This important work can be painstaking, especially in biodiverse areas such as the mountains of Papua New Guinea where, for example, a lamp can be used to attract hundreds of moth species in a single night.

One new technique for identifying such species is called **bar coding**: the use of a well-characterized gene sequence to identify distinct species, using the phylogenetic species concept (see Chapter 27). The idea is that variation in one gene—such as the gene for cytochrome *c* oxidase subunit 1 in the mitochondria of animals—allows biologists to recognize different species in the same way that variations in bar codes enable grocery store scanners to recognize different products. To date, tens of thousands of species have been bar-coded.

It is important to remember, however, that counting species is not like counting beans in a bowl—all species are not equal. Each species has an evolutionary and ecological context.

In terms of evolution, some species occur in lineages that are extremely species rich—they have what biologists call taxonomic diversity. Prominent examples include the beetles, with 360,000 species, and the orchids, with 35,000 species. Other species occur in lineages that are extremely species poor, such as that of the red panda of Asia, which is represented by a single species (**FIGURE 57.2**). When preservation of taxonomic diversity is important, then the preservation of species from species-poor lineages is a priority. If species like the red panda are wiped out, an entire lineage is lost forever.

In terms of ecology, some species have particularly influential roles in a community: they may be key primary producers, important ecosystem engineers creating habitat for other species, keystone predators, or nitrogen fixers (see Chapter 55). When functional diversity is a priority, preservation of species with unique ecological roles is important.

Ecosystem Diversity   Communities vary in the number of species that occur within each trophic level (horizontal diversity) and the number of trophic levels (vertical diversity). **Ecosystem diversity** is a measure of these complex factors plus interactions

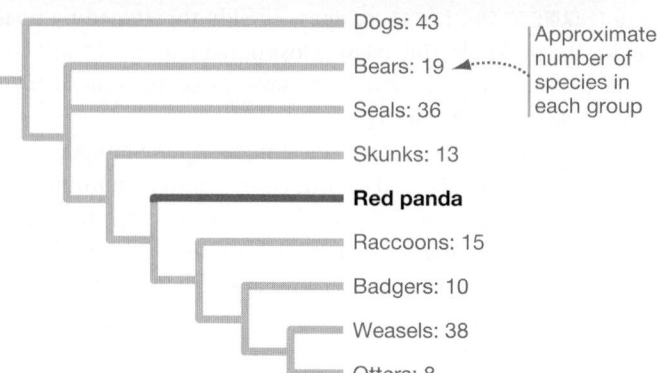

FIGURE 57.2 Phylogenetically Distinct Species May Be High-Priority Targets for Conservation. The red panda has few close relatives and represents a distinct branch on the tree of life. It is a high priority for the preservation of taxonomic diversity.

with the nonliving environment. It is the most difficult level of biodiversity to define and measure because the complexity of ecosystems is so vast.

Intertwined with the concept of ecosystem diversity is **ecosystem function,** which refers to the sum of biological and chemical processes that are characteristic of a given ecosystem—such as primary production, nitrogen cycling, decomposition, and carbon storage. These functions emerge from the sum of the feeding, growing, moving, respiring, excreting, and decomposing processes of the interacting member organisms within their abiotic context. As the field of conservation biology develops, the importance of preserving ecosystem function is gaining more attention.

Note that the three levels of biodiversity are not independent of each other. For example, consider what might happen if the sole species of nitrogen-fixing bacteria disappeared from a grassland community: The genes that code for the nitrogen-fixing enzymes would be lost, followed by loss of the species that depend on that nitrogen as well as loss of important nutrient cycling functions of the ecosystem.

Change through Time   Biodiversity can be recognized and quantified on several distinct levels, but it is also dynamic.

• Mutations that create new alleles increase genetic diversity; natural selection, genetic drift, and gene flow may eliminate certain alleles or change their frequency, leading to an increase or decrease in overall genetic diversity (Chapter 26).

- Speciation increases species diversity; extinction decreases it (Chapter 28).

- Changes in climate or other physical conditions can result in the formation of new ecosystems or ecosystem functions, as can the evolution of new species—such as reef-building corals and salt-tolerant marsh grasses. Disturbances such as volcanic eruptions, human activities, and glaciation can destroy ecosystems or ecosystem functions.

Biodiversity is not static. It has been changing since life on Earth began.

How Many Species Are Living Today?

One of the simplest questions about biodiversity is also one of the most difficult to answer: How many species are there on Earth? The answer is not known, but educated guesses range from 10 to 100 million. Given the massive effort that it would take to document every form of life on Earth, the answer will probably never be known.

Biologists are well aware, though, that the approximately 1.5 million species cataloged to date represent a tiny fraction of the number actually present. The issue is even more acute in poorly studied lineages like bacteria and archaea, where researchers routinely discover dozens of new species in each environmental sequencing or direct sequencing study (see Chapters 21 and 29). For example, a recent direct sequencing study estimated that over 500 species of bacteria live in the human mouth, but only about half had been described and named already. Even among well-studied groups like birds and mammals, new species are discovered every year—including new species of primates.

Given that only a fraction of the organisms alive have been discovered to date, how can biologists go about estimating the total number of species on Earth? One approach is based on intensive surveys of species-rich taxonomic groups in defined sites; a second strategy is based on attempts to identify all of the species present in a particular region.

Taxon-Specific Surveys In a taxon-specific survey, biologists conduct an intensive search for a specific taxon in a well-defined area. For example, in their classic study carried out in 1980, Terry Erwin and Janice Scott fogged several trees with insecticide in a Panamanian jungle and recovered 1000 species of beetles. Many of the species were new to science.

More recently, Philippe Bouchet and colleagues conducted a massive survey of marine mollusks in coral-reef habitats along the west coast of New Caledonia, a tropical island in the southwest Pacific Ocean. The team spent more than a year collecting mollusks at 42 sites over a total area of almost 300 km². The survey represented the most thorough sampling effort ever made to determine the species diversity of mollusks, and produced more than 127,000 individuals representing 2738 species. These numbers far exceed the mollusk diversity recorded for any comparable-sized area.

In reporting their findings, Bouchet and his team emphasized that 20 percent of the species found were represented by a single specimen. This observation suggests that many species are exceedingly rare and thus likely to be missed by less intensive sampling efforts. And when the investigators compared their data with a survey in progress at a second site, different in reef structure but only 200 km away from the original site, they found that only 36 percent of species were shared between the two sites. Taken together, the results support the hypothesis that the 85,000 known mollusk species may represent just a third to a half of the actual total.

All-Taxa Surveys As the name suggests, in an all-taxa survey, biologists attempt to find and catalog all of the species present in an area. The most spectacular example of an all-taxa survey so far is the Census of Marine Life, a massive international effort to record the global diversity of life in the oceans. About 2700 scientists from more than 80 countries launched 540 expeditions over a 10-year period ending in 2010 to survey the number and distribution of marine species around the globe—ranging from deep-sea vents to coral reefs to intertidal mud flats.

The participants collected over 30 million records, including 6000 potentially new species. Even a small sediment core only 7 centimeters in diameter and 10 centimeters deep yielded a surprisingly diverse assemblage of deep-sea animals (**FIGURE 57.3**). Many animal phyla are represented in this sample, most of them miniaturized—a possible adaptation to low nutrient availability in the deep sea. The total number of marine species is currently estimated by the census to be 250,000.

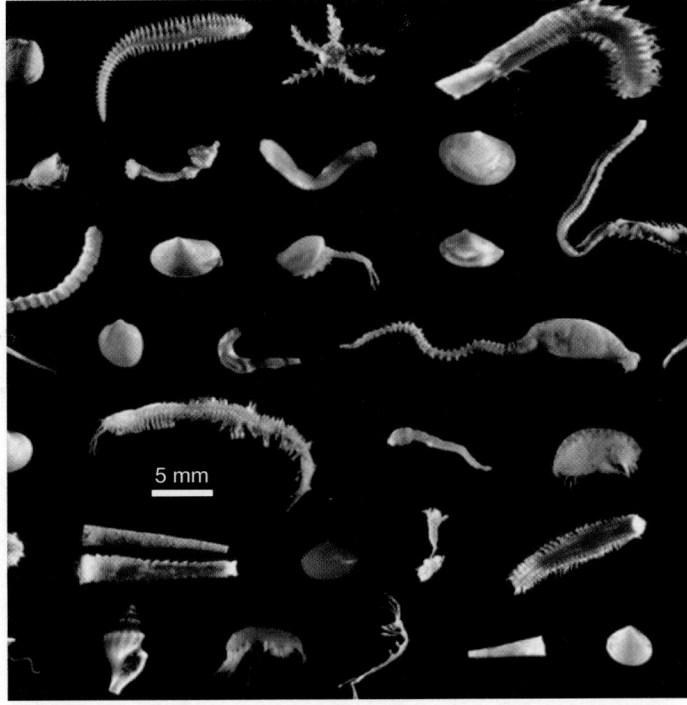

5 mm

FIGURE 57.3 Estimating Species Richness via All-Taxa Surveys. A sample of the diverse, tiny animals sifted from a small core of deep-sea sediments.

Where Is Biodiversity Highest?

Both taxon-specific surveys and all-taxa surveys provide data on the total number of species as well as on their geographic distributions. What patterns of biodiversity emerge?

Mapping Species Richness and Endemism   Patterns of biodiversity are usually measured by dividing geographic areas into a grid of cells and mapping the biodiversity within each cell.

The map in **FIGURE 57.4a**, for example, was constructed by dividing the world's landmasses into a grid of cells 1° latitude by 1° longitude. Using published data, a team of researchers then plotted how many species of birds breed in each cell in the grid, a measure of species richness. The narrow red-and-orange area on the map shows that the number of breeding bird species per area is particularly high in the Andes Mountains of western South America.

Another approach to measuring the distribution of biodiversity is to identify which regions of the world have a high proportion of **endemic species**—meaning, species that are found in a particular area and nowhere else. **FIGURE 57.4b** maps the location of endemic species of breeding birds, based on the same cells and data as the species richness map in Figure 57.4a. Once again, the area of highest species per area occurs in the Andes Mountains. However, the two maps are not identical, so evaluating both species richness and endemism can be more insightful than evaluating either alone.

As you might expect, the areas of highest biodiversity are different for different taxonomic groups. However, some trends do appear.

- The most prominent geographic pattern in the distribution of biodiversity corresponds to latitude: In most, but not all, taxonomic groups, species richness is highest in the tropics and declines toward the poles (see Chapter 55).

- Biodiversity is also higher on land than in the sea. Even though land covers only about 30 percent of Earth's surface, about nine out of 10 species (microbes aside) are terrestrial.

- Areas with greater geographical variation tend to have higher biodiversity than areas that are flat or featureless.

Overall, tropical rain forests are particularly species rich. Even though they represent just 7 percent of Earth's land area, they are thought to contain at least 50 percent of all species present.

Understanding *why* tropical forests are so species rich and *why* the other biodiversity patterns occur is an active research area. The mechanisms that determine patterns of biodiversity are complex because they combine local abiotic factors (solar energy, temperature, precipitation, topography, nutrient availability; Chapters 52 and 56), biotic factors (species interactions; Chapter 55), and historical context (evolutionary history, biogeography; Chapter 52). To complicate matters further, the mechanisms that determine biodiversity differ at the local scale and the regional scale.

(a) Biodiversity distribution in terms of **species richness of birds**

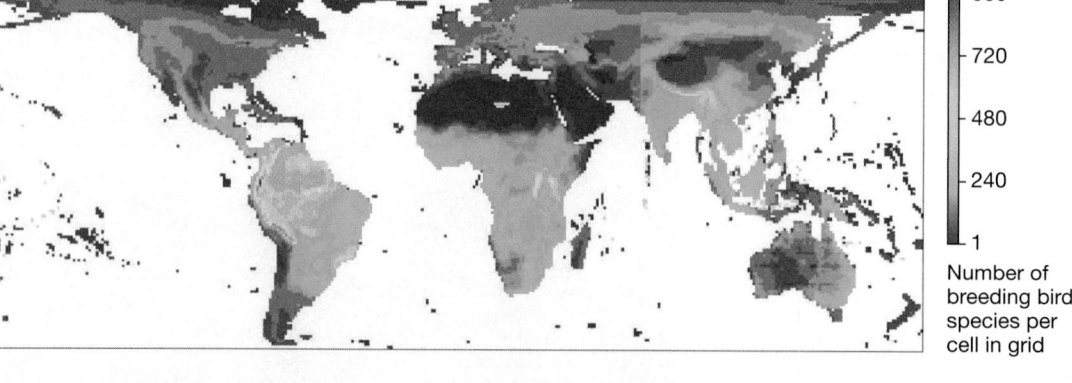

959
720
480
240
1

Number of breeding bird species per cell in grid

FIGURE 57.4 Biodiversity Distribution Maps Show Regions with Extraordinarily High Species Richness and Endemism. Both maps are based on a grid of cells that is 1° longitude by 1° latitude.

DATA: Orme, C. D. L., R. G. Davies, M. Burgess, et al. 2005. *Nature* 436: 1016–1019.

(b) Biodiversity distribution in terms of **endemic species of birds**

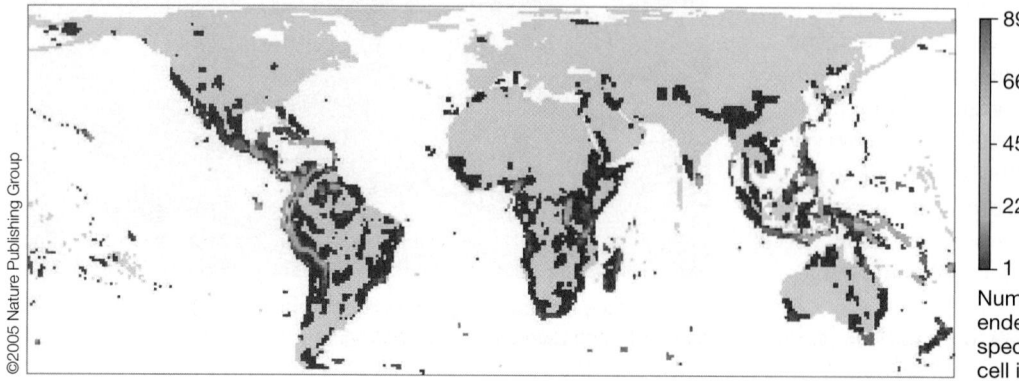

89
66
45
22
1

Number of endemic bird species per cell in grid

Mapping Biodiversity Hotspots In 1988, conservation biologist Norman Myers pioneered the term **biodiversity hotspot** to characterize regions that are in most urgent need of conservation action—areas where efforts to preserve habitat would have the highest return on investment.

Since 1988, Myers and many collaborators, including organizations such as Conservation International, have identified 34 hotspots that meet two criteria:

1. They contain at least 1500 endemic vascular plant species (more than 0.5 percent of the world's total).

2. They have lost at least 70 percent of their traditional or primary vegetation.

Although these areas represent just 2.3 percent of Earth's land area, they contain over 50 percent of all known plant species and 42 percent of all known terrestrial vertebrate species as endemics.

One criticism of the second criterion of biodiversity hotspots is that it considers cumulative past loss, not current rate of loss. Most notably, hotspots do not include the rain forests found in the Amazon, the Congo River basin, and New Guinea—because those environments have not yet lost 70 percent of their primary vegetation. Effectively preserving these areas along with the hotspots would provide protection for over 65 percent of all land plants in just 5 percent of Earth's land surface.

Further, hotspots focus on species diversity rather than ecosystem diversity—but the importance of ecosystem diversity is becoming more apparent every year. In 2002, scientists in collaboration with the World Wildlife Fund created a map of 238 ecoregions—subsets of biomes (see biome maps in Chapter 52). The ecoregion map, shown in **FIGURE 57.5**, highlights particularly

rich areas of biodiversity; it also includes areas with unique communities or abiotic conditions (such as in tundra or desert biomes) or a unique evolutionary history (such as the Galápagos Islands; Chapter 25)—all of which should be priorities for proactive conservation.

Whether global maps of biodiversity hotspots are based on species richness, endemism, level of threat, and/or unique ecological or evolutionary characteristics, they are all useful in their own ways for creating priorities for managing threats to biodiversity.

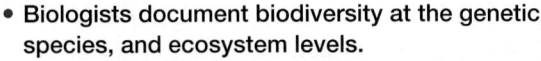

check your understanding

C Y U

If you understand that . . .

- Biologists document biodiversity at the genetic, species, and ecosystem levels.
- On average, biodiversity tends to be higher in the tropics than near the poles, higher on land than in the sea, and higher in areas with geographic diversity. Tropical rain forests are particularly species rich.
- Biodiversity can be mapped according to species richness, endemism, level of threat, and ecological or evolutionary characteristics.

✔ You should be able to . . .

Design a study that would quantify species diversity on your campus.

Answers are available in Appendix A.

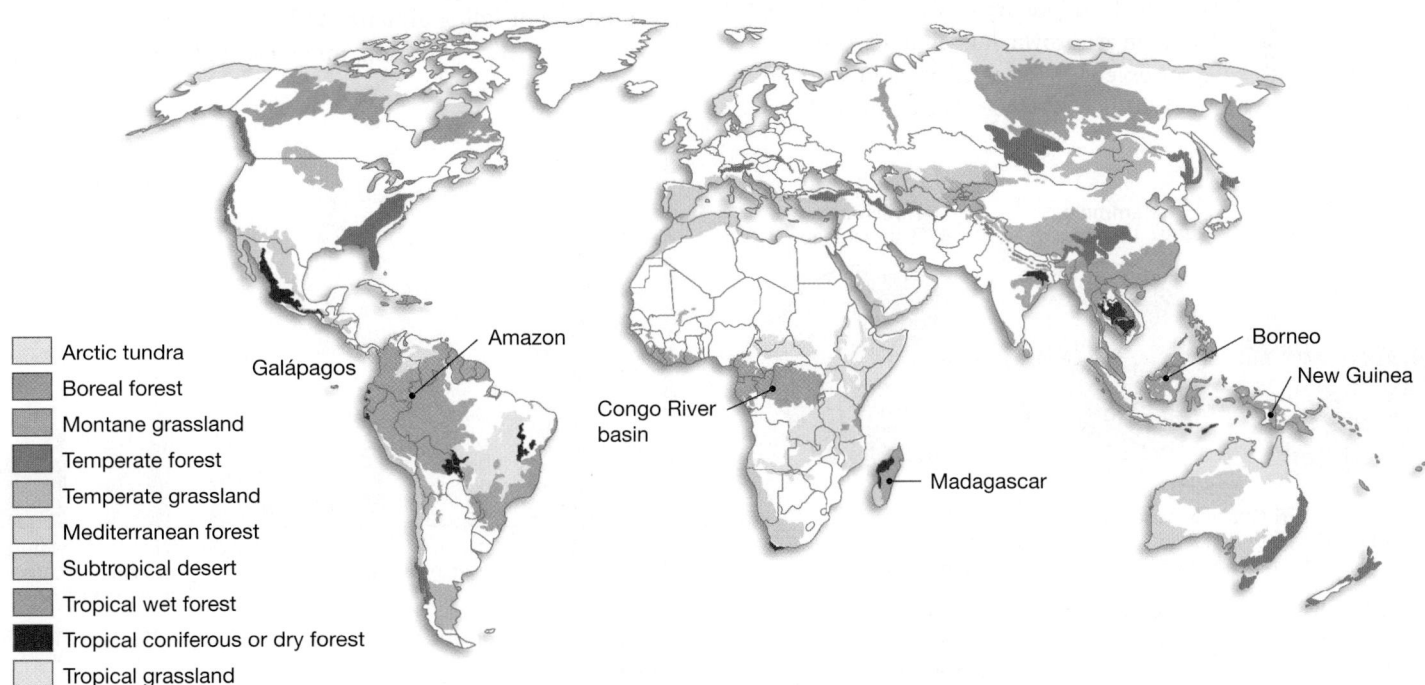

FIGURE 57.5 Ecoregions Designated as High Priority for Conservation. Areas that have the same color belong to the same biome (see Chapter 52).

Legend:
- Arctic tundra
- Boreal forest
- Montane grassland
- Temperate forest
- Temperate grassland
- Mediterranean forest
- Subtropical desert
- Tropical wet forest
- Tropical coniferous or dry forest
- Tropical grassland

Map labels: Galápagos, Amazon, Congo River basin, Madagascar, Borneo, New Guinea

57.2 Threats to Biodiversity

No species lasts forever. Extinction, like death, is a fact of life. If extinction is natural, why are biologists so concerned about habitat and species conservation? The answer is rate.

Today, species are vanishing faster than at virtually any other time in Earth's history. Modern rates of extinction are 100 to 1000 times greater than the average, or "background," rate recorded in the fossil record over the past 550 million years (see Chapter 28). According to data on the rates of extinction in birds and other well-studied groups, 60 percent of all species are likely to be wiped out within 500 years, a blink of an eye on the evolutionary time scale.

Either directly or indirectly, current extinctions are being caused by the demands of a rapidly growing human population—currently at over 7 billion and increasing by about 80 million people per year (Chapter 54). The island that is planet Earth is far more crowded than it used to be.

How do biologists track extinctions that are under way? Beginning in 1994, an organization called the International Union for the Conservation of Nature (IUCN) began compiling the Red List to quantify the risk of extinction of different species. For example, the data in **FIGURE 57.6** show that an astonishing 13 percent of birds, 25 percent of mammals, and 41 percent of amphibians are threatened—that is, they are either Critically Endangered, Endangered, or Vulnerable. An **endangered species** is a species whose numbers have decreased so drastically that it is almost certain to go extinct unless effective conservation programs are put in place.

Based on the data in hand, the vast majority of biologists agree that the sixth mass extinction in the history of multicellular life is occurring. Between the time you are reading this and the time your great-grandchildren are grown, human impacts on the planet promise to equal or exceed those of the gigantic asteroid that smashed into Earth 65 million years ago (see Chapter 28).

Multiple Interacting Threats

What factors are causing so many species to be endangered? Consider a recent study conducted by ecologists Oscar Venter, James Grant, and colleagues. They examined 488 species of endangered plants and animals native to Canada, and they quantified the frequency of different threats based on the categories compiled by the IUCN. The bar graphs in **FIGURE 57.7** show the results. The data reveal several patterns:

- Habitat loss is the single most important factor in the decline of these species. It is a significant issue for over 90 percent of endangered species in terrestrial environments.

- Virtually all of the endangered species are affected by more than one factor. This is important information, because it means that conservation biologists may have to solve more than one problem for any given species to recover.

- Overexploitation is the dominant problem for marine species, while pollution plays a large role for freshwater species.

- Factors beyond human control can be important. These include predation or competition with native species, natural disturbances such as fires or droughts, or the fact that some species have narrow niches and historically are rare. Stated another way, background extinctions will continue to occur.

Similar studies have been conducted for other parts of the world. Do the same conclusions hold up globally? The degree of importance of the different threats varies somewhat according to region, but the take-home message is consistent with the Canadian data. Let's take a closer look at the major threats to species.

Habitat Destruction Humans cause **habitat destruction** (habitat loss) by logging and burning forests, damming rivers, dredging or filling estuaries and wetlands, plowing prairies, grazing livestock, excavating minerals, and building housing

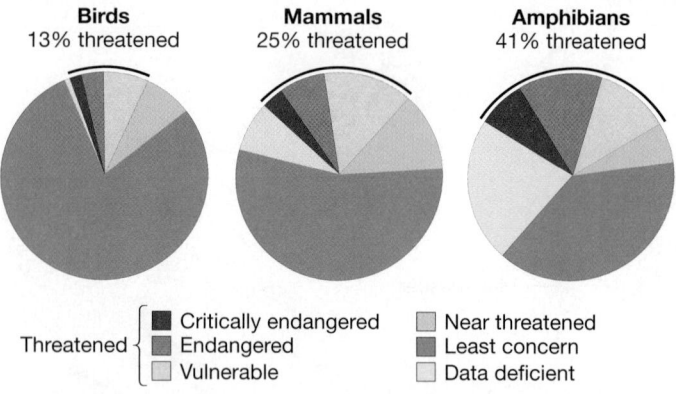

FIGURE 57.6 IUCN Designations of Threatened Species. The total number of threatened species in each group was estimated assuming that the species in the gray (data-deficient) category have the same rate of being threatened as the others. Extinct species are not included.

DATA: IUCN 2012. The IUCN Red List of Threatened Species. Version 2012.2.

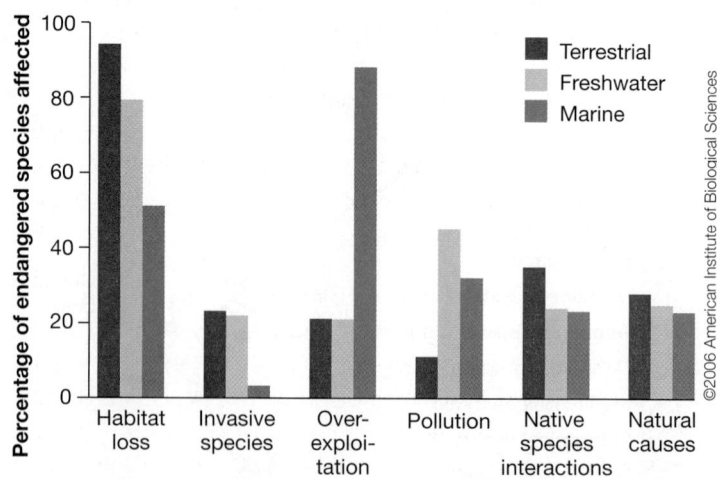

FIGURE 57.7 Endangerment Is Caused by Multiple Factors. These data are from Canada.

✓ **QUANTITATIVE** Why do the data for terrestrial, freshwater, and marine species each add up to more than 100 percent?

DATA: Ventner, O. et al. 2006. *BioScience* 56: 903-910.

developments, golf courses, shopping centers, office complexes, airports, and roads. Today, less than a quarter of Earth's land is in its wild state (see map in Figure 52.10), and most of the wild land is near the poles, where biodiversity is lowest.

On a global scale, one of the most important types of habitat destruction is deforestation—especially the conversion of primary forests to agricultural fields and human settlements. Primary forests—which have never been cut—tend to have high biodiversity due to their complex ecological interactions and deep evolutionary history. Research comparing primary forest and reforested areas shows that primary forest is irreplaceable for sustaining biodiversity.

To appreciate the extent of deforestation in some areas, consider satellite images of wet tropical rain forest in the Brazilian state of Mato Grosso that were made in 1985 and in 2000 (**FIGURE 57.8**). Analyses of satellite photos such as these have shown that as many as 3 million hectares were deforested in the Amazon each year during the 1990s. (A hectare, abbreviated ha, is 100 meters by 100 meters—approximately the size of two football fields.)

More recent analyses suggest that the global rate of deforestation has now slowed compared to an average net loss of 16 million ha/year in the 1990s. The latest report from the United Nations Food and Agriculture Organization, released in 2010, states that the world's forests experienced a net loss of about 13 million ha/year between 2000 and 2010.

But conservationists greeted this report with caution. Net losses dropped because of extensive forest regrowth in China, North America, and Europe, even though over 40 million ha of primary forest have been lost in South America, Southeast Asia, and Africa since 2000. If this rate of tropical deforestation continues, over 28 percent of the wet tropical forest that exists today will be gone in your lifetime, including almost half of the Brazilian Amazon.

Forest losses in the tropics are particularly worrisome because these forests contain many biodiversity hotspots, and they moderate the effects of climate change by storing carbon. Further, when forests are cut, soil and nutrients are often lost from the ecosystem due to erosion, inhibiting future recovery (Chapter 56). Meanwhile, the human population continues to increase rapidly in these areas, putting increased pressure on the forests. Addressing the interaction of these scientific and social variables represents one of the most formidable challenges of our time.

Habitat Degradation Aside from destroying natural areas outright, human activities can also cause **habitat degradation**, a reduction in the quality of a habitat. Humans can degrade habitats in many ways, even by using artificial light at night and creating traffic noise. One of the most pervasive patterns of habitat degradation is **habitat fragmentation,** the parsing of contiguous areas of natural habitats into small, isolated fragments. For some species, a narrow road or pipe intruding into their habitat is a sufficient barrier to limit dispersal.

Habitat fragmentation concerns biologists for several reasons. First, like habitat destruction, fragmentation can reduce habitats to a size that is too small to support some species. This situation is especially true for top predators (see Chapters 55 and 56) such as wolves, mountain lions, and grizzly bears, which need vast natural spaces in which to feed, find mates, and reproduce successfully.

When top predators in food webs are lost, trophic cascades may occur, affecting species throughout the food web in complex ways (see Chapter 56). Trophic cascades often have unexpected impacts on many aspects of ecosystem function, ranging from susceptibility to fire, disease, and invasive species to changes in primary productivity and nutrient cycling.

The second concern about habitat fragmentation is that it can force many species into a metapopulation structure (see Chapter 54). The small, isolated populations that make up a metapopulation are much more likely than large populations to be wiped out, for two reasons:

1. Catastrophic events such as storms, disease outbreaks, or fires exterminate small populations more readily than large populations. Biologists refer to episodes like this as stochastic—meaning, chance—events.

FIGURE 57.8 Rates of Deforestation in Tropical Wet Forests Are High. In the satellite image from Mato Grosso, Brazil, at top, the dark green areas are intact forest. In the lower image of the same location 15 years later, the light areas have been burned or logged and converted to agricultural fields and pastures.

2. Small populations suffer from inbreeding depression and random loss of alleles due to genetic drift.

Why do genetic problems occur? Small populations become inbred simply because so few potential mates are available. Over time, virtually all individuals become related. In most populations, inbreeding leads to lowered fitness due to deleterious recessive alleles being homozygous more frequently—a phenomenon known as inbreeding depression (see Chapter 26).

In addition, genetic drift is much more pronounced in small populations than large populations (see Chapter 26). Because drift leads to the random loss or fixation of alleles, it reduces genetic variation in populations. This result is a concern, because if the environment changes due to the evolution of a new disease or global climate change, the population may lack alleles that confer high fitness in the new environment.

Third, fragmentation creates large amounts of degraded "edge" habitat. The edges of intact habitat are subject to invasion by weedy species and are exposed to more intense sunlight and wind, creating difficult conditions for plants.

The degradation of habitat caused by fragmentation is being documented in a long-term experiment in a tropical wet forest. In an area near Manaus, Brazil, that was slated for clear-cutting, ecologists William Laurance, Thomas Lovejoy, and their colleagues set up 66 square, 1-ha experimental plots that remained uncut. Thirty-nine of these study plots were located in fragments designed to contain 1, 10, or 100 ha of intact forest. Twenty-seven of the plots were set up nearby, in continuous wet forest. As the "Experimental setup" section of **FIGURE 57.9** shows, the distribution of the study plots allowed the research team to monitor changes inside forest fragments of different sizes and to compare these changes with conditions in unfragmented forest. Note that many of the study plots were located on forest edges.

When the research group surveyed the plots at least 10 years after the initial cut, they recorded two predominant effects:

1. A startling drop in aboveground **biomass,** or the total amount of fixed carbon, took place in the study plots located near the edges of logged fragments. As the graph in the "Results" section of Figure 57.9 shows, most of the loss occurred within 2–4 years of the fragmentation event.

2. On average, biomass did not decrease in the interior (control) plots. Further, the rate of biomass loss increased as the distance from the plots to the nearest forest edge decreased.

In tropical wet forests, most of the biomass is concentrated in large trees. Due to exposure to high winds and dry conditions, the edges of the experimental plots contained many downed and dying trees. Follow-up work showed that in edge habitats and in the fragmented patches, large-seeded, slow-growing trees typical of undisturbed forests are being replaced by early successional, "weedy" species (see Chapter 55). A large number of bird and understory plant species have disappeared as well.

The take-home message of this experiment and other research is clear: When habitats are fragmented, the quality *and* quantity of habitats decline drastically. In addition to losing over 8 million ha

RESEARCH

QUESTION: How does fragmentation affect the quality of tropical wet forest habitats?

HYPOTHESIS: Fragmentation reduces the quality of wet forest habitats.

NULL HYPOTHESIS: Fragmentation has no effect on the quality of wet forest habitats.

EXPERIMENTAL SETUP:

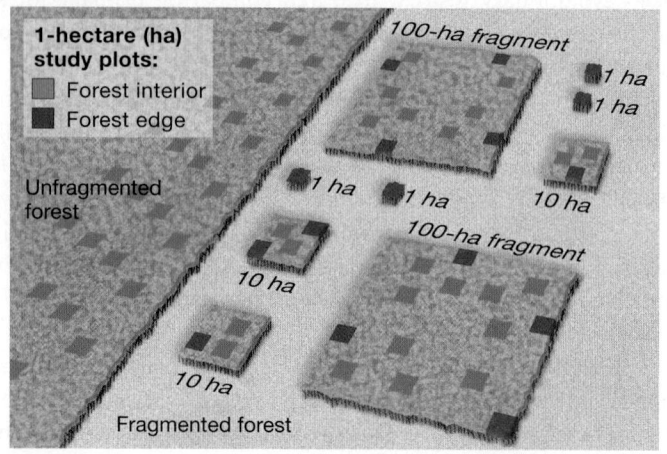

PREDICTION: Biomass will decline in forest fragments compared with those of the forest interior, particularly along edges of fragments.

PREDICTION OF NULL HYPOTHESIS: Biomass will be the same inside forest fragments and along edges as in the forest interior.

RESULTS:

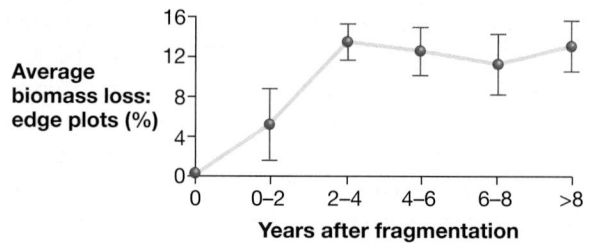

CONCLUSION: Biomass declines sharply along edges of forest fragments.

FIGURE 57.9 Experimental Evidence for Edge Effects in Fragmented Forests. Researchers tracked 66 study plots among four 1-hectare (ha) fragments, three 10-ha fragments, and two 100-ha fragments.

SOURCE: Laurance, W. F., S. G. Laurance, L. V. Ferreira, et al. 1997. Biomass collapse in Amazonian forest fragments. *Science* 278: 1117–1118.

✔**QUESTION** The locations of the fragments and study plots were assigned at random. Why was this step important?

of primary forest in South America, Southeast Asia, and Africa each year to clear-cutting, we are losing large amounts of high-quality habitat throughout the world to fragmentation.

Overexploitation **Overexploitation** refers to any unsustainable removal of wildlife from the natural environment for use by humans. Overexploitation is the dominant threat for marine species.

- Two-thirds of harvestable marine species are now considered fully exploited or depleted, and a recent study of the global fishing industry concluded that 90 percent of individuals in large-bodied fish species from bottom or open-water habitats have already been removed from the world's oceans (see **TABLE 57.1**). The affected species include tuna, swordfish, marlin, cod, halibut, and flounder—and many species of sharks, which are harvested for their fins. Unless stricter fishing limits are established quickly, these popular species will go extinct. Further, removal of these top predators has an immense impact on species at lower trophic levels, causing a cascade of effects on food webs (Chapter 56).

- Overhunting has also emerged recently as a major threat to many mammal populations. Aside from the traditional hunt for game species, the recent growth of the "bushmeat" trade has devastated populations of primates and other mammals in Africa and elsewhere. Adding to the problem, many animals are hunted not for their meat, but for their ivory, skins, horns, or other body parts valued in folk medicines or decorations.

- The capture of exotic animals for the pet trade continues at a rapid rate despite many laws to prevent it. Exotic pets range widely from primates like tamarins and macaques to parrots, iguanas, frogs, corals, and tropical fish.

Overexploitation is a grave concern for marine and terrestrial species worldwide due to the increasing demand and decreasing supply. Ten thousand years ago, only 0.1 percent of Earth's mammalian biomass consisted of humans and their domesticated animals. Today, this number has risen to 90 percent.

Invasive Species A nonnative species that is introduced into a new area is called an **exotic species.** Exotic species do not necessarily pose problems, and can even be beneficial in some circumstances—such as the arrival of coyotes in the eastern United States (from the west) to fill the niche vacated by wolves and other top predators, which were hunted out.

But some exotic species pose a direct threat to native species by eating them, competing with them, causing disease, or other types of interactions. They can also pose indirect threats by changing the local biotic or abiotic resources. You might recall that if an exotic species is introduced to a new area, grows to large population size, and disrupts species native to the area, it is called an **invasive species** (see Chapter 52).

Global trade and global travel have vastly increased the rate of exchange of plants, animals, fungi, and microorganisms—both purposeful and accidental—sometimes with dire consequences. For example, after brown tree snakes were introduced to the Pacific island of Guam 60 years ago (see Table 57.1), the snake population increased exponentially, causing the extinction of 12 native bird species, decreasing populations of small mammals and lizards, and indirectly affecting plant communities. Islands are particularly vulnerable to the effects of invasive species.

SUMMARY TABLE 57.1 Major Threats to Biodiversity

Threat	Example
Habitat destruction and degradation The conversion of primary forest to agricultural fields is a particularly devastating cause of habitat destruction.	
Overexploitation Overexploitation is the dominant threat for marine species, especially for large predators in top trophic levels, like these endangered bluefin tuna.	
Invasive species Invasive species, like this brown tree snake in Guam, are introduced to a new area, multiply rapidly, and threaten native species.	
Pollution Chemical pollutants have reached every corner of the globe, but are particularly threatening to aquatic species.	
Climate change Climate change poses different types of threats to different species. For example, some corals become bleached (lose their symbiotic photosynthetic algae) when water temperature warms.	

Pollution   Humans change both the biotic environment and the abiotic environment by releasing chemicals into ecosystems. Many types of major pollutants pose both direct and indirect threats to species:

- Industrial pollutants include sulfur (causing acid rain), arsenic, mercury, and lead as well as greenhouse gases such as CO_2 and nitric oxide (NO_2).

- Pharmaceutical pollutants (drugs) like oral contraceptives and ibuprofen are excreted in human urine and pass through wastewater-processing facilities into streams and rivers, where they are still biologically active.

- Pesticides and herbicides are a type of persistent organic pollutant (POP) that have a direct effect where they are used, but also affect species far away—sometimes in extremely remote locations—by dispersing in waterways and undergoing biomagnification in food webs (Chapter 56).

- Nutrients like nitrogen and phosphorus run off of farms and fields, causing eutrophication (Chapter 56).

- Garbage fills landfills, waterways, and ocean gyres, posing both physical and chemical threats.

Many pollutants are persistent. They last in organisms and the environment for a long time and may continue to cause harm long after their initial release—and at great distances from their initial release. Freshwater organisms are particularly vulnerable to chemical pollutants.

Climate Change   The acceleration of global climate change due to the burning of fossil fuels, deforestation, and other processes has already had an impact on species, including extinctions (see Chapter 56). Biologists have begun to record how several factors related to climate change influence biodiversity:

- As the global temperature increases, polar and alpine habitats are disappearing, endangering species native to arctic and high-elevation tundra.

- Coastal habitats are threatened by sea-level rise caused by the melting of polar ice caps and glaciers. The most recent analyses project that sea levels will rise an additional 0.8 to 2.0 m by the year 2100, inundating many coastal habitats, including areas with biodiversity hotspots.

- Biomes are becoming redistributed due to changes in both temperature and rainfall, causing mismatches between species distributions and abiotic conditions.

- The species compositions of communities are changing because trees and other slowly dispersing species are unable to keep up with rapid changes in climate.

- Coral reefs are suffering from "bleaching" induced in some species by high temperatures; this change is causing a cascade of effects in those ecosystems.

- Ocean acidification—due to the rise of atmospheric carbon dioxide concentrations and the consequent increase in carbonic acid formation in the oceans—is interfering with many

aspects of organismal physiology, including the ability of corals, crustaceans, and other animals to make calcium carbonate skeletons (see Table 56.1).

Although climate change is a natural phenomenon, the climate is currently changing at an extreme rate.

How Will These Threats Affect Future Extinction Rates?

What impact will the different threats have on biodiversity in the future? The formulation of biodiversity projections is a rapidly growing area of research, due to major advances in computer modeling, increased availability of paleontological and current data to input into the models, and increased demand for projections as the world is faced with important decisions about conservation.

The formulation of biodiversity scenarios is a highly integrative field of ecology because it must use information from multiple levels, including

- the effects of evolutionary processes on genetic variation, adaptation, and extinction in populations (Chapter 26);

- the impact of abiotic factors such as temperature and precipitation on the distribution and abundance of organisms (Chapter 52);

- the behavior of organisms in response to changing conditions (Chapter 53);

- the basic demographic characteristics of populations, such as fecundity and survivorship (Chapter 54);

- the consequences of species interactions at the level of communities (Chapter 55) and ecosystems (Chapter 56); and

- the effect of biodiversity threats at all these levels.

You can see how all these levels fit together in the Big Picture of Ecology on pages 1196–1197. Different models integrate different subsets of this list. Let's consider three examples of models ranging from single populations up to global species distributions.

Estimating the Probability that an Individual Species Will Go Extinct   A population viability analysis, or PVA, is a model that estimates the likelihood that a population will avoid extinction for a given time period. In most cases, a PVA attempts to combine data on

- age-specific survivorship and fecundity (Chapter 54),

- geographic structure (Chapter 52), and

- rate and severity of habitat disturbance (Chapter 55).

Typically, a population is considered viable if the analysis predicts that it has a 95 percent probability of surviving for at least 100 years. This type of analysis is particularly useful for creating management plans for species like the northern spotted owl and island foxes, which are listed as endangered in the IUCN Red List, the U.S. Endangered Species Act, or other metrics.

Estimating the Effect of Habitat Area on Species Richness   One strategy for predicting the future of Earth's biodiversity focuses on

The goal of species–area plots is to quantify the relationship between species richness and habitat area. **FIGURE 57.10** gives an example of a species–area curve—a graph that plots habitat area on the x-axis and the number of species present on the y-axis. In this case, the habitat areas are islands in the Bismarck Archipelago near New Guinea, in the South Pacific, and the species documented are birds. Each data point represents the number of species found on a different island. Note that the relationship between habitat area and the number of species present is linear when both axes are logarithmic. (For help interpreting logarithms, see **BioSkills 6** in Appendix B.)

The "log–linear" relationship turns out to be typical for other habitats and taxonomic groups as well. When biologists have plotted species–area relationships for plants, butterflies, mammals, or birds from islands or continental habitats around the globe, the relationship is consistently described by a function of the form $S = cA^z$, where:

- S is the number of species.
- A represents the habitat area.
- The c term is a constant that scales the data. Its value is high in species-rich areas, such as coral reefs or tropical wet forests, and low in species-poor areas, such as arctic tundra.
- The exponent z represents the slope of the line on a log–log plot. Thus, z describes how rapidly species numbers change with area. Typically, z is about 0.25.

For example, the solid line drawn through the points in Figure 57.10 is described by the function $S = (18.9) A^{0.15}$. This equation can be used to predict the species richness of other areas of a given size, such as an island that has not yet been sampled. If an unsampled island is known to have an area of 8 km², the species richness would be predicted as $S = 18.9 \times 8^{0.15} = 26$.

✔ **QUANTITATIVE** If you understand species–area curves, you should be able to predict the species richness of birds of an island in the Bismarck Archipelago that is 5000 km².

Answer: 68

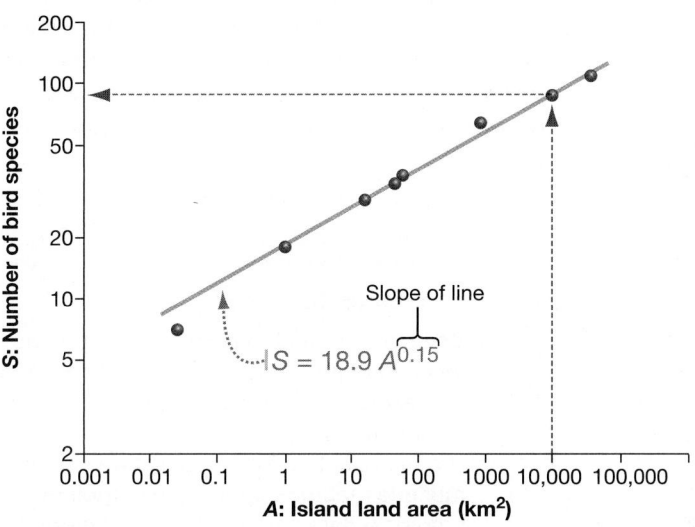

FIGURE 57.10 Species–Area Plots Quantify the Relationship between Species Richness and Habitat Area. The dashed arrows on this graph show the number of bird species that are expected to live on an island with an area of 10,000 km².

DATA: Diamond, J. M. 1975. *Biological Conservation* 7: 129–146.

---

the consequences of the most urgent problem: habitat loss. Given reasonable projections of how much habitat will be lost over a given time period, biologists can estimate rates of extinction based on **species–area relationships.** The concept here is to use species–area curves (see **Quantitative Methods 57.1**) to extrapolate backward to determine the species richness of a given habitat if its area were to decline due to habitat destruction or climate change.

To understand how biologists use the species–area curve in Quantitative Methods 57.1 to project extinction rates, ask yourself the following question: If sea-level rise and deforestation wiped out 90 percent of the habitats in part of the Bismarck Archipelago—for example, if A were reduced from 10,000 km² to 1000 km²—what percentage of species would disappear? According to the graph, the number of bird species present would drop from about 75 to about 53. Thus, the answer is that roughly 30 percent of species would vanish. The slope of the line in species–area curves is species dependent.

The caveat with this method is that using the curve to extrapolate backward can cause overestimation of extinction rates due

to assumptions inherent to the model. Thus, the technique must be used with caution and cross-referenced with other projection methods.

Estimating the Effect of Global Climate Change on Species Distributions How will climate change affect the global distribution of species? Can reliable predictions be made? Bioclimate envelope models, also called niche-based models, use the relationship between current species distributions and abiotic variables such as temperature and precipitation to project the future distributions of species under climate change.

For example, behavioral ecologist Barry Sinervo and an international team of colleagues took a two-step approach. First, they collected extensive data from one part of the globe, measuring extinction rates of 48 species of lizards at 200 sites in Mexico since 1975. The results were shocking: Twelve percent of local populations had already gone extinct because the temperature is changing too rapidly for the lizard populations to adapt (Chapter 56).

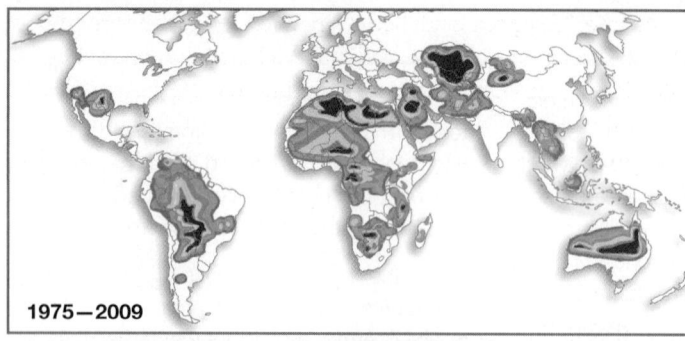

1975–2009

1975–2050

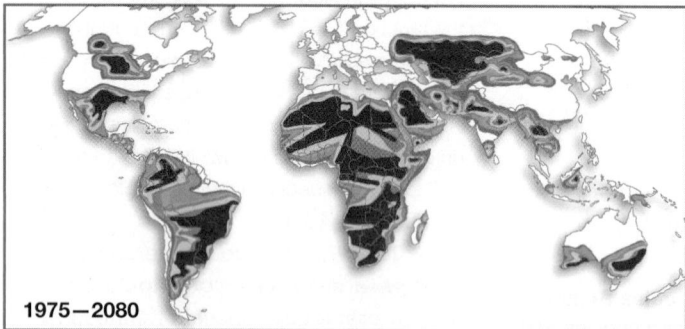

1975–2080

Extinction risk: Low ▢▬▬▬▬▬ High

FIGURE 57.11 Predictions of Global Extinction Rates of Lizards.
The projections are based on current extinction rates in Mexico and projections of global climate change, and were validated using observed extinction rates in regions of four other continents.
DATA: Sinervo, B., F. Mendez-De-La-Cruz, D. B. Miles, et al. 2010. *Science* 328: 894–899.

Sinervo and his team then used their data to predict worldwide responses to global warming given projected climate changes and the relationship between lizard extinctions and climate (**FIGURE 57.11**). They estimated that 4 percent of local lizard populations have already gone extinct worldwide and that lizard species extinctions could reach 20 percent by 2080. These alarming projections were validated with local data, suggesting that climate change is not a problem of the future—it is already a matter of life and death for lizard populations today.

Take-Home Messages about Making Biodiversity Predictions It is important to note that at all levels of study, a great deal of variation exists in projected future biodiversity. And this variation has different sources:

- Projections must make assumptions about future levels of threats to biodiversity, such as the extent of global warming, but these levels are unknown and depend on how rapidly

humans are able to reduce emissions of greenhouse gases and improve land-use practices.

- Variation in some projections is due to knowledge gaps in basic biology. More data collection is greatly needed in these cases for accurate projections.

- Some of the variation is due to differences in how the models are constructed. Scientists address this issue by using multiple models to cross-check projections, and by validating their models with empirical data from the real world.

One of the greatest challenges of making projections is the ability to measure how different factors interact. For example, the precipitous worldwide decline in amphibians is due to a combination of factors ranging from habitat loss, pollution, and climate change to infection by a chytrid fungus. These factors affect each other in complex feedback loops, requiring a systems biology approach that quantifies the network of interactions.

check your understanding

If you understand that . . .

- Species face multiple, interacting threats including habitat destruction and degradation, overexploitation, invasive species, pollution, and climate change.
- Projected extinction rates can be estimated from multiple methods by using a combination of empirical data and mathematical models.

✔ **You should be able to . . .**

1. Explain why fragmentation reduces habitat quality and leads to genetic problems.

2. Describe how you would determine whether each of the threats in Table 57.1 was affecting the orangutans of Borneo.

Answers are available in Appendix A.

57.3 Why Is Biodiversity Important?

One of the questions biologists have to address about biodiversity is one that you may have heard from friends or relatives: Who cares? People are concerned about their grades or job or health or car or love life. Why should they worry about something as abstract as biodiversity?

The answer has two parts, one biological and the other economic and cultural. Let's consider each part in turn.

Biological Benefits of Biodiversity

Do biological communities and ecosystems function better when species diversity is high? The effort to answer this question began in the mid-1980s and has blossomed.

Biodiversity Increases Productivity **FIGURE 57.12** summarizes a classic experiment on how species richness affects one of the most basic aspects of an ecosystem function: the production of biomass. Net primary productivity (NPP) is important because it is the source of chemical energy used by species throughout a food web (see Chapter 56).

As the "Experimental Setup" portion of the figure shows, David Tilman and colleagues divided 32 grassland plant species into five functional categories. The functional groups differ by the timing of their growing season and by whether they allocate most of their resources to manufacturing woody stems or seeds. The researchers planted 13-m by 13-m plots with a mixture of

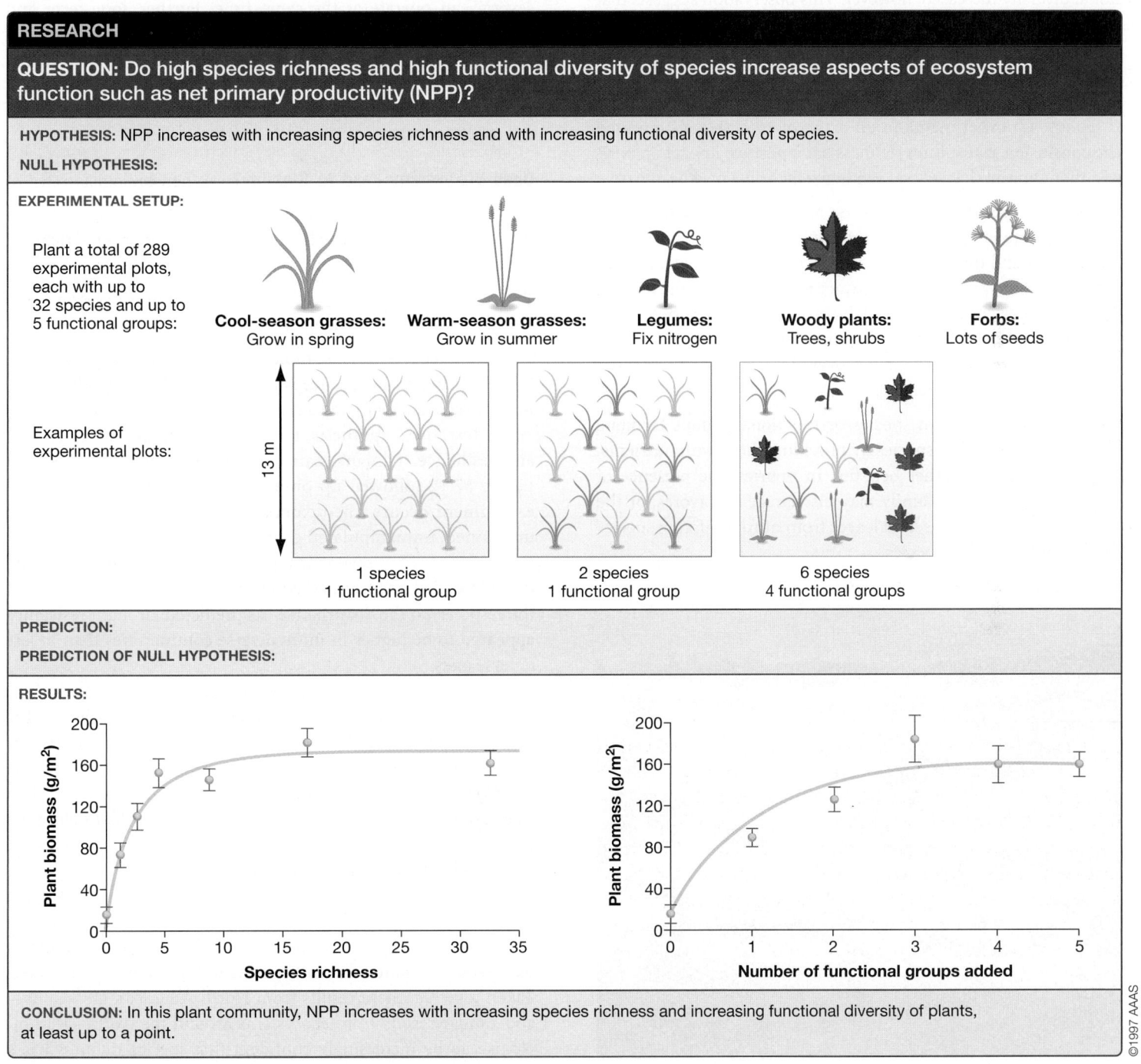

RESEARCH

QUESTION: Do high species richness and high functional diversity of species increase aspects of ecosystem function such as net primary productivity (NPP)?

HYPOTHESIS: NPP increases with increasing species richness and with increasing functional diversity of species.

NULL HYPOTHESIS:

EXPERIMENTAL SETUP:

Plant a total of 289 experimental plots, each with up to 32 species and up to 5 functional groups:

Cool-season grasses: Grow in spring

Warm-season grasses: Grow in summer

Legumes: Fix nitrogen

Woody plants: Trees, shrubs

Forbs: Lots of seeds

Examples of experimental plots:

13 m

1 species
1 functional group

2 species
1 functional group

6 species
4 functional groups

PREDICTION:

PREDICTION OF NULL HYPOTHESIS:

RESULTS:

Plant biomass (g/m²) vs Species richness

Plant biomass (g/m²) vs Number of functional groups added

CONCLUSION: In this plant community, NPP increases with increasing species richness and increasing functional diversity of plants, at least up to a point.

©1997 AAAS

FIGURE 57.12 Evidence that the Productivity of Ecosystems Depends on the Number and Types of Species Present.

SOURCE: Tilman, D., J. Knops, D. Wedin, et al. 1997. The influence of functional diversity and composition on ecosystem processes. *Science* 277: 1300–1302.

✔**EXERCISE** Write in the null hypothesis and both sets of predictions.

between 0 and 32 randomly chosen species, representing from 0 to 5 functional categories, and measured the amount of above-ground biomass produced in each plot (see **FIGURE 57.13**).

The results, graphed in the "Results" section of Figure 57.12, indicate that both the number and type of species present had important effects on biomass production. Experimental plots with more species and with a wider diversity of functional groups were more productive.

Note that total biomass leveled off as species richness and functional diversity increased, however. This observation suggests that increasing species diversity improves ecosystem function only up to a point. This pattern suggests that at least some species in ecosystems are redundant in terms of their ability to contribute to productivity—a characteristic that is important to resilience.

Follow-up experiments in an array of ecosystems supported the conclusion that species richness has a positive impact on NPP, and they showed that several causal mechanisms may be at work.

- *Resource use efficiency* Diverse assemblages of plant species make more efficient use of the sunlight, water, and nutrients available and thus lead to greater overall productivity. For example, some prairie plants extract water near the soil surface, while others use water available a meter or more below the surface, a result of niche differentiation (Chapter 55). When species diversity is high, more overall water is used and more photosynthesis can occur.

- *Facilitation* Certain species or functional groups facilitate the growth of other species by providing them with nutrients, partial shade, or other benefits. In prairies, the presence of plants in the onion family may discourage herbivores, or the decaying roots, stems, and leaves from nitrogen-fixing species may fertilize other species.

FIGURE 57.13 Tilman's classic biodiversity study is among many conducted at the Cedar Creek Natural History Area in Minnesota.

- *Sampling effects* In many habitats, it is common to observe that one or two species are extremely productive. If the number of species in a study plot is low, it is likely that the "big producer" will be missing and NPP will be low. But if the number of species in a study plot is high, then likely the big producer is present and NPP will be high. Simply due to sampling, then, high-species plots will tend to outproduce low-species plots.

The three mechanisms listed are not mutually exclusive—several can operate at the same time. Further, long-term and large-scale experiments have shown that the reason for the pattern can change over time, that the slope of the curve can continue to rise over time, and that the strength of the diversity–productivity relationship depends on the evolutionary history of the species involved.

Does Biodiversity Lead to Stability? When biologists refer to the stability of a community, they mean its ability to

1. recover to former levels of productivity or species richness after a disturbance;

2. maintain productivity and other aspects of ecosystem function as conditions change over time.

Resistance is a measure of how much a community is affected by a disturbance. **Resilience** is a measure of how quickly a community recovers following a disturbance.

To test the hypothesis that diversity increases resistance and resilience, Tilman's team that did the work highlighted in Figure 57.12 followed up on a natural experiment. A **natural experiment** occurs when comparison groups are created by an unplanned, unmanipulated change in conditions.

In 1987–1988, a severe drought hit their study site. When the drought ended, the team asked whether species richness affected the response to the disturbance. As predicted, drought resistance appeared to be higher in more diverse communities than in less diverse ones.

In addition, the group analyzed the change in biomass in each plot four years after the drought versus the biomass before the drought. The data indicated that most plots containing five or fewer species showed a significant lowering of biomass after the disturbance, whereas all of the plots that contained more than five species regained the same biomass that they had before the disturbance. This is strong evidence that species-rich areas are more resilient following a disturbance.

More recent experiments on California grasslands suggest that species-rich ecosystems also may be less prone to invasion by exotic species. The hypothesis here is that species-rich communities use resources more efficiently—leaving less room for invasives. Taken together, these results from North American grasslands—and similar results from ecosystems around the world—lead biologists to be increasingly confident that species richness has a strongly positive effect on how ecosystems function.

Communities that are more diverse appear to be more productive, more resistant to disturbance and invasion, and more resilient than communities that are less diverse. Increased species richness increases the services provided by ecosystems. By implication, biologists can infer that if ecosystems are simplified by extinctions, then productivity and other attributes might decrease.

Ecosystem Services: Economic and Social Benefits of Biodiversity and Ecosystems

The biological benefits of biodiversity—increased productivity, resistance, and resilience—benefit humans too. But this is just the beginning. The prosperity, health, and happiness of humans are completely dependent upon, and intertwined with, the preservation of biodiversity and functioning ecosystems.

Some species provide obvious direct benefits—bluefin tuna are prized as a food source, and tigers are a cultural icon. Other species offer direct benefits that are more difficult to observe—such as the control of crop-eating insects by bats, a service recently valued at $4 billion/year.

Further, many species indirectly benefit humans simply by serving ecological roles in functioning ecosystems, which in turn provide invaluable services such as soil formation and climate moderation. Considered collectively, all the direct and indirect benefits that humans derive from organisms and the ecosystems they compose are called **ecosystem services.**

In 1997 a team of ecologists and economists led by Robert Costanza estimated the value of ecosystem services at $33 trillion—almost twice the gross national product of all nations combined. Let's take a closer look at some of the material and nonmaterial services that nature provides, summarized in **TABLE 57.2**.

Provisioning Services   Wild species have provided the raw material to fuel the development of human societies ever since the first large-scale, highly organized cultures began to develop about 10,000 years ago. People felled vast tracts of forest for building materials and fuel; selectively bred wild plants to yield food and fibers; domesticated animals to provide food, labor, and material goods; harvested the oceans for protein; and processed plants, animals, and fungi as sources of medicines. Today as much as ever, humans rely on other organisms for most of their food and material needs.

The value of provisioning services is relatively easy to quantify because food, fuels, wood, and other materials are traded in the marketplace. For example, mushrooms have an annual food value over $1 billion, the fishing industry has a value of $80–100 billion, and raw materials provided by tropical forests are valued at $300 billion.

Bioprospecting—the exploration of bacteria, archaea, protists, plants, fungi, and animals as novel sources of drugs or ingredients in consumer products—has benefited from the recent explosion of genetic and phylogenetic information. Biologists can now search genomes from a wide array of species to find alleles with desired functions. Recent advances are impressive: researchers have developed a potent painkiller from the paralyzing sting of tropical cone snails and a blood anticoagulant from the saliva of vampire bats.

SUMMARY TABLE 57.2   Ecosystem Services

Type of Service	Example
Provisioning services provide raw materials • Food • Fuel • Fiber and other materials • Medicines • Genetic resources	
Regulating services are part of Earth's life-support system • Climate moderation • Soil formation • Erosion control • O_2 and CO_2 regulation • Water capture • Water purification • Air cleaning • Flood control • Storm mitigation • Waste decomposition • Bioremediation	
Cultural services enrich quality of life • Aesthetics • Recreation • Education • Spiritual value • Human mental and physical health	
Supporting services enable all the other ecosystem services • Primary productivity • Nutrient cycling • Pollination • Biological control	

Regulating Services Services provided by functioning ecosystems—such as water purification, flood control, and waste decomposition—are part of Earth's life-support system. Yet, these regulating services are often taken for granted until human actions disrupt their function.

This was the case in New York City in the 1990s. New York City is known for its excellent drinking water, piped in from the watersheds of the Catskill Mountains. But the water quality declined as those watersheds became developed—and something had to be done. A cost–benefit analysis revealed that it was far cheaper to restore the ecosystem function of the watersheds ($1 billion) than to filter the water artificially ($6–8 billion up front plus $300 million annually). The "free" water purification services that had been provided by the forested Catskill ecosystem suddenly became quite valuable.

Cultural Services People integrate nature in their lives in diverse cultural ways, ranging from artistic celebrations to recreation, education, and spiritual contemplation. Eminent biologist and author Edward O. Wilson coined the term biophilia to refer to the innate, unconscious bond between humans and other organisms as a result of our coevolutionary history. Wilson's hypothesis is that this bond inspires our deep need for, and appreciation of, nature.

An increasing number of studies are showing that humans who are removed from natural environments—such as people who spend most of their time in cities—often suffer both physical and psychological setbacks. Conversely, people who spend more time outdoors in nature, especially as children, tend to have: reduced stress; reduced rates of obesity, diabetes, depression, and attention deficit and hyperactivity disorder; improved problem-solving skills; enhanced creativity and curiosity; improved sleep patterns; and improved vision, among other traits.

On average, people who spend more time surrounded by nature tend to be healthier and happier. (How much time do you spend outdoors?) Government agencies and businesses are taking note because human health and happiness also have economic value.

Supporting Services Some ecosystem services support all the other categories of services. Primary productivity is a key example because the amount of biomass generated affects the provisioning, regulating, and cultural services that ecosystems can provide.

In a similar way, pollination and pest control are critical to multiple ecosystem services. The value of pollination became clear in a region in China where the excessive use of pesticides led to a particularly sharp decline of bees. Human workers must now pollinate apple, pear, and peach orchards by hand, at great expense (**FIGURE 57.14**). In the United States alone, insect-pollinated crops like almonds, apples, and chocolate produce $40 billion worth of products annually.

The air you breathe, the water you drink, and the soils that grow the food you eat are services that ecosystems provide. The quality of your life is tied to the quality of those ecosystems.

FIGURE 57.14 Hand Pollination of Fruit Trees. Following a decline of bees due to pesticide use, workers in Maoxian County, China, hand pollinate entire orchards.

An Ethical Dimension?

The data reviewed in this section suggest that there are sound economic and biological reasons to preserve species and ecosystem function. People live better, and ecosystems work better, when biodiversity is high.

But in addition, an array of biologists, philosophers, and religious leaders argue that humans have an ethical obligation to preserve species and ecosystems. There are at least three arguments for this position:

1. Organisms have intrinsic worth. Humans diminish the world by extinguishing species and destroying ecosystems.

check your understanding

If you understand that . . .

- Experimental plots with higher species richness tend to have higher productivity and show increased stability in response to disturbance or changed abiotic conditions.
- Biodiversity and functioning ecosystems provide many provisioning, regulating, cultural, and supporting services to humans.

✔ **You should be able to . . .**

Design an experiment to test the hypothesis that the rain forest of Borneo is more effective at building soil, retaining soil nutrients, and minimizing soil erosion than the oil palm plantations are.

Answers are available in Appendix A.

2. Industrialized nations are responsible for most of the environmental harm, yet poor countries are disproportionately affected by the consequences of these actions. It is unethical for rich countries to inflict hardship on poor countries.

3. It is unethical for the current generation to deprive future generations of ecosystem services.

Thus, extinction and disruption of ecosystem function are moral issues as well as biological and economic problems. One of the most important reasons to preserve biodiversity and ecosystem function may simply be that it is the right thing to do.

57.4 Preserving Biodiversity and Ecosystem Function

Extinction is irreversible—and biodiversity loss is not slowing down. But there is reason for hope.

Solving any global problem requires a common goal to be defined. In the case of preserving biodiversity and ecosystem function, the objective is to sustain diverse communities in natural landscapes while supporting the extraction of resources required to maintain the health and well-being of the human population. Given the extreme rate of deterioration of present conditions on Earth, conservation biologists often feel like doctors in a busy emergency room: Triage is necessary to maximize positive outcomes. This effort generates sober acceptance of some inevitable losses but vigorous fighting for important gains.

A number of international organizations and agreements are striving to reduce the threats to biodiversity. Central among them is the Convention on Biological Diversity (CBD), which convened in 1992 to establish conservation targets for 2010—many of which were not met. The CBD convened again in 2010 to establish new goals for 2020. These goals serve as an important organizational framework for the international effort. What's working? And what can you do to help?

Addressing the Ultimate Causes of Loss

To begin, most biologists recognize that it will be impossible to preserve high species diversity and high-functioning ecosystems if two things happen:

1. the human population grows to 10 billion or more—from its current 7 billion—over the next half-century, as projected by the "high-fertility" scenario (see Chapter 54); and

2. the rate of consumption of fossil fuels and other resources does not decline, especially in industrialized nations and growing developing nations (see Chapter 56), because consumption threatens biodiversity not only directly but also indirectly via climate change.

While it is complex and controversial to pinpoint a human carrying capacity for the planet (Chapter 54), it is clear that humans are nearing, or have surpassed, some of the recoverable limits of available resources. The way forward involves both reducing population growth and improving **sustainability**—the managed use of resources at a rate only as fast as the rate at which they are replaced.

You may be familiar with efforts to develop renewable sources of energy, promote agricultural and forestry practices that maintain productivity without the addition of synthetic fertilizers or groundwater, limit fishing catches so that populations do not crash irrecoverably, and recycle metals and other raw materials. In essence, the challenge is to create ways for humans to live off the resources that are being produced continuously on Earth rather than using species to extinction and exhausting resources like petroleum, groundwater, and topsoils that have been stored for centuries or millennia.

It may take decades for human population size and resource use to stabilize and for sustainable development practices to become widespread. In the meantime, thousands of species and untold ecosystem services may be lost. What specific approaches do conservationists use to protect natural resources?

Let's consider several examples, while asking: "What level of biodiversity is this targeting?" and "What threats to biodiversity does this approach address?"

Conservation of Genetic Diversity, Populations, and Species

Despite the discouraging persistence of biodiversity loss, conservation efforts have managed a number of successes in preserving genetic diversity, populations, and species in many locations around the globe. What are these successes?

Management Plans for Invasive Species The costs of eradicating some invasive species are so enormous that they outweigh the benefits. However, global awareness of the problem has slowly led to measures that limit the exchange of species, such as requirements that ships exchange ballast water out at sea (an inhospitable environment for coastal species) rather than in port, that cargoes are inspected at border crossings, and that nurseries stop selling invasive plants. Prevention is more effective than treatment, so education is the key.

Seed Banks Agricultural scientists are preserving diverse strains of crop plants in **seed banks**—long-term storage facilities. In addition, they are using wild relatives of domesticated species in breeding programs aimed at improving crop traits. The goal of both efforts is to avoid the loss of valuable genetic diversity.

Ex Situ Conservation and Reintroduction Some species—including Père David's deer, Przewalski's horse, and the Arabian oryx—are extinct in the wild and exist only in captivity. These animals are examples of **ex situ conservation**—the preservation of species in zoos, aquaria, wildlife ranches, botanical gardens, or other artificial settings.

Translated literally, ex situ means out of place. The name was inspired by a contrast with in situ ("in place") conservation—the establishment of protected areas where populations can be maintained in their natural habitat. Ex situ conservation is not ideal, but it is preferred over extinction.

In some cases, captive breeding of ex situ individuals and reintroduction into the wild have succeeded in creating wild populations of endangered species. Recall, for example, the successful effort to establish new flocks of whooping cranes by using eggs and young produced in zoos and other breeding centers (see Chapter 54). The peregrine falcon, California condor, and several other species have been rescued from the brink of extinction by captive breeding and reintroduction to the wild.

Genetic Restoration When habitat fragmentation results in small, isolated populations that are vulnerable to the loss of genetic diversity from chance events, genetic restoration can sometimes be accomplished by artificial gene flow.

Consider the case of a remnant population of the endangered Florida panther in the swampland of southern Florida. By the early 1990s, only about 25 adults remained, and they exhibited several recessive disorders and low genetic diversity—both signs of inbreeding. Using a population viability analysis (PVA; see Section 57.2), conservationists gave the population a 95 percent likelihood of extinction in two decades.

In an effort to prevent this outcome, the U.S. Department of the Interior moved eight panthers (locally called "pumas") from Texas to Florida to introduce new alleles into the Florida population. Geneticist Stephen O'Brien and colleagues followed the genetic consequences, some of which are shown in **FIGURE 57.15** (see inbreeding depression data in Chapter 26). The success of the genetic restoration was evident. The number of panthers quadrupled in 10 years, and the offspring of hybridized animals were larger, stronger, and had reduced incidents of defects.

Note, however, that as time went on, the rate of increase in population size leveled off. Also, the panther habitat continues to deteriorate. Ongoing conservation efforts are needed to successfully protect this panther population from extinction.

Wildlife Corridors One longer-term solution to enabling gene flow among isolated populations is the establishment of **wildlife corridors,** strips of undeveloped habitat that connect preserved areas. In some cases, wildlife corridors are as simple as a walkway under a major highway, so that animals such as salamanders can move from one side to the other without being killed. In other cases, a corridor can consist of 1000 km of makeshift paths across private and public lands, such as the corridors used by isolated populations of tigers in southern India.

By facilitating the movement of individuals, the goals of corridors are to

- allow areas to be recolonized if a species or a population was lost;

- encourage gene flow that introduces new alleles—counteracting the deleterious effects of genetic drift and inbreeding.

Do wildlife corridors actually work? To answer this question experimentally, ecologist Ellen Damschen and colleagues established a series of restored natural areas in the middle of a

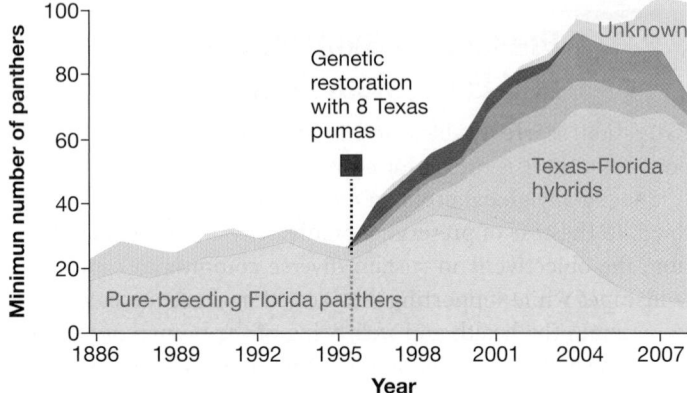

FIGURE 57.15 Experimental Evidence that Gene Flow Can Alleviate Genetic Problems in Small Populations. After eight Texas pumas were introduced to Florida in 1995, the Florida population of panthers quadrupled in size.

DATA: Johnson, W. E., D. P. Onorato, M. E. Roelke, et al. 2010. *Science* 329: 1641–1645.

large, species-poor pine plantation in South Carolina. This project was the equivalent of restoring patches of grassland habitat in the middle of an enormous cornfield. Some of the restoration sites were connected by corridors, while others were isolated. By monitoring the composition of species inside each habitat patch over time, the group was able to show that when compared to unconnected patches, connected patches are steadily gaining more species over time (**FIGURE 57.16**). This is strong evidence that wildlife corridors can increase overall species richness in a metapopulation.

Designing Effective Protected Areas By far the most effective strategy for preserving biodiversity is to protect entire areas of land and sea from exploitation in the first place.

The bad news is that more than two-thirds of the terrestrial biodiversity hotspots lack protection entirely, or have incomplete protection, and less than 0.5 percent of the oceans are protected. Incomplete protection occurs when areas are designated as protected but have insufficient enforcement—such as from poachers.

The good news is that the total amount of protected area is increasing at a steady rate globally. By strategically protecting prime habitat areas, countries worldwide can realize many benefits. For

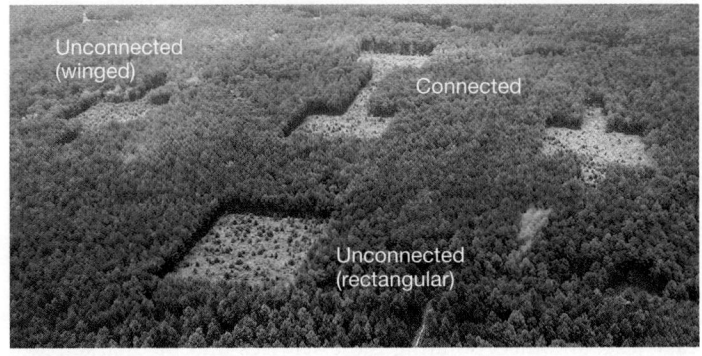

Unconnected (winged)

Connected

Unconnected (rectangular)

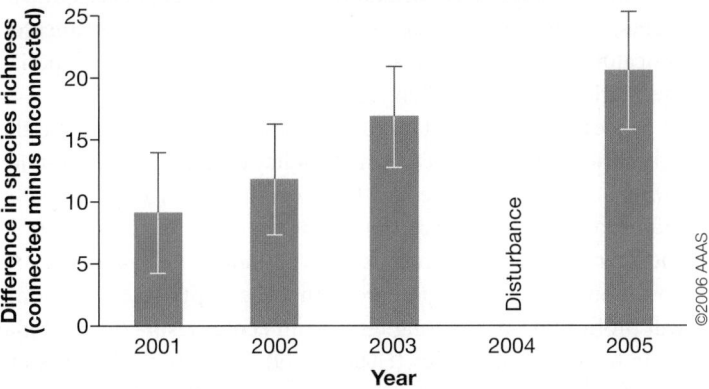

©2006 AAAS

FIGURE 57.16 Experimental Evidence that Wildlife Corridors Work. This graph plots the difference in species richness between connected and unconnected plots. Species richness could not be assessed in 2004 because the study plots were burned as part of a program to restore natural habitats.

DATA: Damschen, E. I., N. M. Haddad, J. L. Orrock, et al. 2006. *Science* 313: 1284–1286.

✔ **QUANTITATIVE** Draw and label bars showing what this graph would look like if corridors had no effect on species richness.

example, preservation of rain forest in Borneo solves many problems: reducing incidence of droughts, floods, and fires, protecting diverse species, and storing carbon.

Researchers are using a geographic approach called the Gap Analysis Program (GAP) to assess the effectiveness of the current system of protected areas. A GAP tries to identify gaps between geographic areas that are particularly rich in biodiversity and areas that are actually managed for the preservation of biodiversity. Mismatches help to identify areas that are high priorities for conservation efforts.

Conservation of Ecosystem Function

As the biological and economic value of the associated services becomes more apparent, many countries are paying more attention to the conservation and restoration of ecosystem function than ever before. Can degraded ecosystems be restored?

Ecosystem Restoration In many areas of the world, ecosystems are already heavily degraded or lost. The state of Illinois, for example, was estimated to contain 22 million acres of prairie when European settlers arrived; today only about 2000 acres of this

biome remain (0.009 percent). When natural areas are badly degraded or gone, biologists turn to restoration.

Thousands of large-scale and small-scale ecosystem restoration and reforestation projects are now occurring around the globe. Recall the dire scenario in Borneo described at the beginning of this chapter. Willie Smits had been working in Borneo for two decades in a desperate attempt to slow the rate of deforestation and save orangutans from the brink of extinction. Frustrated by his lack of progress, he decided to try a new approach—the restoration of 2000 ha of clear-cut rain forest in Samboja.

While Smits would be the first to agree that primary forest is of greater ecological value than restored forest, his progress in just a few years since 2002 is remarkable:

- Over 1600 tree species in a multilayered canopy
- 137 species of birds (up from 5); 9 species of primates
- rainfall up 25 percent; temperature down 3–5°C
- 3000 people earning income

The cornerstone of Smits's Samboja Lestari ("Everlasting Forest") project is integrating the ecological restoration of the forest along with the economic restoration of the local Dayak tribe—a win–win scenario (**FIGURE 57.17**). Although the size of Samboja Lestari is relatively small and the clear-cutting of rain forest continues in Borneo at a rapid pace, Smits's success is reason for hope. How successful are ecosystem restoration projects in general?

Ecologists Holly Jones and Oswald Schmitz analyzed 240 of such restoration projects to determine the resilience of ecosystems after human disturbances. **FIGURE 57.18** (on page 1192) shows some of their surprising results. Recovery of biomass following disturbances was much slower in forests than in other ecosystems, but unexpectedly fast overall—decades rather than centuries. Similarly, biomass recovery was much slower in agricultural and deforested areas, but more rapid overall than anticipated.

FIGURE 57.17 Ecosystem Restoration Effort in Samboja Lestari, Borneo. Rain forest restoration in Borneo is critical to the survival of orangutans. Indonesian workers prepare rescued orangutans for release into forest sanctuaries like Samboja.

(a) Recovery time in different types of restored ecosystems

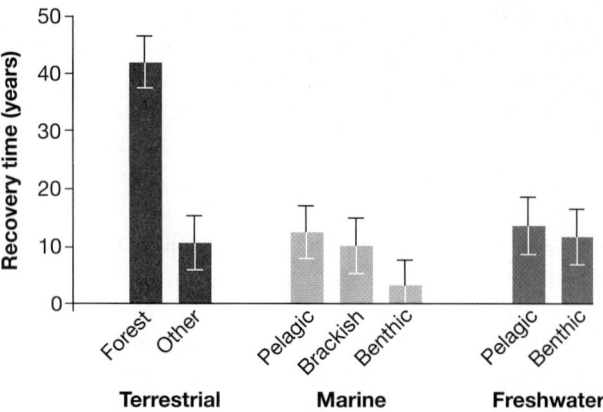

(b) Recovery time following different types of disturbances

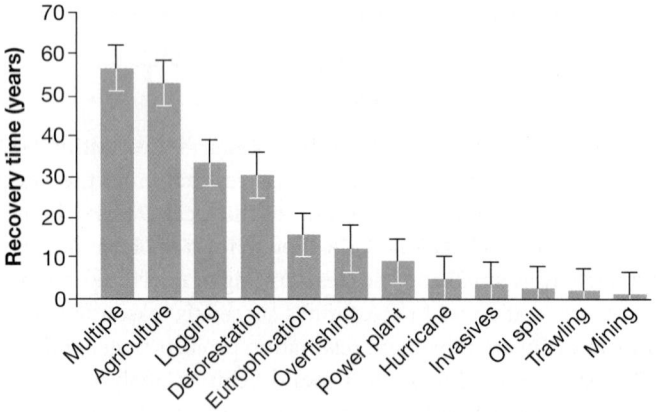

FIGURE 57.18 Average Recovery Times of Ecosystem Biomass.
(a) Recovery of biomass following disturbances was much slower in forests than in other ecosystems, but surprisingly fast overall.
(b) Recovery was much slower in agricultural and deforested areas, but relatively rapid in other areas.

DATA: Jones, H. P., and O. J. Schmitz. 2009. *PLoS ONE* 4: e5653.

Biomass recovery is not equivalent to biodiversity or ecosystem function, but many of the restoration projects show positive outcomes in terms of species richness and other variables. It is not too late to restore many degraded habitats.

In general, restoration success is positively correlated with the size of the area restored, its connectedness, remoteness, and the intactness of surrounding areas. However, another take-home message from the restoration studies is that an accumulation of many small changes can also have a big effect. If you and all your neighbors planted several native trees, shrubs, or other plants in your yards or neighborhoods, you could have a very positive effect on biodiversity and ecosystem function right where you live.

Quantifying Ecosystem Services   Many ecologists are convinced that the traditional approaches to conservation are doomed to fail on their own—nature reserves are too small and isolated to succeed in preserving sufficient biodiversity and ecosystem function. Many of these scientists advocate for an ecosystem services approach to conservation—a quantification of ecosystem services to make explicit in the marketplace that long-term human prosperity depends on the preservation of these services.

The idea is that the cost of replacing ecosystem services is at least 10 to 100 times higher than the cost of maintaining them—if they are replaceable at all (in which case they are infinitely valuable). Therefore, it is an economic imperative to protect these services before they are lost.

The problem is that quantifying ecosystem services is not an easy task, and incorporating these services into the marketplace requires a complicated collaboration among scientists, governments, nongovernmental organizations (NGOs), businesses, educators, communities, and local leadership. Further, policy mechanisms can be challenging to implement.

Successful experiments are under way, however. For example, the Costa Rican government has made policy changes that allow landowners to be compensated for the ecosystem services provided by their properties, such as carbon storage, biodiversity conservation, and scenic beauty. The payments are funded largely from transportation taxes. The logic here is that the Costa Rican economy will profit more in the long term from **ecotourism**—recreational visits to their wild places (**FIGURE 57.19**)—if their wild places are preserved from logging by private owners in the short term. But the private owners must be compensated.

Numerous other similar programs are in development. A global program called REDD (Reducing Emissions from Deforestation

FIGURE 57.19 Ecotourism in Costa Rica. Ecotourists flock to witness the spectacular biodiversity of the Monteverde Cloud Forest Reserve in Costa Rica.

and forest Degradation) is designed to reduce emissions from deforestation by paying landowners *not* to clear-cut—thereby slowing climate change. Other than curbing the production of greenhouse gases in the first place, the preservation of forests is currently the single most effective method of slowing climate change—which is otherwise expected to have enormous economic impacts.

In general, although social and political changes are difficult to implement, outcomes that boost biodiversity and ecosystem function are a boon to humans.

Take-Home Message

Your generation is facing the most serious global environmental crisis in the history of our species. The decisions you make, ranging from how many children you have to how much you drive a car, will have far-reaching consequences. Change happens one person at a time. It can be done.

As someone with a background in biology, you have the intellectual tools to make an important, positive impact. In addition to studying life, biologists have an obligation to help save it.

CHAPTER 57 REVIEW

If you understand . . .

57.1 What Is Biodiversity?

- Biodiversity is quantified at the level of genetic diversity, species diversity, and ecosystem diversity.

- Only about 1.5 million species have been named out of an estimated 10 to 100 million species on Earth.

- On average, biodiversity tends to be higher in the tropics than near the poles, higher on land than in the sea, and higher in areas with geographic diversity.

- Tropical rain forests are particularly species rich, containing at least 50 percent of all species.

- Biodiversity hotspots can be mapped based on species richness, areas of endemism, degree of threat, and/or unique ecological or evolutionary characteristics.

✔ You should be able to explain how you could measure biodiversity in the restored rain forest of Borneo at the genetic, species, and ecosystem levels.

57.2 Threats to Biodiversity

- The sixth mass extinction in the history of life is occurring due to human impacts. Species are vanishing faster than at virtually any other time in Earth's history.

- Habitat destruction and degradation are currently the leading causes of extinction. Conversion of primary forest to agricultural fields and housing developments is particularly devastating.

- Although overexploitation also occurs on land, it is the dominant threat for marine species.

- Invasive species are a long-standing problem on islands, but now they also threaten communities on continents.

- Many types of pollutants are found virtually everywhere on Earth, but they are particularly threatening to aquatic species.

- Climate change is a significant threat to biodiversity due to both direct and indirect impacts.

- To estimate how many species will go extinct in the near future, biologists make calculations based on combinations of variables including current extinction rates, habitat sizes, predicted shifts in climate envelopes, demographics, and species interactions.

✔ You should be able to explain why small, geographically isolated populations of orangutans in isolated islands of forest are at high risk of extinction.

57.3 Why Is Biodiversity Important?

- At the ecosystem level, experiments have shown that high species richness increases aspects of ecosystem function such as productivity, resistance to disturbance, and ability to recover from disturbance.

- Humans gain direct economic benefits from fishing, forestry, agriculture, and other activities that depend on materials provided by ecosystems.

- Humans also benefit from the ability of functioning ecosystems to build and hold soil, moderate local climates, retain and cycle nutrients, retain surface water and recharge groundwater, prevent flooding, and produce oxygen.

- Human happiness and health depend on healthy ecosystems.

✔ You should be able to explain why the high species richness of rain forests leads to high productivity, in terms of the efficiency of resource use, when compared to oil palm plantations.

57.4 Preserving Biodiversity and Ecosystem Function

- The major threats to biodiversity—habitat destruction and climate change—will not lessen until the human population stabilizes and resource use becomes sustainable.

- Solutions to the biodiversity crisis include preserving genetic diversity, protecting key habitats, restoring ecosystems, and integrating conservation with public policy so that protecting ecosystem services is a greater part of both global and local decision-making processes.

- Ecosystem restoration efforts are having a positive impact on biodiversity around the world. Conservation efforts make a difference.

✔ You should be able to evaluate whether the 12 percent of land area protected thus far will be enough to avert a mass extinction.

 MasteringBiology

1. MasteringBiology Assignments

Tutorials and Activities Global Fisheries and Overfishing; Forestation Change; Science, Technology, and Society: DDT; Madagascar and the Biodiversity Crisis; Habitat Fragmentation; Conservation Biology Review; Biodiversity

Questions Reading Quizzes, Blue-Thread Questions, Test Bank

2. eText Read your book online, search, take notes, highlight text, and more.

3. The Study Area Practice Test, Cumulative Test, BioFlix® 3-D Animations, Videos, Activities, Audio Glossary, Word Study Tools, Art

You should be able to . . .

✔ TEST YOUR KNOWLEDGE

1. What does a species–area plot show?
   a. the overall distribution, or area, occupied by a species
   b. the relationship between the body size of a species and the amount of territory or home range it requires
   c. the number of species found, on average, in tropical versus northern areas
   d. the number of species found, on average, in a habitat of a given size

2. What does species richness refer to?
   a. the number of species in an area
   b. the quality of species in an area
   c. the functional importance of a species
   d. the abundance of individuals within a species

3. What is resilience?
   a. resistance to change during a disturbance
   b. the ability to recover from a disturbance
   c. production of aboveground biomass
   d. total biomass production (above- and belowground)

4. What is a biodiversity "hotspot"?
   a. an area where an all-taxon survey is under way
   b. an area where an environmental sequencing study has been completed
   c. a habitat with high NPP
   d. an area with high species richness and high threat

5. Why do small populations become inbred?
   a. They are often part of a metapopulation structure.
   b. Genetic drift becomes a prominent evolutionary force.
   c. Over time, all individuals become increasingly related.
   d. Natural selection does not operate efficiently in small populations.

6. What is the primary cause of endangerment in marine environments?

✔ TEST YOUR UNDERSTANDING

7. If you were asked to evaluate building materials, what would be most "sustainable" about the use of bamboo as a building material?
   a. Bamboo is an invasive species due to its rapid growth rate.
   b. Bamboo is harvested at the same rate at which it grows.
   c. Bamboo is durable so it does not need to be replaced often.
   d. Bamboo plantations create habitat for other species.

8. Biologists claim that the all-taxa survey now under way at the Great Smoky Mountains National Park in the United States will improve their ability to estimate the total number of species living today. Discuss the benefits and limitations that this data set will provide in understanding the extent of global biodiversity.

9. Some biologists prefer to focus efforts on preserving endangered species while others prefer to focus on preserving ecosystems. What is your opinion?

10. What evidence supports the hypothesis that species richness increases ecosystem functions, such as productivity, resistance to disturbance, and resilience?

11. Explain why the construction of wildlife corridors can help maintain biodiversity in a fragmented landscape.

12. What are ecosystem services? Does anyone own them or pay for them—that is, can you make money from them? How does quantifying ecosystem services affect efforts to protect ecosystems?

13. The population size of mountain pine beetles in the American Northwest has long been held in check by freezing temperatures during the winters. As winters warm, populations of pine beetles are increasing, killing whitebark pine forest. Grizzly bears rely on the nuts of whitebark pine trees. Do you think climate change is affecting the fitness of the grizzly bear population?

a. No, because climate change is not occurring yet.

b. No, because climate change is affecting the beetles, not the bears.

c. Yes, because climate change is decreasing the food supply of the bears, which means they will be less strong.

d. Yes, because climate change is decreasing the food supply of the bears, which is likely to reduce their ability to survive or reproduce.

14. You are helping to design a series of reserves in a tropical country. List the steps you would recommend for gathering data and creating a plan that would protect a large amount of biodiversity in a small amount of land.

15. The maps to the right chronicle the loss of old-growth forest (more than 200 years old) that occurred in Warwickshire, England, and in the United States. In your opinion, under what conditions is it ethical for conservationists who live in these countries to lobby government officials in Brazil, Indonesia, and other tropical countries to slow the rate of loss of old-growth forest?

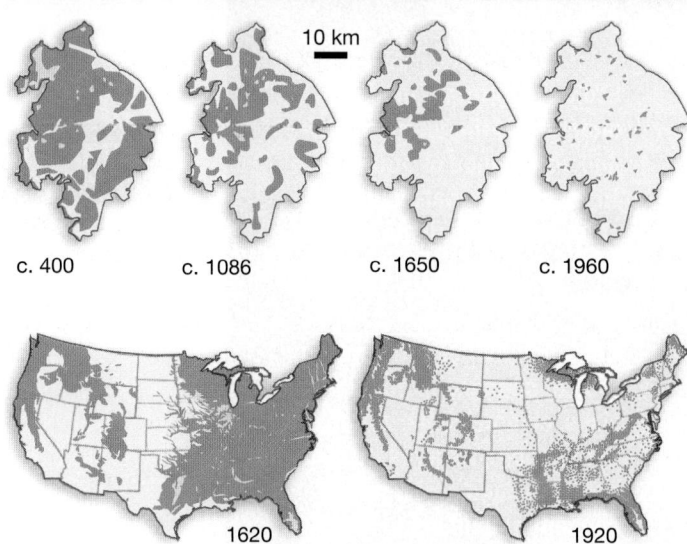

10 km

c. 400   c. 1086   c. 1650   c. 1960

1620   1920

16. Make a list of characteristics that would render a species particularly vulnerable to extinction by humans. Make a list of characteristics that would render a species particularly resistant to pressure from humans. Try to think of an example of each type of species.

The Big Picture

Ecology's primary goal is to understand the distribution and abundance of organisms. To reach this goal, ecologists study how organisms interact with each other and their abiotic environments.

Human activities are causing radical changes in population sizes, climate, soils, water, and atmosphere. As a result, virtually every box and arrow in this concept map is being affected.

In addition to documenting changes, biologists are educating the public and policy makers about the value of biodiversity and healthy ecosystems, and researching the most effective ways to preserve species and keep ecosystems functioning normally.

Note that most boxes in the concept map indicate chapters and sections where you can go for review. Also, be sure to do the blue exercises in the Check Your Understanding box below.

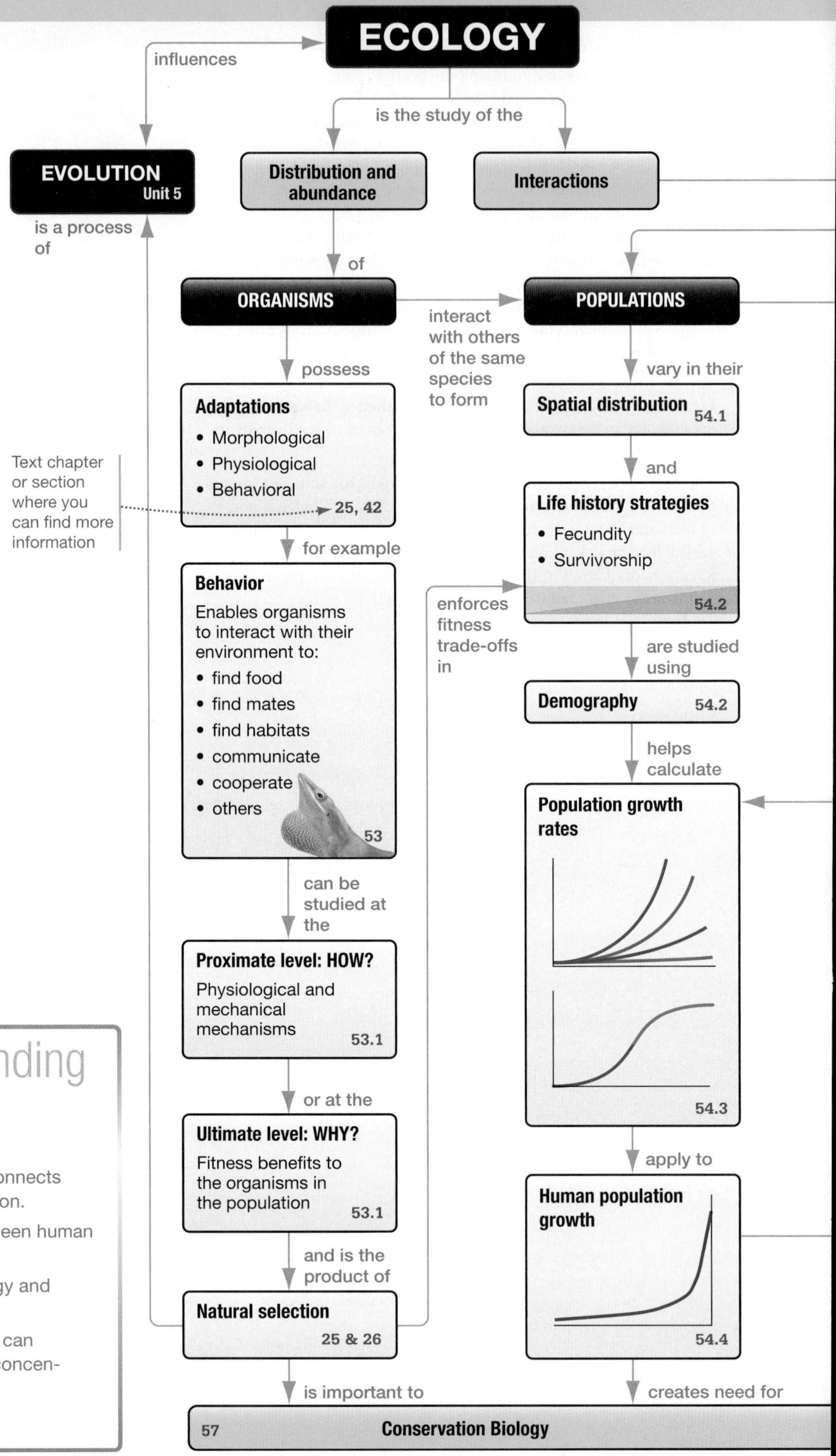

check your understanding

C Y U

If you understand the big picture . . .
✔ You should be able to . . .

1. Add an arrow and linking phrase that connects species interactions with natural selection.

2. Explain why there is a relationship between human population growth and biodiversity.

3. Compare and contrast the flow of energy and nutrients through ecosystems.

4. Give examples of how human activities can increase or decrease the atmospheric concentration of CO_2.

Answers are available in Appendix A.

ECOLOGY

influences → EVOLUTION (Unit 5)

is the study of the

- Distribution and abundance
- Interactions

EVOLUTION — **is a process of** →

Distribution and abundance — **of** →

ORGANISMS — **interact with others of the same species to form** → **POPULATIONS**

ORGANISMS — **possess** →

Adaptations
- Morphological
- Physiological
- Behavioral   → 25, 42

Text chapter or section where you can find more information

for example →

Behavior
Enables organisms to interact with their environment to:
- find food
- find mates
- find habitats
- communicate
- cooperate
- others   53

can be studied at the →

Proximate level: HOW?
Physiological and mechanical mechanisms   53.1

or at the →

Ultimate level: WHY?
Fitness benefits to the organisms in the population   53.1

and is the product of →

Natural selection   25 & 26

is important to →

POPULATIONS — **vary in their** →

Spatial distribution   54.1

and →

Life history strategies
- Fecundity
- Survivorship   54.2

enforces fitness trade-offs in

are studied using →

Demography   54.2

helps calculate →

Population growth rates   54.3

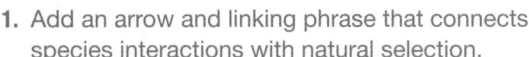

apply to →

Human population growth   54.4

creates need for →

57   **Conservation Biology**

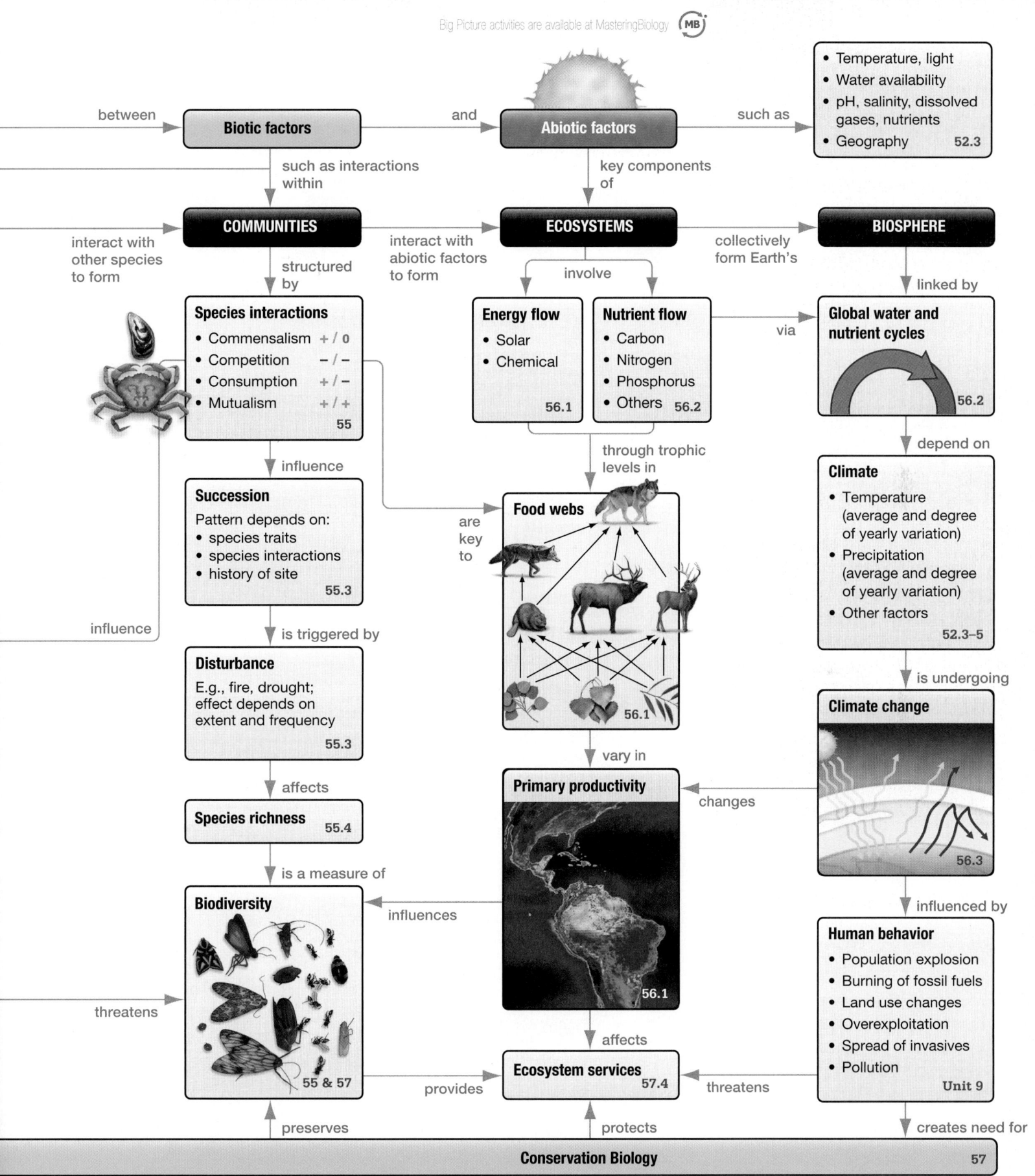

Biotic factors ── *between* ←

Biotic factors ── *and* → **Abiotic factors** ── *such as* →
- Temperature, light
- Water availability
- pH, salinity, dissolved gases, nutrients
- Geography   52.3

Biotic factors ── *such as interactions within* →

Abiotic factors ── *key components of* →

interact with other species to form →

COMMUNITIES ── *interact with abiotic factors to form* → **ECOSYSTEMS** ── *collectively form Earth's* → **BIOSPHERE**

ECOSYSTEMS ── *involve* →

BIOSPHERE ── *linked by* →

COMMUNITIES ── *structured by* →

Species interactions
- Commensalism + / 0
- Competition   − / −
- Consumption   + / −
- Mutualism   + / +
                    55

Energy flow
- Solar
- Chemical
           56.1

Nutrient flow
- Carbon
- Nitrogen
- Phosphorus
- Others   56.2

via →

Global water and nutrient cycles
                        56.2

Nutrient flow ── *through trophic levels in* →

Global water and nutrient cycles ── *depend on* →

Species interactions ── *influence* →

Succession

Pattern depends on:
- species traits
- species interactions
- history of site
                  55.3

are key to →

Food webs
                56.1

Climate
- Temperature (average and degree of yearly variation)
- Precipitation (average and degree of yearly variation)
- Other factors
                52.3–5

Succession ── *is triggered by* →

Disturbance

E.g., fire, drought; effect depends on extent and frequency
                  55.3

Food webs ── *vary in* →

Climate ── *is undergoing* →

Climate change
                56.3

Disturbance ── *affects* →

Species richness   55.4

Primary productivity
                      56.1

changes ← **Climate change**

Species richness ── *is a measure of* →

Primary productivity ── *influences* → **Biodiversity**

Climate change ── *influenced by* →

Biodiversity
                55 & 57

threatens → **Biodiversity**

Primary productivity ── *affects* →

Human behavior
- Population explosion
- Burning of fossil fuels
- Land use changes
- Overexploitation
- Spread of invasives
- Pollution
                  Unit 9

Biodiversity ── *provides* → **Ecosystem services**   57.4 ← *threatens* ── **Human behavior**

Biodiversity ── *preserves* ↑   **Ecosystem services** ── *protects* ↑   **Human behavior** ── *creates need for* ↓

Conservation Biology   57

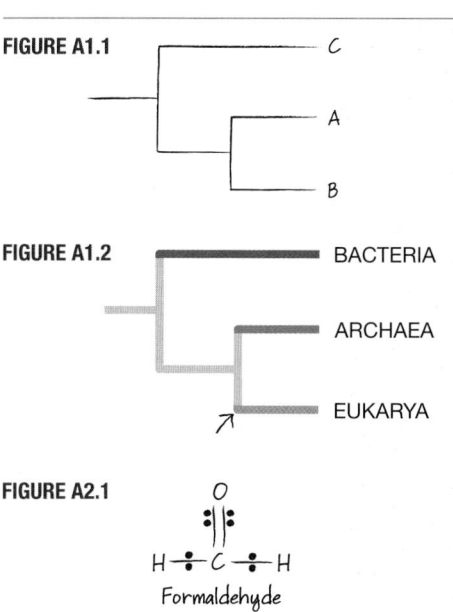

APPENDIX A Answers

CHAPTER 1

IN-TEXT QUESTIONS AND EXERCISES

p. 4 Fig. 1.2 `analyze` If Pasteur had done any of the things listed, he would have had more than one variable in his experiment. This would allow critics to claim that he got different results because of the differences in broth types, heating, or flask types—not the difference in exposure to preexisting cells. The results would not be definitive.
p. 6 `apply` The average kernel protein content would decline, from 11 percent to a much lower value over time.
p. 6 CYU `apply` The data points would all be about 11 percent, indicating no change in average kernel protein content over time.
p. 7 Fig. 1.4 `apply` Molds and other fungi are more closely related to green algae because they differ from plants at two positions (5 and 8 from left) but differ from green algae at only one position (8).
p. 8 Fig. 1.6 `apply` The eukaryotic cell is roughly 10 times the size of the prokaryotic cell.
p. 9 CYU `evaluate` From the sequence data provided, species A and B differ only in one ribonucleotide of the rRNA sequence (position 10 from left). Species C differs from species A and B in four ribonucleotides (positions 1, 2, 9, and 10). A correctly drawn phylogenetic tree would indicate that species A and B appear to be closely related, and species C is more distantly related. See **FIGURE A1.1**.
p. 13 (1) `analyze` You could conclude that the ants weren't navigating normally, because they had been caught and released and transferred to a new channel. **(2)** `analyze` You could conclude that the ants can't navigate normally on their manipulated legs.
p. 13 CYU The key here is to test predation rates during the hottest part of the day (when desert ants actually feed) versus other parts of the day. The experiment would best be done in the field, where natural predators are present. One approach would be to capture a large number of ants, divide the group in two, and measure predation rates (number of ants killed per hour) when they are placed in normal habitat during the hottest part of the day versus an hour before (or after). **(1)** `analyze` The control group here is the normal condition—ants out during the hottest part of the day. If you didn't include a control, a critic could argue that predation did or did not occur because of your experimental setup or manipulation, not because of differences in temperature. **(2)** `analyze` You would need to make sure that there is no difference in body size, walking speed, how they were captured and maintained, or other traits that might make the ants in the two groups more or less susceptible to predators. They should also be put out in the same habitat, so the presence of predators is the same in the two treatments.
p. 13 Fig. 1.10 `analyze` The interpretation of the experiment would not likely change, but your confidence in the conclusions drawn would be reduced if you used just one ant.

IF YOU UNDERSTAND . . .

1.1. `understand` Dead cells cannot regulate the passage of materials between exterior and interior spaces, replicate, use energy, or process information. **1.2** `understand` Observations on thousands of diverse species supported the claim that all organisms consist of cells. The hypothesis that all cells come from preexisting cells was supported when Pasteur showed that new cells do not arise and grow in a boiled liquid unless they are introduced from the air. **1.3** `understand` If seeds with higher protein content leave the most offspring, then individuals with low protein in their seeds will become rare over time. **1.4** `understand` A newly discovered species can be classified as a member of the Bacteria if the sequence of its rRNA contains some features found only in Bacteria. The same logic applies to classifying a new species in the Archaea or Eukarya. **1.5** `understand` (1) A hypothesis is an explanation of how the world works; a prediction is an outcome you should observe if the hypothesis is correct. (2) Experiments are convincing because they measure predictions from two opposing hypotheses. Both predicted actions cannot occur, so one hypothesis will be supported while the other will not.

YOU SHOULD BE ABLE TO . . .

✔ Test Your Knowledge

1. `remember` d **2.** `understand` d **3.** `remember` populations **4.** `understand` b **5.** `understand` An individual's ability to survive and reproduce **6.** `understand` c

✔ Test Your Understanding

7. `evaluate` That the entity they discovered replicates, processes information, acquires and uses energy, is cellular, and that its populations evolve. **8.** `understand` a **9.** `understand` Over time, traits that increased the fitness of individuals in this habitat became increasingly frequent in the population. **10.** `understand` Individuals with certain traits are selected, in the sense that they produce the most offspring. **11.** `analyze` Yes. If evolution is defined as "change in the characteristics of a population over time," then those organisms that are most closely related should have experienced less change over time. On a phylogenetic tree, species with substantially similar rRNA sequences would be diagrammed with a closer common ancestor—one that had the sequences they inherited—than the ancestors shared between species with dissimilar rRNA sequences. **12.** `understand` A null hypothesis specifies what a researcher should observe when the hypothesis being tested isn't correct.

✔ Test Your Problem-Solving Skills

13. `analyze` A scientific theory is not a guess—it is an idea whose validity can be tested with data. Both the cell theory and the theory of evolution have been validated by large bodies of observational and experimental data. **14.** `apply` If all eukaryotes living today have a nucleus, then it is logical to conclude that the nucleus arose in a common ancestor of all eukaryotes, indicated by the arrow you should have added to the figure. See **FIGURE A1.2**. If it had arisen in a common ancestor of Bacteria or Archaea, then species in those groups would have had to lose the trait—an unlikely event. **15.** `evaluate` The data set was so large and diverse that it was no longer reasonable to argue that noncellular life-forms would be discovered. **16.** `apply` b

B**IG** PICTURE Doing Biology

p. 16 CYU (1) `understand` Biologists design and carry out a study, either observational or experimental, to test their ideas. As part of this process, they state their ideas as a hypothesis and null hypothesis and make predictions. They analyze and interpret the data they have gathered, and determine whether the data support their ideas. If not, they revisit their ideas and come up with an alternative hypothesis and design another study to test these new predictions. **(2)** `understand` There are many possible examples. Consider, for example, the experiment on navigation in foraging desert ants (Chapter 1). In addition to testing how the ants use information on stride length and number to calculate how far they are from the nest (multicellular organism and population levels), researchers also could test how the "pedometer" works at the level of cells and molecules. **(3)** `analyze` A hypothesis is a testable statement to explain a specific phenomenon or a set of observations. The word theory refers to proposed explanations for very broad patterns in nature that are supported by a wide body of evidence. A theory serves as a framework for the development of new hypotheses. **(4)** `analyze` The next step is to relate your findings to existing theories and the current scientific literature, and then to communicate your findings to colleagues through informal conversations, presentations at scientific meetings, and eventually publication in peer-reviewed journals.

CHAPTER 2

IN-TEXT QUESTIONS AND EXERCISES

p. 20 Fig. 2.3 `apply` There are 15 electrons in phosphorus, so there must be 15 protons, which is the atomic number. Since the mass number is 31, then the number of neutrons is 16.
p. 23 `understand` *Water:* arrows pointing from hydrogens to oxygen atom; *ammonia:* arrows pointing from hydrogens to nitrogen atom; *methane:* double arrows between carbons and hydrogens; *carbon dioxide:* arrows pointing from carbon to oxygens; *molecular nitrogen:* double arrows between nitrogens.
p. 23 Fig. 2.7 `understand` Oxygen and nitrogen have high electronegativities. They hold shared electrons more tightly than C, H, and many other atoms, resulting in polar bonds.
p. 24 CYU `evaluate` See **FIGURE A2.1**.

FIGURE A1.1

FIGURE A1.2 BACTERIA / ARCHAEA / EUKARYA

FIGURE A2.1 Formaldehyde

p. 25 (1) [evaluate] $\delta^+H{-}O^{\delta-}{-}H^{\delta+}$ **(2)** [apply] If water were linear, the partial negative charge on oxygen would have partial positive charges on either side. Compared to the actual, bent molecule, the partial negative charge would be much less exposed and less able to participate in hydrogen bonding.

p. 26 Fig. 2.14 [understand] Oils are nonpolar. They have long chains of carbon atoms bonded to hydrogen atoms, which share electrons evenly because their electronegativities are similar. When an oil and water are mixed, the polar water molecules interact with each other via hydrogen bonding rather than with the nonpolar oil molecules, which interact with themselves instead.

p. 28 Table 2.2 [understand] "Cause" (Row 1): electrostatic attractions between partial charges on water and opposite charges on ions; hydrogen bonds; water and other polar molecules. "Biological Consequences" (Row 2): ice to float; freezing solid. "Cause" (Row 4): lots of heat energy; break hydrogen bonds and change water to a gas.

p. 29 [apply] The proton concentration would be 3.2×10^{-9} M.

p. 29 Fig. 2.17 [apply] The concentration of protons would decrease because milk is more basic (pH 6.5) than black coffee (pH 5).

p. 30 [apply] The bicarbonate concentration would increase. The protons (H^+) released from carbonic acid would react with the hydroxide ions (OH^-) dissociated from NaOH to form H_2O, leaving fewer protons free to react with bicarbonate to reform carbonic acid.

p. 32 CYU (1) [apply] The reaction would be spontaneous based on the change in potential energy; the reactants have higher chemical energy than the products. The entropy, however, is not increased based on the number of molecules, although heat given off from the reaction still results in increased entropy. **(2)** [understand] The electrons are shifted farther from the nuclei of the carbon and hydrogen atoms and closer to the nuclei of the more electronegative oxygen atoms.

p. 32 Fig. 2.19 [remember] See **FIGURE A2.2**.

p. 33 Fig. 2.21 [analyze] The water-filled flask is the ocean; the gas-filled flask is the atmosphere; the condensed water droplets are rain; the electrical sparks are lightning.

p. 37 Table 2.3 [understand] All the functional groups in Table 2.3, except the sulfhydryl group (—SH), are highly polar. The sulfhydryl group is only very slightly polar.

IF YOU UNDERSTAND . . .

2.1. [understand] The bonds in methane and ammonia are all covalent, but differ in polarity: Methane has nonpolar covalent bonds while ammonia has polar covalent bonds. Sodium chloride does not have covalent bonds; instead, ionic bonds hold the ionized sodium and chloride together. **2.2** [understand] Assuming neutral pH, amino and hydroxyl groups would interact with the partial negative charge on water's oxygen, because they both carry a positive charge (partial for the hydrogen in the hydroxyl). The carboxyl group would interact with the partial positive charges on water's hydrogens, since it would carry a negative charge after losing the proton from its hydroxyl. **2.3** [understand] Like solar radiation, the energy in electricity generates free radicals that would promote the reaction. **2.4** [understand] "Top-down" approach: The reaction responsible for synthesizing acetic acid is observed in cells and can serve as an intermediate for the formation of a more complex molecule (acetyl CoA) that is used by cells throughout the tree of life. "Bottom-up" approach: This reaction can also occur under conditions that mimic the early Earth environment in deep-sea vents. **2.5** [apply] The hydroxyls would increase the solubility of octane by introducing polar covalent bonds, which would make the molecule more hydrophilic. The high electronegativity of oxygen would decrease the potential energy of the modified molecule.

YOU SHOULD BE ABLE TO . . .

✔ **Test Your Knowledge**

1. [understand] b **2.** [remember] a **3.** [remember] c **4.** [remember] d **5.** [remember] Potential energy and entropy. **6.** [remember] The prebiotic soup model and the surface metabolism model.

✔ **Test Your Understanding**

7. [apply] c. Acetic acid has more highly electronegative oxygen atoms than the other molecules. When bonded to carbon or hydrogen, each oxygen will result in a polar covalent bond. **8.** [apply] Relative electronegativities would be $F > O > H > Na$. One bond would form with sodium, and it would be ionic. **9.** [understand] When oxygen is covalently bonded to hydrogen, the difference in electronegativities between the atoms causes the electrons to spend more time near the oxygen. In contrast, the atoms in H_2 and O_2 have the same electronegativities, so they equally share electrons in their covalent bonds. **10.** [apply] See **FIGURE A2.3**. **11.** [apply] The dissociation reaction of carbonic acid lowers the pH of the solution by releasing extra H^+ into the solution. If additional CO_2 is added, the sequence of reactions would be driven to the right, which would make the ocean more acidic. **12.** [understand] The carbon framework determines the overall shape of an organic molecule. The functional groups attached to the carbons determine the molecule's chemical behavior, because these groups are likely to interact with other molecules.

✔ **Test Your Problem-Solving Skills**

13. [analyze] b. **14.** [analyze] No, they don't conflict. Shells that are farther from the protons (positive charges) in the nucleus house electrons that have greater potential energy than shells closer to the nucleus. **15.** [analyze] One possible concept map relating the structure of water to its properties is shown below (see **FIGURE A2.4**). **16.** [create] In hot weather, water absorbs large amounts of heat due to its high specific heat and high heat of vaporization. In cold weather, water releases the large amount of heat that it has absorbed.

FIGURE A2.2

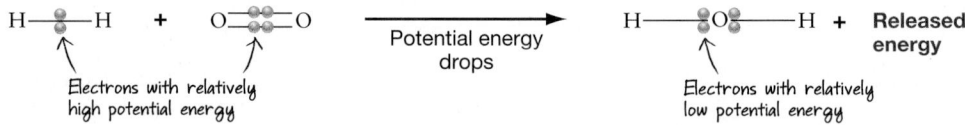

FIGURE A2.3

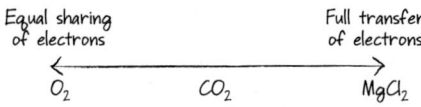

FIGURE A2.4

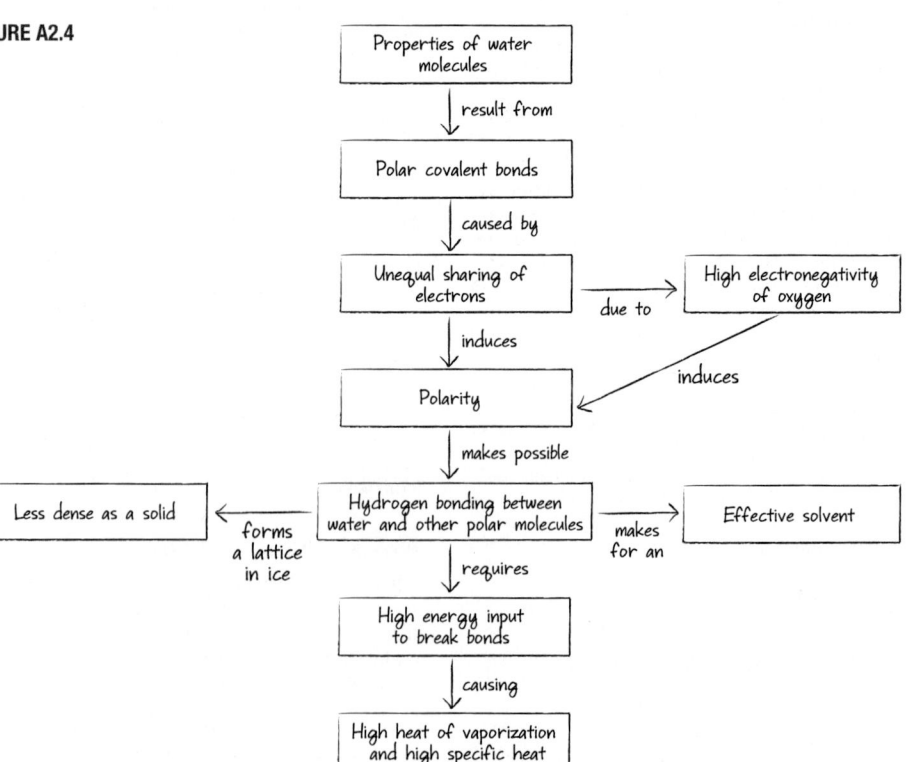

CHAPTER 3

IN-TEXT QUESTIONS AND EXERCISES

p. 43 Fig. 3.2 `understand` The green R-groups contain mostly C and H, which have roughly equal electronegativities. Electrons are evenly shared in C–H bonds and C–S bonds, so the groups are nonpolar. Cysteine has a sulfur that is slightly more electronegative than hydrogen, so it will be less nonpolar than the other green groups. All of the pink R-groups have a highly electronegative oxygen atom with a partial negative charge, making them polar. **p. 44** `apply` From most hydrophilic to most hydrophobic: (1) aspartate, (2) asparagine, (3) tyrosine, (4) valine. The most hydrophilic amino acids will have side chains with full charges (ionized), like aspartate, followed by those with the largest number of highly electronegative atoms, like oxygen or nitrogen. Highly electronegative atoms produce polar covalent bonds with carbon or hydrogen. The most hydrophobic will not have oxygen or nitrogen in their side chains, but instead will have the largest number of C–H bonds, which are nonpolar covalent. **p. 47 CYU** `understand` See **FIGURE A3.1**.
p. 52 CYU (1) `understand` Secondary, tertiary, and quaternary structure all depend on bonds and other interactions between amino acids that are linked in a chain in a specific order (primary structure). **(2)** `apply` There would be 20^5 different peptides, or 3.2×10^6 different primary sequences.
p. 55 CYU `apply` Amino acid changes would be expected to be in the active site or in regions that affect the folded structure of this site. Either of these changes could result in a different active site that either binds a new substrate or catalyzes a different reaction.

IF YOU UNDERSTAND . . .

3.1. `understand` Look at the R-group of the amino acid. If there is a positive charge, then it is basic. If there is a negative charge, then it is acidic. If there is not a charge, but there is an oxygen atom, then it is polar uncharged. If there is no charge or oxygen, then it is nonpolar.
3.2 `apply` Nonpolar amino acid residues would be found in the interior of a globular protein, grouped with other nonpolar residues due to hydrophobic interactions.
3.3 `analyze` Both calmodulin and infectious prions require some form of induction to achieve their active conformations. Calcium ions are required for calmodulin to fold into its functional structure while prions are induced to change their shape by other, improperly folded prion proteins. **3.4** `analyze` *Catalysis:* Proteins are made of amino acids, which have many reactive functional groups, and can fold into different shapes that allow the formation of active sites. *Defense:* Similar to catalysis, the chemical properties and capacity for different shapes allows proteins to be made that can attach to virtually any type of invading virus or cell. *Signaling:* The flexibility in protein structure allows protein activities to be quickly turned on or off based on binding to signal molecules or ions.

YOU SHOULD BE ABLE TO . . .

✔ Test Your Knowledge

1. `remember` d **2.** `remember` The atoms and functional groups found in the side chains. **3.** `remember` b **4.** `remember` b
5. `understand` The order and type of amino acids (i.e., the primary structure) contains the information that directs folding. **6.** `understand` a

✔ Test Your Understanding

7. `understand` Because the nonpolar amino acid residues are not able to interact with the water solvent, they are crowded together in the interior of a protein and surrounded by a network of hydrogen-bonded water molecules. This crowding leads to the development of van der Waals interactions that help glue the nonpolar side chains together. **8.** `understand` No, polymerization is a nonspontaneous reaction because the product molecules have lower entropy than the free form of the reactants and there would be nothing to prevent hydrolysis from reversing the reaction. **9.** `understand` Many possible correct answers, including (1) the presence of an active site in an enzyme that is precisely shaped to fit a substrate or substrates in the correct orientation for a reaction to occur; (2) the doughnut shape of porin that allows certain substances to pass through it; (3) the cable shape of collagen to provide structural support for cells and tissues. **10.** `create` Proteins are highly variable in overall shape and chemical properties due to variation in the composition of R-groups and the array of secondary through quaternary structures that are possible. This variation allows them to fulfill many different roles in the cell. Diversity in the shape and reactivity of active sites also makes them effective catalysts. **11.** `understand` c **12.** `create` In many proteins, especially those involved in cell signaling, their structure is affected by binding to other molecules or ions. Since the shape of the protein is directly involved in its function, the protein's activity is regulated by controlling how it is folded. If proteins were inflexible, this type of control could not occur.

✔ Test Your Problem-Solving Skills

13. `analyze` The side chain of proline is a cyclic structure that is covalently bonded to the nitrogen in the core amino group. This restricts the movement of the side chain relative to the core nitrogen, which further restricts the backbone when the nitrogen participates in a peptide bond with a neighboring amino acid. **14.** `create` See **FIGURE A3.2 15.** `apply` b. Phosphates have a negative charge, so they are most likely to form ionic bonds with the positively-charged side chains of basic amino acid residues. **16.** `apply` The inherited forms likely have some alteration in the primary structure such that the infectious form is spontaneously generated at a higher rate than normal. The amino acid sequence in these prions would likely differ from those transmitted between animals.

CHAPTER 4

IN-TEXT QUESTIONS AND EXERCISES

p. 59 `understand` See **FIGURE A4.1**.

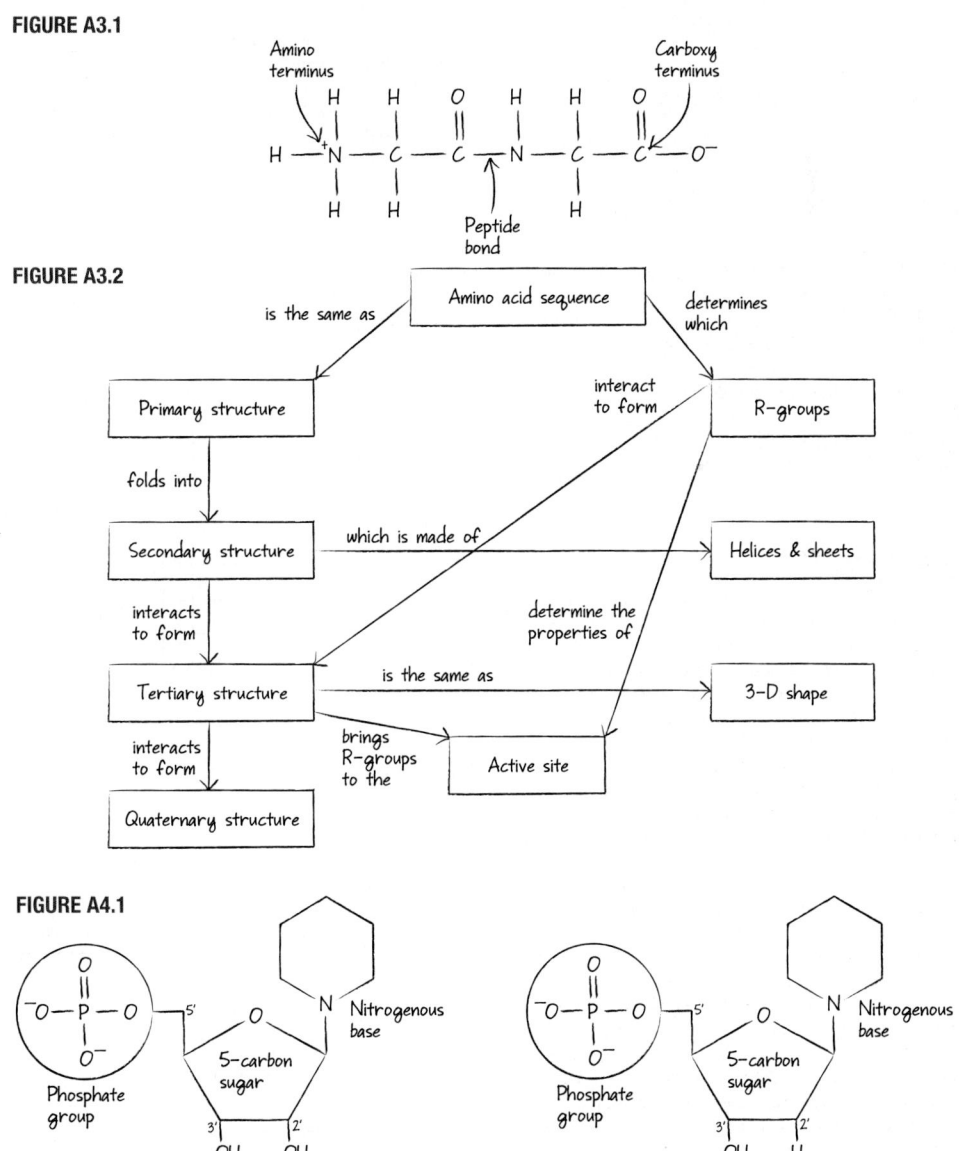

FIGURE A3.1

FIGURE A3.2

FIGURE A4.1

p. 60 Fig. 4.3 [understand] 5'-UAGC-'3
p. 61 CYU [understand] See **FIGURE A4.2**.
p. 63 [understand] If the two strands were parallel, the G-C pairing would align N—H groups together and C=O groups together, which would not allow hydrogen bonds to form.
p. 64 Fig. 4.8 [analyze] It is not spontaneous—energy must be added (as heat) for the reaction to occur.
p. 65 CYU [remember] See **FIGURE A4.3**.

IF YOU UNDERSTAND . . .

4.1. [understand] Cells activate nucleotides by linking additional phosphates to an existing 5' phosphate. Activation increases the chemical energy in the nucleotides enough to offset the decrease in entropy that will result from the polymerization reaction. **4.2** [understand] C-G pairs involve three hydrogen bonds, so they are more stable than A-T pairs with just two hydrogen bonds. **4.3** [analyze] A single-stranded RNA molecule has unpaired bases that can pair with other bases on the same RNA strand, thereby folding the molecule into stem-and-loop configurations. These secondary structures can further fold on themselves, giving the molecule a tertiary structure. Because DNA molecules are double stranded, with no unpaired bases, further internal folding is not possible. **4.4** [understand] Examples would include (1) the production of nucleotides, and (2) polymerization of RNA. It is thought that nucleotides were scarce during chemical evolution, so their catalyzed synthesis by a ribozyme would have been advantageous. Catalysis by an RNA replicase would have dramatically increased the reproductive rate of RNA molecules.

YOU SHOULD BE ABLE TO . . .

✔ Test Your Knowledge

1. [remember] c **2.** [remember] c **3.** [remember] a **4.** [remember] d **5.** [remember] One end has a free phosphate group on the 5' carbon; the other end has a free hydroxyl group bonded to the 3' carbon. **6.** [understand] DNA is a more stable molecule than RNA because it lacks a hydroxyl group on the 2' carbon and is therefore more resistant to cleavage, and because the two sugar-phosphate backbones are held together by many hydrogen bonds between nitrogenous bases.

✔ Test Your Understanding

7. [apply] In DNA, the secondary structure requires that every guanine pairs with a cytosine and every thymine pairs with an adenine, resulting in consistent ratios between the nucleotides. Chargaff's rules do not apply to RNA, since it is single-stranded and the pairing is not consistent throughout the molecules. **8.** [apply] a; if 30 percent is adenine, then 30 percent would be thymine, since they are base-paired together. This means that 40 percent consists of G-C base pairs, which would be equally divided between the two bases. **9.** [apply] The DNA sequence of the new strand would be 5'-ATCGATATC-3'. The RNA sequence would be the same, except each T would be replaced by a U. **10.** [understand] DNA has limited catalytic ability because it (1) lacks functional groups that can participate in catalysis and (2) has a regular structure that is not conducive to forming shapes required for catalysis. RNA molecules can catalyze some reactions because they (1) have exposed hydroxyl functional groups and (2) can fold into shapes that that can function in catalysis. **11.** [apply] No. Catalytic activity in ribozymes depends on the tertiary structure generated from single-stranded molecules. Double-stranded nucleic acids do not form tertiary structures. **12.** [understand] An RNA replicase would undergo replication and be able to evolve. It would process information in the sense of copying itself, and it would use energy to drive polymerization reactions. It would not be bound by a membrane and considered a cell, however, and it would not be able to acquire energy. It would best be considered as an intermediate step between nonlife and true life (as outlined in Chapter 1).

✔ Test Your Problem-Solving Skills

13. [create] See **FIGURE A4.4 14.** [apply] Yes—if the complementary bases lined up over the entire length of the two strands, they would twist into a double helix analogous to a DNA molecule. The same types of hydrogen bonds and hydrophobic interactions would occur as observed in the "stem" portion of hairpins in single-stranded RNA. **15.** [apply] In a triple helix, the bases are unlikely to align properly for hydrogen bonding to occur, so hydrophobic interactions would probably be more important. **16.** [apply] b; the high temperature would make it more likely that the secondary and tertiary structures would be denatured in the ribozymes. To overcome this effect, you would expect the hairpins to possess more G-C pairs, since they consist of three hydrogen bonds compared to the two found in A-T pairs.

CHAPTER 5

IN-TEXT QUESTIONS AND EXERCISES

p. 73 Fig. 5.2 [understand] See the structure of mannose in **FIGURE A5.1**.
p. 75 CYU [apply] See **FIGURE A5.2**.
p. 78 CYU [remember] They could differ in (1) location of linkages (e.g., 1,4 or 1,6); (2) types of linkages (α or β); (3) the sequence of the monomers (e.g., two galactose and then two glucose, versus alternating galactose and glucose); and/or (4) whether the four monomers are linked in a line or whether they branch.
p. 79 Fig. 5.6 [apply] The percentage of inhibition would not change for the intact glycoprotein bar. The purified

---

FIGURE A4.2

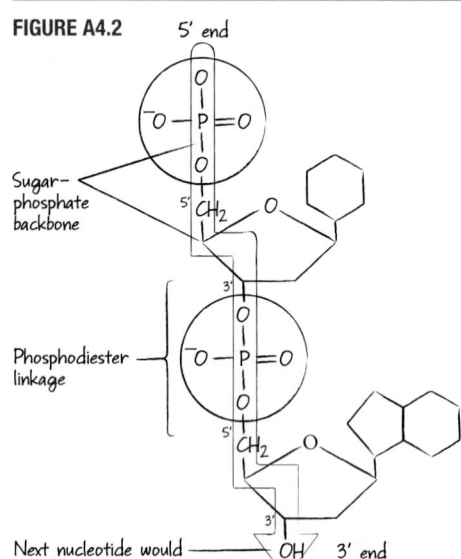

FIGURE A4.3

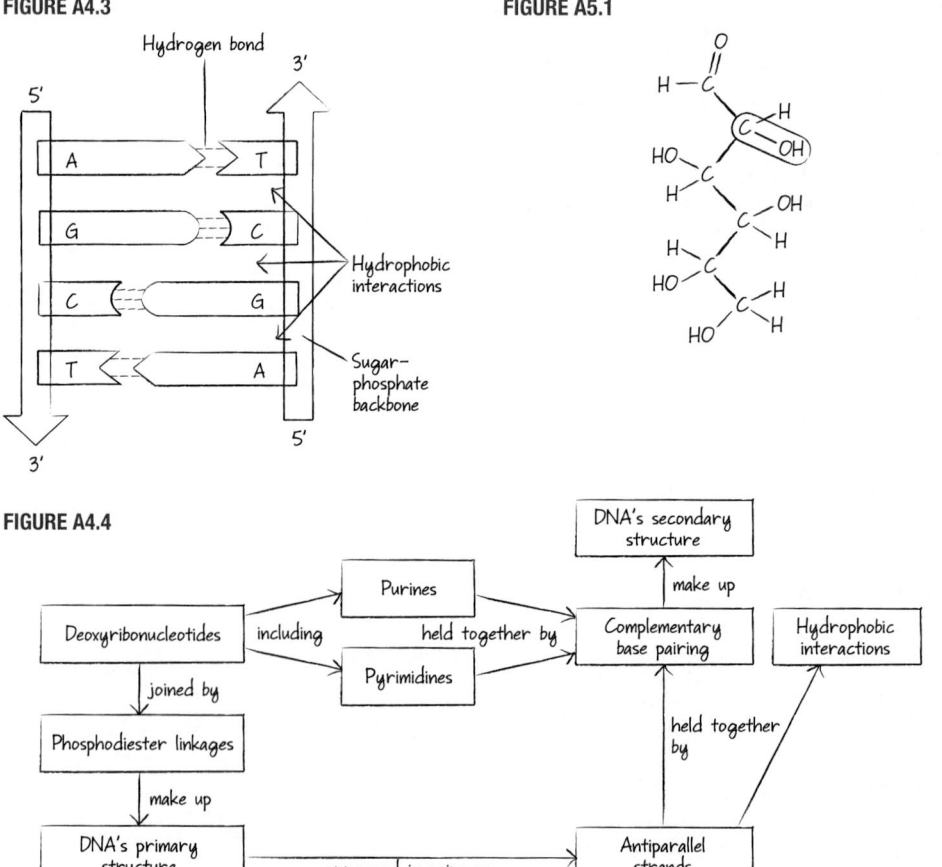

FIGURE A5.1

FIGURE A4.4

carbohydrate bar would be at zero inhibition, and the glycoprotein with digested carbohydrate bar would be similar to the intact glycoprotein bar.

p. 80 Fig. 5.7 understand All of the C—C and C—H bonds should be circled.

p. 81 CYU (1) understand *Aspect 1:* The β-1,4-glycosidic linkages in these molecules result in insoluble fibers that are difficult to degrade. *Aspect 2:* When individual molecules of these carbohydrates align, bonds form between them and produce fibers or sheets that resist pulling and pushing forces. **(2)** apply Most are probably being broken down into glucose, some of which in turn is being broken down in reactions that lead to the synthesis of ATP. Some will be resistant to digestion, such as the insoluble cellulose that makes up dietary fiber. This will help retain water and support the digestion and passage of fecal material.

IF YOU UNDERSTAND . . .

5.1. understand Molecules have to interact in an extremely specific orientation in order for a reaction to occur. Changing the location of a functional group by even one carbon can mean that the molecule will undergo completely different types of reactions. **5.2** analyze Glycosidic linkages can vary more in location and geometry than linkages between amino acids and nucleotides do. This variability increases the structural diversity possible in carbohydrates compared to proteins and nucleic acids. **5.3** understand (1) Polysaccharides used for energy storage are formed entirely from glucose monomers joined by α-glycosidic linkages; structural polysaccharides are made up of glucose or other sugars joined by β-glycosidic linkages. (2) The monomers in energy-storage polysaccharides are linked in a helical arrangement; the monomers in structural polysaccharides are linked in a linear arrangement. (3) Energy-storage polysaccharides may branch; structural polysaccharides do not. (4) Individual chains of energy-storage polysaccharides do not associate with each other; adjacent chains of structural polysaccharides are linked by hydrogen bonds or covalent bonds.

YOU SHOULD BE ABLE TO . . .

✔ Test Your Knowledge

1. remember d **2.** remember Monosaccharides can differ from one another in three ways: (1) the location of their carbonyl group; (2) the number of carbon atoms they contain; and (3) the orientations of their hydroxyl groups. **3.** remember a **4.** remember c **5.** remember a **6.** understand The electrons in the C=O bonds of carbon dioxide molecules are held tightly by the highly electronegative oxygen atoms, so they have low potential energy. The electrons in the C—C and C—H bonds of carbohydrates are shared equally, so they have much higher potential energy.

✔ Test Your Understanding

7. understand c. **8.** apply a; lactose is a disaccharide formed from a β-1,4-glycosidic linkage, so if two glucose molecules were linked with this bond, they would resemble units of cellulose and not be digested by human infants or adults. **9.** understand Carbohydrates are ideal for displaying the identity of the cell because they are so diverse structurally. This diversity enables them to serve as very specific identity tags for cells. **10.** understand When you compare the glucose monomers in an α-1,4-glycosidic linkage versus in a β-1,4-glycosidic linkage, the linkages are located on opposite sides of the plane of the glucose rings, and the glucose monomers are linked in the same orientation versus having every other glucose flipped in orientation. β-1,4-glycosidic linkages are much more likely to form linear fibers and sheets, so they resist degradation. **11.** remember Because (1) no mechanism is known for the prebiotic polymerization of sugars; (2) no catalytic carbohydrates have been discovered that can perform polymerization reactions; and (3) sugar residues in a polysaccharide are not capable of complementary base pairing. **12.** understand Starch and glycogen both consist of glucose monomers joined by α-1,4-glycosidic linkages, and both function as storage carbohydrates. Starch is a mixture of unbranched and branched polysaccharides—called amylose and amylopectin, respectively. All glycogen polysaccharides are branched.

✔ Test Your Problem-Solving Skills

13. analyze Carbohydrates are energy-storage molecules, so minimizing their consumption may reduce total energy intake. Lack of available carbohydrate also forces the body to use fats for energy, reducing the amount of fat that is stored. **14.** apply d; lactose is a disaccharide of glucose and galactose, which can be cleaved by enzymes expressed in the human gut to release galactose. **15.** analyze Amylase breaks down the starch in the cracker into glucose monomers, which stimulate the sweet receptors in your tongue. **16.** apply When bacteria contact lysozyme, the peptidoglycan in their cell walls begins to degrade, leading to the death of the bacteria. Lysozyme therefore helps protect humans against bacterial infections.

CHAPTER 6

IN-TEXT QUESTIONS AND EXERCISES

p. 87 understand Fatty acids are amphipathic because their hydrocarbon tails are hydrophobic but their carboxyl functional groups are hydrophilic.

p. 87 Fig. 6.4 apply At the polar hydroxyl group in cholesterol and the polar head group in phospholipids.

p. 88 CYU (1) analyze Fats consist of three fatty acids linked to glycerol; steroids have a distinctive four-ring structure with variable side groups attached; phospholipids have a hydrophilic, phosphate-containing "head" region and a hydrocarbon tail. **(2)** understand In cholesterol, the hydrocarbon steroid rings and isoprenoid chain are hydrophobic; the hydroxyl group is hydrophilic. In phospholipids, the phosphate-containing head group is hydrophilic; the hydrocarbon chains are hydrophobic.

p. 89 apply Amino acids have amino and carboxyl groups that are ionized in water and nucleotides have negatively charged phosphates. Due to their charge and larger size, both would be placed below the small ions at the bottom of the scale (permeability$<10^{-12}$ cm/sec).

p. 90 CYU create See **TABLE A6.1**.

p. 91 Fig. 6.10 apply Increasing the number of phospholipids with polyunsaturated tails would increase permeability of the liposomes. Starting from the left, the first line (no cholesterol) would represent liposomes with 50% polyunsaturated phospholipids, the second line would be 20% polyunsaturated phospholipids, and the third line would contain only saturated phospholipids.

p. 92 apply If there is a difference in temperature, then there would be a difference in thermal motion. The solute concentration on the side with a higher temperature would decrease because the solute particles would be moving faster and hence be more likely to move to the cooler side of the membrane, where they would slow down.

p. 93 Fig. 6.13 apply Higher, because less water would have to move to the right side to achieve equilibrium.

FIGURE A5.2

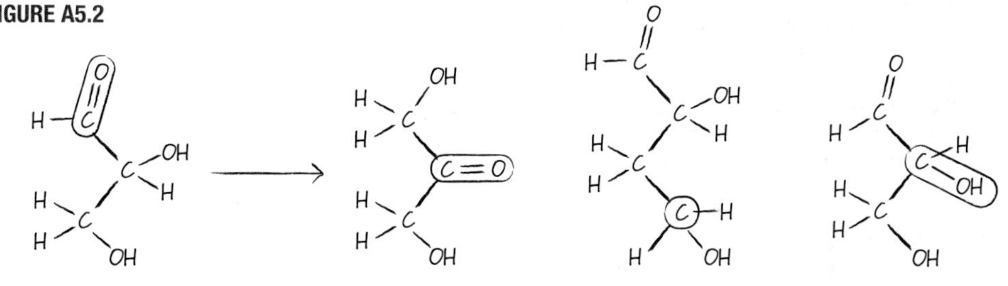

Start with a monosacchride. This one is a 3-carbon aldose (carbonyl group at end)

Variation 1: 3-carbon ketose (carbonyl group in middle)

Variation 2: 4-carbon aldose

Variation 3: 3-carbon aldose with different arrangement of hydroxyl group

TABLE A6.1

Factor	Effect on permeability	Reason
Temperature	Decreases as temperature decreases.	Lower temperature slows movement of hydrocarbon tails, allowing more interactions (membrane is more dense).
Cholesterol	Decreases as cholesterol content increases.	Cholesterol molecules fill in the spaces between the hydrocarbon tails, making the membrane more tightly packed.
Length of hydrocarbon tails	Decreases as length of hydrocarbon tails increases.	Longer hydrocarbon tails have more interactions (membrane is more dense).
Saturation of hydrocarbon tails	Decreases as degree of saturation increases.	Saturated fatty acids have straight hydrocarbon tails that pack together tightly, leaving few gaps.

p. 94 CYU create See **FIGURE A6.1**.

p. 96 Fig. 6.18 create Repeat the procedure using a lipid bilayer that is free of membrane proteins, such as synthetic liposomes constructed from only phospholipids. If proteins were responsible for the pits and mounds, then this control would not show these structures.

p. 97 apply Your arrow should point out of the cell. There is no concentration gradient for chloride, but the outside has a net positive charge, which favors outward movement of negative ions.

p. 97 Fig. 6.21 analyze No—the 10 replicates where no current was recorded probably represent instances where the CFTR protein was damaged and not functioning properly. (In general, no experimental method works "perfectly.")

p. 101 CYU understand Passive transport does not require an input of energy—it happens as a result of energy already present in existing concentration or electrical gradients. Active transport is active in the sense of requiring an input of energy from, for example, ATP. In cotransport, a second ion or molecule is transported against its concentration gradient along with (i.e., "co") an ion that is transported along its concentration gradient.

p. 101 Fig. 6.26 understand *Diffusion*: description as given; no proteins involved. *Facilitated diffusion*: Passive movement of ions or molecules that cannot cross a phospholipid bilayer readily along a concentration gradient; facilitated by channel or carrier proteins. *Active transport*: Active movement of ions or molecules that will build a gradient; facilitated by pump proteins powered by an energy source such as ATP.

IF YOU UNDERSTAND . . .

6.1. analyze Adding H_2 increases the saturation of the oil by converting C=C bonds into C—C bonds with added hydrogens. Lipids with more C—H bonds tend to be solid at room temperature. **6.2** understand Highly permeable and fluid bilayers possess short, unsaturated hydrocarbon tails while those that are highly impermeable and less fluid contain long, saturated hydrocarbon tails. **6.3** apply (1) The solute will diffuse until both sides are at equal concentrations. (2) Water will diffuse toward the side with the higher solute concentration. **6.4** apply See **FIGURE A6.2**.

YOU SHOULD BE ABLE TO . . .

✓ Test Your Knowledge

1. understand c **2.** remember a **3.** understand b **4.** understand d **5.** understand For osmosis to occur, a concentration gradient and membrane that allows water to pass, but not the solute, must be present. **6.** analyze Channel proteins form pores in the membrane and carrier proteins undergo conformational changes to shuttle molecules or ions across the membrane.

✓ Test Your Understanding

7. apply b **8.** analyze No, because they have no polar end to interact with water. Instead, these lipids would float on the surface of water, or collect in droplets suspended in water, reducing their interaction with water to a minimum. **9.** understand Hydrophilic, phosphate-containing head groups interact with water; hydrophobic hydrocarbon tails associate with each other. A bilayer is more stable than independent phospholipids in solution. **10.** apply Ethanol's polar hydroxyl group reduces the speed at which it can cross a membrane, but its small size and lack of charge would allow it to slowly cross membranes—between the rates of water and glucose transport. **11.** understand Only nonpolar, hydrophobic amino acid residues would be found in the portion of the protein that crosses the membrane. In the interior of the bilayer, these residues would be hidden from the water solvent and interact with the nonpolar lipid tails. **12.** apply Chloride ions from sodium chloride will move from the left side to the right through the CFTR. Water

will initially move from the right side to the left by osmosis, but as chloride ions move to the right, water will follow. Na^+ and K^+ ions will not move across the membrane.

✓ Test Your Problem-Solving Skills

13. apply c **14.** create Flip-flops should be rare, because they require a polar head group to pass through the hydrophobic portion of the lipid bilayer. To test this prediction, you could monitor the number of dyed phospholipids that transfer from one side of the membrane to the other in a given period of time. **15.** apply Organisms that live in very cold environments are likely to have highly unsaturated phospholipids. The kinks in unsaturated hydrocarbon tails keep membranes fluid and permeable, even at low temperature. Organisms that live in very hot environments would likely have phospholipids with saturated tails, to prevent membranes from becoming too fluid and permeable. **16.** analyze Adding a methyl group makes a drug more hydrophobic and thus more likely to pass through a lipid bilayer. Adding a charged group makes it hydrophilic and reduces its ability to pass through the lipid bilayer. These modifications would help target the drug to either the inside or outside of cells, respectively.

BIG PICTURE Chemistry of Life

p. 104 CYU (1) understand Oxygen is much more electronegative than hydrogen, so within water, the electrons are unequally shared in the O—H covalent bonds. The resulting partial negative charge around the oxygen and partial positive charges around the hydrogen atoms allow for hydrogen bonds to form among water molecules. **(2)** analyze Unlike other macromolecules, nucleic acids can serve as templates for their own replication. RNA is generally single-stranded and can adopt many different three-dimensional structures. The flexibility in structure, combined with the presence of reactive hydroxyl groups, contribute to the formation of active sites that catalyze chemical reactions. One or more of these catalytic RNA molecules may have evolved the ability to self-replicate. DNA is not likely to have catalyzed its own replication, as it is most often double-stranded, with no clear tertiary structure, and it lacks the reactive hydroxyl groups. **(3)** remember In the amino acid, the nitrogen in the amino (NH_3^+) group and the carbon in the carboxyl (COO^-) group should be circled. In the nucleotide, the oxygen in the hydroxyl (OH) group and the phosphorus in the phosphate (PO_4^{2-}) group on the nucleotide should be circled. **(4)** understand A line representing a protein should be drawn such that it completely crosses the lipid bilayer at least once. The protein could be involved in a variety of different roles, including transport of substances across the membrane in the form of a channel, carrier, or pump.

CHAPTER 7

IN-TEXT QUESTIONS AND EXERCISES

p. 110 CYU remember (1) The nucleoid compacts the chromosome to fit inside the cell via supercoiling while still keeping it accessible for replication and transmission of information. (2) Photosynthetic membranes increase food production by providing a large surface area to hold the pigments and enzymes required for photosynthesis. (3) Flagella propel cells through liquid, often toward a food source. (4) The layer of thick, strong material stiffens the cell wall and provides protection from mechanical damage.

p. 114 Fig. 7.12 create Storing the toxins in vacuoles prevents the toxins from damaging the plant's own organelles and cells.

FIGURE A6.1

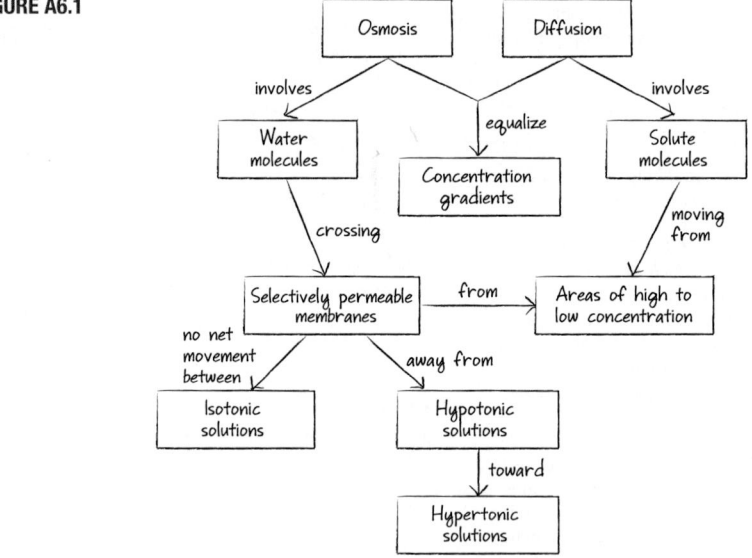

FIGURE A6.2

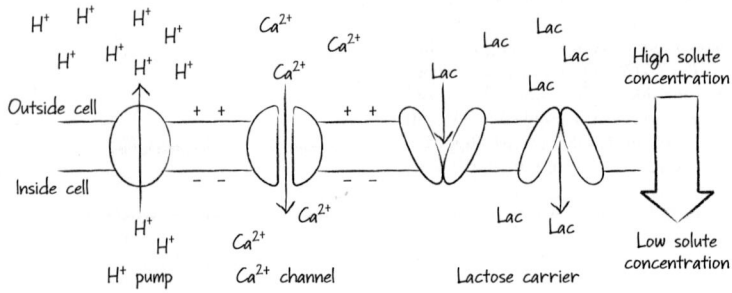

p. 116 CYU (1) `understand` Both organelles contain specific sets of enzymes. Lysosomal enzymes digest macromolecules in the acidic lumen of this organelle, releasing monomers that can be recycled into new macromolecules. Peroxisomes contain catalase and other enzymes that process fatty acids and toxins via oxidation reactions. **(2)** `understand` From top to bottom: administrative/information hub, protein factory, large molecule manufacturing and shipping (protein synthesis and folding center, lipid factory, protein finishing and shipping line, waste processing and recycling center), warehouse, fatty-acid processing and detox center, power station, food-manufacturing facility, support beams, perimeter fencing with secured gates, and leave blank.
p. 118 Fig. 7.16 `remember` See **FIGURE A7.1**.
p. 121 `analyze` (1) Nucleotides are small enough that they would diffuse through the nuclear pore complex along their gradients—a passive process that would not require

energy. (2) Large proteins must be escorted through the nuclear pore complex, which is directional and requires energy, since the protein is concentrated inside the nucleus.
p. 121 Fig. 7.18 `apply` "Prediction": The labeled tail region fragments or the labeled core region fragments of the nucleoplasmin protein will be found in the cell nucleus. "Prediction of null hypothesis": Either both the fragments (no required signal) or neither of them (whole protein signal) will be found in the nucleus of the cell. "Conclusion": The send-to-nucleus signal is in the tail region of the nucleoplasmin protein.
p. 123 `apply` During the chase period, proteins appear to have first entered the Golgi after 7 minutes and then started to move into secretory granules after 37 minutes. This means that in this experiment, it took approximately 30 minutes for the fastest-moving proteins to pass through the Golgi.

p. 126 `apply` In receptor-mediated endocytosis, the conversion of a late endosome to a lysosome is dependent on receiving acid hydrolases from the Golgi. If this receptor is not present, then the enzymes will not be sent and the late endosome will not mature into a lysosome to digest the endocytosed products.
p. 126 CYU (1) `apply` Proteins that enter the nucleus are fully synthesized and have an NLS that interacts with another protein to get it into the organelle. The NLS is not removed. Proteins that enter the ER have a signal sequence that interacts with the SRP during translation. The ribosome is moved to the ER and synthesis continues, moving the protein into the ER. The signal is removed once it enters the organelle. **(2)** `apply` The protein would be in the lysosome. The ER signal would direct the protein into the ER before it is completely synthesized. The M-6-P tag will direct the protein from the Golgi to the late endosome to the lysosome. Thus the complete protein is never free in the cytosol, where the NLS could direct it into the nucleus.
p. 132 CYU `analyze` Actin filaments are made up of two strands of actin monomers, microtubules are made up of tubulin protein dimers that form a tube, and intermediate filaments are made up of a number of different protein subunits. Actin filaments and microtubules exhibit polarity (or directionality), and new subunits are constantly being added or subtracted at either end (but added faster to the plus end). All three elements provide structural support, but only actin filaments and microtubules serve as tracks for motors involved in movement and cell division.
p. 132 Fig. 7.30 `apply` The microtubule doublets of the axoneme would slide past each other completely, but the axoneme would not bend.

IF YOU UNDERSTAND . . .

7.1 `understand` (1) Cells will be unable to synthesize new proteins and will die. (2) In many environments, cells will be unable to resist the osmotic pressure of water entering the cytoplasm and will burst. (3) The cell shape will be different, and cells will not be able to divide.
7.2 `understand` (1) The cell will be unable to produce a sufficient amount of ATP and will die. (2) Reactive molecules, like hydrogen peroxide, will damage the cell, and it will likely die. (3) Nothing will happen, since plants do not have centrioles. **7.3** `analyze` The liver cell would be expected to have more peroxisomes and less rough endoplasmic reticulum than the salivary cells would.
7.4 `create` The addition or removal of phosphates would change the folded structure of the protein, exposing the NLS for nuclear transport. **7.5** `create` As a group, proteins have complex and highly diverse shapes and chemical properties that allow them to recognize a great number of different zip codes in a very specific manner. **7.6** `apply` The Golgi is positioned near the microtubule organizing center, which has microtubules running from the minus end out to the plus end (near the plasma membrane). Kinesin would be used to move these vesicles as it walks toward the plus end.

YOU SHOULD BE ABLE TO . . .
✔ Test Your Knowledge

1. `understand` b **2.** `remember` They have their own small, circular chromosomes; they produce their own ribosomes; and they divide in a manner that is similar to bacterial fission, independent of cellular division. **3.** `remember` c
4. `remember` b **5.** `remember` a **6.** `understand` The phosphate links to the motor protein and causes it to change shape, which results in the protein moving along the filament.

✔ Test Your Understanding

7. `analyze` All cells are bound by a plasma membrane, are filled with cytoplasm, carry their genetic information (DNA) in chromosomes, and contain ribosomes

FIGURE A7.1

(a) Animal pancreatic cell: Exports digestive enzymes.

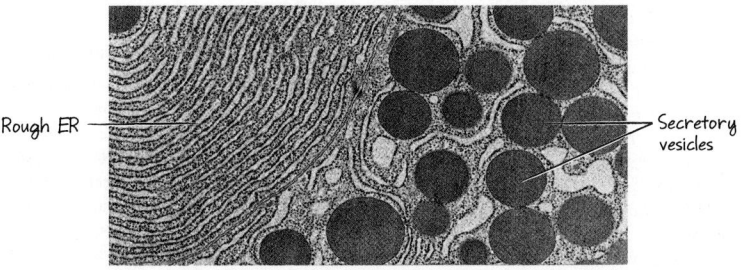

Rough ER — Secretory vesicles

(b) Animal testis cell: Exports lipid-soluble signals.

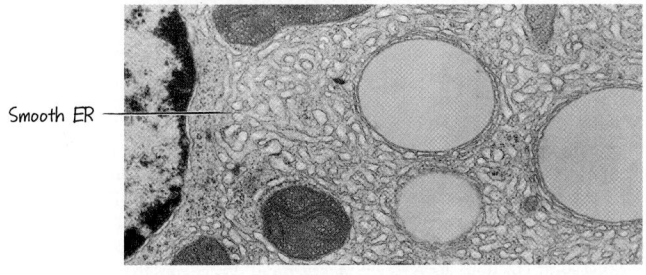

Smooth ER —

(c) Plant leaf cell: Manufactures ATP and sugar.

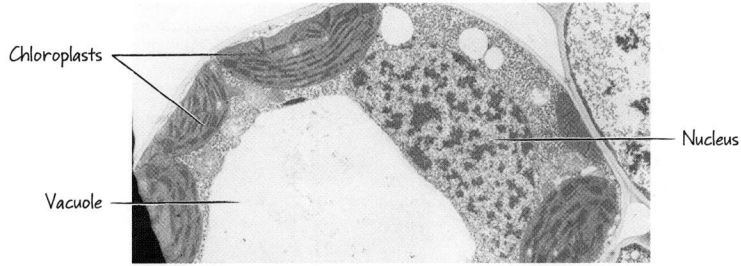

Chloroplasts — Nucleus
Vacuole —

(d) Brown fat cells: Burn fat to generate heat in lieu of ATP.

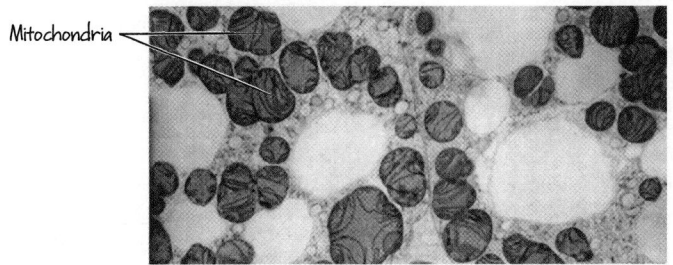

Mitochondria —

(the sites of protein synthesis). Some prokaryotes have organelles not found in plants or animals, such as a magnetite-containing structure. Plant cells have chloroplasts, vacuoles, and a cell wall. Animal cells contain lysosomes and lack a cell wall. **8.** analyze a; the endoplasmic reticulum is responsible for synthesizing the membrane proteins required for the transport of solutes across the plasma membrane. **9.** create The NLS will be used to actively import the protein to the nucleus, leaving very little of the protein in the cytoplasm. Diffusion alone would not drive all the protein into the nucleus. **10.** create Ribosome in cytoplasm (signal is synthesized) → Ribosome at rough ER (protein is completed, folded, and glycosylated) → Transport vesicle → Golgi apparatus (protein is processed; molecular zip code indicating destination) → Transport vesicle → Plasma membrane → Extracellular space. **11.** create This occurs in microfilaments and microtubules because they have ends that differ structurally and functionally—they have different filament growth rates. Intermediate filaments have identical ends, so there is no difference in the rate of assembly between the two ends. **12.** understand Polarized cytoskeletal filaments (microtubules or microfilaments) are present between the organelles. End-directed motor proteins use ATP to move these transport vesicles between them.

✔ Test Your Problem-Solving Skills

13. understand b; fimbriae are involved in bacterial attachment to surfaces and other cells, which would be important in the ability to grow on teeth. **14.** create The proteins must receive a molecular zip code that binds to a receptor on the surface of peroxisomes. They could diffuse randomly to peroxisomes or be transported in a directed way by motor proteins. **15.** create The tails cleaved from nucleoplasmin could be attached to the gold particles that were excluded from crossing the pore complex owing to their size. If these modified particles entered the nucleus, then the tail is not limited to the nucleoplasmin transport alone. **16.** apply The proteins would likely be found in the cytoplasm (e.g., actin and myosin) or imported into the mitochondria. Since there is a high energy demand, you would predict that there are many active mitochondria.

CHAPTER 8

IN-TEXT QUESTIONS AND ANSWERS

p. 139 apply (1) If ΔS is positive (products have more disorder than reactants), then according to the free energy equation, ΔG is more likely to be negative as temperature (T) increases even if ΔH is positive. The increased temperature represents added heat energy that may be used to drive an endothermic reaction to completion, making the reaction spontaneous. (2) Exothermic reactions may be nonspontaneous if they result in a decrease in entropy—meaning that the products are more ordered than the reactants (ΔS is negative).
p. 139 CYU (1) understand Gibbs equation: $\Delta G = \Delta H - T \Delta S$. ΔG symbolizes the change in the Gibbs free energy. ΔH represents the difference in enthalpy (heat, pressure, and volume) between the products and the reactants. T represents the temperature (in degrees Kelvin) at which the reaction is taking place. ΔS symbolizes the change in entropy (amount of disorder). (2) understand When ΔH is negative—meaning that the reactants have lower enthalpy than the products—and when ΔS is positive, meaning that the products have higher entropy (are more disordered) than the reactants.
p. 140 Fig. 8.4 understand Each point represents the data from a single test, not an average of many experiments, so it is not possible to calculate the standard error of the average.
p. 144 analyze Redox reactions transfer energy between molecules or atoms via electrons. When oxidized molecules are reduced, their potential energy increases. ATP

hydrolysis is often coupled with the phosphorylation of another molecule. This phosphorylation increases the potential energy of the molecule.
p. 144 CYU (1) understand Electrons in C−H bonds are not held as tightly as electrons in C−O bonds, so they have higher potential energy. **(2)** understand In part, because its three phosphate groups have four negative charges in close proximity. The electrons repulse each other, raising their potential energy.
p. 144 Fig. 8.9 understand The ΔG in the uncoupled reaction would be positive (>0), and each of the steps in the coupled reaction would have a negative (<0) ΔG.
p. 147 remember (1) binding substrates, (2) transition state, (3) R-groups, (4) structure
p. 146 Fig. 8.12 understand No—a catalyst affects only the activation energy, not the overall change in free energy.
p. 147 Fig. 8.14 analyze See **FIGURE A8.1**.
p. 149 CYU (1) create The rate of the reaction is based primarily on the activity of the enzyme. Once the temperature reaches a level that causes unfolding and inactivation of the enzyme, the rate decreases to the uncatalyzed rate. **(2)** apply The shape change would most likely alter the shape of the active site. If phosphorylation activates catalytic activity, the change to the active site would allow substrates to bind and be brought to their transition state. If phosphorylation inhibits catalytic activity, the shape change to the active site would likely prevent substrates from binding or no longer orient them correctly for the reaction to occur.
p. 150 apply The concentration of A and B would be higher than in the fully functional pathway since they are not depleted to produce C. If D is not depleted by other reactions, then equilibrium would be established between C and D, resulting in lower concentrations of both.

IF YOU UNDERSTAND . . .

8.1 understand Reactions are spontaneous when the free energy in the products is lower than that of the reactants (ΔG is negative). Enthalpy and entropy are measures used to determine free-energy changes. Enthalpy measures the potential energy of the molecules, and entropy measures the disorder. For exergonic, spontaneous reactions, disorder normally increases and the potential energy stored in the products normally decreases relative to the reactants. **8.2** understand Energetic coupling transfers free energy released from exergonic reactions to drive endergonic reactions. Since endergonic reactions are required for sustaining life, without energetic coupling, life would not exist. **8.3** understand Amino acid R-groups lining the active site interact with the substrates, orienting them in a way that stabilizes the transition state, thereby lowering the activation energy needed for the reaction to proceed. **8.4** analyze Allosteric regulation and phosphorylation cause changes in the conformation of the enzyme that affects its catalytic function. Allosteric regulation involves non-covalent bonding, while phosphorylation is a covalent modification of the enzyme's primary structure. **8.5** apply In the first step of the pathway, the rate would increase as the intermediate, which is the product of the first reaction, is removed. In the last step, the rate would decrease due to the loss of the intermediate, which serves as the substrate for the last reaction.

YOU SHOULD BE ABLE TO . . .

✔ Test Your Knowledge

1. remember c **2.** remember a **3.** remember a **4.** remember The enzyme changes shape, but the change is not permanent. The enzyme shape will return to its original conformation after releasing the products. **5.** remember d **6.** remember When the product of a pathway feeds back to interact with an enzyme early in the same pathway to inhibit its function.

✔ Test Your Understanding

7. understand The shape of reactant molecules (the key) fits into the active site of an enzyme (the lock). Fischer's

original model assumed that enzymes were rigid; in fact, enzymes are flexible and dynamic. **8.** understand d. Energy, such as the thermal energy in fire, must be provided to overcome the activation energy barrier before the reaction can proceed. **9.** understand The phosphorylation reaction is exergonic because the electrons in ADP and the phosphate added to the substrate experience less electrical repulsion, and thus have less potential energy, than they did in ATP. A phosphorylated reactant (i.e., an activated intermediate) gains enough potential energy to shift the free energy change for the reaction from endergonic to exergonic. **10.** apply For the coupled reaction, step 1 has a ΔG of about -3 kcal/mol and step 2 has a ΔG of about -3 kcal/mol. The uncoupled reaction has a ΔG of about $+1.3$ kcal/mol. **11.** analyze Both are mechanisms that regulate enzymes; the difference is whether the regulatory molecule binds at the active site (competitive inhibition) or away from the active site (allosteric regulation). **12.** apply Catabolic reactions will often have a negative ΔG based on a decrease in enthalpy and increase in entropy. Anabolic reactions are the opposite—a positive ΔG that is based on an increase in enthalpy and decrease in entropy.

✔ Test Your Problem-Solving Skills

13. create See **FIGURE A8.2. 14.** apply Without the coenzyme, the free-radical-containing transition state would not be stabilized and the reaction rate would drop dramatically. **15.** analyze The data suggest that the enzyme and substrate form a transition state that requires a change in the shape of the active site, and that each movement corresponds to one reaction. **16.** apply b. The sugar likely functions as an allosteric regulator to activate the enzyme.

CHAPTER 9

IN-TEXT QUESTIONS AND EXERCISES

p. 156 Fig. 9.2 remember Glycolysis: "What goes in" = glucose, NAD^+, ADP, inorganic phosphate; "What comes out" = pyruvate, NADH, ATP. Pyruvate processing: "What goes in" = pyruvate, NAD^+; "What comes out" = NADH, CO_2, acetyl CoA. Citric acid cycle: "What goes in" = acetyl CoA, NAD^+, FAD, GDP or ADP, inorganic phosphate; "What comes out" = NADH, $FADH_2$, ATP or GTP, CO_2. Electron transport and oxidative phosphorylation: "What goes in" = NADH, $FADH_2$, O_2, ADP, inorganic phosphate; "What comes out" = ATP, H_2O, NAD^+, FAD.

FIGURE A8.1

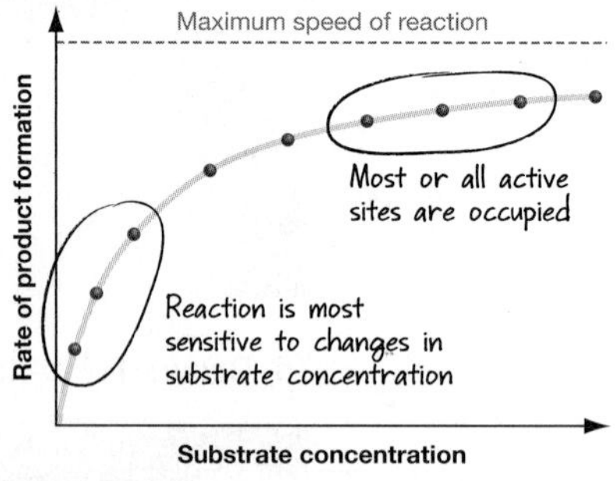

Maximum speed of reaction

Rate of product formation

Most or all active sites are occupied

Reaction is most sensitive to changes in substrate concentration

Substrate concentration

FIGURE A8.2

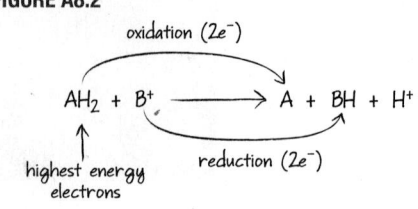

oxidation (2e$^-$)

$AH_2 + B^+ \longrightarrow A + BH + H^+$

highest energy electrons

reduction (2e$^-$)

p. 161 `apply` If the regulatory site had a higher affinity for ATP than the active site, then ATP would always be bound at the regulatory site, and glycolysis would always proceed at a very slow rate.

p. 161 Fig. 9.9 `remember` "Positive control": AMP, NAD$^+$, CoA (reaction substrates). "Negative control by feedback inhibition": acetyl CoA, NADH, ATP (reaction products).

p. 165 CYU `remember` (1) and (2) are combined with the answer to p. 171 CYU (3) Start with 12 triangles on glucose. (These triangles represent the 12 pairs of electrons that will be moved to electron carriers during redox reactions throughout glycolysis and the citric acid cycle.) Move two triangles to the NADH circle generated by glycolysis and the other 10 triangles to the pyruvate circle. Then move these 10 triangles through the pyruvate dehydrogenase square, placing two of them in the NADH circle next to pyruvate dehydrogenase. Add the remaining eight triangles in the acetyl CoA circle. Next move the eight triangles in the acetyl CoA circle through the citric acid cycle, placing six of them in the NADH circle and two in the FADH$_2$ circle generated during the citric acid cycle. (4) These boxes are marked with stars in the diagram.

p. 165 Fig. 9.13 `apply` NADH would be expected to have the highest amount of chemical energy since its production is correlated with the largest drop in free energy in the graph.

p. 167 Fig. 9.15 `understand` The proton gradient arrow should start above in the inner membrane space and point down across the membrane into the mitochondrial matrix. *Complex I:* "What goes in" = NADH; "What comes out" = NAD$^+$, e$^-$, transported H$^+$. *Complex II:* "What goes in" = FADH$_2$; "What comes out" = FAD, e$^-$, H$^+$. *Complex III:* "What goes in" = e$^-$, H$^+$; "What comes out" = e$^-$, transported H$^+$. *Complex IV:* "What goes in" = e$^-$, H$^+$, O$_2$; "What comes out" = H$_2$O, transported H$^+$.

p. 169 `explain` "Indirect" is accurate because most of the energy released during glucose oxidation is not used to produce ATP directly. Instead, this energy is stored in reduced electron carriers that are used by the ETC to generate a proton gradient across a membrane. These protons then diffuse down their concentration gradient across the inner membrane through ATP synthase, which drives ATP synthesis.

p. 169 Fig. 9.17 `create` They could have placed the vesicles in an acidic solution that has a pH below that of the solution in the vesicle. This would set up a proton gradient across the membrane to test for ATP synthesis.

p. 171 CYU `understand` See **FIGURE A9.1**. To illustrate the chemiosmotic mechanism, take the triangles (electrons) piled on the NADH and FADH$_2$ circles and move them through the ETC. While moving these triangles, also move dimes from the mitochondrial matrix to the intermembrane space. As the triangles exit the ETC, add them to the oxygen to water circle. Once all the dimes have been pumped by the ETC into the intermembrane space, move them through ATP synthase back into the mitochondrial matrix to fuel the formation of ATP.

p. 173 CYU `understand` Electron acceptors such as oxygen have a much higher electronegativity than pyruvate. Donating an electron to O$_2$ causes a greater drop in potential energy, making it possible to generate much more ATP per molecule of glucose.

IF YOU UNDERSTAND . . .

9.1 `understand` The radioactive carbons in glucose can be fully oxidized by the central pathways to generate CO$_2$, which would be radiolabeled. Other molecules, like lipids and amino acids, would also be expected to be radiolabeled since they are made using intermediates from the central pathways in other anabolic pathways. **9.2** `apply` See **FIGURE A9.2**. **9.3** `understand` Pyruvate dehydrogenase accomplishes three different tasks that would be expected to require multiple enzymes and active sites: CO$_2$ release, NADH production, and linking of an acetyl group to CoA. **9.4** `apply` NADH would decrease if a drug poisoned the acetyl CoA and oxaloacetate-to-citrate enzyme, since the citric acid cycle would no longer be able to produce NADH in the steps following this reaction in the pathway. **9.5** `apply` The ATP synthase allows protons to reenter the mitochondrial matrix after they have been pumped out by the ETC. By blocking ATP synthase, you would expect the pH of the matrix to increase (decreased proton concentration). **9.6** `understand` Organisms that produce ATP by fermentation would be expected to grow more slowly than those that produce ATP via cellular respiration simply because fermentation produces fewer ATP molecules per glucose molecule than cellular respiration does.

YOU SHOULD BE ABLE TO . . .

✔ Test Your Knowledge

1. `understand` Glycolysis → Pyruvate processing → citric acid cycle → ETC and chemiosmosis. The first three steps are responsible for glucose oxidation; the final step produces the most ATP. **2.** `remember` b **3.** `understand` d **4.** `understand` Most of the energy is stored in the form of NADH. **5.** `remember` c **6.** `remember` a

✔ Test Your Understanding

7. `understand` Stored carbohydrates can be broken down into glucose that enters the glycolytic pathway. If carbohydrates are absent, products from fat and protein catabolism can be used to fuel cellular respiration or fermentation. If ATP is plentiful, anabolic reactions use

FIGURE A9.1

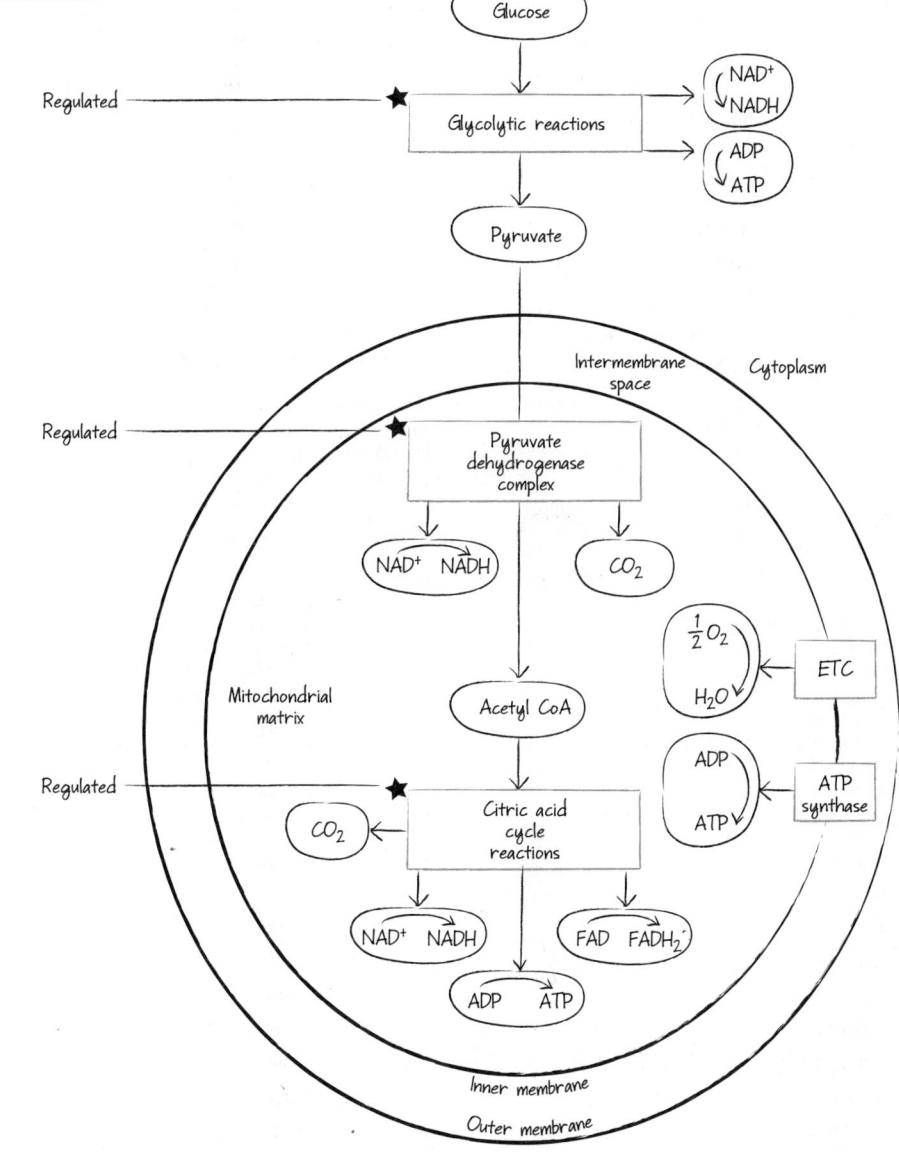

FIGURE A9.2

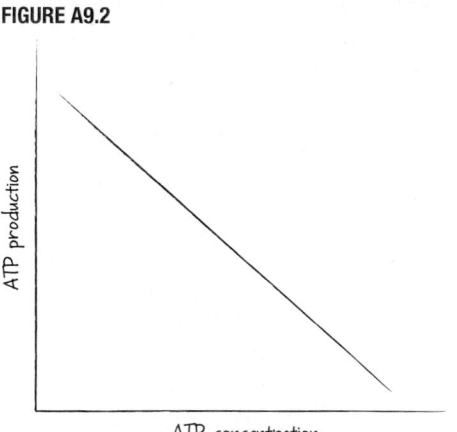

intermediates of the glycolytic pathway and the citric acid cycle to synthesize carbohydrates, fats, and proteins. **8.** analyze Both processes produce ATP from ADP and P_i, but substrate-level phosphorylation occurs when enzymes remove a "high-energy" phosphate from a substrate and directly transfer it to ADP, while oxidative phosphorylation occurs when electrons move through an ETC and produce a proton-motive force that drives ATP synthase. **9.** understand Aerobic respiration is much more productive because oxygen has extremely high electronegativity compared with other electron acceptors, resulting in a greater release of energy during electron transport and more proton pumping. **10.** apply b **11.** analyze Both phosphofructokinase and isocitrate dehydrogenase are regulated by feedback inhibition, where the product of the reaction or series of reactions inhibits the enzyme activity. They differ in that phosphofructokinase is regulated by allosteric inhibition while isocitrate dehydrogenase is controlled by competitive inhibition. **12.** understand Oxidative phosphorylation is possible via a proton gradient that is established by redox reactions in the ETC. ATP synthase consists of a membrane-associated F_o unit and a F_1 unit joined by a rotor shaft. When protons flow through the F_o unit, it spins the rotor shaft within the fixed F_1 unit. This spinning shaft causes structural changes in the F_1 that drives the synthesis of ATP from ADP and P_i.

✔ **Test Your Problem-Solving Skills**

13. create When complex IV is blocked, electrons can no longer be transferred to oxygen, the final acceptor, and cellular respiration stops. Fermentation could keep glycolysis going, but it is inefficient and unlikely to fuel a cell's energy needs over the long term. Cells that lack the enzymes required for fermentation would die first. **14.** apply Because mitochondria with few cristae would have fewer electron transport chains and ATP synthase molecules, they would produce much less ATP than mitochondria with numerous cristae. **15.** apply For each glucose molecule, two ATP are produced in glycolysis and two ATP are produced in the citric acid cycle via substrate-level phosphorylation. A total of 10 NADH and 2 $FADH_2$ molecules are produced from glycolysis, pyruvate oxidation, and the citric acid cycle. If each NADH were to yield 3 ATP, and each $FADH_2$ were to yield 2 ATP, then a total of 34 ATP would be produced via oxidative phosphorylation. Adding these totals would result in 38 ATP. A cell will not produce this much ATP, because the proton-motive force is used in other transport steps and because of other issues that may reduce the overall efficiency. **16.** apply b

CHAPTER 10

IN-TEXT QUESTIONS AND EXERCISES

p. 177 understand See **FIGURE A10.1**. This reaction is endergonic because there are more high-energy chemical bonds in the products compared with the reactants, and there is a decrease in entropy.
p. 177 Fig. 10.1 apply See **FIGURE A10.1**.
p. 180 Fig. 10.6 apply See **FIGURE A10.2**.

p. 182 Fig. 10.9 apply The energy state corresponding to a photon of green light would be located between the energy states corresponding to red and blue photons.
p. 183 CYU apply The outer pigments would be more likely to absorb blue photons (short wavelength, high energy), and interior pigments would absorb red photons (long wavelength, low energy). This establishes a pathway to direct photon energy toward the reaction center since resonance energy is transferred from higher to lower energy levels.
p. 184 Fig. 10.11 understand Yes—otherwise, changes in the production of oxygen could be due to differences in the number of chloroplasts, not differences in the rate of photosynthesis.
p. 186 analyze Light → Antenna complex → Reaction center → Pheophytin → ETC → Proton gradient → ATP synthase. Electrons from water are donated to the reaction center to replace those that were transferred to pheophytin.
p. 188 remember (1) Plastocyanin transfers electrons that move through the cytochrome complex in the ETC to the reaction center of photosystem (PS) I. (2) After they are excited by a photon and donated to the initial electron acceptor.
p. 190 CYU analyze In mitochondria, high-energy electrons are donated by NADH or $FADH_2$ (primary donors) and passed through an ETC to generate a proton-motive force. The low-energy electrons at the end of the chain are accepted by O_2 (terminal acceptor) to form water. In chloroplasts, low-energy electrons are donated by H_2O (primary donor), energized by photons or resonance energy, and passed through an ETC to generate a proton-motive force. These electrons are then excited a second time by photons or resonance energy, and the high-energy electrons are accepted by $NADP^+$ (terminal acceptor) to form NADPH.
p. 190 Fig. 10.18 understand The researchers didn't have any basis on which to predict these intermediates. They needed to perform the experiment to identify them.
p. 192 apply Each complete cycle requires 3 ATP and 2 NADPH molecules. To complete 6 runs through the cycle, a total of 18 ATP and 12 NADPH molecules are needed. By following the number of carbons, it is apparent that only three RuBP molecules are required, since they are fully regenerated every 3 cycles: 3 RuBP (15 carbons) fix and reduce 3 CO_2 to generate 6 G3P (18 carbons), yielding 1 G3P (3 carbons); the other 5 are used to regenerate 3 RuBP (15 carbons). The regeneration of RuBP means that only three would be required for continued runs through the Calvin cycle.
p. 194 Fig. 10.24 apply The morning would have the highest concentration of organic acids in the vacuoles of CAM plants, since these acids are made during the night and used up during the day.
p. 195 CYU (1) understand (a) C_4 plants use PEP carboxylase to fix CO_2 into organic acids in mesophyll cells. These organic acids are then transported into bundle-sheath cells, where they release carbon dioxide to rubisco. (b) CAM plants take in CO_2 at night and have enzymes that fix it into organic acids stored in the central vacuoles of photosynthesizing cells. During the day, the organic acids are processed to release CO_2 to rubisco.

(c) By diffusion through a plant's stomata when they are open. **(2)** apply The concentration of starch would be highest at the end of the day and lowest at the start of the day. Starch is made and stored in the chloroplasts of leaves during periods of high photosynthetic activity during the day. At night, it is broken down to make sucrose, which is transported throughout the plant to drive cellular respiration. (Cellular respiration also occurs during the day, but the impact is minimized due to the photosynthetic production of sugar.)

IF YOU UNDERSTAND . . .

10.1 understand The Calvin cycle depends on the ATP and NADPH produced by the light-capturing reactions, so it is not independent of light. **10.2** understand Most of the energy captured by pigments in chloroplasts is converted into chemical energy by reducing electron acceptors in ETCs. When pigments are extracted, the antenna complexes, reaction centers, and ETCs have been disassembled, so the energy is given off as fluorescence and heat. **10.3** understand Oxygen is produced by a critical step in photosynthesis: splitting water to provide electrons to PS II. If oxygen production increases, it means that more electrons are moving through the photosystems. **10.4** apply Each CO_2 that is fixed and reduced by the Calvin cycle requires 2 NADPH, which means that 12 NADPH molecules are required for a 6-carbon glucose. Each NADPH is made when two high-energy electrons reduce $NADP^+$. Each of these high-energy electrons originates from H_2O only after being excited by 2 photons (one in PS II and one in PS I). This means that 48 photons are required to produce 24 high-energy electrons to reduce 12 $NADP^+$ molecules for the fixation and reduction of 6 CO_2 to make glucose. Photorespiration would increase the number of photons required, since some of the CO_2 that is fixed would be released.

YOU SHOULD BE ABLE TO . . .
✔ **Test Your Knowledge**

1. remember d **2.** understand c **3.** remember c **4.** understand b **5.** remember The conversion of light energy to chemical energy occurs when electrons are transferred from excited pigments to an electron carrier in the photosystems.
6. remember The electron transport chain that accepts electrons from PS II. Plastocyanin is the molecule that transfers electrons from this chain to the PS I reaction center.

✔ **Test Your Understanding**

7. understand The electrons taken from water in PS II are excited twice by either photons or resonance energy. When excited in PS II, the electrons are transferred to PQ and used to build a proton-motive force that makes ATP. After reaching PS I, they are excited a second time and will either be used to reduce $NADP^+$ to make NADPH (noncyclic) or be transported back to PQ to produce more ATP (cyclic). **8.** analyze c **9.** understand The fixation phase is when CO_2 is fixed to RuBP by rubisco to form 3-phosphoglycerate. The reduction phase uses ATP to phosphorylate the carbons and NADPH to

---

FIGURE A10.1

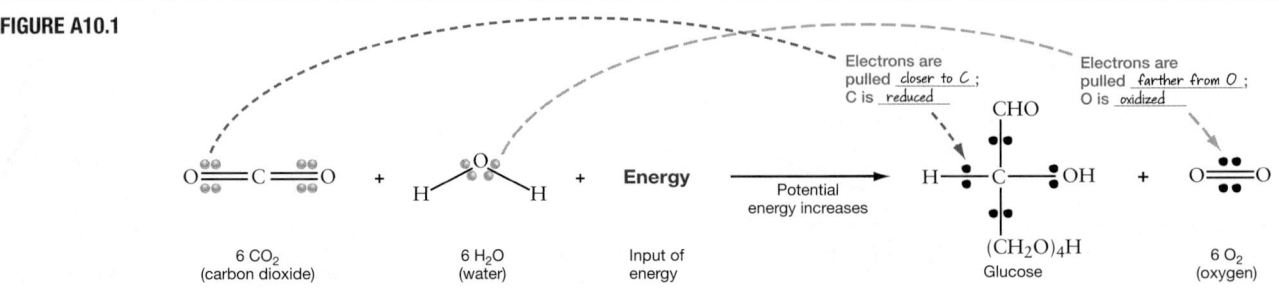

FIGURE A10.2

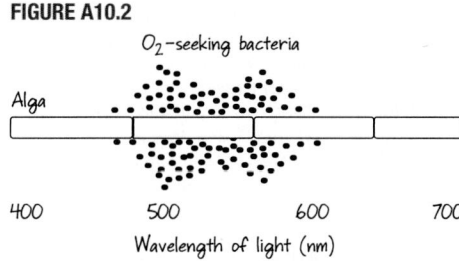

reduce them with high-energy electrons to form G3P. The regeneration phase uses more ATP to convert some of the G3P to RuBP to continue the cycle. **10.** `understand` Photorespiration occurs when levels of CO_2 are low and O_2 are high. Less sugar is produced because (1) CO_2 doesn't participate in the initial reaction catalyzed by rubisco and (2) when rubisco catalyzes the reaction with O_2 instead, one of the products is eventually broken down to CO_2 in a process that uses ATP. **11.** `understand` In both C_4 and CAM plants, atmospheric CO_2 is brought in through stomata and first captured by fixing it to a 3-carbon molecule by PEP carboxylase. The C_4 pathway and CAM differ in the timing of this first fixation step—it occurs during the day in C_4 plants and during the night in CAM plants. They also differ in the location of the Calvin cycle with respect to this first fixation step. In C_4 plants, the two processes occur in different cells, while in CAM plants they occur in the same cell, but at different times (Calvin cycle during the day). **12.** `analyze` Photosynthesis in chloroplasts produces sugar, which is used as a source of carbon for building organic molecules and energy for cellular respiration. Mitochondria harvest the energy stored in sugar to produce ATP, which is used to drive many cellular activities.

✓ Test Your Problem-Solving Skills

13. `apply` (1) O_2, ATP, and NADPH would be formed by noncyclic electron flow. (2) No O_2 or NADPH would be formed, but ATP may be made by cyclic electron flow. (3) Initially, O_2 and NADPH would be formed by noncyclic electron flow, but no ATP would be made. Without ATP, the Calvin cycle would halt and, once all the $NADP^+$ is used up, noncyclic electron flow would switch to cyclic electron flow. **14.** `evaluate` Because rubisco evolved in a high CO_2, low O_2 environment, which would minimize the impact of photorespiration, the hypothesis is credible. But once O_2 levels increased, any change in rubisco that minimized photorespiration would give individuals a huge advantage over organisms with "old" forms of rubisco. There has been plenty of time for such changes to occur, making the "holdover" hypothesis less credible. **15.** `analyze` b; the wavelength of light could excite PS I, but not PS II, resulting in cyclic electron flow since no electrons could be harvested from water by PS II. **16.** `create` No—they are unlikely to have the same complement of photosynthetic pigments. Different wavelengths of light are available in various layers of a forest and water depths. It is logical to predict that plants and algae have pigments that absorb the available wavelengths efficiently. One way to test this hypothesis would be to isolate pigments from species in different locations and test the absorbance spectra of each.

BIG PICTURE Energy

p. 198 CYU (1) `understand` Photosynthesis uses H_2O as a substrate and releases O_2 as a by-product; cellular respiration uses O_2 as a substrate and releases H_2O as a by-product. **(2)** `understand` Photosynthesis uses CO_2 as a substrate; cellular respiration releases CO_2 as a by-product. **(3)** `analyze` CO_2 fixation would essentially stop; CO_2 would continue to be released by cellular respiration. CO_2

levels in the atmosphere would increase rapidly, and production of new plant tissue would cease—meaning that most animals would quickly starve to death. **(4)** `analyze` ATP "is used by" the Calvin cycle; photosystem I "yields" NADPH.

CHAPTER 11

IN-TEXT QUESTIONS AND EXERCISES

p. 204 CYU `analyze` Plant cell walls and animal ECMs are both fiber composites. In plant cell walls the fiber component consists of cross-linked cellulose fibers, and the ground substance is pectin. In animal ECMs the fiber component consists of collagen fibrils, and the ground substance is proteoglycan.
p. 207 `apply` Developing muscle cells could not adhere normally, and muscle tissue would not form properly. The embryo would die.
p. 207 Fig. 11.9 `apply` "Prediction": Cells treated with an antibody that blocks membrane proteins involved in adhesion will not adhere. "Prediction of null hypothesis": All cells will adhere normally.
p. 209 CYU (1) `analyze` The three structures differ in composition, but their function is similar. The middle lamella in plants is composed of pectins that glue adjacent cells together. Tight junctions are made up of membrane proteins that line up and "stitch" adjacent cells together. Desmosomes are "rivet-like" structures composed of proteins that link the cytoskeletons of adjacent cells. **(2)** `understand` The plasma membranes of adjacent plant cells are continuous at plasmodesmata and share portions of the smooth endoplasmic reticulum. Gap junctions connect adjacent animal cells by forming protein-lined pores. Both structures result in openings between the cells that allow cytosol, including ions and small molecules, to be shared.
p. 211 Fig. 11.12 `create` The steroid hormone likely changes the structure of the receptor such that it now exposes a nuclear localization signal, which is required for the protein to be transported into the nucleus.
p. 213 `analyze` The spy is the signaling molecule that arrives at the cell surface (the castle gate). The guard is the G-protein-coupled receptor in the plasma membrane, and the queen is the G protein. The commander of the guard is the enzyme that is activated by the G protein to produce second messengers (the soldiers).
p. 214 `apply` **(1)** The red dominos (RTK components) would be the first two dominos in the chain, followed by the black domino (Ras). This black domino would start two or more new branches, each one represented by a single domino of one color (e.g., green) to represent the activation of one type of kinase. Each of these green dominos would then again branch out, knocking down two or more branches consisting of single dominos of a different color (blue). The same branching pattern would result from each blue domino knocking down two or more yellow dominos. **(2)** Each single black domino (Ras) would require 10 green dominos, 100 blue dominos, and 1000 yellow dominos.
p. 215 CYU (1) `understand` Each cell–cell signaling molecule binds to a specific receptor protein. A cell can respond to a signal only if it has the appropriate receptor. Only certain cell types will have the appropriate receptor for a given signaling molecule. **(2)** `understand` Signals are amplified if one or more steps in a signal transduction pathway, involving either second messengers or a phosphorylation cascade, result in the activation of multiple downstream molecules.
p. 215 Fig. 11.16 `apply` **(1)** cell responses A and C **(2)** cell responses A, B, and C **(3)** cell responses B and C.

IF YOU UNDERSTAND . . .

11.1 `apply` (1) Cells without functional integrin molecules would likely die as a result of not being able to send the

appropriate anchorage-dependent survival signals. (2) Cells would be more sensitive to pulling or shearing forces; both cells and tissues would be weaker and more susceptible to damage. **11.2** `apply` Cells would not be coordinated in their activity, so the heart tissue would not contract in unison and the heart would not beat.
11.3 `analyze` Adrenalin binds to both heart and liver cells, but the activated receptors trigger different signal transduction pathways and lead to different cell responses.
11.4 `analyze` The signal transduction pathways are similarly organized in both unicellular and multicellular organisms—consisting of signaling molecules, receptors, and second messengers. There is more variety in the means of transmitting the signal between cells in multicellular organisms compared to unicellular organisms. For example, there are no direct connections such as gap junctions or plasmodesmata in unicellular organisms.

YOU SHOULD BE ABLE TO . . .

✓ Test Your Knowledge

1. `remember` Fiber composites consist of cross-linked fiber components that withstand tension and a ground substance that withstands compression. The cellulose microfibrils in plants and collagen fibrils in animals functionally resemble the steel rods in reinforced concrete. The pectin in plants and proteoglycan in animals functionally resemble the concrete ground substance. **2.** `remember` b **3.** `remember` a **4.** `understand` b **5.** `remember` d **6.** `remember` Responses that affect which proteins are produced and those that affect the activity of existing proteins.

✓ Test Your Understanding

7. `understand` b **8.** `understand` If each enzyme in the cascade phosphorylates many copies of the enzyme in the next step of the cascade, the initial signal will be amplified many times over. **9.** `analyze` All three are made up of membrane-spanning proteins that directly interact between adjacent cells. Tight junctions seal adjacent animal cells together; gap junctions allow a flow of material from the cytosol of one to the other. Desmosomes firmly secure adjacent cells to one another but do not affect the movement of substances between cells or into the cells. **10.** `understand` When dissociated cells from two sponge species were mixed, the cells sorted themselves into distinct aggregates that contained only cells of the same species. By blocking membrane proteins with antibodies and isolating cells that would not adhere, researchers found that specialized groups of proteins, including cadherins, are responsible for selective adhesion. **11.** `create` Signaling molecule crosses plasma membrane and binds to intracellular receptor (reception) → Receptor changes conformation, and the activated receptor complex moves to target site (processing) → Activated receptor complex binds to a target molecule (e.g., a gene or membrane pump), which changes its activity (response) → Signaling molecule falls off receptor or is destroyed; receptor changes to inactive conformation (deactivation). **12.** `understand` Information from different signals may conflict or be reinforcing. "Crosstalk" between signaling pathways allows cells to integrate information from many signals at the same time instead of responding to each signal in isolation.

✓ Test Your Problem-Solving Skills

13. `apply` d **14.** `analyze` (a) The response would have to be extremely local—the activated receptor complex would have to affect nearby proteins. (b) No amplification could occur, because the number of signaling molecules dictates the amount of the response. (c) The only way to regulate the response would be to block the receptor or make it more responsive to the signaling molecule. **15.** `create` Chitin forms chains that can cross-link with one another. This fungal cell wall would likely need to be either relaxed or destroyed, and new cell wall synthesis

would be coordinated with the directional growth.
16. create Antibody binding to the two parts of the receptor may be causing them to dimerize. Since dimerization normally results after the signaling molecule binds to the receptor, the antibodies could be activating the receptor by mimicking this interaction. The result would be signal transduction even in the absence of the signaling molecule.

CHAPTER 12

IN-TEXT QUESTIONS AND EXERCISES

p. 223 understand **(1)** Genes are segments of chromosomes that code for RNAs and proteins. **(2)** Chromosomes are made of chromatin. **(3)** Sister chromatids are identical copies of the same chromosome, joined together.
p. 224 Fig. 12.5 apply (1) prophase cells would have 4 chromosomes with $2x$ DNA. (2) Anaphase cells would

have 8 chromosomes and $2x$ DNA. (3) Each daughter cell will have 4 chromosomes with x DNA.
p. 226 apply See **TABLE A12.1**.
p. 227 Fig. 12.6 apply The chromosome and black bar would move at the same rate toward the spindle pole.
p. 229 CYU (1) apply See **FIGURE A12.1**. **(2)** apply Loss of the motors would result in two problems: (1) It would reduce the ability of chromosomes to attach to microtubules via their kinetochores; (2) cytokinesis would be inhibited since the Golgi-derived vesicles would not be moved to the center of the spindle to build the cell plate.
p. 230 Fig. 12.10 create Inject cytoplasm from an M-phase frog egg into a somatic cell that is in interphase. If the somatic cell starts mitosis, then the meiotic factor is not limited to gametes.
p. 230 Fig. 12.11 analyze In effect, MPF turns itself off after it is activated. If this didn't happen, the cell might undergo mitosis again before the cell has replicated its DNA.

p. 231 understand MPF activates proteins that get mitosis under way. MPF consists of a cyclin and a Cdk, and it is turned on by phosphorylation at the activating site and dephosphorylation at the inhibitory site. Enzymes that degrade cyclin reduce MPF levels.
p. 232 CYU (1) remember $G_1 \rightarrow S \rightarrow G_2 \rightarrow M$. Checkpoints occur at the end of G_1 and G_2 and during M phase. **(2)** understand Cdk levels are fairly constant throughout the cycle, but cyclin increases during interphase and peaks in M phase. This accumulation of cyclin is a prerequisite for MPF activity, which turns on at the end of G_2, initiating M phase, and declines at the end of M phase.

IF YOU UNDERSTAND . . .

12.1 understand The G_1 and G_2 phases give the cell time to replicate organelles and grow before division as well as perform the normal functions required to stay alive. Chromosomes replicate during S phase and are separated from one another during M phase. Cytokinesis also occurs during M phase, when the parent cell divides into two daughter cells. **12.2** apply In cells that do not dissolve the nuclear envelope, the spindle must be constructed inside the nucleus to attach to the chromosomes and separate them. **12.3** apply Cells would prematurely enter M phase, shortening the length of G_2 and resulting in the daughter cells' being smaller than normal. **12.4** analyze The absence of growth factors in normal cells would cause them to arrest in the G_1 phase—eventually all the cells in the culture would be in G_1. The cancerous cells are not likely to be dependent on these growth factors, so the cells would not arrest and would continue through the cell cycle.

YOU SHOULD BE ABLE TO . . .

✔ Test Your Knowledge

1. remember b **2.** remember d **3.** remember d **4.** remember c
5. understand Daughter chromosomes were observed to move toward the pole faster than do the marked regions of fluorescently labeled kinetochore microtubules.
6. remember The cycle would arrest in M phase, and cytokinesis would not occur.

✔ Test Your Understanding

7. apply For daughter cells to have identical complements of chromosomes, all the chromosomes must be replicated during the S phase, the spindle apparatus must connect with the kinetochores of each sister chromatid in prometaphase, and the sister chromatids of each replicated chromosome must be partitioned in anaphase and fully separated into daughter cells by cytokinesis. **8.** create One possible concept map is shown in **FIGURE A12.2**. **9.** understand Microinjection experiments suggested that something in the cytoplasm of M-phase cells activated the transition from interphase to M phase. The control for this experiment was to inject cytoplasm from a G_2-arrested oocyte into another G_2-arrested oocyte. **10.** apply Protein kinases phosphorylate proteins. Phosphorylation changes a protein's shape, altering its function (activating or inactivating it). As a result, protein kinases regulate the function of proteins. **11.** understand Cyclin concentrations change during the cell cycle. At high concentration, cyclins bind to a specific cyclin-dependent kinase (or Cdk), forming a dimer. This dimer becomes active MPF by changing its shape through the phosphorylation (activating site) and dephosphorylation (inhibitory site) of Cdk. **12.** analyze b.

✔ Test Your Problem-Solving Skills

13. analyze a; adding up each phase allows you to determine that the cell cycle is 8.5 hours long. After 9 hours, the radiolabeled cells would have passed through a full cycle and be in either S phase or G_2—none would have entered M phase. **14.** apply The embryo passes through multiple rounds of the cell cycle, but cytokinesis does

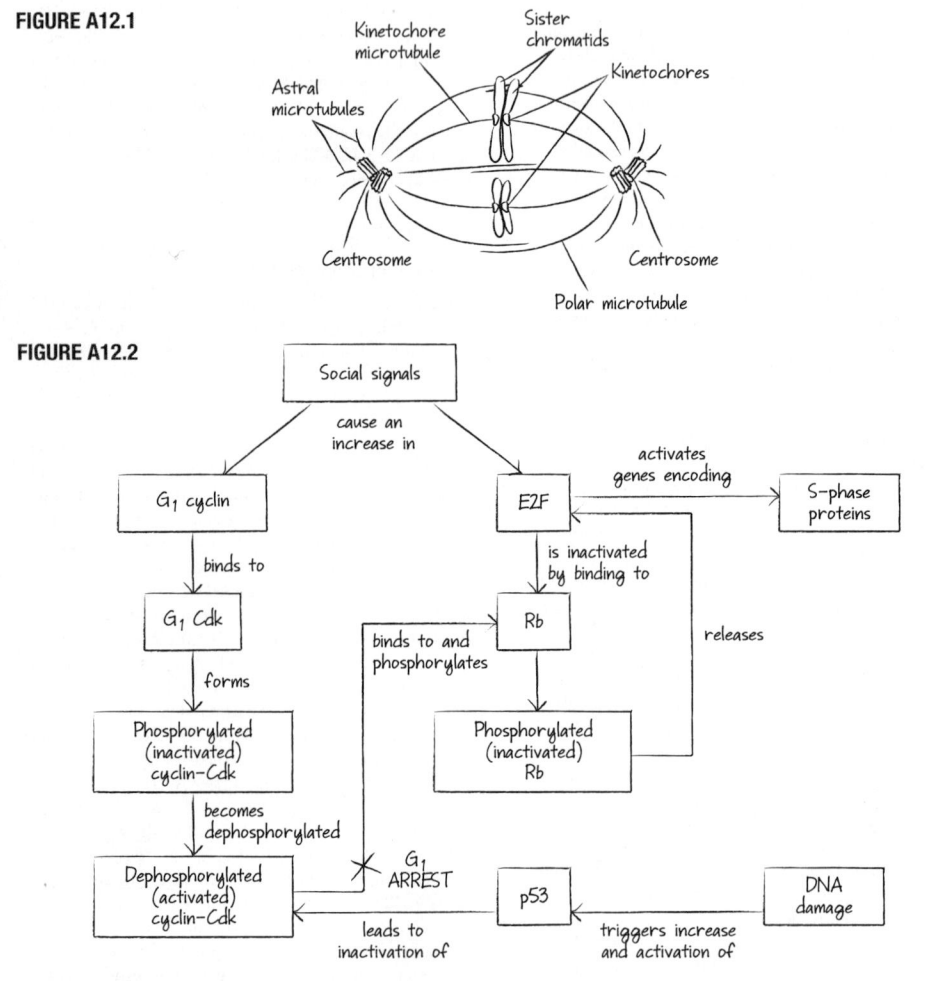

FIGURE A12.1

FIGURE A12.2

TABLE A12.1

	Prophase	Prometaphase	Metaphase	Anaphase	Telophase
Spindle apparatus	Starts to form	Contacts and moves chromosomes	Anchors poles to membrane and produces tension at kinetochores	Pulls chromatids apart	Defines site of cytokinesis
Nuclear envelope	Present	Disintegrates	Nonexistent	Nonexistent	Re-forms
Chromosomes	Condense	Attach to microtubules	Held at metaphase plate	Sister chromatids separate into daughter chromosomes	Collect at opposite poles

not occur during M phases. **15.** `apply` Early detection of cancers leads to a greater likelihood of survival. The widespread implementation of breast and prostate exams allows for the identification and removal of benign tumors before they become malignant. **16.** `analyze` Cancer requires many defects. Older cells have had more time to accumulate defects. Individuals with a genetic predisposition to cancer start out with some cancer-related defects, but this does not mean that the additional defects required for cancer to occur will develop.

CHAPTER 13

IN-TEXT QUESTIONS AND EXERCISES

p. 239 `apply` $n = 4$; the organism is diploid with $2n = 8$.
p. 240 `remember` See **FIGURE A13.1**. Because the two sister chromatids are identical and attached, it is sensible to consider them as parts of a single chromosome.
p. 242 `apply` There will be four DNA molecules in each gamete because the 8 replicated chromosomes in a diploid cell are reduced to 4 replicated chromosomes per cell at the end of meiosis I. In meiosis II, the sister chromatids of each replicated chromosome are separated. Each cell now contains 4 unreplicated chromosomes, each with a single molecule of DNA.
p. 245 `apply` Crossing over would not occur and the daughter cells produced by meiosis would be diploid, not haploid. There would be no reduction division.
p. 246 CYU (1) `understand` Use four long and four short pipe cleaners (or pieces of cooked spaghetti) to represent the chromatids of two replicated homologous chromosomes (four total chromosomes). Mark two long and two short ones with a colored marker pen to distinguish maternal and paternal copies of these chromosomes. Twist identical pipe cleaners (e.g., the two long colored ones) together to simulate replicated chromosomes. Arrange the pipe cleaners to depict the different phases of meiosis I as follows: *Early prophase I:* Align sister chromatids of each homologous pair to form two tetrads. *Late prophase I:* Form one or more crossovers between non-sister chromatids in each tetrad. (This is hard to simulate with pipe cleaners—you'll have to imagine that each chromatid now contains both maternal and paternal segments.) *Metaphase I:* Line up homologous pairs (the two pairs of short pipe cleaners and the two pairs of long pipe cleaners) at the metaphase plate. *Anaphase I:* Separate homologs. Each homolog still consists of sister chromatids joined at the centromere. *Telophase I and cytokinesis:* Move homologs apart to depict formation of two haploid cells, each containing a single replicated copy of two different chromosomes. **(2)** `understand` During anaphase I, homologs (not sister chromatids, as in mitosis) are separated, making the cell products of meiosis I haploid. **(3)** `understand` The pairing of homologs in metaphase I and their separation in anaphase I so that one goes to one daughter cell and the other to the other daughter cell means that each daughter cell obtains precisely one copy of each type of chromosome.
p. 248 `apply` Each gamete would inherit either all maternal or all paternal chromosomes. This would limit genetic variation in the offspring by precluding the many possible gametes containing various combinations of maternal and paternal chromosomes.
p. 248 CYU (1) `apply` See **FIGURE A13.2**. Maternal chromosomes are white and paternal chromosomes are black. Daughter cells with other possible combinations of chromosomes than shown could result from meiosis of this parent cell. **(2)** `understand` Crossing over would increase the genetic diversity of these gametes by creating many different combinations of maternal and paternal alleles along each of the chromosomes. **(3)** `analyze` Asexual reproduction generates no appreciable genetic diversity. Self-fertilization is preceded by meiosis so it generates gametes, through crossing over and independent assortment, that have combinations of alleles not present in the parent. Outcrossing generates the most genetic diversity among offspring because it produces new combinations of alleles from two different individuals.
p. 249 Fig. 13.11 `apply` See **FIGURE A13.3**.
p. 251 Fig. 13.14 `apply` Asexually: 64 (16 individuals from generation three produce 4 offspring per individual). Sexually: 16 (8 individuals from generation three form 4 couples; each couple produces 4 offspring).
p. 252 Fig. 13.15 `apply` The rate of outcrossing is predicted to rise initially, as the pathogen selects for resistant roundworms, and then to fall as the roundworms in the population gain resistance and take advantage of the increased numbers of offspring offered by having more hermaphrodites that can reproduce by self-fertilization.

IF YOU UNDERSTAND . . .

13.1 `remember` See the right panel of Figure 13.7 and note how the cells transition from diploid to haploid in meiosis I. Also note that each chromosome contains two sister chromatids before and after meiosis I.
13.2 `understand` (a) See Figure 13.10 and note how the two different ways of aligning two homologous pairs of chromosomes at metaphase I of meiosis can create four different combinations of maternal and paternal chromosomes in daughter cells. (b) See Figure 13.11. Note that for each homologous pair with one crossover, two chromatids are recombinant and two are unaltered. Since each chromatid will produce a chromosome at the end of meiosis II, your drawing should show that two out of four of these chromosomes are recombinant.
13.3 `apply` (a) Your model should show events similar to those of Figure 13.12. (b) Your model should show two cells with one of each type of chromosome (n), one cell with an extra copy of one chromosome ($n + 1$), and one gamete without any copies of one chromosome ($n - 1$).
13.4 `remember` Sexual reproduction will likely occur during times when conditions are changing rapidly, because genetically diverse offspring may have an advantage in the new conditions.

YOU SHOULD BE ABLE TO . . .

✔ **Test Your Knowledge**

1. `remember` b **2.** `remember` a **3.** `remember` b **4.** `understand` b
5. `remember` 1/2 **6.** `remember` mitosis.

✔ **Test Your Understanding**

7. `remember` Homologous chromosomes are similar in size, shape, and gene content, and originate from different parents. Sister chromatids are exact copies of a chromosome that are generated when chromosomes are replicated (S phase of the cell cycle). **8.** `understand` Refer to Figure 13.7 as a guide for this exercise. The four pens represent the chromatids in one replicated homologous pair; the four pencils, the chromatids in a different homologous pair. To simulate meiosis II, make two "haploid cells"—each with a pair of pens and a pair of pencils representing two replicated chromosomes (one of each type in this species). Line them up in the middle of the cell; then separate the two pens and the two pencils in each cell such that one pen and one pencil go to each of four daughter cells. **9.** `understand` Meiosis I is a reduction division because homologs separate—daughter cells have just one of each type of chromosome instead of two. Meiosis II is not a reduction division because sister chromatids separate— daughter cells have unreplicated chromosomes instead of replicated chromosomes, but still just one of each type. **10.** `apply` b. **11.** `apply` Tetraploids produce diploid gametes, which combine with a haploid gamete from a diploid individual to form a triploid offspring. Mitosis proceeds normally in triploid cells because mitosis doesn't require forming pairs of chromosomes. But during meiosis in a triploid, homologous chromosomes can't pair up correctly. The third set of chromosomes does not have a homologous partner to pair with. **12.** `understand` Asexually produced individuals are genetically identical, so if one is susceptible to a new disease, all are. Sexually produced individuals are genetically unique, so if a new disease strain evolves, at least some plants are likely to be resistant.

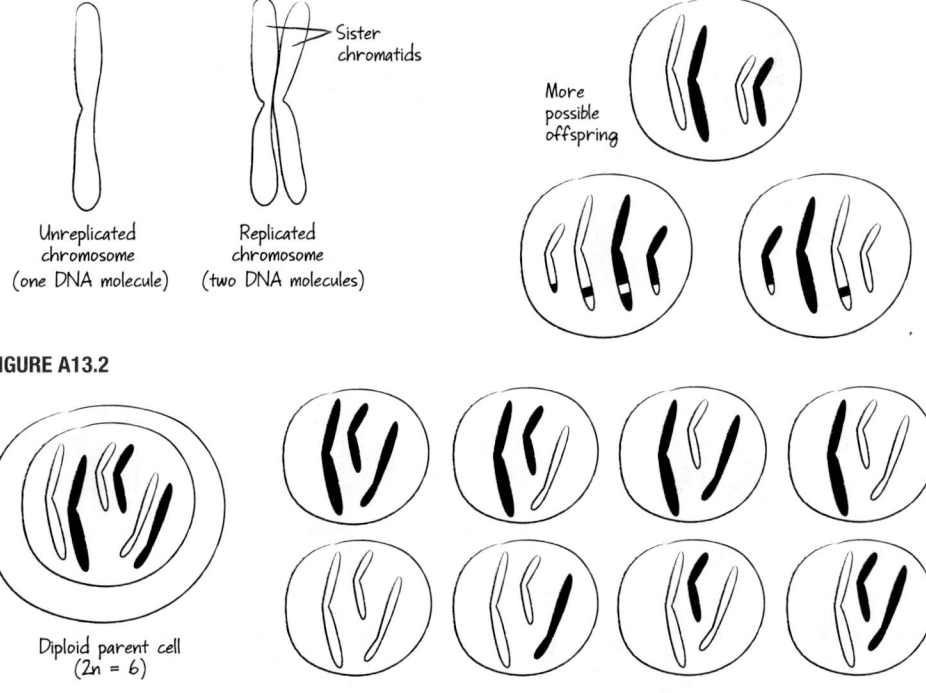

FIGURE A13.1

Sister chromatids

Unreplicated chromosome (one DNA molecule)

Replicated chromosome (two DNA molecules)

FIGURE A13.2

Diploid parent cell ($2n = 6$)

Possible daughter cells ($n = 3$)

FIGURE A13.3

More possible offspring

13. apply The gibbon would have 22 chromosomes in each gamete, and the siamang would have 25. Each somatic cell of the offspring would have 47 chromosomes. The offspring should be sterile because it has some chromosomes that would not form homologous pairs at prophase I of meiosis. **14.** apply c **15.** apply Aneuploidy is the major cause of spontaneous abortion. If spontaneous abortion is rare in older women, it would result in a higher incidence of aneuploid conditions such as Down syndrome in older women, as recorded in the Figure 13.13. **16. (a)** create Such a study might be done in the laboratory, controlling conditions in identical dishes. A population of rotifers infected with fungus could be established in each dish. One dish of rotifers would be kept moist; the other dishes of rotifers would be allowed to dry out. After various periods of time, water would be added to each dish and then the rotifers would be observed to see if fungal infections reappeared. **(b)** create Wind disperses the rotifer to new and often pathogen-free environments. In this case, the ticket to a sex-free existence is not genetic diversity but the evolution of an alternative means of evading pathogens made possible by fungus-infected rotifers ridding themselves of the pathogen when they dry.

CHAPTER 14

IN-TEXT QUESTIONS AND EXERCISES

p. 260 Fig. 14.3 evaluate An experiment is a failure if you didn't learn anything from it. That is not the case here.
p. 261 Table 14.2 analyze *Row 6*: 3.14:1; *Row 7*: 266.
p. 262 apply Filling in the top and side of a Punnett square requires writing out the types and ratios of gametes. For a cross involving one gene (monohybrid cross), this amounts to applying the principle of segregation as one allele is segregated from another. The phenotype ratios are 1:1 round:wrinkled; the genotype ratios are 1:1 *Rr*:*rr*.
p. 262 Fig. 14.4 understand No—the outcome (the expected offspring genotypes that the Punnett square generates) will be the same.

p. 263 CYU (1) apply See answer to Problem 13 in Test Your Problem-Solving Skills, below. **(2)** apply See answer to Problem 15 in Test Your Problem-Solving Skills.
p. 265 apply *AABb* → *AB* and *Ab*. *PpRr* → *PR*, *Pr*, *pR*, and *pr*. *AaPpRr* → *APR*, *APr*, *ApR*, *Apr*, *aPR*, *apR*, *apr*, and *aPr*.
p. 265 CYU (1) apply See answer to Problem 14 in Test Your Problem-Solving Skills. **(2)** apply See answer to Problem 16 in Test Your Problem-Solving Skills.
p. 269 CYU (1) understand See **FIGURE A14.1**. Segregation of alleles occurs when homologs that carry those alleles are separated during anaphase I. One allele ends up in each daughter cell. **(2)** understand See **FIGURE A14.2**. Independent assortment occurs because homologous pairs line up randomly at the metaphase plate during metaphase I. The figure shows two alternative arrangements of homologs in metaphase I. As a result, it is equally possible for a gamete to receive the following four combinations of alleles: *YR*, *Yr*, *yR*, *yr*.
p. 270 Fig. 14.12 apply *XWY*, *XWy*, *XwY*, *Xwy*.

p. 271 Fig. 14.13 analyze Random chance (or perhaps red-eyed, gray-bodied males don't survive well).
p. 273 Fig. 14.16 analyze The gene colored orange is *ruby*; the gene colored blue is *miniature wings*.
p. 275 Fig. 14.19 understand In this case of gene-by-gene interaction, the 9:3:3:1 ratio comes from four different forms of one trait (comb shape), whereas in a standard dihybrid cross, the four different phenotypes come from two different phenotypes for each of two genes.
p. 276 CYU (1) apply The comb phenotype results from interactions between alleles at two different genes, not a single gene. Matings between rose- and pea-comb chickens produce F₂ offspring that may have a new combination of alleles and thus new phenotypes. **(2)** apply Kernel color in wheat is influenced by alleles at many different genes, not a single gene. F₂ offspring have a normal distribution of phenotypes, not a 3:1 ratio.
p. 276 Fig. 14.21 understand Because there are many different genotypes that can produce intermediate coloration and fewer that can produce the extremes of coloration.

FIGURE A14.1

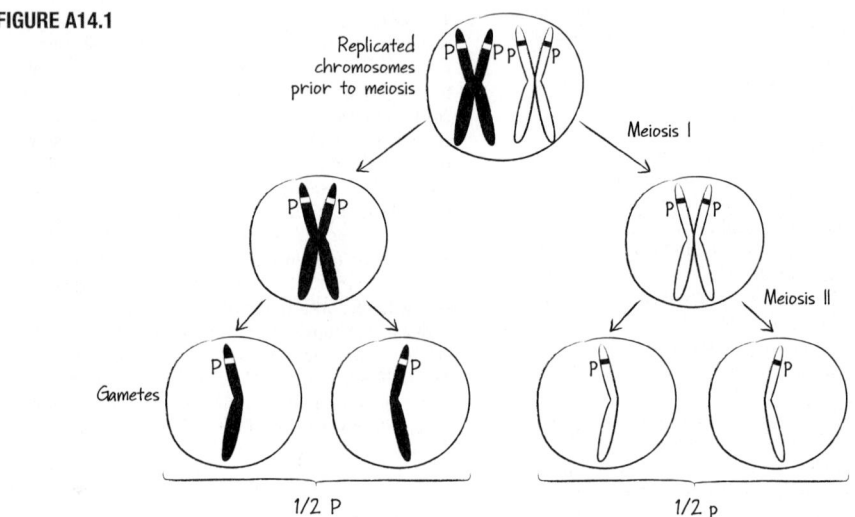

FIGURE A14.2

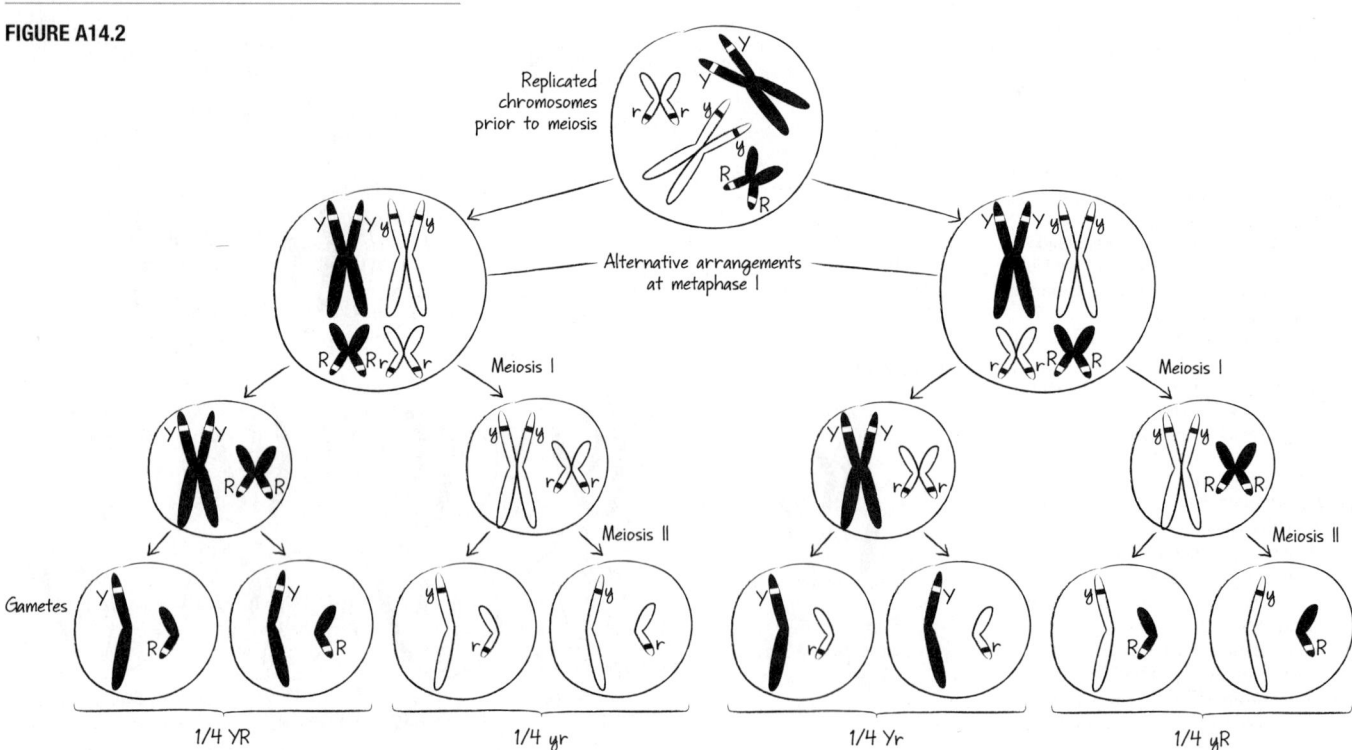

p. 279 Fig. 14.23 apply A heterozygous female and a color-blind male. 1 color-blind male : 1 color-blind female.

IF YOU UNDERSTAND . . .

14.1 understand Because crosses within a pure line never produce individuals with a different phenotype, this indicates that there must be only one allele in pure-line individuals. **14.2** apply B and b. **14.3** apply BR, Br, bR, and br, in equal proportions. The B and b alleles are located on different but homologous chromosomes, which separate into different daughter cells during meiosis I. The $BbRr$ notation indicates that the B and R genes are on different chromosomes. As a result, the chromosomes line up independently of each other in metaphase of meiosis I. The B allele is equally likely to go to a daughter cell with R as with r; likewise, the b allele is equally likely to go to a daughter cell with R as with r. **14.4** apply There are many ways to show this. If eye color in *Drosophila* is chosen as an example, then a pair of Punnett squares like those in Figure 14.11 illustrate how reciprocal crosses involving an X-linked recessive gene give different results. **14.5** understand Linkage refers to the physical connection of two alleles on the same chromosome. Crossing over breaks this linkage between particular alleles as segments of maternal and paternal homologs are exchanged. **14.6** apply See **FIGURE A14.3** for the pedigree. Note that the birth order of daughters and sons is arbitrary; other birth orders are equally valid. A Punnett square will show that all the sons are predicted to be X^+Y and all the daughters X^+X^c, where X^+ shows the dominant, X-linked allele for normal color vision and X^c shows the X-linked recessive color blindness allele.

YOU SHOULD BE ABLE TO . . .

✓ Test Your Understanding

1. understand d **2.** understand b **3.** understand a **4.** understand d
5. understand a **6.** understand a **7.** evaluate d **8.** remember b **9.** apply b
10. understand d

✓ Test Your Problem-Solving Skills

11. apply **Example Solution** Here you're given offspring phenotypes and you're asked to infer parental genotypes. To do this you have to propose hypothetical parental genotypes to test, make a Punnett square to predict the offspring genotypes, and then see if the predicted offspring phenotypes match the data. In this case, coming up with a hypothesis for the parental genotypes is relatively straightforward because the problem states that the trait is due to one gene and two alleles. No information on sex is given, so assume the gene is autosomal (the simplest case). Now look at the second entry in the chart. It shows a 3 : 1 ratio of offspring from a winged individual that self-fertilizes. This result is consistent with the hypothesis that W^+ is dominant and W^- recessive and that this parent's genotype is W^+W^-. Now let's look at the first cross in the chart. If W^+ is dominant, then a wingless parent must be W^-W^-. When you do the Punnett square to predict offspring genotypes from selfing, you find that all the offspring will produce wingless fruits, consistent with the data. In the third cross, all the offspring make winged fruits even though one of the parents produces wingless

FIGURE A14.3

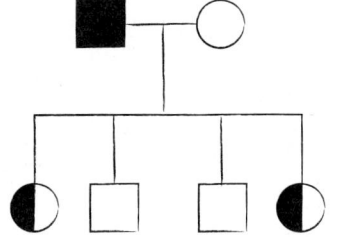

---

fruits and thus is W^-W^-. This would happen only if the winged parent is W^+W^+. (If this reasoning isn't immediately clear to you, work the Punnett square.) In the fourth cross, you could get offspring that all make winged fruits if the parents were W^+W^+ and W^+W^+, or if the parents were W^+W^+ and W^+W^-. Either answer is correct. Again, you can write out the Punnett squares to see that this statement is correct.

12. apply **Example Solution** Here you are given parental and offspring phenotypes and are asked to infer the parental genotypes. As a starting point, assume that the coat colors are due to the simplest genetic system possible: one autosomal gene with two alleles, where one allele is dominant and the other recessive. Because female II produces only black offspring, it's logical to suppose that black is dominant to brown. Let's use B for black and b for brown. Then the male parent is bb. To produce offspring with a 1 : 1 ratio of black : brown coats, female I must be Bb. But to produce all black offspring, female II must be BB. This model explains the data, so you can accept it as correct.

13. apply 3/4; 1/256 (see BioSkills 5 in Appendix B); 1/2 (the probabilities of transmitting the alleles or having sons does not change over time). **14.** apply Your answer to the first three parts should conform to the F_1 and F_2 crosses diagrammed in Figure 14.5b, except that different alleles and traits are being analyzed. The recombinant gametes would be Yi and yI. Yes—there would be some individuals with yellow seeds and constricted pods and with green seeds and inflated pods. **15.** apply Cross 1: non-crested (Cc) × non-crested (Cc) = 22 non-crested ($C\_$); 7 crested (cc). Cross 2: crested (cc) × crested (cc) = 20 crested (cc). Cross 3: non-crested (Cc) × crested (cc) = 7 non-crested (Cc); 6 crested (cc). Non-crested (C) is the dominant allele. **16.** apply This is a dihybrid cross that yields progeny phenotypes in a 9 : 3 : 3 : 1 ratio. Let O stand for the allele for orange petals and o the allele for yellow petals; let S stand for the allele for spotted petals and s the allele for unspotted petals. Start with the hypothesis that O is dominant to o, that S is dominant to s, that the two genes are found on different chromosomes so they assort independently, and that the parent individual's genotype is $OoSs$. If you do a Punnett square for the $OoSs × OoSs$ mating, you'll find that progeny phenotypes should be in the observed 9 : 3 : 3 : 1 proportions. **17.** analyze Let D stand for the normal allele and d for the allele responsible for Duchenne-type muscular dystrophy. The woman's family has no history of the disease, so her genotype is almost certainly DD. The man is not afflicted, so he must be DY. (The trait is X-linked, so he has only one allele; the "Y" stands for the Y chromosome.) Their children are not at risk. The man's sister could be a carrier, however—meaning she has the genotype Dd. If so, then half of the second couple's male children are likely to be affected. **18.** understand Your stages of meiosis should look like a simplified version of Figure 13.7, except with $2n = 4$ instead of $2n = 6$. The A and a alleles could be on the red and blue versions of the longest chromosome, and the B and b alleles could be on the red and blue versions of the smallest chromosomes, similar to the way the hair- and eye-color genes are shown in Figure 13.10. The places you draw them are the locations of the A and B genes, but each chromosome has only one allele. Each pair of red and blue chromosomes is a homologous pair. Sister chromatids bear the same allele (e.g., both sister chromatids of the long blue chromosomes might bear the a allele). Chromatids from the longest and shortest chromosomes are not homologous. To identify the events that result in the principles of segregation and independent assortment, see Figures 14.7 and 14.8 and substitute A, a, and B, b for R, r and Y, y. **19.** apply Half their offspring should have the genotype iI^A and the type A blood phenotype. The other half of their offspring should have the genotype iI^B and the type B blood

---

phenotype. Second case: the genotype and phenotype ratios would be 1 : 1 : 1 : 1 I^AI^B (type AB) : I^Ai (type A) : I^Bi (type B) : ii (type O). **20.** apply Because the children of Tukan and Valco had no eyes and smooth skin, you can conclude that the allele for eyelessness is dominant to eyes and the allele for smooth skin is dominant to hooked skin. E = eyeless, e = two eye sets, S = smooth skin, s = hooked skin. Tukan is $eeSS$; Valco is $EEss$. The children are all $EeSs$. Grandchildren with eyes and smooth skin are eeS-. Assuming that the genes are on different chromosomes, one-fourth of the children's gametes are ee and three-fourths are S-. So 1/4 $ee ×$ 3/4 S- × 32 = 6 children would be expected to have two sets of eyes and smooth skin. **21.** create Although the mothers were treated as children by a reduction of dietary phenylalanine, they would have accumulated phenylalanine and its derivatives once they went off the low-phenylalanine diet as young adults. Children born of such mothers were therefore exposed to high levels of phenylalanine during pregnancy. For this reason, a low-phenylalanine diet is recommended for such mothers throughout the pregnancy. **22.** apply According to Mendel's model, palomino individuals should be heterozygous at the locus for coat color. If you mated palomino individuals, you would expect to see a combination of chestnut, palomino, and cremello offspring. If blending inheritance occurred, however, all the offspring should be palomino. **23.** apply Because this is an X-linked trait, the father who has hemophilia could not have passed the trait on to his son. Thus, the mother in couple 1 must be a carrier and must have passed the recessive allele on to her son, who is XY and affected. To educate a jury about the situation, you should draw what happens to the X and Y during meiosis and then make a drawing showing the chromosomes in couple 1 and couple 2, with a Punnett square showing how these chromosomes are passed to the affected and unaffected children. **24.** apply The curved-wing allele is autosomal recessive; the lozenge-eyed allele is sex-linked (specifically, X-linked) recessive. Let L be the allele for long wings, l be the allele for curved wings; let X^R be the allele for red eyes and X^r the allele for lozenge eyes. The female parent is LlX^RX^r; the male parent is LlX^RY. **25.** apply Albinism indicates the absence of pigment, so let b stand for an allele that gives the absence of blue and y for an allele that gives the absence of yellow pigment. If blue and yellow pigment blend to give green, then both green parents are $BbYy$. The green phenotype is found in $BBYY$, $BBYy$, $BbYY$, and $BbYy$ offspring. The blue phenotype is found in $BByy$ or $Bbyy$ offspring. The yellow phenotype is observed in $bbYY$ or $bbYy$ offspring. Albino offspring are $bbyy$. The phenotypes of the offspring should be in the ratio 9 : 3 : 3 : 1 as green : blue : yellow : albino. Two types of crosses yield $BbYy$ F_1 offspring: $BByy × bbYY$ (blue × yellow) and $BBYY × bbyy$ (green × albino). **26.** apply The chance that their first child will have hemophilia is 1/2. This is because all sons will have the disease and there is a 1/2 chance of having a firstborn son. The chance of having a carrier as their first child is also 1/2. This is because all daughters and none of the sons will be carriers and there is a 1/2 chance of having a firstborn daughter. (Recall that males cannot carry an X-linked recessive trait—with only one X chromosome, males either have the trait or not.) **27.** apply Autosomal dominant.

CHAPTER 15

IN-TEXT QUESTIONS AND EXERCISES

p. 286 Fig. 15.2 analyze The lack of radioactive protein in the pellet (after centrifugation) is strong evidence; they could also make micrographs of infected bacterial cells before and after agitation.
p. 287 Fig. 15.3 apply 5'-TAG-3'.
p. 288 Fig. 15.5 apply The same two bands should appear, but the upper band (DNA containing only ^{14}N) should get bigger and darker and the lower band (hybrid DNA)

should get smaller and lighter in color since each suc-ceeding generation has relatively less heavy DNA.

p. 290 `apply` See **FIGURE A15.1**. The new strands grow in opposite directions, each in the $5' \rightarrow 3'$ direction.

p. 292 `apply` Helicase, topoisomerase, single-strand DNA-binding proteins, primase, and DNA polymerase are all required for leading-strand synthesis. If any one of these proteins is nonfunctional, DNA replication will not occur.

p. 293 `apply` See **FIGURE A15.2**. If DNA ligase were defective, then the leading strand would be continuous, and the lagging strand would have gaps in it where the Okazaki fragments had not been joined.

p. 294 CYU (1) `understand` DNA polymerase adds nucleotides only to the free $3'-$OH on a strand. Primase synthesizes a short RNA sequence that provides the free $3'$ end neces-sary for DNA polymerase to start working. **(2)** `understand` The need to begin DNA synthesis many times on the lagging strand requires many new primers. Since DNA polymerase I is needed to remove primers, it is required predominantly on the primer-rich lagging strand.

p. 296 Fig. 15.13 `apply` As long as the RNA template could bind to the "overhanging" section of single-stranded DNA, any sequence could produce a longer strand. For example, 5'-CCCAUUCCC-3' would work just as well.

p. 297 CYU (1) `understand` This is because telomerase is needed only to replicate one end of a linear DNA and bacterial DNAs lack ends because they are circular. **(2)** `understand` Since telomerase works by extending one strand of DNA without any external template and be-cause DNA synthesis requires a template, telomerase must contain an internal template to allow it to extend a DNA chain.

p. 299 Fig. 15.16 `analyze` They are lower in energy and not absorbed effectively by the DNA bases.

p. 300 CYU (1) `apply` The mutation rate would be pre-dicted to rise because differences in base-pair stability and shape make it possible for DNA polymerase to dis-tinguish correct from incorrect base pairs during DNA replication. **(2)** `apply` The mutation rate should increase because without a way to distinguish which strand to use as a template for repair, about half of mismatches on average would be repaired using the incorrect strand as a template. **(3)** `remember` The enzyme that removes the dimer and surrounding DNA is specific to nucleotide excision repair. DNA polymerase and DNA ligase work in both nucleotide excision repair and normal DNA synthesis.

p. 300 Fig. 15.18 `apply` Exposure to UV radiation can cause formation of thymine dimers. If thymine dimers are not repaired, they represent mutations. If such muta-tions occur in genes controlling the cell cycle, cells can grow abnormally, resulting in cancers.

IF YOU UNDERSTAND . . .

15.1 `apply` These results would not allow distinguishing whether DNA or protein was the genetic material.

15.2 `understand` The bases added during DNA replication are shown in red type.

Original DNA:	**CAATTACGGA**
	GTTAATGCCT
Replicated DNA:	**CAATTACGGA**
	GTTAATGCCT
	CAATTACGGA
	GTTAATGCCT

15.3 `understand` See **FIGURE A15.3**. **15.4** `understand` Because cancer cells divide nearly without limit, it's important for these cells to have active telomerase so that chro-mosomes don't shorten to the point where cell division becomes impossible. **15.5** `understand` If errors in DNA aren't corrected, they represent mutations. When DNA repair systems fail, the mutation rate increases. As the mutation rate increases, the chance that one or more cell-cycle genes will be mutated increases. Mutations in these genes often result in uncontrolled cell division, ultimately leading to cancer.

✔ Test Your Knowledge

1. `remember` d **2.** `understand` a **3.** `remember` topoisomerase
4. `remember` b **5.** `remember` c **6.** `remember` telomerase

✔ Test Your Understanding

7. `remember` Labeling DNA or labeling proteins. **8.** `understand` DNA is constantly damaged, and many pathways have evolved to repair this onslaught of damage. If a DNA re-pair pathway is inactivated by mutation, damage is inef-ficiently repaired. Consequently mutation rates increase, and the increased number of mutations increases the probability that cancer-causing mutations will occur. **9.** `understand` On the lagging strand, DNA polymerase moves away from the replication fork. When helicase unwinds a new section of DNA, primase must build a new primer on the template for the lagging strand (closer to the fork) and another polymerase molecule must begin synthesis at this point. This makes the lagging-strand synthesis discontinuous. On the leading strand, DNA polymerase moves in the same direction as helicase, so synthesis can continue, without interrup-tion, from a single primer (at the origin of replication). **10.** `understand` Telomerase binds to the overhang at the end of a chromosome. Once bound, it begins catalyzing the addition of deoxyribonucleotides to the overhang in the

$5' \rightarrow 3'$ direction, lengthening the overhang. This allows primase, DNA polymerase, and ligase to catalyze the addition of deoxyribonucleotides to the lagging strand in the $5' \rightarrow 3'$ direction, restoring the lagging strand to its original length. **11.** `apply` a (Because the ability to dis-tinguish which strand contains the incorrect base would be lost). **12.** `analyze` d (The regularity of DNA's structure allows enzymes to recognize any type of damage that distorts this regular structure.)

✔ Test Your Problem-Solving Skills

13. `analyze` c (If DNA polymerase could synthesize DNA $3' \rightarrow 5'$ as well as the normal $5' \rightarrow 3'$, then both newly synthesized DNA strands could be extended to follow the replication fork.) **14.** `apply` (a) In **FIGURE A15.4**, the gray lines represent DNA strands containing radioactiv-ity. (b) After one round of replication in radioactive so-lution, one double-stranded DNA would be radioactive in both strands and the other would not be radioactive in either strand. After another round of DNA synthesis, this time in nonradioactive solution, one of the four DNA molecules would be radioactive in both strands and the other three DNA molecules would contain no radioactivity in any strand. **15.** `analyze` (a) The double mu-tant of both *uvrA* and *recA* is most sensitive to UV light; the single mutants are in between; and the wild type is

---

FIGURE A15.1

FIGURE A15.2

FIGURE A15.3

FIGURE A15.4

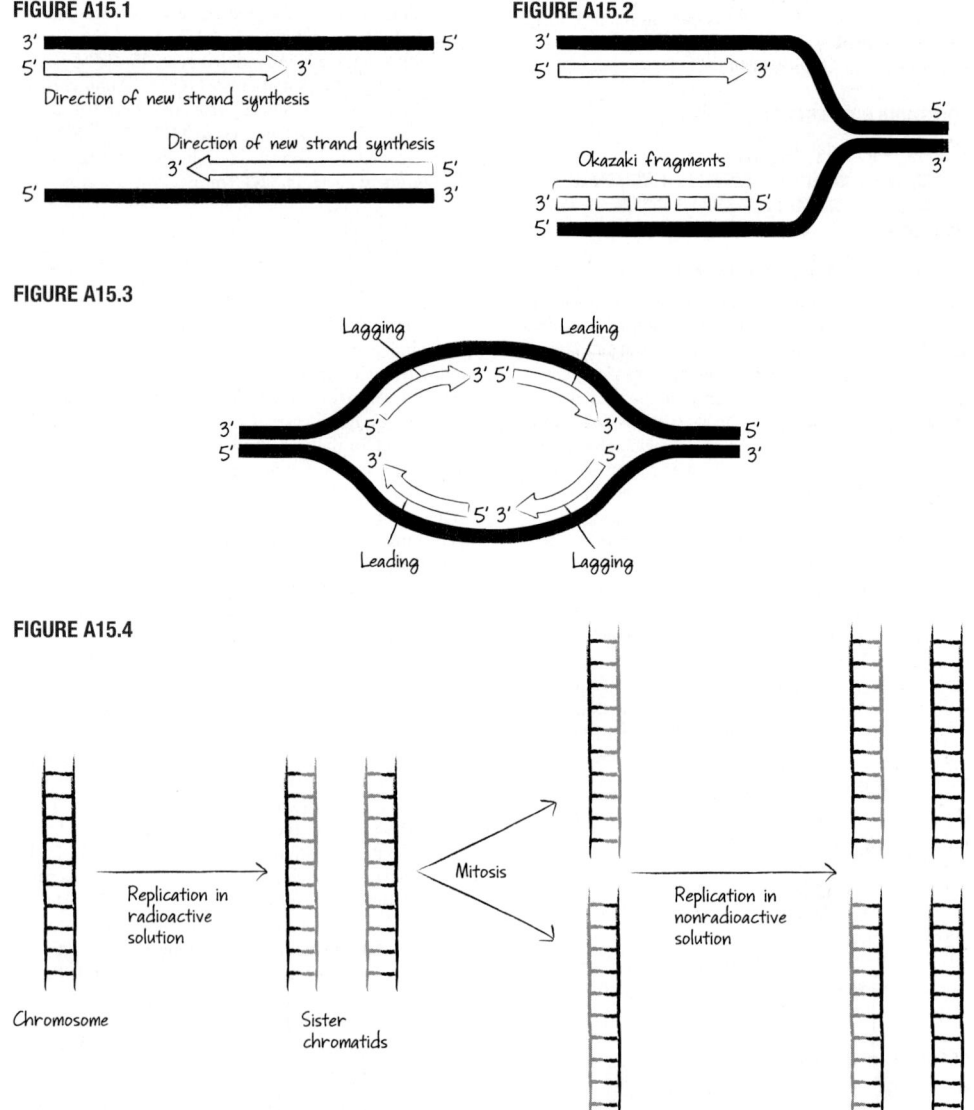

least sensitive. (b) The *recA* gene contributes more to UV repair through most of the UV dose levels. But at very high UV doses, the *uvrA* gene is somewhat more important than the *recA* gene. **16.** apply About 4600 seconds or 77 minutes. This answer comes from knowing that replication proceeds bidirectionally, so replication from each fork is predicted to replicate half the chromosome. This is 4.6 million base pairs/2 = 2.3 million base pairs. At 500 base pairs per second, this requires 2.3 million base pairs/500 base pairs per second = 4600 seconds. To obtain the time in minutes, divide 4600 seconds by 60 seconds per minute.

CHAPTER 16

IN-TEXT QUESTIONS AND EXERCISES

p. 305 Fig. 16.1 apply No, it could not make citrulline from ornithine without enzyme 2. Yes, it would no longer need enzyme 2 to make citrulline.

p. 306 Fig. 16.2 create Many possibilities: strain of fungi used, exact method for creating mutants and harvesting spores to grow, exact growing conditions (temperature, light, recipe for growth medium—including concentrations of supplements), objective criteria for determining growth or no growth.

p. 309 CYU remember Change in DNA sequence, change in sequence of transcribed mRNA, potential change in amino acid sequence of protein, likely altered protein function (if amino acid sequence was altered), likely change in phenotype.

p. 312 analyze (1) The codons in Figure 16.4 are translated correctly. (2) See **FIGURE A16.1**. (3) There are many possibilities (just pick alternative codons for one or more of the amino acids); one is an mRNA sequence (running 5′ → 3′): 5′ GCG-AAC-GAU-UUC-CAG 3′. To get the corresponding DNA sequence, write this sequence but substitute Ts for Us: 5′ GCG-AAC-GAT-TTC-CAG 3′. Now write the complementary bases, which will be in the 3′ → 5′ direction: 3′ CGC-TTG-CTA-AAG-GTC 5′. When this second strand is transcribed by RNA polymerase, it will produce the mRNA given with the proper 5′ → 3′ orientation.

p. 312 CYU (1) apply Note the 3′ → 5′ polarity of the DNA sequences in the accompanying table, and in the subsequent answer. This means that the complementary mRNA codon will read 5′ → 3′. U (rather than T) is the base transcribed from A.

DNA	mRNA Codon	Amino Acid
ATA	UAU	Tyrosine
ATG	UAC	Tyrosine
ATT	UAA	Stop
GCA	UGC	Cysteine

(2) understand The ATA → ATG mutation would have no effect on the protein. The ATA → ATT mutation introduces a stop codon, so the resulting polypeptide would be shortened. This would result in synthesis of a mutant protein much shorter than the original protein. The ATA → GCA mutation might have a profound effect on the protein's conformation because cysteine's structure is different from tyrosine's.

p. 312 Fig. 16.7 apply See **FIGURE A16.1**.

p. 314 Fig. 16.9 analyze Chromosomes 2, 3, 6, 10, 13, 14, 15, 18, 19, 21, 22, and the X chromosome show aneuploidy. Virtually every chromosome has structural rearrangements, and translocations are the most obvious. These are seen when two or more different colors occur on the same chromosome.

IF YOU UNDERSTAND . . .

16.1 understand Ornithine, citrulline or arginine could be added to allow growth. As Figure 16.1 shows, these

compounds are made after the steps catalyzed by the enzymes 1, 2, and 3 that are needed to produce arginine. **16.2** understand An inhibitor of RNA synthesis will eventually prevent the synthesis of new proteins because newly synthesized mRNA is needed for translation. **16.3** apply AUG UGG AAA/AAG CAA/CAG **16.4** understand Since redundancy is having more than one codon specify a particular amino acid, redundancy makes it possible for there to be a point mutation without altering the amino acid. This is a silent mutation. A silent mutation is likely to be neutral because there is no change in amino acid sequence.

YOU SHOULD BE ABLE TO . . .

✔ **Test Your Knowledge**

1. remember d **2.** understand a **3.** understand d **4.** remember Because there is no chemical complementarity between nucleotides and amino acids; and because in eukaryotes, DNA is in the nucleus but translation occurs in the cytoplasm. **5.** understand d **6.** remember A codon that signals the end of translation.

✔ **Test Your Understanding**

7. understand Because the Morse code and genetic code both use simple elements (dots and dashes; 4 different bases) in different orders to encode complex information (words; amino acid sequences).

8.

Substrate 1 $\xrightarrow{A}$ Substrate 2 $\xrightarrow{B}$ Substrate 3 $\xrightarrow{C}$
Substrate 4 $\xrightarrow{D}$ Substrate 5 $\xrightarrow{E}$ Biological Sciazine

apply Substrate 3 would accumulate. Hypothesis: The individuals have a mutation in the gene for enzyme D. **9.** understand They supported an important prediction of the hypothesis: Losing a gene (via mutation) resulted in loss of an enzyme. **10.** understand c **11.** understand In a triplet code, addition or deletion of 1–2 bases disrupts the reading frame "downstream" of the mutation site(s), resulting in a dysfunctional protein. But addition or deletion of 3 bases restores the reading frame—the normal sequence is disrupted only between the first and third mutation. The resulting protein is altered but may still be able to function normally. Only a triplet code would show these patterns. **12.** understand A point mutation changes the nucleotide sequence of an existing allele, creating a new one, so it always changes the genotype. But because the genetic code is redundant, and because point mutations can occur in DNA sequences that do not code for amino acids, these point mutations do not change the protein product and therefore do not change the phenotype.

✔ **Test Your Problem-Solving Skills**

13. apply See **FIGURE A16.2**. **14.** analyze Every copying error would result in a mutation that would change the amino

acid sequence of the protein and would likely affect its function. **15.** analyze Before the central dogma was understood, DNA was known to be the hereditary material, but no one knew how particular sequences of bases resulted in the production of RNA and protein products. The central dogma clarified how genotypes produce phenotypes. **16.** analyze c

CHAPTER 17

IN-TEXT QUESTIONS AND EXERCISES

p. 318 Fig. 17.1 remember RNA is synthesized in the 5′ → 3′ direction; the DNA template is "read" 3′ → 5′.

p. 321 CYU (1) apply Transcription would be reduced or absent because the missing nucleotides are in the −10 region, one of the two critical parts of the promoter. **(2)** understand NTPs are required because the three phosphate groups raise the monomer's potential energy enough to make the polymerization reaction exergonic.

p. 322 Fig. 17.5 apply There would be no loops—the molecules would match up exactly.

p. 323 CYU (1) understand The subunits contain both RNA (the *ribonucleo*– in the name) and proteins. **(2)** understand The cap and tail protect mRNAs from degradation and facilitate translation.

p. 326 Fig. 17.11 analyze If the amino acids stayed attached to the tRNAs, the gray line in the graph would stay high and the green line low. If the amino acids were transferred to some other cell component, the gray line would decline but the green line would be low.

p. 327 understand (1) The amino acid attaches on the top right of the L-shaped structure. (2) The anticodon is antiparallel in orientation to the mRNA codon, and it contains the complementary bases.

p. 332 CYU understand E is for exit—the site where uncharged tRNAs are ejected; P is for peptidyl (or peptide bond)—the site where peptide bond formation takes place; A is for aminoacyl—the site where aminoacyl tRNAs enter.

IF YOU UNDERSTAND . . .

17.1 apply Transcription would continue past the normal point because the insertion of nucleotides would disrupt the structure of the RNA hairpin that functions as a terminator. **17.2** apply The protein-coding segment of the gene is predicted to be longer in eukaryotes because of the presence of introns. **17.3** understand The tRNAs act as adaptors because they couple the information contained in the nucleotides of mRNA to that contained in the amino acid sequence of proteins. **17.4** apply An incorrect amino acid would appear often in proteins. This is because the altered synthetase would sometimes add the correct amino acid for a particular tRNA and at other

---

FIGURE A16.1

mRNA sequence:
5′ AUG-CUG-GAG-GGG-GUU-AGA-CAU 3′

Amino acid sequence:
Met–Leu–Glu–Gly–Val–Arg–His

FIGURE A16.2

Bottom DNA strand:
3′ AACTT–TAC(start)–GGG-CAA-ACC-TCT-AGC-CCA-ATG-TCG-ATC(stop)–AGTTTC 5′

mRNA sequence:
5′ AUG–CCC-GUU-UGG-AGA-UCG-GGU-UAC-AGC-UAG 3′

Amino acid sequence:
Met–Pro–Val–Trp–Arg–Ser–Gly–Tyr–Ser

times add the incorrect amino acid. **17.5** create One possible concept map is shown in **FIGURE A17.1**.

YOU SHOULD BE ABLE TO . . .

✓ Test Your Knowledge

1. understand c **2.** remember c **3.** understand To speed the correct folding of newly synthesized proteins **4.** remember d **5.** remember At the 3′ end **6.** apply b

✓ Test Your Understanding

7. understand Basal transcription factors bind to promoter sequences in eukaryotic DNA and facilitate the binding of RNA polymerase. As part of the RNA polymerase holoenzyme, sigma binds to a promoter sequence in bacterial DNA and to allow RNA polymerase to initiate at the start of genes. **8.** analyze The wobble rules allow a single tRNA to pair with more than one type of mRNA codon. This is distinct from redundancy, in which more than one codon can specify a single amino acid. If the wobble rules did not exist, there would need to be one tRNA for each amino-acid-specifying codon in the redundant genetic code. **9.** apply b **10.** apply After a peptide bond forms between the polypeptide and the amino acid held by the tRNA in the A site, the ribosome moves down the mRNA. As it does, an uncharged tRNA leaves the E site. The now-uncharged tRNA that was in the P site enters the E site; the tRNA holding the polypeptide chain moves from the A site to the P site, and a new aminoacyl tRNA enters the A site. **11.** apply The ribosome's active site is made up of RNA, not protein. **12.** understand The separation allows the aminoacyl tRNA to place the amino acid into the ribosome's active site while reaching to the distant codon on the mRNA.

✓ Test Your Problem-Solving Skills

13. create Ribonucleases degrade mRNAs that are no longer needed by the cell. If an mRNA for a hormone that increased heart rate were never degraded, the hormone would be produced continuously and heart rate would stay elevated—a dangerous situation. **14.** apply d **15.** analyze The regions most crucial to the ribosome's function should be the most highly conserved: the active site, the E, P, and A sites, and the site where mRNAs initially bind. **16.** analyze The most likely locations are one of the grooves or channels where RNA, DNA, and ribonucleotides move through the enzyme—plugging one of them would prevent transcription.

CHAPTER 18

IN-TEXT QUESTIONS AND EXERCISES

p. 338 understand See **TABLE A18.1**.
p. 338 Fig. 18.1 analyze Write "Slowest response, most efficient resource use" next to the transcriptional control label. Write "Fastest response, least efficient resource use" next to the post-translational control label.
p. 339 Fig. 18.2 apply Plates from all three treatments must be identical and contain identical growth medium, except for the presence of the sugars labeled in the figure. Also, all plates must be grown under the same physical conditions (temperature, light) for the same time.
p. 340 Fig. 18.3 apply Use a medium with all 20 amino acids when producing a master plate of mutagenized *E. coli* colonies; then use a replica plate that contains all the amino acids except tryptophan. Choose cells from the master plate that did *not* grow on the replica plate.
p. 341 understand *lacZ* codes for the β-galactosidase enzyme, which breaks the disaccharide lactose into glucose and galactose. *lacY* codes for the galactoside permease enzyme, which transports lactose into the bacterial cell. *lacI* codes for a protein that shuts down production of the other *lac* products. When lactose is absent, the *lacI* product prevents transcription. This is logical because there is no reason for the cell to make β-galactosidase

and galactoside permease if there is no lactose to metabolize. But when lactose is present, it interacts with *lacI* in some way so that *lacZ* and *lacY* are induced (their transcription can occur). When lactose is present, the enzymes that metabolize it are expressed.
p. 343 apply A mutation that prevents lactose binding to repressor is predicted to prevent transcription of the *lac* operon under any condition. This is because the repressor would never come off the operator. A mutation in the operator that prevents repressor binding is predicted to lead to constitutive expression of the *lac* operon.
p. 343 Fig. 18.8 understand Put the "Repressor protein" on the operator. No transcription will take place. Then put the "RNA polymerase" on the promoter. No transcription will take place. Finally, put "lactose" on the repressor protein and then remove the resulting lactose–repressor complex from the operon. Transcription will begin.
p. 344 CYU (1) understand It is logical that the genes for metabolizing lactose should be expressed only when lactose is available. **(2)** understand See **FIGURE A18.1**.
p. 344 apply This mutation is predicted to lower rates of transcription initiation because AraC's binding to

RNA polymerase is essential for AraC to work as an activator.

IF YOU UNDERSTAND . . .

18.1 understand Production of β-galactosidase and galactosidase permease are under transcriptional control—transcription depends on the action of regulatory proteins. The activity of the repressor is under post-translational control. **18.2** apply Treat *E. coli* cells with a mutagen, and create a master plate that is grown at 33°C. Replica-plate this master plate and grow the replica plate at 42°C. Look for colonies that are on the master plate at 33°C but not on the replica plate at 42°C. **18.3** understand The operator is the parking brake; the repressor locks it in place, and the inducer releases it. **18.4** apply *ara* initiator mutations are likely to affect positive and negative control because AraC must bind to the *ara* initiator sequence for both forms of control. **18.5** create Mutations that create operators for the SOS regulon repressor protein would put new genes under control of the repressor and incorporate them into the regulon.

FIGURE A17.1

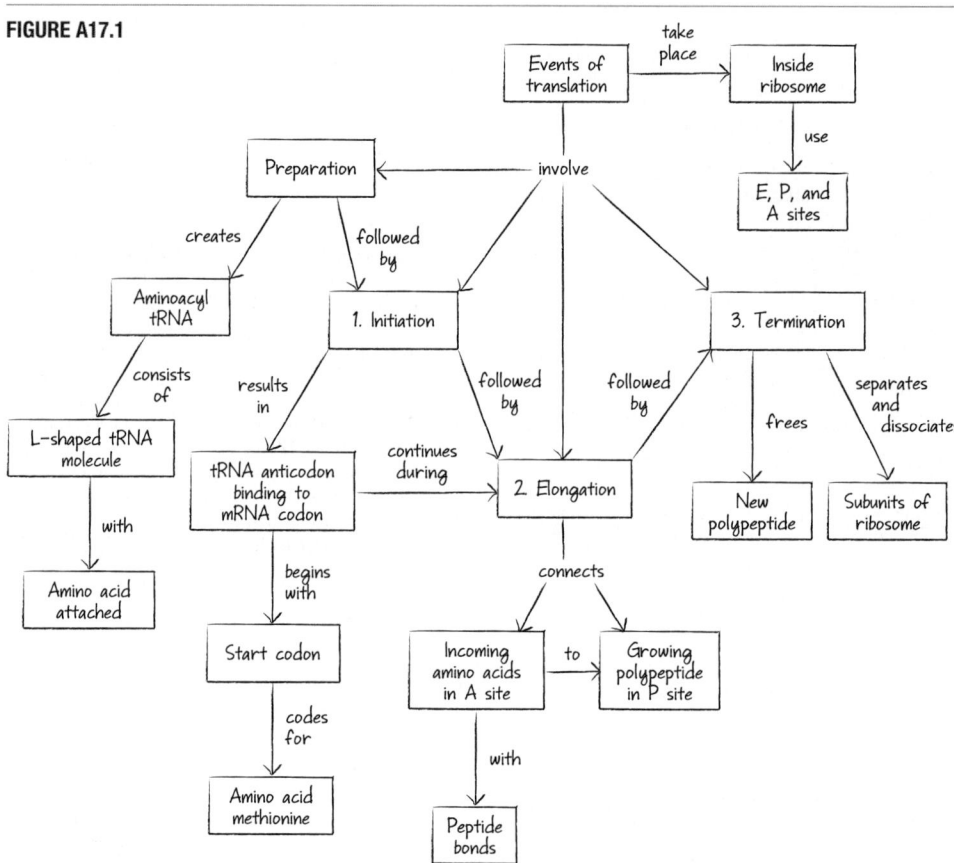

TABLE A18.1

Name	Function
lacZ	Gene for β-galactosidase
lacY	Gene for galactoside permease
Operator	Binding site for repressor
Promoter	Binding site for RNA polymerase
Repressor	Shuts down transcription
Lactose	Binds to repressor and stimulates transcription (removes negative control)
Glucose	At low concentration, increases transcription

✔ **Test Your Knowledge**

1. remember b **2.** understand d **3.** apply b **4.** remember DNA; RNA; **5.** understand When glucose and another sugar are present in the environment, inducer exclusion prevents the use of the other sugar and allows only use of glucose. **6.** remember a

✔ **Test Your Understanding**

7. understand The glycolytic enzymes are always needed in the cell because they are required to produce ATP, and ATP is always needed. **8.** understand Positive control means that a regulatory protein, when present, causes transcription to increase. Negative control means that a regulatory protein, when present, prevents transcription. **9.** apply b. **10.** apply Regulation of the *lac* operon should be normal. The location of the *lacI* gene isn't important, because the gene produces a protein that diffuses within the cell to the operator. **11.** analyze b (since the activator needs to bind to regulatory sequences to activate gene expression, preventing DNA binding would cripple the regulon and prevent cholera).**12.** analyze The *lac* operon would be strongly induced. Once inside the cell, the IPTG will bind to the repressor, causing it to release from DNA. IPTG cannot be broken down, so its concentration will remain high. Finally, since glucose is absent, there will be no inducer exclusion to inhibit IPTG transport through the galactoside permease transporter.

✔ **Test Your Problem-Solving Skills**

13. create Set up cultures with individuals that all come from the same colony of toluene-tolerating bacteria. Half the cultures should have toluene as the only source of carbon; half should have glucose or another common source of carbon. The glucose-containing medium serves as a control to ensure that cells can be grown in the lab. Cells will grow in both cultures if they are able to use toluene as a source of carbon; they will grow only in glucose-containing medium if toluene cannot be used as a carbon source. **14.** apply At a rate of 1×10^{-4} mutants per cell, you would on average find one mutant in every 10,000 (1×10^4) cells. Therefore, you should screen a bit more ($\sim$ 2–3 times more) than 10,000 cells to be reasonably sure of finding at least one mutant. **15.** apply b **16.** analyze Cells with functioning β-galactosidase will produce blue colonies; cells with *lacZ* mutations or *lacY* mutations will not produce β-galactosidase and will produce white colonies. The *lac* promoter could be mutated so that RNA polymerase cannot bind.

CHAPTER 19

IN-TEXT QUESTIONS AND EXERCISES

p. 352 Fig. 19.4 analyze Acetylation of histones decondenses chromatin and allows transcription to begin, so HATs are elements in positive control. Deacetylation condenses chromatin and inactivates transcription, so HDACs are elements in negative control.

p. 353 CYU (1) apply Many more genes than normal are predicted to be expressed because the inability to methylate DNA would lead to more decondensed chromatin. **(2)** understand Addition of acetyl groups to histones or methyl groups to DNA can cause chromatin to decondense or condense, respectively. Different patterns of acetylation or methylation will determine which genes in muscle cells versus brain cells can be transcribed and which are not available for transcription.

p. 353 Fig. 19.5 create They could do something to change histone modifications to see how this affects gene transcription instead of just making the observation that certain histone modifications and low rates of transcription go together.

p. 355 Fig. 19.6 analyze A typical eukaryotic gene usually contains introns and is regulated by multiple enhancers. Bacterial operons lack introns and enhancers. The promoter-proximal element found in some eukaryotic genes is comparable to the *araC* binding site in the *ara* operon of bacteria. Bacterial operons have a single promoter but code for more than one protein; eukaryotic genes code for a single product.

p. 356 analyze The basal transcription factors found in muscle and nerve cells are similar or identical; the regulatory transcription factors found in each cell type are different.

p. 356 understand DNA forms loops when distant regulatory regions, such as silencers and enhancers, are brought close to the promoter through binding of regulatory transcription factors to mediator.

p. 356 CYU (1) analyze Bacterial regulatory sequences are found close to the promoter; eukaryotic regulatory sequences can be close to the promoter or far from it. Bacterial regulatory proteins interact directly with RNA polymerase to initiate or prevent transcription; eukaryotic regulatory proteins influence transcription by altering chromatin structure or binding to the basal transcription complex through mediator proteins. **(2)** understand Certain regulatory proteins decondense chromatin at muscle- or brain-specific genes and then activate or repress the transcription of cell-type-specific genes. Muscle-specific genes are expressed only if muscle-specific regulatory proteins are produced and activated.

p. 358 understand Alternative splicing does not occur in bacteria because bacterial genes do not contain introns—in bacteria, each gene codes for a single product. Alternative splicing is part of step 3 in Figure 19.1.

p. 359 remember Step 4 and step 5. RNA interference either (1) decreases the life span of mRNAs or (2) inhibits translation.

p. 360 CYU (1) understand It became clear that a single gene can code for multiple products instead of a single one. **(2)** understand The miRNAs interfere with mRNAs by targeting them for destruction or preventing them from being translated.

p. 362 CYU (1) understand Many different types of mutations can disrupt control of the cell cycle and initiate cancer. These mutations can affect any of the six levels of control over gene regulation outlined in Figure 19.1. **(2)** understand The p53 protein is responsible for shutting down the cell cycle in cells with damaged DNA. If the protein does not function, then cells with damaged DNA—and thus many mutations—continue to divide. If these cells have mutations in genes that regulate the cell cycle, then they may continue to divide in an uncontrolled fashion.

IF YOU UNDERSTAND . . .

19.1 understand Because eukaryotic RNAs are not translated as soon as transcription occurs, it is possible for RNA processing to occur, which creates variation in the mRNAs produced from a primary RNA transcript and in their life spans. **19.2** understand The default state of eukaryotic genes is "off," because the highly condensed state of the chromatin makes DNA unavailable to RNA polymerase. **19.3** understand See **FIGURE A19.1**. **19.4** understand Alternative splicing makes it possible for a single gene to

FIGURE A18.1

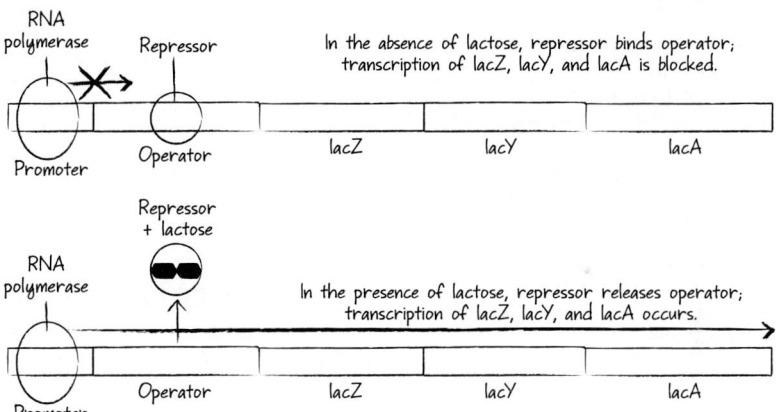

In the absence of lactose, repressor binds operator; transcription of lacZ, lacY, and lacA is blocked.

In the presence of lactose, repressor releases operator; transcription of lacZ, lacY, and lacA occurs.

FIGURE A19.1

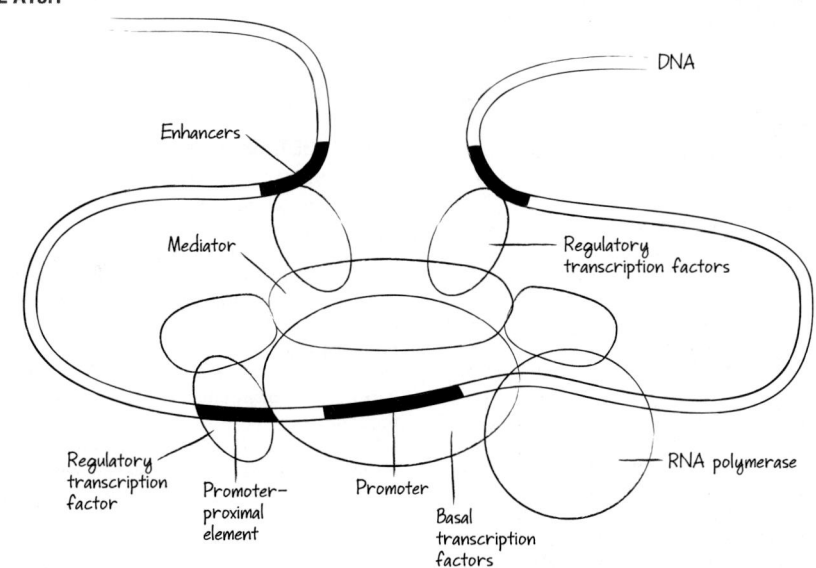

code for multiple products. **19.5** analyze One difference is that, in bacteria, genes that need to be turned on together are often clustered together in operons. Eukaryotes do not use operons. A similarity is in the way eukaryotes turn on many genes together and the strategy of bacterial regulons. Here, genes that are in many different locations share a DNA regulatory sequence and are activated when a regulatory transcription factor binds to the regulatory sequence. **19.6** understand In mutant cells that lack a form of p53 that can bind to enhancers, DNA damage cannot arrest the cell cycle, the cell cannot kill itself, and damaged DNA is replicated. This leads to mutations that can move the cell farther down the road to cancer.

YOU SHOULD BE ABLE TO . . .

✔ Test Your Knowledge

1. remember c **2.** remember A tumor suppressor is a gene or protein that holds cell division in check unless conditions are right for the cell to divide. **3.** understand d **4.** remember The set of regulatory transcription factors present in a particular cell, not differences in DNA sequence, are largely responsible for which genes are expressed. **5.** remember b **6.** remember d

✔ Test Your Understanding

7. analyze (a) Enhancers and the *araC* site are similar because both are sites in DNA where regulatory proteins bind. They are different because enhancers generally are located at great distances from the promoter, whereas the *araC* site is located nearer the promoter. (b) Promoter-proximal elements and the *lac* operon operator are both regulatory sites in DNA located close to the promoter. (c) Basal transcription factors and sigma are proteins that must bind to the promoter before RNA polymerase can initiate transcription. They differ because sigma is part of the RNA polymerase holoenzyme, while the basal transcription complex recruits RNA polymerase to the promoter. **8.** understand If changes in the environment cause changes in how spliceosomes function, then the RNAs and proteins produced from a particular gene could change in a way that helps the cell cope with the new environmental conditions. **9.** analyze (a) Enhancers and silencers are both regulatory sequences located at a distance from the promoter. Enhancers bind regulatory transcription factors that activate transcription; silencers bind regulatory proteins that shut down transcription. (b) Promoter-proximal elements and enhancers are both regulatory sequences that bind positive regulatory transcription factors. Promoter-proximal elements are located close to the promoter; enhancers are far from the promoter. (c) Transcription factors bind to regulatory sites in DNA; mediator does not bind to DNA but instead forms a bridge between regulatory transcription factors and

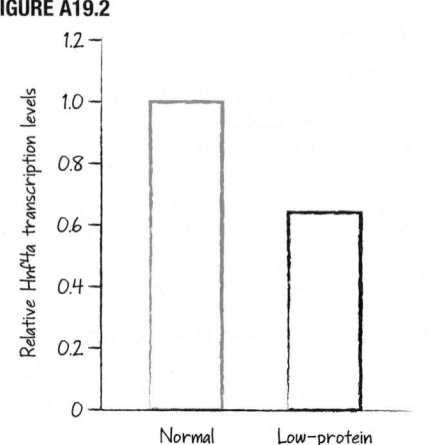

FIGURE A19.2

basal transcription factors. **10.** apply Inhibition of a histone deacetylase is predicted to leave acetyl groups on histones longer than normal. This is predicted to keep gene transcription going longer than normal, leading to higher levels of particular proteins and a change in the phenotype of the cell. **11.** analyze c; this is because there are many more modifications in the histone code relative to the genetic code. **12.** apply The cell is predicted to arrest in the cell cycle and most likely will commit suicide through activation of apoptosis genes because of the continually active *p53*.

✔ Test Your Problem-Solving Skills

13. apply c; This is because promoters and enhancers are brought into close physical proximity when transcription begins (see Figure 19.8). Since rats of malnourished mothers initiate *Hnf4a* gene transcription infrequently, the promoter and enhancer will be together less often in these animals compared with rats born to well-nourished mothers. **14.** understand See **FIGURE A19.2**. The value for the normal diet should be shown as 1.0 and the value for the low-protein diet should be shown as 0.64 (0.64 comes from the ratio of the cpm of the low-protein diet divided by the cpm of the normal diet, or 7368/11,478). **15.** create You could treat a culture of DNA-damaged cells with a drug that stops transcription and then compare them with untreated DNA-damaged cells. If transcriptional control regulates p53 levels, then the p53 level would be lower in the treated cells versus control cells. If control of p53 levels is post-translational, then in both cultures the p53 level would be the same. *Other approaches:* Add labeled NTPs to damaged cells and see if they are incorporated into mRNAs for p53; or add labeled amino acids to damaged cells and see if they are incorporated into completed p53 proteins; or add labeled phosphate groups and see if they are added to p53 proteins. **16.** create The double-stranded RNA could be cut by the same enzyme that creates double-stranded miRNAs from miRNA precursors. This double-stranded RNA may be incorporated into RISC and converted into a single-stranded RNA. The single-stranded RNA held by RISC would work in RNA interference to trigger the destruction of the complementary viral mRNA.

✔ **B I G PICTURE** Genetic Information

p. 366 CYU (1) remember Star = DNA, mRNA, proteins. **(2)** understand RNA "is reverse transcribed by" reverse transcriptase "to form" DNA. **(3)** analyze E = splicing, etc.; E = meiosis and sexual reproduction (along with their links). **(4)** analyze Chromatin "makes up" chromosomes; independent assortment and recombination "contribute to" high genetic diversity.

CHAPTER 20

IN-TEXT QUESTIONS AND EXERCISES

p. 372 understand Isolate the DNA and cut it into small fragments with EcoRI, which leaves sticky ends. Cut copies of a plasmid or other vector with EcoRI. Mix the fragments and plasmids under conditions that promote complementary base pairing by sticky ends of fragments and plasmids. Use DNA ligase to catalyze formation of phosphodiester bonds and seal the sequences.
p. 372 Fig. 20.4 apply No—many times, because many copies of each type of mRNA were present in the pituitary cells, and many pituitary cells were used to prepare the library.
p. 373 apply The probe must be single stranded so that it will bind by complementary base pairing to the target DNA, and it must be labeled so that it can be detected. The probe will base-pair only with fragments that include a sequence complementary to the probe's sequence. A probe with the sequence 5′ AATCG 3′ will

bind to the region of the target DNA that has the sequence 5′ CGATT 3′ as shown here:

5′ AATCG 3′
3′ TCCGGTTAGCATTACCATTTT 5′

p. 374 CYU (1) understand When the endonuclease makes a staggered cut in a palindromic sequence and the strands separate, the single-stranded bases that are left will bind ("stick") to the single-stranded bases left where the endonuclease cut the same palindrome at a different location. **(2)** understand The word probe means to examine thoroughly. A DNA probe "examines" a large set of sequences thoroughly and binds to one—the one that has a complementary base sequence.
p. 375 Fig. 20.7 analyze The polymerase will begin at the 3′ end of each primer. On the top strand in part (b), it will move to the left; on the bottom strand, it will move to the right. As always, synthesis is in the 5′-to-3′ direction.
p. 376 CYU (1) understand Denaturation makes DNA single stranded so the primer can bind to the sequence during the annealing step. Once the primer is in place, *Taq* polymerase can synthesize the rest of the strand during the extension step. It is a "chain reaction" because the products of each reaction cycle are used in the next reaction cycle—this is why the number of copies doubles in each cycle. **(2)** apply One of many possible answers is shown in **FIGURE A20.1**.
p. 378 CYU understand If ddNTPs were present at high concentration, they would almost always be incorporated—meaning that only fragments from the first complementary base in the sequence would be produced.
p. 379 understand Using a map with many markers makes it more likely that there will be one marker that is very tightly linked to the gene of interest—meaning that a form of the marker will almost always be associated with the phenotype you are tracking.
p. 382 CYU create Start with a genetic map with as many polymorphic markers as possible. Determine the genotype at these markers for a large number of individuals who have the same type of alcoholism—one that is thought to have a genetic component—as well as a large number of unaffected individuals. Look for particular versions of a marker that is almost always found in affected individuals. Genes that contribute to a predisposition to alcoholism will be near that marker.
p. 383 Fig. 20.11 apply The insertion will probably disrupt the gene and have serious consequences for the cell and potentially the individual.

IF YOU UNDERSTAND . . .

20.1 understand Special features of DNA are needed to allow replication in a cell. It is unlikely that any DNA fragment generated by cutting the DNA of one species with a restriction enzyme would replicate when inserted into a bacterial cell. By placing DNA fragments within plasmids that normally replicate in a bacterial cell, the inserted DNA can be replicated along with the plasmid.
20.2 analyze Advantages of cloning in cells: No knowledge of the sequence is required. Disadvantages of cloning in cells: It is slower and technically more difficult than PCR. PCR advantages: It is fast and easy, and it can amplify a DNA sequence that is rare in the sample. PCR disadvantage: It requires knowledge of sequences on either side of the target gene, so primers can be designed.
20.3 understand The length of each fragment is dictated by where a ddNTP was incorporated into the growing strand, and each ddNTP corresponds to a base on the template strand. Thus the sequence of fragment sizes corresponds to the sequence on the template DNA.
20.4 understand In this case, both affected and unaffected individuals would have the same marker, so there would be no way to associate a particular form of the marker with the gene and phenotype you are interested in.
20.5 understand An advantage of using retrovirus vectors is that any foreign genes they carry are integrated into

human chromosomes, so delivered genes become a permanent feature of the cell's genome. The concern is that they may integrate at sites that alter gene function in ways that harm cell function, or even worse, lead to cancer. **20.6** `understand` Genes can be inserted into the Ti plasmids carried by *Agrobacterium*, and the recombinant plasmids transferred to plant cells infected by the bacterium.

YOU SHOULD BE ABLE TO . . .

✔ Test Your Knowledge

1. `understand` They cut DNA at specific sites, known as recognition sites, to produce DNA fragments useful for cloning. **2.** `remember` b **3.** `understand` ddNTPs lack the −OH (hydroxyl) group on the 3′ carbon of deoxyribose sugar that is required to extend the DNA chain during synthesis. **4.** `understand` a **5.** `remember` b **6.** `understand` d

✔ Test Your Understanding

7. `understand` When a restriction endonuclease cuts a "foreign gene" sequence and a plasmid, the same sticky ends are created on the excised foreign gene and the cut plasmid. After the sticky ends on the foreign gene and the plasmid anneal, DNA ligase catalyzes closure of the DNA backbone, sealing the foreign gene into the plasmid DNA (see **FIGURE A20.2**). **8.** `apply` If the DNA at each cycle steadily doubles, it is predicted to yield a 33.6 million-fold increase (a 2^{25}-fold increase). **9.** `understand` A cDNA library is a collection of complementary DNAs made from all the mRNAs present in a certain group of cells. A cDNA library from a human nerve cell would be different from one made from a human muscle cell, because nerve cells and muscle cells express many different genes that are specific to their cell type. **10.** `remember` Genetic markers are genes or other loci

that have known locations in the genome. When these locations are diagrammed, they represent the physical relationships between landmarks—in other words, they form a map. **11.** `understand` d; these regulatory sequences promote expression of introduced genes in the endosperm—the rice grain eaten as food. **12.** `analyze` PCR and cellular DNA synthesis are similar in the sense of producing copies of a template DNA. Both rely on primers and DNA polymerase. The major difference between the two is that PCR copies only a specific target sequence, but the entire genome is copied during cellular DNA synthesis.

✔ Test Your Problem-Solving Skills

13. `create` You could use a computer program to identify likely promoter sequences in the sequence data and then look for sequences just downstream that have an AUG start codon and codons that could be part of the protein-coding exons of a potential protein about 500 amino acids (500 codons or 1500 bases) long. **14.** `analyze` False. Since somatic-cell modifications cannot be passed on to future generations, but germ-line modifications may be passed on to offspring, germ-line modification opens a new set of ethical questions. **15.** `analyze` In both techniques, researchers use an indicator to identify either a gene of interest or a colony of bacteria with a particular trait. The problem is the same—picking one particular thing (a certain gene or a cell with a particular mutation) out of a large collection. **16.** `analyze` (a) Primer 1b binds to the top right strand and would allow DNA polymerase to synthesize the top strand across the target gene. Primer 1a, however, binds to the upper left strand and would allow DNA polymerase to synthesize the upper strand *away* from the target gene. Primer 2a binds to the bottom left strand and would allow DNA polymerase to synthesize the bottom strand across the target gene. Primer 2b, however, binds

to the bottom right strand and would allow DNA polymerase to synthesize away from the target gene. (b) She could use primer 1b with primer 2a.

CHAPTER 21

IN-TEXT QUESTIONS AND EXERCISES

p. 391 `understand` If no overlap occurred, there would be no way of ordering the fragments correctly. You would be able to put fragments only in random order—not the correct order.
p. 391 Fig. 21.2 `understand` Shotgun sequencing is based on fragmenting the genome into many small pieces.
p. 393 Fig. 21.3 `understand` Because the mRNA codons that match up with each strand are oriented in the 5′-to-3′ direction.
p. 395 CYU (1) `understand` Parasites don't need genes that code for enzymes required to synthesize molecules they acquire from their hosts. **(2)** `understand` If two closely related species inherited the same gene from a common ancestor, the genes should be similar. But if one species acquired the same gene from a distantly related species via lateral gene transfer, then the genes should be much less similar.
p. 399 Fig. 21.9 `understand`
Chromosome 1:
ψβ2-ε-Gγ-Aγ-ψβ1-δ-ψβ2-ε-Gγ-Aγ-ψβ1-δ-β
Chromosome 2: β
p. 400 CYU (1) `understand` Eukaryotic genomes have vastly different numbers of transposable elements and other repeated sequences that don't directly contribute to phenotype. So genome size can vary widely without changing the number of coding genes or the complexity of the organism. **(2)** `apply` Because most of the genome is transcribed, this number is very close to the percentage of the genome that codes for exons. In humans, the ratio would be roughly 2%.
p. 401 `understand` Start with a microarray containing exons from a large number of human genes. Isolate mRNAs from brain tissue and liver tissue, and make labeled cDNAs from each. Probe the microarray with cDNA made from each type of tissue, and record where binding occurs. Binding events identify exons that are transcribed in each type of tissue. Compare the results to identify genes that are expressed in brain but not liver, or in liver but not brain.
p. 402 `understand` Cell replication is an emergent property because it is due to the interaction of proteins working in a network, yet is a property that could not be predicted from the analysis of any one of these proteins.

IF YOU UNDERSTAND . . .

21.1 `understand` If a search of human gene sequence databases revealed a gene that was similar in base sequence, and if follow-up work confirmed that the mouse and human genes were similar in their pattern of exons and introns and regulatory sequences, then the researchers could claim that they are homologous. **21.2** `understand` The genome of a parasite is predicted to be smaller and to have fewer genes, particularly for metabolism, than the genome of a nonparasite. **21.3** `understand` These features include variable numbers of transposable elements; noncoding repeated sequences; and noncoding, non-repetitive DNA. These add to genome size without adding genes that directly influence phenotype. Additionally, mechanisms such as alternative splicing and the possibility of many noncoding regulatory RNAs can create situations in which few protein-coding genes but many different proteins are expressed in intricately regulated ways—features that would increase morphological complexity without increasing gene number. **21.4** `understand` Color-coded clusters shown in Figure 21.13 represent networks of interacting proteins that work together to carry out a particular cellular function. But it's clear from the figure that smaller, clusters are also connected into larger networks.

FIGURE A20.1

5′ CATGACTATTACGTATCGGGTACTATGCTATCGATCTAGCTACGCTAGCT 3′
3′ GTACTGATAATGCATAGCCCATGATACGATAGCTAGATCGATGCGATCGA 5′

Primer #1, which will anneal to the 3′ end of the top strand:

5′ AGCTAGCGTAGCTAGATCGAT 3′

Primer #2, which will anneal to the 3′ end of the bottom strand:

5′ CATGACTATTACGTATCGGGT 3′

The primers will bind to the separated strands of the parent DNA sequence as follows:

5′ CATGACTATTACGTATCGGGTACTATGCTATCGATCTAGCTACGCTAGCT 3′
                                  3′ TAGCTAGATCGATGCGATCGA 5′

5′ CATGACTATTACGTATCGGGT 3′
3′ GTACTGATAATGCATAGCCCATGATACGATAGCTAGATCGATGCGATCGA 5′

FIGURE A20.2

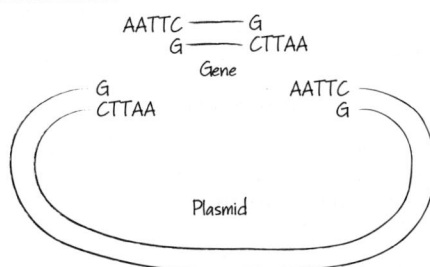

Sticky ends on cut plasmid and gene to be inserted are complementary and can base pair

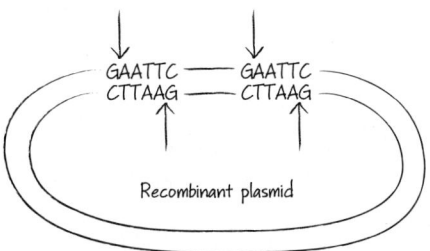

DNA ligase forms four phosphodiester bonds between gene and plasmid DNA at the points indicated by arrows

✔ **Test Your Knowledge**

1. remember c **2.** understand a **3.** remember d **4.** remember b **5.** understand Finding two or more genes of similar sequence in the genome. **6.** remember Having a sequence that is clearly related to a functional gene but with a crippling mutation such as a stop codon or deletion.

✔ **Test Your Understanding**

7. analyze Computer programs are used to scan sequences in both directions to find ATG start codons, a gene-sized logical sequence with recognizable codons, and then a stop codon. One can also look for characteristic promoter, operator, and other regulatory sites. It is more difficult to identify open reading frames in eukaryotes because their genomes are so much larger and because of the presence of introns and repeated sequences. **8.** evaluate c **9.** apply Water flea gene density is about 31,000 genes/200 Mbp = 155 genes/Mbp. In humans, gene density is about 21,000 genes/3000 Mbp = 7 genes/Mbp. The relative gene density of water flea/human = 155/7 = 22. **10.** analyze Because with many different alleles, the chance that a match is coincidental is low relative to the chance that they match by identity (they come from the same person). **11.** understand A DNA microarray experiment identifies which genes are being expressed in a particular cell at a particular time. If a series of experiments shows that different genes are expressed in cells at different times or under different conditions, it implies that expression was turned on or off in response to changes in age or changes in conditions. **12.** understand Homology is a similarity among different species that is due to their inheritance from a common ancestor. If a newly sequenced gene is found to be homologous with a known gene of a different species, it is assumed that the gene products have similar function.

✔ **Test Your Problem-Solving Skills**

13. analyze If "gene A" is not necessary for existence, it can be lost by an event like unequal crossing over (on the chromosome with deleted segments) with no ill effects on the organism. In fact, individuals who have lost unnecessary genes are probably at a competitive advantage, because they no longer have to spend time and energy copying and repairing unused genes. **14.** apply If the grave were authentic, it might include two very different parental patterns along with five children whose patterns each represented a mix between the two parents. The other unrelated individuals would have patterns not shared by anyone else in the grave. **15.** apply d; mutation of such a central gene is likely to influence the phenotypes associated with all the genes it interacts with. If these interacting genes are involved in many different functions, the effects of mutation of a central gene are likely to be widespread, or pleiotropic. **16.** apply You would expect that the livers and blood of chimps and humans would function similarly, but that strong differences occur in brain function. The microarray data support this prediction and suggest chimp and human brains are different because certain genes are turned on or off at different times and expressed in different amounts.

CHAPTER 22

IN-TEXT QUESTIONS AND EXERCISES

p. 410 CYU understand Biologists have been able to grow entire plants from a single differentiated cell taken from an adult. They have also succeeded in producing animals by transferring the nucleus of a fully differentiated cell to an egg whose nucleus has been removed.
p. 411 Fig. 22.4 apply Anterior structures are predicted to be shifted farther back in the embryo, since higher-than-normal levels of the anterior-specifying Bicoid protein would be present in posterior regions.
p. 412 Fig. 22.6 apply Since Bicoid is located in the anterior cytoplasm and works to turn on genes that produce anterior structures, a prediction is that this manipulation would create a concentration gradient of Bicoid with its highest point in the middle of the mutant embryo. This would lead to the most anterior-like structures in the center of the embryo and more posterior structures forming on either side.
p. 414 Fig. 22.10 create Genes are usually considered homologous if they have similar DNA sequence and exon–intron structure.
p. 415 CYU (1) understand Bicoid is a transcription factor, and cells that experience a high versus medium versus low concentrations of Bicoid express Bicoid-regulated genes at different levels. **(2)** understand Bicoid proteins control the expression of gap genes, which—in effect—tell cells which third of the embryo they are in along the anterior–posterior axis. Gap genes control the expression of pair-rule genes, which organize the embryo into individual segments. Pair-rule genes control expression of segment polarity genes, which establish an anterior–posterior polarity to each individual segment. Segment polarity genes turn on homeotic genes, which then trigger genes for producing segment-specific structures like wings or legs.

IF YOU UNDERSTAND . . .

22.1 apply The seedling would be misshapen because controlling the plane of cleavage is essential for formation of normal structures in plants. **22.2.** understand Cells in the stem contain a complete set of genes, so if they can be de-differentiated and receive appropriate signals from other cells, they could differentiate into any cell type and cooperate to form any structure of the plant, including roots. **22.3** analyze Because segmentation genes work in a cascade, mutation of a pair-rule gene is expected to influence the expression of genes in the next level, segment polarity genes, but not affect the expression of earlier acting genes, such as gap genes. **22.4** apply This procedure is predicted to lead to the loss of limbs because the regulatory sequence replacement will cause the regions of expression of *Hoxc8* and *Hoxc6* to overlap.

YOU SHOULD BE ABLE TO . . .

✔ **Test Your Knowledge**

1. remember b **2.** remember stem cells **3.** understand a **4.** remember True **5.** remember a **6.** remember a conserved gene that can be expressed at different times and places during development to produce different types of structures in different organisms.

✔ **Test Your Understanding**

7. understand They occur in both plants and animals, and are responsible for the changes that occur as an embryo develops. **8.** understand If an adult cell used for cloning had some permanent change or loss of genetic information during development, it should not be able to direct the development of a viable adult. But if the adult cell is genetically equivalent to a fertilized egg, then it should be capable of directing the development of a new individual. **9.** understand The researchers exposed adult flies to treatments that induce mutations and then looked for embryos with defects in the anterior–posterior body axis or body segmentation. The embryos had mutations in genes required for body axis formation and segmentation. **10.** understand Development—specifically differentiation—depends on changes in gene expression. Changes in gene expression depend on differences in regulatory transcription factors. **11.** create Differentiation is triggered by the production of transcription factors, which induce other transcription factors, and so on—a sequence that constitutes a regulatory cascade—as development progresses. At each step in the cascade, a new subset of genes is activated—resulting in a step-by-step progression from undifferentiated to fully differentiated cells. **12.** analyze c. This result shows that a mouse tool-kit gene is similar enough (conserved) to a fly gene to be able to take over its function.

✔ **Test Your Problem-Solving Skills**

13. apply There would be more Bicoid protein farther toward the posterior and less in the anterior—meaning that the anterior segments would be "less anterior" in their characteristics and the posterior segments would be "more anterior." **14.** analyze The result means that the human gene was expressed in worms and that its protein product has performed the same function in worms as in humans. This is strong evidence that the genes are homologous and that their function has been evolutionarily conserved. **15.** apply c. Since the morphogen concentration is halved every 100 μm away from the posterior pole, dropping to 1/16 the highest concentration would require 4 "halvings" of the initial concentration ($1/2 \times 1/2 \times 1/2 \times 1/2 = 1/16$), and this would take place in 4 steps of 100 μm each—or 400 μm from the morphogen source. This is where the leg is predicted to be formed. **16.** create Tool-kit genes—such as *Hox* genes—are responsible for triggering the production of structures like wings or legs in particular segments. One hypothesis to explain the variation is that changes in gene expression of tool-kit genes led to different numbers of segments that express leg-forming *Hox* genes.

CHAPTER 23

IN-TEXT QUESTIONS AND EXERCISES

p. 421 understand If the bindin-like protein were blocked, it would not be able to bind to the egg-cell receptor for sperm. Fertilization would not take place.
p. 422 CYU (1) apply This change is predicted to allow the sperm of species A to bind to—and likely fertilize—the eggs of species B. **(2)** understand The entry of a sperm triggers a wave of calcium ion release around the egg-cell membrane. Thus, calcium ions act as a signal that indicates "A sperm has entered the egg." In response to the increase in calcium ion concentration, cortical granules fuse with the egg-cell membrane and release their contents to the exterior, causing the destruction of the sperm receptor and, in sea urchins, the formation of a fertilization envelope.
p. 424 CYU understand If cytoplasmic determinants could not be localized, they would be distributed to all blastomeres. This would specify the same fate for all blastomeres and their descendants instead of different fates.
p. 426 CYU (1) understand Ectoderm and endoderm, because these are the inside and outside layers. **(2)** apply The future anterior–posterior and dorsal–ventral region of the embryo could first be identified as the blastopore forms and cells begin migrating into the embryo. The blastopore marks the future posterior of the embryo, and the region opposite the blastopore is where the anterior will form. Cells that first migrate into the blastopore come from the dorsal side, so this is the future dorsal region; the ventral region is on the opposite side.
p. 427 Fig. 23.9 understand Because neural tube formation depends on rearrangements of the cytoskeleton, a drug that locked the cytoskeleton in place is predicted to prevent formation of the neural tube.
p. 428 Fig. 23.12 analyze It made gene expression possible in non-muscle cells. A "general purpose" promoter can lead to transcription of an inserted cDNA in any type of cell, including fibroblasts.

IF YOU UNDERSTAND . . .

23.1 understand Bindin and the egg-cell receptor for sperm bind to each other. For this to happen, their tertiary structures must fit together somewhat like the structures of a lock and key. **23.2** understand If the cytoplasmic determinant is localized to a certain part of the egg cytoplasm, blastomeres will either contain or not contain the determinant, because the egg's cytoplasm becomes divided between blastomeres. **23.3** analyze Both cells would contain cell–cell signals and transcription factors specific to mesodermal cells, but the anterior cell would contain anterior-specific signals and regulatory transcription factors while the posterior cell would contain posterior-specific signals and regulators. **23.4** understand Cell–cell signals from the notochord direct the formation of the neural tube; subsequently, cell–cell signals from the notochord, neural tube, and ectoderm direct the differentiation of somite cells. Cells expand during neural tube formation. Somite cells move to new positions, proliferate, and differentiate.

YOU SHOULD BE ABLE TO . . .

✔ Test Your Knowledge

1. remember Bindin is a protein on the sperm-cell head that binds to a receptor on the egg plasma membrane to allow species-specific attachment of sperm and egg. **2.** understand c **3.** understand c **4.** understand d **5.** remember ectoderm; **6.** remember a

✔ Test Your Understanding

7. evaluate As mammals, mice form a placenta that allows the embryo to obtain nutrients from the mother. There's no need for lots of nutrient reserves to support development. Frogs develop outside the mother's body, and the embryo is on its own. There is a need for nutrient reserves, and the large egg provides this. Eggs contain stores of nutrients. **8.** apply This treatment is predicted to promote polyspermy, because there would be no release of Ca^{2+} inside the cell to trigger a block to polyspermy. **9.** understand They contain different types and/or concentrations of cytoplasmic determinants. **10.** analyze a; this is because ectodermal cells have to constrict on their dorsal margins and expand on their ventral margin to fold ectoderm into a neural tube. **11.** understand When transplanted early in development, somite cells become the cell type associated with their new location. But when transplanted later in development, somite cells become the cell type associated with the original location. These observations indicate that the same cell is not committed to a particular fate until later in development. **12.** apply More than 50,000 fibroblasts would need to be screened. This is because if 2×10^{-5} of all mRNAs and cDNAs coded for MyoD, then 50,000 cells would need to be screened on average to find one with a MyoD cDNA (50,000 comes from the quotient of $1/2 \times 10^{-5}$ MyoD cDNAs per cDNA). Since this is an average, in some groups of 50,000 cells, there may not be a MyoD cDNA. Therefore, to be reasonably sure of finding at least one MyoD cDNA, more than 50,000 cells should be screened.

✔ Test Your Problem-Solving Skills

13. analyze The interactions that allow proteins to bind to each other are extremely specific—certain amino acids have to line up in certain positions with specific charges and/or chemical groups exposed. Sperm–egg binding is species specific because each form of bindin binds to a different egg-cell receptor for sperm. **14.** evaluate Translation is not needed for cleavage to occur normally—meaning that all the proteins needed for early cleavage are present in the egg. **15.** create The yellow pigment may be a cytoplasmic determinant, or it may be only a molecule that is found in the same cells as the cytoplasmic determinant. Without isolating the pigment and adding it to cells to see if it acts as a cytoplasmic determinant, these two possibilities cannot be distinguished. **16.** apply a (See Figure 23.10 and note how the notochord is closest

to the somite region that forms bone. If an additional notochord were transplanted near the top of the neural tube, signals from the notochord would be expected to convert the nearby somite into bone-producing somite, adding to the amount of bone.)

CHAPTER 24

IN-TEXT QUESTIONS AND EXERCISES

p. 434 Fig. 24.2 remember See **FIGURE A24.1**.
p. 436 CYU (1) understand Cells located on the outside of the embryo become epidermal tissue; cells in the interior become vascular tissue; cells in between become ground tissue. **(2)** apply The *MONOPTEROS* gene would be overexpressed, producing higher levels of the MONOPTEROUS transcription factor. The embryonic cells would get information indicating that they are in the shoot apex, where MONOPTEROS levels are high. As a result, the root would not develop normally.
p. 437 Fig. 24.6 analyze Many animals have three body axes analogous to the three axes of leaves: anterior–posterior (like the leaf's proximal–distal), dorsal–ventral (like the leaf's adaxial–abaxial), and a left–right axis (like the leaf's mediolateral axis). The main plant body, in contrast, has only two axes: apical–basal and radial.
p. 438 CYU (1) create Simply planting a cut stem will often produce a new root meristem. This occurs whenever the cutting produces a new plant. **(2)** apply If the plant could be genetically engineered to express higher levels of the PHAN transcription factor in leaves, the likely result would be a shift in leaf type from simple to compound.
p. 440 apply Mutual inhibition between the *A* and *C* genes helps create separate regions of gene expression. If *C* gene expression is lost, then the inhibition of *A* gene expression is also lost, and *A* genes are expressed in their normal location and where they are normally inhibited by *C* gene expression.
p. 441 CYU (1) analyze Plants generate many reproductive structures—flowers—many times and in many places during the plant's life. Plant reproductive cells are created from vegetative cells, not cells that are set aside early in development as reproductive cells. In contrast, animals form reproductive structures (ovaries and testes) in set locations, produce reproductive structures only once during development and set aside specialized reproductive cells (the germ line) early in their development. **(2)** analyze Like *Hox* genes, *MADS-box* genes code for DNA-binding proteins that function as transcriptional regulators. Both Hox proteins and MADS-domain proteins are involved in regulatory transcription cascades that lead to development of specific structures in specific locations. Although they perform similar functions, *Hox* genes and *MADS-box* genes evolved independently and differ in their base sequences. Thus, these genes and their protein products are not homologous.

IF YOU UNDERSTAND . . .

24.1 apply If there were only one master regulatory gene for patterning, the predicted result of loss of that gene's function by mutation would be disorganization of the entire apical–basal axis, not just removal of parts of it as seen in the actual mutants. **24.2** apply Many things could be imagined, but all should point in the direction of many repeated body parts (for example, 21 wings in a variety of shapes and positions on the body) and a very large overall size (if the fly is old) due to continuous growth. **24.3** apply The prediction is that sepals would develop from every whorl. This is because when the *C* gene product is absent, *A* gene expression spreads across the flower. Since there is no *B* gene expression, only sepals would form—the structure associated with only *A* gene expression.

YOU SHOULD BE ABLE TO . . .

✔ Test Your Knowledge

1. understand This is because mechanisms for development evolved independently in plants and animals. **2.** understand b **3.** remember b **4.** remember d **5.** remember c **6.** understand a.

✔ Test Your Understanding

7. understand c. **8.** understand Animal cells that lead to sperm or egg formation are "sequestered" early in development, so that they are involved only in reproduction and undergo few rounds of mitosis. Because mutations occur in every round of mitosis, plant eggs and sperm should contain many more mutations than animal eggs and sperm. **9.** understand Meristem cells are undifferentiated and divide to produce a cell that remains in the meristem and a cell that can commit to a particular pathway of differentiation. These are properties of stem cells. **10.** understand Just like the three tissue layers in plant embryos, the tissues produced in the SAMs and RAMs of a 300-year-old oak tree can differentiate into all of the specialized cell types found in a mature plant. **11.** evaluate The ABC model predicted that the *A*, *B*, and *C* genes would each be expressed in specific whorls in the developing flower. The experimental data showed that the prediction was correct. **12.** understand Cell proliferation is responsible for expansion of meristems and growth in size. Asymmetric cell division along with cell expansion is responsible for giving the adult plant a particular shape. Differentiation is responsible for generating functional tissues and organs in the adult plant. Cell–cell signals are responsible for setting up the apical–basal axis of the embryo and for guiding the pollen tube to the egg.

✔ Test Your Problem-Solving Skills

13. create Four. This is because there are four states of gene activity (in any cell, *D* can be on or off and *E* can be on or off) and therefore four structures that could be specified using this code. Sets of on–off switches are known as Boolean systems and are the basis for information storage in computers and (sometimes) in cells. **14.** analyze Because vegetative growth is continuous and occurs in all directions (unrestricted), it could be considered indeterminate. In contrast, reproductive growth produces mature reproductive organs and stops when those organs are complete. Because its duration is limited, it could be considered determinate. **15.** apply a; see Figure 24.10 to understand the reasoning. *A* and *B* together create petals, and *B* and *C* together create stamens. **16.** analyze One hundred times more likely. This is because 100 times more cell divisions were required to create the oak egg, and the rate of mutation per cell division is constant. (You don't need to consider the absolute rate of 1×10^{-6}/division, since the question asks about relative chances only).

FIGURE A24.1

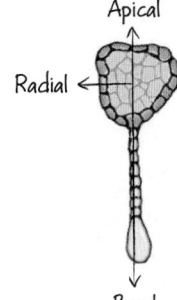

Apical

Radial ←

Basal

CHAPTER 25

IN-TEXT QUESTIONS AND ANSWERS

p. 448 Fig. 25.4 `analyze` Under special creation, fossils with transitional features would be explained as separate types, unrelated to other organisms and just coincidentally having intermediate features.

p. 448 Fig. 25.5 `analyze` If vestigial traits result from inheritance of acquired characteristics, some individuals must have lost the traits during their own lifetimes and passed the reduced traits on to their offspring. For example, a certain monkey's long tail might have been bitten off by a predator. The new traits would then somehow have passed to the individual's eggs or sperm, resulting in shorter-tailed offspring, until humans with a coccyx resulted.

p. 453 CYU (1) `evaluate` If birds evolved from dinosaurs you would expect to find transitional fossil dinosaurs with feathers. Such fossils have been found. **(2)** `analyze` The DNA sequences of chimpanzees and humans are so similar because we share a recent common ancestor.

p. 455 (1) `understand` Relapse occurred because the few bacteria remaining after drug therapy were not eliminated by the patient's weakened immune system and began to reproduce quickly. **(2)** `understand` No—almost all of the cells present at the start of step 3 would have been resistant to the drug.

p. 456 `understand` When antibiotics are overused, some bacteria will be killed, including the bacteria that are harmless or beneficial to us. However, bacteria that are resistant to these antibiotics will flourish and multiply, reducing the likelihood that antibiotic treatment will be effective in the future.

p. 457 Fig. 25.16 `analyze` "Prediction": Beak measurements were different before and after the drought. "Prediction of null hypothesis": No difference was found in beak measurements before and after the drought.

p. 459 CYU (1) `understand` *Postulate 1:* Traits vary within a population. *Postulate 2:* Some of the trait variation is heritable. *Postulate 3:* There is variation in reproductive success (some individuals produce more offspring than others). *Postulate 4:* Individuals with certain heritable traits produce the most offspring. The first two postulates describe heritable variation; the second two describe differential reproductive success. **(2)** `understand` Beak size and shape and body size vary among individual finches, in part because of differences in their genotypes (some alleles lead to deeper or shallower beaks, for example). When a drought hit, individuals with deep beaks survived better and produced more offspring than individuals with shallow beaks.

p. 460 `analyze` In biology, an adaptation is any heritable trait that increases an individual's ability to produce offspring in a particular environment. In everyday English, adaptation is often used to refer to an individual's nonheritable adjustment to meet an environmental challenge, a phenomenon that biologists call acclimatization. The phenotypic changes resulting from acclimatization are not passed on to offspring.

IF YOU UNDERSTAND . . .

25.1 `analyze` Typological thinking is based on the idea that species are unchanging types and that variations within species are unimportant or even misleading. Population thinking recognizes the variation that occurs within a species as critically important. **25.2** `remember` Many answers are possible. For example, the evolution of an antibiotic-resistant TB bacterium demonstrates that species change through time. The homologous sequences in human and fruit fly genes demonstrate that these species are related by common ancestry. **25.3** `analyze` In biology, fitness is the ability of an individual to produce viable, fertile offspring, relative to that ability in other individuals in the population. In everyday English, fitness is a physical attribute

that is acquired as a result of practice or exercise. **25.4** `apply` Under special creation, changes in *Mycobacterium tuberculosis* populations would be explained as individual creative events governed by an intelligent creator. Under the theory of evolution by inheritance of acquired characters, changes in *M. tuberculosis* populations would be explained by the cells trying to transcribe genes in the presence of the drug, and their *rpoB* gene becoming altered as a result. **25.5** `understand` Brain size in *H. sapiens* might be constrained by the need for babies to pass through the mother's birth canal, by the energy required to maintain a large brain as an adult, or by lack of genetic variation for even larger brain size. Flying speed in falcons might be constrained by loss of maneuverability (and thus less success in hunting), the energy demands of extremely rapid flight, or lack of genetic variation for even faster flight.

YOU SHOULD BE ABLE TO . . .

✔ Test Your Knowledge

1. `remember` b **2.** `remember` False **3.** `remember` a **4.** `understand` d
5. `remember` homologous **6.** `remember` c

✔ Test Your Understanding

1. `understand` Mutation produces new genetic variations, at random, without any forethought as to which variations might prove adaptive in the future. Individuals with mutations that are disadvantageous won't produce many offspring, but individuals with beneficial mutations will produce many offspring. The beneficial mutations will thus increase in frequency through selection. **2.** `analyze` Artificial selection is determined by human choice and is goal directed, whereas natural selection is the unplanned differential reproductive success of individuals that vary in their heritable traits. **3.** `analyze` b (In some environments, being big and strong lowers fitness.) **4.** `analyze` There were 751 finches before the drought and 90 finches afterward, so only 90/751 = 0.12, or 12 percent, survived. There were more finches with deep beaks before the drought only because the population was much larger, but the *average* beak depth increased after the drought. **5.** `analyze` The evidence for within-patient evolution is that DNA sequences at the start and end of treatment were identical except for a single nucleotide change in the *rpoB* gene. If rifampin were banned, *rpoB* mutant strains likely would have had lower fitness in the drug-free environment and would not continue to increase in frequency in *M. tuberculosis* populations. **6.** `apply` (1) Traits vary within rabbit populations, such as fur color and hearing ability. (2) Some of this variation is heritable. (3) More rabbits are produced than can survive. (4) Rabbits with certain heritable traits, such as the ability to camouflage in their environment or hear the approach of foxes, produce more offspring.

✔ Test Your Problem-Solving Skills

1. `analyze` The theory of evolution by natural selection predicts that white and brown deer mice are descendants of an ancestral population that varied in color. The white mice had higher fitness in the beach environment, where they were more likely to escape the notice of predators. Likewise, the brown mice had a higher fitness in the shady forest environment. Over time, the average color of each population changed. Special creation claims that the two colors of mice do not change and were created by divine intervention. Evolution by inheritance of acquired characteristic predicts that the mice in the different environments needed to change color, so they did so and then passed their traits onto their offspring. **2.** `analyze` b (No, because changes in height due to nutrition and reduced incidence of disease are not heritable.) **3.** `evaluate` Evidence to support a common ancestor of humans and chimpanzees about 6–7 million years ago includes the following: fossils from Africa that date from 6–7 million years ago with primate characteristics that

are common to chimps and humans, and DNA sequence comparisons showing homologies between chimps and humans (the rate of change of some genes can be used to estimate time of divergence). **4.** `evaluate` The theory of evolution fits the six criteria as follows: (1) and (2): It provides a common underlying mechanism responsible for puzzling observations such as homology, geographic proximity of similar species, the law of succession in the fossil record, vestigial traits, and extinctions. (3) and (4): It suggests new lines of research to test predictions about the outcome of changing environmental conditions in populations, about the presence of transitional forms in the fossil record, and so on. (5): It is a simple idea that explains the tremendous diversity of living and fossil organisms and why species continue to change today. (6): The realization that all organisms are related by common descent and that none are higher or lower than others was a surprise.

CHAPTER 26

IN-TEXT QUESTIONS AND EXERCISES

p. 467 (1) `apply` The frequencies of the three genotypes are shown by $p^2 + 2pq + q^2 = 1$.

$$\text{Freq}(A_1A_1) = p^2 = (0.6)^2 = 0.36.$$
$$\text{Freq}(A_1A_2) = 2pq = 2(0.6)(0.4) = 0.48.$$
$$\text{Freq}(A_2A_2) = q^2 = (0.4)^2 = 0.16.$$

(2) `apply` Allele frequencies in the offspring gene pool:

$$\text{Freq}(A_1) = 0.36 + 1/2(0.48) = 0.60.$$
$$\text{Freq}(A_2) = 1/2(0.48) + 0.16 = 0.40.$$

(3) `apply` Evolution has not occurred, because the allele frequencies have not changed.
p. 469 Table 26.1 `apply` The observed allele frequencies, calculated from the observed genotype frequencies, are 0.43 for *M* and 0.57 for *N*. The expected genotypes, calculated from the observed allele frequencies under the Hardy–Weinberg principle, are 0.185 for *MM*, 0.49 for *MN*, 0.325 for *NN*.
p. 470 CYU `apply` Given the observed genotype frequencies, the observed allele frequencies are

$$\text{Freq}(A_1) = 0.574 + 1/2(0.339) = 0.744.$$
$$\text{Freq}(A_2) = 1/2(0.339) + 0.087 = 0.256.$$

Given these allele frequencies, the genotype frequencies expected under the Hardy–Weinberg principle are A_1A_1: $0.744^2 = 0.554$; A_1A_2: $2(0.744 \times 0.256) = 0.381$; A_2A_2: $0.256^2 = 0.066$. There are 4 percent too few heterozygotes observed, relative to the expected proportion. One of the assumptions of the Hardy–Weinberg principle is not met at this gene in this population, at this time.
p. 470 `apply` The proportions of homozygotes should increase, and the proportions of heterozygotes should decrease, but allele frequencies will not change if no selection is occurring at this locus.
p. 474 `apply` There will be an excess of observed genotypes containing the favored allele compared to the proportion expected under Hardy–Weinberg proportions.
p. 478 (1) `understand` Sperm are small and hence relatively cheap to produce, whereas eggs are large and require a large investment of resources to produce. **(2)** `understand` Sperm are inexpensive to produce, so reproductive success for males depends on their ability to find mates—not on their ability to find resources to produce sperm. The opposite pattern holds for females. Sexual selection is based on variation in ability to find mates, so is more intense in males—leading to more exaggerated traits in males.
p. 479 `apply` If allele frequencies are changing due to drift, the populations in Table 26.1 would behave like the simulated populations in Figure 26.12—frequencies would drift up and down over time, and diverge.
p. 479 Fig. 26.12 `analyze` See **FIGURE A26.1**.
p. 480 Fig. 26.13 `analyze` The smaller the population, the greater the chance that genetic drift will cause alleles to

be fixed or lost in the population in a short time. Starting with a smaller population makes the experiment easier.

p. 481 Fig. 26.14 `analyze` Original population: freq(A_1) = $(9 + 9 + 11)/54 = 0.54$; New population: freq(A_1) = $(2 + 2 + 1)/6 = 0.83$. The frequency of A_1 has increased dramatically.

p. 482 CYU (1) `understand` When allele frequencies fluctuate randomly up and down, sooner or later the frequency of an allele will hit 0. That allele thus is lost from the population, and the other allele at that locus is fixed. **(2)** `understand` In small populations, sampling error is large. For example, the accidental death of a few individuals would have a large impact on allele frequencies.

IF YOU UNDERSTAND . . .

26.1 `apply` The frequencies of two alleles are shown by $p + q = 1$. The frequencies of the three genotypes are shown by $p^2 + 2pq + q^2 = 1$.

$$\text{If freq}(A_1) = 0.2, \text{ then freq}(A_2) = 0.8.$$
$$\text{Freq}(A_1A_1) = p^2 = (0.2)^2 = 0.04.$$
$$\text{Freq}(A_1A_2) = 2pq = 2(0.2)(0.8) = 0.32.$$
$$\text{Freq}(A_2A_2) = q^2 = (0.8)^2 = 0.64.$$

26.2 `apply` It increased homozygosity of recessive alleles associated with diseases such as hemophilia. As a result, the royal families were plagued by genetic diseases. **26.3** `understand` The Hardy–Weinberg principle predicts that allele frequencies will stay the same over time. Natural selection favors some alleles over others, causing the frequency of those alleles to increase. **26.4** `apply` Genetic drift will rapidly reduce genetic variation in small populations. Thus, it is a high priority to keep the populations of endangered animals as large as possible. **26.5** `apply` In some cases, gene flow could be used to increase genetic variation and reduce inbreeding depression. However, gene flow can also be deleterious if the new alleles reduce the fitness of the endangered population. **26.6** `apply` Mutation rates themselves are too slow to affect an endangered population in a short time—unless a deleterious or beneficial mutation exposes the population to selection. (In instances where chemicals cause mutations in organisms, such as pesticides causing abnormal development in frogs, actions could be taken to remove the source of chemicals.)

YOU SHOULD BE ABLE TO . . .

✓ Test Your Knowledge

1. `understand` It defines what genotype and allele frequencies should be expected if evolutionary processes and nonrandom mating are *not* occurring. **2.** `remember` b **3.** `understand` a **4.** `remember` b **5.** `remember` True **6.** `understand` c

✓ Test Your Understanding

1. `apply` d; freq(A_1) = 0.9, so freq(A_2) = 0.1. Freq(A_2A_2) = $q^2 = (0.1)^2 = 0.01$. The number of babies with cystic

FIGURE A26.1

fibrosis = 2500(0.01) = 25. **2.** `apply` Inbreeding causes an increase in the number of homozygotes. Individuals who are homozygous for recessive deleterious alleles are exposed to—and eliminated by—directional natural selection. **3.** `evaluate` This statement is false. Mutations do not occur because an organism wants or needs them. They just happen by accident and can be beneficial, neutral, or deleterious. **4.** `understand` Sexual selection is most intense on the sex that makes the least investment in the offspring, so that sex tends to have exaggerated traits that make individuals successful in competition for mates. In this way, sexual dimorphism evolves. **5.** `analyze` An allele in males, because a successful male can have as many as 100 offspring—10 times more than a successful female. **6.** `create` Your concept map should have linking verbs that relate the following information: Selection may decrease genetic variation or maintain it (e.g., if heterozygotes are favored). Genetic drift reduces it. Gene flow may increase or decrease it (depending on whether immigrants bring new alleles or emigrants remove alleles). Mutation increases it.

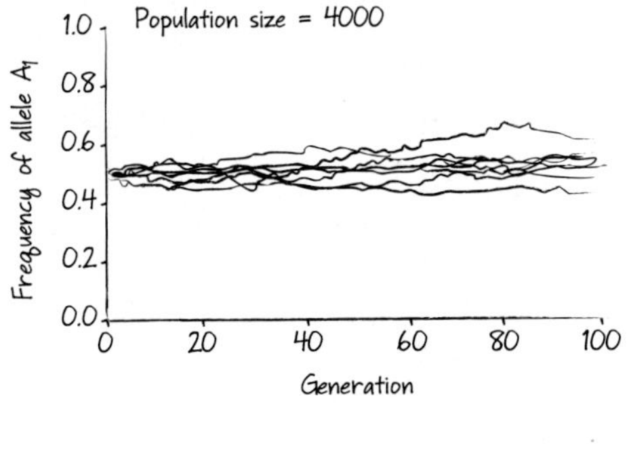

✓ Test Your Problem-Solving Skills

1. `apply` If we let p stand for the frequency of the loss-of-function allele, we know that $p^2 = 0.0001$; therefore $p = \sqrt{0.0001} = 0.01$. By subtraction, the frequency of normal alleles is 0.99. Under the Hardy–Weinberg principle, the frequency of heterozygotes is $2pq$, or $2 \times 0.01 \times 0.99$, which is 0.0198. **2.** `analyze` c **3.** `analyze` If males never lift a finger to help females raise children, the fundamental asymmetry of sex is pronounced and sexual dimorphism should be high. If males invest a great deal in raising offspring, then the fundamental asymmetry of sex is small and sexual dimorphism should be low. **4.** `apply` Captive-bred young, transferred adults, and habitat corridors could be introduced to counteract the damaging effects of drift and inbreeding in the small, isolated populations. It is important, though, to introduce individuals only from similar habitats and connect patches of similar habitat with corridors—to avoid introducing alleles that lower fitness.

CHAPTER 27

IN-TEXT QUESTIONS AND EXERCISES

p. 493 CYU (1) `understand` Biological species concept: Reproductive isolation means that no gene flow is occurring. Morphological species concept: If populations are evolving independently, they may have evolved morphological differences. Phylogenetic species concept: If populations are evolving independently, they should have synapomorphies that identify them as independent twigs on the tree of life. **(2)** `understand` Biological species concept: It cannot be used to evaluate fossils, species that reproduce asexually, or species that do not occur in the same area and therefore never have the opportunity to mate. Morphospecies concept: It cannot identify species that differ in traits other than morphology and is subjective—it can lead to differences of opinion that cannot be resolved by data. Phylogenetic species concept: Reliable phylogenetic information exists only for a small number of organisms.

p. 494 Fig. 27.4 `analyze` The disruption of the salt marsh was a vicariance event that split the range of seaside sparrows into two areas (Gulf and Atlantic coasts), limiting gene flow and eventually resulting in two monophyletic groups.

p. 496 Fig. 27.7 `analyze` It creates genetic isolation—the precondition for divergence—by separating apple maggot flies from hawthorn flies, which mate on hawthorn fruits.

p. 497 `create` The difference in timing of ripening of apples and hawthorns causes divergence by separating apple flies and hawthorn flies not only in space (apple fruits versus hawthorn fruits) but also in time. The

earlier ripening of apples means that apple maggots develop in relatively warm conditions in the fall compared to hawthorn maggots. Natural selection may favor traits in apple maggots that make them adapted to warmer temperatures and traits in hawthorn maggots that make them adapted to cooler temperatures.

p. 498 `apply` A diploid grape plant experiences a defect in meiosis resulting in the formation of diploid gametes. The individual self-fertilizes, producing tetraploid offspring. The tetraploid grape plants self-fertilize or mate with other tetraploid individuals, producing a tetraploid population.

p. 499 `apply` A tetraploid ($4n$) species (in this case, wheat) gives rise to diploid ($2n$) gametes, and a diploid wheat species gives rise to haploid (n) gametes. When a haploid gamete (n) fertilizes a diploid gamete ($2n$), a sterile triploid offspring results. If an error in mitosis occurs and chromosome number doubles before meiosis, then hexaploid wheat is formed.

p. 499 CYU `understand` Use one of the cases given in the chapter to illustrate dispersal (Galápagos finches), vicariance (snapping shrimp), or habitat specialization (apple maggot flies). In each case, drift will cause allele frequencies to change randomly. Differences in the habitats occupied by the isolated populations will cause allele frequencies to change under natural selection. Any mutation that occurs in one population will not occur in the other due to the lack, or low level, of gene flow. Or use polyploidization in maidenhair fern or *Tragopogon* to illustrate how a massive mutation can itself create immediate genetic isolation and divergence.

p. 501 Fig. 27.12 `evaluate` If the same processes were at work in the past that are at work in an experiment with living organisms, then the outcome of the experiment should be a valid replication of the outcome that occurred in the past.

p. 502 `analyze` Gene regions in the experimental hybrid would not align with those of *H. annuus* and *H. petiolaris*, meaning that the gray bar representing the DNA of the experimental hybrid could have had no orange or red regions corresponding to *H. annuus* and *H. petiolaris*.

IF YOU UNDERSTAND THAT . . .

27.1 `apply` Human populations would not be considered separate species under the biological concept, because human populations can successfully interbreed. They would not be considered separate species under the morphological species concept, because all human populations have the same basic morphology. Although human races differ in minor superficial attributes such as skin color and hair texture, they have virtually identical anatomy and physiology in all other regards. Nor would human populations be separate species under the phylogenetic species concept, because they all arose from a very recent common ancestor. (DNA comparisons have revealed that human races are remarkably similar genetically and do not differ enough genetically to qualify even for subspecies status.) **27.2** `create` (Many possibilities) Collect a group of fruit flies that are living in your kitchen, separate the individuals into different cages, and expose the two new populations to dramatically different environmental conditions—heat, lighting, food sources. In essence: Create genetic isolation and then conditions for divergence due to drift, mutation, and selection. **27.3** `evaluate` The sample experiment described above is an example of vicariance, because the experimenter fragmented the habitat into isolated cages. (A different mechanism of speciation may have been involved in your example.) **27.4** `apply` Hybrid individuals will increase in frequency, and natural selection will favor parents that hybridize. The hybrid zone will increase. The two parent populations may go extinct if hybrids live in the same environment.

✔ **Test Your Knowledge**

1. remember a **2.** remember b **3.** understand a **4.** understand a
5. understand d **6.** remember False

✔ **Test Your Understanding**

1. analyze Direct observation: Galápagos finch coloniza-tion, apple maggot flies, *Tragopogon* allopolyploids, maidenhair ferns. Indirect studies: snapping shrimp, sunflowers. **2.** understand a **3.** analyze Based on biological and morphospecies criteria (breeding range and morpholog-ical traits, respectively), one species with six subspecies of seaside sparrows were recognized. However, based on phylogenetic criteria (comparison of gene sequences), biologists concluded that seaside sparrows represent only two monophyletic groups. **4.** evaluate Some flies breed on apple fruits, others breed on hawthorn fruits. This reduces gene flow. Because apple fruits and hawthorn fruits have different scents and other characteristics, selection is causing the populations to diverge. **5.** apply Decreasing. Gene flow tends to equalize allele frequen-cies among populations. **6.** apply c

✔ **Test Your Problem-Solving Skills**

1. analyze b **2.** create Sexual selection can promote diver-gence because it affects gene flow directly. For example, if blond women began mating only with blond men, and dark-haired women began mating only with dark-haired men, gene flow would decrease between these two popu-lations even though they live in the same geographic area. Given enough time, these two populations would diverge due to mutation, genetic drift, and natural selec-tion. **3.** evaluate Two things: (1) Some flies happen to have alleles that allow them to respond to apple scents; others happen to have alleles that allow them to respond to hawthorn scents. There is no "need" involved. (2) The alleles for scent response exist; they are not acquired by spending time on the fruit. **4.** analyze If the isolated populations and habitat fragments are small enough, the species is likely to dwindle to extinction due to in-breeding and loss of genetic variation or catastrophes like a severe storm or a disease outbreak. If the isolated populations survive, they are likely to diverge into new species because they are genetically isolated and because the habitats may differ.

CHAPTER 28

IN-TEXT QUESTIONS AND EXERCISES

p. 507 Fig. 28.1 apply Many trees are possible. See **FIGURE A28.1**.
p. 510 understand Because whales and hippos share SINEs 4–7 and no other species has these four, they are synapo-morphies that define them as a monophyletic group. The similarity is unlikely to be due to homoplasy because the chance of four SINEs inserting in exactly the same place in two different species, independently, is very small.
p. 510 CYU analyze Hair and limb structures in humans and whales are examples of homology because they are traits that can be traced to a common ancestor. All mammals have hair and similar limb bone structure. However, extensive hair loss and advanced social behav-ior in whales and humans are examples of homoplasy. These traits are not common to all mammalian species and likely arose independently during the evolution of specific mammalian lineages.
p. 511 Fig. 28.4 apply See **FIGURE A28.2**. SINEs group (19, 20) identifies peccaries and pigs as a monophyletic group.
p. 513 apply Since some organisms fossilize better than others—such as mollusks with hard shells—a paleontol-ogist would be mistaken to conclude that mollusks were more abundant than other organisms at a certain time just because mollusks are more abundant in the fossil

record. Similarly, a paleontologist would be mistaken to conclude that no organisms occupied an ancient desert based on the lack of fossil evidence; organisms in deserts do not fossilize well. Several other examples are possible.
p. 517 create In habitats on the mainland, complex com-munities have been established for a long time, so tar-weeds in California and *Anolis* lizards on the mainland experience greater competition for resources and are limited to specific niches.
p. 520 CYU understand In terms of ecological opportunity, many resources were available because no other animals (or other types of organisms) existed to exploit them. Also, the evolution of new species in new niches made new niches available for predators. Morphological in-novations like limbs and complex mouthparts were im-portant because they made it possible for animals to live in habitats other than the benthic area, and to consume new types of food.
p. 520 Fig. 28.14 analyze The dinosaurs went extinct during the end-Cretaceous. About 15 percent of families went extinct.

IF YOU UNDERSTAND . . .

28.1 apply Humans are the only mammal that walks up-right on two legs, so this character is a synapomorphy that defines the hominins. **28.2** apply Snail shells are hard structures, and snails often live in marine environments that fossilize readily, so snail fossils are relatively abun-dant. Abundant fossils are more helpful than rare fossils for estimating changes in relative abundance over time. **28.3** analyze They could eat food that was available off the substrate. And once animals lived off (and away from) the bottom, it created an opportunity for other animals who could eat them to evolve. **28.4** evaluate This statement is false. Human-induced extinctions today are not due to poor adaptation to the normal environment. Humans are changing habitats rapidly and in unusual ways (such as clear-cutting forests), so most species cannot adapt to these changes rapidly enough to avoid going extinct. In this way, human-induced extinctions are more similar to mass extinctions than to background extinctions.

YOU SHOULD BE ABLE TO . . .

✔ **Test Your Knowledge**

1. remember c **2.** understand a **3.** understand b **4.** understand False
5. understand d **6.** remember True

✔ **Test Your Understanding**

1. evaluate The fossil record is biased because recent, abun-dant organisms with hard parts that live underground or in environments where sediments are being depos-ited are most likely to fossilize. Even so, it is the only data available on what organisms that lived in the past looked like, and where they lived. **2.** create An enormous amount of animal diversity appeared in a time frame that was short compared to the sweep of earlier Earth history. The most common analogy is to an explosion, referring to the sudden and spectacular radiation of lineages. Another analogy is your hand, where one line (your arm) comes into a node (your palm) and several lineages (your fingers) come out. Many other answers are possible. **3.** understand Homoplasy is rare relative to homology, so phylogenetic trees that minimize the total change required are usually more accurate. In the case of the artiodactyl astragalus, parsimony was misleading because the astragalus was lost when whales evolved limblessness—creating two changes (a loss following a gain) instead of one (a gain). **4.** create (Many answers are possible.) Adaptive radiation of *Anolis* lizards after they colonized new islands in the Caribbean. *Hypothesis:* After lizards arrived on each new island, where there were no predators or competitors, they rapidly diversi-fied to occupy four distinct types of habitats on each island. Adaptive radiation during the Cambrian period

following a morphological innovation. *Hypothesis:* Ad-ditional *Hox* genes made it possible to organize a large, complex body; the evolution of complex mouthparts and limbs made it possible for animals to move and find food in new ways. **5.** apply About 90 percent of spe-cies went extinct during the end-Permian, so about 10 percent survived, written as 0.1 in decimals. $1880 \times 0.1 = 188$ would make it to graduation. **6.** apply d

✔ **Test Your Problem-Solving Skills**

1. understand b **2.** evaluate Many trees are possible. See **FIGURE A28.3**. Sponges are the outgroup. The most par-simonious tree has the fewest number of changes (five in this case). Adding more characters to the matrix can improve the reliability of the results. **3.** understand It would be helpful to have (1) a large crater or other physical evidence from a major impact dated at 251 mya, (2) shocked quartz and microtektites in rock layers dating to the end-Permian era, and (3) high levels of iridium or other elements common in space rocks and rare on Earth, dated to the time of the extinctions. **4.** create Many concept map topologies are possible. Maps should follow this logic: Oxygen is an effective final electron acceptor in cellular respiration because of its high electronegativ-ity. Organisms that use it as a final electron acceptor can produce more usable energy than organisms that do not use oxygen, but only if it is available. With more available energy, aerobic organisms can grow larger and move faster.

BIG PICTURE Evolution

p. 526 CYU (1) understand Circle = inbreeding, sexual selec-tion, nonrandom mating, natural selection, genetic drift,

---

FIGURE A28.1

```
                          Outgroup
                          Species 1
                          Species 2
          A→C    C→A      Species 3
                  C→G      Species 4
```

FIGURE A28.2

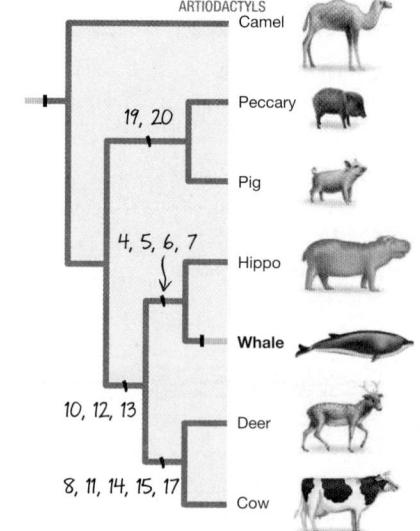

ARTIODACTYLS

Camel
Peccary
Pig
Hippo
Whale
Deer
Cow

19, 20
4, 5, 6, 7
10, 12, 13
8, 11, 14, 15, 17

mutation, and gene flow. **(2)** analyze Adaptation "increases" fitness; synapomorphies "identify branches on" the tree of life. **(3)** analyze Several answers possible, e.g., adaptive radiations and mass extinctions → can be studied using → the fossil record. **(4)** analyze Genetic drift, mutation, and gene flow "are random with respect to" fitness.

CHAPTER 29

IN-TEXT QUESTIONS AND EXERCISES

p. 529 Fig. 29.1 analyze Prokaryotic—it would require just one evolutionary change, the origin of the nuclear envelope in Eukarya. If it were eukaryotic, it would require that the nuclear envelope was lost in both Bacteria and Archaea—two changes are less parsimonious (see Chapter 28).

p. 529 Table 29.1 apply See **FIGURE A29.2**.

p. 534 Fig. 29.5 analyze Weak—different culture conditions may have revealed different species.

p. 535 CYU (1) create Conditions should mimic a spill—sand or stones with a layer of crude oil, or seawater with oil floating on top. Add samples, from sites contaminated with oil, that might contain cells capable of using molecules in oil as electron donors or electron acceptors. Other conditions (temperature, pH, etc.) should be realistic. **(2)** create Use metagenomic analysis. After isolating DNA from a soil sample, fragment and sequence the DNA. Compare the sequences to those from known organisms and use the data to place the species on the tree of life.

p. 538 apply cyanobacteria—photoautotrophs; *Clostridium aceticum*—chemoorganoautotroph; *Nitrosopumilus* sp.—chemolithoautotrophs; helicobacteria—photoheterotrophs; *Escherichia coli*—chemoorganoheterotroph; *Beggitoa*—chemolithotrophic heterotrophs.

p. 539 Fig. 29.10 apply Table 29.5 contains the answers. For example, for organisms called sulfate reducers you would have H_2 as the electron donor, SO_4^{2-} as the electron acceptor, and H_2S as the reduced by-product. For humans the electron donor is glucose ($C_6H_{12}O_6$), the electron acceptor is O_2, and the reduced by-product is water (H_2O).

p. 541 understand Different species of bacteria that have different forms of bacterial chlorophyll will not absorb the same wavelengths of light and will therefore not compete.

p. 542 Fig. 29.12 apply Aerobic respiration. More free energy is released when oxygen is the final electron acceptor than when any other molecule is used, so more ATP can be produced and used for growth.

p. 543 apply If bacteria and archaea did not exist, then (1) the atmosphere would have little or no oxygen, and (2) almost all nitrogen would exist in molecular form (the gas N_2).

p. 543 Fig. 29.13 apply Add a label for "animals" in the upper right quarter of the circle, and draw arrows leading from "Organic compounds with amino groups" to "animals" and from "animals" to "NH_3."

p. 544 CYU evaluate Eukaryotes can only (1) fix carbon via the Calvin–Benson pathway, (2) use aerobic respiration with organic compounds as electron donors, and (3) perform oxygenic photosynthesis. Among bacteria and archaea, there is much more diversity in pathways for carbon fixation, respiration, and photosynthesis, along with many more fermentation pathways.

p. 545a, p. 545b, p. 546a, p. 546b, p. 547 apply See **FIGURE A29.1**.

p. 548 Fig. 29.21 apply About 600,000 *N. maritimus* cells would fit end to end on a meter stick.

IF YOU UNDERSTAND . . .

29.1 analyze Eukaryotes have a nuclear envelope that encloses their chromosomes; bacteria and archaea do not. Bacteria have cell walls that contain peptidoglycan, and archaea have phospholipids containing isoprene subunits in their plasma membranes. Thus, the exteriors of a bacterium and archaeon are radically different. Archaea and eukaryotes also have similar machinery for processing genetic information. **29.2** apply After extracting and sequencing DNA from bacteria that live in an insect gut, analyze the resulting sequences for genes that are similar to known genes involved in nitrogen fixation. **29.3** understand They are compatible, and thrive in each others' presence—one species' waste product is the other species' food.

YOU SHOULD BE ABLE TO . . .

✓ Test Your Knowledge

1. understand c **2.** remember b **3.** remember d **4.** remember d **5.** remember False—the enzyme nitrogenase is poisoned by oxygen. **6.** remember peptidoglycan

✓ Test Your Understanding

1. understand c **2.** understand An electron donor provides the potential energy required to produce ATP. **3.** evaluate Yes. The array of substances that bacteria and archaea can use as electron donors, electron acceptors, and fermentation substrates, along with the diversity of ways that they can fix carbon and perform photosynthesis, allows them to live just about anywhere. **4.** apply They should get energy from reduced organic compounds "stolen" from their hosts. **5.** understand Large amounts of potential energy are released and ATP produced when oxygen is the electron acceptor, because oxygen is so electronegative.

FIGURE A28.3

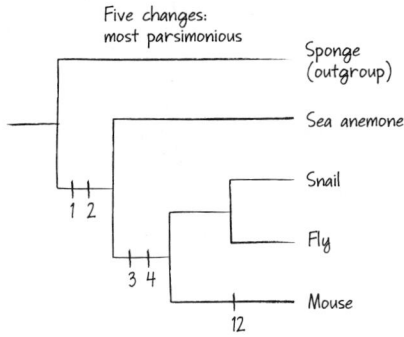
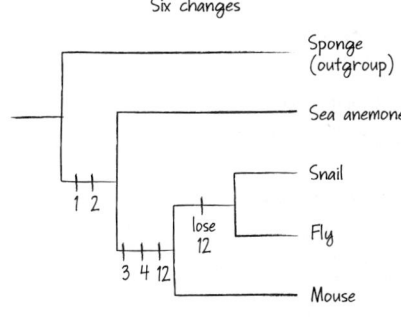
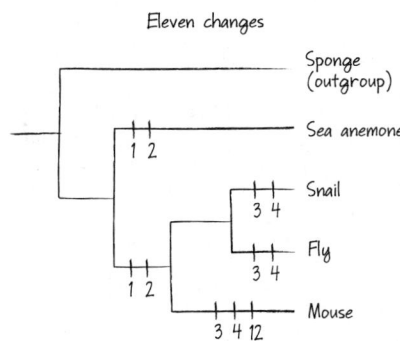

FIGURE A29.1

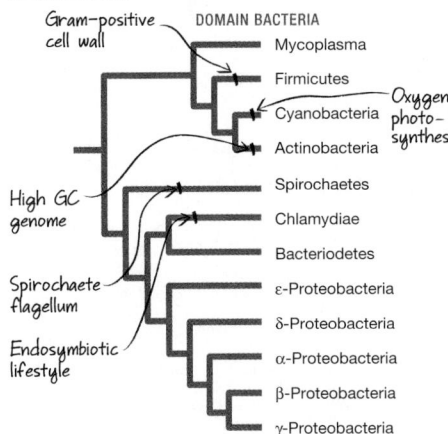

FIGURE A29.2

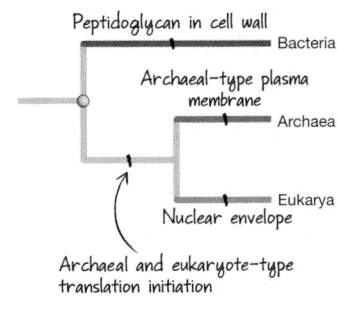

Large body size and high growth rates are not possible without large amounts of ATP. **6.** `understand` They are paraphyletic because the prokaryotes include some (bacteria and archaea) but not all (eukaryotes) groups derived from the common ancestor of all organisms living today.

✔ Test Your Problem-Solving Skills

1. `understand` c (It is pathogenic on a wide variety of organisms). **2.** `analyze` This result supports their hypothesis because the drug poisons the enzymes of the electron transport chain and prevents electron transfer to Fe^{3-}, which is required to drive magnetite synthesis. If magnetite had still formed, another explanation would have been needed. **3.** `create` Hypothesis: A high rate of tooth cavities in Western children is due to an excess of sucrose in the diet, which is absent from the diets of East African children. To test this hypothesis, switch a group of East African children to a diet that contains sucrose, switch a group of Western children to a diet lacking sucrose, and monitor the presence of *S. mutans* in both groups of children. **4.** `create` Look in waters or soil polluted with benzene-containing compounds. Put samples from these environments in culture tubes where benzene is the only source of carbon. Monitor the cultures and study the cells that grow efficiently.

CHAPTER 30

IN-TEXT QUESTIONS AND EXERCISES

p. 529 CYU (1) `understand` Opisthokonts have a flagellum at the base or back of the cell; alveolate cells contain unique support structures called alveoli; stramenopiles—meaning "straw hairs"—have straw-like hairs on their flagella. **(2)** `understand` In direct sequencing, DNA is isolated directly from the environment and analyzed to place species on the tree of life. It is not necessary to actually see the species being studied.

p. 557 CYU (1) `understand` *Plasmodium* species are transmitted to humans by mosquitoes. If mosquitoes can be prevented from biting people, they cannot spread the disease. **(2)** `apply` Iron added → primary producers (photosynthetic protists and bacteria) bloom → more carbon dioxide taken up from atmosphere during photosynthesis → consumers bloom, eat primary producers → bodies of primary producers and consumers fall to bottom of ocean → large deposits of carbon-containing compounds form on ocean floor.

p. 560 `apply` See **FIGURE A30.1**.

p. 560 Fig. 30.8 `analyze` Two—one derived from the original bacterium and one derived from the eukaryotic cell that engulfed the bacterium.

p. 561 `apply` A photosynthetic bacterium (e.g., a cyanobacterium) could have been engulfed by a larger eukaryotic cell. If it was not digested, it could continue to photosynthesize and supply sugars to the host cell.

p. 562 `apply` Membrane infoldings observed in bacterial species today support the hypothesis's plausibility—they confirm that the initial steps actually could have happened. The continuity of the nuclear envelope and ER are consistent with the hypothesis, which predicts that the two structures are derived from the same source (infolded membranes).

p. 564 `apply` The chloroplast genes label should come off of the branch that leads to cyanobacteria. (If you had a phylogeny of just the cyanobacteria, the chloroplast branch would be located somewhere inside.)

p. 565 `analyze` The acquisition of the mitochondrion and the chloroplast represent the transfer of entire genomes, and not just single genes, to a new organism.

p. 566 `apply` Yes—when food is scarce or population density is high, the environment is changing rapidly (deteriorating). Offspring that are genetically unlike their parents may be better able to cope with the new and challenging environment.

p. 569a `understand` (1) Alternation of generations refers to a life cycle in which there are multicellular haploid phases and multicellular diploid phases. A gametophyte is the multicellular haploid phase that produces gametes; the sporophyte is the multicellular diploid phase that produces spores. A spore is a cell that grows into a multicellular individual but is not produced by fusion of two cells. A zygote is a cell that grows into a multicellular individual but *is* produced by fusion of two cells (gametes). Gametes are haploid cells that fuse to form a zygote. (2) See **FIGURE A30.2**.

p. 569b `apply` Most likely, the two types of amoebae evolved independently. The alternative hypothesis is that the common ancestor of alveolates, stramenopiles, rhizarians, plants, opisthokonts, and amoebozoa were amoeboid and that this growth form was lost many times. See **FIGURE A30.1**.

p. 569c, p. 569d, p. 569e, p. 570a, p. 570b, p. 570c `apply` See **FIGURE A30.1**.

p. 571 `analyze` If euglenids could take in food via phagocytosis (ingestive feeding), then that would have provided a mechanism by which a smaller photosynthetic protist could have been engulfed and incorporated into the cell via secondary endosymbiosis.

p. 572a, p. 572b, p. 573, p 574. `apply` See **FIGURE A30.1**.

IF YOU UNDERSTAND . . .

30.1 `understand` Photosynthetic protists use CO_2 and light to produce sugars and other organic compounds, so they furnish the first or primary source of organic material in an ecosystem. **30.2** `remember` Eukaryotic cells contain many organelles, including a nucleus, endomembrane system, and an extensive cytoskeleton. Most prokaryotic cells contain few or no organelles, and no nucleus or endomembrane system and lack a complex cytoskeleton. **30.3** `understand` (1) Outside membrane was from host eukaryote; inside from engulfed cyanobacterium. (2) From the outside in, the four membranes are derived from the eukaryote that engulfed a chloroplast-containing eukaryote, the plasma membrane of the eukaryote that was engulfed, the outer membrane of the engulfed cell's chloroplast, and the inner membrane of its chloroplast.

YOU SHOULD BE ABLE TO . . .

✔ Test Your Knowledge

1. `understand` b **2.** `remember` b **3.** `remember` a **4.** `remember` b **5.** `remember` False. **6.** `remember` diatoms.

✔ Test Your Understanding

1. `understand` a **2.** `understand` Because all eukaryotes living today have cells with a nuclear envelope, it is valid to infer that their common ancestor also had a nuclear envelope. Because bacteria and archaea do not have a nuclear envelope, it is valid to infer that the trait arose in the common ancestor of eukaryotes. **3.** `understand` The host cell provided a protected environment and carbon compounds for the endosymbiont; the endosymbiont provided increased ATP from the carbon compounds **4.** `understand` It confirmed a fundamental prediction made by the hypothesis and could not be explained by any alternative hypothesis. **5.** `understand` All alveolates have alveoli, which are unique structures that function in supporting the cell. Among alveolates there are species that are (1) ingestive feeders, photosynthetic, or parasitic, and that (2) move using cilia, flagella, or a type of amoeboid movement.

✔ Test Your Problem-Solving Skills

1. `analyze` d **2.** `analyze` If the apicoplast that is found in *Plasmodium* (the organism that causes malaria) is genetically similar to chloroplasts, and if glyphosate poisons chloroplasts, it is reasonable to hypothesize that glyphosate will poison the apicoplast and potentially kill the *Plasmodium*. This would be a good treatment strategy

for malaria because humans have no chloroplasts, provided that the glyphosate produces no other effects that would be detrimental to humans. **3.** `evaluate` Primary producers usually grow faster when CO_2 concentration increases, but to date they have not grown fast enough to make CO_2 levels drop—CO_2 levels have been increasing steadily over decades. **4.** `analyze` Given that lateral gene transfer can occur at different points in a phylogenetic history, specific genes can become part of a lineage by a different route from that taken by other genes of the organism. In the case of chlorophyll *a*, its history traces back to a bacterium being engulfed by a protist and forming a chloroplast.

FIGURE A30.1

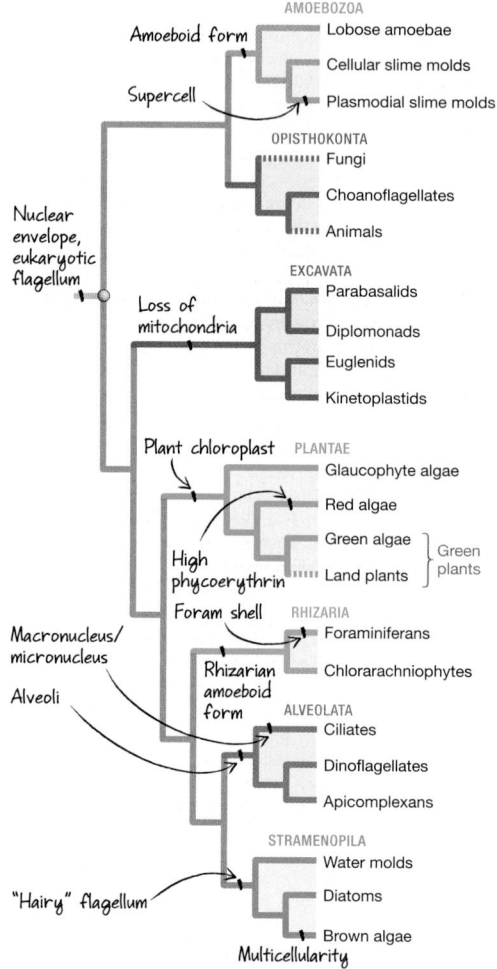

FIGURE A30.2

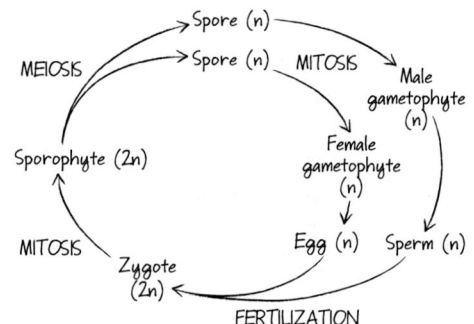

CHAPTER 31

IN-TEXT QUESTIONS AND EXERCISES

p. 582 CYU `understand` (1) Green algae and land plants share an array of morphological traits that are synapomorphies, including the chlorophylls they contain; (2) green algae appear before land plants in the fossil record; and (3) on phylogenetic trees estimated from DNA sequence data, green algae and land plants share a most recent common ancestor—green algae are the initial groups to diverge, and land plants diverge subsequently.

p. 583 Fig. 31.6 `remember` If you find the common ancestor of all the green algae (at the base of the Ulvophyceae), the lineages that are collectively called green algae don't include that common ancestor and all of its descendants—only some of its descendants. The same is true for non-vascular plants (the common ancestor here is at the base of the Hepaticophyta) and seedless vascular plants (the common ancestor here is at the base of the Lycophyta).

p. 586 Fig. 31.8 `understand` Water flows more easily through a short, wide pipe than through a long, skinny one because there is less resistance from the walls of the pipe.

p. 587 Fig. 31.9 `understand` To map an innovation on a phylogenetic tree, biologists determine the location(s) on the tree that is (are) consistent with all descendants from that point having the innovation (unless, in rare cases, the innovation was lost in a few descendants).

p. 589 `understand` Alternation of generations occurs when there are multicellular haploid individuals and multicellular diploid individuals in a life cycle. One hypothesis is that this life cycle evolved because the plants with a large sporophyte were more successful at using the wind to disperse many spores, while a small gametophyte produces gametes that can swim.

p. 590 `apply` In the hornwort photo, the sporophyte is the spike-like green and yellow structure; the gametophyte is the leafy-looking structure underneath. The horsetail gametophyte is the microscopic individual on the left; the sporophyte is the much larger individual on the right.

p. 590 Fig. 31.14 `understand` There are multicellular haploid stages and multicellular diploid stages in these plants.

p. 592 Fig. 31.18 `analyze` Gymnosperm gametophytes are microscopic, so they are even smaller than fern gametophytes. The gymnosperm gametophyte is completely dependent on the sporophyte for nutrition, while fern gametophytes are not. Fern gametophytes are photosynthetic and even supply nutrition to the young sporophytes.

p. 593 Fig. 31.19 `analyze` Consistent—the fossil data suggest that gymnosperms evolved earlier, and gymnosperms have larger gametophytes than angiosperms.

p. 595 Fig. 31.21 `analyze` To eliminate all other variables except for the one being tested, such as the possibility of pollination by other insects in the field. Scales differ because the two insect species visited flowers at very different frequencies.

p. 596 CYU (1) `understand` Cuticle prevents water loss from the plant; UV-absorbing compounds allow plants to be exposed to high light intensities without damage to their DNA; vascular tissue moves water up from the soil and moves photosynthetic products down to the roots.
(2) `apply` See **FIGURE A31.1**.
p. 600a, p. 600b, p. 601, p. 603a, p. 603b, p. 604, p. 605, p. 606 `apply` See **FIGURE A31.1**.
p. 596 Fig. 31.23 `apply` See **FIGURE A31.1**.

IF YOU UNDERSTAND . . .

31.1 `apply` Rates of soil formation will decline; rates of soil loss will increase. **31.2** `understand` All green algae and land plants contain chlorophyll *a* and *b*, have similar arrangements of thylakoid membranes, and store starch. **31.3** `analyze` Both spores and seeds have a tough, protective coat, so they can survive while being dispersed to a new location. Seeds have the advantage of carrying a store of nutrients with them—when a spore germinates, it must make its own food via photosynthesis right away.

YOU SHOULD BE ABLE TO . . .

✓ Test Your Knowledge

1. `remember` c **2.** `remember` c **3.** `remember` a **4.** `remember` c **5.** `remember` (a) Gametangia are found in all land plant groups except angiosperms; (b) pollen is found in gymnosperms and angiosperms; (c) seeds are found in gymnosperms and angiosperms, (d) fruit is found in angiosperms.
6. `understand` In a gametophyte-dominant life cycle, the gametophyte is larger and longer lived than the sporophyte and produces most of the nutrition. In a sporophyte-dominant life cycle, the sporophyte generation is the larger, longer-lived, and photosynthetic phase of the life cycle.

✓ Test Your Understanding

1. `remember` c **2.** `understand` Plants build and hold soils required for human agriculture and forestry, and they increase water supplies that humans can use for drinking, irrigation, or industrial use. Plants release oxygen that we breathe. **3.** `analyze` Cuticle prevents water loss from leaves but also prevents entry of CO_2 required for photosynthesis. Stomata allow CO_2 to diffuse but can close to minimize water loss. Liverwort pores allow gas exchange but cannot be closed if conditions become dry. **4.** `understand` They provided the support needed for plants to grow upright and not fall over in response to wind or gravity. Erect growth allowed plants to compete for light. **5.** `apply` Homosporous plants produce a single type of spore that develops into a gametophyte that produces both egg and sperm. Heterosporous plants produce two different types of spores that develop into two different gametophytes that produce either egg or sperm. In a tulip, the microsporangium is found within the stamen, and the megasporangium is found within the ovule. Microspores divide by mitosis to form male gametophytes (pollen grains); megaspores divide by mitosis to form the female gametophyte. **6.** `understand` Animal pollinators increase the efficiency of pollination so plants can save energy by making less pollen.

✓ Test Your Problem-Solving Skills

1. `understand` d **2.** `analyze` A "reversion" to wind pollination might be favored by natural selection because it is costly to produce a flower that can attract animal pollinators. Because wind-pollinated species grow in dense clusters, they can maximize the chance that the wind will carry pollen from one individual to another (less likely if the individuals are far apart). Wind-pollinated deciduous trees flower in early spring before their developing leaves begin to block the wind. **3.** `create` The combination represents a compromise between efficiency and safety. Tracheids can still transport water if vessels become blocked by air bubbles. **4.** `create` Alter one characteristic of a flower, and present the flower to the normal pollinator. As a control, present the normal (unaltered) flower to the normal pollinator. Record the amount of time the pollinator spends in the flower, the amount of pollen removed, or some other measure of pollination success. Repeat for other altered characteristics. Analyze the data to determine which altered characteristic affects pollination success the most.

CHAPTER 32

IN-TEXT QUESTIONS AND EXERCISES

p. 619 CYU (1) `understand` Because they are made up of a network of thin, branching hyphae, mycelia have a large surface area, which makes absorption efficient.
(2) `remember` Swimming spores and gametes, zygosporangia, basidia, asci.

FIGURE A31.1

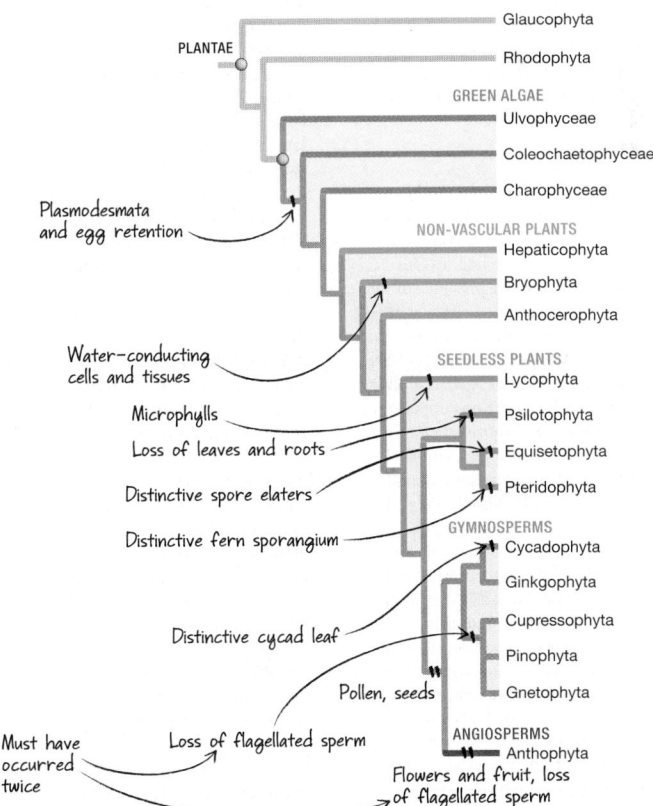

p. 622 Fig. 32.10 `analyze` Labeled nutrient experiments identify which nutrients are exchanged in and in which direction. They explain *why* plants do better in the presence of mycorrhizae, and why the fungus also benefits. **p. 625** `analyze` Human sperm and egg undergo plasmogamy followed by karyogamy during fertilization, but heterokaryotic cells do not occur in humans or other eukaryotes besides fungi. **p. 625 Fig. 32.12** `apply` The haploid mycelium. **p. 626 Fig. 32.13** `analyze` Mycelia produced by asexual reproduction are genetically identical to their parent; mycelia produced by sexual reproduction are genetically different from both parents (each spore has a unique genotype). **p. 628 CYU (1)** `understand` When birch tree seedlings are grown in the presence and absence of EMF, individuals denied their normal EMF are not able to acquire sufficient nitrogen and phosphorus. **(2)** `analyze` Each type of fungus involves meiosis and the production of haploid spores. In a zygosporangium, meiosis occurs in a multinucleate cell; in basidia and asci, this cell has a single diploid nucleus. In asci, meiosis is followed by one round of mitosis. **p. 628, p. 630a, p. 630b, p. 630c, p. 631, p. 632a, p. 632b** `apply` See **FIGURE A32.1**.

IF YOU UNDERSTAND . . .

32.1 `apply` Loss of mycorrhizal fungi would decrease nutrient delivery to plants and reduce their growth inside the experimental plots, compared to control plots with intact fungi. Also, lack of fungi would slow decay of dead plant material, causing a dramatic buildup of dead organic material. **32.2** `understand` Like humans, fungi are composed of eukaryotic cells. Bacteria are prokaryotic and thus have many more unique targets for antibiotics. **32.3** `understand` Haploid hyphae fuse, forming a heterokaryotic mycelium. When karyogamy occurs, a diploid nucleus forms—just as when gametes fuse. **32.4** `understand` See the diversity boxes.

YOU SHOULD BE ABLE TO . . .

✔ Test Your Knowledge

1. `remember` a **2.** `remember` d **3.** `remember` b **4.** `remember` b **5.** `understand` Pores in septa allow nutrients to move from regions of acquisition to regions of mycelial growth; **6.** `remember` commensal.

✔ Test Your Understanding

7. `understand` c **8.** `evaluate` Along with a few bacteria, fungi are the only organisms that can digest wood completely. If the wood is not digested, carbon remains trapped in wood. Without fungi, CO_2 would be tied up and unavailable for photosynthesis, and the presence of undecayed organic matter would reduce the space available for plants to grow. **9.** `analyze` Fungi produce enzymes that degrade cellulose and lignin. **10.** `analyze` Plant roots have much smaller surface area than EMF or AMF. Hyphae are much smaller than the smallest portions of plant roots so can penetrate dead material more efficiently. Extracellular digestion—which plant roots cannot do—allows fungi to break large molecules into small compounds that can be absorbed. **11.** `apply` DNA and RNA contain phosphorus. Mycorrhizal fungi effectively increase the volume of soil mined by root systems and can make adequate amounts of phosphorus available for supporting rapid cell division. **12.** `analyze` Both compounds are processed via extracellular digestion. Different enzymes are involved, however. The degradation of lignin is uncontrolled and does not yield useful products; digestion of cellulose is controlled and produces useful glucose molecules.

✔ Test Your Problem-Solving Skills

13. `apply` c **14.** `create` (1) Confirm that the chytrid fungus is found only in sick frogs and not healthy frogs.

(2) Isolate the chytrid fungus and grow it in a pure culture. (3) Expose healthy frogs to the cultured fungus and see if they become sick. (4) Isolate the fungus from the experimental frogs, grow it in culture, and test whether it is the same as the original fungus. **15.** `create` Each of the different cellulase enzymes attacks cellulose in a different way, so producing all the enzymes together increases the efficiency of the fungus in breaking down cellulose completely. It is likely that lignin peroxidase is produced along with cellulases, so they can act in concert to degrade wood. To test this idea, you could harvest enzymes secreted from a mycelium before and after it contacts wood in a culture dish, and see if the cellulases and lignin peroxidase appear together once the mycelium begins growing on the wood. **16.** `create` Collect a large array of colorful mushrooms that are poisonous; also capture mushroom-eating animals, such as squirrels. Present a hungry squirrel with a choice of mushrooms that have been dyed or painted a drab color versus others treated with a solution that is identical to the dye or paint used but uncolored. Record which mushrooms the squirrel eats. Repeat the test with many squirrels and many mushrooms.

CHAPTER 33

IN-TEXT QUESTIONS AND EXERCISES

p. 639 Fig. 33.2 `apply` The label and bar for "Segmentation?" should go on the branch to the left of the clam shell and "Mollusca" label. The label and bar for "Toolkit genes for segmentation?" should go on the same branch as cephalization, CNS, and coelom. **p. 640** `apply` The genetic tool kit required for multicellularity includes genes that regulate the cell cycle, and thus cell division. When something goes wrong in this process, uncontrolled growth called cancer can result. Thus, cancer may be as ancient as multicellularity itself. **p. 642 Fig. 33.7** `analyze` Stained *Hox* and *dpp* gene products would either not be found in *Nematostella* at all or would not occur in the same anterior–posterior and dorsal–ventral pattern as observed in bilaterians. **p. 646 CYU (1)** `apply` Yes, your body has a tube-within-a-tube design. Humans are vertebrates, vertebrates are chordates, chordates are bilaterians, and all bilaterians have a tube-within-a-tube design, consisting of a bilaterally symmetrical, slender body with an inner tube (gut), an outer tube (mostly skin), muscles and organs derived from mesoderm in between, and one end (head) cephalized with sensory organs and brain. **(2)** `understand` When an animal moves through an environment in one direction, its ability to acquire food and perceive and respond to threats is greater if its feeding, sensing, and information-processing structures are at the leading end. **p. 651 Fig. 33.14** `evaluate` The larva and metamorphosis arrow should be circled in the life cycle. It might be adaptive to skip the feeding larval stage if more food resources are available for the mother (who must expend energy to produce the yolk) than for the larva, if predation pressure on larvae is high, or if other environmental factors for larvae are unfavorable, resulting in high mortality of the larvae. **p. 652 CYU (1)** `understand` The mouthparts of deposit feeders are relatively simple because these animals simply gulp relatively soft material. The mouthparts of mass feeders are more complex because they have to tear off and process chunks of relatively hard material. **(2)** `create` Gametes that are shed into aquatic environments can float or swim. This cannot happen on land, so internal fertilization is more common there. Also, gametes are more at risk of drying out on land, and internal fertilization eliminates that risk. **p. 653** `apply` The origin of cilia-powered swimming should be marked on the Ctenophora branch.

p. 654 `apply` The origin of cnidocytes should be marked on the Cnidaria branch.

IF YOU UNDERSTAND . . .

33.1 `apply` It should increase dramatically—animals would no longer be consuming plant material. **33.2** `understand` Your drawing should show the tube-within-a-tube design of a worm, with one end labeled head and containing the mouth and brain. From the brain, one or two major nerve tracts should run the length of the body. The gut should go from the mouth to the other end, ending in the anus. In between the gut and outer body wall, there should be blocks or layers of muscle. **33.3** `apply` There are many possible answers. For example, within the phylum Arthropoda: The sensory systems have diversified (e.g., eyes are important for sight in flies, antennae are important for smelling in moths); feeding strategies have diversified (e.g., butterflies are fluid feeders but caterpillars are mass feeders); ecological roles have diversified (e.g., butterflies are herbivores but spiders are carnivores); limbs have diversified (e.g., butterflies have legs and wings but caterpillars have lobe-like limbs); and some insects, like grasshoppers, have direct development whereas others, like mosquitoes, have indirect development. **33.4** `remember` If you see a

---

FIGURE A32.1

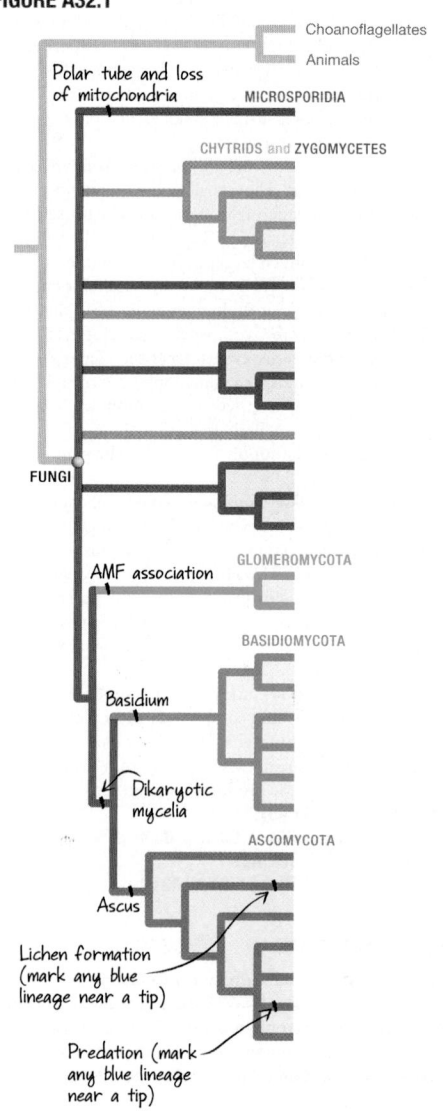

Choanoflagellates
Animals
Polar tube and loss of mitochondria — MICROSPORIDIA
CHYTRIDS and ZYGOMYCETES
FUNGI
AMF association — GLOMEROMYCOTA
BASIDIOMYCOTA
Basidium
Dikaryotic mycelia
ASCOMYCOTA
Ascus
Lichen formation (mark any blue lineage near a tip)
Predation (mark any blue lineage near a tip)

sessile, multicellular animal with no symmetry or radial symmetry, with outcurrent holes, and choanocytes, it is a sponge. If you see a radially symmetric, transparent, planktonic organism moving via rows of cilia, it is a ctenophore. If you see a diploblastic, outwardly radially symmetric animal that is either a single polyp or colony of multiple polyps, or it is a medusa, then the animal is a cnidarian.

YOU SHOULD BE ABLE TO . . .

✔ Test Your Knowledge

1. remember a **2.** remember True **3.** remember protostomes, deuterostomes **4.** understand a **5.** remember d **6.** remember d

✔ Test Your Understanding

1. understand Diploblasts have two types of embryonic tissue (ectoderm, endoderm); triploblasts have three (ectoderm, mesoderm, endoderm). Mesoderm made it possible for an enclosed, mesoderm-lined cavity to develop, creating a coelom. **2.** analyze Many unicellular organisms are heterotrophic, but they can consume only small bits of food. Animals are multicellular, so they are larger and can consume larger amounts of food—making them important consumers in food webs. **3.** apply There are about 18.4 named arthropods for every named species of chordate (1,160,000/63,000), about 1.3 named mollusks for every named species of chordate (85,000/63,000), and about 0.4 named species of nematode worm for every named species of chordate (25,000/63,000). These numbers are likely to be underestimates because there are probably many more arthropods, mollusks, and nematode worms than have been described, yet relatively few of the larger, more familiar chordates (including vertebrates) left to be described. **4.** understand In oviparous species, the mother adds nutrient-rich yolk to the egg that nourishes the developing embryo. In viviparous species, the mother transfers nutrients directly from her body to the growing embryo. **5.** evaluate You could stain the gene products of the segmentation genes in chordates (such as a mouse embryo) and observe whether the expression pattern correlates with the expression pattern observed in arthropods (such as a fly embryo). **6.** apply c

✔ Test Your Problem-Solving Skills

1. analyze Yes—if the same gene is found in nematodes and humans, it was likely found in the common ancestor of protostomes and deuterostomes. If so, then fruit flies should also have this gene. **2.** apply Asexual reproduction is likely to be favored when the sea anemones have plenty of resources such as space and food. Sexual reproduction is likely to be favored when the sea anemones are stressed due to lack of resources, predation, infection, or other factors. In this case, it is adaptive to reproduce in a way that enables dispersal to a new site and/or genetically diverse offspring, some of which will be more likely to survive selection than a clone of identical animals would. **3.** apply Since echinoderms evolved from bilaterally symmetric ancestors with nerve cords, they should have nerve cords too. However, the organization of the nerve cords should be more diffuse, because echinoderms need to take in and process information from multiple directions—not just one. **4.** apply d

CHAPTER 34

IN-TEXT QUESTIONS AND EXERCISES

p. 659 Fig. 34.3 apply We know that aquatic living was the ancestral trait because the outgroups (e.g., Cnidaria, Porifera) were aquatic. We also have fossil evidence that the aquatic lifestyle was ancestral.
p. 660, p. 662, p. 665, p. 666, p. 667, p. 671, p. 674 apply See **FIGURE A34.1**.

p. 662 Fig. 34.6 evaluate Different phyla of worms have different feeding strategies and mouthparts, and they eat different foods, so they are not all in direct competition for food.
p. 664 CYU (1) apply Snails on land would be expected to have gills or lungs inside the body or otherwise protected from drying out, thinner shells so that they are not as heavy, and desiccation-resistant eggs. **(2)** evaluate The tentacles of the octopus could be probed during development for *Hox* genes known to be important in establishing the foot of snails. If the expression pattern is similar, homology would be supported.
p. 664 Fig. 34.10 apply You should have drawn a circle around all the tentacles.

IF YOU UNDERSTAND . . .

34.1 remember Ecological opportunity was important in the diversification of protostomes because many lineages made the transition from water to land, radiating into new niches. Genetic opportunity was important because changes in the expression of developmental tool-kit genes enabled diverse modifications of body plans based on existing genes. **34.2** understand One source of diversification is the loss of ancestral traits; just because a species does not possess a trait does not mean that its ancestor didn't. **34.3** analyze Both arthropods and vertebrates have segmented bodies, which are especially amenable to modularity since a change in the expression pattern of developmental tool-kit genes can change the number, size, and type of segments, exposing morphological diversity to the adaptive forces of natural selection.

YOU SHOULD BE ABLE TO . . .

✔ Test Your Knowledge

1. remember False **2.** remember a **3.** remember c **4.** understand c **5.** remember Both groups are bilaterally symmetric triploblasts with the protostome pattern of development. **6.** remember b

✔ Test Your Understanding

1. remember d **2.** apply Both annelids and arthropods have segmented bodies, unlike other protostomes such as flatworms and nematodes. **3.** evaluate Clams are relatively sessile as adults. Because larvae can swim, they provide a mechanism to disperse to new locations. To test this idea, you could measure the distance traveled by clam larvae and measure their success in recruiting to new habitats compared to larvae that are artificially limited to stay near their parents. **4.** apply Land plants would require cuticles or other mechanisms to keep their tissues from drying out, a mechanism of keeping gametes from drying out, and structural support to retain body shape despite the effects of gravity. **5.** evaluate The ability to fly allowed insects to disperse to new habitats and find new food sources efficiently. **6.** analyze It might be adaptive for juveniles of hemimetabolous individuals to use the same food source as the adults when the food source is plentiful.

✔ Test Your Problem-Solving Skills

1. analyze See **FIGURE A34.2**. **2.** analyze If the ancestors of brachiopods and mollusks lived in similar habitats and experienced natural selection that favored similar traits, then they would have evolved to have similar forms and habitats. This is called convergent evolution (see Chapter 28). **3.** evaluate This hypothesis is reasonable because simple characteristics often occur before more elaborate ones; this could be true for the radula and foot in aplacophorans. An alternate hypothesis would be that these structures' simple forms are derived characteristics that are adaptations for parasitism, but the evidence suggests that aplacophorans are carnivores and detritivores, not parasites. The position of the aplacophorans in the phylogeny could be tested by using DNA sequence data rather than morphological characters in a phylogenetic analysis. **4.** apply b

CHAPTER 35

IN-TEXT QUESTIONS AND EXERCISES

p. 682 Fig. 35.1 apply Invertebrates are a paraphyletic group. The group does not include all the descendants of the common ancestor because vertebrates are excluded.
p. 684 CYU understand If it's an echinoderm, it should have five-part radial symmetry, a calcium carbonate endoskeleton just underneath the skin, and a water vascular system (e.g., visible tube feet).
p. 684, p. 685a, p. 685b apply See **FIGURE A35.1**.

FIGURE A34.2

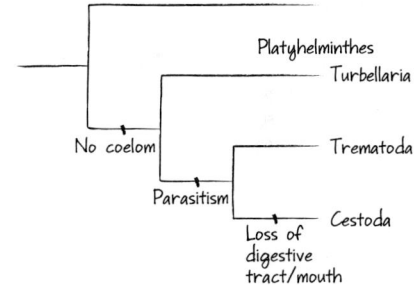

FIGURE A35.1

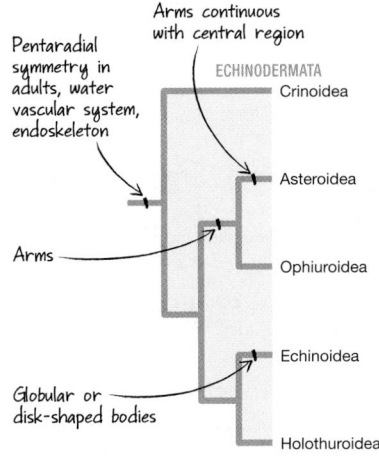

FIGURE A34.1

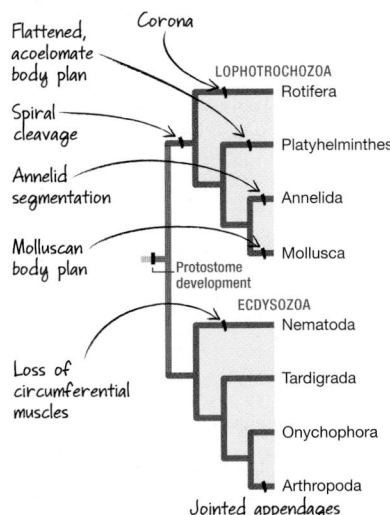

p. 690 Fig. 35.12 (apply) Mammals and birds are equally related to amphibians, because birds and mammals share a common ancestor, and this ancestor shares a common ancestor with amphibians.
p. 691 (apply) See **FIGURE A35.2**
p. 693 Fig. 35.16 (analyze) If this phylogeny were estimated on the basis of limb traits, it would be based on the *assumption* that the limb evolved from fish fins in a series of steps. But because the phylogeny was estimated from other types of data, it is legitimate to use it to analyze the evolution of the limb without making a circular argument.
p. 694 Fig. 35.18 (analyze) The yolk sac is smaller here—their function in an amniotic egg has been taken over by the placenta.
p. 695 CYU (1) (understand) Jaws allow animals to capture food efficiently and process it by crushing or tearing. The ability to process food more efficiently increased the importance of jawed vertebrates as herbivores and carnivores. The tetrapod limb enables vertebrates to locomote on land. The amniotic egg enables vertebrates to lay eggs on land; amniotic eggs are resistant to drying out. The placenta protects the growing fetus and enables the mother to be mobile. Parental care reduces the number of offspring that can be raised, but increases their survivorship. Flight enabled pterosaurs, bats, and birds to expand into new habitats. **(2)** (analyze) Protostomes and deuterostomes

living on land both have mechanisms for preventing their eggs from drying out (e.g., thick membranes in snail and insect eggs; amniotic egg in tetrapods), preventing their skin from drying out (e.g., waxy cuticle in insects, scaly skin in reptiles), and keeping their respiratory surfaces from drying out (e.g., lungs in land snails and tracheal system in insects; lungs in tetrapods).
p. 696 (apply) See **FIGURE A35.3**.
p. 698a, p. 698b, p. 699, p. 700 (apply) See **FIGURE A35.4**.
p. 701a, p. 701b, p. 702a, p. 702b, p. 703a, p. 703b (apply) See **FIGURE A35.3**.
p. 705 Table 35.1 (analyze) 44 percent, 26 percent; relative to body size, *H. floresiensis* has a smaller brain than *H. sapiens*.
p. 706 Fig. 35.37 (apply) Five hominin species existed 1.8 mya, and three existed 100,000 years ago.
p. 706 Fig. 35.38 (analyze) The forehead became much larger, and the face became "flatter"; the brow ridges are less prominent in later skulls than in earlier skulls.

IF YOU UNDERSTAND . . .

35.1 (apply) (1) Mussels and clams will increase dramatically; (2) kelp density will increase.
35.2 (understand) Your cartoon should have a dorsal hollow nerve chord along the top, dorsal edge; another line representing the notochord below the nerve cord; a tail

labeled at one end; and a series of lines or bumps representing pharyngeal gill slits near the other end.
35.3 (understand) Like today's lungfish, the earliest tetrapods could have used their limbs for pulling themselves along the substrate in shallow-water habitats. **35.4** (understand) The earliest fossils of *H. sapiens* are found in Africa, and phylogenetic analyses of living human populations indicate that the most basal groups are all African.

YOU SHOULD BE ABLE TO . . .

✔ **Test Your Knowledge**

1. (remember) True **2.** (remember) a **3.** (understand) d **4.** (remember) mammals **5.** (understand) d **6.** (remember) b

✔ **Test Your Understanding**

1. (understand) c **2.** (understand) Pharyngeal gill slits function in suspension feeding. The notochord furnishes a simple endoskeleton that stiffens the body; electrical signals that coordinate movement are carried by the dorsal hollow nerve cord to the muscles in the post-anal tail, which beats back and forth to make swimming possible. Cephalochordates and Urochordates are chordates, but they are not vertebrates. **3.** (analyze) If jaws are derived forms of gill arches, then the same genes and the same cells should be involved in the development of the jaw and the gill arches. **4.** (understand) Increased parental care allows

FIGURE A35.2

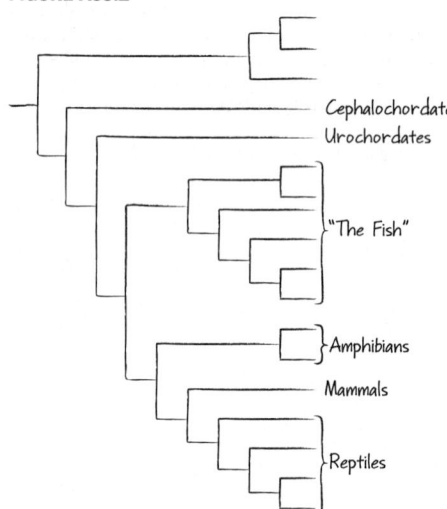

FIGURE A35.3

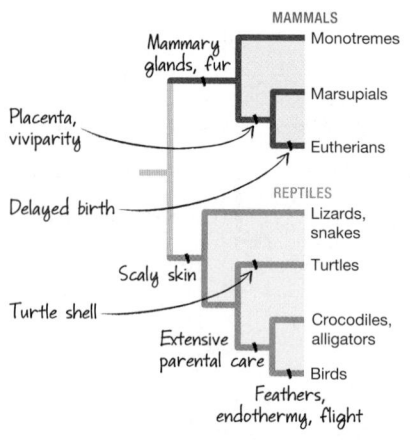

FIGURE A35.4

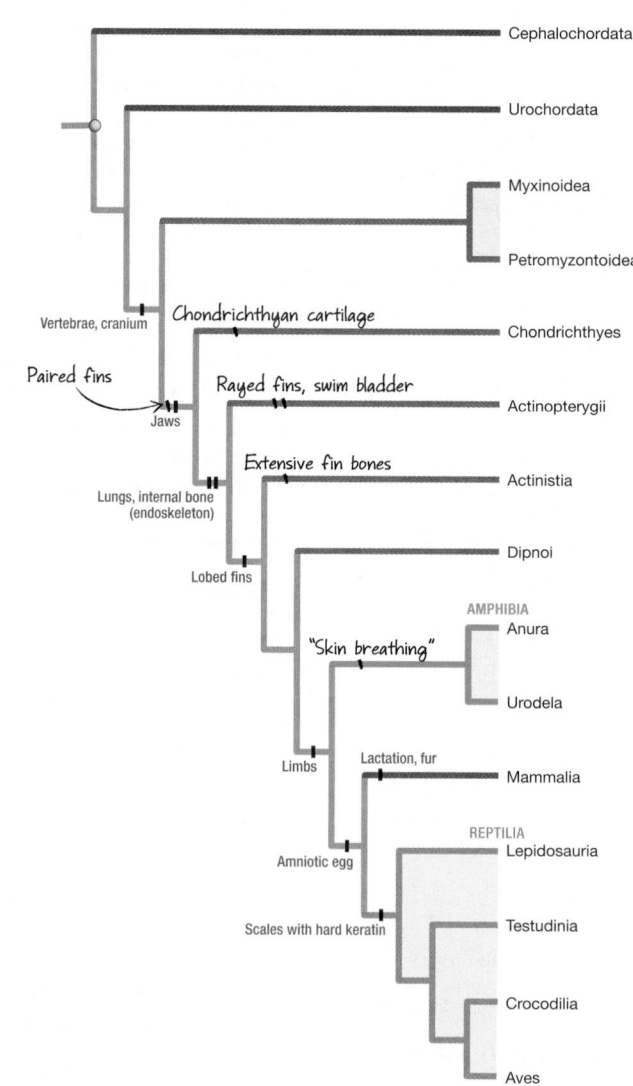

offspring to be better developed, and thus have increased chances of survival, before they have to live on their own. **5.** evaluate Yes: Over a short time interval, many species that occupy an array of foods and habitats evolved. Changes in tooth and jaw structure, tool use, and body size suggest that different hominin species exploited different types of food. **6.** evaluate A phylogeny of modern humans would reject the out-of-Africa hypothesis if the most ancestral lineages in the phylogeny were from any other region but Africa, such as Asia or Europe.

✔ Test Your Problem-Solving Skills

1. apply b **2.** evaluate Xenoturbellidans either retain traits that were present in the common ancestor of all deuterostomes, or they have lost many complex morphological characteristics. **3.** evaluate Vertebrate limbs are diverse in form and function (they can serve as arms, flippers, wings, etc.) but have homologous bones within. Molecular probes could be used to visualize the expression patterns of tool-kit genes like *Hox* in the embryonic limbs of different vertebrate species to show how changes in the expression of homologous genes can cause changes to the structure and function of limbs. **4.** analyze See **FIGURE A35.5**. It is important to use indigenous people for this study to minimize the amount of gene flow (transfer of alleles from one population to another) that occurred during the history of the study populations due to immigration and interbreeding.

CHAPTER 36

IN-TEXT QUESTIONS AND EXERCISES

p. 717 Fig. 36.7 apply Starting with 1 virion, approximately 80 virions were produced after 40 minutes. If all of these virions infected new cells, you expect each of the 80 infected cells to produce 80 virions by 90 minutes, or a total of 6400 virions. The cells appear to replicate once every 30 minutes, which means that they will have replicated 3 times during this period. The total number of cells at 90 minutes would be approximately 2^3, or 8 cells. There would be 800 times as many virions as cells.
p. 717 Fig. 36.8 analyze No—it only shows that CD4 is required. (Subsequent work showed that other proteins are involved as well.)
p. 721 analyze In T4, the viral genome is injected into the host while HIV inserts its genome via membrane fusion. HIV must first convert its genome into double-stranded DNA before replication, which is not necessary for T4, which has a DNA genome. Both viruses use double-stranded DNA to make mRNA that is translated into viral proteins. T4 virions are made in the cytosol of the cell and are released when the cell bursts. HIV virions bud from the cell surface, which does not require cell death.
p. 721 Fig. 36.12 create Budding may disrupt the integrity of the host-cell plasma membrane enough to kill the cell.
p. 722 CYU analyze Like cars in the assembly line of an automobile plant, new virions are assembled from premanufactured parts, and large numbers of progeny are produced per generation. Cells, on the other hand, reproduce by the division of a single integrated unit, resulting in only two progeny per generation.
p. 724 Fig. 36.15 apply You should have the following bars and labels: on branch to HIV-2 (sooty mangabey to human); on branch to HIV-1 strain O (chimp to human); on branch to HIV-1 strain N (chimp to human); on branch to HIV-1 strain M (chimp to human).

p. 725 CYU (1) analyze The escaped-genes and degenerate-cell hypotheses state that viruses originated from cells, while the RNA-world hypothesis suggests that viruses originated in parallel with, maybe even influenced, the origin of cells. The escaped-gene and RNA-world hypotheses state that the viruses originated from parasitic molecules. The degenerate-cell hypothesis states that viruses originated from parasitic organisms. **(2)** apply A mutation that made transmission between humans more efficient would make it more dangerous. Such mutation could occur via genomic reassortment in pigs, where an avian virus and a human virus could co-infect cells and produce recombinants.

IF YOU UNDERSTAND . . .

36.1 understand HIV infects helper T cells. These cells play a central role in the human immune response. As latently infected cells are activated by the immune response, HIV replication causes cell death. Over several years, steady decline in helper T cells leads to a compromised immune response that allows other infections to eventually kill the infected person. **36.2** understand (1) Fusion inhibitors block viral envelope proteins or host cell receptors. (2) Protease inhibitors prevent processing/assembly of viral proteins. (3) Reverse transcriptase inhibitors block reverse transcriptase, preventing replication of the genome. **36.3** apply There is heritable variation among virions, due to random changes that occur as their genomes are copied. There is also differential reproductive success among virions in their ability to successfully infect host cells. This differential success is due to the presence of certain heritable traits. **36.4** analyze Both classes have positive-sense single-stranded RNA genomes. Class IV viruses use an RNA replicase to make a negative-sense copy of the genome, which is then used as a template for producing mRNA and genomic RNAs. Class VI viruses use reverse transcriptase to convert their RNA genome into dsDNA, which is then integrated into the host genome. There it serves as a template to for transcribing viral mRNA and genomic RNAs by host-cell RNA polymerase.

YOU SHOULD BE ABLE TO . . .

✔ Test Your Knowledge

1. remember d **2.** remember b **3.** understand In class VI and VII viruses, reverse transcriptase converts positive-sense single-stranded RNA to double-stranded DNA.
4. remember b **5.** remember c **6.** remember The unique type of genetic material included in the virion (classes I–V) in addition to the manner in which the genome is replicated (compare classes VI and VII with classes IV and I, respectively).

✔ Test Your Understanding

7. analyze A virus with an envelope exits host cells by budding. A virus that lacks an envelope exits host cells by lysis (or other mechanisms that don't involve budding). **8.** analyze (1) Rate of viral genome replication is much higher via the lytic cycle. In lysogeny, the viral genome can replicate only when the host cell replicates. (2) Only the lytic cycle produces virions. (3) The lytic

FIGURE A35.5

cycle results in host cell death, while the host cell continues to survive during lysogeny. **9.** create Viruses rely on host-cell enzymes to replicate, whereas bacteria do not. Therefore, many drugs designed to disrupt the virus life cycle cannot be used, because they would kill host cells as well. Only viral-specific proteins are good targets for drug design. **10.** evaluate Each major hypothesis to explain the origin of viruses is associated with a different type of genome. Escaped genes: single- and double-stranded DNA or possibly single-stranded RNA. Degenerate cells: double-stranded DNA. RNA world: single- or double-stranded RNA. **11.** understand The phylogenies of SIVs and HIVs show that the two shared common ancestors, but that SIVs are ancestral to the HIVs. Also, there are plausible mechanisms for SIVs to be transmitted to humans through butchering or contact with pets, but fewer or no plausible mechanisms for HIVs to be transmitted to monkeys or chimps. **12.** apply a; the single-stranded positive-sense RNA genome of this virus can serve as an mRNA to produce viral proteins, while all of the other possibilities require viral enzymes to transcribe the genome into mRNA.

✔ Test Your Problem-Solving Skills

13. create Culture the *Staphylococcus* strain outside the human host and then add the virus to determine whether the virus kills the bacterium efficiently. Then test the virus on cultured human cells to determine whether the virus harms human cells. Then test the virus on monkeys or other animals to determine if it is safe. Finally, you could test the virus on human volunteers.
14. evaluate Prevention is currently the most cost-effective program, but it does not help people who are already infected. Treatment with effective drugs not only prolongs lives but also reduces virus loads in infected people, so that they have less chance of infecting others. **15.** apply b; by prolonging the infection in this manner, these viruses are more likely to be transmitted to a new host.
16. evaluate Viruses cannot be considered to be alive by the definition given in (1), because viruses are not capable of replicating by themselves. By the definition given in (2), it can be argued that viruses are alive because they store, maintain, replicate, and use genetic information—although they cannot perform all these tasks on their own.

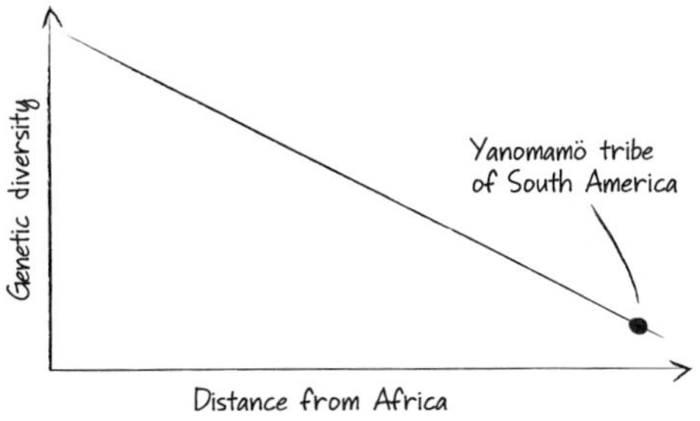

CHAPTER 37

IN-TEXT QUESTIONS AND EXERCISES

p. 732 Fig. 37.1 apply New branches and leaves should develop to the right (the plant will also lean that way); new lateral roots will develop to the left.
p. 734 Fig. 37.3 analyze Lawn grasses are shallow rooted, so when the top part of soil dries out, the plants cannot grow.
p. 735 Table 37.1 understand For aerobic respiration, which generates ATP needed to keep cells alive.
p. 736 Fig. 37.5 evaluate The individuals are genetically identical—thus, any differences in size or shape in the different habitats are due to phenotypic plasticity and not genetic differences.
p. 738 Fig. 37.6 analyze Large surface area provides more area to capture photons, but also more area from which to lose water and be exposed to potentially damaging tearing forces from wind.

p. 739 CYU (1) `remember` See **FIGURE A37.1 (2)** `understand` The generalized body of a plant has a taproot with many lateral roots and broad leaves. Examples of deviations from this generalized body structure follow. Modified roots: Unlike taproots, *fibrous roots* do not have one central root, and *adventitious roots* arise from stems. Modified stems: *Stolons* grow along the soil and grow roots and leaves at each node, and *rhizomes* grow horizontally underground. Both of these modified stems function in asexual reproduction. Modified leaves: The *needle-shaped leaves* of cacti don't lose water to transpiration compared with typical broad leaves, and they protect the plant from herbivory. *Tendrils* on climbing plants are modified leaves that do not photosynthesize, but wrap around trees or other substrates to facilitate climbing.
p. 745 Fig. 37.16 `evaluate` Many possible answers; for example, in leaf cells, genes encoding proteins involved in photosynthesis would be expressed. These genes wouldn't be expressed in the root cells.
p. 747 CYU (1) `understand` The apical meristem gives rise to the three primary meristems. The apical meristem is a single mass of cells localized at the tip of a root or shoot; the primary meristems are localized in distinctive locations behind the apical meristem. **(2)** `understand` Epidermal cells are flattened and lack chloroplasts; they secrete the cuticle (in shoots) or extend water and nutrient-absorbing root hairs (in roots) and protect the plant. Parenchyma cells are "metabolic workhorse cells" found throughout the plant body; they perform photosynthesis and synthesize, store, and/or transport materials. Tracheids are long, slim cells with pits in their secondary cell walls; they are found in xylem and are dead when mature, and they conduct water and solutes up the plant. Sieve-tube elements are long, thin cells found in phloem; they are alive when mature and conduct sugars and other solutes up and down the plant.
p. 750 `understand` Secondary growth adds cells only to the inside and outside of the vascular cambium, so the nail remains the same distance above the ground as the tree grows.

FIGURE A37.1

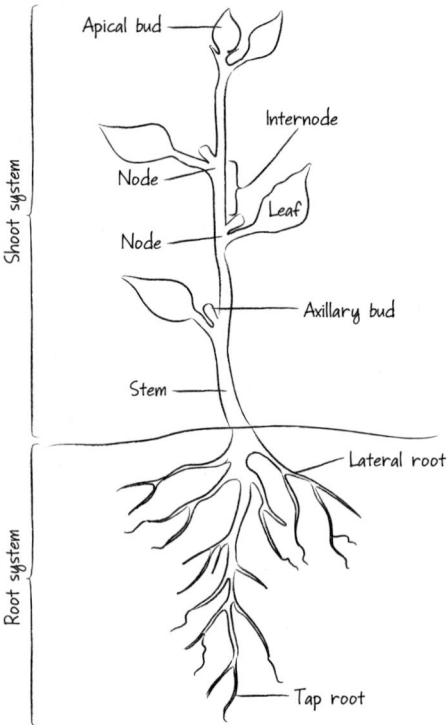

p. 750 Fig. 37.24 `remember` You should have labeled a light band as early wood, and the dark band to its right as late wood; both bands together comprise one growth ring.
p. 751 CYU (1) `apply` The rings are small on the shaded side and large on the sunny side. **(2)** `remember` See **FIGURE A37.2**.

IF YOU UNDERSTAND . . .

37.1 `analyze` Phenotypic plasticity is more important in (1) environments where conditions vary because it gives individuals the ability to change the growth pattern of their roots, shoots, and stems to access sunlight, water, and other nutrients as the environment changes; and (2) in long-lived species because it gives individuals a mechanism to change their growth pattern as the environment changes throughout their lifetime.
37.2 `analyze` The zone of cellular maturation is where root hairs emerge. Root hairs are the site where most of the water and ions are absorbed in the root system. **37.3** `apply` (1) In both cases, the structure would stop growing; (2) the shoot would lack epidermal cells and would die.
37.4 `apply` (1) The plant would not produce secondary xylem and phloem, so its girth and transport ability would be reduced, though it would still produce bark. (2) The trunk would have much wider tree rings on one side than the other side.

YOU SHOULD BE ABLE TO . . .

✓ Test Your Knowledge

1. `remember` b **2.** `understand` a **3.** `remember` c **4.** `remember` a **5.** `remember` Cactus spines are modified leaves; thorns are modified stems. **6.** `analyze` Collenchyma and sclerenchyma cells both have thickened cell walls and function in providing structural support. Unlike collenchyma cells, sclerenchyma cells have lignified cell walls and are dead at maturity.

✓ Test Your Understanding

7. `remember` b **8.** `evaluate` The general function of both systems is to acquire resources: The shoot system captures light and carbon dioxide; the root system absorbs water and nutrients. Vascular tissue is continuous throughout both the shoot and root systems. Diversity in roots and shoots enables plants of different species to live together in the same environment without directly competing for resources. **9.** `analyze` Continuous growth enhances phenotypic plasticity because it allows plants to grow and respond to changes or challenges in their environment (such as changes in light and water availability). **10.** `apply` Cuticle reduces water loss; stomata facilitate gas exchange. Plants from wet habitats should have a relatively large number of stomata and thin cuticle. Plants

FIGURE A37.2

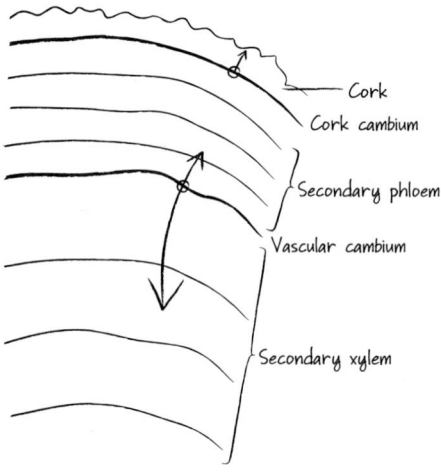

living in dry habitats should have relatively few stomata and thick cuticle. **11.** `analyze` Parenchyma cells in the ground tissue perform photosynthesis and/or synthesize and store materials. In vascular tissue, parenchyma cells in rays conduct water and solutes across the stem; other parenchyma cells differentiate into the sieve-tube elements and companion cells in phloem. **12.** `understand` Cells produced to the inside of the vascular cambium differentiate into secondary xylem; cells produced to the outside of the vascular cambium differentiate into secondary phloem.

✓ Test Your Problem-Solving Skills

13. `apply` c **14.** `remember` Asparagus—stem; Brussels sprouts—lateral buds; celery—petiole; spinach—leaf (petiole and blade); carrots—taproot; potato—modified stem. **15.** `create` They grow continuously, so do not have alternating groups of large and small cells that form rings. **16.** `analyze` Girdling disrupts transport of solutes in secondary phloem. The tree's root system starves.

CHAPTER 38

IN-TEXT QUESTIONS AND EXERCISES

p. 756 `apply` (1) Add a solute (e.g., salt or sugar) to the left side of the U-tube, so that solute concentration is higher than on the right side. (2) Water would move from the right side of the U-tube into the left side.
p. 757 `apply` Pull the plunger up.
p. 757 Fig. 38.2 `apply` It increases (becomes less negative), which reduces the solute potential gradient between the two sides.
p. 758 CYU (1) `apply` When a cell is in equilibrium with pure water, the negative solute potential of the cell will be balanced by the positive pressure potential (turgor pressure). **(2)** `analyze` Irrigation water contains low concentrations of salts so the solute potential will be near zero. As water evaporates, salts remain in the soil, lowering the solute potential.
p. 760 Fig. 38.7 `understand` See **FIGURE A38.1**.
p. 761 `apply` The individual would not be able to exclude toxic solutes from the xylem. If these solutes were transported throughout the plant in high enough concentrations, they could damage tissues. The mutant could also not maintain ions in the xylem and so could not generate root pressure.
p. 764 Fig. 38.12 `analyze` The line on the graph would go up, and large jumps would occur each time the light level was turned up, because increased transpiration rates would increase negative pressure (tension) at the leaf surface and thus increase the water-potential gradient.
p. 765 CYU (1) `understand` At the air–water interface under a stoma, water molecules cannot hydrogen-bond in all directions—they can bond only with water molecules below them. This creates tension at the surface. When transpiration occurs, the curvature of the meniscus increases—increasing tension. **(2)** `apply` When a nearby stoma closes, the humidity inside the air space increases, transpiration decreases, and surface tension on menisci decreases. The same changes occur when a rain shower starts. When air outside the leaf dries, though, the opposite occurs: The humidity inside the air space decreases, transpiration increases, and surface tension on menisci increases.
p. 768 Fig. 38.17 `understand` See **FIGURE A38.2**.
p. 771 CYU (1) `understand` In early spring, growing leaves are sinks and phloem sap moves toward them. In midsummer, leaves are sources and phloem sap moves away from them. **(2)** `create` The movement of water and solutes in sieve elements is by bulk flow, so stationary organelles would provide resistance to the flow.
p. 771 Fig. 38.22 `create` With time the plant will metabolize the radioactive sucrose, and levels of radioactivity will decline.

IF YOU UNDERSTAND . . .

38.1 analyze Because xylem cells are dead at maturity, there are no plasma membranes for water to cross and thus no solute potential. The only significant force acting on water in xylem is pressure. **38.2** evaluate It drops, because transpiration rates from stomata decrease, reducing the water potential gradient between roots and leaves. **38.3** understand CAM plants open their stomata only at night, and thus they transpire when temperatures are cooler. **38.4** analyze To create high pressure in phloem near sources and low pressure near sinks, water has to move from xylem to phloem near sources and from phloem back to xylem near sinks. If xylem were not close to phloem, this water movement could not occur, and the pressure gradient in phloem wouldn't exist.

YOU SHOULD BE ABLE TO . . .

✔ **Test Your Knowledge**

1. understand d **2.** apply b **3.** understand a **4.** remember c **5.** remember Endodermal cells. **6.** remember Companion cells.

✔ **Test Your Understanding**

7. understand c **8.** apply See FIGURE A38.3. **9.** analyze The force responsible for root pressure is the high solute potential of root cells at night, when they continue to accumulate ions from soil and water follows by osmosis. The mechanism is active, in the sense that ions are imported against their concentration gradient. Capillarity is driven by adhesion of water molecules to the sides of xylem cells that creates a pull upward, and by cohesion with water molecules below. Cohesion-tension, generated by transpiration, is driven by the loss of water molecules from menisci in leaves, which creates a large negative pressure (tension). Both capillarity and cohesion-tension are passive processes. **10.** understand Water moves up xylem because of transpiration-induced tension created at the air–water interface under stomata, which is communicated down to the roots by the cohesion of water molecules. Sap flows in phloem because of a pressure gradient that exists between source and sink cells, driven by differences in sucrose concentration and flows of water into or out of nearby xylem. **11.** understand By pumping protons out, companion cells create a strong electrochemical gradient favoring entry of protons. A cotransport protein (a symporter) uses this proton gradient to import sucrose molecules *against* their concentration gradient. **12.** apply When it germinates—sucrose stored inside the seed is released to the growing embryo, which cannot yet make enough sucrose to feed itself.

✔ **Test Your Problem-Solving Skills**

13. apply b **14.** create Plants would not grow as quickly or as tall, because the taller they grew, the more energy they would need to use to transport water and thus the less they would have available for growth. **15.** analyze Because they cannot readily replace water that is lost to transpiration, they have to close stomata. During a heat wave, transpiration cannot cool the plants' tissues. The trees bake to death. **16.** create Closing aquaporins will slow or stop movement of water from cells into the xylem and out of the plant via transpiration. Because water does not leave the cells, they are able to maintain turgor (normal solute potentials).

CHAPTER 39

IN-TEXT QUESTIONS AND EXERCISES

p. 776 Fig. 39.1 evaluate To test the hypothesis that most of a plant's mass comes from CO_2 in the atmosphere, design an experiment where plants are grown in the presence or absence of CO_2 and compare growth, and/or grow plants in the presence of labeled CO_2 and document the presence of labeled carbon in plant tissues. **p. 777 Fig. 39.3** evaluate Because soil is so complex, it would be difficult to defend the assumption that the soils with and without added copper are identical except for the difference in copper concentrations. Also, it would be better to compare treatments that had no copper versus normal amounts of copper, instead of comparing with and without added copper. **p. 783 Fig. 39.10** analyze The proton gradient arrow in (b) should begin above the membrane and cross the

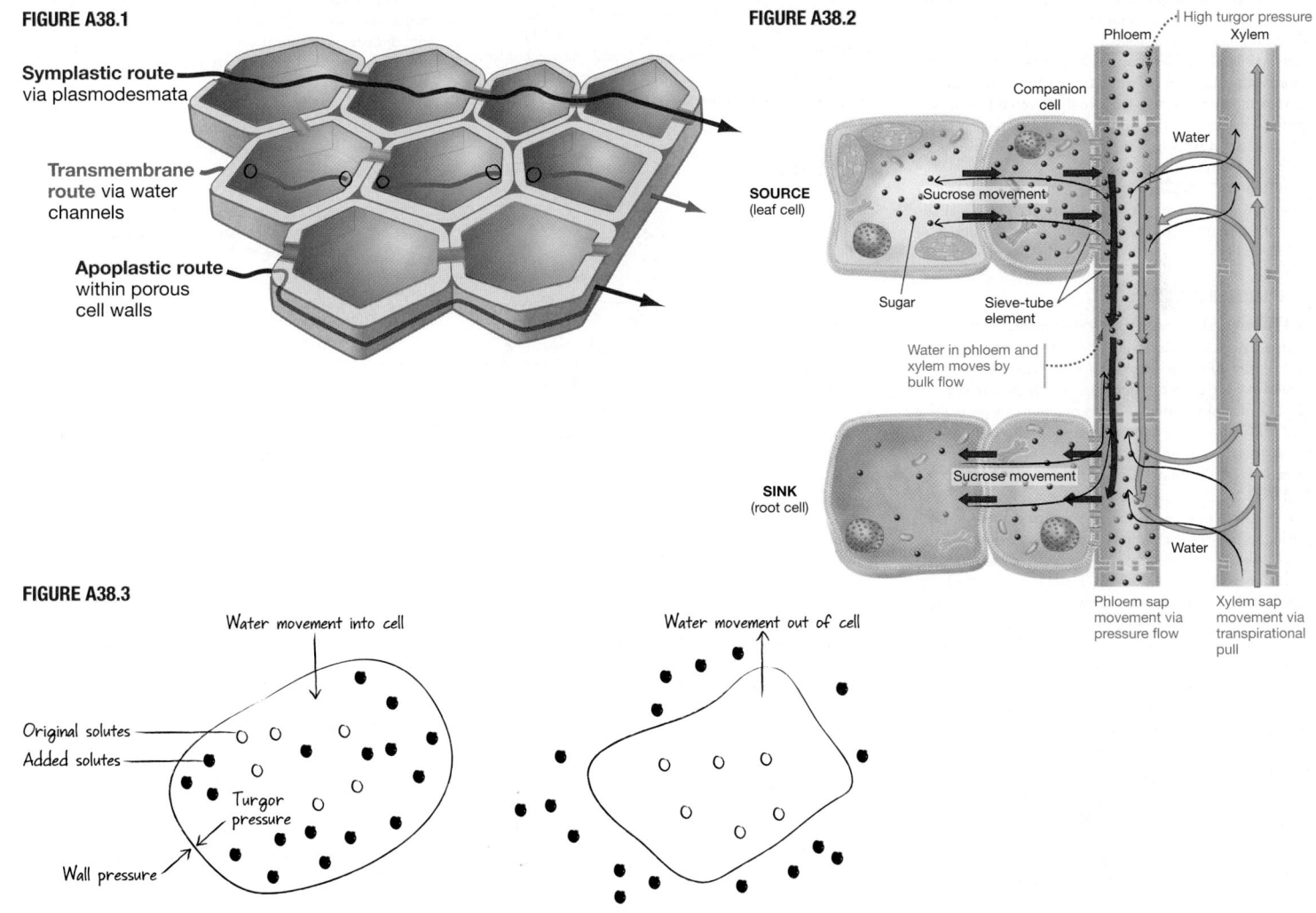

FIGURE A38.1

Symplastic route via plasmodesmata

Transmembrane route via water channels

Apoplastic route within porous cell walls

FIGURE A38.2

High turgor pressure
Phloem   Xylem
Companion cell
Water
SOURCE (leaf cell)
Sucrose movement
Sugar   Sieve-tube element
Water in phloem and xylem moves by bulk flow
SINK (root cell)
Sucrose movement
Water
Phloem sap movement via pressure flow   Xylem sap movement via transpirational pull

FIGURE A38.3

Water movement into cell   Water movement out of cell

Original solutes
Added solutes
Turgor pressure
Wall pressure

membrane pointing down. The electrical gradient arrow in (c) should also begin above the membrane (marked positive) and cross the membrane pointing down (toward negative).

p. 784a apply (1) If there is no voltage across the root-hair membrane, there is no electrical gradient favoring entry of cations, and absorption stops. (2) There is no route for cations to cross the root hair membrane, so absorption stops.

p. 784b apply (1) If there is no proton gradient across the root-hair membrane, there is no gradient favoring entry of protons, so symport of anions stops. (2) There is no route for anions to cross the root-hair membrane, so absorption stops.

p. 786 CYU (1) understand Proton pumps establish an electrical gradient across root-hair membranes, and the inside of the membrane is much more negative than the outside. Cations follow this electrical gradient into the cell. Anions are transported with protons via symporters.

(2) create See **FIGURE A39.1**.

p. 788 CYU (1) analyze Many possible answers, including the following: If N-containing ions are abundant in the soil, the energetic cost of maintaining nitrogen-fixing bacteria may outweigh the benefits in terms of increased nitrogen availability. If mycorrhizal fungi provide a plant with nitrogen-containing ions, the energetic cost of maintaining nitrogen-fixing bacteria may outweigh the benefits in terms of increased nitrogen availability. Plants that grow near species with N-fixing bacteria might gain nitrogen by absorbing nitrogen-containing ions after root nodules die, or by "stealing it" (roots of different species will often grow together). There could be a genetic constraint (see Chapter 25): Plant species may simply lack the alleles required to manage the relationship. Note: To be considered correct, your hypothesis should be plausible (based on the underlying biology) and testable. **(2)** analyze Many possible answers, including: Because the bacterial cell enters root cells, it is important for the plant to have reliable signals that it is not a parasitic bacterium. It is advantageous for the plant to be able to reject nitrogen-fixing bacteria if usable nitrogen is already abundant in the soil.

p. 790 Fig. 39.17 analyze Mistletoe is parasitic (it extracts water and certain nutrients from host trees); bromeliads are not (they are epiphytes and don't affect the fitness of their host plants).

IF YOU UNDERSTAND . . .

39.1 evaluate Add nitrogen to an experimental plot and compare growth within the plot to growth of a similar plot nearby. Do many replicates of the comparison, to convince yourself and others that the results are not due to unusual circumstances in one or a few plots.
39.2 understand Loam is a mixture of sand, clay, and organic matter so it can retain water and nutrients, and provide aeration, all properties that benefit plants.
39.3 apply A higher membrane voltage would allow cations in soil to cross root-hair membranes more readily through membrane channels and would allow anions to enter more readily via symporters. Nutrient absorption should increase. **39.4** evaluate Grow pea plants in the presence of rhizobia, exposed to air with (1) N_2 containing the heavy isotope of nitrogen and (2) radioactive carbon dioxide. Allow the plant to grow and then analyze the rhizobia and plant tissues. If the rhizobia–pea plant interaction is mutualistic, then ^{15}N-containing compounds and radioactive-carbon-containing compounds should be observed in both rhizobia and plant tissues. As a control, grow pea plants in the presence of the labeled compounds but without rhizobia. If the mutualism hypothesis is correct, the plant should contain labeled carbon but no heavy nitrogen. **39.5** analyze Despite the abundance of rain in tropical rain forests, many epiphytes there have limited access to water because their roots don't reach the soil. Like desert plants, these epiphytes are adapted to conserve water.

YOU SHOULD BE ABLE TO . . .

✔ Test Your Knowledge

1. remember b **2.** understand d **3.** remember d **4.** understand c **5.** remember Mycorrhizal fungi obtain most of their nitrogen from proteins in soil organic matter. Rhizobia obtain nitrogen from N_2 gas. **6.** remember Plant

✔ Test Your Understanding

7. remember b **8.** evaluate (1) Grow a large sample of corn plants with a solution containing all essential nutrients needed for normal growth and reproduction. Grow another set of genetically identical (or similar) corn plants in a solution containing all essential nutrients except iron. Compare growth, color, and seed production (and/or other attributes) in the two treatments. (2) Grow corn plants in solutions containing all essential nutrients in normal amounts, but with varying concentrations of iron. Determine the lowest iron concentration at which plants exhibit normal growth rates. **9.** understand If nutrients in soil are scarce, there is intense natural selection favoring alternative ways of obtaining nutrients—for example, by digesting insects or stealing nutrients.
10. apply Higher in soils with both sand and clay. Clay fills spaces between sand grains and slows (1) passage of water through soil and away from roots, and (2) leaching of anions in soil water. Clay particles also provide negative charges that retain cations so they are available to plants. The presence of sand keeps soil loose enough to allow roots to penetrate. **11.** understand Some metal ions are poisonous to plants, and high levels of other ions are toxic. Passive mechanisms of ion exclusion—such as a lack of ion channels that allow passage into root cells—do not require an expenditure of ATP. Active mechanisms of exclusion—such as production of metallothioneins—require an expenditure of ATP.
12. analyze (1) The amounts required and the mechanisms of absorption are different. Much larger amounts of C, H, and O are needed than mineral nutrients. CO_2 enters

plants via stomata, and water is absorbed passively along a water potential gradient; mineral nutrients enter via active uptake in root hairs or via mycorrhizae. (2) Macronutrients are required in relatively large quantities and usually function as components of macromolecules; micronutrients are required in relatively small quantities and usually function as enzyme cofactors.

✔ Test Your Problem-Solving Skills

13. apply The soil lost 60 g while the tree gained 74,000 g. Dividing 60 by 74,000 and converting to a percentage equals 0.08%, far less than 4%. The water that van Helmont used may have contained many of the macro- and micronutrients that were incorporated into plant mass. To test this hypothesis, conduct an experiment of the same design but have one treatment with pure water and one treatment with water containing solutes ("hard water"). Compare the percentage of soil mass incorporated into the willow tree in both treatments. Another explanation: Willow may be unusual. To test this hypothesis, repeat the experiment with willow and several other species under identical conditions, and compare the percentage of soil mass incorporated into the plants. A third hypothesis: Van Helmont's measurements were inaccurate. Repeat the experiment.
14. analyze Acid rain inundates the soil with protons, which bind with the negatively charged clay particles—displacing the cations that are normally bound there. Cations may then be leached away. **15.** evaluate If the hypothesis is correct, vanadate should inhibit phosphorus uptake. To test this hypothesis, grow plants in solution with radioactively labeled phosphorus without vanadate and in solution with vanadate. Measure the amount of labeled phosphorus in root cells with and without vanadate. **16.** apply Alder should decompose much faster, because it provides more nitrogen to the fungi and bacteria that are responsible for decomposition—meaning that they can grow faster.

FIGURE A39.1

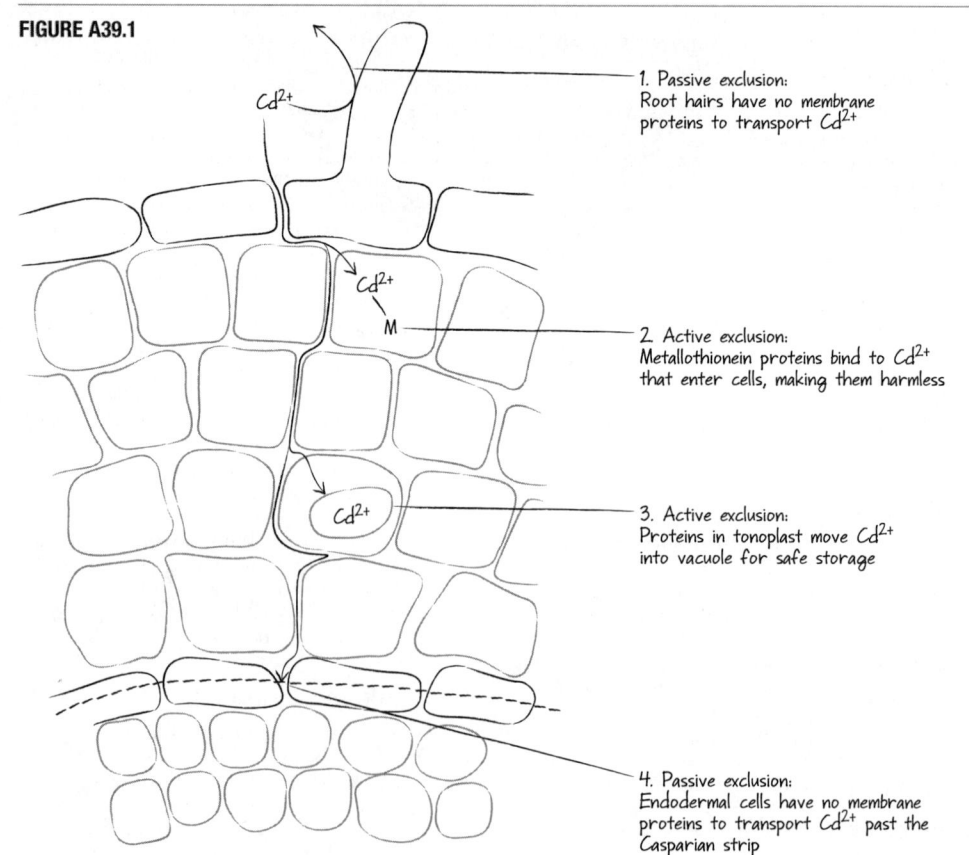

1. Passive exclusion:
Root hairs have no membrane proteins to transport Cd^{2+}

2. Active exclusion:
Metallothionein proteins bind to Cd^{2+} that enter cells, making them harmless

3. Active exclusion:
Proteins in tonoplast move Cd^{2+} into vacuole for safe storage

4. Passive exclusion:
Endodermal cells have no membrane proteins to transport Cd^{2+} past the Casparian strip

CHAPTER 40

IN-TEXT QUESTIONS AND EXERCISES

p. 796 CYU (1) `understand` The only cells that can respond to a specific environmental signal or hormone are cells that have an appropriate receptor for that signal or hormone. **(2)** `remember` Once a receptor is activated, it triggers production of many second messengers or the activation of many proteins in a phosphorylation cascade.

p. 797 Fig. 40.4 `evaluate` Wild-type plants already have a normal *PHOT1* gene and can respond to directional blue light. Inserting a defective gene into a wild-type plant would have no effect.

p. 798 Fig. 40.5 `evaluate` For "Tip removed": Cut the tip and replace it, to control for the hypothesis that the wound itself influences bending (not the loss of the tip). For "Tip covered": Cover the tip with a transparent cover instead of opaque one, to control for the hypothesis that the cover itself affects bending (not excluding light).

p. 799 CYU (1) `apply` Using a syringe or other device, apply auxin to the west side of each stem, just below the tip. **(2)** `apply` Stems would become very long and would not respond to directional light.

p. 801 Table 40.1 `analyze` The average germination rate of lettuce seeds last exposed to red light is about 99 percent. The average for seeds last exposed to far-red light is about 50 percent. Both of these values are much higher than the germination rate of buried seeds that receive no light at all; their germination rate is only 9 percent.

p. 803 CYU (1) `understand` Lettuce grows best in full sunlight. If a seed receives red light, it indicates that the seed is in full sunlight—meaning that conditions for growth are good. But if a seed receives far-red light, it indicates that the seed is in shade—meaning that conditions for growth are poor. **(2)** `understand` Night length changes very predictably throughout the year, so it is a reliable signal for plants to perceive.

p. 803 Fig. 40.10 `apply` An appropriate control treatment would be to graft a leaf from a plant that had never been exposed to a short-night photoperiod onto a new plant that also had never been exposed to the correct conditions for flowering. If the hypothesis is correct, the grafted leaf should not induce flowering.

p. 804 Fig. 40.12 `evaluate` (One possibility) Direct pressure from a statolith changes the shape of the receptor protein. The shape change triggers a signal transduction cascade, leading to a gravitropic cell response.

p. 805 CYU (1) `analyze` In both roots and shoots, auxin is redistributed asymmetrically in response to a signal. In phototropism, auxin is redistributed to the shaded side of the shoot tip in response to blue light. In gravitropism, auxin is redistributed to the lower part of the root or shoot in response to gravity. **(2)** `understand` In root cells, auxin concentrations indicate the direction of gravity. The gravitropic response is triggered by changes in the distribution of auxin in root-tip cells.

p. 809 Fig. 40.19 `analyze` The dwarfed individuals can put more of their available energy into reproduction because they are using less energy for growth than taller individuals.

p. 810 Fig. 40.21 `analyze` Without these data, a critic could argue that the stomata closed in response to water potentials in the leaf, not a signal from the roots.

p. 813 CYU (1) `analyze` ABA: Stomata should close and photosynthesis should stop compared to individuals that do not receive extra ABA. GA: Their stems should elongate rapidly, compared to plants that do not receive extra GA. **(2)** `understand` The rotten apple produces ethylene, which, being a gas, stimulates all of the other apples in the bushel basket to ripen too fast.

p. 816 `analyze` Sacrificing a small number of cells due to the hypersensitive response is far less costly for the plant than death.

p. 819 CYU (1) `analyze` Because specific receptor proteins recognize and match specific proteins produced by

pathogens, having a wide array of these proteins allows individuals to recognize and respond to a wide array of pathogens. **(2)** `understand` If herbivores are not present, synthesizing large quantities of proteinase inhibitors wastes resources (ATP and substrates) that could be used for growth and reproduction.

IF YOU UNDERSTAND . . .

40.1 `analyze` An external signal arrives at a receptor cell/protein (a person sees a barn on fire); the person calls (the receptor transduces the signal); the dispatcher rings a siren (a second messenger is produced or phosphorylation cascade occurs); the fire department members get to the pumper truck (gene expression or some other cell activity changes). **40.2** `apply` Sun-adapted (high-light) plants are likely to have a significant positive phototropic response (i.e., growth toward light), whereas shade-adapted (low-light) plants are less likely to respond phototropically. **40.3** `evaluate` They do not perform photosynthesis and therefore do not need to switch behavior based on being located in full sunlight versus shade. **40.4** `understand` Starch is denser than water, so it falls quickly to the bottom of cells. **40.5** `understand` A reduction in plant height in a windy environment might keep them upright. **40.6** `evaluate` Those hormones need to be transported throughout the plant, because cells throughout the body need to respond to the signal—not in a single direction. **40.7** `understand` In the hypersensitive response, a few cells are sacrificed to prevent a pathogen from spreading throughout the plant.

YOU SHOULD BE ABLE TO . . .

✓ Test Your Knowledge

1. `remember` d **2.** `remember` b **3.** `remember` d **4.** `understand` b **5.** `remember` hypersensitive response **6.** `remember` plasma membrane

✓ Test Your Understanding

7. `remember` a **8.** `remember` The P_{fr}–P_r switch in phytochromes lets plants know whether they are in shade or sunlight, and triggers responses that allow sun-adapted plants to avoid shade. Phototropins are activated by blue light, which indicates full sunlight, and trigger responses that allow plants to photosynthesize at full capacity. **9.** `analyze` Transduce means to convert energy from one form to another. There are many examples: A touch may be converted to an electrochemical potential, light energy or pressure into a shape change in a protein, and so on. **10.** `understand` (Many possibilities) Cells or tissue involved: Auxin directs gravitropism in roots but phototropism in shoots. Developmental stage or age: GA breaks dormancy in seeds but extends stems in older individuals. Concentration: Large concentrations of auxin in stems promote cell elongation; small concentrations do not. Other hormones: The concentration of GA relative to ABA influences whether germination proceeds. **11.** `apply` Ethylene promotes senescence; increased ethylene sensitivity in leaves (combined with reduced auxin concentrations) causes abscission; fruits exposed to ethylene ripen faster than fruits that are not exposed to ethylene. Produce dealers can maximize profits by delaying ripening during transit of fruit by removing ethylene and then adding ethylene just before fruits are displayed. **12.** `apply` When one leaf on the mutant plant is chewed on by a herbivore, the damaged leaf would not be able to signal to the rest of the plant, so it would not be able to fend off future herbivore attacks.

✓ Test Your Problem-Solving Skills

13. `apply` a **14.** `evaluate` Small-seeded plants need to perform photosynthesis early in seedling development or they will starve. Therefore, they need to germinate in direct sunlight. The food reserves in large-seeded plants can support seedling growth for a relatively long time without photosynthesis. Therefore, they do not need to

germinate in direct sunlight. **15.** `evaluate` In these seeds, (a) little or no ABA is present, or (b) ABA is easily leached from the seeds, so its inhibitory effects are eliminated. Test hypothesis (a) by determining whether ABA is present in the seeds. Test hypothesis (b) by comparing the amount of ABA present before and after running water—enough to mimic the amount of rain that falls in a tropical rain forest over a few days or weeks—over the seeds. **16.** `apply` New types of plant resistance provide a selective pressure for pathogens to evolve new mechanisms to evade the resistance.

CHAPTER 41

IN-TEXT QUESTIONS AND EXERCISES

p. 825 `understand` (1) The microsporocyte divides by meiosis to generate male spores; the megasporocyte divides by meiosis to generate female spores. (2) In angiosperms, the female gametophyte stays within the ovary at the base of the flower even after it matures and produces an egg cell. Fertilization and seed development take place in the same location.

p. 829 CYU (1) `analyze` Both are diploid cells that divide by meiosis to produce spores. The megasporocyte produces a megaspore (female spore); a microsporocyte produces a microspore (male spore). **(2)** `analyze` Both are multicellular individuals that produce gametes by mitosis. The female gametophyte is larger than the male gametophyte and produces an egg; the male gametophyte produces sperm.

p. 829 Fig. 41.8 `understand` A gametophyte is the multicellular individual that produces gametes by mitosis. The pollen grain is a gametophyte because it is multicellular and produces sperm by mitosis.

p. 832 CYU (1) `understand` Insects feed on nectar and/or pollen in flowers. Flowers provide food rewards to insects to encourage visitation. The individuals that attract the most pollinators produce the most offspring. **(2)** `understand` One product of double fertilization is the diploid zygote, which will eventually grow by mitosis into a mature sporophyte. The other product is the triploid endosperm nucleus, which will grow by mitosis to form a source of nutrients for the embryo.

p. 833 `remember` The endosperm nucleus in the central cell is triploid; the zygote is diploid; the synergid and other cells remaining from the female gametophyte are haploid.

p. 834 `analyze` A cotyledon and a radicle develop into organs with distinct functions, but both are composed of the same tissues: Epidermis is derived from protoderm, ground tissue is derived from ground meristem, and vascular tissue is derived from the procambium.

p. 835 Fig. 41.14 `evaluate` Hypothesis: Fruit changes color when it ripens as a signal to fruit eaters. The color change is advantageous because seeds are not mature in unripe fruit and are unlikely to survive. Fruit eaters disperse mature seeds in ripe fruit, which are more likely to survive.

p. 836 Fig. 41.16 `evaluate` Hackberries served as a control to demonstrate that animals that avoided chilies were hungry,

p. 837 Fig. 41.17 `understand` Beans—their cotyledons are aboveground.

IF YOU UNDERSTAND . . .

41.1 `remember` The sporophyte is diploid, the gametophyte is haploid, spores and gametes are haploid, and zygotes are diploid. **41.2** `apply` There would be four embryo sacs in each ovule, and thus four eggs. **41.3** `apply` Endosperm tissue is generated when two haploid, female nuclei fuse with one haploid sperm nucleus, forming a cell with three sets of chromosomes. In contrast, a zygote is formed by the fusion of one haploid sperm nucleus and one haploid egg, forming a cell with two sets of chromosomes. **41.4** `analyze` A cold requirement for germination ensures that seeds from plants native to habitats that experience cold winters will not germinate in the fall and fail to become established before winter.

✔ Test Your Knowledge

1. `remember` b 2. `understand` c 3. `remember` a 4. `understand` d
5. `remember` Megasporocytes (female) and microsporocytes (male) undergo meiosis to generate megaspores and microspores, respectively. These divide mitotically to give rise to female and male gametophytes—the embryo sac and pollen grain, respectively. **6.** `remember` Pollination is the transfer of pollen from an anther to a stigma. Fertilization is the fusion of haploid gametes.

✔ Test Your Understanding

7. `understand` c 8. `analyze` Asexual reproduction is a quick, efficient way for a plant to produce large numbers of offspring. The disadvantage is that offspring are genetically identical to the parent and thus are vulnerable to the same diseases and able to thrive only in habitats similar to those inhabited by the parent. **9.** `analyze` Sepals are the outermost structures in a flower and usually protect the structures within; petals are located just inside the sepals and usually function to advertise the flower to pollinators. In wind- and bee-pollinated flowers, sepals probably do not vary much in overall function or general structure. Petals are often lacking in wind-pollinated species; they tend to be broad and flat (to provide a landing site) and colored purple, blue, or yellow in bee-pollinated species (often with ultraviolet markings). **10.** `analyze` Outcrossing increases genetic diversity among offspring, making them more likely to thrive if environmental conditions change from the parental generation. However, outcrossing requires cross-pollination to be successful. Self-fertilization results in relatively low genetic diversity among offspring but ensures that pollination succeeds. **11.** `remember` The female gametophyte develops inside the ovule; the ovary develops into the pericarp of the fruit after fertilization. A mature ovule contains both gametophyte and sporophyte tissue; the ovary and carpel are all sporophyte tissue. **12.** `analyze` The endosperm of a corn seed and the cotyledons of a bean seed both contain nutrients needed for the seed to germinate. The bean cotyledons have absorbed the nutrients of the endosperm.

✔ Test Your Problem-Solving Skills

13. `remember` c 14. `apply` Subtracting the number of used pollen grains per plant (600) from the number of pollen grains generated per plant (10 million) divided by 10 million reveals that 99.994% of the pollen is wasted. **15.** `evaluate` In response to "cheater" plants, insects should be under intense selection pressure to detect and avoid species that do not offer a food reward. Individuals would have to be able to distinguish scents, colors, or other traits that identify cheaters. In response to cheater insects, plants possessing unusually thick or toxic petals or sepals that prevented insects from bypassing the anthers on their way to the nectaries would be more successful at producing seed in the next generation. **16.** `evaluate` Acorns have a large, edible mass and are usually animal dispersed (e.g., by squirrels that store them and forget some). Cherries have an edible fruit and are animal dispersed. Burrs stick to animals and are dispersed as the animals move around. Dandelion seeds float in wind. To estimate the distance that each type of seed is dispersed from the parent, (1) set up "seed traps" to capture seeds at various distances from the parent plant; (2) sample locations at various distances from the parent and analyze young individuals—using techniques introduced in Chapter 21 to determine if they are offspring from the parent being studied; (3) mark seeds, if possible, and find them again after dispersal.

B⃞I⃞G⃞ PICTURE Plant and Animal Form and Function

p. 840 CYU (1) `understand` Plants and animals may accumulate compounds in the epidermis and skin (such as waxes and other lipids) that reduce water loss. Both plants and animals evolved closable openings in this barrier to allow regulated gas exchange. Plants and animals may also change the amount of the body exposed to the sun (in plants by moving leaves, in animals by changing posture or taking shelter) to reduce evaporative water loss.
(2) `understand` Many plants and animals keep their eggs inside their bodies and depend on sperm being deposited near them rather than releasing both to the environment where they could dry out. **(3)** `understand` Very large animals and plants need to build strong supporting structures such as thick bones or trunks to withstand the force of gravity. They also need elaborate internal transport systems to bring nutrients to each cell and carry wastes away.
(4) `analyze` Cellular respiration fits into one of the key functions for survival—Nutrition. Both plants and animals depend on sugar (produced by plants via photosynthesis) as their energy source to make ATP via cellular respiration.

CHAPTER 42

IN-TEXT QUESTIONS AND EXERCISES

p. 844 Fig. 42.2 `create` There would be no difference between experimental and control males.
p. 847 Fig. 42.4 `analyze` Based on the projections called axons that go to specific cells and tissues, neurons transmit signals to specific tissues rather than broadly throughout the body.
p. 850 CYU `analyze` *Connective tissues* consist of cells embedded in a matrix. The density of the matrix determines the rigidity of the connective tissue and its function in padding/protection, structural support, or transport. *Nervous tissue* is composed of neurons and support cells. Neurons have long projections that function as "biological wires" for transmitting electrical signals. *Muscle tissue* has cells that can contract and functions in movement. *Epithelial tissue* has tightly packed layers of cells with distinct apical and basal surfaces. They protect underlying tissues and regulate the entry and exit of materials into these tissues.
p. 850 `apply` A newborn.
p. 851 Fig. 42.10 `apply` The puppy has to eat relatively more because it has a higher mass-specific metabolic rate than an adult dog.
p. 852 Fig. 42.11 `apply` There should be no oxygen uptake on either side of the chamber.
p. 853 CYU (1) `understand` As three-dimensional size increases, volume increases more quickly than does surface area, so the surface area/volume ratio decreases.
(2) `analyze` The squirrel has a higher mass-specific metabolic rate because it is a smaller animal. It therefore needs more energy per unit body mass than a bear.
p. 858 CYU (1) `analyze` Endotherms can remain active during the winter and at night and sustain high levels of aerobic activities such as running or flying, but they require large amounts of food energy. Ectotherms need much less food and can devote a larger proportion of their food intake to reproduction, but they have a hard time maintaining high activity levels at night or in cold weather. **(2)** `apply` The whale's tongue would become colder because not as much heat would be transferred from warm arterial blood to cool venous blood with concurrent flow as with countercurrent flow, and as a result more heat from the arterial blood would be lost to the environment.

IF YOU UNDERSTAND . . .

42.1 `understand` Some adaptations to cold environments in mammals include the evolution of thick coats of hair,

countercurrent heat exchangers in the extremities, and dark coloration to absorb radiation. Acclimatization may include further thickening and darkening of hair upon exposure to cold. **42.2** `apply` Loosening tight junctions would compromise the barrier between the external and internal environments. It would allow water and other substances (including toxins) to pass more readily across the epithelial boundary. **42.3** `create` King Kong is endothermic, and his huge mass would generate a great deal of heat. His relatively small surface area, especially as compared to a normal-sized gorilla, would not be able to dissipate the heat—especially if it were covered with fur. **42.4** `create` Individuals could acclimatize by normal homeostatic mechanisms (increased sweating, seeking shade, etc.). Individuals with alleles that allowed them to function better at higher temperatures would produce more offspring than individuals without those alleles. Over time, this would lead to adaptation via natural selection.
42.5 `create` *Apatosaurus* was likely a homeotherm because of its huge size. Because of its low surface area/volume ratio, metabolically produced heat would be trapped inside the dinosaur's body. Even if exposed to cold or warm temperatures, the dinosaur's temperature would not change much because it takes a long time for it to lose or gain heat across such a relatively small surface area.

✔ Test Your Knowledge

1. `understand` True 2. `remember` Epithelial 3. `understand` b
4. `understand` d 5. `understand` a 6. `analyze` a

✔ Test Your Understanding

7. `understand` It is where environmental changes are sensed first. As an effector, it has a large surface area for losing or gaining heat. **8.** `analyze` A warm frog can move, breathe, and digest much faster than a cold frog, because enzymes work faster (rates of chemical reactions increase) at 35°C than at 5°C. A warm frog needs much more food, though, to support this high metabolic rate. **9.** `understand` *Absorptive sections* have many folds and projections that increase their surface area for absorption. *Capillaries* have a high surface area because they are thin and highly branched, making exchange of fluids and gases efficient. *Beaks of Galápagos finches* have sizes and shapes that correlate with the type of food each species eats. Large beaks are used to crack large seeds; long, thin beaks are used to pick insects off surfaces; etc. *Fish gills* are thin, flattened structures with a large surface area, which facilitates the efficient exchange of gases and wastes. **10.** `analyze` A larger sphere has a relatively smaller surface area for its interior volume than does a smaller sphere, because surface area increases with size at a lower rate than volume does. Diffusion will be more efficient in the smaller sphere. **11.** `apply` See **FIGURE A42.1**. The graph shows that as size (length) increases, the volume increases more rapidly than the surface area. **12.** `understand` c. This is an example of negative feedback because insulin is released in response to an increase in a parameter (blood sugar concentration)

FIGURE A42.1

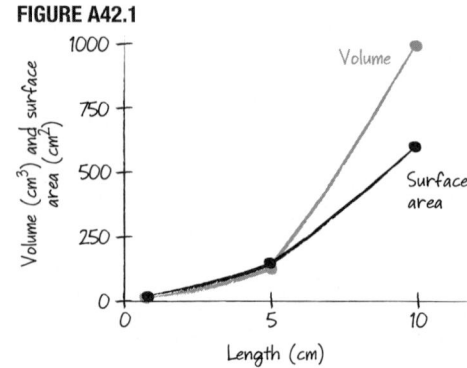

and acts to reduce that parameter so that homeostasis is maintained. In the other examples, homeostasis is not maintained.

✔ Test Your Problem-Solving Skills

13. `create` Large individuals require more food, so are more susceptible to starvation if food sources can't be defended (e.g., when seeds are scattered around). They may also be slower and/or more visible to predators. **14.** `create` You would have to show changes in the frequencies of alleles whose products are affected by temperature—for example, increases in the frequencies of alleles for enzymes that operate best at higher temperature. **15.** `create` The heat-dissipating system should have a very high surface area/volume ratio. Based on biological structures, it should be flattened (thin) and highly folded and branched, and/or contain tubelike projections. **16.** `understand` a. Turtles are ectothermic, so they do not expend much energy to produce body heat. Mice are endothermic, and expend energy to regulate body temperature. Mice will therefore have a higher basal metabolic rate and consume more food.

CHAPTER 43

IN-TEXT QUESTIONS AND EXERCISES

p. 863 Fig. 43.1 `understand` The black and white molecules are moving down their concentration gradients, and the concentration of red molecules does not affect the concentration of black or white molecules.
p. 867 Fig. 43.6 `understand` The sodium–potassium ATPase performs primary active transport; the sodium–chloride–potassium cotransporter performs secondary active transport, the chloride and potassium channels are involved in passive transport.
p. 869 CYU `apply` The osmolarity of the fish's tissue would decrease because the pumps would no longer be able to import sodium.
p. 869 Fig. 43.8 `analyze` The underside of the abdomen is shaded by the grasshopper's body. This location is relatively cool and should reduce the loss of water during respiration.
p. 871 CYU (1) `understand` Uric acid is insoluble in water and can be excreted without much water loss. **(2)** `understand` When electrolytes are reabsorbed in the hindgut epithelia, water follows along an osmotic gradient.
p. 873 `understand` Blood contains cells and large molecules as well as ions, nutrients, and wastes; the filtrate contains only ions, nutrients, and wastes.
p. 876 `apply` (1) Less water reabsorption will occur because the osmotic gradient is not as steep. (2) Filtrate osmolarity will be lower because less water has been reabsorbed. (3) Salt reabsorption will be reduced because the concentration of NaCl in the filtrate is lower.
p. 878 `apply` Ethanol consumption inhibits water reabsorption, leading to a larger volume of less concentrated urine. Nicotine consumption increases water reabsorption, leading to a smaller volume of more concentrated urine.
p. 878 CYU `apply` Water intake increases blood pressure and filtration rate in the renal corpuscle. It leads to lowered electrolyte concentration in the blood and filtrate and the production of large volumes of dilute urine. Eating large amounts of salt results in concentrated, hyperosmotic urine. Water deprivation triggers ADH release and the production of concentrated, hyperosmotic urine.
p. 879 Fig. 43.19 `create` Carrying a large volume of water in the bladder may be energetically costly to the animal. The animal also might not be able to run as quickly and may have an increased predation risk.

IF YOU UNDERSTAND . . .

43.1 `understand` A sodium gradient establishes an osmotic gradient that moves water. It also sets up an electrochemical gradient that can move ions against their electrochemical gradient by secondary active transport through a cotransporter, or with their electrochemical gradient by passive diffusion through a channel.
43.2 `analyze` Cartilaginous fishes such as sharks maintain a high concentration of urea in their blood, which raises their osmolarity to about the same as seawater. Therefore, they lose very little water to the environment. Bony fishes drink seawater and maintain their osmolarity lower than that of seawater by actively transporting excess ions out of the gills. Sharks actively excrete sodium chloride through a rectal gland. **43.3** `understand` When comparing chloride cells in gills of salmon in the ocean versus freshwater, the cells are in different locations on the gills, different forms of Na^+/K^+-ATPase exist, and $Na^+K^+Cl^-$ transporters are located on the basolateral side of the membrane in ocean salmon and on the apical side of the membrane in freshwater salmon. **43.4** `apply` Aquatic insects produce more urea, and especially more ammonia, than terrestrial insects because these nitrogenous wastes are energetically less expensive to produce than uric acid. Ammonia and urea require more water to be excreted, but aquatic insects have access to plenty of water. **43.5** `create` The longer the loop of Henle, the steeper the osmotic gradient in the interstitial fluid in the medulla. Steep osmotic gradients allow a great deal of water to be reabsorbed as urine passes through the collecting duct, resulting in highly concentrated urine.

YOU SHOULD BE ABLE TO . . .
✔ Test Your Knowledge

1. `remember` True **2.** `understand` c **3.** `apply` a **4.** `remember` d **5.** `understand` c **6.** `remember` bladder

✔ Test Your Understanding

7. `create` b. Fishes excrete nitrogenous wastes in the form of ammonia. Although the ammonia can easily diffuse across the fish's gills into the water, freshwater ponds are often small and ammonia can build up to potentially toxic levels. **8.** `analyze` Ammonia is toxic and must be diluted with large amounts of water to be excreted safely. Urea and uric acid are safer and do not have to be excreted with large amounts of water but are more expensive to produce in terms of energy expenditure. Fish excrete ammonia; mammals excrete urea; insects excrete uric acid. You would expect the embryos inside terrestrial eggs to excrete uric acid, since it is the least toxic and is insoluble in water. **9.** `understand` Mitochondria are needed to produce ATP. A key component of salt regulation in fish is the Na^+/K^+-ATPase, which requires ATP to function. Because ATP fuels establishment of the ion gradients necessary for the cotransport mechanisms used by chloride cells as well as other epithelial cells, an abundance of mitochondria would be expected in a chloride cell. **10.** `create` The wax layer should be thicker in insects inhabiting dry habitats, because the animals there are under more osmotic stress (more evaporation due to low humidity; less available drinking water). Insects that live in extremely humid habitats may lack the ability to close their spiracles; animals that live in humid habitats may excrete primarily ammonia or urea and have relatively short loops of Henle. In each case, you could test these predictions by comparing individuals of closely related species that live in dry versus humid habitats. **11.** `create` Without a loop of Henle, there would be no concentration gradient in the medullary interstitial fluid, so water would not be absorbed from the pre-urine in the collecting duct; concentrated urine could not be formed. **12.** `create` Freshwater has such a low osmolarity that if invertebrates osmoconformed to it, they would lack sufficient quantities of ions to conduct electrical currents in their excitable tissues.

✔ Test Your Problem-Solving Skills

13. `apply` c. Without aquaporins in the collecting duct, water cannot be reabsorbed, which would result in increased urine volume and decreased urine osmolarity. **14.** `create` The mammal's urine volume would increase dramatically. Ouabain would poison Na^+/K^+-ATPase pumps in the nephron, which would prevent reabsorption of ions and nutrients. Without the osmotic gradient established by reabsorption of ions and nutrients, the water would remain in the nephron and exit as urine. **15.** `analyze` The observation that the Na^+ concentration decreased by 30 percent, even as volume also decreased, indicates that Na^+ ions were reabsorbed to a greater degree than water. The observation that the urea concentration increased by 50 percent indicates that urea was not reabsorbed. **16.** `analyze` The hypothesis that the mussels are osmoconformers is supported because their hemolymph is very close in osmolarity to that of the wide range of water osmolarities into which they are placed.

CHAPTER 44

IN-TEXT QUESTIONS AND EXERCISES

p. 891 `understand` Nutrient absorption occurs in the small intestine, but not in the esophagus or stomach, and the rate of absorption increases with available surface area.
p. 894 Fig. 44.14 `understand` Frog eggs don't normally make the sodium-glucose cotransporter protein, so the researchers could be confident that if the protein appeared, it was from the injected RNA. They would not be confident of this if the RNAs were injected into rabbit epithelial cells, where the protein was probably already present.
p. 896 `apply` (1) Decreased energy yield from foods due to reduced glucose absorption, (2) increased fecal glucose content due to the passing of unabsorbed glucose into the colon, (3) watery feces due to decreased water reabsorption in the large intestine (lower osmotic gradient). (An additional effect would be increased flatulence due to the metabolism of unabsorbed glucose by bacteria that produce gases as waste products.)
p. 896 CYU (1) `understand` *Mouth:* Food is taken in; teeth physically break down food into smaller particles; salivary amylase begins to break down carbohydrates; lingual lipase initiates the digestion of fats. *Esophagus:* Food is moved to the stomach via peristaltic contractions. *Stomach:* HCl denatures proteins; pepsin begins to digest them. *Small intestine:* Pancreatic enzymes complete the digestion of carbohydrates, proteins, lipids, and nucleic acids. Most of the water and all of the nutrients are absorbed here. *Large intestine:* Water is reabsorbed and feces are formed. *Anus:* Feces accumulate in the rectum and are expelled out the anus. **(2)** `apply` If the release of bile salts is inhibited, fats would not be digested and absorbed efficiently, and they would pass into the large intestine. The individual would likely produce fatty feces and lose weight over time. Trypsin inactivation would inhibit protein digestion and reduce absorption of amino acids, likely leading to weight loss and muscle atrophy due to protein deprivation. If the Na^+-glucose cotransporter is blocked, then glucose would not be absorbed into the bloodstream. The individual would likely become sluggish from lack of energy.
p. 897 Fig. 44.16 `understand` In type 2 diabetes mellitus, the top, black arrow from glucose (in the blood) to glycogen (inside liver and muscle cells) is disrupted. In type 1 diabetes mellitus, the green arrow on the top left, indicating insulin produced by the pancreas, is disrupted.
p. 898 `understand` Individuals with type 1 diabetes mellitus have normal insulin receptors on their liver cells but do not produce and release insulin from the pancreas. Individuals with type 2 diabetes mellitus produce and release insulin, but they have fewer insulin receptors and/or these receptors do not respond as well to insulin. In both types of diseased individuals, glucose concentrations in the blood remain high. In individuals without disease, both insulin production and insulin receptors

normal, so liver cells respond to insulin by taking up glucose from the blood and storing it.

p. 898 Fig. 44.17 analyze The data do not prove that obesity causes type 2 diabetes mellitus. The study provides correlational data and cannot directly assign causation. However, the study provides strong evidence of an association between obesity and type 2 diabetes.

p. 899 CYU (1) understand Type 1 diabetes mellitus is an autoimmune disease, meaning that the body's immune system mistakenly kills the insulin-producing cells of the pancreas. Type 2 diabetes mellitus is caused by a combination of factors, including genetic predisposition, obesity, and poor diet. **(2)** apply When the blood sugar of individuals with type 1 diabetes mellitus gets too high, they should inject themselves with insulin, which will trigger the absorption and storage of glucose by their cells. When their blood glucose gets too low, they should eat something with a high concentration of sugar (such as orange juice or a candy bar) to increase their blood glucose levels quickly. (If glucose is dangerously low, a glucagon shot may be administered.)

IF YOU UNDERSTAND . . .

44.1 analyze One serving of skim milk has approximately 80 kcal, while whole milk has about 152 kcal. **44.2** apply Its mouth or tongue would be modified to make probing of flowers and sucking nectar efficient. It would not require teeth, and its digestive tract would be relatively simple. For example, the stomach would not have to pulverize food or digest large amounts of protein, and the large intestine would not have to process and store large amounts of bulky waste. **44.3** analyze Amylase will work best at a neutral or slightly alkaline pH (typical of mouth and small intestine), whereas pepsin will work best at a low pH (typical of the stomach). **44.4** understand Type 1 diabetes mellitus leads to high glucose because insulin is not produced, and therefore glucose cannot be moved from blood into cells. Type 2 diabetes mellitus leads to high blood glucose because insulin does not bind in sufficient amounts to its receptor, so glucose cannot be transported from blood into cells.

YOU SHOULD BE ABLE TO . . .

✔ Test Your Knowledge

1. remember a **2.** understand d **3.** remember d **4.** remember b **5.** remember insulin, glucagon **6.** remember False

✔ Test Your Understanding

7. understand The bird crop is an enlarged sac that can hold quickly ingested food; in leaf-eating species it is filled with symbiotic organisms and functions as a fermentation vessel. The cow rumen is an enlarged portion of the stomach; the elephant cecum is enlarged relative to other species. Both structures are filled with symbiotic organisms and function as fermentation vessels. **8.** analyze Digestive enzymes break down macromolecules (proteins, carbohydrates, nucleic acids, lipids). If they weren't produced in an inactive form, they would destroy the cells that produce and secrete them. **9.** understand b **10.** understand Pima living in the United States are genetically similar to Pima living in Mexico, but they have higher rates of type 2 diabetes. This suggests that the difference in diabetes rates is due to environmental factors such as diet and exercise. **11.** understand Humans have sharp canines and incisors, which may help to bite and tear meat. Humans also have flat molars, which help in grinding grains and other plant material. **12.** apply In a type 1 diabetic who took too much insulin, the blood glucose would become very low. The person would become sluggish and disoriented, and might slip into a diabetic coma.

✔ Test Your Problem-Solving Skills

13. create A proton pump inhibitor reduces activity of the proton pumps in parietal cells, reducing the amount of HCl produced. The contents of the stomach would therefore become less acidic, helping to prevent acid reflux. **14.** analyze b **15.** evaluate The result is still valid because the injection caused secretion—if it is also true that lack of injection led to no secretion. The criticism is also somewhat valid, however, because the researchers couldn't rule out the hypothesis that signaling from nerves plays some sort of role, too. **16.** create Terrestrial animals are exposed to increased risk of water loss, and the large intestine is where water reabsorption occurs. In most cases, fish do not need to reabsorb large amounts of water from their feces.

CHAPTER 45

IN-TEXT QUESTIONS AND EXERCISES

p. 905 CYU (1) analyze Oxygen partial pressure is high in mountain streams because the water is cold, mixes constantly, and has a high surface area (due to white water). Oxygen partial pressure is low at the ocean bottom because that is far from the surface, where gas exchange takes place, and there is relatively little mixing. **(2)** create *Warm-water species:* Large amount of air, because the oxygen-carrying capacity of warm water is low. *Vigorous algal growth:* Small amount of air, because algae contribute oxygen to the water through photosynthesis. *Sedentary animals:* Small amount of air, because sedentary animals require relatively little oxygen.

p. 906 Fig. 45.4 analyze External gills are ventilated passively and are efficient because they are in direct contact with water. They are exposed to predators and mechanical damage, however. Internal gills are protected but have to be ventilated by some type of active mechanism for water flow.

p. 907 Fig. 45.6 apply If "concurrent flow" occurred, oxygen transfer would stop, because the partial pressure gradient driving diffusion would fall to zero partway along the length of the capillary.

p. 909 Fig. 45.7 analyze Using many animals increases confidence that the results are true for most or all members of the population and are not due to one or a few unusual individuals or circumstances.

p. 912 CYU (1) analyze Common features include large surface area, short diffusion distance (a thin gas exchange membrane), and a mechanism that keeps fresh air or water moving over the gas exchange surface. Only fish gills use a countercurrent exchange mechanism; only tracheae deliver oxygen directly to cells without using a circulatory system; only lungs (mammalian) contain "dead space"—areas that are not involved in gas exchange. **(2)** apply The P_{O_2} would go down as oxygen is used up, the P_{CO_2} would go up as CO_2 diffuses into the blood but cannot be exhaled, and pH would drop as the CO_2 dissolves in blood to form bicarbonate and H^+ ions.

p. 913 create There would be an even larger change in the oxygen saturation of hemoglobin in response to an even smaller change in the partial pressure of oxygen.

p. 914 Fig. 45.16 analyze According to the data in the figure, when the tissue P_{O_2} is 30 mm Hg, the oxygen saturation of hemoglobin is about 15 percent for blood at pH 7.2 and about 25 percent at pH 7.4. Therefore, about 85 percent of the oxygen is released from hemoglobin at pH 7.2, but only about 75 percent of the oxygen is released at pH 7.4.

p. 915 Fig. 45.18 analyze In the lungs, a strong partial pressure gradient favors diffusion of dissolved CO_2 from blood into the alveoli. As the partial pressure of CO_2 in the blood declines, hydrogen ions leave hemoglobin and react with bicarbonate to form more CO_2, which then diffuses into the alveoli and is exhaled from the lungs.

p. 916 CYU analyze The curve for Tibetans should be shifted to the left relative to people adapted to sea level—meaning that their hemoglobin has a higher affinity for oxygen at all partial pressures.

p. 921 Fig. 45.24 analyze Air from the alveoli mixes with air in the "dead space" in the bronchi and trachea on its way out of the body. This dead space air is from the previous inhalation ($P_{O_2} = 160$ mm Hg; $P_{CO_2} = 0.3$), so when the alveolar air mixes with the dead space air, the partial pressures in the exhaled air achieve levels intermediate between that of inhaled and alveolar air.

p. 922 apply Without the delay at the AV node, the ventricles would not have the chance to fully fill with blood from the atria. As a consequence, the volume of blood ejected from the ventricles would decline.

p. 924 CYU (1) understand Blood plasma that leaks out of the capillaries and is not reabsorbed into capillaries enters lymphatic vessels that eventually merge with blood vessels. **(2)** understand See **FIGURE A45.1**.

IF YOU UNDERSTAND . . .

45.1 understand Cellular respiration. **45.2** create It is harder to extract oxygen from water than it is to extract it from air. This limits the metabolic rate of water breathers. **45.3** analyze Large animals have a relatively small body surface area relative to their volume. If they had to rely solely on gas exchange across their skin, the skin surface area would not be large enough to exchange the volume of oxygen needed to meet their metabolic needs. **45.4** create Since O_2 cannot compete as well as CO for binding sites in hemoglobin, O_2 transport decreases. As oxygen levels in the blood drop, tissues (most crucially in the brain) become deprived of oxygen, and suffocation occurs. **45.5** apply Their circulatory systems should have relatively high pressures—achieved by independent systemic and pulmonary circulations powered by a four-chambered heart—to maximize the delivery of oxygenated blood to metabolically active tissues.

YOU SHOULD BE ABLE TO . . .

✔ Test Your Knowledge

1. understand b **2.** remember c **3.** understand d **4.** understand d **5.** remember c **6.** understand An open circulatory system is less efficient than a closed circulatory system because it is not capable of directing the hemolymph toward specific organs, and because the pressure is lower, resulting in slower flow of hemolymph.

✔ Test Your Understanding

7. understand During exercise, P_{O_2} decreases in the tissues and P_{CO_2} increases. The increase in P_{CO_2} lowers tissue pH. The drop in P_{O_2} and pH causes more oxygen to be released from hemoglobin. **8.** understand In blood, CO_2 is converted to carbonic acid, which dissociates into a proton and a bicarbonate ion. Because the protons bind to deoxygenated hemoglobin, they do not cause a dramatic drop in blood pH (hemoglobin acts as a buffer). Also, small decreases in blood pH trigger a homeostatic response that increases breathing rate and expulsion of CO_2. **9.** understand Because airflow through bird lungs is unidirectional, the gas exchange surfaces are

FIGURE A45.1

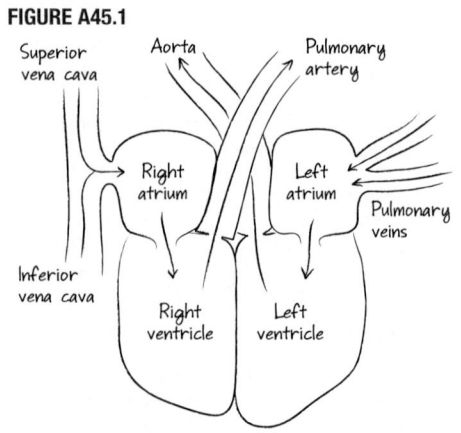

continuously ventilated with fresh, oxygenated air. The trachea and bronchi in a mammalian lung do not have a gas exchange surface, and bidirectional airflow in these species means that "stale" air has to be expelled before "fresh" air can be inhaled. As a result, the alveoli are not ventilated continuously. **10.** understand Lungs increase the temperature of the air and are moist to allow greater solubility of gases (increasing k); alveoli present a large surface area (large A); the epithelium is thin (small D); and constant delivery of deoxygenated blood to alveoli maintains a steep partial pressure gradient, favoring diffusion of oxygen into the body ($P_2 - P_1$ is high). **11.** analyze Yes—the trait compensates for the small P_{O_2} gradient between stagnant water and the blood of the carp. **12.** analyze c

✔ Test Your Problem-Solving Skills

13. create Cold water carries more oxygen than warm water does, so icefish blood can carry enough oxygen to supply the tissues with oxygen even in the absence of hemoglobin. The oxygen and carbon dioxide are simply dissolved in the blood. **14.** analyze The insect tracheal system delivers O_2 directly to respiring cells; in humans, O_2 is first taken up by the blood of the circulatory system, which then delivers it to the cells. The human respiratory system also has specialized muscles devoted to ventilating the lungs. The open circulatory system of the insect is a low-pressure system, which makes use of pumps (hearts) and body movements to circulate hemolymph. The closed circulatory system of humans is a high-pressure system, which can respond to rapid changes in O_2 demand by tissues. **15.** understand If the pulmonary circulation were under pressure as high as that found in the systemic circulation, large amounts of fluid would be forced out of capillaries in the lungs. There is a conflict between the thin surface required for efficient gas exchange and the thickness needed in blood vessels to withstand high pressure. **16.** analyze b

CHAPTER 46

IN-TEXT QUESTIONS AND EXERCISES

p. 931 Fig. 46.3 apply No—as K^+ leaves the cell along its concentration gradient, the interior of the cell becomes more negative. As a result, an electrical gradient favoring movement of K^+ into the cell begins to counteract the concentration gradient favoring movement of K^+ out of the cell. Eventually, the two opposing forces balance out, and there is no net movement of K^+.
p. 932a apply $E_{Ca2+} = 58$ mV $\times \log 1$ mM/0.0001 mM)/ $(+2) = 116$ mV.
p. 932b apply The membrane potential would be near zero because the Na^+ and K^+ would diffuse across the membrane along their concentration gradients.
p. 934 CYU (1) apply The resting potential would fall (be much less negative) because K^+ could no longer leak out of the cell. **(2)** understand The size of an action potential from a particular neuron does not vary, so it cannot contain information. Only the frequency of action potentials from the same neuron varies.
p. 934 understand See **FIGURE A46.1**.
p. 935 (1) understand Na^+ channels open when the cell membrane depolarizes, and as more channels open, more Na^+ flows into the cell, further depolarizing the cell, causing more voltage-gated Na^+ channels to open. **(2)** apply Sodium would flow inward as potassium flowed outward, and their charges would cancel each other out. The membrane would not depolarize.
p. 938 CYU (1) understand Once the threshold level of depolarization is attained, the probability that the voltage-gated sodium channels will open approaches 100 percent. But below threshold, the massive opening of Na^+ channels does not occur. This is why the action potential is all or none. **(2)** apply The membrane would depolarize and stay depolarized.

p. 938 Fig. 46.10 understand Get a solution taken from the synapse between the heart muscle and the vagus nerve *without* the nerve being stimulated. Expose a second heart to this solution. There should be no change in heart rate.
p. 940 analyze The inside of the cell becomes more positive (depolarized). This shifts the membrane potential closer to the threshold, making it more likely that the postsynaptic cell will fire an action potential.
p. 942 CYU apply The EPSP would not be as strong if an enzyme that broke down excitatory neurotransmitter increased in concentration.
p. 943 Fig. 46.16 analyze In the rest-and-digest mode, pupils take in less light stimulation, the heartbeat slows to conserve energy, liver conserves glucose, and digestion is promoted by release of gallbladder products. In the fight-or-flight mode, the pupils open to take in more light, the heartbeat increases to support muscle activity, and the liver releases glucose into the blood to fuel the fight or the flight (for example, from a predator).
p. 944 Fig. 46.18 understand Point to your forehead and top of your head for the frontal lobe, the top and top rear of your head for the parietal lobe, the back of your head for the occipital lobe, and the sides of your head (just above your ear openings) for the temporal lobes.
p. 946 Fig. 46.21 analyze No—for example, part (b) indicates that the size of the brain area devoted to the trunk is no bigger than the size of the brain area devoted to the thumb.

p. 948 CYU (1) understand One method is to study individuals with known brain lesions and correlate the location of the defect with a deficit in mental or physical function. Another is to directly stimulate brain areas in conscious patients during brain surgery and record the response. **(2)** remember Learning and memory involve changes in the number, sensitivity, and placement of synapses and neurons.

IF YOU UNDERSTAND . . .

46.1 understand See **FIGURE A46.2**. **46.2** understand Because the behavior of voltage-gated Na^+ and K^+ channels does not change. Every time a membrane depolarizes past the threshold and all of the voltage-gated Na^+ channels are open, the same amount of Na^+ will enter the cell, and the cell will depolarize to the same extent. In addition, K^+ channels open at the same time and allow the same amount of K^+ to leave the cell, so that it repolarizes to the same extent. **46.3** understand Summation means that EPSPs and IPSPs produced at various sites combine to determine the response or lack of response in a postsynaptic cell. Thus the behavior of a postsynaptic cell depends on the integration of input from many synapses. **46.4** apply The animal would likely die because its basic bodily functions (circulation, breathing, digestion, etc.) would be disrupted.

FIGURE A46.1

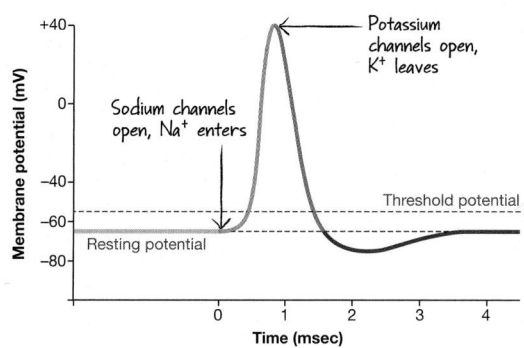

FIGURE A46.2

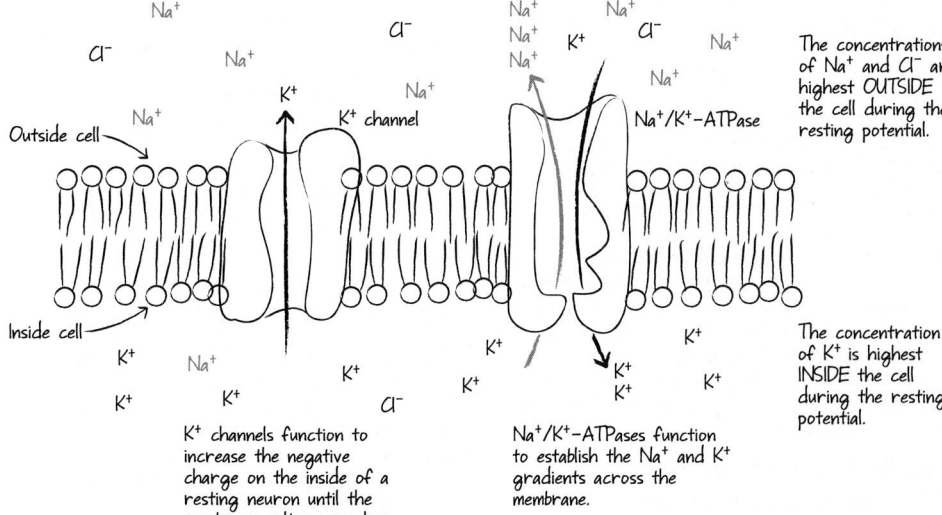

✔ **Test Your Knowledge**

1. remember b 2. remember d 3. understand c 4. remember a 5. remember b 6. understand See **FIGURE A46.3**.

✔ **Test Your Understanding**

7. understand The Na⁺/K⁺-ATPase pumps 3 Na⁺ out of the cell for every 2 K⁺ it brings in. Since more positive charges leave the cell than enter it, there is a difference in charge on the two sides of the membrane and thus a voltage. 8. analyze The ligand-gated channel opens or closes in response to binding by a small molecule (for example, a neurotransmitter); the voltage-gated channel opens or closes in response to changes in membrane potential. 9. understand EPSPs and IPSPs are produced by flow of ions through channels in the membrane, which change the membrane potential in the postsynaptic cell. If a flow of ions at one point depolarizes the cell but a nearby flow of ions hyperpolarizes the cell, then in combination these flows of ions cancel each other out. But if two adjacent flows of ions depolarize the cell, then their effects are additive and sum to produce a total that is more likely to reach threshold. 10. analyze The somatic system responds to external stimuli and controls voluntary skeletal muscle activity, such as movement of arms and legs. The autonomic system controls internal involuntary activities, such as digestion, heart rate, and gland activities. 11. apply A drug that activates the sympathetic nervous system would increase heart rate. 12. understand a

✔ **Test Your Problem-Solving Skills**

13. apply The neurotransmitters will stay in the synaptic cleft longer, so the amount of binding to ligand-gated channels will increase. Ion flows into the postsynaptic cell will increase dramatically, affecting its membrane potential and likelihood of firing action potentials. 14. analyze The benefit of these approaches is that they allow the specific role of the brain region to be assessed in a living person. However, diseased or damaged brains may not respond like healthy and undamaged brains. Also, the extent of lesions and the exact location of electrical stimulation may be difficult to determine, making correlations between the regions affected and the response imprecise. 15. analyze c (Because the current is reduced from normal levels, but not eliminated completely, there must be more than one type of potassium channel present. The venom probably knocks out just one specific type of channel.) 16. create A reasonable hypothesis is that the new neurons arise from stem cells in the brain.

CHAPTER 47

IN-TEXT QUESTIONS AND EXERCISES

p. 959 CYU (1) apply A punctured eardrum wouldn't vibrate correctly and would result in hearing loss at all frequencies in the affected ear. If the stereocilia are too short to come into contact with the tectorial membrane, vibration of the basilar membrane will not cause them to bend, and sound will not be detected. A loss in basilar membrane flexibility would result in the inability to hear lower-pitched sounds, such as human speech. **(2)** understand The lateral line system senses pressure waves in the water, allowing fishes to detect the presence of such objects as prey, predators, or potential mates. **p. 959 Fig. 47.9** create Vision could potentially be used to detect prey in diurnal catfish. To control for this, catfish could be blinded or "blindfolded" with an opaque material over their eyes to permit study of the lateral line system. **p. 961 Fig. 47.12** create The change in retinal's shape would induce a change in opsin's shape. A protein's

function correlates with its shape, so opsin's function is likely to change as well. **p. 962** analyze cGMP is a second messenger in the sense that it signals the sodium channel that transducing is inactive, and it is also a ligand that opens sodium channels. Transducin is like a G protein because it switches from "off" to "on" in response to a receptor protein, activating a key protein as a result. Closing Na⁺ channels stops the entry of positive charges into the cell, making the inside of the cell more negative relative to the outside. **p. 962 Fig. 47.13** understand This ion flow occurs when the photoreceptor cell is not receiving light—when the cell is in the dark. **p. 963 Fig. 47.15** create The S, M, and L opsin proteins are each different in structure. These structural differences affect the ability of retinal to absorb specific frequencies of light and change shape. **p. 964 CYU (1)** understand Retinal acts like an on–off switch that indicates whether light has fallen on a rod cell. When retinal absorbs light, it changes shape. The shape change triggers events that result in a change in action potentials, signaling that light has been absorbed. **(2)** apply Retina detachment from a jarring car accident would separate the retina from the optic nerve, resulting in blindness. The mutation would produce blue–purple color blindness, because the S opsin responds to wavelengths in that region of the spectrum. A clouded lens would reduce the amount of light that reaches the retina, reducing visual sensitivity. **p. 967 CYU (1)** understand Extremely hot food damages taste receptors, so those proteins would no longer be able to respond to their chemical triggers. **(2)** create One hypothesis to explain why the vomeronasal organ is vestigial in most primates is that primates have evolved other complex methods of communication, including gestures, facial expressions, and language, that render sensation with the vomeronasal organ less important. **p. 967** apply (1) A rattlesnake with its eyes covered will strike effectively because its pits will sense the prey. (2) A rattlesnake with its eyes and pits covered with cotton cloth will strike effectively. The cloth will block vision but the heat from the animal will penetrate the cloth to stimulate the thermoreceptors in the pits. (3) A rattlesnake with its eyes and pits covered with an opaque heat-blocking material will not be able to strike effectively, because the material blocks vision and heat energy.

p. 969 CYU (1) create Electrogenic fishes may have a survival advantage over other fishes in murky water, where vision would be compromised. **(2)** create Female sea turtles likely use magnetic cues to locate their natal beaches.

IF YOU UNDERSTAND . . .

47.1 create The sensory neurons that used to serve that limb are cut, but may still fire action potentials that stimulate brain areas associated with awareness of pain in the missing limb. **47.2** analyze Each of these senses involves ion channels that open in response to application of pressure to cell membranes. **47.3.** understand Because different animal species have different sets of opsin molecules that respond to different wavelengths of light absorbed by retinal they see different colors. **47.4** create Smells are part of the sensation of flavor, and people with nasal congestion cannot smell. In addition to smell being important, the small number of taste receptors can send many different frequencies of action potentials to the brain, and the taste receptors can be stimulated in many different combinations. **47.5** understand Capsaicin binds to thermoreceptors that are also activated at high temperatures, leading to the perception of heat. It can also bind to some nociceptors.

YOU SHOULD BE ABLE TO . . .

✔ **Test Your Knowledge**

1. understand a 2. remember d 3. understand c 4. remember b 5. remember false 6. remember magnetoreception

✔ **Test Your Understanding**

7. understand b 8. understand The many possible examples include several blue opsins allowing coelacanths to distinguish the blue wavelengths present in their deep-sea habitat, infrasound hearing allowing elephants to hear over long distances, ultrasonic hearing allowing bats to hunt by echolocation and moths to avoid bat predators, red and yellow opsins allowing fruit-eating mammals to distinguish ripe from unripe fruit. 9. analyze The data provide evidence that a pheromone acted to synchronize the women's menstrual cycles because the average time between onset of a woman's cycle and that of the rest of the women declined over time following move-in to the dorms in fall. It declined most rapidly after they first moved in together. The leveling of the graph over time

FIGURE A46.3

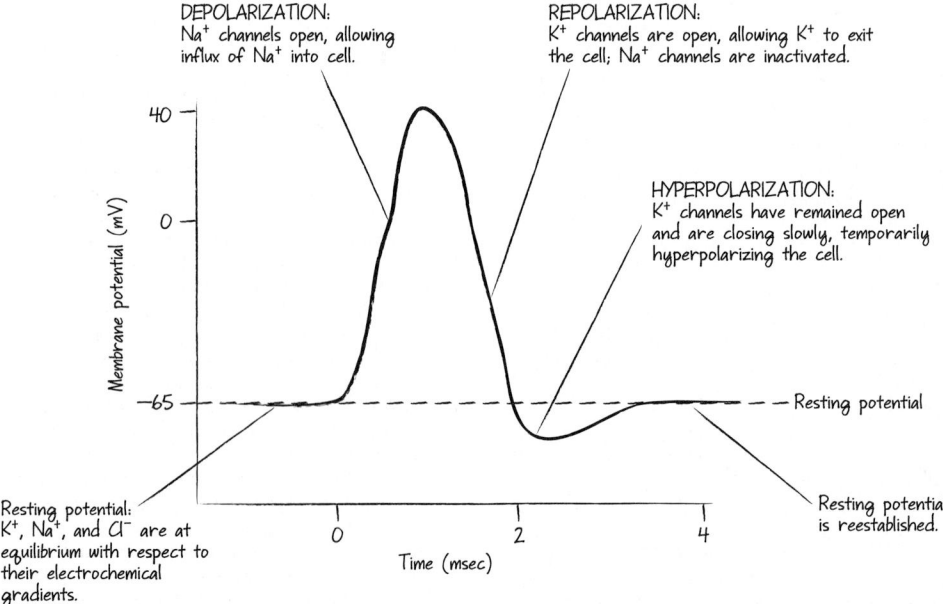

DEPOLARIZATION:
Na⁺ channels open, allowing influx of Na⁺ into cell.

REPOLARIZATION:
K⁺ channels are open, allowing K⁺ to exit the cell; Na⁺ channels are inactivated.

HYPERPOLARIZATION:
K⁺ channels have remained open and are closing slowly, temporarily hyperpolarizing the cell.

Resting potential

Resting potential:
K⁺, Na⁺, and Cl⁻ are at equilibrium with respect to their electrochemical gradients.

Resting potential is reestablished.

Membrane potential (mV)

40

0

-65

Time (msec)

0   2   4

shows that the onset dates became more synchronized.
10. `understand` Ion channels in hair cells are thought to open in response to physical distortion caused by the bending of stereocilia. Ion channels in taste receptors, in contrast, change shape and open when certain bitter-tasting molecules bind to them. **11.** `evaluate` Dalton's hypothesis was reasonable because blue fluid absorbs red light and would prevent it from reaching the retina, and this could be tested by dissecting his eyes. Although the hypothesis was rejected, it inspired a rigorous test and required researchers to think of alternative explanations. **12.** `create` To determine whether electric eels are electrogenic, place an eel in a tank and measure the electricity of water to determine if the eel is emitting an electric field. To determine whether they are electroreceptive, place an eel in a tank in the dark and add an object the size and shape of a prey item that emits no scent or taste. Block the lateral line system with a chemical as in Figure 47.9. If the eel locates the object, it must do so using electroreception.

✓ Test Your Problem-Solving Skills

13. `understand` a; vomeronasal organs evolved in tetrapods, presumably as an adaptation for communication in terrestrial habitats. **14.** `analyze` Both the lateral line system and electroreceptors use hair cells to sense changes in the water. Hair cells in the lateral line system contain mechanoreceptors that are depolarized in response to changes in water pressure. In contrast, electroreceptors cause hair cells to depolarize in response to changes in the electric field. **15.** `evaluate` It is unlikely that a perfume contains human pheromones because they have never actually been characterized, although their presence is suspected. **16.** `create` Block all but a few hundred ommatidia in the center of each compound eye of many dragonflies with an opaque material, and observe any differences that might occur in their ability to detect and pursue prey compared to individuals whose ommatidia had been covered by a transparent material.

CHAPTER 48

IN-TEXT QUESTIONS AND EXERCISES

p. 974 `understand` The trucks are the Z discs, the ropes are the thin filaments, and the burly weightlifters are the thick filaments.
p. 974 Fig. 48.2 `analyze` The dark band includes thin filaments and a dense concentration of bulbous structures extending from the thick filament; the light band consists of thin filaments only.
p. 977 CYU (1) `understand` In a sarcomere, thick myosin filaments are sandwiched between thin actin filaments. When the heads on myosin contact actin and change conformation, they pull the actin filaments toward one another, shortening the whole sarcomere. **(2)** `apply` Increased acetylcholine release would result in an increased rate of muscle-cell contraction. Preventing conformational changes in troponin would prevent muscle contraction. Blocking the uptake of calcium ions into the sarcoplasmic reticulum would lead to sustained muscle contraction.
p. 980 `apply` The breast meat of pigeons is composed of dark meat because these muscles are specialized for endurance. Their dark color is due to a high concentration of myoglobin.
p. 981 `create` When earthworms shorten their longitudinal muscles, the segment containing these muscles shortens and squeezes the internal fluid, which becomes pressurized and pushes in all directions, expanding the circumference of the segment and pushing laterally against the burrow.
p. 983 CYU (1) `analyze` Exoskeletons are made of chitin, proteins, and other substances like calcium carbonate. Endoskeletons are made of calcium phosphate, calcium carbonate, and proteins. Exoskeletons occur on the outsides of animals, whereas endoskeletons occur on the insides. Both types of skeletons are composed of rigid levers separated by joints such that motions are a result of changes in joint angles. Both types of skeletons attach to skeletal muscle and serve to transmit muscle forces. Hydrostatic skeletons also transmit muscle forces but are made of soft tissues that vary from animal to animal. Shape changes in hydrostatic skeletons occur not from changes in joint angles, but from shape changes of the bodies themselves. **(2)** `apply` No movement of the arm would occur, because for these antagonistic muscles to produce movement, one of them must be contracted while the other is relaxed.
p. 986 Fig. 48.16 `apply` Galloping at 3.5 m/s would cost about 75 percent more energy.
p. 987 Fig. 48.18 `apply` About 10 times more costly.

IF YOU UNDERSTAND . . .

48.1 `apply` Paralysis, because acetylcholine has to bind to its receptors on the membrane of postsynaptic muscle fibers for action potentials to propagate in the muscle and cause contraction. **48.2** `analyze` Skeletal muscle is multinucleate, unbranched, striated, and voluntary. Cardiac muscle is branched, striated, and involuntary. Smooth muscle is smooth and involuntary. **48.3** `analyze` When muscles contract in hydrostatic skeletons, the body deforms by becoming longer and thinner, or shorter and narrower, or by bending side to side. In vertebrates and arthropods, muscle contractions cause changes in the joint angles between rigid segments rather than shape changes in the segments themselves. **48.4** `understand` On land, an animal is constrained by its weight; larger animals cannot move as freely and are in greater danger of breaking than are smaller animals. In water, animals are constrained by their shapes. Some shapes would cause high drag, which would inhibit locomotion. In air, animals are constrained by their weight. Larger animals must create more lift to overcome gravity.

YOU SHOULD BE ABLE TO . . .

✓ Test Your Knowledge

1. `remember` a **2.** `remember` c **3.** `understand` d **4.** `remember` True **5.** `understand` a **6.** `understand` False.

✓ Test Your Understanding

7. `create` The key observation was that the banding pattern of sarcomeres changed during contraction. Even though the entire unit became shorter, only some portions moved relative to each other. This observation suggested that some portions of the structure slid past other portions. Muscle fibers have to have many mitochondria because large amounts of ATP are needed to power myosin heads to move along actin filaments; large amounts of calcium stored in smooth ER are needed to initiate contraction by binding to troponin. **8.** `apply` c **9.** `analyze` Acetylcholine reduces the rate and force of contraction of cardiac muscle, whereas it stimulates skeletal muscle to contract. **10.** `create` ATP is no longer produced after death, so calcium cannot be actively transported from the cytosol into the sarcoplasmic reticulum to enable the binding of myosin and actin. The proteins do not unbind, because this process requires ATP. **11.** `apply` The oxygen consumption of the runner would increase because the arches of his or her feet would no longer store as much elastic energy, which normally reduces the energetic cost of running. **12.** `create` Because diverse animals experience the same physical constraints where they live, natural selection favors the same kind of adaptations to those constraints, resulting in convergent evolution. For example, dolphins and ichthyosaurs are distantly related but have a similar body shape due to their similar ability to swim rapidly through water.

✓ Test Your Problem-Solving Skills

13. `apply` In cardiac muscle, the binding of acetylcholine to its receptors causes the heart rate to slow. Ingestion of nightshade would increase heart rate because it blocks acetylcholine receptors. **14.** `apply` If the sarcomere is in a stretched state before stimulation, the initial force production would be reduced because less overlap would occur between actin and myosin; thus, fewer myosin heads could engage in the pull. **15.** `apply` c:

$$0.5 = v^2/(9.8 \text{ m/s}^2 \times 0.9 \text{ m})$$
$$0.5 = v^2/8.8 \text{ m}^2/\text{s}^2$$
$$v^2 = 0.5 \times 8.8 \text{ m}^2/\text{s}^2$$
$$v = \text{square root of } 4.4 \text{ m}^2/\text{s}^2$$
$$v = 2.1 \text{ m/s}$$

16. `evaluate` The bones of *T. rex* are available and can be analyzed to determine their structural properties, such as the mechanical advantage of the leg joints and the ability of the bones to withstand the different forces that the skeleton would experience at different speeds based on data from living animals. A very close estimate could be made, but the exact speed would be hard to determine without observing the dinosaur in action, since behavior affects speed.

CHAPTER 49

IN-TEXT QUESTIONS AND EXERCISES

p. 994 `apply` The current temperature is analogous to the sensory input; the thermostat, to the CNS; the thermostat signal, to the cell-to-cell signal; and the furnace, to the effector. If feedback inhibition fails, hormone production will continually increase.
p. 996 Fig. 49.4 `understand` Injecting the dog with a liquid extract from a different part of the body, or with a solution that is similar in chemical composition to the extract from the pancreas (e.g., in pH or ions present) but lacking molecules produced by the pancreas.
p. 997 CYU (1) `understand` It is difficult to differentiate between the nervous and endocrine systems because some neurons secrete hormones and some endocrine glands respond to neural signals. **(2)** `apply` Thyroid hormone receptors are intracellular because thyroid hormones are lipid soluble.
p. 998 `create` The cells could differ in receptors, signal transduction systems, or response systems (genes or proteins that can be activated).
p. 1001 `understand` Stressed individuals would be less able to heal or to fight pathogens.
p. 1001 Fig. 49.9 `understand` The saline injection controlled for any stress induced by the injection procedure and for introducing additional fluids.
p. 1002 Fig. 49.11 `apply` Same—a defect in both signal and receptor has the same effect—no response—as a defective signal alone or a defective receptor alone.
p. 1003 CYU (1) `analyze` In amphibians, an increase in thyroid hormones stimulates metamorphosis from a tadpole to an adult without a resting stage. In insects, a decrease in juvenile hormone and increase in ecdysone stimulate metamorphosis during a resting (pupal) stage. **(2)** `understand` By reducing immune system function, promoting release of fatty acids from storage cells and use of amino acids from muscles for energy production, and preventing release of glucose in response to signals from insulin, cortisol conserves glucose supplies for use by the brain.
p. 1004 Fig. 49.12 `create` Inject ACTH and monitor cortisol levels. If cortisol does not increase, adrenal failure is likely.

p. 1006 CYU (1) `understand` ACTH triggers the release of cortisol, but cortisol inhibits ACTH release by blocking the release of CRH from the hypothalamus and suppressing ACTH production. High levels of cortisol tend to lower cortisol levels in the future. **(2)** `understand` Processing centers in the brain are responsible for synthesizing a wide array of sensory input. To start a response to this sensory input, they stimulate neurosecretory cells in the hypothalamus. Neurohormones from these cells travel to the anterior pituitary, where they trigger the production and release of other hormones.

p. 1010 CYU (1) `apply` The steroid-hormone–receptor complex would probably fail to bind to the hormone-response element. Then gene expression would not change in response to the hormone—the arrival of the hormone would have little or no effect on the target cell. **(2)** `analyze` The mode of action would be similar to that of a peptide—hormone binding would have to trigger a signal transduction event that resulted in production of a second messenger or a phosphorylation cascade.

IF YOU UNDERSTAND . . .

49.1 `understand` The production and release of hormones are often controlled, directly or indirectly, by electrical or chemical signals from the nervous system. A good example is the hypothalamic–pituitary axis, where neuroendocrine signals regulate endocrine signals. **49.2** `analyze` Electrical signals are short lived and localized. All of the processes listed in the question are long-term changes in the organism that require responses by tissues and organs throughout the body. **49.3** `apply` The hypothalamus would continue to release CRH, stimulating the pituitary to release ACTH, which in turn would stimulate continued release of cortisol from the adrenal cortex. Sustained, high circulating levels of cortisol have negative effects. **49.4** `apply` A cell can respond to more than one hormone at a time, because cells produce different receptors for each hormone.

YOU SHOULD BE ABLE TO . . .

✔ **Test Your Knowledge**

1. `understand` c **2.** `understand` a **3.** `remember` a **4.** `remember` c **5.** `remember` Mammals—eye and pineal gland; lizards—pineal gland; birds—diffuse in brain **6.** `remember` True

✔ **Test Your Understanding**

7. `understand` b **8.** `create` This is one reason that the same hormone can trigger different effects in different tissues. For example, epinephrine binds to four different types of receptors in different tissues—eliciting a different response from each. **9.** `analyze` The posterior pituitary is an extension of the hypothalamus; the anterior pituitary is independent—it communicates with the hypothalamus via chemical signals in blood vessels. The posterior pituitary is a storage area for hypothalamic hormones; the anterior pituitary synthesizes and releases an array of hormones in response to releasing hormones from the hypothalamus. **10.** `analyze` Steroid hormones act directly, and alter gene expression. They usually bind to receptors inside the cell, forming a complex that binds to DNA and activates transcription. Nonsteroid hormones act indirectly, and activate proteins. They bind to receptors on the cell surface and trigger production of a second messenger or a phosphorylation cascade, ending in activation of proteins already present in the cell. **11.** `create` Subject volunteers to travel and jet lag, measure ACTH and cortisol levels, and correlate these results with their perceived level of jet lag symptoms. **12.** `create` Studies have shown that treatment with leptin is largely ineffective in reducing food intake in obese humans.

✔ **Test Your Problem-Solving Skills**

13. `apply` c (dehydration would result, because ADH would not be present) **14.** `create` There are two basic strategies: (1) remove the structure from some individuals

and compare their behavior and condition to untreated individuals in the same environment, or (2) make a liquid extract from the structure, inject it into some individuals, and compare their behavior and condition to individuals in the same environment that were injected with water or a saline solution. **15.** `evaluate` The researcher's hypothesis initially sounds unreasonable, because peptide hormones cannot enter cells to initiate gene transcription. However, a second messenger activated by the hormone could possibly lead to changes in gene transcription. **16.** `evaluate` The graph shows that atrazine treatment reduces the testosterone concentrations in male frogs to near the levels of females. This result supports the hypothesis that atrazine is an endocrine disruptor that feminizes male amphibians.

CHAPTER 50

IN-TEXT QUESTIONS AND EXERCISES

p. 1015 Fig. 50.3 `create` Isolate and identify the molecules found in "crowded" water. Test each molecule by adding it, at the same concentration found in crowded water, to clean water occupied by a single *Daphnia* and recording whether the female produces a male-containing brood. Repeat with many test females, and for each molecule identified. As a control, record the number of male-containing broods produced in clean water.

p. 1016 `apply` Asexual reproduction would be expected in environments where conditions change little over time.

p. 1019 Fig. 50.7 `evaluate` No—the data are consistent with the displacement hypothesis, but there is no direct evidence for it. There are other plausible explanations for the data.

p. 1021 CYU (1) `analyze` Oviparity usually requires less energy input from the mother after egg laying, and mothers do not have to carry eggs around as long—meaning that they can lay more eggs and be more mobile. However, before egg laying, mothers have to produce all the nutrition the embryo will require in the egg, and eggs may not be well protected after laying. Viviparity usually increases the likelihood that the developing offspring will survive until birth, but it limits the number of young that can be produced to the space available in the mother's reproductive tract. If viviparous young can be nourished longer than oviparous young, then they may be larger and more capable of fending for themselves. **(2)** `create` Divide a population of sperm into two groups. Subject one group of sperm to spermkillerene at concentrations observed in the female reproductive tract (the experimental group) but not the other group (the control). Document the number of sperm that are still alive over time in both groups.

p. 1025 Fig. 50.13 `create` (Note: There is more than one possible answer—this is an example.) Having two gonads is an "insurance policy" against loss or damage. To test this idea, surgically remove one gonad from a large number of male and female rats. Do a similar operation on a large number of similar male and female rats, but do not remove either gonad. Once the animals have recovered, place them in a barn or other "natural" setting and let them breed. Compare reproductive success of individuals with paired versus unpaired gonads.

p. 1026 Fig. 50.15 `analyze` It is similar. In both cases, the hypothalamus produces a releasing factor (GnRH or CRH) that acts on the pituitary. The releasing factor stimulates release of regulatory hormones from the anterior pituitary (LH and FSH or ACTH). These hormones travel via the bloodstream and act on the gonads or adrenals to induce the release of hormones from these glands. In both cases, the hormones are involved in negative feedback control of the regulatory hormones from the pituitary.

p. 1029 `apply` (1) There should be an LH spike early in the cycle, followed by early ovulation. (This might not actually occur, though, if the follicle is not yet ready to

rupture.) (2) LH levels should remain low—no mid-cycle spike and no ovulation.

p. 1030 CYU (1) `understand` FSH triggers maturation of an ovarian follicle. Its level rises at the end of a cycle because it is no longer inhibited by progesterone, which stops being produced at high levels when the corpus luteum degenerates. **(2)** `apply` The drug would keep FSH levels low, meaning that follicles would not mature and would not begin producing estradiol and progesterone. The uterine lining would not thicken.

p. 1032 `analyze` (1) The uterine lining may not be maintained adequately—if it degenerates, a miscarriage is likely. (2) Pregnancy tests detect hCG in urine. By the third trimester, hCG is no longer being produced by the placenta.

p. 1034 Fig. 50.23 `apply` The birth will be more difficult, as limbs or other body parts can get caught.

p. 1034 Fig. 50.24 `apply` A mother's chance of dying in childbirth in 1760 was a little over 1000 in 100,000 live births, or close to 1.0 percent. If she gave birth 10 times, she would have a 10 percent chance of dying in childbirth.

IF YOU UNDERSTAND . . .

50.1 `create` Parasites could wipe out a population of asexually produced animals because the animals do not vary genetically. However, a population of sexually produced animals would have enough genetic variability that some individuals could present better resistance to the parasites. **50.2** `create` Parental care demands resources (time, nutrients) that cannot be used to produce more eggs. Sperm competition is not likely when external fertilization occurs, and most fishes use external fertilization. **50.3** `create` The most common methods are surgical ligation (tying off) of the fallopian tubes in a woman to stop the delivery of the egg to the uterus and cutting the vas deferens (vasectomy) in a man to stop the delivery of sperm into the semen. **50.4** `apply` FSH and LH levels would increase following ovariectomy, because sex steroids would no longer be present to exert negative feedback upon the anterior pituitary. **50.5** `understand` The placenta is highly vascularized to provide a large surface area for transfer of oxygen and nutrients from the mother's circulation to the fetus's, and of wastes from the fetus's circulation to the mother's.

YOU SHOULD BE ABLE TO . . .

✔ **Test Your Knowledge**

1. `remember` a **2.** `understand` b **3.** `remember` a **4.** `remember` b **5.** `remember` Luteinizing hormone (LH) and follicle-stimulating hormone (FSH) **6.** `remember` True

✔ **Test Your Understanding**

7. `remember` Every offspring that is produced sexually is genetically unique; offspring that are produced asexually have DNA from their mother only and often are genetically identical to her. **8.** `create` *Daphnia* females produced males only when they were exposed to short day lengths *and* water from crowded populations *and* low food levels. These conditions are likely to occur in the fall. Sexual reproduction could be adaptive in fall if genetically variable offspring are better able to thrive in conditions that occur the following spring. **9.** `analyze` Spermatogenesis generates four haploid sperm cells from each primary spermatocyte; oogenesis produces only one haploid egg cell from each primary oocyte. Egg cells are much larger than sperm cells because they contain more cytoplasm. In males, the second meiotic division occurs right after the first meiotic division, but in females it is delayed until fertilization. **10.** `understand` LH triggers release of estradiol, but at low levels estradiol inhibits further release of LH. LH and FSH trigger release of progesterone, but progesterone inhibits further release of LH and FSH. High levels of estradiol trigger release of more LH. The follicle can produce high levels of estradiol only if it has grown and matured—meaning that it is ready for ovulation to occur. **11.** analyze c **12.** apply Females should be choosier about their second mate because the sperm of the second male has an advantage in sperm competition.

✔ **Test Your Problem-Solving Skills**

13. create If all sheep were genetically identical, it is less likely that any individuals could survive a major adverse event (e.g., a disease outbreak or other environmental challenge) than if the population were composed of genetically diverse individuals, which are likely to vary in their ability to cope with the new conditions. **14.** apply Oviparous populations should produce larger eggs than viviparous populations. Because oviparous species deposit eggs in the environment, the eggs must contain all the nutrients and water required for the entire period of embryonic development. But because embryos in viviparous species receive nourishment directly from the mother, it is likely that their eggs are smaller. **15.** create One hypothesis might be that the shell membrane is a vestigial trait (see Chapter 25)—specifically, an "evolutionary holdover" from an ancestor of today's marsupials that laid eggs. **16.** analyze b

CHAPTER 51

IN-TEXT QUESTIONS AND EXERCISES

p. 1040 apply See **FIGURE A51.1**. Each type of Toll protein is a pattern-recognition receptor that responds to a different group of pathogens. Since bacteria and fungi are very different pathogens, you would not expect a receptor that recognizes fungi to be required for a response against bacteria, so flies lacking Toll would survive as well as the wild-type control.
p. 1041 CYU analyze (1) Applying direct pressure constricts blood vessels, mimicking the effect of histamine released from mast cells. (2) Cleaning the wound to remove dirt and debris mimics the phagocytic activity of macrophages and neutrophils to eliminate foreign cells and material. (3) Applying bandages impregnated with platelet-recruiting compounds mimics the effect of chemokines released from injured tissues and macrophages, which attract circulating leukocytes and platelets to facilitate blood clotting and wound repair.
p. 1041 Fig. 51.3 understand Rubor (reddening) is due to increased blood flow to the infected or injured area. Mast cells trigger dilation of vessels by releasing signaling molecules. Calor (heat) is the result of fever, which is activated by the release of cytokines from macrophages that are in the area of infection. Tumor (swelling) occurs

due to dilation of blood vessels triggered by mast cells and macrophages.
p. 1046 apply (1) In a human, 200 different light chains are possible (see text). The recombination of the V, D, and J segments results in $51 \times 27 \times 6 = 8262$ possible heavy chains. Since these chains combine to produce BCRs, a total of $200 \times 8262 = 1,652,400$ are possible within a human. (2) Only one type of antibody can be produced from a single human B cell.
p. 1047 CYU (1) understand The production of BCRs and TCRs via DNA recombination is required for an effective response. Selection against self-reactive lymphocytes is required for a safe response. **(2)** understand The cell is able to make a larger number of differing variable regions per unit of DNA when the variable region is segmented. For example, 3 segments of V and 3 segments of J can randomly assemble into 8 different variable light chains, whereas only 6 light chains would be possible with the same amount of DNA if the V regions were preassembled.
p. 1048 understand "Clonal" refers to the cloning—producing many exact copies—of cells that are "selected" by the binding of their receptor to a specific antigen (or MHC–peptide complex, which is introduced later).
p. 1050 understand See **TABLE A51.1**.
p. 1051 CYU create Individuals who are heterozygous for MHC genes have greater variability in their MHC proteins than do individuals who are homozygous for these genes. The increased variability in MHC proteins allows a greater variety of peptides to be presented to T cells, which thus would be able to recognize, attack, and eliminate a wider range of pathogens compared with T cells in homozygotes. As a result, heterozygous individuals would likely suffer fewer infections than homozygous individuals.
p. 1054 Fig. 51.17 apply See **FIGURE A51.2**. The response to the new antigen would resemble the primary response that occurred after the first injection of the original antigen.

p. 1055 CYU analyze Attenuated viruses are able to infect the cells of a vaccinated host, but not well enough to cause disease. Inactivated viruses are not able to infect cells at all, and instead remain as extracellular particles. Since attenuated viruses can infect cells, their antigens will be processed and presented by class I MHC complexes, which will promote a stronger cell-mediated response than the inactivated viruses.

IF YOU UNDERSTAND . . .

51.1 understand Toll-like receptors recognize molecular patterns of broad groups of pathogens, such as viruses, fungi, and bacteria. Upon binding to these molecules, the receptors cause signals to be transduced that activate cellular responses directed at eliminating these types of pathogens. **51.2** analyze The BCR locus would be smaller in mature B cells compared to immature cells, since recombination between the gene segments resulted in removing sections of DNA. **51.3** understand Antigen-presenting cells activate the $CD4^+$ T cells that are required for the full activation of B cells to differentiate into plasma cells. **51.4** understand Extracellular viruses are agglutinated and inactivated by antibodies produced in the humoral response. Intracellular viruses are eliminated by the destruction of their host via cytotoxic T cells in the cell-mediated response. **51.5** apply The IgE antibodies generate hypersensitive responses. When these are severe, they can lead to anaphylactic shock. If IgE antibodies were generated against self molecules, you would expect a strong response that will lead to either anaphylactic shock or chronic inflammation.

YOU SHOULD BE ABLE TO . . .

✔ **Test Your Knowledge**

1. remember c **2.** understand a **3.** remember b **4.** understand They are identical except that a B-cell receptor has a transmembrane domain that allows it to be located in the B-cell

TABLE A51.1

MHC type	Origin of Peptide	Type of T cell that binds	Activity stimulated
Class I MHC	Host cell cytosol	CD8+	Activates T cell to secrete molecules that kill infected cell
Class II MHC	Extracellular environment	CD4+	Activates T cell to secrete cytokines that support the immune response of other cells

FIGURE A51.1

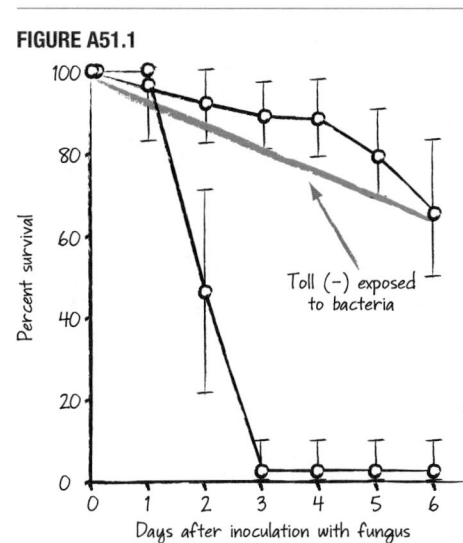

FIGURE A51.2

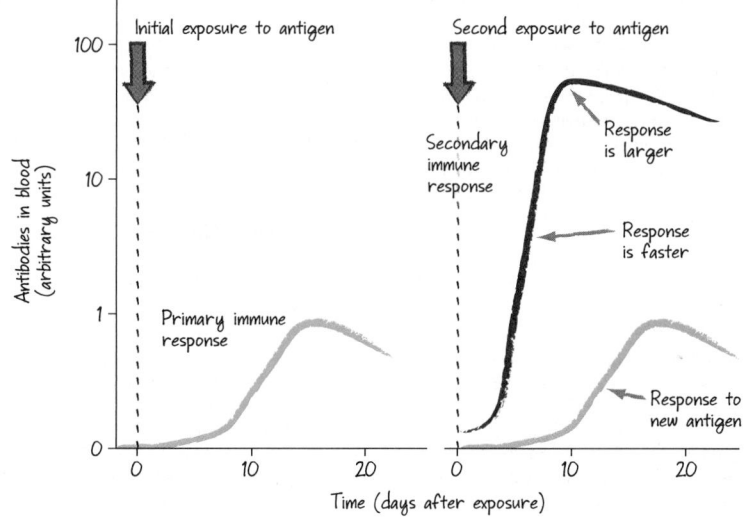

plasma membrane. **5.** `remember` c **6.** `understand` The hygiene
hypothesis claims that immune disorders such as allergies and autoimmunity have increased in frequency due
to the implementation of hygienic practices that
decrease common bacterial, viral, and parasitic
infections.

✔ Test Your Understanding

7. `understand` B cells are first activated to divide via interactions between the BCRs and their complementary
antigens. They are then fully activated to differentiate
and continue to divide when their class II MCH–
peptide complex interacts with an activated CD4$^+$ T cell
that has a complementary TCR. **8.** `analyze` Both the BCR
and TCR interact with epitopes of antigens via binding
sites located in the variable regions of their polypeptide
chains. A TCR is like one "arm" of a BCR. The BCR can
interact with an antigen directly, while the TCR interacts with its epitope in the context of an MHC protein
presented by another cell. **9.** `analyze` Vaccines have to
contain an antigen that can stimulate an appropriate primary immune response. Vaccines have not worked for
HIV, because the antigens on this virus are constantly
changed through mutation, rendering it unrecogniz-
able to memory cells generated following vaccination.
10. `analyze` b **11.** `analyze` Pattern-recognition receptors on
leukocytes bind to surface molecules that are present
on many pathogens. BCRs and TCRs bind to particular
epitopes of antigens that are pathogen specific. Pattern-
recognition-receptor binding directly activates the cell,
while BCRs and TCRs often require additional signals
to become fully activated. **12.** `understand` The mixing and
matching of different combinations of gene segments
from the variable and joining regions of the light-chain
immunoglobulin gene, along with diversity regions of
the heavy-chain gene, gives lymphocytes a unique sequence in both chains of the BCR. The variable regions
in the alpha and beta chains of the TCR are assembled
from segments in a similar manner.

✔ Test Your Problem-Solving Skills

13. `apply` d **14.** `create` If B cells have receptors that bind to
self molecules during maturation, they would become
activated during the maturation process. Instead of the
signal causing the cells to divide, these immature B cells
might respond to the signal by becoming permanently
inactivated or might be induced to undergo apoptosis.
15. `analyze` Natural selection favors individuals that can
create a large array of antibodies, because the high mutation rates and rapid evolution observed in pathogens
means that they will constantly present the immune
system with new antigens. If these antigens were not
recognized by the immune system, the pathogens would
multiply freely and kill the individual. **16.** `analyze` Tissue
from the patient's own body is marked with the major
histocompatibility (MHC) proteins that the patient's immune system recognizes as self. This results in the preservation, rather than the destruction, of the transplanted
tissue. Tissue from a different person is marked with
MHC proteins that are unique to that individual. The
body of the transplant recipient will recognize the MHC
proteins (and other molecules) of the donor as foreign,
resulting in a full immune response and rejection of the
grafted tissue.

CHAPTER 52

IN-TEXT QUESTIONS AND EXERCISES

p. 1062 Fig. 52.1 `apply` It would increase, because cattle
would no longer succumb to the disease carried by tsetse
flies. (In Africa, cattle are limited more by biotic conditions than abiotic conditions.)
p. 1064 Fig. 52.4 `apply` Since there were 28 plots and four
treatments, there were probably about seven replicate

plots per treatment. This is important because the
abiotic and biotic characteristics of individual plots are
likely to vary in natural landscapes.
p. 1066 `apply` See **FIGURE A52.1**.
p. 1067 CYU (1) `apply` There is little rain at the North Pole
because dense, dry air descends and absorbs moisture.
(2) `apply` The highest risk of skin cancer is in the tropics
where the Sun is often directly overhead, resulting in a
large amount of solar radiation per unit area.
p. 1069 Fig. 52.10 `analyze` Most of the remaining wildlands
occur in Canada, at the top of the image.
p. 1070a `understand` Vines and epiphytes increase productivity because they are photosynthetic organisms that fill
space between small trees and large trees—they capture
light and use nutrients that might not be used in a forest
that lacked vines and epiphytes.
p. 1070b `understand` Most leaves have a large surface area to
capture light. Light is abundant in deserts, however, and
plants with a large surface area would be susceptible to
high water loss and/or overheating.
p. 1071a `understand` Crop grains like wheat and corn are
grasses, so they thrive in the grassland biome.
p. 1071b `apply` There is a continuous grass cover and scattered trees. (A biome like this is called a savanna.)
p. 1072a `apply` Boreal forests should move north.
p. 1072b `understand` High elevations present an abiotic environment (precipitation and temperature) that is similar
to the conditions in arctic tundra. As a result, the plant
species present will have similar adaptations to cope
with these physical conditions. These similarities are the
result of convergent evolution.
p. 1073 Fig. 52.17 `analyze` Visible wavelengths enter the
chamber from the top and through the glass, warming
the plants and ground inside the chamber. Some of this
heat energy is retained within the mini-greenhouse.
p. 1074 CYU (1) `analyze` Tropical dry forests are probably
less productive than tropical wet forests because they
have less water available to support photosynthesis
during some periods of the year. **(2)** `apply` Your answer
will depend on where you live. For example, if global
warming continues at the current projected rates, several
effects can be expected. If you live in a coastal area, you
can expect water levels to rise. In general, plant communities will change because average temperature and
moisture—and variation in temperature and moisture—
will change. Also, development is likely to increase in
most areas, transforming the landscape.
p. 1077a `apply` The littoral zone, because light is abundant
and nutrients are available from the substrate.
p. 1077b `apply` Bogs are nitrogen poor, so plants that are
able to capture and digest insects have a large advantage.
This advantage does not exist in marshes and swamps,
where nutrients are more readily available.
p. 1078a `apply` Cold water contains more oxygen than
warm water, so it can support a higher rate of cellular
respiration.
p. 1078b `apply` Species that live in estuaries must be able
to tolerate variable salinity; freshwater marsh species do
not. The abiotic environment (salt concentration) is so
different that few species grow well in both habitats.
p. 1079 `apply` No—the aphotic zone is lightless, so natural
selection favors individuals that do not invest energy in
developing and maintaining eyes.

IF YOU UNDERSTAND . . .

52.1 `understand` Populations are made up of individual
organisms; communities are made up of populations of
different species; ecosystems are made up of groups of
communities along with the abiotic environment; the
biome is made up of all the world's ecosystems. **52.2** `apply`
The Argentine ants would likely increase their range into
the moist agricultural fields, displacing native ants.
52.3 `apply` All latitudes would get equal amounts of
sunlight all year round. There would be no seasons
and no changes in climate with latitude. **52.4** `apply` If

precipitation does not increase, then increased transpiration rates will put plants under water stress and reduce
productivity. **52.5** `apply` Like deserts, open oceans have
plenty of sunlight but very low productivity. Unlike
deserts, which are water limited, open oceans are nutrient limited.

YOU SHOULD BE ABLE TO . . .

✔ Test Your Knowledge

1. `remember` Organismal, population, community, ecosystem, and biosphere/global **2.** `remember` c **3.** `remember` b
4. `remember` d **5.** `remember` c **6.** `remember` a

✔ Test Your Understanding

7. `create` Examples of possible questions: *Organismal:*
How do humans cope with extremely hot (or cold)
weather conditions? *Population:* How large will the total
human population be in 50 years? *Community:* How are
humans affecting ocean species such as cod and tuna?
Ecosystem: How does nitrogen runoff from agricultural
fields affect the availability of oxygen in the Gulf of Mexico? *Global:* How will human-induced changes in global
temperature affect sea level? **8.** `apply` *Biotic:* These ants
compete with native ants. *Abiotic:* Argentine ants require
moist conditions, not too cold, not too hot. These factors
interact; Argentine ants outcompete native ants in moist
environments, but competition slows down their dis-
persal. **9.** `apply` b **10.** `analyze` Productivity in intertidal and
neritic zones is high because sunlight is readily available
and because nutrients are available from estuaries and
deep-ocean currents. Productivity in the oceanic zone
is extremely low—even though light is available at the
surface—because nutrients are scarce. The deepest part
of the oceanic zone may have nutrients available from
the substrate but lacks light to support photosynthesis.
11. `analyze` As elevation increases, biomes often change in
a similar way as increasing in latitude (e.g., you might go
from temperate forest to a boreal-type forest to tundra).
12. `analyze` The natural biome of Eastern North America is
primarily temperate forest. However, most of this area is
now occupied by human-influenced biomes such as cities, suburbs, and farmland.

FIGURE A52.1

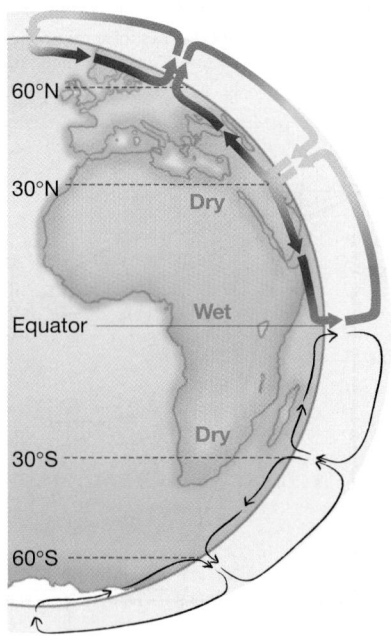

✔ **Test Your Problem-Solving Skills**

13. `apply` a **14.** `create` A single population of lizards was split when South America split from Africa due to continental drift. Fossils are too old to represent animals dispersed by human travel. **15.** `evaluate` Yes. As average global temperatures increase, some organisms may be able to adapt by moving their range. As mountaintops warm, however, organisms adapted to cold conditions will have nowhere to go. Ecologists have good reason to be concerned about these possible extinctions. **16.** `create` Temperature is relatively constant in the tropics all year, whereas temperature fluctuates dramatically at higher latitudes. Organisms that are physiologically adapted to constant temperatures may not have as much genetic variation for traits that enable adaptation to higher temperatures, compared to organisms at higher latitudes, which are adapted to tolerate different temperatures at different times of the year.

CHAPTER 53

IN-TEXT QUESTIONS AND EXERCISES

p. 1086 Fig. 53.4 `analyze` Even though half of each plot serves as a control, the natural landscape varies within plots, so there is a chance that foraging in the two sub-plots could be different due to factors not measured in the experiment. By measuring foraging activity before adding owls and seeds, the researchers established an additional control and a baseline.

p. 1087 CYU `apply` If owls were allowed to hunt within the treatment subplots, the threat of predation would increase. A larger amount of supplementary seed would be required to compensate for the increased risk of foraging. Gerbils do not consciously calculate costs and benefits. One possible proximate explanation is that the presence of hunting owls would increase the stress level in the gerbils (e.g., the hormone cortisol), which could suppress the hunger of the gerbils or reduce their drive to find food.

p. 1088 Fig. 53.6 `analyze` To form a better control, bring females into the lab and give them the same food and housing conditions as the treatment groups, but expose them to artificial lighting that simulates the short daylight conditions of winter.

p. 1089 CYU (1) `analyze` As in lizards, a rise in sex hormones in humans is a proximate cause of sexual readiness. **(2)** `analyze` Humans, like lizards, tend to choose a mate who will increase their fitness. For example, women often prefer men who will be able to help provide for offspring.

p. 1090 Fig. 53.7 `apply` The data would either be distributed randomly around the circles, showing no preference for direction, or the turtles would swim the same direction in both cases, suggesting that they are using compass orientation rather than map orientation.

p. 1093 Fig. 53.10 `apply` See **FIGURE A53.1**.

p. 1095 CYU (1) `analyze` Auditory communication allows individuals to communicate over long distances but can be heard by predators. Olfactory communication is effective in the dark, and scents can continue to carry information long after the signaler has left, but scents do not carry long distances. Visual communication is effective during the day but can be seen by predators. **(2)** `apply` The ability to detect deceit protects the individual from fitness costs (e.g., being eaten, having a mate's eggs fertilized by someone else); avoiding or punishing "liars" should also lower the frequency of deceit (because it becomes less successful). Alleles associated with detecting and avoiding or punishing liars should be favored by natural selection and increase in frequency.

p. 1096 Fig. 53.14 `apply` Between first cousins, $r = 1/2n \times 1/2n \times 1/2n = 1/8$.

p. 1097 `apply` Humans often live near kin, are good at recognizing kin, and in many cases can give kin resources for protection that increase fitness.

p. 1097 Fig. 53.15 `create` The control would be to drag an object of similar size through the colony, at similar times of day. This would test the hypothesis that prairie dogs are reacting to the presence of the experimenter and the disturbance—not to a predator.

IF YOU UNDERSTAND . . .

53.1 `apply` At the proximate level of causation, human teens experience an increase in the sex hormones testosterone and estrogen, which increases sex drive. At the ultimate level of causation, teens with a higher sex drive will, on average, have more offspring, which will increase their fitness and pass along their alleles at a higher rate than individuals with a lower sex drive. **53.2** `apply` You would have to consider variables that provide costs and benefits to the web-building spiders. For example, spinning silk and building webs uses up energy, so the size of the web, the amount of material used to build the web, and the frequency of building webs should be considered, as well as the effect of web-building on vulnerability to predation. But large, well-built webs can increase prey capture, so the rate of capture of prey of various types of webs also should be considered. **53.3** `create` Only a bowerbird with plenty of resources could afford the time and energy to build a highly decorated bower. Thus, the decoration behavior is likely an honest signal of male fitness. **53.4** `create` Migration is condition-dependent in some birds. If enough food is available locally, the costs of migration outweigh the benefits, so the birds' fitness will be higher if they stay put. **53.5** `apply` Deceitful communication should decrease. If large males are gone, few nests are available and more female-mimic males would compete at them. Most female-mimic males would have no (or fewer) eggs to fertilize and no male to take care of the eggs they did fertilize. There would be less natural selection pressure favoring deceitful communication. **53.6** `understand` (1) Long-lived, so that extensive interactions with kin and nonrelatives are possible; (2) good memory, to record reciprocal interactions; (3) kin nearby, to make inclusive fitness gains possible; (4) ability to share fitness benefits (e.g., food) between individuals.

YOU SHOULD BE ABLE TO . . .

✔ **Test Your Knowledge**

1. `understand` c **2.** `understand` Adaptive significance of behavior; how behavior affects fitness **3.** `remember` d **4.** `understand` c **5.** `understand` d **6.** `understand` c

✔ **Test Your Understanding**

7. `understand` The fitness of honeybees depends on how much energy they can bring into the hive by foraging. Hives with honeybees that successfully communicate locations of food resources will be favored by natural

---

FIGURE A53.1

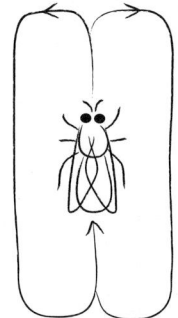

The straight run of the upward waggle dance is twice as long as in Fig. 53.10

---

selection such that the frequency of alleles that determine the communication behavior will increase in the population. **8.** `evaluate` Evolution favors organisms that forage optimally—maximizing benefits and minimizing costs—increasing the frequency of optimal foraging in populations over time. However, individuals in populations vary and environmental circumstances vary. Some individuals will forage suboptimally and suffer reduced fitness. **9.** `apply` Only males that are in good shape (well nourished and free of disease) have the resources required to perform their displays. **10.** `understand` Longer day lengths indicate the arrival of spring, when renewed plant growth and insect activity make more food available. Courtship displays from males indicate the presence of males that can fertilize eggs. **11.** `analyze` This statement is true; sea turtles need both a map and a compass. A map provides information about the spatial relationships of places on a landscape; a compass allows sea turtles to orient the map—so that it aligns with the actual landscape. **12.** `apply` c

✔ **Test Your Problem-Solving Skills**

13. `create` One possibility would be to set up a testing arena with food supplies at various distances from the food source when adult flies are originally released. Identify rover and sitter adults by genotype and test them individually; then record how far each travels during a set time interval. If the rover and sitter alleles affect foraging movements in adults, then rovers should be more likely to move (and to move farther) than sitters. **14.** `analyze` a, b, c, d; the howler monkeys make condition-dependent decisions about what to eat and are likely to prefer high-energy foods, when available, or foods with nutrients or compounds that would increase their fitness. Hamilton's rule could be considered if some monkeys give the highest-quality foods to close relatives rather than consuming it themselves. **15.** `analyze` Individuals have an r of 1/2 with full siblings and an r of 1/8 with first cousins. If the biologist will lose his life, he needs to save two siblings, or eight first cousins, to keep the "lost copy" of his altruism alleles in the population. **16.** `apply` People would be expected to donate blood under two conditions: (1) They either received blood before or expect to have a transfusion in the future, and/or (2) they receive some other benefit in return, such as a good reputation among people who might be able to help them in other ways.

CHAPTER 54

IN-TEXT QUESTIONS AND EXERCISES

p. 1105 Fig. 54.2 `apply` See **FIGURE A54.1** (see A:48)

p. 1106 (1) `apply` Compared to northern populations, fecundity should be high and survivorship low in southern populations. **(2)** `apply` Fecundity can be much higher if females lay eggs instead of retaining them in their bodies and giving birth to live young.

p. 1109 Fig. 54.6 `create` At high population density, competition for food limits the amount of energy available to female song sparrows for egg production.

p. 1112 CYU `create` One possibility is competition for food. To test this idea, compare carrying capacity in identical fenced-in areas that differ only in the amount of food added. (You could also test a hypothesis of space limitation by doubling the size of the enclosure but keeping the amount of food the same.)

p. 1114 Fig. 54.10 `apply` When the population of hares was low, the rate of increase was not density dependent on food. That is, plenty of food was available for the few rabbits, so competition for this resource was low.

p. 1117 CYU (1) `understand` Developed nations have roughly equal numbers of individuals in each age class because the fertility rate has been constant and survivorship high for many years. Developing countries have many more children and young people than older people, because the fertility rate and survivorship have been increasing. **(2)** [understand] Because survivorship is high in most human populations, changes in overall growth rate depend almost entirely on fertility rates.

p. 1118 Fig. 54.14 [apply] The total population size of first-generation females in the third year = 1000; in the fourth year = 240. The total population size of first- and second-generation females in the third year = 2308; in the fourth year = 1230. The number of 1-year-olds created by third-generation females (blue) is $981 \times 0.33 = 324$. The number of newborns (green) created by these 1-year-olds is $324 \times 4.0 = 1296$. Thus, the total population size at the end of the fourth year is expected to be $200 + 792 + 1296 + 324 + 198 + 40 = 2850$.

p. 1119 [apply] The population appears to be increasing over time: From the original 1000 females, there are 2308 females in the third year.

IF YOU UNDERSTAND . . .

54.1 [create] First you would need to observe snowshoe hares to determine how many pellets one individual can produce in one day, on average. Then you could use a quadrat to measure the number of fresh pellets produced in a known area of habitat in one day and use your pellets/hare number to convert this value to hares/area. **54.2** [analyze] In ancient Rome, survivorship was probably low and fecundity high—women started reproducing at a young age and did not live long, on average. In Rome today, the population has high survivorship and low fecundity. **54.3** [apply] See **FIGURE A54.2 54.4** [apply] The population density of mosquitoes would not be constant over time. Ignoring the number of predators that might be present, you would expect high densities of these short-lived insects after rains, followed by low densities during extended dry periods. **54.5** [understand] Fewer children are being born per female, but there are many more females of reproductive age due to high fecundity rates in the previous generation. **54.6** [apply] Small, isolated populations are likely to be wiped out by bad weather, a disease outbreak, or changes in the habitat. But in a metapopulation, migration between the individual small populations helps reestablish populations and maintain the overall size of the metapopulation.

YOU SHOULD BE ABLE TO . . .

✔ **Test Your Knowledge**

1. [remember] b **2.** [remember] births, deaths, immigration, emigration **3.** [remember] a **4.** [remember] True **5.** [understand] d **6.** [understand] c

✔ **Test Your Understanding**

7. [remember] c **8.** [apply] See **TABLE A54.1 9.** [create] The population has undergone near-exponential growth recently because advances in nutrition, sanitation, and medicine have allowed humans to live at high density without suffering from decreased survivorship and fecundity. Eventually, however, growth rates must slow as density-dependent effects such as disease and famine increase death rates and lower birth rates. **10.** [understand] (a) Corridors allow individuals to move between populations, increasing gene flow and making it possible to recolonize habitats where populations have been lost. (b) Maintaining unoccupied habitat makes it possible for the habitat to be recolonized. **11.** [apply] *Lacerta vivipara* is most vulnerable at the southern edge of its ranges, such as in the mountains of Spain. These lizards are cold adapted and may become locally extinct in areas that become too warm **12.** [analyze] $r_{\max}$ gives the population growth rate in the absence of density-dependent limitation; r is the actual growth rate, which is usually affected by density-dependent factors.

✔ **Test Your Problem-Solving Skills**

13. [apply] Fewer older individuals will be left in the population; there will be relatively more young individuals. If too many older individuals are taken, population growth rate may decline sharply as reproduction stops or slows. (But if relatively few older individuals are taken, more resources are available to younger individuals and their survivorship and fecundity, and the population's overall growth rate, may increase.) **14.** [create] The sunflowers and beetles are both metapopulations. To preserve them, you must preserve as many of the sunflower patches as possible (or plant more) and maintain corridors (that may be smaller sunflower patches) along which the beetles can migrate between the patches. **15.** [apply] As a sexually transmitted disease, AIDS will reduce the number of sexually active adults. If the epidemic continues unabated, the numbers of both reproductive-age adults and children will decline, causing a top-heavy age distribution dominated by older adults and the elderly. See **FIGURE A54.3**. **16.** [analyze] c

CHAPTER 55

IN-TEXT QUESTIONS AND EXERCISES

p. 1126 [apply] Your graph should look like Figure 55.4, with *Semibalanus* on the left, *Chthamalus* is on the right, and the *x*-axis indicating height in the intertidal zone. The realized niche of *Chthamalus* is in the upper intertidal zone.

p. 1127 Fig. 55.5 [apply] If Connell had done the treatments on different rocks, a critic could argue that differences in survival were due to differences in the nature of the rocks—not differences in competition.

p. 1128 [apply] Your graph should look like the bottom graph in Figure 55.6, showing that the niches of *Semibalanus* and *Chthamalus* are nonoverlapping with respect to height in the intertidal zone.

p. 1131 Fig. 55.10 [apply] This experimental design tested the hypothesis that mussels can sense the presence of crabs. If the crabs had been fed mussels, a critic could argue that the mussels detect the presence of damaged mussels—not crabs. (As it turns out, the mussels can do both. The experimenters did the same experiment with broken mussel shells in the chamber instead of a fish-fed crab.)

FIGURE A54.1

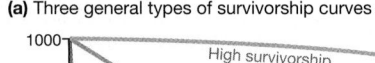
(a) Three general types of survivorship curves

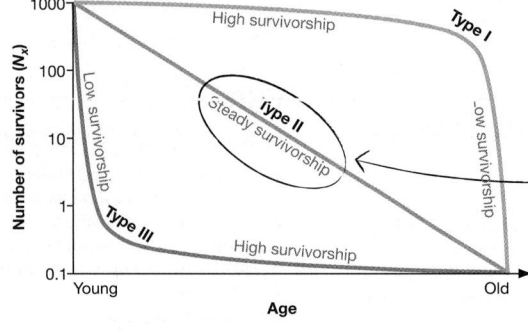

(b) Exercise: Survivorship curve for *Lacerta vivipara*

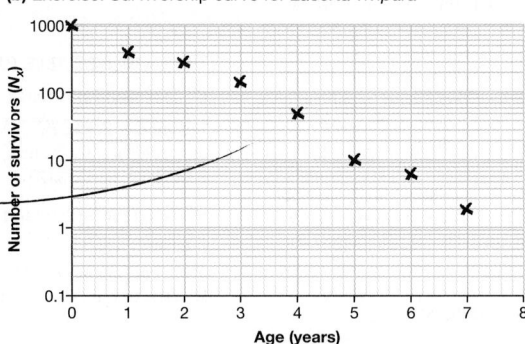

TABLE A54.1

Trait	Life–History Continuum		
	Left	Middle	Right
	High fecundity Low survivorship	Intermediate	Low fecundity High survivorship
Growth habit	Herbaceous	Shrub	Tree
Disease- and predator-fighting ability	Low	Medium	High
Seed size	Small	Medium	Large
Seed number	Many	Moderate	Few
Body size	Small	Medium	Large

FIGURE A54.2

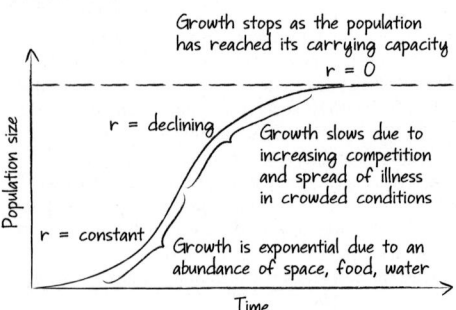

p. 1133 Fig. 55.12 `create` Put equal numbers of infected and uninfected ants in a pen that includes a bird predator. Count how many of each type are eaten over time.

p. 1134 Fig. 55.13 `apply` Many possible answers. For example, acacia trees may expend energy in producing the large bulbs the ants live in and the food they eat. The *Crematogaster* ants may expend energy and risk injury or death in repelling herbivores. The cleaner shrimp may occasionally get injured or eaten by the fish they are cleaning, or their diets may be somewhat restricted by the association. The host fish may miss meals or be at greater risk of predation when they are undergoing cleaning by the shrimp. In both examples of mutualism, however, the overall associations are positive for both parties.

p. 1135 CYU (1) `understand` The individuals do not choose or try to have traits that reduce competition—they simply have those traits (or not). Resource partitioning just happens, because individuals with traits that allow them to exploit different resources produce more offspring, which also have those traits. (Recall from Chapter 25 that natural selection occurs on individuals, but adaptive responses such as resource partitioning are properties of populations.) **(2)** `understand` When species interact via consumption, a trait that gives one species an advantage will exert natural selection on individuals of the other species who have traits that reduce that advantage. This reciprocal adaptation will continue indefinitely. An example is the interaction of *Plasmodium* with the human immune system: The human immune system has evolved the ability to detect proteins from the *Plasmodium* and kill infected cells; in response, *Plasmodium* has evolved different proteins that the immune system does not detect.

p. 1135 Fig. 55.14 `apply` These steps controlled for the alternative hypothesis that differences in treehopper survival were due to differences in the plants they occupied—not the presence or absence of ants.

p. 1137 Fig. 55.15 `analyze` Predictable, at least to a degree.

p. 1142 CYU (1) `understand` The shade provided by early successional species increases humidity, and decomposition of their tissues adds nutrients and organic material to the soil. These conditions favor growth by later successional species, which can outcompete the early successional species. **(2)** `understand` The presence or absence of a plant species where nitrogen fixation occurs would dramatically alter nutrient conditions, and thus the speed of succession and the types of species that could become established. For example, species that require high nitrogen would be favored on sites where alder grew, and species that can tolerate low nitrogen would thrive on sites where alder is absent.

p. 1142 Fig. 55.22 `apply` The effect would be to make the island more remote, which would lower the rate of immigration and move the whole immigration curve downward. The rate of extinction would increase,

shifting the curve upward. The overall effect would be to decrease the number of species, because the island effectively would have become more remote.

p. 1143 `analyze` The calculations for community 1 are as follows. (Note that your final answer is sensitive to the number of digits you use in your calculations; see BioSkills 1 in Appendix B).

Species A	**Species B**
$p_A = 10/18 = 0.555$	$p_B = 1/18 = 0.0555$
$\ln 0.555 = -0.589$	$\ln 0.0555 = -2.891$
$-0.589 \times 0.555 = -0.327$	$-2.891 \times 0.0555 = -0.160$

Species C	**Species D**
$p_C = 1/18 = 0.0555$	$p_D = 3/18 = 0.167$
$\ln 0.0555 = -2.891$	$\ln 0.167 = -1.790$
$-2.891 \times 0.0555 = -0.160$	$-1.790 \times 0.167 = -0.299$

Species E	**Species F**
$p_E = 2/18 = 0.111$	$p_F = 1/18 = 0.0555$
$\ln 0.111 = -2.198$	$\ln 0.0555 = -2.891$
$-2.198 \times 0.111 = -0.244$	$-2.891 \times 0.0555 = -0.160$

Summing the values of $p \times \ln p$ for each species and multiplying by -1 gives the Shannon index of species diversity for community 1:

$$(-1)[(-0.327) + (-0.160) + (-0.160) + (-0.299) + (-0.244) + (-0.160)] = 1.350$$

Similar calculations would give species diversity values of 1.794 for community 2 and 1.610 for community 3.

IF YOU UNDERSTAND . . .

55.1 `understand` A mutualistic relationship becomes a parasitic one if one of the species stops receiving a benefit. The treehopper–ant mutualism becomes parasitic in years when spiders are rare, because the treehoppers no longer derive a benefit but pay a fitness cost (producing honeydew that the ants eat). **55.2** `apply` An invasive species could have a large impact on community structure by replacing the dominant plant species that creates physical structure, which in turn affects other members of the community; by replacing a keystone species that has an indirect effect on many other species; or by outcompeting many species, causing a direct reduction in species richness. **55.3** `apply` All species at Glacier Bay must be able to survive in a cold climate with the local amount of precipitation. The species earliest in the succession must also be able to grow on rock exposed as the glacier melts. But chance, historical differences in seed sources, and presence or absence of alder (and nitrogen fixation) created differences in the species present. **55.4** `apply` Many national parks are surrounded by altered habitats, so they are functionally similar to islands. Species richness in the park is more likely to be preserved if it is large and located nearby other wilderness areas.

YOU SHOULD BE ABLE TO . . .

✔ **Test Your Knowledge**

1. `remember` True **2.** `remember` a **3.** `remember` d **4.** `understand` d **5.** `remember` fecundity **6.** `remember` c

✔ **Test Your Understanding**

7. `evaluate` Yes—the treehopper–ant mutualism is parasitic or mutualistic, depending on conditions; competition can evolve into no competition over time if niche differentiation occurs; arms races mean that the outcome of host–parasite interactions can change over time, and so forth (many other examples). **8.** `understand` b **9.** `evaluate` (a) If community composition is predictable, then the species present should not change over time. But if composition is not predictable, then that species should undergo significant changes over time. (b) If community composition is predictable, then two sites with identical abiotic factors should develop identical communities. If community composition is not predictable, then sites with identical abiotic factors should develop variable communities. In most tests, the data best match the predictions of the "not predictable" hypothesis, though communities show elements of both. **10.** `analyze` Disturbance is any short-lived event that changes the distribution of resources. Compared to low-frequency fires, high-frequency fires would tend to be less severe (less fuel builds up) and would tend to exert more intense natural selection for adaptations to resist the effects of fire. Compared to low-severity fires, high-severity fires would open up more space for pioneering species and would tend to exert more intense natural selection for adaptations to resist the effects of fire. **11.** `apply` Early successional species are adapted to disperse to new environments (small seeds) and grow and reproduce quickly (reproduce at an early age, grow quickly). They can tolerate severe abiotic conditions (high temperature, low humidity, low nutrient availability) but have little competitive ability. These species are able to enter a new environment (with no competitors) and thrive. These attributes are considered adaptations because they increase the fitness of these species. **12.** `analyze` The idea is that high productivity will lead to high population density of consumers, leading to competition and intense natural selection favoring niche differentiation that leads to speciation.

✔ **Test Your Problem-Solving Skills**

13. `analyze` Natural selection will favor orchid individuals that have traits that resist bee attack: thicker flower walls, nectar storage in a different position, a toxin in the flower walls, and so on. Individuals could also be favored if their anthers were in a position that accomplished pollination even if bees eat through the walls of the nectar-storage structure. **14.** `apply` b **15.** `apply` The exact answer will depend on the location of the campus. The first species to appear must possess good dispersal ability, rapid growth, quick reproductive periods, and tolerance for very harsh and severe conditions. The two-acre plot is likely to be colonized first by pioneer species that have very "weedy" characteristics. But once colonization is under way, the course of succession will depend more on how the various species interact with each other. The presence of one species can inhibit or facilitate the arrival and establishment of another. For example, an early-arriving species might provide the shade and nutrients required by a late-arriving species. The site's history and nearby ecosystems may influence which species appear at each stage; for instance, an undisturbed ecosystem nearby could be a source for native species. The pattern and rate of this succession is also influenced by the overall environmental conditions affecting it. Only species with traits appropriate to the local climate are likely to colonize the site. **16.** `create` One reasonable experiment would involve constructing artificial ponds and introducing different numbers of plankton species

FIGURE A54.3

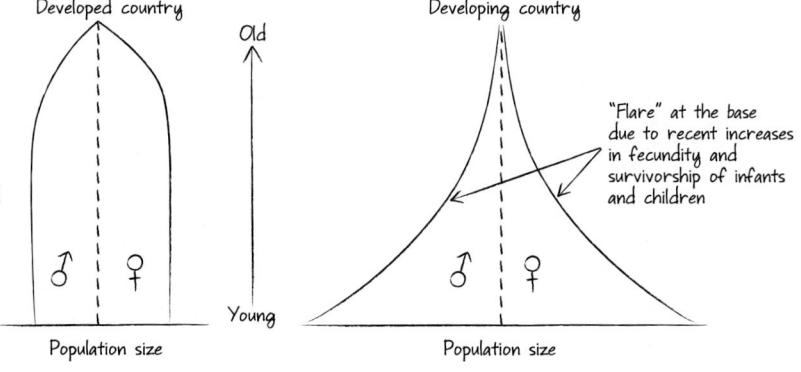

to different ponds, but the same total number of individuals. (Any natural immigration to the ponds would have to be prevented.) After a period of time, remove all of the plankton and measure the biomass present. Make a graph with number of species on the x-axis and total biomass on the y-axis. If the hypothesis is correct, the line of best fit through the data should have a positive slope.

CHAPTER 56

IN-TEXT QUESTIONS AND EXERCISES

p. 1152 apply To grow a kilogram of beef, first you have to grow 10 kg of grain or grass and feed it to the cow. Only 10 percent of this 10 kg will be used for growth and reproduction—the other 9 kg is used for cellular respiration or lost as heat.
p. 1152 apply Crustaceans and fish are ectothermic, so they are much more efficient at converting primary production into the biomass in their bodies than are endothermic birds and mammals.
p. 1153 apply Tuna are top predators in marine food webs. Mercury occurs in low concentrations in the water, but accumulates in higher concentrations at each trophic level. A person who eats a lot of tuna will accumulate mercury over time, potentially causing mercury poisoning. Sardines are primary consumers, a low trophic level, and therefore are not in danger of biomagnification of mercury.
p. 1155 Fig. 56.10 apply Average NPP = 125 g/m^2/year. Area = 65%. Although NPP in the ocean is very small, the ocean is so vast that it ends up being the largest contributor to total NPP.
p. 1156 CYU (1) understand At each trophic level, most of the energy that is consumed is lost to cellular respiration and ultimately heat, metabolism, or other maintenance activities, which leaves only a small percentage of energy for biomass production (growth and reproduction).
(2) apply Since NPP depends on temperature, it is logical to predict that NPP will increase with global warming. (However, Section 56.3 explains that global warming changes precipitation patterns, ocean stratification, and other factors that can limit NPP).
p. 1157 understand When the trees of tropical rain forests are hauled away and the remnants of the forest are burned, the nutrients are exported from the site, leaving only poor soil to nourish the crops.
p. 1158 Fig. 56.13 create One logical hypothesis is that the total amount exported increases as tree roots and other belowground organic material decay and begin to wash into the stream; the amount exported should begin to decline as nitrate reserves become exhausted—eventually, there is no more nitrate to wash away.
p. 1159 Fig. 56.14 apply Much more water should evaporate. One possibility is that more water vapor will be blown over land and increase precipitation over land.
p. 1162 CYU apply **Water**: perhaps the most direct impacts concern rates of groundwater replenishment. Converting biomes into farms or suburbs decreases groundwater recharge and increases runoff. Irrigation pumps groundwater to the surface, where much of it runs off into streams. **Nitrogen**: Humans introduce huge quantities of nitrogen into ecosystems as fertilizer. **Carbon**: Humans are releasing huge quantities of stored carbon, in the form of fossil fuels, into the atmosphere as CO_2. Whereas humans are depleting supplies of freshwater below natural levels, we are increasing quantities of N and C above natural levels.
p. 1164 Fig. 56.22 analyze Photosynthesis increases in summer, resulting in removal of CO_2 from the atmosphere.
p. 1165 Fig. 56.23 apply $2 \times 7000 = 14,000$ million metric tons of CO_2, double the current total.

p. 1168 apply The pH of the water will fall (acidity will increase) because CO_2 from your breath reacts with water to form carbonic acid.

IF YOU UNDERSTAND . . .

56.1 apply Tofu is a plant, a primary producer at the base of the productivity pyramid. Cows are primary consumers, one level up. It would be about 10 times more efficient to eat tofu, because 10 times the quantity of plants would be necessary to provide one pound of beef.
56.2 apply Decomposition is slowest in cold, wet temperatures on land and in anoxic conditions in the ocean (or in freshwater). Stagnant water often becomes anoxic as decomposers use up available oxygen that is not replenished by diffusion from the atmosphere. **56.3** apply Many possible answers, including (1) reducing carbon dioxide emissions by finding alternatives to fossil fuels should decrease the amount of carbon dioxide released into the atmosphere, (2) recycling programs decrease carbon dioxide emissions from manufacturing plants and power plants, and (3) reforestation can tie up carbon dioxide in wood.

YOU SHOULD BE ABLE TO . . .

✔ Test Your Knowledge

1. remember True **2.** understand d **3.** understand a **4.** remember c
5. remember c **6.** remember b

✔ Test Your Understanding

7. apply a **8.** analyze Sharks normally eat fish that eat the herbivorous fish that keep the growth of algae on corals in check. When sharks are absent, more of the intermediate fish survive, which decreases the number of herbivorous fish, causing algae to thrive on the coral reef. **9.** analyze Warmer, wetter climates speed decomposition; cool temperatures slow it. Decomposition regulates nutrient availability because it releases nutrients from detritus and allows them to reenter the food web. **10.** evaluate Yes, corals would be a good choice. Corals are important ecologically because they have a high NPP, providing energy to diverse animals in reef food webs. Corals are also vulnerable to climate change from the increase in water temperature, which can cause bleaching (loss of photosynthetic algae), as well as from the effects of ocean acidification, which slows the rate of coral growth. **11.** analyze Forest fires are an example of positive feedback; drier, hotter summers increase the risk of fires, and fires release CO_2 to the atmosphere, increasing global warming and the risk of fires. Forest growth is an example of negative feedback; as CO_2 increases in the atmosphere, the rate of forest growth in some areas may increase (due to increased availability of CO_2, increased temperature, and/or increased precipitation), reducing the concentration of CO_2 in the atmosphere. **12.** analyze The open ocean has almost no nutrient input from the land and has little upwelling to supply nutrients from the deep ocean. In contrast, intertidal and coastal areas receive large inputs of nutrients from rivers as well as from upwellings from ocean depths.

✔ Test Your Problem-Solving Skills

13. create One possibility is to add radioactive nitrogen to the water of the ponds in natural concentrations (control ponds) and elevated nitrogen concentrations (treatment ponds) and then follow this isotope through the primary producers, primary consumers, and secondary consumers in the system by measuring the amount of radioisotope in the tissues of organisms, and quantity of biomass, at each trophic level in the treatment versus control ponds **14.** evaluate c **15.** apply Without herbivores, there is no link in the nitrogen cycle between primary producers and secondary consumers. All of the plant nitrogen would go to the primary decomposers and back

into the soil. If decomposition is rapid enough, nitrogen would cycle quickly between primary producers, decomposers, the soil, and back to primary producers.
16. apply Atmospheric oxygen would increase due to extensive photosynthesis, but carbon dioxide levels would decrease because little decomposition was occurring. The temperature would drop because fewer greenhouse gases would be trapping heat reflected from the Earth's surface.

CHAPTER 57

IN-TEXT QUESTIONS AND EXERCISES

p. 1177 CYU create Do an all-taxon survey: organize experts and volunteers to collect, examine, and identify all species present. This could include direct sequencing studies (see Chapter 29) to document the bacteria and archaea present in different habitats on campus.
p. 1178 Fig. 57.7 apply Nearly all endangered species are threatened by more than one factor and therefore appear on the graph more than once, leading to species totals greater than 100 percent.
p. 1180 Fig. 57.9 analyze Because the treatments and study areas were assigned randomly, there is no bias involved in picking certain areas that are unusual in terms of their biomass or species diversity. They should represent a random sample of biomass and species diversity in fragments of various sizes versus intact forest.
p. 1183 Using the equation $S = (18.9)A^{0.15}$, plug in 5000 km^2 for A and solve for S such that $S = 18.9 \times 5000^{0.15} = 68$ species.
p. 1184 CYU (1) understand Fragmentation reduces habitat quality by creating edges that are susceptible to invasion and loss of species, due to changed abiotic conditions. Genetic problems occur inside fragments as species become inbred and/or lose genetic diversity via genetic drift. **(2)** create Habitat destruction and fragmentation in Borneo are obvious from the clear-cut areas. Overexploitation could be assessed by interviewing local citizens and law enforcement officers to the history of poaching orangutans for the pet trade or for other uses. Effects from invasive species could be assessed by published or local accounts of community structure over time. Pollution could be assessed by direct measurement of soil, water, air, and possibly tissue samples (in dead orangutans, feces, or blood samples). The effects of climate change could be assessed based on local weather records.
p. 1185 Fig. 57.12 analyze *Null hypothesis:* NPP is not affected by species richness or functional diversity of species. *Prediction:* NPP will be greater in plots with more species and more functional groups. *Prediction of null:* There will be no difference in NPP based on species richness or number of functional groups.
p. 1188 CYU create Establish several study plots in an area of Borneo where some of the forest has been converted to palm plantation but the adjacent forest (of similar elevation and topography) has rain forest still intact—half of the plots in the plantation and half in the rain forest. Monitor soil depth, soil nutrient levels, precipitation, and stream sediment levels regularly, and compare after several years.
p. 1191 Fig. 57.16 apply See **FIGURE A57.1**.

IF YOU UNDERSTAND . . .

57.1 apply At the genetic level, you could measure the genetic diversity of the orangutan population by comparing DNA sequences obtained from blood or feces samples. This information could help you to determine if a genetic bottleneck had occurred during the clear-cutting period. At the species level, you could monitor the increase in species richness of trees, birds, mammals, and other groups as the restoration progressed. At the ecosystem level, you could measure the ability of the growing forest to retain nutrients in the soil, prevent erosion, trap carbon, increase the prevalence of rain, and other functions **57.2** apply Small, geographically isolated populations of orangutans have low genetic diversity due to lack of gene flow and genetic drift and can suffer from inbreeding depression. Low genetic diversity makes them vulnerable to changes in their habitat, and small population size makes them susceptible to catastrophic events such as storms, disease outbreaks, or fires. **57.3** apply If many species are present, and each uses the available resources in a unique way, then a larger proportion of all available resources should be used—leading to higher biomass production. **57.4** evaluate Twelve percent of land area is already protected, but a mass extinction event is still going on. This is because 12 percent is not sufficient to save most species, and because many or most protected areas are not located in biodiversity hotspots.

YOU SHOULD BE ABLE TO . . .

✔ Test Your Knowledge

1. remember d **2.** remember a **3.** remember b **4.** remember d
5. understand c **6.** understand overexploitation

✔ Test Your Understanding

7. analyze b **8.** analyze By comparing the number of species estimated to reside in the park before and after the survey, biologists will have an idea of how much actual species diversity is being underestimated in other parts of the world—where research was at the level of Great Smoky Mountains National Park before the survey. The limitation is that it may be difficult to extrapolate from the data—no one knows if the situation at this park is typical of other habitats. **9.** evaluate No one correct answer; there are good arguments for both. In a case like the orangutans of Borneo, it is not possible to save the orangutans without saving their ecosystem. On the other hand, it is much easier to attract money and volunteers to help save a charismatic primate like an orangutan than to save a patch of forest. Both types of efforts are needed, their relative importance depends on the circumstance. **10.** evaluate In experiments with species native to North American grasslands, study plots that have more species produce more biomass, change less during a disturbance (drought), and recover from a disturbance faster than study plots with fewer species. **11.** understand Wildlife corridors facilitate the movement of individuals. Corridors allow areas to be recolonized if a species is lost and enable the introduction of new alleles that can counteract genetic drift and inbreeding in small, isolated populations. **12.** apply Ecosystem services are beneficial effects that ecosystems have on humans. It is rare to own or pay for these services, so no one has a vested interest in maintaining them. This is a primary reason that ecosystems are destroyed, even though the services they offer are valuable. The exception would be landowners who are paid *not* to log their forests, to maintain ecosystem functions like carbon storage.

✔ Test Your Problem-Solving Skills

13. apply d **14.** create Catalog existing biodiversity at the genetic, species, and ecosystem levels. Using these data, find areas that have the highest concentration of biodiversity at each level. Protect as many of these areas as possible, and connect them with corridors of habitat. **15.** evaluate There is no "correct" answer. One argument is that conservationists could lobby officials in Brazil and Indonesia to learn from the mistakes made in developed nations, and preserve enough forests to maintain biodiversity and keep ecosystem services intact—avoiding the expenditures that developed countries had to make to clean up pollution and restore ecosystems and endangered species. **16.** analyze Species with specialized food or habitat requirements, large size (and thus large requirements for land area), small population size, and low reproductive rate are vulnerable to extinction. A good example is the koala bear. Species that are particularly resistant to pressure from humans possess the opposite of many of these traits. A good example of a resistant species is the Norway rat.

BIG PICTURE Ecology

p. 1196 CYU (1) analyze Species interactions "are agents of" natural selection. **(2)** analyze As the human population climbs, more resources are needed to support the population. The harvesting of resources affects other species both directly (e.g. overfishing) and indirectly (e.g. climate change), resulting in decreased biodiversity. **(3)** analyze Both energy and nutrients flow through trophic levels in food webs. However, energy flows in one direction, ultimately dissipating as heat, whereas nutrients continue to cycle through ecosystems. **(4)** apply Human behavior "increases emissions of" CO_2 "causes" greenhouse effect "causes" climate change.

Bioskills

BIOSKILLS 1; p. B:2 CYU (1) apply 3.1 miles **(2)** apply 37°C
(3) apply Multiply your weight in pounds by 1/2.2 (0.45).

(4) apply 4 **(5)** apply 4 **(6)** Two significant figures. When you multiply, the answer can have no more significant figures than the least accurate measurement—in this case, 1.6.

BIOSKILLS 2; p. B:3 CYU (1) apply "different-yoked-together" **(2)** apply "sugary-loosened" **(3)** apply "study-of-form" **(4)** apply "three-bodies"

BIOSKILLS 3; p. B:6 CYU (1) apply about 18% **(2)** apply a dramatic drop (almost 10%) **(3)** understand No—the order of presentation in a bar chart does not matter (though it's convenient to arrange the bars in a way that reinforces the overall message). **(4)** apply 11 **(5)** apply 68 inches

BIOSKILLS 4; p. B:7 CYU apply Test 2, the estimate based on the larger sample—the more replicates or observations you have, the more precise your estimate of the average should be.

BIOSKILLS 5; p. B:8 CYU (1) apply $1/2 \times 1/2 \times 1/2 \times 1/2 = 1/16$ **(2)** apply $1/6 + 1/6 + 1/6 = 1/2$

BIOSKILLS 6; p. B:9 CYU (1) understand exponential **(2)** apply $\ln N_t = \ln N_0 + rt$

BIOSKILLS 7; p. B:9 Fig. B7.1. analyze See **FIGURE BA.1**.
p. B:10 Fig. B7.2. analyze See **FIGURE BA.2**.
p. B:11 Fig. B7.3. analyze Figure B7.3d is different.

FIGURE BA.1

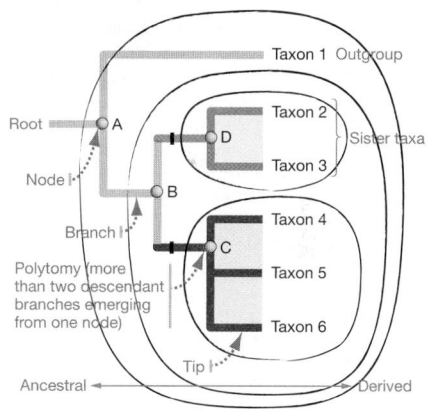

FIGURE BA.2

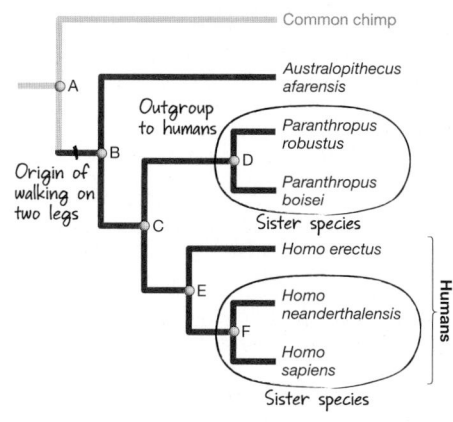

FIGURE A57.1

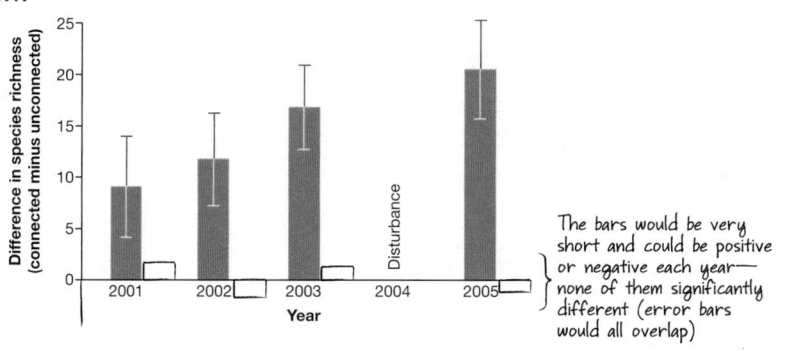

BIOSKILLS 8; p. B:12 Fig. B8.1. *apply* See **FIGURE BA.3**.

BIOSKILLS 9; p. B:13 Fig. B9.1 *understand* DNA and RNA are acids that tend to drop a proton in solution, giving them a negative charge.
p. B:15 CYU *analyze* The lane with no band comes from a sample where RNA X is not present. The same size RNA X is present in the next two lanes, but the faint band has very few copies while the dark band has many. In the fourth lane, the band is formed by a smaller version of RNA X, and relatively few copies are present.

BIOSKILLS 10; p. B:17 CYU (1) *understand* size and/or density. **(2)** *apply* Mitochondria, because they are larger in size compared with ribosomes.

BIOSKILLS 11; p. B:19 CYU (1) *analyze* No—it's just that no mitochondria happened to be present in this section

sliced through the cell. **(2)** *explain* Understanding a molecule's structure is often critical to understanding how it functions in cells.

BIOSKILLS 12; p. B:21 CYU (1) *analyze* It may not be clear that the results are relevant to noncancerous cells that are not growing in cell culture—that is, that the artificial conditions mimic natural conditions. **(2)** *analyze* It may not be clear that the results are relevant to individuals that developed normally, from an embryo—that is, that the artificial conditions mimic natural conditions.

BIOSKILLS 13; p. B:23 Fig. B13.1 *analyze* This is human body temperature—the natural habitat of *E. coli*.
p. B:24 CYU (1) *analyze* *Caenorhabditis elegans* would be a good possibility, because the cells that normally die have already been identified. You could find mutant individuals that lacked normal cell death; you could compare

the resulting embryos to normal embryos and be able to identify exactly which cells change as a result. **(2)** *analyze* Any of the multicellular organisms in the list would be a candidate, but *Dictyostelium discoideum* might be particularly interesting because cells stick to each other only during certain points in the life cycle. **(3)** *analyze* *Mus musculus*—as the only mammal in the list, it is the organism most likely to have a gene similar to the one you want to study.

BIOSKILLS 14; p. B:26 CYU *synthesize* Many examples are possible. See Figure 1.9 in Chapter 1, as an example of the format to use for your Research Box.

BIOSKILLS 15; p. B:27 CYU (1) *analyze* See **FIGURE BA.4**.
(2) *analyze* See **FIGURE BA.4**.

FIGURE BA.3

Molecular formula: CO_2

Structural formula: $O=C=O$

Ball-and-stick model:

Space-filling model:

FIGURE BA.4

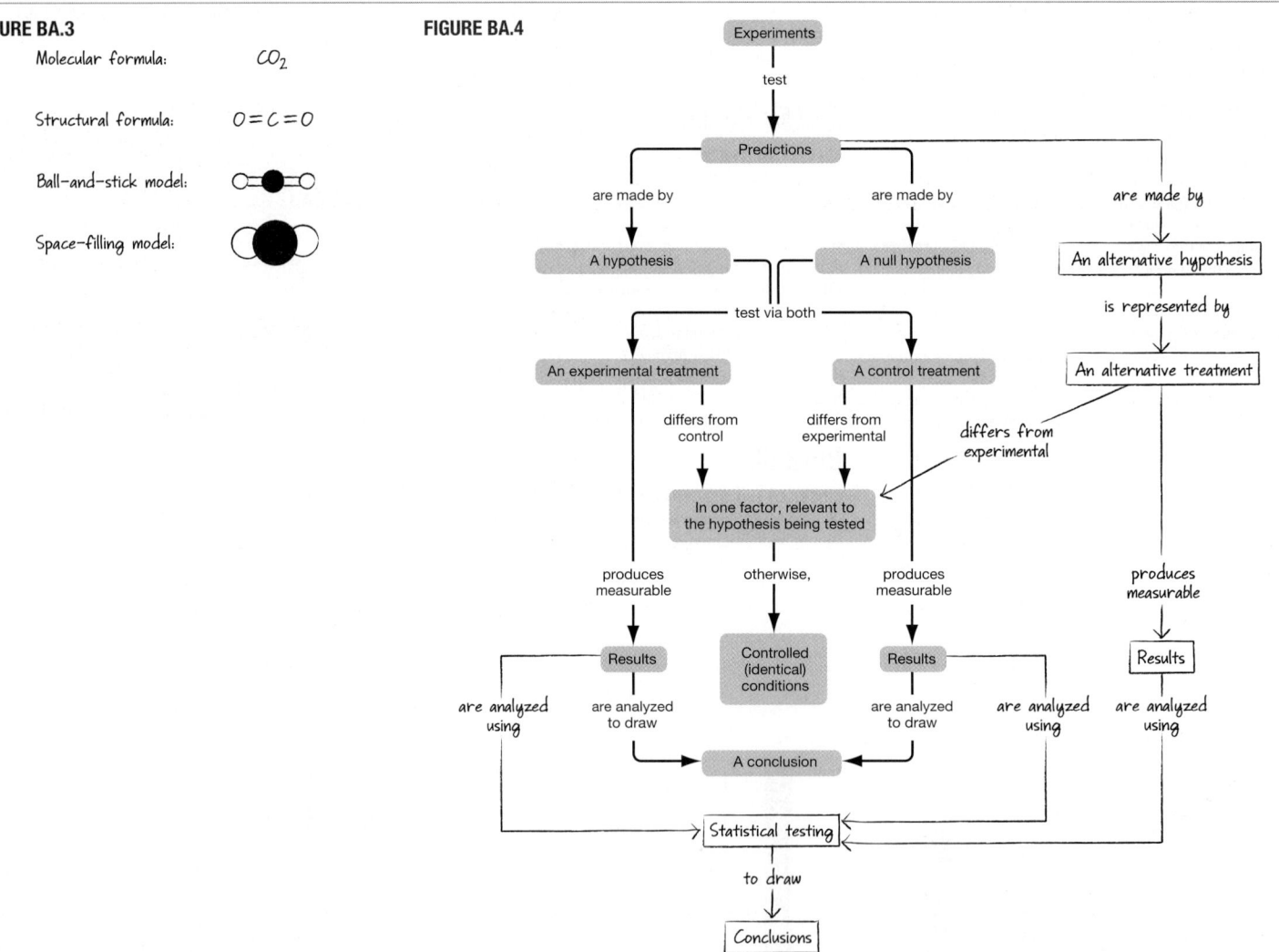

BIOSKILL 1 the metric system and significant figures

Scientists ask questions that can be answered by observing or measuring things—by collecting data. What units are used to make measurements? When measurements are reported, how can you tell how reliable the data are?

The Metric System

The metric system is the system of units of measure used in every country of the world but three (Liberia, Myanmar, and the United States). It is also the basis of the SI system—the International System of Units (abbreviated from the French, *Système international d'unités*)—used in scientific publications.

The popularity of the metric system is based on its consistency and ease of use. These attributes, in turn, arise from the system's use of the base 10. For example, each unit of length in the system is related to all other measures of length in the system by a multiple of 10. There are 10 millimeters in a centimeter; 100 centimeters in a meter; 1000 meters in a kilometer.

Measures of length in the English system, in contrast, do not relate to each other in a regular way. Inches are routinely divided into 16ths; there are 12 inches in a foot; 3 feet in a yard; 5280 feet (or 1760 yards) in a mile.

If you have grown up in the United States and are accustomed to using the English system, it is extremely important to begin developing a working familiarity with metric units and values. **Tables B1.1** and **B1.2** (see B:2) should help you get started with this process.

As an example, consider the following question: An American football field is 120 yards long, while rugby fields are 144 meters

TABLE B1.1 **Metric System Units and Conversions**

Measurement	Unit of Measurement and Abbreviation	Metric System Equivalent	Converting Metric Units to English Units
Length	kilometer (km)	1 km = 1000 m = 10^3 m	1 km = 0.62 mile
	meter (m)	1 m = 100 cm	1 m = 1.09 yards = 3.28 feet = 39.37 inches
	centimeter (cm)	1 cm = 0.01 m = 10^{-2} m	1 cm = 0.3937 inch
	millimeter (mm)	1 mm = 0.001 m = 10^{-3} m	1 mm = 0.039 inch
	micrometer (μm)	1 μm = 10^{-6} m = 10^{-3} mm	
	nanometer (nm)	1 nm = 10^{-9} m = 10^{-3} μm	
Area	hectare (ha)	1 ha = 10,000 m^2	1 ha = 2.47 acres
	square meter (m^2)	1 m^2 = 10,000 cm^2	1 m^2 = 1.196 square yards
	square centimeter (cm^2)	1 cm^2 = 100 mm^2 = 10^{-4} m^2	1 cm^2 = 0.155 square inch
Volume	liter (L)	1 L = 1000 mL	1 L = 1.06 quarts
	milliliter (mL)	1 mL = 1000 μL = 10^{-3} L	1 mL = 0.034 fluid ounce
	microliter (μL)	1 μL = 10^{-6} L	
Mass	kilogram (kg)	1 kg = 1000 g	1 kg = 2.20 pounds
	gram (g)	1 g = 1000 mg	1 g = 0.035 ounce
	milligram (mg)	1 mg = 1000 μg = 10^{-3} g	
	microgram (μg)	1 μg = 10^{-6} g	
Temperature	Kelvin (K)*		K = °C + 273.15
	degrees Celsius (°C)		°C = $\frac{5}{9}$ (°F − 32)
	degrees Fahrenheit (°F)		°F = $\frac{9}{5}$°C + 32

*Absolute zero is −273.15 °C = 0 K.

Prefix	Abbreviation	Definition
nano–	n	$0.000\,000\,001 = 10^{-9}$
micro–	μ	$0.000\,001 = 10^{-6}$
milli–	m	$0.001 = 10^{-3}$
centi–	c	$0.01 = 10^{-2}$
deci–	d	$0.1 = 10^{-1}$
–	–	$1 = 10^{0}$
kilo–	k	$1000 = 10^{3}$
mega–	M	$1\,000\,000 = 10^{6}$
giga–	G	$1\,000\,000\,000 = 10^{9}$

long. In yards, how much longer is a rugby field than an American football field? To solve this problem, first convert meters to yards: 144 m × 1.09 yards/m = 157 yards (note that the unit "m" cancels out). The difference in yards is thus: 157 – 120 = 37 yards. If you did these calculations on a calculator, you might have come up with 36.96 yards. Why has the number of yards been rounded off? The answer lies in significant figures. Let's take a closer look.

Significant Figures

Significant figures or "sig figs"—the number of digits used to report the measurement—are critical when reporting scientific data. The number of significant figures in a measurement, such as 3.524, is the number of digits that are known with some degree of confidence (3, 5, and 2) plus the last digit (4), which is an estimate or approximation. How do scientists know how many digits to report?

Rules for Working with Significant Figures

The rules for counting significant figures are summarized here:

- All nonzero numbers are always significant.
- Leading zeros are never significant; these zeros do nothing but set the decimal point.
- Embedded zeros are always significant.

- Trailing zeros are significant *only* if the decimal point is specified (Hint: Change the number to scientific notation. It is easier to see the "trailing" zeros.)

Table B1.3 provides examples of how to apply these rules. The bottom line is that significant figures indicate the precision of measurements.

Precision versus Accuracy

If biologists count the number of bird eggs in a nest, they report the data as an exact number—say, 3 eggs. But if the same biologists are measuring the diameter of the eggs, the numbers will be inexact. Just how inexact they are depends on the equipment used to make the measurements. For example, if you measure the width of your textbook with a ruler several times, you'll get essentially the same measurement again and again. Precision refers to how closely individual measurements agree with each other. So, you have determined the length with precision, but how do you know if the ruler was accurate to begin with?

Accuracy refers to how closely a measured value agrees with the correct value. You don't know the accuracy of a measuring device unless you calibrate it, by comparing it against a ruler that is known to be accurate. As the sensitivity of equipment used to

TABLE B1.3 Rules for Working with Significant Figures

Example	Number of Significant Figures	Scientific Notation	Rule
35,200	5	3.52×10^{4}	All nonzero numbers are always significant
0.00352	3	3.52×10^{-3}	Leading zeros are not significant
1.035	4	$1.035\ (\times 10^{0})$	Imbedded zeros are always significant
200	1	2×10^{2}	Trailing zeros are significant only if the decimal point is specified
200.0	4	2.000×10^{2}	Trailing zeros are significant only if the decimal point is specified

make a measurement increases, the number of significant figures increases. For example, if you used a kitchen scale to weigh out some sodium chloride, it might be accurate to 3 ± 1 g (1 significant figure); but an analytical balance in the lab might be accurate to 3.524 ± 0.001 g (4 significant figures).

In science, only the numbers that have significance—that are obtained from measurement—are reported. It is important to follow the "sig fig rules" when reporting a measurement, so that data do not appear to be more accurate than the equipment allows.

Combining Measurements

How do you deal with combining measurements with different degrees of accuracy and precision? The simple rule to follow is that the accuracy of the final answer can be no greater than the least accurate measurement. So, when you multiply or divide measurements, the answer can have no more significant figures than the least accurate measurement. When you add or subtract measurements, the answer can have no more decimal places than the least accurate measurement.

As an example, consider that you are adding the following measurements: 5.9522, 2.065, and 1.06. If you plug these numbers into your calculator, the answer your calculator will give you is 9.0772. However, this is incorrect—you must round your answer off to the nearest value, 9.08, to the least number of decimal places in your data.

It is important to nail down the concept of significant figures and to practice working with metric units and values. The Check Your Understanding questions in this BioSkill should help you get started with this process.

BIOSKILL 2   some common Latin and Greek roots used in biology

Greek or Latin Root	English Translation	Example Term
a, an	not	anaerobic
aero	air	aerobic
allo	other	allopatric
amphi	on both sides	amphipathic
anti	against	antibody
auto	self	autotroph
bi	two	bilateral symmetry
bio	life, living	bioinformatics
blast	bud, sprout	blastula
co	with	cofactor
cyto	cell	cytoplasm
di	two	diploid
ecto	outer	ectoparasite
endo	inner, within	endoparasite
epi	outer, upon	epidermis
exo	outside	exothermic
glyco	sugary	glycolysis
hetero	different	heterozygous
homo	alike	homozygous
hydro	water	hydrolysis
hyper	over, more than	hypertonic
hypo	under, less than	hypotonic
inter	between	interspecific
intra	within	intraspecific
iso	same	isotonic
logo, logy	study of	morphology
lyse, lysis	loosen, burst	glycolysis
macro	large	macromolecule

Greek or Latin Root	English Translation	Example Term
meta	change, turning point	metamorphosis
micro	small	microfilament
morph	form	morphology
oligo	few	oligopeptide
para	beside	parathyroid gland
photo	light	photosynthesis
poly	many	polymer
soma	body	somatic cells
sym, syn	together	symbiotic, synapsis
trans	across	translation
tri	three	trisomy
zygo	yoked together	zygote

check your understanding

If you understand BioSkill 2

✔ **You should be able to . . .**

Provide literal translations of the following terms:

1. heterozygote
2. glycolysis
3. morphology
4. trisomy

Graphs are the most common way to report data, for a simple reason. Compared to reading raw numerical values in a table or list, a graph makes it much easier to understand what the data mean.

Learning how to read and interpret graphs is one of the most basic skills you'll need to acquire as a biology student. As when learning piano or soccer or anything else, you need to understand a few key ideas to get started and then have a chance to practice—a lot—with some guidance and feedback.

Getting Started

To start reading a graph, you need to do three things: read the axes, figure out what the data points represent—that is, where they came from—and think about the overall message of the data. Let's consider each step in turn.

What Do the Axes Represent?

Graphs have two axes: one horizontal and one vertical. The horizontal axis of a graph is also called the x-axis or the abscissa. The vertical axis of a graph is also called the y-axis or the ordinate. Each axis represents a variable that takes on a range of values. These values are indicated by the ticks and labels on the axis. Note that each axis should *always* be clearly labeled with the unit or treatment it represents.

FIGURE B3.1 shows a scatterplot—a type of graph where continuous data are graphed on each axis. Continuous data can take an array of values over a range. In contrast, discrete data can take only a restricted set of values. If you were graphing the average height of men and women in your class, height is a continuous variable, but gender is a discrete variable.

For the example in this figure, the x-axis represents time in units of generations of maize; the y-axis represents the average percentage of the dry weight of a maize kernel that is protein.

To create a graph, researchers plot the independent variable on the x-axis and the dependent variable on the y-axis (Figure B3.1a). The terms independent and dependent are used because the values on the y-axis depend on the x-axis values. In our example, the researchers wanted to show how the protein content of maize kernels in a study population changed over time. Thus, the protein concentration plotted on the y-axis depended on the year (generation) plotted on the x-axis. The value on the y-axis always depends on the value on the x-axis, but not vice versa.

In many graphs in biology, the independent variable is either time or the various treatments used in an experiment. In these cases, the y-axis records how some quantity changes as a function of time or as the outcome of the treatments applied to the experimental cells or organisms.

(a) Read the axes—what is being plotted?

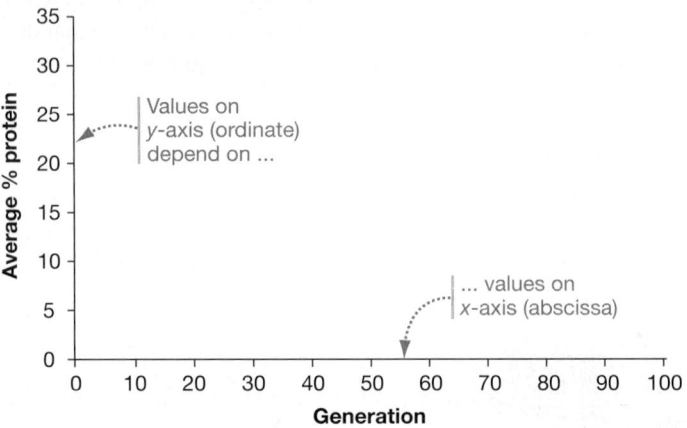

(b) Look at the bars or data points—what do they represent?

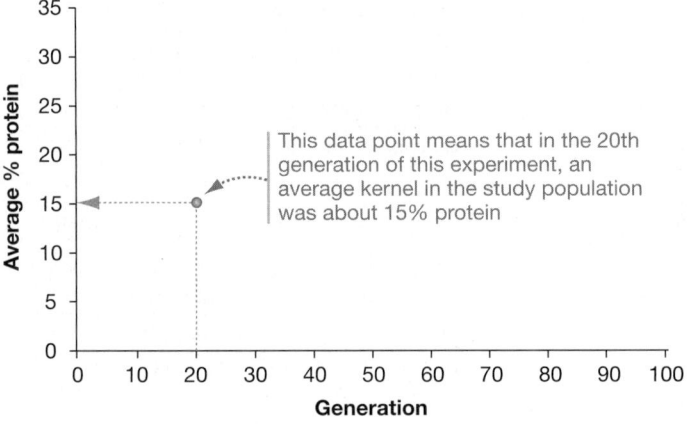

(c) What's the punchline?

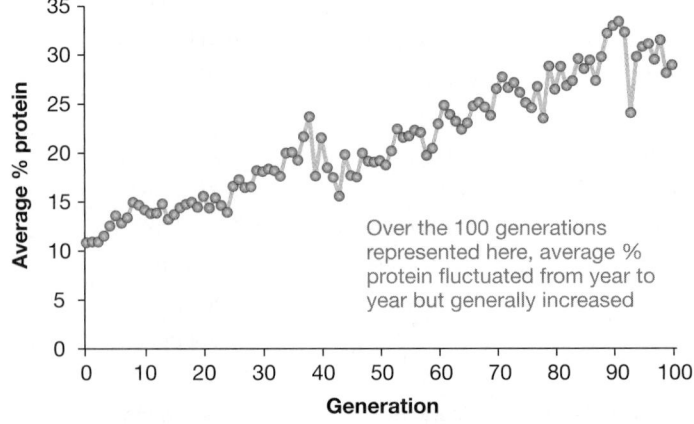

FIGURE B3.1 Scatterplots Are Used to Graph Continuous Data.

What Do the Data Points Represent?

Once you've read the axes, you need to figure out what each data point is. In our maize kernel example, the data point in Figure B3.1b represents the average percentage of protein found in a sample of kernels from a study population in a particular generation.

If it's difficult to figure out what the data points are, ask yourself where they came from—meaning, how the researchers got them. You can do this by understanding how the study was done and by understanding what is being plotted on each axis. The *y*-axis will tell you what they measured; the *x*-axis will usually tell you when they measured it or what group was measured. In some cases—for example, in a plot of average body size versus average brain size in primates—the *x*-axis will report a second variable that was measured.

In other cases, a data point on a graph may represent a relative or arbitrary unit of measurement. The data point shows the ratio of the amount of a substance, intensity, or other quantities, relative to a predetermined reference measurement. For example, the *y*-axis might show the percentage of relative activity of an enzyme—the rate of the enzyme-catalyzed reaction, scaled to the highest rate of activity observed (100 percent)—in experiments conducted under conditions that are identical except for one variable, such as pH or temperature (see Figure 8.14).

What Is the Overall Trend or Message?

Look at the data as a whole, and figure out what they mean. Figure B3.1c suggests an interpretation of the maize kernel example. If the graph shows how some quantity changes over time, ask yourself if that quantity is increasing, decreasing, fluctuating up and down, or staying the same. Then ask whether the pattern is the same over time or whether it changes over time.

When you're interpreting a graph, it's extremely important to limit your conclusions to the data presented. Don't extrapolate beyond the data, unless you are explicitly making a prediction based on the assumption that present trends will continue. For example, you can't say that the average percentage of protein content was increasing in the population before the experiment started, or that it will continue to increase in the future. You can say only what the data tell you.

Types of Graphs

Many of the graphs in this text are scatterplots like the one shown in Figure B3.1c, where individual data points are plotted. But you will also come across other types of graphs in this text.

Scatterplots, Lines, and Curves

Scatterplots sometimes have data points that are by themselves, but at other times data points will be connected by dot-to-dot lines to help make the overall trend clearer, as in Figure B3.1c, or may have a smooth line through them.

A *smooth line* through data points—sometimes straight, sometimes curved—is a mathematical "line of best fit." A line of best fit represents a mathematical function that summarizes the relationship between the *x* and *y* variables. It is "best" in the sense of fitting the data points most precisely. The line may pass through some of the points, none of the points, or all of the points.

Curved lines often take on characteristic shapes depending on the relationships between the *x* and *y* variable. For example, a bell-shaped curve depicts a normal distribution in which most data points are clumped near the middle, while a sigmoid or S-shaped curve exhibits small changes at first, which then accelerate and approach maximal value over time. Data from studies on population growth (see Chapter 54), enzyme kinetics (see Chapter 8), and oxygen–hemoglobin dissociation (see Chapter 45) typically fall on a curved line.

Bar Charts, Histograms, and Box-and-Whisker Plots

Scatterplots, or line-of-best-fit graphs, are the most appropriate type of graph when the data have a continuous range of values and you want to show individual data points. But other types of graphs are used to represent different types of distributions:

- *Bar charts* plot data that have discrete or categorical values instead of a continuous range of values. In many cases the bars might represent different treatment groups in an experiment, as in **FIGURE B3.2a** (see B:6). In this graph, the height of the bar indicates the average value. Statistical tests can be used to determine whether a difference between treatment groups is significant (see **BIOSKILLS 4**).

- *Histograms* illustrate frequency data and can be plotted as numbers or percentages. **FIGURE B3.2b** shows an example where height is plotted on the *x*-axis, and the number of students in a population is plotted on the *y*-axis. Each rectangle indicates the number of individuals in each interval of height, which reflects the relative frequency, in this population, of people whose heights are in that interval. The measurements could also be recalculated so that the *y*-axis would report the proportion of people in each interval. Then the sum of all the bars would equal 100 percent. Note that if you were to draw a smooth curve connecting the top of the bars on this histogram, the smooth curve would represent the shape of a bell.

- *Box-and-whisker plots* allow you to easily see where most of the data fall. Each box indicates where half of the data numbers are. The whiskers indicate the lower extreme and the upper extreme of the data. The vertical line inside each box indicates the median—meaning that half of the data are above this value and half are below (see Figure 1.9 for an example).

When you are looking at a bar chart that plots values from different treatments in an experiment, ask yourself if these values are the same or different. If the bar chart reports averages over discrete ranges of values, ask what trend is implied—as you would for a scatterplot.

(a) Bar chart

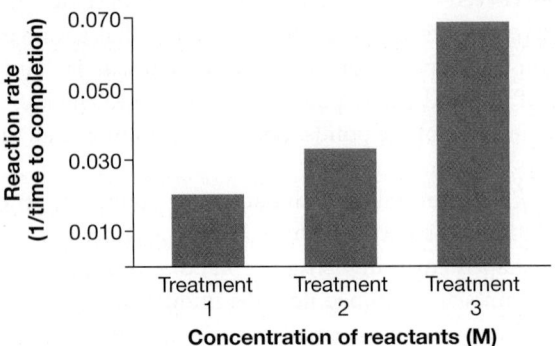

(b) Histogram

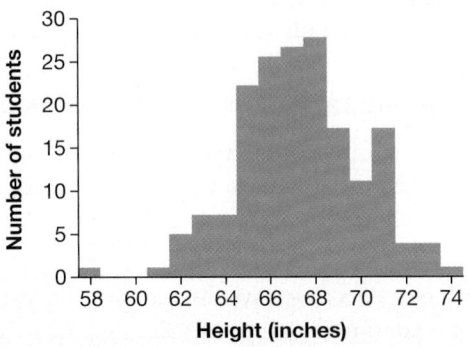

FIGURE B3.2 Bar Charts and Histograms. (a) Bar charts are used to graph data that are discontinuous or categorical. **(b)** Histograms show the distribution of frequencies or values in a population.

When you are looking at a histogram, ask whether there is a "hump" in the data—indicating a group of values that are more frequent than others. Is the hump in the center of the distribution of values, toward the left, or toward the right? If so, what does it mean?

Similarly, when you are looking at a box-and-whisker plot, ask yourself what information the graph gives you. What is the

range of values for the data? Where are half the data points? Below what value is three quarters of the data?

Getting Practice

Working with this text will give you lots of practice with reading graphs—they appear in almost every chapter. In many cases we've inserted an arrow to represent your instructor's hand at the whiteboard, with a label that suggests an interpretation or draws your attention to an important point on the graph. In other cases, you should be able to figure out what the data mean on your own or with the help of other students or your instructor.

check your understanding

If you understand BioSkill 3

✔ **You should be able to . . .**

1. **QUANTITATIVE** Determine the total change in average percentage of protein in maize kernels, from the start of the experiment until the end.

2. **QUANTITATIVE** Determine the trend in average percentage of protein in maize kernels between generation 37 and generation 42.

3. Explain whether the conclusions from the bar chart in Figure B3.2a would be different if the data and label for Treatment 3 were put on the far left and the data and label for Treatment 1 on the far right.

4. **QUANTITATIVE** Determine approximately how many students in this class are 70 inches tall, by using Figure B3.2b.

5. **QUANTITATIVE** Determine the most common height in the class graphed in Figure B3.2b.

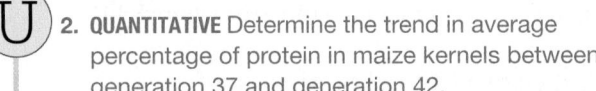

BIOSKILL 4 using statistical tests and interpreting standard error bars

When biologists do an experiment, they collect data on individuals in a treatment group and a control group, or several such comparison groups. Then they want to know whether the individuals in the two (or more) groups are different. For example, in one experiment student researchers measured how fast a product formed when they set up a reaction with three different concentrations of reactants (introduced in Chapter 8). Each treatment—meaning, each combination of reactant concentrations—was replicated many times.

FIGURE B4.1 graphs the average reaction rate for each of the three treatments in the experiment. Note that Treatments 1, 2, and 3 represent increasing concentrations of reactants. The thin

"I-beams" on each bar indicate the standard error of each average. The standard error is a quantity that indicates the uncertainty in the calculation of an average.

For example, if two trials with the same concentration of reactants had a reaction rate of 0.075 and two trials had a reaction rate of 0.025, then the average reaction rate would be 0.050. In this case, the standard error would be large. But if two trials had a reaction rate of 0.051 and two had a reaction rate of 0.049, the average would still be 0.050, but the standard error would be small.

In effect, the standard error quantifies how confident you are that the average you've calculated is the average you'd observe if

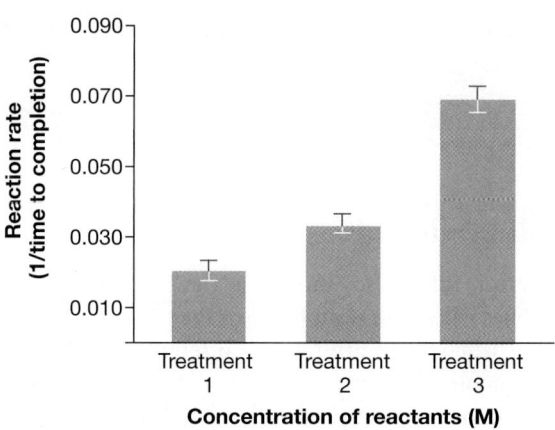

FIGURE B4.1 **Standard Error Bars Indicate the Uncertainty in an Average.**

you did the experiment under the same conditions an extremely large number of times. It is a measure of precision (see **BIOSKILLS 1**).

Once they had calculated these averages and standard errors, the students wanted to answer a question: Does reaction rate increase when reactant concentration increases?

After looking at the data, you might conclude that the answer is yes. But how could you come to a conclusion like this objectively, instead of subjectively?

The answer is to use a statistical test. This can be thought of as a three-step process.

1. Specify the null hypothesis, which is that reactant concentration has no effect on reaction rate.

2. Calculate a test statistic, which is a number that characterizes the size of the difference among the treatments. In this case, the test statistic compares the actual differences in reaction rates among treatments to the difference predicted by the null hypothesis. The null hypothesis predicts that there should be no difference.

3. The third step is to determine the probability of getting a test statistic at least as large as the one calculated just by chance. The answer comes from a reference distribution—a mathematical function that specifies the probability of getting various values of the test statistic if the null hypothesis is correct. (If you take a statistics course, you'll learn which test

statistics and reference distributions are relevant to different types of data.)

You are very likely to see small differences among treatment groups just by chance—even if no differences actually exist. If you flipped a coin 10 times, for example, you are unlikely to get exactly five heads and five tails, even if the coin is fair. A reference distribution tells you how likely you are to get each of the possible outcomes of the 10 flips if the coin is fair, just by chance.

In this case, the reference distribution indicated that if the null hypothesis of no actual difference in reaction rates is correct, you would see differences at least as large as those observed only 0.01 percent of the time just by chance. By convention, biologists consider a difference among treatment groups to be statistically significant if there is less than a 5 percent probability of observing it just by chance. Based on this convention, the student researchers were able to claim that the null hypothesis is not correct for reactant concentration. According to their data, the reaction they studied really does happen faster when reactant concentration increases.

You'll likely be doing actual statistical tests early in your undergraduate career. To use this text, though, you only need to know what statistical testing does. And you should take care to inspect the standard error bars on graphs in this book. As a *very* rough rule of thumb, averages often turn out to be significantly different, according to an appropriate statistical test, if there is no overlap between two times the standard errors.

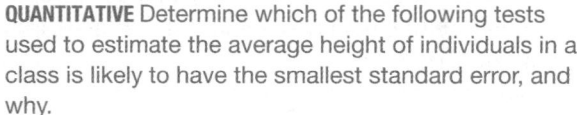

If you understand BioSkill 4

✔ **You should be able to . . .**

QUANTITATIVE Determine which of the following tests used to estimate the average height of individuals in a class is likely to have the smallest standard error, and why.

- Measuring the height of two individuals chosen at random to estimate the average.

- Measuring the height of every student who showed up for class on a particular day to estimate the average.

In several cases in this text, you'll need to combine probabilities from different events in order to solve a problem. One of the most common applications is in genetics problems. For example, Punnett squares work because they are based on two fundamental rules of probability. Each rule pertains to a distinct situation.

The Both-And Rule

The both-and rule—also known as the product rule or multiplication rule—applies when you want to know the probability that two or more independent events occur together. Let's use the rolling of two dice as an example. What is the probability of rolling two sixes? These two events are independent, because the probability of rolling a six on one die has no effect on the probability of rolling a six on the other die. (In the same way, the probability of getting a gamete with allele R from one parent has no effect on the probability of getting a gamete with allele R from the other parent. Gametes fuse randomly.)

The probability of rolling a six on the first die is 1/6. The probability of rolling a six on the second die is also 1/6. The probability of rolling a six on *both* dice, then, is $1/6 \times 1/6 = 1/36$. In other words, if you rolled two dice 36 times, on average you would expect to roll two sixes once.

In the case of a cross between two parents heterozygous at the R gene, the probability of getting allele R from the father is 1/2 and the probability of getting R from the mother is 1/2. Thus, the probability of getting both alleles and creating an offspring with genotype RR is $1/2 \times 1/2 = 1/4$.

The Either-Or Rule

The either-or rule—also known as the sum rule or addition rule—applies when you want to know the probability of an event happening when there are several different ways for the same event or outcome to occur. In this case, the probability that the event will occur is the sum of the probabilities of each way that it can occur.

For example, suppose you wanted to know the probability of rolling either a one or a six when you toss a die. The probability of drawing each is 1/6, so the probability of getting one or the other is $1/6 + 1/6 = 1/3$. If you rolled a die three times, on average you'd expect to get a one or a six once.

In the case of a cross between two parents heterozygous at the R gene, the probability of getting an R allele from the father and an r allele from the mother is $1/2 \times 1/2 = 1/4$. Similarly, the probability of getting an r allele from the father and an R allele from the mother is $1/2 \times 1/2 = 1/4$. Thus, the combined probability of getting the Rr genotype in either of the two ways is $1/4 + 1/4 = 1/2$.

check your understanding

If you understand BioSkill 5

✔ **You should be able to . . .**

1. **QUANTITATIVE** Calculate the probability of getting four "tails" if four students each toss a coin.

2. **QUANTITATIVE** Calculate the probability of getting a two, a three, or a six after a single roll of a die.

You have probably been introduced to logarithms and logarithmic notation in algebra courses, and you will encounter logarithms at several points in this course. Logarithms are a way of working with powers—meaning, numbers that are multiplied by themselves one or more times.

Scientists use exponential notation to represent powers. For example,

$$a^x = y$$

means that if you multiply a by itself x times, you get y. In exponential notation, a is called the base and x is called the exponent. The entire expression is called an exponential function.

What if you know y and a, and you want to know x? This is where logarithms come in. You can solve for exponents using logarithms. For example,

$$x = \log_a y$$

This equation reads, x is equal to the logarithm of y to the base a. Logarithms are a way of working with exponential functions. They are important because so many processes in biology (and chemistry and physics, for that matter) are exponential. To understand what's going on, you have to describe the process with an exponential function and then use logarithms to work with that function.

Although a base can be any number, most scientists use just two bases when they employ logarithmic notation: 10 and e (sometimes called Euler's number after Swiss mathematician Leonhard Euler). What is e? It is a rate of exponential growth shared by many natural processes, where e is the limit of $(1 + \frac{1}{n})^n$ (as n tends to infinity). Mathematicians have shown that the base e is an irrational number (like π) that is approximately equal to 2.718. Like 10, e is just a number; $10^0 = 1$ and, likewise, $e^0 = 1$. But both 10 and e have qualities that make them convenient to use in biology (as well as chemistry and physics).

Logarithms to the base 10 are so common that they are usually symbolized in the form log y instead of $\log_{10} y$. A logarithm to the base e is called a natural logarithm and is symbolized ln (pronounced *EL-EN*) instead of log. You write "the natural logarithm of y" as ln y.

Most scientific calculators have keys that allow you to solve problems involving base 10 and base e. For example, if you know y, they'll tell you what log y or ln y are—meaning that they'll solve for x in our first example equation. They'll also allow you to find a number when you know its logarithm to base 10 or base

e. Stated another way, they'll tell you what y is if you know x, and y is equal to e^x or 10^x. This is called taking an antilog. In most cases, you'll use the inverse or second function button on your calculator to find an antilog (above the log or ln key).

To get some practice with your calculator, consider this equation:

$$10^2 = 100$$

If you enter 100 in your calculator and then press the log key, the screen should say 2. The logarithm tells you what the exponent is. Now press the antilog key while 2 is on the screen. The calculator screen should return to 100. The antilog solves the exponential function, given the base and the exponent.

If your background in algebra isn't strong, you'll want to get more practice working with logarithms—you'll see them frequently during your undergraduate career. Remember that once you understand the basic notation, there's nothing mysterious about logarithms. They are simply a way of working with exponential functions, which describe what happens when something is multiplied by itself a number of times—like cells that divide and then divide again and then again.

Using logarithms will also come up when you are studying something that can have a large range of values, like the concentration of hydrogen ions in a solution or the intensity of sound that the human ear can detect. In cases like this, it's convenient to express the numbers involved as exponents. Using exponents makes a large range of numbers smaller and more manageable. For example, instead of saying that hydrogen ion concentration in a solution can range from 1 to 10^{-14}, the pH scale allows you to simply say that it ranges from 1 to 14. Instead of giving the actual value, you're expressing it as an exponent. It just simplifies things.

check your understanding

If you understand BioSkill 6

✔ **You should be able to . . .**

Use the equation $N_t = N_0 e^{rt}$ (see Chapter 54).

1. Explain what type of function this equation describes.

2. **QUANTITATIVE** Determine how you would write the equation, after taking the natural logarithm of both sides.

Phylogenetic trees show the evolutionary relationships among species, just as a genealogy shows the relationships among people in your family. They are unusual diagrams, however, and it can take practice to interpret them correctly.

To understand how evolutionary trees work, consider **FIGURE B7.1**. Notice that a phylogenetic tree consists of a root (the most ancestral branch in the tree), branches, nodes, and tips.

- Branches represent populations through time. In this text, branches are drawn as horizontal lines. In most cases the length of the branch is arbitrary and has no meaning, but in some cases branch lengths are proportional to time or the extent of genetic difference among populations (if so, there will be a scale at the bottom of the tree). The vertical lines on the tree represent splitting events, where one group broke into two independent groups. Their length is arbitrary—chosen simply to make the tree more readable.

- Nodes (also called forks) occur where an ancestral group splits into two or more descendant groups (see point A in Figure B7.1). Thus, each node represents the most recent common ancestor of the two or more descendant populations that emerge from it. If more than two descendant groups emerge from a node, the node is called a polytomy (see node C). A polytomy usually means that the populations split from one another so quickly that it is not possible to tell which split off earlier or later.

- Tips (also called terminal nodes) are the tree's endpoints, which represent groups living today or a dead end—a branch

ending in extinction. The names at the tips can represent species or larger groups such as mammals or conifers.

Recall that a taxon (plural: taxa) is any named group of organisms (see Chapter 1). A taxon could be a single species, such as *Homo sapiens*, or a large group of species, such as Primates. Tips connected by a single node on a tree are called sister taxa.

The phylogenetic trees used in this text are all rooted. This means that the first, or most basal, node on the tree—the one on the far left in this book—is the most ancient. To determine where the root on a tree occurs, biologists include one or more outgroup species when they are collecting data to estimate a particular phylogeny. An outgroup is a taxonomic group that is known to have diverged before the rest of the taxa in the study. Outgroups are used to establish whether a trait is ancestral or derived. An ancestral trait is a characteristic that existed in an ancestor; a derived trait is a characteristic that is a modified form of the ancestral trait, found in a descendant.

In Figure B7.1, Taxon 1 is an outgroup to the monophyletic group consisting of taxa 2–6. A monophyletic group consists of an ancestral species and all of its descendants. The root of a tree is placed between the outgroup and the monophyletic group being studied. This position in Figure B7.1 is node A. Note that black hash marks are used to indicate a derived trait that is shared among the red branches, and another derived trait that is shared among the orange branches.

Understanding monophyletic groups is fundamental to reading and estimating phylogenetic trees. Monophyletic groups may also be called lineages or clades and can be identified using the "one-snip test": If you cut any branch on a phylogenetic tree, all of the branches and tips that fall off represent a monophyletic group. Using the one-snip test, you should be able to convince yourself that the monophyletic groups on a tree are nested. In Figure B7.1, for example, the monophyletic group comprising node A and taxa 1–6 contains a monophyletic group consisting of node B and taxa 2–6, which includes the monophyletic group represented by node C and taxa 4–6.

To put all these new terms and concepts to work, consider the phylogenetic tree in **FIGURE B7.2**, which shows the relationships between common chimpanzees and six human and humanlike species that lived over the past 5–6 million years. Chimps functioned as an outgroup in the analysis that led to this tree, so the root was placed at node A. The branches marked in red identify a monophyletic group called the hominins.

To practice how to read a tree, put your finger at the tree's root, at the far left, and work your way to the right. At node A, the ancestral population split into two descendant populations. One of these populations eventually evolved into today's chimps; the other gave rise to the six species of hominins pictured. Now

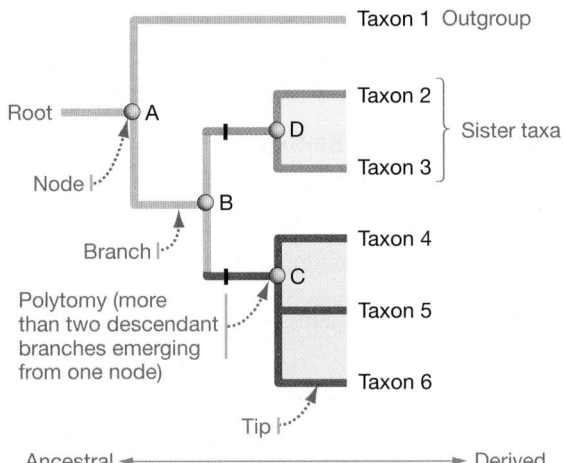

FIGURE B7.1 Phylogenetic Trees Have Roots, Branches, Nodes, and Tips.

✔**EXERCISE** Circle all four monophyletic groups present.

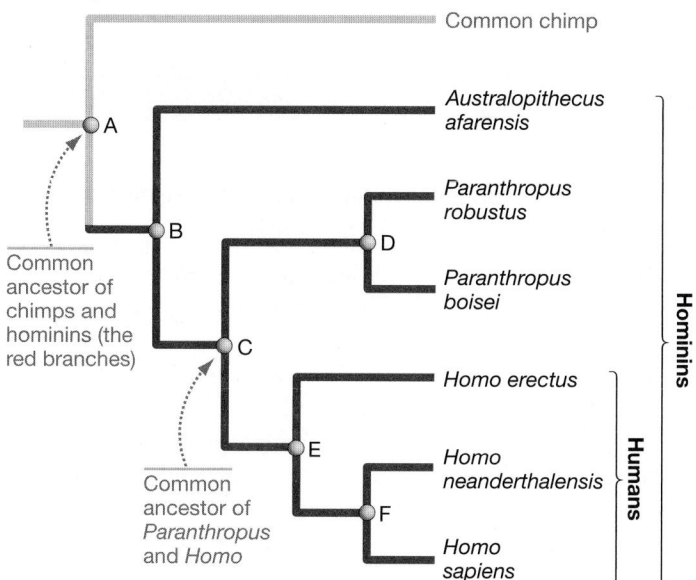

FIGURE B7.2 An Example of a Phylogenetic Tree. A phylogenetic tree showing the relationships of species in the monophyletic group called hominins.

✔**EXERCISE** All of the hominins walked on two legs—unlike chimps and all of the other primates. Add a mark on the phylogeny to show where upright posture evolved, and label it "origin of walking on two legs." Circle and label a pair of sister species. Label an outgroup to the monophyletic group called humans (species in the genus *Homo*).

continue moving your finger toward the tips of the tree until you hit node C. It should make sense to you that at this splitting event, one descendant population eventually gave rise to two *Paranthropus* species, while the other became the ancestor of humans—species in the genus *Homo*. As you study Figure B7.2, consider these two important points:

1. There are many equivalent ways of drawing this tree. For example, this version shows *Homo sapiens* on the bottom. But the tree would be identical if the two branches emerging from node E were rotated 180°, so that the species appeared in the order *Homo sapiens*, *Homo neanderthalensis*, *Homo erectus*. Trees are read from root to tips, not from top to bottom or bottom to top.

2. No species on any tree is any higher or lower than any other. Chimps and *Homo sapiens* have been evolving exactly the same amount of time since their divergence from a common ancestor—neither species is higher or lower than the other. It is legitimate to say that more ancient groups like *Australopithecus afarensis* have traits that are ancestral or more basal—meaning, that appeared earlier in evolution—compared to traits that appear in *Homo sapiens*, which are referred to as more derived.

FIGURE B7.3 presents a chance to test your tree-reading ability. Five of the six trees shown in this diagram are identical in terms of the evolutionary relationships they represent. One differs. The key to understanding the difference is to recognize that the ordering of tips does not matter in a tree—only the ordering of nodes (branch points) matters. You can think of a tree as being like a mobile: The tips can rotate without changing the underlying relationships.

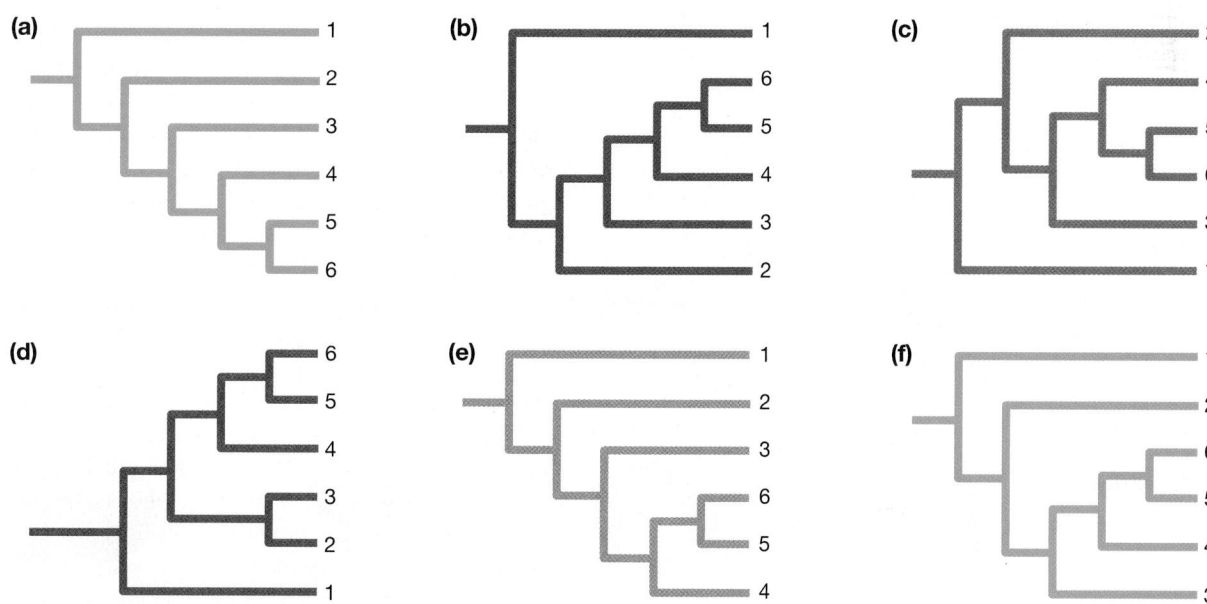

FIGURE B7.3 Alternative Ways of Drawing the Same Tree.

✔**QUESTION** Five of these six trees describe exactly the same relationships among taxa 1 through 6. Identify the tree that is different from the other five.

If you haven't had much chemistry yet, learning basic biological chemistry can be a challenge. One stumbling block is simply being able to read chemical structures efficiently and understand what they mean. This skill will come much easier once you have a little notation under your belt and you understand some basic symbols.

Atoms are the basic building blocks of everything in the universe, just as cells are the basic building blocks of your body. Every atom has a one- or two-letter symbol. **Table B8.1** shows the symbols for most of the atoms you'll encounter in this book. You should memorize these. The table also offers details on how the atoms form bonds as well as how they are represented in some visual models.

When atoms attach to each other by covalent bonding, a molecule forms. Biologists have a couple of different ways of representing molecules—you'll see each of these in the book and in class.

- Molecular formulas like those in **FIGURE B8.1a** simply list the atoms present in a molecule. Subscripts indicate how many of each atom are present. If the formula has no subscript, only one atom of that type is present. A methane (natural gas) molecule, for example, can be written as CH_4. It consists of one carbon atom and four hydrogen atoms.

TABLE B8.1 Some Attributes of Atoms Found in Organisms

Atom	Symbol	Number of Bonds It Can Form	Standard Color Code*
Hydrogen	H	1	white
Carbon	C	4	black
Nitrogen	N	3	blue
Oxygen	O	2	red
Sodium	Na	1	—
Magnesium	Mg	2	—
Phosphorus	P	5	orange or purple
Sulfur	S	2	yellow
Chlorine	Cl	1	—
Potassium	K	1	—
Calcium	Ca	2	—

*In ball-and-stick or space-filling models.

- Structural formulas like those in **FIGURE B8.1b** show which atoms in the molecule are bonded to each other. Each bond is indicated by a dash. The structural formula for methane in-

	Methane	Ammonia	Water	Oxygen
(a) Molecular formulas:	CH_4	NH_3	H_2O	O_2

(b) Structural formulas:

(c) Ball-and-stick models:

(d) Space-filling models:

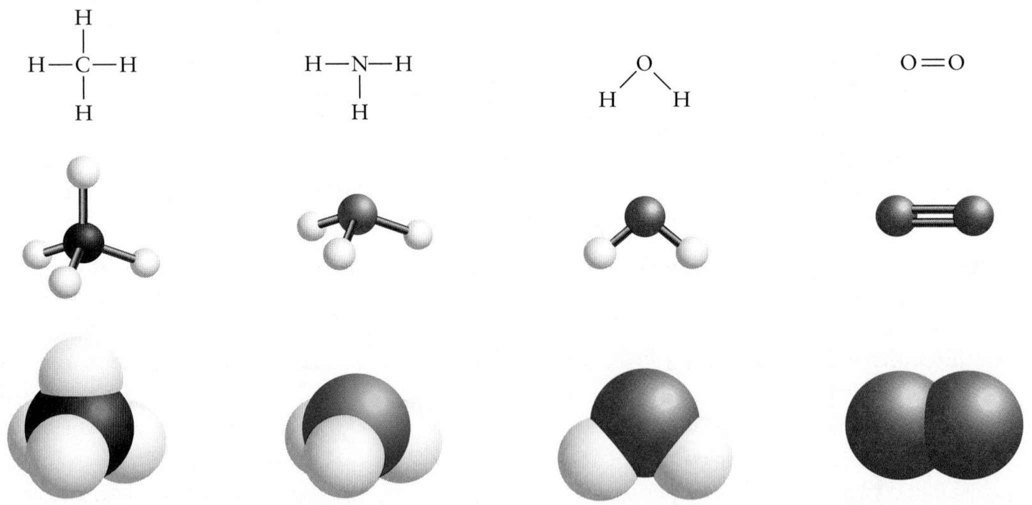

FIGURE B8.1 Molecules Can Be Represented in Several Different Ways.

✓**EXERCISE** Carbon dioxide consists of a carbon atom that forms a double bond with each of two oxygen atoms, for a total of four bonds. It is a linear molecule. Write carbon dioxide's molecular formula and then draw its structural formula, a ball-and-stick model, and a space-filling model.

dicates that each of the four hydrogen atoms forms one covalent bond with carbon, and that carbon makes a total of four covalent bonds. Single covalent bonds are symbolized by a single dash; double bonds are indicated by two dashes.

Even simple molecules have distinctive shapes, because different atoms make covalent bonds at different angles. Ball-and-stick and space-filling models show the geometry of the bonds accurately.

- In a ball-and-stick model, a stick is used to represent each covalent bond (see **FIGURE B8.1c**).
- In space-filling models, the atoms are simply stuck onto each other in their proper places (see **FIGURE B8.1d**).

To learn more about a molecule when you look at a chemical structure, ask yourself three questions:

1. *Is the molecule polar—meaning that some parts are more negatively or positively charged than others?* Molecules that contain nitrogen or oxygen atoms are often polar, because these atoms have such high electronegativity (see Chapter 2). This trait is important because polar molecules dissolve in water.

2. *Does the structural formula show atoms that might participate in chemical reactions?* For example, are there charged atoms or amino or carboxyl ($-COOH$) groups that might act as a base or an acid?

3. *In ball-and-stick and especially space-filling models of large molecules, are there interesting aspects of overall shape?* For example, is there a groove where a protein might bind to DNA, or a cleft where a substrate might undergo a reaction in an enzyme?

BIOSKILL 9 separating and visualizing molecules

To study a molecule, you have to be able to isolate it. Isolating a molecule is a two-step process: the molecule has to be separated from other molecules in a mixture and then physically picked out or located in a purified form. **BIOSKILLS 9** focuses on the techniques that biologists use to separate nucleic acids and proteins and then find the particular one they are interested in.

Using Electrophoresis to Separate Molecules

In molecular biology, the standard technique for separating proteins and nucleic acids is called gel electrophoresis or, simply, electrophoresis (literally, "electricity-moving"). You may be using electrophoresis in a lab for this course, and you will certainly be analyzing data derived from electrophoresis in this text.

The principle behind electrophoresis is simple. Proteins (when denatured and coated with a special detergent) and nucleic acids carry a charge. As a result, these molecules move when placed in an electric field. Negatively charged molecules move toward the positive electrode; positively charged molecules move toward the negative electrode.

To separate a mixture of macromolecules so that each one can be isolated and analyzed, researchers place the sample in a gelatinous substance. More specifically, the sample is placed in a "well"—a slot in a sheet or slab of the gelatinous substance. The "gel" itself consists of long molecules that form a matrix of fibers. The gelatinous matrix has pores that act like a sieve through which the molecules can pass.

When an electrical field is applied across the gel, the molecules in the well move through the gel toward an electrode. Molecules that are smaller or more highly charged for their size move faster than do larger or less highly charged molecules. As they move, then, the molecules separate by size and by charge. Small and highly charged molecules end up at the bottom of the gel; large, less-charged molecules remain near the top.

An Example "Run"

FIGURE B9.1 (see B:14) shows the electrophoresis setup used in an experiment investigating how RNA molecules polymerize. In this case, the investigators wanted to document how long RNA molecules became over time, when ribonucleoside triphosphates were present in a particular type of solution.

Step 1 shows how they loaded samples of macromolecules, taken on different days during the experiment, into wells at the top of the gel slab. This is a general observation: Each well holds a different sample. In this and many other cases, the researchers also filled a well with a sample containing fragments of known size, called a size standard or "ladder."

In step 2, the researchers immersed the gel in a solution that conducts electricity and applied a voltage across the gel. The molecules in each well started to run down the gel, forming a lane. After several hours of allowing the molecules to move, the researchers removed the electric field (step 3). By then, molecules of different size and charge had separated from one another. In this case, small RNA molecules had reached the bottom of the gel. Above them were larger RNA molecules, which had run more slowly.

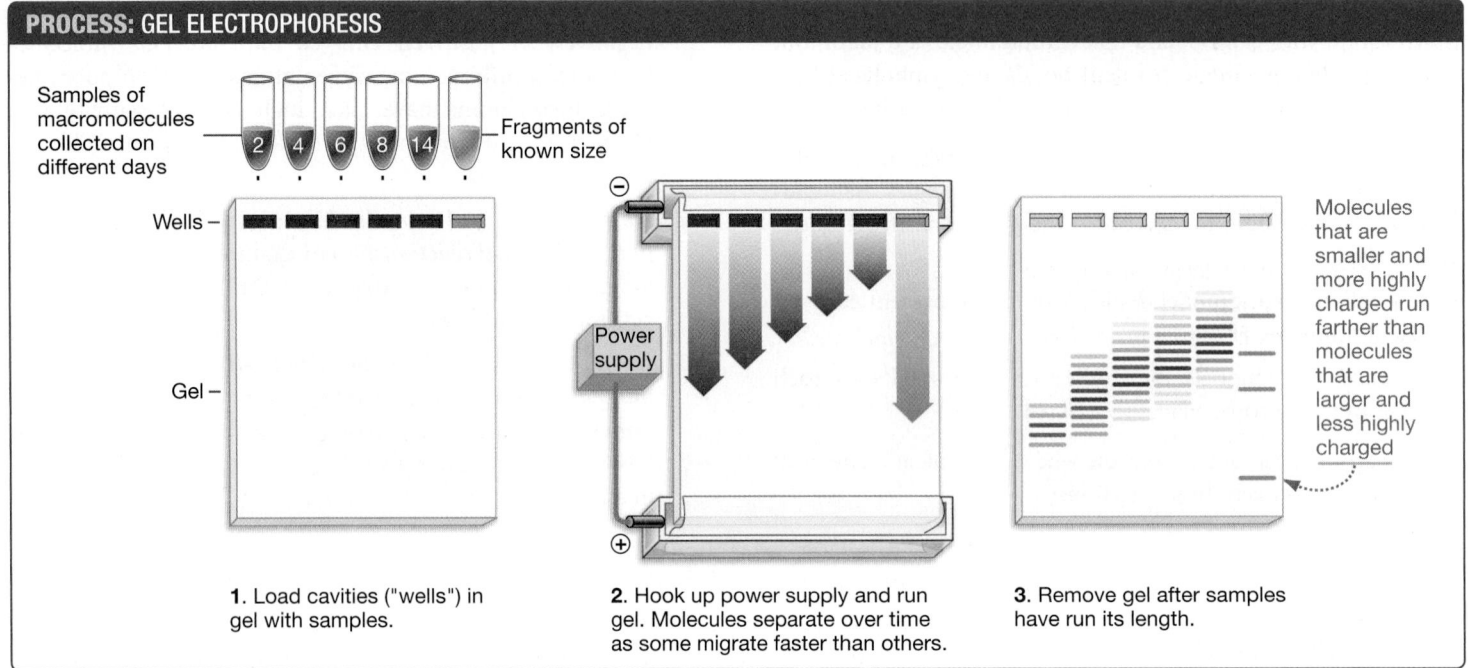

Samples of macromolecules collected on different days

Fragments of known size

Wells —

Gel —

Power supply

Molecules that are smaller and more highly charged run farther than molecules that are larger and less highly charged

1. Load cavities ("wells") in gel with samples.

2. Hook up power supply and run gel. Molecules separate over time as some migrate faster than others.

3. Remove gel after samples have run its length.

FIGURE B9.1 Macromolecules Can Be Separated via Gel Electrophoresis.

✔**QUESTION** DNA and RNA run toward the positive electrode. Why are these molecules negatively charged?

Why Do Separated Molecules Form Bands?

When researchers visualize a particular molecule on a gel, using techniques described in this section, the image that results consists of bands: shallow lines that are as wide as a lane in the gel. Why?

To understand the answer, study **FIGURE B9.2**. The left panel shows the original mixture of molecules. In this cartoon, the size of each dot represents the size of each molecule. The key is to realize that the original sample contains many copies of each specific molecule, and that these copies run down the length of the gel together—meaning, at the same rate—because they have the same size and charge.

It's that simple: Molecules that are alike form a band because they stay together.

Using Thin Layer Chromatography to Separate Molecules

Gel electrophoresis is not the only way to separate molecules. Researchers also use a method called thin layer chromatography. This method was developed in the early 1900s by botanists who were analyzing the different-colored pigments from leaves of a plant (see Chapter 10), hence the name chromatography from the Greek words khroma for "color" and graphein, "to write."

In this method, rather than loading the sample into the well of a gel, the samples are deposited or "spotted" near the bottom of a stiff support, either glass or plastic, that is coated with a thin layer of silica gel, cellulose, or a similar porous material. The coated support is placed in a solvent solution. As the solvent

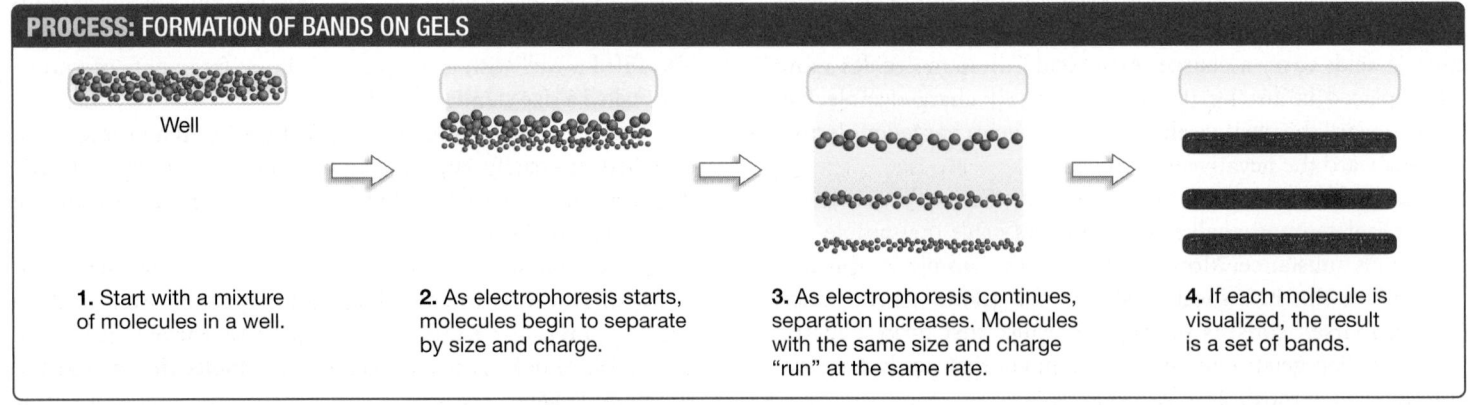

Well

1. Start with a mixture of molecules in a well.

2. As electrophoresis starts, molecules begin to separate by size and charge.

3. As electrophoresis continues, separation increases. Molecules with the same size and charge "run" at the same rate.

4. If each molecule is visualized, the result is a set of bands.

FIGURE B9.2 On a Gel, Molecules That Are Alike Form Bands.

wicks upward through the coating by capillary action, it carries the molecules in the mixture with it. Molecules are carried at different rates, based on their size and solubility in the solvent.

Visualizing Molecules

Once molecules have been separated using electrophoresis or thin layer chromatography, they have to be detected. Unfortunately, although plant pigments are colored, proteins and nucleic acids are invisible unless they are tagged in some way. Let's first look at two of the most common tagging systems and then consider how researchers can tag and visualize specific molecules of interest and not others.

Using Radioactive Isotopes and Autoradiography

When molecular biology was getting under way, the first types of tags in common use were radioactive isotopes—forms of atoms that are unstable and release energy in the form of radiation.

In the polymerization experiment diagrammed in Figure B9.1, for example, the researchers had attached a radioactive phosphorus atom to the monomers—ribonucleoside triphosphates—used in the original reaction mix. Once polymers formed, they contained radioactive atoms. When electrophoresis was complete, the investigators visualized the polymers by laying X-ray film over the gel. Because radioactive emissions expose film, a black dot appears wherever a radioactive atom is located in the gel. So many black dots occur so close together that the collection forms a dark band.

This technique for visualizing macromolecules is called autoradiography. The autoradiograph that resulted from the polymerization experiment is shown in **FIGURE B9.3**. The samples, taken on days 2, 4, 6, 8, and 14 of the experiment, are noted along the bottom. The far right lane contains macromolecules of known size; this lane is used to estimate the size of the molecules in the experimental samples. The bands that appear in each sample lane represent the different polymers that had formed.

Reading a Gel

One of the keys to interpreting or "reading" a gel, or the corresponding autoradiograph, is to realize that darker bands contain more radioactive markers, indicating the presence of many radioactive molecules. Lighter bands contain fewer molecules.

To read a gel, then, you look for (1) the presence or absence of bands in some lanes—meaning, some experimental samples—versus others, and (2) contrasts in the darkness of the bands—meaning, differences in the number of molecules present.

For example, several conclusions can be drawn from the data in Figure B9.3. First, a variety of polymers formed at each stage. After the second day, for example, polymers from 12 to 18 monomers long had formed. Second, the overall length of polymers produced increased with time. At the end of the fourteenth day, most of the RNA molecules were between 20 and 40 monomers long.

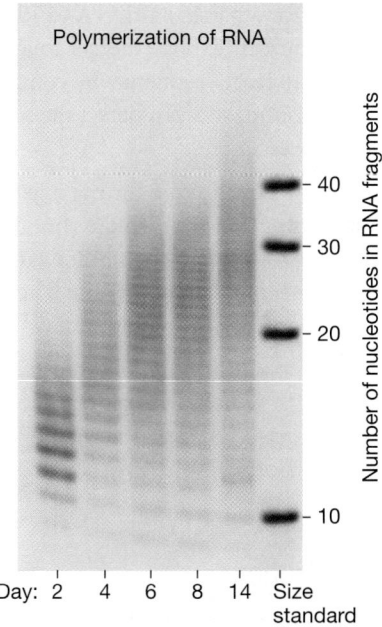

FIGURE B9.3 Autoradiography Is a Technique for Visualizing Macromolecules. The molecules in a gel can be visualized in a number of ways. In this case, the RNA molecules in the gel exposed an X-ray film because they had radioactive atoms attached. When developed, the film is called an autoradiograph.

Starting in the late 1990s and early 2000s, it became much more common to tag nucleic acids with fluorescent tags. Once electrophoresis is complete, fluorescence can be detected by exposing the gel to an appropriate wavelength of light; the fluorescent tag fluoresces or glows in response (fluorescence is explained in Chapter 10).

Fluorescent tags have important advantages over radioactive isotopes: (1) They are safer to handle. (2) They are faster—you don't have to wait hours or days for the radioactive isotope to expose a film. (3) They come in multiple colors, so you can tag several different molecules in the same experiment and detect them independently.

Using Nucleic Acid Probes

In many cases, researchers want to find one specific molecule—a certain DNA sequence, for example—in the collection of molecules on a gel. How is this possible? The answer hinges on using a particular molecule as a probe.

You'll learn in more detail about how probes work in this text (Chapter 20). Here it's enough to get the general idea: A probe is a marked molecule that binds specifically to your molecule of interest. The "mark" is often a radioactive atom, a fluorescent tag, or an enzyme that catalyzes a color-forming or light-emitting reaction.

If you are looking for a particular DNA or RNA sequence on a gel, for example, you can expose the gel to a single-stranded probe that binds to the target sequence by complementary base pairing. Once it has bound, you can detect the band through autoradiography or fluorescence.

- *Southern blotting* is a technique for making DNA fragments that have been run out on a gel single stranded, transferring them from the gel to a nylon membrane, and then probing them to identify segments of interest. The technique was named after its inventor, Edwin Southern.

- *Northern blotting* is a technique for transferring RNA fragments from a gel to a nylon membrane and then probing them to detect target segments. The name is a lighthearted play on Southern blotting—the protocol from which it was derived.

Using Antibody Probes

How can researchers find a particular protein out of a large collection of different proteins? The answer is to use an antibody. An antibody is a protein that binds specifically to a section of a different protein (see Chapter 51 if you want more detail on the structure of antibodies and their function in the immune system).

To use an antibody as a probe, investigators attach a tag molecule—often an enzyme that catalyzes a color-forming reaction—to the antibody and allow it to react with proteins in a mixture. The antibody will stick to the specific protein that it binds to and then can be visualized thanks to the tag it carries.

If the proteins in question have been separated by gel electrophoresis and transferred to a membrane, the result is called a western blot. The name western is an extension of the Southern and northern patterns.

Using Radioimmunoassay and ELISA to Measure Amounts of Molecules

Another important method that makes use of antibodies is called a radioimmunoassay. This method is used when investigators want to measure tiny amounts of a molecule, such as a hormone in the blood. In this case, a known quantity of a hormone is labeled with a radioactive tag. This tagged hormone is then mixed with a known amount of antibody, and the two bind to one another. Next, a sample of blood, containing an unknown quantity of that same hormone, is added. The hormone from the blood and the radiolabeled hormone compete for antibody binding sites. As the concentration of unlabeled hormone increases, more of it binds to the antibody, displacing more of the radiolabeled hormone. The amount of unbound radiolabeled hormone is then measured. Using known standards as a reference, the amount of hormone in the blood can be determined.

Another commonly used technique based on similar principles is called ELISA (enzyme-linked immunosorbent assay). In this case, the amount of a particular molecule is measured using colorimetric signals instead of a radioactive signal.

check your understanding

If you understand BioSkill 9

✔**You should be able to . . .**

Interpret a gel that has been stained for "RNA X." One lane contains no bands. Two lanes have a band in the same location, even though one of the bands is barely visible and the other is extremely dark. The fourth lane has a faint band located below the bands in the other lanes.

Biologists use a technique called differential centrifugation to isolate specific cell components. Differential centrifugation is based on breaking cells apart to create a complex mixture and then separating components in a centrifuge. A centrifuge accomplishes this task by spinning cells in a solution that allows molecules and other cell components to separate according to their density or size and shape. The individual parts of the cell can then be purified and studied in detail, in isolation from other parts of the cell.

The first step in preparing a cell sample for centrifugation is to release the cell components by breaking the cells apart. This can be done by putting them in a hypotonic solution, by exposing them to high-frequency vibration, by treating cells with a detergent, or by grinding them up. Each of these methods breaks apart plasma membranes and releases the contents of the cells.

The resulting pieces of plasma membrane quickly reseal to form small vesicles, often trapping cell components inside. The solution that results from the homogenization step is a mixture of these vesicles, free-floating macromolecules released from the cells, and organelles. A solution like this is called a cell extract or cell homogenate.

When a cell homogenate is placed in a centrifuge tube and spun at high speed, the components that are in solution tend to move outward, along the red arrow in **FIGURE B10.1a**. The effect is similar to a merry-go-round, which seems to push you outward in a straight line away from the spinning platform. In response to this outward-directed force, the cell homogenate exerts a centripetal (literally, "center-seeking") force that pushes the homogenate away from the bottom of the tube. Larger, denser molecules or particles resist this inward force more readily than do smaller, less dense ones and so reach the bottom of the centrifuge tube faster.

To separate the components of a cell extract, researchers often perform a series of centrifuge runs. Steps 1 and 2 of

(a) How a centrifuge works

When the centrifuge spins, the macromolecules tend to move toward the bottom of the centrifuge tube (red arrow)

The solution in the tube exerts a centripetal force, which resists movement of the molecules to the bottom of the tube (blue arrow)

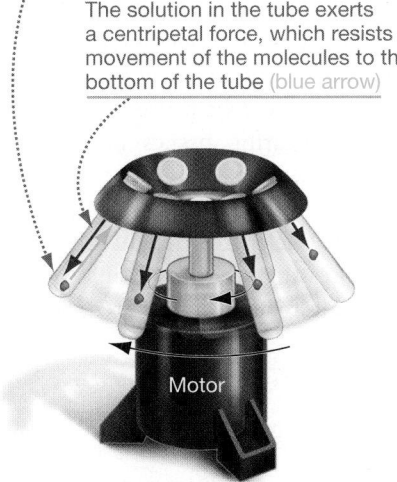

Very large or dense molecules overcome the centripetal force more readily than smaller, less dense ones. As a result, larger, denser molecules move toward the bottom of the tube faster.

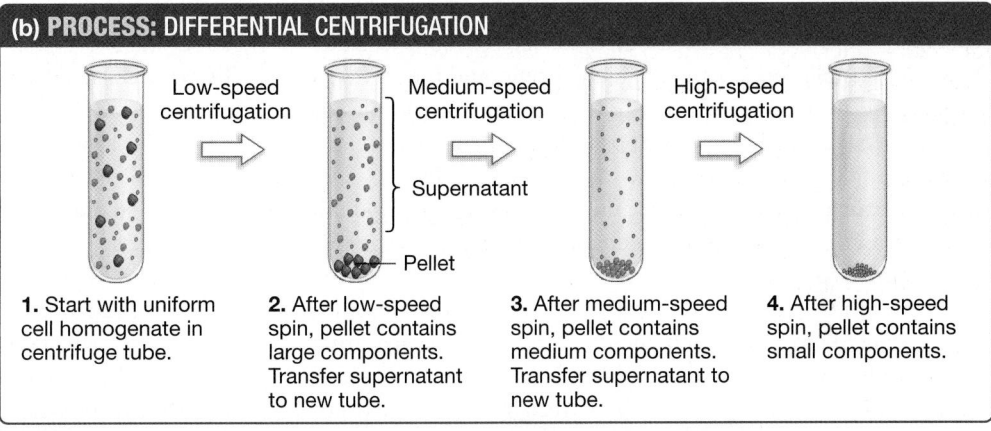

(b) PROCESS: DIFFERENTIAL CENTRIFUGATION

Low-speed centrifugation → Medium-speed centrifugation → High-speed centrifugation

Supernatant

Pellet

1. Start with uniform cell homogenate in centrifuge tube.

2. After low-speed spin, pellet contains large components. Transfer supernatant to new tube.

3. After medium-speed spin, pellet contains medium components. Transfer supernatant to new tube.

4. After high-speed spin, pellet contains small components.

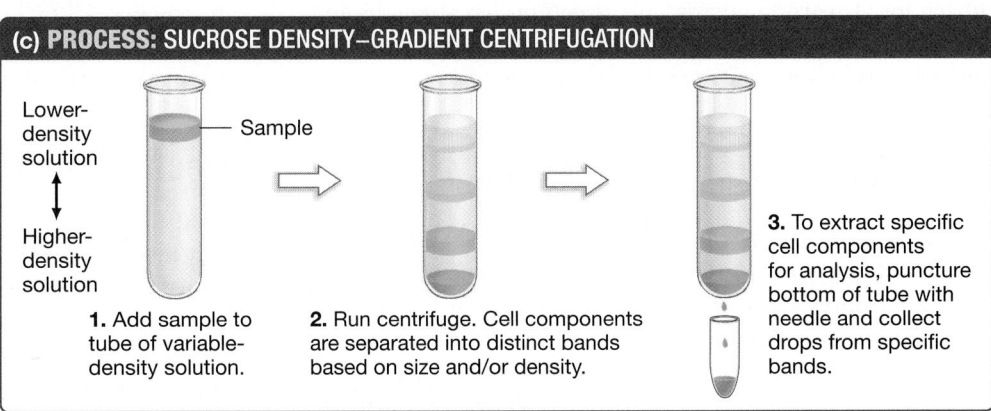

(c) PROCESS: SUCROSE DENSITY–GRADIENT CENTRIFUGATION

Lower-density solution

Higher-density solution

Sample

1. Add sample to tube of variable-density solution.

2. Run centrifuge. Cell components are separated into distinct bands based on size and/or density.

3. To extract specific cell components for analysis, puncture bottom of tube with needle and collect drops from specific bands.

FIGURE B10.1 Cell Components Can Be Separated by Centrifugation. (a) The forces inside a centrifuge tube allow cell components to be separated. **(b)** Through a series of centrifuge runs made at increasingly higher speeds, an investigator can separate fractions of a cell homogenate by size via differential centrifugation. **(c)** A high-speed centrifuge run can achieve extremely fine separation among cell components by sucrose density–gradient centrifugation.

FIGURE B10.1b illustrate how an initial treatment at low speed causes larger, heavier parts of the homogenate to move below smaller, lighter parts. The material that collects at the bottom of the tube is called the pellet, and the solution and solutes left behind form the supernatant ("above-swimming"). The supernatant is placed in a fresh tube and centrifuged at increasingly higher speeds and longer durations. Each centrifuge run continues to separate cell components based on their size and density.

To accomplish separation of macromolecules or organelles, researchers frequently follow up with centrifugation at extremely high speeds. One strategy is based on filling the centrifuge tube with a series of sucrose solutions of increasing density (**FIGURE B10.1c**). The density gradient allows cell components to separate on the basis of small differences in size, shape, and density. When the centrifuge run is complete, each cell component occupies a distinct band of material in the tube, based on how quickly each component moves through the increasingly

dense gradient of sucrose solution during the centrifuge run. A researcher can then collect the material in each band for further study.

BIOSKILL 11 — biological imaging: microscopy and x-ray crystallography

A lot of biology happens at levels that can't be detected with the naked eye. Biologists use an array of microscopes to study small multicellular organisms, individual cells, and the contents of cells. And to understand what individual macromolecules or macromolecular machines like ribosomes look like, researchers use data from a technique called X-ray crystallography.

You'll probably use dissecting microscopes and compound light microscopes to view specimens during your labs for this course, and throughout this text you'll be seeing images generated from other types of microscopy and from X-ray crystallographic data. Among the fundamental skills you'll be acquiring as an introductory student, then, is a basic understanding of how these techniques work. The key is to recognize that each approach for visualizing microscopic structures has strengths and weaknesses. As a result, each technique is appropriate for studying certain types or aspects of cells or molecules.

Light and Fluorescence Microscopy

If you use a dissecting microscope during labs, you'll recognize that it works by magnifying light that bounces off a whole specimen—often a live organism. You'll be able to view the specimen in three dimensions, which is why these instruments are sometimes called stereomicroscopes, but the maximum magnification possible is only about 20 to 40 times normal size (20× to 40×).

To view smaller objects, you'll probably use a compound microscope. Compound microscopes magnify light that is passed *through* a specimen. The instruments used in introductory labs are usually capable of 400× magnifications; the most sophisticated

compound microscopes available can achieve magnifications of about 2000×. This is enough to view individual bacterial or eukaryotic cells and see large structures inside cells, like condensed chromosomes (see Chapter 12). To prepare a specimen for viewing under a compound light microscope, the tissues or cells are usually sliced to create a section thin enough for light to pass through efficiently. The section is then dyed to increase contrast and make structures visible. In many cases, different types of dyes are used to highlight different types of structures.

To visualize specific proteins, researchers use a technique called immunostaining. After preparing tissues or cells for viewing, the specimen is stained with fluorescently tagged antibodies. In this case, the cells are viewed under a fluorescence microscope. Ultraviolet or other wavelengths of light are passed through the specimen. The fluorescing tag emits visible light in response. The result? Beautiful cells that glow green, red, or blue.

Electron Microscopy

Until the 1950s, the compound microscope was the biologist's only tool for viewing cells directly. But the invention of the electron microscope provided a new way to view specimens. Two basic types of electron microscopy are now available: one that allows researchers to examine cross sections of cells at extremely high magnification, and one that offers a view of surfaces at somewhat lower magnification.

Transmission Electron Microscopy

The transmission electron microscope (TEM) is an extraordinarily effective tool for viewing cell structure at high magnification. TEM forms an image from electrons that pass through a specimen, just as a light microscope forms an image from light rays that pass through a specimen.

Biologists who want to view a cell under a transmission electron microscope begin by "fixing" the cell, meaning that they treat it with a chemical agent that stabilizes the cell's structure and contents while disturbing them as little as possible. Then the researcher permeates the cell with an epoxy plastic that stiffens the structure. Once this epoxy hardens, the cell can be cut into extremely thin sections with a glass or diamond knife. Finally, the sectioned specimens are impregnated with a metal—often lead. (The reason for this last step is explained shortly.)

FIGURE B11.1a outlines how the transmission electron microscope works. A beam of electrons is produced by a tungsten filament at the top of a column and directed downward. (All of the air is pumped out of the column, so that the electron beam isn't scattered by collisions with air molecules.) The electron beam passes through a series of lenses and through the specimen. The lenses are actually electromagnets, which alter the path of the beam much like a glass lens in a dissecting or compound microscope bends light. The electromagnet lenses magnify and focus the image on a screen at the bottom of the column. There the electrons strike a coating of fluorescent crystals, which emit visible light in response—just like a television screen. When the microscopist moves the screen out of the way and allows the electrons to expose a sheet of black-and-white film or to be detected by a digital camera, the result is a micrograph—a photograph of an image produced by microscopy.

The image itself is created by electrons that pass through the specimen. If no specimen were in place, all the electrons would pass through and the screen (and micrograph) would be uniformly bright. Unfortunately, cell materials by themselves would also appear fairly uniform and bright. This is because an atom's ability to deflect an electron depends on its mass. In turn, an atom's mass is a function of its atomic number. The hydrogen, carbon, oxygen, and nitrogen atoms that dominate biological molecules have low atomic numbers. This is why cell biologists must saturate cell sections with lead solutions. Lead has a high atomic number and scatters electrons effectively. Different macromolecules take up lead atoms in different amounts, so the metal acts as a "stain" that produces contrast. With TEM, areas of dense metal scatter the electron beam most, producing dark areas in micrographs.

The advantage of TEM is that it can magnify objects up to 250,000×—meaning that intracellular structures are clearly visible. The downsides are that researchers are restricted to observing dead, sectioned material, and they must take care that the preparation process does not distort the specimen.

Scanning Electron Microscopy

The scanning electron microscope (SEM) is the most useful tool biologists have for looking at the surfaces of structures. Materials are prepared for scanning electron microscopy by coating their surfaces with a layer of metal atoms. To create an image of this surface, the microscope scans the surface with a narrow beam of electrons. Electrons that are reflected back from the surface or that are emitted by the metal atoms in response to the beam then strike a detector. The signal from the detector controls a second electron beam, which scans a TV-like screen and forms an image magnified up to 50,000 times the object's size.

Because SEM records shadows and highlights, it provides images with a three-dimensional appearance (**FIGURE B11.1b**). It cannot magnify objects nearly as much as TEM can, however.

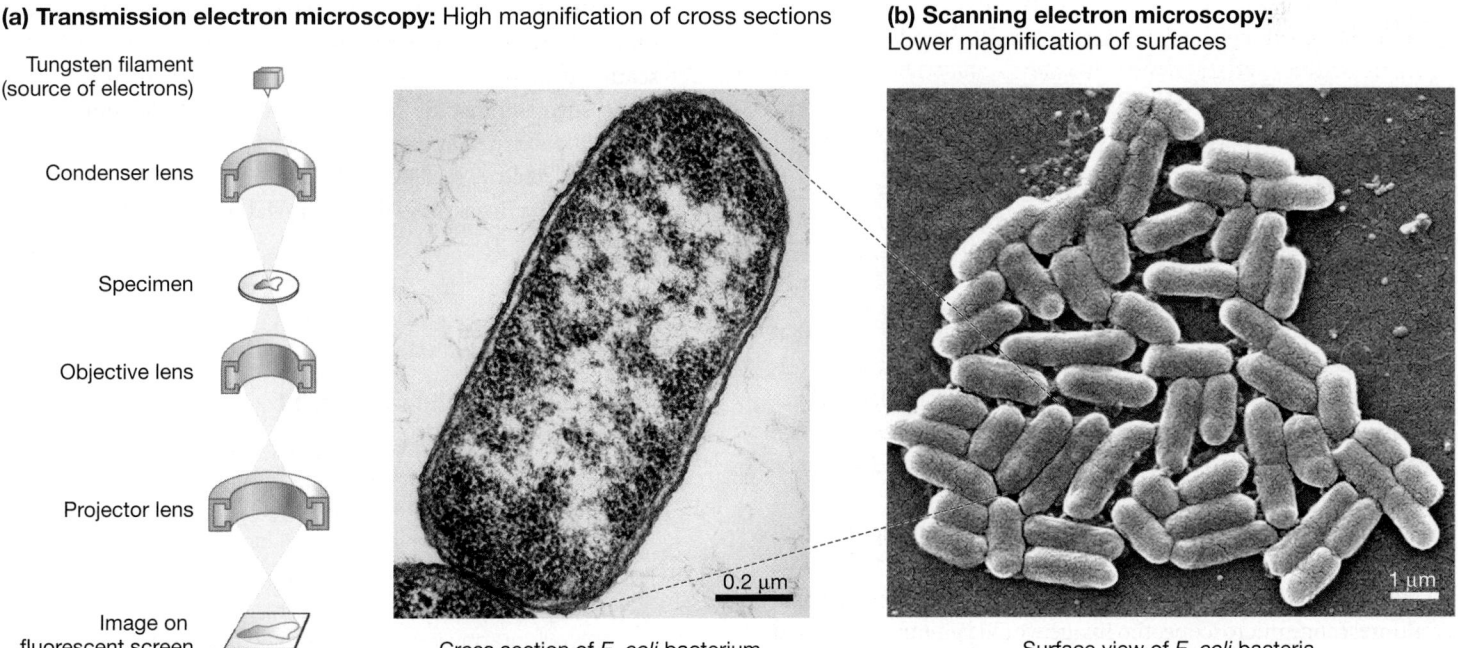

(a) Transmission electron microscopy: High magnification of cross sections

Tungsten filament (source of electrons)

Condenser lens

Specimen

Objective lens

Projector lens

Image on fluorescent screen

Cross section of *E. coli* bacterium

0.2 μm

(b) Scanning electron microscopy: Lower magnification of surfaces

Surface view of *E. coli* bacteria

1 μm

FIGURE B11.1 There Are Two Basic Types of Electron Microscopy.

Studying Live Cells and Real-Time Processes

Until the 1960s, biologists were unable to get clear, high-magnification images of living cells. But a series of innovations over the past 50 years has made it possible to observe organelles and subcellular structures in action.

The development of video microscopy, where the image from a light microscope is captured by a video camera instead of by an eye or a film camera, proved revolutionary. It allowed specimens to be viewed at higher magnification, because video cameras are more sensitive to small differences in contrast than are the human eye or still cameras. It also made it easier to keep live specimens functioning normally, because the increased light sensitivity of video cameras allows them to be used with low illumination, so specimens don't overheat. And when it became possible to digitize video images, researchers began using computers to remove out-of-focus background material and increase image clarity.

A more recent innovation was the use of a fluorescent molecule called green fluorescent protein, or GFP, which allows researchers to tag specific molecules or structures and follow their movement in live cells over time. This was a major advance over immunostaining, in which cells have to be fixed. GFP is naturally synthesized in jellyfish that fluoresce, or emit light. By affixing GFP to another protein (using genetic engineering techniques described in Chapter 20) and then inserting it into a live cell, investigators can follow the protein's fate over time and even videotape its movement. For example, researchers have videotaped GFP-tagged proteins being transported from the rough ER through the Golgi apparatus and out to the plasma membrane. This is cell biology: the movie.

GFP's influence has been so profound that the researchers who developed its use in microscopy were awarded the 2008 Nobel Prize in Chemistry.

Visualizing Structures in 3-D

The world is three-dimensional. To understand how microscopic structures and macromolecules work, it is essential to understand their shape and spatial relationships. Consider three techniques currently being used to reconstruct the 3-D structure of cells, organelles, and macromolecules.

- *Confocal microscopy* is carried out by mounting cells that have been treated with one or more fluorescing tags on a microscope slide and then focusing a beam of ultraviolet or other wavelengths of light at a specific depth within the specimen. The fluorescing tag emits visible light in response. A detector for this light is then set up at exactly the position where the emitted light comes into focus. The result is a sharp image of a precise plane in the cell being studied (**FIGURE B11.2a**). Note that if you viewed the same specimen under a conventional fluorescence microscope, the image would be blurry because it results from light emitted by the entire cell (**FIGURE B11.2b**). By altering the focal plane, a researcher can record images from

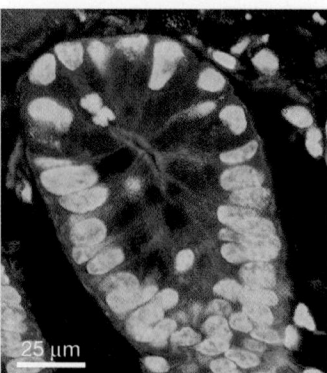 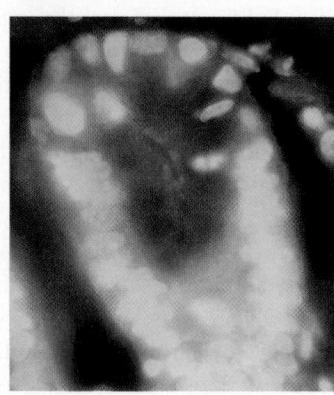

(a) Confocal fluorescence image of single cell

(b) Conventional fluorescence image of same cell

25 µm

FIGURE B11.2 Confocal Microscopy Provides Sharp Images of Living Cells. (a) The confocal image of this mouse intestinal cell is sharp, because it results from light emitted at a single plane inside the cell. **(b)** The conventional image of this same cell is blurred, because it results from light emitted by the entire cell.

an array of depths in the specimen; a computer can then be used to generate a 3-D image of the cell.

- *Electron tomography* uses a transmission electron microscope to generate a 3-D image of an organelle or other subcellular structure. The specimen is rotated around a single axis while the researcher takes many "snapshots." The individual images are then pieced together with a computer. This technique has provided a much more accurate view of mitochondrial structure than was possible using traditional TEM (see Chapter 7).

- *X-ray crystallography, or X-ray diffraction analysis*, is the most widely used technique for reconstructing the 3-D structure of molecules. As its name implies, the procedure is based on bombarding crystals of a molecule with X-rays. X-rays are scattered in precise ways when they interact with the electrons surrounding the atoms in a crystal, producing a diffraction pattern that can be recorded on X-ray film or other types of detectors (**FIGURE B11.3**). By varying the orientation of the X-ray beam as it strikes a crystal and documenting the

check your understanding

If you understand BioSkill 11

✓ **You should be able to . . .**

1. Interpret whether the absence of mitochondria in a transmission electron micrograph of a cancerous human liver means that the cell lacks mitochondria.

2. Explain why the effort to understand the structure of biological molecules is worthwhile even though X-ray crystallography is time consuming and technically difficult. What's the payoff?

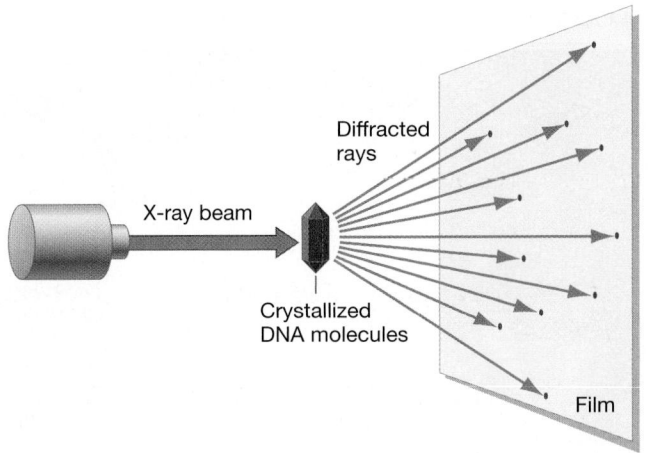

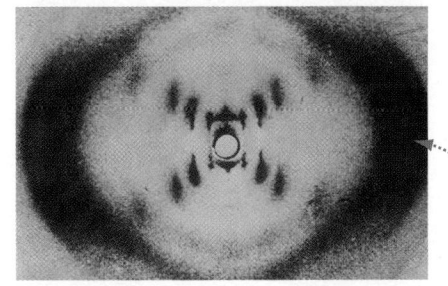

The patterns are determined by the structure of the molecules within the crystal

FIGURE B11.3 X-Ray Crystallography. When crystallized molecules are bombarded with X-rays, the radiation is scattered in distinctive patterns. The photograph at the right shows an X-ray film that recorded the pattern of scattered radiation from DNA molecules.

diffraction patterns that result, researchers can construct a map representing the density of electrons in the crystal. By relating these electron-density maps to information about the primary structure of the nucleic acid or protein, a 3-D model of the molecule can be built. Virtually all of the molecular models used in this book were built from X-ray crystallographic data.

BIOSKILL 12 cell and tissue culture methods

For researchers, there are important advantages to growing plant and animal cells and tissues outside the organism itself. Cell and tissue cultures provide large populations of a single type of cell or tissue and the opportunity to control experimental conditions precisely.

Animal Cell Culture

The first successful attempt to culture animal cells occurred in 1907, when a researcher cultivated amphibian nerve cells in a drop of fluid from the spinal cord. But it was not until the 1950s and 1960s that biologists could routinely culture plant and animal cells in the laboratory. The long lag time was due to the difficulty of re-creating conditions that exist in the intact organism precisely enough for cells to grow normally.

To grow in culture, animal cells must be provided with a liquid mixture containing the nutrients, vitamins, and hormones that stimulate growth. Initially, this mixture was serum, the liquid portion of blood; now, serum-free media are available for certain cell types. Serum-free media are preferred because they are much more precisely defined chemically than serum.

In addition, many types of animal cells will not grow in culture unless they are provided with a solid surface that mimics the types of surfaces that enable cells in the intact organisms to adhere. As a result, cells are typically cultured in flasks (**FIGURE B12.1a**, left; see B:22).

Even under optimal conditions, though, normal cells display a finite life span in culture. In contrast, many cultured cancerous cells grow indefinitely. This characteristic correlates with a key feature of cancerous cells in organisms: Their growth is continuous and uncontrolled.

Because of their immortality and relative ease of growth, cultured cancer cells are commonly used in research on basic aspects of cell structure and function. For example, the first human cell type to be grown in culture was isolated in 1951 from a malignant tumor of the uterine cervix. These cells are called HeLa cells in honor of their donor, Henrietta Lacks, who died soon thereafter from her cancer. HeLa cells continue to grow in laboratories around the world (Figure B12.1a, right).

Plant Tissue Culture

Certain cells found in plants are totipotent—meaning that they retain the ability to divide and differentiate into a complete, mature plant, including new types of tissue. These cells, called parenchyma cells, are important in wound healing and asexual reproduction. But they also allow researchers to grow complete adult plants in the laboratory, starting with a small number of parenchyma cells.

Biologists who grow plants in tissue culture begin by placing parenchyma cells in a liquid or solid medium containing all the

(a) Animal cell culture: immortal HeLa cancer cells

 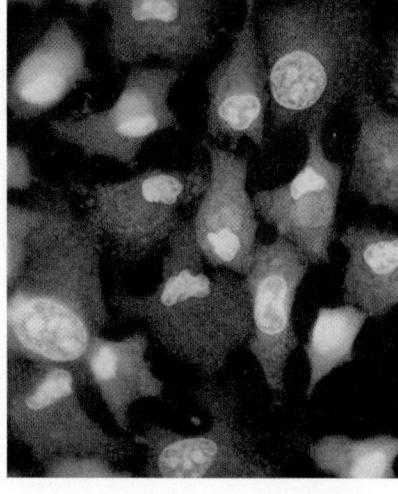

(b) Plant tissue culture: tobacco callus

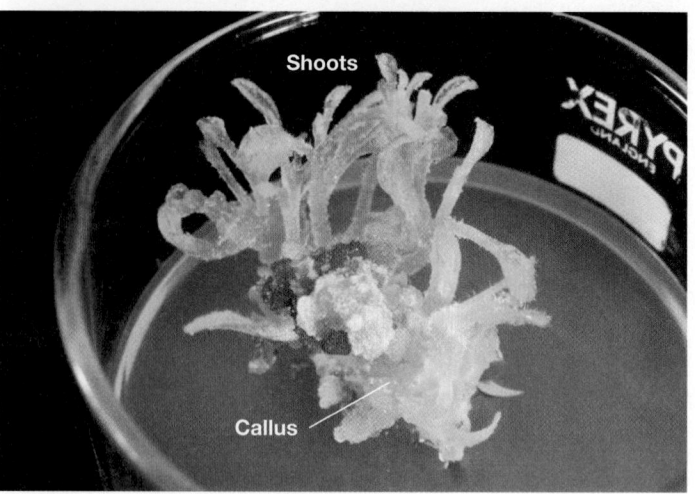

Shoots

Callus

FIGURE B12.1 Animal and Plant Cells Can Be Grown in the Lab.

nutrients required for cell maintenance and growth. In the early days of plant tissue culture, investigators found not only that specific growth signals called hormones were required for successful growth and differentiation but also that the relative abundance of hormones present was critical to success.

The earliest experiments on hormone interactions in tissue cultures were done with tobacco cells in the 1950s by Folke Skoog and co-workers. These researchers found that when the hormone called auxin was added to the culture by itself, the cells enlarged but did not divide. But if the team added roughly equal amounts of auxin and another growth signal called cytokinin to the cells, the cells began to divide and eventually formed a callus, or an undifferentiated mass of parenchyma cells.

By varying the proportion of auxin to cytokinins in different parts of the callus and through time, the team could stimulate the growth and differentiation of root and shoot systems and produce whole new plants (**FIGURE B12.1b**). A high ratio of auxin to cytokinin led to the differentiation of a root system, while a high ratio of cytokinin to auxin led to the development of a shoot system. Eventually Skoog's team was able to produce a complete plant from just one parenchyma cell.

The ability to grow whole new plants in tissue culture from just one cell has been instrumental in the development of genetic engineering (see Chapter 20). Researchers insert recombinant genes into target cells, test the cells to identify those that successfully express the recombinant genes, and then use tissue culture techniques to grow those cells into adult individuals with a novel genotype and phenotype.

check your understanding

If you understand BioSkill 12

✔ **You should be able to . . .**

1. Identify a limitation of how experiments on HeLa cells are interpreted.

2. State a disadvantage of doing experiments on plants that have been propagated from single cells growing in tissue culture.

Research in biological science starts with a question. In most cases, the question is inspired by an observation about a cell or an organism. To answer it, biologists have to study a particular species. Study organisms are often called model organisms, because investigators hope that they serve as a model for what is going on in a wide array of species.

Model organisms are chosen because they are convenient to study and because they have attributes that make them appropriate for the particular research proposed. They tend to have some common characteristics:

- *Short generation time and rapid reproduction* This trait is important because it makes it possible to produce offspring quickly and perform many experiments in a short amount of time—you don't have to wait long for individuals to grow.

- *Large numbers of offspring* This trait is particularly important in genetics, where many offspring phenotypes and genotypes need to be assessed to get a large sample size.

- *Small size, simple feeding and habitat requirements* These attributes make it relatively cheap and easy to maintain individuals in the lab.

The following notes highlight just a few model organisms supporting current work in biological science.

Escherichia coli

Of all model organisms in biology, perhaps none has been more important than the bacterium *Escherichia coli*—a common inhabitant of the human gut. The strain that is most commonly worked on today, called K-12 (**FIGURE B13.1a**; see B:24), was originally isolated from a hospital patient in 1922.

During the last half of the twentieth century, key results in molecular biology originated in studies of *E. coli*. These results include the discovery of enzymes such as DNA polymerase, RNA polymerase, DNA repair enzymes, and restriction endonucleases; the elucidation of ribosome structure and function; and the initial characterization of promoters, regulatory transcription factors, regulatory sites in DNA, and operons. In many cases, initial discoveries made in *E. coli* allowed researchers to confirm that homologous enzymes and processes existed in an array of organisms, often ranging from other bacteria to yeast, mice, and humans.

The success of *E. coli* as a model for other species inspired Jacques Monod's claim that "Once we understand the biology of *Escherichia coli*, we will understand the biology of an elephant." The genome of *E. coli* K-12 was sequenced in 1997, and the strain continues to be a workhorse in studies of gene function, biochemistry, and particularly biotechnology. Much remains to be learned, however. Despite over 60 years of intensive study, the function of about a third of the *E. coli* genome is still unknown.

In the lab, *E. coli* is usually grown in suspension culture, where cells are introduced to a liquid nutrient medium, or on plates containing agar—a gelatinous mix of polysaccharides. Under optimal growing conditions—meaning before cells begin to get crowded and compete for space and nutrients—a cell takes just 30 minutes on average to grow and divide. At this rate, a single cell can produce a population of over a million descendants in just 10 hours. Except for new mutations, all of the descendant cells are genetically identical.

Dictyostelium discoideum

The cellular slime mold *Dictyostelium discoideum* is not always slimy, and it is not a mold—meaning a type of fungus. Instead, it is an amoeba. Amoeba is a general term that biologists use to characterize a unicellular eukaryote that lacks a cell wall and is extremely flexible in shape. *Dictyostelium* has long fascinated biologists because it is a social organism. Independent cells sometimes aggregate to form a multicellular structure.

Under most conditions, *Dictyostelium* cells are haploid (*n*) and move about in decaying vegetation on forest floors or other habitats. They feed on bacteria by engulfing them whole. When these cells reproduce, they can do so sexually by fusing with another cell then undergoing meiosis, or asexually by mitosis, which is more common. If food begins to run out, the cells begin to aggregate. In many cases, tens of thousands of cells cohere to form a 2-mm-long mass called a slug (**FIGURE B13.1b**). (This is not the slug that is related to snails.)

After migrating to a sunlit location, the slug stops and individual cells differentiate according to their position in the slug. Some form a stalk; others form a mass of spores at the tip of the stalk. (A spore is a single cell that develops into an adult organism, but it is not formed from gamete fusion like a zygote is.) The entire structure, stalk plus mass of spores, is called a fruiting body. Cells that form spores secrete a tough coat and represent a durable resting stage. The fruiting body eventually dries out, and the wind disperses the spores to new locations, where more food might be available.

Dictyostelium has been an important model organism for investigating questions about eukaryotes:

- Cells in a slug are initially identical in morphology but then differentiate into distinctive stalk cells and spores. Studying this process helped biologists better understand how cells in plant and animal embryos differentiate into distinct cell types.

(a) Bacterium *Escherichia coli* (strain K-12)

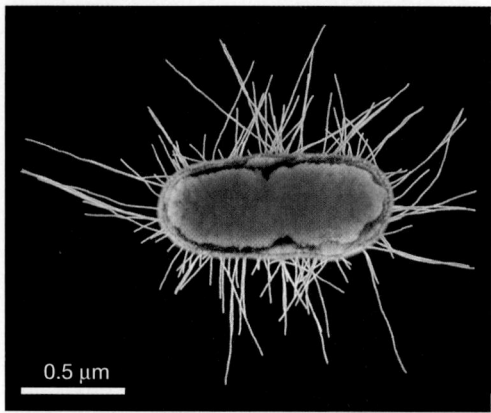

0.5 µm

(b) Slime mold *Dictyostelium discoideum*

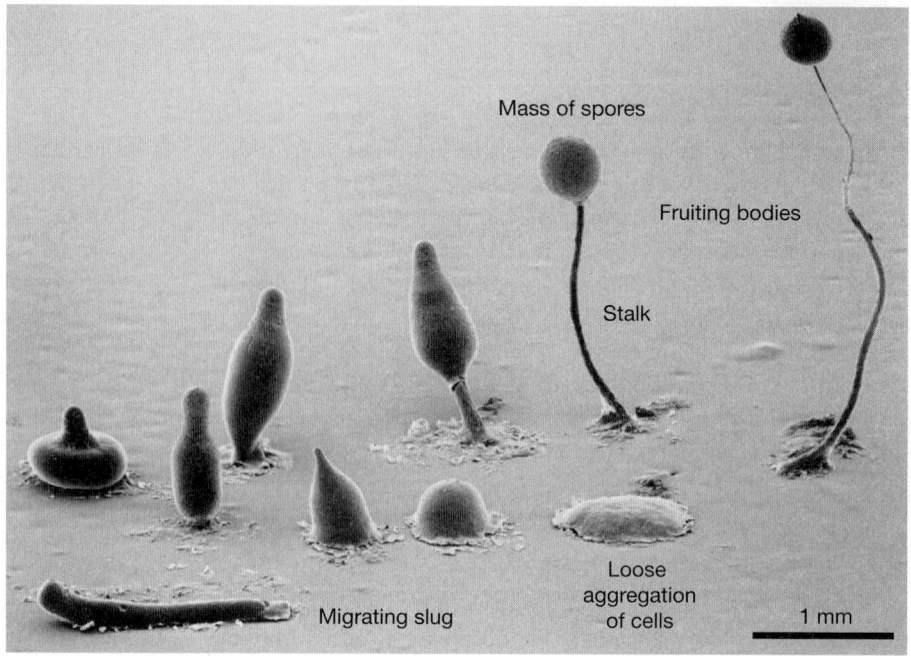

Mass of spores

Fruiting bodies

Stalk

Migrating slug

Loose aggregation of cells

1 mm

(c) Thale cress *Arabidopsis thaliana*

5 cm

(e) Fruit fly *Drosophila melanogaster*

0.5 mm

(f) Roundworm *Caenorhabditis elegans*

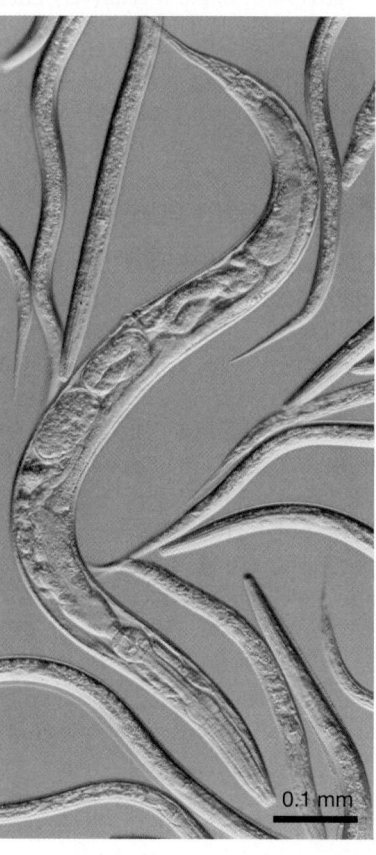

0.1 mm

(d) Yeast *Saccharomyces cerevisiae*

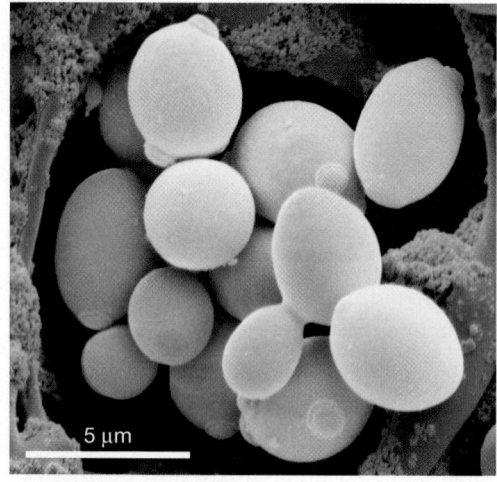

5 µm

(g) Mouse *Mus musculus*

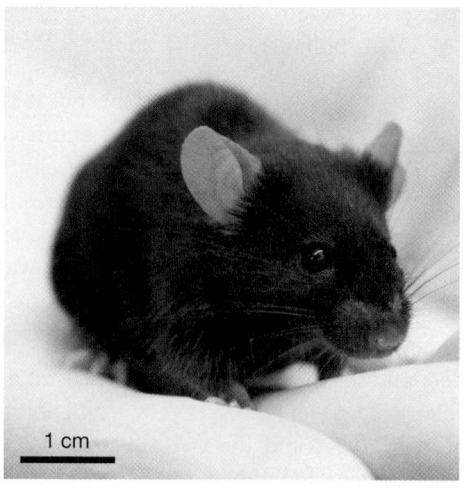

1 cm

FIGURE B13.1 Model Organisms.

✔ **QUESTION** *E. coli* is grown at a temperature of 37°C. Why?

- The process of slug formation has helped biologists study how animal cells move and how they aggregate as they form specific types of tissues.

- When *Dictyostelium* cells aggregate to form a slug, they stick to each other. The discovery of membrane proteins responsible for cell–cell adhesion helped biologists understand some of the general principles of multicellular life (highlighted in Chapter 11).

Arabidopsis thaliana

In the early days of biology, the best-studied plants were agricultural varieties such as maize (corn), rice, and garden peas. When biologists began to unravel the mechanisms responsible for oxygenic photosynthesis in the early to mid-1900s, they relied on green algae that were relatively easy to grow and manipulate in the lab—often the unicellular species *Chlamydomonas reinhardii*—as an experimental subject.

Although crop plants and green algae continue to be the subject of considerable research, a new model organism emerged in the 1980s and now serves as the preeminent experimental subject in plant biology. That organism is *Arabidopsis thaliana*, commonly known as thale cress or wall cress (**FIGURE B13.1c**).

Arabidopsis is a member of the mustard family, or Brassicaceae, so it is closely related to radishes and broccoli. In nature it is a weed—meaning a species that is adapted to thrive in habitats where soils have been disturbed.

One of the most attractive aspects of working with *Arabidopsis* is that individuals can grow from a seed into a mature, seed-producing plant in just four to six weeks. Several other attributes make it an effective subject for study: It has just five chromosomes, has a relatively small genome with limited numbers of repetitive sequences, can self-fertilize as well as undergo cross-fertilization, can be grown in a relatively small amount of space and with a minimum of care in the greenhouse, and produces up to 10,000 seeds per individual per generation.

Arabidopsis has been instrumental in a variety of studies in plant molecular genetics and development, and it is increasingly popular in ecological and evolutionary studies. In addition, the entire genome of the species has now been sequenced, and studies have benefited from the development of an international "*Arabidopsis* community"—a combination of informal and formal associations of investigators who work on *Arabidopsis* and use regular meetings, e-mail, and the Internet to share data, techniques, and seed stocks.

Saccharomyces cerevisiae

When biologists want to answer basic questions about how eukaryotic cells work, they often turn to the yeast *Saccharomyces cerevisiae*.

S. cerevisiae is unicellular and relatively easy to culture and manipulate in the lab (**FIGURE B13.1d**). In good conditions, yeast cells grow and divide almost as rapidly as bacteria. As a result,

the species has become the organism of choice for experiments on control of the cell cycle and regulation of gene expression in eukaryotes. For example, research has confirmed that several of the genes controlling cell division and DNA repair in yeast have homologs in humans; and when mutated, these genes contribute to cancer. Strains of yeast that carry these mutations are now being used to test drugs that might be effective against cancer.

S. cerevisiae has become even more important in efforts to interpret the genomes of organisms like rice, mice, zebrafish, and humans. It is much easier to investigate the function of particular genes in *S. cerevisiae* by creating mutants or transferring specific alleles among individuals than it is to do the same experiments in mice or zebrafish. Once the function of a gene has been established in yeast, biologists can look for the homologous gene in other eukaryotes. If such a gene exists, they can usually infer that it has a function similar to its role in *S. cerevisiae*. It was also the first eukaryote with a completely sequenced genome.

Drosophila melanogaster

If you walk into a biology building on any university campus around the world, you are almost certain to find at least one lab where the fruit fly *Drosophila melanogaster* is being studied (**FIGURE B13.1e**).

Drosophila has been a key experimental subject in genetics since the early 1900s. It was initially chosen as a focus for study by T. H. Morgan, because it can be reared in the laboratory easily and inexpensively, matings can be arranged, the life cycle is completed in less than two weeks, and females lay a large number of eggs. These traits made fruit flies valuable subjects for breeding experiments designed to test hypotheses about how traits are transmitted from parents to offspring (see Chapter 14).

More recently, *Drosophila* has also become a key model organism in the field of developmental biology. The use of flies in developmental studies was inspired largely by the work of Christianne Nüsslein-Volhard and Eric Wieschaus, who in the 1980s isolated flies with genetic defects in early embryonic development. By investigating the nature of these defects, researchers have gained valuable insights into how various gene products influence the development of eukaryotes (see Chapter 22). The complete genome sequence of *Drosophila* has been available to investigators since the year 2000.

Caenorhabditis elegans

The roundworm *Caenorhabditis elegans* emerged as a model organism in developmental biology in the 1970s, due largely to work by Sydney Brenner and colleagues. (*Caenorhabditis* is pronounced *see-no-rab-DIE-tiss*.)

C. elegans was chosen for three reasons: (**1**) Its cuticle (soft outer layer) is transparent, making individual cells relatively easy to observe (**FIGURE B13.1f**); (**2**) adults have exactly 959 nonreproductive cells; and, most important, (**3**) the fate of each cell in an embryo can be predicted because cell fates are invariant among individuals. For example, when researchers examine a 33-cell *C. elegans* embryo, they know exactly which of the 959 cells in the adult will be derived from each of those 33 embryonic cells.

In addition, *C. elegans* are small (less than 1 mm long), are able to self-fertilize or cross-fertilize, and undergo early development in just 16 hours. The entire genome of *C. elegans* has now been sequenced.

Mus musculus

The house mouse *Mus musculus* is the most important model organism among mammals. For this reason, it is especially prominent in biomedical research, where researchers need to work on individuals with strong genetic and developmental similarities to humans.

The house mouse was an intelligent choice of model organism in mammals because it is small and thus relatively inexpensive to maintain in captivity, and because it breeds rapidly. A litter can contain 10 offspring, and generation time is only 12 weeks—meaning that several generations can be produced in a year. Descendants of wild house mice have been selected for docility and other traits that make them easy to handle and rear; these populations are referred to as laboratory mice (**FIGURE B13.1g**).

Some of the most valuable laboratory mice are strains with distinctive, well-characterized genotypes. Inbred strains are

virtually homogenous genetically (see Chapter 26) and are useful in experiments where gene-by-gene or gene-by-environment interactions have to be controlled. Other populations carry mutations that knock out genes and cause diseases similar to those observed in humans. These individuals are useful for identifying the cause of genetic diseases and testing drugs or other types of therapies.

check your understanding

If you understand BioSkill 13
✔ **You should be able to . . .**

Determine which model organisms described here would be the best choice for the following studies. In each case, explain your reasoning.

1. A study of why specific cells in an embryo die at certain points in normal development. One goal is to understand the consequences for the individual when programmed cell death does not occur when it should.

2. A study of proteins that are required for cell–cell adhesion.

3. Research on a gene suspected to be involved in the formation of breast cancer in humans.

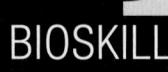

BIOSKILL 14   primary literature and peer review

As part of the process of doing science, biologists communicate their results to the scientific community through publications in scientific journals that report on their original research discoveries (see Chapter 1). These published reports are referred to, interchangeably, as the primary literature, research papers, or primary research articles.

What Is the Primary Literature?

Scientists publish "peer-reviewed" papers. This means that several experts in the field have carefully read the paper and considered its strengths and weaknesses. Reviewers write a critique of the paper and make a recommendation to the journal editor as to whether the paper should be published. Often reviewers will suggest additional experiments that need to be completed before a paper is considered acceptable for publication. The peer review process means that research discoveries are carefully vetted before they go to press.

A primary research paper can be distinguished from secondary sources—such as review articles, textbooks, and magazine articles—by looking for key characteristics. A primary research paper includes a detailed description of methods and results,

written by the researchers who did the work. A typical paper contains a Title, Abstract, Introduction, Materials and Methods (or Experimental Design), Results and Discussion (**Table B14.1**), although the order and name of the sections varies among journals.

Getting Started

At first, trying to read the primary literature may seem like a daunting task. A paper may be peppered with unfamiliar terms and acronyms. If you tried to read a research paper from start to finish, like you might read a chapter in this textbook, it would be a frustrating experience. But, with practice, the scientific literature becomes approachable, and it is well worth the effort. The primary literature is the cutting edge, the place to read firsthand about the process of doing science. Becoming skilled at reading and evaluating scientific reports is a powerful way to learn how to think critically—to think like a biologist.

To get started, try breaking down reading the primary research article into a series of steps:

1. Read the authors' names. Where are they from? Are they working as a team or alone? After delving into the literature,

TABLE B14.1 Sections of a Primary Research Paper

Section	Characteristics
Title	Short, succinct, eye-catching
Abstract	Summary of Methods, Results, Discussion. Explains why the research was done and why the results are significant.
Introduction	Background information (what past work was done, why the work was important). States the objectives and hypotheses of the study and explains why the study is important.
Materials and Methods	Explains how the work was done and where it was done.
Results	Explains what the data show.
Discussion	Explains why the data show what they show, how the analysis relates to the objectives from the Introduction, and the significance of findings and how they advance the field.

certain familiar names will crop up again and again. You'll begin to recognize the experts in a particular field.

2. Read the title. It should summarize the key finding of the paper and tell you what you can expect to learn from the paper.

3. Read the abstract. The abstract summarizes the entire paper in a short paragraph. At this point, it might be tempting to stop reading. But sometimes authors understate or overstate the significance and conclusions of their work. You should never cite an article as a reference after having read only the abstract.

4. Read the Introduction. The first couple of paragraphs should make it clear what the objectives or hypotheses of the paper are; the remaining paragraphs will give you the background information you need to understand the point of the paper.

5. Flip through the article and look at the figures, illustrations, and tables, including reading the legends.

6. Read the Results section carefully. Ask yourself these questions: Do the results accurately describe the data presented in the paper? Were all the appropriate controls carried out in an experiment? Are there additional experiments that you think should have been performed? Are the figures and tables clearly labeled?

7. Consult the Materials and Methods section to help understand the research design and the techniques used.

8. Read the Discussion. The first and last paragraphs usually summarize the key findings and state their significance. The Discussion is the part of the paper where the results are explained in the context of the scientific literature. The authors should explain what their results mean.

Getting Practice

The best way to get practice is to start reading the scientific literature as often as possible. You could begin by reading some of the references cited in this textbook. You can get an electronic copy of most articles through online databases such as PubMed, ScienceDirect, or Google Scholar, or through your institution's library.

After reading a primary research paper, you should be able to paraphrase the significance of the paper in a few sentences, free of technical jargon. You should also be able to both praise and criticize several points of the paper. As you become more familiar with reading the scientific literature, you're likely to start thinking about what questions remain to be answered. And, you may even come up with "the next experiment."

check your understanding

If you understand BioSkill 14

✔ **You should be able to . . .**

After choosing a primary research paper on a topic in biology that you would like to know more about, select one figure in the Results section that reports on the experiment and construct a Research box (like the ones in this textbook) that depicts this experiment.

A concept map is a graphical device for organizing and expressing what you know about a topic. It has two main elements: (1) concepts that are identified by words or short phrases and placed in a box or circle, and (2) labeled arrows that physically link two concepts and explain the relationship between them. The concepts are arranged hierarchically on a page with the most general concepts at the top and the most specific ideas at the bottom.

The combination of a concept, a linking word, and a second concept is called a proposition. Good concept maps also have cross-links—meaning, labeled arrows that connect different elements in the hierarchy as you read down the page.

Concept maps, initially developed by Joseph Novak in the early 1970s, have proven to be an effective studying and learning tool. They can be particularly valuable if constructed by a group, or when different individuals exchange and critique concept maps they have created independently. Although concept maps vary widely in quality and can be graded using objective criteria, there are many equally valid ways of making a high-quality concept map on a particular topic.

When you are asked to make a concept map in this text, you will usually be given at least a partial list of concepts to use. As an example, suppose you were asked to create a concept map on experimental design and were given the following concepts: results, predictions, control treatment, experimental treatment, controlled (identical) conditions, conclusions, experiment, hypothesis to be tested, null hypothesis. One possible concept map is shown in **FIGURE B15.1**.

Good concept maps have four qualities:

1. They exhibit an organized hierarchy, indicating how each concept on the map relates to larger and smaller concepts.

2. The concept words are specific—not vague.

3. The propositions are accurate.

4. There is cross-linking between different elements in the hierarchy of concepts.

As you practice making concept maps, go through these criteria and use them to evaluate your own work as well as the work of fellow students.

check your understanding

If you understand BioSkill 15

✔ **You should be able to . . .**

1. Add an "Alternative hypothesis" concept to the map in Figure B15.1, along with other concepts and labeled linking arrows needed to indicate its relationship to other information on the map. (Hint: Recall that investigators often contrast a hypothesis being tested with an alternative hypothesis that does not qualify as a null hypothesis.)

2. Add a box for the concept "Statistical testing" (see **BIOSKILLS 4**) along with appropriately labeled linking arrows.

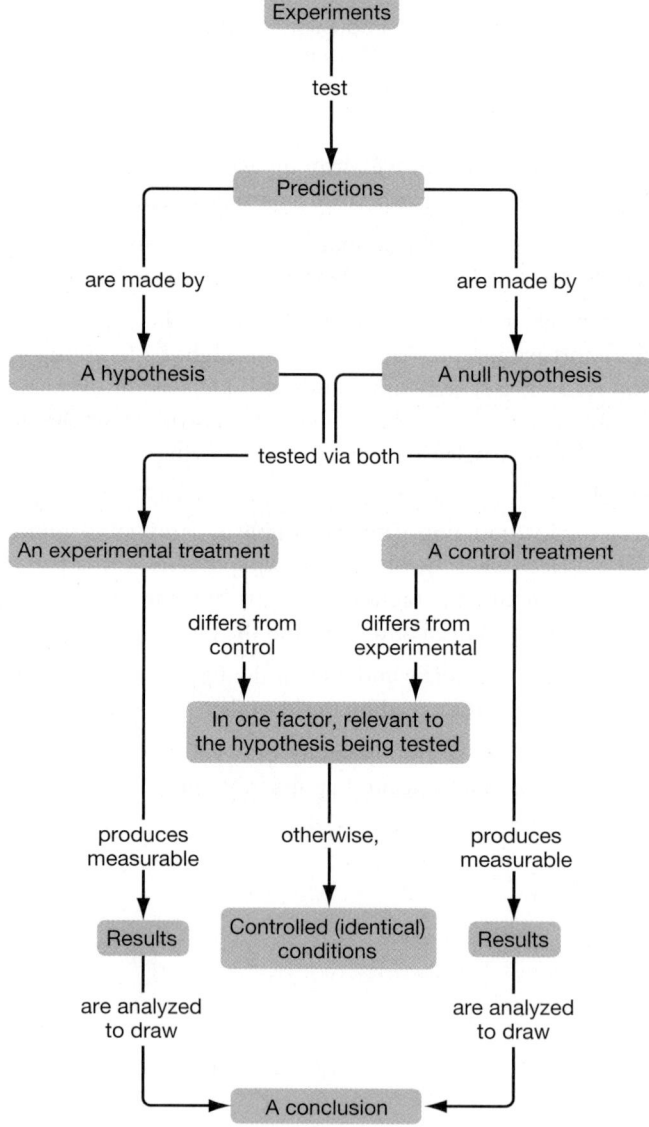

FIGURE B15.1 A Concept Map on Principles of Experimental Design.

Most students have at one time or another wondered why a particular question on an exam seemed so hard, while others seemed easy. The explanation lies in the type of cognitive skills required to answer the question. Let's take a closer look.

Categories of Human Cognition

Bloom's Taxonomy is a classification system that instructors use to identify the cognitive skill levels at which they are asking students to work, particularly on practice problems and exams. Bloom's Taxonomy is also a very useful tool for students to know—it can help you to figure out the appropriate level at which you should be studying to succeed in a course.

Bloom's Taxonomy distinguishes six different categories of human thinking: remember, understand, apply, analyze, evaluate, and create. One of the most useful distinctions lies not in the differences among the six categories, but rather in the difference between high-order cognitive (HOC) and low-order cognitive (LOC) skills. **FIGURE B16.1** shows how the different levels of the taxonomy can be broken into HOC and LOC skills.

Skills that hallmark LOCs include recall, explanation, or application of knowledge in the exact way that you have before (remember, understand, apply), while skills that typify HOCs include the application of knowledge in a new way, as well as the breakdown, critique, or creation of information (analyze, evaluate, and create). Most college instructors will assume students are proficient at solving LOC questions and will expect you to frequently work at the HOC levels. HOC problems usually require use of basic vocabulary and applying knowledge—working at this level helps students to master the LOC levels.

Six Study Steps to Success

Bloom's Taxonomy provides a useful guide for preparing for an exam, using the following six steps:

1. *Answer in-chapter questions while reading the chapter.* All questions in this book have been assigned Bloom's levels, so you can review the question answers and the Bloom's level while you study.

2. *Identify the Bloom's level(s) of the questions that you are having greatest difficulty answering.* While working through

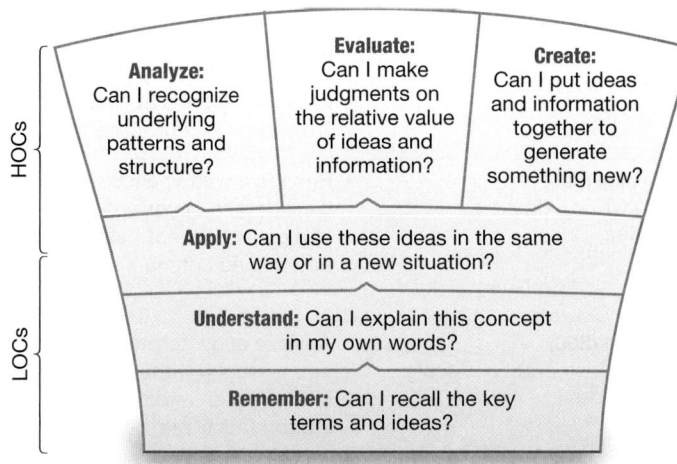

FIGURE B16.1 Bloom's Taxonomy.

the text, take note of the content and Bloom's level(s) that you find the most challenging.

3. *Use the Bloom's Taxonomy Study Guide* (**Table B16.1;** see **B:30**) *to focus your study efforts at the appropriate Bloom's level.* Table B16.1 lists specific study methods that can help you practice your understanding of the material at both the LOC and HOC levels, whether you are studying alone or with a study group.

4. *Complete the end-of-chapter questions as if you're taking an exam, without looking for the answers.* If you look at the chapter text or jump to the answers, then you really aren't testing your ability to work with the content and have reduced the questions to the lowest Bloom's level of remember.

5. *Grade your answers and note the Bloom's level of the questions you got wrong.* At what level of Bloom's Taxonomy were the questions you missed?

6. *Use the Bloom's Taxonomy Study Guide to focus your study efforts at the appropriate Bloom's level.* If you missed a lot of questions, then spend more time studying the material and find other resources for quizzing yourself.

By following these steps and studying at the HOC levels, you should succeed in answering questions on in-class exams.

	Individual Study Activities	Group Study Activities
Create (HOC) Generate something new	• Generate a hypothesis or design an experiment based on information you are studying • Create a model based on a given data set • Create summary sheets that show how facts and concepts relate to each other • Create questions at each level of Bloom's Taxonomy as a practice test and then take the test	• Each student puts forward a hypothesis about biological process and designs an experiment to test it. Peers critique the hypotheses and experiments • Create a new model/summary sheet/concept map that integrates each group member's ideas
Evaluate (HOC) Defend or judge a concept or idea	• Provide a written assessment of the strengths and weaknesses of your peers' work or understanding of a given concept based on previously determined criteria	• Provide a verbal assessment of the strengths and weaknesses of your peers' work or understanding of a given concept based on previously described criteria, and have your peers critique your assessment
Analyze (HOC) Distinguish parts and make inferences	• Analyze and interpret data in primary literature or a textbook without reading the author's interpretation and then compare the authors' interpretation with your own • Analyze a situation and then identify the assumptions and principles of the argument • Compare and contrast two ideas or concepts • Construct a map of the main concepts by defining the relationships of the concepts using one- or two-way arrows	• Work together to analyze and interpret data in primary literature or a textbook without reading the author's interpretation, and defend your analysis to your peers • Work together to identify all of the concepts in a paper or textbook chapter, construct individual maps linking the concepts together with arrows and words that relate the concepts, and then grade each other's concept maps
Apply (HOC or LOC) Use information or concepts in new ways (HOC) or in the same ways (LOC)	• Review each process you have learned and then ask yourself: What would happen if you increase or decrease a component in the system, or what would happen if you alter the activity of a component in the system? • If possible, graph a biological process and create scenarios that change the shape or slope of the graph	• Practice writing out answers to old exam questions on the board, and have your peers check to make sure you don't have too much or too little information in your answer • Take turns teaching your peers a biological process while the group critiques the content
Understand (LOC) Explain information or concepts	• Describe a biological process in your own words without copying it from a book or another source • Provide examples of a process • Write a sentence using the word • Give examples of a process	• Discuss content with peers • Take turns quizzing each other about definitions, and have your peers check your answer
Remember (LOC) Recall information	• Practice labeling diagrams • List characteristics • Identify biological objects or components from flash cards • Quiz yourself with flash cards • Take a self-made quiz on vocabulary • Draw, classify, select, or match items • Write out the textbook definitions	• Check a drawing that another student labeled • Create lists of concepts and processes that your peers can match • Place flash cards in a bag and take turns selecting one for which you must define a term • Do the preceding activities, and have peers check your answers

APPENDIX C Periodic Table of Elements

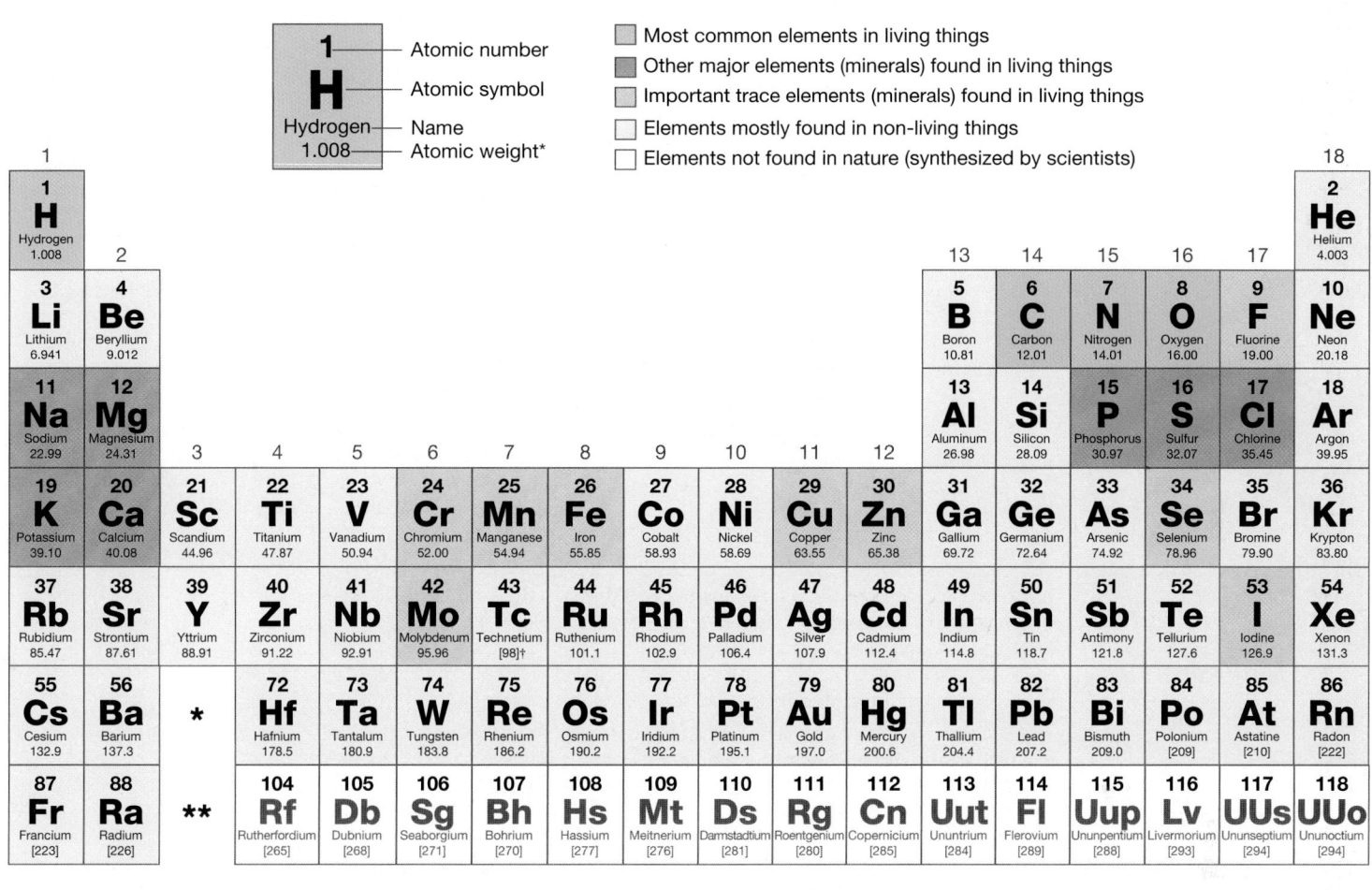

DATA: Wieser, M. E., and M. Berglund. 2009. *Pure and Applied Chemistry* 81: 2131–2156.

*Atomic weights are reported to four significant figures.

†For elements with a variable number of protons and/or neutrons, the mass number of the longest-lived isotope of the element is reported in brackets.

Glossary

5′ cap A modified guanine (G) nucleotide (7-methylguanylate) added to the 5′ end of eukaryotic mRNAs. Helps protect the mRNA from being degraded and promotes efficient initiation of translation.

abdomen A region of the body; in insects, one of the three prominent body regions called tagmata.

abiotic Not alive (e.g., air, water, and some components of soil). Compare with **biotic**.

aboveground biomass The total mass of living plants in an area, excluding roots.

abscisic acid (ABA) A plant hormone that inhibits growth; it stimulates stomatal closure and triggers dormancy.

abscission In plants, the normal (often seasonal) shedding of leaves, fruits, or flowers.

abscission zone The region at the base of a petiole where cell wall degradation occurs; results in the dropping of leaves.

absorption In animals, the uptake of ions and small molecules, derived from food, across the lining of the intestine and into the bloodstream.

absorption spectrum The amount of light of different wavelengths absorbed by a pigment. Usually depicted as a graph of light absorbed versus wavelength. Compare with **action spectrum**.

acclimation A change in a study organism's phenotype that occurs in response to laboratory conditions.

acclimatization A change in an individual's phenotype that occurs in response to a change in natural environmental conditions.

acetyl CoA A molecule produced by oxidation of pyruvate (the final product of glycolysis) in a reaction catalyzed by pyruvate dehydrogenase. Can enter the citric acid cycle and is used as a carbon source in the synthesis of fatty acids, steroids, and other compounds.

acetylation Addition of an acetyl group ($-COCH_3$) to a molecule. Acetylation of histone proteins is important in controlling chromatin condensation.

acetylcholine (Ach) A neurotransmitter, released by nerve cells at neuromuscular junctions, that triggers contraction of skeletal muscle cells but slows the rate of contraction in cardiac muscle cells. Also used as a neurotransmitter between neurons.

acid Any compound that gives up protons or accepts electrons during a chemical reaction or that releases hydrogen ions when dissolved in water.

acid-growth hypothesis The hypothesis that auxin triggers elongation of plant cells by increasing the activity of proton pumps, making the cell wall more acidic and leading to expansion of the cell wall and an influx of water.

acoelomate A bilaterian animal that lacks an internal body cavity (coelom). Compare with **coelomate** and **pseudocoelomate**.

acquired immune deficiency syndrome (AIDS) A human disease characterized by death of immune system cells (in particular helper T cells) and subsequent vulnerability to other infections. Caused by the human immunodeficiency virus (HIV).

acrosome A caplike structure, located on the head of a sperm cell, that contains enzymes capable of dissolving the outer coverings of an egg.

ACTH See **adrenocorticotropic hormone**.

actin A globular protein that can be polymerized to form filaments. Actin filaments are part of the cytoskeleton and constitute the thin filaments in skeletal muscle cells.

actin filament A long fiber, about 7 nm in diameter, composed of two intertwined strands of polymerized actin protein; one of the three types of cytoskeletal fibers. Involved in cell movement. Also called a *microfilament*. Compare with **intermediate filament** and **microtubule**.

action potential A rapid, temporary change in electrical potential across a membrane, from negative to positive and back to negative. Occurs in cells, such as neurons and muscle cells, that have an excitable membrane.

action spectrum The relative effectiveness of different wavelengths of light in driving a light-dependent process such as photosynthesis. Usually depicted as a graph of some measure of the process versus wavelength. Compare with **absorption spectrum**.

activation energy The amount of energy required to initiate a chemical reaction; specifically, the energy required to reach the transition state.

activator A protein that binds to a DNA regulatory sequence to increase the frequency of transcription initiation by RNA polymerase.

active site The location in an enzyme molecule where substrates (reactant molecules) bind and react.

active transport The movement of ions or molecules across a membrane against an electrochemical gradient. Requires energy (e.g., from hydrolysis of ATP) and assistance of a transport protein (e.g., pump).

adaptation Any heritable trait that increases the fitness of an individual with that trait, compared with individuals without that trait, in a particular environment.

adaptive immune response See **adaptive immunity**.

adaptive immunity Immunity to a particular pathogen or other antigen conferred by activated B and T cells in vertebrates. Characterized by specificity, diversity, memory, and self–nonself recognition. Also called *adaptive immune response*. Compare with **innate immunity**.

adaptive radiation Rapid evolutionary diversification within one lineage, producing many descendant species with a wide range of adaptive forms.

adenosine triphosphate (ATP) A molecule consisting of an adenine base, a sugar, and three phosphate groups that can be hydrolyzed to release energy. Universally used by cells to store and transfer energy.

adhesion The tendency of certain dissimilar molecules to cling together due to attractive forces. Compare with **cohesion**.

adipocyte A fat cell.

adrenal gland Either of two small endocrine glands, one above each kidney. The outer portion (cortex) secretes several steroid hormones; the inner portion (medulla) secretes epinephrine and norepinephrine.

adrenaline See **epinephrine**.

adrenocorticotropic hormone (ACTH) A peptide hormone, produced and secreted by the anterior pituitary, that stimulates release of steroid hormones (e.g., cortisol, aldosterone) from the adrenal cortex.

adult A sexually mature individual.

adventitious root A root that develops from a plant's shoot system instead of from the plant's root system.

aerobic Referring to any metabolic process, cell, or organism that uses oxygen as an electron acceptor. Compare with **anaerobic**.

afferent division The part of the nervous system that transmits information about the internal and external environment to the central nervous system. Consists mainly of sensory neurons. Compare with **efferent division**.

age class All the individuals of a specific age in a population.

age-specific fecundity The average number of female offspring produced by a female in a certain age class.

age structure The proportion of individuals in a population that are of each possible age.

agglutination Clumping together of cells or viruses by antibodies or other cross-linking molecules.

aggregate fruit A fruit (e.g., raspberry) that develops from a single flower that has many separate carpels. Compare with **multiple** and **simple fruit**.

AIDS See **acquired immune deficiency syndrome**.

albumen A solution of water and protein (particularly albumins), found in amniotic eggs, that nourishes the growing embryo. Also called *egg white*.

alcohol fermentation Catabolic pathway in which pyruvate produced by glycolysis is converted to ethanol in the absence of oxygen.

aldosterone A hormone that stimulates the kidney to conserve salt and water and promotes retention of sodium; produced in the adrenal cortex.

allele A particular version of a gene.

allergen Any molecule (antigen) that triggers an allergic response (an allergy).

allergy An IgE-mediated abnormal response to an antigen, usually characterized by dilation of blood vessels, contraction of smooth muscle cells, and increased activity of mucus-secreting cells.

allopatric speciation Speciation that occurs when populations of the same species become geographically isolated, often due to dispersal or vicariance. Compare with **sympatric speciation**.

allopatry Condition in which two or more populations live in different geographic areas. Compare with **sympatry**.

allopolyploidy (adjective: allopolyploid) The state of having more than two full sets of chromosomes (polyploidy) due to hybridization between different species. Compare with **autopolyploidy**.

allosteric regulation Regulation of a protein's function by binding of a regulatory molecule, usually to a specific site distinct from the active site, that causes a change in the protein's shape.

α-amylase See **amylase**.

α-helix (alpha-helix) A protein secondary structure in which the polypeptide backbone coils into a spiral shape stabilized by hydrogen bonding.

alternation of generations A life cycle involving alternation of a multicellular haploid stage (gametophyte) with a multicellular diploid stage (sporophyte). Occurs in most plants and some protists.

alternative splicing In eukaryotes, the splicing of primary RNA transcripts from a single gene in different ways to produce different mature mRNAs and thus different polypeptides.

altruism Any behavior that has a fitness cost to the individual (lowered survival and/or reproduction) and a fitness benefit to the recipient. See **reciprocal altruism**.

alveolus (plural: alveoli) Any of the tiny air-filled sacs of a mammalian lung.

ambisense RNA virus A ssRNA virus whose genome consists of at least one strand that contains both positive-sense and negative-sense regions.

aminoacyl tRNA A transfer RNA molecule that is covalently bound to an amino acid.

aminoacyl-tRNA synthetase An enzyme that catalyzes the addition of a particular amino acid to its corresponding tRNA molecule.

ammonia (NH_3) A small molecule, produced by the breakdown of proteins and nucleic acids, that is very toxic to cells. Is a strong base that gains a proton to form the ammonium ion (NH_4^+).

amnion The innermost of the membranes surrounding the embryo in an amniotic egg.

amniotes A major lineage of vertebrates (Amniota) that reproduce with amniotic eggs. Includes all reptiles (including birds) and mammals—all tetrapods except amphibians.

amniotic egg An egg that has a watertight shell or case enclosing a membrane-bound water supply (the amnion), food supply (yolk sac), and waste sac (allantois).

amoeboid motion See **cell crawling**.

amphibians A lineage of vertebrates, many of which breathe through their skin and feed on land but lay their eggs in water. Represent the earliest tetrapods; include frogs, salamanders, and caecilians.

amphipathic Containing hydrophilic and hydrophobic regions.

ampullae of Lorenzini Structures on the heads of sharks that contain cells with electroreceptors.

amylase Any enzyme that can break down starch by catalyzing hydrolysis of the glycosidic linkages between the glucose residues.

amyloplasts Starch-storing organelles (plastids) in plants. In root cap cells, they settle to the bottom of the cell and may be used as gravity detectors.

anabolic pathway Any set of chemical reactions that synthesizes large molecules from smaller ones. Generally requires an input of energy. Compare with **catabolic pathway**.

anadromous Having a life cycle in which adults live in the ocean (or large lakes) but migrate up freshwater streams to breed and lay eggs.

anaerobic Referring to any metabolic process, cell, or organism that uses an electron acceptor other than oxygen, including fermentation or anaerobic respiration. Compare with **aerobic**.

anaphase A stage in mitosis or meiosis during which chromosomes are moved to opposite poles of the spindle apparatus.

anatomy The study of the physical structure of organisms.

ancestral trait A trait found in the ancestors of a particular group.

aneuploidy (adjective: aneuploid) The state of having an abnormal number of copies of a certain chromosome.

angiosperm A flowering vascular plant that produces seeds within mature ovaries (fruits). The angiosperms form a single lineage. Compare with **gymnosperm**.

animal A member of a major lineage of eukaryotes (Animalia) whose members typically have a complex, large, multicellular body; eat other organisms; and are mobile.

animal model Any disease that occurs in a non-human animal and has parallels to a similar disease of humans. Studied by medical researchers in hopes that findings may apply to human disease.

anion A negatively charged ion.

annelids Members of the phylum Annelida (segmented worms). Distinguished by a segmented body and a coelom that functions as a hydrostatic skeleton. Annelids belong to the lophotrochozoan branch of the protostomes.

annual Referring to a plant whose life cycle normally lasts only one growing season—less than one year. Compare with **perennial**.

anoxygenic Referring to any process or reaction that does not produce oxygen. Photosynthesis in purple sulfur and purple nonsulfur bacteria, which does not involve photosystem II, is anoxygenic. Compare with **oxygenic**.

antagonistic muscle group A set of two or more muscles that reextend one another by transmitting their forces via the skeleton.

antenna (plural: antennae) A long appendage of the head that is used to touch or smell.

antenna complex Part of a photosystem, containing an array of chlorophyll molecules and accessory pigments, that receives energy from light and directs the energy to a central reaction center during photosynthesis.

anterior Toward an animal's head and away from its tail. The opposite of posterior.

anterior pituitary The part of the pituitary gland containing endocrine cells that produce and release a variety of peptide hormones in response to other hormones from the hypothalamus. Compare with **posterior pituitary**.

anther The pollen-producing structure at the end of a stamen in flowering plants (angiosperms).

antheridium (plural: antheridia) The sperm-producing structure in most land plants except angiosperms.

anthropoids One of the two major lineages of primates, including apes, humans, and all monkeys. Compare with **prosimians**.

antibiotic Any substance, such as penicillin, that can kill or inhibit the growth of bacteria.

antibody A protein produced by B cells that can bind to a specific part of an antigen, tagging it for removal by the immune system. All monomeric forms of antibodies consist of two light chains and two heavy chains, which vary between different antibodies. Also called *immunoglobulin*.

anticodon The sequence of three bases (a triplet) in a transfer RNA molecule that can bind to an mRNA codon with a complementary sequence.

antidiuretic hormone (ADH) A peptide hormone, secreted from the posterior pituitary gland, that stimulates water retention by the kidney. Also called *vasopressin*.

antigen Any foreign molecule, often a protein, that can stimulate an innate or adaptive response by the immune system.

antigen presentation Process by which small peptides, derived from ingested particulate antigens (e.g., bacteria) or intracellular antigens (e.g., viruses in infected cell) are complexed with MHC proteins and transported to the cell surface, where they are displayed and can be recognized by T cells.

antiparallel Describing the opposite orientation of nucleic acid strands that are hydrogen bonded to one another, with one strand running in the $5' \rightarrow 3'$ direction and the other in the $3' \rightarrow 5'$ direction.

antiporter A carrier protein that allows an ion to diffuse down an electrochemical gradient, using the energy of that process to transport a different substance in the opposite direction *against* its concentration gradient. Compare with **symporter**.

antiviral Any drug or other agent that can interfere with the transmission or replication of viruses.

aorta In terrestrial vertebrates, the major artery carrying oxygenated blood away from the heart.

aphotic zone Deep water receiving no sunlight. Compare with **photic zone**.

apical Toward the top. In plants, at the tip of a branch. In animals, on the side of an epithelial layer that faces the environment and not other body tissues. Compare with **basal**.

apical–basal axis The shoot-to-root axis of a plant.

apical bud A bud at the tip of a stem or branch, where growth occurs to lengthen the stem or branch.

apical dominance Inhibition of lateral bud growth by the apical meristem at the tip of a plant branch.

apical meristem A group of undifferentiated plant cells, at the tip of a shoot or root, that is responsible for primary growth. Compare with **cambium**.

apodeme Any of the chitinous ingrowths of the exoskeleton to which muscles attach.

apomixis The formation of mature seeds without fertilization occurring; a type of asexual reproduction.

apoplast In plants, the region outside plasma membranes consisting of the porous cell walls and the intervening extracellular air space. Compare with **symplast**.

apoptosis Series of tightly controlled changes that lead to the self-destruction of a cell. Occurs frequently during embryological development and as part of the immune response to remove infected or cancerous cells. Also called *programmed cell death*.

appendix A blind sac (having only one opening) that extends from the cecum in some mammals.

aquaporin A type of channel protein through which water can move by osmosis across a plasma membrane.

aquifer An underground layer of porous rock, sand, or gravel that is saturated with water.

ara operon A set of three genes in *E. coli* that are transcribed into a single mRNA and required for metabolism of the sugar arabinose. Transcription of the *ara* operon is controlled by the AraC regulatory protein.

araC The regulatory gene (written as *araC*) or regulatory protein (when written as AraC) of the *E. coli* *ara* operon.

arbuscular mycorrhizal fungi (AMF) Fungi from the Glomeromycota lineage whose hyphae enter the root cells of their host plants. Also called *endomycorrhizal fungi*.

Archaea One of the three taxonomic domains of life, consisting of unicellular prokaryotes distinguished by cell walls made of certain polysaccharides not found in bacterial or eukaryotic cell walls, plasma membranes composed of unique isoprene-containing phospholipids, and ribosomes and RNA polymerase similar to those of eukaryotes. Compare with **Bacteria** and **Eukarya**.

archegonium (plural: archegonia) The egg-producing structure in most land plants except angiosperms.

arteriole Any of the many tiny vessels that carry blood from arteries to capillaries.

arteriosclerosis Hardening and loss of elasticity of arteries.

artery Any thick-walled blood vessel that carries blood (oxygenated or not) under relatively high pressure away from the heart to organs throughout the body. Compare with **vein**.

arthropods Members of the phylum Arthropoda. Distinguished by a segmented body; a hard, jointed exoskeleton; paired, jointed appendages; and an extensive body cavity called a hemocoel. Arthropods belong to the ecdysozoan branch of the protostomes.

articulation A movable point of contact between two rigid components of a skeleton, such as between bones of a vertebrate endoskeleton or between segments of cuticle in an arthropod exoskeleton. See **joint**.

artificial selection Deliberate manipulation by humans, as in animal and plant breeding, of the genetic composition of a population by allowing only individuals with desirable traits to reproduce.

ascus (plural: asci) Specialized spore-producing cell found at the ends of hyphae in "sac fungi" (Ascomycota).

asexual reproduction Any form of reproduction where offspring inherit DNA from only one parent. Includes binary fission, budding, and parthenogenesis. Compare with **sexual reproduction**.

astral microtubules Mitotic and meiotic microtubules that have arisen from the two spindle poles and interact with proteins on the plasma membrane.

asymmetric competition Ecological competition between two species in which one species suffers a much greater fitness decline than the other. Compare with **symmetric competition**.

atomic number The number of protons in the nucleus of an atom, giving the atom its identity as a particular chemical element.

atomic weight The average mass of an element that is based on the relative proportions of all the naturally occurring isotopes.

ATP synthase A large membrane-bound protein complex that uses the energy of protons flowing through it to synthesize ATP.

ATP See **adenosine triphosphate**.

atrioventricular (AV) node A region of the heart between the right atrium and right ventricle where electrical signals from the atrium are slowed briefly before spreading to the ventricle. This delay allows the ventricle to fill with blood before contracting. Compare with **sinoatrial (SA) node**.

atrium (plural: atria) A thin-walled chamber of the heart that receives blood from veins and pumps it to a neighboring chamber (the ventricle).

autocrine Relating to a chemical signal that affects the same cell that produced and released it.

autoimmunity A pathological condition in which the immune system attacks self cells or tissues of an individual's own body.

autonomic nervous system The part of the peripheral nervous system that controls internal organs and involuntary processes, such as stomach contraction, hormone release, and heart rate. Includes parasympathetic and sympathetic nerves. Compare with **somatic nervous system**.

autophagy The process by which damaged organelles are surrounded by a membrane and delivered to a lysosome to be recycled.

autopolyploidy (adjective: autopolyploid) The state of having more than two full sets of chromosomes

(polyploidy) due to a mutation that doubled the chromosome number. All the chromosomes come from the same species. Compare with **allopolyploidy**.

autosomal inheritance The inheritance patterns that occur when genes are located on autosomes rather than on sex chromosomes.

autosome Any chromosome other than a sex chromosome (i.e,. any chromosome other than the X or Y in mammals).

autotroph Any organism that can synthesize reduced organic compounds from simple inorganic sources such as CO_2 or CH_4. Most plants and some bacteria and archaea are autotrophs. Also called *primary producer*. Compare with **heterotroph**.

auxin Indoleacetic acid (IAA), a plant hormone that stimulates phototropism and cell elongation.

axillary bud A bud that forms at a node and may develop into a lateral (side) branch. Also called *lateral bud*.

axon A long projection of a neuron that can propagate an action potential.

axon hillock The site in a neuron where an axon joins the cell body and where action potentials are first triggered.

axoneme A structure found in eukaryotic cilia and flagella and responsible for their motion; composed of two central microtubules surrounded by nine doublet microtubules (9 + 2 arrangement).

B cell A type of lymphocyte that matures in the bone marrow and, with T cells, is responsible for adaptive immunity. Produces antibodies and also functions in antigen presentation. Also called *B lymphocyte*.

B-cell receptor (BCR) An immunoglobulin protein embedded in the plasma membrane of mature B cells and to which antigens bind. Apart from the transmembrane domain, it is identical in structure to antibodies.

BAC library A collection of all the sequences found in the genome of a species, inserted into **bacterial artificial chromosomes (BACs)**.

background extinction The average rate of low-level extinction that has occurred continuously throughout much of evolutionary history. Compare with **mass extinction**.

Bacteria One of the three taxonomic domains of life, consisting of unicellular prokaryotes distinguished by cell walls composed largely of peptidoglycan, plasma membranes similar to those of eukaryotic cells, and ribosomes and RNA polymerase that differ from those in archaeans or eukaryotes. Compare with **Archaea** and **Eukarya**.

bacterial artificial chromosome (BAC) An artificial version of a bacterial chromosome that can be used as a cloning vector to produce many copies of large DNA fragments.

bacteriophage Any virus that infects bacteria.

baculum A bone inside the penis; usually present in mammals with a penis that lacks erectile tissue.

balancing selection A mode of natural selection in which no single allele is favored in all populations of a species at all times. Instead, there is a balance among alleles in terms of fitness and frequency.

ball-and-stick model A representation of a molecule where atoms are shown as balls—colored and scaled to indicate the atom's identity—and covalent bonds are shown as rods or sticks connecting the balls in the correct geometry.

bar coding The use of well-characterized gene sequences to identify species.

bark The protective outer layer of woody plants, composed of cork cells, cork cambium, and secondary phloem.

baroreceptors Specialized nerve cells in the walls of the heart and certain major arteries that detect changes in blood pressure and trigger appropriate responses by the brain.

basal Toward the base. In plants, toward the root or at the base of a branch where it joins the stem. In animals, on the side of an epithelial layer that abuts underlying body tissues. Compare with **apical**.

basal body The microtubule organizing center for cilia and flagella in eukaryotic cells. Consists of nine triplets of microtubules arranged in a circle and establishes the structure of axonemes. Structurally identical with a centriole.

basal lamina A thick, collagen-rich extracellular matrix that underlies most epithelial tissues in animals and connects it to connective tissue.

basal metabolic rate (BMR) The total energy consumption by an organism at rest in a comfortable environment. For aerobes, often measured as the amount of oxygen consumed per hour.

basal transcription factor Proteins, present in all eukaryotic cells, that bind to promoters and help initiate transcription. Compare with **regulatory transcription factor**.

base Any compound that acquires protons or gives up electrons during a chemical reaction or accepts hydrogen ions when dissolved in water.

basidium (plural: basidia) Specialized spore-producing cell at the ends of hyphae in club fungi, members of the Basidiomycota.

basilar membrane The membrane on which the bottom portion of hair cells sits in the vertebrate cochlea.

basolateral Toward the bottom and sides. In animals, the side of an epithelial layer that faces other body tissues and not the environment.

Batesian mimicry A type of mimicry in which a harmless or palatable species resembles a dangerous or poisonous species. Compare with **Müllerian mimicry**.

beak A structure that exerts biting forces and is associated with the mouth; found in birds, cephalopods, and some insects.

behavior Any action by an organism, often in response to a stimulus.

behavioral ecology The study of how organisms respond to particular abiotic and biotic stimuli from their environment.

beneficial In genetics, referring to any mutation, allele, or trait that increases an individual's fitness.

benign tumor A mass of abnormal tissue that appears due to unregulated growth but does not spread to other organs. Benign tumors are not cancers. Compare with **malignant tumor**.

benthic Living at the bottom of an aquatic environment.

benthic zone The area along the bottom of an aquatic environment.

β-pleated sheet (beta-pleated sheet) A protein secondary structure in which the polypeptide backbone folds into a sheetlike shape stabilized by hydrogen bonding.

bilateral symmetry An animal body pattern in which one plane of symmetry divides the body into a left side and a right side. Typically, the body is long and narrow, with a distinct head end and tail end. Compare with **radial symmetry**.

bilaterian A member of a major lineage of animals (Bilateria) that are bilaterally symmetrical at some point in their life cycle, have three embryonic germ layers, and have a coelom. All protostomes and deuterostomes are bilaterians.

bile A complex solution produced by the liver, stored in the gallbladder, and secreted into the intestine. Contains steroid derivatives called bile salts that are responsible for emulsification of fats during digestion.

binary fission The process of cell division used for asexual reproduction of many prokaryotic cells. The genetic material is replicated and partitioned to opposite sides of a growing cell, which is then divided in half to create two genetically identical cells.

biodiversity The diversity of life considered at three levels: genetic diversity (variety of alleles and/or genes in a population, species, or group of species); species diversity (variety and relative abundance of species present in a certain area); and ecosystem diversity (variety of communities and abiotic components in a region).

biodiversity hotspot A region that is extraordinarily rich in species.

biofilm A complex community of bacteria enmeshed in a polysaccharide-rich, extracellular matrix that allows them to attach to a surface.

biogeochemical cycle The pattern of circulation of an element or molecule among living organisms and the environment.

biogeography The study of how species and populations are distributed geographically.

bioinformatics The field of study concerned with managing, analyzing, and interpreting biological information, particularly DNA sequences.

biological species concept The definition of a species as a population or group of populations that are reproductively isolated from other groups. Members of a species have the potential to interbreed in nature to produce viable, fertile offspring but cannot interbreed successfully with members of other species. Compare with **morphospecies** and **phylogenetic species concepts**.

bioluminescence The emission of light by a living organism via an enzyme-catalyzed reaction.

biomagnification In animal tissues, an increase in the concentration of particular molecules that may occur as those molecules are passed up a food chain.

biomass The total mass of all organisms in a given population or geographical area; usually expressed as total dry weight.

biome A large terrestrial or marine region characterized by distinct abiotic characteristics and dominant types of vegetation.

biomechanics A field of biology that applies the principles of physics and engineering to analyze the mechanical structure and function of organisms.

bioprospecting The effort to find commercially useful compounds by studying organisms—especially species that are poorly studied to date.

bioremediation The use of living organisms, usually bacteria or archaea, to degrade environmental pollutants.

biosphere The thin zone surrounding the Earth where all life exists; the sum of all terrestrial and aquatic ecosystems.

biotechnology The application of biological techniques and discoveries to medicine, industry, and agriculture.

biotic Living, or produced by a living organism. Compare with **abiotic**.

bipedal Walking primarily on two legs; characteristic of hominins.

bipolar cell A type of cell in the vertebrate retina that receives information from one or more photoreceptors and passes it to other bipolar cells or ganglion cells.

bivalent The structure formed by synapsed homologous chromosomes during prophase of meiosis I. Also known as a *tetrad*.

bivalves A lineage of mollusks that have shells made of two parts, or valves, such as clams and mussels.

bladder A mammalian organ that holds urine until it can be excreted.

blade The wide, flat part of a plant leaf.

blastocoel Fluid-filled cavity in the blastula of many animal species.

blastocyst The mammalian blastula. A roughly spherical structure composed of trophoblast cells on the exterior and a cluster of cells (the inner cell mass) that fills part of the interior space.

blastomere A cell created by cleavage divisions in early animal embryos.

blastopore An opening (pore) in the surface of some early embryos, through which cells move during gastrulation.

blastula In vertebrate development, a ball of cells (blastomere cells) typically surrounding a fluid-filled cavity (the blastocoel). The blastula is formed by cleavage of a zygote and undergoes gastrulation. See **blastocyst**.

blood A type of connective tissue consisting of red blood cells and leukocytes suspended in a fluid portion, an extracellular matrix called plasma. Transports materials throughout the vertebrate body.

body mass index (BMI) A mathematical relationship of weight and height used to assess obesity in humans. Calculated as weight (kg) divided by the square of height (m^2).

body plan The basic architecture of an animal's body, including the number and arrangement of limbs, body segments, and major tissue layers.

bog A freshwater wetland that has no or almost no water flow, resulting in very low oxygen levels and acidic conditions.

Bohr shift The rightward shift of the oxygen–hemoglobin equilibrium curve that occurs with decreasing pH. It results in hemoglobin being more likely to release oxygen in the acidic environment of exercising muscle.

bone A type of vertebrate connective tissue consisting of living cells and blood vessels within a hard extracellular matrix composed of calcium phosphate ($CaPO_4$) and small amounts of calcium carbonate ($CaCO_3$) and protein fibers.

bone marrow The soft tissue filling the inside of large bones; contains stem cells that develop into red blood cells and leukocytes throughout life.

Bowman's capsule The hollow, double-walled, cup-shaped portion of a nephron that surrounds a glomerulus in the vertebrate kidney.

brain A large mass of neurons, located in the head region of an animal, that is involved in information processing; may also be called the cerebral ganglion.

brain stem The most posterior portion of the vertebrate brain, connecting to the spinal cord and responsible for autonomic body functions such as heart rate, respiration, and digestion.

braincase See **cranium**.

branch (1) A part of a phylogenetic tree that represents populations through time. (2) Any extension of a plant's shoot system.

brassinosteroids A family of steroid hormones found in plants; stimulate growth.

bronchiole Any of the small tubes in mammalian lungs that carry air from the bronchi to the alveoli.

bronchus (plural: bronchi) In mammals, one of a pair of large tubes that lead from the trachea to each lung.

bryophyte See **non-vascular plants**.

budding Asexual reproduction in which an outgrowth from the parent breaks free as an independent individual; occurs in yeasts and some invertebrates.

buffer A substance that, in solution, acts to minimize changes in the pH of that solution when acid or base is added.

bulbourethral gland In male mammals, either of a small pair of glands at the base of the urethra that secrete an alkaline mucus (part of semen), which lubricates the tip of the penis and neutralizes acids in the urethra during copulation. In humans, also called *Cowper's glands*.

bulk flow The directional mass movement of a fluid due to pressure differences, such as movement of water through plant xylem and phloem, and movement of blood in animals.

bulk-phase endocytosis Nonspecific uptake of extracellular fluid by pinching off the plasma membrane to form small membrane-bound vesicles; considered to be a means of retrieving membrane

from the surface following exocytosis. Compare with **receptor-mediated endocytosis**.

bundle-sheath cell A type of cell found around the vascular tissue (veins) of plant leaves.

C_3 pathway The most common form of photosynthesis in which atmospheric CO_2 is fixed by rubisco to form 3-phosphoglycerate, a three-carbon molecule. Used in first phase of the Calvin cycle.

C_4 pathway A variant type of photosynthesis in which atmospheric CO_2 is first fixed by PEP carboxylase into four-carbon acids, rather than the three-carbon molecules of the classic C_3 pathway. Used to concentrate CO_2 to reduce photorespiration in the Calvin cycle while stomata are closed.

C_4 photosynthesis A variant type of photosynthesis in which atmospheric CO_2 is first fixed into four-carbon sugars, rather than the three-carbon sugars of classic C_3 photosynthesis. Enhances photosynthetic efficiency in hot, dry environments by reducing loss of oxygen due to photorespiration.

cadherin Any of a class of cell-surface proteins involved in selective cell–cell adhesion. Important for coordinating movements of cells and the establishment of tissues during embryological development.

callus In plants, a mass of undifferentiated cells that can generate roots and other tissues necessary to create a mature plant.

Calvin cycle In photosynthesis, the set of reactions that use NADPH and ATP formed in the light-dependent reactions to drive the fixation of CO_2, reduction of the fixed carbon to produce sugar, and the regeneration of the substrate used to fix CO_2. Also called *light-independent reactions*.

calyx All of the sepals of a flower.

CAM See **crassulacean acid metabolism**.

cambium (plural: cambia) In woody plants, tissue that consists of two types of cylindrical meristems that increase the width of roots and shoots through the process of secondary growth. See **vascular cambium** and **cork cambium**.

Cambrian explosion The rapid diversification of animal body types and lineages that occured between the species present in the Doushantuo faunas (around 570 mya), Ediacaran faunas (565–542 mya), and the Early Cambrian faunas (525–515 mya).

cancer General term for any tumor whose cells grow in an uncontrolled fashion, invade nearby tissues, and spread to other sites in the body.

capillarity The tendency of water to move up a narrow tube due to adhesion, cohesion, and surface tension (also called capillary action).

capillary Any of the many small, thin-walled blood vessels that permeate all tissues and organs, and allow exchange of gases and other molecules between blood and body cells.

capillary bed A thick network of capillaries.

capsid A shell of protein enclosing the genome of a virus particle.

carapace In crustaceans, a large platelike section of the exoskeleton that covers and protects the cephalo-thorax (e.g., a crab's "shell").

carbohydrate Any of a class of molecules that contain a carbonyl group, several hydroxyl groups, and

several to many carbon-hydrogen bonds. See **monosaccharide** and **polysaccharide**.

carbon cycle, global The worldwide movement of carbon among terrestrial ecosystems, the oceans, and the atmosphere.

carbon fixation The process of converting gaseous carbon dioxide into an organic molecule, often associated with photosynthesis. See also **PEP carboxylase** and **rubisco**.

carbonic anhydrase An enzyme that catalyzes the formation of carbonic acid (H_2CO_3) from carbon dioxide and water.

carboxylic acids Organic acids with the form R-COOH (a carboxyl group).

cardiac cycle One complete heartbeat cycle, including systole and diastole.

cardiac muscle The muscle tissue of the vertebrate heart; responsible for pumping blood. Consists of long, branched fibers that are electrically connected and that initiate their own contractions; not under voluntary control. Compare with **skeletal** and **smooth muscle**.

cardiovascular disease A group of diseases of the heart and blood vessels caused by poor diet, obesity, inactivity, genetics, tobacco use, age, and other factors.

carnivore (adjective: carnivorous) An animal whose diet consists predominantly of meat, or other animals. Most members of the mammalian taxon Carnivora are carnivores. Some plants are carnivorous, trapping and killing small animals and then absorbing nutrients from the prey's body. Compare with **herbivore** and **omnivore**.

carotenoid Any of a class of accessory pigments, found in chloroplasts, that absorb wavelengths of light not absorbed by chlorophyll; typically appear yellow, orange, or red. Includes carotenes and xanthophylls.

carpel The female reproductive organ in a flower. Consists of the stigma, to which pollen grains adhere; the style, through which the pollen tube grows; and the ovary, which houses the ovule. Compare with **stamen**.

carrier protein A membrane protein that facilitates diffusion of a small molecule (e.g., glucose) across a membrane by a process involving a reversible change in the shape of the protein. Also called *carrier* or *transporter*. Compare with **channel protein**.

carrier A heterozygous individual carrying a normal allele and a recessive allele for an inherited trait; does not display the phenotype of the recessive trait but can pass the recessive allele to offspring.

carrying capacity (*K*) The maximum population size of a certain species that a given habitat can support.

cartilage A type of vertebrate connective tissue that consists of relatively few cells scattered in a stiff matrix of polysaccharides and protein fibers. Provides structural support.

Casparian strip In plant roots, a waxy layer containing suberin, a water-repellent substance that prevents movement of water through the walls of endodermal cells, thus blocking the apoplastic pathway of water and ion movement into the vascular tissue.

cast A type of fossil, formed when the decay of a body part leaves a void that is then filled with minerals that later harden.

catabolic pathway Any set of chemical reactions that breaks down large, complex molecules into smaller ones, releasing energy in the process. Compare with **anabolic pathway**.

catalysis (verb: catalyze) Acceleration of the rate of a chemical reaction due to a decrease in the free energy of the transition state, called the activation energy.

catalyst Any substance that increases the rate of a chemical reaction without itself undergoing any permanent chemical change.

catecholamines A class of small compounds, derived from the amino acid tyrosine, that are used as hormones or neurotransmitters. Include epinephrine, norepinephrine, and dopamine.

cation A positively charged ion.

cation exchange The release (displacement) of cations, such as magnesium and calcium from soil particles, by protons in acidic soil water. The released cations are available for uptake by plants.

CD4 A membrane protein on the surface of some T cells in humans. $CD4^+$ T cells can give rise to helper T cells.

CD8 A membrane protein on the surface of some T cells in humans. $CD8^+$ T cells can give rise to cytotoxic T cells.

Cdk See **cyclin-dependent kinase**.

cDNA See **complementary DNA**.

cDNA library A set of cDNAs from a particular cell type or stage of development. Each cDNA is carried by a plasmid or other cloning vector and can be separated from other cDNAs. Compare with **genomic library**.

cecum A blind sac between the small intestine and the colon. Is enlarged in some species (e.g., rabbits) that use it as a fermentation vat for digestion of cellulose.

cell A highly organized compartment bounded by a thin, flexible structure (plasma membrane) and containing concentrated chemicals in an aqueous (watery) solution. The basic structural and functional unit of all organisms.

cell body The part of a neuron that contains the nucleus; where incoming signals are integrated. Also called the *soma*.

cell crawling A form of cellular movement involving actin filaments in which the cell produces bulges (pseudopodia) that stick to the substrate and pull the cell forward. Also called *amoeboid motion*.

cell cycle Ordered sequence of events in which a eukaryotic cell replicates its chromosomes, evenly partitions the chromosomes to two daughter cells, and then undergoes division of the cytoplasm.

cell-cycle checkpoint Any of several points in the cell cycle at which progression of a cell through the cycle can be regulated.

cell division Creation of new cells by division of preexisting cells.

cell-mediated (immune) response The type of immune response that involves generation of cytotoxic

T cells from $CD8^+$ T cells. Defends against pathogen-infected cells, cancer cells, and transplanted cells. Compare with **humoral (immune) response**.

cell membrane See **plasma membrane**.

cell plate A flattened sac-like structure formed in the middle of a dividing plant cell from Golgi-derived vesicles containing cell wall material; ultimately divides the cytoplasm into two separate cells.

cell sap An aqueous solution found in the vacuoles of plant cells.

cell theory The theory that all organisms are made of cells and that all cells come from preexisting cells.

cell wall A fibrous layer found outside the plasma membrane of most bacteria and archaea and many eukaryotes.

cellular respiration A common pathway for production of ATP, involving transfer of electrons from compounds with high potential energy through an electron transport chain and ultimately to an electron acceptor (often oxygen).

cellulase An enzyme that can break down cellulose by catalyzing hydrolysis of the glycosidic linkages between the glucose residues.

cellulose A structural polysaccharide composed of glucose monomers joined by β-1,4-glycosidic linkages. Found in the cell wall of algae, plants, and some bacteria and fungi.

Cenozoic era The most recent interval of geologic time, beginning 65.5 million years ago, during which mammals became the dominant vertebrates and angiosperms became the dominant plants.

central dogma The scheme for information flow in the cell: DNA → RNA → protein.

central nervous system (CNS) Large numbers of neurons aggregated into clusters called ganglia in bilaterian animals. In vertebrates, the central nervous system consists of the brain and spinal cord. Compare with **nerve net** and **peripheral nervous system (PNS)**.

centriole One of two small cylindrical structures found together within the centrosome near the nucleus of a eukaryotic cell (not found in plants). Consists of microtubule triplets and is structurally identical with a basal body.

centromere Constricted region of a replicated chromosome where the two sister chromatids are joined and the kinetochore is located.

centrosome Structure in animal and fungal cells, containing two centrioles, that serves as a microtubule organizing center for the cell's cytoskeleton and for the spindle apparatus during cell division.

cephalization The formation in animals of a distinct anterior region (the head) where sense organs and a mouth are clustered.

cephalochordates One of the three major chordate lineages (Cephalochordata), comprising small, mobile organisms that live in marine sands and suspension feed; also called *lancelets* or *amphioxus*. Compare with **urochordates** and **vertebrates**.

cephalopods A lineage of mollusks including the squid, octopuses, and nautiluses. Distinguished by large brains, excellent vision, tentacles, and a reduced or absent shell.

cerebellum Posterior section of the vertebrate brain; involved in coordination of complex muscle movements, such as those required for locomotion and maintaining balance.

cerebrum The most anterior section of the vertebrate brain. Divided into left and right hemispheres and four lobes: frontal lobe, involved in complex decision making (in humans); occipital lobe, receives and interprets visual information; parietal lobe, involved in integrating sensory and motor functions; and temporal lobe, functions in memory, speech (in humans), and interpreting auditory information.

cervix The bottom portion of the uterus, containing a canal that leads to the vagina.

chaetae (singular: chaeta) Bristle-like extensions found in some annelids.

channel protein A protein that forms a pore in a cell membrane. The structure of most channels allows them to admit just one or a few types of ions or molecules. Compare with **carrier protein**.

character Any genetic, morphological, physiological, or behavioral characteristic of an organism to be studied.

character displacement The evolutionary tendency for the traits of similar species that occupy overlapping ranges to change in a way that reduces interspecific competition.

chelicerae A pair of clawlike appendages found around the mouth of certain arthropods called chelicerates (spiders, mites, and allies).

chemical bond An attractive force binding two atoms together. Covalent bonds, ionic bonds, and hydrogen bonds are types of chemical bonds.

chemical energy The potential energy stored in covalent bonds between atoms.

chemical equilibrium A dynamic but stable state of a reversible chemical reaction in which the forward reaction and reverse reactions proceed at the same rate, so that the concentrations of reactants and products remain constant.

chemical evolution A The theory that simple chemical compounds in the early atmosphere and ocean combined via chemical reactions to form larger, more complex substances, eventually leading to the origin of life and the start of biological evolution.

chemical reaction Any process in which one compound or element is combined with others or is broken down; involves the making and/or breaking of chemical bonds.

chemiosmosis An energetic coupling mechanism whereby energy stored in an electrochemical proton gradient is used to drive an energy-requiring process such as production of ATP.

chemokine Any of a subset of cytokines that acts as a chemical signal attracting leukocytes to a site of tissue injury or infection.

chemolithotroph An organism (bacteria or archaea) that produces ATP by oxidizing inorganic molecules with high potential energy such as ammonia (NH_3) or methane (CH_4). Also called *lithotroph*. Compare with **chemoorganotroph**.

chemoorganotroph An organism that produces ATP by oxidizing organic molecules with high potential energy such as sugars. Also called *organotroph*. Compare with **chemolithotroph**.

chemoreception A sensory system in which receptors are activated in response to the binding of chemicals.

chemoreceptor A sensory cell or organ specialized for detection of specific molecules or classes of molecules.

chiasma (plural: chiasmata) The X-shaped structure formed during meiosis by crossing over between non-sister chromatids in a pair of homologous chromosomes.

chitin A structural polysaccharide composed of *N*-acetyl-glucosamine (NAG) monomers joined end to end by β-1,4-glycosidic linkages. Found in cell walls of fungi and many algae, and in external skeletons of insects and crustaceans.

chitons A lineage of marine mollusks that have a protective shell formed of eight calcium carbonate plates.

chlorophyll Any of several closely related green pigments, found in chloroplasts, that absorb light during photosynthesis.

chloroplast A chlorophyll-containing organelle, bounded by a double membrane, in which photosynthesis occurs; found in plants and photosynthetic protists. Also the location of starch, amino acid, fatty acid, purine, and pyrimidine synthesis.

choanocyte A specialized, flagellated feeding cell found in choanoflagellates (protists that are the closest living relatives of animals) and sponges (the most ancient animal phylum).

cholecystokinin A peptide hormone secreted by cells in the lining of the small intestine. Stimulates the secretion of digestive enzymes from the pancreas and of bile from the liver and gallbladder.

chordate Any member of the phylum Chordata. Chordates are deuterostomes distinguished by a dorsal hollow nerve cord, pharyngeal gill slits, a notochord, and a post-anal tail. Includes vertebrates, cephalochordata, and urochordata.

chromatid One of the two identical double-stranded DNAs composing a replicated chromosome that is connected at the centromere to the other strand.

chromatin The complex of DNA and proteins, mainly histones, that compose eukaryotic chromosomes. Can be highly compact (heterochromatin) or loosely coiled (euchromatin).

chromatin remodeling The process by which the DNA in chromatin is unwound from its associated proteins to allow transcription or replication. May involve chemical modification of histone proteins or reshaping of the chromatin by large multiprotein complexes in an ATP-requiring process.

chromatin remodeling complexes Sets of enzymes that use energy from ATP hydrolysis shift nucleosomes on DNA to expose regulatory sequences to transcription factors.

chromosome theory of inheritance The principle that genes are located on chromosomes and that patterns of inheritance are determined by the behavior of chromosomes during meiosis.

chromosome Gene-carrying structure consisting of a single long molecule of double-stranded DNA and

associated proteins (e.g., histones). Most prokaryotic cells contain a single, circular chromosome; eukaryotic cells contain multiple noncircular (linear) chromosomes located in the nucleus.

cilium (plural: cilia) One of many short, filamentous projections of some eukaryotic cells, containing a core of microtubules. Used to move the cell as well as to circulate fluid or particles around the surface of a stationary cell. See **axoneme**.

circadian clock An internal mechanism found in most organisms that regulates many body processes (sleep–wake cycles, hormonal patterns, etc.) in a roughly 24-hour cycle.

circulatory system The system responsible for moving oxygen, carbon dioxide, and other materials (hormones, nutrients, wastes) within the animal body.

cisternae (singular: cisterna) Flattened, membrane-bound compartments that make up the Golgi apparatus.

cisternal maturation The process of cargo movement through the Golgi apparatus by residing in cisternae that mature from *cis* to *trans* via the import and export of different Golgi enzymes.

citric acid cycle A series of eight chemical reactions that start with citrate (deprotonated citric acid) and ends with oxaloacetate, which reacts with acetyl CoA to form citrate—forming a cycle that is part of the pathway that oxidizes glucose to CO_2. Also known as the *Krebs cycle* or *tricarboxylic acid (TCA) cycle*.

clade See **monophyletic group**.

cladistic approach A method for constructing a phylogenetic tree that is based on identifying the unique traits (shared, derived characters, called synapomorphies) of each monophyletic group.

Class I MHC protein A type of MHC protein that is present on the plasma membrane of virtually all nucleated cells and functions in presenting antigen to $CD8^+$ T cells.

Class II MHC protein A type of MHC protein that is present only on the plasma membrane of certain cells in the immune response, such as dendritic cells, macrophages, and B cells. It functions in presenting epitopes of antigens to $CD4^+$ T cells.

cleavage In animal development, the series of rapid mitotic cell divisions, with little cell growth, that produces successively smaller cells (blastomeres) and transforms a zygote into a multicellular blastula.

cleavage furrow A pinching in of the plasma membrane that occurs as cytokinesis begins in animal cells and deepens until the cytoplasm is divided into two daughter cells.

climate The prevailing, long-term weather conditions in a particular region.

climax community The stable, final community that develops from ecological succession.

clitoris A rod of erectile tissue in the external genitalia of female mammals. Is formed from the same embryonic tissue as the male penis and has a similar function in sexual arousal.

cloaca In a few mammals and many nonmammalian vertebrates, a body cavity opening to the outside and used by both the excretory and reproductive systems.

clonal selection theory The dominant explanation of the generation of an adaptive immune response. According to the theory, the immune system retains a vast pool of inactive lymphocytes, each with a unique receptor for a unique epitope. Lymphocytes that encounter their complementary epitopes are stimulated to divide (selected and cloned), producing daughter cells that combat infection and confer immunity.

clone (1) An individual that is genetically identical to another individual. (2) A lineage of genetically identical individuals or cells. (3) As a verb, to make one or more genetic replicas of a cell or individual.

cloning vector A plasmid or other agent used to transfer recombinant genes into cultured host cells. Also called simply *vector*.

closed circulatory system A circulatory system in which the circulating fluid (blood) is confined to blood vessels and flows in a continuous circuit. Compare with **open circulatory system**.

cnidocyte A specialized stinging cell found in cnidarians (e.g., jellyfish, corals, and anemones) and used in capturing prey.

co-receptor Any membrane protein that acts with some other membrane protein in a cell interaction or cell response.

cochlea The organ of hearing in the inner ear of mammals, birds, and crocodilians. A coiled, fluid-filled tube containing specialized pressure-sensing neurons (hair cells) that detect sounds of different pitches.

coding strand See **non-template strand**.

codominance An inheritance pattern in which heterozygotes exhibit both of the traits seen in each type of homozygous individual.

codon A sequence of three nucleotides in DNA or RNA that codes for an amino acid or a start or stop signal for protein synthesis.

coefficient of relatedness (*r*) A measure of how closely two individuals are related. Calculated as the probability that an allele in two individuals is inherited from the same ancestor.

coelom An internal, usually fluid-filled body cavity that is completely or partially lined with mesoderm.

coelomate An animal that has a true coelom, completely lined with mesoderm. Compare with **acoelomate** and **pseudocoelomate**.

coenocytic Containing many nuclei and a continuous cytoplasm through a filamentous body, without the body being divided into distinct cells. Some fungi are coenocytic.

coenzyme A small organic molecule that is a required cofactor for an enzyme-catalyzed reaction. Often donates or receives electrons or functional groups during the reaction.

coenzyme A (CoA) A molecule that is required for many cellular reactions and that is often transiently linked to other molecules, such as acetyl groups (see **acetyl CoA**).

coenzyme Q A nonprotein molecule that shuttles electrons between membrane-bound complexes in the mitochondrial electron transport chain. Also called **ubiquinone** or Q.

coevolution A pattern of evolution in which two interacting species reciprocally influence each other's adaptations over time.

coevolutionary arms race A series of adaptations and counter-adaptations observed in species that interact closely over time and affect each other's fitness.

cofactor An inorganic ion, such as a metal ion, that is required for an enzyme to function normally. May be bound tightly to an enzyme or associate with it transiently during catalysis.

cohesion The tendency of certain like molecules (e.g., water molecules) to cling together due to attractive forces. Compare with **adhesion**.

cohesion-tension theory The theory that water movement upward through plant vascular tissues is due to loss of water from leaves (transpiration), which pulls a cohesive column of water upward.

cohort A group of individuals that are the same age and can be followed through time.

coleoptile A modified leaf that covers and protects the stems and leaves of grass seedlings.

collagen A fibrous, pliable, cable-like glycoprotein that is a major component of the extracellular matrix of animal cells. Various subtypes differ in their tissue distribution, some of which are assembled into large fibrils in the extracellular space.

collecting duct In the vertebrate kidney, a large straight tube that receives filtrate from the distal tubules of several nephrons. Involved in the regulated reabsorption of water.

collenchyma cell In plants, an elongated cell with cell walls thickened at the corners that provides support to growing plant parts; usually found in strands along leaf veins and stalks. Compare with **parenchyma cell** and **sclerenchyma cell**.

colon The portion of the large intestine where feces are formed by compaction of wastes and reabsorption of water.

colony An assemblage of individuals. May refer to an assemblage of semi-independent cells or to a breeding population of multicellular organisms.

commensalism (adjective: commensal) A symbiotic relationship in which one organism (the commensal) benefits and the other (the host) is not harmed. Compare with **mutualism** and **parasitism**.

communication In ecology, any process in which a signal from one individual modifies the behavior of another individual.

community All of the species that interact with each other in a certain area.

companion cell In plants, a cell in the phloem that is connected via many plasmodesmata to adjacent sieve-tube elements. Companion cells provide materials to maintain sieve-tube elements and function in the loading and unloading of sugars into sieve-tube elements.

compass orientation A type of navigation in which movement occurs in a specific direction.

competition In ecology, the interaction of two species or two individuals trying to use the same limited resource (e.g., water, food, living space). May occur between individuals of the same species (intraspecific competition) or different species (interspecific competition).

competitive exclusion principle The principle that two species cannot coexist in the same ecological niche in the same area because one species will outcompete the other.

competitive inhibition Inhibition of an enzyme's ability to catalyze a chemical reaction via binding of a nonreactant molecule that competes with the substrate(s) for access to the active site.

complement system A set of proteins that circulate in the bloodstream and can destroy bacteria by forming holes in the bacterial plasma membrane.

complementary base pairing The association between specific nitrogenous bases of nucleic acids stabilized by hydrogen bonding. Adenine pairs only with thymine (in DNA) or uracil (in RNA), and guanine pairs only with cytosine.

complementary DNA (cDNA) DNA produced in the laboratory using an RNA transcript as a template and reverse transcriptase; corresponds to a gene but lacks introns. Also produced naturally by retroviruses.

complementary strand A newly synthesized strand of RNA or DNA that has a base sequence complementary to that of the template strand.

complete digestive tract A digestive tract with two openings, usually called a mouth and an anus.

complete metamorphosis See **holometabolous metamorphosis**.

compound eye An eye formed of many independent light-sensing columns (ommatidia); occurs in arthropods. Compare with **simple eye**.

concentration gradient Difference across space (e.g., across a membrane) in the concentration of a dissolved substance.

condensation reaction A chemical reaction in which two molecules are joined covalently with the removal of an $-OH$ from one and an $-H$ from another to form water. Also called a *dehydration reaction*. Compare with **hydrolysis**.

conduction (1) Direct transfer of heat between two objects that are in physical contact. Compare with **convection**. (2) Transmission of an electrical impulse along the axon of a nerve cell.

cone cell A photoreceptor cell with a cone-shaped outer portion that is particularly sensitive to bright light of a certain color. Also called simply *cone*. Compare with **rod cell**.

connective tissue An animal tissue consisting of scattered cells in a liquid, jellylike, or solid extracellular matrix. Includes bone, cartilage, tendons, ligaments, and blood.

conservation biology The effort to study, preserve, and restore threatened genetic diversity, populations, communities, and ecosystems.

constant (C) region The invariant amino acid sequence in polypeptides that are used to make antibodies, B-cell receptors, and T-cell receptors. Apart from antibody class types (IgG, IgM, etc.), this region remains constant within an individual. Compare with **variable (V) region**.

constitutive Always occurring; always present. Commonly used to describe enzymes and other proteins that are synthesized continuously or mutants in which one or more genetic loci are constantly expressed due to defects in gene control.

constitutive defense A defensive trait that is always manifested even in the absence of a predator or pathogen. Also called *standing defense*. Compare with **inducible defenses**.

constitutive mutant An abnormal (mutated) strain that produces a product at all times, instead of under certain conditions only.

consumer See **heterotroph**.

consumption Predation or herbivory.

continental shelf The portion of a geologic plate that extends from a continent under seawater.

continuous strand See **leading strand**.

contraception Any of several methods to prevent pregnancy.

control In a scientific experiment, a group of organisms or samples that do not receive the experimental treatment but are otherwise identical to the group that does.

convection Transfer of heat by movement of large volumes of a gas or liquid. Compare with **conduction**.

convergent evolution The independent evolution of similar traits in distantly related organisms due to adaptation to similar environments and a similar way of life.

cooperative binding The tendency of the protein subunits of hemoglobin to affect each other's oxygen binding such that each bound oxygen molecule increases the likelihood of further oxygen binding.

coprophagy The eating of feces.

coral reef A large assemblage of colonial marine corals that usually serves as shallow-water, sunlit habitat for many other species as well.

co-receptor Any membrane protein that acts with some other membrane protein in a cell interaction or cell response.

core enzyme The enzyme responsible for catalysis in a multipart holoenzyme.

cork cambium One of two types of cylindrical meristem, consisting of a ring of undifferentiated plant cells found just under the cork layer of woody stems and roots; produces new cork cells on its outer side. Compare with **vascular cambium**.

cork cell A cell in the protective outermost layer of a woody stem and root that produces and accumulates waxes that make the cell less permeable to water and gases.

corm A rounded, thick underground stem that can produce new plants via asexual reproduction.

cornea The transparent sheet of connective tissue at the very front of the eye in vertebrates and some other animals. Protects the eye and helps focus light.

corolla All of the petals of a flower.

corona The cluster of cilia at the anterior end of a rotifer; in many species it facilitates suspension feeding.

corpus callosum A thick band of neurons that connects the two hemispheres of the cerebrum in the mammalian brain.

corpus luteum A yellowish structure that secretes progesterone in an ovary. Is formed from a follicle that has recently ovulated.

cortex (1) In animals, the outermost region of an organ, such as the kidney or adrenal gland. (2) In plants, a layer of ground tissue found outside the vascular bundles of roots and outside the pith of a stem.

corticotropin-releasing hormone (CRH) A peptide hormone, produced and secreted by the hypothalamus, that stimulates the anterior pituitary to release ACTH.

cortisol A steroid hormone, produced and secreted by the adrenal cortex, that increases blood glucose and prepares the body for stress. The major glucocorticoid hormone in some mammals. Also called *hydrocortisone*.

cost–benefit analysis Decisions or analyses that weigh the fitness costs and benefits of a particular action.

cotransporter A transmembrane protein that facilitates diffusion of an ion down its previously established electrochemical gradient and uses the energy of that process to transport some other substance, in the same or opposite direction, *against* its concentration gradient. Also called *secondary active transporter*. See **antiporter** and **symporter**.

cotyledon The first leaf, or seed leaf, of a plant embryo. Used for storing and digesting nutrients and/or for early photosynthesis.

countercurrent exchanger In animals, any anatomical arrangement that allows the maximum transfer of heat or a soluble substance from one fluid to another. The two fluids must be flowing in opposite directions and have a heat or concentration gradient between them.

covalent bond A type of chemical bond in which two atoms share one or more pairs of electrons. Compare with **hydrogen bond** and **ionic bond**.

cranium A bony, cartilaginous, or fibrous case that encloses and protects the brain of vertebrates. Forms part of the skull. Also called *braincase*.

crassulacean acid metabolism (CAM) A variant type of photosynthesis in which CO_2 is fixed and stored in organic acids at night when stomata are open and then released to feed the Calvin cycle during the day when stomata are closed. Helps reduce water loss and CO_2 loss by photorespiration.

cristae (singular: crista) Sac-like invaginations of the inner membrane of a mitochondrion. Location of the electron transport chain and ATP synthase.

Cro-Magnon A prehistoric European population of modern humans (*Homo sapiens*) known from fossils, paintings, sculptures, and other artifacts.

crop A storage organ in the digestive system of certain vertebrates.

cross A mating between two individuals that is used for genetic analysis.

cross-pollination Pollination of a flower by pollen from another individual, rather than by self-fertilization. Also called *crossing*.

cross-talk Interactions among signaling pathways, triggered by different signals, that modify a cellular response.

crossing over The exchange of segments of non-sister chromatids between a pair of homologous chromosomes that occurs during meiosis I.

crosstalk Interactions among signaling pathways that modify a cellular response.

crustaceans A lineage of arthropods that includes shrimp, lobster, and crabs. Many have a carapace (a platelike portion of the exoskeleton covering the cephalothorax) and mandibles for biting or chewing.

cryptic species A species that cannot be distinguished from similar species by easily identifiable morphological traits.

culture In cell biology, a collection of cells or a tissue growing under controlled conditions, usually in suspension or on the surface of a dish of solid growth medium.

Cushing's disease A human endocrine disorder caused by loss of feedback inhibition of cortisol on ACTH secretion. Characterized by high ACTH and cortisol levels and wasting of body protein reserves.

cuticle A protective coating secreted by the outermost layer of cells of an animal or a plant; often functioning to reduce evaporative water loss.

cyanobacteria A lineage of photosynthetic bacteria formerly known as blue-green algae. Likely the first life-forms to carry out oxygenic photosynthesis.

cyclic AMP (cAMP) Cyclic adenosine monophosphate; a small molecule, derived from ATP, that is widely used by cells in signal transduction and transcriptional control.

cyclic electron flow Path of electrons in which excited electrons of photosystem I are transferred back to plastoquinone (PQ), the start of the electron transport chain normally associated with photosystem II. Instead of reducing $NADP^+$ to make NADPH, the electron energy is used to make ATP via photophosphorylation. Compare with **noncyclic electron flow**.

cyclin One of several regulatory proteins whose concentrations fluctuate cyclically throughout the cell cycle. Involved in the control of the cell cycle via cyclin-dependent kinases.

cyclin-dependent kinase (Cdk) Any of several related protein kinases that are functional only when bound to a cyclin and are activated by other modifications. Involved in control of the cell cycle.

cytochrome *c* (cyt *c*) A soluble protein that shuttles electrons between membrane-bound complexes in the mitochondrial electron transport chain.

cytokine Any of a diverse group of signaling proteins, secreted largely by cells of the immune system, whose effects include stimulating leukocyte production, recruiting cells to the site of infection, tissue repair, and fever. Generally function to regulate the type, intensity, and duration of an immune response.

cytokinesis Division of the cytoplasm to form two daughter cells. Typically occurs immediately after division of the nucleus by mitosis or meiosis.

cytokinins A class of plant hormones that stimulate cell division and retard aging.

cytoplasm All the contents of a cell, excluding the nucleus, bounded by the plasma membrane.

cytoplasmic determinant A regulatory transcription factor or signaling molecule that is distributed unevenly in the cytoplasm of the egg and that directs early pattern formation in an embryo.

cytoplasmic streaming The directed flow of cytosol and organelles that facilitates distribution of materials within some large plant and fungal cells. Occurs along actin filaments and is powered by myosin.

cytoskeleton In eukaryotic cells, a network of protein fibers in the cytoplasm that are involved in cell shape, support, locomotion, and transport of materials within the cell. Prokaryotic cells have a similar but much less extensive network of fibers.

cytosol The fluid portion of the cytoplasm, excluding the contents of membrane-enclosed organelles.

cytotoxic T cell A type of $CD8^+$ effector T cell that induces apoptosis in infected and cancerous cells. Recognizes target cells via interactions with complementary class I MHC–peptide complexes. Also called *cytotoxic T lymphocyte (CTL)* and *killer T cell*. Compare with **helper T cell**.

dalton (Da) A unit of mass equal to 1/12 the mass of one carbon-12 atom; about the mass of 1 proton or 1 neutron.

daughter strand The strand of DNA that is newly replicated from an existing template strand of DNA.

day-neutral plant A plant whose flowering time is not affected by the relative length of day and night (the photoperiod). Compare with **long-day** and **short-day plant**.

dead space Air passages that are not involved in gas exchange with the blood; examples are the trachea and bronchi.

deciduous Describing a plant that sheds leaves or other structures at regular intervals (e.g., each autumn).

decomposer See **detritivore**.

decomposer food chain An ecological network of detritus, decomposers that eat detritus, and predators and parasites of the decomposers.

deep sequencing A method to learn the types of mRNAs or DNA sequences present in cells, and their relative amounts, involving the preparation and sequencing of cDNA libraries.

definitive host The host species in which a parasite reproduces sexually. Compare with **intermediate host**.

dehydration reaction See **condensation reaction**.

deleterious In genetics, referring to any mutation, allele, or trait that reduces an individual's fitness.

deletion In genetics, refers to the loss of part of a chromosome.

demography The study of factors that determine the size and structure of populations through time.

dendrite A short extension from a neuron's cell body that receives signals from other neurons.

dendritic cell A type of leukocyte that ingests and digests foreign antigens, moves to a lymph node, and presents the antigens' epitopes, in the context of MHC proteins on its membrane, to $CD4^+$ and $CD8^+$ T cells.

dense connective tissue A type of connective tissue, distinguished by having an extracellular matrix dominated by collagen fibers. Found in tendons and ligaments.

density dependent In population ecology, referring to any characteristic that varies depending on population density.

density independent In population ecology, referring to any characteristic that does not vary with population density.

deoxyribonucleic acid (DNA) A nucleic acid composed of deoxyribonucleotides that carries the genetic information of a cell. Generally occurs as two intertwined strands, but these can be separated. See also **double helix**.

deoxyribonucleoside triphosphate (dNTP) A monomer used by DNA polymerase to polymerize DNA. Consists of the sugar deoxyribose, a base (A, T, G, or C), and three phosphate groups.

deoxyribonucleotide See **nucleotide**.

depolarization A change in membrane potential from its resting negative state to a less negative or a positive state; a normal phase in an action potential. Compare with **hyperpolarization**.

depolarization Change in membrane potential from its resting negative state to a less negative or to a positive state; a normal phase in an action potential. Compare with **hyperpolarization**.

deposit feeder An animal that eats its way through a food-containing substrate.

derived trait A trait that is clearly homologous with a trait found in an ancestor of a particular group, but that has a new form.

dermal tissue system The tissue forming the outer layer of a plant; also called *epidermis*.

descent with modification The phrase used by Darwin to describe his hypothesis of evolution by natural selection.

desmosome A type of cell–cell attachment structure, consisting of cadherin proteins, that is anchored to intermediate filaments. Serves to link the cytoskeletons of adjacent animal cells and form strong cell–cell attachments throughout a tissue. Compare with **gap junction** and **tight junction**.

detergent A type of small amphipathic molecule used to solubilize hydrophobic molecules in aqueous solution.

determination In development, the commitment of a cell to a particular differentiated fate. Once a cell is fully determined, it can differentiate only into a particular cell type (e.g., liver cell, brain cell).

detritivore An organism whose diet consists mainly of dead organic matter (detritus). Various bacteria, fungi, protists, and animals are detritivores. Also called *decomposer*.

detritus A layer of dead organic matter that accumulates at ground level or on seafloors and lake bottoms.

deuterostomes A major lineage of bilaterian animals that share a pattern of embryological development, including formation of the anus earlier than the mouth, and formation of the coelom by pinching off of layers of mesoderm from the gut. Includes echinoderms and chordates. Compare with **protostomes**.

developmental homology A similarity in embryonic form, or in the fate of embryonic tissues, that is due to inheritance from a common ancestor.

diabetes mellitus A disease caused by defects in insulin production (type 1) or in the response of cells to insulin (type 2). Characterized by abnormally high blood glucose levels and huge amounts of glucose-containing urine.

diaphragm An elastic, sheetlike structure. In mammals, the muscular sheet of tissue that separates the chest and abdominal cavities. It contracts and moves downward during inhalation, expanding the chest cavity.

diastole The portion of the cardiac cycle during which the atria or ventricles of the heart are relaxed. Compare with **systole**.

diastolic blood pressure The force exerted by blood against artery walls during relaxation of the heart's left ventricle. Compare with **systolic blood pressure**.

dicot Any flowering plant (angiosperm) that has two cotyledons (embryonic leaves) upon germination. The dicots do not form a monophyletic group. Also called *dicotyledonous plant*. Compare with **eudicot** and **monocot**.

dideoxy sequencing A laboratory technique for determining the exact nucleotide sequence of DNA. Relies on the use of dideoxynucleoside triphosphates (ddNTPs), which terminate DNA replication.

diencephalon The part of the mammalian brain that relays sensory information to the cerebellum and functions in maintaining homeostasis.

differential centrifugation Procedure for separating cellular components according to their size and density by spinning a cell homogenate in a series of centrifuge runs. After each run, the supernatant is removed from the deposited material (pellet) and spun again at progressively higher speeds.

differential gene expression Expression of different sets of genes in cells with the same genome. Responsible for creating different cell types.

differentiation The process by which any unspecialized cell becomes a distinct specialized cell type (e.g., liver cell, brain cell), usually by changes in gene expression. Also called *cell differentiation*.

diffusion Spontaneous movement of a substance from one region to another, often with a net movement from a region of high concentration to one of low concentration (i.e., down a concentration gradient).

digestion The physical and chemical breakdown of food into molecules that can be absorbed into the body of an animal.

digestive tract The long tube that begins at the mouth and ends at the anus. Also called *alimentary canal* or *gastrointestinal (GI) tract*.

dihybrid cross A mating between two parents that are heterozygous for two different genes.

dikaryotic Describing a cell or fungal mycelium having two haploid nuclei that are genetically distinct.

dimer An association of two molecules that may be identical (homodimer) or different (heterodimer).

dioecious Describing an angiosperm species that has male and female reproductive structures on separate plants. Compare with **monoecious**.

diploblast (adjective: diploblastic) An animal whose body develops from two basic embryonic cell layers or tissues—ectoderm and endoderm. Compare with **triploblast**.

diploid (1) Having two sets of chromosomes ($2n$). (2) A cell or an individual organism with two sets of chromosomes, one set inherited from the mother and one set from the father. Compare with **haploid**.

direct sequencing A technique for identifying and studying microorganisms that cannot be grown in culture. Involves detecting and amplifying copies of certain specific genes in the microorganisms' DNA, sequencing these genes, and then comparing the sequences with the known sequences from other organisms.

directional selection A mode of natural selection that favors one extreme phenotype with the result that the average phenotype of a population changes in one direction. Generally reduces overall genetic variation in a population. Compare with **disruptive selection** and **stabilizing selection**.

disaccharide A carbohydrate consisting of two monosaccharides (sugar residues) linked together.

discontinuous strand See **lagging strand**.

discrete trait An inherited trait that exhibits distinct phenotypes rather than the continuous variation characteristic of a quantitative trait such as body height.

dispersal The movement of individuals from their place of origin (birth, hatching) to a new location.

disruptive selection A mode of natural selection that favors extreme phenotypes at both ends of the range of phenotypic variation. Maintains overall genetic variation in a population. Compare with **stabilizing selection** and **directional selection**.

distal tubule In the vertebrate kidney, the convoluted portion of a nephron into which filtrate moves from the loop of Henle. Involved in the regulated reabsorption of sodium and water. Compare with **proximal tubule**.

disturbance In ecology, any strong, short-lived disruption to a community that changes the distribution of living and/or nonliving resources.

disturbance regime The characteristic disturbances that affect a given ecological community.

disulfide bond A covalent bond between two sulfur atoms, typically in the side chains of certain amino acids (e.g., cysteine). Often contributes to tertiary and quaternary levels of protein structure.

DNA See **deoxyribonucleic acid**.

DNA cloning Any of several techniques for producing many identical copies of a particular gene or other DNA sequence.

DNA fingerprinting Any of several methods for identifying individuals by unique features of their genomes. Commonly involves PCR to produce many copies of certain short tandem repeats (microsatellites) and then analyzing their lengths.

DNA helicase An enzyme that breaks hydrogen bonds between nucleotides of DNA, "unzipping" a double-stranded DNA molecule.

DNA library See **cDNA library** and **genomic library**.

DNA ligase An enzyme that joins pieces of DNA by catalyzing the formation of a phosphodiester bond between the pieces.

DNA methylation The addition of a methyl group ($-CH_3$) to a DNA molecule.

DNA methyltransferase A class of eukaryotic enzymes that add a methyl group to cytosines in DNA. Methylation of DNA leads to chromatin condensation and is an important means of regulating gene expression in eukaryotes.

DNA microarray A set of single-stranded DNA fragments, representing thousands of different genes that are permanently fixed to a small glass slide. Can be used to determine which genes are expressed in different cell types, under different conditions, or at different developmental stages.

DNA polymerase Any enzyme that catalyzes synthesis of DNA from deoxyribonucleotide triphosphates (dNTPs).

domain (1) A taxonomic category, based on similarities in basic cellular biochemistry, above the kingdom level. The three recognized domains are Bacteria, Archaea, and Eukarya. (2) A section of a protein that has a distinctive tertiary structure and function.

dominant Referring to an allele that determines the same phenotype when it is present in homozygous or heterozygous form.. Compare with **recessive**.

dormancy A temporary state of greatly reduced metabolic activity and growth in plants or plant parts (e.g., seeds, spores, bulbs, and buds).

dorsal Toward an animal's back and away from its belly. The opposite of ventral.

dorsal hollow nerve chord See **nerve chord**.

double fertilization An unusual form of reproduction seen in flowering plants, in which one sperm cell fuses with an egg to form a zygote and the other sperm cell fuses with two polar nuclei to form the triploid endosperm.

double helix The secondary structure of DNA, consisting of two antiparallel DNA strands wound around each other.

Down syndrome A human developmental disorder caused by trisomy of chromosome 21.

downstream In genetics, the direction in which RNA polymerase moves along a DNA strand. Compare with **upstream**.

duplication In genetics, refers to an additional copy of part of a chromosome.

dynein A class of motor proteins that uses the chemical energy of ATP to "walk" toward the minus end of a microtubule. Dyneins are responsible for bending of cilia and flagella, play a role in chromosome movement during mitosis, and can transport vesicles and organelles.

early endosome A small transient organelle that is formed by the accumulation of vesicles from receptor-mediated endocytosis and is an early stage in the formation of a lysosome.

ecdysone An insect hormone that triggers either molting (to a larger larval form) or metamorphosis (to the adult form), depending on the level of juvenile hormone.

ecdysozoans A major lineage of protostomes (Ecdysozoa) that grow by shedding their external skeletons (molting) and expanding their bodies. Includes arthropods, nematodes, and other groups. Compare with **lophotrochozoans**.

echinoderms A major lineage of deuterostomes (Echinodermata) distinguished by adult bodies with five-sided radial symmetry, a water vascular system, and tube feet. Includes sea urchins, sand dollars, and sea stars.

echolocation The use of echoes from vocalizations to obtain information about locations of objects in the environment.

ecological selection Also known as environmental selection. A type of natural selection that favors individuals with heritable traits that enhance their ability to survive and reproduce in a certain physical and/or biological environment, excluding their ability to obtain a mate. Compare with **sexual selection**.

ecology The study of how organisms interact with each other and with their surrounding environment.

ecosystem All the organisms that live in a geographic area, together with the nonliving (abiotic) components that affect or exchange materials with the organisms; a community and its physical environment.

ecosystem diversity The variety of biotic components in a region along with abiotic components, such as soil, water, and nutrients.

ecosystem function The sum of biological and chemical processes that are characteristic of a given ecosystem—such as primary production, nitrogen cycling, and carbon storage.

ecosystem services All of the benefits that humans derive, directly or indirectly, from ecosystem functions.

ecotourism Tourism that is based on observing wildlife or experiencing other aspects of natural areas.

ectoderm The outermost of the three basic cell layers (germ layers) in most animal embryos; gives rise to the outer covering and nervous system. Compare with **endoderm** and **mesoderm**.

ectomycorrhizal fungi (EMF) Fungi whose hyphae form a dense network that covers their host plant's roots but do not enter the root cells.

ectoparasite A parasite that lives on the outer surface of the host's body.

ectotherm An animal that gains most of its body heat from external sources as opposed to metabolic processes. Compare with **endotherm**.

effector Any cell, organ, or structure with which an animal can respond to external or internal stimuli. Usually functions, along with a sensor and integrator, as part of a homeostatic system.

efferent division The part of the nervous system that carries commands from the central nervous system to the body. Consists primarily of motor neurons.

egg A mature female gamete and any associated external layers (such as a shell). Larger and less mobile than the male gamete. In animals, also called *ovum*.

ejaculation The release of semen from the copulatory organ of a male animal.

ejaculatory duct A short duct through which sperm move during ejaculation; connects the vas deferens to the urethra.

electric current A flow of electrical charge past a point. Also called *current*.

electrical potential Potential energy created by a separation of electrical charges between two points. Also called *voltage*.

electrocardiogram (EKG) A recording of the electrical activity of the heart, as measured through electrodes on the skin.

electrochemical gradient The combined effect of an ion's concentration gradient and electrical (charge) gradient across a membrane that affects the diffusion of ions across the membrane.

electrogenic fish Any of various kinds of fishes having specialized electric organs that emit a current into the water to detect objects.

electrolyte Any compound that dissociates into ions when dissolved in water. In nutrition, any of the major ions necessary for normal cell function.

electromagnetic spectrum The entire range of wavelengths of radiation extending from short wavelengths (high energy) to long wavelengths (low energy). Includes gamma rays, X-rays, ultraviolet, visible light, infrared, microwaves, and radio waves (from short to long wavelengths).

electron acceptor A reactant that gains an electron and is reduced in a reduction–oxidation reaction.

electron carrier Any molecule that readily accepts electrons from and donates electrons to other molecules. Protons may be transferred with the electrons in the form of hydrogen atoms.

electron donor A reactant that loses an electron and is oxidized in a reduction–oxidation reaction.

electron shell A group of orbitals of electrons with similar energies. Electron shells are arranged in roughly concentric layers around the nucleus of an atom, and electrons in outer shells have more energy than those in inner shells. Electrons in the outermost shell, the valence shell, often are involved in chemical bonding.

electron transport chain (ETC) Any set of membrane-bound protein complexes and mobile electron carriers involved in a coordinated series of redox reactions in which the potential energy of electrons is successively decreased and used to pump protons from one side of a membrane to the other.

electronegativity A measure of the ability of an atom to attract electrons toward itself from an atom to which it is bonded.

electroreception A sensory system in which receptors are activated by electric fields.

electroreceptor A sensory cell or organ specialized to detect electric fields.

element A substance, consisting of atoms with a specific number of protons. Elements preserve their identity in chemical reactions.

elongation (1) The process by which RNA lengthens during transcription. (2) The process by which a polypeptide chain lengthens during translation.

elongation factors Proteins involved in the elongation phase of translation, assisting ribosomes in the synthesis of the growing peptide chain.

embryo A young, developing organism; the stage after fertilization and zygote formation.

embryo sac The female gametophyte in flowering plants.

embryogenesis The production of an embryo from a zygote. Embryogenesis is an early event in development of animals and plants.

Embryophyta An increasingly popular name for the lineage called land plants; reflects their retention of a fertilized egg.

embryophyte A plant that nourishes its embryos inside its own body. All land plants are embryophytes.

emergent property A property that stems from the interaction of simpler elements and that is impossible to predict from the study of individual elements.

emergent vegetation Any plants in an aquatic habitat that extend above the surface of the water.

emerging disease Any infectious disease, often a viral disease, that suddenly afflicts significant numbers of humans for the first time; often due to changes in the host species for a pathogen or to radical changes in the genetic material of the pathogen.

emigration The migration of individuals away from one population to other populations. Compare with **immigration**.

emulsification (verb: emulsify) The dispersion of fat into an aqueous solution. Usually requires the aid of an amphipathic substance such as a detergent or bile salts, which can break large fat globules into microscopic fat droplets.

endangered species A species whose numbers have decreased so much that it is in danger of extinction throughout all or part of its range.

endemic species A species that lives in one geographic area and nowhere else.

endergonic Referring to a chemical reaction that requires an input of energy to occur and for which the Gibbs free-energy change (ΔG) is greater than zero. Compare with **exergonic**.

endocrine Relating to a chemical signal (hormone) that is released into the bloodstream by a producing cell and acts on a distant target cell.

endocrine disruptor An exogenous chemical that interferes with normal hormonal signaling.

endocrine gland A gland that secretes hormones directly into the bloodstream or interstitial fluid instead of into ducts. Compare with **exocrine gland**.

endocrine system All of the glands and tissues that produce and secrete hormones into the bloodstream.

endocytosis General term for any pinching off of the plasma membrane that results in the uptake of material from outside the cell. Includes phagocytosis, pinocytosis, and receptor-mediated endocytosis. Compare with **exocytosis**.

endoderm The innermost of the three basic cell layers (germ layers) in most animal embryos; gives rise to the digestive tract and organs that connect to it (liver, lungs, etc.). Compare with **ectoderm** and **mesoderm**.

endodermis In plant roots, a cylindrical layer of cells that separates the cortex from the vascular tissue and location of the Casparian strip.

endomembrane system A system of organelles in eukaryotic cells that synthesizes, processes, transports, and recycles proteins and lipids. Includes the endoplasmic reticulum (ER), Golgi apparatus, and lysosomes.

endomycorrhizal fungi See **arbuscular mycorrhizal fungi (AMF)**.

endoparasite A parasite that lives inside the host's body.

endophyte (adjective: endophytic) A fungus that lives inside the aboveground parts of a plant in a symbiotic relationship. Compare with **epiphyte**.

endoplasmic reticulum (ER) A network of interconnected membranous sacs and tubules found inside eukaryotic cells. See **rough** and **smooth endoplasmic reticulum**.

endoskeleton Bony and/or cartilaginous structures within the body that provide support. Examples are the spicules of sponges, the plates in echinoderms, and the bony skeleton of vertebrates. Compare with **exoskeleton**.

endosperm A triploid ($3n$) tissue in the seed of a flowering plant (angiosperm) that serves as food for the plant embryo. Functionally analogous to the yolk in some animal eggs.

endosymbiont An organism that lives in a symbiotic relationship inside the body of its host.

endosymbiosis An association between organisms of two different species in which one lives inside the cell or cells of the other.

endosymbiosis theory The theory that mitochondria and chloroplasts evolved from prokaryotes that were engulfed by host cells and took up a symbiotic existence within those cells, a process termed primary endosymbiosis. In some eukaryotes, chloroplasts may have originated by secondary endosymbiosis; that is, when a cell engulfed a chloroplast-containing protist and retained its chloroplasts.

endotherm An animal whose primary source of body heat is internally generated. Compare with **ectotherm**.

endothermic Referring to a chemical reaction that absorbs heat. Compare with **exothermic**.

energetic coupling In cellular metabolism, the mechanism by which energy released from an exergonic reaction (commonly, hydrolysis of ATP) is used to drive an endergonic reaction.

energy The capacity to do work or to supply heat. May be stored (potential energy) or available in the form of motion (kinetic energy).

enhancer A regulatory sequence in eukaryotic DNA that may be located far from the gene it controls or within introns of the gene. Binding of specific proteins to an enhancer enhances the transcription of certain genes.

enrichment culture A method of detecting and obtaining cells with specific characteristics by placing a sample, containing many types of cells, under a specific set of conditions (e.g., temperature, salt concentration, available nutrients) and isolating those cells that grow rapidly in response.

enthalpy (*H*) A quantitative measure of the amount of potential energy, or heat content, of a system plus the pressure and volume it exerts on its surroundings.

entropy (*S*) A quantitative measure of the amount of disorder of any system, such as a group of molecules.

envelope (viral) A membrane that encloses the capsids of some viruses. Normally includes specialized proteins that attach to host-cell surfaces.

environmental sequencing See **metagenomics**.

enzyme A protein catalyst used by living organisms to speed up and control biological reactions.

epicotyl In some embryonic plants, a portion of the embryonic stem that extends above the cotyledons.

epidemic The spread of an infectious disease throughout a population in a short time period. Compare with **pandemic**.

epidermis The outermost layer of cells of any multicellular organism.

epididymis A coiled tube wrapped around each testis in reptiles, birds, and mammals. The site of the final stages of sperm maturation.

epigenetic inheritance Pattern of inheritance involving differences in phenotype that are not due to differences in the nucleotide sequence of genes.

epinephrine A catecholamine hormone, produced and secreted by the adrenal medulla, that triggers rapid responses related to the fight-or-flight response. Also called *adrenaline*.

epiphyte (adjective: epiphytic) A nonparasitic plant that grows on the trunks or branches of other plants and is not rooted in soil.

epithelial tissues See **epithelium**.

epithelium (plural: epithelia) An animal tissue consisting of sheetlike layers of tightly packed cells that line an organ, a gland, a duct, or a body surface. Also called *epithelial tissue*.

epitope A small region of a particular antigen to which an antibody, B-cell receptor, or T-cell receptor binds.

equilibrium potential The membrane potential at which there is no net movement of a particular ion into or out of a cell.

ER signal sequence A short amino acid sequence that marks a polypeptide for transport to the endoplasmic reticulum, where synthesis of the polypeptide chain is completed and the signal sequence removed. See **signal recognition particle**.

erythropoietin (EPO) A peptide hormone, released by the kidney in response to low blood-oxygen levels, that stimulates the bone marrow to produce more red blood cells.

esophagus The muscular tube that connects the mouth to the stomach.

essential amino acid Any amino acid that an animal cannot synthesize and must obtain from the diet. May refer specifically to one of the eight essential amino acids of adult humans: isoleucine, leucine, lysine, methionine, phenylalanine, threonine, tryptophan, and valine.

essential nutrient Any chemical element, ion, or compound that is required for normal growth, reproduction, and maintenance of a living organism and that cannot be synthesized by the organism.

EST See **expressed sequence tag**.

ester linkage The covalent bond formed by a condensation reaction between a carboxyl group and a hydroxyl group. Ester linkages join fatty acids to glycerol to form a fat or phospholipid.

estradiol The major estrogen produced by the ovaries of female mammals and many other vertebrates. Stimulates development of the female reproductive tract, growth of ovarian follicles, and growth of breast tissue in mammals.

estrogens A class of steroid hormones, including estradiol, estrone, and estriol, that generally promote female-like traits. Secreted by the gonads, fat tissue, and some other organs.

estrous cycle A female reproductive cycle in which the uterine lining is reabsorbed rather than shed in the absence of pregnancy and the female is sexually receptive only briefly during mid-cycle (estrus). It is seen in all mammals except Old World monkeys and apes (including humans). Compare with **menstrual cycle**.

ethylene A gaseous plant hormone associated with senescence that induces fruits to ripen and flowers to fade.

eudicot A member of a monophyletic group (lineage) of angiosperms that includes complex flowering plants and trees (e.g., roses, daisies, maples). All eudicots have two cotyledons, but not all dicots are members of this lineage. Compare with **dicot** and **monocot**.

Eukarya One of the three taxonomic domains of life, consisting of unicellular organisms (most protists, yeasts) and multicellular organisms (fungi, plants, animals) distinguished by a membrane-bound cell nucleus, numerous organelles, and an extensive cytoskeleton. Compare with **Archaea** and **Bacteria**.

eukaryote A member of the domain Eukarya; an organism whose cells contain a nucleus, numerous membrane-bound organelles, and an extensive cytoskeleton. May be unicellular or multicellular. Compare with **prokaryote**.

eusociality A complex social structure in which workers sacrifice most or all of their direct reproduction to help rear the queen's offspring. Common in insects such as ants, bees, wasps, and termites.

eutherians A lineage of mammals (Eutheria) whose young develop in the uterus and are not housed in an abdominal pouch. Also called *placental mammals*.

evaporation The energy-absorbing phase change from a liquid state to a gaseous state. Many organisms evaporate water as a means of heat loss.

evo-devo Popular term for evolutionary developmental biology, a research field focused on how changes in developmentally important genes have led to the evolution of new phenotypes.

evolution (1) The theory that all organisms on Earth are related by common ancestry and that they have changed over time, and continue to change, via natural selection and other processes. (2) Any change in the genetic characteristics of a population over time, especially, a change in allele frequencies.

ex situ conservation Preserving species outside of natural areas (e.g., in zoos, aquaria, or botanical gardens).

excitable membrane A plasma membrane that is capable of generating an action potential. Neurons, muscle cells, and some other cells have excitable membranes.

excitatory postsynaptic potential (EPSP) A change in membrane potential, usually depolarization, at a neuron dendrite that makes an action potential more likely.

exergonic Referring to a chemical reaction that can occur spontaneously, releasing heat and/or increasing entropy, and for which the Gibbs free-energy change (ΔG) is less than zero. Compare with **endergonic**.

exocrine gland A gland that secretes some substance through a duct into a space other than the circulatory system, such as the digestive tract or the skin surface. Compare with **endocrine gland**.

exocytosis Secretion of intracellular molecules (e.g., hormones, collagen), contained within membrane-bound vesicles, to the outside of the cell by fusion of vesicles to the plasma membrane. Compare with **endocytosis**.

exon A transcribed region of a eukaryotic gene or region of a primary transcript that is retained in the mature RNA. Except for 5′ and 3′ UTRs, mRNA exons code for amino acids. Compare with **intron**.

exoskeleton A hard covering secreted on the outside of the body, used for body support, protection, and muscle attachment. Prominent in arthropods. Compare with **endoskeleton**.

exothermic Referring to a chemical reaction that releases heat. Compare with **endothermic**.

exotic species A nonnative species that is introduced into a new area. Exotic species often are competitors, pathogens, or predators of native species.

expansins A class of plant proteins that break hydrogen bonds between components in the primary cell wall to allow it to expand for cell growth.

exponential population growth The accelerating increase in the size of a population that occurs when the growth rate is constant and density independent. Compare with **logistic population growth**.

expressed sequence tag (EST) A portion of a transcribed gene (synthesized from an mRNA in a cell), used to find the physical location of that gene in the genome.

extant species A species that is living today.

extensor A muscle that pulls two bones farther apart from each other, increasing the angle of the joint, as in the extension of a limb or the spine. Compare with **flexor**.

extinct species A species that has died out.

extracellular digestion Digestion that takes place outside of an organism, as occurs in many fungi that make and secrete digestive enzymes.

extracellular matrix (ECM) A complex meshwork in which animal cells are embedded, consisting of proteins (e.g., collagen, proteoglycan, laminin) and polysaccharides produced by the cells.

extremophile A bacterium or archaean that thrives in an "extreme" environment (e.g., high-salt, high-temperature, low-temperature, or low-pressure).

F_1 generation First filial generation. The first generation of offspring produced from a mating (i.e., the offspring of the parental generation).

facilitated diffusion Passive movement of a substance across a membrane with the assistance of transmembrane carrier proteins or channel proteins.

facilitation In ecological succession, the phenomenon in which early-arriving species make conditions more favorable for later-arriving species. Compare with **inhibition** and **tolerance**.

facultative anaerobe Any organism that can survive and reproduce by performing aerobic respiration when oxygen is available or fermentation when it is not.

FAD/FADH$_2$ Oxidized and reduced forms, respectively, of flavin adenine dinucleotide. A nonprotein electron carrier that functions in the citric acid cycle and oxidative phosphorylation.

fallopian tube A narrow tube connecting the uterus to the ovary in humans, through which the egg travels after ovulation. Site of fertilization and cleavage. In nonhuman animals, called *oviduct*.

fast muscle fiber Type of skeletal muscle fiber that is white in color, generates ATP by glycolysis, and contracts rapidly but fatigues easily. Also called *fast glycolytic*, or *Type IIb*, *fiber*.

fat A lipid consisting of three fatty acid molecules joined by ester linkages to a glycerol molecule. Also called *triacylglycerol* or *triglyceride*.

fatty acid A lipid consisting of a hydrocarbon chain bonded at one end to a carboxyl group. Used by many organisms to store chemical energy; a major component of animal and plant fats and phospholipids.

fauna All the animal species characteristic of a particular region, period, or environment.

feather A specialized skin outgrowth, composed of β-keratin, present in all birds as well as in some non-avian dinosaurs. Used for flight, insulation, display, and other purposes.

feces The waste products of digestion.

fecundity The average number of female offspring produced by a single female in the course of her lifetime.

feedback inhibition A type of control in which high concentrations of the product of a metabolic pathway inhibit one of the enzymes early in the pathway. A form of negative feedback.

fermentation Any of several metabolic pathways that regenerate oxidizing agents, such as NAD$^+$, by transferring electrons to a final electron acceptor in the absence of an electron transport chain. Allows pathways such as glycolysis to continue to make ATP.

ferredoxin In photosynthetic organisms, an iron- and sulfur-containing protein in the electron transport chain of photosystem I. Can transfer electrons to the enzyme NADP$^+$ reductase, which catalyzes formation of NADPH.

fertility The average number of surviving children that each woman has during her lifetime.

fertilization Fusion of the nuclei of two haploid gametes to form a zygote with a diploid nucleus.

fertilization envelope A physical barrier that forms around a fertilized egg in many animals. The fertilization envelope prevents fertilization by more than one sperm (polyspermy).

fetal alcohol syndrome (FAS) A condition marked by hyperactivity, severe learning disabilities, and depression. Thought to be caused by exposure of an individual to high blood alcohol concentrations during embryonic development.

fetus In live-bearing animals, the unborn offspring after the embryonic stage. It usually is developed sufficiently to be recognizable as belonging to a certain species.

fiber In plants, a type of elongated sclerenchyma cell that provides support to vascular tissue. Compare with **sclereid**.

Fick's law of diffusion A mathematical relationship that describes the rates of diffusion of gases.

fight-or-flight response Rapid physiological changes that prepare the body for emergencies. Includes increased heart rate, increased blood pressure, and decreased digestion.

filament Any thin, threadlike structure, particularly (1) the threadlike extensions of a fish's gills or (2) part of a stamen: the slender stalk that bears the anthers in a flower.

filter feeder See **suspension feeder**.

filtrate Any fluid produced by filtration, in particular the fluid ("pre-urine") in the Malpighian tubules of insects and the nephrons of vertebrate kidneys.

filtration A process of removing large components from a fluid by forcing it through a filter. Occurs in a renal corpuscle of the vertebrate kidney, allowing water and small solutes to pass from the blood into the nephron.

fimbria (plural: fimbriae) A long, needlelike projection from the cell membrane of prokaryotes that is involved in attachment to nonliving surfaces or other cells.

finite rate of increase (λ) The rate of increase of a population over a given period of time. Calculated as the ending population size divided by the starting population size. Compare with **intrinsic rate of increase**.

first law of thermodynamics The principle of physics that energy is conserved in any process. Energy can be transferred and converted into different forms, but it cannot be created or destroyed.

fission (1) A form of asexual reproduction in which a prokaryotic cell divides to produce two genetically similar daughter cells by a process similar to mitosis of eukaryotic cells. Also called *binary fission*. (2) A form of asexual reproduction in which an animal splits into two or more individuals of approximately equal size; common among invertebrates.

fitness The ability of an individual to produce viable offspring relative to others of the same species.

fitness trade-off See **trade-off**.

fixed action pattern (FAP) Highly stereotyped behavior pattern that occurs in a certain invariant way in a certain species. A form of innate behavior.

flaccid Limp as a result of low internal (turgor) pressure (e.g., a wilted plant leaf). Compare with **turgid**.

flagellum (plural: flagella) A long, cellular projection that undulates (in eukaryotes) or rotates (in prokaryotes) to move the cell through an aqueous environment. See **axoneme**.

flatworms Members of the phylum Platyhelminthes. Distinguished by a broad, flat, unsegmented body that lacks a coelom. Flatworms belong to the lophotrochozoan branch of the protostomes.

flavin adenine dinucleotide See **FAD/FADH₂**.

flexor A muscle that pulls two bones closer together, decreasing the joint angle, as in the flexing of a limb or the spine. Compare with **extensor**.

floral meristem A group of undifferentiated plant stem cells that can give rise to the four organs making up a flower.

florigen In plants, a protein hormone that is synthesized in leaves and transported to the shoot apical meristem, where it stimulates flowering.

flower In angiosperms, the part of a plant that contains reproductive structures. Typically includes a calyx, a corolla, and one or more stamens and/or carpels. See **perfect** and **imperfect flower**.

fluid connective tissue A type of connective tissue, distinguished by having a liquid extracellular matrix; includes blood.

fluid feeder Any animal that feeds by sucking or mopping up liquids such as nectar, plant sap, or blood.

fluid-mosaic model The widely accepted hypothesis that the plasma membrane and organelle membranes consist of proteins embedded in a fluid phospholipid bilayer.

fluorescence The spontaneous emission of light from an excited electron falling back to its normal (ground) state.

follicle In a mammalian ovary, a sac of supportive cells containing an egg cell.

follicle-stimulating hormone (FSH) A peptide hormone, produced and secreted by the anterior pituitary; it stimulates (in females) growth of eggs and follicles in the ovaries or (in males) sperm production in the testes.

follicular phase In a menstrual cycle, the first major phase, during which follicles grow and estradiol levels increase; ends with ovulation.

food Any nutrient-containing material that can be consumed and digested by animals.

food chain A relatively simple pathway of energy flow through a few species, each at a different trophic level, in an ecosystem. Might include, for example, a primary producer, a primary consumer, a secondary consumer, and a decomposer. A subset of a **food web**.

food web The complex network of interactions among species in an ecosystem formed by the transfer of energy and nutrients among trophic levels. Consists of many food chains.

foot One of the three main parts of the mollusk body; a muscular appendage, used for movements such as crawling and/or burrowing into sediment.

foraging Searching for food.

forebrain One of the three main regions of the vertebrate brain; includes the cerebrum, thalamus, and hypothalamus. Compare with **hindbrain** and **midbrain**.

fossil Any physical trace of an organism that existed in the past. Includes tracks, burrows, fossilized bones, casts, and so on.

fossil record All of the fossils that have been found anywhere on Earth and that have been formally described in the scientific literature.

founder effect A change in allele frequencies that often occurs when a new population is established from a small group of individuals (founder event) due to sampling error (i.e., the small group is not a representative sample of the source population).

fovea In the vertebrate eye, a portion of the retina where incoming light is focused; contains a high proportion of cone cells.

frameshift mutation The addition or deletion of a nucleotide in a coding sequence that shifts the reading frame of the mRNA.

free energy The energy of a system that can be converted into work. It may be measured only through the change in free energy in a reaction. See **Gibbs free-energy change**.

free radicals Any substance containing one or more atoms with an unpaired electron. Unstable and highly reactive.

frequency The number of wave crests per second traveling past a stationary point. Determines the pitch of sound and the color of light.

frequency-dependent selection A pattern of selection in which certain alleles are favored only when they are rare; a form of balancing selection.

fronds The large leaves of ferns.

frontal lobe In the vertebrate brain, one of the four major areas in the cerebrum.

fruit In flowering plants (angiosperms), a mature, ripened plant ovary (or group of ovaries), along with the seeds it contains and any adjacent fused parts; often functions in seed dispersal. See **aggregate**, **multiple**, and **simple fruit**.

fruiting body A structure formed in some prokaryotes, fungi, and protists for spore dispersal; usually consists of a base, a stalk, and a mass of spores at the top.

functional genomics The study of how a genome works; that is, when and where specific genes are expressed and how their products interact to produce a functional organism.

functional group A small group of atoms bonded together in a precise configuration and exhibiting particular chemical properties that it imparts to any organic molecule in which it occurs.

fundamental niche The total theoretical range of environmental conditions that a species can tolerate. Compare with **realized niche**.

fungi A lineage of eukaryotes that typically have a filamentous body (mycelium) and obtain nutrients by absorption.

fungicide Any substance that can kill fungi or slow their growth.

G protein Any of various proteins that are activated by binding to guanosine triphosphate (GTP) and inactivated when GTP is hydrolyzed to GDP. In G-protein-coupled receptors, signal binding directly triggers the activation of a G protein, leading to production of a second messenger or initiation of a phosphorylation cascade.

G₁ phase The phase of the cell cycle that constitutes the first part of interphase before DNA synthesis (S phase).

G₂ phase The phase of the cell cycle between synthesis of DNA (S phase) and mitosis (M phase); the last part of interphase.

gallbladder A small pouch that stores bile from the liver and releases it as needed into the small intestine during digestion of fats.

gametangium (plural: gametangia) (1) The gamete-forming structure found in all land plants except angiosperms. Contains a sperm-producing antheridium and an egg-producing archegonium. (2) The gamete-forming structure of some chytrid fungi.

gamete A haploid reproductive cell that can fuse with another haploid cell to form a zygote. Most multicellular eukaryotes have two distinct forms of gametes: egg cells (ova) and sperm cells.

gametogenesis The production of gametes (eggs or sperm).

gametophyte In organisms undergoing alternation of generations, the multicellular haploid form that arises from a single haploid spore and produces gametes. Compare with **sporophyte**.

ganglion (plural: ganglia) A mass of neurons in a centralized nervous system.

ganglion cell In the retina, a type of neuron whose axons form the optic nerves.

gap junction A type of cell–cell attachment structure that directly connects the cytosolic components of adjacent animal cells, allowing passage of water, ions, and small molecules between the cells. Compare with **desmosome** and **tight junction**.

gastrin A hormone produced by cells in the stomach lining in response to the arrival of food or to a neural signal from the brain. Stimulates other stomach cells to release hydrochloric acid.

gastropods A lineage of mollusks distinguished by a large, muscular foot and a unique feeding structure, the radula. Include slugs and snails.

gastrulation The process of coordinated cell movements, including the moving of some cells from the outer surface of the embryo to the interior, that results in the formation of three germ layers (endoderm, mesoderm, and ectoderm) and the axes of the embryo.

gated channel A channel protein that opens and closes in response to a specific stimulus, such as the binding of a particular molecule or a change in voltage across the membrane.

gemma (plural: gemmae) A small reproductive structure that is produced asexually in some plants during the gametophyte phase and can grow into a mature gametophyte; most common in non-vascular plants, particularly liverworts and club mosses, and in ferns

gene A section of DNA (or RNA, for some viruses) that encodes information for building one or more related polypeptides or functional RNA molecules along with the regulatory sequences required for its transcription.

gene duplication The formation of an additional copy of a gene, typically by misalignment of

chromosomes during crossing over. Thought to be an important evolutionary process in creating new genes.

gene expression The set of processes, including transcription and translation, that convert information in DNA into a product of a gene, most commonly a protein.

gene family A set of genetic loci whose DNA sequences are extremely similar. Thought to have arisen by duplication of a single ancestral gene and subsequent mutations in the duplicated sequences.

gene flow The movement of alleles between populations; occurs when individuals leave one population, join another, and breed.

gene pool All the alleles of all the genes in a certain population.

gene therapy The treatment of an inherited disease by introducing a normal form of the gene.

generation The average time between a mother's first offspring and her daughter's first offspring.

genetic bottleneck A reduction in allelic diversity resulting from a sudden reduction in the size of a large population (population bottleneck) due to a random event.

genetic code The set of all codons and their meaning.

genetic correlation A type of evolutionary constraint in which selection on one trait causes a change in another trait as well; may occur when the same gene(s) affect both traits.

genetic diversity The diversity of alleles or genes in a population, species, or group of species.

genetic drift Any change in allele frequencies due to random events. Causes allele frequencies to drift up and down randomly over time, and eventually can lead to the fixation or loss of alleles.

genetic equivalence Having all different cell types of a multicellular individual possess the same genome.

genetic homology Similarity in DNA nucleotide sequences, RNA nucleotide sequences, or amino acid sequences due to inheritance from a common ancestor.

genetic map A list of genes on a chromosome that indicates their position and relative distances from one another. Also called a *linkage map*. Compare with **physical map**.

genetic marker A genetic locus that can be identified and traced in populations by laboratory techniques or by a distinctive visible phenotype.

genetic recombination A change in the combination of alleles on a given chromosome or in a given individual. Also called *recombination*.

genetic screen Any technique that identifies individuals with a particular type of mutation.

genetic variation (1) The number and relative frequency of alleles present in a particular population. (2) The proportion of phenotypic variation in a trait that is due to genetic rather than environmental influences in a certain population in a certain environment.

genetics The study of the inheritance of traits.

genital (plural: genitalia) Any external copulatory organ.

genome All the hereditary information in an organism, including not only genes but also stretches of DNA that do not contain genes.

genome annotation The process of analyzing a genome sequence to identify key features such as genes, regulatory sequences, and splice sites.

genomic library A set of DNA segments representing the entire genome of a particular organism. Each segment is carried by a plasmid or other cloning vector and can be separated from other segments. Compare with **cDNA library**.

genomics The field of study concerned with sequencing, interpreting, and comparing whole genomes from different organisms.

genotype All the alleles of every gene present in a given individual. Often specified only for the alleles of a particular set of genes under study. Compare with **phenotype**.

genus (plural: genera) In Linnaeus' system, a taxonomic category of closely related species. Always italicized and capitalized to indicate that it is a recognized scientific genus.

geologic time scale The sequence of eons, eras, and periods used to describe the geologic history of Earth.

germ cell In animals, any cell that can potentially give rise to gametes. Also called *germ-line cells*.

germ layer In animals, one of the three basic types of tissue formed during gastrulation; gives rise to all other tissues. See **endoderm**, **mesoderm**, and **ectoderm**.

germ line In animals, any of the cells that are capable of giving rise to reproductive cells (sperm or egg). Compare with **germ cell**.

germ theory of disease The theory that infectious diseases are caused by bacteria, viruses, and other microorganisms.

germination The process by which a seed becomes a young plant.

gestation The period of development inside the mother, from implantation to birth, in those species that have live birth.

gibberellins A class of hormones, found in plants and fungi, that stimulate growth. Gibberellic acid (GA) is one of the major gibberellins.

Gibbs free-energy change (ΔG) A measure of the change in enthalpy and entropy that occurs in a given chemical reaction. ΔG is less than 0 for spontaneous reactions and greater than 0 for nonspontaneous reactions.

gill Any organ in aquatic animals that exchanges gases and other dissolved substances between the blood and the surrounding water. Typically, a filamentous outgrowth of a body surface.

gill arch In aquatic vertebrates, a curved region of tissue between the gills. Gills are suspended from the gill arches.

gill filament In fish, any of the many long, thin structures that extend from gill arches into the water and across which gas exchange occurs.

gill lamella (plural: gill lamellae) Any of hundreds to thousands of sheetlike structures, each containing a capillary bed, that make up a gill filament.

gland An organ whose primary function is to secrete some substance, either into the blood (endocrine gland) or into some other space such as the gut or skin (exocrine gland).

glia Collective term for several types of cells in nervous tissue that are not neurons and do not conduct electrical signals but perform other functions, such as providing support, nourishment, or electrical insulation. Also called *glial cells*.

global carbon cycle See **carbon cycle, global**.

global climate change The global sum of all the local changes in temperature and precipitation patterns that accompany global warming (or in some past events, global cooling).

global gene regulation The regulation of multiple bacterial genes that are not part of one operon.

global nitrogen cycle See **nitrogen cycle, global**.

global warming A sustained increase in Earth's average surface temperature.

global water cycle See **water cycle, global**.

glomalin A glycoprotein that is abundant in the hyphae of arbuscular mycorrhizal fungi; when hyphae decay, it is an important component of soil.

glomerulus (plural: glomeruli) (1) In the vertebrate kidney, a ball-like cluster of capillaries, surrounded by Bowman's capsule, at the beginning of a nephron. (2) In the brain, a ball-shaped cluster of neurons in the olfactory bulb.

glucagon A peptide hormone produced by the pancreas in response to low blood glucose. Raises blood glucose by triggering breakdown of glycogen and stimulating gluconeogenesis. Compare with **insulin**.

glucocorticoids A class of steroid hormones, produced and secreted by the adrenal cortex, that increase blood glucose and prepare the body for stress. Include cortisol and corticosterone. Compare with **mineralocorticoids**.

gluconeogenesis Synthesis of glucose, often from non-carbohydrate sources (e.g., proteins and fatty acids). In plants, used to produce glucose from products of the Calvin cycle. In animals, occurs in the liver in response to low insulin levels and high glucagon levels.

glucose Six-carbon monosaccharide whose oxidation in cellular respiration is the major source of ATP in animal cells.

glyceraldehyde-3-phosphate (G3P) The phosphorylated three-carbon compound formed as the result of carbon fixation in the first step of the Calvin cycle.

glycerol A three-carbon molecule that forms the "backbone" of phospholipids and most fats.

glycogen A highly branched storage polysaccharide composed of glucose monomers joined by α-1,4- and α-1,6-glycosidic linkages. The major form of stored carbohydrate in animals.

glycolipid Any lipid molecule that is covalently bonded to a carbohydrate group.

glycolysis A series of 10 chemical reactions that oxidize glucose to produce pyruvate, NADH, and ATP.

Used by organisms as part of fermentation or cellular respiration.

glycoprotein Any protein with one or more covalently bonded carbohydrates, typically oligosaccharides.

glycosidic linkage The covalent linkage formed by a condensation reaction between two sugar monomers; joins the residues of a polysaccharide.

glycosylation Addition of a carbohydrate group to a molecule.

glyoxysome Specialized type of peroxisome found in plant cells and packed with enzymes for processing the products of photosynthesis.

gnathostomes Animals with jaws. Most vertebrates are gnathostomes.

Golgi apparatus A eukaryotic organelle, consisting of stacks of flattened membranous sacs (cisternae), that functions in processing and sorting proteins and lipids destined to be secreted or directed to other organelles. Also called *Golgi complex*.

gonad An organ, such as a testis or an ovary, that produces reproductive cells.

gonadotropin-releasing hormone (GnRH) A peptide hormone, produced and secreted by the hypothalamus, that stimulates release of follicle-stimulating hormone (FSH) and luteinizing hormone (LH) from the anterior pituitary.

grade In taxonomy, a group of species that share some, but not all, of the descendants of a common ancestor. Also called a *paraphyletic group*.

Gram-negative Describing bacteria that look pink when treated with a Gram stain. These bacteria have a cell wall composed of a thin layer of peptidoglycan and an outer phospholipid layer. Compare with **Gram-positive**.

Gram-positive Describing bacteria that look purple when treated with a Gram stain. These bacteria have cell walls composed of a thick layer of peptidoglycan and no outer phospholipid later. Compare with **Gram-negative**.

Gram stain A dye that distinguishes the two general types of cell walls found in bacteria. Used to routinely classify bacteria as Gram-negative or Gram-positive.

granum (plural: grana) In chloroplasts, a stack of flattened, membrane-bound thylakoid discs where the light reactions of photosynthesis occur.

gravitropism The growth or movement of a plant in a particular direction in response to gravity.

grazing food chain The ecological network of herbivores and the predators and parasites that consume them.

great apes See **hominids**.

green algae A paraphyletic group of photosynthetic organisms that contain chloroplasts similar to those in green plants. Often classified as protists, green algae are the closest living relatives of land plants and form a monophyletic group with them.

greenhouse gas An atmospheric gas that absorbs and reflects infrared radiation, so that heat radiated from Earth is retained in the atmosphere instead of being lost to space.

gross primary productivity In an ecosystem, the total amount of carbon fixed by photosynthesis (or more

rarely, chemosynthesis), including that used for cellular respiration, over a given time period. Compare with **net primary productivity**.

ground meristem The middle layer of a young plant embryo. Gives rise to the ground tissue system.

ground tissue An embryonic tissue layer that gives rise to parenchyma, collenchyma, and sclerenchyma—tissues other than the epidermis and vascular tissue. Also called *ground tissue system*.

groundwater Any water below the land surface.

growth factor Any of a large number of signaling molecules that are secreted by certain cells and that stimulate other cells to grow, divide, or differentiate.

growth hormone (GH) A peptide hormone, produced and secreted by the mammalian anterior pituitary, that promotes lengthening of the long bones in children and muscle growth, tissue repair, and lactation in adults. Also called *somatotropin*.

guanosine triphosphate (GTP) A nucleotide consisting of guanine, a ribose sugar, and three phosphate groups. Can be hydrolyzed to release free energy. Commonly used in RNA synthesis and also functions in signal transduction in association with G proteins.

guard cell One of two specialized, crescent-shaped cells forming the border of a plant stoma. Guard cells can change shape to open or close the stoma. See also **stoma**.

gustation The perception of taste.

guttation Excretion of water droplets from plant leaves; visible in the early morning. Caused by root pressure.

gymnosperm A vascular plant that makes seeds but does not produce flowers. The gymnosperms include five lineages of green plants (cycads, ginkgoes, conifers, redwoods, and gnetophytes). Compare with **angiosperm**.

H⁺-ATPase See **proton pump**.

habitat degradation The reduction of the quality of a habitat.

habitat destruction Human-caused destruction of a natural habitat, replaced by an urban, suburban, or agricultural landscape.

habitat fragmentation The breakup of a large region of a habitat into many smaller regions, separated from others by a different type of habitat.

Hadley cell An atmospheric cycle of large-scale air movement in which warm equatorial air rises, moves north or south, and then descends at approximately 30° N or 30° S latitude.

hair cell A pressure-detecting sensory cell that has tiny "hairs" (stereocilia) jutting from its surface. Found in the inner ear, lateral line system, and ampullae of Lorenzini.

hairpin A secondary structure in RNA consisting of a stable loop formed by hydrogen bonding between purine and pyrimidine bases on the same strand.

Hamilton's rule The proposition that an allele for altruistic behavior will be favored by natural selection only if $Br C$, where B = the fitness benefit to the recipient, C = the fitness cost to the actor, and r = the coefficient of relatedness between recipient and actor.

haploid (1) Having one set of chromosomes (1*n* or *n* for short). (2) A cell or an individual organism with one set of chromosomes. Compare with **diploid**.

haploid number The number of different types of chromosomes in a cell. Symbolized as *n*.

Hardy–Weinberg principle A principle of population genetics stating that genotype frequencies in a large population do not change from generation to generation in the absence of evolutionary processes (e.g., mutation, gene flow, genetic drift, and selection), and nonrandom mating.

haustorium (plural: haustoria) Highly modified stem or root of a parasitic plant. The haustorium penetrates the tissues of a host and absorbs nutrients and water.

hearing The sensation of the wavelike changes in air pressure called sound.

heart A muscular pump that circulates blood throughout the body.

heart murmur A distinctive sound caused by back-flow of blood through a defective heart valve.

heartwood The older xylem in the center of an older stem or root, containing protective compounds and no longer functioning in water transport.

heat Thermal energy that is transferred from an object at higher temperature to one at lower temperature.

heat of vaporization The energy required to vaporize 1 gram of a liquid into a gas.

heat-shock proteins Proteins that facilitate refolding of proteins that have been denatured by heat or other agents.

heavy chain The larger of the two types of polypeptide chains in an antibody or B-cell receptor molecule; composed of a variable (*V*) region, which contributes to the antigen-binding site, and a constant (*C*) region. Differences in heavy-chain constant regions determine the different classes of immunoglobulins (IgA, IgE, etc.). Compare with **light chain**.

helper T cell A CD4$^+$ effector T cell that secretes cytokines and in other ways promotes the activation of other lymphocytes. Activated by interacting with complementary class II MHC–peptide complexes on the surface of antigen-presenting cells such as dendritic cells.

heme A small molecule that binds to each of the four polypeptides that combine to form hemoglobin; contains an iron ion that can bind oxygen.

hemimetabolous metamorphosis A type of metamorphosis in which the animal increases in size from one stage to the next, but does not dramatically change its body form. Also called *incomplete metamorphosis*.

hemocoel A body cavity, present in arthropods and some mollusks, containing a pool of circulatory fluid (hemolymph) bathing the internal organs. Unlike a coelom, a hemocoel is not lined in mesoderm.

hemoglobin An oxygen-binding protein consisting of four polypeptide subunits, each containing an oxygen-binding heme group. The major oxygen carrier in mammalian blood.

hemolymph The circulatory fluid of animals with open circulatory systems (e.g., insects) in which the fluid is not confined to blood vessels.

herbaceous Referring to a plant that is not woody.

herbivore (adjective: herbivorous) An animal that eats primarily plants and rarely or never eats meat. Compare with **carnivore** and **omnivore**.

herbivory The practice of eating plant tissues.

heredity The transmission of traits from parents to offspring via genetic information.

heritable Referring to traits that can be transmitted from one generation to the next.

hermaphrodite An organism that produces both male and female gametes.

heterokaryotic Describing a cell or fungal mycelium containing two or more haploid nuclei that are genetically distinct.

heterospory (adjective: heterosporous) In seed plants, the production of two distinct types of spores: microspores, which become the male gametophyte, and megaspores, which become the female gametophyte. Compare with **homospory**.

heterotherm An animal whose body temperature varies markedly. Compare with **homeotherm**.

heterotroph Any organism that cannot synthesize reduced organic compounds from inorganic sources and that must obtain them from other organisms. Some bacteria, some archaea, and virtually all fungi and animals are heterotrophs. Also called *consumer*. Compare with **autotroph**.

heterozygote advantage A pattern of natural selection that favors heterozygous individuals compared with homozygotes. Tends to maintain genetic variation in a population, thus is a form of balancing selection.

heterozygous Having two different alleles of a gene.

hexose A monosaccharide (simple sugar) containing six carbon atoms.

hibernation An energy-conserving physiological state, marked by a decrease in metabolic rate, body temperature, and activity, that lasts for a prolonged period (weeks to months). Occurs in some animals in response to winter cold and scarcity of food. Compare with **torpor**.

hindbrain One of the three main regions of the vertebrate brain, responsible for balance and sometimes hearing; includes the cerebellum and medulla oblongata. Compare with **forebrain** and **midbrain**.

histamine A molecule released from mast cells during an inflammatory response that, at high concentrations, causes blood vessels to constrict to reduce blood loss from tissue damage.

histone One of several positively charged (basic) proteins associated with DNA in the chromatin of eukaryotic cells.

histone acetyltransferases (HATs) A class of eukaryotic enzymes that loosen chromatin structure by adding acetyl groups to histone proteins.

histone code The hypothesis that specific combinations of chemical modifications of histone proteins contain information that influences chromatin condensation and gene expression.

histone deacetylases (HDACs) A class of eukaryotic enzymes that condense chromatin by removing acetyl groups from histone proteins.

holoenzyme A multipart enzyme consisting of a core enzyme (containing the active site for catalysis) along with other required proteins.

holometabolous metamorphosis A type of metamorphosis in which the animal completely changes its form; includes a distinct larval stage. Also called *complete metamorphosis*.

homeobox A DNA sequence of about 180 base pairs that codes for a DNA binding motif called the homeodomain in the resulting protein. Genes containing a homeobox usually play a role in controlling development of organisms from fruit flies to humans.

homeostasis (adjective: homeostatic) The array of relatively stable chemical and physical conditions in an animal's cells, tissues, and organs. May be achieved by the body's passively matching the conditions of a stable external environment (conformational homeostasis) or by active physiological processes (regulatory homeostasis) triggered by variations in the external or internal environment.

homeotherm An animal that has a constant or relatively constant body temperature. Compare with **heterotherm**.

homeotic mutation A mutation that causes one body part to be substituted for another.

hominids Members of the family Hominidae, which includes humans and extinct related forms, chimpanzees, gorillas, and orangutans. Distinguished by large body size, no tail, and an exceptionally large brain. Also called *great apes*.

hominins Any extinct or living species of bipedal apes, such as *Australopithecus africanus*, *Homo erectus*, and *Homo sapiens*.

homologous See **homology**.

homologous chromosomes In a diploid organism, chromosomes that are similar in size, shape, and gene content. Also called *homologs*.

homology (adjective: homologous) Similarity among organisms of different species due to their inheritance from a common ancestor. Features that exhibit such similarity (e.g., DNA sequences, proteins, body parts) are said to be homologous. Compare with **homoplasy**.

homoplasy (adjective: homoplastic) Similarity among organisms of different species due to reasons other than common ancestry, such as convergent evolution. Features that exhibit such similarity (e.g., the wings of birds and bats) are said to be homoplastic, or convergent. Compare with **homology**.

homospory (adjective: homosporous) In seedless vascular plants, the production of just one type of spore. Compare with **heterospory**.

homozygous Having two identical alleles of a gene.

hormone Any of many different signaling molecules that circulate throughout the plant or animal body and can trigger characteristic responses in distant target cells at very low concentrations.

hormone-response element A specific sequence in DNA to which a steroid hormone–receptor complex can bind and affect gene transcription.

host An individual that has been invaded by an organism such as a parasite or a virus, or that provides habitat or resources to a commensal organism.

host cell A cell that has been invaded by a parasitic organism or a virus and provides an environment that is conducive to the pathogen's growth and reproduction.

***Hox* genes** A class of genes found in several animal phyla, including vertebrates, that are expressed in a distinctive pattern along the anterior–posterior axis in early embryos and control formation of specific structures. *Hox* genes code for transcription factors with a DNA-binding sequence called a homeobox.

human Any member of the genus *Homo,* which includes modern humans (*Homo sapiens*) and several extinct species.

human chorionic gonadotropin (hCG) A glycoprotein hormone produced by a human embryo and placenta from about week 3 to week 14 of pregnancy. Maintains the corpus luteum, which produces hormones that preserve the uterine lining.

Human Genome Project The multinational research project that sequenced the human genome.

human immunodeficiency virus (HIV) A retrovirus that causes acquired immune deficiency syndrome (AIDS) in humans.

humoral (immune) response The type of immune response that is mediated through the production and secretion of antibodies, complement proteins, and other soluble factors that eliminate extracellular pathogens. Compare with **cell-mediated (immune) response**.

humus The decayed organic matter in soils.

hybrid The offspring of parents from two different strains, populations, or species.

hybrid zone A geographic area where interbreeding occurs between two species, sometimes producing fertile hybrid offspring.

hydrocarbon An organic molecule that contains only hydrogen and carbon atoms.

hydrogen bond A weak interaction between two molecules or different parts of the same molecule resulting from the attraction between a hydrogen atom with a partial positive charge and another atom (usually O or N) with a partial negative charge. Compare with **covalent bond** and **ionic bond**.

hydrogen ion (H$^+$) A single proton with a charge of 1+; typically, one that is dissolved in solution or that is being transferred from one atom to another in a chemical reaction.

hydrolysis A chemical reaction in which a molecule is split into smaller molecules by reacting with water. In biology, most hydrolysis reactions involve the splitting of polymers into monomers. Compare with **condensation reaction**.

hydrophilic Interacting readily with water. Hydrophilic compounds are typically polar compounds containing partially or fully charged atoms. Compare with **hydrophobic**.

hydrophobic Not readily interacting with water. Hydrophobic compounds are typically nonpolar compounds that lack partially or fully charged atoms. Compare with **hydrophilic**.

hydrophobic interactions Very weak interactions between nonpolar molecules, or nonpolar regions of the same molecule, when exposed to an aqueous solvent. The surrounding water molecules support these interactions by interacting with one another and encapsulating the nonpolar molecules.

hydroponic growth Growth of plants in liquid cultures instead of soil.

hydrostatic skeleton A system of body support involving a body wall in tension surrounding a fluid or soft tissue under compression.

hydroxide ion (OH$^-$) An oxygen atom and a hydrogen atom joined by a single covalent bond and carrying a negative charge; formed by dissociation of water.

hygiene hypothesis The claim that immune disorders arise in individuals less likely to have been exposed to pathogens and parasites, especially in early childhood. Provides an explanation for the increased risk of allergies and autoimmune disease in countries with high levels of sanitation.

hyperosmotic Comparative term designating a solution that has a greater solute concentration, and therefore a lower water concentration, than another solution. Compare with **hyposmotic** and **isosmotic**.

hyperpolarization A change in membrane potential from its resting negative state to an even more negative state; a normal phase in an action potential. Compare with **depolarization**.

hypersensitive reaction An intense allergic response by cells that have been sensitized by previous exposure to an allergen.

hypersensitive response In plants, the rapid death of a cell that has been infected by a pathogen, thereby reducing the potential for infection to spread throughout a plant. Compare with **systemic acquired resistance**.

hypertension Abnormally high blood pressure.

hypertonic Comparative term designating a solution that, if inside a cell or vesicle, results in the uptake of water and swelling or even bursting of the membrane-bound structure. This solution has a greater solute concentration than the solution on the other side of the membrane. Used when the solute is unable to pass through the membrane. Compare with **hypotonic** and **isotonic**.

hypha (plural: hyphae) One of the long, branching strands of a fungal mycelium (the mesh-like body of a fungus). Also found in some protists.

hypocotyl The stem of a very young plant; the region between the cotyledon (embryonic leaf) and the radicle (embryonic root).

hyposmotic Comparative term designating a solution that has a lower solute concentration, and therefore a higher water concentration, than another solution. Compare with **hyperosmotic** and **isosmotic**.

hypothalamic–pituitary axis The functional interaction of the hypothalamus and the anterior pituitary gland, which are anatomically distinct but work together to regulate most of the other endocrine glands in the body.

hypothalamus A part of the brain that functions in maintaining the body's internal physiological state by regulating the autonomic nervous system, endocrine system, body temperature, water balance, and appetite.

hypothesis A testable statement that explains a phenomenon or a set of observations.

hypotonic Comparative term designating a solution that, if inside a cell or vesicle, results in the loss of water and shrinkage of the membrane-bound structure. This solution has a lower solute concentration than the solution on the other side of the membrane. Used when the solute is unable to pass through the membrane. Compare with **hypertonic** and **isotonic**.

immigration The migration of individuals into a particular population from other populations. Compare with **emigration**.

immune system The system whose primary function is to defend the host organism against pathogens. Includes several types of cells (e.g., leukocytes). In vertebrates, several organs are also involved where specialized cells develop or reside (e.g., lymph nodes and thymus).

immunity (adjective: immune) State of being protected against infection by disease-causing pathogens.

immunization The conferring of immunity to a particular disease by artificial means.

immunoglobulin (Ig) Any of the class of proteins that are structurally related to antibodies.

immunological memory The ability of the immune system to "remember" an antigen and mount a rapid, effective adaptive immune response to a pathogen encountered years or decades earlier. Based on the formation of memory lymphocytes.

impact hypothesis The hypothesis that a collision between the Earth and an asteroid caused the mass extinction at the K–P boundary, 65 million years ago.

imperfect flower A flower that contains male parts (stamens) *or* female parts (carpels) but not both. Compare with **perfect flower**.

implantation The process by which an embryo buries itself in the uterine or oviductal wall and forms a placenta. Occurs in mammals and some other viviparous vertebrates.

in situ hybridization A technique for detecting specific DNAs and mRNAs in cells and tissues by use of labeled complementary probes. Can be used to determine where and when particular genes are expressed in embryos.

inbreeding Mating between closely related individuals. Increases homozygosity of a population and often leads to a decline in the average fitness via selection (inbreeding depression).

inbreeding depression In inbred offspring, fitness declines due to deleterious recessive alleles that are homozygous, thus exposed to selection.

inclusive fitness The combination of (1) direct production of offspring (direct fitness) and (2) extra production of offspring by relatives in response to help provided by the individual in question (indirect fitness).

incomplete digestive tract A digestive tract that has just one opening.

incomplete dominance An inheritance pattern in which the heterozygote phenotype is in between the homozygote phenotypes.

incomplete metamorphosis See **hemimetabolous metamorphosis**.

independent assortment, principle of The concept that each pair of hereditary elements (alleles of the same gene) segregates (separates) independently of alleles of other genes during meiosis. One of Mendel's two principles of genetics.

indeterminate growth A pattern of growth in which an individual continues to increase its overall body size throughout its life.

induced fit Change in the shape of the active site of an enzyme, as the result of initial weak binding of a substrate, so that it binds substrate more tightly.

inducer A small molecule that triggers transcription of a specific gene, often by binding to and inactivating a repressor protein.

inducible defense A defensive trait that is manifested only in response to the presence of a consumer (predator or herbivore) or pathogen. Compare with **constitutive defense**.

infection thread An invagination of the plasma membrane of a root hair through which beneficial nitrogen-fixing bacteria enter the roots of their host plants (legumes).

inflammatory response An aspect of the innate immune response, seen in most cases of infection or tissue injury, in which the affected tissue becomes swollen, red, warm, and painful.

ingestion The act of bringing food into the digestive tract.

inhibition In ecological succession, the phenomenon in which early-arriving species make conditions less favorable for the establishment of certain later-arriving species. Compare with **facilitation** and **tolerance**.

inhibitory postsynaptic potential (IPSP) A change in membrane potential, usually hyperpolarization, at a neuron dendrite that makes an action potential less likely.

initiation (1) In an enzyme-catalyzed reaction, the stage during which enzymes orient reactants precisely as they bind at specific locations within the enzyme's active site. (2) In DNA transcription, the stage during which RNA polymerase and other proteins assemble at the promoter sequence and open the strands of DNA to start transcription. (3) In translation, the stage during which a complex consisting of initiation factor proteins, a ribosome, an mRNA, and an aminoacyl tRNA corresponding to the start codon is formed.

initiation factors A class of proteins that assist ribosomes in binding to a messenger RNA molecule to begin translation.

innate behavior Behavior that is inherited genetically, does not have to be learned, and is typical of a species.

innate immune response See **innate immunity**.

innate immunity A set of barriers to infection and generic defenses against broad types of pathogens. Produces an immediate response that involves many different leukocytes, which often activate an inflammatory response. Compare with **acquired immunity**.

inner cell mass (ICM) A cluster of cells, in the interior of a mammalian blastocyst, that eventually develop into the embryo. Contrast with **trophoblast**.

inner ear The innermost portion of the mammalian ear, consisting of a fluid-filled system of tubes that includes the cochlea (which receives sound vibrations from the middle ear) and the semicircular canals (which function in balance).

insects A terrestrial lineage of arthropods distinguished by three tagmata (head, thorax, abdomen), a single pair of antennae, and unbranched appendages.

insulin A peptide hormone produced by the pancreas in response to high levels of glucose (or amino acids) in blood. Enables cells to absorb glucose and coordinates synthesis of fats, proteins, and glycogen. Compare with **glucagon**.

integral membrane protein Any membrane protein that spans the entire lipid bilayer. Also called *transmembrane protein*. Compare with **peripheral membrane protein**.

integrated pest management In agriculture or forestry, systems for managing insects or other pests that include carefully controlled applications of toxins, introduction of species that prey on pests, planting schemes that reduce the chance of a severe pest outbreak, and other techniques.

integrator A component of an animal's nervous system that functions as part of a homeostatic system by evaluating sensory information and triggering appropriate responses. See **effector** and **sensor**.

integrin Any of a class of cell-surface proteins that bind to laminins and other proteins in the extracellular matrix, thus holding cells in place.

intercalated disc A type of specialized connection between adjacent heart muscle cells that contains gap junctions, allowing electrical signals to pass between the cells.

intermediate filament A long fiber, about 10 nm in diameter, composed of one of various proteins (e.g., keratins, lamins); one of the three types of cytoskeletal fibers. Used to form networks that help maintain cell shape and hold the nucleus in place. Compare with **actin filament** and **microtubule**.

intermediate host The host species in which a parasite reproduces asexually. Compare with **definitive host**.

intermediate muscle fiber Type of skeletal muscle fiber that is pink in color, generates ATP by both glycolysis and aerobic respiration, and has contractile properties that are intermediate between slow fibers and fast fibers. Also called fast oxidative fiber.

interneuron A neuron that passes signals from one neuron to another. Compare with **motor neuron** and **sensory neuron**.

internode The section of a plant stem between two nodes (sites where leaves attach).

interphase The portion of the cell cycle between one M phase and the next. Includes the G_1 phase, S phase, and G_2 phase.

intersexual selection The sexual selection of an individual of one gender for mating by an individual of the other gender (usually by female choice).

interspecific competition Competition between members of different species for the same limited resource. Compare with **intraspecific competition**.

interstitial fluid The plasma-like fluid found in the region (interstitial space) between cells.

intertidal zone The region between the low-tide and high-tide marks on a seashore.

intrasexual selection Competition among members of one gender for an opportunity to mate (usually male–male competition).

intraspecific competition Competition between members of the same species for the same limited resource. Compare with **interspecific competition**.

intrinsic rate of increase (r_{max}) The rate at which a population will grow under optimal conditions (i.e., when birthrates are as high as possible and death rates are as low as possible). Compare with **finite rate of increase**.

intron A region of a eukaryotic gene that is transcribed into RNA but is later removed. Compare with **exon**.

invasive species An exotic (nonnative) species that, upon introduction to a new area, spreads rapidly and competes successfully with native species.

inversion A mutation in which a segment of a chromosome breaks from the rest of the chromosome, flips, and rejoins in reversed orientation.

invertebrates A paraphyletic group composed of animals without a backbone; includes about 95 percent of all animal species. Compare with **vertebrates**.

involuntary muscle Muscle that contracts in response to stimulation by involuntary (parasympathetic or sympathetic), but not voluntary (somatic), neural stimulation.

ion An atom or a molecule that has lost or gained electrons and thus carries an electric charge, either positive (cation) or negative (anion), respectively.

ion channel A type of channel protein that allows certain ions to diffuse across a plasma membrane down an electrochemical gradient.

ionic bond A chemical bond that is formed when an electron is completely transferred from one atom to another so that the atoms remain associated due to their opposite electric charges. Compare with **covalent bond** and **hydrogen bond**.

iris A ring of pigmented muscle just behind the cornea in the vertebrate eye that contracts or expands to control the amount of light entering the eye through the pupil.

isosmotic Comparative term designating a solution that has the same solute concentration and water concentration as another solution. Compare with **hyperosmotic** and **hyposmotic**.

isotonic Comparative term designating a solution that, if inside a cell or vesicle, results in no net uptake or loss of water and thus no effect on the volume of the membrane-bound structure. This solution has the same solute concentration as the solution on the other side of the membrane. Compare with **hypertonic** and **hypotonic**.

isotope Any of several forms of an element that have the same number of protons but differ in the number of neutrons.

joint A place where two components (bones, cartilages, etc.) of a skeleton meet. May be movable (an articulated joint) or immovable (e.g., skull sutures).

juvenile An individual that has adult-like morphology but is not sexually mature.

juvenile hormone An insect hormone that prevents larvae from metamorphosing into adults.

karyogamy Fusion of two haploid nuclei to form a diploid nucleus. Occurs in many fungi, and in animals and plants during fertilization of gametes.

karyotype The distinctive appearance of all of the chromosomes in an individual, including the number of chromosomes and their length and banding patterns (after staining with dyes).

keystone species A species that has an exceptionally great impact on the other species in its ecosystem relative to its abundance.

kidney In terrestrial vertebrates, one of a paired organ situated at the back of the abdominal cavity that filters the blood, produces urine, and secretes several hormones.

kilocalorie (kcal) A unit of energy often used to measure the energy content of food. A kcal of energy raises 1 kg of water 1°C.

kin selection A form of natural selection that favors traits that increase survival or reproduction of an individual's kin at the expense of the individual.

kinesin A class of motor proteins that uses the chemical energy of ATP to "walk" toward the plus end of a microtubule. Used to transport vesicles, particles, organelles and chromosomes.

kinetic energy The energy of motion. Compare with **potential energy**.

kinetochore A protein complex at the centromere where microtubules attach to the chromosome. Contains motor proteins and microtubule-binding proteins that are involved in chromosome segregation during M phase.

kinetochore microtubules Microtubules in the spindle formed during mitosis and meiosis that are attached to the kinetochore on a chromosome.

kinocilium (plural: kinocilia) A single cilium that juts from the surface of many hair cells and functions in detection of sound or pressure.

knock-out allele A mutant allele that does not produce a functional product. Also called a *null allele* or *loss-of-function allele*.

Koch's postulates Four criteria used to determine whether a suspected infectious agent causes a particular disease.

labium majus (plural: labia majora) One of two outer folds of skin that surround the labia minora, clitoris, and vaginal opening of female mammals.

labium minus (plural: labia minora) One of two folds of skin inside the labia majora and surrounding the opening of the urethra and vagina.

labor The strong muscular contractions of the uterus that expel the fetus during birth.

lac operon A set of three genes in *E. coli* that are transcribed into a single mRNA and required for lactose metabolism. Studies of the *lac* operon revealed many insights about gene regulation.

lactation (verb: lactate) Production of milk to feed offspring, from mammary glands of mammals.

lacteal A small lymphatic vessel extending into the center of a villus in the small intestine. Receives chylomicrons containing fat absorbed from food.

lactic acid fermentation Catabolic pathway in which pyruvate produced by glycolysis is converted to lactic acid in the absence of oxygen.

lagging strand In DNA replication, the strand of new DNA that is synthesized discontinuously in a series of short pieces that are later joined. Also called *discontinuous strand*. Compare with **leading strand**.

laminins An abundant protein in the extracellular matrix that binds to other ECM components and to integrins in plasma membranes; helps anchor cells in place. Predominantly found in the basal lamina; many subtypes function in different tissues.

large intestine The distal portion of the digestive tract, consisting of the cecum, colon, and rectum. Its primary function is to compact the wastes delivered from the small intestine and absorb enough water to form feces.

larva (plural: larvae) An immature stage of an animal species in which the immature and adult stages have different body forms.

late endosome A membrane-bound vesicle that arises from an early endosome, accepts lysosomal enzymes from the Golgi, and matures into a lysosome.

latency In viruses that infect animals, the ability to coexist with the host cell in a dormant state without producing new virions. The viral genetic material is replicated as the host cell replicates. Genetic material may or may not be integrated in the host genome, depending on the virus.

lateral bud A bud that forms at the nodes of a stem and may develop into a lateral (side) branch. Also called *axillary bud*.

lateral gene transfer Transfer of DNA between two different species.

lateral line system A pressure-sensitive sensory organ found in many aquatic vertebrates.

lateral root A plant root that extends horizontally from another root.

leaching Loss of nutrients from soil via percolating water.

leading strand In DNA replication, the strand of new DNA that is synthesized in one continuous piece. Also called *continuous strand*. Compare with **lagging strand**.

leaf The main photosynthetic organ of vascular plants.

leak channel Ion channel that allows ions to leak across the membrane of a neuron in its resting state.

learning An enduring change in an individual's behavior that results from specific experience(s).

leghemoglobin An iron-containing protein similar to hemoglobin. Found in infected cells of legume root nodules where it binds oxygen, preventing it from poisoning a bacterial enzyme needed for nitrogen fixation.

legumes Members of the pea plant family. Many form symbiotic associations with nitrogen-fixing bacteria in their roots.

lens A transparent structure that focuses incoming light onto a retina or other light-sensing apparatus of an eye.

lenticel Spongy segment in bark that allows gas exchange between cells in a woody stem and the atmosphere.

leptin A hormone produced and secreted by fat cells (adipocytes) that acts to stabilize fat-tissue mass partly by inhibiting appetite and increasing energy expenditure.

leukocyte Any of several types of blood cells, including neutrophils, macrophages, and lymphocytes, that reside in tissues and circulate in blood and lymph. Functions in tissue repair and defense against pathogens. Also called *white blood cell*.

lichen A symbiotic association of a fungus, often in the Ascomycota lineage, and a photosynthetic alga or cyanobacterium.

life cycle The sequence of developmental events and phases that occurs during the life span of an organism, from fertilization to offspring production.

life history The sequence of events in an individual's life from birth to reproduction to death, including how an individual allocates resources to growth, reproduction, and activities or structures that are related to survival.

life table A data set that summarizes the probability that an individual in a certain population will survive and reproduce in any given year over the course of its lifetime.

ligament Connective tissue that joins bones of an endoskeleton together.

ligand Any molecule that binds to a specific site on a receptor molecule.

ligand-gated channel An ion channel that opens or closes in response to binding by a certain molecule. Compare with **voltage-gated channel**.

light chain The smaller of the two types of polypeptide chains in an antibody or B-cell receptor molecule; composed of a variable (*V*) region, which contributes to the antigen-binding site, and a constant (*C*) region. Compare with **heavy chain**.

lignin A substance, found in the secondary cell walls of some plants, that is exceptionally stiff and strong; a complex polymer built from six-carbon rings. Most abundant in woody plant parts.

limiting nutrient Any essential nutrient whose scarcity in the environment significantly reduces growth and reproduction of organisms.

limnetic zone Open water (not near shore) that receives enough sunlight to support photosynthesis.

lineage See **monophyletic group**.

LINEs (long interspersed nuclear elements) The most abundant class of transposable elements in human genomes; can create copies of itself and insert them elsewhere in the genome. Compare with **SINEs**.

lingual lipase An enzyme produced by glands in the tongue. It breaks down fat molecules into fatty acids and monoglycerides.

linkage In genetics, a physical association between two genes because they are on the same chromosome; the inheritance patterns resulting from this association.

linkage map See **genetic map**.

lipid Any organic substance that does not dissolve in water, but dissolves well in nonpolar organic solvents. Lipids include fatty acids, fats, oils, waxes, steroids, and phospholipids.

lipid bilayer The basic structural element of all cellular membranes; consists of a two-layer sheet of phospholipid molecules with their hydrophobic tails oriented toward the inside and their hydrophilic heads toward the outside. Also called *phospholipid bilayer*.

littoral zone Shallow water near shore that receives enough sunlight to support photosynthesis. May be marine or freshwater; often flowering plants are present.

liver A large, complex organ of vertebrates that performs many functions, including storage of glycogen, processing and conversion of food and wastes, and production of bile.

lobe-finned fish Fish with fins supported by bony elements that extend down the length of the structure.

locomotion Movement of an organism under its own power.

locus (plural: loci) A gene's physical location on a chromosome.

logistic population growth The density-dependent decrease in growth rate as population size approaches the carrying capacity. Compare with **exponential population growth**.

long interspersed nuclear elements See **LINEs**.

long-day plant A plant that blooms in response to short nights (usually in late spring or early summer in the Northern Hemisphere). Compare with **day-neutral** and **short-day plant**.

loop of Henle In the vertebrate kidney, a long U-shaped loop in a nephron that extends into the medulla. Functions as a countercurrent exchanger to set up an osmotic gradient that allows reabsorption of water from the collecting duct.

loose connective tissue A type of connective tissue consisting of fibrous proteins in a soft matrix. Often functions as padding for organs.

lophophore A specialized feeding structure found in some lophotrochozoans and used in suspension (filter) feeding.

lophotrochozoans A major lineage of protostomes (Lophotrochozoa) that grow by extending the size of their skeletons rather than by molting. Many phyla have a specialized feeding structure (lophophore) and/or ciliated larvae (trochophore). Includes rotifers, flatworms, segmented worms, and mollusks. Compare with **ecdysozoans**.

loss-of-function allele See **knock-out allele**.

LUCA The *l*ast *u*niversal *c*ommon *a*ncestor of cells. This theoretical entity is proposed to be the product of chemical evolution and provided characteristics of life that are shared by all living organisms on Earth today.

lumen The interior space of any hollow structure (e.g., the rough ER) or organ (e.g., the stomach).

lung Any respiratory organ used for gas exchange between blood and air.

luteal phase The second major phase of a menstrual cycle, after ovulation, when the progesterone levels are high and the body is preparing for a possible pregnancy.

luteinizing hormone (LH) A peptide hormone, produced and secreted by the anterior pituitary, that stimulates estrogen production, ovulation, and formation of the corpus luteum in females and testosterone production in males.

lymph The mixture of fluid and white blood cells that circulates through the ducts and lymph nodes of the lymphatic system in vertebrates.

lymph node Any of many small, oval structures that lymph moves through in the lymphatic system. Filters the lymph and screens it for pathogens and other antigens. Major sites of lymphocyte activation.

lymphatic system In vertebrates, a body-wide network of thin-walled ducts (or vessels) and lymph nodes, separate from the circulatory system. Collects excess fluid from body tissues and returns it to the blood; also functions as part of the immune system.

lymphocyte A cell that circulates through the bloodstream and lymphatic system and is responsible for the development of adaptive immunity. In most cases belongs to one type of leukocyte—either B cells or T cells.

lysogeny In viruses that infect bacteria (bacteriophages), the ability to coexist with the host cell in a dormant state without producing new virions. The viral genetic material is integrated in the host chromosome and is replicated as the host cell replicates. Compare with **lytic cycle**.

lysosome A small, acidified organelle in an animal cell containing enzymes that catalyze hydrolysis reactions and can digest large molecules. Compare with **vacuole**.

lysozyme An enzyme that functions in innate immunity by digesting bacterial cell walls. Occurs in lysosomes of phagocytes and is secreted in saliva, tears, mucus, and egg white.

lytic cycle A type of viral replicative growth in which the production and release of virions kills the host cell. Compare with **lysogeny**.

M phase The phase of the cell cycle during which cell division occurs. Includes mitosis or meiosis and often cytokinesis.

M-phase-promoting factor (MPF) A complex of a cyclin and cyclin-dependent kinase that, when activated, phosphorylates a number of specific proteins needed to initiate mitosis in eukaryotic cells.

macromolecular machine A group of proteins that assemble to carry out a particular function.

macromolecule Any very large organic molecule, usually made up of smaller molecules (monomers) joined together into a polymer. The main biological macromolecules are proteins, nucleic acids, and polysaccharides.

macronutrient Any element (e.g., nitrogen) that is required in large quantities for normal growth, reproduction, and maintenance of a living organism. Compare with **micronutrient**.

macrophage A type of leukocyte in the innate immune system that participates in the inflammatory response by secreting cytokines and phagocytizing invading pathogens and apoptotic cells. Also serves as an antigen-presenting cell to activate lymphocytes.

MADS box A DNA sequence that codes for a DNA-binding motif in proteins; present in floral organ identity genes in plants. Functionally similar sequences are found in some fungal and animal genes.

magnetoreception A sensory system in which receptors are activated in response to magnetic fields.

magnetoreceptor A sensory cell or organ specialized for detecting magnetic fields.

major histocompatibility protein See **MHC protein**.

maladaptive Describing a trait that lowers fitness.

malaria A human disease caused by five species of the protist *Plasmodium* and passed to humans by mosquitoes.

malignant tumor A tumor that is actively growing and disrupting local tissues or is spreading to other organs. Cancer consists of one or more malignant tumors. Compare with **benign tumor**.

Malpighian tubules A major excretory organ of insects, consisting of blind-ended tubes that extend from the gut into the hemocoel. Filter hemolymph to form "pre-urine" and then send it to the hindgut for further processing.

mammals One of the two lineages of amniotes (vertebrates that produce amniotic eggs) distinguished by hair (or fur) and mammary glands. Includes the monotremes (platypuses), marsupials, and eutherians (placental mammals).

mammary glands Specialized exocrine glands that produce and secrete milk for nursing offspring. A diagnostic feature of mammals.

mandibles Any mouthpart used in chewing. In vertebrates, the lower jaw. In insects, crustaceans, and myriapods, the first pair of mouthparts.

mantle One of the three main parts of the mollusk body; the thick outer tissue that protects the visceral mass and may secrete a calcium carbonate shell.

marsh A wetland dominated by grasses and other nonwoody plants.

marsupials A lineage of mammals (Marsupiala) that nourish their young in an abdominal pouch after a very short period of development in the uterus.

mass extinction The extinction of a large number of diverse evolutionary groups during a relatively short period of geologic time (about 1 million years). May occur due to sudden and extraordinary environmental changes. Compare with **background extinction**.

mass feeder An animal that ingests chunks of food.

mass number The total number of protons and neutrons in an atom.

mast cell A type of leukocyte that is stationary (embedded in tissue) and helps trigger the inflammatory response, including secretion of histamine, to infection or injury. Particularly important in allergic responses and defense against parasites.

maternal chromosome A chromosome inherited from the mother.

mechanical advantage The ratio of force exerted on a load to the muscle force of the effort. A measure of the force efficiency of a mechanical system.

mechanoreception A sensory system in which receptors are activated in response to changes in pressure.

mechanoreceptor A sensory cell or organ specialized for detecting distortions caused by touch or pressure. One example is hair cells in the cochlea.

mediator Regulatory proteins in eukaryotes that form a physical link between regulatory transcription factors that are bound to DNA, the basal transcription complex, and RNA polymerase.

medium A liquid or solid that supports the growth of cells.

medulla The innermost part of an organ (e.g., kidney or adrenal gland).

medulla oblongata In vertebrates, a region of the brain stem that along with the cerebellum forms the hindbrain.

medusa (plural: medusae) The free-floating stage in the life cycle of some cnidarians (e.g., jellyfish). Compare with **polyp**.

megapascal (MPa) A unit of pressure (force per unit area) equivalent to 1 million pascals (Pa).

megasporangium (plural: megasporangia) In heterosporous species of plants, a spore-producing structure that produces megaspores, which go on to develop into female gametophytes.

megaspore In seed plants, a haploid (*n*) spore that is produced in a megasporangium by meiosis of a diploid (*2n*) megasporocyte; develops into a female gametophyte. Compare with **microspore**.

meiosis In sexually reproducing organisms, a special two-stage type of cell division in which one diploid (*2n*) parent cell produces haploid (*n*) cells (gametes); results in halving of the chromosome number. Also called *reduction division*.

meiosis I The first cell division of meiosis, in which synapsis and crossing over occur and homologous chromosomes are separated from each other, producing daughter cells with half as many chromosomes (each composed of two sister chromatids) as the parent cell.

meiosis II The second cell division of meiosis, in which sister chromatids are separated from each other. Similar to mitosis.

melatonin A hormone, produced by the pineal gland, that regulates sleep–wake cycles and seasonal reproduction in vertebrates.

membrane potential A difference in electric charge across a cell membrane; a form of potential energy. Also called *membrane voltage*.

memory Retention of learned information.

memory cells A type of lymphocyte responsible for maintaining immunity for years or decades after an infection. Descended from a B cell or T cell activated during a previous infection or vaccination.

meniscus (plural: menisci) The concave boundary layer formed at most air–water interfaces due to adhesion and surface tension.

menstrual cycle A female reproductive cycle seen in Old World monkeys and apes (including humans) in which the uterine lining is shed (menstruation) if no pregnancy occurs. Compare with **estrous cycle**.

menstruation The periodic shedding of the uterine lining through the vagina that occurs in females of Old World monkeys and apes, including humans.

meristem (adjective: meristematic) In plants, a group of undifferentiated cells, including stem cells, which can divide and develop into various adult tissues throughout the life of a plant. See also **apical meristem** and **ground meristem**.

mesoderm The middle of the three basic cell layers (germ layers) in most animal embryos; gives rise to muscles, bones, blood, and some internal organs (kidney, spleen, etc.). Compare with **ectoderm** and **endoderm**.

mesoglea A gelatinous material, containing scattered ectodermal cells, that is located between the ectoderm and endoderm of cnidarians (e.g., jellyfish, corals, and anemones).

mesophyll cell A type of cell, found near the surfaces of plant leaves, that is specialized for the light-dependent reactions of photosynthesis.

Mesozoic era The interval of geologic time, from 251 million to 65.5 million years ago, during which gymnosperms were the dominant plants and dinosaurs the dominant vertebrates. Ended with extinction of the dinosaurs (except birds).

messenger RNA (mRNA) An RNA molecule transcribed from DNA that carries information (in codons) that specifies the amino acid sequence of a polypeptide.

meta-analysis A comparative analysis of the results of many smaller, previously published studies.

metabolic pathway A linked series of biochemical reactions that build up or break down a particular molecule; the product of one reaction is the substrate of the next reaction.

metabolic rate The total energy use by all the cells of an individual. For aerobic organisms, often measured as the amount of oxygen consumed per hour.

metabolic water The water that is produced as a by-product of cellular respiration.

metagenomics The inventory of all the genes in a community or ecosystem by sequencing, analyzing, and comparing the genomes of the component organisms. Also called *environmental sequencing*.

metagenomics The inventory of all the genes in a community or ecosystem by sequencing, analyzing, and comparing the genomes of the component organisms. Sequencing of all or most of the genes present in an environment directly (also called *environmental sequencing*).

metallothioneins Small plant proteins that bind to and prevent excess metal ions from acting as toxins.

metamorphosis Transition from one developmental stage to another, such as from the larval to the adult form of an animal.

metaphase A stage in mitosis or meiosis during which chromosomes line up in the middle of the cell.

metaphase plate The plane along which chromosomes line up in the middle of the spindle during metaphase of mitosis or meiosis; not an actual structure.

metapopulation A population made up of many small, physically isolated populations connected by migration.

metastasis The spread of cancerous cells from their site of origin to distant sites in the body where they may establish additional tumors.

methanogen A prokaryote that produces methane (CH_4) as a by-product of cellular respiration.

methanotroph An organism (bacteria or archaea) that uses methane (CH_4) as its primary electron donor and source of carbon.

methyl salicylate (MeSA) A molecule that is hypothesized to function as a signal, transported among tissues, that triggers systematic acquired resistance in plants—a response to pathogen attack.

MHC protein Any of a large set of mammalian cell-surface glycoproteins involved in marking cells as self and in antigen presentation to T cells. Also called *MHC molecule*. Compare with **Class I** and **Class II MHC protein**.

microbe Any microscopic organism, including bacteria, archaea, and various tiny eukaryotes.

microbiology The field of study concerned with microscopic organisms.

microfibril Bundled strands of cellulose that serve as the fibrous component in plant cell walls.

microfilament See **actin filament**.

micronutrient Any element (e.g., iron, molybdenum, magnesium) that is required in very small quantities for normal growth, reproduction, and maintenance of a living organism. Compare with **macronutrient**.

micropyle The tiny pore in a plant ovule through which the pollen tube reaches the embryo sac.

microRNA (miRNA) A small, single-stranded RNA associated with proteins in an RNA-induced silencing complex (RISC). Processed from a longer premiRNA gene transcript. Can bind to complementary sequences in mRNA molecules, allowing the associated proteins of RISC to degrade the bound mRNA or inhibit its translation. See **RNA interference**.

microsatellite A noncoding stretch of eukaryotic DNA consisting of a repeating sequence 2 to 6 base pairs long. Also called *short tandem repeat* or *simple sequence repeat*.

microsporangium (plural: microsporangia) In heterosporous species of plants, a spore-producing structure that produces microspores, which go on to develop into male gametophytes.

microspore In seed plants, a haploid (*n*) spore that is produced in a microsporangium by meiosis of a diploid (*2n*) microsporocyte; develops into a male gametophyte. Compare with **megaspore**.

microtubule A long, tubular fiber, about 25 nm in diameter, formed by polymerization of tubulin protein dimers; one of the three types of cytoskeletal fibers. Involved in cell movement and transport of materials within the cell. Compare with **actin filament** and **intermediate filament**.

microtubule organizing center (MTOC) General term for any structure (e.g., centrosome and basal body) that organizes microtubules in cells.

microvilli (singular: microvillus) Tiny protrusions from the surface of an epithelial cell that increase the surface area for absorption of substances.

midbrain One of the three main regions of the vertebrate brain; includes sensory integrating and relay centers. Compare with **forebrain** and **hindbrain**.

middle ear The air-filled middle portion of the mammalian ear, which contains three small bones (ossicles) that transmit and amplify sound from the tympanic membrane to the inner ear. Is connected to the throat via the eustachian tube.

migration (1) In ecology, a seasonal movement of large numbers of organisms from one geographic location or habitat to another. (2) In population genetics, movement of individuals from one population to another.

millivolt (mV) A unit of voltage equal to 1/1000 of a volt.

mimicry A phenomenon in which one species has evolved (or learns) to look or sound like another species. See **Batesian mimicry** and **Müllerian mimicry**.

mineral One of various inorganic substances that are important components of enzyme cofactors or of structural materials in an organism.

mineralocorticoids A class of steroid hormones, produced and secreted by the adrenal cortex, that regulate electrolyte levels and the overall volume of body fluids. Aldosterone is the principal one in humans. Compare with **glucocorticoids**.

minisatellite A noncoding stretch of eukaryotic DNA consisting of a repeating sequence that is 6 to 100 base pairs long. Also called *variable number tandem repeat (VNTR)*.

mismatch repair The process by which mismatched base pairs in DNA are fixed.

missense mutation A point mutation (change in a single base pair) that changes one amino acid for another within the sequence of a protein.

mitochondrial matrix Central compartment of a mitochondrion, which is lined by the inner membrane; contains mitochondrial DNA, ribosomes, and the enzymes for pyruvate processing and the citric acid cycle.

mitochondrion (plural: mitochondria) A eukaryotic organelle that is bounded by a double membrane and is the site of aerobic respiration and ATP synthesis.

mitogen-activated protein kinases (MAPK) Enzymes that are involved in signal transduction pathways that often lead to the induction of cell replication. Different types are organized in a series, where one kinase activates another via phosphorylation. See also **phosphorylation cascade**.

mitosis In eukaryotic cells, the process of nuclear division that results in two daughter nuclei genetically identical with the parent nucleus. Subsequent cytokinesis (division of the cytoplasm) yields two daughter cells.

mode of transmission The type of inheritance observed as a trait is passed from parent to offspring. Some common types are autosomal recessive, autosomal dominant, and X-linked recessive.

model organism An organism selected for intensive scientific study based on features that make it easy to work with (e.g., body size, life span), in the hope that findings will apply to other species.

molarity A common unit of solute concentration equal to the number of moles of a dissolved solute in 1 liter of solution.

mole The amount of a substance that contains 6.022×10^{23} of its elemental entities (e.g., atoms, ions, or molecules). This number of molecules of a compound will have a mass equal to the molecular weight of that compound expressed in grams.

molecular chaperone A protein that facilitates the folding of newly synthesized proteins into their correct three-dimensional shape. Usually works by an ATP-dependent mechanism.

molecular formula A notation that indicates only the numbers and types of atoms in a molecule, such as H_2O for the water molecule. Compare with **structural formula**.

molecular weight The sum of the atomic weights of all of the atoms in a molecule; roughly, the total number of protons and neutrons in the molecule.

molecule A combination of two or more atoms held together by covalent bonds.

molting A method of body growth, used by ecdysozoans, that involves the shedding of an external protective cuticle or skeleton, expansion of the soft body, and growth of a new external layer.

monocot Any flowering plant (angiosperm) that has a single cotyledon (embryonic leaf) upon germination. Monocots form a monophyletic group. Also called a monocotyledonous plant. Compare with **dicot**.

monoecious Describing an angiosperm species that has both male and female reproductive structures on each plant. Compare with **dioecious**.

monohybrid cross A mating between two parents that are both heterozygous for one given gene.

monomer A small molecule that can covalently bind to other similar molecules to form a larger macromolecule. Compare with **polymer**.

monophyletic group An evolutionary unit that includes an ancestral population and all of its descendants but no others. Also called a *clade* or *lineage*. Compare with **paraphyletic group** and **polyphyletic group**.

monosaccharide A molecule that has the molecular formula $(CH_2O)_n$ and cannot be hydrolyzed to form any smaller carbohydrates. Also called *simple sugar*. Compare with **oligosaccharide** and **polysaccharide**.

monosomy The state of having only one copy of a particular type of chromosome in an otherwise diploid cell.

monotremes A lineage of mammals (Monotremata) that lay eggs and then nourish the young with milk. Includes just five living species: the platypus and four species of echidna, all with leathery beaks or bills.

morphogen A molecule that exists in a concentration gradient and provides spatial information to embryonic cells.

morphology The shape and appearance of an organism's body and its component parts.

morphospecies concept The definition of a species as a population or group of populations that have measurably different anatomical features from other groups. Also called *morphological species concept*. Compare with **biological species concept** and **phylogenetic species concept**.

motor neuron A nerve cell that carries signals from the central nervous system (brain and spinal cord) to an effector, such as a muscle or gland. Compare with **interneuron** and **sensory neuron**.

motor protein A class of proteins whose major function is to convert the chemical energy of ATP into motion. Includes dynein, kinesin, and myosin.

MPF See **M-phase-promoting factor**.

mRNA See **messenger RNA**.

mucosal-associated lymphoid tissue (MALT) Collective term for lymphocytes and other leukocytes associated with skin cells and mucus-secreting epithelial tissues in the gut and respiratory tract. Plays an important role in preventing entry of pathogens into the body.

mucous cell A type of cell found in the epithelial layer of the stomach; responsible for secreting mucus into the stomach.

mucus (adjective: mucous) A slimy mixture of glycoproteins (called mucins) and water that is secreted in many animal organs for lubrication. Serves as a barrier to protect surfaces from infection.

Müllerian inhibitory substance A peptide hormone, secreted by the embryonic testis, that causes regression (withering away) of the female reproductive ducts.

Müllerian mimicry A type of mimicry in which two (or more) harmful species resemble each other. Compare with **Batesian mimicry**.

multicellularity The state of being composed of many cells that adhere to each other and do not all express the same genes, with the result that some cells have specialized functions.

multiple allelism The existence of more than two alleles of the same gene.

multiple fruit A fruit (e.g., pineapple) that develops from many separate flowers and thus many carpels. Compare with **aggregate** and **simple fruit**.

multiple sclerosis (MS) A human autoimmune disease in which the immune system attacks the myelin sheaths that insulate axons of neurons.

muscle fiber A single muscle cell.

muscle tissue An animal tissue consisting of bundles of long, thin, contractile cells (muscle fibers). Functions primarily in movement.

mutagen Any physical or chemical agent that increases the rate of mutation.

mutant An individual that carries a mutation, particularly a new or rare mutation.

mutation Any change in the hereditary material of an organism (DNA in most organisms, RNA in some viruses). The only source of new alleles in populations.

mutualism (adjective: mutualistic) A symbiotic relationship between two organisms (mutualists) that benefits both. Compare with **commensalism** and **parasitism**.

mutualist Organism that is a participant and partner in a mutualistic relationship. See **mutualism**.

mycelium (plural: mycelia) A mass of underground filaments (hyphae) that form the body of a fungus. Also found in some protists and bacteria.

mycorrhiza (plural: mycorrhizae) A mutualistic association between certain fungi and the roots of most vascular plants, sometimes visible as nodules or nets in or around plant roots.

myelin sheath Multiple layers of myelin, derived from the cell membranes of certain glial cells, wrapped around the axon of a neuron and providing electrical insulation.

myocardial infarction Death of cardiac muscle cells when deprived of oxygen.

myoD A transcription factor that is a master regulator of muscle cell differentiation (short for "*myo*blast *determination*").

myofibril Long, slender structure composed of contractile proteins organized into repeating units (sarcomeres) in vertebrate heart muscle and skeletal muscle.

myosin Any one of a class of motor proteins that use the chemical energy of ATP to move along actin filaments in muscle contraction, cytokinesis, and vesicle transport.

myriapods A lineage of arthropods with long segmented trunks, each segment bearing one or two pairs of legs. Includes millipedes and centipedes.

NAD⁺/NADH Oxidized and reduced forms, respectively, of nicotinamide adenine dinucleotide. A nonprotein electron carrier that functions in many of the redox reactions of metabolism.

NADP⁺/NADPH Oxidized and reduced forms, respectively, of nicotinamide adenine dinucleotide phosphate. A nonprotein electron carrier that is reduced during the light-dependent reactions in photosynthesis and extensively used in biosynthetic reactions.

natural experiment A situation in which a natural change in conditions enables comparisons of groups, rather than a manipulation of conditions by researchers.

natural selection The process by which individuals with certain heritable traits tend to produce more surviving offspring than do individuals without those traits, often leading to a change in the genetic makeup of the population. A major mechanism of evolution.

nauplius A distinct planktonic larval stage seen in many crustaceans.

Neanderthal A recently extinct European species of hominin, *Homo neanderthalensis*, closely related to but distinct from modern humans.

nectar The sugary fluid produced by flowers to attract and reward pollinating animals.

nectary A nectar-producing structure in a flower.

negative control Of genes, when a regulatory protein shuts down expression by binding to DNA on or near the gene.

negative feedback A self-limiting, corrective response in which a deviation in some variable (e.g., concentration of some compound) triggers responses aimed at returning the variable to normal. Represents a means of maintaining homeostasis. Compare with **positive feedback**.

negative pressure ventilation Ventilation of the lungs by expanding the rib cage so as to "pull" air into the lungs. Compare with **positive pressure ventilation**.

negative-sense RNA virus An ssRNA virus whose genome contains sequences complementary to those in the mRNA required to produce viral proteins. Compare with **ambisense virus** and **positive-sense virus**.

nematodes See **roundworms**.

nephron One of many tiny tubules inside the kidney that function in the formation of urine.

neritic zone Shallow marine waters beyond the intertidal zone, extending down to about 200 meters, where the continental shelf ends.

nerve A long, tough strand of nervous tissue, typically containing thousands of axons, wrapped in connective tissue; carries impulses between the central nervous system and some other part of the body.

nerve cord In chordate animals, a hollow bundle of nerves extending from the brain along the dorsal (back) side of the animal, with cerebrospinal fluid inside a central channel. One of the defining features of chordates.

nerve net A nervous system in which neurons are diffuse instead of being clustered into large ganglia or tracts; found in cnidarians and ctenophores.

nervous tissue An animal tissue consisting of nerve cells (neurons) and various supporting cells.

net primary productivity (NPP) In an ecosystem, the total amount of carbon fixed by photosynthesis over a given time period minus the amount oxidized during cellular respiration. Compare with **gross primary productivity**.

net reproductive rate (R_0) The growth rate of a population per generation; equivalent to the average number of female offspring that each female produces over her lifetime.

neural tube A folded tube of ectoderm that forms along the dorsal side of a young vertebrate embryo; gives rise to the brain and spinal cord.

neuroendocrine Referring to nerve cells (neurons) that release hormones into the blood or to such hormones themselves.

neurogenesis The birth of new neurons from central nervous system stem cells.

neurohormones Hormones produced by neurons.

neuron A cell that is specialized for the transmission of nerve impulses. Typically has dendrites, a cell body, and a long axon that forms synapses with other neurons. Also called *nerve cell*.

neurosecretory cell A nerve cell (neuron) that produces and secretes hormones into the bloodstream. Principally found in the hypothalamus. Also called *neuroendocrine cell*.

neurotoxin Any substance that specifically destroys or blocks the normal functioning of neurons.

neurotransmitter A molecule that transmits signals from one neuron to another or from a neuron to a muscle or gland. Examples are acetylcholine, dopamine, serotonin, and norepinephrine.

neutral In genetics, referring to any mutation or mutant allele that has no effect on an individual's fitness.

neutrophil A type of leukocyte, capable of moving through body tissues, that engulfs and digests pathogens and other foreign particles; also secretes various compounds that attack bacteria and fungi.

niche The range of resources that a species can use and the range of conditions that it can tolerate. More broadly, the role that species plays in its ecosystem.

niche differentiation The evolutionary change in resource use by competing species that occurs as the result of character displacement.

nicotinamide adenine dinucleotide See **NAD⁺/NADH**.

nitrogen cycle, global The movement of nitrogen among terrestrial ecosystems, the oceans, and the atmosphere.

nitrogen fixation The incorporation of atmospheric nitrogen (N_2) into ammonia (NH_3), which can be used to make many organic compounds. Occurs in only a few lineages of bacteria and archaea.

nociceptor A sensory cell or organ specialized to detect tissue damage, usually producing the sensation of pain.

Nod factors Molecules produced by nitrogen-fixing bacteria that help them recognize and bind to roots of legumes.

node (1) In animals, any small thickening (e.g., a lymph node). (2) In plants, the part of a stem where leaves or leaf buds are attached. (3) In a phylogenetic tree, the point where two branches diverge, representing the point in time when an ancestral group split into two or more descendant groups. Also called *fork*.

node of Ranvier One of the unmyelinated sections that occurs periodically along a neuron's axon and serves as a site where an action potential can be regenerated.

nodule Globular structure on roots of legume plants that contain symbiotic nitrogen-fixing bacteria.

noncyclic electron flow Path of electron flow in which electrons pass from photosystem II, through an electron transport chain, to photosystem I, and ultimately to NADP⁺ during the light-dependent reactions of photosynthesis. See also **Z scheme**.

nondisjunction An error that can occur during meiosis or mitosis, in which one daughter cell receives two copies of a particular chromosome and the other daughter cell receives none.

nonpolar covalent bond A covalent bond in which electrons are equally shared between two atoms of the same or similar electronegativity. Compare with **polar covalent bond**.

nonsense mutation A point mutation (change in a single base pair) that converts an amino-acid-specifying codon into a stop codon.

non-sister chromatids The chromatids of a particular type of chromosome (after replication) with respect to the chromatids of its homologous chromosome. Crossing over occurs between non-sister chromatids. Compare with **sister chromatids**.

non-template strand The strand of DNA that is not transcribed during synthesis of RNA. Its sequence corresponds to that of the mRNA produced from the other strand. Also called *coding strand*.

non-vascular plants A paraphyletic group of land plants that lack vascular tissue and reproduce using spores. The non-vascular plants include three lineages of green plants (liverworts, mosses, and hornworts). These lineages are sometimes called *bryophytes*.

norepinephrine A catecholamine used as a neurotransmitter in the sympathetic nervous system. Also is produced by the adrenal medulla and functions as a hormone that triggers rapid responses relating to the fight-or-flight response.

notochord A supportive but flexible rod that occurs in the back of a chordate embryo, ventral to the developing spinal cord. Replaced by vertebrae in most adult vertebrates. A defining feature of chordates.

nuclear envelope The double-layered membrane enclosing the nucleus of a eukaryotic cell.

nuclear lamina A lattice-like sheet of fibrous nuclear lamins, which are one type of intermediate filament. Lines the inner membrane of the nuclear envelope, stiffening the envelope and helping to organize the chromosomes.

nuclear lamins Intermediate filaments that make up the nuclear lamina layer—a lattice-like layer inside the nuclear envelope that stiffens the structure.

nuclear localization signal (NLS) A short amino acid sequence that marks a protein for delivery to the nucleus.

nuclear pore An opening in the nuclear envelope that connects the inside of the nucleus with the cytoplasm and through which molecules such as mRNA and some proteins can pass.

nuclear pore complex A large complex of dozens of proteins lining a nuclear pore, defining its shape and regulating transport through the pore.

nuclease Any enzyme that can break down RNA or DNA molecules.

nucleic acid A macromolecule composed of nucleotide monomers. Generally used by cells to store or transmit hereditary information. Includes ribonucleic acid and deoxyribonucleic acid.

nucleoid In prokaryotic cells, a dense, centrally located region that contains DNA but is not surrounded by a membrane.

nucleolus In eukaryotic cells, a specialized structure in the nucleus where ribosomal RNA processing occurs and ribosomal subunits are assembled.

nucleosome A repeating, bead-like unit of eukaryotic chromatin, consisting of about 200 nucleotides of DNA wrapped twice around eight histone proteins.

nucleotide excision repair The process of removing a damaged region in one strand of DNA and replacing it with the correct sequence using the undamaged strand as a template.

nucleotide A molecule consisting of a five-carbon sugar (ribose or deoxyribose), a phosphate group, and one of several nitrogen-containing bases. DNA and RNA are polymers of nucleotides containing deoxyribose (deoxyribonucleotides) and ribose (ribonucleotides), respectively. Equivalent to a nucleoside plus one phosphate group.

nucleus (1) The center of an atom, containing protons and neutrons. (2) In eukaryotic cells, the large organelle containing the chromosomes and surrounded by a double membrane. (3) A discrete clump of neuron cell bodies in the brain, usually sharing a distinct function.

null allele See **knock-out allele**.

null hypothesis A hypothesis that specifies what the results of an experiment will be if the main hypothesis being tested is wrong. Often states that there will be no difference between experimental groups.

nutrient Any substance that an organism requires for normal growth, maintenance, or reproduction.

occipital lobe In the vertebrate brain, one of the four major areas in the cerebrum.

oceanic zone The waters of the open ocean beyond the continental shelf.

odorant Any volatile molecule that conveys information about food or the environment.

oil An unsaturated fat that is liquid at room temperature.

Okazaki fragment Short segment of DNA produced during replication of the lagging strand template. Many Okazaki fragments make up the lagging strand in newly synthesized DNA.

olfaction The perception of odors.

olfactory bulb A bulb-shaped projection of the brain just above the nose. Receives and interprets odor information from the nose.

oligodendrocyte A type of glial cell that wraps around axons of some neurons in the central nervous system, forming a myelin sheath that provides electrical insulation. Compare with **Schwann cell**.

oligopeptide A chain composed of fewer than 50 amino acids linked together by peptide bonds. Often referred to simply as *peptide*.

oligosaccharide A linear or branched polymer consisting of less than 50 monosaccharides joined by glycosidic linkages. Compare with **monosaccharide** and **polysaccharide**.

ommatidium (plural: ommatidia) A light-sensing column in an arthropod's compound eye.

omnivore (adjective: omnivorous) An animal whose diet regularly includes both meat and plants. Compare with **carnivore** and **herbivore**.

oncogene Any gene whose protein product stimulates cell division at all times and thus promotes cancer development. Often is a mutated form of a gene involved in regulating the cell cycle. See **proto-oncogene**.

one-gene, one-enzyme hypothesis The hypothesis that each gene is responsible for making one enzyme. This hypothesis has expanded to include genes that produce proteins other than enzymes or that produce RNAs as final products.

oogenesis The production of egg cells (ova).

oogonium (plural: oogonia) In an ovary, any of the diploid cells that can divide by mitosis to create primary oocytes (which can undergo meiosis) and more oogonia.

open circulatory system A circulatory system in which the circulating fluid (hemolymph) is not confined to blood vessels. Compare with **closed circulatory system**.

open reading frame (ORF) Any DNA sequence, ranging in length from several hundred to thousands of base pairs long, that is flanked by a start codon and a stop codon. ORFs identified by computer analysis of DNA may be functional genes, especially if they have other features characteristic of genes (e.g., promoter sequence).

operator In prokaryotic DNA, a binding site for a repressor protein; located near the start of an operon.

operculum The stiff flap of tissue that covers the gills of teleost fishes.

operon A region of prokaryotic DNA that codes for a series of functionally related genes and is transcribed from a single promoter into one mRNA.

opsin A transmembrane protein that is covalently linked to retinal, the light-detecting pigment in rod and cone cells.

optic nerve A bundle of neurons that runs from the eye to the brain.

optimal foraging The concept that animals forage in a way that maximizes the amount of usable energy they take in, given the costs of finding and ingesting their food and the risk of being eaten while they're at it.

orbital The region of space around an atomic nucleus in which an electron is present most of the time.

organ A group of tissues organized into a functional and structural unit.

organ system Groups of tissues and organs that work together to perform a function.

organelle Any discrete, membrane-bound structure within a cell (e.g., mitochondrion) that has a characteristic structure and function.

organic For a compound, containing carbon and hydrogen and usually containing carbon–carbon bonds. Organic compounds are widely used by living organisms.

organism Any living entity that contains one or more cells.

organogenesis A stage of embryonic development that follows gastrulation and that creates organs from the three germ layers.

origin of replication The site on a chromosome at which DNA replication begins.

osmoconformer An animal that does not actively regulate the osmolarity of its tissues but conforms to the osmolarity of the surrounding environment.

osmolarity The concentration of dissolved substances in a solution, measured in osmoles per liter.

osmoregulation The process by which a living organism controls the concentration of water and salts in its body.

osmoregulator An animal that actively regulates the osmolarity of its tissues.

osmosis Diffusion of water across a selectively permeable membrane from a region of low solute concentration (high water concentration) to a region of high solute concentration (low water concentration).

ossicles, ear In mammals, three bones found in the middle ear that function in transferring and amplifying sound from the outer ear to the inner ear.

ouabain A plant toxin that poisons the sodium–potassium pumps of animals.

out-of-Africa hypothesis The hypothesis that modern humans (*Homo sapiens*) evolved in Africa and spread to other continents, replacing other *Homo* species without interbreeding with them.

outcrossing Reproduction by fusion of the gametes of different individuals, rather than by self-fertilization.

outer ear The outermost portion of the mammalian ear, consisting of the pinna (ear flap) and the ear canal. Funnels sound to the tympanic membrane.

outgroup A taxon that is closely related to a particular monophyletic group but is not part of it.

oval window A membrane separating the fluid-filled cochlea from the air-filled middle ear; sound vibrations pass through it from the middle ear to the inner ear in mammals.

ovary The egg-producing organ of a female animal, or the fruit- and seed-producing structure in the female part of a flower.

overexploitation Unsustainable removal of wildlife from the natural environment for use by humans.

oviduct See **fallopian tube**.

oviparous In animals, producing eggs that are laid outside the body where they develop and hatch. Compare with **ovoviviparous** and **viviparous**.

ovoviviparous In animals, producing eggs that are retained inside the body until they are ready to hatch. Compare with **oviparous** and **viviparous**.

ovulation The release of an egg from an ovary of a female vertebrate. In humans, an ovarian follicle releases an egg at the end of the follicular phase of the menstrual cycle.

ovule In flowering plants, the structure inside an ovary that contains the female gametophyte and eventually (if fertilized) becomes a seed.

ovum (plural: ova) See **egg**.

oxidation The loss of electrons from an atom or molecule during a redox reaction, either by donation of an electron to another atom or molecule, or by the shared electrons in covalent bonds moving farther from the atomic nucleus.

oxidative phosphorylation Production of ATP molecules by ATP synthase using the proton gradient established via redox reactions of an electron transport chain.

oxygen–hemoglobin equilibrium curve The graphed depiction of the percentage of hemoglobin in the blood that is bound to oxygen at various partial pressures of oxygen.

oxygenic Referring to any process or reaction that produces oxygen. Photosynthesis in plants, algae, and cyanobacteria, which involves photosystem II, is oxygenic. Compare with **anoxygenic**.

oxytocin A peptide hormone, secreted by the posterior pituitary, that triggers labor and milk production in females and that stimulates pair bonding, parental care, and affiliative behavior in both sexes.

p53 A tumor-suppressor protein (molecular weight of 53 kilodaltons) that responds to DNA damage by stopping the cell cycle, turning on DNA repair

machinery, and, if necessary, triggering apoptosis. Encoded by the *p53* gene.

pacemaker cell Any of a group of specialized cardiac muscle cells in the sinoatrial (SA) node of the vertebrate heart that have an inherent rhythm and can generate an electrical impulse that spreads to other heart cells.

paleontologists Scientists who study the fossil record and the history of life.

Paleozoic era The interval of geologic time, from 542 million to 251 million years ago, during which fungi, land plants, and animals first appeared and diversified. Began with the Cambrian explosion and ended with the extinction of many invertebrates and vertebrates at the end of the Permian period.

pancreas A large gland in vertebrates that has both exocrine and endocrine functions. Secretes digestive enzymes into a duct connected to the intestine and secretes several hormones (notably, insulin and glucagon) into the bloodstream.

pancreatic amylase An enzyme produced by the pancreas that breaks down glucose chains by catalyzing hydrolysis of the glycosidic linkages between the glucose residues.

pancreatic lipase An enzyme that is produced in the pancreas and acts in the small intestine to break bonds in complex fats, releasing small lipids.

pandemic The spread of an infectious disease in a short time period over a wide geographic area and affecting a very high proportion of the population. Compare with **epidemic**.

parabiosis An experimental technique for determining whether a certain physiological phenomenon is regulated by a hormone; consists of surgically uniting two individuals so that hormones can pass between them.

paracrine Relating to a chemical signal that is released by one cell and affects neighboring cells.

paraphyletic group A group that includes an ancestral population and *some* but not all of its descendants. Compare with **monophyletic group**.

parapodia (singular: parapodium) Appendages found in some annelids from which bristle-like structures (chaetae) extend.

parasite An organism that lives on a host species (ectoparasite) or in a host species (endoparasite) and that damages its host.

parasitism (adjective: parasitic) A symbiotic relationship between two organisms that is beneficial to one organism (the parasite) but detrimental to the other (the host). Compare with **commensalism** and **mutualism**.

parasitoid An organism that has a parasitic larval stage and a free-living adult stage. Most parasitoids are insects that lay eggs in the bodies of other insects.

parasympathetic nervous system The part of the autonomic nervous system that stimulates responses for conserving or restoring energy, such as reduced heart rate and increased digestion. Compare with **sympathetic nervous system**.

parenchyma cell In plants, a general type of cell with a relatively thin primary cell wall. These cells, found in leaves, the centers of stems and roots, and fruits, are involved in photosynthesis, storage, and

transport. Compare with **collenchyma cell** and **sclerenchyma cell**.

parental care Any action by which an animal expends energy or assumes risks to benefit its offspring (e.g., nest building, feeding of young, defense).

parental generation The adults used in the first experimental cross of a breeding experiment.

parental strand A strand of DNA that is used as a template during DNA synthesis.

parietal cell A cell in the stomach lining that secretes hydrochloric acid.

parietal lobe In the vertebrate brain, one of the four major areas in the cerebrum.

parsimony The logical principle that the most likely explanation of a phenomenon is the most economical or simplest. When applied to comparison of alternative phylogenetic trees, it suggests that the one requiring the fewest evolutionary changes is most likely to be correct.

parthenogenesis Development of offspring from unfertilized eggs; a type of asexual reproduction.

partial pressure The pressure of one particular gas in a mixture of gases; the contribution of that gas to the overall pressure.

particulate inheritance The observation that genes from two parents do not blend together in offspring, but instead remain separate or particle-like.

pascal (Pa) A unit of pressure (force per unit area).

passive transport Diffusion of a substance across a membrane. When this event occurs with the assistance of membrane proteins, it is called *facilitated diffusion*.

patch clamping A technique for studying the electrical currents that flow through individual ion channels by sucking a tiny patch of membrane to the hollow tip of a microelectrode.

paternal chromosome A chromosome inherited from the father.

pathogen (adjective: pathogenic) Any entity capable of causing disease, such as a microbe, virus, or prion.

pattern formation The series of events that determines the spatial organization of an entire embryo or parts of an embryo, for example, setting the major body axes early in development.

pattern-recognition receptor On leukocytes, a class of membrane proteins that bind to molecules commonly associated with foreign cells and viruses and signal responses against broad types of pathogens. Part of the innate immune response.

peat Semi-decayed organic matter that accumulates in moist, low-oxygen environments such as *Sphagnum* (moss) bogs.

pectin A gelatinous polysaccharide found in the primary cell wall of plant cells. Attracts and holds water, forming a gel that resists compression forces and helps keep the cell wall moist.

pedigree A family tree of parents and offspring, showing inheritance of particular traits of interest.

penis The copulatory organ of male mammals, used to insert sperm into a female.

pentose A monosaccharide (simple sugar) containing five carbon atoms.

PEP carboxylase An enzyme that catalyzes addition of CO_2 to phosphoenolpyruvate, a three-carbon compound, forming a four-carbon organic acid. See also **C_4 pathway** and **crassulacean acid metabolism (CAM)**.

pepsin A protein-digesting enzyme present in the stomach.

peptide See **oligopeptide**.

peptide bond The covalent bond formed by a condensation reaction between two amino acids; links the residues in peptides and proteins.

peptidoglycan A complex structural polysaccharide found in bacterial cell walls.

perennial Describing a plant whose life cycle normally lasts for more than one year. Compare with **annual**.

perfect flower A flower that contains both male parts (stamens) and female parts (carpels). Compare with **imperfect flower**.

perforation In plants, a small hole in the primary and secondary cell walls of vessel elements that allows passage of water.

pericarp The part of a fruit, formed from the ovary wall, that surrounds the seeds and protects them. Corresponds to the flesh of most edible fruits and the hard shells of most nuts.

pericycle In plant roots, a layer of cells just inside the endodermis that give rise to lateral roots.

peripheral membrane protein Any membrane protein that does not span the entire lipid bilayer and associates with only one side of the bilayer. Compare with **integral membrane protein**.

peripheral nervous system (PNS) All the components of the nervous system that are outside the central nervous system (the brain and spinal cord). Includes the somatic nervous system and the autonomic nervous system.

peristalsis Rhythmic waves of muscular contraction. In the digestive tract, pushes food along. In animals with hydrostatic skeletons, enables crawling.

permafrost A permanently frozen layer of icy soil found in most tundra and some taiga.

permeability The tendency of a structure, such as a membrane, to allow a given substance to diffuse across it.

peroxisome An organelle found in most eukaryotic cells that contains enzymes for oxidizing fatty acids and other compounds, including many toxins, rendering them harmless. See **glyoxysome**.

petal Any of the leaflike organs arranged around the reproductive organs of a flower. Often colored and scented to attract pollinators.

petiole The stalk of a leaf.

pH A measure of the concentration of protons in a solution and thus of acidity or alkalinity. Defined as the negative of the base-10 logarithm of the proton concentration: $pH = -\log[H^+]$.

phagocytosis Uptake by a cell of small particles or cells by invagination and pinching off of the plasma membrane to form small, membrane-bound vesicles; one type of **endocytosis**.

pharyngeal gill slits A set of parallel openings from the throat to the outside that function in both feeding and gas exchange. A diagnostic trait of chordates.

pharyngeal jaw A secondary jaw in the back of the throat; found in some fishes, it aids in food processing. Derived from modified gill arches.

phenology The timing of events during the year, in environments where seasonal changes occur.

phenotype The detectable traits of an individual. Compare with **genotype**.

phenotypic plasticity Within-species variation in phenotype that is due to differences in environmental conditions. Occurs more commonly in plants than animals.

pheophytin The molecule in photosystem II that accepts excited electrons from the reaction center chlorophyll and passes them to an electron transport chain.

pheromone A chemical signal, released by an individual into the external environment, that can trigger changes in the behavior or physiology or both of another member of the same species.

phloem A plant vascular tissue that conducts sugars between roots and shoots; contains sieve-tube elements and companion cells. Primary phloem develops from the procambium of apical meristems; secondary phloem, from the vascular cambium. Compare with **xylem**.

phosphatase An enzyme that removes phosphate groups from proteins or other molecules. Phosphatases are often used in the inactivation of signaling pathways that involve the phosphorylation and activation of proteins.

phosphodiester linkage Chemical linkage between adjacent nucleotide residues in DNA and RNA. Forms when the phosphate group of one nucleotide condenses with the hydroxyl group on the sugar of another nucleotide. Also known as *phosphodiester bond*.

phosphofructokinase The enzyme that catalyzes synthesis of fructose-1,6-bisphosphate from fructose-6-phosphate, a key reaction in glycolysis (step 3). Also called *6-phosphofructokinase*.

phospholipid A class of lipid having a hydrophilic head (including a phosphate group) and a hydrophobic tail (consisting of two hydrocarbon chains). Major components of the plasma membrane and organelle membranes.

phosphorylase An enzyme that breaks down glycogen by catalyzing hydrolysis of the α-glycosidic linkages between the glucose residues.

phosphorylation (verb: phosphorylate) The addition of a phosphate group to a molecule.

phosphorylation cascade A series of enzyme-catalyzed phosphorylation reactions commonly used in signal transduction pathways to amplify and convey a signal inward from the plasma membrane.

photic zone In an aquatic habitat, water that is shallow enough to receive some sunlight (whether or not it is enough to support photosynthesis). Compare with **aphotic zone**.

photon A discrete packet of light energy; a particle of light.

photoperiod The amount of time per day (usually in hours) that an organism is exposed to light.

photoperiodism Any response by an organism to the relative lengths of day and night (i.e., photoperiod).

photophosphorylation Production of ATP molecules by ATP synthase using the proton-motive force generated as light-excited electrons flow through an electron transport chain during photosynthesis.

photoreception A sensory system in which receptors are activated by light.

photoreceptor A molecule, a cell, or an organ that is specialized to detect light.

photorespiration A series of light-driven chemical reactions that consumes oxygen and releases carbon dioxide, basically undoing photosynthesis. Usually occurs when there are high O_2 and low CO_2 concentrations inside plant cells; often occurs when stomata must be kept closed to prevent dehydration.

photoreversibility A change in conformation that occurs in certain plant pigments when they are exposed to the particular wavelengths of light that they absorb; triggers responses by the plant.

photosynthesis The complex biological process that converts the energy of light into chemical energy stored in glucose and other organic molecules. Occurs in most plants, algae, and some bacteria.

photosystem One of two types of units, consisting of a central reaction center surrounded by antenna complexes, that is responsible for the light-dependent reactions of photosynthesis.

photosystem I A photosystem that contains a pair of P700 chlorophyll molecules and uses absorbed light energy to reduce $NADP^+$ to NADPH.

photosystem II A photosystem that contains a pair of P680 chlorophyll molecules and uses absorbed light energy to produce a proton-motive force for the synthesis of ATP. Oxygen is produced as a by-product when water is split to obtain electrons.

phototroph An organism (most plants, algae, and some bacteria) that produces ATP through photosynthesis.

phototropins A class of plant photoreceptors that detect blue light and initiate various responses.

phototropism Growth or movement of an organism in a particular direction in response to light.

phylogenetic species concept The definition of a species as the smallest monophyletic group in a phylogenetic tree. Compare with **biological species concept** and **morphospecies concept**.

phylogenetic tree A branching diagram that depicts the evolutionary relationships among species or other taxa.

phylogeny The evolutionary history of a group of organisms.

phylum (plural: phyla) In Linnaeus' system, a taxonomic category above the class level and below the kingdom level. In plants, sometimes called a *division*.

physical map A map of a chromosome that shows the number of base pairs between various genetic markers. Compare with **genetic map**.

physiology The study of how an organism's body functions.

phytochrome A specialized plant photoreceptor that exists in two shapes depending on the ratio of red to far-red light and is involved in the timing of certain physiological processes, such as flowering, stem elongation, and germination.

pigment Any molecule that absorbs certain wavelengths of visible light and reflects or transmits other wavelengths.

piloting A type of navigation in which animals use familiar landmarks to find their way.

pineal gland An endocrine gland, located in the brain, that secretes the hormone melatonin.

pioneering species Those species that appear first in recently disturbed areas.

pit In plants, a small hole in the secondary cell walls of tracheids and vessel elements that allows passage of water.

pitch The sensation produced by a particular frequency of sound. Low frequencies are perceived as low pitches; high frequencies, as high pitches.

pith In the shoot systems of plants, ground tissue located to the inside of the vascular bundles.

pituitary gland A small gland located directly under the brain and physically and functionally connected to the hypothalamus. Produces and secretes an array of hormones that affect many other glands and organs.

placenta A structure that forms in the pregnant uterus from maternal and fetal tissues. Delivers oxygen to the fetus, exchanges nutrients and wastes between mother and fetus, anchors the fetus to the uterine wall, and produces some hormones. Occurs in most mammals and in a few other vertebrates.

placental mammals See **eutherians**.

plankton Drifting organisms (animals, plants, archaea, or bacteria) in aquatic environments.

Plantae The monophyletic group that includes red, green, and glaucophyte algae, and land plants.

plasma The non-cellular portion of blood.

plasma cell A B cell that produces large quantities of antibodies after being activated by interacting with antigen and a CD4+ T cell via peptide presentation. Also called an *effector B cell*.

plasma membrane A membrane that surrounds a cell, separating it from the external environment and selectively regulating passage of molecules and ions into and out of the cell. Also called *cell membrane*.

plasmid A small, usually circular, supercoiled DNA molecule independent of the cell's main chromosome(s) in prokaryotes and some eukaryotes.

plasmodesmata (singular: plasmodesma) Physical connections between two plant cells, consisting of membrane-lined gaps in the cell walls through which the two cells' plasma membranes, cytoplasm, and smooth ER can connect directly. Functionally similar to gap junctions in animal cells.

plasmogamy Fusion of the cytoplasm of two individuals. Occurs in many fungi.

plastocyanin A small protein that shuttles electrons originating from photosystem II to the reaction center of photosystem I during photosynthesis.

plastoquinone (PQ) A nonprotein electron carrier in the chloroplast electron transport chain. Receives excited electrons from photosystem II (noncyclic) or photosystem I (cyclic) and passes them to more

electronegative molecules in the chain. Also transports protons from the stroma to the thylakoid lumen, generating a proton-motive force.

platelet A small membrane-bound cell fragment in vertebrate blood that functions in blood clotting. Derived from large cells in the bone marrow.

pleiotropy (adjective: pleiotropic) The ability of a single gene to affect more than one trait.

ploidy The number of complete chromosome sets present. *Haploid* refers to a ploidy of 1; *diploid,* a ploidy of 2; *triploid,* a ploidy of 3; and *tetraploid,* a ploidy of 4.

point mutation A mutation that results in a change in a single base pair in DNA.

polar (1) Asymmetrical or unidirectional. (2) Carrying a partial positive charge on one side of a molecule and a partial negative charge on the other. Polar molecules are generally hydrophilic.

polar body Any of the tiny, nonfunctional cells that are made as a by-product during meiosis of a primary oocyte, due to most of the cytoplasm going to the ovum.

polar covalent bond A covalent bond in which electrons are shared unequally between atoms differing in electronegativity, resulting in the more electronegative atom having a partial negative charge and the other atom, a partial positive charge. Compare with **nonpolar covalent bond**.

polar microtubules Mitotic and meiotic microtubules that have arisen from the two spindle poles and overlap with each other in the middle of the spindle apparatus.

polar nuclei In flowering plants, the nuclei in the female gametophyte that fuse with one sperm nucleus to produce the endosperm. Most species have two.

pollen grain In seed plants, a male gametophyte enclosed within a protective coat of sporopollenin.

pollen tube In flowering plants, a structure that grows out of a pollen grain after it reaches the stigma, extends down the style, and through which two sperm cells are delivered to the ovule.

pollination The process by which pollen reaches the carpel of a flower (in flowering plants), transferred from anther to stigma, or reaches the ovule directly (in conifers and their relatives).

pollination syndrome Suites of flower characters that are associated with certain types of pollinators and that have evolved through natural selection imposed by the interaction between flowers and pollinators.

poly(A) signal In eukaryotes, a short sequence of nucleotides near the 3′ end of pre-mRNAs that signals cleavage of the RNA and addition of the poly(A) tail.

poly(A) tail In eukaryotes, a sequence of about 100–250 adenine nucleotides added to the 3′ end of newly transcribed messenger RNA molecules.

polygenic inheritance Having many genes influence one trait.

polymer Any long molecule composed of small repeating units (monomers) bonded together. The main biological polymers are proteins, nucleic acids, and polysaccharides.

polymerase chain reaction (PCR) A laboratory technique for rapidly generating millions of identical

copies of a specific stretch of DNA. Works by incubating the original DNA sequence of interest with primers, nucleotides, and DNA polymerase.

polymerization (verb: polymerize) The process by which many identical or similar small molecules (monomers) are covalently bonded to form a large molecule (polymer).

polymorphic species A species that has two or more distinct phenotypes in the same interbreeding population at the same time.

polymorphism (adjective: polymorphic) (1) The occurrence of more than one allele at a genetic locus in a population. (2) The occurrence of more than two distinct phenotypes of a trait in a population.

polyp The immotile (sessile) stage in the life cycle of some cnidarians (e.g., jellyfish). Compare with **medusa**.

polypeptide A chain of 50 or more amino acids linked together by peptide bonds. Compare with **oligopeptide** and **protein**.

polyphyletic group An unnatural group based on convergent (homoplastic) characteristics that are not present in a common ancestor. Compare with **monophyletic group**.

polyploidy (adjective: polyploid) The state of having more than two full sets of chromosomes, either from the same species (autopolyploidy) or from different species (allopolyploidy).

polyribosome A messenger RNA molecule along with more than one attached ribosome and their growing peptide strands.

polysaccharide A linear or branched polymer consisting of many monosaccharides joined by glycosidic linkages. Compare with **monosaccharide** and **oligosaccharide**.

polytomy A node in a phylogenetic tree that depicts an ancestral branch dividing into three or more descendant branches; usually indicates that insufficient data were available to resolve which taxa are more closely related.

population A group of individuals of the same species living in the same geographic area at the same time.

population density The number of individuals of a population per unit area.

population dynamics Changes in the size and other characteristics of populations through time and space.

population ecology The study of how and why the number of individuals in a population changes over time and space.

population thinking The ability to analyze trait frequencies, event probabilities, and other attributes of populations of molecules, cells, or organisms.

pore In land plants, an opening in the epithelium that allows gas exchange. See also **stoma**.

positive control Of genes, when a regulatory protein triggers expression by binding to DNA on or near the gene.

positive feedback A physiological mechanism in which a change in some variable stimulates a response that increases the change. Relatively rare in organisms but is important in generation of the action potential. Compare with **negative feedback**.

positive pressure ventilation Ventilation of the lungs by using positive pressure in the mouth to "push" air into the lungs. Compare with **negative pressure ventilation**.

positive-sense RNA virus An ssRNA virus whose genome contains the same sequences as the mRNA required to produce viral proteins. Compare with **ambisense virus** and **negative-sense virus**.

posterior Toward an animal's tail and away from its head. The opposite of anterior.

posterior pituitary The part of the pituitary gland that contains the ends of hypothalamic neurosecretory cells and from which oxytocin and antidiuretic hormone are secreted. Compare with **anterior pituitary**.

postsynaptic neuron A neuron that receives signals, usually via neurotransmitters, from another neuron at a synapse. Compare with **presynaptic neuron**.

post-translational control Regulation of gene expression by modification of proteins (e.g., addition of a phosphate group or sugar residues) after translation.

postzygotic isolation Reproductive isolation resulting from mechanisms that operate after mating of individuals of two different species occurs. The most common mechanisms are the death of hybrid embryos or reduced fitness of hybrids.

potential energy Energy stored in matter as a result of its position or molecular arrangement. Compare with **kinetic energy**.

prebiotic soup model Hypothetical explanation for chemical evolution whereby small molecules reacted with one another in a mixture of organic molecules condensed into a body of water, typically in reference to the early oceans.

Precambrian The interval between the formation of the Earth, about 4.6 billion years ago, and the appearance of most animal groups about 542 million years ago. Unicellular organisms were dominant for most of this era, and oxygen was virtually absent for the first 2 billion years.

pre-mRNA In eukaryotes, the primary transcript of protein-coding genes. Pre-mRNA is processed to form mRNA.

predation The killing and eating of one organism (the prey) by another (the predator).

predator Any organism that kills other organisms for food.

prediction A measurable or observable result of an experiment based on a particular hypothesis. A correct prediction provides support for the hypothesis being tested.

pressure-flow hypothesis The hypothesis that sugar movement through phloem tissue is due to differences in the turgor pressure of phloem sap.

pressure potential (ψ_p) A component of the potential energy of water caused by physical pressures on a solution. It can be positive or negative. Compare with **solute potential (ψ_S)**.

presynaptic neuron A neuron that transmits signals, usually by releasing neurotransmitters, to another neuron or to an effector cell at a synapse.

prezygotic isolation Reproductive isolation resulting from any one of several mechanisms that prevent individuals of two different species from mating.

primary active transport A form of active transport in which a source of energy like ATP is directly used to move ions against their electrochemical gradients.

primary cell wall The outermost layer of a plant cell wall, made of cellulose fibers and gelatinous polysaccharides, that defines the shape of the cell and withstands the turgor pressure of the plasma membrane.

primary consumer An herbivore; an organism that eats plants, algae, or other primary producers. Compare with **secondary consumer**.

primary decomposer A decomposer (detritivore) that consumes detritus from plants.

primary growth In plants, an increase in the length of stems and roots due to the activity of apical meristems. Compare with **secondary growth**.

primary immune response An adaptive immune response to a pathogen that the immune system has not encountered before. Compare with **secondary immune response**.

primary meristem In plants, three types of partially differentiated cells that are produced by apical meristems, including protoderm, ground meristem, and procambium. Compare with **apical meristem** and **cambium**.

primary oocyte Any of the large diploid cells in an ovarian follicle that can initiate meiosis to produce a haploid secondary oocyte and a polar body.

primary producer Any organism that creates its own food by photosynthesis or from reduced inorganic compounds and that is a food source for other species in its ecosystem. Also called *autotroph*.

primary spermatocyte Any of the diploid cells in the testis that can initiate meiosis I to produce two secondary spermatocytes.

primary structure The sequence of amino acid residues in a peptide or protein; also the sequence of nucleotides in a nucleic acid. Compare with **secondary**, **tertiary**, and **quaternary structure**.

primary succession The gradual colonization of a habitat of bare rock or gravel, usually after an environmental disturbance that removes all soil and previous organisms. Compare with **secondary succession**.

primary transcript In eukaryotes, a newly transcribed RNA molecule that has not yet been processed to a mature RNA. Called *pre-mRNA* when the final product is a protein.

primase An enzyme that synthesizes a short stretch of RNA to use as a primer during DNA replication.

primates The lineage of mammals that includes prosimians (lemurs, lorises, etc.), monkeys, and great apes (including humans).

primer A short, single-stranded RNA molecule that base-pairs with a DNA template strand and is elongated by DNA polymerase during DNA replication.

probe A radioactively or chemically labeled single-stranded fragment of a known DNA or RNA sequence that can bind to and detect its complementary sequence in a sample containing many different sequences.

proboscis A tubular, often extensible feeding appendage with which food can be obtained.

procambium A primary meristem tissue that gives rise to the vascular tissue.

product Any of the final materials formed in a chemical reaction.

progesterone A steroid hormone produced and secreted by the corpus luteum in the ovaries after ovulation and by the placenta during gestation; protects the uterine lining.

programmed cell death Regulated cell death that is used in development, tissue maintainance, and destruction of infected cells. Can occur in different ways; apoptosis is the best-known mechanism.

prokaryote A member of the domain Bacteria or Archaea; a unicellular organism lacking a nucleus and containing relatively few organelles or cytoskeletal components. Compare with **eukaryote**.

prolactin A peptide hormone, produced and secreted by the anterior pituitary, that promotes milk production in female mammals and has a variety of effects on parental behavior and seasonal reproduction in other vertebrates.

prometaphase A stage in mitosis or meiosis during which the nuclear envelope breaks down and microtubules attach to kinetochores.

promoter A short nucleotide sequence in DNA that binds a sigma factor (in bacteria) or basal transcription factors (in eukaryotes) to enable RNA polymerase to begin transcription. In bacteria, several contiguous genes are often transcribed from a single promoter. In eukaryotes, each gene generally has its own promoter.

promoter-proximal element In eukaryotes, regulatory sequences in DNA that are close to a promoter and that can bind regulatory transcription factors.

proofreading The process by which a DNA polymerase recognizes and removes a wrong base added during DNA replication and then continues synthesis.

prophase The first stage in mitosis or meiosis during which chromosomes become visible and the spindle apparatus forms. Synapsis and crossing over occur during prophase of meiosis I.

prosimians One of the two major lineages of primates, a paraphyletic group including lemurs, pottos, and lorises. Compare with **anthropoids**.

prostate gland A gland in male mammals that surrounds the base of the urethra and secretes a fluid that is a component of semen.

prosthetic group A non-amino acid atom or molecule that is permanently attached to an enzyme or other protein and is required for its function.

protease An enzyme that can break up proteins by cleaving the peptide bonds between amino acid residues.

proteasome A macromolecular machine that destroys proteins that have been marked by the addition of ubiquitin.

protein A macromolecule consisting of one or more polypeptide chains composed of 50 or more amino acids linked together. Each protein has a unique sequence of amino acids and generally possesses a characteristic three-dimensional shape.

protein kinase An enzyme that catalyzes the addition of a phosphate group to another protein, typically activating or inactivating the substrate protein.

proteinase inhibitors Defense compounds, produced by plants, that induce illness in herbivores by inhibiting digestive enzymes.

proteoglycan A type of highly glycosylated protein found in the extracellular matrix of animal cells that attracts and holds water, forming a gel that resists compression forces.

proteome The complete set of proteins produced by a particular cell type.

proteomics The systematic study of the interactions, localization, functions, regulation, and other features of the full protein set (proteome) in a particular cell type.

protist Any eukaryote that is not a green plant, animal, or fungus. Protists are a diverse paraphyletic group. Most are unicellular, but some are multicellular or form aggregations called colonies.

protocell A hypothetical pre-cell structure consisting of a membrane compartment that encloses replicating macromolecules, such as ribozymes.

protoderm The exterior layer of a young plant embryo that gives rise to the epidermis.

proto-oncogene Any gene that encourages cell division in a regulated manner, typically by triggering specific phases in the cell cycle. Mutation may convert it into an oncogene. See **oncogene**.

proton pump A membrane protein that can hydrolyze ATP to power active transport of protons (H^+ ions) across a membrane against an electrochemical gradient. Also called H^+-ATPase.

proton-motive force The combined effect of a proton gradient and an electric potential gradient across a membrane, which can drive protons across the membrane. Used by mitochondria and chloroplasts to power ATP synthesis via the mechanism of chemiosmosis.

protostomes A major lineage of animals that share a pattern of embryological development, including formation of the mouth earlier than the anus, and formation of the coelom by splitting of a block of mesoderm. Includes arthropods, mollusks, and annelids. Compare with **deuterostomes**.

proximal tubule In the vertebrate kidney, the convoluted section of a nephron into which filtrate moves from Bowman's capsule. Involved in the largely unregulated reabsorption of electrolytes, nutrients, and water. Compare with **distal tubule**.

proximate causation In biology, the immediate, mechanistic cause of a phenomenon (how it happens), as opposed to why it evolved. Also called *proximate explanation*. Compare with **ultimate causation**.

pseudocoelomate An animal that has a coelom that is only partially lined with mesoderm. Compare with **acoelomate** and **coelomate**.

pseudogene A DNA sequence that closely resembles a functional gene but is not transcribed. Thought to have arisen by duplication of the functional gene followed by inactivation due to a mutation.

pseudopodium (plural: pseudopodia) A temporary bulge-like extension of certain protist cells used in cell crawling and ingestion of food.

puberty The various physical and emotional changes that an immature human undergoes in reaching reproductive maturity. Also the period when such changes occur.

pulmonary artery A short, thick-walled artery that carries oxygen-poor blood from the heart to the lungs.

pulmonary circulation The part of the circulatory system that sends oxygen-poor blood to the lungs. It is separate from the rest of the circulatory system (the systemic circulation) in mammals and birds.

pulmonary vein A short, thin-walled vein that carries oxygen-rich blood from the lungs to the heart. Humans have four such veins.

pulse–chase experiment A type of experiment in which a population of cells or molecules at a particular moment in time is marked by means of a labeled molecule (pulse) and then their fate is followed over time (chase).

pump Any membrane protein that can hydrolyze ATP and change shape to power active transport of a specific ion or small molecule across a plasma membrane against its electrochemical gradient. See **proton pump**.

pupa (plural: pupae) A metamorphosing insect that is enclosed in a protective case.

pupil The hole in the center of the iris through which light enters a vertebrate or cephalopod eye.

pure line In animal or plant breeding, a strain that produces offspring identical with themselves when self-fertilized or crossed to another member of the same population. Pure lines are homozygous for most, if not all, genetic loci.

purifying selection Selection that lowers the frequency of or even eliminates deleterious alleles.

purines A class of small, nitrogen-containing, double-ringed bases (guanine, adenine) found in nucleotides. Compare with **pyrimidines**.

pyrimidines A class of small, nitrogen-containing, single-ringed bases (cytosine, uracil, thymine) found in nucleotides. Compare with **purines**.

pyruvate dehydrogenase A large enzyme complex, located in the mitochondrial matix, that is responsible for converting pyruvate to acetyl CoA during cellular respiration.

quantitative trait A trait that exhibits continuous phenotypic variation (e.g., human height), rather than the distinct forms characteristic of discrete traits.

quaternary structure In proteins, the overall three-dimensional shape formed from two or more polypeptide chains (subunits); determined by the number, relative positions, and interactions of the subunits. In single stranded nucleic acids, the hydrogen bonding between two or more distinct strands will form this level of structure through hydrophobic interactions between complementary bases. Compare with **primary**, **secondary**, and **tertiary structures**.

quorum sensing Cell–cell signaling in unicellular organisms, in which cells of the same species communicate via chemical signals. It is often observed that cell activity changes dramatically when the population reaches a threshold size, or quorum.

radial symmetry An animal body pattern that has at least two planes of symmetry. Typically, the body is in the form of a cylinder or disk, and the body parts radiate from a central hub. Compare with **bilateral symmetry**.

radiation Transfer of heat between two bodies that are not in direct physical contact. More generally, the emission of electromagnetic energy of any wavelength.

radicle The root of a plant embryo.

radioactive isotope A version of an element that has an unstable nucleus, which will release radiation energy as it decays to a more stable form. Decay often results in the radioisotope becoming a different element.

radioimmunoassay A competitive binding assay in which the quantity of hormone in a sample can be estimated. Uses radioactively labeled hormones that compete with the unknown hormone to bind with an antibody.

radula A rasping feeding appendage in mollusks such as gastropods (snails, slugs).

rain shadow The dry region on the side of a mountain range away from the prevailing wind.

range The geographic distribution of a species.

Ras protein A type of G protein that is activated by enzyme-linked cell-surface receptors, including receptor tyrosine kinases. Activated Ras then initiates a phosphorylation cascade, culminating in a cell response.

ray-finned fishes Members of the Actinopterygii, a diverse group of fishes with fins supported by bony rods arranged in a ray pattern.

rays In plant shoot systems with secondary growth, a lateral row of parenchyma cells produced by vascular cambium. Transport water and nutrients laterally across the stem.

Rb protein A tumor suppressor protein that helps regulate progression of a cell from the G_1 phase to the S phase of the cell cycle. Defects in Rb protein are found in many types of cancer.

reactant Any of the starting materials in a chemical reaction.

reaction center Centrally located component of a photosystem containing proteins and a pair of specialized chlorophyll molecules. It is surrounded by antenna complexes that transmit resonance energy to excite the reaction center pigments.

reading frame A series of non-overlapping, three-base-long sequences (potential codons) in DNA or RNA. The reading frame for a polypeptide is set by the start codon.

realized niche The portion of the fundamental niche that a species actually occupies given limiting factors such as competition with other species. Compare with **fundamental niche**.

receptor-mediated endocytosis Uptake by a cell of certain extracellular macromolecules, bound to specific receptors in the plasma membrane, by pinching off the membrane to form small membrane-bound vesicles.

receptor tyrosine kinase (RTK) Any of a class of enzyme-linked cell-surface signal receptors that undergo phosphorylation after binding a signaling molecule. The activated, phosphorylated receptor then triggers a signal transduction pathway inside the cell.

recessive Referring to an allele whose phenotypic effect is observed only in homozygous individuals. Compare with **dominant**.

reciprocal altruism Altruistic behavior that is exchanged between a pair of individuals at different times (i.e., sometimes individual A helps individual B, and sometimes B helps A).

reciprocal cross A cross in which the mother's and father's phenotypes are the reverse of that examined in a previous cross.

recombinant Possessing a new combination of alleles. May refer to a single chromosome or DNA molecule, or to an entire organism.

recombinant DNA technology A variety of techniques for isolating specific DNA fragments and introducing them into different regions of DNA or a different host organism.

rectal gland A salt-excreting gland in the digestive system of sharks, skates, and rays.

rectum The last portion of the digestive tract. It is where feces are held until they are expelled.

red blood cell A hemoglobin-containing cell that circulates in the blood and delivers oxygen from the lungs to the tissues.

redox reaction Any chemical reaction that involves either the complete transfer of one or more electrons from one reactant to another, or a reciprocal shift in the position of shared electrons within one or more of the covalent bonds of two reactants. Also called *reduction–oxidation reaction*.

reduction The gain of electrons by an atom or molecule during a redox reaction, either by acceptance of an electron from another atom or molecule, or by the shared electrons in covalent bonds moving closer to the atomic nucleus.

reduction–oxidation reaction See **redox reaction**.

reflex An involuntary response to environmental stimulation. May involve the brain (conditioned reflex) or not (spinal reflex).

refractory No longer responding to stimuli that previously elicited a response. An example is the tendency of voltage-gated sodium channels to remain closed immediately after an action potential.

regulatory sequence Any segment of DNA that is involved in controlling transcription of a specific gene by binding a regulatory transcription factor protein.

regulatory transcription factor General term for proteins that bind to DNA regulatory sequences (eukaryotic enhancers, silencers, and promoter-proximal elements), but not to the promoter itself, leading to an increase or decrease in transcription of specific genes. Compare with **basal transcription factor**.

regulon A large set of genes in bacteria that are controlled by a single type of regulatory molecule. Regulon genes are transcribed in response to environmental cues and allow cells to respond to changing environments.

reinforcement In evolutionary biology, the natural selection for traits that prevent interbreeding between recently diverged species.

release factors Proteins that trigger termination of translation when a ribosome reaches a stop codon.

renal corpuscle In the vertebrate kidney, the ball-like structure at the beginning of a nephron, consisting of a glomerulus and the surrounding Bowman's capsule. Acts as a filtration device.

replacement rate The number of offspring each female must produce over her entire life to "replace" herself and her mate, resulting in zero population growth. The actual number is slightly more than 2 because some offspring die before reproducing.

replica plating A method of identifying bacterial colonies that have certain mutations by transferring cells from each colony on a master plate to a second (replica) plate and observing their growth when exposed to different conditions.

replication fork The Y-shaped site at which a double-stranded molecule of DNA is separated into two single strands for replication.

replicative growth The process by which viruses produce new virions.

replisome The macromolecular machine that copies DNA; includes DNA polymerase, helicase, primase, and other enzymes.

repolarization Return to a resting potential after a membrane potential has changed; a normal phase in an action potential.

repressor (1) In bacteria, a protein that binds to an operator sequence in DNA to prevent transcription when an inducer is not present and that comes off DNA to allow transcription when an inducer binds to the repressor protein. (2) In eukaryotes, a protein that binds to a silencer sequence in DNA to prevent or reduce gene transcription.

reproductive development The phase of plant development that involves development of the flower and reproductive cells. Follows vegetative development and occurs when a shoot apical meristem (SAM) transitions to a flower-producing meristem.

reptiles One of the two lineages of amniotes (vertebrates that produce amniotic eggs) distinguished by adaptations for life and reproduction on land. Living reptiles include turtles, snakes and lizards, crocodiles and alligators, and birds. Except for birds, all are ectotherms.

resilience, community A measure of how quickly a community recovers following a disturbance.

resistance, community A measure of how much a community is affected by a disturbance.

respiratory system The collection of cells, tissues, and organs responsible for gas exchange between an animal and its environment.

resting potential The membrane potential of a cell in its resting, or normal, state.

restriction endonucleases Bacterial enzymes that cut DNA at a specific base-pair sequence (restriction site). Also called *restriction enzymes*.

retina A thin layer of light-sensitive cells (rods and cones) and neurons at the back of a simple eye, such as that of cephalopods and vertebrates.

retinal A light-absorbing pigment that is linked to the protein opsin in rods and cones of the vertebrate eye.

retrovirus A virus with an RNA genome that reverse-transcribes its RNA into a double-stranded DNA sequence, which is then inserted into the host's genome as part of its replicative cycle.

reverse transcriptase An enzyme that can synthesize double-stranded DNA from a single-stranded RNA template.

rhizobia (singular: rhizobium) Members of the bacterial genus *Rhizobium*; nitrogen-fixing bacteria that live in root nodules of members of the pea family (legumes).

rhizoid The hairlike structure that anchors a non-vascular plant to the substrate.

rhizome A modified stem that runs horizontally underground and produces new plants at the nodes (a form of asexual reproduction). Compare with **stolon**.

rhodopsin A transmembrane complex that is instrumental in detection of light by rods of the vertebrate eye. Is composed of the transmembrane protein opsin covalently linked to retinal, a light-absorbing pigment.

ribonucleic acid (RNA) A nucleic acid composed of ribonucleotides that usually is single stranded. Functions include structural components of ribosomes (rRNA), transporters of amino acids (tRNA), and messages of the DNA code required for protein synthesis (mRNA), among others.

ribonucleotide See **nucleotide**.

ribosomal RNA (rRNA) An RNA molecule that forms part of the ribosome.

ribosome A large macromolecular machine that synthesizes proteins by using the genetic information encoded in messenger RNA. Consists of two subunits, each composed of ribosomal RNA and proteins.

ribosome binding site In a bacterial mRNA molecule, the sequence just upstream of the start codon to which a ribosome binds to initiate translation. Also called the *Shine–Dalgarno sequence*.

ribozyme Any RNA molecule that can act as a catalyst, that is, speed up a chemical reaction.

ribulose bisphosphate (RuBP) A five-carbon compound that combines with CO_2 in the first step of the Calvin cycle during photosynthesis.

RNA See **ribonucleic acid**.

RNA interference (RNAi) Degradation of an RNA molecule or inhibition of its translation following its binding by a short RNA (microRNA) whose sequence is complementary to a portion of the mRNA.

RNA polymerase An enzyme that catalyzes the synthesis of RNA from ribonucleotides using a DNA template.

RNA processing In eukaryotes, the changes that a primary RNA transcript undergoes to become a mature RNA molecule. For pre-mRNA it includes the addition of a 5′ cap and poly(A) tail and splicing to remove introns.

RNA replicase A viral enzyme that can synthesize RNA from an RNA template. Also called an *RNA-dependent RNA polymerase*.

RNA world hypothesis Proposal that chemical evolution produced RNAs that could catalyze key reactions involved in their own replication and basic metabolism, which led to the evolution of proteins and the first life-form.

rod cell A photoreceptor cell with a rod-shaped outer portion that is particularly sensitive to dim light but not used to distinguish colors. Also called simply *rod*. Compare with **cone cell**.

root (1) An underground part of a plant that anchors the plant and absorbs water and nutrients. (2) The most ancestral branch in a phylogenetic tree.

root apical meristem (RAM) A group of undifferentiated plant stem cells at the tip of a plant root that can differentiate into mature root tissue.

root cap A small group of cells that covers and protects the root apical meristem. Senses gravity and determines the direction of root growth.

root hair A long, thin outgrowth of the epidermal cells of plant roots, providing increased surface area for absorption of water and nutrients.

root pressure Positive pressure of xylem sap in the vascular tissue of roots. Generated during the night as a result of the accumulation of ions from the soil and subsequent osmotic movement of water into the xylem.

root system The belowground part of a plant.

rotifer Member of the phylum Rotifera. Distinguished by a cluster of cilia, called a corona, used in suspension feeding in marine and freshwater environments. Rotifers belong to the lophotrochozoan branch of the protostomes.

rough endoplasmic reticulum (rough ER) The portion of the endoplasmic reticulum that is dotted with ribosomes. Involved in synthesis of plasma membrane proteins, secreted proteins, and proteins localized to the ER, Golgi apparatus, and lysosomes. Compare with **smooth endoplasmic reticulum**.

roundworms Members of the phylum Nematoda. Distinguished by an unsegmented body with a pseudocoelom and no appendages. Roundworms belong to the ecdysozoan branch of the protostomes. Also called *nematodes*.

rubisco The enzyme that catalyzes the first step of the Calvin cycle during photosynthesis: the addition of a molecule of CO_2 to ribulose bisphosphate. See also **carbon fixation**.

ruminant Member of a group of hoofed mammals (e.g., cattle, sheep, deer) that have a four-chambered stomach specialized for digestion of plant cellulose. Ruminants regurgitate cud, a mixture of partially digested food and cellulose-digesting bacteria, from the largest chamber (the rumen) for further chewing.

salinity The proportion of solutes dissolved in water in natural environments, often designated in grams of solute per kilogram of water (cited as parts per thousand).

salivary amylase An enzyme that is produced by the salivary glands and that can break down starch by catalyzing hydrolysis of the glycosidic linkages between the glucose residues.

salivary gland A type of vertebrate gland that secretes saliva (a mixture of water, mucus-forming glycoproteins, and digestive enzymes) into the mouth.

sampling error The selection of a nonrepresentative sample from some larger population, due to chance.

saprophyte An organism that feeds primarily on dead plant material.

sapwood The younger xylem in the outer layer of wood of a stem or root, functioning primarily in water transport.

sarcomere The repeating contractile unit of a skeletal muscle cell; the portion of a myofibril located between adjacent Z disks.

sarcoplasmic reticulum Sheets of smooth endoplasmic reticulum in a muscle cell. Contains high concentrations of calcium, which can be released into the cytoplasm to trigger contraction.

saturated Referring to lipids in which all the carbon-carbon bonds are single bonds. Such compounds have relatively high melting points. Compare with **unsaturated**.

scanning electron microscope (SEM) A microscope that produces images of the surfaces of objects by reflecting electrons from a specimen coated with a layer of metal atoms. Compare with **transmission electron microscope**.

scarify To scrape, rasp, cut, or otherwise damage the coat of a seed. Necessary in some species to trigger germination.

Schwann cell A type of glial cell that wraps around axons of some neurons outside the brain and spinal cord, forming a myelin sheath that provides electrical insulation. Compare with **oligodendrocyte**.

scientific name The unique, two-part name given to each species, with a genus name followed by a species name—as in *Homo sapiens*. Scientific names are always italicized, and are also known as Latin names.

sclereid In plants, a relatively short type of sclerenchyma cell that usually functions in protection, such as in seed coats and nutshells. Compare with **fiber**.

sclerenchyma cell In plants, a cell that has a thick secondary cell wall and provides support; typically contains the tough structural polymer lignin and usually is dead at maturity. Includes fibers and sclereids. Compare with **collenchyma cell** and **parenchyma cell**.

scrotum A sac of skin containing the testes and suspended just outside the abdominal body cavity of many male mammals.

second law of thermodynamics The principle of physics that the entropy of the universe or any closed system always increases.

second-male advantage The reproductive advantage, in some species, of a male who mates with a female last, after other males have mated with her.

second messenger A nonprotein signaling molecule produced or activated inside a cell in response to stimulation at the cell surface. Commonly used to relay the message of a hormone or other extracellular signaling molecule.

secondary active transport Transport of an ion or molecule in a defined direction that is often against its electrochemical gradient, in company with an ion or molecule being transported along its electrochemical gradient. Also called *cotransport*.

secondary cell wall The thickened inner layer of a plant cell wall formed by certain cells as they mature and have stopped growing; in water-conducting cells, contains lignin. Provides support or protection.

secondary consumer A carnivore; an organism that eats herbivores. Compare with **primary consumer**.

secondary growth In plants, an increase in the width of stems and roots due to the activity of cambium. Compare with **primary growth**.

secondary immune response The adaptive immune response to a pathogen that the immune system has encountered before. Normally much faster and more efficient than the primary response, due to immunological memory. Compare with **primary immune response**.

secondary metabolites Molecules that are closely related to compounds in key synthetic pathways and that often function in defense.

secondary oocyte A cell produced by meiosis I of a primary oocyte in the ovary. If fertilized, will complete meiosis II to produce an ootid (which develops into an ovum) and a polar body.

secondary spermatocyte A cell produced by meiosis I of a primary spermatocyte in the testis. Can undergo meiosis II to produce spermatids.

secondary structure In proteins, localized folding of a polypeptide chain into regular structures (i.e., alpha-helix and beta-pleated sheet) stabilized by hydrogen bonding between atoms of the peptide backbone. In nucleic acids, elements of structure (e.g., helices and hairpins) stabilized by hydrogen bonding and hydrophobic interactions between complementary bases. Compare with **primary**, **tertiary**, and **quaternary structures**.

secondary succession Gradual colonization of a habitat after an environmental disturbance (e.g., fire, windstorm, logging) that removes some or all previous organisms but leaves the soil intact. Compare with **primary succession**.

secretin A peptide hormone produced by cells in the small intestine in response to the arrival of food from the stomach. Stimulates secretion of bicarbonate (HCO_3^-) from the pancreas.

sedimentary rock A type of rock formed by gradual accumulation of sediment, particularly sand and mud, as in riverbeds and on the ocean floor. Most fossils are found in sedimentary rocks.

seed A plant reproductive structure consisting of an embryo, associated nutritive tissue (endosperm), and an outer protective layer (seed coat). In angiosperms, develops from the fertilized ovule of a flower.

seed bank A repository where seeds, representing many different varieties of domestic crops or other species, are preserved.

seed coat A protective layer around a seed that encases both the embryo and the endosperm.

segment A well-defined, repeated region of the body along the anterior–posterior body axis, containing structures similar to other nearby segments.

segmentation Division of the body or a part of it into a series of similar structures; exemplified by the body segments of insects and worms and by the somites of vertebrates.

segmentation genes A group of genes that control the formation and patterning of body segmentation in embryonic development. Includes maternal genes, gap genes, pair-rule genes, and segment polarity genes.

segregation, principle of The concept that each pair of hereditary elements (alleles of the same gene) separate from each other during meiosis. One of Mendel's two principles of genetics.

selective adhesion The tendency of cells of one tissue type to adhere to other cells of the same type.

selective permeability The property of a membrane that allows some substances to diffuse across it much more readily than other substances.

selectively permeable membrane Any membrane across which some solutes can move more readily than others.

self-fertilization The fusion of two gametes from the same individual to form offspring. Also called *selfing*.

self molecule A molecule that is synthesized by an organism and is a normal part of its cells and/or body; as opposed to nonself, or foreign, molecules.

semen The combination of sperm and accessory fluids that is released by male mammals and reptiles during ejaculation.

semiconservative replication The way DNA replicates, in which each strand of an existing DNA molecule serves as a template to create a new complementary DNA strand. It is called semiconservative because each newly replicated DNA molecule conserves one of the parental strands and contains another, newly replicated strand.

seminal vesicle In male mammals, either of a pair of reproductive glands that secrete a sugar-containing fluid into semen, which provides energy for sperm movement.

senescence The genetically programmed, active process of aging.

sensor Any cell, organ, or structure with which an animal can sense some aspect of the external or internal environment. Usually functions, along with an integrator and effector, as part of a homeostatic system.

sensory neuron A nerve cell that carries signals from sensory receptors to the central nervous system. Compare with **interneuron** and **motor neuron**.

sepal One of the protective leaflike organs enclosing a flower bud and later part of the outermost portion of the flower.

septum (plural: septa) Any wall-like structure. In fungi, septa divide the filaments (hyphae) of mycelia into cell-like compartments.

serotonin A neurotransmitter involved in many brain functions, including sleep, pleasure, and mood.

serum The liquid that remains when cells and clot material are removed from clotted blood. Contains water, dissolved gases, growth factors, nutrients, and other soluble substances. Compare with **plasma**.

sessile Permanently attached to a substrate; not capable of moving to another location.

set point A normal or target value for a regulated internal variable, such as body heat or blood pH.

sex chromosome Chromosomes that differ in shape or in number in males and females. For example, the X and Y chromosomes of many animals. Compare with **autosome**.

sex-linked inheritance Inheritance patterns observed in genes carried on sex chromosomes. In this case, females and males have different numbers of alleles of a gene. Often creates situations in which a trait appears more often in one sex. Also called *sex-linkage*.

sexual dimorphism Any trait that differs between males and females.

sexual reproduction Any form of reproduction in which genes from two parents are combined via fusion of gametes, producing offspring that are genetically distinct from both parents. Compare with **asexual reproduction**.

sexual selection A type of natural selection that favors individuals with traits that increase their ability to obtain mates. Acts more strongly on males than females. (Compare with **ecological selection**.)

shell A hard, protective outer structure.

Shine–Dalgarno sequence See **ribosome binding sequence**.

shoot In a plant embryo, the combination of hypocotyl and cotyledons, which will become the aboveground portions of the plant.

shoot apical meristem (SAM) A group of undifferentiated plant stem cells at the tip of a plant stem that can differentiate into mature shoot tissues.

shoot system The aboveground part of a plant comprising stems, leaves, and flowers (in angiosperms).

short-day plant A plant that blooms in response to long nights (usually in late summer or fall in the Northern Hemisphere). Compare with **day-neutral** and **long-day plant**.

short tandem repeats (STRs) Relatively short DNA sequences that are repeated, one after another, down the length of a chromosome. See **microsatellite**.

shotgun sequencing A method of sequencing genomes that is based on breaking the genome into small pieces, sequencing each piece separately, and then figuring out how the pieces are connected.

sieve plate In plants, a pore-containing structure at each end of a sieve-tube element in phloem.

sieve-tube element In plants, an elongated sugar-conducting cell in phloem that lacks nuclei and has sieve plates at both ends, allowing sap to flow to adjacent cells.

sigma A bacterial protein that associates with the core RNA polymerase to allow recognition of promoters.

signal In behavioral ecology, any information-containing behavior or characteristic.

signal receptor Any cellular protein that binds to a particular signaling molecule (e.g., a hormone or neurotransmitter) and triggers a response by the cell. Receptors for lipid-insoluble signals are transmembrane proteins in the plasma membrane; those for many lipid-soluble signals (e.g., steroid hormones) are located inside the cell.

signal recognition particle (SRP) An RNA–protein complex that binds to the ER signal sequence in a polypeptide as it emerges from a ribosome and transports the ribosome–polypeptide complex to the ER membrane, where synthesis of the polypeptide is completed.

signal transduction The process by which a stimulus (e.g., a hormone, a neurotransmitter, or sensory information) outside a cell is converted into an intracellular signal required for a cellular response. Usually involves a specific sequence of molecular events, or signal transduction pathway, that may lead to amplification of the signal.

signal transduction cascade See **phosphorylation cascade**.

silencer A regulatory sequence in eukaryotic DNA to which repressor proteins can bind, inhibiting gene transcription.

silent mutation A point mutation that changes the sequence of a codon without changing the amino acid that is specified.

simple eye An eye with only one light-collecting apparatus (e.g., one lens), as in vertebrates and cephalopods. Compare with **compound eye**.

simple fruit A fruit (e.g., apricot) that develops from a single flower that has a single carpel or several fused carpels. Compare with **aggregate** and **multiple fruit**.

simple sequence repeat See **microsatellite**.

SINEs (short interspersed nuclear elements) The second most abundant class of transposable elements in human genomes; can create copies of itself and insert them elsewhere in the genome. Compare with **LINEs**.

single nucleotide polymorphism (SNP) A site on a chromosome where individuals in a population have different nucleotides. Can be used as a genetic marker to help track the inheritance of nearby genes.

single-strand DNA-binding proteins (SSBP) A protein that attaches to separated strands of DNA during replication or transcription, preventing them from re-forming a double helix.

sink Any tissue, site, or location where an element or a molecule is consumed or taken out of circulation (e.g., in plants, a tissue where sugar exits the phloem). Compare with **source**.

sinoatrial (SA) node In the right atrium of the vertebrate heart, a cluster of cardiac muscle cells that initiates the heartbeat and determines the heart rate. Compare with **atrioventricular (AV) node**.

siphon A tubelike appendage of many mollusks, often used for feeding or propulsion.

sister chromatids The paired strands of a recently replicated chromosome, which are connected at the centromere and eventually separate during anaphase of mitosis and meiosis II. Compare with **non-sister chromatids**.

sister species Closely related species that occupy adjacent branches in a phylogenetic tree.

skeletal muscle The muscle tissue attached to the bones of the vertebrate skeleton. Consists of long, unbranched muscle fibers with a characteristic striped (striated) appearance; controlled voluntarily. Compare with **cardiac** and **smooth muscle**.

sliding-filament model The hypothesis that thin (actin) filaments and thick (myosin) filaments slide past each other, thereby shortening the sarcomere. Shortening of all the sarcomeres in a myofibril results in contraction of the entire myofibril.

slow muscle fiber Type of skeletal muscle fiber that is red in color due to the abundance of myoglobin, generates ATP by oxidative phosphorylation, and contracts slowly but does not fatigue easily. Also called *slow oxidative*, or *Type I, fiber*.

small intestine The portion of the digestive tract between the stomach and the large intestine. The site of the final stages of digestion and of most nutrient absorption.

small nuclear ribonucleoproteins See **snRNPs**.

smooth endoplasmic reticulum (smooth ER) The portion of the endoplasmic reticulum that does not have ribosomes attached to it. Involved in synthesis and secretion of lipids. Compare with **rough endoplasmic reticulum**.

smooth muscle The unstriated muscle tissue that lines the intestine, blood vessels, and some other organs. Consists of tapered, unbranched cells that can sustain long contractions. Not voluntarily controlled. Compare with **cardiac** and **skeletal muscle**.

snRNPs (small nuclear ribonucleoproteins) Complexes of proteins and small RNA molecules that function as components of spliceosomes during splicing (removal of introns from pre-mRNAs).

sodium–potassium pump A transmembrane protein that uses the energy of ATP to move sodium ions out of the cell and potassium ions in. Also called Na^+/K^+-ATPase.

soil organic matter Organic (carbon-containing) compounds found in soil.

solute Any substance that is dissolved in a liquid.

solute potential (ψ_S) A component of the potential energy of water caused by a difference in solute concentrations at two locations. Can be zero (pure water) or negative. Compare with **pressure potential** (ψ_P).

solution A liquid containing one or more dissolved solids or gases in a homogeneous mixture.

solvent Any liquid in which one or more solids or gases can dissolve.

soma See **cell body**.

somatic cell Any type of cell in a multicellular organism except eggs, sperm, and their precursor cells. Also called *body cells*.

somatic hypermutation Mutation that occurs in the variable regions of immunoglobulin genes when B cells are first activated and in memory cells, resulting in novel variation in the receptors that bind to antigens.

somatic nervous system The part of the peripheral nervous system (outside the brain and spinal cord) that controls skeletal muscles and is under voluntary control. Compare with **autonomic nervous system**.

somatostatin A hormone secreted by the pancreas and hypothalamus that inhibits the release of several other hormones.

somite A block of mesoderm that occurs in pairs along both sides of the developing neural tube in a vertebrate embryo. Gives rise to muscle, vertebrae, ribs, and the dermis of the skin.

sori In ferns, a cluster of spore-producing structures (sporangia) usually found on the underside of fronds.

source Any tissue, site, or location where a substance is produced or enters circulation (e.g., in plants, the tissue where sugar enters the phloem). Compare with **sink**.

space-filling model A representation of a molecule where atoms are shown as balls—color-coded and scaled to indicate the atom's identify—attached to each other in the correct geometry.

speciation The evolution of two or more distinct species from a single ancestral species.

species An evolutionarily independent population or group of populations. Generally distinct from other species in appearance, behavior, habitat, ecology, genetic characteristics, and so on.

species–area relationship The mathematical relationship between the area of a certain habitat and the number of species that it can support.

species diversity The variety and relative abundance of the species present in a given ecological community.

species richness The number of species present in a given ecological community.

specific heat The amount of energy required to raise the temperature of 1 gram of a substance by 1°C; a measure of the capacity of a substance to absorb energy.

sperm A mature male gamete; smaller and more mobile than the female gamete.

sperm competition Competition between the sperm of different males to fertilize eggs inside the same female.

spermatid An immature sperm cell.

spermatogenesis The production of sperm. Occurs continuously in a testis.

spermatogonium (plural: spermatogonia) Any of the diploid cells in a testis that can give rise to primary spermatocytes.

spermatophore A gelatinous package containing sperm cells that is produced by males of species that have internal fertilization without copulation.

sphincter A muscular valve that can close off a tube, as in a blood vessel or a part of the digestive tract.

spicule Stiff spike of silica or calcium carbonate that provides structural support in the body of many sponges.

spindle apparatus The array of microtubules responsible for moving chromosomes during mitosis and meiosis; includes kinetochore microtubules, polar microtubules, and astral microtubules.

spines In plants, modified leaves that are stiff and sharp and that function in defense.

spiracle In insects, a small opening that connects air-filled tracheae to the external environment, allowing for gas exchange.

spleen A dark red organ, found near the stomach of most vertebrates, that filters blood, stores extra red blood cells in case of emergency, and plays a role in immunity.

spliceosome In eukaryotes, a large, complex assembly of snRNPs (small nuclear ribonucleoproteins) that catalyzes removal of introns from primary RNA transcripts.

splicing The process by which introns are removed from primary RNA transcripts and the remaining exons are connected together.

sporangium (plural: sporangia) A spore-producing structure found in seed plants, some protists, and some fungi (e.g., chytrids).

spore (1) In bacteria, a dormant form that generally is resistant to extreme conditions. (2) In eukaryotes, a single haploid cell produced by mitosis or meiosis (not by fusion of gametes) that is capable of developing into an adult organism.

sporophyte In organisms undergoing alternation of generations, the multicellular diploid form that arises from two fused gametes and produces haploid spores. Compare with **gametophyte**.

sporopollenin A watertight material that encases spores and pollen of modern land plants.

stabilizing selection A mode of natural selection that favors phenotypes near the middle of the range of phenotypic variation. Reduces overall genetic variation in a population. Compare with **disruptive selection** and **directional selection**.

stamen The male reproductive structure of a flower. Consists of an anther, in which pollen grains are produced, and a filament, which supports the anther. Compare with **carpel**.

standing defense See **constitutive defense**.

stapes The last of three small bones (ossicles) in the middle ear of vertebrates. Receives vibrations from the tympanic membrane and by vibrating against the oval window passes them to the cochlea.

starch A mixture of two storage polysaccharides, amylose and amylopectin, both formed from α-glucose monomers. Amylopectin is branched, and amylose is unbranched. The major form of stored carbohydrate in plants.

start codon The AUG triplet in mRNA at which protein synthesis begins; codes for the amino acid methionine.

statocyst A sensory organ of many arthropods that detects the animal's orientation in space (e.g., whether the animal is flipped upside down).

statolith A tiny stone or dense particle found in specialized gravity-sensing organs in some animals such as lobsters, and in gravity-sensing tissues of plants.

statolith hypothesis The hypothesis that amyloplasts (dense, starch-storing plant organelles) serve as statoliths in gravity detection by plants.

stem cell Any relatively undifferentiated cell that can divide to produce a daughter cell that remains a stem cell and a daughter cell that can differentiate into specific cell types.

stems Vertical, aboveground structures that make up the shoot system of plants.

stereocilium (plural: stereocilia) One of many stiff outgrowths from the surface of a hair cell that are involved in detection of sound by terrestrial vertebrates or of waterborne vibrations by fishes.

steroid A class of lipid with a characteristic four-ring hydrocarbon structure.

sticky ends The short, single-stranded ends of a DNA molecule cut by a restriction endonuclease. Tend to form hydrogen bonds with other sticky ends that have complementary sequences.

stigma The sticky tip at the end of a flower carpel to which pollen grains adhere.

stolon A modified stem that runs horizontally over the soil surface and produces new plants at the nodes (a form of asexual reproduction). Compare with **rhizome**.

stoma (plural: stomata) Generally, a pore or opening. In plants, a microscopic pore, surrounded by specialized cells that open the pore, on the surface of a leaf or stem through which gas exchange occurs. See also **guard cells**.

stomach A tough, muscular pouch in the vertebrate digestive tract between the esophagus and small intestine. Physically breaks up food and begins digestion of proteins.

stop codon Any of three mRNA triplets (UAG, UGA, or UAA) that cause termination of protein synthesis. Also called a *termination codon*.

strain The lowest, most specific level of taxonomy that refers to a population of individuals that are genetically very similar or identical.

striated muscle Muscle tissue containing protein filaments organized into repeating structures that give the cells and tissues a banded appearance.

stroma The fluid matrix of a chloroplast in which the thylakoids are embedded. Site where the Calvin cycle reactions occur.

structural formula A two-dimensional notation in which the chemical symbols for the constituent atoms are joined by straight lines representing single (–), double (=), or triple (≡) covalent bonds. Compare with **molecular formula**.

structural homology Similarities in adult organismal structures (e.g., limbs, shells, flowers) that are due to inheritance from a common ancestor.

style The slender stalk of a flower carpel connecting the stigma and the ovary.

suberin Waxy substance found in the cell walls of cork tissue and in the Casparian strip of endodermal cells.

subspecies A population that has distinctive traits and some genetic differences relative to other populations of the same species but that is not distinct enough to be classified as a separate species.

substrate (1) A reactant that interacts with a catalyst, such as an enzyme or ribozyme, in a chemical reaction. (2) A surface on which a cell or organism sits.

substrate-level phosphorylation Production of ATP or GTP by the transfer of a phosphate group from an intermediate substrate directly to ADP or GDP. Occurs in glycolysis and in the citric acid cycle.

succession In ecology, the gradual colonization of a habitat after an environmental disturbance (e.g., fire, flood), usually by a series of species. See **primary** and **secondary succession**.

sucrose A disaccharide formed from glucose and fructose. One of the two main products of photosynthesis.

sugar Synonymous with carbohydrate, though usually used in an informal sense to refer to small carbohydrates (monosaccharides and disaccharides).

summation The additive effect of different postsynaptic potentials on a nerve or muscle cell, such that several subthreshold stimulations can cause or inhibit an action potential.

supporting connective tissue A type of connective tissue distinguished by having a firm extracellular matrix. Includes bone and cartilage.

surface metabolism model Hypothetical explanation for chemical evolution whereby small molecules reacted with one another through catalytic activity associated with a surface, such as the mineral deposits found in deep-sea hydrothermal vents.

surface tension The cohesive force that causes molecules at the surface of a liquid to stick together, thereby resisting deformation of the liquid's surface and minimizing its surface area.

survivorship On average, the proportion of offspring that survive to a particular age.

survivorship curve A graph depicting the percentage of a population that survives to different ages.

suspension feeder Any organism that obtains food by filtering small particles or small organisms out of water or air. Also called *filter feeder*.

sustainability The planned use of environmental resources at a rate no faster than the rate at which they are naturally replaced.

sustainable agriculture Agricultural techniques that are designed to maintain long-term soil quality and productivity.

swamp A wetland that has a steady rate of water flow and is dominated by trees and shrubs.

swim bladder A gas-filled organ of many ray-finned fishes that regulates buoyancy.

symbiosis (adjective: symbiotic) Any close and prolonged physical relationship between individuals of two different species. See **commensalism**, **mutualism**, and **parasitism**.

symmetric competition Ecological competition between two species in which both suffer similar declines in fitness. Compare with **asymmetric competition**.

sympathetic nervous system The part of the autonomic nervous system that stimulates fight-or-flight responses, such as increased heart rate, increased blood pressure, and decreased digestion. Compare with **parasympathetic nervous system**.

sympatric speciation The divergence of populations living within the same geographic area into different species as the result of their genetic (not physical) isolation. Compare with **allopatric speciation**.

sympatry Condition in which two or more populations live in the same geographic area, or close enough to permit interbreeding. Compare with **allopatry**.

symplast In plants, the space inside the plasma membranes. The symplast of adjacent cells is often connected through plasmodesmata. Compare with **apoplast**.

symporter A cotransport protein that allows an ion to diffuse down an electrochemical gradient, using the energy of that process to transport a different substance in the same direction *against* its concentration gradient. Compare with **antiporter**.

synapomorphy A shared, derived trait found in two or more taxa that is present in their most recent common ancestor but is missing in more distant ancestors. Useful for inferring evolutionary relationships.

synapse The interface between two neurons or between a neuron and an effector cell.

synapsis The physical pairing of two homologous chromosomes during prophase I of meiosis. Crossing over is observed during synapsis.

synaptic cleft The space between two communicating nerve cells (or between a neuron and effector cell) at a synapse, across which neurotransmitters diffuse.

synaptic plasticity Long-term changes in the responsiveness or physical structure of a synapse that can occur after particular stimulation patterns. Thought to be the basis of learning and memory.

synaptic vesicle A small neurotransmitter-containing vesicle at the end of an axon that releases neurotransmitter into the synaptic cleft by exocytosis.

synaptonemal complex A network of proteins that holds non-sister chromatids together during synapsis in meiosis I.

synthesis (S) phase The phase of the cell cycle during which DNA is synthesized and chromosomes are replicated.

system A defined set of interacting chemical components under observation.

systemic acquired resistance (SAR) A slow, widespread response of plants to a localized infection that protects healthy tissue from invasion by pathogens. Compare with **hypersensitive response**.

systemic circulation The part of the circulatory system that sends oxygen-rich blood from the lungs out to the rest of the body. It is separate from the pulmonary circulation in mammals and birds.

systemin A peptide hormone, produced by plant cells damaged by herbivores, that initiates a protective response in undamaged cells.

systems biology The study of the structure of networks and how interactions between individual network components such as genes or proteins can lead to emergent biological properties.

systole The portion of the cardiac cycle during which the heart muscles are contracting. Compare with **diastole**.

systolic blood pressure The force exerted by blood against artery walls during contraction of the heart's left ventricle. Compare with **diastolic blood pressure**.

T cell A type of lymphocyte that matures in the thymus and, with B cells, is responsible for adaptive immunity. Involved in activation of B cells (CD4⁺ helper T cells) and destruction of infected cells (CD8⁺ cytotoxic T cells). Also called *T lymphocytes*.

T-cell receptor (TCR) A type of transmembrane protein found on T cells that can bind to antigens displayed on the surfaces of other cells. Composed of two polypeptides, called the alpha chain and beta chain, that consist of variable and constant regions. See **antigen presentation**.

T tubule Any of the membranous tubes that extend into the interior of muscle cells, propagating action potentials throughout the cell and triggering release of calcium from the sarcoplasmic reticulum.

tagmata (singular: tagma) Prominent body regions in arthropods, such as the head, thorax, and abdomen in insects.

taiga A vast forest biome throughout subarctic regions, consisting primarily of short coniferous trees. Characterized by intensely cold winters, short summers, and high annual variation in temperature.

taproot A large, vertical main root of a plant's root system.

taste bud A sensory structure, found chiefly in the mammalian tongue, containing spindle-shaped cells that respond to chemical stimuli.

TATA-binding protein (TBP) A protein that binds to the TATA box in eukaryotic promoters and is a component of the basal transcription complex.

TATA box A short DNA sequence in many eukaryotic promoters about 30 base pairs upstream from the transcription start site.

taxon (plural: taxa) Any named group of organisms at any level of a classification system.

taxonomy The branch of biology concerned with the classification and naming of organisms.

tectorial membrane A membrane, located in the vertebrate cochlea, that takes part in the transduction of sound by bending the stereocilia of hair cells in response to sonic vibrations.

telomerase An enzyme that adds DNA to the ends of chromosomes (telomeres) by catalyzing DNA synthesis from an RNA template that is part of the enzyme.

telomere The end of a linear chromosome that contains a repeated sequence of DNA.

telophase The final stage in mitosis or meiosis during which daughter chromosomes (homologous chromosomes in meiosis I) have separated and new nuclear envelopes begin to form around each set of chromosomes.

temperate Having a climate with pronounced annual fluctuations in temperature (i.e., warm summers and cold winters) but typically neither as hot as the tropics nor as cold as the poles.

temperature A measurement of thermal energy present in an object or substance, reflecting how much the constituent molecules are moving.

template strand An original nucleic acid strand used to make a new, complementary copy based on hydrogen bonding between nitrogeneous bases.

temporal lobe In the vertebrate brain, one of the four major areas in the cerebrum.

tendon A band of tough, fibrous connective tissue that connects a muscle to a bone.

tentacle A long, thin, muscular appendage typically used for feeling and feeding. Occurs in different forms in diverse animals such as cephalopod mollusks and sea anemones.

termination (1) In enzyme-catalyzed reactions, the final stage in which the enzyme returns to its original conformation and products are released. (2) In transcription, the dissociation of RNA polymerase from DNA. (3) In translation, the dissociation of a ribosome from mRNA when it reaches a stop codon.

territory An area that is actively defended by an animal from others of its species and that provides exclusive or semi-exclusive use of its resources by the owner.

tertiary consumers In a food chain or food web, organisms that feed on secondary consumers. Compare with **primary consumer** and **secondary consumer**.

tertiary structure The overall three-dimensional shape of a single polypeptide chain, resulting from multiple interactions among the amino acid side chains and the peptide backbone. In single-stranded nucleic acids, the three-dimensional shape is formed by hydrogen bonding and hydrophobic interactions between complementary bases. Compare with **primary**, **secondary**, and **quaternary structure**.

testcross The breeding of an individual that expresses a dominant phenotype but has an unknown genotype with an individual having only recessive alleles for the traits of interest. Used to order to infer the unknown genotype from observation of the phenotypes seen in offspring.

testis (plural: testes) The sperm-producing organ of a male animal.

testosterone A steroid hormone, produced and secreted by the testes, that stimulates sperm production and various male traits and reproductive behaviors.

tetrad The structure formed by synapsed homologous chromosomes during prophase of meiosis I. Also known as a *bivalent*.

tetrapod Any member of the lineage that includes all vertebrates with two pairs of limbs (amphibians, mammals, and reptiles, including birds).

texture A quality of soil, resulting from the relative abundance of different-sized particles.

theory An explanation for a broad class of phenomena that is supported by a wide body of evidence. A theory serves as a framework for the development of new hypotheses.

thermal energy The kinetic energy of molecular motion.

thermocline A steep gradient (cline) in environmental temperature, such as occurs in a thermally stratified lake or ocean.

thermophile A bacterium or archaean that thrives in very hot environments.

thermoreception A sensory system in which receptors are activated by changes in heat energy.

thermoreceptor A sensory cell or an organ specialized for detection of changes in temperature.

thermoregulation Regulation of body temperature.

thick filament A filament composed of bundles of the motor protein myosin; anchored to the center of the sarcomere. Compare with **thin filament**.

thigmotropism Growth or movement of an organism in response to contact with a solid object.

thin filament A filament composed of two coiled chains of actin and associated regulatory proteins; anchored at the Z disk of the sarcomere. Compare with **thick filament**.

thorax A region of the body; in insects, one of the three prominent body regions, along with the head and abdomen, called tagmata.

thorn A modified plant stem shaped as a sharp, protective structure. Helps protect a plant against feeding by herbivores.

threshold potential The membrane potential that will trigger an action potential in a neuron or other excitable cell. Also called simply *threshold*.

thylakoid A membrane-bound network of flattened sac-like structures inside a plant chloroplast that functions in converting light energy to chemical energy. A stack of thylakoid discs is a granum.

thymus An organ, located in the anterior chest or neck of vertebrates, in which immature T cells that originated in the bone marrow undergo maturation.

thyroid gland A gland in the neck that releases thyroid hormone (which increases metabolic rate) and calcitonin (which lowers blood calcium).

thyroid hormones Either of two hormones, triiodothyronine (T_3) or thyroxine (T_4), produced by the thyroid gland. See **triiodothyronine** and **thyroxine**.

thyroid-stimulating hormone (TSH) A peptide hormone, produced and secreted by the anterior pituitary, that stimulates release of thyroid hormones from the thyroid gland.

thyroxine (T_4) A lipid-soluble hormone, derived from the amino acid tyrosine, containing four iodine atoms and produced and secreted by the thyroid gland. Acts primarily to increase cellular metabolism. In mammals, T_4 is converted to the more active hormone triiodothyronine (T_3) in the liver.

Ti plasmid A plasmid carried by *Agrobacterium* (a bacterium that infects plants) that can integrate into a plant cell's chromosomes and induce formation of a gall.

tight junction A type of cell–cell attachment structure that links the plasma membranes of adjacent animal cells, forming a barrier that restricts movement of substances in the space between the cells. Most abundant in epithelia (e.g., the intestinal lining). Compare with **desmosome** and **gap junction**.

tip The end of a branch on a phylogenetic tree. Represents a specific species or larger taxon that has not (yet) produced descendants—either a group living today or a group that ended in extinction. Also called *terminal node*.

tissue A group of cells that function as a unit, such as muscle tissue in an animal or xylem tissue in a plant.

tolerance In ecological succession, the phenomenon in which early-arriving species do not affect the probability that subsequent species will become established. Compare with **facilitation** and **inhibition**.

tonoplast The membrane surrounding a plant vacuole.

tool-kit genes A set of key developmental genes that establishes the body plan of animals and plants; present at the origin of the multicellular lineages and elaborated upon over evolutionary time by a process of duplication and divergence. Includes *Hox* genes.

top-down control The hypothesis that population size is limited by predators or herbivores (consumers).

topoisomerase An enzyme that prevents the twisting of DNA ahead of the advancing replication fork by cutting the DNA, allowing it to unwind, and rejoining it.

torpor An energy-conserving physiological state, marked by a decrease in metabolic rate, body temperature, and activity, that lasts for a short period (overnight to a few days or weeks). Occurs in some small mammals when the ambient temperature drops significantly. Compare with **hibernation**.

totipotent Capable of dividing and developing to form a complete, mature organism.

toxin A poison produced by a living organism, such as a plant, animal, or microorganism.

trachea (plural: tracheae) (1) In insects, any of the small air-filled tubes that extend throughout the body and function in gas exchange. (2) In terrestrial vertebrates, the airway connecting the larynx to the bronchi. Also called *windpipe*.

tracheid In vascular plants, a long, thin, water-conducting cell that has pits where its lignin-containing secondary cell wall is absent, allowing water movement between adjacent cells. Compare with **vessel element**.

trade-off In evolutionary biology, an inescapable compromise between two traits that cannot be optimized simultaneously. Also called *fitness trade-off*.

trait Any observable characteristic of an individual.

transcription The process that uses a DNA template to produce a complementary RNA.

transcription factor General term for a protein that binds to a DNA regulatory sequence to influence transcription. It includes both regulatory and basal transcription factors.

transcriptional activator A eukaryotic regulatory transcription factor that binds to regulatory DNA sequences in enhancers or promoter-proximal elements to promote the initiation of transcription.

transcriptional control Regulation of gene expression by various mechanisms that change the rate at which genes are transcribed to form messenger RNA. In negative control, binding of a regulatory protein to DNA represses transcription; in positive control, binding of a regulatory protein to DNA promotes transcription.

transcriptome The complete set of genes transcribed in a particular cell.

transduction The conversion of information from one mode to another. For example, the process by which a stimulus outside a cell is converted into a response by the cell.

transfer RNA (tRNA) An RNA molecule that has an anticodon at one end and an amino acid attachment site at the other. Each tRNA carries a specific amino acid and binds to the corresponding codon in messenger RNA during translation.

transformation (1) Incorporation of external DNA into a cell. Occurs naturally in some bacteria; can be induced in the laboratory. (2) Conversion of a normal mammalian cell to one that divides uncontrollably.

transgenic A plant or animal whose genome contains DNA introduced from another individual, often from a different species.

transition state A high-energy intermediate state of the reactants during a chemical reaction that must be achieved for the reaction to proceed. Compare with **activation energy**.

transitional feature A trait that is intermediate between a condition observed in ancestral (older) species and the condition observed in derived (younger) species.

translation The process by which a polypeptide (a string of amino acids joined by peptide bonds) is synthesized from information in codons of messenger RNA.

translational control Regulation of gene expression by various mechanisms that alter the life span of messenger RNA or the efficiency of translation.

translocation (1) In plants, the movement of sugars and other organic nutrients through the phloem by bulk flow. (2) A type of mutation in which a piece of a chromosome moves to a nonhomologous chromosome. (3) The movement of a ribosome down a messenger RNA during translation.

transmembrane protein See **integral membrane protein**.

transmission The passage or transfer of (1) a disease from one individual to another or (2) electrical impulses from one neuron to another.

transpiration Loss of water vapor from aboveground plant parts. Occurs primarily through stomata.

transporter See **carrier protein**.

transposable elements Any of several kinds of DNA sequences that are capable of moving themselves, or copies of themselves, to other locations in the genome. Include LINEs and SINEs.

tree of life The phylogenetic tree that includes all organisms.

trichome A hairlike appendage that grows from epidermal cells in the shoot system of some plants. Trichomes exhibit a variety of shapes, sizes, and functions depending on species.

triiodothyronine (T₃) A lipid-soluble hormone, derived from the amino acid tyrosine, containing three iodine atoms and produced and secreted by the thyroid gland. Acts primarily to increase cellular metabolism. In mammals, T_3 has a stronger effect than does the related hormone thyroxine (T_4).

triose A monosaccharide (simple sugar) containing three carbon atoms.

triplet code A code in which a "word" of three letters encodes one piece of information. The genetic code is a triplet code because a codon is three nucleotides long and encodes one amino acid.

triploblast (adjective: triploblastic) An animal whose body develops from three basic embryonic cell layers or tissues: ectoderm, mesoderm, and endoderm. Compare with **diploblast**.

trisomy The state of having three copies of one particular type of chromosome in an otherwise diploid cell.

tRNA See **transfer RNA**.

trochophore A larva with a ring of cilia around its middle that is found in some lophotrochozoans.

trophic cascade A series of changes in the abundance of species in a food web, usually caused by the addition or removal of a key predator.

trophic level A feeding level in an ecosystem.

trophoblast The exterior of a blastocyst (the structure that results from cleavage in embryonic development of mammals).

tropomyosin A regulatory protein present in thin (actin) filaments that blocks the myosin-binding sites on these filaments in resting muscles, thereby preventing muscle contraction.

troponin A regulatory protein, present in thin (actin) filaments, that can move tropomyosin off the myosin-binding sites on these filaments, thereby triggering muscle contraction. Activated by high intracellular calcium.

true navigation The type of navigation by which an animal can reach a specific point on Earth's surface. Also called *map orientation*.

trypsin A protein-digesting enzyme present in the small intestine that activates several other protein-digesting enzymes.

tube foot One of the many small, mobile, fluid-filled extensions of the water vascular system of echinoderms; the part extending outside the body is called a podium, while the bulb within the body is the ampulla. Used in locomotion, feeding, and respiration.

tuber A modified plant rhizome that functions in storage of carbohydrates.

tuberculosis A disease of the lungs caused by infection with the bacterium *Mycobacterium tuberculosis*.

tumor A mass of cells formed by uncontrolled cell division. Can be benign or malignant.

tumor suppressor A protein (e.g., p53 or Rb) that prevents cell division, such as when the cell has DNA damage. Mutant genes that code for tumor suppressors are associated with cancer.

tundra The treeless biome in polar and alpine regions, characterized by short, slow-growing vegetation, permafrost, and a climate of long, intensely cold winters and very short summers.

turbidity Cloudiness of water caused by sediments and/or microscopic organisms.

turgid Swollen and firm as a result of high internal pressure (e.g., a plant cell containing enough water for the cytoplasm to press against the cell wall). Compare with **flaccid**.

turgor pressure The outward pressure exerted by the fluid contents of a living plant cell against its cell wall.

turnover In lake ecology, the complete mixing of upper and lower layers of water of different temperatures; occurs each spring and fall in temperate-zone lakes.

tympanic membrane The membrane separating the middle ear from the outer ear in terrestrial vertebrates, or similar structures in insects. Also called the *eardrum*.

ubiquinone See **coenzyme Q**.

ulcer A hole in an epithelial layer, exposing the underlying tissues to damage.

ultimate causation In biology, the reason that a trait or phenomenon is thought to have evolved; the adaptive advantage of that trait. Also called *ultimate explanation*. Compare with **proximate causation**.

umami The taste of glutamate, responsible for the "meaty" taste of most proteins and of monosodium glutamate.

umbilical cord The cord that connects a developing mammalian embryo or fetus to the placenta and through which the embryo or fetus receives oxygen and nutrients.

unequal crossover An error in crossing over during meiosis I in which the two non-sister chromatids match up at different sites. Results in gene duplication in one chromatid and gene loss in the other.

unsaturated Referring to lipids in which at least one carbon-carbon bond is a double bond. Double bonds produce kinks in hydrocarbon chains and decrease the compound's melting point. Compare with **saturated**.

upstream In genetics, opposite to the direction in which RNA polymerase moves along a DNA strand. Compare with **downstream**.

urea The major nitrogenous waste of mammals, adult amphibains, and cartilaginous fishes. Compare with **ammonia** and **uric acid**.

ureter In vertebrates, a tube that transports urine from one kidney to the bladder.

urethra The tube that drains urine from the bladder to the outside environment. In male vertebrates, also used for passage of semen during ejaculation.

uric acid A whitish excretory product of birds, reptiles, and terrestrial arthropods. Used to remove from the body excess nitrogen derived from the breakdown of amino acids. Compare with **ammonia** and **urea**.

urochordates One of the three major chordate lineages (Urochordata), comprising sessile or floating, filter-feeding animals that have a polysaccharide covering (tunic) and two siphons through which water enters and exits; include tunicates and salps. Compare with **cephalochordates** and **vertebrates**.

uterus The organ in which developing embryos are housed in mammals and some other viviparous vertebrates.

vaccination Artificially producing immunological memory against a pathogen by using isolated antigens or altered versions of the pathogen to stimulate an adaptive immune response in the absence of disease.

vaccine A preparation designed to stimulate an immune response against a particular pathogen without causing illness. Vaccines consist of inactivated (killed) pathogens, live but weakened (attenuated) pathogens, or parts of a pathogen (subunit vaccine).

vacuole A large organelle in plant and fungal cells that usually is used for bulk storage of water, pigments, oils, or other substances. Some vacuoles contain enzymes and have a digestive function similar to lysosomes in animal cells.

vagina The birth canal of female mammals; a muscular tube that extends from the uterus through the pelvis to the exterior.

valence The number of unpaired electrons in the outermost electron shell of an atom; often determines how many covalent bonds the atom can form.

valence electron An electron in the outermost electron shell, the valence shell, of an atom. Valence electrons tend to be involved in chemical bonding.

valence shell The outermost electron shell of an atom.

valve In circulatory systems, any of the flaps of tissue that prevent backward flow of blood, particularly in veins and in the heart.

van der Waals interactions A weak electrical attraction between two nonpolar molecules that have been brought together through hydrophobic interactions. Often contributes to tertiary and quaternary structures in proteins.

variable number tandem repeat See **minisatellite**.

variable (V) region The amino acid sequence that changes in polypeptides used to make antibodies, B-cell receptors, and T-cell receptors, This portion of the protein is highly variable within an individual and forms the epitope-binding site. Compare with **constant (C) region**.

vas deferens (plural: vasa deferentia) A muscular tube that stores and transports semen from the epididymis to the ejaculatory duct. Also called the *ductus deferens*.

vasa recta In the vertebrate kidney, a network of blood vessels that runs alongside the loop of Henle of a nephron. Functions in reabsorption of water and solutes from the filtrate.

vascular bundle In a plant stem, a cluster of xylem and phloem strands that run the length of the stem.

vascular cambium One of two types of cylindrical meristem, consisting of a ring of undifferentiated plant cells in the stem and root of woody plants; produces secondary xylem (wood) and secondary phloem. Compare with **cork cambium**.

vascular tissue In plants, tissue that transports water, nutrients, and sugars. Made up of the complex tissues xylem and phloem, each of which contains several cell types. Also called *vascular tissue system*.

vector A biting insect or other organism that transfers pathogens from one species to another. See also **cloning vector**.

vegetative development The phase of plant development that involves growth and the the production of all plant structures except the flower.

vein Any blood vessel that carries blood (oxygenated or not) under relatively low pressure from the tissues toward the heart. Compare with **artery**.

veliger A distinctive type of larva, found in mollusks.

vena cava (plural: venae cavae) Either of two large veins that return oxygen-poor blood to the heart.

ventral Toward an animal's belly and away from its back. The opposite of dorsal.

ventricle (1) A thick-walled chamber of the heart that receives blood from an atrium and pumps it to the body or to the lungs. (2) Any of several small fluid-filled chambers in the vertebrate brain.

venule Any of the body's many small veins (blood vessels that return blood to the heart).

vertebrae (singular: vertebra) The cartilaginous or bony elements that form the backbones of vertebrate animals.

vertebrates One of the three major chordate lineages (Vertebrata), comprising animals with a dorsal column of cartilaginous or bony structures (vertebrae) and a skull enclosing the brain. Includes fishes, amphibians, mammals, and reptiles (including birds). Compare with **cephalochordates** and **urochordates**.

vessel element In vascular plants, a short, wide, water-conducting cell that has gaps through both the primary and secondary cell walls, allowing unimpeded passage of water between adjacent cells. Compare with **tracheid**.

vestigial trait A reduced or incompletely developed structure that has no function or reduced function, but is clearly similar to functioning organs or structures in closely related species.

vicariance The physical splitting of a population into smaller, isolated populations by a geographic barrier.

villi (singular: villus) Small, fingerlike projections (1) of the lining of the small intestine or (2) of the fetal portion of the placenta adjacent to maternal arteries. Function to increase the surface area available for absorption of nutrients and gas exchange.

virion The infectious extracellular particle that is produced from a viral infection; used for transmitting the virus between hosts. It consists of a DNA or RNA genome enclosed within a protein shell (capsid) that may be further enveloped in a phospholipid bilayer. Compare with **virus**.

virulence (adjective: virulent) Referring to the ability of pathogens to cause severe disease in susceptible hosts.

virus An obligate, intracellular parasite that is acellular, but uses host-cell biosynthetic machinery to replicate. Compare with **virion**.

visceral mass One of the three main parts of the mollusk body; contains most of the internal organs and external gill.

visible light The range of wavelengths of electromagnetic radiation that humans can see, from about 400 to 700 nanometers.

vitamin Any of various organic micronutrients that usually function as coenzymes.

vitelline envelope A fibrous sheet of glycoproteins that surrounds mature egg cells in many vertebrates. Surrounded by a thick gelatinous matrix (the jelly layer) in some aquatic species. In mammals, called the *zona pellucida*.

viviparous Producing live young (instead of eggs) that develop within the body of the mother before birth. Compare with **oviparous** and **ovoviviparous**.

volt (V) A unit of electrical potential (voltage).

voltage Potential energy created by a separation of electric charges between two points. Also called *electrical potential*.

voltage clamping A technique for imposing a constant membrane potential on a cell. Widely used to investigate ion channels.

voltage-gated channel An ion channel that opens or closes in response to changes in membrane voltage. Compare with **ligand-gated channel**.

voluntary muscle Muscle that contracts in response to stimulation by voluntary (somatic), but not involuntary (parasympathetic or sympathetic), neural stimulation.

vomeronasal organ A paired sensory organ, located in the nasal region, containing chemoreceptors that bind odorants and pheromones.

wall pressure The inward pressure exerted by a cell wall against the fluid contents of a living plant cell.

Wallace line A line in the Indonesian region that demarcates two areas, each of which is characterized by a distinct set of animal species.

water potential (ψ) The potential energy of water in a certain environment compared with the potential energy of pure water at room temperature and atmospheric pressure. In living organisms, ψ equals the solute potential (ψ_S) plus the pressure potential (ψ_P).

water-potential gradient A difference in water potential in one region compared with that in another region. Determines the direction that water moves, always from regions of higher water potential to regions of lower water potential.

water table The upper limit of the underground layer of soil that is saturated with water.

water vascular system In echinoderms, a system of fluid-filled tubes and chambers that functions as a hydrostatic skeleton.

watershed The area drained by a single stream or river.

Watson–Crick pairing See **complementary base pairing**.

wavelength The distance between two successive crests in any regular wave, such as light waves, sound waves, or waves in water.

wax A class of lipid with extremely long, saturated hydrocarbon tails. Harder and less greasy than fats.

weather The specific short-term atmospheric conditions of temperature, moisture, sunlight, and wind in a certain area.

weathering The gradual wearing down of large rocks by rain, running water, temperature changes, and wind; one of the processes that transform rocks into soil.

weed Any plant that is adapted for growth in disturbed soils.

white blood cell Any of several types of blood cells, including neutrophils, macrophages, and lymphocytes, that circulate in blood and lymph and function in defense against pathogens.

wild type The most common phenotype seen in a wild population.

wildlife corridor Strips of wildlife habitat connecting populations that otherwise would be isolated by human-made development.

wilt To lose turgor pressure in a plant tissue.

wobble hypothesis The hypothesis that some tRNA molecules can pair with more than one mRNA codon by tolerating some non-standard base pairing in the third base, so long as the first and second bases are correctly matched.

wood Xylem resulting from secondary growth; forms strong supporting material. Also called *secondary xylem*.

worm An animal with a long, thin, tubelike body lacking limbs.

Woronin body A dense organelle in certain fungi that plugs pores in damaged septa to prevent leakage of cytoplasm.

X-linked inheritance Inheritance patterns for genes located on the mammalian X chromosome. Also called *X-linkage*.

X-ray crystallography A technique for determining the three-dimensional structure of large molecules, including proteins and nucleic acids, by analysis of the diffraction patterns produced by X-rays beamed at crystals of the molecule.

xenoestrogens Foreign chemicals that bind to estrogen receptors or otherwise induce estrogen-like effects.

xeroderma pigmentosum (XP) A human disease characterized by extreme sensitivity to ultraviolet light. Caused by an autosomal recessive allele that inactivates the nucleotide excision DNA repair system.

xylem A plant vascular tissue that conducts water and ions; contains tracheids and/or vessel elements.

Primary xylem develops from the procambium of apical meristems; secondary xylem, or wood, from the vascular cambium. Compare with **phloem**.

Y-linked inheritance Inheritance patterns for genes located on the mammalian Y chromosome. Also called *Y-linkage*.

yeast Any fungus growing as a single-celled form. Also, a specific lineage of Ascomycota.

yolk The nutrient-rich cytoplasm inside an egg cell; used as food for the growing embryo.

Z disk The structure that forms each end of a sarcomere. Contains a protein that binds tightly to actin, thereby anchoring thin filaments.

Z scheme Model for changes in the potential energy of electrons as they pass from photosystem II to photosystem I and ultimately to NADP$^+$ during the light-dependent reactions of photosynthesis. See also **noncyclic electron flow**.

zero population growth (ZPG) A state of stable population size due to fertility staying at the replacement rate for at least one generation.

zona pellucida The gelatinous layer around a mammalian egg cell. In other vertebrates, called the *vitelline envelope*.

zone of (cellular) division In plant roots, a group of apical meristematic cells just behind the root cap where cells are actively dividing.

zone of (cellular) elongation In plant roots, a group of young cells, derived from primary meristem tissues and located behind the apical meristem, that are increasing in length.

zone of (cellular) maturation In plant roots, a group of plant cells, located several millimeters behind the root cap, that are differentiating into mature tissues.

zygosporangium (plural: zygosporangia) The distinctive spore-producing structure in fungi that are members of the Zygomycota.

zygote The cell formed by the union of two gametes; a fertilized egg.

Credits

Photo Credits

Frontmatter Title page Eric Isselée/Fotolia **p. iii** IFE, URI-IAO, UW, Lost City Science Party; NOAA/OAR/OER; The Lost City 2005 Expedition **p. v** R. B. Taylor/Photo Researchers, Inc. **p. vi** Brian Johnston **p. viii** Anthony Bannister/Photo Researchers, Inc. **p. x** Tim Laman/National Geographic Stock **p. xiii** Lee W. Wilcox **p. xiv** Steve Winter/National Geographic/Getty Images **p. xvi** Associated Press/Gerald Herbert **p. xviii (bottom left)** Natalie B. Fobes Photography

Chapter 1 Opener Thierry Montford/Biosphoto/Photo Researchers Inc. **1.1a** Biophoto Associates/Photo Researchers, Inc. **1.1b** Brian J. Ford **1.1c** The Print Collector/Alamy **1.6a** Steve Gschmeissner/Photo Researchers, Inc. **1.6b** Kwangshin Kim/Photo Researchers, Inc. **1.7** Corbis Premium RF/Alamy **1.10** Reproduced with permission from H. Wolf. 2011. Odometry and Insect Navigation. *Journal of Experimental Biology* 214: 1629–1641.

Chapter 2 Opener IFE, URI-IAO, UW, Lost City Science Party; NOAA/OAR/OER; The Lost City 2005 Expedition **2.1** Dragan Trifunovic/Shutterstock **2.6c** Zedcor Wholly Owned/Photos.com **2.11** LianeM/Alamy **2.15b** Dietmar Nill/Getty Images **2.15c** John Sylvester/First Light/AGE fotostock **2.29** picture-alliance/Judaica-Samml/Newscom

Chapter 3 3.9ab Janice Carr, CDC

Chapter 4 Opener SSPL/The Image Works **4.5** National Cancer Institute

Chapter 5 Opener Peter Arnold/Alamy

Chapter 6 6.2a Multiart/iStockphoto **6.2b** lepas2004/iStockphoto **6.2c** ilker canikligil/Shutterstock **6.6L** Dr. rer. nat. Markus Drechsler **6.15** Jack W. Szostak **6.18** Don W. Fawcett/Photo Researchers, Inc.

Chapter 7 Opener Dr. Torsten Wittmann/SPL/Photo Researchers, Inc. **7.1** SPL/Photo Researchers, Inc. **7.2a** Gopal Murti/Photo Researchers, Inc. **7.2b** Biology Pics/Photo Researchers, Inc. **7.3** Eye of Science/Photo Researchers, Inc. **7.4** Eye of Science/Photo Researchers, Inc. **7.7** Don W. Fawcett/Photo Researchers, Inc. **7.8** Omikron/Photo Researchers, Inc. **7.9LR** Don W. Fawcett/Photo Researchers, Inc. **7.10** Biophoto Associates/Photo Researchers, Inc. **7.11** Don W. Fawcett/Photo Researchers, Inc. **7.12** E.H. Newcomb & W.P Wergin/Biological Photo Service **7.13** Fawcett/Friend/Photo Researchers, Inc. **7.14** Don W. Fawcett/Photo Researchers, Inc **7.15** E.H. Newcomb & W.P. Wergin/Biological Photo Service **7.16a** Dr. Don Fawcett/S. Ito & A. Like/Photo Researchers, Inc. **7.16b** Don W. Fawcett/Photo Researchers, Inc. **7.16c** Biophoto Associates/Photo Researchers, Inc. **7.16d** From "*Caveolin-1* expression is essential for proper nonshivering thermogenesis in brown adipose tissue." Cohen, A. W., Schubert, W., Brasaemle, D. L., Scherer, P. E., Lisanti, M. P. *Diabetes.* 2005 Mar; 54(3): 679–86, Fig. 6. **7.17** Don W. Fawcett/Photo Researchers, Inc. **7.26** Micrograph by Dr. Conly L. Rieder, Wadsworth Center, Albany, New York 12201-0509 **7.27ab** Reproduced by permission of the American Society for Cell Biology from *Molecular Biology of the Cell* 9 (12), December 1998, cover. ©1999 by the American Society for Cell Biology. Image courtesy of Bruce J. Schnapp, Oregon Health Sciences University. **7.28** John E. Heuser, M.D., Washington University School of Medicine, St. Louis, Missouri **7.29a** SPL/Photo Researchers, Inc. **7.29b** Biomedical Imaging Unit, Southampton General Hospital/Photo Researchers, Inc. **7.30** Don W. Fawcett/Photo Researchers, Inc.

Chapter 8 Opener Richard Megna/Fundamental Photographs **8.3** Thomas Eisner

Chapter 9 Opener Raiden32/Used under the Creative Commons license. http://creativecommons.org/licenses/by-sa/3.0/deed.en **9.1a** Oliver Hoffmann/Shutterstock **9.1b** Gergo Orban/Shutterstock **9.8** Terry Frey **9.16** Michael Delannoy

Chapter 10 Opener R.B. Taylor/Photo Researchers, Inc. **10.3T** John Durham/Science Photo Library/Photo Researchers, Inc. **10.3MB** Barbara Erienborn/Electron micrograph by Wm. P. Wergin, courtesy of E.H. Newcomb, University of Wisconsin **10.5b** Sinclair Stammers/Photo Researchers, Inc. **10.18LR** James A. Bassham, Lawrence Berkeley Laboratory, UCB (retired) **10.21ab** Dr. Jeremy Burgess/Science Photo Library/Photo Researchers, Inc.

Chapter 11 Opener Roger J. Bick & Brian J. Poindexter/UT-Houston Medical School/Photo Researchers, Inc. **11.1** Jochen Tack/imagebroker/Alamy **11.2** Biophoto Associates/Photo Researchers, Inc. **11.3b** ©ROCKWATER, Inc./*Journal of Bone and Joint Surgery*, 1997, 79, Instructional Course Lectures, The American Academy of Orthopaedic Surgeons—Articular Cartilage. Part I: Tissue Design and Chondrocyte-Matrix Interactions, Buckwalter, 600–611 **11.5** Biophoto Associates/Photo Researchers, Inc. **11.7aLR** Don W. Fawcett/Photo Researchers, Inc. **11.8** Don W. Fawcett/Photo Researchers **11.10a** Dr. Don Fawcett/Photo Researcher's, Inc. **11.10b** E. H. Newcomb & W.P. Wergin/Biological Photo Service **11.17a** A. Nern and R. A. Arkowitz. 2000. "G proteins mediate changes in cell shape by stabilizing the axis of polarity." *Molecular Cell* 5(5): 853–864, Fig. 2C. **11.17b** R. H. Kessin. 2003. "Cell motility: Making streams." *Nature* 422: 481–482. ©2003 Nature Publishing Group. Used with permission.

Chapter 12 Opener G. Gimenz-Martin/SPL/Photo Researchers, Inc. **12.1T** Gopal Murti/Photo Researchers, Inc. **12.1B** Biophoto Associates/Photo Researchers, Inc. **12.5(1–7)** Micrographs by Conly L. Rieder, Department of Biology, Rensselaer Polytechnic Institute, Troy, New York. **12.8a** Ed Reschke/Getty **12.8b** Michael Danilchik

Chapter 13 Opener David Phillips/The Population Council/Photo Researchers, Inc. **13.1** Look at Sciences/Photo Researchers, Inc. **13.6** Jupiterimages/Photos.com **13.7(1–9)** Warren Rosenberg/Fundamental Photographs **13.15** Dr. Rebecca Shulte

Chapter 14 Opener Brian Johnston **14.9a** Benjamin Prud'homme/Nicolas Gompel **14.9bLR** From "Learning to Fly: Phenotypic Markers in Drosophila." A poster of common phenotypic markers used in Drosophila genetics. Jennifer Childress, Richard Behringer, and Georg Halder. 2005. *Genesis* 43(1). Cover illustration. **14.16** Carla Fernanda Reis

Chapter 15 Opener Dr. Gopal Murti/Science Photo Library/Photo Researchers, Inc. **15.1** Eye of Science/Photo Researchers, Inc. **15.5** Reproduced by permission of Matthew S. Meselson, Harvard University, from M. Meselson and F.W. Stahl, "The replication of DNA in Escherichia coli." *PNAS* 44(7): 671–682 (July 1958), p. 675, Fig. 4. **15.7** Dr. Gopal Murti/Science Photo Library/Photo Researchers, Inc.

Chapter 16 Opener J. Craig Venter Institute **16.4a** Rod Williams/Nature Picture Library **16.4b** Clint Cook & Janet P. Crossland, The Peromyscu Genetic Stock Center at the U. of South Carolina **16.9** Courtesy of Hesed M. Padilla-Nash and Thomas Ried. Affiliation is Section of Cancer Genomics, Genetics Branch, Center for Research, National Cancer Institute, National Institutes of Health, Bethesda, MD 20892

Chapter 17 Opener Oscar Miller/SPL/Photo Researchers, Inc. **17.5** Bert W. O'Malley, M.D., Baylor College of Medicine **17.8** From Hamkalo and Miller, "Electronmicroscopy of Genetic Material," Figure 6a, p. 379. Reproduced with permission, from *Annual Review of Biochemistry*, Volume 42. ©1973 by Annual Reviews, Inc. **17.9** Dr. Elena Kiseleva/Photo Researchers, Inc.

Chapter 18 Opener Stephanie Schuller/Photo Researchers, Inc. **18.UN1** EDVOTEK, The Biotechnology Education Company (www.edvotek.com)/Jeff Chirikjian

Chapter 19 19.2 Ada Olins/Don Fawcett/Photo Researchers, Inc. **19.4TB** ©2002 from *Molecular Biology of the Cell* 4/e Fig. 4-23 by Bruce Alberts et al. Reproduced by Permission of Garland Science Taylor & Francis Books, Inc., and Dr. Barbara Hamkalo.

Chapter 20 Opener International Rice Research Institute (IRRI) used under Creative Commons 2.0 license. http://creativecommons.org/licenses/by-nc-sa/2.0/deed.en **20.1** Brady-Handy Photograph Collection, Library of Congress. Reproduction number: LC-DIG-cwpbh-02977. **20.12** Science Source/Photo Researchers, Inc. **20.15BLR** Golden Rice

Chapter 21 Opener Martin Krzywinski/Science Photo Libary/Alamy **21.8BLR** Test results provided by GENDIA (www.gendia.net) **21.11** Agilent Technologies, Inc. **21.12** Camilla M. Kao and Patrick O. Brown, Stanford University.

Chapter 22 Opener Anthony Bannister/Photo Researchers, Inc. **22.1a** From "The BMP antagonist Gremlin regulates outgrowth, chondrogenesis and programmed cell death in the developing limb." R. Merino, J. Rodriguez-Leon, D. Macias, Y. Ganan, A.N. Economides and J. M. Hurle. *Development* 126: 5515-5522 (1999), Fig. 6 E & F / Juan M. Hurle. **22.1b** Reproduced by permission of Elsevier Science from K. Kuida et al., "Reduced apoptosis and cytochrome c–mediated caspase activation in mice lacking caspase 9. *Cell* 94:325-337 (1998), Figs. 2E and 2F. ©1998 by Elsevier Science Ltd. Image courtesy of Keisuke Kuida, Vertex Pharmaceuticals. **22.2** John Chadwick/Agence France Presse/Newscom **22.3a** Richard Hutchings/Photo Researchers, Inc. **22.3b** From Drs. Adamson and Walls, "Embryonic and Neonatal Phenotyping of Gentically Engineered Mice." *ILAR Journal* 47(2), Fig. 12A, Optical projection tomographhy (OPT) image of a day 12.5 mouse embryo. (The rotation of the embryo

Illustration and Text Credits

The three-dimensional structure of the first EGF-like module of human factor IX: Comparison with EGF and TGF-alpha. *Protein Science* 1: 81–90. **3.12a** PDB ID: 1D1L. P. B. Rupert, A. K. Mollah, M. C. Mossing, et al. 2000. The structural basis for enhanced stability and reduced DNA binding seen in engineered second-generation Cro monomers and dimers. *Journal of Molecular Biology* 296: 1079–1090. **3.12b** PDB ID: 2DN2. S.Y. Park, T. Yokoyama, N. Shibayama, et al. 2006. 1.25 Å resolution crystal structures of human hemoglobin in the oxy, deoxy and carbonmonoxy forms. *Journal of Molecular Biology* 360: 690–701. **3.14a** PDB ID: 1DMO. M. Zhang, T. Tanaka, and M. Ikura. 1995. Calcium-induced conformational transition revealed by the solution structure of apo calmodulin. *Nature Structural Biology* 2: 758–767. **3.14b** PDB ID: 3CLN. Y. S. Babu, C. E. Bugg, and W. J. Cook. 1988. Structure of calmodulin refined at 2.2 A resolution. *Journal of Molecular Biology* 204: 191–204. **3.16** PDB ID: 2PTC. M. Marquart, J. Walter, J. Deisenhofer, et al. 1983. The geometry of the reactive site and of the peptide groups in trypsin, trypsinogen and its complexes with inhibitors. *Acta Crystallographica* Section B39: 480–490.

Chapter 4 **T4-1** PDB ID: 1EHZ. H. Shi and P. B. Moore. 2000. The crystal structure of yeast phenylalanine tRNA at 1.93 Å resolution: A classic structure revisited. *RNA* 6: 1091–1105. **4.11** PDB ID: 1X8W. F. Guo, A. R. Gooding, and T. R. Cech. 2004. Structure of the *Tetrahymena* ribozyme: Base triple sandwich and metal ion at the active site. *Molecular Cell* 16: 351–362.

Chapter 5 **5.6** H. M. Florman, K. B. Bechtol, and P. M. Wassarman. 1984. Enzymatic dissection of the functions of the mouse egg's receptor for sperm. *Developmental Biology* 106: 243–255. Also H. M. Florman and P. M. Wassarman. 1985. O-linked oligosaccharides of mouse egg ZP3 account for its sperm receptor activity. *Cell* 41: 313–324; J. D. Bleil and P. M. Wassarman. 1988. Galactose at the nonreducing terminus of O-linked oligosaccharides of mouse egg zona pellucida glycoprotein ZP3 is essential for the glycoprotein's sperm receptor activity. *PNAS* 85: 6778–6782.

Chapter 6 **Opener** CHARMM-GUI Archive—Library of Pure Lipid Bilayer (www.charmm-gui.org/?doc=archive&lib=lipid_pure), POPE Bilayer Library (pope_n256.pdb). Reference: S. Jo, T. Kim, and W. Im. 2007. Automated builder and database of protein/membrane complexes for molecular dynamics simulations. *PLoS ONE* 2 (9): e880. **6.10** Data: J. de Gier, J. G. Mandersloot, and L. L. van Deenen. 1968. Lipid composition and permeability of liposomes. *Biochimica et Biophysica Acta* 150: 666–675. **6.21** Data: C. E. Bear, F. Duguay, A. L. Naismith, et al. 1992. Purification and functional reconstitution of the cystic fibrosis transmembrane conductance regulator (CFTR). *Cell* 68: 809–818. **6.22** PDB ID: 2ZZ9. K. Tani, T. Mitsuma, Y. Hiroaki, et al. 2009. Mechanism of aquaporin-4's fast and highly selective water conduction and proton exclusion. *Journal of Molecular Biology* 389: 694–706. **6.23** PDB ID: 1K4C. Y. Zhou, J. H. Morais-Cabral, A. Kaufman, et al. 2001. Chemistry of ion coordination and hydration revealed by a K^+ channel–Fab complex at 2.0 Å resolution. *Nature* 414: 43–48. **6.23** PDB ID: 3FB7. L. G. Cuello, V. Jogini, D. M. Cortes, et al. Open KcsA potassium channel in the presence of Rb^+ ion. (To be published.) **7.18** A. D. Mills, R. A. Laskey, P. Black, et al. 1980. An acidic protein which assembles nucleosomes in vitro is the most abundant protein in *Xenopus* oocyte nuclei. *Journal of Molecular Biology* 139: 561–568. Also C. Dingwall, S. V. Sharnick, and R. A. Laskey. 1982. A polypeptide domain that specifies migration of nucleoplasmin into the nucleus. *Cell* 30: 449–458.

Chapter 7 **7.5** Reprinted with kind permission from Springer Science+Business Media B.V. from David S. Goodsell, *The Machinery of Life*. 2nd ed., 2009.

Chapter 8 **8.8** PDB ID: 1Q18. V. V. Lunin, Y. Li, J. D. Schrag, et al. 2004. Crystal structures of *Escherichia coli* ATP-dependent glucokinase and its complex with glucose. *Journal of Bacteriology* 186: 6915–6927. **8.8** PDB ID: 2Q2R. A. T. Cordeiro, A. J. Caceres, D. Vertommen, et al. 2007. The crystal structure of *Trypanosoma cruzi* glucokinase reveals features determining oligomerization and anomer specificity of hexose-phosphorylating enzymes. *Journal of Molecular Biology* 372: 1215–1226. **8.14a&b** Data: N. N. Nawani, B. P. Kapadnis, A. D. Das, et al. 2002. Purification and characterization of a thermophilic and acidophilic chitinase from Microbispora sp. V2. *Journal of Applied Microbiology* 93: 865–975, Figs. 7, 8a. Also N. N. Nawani and B. P. Kapadnis. 2001. One-step purification of chitinase from *Serratia marcescens* NK1, a soil isolate. *Journal of Applied Microbiology* 90: 803–808, Figs. 3, 4. **8.15** PDB ID: 2ERK. B. J. Canagarajah, A. Khokhlatchev, M. H. Cobb, et al. 1997. Activation mechanism of the MAP kinase ERK2 by dual phosphorylation. *Cell* (Cambridge, MA) 90: 859–869. **8.15** PDB ID 3ERK. Z. Wang, B. J. Canagarajah, J. C. Boehm, et al. 1998. Structural basis of inhibitor selectivity in MAP kinases. *Structure* 6: 1117–1128.

Chapter 9 **9.7** PDB ID: 4PFK. P. R. Evans and P. J. Hudson. 1981. Phosphofructokinase: Structure and control. *Philosophical Transactions R. Soc. Lond. B: Biol. Sci.* 293: 53–62. **9.13** Data: X. Li, R. K. Dash, R. K. Pradhan, et al. 2010. A database of thermodynamic quantities for the reactions of glycolysis and the tricarboxylic acid cycle. *Journal of Physical Chemistry B* 114: 16068–16082, Table 4. **9.14** Data: D.F . Wilson, M. Erecinska, and P. L. Dutton. 1974. Thermodynamic relationships in mitochondrial oxidative phosphorylation. *Annual Review of Biophysics and Bioengineering* 3: 203–230, Tables 1, 3. Also V. D. Sled, N. I. Rudnitzky, Y. Hatefi, et al. 1994. Thermodynamic analysis of flavin in mitochondrial NADH: Ubiquinone oxidoreductase (complex I). *Biochemistry* 33: 10069–10075.

9.17 E. Racker and W. Stoeckenius. 1974. Reconstitution of purple membrane vesicles catalyzing light-driven proton uptake and adenosine triphosphate formation. *Journal of Biological Chemistry* 249: 662–663.

Chapter 10 **10.6** T. W. Engelmann. 1882. Oxygen excretion from plant cells in a microspectrum. *Botanische Zeitung* 40: 419–426. **10.7** Reprinted with kind permission from Springer Science+Business Media B.V. from The Photosynthetic Process. In *Concepts in Photobiology: Photosynthesis and Photomorphogenesis*, edited by G. S. Singhal, G. Renger, S. K. Sopory et al. Dordrecht: Kluwer Academic; co-published with Narosa Publishing House (New Delhi), pp. 11–51, Fig. 5. **10.11** R. Govindjee, Govindjee, and G. Hoch. 1964. Emerson enhancement effect in chloroplast reactions. *Plant Physiology* 39: 10–14. **10.18** A. A. Benson, J. A. Bassham, M. Calvin, et al. 1950. The path of carbon in photosynthesis. V. Paper chromatography and radioautography of the products. *Journal of the American Chemistry Society* 72: 1710–1718. **10.20a** PDB ID: 1RCX. T. C. Taylor and I. Andersson. 1997. The structure of the complex between rubisco and its natural substrate ribulose 1,5-bisphosphate. *Journal of Molecular Biology* 265: 432–444.

Chapter 11 **11.4** Reprinted with kind permission from Springer Science+Business Media B.V. from David S. Goodsell, *The Machinery of Life*. 2nd ed., 2009. **11.7b** B. Alberts, A. Johnson, J. Lewis, et al. 2002. *Molecular Biology of the Cell.* 4th ed., Fig. 19.5, p. 1069. **11.9** K. Hatta and M. Takeichi. 1986. Expression of N-cadherin adhesion molecules associated with early morphogenetic events in chick development. *Nature* 320: 447–449. Also M. Takeichi. 1988. The cadherins: Cell–cell adhesion molecules controlling animal morphogenesis. *Development* 102: 639–655.

Chapter 12 **12.6** G. J. Gorbsky, P. J. Sammack, and G. G. Borisey. 1987. Chromosomes move poleward during anaphase along stationary microtubules that coordinately dissemble from their kinetochore ends. *Journal of Cell Biology* 104: 9–18. **12.9** Based on J. L. Ptacin, S. F. Lee, E. C. Garner, et al. 2010. A spindle-like apparatus guides bacterial chromosome segregation. *Nature Cell Biology* 12: 791–798, Fig. 5. **12.10** Y. Masui and C. L. Markert. 1971. Cytoplasmic control of nuclear behavior during meiotic maturation of frog oocytes. *Journal of Experimental Zoology* 177: 129–145. **12.13** Data: the website of the National Cancer Institute (www.cancer.gov), Common Cancer Types, November 2010.

Chapter 13 **13.13** Data: National Down Syndrome Society, 2012. **13.15** L. T. Morran, O. G. Schmidt, I. A. Gelarden, et al. 2011. Running with the red queen: Host-parasite coevolution selects for biparental sex. *Science* 333: 216–218.

Chapter 14 **14.3** G. Mendel. 1866. Versuche über Pflanzen-hybriden. *Verhandlungen des naturforschenden Vereines in Brünn* 4: 3–47. English translation available from Electronic Scholarly Publishing (www.esp.org). **T14.2** Data: G. Mendel. 1866. Versuche über Pflanzen-hybriden. *Verhandlungen des naturforschenden Vereines in Brünn* 4: 3–47. **14.13** Data: T. H. Morgan. 1911. An attempt to analyze the constitution of the chromosomes on the basis of sex-limited inheritance in *Drosophila*. *Journal of Experimental Zoology* 11: 365–414. **14.20** Data: Pan American Health Organization/WHO. 2004. *Epidemiological Bulletin* 25: 9–13, Graph 1.

Chapter 15 **15.2** A. D. Hershey and M. Chase. 1952. Independent functions of viral protein and nucleic acid in growth of bacteriophage. *Journal of General Physiology* 36: 39–56. **15.5** Adapted by permission of Dr. Matthew Meselson after M. Meselson and F. W. Stahl. 1958. The replication of DNA in *Escherichia coli*. *PNAS* 44: 671–682, Fig. 6. **15.14** Data: R. C. Allsopp, H. Vaziri, C. Patterson, et al. 1992. Telomere length predicts replicative capacity of human fibroblasts. *PNAS* 89: 10114–10118. **15.18a** Data: J. E. Cleaver. 1970. DNA repair and radiation sensitivity in human (xeroderma pigmentosum) cells. *International Journal of Radiation Biology* 18: 557–565, Fig. 3. **15.18b** Data: J .E. Cleaver. 1972. Xeroderma pigmentosum: Variants with normal DNA repair and normal sensitivity to ultraviolet light. *Journal of Investigative Dermatology* 58: 124–128, Fig. 1. **15.UN2** Graph adapted by permission of the Radiation Research Society from P. Howard-Flanders and R. P. Boyce. 1966. DNA repair and genetic recombination: Studies on mutants of *Escherichia coli* defective in these processes. *Radiation Research Supplement* 6: 156–184, Fig. 8.

Chapter 16 **16.2** Data: A. M. Srb and N. H. Horowitz. 1944. The ornithine cycle in *Neurospora* and its genetic control. *Journal of Biological Chemistry* 154: 129–139.

Chapter 17 **17.2a** PDB ID: 3IYD. B. P. Hudson, J. Quispe, S. Lara-Gonzalez, et al. 2009. Three-dimensional EM structure of an intact activator-dependent transcription initiation complex. *PNAS* 106: 19830–19835. **17.11** Based on M. B. Hoagland, M. L. Stephenson, J. F. Scott, et al. 1958. A soluble ribonucleic acid intermediate in protein synthesis. *Journal of Biological Chemistry* 231: 241–257, Fig. 6. **17.13** PDB ID: 1ZJW: I. Gruic-Sovulj, N. Uter, T. Bullock, et al. 2005. tRNA-dependent aminoacyl-adenylate hydrolysis by a nonediting class I aminoacyl-tRNA synthetase. *Journal of Biological Chemistry* 280: 23978–23986. **17.14b** PDB IDs: 3FIK, 3FIH. E. Villa, J. Sengupta, L. G. Trabuco, et al. 2009. Ribosome-induced changes in elongation factor Tu conformation control GTP hydrolysis. *PNAS* 106: 1063–1068.

Chapter 18 **18.2** Data: A. B. Pardee, F. Jacob, and J. Monod. 1959. The genetic control and cytoplasmic expression of "inducibility" in the synthesis of ß-glactosidase by *E. coli*. *Journal of Molecular Biology* 1: 165–178.

Chapter 19 **Opener** PDB ID: 1ZBB. T. Schalch, S. Duda, D. F. Sargent, et al. 2005. X-ray structure of a tetranucleosome and its implications for the chromatin fibre. *Nature* 436: 138–141. **19.5** I. Sandovici, N. H. Smith, and M. D. Nitert. 2011. Maternal diet and aging alter the epigenetic control of a promoter–enhancer interaction at the *Hnf4a* gene in rat pancreatic islets. *PNAS* 108: 5449–5454. **19.7b** PDB ID: 1MDY. P. C. Ma, M. A. Rould, H. Weintraub, et al. 1994. Crystal structure of MyoD bHLH domain–DNA complex: Perspectives on DNA recognition and implications for transcriptional activation. *Cell* (Cambridge, MA) 77: 451–459.

Chapter 21 **21.1** Data: European Nucleotide Archive/EMBL–Bank Release Notes. Release 110, December 2011 (www.ebi.ac.uk/embl/). **21.4a&b** Data: (a) Y. Hou and S. Lin. 2009. *PLoS ONE* 4 (9): e6978, Supplemental Table S1. (b) KEGG: *Kyoto Encyclopedia of Genes and Genomes,* KEGG Organisms: Complete genomes (www.genome.jp/kegg/). **21.8** Reproduced by permission of GENDIA from www.paternity.be/information_EN.html#identitytest, Examples 1 and 2. **21.10** Based on G. T. Ryan. 2005. Synergy between sequence and size in large-scale genomics. *Nature Reviews Genetics* 6, Box 3, p. 702. **21.13** V. Pancaldi, O. S. Sarac, C. Rallis, et al. 2012. Predicting the fission yeast protein interaction network. *G3: Genes, Genomes, Genetics* 2 (4): 453–567, Fig. 5.

Chapter 22 **22.10a&b** Reproduced by permission from Macmillan Publishers Ltd after J. C. Pearson, D. Lemons, and W. McGinnis. 2005. Modulating *Hox* gene functions during animal body patterning. *Nature Reviews Genetics* 6: 893–904, Fig. 1.

Chapter 23 **23.12** H. Weintraub, S. J. Tapscott, R. L. Davis, et al. 1989. Activation of muscle-specific genes in pigment, nerve, fat, liver, and fibroblast cell lines by forced expression of MyoD. *PNAS* 86: 5434–5438.

Chapter 25 **25.4** Based on E. B. Daeschler et al. 2006. A Devonian tetrapod-like fish and the evolution of the tetrapod body plan. *Nature* 440: 757–763, Fig. 6; P. E. Ahlberg and J. A. Clack. 2006. A firm step from water to land. *Nature* 440: 747–749, Fig. 1; N. H. Shubin et al. 2006. The pectoral fin of *Tiktaalik roseae* and the origin of the tetrapod limb. *Nature* 440: 764–771, Fig. 4; M. Hildebrand and G. Goslow. 2001. *Analysis of Vertebrate Structures,* 5th ed. John Wiley and Sons, Inc. **25.11** *Indohyus* reproduced by permission from Macmillan Publishers Ltd after J. G. M. Thewissen, L. N. Cooper, M. T. Clementz, et al. 2007. Whales originated from aquatic artiodactyls in the Eocene epoch of India. *Nature* 450: 1190–1194, Fig. 5. *Rhodhocetus* reproduced by permission of AAAS after P. D. Gingerich, M. ul Haq, I. S. Zalmout, et al. 2001. Origin of whales from early Artiodactyls: Hands and feet of Eocene Protocetidae from Pakistan. *Science* 293: 2239–2242, Fig. 3. (http://www.sciencemag.org/content/293/5538/2239.abstract). *Dorudon* reproduced by permission after P. D. Gingerich, M. ul Haq, W. Von Koenigswald, et al. 2009. New protocetid whale from the middle Eocene of Pakistan: Birth on land, precocial development, and sexual dimorphism. *PLoS ONE* 4 (2): e4366, Fig. 1B. New data for Dorudon from M. D. Uhen. 2004. Form, function, and anatomy of *Dorudon atrox* (Mammalia, Cetacea): An Archaeocete from the Middle to Late Eocene of Egypt. University of Michigan Papesrs on Paleontology No. 34. *Delphinapterus* reproduced by permission of Skulls Unlimited International, Inc. (www.SkullsUnlimited.com). **25.14** Adapted by permission from D. P. Genereux and C. T. Bergstrom. 2005. Evolution in action: Understanding antibiotic resistance. In *Evolutionary Science and Society: Educating a New Generation*, edited by J. Cracraft and R. W. Bybee. Colorado Springs, CO: BSCS, 145–153, Fig 3. Data: Centers for Disease Control, 2004. **25.16** Data: P. T. Boag and P. R. Grant. 1981. Intense natural selection in a population of Darwin's finches (*Geospizinae*) in the Galápagos. *Science New Series* 214 (4516): 82–85, Table 1. **25.17** Body size and beak shape graph reproduced by permission of AAAS from P. R. Grant and B. R. Grant. 2002. *Science* 296:707–711, Fig. 1. Beak size graph reprinted by permission of AAAS from P. R. Grant and B. R. Grant. 2006. Evolution of character displacement in Darwin's finches. *Science* 313: 224–226, Fig. 2. (http://www.sciencemag.org/content/296/5568/707.abstract; http://www.sciencemag.org/content/313/5784/224.abstract).

Chapter 26 **T26.1** Data: W. C. Boyd. 1950. *Genetics and the Races of Man.* Boston: Little, Brown. Data are from Chapter 8, Blood Groups: Table 27, p. 238. **T26.2** Data: T. Markow, P. H. Hedrick, K. Zuerlein, et al. 1993. HLA polymorphism in the Havasupai: Evidence for balancing selection. *American Journal of Human Genetics* 53: 943–952, Table 3. **26.4** Data: W. E. Johnson, E. P. Onorato, M. E. Roelke, et al. 2010. Genetic restoration of the Florida panther. *Science* 329: 1641–1645. **26.5b** Data: C. R. Brown and M. B. Brown. 1998. Intense natural selection on body size and wing and tail asymmetry in cliff swallows during severe weather. *Evolution* 52: 1461–1475. **26.6b** Data: M. N. Karn, H. Lang-Brown, H. J. MacKenzie, et al. 1951. Birth weight, gestation time and survival in sibs. *Annals of Eugenics* 15: 306–322. **26.7b** Data: T. B. Smith. 1987. Bill size polymorphism and intraspecific niche utilization in an African finch. *Nature* 329: 717–719. **26.8b** Data: J. D. Blount, N. B. Metcalf, T. R. Birkhead, et al. 2003. Carotenoid modulation of immune function and sexual attractiveness in zebra finches. *Science* 300: 125–127, Fig. 1b. **26.10 b&c** Data: B. J. Le Boeuf and R. S. Peterson. 1969. Social status and mating activity in elephant seals. *Science* 163: 91–93. **26.12** Reproduced by permission of Pearson Education, Inc., from S. Freeman and J. Herron. 2004. *Evolutionary Analysis.* 3rd ed., Figs. 6.15a, 6.15c. ©2004. **26.13** W. E. Kerr and S. Wright. 1954. Experimental studies of the distribution of gene frequencies in very small populations of *Drosophilia melanogaster*: I. Forked. *Evolution* 8: 172–177. **26.16b** Data: H. Araki, B. Cooper, and M. S. Blouin. 2009. Carry-over effect of captive breeding reduces reproductive fitness of wild-born descendants in the wild. *Biology Letters* 5 (5): 621–624, Fig. 2b. **26.17** Data: S. F. Elena, V. S. Cooper, and R. E. Lenski. 1996. Punctuated evolution caused by selection of rare beneficial mutations. *Science New Series* 272: 1802–1804. **26.18b** N. Moran and T. Jarvik. 2010. Lateral transfer of genes from fungi underlies carotenoid production in aphids. *Science* 328: 624–627. **27.3a** Based on J. C. Avise and W. S. Nelson. 1989. Molecular genetic relationships of the extinct dusky seaside sparrow. *Science* 243 (4891): 646–648, Fig. 2.

Chapter 27 **27.3b** J. C. Avise and W. S. Nelson. 1989. Molecular genetic relationships of the extinct dusky seaside sparrow. *Science* 243 (4891): 646–648. **27.5b** Data: N. Knowlton, L. A. Weigt, L. A. Solorzano, et al. 1993. Divergence in proteins, mitochondrial DNA, and reproductive compatibility across the Isthmus of Panama. *Science* 260 (5114): 1629–1632, Fig. 1. **27.7** Adapted by permission of Wiley-Blackwell from H. R. Dambroski, C. Linn Jr., S. H. Berlocher, et al. 2005. The genetic basis for fruit odor discrimination in *Rhagoletis* flies and its significance for sympatric host shifts. *Evolution* 59 (9): 1953–1964, Figs. 1A, 1B. **27.11b** Data: S. E. Rohwer, E. Bermingham, and C. Wood. 2001. Plumage and mitochondrial DNA haplotype variation across a moving hybrid zone. *Evolution* 55: 405–422. **27.12** Data: L. H. Rieseberg, B. Sinervo, C. R. Linder, et al. 1996. Role of gene interactions in hybrid speciation: Evidence from ancient and experimental hybrids. *Science* 272 (5262): 741–745.

Chapter 28 **28.4c** Data adapted by permission of the publisher from M. A. Nikaido, P. Rooney, and N. Okada. 1999. Phylogenetic relationships among cetartiodactyls based on insertions of short and long interspersed elements: Hippopotamuses are the closest extant relatives of whales. *PNAS* 96: 10261–10266, Figs. 1, 7. ©1999 National Academy of Sciences, USA. **28.6, 28.7, 28.8** Data: International Commission on Stratigraphy, "International Stratigraphic Chart, 2009" (www.stratigraphy.org/column.php?id=Chart/TimeScale). The data on this site are modified from F. M. Gradstein and J. C. Ogg (eds.). 2004. *A Geologic Time Scale 2004*. Cambridge, UK: Cambridge University Press; and J. G. Ogg, G. Ogg, and F. M. Gradstein. 2008. *The Concise Geologic Time Scale.* Cambridge, UK: Cambridge University Press. **28.9** B. G. Baldwin, D. W. Kyhos, and J. Dvorak. 1990. Chloroplast DNA evolution and adaptive radiation in the Hawaiian Silversword Alliance (*Asteraceae-Madiinae*). *Annals of the Missouri Botanical Garden* 77 (1): 96–109, Fig. 2. **28.10c** J. B Losos, K. I. Warheit, and T. W. Schoener. 1997. Adaptive differentiation following experimental island colonization in Anolis lizards. *Nature* 387: 70–73. **28.13** Data: J. W. Valentine, D. H. Erwin, and D. Jablonski. 1996. Developmental evolution of metazoan body plans: The fossil evidence. *Developmental Biology* 173: 373–381, Fig. 3. Also R. de Rosa et al. 1999. *Hox* genes in brachiopods and priapulids and protostome evolution. *Nature* 399: 772–776; D. Chourrout, F. Delsuc, P. Chourrout, et al. 2006. Minimal *ProtoHox* cluster inferred from bilaterian and cnidarian *Hox* complements. *Nature* 442: 684–687. **28.14** Data: M. J. Benton. 1995. Diversification and extinction in the history of life. *Science* 268: 52–58.

Chapter 29 **29.12** Data: D. F. Wilson, M. Erecinska, and P. L. Dutton. 1974. Thermodynamic relationships in mitochondrial oxidative phosphorylation. *Annual Review of Biophysics and Bioengineering* 3: 203–230, Tables 1, 3. **30.1** S. M. Adl, A. G. Simpson, M. A. Farmer, et al. 2005. The new higher level classification of eukaryotes with emphasis on the taxonomy of protists. *Journal of Eukaryotic Microbiology* 52: 399–451. Also N. Arisue, M. Hasegawa, and T. Hashimoto. 2005. Root of the eukaryota tree as inferred from combined maximum likelihood analyses of multiple molecular sequence data. *Society for Molecular Biology and Evolution, Molecular Biology and Evolution* 22: 409–420; V. Hampl, L. Hug, J. W. Leigh, et al. 2009. Phylogenomic analyses support the monophyly of Excavata and resolve relationships among eukaryotic "supergroups." *PNAS* 106 (10): 3859–3864, Figs. 1, 2, 3; J. D. Hackett, H. S. Yoon, S. Li, et al. 2007. Phylogenomic analysis supports the monophyly of cryptophytes and haptophytes and the association of Rhizaria with chromalveolates. *Molecular Biology and Evolution* 24: 1702–1713, Fig. 1; P. Schaap, T. Winckler, M. Nelson, et al. 2006. Molecular phylogeny and evolution of morphology in the social amoebas. *Science* 314: 661–663.

Chapter 30 **30.1, 30.7** S. M. Adl, A. G. Simpson, M. A. Farmer, et al. 2005. The new higher level classification of eukaryotes with emphasis on the taxonomy of protists. *Journal of Eukaryotic Microbiology* 52: 399–451. Also N. Arisue, M. Hasegawa, and T. Hashimoto. 2005. Root of the eukaryota tree as inferred from combined maximum likelihood analyses of multiple molecular sequence data. *Society for Molecular Biology and Evolution, Molecular Biology and Evolution* 22: 409–420; V. Hampl, L. Hug, J. W. Leigh, et al. 2009. Phylogenomic analyses support the monophyly of Excavata and resolve relationships among eukaryotic "supergroups." *PNAS* 106 (10): 3859–3864, Figs. 1, 2, 3; J. D. Hackett, H. S. Yoon, S. Li, et al. 2007. Phylogenomic analysis supports the monophyly of cryptophytes and haptophytes and the association of Rhizaria with chromalveolates. *Molecular Biology and Evolution* 24: 1702–1713, Fig. 1; P. Schaap, T. Winckler, M. Nelson, et al. 2006. Molecular phylogeny and evolution of morphology in the social amoebas. *Science* 314: 661–663.

Chapter 31 **31.6, 31.9, 31.23** Y.-L. Qiu, L. Li, B. Wang, et al. 2006. The deepest divergences in land plants inferred from phylogenomic evidence. *PNAS* 103 (42): 15511–15516, Fig. 1. Also K. S. Renzaglia, S. Schuette, R. J. Duff, et al. 2007. Bryophyte phylogeny: Advancing the molecular and morphological frontiers. American Bryological and Lichenological Society, *The Bryologist* 110 (2): 179–213; J.-F. Pombert, C. Otis, C. Lemieux, et al. 2005. The chloroplast genome sequence of the green alga *Pseudendoclonium akinetum* (*Ulvophyceae*) reveals unusual structural features and new insights into the branching order of chlorophyte lineages. *Molecular Biology and Evolution* 22: 1903–1918. **31.21** Based on M. E. Hoballah, T. Gubitz, J. Stuurman, et al. 2007. Single gene-mediated shift in pollinator attraction in Petunia. *Plant Cell* 19: 779–790, Fig. 6g, 6h. **31.25** P. S. Soltis and D. E. Soltis. 2004. The origin and diversification of angiosperms. *American Journal of Botany* 91 (10): 1614–1626, Figs. 1, 2, 3.

Chapter 32 **32.7** S. M. Adl, A. G. Simpson, M. A. Farmer, et al. 2005. The new higher level classification of eukaryotes with emphasis on the taxonomy of protists. *Journal of Eukaryotic Microbiology* 52: 399–451. Also N. Arisue, M. Hasegawa, and T. Hashimoto. 2005. Root of the Eukaryota tree as inferred from combined maximum likelihood analyses of multiple molecular sequence data. *Molecular Biology and Evolution* 22: 409–420; V. Hampl, L. Hug, J. W. Leigh, et al. 2009. Phylogenomic analyses support the monophyly of Excavata and resolve relationships among eukaryotic "supergroups." *PNAS* 106: 3859–3864, Figs. 1, 2, 3; J. D. Hackett, H. S. Yoon, S. Li, et al. 2007. Phylogenomic analysis supports the monophyly of cryptophytes and haptophytes and the association of Rhizaria with chromalveolates. *Molecular Biology and Evolution* 24: 1702–1713, Fig 1; P. Schaap, T. Winckler, M. Nelson, et al. 2006. Molecular phylogeny and evolution of morphology in the social amoebas. *Science* 314: 661–663. **32.8** T. Y. James, F. Kauff, C. L. Schoch, et. al. 2006. Reconstructing the early evolution of fungi using a six-gene phylogeny. *Nature* 443: 818–822, Fig. 1. Also D. J. McLaughlin, D. S. Hibbett, F. Lutzoni, et al. 2009. The search for the fungal tree of life. *Trends in Microbiology* 17: 488–497; D. S. Hibbett, M. Binder, J. F. Bischoff, et al. 2007. A higher-level phylogenetic classification of the Fungi. *Mycological Research* 111: 509–547; H. Wang, X. Zhao, L. Gao, et al. 2009. A fungal phylogeny based on 82 complete genomes using the composition vector method. *BMC Evolutionary Biology* 9: 195. **32.10** After H. Bücking and W. Heyser. 2001. Microautoradiographic localization of phosphate and carbohydrates in mycorrhizal roots of *Populus tremula* x *Populus alba* and the implications for transfer processes in ectmycorrhizal associations. *Tree Physiology* 21: 101–107.

Chapter 33 **33.2** A. M. A. Aguinaldo, J. M. Turbeville, L. S. Linford, et al. 1997. Evidence for a clade of nematodes, arthropods, and other moulting animals. *Nature* 387: 489–493, Figs. 1, 2, 3. Also C. W. Dunn, A. Hejnol, D. Q. Matus, et al. 2008. Broad phylogenomic sampling improves resolution of the animal tree of life. *Nature* 452: 745–750, Figs. 1, 2. **33.7** Based on J. R. Finnerty, K. Pang, P. Burton, et al. 2004. Origins of bilateral symmetry: *Hox* and *Dpp* expression in a sea anemone. *Science* 304: 1335–1337.

Chapter 34 **34.1** A. D. Chapman. 2009. *Numbers of Living Species in Australia and the World*. 2nd ed. Canberra: Australian Biological Resources Study. **34.2** A. M. A. Aguinaldo, J. M. Turbeville, L. S. Linford, et al. 1997. Evidence for a clade of nematodes, arthropods, and other moulting animals. *Nature* 387: 489–493, Figs. 1, 2, 3. Also C. W. Dunn, A. Hejnol, D. Q. Matus, et al. 2008. Broad phylogenomic sampling improves resolution of the animal tree of life. *Nature* 452: 745–750, Figs. 1, 2. **34.28, 34.30** Based on data in J. Regier, J. W. Schultz, A. Zwick, et al. 2010. Arthropod relationships revealed by phylogenomic analysis of nuclear protein-coding sequences. *Nature* 463: 1079–1083. **34.UN1** Based on M. Averof and N. H. Patel. 1997. Crustacean appendage evolution associated with changes in *Hox* gene expression. *Nature* 388: 682–686, Fig. 4.

Chapter 35 **35.1** After C. W. Dunn, A. Hejnol. D. Q. Matus, et al. 2008. Broad phylogenomic sampling improves resolution of the animal tree of life. *Nature* 452: 745–750, Figs. 1, 2. Also F. Delsuc, H. Brinkmann, D. Chourrout, et al. 2006. Tunicates and not cephalochordates are the closest living relatives of vertebrates. *Nature* 439: 965–968, Fig. 1. **35.12** J. E. Blair and S. B. Hedges. 2005. Molecular phylogeny and divergence times of deuterostome animals. *Molecular Biology and Evolution* 22: 2275–2284, Figs. 1, 3, 4. **35.13** Data: American Museum of Natural History; BirdLife International (birdlife.org); FishBase (www.fishbase.org); IUCN 2010. *The IUCN Red List of Threatened Species* (http://www.iucnredlist.org); The Reptile Database (reptile-database.org). **35.16** Based on E. B. Daeschler, N. H. Shubin, F. A. Jenkins, et al. 2006. A Devonian tetrapod-like fish and the evolution of the tetrapod body plan. *Nature* 440: 757–763, Fig. 6. Also N. H. Shubin, E. B. Daeschler, F. A. Jenkins, et al. 2006. The pectoral fin of *Tiktaalik roseae* and the origin of the tetrapod limb. *Nature* 440:764–771, Fig. 4. **35.20** Color plate by Michael DiGiorgio. From Q. Li, K.-Q. Gao, J. Vinther, et al. 2010. Plumage color patterns of an extinct dinosaur. *Science* 327, Fig. 4, p. 1371. Reprinted by permission of the artist, Michael DiGiorgio. **35.22** J. E. Blair and S. B. Hedges. 2005. Molecular phylogeny and divergence times of deuterostome animals. *Molecular Biology and Evolution*, 22: 2275–2284, Figs. 1, 3, 4. Also F. R. Liu, M. M. Miyamoto, N. P. Freire, et al. 2001. Molecular and morphological supertrees for eutherian (placental) mammals. *Science*

291: 1786–1789; A. B. Prasad, M. W. Allard, E. D. Green, et al. 2008. Confirming the phylogeny of mammals by use of large comparative sequence data sets. Oxford University Press, *Molecular Biology and Evolution* 25: 1795–1808. **35.36** A. B. Prasad, M. W. Allard, E. D. Green, et al. 2008. Confirming the phylogeny of mammals by use of large comparative sequence data sets. Oxford University Press, *Molecular Biology and Evolution* 25 (9):1795–1808, Figs. 1, 2, 3. **35.37** Data: D. E. Lieberman. 2009. Palaeoanthropology: *Homo floresiensis* from head to toe. *Nature* 459: 41–42. **T35.1** S. L. Robson and B. Wood. 2008. Hominin life history: Reconstruction and evolution. *Journal of Anatomy* 212: 394–425. **35.39** J. Z. Li, D. M. Absher, H. Tang, et al. 2008. Worldwide human relationships inferred from genome-wide patterns of variation. *Science* 319: 1100–1104, Fig. 1. **35.40** After L. L. Cavalli-Sforza and M. W. Feldman. 2003. The application of molecular genetic approaches to the study of human evolution. *Nature Genetics Supplement* 33: 266–275, Fig. 3; Data: Genographic Project at National Geographic: https://genographic.nationalgeographic.com/genographic/lan/en/atlas.html; dates from August 2009 summary map: http://ngm.nationalgeographic.com/big-idea/02/queens-genes; M. Rasmussen, X. Guo, Y. Wang, et al. 2011. An Aboriginal Australian genome reveals separate human dispersals into Asia. *Science* 334: 94–98.

Chapter 36 **36.2a** Data: E. Arias. 2010. United States life tables, 2006. *National Vital Statistics Reports* 58 (21): 1–40, Table 10. Hyattsville, MD: National Center for Health Statistics. **36.3** Adapted with permission of Annual Reviews from G. Pantaleo and A. S. Fauci. 1996. Immunopathogensis of HIV infection. *Annual Review of Microbiology* 50: 825–854. Fig. 1a, p. 829. **36.7** Data courtesy of the Undergraduate Biotechnology Laboratory, California Polytechnic State University, San Luis Obispo. **36.8** A. G. Dalgleish, P. C. Beverley, P. R. Clapham, et al. 1984. The CD4 (T4) antigen is an essential component of the receptor for the AIDS retrovirus. *Nature* 312: 763–767. Also D. Klatzmann, E. Champagne, S. Chamaret, et al. 1984. T-lymphocyte T4 molecule behaves as the receptor for human retrovirus LAV. *Nature* 312: 767–768. **36.15** F. Gao, E. Bailes, D. L. Robertson, et al. 1999. Origin of HIV-1 in the chimpanzee *Pan troglodytes troglodytes*. *Nature* 397: 436–441, Fig. 2.

Chapter 37 **37.3** Illustration adapted by permission from "Root Systems of Prairie Plants," by Heidi Natura. ©1995 Conservation Research Institute. **37.5** After J. Clausen, D. D. Keck, and M. Hiesey. 1945. Experimental studies on the nature of species II: Plant evolution through amphiploidy and autoploidy, with examples from the *Madiinae*. Publication No. 564. Washington, DC: Carnegie Institute of Washington.

Chapter 38 **38.3** Data: R. G. Cline and W. S. Campbell. 1976. Seasonal and diurnal water relations of selected forest species. *Ecology* 57: 367–373, Fig. 1. **38.12** Adapted by permission of the American Society of Plant Biologists from C. M. Wei, M. T. Tyree, and E. Steudle. 1999. Direct measurement of xylem pressure in leaves of intact maize plants. A test of the cohesion-tension theory taking hydraulic architecture into consideration. *Plant Physiology* 121: 1191–1205, Fig. 6. ©1999 by American Society of Plant Biologists. **38.22** J. R. Gottwald, P. J. Krysan, J. C. Young, et al. 2000. Genetic evidence for the in planta role of phloem-specific plasma membrane sucrose transporters. *PNAS* 97: 13979–13984.

Chapter 39 **39.3** D. I. Arnon and P. R. Stout. 1939. The essentiality of certain elements in minute quantity for plants with special reference to copper. *Plant Physiology* 14: 371–375.

Chapter 40 **40.4** Based on S. Inoue, T. Kinoshita, M. Matsumoto, et al. 2008. Blue light-induced autophosphorylation of phototropin is a primary step for signaling. *PNAS* 105: 5626–5631. **40.5** After C. Darwin and F. Darwin. 1897. *The Power of Movement in Plants*. New York: D. Appleton. **40.10** After A. Lang, M. K. Chailakhyan, and I. A. Frolova. 1977. Promotion and inhibition of flower formation in a day-neutral plant in grafts with a short-day plant and a long-day plant. *PNAS* 74: 2412–2416. **T40.1** H. A. Borthwick, S. B. Hendricks, M. W. Parker, et al. 1952. A reversible photoreaction controlling seed germination. *PNAS* 38: 662–666, Table 1. **40.21** After P. G. Blackman and W. J. Davies. 1985. Root to shoot communication in maize plants of the effect of soil drying. *Journal of Experimental Botany* 36: 39–48.

Chapter 41 **41.10** Y.-L. Qiu, L. Li, B. Wang, et al. 2006. The deepest divergences in land plants inferred from phylogenomic evidence. *PNAS* 103 (42): 15511–15516, Fig. 1. Also K. S. Renzaglia, S. Schuette, R. J. Duff, et al. 2007. Bryophyte phylogeny: Advancing the molecular and morphological frontiers. American Bryological and Lichenological Society, *The Bryologist* 110 (2): 179–213; J.-F. Pombert, C. Otis, C. Lemieux, et al. 2005. The chloroplast genome sequence of the green alga *Pseudendoclonium akinetum* (*Ulvophyceae*) reveals unusual structural features and new insights into the branching order of chlorophyte lineages. *Molecular Biology and Evolution* 22: 1903–1918. **41.17** J. J. Tewksbury and G. P. Nabhan. 2001. Directed deterrence by capsaicin in chilies. *Nature* 412: 403–404.

Chapter 42 **42.2** A. M. Kerr, S. N. Gershman, and S. K. Sakaluk. 2010. Experimentally induced spermatophore production and immune responses reveal a trade-off in crickets. *Behavioral Ecology* 21: 647–654. **42.10** Data: K. Schmidt-Nielsen. 1984. *Scaling: Why Is Animal Size So Important?* Cambridge, UK: Cambridge University Press. **42.11** Adapted by permission of The Company of Biologists from

P. R. Wells and A. W. Pinder. 1996. The respiratory development of Atlantic salmon. II. Partitioning of oxygen uptake among gills, yolk sac and body surfaces. *Journal of Experimental Biology* 199: 2737–2744, Figs. 1A, 3.

Chapter 43 **43.13** Anterolateral view reproduced by permission from Macmillan Publishers Ltd after S. E. Pierce, J. A. Clack, and J. R. Hutchinson. 2012. Three-dimensional limb joint mobility in the early tetrapod *Ichthyostega*. *Nature* 486: Fig. 1, p. 523. **43.14** Adapted by permission of Elsevier from K. J. Ullrich, K. Kramer, and J. W. Boyer. 1961. Present knowledge of the counter-current system in the mammalian kidney. *Progress in Cardiovascular Diseases* 3: 395–431, Figs. 5, 7. ©1961. **43.19** J. R. Davis and D. F. DeNardo. 2007. The urinary bladder as a physiological reservoir that moderates dehydration in a large desert lizard, the Gila monster *Heloderma suspectum*. *Journal of Experimental Biology* 210 (8): 1472–1480. Reprinted with permission from The Company of Biologists.

Chapter 44 **44.14** After E. M. Wright. 1993. The intestinal Na$^+$/glucose cotransporter. *Annual Review of Physiology* 55: 575–589. Also K. Takata, T. Kasahara, M. Kasahar, et al. 1992. Immunohistochemical localization of Na$^+$-dependent glucose transporter in rat jejunum. *Cell and Tissue Research* 267: 3–9. **44.17a** Graph reproduced with permission from The American Diabetes Association from L. O. Schulz, P. H. Bennett, E. Ravussin, et al. 2006. Effects of traditional and Western environments on prevalence of type 2 diabetes in Pima Indians in Mexico and the U.S. *Diabetes Care* 29: 1866–1871, Fig. 1. ©2006 American Diabetes Association. **44.17b** Data: L. O. Schulz, P. H. Bennett, E. Ravussin, et al. 2006. Effects of traditional and Western environments on prevalence of type 2 diabetes in Pima Indians in Mexico and the U.S. *Diabetes Care* 29: 1866–1871, Table 2.

Chapter 45 **45.7** Based on Y. Komai. 1998. Augmented respiration in a flying insect. *Journal of Experimental Biology* 201: 2359–2366, Fig. 7. **45.9** R. A. Berner. 1999. Atmospheric oxygen over Phanerozoic time. *PNAS* 96: 10955–10957. Fig. 2. Copyright ©1999 National Academy of Sciences, USA. Reprinted with permission. **45.12** W. Bretz and K. Schmidt Nielsen. 1971. Bird respiration: Flow patterns in the duck lung. *Journal of Experimental Biology* 54: 103–118. **45.21** Data: E. H. Starling. 1896. On the absorption of fluids from the connective tissue spaces. *Journal of Physiology* 19: 312–326. **45.23** J. E. Blair and S. B. Hedges. 2005. Molecular phylogeny and divergence times of deuterostome animals. *Molecular Biology and Evolution* 22: 2275–2284, Figs. 1, 3, 4.

Chapter 46 **46.10** Data: O. Loewi. 1921. Über humorale Übertragbarkeit der Herznervenwirkung. *Pflügers Archiv European Journal of Physiology* 189: 239–242. **46.22** Data: R. Q. Quiroga, L. Reddy, G. Kreiman, et al. 2005. Invariant visual representation by single neurons in the human brain. *Nature* 435: 1102–1107, Fig. 1B. **46.24a&c** Based on K. C. Martin, A. Casadio, H. Zhu, et al. 1997. Synapse-specific, long-term facilitation of *Aplysia* sensory to motor synapses: A function for local protein synthesis in memory storage. *Cell* 91: 927–938.

Chapter 47 **47.2a&b** Data: J. E. Rose, J. E. Hind, D. J. Anderson, et al. 1971. Some effects of stimulus intensity on response of auditory nerve fibers in the squirrel monkey. *Journal of Neurophysiology* 34: 685–699. **47.9** K. Pohlmann, J. Atema, and T. Breithaupt. 2004. The importance of the lateral line in nocturnal predation of piscivorous catfish. *Journal of Experimental Biology* 207: 2971–2978. **47.15** Based on D. M. Hunt, K. S. Dulai, J. K. Bowmaker, et al. 1995. The chemistry of John Dalton's color blindness. *Science* 267: 984–988, Fig. 3. **47.UN1** Reprinted by permission from Macmillan Publishers Ltd. from M. K. McClintock. 1971. Menstrual synchrony and suppression. *Nature* 229: 244–245, Fig. 1.

Chapter 48 **48.3** PDB ID: 1KWO. D. M. Himmel, S. Gourinath, L. Reshetnikova, et al. 2002. Crystallographic findings on the internally uncoupled and near-rigor states of myosin: Further insights into the mechanics of the motor. *PNAS* 99: 12645–12650. **48.7** After Exercise Information—Interactive Muscle Map (www.askthetrainer.com/exercise-information/). **48.15** Reprinted by permission from Macmillan Publishers Ltd from D. F. Hoyt and C. R. Taylor. 1981. Gait and the energetics of locomotion in horses. *Nature* 292: 239–240, Fig. 2, p. 240. **48.16** S. Vogel. 2003. *Comparative Biomechanics: Life's Physical World*. 2nd ed. Princeton, NJ: Princeton University Press, Fig. 3.5, p. 60. Reprinted by permission of Princeton University Press. **48.17** Reprinted with permission from AAAS from K. Schmidt-Nielsen. 1972. Locomotion: Energy cost of swimming, flying, and running. *Science* 177: 222–228. (http://www.sciencemag.org/content/177/4045/222.extract).

Chapter 49 **49.4** W. M. Bayliss and E. H. Starling. 1902. The mechanism of pancreatic secretion. *Journal of Physiology* 28: 325–353. **49.7** Graph from J. Vanecek. 1998. Cellular mechanisms of melatonin action. *Psychological Reviews* 78: 687–721, Fig. 1, p. 688. Original Data: H. Illnerová, K. Hoffmann, J. Vanácek, et al. 1984. Adjustment of pineal melatonin and N-acetyltransferase rhythms to change from long to short photoperiod in the Djungarian hamster Phodopus sungorus. *Neuroendocrinology* 38: 226–231. **49.9** Data: B. D. King, L. Sokoloff, and R. L. Wechsler. 1952. The effects of l-epinephrine and l-nor-epinephrine upon cerebral circulation and metabolism in man. *Journal of Clinical Investigation* 31: 273–279. Also P. S. Mueller and D. Horwitz. 1962. Plasma free fatty acid and blood glucose responses to analogues of norepinephrine in man. *Journal of Lipid Research* 3: 251–255. **49.11** Reprinted by permission from Macmillan Publishers Ltd. from D. L. Coleman. 2010. A historical perspective on Leptin. *Nature Medicine* 16: 1097–1099, Fig. 2. **49.12a&b** Data: G. V. Upton, A. Corbin, C. C. Mabry, et al. 1973. Evidence for the internal feedback phenomenon in human subjects: Effects of ACTH on plasma CRF. *Acta Endocrinologica* 73: 437–443. **48.13** Anterolateral view reproduced by permission from Macmillan Publishers Ltd after S. E. Pierce, J. A. Clack, and J. R. Hutchinson. 2012. Three-dimensional limb joint mobility in the early tetrapod Ichthyostega. *Nature* 486: Fig. 1, p. 523. **49.14** Data: D. Toft and J. Gorski. 1966. A receptor molecule for estrogens: Isolation from the rat uterus and preliminary characterization. *PNAS* 55: 1574–1581. **49.16b** Data: T. W. Rall, E. W. Sutherland, and J. Berthet, et al. 1957. The relationship of epinephrine and glucagon to liver phosphorylase. *Journal of Biological Chemistry* 224: 463–475. **49.17** Reproduced from E. W. Sutherland. 1971. Studies on the mechanism of hormone action. (A lecture delivered in Stockholm, December 11, 1971.) ©1971 Nobel Foundation; E. W. Sutherland. 1972. Studies on the mechanism of hormone action. *Science* 177: 401–408, Fig. 6. Also G. A. Robison, R. W. Butcher, I. Oye, et al. 1965. The effect of epinephrine on adenosine 3′,5′-phosphate levels in the isolated perfused rat heart. *Molecular Pharmacology* 1: 168–177. **49.UN1** Data: T. B. Hayes, A. Collins, M. Lee, et al. 2002. Hermaphroditic, demasculinized frogs after exposure to the herbicide atrazine at low ecologically relevant doses. *PNAS* 99: 5476–5480.

Chapter 50 **50.3a** Based on R. G. Stross and J. C. Hill. 1965. Diapause induction in *Daphnia* requires two stimuli. *Science* 150: 1462–1464, Fig. 3. **50.3b** Data: O. T. Kleiven, P. Larsson, and A. Hobæk. 1992. Sexual reproduction in *Daphnia magna* requires three stimuli. *Oikos* 65: 197–206, Table 4. **50.7** Based on C. S. C. Price, K. A. Dyer, and J. A. Coyne. 1999. Sperm competition between *Drosophila* males involves both displacement and incapacitation. *Nature* 400: 449–452, Fig. 3c. **50.9** R. Shine and M. S. Y. Lee. 1999. A reanalysis of the evolution of viviparity and egg-guarding in squamate reptiles. The Herpetologists' League, Inc., *Herpetologica* 55 (4): 538–549, Figs. 1, 2, 3. **50.10b** Reproduced by permission of Elsevier from C. Hotzy and G. Arnqvist. 2009. Sperm competition favors harmful males in seed beetles. *Current Biology* 19: 404–407, Fig. 2. ©2009. **50.17, 50.18** Data: R. Stricker, R. Eberhart, M.-C. Chevallier, et al. 2006. Establishment of detailed reference values for luteinizing hormone, follicle stimulating hormone, estradiol, and progesterone during different phases of the menstrual cycle on the Abbott ARCHITECT* analyzer. *Clinical Chemistry and Laboratory Medicine* 44: 883–887, Tables 1A, 1B. **50.22b** Data: V. W. Swayze II., V. P. Johnson, J. W. Hanson, et al. 1997. Magnetic resonance imaging of brain anomalies in fetal alcohol syndrome. *Pediatrics* 99: 232–240. **50.24** Adapted by permission of the publisher and author from U. Högberg and I. Joelsson. 1985. The decline in maternal mortality in Sweden, 1931–1980. *Acta Obstetricia et Gynecologica Scandinavica* 64: 583–592, Fig. 1. Taylor & Francis Group, http://www.informaworld.com.

Chapter 51 **51.2** Based on B. Lemaitre, E. Nicolas, L. Michaut, et al. 1996. The dorsoventral regulatory gene cassette *spätzle/Toll/cactus* controls the potent antifungal response in *Drosophila* adults. *Cell* 86: 973–983. **51.5a** PDB ID: 1IGT. L. J. Harris, S. B. Larson, K. W. Hasel, et al. 1997. Refined structure of an intact IgG2a monoclonal antibody. *Biochemistry* 36: 1581–1597. **51.5b** PDB ID: 1TCR. K. C. Garcia, M. Degano, R. L. Stanfield, et al. 1996. An aß T cell receptor structure at 2.5 Å and its orientation in the TCR–MHC complex. *Science* 274: 209–219. **51.6** PDB ID: 1HMG. M. Knossow, R. S. Daniels, A. R. Douglas, et al. 1984. Three-dimensional structure of an antigenic mutant of the influenza virus haemagglutinin. *Nature* 311 (5987): 678–680. **51.10** PDB ID: 1DLH. L. J. Stern, J. H. Brown, T. S. Jardetzky, et al. 1994. Crystal structure of the human class II MHC protein HLA–DR1 complexed with an influenza virus peptide. *Nature* 368: 215–221. **51.17** Based on B. Alberts, A. Johnson, J. Lewis, et al. 2002. *Molecular Biology of the Cell*. 4th ed. New York: Garland Science, Taylor & Francis Group. Modified from Fig. 24-10. **52.1** Data: Food and Agricultural Organization of the United Nations.

Chapter 52 **52.4** S. B. Menke, R. N. Fisher, W. Jetz, et al. 2007. Biotic and abiotic controls of Argentine ant invasion success at local and landscape scales. *Ecology* 88 (12): 3164–3173. **52.9** D. M. Olson, E. Dinerstein, E. D. Wikramanayake, et al. 2001. Terrestrial ecoregions of the world: A new map of life on Earth. *BioScience* 51: 933–938. **52.11, 52.12, 52.13, 52.14, 52.15, 52.16** Data: E. Aguado and J. E. Burt. 1999. *Understanding Weather and Climate*. Upper Saddle River, NJ: Prentice Hall. **52.18a** Based on M. B. Saffo. 1987. New light on seaweeds. *BioScience* 37 (9): 654–664, Fig. 1. ©1987 American Institute of Biological Sciences.

Chapter 53 **53.3** After M. Sokolowski. 2001. *Drosophila*: Genetics meets behavior. *Nature Reviews Genetics* 2: 879–890, Fig. 2. **53.4** Z. Abramsky, M. L. Rosenzweig, and A. Subach. 2002. The costs of apprehensive foraging. *Ecology* 85: 1330–1340. **53.5b** Adapted by permission of the author from D. Crews. 1975. Psychobiology of reptilian reproduction. *Science* 189: 1059–1065, Fig. 1. **53.6** Adapted by permission of the author from D. Crews. 1975. Psychobiology of reptilian reproduction. *Science* 189: 1059–1065, Fig. 2. **53.7** Reprinted by permission from Macmillan Publishers Ltd from K. J. Lohmann, C. M. F. Lohmann, L. M. Ehrhart, et al. 2004. Geomagnetic map used in sea-turtle navigation. *Nature* 428: 909. **53.8** C. Egevang, et al. 2010. Tracking of

Arctic terns *Sterna paradisaea* reveals longest animal migration. *PNAS* 107(5): 2078–2081, Fig. 1B. ©2010 National Academy of Sciences, USA. Reprinted with permission. **53.15** Data: J. L. Hoogland. 1983. Nepotism and alarm calling in the black-tailed prairie dog (*Cynomys ludovicianus*). *Animal Behavior* 31: 472–479.

Chapter 54 **T54.1** Data reproduced from H. Strijbosch and R. C. M. Creemers. 1988. Comparative demography of sympatric populations of *Lacerta vivipara* and *Lacerta agilis*. *Oecologia* 76: 20–26. **54.5b** Data: G. F. Gause. 1934. *The Struggle for Existence*. Baltimore, MD: Williams & Wilkins. **54.6a** G. E. Forrester and M. A. Steele. 2000. Variation in the presence and cause of density-dependent mortality in three species of reef fishes. *Ecology* 81 (9): 2416–2427, Fig. 5. Copyright © 2000 by the Ecological Society of America. Reprinted with permission. **54.6b** Data: P. Arcese and J. N. M. Smith. 1988. Effects of population density and supplemental food on reproduction in song sparrows. *Journal of Animal Ecology* 57: 119–136, Fig. 2. **54.9** Data: D. A. MacLulich. 1937. Fluctuations in the numbers of the varying hare (*Lepus americanus*). *University of Toronto Studies Biological Series* 43. **54.10** Graphs reproduced by permission of AAAS from C. J. Krebs, S. Boutin, R. Boonstra, et al. 1995. Impact of food and predation on the snowshoe hare cycle. *Science* 269: 1112–1115, Fig. 3. (http://www.sciencemag.org/content/269/5227/1112). **54.11a** Data: United Nations, Department of Economic and Social Affairs, Population Division. 2011. *World Population Prospects: The 2010 Revision*. **54.11b** Data: United Nations, Department of Economic and Social Affairs, Population Division. 2011. *World Population Prospects: The 2010 Revision*. **54.13** United Nations, Department of Economic and Social Affairs, Population Division. 2011. *World Population Prospects: The 2010 Revision*.

Chapter 55 **55.2a&b, 55.3a&b** G. F. Gause. 1934. *The Struggle for Existence*. Baltimore. MD: Williams & Wilkins. **55.5** Data: J. H. Connell. 1961. The influence of interspecific competition and other factors on the distribution of the barnacle *Chthamalus stellatus*. *Ecology* 42: 710–723. **55.7** P. R. Grant. 1999. *Ecology and Evolution of Darwin's Finches*. ©1986 Princeton University Press. Fig. 15, p. 78. Reprinted by permission of Princeton University Press. **55.10** Adapted by permission of Ecological Society of America from G. H. Leonard, M. D. Bertness, and P. O. Yund. 1999. Crab predation, waterborne cues, and inducible defenses in the blue mussel, *Mytilus edulis*. *Ecology* 80: 1–14. **55.14** Based on J. H. Cushman and T. G. Whitham. 1989. Conditional mutualism in a membracid–ant association: temporal, age-specific, and density-dependent effects. *Ecology* 70 (4): 1040–1047. **55.15** Based on D. G. Jenkins and A. L. Buikema. 1998. Do similar communities develop in similar sites? A test with zooplankton structure and function. *Ecological Monographs* 68: 421–443. **55.16** Adapted with kind permission from Springer Science+Business Media B.V. from R. T. Paine. 1974. Intertidal community structure. *Oecologia* 15: 93–120. ©1974 Springer-Verlag. **55.18b** Data: T. W. Swetnam. 1993. Fire history and climate change in giant sequoia groves. *Science* 262: 885–889. **55.21a** Adapted by permission of Wiley-Blackwell from R. H. MacArthur and E. O. Wilson. 1963. An equilibrium theory of insular zoogeography. *Evolution* 17 (4): 373–387, Fig. 4. **55.21 b & c** Data: R. H. MacArthur and E. O. Wilson. 1963. An equilibrium theory of insular zoogeography. *Evolution* 17 (4): 373–387, Fig. 5. **55.23** Data: W. V. Reid and K. R. Miller. 1989. *Keeping Options Alive: The Scientific Basis for Conserving Biodiversity*. Washington, DC: World Resources Institute. **56.10** Data: H. Leith and R. H. Whittaker (eds.). 1975. *Primary Productivity in the Biosphere. Ecological Studies and Synthesis*, Vol. 14. New York: Springer-Verlag.

Chapter 56 **56.5** Based on D. E. Canfield, A. N. Glazer, and P. G. Falkowski. 2010. The evolution and future of Earth's nitrogen cycle. *Science* 330: 192–196. **56.13** Data: G. E. Likens, F. H. Bormann, N. M. Johnson, et al. 1970. Effects of forest cutting and herbicide treatment on nutrient budgets in the Hubbard Brook watershed-ecosystem. *Ecological Monographs* 40: 23–47. **56.14** T. Oki and S. Kanae. 2006. Global hydrological cycles and world water resources. *Science* 313: 1068–1072. **56.15** U.S Geological Survey: Changes in water levels and storage in the High Plains Aquifer, predevelopment to 2009. Fact Sheet, July 2011, Fig. 1. **56.16** N. Gruber and J. N. Galloway. 2008. An Earth-system perspective of the global nitrogen cycle. *Nature* 451: 293–296. **56.17** Based on D. E. Canfield, A. N. Glazer, and P. G. Falkowski. 2010. The evolution and future of Earth's nitrogen cycle. *Science* 330: 192–196.

56.19 B. Hönisch, A. Ridgwell, D. N. Schmidt, et al. 2012. The geological record of ocean acidification. *Science* 335: 1058–1063. **56.20** Carbon Dioxide Information Analysis Center (http://cdiac.ornl.gov/). **56.22, top** NOAA. **56.22, middle & bottom** T. R. Karl, J. M. Melillo, and T. C. Peterson (eds.). 2009. *Global Climate Change Impacts in the United States*. New York: Cambridge University Press, p. 6. **56.23** Carbon Dioxide Information Analysis Center (http://cdiac.ornl.gov/). **56.24** T. R. Karl, J. M. Melillo, and T. C. Peterson (eds). 2009. *Global Climate Change Impacts in the United States*. New York: Cambridge University Press, p. 34. **56.25** Reproduced by permission of AAAS from M. Zhao and S. W. Running. 2010. Drought-induced reduction in global terrestrial net primary production from 2000 through 2009. *Science* 329: 940–943, Fig. 2. (http://www.sciencemag.org/content/329/5994/940.abstract).

Chapter 57 **57.2** F.-R. Liu, M. M. Miyamoto, N. P. Freire, et al. 2001. Molecular and morphological supertrees for eutherian (placental) mammals. *Science* 291: 1786–1789, Fig. 1. Also A. B. Prasad, M. W. Mallard, E. D. Green, et al. 2008. Confirming the phylogeny of mammals by use of large comparative sequence data sets. Oxford University Press, *Molecular Biology and Evolution* 25 (9):1795–1808; U. Arnason, A. Gullberg, and A. Janke. 2004. Mitogenomic analyses provide new insights into cetacean origin and evolution. Elsevier, *Gene* 333: 27–34, Fig. 1. **57.4a&b** Adapted by permission of Macmillan Publishers Ltd after C. D. L. Orme, R. G. Davies, M. Burgess, et al. 2005. Global hotspots of species richness are not congruent with endemism or threat. *Nature* 436: 1016–1019, Figs. 1a, 1c. **57.5** D. M. Olson and E. Einerstein. 2002. The Global 200: Priority ecoregions for global conservation. *Annals of the Missouri Botanical Garden* 89: 199–224. Fig. 2, p. 201. Reprinted by permission of the Missouri Botanical Garden Press. **57.6** Data: IUCN 2012. *The IUCN Red List of Threatened Species. Version 2012.2.* (http://www.iucnredlist.org). **57.7** Reproduced by permission of the University of California Press Journals and the author from O. Venter et al. 2006. Threats to endangered species in Canada. *BioScience* 56 (11): 903–910, Fig 2. ©2006 American Institute of Biological Sciences. **57.9** Data: W. F. Laurance, S. G. Laurance, L. V. Ferreira, et al. 2000. Conservation: Rainforest fragmentation kills big trees. *Nature* 404: 836. **57.9** Based on W. F. Laurance, S. G. Laurance, L.V . Ferreira, et al. 1997. Biomass collapse in Amazonian forest fragments. *Science* 278: 1117–1118, Fig. 2. **57.10** Adapted by permission of Elsevier from J. M. Diamond. 1975. The island dilemma: Lessons of modern biogeographic studies for the design of natural reserves. *Biological Conservation* 7: 129–146. ©1975. **57.11** Reproduced by permission of AAAS from B. Sinervo, F. Mendez de la Cruz, D. B. Miles, et al. 2010. Erosion of lizard diversity by climate change and altered thermal niches. *Science* 328: 894–899, Fig. 3. (http://www.sciencemag.org/content/328/5980/894.abstract). **57.12** Graphs reproduced by permission of AAAS from D. Tilman, J. Knops, D. Wedin, et al. 1997. The influence of functional diversity and composition on ecosystem processes. *Science* 277: 1300–1302, Fig. 1. (http://www.sciencemag.org/content/277/5330/1300.short). **57.14** Data: W. E. Johnson, D. P. Onorato, M. E. Roelke, et al. 2010. Genetic restoration of the Florida panther. *Science* 329: 1641–1645. **57.15** Reproduced by permission of AAAS from E. I. Damschen, N. M. Haddad, J. L. Orrock, et al. 2006. Corridors increase plant species richness at large scales. *Science* 313: 1284–1286, Fig. 2B. (http://www.sciencemag.org/content/313/5791/1284.abstract). **57.17** Data: H. P. Jones and O. J. Schmitz. 2009. Rapid recovery of damaged ecosystems. *PLoS One* 4 (5): e5653, Fig. 4. **57.UN1, top** Data: A. P. Dobson. 1996. *Conservation and Biodiversity*. Scientific American Library. New York: Freeman. Original source of England maps: Reproduced by permission of the author from D. S. Wilcove, C. H. McLellan, and A. P. Dobson. 1986. Habitat fragmentation in the temperate zone. In *Conservation Biology: The Science of Scarcity and Diversity*, edited by M. E. Soule. Sunderland, MA: Sinauer Associates, pp. 237–256, Fig. 1. **57.UN1, bottom** Data: A. P. Dobson. 1996. *Conservation and Biodiversity*. Scientific American Library. New York: Freeman. Original source of U.S. Maps: W. B. Greeley, Chief, U.S. Forest Service. 1925. Relation of geography to timber supply. *Economic Geography* 1: 1–11.

Appendix B: BioSkills **TB16.1** A. Crowe, C. Dicks, and M. P. Wenderoth. 2008. Biology in bloom: Implementing Bloom's Taxonomy to enhance student learning in Biology. *CBE–Life Sciences Education* 7: 368–381, Table 3.

Index

Boldface page numbers indicate a glossary entry; page numbers followed by an *f* indicate a figure; those followed by *t* indicate a table.

Animal form and function, 842–60
  adaptations of, 843–45
  anatomy and physiology in, 843, 849f
  body size effects on physiology in, 850–53
  body temperature regulation in, 854–58
  dry habitats and, 842–43
  fitness trade-offs in, 843–45
  homeostasis in, 853–54
  organs and organ systems in, 849–50
  structure-function relationships in, 845–46
  tissues in, 845–49
Animal models, disease, **382**
Animal movement
  bipedalism, 705–7
  diversification of limbs for, 649–50f
  ecdysozoan, 674–78
  echinoderm, 685–86
  evolution of muscle-generated, 972–73
  hydrostatic skeletons and, 644
  invertebrate, 687–88
  locomotion in, 983–88
  lophotrochozoan, 665–69
  mollusk, 663–64f, 669
  muscle contraction in, 973–77
  muscle tissue and, 847–48, 977–80
  neurons and muscle cells of, 637
  skeletal systems in, 980–83
  types of, 973
  vertebrate, 697–703
Animal nutrition, 882–901
  cellulose as dietary fiber for, 78
  digestion and absorption of nutrients in, 886–97
  fetal, during pregnancy, 1032–33
  food capture and mouthparts in, 884–86
  food types in, 892–93
  four steps of, 883f
  glucose and nutritional homeostasis in, 897–99
  glycogen and starch hydrolysis in, 76, 80–81
  herbivore limitation by poor, 1131–32
  human digestive tract, 887f
  human mineral and electrolyte requirements, 885t
  human vitamin requirements, 884t
  mammalian digestive enzymes in, 895t
  nutritional homeostasis in, 897–99
  nutritional requirements and, 883–84
  waste elimination in, 896–97
Animal reproduction, 1013–36
  asexual vs. sexual, 1014
  deceitful communication in, 1094–95
  diversity in, 650–51
  ecdysozoan, 674–78
  echinoderm, 685–86
  in energy flow, 1149
  fertilization and egg development in, 1018–21
  fitness trade-offs in, 843–45
  gametogenesis in, 1016–18f
  hormones in, 998–99
  invertebrate, 687–88
  life cycles of, 651–52
  life tables and, 1104–5, 1106f
  lophotrochozoan, 665–69
  mammalian pregnancy and birth in, 1030–34
  mammalian reproductive systems in, 1021–25
  mammalian sex hormones in, 1025–30
  mating behaviors and (see Mating)

  migration and, 1091
  net reproductive rate in, 1105
  parthenogenesis, 665
  sexual selection and, 475–78
  switching modes of, between asexual and sexual, 1014–16
  vertebrate, 693–94, 697–703
Anion(s), **22f**, 781–83
Annelids, 638t, 657–58, 667
Aniston, Jennifer, 946, 947f
Annual growth rings, tree, 751, 1139
Annuals, **599**
Anolis lizards, 516, 1087–89, 1125
Anoxic oceans, 521
Anoxygenic photosynthesis, **186**, 539t, **541**, 545
Antagonistic muscle groups, **981**
Antagonistic-pair feedback systems, 857
Antarctic fur seals, 1086
Antbirds, 1124–25
Antenna complex, photosystem, **182**–83
Antennae, insect, **675**
Anterior body axis, **410**
Anterior pituitary gland, **1004**, 1006
Antheridia, **587**
Anthers, **827**
Anthocerophyta, 602
Anthophyta (angiosperms), 608–9f. *See also* Angiosperms
Anthrax, 532
Anthropocene epoch, 515
Anthropogenic biomes, 1069
Anthropoids, 704f–**5**
Antibacterial compounds, 815
Antibiotic resistance, 395, 454–56
Antibiotics, **532**, **1039**
  actinobacteria and, 546
  bacterial ribosomes and, 529
  fungi and, 613, 632
  in human semen, 1023t
  lysozyme as, 1039
  pathogenic bacteria and, 532–33
Antibodies, **206**, **717**, **1042**
  adaptive immune response and, 1042 (*see also* Adaptive immune response)
  B-cell activation and secretion of, 1050–51
  binding of, to epitopes, 1045
  cell adhesion proteins and, 206–7
  five classes of immunoglobulins and, 1045t (*see also* Immunoglobulins)
  in fluorescent microscopy, B:18
  in HIV research, 717–18
  humoral response and, 1051–52
  as probes, B:16
  proteins as, 54
  specificity and diversity of, 1045–46
Anticodons, **327**, 328–29f
Antidiuretic hormone (ADH), **877**–78, **1003**, 1005–6
Antifungal compounds, 815
Antigen presentation, **1045**, 1048–49
Antigens, **1040**, 1044–50
Antilogs, logarithms and, B:9
Antiparallel DNA strands, **62**, 286–87
Antiporters, **770**, 786–**87**, **865**
Antiviral drugs, 714
Ants, 11–14, 623, 1063–64, 1083–85, 1124–25, 1133–35
Anus, 644–45, 658, 886, 887f
Aorta, **917**
Aphids, 485, 672f, 768–69f
Aphotic zone, lake and pond, **1077**
Aphotic zone, ocean, **1079**
Apical-basal axis, **434**, 435–36

Apical buds, **735**
Apical cells, **434**
Apical dominance, **806**–7
Apical meristems, 436, 437f, **740**–42, 748
Apical mutants, 435
Apical side, epithelium, **849**
Apicomplexans, 573
Apodemes, **983**
Apomixis, **825**
Apoplast, 208–9f, **760**
Apoplastic pathway, 760–61, 785
Apoptosis, **232**, **407**, **1053**
  in cancer, 362–63
  in cell-mediated response, 1053–54
  in Huntington's disease, 382
  in organogenesis, 426
  as programmed cell death, 406t, 407–8
  tumor suppressors and, 232
Appendix, 887f, **896**
Apple maggot flies, 496–97
Apply (Bloom's taxonomy skill), B:29f–B:30t
Aquaporins, **98**, **865**, **896**
  large intestine water absorption and, 896
  in osmoregulation, 865, 877–78
  in plant water movement, 760
  selectivity and structure of, 98
Aquatic biomes
  animal locomotion in, 984, 985, 988
  animal transition to terrestrial environments from, 659–60
  behavior of oxygen and carbon dioxide in, 904–5
  direct and indirect human impacts on, 1076
  estuaries as freshwater and marine, 1078
  external fertilization in, 1018
  freshwater lakes, ponds, wetlands, and streams, 1077–78
  gas exchange by gills in, 906–7
  global net primary productivity of, 1154–56
  global water cycle and, 1159–60
  nitrogen pollution in, 1160
  nutrient availability in, 1075–76
  oceans as marine, 1079
  osmoregulation in, 862–64, 866–68
  plant transition to terrestrial environments from, 584–95
  protists in, 553, 556
  salinity in, 1074
  water depth and water flow in, 1074–75
Aqueous solutions
  acid-base reactions in, 28–30
  chemical evolution in, 24
  molarity of, 29
  properties of water and, 25–28
Aquifers, 1159–**60**
ara operon, 344, 345f
Arabian oryx, 842–43, 855, 864
Arabidopsis thaliana. *See* Mustard plant
Arabinose, 344, 345f
araC gene, **344**
AraC protein, 344, 345f
Arachnids, 678
Arbuscular mycorrhizal fungi (AMF), **620**–22, 630
Arcese, Peter, 1111–12
Archaea, **529**. *See also* Archaea domain
  aerobic vs. anaerobic respiration in, 171
  ammonia-oxidizing, 539t
  bacteria vs., as prokaryotes, 528–29 (*see also* Bacteria; Prokaryotes)

  citric acid cycle in, 162
  human diseases and, 531
  phospholipids in plasma membranes of, 529
  phylogeny of, 536f
  as prokaryotes, 107
Archaea domain. *See also* Archaea
  characteristics of, 529t
  key lineages of, 544, 548–49
  phospholipids in, 87
  phylogeny of, 7f–8, 536f
  as prokaryotic, 107, 528–29 (*see also* Prokaryotes)
Archaeopteryx, 512
Archegonia, **587**
Arctic terns, 1091
Arctic tundra, **1072**, 1073f, 1165, 1167
Ardipithecus genus, 705t
Area, metric units and conversions for, B:1t
Area and age hypothesis on species richness, 1145
Argentine ant, 1063–64, 1083–84, 1085. *See also* Ants
Arginine, 305–6
Aristotle, 445
Armor, defensive, 1129f
Arms races, coevolutionary, 1124, 1132
Arrangement, leaf, 738
Arrhenius, Svante, 1163
Arteries, 871, **917**–18, 921
Arterioles, **917**–18, 924
Arteriosclerosis, **924**
Arthrobacter, 546
Arthropods, **670**
  body plan of, 670–71
  as ecdysozoans, 673
  exoskeletons of, 983
  key lineages of, 638t, 675–78
  metamorphosis in, 672
  origin of wings in evolution of, 671–72
  as protostomes, 657–58
Articulations (joints), **981**–82
Artificial membranes, 88–89
Artificial selection, 5–6, 257, **453**, **579**. *See also* Crosses
Artiodactyls, 510–11f
Ascaris (roundworms), 238. *See also* Roundworms
Asci, **617**–18
Ascidians, 688
Ascomycetes, 617–18, 627, 631–32
Ascorbic acid, 884t
Asexual reproduction, **220**, **825**, **1014**. *See also* Animal reproduction
  animal, 650, 1014
  by fragmentation, 601, 603
  fungal, 617, 625
  mitosis and, 220 (*see also* Mitosis)
  plant, 825–26
  prokaryotic vs. eukaryotic, 566
  sexual reproduction vs., 247, 566, 1014 (*see also* Sexual reproduction)
  switching between sexual and, 1014–16
Asteroidea, 685
Asteroid impact hypothesis, 521–23
Asthma, 1056
Astragalus, 510
Astral microtubules, **225**
Astrobiologists, 531
Asymmetric cell division, plant, 433–34
Asymmetric competition, **1125**–26
Athletes
  EPO abuse by, 1003
  human growth hormone and, 374

Boldface page numbers indicate a glossary entry; page numbers followed by *f* indicate a figure; those followed by *t* indicate a table.

Bogs, **1077**
Bohr shift, **914**, 915
Boluses, 884
Bombadier beetles, 138, 139*f*
Bond saturation, lipid, 85–86, 90
Bonds, B:12–B:13. *See also* Covalent bonds; Hydrogen bonds; Ionic bonds
Bone, 427–29, **689**, 846*f*, **847**, **981**–82
Bone marrow, 982–83, **1043**
Bone marrow transplants, 384
Bony fishes, osmoregulation by, 862–63
Boreal forests, **1072**, 1157
Bormann, Frank, 1149–50
Borneo, 1172–73, 1191
Borthwick, H. A., 800–801
Both-and rule, probability, B:8
Bottom-up limitation hypothesis, herbivore population, 1131–32
Bottom-up research approach, chemical evolution, 32–33
Bouchet Philippe, 1175
Boveri, Theodor, 257, 266–67
Bowerbird, 1082*f*
Bowman's capsule, **873**, 878*t*
Box-and-whisker plots, B:5–B:6
Bracket fungi, 617, 626–27
Brain stem, **944**
Braincase, **707**
Brains, **643**
    anatomy of, 944–46
    central nervous system and, 929
    diffusion spectrum imaging of neurons in, 928*f*
    in hormonal control of urine formation, 877
    Huntington's disease and degeneration of, 278, 378, 382
    in initiation of muscle contraction, 976–77
    neural tubes and, 427
    origin of, in animal evolution, 643
    primate, 705
    prions and diseases of, 53, 369–70
    sensory organ transmission of information to, 954
    in thermoregulation, 856–57
    vertebrate, 688
Brakefield, Paul, 1167
Branches, phylogenetic tree, **506**, B:10–B:11
Branches, plant, **735**
Branching adaptations, animal surface-area, 852–53
Brassinosteroids, 210*t*, **812**, 813*t*
Bread mold, 305, 629. *See also* *Neurospora crassa*
Bread yeast. *See* Saccharomyces cerevisiae
Breathing. *See* Ventilation
Breeding. *See also* Artificial selection
    captive, 482–83, 1189–90
    selective, 257 (*see also* Crosses)
Brenner, Sydney, 310–11, B:25
Brewer's yeast. *See* Saccharomyces cerevisiae
Bridled goby fish, 1109–11
Briggs, Winslow, 799
Brine shrimp, 671
Bristlecone pine trees, 432*f*, 607, 732
Brittle stars, 684
Britton, Roy, 324

Broca, Paul, 944–45
Bronchi, **909**
Bronchioles, **909**
Bronstein, Judith, 1134
Brown algae, 7, 552*f*, 557, 568–69, 574. *See also* Algae
Brown fat cells, 118
Brown kiwis, 476
Brown tree snake, 1181
Brushtail possums, 1031*f*
Bryophytes (mosses), 580–81, 601–2, 824*f*–25
Bryozoans, 638*t*, 665
Buchner, Hans and Edward, 158
Buck, Linda, 966
Budding, animal, 654, 685, 688, **1014**
Budding, viral, 720, 721*f*
Buds, axillary and apical, 735
Buds, flower, 826
Buffers, pH, **30**, **916**
Buikema, Arthur, 1136–37
Building materials, 579–80
Bulbourethral gland, **1022**–23*f*
Bulbs, 739
Bulk flow, **762**
Bulk-phase endocytosis, **127**
Bundle-sheath cells, **193**–94, **766**
Burials, fossils and, 512
Bursa-dependent lymphocytes, 1043. *See also* B cells
Bursting, viral, 720, 721*f*
Butter, 86*f*
Buttercup root, 72*f*
Butterflies, 1091, 1112–13, 1129–30
Bypass vessels, 920–21

C

C-terminus, 46, 328
C_3 pathway, **193**
C_4 pathway, **193**–94, **766**
Cacti, 194, 736–37
Cactus mice, 835–36
Cadherins, **207**
Caecilians, 700
Caenorhabditis elegans. *See* Roundworms
Caffeine, 815
Calcium. *See also* Calcium ions
    animal bones and, 982–83
    in ecosystem nutrient cycles, 1156
    as human nutrient, 885*t*
    as plant nutrient, 780*t*
Calcium carbonate, 556–57, 572, 600, 981
Calcium ions
    as electrolytes, 862
    in polyspermy prevention, 421–23
    protein folding and, 53
    as second messengers, 213*t*
Calcium phosphate, 981
Calculators, scientific, B:9
Calibration, accuracy and, B:2–B:3
California mussel, 1137–38
Callus, **744**, B:22
Calmodulin, 53, 458
Calvin, Melvin, 178, 190–91
Calvin cycle, **178**, 190–95, 541, 766. *See also* Photosynthesis
Calyx, **826**
CAM (crassulacean acid metabolism), **194**–95, 766, **766**, 1070
Cambium, **748**–50

Cambrian explosion, **518**–20, **636**–37
cAMP (cyclic adenosine monophosphate; cyclic AMP), 213*t* **1009**–10
Cancer, **232**
    anticancer drugs, 607
    cell cultures of, B:21, B:22*f*
    chromosome mutations and, 314–15
    defective Ras proteins and, 214
    as family of diseases, 234
    from gene therapy for severe combined immunodeficiency (SCID), 384–85
    genetic basis of uncontrolled cell growth in, 361–63
    genetic testing for, 382
    homology and, 451
    loss of cell-cycle control and, 233–34
    mutations and, 298–301
    properties of cancer cells, 233
    retroviruses and, 728
    telomerase and, 297
    types and severity of, 232*t*
    types of cell division defects in, 232–33
Cannabis sativa, 828*f*
Cannibalization, mating and, 1020
Capillaries, **852**–53*f*, 873, **917**–18, 923
Capillarity, **761**
Capillary action, plant, 759, 761–62
Capillary beds, **917**–18
Caps, mRNA, 323–24
Capsaicin, 835–36
Capsids, 285–86, **715**, 720
Captive breeding, 482–83, 1189–90
Carapace, **677**
Carbohydrates, **72**
    animal digestion and transportation of, 883, 888, 893–94
    cell characteristics and, 107
    cell identity functions of, 78–80
    energy storage functions of, 76, 80–81
    fermentation of, 540
    glucose production from, 155
    glycosylation of, to form glycoproteins, 124
    in metabolic pathways, 157–58
    monosaccharides, 73–75
    photosynthetic production of, 176–78 (*see also* Photosynthesis)
    polysaccharide types and structures, 77*t*
    polysaccharides, 75–78
    processing of, in photosynthesis, 195
    reduction of carbon dioxide to produce (*see* Calvin cycle)
    structural functions of, 76, 78, 117
    types of, 72–73
Carbon
    in amino acid structure, 42
    Calvin cycle fixation of, during photosynthesis (*see* Calvin cycle)
    in citric acid cycle, 162
    in ecosystem nutrient cycles, 1156
    electronegativity of, 21
    functional groups and atoms of, 37*t*
    lipids, hydrocarbons, and, 85–86
    in living organisms, 19
    in metabolic pathways, 157
    in net primary productivity, 1069
    in organic molecules, 36–38
    prokaryotic strategies for obtaining and fixing, 538–41
    in protein secondary structure, 48–49
    in redox reactions, 141–42
    simple molecules from, 23
Carbon–carbon bonds, 36

Carbon cycle, **614**
    global, 1161–62
    plants and, 579
    saprophytic fungi and, 614–15
    Sphagnum moss and, 602
Carbon dioxide
    animal gas exchange and, 902–3
    atmospheric, in global carbon cycle, 1161
    behavior of, in air, 903–4
    behavior of, in water, 904–5
    carbonic anhydrase protein and, 54
    covalent bonds of, 23
    end-Permian atmospheric, 521
    global climate change and, 556–57, 579, 1073
    land plants and, 584
    mechanisms for increasing concentration of, 193–95
    in origin-of-life experiments, 34–36
    oxidation of acetyl CoA to, in citric acid cycle, 162–65
    passage of, through stomata of leaves, 192–93
    photosynthetic carbohydrate production from, 177–78 (*see also* Photosynthesis)
    plant pathways to increase, in dry habitats, 766
    plant water loss and requirements for, 754–55
    in prokaryotic metabolism, 539*t*, 541
    in pyruvate processing, 161–62
    reduction of, in Calvin cycle (*see* Calvin cycle)
    transport of, in blood, 915–16
    in volcanic gases, 30
Carbon fixation, **190**, 541. *See also* Calvin cycle
Carbonic acid, 30
Carbonic anhydrase, 54, **890**, **915**–16
Carboniferous period, 582
Carbonyl functional group, 37*t*–38, 45–46, 48–49*f*, 73–74
Carboxy-terminus, 46, 328
Carboxyl functional group, 37*t*–38, 42, 45–46, 85–87, 162
Carboxylic acids, 37*t*, **162**
Carboxypeptidase, 895*t*
Cardiac cycle, **923**. *See also* Hearts
Cardiac muscle, 847*f*–48, **978**
Cardiac tissue, 200*f*
Cardiovascular disease, **924**
Carnauba palms, 743
Carnivores, 579, **647**–48, **1128**
Carnivorous plants, **789**–90
Carotenes, 180–82
Carotenoids, **180**. *See also* Pigments, photosynthetic
    in photosynthesis, 180–82
    in sexual selection, 475–76
    synthesis of, by aphids, 485
Carpellate flowers, 827(*footnote*)
Carpels, **439**, **592**, 827–28
Carrier proteins, **99**, **769**, **864**
    facilitated diffusion via, 99
    in osmoregulation, 864–65
    in translocation, 769–72
Carriers, recessive trait, **277**–78
Carrion flowers, 594
Carroll, Sean, 419, 649–50*f*, 660
Carrying capacity, **1108**–9*f*, 1112, 1130–31, 1189
Carson, Rachel, 999–1000, 1152–53
Cartilage, **689**, 846*f*, **847**, **981**–82
Cascades, regulatory gene, 413–14

Boldface page numbers indicate a glossary entry; page numbers followed by an *f* indicate a figure; those followed by *t* indicate a table.

Genetic engineering, 368–88
  agricultural biotechnology and
    development of transgenic golden
    rice in, 385–86
  in agriculture, 579
  amplification of DNA with polymerase
    chain reaction in, 374–76
  common techniques used in, 381t
  common tools used in, 380t
  dideoxy DNA sequencing in, 376–78
  gene therapies in, 383–85
  genetic mapping in, 378–82
  of human growth hormone for pituitary
    dwarfism using recombinant DNA
    technology, 369–74
  molecular biology techniques and
    recombinant DNA technology in,
    368–69
  plant cell and tissue cultures in, B:22
  in video microscopy, B:20
Genetic equivalence, 408, 409
Genetic homologies, 450–51
Genetic isolation, 490, 494–99
Genetic libraries. See DNA libraries
Genetic maps, 271, 379
  benefits of finding disease genes
    with, 382
  crossing over and, in chromosome
    theory of inheritance, 271
  development of first, 274f
  ethical concerns over genetic testing
    and, 382
  finding Huntington's disease gene
    with, 378–82
  in genetic engineering, 381t
Genetic markers, 379, 479
  finding Huntington's disease gene
    with, 379
  in hybridization research, 502
  in studies of genetic drift, 479–80
Genetic recombination, 247–48,
    270–71, 274f
Genetic restoration, 1190
Genetic screens, 305–6, 339
Genetic testing, 382
Genetic variation, 472. See also Genetic
    diversity
  asexual vs. sexual reproduction
    and, 247
  from crossing over, 248
  drug resistance and, 455–56
  effects of evolutionary processes on,
    466, 486t
  effects of modes of natural selection
    on, 472, 475t
  genetic drift and, 479
  from independent assortment, 247–48
  lack of, as genetic constraint, 462
  mutations and, 485
  natural selection and, 454, 456
  sexual reproduction and, 566
  types of fertilization and, 248–49
Genetically modified food, 385–86
Genetics, 257. See also Genes
  behavior causation in (see Proximate
    causation)
  chromosome theory of inheritance
    and, 256–57 (see also Chromosome
    theory of inheritance)
  combining probabilities in, B:8
  developing tree of life from rRNA
    sequences in, 6–8

DNA in, 64 (see also DNA
    [deoxyribonucleic acid])
  genetic disorders (see Genetic
    disorders, human)
  Mendelian (see Mendelian genetics)
  model organisms in, B:23
  protostome, 660
  RNA in, 67 (see also RNA [ribonucleic
    acid])
  RNA world hypothesis and, 69
Genitalia, 1022–23
Gennett, J. Claude, 1046
Genome annotation, 392–93
Genome sequencing, 389f, 391, 725, 1173
Genomes, 289, 389
  emerging viruses from reassortment
    of, 723–24
  eukaryotic, 393, 395–400, 628 (see also
    Eukaryotic genomes)
  genetic homology and, 451
  genomics and, 389–90 (see also
    Genomics)
  identifying genes in, 392–93
  in Meselson–Stahl experiment, 289
  model organism, B:23–B:26
  origin of multicellularity in, 640
  prokaryotic, 392–95
  Saccharomyces cerevisiae, 618
  selecting organisms for sequencing
    of, 392
  technologies for sequencing, 376, 378
  viral, 715–16, 719–22
  whole-genome sequencing of, 390–93
Genomic libraries, 372, 381t
Genomic reassortment, viruses and,
    723–24
Genomics, 389–404
  bacterial and archaeal genomes in,
    393–95
  comparative, in studying evolutionary
    innovations of animals, 638
  development of, 389–90
  eukaryotic genomes in, 395–400
  functional genomics and, 400–402
  genome sequencers in, 389f, 391
  Human Genome Project in, 389–90,
    399–400
  metagenomics (environmental
    sequencing) in, 395, 534–35
  proteomics and, 402
  systems biology and, 402
  whole-genome sequencing in, 390–93
Genotypes, 261
  central dogma and linking phenotypes
    and, 308, 309f
  effect of inbreeding on frequencies of,
    470–71f
  exceptions and extensions to
    Mendelian rules on, 277t
  frequencies of, for human MN blood
    group, 468–69
  Hardy–Weinberg principle and
    frequencies of, 466–70
  mating, body odor, and, 470
  in Mendelian genetics, 258t
  mutations as changes in, 313–15 (see
    also Mutation[s])
  in particulate inheritance, 261–63
  phenotypes and, 261, 273–75 (see also
    Phenotypes)
  predicting, with Punnett square,
    262–63

Genus, taxonomic, 8
Geographic distribution and abundance
    of organisms
  abiotic factors in, 1062
  allopatric speciation and, 494–95
  biodiversity hotspots and, 1176–77
  biogeography and, 1079
  biotic factors in, 1062
  ecology and, 1059–65 (see also
    Ecology)
  estimating effects of global climate
    change on, 1183–84
  evidence for evolutionary change in,
    449–50
  global climate change and, 1073–74,
    1166f, 1167
  historical factors in, 1062–63
  interaction of biotic and abiotic factors
    in, 1063–65
  island biogeography and, 1144
  mapping current and past, 1137
  in population ecology, 1102–3
  protected areas and, 1190–91
  species interactions and, 1124
Geologic time scale, 446–47
Geothermal radiation, 541
Gerbils, 1086–87
Germ cells, 438
Germ layers, 424f–25, 639, 640–41
Germ line, 433
Germ theory of disease, 532
Germination, 433, 800–801, 809–10,
    832, 836–37
Gestation, 694, 1031–33
GH1 gene, pituitary dwarfism, 369, 373
Giant axon, squid, 130
Giant sequoias, 1138–39
Giant sperm, 1020
Giardia, 555t, 557, 571
Giardiasis, 555t, 571
Gibberellins, 808–10, 813t, 836–37
Gibbons, Ian, 132–33
Gibbs free-energy change, 138–39
Gila monsters, 878f, 879
Gilbert, Walter, 322
Gill arches, 692
Gill filaments, 906–7f
Gill lamellae, 852–53f, 906–7
Gill pouches, 450
Gills, 668, 687, 852, 906
  developmental changes in gas
    exchange with, 851–52
  mollusk, 662, 668
  in origin of insect wings, 671–72
  in osmoregulation, 862–64, 868
  protostome, 659
  structure and function of, 906–7
  vertebrate, 686–87
Ginkgoes, 606
Giraffe neck hypothesis testing, 9–11
Gizzards, avian, 891–92
Glabe, Charles, 421
Glacier Bay succession case history,
    1141–42
Glands, 848, 993, 994f–95
Glanville fritillaries (butterflies), 1112–
    13, 1118–20
Glaucophyte algae, 569
Gleason, Henry, 1136–37
Glia, 937
Global air circulation patterns, 1065–66
Global biogeochemical cycles, 1156,
    1159–62. See also Nutrient cycling
Global carbon cycle, 556, 1161–62
Global climate change, 1163–69
  causes of, 1163–64

devegetation and, 579
  effects of, on aquatic biomes, 1076
  effects of, on terrestrial biomes,
    1073–74
  estimating effects of, on species
    distributions, 1183–84
  global warming and, 1163
  impact of, on ecosystem net primary
    productivity, 1168–69
  impact of, on organisms, 1166–68
  life-history traits of endangered species
    and, 1118
  local consequences of, 1169
  positive and negative feedback in,
    1164–66
  protists and limitation of, 556–57
  Sphagnum moss and, 602
  summary of impacts on organisms
    of, 1166t
  temperature and precipitation changes
    from, 1164, 1166
  as threat to biodiversity, 1181t, 1182
Global ecology, 1159–69
  biosphere and ecosystems in, 1060t,
    1061, 1148–49 (see also Ecosystem
    ecology)
  global biogeochemical cycles in,
    1159–62
  global climate change in, 1163–69
  global net primary productivity
    patterns in, 1153–56
Global gene regulation, 344–45
Global human population, 1101–2,
    1117–18
Global net primary productivity
    patterns, 1153–56
Global nitrogen cycle, 1160–61
Global warming, global climate change
    and, 1163. See also Global climate
    change
Global water cycle, 1159–60, 1166
Globin genes, 399
Globular proteins, 47
Glomalin, 621
Glomeromycota, 618, 630
Glomeruli, 873, 878t, 966
Glucagon, 54, 897, 992
Glucocorticoids, 1000–1001, 1004–6
Glucokinase, 145
Gluconeogenesis, 195, 897
Glucose, 154
  animal absorption of, 893–94
  ATP synthesis from, 81
  ATP synthesis from oxidation of, in
    cellular respiration, 154–56, 170–72
  ATP synthesis from oxidation of, in
    fermentation, 155, 172–73
  carrier protein for, in human red blood
    cells of, 99
  configurations of, 73–74
  cortisol and availability of, 1000–1001
  in diabetes mellitus and nutritional
    homeostasis, 897–99
  enzyme catalysis of phosphorylation
    of, 145
  free energy changes in oxidation
    of, 165f
  glycogen and, 76
  glycolysis as processing of, to pyruvate,
    158–61
  hydrolysis of carbohydrates by
    enzymes to release, 80–81
  lac operon regulation by, 343
  lactose metabolism and preference for,
    338–39f
  in metabolic pathways, 157

Boldface page numbers indicate a glossary entry; page numbers followed by an *f* indicate a figure; those followed by *t* indicate a table.

Boldface page numbers indicate a glossary entry; page numbers followed by an *f* indicate a figure; those followed by *t* indicate a table.

Boldface page numbers indicate a glossary entry; page numbers followed by an *f* indicate a figure; those followed by *t* indicate a table.

Boldface page numbers indicate a glossary entry; page numbers followed by an *f* indicate a figure; those followed by *t* indicate a table.

I:30   INDEX

Boldface page numbers indicate a glossary entry; page numbers followed by an *f* indicate a figure; those followed by *t* indicate a table.

Random distribution, 1102
Random mating, 468–70
Random mutations, 312
Ranges, species, **1062**, **1102**. *See also* Geographic distribution and abundance of organisms
for human population, 1116
impacts of global climate change on, 1166t, 1167
mapping, 1137
in population ecology, 1102–3
in preservation of metapopulations, 1120
Ras protein, **213**, 214, 232–33
Ray-finned fishes, 692, **698**–99
Rayment, Ivan, 974–76
Rays, cambium, **749**
Rb protein, **234**
Reabsorption, animal, 870–71, 873–74, 877–78
Reactants, chemical reaction, **30**–32, 139–40f, 143–47
Reaction center, photosystem, 183–84
Reaction rates, chemical reaction, 139–40f, 145–47
Reactivity, amino acid side chain, 44
Reading frame, DNA sequence, **310**–11, 313, 392–93
Realized niches, **1126**
Receptacle, plant flower, 826
Receptor-mediated endocytosis, **126**
Receptor tyrosine kinases (RTKs), **213**–14
Receptors. *See also* B-cell receptors (BCRs); Enzyme-linked receptors; G-protein-coupled receptors; Signal receptors; T-cell receptors (TCRs)
animal hormone, 1006–10
animal sensory, 856–57, 953
cell–cell signal, 210
plant sensory system, 794–97, 800–803
Recessive alleles, 258t, 471
Recessive traits, **259**–60, 277–79, 299, 369. *See also* Dominant traits
Reciprocal altruism, **1098**
Reciprocal crosses, 258t, **260**–61, 268–69
Recognition phase, adaptive immune response, 1041–47
Recolonization, metapopulation, 1112f–13
Recombinant DNA technology, **369**–74. *See also* Genetic engineering
agricultural biotechnology, transgenic crops, and, 385–86
creating animal models of disease using, 382
discovery of gene recombination, 1046, 1047f
engineering human growth hormone using, 370–74
ethical concerns about, 374
gene therapy and, 383–85
in genetic engineering, 368–69, 381t
pituitary dwarfism, human growth hormone, and, 369–70
plant cell and tissue cultures in, B:22
Recombinant organisms, **270**–71, 274f
Recommended Dietary Allowances (RDAs), 883
Rectal glands, shark, **866**–88
Rectum, **896**
Red algae, 569, 572. *See also* Algae
Red blood cells, 48, 54, 99, **912**
Red/far-red light responses, plant, 179–81f, 800–803, 818t, 837

Red-green color blindness, 278–79, 963–64
REDD (Reducing Emissions from Deforestation and Forest Degradation), 1192–93
Re-differentiation, cell, 407
Redox (reduction-oxidation) reactions
in Calvin cycle, 190–92
in electron transport chain, 166–67
peroxisomes and, 115
in photosystems, 183–85
prokaryotic nitrogen fixation and, 542–43
transfer of energy via electrons in, 141–43
Reduction, **141**, 191, 241–42. *See also* Redox (reduction-oxidation) reactions
Re-extension, muscle, 980
Reference distributions, statistical tests and, B:7
Reflex, spinal, **944**
Refractory state, **936**
Regeneration phase, Calvin cycle, 191
Regulating services, ecosystem, 1187t, 1188
Regulation
of animal body temperature, 854–58
of animal homeostasis, 853–54
of animal hormone production, 1003–6
of blood pH, 915–16
of blood pressure and blood flow, 924
cancer as loss of cell-cycle, 233–34
of cell cycle, 229–31
of citric acid cycle, 162–63f
of enzymes, 149–50
of flower structures, 439–41
of gene expression (*see* Gene regulation, bacterial; Gene regulation, eukaryotic)
of glycolysis, 159–61
homeostatic, of ventilation, 911–12
kidney, 872, 877–78
of mammalian menstrual cycle, 1027–30
of mammalian puberty, 1026–27
of metabolic pathways, 150
of pancreatic enzymes, 893
of photosynthesis, 195
of plant body axes, 435–36
of plant leaf shape, 436–38
of protein folding, 53
of pyruvate processing, 161–62
of telomerase, 297
Regulatory cascades, 413–14, 436–38, 439–41
Regulatory genes
evolutionary change and, 416
evolutionary conservation of, 414–15
Hox genes as, for developmental positional information, 412–14
morphogens, 410–12
tool-kit genes as, 415
transcription factors, 410, 411–15 (*see also* Transcription factors)
Regulatory homeostasis, 853–58
Regulatory hormones, regulation of, 1026–27
Regulatory molecules, cell-cycle, 229–31
Regulatory proteins
in bacterial gene regulation, 337
in control of plant body axes, 435–36
in determination of leaf shape, 436–38
in differentiation of muscle cells, 428f–29

in eukaryotic transcription initiation, 353–56, 357f
in negative control, 341–44
in positive control, 341, 344, 345f
regulatory DNA sequences and, 353–54, 355f
as transcription factors in differential gene expression, 354–56
Regulatory sequences, 353–54, 355f, 356, 357f, 363
Regulatory transcription factors, **410**. *See also* Transcription factors
Regulons, **345**, 360
Reinforcement, **499**–500
Reintroduction programs, species, 1189–90
Relative dating, 446–47, 453
Release factors, **331**, 332f
Religious faith vs. science, 9
Remember (Bloom's taxonomy skill), B:29f–B:30t
Remodeling, chromatin. *See* Chromatin remodeling
Renal blood vessels, 871
Renal corpuscle, 872, **873**, 878t
Renewable energy, 1189
Repair, DNA, 297–301
Repetition, scientific experiments and, 13
Replacement rate, **1117**
Replica plating, **339**–41
Replicated chromosomes, 239–40
Replication, chromosome, 220–21
Replication, DNA, 287–90. *See also* DNA synthesis
Replication fork, **290**–95
Replicative growth, viral, **716**–21
Replisomes, **294**–95
Repolarization, **933**–34
Reports, scientific, B:26–B:27
Repressors, **341**–44, 345f, **354**
Reproduction
animal (*see* Animal reproduction)
cell cycle and, 219–20 (*see also* Cell cycle)
evolution by natural selection and, 5–6, 454, 456
evolution of giraffe necks and, 9–11
fungal, 616f, 617–18, 624–25
as goal of living organisms, 2
mitosis and, 220 (*see also* Mitosis)
in model organisms, B:23
plant (*see* Plant reproduction)
protist, 566
Reproductive development, plant, **433**, 438–41
Reproductive isolation
allopatric speciation and, 494–95
biological species concept and, 490
ecological niches and, 496–97
mechanisms of, 491t
sympatric speciation and, 495–99
Reptiles, **696**
birds as, 694–95
lineages of, 696–97, 702–3
nitrogenous wastes of, 865
urine formation in, 878–79
in vertebrate phylogeny, 691
Research papers, B:26–B:27
Reservoirs, biogeochemical, 1159
Residues, amino acids as, 45–46
Resilience, community, **1186**–87, 1191–92
Resistance, community, **1186**–87
Resource partitioning, 1127–28
Resources
as ecosystem provisioning services, 1187

human consumption of, 1164, 1189
species richness and efficient use of, 1186
Respiratory systems, **903**–16
air and water as respiratory media in, 903–5
circulatory systems and, in gas exchange and circulation, 902–3 (*see also* Circulatory systems)
Fick's law of diffusion and, 905–6
fish gills in, 906–7
homeostatic control of ventilation in, 911–12
insect trachaea in, 907–9
organs of, 905–12
pulmonary circulation and, 920f
vertebrate lungs in, 909–11f
Responder cells, plant, 794–96
Response, cell–cell signal, 214
Resting potentials, **931**–32
Restoration, ecosystem, 1191–92
Restriction endonucleases, **370**–72, 379, 380t
Results, primary literature, B:27t
Reticulum, 891
Retina, eye, **960**, 961f
Retinal, 147, 961f–62
Retinoblastoma, 234
Retro-evolution hypothesis, 150–51f
Retroviruses, **383**–85, 720, 726f, 728
Revelle, Roger, 1163
Reverse transcriptase, **309**, **720**
in gene therapy, 383
in genetic engineering, 380t
inhibitors, 720
in long interspersed nuclear elements (LINEs), 396, 397f
producing cDNAs with, 370
Rheumatoid arthritis, 1056
Rhinovirus, 727
Rhizaria, 558t, 569, 572
Rhizobia, **787**–89
Rhizoids, **598**
Rhizomes, **603**, **737**, 825
Rhodopsin, 395, 961f–62
Rhynie Chert, 585
Ribbon diagrams, 48–49f
Ribonuclease, 52
Ribonucleic acid (RNA). *See* RNA (ribonucleic acid)
Ribonucleotides, 6–7, **58**–59, 93–94. *See also* Nucleotides
Ribonucleotide triphosphate (NTP), 318–19
Ribose, 58–60, 65, 68–69, 73, 78
Ribosomal RNAs (rRNAs), **119**, **328**
developing tree of life from sequences of, 6–8
in gene families, 398
phylogenies based on, 535–36
in ribosomes, 328, 330
Ribosome binding sites, **329**–30
Ribosomes, **108**, **324**
in cracking of genetic code, 311
elongation of polypeptides during translation in, 330–31
eukaryotic, 112, 113
initiation of translation in, 329–30
mitochondria and, 115–16
polypeptide synthesis by, 119
post-translational modifications of proteins in, 331–32
prokaryotic, 108
protein synthesis by free, 123–24
ribosomal RNA and structure of, 328–29f

Boldface page numbers indicate a glossary entry; page numbers followed by an
f indicate a figure; those followed by *t* indicate a table.

I:34   INDEX

hemoglobin, 51t
  protein, 48–49
  RNA, 65–66
  tRNA, 327
Secondary succession, **1139**–40
Secondary xylem, 748–50
Secretin, **893**, 995, **996**–97
Secretions, barrier, 1038–39
Secretory pathway, endomembrane
    system, 121–26
Secretory vesicles, 123
Sedimentary rocks, **446**, 447f
Sediments, fossils and, 512
Seed banks, **1189**
Seed coat, **833**
Seed plants. *See also* Angiosperms;
    Gymnosperms
  characteristics of, 581
  evolution of, 831
  heterospory in, 590, 591f, 592f
  lineages of, 599, 606–9f
  molecular phylogeny of, 583
  pollen grains in, 590–91
  seeds in, 591–92
Seeding, bioremediation with, 533
Seedless vascular plants
  characteristics of, 581
  homospory in, 590, 591f
  lineages of, 598–99, 603–5
  molecular phylogeny of, 583
Seeds, **433**, **581**, **823**, 833–37
  creation of, in embryogenesis, 433–36
  dormancy of, 836–37
  embryogenesis and, 833–34
  fruit development and dispersal of,
    834–36
  germination of, 832, 836–37
  gibberellins and abscisic acid in
    dormancy and germination of,
    809–10
  ground tissue, sclereids, and, 746
  in plant reproduction, 591–92, 823
  red/far-red light reception and
    germination of, 800–801
  role of drying in maturation of, 834
  seed coat, 833
  seed plants and, 581, 599
  vacuoles in, 114
Segment identity genes. *See Hox* genes
Segment polarity genes, 412, 413f, 414f
Segmentation, **645**–46
Segmentation genes, **412**, 413f
Segmented bodies, arthropod, 670–71
Segmented worms, 638t, 662, 667
Segments, embryo, **412**
Segregation, principle of, **262**
  meiosis as explanation for, 266–67
  particulate inheritance and, 261–63
  Punnett squares and, 262–63
Selective adhesion, **206**–7
Selective breeding, 257. *See also* Crosses
Selective permeability, plasma
    membrane, **89**, **862**. *See also*
    Permeability, plasma membrane
  importance of, to cells, 84, 107
  intracellular environments and,
    100–101
  ion channel proteins and, 98
  lipid bilayers and, 89
  osmoregulation and, 862
  osmosis and, 92–93
Self-fertilization, **248**–49, **258**, 470–71f,
    497–98, **830**, B:25–B:26
Self molecule, **1046**–47
Self recognition, adaptive immune
    response and, 1042, 1046–47

Self-replicating molecules
  DNA as, 64–65
  lipid bilayers and, 93–94
  research on proteins as first, 42,
    44–45, 55
  RNA world hypothesis on RNA as
    first, 57–58, 68–69
Self-sacrificing behavior, 461. *See also*
    Altruism
Selfish DNA sequences, 396, 397f
Selfish genes, 461
Semen, **1022**–23t
Semiconservative replication, DNA,
    287–89
Seminal vesicles, **1022**–23f
Senescence, plant, 812–13
Sensors, **854**
Sensory neurons, **929**
Sensory organs, animal, 646, 647t
Sensory receptors, homeostatic, 1001
Sensory systems, animal, 952–71
  chemoreception of chemicals in,
    964–67
  electroreception of electric fields in,
    967–68
  information transmission to brains
    in, 954
  magnetoreception of magnetic fields
    in, 968–69
  mechanoreception of pressure changes
    in, 954–59
  photoreception of light in, 959–64
  processes and receptors in, 953
  senses and sensory organs in, 646, 647t
  in sensing of environmental changes,
    952–53
  sensory transduction in, 953–54
  thermoreception of temperature
    in, 967
Sensory systems, plant, 793–821
  defense responses to pathogens and
    herbivores, 815–19
  environments and, 793–94
  germination, stem elongation, and
    flowering responses to red and far-
    red light, 800–803
  gravitropic responses to gravity, 803–5
  growth regulators, 814t
  information processing and, 794–96
  phototropic responses to blue light,
    796–800
  responses to wind and touch, 805–6
  selected, 818t
  youth, maturity, and aging as growth
    responses, 806–15
Sepals, **439**, **593**, **826**
Separation
  of cell components, B:17–B:18
  of molecules, B:13–B:15
Septa, **616**
Sequences, amino acid, 46, 47–48
Sequences, DNA. *See* DNA sequences
Sequences, rRNA, 6–8
Sequencing technology, 389f, 391,
    1173. *See also* Direct sequencing;
    DNA sequencing; Metagenomics
    (environmental sequencing);
    Whole-genome sequencing
Serotonin, 941t, **947**
Serum, blood, 233, B:21
Services, ecosystem. *See* Ecosystem
    services
Sessile organisms, **574**, **638**
Set point, **854**
Severe combined immunodeficiency
    (SCID), 383–84, 1056

Sex chromosomes, **238**, 240t, 268–69, 277t
Sex hormones
  binding of, to intracellular receptors,
    1006–8f
  contraception and, 1029–30
  control of menstrual cycle by,
    1027–30
  control of puberty by, 1026–27
  mammalian, 1025–30
  in mating, 1087–88
  testosterone and estradiol as steroids
    and, 1025–26
  in vertebrate sexual development and
    activity, 998–99
Sex-linked inheritance (sex-linkage), **268**
  in chromosome theory of inheritance,
    268–69
  exceptions and extensions to
    Mendelian rules on, 277t
  human, 278–79
  linkage vs., 270
Sex pheromones, 215–16
Sexual abstinence, 722
Sexual activity. *See* Mating
Sexual competition, 9–11
Sexual dimorphism, **477**–78
Sexual reproduction, **247**, **823**, **1014**
  animal, 650–51, 998–99, 1016–18f (*see
    also* Animal reproduction)
  asexual reproduction vs., 247, 251–52,
    566, 1014
  changing-environment hypothesis
    on, 253
  in eukaryotes, 553
  fertilization in (*see* Fertilization)
  fungal, 617–18, 624–27
  genetic variation from, 247–49
  hormones in, 998–99
  meiosis and, 219, 237–38 (*see also*
    Meiosis)
  paradox of, 251–52
  plant, 586–87, 823–25 (*see also* Plant
    reproduction)
  protist, 566
  purifying selection hypothesis on,
    252–53
  switching between asexual and, by
    Daphnia, 1014–16
Sexual selection, **472**, 475–78, **1088**–89
Sexually transmitted diseases, 546, 547,
    571. *See also* AIDS (acquired
    immune deficiency syndrome);
    HIV (human immunodeficiency
    virus)
Shade, far-red light as, 800–801
Shade leaves, 738
Shape
  active transport pump, 99–100
  animal locomotion and body, 985, 986
  enzyme, 54, 149–50
  flowers, 593–94
  genetic control of leaf, 436–38
  membrane carrier protein, 99
  organic molecule, 36f, 38
  plant receptor cell, 794–95
  prokaryotic diversity in, 536, 537f
  protein, 47–53, 331–32
  retinal change in, 962
  shoot system, 735–36
  signal receptor, 210
  simple molecule, 23–24, B:12–B:13
  transcription factors, 355–56
  viral, 715
Shared ancestry. *See* Common ancestry
Sharks, 512, 698, 866–68, 967–68
Sharp, Phillip, 321–22

Sheep, cloned, 409
Shells, **562**
  eggs, 693–94
  mollusk, 663, 668
  turtle, 702
Shine-Dagarno sequences, 329–30
Shocked quartz, 521–22f
Shoot apical meristem (SAM), **434**
  in reproductive development, 438–39
  in vegetative development, 436, 437f
Shoot systems, **732**
  adventitious roots and, 734–35
  characteristics of, 735
  collenchyma cells and shoot support
    in, 744–45
  formation of, 434
  functions of, 732
  modified stems in, 736–37
  morphological diversity of, 735–36
  organization of primary, 742
  phenotypic plasticity in, 736
  water transport in (*see* Water
    transport, plant)
Shoots, **434**
Short-day plants, **802**
Short tandem repeats (STRs), 396–98f
Shotgun sequencing, **390**–91
Shrimp, 495, 671, 677, 1133, 1134f
SI system, B:1
Sickle-cell disease, 48, 278
Side chains, amino acid
  amphipathic proteins and, 94
  in enzyme catalysis, 144–47
  function of, 42–44
  in major amino acids, 43f
  peptide bonds and, 45–46
  polarity of, and water solubility, 44
  protein structure and, 48–50
  of steroids, 87
Sieve plates, **747**, **767**
Sieve-tube elements, **747**, **767**, 770–71
Sig figs, metric system and, B:2–B:3
Sight, 481–82, 646, 647t
Sigma proteins, **319**–20f, 325t
Sigmoid curves, scatterplot, B:5
Signal hypothesis, endomembrane
    system, 123–24
Signal processing, cell–cell, 210–14
Signal receptors, **210**. *See also* Enzyme-
    linked receptors; G-protein-
    coupled receptors
Signal recognition particle (SRP), **124**
Signal response, cell–cell, 214
Signal transduction, **210**, **794**, **1008**
  cross-talk in, 813–14
  via G-protein-coupled receptors,
    211–13
  in hormone binding to cell-surface
    receptors, 1008–10
  of lipid-insoluble signals, 210–11
  in mechanoreception, 955
  in photoreception, 962–63
  in plant cells, 794–96
  via enzyme-linked receptors, 211,
    213–14
Signal transduction cascades, **1009**–10
Signaling, cell–cell. *See* Cell–cell
    signaling
Signaling, chemical. *See* Chemical
    signaling
Signaling pathways, animal hormone,
    993–94
Signals, communication, **1092**
Significant figures, metric system and,
    B:2–B:3
Silencers, **354**

Boldface page numbers indicate a glossary entry; page numbers followed by an
f indicate a figure; those followed by *t* indicate a table.

Boldface page numbers indicate a glossary entry; page numbers followed by an *f* indicate a figure; those followed by *t* indicate a table.

Boldface page numbers indicate a glossary entry; page numbers followed by an *f* indicate a figure; those followed by *t* indicate a table.